1994 SAE Handbook

Volume 2
Parts and Components

Cooperative Engineering Program

Published by:
Society of Automotive Engineers, Inc.
400 Commonwealth Drive
Warrendale, PA 15096-0001
U.S.A.
Phone: (412) 776-4841
Fax: (412) 776-5760

All SAE Standards are submitted to the American National Standards Institute for recognition as American National Standards. The second printing of the standard will indicate the recognition.

The 1994 SAE Handbook is comprised of three volumes, the titles of which are:
- Volume 1—Materials, Fuels, Emissions, and Noise
- Volume 2—Parts and Components
- Volume 3—On-Highway Vehicles and Off-Highway Machinery

Each volume contains a complete Subject Index in addition to a Numerical Index of Standards.

Page numbers shown in the indexes include the volume number followed by the section number and the page number within that section.

ISBN 1-56091-461-0
ISSN 0362-8205
Library of Congress Catalog Card Number: 25-16527

Copyright © 1994 Society of Automotive Engineers, Inc.

All rights reserved. Printed in the United States of America.

Permission to photocopy for internal or personal use, or the internal or personal use of specific clients, is granted by SAE for libraries and other users registered with the Copyright Clearance Center (CCC), provided that the base fee of $5.00 per report is paid directly to CCC, 27 Congress St., Salem, MA 01970. Special requests should be addressed to the SAE Publications Group. 1-56091-(set) /93 $5.00.

HOW TO USE THE 1994 SAE HANDBOOK

Each surface vehicle Standard, Recommended Practice, or Information Report has a designation consisting of the letter "J" combined with a number. The letter "J" is combined with a nonsignificant number to eliminate any possible confusion between the report number and the SAE numbers within the report.

In SAE usage, the term "technical report" includes standards, recommended practices, and information reports.

Effective with the 1981 SAE Handbook, revisions of reports are indicated by the month and year of revision; for example, J300 MAR92. A lower case "a," "b," etc., appended to the report designation number indicates successive revisions of older reports.

Because the SAE Handbook is published on an annual basis, certain reports in this book may not be the latest issue of the document. Therefore, users are cautioned to contact SAE Headquarters to determine the status of specific documents.

All new and revised reports contain SI (metric) equivalents of all dimensions. The SAE Editorial Advisory Committee appreciates receiving any comments and/or suggestions regarding the conversions.

The φ or (R) symbol next to a section or line of a report indicates areas where technical revisions have been made since the previous issue of the report. If the symbol is next to the report title, it indicates a complete revision of the report. The notation "ed." was used on older reports to indicate editorial changes.

Reports not included in the Handbook are listed under "Related Technical Reports" in Volume 1.

A "†" or "*" next to a report title in the index indicates the report is new or has been revised in the past year.

Page numbers shown in the index include the volume number followed by the section number and the page number within that section.

NOTE

All technical reports, including standards approved and practices recommended, are advisory only. Their use by anyone engaged in industry or trade or their use by governmental agencies is entirely voluntary. There is no agreement to adhere to any SAE Standard or Recommended Practice, and no commitment to conform to or be guided by any technical report. In formulating and approving technical reports, the Technical Board, its councils, and committees will not investigate or consider patents which may apply to the subject matter. Prospective users of the report are responsible for protecting themselves against liability for infringement of patents, trademarks, and copyrights.

1994 SAE HANDBOOK CONTENTS—VOLUME 2

Numerical Index 1
Subject Index 27

Section 15—Threads
Section 16—Fasteners
Section 17—Ball Studs and Joints
Section 18—Splines
Section 19—V-Belts
Section 20—Springs
Section 21—Speedometers and Tachometers
Section 22—Tubing, Hose and Fittings
Section 23—Electrical/Electronic Equipment
Section 24—Lighting
Section 25—Brakes
Section 26—Powerplant Components and Accessories

Numerical Index

Document	Title	Reference
SAE TSB001 JUN93 †	SAE Technical Standards Board Rules and Regulations	1:A.01
SAE TSB002 JUN92	Preparation of SAE Technical Reports	1:A.05
SAE TSB003 JUN92	Rules for SAE Use of SI (Metric) Units	1:A.08
SAE TSB004 JUN92	Technical Committee Guidelines	1:A.30
SAE HS-3	Surface Rolling and Other Methods for Mechanical Prestressing of Metals	Cancelled (Superseded by J811)
SAE J4	Motor Vehicle Seat Belt Assemblies	Cancelled (Superseded by J114, J117, J140, J141, J339, J800)
SAE J10 OCT90	Automotive and Off-Highway Air Brake Reservoir Performance and Identification Requirements (A)	3:36.132
SAE HS-13 MAY93	*Vehicle Occupant Restraint Systems and Components Standards Manual	Available as a separate publication (See Related Technical Reports Section)
SAE J14 MAY80	Specifications for Elastomer Compounds for Automotive Applications	Cancelled 1980 (Superseded by J200)
SAE J15 MAY78	Flexible Foams Made from Polymers or Copolymers of Vinyl Chloride	Cancelled 1978
SAE J17 JAN85	Latex Foam Rubbers	1:11.53
SAE J18 JUL92	*Sponge and Expanded Cellular Rubber Products	1:11.56
SAE J19 JAN85	Latex Dipped Goods and Coatings for Automotive Applications	1:11.63
SAE HS-19 JAN87	SAE Documents Referenced in Federal Motor Vehicle Safety Standards	Cancelled 1987
SAE J20 MAR88	Coolant System Hoses (A)	1:11.222
SAE HS-23 FEB93	*Fuels and Lubricants Standards Manual	Available as a separate publication (See Related Technical Reports Section)
SAE HS-24 JUL92	*Surface Vehicle Brake Manual	Available as a separate publication (See Related Technical Reports Section)
ANSI SAE Z26.1-1990	American National Standard for Safety Glazing Materials for Glazing Motor Vehicles and Motor Vehicle Equipment Operating on Land Highways—Safety Code	Available as a separate publication (See Related Technical Reports Section)
SAE J29 JUL83	Plastic Material for Use in Housings of Motor Vehicle Lighting Devices	Cancelled 1983
SAE J30 MAY93	*Fuel and Oil Hoses (A)	1:11.226
SAE J31 MAR86	Hydraulic Backhoe Lift Capacity (A)	3:40.331
SAE J32 MAR88	Service Performance Requirements for Sealed Beam Headlamp Units for Motor Vehicles	Cancelled 1988
SAE J33 DEC91	Snowmobile Definitions and Nomenclature—General	3:38.01
SAE J34 DEC91	Exterior Sound Level Measurement Procedure for Pleasure Motorboats	1:14.42
SAE HS-34 JAN93	*SAE Ground Vehicle Lighting Standards Manual	Available as a separate publication (See Related Technical Reports Section)
SAE J35 SEP88	Diesel Smoke Measurement Procedure (A)	1:13.64
SAE J38 AUG91	Lift Arm Support Device for Loaders	3:40.353
SAE J39 JUN93	*T-Hook Slots for Securement in Shipment of Agricultural Equipment (A)	3:39.58
SAE J40 JUN69	Automotive Brake Hoses	Cancelled 1969 (Superseded by J1401, J1402, J1403, J1406)
SAE HS-40 JAN91	*Principles of Engine Cooling Systems, Components and Maintenance	Available as a separate publication (See Related Technical Reports Section)
SAE J43 FEB88	Axle Application Load Rating for Industrial Wheel Loaders and Backhoe Loaders	3:40.333
SAE J44 DEC91	Service Brake System Performance Requirements—Snowmobiles	3:38.11
SAE J45 OCT84	Brake System Test Procedure—Snowmobiles	3:38.11
SAE J46 JUN80	Wheel Slip Brake Control System Road Test Code	2:25.158
SAE J47 JUN86	Maximum Sound Level Potential for Motorcycles (A)	1:14.01
SAE J48 MAY93	*Guidelines for Liquid Level Indicators	3:40.103
SAE J49 APR80	Specification Definitions—Hydraulic Backhoes (A)	3:40.70
SAE J50 APR87	Windshield Wiper Hose	Cancelled 1987
SAE J51 MAY89	Automotive Air Conditioning Hose	1:11.249
SAE J52 MAY81	Steering Wheel Rim Faceform Impact Test Procedure	Cancelled 1981
SAE J53 OCT84	Minimum Performance Criteria for Emergency Steering of Wheeled Earthmoving Construction Machines	3:40.515
SAE J56 JUN83	Electrical Generating System (Alternator Type) Performance Curve and Test Procedure (A)	2:23.82
SAE J57 FEB87	Sound Level of Highway Truck Tires (A)	1:14.31
SAE J58 FEB90	Flanged 12-Point Screws	Cancelled 1990
SAE HS-58	Diesel Fuel Injection Equipment and Test Methods	Available as a separate publication (See Related Technical Reports Section)
SAE J59 FEB86	Fuel Injection Nozzle Gaskets	Cancelled 1986
SAE J60 FEB77	Rubber Cups for Hydraulic Actuating Cylinders	Cancelled 1977 (Superseded by J1601)
SAE J62 FEB76	Rubber Seals for Hydraulic Disc Brake Cylinders	Cancelled 1976 (Superseded by J1603)
SAE J63 MAY86	Hole Placement on Dozer End Bits (A)	3:40.231
SAE HS-63 MAY88	Manual on Design and Manufacture of Coned Disk Springs or Belleville Springs	Cancelled 1988 (Superseded by HS-1582)
SAE J64 APR88	Vehicle Identification Numbers—Snowmobiles (A)	3:38.03
SAE J65 APR76	Rubber Boots for Use on Hydraulic Brake Actuating Cylinders	Cancelled 1976 (Superseded by J1604)
SAE J66 APR76	Brake Master Cylinder Reservoir Diaphragm Gasket	Cancelled 1976 (Superseded by J1605)
SAE J67 NOV84	Shovel Dipper, Clam Bucket, and Dragline Bucket Rating	3:40.323
SAE J68 DEC91	Tests for Snowmobile Switching Devices and Components	3:38.08
SAE J70 DEC68	Hydraulic Brake Fluid	Cancelled 1968 (Superseded by J1703)
SAE J71 DEC89	Central Fluid Systems	Cancelled 1989
SAE J72	Test for Central System Fluids	Cancelled
SAE HS-J73 DEC86	Multipurpose Petroleum Base Fluids	Cancelled 1986
SAE J75 OCT88	Motor Vehicle Brake Fluid Container Compatibility	2:25.66

* Technical Revision   † New Document   (D)—DODISS Adopted   (A)—ANSI Recognized

Numerical Index

Document	Title	Reference
SAE J76 OCT89	Handling and Dispensing of Motor Vehicle Brake Fluid	Cancelled 1989
SAE J77 OCT89	Service Maintenance of Motor Vehicle Brake Fluid in Motor Vehicle Brake Actuating Systems	Cancelled 1989
SAE J78 JUN79	Steel Self-Drilling Tapping Screws	1:4.21
SAE J79	Brake Disc and Drum Thermocouple Installation	2:25.157
SAE J80 JAN88	Automotive Rubber Mats (A)	1:11.185
SAE J81 JUN79	Thread Rolling Screws	1:4.26
SAE J82 JUN79	Mechanical and Quality Requirements for Machine Screws	1:4.14
SAE HS-82	Truck Ability Prediction Procedure—SAE J688	Available as a separate publication (See Related Technical Reports Section)
SAE HS-83	Truck Ability Work Sheet Pad	Cancelled (Superseded by HS-82)
SAE HS-83	SAE Commercial Vehicle Ability Report Form	Cancelled (Superseded by HS-82)
SAE HS-84	Manual on Shot Peening	Available as a separate publication (See Related Technical Reports Section)
SAE HS-87	Fiberboards	Available as a separate publication (See Related Technical Reports Section)
SAE J88 JUN86	Sound Measurement—Earthmoving Machinery—Exterior (A) (D)	1:14.106
SAE J89 JAN85	Dynamic Cushioning Performance Criteria for Snowmobile Seats (A)	3:38.13
SAE J90 JUN90	Standard Classification System for Nonmetallic Automotive Gasket Materials (A)	1:11.197
SAE J91	Comments on Traffic Noise Reduction	Cancelled
ANSI B92.1, 1a	Involute Splines and Inspection Standard	Available only from American National Standards Institute 1430 Broadway New York, NY 10018 (212) 642-4991
ANSI B92.2M	Metric Involute Splines and Inspection Standard	Available only from American National Standards Institute 1430 Broadway New York, NY 10018 (212) 642-4991
ANSI B92.2Ma	Part III: Inspection—Metric Module Involute Splines	Available only from American National Standards Institute 1430 Broadway New York, NY 10018 (212) 642-4991
SAE J92 OCT84	Snowmobile Throttle Control Systems (A)	3:38.15
SAE J93 OCT87	Snowmobile and All-Terrain Vehicle Pulley Guards and Shields	Cancelled 1987
SAE J94 OCT88	Combination Tail and Floodlamp for Industrial Equipment	Cancelled 1988
SAE J95 MAR86	Headlamps for Industrial Equipment (A)	3:40.262
SAE J96 MAR86	Flashing Warning Lamp for Industrial Equipment (A)	3:40.261
SAE J97 MAR87	Nomenclature for Exhaust System Parts	Cancelled 1987
SAE J98 NOV92	*Personnel Protection for General Purpose Industrial Machines (A)	3:40.472
SAE J99 MAR86	Lighting and Marking of Industrial Equipment on Highways (A)	3:40.263
SAE J100 MAR88	Passenger Car Glazing Shade Bands (A)	3:34.104
SAE J101 DEC89	Hydraulic Wheel Cylinders for Automotive Drum Brakes (A)	2:25.25
SAE J102 DEC90	Square Head Setscrews	Cancelled 1990 (Superseded by ANSI/B 18.6.2)
SAE J103 DEC90	Lag Screws	Cancelled 1990 (Superseded by ANSI/B 18.2.1)
SAE J104 DEC90	Square and Hex Nuts	Cancelled 1990 (Superseded by ANSI/ASME/B 18.2.2, ANSI/B 18.6.3)
SAE J105 DEC90	Hex Bolts	Cancelled 1990 (Superseded by ANSI/B 18.2.1)
SAE J106 DEC82	Soil Type and Strength Classification	Cancelled 1982
SAE J107 APR80	Operator Controls and Displays on Motorcycles	3:37.05
SAE J108 MAR87	Brake System Road Test Code—Motorcycles (A)	2:25.144
SAE J109 MAR87	Service Brake System Performance Requirements—Motorcycles and Motor-Driven Cycles (A)	2:25.146
SAE J110 DEC91	Seals—Testing of Radial Lip	3:29.140
SAE J111 JUN88	Seals—Terminology of Radial Lip (A)	3:29.103
SAE J112a	Electric Windshield Wiper Switch	2:23.188
SAE J113 DEC88	Hard Drawn Mechanical Spring Wire and Springs	1:5.04
SAE J114 MAR86	Seat Belt Assembly Webbing Abrasion Performance Requirements (A)	3:33.11
SAE J115 JAN87	Safety Signs (A)	3:40.494
SAE J116 JAN87	Gas Turbine Engine Test Code	Cancelled 1987
SAE J117	Dynamic Test Procedure—Type 1 and Type 2 Seat Belt Assemblies	Cancelled 1993.06
SAE J118 JAN84	Formerly Standard SAE Carbon Steels	Cancelled 1984 (Superseded by J1249)
SAE J119 FEB87	Fiberboard Crease Bending Test (A)	1:11.166
SAE J120a	Rubber Rings for Automotive Applications	1:11.203
SAE J121 AUG83	Decarburization in Hardened and Tempered Threaded Fasteners (A)	1:4.44
SAE J122a	Surface Discontinuities on Nuts	1:4.40
SAE J123c	Surface Discontinuities on Bolts, Screws, and Studs	1:4.37
SAE HS-124	SAE Manual on Blast Cleaning—SAE J792a	Cancelled (Superseded by J792)
SAE J125 MAY88	Elevated Temperature Properties of Cast Irons	1:6.12
SAE J126 JUN86	Selecting and Specifying Hot and Cold Rolled Steel Sheet and Strip (A)	1:3.27
SAE J127	Selection of Tires for Light Duty Trucks for Normal Highway Service	Cancelled
SAE J128	Occupant Restraint System Evaluation—Passenger Cars	3:33.12
SAE J129 DEC81	Engine and Transmission Identification Numbers	3:27.03
SAE J130 DEC77	Nonrigid Thermoplastic Compounds for Automotive Applications	Cancelled 1977
SAE J131 MAR83	Motorcycle Turn Signal Lamps (A)	2:24.136
SAE J132 DEC88	Oil Tempered Chromium-Vanadium Valve Spring Quality Wire and Springs	1:5.03
SAE J133 OCT87	Fifth Wheel Kingpin Performance—Commercial Trailers and Semitrailers	3:36.86
SAE J134 MAY85	Brake System Road Test Code—Passenger Car and Light Duty Truck-Trailer Combinations (A)	2:25.134

* Technical Revision † New Document (D)—DODISS Adopted (A)—ANSI Recognized

Numerical Index

SAE J135 MAY85	Service Brake System Performance Requirements—Passenger Car—Trailer Combinations	2:25.143
SAE J136 MAY81	Simplified Method for Simulating Glancing Blow Impacts—Motor Vehicles	Cancelled 1981
SAE J137 JUN89	Lighting and Marking of Agricultural Equipment on Highways	3:39.60
SAE J138	Film Analysis Guides for Dynamic Studies of Test Subjects	3:34.245
SAE J139 JUN90	Ignition System Nomenclature and Terminology (A)	2:23.58
SAE J140a	Seat Belt Hardware Test Procedure	3:33.01
SAE J141	Seat Belt Hardware Performance Requirements	3:33.03
SAE HS-150 JUN92	Fluid Conductors and Connectors	Available as a separate publication (See Related Technical Reports Section)
SAE J151 JUN91	Pressure Relief for Cooling System (A)	2:26.350
SAE J152 JUN80	Engine Specification Tag	Cancelled 1980
SAE J153 MAY87	Operator Precautions (A)	3:40.470
SAE J154 JUN92	Operator Space Envelope Dimensions for Off-Road Machines	3:40.419
SAE J155 JUN89	Service Brake System Performance Requirements—Light-Duty Truck	Cancelled 1989
SAE J156 APR86	Fusible Links (A)	2:23.171
SAE J157 DEC88	Oil Tempered Chromium—Silicon Alloy Steel Wire and Springs	1:5.07
SAE J158 JUN86	Automotive Malleable Iron Castings (A)	1:6.07
SAE J159 APR85	Load Moment System (A)	3:40.216
SAE J160 JUN80	Swell, Growth, and Dimensional Stability of Brake Linings (A)	2:25.90
SAE J161 JUN87	Selection of Tires for Multipurpose Passenger Vehicles, Light-, Medium-, and Heavy-Duty Trucks, Buses, Trailers, and Semitrailers for Normal Highway Service	Cancelled 1987
SAE J162 JUN88	Flywheels for Single Bearing Engine Mounted Power Generators	Cancelled 1988
SAE J163	Low Tension Wiring and Cable Terminals and Splice Clips	2:23.157
SAE J164 JUN91	Radiator Caps and Filler Necks (A)	2:26.373
SAE J165 JUN82	Fan Blast Deflectors for Earthmoving Machines	Cancelled 1982
SAE J166 JUN77	Minimum Performance Criteria for Brake Systems for Off-Highway Trucks and Wagons	Cancelled 1977
SAE J167 APR92	Overhead Protection for Agricultural Tractors—Test Procedures and Performance Requirements	3:39.89
SAE J168 APR78	Protective Enclosures for Agricultural Tractors—Test Procedures and Performance Requirements	Cancelled 1978 (Superseded by J1194)
SAE J169 MAR85	Design Guidelines for Air Conditioning Systems for Off-Road Operator Enclosures	3:40.510
SAE J170 MAR90	Measurement of Fuel Evaporative Emissions from Gasoline Powered Passenger Cars and Light Trucks by the Trap Method	Cancelled 1990 (Superseded by J170)
SAE J171 APR91	Measurement of Fuel Evaporative Emissions from Gasoline Powered Passenger Cars and Light Trucks Using the Enclosure Technique (A)	1:13.95
SAE J172 DEC88	Hard Drawn Carbon Steel Valve Spring Quality Wire and Springs	1:5.08
SAE J173 DEC87	Specification Definitions—Dozers	Cancelled 1987 (Superseded by J729)
SAE J174	Torque-Tension Test Procedure for Steel Threaded Fasteners	2:16.22
SAE J175 JUN88	Wheels—Impact Test Procedures—Road Vehicles (A)	3:31.07
SAE J176 DEC89	Fast Fill Fueling Installation for Off-Road Work Machines (A)	3:40.103
SAE J177 APR82	Measurement of Carbon Dioxide, Carbon Monoxide, and Oxides of Nitrogen in Diesel Exhaust	1:13.01
SAE J178 DEC88	Music Steel Spring Wire and Springs	1:5.10
SAE J179 JUL84	Labeling—Disc Wheel and Demountable Rims—Trucks (A)	3:31.30
SAE J180 MAY87	Electrical Charging Systems for Construction and Industrial Machinery (A)	3:40.270
SAE J181 MAY74	Brake Test Procedure for Industrial Wheel Tractors and Equipment	Cancelled 1974
SAE J182 NOV90	Motor Vehicle Fiducial Marks and Three Dimensional Reference System (A)	3:34.229
SAE J183 JUN91	Engine Oil Performance and Engine Service Classification (Other Than "Energy-Conserving") (A)	1:12.01
SAE J184 AUG87	Qualifying a Sound Data Acquisition System (D)	1:14.09
SAE HS-184 NOV89	Surface Vehicle Sound Measurement Manual	Available as a separate publication (See Related Technical Reports Section)
SAE J185 JUN88	Access Systems for Off-Road Machines (D)	3:40.383
SAE J186 DEC89	Supplemental High Mounted Stop and Rear Turn Signal Lamps for Use on Vehicles Less Than 2032 mm in Overall Width (A)	2:24.155
SAE J187	Truck Vehicle Identification Numbers	3:27.05
SAE J188 OCT89	Power Steering Pressure Hose—High Volumetric Expansion Type (A)	1:11.241
SAE J189 DEC89	Power Steering Return Hose—Low Pressure (A)	1:11.243
SAE J190	Power Steering Pressure Hose—Wire Braid	1:11.248
SAE J191 OCT89	Power Steering Pressure Hose—Low Volumetric Expansion Type (A)	1:11.244
SAE J192 MAR85	Exterior Sound Level for Snowmobiles (A)	1:14.36
SAE J193 FEB87	Ball Stud and Socket Assembly—Test Procedures (A)	2:17.04
SAE J194 FEB78	Drawbar for Forestry Tractors	Cancelled 1978
SAE J195 DEC88	Automatic Vehicle Speed Control—Motor Vehicles	2:21.16
SAE J196 DEC84	Electric Tachometer Specification—On Road	Cancelled 1984 (Superseded by J1399)
SAE J197 DEC84	Electric Tachometer Specification—Off-Road	Cancelled 1984 (Superseded by J1399)
SAE J198 JUN93	*Windshield Wiper Systems—Trucks, Buses, and Multipurpose Vehicles	3:34.12
SAE J199	Track Type Tractor Availability Test Code	Cancelled
SAE J200 MAY93	*Classification System for Rubber Materials	1:11.01
SAE J201 MAY89	In-Service Brake Performance Test Procedure Passenger-Car and Light-Duty Truck	2:25.85
SAE J202 MAY92	Synthetic Skins for Automotive Testing	Cancelled 1992
SAE J207 FEB85	Electroplating of Nickel and Chromium on Metal Parts—Automotive Ornamentation and Hardware	1:10.150
SAE J208 JUN93	*Safety for Agricultural Tractors	3:39.69

* Technical Revision   † New Document   (D)—DODISS Adopted   (A)—ANSI Recognized

Numerical Index

SAE J209 JAN87	Instrument Face Design and Location for Construction and Industrial Equipment (A)	3:40.421
SAE J210 JAN87	Wiring Identification System for Industrial and Construction Equipment	Cancelled 1987
SAE HS-210	Laboratory Testing Machines and Procedures for Measuring the Steady State Force and Movement Properties of Passenger Car Tires—SAE J1106 and SAE J1107	Cancelled (Superseded by J1106, J1107)
SAE J211 OCT88	Instrumentation for Impact Test (A)	3:34.270
SAE J212 JUN80	Brake System Dynamometer Test Procedure—Passenger Car	2:25.129
SAE J213a	Definitions—Motorcycles	3:37.01
SAE J214 MAR86	Hydraulic Cylinder Test Procedure (A)	3:40.138
SAE J215 JUN88	Continuous Hydrocarbon Analysis of Diesel Emissions (A)	1:13.29
SAE HS-215	Handbook of Motor Vehicle, Safety and Environmental Terminology	Available as a separate publication (See Related Technical Reports Section)
SAE J216 DEC89	Passenger Car Glazing—Electrical Circuits (A)	3:34.104
SAE J217 DEC88	Stainless Steel 17-7 PH Spring Wire and Springs	1:5.08
SAE J218 DEC81	Passenger Car Identification Terminology	3:27.02
SAE J220 MAR91	Crane Boomstop (A)	3:40.212
SAE J222 DEC91	Parking Lamps (Front Position Lamps)	2:24.167
SAE J223 APR80	Symbols and Color Codes for Maintenance Instructions, Container and Filler Identification (A)	3:40.104
SAE J224 MAR80	Collision Deformation Classification	3:34.235
SAE J225 JUN93	*Brake System Torque Balance Test Code Commercial Vehicles (A)	3:36.133
SAE J226 JAN86	Engine Preheaters (A)	2:26.370
SAE J227a	Electric Vehicle Test Procedure	Cancelled 1993.05
SAE J228 JUN90	Airflow Reference Standards (A)	2:26.372
SAE J229 JUN80	Service Brake Structural Integrity Test Procedure—Passenger Car	2:25.133
SAE J230 DEC88	Stainless Steel, SAE 30302, Spring Wire and Springs	1:5.09
SAE J231 JAN81	Minimum Performance Criteria for Falling Object Protective Structure (FOPS) (A)	3:40.364
SAE J232 DEC84	Industrial Rotary Mowers (A)	3:40.477
SAE J233 DEC87	Marine Engine Mountings Direct Drive Transmission	Cancelled 1987
SAE J234	Electric Windshield Washer Switch	2:23.188
SAE J235	Electric Blower Motor Switch	2:23.189
SAE J236 DEC76	Minimum Performance Criteria for Brake Systems for Rubber Tired, Self-Propelled Graders	Cancelled 1976 (Superseded by J1152)
SAE J237 DEC76	Minimum Performance Criteria for Brake Systems for Off-Highway, Rubber Tired, Front End Loaders and Dozers	Cancelled 1976 (Superseded by J1152)
SAE J238	Nut and Conical Spring Washer Assemblies	2:16.20
SAE J240 JUN93	*Life Test for Automotive Storage Batteries	2:23.71
SAE J242a	Metric Thread Fuel Injection Tubing Connections	Cancelled 1987
SAE J243	Methods of Tests for Automotive-Type Sealers, Adhesives, and Deadeners	1:11.68
SAE J244 AUG92	*Measurement of Intake Air or Exhaust Gas Flow of Diesel Engines (A)	1:13.11
SAE J245 AUG82	Engine-Rating Code—Spark Ignition	Cancelled 1982 (Superseded by J1349)
SAE J246 JUN93	*Spherical and Flanged Sleeve (Compression) Tube Fittings	2:22.21
SAE J247 FEB87	Instrumentation for Measuring Acoustic Impulses Within Vehicles (A)	1:14.07
SAE J248 FEB78	Crane Overload Indicating System Test Procedure	Cancelled 1978
SAE J249 JUN88	Mechanical Stop Lamp Switch (A)	2:24.150
SAE J250	Synthetic Resin Plastic Sealers, Nondrying Type	1:11.65
SAE J253 DEC89	Headlamp Switch (A)	2:24.176
SAE J254 AUG84	Instrumentation and Techniques for Exhaust Gas Emissions Measurement (A)	1:13.104
SAE J255a	Diesel Engine Smoke Measurement	1:13.56
SAE J256 AUG87	Service Performance Requirements for Motor Vehicle Lighting Devices and Components	Cancelled 1987
SAE J257 MAR85	Brake Rating Horsepower Requirements—Commercial Vehicles	3:36.137
SAE J258	Circuit Breaker—Internal Mounted—Automatic Reset	2:23.141
SAE J259	Ignition Switch	2:23.141
SAE J260 JUN90	Rear Underride Guard Test Procedure (A)	3:34.287
SAE J261 JUN87	Muffler Parts Nomenclature	Cancelled 1987
SAE J262 JUN87	Resonator Parts Nomenclature	Cancelled 1987
SAE J263 JUN85	Emergency Air Brake Systems—Motor Vehicles and Vehicle Combinations	Cancelled 1985
SAE J264 JUN88	Vision Glossary	3:34.227
SAE J265 APR91	Diesel Fuel Injector Assembly—Types 8, 9, 10, and 11 (A)	2:26.306
SAE J267 JAN91	Wheels/Rims—Trucks—Test Procedures and Performance Requirements (A)	3:31.14
SAE J268 MAY89	Rear View Mirrors—Motorcycles	3:37.06
SAE J270 MAY82	Engine Rating Code—Diesel	Cancelled 1982 (Superseded by J1349)
SAE J271 DEC88	Special Quality High Tensile, Hard Drawn Mechanical Spring Wire and Springs	1:5.05
SAE J272 DEC81	Vehicle Identification Number Systems	3:27.01
SAE J273 DEC81	Passenger Car Vehicle Identification Number System	3:27.02
SAE J274 JUN89	Rated Suspension Spring Capacity	2:20.30
SAE J275 FEB85	Test Method for Determining Window Fogging Resistance of Interior Trim Materials (A)	1:11.117
SAE J276 NOV92	*Steering Frame Lock for Articulated Loaders and Tractors	3:40.353
SAE J277 JUN90	Maintenance of Design Voltage—Snowmobile Electrical Systems (A)	3:38.03
SAE J278 OCT84	Snowmobile Stop Lamp (A)	3:38.05
SAE J279 OCT84	Snowmobile Tail Lamp (Rear Position Lamp) (A)	3:38.06
SAE J280 JUN84	Snowmobile Headlamps (A)	3:38.04
SAE J281 SEP80	Cast Iron Sealing Rings	3:29.148
SAE J282 SEP79	Automotive Gasoline Performance and Information System	Cancelled 1979
SAE J283 JUN93	*Test Procedure for Measuring Hydraulic Lift Capacity on Agricultural Tractors Equipped with Three-Point Hitch	3:39.41
SAE J284 JAN91	Safety Alert Symbol for Agricultural, Construction and Industrial Equipment (A)	3:39.68
SAE J285 SEP92	*Gasoline Dispenser Nozzle Spouts (A)	2:26.441
SAE J286 NOV83	SAE No. 2 Clutch Friction Test Machine Test Procedure (A)	3:29.95

* Technical Revision † New Document (D)—DODISS Adopted (A)—ANSI Recognized

Numerical Index

SAE J287 JUN88	Driver Hand Control Reach (A)	3:34.157
SAE J288 NOV83	Snowmobile Fuel Tanks (A)	3:38.15
SAE J291 JUN80	Determination of Brake Fluid Temperature	2:25.157
SAE J292 OCT84	Snowmobile and Snowmobile Cutter Lamps, Reflective Devices, and Associated Equipment (A)	3:38.07
SAE J293 OCT88	Vehicle Grade Parking Performance Requirements (A)	2:25.102
SAE J294	*Service Brake Structural Integrity Test Procedure—Vehicles Over 4500 kg (10 000 lb) GVWR	3:36.138
SAE J296 JUN93	*Excavator, Mini-Excavator, and Backhoe Hoe Bucket Volumetrtic Rating	3:40.329
SAE J297 JUN85	Operator Controls on Industrial Equipment	3:40.469
SAE J298 JUN87	Universal Symbols for Operator Controls on Industrial Equipment	Cancelled 1987 (Superseded by J1362)
SAE J299 AUG87	Stopping Distance Test Procedure (A)	2:25.156
SAE J300 MAR93	*Engine Oil Viscosity Classification	1:12.16
SAE J301 MAR93	*Effective Dates of New or Revised Technical Reports (A)	1:12.01
SAE J302 MAR67	Temperatures for Recommending Crankcase Oil	Cancelled 1967
SAE J303 MAR72	Internal Combustion Engine Service Classifications	Cancelled 1972
SAE J304 JUN93	*Engine Oil Tests (A)	1:12.19
SAE J305 JUN73	Extrapolated Oil Viscosities	Cancelled 1973
SAE J306 OCT91	Axle and Manual Transmission Lubricant Viscosity Classification	1:12.37
SAE J307 OCT73	Temperatures for Recommending Gear Lubricants	Cancelled 1973
SAE J308 JUN89	Axle and Manual Transmission Lubricants	1:12.36
SAE J309 JUN73	Corresponding Ranges in Kinematic, Redwood, and Engler Viscosities	Cancelled 1973
SAE J310 JUN93	*Automotive Lubricating Greases	1:12.39
SAE J311 APR86	Fluid for Passenger Car Type Automatic Transmissions	1:12.44
SAE J312 JAN93	*Automotive Gasolines	1:12.47
SAE J313 MAR92	Diesel Fuels	1:12.60
SAE J314 MAY81	Felts—Wool and Part Wool (A)	1:11.183
SAE J315 JAN85	Fiberboard Test Procedure (A)	1:11.158
SAE J316 DEC88	Oil Tempered Carbon Steel Spring Wire and Springs	1:5.01
SAE J318 SEP80	Air Brake Gladhand Service (Control) and Emergency (Supply) Line Couplers—Trucks, Truck-Tractors, and Trailers	3:36.139
SAE J319 SEP76	Minimum Performance Criteria for Brake Systems for Off-Highway, Rubber Tired, Self-Propelled Scrapers	Cancelled 1976 (Superseded by J1152)
SAE J320 SEP74	Minimum Performance Criteria for Roll-Over Protective System for Rubber-Tired, Self-Propelled Scrapers	Cancelled 1974 (Superseded by J1040)
SAE J321b	Tire Guards for Protection of Operator of Earthmoving Haulage Machines (A)	3:40.352
SAE J322 DEC85	Nonmetallic Trim Materials—Test Method for Determining the Staining Resistance to Hydrogen Sulfide Gas (A)	1:11.116
SAE J323	Test Method for Determining Cold Cracking of Flexible Plastic Materials	1:11.115
SAE J326 MAR86	Nomenclature—Hydraulic Backhoes (A)	3:40.326
SAE J327 MAR87	Ignition Timing	Cancelled 1987
SAE J328 MAR90	Wheels—Passenger Cars—Performance Requirements and Test Procedures (A)	3:31.06
SAE J331 OCT92	*Sound Levels for Motorcycles	1:14.03
SAE J332 AUG81	Testing Machines for Measuring the Uniformity of Passenger Car and Light Truck Tires (A)	3:30.32
SAE J333 AUG78	Operator Protection for Wheel Type Agricultural Tractors	Cancelled 1978 (Superseded by J1194)
SAE J334 AUG78	Protective Frame for Agricultural Tractors—Test Procedures and Performance Requirements	Cancelled 1978 (Superseded by J1194)
SAE J335 SEP90	Multiposition Small Engine Exhaust System Fire Ignition Suppression (A)	2:26.351
SAE J336 OCT88	Sound Level for Truck Cab Interior	1:14.29
SAE J337 OCT87	Brake Flange Mounting—On-Highway Vehicles	Cancelled 1987
SAE J338 OCT81	Motor Vehicle Instrument Panel Laboratory Impact Test Procedure—Knee-Leg Area	Cancelled 1981
SAE J339 MAR86	Seat Belt Assembly Webbing Abrasion Test Procedure (A)	3:33.11
SAE J341 MAR87	Truck and Bus Tire Performance Requirements and Test Procedures	Cancelled 1987
SAE J342 JAN91	Spark Arrester Test Procedure for Large Size Engines (A)	2:26.360
SAE J343 JUN93	*Tests and Procedures for SAE 100R Series Hydraulic Hose and Hose Assemblies (A)	2:22.185
SAE J344 JUN86	Lawn and Garden Tractors Load and Inflation Pressures and Tire Selection Table for Future Design	Cancelled 1986
SAE J345a	Wet or Dry Pavement Passenger Car Tire Peak and Locked Wheel Braking Traction	3:30.24
SAE J346 JUN81	Motor Vehicle Seatback Assembly Laboratory Impact Test Procedure—Head Area	Cancelled 1981
SAE J347 JUL88	Diesel Fuel Injector Assembly Type 7 (9.5 mm) (A)	2:26.305
SAE J348 JUN90	Wheel Chocks	3:36.122
SAE J349 FEB91	Detection of Surface Imperfections in Ferrous Rods, Bars, Tubes, and Wires	1:3.63
SAE J350 JAN91	Spark Arrester Test Procedure for Medium Size Engines (A)	2:26.365
SAE J351 DEC88	Oil Tempered Carbon Steel Valve Spring Quality Wire and Springs	1:5.02
SAE J352 DEC79	External Ignition-Proofing of Marine Engine Alternators	Cancelled 1979
SAE J353 DEC79	External Ignition-Proofing of Marine Engine Regulators	Cancelled 1979
SAE J354 DEC79	External Ignition-Proofing of Marine Engine Distributors	Cancelled 1979
SAE J355 DEC79	External Ignition-Proofing of Marine Engine Cranking Motors	Cancelled 1979
SAE J356 JUN91	Welded Flash Controlled Low Carbon Steel Tubing Normalized for Bending, Double Flaring, and Beading (A)	2:22.213
SAE J357 JUN91	Physical and Chemical Properties of Engine Oils (A)	1:12.24
SAE J358 FEB91	Nondestructive Tests (A)	1:3.67
SAE J359 FEB91	Infrared Testing	1:3.69
SAE J360 OCT88	Truck and Bus Grade Parking Performance Test Procedure	2:25.101
SAE J361 MAR85	Test Method for Determining Visual Color Match to Master Specimen for Fabrics, Vinyls, Coated Fiberboards, and Other Automotive Trim Materials (A)	1:11.180

* Technical Revision   † New Document   (D)—DODISS Adopted   (A)—ANSI Recognized

Numerical Index

SAE J362 JUL82	Vehicle Hood Latch Systems (A)	3:34.06
SAE J363 FEB87	Filter Base Mounting (A)	2:26.246
SAE J364 FEB85	Vehicle Identification Numbers—Motorcycles	Cancelled 1985
SAE J365 FEB85	Method of Testing Resistance to Scuffing of Trim Materials (A)	1:11.163
SAE J366 FEB87	Exterior Sound Level for Heavy Trucks and Buses (A) (D)	1:14.27
SAE J367 JUN80	Passenger Car Door System Crush Test Procedure	3:34.286
SAE J368 MAR93	*High-Strength, Quenched, and Tempered Structural Steels	1:1.164
SAE J369 JUN89	Flammability of Automotive Interior Materials—Horizontal Test Method	1:11.176
SAE J370 JUN90	Bolt and Capscrew Sizes for Use in Construction and Industrial Machinery	Cancelled 1990
SAE J371 MAY93	*Drain, Fill, and Level Plugs for Off-Road, Self-Propelled Work Machines	3:40.257
SAE J372 MAY87	LP-Gas Fuel Systems Components	Cancelled 1987
SAE J373 APR93	*Housing Internal Dimensions for Single- and Two-Plate Spring-Loaded Clutches (A)	2:26.241
SAE J374 MAY91	Vehicle Roof Strength Test Procedure (A)	3:34.285
SAE J375 APR85	Radius-of-Load or Boom Angle Indicating Systems	3:40.214
SAE J376 APR85	Load Indicating Devices in Lifting Crane Service (A)	3:40.215
SAE J377 FEB87	Performance of Vehicle Traffic Horns (A)	1:14.35
SAE J378 JAN88	Marine Engine Wiring (A)	3:41.22
SAE J379a	Gogan Hardness of Brake Lining	2:25.88
SAE J380 FEB93	*Specific Gravity of Brake Lining	2:25.89
SAE J381 JUN84	Windshield Defrosting Systems Test Procedure—Trucks, Buses, and Multipurpose Vehicles (A) (D)	3:34.21
SAE J382 OCT84	Windshield Defrosting Systems Performance Guidelines—Trucks, Buses, and Multi-Purpose Vehicles (A)	3:34.23
SAE J383 APR86	Motor Vehicle Seat Belt Anchorages—Design Recommendations (A)	3:33.05
SAE J384 APR86	Motor Vehicle Seat Belt Anchorages—Test Procedure (A)	3:33.07
SAE J385 APR86	Motor Vehicle Seat Belt Anchorages—Performance Requirements (A)	3:33.08
SAE J386 JUN93	*Operator Restraint System for Off-Road Work Machines	3:40.491
SAE J387 OCT88	Terminology—Motor Vehicle Lighting	2:24.01
SAE J388	Dynamic Flex Fatigue Test for Slab Polyurethane Foam	3:34.98
SAE J389 OCT87	Universal Symbols for Operator Controls on Agricultural Equipment	Cancelled (Superseded by J1362)
SAE J390 JUN82	Dual Dimensioning (A)	1:A.35
SAE HS-J390	Dual Dimensioning	Cancelled (Superseded by J390)
SAE J391 JUL81	Definition for Particle Size	1:3.81
SAE J392 FEB92	Motorcycle and Motor Driven Cycle Electrical System—Maintenance of Design Voltage	2:24.135
SAE J393 OCT91	Nomenclature—Wheels, Hubs, and Rims for Commercial Vehicles	3:31.32
SAE J394 OCT76	Minimum Performance Criteria for Roll-Over Protective Structure for Rubber-Tired Front End Loaders and Rubber-Tired Dozers	Cancelled 1976 (Superseded by J1040)
SAE J395 OCT74	Minimum Performance Criteria for Roll-Over Protective Structure for Crawler Tractors and Crawler-Type Loaders	Cancelled 1974 (Superseded by J1040)
SAE J396 OCT74	Minimum Performance Criteria for Roll-Over Protective Structures for Motor Graders	Cancelled 1974 (Superseded by J1040)
SAE J397 APR88	Deflection Limiting Volume—ROPS/FOPS Laboratory Evaluation (A)	3:40.363
SAE J398 FEB88	Fuel Tank Filler Conditions—Passenger Car, Multi-Purpose Passenger Vehicles, and Light Duty Trucks	2:26.436
SAE J399 FEB85	Anodized Aluminum Automotive Parts	1:10.38
SAE J400 JAN85	Test for Chip Resistance of Surface Coatings	1:11.126
SAE J401 NOV92	*Selection and Use of Steels	1:1.06
SAE J402 DEC88	SAE Numbering System for Wrought or Rolled Steel	1:1.08
SAE J403 MAY92	Chemical Compositions of SAE Carbon Steels	1:1.09
SAE J404 FEB91	Chemical Compositions of SAE Alloy Steels (A)	1:1.14
SAE J405 JAN89	Chemical Compositions of SAE Wrought Stainless Steels	1:1.16
SAE J406 JUN93	*Methods of Determining Hardenability of Steels	1:1.23
SAE J407 JUN81	Hardenability Bands for Alloy H Steels	Cancelled 1981 (Superseded by J1268)
SAE J408 DEC83	Methods of Sampling Steel for Chemical Analysis (A)	1:1.152
SAE J409 DEC90	Product Analysis—Permissible Variations from Specified Chemical Analysis of a Heat or Cast of Steel (A)	1:1.153
SAE J410 DEC92	High Strength, Low Alloy Steel	Cancelled 1992 (Superseded by J1392, J1442)
SAE J411 NOV89	Carbon and Alloy Steels	1:2.01
SAE J412 JUN89	General Characteristics and Heat Treatments of Steels	1:2.05
SAE J413 JUN90	Mechanical Properties of Heat Treated Wrought Steels	1:2.24
SAE J414 JUN84	Estimated Mechanical Properties and Machinability of Hot Rolled and Cold Drawn Carbon Steel Bars	Cancelled 1984 (Superseded by J1397)
SAE J415 JUN83	Definitions of Heat Treating Terms	1:2.37
SAE J416 DEC83	Tensile Test Specimens (A)	1:3.01
SAE J417 DEC83	Hardness Tests and Hardness Number Conversions	1:3.04
SAE J418 DEC83	Grain Size Determination of Steels (A)	1:3.10
SAE J419 DEC83	Methods of Measuring Decarburization (A)	1:3.12
SAE J420 MAR91	Magnetic Particle Inspection	1:3.70
SAE J421 MAY93	*Cleanliness Rating of Steels by the Magnetic Particle Method	1:3.57
SAE J422 DEC83	Microscopic Determination of Inclusions in Steels (A)	1:3.59
SAE J423 DEC83	Methods of Measuring Case Depth (A)	1:3.64
SAE J424 JUN86	Method for Determining Breakage Allowances for Sheet Steel (A)	1:3.66
SAE J425 MAR91	Electromagnetic Testing by Eddy Current Methods	1:3.72
SAE J426 MAR91	Liquid Penetrant Test Methods	1:3.75
SAE J427 MAR91	Penetrating Radiation Inspection (A)	1:3.76
SAE J428 MAR91	Ultrasonic Inspection (A)	1:3.79
SAE J429 AUG83	Mechanical and Material Requirements for Externally Threaded Fasteners (A)	1:4.01
SAE J430	Mechanical and Chemical Requirements for Nonthreaded Fasteners	1:4.18
SAE J431 MAR93	*Automotive Gray Iron Castings	1:6.01
SAE J432 MAR70	Automotive Malleable Iron Castings	Cancelled 1970 (Superseded by J158)

* Technical Revision † New Document (D)—DODISS Adopted (A)—ANSI Recognized

Numerical Index

SAE J433 MAR70	Pearlitic Malleable Iron Castings	Cancelled 1970 (Superseded by J158)	
SAE J434 JUN86	Automotive Ductile (Nodular) Iron Castings (A)	1:6.09	
SAE J435c	Automotive Steel Castings	1:6.17	
SAE J436 JUN66	Chemical Compositions of SAE Corrosion and Heat Resistant Ferrous Castings	Cancelled 1966	
SAE J437a	Selection and Heat Treatment of Tool and Die Steels	1:7.02	
SAE J438b	Tool and Die Steels	1:7.01	
SAE J439a	Sintered Carbide Tools	1:7.07	
SAE J441 JUN93	*Cut Wire Shot	1:8.05	
SAE J442 AUG79	Test Strip, Holder and Gage for Shot Peening	1:8.06	
SAE J443 JAN84	Procedures for Using Standard Shot Peening Test Strip (A)	1:8.08	
SAE J444 MAY93	*Cast Shot and Grit Size Specifications for Peening and Cleaning	1:8.11	
SAE J445 AUG84	Metallic Shot and Grit Mechanical Testing	1:8.12	
SAE J446 JUN84	Preferred Thicknesses for Uncoated Flat Metals (Thru 12 mm) (A)	1:9.07	
SAE HS-J447 JUN81	Prevention of Corrosion of Motor Vehicle Body and Chassis Components	Available as a separate publication (See Related Technical Reports Section)	
SAE J448a	Surface Texture	1:9.03	
SAE J449a	Surface Texture Control	1:9.05	
SAE J450 JUN91	Use of Terms Yield Strength and Yield Point (A)	1:9.01	
SAE J451 JAN89	Aluminum Alloys—Fundamentals	1:10.01	
SAE J452 JAN89	General Information—Chemical Compositions, Mechanical and Physical Properties of SAE Aluminum Casting Alloys	1:10.06	
SAE J453 JAN86	Chemical Compositions and Mechanical and Physical Properties of SAE Aluminum Casting Alloys	Cancelled 1986 (Superseded by J452)	
SAE J454 FEB91	General Data on Wrought Aluminum Alloys (A)	1:10.26	
SAE J455 FEB65	Chemical Compositions of SAE Wrought Aluminum Alloys	Cancelled 1965 (Superseded by J457)	
SAE J456 FEB65	Typical Physical and Mechanical Properties of SAE Wrought Aluminum Alloys	Cancelled 1965 (Superseded by J454)	
SAE J457 FEB91	Chemical Compositions, Mechanical Property Limits, and Dimensional Tolerances of SAE Wrought Aluminum Alloys (A)	1:10.38	
SAE J458 FEB65	Dimensional Tolerances of SAE Wrought Aluminum Alloys	Cancelled 1965 (Superseded by J457)	
SAE J459 OCT91	Bearing and Bushing Alloys	1:10.43	
SAE J460 OCT91	Bearing and Bushing Alloys—Chemical Composition of SAE Bearing and Bushing Alloys	1:10.41	
SAE J461 SEP81	Wrought and Cast Copper Alloys	1:10.45	
SAE J462 SEP81	Cast Copper Alloys	1:10.78	
SAE J463 SEP81	Wrought Copper and Copper Alloys	1:10.81	
SAE J464 JAN89	Magnesium Alloys	1:10.120	
SAE J465 JAN89	Magnesium Casting Alloys	1:10.120	
SAE J466 DEC89	Magnesium Wrought Alloys	1:10.125	
SAE J467b	Special Purpose Alloys ("Superalloys")	1:10.128	
SAE J468 DEC88	Zinc Alloy Ingot and Die Casting Compositions	1:10.139	
SAE J469 JAN89	Zinc Die Casting Alloys	1:10.139	
SAE J470c	Wrought Nickel and Nickel-Related Alloys	1:10.140	
SAE J471d	Sintered Powder Metal Parts: Ferrous	1:8.01	
SAE J472	Copper and Silver Base Brazing Alloys	Cancelled	
SAE J473a	Solders	1:10.157	
SAE J474 FEB85	Electroplating and Related Finishes	1:10.157	
SAE J475a	Screw Threads (ANSI B1.1)	Available only from American National Standards Institute 1430 Broadway New York, NY 10018 (212) 642-4991	
SAE J476a	Dryseal Pipe Threads	2:15.01	
SAE J477 FEB69	Square and Hexagon Bolts and Nuts	Cancelled 1969 (Superseded by J102, J103, J104, J105)	
SAE J478a	Slotted and Recessed Head Screws	Cancelled 1990 (Superseded by ANSI/B 18.6.1, ANSI/B 18.6.2)	
SAE J479 FEB90	Slotted Headless Setscrews	Cancelled 1990 (Superseded by ANSI/B 18.6.2)	
SAE J480 FEB68	Socket Head Capscrews and Setscrews	Cancelled 1968	
SAE J481 FEB90	Round Head Bolts	Cancelled 1990	
SAE J482 FEB90	Hexagon High Nuts	Cancelled 1990	
SAE J483 FEB90	Crown (Blind, Acorn) Nuts	Cancelled 1990	
SAE J484 FEB90	Alignment of Nut Slots	Cancelled 1990 (Superseded by ANSI/ASME/B 18.2.2)	
SAE J485 FEB90	Holes in Bolt and Capscrew Shanks and Slots in Nuts for Cotter Pins	Cancelled 1990	
SAE J486 FEB70	Slots in Bolt Heads	Cancelled 1970	
SAE J487 FEB90	Cotter Pins	Cancelled 1990 (Superseded by ANSI/B 18.8.1)	
SAE J488 FEB90	Plain Washers	Cancelled 1990 (Superseded by ANSI/B 18.22.1)	
SAE J489 FEB90	Lock Washers	Cancelled 1990 (Superseded by ANSI/ASME/B 18.21)	
SAE J490 OCT81	Ball Joints	2:17.08	
SAE J491 NOV87	Steering Ball Studs and Socket Assemblies (A)	2:17.01	
SAE J492	Rivets and Riveting	2:16.29	
SAE J493	Rod Ends and Clevis Pins	2:16.28	
SAE J494	Grooved Straight Pins	2:16.23	
SAE J495	Straight Pins (Solid)	2:16.23	
SAE J496	Spring Type Straight Pins	2:16.26	
SAE J497	Unhardened Ground Dowel Pins	2:16.28	
SAE J498 NOV87	Involute Splines and Inspections	Cancelled 1987 (Superseded by ANSI B92.1)	
SAE J499a	Parallel Side Splines for Soft Broached Holes in Fittings	2:18.01	
SAE J500 AUG89	Serrated Shaft Ends (A)	2:18.03	
SAE J501	Shaft Ends	2:18.04	
SAE J502	Woodruff Keys	2:18.05	
SAE J503	Woodruff Key Slots and Keyways	2:18.06	
SAE J506b	Sleeve Type Half Bearings	2:26.194	
SAE J507 AUG77	Helical Compression Springs Hot-Coiled for General Automotive Use	Cancelled 1977 (Superseded by J1121)	
SAE J508 AUG77	Helical Compression and Extension Springs Cold-Coiled for General Automotive Use	Cancelled 1977 (Superseded by J1121)	
SAE J509 AUG71	Helical Springs for Motor-Vehicle Suspension	Cancelled 1971 (Superseded by J1121)	

* Technical Revision   † New Document   (D)—DODISS Adopted   (A)—ANSI Recognized

Numerical Index

SAE J510 NOV92	*Leaf Springs for Motor Vehicle Suspension—Made to Customary U.S. Units	2:20.09
SAE J511 JUN89	Pneumatic Spring Terminology	2:20.16
SAE J512 JUN93	*Automotive Tube Fittings	2:22.06
SAE J513 JUN93	*Refrigeration Tube Fittings—General Specifications	2:22.33
SAE J514 JUN93	*Hydraulic Tube Fittings	2:22.59
SAE J515 JUN92	Hydraulic O-Ring	2:22.200
SAE J516 JUN93	*Hydraulic Hose Fittings	2:22.114
SAE J517 JUN93	*Hydraulic Hose	2:22.161
SAE J518 JUN93	*Hydraulic Flanged Tube, Pipe, and Hose Connections, Four-Bolt Split Flange Type	2:22.190
SAE J519 JUN71	Hydraulic Hose Flanged "O" Ring Connections	Cancelled 1971 (Superseded by J517)
SAE J520 JUN78	Fuel Supply Connections	Cancelled 1978
SAE J521 JUN87	Fuel Injection Tubing Connections	Cancelled 1987
SAE J524 JUN91	Seamless Low Carbon Steel Tubing Annealed for Bending and Flaring (A)	2:22.206
SAE J525 JUN91	Welded and Cold Drawn Low Carbon Steel Tubing Annealed for Bending and Flaring (A)	2:22.207
SAE J526 JUN91	Welded Low Carbon Steel Tubing (A)	2:22.208
SAE J527 JUN91	Brazed Double Wall Low Carbon Steel Tubing (A)	2:22.209
SAE J528 JUN91	Seamless Copper Tube (A)	2:22.218
SAE J529 JUN88	Fuel Injection Tubing	Cancelled 1988
SAE J530 JUN93	*Automotive Pipe Fittings	2:22.232
SAE J531 JUN93	*Automotive Pipe, Filler, and Drain Plugs	2:22.241
SAE J532 JUN93	*Automotive Straight Thread Filler and Drain Plugs	2:22.246
SAE J533 JUN92	Flares for Tubing	2:22.203
SAE J534 JUN93	*Lubrication Fittings (A)	2:22.252
SAE J535 JUN86	Water Connection Flanges	Cancelled 1986
SAE J536 JUN92	Hose Clamps	Cancelled 1992
SAE J537 JUN92	Storage Batteries	2:23.62
SAE J538 JUL89	Grounding of Storage Batteries (A)	2:23.74
SAE J539 MAR87	Voltages for Diesel Electrical Systems (A)	2:23.75
SAE J540 MAR69	Ignition Coil Mounting	Cancelled 1969
SAE J541 FEB89	Voltage Drop for Starting Motor Circuits	2:23.74
SAE J542 JUN91	Starting Motor Mountings (A)	2:23.76
SAE J543 JUN88	Starting Motor Pinions and Ring Gears (A)	2:23.78
SAE J544 MAR88	Electric Starting Motor Test Procedure (A)	2:23.80
SAE J545 MAR84	Generating Mountings	Cancelled 1984
SAE J546 MAR85	Magneto Mountings	Cancelled 1985
SAE J548/1 JUN92	Spark Plugs	2:23.83
SAE J548/2 JUN92	Spark Plug Installation Sockets	2:23.95
SAE J549 JUN90	Preignition Rating of Spark Plugs	2:23.96
SAE J550 JUN74	Torque Requirements for Spark-Plug Installation	Cancelled 1974 (Superseded by J548)
SAE J551 MAR90	Performance Levels and Methods of Measurement of Electromagnetic Radiation from Vehicles and Devices (30 to 1000 MHz)	2:23.190
SAE J552 MAR82	External Electromagnetic Radiation Suppressor	Cancelled 1982
SAE J553 JUN92	*Circuit Breakers	2:23.135
SAE J554 AUG87	Electric Fuses (Cartridge Type) (A)	2:23.123
SAE J555 AUG81	Truck, Truck-Tractor, Trailer, and Motor-Coach Wiring	Cancelled 1981 (Superseded by J1292)
SAE J556 AUG81	Automobile Wiring	Cancelled 1981 (Superseded by J1292)
SAE J557 AUG90	High Tension Ignition Cable	Cancelled 1990 (Superseded by J2031)
SAE J558 AUG78	Low Tension Cable	Cancelled 1978 (Superseded by J1127, J1128)
SAE J559	Seven Conductor Jacketed Cable for Truck and Trailer Connections	Cancelled
SAE J560 JUN93	*Seven Conductor Electrical Connector for Truck-Trailer Jumper Cable	2:23.175
SAE J561 JUN93	*Electrical Terminals—Eyelet and Spade Type	2:23.182
SAE J562 APR86	Nonmetallic Loom (A)	2:23.187
SAE J563 MAR90	Six- and Twelve-Volt Cigar Lighter Receptacles (A)	2:23.189
SAE J564 MAR90	Headlamp Beam Switching (A)	2:24.08
SAE J565 JUN89	Semiautomatic Headlamp Beam Switching Devices (D)	2:24.09
SAE J566 JUN71	Headlamp Mountings	Cancelled 1971
SAE J567 NOV87	Lamp Bulb Retention System (A)	2:24.11
SAE J568 NOV78	Sockets Receiving Prefocus Base Lamps	Cancelled 1978
SAE J569 NOV70	Lamp Bulbs and Bases	Cancelled 1970
SAE J570	Wedge Base Bulbs	Cancelled
SAE J571 NOV88	Dimensional Specifications for Sealed Beam Headlamp Units	Cancelled 1988
SAE J572 MAY93	*Requirements for Sealed Lighting Unit for Construction and Industrial Machines	3:40.260
SAE J573 DEC89	Miniature Lamp Bulbs (A) (D)	2:24.21
SAE J574 DEC64	12-Volt Lamp Bulbs and Sealed Units for Heavy Duty Commercial Vehicles	Cancelled 1964 (Superseded by J573)
SAE J575 JUN92	Test Methods and Equipment for Lighting Devices and Components for Use on Vehicles Less Than 2032 mm in Overall Width	2:24.30
SAE J576 JUL91	Plastic Materials for Use in Optical Parts Such as Lenses and Reflectors of Motor Vehicle Lighting Devices	2:24.33
SAE J577 JUL80	Vibration Test Machine	Cancelled 1980
SAE J578 MAY88	Color Specification (A) (D)	2:24.34
SAE J579 MAY92	Sealed Beam Headlamp Units for Motor Vehicles	Cancelled 1992
SAE J580 MAY92	Sealed Beam Headlamp Assembly	Cancelled 1992
SAE J581 JUN89	Auxiliary Driving Lamps	2:24.117
SAE J582 SEP84	Auxiliary Low Beam Lamps (A)	2:24.118
SAE J583 JUN93	*Front Fog Lamps	2:24.119
SAE J584 DEC83	Motorcycle Headlamps (A)	2:24.120
SAE J585 DEC91	Tail Lamps (Rear Position Lamps) for Use on Motor Vehicles Less Than 2032 mm in Overall Width	2:24.137
SAE J586 DEC89	Stop Lamps for Use on Motor Vehicles Less Than 2032 mm in Overall Width (A)	2:24.143
SAE J587 MAR93	*License Plate Illumination Devices (Rear Registration Plate Illumination Devices)	2:24.151
SAE J588 JUN91	Turn Signal Lamps for Use on Motor Vehicles Less Than 2032 mm in Overall Width (A)	2:24.153
SAE J589b	Turn Signal Switch (D)	2:24.159
SAE J590 APR93	*Turn Signal Flashers (A)	2:24.164
SAE J591 MAY89	Spot Lamps	2:24.167
SAE J592 JUN92	Clearance, Side Marker, and Identification Lamps	2:24.170
SAE J593 JUN87	Backup Lamps (Reversing Lamps) (A) (D)	2:24.174
SAE J594 MAY89	Reflex Reflectors (D)	2:24.177

* Technical Revision   † New Document   (D)—DODISS Adopted   (A)—ANSI Recognized

Numerical Index

Number	Date	Title	Reference
SAE J595	JAN90	Flashing Warning Lamps for Authorized Emergency, Maintenance and Service Vehicles (A)	2:24.183
SAE J596	JAN79	Electric Emergency Lanterns	Cancelled 1979
SAE J597	JAN79	Liquid Burning Emergency Flares	Cancelled 1979
SAE J598	MAY87	Sealed Lighting Units for Construction, Industrial and Forest Machinery (A)	3:40.259
SAE J599	MAY81	Lighting Inspection Code	2:24.202
SAE J600	FEB93	*Headlamp Aim Testing Machines	2:24.205
SAE J601	FEB65	Light Output Meter	Cancelled 1965
SAE J602	DEC89	Headlamp Aiming Device for Mechanically Aimable Headlamp Units (A)	2:24.206
SAE J603	DEC81	Incandescent Lamp Impact Test	Cancelled 1981
SAE J604	JAN86	Engine Terminology and Nomenclature—General (A)	2:26.01
SAE J605	JAN73	Free Piston Engine Nomenclature	Cancelled 1973
SAE J607	AUG88	Small Spark Ignition Engine Test Code (A)	2:26.22
SAE J608		Minimum Identification Markings for Small Air-Cooled Engines	Cancelled
SAE J609a		Mounting Flanges and Power Take-Off Shafts for Small Engines	2:26.26
SAE J610	AUG87	Valve Seat Inserts—Engine	Cancelled 1987
SAE J611	AUG65	Piston Rings and Pistons	Cancelled 1965 (Superseded by J929)
SAE J612	AUG65	Piston and Piston Ring Nomenclature	Cancelled 1965 (Superseded by J929)
SAE J613	AUG76	Turbocharger Connections	Cancelled 1976 (Superseded by J1135)
SAE J614b		Engine and Transmission Dipstick Marking (A)	2:26.233
SAE J615	APR91	Engine Mountings (A)	2:26.234
SAE J616	APR90	Engine Foot Mounting (Front and Rear)	3:40.252
SAE J617	MAY92	Engine Flywheel Housing and Mating Transmission Housing Flanges	3:40.247
SAE J618	JAN91	Flywheels for Single-Plate Spring-Loaded Clutches (A)	2:26.238
SAE J619	JUN80	Flywheels for Two-Plate Spring-Loaded Clutches	2:26.240
SAE J620	MAY93	*Flywheels for Industrial Engines Used with Industrial Power Take-Offs Equipped with Driving-Ring Type Overcenter Clutches and Engine-Mounted Marine Gears and Single Bearing Engine-Mounted Power Generators	3:40.256
SAE J621	APR89	Industrial Power Takeoffs with Driving Ring-Type Overcenter Clutches	3:40.243
SAE J622	APR87	Airflow or Vacuum Governor Flanges	Cancelled 1987
SAE J623	APR83	Automotive Carburetor Flanges	Cancelled 1983
SAE J624	SEP89	Tapped and Flanged Exhaust Connections for Small Engines (A)	3:40.255
SAE J625	SEP87	Fuel Pump Mountings for Diaphragm Type Pumps	Cancelled 1987
SAE J626	JUN88	Diesel Fuel Injection—End Mounting Flanges for Fuel Injection Pumps (A)	2:26.297
SAE J627	JUN65	Diesel Fuel Injection Pump Mounting, Flange No. 5	Cancelled 1965 (Superseded by J626)
SAE J628	JUN65	Diesel Fuel Injection Pump Mounting, Flange No. 6	Cancelled 1965 (Superseded by J626)
SAE J629	APR91	Diesel Fuel Injector Assembly—Flange Mounted Types 5 and 6 (A)	2:26.304
SAE J631	APR93	*Radiator Nomenclature (A)	2:26.385
SAE J632		Radiator Filler Necks	Cancelled
SAE J633		Large Size Radiator Filler Necks	Cancelled
SAE J634	APR83	Water Thermostat Pockets	Cancelled 1983
SAE J635	JUL84	Fan Hub Bolt Circles and Pilot Holes (A)	2:26.414
SAE J636	MAY92	V-Belts and Pulleys	2:19.01
SAE J637	FEB89	Automotive V-Belt Drives	2:19.04
SAE J638	APR93	*Motor Vehicle Heater Test Procedure	3:34.34
SAE J639	NOV91	Safety and Containment of Refrigerant for Mechanical Vapor Compression Systems Used for Mobile Air-Conditioning Systems	3:34.40
SAE J640	APR93	*Symbols for Hydrodynamic Drives (A)	3:29.01
SAE J641	JUL88	Hydrodynamic Drives Terminology (A)	3:29.02
SAE J642	JUL72	Color Code for Hydrodynamic Drive Illustrations	Cancelled 1972
SAE J643	JUN89	Hydrodynamic Drive Test Code	3:29.05
SAE J644	JUN72	Performance Curve Sheets for Hydrodynamic Drive Vehicles	Cancelled 1972
SAE J645	OCT92	*Automotive Transmission Terminology	3:29.23
SAE J646	APR93	*Planetary Gear(s)—Terminology	3:29.24
SAE J647	APR90	Transmissions—Schematic Diagrams (A)	3:29.24
SAE J648	APR93	*Automatic Transmission Hydraulic Control Systems—Terminology	3:29.27
SAE J649	JUL88	Automatic Transmission Functions—Terminology (A)	3:29.27
SAE J650	JUL72	Hydraulic Control Systems for Automatic Transmissions	Cancelled 1972
SAE J651	JAN91	Passenger Car and Truck Automatic Transmission Test Code (A)	3:29.28
SAE J652	JAN80	Truck Transmissions—Test Code	Cancelled 1980
SAE J653	JAN76	Form for Transmission Compression Spring	Cancelled 1976 (Superseded by J1122)
SAE J654	MAR93	*Static and Reciprocating Elastomeric Trnamission Seals	3:29.145
SAE J655	MAR77	Temperature Instrument Mounting	Cancelled 1977
SAE J656	APR88	Automotive Brake Definitions and Nomenclature (A)	2:25.70
SAE J657	APR74	Definitions for Braking Terminology and Brake Operation Terminology	Cancelled 1974 (Superseded by J656)
SAE J658	APR77	Service Brake Performance	Cancelled 1977 (Superseded by J201)
SAE J659	APR91	Color Code for Location Identification of Combination Linings of Two Different Materials or Two Shoe Brakes	Cancelled 1991
SAE J660	APR80	Brake Linings	Cancelled 1980
SAE J661	AUG87	Brake Lining Quality Control Test Procedure (A)	2:25.90
SAE J662	NOV90	Brake Block Chamfer (A)	2:25.93
SAE J663b		Rivets for Brake Linings and Bolts for Brake Blocks	2:25.94
SAE J664	NOV75	Brake Spider Mounting	Cancelled 1975
SAE J665	NOV75	Brake Drum Mountings	Cancelled 1975
SAE J666	NOV87	Brake Diaphragm Maintenance—Tractor and Semitrailer	Cancelled 1987
SAE J667		Brake Test Code—Inertia Dynamometer	2:25.103
SAE J668	NOV72	Air Brake Build-Up Time Instrumentation	Cancelled 1972
SAE J670e		Vehicle Dynamics Terminology	3:34.297
SAE HS-J670		Vehicle Dynamics Terminology	Cancelled (Superseded by J670)
SAE J671	APR82	Vibration Damping Materials and Underbody Coatings (A)	1:11.111
SAE J672	APR75	Exterior Loudness Evaluation of Heavy Trucks and Buses	Cancelled 1975
SAE J673	APR93	*Automotive Safety Glasses	3:34.100
SAE J674	NOV90	Safety Glazing Materials—Motor Vehicles (A)	3:34.103
SAE J678	DEC88	Speedometers and Tachometers—Automotive	2:21.03
SAE J679	DEC69	Panel Mountings for Diesel Fuel Pump or Engine Primers	Cancelled 1969

* Technical Revision † New Document (D)—DODISS Adopted (A)—ANSI Recognized

Numerical Index

SAE J680 SEP88	Location and Operation of Instruments and Controls in Motor Truck Cabs (A)	3:36.121
SAE J681 SEP75	Bumper Location	Cancelled 1975
SAE J682 OCT84	Rear Wheel Splash and Stone Throw Protection	3:36.120
SAE J683 AUG85	Tire Chain Clearance—Trucks, Buses (Except Suburban, Intercity, and Transit Buses), and Combinations of Vehicles	3:36.115
SAE J684 MAY87	Trailer Couplings, Hitches and Safety Chains—Automotive Type (A)	3:35.01
SAE J685 MAY81	Data Plate—Automotive Type Trailers	Cancelled 1981
SAE J686 JUL81	Motor Vehicle License Plates	3:27.06
SAE J687 JUN88	Nomenclature—Truck, Bus, Trailer (A)	3:36.01
SAE J688 AUG87	Truck Ability Prediction Procedure (A)	3:36.03
SAE J689 DEC89	Curb Clearance Approach, Departure, and Ramp Breakover Angles—Passenger Car and Light Truck (A)	3:32.03
SAE J690	Certificates of Maximum Net Horsepower for Motor Trucks and Truck Tractors	3:36.65
SAE J691 SEP90	Motor Truck CA Dimensions (A)	3:36.67
SAE J692 SEP68	Truck Tractor Trailer Clearances	Cancelled 1968
SAE J693 APR89	Truck Overall Widths Across Dual Tires	3:36.111
SAE J694 SEP88	Disc Wheel/Hub or Drum Interface Dimensions—Commercial Vehicles (A) (D)	3:36.100
SAE J695 DEC89	Turning Ability and Off Tracking—Motor Vehicles (A)	3:36.78
SAE J696	Truck Tractor Fifth Wheel Heights	Cancelled
SAE J697 MAY88	Safety Chain of Full Trailers or Converter Dollies (A)	3:36.98
SAE J699 NOV85	Average Vehicle Dimensions for Use in Designing Docking Facilities for Motor Vehicles (A)	3:36.67
SAE J700 JUN93	*Upper Coupler Kingpin—Commercial Trailers and Semitrailers (A)	3:36.90
SAE J701 AUG84	Truck Tractor Semitrailer Interchange Coupling Dimensions (A)	3:36.83
SAE J702 MAY85	Brake and Electrical Connection Locations—Truck-Tractor and Truck-Trailer (A)	3:36.114
SAE J703 MAY80	Fuel Systems—Truck and Truck Tractors	Cancelled 1980
SAE J704 DEC92	*Openings for Six- and Eight-Bolt Truck Transmission Mounted Power Take-Offs	3:29.41
SAE J705 DEC87	Rating of Truck Power Take-Offs	Cancelled 1987
SAE J706 NOV90	*Rating of Winches (A)	3:36.117
SAE J707 NOV73	Spark Arresters for Internal Combustion Engines	Cancelled 1973 (Superseded by J350)
SAE J708 DEC84	Agricultural Tractor Test Code	3:39.01
SAE J709d	Agricultural Tractor Tire Loadings, Torque Factors, and Inflation Pressures (A)	3:39.20
SAE J711 MAR91	Tire Selection Tables for Agricultural Tractors of Future Design (A)	3:39.26
SAE J712 APR93	*Industrial and Agricultural Disc Wheels (A)	3:39.54
SAE J713 APR67	Interchangeability of Disc Halves for Agricultural Press and Gage Wheels	Cancelled 1967 (Superseded by ASAE R221)
SAE J714 APR93	*Wheel Mounting Elements for Industrial and Agricultural Disc Wheels (A)	3:39.57
SAE J715 JUN93	*Three-Point Free-Link Hitch Attachment of Implements to Agricultural Wheeled Tractors	3:39.37
SAE J716 MAR91	Application of Hydraulic Remote Control to Agricultural Tractors and Trailing Type Agricultural Implements (A)	3:39.45
SAE J717 APR93	*Auxiliary Power Take-Off Drives for Agricultural Tractors	3:39.53
SAE J718 APR78	540-RPM Power Take-Off for Farm Tractors	Cancelled 1978
SAE J719 APR78	1000-RPM Power Take-Off Farm Tractors	Cancelled 1978
SAE J720 APR86	Tractor Belt Speed and Pulley Width	Cancelled 1986
SAE J721 FEB93	*Operating Requirements for Tractors and Power Take-Off Driven Implements (A)	3:39.52
SAE J722 MAR91	Power Take-Off Definitions and Terminology for Agricultural Tractors (A)	3:39.53
SAE J723 MAR77	Farm Equipment Breakaway Connector	Cancelled 1977
SAE J724 MAR77	Farm Equipment Two Conductor Breakaway Connector Cable	Cancelled 1977
SAE J725 MAR91	Mounting Brackets and Socket for Warning Lamp and Slow-Moving Vehicle (SMV) Identification Emblem (A)	3:39.65
SAE J726 JUN93	*Air Cleaner Test Code	2:26.214
SAE J727 JAN86	Nomenclature—Crawler Tractor (A)	3:40.30
SAE J728 JUL90	Component Nomenclature—Scrapers (A)	3:40.34
SAE J729 SEP86	Nomenclature and Specification Definitions—Dozers (A)	3:40.37
SAE J730 SEP74	Nomenclature—Cable Control Units	Cancelled 1974
SAE J731 FEB85	Component Nomenclature—Loader	3:40.39
SAE J732 JUN92	Specification Definitions—Loaders (A)	3:40.65
SAE J733 JUN86	Nomenclature—Rippers and Scarifiers (A)	3:40.45
SAE J734 JUL90	Component Nomenclature—Dumper Trailer (A)	3:40.47
SAE J736 JUL79	Mechanical Power Outlet Test Code	Cancelled 1979
SAE J737 AUG89	Hole Spacing for Scraper and Dozer Cutting Edges (A)	3:40.231
SAE J738 JUN86	Cutting Edge—Double Bevel Cross Sections (A)	3:40.235
SAE J739 JUN91	Cutting Edge—Curved Grader (A)	3:40.233
SAE J740 JUL86	Countersunk Square Holes for Cutting Edges and End Bits (A)	3:40.237
SAE J741 JUN93	*Capacity Rating—Scraper, Open Bowl (A)	3:40.319
SAE J742 FEB85	Capacity Rating—Loader Bucket	3:40.324
SAE J743 MAR92	Lift Capacity Calculation and Test Procedure—Pipelayer and Sideboom	3:40.223
SAE J744 JUL88	Hydraulic Pump and Motor Mounting and Drive Dimensions (A)	3:40.116
SAE J745 APR87	Hydraulic Power Pump Test Procedure (A)	3:40.111
SAE J746 MAR86	Hydraulic Motor Test Procedure (A)	3:40.128
SAE J747 MAY90	Control Valve Test Procedure (A)	3:40.130
SAE J748 MAR86	Hydraulic Directional Control Valves, 3000 psi Maximum (A)	3:40.120
SAE J749 AUG92	*Drawbars—Crawler Tractor (A)	3:40.318
SAE J750 AUG71	Industrial (Track Type) Tractor Equipment Mounting	Cancelled 1971
SAE J751 APR86	Off-Road Tire and Rim Classification—Construction Machines (A)	3:40.301
SAE J752 APR91	Maintenance Interval—Construction Equipment	Cancelled 1991 (Superseded by J753)
SAE J753 MAY91	Maintenance Interval Chart (A)	3:40.101
SAE J754a	Lubricant Types—Construction and Industrial Machinery	3:40.102
SAE J755 JUN80	Marine Propeller-Shaft Ends and Hubs	3:41.06

*Technical Revision   †New Document   (D)—DODISS Adopted   (A)—ANSI Recognized

Numerical Index

SAE J756 AUG87	Marine Propeller-Shaft Couplings (A)	3:41.01
SAE J759 JUN91	Lighting Identification Code (A)	2:24.03
SAE J760a	Dimensional Specifications for General Service Sealed Lighting Units	2:24.17
SAE J763 APR91	Aging of Carbon Steel Sheet and Strip (A)	1:2.36
SAE J764 NOV85	Loading Ability Test Code—Scrapers (A)	3:40.159
SAE J765 OCT90	Crane Load Stability Test Code (A)	3:40.210
SAE J766 OCT68	SAE Temper Designation System, Cast and Wrought Aluminum and Aluminum Base Alloys	Cancelled 1968
SAE J767 OCT73	Mechanical and Physical Properties of SAE Aluminum Casting Alloys	Cancelled 1973
SAE J768 OCT68	SAE Numbering System for Designation of Wrought Aluminum Alloys	Cancelled 1968
SAE J770 OCT84	Alloy Steel Machinability Ratings	Cancelled 1984 (Superseded by J1397)
SAE J771 APR86	Automotive Printed Circuits (A)	2:23.145
SAE J772 DEC92	*Clearance Envelopes for Six-Bolt, Eight-Bolt, and Rear Truck Transmission Mounted Power Takeoffs (A)	3:29.33
SAE J773b	Conical Spring Washers	2:16.19
SAE J774 DEC89	Emergency Warning Device (Triangular Shape) (A)	2:24.181
SAE J775 JAN88	Engine Poppet Valve Information Report	1:1.165
SAE J776 JAN81	Hardenability Bands for Carbon H Steels	Cancelled 1981 (Superseded by J1268)
SAE J778 JAN84	Formerly Standard SAE Alloy Steels	Cancelled 1984 (Superseded by J1249)
SAE J779 JAN83	Tractor Protection Valve Control	Cancelled 1983
SAE J780 JUN90	Engine Coolant Pump Seals	2:26.415
SAE J781 JUN68	Battery Identification Selection Chart	Cancelled 1968 (Superseded by J537)
SAE J782b	Motor Vehicle Seating Manual	3:34.73
SAE HS-J782	Motor Vehicle Seating Manual	Cancelled (Superseded by J782)
SAE HS-J784a	Residual Stress Measurements by X-Ray Diffraction	Available as a separate publication (See Related Technical Reports Section)
SAE J786a	Brake System Road Test Code—Truck, Bus, and Combination of Vehicles	3:36.140
SAE J787 JUN77	Motor Vehicle Seat Belt Anchorage	Cancelled 1977 (Superseded by J383, J384, J385)
SAE HS-788 APR80	Manual on Design and Application of Leaf Springs	Available as a separate publication (See Related Technical Reports Section)
SAE J792a	SAE Manual on Blast Cleaning	1:8.14
SAE HS-J795 APR90	Manual on Design and Application of Helical and Spiral Springs	Available as a separate publication (See Related Technical Reports Section)
SAE HS-796 JUL90	Manual on Design and Manufacture of Torsion Bar Springs	Available as a separate publication (See Related Technical Reports Section)
SAE J800 APR86	Motor Vehicle Seat Belt Assembly Installations (A)	3:33.09
SAE HS-806 JUN83	Oil Filter Test Procedure	Available as a separate publication (See Related Technical Reports Section)
SAE J808 JUN67	Manual on Shot Peening	Cancelled (Superseded by HS-84)
SAE J810 MAR87	Classification of Common Imperfections in Sheet Steel	1:3.30
SAE J811 AUG81	Surface Rolling and Other Methods for Mechanical Prestressing of Metals	1:8.32
SAE J813 AUG85	Automotive Air Brake Reservoir Volume	Cancelled 1985
SAE J814 JUL88	Engine Coolants	1:11.209
SAE J815 MAY81	Load Deflection Testing of Urethane Foams for Automotive Seating	3:34.100
SAE J816 MAY82	Engine Test Code—Spark Ignition and Diesel	Cancelled 1982 (Superseded by J1349)
SAE J817/1 MAR91	Engineering Design Serviceability Guidelines—Construction and Industrial Machinery—Serviceability Definitions—Off-Road Work Machines (A)	3:40.521
SAE J817/2 MAR91	Engineering Design Serviceability Guidelines—Construction and Industrial Machinery—Maintainability Index—Off-Road Work Machines (A)	3:40.522
SAE J818 MAY87	Rated Operating Load for Loaders (A)	3:40.100
SAE J819 MAR87	Engine Cooling System Field Test (Air-to-Boil) (A)	3:40.157
SAE J820 OCT80	Crane Hoist Line Speed and Power Test Code (A)	3:40.208
SAE J821 MAY85	Electrical Wiring Systems for Construction, Agricultural and Off-Road Machines	3:40.264
SAE J822 MAY72	Wedge Base Type Socket	Cancelled 1972
SAE J823 JUN91	Flasher Test (A)	2:24.210
SAE J824 JAN86	Engine Rotation and Cylinder Numbering (A)	2:26.28
SAE J826 JUN92	Devices for Use in Defining and Measuring Vehicle Seating Accommodation	3:34.90
SAE J827 MAR90	Cast Steel Shot (A)	1:8.09
SAE J828 MAR87	Gas Turbine Engine Nomenclature	Cancelled 1987
SAE J829 FEB88	Fuel Tank Filler Cap and Cap Retainer (A)	2:26.432
SAE J830 OCT92	*Fuel Injection Equipment Nomenclature	2:26.254
SAE J831	Nomenclature—Automotive Electrical Systems	2:23.01
SAE J832 OCT87	Locating Ledge for Flat Pad Engine Mountings	Cancelled 1987
SAE J833 MAY89	Human Physical Dimensions	3:40.377
SAE J834a	Passenger Car Rear Vision	3:34.114
SAE J835 DEC81	Split Type Bushings—Design and Application (A)	2:26.201
SAE J836a	Automotive Metallurgical Joining	1:9.08
SAE HS-J836	Automotive Metallurgical Joining	Cancelled (Superseded by J836)
SAE J839 JUN91	Passenger Car Side Door Latch Systems (A)	3:34.02
SAE J840 AUG82	Test Procedures for Brake Shoe and Lining Adhesives and Bonds	2:25.95
SAE J841 DEC84	Operator Controls for Agricultural Wheeled Tractors	3:39.70
SAE J842	Restraining Devices for Children (8 Months to 6 Years) for Use in Motor Vehicles	Cancelled
SAE J843 NOV90	Brake System Road Test Code—Passenger Car and Light-Duty Truck (A)	2:25.108
SAE J844 JUN90	Nonmetallic Air Brake System Tubing (A)	2:22.221
SAE J845 MAR92	360 Degree Warning Devices for Authorized Emergency, Maintenance, and Service Vehicles	2:24.185

* Technical Revision   † New Document   (D)—DODISS Adopted   (A)—ANSI Recognized

Numerical Index

SAE J846 JUN89	Coding Systems for Identification of Fluid Conductors and Connectors (D)	2:22.01
SAE J847 NOV87	Trailer Tow Bar Eye and Pintle Hook/Coupler Performance (A)	3:36.95
SAE J848 JAN91	Fifth Wheel Kingpin, Heavy-Duty—Commercial Trailers and Semitrailers (A)	3:36.92
SAE J849 NOV85	Connection and Accessory Locations for Towing Multiple Trailers (A)	3:36.93
SAE J850 NOV88	Fixed Rigid Barrier Collision Tests	3:34.281
SAE J851 JUN87	Dimensions—Wheels for Demountable Rims, Demountable Rims and Rim Spacers—Commercial Vehicles (A)	3:36.98
SAE J852 NOV87	Front Cornering Lamps for Use on Motor Vehicles (A)	2:24.162
SAE J853 DEC81	Vehicle Identification Numbers	3:27.03
SAE J854	Harness Type Restraint Assemblies for Use in Motor Vehicles	Cancelled
SAE J855 MAR87	Test Method of Stretch and Set of Textiles and Plastics	1:11.181
SAE J856	Connectors and Plugs	2:24.15
SAE J857 MAR91	Roll-Over Tests without Collision	Cancelled 1991
SAE J858a	Electrical Terminals—Blade Type	2:23.179
SAE J859 OCT92	*Numbering System for Designating Grades of Cast Ferrous Materials	1:6.01
SAE J860 JAN85	Test Method for Measuring Weight of Organic Trim Materials (A)	1:11.114
SAE J861	Method of Testing Resistance to Crocking of Organic Trim Materials	1:11.114
SAE J862 JAN89	Factors Affecting Accuracy of Mechanically Driven Automotive Speedometer and Odometers	2:21.01
SAE J863 JUN86	Methods for Determining Plastic Deformation in Sheet Metal Stampings (A)	1:3.17
SAE J864 MAY93	*Surface Hardness Testing with Files	1:3.03
SAE J866 NOV90	Friction Coefficient Identification System for Brake Linings (A)	2:25.89
SAE J868 FEB86	Large Size Radiator Filler Caps and Necks	Cancelled 1993.05
SAE J869 JUL90	Component Nomenclature—Construction Two- and Four-Wheel Tractors (A)	3:40.32
SAE J870 JUL90	Component Nomenclature—Graders (A)	3:40.36
SAE J871 JUL90	Hi-Head Finished Hex Bolts	Cancelled 1990
SAE J872 MAY86	Drawbar Test Procedure for Construction, Forestry and Industrial Machines (A)	3:40.146
SAE J873 FEB86	Drag Force Test Procedure for Construction, Forestry, and Industrial Machines (A)	3:40.150
SAE J874 OCT85	Center of Gravity Test Code (A)	3:40.152
SAE J875 JUN85	Trailer Axle Alignment (A)	3:36.112
SAE J876 JAN91	Wide Base Tire Rims and Wheels (A)	3:36.109
SAE J877 JUN84	Properties of Low Carbon Sheet Steel and Their Relationship to Formability	1:3.20
SAE J878 JUN78	Low Tension Cable Thermosetting Insulation	Cancelled 1978 (Superseded by J1128)
SAE J879b	Motor Vehicle Seating Systems	3:34.94
SAE J880 MAR85	Brake System Rating Test Code—Commercial Vehicles	3:36.147
SAE J881 FEB85	Lifting Crane Sheave and Drum Sizes	3:40.230
SAE J882 FEB85	Test Method for Measuring Thickness of Automotive Textiles and Plastics (A)	1:11.115
SAE J883 FEB86	Test Method for Determining Dimensional Stability of Automotive Textile Materials (A)	1:11.114
SAE J884 MAR91	Liquid Ballast Table for Drive Tires of Agricultural Tractors (A)	3:39.25
SAE J885 JUL86	Human Tolerance to Impact Conditions as Related to Motor Vehicle Design	3:34.306
SAE HS-J885	Human Tolerance to Impact Conditions as Related to Motor Vehicle Design	Cancelled (Superseded by J885)
SAE J887 AUG87	School Bus Warning Lamps (A)	2:24.192
SAE J890 AUG72	Air Brake System Test Procedure for Vehicles in Service	Cancelled 1972
SAE J891 JUN93	*Spring Nuts	2:16.01
SAE J892a	Push-On Spring Nuts	2:16.13
SAE J893 JUN79	Vehicle Fuel Consumption Test Code	Cancelled 1979
SAE J894 JUN78	Terminology—Construction and Industrial Machinery	Cancelled 1978 (Superseded by J1234)
SAE J895 APR86	Five Conductor Electrical Connectors for Automotive Type Trailers (A)	2:23.171
SAE J896 MAY83	Mounting Flanges for Engine Accessory Drives (A)	3:40.253
SAE J897 OCT85	Machine Slope Operation Test Code (A)	3:40.158
SAE J898 OCT87	Control Locations for Off-Road Work Machines (A)	3:40.404
SAE J899 DEC88	Operator's Seat Dimensions for Off-Road Self-Propelled Work Machines	3:40.395
SAE J900 AUG85	Crankcase Emission Control Test Code (A)	1:13.69
SAE J901 OCT90	Universal Joints and Driveshafts—Nomenclature—Terminology—Application (A)	3:29.42
SAE J902 APR93	*Passenger Car Windshield Defrosting Systems	3:34.18
SAE J903c	Passenger Car Windshield Wiper Systems	3:34.07
SAE J905 JAN87	Fuel Filter Test Methods (A)	2:26.329
SAE HS-J905	Fuel Filter Test Method	Cancelled (Superseded by J905)
SAE J907 JAN71	Recommendations for Improving Safety on Farm Tractors	Cancelled 1971 (Superseded by J208)
SAE J908 JAN71	Lighting and Marking of Farm and Light Industrial Equipment on Public Roads	Cancelled 1971 (Superseded by J137)
SAE J909 JUN93	*Three-Point Hitch, Implement Quick-Attaching Coupler, Agricultural Tractors	3:39.34
SAE J910 OCT88	Hazard Warning Signal Switch (A) (D)	2:24.184
SAE J911 JUN86	Surface Texture Measurement of Cold Rolled Sheet Steel (A)	1:9.01
SAE J912a	Test Method for Determining Blocking Resistance and Associated Characteristics of Automotive Trim Materials	1:11.165
SAE J913 FEB85	Test Method for Wicking of Automotive Fabrics and Fibrous Materials (A)	1:11.163
SAE J914 NOV87	Side Turn Signal Lamps (A)	2:24.160
SAE J915 APR93	*Automatic Transmissions—Manual Control Sequence	3:29.32
SAE J916 APR92	Rules for SAE Use of SI (Metric) Units	Cancelled 1992 (Superseded by TSB 003)
SAE J917 JUN80	Marine Push-Pull Control Cables	3:41.03
SAE J918c	Passenger Car Tire Performance Requirements and Test Procedures	3:30.01
SAE J919 JUN86	Sound Measurement—Earthmoving Machinery—Operator—Singular Type (A) (D)	1:14.53
SAE J920 SEP85	Technical Publications for Off-Road Work Machines (A)	3:40.526
SAE J921 SEP81	Motor Vehicle Instrument Panel Laboratory Impact Test Procedure—Head Area	Cancelled 1981

* Technical Revision † New Document (D)—DODISS Adopted (A)—ANSI Recognized

Numerical Index

Document	Date	Title	Reference
SAE J922	JUL88	Turbocharger Nomenclature and Terminology	2:26.04
SAE J923	APR91	Nomenclature and Terminology for Truck and Bus Drive Axles (A)	3:29.72
SAE J924	JAN81	Thrust Washers—Design and Application	2:26.206
SAE J925	JUN93	*Minimum Service Access Dimensions for Off-Road Machines (A)	3:40.380
SAE J926	JUN79	Hydraulic Pipe Fittings	Cancelled 1979 (Superseded by J514, Section IV)
SAE J927	NOV88	Flywheels for Engine Mounted Torque Converters	3:40.254
SAE J928	JUL89	Electrical Terminals—Pin and Receptacle Type (A)	2:23.180
SAE J929	JUL90	Piston Rings and Pistons	Cancelled 1990
SAE J930	AUG84	Storage Batteries for Off-Road Work Machines (A)	3:40.528
SAE J931	MAR86	Hydraulic Power Circuit Filtration (D)	3:40.137
SAE J932	AUG85	Definitions for Macrostrain and Microstrain (A)	1:3.81
SAE J933	JUN79	Mechanical and Quality Requirements for Tapping Screws	1:4.19
SAE J934	JUL82	Vehicle Passenger Door Hinge Systems	3:34.01
SAE J935	JUN90	High Strength Carbon and Alloy Die Drawn Steels	1:2.04
SAE J937	JUN89	Service Brake System Performance Requirements—Passenger Car	Cancelled 1989
SAE J938	JUN92	Drop Test for Evaluating Laminated Safety Glass for Use in Automotive Windshields	Cancelled 1992
SAE J940	DEC88	Glossary of Carbon Steel Sheet and Strip Terms	1:2.34
SAE J941	JUN92	Motor Vehicle Drivers' Eye Locations	3:34.204
SAE J942b		Passenger Car Windshield Washer Systems (D)	3:34.16
SAE J943	JUN93	*Slow-Moving Vehicle Identification Emblem	3:39.62
SAE J944	JUN92	Steering Control System—Passenger Car—Laboratory Test Procedure	Cancelled 1992
SAE J945	JUN93	*Vehicular Hazard Warning Signal Flashers (A)	2:24.164
SAE J946	OCT91	Application Guide to Radial Lip Seals	3:29.109
SAE J947	JAN85	Glossary of Fiberboard Terminology	1:11.156
SAE J948	FEB86	Test Method for Determining Resistance to Abrasion of Automotive Bodycloth, Vinyl, and Leather, and the Snagging of Automotive Bodycloth (A)	1:11.178
SAE J949	AUG81	Test Method for Determining Stiffness (Modulus of Bending) of Fiberboards	1:11.178
SAE J950	OCT85	Gradeability Test Code	Cancelled 1993.06
SAE J951	JAN85	Florida Exposure of Automotive Finishes	1:11.130
SAE J952	JAN77	Sound Levels for Engine Powered Equipment	Cancelled 1977
SAE J953	APR93	*Passenger Car Backlight Defogging System (A)	3:34.28
SAE J954		Urethane for Automotive Seating	3:34.99
SAE J955	APR75	Full Shielding of Power Drivelines for Agricultural Implements and Tractors	Cancelled 1975
SAE J956	APR85	Remote and Automatic Control Systems for Construction and Industrial Machinery	Cancelled 1985
SAE J957 ISO 6484-D		*Capacity Rating—Elevating Scrapers	3:40.322
SAE J958	JUN89	Nomenclature and Dimensions for Crane Shovels	3:40.53
SAE J959	MAY91	Lifting Crane, Wire-Rope Strength Factors (A)	3:40.217
SAE J960	JUN91	Marine Control Cable Connection—Engine Clutch Lever (A)	3:41.04
SAE J961	JUN91	Marine Control Cable Connection—Engine Throttle Lever (A)	3:41.05
SAE J962	MAY86	Formed Tube Ends for Hose Connections (A)	2:22.202
SAE J963	MAY79	Anthropomorphic Test Device for Use in Dynamic Testing of Motor Vehicles	Cancelled 1979
SAE J964	JUN92	Test Procedure for Determining Reflectivity of Rear View Mirrors	3:34.107
SAE J965	AUG66	Abrasive Wear	1:9.22
SAE HS-J965		Abrasive Wear	Cancelled (Superseded by J965)
SAE J966		Test Procedure for Measuring Passenger Car Tire Revolutions per Mile	3:30.36
SAE J967	SEP88	Calibration Fluid for Diesel Injection Equipment (A)	2:26.328
SAE J968/1	MAY91	Diesel Injection Pump Testing—Part 1: Calibrating Nozzle and Holder Assemblies (A)	2:26.309
SAE J968/2	MAY91	Diesel Injection Pump Testing—Part 2: Orifice Plate Flow Measurement (A)	2:26.318
SAE J969	MAY89	Testing Techniques for Diesel Fuel Injection Systems	Cancelled 1989
SAE J970	MAY89	Test Stands for Diesel Fuel Injection Systems	Cancelled 1989
SAE J971	JUN91	Brake Power Rating Test Code—Commercial Vehicle Inertia Dynamometer (A)	2:25.152
SAE J972	DEC88	Moving Rigid Barrier Collision Tests	3:34.278
SAE J973	JUN93	*Ignition System Measurements Procedure	2:23.59
SAE J974	JUN93	*Flashing Warning Lamp for Agricultural Equipment	3:39.59
SAE J975	JUN93	*Headlamps for Agricultural Equipment	3:39.59
SAE J976	JUN86	Combination Tail and Floodlamp for Agricultural Equipment	Cancelled 1986
SAE J977	JUN73	Instrumentation for Laboratory Impact Tests	Cancelled 1973 (Superseded by J211)
SAE J978		Bumper Jacking Test Procedure—Motor Vehicles	3:32.01
SAE J979		Bumper Jack Requirements—Motor Vehicles	3:32.01
SAE J980a		Bumper Evaluation Test Procedure—Passenger Cars	3:32.02
SAE J981	JUN87	Master Calibrating Fuel Injection Pump	Cancelled 1987
SAE J982	JUN90	Test Code—Truck, Truck-Tractor, and Trailer Air Service Brake System Pneumatic Pressure and Time Levels	Cancelled 1990
SAE J983	OCT80	Crane and Cable Excavator Basic Operating Control Arrangements (A)	3:40.220
SAE J984	OCT92	Bodyforms for Use in Motor Vehicle Passenger Compartment Impact Development	Cancelled 1992
SAE J985	OCT88	Vision Factors Considerations in Rear View Mirror Design	3:34.110
SAE J986	OCT88	Sound Level for Passenger Cars and Light Trucks (D)	1:14.14
SAE J987	APR85	Rope Supported Lattice-Type Boom Crane Structures—Method of Test	3:40.161
SAE J988	APR88	Labeling of Motor Vehicle Brake Fluid Containers	Cancelled 1988
SAE J989	APR87	Test Procedure for Measuring Carbon Monoxide Concentrations in Vehicle Passenger Compartments	Cancelled 1987
SAE J990	MAR87	Nomenclature—Industrial and Agricultural Mowers (A)	3:40.475
SAE J991	MAR82	Soil-Machine Terminology	Cancelled 1982

* Technical Revision   † New Document   (D)—DODISS Adopted   (A)—ANSI Recognized

Numerical Index

SAE J992 MAR85	Brake System Performance Requirements—Truck, Bus, and Combination of Vehicles (A)	3:36.164
SAE J993 JAN89	Alloy and Temper Designation Systems for Aluminum	1:10.02
SAE J994 MAR85	Alarm—Backup—Electric—Performance, Test, and Application (A)	1:14.45
SAE J995 JUN79	Mechanical and Material Requirements for Steel Nuts	1:4.15
SAE J996 JUN91	Inverted Vehicle Drop Test Procedure	Cancelled 1991
SAE J997 SEP90	Spark Arrester Test Carbon (A)	2:26.368
SAE J998 SEP86	Minimum Requirements for Motor Vehicle Brake Linings	Cancelled 1986
SAE J999 FEB85	Crane Boom Hoist Disengaging Device	3:40.216
SAE J1001 MAR87	Industrial Flail Mowers and Power Rakes (A)	3:40.488
SAE J1002 JUN87	Seals—Evaluation of Elastohydrodynamic	3:29.137
SAE J1003 JUN90	Diesel Engine Emission Measurement Procedure (A)	1:13.33
SAE J1004 APR93	*Glossary of Engine Cooling System Terms (A)	2:26.380
SAE J1006 APR87	Performance Test for Air Conditioned Agricultural Equipment	Cancelled 1987 (Superseded by J1503)
SAE J1008 JAN87	Sound Measurement—Self-Propelled Agricultural Equipment—Exterior (A)	1:14.113
SAE J1010 JAN83	Emission Control Hose	Cancelled 1983
SAE J1011 JAN74	Performance Criteria for Roll-Over Protective Structures for Rubber Tired, Off-Highway, Non-Trailed Hauling Units with Rear or Side Dump Bodies	Cancelled 1974 (Superseded by J1040)
SAE J1012 JUN93	*Operator Enclosure Pressurization System Test Procedure	3:39.67
SAE J1013 AUG92	*Measurement of Whole Body Vibration of the Seated Operator of Off-Highway Work Machines	3:39.92
SAE J1014 JAN85	Classification and Nomenclature Towing Winch for Skidders and Crawler Tractors	3:40.51
SAE J1015 MAR91	Ton-Kilometer Per Hour Test Procedure (A)	3:40.314
SAE J1016 JUL90	Component Nomenclature—Dumpers (A)	3:40.48
SAE J1017 JAN86	Nomenclature—Rollers and Compactors (A)	3:40.49
SAE J1018 JAN81	Recreational Trailer Vehicle Identification Number System	Cancelled 1981
SAE J1019 JUN90	Tests and Procedures for High Temperature Transmission Oil Hose, Engine Lubricating Oil Hose, and Hose Assemblies (A)	2:22.189
SAE J1024 DEC89	Fuel-Fired Heaters—Air Heating—For Construction and Industrial Machinery	3:40.496
SAE J1025	Test Procedures for Measuring Truck Tire Revolutions per Mile	3:30.36
SAE J1026 APR90	Braking Performance—Crawler Tractors and Crawler Loaders	3:40.281
SAE J1028 AUG89	Mobile Crane Working Area Definitions (A)	3:40.222
SAE J1029 MAR86	Lighting and Marking of Construction and Industrial Machinery	3:40.262
SAE J1030 FEB87	Maximum Sound Level for Passenger Cars and Light Trucks (A)	1:14.17
SAE J1032 APR87	Definitions for Machine Availability (Off-Road Work Machines) (A)	3:40.98
SAE J1033 APR93	*Procedure for Measuring Bore and Face Runout of Flywheels, Flywheel Housings and Flywheel Housing Adapters	3:40.245
SAE J1034 APR91	Automotive and Light Truck Engine Coolant Concentrate—Ethylene Glycol Type (A)	1:11.214
SAE J1035 APR86	Technical Publications for Agricultural Equipment	Cancelled 1986 (Superseded by J920)
SAE J1036 APR93	*Dimensional Standard for Cylindrical Hydraulic Couplers for Agricultural Tractors (A)	3:39.48
SAE J1037 DEC87	Windshield Washer Tubing (A)	1:11.254
SAE J1038 DEC92	*Recommendations for Children's Snowmobile	3:38.18
SAE J1039 DEC80	Size Classification for Crawler Tractors	Cancelled 1980
SAE J1040 APR88	Performance Criteria for Rollover Protective Structures (ROPS) for Construction, Earthmoving, Forestry, and Mining Machines (A)	3:40.356
SAE J1041 OCT91	Braking System Test Procedures and Braking Performance Criteria for Agricultural Tractors	3:39.29
SAE J1042 JUN93	*Operator Protection for General-Purpose Industrial Machines	3:40.363
SAE J1043 SEP87	Performance Criteria for Falling Object Protective Structure (FOPS) for Industrial Machines	3:40.365
SAE J1044 DEC81	World Manufacturer Identifier	3:27.05
SAE J1045 MAY93	*Instrumentation and Techniques for Vehicle Refueling Emissions Measurement	1:13.137
SAE J1046 MAY82	Exterior Sound Level Measurement Procedure for Small Engine Powered Equipment	Cancelled 1982
SAE J1047 JUN90	Tubing—Motor Vehicle Brake System Hydraulic (A)	2:25.162
SAE J1048 MAR80	Symbols for Motor Vehicle Controls, Indicators and Tell-Tales	3:34.288
SAE J1049 MAR90	Service Performance Requirements and Test Procedures for Motor Vehicle Lamp Bulbs	Cancelled 1990
SAE J1050a	Describing and Measuring the Driver's Field of View	3:34.214
SAE J1051 DEC88	Force-Deflection Measurements of Cushioned Components of Seats for Off-Road Work Machines	3:40.388
SAE J1052 MAY87	Motor Vehicle Driver and Passenger Head Position (A)	3:34.231
SAE J1053	Steel Stamped Nuts of One Pitch Thread Design	2:16.14
SAE J1054 OCT89	Warning Lamp Alternating Flashers (A)	2:24.166
SAE J1055 OCT87	Service Performance Requirements for Turn Signal Flashers	Cancelled 1987
SAE J1056 OCT87	Service Performance Requirements for Vehicular Hazard Warning Flashers	Cancelled 1987
SAE J1057 SEP88	Identification Terminology of Earthmoving Machines (A)	3:40.05
SAE J1058 APR91	Sheet Steel Thickness and Profile (A)	1:3.19
SAE J1059 JUN84	Speedometer Test Procedure (A)	2:21.08
SAE J1060	Subjective Rating Scale for Evaluation of Noise and Ride Comfort Characteristics Related to Motor Vehicle Tires	3:30.37
SAE J1061 OCT92	*Surface Discontinuities on General Application Bolts, Screws, and Studs	1:4.33
SAE J1062 OCT84	Snowmobile Passenger Handgrips (A)	3:38.09
SAE J1063 OCT80	Cantilevered Boom Crane Structures—Method of Test (A) (D)	3:40.170
SAE J1065 MAR92	Pressure Ratings for Hydraulic Tubing and Fittings (D)	2:22.215
SAE HS-J1066 MAR87	Recommended Guidelines for Company Metrication Programs in the Metalworking Industry	Cancelled 1987

* Technical Revision † New Document (D)—DODISS Adopted (A)—ANSI Recognized

Numerical Index

Document	Title	Reference
SAE J1067	Seven Conductor Jacketed Cable for Truck Trailer Connections	2:23.178
SAE J1069 JUN81	Oil Change System for Quick Service of Off-Road Self-Propelled Work Machines (A)	3:40.102
SAE J1071 JUN85	Operator Controls for Graders (A)	3:40.470
SAE J1072	Sintered Tool Materials	1:7.12
SAE J1073 JUN90	Spring-Loaded Clutch Spin Test Procedure (A)	2:26.242
SAE J1074 FEB87	Engine Sound Level Measurement Procedure (A) (D)	1:14.114
SAE J1075 JUN93	*Sound Measurement—Construction Site (A)	1:14.99
SAE J1076 MAR90	Backup Lamp Switch	2:24.175
SAE J1077 MAR87	Measurement of Exterior Sound Level of Trucks with Auxiliary Equipment	Cancelled 1987
SAE J1078 APR86	A Recommended Method of Analytically Determining the Competence of Hydraulic Telescopic Cantilevered Crane Booms	3:40.196
SAE J1079 SEP88	Overcenter Clutch Spin Test Procedure	3:40.243
SAE J1080 SEP76	Minimum Performance Criteria for Braking Systems for Rubber Tired Cranes and Excavators	Cancelled 1976 (Superseded by J1152, J1093)
SAE J1081 DEC88	Potential Standard Steels	1:1.18
SAE J1082 JAN89	Fuel Economy Measurement Road Test Procedure	2:26.459
SAE J1083 JUL85	Unauthorized Starting or Movement of Machines (A)	3:40.96
SAE J1084 APR80	Operator Protective Structure Performance Criteria for Certain Forestry Equipment (A)	3:40.370
SAE J1085a	Test for Dynamic Properties of Elastomeric Isolators	1:11.50
SAE J1086 DEC92	*Numbering Metals and Alloys	1:1.01
SAE HS-1086 DEC93	Fifth Edition Unified Numbering System Handbook for Metals and Alloys (D)	Available as a separate publication (See Related Technical Reports Section)
SAE J1087 OCT90	One-Way Clutches—Nomenclature and Terminology	3:29.93
SAE J1088 FEB93	*Test Procedure for the Measurement of Gaseous Exhaust Emissions from Small Utility Engines (A)	1:13.151
SAE J1089 FEB81	Lateral Impact Test Procedure for Vehicle Interiors	Cancelled 1981
SAE J1090	Preparation of Diesel Engines in Construction and Industrial Machinery for Operation in Cold Climate	Cancelled (Superseded by SP-346)
SAE J1092 JUN86	Nomenclature—Industrial Tractors (Wheel) (A)	3:40.31
SAE J1093 JUN82	Latticed Crane Boom Systems—Analytical Procedure	3:40.167
SAE HS-J1093	Latticed Crane Boom Systems—Analytical Procedure	Cancelled (Superseded by J1093)
SAE J1094 JUN92	Constant Volume Sampler System for Exhaust Emissions Measurement	1:13.117
SAE J1095 JAN91	Spoke Wheels and Hub Fatigue Test Procedures (A)	3:31.20
SAE J1096 FEB93	*Measurement of Exterior Sound Levels for Heavy Trucks Under Stationary Conditions	1:14.29
SAE J1097 NOV88	Hydraulic Excavator Lift Capacity Calculation and Test Procedure	3:40.334
SAE J1098 MAR91	Ton Kilometer Per Hour Application (A)	3:40.317
SAE J1099	Technical Report on Fatigue Properties	1:3.81
SAE J1100 JUN93	*Motor Vehicle Dimensions (A)	3:34.118
SAE J1101 JUN92	Test Procedure for Parking Stability of Motorcycles	3:37.08
SAE J1102	Mechanical and Material Requirements for Wheel Bolts	1:4.18
SAE J1104 JUN89	Service Performance Requirements for Warning Lamp Alternating Flashers	Cancelled 1989
SAE J1105 SEP89	Horn—Forward Warning—Electric—Performance, Test, and Application (A)	1:14.47
SAE J1106 SEP75	Laboratory Testing Machines for Measuring the Steady State Force and Moment Properties of Passenger Car Tires	3:30.40
SAE J1107 SEP75	Laboratory Testing Machines and Procedures for Measuring the Steady State Force and Moment Properties of Passenger Car Tires	3:30.43
SAE J1108 DEC81	Truck and Truck Tractor Vehicle Identification Number Systems	3:27.07
SAE J1109 JUN93	*Component Nomenclature—Articulated Log Skidder, Rubber-Tired (A)	3:40.62
SAE J1110 JUN93	*Specification Definitions—Articulated Rubber-Tired Log Skidder (A)	3:40.73
SAE J1111 MAR81	Component Nomenclature—Skidder-Grapple (A)	3:40.64
SAE J1112 APR80	Specification Definitions—Skidder-Grapple (A)	3:40.75
SAE J1113 AUG87	Electromagnetic Susceptibility Measurement Procedures for Vehicle Components (Except Aircraft) (A)	2:23.207
SAE J1114 NOV89	Fuel Tank Filler Cap and Cap Retainer—Threaded (A)	2:26.430
SAE J1115	Guidelines for Developing and Revising SAE Nomenclature and Definitions	1:A.35
SAE J1116 JUN86	Categories of Off-Road Self-Propelled Work Machines (A)	3:40.01
SAE J1117 MAR86	Method of Measuring and Reporting the Pressure Differential-Flow Characteristics of a Hydraulic Fluid Power Valve (A)	3:40.120
SAE J1118 JUN93	*Hydraulic Valves for Motor Vehicle Brake Systems Test Procedure	2:25.35
SAE J1119 DEC88	Steel Products for Rollover Protective Structures (ROPS) and Falling Object Protective Structures (FOPS)	3:40.371
SAE J1120 JUN89	Spherical Rod Ends	2:17.14
SAE J1121 JUL88	Helical Compression and Extension Spring Terminology (A)	2:20.17
SAE J1122 JUL88	Helical Springs: Specification Check Lists (A)	2:20.27
SAE J1123 NOV92	*Leaf Springs for Motor Vehicle Suspension—Made to Metric Units	2:20.01
SAE J1124 MAR87	Glossary of Terms Related to Fluid Filters and Filter Testing (A)	2:26.342
SAE J1127 JUN88	Battery Cable (A)	2:23.158
SAE J1128 JUN88	Low Tension Primary Cable (A)	2:23.162
SAE J1129 JUN87	Operator Cab Environment for Heated, Ventilated, and Air Conditioned Construction and Industrial Equipment	Cancelled 1987 (Superseded by J1503)
SAE J1130 MAY83	Determination of Emissions from Gas Turbine Powered Light Duty Surface Vehicles (A)	1:13.156
SAE J1131 DEC87	Performance Requirements for SAE J844 Nonmetallic Tubing and Fitting Assemblies Used in Automotive Air Brake Systems (A)	2:22.231
SAE J1132 DEC92	142 mm X 200 mm Sealed Beam Headlamp Unit	Cancelled 1992
SAE J1133 JUL89	School Bus Stop Arm (A)	2:24.196

* Technical Revision   † New Document   (D)—DODISS Adopted   (A)—ANSI Recognized

Numerical Index

SAE J1134 FEB92	SAE Nodal Mount	3:36.112
SAE J1135 FEB86	Turbocharger Connections	Cancelled 1986
SAE J1136 APR87	Braking Performance—Roller/Compactors (A)	3:40.300
SAE J1137 MAR85	Hydraulic Valves for Motor Vehicle Brake Systems—Performance Requirements	2:25.39
SAE J1138	Design Criteria—Driver Hand Controls Location for Passenger Cars, Multi-Purpose Passenger Vehicles, and Trucks (10 000 GVW and Under)	3:34.191
SAE J1139	Supplemental Information—Driver Hand Controls Location for Passenger Cars, Multi-Purpose Passenger Vehicles, and Trucks (10 000 GVW and Under)	3:34.193
SAE J1140 FEB88	Filler Pipes and Openings of Motor Vehicle Fuel Tanks (A)	2:26.433
SAE J1141 JUL87	Air Cleaner Elements	2:26.213
SAE J1142 JUN91	Towability Design Criteria and Equipment Use—Passenger Cars, Vans, and Light Duty Trucks (A)	3:42.01
SAE J1143 JUN91	Towed Vehicle/Tow Equipment Attachment Test Procedure—Passenger Cars, Vans, and Light Duty Trucks (A)	3:42.12
SAE J1144 JUN91	Towed Vehicle Drivetrain Test Procedure—Passenger Cars, Vans, and Light Duty Trucks (A)	3:42.16
SAE J1145 FEB93	*Emissions Terminology and Nomenclature	1:13.166
SAE J1146 JUN86	The Automotive Lubricant Performance and Service Classification Maintenance Procedure (A)	1:12.87
SAE J1147 JUN83	Welding, Brazing, and Soldering—Materials and Practices	1:9.07
SAE J1148 JUN90	Engine Charge Air Cooler Nomenclature (A)	2:26.392
SAE J1149 JUN91	Metallic Air Brake System Tubing and Pipe (A)	2:22.219
SAE J1150 OCT92	*Terminology for Agricultural Equipment	3:39.01
SAE J1151 DEC91	Methane Measurement Using Gas Chromatography	1:13.170
SAE J1152 APR80	Braking Performance—Rubber-Tired Construction Machines (A)	3:40.289
SAE J1153 JUN91	Hydraulic Master Cylinders for Motor Vehicle Brakes—Test Procedure (A)	2:25.29
SAE J1154 JUN91	Hydraulic Master Cylinders for Motor Vehicle Brakes—Performance Requirements (A)	2:25.34
SAE J1155 JUN88	Displacement of Pneumatic Service Brake Actuators	Cancelled 1988
SAE HS-J1156	Automotive Resistance Spot Welding Electrodes (AWS D8.6-77)	Available only from American Welding Society 550 N.W. LeJeune Rd. Miami, FL 33135 (305) 443-9353
SAE J1157 JUN88	Measurement Procedure for Evaluation of Full-Flow, Light Extinction Smokemeter Performance	Cancelled 1988
SAE J1158 JAN85	Specification Definitions—Winches for Crawler Tractors and Skidders	3:40.79
SAE J1159 JAN92	Preparation of SAE Technical Reports—Surface Vehicles and Machines: Standards, Recommended Practices, Information Reports	Cancelled 1992 (Superseded by TSB 002)
SAE J1160 MAR83	Operator Ear Sound Level Measurement Procedure for Snow Vehicles (A)	1:14.37
SAE J1161 MAR83	Operational Sound Level Measurement Procedure for Snow Vehicles (A)	1:14.39
SAE J1163 JUN91	Determining Seat Index Point (A)	3:40.391
SAE J1164 JAN91	Labeling of ROPS and FOPS and OPS (A)	3:40.362
SAE J1165 MAR86	Reporting Cleanliness Levels of Hydraulic Fluids (A)	3:40.124
SAE J1166 MAY90	Sound Measurement—Off-Road Self-Propelled Work Machines—Operator—Work Cycle (A)	1:14.55
SAE J1167 JUN89	Motorcycle Stop Lamp Switch	3:37.07
SAE J1168 MAY89	Motorcycle Bank Angle Measurement Procedure	3:37.07
SAE J1169 MAR92	Measurement of Light Vehicle Exhaust Sound Level Under Stationary Conditions	1:14.21
SAE J1170 FEB93	*Front and Rear Power Take-Off for Agricultural Tractors (A)	3:39.49
SAE J1171 JAN86	External Ignition Protection of Marine Electrical Devices	3:41.09
SAE J1172 MAY93	*Engine Flywheel Housings with Sealed Flanges	2:26.242
SAE J1173 SEP88	Size Classification and Characteristics of Glass Beads for Peening (A)	1:8.30
SAE J1174 MAR85	Operator Ear Sound Level Measurement Procedure for Small Engine Powered Equipment (A)	1:14.95
SAE J1175 MAR85	Bystander Sound Level Measurement Procedure for Small Engine Powered Equipment (A)	1:14.97
SAE J1176 MAR86	External Leakage Classifications for Hydraulic Systems (A)	3:40.146
SAE J1177 JUN88	Hydraulic Excavator Operator Controls	3:40.341
SAE J1178 JUN87	Braking Performance—Rubber-Tired Skidders (A)	3:40.97
SAE J1179 FEB90	Hydraulic Excavator and Backhoe Digging Forces (A)	3:40.340
SAE J1180 OCT80	Telescopic Boom Length Indicating System (A)	3:40.213
SAE J1183 FEB85	Recommended Guidelines for Fatigue Testing of Elastomeric Materials and Components	1:11.35
SAE J1184 FEB86	Definitions of Acoustical Terms	Cancelled 1986
SAE HS-J1188	Specification for Automotive Weld Quality—Resistance Spot Welding (AWS D8.7-78)	Available only from American Welding Society 550 N.W. LeJeune Rd. Miami, FL 33135 (305) 443-9353
SAE J1191 JAN86	High Tension Ignition Cable Assemblies—Marine (A)	3:41.26
SAE J1193 NOV84	Nomenclature and Dimensions for Hydraulic Excavators	3:40.59
SAE J1194 MAY89	Rollover Protective Structures (ROPS) for Wheeled Agricultural Tractors (A)	3:39.71
SAE J1195 MAR86	Cylinder Rod Wiper Seal Ingression Test (A)	3:40.141
SAE HS-J1196	Specification for Automotive Frame Weld Quality—Arc Welding (AWS D8.8-79)	Available only from American Welding Society 550 N.W. LeJeune Rd. Miami, FL 33135 (305) 443-9353
SAE J1197 FEB91	Rated Operating Load for Loaders Equipped with Log or Material Forks without Vertical Mast (A)	3:40.101
SAE J1199 SEP83	Mechanical and Material Requirements for Metric Externally Threaded Steel Fasteners (A)	1:4.06
SAE J1200	Blind Rivets—Break Mandrel Type	2:16.39

* Technical Revision † New Document (D)—DODISS Adopted (A)—ANSI Recognized

Numerical Index

Document	Title	Reference
SAE J1201 SEP90	Piston Rings and Grooves (Metric)	Cancelled 1990
SAE J1203	Light Transmittance of Automotive Windshields Safety Glazing Materials	3:34.102
SAE J1204 DEC89	Wheels—Recreational and Utility Trailer Test Procedure (A)	3:31.08
SAE J1205	Performance Requirements for Snap-In Tubeless Tire Valves	3:30.40
SAE J1206	Methods for Testing Snap-In Tubeless Tire Valves	3:30.38
SAE J1207 FEB87	Measurement Procedure for Determination of Silencer Effectiveness in Reducing Engine Intake or Exhaust Sound Level (A)	1:14.115
SAE J1209 JAN85	Identification Terminology of Mobile Forestry Machines	3:40.25
SAE J1211	Recommended Environmental Practices for Electronic Equipment Design	2:23.228
SAE J1212 JAN85	Fire Prevention on Forestry Equipment	3:40.95
SAE J1213 NOV82	Glossary of Automotive Electronic Terms	2:23.01
SAE J1213/1 JUN91	Glossary of Vehicle Networks for Multiplexing and Data Communications (A)	2:23.19
SAE J1213/2 OCT88	Glossary of Reliability Terminology Associated with Automotive Electronics	2:23.22
SAE HS-J1213	Glossary of Automotive Electronic Terms	Cancelled (Superseded by J1213)
SAE J1214 JAN92	Tire to Body Clearance Check for Recreational Vehicles	3:35.04
SAE J1215	Performance Prediction of Roll-Over Protective Structures (ROPS) Through Analytical Methods	Cancelled 1993.06
SAE J1216	Test Methods for Metric Threaded Fasteners (A)	1:4.10
SAE J1220 JAN87	Rotary-Trochoidal Engine Nomenclature and Terminology	Cancelled 1987
SAE J1221 JAN88	Headlamp—Turn Signal Spacing	Cancelled 1988
SAE J1222 OCT84	Speed Control Assurance for Snowmobiles (A)	3:38.17
SAE J1223 JUN93	*Marine Carburetors and Fuel Injection Throttle Bodies	3:41.28
SAE J1224 JUN86	Braking Performance—New Off-Highway Dumpers	Cancelled 1986 (Superseded by J1473)
SAE J1225 JUN87	Development of a Frequency Weighted Portable Ride Meter	Cancelled 1987
SAE J1226 FEB83	Electric Speedometer Specification—On Road (A)	2:21.09
SAE J1227 MAR86	Assessing Cleanliness of Hydraulic Fluid Power Components and Systems (A)	3:40.104
SAE J1228 NOV91	Small Craft—Marine Propulsion Engine and Systems—Power Measurements and Declarations	3:41.12
SAE J1229 DEC81	Truck Identification Terminology (A)	3:27.04
SAE J1230 OCT79	Minimum Requirements for Wheel Slip Brake Control System Malfunction Signals	2:25.162
SAE J1231 JUN93	*Beaded Tube Hose Fittings	2:22.158
SAE J1232	Passenger and Light Truck Tire Traction Device Profile Determination and Classification	3:36.115
SAE J1233 JAN85	Commercial Literature Specifications—Off-Road Work Machines (A)	3:40.28
SAE J1234 JAN85	Specification Definitions—Off-Road Work Machines (A)	3:40.22
SAE J1235 MAR86	Measuring and Reporting the Internal Leakage of a Hydraulic Fluid Power Valve (A)	3:40.123
SAE J1236 APR93	*Cast Iron Sealing Rings (Metric)	3:29.156
SAE J1237	Metric Thread Rolling Screws	1:4.29
SAE J1238 OCT78	Rating Lift Cranes on Fixed Platforms Operating in the Ocean Environment	3:40.206
SAE J1239	Four- and Eight-Conductor Rectangular Electrical Connectors for Automotive Type Trailers	2:23.173
SAE J1240 DEC91	Flywheel Spin Test Procedure	2:26.242
SAE J1241	Fuel and Lubricant Tanks for Motorcycles	3:37.11
SAE J1242 MAR91	Acoustic Emission Test Methods (A)	1:3.80
SAE J1243 MAY88	Diesel Emission Production Audit Test Procedure (A)	1:13.50
SAE J1244 AUG88	Oil Cooler Nomenclature and Glossary (A)	2:26.405
SAE J1245 JUN82	Guide to the Application and Use of Engine Coolant Pump Face Seals (A)	2:26.417
SAE J1246 MAR90	Measuring the Radius of Curvature of Convex Mirrors (A)	3:34.109
SAE J1247 APR80	Simulated Mountain Brake Performance Test Procedure	2:25.123
SAE J1248 JUN91	Performance Requirements for Parking Stability of Motorcycles (A)	3:37.09
SAE J1249 JAN89	Former SAE Standard and Former SAE EX-Steels	1:1.19
SAE J1250 NOV92	*In-Service Brake Performance Test Procedure—Vehicles Over 4500 kg (10 000 lb)	2:25.86
SAE J1252 JUL81	SAE Wind Tunnel Test Procedure for Trucks and Buses (A)	3:36.61
SAE J1253 JUN93	*Low-Temperature Cranking Load Requirements of an Engine	2:23.81
SAE J1254 JAN85	Component Nomenclature—Feller Buncher	3:40.63
SAE J1255 JAN85	Specification Definitions—Feller/Buncher	3:40.93
SAE J1256 OCT88	Fuel Economy Measurement—Road Test Procedure—Cold Start and Warm-Up Fuel Economy	2:26.506
SAE J1257 OCT80	Rating Chart for Cantilevered Boom Cranes	3:40.176
SAE J1258 OCT80	Automotive Hydraulic Brake System Metric Connections	Cancelled 1980 (Superseded by J1290)
SAE J1259 JUN89	Metric Spherical Rod Ends	2:17.16
SAE J1260 APR89	Standard Oil Filter Test Oil	2:26.350
SAE J1261 NOV84	Agricultural Tractor Tire Dynamic Indices	3:39.28
SAE J1262 OCT90	Sound Measurement—Trenching Machines	1:14.109
SAE J1263 JUN91	Road Load Measurement and Dynamometer Simulation Using Coastdown Techniques	2:26.470
SAE J1264 OCT86	Joint RCCC/SAE Fuel Consumption Test Procedure (Short Term In-Service Vehicle) Type I (A)	3:36.06
SAE J1265 MAR88	Capacity Rating—Dozer Blades (A)	3:40.236
SAE J1266 JUN90	Axle Efficiency Test Procedure (A)	3:29.85
SAE J1267 DEC88	Leakage Testing	1:3.73
SAE J1268 JUN93	*Hardenability Bands for Carbon and Alloy H Steels	1:1.48
SAE J1269 MAR87	Rolling Resistance Measurement Procedure for Passenger Car, Light Truck, and Highway Truck and Bus Tires (A)	3:30.18
SAE J1270 MAR87	Measurement of Passenger Car, Light Truck, and Highway Truck and Bus Tire Rolling Resistance	3:30.21
SAE J1271 MAR92	Technical Committee Guideposts	Cancelled 1992 (Superseded by TSB 004)

* Technical Revision   † New Document   (D)—DODISS Adopted   (A)—ANSI Recognized

Numerical Index

Document	Date	Title	Reference
SAE J1272	JAN85	Felling Head Terminology and Nomenclature	3:40.77
SAE J1273	NOV91	Selection, Installation, and Maintenance of Hose and Hose Assemblies	2:22.184
SAE J1274	NOV91	Performance Requirements for the Automotive Audio Cassette	Cancelled 1991
SAE J1275	NOV91	Testing Methods for Audio Cassettes	Cancelled 1991
SAE J1276	MAR86	Standardized Fluid for Hydraulic Component Tests (A)	3:40.124
SAE J1277	JUL90	Method for Assessing the Cleanliness Level of New Hydraulic Fluid (A)	3:40.127
SAE J1278	MAR93	*SI (Metric) Synchronous Belts and Pulleys	2:19.09
SAE J1279	DEC88	Snowmobile Drive Mechanisms	3:38.18
SAE J1280	JUN92	Determination of Sulfur Compounds in Automotive Exhaust	1:13.72
SAE J1281	MAR85	Operator Sound Level Exposure Assessment Procedure for Pleasure Motorboats (A)	1:14.43
SAE J1282	OCT84	Snowmobile Brake Control Systems (A)	3:38.09
SAE J1283	MAR86	Electrical Connector for Auxiliary Starting of Construction, Agricultural, and Off-Road Machinery (A)	3:40.267
SAE J1284	APR88	Blade Type Electric Fuses (A)	3:41.40
SAE J1285	JAN85	Powershift Transmission Fluid Classification	1:12.45
SAE J1286	APR87	Thrust Test Device	3:41.05
SAE J1287	JUN93	*Measurement of Exhaust Sound Levels of Stationary Motorcycles	1:14.05
SAE J1288	MAR90	Packaging, Storage, and Shelf Life of Hydraulic Brake Hose Assemblies	2:25.13
SAE J1289	APR81	Mobile Crane Stability Ratings (A)	3:40.223
SAE J1290	MAY89	Automotive Hydraulic Brake System—Metric Tube Connections	2:25.165
SAE J1291	MAY89	Automotive Hydraulic Brake System—Metric Banjo Bolt Connections	2:25.170
SAE J1292	OCT81	Automobile, Truck, Truck-Tractor, Trailer, and Motor Coach Wiring	2:23.142
SAE J1293	JAN90	Undervehicle Coupon Corrosion Tests	1:3.48
SAE J1294	JAN86	Ignition Distributors—Marine	3:41.19
SAE J1295	JUN89	Identification Terminology and Specification Definitions—Pipelayers and Side Booms, Tractor or Loader Mounted	3:40.55
SAE J1297	MAR93	*Alternative Automotive Fuels	1:12.74
SAE J1298	SEP88	Hydraulic Systems Diagnostic Port Sizes and Locations (A)	3:40.110
SAE J1299	JAN91	Electrical Propulsion Control—Off-Road Dumpers (A)	3:40.282
SAE J1303	FEB85	Cutting Edge—Cross Sections Loader Straight	3:40.239
SAE J1304	FEB85	Cutting Edge—Cross Sections Loader Straight with Bolt Holes (A)	3:40.241
SAE J1305	JUN87	Two-Block Warning and Limit Systems in Lifting Crane Service	3:40.216
SAE J1306	JUN89	Motorcycle Auxiliary Front Lamps	2:24.122
SAE J1307	NOV86	Excavator and Backhoe Hand Signals (A)	3:40.348
SAE J1308	SEP85	Fan Guard for Off-Road Machines	3:40.257
SAE J1309	MAY91	Crawler Mounted Hydraulic Excavator Travel Performance (A)	3:40.328
SAE J1310	JUN93	*Electric Engine Preheaters and Battery Warmers for Diesel Engines	3:40.500
SAE J1312	JAN90	Procedure for Mapping Engine Performance—Spark Ignition and Compression Ignition Engines (A)	2:26.22
SAE J1313	MAR93	*Automotive Synchronous Belt Drives (A)	2:19.13
SAE J1315	JAN91	Off-Road Tire and Rim Selection and Application (A)	3:40.307
SAE J1317	JUN82	Electrical Propulsion Rotating Equipment—Off-Road Dumper (A)	3:40.279
SAE J1318	APR86	Gaseous Discharge Warning Lamp for Authorized Emergency, Maintenance, and Service Vehicles (A)	2:24.188
SAE J1319	JUN93	*Fog Tail Lamp (Rear Fog Light) Systems (A)	2:24.141
SAE J1320	APR87	Marine Electrical Switches	3:41.28
SAE J1321	OCT86	Joint TMC/SAE Fuel Consumption Test Procedure—Type II (A)	3:36.11
SAE J1322	JUN85	Preferred Conversion Values for Dimensions in Lighting—Inch-Pound Units/SI	2:24.01
SAE J1323	SEP90	Standard Classification System for Fiberboards (A)	1:11.161
SAE J1324	OCT89	Acoustical and Thermal Materials Test Procedure (A)	1:11.111
SAE J1325	FEB85	Test Method for Measuring the Relative Drapeability of Flexible Insulation Materials	1:11.169
SAE J1326	FEB85	Test Method for Measuring Wet Color Transfer Characteristics (A)	1:11.169
SAE J1327	FEB80	Fuel Fired Engine Preheaters and Their Application	Cancelled (Superseded by J1350)
SAE J1329	JUL89	Minimum Performance Criteria for Braking Systems for Specialized Rubber-Tired, Self-Propelled Underground Mining Machines (A)	3:40.295
SAE J1330	JUN88	Photometry Laboratory Accuracy Guidelines	2:24.04
SAE J1332	OCT81	Rope Drum Rotation Indicating Device (A)	3:40.229
SAE J1333	MAR90	Hydraulic Cylinder Rod Corrosion Test (A)	3:40.343
SAE J1334	JUN87	Hydraulic Cylinder Integrity Test	3:40.344
SAE J1335	APR90	Hydraulic Cylinder No-Load Friction Test (A)	3:40.346
SAE J1336	JUN87	Hydraulic Cylinder Leakage Test	3:40.346
SAE J1338	JUN81	Open Field Whole-Vehicle Radiated Susceptibility 10 kHz–18 GHz, Electric Field	2:23.447
SAE J1339	AUG89	Test Method for Measuring Power Consumption of Truck and Bus Engine Cooling Fans (A)	3:36.52
SAE J1340	APR90	Test Method for Measuring Power Consumption of Air Conditioning and Brake Compressors for Trucks and Buses (A)	3:36.53
SAE J1341	APR90	Test Method for Measuring Power Consumption of Hydraulic Pumps for Trucks and Buses (A)	3:36.54
SAE J1342	AUG89	Method for Determining Power Consumption of Engine Cooling Fan-Drive Systems (A)	3:36.55
SAE J1343	OCT88	Information Relating to Duty Cycles and Average Power Requirements of Truck and Bus Engine Accessories	3:36.57
SAE J1344	APR93	*Marking of Plastic Parts	1:11.131
SAE J1345	FEB82	Automotive Plastic Parts Specification (A)	1:11.148
SAE J1346	JUN81	Guide to Manifold Absolute Pressure Transducer Representative Test Method	2:23.510
SAE J1347	JUN81	Guide to Manifold Absolute Pressure Transducer Representative Specification	2:23.512
SAE J1349	JUN90	Engine Power Test Code—Spark Ignition and Compression Ignition—Net Power Rating	2:26.16

* Technical Revision † New Document (D)—DODISS Adopted (A)—ANSI Recognized

Numerical Index

Document	Title	Reference
SAE J1350 MAR88	Selection and Application Guidelines for Diesel, Gasoline, and Propane Fired Liquid Cooled Engine Pre-Heaters	3:40.496
SAE J1351 JUN93	*Hot Odor Test for Insulation Materials	1:11.172
SAE J1352 APR87	Compression and Recovery of Insulation Paddings	1:11.172
SAE J1353 APR87	Nomenclature—Clam Bunk Skidder (A)	3:40.80
SAE J1354 APR87	Nomenclature—Forwarder (A)	3:40.88
SAE J1355 MAY93	*Test Method for Measuring Thickness of Resilient Insulating Paddings	1:11.170
SAE J1356 FEB88	Performance Criteria for Falling Object Guards for Excavators	3:40.367
SAE J1360 SEP87	Product Identification Numbering System of Off-Road Work Machines (A)	3:40.04
SAE J1361 APR87	Hot Plate Method for Evaluating Heat Resistance and Thermal Insulation Properties of Materials	1:11.171
SAE J1362 JUN92	Graphical Symbols for Operator Controls and Displays on Off-Road Self-Propelled Work Machines	3:40.423
SAE J1363 JAN85	Capacity Rating—Dumper Body and Trailer Body (A)	3:40.321
SAE J1367 SEP81	Performance Test Procedure—Ball Joints and Spherical Rod Ends (A)	2:17.18
SAE J1368 JUN93	*Child Restraint Anchorages and Attachment Hardware (A)	3:33.14
SAE J1369 MAR92	Anchorage Provisions for Installation of Child Restraint Tether Straps in Rear Seating Positions	3:33.16
SAE J1371 JUN93	*Hydraulic Excavator Swing Minimum Performance and Rating Procedure	3:40.342
SAE J1372 JUL88	Sound Power Determination—Earthmoving Machinery—Static Condition (A)	1:14.102
SAE J1373 OCT87	Rear Cornering Lamp (A)	2:24.163
SAE J1374 MAY85	Hydraulic Cylinder Rod Seal Endurance Test Procedure (A)	3:40.143
SAE J1375 MAR93	*Cranking Motor Application Considerations (A)	2:23.81
SAE J1376 JUL82	Fuel Economy Measurement Test (Engineering Type) for Trucks and Buses (A)	3:36.31
SAE J1377 JAN89	Transmission Mounted Vehicle Speed Signal Rotor Specification	2:21.17
SAE J1378 MAR83	Electric Hourmeter Specification	2:21.14
SAE J1379 MAR88	Rolling Resistance Measurement Procedure for Highway Truck and Bus Tires	Cancelled 1988
SAE J1380 MAR88	The Measurement of Highway Truck and Bus Tire Rolling Resistance	Cancelled 1988
SAE J1382 JUN93	*Classification, Nomenclature, and Specification Definitions for Trenching Machines	3:40.41
SAE J1383 JUN90	Performance Requirements for Motor Vehicle Headlamps (A)	2:24.36
SAE J1384 JUN93	*Vibration Performance Evaluation of Operator Seats (A)	3:40.396
SAE J1385 JUN83	Classification of Earthmoving Machines for Vibration Tests of Operator Seats (A)	3:40.398
SAE J1386 JAN86	Classification of Agricultural Wheeled Tractors for Vibration Tests of Operator Seats (A)	3:40.401
SAE J1388 JUN85	Personnel Protection—Skid Steer Loaders (A)	3:40.474
SAE J1389 MAY93	*Corrosion Test for Insulation Materials	1:11.173
SAE J1390 APR82	Engine Cooling Fan Structural Analysis (A)	2:26.400
SAE J1392 JUN84	Steel, High Strength, Hot Rolled Sheet and Strip, Cold Rolled Sheet, and Coated Sheet	1:1.159
SAE J1393 JUN84	On-Highway Truck Cooling Test Code (A)	2:26.388
SAE J1394 APR91	Metric Nonmetallic Air Brake System Tubing (A)	2:22.227
SAE J1395 JUN91	Front and Rear Turn Signal Lamps for Use on Motor Vehicles 2032 mm or More in Overall Width (A)	2:24.156
SAE J1397 MAY92	Estimated Mechanical Properties and Machinability of Steel Bars	1:2.26
SAE J1398 JUN91	Stop Lamps for Use on Motor Vehicles 2032 mm or More in Overall Width (A)	2:24.145
SAE J1399 JUN84	Electric Tachometer Specification	2:21.12
SAE J1400 MAY90	Laboratory Measurement of the Airborne Sound Barrier Performance of Automotive Materials and Assemblies (A)	1:11.174
SAE J1401 JUN93	*Road Vehicle—Hydraulic Brake Hose Assemblies for Use with Nonpetroleum-Base Hydraulic Fluids	2:25.01
SAE J1402 JUN85	Automotive Air Brake Hose and Hose Assemblies (A)	2:25.11
SAE J1403 JUL89	Vacuum Brake Hose (A)	2:25.14
SAE J1404 JUN93	*Service Brake Structural Integrity Requirements—Vehicles Over 10 000 lb (4500 kg) GVWR (A)	3:36.171
SAE J1405 JUN90	Optional Impulse Test Procedures for Hydraulic Hose Assemblies	2:22.188
SAE J1406 JUN93	*Application of Hydraulic Brake Hose to Motor Vehicles	2:25.16
SAE J1407 MAR88	Vehicle Electromagnetic Radiated Susceptibility Testing Using a Large TEM Cell	2:23.460
SAE J1409 JUN88	Air Brake Valves Test Procedure (A)	3:36.171
SAE J1410 OCT86	Air Brake Valve—Performance Requirements	3:36.173
SAE HS-1417	Fluid Sealing Handbook—Radial Lip Seals	Available as a separate publication (See Related Technical Reports Section)
SAE J1418 DEC87	Fuel Injection Pumps — High Pressure Pipes (Tubing) for Testing (A)	2:26.263
SAE J1419 FEB88	Tapers for Shaft Ends and Hubs for Fuel Injection Pumps (A)	2:26.245
SAE J1420 AUG88	Supportive Information Report for the Fuel Economy Measurement Test (Engineering Type) for Trucks and Buses	3:36.44
SAE J1422 NOV89	Fuel Warmer—Diesel Engines (A)	3:40.509
SAE J1423 FEB92	Classification of Energy-Conserving Engine Oil for Passenger Cars, Vans, and Light-Duty Trucks	1:12.29
SAE J1424 JUN93	*Cargo Lamps for Use on Vehicles Under 12 000 lb GVWR	2:24.120
SAE J1428 MAY85	Marine Circuit Breakers (A)	3:41.20
SAE J1430 FEB93	*Retardation Capability of Off-Highway Dumpers and Scrapers	3:40.298
SAE J1432 OCT88	High Mounted Stop Lamps for Use on Vehicles 2032 mm or More in Overall Width	2:24.147
SAE J1434 JAN89	Wrought Aluminum Applications Guidelines	1:10.15
SAE J1436 JUN89	Requirements for Engine Cooling System Filling, Deaeration and Drawdown Tests	2:26.379
SAE J1437 JUN84	Lead-Free Replacement Paints (A)	1:11.129
SAE J1440 APR86	Off-Road Tire and Rim Classification—Forestry Machines (A)	3:40.310
SAE J1441 FEB92	Subjective Rating Scale for Vehicle Handling	3:28.16
SAE J1442 DEC88	High Strength, Hot Rolled Steel Plates, Bars and Shapes	1:1.162

* Technical Revision   † New Document   (D)—DODISS Adopted   (A)—ANSI Recognized

Numerical Index

Document	Title	Reference
SAE J1444 JUN91	Procedure for Evaluating Transient Response of Small Engine Driven Generator Sets (A)	2:26.23
SAE J1446 MAY89	On-Machine Alarm Test and Evaluation Procedure for Construction and General Purpose Industrial Machinery	1:14.50
SAE J1447 MAY86	Fire-Resistant Fluid Usage in Hydraulic Systems of Off-Road Work Machines	3:40.126
SAE J1448 JAN84	Electromagnetic Susceptibility Measurements of Vehicle Components Using TEM Cells (14 kHz-200 MHz)	2:23.471
SAE J1449 FEB87	Graphic Symbols for Boats (A)	3:41.31
SAE J1450 MAY88	Air Brake Actuator Diaphragm Test Procedure (A)	3:36.166
SAE J1451 FEB85	A Dictionary of Terms for the Dynamics and Handling of Single Track Vehicles (Motorcycles, Mopeds, and Bicycles)	3:37.01
SAE HS-J1451	A Dictionary of Terms for the Dynamics and Handling of Single Track Vehicles (Motorcycles, Mopeds, and Bicycles)	Cancelled (Superseded by J1451)
SAE J1452 OCT88	Trailer Grade Parking Performance Test Procedure	2:25.155
SAE J1453 JUN93	*Fitting—O-Ring Face Seal	2:22.256
SAE J1454 JAN86	Dynamic Durability Testing of Seat Cushions for Off-Road Work Machines (A)	3:40.389
SAE J1455 JAN88	Joint SAE/TMC Recommended Environmental Practices for Electronic Equipment Design (Heavy-Duty Trucks)	2:23.556
SAE J1456 JUN90	Maximum Allowable Rotational Speed for Internal Combustion Engine Flywheels (A)	2:26.243
SAE J1459 AUG88	V-Ribbed Belts and Pulleys	2:19.15
SAE J1460 MAR85	Human Mechanical Response Characteristics	3:34.292
SAE J1461 APR91	Manual Slack Adjuster Test Procedure (A)	3:36.174
SAE J1462 MAY87	Automatic Slack Adjuster Test Procedure (A)	3:36.175
SAE J1463 NOV84	Pull-Type Clutch—Transmission Installation Dimensions (A)	3:36.120
SAE J1464 APR88	Identification Terminology of Loaders/Tractors with Forks and Rough Terrain Forklifts	3:40.58
SAE J1466 OCT85	Passenger Car and Light Truck Tire Dynamic Driving Traction in Snow (A)	3:30.26
SAE J1467 JUN93	*Clip Fastener Fitting	2:22.197
SAE J1468 MAY93	*Application Testing of Oil-to-Air Oil Coolers for Cooling Performance	2:26.398
SAE J1469 JUN88	Air Brake Actuator Test Procedure—Truck-Tractor, Bus and Trailer (A)	3:36.157
SAE J1470 MAR92	Measurement of Noise Emitted by Accelerating Highway Vehicles	1:14.32
SAE J1472 JUN87	Braking Performance—Roller Compactors (A)	3:40.288
SAE J1473 OCT90	Brake Performance—Rubber-Tired Earthmoving Machines (A)	3:40.285
SAE J1474 JAN85	Heavy Duty Non-Metallic Engine Cooling Fans—Material, Manufacturing and Test Considerations	2:26.403
SAE J1475 JUN84	Hydraulic Hose Fittings for Marine Applications (A)	3:41.39
SAE J1476 JUN87	Parking Brake Structural Integrity Test Procedure—Vehicles Over 10,000 Pounds (4500 KG) GVWR (A)	3:36.123
SAE J1477 JAN86	Measurement of Interior Sound Levels of Light Vehicles (A)	1:14.19
SAE J1479 APR91	Automotive Pull Type Clutch Terminology (A)	3:36.02
SAE J1487 MAY85	Rating Air Conditioner Evaporator Air Delivery and Cooling Capacities (A)	3:36.58
SAE J1488 MAY90	Emulsified Water/Fuel Separation Test Procedure	2:26.345
SAE J1489 JUN87	Heavy Truck and Bus Retarder Downhill Performance Mapping Procedure (A)	3:36.124
SAE J1490 JAN87	Measurement and Presentation of Truck Ride Vibrations (A)	3:34.252
SAE J1491 JUN90	Vehicle Acceleration Measurement	2:26.503
SAE J1492 MAR92	Measurement of Light Vehicle Stationary Exhaust System Sound Level Engine Speed Sweep Method	1:14.24
SAE J1493 OCT91	Shielding of Starter System Energization	3:40.266
SAE J1494 JUN89	Battery Booster Cables	2:23.159
SAE J1495 MAR92	Test Procedure for Battery Flame Retardant Venting Systems	2:23.72
SAE J1496 MAR92	Replaceable Bulbs for Headlamps	Cancelled 1992
SAE J1497 DEC88	Design Guide for Formed-In-Place Gaskets (A)	1:11.187
SAE J1498 MAY90	Heating Value of Fuels	1:12.82
SAE J1499 FEB87	Band Friction Test Machine (SAE) Test Procedure	3:29.99
SAE J1500 FEB87	Universal Symbols for Operator Controls	Cancelled 1987 (Superseded by J107, J389, J1048, J1362, J1449)
SAE J1502 APR86	Hydraulic Diagnostic Couplings (A)	3:40.110
SAE J1503 JUL86	Performance Test for Air-Conditioned, Heated, and Ventilated Off-Road Self-Propelled Work Machines (A)	3:40.511
SAE J1505 MAY85	Brake Force Distribution Test Code—Commercial Vehicles (A)	2:25.147
SAE J1506 APR93	*Emission Test Driving Schedules (A)	1:13.141
SAE J1507 JAN87	Anechoic Test Facility Radiated Susceptibility 20 MHz - 18 GHz Electromagnetic Field	2:23.477
SAE J1508 JUN93	*Hose Clamp Specifications	2:16.45
SAE J1510 OCT88	Lubricants for Two-Stroke-Cycle Gasoline Engines	1:12.32
SAE J1511 OCT90	Steering for Off-Road, Rubber-Tired Machines (A)	3:40.517
SAE J1512 APR90	Manual Slack Adjuster Performance Requirements (A)	3:36.176
SAE J1513 APR90	Automatic Slack Adjuster Performance Requirements (A)	3:36.176
SAE J1515 MAR88	Impact of Alternative Fuels on Engine Test and Reporting Procedures	1:12.89
SAE J1516 MAR90	Accommodation Tool Reference Point	3:34.247
SAE J1517 MAR90	Driver Selected Seat Position (A)	3:34.250
SAE J1518 JUN86	Three Point Hitch (Type A) Backhoe Personnel Protection	3:40.353
SAE J1521 MAR90	Truck Driver Shin-Knee Position for Clutch and Accelerator (A)	3:34.263
SAE J1522 MAR90	Truck Driver Stomach Position (A)	3:34.265
SAE J1523 JUN85	Metal to Metal Overlap Shear Strength Test for Automotive Type Adhesives (A)	1:11.78
SAE J1524 NOV88	Method of Viscosity Test for Automotive Type Adhesives, Sealers, and Deadeners	1:11.78
SAE J1525 JUN85	Lap Shear Test for Automotive Type Adhesives for Fiber Reinforced Plastic (FRP) Bonding (A)	1:11.81
SAE J1526 JUN87	Joint TMC/SAE Fuel Consumption In-Service Test Procedure Type III (A)	2:26.450
SAE J1527 FEB93	*Marine Fuel Hoses (A)	3:41.16
SAE J1528 JUN90	Fatigue Testing Procedure for Suspension-Leaf Springs (A)	2:20.08

* Technical Revision   † New Document   (D)—DODISS Adopted   (A)—ANSI Recognized

Numerical Index

SAE J1529 MAY86	Overlap Shear Test for Automotive Type Sealant for Stationary Glass Bonding (A)	1:11.95
SAE J1530 JUN85	Test Method for Determining Resistance to Abrasion; Bearding; and Fiber Loss of Automotive Carpet Materials (A)	1:11.107
SAE J1532 JUN93	*Transmission Oil Cooler Hose	1:11.252
SAE J1533 JUN93	*Operator Enclosure Air Filter Element Test Procedure	3:40.513
SAE J1536 OCT88	Two-Stroke-Cycle Engine Oil Miscibility/Fluidity Classification	1:12.33
SAE J1537 JUN90	Validation Testing of Electric Fuel Pumps for Gasoline Fuel Injection Systems (A)	2:26.291
SAE J1538 APR88	Glossary of Automotive Inflatable Restraint Terms	3:33.31
SAE J1540 APR92	Manual Transmission and Transaxle Efficiency and Parasitic Loss Measurement	3:29.61
SAE J1541 FEB88	Fuel Injection Nomenclature — Spark Ignition Engines (A)	2:26.290
SAE J1542 JAN89	Laboratory Testing of Vehicle and Industrial Heat Exchangers for Thermal Cycle Durability	2:26.382
SAE J1544 JAN88	Revolutions per Mile and Static Loaded Radius for Off-Road Tires	3:40.307
SAE J1545 JUN86	Instrumental Color Difference Measurement for Exterior Finishes, Textiles, and Colored Trim (A)	1:11.279
SAE J1547 OCT88	Electromagnetic Susceptibility Procedures for Common Mode Injection (1—400 MHz), Module Testing	2:23.458
SAE J1548 FEB93	*Drawbars—Agricultural Wheel Tractors	3:39.90
SAE J1549 APR88	Diesel Fuel Injection Pump—Validation of Calibrating Nozzle Holder Assemblies (A)	2:26.321
SAE J1553 MAY86	Cross Peel Test for Automotive Type Adhesives for Fiber Reinforced Plastic (FRP) Bonding (A)	1:11.100
SAE J1554 NOV85	Identifying and Repairing High Strength Steel Vehicle Components	3:42.18
SAE J1555 NOV85	Design Guidelines for Optimizing Automobile Collision Damage Resistance, Repairability, and Serviceability	3:42.19
SAE J1556 DEC85	Stationary Glass Replacement	3:42.21
SAE J1560 JAN92	Low Tension Thin Wall Primary Cable	2:23.167
SAE J1561 FEB93	*Laboratory Speed Test Procedure for Passenger Car Tires	3:30.34
SAE J1563 MAY87	Guidelines and Limitations of Laboratory Cyclic Corrosion Test Procedures for Exterior, Painted, Automotive Body Panels	1:3.51
SAE HS-1566	Aerodynamic Flow Visualization Techniques and Procedures	Available as a separate publication (See Related Technical Reports Section)
SAE J1567 AUG87	Collision Detection Serial Data Communications Multiplex Bus	2:23.481
SAE J1568 JUN93	† Materials for Plastic Pistons for Hydraulic Disc Brake Cylinders	2:25.18
SAE J1570 SEP91	Rubber Dust Boots for the Hydraulic Disc Brake Piston	2:25.181
SAE J1571 MAY87	Inspection of Energy Absorbing Bumper Mounts (A)	3:42.25
SAE J1573 FEB87	Flexible Bumper Repair	3:42.22
SAE HS-1576	Manual for Incorporating Pneumatic Springs in Vehicle Suspension Designs	Available as a separate publication (See Related Technical Reports Section)
SAE J1577 JUN91	Replaceable Motorcycle Headlamp Bulbs (A)	2:24.123
SAE J1578 JUN89	Motorcycle Side Stand Retraction Test Procedure	3:37.14
SAE J1579 JUN89	Motorcycle Side Stand Retraction Performance Requirements	3:37.17
SAE J1580 DEC89	Metric Countersunk Holes for Cutting Edges and End Bits (A)	3:40.238
SAE J1581 SEP89	Cutting Edge—Optional Cross-Sections and Dimensions Loader Straight (A)	3:40.242
SAE HS-1582	Manual on Design and Manufacture of Coned Disk Springs (Belleville Springs) and Spring Washers	Available as a separate publication (See Related Technical Reports Section)
SAE J1583 MAR90	Controller Area Network (CAN), An In-Vehicle Serial Communication Protocol	2:23.486
SAE J1587 AUG92	*Joint SAE/TMC Electronic Data Interchange Between Microcomputer Systems in Heavy-Duty Vehicle Applications	2:23.531
SAE J1588 OCT92	*Internal Combustion Engines—Piston Rings—Vocabulary	2:26.28
SAE J1589 OCT92	*Internal Combustion Engines—Piston Rings—Inspection Measuring Principles	2:26.52
SAE J1590 OCT92	*Internal Combustion Engines—Piston Rings—Material Specifications (A)	2:26.72
SAE J1591 OCT92	*Internal Combustion Engines—Piston Rings—General Specifications (A)	2:26.186
SAE J1594 JUN87	Vehicle Aerodynamics Terminology (A)	3:28.17
SAE J1595 OCT88	Electrostatic Discharge Test for Vehicles	2:23.464
SAE J1596 JUN89	Automotive V-Ribbed Belt Drives and Test Methods	2:19.17
SAE J1597 JUL88	Laboratory Testing of Vehicle and Industrial Heat Exchangers for Pressure Cycle Durability (A)	2:26.390
SAE J1598 JUL88	Laboratory Testing of Vehicle and Industrial Heat Exchangers for Durability Under Vibration Induced Loading (A)	2:26.389
SAE J1600 APR87	Method for Evaluating the Adhesion Characteristics of Automotive Sealers	1:11.82
SAE J1601 NOV90	Rubber Cups for Hydraulic Actuating Cylinders (A)	2:25.175
SAE J1603 JUN90	Rubber Seals for Hydraulic Disc Brake Cylinders (A)	2:25.19
SAE J1604 OCT89	Rubber Boots for Drum-Type Hydraulic Brake Wheel Cylinders (A)	2:25.24
SAE J1605 MAR92	Brake Master Cylinder Reservoir Diaphragm Gasket	2:25.22
SAE J1606 MAR93	† Headlamp Design Guidelines for Mature Drivers	3:34.225
SAE J1610 APR93	*Test Method for Evaluating the Sealing Capability of Hose Connections with a PVT Test Facility	2:16.80
SAE J1611 JUN93	† Trenchless Equipment - (Horizontal Earthboring Machines) Operator Control Definitions	3:40.21
SAE J1615 JUN93	† Thread Sealants	2:22.240
SAE J1626 MAR93	† Braking, Stability, and Control Performance Test Procedures for Air-Brake- Equiped Truck Tractors	3:36.130
SAE J1627 JUN93	† Rating Criteria for Electronic Refrigerant Leak Detectors	3:34.71

* Technical Revision † New Document (D)—DODISS Adopted (A)—ANSI Recognized

Numerical Index

Document	Date	Title	Reference
SAE J1628	JUN93	† Technican Procedure for Using Electronic Refrigerant Leak Detectors for Service of Mobile Air Conditioning Systems	3:34.72
SAE J1629	JUN93	† Cautionary Statements for Handling HFC-134a During Mobile Air Conditioning Service	3:34.60
SAE J1634	MAY93	† Electric Vehicle Energy Consumption and Range Test Procedure	3:28.01
SAE J1635	MAY93	† SAE Cold Start and Driveability Procedure	2:26.498
SAE J1636	FEB93	† Recommended Guidelines for Load/Deformation Testing of Elastomeric Components	1:11.46
SAE J1637	FEB93	† Laboratory Measurement of the Composite Vibration Damping Properties of Materials on a Supporting Steel Bar	1:2.30
SAE J1638	MAY93	† Compression Set of Hoses or Solid Discs	1:11.33
SAE J1639	MAR93	† Classification System for Automotive Polyamide (PA) Plastics	1:11.145
SAE J1644	MAY93	† Metallic Tube Connections for Fluid Power and General Use-Test Methods for Threaded Hydraulic Fluid Power Connectors	2:22.285
SAE J1649/1	JUN93	*Compressed Air for General Use—Part 1: Contaminants and Quality Classes	3:34.47
SAE J1650	MAR93	† Seamless Copper-Nickel 90-10 Tubing	2:22.211
SAE J1657	JUN93	† Selection Criteria for Retrofit Refrigerants to Replace R-12 in Mobile Air Conditioning Systems	3:34.61
SAE J658	JUN93	† Alternate Refrigerant Consistency Criteria for use in Mobile Air Conditioning Systems	3:34.63
SAE J659	JUN93	† Vehicle Testing Requirements for Replacement Refrigerants for CFC-12 (R12) Mobile Air Conditioning Systems	3:34.64
SAE J1660	JUN93	† Fittings and Labels for Retrofit of CFC-12 (R12) Mobile Air Conditioning Systems to HFC-134a (R134a)	3:34.66
SAE J1661	JUN93	† Procedure for Retrofitting CFC-12 (R12) Mobile Air Conditioning Systems to HFC-134a (R134a)	3:34.67
SAE J1662	JUN93	† Compatibility of Retrofit Refrigerants with Air Conditioning System Materials	3:34.69
SAE J1666	MAY93	† Electric Vehicle Acceleration, Gradeability, and Deceleration Test Procedure	3:28.12
SAE J1668	JUN93	† Diesel Engines—Fuel Injection Pump Testing	2:26.322
SAE J1670	MAY93	† Type 'F' Clamps for Plumbing Applications	2:16.79
SAE J1702	MAY87	Motor Vehicle Brake Fluid—Arctic	Cancelled 1987
SAE J1703	JUN91	Motor Vehicle Brake Fluid (A)	2:25.41
SAE J1705	OCT88	Low Water Tolerant Brake Fluids (A)	2:25.56
SAE J1706	OCT88	Production, Handling and Dispensing of SAE J1703 Motor Vehicle Brake Fluids (A)	2:25.68
SAE J1707	NOV91	Service Maintenance of SAE J1703 Brake Fluids in Motor Vehicle Brake Systems	2:25.53
SAE J1708	OCT90	Serial Data Communications Between Microcomputer Systems in Heavy Duty Vehicle Applications (A)	2:23.553
SAE J1709		† European Brake Fluid Technology	2:25.64
SAE J1800	APR87	Method for Evaluating the Paintable Characteristics of Automotive Sealers	1:11.83
SAE J1801	JUN93	† Brake Effectiveness Marking for Brake Blocks	2:25.118
SAE J1802	JUN93	† Brake Block Effectiveness Rating	2:25.119
SAE J1804	JUN89	Corrosion Preventive Compound, Topside Vehicle Corrosion Protection	1:3.52
SAE J1805	APR93	*Sound Power Level Measurements of Earthmoving Machinery - Static and In-Place Dynamic Methods	1:14.90
SAE J1806	NOV92	*Clutch Dimensions for Truck and Bus Applications (A)	3:36.119
SAE J1808	OCT89	Vacuum Power Assist Brake Booster Test Procedure (A)	2:25.172
SAE J1810	JAN93	† Electrical Indicating System Specification	3:40.275
SAE J1812	OCT88	Function Performance Status Classification for EMC Susceptibility Testing of Automotive Electronic and Electrical Devices	2:23.226
SAE J1813	AUG87	A Vehicle Network Protocol with a Fault Tolerant Multiplex Signal Bus	2:23.499
SAE J1814	FEB93	† Operator Controls--Off-Road Machines	3:40.414
SAE J1816	OCT87	Performance Levels and Methods of Measurement of Electromagnetic Radiation from Vehicles and Devices, Narrowband, 10 kHz to 1000 MHz (A)	2:23.508
SAE J1817	JUN91	Long Stroke Air Brake Actuator Marking (A)	3:36.167
SAE J1819	NOV90	Securing Child Restraint Systems in Motor Vehicle Rear Seats (A)	3:33.18
SAE J1823	JUN91	Specification Definitions—Articulated Rubber-Tired Forwarder (A)	3:40.89
SAE J1824	JUN91	Specification Definitions—Clam Bunk Skidder (A)	3:40.81
SAE J1825	APR88	Shelf Storage of Hydraulic Brake Components	2:25.35
SAE J1826	APR89	Turbocharger Gas Stand Test Code	2:26.355
SAE J1827	MAY87	Unibody Weld Quality Testing (A)	3:42.23
SAE J1828	MAY87	Uniform Reference and Dimensional Guidelines for Unibody Vehicles	3:42.28
SAE J1829	MAY92	Stoichiometric Air/Fuel Ratios of Automotive Fuels	1:12.80
SAE J1830	MAY87	Size Classification and Characteristics of Ceramic Shot for Peening (A)	1:8.32
SAE J1832	NOV89	Gasoline Fuel Injector (A)	2:26.263
SAE J1833	NOV88	Hot Impulse Test for Hydraulic Brake Hose Assemblies	2:25.17
SAE J1834	JUN91	Seat Belt Comfort, Fit, and Convenience (A)	3:34.267
SAE J1835	MAR90	Fastener Hardware for Wheels for Demountable Rims (A)	3:31.26
SAE J1836	OCT88	Overlap Shear Test for Sealant Adhesive Bonding of Automotive Glass Encapsulating Material to Body Opening	1:11.97
SAE J1837	JUN91	Electroplate Requirements for Decorative Chromium Deposits on Zinc Base Materials Used for Exterior Ornamentation (A)	1:10.155
SAE J1839	MAY90	Coarse Droplet Water/Fuel Separation Test Procedure (A)	2:26.437
SAE J1843	APR93	*Accelerator Pedal Position Sensor for Use with Electronic Controls in Medium- and Heavy-Duty Vehicle Applications	2:23.515
SAE J1846	JUN89	Characterizing a Test Surface for Motorcycle Side Stand Retraction Performance Testing	3:37.17
SAE J1847	JUN89	Abrasion Resistance Testing—Vehicle Exterior Graphics and Pin Striping	1:11.262
SAE J1849	JUL89	Emergency Vehicle Sirens (A)	2:24.198
SAE J1850	AUG91	Class B Data Communication Network Interface	2:23.268
SAE J1851	MAY87	Induction Cure Test for Metal Bonding Adhesives (A)	1:11.109

* Technical Revision   † New Document   (D)—DODISS Adopted   (A)—ANSI Recognized

Numerical Index

Document	Title	Location
SAE J1852 SEP87	Properties of Galvanized Low Carbon Sheet Steels and Their Relation to Formability	1:3.22
SAE J1853 APR87	Hand Winches—Boat Trailer Type	3:35.05
SAE J1854 JUN88	Brake Force/Distribution Performance Guide—Commercial Vehicles (A)	3:36.170
SAE J1855 MAR92	Deployment of Electrically Activated Automotive Air Bags for Automobile Reclamation	3:33.29
SAE J1856 MAY89	Identification of Automotive Air Bags	3:33.29
SAE J1857 MAY87	Flywheel Dimensions for Truck and Bus Applications (A)	2:26.237
SAE J1858 JUN88	Full Flow Lubricating Oil Filters Multipass Method for Evaluating Filtration Performance (A)	2:26.248
SAE J1859 APR89	Test Procedures for Determining Air Brake Valve Input-Output Characteristics	3:36.162
SAE J1860 AUG90	Labeling Air Brake Valves with Their Performance (Input-Output) Characteristics (A)	3:36.164
SAE J1862 FEB90	Fuel Injection System Fuel Pressure Regulator and Pressure Damper	2:26.279
SAE J1863 APR87	Coach Joint Fracture Test	1:11.85
SAE J1864 APR87	Method for Evaluating Material Separation in Automotive Sealers Under Pressure in Static Conditions	1:11.92
SAE J1868 JUN90	Restricted Hardenability Bands for Selected Alloy Steels (A)	1:1.139
SAE J1873 JUN93	*Moisture Transmission Test Procedure—Hydraulic Brake Hose Assemblies	2:25.09
SAE J1875 JUN93	†Materials for Plastic Check Valves for Vacuum Booster System	2:25.40
SAE J1876 MAR90	Plastic Dust Shield for Hydraulic Disc Brakes (A)	2:25.67
SAE J1877 MAY88	Recommended Practice for Bar-Coded Vehicle Identification Number Label (A)	3:42.37
SAE J1879 OCT88	General Qualification and Production Acceptance Criteria for Integrated Circuits in Automotive Applications	2:23.583
SAE J1881 FEB93	*Initial Graphics Exchange Specification (A)	1:A.38
SAE J1882 AUG87	Method for Evaluating the Cleavage Strength of Structurally Bonded Fiberglass Reinforced Plastic (Wedge Test) (A)	1:11.102
SAE J1883 MAR88	Elastomeric Bushing "TRAC" Application Code (A)	1:11.40
SAE J1885 MAR92	Accelerated Exposure of Automotive Interior Trim Components Using a Controlled Irradiance Water Cooled Xenon-Arc Apparatus	1:11.255
SAE J1888 NOV90	High Current Time Lag Electric Fuses (A)	2:23.128
SAE J1889 JUN88	L.E.D. Lighting Devices (A)	2:24.19
SAE J1890 JUN88	Performance Assurance of Remanufactured, Hydraulically-Operated Rack and Pinion Steering Gears (A)	3:42.33
SAE J1892 MAY88	Recommended Practice for Bar-Coded Vehicle Emission Configuration Label (A)	3:42.34
SAE J1899 JUN91	Lubricating Oil, Aircraft Piston Engine (Ashless Dispersant) (A)	1:12.98
SAE J1900 OCT90	Seals—Bond Test Fixture and Procedure (A)	3:29.128
SAE J1901 SEP90	Lip Force Measurement, Radial Lip Seals (A)	3:29.135
SAE J1903 AUG90	Automotive Adaptive Driver Controls, Manual (A)	3:34.198
SAE J1907 OCT88	Peel Adhesion Test for Glass to Elastomeric Material for Automotive Glass Encapsulation	1:11.99
SAE J1908 AUG90	Electrical Grounding Practice (A)	3:40.265
SAE J1911 SEP90	Test Procedure for Air Reservoir Capacity—Highway Type Vehicles (A)	3:36.153
SAE J1914 MAR88	Dynamic Ozone Test Procedure—Hydraulic Brake Hose (A)	2:25.10
SAE J1915 FEB90	Recommended Remanufacturing Procedures for Manual Transmission Clutch Assemblies	3:42.39
SAE J1916 MAY89	Engine Water Pump Remanufacture Procedures and Acceptance Criteria	3:42.41
SAE J1918 MAY88	Method for the Determination of Expansion and Water Absorption of Automotive Sealers (A)	1:11.86
SAE J1922 DEC89	Powertrain Control Interface for Electronic Controls Used in Medium and Heavy Duty Diesel On-Highway Vehicle Applications (A)	3:36.177
SAE J1924 DEC92	†OEM/Vendor's Interface Specification for Vehicle Electronic Programming Station	2:23.522
SAE J1926/1 ISO 119 MAR93	*Connections for General Use and Fluid Power-Ports and Stud Ends with ISO 725 Threads and O-ring Sealing—Part 1: Threaded Port with O-Ring Seal in Truncated Housing	2:22.99
SAE J1926/2 ISO 119 MAR93	†Connections for General use and Fluid Power-Ports and Stud Ends with ISO 725 Threads and O-Ring Sealing—Part 2: Heavy-Duty (S Series) Stud Ends	2:22.102
SAE J1926/3 ISO 119 MAR93	†Connections for General use and Fluid Power-Ports and Stud Ends with ISO 725 Threads and O-Ring Sealing—Part 3: Light-Duty (L Series) Stud Ends	2:22.108
SAE J1927 OCT88	Cumulative Damage Analysis for Hydraulic Hose Assemblies	2:22.180
SAE J1928 APR93	*Devices Providing Backfire Flame Control for Gasoline Engines in Marine Applications	3:41.44
SAE J1930 JUN93	*Electrical/Electronic Systems Diagnostic Terms, Definitions, Abbreviations and Acronyms	2:23.288
SAE J1936 OCT89	Chemical Methods for the Measurement of Nonregulated Diesel Emissions (A)	1:13.38
SAE J1937 NOV89	Engine Testing with Low Temperature Charge Air Cooler Systems in a Dynamometer Test Cell (A)	1:13.23
SAE J1938 OCT88	Design/Process Checklist for Vehicle Electronic Systems	2:23.578
SAE J1940 JUN89	Small Engine Power Rating Procedure	2:26.23
SAE J1941 APR90	Coolant Concentrate (Low Silicate, Ethylene Glycol Type Requiring an Initial Charge of Supplemental Coolant Additive) for Heavy-Duty Engines (A)	1:11.216
SAE J1942 JUN93	*Hose and Hose Assemblies for Marine Applications	2:22.156
SAE J1944 MAY91	Truck & Bus Multipurpose Windshield Washer System (A)	3:34.10
SAE J1945 JAN90	Cross-Tooth Companion Flanges, Type T (A)	3:29.71
SAE J1946 JAN91	Companion Flanges, Type A (External Pilot) and Type S (Internal Pilot) (A)	3:29.65
SAE J1947 AUG89	O.D. Coatings for Radial Lip-Type Shaft Seals	1:11.65
SAE J1948 MAR89	Cab Sleeper Occupant Restraint System Test	3:34.267

* Technical Revision † New Document (D)—DODISS Adopted (A)—ANSI Recognized

Numerical Index

SAE J1949 OCT88	Road Vehicles—High Pressure Fuel Injection Pipe End-Connections with 60 Deg Female Cone	2:26.302
SAE J1950 MAY89	Proving Ground Vehicle Corrosion Testing	1:3.41
SAE J1952 JAN91	All Wheel Drive Systems Classification (A)	3:29.15
SAE J1954 MAY90	Guide to the Application and Use of Passenger Car Air-Conditioning Compressor Face Seals (A)	2:26.423
SAE J1957 JUN93	† Central High Mounted Stop Lamp Standard for Use on Vehicles Less than 2032 mm Overall Width	2:24.149
SAE J1958 APR89	Diesel Engines—Steel Tubes for High Pressure Fuel Injection Pipes (Tubing)	2:26.260
SAE J1959 JUN89	Corrosion Preventive Compound, Underbody Vehicle Corrosion Protection	1:3.55
SAE J1960 JUN89	Accelerated Exposure of Automotive Exterior Materials Using a Controlled Irradiance Water Cooled Xenon Arc Apparatus	1:11.268
SAE J1961 DEC88	Accelerated Exposure of Automotive Exterior Materials Using a Solar Fresnel-Reflective Apparatus	1:11.286
SAE J1962 JUN93	* Diagnostic Connector	2:23.245
SAE J1965 FEB93	† Road Vehicles--Wheels for Commercial Vehicles and Multipurpose Passenger Vehicles--Fixing Nuts--Test Methods	3:36.107
SAE J1966 JUN91	Lubricating Oil, Aircraft Piston Engine (Nondispersant Mineral Oil) (A)	1:12.110
SAE J1967 MAR91	Retroreflective Materials for Vehicle Conspicuity (A)	3:34.112
SAE J1969 OCT88	Electrocoat Compatibilities of Automotive Sealers	1:11.90
SAE J1970 DEC91	Shoreline Sound Level Measurement Procedure	1:14.40
SAE J1971 JUN89	Radial Lip Seal Torque: Measurement Methods and Results	3:29.151
SAE J1973 APR91	Personal Watercraft—Flotation (A)	3:41.49
SAE J1974 JUN93	† Decorative Anodizing Specification for Automotive Applications	1:10.39
SAE J1975 JUN91	Case Hardenability of Carburized Steels (A)	1:2.15
SAE J1976 JUN89	Outdoor Weathering of Exterior Materials	1:11.275
SAE J1978 MAR92	† OBD II Scan Tool	2:23.325
SAE J1979 DEC91	E/E Diagnostic Test Modes	2:23.30
SAE J1980 NOV90	Guidelines for Evaluating Out-of-Position Vehicle Occupant Interactions with Deploying Airbags (A)	3:33.33
SAE J1982 DEC91	Nomenclature—Wheels for Passenger Cars, Light Trucks, and Multipurpose Vehicles	3:31.01
SAE J1984 NOV89	Diesel Fuel Injector Assembly Type 27 (9.5 mm) (A)	2:26.308
SAE J1986 FEB93	† Balance Weight and Rim Flange Design Specifications, Test Procedures, and Performance Recommendations	3:31.10
SAE J1989 OCT89	Recommended Service Procedure for the Containment of R-12	3:34.42
SAE J1990 MAR92	Extraction and Recycle Equipment for Mobile Automotive Air-Conditioning Systems	3:34.43
SAE J1991 OCT89	Standard of Purity for Use in Mobile Air-Conditioning Systems	3:34.47
SAE J1993 MAR93	† Cast Steel Grit	1:8.13
SAE J1995 JUN90	Engine Power Test Code—Spark Ignition and Compression Ignition—Gross Power Rating	2:26.11
SAE J1996 OCT92	* Internal Combustion Engines—Piston Rings—Quality Requirements	2:26.74
SAE J1997 OCT92	* Internal Combustion Engines—Piston Rings—Rectangular Rings	2:26.79
SAE J1998 OCT92	* Internal Combustion Engines—Piston Rings—Rectangular Rings with Narrow Ring Width (A)	2:26.43
SAE J1999 OCT92	* Internal Combustion Engines—Piston Rings—Scraper Rings	2:26.90
SAE J2000 OCT92	* Internal Combustion Engines—Piston Rings—Keystone Rings	2:26.100
SAE J2001 OCT92	* Internal Combustion Engines—Piston Rings—Half Keystone Rings (A)	2:26.111
SAE J2002 OCT92	* Internal Combustion Engines—Piston Rings—Oil Control Rings	2:26.117
SAE J2003 OCT92	* Internal Combustion Engines—Piston Rings Coil Spring Loaded Oil Control Rings	2:26.131
SAE J2004 OCT92	* Internal Combustion Engines—Piston Rings Expander/Segment Oil Control Rings (A)	2:26.175
SAE J2005 DEC91	Stationary Sound Level Measurement Procedure for Pleasure Motorboats	1:14.41
SAE J2009 FEB93	† Discharge Forward Lighting System	2:24.113
SAE J2012 MAR92	Diagnostic Trouble Code Definitions	2:23.284
SAE J2013 MAY91	Military Tire Glossary (A)	3:30.62
SAE J2014 APR91	Pneumatic Tires for Military Tactical Wheeled Vehicles (A)	3:30.04
SAE J2016 JUN89	Chemical Stress Resistance of Polymers	1:11.154
SAE J2017 AUG89	Developing Technician Training (A)	3:42.48
SAE J2018 JUL89	Assessing Technician Training (A)	3:42.50
SAE J2020 JUN89	Accelerated Exposure of Automotive Exterior Materials Using a Fluorescent UV and Condensation Apparatus	1:11.288
SAE J2022 JUN91	Classification, Nomenclature, and Specification Definitions for Horizontal Earthboring Machines (A)	3:40.14
SAE J2023 JAN93	† Operating Precautions for Horizontal Earthboring Machines	3:40.20
SAE J2024 JUN93	* Contaminants for Testing Air Brake Components and Auxiliary Pneumatic Devices	3:36.131
SAE J2025 JUN89	Method for Evaluating the Flow Properties of Pumpable Sealers	1:11.66
SAE J2026 NOV91	Recommended Practice for the Selection of Engineering Workstations	3:43.01
SAE J2028 JUN92	Front-Wheel-Drive Constant Velocity Joint Boot Seals	3:29.07
SAE J2031 JAN90	High Tension Ignition Cable (A)	2:23.149
SAE J2032 NOV91	Ignition Cable Assemblies	2:23.154
SAE J2034 JAN93	† Personal Water craft Ventilation Systems	3:41.49
SAE J2037 NOV90	Off-Board Diagnostic Message Formats (A)	2:23.304
SAE J2038 APR92	Engine Weight, Dimensions, Center of Gravity, and Moment of Inertia	2:26.02
SAE J2040 JUN91	Tail Lamps (Rear Position Lamps) for Use on Vehicles 2032 mm or More in Overall Width (A)	2:24.139
SAE J2041 JUN92	† Relfex Reflectors for use on Vehicles 2032 mm or More in Overall Width	2:24.178
SAE J2042 JUN91	Clearance, Sidemarker, and Identification Lamps for Use on Motor Vehicles 2032 mm or More in Overall Width (A)	2:24.172
SAE J2044 JUN92	SAE Quick Connector Specifications for Liquid Fuel Systems	2:26.442
SAE J2045 OCT92	† Tube/Hose Assemblies	1:12.85
SAE J2046 JAN93	† Personal Watercraft Fuel Systems	3:41.46

* Technical Revision   † New Document   (D)—DODISS Adopted   (A)—ANSI Recognized

Numerical Index

SAE J2050 FEB93	† High-Temperature Power Steering Pressure Hose	1:11.245
SAE J2051 JUN92	Qualifications for Four-Way Subbase Mounted Air Valves for Automotive Manufacturing Applications	3:43.21
SAE J2052 MAR90	Test Device Head Contact Duration Analysis	3:34.276
SAE J2054 NOV90	E/E Diagnostic Data Communications (A)	2:23.329
SAE J2055 JUN91	Identification Terminology & Component Nomenclature—Knuckle Boom Log Loader (A)	3:40.67
SAE J2056/1 JUN93	† Class C Application Requirement Considerations	2:23.366
SAE J2056/2 APR93	† Survey of Known Protocols	2:23.372
SAE J2056/3 JUN91	Selection of Transmission Media (A)	2:23.391
SAE J2057/1 JUN91	Class A Application/Definition (A)	2:23.407
SAE J2057/3 JUN93	† Class A Multiplexing Sensors	2:23.413
SAE J2057/4 JUN93	† Class A Multiplexing Architecture Strategies	2:23.415
SAE J2058 JUN90	Chrysler Sensor and Control (CSC) Bus Multiplexing Network for Class 'A' Applications	2:23.422
SAE J2059 AUG92	All-Wheel-Drive Drivetrain Schematic Symbol Standards	3:29.19
SAE J2064 JUN93	† R134a Refrigerant Automotive Air Conditioning Hose	1:11.219
SAE J2068 JAN90	Combination Turn Signal Hazard Warning Signal Flashers (A)	2:24.165
SAE J2069 JAN93	† Recovery Attachment Points for Passenger Cars, Vans, and Light Trucks	3:42.21
SAE J2071 MAR90	Aerodynamic Testing of Road Vehicles	3:28.18
SAE J2073 MAR93	† Automotive Starter Remanufactured Procedures	3:42.42
SAE J2074 JUN93	† The Air Bag Systems in Your Cars 'What the Public Needs to Know'	3:33.42
SAE J2076 FEB93	† High-Temperature Power Steering Return Hose--Low Pressure	1:11.247
SAE J2077 NOV90	Miniature Blade Type Electrical Fuses (A)	2:23.125
SAE J2079 MAY92	Location of Atomizer of Ether Systems for Diesel Engines	2:26.329
SAE J2081 JUN91	Test Procedure for Determining the Resistance of Safety Glazing Materials, of Which One Surface is Plastic, to Simulated Weathering (A)	3:34.106
SAE J2082 JUN92	Cooling Flow Measurement Techniques	3:28.46
SAE J2084 JAN93	† Aerodynamic Testing of Road Vehicles--Testing Methods and Procedures	3:28.61
SAE J2087 AUG91	Daytime Running Lamps for Use on Motor Vehicles	2:24.168
SAE J2094 JUN92	Vehicle and Control Modifications for Drivers with Physical Disabilities Terminology	3:34.196
SAE J2096 JUL92	† Categorization of Low Carbon Automotive Sheet Steel	1:3.25
SAE J2099 DEC91	Standard of Purity for Recycled HFC-134a for Use in Mobile Air-Conditioning Systems	3:34.60
SAE HS-2100 NOV90	Numbering System for Standard Drills, Standard Taps, and Reamers	Available as a separate publication (See Related Technical Reports Section)
SAE J2100 AUG92	† Accelerated Environmental Testing for Bonded Automotive Assemblies	1:11.63
SAE J2101 JUN91	Acoustics—Measurement of Airborne Noise Emitted by Construction Equipment Intended for Outdoor Use—Method for Determining Compliance with Noise Limits (A)	1:14.68
SAE J2102 JUN91	Acoustics—Measurement of Airborne Noise Emitted by Earthmoving Machinery—Method for Determining Compliance with Limits for Exterior Noise—Stationary Test Condition (A)	1:14.74
SAE J2103 JUN91	Acoustics—Measurement of Airborne Noise Emitted by Earthmoving Machinery—Operator's Position—Stationary Testing Condition (A)	1:14.78
SAE J2104 JUN91	Acoustics—Measurement of Exterior Noise Emitted by Earthmoving Machinery—Dynamic Test Conditions (A)	1:14.81
SAE J2105 JUN91	Acoustics—Measurement of Noise Emitted by Earthmoving Machinery at the Operator's Position—Simulated Work Cycle Test Conditions (A)	1:14.87
SAE J2106 APR91	Token Slot Network for Automotive Control (A)	2:23.431
SAE J2108 DEC91	Door Courtesy Switch (A)	2:23.141
SAE J2114 APR93	† Dolly Rollover Recommended Test Procedure	3:34.283
SAE J2115 JUN93	† Brake Performance and Wear Test Code Commercial Vehicle Inertia Dynamometer	2:25.106
SAE J2116 JUN93	* Two-Stroke-Cycle Gasoline Engine Lubricants Performance and Service Classification (A)	1:12.33
SAE J2119 JUN93	† Manual Controls for Mature Drivers	3:34.194
SAE J2120 JUL92	† Personal Watercraft--Electrical Systems	3:41.30
SAE J2122 NOV90	Numbering System for Standard Drills (A)	3:43.11
SAE J2123 NOV90	Numbering System for Standard Taps (A)	3:43.15
SAE J2124 NOV90	Numbering System for Reamers (A)	3:43.18
SAE J2129 NOV90	Guidelines for Requests Received from Outside Sources for the ConAg Council to Originate or Review Technical Reports	3:40.03
SAE J2130 APR93	† Self-Propelled Sweepers	3:40.23
SAE J2131 OCT92	* Front-Mounted Linkage for Agricultural Wheeled Tractors	3:39.43
SAE J2133 JUN93	* Disc Wheel Radial Runout Low Point Marking	3:31.38
SAE J2162 JUN92	Spark Plug Heat Rating Classifications	2:23.97
SAE J2175 JUN91	Specifications for Low Carbon Cast Steel Shot (A)	1:8.10
SAE J2177 APR92	Chassis Dynamometer Test Procedure—Heavy-Duty Road Vehicles	2:26.06
SAE J2178/1 JUN92	Class B Data Communication Network Messages: Detailed Header Formats and Physical Address Assignments	2:23.334
SAE J2178/2 JUN93	† Class B Data Communication Network Messages Part 2: Data Parameter Definitions	2:23.345
SAE J2180 APR93	† A Tilt-table Procedure for Measuring the Static Rollover Threshold for Heavy Trucks	2:23.528
SAE J2181 JUN93	† Steady-State Circular Test Procedure for Trucks and Buses	3:36.68
SAE J2184 OCT92	† Vehicle Lift Points for Service Garage Lifting	3:42.53
SAE J2185 NOV91	Life Test for Heavy-Duty Storage Batteries	2:23.72
SAE J2186 SEP91	E/E Data Link Security	2:23.507

* Technical Revision   † New Document   (D)—DODISS Adopted   (A)—ANSI Recognized

Numerical Index

Document	Title	Reference
SAE J2189 FEB93	† Guidelines for Evaluating Child Restraint System Interactions with Deploying Airbags	3:33.20
SAE J2190 JUN93	† Enhanced E/E Diagnostic Test Modes	2:23.41
SAE J2194 JUN93	* Roll-Over Protective Structures (ROPS) for Wheeled Agricultural Tractors (A)	3:39.78
SAE J2196 JUN92	Service Hose for Automotive Air Conditioning	3:34.51
SAE J2197 JUN92	HFC-134a (R-134a) Service Hose Fittings for Automotive Air-Conditioning Service Equipment	3:34.53
SAE J2198 JUL92	† Glossary Automatic Belt Tensioner	2:19.19
SAE J2200 JAN91	Passenger Cars and Light Truck Axles (A)	3:29.88
SAE J2201 JUN93	† Universal Interface for OBD II Scan	2:23.250
SAE J2203 JUN91	SAE 17.6 Cubic Inch Spark Plug Rating Engine (A)	2:23.98
SAE J2208 JUN93	† Park Standard for Automatic Transmissions	3:29.42
SAE J2209 JUN92	CFC-12 (R-12) Extraction Equipment for Mobile Automotive Air-Conditioning Systems	3:34.46
SAE J2210 DEC91	HFC-134a Recycling Equipment for Mobile Air-Conditioning Systems	3:34.58
SAE J2211 DEC91	Recommended Service Procedure for the Containment of HFC-134a	3:34.57
SAE J2212 MAR92	† Accelerated Exposure of Automotive Interior Trim Components Using a Controlled Irradiance Air-Cooled Xenon-Arc Apparatus	1:11.263
SAE J2213 JUN91	Metric Ball Joints (A)	2:17.12
SAE J2215 DEC91	Surface Match Verification Method for Pressure Sensitive Adhesively Attached Components	1:11.267
SAE J2217 OCT91	Photometric Guidelines for Instrument Panel Displays That Accomodate Older Drivers	3:34.226
SAE J2226 OCT92	† Internal Combustion Engines—Piston Rings—Steel Rectangular Rings	2:26.180
SAE J2227 MAY93	* International Tests and Specifications for Automotive Engine Oils	1:12.10
SAE J2228 JUN93	† Kingpin Wear Limits Commercial Trailers and Semi-Trailers	3:36.89
SAE J2229 FEB93	† Accelerated Exposure of Automotive Interior Trim Materials Using an Outdoor Under Glass Variable Angled Controlled Temperature Apparatus	1:11.119
SAE J2230 FEB93	† Accelerated Exposure of Automotive Interior Trim Material Using Outdoor Under-Glass Controlled Sun-Tracking Temperature and Humidity Apparatus	1:11.123
SAE J2232 JUN92	Vehicle System Voltage—Initial Recommendations	2:23.75
SAE J2234 JAN93	† Equivalent Temperature	3:34.25
SAE J2235 JUN93	† Paint and Trim Code Location	3:33.44
SAE J2236 JUN92	Standard Method for Determining Continuous Upper Temperature Resistance of Elastomers	1:11.32
SAE J2240 MAR93	† Starter Armature Remanufacturing Procedures	3:42.45
SAE J2241 MAR93	† Automotive Starter Drive Assembly Remanufacturing Procedures	3:42.46
SAE J2242 MAR93	† Automotive Starter Solenoid Remanufacturing Procedures	3:42.47
SAE J2244/1 DEC91	Connections for Fluid Power and General Use—Ports and Stud Ends with ISO 261 Threads and O-Ring Sealing Part 1: Port with O-Ring Seal in Truncated Housing	2:22.290
SAE J2244/2 DEC91	Connections for Fluid Power and General Use—Ports and Stud Ends with ISO 261 Threads and O-Ring Sealing Part 2: Heavy-Duty (S Series) Stud Ends— Dimensions, Design, Test Methods, and Requirements	2:22.293
SAE J2246 JUN92	Antilock Brake System Review	2:25.72
SAE HS-2300 JUN93	† Introduction to All-Wheel Drive	Available as a separate publication (See Related Technical Reports Section)
SAE HS-2400 FEB93	† SAE Surface Vehicle Emissions Standards Manual	Available as a separate publication (See Related Technical Reports Section)
SAE HS-2500 APR93	† SAE Motorcycle Standards Manual	Available as a separate publication (See Related Technical Reports Section)
SAE HS-2700 APR93	† SAE Automotive Textiles and Trim Standards Manual	Available as a separate publication (See Related Technical Reports Section)
SAE J2708 APR93	* Agricultural Tractor Test Code (OECD)	3:39.05
SAE J3000 JUN93	* Thermoplastic Elastomer Classification System	1:11.27
SAE J4153 JUN87	Human Engineering Considerations for TCAS	Available as a separate publication (See Related Technical Reports Section)

* Technical Revision   † New Document   (D)—DODISS Adopted   (A)—ANSI Recognized

Subject Index

ABBREVIATIONS
See: TERMINOLOGY

ABDOMEN
Impact Tolerance
Human Tolerance to Impact Conditions as Related to Motor Vehicle Design—SAE **J885** JUL86 .... 3:34.306

ABRASION
Abrasive Wear—SAE **J965** AUG66 .... 1:9.22
Carpeting
Test Method for Determining Resistance to Abrasion; Bearding; and Fiber Loss of Automotive Carpet Materials (A)—SAE **J1530** JUN85 .... 1:11.107
Materials
Abrasion Resistance Testing—Vehicle Exterior Graphics and Pin Striping—SAE **J1847** JUN89 .... 1:11.262
Seat Belt Assembly Webbing Abrasion Performance Requirements (A)—SAE **J114** MAR86 .... 3:33.11
Seat Belt Assembly Webbing Abrasion Test Procedure (A)—SAE **J339** MAR86 .... 3:33.11
Test Method for Determining Resistance to Abrasion of Automotive Bodycloth, Vinyl, and Leather, and the Snagging of Automotive Bodycloth (A)—SAE **J948** FEB86 .... 1:11.178

ACCELERATED TESTING
Accelerated Exposure of Automotive Exterior Materials Using a Solar Fresnel-Reflective Apparatus—SAE **J1961** DEC88 .... 1:11.286
Accelerated Exposure of Automotive Exterior Materials Using a Controlled Irradiance Water Cooled Xenon Arc Apparatus—SAE **J1960** JUN89 .... 1:11.268
Accelerated Exposure of Automotive Exterior Materials Using a Fluorescent UV and Condensation Apparatus—SAE **J2020** JUN89 .... 1:11.288
Accelerated Exposure of Automotive Interior Trim Components Using a Controlled Irradiance Water Cooled Xenon-Arc Apparatus—SAE **J1885** MAR92 .... 1:11.255
High Current Time Lag Electric Fuses (A)—SAE **J1888** NOV90 .... 2:23.128
Hydraulic Cylinder Rod Corrosion Test (A)—SAE **J1333** MAR90 .... 3:40.343
† Life Test for Automotive Storage Batteries—SAE **J240** JUN93 .... 2:23.71
Life Test for Heavy-Duty Storage Batteries—SAE **J2185** NOV91 .... 2:23.72
Proving Ground Vehicle Corrosion Testing—SAE **J1950** MAY89 .... 1:3.41
† Sponge and Expanded Cellular Rubber Products—SAE **J18** JUL92 .... 1:11.56
Test Procedure for Determining the Resistance of Safety Glazing Materials, of Which One Surface is Plastic, to Simulated Weathering (A)—SAE **J2081** JUN91 .... 3:34.106
Automobile Trim
*Accelerated Exposure of Automotive Interior Trim Components Using a Controlled Irradiance Air-Cooled Xenon-Arc Apparatus—SAE **J2212** MAR93 .... 1:11.263
*Accelerated Exposure of Automotive Interior Trim Materials Using an Outdoor Under Glass Variable Angled Controlled Temperature Apparatus—SAE **J2229** FEB93 . 1:11.119
*Accelerated Exposure of Automotive Interior Trim Material Using Outdoor Under-Glass Controlled Sun-Tracking Temperature and Humidity Apparatus—SAE **J2230** FEB93 .... 1:11.123
Bonded Assemblies
*Accelerated Environmental Testing for Bonded Automotive Assemblies—SAE **J2100** AUG92 .... 1:11.63

ACCELERATION
See also: PERFORMANCE
Automobile
Fixed Rigid Barrier Collision Tests—SAE **J850** NOV88 . 3:34.281
Sound Level for Passenger Cars and Light Trucks (D)—SAE **J986** OCT88 .... 1:14.14
Vehicle Acceleration Measurement—SAE **J1491** JUN90 2:26.503
Electric Vehicles
*Electric Vehicle Acceleration, Gradeability, and Deceleration Test Procedure—SAE **J1666** MAY93 ..... 3:28.12
Trucks and Buses
Vehicle Acceleration Measurement—SAE **J1491** JUN90 2:26.503

ACCELERATOR PEDALS
† Accelerator Pedal Position Sensor for Use with Electronic Controls in Medium- and Heavy-Duty Vehicle Applications—SAE **J1843** APR93 .... 2:23.515
Accommodation Tool Reference Point—SAE **J1516** MAR90 .... 3:34.247
Driver Selected Seat Position (A)—SAE **J1517** MAR90 . 3:34.250
Truck Driver Shin-Knee Position for Clutch and Accelerator (A)—SAE **J1521** MAR90 .... 3:34.263

ACCESSORY DRIVES
Automotive V-Ribbed Belt Drives and Test Methods—SAE **J1596** JUN89 .... 2:19.17
*Glossary Automatic Belt Tensioner—SAE **J2198** JUL92 . 2:19.19
Mounting Flanges for Engine Accessory Drives (A)—SAE **J896** MAY83 .... 3:40.253

ACCIDENT INVESTIGATION
See also: CRASHWORTHINESS
*Guidelines for Evaluating Child Restraint System Interactions with Deploying Airbags—SAE **J2189** FEB93 3:33.20
Deformation Classification
Collision Deformation Classification—SAE **J224** MAR80 . 3:34.235

ACOUSTIC EMISSION ANALYSIS
See also: NONDESTRUCTIVE TESTS
Acoustic Emission Test Methods (A)—SAE **J1242** MAR91 .... 1:3.80
Acoustics—Measurement of Airborne Noise Emitted by Earthmoving Machinery—Method for Determining Compliance with Limits for Exterior Noise—Stationary Test Condition (A)—SAE **J2102** JUN91 .... 1:14.74
Acoustics—Measurement of Airborne Noise Emitted by Earthmoving Machinery—Operator's Position—Stationary Testing Condition (A)—SAE **J2103** JUN91 .... 1:14.78
Acoustics—Measurement of Airborne Noise Emitted by Construction Equipment Intended for Outdoor Use—Method for Determining Compliance with Noise Limits (A)—SAE **J2101** JUN91 .... 1:14.68
Acoustics—Measurement of Exterior Noise Emitted by Earthmoving Machinery—Dynamic Test Conditions (A)—SAE **J2104** JUN91 .... 1:14.81
Acoustics—Measurement of Noise Emitted by Earthmoving Machinery at the Operator's Position—Simulated Work Cycle Test Conditions (A)—SAE **J2105** JUN91 .... 1:14.87
Measurement of Noise Emitted by Accelerating Highway Vehicles—SAE **J1470** MAR92 .... 1:14.32
Nondestructive Tests (A)—SAE **J358** FEB91 .... 1:3.67
† Sound Measurement—Construction Site (A)—SAE **J1075** JUN93 .... 1:14.99

ACOUSTIC INSULATION
See: INSULATION

ACOUSTICS
See also: ACOUSTIC EMISSION ANALYSIS, NOISE
Acoustical and Thermal Materials Test Procedure (A)—SAE **J1324** OCT89 .... 1:11.111
Instrumentation for Measuring Acoustic Impulses Within Vehicles (A)—SAE **J247** FEB87 .... 1:14.07
Laboratory Measurement of the Airborne Sound Barrier Performance of Automotive Materials and Assemblies (A)—SAE **J1400** MAY90 .... 1:11.174

---
* Technical Revision   † New Document   (D)—DODISS Adopted   (A)—ANSI Recognized

Subject Index

ACOUSTICS (continued)
Surface Vehicle Sound Measurement Manual—SAE HS-184 NOV89 .... Available as a separate publication (See Related Technical Reports Section)

ACRONYMS
See: TERMINOLOGY

ACTUATORS
See also: STARTERS
Air Brake Actuator Diaphragm Test Procedure (A)—SAE J1450 MAY88 .... 3:36.166
Air Brake Actuator Test Procedure—Truck-Tractor, Bus and Trailer (A)—SAE J1469 JUN88 .... 3:36.157
Long Stroke Air Brake Actuator Marking (A)—SAE J1817 JUN91 .... 3:36.167

ADAPTIVE PRODUCTS
Automotive Adaptive Driver Controls, Manual (A)—SAE J1903 AUG90 .... 3:34.198
Vehicle and Control Modifications for Drivers with Physical Disabilities Terminology—SAE J2094 JUN92 .. 3:34.196

ADDITIVES
Coolants
Coolant Concentrate (Low Silicate, Ethylene Glycol Type Requiring an Initial Charge of Supplemental Coolant Additive) for Heavy-Duty Engines (A)—SAE J1941 APR90 .... 1:11.216
Diesel Fuels
Diesel Fuels—SAE J313 MAR92 .... 1:12.60
Gasoline
† Automotive Gasolines—SAE J312 JAN93 .... 1:12.47
Lubricants
Lubricating Oil, Aircraft Piston Engine (Ashless Dispersant) (A)—SAE J1899 JUN91 .... 1:12.98
Physical and Chemical Properties of Engine Oils (A)—SAE J357 JUN91 .... 1:12.24

ADHESION
Method for Evaluating the Adhesion Characteristics of Automotive Sealers—SAE J1600 APR87 .... 1:11.82
Peel Adhesion Test for Glass to Elastomeric Material for Automotive Glass Encapsulation—SAE J1907 OCT88 .. 1:11.99
Seals—Bond Test Fixture and Procedure (A)—SAE J1900 OCT90 .... 3:29.128
Surface Match Verification Method for Pressure Sensitive Adhesively Attached Components—SAE J2215 DEC91 . 1:11.267
Coatings
Electroplating of Nickel and Chromium on Metal Parts—Automotive Ornamentation and Hardware—SAE J207 FEB85 .... 1:10.150

ADHESIVES
See also: SEALANTS
Coach Joint Fracture Test—SAE J1863 APR87 ..... 1:11.85
Cross Peel Test for Automotive Type Adhesives for Fiber Reinforced Plastic (FRP) Bonding (A)—SAE J1553 MAY86 .... 1:11.100
Induction Cure Test for Metal Bonding Adhesives (A)—SAE J1851 MAY87 .... 1:11.109
Lap Shear Test for Automotive Type Adhesives for Fiber Reinforced Plastic (FRP) Bonding (A)—SAE J1525 JUN85 .... 1:11.81
Metal to Metal Overlap Shear Strength Test for Automotive Type Adhesives (A)—SAE J1523 JUN85 ... 1:11.78
Method for Evaluating the Cleavage Strength of Structurally Bonded Fiberglass Reinforced Plastic (Wedge Test) (A)—SAE J1882 AUG87 .... 1:11.102
Method of Viscosity Test for Automotive Type Adhesives, Sealers, and Deadeners—SAE J1524 NOV88 ...... 1:11.78
Methods of Tests for Automotive-Type Sealers, Adhesives, and Deadeners—SAE J243 .... 1:11.68
Overlap Shear Test for Automotive Type Sealant for Stationary Glass Bonding (A)—SAE J1529 MAY86 ... 1:11.95
Overlap Shear Test for Sealant Adhesive Bonding of Automotive Glass Encapsulating Material to Body Opening—SAE J1836 OCT88 .... 1:11.97
Stationary Glass Replacement—SAE J1556 DEC85 ... 3:42.21
Test Procedures for Brake Shoe and Lining Adhesives and Bonds—SAE J840 AUG82 .... 2:25.95

AERODYNAMICS
See also: WIND TUNNEL TESTING
Aerodynamic Testing of Road Vehicles—SAE J2071 MAR90 .... 3:28.18
Automobiles
Aerodynamic Flow Visualization Techniques and Procedures—SAE HS-1566 .... Available as a separate publication (See Related Technical Reports Section)
*Aerodynamic Testing of Road Vehicles--Testing Methods and Procedures—SAE J2084 JAN93 .... 3:28.61
Vehicle Aerodynamics Terminology (A)—SAE J1594 JUN87 .... 3:28.17
Construction and Industrial Equipment
Drag Force Test Procedure for Construction, Forestry, and Industrial Machines (A)—SAE J873 FEB86 ...... 3:40.150
Terminology
Vehicle Aerodynamics Terminology (A)—SAE J1594 JUN87 .... 3:28.17
Vehicle Dynamics Terminology—SAE J670e JUN76 ... 3:34.297
Trucks and Buses
Aerodynamic Flow Visualization Techniques and Procedures—SAE HS-1566 .... Available as a separate publication (See Related Technical Reports Section)
Heavy Truck and Bus Retarder Downhill Performance Mapping Procedure (A)—SAE J1489 JUN87 .... 3:36.124
SAE Wind Tunnel Test Procedure for Trucks and Buses (A)—SAE J1252 JUL81 .... 3:36.61
Supportive Information Report for the Fuel Economy Measurement Test (Engineering Type) for Trucks and Buses—SAE J1420 AUG88 .... 3:36.44

AFTERCOOLERS
Engine Charge Air Cooler Nomenclature (A)—SAE J1148 JUN90 .... 2:26.392

AGRICULTURAL EQUIPMENT
See also: BACKHOES, EARTHMOVING EQUIPMENT, OFF-ROAD VEHICLES, TILLERS, TRENCHING EQUIPMENT
Categories of Off-Road Self-Propelled Work Machines (A)—SAE J1116 JUN86 .... 3:40.01
Guidelines for Requests Received from Outside Sources for the ConAg Council to Originate or Review Technical Reports—SAE J2129 NOV90 .... 3:40.03
† Steering Frame Lock for Articulated Loaders and Tractors—SAE J276 NOV92 .... 3:40.353
† T-Hook Slots for Securement in Shipment of Agricultural Equipment (A)—SAE J39 JUN93 .... 3:39.58
Brakes
Braking System Test Procedures and Braking Performance Criteria for Agricultural Tractors—SAE J1041 OCT91 .... 3:39.29
Electric Equipment
Electrical Connector for Auxiliary Starting of Construction, Agricultural, and Off-Road Machinery (A)—SAE J1283 MAR86 .... 3:40.267
Hitches
† Drawbars—Agricultural Wheel Tractors—SAE J1548 FEB93 .... 3:39.90
† Front-Mounted Linkage for Agricultural Wheeled Tractors—SAE J2131 OCT92 .... 3:39.43

* Technical Revision   † New Document   (D)—DODISS Adopted   (A)—ANSI Recognized

Subject Index

AGRICULTURAL EQUIPMENT (continued)
Hitches (continued)
† Three-Point Free-Link Hitch Attachment of Implements to Agricultural Wheeled Tractors—SAE **J715** JUN93 ..... 3:39.37

† Three-Point Hitch, Implement Quick-Attaching Coupler, Agricultural Tractors—SAE **J909** JUN93 ..... 3:39.34

Hydraulic Systems
Application of Hydraulic Remote Control to Agricultural Tractors and Trailing Type Agricultural Implements (A)—SAE **J716** MAR91 ..... 3:39.45

† Dimensional Standard for Cylindrical Hydraulic Couplers for Agricultural Tractors (A)—SAE **J1036** APR93 ..... 3:39.48

† Test Procedure for Measuring Hydraulic Lift Capacity on Agricultural Tractors Equipped with Three-Point Hitch—SAE **J283** JUN93 ..... 3:39.41

Implements
† Front-Mounted Linkage for Agricultural Wheeled Tractors—SAE **J2131** OCT92 ..... 3:39.43

Lamps
† Flashing Warning Lamp for Agricultural Equipment—SAE **J974** JUN93 ..... 3:39.59

† Headlamps for Agricultural Equipment—SAE **J975** JUN93 ..... 3:39.59

Lighting and Marking of Agricultural Equipment on Highways—SAE **J137** JUN89 ..... 3:39.60

† SAE Ground Vehicle Lighting Standards Manual—SAE **HS-34** JAN93 ..... Available as a separate publication (See Related Technical Reports Section)

Noise
Sound Measurement—Self-Propelled Agricultural Equipment—Exterior (A)—SAE **J1008** JAN87 ..... 1:14.113

Operator Controls
Operator Controls for Agricultural Wheeled Tractors—SAE **J841** DEC84 ..... 3:39.70

* Operator Controls--Off-Road Machines—SAE **J1814** FEB93 ..... 3:40.414

Operator Protective Structures
Overhead Protection for Agricultural Tractors—Test Procedures and Performance Requirements—SAE **J167** APR92 ..... 3:39.89

† Roll-Over Protective Structures (ROPS) for Wheeled Agricultural Tractors (A)—SAE **J2194** JUN93 ..... 3:39.78

Rollover Protective Structures (ROPS) for Wheeled Agricultural Tractors (A)—SAE **J1194** MAY89 ..... 3:39.71

Performance
Agricultural Tractor Test Code—SAE **J708** DEC84 ..... 3:39.01

† Agricultural Tractor Test Code (OECD)—SAE **J2708** APR93 ..... 3:39.05

Power Take-Off
† Auxiliary Power Take-Off Drives for Agricultural Tractors—SAE **J717** APR93 ..... 3:39.53

† Front and Rear Power Take-Off for Agricultural Tractors (A)—SAE **J1170** FEB93 ..... 3:39.49

† Operating Requirements for Tractors and Power Take-Off Driven Implements (A)—SAE **J721** FEB93 ..... 3:39.52

Power Take-Off Definitions and Terminology for Agricultural Tractors (A)—SAE **J722** MAR91 ..... 3:39.53

Product Identification Number
Product Identification Numbering System of Off-Road Work Machines (A)—SAE **J1360** SEP87 ..... 3:40.04

Safety
Overhead Protection for Agricultural Tractors—Test Procedures and Performance Requirements—SAE **J167** APR92 ..... 3:39.89

Safety Alert Symbol for Agricultural, Construction and Industrial Equipment (A)—SAE **J284** JAN91 ..... 3:39.68

† Safety for Agricultural Tractors—SAE **J208** JUN93 ..... 3:39.69

Seats
Classification of Agricultural Wheeled Tractors for Vibration Tests of Operator Seats (A)—SAE **J1386** JAN86 ..... 3:40.401

Force-Deflection Measurements of Cushioned Components of Seats for Off-Road Work Machines—SAE **J1051** DEC88 ..... 3:40.388

† Vibration Performance Evaluation of Operator Seats (A)—SAE **J1384** JUN93 ..... 3:40.396

Terminology
Identification Terminology of Loaders/Tractors with Forks and Rough Terrain Forklifts—SAE **J1464** APR88 ..... 3:40.58

† Terminology for Agricultural Equipment—SAE **J1150** OCT92 ..... 3:39.01

Tires
Agricultural Tractor Tire Dynamic Indices—SAE **J1261** NOV84 ..... 3:39.28

Agricultural Tractor Tire Loadings, Torque Factors, and Inflation Pressures (A)—SAE **J709d** 3:39.20

Liquid Ballast Table for Drive Tires of Agricultural Tractors (A)—SAE **J884** MAR91 ..... 3:39.25

Tire Selection Tables for Agricultural Tractors of Future Design (A)—SAE **J711** MAR91 ..... 3:39.26

Vibration
† Measurement of Whole Body Vibration of the Seated Operator of Off-Highway Work Machines—SAE **J1013** AUG92 ..... 3:39.92

† Vibration Performance Evaluation of Operator Seats (A)—SAE **J1384** JUN93 ..... 3:40.396

Wheels
† Industrial and Agricultural Disc Wheels (A)—SAE **J712** APR93 ..... 3:39.54

† Wheel Mounting Elements for Industrial and Agricultural Disc Wheels (A)—SAE **J714** APR93 ..... 3:39.57

Wiring Systems
Electrical Wiring Systems for Construction, Agricultural and Off-Road Machines—SAE **J821** MAY85 ..... 3:40.264

AGRICULTURAL TRACTORS
See: AGRICULTURAL EQUIPMENT

AIR BAGS
See: INFLATABLE RESTRAINTS

AIR BRAKES
See also: BRAKES

Air Brake Actuator Diaphragm Test Procedure (A)—SAE **J1450** MAY88 ..... 3:36.166

Air Brake Actuator Test Procedure—Truck-Tractor, Bus and Trailer (A)—SAE **J1469** JUN88 ..... 3:36.157

Air Brake Gladhand Service (Control) and Emergency (Supply) Line Couplers—Trucks, Truck-Tractors, and Trailers—SAE **J318** SEP80 ..... 3:36.139

Automotive and Off-Highway Air Brake Reservoir Performance and Identification Requirements (A)—SAE **J10** OCT90 ..... 3:36.132

Brake Force/Distribution Performance Guide—Commercial Vehicles (A)—SAE **J1854** JUN88 ..... 3:36.170

* Braking, Stability, and Control Performance Test Procedures for Air-Brake- Equiped Truck Tractors—SAE **J1626** MAR93 ..... 3:36.130

† Contaminants for Testing Air Brake Components and Auxiliary Pneumatic Devices—SAE **J2024** JUN93 ..... 3:36.131

Information Relating to Duty Cycles and Average Power Requirements of Truck and Bus Engine Accessories—SAE **J1343** OCT88 ..... 3:36.57

Long Stroke Air Brake Actuator Marking (A)—SAE **J1817** JUN91 ..... 3:36.167

† Surface Vehicle Brake Manual—SAE **HS-24** JUL92 ..... Available as a separate publication (See Related Technical Reports Section)

---

* Technical Revision † New Document (D)—DODISS Adopted (A)—ANSI Recognized

Subject Index

AIR BRAKES (continued)

Test Method for Measuring Power Consumption of Air Conditioning and Brake Compressors for Trucks and Buses (A)—SAE **J1340** APR90 .... 3:36.53

Test Procedure for Air Reservoir Capacity—Highway Type Vehicles (A)—SAE **J1911** SEP90 .... 3:36.153

Tubing

Metallic Air Brake System Tubing and Pipe (A)—SAE **J1149** JUN91 .... 2:22.219

Metric Nonmetallic Air Brake System Tubing (A)—SAE **J1394** APR91 .... 2:22.227

Nonmetallic Air Brake System Tubing (A)—SAE **J844** JUN90 .... 2:22.221

Performance Requirements for SAE J844 Nonmetallic Tubing and Fitting Assemblies Used in Automotive Air Brake Systems (A)—SAE **J1131** DEC87 .... 2:22.231

Valves

Air Brake Valves Test Procedure (A)—SAE **J1409** JUN88 3:36.171

Air Brake Valve—Performance Requirements—SAE **J1410** OCT86 .... 3:36.173

Labeling Air Brake Valves with Their Performance (Input-Output) Characteristics (A)—SAE **J1860** AUG90 ...... 3:36.164

Test Procedures for Determining Air Brake Valve Input-Output Characteristics—SAE **J1859** APR89 .... 3:36.162

AIR CLEANERS
See also: FILTERS

Air Cleaner Elements—SAE **J1141** JUL87 .... 2:26.213

†Air Cleaner Test Code—SAE **J726** JUN93 .... 2:26.214

AIR CONDITIONING
See also: VENTILATION

*Alternate Refrigerant Consistency Criteria for use in Mobile Air Conditioning Systems—SAE **J658** JUN93 ... 3:34.63

*Cautionary Statements for Handling HFC-134a During Mobile Air Conditioning Service—SAE **J1629** JUN93 .. 3:34.60

CFC-12 (R-12) Extraction Equipment for Mobile Automotive Air-Conditioning Systems—SAE **J2209** JUN92 .... 3:34.46

*Compatibility of Retrofit Refrigerants with Air Conditioning System Materials—SAE **J1662** JUN93 .... 3:34.69

*Equivalent Temperature—SAE **J2234** JAN93 .... 3:34.25

Extraction and Recycle Equipment for Mobile Automotive Air-Conditioning Systems—SAE **J1990** MAR92 ...... 3:34.43

*Fittings and Labels for Retrofit of CFC-12 (R12) Mobile Air Conditioning Systems to HFC-134a (R134a)—SAE **J1660** JUN93 .... 3:34.66

HFC-134a Recycling Equipment for Mobile Air-Conditioning Systems—SAE **J2210** DEC91 .... 3:34.58

*Procedure for Retrofitting CFC-12 (R12) Mobile Air Conditioning Systems to HFC-134a (R134a)—SAE **J1661** JUN93 .... 3:34.67

*Rating Criteria for Electronic Refrigerant Leak Detectors—SAE **J1627** JUN93 .... 3:34.71

Recommended Service Procedure for the Containment of HFC-134a—SAE **J2211** DEC91 .... 3:34.57

Recommended Service Procedure for the Containment of R-12—SAE **J1989** OCT89 .... 3:34.42

Safety and Containment of Refrigerant for Mechanical Vapor Compression Systems Used for Mobile Air-Conditioning Systems—SAE **J639** NOV91 .... 3:34.40

*Selection Criteria for Retrofit Refrigerants to Replace R-12 in Mobile Air Conditioning Systems—SAE **J1657** JUN93 .... 3:34.61

Standard of Purity for Recycled HFC-134a for Use in Mobile Air-Conditioning Systems—SAE **J2099** DEC91 .. 3:34.60

Standard of Purity for Use in Mobile Air-Conditioning Systems—SAE **J1991** OCT89 .... 3:34.47

*Technican Procedure for Using Electronic Refrigerant Leak Detectors for Service of Mobile Air Conditioning Systems—SAE **J1628** JUN93 .... 3:34.72

*Vehicle Testing Requirements for Replacement Refrigerants for CFC-12 (R12) Mobile Air Conditioning Systems—SAE **J659** JUN93 .... 3:34.64

Hoses

Automotive Air Conditioning Hose—SAE **J51** MAY89 ... 1:11.249

HFC-134a (R-134a) Service Hose Fittings for Automotive Air-Conditioning Service Equipment—SAE **J2197** JUN92 3:34.53

*R134a Refrigerant Automotive Air Conditioning Hose—SAE **J2064** JUN93 .... 1:11.219

Service Hose for Automotive Air Conditioning—SAE **J2196** JUN92 .... 3:34.51

Off-Road Vehicles

Design Guidelines for Air Conditioning Systems for Off-Road Operator Enclosures—SAE **J169** MAR85 .... 3:40.510

Performance Test for Air-Conditioned, Heated, and Ventilated Off-Road Self-Propelled Work Machines (A)—SAE **J1503** JUL86 .... 3:40.511

Trucks and Buses

Information Relating to Duty Cycles and Average Power Requirements of Truck and Bus Engine Accessories—SAE **J1343** OCT88 .... 3:36.57

Rating Air Conditioner Evaporator Air Delivery and Cooling Capacities (A)—SAE **J1487** MAY85 .... 3:36.58

Test Method for Measuring Power Consumption of Air Conditioning and Brake Compressors for Trucks and Buses (A)—SAE **J1340** APR90 .... 3:36.53

AIR FILTERS

Air Cleaner Elements—SAE **J1141** JUL87 .... 2:26.213

†Air Cleaner Test Code—SAE **J726** JUN93 .... 2:26.214

†Operator Enclosure Air Filter Element Test Procedure—SAE **J1533** JUN93 .... 3:40.513

AIR FLOW

Airflow Reference Standards (A)—SAE **J228** JUN90 .. 2:26.372

Cooling Flow Measurement Techniques—SAE **J2082** JUN92 .... 3:28.46

Diesel Engines

†Measurement of Intake Air or Exhaust Gas Flow of Diesel Engines (A)—SAE **J244** AUG92 .... 1:13.11

AIR-FUEL RATIO

Impact of Alternative Fuels on Engine Test and Reporting Procedures—SAE **J1515** MAR88 .... 1:12.89

Stoichiometric Air/Fuel Ratios of Automotive Fuels—SAE **J1829** MAY92 .... 1:12.80

ALARMS
See also: WARNING SYSTEMS

Alarm—Backup—Electric—Performance, Test, and Application (A)—SAE **J994** MAR85 .... 1:14.45

On-Machine Alarm Test and Evaluation Procedure for Construction and General Purpose Industrial Machinery—SAE **J1446** MAY89 .... 1:14.50

ALL WHEEL DRIVE

All Wheel Drive Systems Classification (A)—SAE **J1952** JAN91 .... 3:29.15

All-Wheel-Drive Drivetrain Schematic Symbol Standards—SAE **J2059** AUG92 .... 3:29.19

*Introduction to All-Wheel Drive—SAE **HS-2300** AUG93 Available as a separate publication (See Related Technical Reports Section)

ALLOYS
See also: ALUMINUM ALLOYS, COBALT ALLOYS, COPPER ALLOYS, IRON ALLOYS, LEAD ALLOYS, MAGNESIUM ALLOYS, NICKEL ALLOYS, STEELS, ZINC ALLOYS

Fifth Edition Unified Numbering System Handbook for Metals and Alloys (D)—SAE **HS-1086** AUG93 .... Available as a separate publication (See Related Technical Reports Section)

Numbering Systems

†Numbering Metals and Alloys—SAE **J1086** DEC92 .... 1:1.01

---

* Technical Revision   † New Document   (D)—DODISS Adopted   (A)—ANSI Recognized

Subject Index

ALTERNATE FUELS
See also: CHEMICAL PROPERTIES, EMULSIFIED FUELS
† Alternative Automotive Fuels—SAE **J1297** MAR93 .... 1:12.74
† Fuels and Lubricants Standards Manual—SAE **HS-23** FEB93 .... Available as a separate publication (See Related Technical Reports Section)
Impact of Alternative Fuels on Engine Test and Reporting Procedures—SAE **J1515** MAR88 .... 1:12.89

ALTERNATORS
Electrical Charging Systems for Construction and Industrial Machinery (A)—SAE **J180** MAY87 .... 3:40.270
Electrical Generating System (Alternator Type) Performance Curve and Test Procedure (A)—SAE **J56** JUN83 .... 2:23.82
Information Relating to Duty Cycles and Average Power Requirements of Truck and Bus Engine Accessories—SAE **J1343** OCT88 .... 3:36.57

ALTITUDE
† Automotive Gasolines—SAE **J312** JAN93 .... 1:12.47
Recommended Environmental Practices for Electronic Equipment Design—SAE **J1211** 2:23.228

ALUMINUM ALLOYS
See also: CHEMICAL COMPOSITION, MECHANICAL PROPERTIES, PHYSICAL PROPERTIES
Aluminum Alloys—Fundamentals—SAE **J451** JAN89 ... 1:10.01
Bearing and Bushing Alloys—SAE **J459** OCT91 ...... 1:10.43
Bearing and Bushing Alloys—Chemical Composition of SAE Bearing and Bushing Alloys—SAE **J460** OCT91 ... 1:10.41
Technical Report on Fatigue Properties—SAE **J1099** ... 1:3.81
Cast
Alloy and Temper Designation Systems for Aluminum—SAE **J993** JAN89 .... 1:10.02
General Information—Chemical Compositions, Mechanical and Physical Properties of SAE Aluminum Casting Alloys—SAE **J452** JAN89 .... 1:10.06
Coating
Anodized Aluminum Automotive Parts—SAE **J399** FEB85 1:10.38
*Decorative Anodizing Specification for Automotive Applications—SAE **J1974** JUN93 .... 1:10.39
Electroplating of Nickel and Chromium on Metal Parts—Automotive Ornamentation and Hardware—SAE **J207** FEB85 .... 1:10.150
Wrought
Alloy and Temper Designation Systems for Aluminum—SAE **J993** JAN89 .... 1:10.02
Chemical Compositions, Mechanical Property Limits, and Dimensional Tolerances of SAE Wrought Aluminum Alloys (A)—SAE **J457** FEB91 .... 1:10.38
General Data on Wrought Aluminum Alloys (A)—SAE **J454** FEB91 .... 1:10.26
Wrought Aluminum Applications Guidelines—SAE **J1434** JAN89 .... 1:10.15

ANECHOIC CHAMBERS
Anechoic Test Facility Radiated Susceptibility 20 MHz - 18 GHz Electromagnetic Field—SAE **J1507** JAN87 .... 2:23.477

ANEMOMETERS
Cooling Flow Measurement Techniques—SAE **J2082** JUN92 .... 3:28.46

ANNEALING
Copper Alloys
Wrought and Cast Copper Alloys—SAE **J461** SEP81 ... 1:10.45
Steel
General Characteristics and Heat Treatments of Steels—SAE **J412** JUN89 .... 1:2.05
Selection and Heat Treatment of Tool and Die Steels—SAE **J437a** 1:7.02

ANTHROPOMETRY
See also: HUMAN BODY, HUMAN ENGINEERING
Accommodation Tool Reference Point—SAE **J1516** MAR90 .... 3:34.247
Control Locations for Off-Road Work Machines (A)—SAE **J898** OCT87 .... 3:40.404
Devices for Use in Defining and Measuring Vehicle Seating Accommodation—SAE **J826** JUN92 .... 3:34.90
Driver Selected Seat Position (A)—SAE **J1517** MAR90 . 3:34.250
Human Mechanical Response Characteristics—SAE **J1460** MAR85 .... 3:34.292
Human Physical Dimensions—SAE **J833** MAY89 ...... 3:40.377
† Minimum Service Access Dimensions for Off-Road Machines (A)—SAE **J925** JUN93 .... 3:40.380
Truck Driver Shin-Knee Position for Clutch and Accelerator (A)—SAE **J1521** MAR90 .... 3:34.263
Truck Driver Stomach Position (A)—SAE **J1522** MAR90   3:34.265
Test Devices
Occupant Restraint System Evaluation—Passenger Cars—SAE **J128** 3:33.12

ANTIFREEZE
See: COOLANTS

ANTIKNOCK RATINGS
† Automotive Gasolines—SAE **J312** JAN93 .... 1:12.47

ARMATURE
*Starter Armature Remanufacturing Procedures—SAE **J2240** MAR93 .... 3:42.45

ASBESTOS
Gaskets
Standard Classification System for Nonmetallic Automotive Gasket Materials (A)—SAE **J90** JUN90 ..... 1:11.197

AUTOMOBILE DIMENSIONS
Curb Clearance Approach, Departure, and Ramp Breakover Angles—Passenger Car and Light Truck (A)—SAE **J689** DEC89 .... 3:32.03
† Motor Vehicle Dimensions (A)—SAE **J1100** JUN93 .... 3:34.118
Motor Vehicle Fiducial Marks and Three Dimensional Reference System (A)—SAE **J182** NOV90 .... 3:34.229
Uniform Reference and Dimensional Guidelines for Unibody Vehicles—SAE **J1828** MAY87 .... 3:42.28

AUTOMOBILE INTERIORS
See: PASSENGER COMPARTMENTS

AUTOMOBILE TRIM
Accelerated Testing
*Accelerated Exposure of Automotive Interior Trim Components Using a Controlled Irradiance Air-Cooled Xenon-Arc Apparatus—SAE **J2212** MAR93 .... 1:11.263
*Accelerated Exposure of Automotive Interior Trim Materials Using an Outdoor Under Glass Variable Angled Controlled Temperature Apparatus—SAE **J2229** FEB93 . 1:11.119
*Accelerated Exposure of Automotive Interior Trim Material Using Outdoor Under-Glass Controlled Sun-Tracking Temperature and Humidity Apparatus—SAE **J2230** FEB93 .... 1:11.123
Color
Instrumental Color Difference Measurement for Exterior Finishes, Textiles, and Colored Trim (A)—SAE **J1545** JUN86 .... 1:11.279
Exterior
Accelerated Exposure of Automotive Exterior Materials Using a Fluorescent UV and Condensation Apparatus—SAE **J2020** JUN89 .... 1:11.288
Accelerated Exposure of Automotive Exterior Materials Using a Solar Fresnel-Reflective Apparatus—SAE **J1961** DEC88 .... 1:11.286
Accelerated Exposure of Automotive Exterior Materials Using a Controlled Irradiance Water Cooled Xenon Arc Apparatus—SAE **J1960** JUN89 .... 1:11.268
Anodized Aluminum Automotive Parts—SAE **J399** FEB85 1:10.38

---
* Technical Revision   † New Document   (D)—DODISS Adopted   (A)—ANSI Recognized

Subject Index

AUTOMOBILE TRIM (continued)
  Exterior (continued)
  *Decorative Anodizing Specification for Automotive Applications—SAE **J1974** JUN93 ... 1:10.39
  Electroplate Requirements for Decorative Chromium Deposits on Zinc Base Materials Used for Exterior Ornamentation (A)—SAE **J1837** JUN91 ... 1:10.155
  Electroplating of Nickel and Chromium on Metal Parts—Automotive Ornamentation and Hardware—SAE **J207** FEB85 ... 1:10.150
  Outdoor Weathering of Exterior Materials—SAE **J1976** JUN89 ... 1:11.275
  *SAE Automotive Textiles and Trim Standards Manual—SAE **HS-2700** JUN93 ... Available as a separate publication *(See Related Technical Reports Section)*
  Surface Match Verification Method for Pressure Sensitive Adhesively Attached Components—SAE **J2215** DEC91 ... 1:11.267

  Interior
  Accelerated Exposure of Automotive Interior Trim Components Using a Controlled Irradiance Water Cooled Xenon-Arc Apparatus—SAE **J1885** MAR92 ... 1:11.255
  *Decorative Anodizing Specification for Automotive Applications—SAE **J1974** JUN93 ... 1:10.39
  Flammability of Automotive Interior Materials—Horizontal Test Method—SAE **J369** JUN89 ... 1:11.176
  † Hot Odor Test for Insulation Materials—SAE **J1351** JUN93 ... 1:11.172
  Hot Plate Method for Evaluating Heat Resistance and Thermal Insulation Properties of Materials—SAE **J1361** APR87 ... 1:11.171
  Method of Testing Resistance to Crocking of Organic Trim Materials—SAE **J861** ... 1:11.114
  Method of Testing Resistance to Scuffing of Trim Materials (A)—SAE **J365** FEB85 ... 1:11.163
  Nonmetallic Trim Materials—Test Method for Determining the Staining Resistance to Hydrogen Sulfide Gas (A)—SAE **J322** DEC85 ... 1:11.116
  *SAE Automotive Textiles and Trim Standards Manual—SAE **HS-2700** DEC93 ... Available as a separate publication *(See Related Technical Reports Section)*
  Test Method for Determining Blocking Resistance and Associated Characteristics of Automotive Trim Materials—SAE **J912a** ... 1:11.165
  Test Method for Determining Dimensional Stability of Automotive Textile Materials (A)—SAE **J883** FEB86 ... 1:11.114
  Test Method for Determining Resistance to Abrasion of Automotive Bodycloth, Vinyl, and Leather, and the Snagging of Automotive Bodycloth (A)—SAE **J948** FEB86 ... 1:11.178
  Test Method for Determining Visual Color Match to Master Specimen for Fabrics, Vinyls, Coated Fiberboards, and Other Automotive Trim Materials (A)—SAE **J361** MAR85 ... 1:11.180
  Test Method for Determining Window Fogging Resistance of Interior Trim Materials (A)—SAE **J275** FEB85 ... 1:11.117
  Test Method for Measuring Weight of Organic Trim Materials (A)—SAE **J860** JAN85 ... 1:11.114
  Test Method for Wicking of Automotive Fabrics and Fibrous Materials (A)—SAE **J913** FEB85 ... 1:11.163

AUTOMOBILES
  See also: AUTOMOBILE DIMENSIONS, AUTOMOBILE TRIM
  Acceleration
  Vehicle Acceleration Measurement—SAE **J1491** JUN90 ... 2:26.503

  Aerodynamics
  Aerodynamic Flow Visualization Techniques and Procedures—SAE **HS-1566** ... Available as a separate publication *(See Related Technical Reports Section)*
  *Aerodynamic Testing of Road Vehicles--Testing Methods and Procedures—SAE **J2084** JAN93 ... 3:28.61
  Vehicle Aerodynamics Terminology (A)—SAE **J1594** JUN87 ... 3:28.17

  Brakes
  Antilock Brake System Review—SAE **J2246** JUN92 ... 2:25.72
  Automotive and Off-Highway Air Brake Reservoir Performance and Identification Requirements (A)—SAE **J10** OCT90 ... 3:36.132
  Automotive Brake Definitions and Nomenclature (A)—SAE **J656** APR88 ... 2:25.70
  Automotive Hydraulic Brake System—Metric Banjo Bolt Connections—SAE **J1291** MAY89 ... 2:25.170
  Automotive Hydraulic Brake System—Metric Tube Connections—SAE **J1290** MAY89 ... 2:25.165
  Brake System Dynamometer Test Procedure—Passenger Car—SAE **J212** JUN80 ... 2:25.129
  Brake System Road Test Code—Passenger Car and Light Duty Truck-Trailer Combinations (A)—SAE **J134** MAY85 ... 2:25.134
  Brake System Road Test Code—Passenger Car and Light-Duty Truck (A)—SAE **J843** NOV90 ... 2:25.108
  Brake Test Code—Inertia Dynamometer—SAE **J667** ... 2:25.103
  Hydraulic Master Cylinders for Motor Vehicle Brakes—Performance Requirements (A)—SAE **J1154** JUN91 ... 2:25.34
  Hydraulic Master Cylinders for Motor Vehicle Brakes—Test Procedure (A)—SAE **J1153** JUN91 ... 2:25.29
  Hydraulic Valves for Motor Vehicle Brake Systems—Performance Requirements—SAE **J1137** MAR85 ... 2:25.39
  Hydraulic Wheel Cylinders for Automotive Drum Brakes (A)—SAE **J101** DEC89 ... 2:25.25
  In-Service Brake Performance Test Procedure Passenger-Car and Light-Duty Truck—SAE **J201** MAY89 ... 2:25.85
  Service Brake Structural Integrity Test Procedure—Passenger Car—SAE **J229** JUN80 ... 2:25.133
  Service Brake System Performance Requirements—Passenger Car—Trailer Combinations—SAE **J135** MAY85 ... 2:25.143
  Simulated Mountain Brake Performance Test Procedure—SAE **J1247** APR80 ... 2:25.123
  Stopping Distance Test Procedure (A)—SAE **J299** AUG87 ... 2:25.156
  † Surface Vehicle Brake Manual—SAE **HS-24** JUL92 ... Available as a separate publication *(See Related Technical Reports Section)*
  Wheel Slip Brake Control System Road Test Code—SAE **J46** JUN80 ... 2:25.158

  Bumpers
  Bumper Evaluation Test Procedure—Passenger Cars—SAE **J980a** ... 3:32.02

  Defoggers
  † Passenger Car Backlight Defogging System (A)—SAE **J953** APR93 ... 3:34.28

  Defrosters
  † Passenger Car Windshield Defrosting Systems—SAE **J902** APR93 ... 3:34.18

  Electric Equipment
  Automobile, Truck, Truck-Tractor, Trailer, and Motor Coach Wiring—SAE **J1292** OCT81 ... 2:23.142
  Automotive Printed Circuits (A)—SAE **J771** APR86 ... 2:23.145
  Electric Hourmeter Specification—SAE **J1378** MAR83 ... 2:21.14
  Fusible Links (A)—SAE **J156** APR86 ... 2:23.171
  Ignition Switch—SAE **J259** ... 2:23.141

* Technical Revision † New Document (D)—DODISS Adopted (A)—ANSI Recognized

AUTOMOBILES (continued)

Fiducial Marks
† Motor Vehicle Dimensions (A)—SAE J1100 JUN93 .... 3:34.118
Motor Vehicle Fiducial Marks and Three Dimensional Reference System (A)—SAE J182 NOV90 .... 3:34.229

Fuel Tanks
Filler Pipes and Openings of Motor Vehicle Fuel Tanks (A)—SAE J1140 FEB88 .... 2:26.433
Fuel Tank Filler Conditions—Passenger Car, Multi-Purpose Passenger Vehicles, and Light Duty Trucks—SAE J398 FEB88 .... 2:26.436

Heating
† Motor Vehicle Heater Test Procedure—SAE J638 APR93 3:34.34

Lift Points
* Vehicle Lift Points for Service Garage Lifting—SAE J2184 OCT92 .... 3:42.53

Lubricants
Classification of Energy-Conserving Engine Oil for Passenger Cars, Vans, and Light-Duty Trucks—SAE J1423 FEB92 .... 1:12.29

Noise
Maximum Sound Level for Passenger Cars and Light Trucks (A)—SAE J1030 FEB87 .... 1:14.17
Measurement of Light Vehicle Exhaust Sound Level Under Stationary Conditions—SAE J1169 MAR92 ..... 1:14.21
Measurement of Light Vehicle Stationary Exhaust System Sound Level Engine Speed Sweep Method—SAE J1492 MAR92 .... 1:14.24
Measurement of Noise Emitted by Accelerating Highway Vehicles—SAE J1470 MAR92 .... 1:14.32
Sound Level for Passenger Cars and Light Trucks (D)—SAE J986 OCT88 .... 1:14.14

Operator Controls
Automotive Adaptive Driver Controls, Manual (A)—SAE J1903 AUG90 .... 3:34.198
Design Criteria—Driver Hand Controls Location for Passenger Cars, Multi-Purpose Passenger Vehicles, and Trucks (10 000 GVW and Under)—SAE J1138 .... 3:34.191
Supplemental Information—Driver Hand Controls Location for Passenger Cars, Multi-Purpose Passenger Vehicles, and Trucks (10 000 GVW and Under)—SAE J1139 .... 3:34.193

Restraint Systems
† Vehicle Occupant Restraint Systems and Components Standards Manual—SAE HS-13 MAY93 .... Available as a separate publication (See Related Technical Reports Section)

Safety
Anchorage Provisions for Installation of Child Restraint Tether Straps in Rear Seating Positions—SAE J1369 MAR92 .... 3:33.16
† Automotive Safety Glasses—SAE J673 APR93 .... 3:34.100
Collision Deformation Classification—SAE J224 MAR80 3:34.235
Fixed Rigid Barrier Collision Tests—SAE J850 NOV88 . 3:34.281
Human Tolerance to Impact Conditions as Related to Motor Vehicle Design—SAE J885 JUL86 .... 3:34.306
Instrumentation for Impact Test (A)—SAE J211 OCT88 . 3:34.270
Moving Rigid Barrier Collision Tests—SAE J972 DEC88 3:34.278
Passenger Car Door System Crush Test Procedure—SAE J367 JUN80 .... 3:34.286
Safety Glazing Materials—Motor Vehicles (A)—SAE J674 NOV90 .... 3:34.103
* The Air Bag Systems in Your Cars 'What the Public Needs to Know'—SAE J2074 JUN93 .... 3:33.42

Seats
Accommodation Tool Reference Point—SAE J1516 MAR90 .... 3:34.247
Driver Selected Seat Position (A)—SAE J1517 MAR90 . 3:34.250

Speed Control
Automatic Vehicle Speed Control—Motor Vehicles—SAE J195 DEC88 .... 2:21.16
Transmission Mounted Vehicle Speed Signal Rotor Specification—SAE J1377 JAN89 .... 2:21.17

Tires
† Laboratory Speed Test Procedure for Passenger Car Tires—SAE J1561 FEB93 .... 3:30.34
Laboratory Testing Machines and Procedures for Measuring the Steady State Force and Moment Properties of Passenger Car Tires—SAE J1107 FEB75 . 3:30.43
Laboratory Testing Machines for Measuring the Steady State Force and Moment Properties of Passenger Car Tires—SAE J1106 FEB75 .... 3:30.40
Measurement of Passenger Car, Light Truck, and Highway Truck and Bus Tire Rolling Resistance—SAE J1270 MAR87 .... 3:30.21
Passenger and Light Truck Tire Traction Device Profile Determination and Classification—SAE J1232 .... 3:36.115
Passenger Car and Light Truck Tire Dynamic Driving Traction in Snow (A)—SAE J1466 OCT85 .... 3:30.26
Passenger Car Tire Performance Requirements and Test Procedures—SAE J918c 3:30.01
Rolling Resistance Measurement Procedure for Passenger Car, Light Truck, and Highway Truck and Bus Tires (A)—SAE J1269 MAR87 .... 3:30.18
Test Procedure for Measuring Passenger Car Tire Revolutions per Mile—SAE J966 .... 3:30.36
Testing Machines for Measuring the Uniformity of Passenger Car and Light Truck Tires (A)—SAE J332 AUG81 .... 3:30.32
Wet or Dry Pavement Passenger Car Tire Peak and Locked Wheel Braking Traction—SAE J345a 3:30.24

Towing
* Recovery Attachment Points for Passenger Cars, Vans, and Light Trucks—SAE J2069 JAN93 .... 3:42.21

Transmissions
Automatic Transmission Functions—Terminology (A)—SAE J649 JUL88 .... 3:29.27
† Automatic Transmissions—Manual Control Sequence—SAE J915 APR93 .... 3:29.32
† Automotive Transmission Terminology—SAE J645 OCT92 .... 3:29.23
Passenger Car and Truck Automatic Transmission Test Code (A)—SAE J651 JAN91 .... 3:29.28
Transmissions—Schematic Diagrams (A)—SAE J647 APR90 .... 3:29.24

Vehicle Identification Number (VIN)
Passenger Car Identification Terminology—SAE J218 DEC81 .... 3:27.02
Passenger Car Vehicle Identification Number System—SAE J273 DEC81 .... 3:27.02
Vehicle Identification Numbers—SAE J853 DEC81 .... 3:27.03

Vision
Passenger Car Rear Vision—SAE J834a 3:34.114

Wheels
Wheels—Impact Test Procedures—Road Vehicles (A)—SAE J175 JUN88 .... 3:31.07
Wheels—Passenger Cars—Performance Requirements and Test Procedures (A)—SAE J328 MAR90 .... 3:31.06

Windshields
† Passenger Car Windshield Defrosting Systems—SAE J902 APR93 .... 3:34.18
Passenger Car Windshield Washer Systems (D)—SAE J942b 3:34.16
Passenger Car Windshield Wiper Systems—SAE J903c 3:34.07

AXLES
Axle Application Load Rating for Industrial Wheel Loaders and Backhoe Loaders—SAE J43 FEB88 ..... 3:40.333
Axle Efficiency Test Procedure (A)—SAE J1266 JUN90 . 3:29.85

---

* Technical Revision   † New Document   (D)—DODISS Adopted   (A)—ANSI Recognized

Subject Index

AXLES (continued)
Passenger Cars and Light Truck Axles (A)—SAE **J2200** JAN91 .... 3:29.88
Trailer Axle Alignment (A)—SAE **J875** JUN85 .... 3:36.112
Lubricants
Axle and Manual Transmission Lubricant Viscosity Classification—SAE **J306** OCT91 .... 1:12.37
Axle and Manual Transmission Lubricants—SAE **J308** JUN89 .... 1:12.36
The Automotive Lubricant Performance and Service Classification Maintenance Procedure (A)—SAE **J1146** JUN86 .... 1:12.87
Terminology
Nomenclature and Terminology for Truck and Bus Drive Axles (A)—SAE **J923** APR91 .... 3:29.72

BACK LAMPS
See: BACKUP LAMPS, STOP LAMPS, TAIL LAMPS

BACKHOES
See also: CONSTRUCTION AND INDUSTRIAL EQUIPMENT, OFF-ROAD VEHICLES
† Excavator, Mini-Excavator, and Backhoe Hoe Bucket Volumetric Rating—SAE **J296** JUN93 .... 3:40.329
Hydraulic Backhoe Lift Capacity (A)—SAE **J31** MAR86 . 3:40.331
Hydraulic Excavator and Backhoe Digging Forces (A)—SAE **J1179** FEB90 .... 3:40.340
Axles
Axle Application Load Rating for Industrial Wheel Loaders and Backhoe Loaders—SAE **J43** FEB88 ..... 3:40.333
Noise
Sound Measurement—Off-Road Self-Propelled Work Machines—Operator—Work Cycle (A)—SAE **J1166** MAY90 .... 1:14.55
† Sound Power Level Measurements of Earthmoving Machinery - Static and In-Place Dynamic Methods—SAE **J1805** APR93 .... 1:14.90
Safety
Three Point Hitch (Type A) Backhoe Personnel Protection—SAE **J1518** JUN86 .... 3:40.353
Terminology
Identification Terminology of Earthmoving Machines (A)—SAE **J1057** SEP88 .... 3:40.05
Identification Terminology of Loaders/Tractors with Forks and Rough Terrain Forklifts—SAE **J1464** APR88 ...... 3:40.58
Nomenclature—Hydraulic Backhoes (A)—SAE **J326** MAR86 .... 3:40.326
Specification Definitions—Hydraulic Backhoes (A)—SAE **J49** APR80 .... 3:40.70

BACKLIGHTS
See: REAR WINDOWS

BACKUP ALARMS
See: ALARMS, WARNING SYSTEMS

BACKUP LAMPS
See also: TAIL LAMPS
Backup Lamp Switch—SAE **J1076** MAR90 .... 2:24.175
Backup Lamps (Reversing Lamps) (A) (D)—SAE **J593** JUN87 .... 2:24.174
† SAE Ground Vehicle Lighting Standards Manual—SAE **HS-34** JAN93 .... Available as a separate publication *(See Related Technical Reports Section)*

BALANCE WEIGHTS
*Balance Weight and Rim Flange Design Specifications, Test Procedures, and Performance Recommendations—SAE **J1986** FEB93 .... 3:31.10

BALL JOINTS
See also: SUSPENSION SYSTEMS
Ball Joints—SAE **J490** OCT81 .... 2:17.08
Metric Ball Joints (A)—SAE **J2213** JUN91 .... 2:17.12

Metric Spherical Rod Ends—SAE **J1259** JUN89 .... 2:17.16
Performance Test Procedure—Ball Joints and Spherical Rod Ends (A)—SAE **J1367** SEP81 .... 2:17.18
Spherical Rod Ends—SAE **J1120** JUN89 .... 2:17.14

BALL STUDS
Ball Stud and Socket Assembly—Test Procedures (A)—SAE **J193** FEB87 .... 2:17.04
Steering Ball Studs and Socket Assemblies (A)—SAE **J491** NOV87 .... 2:17.01

BARS
Copper Alloys
Wrought Copper and Copper Alloys—SAE **J463** SEP81 . 1:10.81
Ferrous Materials
Detection of Surface Imperfections in Ferrous Rods, Bars, Tubes, and Wires—SAE **J349** FEB91 .... 1:3.63
Magnesium Alloys
Magnesium Casting Alloys—SAE **J465** JAN89 .... 1:10.120
Steel
Estimated Mechanical Properties and Machinability of Steel Bars—SAE **J1397** MAY92 .... 1:2.26
High Strength, Hot Rolled Steel Plates, Bars and Shapes—SAE **J1442** DEC88 .... 1:1.162
Selection and Heat Treatment of Tool and Die Steels—SAE **J437a** 1:7.02

BATTERIES
Vehicle System Voltage—Initial Recommendations—SAE **J2232** JUN92 .... 2:23.75
Cables
Battery Booster Cables—SAE **J1494** JUN89 .... 2:23.159
Battery Cable (A)—SAE **J1127** JUN88 .... 2:23.158
Storage
† Circuit Breakers—SAE **J553** JUN92 .... 2:23.135
Grounding of Storage Batteries (A)—SAE **J538** JUL89 . 2:23.74
† Life Test for Automotive Storage Batteries—SAE **J240** JUN93 .... 2:23.71
Life Test for Heavy-Duty Storage Batteries—SAE **J2185** NOV91 .... 2:23.72
Storage Batteries—SAE **J537** JUN92 .... 2:23.62
Storage Batteries for Off-Road Work Machines (A)—SAE **J930** AUG84 .... 3:40.528
Venting System
Test Procedure for Battery Flame Retardant Venting Systems—SAE **J1495** MAR92 .... 2:23.72
Warmers
† Electric Engine Preheaters and Battery Warmers for Diesel Engines—SAE **J1310** JUN93 .... 3:40.500

BEARINGS
Bearing and Bushing Alloys—SAE **J459** OCT91 .... 1:10.43
Bearing and Bushing Alloys—Chemical Composition of SAE Bearing and Bushing Alloys—SAE **J460** OCT91 ... 1:10.41
Metric Spherical Rod Ends—SAE **J1259** JUN89 .... 2:17.16
Performance Test Procedure—Ball Joints and Spherical Rod Ends (A)—SAE **J1367** SEP81 .... 2:17.18
Sintered Powder Metal Parts: Ferrous—SAE **J471d** 1:8.01
Sleeve Type Half Bearings—SAE **J506b** 2:26.194
Spherical Rod Ends—SAE **J1120** JUN89 .... 2:17.14

BELTS
See also: SAFETY BELTS
† Automotive Synchronous Belt Drives (A)—SAE **J1313** MAR93 .... 2:19.13
Automotive V-Belt Drives—SAE **J637** FEB89 .... 2:19.04
Automotive V-Ribbed Belt Drives and Test Methods—SAE **J1596** JUN89 .... 2:19.17
*Glossary Automatic Belt Tensioner—SAE **J2198** JUL92 . 2:19.19
† SI (Metric) Synchronous Belts and Pulleys—SAE **J1278** MAR93 .... 2:19.09
V-Belts and Pulleys—SAE **J636** MAY92 .... 2:19.01
V-Ribbed Belts and Pulleys—SAE **J1459** AUG88 ...... 2:19.15

---

* Technical Revision   † New Document   (D)—DODISS Adopted   (A)—ANSI Recognized

BICYCLES
Terminology
A Dictionary of Terms for the Dynamics and Handling of Single Track Vehicles (Motorcycles, Mopeds, and Bicycles)—SAE **J1451** FEB85 . . . . 3:37.01

BIOMECHANICS
Human Mechanical Response Characteristics—SAE **J1460** MAR85 . . . . 3:34.292

Human Tolerance to Impact Conditions as Related to Motor Vehicle Design—SAE **J885** JUL86 . . . . 3:34.306

BLADES
Turbine
Hydrodynamic Drives Terminology (A)—SAE **J641** JUL88 3:29.02

BLOWBY TESTS
Crankcase Emission Control Test Code (A)—SAE **J900** AUG85 . . . . 1:13.69

BLOWERS
Switches
Electric Blower Motor Switch—SAE **J235** 2:23.189

BOATS
See: MARINE EQUIPMENT, MOTORBOATS

BODIES (Automobile)
See also: DOORS, FRAMES, PASSENGER COMPARTMENTS, ROOFS

Accelerated Exposure of Automotive Exterior Materials Using a Fluorescent UV and Condensation Apparatus—SAE **J2020** JUN89 . . . . 1:11.288

Accelerated Exposure of Automotive Exterior Materials Using a Solar Fresnel-Reflective Apparatus—SAE **J1961** DEC88 . . . . 1:11.286

Accelerated Exposure of Automotive Exterior Materials Using a Controlled Irradiance Water Cooled Xenon Arc Apparatus—SAE **J1960** JUN89 . . . . 1:11.268

Outdoor Weathering of Exterior Materials—SAE **J1976** JUN89 . . . . 1:11.275

Aluminum
Wrought Aluminum Applications Guidelines—SAE **J1434** JAN89 . . . . 1:10.15

Corrosion
Corrosion Preventive Compound, Topside Vehicle Corrosion Protection—SAE **J1804** JUN89 . . . . 1:3.52

Corrosion Preventive Compound, Underbody Vehicle Corrosion Protection—SAE **J1959** JUN89 . . . . 1:3.55

Guidelines and Limitations of Laboratory Cyclic Corrosion Test Procedures for Exterior, Painted, Automotive Body Panels—SAE **J1563** MAY87 . . . . 1:3.51

Prevention of Corrosion of Motor Vehicle Body and Chassis Components—SAE **HS-J447** JUN81 . . . . Available as a separate publication (See Related Technical Reports Section)

Undervehicle Coupon Corrosion Tests—SAE **J1293** JAN90 . . . . 1:3.48

Lift Points
*Vehicle Lift Points for Service Garage Lifting—SAE **J2184** OCT92 . . . . 3:42.53

Repair
Uniform Reference and Dimensional Guidelines for Unibody Vehicles—SAE **J1828** MAY87 . . . . 3:42.28

Welding
Unibody Weld Quality Testing (A)—SAE **J1827** MAY87 . 3:42.23

BOLTS
See also: FASTENERS

Automotive Hydraulic Brake System—Metric Banjo Bolt Connections—SAE **J1291** MAY89 . . . . 2:25.170

Decarburization in Hardened and Tempered Threaded Fasteners (A)—SAE **J121** AUG83 . . . . 1:4.44

Fan Hub Bolt Circles and Pilot Holes (A)—SAE **J635** JUL84 . . . . 2:26.414

Mechanical and Material Requirements for Externally Threaded Fasteners (A)—SAE **J429** AUG83 . . . . 1:4.01

Mechanical and Material Requirements for Metric Externally Threaded Steel Fasteners (A)—SAE **J1199** SEP83 . . . . 1:4.06

Mechanical and Material Requirements for Wheel Bolts—SAE **J1102** 1:4.18

Metric Countersunk Holes for Cutting Edges and End Bits (A)—SAE **J1580** DEC89 . . . . 3:40.238

Rivets for Brake Linings and Bolts for Brake Blocks—SAE **J663b** 2:25.94

Surface Discontinuities on Bolts, Screws, and Studs—SAE **J123c** 1:4.37

†Surface Discontinuities on General Application Bolts, Screws, and Studs—SAE **J1061** OCT92 . . . . 1:4.33

Test Methods for Metric Threaded Fasteners (A)—SAE **J1216** 1:4.10

Torque-Tension Test Procedure for Steel Threaded Fasteners—SAE **J174** 2:16.22

BONDING
Metal to Metal Overlap Shear Strength Test for Automotive Type Adhesives (A)—SAE **J1523** JUN85 . . . 1:11.78

Seals—Bond Test Fixture and Procedure (A)—SAE **J1900** OCT90 . . . . 3:29.128

Accelerated Testing
*Accelerated Environmental Testing for Bonded Automotive Assemblies—SAE **J2100** AUG92 . . . . 1:11.63

Aluminum Alloys
Wrought Aluminum Applications Guidelines—SAE **J1434** JAN89 . . . . 1:10.15

Brakes
Test Procedures for Brake Shoe and Lining Adhesives and Bonds—SAE **J840** AUG82 . . . . 2:25.95

Glass
Overlap Shear Test for Automotive Type Sealant for Stationary Glass Bonding (A)—SAE **J1529** MAY86 . . . 1:11.95

Overlap Shear Test for Sealant Adhesive Bonding of Automotive Glass Encapsulating Material to Body Opening—SAE **J1836** OCT88 . . . . 1:11.97

Peel Adhesion Test for Glass to Elastomeric Material for Automotive Glass Encapsulation—SAE **J1907** OCT88 . 1:11.99

Stationary Glass Replacement—SAE **J1556** DEC85 . . . 3:42.21

Metal
Induction Cure Test for Metal Bonding Adhesives (A)—SAE **J1851** MAY87 . . . . 1:11.109

Plastics
Cross Peel Test for Automotive Type Adhesives for Fiber Reinforced Plastic (FRP) Bonding (A)—SAE **J1553** MAY86 . . . . 1:11.100

Lap Shear Test for Automotive Type Adhesives for Fiber Reinforced Plastic (FRP) Bonding (A)—SAE **J1525** JUN85 . . . . 1:11.81

Method for Evaluating the Cleavage Strength of Structurally Bonded Fiberglass Reinforced Plastic (Wedge Test) (A)—SAE **J1882** AUG87 . . . . 1:11.102

BOOMS
See also: CRANES

A Recommended Method of Analytically Determining the Competence of Hydraulic Telescopic Cantilevered Crane Booms—SAE **J1078** APR86 . . . . 3:40.196

Crane Boom Hoist Disengaging Device—SAE **J999** FEB85 . . . . 3:40.216

Crane Boomstop (A)—SAE **J220** MAR91 . . . . 3:40.212

Identification Terminology & Component Nomenclature—Knuckle Boom Log Loader (A)—SAE **J2055** JUN91 . . . 3:40.67

Latticed Crane Boom Systems—Analytical Procedure—SAE **J1093** JUN82 . . . . 3:40.167

Lift Capacity Calculation and Test Procedure—Pipelayer and Sideboom—SAE **J743** MAR92 . . . . 3:40.223

---

*Technical Revision   † New Document   (D)—DODISS Adopted   (A)—ANSI Recognized

Subject Index

BOOMS (continued)
Radius-of-Load or Boom Angle Indicating Systems—SAE J375 APR85 .... 3:40.214
Telescopic Boom Length Indicating System (A)—SAE J1180 OCT80 .... 3:40.213
Two-Block Warning and Limit Systems in Lifting Crane Service—SAE J1305 JUN87 .... 3:40.216

BOOTS
Front-Wheel-Drive Constant Velocity Joint Boot Seals—SAE J2028 JUN92 .... 3:29.07

BRAKE BLOCKS
Brake Block Chamfer (A)—SAE J662 NOV90 .... 2:25.93
*Brake Block Effectiveness Rating—SAE J1802 JUN93 .... 2:25.119
*Brake Effectiveness Marking for Brake Blocks—SAE J1801 JUN93 .... 2:25.118
Rivets for Brake Linings and Bolts for Brake Blocks—SAE J663b 2:25.94
†Surface Vehicle Brake Manual—SAE HS-24 JUL92 .... Available as a separate publication (See Related Technical Reports Section)

BRAKE CYLINDERS
Hydraulic Master Cylinders for Motor Vehicle Brakes—Performance Requirements (A)—SAE J1154 JUN91 .... 2:25.34
Hydraulic Master Cylinders for Motor Vehicle Brakes—Test Procedure (A)—SAE J1153 JUN91 .... 2:25.29
Hydraulic Wheel Cylinders for Automotive Drum Brakes (A)—SAE J101 DEC89 .... 2:25.25
*Materials for Plastic Pistons for Hydraulic Disc Brake Cylinders—SAE J1568 JUN93 .... 2:25.18
Rubber Boots for Drum-Type Hydraulic Brake Wheel Cylinders (A)—SAE J1604 OCT89 .... 2:25.24
Rubber Cups for Hydraulic Actuating Cylinders (A)—SAE J1601 NOV90 .... 2:25.175
Rubber Seals for Hydraulic Disc Brake Cylinders (A)—SAE J1603 JUN90 .... 2:25.19
†Surface Vehicle Brake Manual—SAE HS-24 JUL92 .... Available as a separate publication (See Related Technical Reports Section)

BRAKE DRUMS
†Automotive Gray Iron Castings—SAE J431 MAR93 .... 1:6.01
*Brake Performance and Wear Test Code Commercial Vehicle Inertia Dynamometer—SAE J2115 JUN93 ...... 2:25.106
Brake Test Code—Inertia Dynamometer—SAE J667 ... 2:25.103
Rubber Boots for Drum-Type Hydraulic Brake Wheel Cylinders (A)—SAE J1604 OCT89 .... 2:25.24
†Surface Vehicle Brake Manual—SAE HS-24 JUL92 .... Available as a separate publication (See Related Technical Reports Section)

BRAKE FLUIDS
See also: LUBRICANTS
†Contaminants for Testing Air Brake Components and Auxiliary Pneumatic Devices—SAE J2024 JUN93 ..... 3:36.131
Determination of Brake Fluid Temperature—SAE J291 JUN80 .... 2:25.157
*European Brake Fluid Technology—SAE J1709 .... 2:25.64
Hot Impulse Test for Hydraulic Brake Hose Assemblies—SAE J1833 NOV88 .... 2:25.17
Low Water Tolerant Brake Fluids (A)—SAE J1705 OCT88 .... 2:25.56
Motor Vehicle Brake Fluid (A)—SAE J1703 JUN91 .... 2:25.41
Motor Vehicle Brake Fluid Container Compatibility—SAE J75 OCT88 .... 2:25.66
Production, Handling and Dispensing of SAE J1703 Motor Vehicle Brake Fluids (A)—SAE J1706 OCT88 ... 2:25.68
†Road Vehicle—Hydraulic Brake Hose Assemblies for Use with Nonpetroleum-Base Hydraulic Fluids—SAE J1401 JUN93 .... 2:25.01
Service Maintenance of SAE J1703 Brake Fluids in Motor Vehicle Brake Systems—SAE J1707 NOV91 .... 2:25.53
†Surface Vehicle Brake Manual—SAE HS-24 JUL92 .... Available as a separate publication (See Related Technical Reports Section)

BRAKE HOSES
See also: HOSES
†Application of Hydraulic Brake Hose to Motor Vehicles—SAE J1406 JUN93 .... 2:25.16
Automotive Air Brake Hose and Hose Assemblies (A)—SAE J1402 JUN85 .... 2:25.11
Dynamic Ozone Test Procedure—Hydraulic Brake Hose (A)—SAE J1914 MAR88 .... 2:25.10
Hot Impulse Test for Hydraulic Brake Hose Assemblies—SAE J1833 NOV88 .... 2:25.17
†Moisture Transmission Test Procedure—Hydraulic Brake Hose Assemblies—SAE J1873 JUN93 .... 2:25.09
Packaging, Storage, and Shelf Life of Hydraulic Brake Hose Assemblies—SAE J1288 MAR90 .... 2:25.13
†Road Vehicle—Hydraulic Brake Hose Assemblies for Use with Nonpetroleum-Base Hydraulic Fluids—SAE J1401 JUN93 .... 2:25.01
†Surface Vehicle Brake Manual—SAE HS-24 JUL92 .... Available as a separate publication (See Related Technical Reports Section)
Vacuum Brake Hose (A)—SAE J1403 JUL89 .... 2:25.14

BRAKE LININGS
Brake Lining Quality Control Test Procedure (A)—SAE J661 AUG87 .... 2:25.90
Friction Coefficient Identification System for Brake Linings (A)—SAE J866 NOV90 .... 2:25.89
Gogan Hardness of Brake Lining—SAE J379a 2:25.88
Parking Brake Structural Integrity Test Procedure—Vehicles Over 10,000 Pounds (4500 KG) GVWR (A)—SAE J1476 JUN87 .... 3:36.123
Rivets for Brake Linings and Bolts for Brake Blocks—SAE J663b 2:25.94
†Specific Gravity of Brake Lining—SAE J380 FEB93 .... 2:25.89
†Surface Vehicle Brake Manual—SAE HS-24 JUL92 .... Available as a separate publication (See Related Technical Reports Section)
Swell, Growth, and Dimensional Stability of Brake Linings (A)—SAE J160 JUN80 .... 2:25.90
Test Procedures for Brake Shoe and Lining Adhesives and Bonds—SAE J840 AUG82 .... 2:25.95

BRAKE SEALS
Low Water Tolerant Brake Fluids (A)—SAE J1705 OCT88 .... 2:25.56
Motor Vehicle Brake Fluid (A)—SAE J1703 JUN91 .... 2:25.41
Rubber Boots for Drum-Type Hydraulic Brake Wheel Cylinders (A)—SAE J1604 OCT89 .... 2:25.24
Rubber Dust Boots for the Hydraulic Disc Brake Piston—SAE J1570 SEP91 .... 2:25.181
Rubber Seals for Hydraulic Disc Brake Cylinders (A)—SAE J1603 JUN90 .... 2:25.19

BRAKE SHOES
†Surface Vehicle Brake Manual—SAE HS-24 JUL92 .... Available as a separate publication (See Related Technical Reports Section)
Test Procedures for Brake Shoe and Lining Adhesives and Bonds—SAE J840 AUG82 .... 2:25.95

---
* Technical Revision   † New Document   (D)—DODISS Adopted   (A)—ANSI Recognized

Subject Index

BRAKES
See also: AIR BRAKES, BRAKE BLOCKS, BRAKE CYLINDERS, BRAKE DRUMS, BRAKE FLUIDS, BRAKE HOSES, BRAKE LININGS, BRAKE SEALS, BRAKE SHOES, HYDRAULIC SYSTEMS

Shelf Storage of Hydraulic Brake Components—SAE J1825 APR88 .... 2:25.35

† Surface Vehicle Brake Manual—SAE **HS-24** JUL92 .... Available as a separate publication *(See Related Technical Reports Section)*

Actuators
Air Brake Actuator Diaphragm Test Procedure (A)—SAE **J1450** MAY88 .... 3:36.166

Air Brake Actuator Test Procedure—Truck-Tractor, Bus and Trailer (A)—SAE **J1469** JUN88 .... 3:36.157

Long Stroke Air Brake Actuator Marking (A)—SAE **J1817** JUN91 .... 3:36.167

Agricultural Equipment
Braking System Test Procedures and Braking Performance Criteria for Agricultural Tractors—SAE **J1041** OCT91 .... 3:39.29

Automobile
Antilock Brake System Review—SAE **J2246** JUN92 .... 2:25.72

Automotive and Off-Highway Air Brake Reservoir Performance and Identification Requirements (A)—SAE **J10** OCT90 .... 3:36.132

Automotive Brake Definitions and Nomenclature (A)—SAE **J656** APR88 .... 2:25.70

Automotive Hydraulic Brake System—Metric Banjo Bolt Connections—SAE **J1291** MAY89 .... 2:25.170

Automotive Hydraulic Brake System—Metric Tube Connections—SAE **J1290** MAY89 .... 2:25.165

Brake System Dynamometer Test Procedure—Passenger Car—SAE **J212** JUN80 .... 2:25.129

Brake System Road Test Code—Passenger Car and Light Duty Truck-Trailer Combinations (A)—SAE **J134** MAY85 .... 2:25.134

Brake System Road Test Code—Passenger Car and Light-Duty Truck (A)—SAE **J843** NOV90 .... 2:25.108

Brake Test Code—Inertia Dynamometer—SAE **J667** ... 2:25.103

Hydraulic Wheel Cylinders for Automotive Drum Brakes (A)—SAE **J101** DEC89 .... 2:25.25

In-Service Brake Performance Test Procedure Passenger-Car and Light-Duty Truck—SAE **J201** MAY89 2:25.85

Service Brake Structural Integrity Test Procedure—Passenger Car—SAE **J229** JUN80 .... 2:25.133

Service Brake System Performance Requirements—Passenger Car—Trailer Combinations—SAE **J135** MAY85 .... 2:25.143

Simulated Mountain Brake Performance Test Procedure—SAE **J1247** APR80 .... 2:25.123

Stopping Distance Test Procedure (A)—SAE **J299** AUG87 .... 2:25.156

Vacuum Power Assist Brake Booster Test Procedure (A)—SAE **J1808** OCT89 .... 2:25.172

Wheel Slip Brake Control System Road Test Code—SAE **J46** JUN80 .... 2:25.158

Cold Weather Operation
Air Brake Actuator Diaphragm Test Procedure (A)—SAE **J1450** MAY88 .... 3:36.166

Construction and Industrial Equipment
Braking Performance—Crawler Tractors and Crawler Loaders—SAE **J1026** APR90 .... 3:40.281

Braking Performance—Roller Compactors (A)—SAE **J1472** JUN87 .... 3:40.288

Braking Performance—Roller/Compactors (A)—SAE **J1136** APR87 .... 3:40.300

Braking Performance—Rubber-Tired Construction Machines (A)—SAE **J1152** APR80 .... 3:40.289

Braking Performance—Rubber-Tired Skidders (A)—SAE **J1178** JUN87 .... 3:40.97

Minimum Performance Criteria for Braking Systems for Specialized Rubber-Tired, Self-Propelled Underground Mining Machines (A)—SAE **J1329** JUL89 .... 3:40.295

Control Systems
Minimum Requirements for Wheel Slip Brake Control System Malfunction Signals—SAE **J1230** OCT79 ..... 2:25.162

Snowmobile Brake Control Systems (A)—SAE **J1282** OCT84 .... 3:38.09

Test Procedures for Determining Air Brake Valve Input-Output Characteristics—SAE **J1859** APR89 .... 3:36.162

Disc
Brake Disc and Drum Thermocouple Installation—SAE **J79** 2:25.157

* Brake Performance and Wear Test Code Commercial Vehicle Inertia Dynamometer—SAE **J2115** JUN93 ..... 2:25.106

Determination of Brake Fluid Temperature—SAE **J291** JUN80 .... 2:25.157

* Materials for Plastic Pistons for Hydraulic Disc Brake Cylinders—SAE **J1568** JUN93 .... 2:25.18

Plastic Dust Shield for Hydraulic Disc Brakes (A)—SAE **J1876** MAR90 .... 2:25.67

Rubber Dust Boots for the Hydraulic Disc Brake Piston—SAE **J1570** SEP91 .... 2:25.181

Rubber Seals for Hydraulic Disc Brake Cylinders (A)—SAE **J1603** JUN90 .... 2:25.19

Earthmoving Equipment
Brake Performance—Rubber-Tired Earthmoving Machines (A)—SAE **J1473** OCT90 .... 3:40.285

Braking Performance—Crawler Tractors and Crawler Loaders—SAE **J1026** APR90 .... 3:40.281

Braking Performance—Rubber-Tired Skidders (A)—SAE **J1178** JUN87 .... 3:40.97

Minimum Performance Criteria for Braking Systems for Specialized Rubber-Tired, Self-Propelled Underground Mining Machines (A)—SAE **J1329** JUL89 .... 3:40.295

Gaskets
Brake Master Cylinder Reservoir Diaphragm Gasket—SAE **J1605** MAR92 .... 2:25.22

Motorcycle
Brake System Road Test Code—Motorcycles (A)—SAE **J108** MAR87 .... 2:25.144

Service Brake System Performance Requirements—Motorcycles and Motor-Driven Cycles (A)—SAE **J109** MAR87 .... 2:25.146

Off-Road Vehicles
Automotive and Off-Highway Air Brake Reservoir Performance and Identification Requirements (A)—SAE **J10** OCT90 .... 3:36.132

† Retardation Capability of Off-Highway Dumpers and Scrapers—SAE **J1430** FEB93 .... 3:40.298

Parking
Manual Slack Adjuster Test Procedure (A)—SAE **J1461** APR91 .... 3:36.174

Parking Brake Structural Integrity Test Procedure—Vehicles Over 10,000 Pounds (4500 KG) GVWR (A)—SAE **J1476** JUN87 .... 3:36.123

Trailer Grade Parking Performance Test Procedure—SAE **J1452** OCT88 .... 2:25.155

Truck and Bus Grade Parking Performance Test Procedure—SAE **J360** OCT88 .... 2:25.101

Vehicle Grade Parking Performance Requirements (A)—SAE **J293** OCT88 .... 2:25.102

Wheel Chocks—SAE **J348** JUN90 .... 3:36.122

Power Assisted
Vacuum Power Assist Brake Booster Test Procedure (A)—SAE **J1808** OCT89 .... 2:25.172

Slack Adjuster
Automatic Slack Adjuster Performance Requirements—SAE **J1513** APR90 .... 3:36.176

---

* Technical Revision   † New Document   (D)—DODISS Adopted   (A)—ANSI Recognized

Subject Index

BRAKES (continued)

Slack Adjuster (continued)
Automatic Slack Adjuster Test Procedure (A)—SAE J1462 MAY87 .... 3:36.175
Manual Slack Adjuster Performance Requirements (A)—SAE J1512 APR90 .... 3:36.176
Manual Slack Adjuster Test Procedure (A)—SAE J1461 APR91 .... 3:36.174

Snowmobile
Brake System Test Procedure—Snowmobiles (A)—SAE J45 OCT84 .... 3:38.11
Service Brake System Performance Requirements—Snowmobiles—SAE J44 DEC91 .... 3:38.11
Snowmobile Brake Control Systems (A)—SAE J1282 OCT84 .... 3:38.09

Terminology
Antilock Brake System Review—SAE J2246 JUN92 .... 2:25.72
Automotive Brake Definitions and Nomenclature (A)—SAE J656 APR88 .... 2:25.70

Trailer
Service Brake System Performance Requirements—Passenger Car—Trailer Combinations—SAE J135 MAY85 .... 2:25.143
Trailer Grade Parking Performance Test Procedure—SAE J1452 OCT88 .... 2:25.155

Truck Trailer
Brake and Electrical Connection Locations—Truck-Tractor and Truck-Trailer (A)—SAE J702 MAY85 . 3:36.114
Brake Force/Distribution Performance Guide—Commercial Vehicles (A)—SAE J1854 JUN88 .... 3:36.170

Trucks and Buses
Air Brake Gladhand Service (Control) and Emergency (Supply) Line Couplers—Trucks, Truck-Tractors, and Trailers—SAE J318 SEP80 .... 3:36.139
Antilock Brake System Review—SAE J2246 JUN92 .... 2:25.72
Automotive and Off-Highway Air Brake Reservoir Performance and Identification Requirements (A)—SAE J10 OCT90 .... 3:36.132
Brake Force Distribution Test Code—Commercial Vehicles (A)—SAE J1505 MAY85 .... 2:25.147
Brake Force/Distribution Performance Guide—Commercial Vehicles (A)—SAE J1854 JUN88 .... 3:36.170
*Brake Performance and Wear Test Code Commercial Vehicle Inertia Dynamometer—SAE J2115 JUN93 ..... 2:25.106
Brake Power Rating Test Code—Commercial Vehicle Inertia Dynamometer (A)—SAE J971 JUN91 .... 2:25.152
Brake Rating Horsepower Requirements—Commercial Vehicles—SAE J257 MAR85 .... 3:36.137
Brake System Performance Requirements—Truck, Bus, and Combination of Vehicles (A)—SAE J992 MAR85 ... 3:36.164
Brake System Rating Test Code—Commercial Vehicles—SAE J880 MAR85 .... 3:36.147
Brake System Road Test Code—Passenger Car and Light Duty Truck-Trailer Combinations (A)—SAE J134 MAY85 .... 2:25.134
Brake System Road Test Code—Passenger Car and Light-Duty Truck (A)—SAE J843 NOV90 .... 2:25.108
Brake System Road Test Code—Truck, Bus, and Combination of Vehicles—SAE J786a 3:36.140
†Brake System Torque Balance Test Code Commercial Vehicles (A)—SAE J225 JUN93 .... 3:36.133
*Braking, Stability, and Control Performance Test Procedures for Air-Brake- Equiped Truck Tractors—SAE J1626 MAR93 .... 3:36.130
Hydraulic Wheel Cylinders for Automotive Drum Brakes (A)—SAE J101 DEC89 .... 2:25.25
In-Service Brake Performance Test Procedure Passenger-Car and Light-Duty Truck—SAE J201 MAY89 ... 2:25.85
†In-Service Brake Performance Test Procedure—Vehicles Over 4500 kg (10 000 lb)—SAE J1250 NOV92 .... 2:25.86
†Service Brake Structural Integrity Requirements—Vehicles Over 10 000 lb (4500 kg) GVWR (A)—SAE J1404 JUN93 .... 3:36.171
†Service Brake Structural Integrity Test Procedure—Vehicles Over 4500 kg (10 000 lb) GVWR—SAE J294 .. 3:36.138
Simulated Mountain Brake Performance Test Procedure—SAE J1247 APR80 .... 2:25.123
Stopping Distance Test Procedure (A)—SAE J299 AUG87 .... 2:25.156
Test Method for Measuring Power Consumption of Air Conditioning and Brake Compressors for Trucks and Buses (A)—SAE J1340 APR90 .... 3:36.53
Test Procedure for Air Reservoir Capacity—Highway Type Vehicles (A)—SAE J1911 SEP90 .... 3:36.153
Truck and Bus Grade Parking Performance Test Procedure—SAE J360 OCT88 .... 2:25.101
Wheel Slip Brake Control System Road Test Code—SAE J46 JUN80 .... 2:25.158

Tubing
Automotive Hydraulic Brake System—Metric Tube Connections—SAE J1290 MAY89 .... 2:25.165
Tubing—Motor Vehicle Brake System Hydraulic (A)—SAE J1047 JUN90 .... 2:25.162

Valves
†Hydraulic Valves for Motor Vehicle Brake Systems Test Procedure—SAE J1118 JUN93 .... 2:25.35
Hydraulic Valves for Motor Vehicle Brake Systems—Performance Requirements—SAE J1137 MAR85 ..... 2:25.39
*Materials for Plastic Check Valves for Vacuum Booster System—SAE J1875 JUN93 .... 2:25.40

BRASS
Manual on Design and Application of Helical and Spiral Springs—SAE HS-J795 APR90 .... Available as a separate publication (See Related Technical Reports Section)

BRAZING
Automotive Metallurgical Joining—SAE J836a APR70 .. 1:9.08
Welding, Brazing, and Soldering—Materials and Practices—SAE J1147 JUN83 .... 1:9.07

Copper Alloys
Wrought and Cast Copper Alloys—SAE J461 SEP81 ... 1:10.45

BUCKETS
Capacity Rating—Loader Bucket—SAE J742 FEB85 ... 3:40.324
Cutting Edge—Cross Sections Loader Straight (A)—SAE J1303 FEB85 .... 3:40.239
Cutting Edge—Cross Sections Loader Straight with Bolt Holes (A)—SAE J1304 FEB85 .... 3:40.241
Cutting Edge—Optional Cross-Sections and Dimensions Loader Straight (A)—SAE J1581 SEP89 .... 3:40.242
†Excavator, Mini-Excavator, and Backhoe Hoe Bucket Volumetrtic Rating—SAE J296 JUN93 .... 3:40.329
Shovel Dipper, Clam Bucket, and Dragline Bucket Rating—SAE J67 NOV84 .... 3:40.323

BULLDOZERS
See also: CONSTRUCTION AND INDUSTRIAL EQUIPMENT, EARTHMOVING EQUIPMENT

Capacity Rating—Dozer Blades (A)—SAE J1265 MAR88 3:40.236
Cutting Edge—Double Bevel Cross Sections (A)—SAE J738 JUN86 .... 3:40.235
Hole Placement on Dozer End Bits (A)—SAE J63 MAY86 .... 3:40.231
Hole Spacing for Scraper and Dozer Cutting Edges (A)—SAE J737 AUG89 .... 3:40.231

Brakes
Braking Performance—Rubber-Tired Construction Machines (A)—SAE J1152 APR80 .... 3:40.289

---

* Technical Revision † New Document (D)—DODISS Adopted (A)—ANSI Recognized

BULLDOZERS (continued)
Noise
Sound Measurement—Off-Road Self-Propelled Work Machines—Operator—Work Cycle (A)—SAE J1166 MAY90 .... 1:14.55

†Sound Power Level Measurements of Earthmoving Machinery - Static and In-Place Dynamic Methods—SAE J1805 APR93 .... 1:14.90

Terminology
Nomenclature and Specification Definitions—Dozers (A)—SAE J729 SEP86 .... 3:40.37

BULLET RESISTANCE
American National Standard for Safety Glazing Materials for Glazing Motor Vehicles and Motor Vehicle Equipment Operating on Land Highways—Safety Code—ANSI SAE Z26.1-1990 .... Available as a separate publication (See Related Technical Reports Section)

BUMPERS
Bumper Evaluation Test Procedure—Passenger Cars—SAE J980a 3:32.02

Curb Clearance Approach, Departure, and Ramp Breakover Angles—Passenger Car and Light Truck (A)—SAE J689 DEC89 .... 3:32.03

Aluminum
Anodized Aluminum Automotive Parts—SAE J399 FEB85 1:10.38

Jacks
Bumper Jack Requirements—Motor Vehicles—SAE J979 3:32.01

Bumper Jacking Test Procedure—Motor Vehicles—SAE J978 .... 3:32.01

Mounts
Inspection of Energy Absorbing Bumper Mounts (A)—SAE J1571 MAY87 .... 3:42.25

Plastic
Flexible Bumper Repair—SAE J1573 FEB87 .... 3:42.22

BUSES
See: TRUCKS AND BUSES

BUSHINGS
Bearing and Bushing Alloys—SAE J459 OCT91 .... 1:10.43

Bearing and Bushing Alloys—Chemical Composition of SAE Bearing and Bushing Alloys—SAE J460 OCT91 .... 1:10.41

Elastomeric Bushing "TRAC" Application Code (A)—SAE J1883 MAR88 .... 1:11.40

Split Type Bushings—Design and Application (A)—SAE J835 DEC81 .... 2:26.201

CABLES
See also: BATTERIES, ELECTRIC CABLES, IGNITION SYSTEMS, WIRING SYSTEMS

Ignition Cable Assemblies—SAE J2032 NOV91 .... 2:23.154

Selection of Transmission Media (A)—SAE J2056/3 JUN91 .... 2:23.391

Marine Engines
Marine Control Cable Connection—Engine Clutch Lever (A)—SAE J960 JUN91 .... 3:41.04

Marine Control Cable Connection—Engine Throttle Lever (A)—SAE J961 JUN91 .... 3:41.05

Marine Push-Pull Control Cables—SAE J917 JUN80 ... 3:41.03

CABS
See also: TRUCKS AND BUSES

Human Engineering
Operator Space Envelope Dimensions for Off-Road Machines—SAE J154 JUN92 .... 3:40.419

Trucks and Buses
Accommodation Tool Reference Point—SAE J1516 MAR90 .... 3:34.247

Cab Sleeper Occupant Restraint System Test—SAE J1948 MAR89 .... 3:34.267

Driver Selected Seat Position (A)—SAE J1517 MAR90 . 3:34.250

Rating Air Conditioner Evaporator Air Delivery and Cooling Capacities (A)—SAE J1487 MAY85 .... 3:36.58

Sound Level for Truck Cab Interior—SAE J336 OCT88 . 1:14.29

Truck Driver Stomach Position (A)—SAE J1522 MAR90 3:34.265

CALIBRATION FLUIDS
Calibration Fluid for Diesel Injection Equipment (A)—SAE J967 SEP88 .... 2:26.328

CAMSHAFTS
†Automotive Gray Iron Castings—SAE J431 MAR93 .... 1:6.01

Drives
†Automotive Synchronous Belt Drives (A)—SAE J1313 MAR93 .... 2:19.13

†SI (Metric) Synchronous Belts and Pulleys—SAE J1278 MAR93 .... 2:19.09

CARBIDE TOOLS
Sintered Carbide Tools—SAE J439a 1:7.07

Sintered Tool Materials—SAE J1072 .... 1:7.12

CARBON
Spark Arrester Test Carbon (A)—SAE J997 SEP90 .... 2:26.368

CARBON DIOXIDE
Exhaust Emissions
Determination of Emissions from Gas Turbine Powered Light Duty Surface Vehicles (A)—SAE J1130 MAY83 .. 1:13.156

Diesel Engine Emission Measurement Procedure (A)—SAE J1003 JUN90 .... 1:13.33

Instrumentation and Techniques for Exhaust Gas Emissions Measurement (A)—SAE J254 AUG84 ...... 1:13.104

Measurement of Carbon Dioxide, Carbon Monoxide, and Oxides of Nitrogen in Diesel Exhaust—SAE J177 APR82 1:13.01

CARBON MONOXIDE
Exhaust Emissions
Determination of Emissions from Gas Turbine Powered Light Duty Surface Vehicles (A)—SAE J1130 MAY83 .. 1:13.156

Diesel Engine Emission Measurement Procedure (A)—SAE J1003 JUN90 .... 1:13.33

Instrumentation and Techniques for Exhaust Gas Emissions Measurement (A)—SAE J254 AUG84 ...... 1:13.104

Measurement of Carbon Dioxide, Carbon Monoxide, and Oxides of Nitrogen in Diesel Exhaust—SAE J177 APR82 1:13.01

CARBURETORS
Marine Engines
†Devices Providing Backfire Flame Control for Gasoline Engines in Marine Applications—SAE J1928 APR93 ... 3:41.44

†Marine Carburetors and Fuel Injection Throttle Bodies—SAE J1223 JUN93 .... 3:41.28

CARBURIZING
See also: CASE HARDENING

Steel
Case Hardenability of Carburized Steels (A)—SAE J1975 JUN91 .... 1:2.15

General Characteristics and Heat Treatments of Steels—SAE J412 JUN89 .... 1:2.05

†Methods of Determining Hardenability of Steels—SAE J406 JUN93 .... 1:1.23

CARPETING
Method of Testing Resistance to Crocking of Organic Trim Materials—SAE J861 .... 1:11.114

Test Method for Determining Resistance to Abrasion; Bearding; and Fiber Loss of Automotive Carpet Materials (A)—SAE J1530 JUN85 .... 1:11.107

CASE HARDENING
See also: CARBURIZING

Screws
Steel Self-Drilling Tapping Screws—SAE J78 JUN79 ... 1:4.21

Steel
Case Hardenability of Carburized Steels (A)—SAE J1975 JUN91 .... 1:2.15

---

* Technical Revision   † New Document   (D)—DODISS Adopted   (A)—ANSI Recognized

Subject Index

CASE HARDENING (continued)
  Steel (continued)
    General Characteristics and Heat Treatments of Steels—SAE **J412** JUN89 ... 1:2.05
    Methods of Measuring Case Depth **(A)**—SAE **J423** DEC83 ... 1:3.64

CAST IRON
  See: IRON

CASTINGS
  Aluminum
    General Information—Chemical Compositions, Mechanical and Physical Properties of SAE Aluminum Casting Alloys—SAE **J452** JAN89 ... 1:10.06
  Iron
    Automotive Ductile (Nodular) Iron Castings **(A)**—SAE **J434** JUN86 ... 1:6.09
    † Automotive Gray Iron Castings—SAE **J431** MAR93 ... 1:6.01
    Automotive Malleable Iron Castings **(A)**—SAE **J158** JUN86 ... 1:6.07
  Magnesium
    Magnesium Casting Alloys—SAE **J465** JAN89 ... 1:10.120
  Steel
    Automotive Steel Castings—SAE **J435c** ... 1:6.17
  Zinc
    Zinc Alloy Ingot and Die Casting Compositions—SAE **J468** DEC88 ... 1:10.139
    Zinc Die Casting Alloys—SAE **J469** JAN89 ... 1:10.139

CATALYST
  Terminology
    † Emissions Terminology and Nomenclature—SAE **J1145** FEB93 ... 1:13.166

CELLULOSE
  Gaskets
    Standard Classification System for Nonmetallic Automotive Gasket Materials **(A)**—SAE **J90** JUN90 ... 1:11.197

CENTER OF GRAVITY
  Center of Gravity Test Code **(A)**—SAE **J874** OCT85 ... 3:40.152

CERAMICS
  Sintered Tool Materials—SAE **J1072** ... 1:7.12
  Shot
    Size Classification and Characteristics of Ceramic Shot for Peening **(A)**—SAE **J1830** MAY87 ... 1:8.32

CETANE NUMBER
  Diesel Fuels—SAE **J313** MAR92 ... 1:12.60

CHAINS
  Safety Chain of Full Trailers or Converter Dollies **(A)**—SAE **J697** MAY88 ... 3:36.98
  Trailer Couplings, Hitches and Safety Chains—Automotive Type **(A)**—SAE **J684** MAY87 ... 3:35.01
  Tire
    Passenger and Light Truck Tire Traction Device Profile Determination and Classification—SAE **J1232** ... 3:36.115
    Tire Chain Clearance—Trucks, Buses (Except Suburban, Intercity, and Transit Buses), and Combinations of Vehicles—SAE **J683** AUG85 ... 3:36.115

CHASSIS
  See also: BODIES (Automobile), FRAMES
  Corrosion
    Prevention of Corrosion of Motor Vehicle Body and Chassis Components—SAE **HS-J447** JUN81 ... Available as a separate publication *(See Related Technical Reports Section)*

CHEMICAL ANALYSIS
  Steel
    Methods of Sampling Steel for Chemical Analysis **(A)**—SAE **J408** DEC83 ... 1:1.152
    Product Analysis—Permissible Variations from Specified Chemical Analysis of a Heat or Cast of Steel **(A)**—SAE **J409** DEC90 ... 1:1.153

CHEMICAL COMPOSITION
  Alloys
    Fifth Edition Unified Numbering System Handbook for Metals and Alloys **(D)**—SAE **HS-1086** DEC93 ... Available as a separate publication *(See Related Technical Reports Section)*
  Alternate Fuels
    † Alternative Automotive Fuels—SAE **J1297** MAR93 ... 1:12.74
  Aluminum Alloys
    Bearing and Bushing Alloys—Chemical Composition of SAE Bearing and Bushing Alloys—SAE **J460** OCT91 ... 1:10.41
    Chemical Compositions, Mechanical Property Limits, and Dimensional Tolerances of SAE Wrought Aluminum Alloys **(A)**—SAE **J457** FEB91 ... 1:10.38
    Fifth Edition Unified Numbering System Handbook for Metals and Alloys **(D)**—SAE **HS-1086** FEB93 ... Available as a separate publication *(See Related Technical Reports Section)*
    General Data on Wrought Aluminum Alloys **(A)**—SAE **J454** FEB91 ... 1:10.26
    General Information—Chemical Compositions, Mechanical and Physical Properties of SAE Aluminum Casting Alloys—SAE **J452** JAN89 ... 1:10.06
  Cobalt Alloys
    Engine Poppet Valve Information Report—SAE **J775** JAN88 ... 1:1.165
    Special Purpose Alloys ("Superalloys")—SAE **J467b** ... 1:10.128
  Copper Alloys
    Bearing and Bushing Alloys—Chemical Composition of SAE Bearing and Bushing Alloys—SAE **J460** OCT91 ... 1:10.41
    Cast Copper Alloys—SAE **J462** SEP81 ... 1:10.78
    Fifth Edition Unified Numbering System Handbook for Metals and Alloys **(D)**—SAE **HS-1086** SEP93 ... Available as a separate publication *(See Related Technical Reports Section)*
    Wrought Copper and Copper Alloys—SAE **J463** SEP81 ... 1:10.81
  Copper Tubing
    Metallic Air Brake System Tubing and Pipe **(A)**—SAE **J1149** JUN91 ... 2:22.219
    Seamless Copper Tube **(A)**—SAE **J528** JUN91 ... 2:22.218
  Exhaust Emissions
    Chemical Methods for the Measurement of Nonregulated Diesel Emissions **(A)**—SAE **J1936** OCT89 ... 1:13.38
  Felt
    Felts—Wool and Part Wool **(A)**—SAE **J314** MAY81 ... 1:11.183
  Fuels
    Stoichiometric Air/Fuel Ratios of Automotive Fuels—SAE **J1829** MAY92 ... 1:12.80
  Glass Beads
    Size Classification and Characteristics of Glass Beads for Peening **(A)**—SAE **J1173** SEP88 ... 1:8.30
  Iron Alloys
    Automotive Ductile (Nodular) Iron Castings **(A)**—SAE **J434** JUN86 ... 1:6.09
    † Automotive Gray Iron Castings—SAE **J431** MAR93 ... 1:6.01
    Automotive Malleable Iron Castings **(A)**—SAE **J158** JUN86 ... 1:6.07
    Elevated Temperature Properties of Cast Irons—SAE **J125** MAY88 ... 1:6.12
    Engine Poppet Valve Information Report—SAE **J775** JAN88 ... 1:1.165
    Fifth Edition Unified Numbering System Handbook for Metals and Alloys **(D)**—SAE **HS-1086** JAN93 ... Available as a separate publication *(See Related Technical Reports Section)*

---

* Technical Revision   † New Document   **(D)**—DODISS Adopted   **(A)**—ANSI Recognized

Subject Index

CHEMICAL COMPOSITION (continued)

Iron Alloys (continued)
Special Purpose Alloys ("Superalloys")—SAE **J467b** ... 1:10.128

Lead Alloys
Bearing and Bushing Alloys—Chemical Composition of SAE Bearing and Bushing Alloys—SAE **J460** OCT91 ... 1:10.41
Fifth Edition Unified Numbering System Handbook for Metals and Alloys (D)—SAE **HS-1086** OCT93 .... Available as a separate publication (See Related Technical Reports Section)

Low Melting Metals and Alloys
Fifth Edition Unified Numbering System Handbook for Metals and Alloys (D)—SAE **HS-1086** OCT93 .... Available as a separate publication (See Related Technical Reports Section)

Lubricants
*European Brake Fluid Technology—SAE **J1709** 2:25.64

Magnesium Alloys
Fifth Edition Unified Numbering System Handbook for Metals and Alloys (D)—SAE **HS-1086** OCT93 .... Available as a separate publication (See Related Technical Reports Section)

Magnesium Casting Alloys—SAE **J465** JAN89 .... 1:10.120
Magnesium Wrought Alloys—SAE **J466** DEC89 .... 1:10.125

Metals and Alloys
Fifth Edition Unified Numbering System Handbook for Metals and Alloys (D)—SAE **HS-1086** DEC93 .... Available as a separate publication (See Related Technical Reports Section)

Nickel Alloys
Engine Poppet Valve Information Report—SAE **J775** JAN88 .... 1:1.165
Fifth Edition Unified Numbering System Handbook for Metals and Alloys (D)—SAE **HS-1086** JAN93 .... Available as a separate publication (See Related Technical Reports Section)

Special Purpose Alloys ("Superalloys")—SAE **J467b** ... 1:10.128
Wrought Nickel and Nickel-Related Alloys—SAE **J470c** . 1:10.140

Precious Metals and Alloys
Fifth Edition Unified Numbering System Handbook for Metals and Alloys (D)—SAE **HS-1086** JAN93 .... Available as a separate publication (See Related Technical Reports Section)

Rare Earth Metals and Alloys
Fifth Edition Unified Numbering System Handbook for Metals and Alloys (D)—SAE **HS-1086** JAN93 .... Available as a separate publication (See Related Technical Reports Section)

Reactive and Refractory Metals and Alloys
Fifth Edition Unified Numbering System Handbook for Metals and Alloys (D)—SAE **HS-1086** JAN93 .... Available as a separate publication (See Related Technical Reports Section)

Sintered Powder Materials
Sintered Powder Metal Parts: Ferrous—SAE **J471d** 1:8.01

Steel
Automotive Steel Castings—SAE **J435c** 1:6.17
Chemical Compositions of SAE Alloy Steels (A)—SAE **J404** FEB91 .... 1:1.14
Chemical Compositions of SAE Carbon Steels—SAE **J403** MAY92 .... 1:1.09
Chemical Compositions of SAE Wrought Stainless Steels—SAE **J405** JAN89 .... 1:1.16
Engine Poppet Valve Information Report—SAE **J775** JAN88 .... 1:1.165
Fifth Edition Unified Numbering System Handbook for Metals and Alloys (D)—SAE **HS-1086** JAN93 .... Available as a separate publication (See Related Technical Reports Section)

Former SAE Standard and Former SAE EX-Steels—SAE **J1249** JAN89 .... 1:1.19
General Characteristics and Heat Treatments of Steels—SAE **J412** JUN89 .... 1:2.05
† Hardenability Bands for Carbon and Alloy H Steels (A)—SAE **J1268** JUN93 .... 1:1.48
High Strength, Hot Rolled Steel Plates, Bars and Shapes—SAE **J1442** DEC88 .... 1:1.162
† High-Strength, Quenched, and Tempered Structural Steels—SAE **J368** MAR93 .... 1:1.164
Mechanical and Chemical Requirements for Nonthreaded Fasteners—SAE **J430** 1:4.18
Mechanical and Material Requirements for Externally Threaded Fasteners (A)—SAE **J429** AUG83 ... 1:4.01
Mechanical and Material Requirements for Metric Externally Threaded Steel Fasteners (A)—SAE **J1199** SEP83 .... 1:4.06
Mechanical and Material Requirements for Steel Nuts—SAE **J995** JUN79 .... 1:4.15
Mechanical and Material Requirements for Wheel Bolts—SAE **J1102** 1:4.18
Mechanical and Quality Requirements for Machine Screws—SAE **J82** JUN79 .... 1:4.14
Mechanical and Quality Requirements for Tapping Screws—SAE **J933** JUN79 .... 1:4.19
† Methods of Determining Hardenability of Steels—SAE **J406** JUN93 .... 1:1.23
Metric Thread Rolling Screws—SAE **J1237** 1:4.29
Potential Standard Steels—SAE **J1081** DEC88 ..... 1:1.18
Product Analysis—Permissible Variations from Specified Chemical Analysis of a Heat or Cast of Steel (A)—SAE **J409** DEC90 .... 1:1.153
Restricted Hardenability Bands for Selected Alloy Steels (A)—SAE **J1868** JUN90 .... 1:1.139
Special Purpose Alloys ("Superalloys")—SAE **J467b** ... 1:10.128
Steel Self-Drilling Tapping Screws—SAE **J78** JUN79 ... 1:4.21
Steel, High Strength, Hot Rolled Sheet and Strip, Cold Rolled Sheet, and Coated Sheet—SAE **J1392** JUN84 .. 1:1.159
Thread Rolling Screws—SAE **J81** JUN79 .... 1:4.26
Tool and Die Steels—SAE **J438b** 1:7.01

Steel Alloys
Fifth Edition Unified Numbering System Handbook for Metals and Alloys (D)—SAE **HS-1086** JUN93 .... Available as a separate publication (See Related Technical Reports Section)

Steel Shot
*Cast Steel Grit—SAE **J1993** MAR93 .... 1:8.13
Cast Steel Shot (A)—SAE **J827** MAR90 .... 1:8.09
Specifications for Low Carbon Cast Steel Shot (A)—SAE **J2175** JUN91 .... 1:8.10

Steel Springs and Wires
† Cut Wire Shot—SAE **J441** JUN93 .... 1:8.05
Hard Drawn Carbon Steel Valve Spring Quality Wire and Springs—SAE **J172** DEC88 .... 1:5.08
Hard Drawn Mechanical Spring Wire and Springs—SAE **J113** DEC88 .... 1:5.04
Music Steel Spring Wire and Springs—SAE **J178** DEC88 1:5.10
Oil Tempered Carbon Steel Spring Wire and Springs—SAE **J316** DEC88 .... 1:5.01
Oil Tempered Carbon Steel Valve Spring Quality Wire and Springs—SAE **J351** DEC88 .... 1:5.02
Oil Tempered Chromium-Vanadium Valve Spring Quality Wire and Springs—SAE **J132** DEC88 .... 1:5.03
Oil Tempered Chromium—Silicon Alloy Steel Wire and Springs—SAE **J157** DEC88 .... 1:5.07

*Technical Revision   † New Document   (D)—DODISS Adopted   (A)—ANSI Recognized

Subject Index

CHEMICAL COMPOSITION (continued)
Steel Springs and Wires (continued)
Special Quality High Tensile, Hard Drawn Mechanical Spring Wire and Springs—SAE **J271** DEC88 .... 1:5.05

Stainless Steel 17-7 PH Spring Wire and Springs—SAE **J217** DEC88 .... 1:5.08

Stainless Steel, SAE 30302, Spring Wire and Springs—SAE **J230** DEC88 .... 1:5.09

Steel Tubing
Seamless Low Carbon Steel Tubing Annealed for Bending and Flaring (A)—SAE **J524** JUN91 .... 2:22.206

Welded and Cold Drawn Low Carbon Steel Tubing Annealed for Bending and Flaring (A)—SAE **J525** JUN91 ... 2:22.207

Welded Flash Controlled Low Carbon Steel Tubing Normalized for Bending, Double Flaring, and Beading (A)—SAE **J356** JUN91 .... 2:22.213

Superalloys
Special Purpose Alloys ("Superalloys")—SAE **J467b** ... 1:10.128

Tin Alloys
Bearing and Bushing Alloys—Chemical Composition of SAE Bearing and Bushing Alloys—SAE **J460** OCT91 ... 1:10.41

Titanium Alloys
Special Purpose Alloys ("Superalloys")—SAE **J467b** ... 1:10.128

Zinc Alloys
Fifth Edition Unified Numbering System Handbook for Metals and Alloys (D)—SAE **HS-1086** OCT93 .... Available as a separate publication (See Related Technical Reports Section)

Zinc Alloy Ingot and Die Casting Compositions—SAE **J468** DEC88 .... 1:10.139

CHEMICAL PROPERTIES
Alternate Fuels
Impact of Alternative Fuels on Engine Test and Reporting Procedures—SAE **J1515** MAR88 .... 1:12.89

Coolants
Automotive and Light Truck Engine Coolant Concentrate—Ethylene Glycol Type (A)—SAE **J1034** APR91 .... 1:11.214

Diesel Fuels
Coarse Droplet Water/Fuel Separation Test Procedure (A)—SAE **J1839** MAY90 .... 2:26.437

Diesel Fuels—SAE **J313** MAR92 .... 1:12.60

Emulsified Water/Fuel Separation Test Procedure—SAE **J1488** MAY90 .... 2:26.345

Fuels
† Fuels and Lubricants Standards Manual—SAE **HS-23** FEB93 .... Available as a separate publication (See Related Technical Reports Section)

Heating Value of Fuels—SAE **J1498** MAY90 .... 1:12.82

Gasoline
† Automotive Gasolines—SAE **J312** JAN93 .... 1:12.47

Lubricants
*European Brake Fluid Technology—SAE **J1709** 2:25.64

† Fuels and Lubricants Standards Manual—SAE **HS-23** FEB93 .... Available as a separate publication (See Related Technical Reports Section)

Lubricants for Two-Stroke-Cycle Gasoline Engines—SAE **J1510** OCT88 .... 1:12.32

Lubricating Oil, Aircraft Piston Engine (Ashless Dispersant) (A)—SAE **J1899** JUN91 .... 1:12.98

Lubricating Oil, Aircraft Piston Engine (Nondispersant Mineral Oil) (A)—SAE **J1966** JUN91 .... 1:12.110

Physical and Chemical Properties of Engine Oils (A)—SAE **J357** JUN91 .... 1:12.24

CHILD RESTRAINT SYSTEMS
Anchorage Provisions for Installation of Child Restraint Tether Straps in Rear Seating Positions—SAE **J1369** MAR92 .... 3:33.16

† Child Restraint Anchorages and Attachment Hardware (A)—SAE **J1368** JUN93 .... 3:33.14

*Guidelines for Evaluating Child Restraint System Interactions with Deploying Airbags—SAE **J2189** FEB93 ... 3:33.20

Securing Child Restraint Systems in Motor Vehicle Rear Seats (A)—SAE **J1819** NOV90 .... 3:33.18

† Vehicle Occupant Restraint Systems and Components Standards Manual—SAE **HS-13** MAY93 .... Available as a separate publication (See Related Technical Reports Section)

CHROMIUM
Coatings
Electroplate Requirements for Decorative Chromium Deposits on Zinc Base Materials Used for Exterior Ornamentation (A)—SAE **J1837** JUN91 .... 1:10.155

Electroplating of Nickel and Chromium on Metal Parts—Automotive Ornamentation and Hardware—SAE **J207** FEB85 .... 1:10.150

CIGAR LIGHTERS
Six- and Twelve-Volt Cigar Lighter Receptacles (A)—SAE **J563** MAR90 .... 2:23.189

CIRCUIT BREAKERS
See: ELECTRIC CIRCUIT BREAKERS

CIRCUITS
See: ELECTRIC CIRCUITS

CLAM BUCKETS
See: BUCKETS

CLAMPS
See also: FASTENERS

† Hose Clamp Specifications—SAE **J1508** JUN93 ...... 2:16.45

*Type 'F' Clamps for Plumbing Applications—SAE **J1670** MAY93 .... 2:16.79

CLASSIFICATION
See also: NUMBERING SYSTEMS

All Wheel Drive Systems Classification (A)—SAE **J1952** JAN91 .... 3:29.15

*Introduction to All-Wheel Drive—SAE **HS-2300** JAN93 ... Available as a separate publication (See Related Technical Reports Section)

Preparation of SAE Technical Reports—SAE **TSB 002** JUN92 .... 1:A.05

Collision Deformation
Collision Deformation Classification—SAE **J224** MAR80 ... 3:34.235

Coupling
† Classification, Nomenclature, and Specification Definitions for Trenching Machines—SAE **J1382** JUN93 ... 3:40.41

Elastomers
† Thermoplastic Elastomer Classification System—SAE **J3000** JUN93 .... 1:11.27

Fiberboards
Standard Classification System for Fiberboards (A)—SAE **J1323** SEP90 .... 1:11.161

Gasket Materials
Standard Classification System for Nonmetallic Automotive Gasket Materials (A)—SAE **J90** JUN90 .... 1:11.197

Heat Rating
Spark Plug Heat Rating Classifications—SAE **J2162** JUN92 .... 2:23.97

Lubricants
† Automotive Lubricating Greases—SAE **J310** JUN93 ... 1:12.39

Axle and Manual Transmission Lubricant Viscosity Classification—SAE **J306** OCT91 .... 1:12.37

Axle and Manual Transmission Lubricants—SAE **J308** JUN89 .... 1:12.36

---
* Technical Revision † New Document (D)—DODISS Adopted (A)—ANSI Recognized

CLASSIFICATION (continued)
Lubricants (continued)
Classification of Energy-Conserving Engine Oil for Passenger Cars, Vans, and Light-Duty Trucks—SAE J1423 FEB92 .... **1:12.29**

Engine Oil Performance and Engine Service Classification (Other Than "Energy-Conserving") (A)—SAE J183 JUN91 .... **1:12.01**

† Engine Oil Viscosity Classification—SAE J300 MAR93 .. **1:12.16**

† Fuels and Lubricants Standards Manual—SAE HS-23 FEB93 .... Available as a separate publication (See Related Technical Reports Section)

The Automotive Lubricant Performance and Service Classification Maintenance Procedure (A)—SAE J1146 JUN86 .... **1:12.87**

Two-Stroke-Cycle Engine Oil Miscibility/Fluidity Classification—SAE J1536 OCT88 .... **1:12.33**

† Two-Stroke-Cycle Gasoline Engine Lubricants Performance and Service Classification (A)—SAE J2116 JUN93 .... **1:12.33**

Piston Ring Materials
† Internal Combustion Engines—Piston Rings—Material Specifications (A)—SAE J1590 OCT92 .... **2:26.72**

Plastics
*Classification System for Automotive Polyamide (PA) Plastics—SAE J1639 MAR93 .... **1:11.145**

Rubber
† Classification System for Rubber Materials—SAE J200 MAY93 .... **1:11.01**

Steels
*Categorization of Low Carbon Automotive Sheet Steel—SAE J2096 JUL92 .... **1:3.25**

CLEANING
† Cast Shot and Grit Size Specifications for Peening and Cleaning—SAE J444 MAY93 .... **1:8.11**

Metallic Shot and Grit Mechanical Testing—SAE J445 AUG84 .... **1:8.12**

SAE Manual on Blast Cleaning—SAE J792a AUG68 .... **1:8.14**

Specifications for Low Carbon Cast Steel Shot (A)—SAE J2175 JUN91 .... **1:8.10**

CLEANLINESS
Hydraulic Fluids
Assessing Cleanliness of Hydraulic Fluid Power Components and Systems (A)—SAE J1227 MAR86 ... **3:40.104**

Method for Assessing the Cleanliness Level of New Hydraulic Fluid (A)—SAE J1277 JUL90 .... **3:40.127**

Reporting Cleanliness Levels of Hydraulic Fluids (A)—SAE J1165 MAR86 .... **3:40.124**

The Automotive Lubricant Performance and Service Classification Maintenance Procedure (A)—SAE J1146 JUN86 .... **1:12.87**

Steel
† Cleanliness Rating of Steels by the Magnetic Particle Method—SAE J421 MAY93 .... **1:3.57**

CLIMATE CONTROL
See: AIR CONDITIONING, HEATING, VENTILATION

CLOCKS
Electric Hourmeter Specification—SAE J1378 MAR83 .. **2:21.14**

CLUTCH PEDALS
Accommodation Tool Reference Point—SAE J1516 MAR90 .... **3:34.247**

Driver Selected Seat Position (A)—SAE J1517 MAR90 . **3:34.250**

Truck Driver Shin-Knee Position for Clutch and Accelerator (A)—SAE J1521 MAR90 .... **3:34.263**

CLUTCH PLATES
† Automotive Gray Iron Castings—SAE J431 MAR93 .... **1:6.01**

CLUTCHES
See also: CLUTCH PEDALS, CLUTCH PLATES
† Clutch Dimensions for Truck and Bus Applications (A)—SAE J1806 NOV92 .... **3:36.119**

Flywheels for Single-Plate Spring-Loaded Clutches (A)—SAE J618 JAN91 .... **2:26.238**

Flywheels for Two-Plate Spring-Loaded Clutches—SAE J619 JUN80 .... **2:26.240**

† Housing Internal Dimensions for Single- and Two-Plate Spring-Loaded Clutches (A)—SAE J373 APR93 .... **2:26.241**

Overcenter Clutch Spin Test Procedure—SAE J1079 SEP88 .... **3:40.243**

Pull-Type Clutch—Transmission Installation Dimensions (A)—SAE J1463 NOV84 .... **3:36.120**

Recommended Remanufacturing Procedures for Manual Transmission Clutch Assemblies—SAE J1915 FEB90 .. **3:42.39**

SAE No. 2 Clutch Friction Test Machine Test Procedure (A)—SAE J286 NOV83 .... **3:29.95**

Spring-Loaded Clutch Spin Test Procedure (A)—SAE J1073 JUN90 .... **2:26.242**

Marine Engines
Marine Control Cable Connection—Engine Clutch Lever (A)—SAE J960 JUN91 .... **3:41.04**

Mounts
SAE Nodal Mount—SAE J1134 FEB92 .... **3:36.112**

Terminology
Automotive Pull Type Clutch Terminology (A)—SAE J1479 APR91 .... **3:36.02**

One-Way Clutches—Nomenclature and Terminology—SAE J1087 OCT90 .... **3:29.93**

COASTDOWN TESTS
Road Load Measurement and Dynamometer Simulation Using Coastdown Techniques—SAE J1263 JUN91 .... **2:26.470**

Supportive Information Report for the Fuel Economy Measurement Test (Engineering Type) for Trucks and Buses—SAE J1420 AUG88 .... **3:36.44**

COATINGS
See also: FINISHES, PAINTS
Electrocoat Compatibilities of Automotive Sealers—SAE J1969 OCT88 .... **1:11.90**

Method for the Determination of Expansion and Water Absorption of Automotive Sealers (A)—SAE J1918 MAY88 .... **1:11.86**

O.D. Coatings for Radial Lip-Type Shaft Seals—SAE J1947 AUG89 .... **1:11.65**

Outdoor Weathering of Exterior Materials—SAE J1976 JUN89 .... **1:11.275**

Test for Chip Resistance of Surface Coatings—SAE J400 JAN85 .... **1:11.126**

Vibration Damping Materials and Underbody Coatings (A)—SAE J671 APR82 .... **1:11.111**

Chemical
Prevention of Corrosion of Motor Vehicle Body and Chassis Components—SAE HS-J447 JUN81 .... Available as a separate publication (See Related Technical Reports Section)

Chromium
Electroplate Requirements for Decorative Chromium Deposits on Zinc Base Materials Used for Exterior Ornamentation (A)—SAE J1837 JUN91 .... **1:10.155**

Electroplating of Nickel and Chromium on Metal Parts—Automotive Ornamentation and Hardware—SAE J207 FEB85 .... **1:10.150**

Corrosion Resistant
Corrosion Preventive Compound, Topside Vehicle Corrosion Protection—SAE J1804 JUN89 .... **1:3.52**

*Decorative Anodizing Specification for Automotive Applications—SAE J1974 JUN93 .... **1:10.39**

---

*Technical Revision † New Document (D)—DODISS Adopted (A)—ANSI Recognized

Subject Index

COATINGS (continued)
 Corrosion Resistant (continued)
  Electroplate Requirements for Decorative Chromium Deposits on Zinc Base Materials Used for Exterior Ornamentation (A)—SAE J1837 JUN91 .... 1:10.155
  Guidelines and Limitations of Laboratory Cyclic Corrosion Test Procedures for Exterior, Painted, Automotive Body Panels—SAE J1563 MAY87 .... 1:3.51
  Prevention of Corrosion of Motor Vehicle Body and Chassis Components—SAE HS-J447 JUN81 .... Available as a separate publication (See Related Technical Reports Section)
  Properties of Galvanized Low Carbon Sheet Steels and Their Relation to Formability—SAE J1852 SEP87 ..... 1:3.22
  Undervehicle Coupon Corrosion Tests—SAE J1293 JAN90 .... 1:3.48
 Metallic
  Prevention of Corrosion of Motor Vehicle Body and Chassis Components—SAE HS-J447 JUN81 .... Available as a separate publication (See Related Technical Reports Section)
 Nickel
  Electroplating of Nickel and Chromium on Metal Parts—Automotive Ornamentation and Hardware—SAE J207 FEB85 .... 1:10.150
 Vinyl
  Test Method for Determining Cold Cracking of Flexible Plastic Materials—SAE J323 .... 1:11.115

COBALT ALLOYS
 Special Purpose Alloys ("Superalloys")—SAE J467b ... 1:10.128

CODING
 *Class B Data Communication Network Messages Part 2: Data Parameter Definitions—SAE J2178/2 JUN93 ... 2:23.345
 Diagnostic Trouble Code Definitions—SAE J2012 MAR92 ... 2:23.284
 E/E Diagnostic Test Modes—SAE J1979 DEC91 ...... 2:23.30
 *Enhanced E/E Diagnostic Test Modes—SAE J2190 JUN93 .... 2:23.41
 †Internal Combustion Engines—Piston Rings—General Specifications (A)—SAE J1591 OCT92 .... 2:26.186

COLD WEATHER OPERATION
 Brakes
  Hydraulic Valves for Motor Vehicle Brake Systems—Performance Requirements—SAE J1137 MAR85 ..... 2:25.39
 Diesel Engines
  †Electric Engine Preheaters and Battery Warmers for Diesel Engines—SAE J1310 JUN93 .... 3:40.500
  Location of Atomizer of Ether Systems for Diesel Engines—SAE J2079 MAY92 .... 2:26.329
  *SAE Cold Start and Driveability Procedure—SAE J1635 MAY93 .... 2:26.498
 Seals
  Seals—Testing of Radial Lip—SAE J110 DEC91 ..... 3:29.140
 Spark Ignition Engines
  Fuel Economy Measurement—Road Test Procedure—Cold Start and Warm-Up Fuel Economy—SAE J1256 OCT88 .... 2:26.506
  †Low-Temperature Cranking Load Requirements of an Engine—SAE J1253 JUN93 .... 2:23.81
  *SAE Cold Start and Driveability Procedure—SAE J1635 MAY93 .... 2:26.498

COLLISION DEFORMATION
 See: CRASHWORTHINESS, IMPACT TESTS

COLLISION DETECTION
 Collision Detection Serial Data Communications Multiplex Bus—SAE J1567 AUG87 .... 2:23.481

COLORS
 *Paint and Trim Code Location—SAE J2235 JUN93 .... 3:33.44
 *SAE Automotive Textiles and Trim Standards Manual—SAE HS-2700 JUN93 .... Available as a separate publication (See Related Technical Reports Section)
 Test Method for Determining Visual Color Match to Master Specimen for Fabrics, Vinyls, Coated Fiberboards, and Other Automotive Trim Materials (A)—SAE J361 MAR85 .... 1:11.180
 Test Method for Measuring Wet Color Transfer Characteristics (A)—SAE J1326 FEB85 .... 1:11.169
 Fading
  *Accelerated Exposure of Automotive Interior Trim Components Using a Controlled Irradiance Air-Cooled Xenon-Arc Apparatus—SAE J2212 MAR93 .... 1:11.263
  *Accelerated Exposure of Automotive Interior Trim Material Using Outdoor Under-Glass Controlled Sun-Tracking Temperature and Humidity Apparatus—SAE J2230 FEB93 .... 1:11.123
  *Accelerated Exposure of Automotive Interior Trim Materials Using an Outdoor Under Glass Variable Angled Controlled Temperature Apparatus—SAE J2229 FEB93 ... 1:11.119
 Lighting Devices
  Color Specification (A) (D)—SAE J578 MAY88 .... 2:24.34
 Measurement
  Instrumental Color Difference Measurement for Exterior Finishes, Textiles, and Colored Trim (A)—SAE J1545 JUN86 .... 1:11.279

COMBUSTION
 Fuels
  Heating Value of Fuels—SAE J1498 MAY90 .... 1:12.82

COMMUNICATIONS SYSTEM
 A Vehicle Network Protocol with a Fault Tolerant Multiplex Signal Bus—SAE J1813 AUG87 .... 2:23.499
 Chrysler Sensor and Control (CSC) Bus Multiplexing Network for Class 'A' Applications—SAE J2058 JUN90 . 2:23.422
 *Class C Application Requirement Considerations—SAE J2056/1 JUN93 .... 2:23.366
 Collision Detection Serial Data Communications Multiplex Bus—SAE J1567 AUG87 .... 2:23.481
 Controller Area Network (CAN), An In-Vehicle Serial Communication Protocol—SAE J1583 MAR90 .... 2:23.486
 E/E Diagnostic Test Modes—SAE J1979 DEC91 ...... 2:23.30
 *OBD II Scan Tool—SAE J1978 MAR92 .... 2:23.325
 *OEM/Vendor's Interface Specification for Vehicle Electronic Programming Station—SAE J1924 DEC92 .. 2:23.522
 Powertrain Control Interface for Electronic Controls Used in Medium and Heavy Duty Diesel On-Highway Vehicle Applications (A)—SAE J1922 DEC89 .... 3:36.177
 Token Slot Network for Automotive Control (A)—SAE J2106 APR91 .... 2:23.431

COMPACTORS
 †Steering Frame Lock for Articulated Loaders and Tractors—SAE J276 NOV92 .... 3:40.353
 Brakes
  Braking Performance—Roller Compactors (A)—SAE J1472 JUN87 .... 3:40.288
  Braking Performance—Roller/Compactors (A)—SAE J1136 APR87 .... 3:40.300
 Noise
  Sound Measurement—Off-Road Self-Propelled Work Machines—Operator—Work Cycle (A)—SAE J1166 MAY90 .... 1:14.55
 Terminology
  Nomenclature—Rollers and Compactors (A)—SAE J1017 JAN86 .... 3:40.49

COMPOSITE MATERIALS
 See also: FIBER REINFORCED PLASTICS
 Compression and Recovery of Insulation Paddings—SAE J1352 APR87 .... 1:11.172

---

* Technical Revision   † New Document   (D)—DODISS Adopted   (A)—ANSI Recognized

COMPOSITE MATERIALS (continued)

Hot Plate Method for Evaluating Heat Resistance and Thermal Insulation Properties of Materials—SAE J1361 APR87 .... 1:11.171

Test Method for Measuring the Relative Drapeability of Flexible Insulation Materials—SAE J1325 FEB85 ...... 1:11.169

COMPRESSED AIR

† Compressed Air for General Use—Part 1: Contaminants and Quality Classes—SAE J1649/1 JUN93 .... 3:34.47

COMPRESSORS

Guide to the Application and Use of Passenger Car Air-Conditioning Compressor Face Seals (A)—SAE J1954 MAY90 .... 2:26.423

Information Relating to Duty Cycles and Average Power Requirements of Truck and Bus Engine Accessories—SAE J1343 OCT88 .... 3:36.57

Test Method for Measuring Power Consumption of Air Conditioning and Brake Compressors for Trucks and Buses (A)—SAE J1340 APR90 .... 3:36.53

Performance

Turbocharger Gas Stand Test Code—SAE J1826 APR89 2:26.355

Turbocharger Nomenclature and Terminology—SAE J922 JUL88 .... 2:26.04

COMPUTER EQUIPMENT

Recommended Practice for the Selection of Engineering Workstations—SAE J2026 NOV91 .... 3:43.01

COMPUTER PROGRAMS

Road Load Measurement and Dynamometer Simulation Using Coastdown Techniques—SAE J1263 JUN91 .... 2:26.470

CONNECTORS

See also: COUPLINGS, FASTENERS, FITTINGS, FLANGES

Automotive Hydraulic Brake System—Metric Banjo Bolt Connections—SAE J1291 MAY89 .... 2:25.170

Automotive Hydraulic Brake System—Metric Tube Connections—SAE J1290 MAY89 .... 2:25.165

† Automotive Tube Fittings—SAE J512 JUN93 .... 2:22.06

† Beaded Tube Hose Fittings—SAE J1231 JUN93 ...... 2:22.158

Brake and Electrical Connection Locations—Truck-Tractor and Truck-Trailer (A)—SAE J702 MAY85 . 3:36.114

Coding Systems for Identification of Fluid Conductors and Connectors (D)—SAE J846 JUN89 .... 2:22.01

Connection and Accessory Locations for Towing Multiple Trailers (A)—SAE J849 NOV85 .... 3:36.93

Connections for Fluid Power and General Use—Ports and Stud Ends with ISO 261 Threads and O-Ring Sealing Part 2: Heavy-Duty (S Series) Stud Ends—Dimensions, Design, Test Methods, and Requirements—SAE J2244/2 DEC91 .... 2:22.293

Connections for Fluid Power and General Use—Ports and Stud Ends with ISO 261 Threads and O-Ring Sealing Part 1: Port with O-Ring Seal in Truncated Housing—SAE J2244/1 DEC91 .... 2:22.290

*Connections for General use and Fluid Power-Ports and Stud Ends with ISO 725 Trheads and O-Ring Sealing—Part 2: Heavy-Duty (S Series) Stud Ends—SAE J1926/2 ISO 11 MAR93 .... 2:22.102

† Connections for General Use and Fluid Power-Ports and Stud Ends with ISO 725 Threads and O-ring Sealing—Part 1: Threaded Port with O-Ring Seal in Truncated Housing—SAE J1926/1 ISO 11 MAR93 .... 2:22.99

*Connections for General use and Fluid Power-Ports and Stud Ends with ISO 725 Threads and O-Ring Sealing—Part 3: Light-Duty (L Series) Stud Ends—SAE J1926/3 ISO 11 MAR93 .... 2:22.108

Connectors and Plugs—SAE J856 .... 2:24.15

† Diagnostic Connector—SAE J1962 JUN93 .... 2:23.245

Fluid Conductors and Connectors—SAE HS-150 JUN92 Available as a separate publication *(See Related Technical Reports Section)*

Formed Tube Ends for Hose Connections (A)—SAE J962 MAY86 .... 2:22.202

† Hydraulic Flanged Tube, Pipe, and Hose Connections, Four-Bolt Split Flange Type—SAE J518 JUN93 .... 2:22.190

Ignition Cable Assemblies—SAE J2032 NOV91 .... 2:23.154

*Metallic Tube Connections for Fluid Power and General Use-Test Methods for Threaded Hydraulic Fluid Power Connectors—SAE J1644 MAY93 .... 2:22.285

† Refrigeration Tube Fittings—General Specifications—SAE J513 JUN93 .... 2:22.33

Road Vehicles—High Pressure Fuel Injection Pipe End-Connections with 60 Deg Female Cone—SAE J1949 OCT88 .... 2:26.302

SAE Quick Connector Specifications for Liquid Fuel Systems—SAE J2044 JUN92 .... 2:26.442

† Seven Conductor Electrical Connector for Truck-Trailer Jumper Cable—SAE J560 JUN93 .... 2:23.175

† Test Method for Evaluating the Sealing Capability of Hose Connections with a PVT Test Facility—SAE J1610 APR93 .... 2:16.80

CONSTRUCTION AND INDUSTRIAL EQUIPMENT

See also: BACKHOES, BULLDOZERS, CRANES, CRAWLERS, DUMPERS, EARTHMOVING EQUIPMENT, EXCAVATORS, GRADERS, LOADERS, OFF-ROAD VEHICLES, RIPPERS, ROLLERS, SCARIFIERS, SCRAPERS

Categories of Off-Road Self-Propelled Work Machines (A)—SAE J1116 JUN86 .... 3:40.01

Countersunk Square Holes for Cutting Edges and End Bits (A)—SAE J740 JUL86 .... 3:40.237

Definitions for Machine Availability (Off-Road Work Machines) (A)—SAE J1032 APR87 .... 3:40.98

† Flywheels for Industrial Engines Used with Industrial Power Take-Offs Equipped with Driving-Ring Type Overcenter Clutches and Engine-Mounted Marine Gears and Single Bearing Engine-Mounted Power Generators—SAE J620 MAY93 .... 3:40.256

Guidelines for Requests Received from Outside Sources for the ConAg Council to Originate or Review Technical Reports—SAE J2129 NOV90 .... 3:40.03

Machine Slope Operation Test Code (A)—SAE J897 OCT85 .... 3:40.158

Aerodynamics

Drag Force Test Procedure for Construction, Forestry, and Industrial Machines (A)—SAE J873 FEB86 .... 3:40.150

Alarms

Alarm—Backup—Electric—Performance, Test, and Application (A)—SAE J994 MAR85 .... 1:14.45

On-Machine Alarm Test and Evaluation Procedure for Construction and General Purpose Industrial Machinery—SAE J1446 MAY89 .... 1:14.50

Brakes

Braking Performance—Rubber-Tired Construction Machines (A)—SAE J1152 APR80 .... 3:40.289

Minimum Performance Criteria for Braking Systems for Specialized Rubber-Tired, Self-Propelled Underground Mining Machines (A)—SAE J1329 JUL89 .... 3:40.295

Center of Gravity

Center of Gravity Test Code (A)—SAE J874 OCT85 ... 3:40.152

Electric Equipment

Electrical Charging Systems for Construction and Industrial Machinery (A)—SAE J180 MAY87 .... 3:40.270

Electrical Connector for Auxiliary Starting of Construction, Agricultural, and Off-Road Machinery (A)—SAE J1283 MAR86 .... 3:40.267

Electrical Propulsion Rotating Equipment—Off-Road Dumper (A)—SAE J1317 JUN82 .... 3:40.279

Electrical Wiring Systems for Construction, Agricultural and Off-Road Machines—SAE J821 MAY85 .... 3:40.264

Engine Cooling Systems

Requirements for Engine Cooling System Filling, Deaeration and Drawdown Tests—SAE J1436 JUN89 .. 2:26.379

---

* Technical Revision   † New Document   (D)—DODISS Adopted   (A)—ANSI Recognized

Subject Index

CONSTRUCTION AND INDUSTRIAL EQUIPMENT (continued)

Fuel Systems

Fast Fill Fueling Installation for Off-Road Work Machines (A)—SAE **J176** DEC89 .... 3:40.103

† Guidelines for Liquid Level Indicators—SAE **J48** MAY93 ... 3:40.103

Heating

Fuel-Fired Heaters—Air Heating—For Construction and Industrial Machinery—SAE **J1024** DEC89 .... 3:40.496

Horns

Horn—Forward Warning—Electric—Performance, Test, and Application (A)—SAE **J1105** SEP89 .... 1:14.47

Instrument Panels

Instrument Face Design and Location for Construction and Industrial Equipment (A)—SAE **J209** JAN87 ...... 3:40.421

Lamps

Flashing Warning Lamp for Industrial Equipment (A)—SAE **J96** MAR86 .... 3:40.261

Headlamps for Industrial Equipment (A)—SAE **J95** MAR86 .... 3:40.262

Lighting and Marking of Construction and Industrial Machinery—SAE **J1029** MAR86 .... 3:40.262

Lighting and Marking of Industrial Equipment on Highways (A)—SAE **J99** MAR86 .... 3:40.263

† Requirements for Sealed Lighting Unit for Construction and Industrial Machines—SAE **J572** MAY93 .... 3:40.260

† SAE Ground Vehicle Lighting Standards Manual—SAE **HS-34** JAN93 .... Available as a separate publication (See Related Technical Reports Section)

Sealed Lighting Units for Construction, Industrial and Forest Machinery (A)—SAE **J598** MAY87 .... 3:40.259

Lift Capacity

Lift Capacity Calculation and Test Procedure—Pipelayer and Sideboom—SAE **J743** MAR92 .... 3:40.223

Lubrication

Lubricant Types—Construction and Industrial Machinery—SAE **J754a** 3:40.102

Maintenance

Engineering Design Serviceability Guidelines—Construction and Industrial Machinery—Serviceability Definitions—Off-Road Work Machines (A)—SAE **J817/1** MAR91 .... 3:40.521

Engineering Design Serviceability Guidelines—Construction and Industrial Machinery—Maintainability Index—Off-Road Work Machines (A)—SAE **J817/2** MAR91 .... 3:40.522

Lift Arm Support Device for Loaders—SAE **J38** AUG91 . 3:40.353

Maintenance Interval Chart (A)—SAE **J753** MAY91 .... 3:40.101

† Personnel Protection for General Purpose Industrial Machines (A)—SAE **J98** NOV92 .... 3:40.472

Technical Publications for Off-Road Work Machines (A)—SAE **J920** SEP85 .... 3:40.526

Noise

Acoustics—Measurement of Airborne Noise Emitted by Construction Equipment Intended for Outdoor Use—Method for Determining Compliance with Noise Limits (A)—SAE **J2101** JUN91 .... 1:14.68

† Sound Measurement—Construction Site (A)—SAE **J1075** JUN93 .... 1:14.99

Sound Measurement—Off-Road Self-Propelled Work Machines—Operator—Work Cycle (A)—SAE **J1166** MAY90 .... 1:14.55

Operator Controls

Operator Controls on Industrial Equipment—SAE **J297** JUN85 .... 3:40.469

*Operator Controls--Off-Road Machines—SAE **J1814** FEB93 .... 3:40.414

† Personnel Protection for General Purpose Industrial Machines (A)—SAE **J98** NOV92 .... 3:40.472

Operator Protective Structures

Deflection Limiting Volume—ROPS/FOPS Laboratory Evaluation (A)—SAE **J397** APR88 .... 3:40.363

Labeling of ROPS and FOPS and OPS (A)—SAE **J1164** JAN91 .... 3:40.362

Minimum Performance Criteria for Falling Object Protective Structure (FOPS) (A)—SAE **J231** JAN81 .... 3:40.364

† Operator Protection for General-Purpose Industrial Machines—SAE **J1042** JUN93 .... 3:40.363

Performance Criteria for Falling Object Protective Structure (FOPS) for Industrial Machines—SAE **J1043** SEP87 .... 3:40.365

Performance Criteria for Rollover Protective Structures (ROPS) for Construction, Earthmoving, Forestry, and Mining Machines (A)—SAE **J1040** APR88 .... 3:40.356

† Personnel Protection for General Purpose Industrial Machines (A)—SAE **J98** NOV92 .... 3:40.472

Steel Products for Rollover Protective Structures (ROPS) and Falling Object Protective Structures (FOPS)—SAE **J1119** DEC88 .... 3:40.371

Safety

Operator Precautions (A)—SAE **J153** MAY87 .... 3:40.470

† Personnel Protection for General Purpose Industrial Machines (A)—SAE **J98** NOV92 .... 3:40.472

Safety Alert Symbol for Agricultural, Construction and Industrial Equipment (A)—SAE **J284** JAN91 .... 3:39.68

Steering Systems

Minimum Performance Criteria for Emergency Steering of Wheeled Earthmoving Construction Machines—SAE **J53** OCT84 .... 3:40.515

Terminology

Component Nomenclature—Construction Two- and Four-Wheel Tractors (A)—SAE **J869** JUL90 .... 3:40.32

Identification Terminology of Loaders/Tractors with Forks and Rough Terrain Forklifts—SAE **J1464** APR88 ...... 3:40.58

Nomenclature—Industrial Tractors (Wheel) (A)—SAE **J1092** JUN86 .... 3:40.31

Tractive Ability

Drawbar Test Procedure for Construction, Forestry and Industrial Machines (A)—SAE **J872** MAY86 .... 3:40.146

Revolutions per Mile and Static Loaded Radius for Off-Road Tires—SAE **J1544** JAN88 .... 3:40.307

Wheels

† Industrial and Agricultural Disc Wheels (A)—SAE **J712** APR93 .... 3:39.54

† Wheel Mounting Elements for Industrial and Agricultural Disc Wheels (A)—SAE **J714** APR93 .... 3:39.57

CONTAINMENT

Recommended Service Procedure for the Containment of HFC-134a—SAE **J2211** DEC91 .... 3:34.57

Recommended Service Procedure for the Containment of R-12—SAE **J1989** OCT89 .... 3:34.42

Safety and Containment of Refrigerant for Mechanical Vapor Compression Systems Used for Mobile Air-Conditioning Systems—SAE **J639** NOV91 .... 3:34.40

CONTAMINANTS

† Compressed Air for General Use—Part 1: Contaminants and Quality Classes—SAE **J1649/1** JUN93 .... 3:34.47

† Contaminants for Testing Air Brake Components and Auxiliary Pneumatic Devices—SAE **J2024** JUN93 ...... 3:36.131

HFC-134a Recycling Equipment for Mobile Air-Conditioning Systems—SAE **J2210** DEC91 .... 3:34.58

Standard of Purity for Use in Mobile Air-Conditioning Systems—SAE **J1991** OCT89 .... 3:34.47

CONTROL SYSTEMS

See also: **SPEED CONTROL**

Automotive Adaptive Driver Controls, Manual (A)—SAE **J1903** AUG90 .... 3:34.198

*Park Standard for Automatic Transmissions—SAE **J2208** JUN93 .... 3:29.42

---

* Technical Revision   † New Document   (D)—DODISS Adopted   (A)—ANSI Recognized

CONTROL SYSTEMS (continued)

Brakes

Braking Performance—Roller Compactors (A)—SAE J1472 JUN87 .... 3:40.288

Minimum Requirements for Wheel Slip Brake Control System Malfunction Signals—SAE J1230 OCT79 ..... 2:25.162

Snowmobile Brake Control Systems (A)—SAE J1282 OCT84 .... 3:38.09

Marine Engines

Marine Control Cable Connection—Engine Clutch Lever (A)—SAE J960 JUN91 .... 3:41.04

Marine Control Cable Connection—Engine Throttle Lever (A)—SAE J961 JUN91 .... 3:41.05

Marine Push-Pull Control Cables—SAE J917 JUN80 ... 3:41.03

Off-Road Vehicles

Electrical Propulsion Control—Off-Road Dumpers (A)— SAE J1299 JAN91 .... 3:40.282

Powertrains

Powertrain Control Interface for Electronic Controls Used in Medium and Heavy Duty Diesel On-Highway Vehicle Applications (A)—SAE J1922 DEC89 .... 3:36.177

Real Time

Chrysler Sensor and Control (CSC) Bus Multiplexing Network for Class 'A' Applications—SAE J2058 JUN90 . 2:23.422

*Class C Application Requirement Considerations—SAE J2056/1 JUN93 .... 2:23.366

*Survey of Known Protocols—SAE J2056/2 APR93 ..... 2:23.372

Token Slot Network for Automotive Control (A)—SAE J2106 APR91 .... 2:23.431

Snowmobiles

Snowmobile Brake Control Systems (A)—SAE J1282 OCT84 .... 3:38.09

Snowmobile Throttle Control Systems (A)—SAE J92 OCT84 .... 3:38.15

Transmissions

†Automatic Transmission Hydraulic Control Systems— Terminology—SAE J648 APR93 .... 3:29.27

CONTROLS
See: OPERATOR CONTROLS

CONVERTIBLE TOPS

Test Method for Wicking of Automotive Fabrics and Fibrous Materials (A)—SAE J913 FEB85 .... 1:11.163

COOLANTS

Automotive and Light Truck Engine Coolant Concentrate—Ethylene Glycol Type (A)—SAE J1034 APR91 .... 1:11.214

Coolant Concentrate (Low Silicate, Ethylene Glycol Type Requiring an Initial Charge of Supplemental Coolant Additive) for Heavy-Duty Engines (A)—SAE J1941 APR90 .... 1:11.216

Engine Coolants—SAE J814 JUL88 .... 1:11.209

†Principles of Engine Cooling Systems, Components and Maintenance—SAE HS-40 JAN91 .... Available as a separate publication (See Related Technical Reports Section)

†Test Method for Evaluating the Sealing Capability of Hose Connections with a PVT Test Facility—SAE J1610 APR93 .... 2:16.80

Preheaters

Engine Preheaters (A)—SAE J226 JAN86 .... 2:26.370

COOLING SYSTEMS
See also: AIR CONDITIONING, COOLANTS, FANS, OIL COOLERS, RADIATORS, REFRIGERATION

†Application Testing of Oil-to-Air Oil Coolers for Cooling Performance—SAE J1468 MAY93 .... 2:26.398

Cooling Flow Measurement Techniques—SAE J2082 JUN92 .... 3:28.46

Engine Cooling Fan Structural Analysis (A)—SAE J1390 APR82 .... 2:26.400

Engine Cooling System Field Test (Air-to-Boil) (A)—SAE J819 MAR87 .... 3:40.157

Engine Testing with Low Temperature Charge Air Cooler Systems in a Dynamometer Test Cell (A)—SAE J1937 NOV89 .... 1:13.23

Fan Hub Bolt Circles and Pilot Holes (A)—SAE J635 JUL84 .... 2:26.414

Guide to the Application and Use of Engine Coolant Pump Face Seals (A)—SAE J1245 JUN82 .... 2:26.417

Heavy Duty Non-Metallic Engine Cooling Fans—Material, Manufacturing and Test Considerations—SAE J1474 JAN85 .... 2:26.403

Information Relating to Duty Cycles and Average Power Requirements of Truck and Bus Engine Accessories— SAE J1343 OCT88 .... 3:36.57

Laboratory Testing of Vehicle and Industrial Heat Exchangers for Durability Under Vibration Induced Loading (A)—SAE J1598 JUL88 .... 2:26.389

Laboratory Testing of Vehicle and Industrial Heat Exchangers for Pressure Cycle Durability (A)—SAE J1597 JUL88 .... 2:26.390

Laboratory Testing of Vehicle and Industrial Heat Exchangers for Thermal Cycle Durability—SAE J1542 JAN89 .... 2:26.382

Method for Determining Power Consumption of Engine Cooling Fan-Drive Systems (A)—SAE J1342 AUG89 ... 3:36.55

On-Highway Truck Cooling Test Code (A)—SAE J1393 JUN84 .... 2:26.388

Pressure Relief for Cooling System (A)—SAE J151 JUN91 .... 2:26.350

†Principles of Engine Cooling Systems, Components and Maintenance—SAE HS-40 JAN91 .... Available as a separate publication (See Related Technical Reports Section)

Requirements for Engine Cooling System Filling, Deaeration and Drawdown Tests—SAE J1436 JUN89 .. 2:26.379

Test Method for Measuring Power Consumption of Truck and Bus Engine Cooling Fans (A)—SAE J1339 AUG89 . 3:36.52

Hoses

*Compression Set of Hoses or Solid Discs—SAE J1638 MAY93 .... 1:11.33

Coolant System Hoses (A)—SAE J20 MAR88 .... 1:11.222

Terminology

Engine Charge Air Cooler Nomenclature (A)—SAE J1148 JUN90 .... 2:26.392

Engine Coolant Pump Seals—SAE J780 JUN90 .... 2:26.415

†Glossary of Engine Cooling System Terms (A)—SAE J1004 APR93 .... 2:26.380

COPPER ALLOYS
See also: CHEMICAL COMPOSITION, MECHANICAL PROPERTIES, PHYSICAL PROPERTIES

Bearing and Bushing Alloys—SAE J459 OCT91 ...... 1:10.43

Bearing and Bushing Alloys—Chemical Composition of SAE Bearing and Bushing Alloys—SAE J460 OCT91 ... 1:10.41

Manual on Design and Application of Helical and Spiral Springs—SAE HS-J795 APR90 .... Available as a separate publication (See Related Technical Reports Section)

Cast

Cast Copper Alloys—SAE J462 SEP81 .... 1:10.78

Wrought and Cast Copper Alloys—SAE J461 SEP81 ... 1:10.45

Coated

Electroplating of Nickel and Chromium on Metal Parts— Automotive Ornamentation and Hardware—SAE J207 FEB85 .... 1:10.150

Wrought

Wrought and Cast Copper Alloys—SAE J461 SEP81 ... 1:10.45

Wrought Copper and Copper Alloys—SAE J463 SEP81 . 1:10.81

---

* Technical Revision † New Document (D)—DODISS Adopted (A)—ANSI Recognized

Subject Index

CORK
Gaskets
Standard Classification System for Nonmetallic Automotive Gasket Materials (A)—SAE **J90** JUN90 .... 1:11.197

CORNERING LAMPS
Front Cornering Lamps for Use on Motor Vehicles (A)—SAE **J852** NOV87 .... 2:24.162
Rear Cornering Lamp (A)—SAE **J1373** OCT87 .... 2:24.163
† SAE Ground Vehicle Lighting Standards Manual—SAE **HS-34** JAN93 .... Available as a separate publication (See Related Technical Reports Section)

CORROSION
Aluminum Alloys
*Decorative Anodizing Specification for Automotive Applications—SAE **J1974** JUN93 .... 1:10.39
Wrought Aluminum Applications Guidelines—SAE **J1434** JAN89 .... 1:10.15

Bodies (Automobile)
Corrosion Preventive Compound, Topside Vehicle Corrosion Protection—SAE **J1804** JUN89 .... 1:3.52
Corrosion Preventive Compound, Underbody Vehicle Corrosion Protection—SAE **J1959** JUN89 .... 1:3.55
Guidelines and Limitations of Laboratory Cyclic Corrosion Test Procedures for Exterior, Painted, Automotive Body Panels—SAE **J1563** MAY87 .... 1:3.51
Prevention of Corrosion of Motor Vehicle Body and Chassis Components—SAE **HS-J447** JUN81 .... Available as a separate publication (See Related Technical Reports Section)
Proving Ground Vehicle Corrosion Testing—SAE **J1950** MAY89 .... 1:3.41
Undervehicle Coupon Corrosion Tests—SAE **J1293** JAN90 .... 1:3.48

Brake Fluids
Low Water Tolerant Brake Fluids (A)—SAE **J1705** OCT88 .... 2:25.56
Motor Vehicle Brake Fluid (A)—SAE **J1703** JUN91 .... 2:25.41

Chassis
Prevention of Corrosion of Motor Vehicle Body and Chassis Components—SAE **HS-J447** JUN81 .... Available as a separate publication (See Related Technical Reports Section)

Cooling Systems
Automotive and Light Truck Engine Coolant Concentrate—Ethylene Glycol Type (A)—SAE **J1034** APR91 .... 1:11.214
Engine Coolants—SAE **J814** JUL88 .... 1:11.209

Copper Alloys
Wrought and Cast Copper Alloys—SAE **J461** SEP81 ... 1:10.45

Hydraulic Cylinders
Hydraulic Cylinder Rod Corrosion Test (A)—SAE **J1333** MAR90 .... 3:40.343

Insulation Materials
† Corrosion Test for Insulation Materials—SAE **J1389** MAY93 .... 1:11.173

COTTER PINS
See: PINS

COUPLINGS
See also: CONNECTORS, FASTENERS, FLUID COUPLINGS
Agricultural Equipment
Application of Hydraulic Remote Control to Agricultural Tractors and Trailing Type Agricultural Implements (A)—SAE **J716** MAR91 .... 3:39.45
† Dimensional Standard for Cylindrical Hydraulic Couplers for Agricultural Tractors (A)—SAE **J1036** APR93 .... 3:39.48
† Three-Point Hitch, Implement Quick-Attaching Coupler, Agricultural Tractors—SAE **J909** JUN93 .... 3:39.34

Drive Shafts
Marine Propeller-Shaft Couplings (A)—SAE **J756** AUG87 3:41.01
Hydraulic Systems
Hydraulic Diagnostic Couplings (A)—SAE **J1502** APR86 3:40.110
Trailers
*Kingpin Wear Limits Commercial Trailers and Semi-Trailers—SAE **J2228** JUN93 .... 3:36.89
Trailer Couplings, Hitches and Safety Chains—Automotive Type (A)—SAE **J684** MAY87 .... 3:35.01
Truck Trailers
Air Brake Gladhand Service (Control) and Emergency (Supply) Line Couplers—Trucks, Truck-Tractors, and Trailers—SAE **J318** SEP80 .... 3:36.139
Connection and Accessory Locations for Towing Multiple Trailers (A)—SAE **J849** NOV85 .... 3:36.93
Fifth Wheel Kingpin Performance—Commercial Trailers and Semitrailers—SAE **J133** OCT87 .... 3:36.86
Fifth Wheel Kingpin, Heavy-Duty—Commercial Trailers and Semitrailers (A)—SAE **J848** JAN91 .... 3:36.92
*Kingpin Wear Limits Commercial Trailers and Semi-Trailers—SAE **J2228** JUN93 .... 3:36.89
† Upper Coupler Kingpin—Commercial Trailers and Semitrailers (A)—SAE **J700** JUN93 .... 3:36.90
Trucks
Truck Tractor Semitrailer Interchange Coupling Dimensions (A)—SAE **J701** AUG84 .... 3:36.83

CRANES
See also: BOOMS
Crane Hoist Line Speed and Power Test Code (A)—SAE **J820** OCT80 .... 3:40.208
Crane Load Stability Test Code (A)—SAE **J765** OCT90 . 3:40.210
Lifting Crane Sheave and Drum Sizes—SAE **J881** FEB85 3:40.230
Lifting Crane, Wire-Rope Strength Factors (A)—SAE **J959** MAY91 .... 3:40.217
Load Indicating Devices in Lifting Crane Service (A)—SAE **J376** APR85 .... 3:40.215
Load Moment System (A)—SAE **J159** APR85 .... 3:40.216
Mobile Crane Working Area Definitions (A)—SAE **J1028** AUG89 .... 3:40.222
Rating Chart for Cantilevered Boom Cranes (A)—SAE **J1257** OCT80 .... 3:40.176
Rating Lift Cranes on Fixed Platforms Operating in the Ocean Environment—SAE **J1238** OCT78 .... 3:40.206
Rope Drum Rotation Indicating Device (A)—SAE **J1332** OCT81 .... 3:40.229
Two-Block Warning and Limit Systems in Lifting Crane Service—SAE **J1305** JUN87 .... 3:40.216
Brakes
Braking Performance—Rubber-Tired Construction Machines (A)—SAE **J1152** APR80 .... 3:40.289
Operator Controls
Crane and Cable Excavator Basic Operating Control Arrangements (A)—SAE **J983** OCT80 .... 3:40.220
Shovels
Nomenclature and Dimensions for Crane Shovels—SAE **J958** JUN89 .... 3:40.53
Stability
Mobile Crane Stability Ratings (A)—SAE **J1289** APR81 . 3:40.223
Stresses
Cantilevered Boom Crane Structures—Method of Test (A) (D)—SAE **J1063** OCT80 .... 3:40.170
Rope Supported Lattice-Type Boom Crane Structures—Method of Test—SAE **J987** APR85 .... 3:40.161

CRANKCASE EMISSIONS
Crankcase Emission Control Test Code (A)—SAE **J900** AUG85 .... 1:13.69
Recommended Practice for Bar-Coded Vehicle Emission Configuration Label (A)—SAE **J1892** MAY88 .... 3:42.34

---

* Technical Revision † New Document (D)—DODISS Adopted (A)—ANSI Recognized

Subject Index

CRANKCASE EMISSIONS (continued)
  *SAE Surface Vehicle Emissions Standards Manual—SAE HS-2400 FEB93 .... Available as a separate publication (See Related Technical Reports Section)

CRANKING MOTORS
  See: STARTERS

CRANKSHAFTS
  Mounting Flanges and Power Take-Off Shafts for Small Engines—SAE J609a 2:26.26

CRASH BARRIERS
  Fixed Rigid Barrier Collision Tests—SAE J850 NOV88 . 3:34.281
  Moving Rigid Barrier Collision Tests—SAE J972 DEC88 3:34.278

CRASHWORTHINESS
  See also: DROP TESTS, IMPACT TESTS
  Collision Deformation Classification—SAE J224 MAR80 3:34.235
  Design Guidelines for Optimizing Automobile Collision Damage Resistance, Repairability, and Serviceability—SAE J1555 NOV85 .... 3:42.19
  Fixed Rigid Barrier Collision Tests—SAE J850 NOV88 . 3:34.281
  Fuel and Lubricant Tanks for Motorcycles—SAE J1241 . 3:37.11
  Identifying and Repairing High Strength Steel Vehicle Components—SAE J1554 NOV85 .... 3:42.18
  Moving Rigid Barrier Collision Tests—SAE J972 DEC88 3:34.278
  Passenger Car Door System Crush Test Procedure—SAE J367 JUN80 .... 3:34.286

CRAWLERS
  See also: CONSTRUCTION AND INDUSTRIAL EQUIPMENT
  Classification and Nomenclature Towing Winch for Skidders and Crawler Tractors—SAE J1014 JAN85 .... 3:40.51
  Crawler Mounted Hydraulic Excavator Travel Performance (A)—SAE J1309 MAY91 .... 3:40.328
  Rated Operating Load for Loaders Equipped with Log or Material Forks without Vertical Mast (A)—SAE J1197 FEB91 .... 3:40.101
  Specification Definitions—Winches for Crawler Tractors and Skidders—SAE J1158 JAN85 .... 3:40.79
  Brakes
  Braking Performance—Crawler Tractors and Crawler Loaders—SAE J1026 APR90 .... 3:40.281
  Drawbars
  †Drawbars—Crawler Tractor (A)—SAE J749 AUG92 .... 3:40.318
  Noise
  Sound Measurement—Off-Road Self-Propelled Work Machines—Operator—Work Cycle (A)—SAE J1166 MAY90 .... 1:14.55
  Terminology
  Nomenclature—Crawler Tractor (A)—SAE J727 JAN86 . 3:40.30

CROCKING TESTS
  Method of Testing Resistance to Crocking of Organic Trim Materials—SAE J861 .... 1:11.114

CUPS
  Rubber
  Motor Vehicle Brake Fluid (A)—SAE J1703 JUN91 .... 2:25.41
  Rubber Cups for Hydraulic Actuating Cylinders (A)—SAE J1601 NOV90 .... 2:25.175

CUTTING EDGES
  Countersunk Square Holes for Cutting Edges and End Bits (A)—SAE J740 JUL86 .... 3:40.237
  Cutting Edge—Cross Sections Loader Straight (A)—SAE J1303 FEB85 .... 3:40.239
  Cutting Edge—Cross Sections Loader Straight with Bolt Holes (A)—SAE J1304 FEB85 .... 3:40.241
  Cutting Edge—Curved Grader (A)—SAE J739 JUN91 . 3:40.233
  Cutting Edge—Double Bevel Cross Sections (A)—SAE J738 JUN86 .... 3:40.235
  Cutting Edge—Optional Cross-Sections and Dimensions Loader Straight (A)—SAE J1581 SEP89 .... 3:40.242
  Hole Spacing for Scraper and Dozer Cutting Edges (A)—SAE J737 AUG89 .... 3:40.231
  Metric Countersunk Holes for Cutting Edges and End Bits (A)—SAE J1580 DEC89 .... 3:40.238

CUTTING TOOLS
  See: TOOLS

CYLINDERS
  See also: BRAKE CYLINDERS
  Engine
  Engine Rotation and Cylinder Numbering (A)—SAE J824 JAN86 .... 2:26.28
  Hydraulic
  Application of Hydraulic Remote Control to Agricultural Tractors and Trailing Type Agricultural Implements (A)—SAE J716 MAR91 .... 3:39.45
  Cylinder Rod Wiper Seal Ingression Test (A)—SAE J1195 MAR86 .... 3:40.141
  Hydraulic Cylinder Integrity Test—SAE J1334 JUN87 .. 3:40.344
  Hydraulic Cylinder Leakage Test—SAE J1336 JUN87 .. 3:40.346
  Hydraulic Cylinder No-Load Friction Test (A)—SAE J1335 APR90 .... 3:40.346
  Hydraulic Cylinder Rod Corrosion Test (A)—SAE J1333 MAR90 .... 3:40.343
  Hydraulic Cylinder Rod Seal Endurance Test Procedure (A)—SAE J1374 MAY85 .... 3:40.143
  Hydraulic Cylinder Test Procedure (A)—SAE J214 MAR86 .... 3:40.138
  Hydraulic Master Cylinders for Motor Vehicle Brakes—Test Procedure (A)—SAE J1153 JUN91 .... 2:25.29
  Hydraulic Master Cylinders for Motor Vehicle Brakes—Performance Requirements (A)—SAE J1154 JUN91 ... 2:25.34
  Hydraulic Wheel Cylinders for Automotive Drum Brakes (A)—SAE J101 DEC89 .... 2:25.25
  Rubber Cups for Hydraulic Actuating Cylinders (A)—SAE J1601 NOV90 .... 2:25.175
  Shelf Storage of Hydraulic Brake Components—SAE J1825 APR88 .... 2:25.35

DAMAGEABILITY
  See: CRASHWORTHINESS

DAMPING
  *Laboratory Measurement of the Composite Vibration Damping Properties of Materials on a Supporting Steel Bar—SAE J1637 FEB93 .... 1:2.30

DATA ACQUISITION
  Qualifying a Sound Data Acquisition System (D)—SAE J184 AUG87 .... 1:14.09

DATA BUS
  A Vehicle Network Protocol with a Fault Tolerant Multiplex Signal Bus—SAE J1813 AUG87 .... 2:23.499
  Chrysler Sensor and Control (CSC) Bus Multiplexing Network for Class 'A' Applications—SAE J2058 JUN90 . 2:23.422
  *Class A Multiplexing Architecture Strategies—SAE J2057/4 JUN93 .... 2:23.415
  *Class A Multiplexing Sensors—SAE J2057/3 JUN93 ... 2:23.413
  Class B Data Communication Network Interface—SAE J1850 AUG91 .... 2:23.268
  Collision Detection Serial Data Communications Multiplex Bus—SAE J1567 AUG87 .... 2:23.481
  Controller Area Network (CAN), An In-Vehicle Serial Communication Protocol—SAE J1583 MAR90 .... 2:23.486
  Serial Data Communications Between Microcomputer Systems in Heavy Duty Vehicle Applications (A)—SAE J1708 OCT90 .... 2:23.553
  *Survey of Known Protocols—SAE J2056/2 APR93 ..... 2:23.372
  Token Slot Network for Automotive Control (A)—SAE J2106 APR91 .... 2:23.431

---

* Technical Revision   † New Document   (D)—DODISS Adopted   (A)—ANSI Recognized

Subject Index

DATA COMMUNICATION
   Chrysler Sensor and Control (CSC) Bus Multiplexing
   Network for Class 'A' Applications—SAE **J2058** JUN90 . 2:23.422
   Class A Application/Definition **(A)**—SAE **J2057/1** JUN91   2:23.407
   *Class A Multiplexing Architecture Strategies—SAE
   J2057/4 JUN93 .... 2:23.415
   *Class A Multiplexing Sensors—SAE **J2057/3** JUN93 ... 2:23.413
   Class B Data Communication Network Interface—SAE
   J1850 AUG91 .... 2:23.268
   *Class B Data Communication Network Messages Part 2:
   Data Parameter Definitions—SAE **J2178/2** JUN93 ..... 2:23.345
   Class B Data Communication Network Messages:
   Detailed Header Formats and Physical Address
   Assignments—SAE **J2178/1** JUN92 .... 2:23.334
   *Class C Application Requirement Considerations—SAE
   J2056/1 JUN93 .... 2:23.366
   Controller Area Network (CAN), An In-Vehicle Serial
   Communication Protocol—SAE **J1583** MAR90 .... 2:23.486
   E/E Data Link Security—SAE **J2186** SEP91 .... 2:23.507
   E/E Diagnostic Data Communications **(A)**—SAE **J2054**
   NOV90 .... 2:23.329
   Glossary of Vehicle Networks for Multiplexing and Data
   Communications **(A)**—SAE **J1213/1** JUN91 .... 2:23.19
   † Initial Graphics Exchange Specification **(A)**—SAE **J1881**
   FEB93 .... 1:A.38
   † Joint SAE/TMC Electronic Data Interchange Between
   Microcomputer Systems in Heavy-Duty Vehicle
   Applications—SAE **J1587** AUG92 .... 2:23.531
   Off-Board Diagnostic Message Formats **(A)**—SAE **J2037**
   NOV90 .... 2:23.304
   Powertrain Control Interface for Electronic Controls Used
   in Medium and Heavy Duty Diesel On-Highway Vehicle
   Applications **(A)**—SAE **J1922** DEC89 .... 3:36.177
   Selection of Transmission Media **(A)**—SAE **J2056/3**
   JUN91 .... 2:23.391
   Serial Data Communications Between Microcomputer
   Systems in Heavy Duty Vehicle Applications **(A)**—SAE
   J1708 OCT90 .... 2:23.553
   *Survey of Known Protocols—SAE **J2056/2** APR93 ..... 2:23.372
   Token Slot Network for Automotive Control **(A)**—SAE
   J2106 APR91 .... 2:23.431
   *Universal Interface for OBD II Scan—SAE **J2201** JUN93   2:23.250

DATA TRANSMISSION
   Collision Detection Serial Data Communications Multiplex
   Bus—SAE **J1567** AUG87 .... 2:23.481
   *Enhanced E/E Diagnostic Test Modes—SAE **J2190**
   JUN93 .... 2:23.41
   † Initial Graphics Exchange Specification **(A)**—SAE **J1881**
   FEB93 .... 1:A.38
   † Joint SAE/TMC Electronic Data Interchange Between
   Microcomputer Systems in Heavy-Duty Vehicle
   Applications—SAE **J1587** AUG92 .... 2:23.531

DECARBURIZATION
  Steel
   Decarburization in Hardened and Tempered Threaded
   Fasteners **(A)**—SAE **J121** AUG83 .... 1:4.44
   † Leaf Springs for Motor Vehicle Suspension—Made to
   Metric Units—SAE **J1123** NOV92 .... 2:20.01
   † Leaf Springs for Motor Vehicle Suspension—Made to
   Customary U.S. Units—SAE **J510** NOV92 .... 2:20.09
   Methods of Measuring Decarburization **(A)**—SAE **J419**
   DEC83 .... 1:3.12

DECELERATION
   Sound Level for Passenger Cars and Light Trucks **(D)**—
   SAE **J986** OCT88 .... 1:14.14
  Electric Vehicles
   *Electric Vehicle Acceleration, Gradeability, and
   Deceleration Test Procedure—SAE **J1666** MAY93 ..... 3:28.12

DEFECTS
   Acoustic Emission Test Methods **(A)**—SAE **J1242**
   MAR91 .... 1:3.80
   Nondestructive Tests **(A)**—SAE **J358** FEB91 .... 1:3.67
   Ultrasonic Inspection **(A)**—SAE **J428** MAR91 .... 1:3.79
  Ferrous Materials
   Detection of Surface Imperfections in Ferrous Rods,
   Bars, Tubes, and Wires—SAE **J349** FEB91 .... 1:3.63
  Piston Rings
   † Internal Combustion Engines—Piston Rings—Quality
   Requirements—SAE **J1996** OCT92 .... 2:26.74
  Steel
   Classification of Common Imperfections in Sheet Steel—
   SAE **J810** MAR87 .... 1:3.30

DEFOGGERS
   † Passenger Car Backlight Defogging System **(A)**—SAE
   J953 APR93 .... 3:34.28
   Test Method for Determining Window Fogging
   Resistance of Interior Trim Materials **(A)**—SAE **J275**
   FEB85 .... 1:11.117

DEFROSTERS
   † Passenger Car Backlight Defogging System **(A)**—SAE
   J953 APR93 .... 3:34.28
   † Passenger Car Windshield Defrosting Systems—SAE
   J902 APR93 .... 3:34.18
   Windshield Defrosting Systems Performance
   Guidelines—Trucks, Buses, and Multi-Purpose
   Vehicles **(A)**—SAE **J382** OCT84 .... 3:34.23
   Windshield Defrosting Systems Test Procedure—Trucks,
   Buses, and Multipurpose Vehicles **(A) (D)**—SAE **J381**
   JUN84 .... 3:34.21

DESSICANTS
   *Compatibility of Retrofit Refrigerants with Air Conditioning
   System Materials—SAE **J1662** JUN93 .... 3:34.69

DIAGNOSTICS
   † Diagnostic Connector—SAE **J1962** JUN93 .... 2:23.245
   Diagnostic Trouble Code Definitions—SAE **J2012** MAR92   2:23.284
   E/E Diagnostic Test Modes—SAE **J1979** DEC91 ...... 2:23.30
   † Electrical/Electronic Systems Diagnostic Terms,
   Definitions, Abbreviations and Acronyms—SAE **J1930**
   JUN93 .... 2:23.288
   *Enhanced E/E Diagnostic Test Modes—SAE **J2190**
   JUN93 .... 2:23.41
   *OBD II Scan Tool—SAE **J1978** MAR92 .... 2:23.325
   Off-Board Diagnostic Message Formats **(A)**—SAE **J2037**
   NOV90 .... 2:23.304

DIESEL ENGINES
  Cold Weather Operation
   *SAE Cold Start and Driveability Procedure—SAE **J1635**
   MAY93 .... 2:26.498
  Cooling Systems
   Engine Testing with Low Temperature Charge Air Cooler
   Systems in a Dynamometer Test Cell **(A)**—SAE **J1937**
   NOV89 .... 1:13.23
   Fan Hub Bolt Circles and Pilot Holes **(A)**—SAE **J635**
   JUL84 .... 2:26.414
   On-Highway Truck Cooling Test Code **(A)**—SAE **J1393**
   JUN84 .... 2:26.388
  Electric Equipment
   Voltages for Diesel Electrical Systems **(A)**—SAE **J539**
   MAR87 .... 2:23.75
  Ether Systems
   Location of Atomizer of Ether Systems for Diesel
   Engines—SAE **J2079** MAY92 .... 2:26.329
  Exhaust Emissions
   Chemical Methods for the Measurement of Nonregulated
   Diesel Emissions **(A)**—SAE **J1936** OCT89 .... 1:13.38
   Continuous Hydrocarbon Analysis of Diesel
   Emissions **(A)**—SAE **J215** JUN88 .... 1:13.29

---

* Technical Revision   † New Document   (D)—DODISS Adopted   (A)—ANSI Recognized

DIESEL ENGINES (continued)
Exhaust Emissions (continued)
Diesel Emission Production Audit Test Procedure (A)—SAE **J1243** MAY88 .... 1:13.50

Diesel Engine Emission Measurement Procedure (A)—SAE **J1003** JUN90 .... 1:13.33

Measurement of Carbon Dioxide, Carbon Monoxide, and Oxides of Nitrogen in Diesel Exhaust—SAE **J177** APR82 1:13.01

*SAE Surface Vehicle Emissions Standards Manual—SAE **HS-2400** FEB93 .... Available as a separate publication (See Related Technical Reports Section)

Exhaust Gas Flow
†Measurement of Intake Air or Exhaust Gas Flow of Diesel Engines (A)—SAE **J244** AUG92 .... 1:13.11

Flywheels
Maximum Allowable Rotational Speed for Internal Combustion Engine Flywheels (A)—SAE **J1456** JUN90 . 2:26.243

Fuel Injection
Calibration Fluid for Diesel Injection Equipment (A)—SAE **J967** SEP88 .... 2:26.328

*Diesel Engines—Fuel Injection Pump Testing—SAE **J1668** JUN93 .... 2:26.322

Diesel Fuel Injection Equipment and Test Methods—SAE **HS-58** .... Available as a separate publication (See Related Technical Reports Section)

Diesel Fuel Injection Pump—Validation of Calibrating Nozzle Holder Assemblies (A)—SAE **J1549** APR88 .... 2:26.321

Diesel Fuel Injection—End Mounting Flanges for Fuel Injection Pumps (A)—SAE **J626** JUN88 .... 2:26.297

Diesel Fuel Injector Assembly Type 27 (9.5 mm) (A)—SAE **J1984** NOV89 .... 2:26.308

Diesel Fuel Injector Assembly Type 7 (9.5 mm) (A)—SAE **J347** JUL88 .... 2:26.305

Diesel Fuel Injector Assembly—Types 8, 9, 10, and 11 (A)—SAE **J265** APR91 .... 2:26.306

Diesel Injection Pump Testing—Part 1: Calibrating Nozzle and Holder Assemblies (A)—SAE **J968/1** MAY91 ...... 2:26.309

Road Vehicles—High Pressure Fuel Injection Pipe End-Connections with 60 Deg Female Cone—SAE **J1949** OCT88 .... 2:26.302

Tapers for Shaft Ends and Hubs for Fuel Injection Pumps (A)—SAE **J1419** FEB88 .... 2:26.245

Fuel Warmers
Fuel Warmer—Diesel Engines (A)—SAE **J1422** NOV89 . 3:40.509

Identification Numbers
Engine and Transmission Identification Numbers—SAE **J129** DEC81 .... 3:27.03

Lubricants
Engine Oil Performance and Engine Service Classification (Other Than "Energy-Conserving") (A)—SAE **J183** JUN91 .... 1:12.01

†Engine Oil Tests (A)—SAE **J304** JUN93 .... 1:12.19

†International Tests and Specifications for Automotive Engine Oils—SAE **J2227** MAY93 .... 1:12.10

Noise
Engine Sound Level Measurement Procedure (A) (D)—SAE **J1074** FEB87 .... 1:14.114

Measurement Procedure for Determination of Silencer Effectiveness in Reducing Engine Intake or Exhaust Sound Level (A)—SAE **J1207** FEB87 .... 1:14.115

Performance
Certificates of Maximum Net Horsepower for Motor Trucks and Truck Tractors—SAE **J690** 3:36.65

Engine Power Test Code—Spark Ignition and Compression Ignition— Gross Power Rating—SAE **J1995** JUN90 .... 2:26.11

Engine Power Test Code—Spark Ignition and Compression Ignition—Net Power Rating—SAE **J1349** JUN90 .... 2:26.16

Information Relating to Duty Cycles and Average Power Requirements of Truck and Bus Engine Accessories—SAE **J1343** OCT88 .... 3:36.57

Procedure for Mapping Engine Performance—Spark Ignition and Compression Ignition Engines (A)—SAE **J1312** JAN90 .... 2:26.22

Preheaters
†Electric Engine Preheaters and Battery Warmers for Diesel Engines—SAE **J1310** JUN93 .... 3:40.500

Engine Preheaters (A)—SAE **J226** JAN86 .... 2:26.370

Selection and Application Guidelines for Diesel, Gasoline, and Propane Fired Liquid Cooled Engine Pre-Heaters—SAE **J1350** MAR88 .... 3:40.496

Small
Small Engine Power Rating Procedure—SAE **J1940** JUN89 .... 2:26.23

Smoke
Diesel Emission Production Audit Test Procedure (A)—SAE **J1243** MAY88 .... 1:13.50

Diesel Engine Smoke Measurement—SAE **J255a** 1:13.56

Diesel Smoke Measurement Procedure (A)—SAE **J35** SEP88 .... 1:13.64

*SAE Surface Vehicle Emissions Standards Manual—SAE **HS-2400** FEB93 .... Available as a separate publication (See Related Technical Reports Section)

Spark Arresters
Spark Arrester Test Procedure for Large Size Engines (A)—SAE **J342** JAN91 .... 2:26.360

Spark Arrester Test Procedure for Medium Size Engines (A)—SAE **J350** JAN91 .... 2:26.365

Terminology
Engine Terminology and Nomenclature—General (A)—SAE **J604** JAN86 .... 2:26.01

Thermal Efficiency
Heating Value of Fuels—SAE **J1498** MAY90 .... 1:12.82

DIESEL FUELS
See also: CHEMICAL PROPERTIES

†Alternative Automotive Fuels—SAE **J1297** MAR93 .... 1:12.74

Coarse Droplet Water/Fuel Separation Test Procedure (A)—SAE **J1839** MAY90 .... 2:26.437

Diesel Fuels—SAE **J313** MAR92 .... 1:12.60

Emulsified Water/Fuel Separation Test Procedure—SAE **J1488** MAY90 .... 2:26.345

Fuel Economy Measurement—Road Test Procedure—Cold Start and Warm-Up Fuel Economy—SAE **J1256** OCT88 .... 2:26.506

Fuel Warmer—Diesel Engines (A)—SAE **J1422** NOV89 . 3:40.509

†Fuels and Lubricants Standards Manual—SAE **HS-23** FEB93 .... Available as a separate publication (See Related Technical Reports Section)

Heating Value of Fuels—SAE **J1498** MAY90 .... 1:12.82

DIMENSIONS
See also: ANTHROPOMETRY

†Connections for General Use and Fluid Power-Ports and Stud Ends with ISO 725 Threads and O-ring Sealing—Part 1: Threaded Port with O-Ring Seal in Truncated Housing—SAE **J1926/1 ISO 11** MAR93 .... 2:22.99

Dual Dimensioning (A)—SAE **J390** JUN82 .... 1:A.35

†Electrical Terminals—Eyelet and Spade Type—SAE **J561** JUN93 .... 2:23.182

Electrical Terminals—Pin and Receptacle Type (A)—SAE **J928** JUL89 .... 2:23.180

Engine Weight, Dimensions, Center of Gravity, and Moment of Inertia—SAE **J2038** APR92 .... 2:26.02

---

* Technical Revision † New Document (D)—DODISS Adopted (A)—ANSI Recognized

Subject Index

DIMENSIONS (continued)

Low Tension Thin Wall Primary Cable—SAE **J1560** JAN92 .... 2:23.167

Preparation of SAE Technical Reports—SAE **TSB 002** JUN92 .... 1:A.05

Radiator Caps and Filler Necks (A)—SAE **J164** JUN91 . 2:26.373

† Spring Nuts—SAE **J891** JUN93 .... 2:16.01

Agricultural Equipment

† Drawbars—Crawler Tractor (A)—SAE **J749** AUG92 .... 3:40.318

Automobile

Curb Clearance Approach, Departure, and Ramp Breakover Angles—Passenger Car and Light Truck (A)—SAE **J689** DEC89 .... 3:32.03

† Motor Vehicle Dimensions (A)—SAE **J1100** JUN93 .... 3:34.118

Motor Vehicle Fiducial Marks and Three Dimensional Reference System (A)—SAE **J182** NOV90 .... 3:34.229

Ball Joints

Metric Ball Joints (A)—SAE **J2213** JUN91 .... 2:17.12

Belts and Pulleys

† SI (Metric) Synchronous Belts and Pulleys—SAE **J1278** MAR93 .... 2:19.09

Clamps

† Hose Clamp Specifications—SAE **J1508** JUN93 ...... 2:16.45

* Type 'F' Clamps for Plumbing Applications—SAE **J1670** MAY93 .... 2:16.79

Clutch Housings

† Housing Internal Dimensions for Single- and Two-Plate Spring-Loaded Clutches (A)—SAE **J373** APR93 ..... 2:26.241

Clutches

† Clutch Dimensions for Truck and Bus Applications (A)—SAE **J1806** NOV92 .... 3:36.119

Connections

† Automotive Tube Fittings—SAE **J512** JUN93 .... 2:22.06

Connections for Fluid Power and General Use—Ports and Stud Ends with ISO 261 Threads and O-Ring Sealing Part 1: Port with O-Ring Seal in Truncated Housing—SAE **J2244/1** DEC91 .... 2:22.290

Connections for Fluid Power and General Use—Ports and Stud Ends with ISO 261 Threads and O-Ring Sealing Part 2: Heavy-Duty (S Series) Stud Ends—Dimensions, Design, Test Methods, and Requirements—SAE **J2244/2** DEC91 .... 2:22.293

* Connections for General use and Fluid Power-Ports and Stud Ends with ISO 725 Threads and O-Ring Sealing—Part 2: Heavy-Duty (S Series) Stud Ends—SAE **J1926/2** ISO 11 MAR93 .... 2:22.102

* Connections for General use and Fluid Power-Ports and Stud Ends with ISO 725 Threads and O-Ring Sealing—Part 3: Light-Duty (L Series) Stud Ends—SAE **J1926/3** ISO 11 MAR93 .... 2:22.108

† Connections for General Use and Fluid Power-Ports and Stud Ends with ISO 725 Threads and O-ring Sealing—Part 1: Threaded Port with O-Ring Seal in Truncated Housing—SAE **J1926/1 ISO 11** MAR93 .... 2:22.99

† Hydraulic Flanged Tube, Pipe, and Hose Connections, Four-Bolt Split Flange Type—SAE **J518** JUN93 .... 2:22.190

Ignition Cable Assemblies—SAE **J2032** NOV91 .... 2:23.154

† Refrigeration Tube Fittings—General Specifications—SAE **J513** JUN93 .... 2:22.33

Couplers

† Dimensional Standard for Cylindrical Hydraulic Couplers for Agricultural Tractors (A)—SAE **J1036** APR93 ..... 3:39.48

† Three-Point Hitch, Implement Quick-Attaching Coupler, Agricultural Tractors—SAE **J909** JUN93 .... 3:39.34

Crane Shovels

Nomenclature and Dimensions for Crane Shovels—SAE **J958** JUN89 .... 3:40.53

Excavators

Nomenclature and Dimensions for Hydraulic Excavators—SAE **J1193** NOV84 .... 3:40.59

Fabrics

Test Method for Determining Dimensional Stability of Automotive Textile Materials (A)—SAE **J883** FEB86 .... 1:11.114

Fittings

† Automotive Pipe Fittings—SAE **J530** JUN93 .... 2:22.232

† Automotive Tube Fittings—SAE **J512** JUN93 .... 2:22.06

† Beaded Tube Hose Fittings—SAE **J1231** JUN93 ...... 2:22.158

† Clip Fastener Fitting—SAE **J1467** JUN93 .... 2:22.197

† Fitting—O-Ring Face Seal—SAE **J1453** JUN93 .... 2:22.256

† Hydraulic Hose Fittings—SAE **J516** JUN93 .... 2:22.114

† Hydraulic Tube Fittings—SAE **J514** JUN93 .... 2:22.59

† Lubrication Fittings (A)—SAE **J534** JUN93 .... 2:22.252

† Refrigeration Tube Fittings—General Specifications—SAE **J513** JUN93 .... 2:22.33

† Spherical and Flanged Sleeve (Compression) Tube Fittings—SAE **J246** JUN93 .... 2:22.21

Flanges

Companion Flanges, Type A (External Pilot) and Type S (Internal Pilot) (A)—SAE **J1946** JAN91 .... 3:29.65

Starting Motor Mountings (A)—SAE **J542** JUN91 ...... 2:23.76

Flywheels

Engine Flywheel Housing and Mating Transmission Housing Flanges—SAE **J617** MAY92 .... 3:40.247

† Engine Flywheel Housings with Sealed Flanges—SAE **J1172** MAY93 .... 2:26.242

† Flywheels for Industrial Engines Used with Industrial Power Take-Offs Equipped with Driving-Ring Type Overcenter Clutches and Engine-Mounted Marine Gears and Single Bearing Engine-Mounted Power Generators—SAE **J620** MAY93 .... 3:40.256

Flywheels for Single-Plate Spring-Loaded Clutches (A)—SAE **J618** JAN91 .... 2:26.238

Forestry Equipment

Specification Definitions—Articulated Rubber-Tired Forwarder (A)—SAE **J1823** JUN91 .... 3:40.89

Specification Definitions—Clam Bunk Skidder (A)—SAE **J1824** JUN91 .... 3:40.81

Frames

Motor Truck CA Dimensions (A)—SAE **J691** SEP90 ... 3:36.67

Fuel Injection Nozzles

Diesel Fuel Injector Assembly—Types 8, 9, 10, and 11 (A)—SAE **J265** APR91 .... 2:26.306

Gasoline Dispenser Nozzles

† Gasoline Dispenser Nozzle Spouts (A)—SAE **J285** SEP92 .... 2:26.441

Headlamps

Dimensional Specifications for General Service Sealed Lighting Units—SAE **J760a** 2:24.17

Replaceable Motorcycle Headlamp Bulbs (A)—SAE **J1577** JUN91 .... 2:24.123

Hitches

† Three-Point Free-Link Hitch Attachment of Implements to Agricultural Wheeled Tractors—SAE **J715** JUN93 ..... 3:39.37

† Three-Point Hitch, Implement Quick-Attaching Coupler, Agricultural Tractors—SAE **J909** JUN93 .... 3:39.34

Hoses

† Fuel and Oil Hoses (A)—SAE **J30** MAY93 .... 1:11.226

† Hydraulic Hose—SAE **J517** JUN93 .... 2:22.161

† Marine Fuel Hoses (A)—SAE **J1527** FEB93 .... 3:41.16

† Transmission Oil Cooler Hose—SAE **J1532** JUN93 .... 1:11.252

Lamp Bulbs

Miniature Lamp Bulbs (A) (D)—SAE **J573** DEC89 ..... 2:24.21

Replaceable Motorcycle Headlamp Bulbs (A)—SAE **J1577** JUN91 .... 2:24.123

† Requirements for Sealed Lighting Unit for Construction and Industrial Machines—SAE **J572** MAY93 .... 3:40.260

Mounts

Engine Foot Mounting (Front and Rear)—SAE **J616** APR90 .... 3:40.252

---

* Technical Revision † New Document (D)—DODISS Adopted (A)—ANSI Recognized

Subject Index

DIMENSIONS (continued)
Mounts (continued)
Engine Mountings (A)—SAE J615 APR91 .... 2:26.234
Fastener Hardware for Wheels for Demountable
Rims (A)—SAE J1835 MAR90 .... 3:31.26
Hydraulic Pump and Motor Mounting and Drive
Dimensions (A)—SAE J744 JUL88 .... 3:40.116
Starting Motor Mountings (A)—SAE J542 JUN91 ...... 2:23.76

O-Rings
Hydraulic O-Ring—SAE J515 JUN92 .... 2:22.200

Off-Road Vehicles
† Minimum Service Access Dimensions for Off-Road
Machines (A)—SAE J925 JUN93 .... 3:40.380
Operator Space Envelope Dimensions for Off-Road
Machines—SAE J154 JUN92 .... 3:40.419
Operator's Seat Dimensions for Off-Road Self-Propelled
Work Machines—SAE J899 DEC88 .... 3:40.395
Specification Definitions—Articulated Rubber-Tired
Forwarder (A)—SAE J1823 JUN91 .... 3:40.89
Specification Definitions—Clam Bunk Skidder (A)—SAE
J1824 JUN91 .... 3:40.81
Specification Definitions—Off-Road Work Machines (A)—
SAE J1234 JAN85 .... 3:40.22

Passenger Compartments
Devices for Use in Defining and Measuring Vehicle
Seating Accommodation—SAE J826 JUN92 .... 3:34.90

Piston Rings
† Internal Combustion Engines—Piston Rings Coil Spring
Loaded Oil Control Rings—SAE J2003 OCT92 .... 2:26.131
† Internal Combustion Engines—Piston Rings Expander/
Segment Oil Control Rings—SAE J2004 OCT92 ... 2:26.175
† Internal Combustion Engines—Piston Rings—General
Specifications (A)—SAE J1591 OCT92 .... 2:26.186
† Internal Combustion Engines—Piston Rings—Half
Keystone Rings (A)—SAE J2001 OCT92 .... 2:26.111
† Internal Combustion Engines—Piston Rings—Inspection
Measuring Principles—SAE J1589 OCT92 .... 2:26.52
† Internal Combustion Engines—Piston Rings—Keystone
Rings—SAE J2000 OCT92 .... 2:26.100
† Internal Combustion Engines—Piston Rings—Oil Control
Rings—SAE J2002 OCT92 .... 2:26.117
† Internal Combustion Engines—Piston Rings—Quality
Requirements—SAE J1996 OCT92 .... 2:26.74
† Internal Combustion Engines—Piston Rings—
Rectangular Rings—SAE J1997 OCT92 .... 2:26.79
† Internal Combustion Engines—Piston Rings—
Rectangular Rings with Narrow Ring Width (A)—SAE
J1998 OCT92 .... 2:26.43
† Internal Combustion Engines—Piston Rings—Scraper
Rings—SAE J1999 OCT92 .... 2:26.90
* Internal Combustion Engines—Piston Rings—Steel
Rectangular Rings—SAE J2226 OCT92 .... 2:26.180

Plugs
† Automotive Pipe, Filler, and Drain Plugs—SAE J531
JUN93 .... 2:22.241
† Automotive Straight Thread Filler and Drain Plugs—SAE
J532 JUN93 .... 2:22.246

Power Take-Off
† Front and Rear Power Take-Off for Agricultural
Tractors (A)—SAE J1170 FEB93 .... 3:39.49

Pulleys
V-Belts and Pulleys—SAE J636 MAY92 .... 2:19.01

Recreational Vehicles
Tire to Body Clearance Check for Recreational
Vehicles—SAE J1214 JAN92 .... 3:35.04

Rims
Wide Base Tire Rims and Wheels (A)—SAE J876 JAN91 3:36.109

Seals
Application Guide to Radial Lip Seals—SAE J946 OCT91 3:29.109
Engine Coolant Pump Seals—SAE J780 JUN90 ...... 2:26.415

Sockets
Spark Plug Installation Sockets—SAE J548/2 JUN92 ... 2:23.95

Spark Plugs
Spark Plugs—SAE J548/1 JUN92 .... 2:23.83

Trucks and Buses
Average Vehicle Dimensions for Use in Designing
Docking Facilities for Motor Vehicles (A)—SAE J699
NOV85 .... 3:36.67
Motor Truck CA Dimensions (A)—SAE J691 SEP90 ... 3:36.67
† Motor Vehicle Dimensions (A)—SAE J1100 JUN93 .... 3:34.118
Motor Vehicle Fiducial Marks and Three Dimensional
Reference System (A)—SAE J182 NOV90 .... 3:34.229
Truck Overall Widths Across Dual Tires—SAE J693
APR89 .... 3:36.111
Truck Tractor Semitrailer Interchange Coupling
Dimensions (A)—SAE J701 AUG84 .... 3:36.83

Tubing
Brazed Double Wall Low Carbon Steel Tubing (A)—SAE
J527 JUN91 .... 2:22.209
Flares for Tubing—SAE J533 JUN92 .... 2:22.203
Metallic Air Brake System Tubing and Pipe (A)—SAE
J1149 JUN91 .... 2:22.219
Metric Nonmetallic Air Brake System Tubing (A)—SAE
J1394 APR91 .... 2:22.227
Nonmetallic Air Brake System Tubing (A)—SAE J844
JUN90 .... 2:22.221
Seamless Copper Tube (A)—SAE J528 JUN91 ...... 2:22.218
* Seamless Copper-Nickel 90-10 Tubing—SAE J1650
MAR93 .... 2:22.211
Seamless Low Carbon Steel Tubing Annealed for
Bending and Flaring (A)—SAE J524 JUN91 .... 2:22.206
Welded and Cold Drawn Low Carbon Steel Tubing
Annealed for Bending and Flaring (A)—SAE J525 JUN91 2:22.207
Welded Flash Controlled Low Carbon Steel Tubing
Normalized for Bending, Double Flaring, and
Beading (A)—SAE J356 JUN91 .... 2:22.213
Welded Low Carbon Steel Tubing (A)—SAE J526 JUN91 2:22.208

DIPPERS
See also: OFF-ROAD VEHICLES
Shovel Dipper, Clam Bucket, and Dragline Bucket
Rating—SAE J67 NOV84 .... 3:40.323

DIPSTICKS
Engine and Transmission Dipstick Marking (A)—SAE
J614b 2:26.233

DIRECTIONAL CONTROL
See also: HANDLING
Control Valve Test Procedure (A)—SAE J747 MAY90 .. 3:40.130
* Steady-State Circular Test Procedure for Trucks and
Buses—SAE J2181 JUN93 .... 3:36.68
Terminology
Vehicle Dynamics Terminology—SAE J670e JUN76 ... 3:34.297

DISC WHEELS
See: WHEELS

DISTRIBUTORS
See also: ELECTRIC EQUIPMENT, HANDLING
Drives
† Automotive Synchronous Belt Drives (A)—SAE J1313
MAR93 .... 2:19.13
† SI (Metric) Synchronous Belts and Pulleys—SAE J1278
MAR93 .... 2:19.09
Marine Engines
Ignition Distributors—Marine—SAE J1294 JAN86 ..... 3:41.19

DOORS
Courtesy Switches
Door Courtesy Switch (A)—SAE J2108 DEC91 .... 2:23.141
Crashworthiness
Passenger Car Door System Crush Test Procedure—
SAE J367 JUN80 .... 3:34.286

---

* Technical Revision   † New Document   (D)—DODISS Adopted   (A)—ANSI Recognized

Subject Index

DOORS (continued)
  Hinges
  Vehicle Passenger Door Hinge Systems—SAE **J934** JUL82 .... 3:34.01
  Latches
  Passenger Car Side Door Latch Systems (A)—SAE **J839** JUN91 .... 3:34.02

DOZERS
See: BULLDOZERS

DRAIN PLUGS
  † Automotive Pipe, Filler, and Drain Plugs—SAE **J531** JUN93 .... 2:22.241
  † Automotive Straight Thread Filler and Drain Plugs—SAE **J532** JUN93 .... 2:22.246
  † Drain, Fill, and Level Plugs for Off-Road, Self-Propelled Work Machines—SAE **J371** MAY93 .... 3:40.257

DRAWBARS
  Drawbar Test Procedure for Construction, Forestry and Industrial Machines (A)—SAE **J872** MAY86 .... 3:40.146
  † Drawbars—Agricultural Wheel Tractors—SAE **J1548** FEB93 .... 3:39.90
  † Drawbars—Crawler Tractor (A)—SAE **J749** AUG92 .... 3:40.318

DRAWINGS
  Dual Dimensioning (A)—SAE **J390** JUN82 .... 1:A.35

DRILLS
  Numbering System for Standard Drills (A)—SAE **J2122** NOV90 .... 3:43.11

DRIVEABILITY
  † Automotive Gasolines—SAE **J312** JAN93 .... 1:12.47
  * SAE Cold Start and Driveability Procedure—SAE **J1635** MAY93 .... 2:26.498

DRIVELINES
See also: DRIVESHAFTS
  All Wheel Drive Systems Classification (A)—SAE **J1952** JAN91 .... 3:29.15
  Universal Joints and Driveshafts—Nomenclature—Terminology—Application (A)—SAE **J901** OCT90 .... 3:29.42

DRIVERS
  Head Position
  Guidelines for Evaluating Out-of-Position Vehicle Occupant Interactions with Deploying Airbags (A)—SAE **J1980** NOV90 .... 3:33.33
  Motor Vehicle Driver and Passenger Head Position (A)—SAE **J1052** MAY87 .... 3:34.231
  Older
  * Headlamp Design Guidelines for Mature Drivers—SAE **J1606** MAR93 .... 3:34.225
  * Manual Controls for Mature Drivers—SAE **J2119** JUN93 .... 3:34.194
  Operator Controls
  Automotive Adaptive Driver Controls, Manual (A)—SAE **J1903** AUG90 .... 3:34.198
  Design Criteria—Driver Hand Controls Location for Passenger Cars, Multi-Purpose Passenger Vehicles, and Trucks (10 000 GVW and Under)—SAE **J1138** 3:34.191
  Driver Hand Control Reach (A)—SAE **J287** JUN88 .... 3:34.157
  Supplemental Information—Driver Hand Controls Location for Passenger Cars, Multi-Purpose Passenger Vehicles, and Trucks (10 000 GVW and Under)—SAE **J1139** 3:34.193
  Operator Contros.
  * Manual Controls for Mature Drivers—SAE **J2119** JUN93 .... 3:34.194
  Seat Position
  Accommodation Tool Reference Point—SAE **J1516** MAR90 .... 3:34.247
  Driver Selected Seat Position (A)—SAE **J1517** MAR90 .... 3:34.250
  Guidelines for Evaluating Out-of-Position Vehicle Occupant Interactions with Deploying Airbags (A)—SAE **J1980** NOV90 .... 3:33.33
  Truck Driver Shin-Knee Position for Clutch and Accelerator (A)—SAE **J1521** MAR90 .... 3:34.263
  Truck Driver Stomach Position (A)—SAE **J1522** MAR90 .... 3:34.265
  Vision
  Describing and Measuring the Driver's Field of View—SAE **J1050a** 3:34.214
  * Headlamp Design Guidelines for Mature Drivers—SAE **J1606** MAR93 .... 3:34.225
  * Manual Controls for Mature Drivers—SAE **J2119** JUN93 .... 3:34.194
  Motor Vehicle Drivers' Eye Locations—SAE **J941** JUN92 .... 3:34.204
  Passenger Car Rear Vision—SAE **J834a** 3:34.114
  Photometric Guidelines for Instrument Panel Displays That Accomodate Older Drivers—SAE **J2217** OCT91 .... 3:34.226
  Vision Factors Considerations in Rear View Mirror Design—SAE **J985** OCT88 .... 3:34.110

DRIVES
See also: TORQUE CONVERTERS, TRANSMISSIONS
  * Automotive Starter Drive Assembly Remanufacturing Procedures—SAE **J2241** MAR93 .... 3:42.46
  * Automotive Starter Solenoid Remanufacturing Procedures—SAE **J2242** MAR93 .... 3:42.47
  Passenger Cars and Light Truck Axles (A)—SAE **J2200** JAN91 .... 3:29.88
  Snowmobile Drive Mechanisms—SAE **J1279** DEC88 .... 3:38.18
  Accessory
  Mounting Flanges for Engine Accessory Drives (A)—SAE **J896** MAY83 .... 3:40.253
  Belt
  † Automotive Synchronous Belt Drives (A)—SAE **J1313** MAR93 .... 2:19.13
  Automotive V-Belt Drives—SAE **J637** FEB89 .... 2:19.04
  Automotive V-Ribbed Belt Drives and Test Methods—SAE **J1596** JUN89 .... 2:19.17
  * Glossary Automatic Belt Tensioner—SAE **J2198** JUL92 . 2:19.19
  † SI (Metric) Synchronous Belts and Pulleys—SAE **J1278** MAR93 .... 2:19.09
  V-Belts and Pulleys—SAE **J636** MAY92 .... 2:19.01
  V-Ribbed Belts and Pulleys—SAE **J1459** AUG88 .... 2:19.15
  Fan
  Method for Determining Power Consumption of Engine Cooling Fan-Drive Systems (A)—SAE **J1342** AUG89 .... 3:36.55
  Hydrodynamic
  Hydrodynamic Drive Test Code—SAE **J643** JUN89 .... 3:29.05
  Hydrodynamic Drives Terminology (A)—SAE **J641** JUL88 3:29.02
  † Symbols for Hydrodynamic Drives (A)—SAE **J640** APR93 .... 3:29.01

DRIVESHAFTS
  Marine Engines
  Marine Propeller-Shaft Couplings (A)—SAE **J756** AUG87 3:41.01
  Marine Propeller-Shaft Ends and Hubs—SAE **J755** JUN80 .... 3:41.06
  Terminology
  All Wheel Drive Systems Classification (A)—SAE **J1952** JAN91 .... 3:29.15
  Universal Joints and Driveshafts—Nomenclature—Terminology—Application (A)—SAE **J901** OCT90 .... 3:29.42

DRIVETRAINS
See: POWERTRAINS

DRIVING CYCLES
  * Electric Vehicle Energy Consumption and Range Test Procedure—SAE **J1634** MAY93 .... 3:28.01
  † Emission Test Driving Schedules (A)—SAE **J1506** APR93 .... 1:13.141
  Fuel Economy Measurement Road Test Procedure—SAE **J1082** JAN89 .... 2:26.459
  Fuel Economy Measurement—Road Test Procedure—Cold Start and Warm-Up Fuel Economy—SAE **J1256** OCT88 .... 2:26.506

---

* Technical Revision   † New Document   (D)—DODISS Adopted   (A)—ANSI Recognized

DROP TESTS
Operator Protective Structures
Overhead Protection for Agricultural Tractors—Test Procedures and Performance Requirements—SAE **J167** APR92 .... 3:39.89

DUMPERS
See also: **CONSTRUCTION AND INDUSTRIAL EQUIPMENT, EARTHMOVING EQUIPMENT, OFF-ROAD VEHICLES**

Capacity Rating—Dumper Body and Trailer Body (A)—SAE **J1363** JAN85 .... 3:40.321

Electrical Propulsion Control—Off-Road Dumpers (A)—SAE **J1299** JAN91 .... 3:40.282

Electrical Propulsion Rotating Equipment—Off-Road Dumper (A)—SAE **J1317** JUN82 .... 3:40.279

Brakes
Braking Performance—Rubber-Tired Construction Machines (A)—SAE **J1152** APR80 .... 3:40.289

Noise
Sound Measurement—Off-Road Self-Propelled Work Machines—Operator—Work Cycle (A)—SAE **J1166** MAY90 .... 1:14.55

Retarders
† Retardation Capability of Off-Highway Dumpers and Scrapers—SAE **J1430** FEB93 .... 3:40.298

Steering
Minimum Performance Criteria for Emergency Steering of Wheeled Earthmoving Construction Machines—SAE **J53** OCT84 .... 3:40.515

Terminology
Component Nomenclature—Dumper Trailer (A)—SAE **J734** JUL90 .... 3:40.47

Component Nomenclature—Dumpers (A)—SAE **J1016** JUL90 .... 3:40.48

Identification Terminology of Earthmoving Machines (A)—SAE **J1057** SEP88 .... 3:40.05

Tire Guards
Tire Guards for Protection of Operator of Earthmoving Haulage Machines (A)—SAE **J321b** 3:40.352

DUST SHIELDS
Plastic Dust Shield for Hydraulic Disc Brakes (A)—SAE **J1876** MAR90 .... 2:25.67

DYNAMICS
A Dictionary of Terms for the Dynamics and Handling of Single Track Vehicles (Motorcycles, Mopeds, and Bicycles)—SAE **J1451** FEB85 .... 3:37.01

Vehicle Aerodynamics Terminology (A)—SAE **J1594** JUN87 .... 3:28.17

Vehicle Dynamics Terminology—SAE **J670e** JUN76 .... 3:34.297

DYNAMOMETER TESTING
Acoustics—Measurement of Airborne Noise Emitted by Earthmoving Machinery—Method for Determining Compliance with Limits for Exterior Noise—Stationary Test Condition (A)—SAE **J2102** JUN91 .... 1:14.74

Acoustics—Measurement of Airborne Noise Emitted by Earthmoving Machinery—Operator's Position—Stationary Testing Condition (A)—SAE **J2103** JUN91 .... 1:14.78

Axle Efficiency Test Procedure (A)—SAE **J1266** JUN90 .... 3:29.85

*Brake Block Effectiveness Rating—SAE **J1802** JUN93 .... 2:25.119

*Brake Performance and Wear Test Code Commercial Vehicle Inertia Dynamometer—SAE **J2115** JUN93 .... 2:25.106

Brake Power Rating Test Code—Commercial Vehicle Inertia Dynamometer (A)—SAE **J971** JUN91 .... 2:25.152

Brake System Dynamometer Test Procedure—Passenger Car—SAE **J212** JUN80 .... 2:25.129

Brake Test Code—Inertia Dynamometer—SAE **J667** 2:25.103

Chassis Dynamometer Test Procedure—Heavy-Duty Road Vehicles—SAE **J2177** APR92 .... 2:26.06

Constant Volume Sampler System for Exhaust Emissions Measurement—SAE **J1094** JUN92 .... 1:13.117

Determination of Emissions from Gas Turbine Powered Light Duty Surface Vehicles (A)—SAE **J1130** MAY83 .... 1:13.156

Diesel Engine Emission Measurement Procedure (A)—SAE **J1003** JUN90 .... 1:13.33

*Electric Vehicle Acceleration, Gradeability, and Deceleration Test Procedure—SAE **J1666** MAY93 .... 3:28.12

*Electric Vehicle Energy Consumption and Range Test Procedure—SAE **J1634** MAY93 .... 3:28.01

† Emission Test Driving Schedules (A)—SAE **J1506** APR93 .... 1:13.141

Engine Power Test Code—Spark Ignition and Compression Ignition—Net Power Rating—SAE **J1349** JUN90 .... 2:26.16

Engine Power Test Code—Spark Ignition and Compression Ignition— Gross Power Rating—SAE **J1995** JUN90 .... 2:26.11

Engine Testing with Low Temperature Charge Air Cooler Systems in a Dynamometer Test Cell (A)—SAE **J1937** NOV89 .... 1:13.23

Impact of Alternative Fuels on Engine Test and Reporting Procedures—SAE **J1515** MAR88 .... 1:12.89

Passenger Car and Truck Automatic Transmission Test Code (A)—SAE **J651** JAN91 .... 3:29.28

Road Load Measurement and Dynamometer Simulation Using Coastdown Techniques—SAE **J1263** JUN91 .... 2:26.470

Spark Arrester Test Procedure for Medium Size Engines (A)—SAE **J350** JAN91 .... 2:26.365

† Test Procedure for the Measurement of Gaseous Exhaust Emissions from Small Utility Engines (A)—SAE **J1088** FEB93 .... 1:13.151

EARTHBORING MACHINES
Operator Controls
*Trenchless Equipment - (Horizontal Earthboring Machines) Operator Control Definitions—SAE **J1611** JUN93 .... 3:40.21

Precautions
*Operating Precautions for Horizontal Earthboring Machines—SAE **J2023** JAN93 .... 3:40.20

Terminology
Classification, Nomenclature, and Specification Definitions for Horizontal Earthboring Machines (A)—SAE **J2022** JUN91 .... 3:40.14

*Trenchless Equipment - (Horizontal Earthboring Machines) Operator Control Definitions—SAE **J1611** JUN93 .... 3:40.21

EARTHMOVING EQUIPMENT
See also: **AGRICULTURAL EQUIPMENT, BACKHOES, BULLDOZERS, CONSTRUCTION AND INDUSTRIAL EQUIPMENT, DUMPERS, EXCAVATORS, GRADERS, OFF-ROAD VEHICLES, SCRAPERS, TILLERS, TRENCHING EQUIPMENT**

Brakes
Brake Performance—Rubber-Tired Earthmoving Machines (A)—SAE **J1473** OCT90 .... 3:40.285

Braking Performance—Crawler Tractors and Crawler Loaders—SAE **J1026** APR90 .... 3:40.281

Minimum Performance Criteria for Braking Systems for Specialized Rubber-Tired, Self-Propelled Underground Mining Machines (A)—SAE **J1329** JUL89 .... 3:40.295

Noise
Acoustics—Measurement of Airborne Noise Emitted by Earthmoving Machinery—Operator's Position—Stationary Testing Condition (A)—SAE **J2103** JUN91 .... 1:14.78

Acoustics—Measurement of Airborne Noise Emitted by Earthmoving Machinery—Method for Determining Compliance with Limits for Exterior Noise—Stationary Test Condition (A)—SAE **J2102** JUN91 .... 1:14.74

Acoustics—Measurement of Exterior Noise Emitted by Earthmoving Machinery—Dynamic Test Conditions (A)—SAE **J2104** JUN91 .... 1:14.81

---

* Technical Revision † New Document (D)—DODISS Adopted (A)—ANSI Recognized

Subject Index

EARTHMOVING EQUIPMENT (continued)
Noise (continued)
Acoustics—Measurement of Noise Emitted by Earthmoving Machinery at the Operator's Position—Simulated Work Cycle Test Conditions (A)—SAE J2105 JUN91 .... 1:14.87

Sound Measurement—Earthmoving Machinery—Exterior (A) (D)—SAE J88 JUN86 .... 1:14.106

Sound Measurement—Earthmoving Machinery—Operator—Singular Type (A) (D)—SAE J919 JUN86 ... 1:14.53

Sound Measurement—Off-Road Self-Propelled Work Machines—Operator—Work Cycle (A)—SAE J1166 MAY90 .... 1:14.55

Sound Power Determination—Earthmoving Machinery—Static Condition (A)—SAE J1372 JUL88 .... 1:14.102

† Sound Power Level Measurements of Earthmoving Machinery - Static and In-Place Dynamic Methods—SAE J1805 APR93 .... 1:14.90

Operator Protective Structures
Performance Criteria for Falling Object Guards for Excavators—SAE J1356 FEB88 .... 3:40.367

Performance Criteria for Rollover Protective Structures (ROPS) for Construction, Earthmoving, Forestry, and Mining Machines (A)—SAE J1040 APR88 .... 3:40.356

Seats
Classification of Earthmoving Machines for Vibration Tests of Operator Seats (A)—SAE J1385 JUN83 ...... 3:40.398

† Vibration Performance Evaluation of Operator Seats (A)—SAE J1384 JUN93 .... 3:40.396

Steering
Minimum Performance Criteria for Emergency Steering of Wheeled Earthmoving Construction Machines—SAE J53 OCT84 .... 3:40.515

Terminology
Identification Terminology of Earthmoving Machines (A)—SAE J1057 SEP88 .... 3:40.05

Tires
Off-Road Tire and Rim Selection and Application (A)—SAE J1315 JAN91 .... 3:40.307

Revolutions per Mile and Static Loaded Radius for Off-Road Tires—SAE J1544 JAN88 .... 3:40.307

Tire Guards for Protection of Operator of Earthmoving Haulage Machines (A)—SAE J321b 3:40.352

Ton Kilometer Per Hour Application (A)—SAE J1098 MAR91 .... 3:40.317

Ton-Kilometer Per Hour Test Procedure (A)—SAE J1015 MAR91 .... 3:40.314

Vibration
† Vibration Performance Evaluation of Operator Seats (A)—SAE J1384 JUN93 .... 3:40.396

EDDY CURRENT TESTING
See also: NONDESTRUCTIVE TESTS

Detection of Surface Imperfections in Ferrous Rods, Bars, Tubes, and Wires—SAE J349 FEB91 .... 1:3.63

Electromagnetic Testing by Eddy Current Methods—SAE J425 MAR91 .... 1:3.72

Nondestructive Tests (A)—SAE J358 FEB91 .... 1:3.67

ELASTOMERS
See also: PLASTICS, RUBBER

*Compatibility of Retrofit Refrigerants with Air Conditioning System Materials—SAE J1662 JUN93 .... 3:34.69

Elastomeric Bushing "TRAC" Application Code (A)—SAE J1883 MAR88 .... 1:11.40

Peel Adhesion Test for Glass to Elastomeric Material for Automotive Glass Encapsulation—SAE J1907 OCT88 .. 1:11.99

*Recommended Guidelines for Load/Deformation Testing of Elastomeric Components—SAE J1636 FEB93 ...... 1:11.46

Standard Method for Determining Continuous Upper Temperature Resistance of Elastomers—SAE J2236 JUN92 .... 1:11.32

Test for Dynamic Properties of Elastomeric Isolators—SAE J1085a 1:11.50

† Thermoplastic Elastomer Classification System—SAE J3000 JUN93 .... 1:11.27

Fatigue
Recommended Guidelines for Fatigue Testing of Elastomeric Materials and Components—SAE J1183 FEB85 .... 1:11.35

Seals
† Static and Reciprocating Elastomeric Trnamission Seals—SAE J654 MAR93 .... 3:29.145

Tubing
Windshield Washer Tubing (A)—SAE J1037 DEC87 ... 1:11.254

ELECTRIC CABLES
Automobile, Truck, Truck-Tractor, Trailer, and Motor Coach Wiring—SAE J1292 OCT81 .... 2:23.142

Battery Cable (A)—SAE J1127 JUN88 .... 2:23.158

Fusible Links (A)—SAE J156 APR86 .... 2:23.171

High Tension Ignition Cable (A)—SAE J2031 JAN90 ... 2:23.149

High Tension Ignition Cable Assemblies—Marine (A)—SAE J1191 JAN86 .... 3:41.26

Low Tension Primary Cable (A)—SAE J1128 JUN88 ... 2:23.162

Low Tension Thin Wall Primary Cable—SAE J1560 JAN92 .... 2:23.167

Low Tension Wiring and Cable Terminals and Splice Clips—SAE J163 .... 2:23.157

Seven Conductor Jacketed Cable for Truck Trailer Connections—SAE J1067 .... 2:23.178

Jumper
Battery Booster Cables—SAE J1494 JUN89 .... 2:23.159

† Seven Conductor Electrical Connector for Truck-Trailer Jumper Cable—SAE J560 JUN93 .... 2:23.175

ELECTRIC CIRCUIT BREAKERS
† Circuit Breakers—SAE J553 JUN92 .... 2:23.135

Circuit Breaker—Internal Mounted—Automatic Reset—SAE J258 .... 2:23.141

Marine Circuit Breakers (A)—SAE J1428 MAY85 ..... 3:41.20

Miniature Blade Type Electrical Fuses (A)—SAE J2077 NOV90 .... 2:23.125

ELECTRIC CIRCUITS
Automotive Printed Circuits (A)—SAE J771 APR86 .... 2:23.145

Passenger Car Glazing—Electrical Circuits (A)—SAE J216 DEC89 .... 3:34.104

ELECTRIC EQUIPMENT
See also: BATTERIES, DISTRIBUTORS, ELECTRIC CABLES, ELECTRIC CIRCUIT BREAKERS, ELECTRIC CIRCUITS, ELECTRIC GENERATORS, ELECTRIC TERMINALS, ELECTRONIC EQUIPMENT, IGNITION SYSTEMS, LAMPS, SPEEDOMETERS, STARTERS, SWITCHES, TACHOMETERS, WIRING SYSTEMS

Blade Type Electric Fuses (A)—SAE J1284 APR88 .... 3:41.40

† Cranking Motor Application Considerations (A)—SAE J1375 MAR93 .... 2:23.81

Electric Starting Motor Test Procedure (A)—SAE J544 MAR88 .... 2:23.80

Electrical Generating System (Alternator Type) Performance Curve and Test Procedure (A)—SAE J56 JUN83 .... 2:23.82

Function Performance Status Classification for EMC Susceptibility Testing of Automotive Electronic and Electrical Devices—SAE J1812 OCT88 .... 2:23.226

High Current Time Lag Electric Fuses (A)—SAE J1888 NOV90 .... 2:23.128

† Low-Temperature Cranking Load Requirements of an Engine—SAE J1253 JUN93 .... 2:23.81

† SAE Ground Vehicle Lighting Standards Manual—SAE HS-34 JAN93 .... Available as a separate publication (See Related Technical Reports Section)

---

*Technical Revision   † New Document   (D)—DODISS Adopted   (A)—ANSI Recognized

Subject Index

ELECTRIC EQUIPMENT (continued)
- Starting Motor Pinions and Ring Gears (A)—SAE J543 JUN88 .... 2:23.78
- Vehicle System Voltage—Initial Recommendations—SAE J2232 JUN92 .... 2:23.75

Agricultural Equipment
- Electrical Connector for Auxiliary Starting of Construction, Agricultural, and Off-Road Machinery (A)—SAE J1283 MAR86 .... 3:40.267
- Electrical Wiring Systems for Construction, Agricultural and Off-Road Machines—SAE J821 MAY85 .... 3:40.264

Automobiles
- Automobile, Truck, Truck-Tractor, Trailer, and Motor Coach Wiring—SAE J1292 OCT81 .... 2:23.142
- Automotive Printed Circuits (A)—SAE J771 APR86 .... 2:23.145
- Electric Hourmeter Specification—SAE J1378 MAR83 .... 2:21.14
- Fusible Links (A)—SAE J156 APR86 .... 2:23.171
- Ignition Switch—SAE J259 .... 2:23.141
- Voltage Drop for Starting Motor Circuits—SAE J541 FEB89 .... 2:23.74
- Voltages for Diesel Electrical Systems (A)—SAE J539 MAR87 .... 2:23.75

Construction and Industrial Equipment
- Electrical Charging Systems for Construction and Industrial Machinery (A)—SAE J180 MAY87 .... 3:40.270
- Electrical Connector for Auxiliary Starting of Construction, Agricultural, and Off-Road Machinery (A)—SAE J1283 MAR86 .... 3:40.267
- Electrical Propulsion Control—Off-Road Dumpers (A)—SAE J1299 JAN91 .... 3:40.282
- Electrical Propulsion Rotating Equipment—Off-Road Dumper (A)—SAE J1317 JUN82 .... 3:40.279
- Electrical Wiring Systems for Construction, Agricultural and Off-Road Machines—SAE J821 MAY85 .... 3:40.264

Diagnostics
- † Diagnostic Connector—SAE J1962 JUN93 .... 2:23.245
- † Electrical/Electronic Systems Diagnostic Terms, Definitions, Abbreviations and Acronyms—SAE J1930 JUN93 .... 2:23.288
- *Enhanced E/E Diagnostic Test Modes—SAE J2190 JUN93 .... 2:23.41

Environmental Effects
- Miniature Blade Type Electrical Fuses (A)—SAE J2077 NOV90 .... 2:23.125
- Recommended Environmental Practices for Electronic Equipment Design—SAE J1211 .... 2:23.228

Grounds
- Electrical Grounding Practice (A)—SAE J1908 AUG90 .... 3:40.265

Marine
- External Ignition Protection of Marine Electrical Devices—SAE J1171 JAN86 .... 3:41.09
- Ignition Distributors—Marine—SAE J1294 JAN86 .... 3:41.19
- Marine Circuit Breakers (A)—SAE J1428 MAY85 .... 3:41.20

Motorcycles
- Motorcycle and Motor Driven Cycle Electrical System—Maintenance of Design Voltage—SAE J392 FEB92 .... 2:24.135

Off-Road Vehicles
- Electrical Grounding Practice (A)—SAE J1908 AUG90 .... 3:40.265
- *Electrical Indicating System Specification—SAE J1810 JAN93 .... 3:40.275

Personal Watercraft
- *Personal Watercraft--Electrical Systems—SAE J2120 JUL92 .... 3:41.30

Snowmobiles
- Maintenance of Design Voltage—Snowmobile Electrical Systems (A)—SAE J277 JUN90 .... 3:38.03

Terminology
- † Electrical/Electronic Systems Diagnostic Terms, Definitions, Abbreviations and Acronyms—SAE J1930 JUN93 .... 2:23.288

- Nomenclature—Automotive Electrical Systems—SAE J831 .... 2:23.01

Trailers
- Five Conductor Electrical Connectors for Automotive Type Trailers (A)—SAE J895 APR86 .... 2:23.171
- Four- and Eight-Conductor Rectangular Electrical Connectors for Automotive Type Trailers—SAE J1239 .... 2:23.173

Truck Trailers
- Automobile, Truck, Truck-Tractor, Trailer, and Motor Coach Wiring—SAE J1292 OCT81 .... 2:23.142
- Brake and Electrical Connection Locations—Truck-Tractor and Truck-Trailer (A)—SAE J702 MAY85 .... 3:36.114
- † Seven Conductor Electrical Connector for Truck-Trailer Jumper Cable—SAE J560 JUN93 .... 2:23.175

Trucks and Buses
- Automobile, Truck, Truck-Tractor, Trailer, and Motor Coach Wiring—SAE J1292 OCT81 .... 2:23.142
- Information Relating to Duty Cycles and Average Power Requirements of Truck and Bus Engine Accessories—SAE J1343 OCT88 .... 3:36.57

ELECTRIC FUSES
See: FUSES

ELECTRIC GENERATORS
- Electrical Generating System (Alternator Type) Performance Curve and Test Procedure (A)—SAE J56 JUN83 .... 2:23.82
- Nomenclature—Automotive Electrical Systems—SAE J831 .... 2:23.01
- Procedure for Evaluating Transient Response of Small Engine Driven Generator Sets (A)—SAE J1444 JUN91 .... 2:26.23

ELECTRIC MOTORS
See also: STARTERS
- Electric Blower Motor Switch—SAE J235 .... 2:23.189

ELECTRIC TERMINALS
- Electrical Terminals—Blade Type—SAE J858a 2:23.179
- † Electrical Terminals—Eyelet and Spade Type—SAE J561 JUN93 .... 2:23.182
- Electrical Terminals—Pin and Receptacle Type (A)—SAE J928 JUL89 .... 2:23.180
- Low Tension Wiring and Cable Terminals and Splice Clips—SAE J163 .... 2:23.157

ELECTRIC VEHICLES
- *Electric Vehicle Acceleration, Gradeability, and Deceleration Test Procedure—SAE J1666 MAY93 .... 3:28.12
- *Electric Vehicle Energy Consumption and Range Test Procedure—SAE J1634 MAY93 .... 3:28.01
- Performance Levels and Methods of Measurement of Electromagnetic Radiation from Vehicles and Devices (30 to 1000 MHz)—SAE J551 MAR90 .... 2:23.190

ELECTROMAGNETIC COMPATIBILITY (EMC)
- Anechoic Test Facility Radiated Susceptibility 20 MHz - 18 GHz Electromagnetic Field—SAE J1507 JAN87 .... 2:23.477
- Electromagnetic Susceptibility Measurements of Vehicle Components Using TEM Cells (14 kHz-200 MHz)—SAE J1448 JAN84 .... 2:23.471
- Function Performance Status Classification for EMC Susceptibility Testing of Automotive Electronic and Electrical Devices—SAE J1812 OCT88 .... 2:23.226
- Open Field Whole-Vehicle Radiated Susceptibility 10 kHz—18 GHz, Electric Field—SAE J1338 JUN81 .... 2:23.447
- Serial Data Communications Between Microcomputer Systems in Heavy Duty Vehicle Applications (A)—SAE J1708 OCT90 .... 2:23.553

ELECTROMAGNETIC INTERFERENCE (EMI)
- Collision Detection Serial Data Communications Multiplex Bus—SAE J1567 AUG87 .... 2:23.481
- Electromagnetic Susceptibility Measurement Procedures for Vehicle Components (Except Aircraft) (A)—SAE J1113 AUG87 .... 2:23.207

---

* Technical Revision   † New Document   (D)—DODISS Adopted   (A)—ANSI Recognized

Subject Index

ELECTROMAGNETIC INTERFERENCE (EMI) (continued)
Electromagnetic Susceptibility Procedures for Common Mode Injection (1—400 MHz), Module Testing—SAE **J1547** OCT88 .... 2:23.458

Function Performance Status Classification for EMC Susceptibility Testing of Automotive Electronic and Electrical Devices—SAE **J1812** OCT88 .... 2:23.226

Open Field Whole-Vehicle Radiated Susceptibility 10 kHz—18 GHz, Electric Field—SAE **J1338** JUN81 ..... 2:23.447

Performance Levels and Methods of Measurement of Electromagnetic Radiation from Vehicles and Devices, Narrowband, 10 kHz to 1000 MHz (A)—SAE **J1816** OCT87 .... 2:23.508

Performance Levels and Methods of Measurement of Electromagnetic Radiation from Vehicles and Devices (30 to 1000 MHz)—SAE **J551** MAR90 .... 2:23.190

Selection of Transmission Media (A)—SAE **J2056/3** JUN91 .... 2:23.391

Vehicle Electromagnetic Radiated Susceptibility Testing Using a Large TEM Cell—SAE **J1407** MAR88 .... 2:23.460

ELECTROMAGNETIC TESTS
See also: NONDESTRUCTIVE TESTS

Detection of Surface Imperfections in Ferrous Rods, Bars, Tubes, and Wires—SAE **J349** FEB91 .... 1:3.63

Electromagnetic Susceptibility Procedures for Common Mode Injection (1—400 MHz), Module Testing—SAE **J1547** OCT88 .... 2:23.458

Electromagnetic Testing by Eddy Current Methods—SAE **J425** MAR91 .... 1:3.72

Nondestructive Tests (A)—SAE **J358** FEB91 .... 1:3.67

Performance Levels and Methods of Measurement of Electromagnetic Radiation from Vehicles and Devices, Narrowband, 10 kHz to 1000 MHz (A)—SAE **J1816** OCT87 .... 2:23.508

Performance Levels and Methods of Measurement of Electromagnetic Radiation from Vehicles and Devices (30 to 1000 MHz)—SAE **J551** MAR90 .... 2:23.190

ELECTRONIC EQUIPMENT
See also: ELECTRIC EQUIPMENT, TRANSDUCERS

Chrysler Sensor and Control (CSC) Bus Multiplexing Network for Class 'A' Applications—SAE **J2058** JUN90 . 2:23.422

Collision Detection Serial Data Communications Multiplex Bus—SAE **J1567** AUG87 .... 2:23.481

Design/Process Checklist for Vehicle Electronic Systems—SAE **J1938** OCT88 .... 2:23.578

Electromagnetic Susceptibility Measurements of Vehicle Components Using TEM Cells (14 kHz-200 MHz)—SAE **J1448** JAN84 .... 2:23.471

Electromagnetic Susceptibility Procedures for Common Mode Injection (1—400 MHz), Module Testing—SAE **J1547** OCT88 .... 2:23.458

Function Performance Status Classification for EMC Susceptibility Testing of Automotive Electronic and Electrical Devices—SAE **J1812** OCT88 .... 2:23.226

Gasoline Fuel Injector (A)—SAE **J1832** NOV89 .... 2:26.263

General Qualification and Production Acceptance Criteria for Integrated Circuits in Automotive Applications—SAE **J1879** OCT88 .... 2:23.583

Performance Levels and Methods of Measurement of Electromagnetic Radiation from Vehicles and Devices, Narrowband, 10 kHz to 1000 MHz (A)—SAE **J1816** OCT87 .... 2:23.508

* Rating Criteria for Electronic Refrigerant Leak Detectors—SAE **J1627** JUN93 .... 3:34.71

* Technican Procedure for Using Electronic Refrigerant Leak Detectors for Service of Mobile Air Conditioning Systems—SAE **J1628** JUN93 .... 3:34.72

Diagnostics
† Diagnostic Connector—SAE **J1962** JUN93 .... 2:23.245

E/E Diagnostic Data Communications (A)—SAE **J2054** NOV90 .... 2:23.329

† Electrical/Electronic Systems Diagnostic Terms, Definitions, Abbreviations and Acronyms—SAE **J1930** JUN93 .... 2:23.288

* OBD II Scan Tool—SAE **J1978** MAR92 .... 2:23.325

Environmental Effects
Joint SAE/TMC Recommended Environmental Practices for Electronic Equipment Design (Heavy-Duty Trucks)—SAE **J1455** JAN88 .... 2:23.556

Recommended Environmental Practices for Electronic Equipment Design—SAE **J1211** 2:23.228

Glossary
Glossary of Automotive Electronic Terms—SAE **J1213** NOV82 .... 2:23.01

Glossary of Reliability Terminology Associated with Automotive Electronics—SAE **J1213/2** OCT88 .... 2:23.22

Glossary of Vehicle Networks for Multiplexing and Data Communications (A)—SAE **J1213/1** JUN91 .... 2:23.19

Security
E/E Data Link Security—SAE **J2186** SEP91 .... 2:23.507

Speed Sensors
Transmission Mounted Vehicle Speed Signal Rotor Specification—SAE **J1377** JAN89 .... 2:21.17

Terminology
† Electrical/Electronic Systems Diagnostic Terms, Definitions, Abbreviations and Acronyms—SAE **J1930** JUN93 .... 2:23.288

Trucks
Joint SAE/TMC Recommended Environmental Practices for Electronic Equipment Design (Heavy-Duty Trucks)—SAE **J1455** JAN88 .... 2:23.556

Trucks and Buses
† Accelerator Pedal Position Sensor for Use with Electronic Controls in Medium- and Heavy-Duty Vehicle Applications—SAE **J1843** APR93 .... 2:23.515

† Joint SAE/TMC Electronic Data Interchange Between Microcomputer Systems in Heavy-Duty Vehicle Applications—SAE **J1587** AUG92 .... 2:23.531

* OEM/Vendor's Interface Specification for Vehicle Electronic Programming Station—SAE **J1924** DEC92 .. 2:23.522

Powertrain Control Interface for Electronic Controls Used in Medium and Heavy Duty Diesel On-Highway Vehicle Applications (A)—SAE **J1922** DEC89 .... 3:36.177

Serial Data Communications Between Microcomputer Systems in Heavy Duty Vehicle Applications (A)—SAE **J1708** OCT90 .... 2:23.553

ELECTROPLATING
Electroplate Requirements for Decorative Chromium Deposits on Zinc Base Materials Used for Exterior Ornamentation (A)—SAE **J1837** JUN91 .... 1:10.155

Electroplating and Related Finishes—SAE **J474** FEB85 . 1:10.157

Electroplating of Nickel and Chromium on Metal Parts—Automotive Ornamentation and Hardware—SAE **J207** FEB85 .... 1:10.150

ELECTROSTATIC DISCHARGE
Electrostatic Discharge Test for Vehicles—SAE **J1595** OCT88 .... 2:23.464

EMERGENCY VEHICLES
Lamps
360 Degree Warning Devices for Authorized Emergency, Maintenance, and Service Vehicles—SAE **J845** MAR92 . 2:24.185

Flashing Warning Lamps for Authorized Emergency, Maintenance and Service Vehicles (A)—SAE **J595** JAN90 .... 2:24.183

Gaseous Discharge Warning Lamp for Authorized Emergency, Maintenance, and Service Vehicles (A)—SAE **J1318** APR86 .... 2:24.188

† SAE Ground Vehicle Lighting Standards Manual—SAE **HS-34** JAN93 .... Available as a separate publication *(See Related Technical Reports Section)*

---

* Technical Revision   † New Document   (D)—DODISS Adopted   (A)—ANSI Recognized

Subject Index

EMERGENCY VEHICLES (continued)
Sirens
Emergency Vehicle Sirens (A)—SAE J1849 JUL89 ..... 2:24.198

EMISSIONS
See: ACOUSTIC EMISSION ANALYSIS, CRANKCASE EMISSIONS, EVAPORATIVE EMISSIONS, EXHAUST EMISSIONS, REFUELING EMISSIONS, SMOKE

EMULSIFIED FUELS
See also: ALTERNATE FUELS
Emulsified Water/Fuel Separation Test Procedure—SAE J1488 MAY90 ..... 2:26.345
Heating Value of Fuels—SAE J1498 MAY90 ..... 1:12.82

EMULSIFIED WATER
Coarse Droplet Water/Fuel Separation Test Procedure (A)—SAE J1839 MAY90 ..... 2:26.437
Emulsified Water/Fuel Separation Test Procedure—SAE J1488 MAY90 ..... 2:26.345

ENERGY CONSUMPTION
See also: FUEL CONSUMPTION
Axles
Axle Efficiency Test Procedure (A)—SAE J1266 JUN90 . 3:29.85
Electric Vehicles
*Electric Vehicle Energy Consumption and Range Test Procedure—SAE J1634 MAY93 ..... 3:28.01

ENGINE IDENTIFICATION NUMBER (EIN)
Engine and Transmission Identification Numbers—SAE J129 DEC81 ..... 3:27.03

ENGINES
See: DIESEL ENGINES, GAS TURBINES, MARINE ENGINES, SPARK IGNITION ENGINES, TWO STROKE CYCLE ENGINES

ENVIRONMENTAL EFFECTS
See also: CORROSION
Accelerated Exposure of Automotive Exterior Materials Using a Solar Fresnel-Reflective Apparatus—SAE J1961 DEC88 ..... 1:11.286
Joint SAE/TMC Recommended Environmental Practices for Electronic Equipment Design (Heavy-Duty Trucks)—SAE J1455 JAN88 ..... 2:23.556
Bodies (Automobile)
Accelerated Exposure of Automotive Exterior Materials Using a Fluorescent UV and Condensation Apparatus—SAE J2020 JUN89 ..... 1:11.288
Accelerated Exposure of Automotive Exterior Materials Using a Controlled Irradiance Water Cooled Xenon Arc Apparatus—SAE J1960 JUN89 ..... 1:11.268
Guidelines and Limitations of Laboratory Cyclic Corrosion Test Procedures for Exterior, Painted, Automotive Body Panels—SAE J1563 MAY87 ..... 1:3.51
Outdoor Weathering of Exterior Materials—SAE J1976 JUN89 ..... 1:11.275
Proving Ground Vehicle Corrosion Testing—SAE J1950 MAY89 ..... 1:3.41
Bonded Assemblies
*Accelerated Environmental Testing for Bonded Automotive Assemblies—SAE J2100 AUG92 ..... 1:11.63
Electronic Equipment
General Qualification and Production Acceptance Criteria for Integrated Circuits in Automotive Applications—SAE J1879 OCT88 ..... 2:23.583
Recommended Environmental Practices for Electronic Equipment Design—SAE J1211 ..... 2:23.228
Trim
Accelerated Exposure of Automotive Interior Trim Components Using a Controlled Irradiance Water Cooled Xenon-Arc Apparatus—SAE J1885 MAR92 ..... 1:11.255

ETHYLENE-GLYCOL
Automotive and Light Truck Engine Coolant Concentrate—Ethylene Glycol Type (A)—SAE J1034 APR91 ..... 1:11.214
Coolant Concentrate (Low Silicate, Ethylene Glycol Type Requiring an Initial Charge of Supplemental Coolant Additive) for Heavy-Duty Engines (A)—SAE J1941 APR90 ..... 1:11.216
Engine Coolants—SAE J814 JUL88 ..... 1:11.209

EVAPORATIVE EMISSIONS
† Automotive Gasolines—SAE J312 JAN93 ..... 1:12.47
† Instrumentation and Techniques for Vehicle Refueling Emissions Measurement—SAE J1045 MAY93 ..... 1:13.137
Measurement of Fuel Evaporative Emissions from Gasoline Powered Passenger Cars and Light Trucks Using the Enclosure Technique (A)—SAE J171 APR91 . 1:13.95
*SAE Surface Vehicle Emissions Standards Manual—SAE HS-2400 FEB93 ..... Available as a separate publication (See Related Technical Reports Section)

EVAPORATORS
Rating Air Conditioner Evaporator Air Delivery and Cooling Capacities (A)—SAE J1487 MAY85 ..... 3:36.58

EXCAVATORS
See also: CONSTRUCTION AND INDUSTRIAL EQUIPMENT, EARTHMOVING EQUIPMENT, OFF-ROAD VEHICLES
Crawler Mounted Hydraulic Excavator Travel Performance (A)—SAE J1309 MAY91 ..... 3:40.328
† Excavator, Mini-Excavator, and Backhoe Hoe Bucket Volumetrtic Rating—SAE J296 JUN93 ..... 3:40.329
Hydraulic Excavator and Backhoe Digging Forces (A)—SAE J1179 FEB90 ..... 3:40.340
Hydraulic Excavator Lift Capacity Calculation and Test Procedure—SAE J1097 NOV88 ..... 3:40.334
† Hydraulic Excavator Swing Minimum Performance and Rating Procedure—SAE J1371 JUN93 ..... 3:40.342
Brakes
Braking Performance—Rubber-Tired Construction Machines (A)—SAE J1152 APR80 ..... 3:40.289
Dimensions
Nomenclature and Dimensions for Hydraulic Excavators—SAE J1193 NOV84 ..... 3:40.59
Noise
Sound Measurement—Off-Road Self-Propelled Work Machines—Operator—Work Cycle (A)—SAE J1166 MAY90 ..... 1:14.55
† Sound Power Level Measurements of Earthmoving Machinery - Static and In-Place Dynamic Methods—SAE J1805 APR93 ..... 1:14.90
Operator Controls
Crane and Cable Excavator Basic Operating Control Arrangements (A)—SAE J983 OCT80 ..... 3:40.220
Excavator and Backhoe Hand Signals—SAE J1307 NOV86 ..... 3:40.348
Hydraulic Excavator Operator Controls—SAE J1177 JUN88 ..... 3:40.341
Operator Protective Structures
Performance Criteria for Falling Object Guards for Excavators—SAE J1356 FEB88 ..... 3:40.367
Terminology
Identification Terminology of Earthmoving Machines (A)—SAE J1057 SEP88 ..... 3:40.05
Nomenclature and Dimensions for Hydraulic Excavators—SAE J1193 NOV84 ..... 3:40.59

EXHAUST EMISSIONS
See also: EXHAUST GASES, INDIVIDUAL POLLUTANTS, SMOKE
E/E Diagnostic Test Modes—SAE J1979 DEC91 ..... 2:23.30
† Emission Test Driving Schedules (A)—SAE J1506 APR93 ..... 1:13.141
Instrumentation and Techniques for Exhaust Gas Emissions Measurement (A)—SAE J254 AUG84 ..... 1:13.104

---

* Technical Revision   † New Document   (D)—DODISS Adopted   (A)—ANSI Recognized

Subject Index

EXHAUST EMISSIONS (continued)
† Instrumentation and Techniques for Vehicle Refueling Emissions Measurement—SAE **J1045** MAY93 .... **1:13.137**

Diesel Engine
Chemical Methods for the Measurement of Nonregulated Diesel Emissions (A)—SAE **J1936** OCT89 .... **1:13.38**

Continuous Hydrocarbon Analysis of Diesel Emissions (A)—SAE **J215** JUN88 .... **1:13.29**

Diesel Emission Production Audit Test Procedure (A)—SAE **J1243** MAY88 .... **1:13.50**

Diesel Engine Emission Measurement Procedure (A)—SAE **J1003** JUN90 .... **1:13.33**

Diesel Engine Smoke Measurement—SAE **J255a** **1:13.56**

Diesel Fuels—SAE **J313** MAR92 .... **1:12.60**

Diesel Smoke Measurement Procedure (A)—SAE **J35** SEP88 .... **1:13.64**

Engine Testing with Low Temperature Charge Air Cooler Systems in a Dynamometer Test Cell (A)—SAE **J1937** NOV89 .... **1:13.23**

Measurement of Carbon Dioxide, Carbon Monoxide, and Oxides of Nitrogen in Diesel Exhaust—SAE **J177** APR82 **1:13.01**

*SAE Surface Vehicle Emissions Standards Manual—SAE **HS-2400** FEB93 .... *Available as a separate publication (See Related Technical Reports Section)*

Gas Turbine
Determination of Emissions from Gas Turbine Powered Light Duty Surface Vehicles (A)—SAE **J1130** MAY83 .. **1:13.156**

Measurement
*SAE Surface Vehicle Emissions Standards Manual—SAE **HS-2400** FEB93 .... *Available as a separate publication (See Related Technical Reports Section)*

Sampling
Constant Volume Sampler System for Exhaust Emissions Measurement—SAE **J1094** JUN92 .... **1:13.117**

Determination of Sulfur Compounds in Automotive Exhaust—SAE **J1280** JUN92 .... **1:13.72**

† Test Procedure for the Measurement of Gaseous Exhaust Emissions from Small Utility Engines (A)—SAE **J1088** FEB93 .... **1:13.151**

Spark Ignition Engine
Constant Volume Sampler System for Exhaust Emissions Measurement—SAE **J1094** JUN92 .... **1:13.117**

Determination of Sulfur Compounds in Automotive Exhaust—SAE **J1280** JUN92 .... **1:13.72**

Instrumentation and Techniques for Exhaust Gas Emissions Measurement (A)—SAE **J254** AUG84 ...... **1:13.104**

Methane Measurement Using Gas Chromatography—SAE **J1151** DEC91 .... **1:13.170**

*SAE Surface Vehicle Emissions Standards Manual—SAE **HS-2400** FEB93 .... *Available as a separate publication (See Related Technical Reports Section)*

† Test Procedure for the Measurement of Gaseous Exhaust Emissions from Small Utility Engines (A)—SAE **J1088** FEB93 .... **1:13.151**

Spark Ignition Engines
† Automotive Gasolines—SAE **J312** JAN93 .... **1:12.47**

Terminology
† Emissions Terminology and Nomenclature—SAE **J1145** FEB93 .... **1:13.166**

Handbook of Motor Vehicle, Safety and Environmental Terminology—SAE **HS-215** .... *Available as a separate publication (See Related Technical Reports Section)*

EXHAUST GASES
See also: EXHAUST EMISSIONS

† Test Procedure for the Measurement of Gaseous Exhaust Emissions from Small Utility Engines (A)—SAE **J1088** FEB93 .... **1:13.151**

Flow
† Measurement of Intake Air or Exhaust Gas Flow of Diesel Engines (A)—SAE **J244** AUG92 .... **1:13.11**

EXHAUST SYSTEMS
Determination of Sulfur Compounds in Automotive Exhaust—SAE **J1280** JUN92 .... **1:13.72**

Tapped and Flanged Exhaust Connections for Small Engines (A)—SAE **J624** SEP89 .... **3:40.255**

Fires
Multiposition Small Engine Exhaust System Fire Ignition Suppression (A)—SAE **J335** SEP90 .... **2:26.351**

Noise
† Measurement of Exhaust Sound Levels of Stationary Motorcycles—SAE **J1287** JUN93 .... **1:14.05**

Measurement of Light Vehicle Exhaust Sound Level Under Stationary Conditions—SAE **J1169** MAR92 ..... **1:14.21**

Measurement of Light Vehicle Stationary Exhaust System Sound Level Engine Speed Sweep Method—SAE **J1492** MAR92 .... **1:14.24**

Measurement Procedure for Determination of Silencer Effectiveness in Reducing Engine Intake or Exhaust Sound Level (A)—SAE **J1207** FEB87 .... **1:14.115**

EXTERIORS
See: AUTOMOBILE DIMENSIONS

EYES
See also: VISION

Motor Vehicle Drivers' Eye Locations—SAE **J941** JUN92 **3:34.204**

Movement
Describing and Measuring the Driver's Field of View—SAE **J1050a** **3:34.214**

Vision Factors Considerations in Rear View Mirror Design—SAE **J985** OCT88 .... **3:34.110**

FABRICS
See also: TRIM MATERIALS

Flammability of Automotive Interior Materials—Horizontal Test Method—SAE **J369** JUN89 .... **1:11.176**

Method of Testing Resistance to Crocking of Organic Trim Materials—SAE **J861** **1:11.114**

Method of Testing Resistance to Scuffing of Trim Materials (A)—SAE **J365** FEB85 .... **1:11.163**

*SAE Automotive Textiles and Trim Standards Manual—SAE **HS-2700** FEB93 .... *Available as a separate publication (See Related Technical Reports Section)*

Test Method for Determining Dimensional Stability of Automotive Textile Materials (A)—SAE **J883** FEB86 ... **1:11.114**

Test Method for Determining Resistance to Abrasion of Automotive Bodycloth, Vinyl, and Leather, and the Snagging of Automotive Bodycloth (A)—SAE **J948** FEB86 .... **1:11.178**

Test Method for Determining Resistance to Abrasion; Bearding; and Fiber Loss of Automotive Carpet Materials (A)—SAE **J1530** JUN85 .... **1:11.107**

Test Method for Measuring Thickness of Automotive Textiles and Plastics (A)—SAE **J882** FEB85 .... **1:11.115**

Test Method for Wicking of Automotive Fabrics and Fibrous Materials (A)—SAE **J913** FEB85 .... **1:11.163**

Test Method of Stretch and Set of Textiles and Plastics—SAE **J855** MAR87 .... **1:11.181**

Color
Instrumental Color Difference Measurement for Exterior Finishes, Textiles, and Colored Trim (A)—SAE **J1545** JUN86 .... **1:11.279**

---

* Technical Revision   † New Document   (D)—DODISS Adopted   (A)—ANSI Recognized

Subject Index

FABRICS (continued)
Color (continued)
Test Method for Determining Visual Color Match to Master Specimen for Fabrics, Vinyls, Coated Fiberboards, and Other Automotive Trim Materials (A)—SAE **J361** MAR85 .... 1:11.180

Test Method for Measuring Wet Color Transfer Characteristics (A)—SAE **J1326** FEB85 .... 1:11.169

Wear
Accelerated Exposure of Automotive Interior Trim Components Using a Controlled Irradiance Water Cooled Xenon-Arc Apparatus—SAE **J1885** MAR92 .... 1:11.255

Dynamic Durability Testing of Seat Cushions for Off-Road Work Machines (A)—SAE **J1454** JAN86 .... 3:40.389

FACE
Impact Tolerance
Human Mechanical Response Characteristics—SAE **J1460** MAR85 .... 3:34.292

FAILURE
See also: FATIGUE, FRACTURE
Fiber Reinforced Plastics
Method for Evaluating the Cleavage Strength of Structurally Bonded Fiberglass Reinforced Plastic (Wedge Test) (A)—SAE **J1882** AUG87 .... 1:11.102

Sealers
Method for Evaluating Material Separation in Automotive Sealers Under Pressure in Static Conditions—SAE **J1864** APR87 .... 1:11.92

Valves
Engine Poppet Valve Information Report—SAE **J775** JAN88 .... 1:1.165

FALLING OBJECT PROTECTIVE STRUCTURES (FOPS)
See: OPERATOR PROTECTIVE STRUCTURES

FANS
See also: COOLING SYSTEMS

Fan Hub Bolt Circles and Pilot Holes (A)—SAE **J635** JUL84 .... 2:26.414

Heavy Duty Non-Metallic Engine Cooling Fans—Material, Manufacturing and Test Considerations—SAE **J1474** JAN85 .... 2:26.403

† Principles of Engine Cooling Systems, Components and Maintenance—SAE **HS-40** JAN91 .... Available as a separate publication *(See Related Technical Reports Section)*

Guards
Fan Guard for Off-Road Machines—SAE **J1308** SEP85 . 3:40.257

Power Consumption
Information Relating to Duty Cycles and Average Power Requirements of Truck and Bus Engine Accessories—SAE **J1343** OCT88 .... 3:36.57

Method for Determining Power Consumption of Engine Cooling Fan-Drive Systems (A)—SAE **J1342** AUG89 ... 3:36.55

Test Method for Measuring Power Consumption of Truck and Bus Engine Cooling Fans (A)—SAE **J1339** AUG89 . 3:36.52

Structural Analysis
Engine Cooling Fan Structural Analysis (A)—SAE **J1390** APR82 .... 2:26.400

FASTENERS
See also: BOLTS, CLAMPS, CONNECTORS, FITTINGS, NUTS, PINS, RIVETS, SCREWS, STUDS, WASHERS

Fastener Hardware for Wheels for Demountable Rims (A)—SAE **J1835** MAR90 .... 3:31.26

Aluminum
Wrought Aluminum Applications Guidelines—SAE **J1434** JAN89 .... 1:10.15

Steel
† Clip Fastener Fitting—SAE **J1467** JUN93 .... 2:22.197

Decarburization in Hardened and Tempered Threaded Fasteners (A)—SAE **J121** AUG83 .... 1:4.44

Mechanical and Chemical Requirements for Nonthreaded Fasteners—SAE **J430** 1:4.18

Mechanical and Material Requirements for Externally Threaded Fasteners (A)—SAE **J429** AUG83 .... 1:4.01

Mechanical and Material Requirements for Metric Externally Threaded Steel Fasteners (A)—SAE **J1199** SEP83 .... 1:4.06

Test Methods for Metric Threaded Fasteners (A)—SAE **J1216** 1:4.10

Torque-Tension Test Procedure for Steel Threaded Fasteners—SAE **J174** 2:16.22

FATIGUE
See also: FAILURE, FRACTURE, WEAR

Cumulative Damage Analysis for Hydraulic Hose Assemblies—SAE **J1927** OCT88 .... 2:22.180

Manual on Shot Peening—SAE **HS-84** .... Available as a separate publication *(See Related Technical Reports Section)*

Aluminum Alloys
Technical Report on Fatigue Properties—SAE **J1099** ... 1:3.81

Wrought Aluminum Applications Guidelines—SAE **J1434** JAN89 .... 1:10.15

Belts
Automotive V-Belt Drives—SAE **J637** FEB89 .... 2:19.04

Copper Alloys
Wrought and Cast Copper Alloys—SAE **J461** SEP81 ... 1:10.45

Elastomers
Recommended Guidelines for Fatigue Testing of Elastomeric Materials and Components—SAE **J1183** FEB85 .... 1:11.35

Polyurethane
Dynamic Flex Fatigue Test for Slab Polyurethane Foam—SAE **J388** 3:34.98

Springs
Fatigue Testing Procedure for Suspension-Leaf Springs (A)—SAE **J1528** JUN90 .... 2:20.08

Manual on Design and Application of Helical and Spiral Springs—SAE **HS-J795** APR90 .... Available as a separate publication *(See Related Technical Reports Section)*

Manual on Design and Application of Leaf Springs—SAE **HS-788** APR80 .... Available as a separate publication *(See Related Technical Reports Section)*

Manual on Design and Manufacture of Torsion Bar Springs—SAE **HS-796** JUL90 .... Available as a separate publication *(See Related Technical Reports Section)*

Steel
† Selection and Use of Steels—SAE **J401** NOV92 .... 1:1.06

Technical Report on Fatigue Properties—SAE **J1099** ... 1:3.81

Wheels
Spoke Wheels and Hub Fatigue Test Procedures (A)—SAE **J1095** JAN91 .... 3:31.20

Wheels/Rims—Trucks—Test Procedures and Performance Requirements (A)—SAE **J267** JAN91 .... 3:31.14

Wheels—Passenger Cars—Performance Requirements and Test Procedures (A)—SAE **J328** MAR90 ...... 3:31.06

Wheels—Recreational and Utility Trailer Test Procedure (A)—SAE **J1204** DEC89 .... 3:31.08

FELLER BUNCHER
See also: FORESTRY EQUIPMENT

Component Nomenclature—Feller Buncher—SAE **J1254** JAN85 .... 3:40.63

Specification Definitions—Feller/Buncher—SAE **J1255** JAN85 .... 3:40.93

---

* Technical Revision   † New Document   (D)—DODISS Adopted   (A)—ANSI Recognized

Subject Index

FELLER BUNCHER (continued)
Noise
Sound Measurement—Off-Road Self-Propelled Work Machines—Operator—Work Cycle (A)—SAE **J1166** MAY90 .... 1:14.55

FELLING HEAD
See also: FORESTRY EQUIPMENT
Terminology
Felling Head Terminology and Nomenclature—SAE **J1272** JAN85 .... 3:40.77

FELTS
Felts—Wool and Part Wool (A)—SAE **J314** MAY81 .... 1:11.183

FERROUS MATERIALS
See also: IRON ALLOYS, STEELS
Detection of Surface Imperfections in Ferrous Rods, Bars, Tubes, and Wires—SAE **J349** FEB91 .... 1:3.63
Methods of Measuring Decarburization (A)—SAE **J419** DEC83 .... 1:3.12
Sintered Powder Metal Parts: Ferrous—SAE **J471d** 1:8.01
Use of Terms Yield Strength and Yield Point (A)—SAE **J450** JUN91 .... 1:9.01
Numbering Systems
† Numbering System for Designating Grades of Cast Ferrous Materials—SAE **J859** OCT92 .... 1:6.01

FIBER OPTICS
Selection of Transmission Media (A)—SAE **J2056/3** JUN91 .... 2:23.391

FIBER REINFORCED PLASTICS
See also: COMPOSITE MATERIALS, PLASTICS
Lap Shear Test for Automotive Type Adhesives for Fiber Reinforced Plastic (FRP) Bonding (A)—SAE **J1525** JUN85 .... 1:11.81
*Materials for Plastic Pistons for Hydraulic Disc Brake Cylinders—SAE **J1568** JUN93 .... 2:25.18
Overlap Shear Test for Automotive Type Sealant for Stationary Glass Bonding (A)—SAE **J1529** MAY86 .... 1:11.95
Bonding
Cross Peel Test for Automotive Type Adhesives for Fiber Reinforced Plastic (FRP) Bonding (A)—SAE **J1553** MAY86 .... 1:11.100
Method for Evaluating the Cleavage Strength of Structurally Bonded Fiberglass Reinforced Plastic (Wedge Test) (A)—SAE **J1882** AUG87 .... 1:11.102

FIBERBOARD
See also: TRIM MATERIALS
Fiberboard Crease Bending Test (A)—SAE **J119** FEB87 1:11.166
Fiberboard Test Procedure (A)—SAE **J315** JAN85 .... 1:11.158
Fiberboards—SAE **HS-87** .... Available as a separate publication *(See Related Technical Reports Section)*
Method of Testing Resistance to Crocking of Organic Trim Materials—SAE **J861** 1:11.114
Method of Testing Resistance to Scuffing of Trim Materials (A)—SAE **J365** FEB85 .... 1:11.163
Classification
Standard Classification System for Fiberboards (A)—SAE **J1323** SEP90 .... 1:11.161
Color
Test Method for Determining Visual Color Match to Master Specimen for Fabrics, Vinyls, Coated Fiberboards, and Other Automotive Trim Materials (A)—SAE **J361** MAR85 .... 1:11.180
Stiffness
Test Method for Determining Stiffness (Modulus of Bending) of Fiberboards—SAE **J949** AUG81 .... 1:11.178
Terminology
Glossary of Fiberboard Terminology—SAE **J947** JAN85 . 1:11.156

FIBERGLASS
See: FIBER REINFORCED PLASTICS

FIDUCIAL MARKS
† Motor Vehicle Dimensions (A)—SAE **J1100** JUN93 .... 3:34.118
Motor Vehicle Fiducial Marks and Three Dimensional Reference System (A)—SAE **J182** NOV90 .... 3:34.229

FIFTH WHEEL
See: TRUCK TRAILERS

FILLER PLUGS
† Automotive Pipe, Filler, and Drain Plugs—SAE **J531** JUN93 .... 2:22.241
† Automotive Straight Thread Filler and Drain Plugs—SAE **J532** JUN93 .... 2:22.246
† Drain, Fill, and Level Plugs for Off-Road, Self-Propelled Work Machines—SAE **J371** MAY93 .... 3:40.257

FILTERS
Air
Air Cleaner Elements—SAE **J1141** JUL87 .... 2:26.213
† Air Cleaner Test Code—SAE **J726** JUN93 .... 2:26.214
† Operator Enclosure Air Filter Element Test Procedure—SAE **J1533** JUN93 .... 3:40.513
Fuel
Fuel Filter Test Methods (A)—SAE **J905** JAN87 ...... 2:26.329
Glossary
Glossary of Terms Related to Fluid Filters and Filter Testing (A)—SAE **J1124** MAR87 .... 2:26.342
Hydraulic System
Hydraulic Power Circuit Filtration (D)—SAE **J931** MAR86 3:40.137
Oil
Filter Base Mounting (A)—SAE **J363** FEB87 .... 2:26.246
Full Flow Lubricating Oil Filters Multipass Method for Evaluating Filtration Performance (A)—SAE **J1858** JUN88 .... 2:26.248
Oil Filter Test Procedure—SAE **HS-806** JUN83 ...... Available as a separate publication *(See Related Technical Reports Section)*
Standard Oil Filter Test Oil—SAE **J1260** APR89 ...... 2:26.350
Particulates
Determination of Sulfur Compounds in Automotive Exhaust—SAE **J1280** JUN92 .... 1:13.72
† Operator Enclosure Air Filter Element Test Procedure—SAE **J1533** JUN93 .... 3:40.513

FINISHES
See also: CLEANING, COATINGS, PAINTS
Automotive Plastic Parts Specification (A)—SAE **J1345** FEB82 .... 1:11.148
Florida Exposure of Automotive Finishes—SAE **J951** JAN85 .... 1:11.130
Outdoor Weathering of Exterior Materials—SAE **J1976** JUN89 .... 1:11.275
Color
Instrumental Color Difference Measurement for Exterior Finishes, Textiles, and Colored Trim (A)—SAE **J1545** JUN86 .... 1:11.279

FINISHING
Aluminum Alloys
Aluminum Alloys—Fundamentals—SAE **J451** JAN89 ... 1:10.01
Wrought Aluminum Applications Guidelines—SAE **J1434** JAN89 .... 1:10.15

FIRE RESISTANT ALLOYS
See: HYDRAULIC FLUIDS

FIRES
See also: FLAMMABILITY
Exhaust Systems
Multiposition Small Engine Exhaust System Fire Ignition Suppression (A)—SAE **J335** SEP90 .... 2:26.351

---

* Technical Revision   † New Document   (D)—DODISS Adopted   (A)—ANSI Recognized

Subject Index

FIRES (continued)
Forestry Equipment
Fire Prevention on Forestry Equipment—SAE J1212 JAN85 .... 3:40.95

FITTINGS
See also: CONNECTORS

*Fittings and Labels for Retrofit of CFC-12 (R12) Mobile Air Conditioning Systems to HFC-134a (R134a)—SAE J1660 JUN93 .... 3:34.66

Fluid Conductors and Connectors—SAE **HS-150** JUN92 — Available as a separate publication *(See Related Technical Reports Section)*

Hose
†Beaded Tube Hose Fittings—SAE J1231 JUN93 ...... 2:22.158
†Clip Fastener Fitting—SAE J1467 JUN93 .... 2:22.197
†Fitting—O-Ring Face Seal—SAE J1453 JUN93 ...... 2:22.256
HFC-134a (R-134a) Service Hose Fittings for Automotive Air-Conditioning Service Equipment—SAE J2197 JUN92 ...... 3:34.53
†Hydraulic Hose Fittings—SAE J516 JUN93 ...... 2:22.114
Hydraulic Hose Fittings for Marine Applications (A)—SAE J1475 JUN84 .... 3:41.39

Numbering Systems
Coding Systems for Identification of Fluid Conductors and Connectors (D)—SAE J846 JUN89 .... 2:22.01

Tube and Pipe
†Automotive Pipe Fittings—SAE J530 JUN93 .... 2:22.232
†Automotive Tube Fittings—SAE J512 JUN93 .... 2:22.06
†Fitting—O-Ring Face Seal—SAE J1453 JUN93 ...... 2:22.256
Flares for Tubing—SAE J533 JUN92 .... 2:22.203
†Hydraulic Tube Fittings—SAE J514 JUN93 .... 2:22.59
†Lubrication Fittings (A)—SAE J534 JUN93 .... 2:22.252
Performance Requirements for SAE J844 Nonmetallic Tubing and Fitting Assemblies Used in Automotive Air Brake Systems (A)—SAE J1131 DEC87 .... 2:22.231
Pressure Ratings for Hydraulic Tubing and Fittings (D)—SAE J1065 MAR92 .... 2:22.215
†Refrigeration Tube Fittings—General Specifications—SAE J513 JUN93 .... 2:22.33
SAE Quick Connector Specifications for Liquid Fuel Systems—SAE J2044 JUN92 .... 2:26.442
†Spherical and Flanged Sleeve (Compression) Tube Fittings—SAE J246 JUN93 .... 2:22.21

FLAME ARRESTERS
†Devices Providing Backfire Flame Control for Gasoline Engines in Marine Applications—SAE J1928 APR93 ... 3:41.44

FLAMMABILITY
American National Standard for Safety Glazing Materials for Glazing Motor Vehicles and Motor Vehicle Equipment Operating on Land Highways—Safety Code—ANSI **SAE Z26.1-1990** .... Available as a separate publication *(See Related Technical Reports Section)*

Flammability of Automotive Interior Materials—Horizontal Test Method—SAE J369 JUN89 .... 1:11.176

Refrigerants
*Selection Criteria for Retrofit Refrigerants to Replace R-12 in Mobile Air Conditioning Systems—SAE J1657 JUN93 .... 3:34.61

FLANGES
*Balance Weight and Rim Flange Design Specifications, Test Procedures, and Performance Recommendations—SAE J1986 FEB93 .... 3:31.10

Companion Flanges, Type A (External Pilot) and Type S (Internal Pilot) (A)—SAE J1946 JAN91 .... 3:29.65
Cross-Tooth Companion Flanges, Type T (A)—SAE J1945 JAN90 .... 3:29.71
Engine Flywheel Housing and Mating Transmission Housing Flanges—SAE J617 MAY92 .... 3:40.247
†Engine Flywheel Housings with Sealed Flanges—SAE J1172 MAY93 .... 2:26.242
Mounting Flanges and Power Take-Off Shafts for Small Engines—SAE J609a 2:26.26
Mounting Flanges for Engine Accessory Drives (A)—SAE J896 MAY83 .... 3:40.253
Starting Motor Mountings (A)—SAE J542 JUN91 ...... 2:23.76
Tapped and Flanged Exhaust Connections for Small Engines (A)—SAE J624 SEP89 .... 3:40.255

FLASHERS
See: HAZARD WARNING SYSTEMS, TURN INDICATORS, WARNING SYSTEMS

FLOOD LAMPS
†Requirements for Sealed Lighting Unit for Construction and Industrial Machines—SAE J572 MAY93 .... 3:40.260
Sealed Lighting Units for Construction, Industrial and Forest Machinery (A)—SAE J598 MAY87 .... 3:40.259

FLOOR MATS
Automotive Rubber Mats (A)—SAE J80 JAN88 ...... 1:11.185

FLOTATION
Personal Watercraft—Flotation (A)—SAE J1973 APR91 .... 3:41.49

FLOWMETERS
Airflow Reference Standards (A)—SAE J228 JUN90 ... 2:26.372
Diesel Injection Pump Testing—Part 2: Orifice Plate Flow Measurement (A)—SAE J968/2 MAY91 .... 2:26.318
†Measurement of Intake Air or Exhaust Gas Flow of Diesel Engines (A)—SAE J244 AUG92 .... 1:13.11
Method for Evaluating the Flow Properties of Pumpable Sealers—SAE J2025 JUN89 .... 1:11.66

FLUID CONNECTORS
See: CONNECTORS, FITTINGS, HOSES, TUBING

FLUID COUPLINGS
See also: TRANSMISSIONS

Hydrodynamic Drives Terminology (A)—SAE J641 JUL88 .... 3:29.02

FLUID POWER SYSTEMS
See: HYDRAULIC SYSTEMS

FLUIDS
See: BRAKE FLUIDS, CALIBRATION FLUIDS, COOLANTS, HYDRAULIC FLUIDS, TRANSMISSION FLUIDS

FLUOROSCOPIC INSPECTION
See also: NONDESTRUCTIVE TESTS

Penetrating Radiation Inspection (A)—SAE J427 MAR91 .... 1:3.76

FLUX TESTING
Detection of Surface Imperfections in Ferrous Rods, Bars, Tubes, and Wires—SAE J349 FEB91 .... 1:3.63

FLYWHEELS
Engine Flywheel Housing and Mating Transmission Housing Flanges—SAE J617 MAY92 .... 3:40.247
†Engine Flywheel Housings with Sealed Flanges—SAE J1172 MAY93 .... 2:26.242
Flywheel Dimensions for Truck and Bus Applications (A)—SAE J1857 MAY87 .... 2:26.237
Flywheel Spin Test Procedure—SAE J1240 DEC91 .... 2:26.242
Flywheels for Engine Mounted Torque Converters—SAE J927 NOV88 .... 3:40.254
†Flywheels for Industrial Engines Used with Industrial Power Take-Offs Equipped with Driving-Ring Type Overcenter Clutches and Engine-Mounted Marine Gears and Single Bearing Engine-Mounted Power Generators—SAE J620 MAY93 .... 3:40.256
Flywheels for Single-Plate Spring-Loaded Clutches (A)—SAE J618 JAN91 .... 2:26.238
Flywheels for Two-Plate Spring-Loaded Clutches—SAE J619 JUN80 .... 2:26.240
Maximum Allowable Rotational Speed for Internal Combustion Engine Flywheels (A)—SAE J1456 JUN90 . 2:26.243

---

* Technical Revision   † New Document   (D)—DODISS Adopted   (A)—ANSI Recognized

Subject Index

FLYWHEELS (continued)
† Procedure for Measuring Bore and Face Runout of Flywheels, Flywheel Housings and Flywheel Housing Adapters—SAE J1033 APR93 .... 3:40.245

FOAM RUBBER
See: RUBBER

FOAMS
Urethane
Dynamic Flex Fatigue Test for Slab Polyurethane Foam—SAE J388 .... 3:34.98
Load Deflection Testing of Urethane Foams for Automotive Seating—SAE J815 MAY81 .... 3:34.100
Urethane for Automotive Seating—SAE J954 .... 3:34.99

FOG LAMPS
† Fog Tail Lamp (Rear Fog Light) Systems (A)—SAE J1319 JUN93 .... 2:24.141
† Front Fog Lamps—SAE J583 JUN93 .... 2:24.119
† Headlamp Aim Testing Machines—SAE J600 FEB93 .... 2:24.205
Lighting Inspection Code—SAE J599 MAY81 .... 2:24.202
† SAE Ground Vehicle Lighting Standards Manual—SAE HS-34 JAN93 .... Available as a separate publication (See Related Technical Reports Section)

FOPS
See: OPERATOR PROTECTIVE STRUCTURES

FORESTRY EQUIPMENT
See also: FELLER BUNCHER, FELLING HEAD, OFF-ROAD VEHICLES, SKIDDERS
Dimensions
Specification Definitions—Articulated Rubber-Tired Forwarder (A)—SAE J1823 JUN91 .... 3:40.89
Specification Definitions—Clam Bunk Skidder (A)—SAE J1824 JUN91 .... 3:40.81
Fires
Fire Prevention on Forestry Equipment—SAE J1212 JAN85 .... 3:40.95
Lamps
Sealed Lighting Units for Construction, Industrial and Forest Machinery (A)—SAE J598 MAY87 .... 3:40.259
Operator Protective Structures
Labeling of ROPS and FOPS and OPS (A)—SAE J1164 JAN91 .... 3:40.362
Minimum Performance Criteria for Falling Object Protective Structure (FOPS) (A)—SAE J231 JAN81 .... 3:40.364
Operator Protective Structure Performance Criteria for Certain Forestry Equipment (A)—SAE J1084 APR80 .... 3:40.370
Performance Criteria for Rollover Protective Structures (ROPS) for Construction, Earthmoving, Forestry, and Mining Machines (A)—SAE J1040 APR88 .... 3:40.356
Terminology
Identification Terminology & Component Nomenclature—Knuckle Boom Log Loader (A)—SAE J2055 JUN91 .... 3:40.67
Identification Terminology of Mobile Forestry Machines—SAE J1209 JAN85 .... 3:40.25
Nomenclature—Clam Bunk Skidder (A)—SAE J1353 APR87 .... 3:40.80
Nomenclature—Forwarder (A)—SAE J1354 APR87 .... 3:40.88
Tires
Off-Road Tire and Rim Classification—Forestry Machines (A)—SAE J1440 APR86 .... 3:40.310

FORGING
Steel
Selection and Heat Treatment of Tool and Die Steels—SAE J437a 1:7.02

FORKS
Identification Terminology of Loaders/Tractors with Forks and Rough Terrain Forklifts—SAE J1464 APR88 .... 3:40.58

Rated Operating Load for Loaders Equipped with Log or Material Forks without Vertical Mast (A)—SAE J1197 FEB91 .... 3:40.101

FORMABILITY
Aluminum Alloys
Wrought Aluminum Applications Guidelines—SAE J1434 JAN89 .... 1:10.15
Steel
Properties of Galvanized Low Carbon Sheet Steels and Their Relation to Formability—SAE J1852 SEP87 .... 1:3.22
Properties of Low Carbon Sheet Steel and Their Relationship to Formability—SAE J877 JUN84 .... 1:3.20
Selecting and Specifying Hot and Cold Rolled Steel Sheet and Strip (A)—SAE J126 JUN86 .... 1:3.27

FORWARDERS
See also: OFF-ROAD VEHICLES
Specification Definitions—Articulated Rubber-Tired Forwarder (A)—SAE J1823 JUN91 .... 3:40.89
Terminology
Nomenclature—Forwarder (A)—SAE J1354 APR87 .... 3:40.88

FRACTURE
See also: FAILURE, FATIGUE
Coach Joint Fracture Test—SAE J1863 APR87 .... 1:11.85
Steel
† Selection and Use of Steels—SAE J401 NOV92 .... 1:1.06

FRAMES
See also: CHASSIS
Motor Truck CA Dimensions (A)—SAE J691 SEP90 .... 3:36.67
Lift Points
*Vehicle Lift Points for Service Garage Lifting—SAE J2184 OCT92 .... 3:42.53
Seat
Motor Vehicle Seating Manual—SAE J782b OCT80 .... 3:34.73
Welding
Specification for Automotive Frame Weld Quality—Arc Welding (AWS D8.8-79)—SAE HS-J1196 .... Available as a separate publication (See Related Technical Reports Section)

FREE PISTON ENGINES
Terminology
Engine Terminology and Nomenclature—General (A)—SAE J604 JAN86 .... 2:26.01

FRICTION
See also: WEAR
Band Friction Test Machine (SAE) Test Procedure—SAE J1499 FEB87 .... 3:29.99
Characterizing a Test Surface for Motorcycle Side Stand Retraction Performance Testing—SAE J1846 JUN89 .... 3:37.17
Brake Linings
Friction Coefficient Identification System for Brake Linings (A)—SAE J866 NOV90 .... 2:25.89
Clutches
SAE No. 2 Clutch Friction Test Machine Test Procedure (A)—SAE J286 NOV83 .... 3:29.95
Hydraulic Cylinders
Hydraulic Cylinder No-Load Friction Test (A)—SAE J1335 APR90 .... 3:40.346

FRICTION MATERIALS
Band Friction Test Machine (SAE) Test Procedure—SAE J1499 FEB87 .... 3:29.99
SAE No. 2 Clutch Friction Test Machine Test Procedure (A)—SAE J286 NOV83 .... 3:29.95

FUEL BLENDS
† Alternative Automotive Fuels—SAE J1297 MAR93 .... 1:12.74

---

* Technical Revision † New Document (D)—DODISS Adopted (A)—ANSI Recognized

FUEL CONSUMPTION

Classification of Energy-Conserving Engine Oil for Passenger Cars, Vans, and Light-Duty Trucks—SAE **J1423** FEB92 .... 1:12.29

Automobiles

† Emission Test Driving Schedules (A)—SAE **J1506** APR93 .... 1:13.141

Fuel Economy Measurement Road Test Procedure—SAE **J1082** JAN89 .... 2:26.459

Fuel Economy Measurement—Road Test Procedure—Cold Start and Warm-Up Fuel Economy—SAE **J1256** OCT88 .... 2:26.506

Procedure for Mapping Engine Performance—Spark Ignition and Compression Ignition Engines (A)—SAE **J1312** JAN90 .... 2:26.22

Trucks and Buses

† Emission Test Driving Schedules (A)—SAE **J1506** APR93 .... 1:13.141

Fuel Economy Measurement Test (Engineering Type) for Trucks and Buses (A)—SAE **J1376** JUL82 .... 3:36.31

Joint RCCC/SAE Fuel Consumption Test Procedure (Short Term In-Service Vehicle) Type I (A)—SAE **J1264** OCT86 .... 3:36.06

Joint TMC/SAE Fuel Consumption In-Service Test Procedure Type III (A)—SAE **J1526** JUN87 .... 2:26.450

Joint TMC/SAE Fuel Consumption Test Procedure—Type II (A)—SAE **J1321** OCT86 .... 3:36.11

Method for Determining Power Consumption of Engine Cooling Fan-Drive Systems (A)—SAE **J1342** AUG89 ... 3:36.55

Supportive Information Report for the Fuel Economy Measurement Test (Engineering Type) for Trucks and Buses—SAE **J1420** AUG88 .... 3:36.44

Test Method for Measuring Power Consumption of Air Conditioning and Brake Compressors for Trucks and Buses (A)—SAE **J1340** APR90 .... 3:36.53

Test Method for Measuring Power Consumption of Hydraulic Pumps for Trucks and Buses (A)—SAE **J1341** APR90 .... 3:36.54

FUEL ECONOMY
See: FUEL CONSUMPTION

FUEL FILLER CAPS
See also: FUEL TANKS

Fuel Tank Filler Cap and Cap Retainer (A)—SAE **J829** FEB88 .... 2:26.432

Fuel Tank Filler Cap and Cap Retainer—Threaded (A)—SAE **J1114** NOV89 .... 2:26.430

FUEL FILTERS
See: FILTERS

FUEL INJECTION
See also: FUEL INJECTION NOZZLES, FUEL INJECTION PUMPS

Calibration Fluid for Diesel Injection Equipment (A)—SAE **J967** SEP88 .... 2:26.328

Diesel Fuel Injection Pump—Validation of Calibrating Nozzle Holder Assemblies (A)—SAE **J1549** APR88 .... 2:26.321

Fuel Injection Pumps — High Pressure Pipes (Tubing) for Testing (A)—SAE **J1418** DEC87 .... 2:26.263

Fuel Injection System Fuel Pressure Regulator and Pressure Damper—SAE **J1862** FEB90 .... 2:26.279

Gasoline Fuel Injector (A)—SAE **J1832** NOV89 ...... 2:26.263

† Marine Carburetors and Fuel Injection Throttle Bodies—SAE **J1223** JUN93 .... 3:41.28

Road Vehicles—High Pressure Fuel Injection Pipe End-Connections with 60 Deg Female Cone—SAE **J1949** OCT88 .... 2:26.302

Hoses

*Tube/Hose Assemblies—SAE **J2045** OCT92 .... 1:12.85

Terminology

† Fuel Injection Equipment Nomenclature—SAE **J830** OCT92 .... 2:26.254

Fuel Injection Nomenclature — Spark Ignition Engines (A)—SAE **J1541** FEB88 .... 2:26.290

FUEL INJECTION NOZZLES

Diesel Fuel Injection Equipment and Test Methods—SAE **HS-58** .... Available as a separate publication (See Related Technical Reports Section)

Diesel Fuel Injection Pump—Validation of Calibrating Nozzle Holder Assemblies (A)—SAE **J1549** APR88 .... 2:26.321

Diesel Fuel Injector Assembly Type 27 (9.5 mm) (A)—SAE **J1984** NOV89 .... 2:26.308

Diesel Fuel Injector Assembly Type 7 (9.5 mm) (A)—SAE **J347** JUL88 .... 2:26.305

Diesel Fuel Injector Assembly—Flange Mounted Types 5 and 6 (A)—SAE **J629** APR91 .... 2:26.304

Diesel Fuel Injector Assembly—Types 8, 9, 10, and 11 (A)—SAE **J265** APR91 .... 2:26.306

Diesel Injection Pump Testing—Part 1: Calibrating Nozzle and Holder Assemblies (A)—SAE **J968/1** MAY91 ..... 2:26.309

† Fuel Injection Equipment Nomenclature—SAE **J830** OCT92 .... 2:26.254

FUEL INJECTION PUMPS

*Diesel Engines—Fuel Injection Pump Testing—SAE **J1668** JUN93 .... 2:26.322

Diesel Fuel Injection Equipment and Test Methods—SAE **HS-58** .... Available as a separate publication (See Related Technical Reports Section)

Diesel Fuel Injection Pump—Validation of Calibrating Nozzle Holder Assemblies (A)—SAE **J1549** APR88 .... 2:26.321

Diesel Injection Pump Testing—Part 2: Orifice Plate Flow Measurement (A)—SAE **J968/2** MAY91 .... 2:26.318

† Fuel Injection Equipment Nomenclature—SAE **J830** OCT92 .... 2:26.254

Validation Testing of Electric Fuel Pumps for Gasoline Fuel Injection Systems (A)—SAE **J1537** JUN90 .... 2:26.291

Mounts

Diesel Fuel Injection—End Mounting Flanges for Fuel Injection Pumps (A)—SAE **J626** JUN88 .... 2:26.297

Pipes

Diesel Engines—Steel Tubes for High Pressure Fuel Injection Pipes (Tubing)—SAE **J1958** APR89 .... 2:26.260

Fuel Injection Pumps — High Pressure Pipes (Tubing) for Testing (A)—SAE **J1418** DEC87 .... 2:26.263

Road Vehicles—High Pressure Fuel Injection Pipe End-Connections with 60 Deg Female Cone—SAE **J1949** OCT88 .... 2:26.302

Shafts

Tapers for Shaft Ends and Hubs for Fuel Injection Pumps (A)—SAE **J1419** FEB88 .... 2:26.245

FUEL SYSTEMS
See also: FUEL INJECTION, FUEL INJECTION NOZZLES, FUEL INJECTION PUMPS, FUEL TANKS

Fuel Filter Test Methods (A)—SAE **J905** JAN87 ...... 2:26.329

Fuel Injection System Fuel Pressure Regulator and Pressure Damper—SAE **J1862** FEB90 .... 2:26.279

*Personal Watercraft Fuel Systems—SAE **J2046** JAN93 . 3:41.46

SAE Quick Connector Specifications for Liquid Fuel Systems—SAE **J2044** JUN92 .... 2:26.442

Construction and Industrial Equipment

Fast Fill Fueling Installation for Off-Road Work Machines (A)—SAE **J1176** DEC89 .... 3:40.103

† Guidelines for Liquid Level Indicators—SAE **J48** MAY93 3:40.103

Gasoline

Validation Testing of Electric Fuel Pumps for Gasoline Fuel Injection Systems (A)—SAE **J1537** JUN90 .... 2:26.291

Hoses

† Fuel and Oil Hoses (A)—SAE **J30** MAY93 .... 1:11.226

---

* Technical Revision   † New Document   (D)—DODISS Adopted   (A)—ANSI Recognized

Subject Index

FUEL SYSTEMS (continued)
Hoses (continued)
*Tube/Hose Assemblies—SAE **J2045** OCT92 .... 1:12.85

FUEL TANKS
See also: **FUEL FILLER CAPS**

Fast Fill Fueling Installation for Off-Road Work Machines (A)—SAE **J176** DEC89 .... 3:40.103

Filler Pipes and Openings of Motor Vehicle Fuel Tanks (A)—SAE **J1140** FEB88 .... 2:26.433

Fuel and Lubricant Tanks for Motorcycles—SAE **J1241** . 3:37.11

Fuel Tank Filler Cap and Cap Retainer (A)—SAE **J829** FEB88 .... 2:26.432

Fuel Tank Filler Cap and Cap Retainer—Threaded (A)—SAE **J1114** NOV89 .... 2:26.430

Fuel Tank Filler Conditions—Passenger Car, Multi-Purpose Passenger Vehicles, and Light Duty Trucks—SAE **J398** FEB88 .... 2:26.436

*Personal Watercraft Fuel Systems—SAE **J2046** JAN93 . 3:41.46

Snowmobile Fuel Tanks (A)—SAE **J288** NOV83 ...... 3:38.15

FUEL/WATER SEPARATORS
Coarse Droplet Water/Fuel Separation Test Procedure (A)—SAE **J1839** MAY90 .... 2:26.437

Emulsified Water/Fuel Separation Test Procedure—SAE **J1488** MAY90 .... 2:26.345

FUELS
See also: **ALTERNATE FUELS, DIESEL FUELS, GASOLINE**

†Effective Dates of New or Revised Technical Reports (A)—SAE **J301** MAR93 .... 1:12.01

†Fuels and Lubricants Standards Manual—SAE **HS-23** FEB93 .... Available as a separate publication (See Related Technical Reports Section)

Heating Value of Fuels—SAE **J1498** MAY90 .... 1:12.82

Impact of Alternative Fuels on Engine Test and Reporting Procedures—SAE **J1515** MAR88 .... 1:12.89

Stoichiometric Air/Fuel Ratios of Automotive Fuels—SAE **J1829** MAY92 .... 1:12.80

FUSES
Blade Type Electric Fuses (A)—SAE **J1284** APR88 .... 3:41.40

Electric Fuses (Cartridge Type) (A)—SAE **J554** AUG87 . 2:23.123

Fusible Links (A)—SAE **J156** APR86 .... 2:23.171

High Current Time Lag Electric Fuses (A)—SAE **J1888** NOV90 .... 2:23.128

Miniature Blade Type Electrical Fuses (A)—SAE **J2077** NOV90 .... 2:23.125

GAS CHROMATOGRAPHY
Methane Measurement Using Gas Chromatography—SAE **J1151** DEC91 .... 1:13.170

GAS TURBINES
Exhaust Emissions
Determination of Emissions from Gas Turbine Powered Light Duty Surface Vehicles (A)—SAE **J1130** MAY83 .. 1:13.156

GASKETS
Brake Master Cylinder Reservoir Diaphragm Gasket—SAE **J1605** MAR92 .... 2:25.22

Design Guide for Formed-In-Place Gaskets (A)—SAE **J1497** DEC88 .... 1:11.187

Standard Classification System for Nonmetallic Automotive Gasket Materials (A)—SAE **J90** JUN90 .... 1:11.197

GASOLINE
See also: **CHEMICAL PROPERTIES**

†Automotive Gasolines—SAE **J312** JAN93 .... 1:12.47

†Fuels and Lubricants Standards Manual—SAE **HS-23** FEB93 .... Available as a separate publication (See Related Technical Reports Section)

Nozzles
†Gasoline Dispenser Nozzle Spouts (A)—SAE **J285** SEP92 .... 2:26.441

GEARS
See also: **AXLES, TRANSMISSIONS**

Performance Assurance of Remanufactured, Hydraulically-Operated Rack and Pinion Steering Gears (A)—SAE **J1890** JUN88 .... 3:42.33

†Planetary Gear(s)—Terminology—SAE **J646** APR93 ... 3:29.24

Starting Motor Pinions and Ring Gears (A)—SAE **J543** JUN88 .... 2:23.78

GENERATORS
See: **ELECTRIC GENERATORS**

GLASS
See also: **WINDOWS, WINDSHIELDS**
Bonding
Overlap Shear Test for Automotive Type Sealant for Stationary Glass Bonding (A)—SAE **J1529** MAY86 .... 1:11.95

Overlap Shear Test for Sealant Adhesive Bonding of Automotive Glass Encapsulating Material to Body Opening—SAE **J1836** OCT88 .... 1:11.97

Peel Adhesion Test for Glass to Elastomeric Material for Automotive Glass Encapsulation—SAE **J1907** OCT88 .. 1:11.99

Stationary Glass Replacement—SAE **J1556** DEC85 ... 3:42.21

Glazing
American National Standard for Safety Glazing Materials for Glazing Motor Vehicles and Motor Vehicle Equipment Operating on Land Highways—Safety Code—ANSI SAE **Z26.1-1990** .... Available as a separate publication (See Related Technical Reports Section)

†Automotive Safety Glasses—SAE **J673** APR93 .... 3:34.100

Light Transmittance of Automotive Windshields Safety Glazing Materials—SAE **J1203** 3:34.102

Passenger Car Glazing Shade Bands (A)—SAE **J100** MAR88 .... 3:34.104

Passenger Car Glazing—Electrical Circuits (A)—SAE **J216** DEC89 .... 3:34.104

Safety Glazing Materials—Motor Vehicles (A)—SAE **J674** NOV90 .... 3:34.103

Test Procedure for Determining the Resistance of Safety Glazing Materials, of Which One Surface is Plastic, to Simulated Weathering (A)—SAE **J2081** JUN91 .... 3:34.106

GLASS BEADS
Size Classification and Characteristics of Glass Beads for Peening (A)—SAE **J1173** SEP88 .... 1:8.30

GLAZING
See: **GLASS**

GLOSSARY
See: **TERMINOLOGY**

GOVERNORS
Maximum Allowable Rotational Speed for Internal Combustion Engine Flywheels (A)—SAE **J1456** JUN90 . 2:26.243

Procedure for Evaluating Transient Response of Small Engine Driven Generator Sets (A)—SAE **J1444** JUN91 . 2:26.23

Terminology
†Fuel Injection Equipment Nomenclature—SAE **J830** OCT92 .... 2:26.254

GRADEABILITY
Electric Vehicles
*Electric Vehicle Acceleration, Gradeability, and Deceleration Test Procedure—SAE **J1666** MAY93 ..... 3:28.12

Trucks and Buses
Truck Ability Prediction Procedure (A)—SAE **J688** AUG87 .... 3:36.03

---

* Technical Revision   † New Document   (D)—DODISS Adopted   (A)—ANSI Recognized

Subject Index

GRADEABILITY (continued)
Trucks and Buses (continued)
Truck Ability Prediction Procedure—SAE J688—SAE HS-82 .... Available as a separate publication *(See Related Technical Reports Section)*

GRADERS
See also: CONSTRUCTION AND INDUSTRIAL EQUIPMENT, EARTHMOVING EQUIPMENT
Cutting Edge—Curved Grader (A)—SAE J739 JUN91 .. 3:40.233
Revolutions per Mile and Static Loaded Radius for Off-Road Tires—SAE J1544 JAN88 .... 3:40.307
Brakes
Braking Performance—Rubber-Tired Construction Machines (A)—SAE J1152 APR80 .... 3:40.289
Noise
Sound Measurement—Off-Road Self-Propelled Work Machines—Operator—Work Cycle (A)—SAE J1166 MAY90 .... 1:14.55
Operator Controls
Operator Controls for Graders (A)—SAE J1071 JUN85 . 3:40.470
Steering
Minimum Performance Criteria for Emergency Steering of Wheeled Earthmoving Construction Machines—SAE J53 OCT84 .... 3:40.515
Terminology
Component Nomenclature—Graders (A)—SAE J870 JUL90 .... 3:40.36

GRADES (Road)
See: PARKING PERFORMANCE

GRAIN SIZE
Steel
Grain Size Determination of Steels (A)—SAE J418 DEC83 .... 1:3.10

GRAPHICS
† Initial Graphics Exchange Specification (A)—SAE J1881 FEB93 .... 1:A.38

GREASES
See also: LUBRICANTS
† Automotive Lubricating Greases—SAE J310 JUN93 ... 1:12.39
† Fuels and Lubricants Standards Manual—SAE HS-23 FEB93 .... Available as a separate publication *(See Related Technical Reports Section)*

Prevention of Corrosion of Motor Vehicle Body and Chassis Components—SAE HS-J447 JUN81 .... Available as a separate publication *(See Related Technical Reports Section)*

The Automotive Lubricant Performance and Service Classification Maintenance Procedure (A)—SAE J1146 JUN86 .... 1:12.87

GRIT
† Cast Shot and Grit Size Specifications for Peening and Cleaning—SAE J444 MAY93 .... 1:8.11
Metallic Shot and Grit Mechanical Testing—SAE J445 AUG84 .... 1:8.12

HAND GRIPS
Snowmobile Passenger Handgrips (A)—SAE J1062 OCT84 .... 3:38.09

HAND SIGNALS
Excavator and Backhoe Hand Signals (A)—SAE J1307 NOV86 .... 3:40.348

HANDLING
See also: DIRECTIONAL CONTROL, STABILITY
A Dictionary of Terms for the Dynamics and Handling of Single Track Vehicles (Motorcycles, Mopeds, and Bicycles)—SAE J1451 FEB85 .... 3:37.01

Motorcycle Bank Angle Measurement Procedure—SAE J1168 MAY89 .... 3:37.07
*Steady-State Circular Test Procedure for Trucks and Buses—SAE J2181 JUN93 .... 3:36.68
Subjective Rating Scale for Vehicle Handling—SAE J1441 FEB92 .... 3:28.16

HANDS
Driver Hand Control Reach (A)—SAE J287 JUN88 .... 3:34.157

HARDENABILITY
See also: SURFACE HARDNESS
Steel
General Characteristics and Heat Treatments of Steels—SAE J412 JUN89 .... 1:2.05
† Hardenability Bands for Carbon and Alloy H Steels (A)—SAE J1268 JUN93 .... 1:1.48
Mechanical Properties of Heat Treated Wrought Steels—SAE J413 JUN90 .... 1:2.24
† Methods of Determining Hardenability of Steels—SAE J406 JUN93 .... 1:1.23
Restricted Hardenability Bands for Selected Alloy Steels (A)—SAE J1868 JUN90 .... 1:1.139
Selection and Heat Treatment of Tool and Die Steels—SAE J437a 1:7.02

HAZARD WARNING SYSTEMS
See also: ALARMS, WARNING SYSTEMS
Combination Turn Signal Hazard Warning Signal Flashers (A)—SAE J2068 JAN90 .... 2:24.165
Emergency Warning Device (Triangular Shape) (A)—SAE J774 DEC89 .... 2:24.181
Flasher Test (A)—SAE J823 JUN91 .... 2:24.210
Hazard Warning Signal Switch (A) (D)—SAE J910 OCT88 .... 2:24.184
† SAE Ground Vehicle Lighting Standards Manual—SAE HS-34 JAN93 .... Available as a separate publication *(See Related Technical Reports Section)*

Two-Block Warning and Limit Systems in Lifting Crane Service—SAE J1305 JUN87 .... 3:40.216
† Vehicular Hazard Warning Signal Flashers (A)—SAE J945 JUN93 .... 2:24.164

HEAD
Impact Tolerance
Human Mechanical Response Characteristics—SAE J1460 MAR85 .... 3:34.292
Human Tolerance to Impact Conditions as Related to Motor Vehicle Design—SAE J885 JUL86 .... 3:34.306
Movement
Motor Vehicle Driver and Passenger Head Position (A)—SAE J1052 MAY87 .... 3:34.231
Test Device Head Contact Duration Analysis—SAE J2052 MAR90 .... 3:34.276
Vision Factors Considerations in Rear View Mirror Design—SAE J985 OCT88 .... 3:34.110

HEAD RESTRAINTS
See also: RESTRAINT SYSTEMS
Motor Vehicle Seating Manual—SAE J782b OCT80 ... 3:34.73

HEADLAMPS
See also: LAMPS
*Discharge Forward Lighting System—SAE J2009 FEB93 2:24.113
† Headlamp Aim Testing Machines—SAE J600 FEB93 ... 2:24.205
Headlamp Aiming Device for Mechanically Aimable Headlamp Units (A)—SAE J602 DEC89 .... 2:24.206
*Headlamp Design Guidelines for Mature Drivers—SAE J1606 MAR93 .... 3:34.225
Lighting Inspection Code—SAE J599 MAY81 .... 2:24.202
Performance Requirements for Motor Vehicle Headlamps (A)—SAE J1383 JUN90 .... 2:24.36

---
* Technical Revision   † New Document   (D)—DODISS Adopted   (A)—ANSI Recognized

Subject Index

HEADLAMPS (continued)
† SAE Ground Vehicle Lighting Standards Manual—SAE
   HS-34 JAN93 . . . . Available as a separate publication (See Related Technical Reports Section)

Agricultural Equipment
† Headlamps for Agricultural Equipment—SAE **J975** JUN93   3:39.59

Construction and Industrial Equipment
Headlamps for Industrial Equipment (A)—SAE **J95**
MAR86 . . . . 3:40.262
Lighting and Marking of Construction and Industrial
Machinery—SAE **J1029** MAR86 . . . . 3:40.262
Lighting and Marking of Industrial Equipment on
Highways (A)—SAE **J99** MAR86 . . . . 3:40.263
† Requirements for Sealed Lighting Unit for Construction
and Industrial Machines—SAE **J572** MAY93 . . . . 3:40.260
Sealed Lighting Units for Construction, Industrial and
Forest Machinery (A)—SAE **J598** MAY87 . . . . 3:40.259

Dimensions
Dimensional Specifications for General Service Sealed
Lighting Units—SAE **J760a** 2:24.17

Motorcycles
Motorcycle Headlamps (A)—SAE **J584** DEC83 . . . . 2:24.120
Replaceable Motorcycle Headlamp Bulbs (A)—SAE
J1577 JUN91 . . . . 2:24.123

Snowmobiles
Snowmobile Headlamps (A)—SAE **J280** JUN84 . . . . . . 3:38.04

Switches
Headlamp Beam Switching (A)—SAE **J564** MAR90 . . . . 2:24.08
Headlamp Switch (A)—SAE **J253** DEC89 . . . . 2:24.176
Semiautomatic Headlamp Beam Switching Devices (D)—
SAE **J565** JUN89 . . . . 2:24.09
Tests for Snowmobile Switching Devices and
Components—SAE **J68** DEC91 . . . . 3:38.08

HEAT EXCHANGERS
Engine Charge Air Cooler Nomenclature (A)—SAE
J1148 JUN90 . . . . 2:26.392
Laboratory Testing of Vehicle and Industrial Heat
Exchangers for Pressure Cycle Durability (A)—SAE
J1597 JUL88 . . . . 2:26.390
Laboratory Testing of Vehicle and Industrial Heat
Exchangers for Durability Under Vibration Induced
Loading (A)—SAE **J1598** JUL88 . . . . 2:26.389
Laboratory Testing of Vehicle and Industrial Heat
Exchangers for Thermal Cycle Durability—SAE **J1542**
JAN89 . . . . 2:26.382

HEAT INSULATION
See: INSULATION

HEAT RESISTANCE
Standard Method for Determining Continuous Upper
Temperature Resistance of Elastomers—SAE **J2236**
JUN92 . . . . 1:11.32

HEAT TRANSFER
Acoustical and Thermal Materials Test Procedure (A)—
SAE **J1324** OCT89 . . . . 1:11.111

HEAT TREATMENT
Adhesives
Induction Cure Test for Metal Bonding Adhesives (A)—
SAE **J1851** MAY87 . . . . 1:11.109

Aluminum Alloys
Aluminum Alloys—Fundamentals—SAE **J451** JAN89 . . . 1:10.01
General Data on Wrought Aluminum Alloys (A)—SAE
J454 FEB91 . . . . 1:10.26
General Information—Chemical Compositions,
Mechanical and Physical Properties of SAE Aluminum
Casting Alloys—SAE **J452** JAN89 . . . . 1:10.06

Bolts
Mechanical and Material Requirements for Wheel Bolts—
SAE **J1102** 1:4.18

Fasteners
Mechanical and Material Requirements for Externally
Threaded Fasteners (A)—SAE **J429** AUG83 . . . . 1:4.01
Mechanical and Material Requirements for Metric
Externally Threaded Steel Fasteners (A)—SAE **J1199**
SEP83 . . . . 1:4.06

Glossary
Definitions of Heat Treating Terms—SAE **J415** JUN83 . . 1:2.37

Screws
Metric Thread Rolling Screws—SAE **J1237** 1:4.29

Steel
General Characteristics and Heat Treatments of Steels—
SAE **J412** JUN89 . . . . 1:2.05
† High-Strength, Quenched, and Tempered Structural
Steels—SAE **J368** MAR93 . . . . 1:1.164
Methods of Measuring Case Depth (A)—SAE **J423**
DEC83 . . . . 1:3.64
Selection and Heat Treatment of Tool and Die Steels—
SAE **J437a** 1:7.02

Valve Materials
Engine Poppet Valve Information Report—SAE **J775**
JAN88 . . . . 1:1.165

HEATING
Automobiles
* Equivalent Temperature—SAE **J2234** JAN93 . . . . 3:34.25
† Motor Vehicle Heater Test Procedure—SAE **J638** APR93   3:34.34

Construction and Industrial Equipment
Fuel-Fired Heaters—Air Heating—For Construction and
Industrial Machinery—SAE **J1024** DEC89 . . . . 3:40.496

Diesel Engines
† Electric Engine Preheaters and Battery Warmers for
Diesel Engines—SAE **J1310** JUN93 . . . . 3:40.500
Selection and Application Guidelines for Diesel, Gasoline,
and Propane Fired Liquid Cooled Engine Pre-Heaters—
SAE **J1350** MAR88 . . . . 3:40.496

Fuels
Fuel Warmer—Diesel Engines (A)—SAE **J1422** NOV89 . 3:40.509
Heating Value of Fuels—SAE **J1498** MAY90 . . . . 1:12.82

Off-Road Vehicles
Performance Test for Air-Conditioned, Heated, and
Ventilated Off-Road Self-Propelled Work Machines (A)—
SAE **J1503** JUL86 . . . . 3:40.511

HINGES
Door
Vehicle Passenger Door Hinge Systems—SAE **J934**
JUL82 . . . . 3:34.01

HITCHES
† Drawbars—Agricultural Wheel Tractors—SAE **J1548**
FEB93 . . . . 3:39.90
Three Point Hitch (Type A) Backhoe Personnel
Protection—SAE **J1518** JUN86 . . . . 3:40.353
† Three-Point Free-Link Hitch Attachment of Implements to
Agricultural Wheeled Tractors—SAE **J715** JUN93 . . . 3:39.37
† Three-Point Hitch, Implement Quick-Attaching Coupler,
Agricultural Tractors—SAE **J909** JUN93 . . . . 3:39.34
Towability Design Criteria and Equipment Use—
Passenger Cars, Vans, and Light Duty Trucks (A)—SAE
J1142 JUN91 . . . . 3:42.01
Trailer Couplings, Hitches and Safety Chains—
Automotive Type (A)—SAE **J684** MAY87 . . . . 3:35.01

Agricultural Equipment
† Front-Mounted Linkage for Agricultural Wheeled
Tractors—SAE **J2131** OCT92 . . . . 3:39.43

HOODS
Vehicle Hood Latch Systems (A)—SAE **J362** JUL82 . . . 3:34.06

HORNS
Horn—Forward Warning—Electric—Performance, Test,
and Application (A)—SAE **J1105** SEP89 . . . . 1:14.47

---

* Technical Revision   † New Document   (D)—DODISS Adopted   (A)—ANSI Recognized

HORNS (continued)
Performance of Vehicle Traffic Horns (A)—SAE J377
FEB87 .... 1:14.35

HORSEPOWER
See also: PERFORMANCE, POWER CONSUMPTION

Brake Rating Horsepower Requirements—Commercial
Vehicles—SAE J257 MAR85 .... 3:36.137

Certificates of Maximum Net Horsepower for Motor
Trucks and Truck Tractors—SAE J690 .... 3:36.65

Truck Ability Prediction Procedure (A)—SAE J688
AUG87 .... 3:36.03

HOSES
Cumulative Damage Analysis for Hydraulic Hose
Assemblies—SAE J1927 OCT88 .... 2:22.180

Fluid Conductors and Connectors—SAE HS-150 JUN92 — Available as a separate publication (See Related Technical Reports Section)

† Hose and Hose Assemblies for Marine Applications—
SAE J1942 JUN93 .... 2:22.156

† Hydraulic Hose—SAE J517 JUN93 .... 2:22.161

Optional Impulse Test Procedures for Hydraulic Hose
Assemblies—SAE J1405 JUN90 .... 2:22.188

Selection, Installation, and Maintenance of Hose and
Hose Assemblies—SAE J1273 NOV91 .... 2:22.184

† Tests and Procedures for SAE 100R Series Hydraulic
Hose and Hose Assemblies (A)—SAE J343 JUN93 .... 2:22.185

Air Conditioning
Automotive Air Conditioning Hose—SAE J51 MAY89 ... 1:11.249

HFC-134a (R-134a) Service Hose Fittings for Automotive
Air-Conditioning Service Equipment—SAE J2197 JUN92 .... 3:34.53

*R134a Refrigerant Automotive Air Conditioning Hose—
SAE J2064 JUN93 .... 1:11.219

Service Hose for Automotive Air Conditioning—SAE
J2196 JUN92 .... 3:34.51

Brake
† Application of Hydraulic Brake Hose to Motor Vehicles—
SAE J1406 JUN93 .... 2:25.16

Automotive Air Brake Hose and Hose Assemblies (A)—
SAE J1402 JUN85 .... 2:25.11

Dynamic Ozone Test Procedure—Hydraulic Brake
Hose (A)—SAE J1914 MAR88 .... 2:25.10

Hot Impulse Test for Hydraulic Brake Hose Assemblies—
SAE J1833 NOV88 .... 2:25.17

† Moisture Transmission Test Procedure—Hydraulic Brake
Hose Assemblies—SAE J1873 JUN93 .... 2:25.09

Packaging, Storage, and Shelf Life of Hydraulic Brake
Hose Assemblies—SAE J1288 MAR90 .... 2:25.13

† Road Vehicle—Hydraulic Brake Hose Assemblies for Use
with Nonpetroleum-Base Hydraulic Fluids—SAE J1401
JUN93 .... 2:25.01

† Surface Vehicle Brake Manual—SAE HS-24 JUL92 .... Available as a separate publication (See Related Technical Reports Section)

Vacuum Brake Hose (A)—SAE J1403 JUL89 .... 2:25.14

Clamps
† Hose Clamp Specifications—SAE J1508 JUN93 ...... 2:16.45

*Type 'F' Clamps for Plumbing Applications—SAE J1670
MAY93 .... 2:16.79

Connections
Connections for Fluid Power and General Use—Ports
and Stud Ends with ISO 261 Threads and O-Ring
Sealing Part 2: Heavy-Duty (S Series) Stud Ends—
Dimensions, Design, Test Methods, and Requirements—
SAE J2244/2 DEC91 .... 2:22.293

Connections for Fluid Power and General Use—Ports
and Stud Ends with ISO 261 Threads and O-Ring
Sealing Part 1: Port with O-Ring Seal in Truncated
Housing—SAE J2244/1 DEC91 .... 2:22.290

† Hydraulic Flanged Tube, Pipe, and Hose Connections,
Four-Bolt Split Flange Type—SAE J518 JUN93 .... 2:22.190

† Test Method for Evaluating the Sealing Capability of
Hose Connections with a PVT Test Facility—SAE J1610
APR93 .... 2:16.80

Connectors
Formed Tube Ends for Hose Connections (A)—SAE
J962 MAY86 .... 2:22.202

Cooling Systems
*Compression Set of Hoses or Solid Discs—SAE J1638
MAY93 .... 1:11.33

Coolant System Hoses (A)—SAE J20 MAR88 .... 1:11.222

Fittings
† Beaded Tube Hose Fittings—SAE J1231 JUN93 ...... 2:22.158

† Clip Fastener Fitting—SAE J1467 JUN93 .... 2:22.197

Coding Systems for Identification of Fluid Conductors and
Connectors (D)—SAE J846 JUN89 .... 2:22.01

† Fitting—O-Ring Face Seal—SAE J1453 JUN93 ...... 2:22.256

HFC-134a (R-134a) Service Hose Fittings for Automotive
Air-Conditioning Service Equipment—SAE J2197 JUN92 .... 3:34.53

Hot Impulse Test for Hydraulic Brake Hose Assemblies—
SAE J1833 NOV88 .... 2:25.17

† Hydraulic Hose Fittings—SAE J516 JUN93 .... 2:22.114

Hydraulic Hose Fittings for Marine Applications (A)—SAE
J1475 JUN84 .... 3:41.39

Fuel
† Fuel and Oil Hoses (A)—SAE J30 MAY93 .... 1:11.226

† Marine Fuel Hoses (A)—SAE J1527 FEB93 .... 3:41.16

Fuels Systems
*Tube/Hose Assemblies—SAE J2045 OCT92 .... 1:12.85

Lubrication
† Fuel and Oil Hoses (A)—SAE J30 MAY93 .... 1:11.226

Tests and Procedures for High Temperature
Transmission Oil Hose, Engine Lubricating Oil Hose, and
Hose Assemblies (A)—SAE J1019 JUN90 .... 2:22.189

† Transmission Oil Cooler Hose—SAE J1532 JUN93 .... 1:11.252

Power Steering
*High-Temperature Power Steering Pressure Hose—SAE
J2050 FEB93 .... 1:11.245

*High-Temperature Power Steering Return Hose--Low
Pressure—SAE J2076 FEB93 .... 1:11.247

Information Relating to Duty Cycles and Average Power
Requirements of Truck and Bus Engine Accessories—
SAE J1343 OCT88 .... 3:36.57

Power Steering Pressure Hose—High Volumetric
Expansion Type (A)—SAE J188 OCT89 .... 1:11.241

Power Steering Pressure Hose—Low Volumetric
Expansion Type (A)—SAE J191 OCT89 .... 1:11.244

Power Steering Pressure Hose—Wire Braid—SAE J190 .... 1:11.248

Power Steering Return Hose—Low Pressure (A)—SAE
J189 DEC89 .... 1:11.243

HOURMETERS
Electric Hourmeter Specification—SAE J1378 MAR83 .... 2:21.14

HOUSINGS
Engine Flywheel Housing and Mating Transmission
Housing Flanges—SAE J617 MAY92 .... 3:40.247

† Engine Flywheel Housings with Sealed Flanges—SAE
J1172 MAY93 .... 2:26.242

† Housing Internal Dimensions for Single- and Two-Plate
Spring-Loaded Clutches (A)—SAE J373 APR93 .... 2:26.241

† Procedure for Measuring Bore and Face Runout of
Flywheels, Flywheel Housings and Flywheel Housing
Adapters—SAE J1033 APR93 .... 3:40.245

SAE Nodal Mount—SAE J1134 FEB92 .... 3:36.112

HUBS
See also: WHEELS

Nomenclature—Wheels, Hubs, and Rims for Commercial
Vehicles—SAE J393 OCT91 .... 3:31.32

---

* Technical Revision   † New Document   (D)—DODISS Adopted   (A)—ANSI Recognized

Subject Index

HUBS (continued)
Spoke Wheels and Hub Fatigue Test Procedures (A)—
SAE **J1095** JAN91 .... 3:31.20

HUMAN BODY
  Dimensions
  Human Physical Dimensions—SAE **J833** MAY89 ...... 3:40.377
  † Minimum Service Access Dimensions for Off-Road
  Machines (A)—SAE **J925** JUN93 .... 3:40.380
  Impact Response
  Film Analysis Guides for Dynamic Studies of Test
  Subjects—SAE **J138** 3:34.245
  Guidelines for Evaluating Out-of-Position Vehicle
  Occupant Interactions with Deploying Airbags (A)—SAE
  J1980 NOV90 .... 3:33.33
  Human Mechanical Response Characteristics—SAE
  J1460 MAR85 .... 3:34.292
  Human Tolerance to Impact Conditions as Related to
  Motor Vehicle Design—SAE **J885** JUL86 .... 3:34.306
  Vibration
  † Measurement of Whole Body Vibration of the Seated
  Operator of Off-Highway Work Machines—SAE **J1013**
  AUG92 .... 3:39.92

HUMAN ENGINEERING
  See also: OPERATOR CONTROLS, VISION
  Human Mechanical Response Characteristics—SAE
  J1460 MAR85 .... 3:34.292
  Agricultural Equipment
  † Vibration Performance Evaluation of Operator
  Seats (A)—SAE **J1384** JUN93 .... 3:40.396
  Automobiles and Trucks
  Accommodation Tool Reference Point—SAE **J1516**
  MAR90 .... 3:34.247
  Automotive Adaptive Driver Controls, Manual (A)—SAE
  J1903 AUG90 .... 3:34.198
  Describing and Measuring the Driver's Field of View—
  SAE **J1050a** 3:34.214
  Design Criteria—Driver Hand Controls Location for
  Passenger Cars, Multi-Purpose Passenger Vehicles, and
  Trucks (10 000 GVW and Under)—SAE **J1138** 3:34.191
  Devices for Use in Defining and Measuring Vehicle
  Seating Accommodation—SAE **J826** JUN92 .... 3:34.90
  Driver Hand Control Reach (A)—SAE **J287** JUN88 .... 3:34.157
  Driver Selected Seat Position (A)—SAE **J1517** MAR90 . 3:34.250
  *Headlamp Design Guidelines for Mature Drivers—SAE
  J1606 MAR93 .... 3:34.225
  *Manual Controls for Mature Drivers—SAE **J2119** JUN93 3:34.194
  Motor Vehicle Driver and Passenger Head Position (A)—
  SAE **J1052** MAY87 .... 3:34.231
  Photometric Guidelines for Instrument Panel Displays
  That Accomodate Older Drivers—SAE **J2217** OCT91 .. 3:34.226
  Seat Belt Comfort, Fit, and Convenience (A)—SAE
  J1834 JUN91 .... 3:34.267
  Sound Level for Truck Cab Interior—SAE **J336** OCT88 . 1:14.29
  Supplemental Information—Driver Hand Controls
  Location for Passenger Cars, Multi-Purpose Passenger
  Vehicles, and Trucks (10 000 GVW and Under)—SAE
  J1139 .... 3:34.193
  Truck Driver Shin-Knee Position for Clutch and
  Accelerator (A)—SAE **J1521** MAR90 .... 3:34.263
  Truck Driver Stomach Position (A)—SAE **J1522** MAR90 3:34.265
  Earthmoving Equipment
  Acoustics—Measurement of Airborne Noise Emitted by
  Earthmoving Machinery—Operator's Position—Stationary
  Testing Condition (A)—SAE **J2103** JUN91 .... 1:14.78
  Acoustics—Measurement of Exterior Noise Emitted by
  Earthmoving Machinery—Dynamic Test Conditions (A)—
  SAE **J2104** JUN91 .... 1:14.81
  Acoustics—Measurement of Noise Emitted by
  Earthmoving Machinery at the Operator's Position—
  Simulated Work Cycle Test Conditions (A)—SAE **J2105**
  JUN91 .... 1:14.87
  Sound Measurement—Earthmoving Machinery—
  Operator—Singular Type (A) (D)—SAE **J919** JUN86 .. 1:14.53
  Sound Measurement—Off-Road Self-Propelled Work
  Machines—Operator—Work Cycle (A)—SAE **J1166**
  MAY90 .... 1:14.55
  † Vibration Performance Evaluation of Operator
  Seats (A)—SAE **J1384** JUN93 .... 3:40.396
  Motorboats
  Operator Sound Level Exposure Assessment Procedure
  for Pleasure Motorboats (A)—SAE **J1281** MAR85 .... 1:14.43
  Off-Road Vehicles
  Access Systems for Off-Road Machines (D)—SAE **J185**
  JUN88 .... 3:40.383
  Control Locations for Off-Road Work Machines (A)—SAE
  J898 OCT87 .... 3:40.404
  † Measurement of Whole Body Vibration of the Seated
  Operator of Off-Highway Work Machines—SAE **J1013**
  AUG92 .... 3:39.92
  † Minimum Service Access Dimensions for Off-Road
  Machines (A)—SAE **J925** JUN93 .... 3:40.380
  *Operator Controls--Off-Road Machines—SAE **J1814**
  FEB93 .... 3:40.414
  Operator Space Envelope Dimensions for Off-Road
  Machines—SAE **J154** JUN92 .... 3:40.419
  Operator's Seat Dimensions for Off-Road Self-Propelled
  Work Machines—SAE **J899** DEC88 .... 3:40.395
  Rear View Mirrors
  Vision Factors Considerations in Rear View Mirror
  Design—SAE **J985** OCT88 .... 3:34.110
  Snow Blowers
  Bystander Sound Level Measurement Procedure for
  Small Engine Powered Equipment (A)—SAE **J1175**
  MAR85 .... 1:14.97
  Operator Ear Sound Level Measurement Procedure for
  Small Engine Powered Equipment (A)—SAE **J1174**
  MAR85 .... 1:14.95
  Snowmobiles
  Operator Ear Sound Level Measurement Procedure for
  Snow Vehicles (A)—SAE **J1160** MAR83 .... 1:14.37

HUMIDITY
  Guidelines and Limitations of Laboratory Cyclic Corrosion
  Test Procedures for Exterior, Painted, Automotive Body
  Panels—SAE **J1563** MAY87 .... 1:3.51
  Recommended Environmental Practices for Electronic
  Equipment Design—SAE **J1211** 2:23.228

HYDRAULIC FLUIDS
  See also: BRAKE FLUIDS, TRANSMISSION FLUIDS
  Hydraulic O-Ring—SAE **J515** JUN92 .... 2:22.200
  Hydraulic Power Circuit Filtration (D)—SAE **J931** MAR86 3:40.137
  Low Water Tolerant Brake Fluids (A)—SAE **J1705**
  OCT88 .... 2:25.56
  Standardized Fluid for Hydraulic Component Tests (A)—
  SAE **J1276** MAR86 .... 3:40.124
  Cleanliness
  Assessing Cleanliness of Hydraulic Fluid Power
  Components and Systems (A)—SAE **J1227** MAR86 ... 3:40.104
  Method for Assessing the Cleanliness Level of New
  Hydraulic Fluid (A)—SAE **J1277** JUL90 .... 3:40.127
  Reporting Cleanliness Levels of Hydraulic Fluids (A)—
  SAE **J1165** MAR86 .... 3:40.124
  Fire Resistant
  Fire-Resistant Fluid Usage in Hydraulic Systems of Off-
  Road Work Machines—SAE **J1447** MAY86 .... 3:40.126

HYDRAULIC SYSTEMS
  Cumulative Damage Analysis for Hydraulic Hose
  Assemblies—SAE **J1927** OCT88 .... 2:22.180

---

* Technical Revision † New Document (D)—DODISS Adopted (A)—ANSI Recognized

Subject Index

HYDRAULIC SYSTEMS (continued)

† Hydraulic Hose Fittings—SAE J516 JUN93 .... **2:22.114**

Hydraulic Hose Fittings for Marine Applications (A)—SAE J1475 JUN84 .... **3:41.39**

Hydraulic Systems Diagnostic Port Sizes and Locations (A)—SAE J1298 SEP88 .... **3:40.110**

Optional Impulse Test Procedures for Hydraulic Hose Assemblies—SAE J1405 JUN90 .... **2:22.188**

Performance Assurance of Remanufactured, Hydraulically-Operated Rack and Pinion Steering Gears (A)—SAE J1890 JUN88 .... **3:42.33**

Agricultural Equipment

Application of Hydraulic Remote Control to Agricultural Tractors and Trailing Type Agricultural Implements (A)—SAE J716 MAR91 .... **3:39.45**

† Dimensional Standard for Cylindrical Hydraulic Couplers for Agricultural Tractors (A)—SAE J1036 APR93 ...... **3:39.48**

† Test Procedure for Measuring Hydraulic Lift Capacity on Agricultural Tractors Equipped with Three-Point Hitch—SAE J283 JUN93 .... **3:39.41**

Brakes

† Application of Hydraulic Brake Hose to Motor Vehicles—SAE J1406 JUN93 .... **2:25.16**

Automotive Hydraulic Brake System—Metric Banjo Bolt Connections—SAE J1291 MAY89 .... **2:25.170**

Automotive Hydraulic Brake System—Metric Tube Connections—SAE J1290 MAY89 .... **2:25.165**

Dynamic Ozone Test Procedure—Hydraulic Brake Hose (A)—SAE J1914 MAR88 .... **2:25.10**

Hot Impulse Test for Hydraulic Brake Hose Assemblies—SAE J1833 NOV88 .... **2:25.17**

Hydraulic Master Cylinders for Motor Vehicle Brakes—Test Procedure (A)—SAE J1153 JUN91 .... **2:25.29**

Hydraulic Master Cylinders for Motor Vehicle Brakes—Performance Requirements (A)—SAE J1154 JUN91 ... **2:25.34**

Hydraulic Wheel Cylinders for Automotive Drum Brakes (A)—SAE J101 DEC89 .... **2:25.25**

* Materials for Plastic Check Valves for Vacuum Booster System—SAE J1875 JUN93 .... **2:25.40**

† Moisture Transmission Test Procedure—Hydraulic Brake Hose Assemblies—SAE J1873 JUN93 .... **2:25.09**

Packaging, Storage, and Shelf Life of Hydraulic Brake Hose Assemblies—SAE J1288 MAR90 .... **2:25.13**

Plastic Dust Shield for Hydraulic Disc Brakes (A)—SAE J1876 MAR90 .... **2:25.67**

† Road Vehicle—Hydraulic Brake Hose Assemblies for Use with Nonpetroleum-Base Hydraulic Fluids—SAE J1401 JUN93 .... **2:25.01**

Rubber Boots for Drum-Type Hydraulic Brake Wheel Cylinders (A)—SAE J1604 OCT89 .... **2:25.24**

Rubber Cups for Hydraulic Actuating Cylinders (A)—SAE J1601 NOV90 .... **2:25.175**

Service Maintenance of SAE J1703 Brake Fluids in Motor Vehicle Brake Systems—SAE J1707 NOV91 .... **2:25.53**

Shelf Storage of Hydraulic Brake Components—SAE J1825 APR88 .... **2:25.35**

† Surface Vehicle Brake Manual—SAE HS-24 JUL92 .... Available as a separate publication *(See Related Technical Reports Section)*

Cleanliness

Assessing Cleanliness of Hydraulic Fluid Power Components and Systems (A)—SAE J1227 MAR86 ... **3:40.104**

Connections

† Connections for General Use and Fluid Power-Ports and Stud Ends with ISO 725 Threads and O-ring Sealing—Part 1: Threaded Port with O-Ring Seal in Truncated Housing—SAE J1926/1 ISO 11 MAR93 .... **2:22.99**

Coupling

Connections for Fluid Power and General Use—Ports and Stud Ends with ISO 261 Threads and O-Ring Sealing Part 2: Heavy-Duty (S Series) Stud Ends—Dimensions, Design, Test Methods, and Requirements—SAE J2244/2 DEC91 .... **2:22.293**

Connections for Fluid Power and General Use—Ports and Stud Ends with ISO 261 Threads and O-Ring Sealing Part 1: Port with O-Ring Seal in Truncated Housing—SAE J2244/1 DEC91 .... **2:22.290**

* Metallic Tube Connections for Fluid Power and General Use-Test Methods for Threaded Hydraulic Fluid Power Connectors—SAE J1644 MAY93 .... **2:22.285**

Couplings

† Dimensional Standard for Cylindrical Hydraulic Couplers for Agricultural Tractors (A)—SAE J1036 APR93 ...... **3:39.48**

Hydraulic Diagnostic Couplings (A)—SAE J1502 APR86 **3:40.110**

Cylinders

Cylinder Rod Wiper Seal Ingression Test (A)—SAE J1195 MAR86 .... **3:40.141**

Hydraulic Cylinder Integrity Test—SAE J1334 JUN87 .. **3:40.344**

Hydraulic Cylinder Leakage Test—SAE J1336 JUN87 .. **3:40.346**

Hydraulic Cylinder No-Load Friction Test (A)—SAE J1335 APR90 .... **3:40.346**

Hydraulic Cylinder Rod Corrosion Test (A)—SAE J1333 MAR90 .... **3:40.343**

Hydraulic Cylinder Rod Seal Endurance Test Procedure (A)—SAE J1374 MAY85 .... **3:40.143**

Hydraulic Cylinder Test Procedure (A)—SAE J214 MAR86 .... **3:40.138**

Hydraulic Master Cylinders for Motor Vehicle Brakes—Performance Requirements (A)—SAE J1154 JUN91 ... **2:25.34**

Hydraulic Master Cylinders for Motor Vehicle Brakes—Test Procedure (A)—SAE J1153 JUN91 .... **2:25.29**

Shelf Storage of Hydraulic Brake Components—SAE J1825 APR88 .... **2:25.35**

Filters

Hydraulic Power Circuit Filtration (D)—SAE J931 MAR86 **3:40.137**

Hoses

† Application of Hydraulic Brake Hose to Motor Vehicles—SAE J1406 JUN93 .... **2:25.16**

† Hydraulic Hose—SAE J517 JUN93 .... **2:22.161**

† Tests and Procedures for SAE 100R Series Hydraulic Hose and Hose Assemblies (A)—SAE J343 JUN93 .... **2:22.185**

Leakage

External Leakage Classifications for Hydraulic Systems (A)—SAE J1176 MAR86 .... **3:40.146**

Hydraulic Cylinder Leakage Test—SAE J1336 JUN87 .. **3:40.346**

Hydraulic Cylinder Test Procedure (A)—SAE J214 MAR86 .... **3:40.138**

Measuring and Reporting the Internal Leakage of a Hydraulic Fluid Power Valve (A)—SAE J1235 MAR86 .. **3:40.123**

Motors

Hydraulic Motor Test Procedure (A)—SAE J746 MAR86 **3:40.128**

Off-Road Vehicles

Fire-Resistant Fluid Usage in Hydraulic Systems of Off-Road Work Machines—SAE J1447 MAY86 .... **3:40.126**

Pumps

Hydraulic Power Pump Test Procedure (A)—SAE J745 APR87 .... **3:40.111**

Hydraulic Pump and Motor Mounting and Drive Dimensions (A)—SAE J744 JUL88 .... **3:40.116**

Test Method for Measuring Power Consumption of Hydraulic Pumps for Trucks and Buses (A)—SAE J1341 APR90 .... **3:36.54**

Seals

† Cast Iron Sealing Rings (Metric)—SAE J1236 APR93 ... **3:29.156**

Transmissions

† Automatic Transmission Hydraulic Control Systems—Terminology—SAE J648 APR93 .... **3:29.27**

---

* Technical Revision † New Document (D)—DODISS Adopted (A)—ANSI Recognized

Subject Index

HYDRAULIC SYSTEMS (continued)
Transmissions (continued)
Fluid for Passenger Car Type Automatic Transmissions—SAE J311 APR86 .... 1:12.44

Tubing
Connections for Fluid Power and General Use—Ports and Stud Ends with ISO 261 Threads and O-Ring Sealing Part 1: Port with O-Ring Seal in Truncated Housing—SAE J2244/1 DEC91 .... 2:22.290

Connections for Fluid Power and General Use—Ports and Stud Ends with ISO 261 Threads and O-Ring Sealing Part 2: Heavy-Duty (S Series) Stud Ends—Dimensions, Design, Test Methods, and Requirements—SAE J2244/2 DEC91 .... 2:22.293

Fuel Injection Pumps — High Pressure Pipes (Tubing) for Testing (A)—SAE J1418 DEC87 .... 2:26.263

† Hydraulic Tube Fittings—SAE J514 JUN93 .... 2:22.59

Pressure Ratings for Hydraulic Tubing and Fittings (D)—SAE J1065 MAR92 .... 2:22.215

* Seamless Copper-Nickel 90-10 Tubing—SAE J1650 MAR93 .... 2:22.211

Tubing—Motor Vehicle Brake System Hydraulic (A)—SAE J1047 JUN90 .... 2:25.162

Valves
Control Valve Test Procedure (A)—SAE J747 MAY90 .. 3:40.130

Hydraulic Directional Control Valves, 3000 psi Maximum (A)—SAE J748 MAR86 .... 3:40.120

† Hydraulic Valves for Motor Vehicle Brake Systems Test Procedure—SAE J1118 JUN93 .... 2:25.35

Hydraulic Valves for Motor Vehicle Brake Systems—Performance Requirements—SAE J1137 MAR85 ..... 2:25.39

Measuring and Reporting the Internal Leakage of a Hydraulic Fluid Power Valve (A)—SAE J1235 MAR86 .. 3:40.123

Method of Measuring and Reporting the Pressure Differential-Flow Characteristics of a Hydraulic Fluid Power Valve (A)—SAE J1117 MAR86 .... 3:40.120

Shelf Storage of Hydraulic Brake Components—SAE J1825 APR88 .... 2:25.35

HYDROCARBONS
Exhaust Emissions
Continuous Hydrocarbon Analysis of Diesel Emissions (A)—SAE J215 JUN88 .... 1:13.29

Diesel Engine Emission Measurement Procedure (A)—SAE J1003 JUN90 .... 1:13.33

Instrumentation and Techniques for Exhaust Gas Emissions Measurement (A)—SAE J254 AUG84 ...... 1:13.104

† Instrumentation and Techniques for Vehicle Refueling Emissions Measurement—SAE J1045 MAY93 .... 1:13.137

HYDRODYNAMIC DRIVES
See: DRIVES

HYDROGEN SULFIDES
Nonmetallic Trim Materials—Test Method for Determining the Staining Resistance to Hydrogen Sulfide Gas (A)—SAE J322 DEC85 .... 1:11.116

IGNITION SYSTEMS
See also: CABLES, DISTRIBUTORS, ELECTRIC EQUIPMENT, ELECTRONIC EQUIPMENT, SPARK PLUGS, STARTERS

† Ignition System Measurements Procedure—SAE J973 JUN93 .... 2:23.59

Cables
High Tension Ignition Cable (A)—SAE J2031 JAN90 ... 2:23.149

High Tension Ignition Cable Assemblies—Marine (A)—SAE J1191 JAN86 .... 3:41.26

Ignition Cable Assemblies—SAE J2032 NOV91 .... 2:23.154

Switches
Ignition Switch—SAE J259 .... 2:23.141

Tests for Snowmobile Switching Devices and Components—SAE J68 DEC91 .... 3:38.08

Terminology
Ignition System Nomenclature and Terminology (A)—SAE J139 JUN90 .... 2:23.58

IMPACT TESTS
See also: CRASHWORTHINESS, DROP TESTS

Fixed Rigid Barrier Collision Tests—SAE J850 NOV88 . 3:34.281

Moving Rigid Barrier Collision Tests—SAE J972 DEC88 3:34.278

Test Device Head Contact Duration Analysis—SAE J2052 MAR90 .... 3:34.276

Classification
Collision Deformation Classification—SAE J224 MAR80 3:34.235

Doors
Passenger Car Door System Crush Test Procedure—SAE J367 JUN80 .... 3:34.286

Fuel Tanks
Fuel and Lubricant Tanks for Motorcycles—SAE J1241 . 3:37.11

Snowmobile Fuel Tanks (A)—SAE J288 NOV83 ...... 3:38.15

Glazing
American National Standard for Safety Glazing Materials for Glazing Motor Vehicles and Motor Vehicle Equipment Operating on Land Highways—Safety Code—ANSI SAE Z26.1-1990 .... Available as a separate publication (See Related Technical Reports Section)

Instrumentation
Instrumentation for Impact Test (A)—SAE J211 OCT88 . 3:34.270

Operator Protective Structures
Overhead Protection for Agricultural Tractors—Test Procedures and Performance Requirements—SAE J167 APR92 .... 3:39.89

Rear Underride Guard Test Procedure (A)—SAE J260 JUN90 .... 3:34.287

† Roll-Over Protective Structures (ROPS) for Wheeled Agricultural Tractors (A)—SAE J2194 JUN93 .... 3:39.78

Rollover Protective Structures (ROPS) for Wheeled Agricultural Tractors (A)—SAE J1194 MAY89 .... 3:39.71

Restraint Systems
* Guidelines for Evaluating Child Restraint System Interactions with Deploying Airbags—SAE J2189 FEB93 3:33.20

Guidelines for Evaluating Out-of-Position Vehicle Occupant Interactions with Deploying Airbags (A)—SAE J1980 NOV90 .... 3:33.33

Occupant Restraint System Evaluation—Passenger Cars—SAE J128 .... 3:33.12

† Vehicle Occupant Restraint Systems and Components Standards Manual—SAE HS-13 MAY93 .... Available as a separate publication (See Related Technical Reports Section)

Seats
Dynamic Cushioning Performance Criteria for Snowmobile Seats (A)—SAE J89 JAN85 .... 3:38.13

Wheels
Wheels—Impact Test Procedures—Road Vehicles (A)—SAE J175 JUN88 .... 3:31.07

IMPACT TOLERANCE
Human Body
Film Analysis Guides for Dynamic Studies of Test Subjects—SAE J138 .... 3:34.245

Human Mechanical Response Characteristics—SAE J1460 MAR85 .... 3:34.292

Human Tolerance to Impact Conditions as Related to Motor Vehicle Design—SAE J885 JUL86 .... 3:34.306

INCLUSIONS
Steels
Microscopic Determination of Inclusions in Steels (A)—SAE J422 DEC83 .... 1:3.59

---

* Technical Revision † New Document (D)—DODISS Adopted (A)—ANSI Recognized

Subject Index

INDICATORS
  *Electrical Indicating System Specification—SAE **J1810**
  JAN93 .... 3:40.275

INDUSTRIAL EQUIPMENT
  See: CONSTRUCTION AND INDUSTRIAL EQUIPMENT

INDUSTRIAL TRACTORS
  See: CONSTRUCTION AND INDUSTRIAL EQUIPMENT

INFLATABLE RESTRAINTS
  Deployment of Electrically Activated Automotive Air Bags
  for Automobile Reclamation—SAE **J1855** MAR92 ..... 3:33.29
  Glossary of Automotive Inflatable Restraint Terms—SAE
  J1538 APR88 .... 3:33.31
  *Guidelines for Evaluating Child Restraint System
  Interactions with Deploying Airbags—SAE **J2189** FEB93 3:33.20
  Guidelines for Evaluating Out-of-Position Vehicle
  Occupant Interactions with Deploying Airbags (A)—SAE
  J1980 NOV90 .... 3:33.33
  Identification of Automotive Air Bags—SAE **J1856**
  MAY89 .... 3:33.29
  *SAE Automotive Textiles and Trim Standards Manual—
  SAE **HS-2700** MAY93 .... Available as a separate publication *(See Related Technical Reports Section)*
  *The Air Bag Systems in Your Cars 'What the Public
  Needs to Know'—SAE **J2074** JUN93 .... 3:33.42
  †Vehicle Occupant Restraint Systems and Components
  Standards Manual—SAE **HS-13** MAY93 .... Available as a separate publication *(See Related Technical Reports Section)*

INFRARED TESTING
  See also: NONDESTRUCTIVE TESTS
  Infrared Testing—SAE **J359** FEB91 .... 1:3.69
  Nondestructive Tests (A)—SAE **J358** FEB91 .... 1:3.67

INJURIES
  Human Mechanical Response Characteristics—SAE
  J1460 MAR85 .... 3:34.292
  Human Tolerance to Impact Conditions as Related to
  Motor Vehicle Design—SAE **J885** JUL86 .... 3:34.306
  Test Device Head Contact Duration Analysis—SAE
  J2052 MAR90 .... 3:34.276

INSPECTION
  See also: MAGNETIC PARTICLE INSPECTION,
  NONDESTRUCTIVE TESTS, ULTRASONIC TESTS
  Acoustic Emission Test Methods (A)—SAE **J1242**
  MAR91 .... 1:3.80
  Detection of Surface Imperfections in Ferrous Rods,
  Bars, Tubes, and Wires—SAE **J349** FEB91 .... 1:3.63
  Engine Water Pump Remanufacture Procedures and
  Acceptance Criteria—SAE **J1916** MAY89 .... 3:42.41
  Liquid Penetrant Test Methods—SAE **J426** MAR91 .... 1:3.75
  Microscopic Determination of Inclusions in Steels (A)—
  SAE **J422** DEC83 .... 1:3.59
  Penetrating Radiation Inspection (A)—SAE **J427** MAR91 1:3.76
  Pneumatic Tires for Military Tactical Wheeled
  Vehicles (A)—SAE **J2014** APR91 .... 3:30.04
  †Surface Discontinuities on General Application Bolts,
  Screws, and Studs—SAE **J1061** OCT92 .... 1:4.33
  Lighting Equipment
  Headlamp Aiming Device for Mechanically Aimable
  Headlamp Units (A)—SAE **J602** DEC89 .... 2:24.206
  Lighting Inspection Code—SAE **J599** MAY81 .... 2:24.202
  Steel Castings
  Automotive Steel Castings—SAE **J435c** 1:6.17

INSTRUMENT PANELS
  Automobiles and Trucks
  Design Criteria—Driver Hand Controls Location for
  Passenger Cars, Multi-Purpose Passenger Vehicles, and
  Trucks (10 000 GVW and Under)—SAE **J1138** 3:34.191
  Location and Operation of Instruments and Controls in
  Motor Truck Cabs (A)—SAE **J680** SEP88 .... 3:36.121
  Photometric Guidelines for Instrument Panel Displays
  That Accomodate Older Drivers—SAE **J2217** OCT91 .. 3:34.226
  Supplemental Information—Driver Hand Controls
  Location for Passenger Cars, Multi-Purpose Passenger
  Vehicles, and Trucks (10 000 GVW and Under)—SAE
  J1139 .... 3:34.193
  Construction and Industrial Equipment
  Instrument Face Design and Location for Construction
  and Industrial Equipment (A)—SAE **J209** JAN87 ..... 3:40.421
  Motorcycles
  Operator Controls and Displays on Motorcycles—SAE
  J107 APR80 .... 3:37.05

INSULATION
  Acoustical and Thermal Materials Test Procedure (A)—
  SAE **J1324** OCT89 .... 1:11.111
  Compression and Recovery of Insulation Paddings—SAE
  J1352 APR87 .... 1:11.172
  †Corrosion Test for Insulation Materials—SAE **J1389**
  MAY93 .... 1:11.173
  †Hot Odor Test for Insulation Materials—SAE **J1351**
  JUN93 .... 1:11.172
  Hot Plate Method for Evaluating Heat Resistance and
  Thermal Insulation Properties of Materials—SAE **J1361**
  APR87 .... 1:11.171
  Laboratory Measurement of the Airborne Sound Barrier
  Performance of Automotive Materials and
  Assemblies (A)—SAE **J1400** MAY90 .... 1:11.174
  *Laboratory Measurement of the Composite Vibration
  Damping Properties of Materials on a Supporting Steel
  Bar—SAE **J1637** FEB93 .... 1:2.30
  Method of Viscosity Test for Automotive Type Adhesives,
  Sealers, and Deadeners—SAE **J1524** NOV88 .... 1:11.78
  Methods of Tests for Automotive-Type Sealers,
  Adhesives, and Deadeners—SAE **J243** 1:11.68
  Nomenclature—Clam Bunk Skidder (A)—SAE **J1353**
  APR87 .... 3:40.80
  Nonmetallic Loom (A)—SAE **J562** APR86 .... 2:23.187
  *SAE Automotive Textiles and Trim Standards Manual—
  SAE **HS-2700** APR93 .... Available as a separate publication *(See Related Technical Reports Section)*
  Test Method for Measuring the Relative Drapeability of
  Flexible Insulation Materials—SAE **J1325** FEB85 ..... 1:11.169
  †Test Method for Measuring Thickness of Resilient
  Insulating Paddings—SAE **J1355** MAY93 .... 1:11.170
  Vibration Damping Materials and Underbody
  Coatings (A)—SAE **J671** APR82 .... 1:11.111

INTAKE SYSTEMS
  Airflow Reference Standards (A)—SAE **J228** JUN90 ... 2:26.372
  †Devices Providing Backfire Flame Control for Gasoline
  Engines in Marine Applications—SAE **J1928** APR93 ... 3:41.44
  Location of Atomizer of Ether Systems for Diesel
  Engines—SAE **J2079** MAY92 .... 2:26.329
  Noise
  Measurement Procedure for Determination of Silencer
  Effectiveness in Reducing Engine Intake or Exhaust
  Sound Level (A)—SAE **J1207** FEB87 .... 1:14.115

INTEGRATED CIRCUITS
  Collision Detection Serial Data Communications Multiplex
  Bus—SAE **J1567** AUG87 .... 2:23.481
  General Qualification and Production Acceptance Criteria
  for Integrated Circuits in Automotive Applications—SAE
  J1879 OCT88 .... 2:23.583

INTERFACE PROTOCOL
  *OBD II Scan Tool—SAE **J1978** MAR92 .... 2:23.325
  *OEM/Vendor's Interface Specification for Vehicle
  Electronic Programming Station—SAE **J1924** DEC92 .. 2:23.522

---

* Technical Revision   † New Document   (D)—DODISS Adopted   (A)—ANSI Recognized

Subject Index

INTERFACE PROTOCOL (continued)
* Survey of Known Protocols—SAE **J2056/2** APR93 ..... 2:23.372
* Universal Interface for OBD II Scan—SAE **J2201** JUN93 ..... 2:23.250

IRON
Cast
Automotive Ductile (Nodular) Iron Castings (A)—SAE **J434** JUN86 ..... 1:6.09
† Automotive Gray Iron Castings—SAE **J431** MAR93 .... 1:6.01
Automotive Malleable Iron Castings (A)—SAE **J158** JUN86 ..... 1:6.07
Cast Iron Sealing Rings—SAE **J281** SEP80 ..... 3:29.148
† Cast Iron Sealing Rings (Metric)—SAE **J1236** APR93 .. 3:29.156
Elevated Temperature Properties of Cast Irons—SAE **J125** MAY88 ..... 1:6.12
† Internal Combustion Engines—Piston Rings—Half Keystone Rings (A)—SAE **J2001** OCT92 ..... 2:26.111
† Numbering System for Designating Grades of Cast Ferrous Materials—SAE **J859** OCT92 ..... 1:6.01
Surface Hardness
† Surface Hardness Testing with Files—SAE **J864** MAY93 ..... 1:3.03

IRON ALLOYS
See also: FERROUS MATERIALS, STEELS
Engine Poppet Valve Information Report—SAE **J775** JAN88 ..... 1:1.165
Special Purpose Alloys ("Superalloys")—SAE **J467b** ... 1:10.128

JACKS
Bumper Jack Requirements—Motor Vehicles—SAE **J979** 3:32.01
Bumper Jacking Test Procedure—Motor Vehicles—SAE **J978** 3:32.01

JOINING
See also: WELDING
Automotive Metallurgical Joining—SAE **J836a** JAN70 .. 1:9.08
Unibody Weld Quality Testing (A)—SAE **J1827** MAY87 . 3:42.23
Aluminum Alloys
Aluminum Alloys—Fundamentals—SAE **J451** JAN89 .... 1:10.01
Wrought Aluminum Applications Guidelines—SAE **J1434** JAN89 ..... 1:10.15
Copper Alloys
Wrought and Cast Copper Alloys—SAE **J461** SEP81 ... 1:10.45
Fracture
Coach Joint Fracture Test—SAE **J1863** APR87 .... 1:11.85

KEYS
Woodruff Key Slots and Keyways—SAE **J503** 2:18.06
Woodruff Keys—SAE **J502** 2:18.05

KINEMATICS
Automobile Occupants
* Dolly Rollover Recommended Test Procedure—SAE **J2114** APR93 ..... 3:34.283
Guidelines for Evaluating Out-of-Position Vehicle Occupant Interactions with Deploying Airbags (A)—SAE **J1980** NOV90 ..... 3:33.33
Human Mechanical Response Characteristics—SAE **J1460** MAR85 ..... 3:34.292
Terminology
Vehicle Dynamics Terminology—SAE **J670e** MAR76 ... 3:34.297

KINGPINS
Fifth Wheel Kingpin Performance—Commercial Trailers and Semitrailers—SAE **J133** OCT87 ..... 3:36.86
Fifth Wheel Kingpin, Heavy-Duty—Commercial Trailers and Semitrailers (A)—SAE **J848** JAN91 ..... 3:36.92
* Kingpin Wear Limits Commercial Trailers and Semi-Trailers—SAE **J2228** JUN93 ..... 3:36.89
† Upper Coupler Kingpin—Commercial Trailers and Semitrailers (A)—SAE **J700** JUN93 ..... 3:36.90

KNEE RESTRAINTS
Glossary of Automotive Inflatable Restraint Terms—SAE **J1538** APR88 ..... 3:33.31

LAMP BULBS
Lamp Bulb Retention System (A)—SAE **J567** NOV87 .. 2:24.11
Miniature Lamp Bulbs (A) (D)—SAE **J573** DEC89 ..... 2:24.21
Replaceable Motorcycle Headlamp Bulbs (A)—SAE **J1577** JUN91 ..... 2:24.123
† Requirements for Sealed Lighting Unit for Construction and Industrial Machines—SAE **J572** MAY93 .... 3:40.260
† SAE Ground Vehicle Lighting Standards Manual—SAE **HS-34** JAN93 ..... Available as a separate publication (See Related Technical Reports Section)

LAMPS
See also: BACKUP LAMPS, CORNERING LAMPS, FLOOD LAMPS, FOG LAMPS, HEADLAMPS, LAMP BULBS, LICENSE PLATE LAMPS, PARKING LAMPS, REFLECTORS, SIDE MARKER LAMPS, SPOT LAMPS, STOP LAMPS, TAIL LAMPS, TURN INDICATORS, WARNING SYSTEMS
360 Degree Warning Devices for Authorized Emergency, Maintenance, and Service Vehicles—SAE **J845** MAR92 . 2:24.185
Accelerated Exposure of Automotive Exterior Materials Using a Fluorescent UV and Condensation Apparatus—SAE **J2020** JUN89 ..... 1:11.288
Auxiliary Driving Lamps—SAE **J581** JUN89 .... 2:24.117
Auxiliary Low Beam Lamps (A)—SAE **J582** SEP84 .... 2:24.118
† Cargo Lamps for Use on Vehicles Under 12 000 lb GVWR—SAE **J1424** JUN93 ..... 2:24.120
* Central High Mounted Stop Lamp Standard for Use on Vehicles Less than 2032 mm Overall Width—SAE **J1957** JUN93 ..... 2:24.149
Clearance, Side Marker, and Identification Lamps—SAE **J592** JUN92 ..... 2:24.170
Clearance, Sidemarker, and Identification Lamps for Use on Motor Vehicles 2032 mm or More in Overall Width (A)—SAE **J2042** JUN91 ..... 2:24.172
Connectors and Plugs—SAE **J856** 2:24.15
Daytime Running Lamps for Use on Motor Vehicles—SAE **J2087** AUG91 ..... 2:24.168
* Discharge Forward Lighting System—SAE **J2009** FEB93 . 2:24.113
† Flashing Warning Lamp for Agricultural Equipment—SAE **J974** JUN93 ..... 3:39.59
Flashing Warning Lamp for Industrial Equipment (A)—SAE **J96** MAR86 ..... 3:40.261
Flashing Warning Lamps for Authorized Emergency, Maintenance and Service Vehicles (A)—SAE **J595** JAN90 ..... 2:24.183
Gaseous Discharge Warning Lamp for Authorized Emergency, Maintenance, and Service Vehicles (A)—SAE **J1318** APR86 ..... 2:24.188
High Mounted Stop Lamps for Use on Vehicles 2032 mm or More in Overall Width—SAE **J1432** OCT88 ..... 2:24.147
L.E.D. Lighting Devices (A)—SAE **J1889** JUN88 ...... 2:24.19
Lighting and Marking of Agricultural Equipment on Highways—SAE **J137** JUN89 ..... 3:39.60
Lighting Identification Code (A)—SAE **J759** JUN91 .... 2:24.03
Lighting Inspection Code—SAE **J599** MAY81 .... 2:24.202
Motorcycle Auxiliary Front Lamps—SAE **J1306** JUN89 . 2:24.122
Mounting Brackets and Socket for Warning Lamp and Slow-Moving Vehicle (SMV) Identification Emblem (A)—SAE **J725** MAR91 ..... 3:39.65
Photometry Laboratory Accuracy Guidelines—SAE **J1330** JUN88 ..... 2:24.04
Preferred Conversion Values for Dimensions in Lighting—Inch-Pound Units/SI—SAE **J1322** JUN85 ... 2:24.01
† SAE Ground Vehicle Lighting Standards Manual—SAE **HS-34** JAN93 ..... Available as a separate publication (See Related Technical Reports Section)
School Bus Warning Lamps (A)—SAE **J887** AUG87 ... 2:24.192

---

* Technical Revision   † New Document   (D)—DODISS Adopted   (A)—ANSI Recognized

Subject Index

LAMPS (continued)
Sealed Lighting Units for Construction, Industrial and Forest Machinery (A)—SAE **J598** MAY87 .... 3:40.259

Snowmobile and Snowmobile Cutter Lamps, Reflective Devices, and Associated Equipment (A)—SAE **J292** OCT84 .... 3:38.07

Test Methods and Equipment for Lighting Devices and Components for Use on Vehicles Less Than 2032 mm in Overall Width—SAE **J575** JUN92 .... 2:24.30

Turn Signal Lamps for Use on Motor Vehicles Less Than 2032 mm in Overall Width (A)—SAE **J588** JUN91 ...... 2:24.153

Color
Color Specification (A) (D)—SAE **J578** MAY88 .... 2:24.34

Lenses
Plastic Materials for Use in Optical Parts Such as Lenses and Reflectors of Motor Vehicle Lighting Devices—SAE **J576** JUL91 .... 2:24.33

Terminology
Terminology—Motor Vehicle Lighting—SAE **J387** OCT88 2:24.01

LATCHES
Door
Passenger Car Side Door Latch Systems (A)—SAE **J839** JUN91 .... 3:34.02

Hood
Vehicle Hood Latch Systems (A)—SAE **J362** JUL82 ... 3:34.06

LATEX
See: RUBBER

LAWNMOWERS
Noise
Bystander Sound Level Measurement Procedure for Small Engine Powered Equipment (A)—SAE **J1175** MAR85 .... 1:14.97

Operator Ear Sound Level Measurement Procedure for Small Engine Powered Equipment (A)—SAE **J1174** MAR85 .... 1:14.95

Seals
Hydraulic Cylinder Rod Seal Endurance Test Procedure (A)—SAE **J1374** MAY85 .... 3:40.143

LEAD ALLOYS
Bearing and Bushing Alloys—SAE **J459** OCT91 ...... 1:10.43

Bearing and Bushing Alloys—Chemical Composition of SAE Bearing and Bushing Alloys—SAE **J460** OCT91 ... 1:10.41

LEAKAGES
Leakage Testing—SAE **J1267** DEC88 .... 1:3.73

† Marine Carburetors and Fuel Injection Throttle Bodies—SAE **J1223** JUN93 .... 3:41.28

Nondestructive Tests (A)—SAE **J358** FEB91 .... 1:3.67

Hydraulic Systems
External Leakage Classifications for Hydraulic Systems (A)—SAE **J1176** MAR86 .... 3:40.146

Hydraulic Cylinder Leakage Test—SAE **J1336** JUN87 . 3:40.346

Hydraulic Cylinder Test Procedure (A)—SAE **J214** MAR86 .... 3:40.138

Refrigerants
*Rating Criteria for Electronic Refrigerant Leak Detectors—SAE **J1627** JUN93 .... 3:34.71

*Technican Procedure for Using Electronic Refrigerant Leak Detectors for Service of Mobile Air Conditioning Systems—SAE **J1628** JUN93 .... 3:34.72

Valves
Measuring and Reporting the Internal Leakage of a Hydraulic Fluid Power Valve (A)—SAE **J1235** MAR86 . 3:40.123

LEATHER
Method of Testing Resistance to Crocking of Organic Trim Materials—SAE **J861** 1:11.114

Method of Testing Resistance to Scuffing of Trim Materials (A)—SAE **J365** FEB85 .... 1:11.163

Test Method for Determining Resistance to Abrasion of Automotive Bodycloth, Vinyl, and Leather, and the Snagging of Automotive Bodycloth (A)—SAE **J948** FEB86 .... 1:11.178

LENSES
† SAE Ground Vehicle Lighting Standards Manual—SAE **HS-34** JAN93 .... Available as a separate publication (See Related Technical Reports Section)

Plastic
Plastic Materials for Use in Optical Parts Such as Lenses and Reflectors of Motor Vehicle Lighting Devices—SAE **J576** JUL91 .... 2:24.33

LICENSE PLATE LAMPS
† License Plate Illumination Devices (Rear Registration Plate Illumination Devices)—SAE **J587** MAR93 .... 2:24.151

LICENSE PLATES
Motor Vehicle License Plates—SAE **J686** JUL81 ...... 3:27.06

LIFT POINTS
Automobiles
*Vehicle Lift Points for Service Garage Lifting—SAE **J2184** OCT92 .... 3:42.53

LIGHTERS
Six- and Twelve-Volt Cigar Lighter Receptacles (A)—SAE **J563** MAR90 .... 2:23.189

LIGHTING
See: LAMPS

LINKAGES
Agricultural Equipment
† Front-Mounted Linkage for Agricultural Wheeled Tractors—SAE **J2131** OCT92 .... 3:39.43

LIQUID LEVEL INDICATORS
† Guidelines for Liquid Level Indicators—SAE **J48** MAY93 3:40.103

LIQUID PENETRANT TESTING
See: NONDESTRUCTIVE TESTS

Detection of Surface Imperfections in Ferrous Rods, Bars, Tubes, and Wires—SAE **J349** FEB91 .... 1:3.63

Liquid Penetrant Test Methods—SAE **J426** MAR91 ... 1:3.75

Nondestructive Tests (A)—SAE **J358** FEB91 .... 1:3.67

LOADERS
See also: CONSTRUCTION AND INDUSTRIAL EQUIPMENT, CRAWLERS

Capacity Rating—Loader Bucket—SAE **J742** FEB85 ... 3:40.324

Cutting Edge—Cross Sections Loader Straight (A)—SAE **J1303** FEB85 .... 3:40.239

Cutting Edge—Cross Sections Loader Straight with Bolt Holes (A)—SAE **J1304** FEB85 .... 3:40.241

Cutting Edge—Optional Cross-Sections and Dimensions Loader Straight (A)—SAE **J1581** SEP89 .... 3:40.242

Lift Arm Support Device for Loaders—SAE **J38** AUG91 . 3:40.353

Lift Capacity Calculation and Test Procedure—Pipelayer and Sideboom—SAE **J743** MAR92 .... 3:40.223

Rated Operating Load for Loaders (A)—SAE **J818** MAY87 .... 3:40.100

Rated Operating Load for Loaders Equipped with Log or Material Forks without Vertical Mast (A)—SAE **J1197** FEB91 .... 3:40.101

† Steering Frame Lock for Articulated Loaders and Tractors—SAE **J276** NOV92 .... 3:40.353

Axles
Axle Application Load Rating for Industrial Wheel Loaders and Backhoe Loaders—SAE **J43** FEB88 ..... 3:40.333

Brakes
Braking Performance—Crawler Tractors and Crawler Loaders—SAE **J1026** APR90 .... 3:40.281

Braking Performance—Rubber-Tired Construction Machines (A)—SAE **J1152** APR80 .... 3:40.289

---

* Technical Revision   † New Document   (D)—DODISS Adopted   (A)—ANSI Recognized

Subject Index

LOADERS (continued)
Noise
Sound Measurement—Off-Road Self-Propelled Work Machines—Operator—Work Cycle (A)—SAE J1166 MAY90 .... 1:14.55

† Sound Power Level Measurements of Earthmoving Machinery - Static and In-Place Dynamic Methods—SAE J1805 APR93 .... 1:14.90

Operator Protective Structures
Personnel Protection—Skid Steer Loaders (A)—SAE J1388 JUN85 .... 3:40.474

Steering
Minimum Performance Criteria for Emergency Steering of Wheeled Earthmoving Construction Machines—SAE J53 OCT84 .... 3:40.515

Terminology
Component Nomenclature—Loader—SAE J731 FEB85 . 3:40.39

Identification Terminology & Component Nomenclature—Knuckle Boom Log Loader (A)—SAE J2055 JUN91 .... 3:40.67

Identification Terminology of Earthmoving Machines (A)—SAE J1057 SEP88 .... 3:40.05

Identification Terminology of Loaders/Tractors with Forks and Rough Terrain Forklifts—SAE J1464 APR88 ...... 3:40.58

Specification Definitions—Loaders—SAE J732 JUN92 .. 3:40.65

LOCKS
† Steering Frame Lock for Articulated Loaders and Tractors—SAE J276 NOV92 .... 3:40.353

LOGGING EQUIPMENT
See: FORESTRY EQUIPMENT

LOOMS
Nonmetallic Loom (A)—SAE J562 APR86 .... 2:23.187

LUBRICANTS
See also: BRAKE FLUIDS, CHEMICAL PROPERTIES, GREASES, TRANSMISSION FLUIDS

† Automotive Lubricating Greases—SAE J310 JUN93 ... 1:12.39

† Effective Dates of New or Revised Technical Reports (A)—SAE J301 MAR93 .... 1:12.01

† Fuels and Lubricants Standards Manual—SAE HS-23 FEB93 .... Available as a separate publication *(See Related Technical Reports Section)*

Air Conditioning
*Compatibility of Retrofit Refrigerants with Air Conditioning System Materials—SAE J1662 JUN93 .... 3:34.69

Construction and Industrial Equipment
Lubricant Types—Construction and Industrial Machinery—SAE J754a 3:40.102

Corrosion Prevention
Prevention of Corrosion of Motor Vehicle Body and Chassis Components—SAE HS-J447 JUN81 .... Available as a separate publication *(See Related Technical Reports Section)*

Engine
Classification of Energy-Conserving Engine Oil for Passenger Cars, Vans, and Light-Duty Trucks—SAE J1423 FEB92 .... 1:12.29

Engine Oil Performance and Engine Service Classification (Other Than "Energy-Conserving") (A)—SAE J183 JUN91 .... 1:12.01

† Engine Oil Tests (A)—SAE J304 JUN93 .... 1:12.19

† Engine Oil Viscosity Classification—SAE J300 MAR93 . 1:12.16

Fuel and Lubricant Tanks for Motorcycles—SAE J1241 . 3:37.11

Full Flow Lubricating Oil Filters Multipass Method for Evaluating Filtration Performance (A)—SAE J1858 JUN88 .... 2:26.248

† International Tests and Specifications for Automotive Engine Oils—SAE J2227 MAY93 .... 1:12.10

Lubricants for Two-Stroke-Cycle Gasoline Engines—SAE J1510 OCT88 .... 1:12.32

Lubricating Oil, Aircraft Piston Engine (Ashless Dispersant) (A)—SAE J1899 JUN91 .... 1:12.98

Lubricating Oil, Aircraft Piston Engine (Nondispersant Mineral Oil) (A)—SAE J1966 JUN91 .... 1:12.110

Physical and Chemical Properties of Engine Oils (A)—SAE J357 JUN91 .... 1:12.24

Standard Oil Filter Test Oil—SAE J1260 APR89 ...... 2:26.350

The Automotive Lubricant Performance and Service Classification Maintenance Procedure (A)—SAE J1146 JUN86 .... 1:12.87

Two-Stroke-Cycle Engine Oil Miscibility/Fluidity Classification—SAE J1536 OCT88 .... 1:12.33

† Two-Stroke-Cycle Gasoline Engine Lubricants Performance and Service Classification (A)—SAE J2116 JUN93 .... 1:12.33

Off-Road Vehicles
Oil Change System for Quick Service of Off-Road Self-Propelled Work Machines (A)—SAE J1069 JUN81 .... 3:40.102

Synthetic
Physical and Chemical Properties of Engine Oils (A)—SAE J357 JUN91 .... 1:12.24

Transmission and Axle
Axle and Manual Transmission Lubricant Viscosity Classification—SAE J306 OCT91 .... 1:12.37

Axle and Manual Transmission Lubricants—SAE J308 JUN89 .... 1:12.36

Fluid for Passenger Car Type Automatic Transmissions—SAE J311 APR86 .... 1:12.44

Powershift Transmission Fluid Classification—SAE J1285 JAN85 .... 1:12.45

The Automotive Lubricant Performance and Service Classification Maintenance Procedure (A)—SAE J1146 JUN86 .... 1:12.87

MACHINABILITY
Copper Alloys
Wrought and Cast Copper Alloys—SAE J461 SEP81 ... 1:10.45

Steel
Estimated Mechanical Properties and Machinability of Steel Bars—SAE J1397 MAY92 .... 1:2.26

High Strength Carbon and Alloy Die Drawn Steels—SAE J935 JUN90 .... 1:2.04

MACHINE VISION
Recommended Practice for Bar-Coded Vehicle Identification Number Label (A)—SAE J1877 MAY88 ... 3:42.37

MACROSTRAIN
Definitions for Macrostrain and Microstrain (A)—SAE J932 AUG85 .... 1:3.81

MAGNESIUM ALLOYS
See also: CHEMICAL COMPOSITION, MECHANICAL PROPERTIES, PHYSICAL PROPERTIES

Magnesium Alloys—SAE J464 JAN89 .... 1:10.120

Magnesium Casting Alloys—SAE J465 JAN89 .... 1:10.120

Magnesium Wrought Alloys—SAE J466 DEC89 ...... 1:10.125

MAGNETIC PARTICLE INSPECTION
See also: NONDESTRUCTIVE TESTS

† Cleanliness Rating of Steels by the Magnetic Particle Method—SAE J421 MAY93 .... 1:3.57

Detection of Surface Imperfections in Ferrous Rods, Bars, Tubes, and Wires—SAE J349 FEB91 .... 1:3.63

Magnetic Particle Inspection—SAE J420 MAR91 ...... 1:3.70

Nondestructive Tests (A)—SAE J358 FEB91 .... 1:3.67

MAINTENANCE
See also: REPAIR, SERVICEABILITY

Brakes
Service Maintenance of SAE J1703 Brake Fluids in Motor Vehicle Brake Systems—SAE J1707 NOV91 .... 2:25.53

---
* Technical Revision † New Document (D)—DODISS Adopted (A)—ANSI Recognized

Subject Index

MAINTENANCE (continued)
Construction and Industrial Equipment
Engineering Design Serviceability Guidelines—Construction and Industrial Machinery—Maintainability Index—Off-Road Work Machines (A)—SAE J817/2 MAR91 .... 3:40.522
Lift Arm Support Device for Loaders—SAE J38 AUG91 . 3:40.353
Maintenance Interval Chart (A)—SAE J753 MAY91 .... 3:40.101
† Personnel Protection for General Purpose Industrial Machines (A)—SAE J98 NOV92 .... 3:40.472
Technical Publications for Off-Road Work Machines (A)—SAE J920 SEP85 .... 3:40.526

Off-Road Vehicles
Maintenance Interval Chart (A)—SAE J753 MAY91 .... 3:40.101

Preventive
† Principles of Engine Cooling Systems, Components and Maintenance—SAE HS-40 JAN91 .... Available as a separate publication (See Related Technical Reports Section)

Service Technicians
Assessing Technician Training (A)—SAE J2018 JUL89 . 3:42.50
Developing Technician Training (A)—SAE J2017 AUG89 3:42.48

Symbols
Symbols and Color Codes for Maintenance Instructions, Container and Filler Identification (A)—SAE J223 APR80 3:40.104

MAINTENANCE VEHICLES
Lamps
360 Degree Warning Devices for Authorized Emergency, Maintenance, and Service Vehicles—SAE J845 MAR92 . 2:24.185
Flashing Warning Lamps for Authorized Emergency, Maintenance and Service Vehicles (A)—SAE J595 JAN90 .... 2:24.183
Gaseous Discharge Warning Lamp for Authorized Emergency, Maintenance, and Service Vehicles (A)—SAE J1318 APR86 .... 2:24.188
† SAE Ground Vehicle Lighting Standards Manual—SAE HS-34 JAN93 .... Available as a separate publication (See Related Technical Reports Section)

MANIKINS
Guidelines for Evaluating Out-of-Position Vehicle Occupant Interactions with Deploying Airbags (A)—SAE J1980 NOV90 .... 3:33.33
Test Device Head Contact Duration Analysis—SAE J2052 MAR90 .... 3:34.276

Impact Response
*Guidelines for Evaluating Child Restraint System Interactions with Deploying Airbags—SAE J2189 FEB93 3:33.20
Human Mechanical Response Characteristics—SAE J1460 MAR85 .... 3:34.292

MARINE ENGINES
See also: MARINE EQUIPMENT
Small Craft—Marine Propulsion Engine and Systems—Power Measurements and Declarations—SAE J1228 NOV91 .... 3:41.12
Thrust Test Device—SAE J1286 APR87 .... 3:41.05

Cables
Marine Control Cable Connection—Engine Clutch Lever (A)—SAE J960 JUN91 .... 3:41.04
Marine Control Cable Connection—Engine Throttle Lever (A)—SAE J961 JUN91 .... 3:41.05
Marine Push-Pull Control Cables—SAE J917 JUN80 ... 3:41.03

Carburetors
† Devices Providing Backfire Flame Control for Gasoline Engines in Marine Applications—SAE J1928 APR93 ... 3:41.44
† Marine Carburetors and Fuel Injection Throttle Bodies—SAE J1223 JUN93 .... 3:41.28

Distributors
Blade Type Electric Fuses (A)—SAE J1284 APR88 .... 3:41.40

Electric Equipment
External Ignition Protection of Marine Electrical Devices—SAE J1171 JAN86 .... 3:41.09
Marine Circuit Breakers (A)—SAE J1428 MAY85 ..... 3:41.20

Flame Arresters
† Devices Providing Backfire Flame Control for Gasoline Engines in Marine Applications—SAE J1928 APR93 ... 3:41.44

Hoses
Hydraulic Hose Fittings for Marine Applications (A)—SAE J1475 JUN84 .... 3:41.39

Ignition Systems
High Tension Ignition Cable Assemblies—Marine (A)—SAE J1191 JAN86 .... 3:41.26
Ignition Distributors—Marine—SAE J1294 JAN86 ..... 3:41.19

Throttles
† Marine Carburetors and Fuel Injection Throttle Bodies—SAE J1223 JUN93 .... 3:41.28

Wiring Systems
Marine Engine Wiring (A)—SAE J378 JAN88 .... 3:41.22

MARINE EQUIPMENT
See also: MARINE ENGINES
† Hose and Hose Assemblies for Marine Applications—SAE J1942 JUN93 .... 2:22.156
Marine Electrical Switches—SAE J1320 APR87 .... 3:41.28
† Marine Fuel Hoses (A)—SAE J1527 FEB93 .... 3:41.16
Marine Propeller-Shaft Couplings (A)—SAE J756 AUG87 3:41.01
Marine Propeller-Shaft Ends and Hubs—SAE J755 JUN80 .... 3:41.06

MASTER CYLINDERS
See: BRAKE CYLINDERS

MATERIALS
See: SPECIFIC MATERIALS

MEASUREMENT
Acoustics—Measurement of Airborne Noise Emitted by Earthmoving Machinery—Operator's Position—Stationary Testing Condition (A)—SAE J2103 JUN91 .... 1:14.78
Acoustics—Measurement of Airborne Noise Emitted by Construction Equipment Intended for Outdoor Use—Method for Determining Compliance with Noise Limits (A)—SAE J2101 JUN91 .... 1:14.68
Acoustics—Measurement of Airborne Noise Emitted by Earthmoving Machinery—Method for Determining Compliance with Limits for Exterior Noise—Stationary Test Condition (A)—SAE J2102 JUN91 .... 1:14.74
Acoustics—Measurement of Exterior Noise Emitted by Earthmoving Machinery—Dynamic Test Conditions (A)—SAE J2104 JUN91 .... 1:14.81
Acoustics—Measurement of Noise Emitted by Earthmoving Machinery at the Operator's Position—Simulated Work Cycle Test Conditions (A)—SAE J2105 JUN91 .... 1:14.87
Chemical Methods for the Measurement of Nonregulated Diesel Emissions (A)—SAE J1936 OCT89 .... 1:13.38
Constant Volume Sampler System for Exhaust Emissions Measurement—SAE J1094 JUN92 .... 1:13.117
Cooling Flow Measurement Techniques—SAE J2082 JUN92 .... 3:28.46
Detection of Surface Imperfections in Ferrous Rods, Bars, Tubes, and Wires—SAE J349 FEB91 .... 1:3.63
Devices for Use in Defining and Measuring Vehicle Seating Accommodation—SAE J826 JUN92 .... 3:34.90
Engine Weight, Dimensions, Center of Gravity, and Moment of Inertia—SAE J2038 APR92 .... 2:26.02
*Laboratory Measurement of the Composite Vibration Damping Properties of Materials on a Supporting Steel Bar—SAE J1637 FEB93 .... 1:2.30
Lip Force Measurement, Radial Lip Seals (A)—SAE J1901 SEP90 .... 3:29.135

* Technical Revision   † New Document   (D)—DODISS Adopted   (A)—ANSI Recognized

Subject Index

MEASUREMENT (continued)

Manual Transmission and Transaxle Efficiency and Parasitic Loss Measurement—SAE **J1540** APR92 ..... 3:29.61

† Measurement of Exhaust Sound Levels of Stationary Motorcycles—SAE **J1287** JUN93 ..... 1:14.05

Measurement of Fuel Evaporative Emissions from Gasoline Powered Passenger Cars and Light Trucks Using the Enclosure Technique (A)—SAE **J171** APR91 ..... 1:13.95

† Measurement of Intake Air or Exhaust Gas Flow of Diesel Engines (A)—SAE **J244** AUG92 ..... 1:13.11

Measurement of Light Vehicle Exhaust Sound Level Under Stationary Conditions—SAE **J1169** MAR92 ..... 1:14.21

Measurement of Light Vehicle Stationary Exhaust System Sound Level Engine Speed Sweep Method—SAE **J1492** MAR92 ..... 1:14.24

Measurement of Noise Emitted by Accelerating Highway Vehicles—SAE **J1470** MAR92 ..... 1:14.32

Methane Measurement Using Gas Chromatography—SAE **J1151** DEC91 ..... 1:13.170

Preparation of SAE Technical Reports—SAE **TSB 002** JUN92 ..... 1:A.05

† Procedure for Measuring Bore and Face Runout of Flywheels, Flywheel Housings and Flywheel Housing Adapters—SAE **J1033** APR93 ..... 3:40.245

Road Load Measurement and Dynamometer Simulation Using Coastdown Techniques—SAE **J1263** JUN91 ..... 2:26.470

Rules for SAE Use of SI (Metric) Units—SAE **TSB 003** JUN92 ..... 1:A.08

* SAE Surface Vehicle Emissions Standards Manual—SAE **HS-2400** FEB93 ..... Available as a separate publication (See Related Technical Reports Section)

Shoreline Sound Level Measurement Procedure—SAE **J1970** DEC91 ..... 1:14.40

Small Craft—Marine Propulsion Engine and Systems—Power Measurements and Declarations—SAE **J1228** NOV91 ..... 3:41.12

† Sound Measurement—Construction Site (A)—SAE **J1075** JUN93 ..... 1:14.99

† Sound Power Level Measurements of Earthmoving Machinery - Static and In-Place Dynamic Methods—SAE **J1805** APR93 ..... 1:14.90

Stationary Sound Level Measurement Procedure for Pleasure Motorboats—SAE **J2005** DEC91 ..... 1:14.41

† Test Method for Measuring Thickness of Resilient Insulating Paddings—SAE **J1355** MAY93 ..... 1:11.170

Test Procedure for Determining Reflectivity of Rear View Mirrors—SAE **J964** JUN92 ..... 3:34.107

† Test Procedure for Measuring Hydraulic Lift Capacity on Agricultural Tractors Equipped with Three-Point Hitch—SAE **J283** JUN93 ..... 3:39.41

Aerodynamics

* Aerodynamic Testing of Road Vehicles--Testing Methods and Procedures—SAE **J2084** JAN93 ..... 3:28.61

Piston Rings

† Internal Combustion Engines—Piston Rings—Inspection Measuring Principles—SAE **J1589** OCT92 ..... 2:26.52

† Internal Combustion Engines—Piston Rings—Keystone Rings—SAE **J2000** OCT92 ..... 2:26.100

† Internal Combustion Engines—Piston Rings—Oil Control Rings—SAE **J2002** OCT92 ..... 2:26.117

† Internal Combustion Engines—Piston Rings—Quality Requirements—SAE **J1996** OCT92 ..... 2:26.74

† Internal Combustion Engines—Piston Rings—Rectangular Rings—SAE **J1997** OCT92 ..... 2:26.79

† Internal Combustion Engines—Piston Rings—Scraper Rings—SAE **J1999** OCT92 ..... 2:26.90

MECHANICAL PROPERTIES

General Qualification and Production Acceptance Criteria for Integrated Circuits in Automotive Applications—SAE **J1879** OCT88 ..... 2:23.583

Use of Terms Yield Strength and Yield Point (A)—SAE **J450** JUN91 ..... 1:9.01

Aluminum Alloys

Bearing and Bushing Alloys—SAE **J459** OCT91 ..... 1:10.43

Chemical Compositions, Mechanical Property Limits, and Dimensional Tolerances of SAE Wrought Aluminum Alloys (A)—SAE **J457** FEB91 ..... 1:10.38

General Data on Wrought Aluminum Alloys (A)—SAE **J454** FEB91 ..... 1:10.26

General Information—Chemical Compositions, Mechanical and Physical Properties of SAE Aluminum Casting Alloys—SAE **J452** JAN89 ..... 1:10.06

Technical Report on Fatigue Properties—SAE **J1099** 1:3.81

Wrought Aluminum Applications Guidelines—SAE **J1434** JAN89 ..... 1:10.15

Bolts

Mechanical and Material Requirements for Wheel Bolts—SAE **J1102** 1:4.18

Cobalt Alloys

Engine Poppet Valve Information Report—SAE **J775** JAN88 ..... 1:1.165

Special Purpose Alloys ("Superalloys")—SAE **J467b** 1:10.128

Copper Alloys

Bearing and Bushing Alloys—SAE **J459** OCT91 ..... 1:10.43

Cast Copper Alloys—SAE **J462** SEP81 ..... 1:10.78

Manual on Design and Application of Helical and Spiral Springs—SAE **HS-J795** APR90 ..... Available as a separate publication (See Related Technical Reports Section)

Wrought and Cast Copper Alloys—SAE **J461** SEP81 ..... 1:10.45

Wrought Copper and Copper Alloys—SAE **J463** SEP81 ..... 1:10.81

Copper Tubing

Metallic Air Brake System Tubing and Pipe (A)—SAE **J1149** JUN91 ..... 2:22.219

Seamless Copper Tube (A)—SAE **J528** JUN91 ..... 2:22.218

Copper-Nickel Tubing

* Seamless Copper-Nickel 90-10 Tubing—SAE **J1650** MAR93 ..... 2:22.211

Elastomers

* Recommended Guidelines for Load/Deformation Testing of Elastomeric Components—SAE **J1636** FEB93 ..... 1:11.46

† Thermoplastic Elastomer Classification System—SAE **J3000** JUN93 ..... 1:11.27

Fasteners

Mechanical and Material Requirements for Externally Threaded Fasteners (A)—SAE **J429** AUG83 ..... 1:4.01

Mechanical and Material Requirements for Metric Externally Threaded Steel Fasteners (A)—SAE **J1199** SEP83 ..... 1:4.06

Felt

Felts—Wool and Part Wool (A)—SAE **J314** MAY81 ..... 1:11.183

Gasket Materials

Standard Classification System for Nonmetallic Automotive Gasket Materials (A)—SAE **J90** JUN90 ..... 1:11.197

Iron Alloys

Automotive Ductile (Nodular) Iron Castings (A)—SAE **J434** JUN86 ..... 1:6.09

† Automotive Gray Iron Castings—SAE **J431** MAR93 ..... 1:6.01

Automotive Malleable Iron Castings (A)—SAE **J158** JUN86 ..... 1:6.07

Elevated Temperature Properties of Cast Irons—SAE **J125** MAY88 ..... 1:6.12

Engine Poppet Valve Information Report—SAE **J775** JAN88 ..... 1:1.165

---

* Technical Revision   † New Document   (D)—DODISS Adopted   (A)—ANSI Recognized

MECHANICAL PROPERTIES (continued)
Iron Alloys (continued)
† Internal Combustion Engines—Piston Rings—Material Specifications (A)—SAE **J1590** OCT92 .... **2:26.72**
Special Purpose Alloys ("Superalloys")—SAE **J467b** ... **1:10.128**

Lead Alloys
Bearing and Bushing Alloys—SAE **J459** OCT91 ...... **1:10.43**

Magnesium Alloys
Magnesium Casting Alloys—SAE **J465** JAN89 .... **1:10.120**
Magnesium Wrought Alloys—SAE **J466** DEC89 ...... **1:10.125**

Nickel Alloys
Engine Poppet Valve Information Report—SAE **J775** JAN88 .... **1:1.165**
Manual on Design and Application of Helical and Spiral Springs—SAE **HS-J795** APR90 .... Available as a separate publication (See Related Technical Reports Section)
Special Purpose Alloys ("Superalloys")—SAE **J467b** ... **1:10.128**
Wrought Nickel and Nickel-Related Alloys—SAE **J470c** . **1:10.140**

Nylon Tubing
Metric Nonmetallic Air Brake System Tubing (A)—SAE **J1394** APR91 .... **2:22.227**
Nonmetallic Air Brake System Tubing (A)—SAE **J844** JUN90 .... **2:22.221**

Plastics
Automotive Plastic Parts Specification (A)—SAE **J1345** FEB82 .... **1:11.148**
* Classification System for Automotive Polyamide (PA) Plastics—SAE **J1639** MAR93 .... **1:11.145**
* Materials for Plastic Check Valves for Vacuum Booster System—SAE **J1875** JUN93 .... **2:25.40**
* Materials for Plastic Pistons for Hydraulic Disc Brake Cylinders—SAE **J1568** JUN93 .... **2:25.18**
Plastic Dust Shield for Hydraulic Disc Brakes (A)—SAE **J1876** MAR90 .... **2:25.67**

Rubber
† Classification System for Rubber Materials—SAE **J200** MAY93 .... **1:11.01**

Screws
Mechanical and Quality Requirements for Machine Screws—SAE **J82** JUN79 .... **1:4.14**
Mechanical and Quality Requirements for Tapping Screws—SAE **J933** JUN79 .... **1:4.19**
Steel Self-Drilling Tapping Screws—SAE **J78** JUN79 ... **1:4.21**
Thread Rolling Screws—SAE **J81** JUN79 .... **1:4.26**

Sintered Powder Metals
Sintered Powder Metal Parts: Ferrous—SAE **J471d** **1:8.01**

Spring Materials
Manual on Design and Application of Helical and Spiral Springs—SAE **HS-J795** APR90 .... Available as a separate publication (See Related Technical Reports Section)

Manual on Design and Manufacture of Coned Disk Springs (Belleville Springs) and Spring Washers—SAE **HS-1582** .... Available as a separate publication (See Related Technical Reports Section)

Steel
Automotive Steel Castings—SAE **J435c** **1:6.17**
Case Hardenability of Carburized Steels (A)—SAE **J1975** JUN91 .... **1:2.15**
Engine Poppet Valve Information Report—SAE **J775** JAN88 .... **1:1.165**
Estimated Mechanical Properties and Machinability of Steel Bars—SAE **J1397** MAY92 .... **1:2.26**
Hardness Tests and Hardness Number Conversions—SAE **J417** DEC83 .... **1:3.04**

High Strength Carbon and Alloy Die Drawn Steels—SAE **J935** JUN90 .... **1:2.04**
High Strength, Hot Rolled Steel Plates, Bars and Shapes—SAE **J1442** DEC88 .... **1:1.162**
† High-Strength, Quenched, and Tempered Structural Steels—SAE **J368** MAR93 .... **1:1.164**
† Internal Combustion Engines—Piston Rings—Material Specifications (A)—SAE **J1590** OCT92 .... **2:26.72**
Mechanical and Chemical Requirements for Nonthreaded Fasteners—SAE **J430** **1:4.18**
Mechanical and Material Requirements for Steel Nuts—SAE **J995** JUN79 .... **1:4.15**
Mechanical Properties of Heat Treated Wrought Steels—SAE **J413** JUN90 .... **1:2.24**
Methods for Determining Plastic Deformation in Sheet Metal Stampings (A)—SAE **J863** JUN86 .... **1:3.17**
Properties of Galvanized Low Carbon Sheet Steels and Their Relation to Formability—SAE **J1852** SEP87 ..... **1:3.22**
Properties of Low Carbon Sheet Steel and Their Relationship to Formability—SAE **J877** JUN84 .... **1:3.20**
Selection and Heat Treatment of Tool and Die Steels—SAE **J437a** **1:7.02**
† Selection and Use of Steels—SAE **J401** NOV92 ...... **1:1.06**
Special Purpose Alloys ("Superalloys")—SAE **J467b** ... **1:10.128**
Steel, High Strength, Hot Rolled Sheet and Strip, Cold Rolled Sheet, and Coated Sheet—SAE **J1392** JUN84 ... **1:1.159**
Technical Report on Fatigue Properties—SAE **J1099** ... **1:3.81**

Steel Pipe
Metallic Air Brake System Tubing and Pipe (A)—SAE **J1149** JUN91 .... **2:22.219**

Steel Shot
Cast Steel Shot (A)—SAE **J827** MAR90 .... **1:8.09**
Specifications for Low Carbon Cast Steel Shot (A)—SAE **J2175** JUN91 .... **1:8.10**

Steel Springs and Wires
Hard Drawn Carbon Steel Valve Spring Quality Wire and Springs—SAE **J172** DEC88 .... **1:5.08**
Hard Drawn Mechanical Spring Wire and Springs—SAE **J113** DEC88 .... **1:5.04**
Manual on Design and Application of Helical and Spiral Springs—SAE **HS-J795** APR90 .... Available as a separate publication (See Related Technical Reports Section)

Music Steel Spring Wire and Springs—SAE **J178** DEC88 **1:5.10**
Oil Tempered Carbon Steel Spring Wire and Springs—SAE **J316** DEC88 .... **1:5.01**
Oil Tempered Carbon Steel Valve Spring Quality Wire and Springs—SAE **J351** DEC88 .... **1:5.02**
Oil Tempered Chromium-Vanadium Valve Spring Quality Wire and Springs—SAE **J132** DEC88 .... **1:5.03**
Oil Tempered Chromium—Silicon Alloy Steel Wire and Springs—SAE **J157** DEC88 .... **1:5.07**
Special Quality High Tensile, Hard Drawn Mechanical Spring Wire and Springs—SAE **J271** DEC88 .... **1:5.05**
Stainless Steel 17-7 PH Spring Wire and Springs—SAE **J217** DEC88 .... **1:5.08**
Stainless Steel, SAE 30302, Spring Wire and Springs—SAE **J230** DEC88 .... **1:5.09**

Steel Tubing
Brazed Double Wall Low Carbon Steel Tubing (A)—SAE **J527** JUN91 .... **2:22.209**
Diesel Engines—Steel Tubes for High Pressure Fuel Injection Pipes (Tubing)—SAE **J1958** APR89 .... **2:26.260**
Seamless Low Carbon Steel Tubing Annealed for Bending and Flaring (A)—SAE **J524** JUN91 .... **2:22.206**
Welded and Cold Drawn Low Carbon Steel Tubing Annealed for Bending and Flaring (A)—SAE **J525** JUN91 **2:22.207**
Welded Flash Controlled Low Carbon Steel Tubing Normalized for Bending, Double Flaring, and Beading (A)—SAE **J356** JUN91 .... **2:22.213**

---

* Technical Revision   † New Document   (D)—DODISS Adopted   (A)—ANSI Recognized

Subject Index

MECHANICAL PROPERTIES (continued)
  Steel Tubing (continued)
    Welded Low Carbon Steel Tubing (A)—SAE **J526** JUN91   **2:22.208**
  Steel Wire Shot
    † Cut Wire Shot—SAE **J441** JUN93 .... **1:8.05**
  Steels
    *Categorization of Low Carbon Automotive Sheet Steel—SAE **J2096** JUL92 .... **1:3.25**
  Tin Alloys
    Bearing and Bushing Alloys—SAE **J459** OCT91 ...... **1:10.43**
  Titanium Alloys
    Special Purpose Alloys ("Superalloys")—SAE **J467b** .. **1:10.128**
  Zinc Alloys
    Zinc Die Casting Alloys—SAE **J469** JAN89 .... **1:10.139**

METAL POWDERS
  Sintered Powder Metal Parts: Ferrous—SAE **J471d** **1:8.01**

METALS
  See also: **Specific Metal (Aluminum, etc.)**
  Preferred Thicknesses for Uncoated Flat Metals (Thru 12 mm) (A)—SAE **J446** JUN84 .... **1:9.07**
  Surface Rolling and Other Methods for Mechanical Prestressing of Metals—SAE **J811** AUG81 .... **1:8.32**
  Cleaning
    SAE Manual on Blast Cleaning—SAE **J792a** AUG68 .... **1:8.14**
  Numbering System
    Fifth Edition Unified Numbering System Handbook for Metals and Alloys (D)—SAE **HS-1086** AUG93 .... Available as a separate publication *(See Related Technical Reports Section)*
    † Numbering Metals and Alloys—SAE **J1086** DEC92 .... **1:1.01**

METALS FINISHING
  See: **FINISHING**

METALS JOINING
  See: **JOINING**

METALWORKING
  See also: **FORGING, FORMABILITY, ROLLING**
  Aluminum Alloys
    Aluminum Alloys—Fundamentals—SAE **J451** JAN89 ... **1:10.01**

METHANE
  Exhaust Emissions
    Methane Measurement Using Gas Chromatography—SAE **J1151** DEC91 .... **1:13.170**

METRIC SYSTEM
  Dual Dimensioning (A)—SAE **J390** JUN82 .... **1:A.35**
  Preparation of SAE Technical Reports—SAE **TSB 002** JUN92 .... **1:A.05**
  Rules for SAE Use of SI (Metric) Units—SAE **TSB 003** JUN92 .... **1:A.08**
  Bearings
    Metric Spherical Rod Ends—SAE **J1259** JUN89 ...... **2:17.16**
  Belts and Pulleys
    † SI (Metric) Synchronous Belts and Pulleys—SAE **J1278** MAR93 .... **2:19.09**
  Bolts
    Automotive Hydraulic Brake System—Metric Banjo Bolt Connections—SAE **J1291** MAY89 .... **2:25.170**
    Metric Countersunk Holes for Cutting Edges and End Bits (A)—SAE **J1580** DEC89 .... **3:40.238**
  Brake Tubing
    Automotive Hydraulic Brake System—Metric Tube Connections—SAE **J1290** MAY89 .... **2:25.165**
    Metric Nonmetallic Air Brake System Tubing (A)—SAE **J1394** APR91 .... **2:22.227**
  Fasteners
    Decarburization in Hardened and Tempered Threaded Fasteners (A)—SAE **J121** AUG83 .... **1:4.44**
    Mechanical and Material Requirements for Metric Externally Threaded Steel Fasteners (A)—SAE **J1199** SEP83 .... **1:4.06**
    Metric Thread Rolling Screws—SAE **J1237** **1:4.29**
    Test Methods for Metric Threaded Fasteners (A)—SAE **J1216** **1:4.10**
  Lamps
    Preferred Conversion Values for Dimensions in Lighting—Inch-Pound Units/SI—SAE **J1322** JUN85 .. **2:24.01**
  Seals
    † Cast Iron Sealing Rings (Metric)—SAE **J1236** APR93 .. **3:29.156**
  Splines
    Metric Involute Splines and Inspection Standard—ANSI **B92.2M** .... Available as a separate publication *(See Related Technical Reports Section)*
    Part III: Inspection—Metric Module Involute Splines—ANSI **B92.2Ma** .... Available as a separate publication *(See Related Technical Reports Section)*
  Springs
    † Leaf Springs for Motor Vehicle Suspension—Made to Metric Units—SAE **J1123** NOV92 .... **2:20.01**

MICROCOMPUTERS
  See: **MICROPROCESSORS**

MICROPROCESSORS
  See also: **ELECTRONIC EQUIPMENT**
  Collision Detection Serial Data Communications Multiplex Bus—SAE **J1567** AUG87 .... **2:23.481**
  † Joint SAE/TMC Electronic Data Interchange Between Microcomputer Systems in Heavy-Duty Vehicle Applications—SAE **J1587** AUG92 .... **2:23.531**
  Serial Data Communications Between Microcomputer Systems in Heavy Duty Vehicle Applications (A)—SAE **J1708** OCT90 .... **2:23.553**

MICROSTRAIN
  Definitions for Macrostrain and Microstrain (A)—SAE **J932** AUG85 .... **1:3.81**

MINIBIKES
  Operator Controls and Displays on Motorcycles—SAE **J107** APR80 .... **3:37.05**

MINING EQUIPMENT
  Minimum Performance Criteria for Braking Systems for Specialized Rubber-Tired, Self-Propelled Underground Mining Machines (A)—SAE **J1329** JUL89 .... **3:40.295**
  Unauthorized Starting or Movement of Machines (A)—SAE **J1083** JUL85 .... **3:40.96**
  Operator Controls
    *Operator Controls--Off-Road Machines—SAE **J1814** FEB93 .... **3:40.414**
  Operator Protective Structures
    Labeling of ROPS and FOPS and OPS (A)—SAE **J1164** JAN91 .... **3:40.362**
    Minimum Performance Criteria for Falling Object Protective Structure (FOPS) (A)—SAE **J231** JAN81 .... **3:40.364**
    Performance Criteria for Rollover Protective Structures (ROPS) for Construction, Earthmoving, Forestry, and Mining Machines (A)—SAE **J1040** APR88 .... **3:40.356**
  Product Identification Number
    Product Identification Numbering System of Off-Road Work Machines (A)—SAE **J1360** SEP87 .... **3:40.04**

MIRRORS
  See also: **REAR VIEW MIRRORS**
  Measuring the Radius of Curvature of Convex Mirrors (A)—SAE **J1246** MAR90 .... **3:34.109**

MODELS
Wind Tunnel Testing
Aerodynamic Testing of Road Vehicles—SAE **J2071** MAR90 .... 3:28.18

SAE Wind Tunnel Test Procedure for Trucks and Buses (A)—SAE **J1252** JUL81 .... 3:36.61

MOPEDS
Operator Controls and Displays on Motorcycles—SAE **J107** APR80 .... 3:37.05

Terminology
A Dictionary of Terms for the Dynamics and Handling of Single Track Vehicles (Motorcycles, Mopeds, and Bicycles)—SAE **J1451** FEB85 .... 3:37.01

MOTORBOATS
See also: MARINE EQUIPMENT
Engines
Thrust Test Device—SAE **J1286** APR87 .... 3:41.05
Fuel Hoses
† Marine Fuel Hoses (A)—SAE **J1527** FEB93 .... 3:41.16
Fuses
Blade Type Electric Fuses (A)—SAE **J1284** APR88 .... 3:41.40
Noise
Exterior Sound Level Measurement Procedure for Pleasure Motorboats—SAE **J34** DEC91 .... 1:14.42

Operator Sound Level Exposure Assessment Procedure for Pleasure Motorboats (A)—SAE **J1281** MAR85 ..... 1:14.43

Shoreline Sound Level Measurement Procedure—SAE **J1970** DEC91 .... 1:14.40

Stationary Sound Level Measurement Procedure for Pleasure Motorboats—SAE **J2005** DEC91 .... 1:14.41

Symbols
Graphic Symbols for Boats (A)—SAE **J1449** FEB87 .... 3:41.31

MOTORCYCLES
*SAE Motorcycle Standards Manual—SAE **HS-2500** APR93 — Available as a separate publication (See Related Technical Reports Section)

Brakes
Brake System Road Test Code—Motorcycles (A)—SAE **J108** MAR87 .... 2:25.144

Service Brake System Performance Requirements—Motorcycles and Motor-Driven Cycles (A)—SAE **J109** MAR87 .... 2:25.146

† Surface Vehicle Brake Manual—SAE **HS-24** JUL92 .... Available as a separate publication (See Related Technical Reports Section)

Electric Equipment
Motorcycle and Motor Driven Cycle Electrical System—Maintenance of Design Voltage—SAE **J392** FEB92 .... 2:24.135

Fuel Tanks
Fuel and Lubricant Tanks for Motorcycles—SAE **J1241** . 3:37.11

Handling
A Dictionary of Terms for the Dynamics and Handling of Single Track Vehicles (Motorcycles, Mopeds, and Bicycles)—SAE **J1451** FEB85 .... 3:37.01

Motorcycle Bank Angle Measurement Procedure—SAE **J1168** MAY89 .... 3:37.07

Lamps
Motorcycle Auxiliary Front Lamps—SAE **J1306** JUN89 .. 2:24.122

Motorcycle Headlamps (A)—SAE **J584** DEC83 .... 2:24.120

Motorcycle Stop Lamp Switch—SAE **J1167** JUN89 ..... 3:37.07

Replaceable Motorcycle Headlamp Bulbs (A)—SAE **J1577** JUN91 .... 2:24.123

† SAE Ground Vehicle Lighting Standards Manual—SAE **HS-34** JAN93 .... Available as a separate publication (See Related Technical Reports Section)

Mirrors
Rear View Mirrors—Motorcycles—SAE **J268** MAY89 ... 3:37.06

Noise
Maximum Sound Level Potential for Motorcycles (A)—SAE **J47** JUN86 .... 1:14.01

† Measurement of Exhaust Sound Levels of Stationary Motorcycles—SAE **J1287** JUN93 .... 1:14.05

† Sound Levels for Motorcycles—SAE **J331** OCT92 ..... 1:14.03

Operator Controls
Operator Controls and Displays on Motorcycles—SAE **J107** APR80 .... 3:37.05

Parking Stability
Performance Requirements for Parking Stability of Motorcycles (A)—SAE **J1248** JUN91 .... 3:37.09

Test Procedure for Parking Stability of Motorcycles—SAE **J1101** JUN92 .... 3:37.08

Side Stands
Characterizing a Test Surface for Motorcycle Side Stand Retraction Performance Testing—SAE **J1846** JUN89 ... 3:37.17

Motorcycle Side Stand Retraction Performance Requirements—SAE **J1579** JUN89 .... 3:37.17

Motorcycle Side Stand Retraction Test Procedure—SAE **J1578** JUN89 .... 3:37.14

Terminology
A Dictionary of Terms for the Dynamics and Handling of Single Track Vehicles (Motorcycles, Mopeds, and Bicycles)—SAE **J1451** FEB85 .... 3:37.01

Definitions—Motorcycles—SAE **J213a** 3:37.01

Turn Indicators
Motorcycle Turn Signal Lamps (A)—SAE **J131** MAR83 . 2:24.136

MOUNTS
Qualifications for Four-Way Subbase Mounted Air Valves for Automotive Manufacturing Applications—SAE **J2051** JUN92 .... 3:43.21

Test for Dynamic Properties of Elastomeric Isolators—SAE **J1085a** 1:11.50

Accessory Drive
Mounting Flanges for Engine Accessory Drives (A)—SAE **J896** MAY83 .... 3:40.253

Alternator
Electrical Charging Systems for Construction and Industrial Machinery (A)—SAE **J180** MAY87 .... 3:40.270

Bumper
Inspection of Energy Absorbing Bumper Mounts (A)—SAE **J1571** MAY87 .... 3:42.25

Clutch
SAE Nodal Mount—SAE **J1134** FEB92 .... 3:36.112

Engine
Engine Foot Mounting (Front and Rear)—SAE **J616** APR90 .... 3:40.252

Engine Mountings (A)—SAE **J615** APR91 .... 2:26.234

Fan
Fan Hub Bolt Circles and Pilot Holes (A)—SAE **J635** JUL84 .... 2:26.414

Filter
Filter Base Mounting (A)—SAE **J363** FEB87 .... 2:26.246

Fuel Systems
Diesel Fuel Injection—End Mounting Flanges for Fuel Injection Pumps (A)—SAE **J626** JUN88 .... 2:26.297

Diesel Fuel Injector Assembly Type 27 (9.5 mm) (A)—SAE **J1984** NOV89 .... 2:26.308

Diesel Fuel Injector Assembly—Flange Mounted Types 5 and 6 (A)—SAE **J629** APR91 .... 2:26.304

Hydraulic Systems
Hydraulic Pump and Motor Mounting and Drive Dimensions (A)—SAE **J744** JUL88 .... 3:40.116

Lamp
High Mounted Stop Lamps for Use on Vehicles 2032 mm or More in Overall Width—SAE **J1432** OCT88 .... 2:24.147

---
* Technical Revision † New Document (D)—DODISS Adopted (A)—ANSI Recognized

Subject Index

MOUNTS (continued)
  Lamp (continued)
    Mounting Brackets and Socket for Warning Lamp and Slow-Moving Vehicle (SMV) Identification Emblem (A)—SAE J725 MAR91 .... 3:39.65
  Starter
    Starting Motor Mountings (A)—SAE J542 JUN91 ...... 2:23.76
  Transmissions
    † Clearance Envelopes for Six-Bolt, Eight-Bolt, and Rear Truck Transmission Mounted Power Takeoffs (A)—SAE J772 DEC92 .... 3:29.33
    Engine Flywheel Housing and Mating Transmission Housing Flanges—SAE J617 MAY92 .... 3:40.247
    † Openings for Six- and Eight-Bolt Truck Transmission Mounted Power Take-Offs—SAE J704 DEC92 .... 3:29.41
  Wheel
    Fastener Hardware for Wheels for Demountable Rims (A)—SAE J1835 MAR90 .... 3:31.26
    † Wheel Mounting Elements for Industrial and Agricultural Disc Wheels (A)—SAE J714 APR93 .... 3:39.57

MOWERS
  See also: CONSTRUCTION AND INDUSTRIAL EQUIPMENT, LAWNMOWERS
  Safety
    Industrial Flail Mowers and Power Rakes (A)—SAE J1001 MAR87 .... 3:40.488
    Industrial Rotary Mowers (A)—SAE J232 DEC84 ..... 3:40.477
  Terminology
    Nomenclature—Industrial and Agricultural Mowers (A)—SAE J990 MAR87 .... 3:40.475

MULTIPLEXING
    A Vehicle Network Protocol with a Fault Tolerant Multiplex Signal Bus—SAE J1813 AUG87 .... 2:23.499
    Chrysler Sensor and Control (CSC) Bus Multiplexing Network for Class 'A' Applications—SAE J2058 JUN90 . 2:23.422
    Class A Application/Definition (A)—SAE J2057/1 JUN91 2:23.407
    *Class A Multiplexing Architecture Strategies—SAE J2057/4 JUN93 .... 2:23.415
    *Class A Multiplexing Sensors—SAE J2057/3 JUN93 ... 2:23.413
    Class B Data Communication Network Interface—SAE J1850 AUG91 .... 2:23.268
    *Class C Application Requirement Considerations—SAE J2056/1 JUN93 .... 2:23.366
    Collision Detection Serial Data Communications Multiplex Bus—SAE J1567 AUG87 .... 2:23.481
    Glossary of Vehicle Networks for Multiplexing and Data Communications (A)—SAE J1213/1 JUN91 .... 2:23.19
    Selection of Transmission Media (A)—SAE J2056/3 JUN91 .... 2:23.391
    *Survey of Known Protocols—SAE J2056/2 APR93 ..... 2:23.372

NECK
  Impact Tolerance
    Human Mechanical Response Characteristics—SAE J1460 MAR85 .... 3:34.292
    Human Tolerance to Impact Conditions as Related to Motor Vehicle Design—SAE J885 JUL86 .... 3:34.306

NETWORK
    A Vehicle Network Protocol with a Fault Tolerant Multiplex Signal Bus—SAE J1813 AUG87 .... 2:23.499
    Chrysler Sensor and Control (CSC) Bus Multiplexing Network for Class 'A' Applications—SAE J2058 JUN90 . 2:23.422
    Class A Application/Definition (A)—SAE J2057/1 JUN91 2:23.407
    *Class A Multiplexing Architecture Strategies—SAE J2057/4 JUN93 .... 2:23.415
    *Class A Multiplexing Sensors—SAE J2057/3 JUN93 ... 2:23.413
    Class B Data Communication Network Interface—SAE J1850 AUG91 .... 2:23.268
    *Class B Data Communication Network Messages Part 2: Data Parameter Definitions—SAE J2178/2 JUN93 ..... 2:23.345
    Class B Data Communication Network Messages: Detailed Header Formats and Physical Address Assignments—SAE J2178/1 JUN92 .... 2:23.334
    Controller Area Network (CAN), An In-Vehicle Serial Communication Protocol—SAE J1583 MAR90 .... 2:23.486
    Glossary of Vehicle Networks for Multiplexing and Data Communications (A)—SAE J1213/1 JUN91 .... 2:23.19
    Selection of Transmission Media (A)—SAE J2056/3 JUN91 .... 2:23.391
    Serial Data Communications Between Microcomputer Systems in Heavy Duty Vehicle Applications (A)—SAE J1708 OCT90 .... 2:23.553
    *Survey of Known Protocols—SAE J2056/2 APR93 ..... 2:23.372
    Token Slot Network for Automotive Control (A)—SAE J2106 APR91 .... 2:23.431

NICKEL
  Coatings
    Electroplating of Nickel and Chromium on Metal Parts—Automotive Ornamentation and Hardware—SAE J207 FEB85 .... 1:10.150

NICKEL ALLOYS
  See also: CHEMICAL COMPOSITION, MECHANICAL PROPERTIES, PHYSICAL PROPERTIES
    Engine Poppet Valve Information Report—SAE J775 JAN88 .... 1:1.165
    Special Purpose Alloys ("Superalloys")—SAE J467b ... 1:10.128
    Wrought Nickel and Nickel-Related Alloys—SAE J470c . 1:10.140

NITROGEN OXIDES
  Exhaust Emissions
    Determination of Emissions from Gas Turbine Powered Light Duty Surface Vehicles (A)—SAE J1130 MAY83 .. 1:13.156
    Diesel Engine Emission Measurement Procedure (A)—SAE J1003 JUN90 .... 1:13.33
    Instrumentation and Techniques for Exhaust Gas Emissions Measurement (A)—SAE J254 AUG84 ...... 1:13.104
    Measurement of Carbon Dioxide, Carbon Monoxide, and Oxides of Nitrogen in Diesel Exhaust—SAE J177 APR82 1:13.01

NOISE
  See also: ACOUSTICS
    Laboratory Measurement of the Airborne Sound Barrier Performance of Automotive Materials and Assemblies (A)—SAE J1400 MAY90 .... 1:11.174
    Qualifying a Sound Data Acquisition System (D)—SAE J184 AUG87 .... 1:14.09
  Agricultural Equipment
    Sound Measurement—Self-Propelled Agricultural Equipment—Exterior (A)—SAE J1008 JAN87 .... 1:14.113
  Automobile
    Maximum Sound Level for Passenger Cars and Light Trucks (A)—SAE J1030 FEB87 .... 1:14.17
    Measurement of Light Vehicle Exhaust Sound Level Under Stationary Conditions—SAE J1169 MAR92 .... 1:14.21
    Measurement of Light Vehicle Stationary Exhaust System Sound Level Engine Speed Sweep Method—SAE J1492 MAR92 .... 1:14.24
    Measurement of Noise Emitted by Accelerating Highway Vehicles—SAE J1470 MAR92 .... 1:14.32
    Sound Level for Passenger Cars and Light Trucks (D)—SAE J986 OCT88 .... 1:14.14
  Construction and Industrial Equipment
    Acoustics—Measurement of Airborne Noise Emitted by Construction Equipment Intended for Outdoor Use—Method for Determining Compliance with Noise Limits (A)—SAE J2101 JUN91 .... 1:14.68
    † Sound Measurement—Construction Site (A)—SAE J1075 JUN93 .... 1:14.99
    Sound Measurement—Off-Road Self-Propelled Work Machines—Operator—Work Cycle (A)—SAE J1166 MAY90 .... 1:14.55

---

* Technical Revision   † New Document   (D)—DODISS Adopted   (A)—ANSI Recognized

Subject Index

NOISE (continued)
Earthmoving Equipment
Acoustics—Measurement of Airborne Noise Emitted by Earthmoving Machinery—Operator's Position—Stationary Testing Condition (A)—SAE J2103 JUN91 .... 1:14.78

Acoustics—Measurement of Airborne Noise Emitted by Earthmoving Machinery—Method for Determining Compliance with Limits for Exterior Noise—Stationary Test Condition (A)—SAE J2102 JUN91 .... 1:14.74

Acoustics—Measurement of Exterior Noise Emitted by Earthmoving Machinery—Dynamic Test Conditions (A)—SAE J2104 JUN91 .... 1:14.81

Acoustics—Measurement of Noise Emitted by Earthmoving Machinery at the Operator's Position—Simulated Work Cycle Test Conditions (A)—SAE J2105 JUN91 .... 1:14.87

Sound Measurement—Earthmoving Machinery—Exterior (A) (D)—SAE J88 JUN86 .... 1:14.106

Sound Measurement—Earthmoving Machinery—Operator—Singular Type (A) (D)—SAE J919 JUN86 .... 1:14.53

Sound Measurement—Off-Road Self-Propelled Work Machines—Operator—Work Cycle (A)—SAE J1166 MAY90 .... 1:14.55

Sound Power Determination—Earthmoving Machinery—Static Condition (A)—SAE J1372 JUL88 .... 1:14.102

† Sound Power Level Measurements of Earthmoving Machinery - Static and In-Place Dynamic Methods—SAE J1805 APR93 .... 1:14.90

Engines
Engine Sound Level Measurement Procedure (A) (D)—SAE J1074 FEB87 .... 1:14.114

Measurement Procedure for Determination of Silencer Effectiveness in Reducing Engine Intake or Exhaust Sound Level (A)—SAE J1207 FEB87 .... 1:14.115

Lawnmower
Bystander Sound Level Measurement Procedure for Small Engine Powered Equipment (A)—SAE J1175 MAR85 .... 1:14.97

Motorboats
Exterior Sound Level Measurement Procedure for Pleasure Motorboats—SAE J34 DEC91 .... 1:14.42

Operator Sound Level Exposure Assessment Procedure for Pleasure Motorboats (A)—SAE J1281 MAR85 .... 1:14.43

Shoreline Sound Level Measurement Procedure—SAE J1970 DEC91 .... 1:14.40

Stationary Sound Level Measurement Procedure for Pleasure Motorboats—SAE J2005 DEC91 .... 1:14.41

Motorcycle
Maximum Sound Level Potential for Motorcycles (A)—SAE J47 JUN86 .... 1:14.01

† Measurement of Exhaust Sound Levels of Stationary Motorcycles—SAE J1287 JUN93 .... 1:14.05

† Sound Levels for Motorcycles—SAE J331 OCT92 .... 1:14.03

Passenger Compartment
Instrumentation for Measuring Acoustic Impulses Within Vehicles (A)—SAE J247 FEB87 .... 1:14.07

Measurement of Interior Sound Levels of Light Vehicles (A)—SAE J1477 JAN86 .... 1:14.19

Small Engine Powered Equipment
Bystander Sound Level Measurement Procedure for Small Engine Powered Equipment (A)—SAE J1175 MAR85 .... 1:14.97

Operator Ear Sound Level Measurement Procedure for Small Engine Powered Equipment (A)—SAE J1174 MAR85 .... 1:14.95

Surface Vehicle Sound Measurement Manual—SAE HS-184 NOV89 .... Available as a separate publication (See Related Technical Reports Section)

Snowmobile
Exterior Sound Level for Snowmobiles (A)—SAE J192 MAR85 .... 1:14.36

Operational Sound Level Measurement Procedure for Snow Vehicles (A)—SAE J1161 MAR83 .... 1:14.39

Operator Ear Sound Level Measurement Procedure for Snow Vehicles (A)—SAE J1160 MAR83 .... 1:14.37

Tires
Sound Level of Highway Truck Tires (A)—SAE J57 FEB87 .... 1:14.31

Subjective Rating Scale for Evaluation of Noise and Ride Comfort Characteristics Related to Motor Vehicle Tires—SAE J1060 .... 3:30.37

Trenching Equipment
Sound Measurement—Trenching Machines—SAE J1262 OCT90 .... 1:14.109

Trucks and Buses
Exterior Sound Level for Heavy Trucks and Buses (A) (D)—SAE J366 FEB87 .... 1:14.27

Maximum Sound Level for Passenger Cars and Light Trucks (A)—SAE J1030 FEB87 .... 1:14.17

† Measurement of Exterior Sound Levels for Heavy Trucks Under Stationary Conditions—SAE J1096 FEB93 .... 1:14.29

Measurement of Light Vehicle Exhaust Sound Level Under Stationary Conditions—SAE J1169 MAR92 .... 1:14.21

Measurement of Light Vehicle Stationary Exhaust System Sound Level Engine Speed Sweep Method—SAE J1492 MAR92 .... 1:14.24

Measurement of Noise Emitted by Accelerating Highway Vehicles—SAE J1470 MAR92 .... 1:14.32

Sound Level for Passenger Cars and Light Trucks (D)—SAE J986 OCT88 .... 1:14.14

Sound Level for Truck Cab Interior—SAE J336 OCT88 .... 1:14.29

NOMENCLATURE
See: TERMINOLOGY

NONDESTRUCTIVE TESTS
See also: EDDY CURRENT TESTING, ELECTROMAGNETIC TESTS, FLUOROSCOPIC INSPECTION, MAGNETIC PARTICLE INSPECTION, RADIATION INSPECTION, ULTRASONIC TESTS

Acoustic Emission Test Methods (A)—SAE J1242 MAR91 .... 1:3.80

Detection of Surface Imperfections in Ferrous Rods, Bars, Tubes, and Wires—SAE J349 FEB91 .... 1:3.63

Electromagnetic Testing by Eddy Current Methods—SAE J425 MAR91 .... 1:3.72

Infrared Testing—SAE J359 FEB91 .... 1:3.69

Leakage Testing—SAE J1267 DEC88 .... 1:3.73

Liquid Penetrant Test Methods—SAE J426 MAR91 .... 1:3.75

Magnetic Particle Inspection—SAE J420 MAR91 .... 1:3.70

Method for Determining Breakage Allowances for Sheet Steel (A)—SAE J424 JUN86 .... 1:3.66

Nondestructive Tests (A)—SAE J358 FEB91 .... 1:3.67

Penetrating Radiation Inspection (A)—SAE J427 MAR91 .... 1:3.76

† Specific Gravity of Brake Lining—SAE J380 FEB93 .... 2:25.89

Ultrasonic Inspection (A)—SAE J428 MAR91 .... 1:3.79

NOZZLES
Airflow Reference Standards (A)—SAE J228 JUN90 .... 2:26.372

† Measurement of Intake Air or Exhaust Gas Flow of Diesel Engines (A)—SAE J244 AUG92 .... 1:13.11

Fuel Injection
Diesel Fuel Injection Pump—Validation of Calibrating Nozzle Holder Assemblies (A)—SAE J1549 APR88 .... 2:26.321

Diesel Fuel Injector Assembly Type 27 (9.5 mm) (A)—SAE J1984 NOV89 .... 2:26.308

Diesel Fuel Injector Assembly Type 7 (9.5 mm) (A)—SAE J347 JUL88 .... 2:26.305

Diesel Fuel Injector Assembly—Flange Mounted Types 5 and 6 (A)—SAE J629 APR91 .... 2:26.304

---
* Technical Revision   † New Document   (D)—DODISS Adopted   (A)—ANSI Recognized

Subject Index

NOZZLES (continued)
  Fuel Injection (continued)
    Diesel Fuel Injector Assembly—Types 8, 9, 10, and 11 (A)—SAE **J265** APR91 .... 2:26.306
    Diesel Injection Pump Testing—Part 1: Calibrating Nozzle and Holder Assemblies (A)—SAE **J968/1** MAY91 ..... 2:26.309
    † Fuel Injection Equipment Nomenclature—SAE **J830** OCT92 .... 2:26.254
  Gasoline Dispenser
    † Gasoline Dispenser Nozzle Spouts (A)—SAE **J285** SEP92 .... 2:26.441

NUMBERING SYSTEMS
See also: CLASSIFICATION, ENGINE IDENTIFICATION NUMBER (EIN), PRODUCT IDENTIFICATION NUMBER, TRANSAXLE IDENTIFICATION NUMBER (TIN), TRANSMISSION IDENTIFICATION NUMBER (TIN), VEHICLE IDENTIFICATION NUMBER (VIN), WORLD MANUFACTURING SYSTEMS (WMI)
  Numbering System for Standard Drills, Standard Taps, and Reamers—SAE **HS-2100** NOV90 .... Available as a separate publication (See Related Technical Reports Section)
  Preparation of SAE Technical Reports—SAE **TSB 002** JUN92 .... 1:A.05
  Aluminum Alloys
    Alloy and Temper Designation Systems for Aluminum—SAE **J993** JAN89 .... 1:10.02
    General Information—Chemical Compositions, Mechanical and Physical Properties of SAE Aluminum Casting Alloys—SAE **J452** JAN89 .... 1:10.06
    Wrought Aluminum Applications Guidelines—SAE **J1434** JAN89 .... 1:10.15
  Cobolt Alloys
    Special Purpose Alloys ("Superalloys")—SAE **J467b** ... 1:10.128
  Diagnostics
    Diagnostic Trouble Code Definitions—SAE **J2012** MAR92 2:23.284
  Drills
    Numbering System for Standard Drills (A)—SAE **J2122** NOV90 .... 3:43.11
  Fiberboard
    Standard Classification System for Fiberboards (A)—SAE **J1323** SEP90 .... 1:11.161
  Fittings
    Coding Systems for Identification of Fluid Conductors and Connectors (D)—SAE **J846** JUN89 .... 2:22.01
  Gasket Materials
    Standard Classification System for Nonmetallic Automotive Gasket Materials (A)—SAE **J90** JUN90 .... 1:11.197
  Iron
    † Numbering System for Designating Grades of Cast Ferrous Materials—SAE **J859** OCT92 .... 1:6.01
    Special Purpose Alloys ("Superalloys")—SAE **J467b** ... 1:10.128
  Metals and Alloys
    Fifth Edition Unified Numbering System Handbook for Metals and Alloys (D)—SAE **HS-1086** OCT93 .... Available as a separate publication (See Related Technical Reports Section)
    † Numbering Metals and Alloys—SAE **J1086** DEC92 .... 1:1.01
  Nickel Alloys
    Special Purpose Alloys ("Superalloys")—SAE **J467b** ... 1:10.128
  Plastics
    Automotive Plastic Parts Specification (A)—SAE **J1345** FEB82 .... 1:11.148
    * Classification System for Automotive Polyamide (PA) Plastics—SAE **J1639** MAR93 .... 1:11.145
  Reamers
    Numbering System for Reamers (A)—SAE **J2124** NOV90 3:43.18
  Rubber
    † Classification System for Rubber Materials—SAE **J200** MAY93 .... 1:11.01
  Steels
    * Categorization of Low Carbon Automotive Sheet Steel—SAE **J2096** JUL92 .... 1:3.25
    Engine Poppet Valve Information Report—SAE **J775** JAN88 .... 1:1.165
    Potential Standard Steels—SAE **J1081** DEC88 .... 1:1.18
    SAE Numbering System for Wrought or Rolled Steel—SAE **J402** DEC88 .... 1:1.08
    Special Purpose Alloys ("Superalloys")—SAE **J467b** ... 1:10.128
  Superalloys
    Special Purpose Alloys ("Superalloys")—SAE **J467b** ... 1:10.128
  Taps
    Numbering System for Standard Taps (A)—SAE **J2123** NOV90 .... 3:43.15
  Undercoatings
    Vibration Damping Materials and Underbody Coatings (A)—SAE **J671** APR82 .... 1:11.111
  Vibration Damping Materials
    Vibration Damping Materials and Underbody Coatings (A)—SAE **J671** APR82 .... 1:11.111

NUTS
See also: FASTENERS
  Automotive Hydraulic Brake System—Metric Tube Connections—SAE **J1290** MAY89 .... 2:25.165
  Mechanical and Material Requirements for Steel Nuts—SAE **J995** JUN79 .... 1:4.15
  Nut and Conical Spring Washer Assemblies—SAE **J238** 2:16.20
  Push-On Spring Nuts—SAE **J892a** 2:16.13
  * Road Vehicles--Wheels for Commercial Vehicles and Multipurpose Passenger Vehicles--Fixing Nuts--Test Methods—SAE **J1965** FEB93 .... 3:36.107
  † Spring Nuts—SAE **J891** JUN93 .... 2:16.01
  Steel Stamped Nuts of One Pitch Thread Design—SAE **J1053** 2:16.14
  Surface Discontinuities on Nuts—SAE **J122a** 1:4.40
  Torque-Tension Test Procedure for Steel Threaded Fasteners—SAE **J174** 2:16.22

NYLON
  * Classification System for Automotive Polyamide (PA) Plastics—SAE **J1639** MAR93 .... 1:11.145
  Metric Nonmetallic Air Brake System Tubing (A)—SAE **J1394** APR91 .... 2:22.227
  Nonmetallic Air Brake System Tubing (A)—SAE **J844** JUN90 .... 2:22.221
  Performance Requirements for SAE J844 Nonmetallic Tubing and Fitting Assemblies Used in Automotive Air Brake Systems (A)—SAE **J1131** DEC87 .... 2:22.231

O-RINGS
  * Connections for General use and Fluid Power-Ports and Stud Ends with ISO 725 Trheads and O-Ring Sealing—Part 2: Heavy-Duty (S Series) Stud Ends—SAE **J1926/2 ISO 11** MAR93 .... 2:22.102
  * Connections for General use and Fluid Power-Ports and Stud Ends with ISO 725 Threads and O-Ring Sealing—Part 3: Light-Duty (L Series) Stud Ends—SAE **J1926/3 ISO 11** MAR93 .... 2:22.108
  † Connections for General Use and Fluid Power-Ports and Stud Ends with ISO 725 Threads and O-ring Sealing—Part 1: Threaded Port with O-Ring Seal in Truncated Housing—SAE **J1926/1 ISO 11** MAR93 .... 2:22.99
  † Fitting—O-Ring Face Seal—SAE **J1453** JUN93 ..... 2:22.256
  Fluid Conductors and Connectors—SAE **HS-150** JUN92 Available as a separate publication (See Related Technical Reports Section)
  Hydraulic O-Ring—SAE **J515** JUN92 .... 2:22.200
  † Hydraulic Tube Fittings—SAE **J514** JUN93 .... 2:22.59
  Rubber Rings for Automotive Applications—SAE **J120a** . 1:11.203

---

* Technical Revision   † New Document   (D)—DODISS Adopted   (A)—ANSI Recognized

Subject Index

OCCUPANT PROTECTION
See: OPERATOR PROTECTIVE STRUCTURES, RESTRAINT SYSTEMS, SAFETY BELTS

OCCUPANTS

Impact Kinematics

*Dolly Rollover Recommended Test Procedure—SAE
J2114 APR93 .... 3:34.283

Guidelines for Evaluating Out-of-Position Vehicle Occupant Interactions with Deploying Airbags (A)—SAE J1980 NOV90 .... 3:33.33

Human Mechanical Response Characteristics—SAE
J1460 MAR85 .... 3:34.292

OCTANE NUMBER

†Automotive Gasolines—SAE J312 JAN93 .... 1:12.47

ODOMETERS
See also: SPEEDOMETERS, TACHOMETERS

Electric Speedometer Specification—On Road (A)—SAE
J1226 FEB83 .... 2:21.09

Factors Affecting Accuracy of Mechanically Driven Automotive Speedometer and Odometers—SAE J862 JAN89 .... 2:21.01

Speedometers and Tachometers—Automotive—SAE
J678 DEC88 .... 2:21.03

ODORS

†Hot Odor Test for Insulation Materials—SAE J1351
JUN93 .... 1:11.172

OFF-ROAD VEHICLES
See also: AGRICULTURAL EQUIPMENT, BACKHOES, CONSTRUCTION AND INDUSTRIAL EQUIPMENT, DIPPERS, DUMPERS, EARTHMOVING EQUIPMENT, EXCAVATORS, FORESTRY EQUIPMENT, FORWARDERS, SCRAPERS, TRENCHING EQUIPMENT

Categories of Off-Road Self-Propelled Work Machines (A)—SAE J1116 JUN86 .... 3:40.01

Commercial Literature Specifications—Off-Road Work Machines (A)—SAE J1233 JAN85 .... 3:40.28

Definitions for Machine Availability (Off-Road Work Machines)—SAE J1032 APR87 .... 3:40.98

†Drain, Fill, and Level Plugs for Off-Road, Self-Propelled Work Machines—SAE J371 MAY93 .... 3:40.257

Engineering Design Serviceability Guidelines—Construction and Industrial Machinery—Maintainability Index—Off-Road Work Machines (A)—SAE J817/2 MAR91 .... 3:40.522

†Guidelines for Liquid Level Indicators—SAE J48 MAY93   3:40.103

Guidelines for Requests Received from Outside Sources for the ConAg Council to Originate or Review Technical Reports—SAE J2129 NOV90 .... 3:40.03

Identification Terminology of Loaders/Tractors with Forks and Rough Terrain Forklifts—SAE J1464 APR88 ...... 3:40.58

†Operator Enclosure Air Filter Element Test Procedure—SAE J1533 JUN93 .... 3:40.513

†Operator Enclosure Pressurization System Test Procedure—SAE J1012 JUN93 .... 3:39.67

Performance Test for Air-Conditioned, Heated, and Ventilated Off-Road Self-Propelled Work Machines (A)—SAE J1503 JUL86 .... 3:40.511

Shielding of Starter System Energization—SAE J1493 OCT91 .... 3:40.266

Technical Publications for Off-Road Work Machines (A)—SAE J920 SEP85 .... 3:40.526

Access Systems

Access Systems for Off-Road Machines (D)—SAE J185 JUN88 .... 3:40.383

†Minimum Service Access Dimensions for Off-Road Machines (A)—SAE J925 JUN93 .... 3:40.380

Air Conditioning

Design Guidelines for Air Conditioning Systems for Off-Road Operator Enclosures—SAE J169 MAR85 .... 3:40.510

Batteries

Storage Batteries for Off-Road Work Machines (A)—SAE J930 AUG84 .... 3:40.528

Brakes

Automotive and Off-Highway Air Brake Reservoir Performance and Identification Requirements (A)—SAE J10 OCT90 .... 3:36.132

Braking Performance—Roller Compactors (A)—SAE J1472 JUN87 .... 3:40.288

Braking Performance—Roller/Compactors (A)—SAE J1136 APR87 .... 3:40.300

†Surface Vehicle Brake Manual—SAE HS-24 JUL92 .... Available as a separate publication (See Related Technical Reports Section)

Dimensions

†Minimum Service Access Dimensions for Off-Road Machines (A)—SAE J925 JUN93 .... 3:40.380

Operator Space Envelope Dimensions for Off-Road Machines—SAE J154 JUN92 .... 3:40.419

Specification Definitions—Articulated Rubber-Tired Forwarder (A)—SAE J1823 JUN91 .... 3:40.89

Specification Definitions—Clam Bunk Skidder (A)—SAE J1824 JUN91 .... 3:40.81

Specification Definitions—Off-Road Work Machines (A)—SAE J1234 JAN85 .... 3:40.22

Electric Drives

Electrical Propulsion Control—Off-Road Dumpers (A)—SAE J1299 JAN91 .... 3:40.282

Electrical Propulsion Rotating Equipment—Off-Road Dumper (A)—SAE J1317 JUN82 .... 3:40.279

Electric Equipment

Electrical Charging Systems for Construction and Industrial Machinery (A)—SAE J180 MAY87 .... 3:40.270

Electrical Connector for Auxiliary Starting of Construction, Agricultural, and Off-Road Machinery (A)—SAE J1283 MAR86 .... 3:40.267

Electrical Grounding Practice (A)—SAE J1908 AUG90 . 3:40.265

*Electrical Indicating System Specification—SAE J1810 JAN93 .... 3:40.275

Electrical Wiring Systems for Construction, Agricultural and Off-Road Machines—SAE J821 MAY85 .... 3:40.264

Human Engineering

†Minimum Service Access Dimensions for Off-Road Machines (A)—SAE J925 JUN93 .... 3:40.380

Operator Space Envelope Dimensions for Off-Road Machines—SAE J154 JUN92 .... 3:40.419

Humand Engineering

*Operator Controls--Off-Road Machines—SAE J1814 FEB93 .... 3:40.414

Hydraulic Systems

Fire-Resistant Fluid Usage in Hydraulic Systems of Off-Road Work Machines—SAE J1447 MAY86 .... 3:40.126

Hydraulic Cylinder No-Load Friction Test (A)—SAE J1335 APR90 .... 3:40.346

Hydraulic Cylinder Rod Corrosion Test (A)—SAE J1333 MAR90 .... 3:40.343

Indicators

*Electrical Indicating System Specification—SAE J1810 JAN93 .... 3:40.275

Lubrication

Oil Change System for Quick Service of Off-Road Self-Propelled Work Machines (A)—SAE J1069 JUN81 .... 3:40.102

Maintenance

Engineering Design Serviceability Guidelines—Construction and Industrial Machinery—Serviceability Definitions—Off-Road Work Machines (A)—SAE J817/1 MAR91 .... 3:40.521

Maintenance Interval Chart (A)—SAE J753 MAY91 .... 3:40.101

---

* Technical Revision   † New Document   (D)—DODISS Adopted   (A)—ANSI Recognized

Subject Index

OFF-ROAD VEHICLES (continued)
Noise
Sound Measurement—Trenching Machines—SAE J1262 OCT90 .... 1:14.109

Operator Controls
Control Locations for Off-Road Work Machines (A)—SAE J898 OCT87 .... 3:40.404

*Electrical Indicating System Specification—SAE J1810 JAN93 .... 3:40.275

Graphical Symbols for Operator Controls and Displays on Off-Road Self-Propelled Work Machines—SAE J1362 JUN92 .... 3:40.423

*Operator Controls--Off-Road Machines—SAE J1814 FEB93 .... 3:40.414

Product Identification Number
Product Identification Numbering System of Off-Road Work Machines (A)—SAE J1360 SEP87 .... 3:40.04

Restraint Systems
†Operator Restraint System for Off-Road Work Machines—SAE J386 JUN93 .... 3:40.491

†Vehicle Occupant Restraint Systems and Components Standards Manual—SAE HS-13 MAY93 .... Available as a separate publication (See Related Technical Reports Section)

Retarders
†Retardation Capability of Off-Highway Dumpers and Scrapers—SAE J1430 FEB93 .... 3:40.298

Safety
Access Systems for Off-Road Machines (D)—SAE J185 JUN88 .... 3:40.383

Fan Guard for Off-Road Machines—SAE J1308 SEP85 . 3:40.257

Safety Signs (A)—SAE J115 JAN87 .... 3:40.494

Unauthorized Starting or Movement of Machines (A)—SAE J1083 JUL85 .... 3:40.96

Seats
Classification of Earthmoving Machines for Vibration Tests of Operator Seats (A)—SAE J1385 JUN83 ...... 3:40.398

Determining Seat Index Point (A)—SAE J1163 JUN91 .. 3:40.391

Dynamic Durability Testing of Seat Cushions for Off-Road Work Machines (A)—SAE J1454 JAN86 .... 3:40.389

Force-Deflection Measurements of Cushioned Components of Seats for Off-Road Work Machines—SAE J1051 DEC88 .... 3:40.388

Operator's Seat Dimensions for Off-Road Self-Propelled Work Machines—SAE J899 DEC88 .... 3:40.395

Steering Systems
Steering for Off-Road, Rubber-Tired Machines (A)—SAE J1511 OCT90 .... 3:40.517

Tires
Off-Road Tire and Rim Classification—Construction Machines (A)—SAE J751 APR86 .... 3:40.301

Off-Road Tire and Rim Classification—Forestry Machines (A)—SAE J1440 APR86 .... 3:40.310

Off-Road Tire and Rim Selection and Application (A)—SAE J1315 JAN91 .... 3:40.307

Revolutions per Mile and Static Loaded Radius for Off-Road Tires—SAE J1544 JAN88 .... 3:40.307

Ton Kilometer Per Hour Application (A)—SAE J1098 MAR91 .... 3:40.317

Ton-Kilometer Per Hour Test Procedure (A)—SAE J1015 MAR91 .... 3:40.314

Vibrations
†Measurement of Whole Body Vibration of the Seated Operator of Off-Highway Work Machines—SAE J1013 AUG92 .... 3:39.92

OFF-TRACKING
Turning Ability and Off Tracking—Motor Vehicles (A)—SAE J695 DEC89 .... 3:36.78

OIL COOLERS
See also: LUBRICANTS

†Application Testing of Oil-to-Air Oil Coolers for Cooling Performance—SAE J1468 MAY93 .... 2:26.398

Oil Cooler Nomenclature and Glossary (A)—SAE J1244 AUG88 .... 2:26.405

†Principles of Engine Cooling Systems, Components and Maintenance—SAE HS-40 JAN91 .... Available as a separate publication (See Related Technical Reports Section)

OIL FILTERS
Filter Base Mounting (A)—SAE J363 FEB87 .... 2:26.246

Full Flow Lubricating Oil Filters Multipass Method for Evaluating Filtration Performance (A)—SAE J1858 JUN88 .... 2:26.248

Hydraulic Power Circuit Filtration (D)—SAE J931 MAR86 3:40.137

Oil Filter Test Procedure—SAE HS-806 JUN83 .... Available as a separate publication (See Related Technical Reports Section)

Standard Oil Filter Test Oil—SAE J1260 APR89 ...... 2:26.350

OIL INDICATORS
Qualifications for Four-Way Subbase Mounted Air Valves for Automotive Manufacturing Applications—SAE J2051 JUN92 .... 3:43.21

OILS
See: LUBRICANTS

OPERATOR CONTROLS
Agricultural Equipment
Operator Controls for Agricultural Wheeled Tractors—SAE J841 DEC84 .... 3:39.70

*Operator Controls--Off-Road Machines—SAE J1814 FEB93 .... 3:40.414

Automobiles and Trucks
Automotive Adaptive Driver Controls, Manual (A)—SAE J1903 AUG90 .... 3:34.198

Design Criteria—Driver Hand Controls Location for Passenger Cars, Multi-Purpose Passenger Vehicles, and Trucks (10 000 GVW and Under)—SAE J1138 .... 3:34.191

Driver Hand Control Reach (A)—SAE J287 JUN88 .... 3:34.157

Location and Operation of Instruments and Controls in Motor Truck Cabs (A)—SAE J680 SEP88 .... 3:36.121

*Manual Controls for Mature Drivers—SAE J2119 JUN93 3:34.194

Semiautomatic Headlamp Beam Switching Devices (D)—SAE J565 JUN89 .... 2:24.09

Supplemental Information—Driver Hand Controls Location for Passenger Cars, Multi-Purpose Passenger Vehicles, and Trucks (10 000 GVW and Under)—SAE J1139 .... 3:34.193

Construction and Industrial Equipment
Crane and Cable Excavator Basic Operating Control Arrangements (A)—SAE J983 OCT80 .... 3:40.220

Hydraulic Excavator Operator Controls—SAE J1177 JUN88 .... 3:40.341

Operator Controls for Graders (A)—SAE J1071 JUN85 . 3:40.470

Operator Controls on Industrial Equipment—SAE J297 JUN85 .... 3:40.469

*Operator Controls--Off-Road Machines—SAE J1814 FEB93 .... 3:40.414

†Personnel Protection for General Purpose Industrial Machines (A)—SAE J98 NOV92 .... 3:40.472

Earthboring Machines
*Trenchless Equipment - (Horizontal Earthboring Machines) Operator Control Definitions—SAE J1611 JUN93 .... 3:40.21

Indicators
*Electrical Indicating System Specification—SAE J1810 JAN93 .... 3:40.275

---

* Technical Revision   † New Document   (D)—DODISS Adopted   (A)—ANSI Recognized

Subject Index

OPERATOR CONTROLS (continued)
  Mining Equipment
    *Operator Controls--Off-Road Machines—SAE **J1814**
    FEB93 .... 3:40.414
  Motorcycles
    Operator Controls and Displays on Motorcycles—SAE
    J107 APR80 .... 3:37.05
  Off-Road Vehicles
    Control Locations for Off-Road Work Machines (A)—SAE
    J898 OCT87 .... 3:40.404
    *Electrical Indicating System Specification—SAE **J1810**
    JAN93 .... 3:40.275
    Graphical Symbols for Operator Controls and Displays on
    Off-Road Self-Propelled Work Machines—SAE **J1362**
    JUN92 .... 3:40.423
  Off-road vehicles
    *Operator Controls--Off-Road Machines—SAE **J1814**
    FEB93 .... 3:40.414
  Snowmobiles
    Snowmobile Brake Control Systems (A)—SAE **J1282**
    OCT84 .... 3:38.09
  Symbols
    Graphic Symbols for Boats (A)—SAE **J1449** FEB87 ... 3:41.31
    Graphical Symbols for Operator Controls and Displays on
    Off-Road Self-Propelled Work Machines—SAE **J1362**
    JUN92 .... 3:40.423
    Symbols for Motor Vehicle Controls, Indicators and
    Tell-Tales—SAE **J1048** MAR80 .... 3:34.288
  Terminology
    *Trenchless Equipment - (Horizontal Earthboring
    Machines) Operator Control Definitions—SAE **J1611**
    JUN93 .... 3:40.21
  Transmissions
    † Automatic Transmissions—Manual Control Sequence—
    SAE **J915** APR93 .... 3:29.32

OPERATOR PROTECTIVE STRUCTURES
  Deflection Limiting Volume—ROPS/FOPS Laboratory
  Evaluation (A)—SAE **J397** APR88 .... 3:40.363
  Labeling of ROPS and FOPS and OPS (A)—SAE **J1164**
  JAN91 .... 3:40.362
  Minimum Performance Criteria for Falling Object
  Protective Structure (FOPS) (A)—SAE **J231** JAN81 .... 3:40.364
  † Operator Enclosure Air Filter Element Test Procedure—
  SAE **J1533** JUN93 .... 3:40.513
  † Operator Enclosure Pressurization System Test
  Procedure—SAE **J1012** JUN93 .... 3:39.67
  † Operator Protection for General-Purpose Industrial
  Machines—SAE **J1042** JUN93 .... 3:40.363
  Operator Protective Structure Performance Criteria for
  Certain Forestry Equipment (A)—SAE **J1084** APR80 ... 3:40.370
  † Operator Restraint System for Off-Road Work
  Machines—SAE **J386** JUN93 .... 3:40.491
  Operator Space Envelope Dimensions for Off-Road
  Machines—SAE **J154** JUN92 .... 3:40.419
  Overhead Protection for Agricultural Tractors—Test
  Procedures and Performance Requirements—SAE **J167**
  APR92 .... 3:39.89
  Performance Criteria for Falling Object Guards for
  Excavators—SAE **J1356** FEB88 .... 3:40.367
  Performance Criteria for Falling Object Protective
  Structure (FOPS) for Industrial Machines—SAE **J1043**
  SEP87 .... 3:40.365
  Performance Criteria for Rollover Protective Structures
  (ROPS) for Construction, Earthmoving, Forestry, and
  Mining Machines (A)—SAE **J1040** APR88 .... 3:40.356
  † Personnel Protection for General Purpose Industrial
  Machines (A)—SAE **J98** NOV92 .... 3:40.472
  Personnel Protection—Skid Steer Loaders (A)—SAE
  J1388 JUN85 .... 3:40.474
  Rear Underride Guard Test Procedure (A)—SAE **J260**
  JUN90 .... 3:34.287

  † Roll-Over Protective Structures (ROPS) for Wheeled
  Agricultural Tractors (A)—SAE **J2194** JUN93 .... 3:39.78
  Rollover Protective Structures (ROPS) for Wheeled
  Agricultural Tractors (A)—SAE **J1194** MAY89 .... 3:39.71
  Steel Products for Rollover Protective Structures (ROPS)
  and Falling Object Protective Structures (FOPS)—SAE
  J1119 DEC88 .... 3:40.371
  Steel
    † Selection and Use of Steels—SAE **J401** NOV92 ...... 1:1.06

OPTICAL SYSTEMS
  Recommended Practice for Bar-Coded Vehicle
  Identification Number Label (A)—SAE **J1877** MAY88 ... 3:42.37

OXYGEN
  Exhaust Gases
    Instrumentation and Techniques for Exhaust Gas
    Emissions Measurement (A)—SAE **J254** AUG84 ...... 1:13.104

OXYGENATES
  † Alternative Automotive Fuels—SAE **J1297** MAR93 .... 1:12.74

PAINTS
  See also: **COATINGS, FINISHES**
  Electrocoat Compatibilities of Automotive Sealers—SAE
  J1969 OCT88 .... 1:11.90
  Florida Exposure of Automotive Finishes—SAE **J951**
  JAN85 .... 1:11.130
  Lead-Free Replacement Paints (A)—SAE **J1437** JUN84 1:11.129
  Method for Evaluating the Paintable Characteristics of
  Automotive Sealers—SAE **J1800** APR87 .... 1:11.83
  Color
    Instrumental Color Difference Measurement for Exterior
    Finishes, Textiles, and Colored Trim (A)—SAE **J1545**
    JUN86 .... 1:11.279
    *Paint and Trim Code Location—SAE **J2235** JUN93 .... 3:33.44
  Corrosion
    Guidelines and Limitations of Laboratory Cyclic Corrosion
    Test Procedures for Exterior, Painted, Automotive Body
    Panels—SAE **J1563** MAY87 .... 1:3.51

PARAMETERS
  *Class B Data Communication Network Messages Part 2:
  Data Parameter Definitions—SAE **J2178/2** JUN93 ..... 2:23.345

PARKING BRAKES
  See: **BRAKES**

PARKING LAMPS
  Parking Lamps (Front Position Lamps)—SAE **J222**
  DEC91 .... 2:24.167
  † SAE Ground Vehicle Lighting Standards Manual—SAE
  HS-34 JAN93 .... Available as a separate publication (See Related Technical Reports Section)

PARKING PERFORMANCE
  Motor Vehicles
    Truck and Bus Grade Parking Performance Test
    Procedure—SAE **J360** OCT88 .... 2:25.101
    Vehicle Grade Parking Performance Requirements (A)—
    SAE **J293** OCT88 .... 2:25.102
  Motorcycles
    Performance Requirements for Parking Stability of
    Motorcycles (A)—SAE **J1248** JUN91 .... 3:37.09
    Test Procedure for Parking Stability of Motorcycles—SAE
    J1101 JUN92 .... 3:37.08
  Trailers
    Trailer Grade Parking Performance Test Procedure—SAE
    J1452 OCT88 .... 2:25.155

PARTICLE SIZE
  † Compressed Air for General Use—Part 1: Contaminants
  and Quality Classes—SAE **J1649/1** JUN93 .... 3:34.47
  Definition for Particle Size—SAE **J391** JUL81 .... 1:3.81

---

* Technical Revision   † New Document   (D)—DODISS Adopted   (A)—ANSI Recognized

Subject Index

PARTICULATES
See also: EXHAUST EMISSIONS
Chemical Methods for the Measurement of Nonregulated Diesel Emissions (A)—SAE J1936 OCT89 .... 1:13.38
Filters
Determination of Sulfur Compounds in Automotive Exhaust—SAE J1280 JUN92 ...... 1:13.72
† Operator Enclosure Air Filter Element Test Procedure—SAE J1533 JUN93 .... 3:40.513

PASSENGER CARS
See: AUTOMOBILES

PASSENGER COMPARTMENTS
See also: FABRICS, HUMAN ENGINEERING, RESTRAINT SYSTEMS, SEATS
Accommodation Tool Reference Point—SAE J1516 MAR90 .... 3:34.247
Devices for Use in Defining and Measuring Vehicle Seating Accommodation—SAE J826 JUN92 .... 3:34.90
Driver Selected Seat Position (A)—SAE J1517 MAR90 . 3:34.250
† Motor Vehicle Dimensions (A)—SAE J1100 JUN93 .... 3:34.118
Accelerated Testing
*Accelerated Exposure of Automotive Interior Trim Components Using a Controlled Irradiance Air-Cooled Xenon-Arc Apparatus—SAE J2212 MAR93 ...... 1:11.263
*Accelerated Exposure of Automotive Interior Trim Material Using Outdoor Under-Glass Controlled Sun-Tracking Temperature and Humidity Apparatus—SAE J2230 FEB93 .... 1:11.123
*Accelerated Exposure of Automotive Interior Trim Materials Using an Outdoor Under Glass Variable Angled Controlled Temperature Apparatus—SAE J2229 FEB93 . 1:11.119
Head Clearance
Motor Vehicle Driver and Passenger Head Position (A)—SAE J1052 MAY87 .... 3:34.231
Noise
Instrumentation for Measuring Acoustic Impulses Within Vehicles (A)—SAE J247 FEB87 .... 1:14.07
Measurement of Interior Sound Levels of Light Vehicles (A)—SAE J1477 JAN86 .... 1:14.19

PASSIVE RESTRAINT SYSTEMS
See also: OCCUPANTS, RESTRAINT SYSTEMS, SAFETY BELTS
Glossary of Automotive Inflatable Restraint Terms—SAE J1538 APR88 .... 3:33.31
*The Air Bag Systems in Your Cars 'What the Public Needs to Know'—SAE J2074 JUN93 ...... 3:33.42
† Vehicle Occupant Restraint Systems and Components Standards Manual—SAE HS-13 MAY93 .... Available as a separate publication *(See Related Technical Reports Section)*

Noise
Instrumentation for Measuring Acoustic Impulses Within Vehicles (A)—SAE J247 FEB87 .... 1:14.07

PEDALS
Accommodation Tool Reference Point—SAE J1516 MAR90 .... 3:34.247
Driver Selected Seat Position (A)—SAE J1517 MAR90 . 3:34.250
Truck Driver Shin-Knee Position for Clutch and Accelerator (A)—SAE J1521 MAR90 .... 3:34.263

PERFORMANCE
See also: ACCELERATION, HORSEPOWER
Agricultural Equipment
Agricultural Tractor Test Code—SAE J708 DEC84 .... 3:39.01
† Agricultural Tractor Test Code (OECD)—SAE J2708 APR93 .... 3:39.05
Automobile
Road Load Measurement and Dynamometer Simulation Using Coastdown Techniques—SAE J1263 JUN91 ... 2:26.470
Vehicle Acceleration Measurement—SAE J1491 JUN90 . 2:26.503

Brakes
Brake Power Rating Test Code—Commercial Vehicle Inertia Dynamometer (A)—SAE J971 JUN91 .... 2:25.152
Electric VehicLes
*Electric Vehicle Acceleration, Gradeability, and Deceleration Test Procedure—SAE J1666 MAY93 ..... 3:28.12
Engines
Certificates of Maximum Net Horsepower for Motor Trucks and Truck Tractors—SAE J690 .... 3:36.65
Engine Power Test Code—Spark Ignition and Compression Ignition— Gross Power Rating—SAE J1995 JUN90 .... 2:26.11
Engine Power Test Code—Spark Ignition and Compression Ignition—Net Power Rating—SAE J1349 JUN90 .... 2:26.16
Procedure for Mapping Engine Performance—Spark Ignition and Compression Ignition Engines (A)—SAE J1312 JAN90 .... 2:26.22
Small Craft—Marine Propulsion Engine and Systems—Power Measurements and Declarations—SAE J1228 NOV91 .... 3:41.12
Testing
† Agricultural Tractor Test Code (OECD)—SAE J2708 APR93 .... 3:39.05
† Application Testing of Oil-to-Air Oil Coolers for Cooling Performance—SAE J1468 MAY93 .... 2:26.398
Automotive and Off-Highway Air Brake Reservoir Performance and Identification Requirements (A)—SAE J10 OCT90 .... 3:36.132
Backup Lamp Switch—SAE J1076 MAR90 .... 2:24.175
*Brake Block Effectiveness Rating—SAE J1802 JUN93 . 2:25.119
*Brake Performance and Wear Test Code Commercial Vehicle Inertia Dynamometer—SAE J2115 JUN93 ..... 2:25.106
Braking System Test Procedures and Braking Performance Criteria for Agricultural Tractors—SAE J1041 OCT91 .... 3:39.29
*Braking, Stability, and Control Performance Test Procedures for Air-Brake- Equiped Truck Tractors—SAE J1626 MAR93 .... 3:36.130
Brazed Double Wall Low Carbon Steel Tubing (A)—SAE J527 JUN91 .... 2:22.209
Daytime Running Lamps for Use on Motor Vehicles—SAE J2087 AUG91 .... 2:24.168
*Discharge Forward Lighting System—SAE J2009 FEB93 . 2:24.113
Door Courtesy Switch (A)—SAE J2108 DEC91 ...... 2:23.141
Engine Oil Performance and Engine Service Classification (Other Than "Energy-Conserving") (A)—SAE J183 JUN91 .... 1:12.01
† Engine Oil Tests (A)—SAE J304 JUN93 .... 1:12.19
Flasher Test (A)—SAE J823 JUN91 .... 2:24.210
Front and Rear Turn Signal Lamps for Use on Motor Vehicles 2032 mm or More in Overall Width (A)—SAE J1395 JUN91 .... 2:24.156
Front-Wheel-Drive Constant Velocity Joint Boot Seals—SAE J2028 JUN92 .... 3:29.07
† Fuel and Oil Hoses (A)—SAE J30 MAY93 .... 1:11.226
Fuel Injection System Fuel Pressure Regulator and Pressure Damper—SAE J1862 FEB90 .... 2:26.279
† Fuels and Lubricants Standards Manual—SAE HS-23 FEB93 .... Available as a separate publication *(See Related Technical Reports Section)*
Headlamp Beam Switching (A)—SAE J564 MAR90 .... 2:24.08
High Current Time Lag Electric Fuses (A)—SAE J1888 NOV90 .... 2:23.128
*High-Temperature Power Steering Pressure Hose—SAE J2050 FEB93 .... 1:11.245
*High-Temperature Power Steering Return Hose--Low Pressure—SAE J2076 FEB93 .... 1:11.247

---

* Technical Revision † New Document (D)—DODISS Adopted (A)—ANSI Recognized

PERFORMANCE (continued)
Testing (continued)

Hydraulic Master Cylinders for Motor Vehicle Brakes—Test Procedure (A)—SAE J1153 JUN91 .... 2:25.29

Hydraulic Wheel Cylinders for Automotive Drum Brakes (A)—SAE J101 DEC89 .... 2:25.25

Ignition Cable Assemblies—SAE J2032 NOV91 .... 2:23.154

† International Tests and Specifications for Automotive Engine Oils—SAE J2227 MAY93 .... 1:12.10

† Laboratory Speed Test Procedure for Passenger Car Tires—SAE J1561 FEB93 .... 3:30.34

Manual Slack Adjuster Test Procedure (A)—SAE J1461 APR91 .... 3:36.174

Metallic Air Brake System Tubing and Pipe (A)—SAE J1149 JUN91 .... 2:22.219

*Metallic Tube Connections for Fluid Power and General Use-Test Methods for Threaded Hydraulic Fluid Power Connectors—SAE J1644 MAY93 .... 2:22.285

Metric Nonmetallic Air Brake System Tubing (A)—SAE J1394 APR91 .... 2:22.227

† Operator Enclosure Air Filter Element Test Procedure—SAE J1533 JUN93 .... 3:40.513

Overhead Protection for Agricultural Tractors—Test Procedures and Performance Requirements—SAE J167 APR92 .... 3:39.89

Passenger Car and Truck Automatic Transmission Test Code (A)—SAE J651 JAN91 .... 3:29.28

Personal Watercraft—Flotation (A)—SAE J1973 APR91 3:41.49

Pneumatic Tires for Military Tactical Wheeled Vehicles (A)—SAE J2014 APR91 .... 3:30.04

Qualifications for Four-Way Subbase Mounted Air Valves for Automotive Manufacturing Applications—SAE J2051 JUN92 .... 3:43.21

Retroreflective Materials for Vehicle Conspicuity (A)—SAE J1967 MAR91 .... 3:34.112

† Road Vehicle—Hydraulic Brake Hose Assemblies for Use with Nonpetroleum-Base Hydraulic Fluids—SAE J1401 JUN93 .... 2:25.01

† Roll-Over Protective Structures (ROPS) for Wheeled Agricultural Tractors (A)—SAE J2194 JUN93 .... 3:39.78

Rubber Cups for Hydraulic Actuating Cylinders (A)—SAE J1601 NOV90 .... 2:25.175

Rubber Dust Boots for the Hydraulic Disc Brake Piston—SAE J1570 SEP91 .... 2:25.181

*SAE Motorcycle Standards Manual—SAE HS-2500 APR93 .... Available as a separate publication (See Related Technical Reports Section)

Seals—Testing of Radial Lip—SAE J110 DEC91 ...... 3:29.140

Seamless Copper Tube (A)—SAE J528 JUN91 .... 2:22.218

Seamless Low Carbon Steel Tubing Annealed for Bending and Flaring (A)—SAE J524 JUN91 .... 2:22.206

† Service Brake Structural Integrity Test Procedure—Vehicles Over 4500 kg (10 000 lb) GVWR—SAE J294 .. 3:36.138

Service Hose for Automotive Air Conditioning—SAE J2196 JUN92 .... 3:34.51

Standard Method for Determining Continuous Upper Temperature Resistance of Elastomers—SAE J2236 JUN92 .... 1:11.32

*Steady-State Circular Test Procedure for Trucks and Buses—SAE J2181 JUN93 .... 3:36.68

Steering for Off-Road, Rubber-Tired Machines (A)—SAE J1511 OCT90 .... 3:40.517

Stop Lamps for Use on Motor Vehicles 2032 mm or More in Overall Width (A)—SAE J1398 JUN91 .... 2:24.145

Subjective Rating Scale for Vehicle Handling—SAE J1441 FEB92 .... 3:28.16

Tail Lamps (Rear Position Lamps) for Use on Vehicles 2032 mm or More in Overall Width (A)—SAE J2040 JUN91 .... 2:24.139

† Test Method for Evaluating the Sealing Capability of Hose Connections with a PVT Test Facility—SAE J1610 APR93 .... 2:16.80

Test Methods and Equipment for Lighting Devices and Components for Use on Vehicles Less Than 2032 mm in Overall Width—SAE J575 JUN92 .... 2:24.30

Test Procedure for Air Reservoir Capacity—Highway Type Vehicles (A)—SAE J1911 SEP90 .... 3:36.153

Tests and Procedures for High Temperature Transmission Oil Hose, Engine Lubricating Oil Hose, and Hose Assemblies (A)—SAE J1019 JUN90 .... 2:22.189

† Tests and Procedures for SAE 100R Series Hydraulic Hose and Hose Assemblies (A)—SAE J343 JUN93 .... 2:22.185

Tests for Snowmobile Switching Devices and Components—SAE J68 DEC91 .... 3:38.08

† Two-Stroke-Cycle Gasoline Engine Lubricants Performance and Service Classification (A)—SAE J2116 JUN93 .... 1:12.33

Vacuum Power Assist Brake Booster Test Procedure (A)—SAE J1808 OCT89 .... 2:25.172

Validation Testing of Electric Fuel Pumps for Gasoline Fuel Injection Systems (A)—SAE J1537 JUN90 .... 2:26.291

*Vehicle Testing Requirements for Replacement Refrigerants for CFC-12 (R12) Mobile Air Conditioning Systems—SAE J659 JUN93 .... 3:34.64

Welded and Cold Drawn Low Carbon Steel Tubing Annealed for Bending and Flaring (A)—SAE J525 JUN91 2:22.207

Welded Flash Controlled Low Carbon Steel Tubing Normalized for Bending, Double Flaring, and Beading (A)—SAE J356 JUN91 .... 2:22.213

Welded Low Carbon Steel Tubing (A)—SAE J526 JUN91 2:22.208

Wheels—Passenger Cars—Performance Requirements and Test Procedures (A)—SAE J328 MAR90 .... 3:31.06

Trucks and Buses

Certificates of Maximum Net Horsepower for Motor Trucks and Truck Tractors—SAE J690 .... 3:36.65

Heavy Truck and Bus Retarder Downhill Performance Mapping Procedure (A)—SAE J1489 JUN87 .... 3:36.124

Information Relating to Duty Cycles and Average Power Requirements of Truck and Bus Engine Accessories—SAE J1343 OCT88 .... 3:36.57

Truck Ability Prediction Procedure—SAE J688—SAE HS-82 .... Available as a separate publication (See Related Technical Reports Section)

Vehicle Acceleration Measurement—SAE J1491 JUN90 2:26.503

Turbochargers

Turbocharger Gas Stand Test Code—SAE J1826 APR89 2:26.355

Turbocharger Nomenclature and Terminology—SAE J922 JUL88 .... 2:26.04

Valves

Labeling Air Brake Valves with Their Performance (Input-Output) Characteristics (A)—SAE J1860 AUG90 ...... 3:36.164

PERSONAL WATERCRAFT

*Personal Water craft Ventilation Systems—SAE J2034 JAN93 .... 3:41.49

*Personal Watercraft Fuel Systems—SAE J2046 JAN93 . 3:41.46

*Personal Watercraft--Electrical Systems—SAE J2120 JUL92 .... 3:41.30

Personal Watercraft—Flotation (A)—SAE J1973 APR91 3:41.49

PHOTOGRAPHY

Film Analysis Guides for Dynamic Studies of Test Subjects—SAE J138 .... 3:34.245

PHOTOMETRY

360 Degree Warning Devices for Authorized Emergency, Maintenance, and Service Vehicles—SAE J845 MAR92 . 2:24.185

† Cargo Lamps for Use on Vehicles Under 12 000 lb GVWR—SAE J1424 JUN93 .... 2:24.120

---

* Technical Revision   † New Document   (D)—DODISS Adopted   (A)—ANSI Recognized

Subject Index

PHOTOMETRY (continued)

*Central High Mounted Stop Lamp Standard for Use on Vehicles Less than 2032 mm Overall Width—SAE **J1957** JUN93 .... 2:24.149

Clearance, Side Marker, and Identification Lamps—SAE **J592** JUN92 .... 2:24.170

Clearance, Sidemarker, and Identification Lamps for Use on Motor Vehicles 2032 mm or More in Overall Width (A)—SAE **J2042** JUN91 .... 2:24.172

Daytime Running Lamps for Use on Motor Vehicles—SAE **J2087** AUG91 .... 2:24.168

†Fog Tail Lamp (Rear Fog Light) Systems (A)—SAE **J1319** JUN93 .... 2:24.141

Front and Rear Turn Signal Lamps for Use on Motor Vehicles 2032 mm or More in Overall Width (A)—SAE **J1395** JUN91 .... 2:24.156

†Front Fog Lamps—SAE **J583** JUN93 .... 2:24.119

L.E.D. Lighting Devices (A)—SAE **J1889** JUN88 .... 2:24.19

Parking Lamps (Front Position Lamps)—SAE **J222** DEC91 .... 2:24.167

Photometric Guidelines for Instrument Panel Displays That Accomodate Older Drivers—SAE **J2217** OCT91 .. 3:34.226

Photometry Laboratory Accuracy Guidelines—SAE **J1330** JUN88 .... 2:24.04

*Relfex Reflectors for use on Vehicles 2032 mm or More in Overall Width—SAE **J2041** JUN92 .... 2:24.178

School Bus Stop Arm (A)—SAE **J1133** JUL89 .... 2:24.196

Stop Lamps for Use on Motor Vehicles 2032 mm or More in Overall Width (A)—SAE **J1398** JUN91 .... 2:24.145

Tail Lamps (Rear Position Lamps) for Use on Vehicles 2032 mm or More in Overall Width (A)—SAE **J2040** JUN91 .... 2:24.139

Tail Lamps (Rear Position Lamps) for Use on Motor Vehicles Less Than 2032 mm in Overall Width—SAE **J585** DEC91 .... 2:24.137

Turn Signal Lamps for Use on Motor Vehicles Less Than 2032 mm in Overall Width (A)—SAE **J588** JUN91 ..... 2:24.153

PHYSICAL PROPERTIES

Alternate Fuels
†Alternative Automotive Fuels—SAE **J1297** MAR93 .... 1:12.74

Aluminum Alloys
General Data on Wrought Aluminum Alloys (A)—SAE **J454** FEB91 .... 1:10.26

General Information—Chemical Compositions, Mechanical and Physical Properties of SAE Aluminum Casting Alloys—SAE **J452** JAN89 .... 1:10.06

Wrought Aluminum Applications Guidelines—SAE **J1434** JAN89 .... 1:10.15

Brake Linings
†Specific Gravity of Brake Lining—SAE **J380** FEB93 .... 2:25.89

Ceramic Shot
Size Classification and Characteristics of Ceramic Shot for Peening (A)—SAE **J1830** MAY87 .... 1:8.32

Cobalt Alloys
Engine Poppet Valve Information Report—SAE **J775** JAN88 .... 1:1.165

Special Purpose Alloys ("Superalloys")—SAE **J467b** ... 1:10.128

Coolants
Automotive and Light Truck Engine Coolant Concentrate—Ethylene Glycol Type (A)—SAE **J1034** APR91 .... 1:11.214

Engine Coolants—SAE **J814** JUL88 .... 1:11.209

Copper Alloys
Wrought and Cast Copper Alloys—SAE **J461** SEP81 ... 1:10.45

Diesel Fuels
Diesel Fuels—SAE **J313** MAR92 .... 1:12.60

Elastomers
†Thermoplastic Elastomer Classification System—SAE **J3000** JUN93 .... 1:11.27

Fuels
†Fuels and Lubricants Standards Manual—SAE **HS-23** FEB93 .... Available as a separate publication *(See Related Technical Reports Section)*

Gasket Materials
Design Guide for Formed-In-Place Gaskets (A)—SAE **J1497** DEC88 .... 1:11.187

Standard Classification System for Nonmetallic Automotive Gasket Materials (A)—SAE **J90** JUN90 .... 1:11.197

Gasoline
†Alternative Automotive Fuels—SAE **J1297** MAR93 ... 1:12.74

†Automotive Gasolines—SAE **J312** JAN93 .... 1:12.47

Greases
†Automotive Lubricating Greases—SAE **J310** JUN93 ... 1:12.39

Hoses
†Fuel and Oil Hoses (A)—SAE **J30** MAY93 .... 1:11.226

Insulation
Low Tension Thin Wall Primary Cable—SAE **J1560** JAN92 .... 2:23.167

Iron Alloys
Elevated Temperature Properties of Cast Irons—SAE **J125** MAY88 .... 1:6.12

Engine Poppet Valve Information Report—SAE **J775** JAN88 .... 1:1.165

Special Purpose Alloys ("Superalloys")—SAE **J467b** ... 1:10.128

Lubricants
†Engine Oil Viscosity Classification—SAE **J300** MAR93 .. 1:12.16

†Fuels and Lubricants Standards Manual—SAE **HS-23** FEB93 .... Available as a separate publication *(See Related Technical Reports Section)*

Lubricants for Two-Stroke-Cycle Gasoline Engines—SAE **J1510** OCT88 .... 1:12.32

Lubricating Oil, Aircraft Piston Engine (Ashless Dispersant) (A)—SAE **J1899** JUN91 .... 1:12.98

Lubricating Oil, Aircraft Piston Engine (Nondispersant Mineral Oil) (A)—SAE **J1966** JUN91 .... 1:12.110

Physical and Chemical Properties of Engine Oils (A)—SAE **J357** JUN91 .... 1:12.24

Magnesium Alloys
Magnesium Alloys—SAE **J464** JAN89 .... 1:10.120

Magnesium Casting Alloys—SAE **J465** JAN89 ...... 1:10.120

Nickel Alloys
Special Purpose Alloys ("Superalloys")—SAE **J467b** ... 1:10.128

Wrought Nickel and Nickel-Related Alloys—SAE **J470c** . 1:10.140

NickelAlloys
Engine Poppet Valve Information Report—SAE **J775** JAN88 .... 1:1.165

O-Rings
Rubber Rings for Automotive Applications—SAE **J120a** . 1:11.203

Plastics
Automotive Plastic Parts Specification (A)—SAE **J1345** FEB82 .... 1:11.148

*Materials for Plastic Check Valves for Vacuum Booster System—SAE **J1875** JUN93 .... 2:25.40

Synthetic Resin Plastic Sealers, Nondrying Type—SAE **J250** 1:11.65

Rubber
Brake Master Cylinder Reservoir Diaphragm Gasket—SAE **J1605** MAR92 .... 2:25.22

†Classification System for Rubber Materials—SAE **J200** MAY93 .... 1:11.01

Latex Dipped Goods and Coatings for Automotive Applications—SAE **J19** JAN85 .... 1:11.63

Latex Foam Rubbers—SAE **J17** JAN85 .... 1:11.53

†Sponge and Expanded Cellular Rubber Products—SAE **J18** JUL92 .... 1:11.56

---

* Technical Revision   † New Document   (D)—DODISS Adopted   (A)—ANSI Recognized

Subject Index

PHYSICAL PROPERTIES (continued)
Seals
Front-Wheel-Drive Constant Velocity Joint Boot Seals—SAE J2028 JUN92 .... 3:29.07
Sintered Carbide
Sintered Carbide Tools—SAE J439a 1:7.07
Sintered Powder Materials
Sintered Powder Metal Parts: Ferrous—SAE J471d 1:8.01
Steel Wire Shot
† Cut Wire Shot—SAE J441 JUN93 .... 1:8.05
Steels
*Cast Steel Grit—SAE J1993 MAR93 .... 1:8.13
Engine Poppet Valve Information Report—SAE J775 JAN88 .... 1:1.165
† Hardenability Bands for Carbon and Alloy H Steels (A)—SAE J1268 JUN93 .... 1:1.48
Sheet Steel Thickness and Profile (A)—SAE J1058 APR91 .... 1:3.19
Special Purpose Alloys ("Superalloys")—SAE J467b ... 1:10.128
Superalloys
Special Purpose Alloys ("Superalloys")—SAE J467b ... 1:10.128
Titanium Alloys
Special Purpose Alloys ("Superalloys")—SAE J467b ... 1:10.128
Transmission Fluids
Fluid for Passenger Car Type Automatic Transmissions—SAE J311 APR86 .... 1:12.44
Powershift Transmission Fluid Classification—SAE J1285 JAN85 .... 1:12.45
Vibration Damping Materials
Vibration Damping Materials and Underbody Coatings (A)—SAE J671 APR82 .... 1:11.111
Zinc Alloys
Zinc Die Casting Alloys—SAE J469 JAN89 .... 1:10.139

PILOT HOLES
Companion Flanges, Type A (External Pilot) and Type S (Internal Pilot) (A)—SAE J1946 JAN91 .... 3:29.65
Fan Hub Bolt Circles and Pilot Holes (A)—SAE J635 JUL84 .... 2:26.414

PINS
See also: FASTENERS, KINGPINS
Clevis
Rod Ends and Clevis Pins—SAE J493 .... 2:16.28
Dowel
Unhardened Ground Dowel Pins—SAE J497 .... 2:16.28
Straight
Grooved Straight Pins—SAE J494 .... 2:16.23
Spring Type Straight Pins—SAE J496 .... 2:16.26
Straight Pins (Solid)—SAE J495 .... 2:16.23

PIPELAYERS
See also: CONSTRUCTION AND INDUSTRIAL EQUIPMENT, OFF-ROAD VEHICLES
Identification Terminology and Specification Definitions—Pipelayers and Side Booms, Tractor or Loader Mounted—SAE J1295 JUN89 .... 3:40.55
Lift Capacity Calculation and Test Procedure—Pipelayer and Sideboom—SAE J743 MAR92 .... 3:40.223

PIPES
Diesel Engines—Steel Tubes for High Pressure Fuel Injection Pipes (Tubing)—SAE J1958 APR89 .... 2:26.260
Fuel Injection Pumps — High Pressure Pipes (Tubing) for Testing (A)—SAE J1418 DEC87 .... 2:26.263
Connections
Connections for Fluid Power and General Use—Ports and Stud Ends with ISO 261 Threads and O-Ring Sealing Part 1: Port with O-Ring Seal in Truncated Housing—SAE J2244/1 DEC91 .... 2:22.290
Connections for Fluid Power and General Use—Ports and Stud Ends with ISO 261 Threads and O-Ring Sealing Part 2: Heavy-Duty (S Series) Stud Ends—Dimensions, Design, Test Methods, and Requirements—SAE J2244/2 DEC91 .... 2:22.293
† Hydraulic Flanged Tube, Pipe, and Hose Connections, Four-Bolt Split Flange Type—SAE J518 JUN93 ...... 2:22.190
Road Vehicles—High Pressure Fuel Injection Pipe End-Connections with 60 Deg Female Cone—SAE J1949 OCT88 .... 2:26.302
Fittings
† Automotive Pipe Fittings—SAE J530 JUN93 .... 2:22.232
Coding Systems for Identification of Fluid Conductors and Connectors (D)—SAE J846 JUN89 .... 2:22.01
† Hydraulic Tube Fittings—SAE J514 JUN93 .... 2:22.59
Plugs
† Automotive Pipe, Filler, and Drain Plugs—SAE J531 JUN93 .... 2:22.241
Steel
Metallic Air Brake System Tubing and Pipe (A)—SAE J1149 JUN91 .... 2:22.219
Threads
† Beaded Tube Hose Fittings—SAE J1231 JUN93 ...... 2:22.158
Dryseal Pipe Threads—SAE J476a 2:15.01
*Thread Sealants—SAE J1615 JUN93 .... 2:22.240

PISTON RINGS
† Internal Combustion Engines—Piston Rings Expander/Segment Oil Control Rings (A)—SAE J2004 OCT92 ... 2:26.175
† Internal Combustion Engines—Piston Rings—General Specifications (A)—SAE J1591 OCT92 .... 2:26.186
† Internal Combustion Engines—Piston Rings—Half Keystone Rings (A)—SAE J2001 OCT92 .... 2:26.111
† Internal Combustion Engines—Piston Rings—Material Specifications (A)—SAE J1590 OCT92 .... 2:26.72
† Internal Combustion Engines—Piston Rings—Rectangular Rings with Narrow Ring Width (A)—SAE J1998 OCT92 .... 2:26.43
*Internal Combustion Engines—Piston Rings—Steel Rectangular Rings—SAE J2226 OCT92 .... 2:26.180
Defects
† Internal Combustion Engines—Piston Rings—Quality Requirements—SAE J1996 OCT92 .... 2:26.74
Dimensions
† Internal Combustion Engines—Piston Rings Coil Spring Loaded Oil Control Rings—SAE J2003 OCT92 ...... 2:26.131
† Internal Combustion Engines—Piston Rings—Keystone Rings—SAE J2000 OCT92 .... 2:26.100
† Internal Combustion Engines—Piston Rings—Oil Control Rings—SAE J2002 OCT92 .... 2:26.117
† Internal Combustion Engines—Piston Rings—Rectangular Rings—SAE J1997 OCT92 .... 2:26.79
† Internal Combustion Engines—Piston Rings—Scraper Rings—SAE J1999 OCT92 .... 2:26.90
Foreign Equivalents
† Internal Combustion Engines—Piston Rings—Vocabulary—SAE J1588 OCT92 .... 2:26.28
Measurement
† Internal Combustion Engines—Piston Rings—Inspection Measuring Principles—SAE J1589 OCT92 .... 2:26.52
Terminology
† Internal Combustion Engines—Piston Rings—Vocabulary—SAE J1588 OCT92 .... 2:26.28

PLASTIC DEFORMATION
Steel
Methods for Determining Plastic Deformation in Sheet Metal Stampings (A)—SAE J863 JUN86 .... 1:3.17

---

*Technical Revision   † New Document   (D)—DODISS Adopted   (A)—ANSI Recognized

Subject Index

PLASTICS
See also: ELASTOMERS, FIBER REINFORCED PLASTICS, URETHANE, VINYL

Abrasion Resistance Testing—Vehicle Exterior Graphics and Pin Striping—SAE J1847 JUN89 .... 1:11.262

Chemical Stress Resistance of Polymers—SAE J2016 JUN89 .... 1:11.154

*Classification System for Automotive Polyamide (PA) Plastics—SAE J1639 MAR93 .... 1:11.145

*Compatibility of Retrofit Refrigerants with Air Conditioning System Materials—SAE J1662 JUN93 .... 3:34.69

*SAE Automotive Textiles and Trim Standards Manual—SAE HS-2700 JUN93 .... Available as a separate publication (See Related Technical Reports Section)

Synthetic Resin Plastic Sealers, Nondrying Type—SAE J250 .... 1:11.65

Test Method for Determining Cold Cracking of Flexible Plastic Materials—SAE J323 .... 1:11.115

Test Method for Measuring Thickness of Automotive Textiles and Plastics (A)—SAE J882 FEB85 .... 1:11.115

Test Method of Stretch and Set of Textiles and Plastics—SAE J855 MAR87 .... 1:11.181

Test Procedure for Determining the Resistance of Safety Glazing Materials, of Which One Surface is Plastic, to Simulated Weathering (A)—SAE J2081 JUN91 .... 3:34.106

Bonding
Cross Peel Test for Automotive Type Adhesives for Fiber Reinforced Plastic (FRP) Bonding (A)—SAE J1553 MAY86 .... 1:11.100

Lap Shear Test for Automotive Type Adhesives for Fiber Reinforced Plastic (FRP) Bonding (A)—SAE J1525 JUN85 .... 1:11.81

Overlap Shear Test for Automotive Type Sealant for Stationary Glass Bonding (A)—SAE J1529 MAY86 .... 1:11.95

Bumpers
Flexible Bumper Repair—SAE J1573 FEB87 .... 3:42.22

Fading
*Accelerated Exposure of Automotive Interior Trim Components Using a Controlled Irradiance Air-Cooled Xenon-Arc Apparatus—SAE J2212 MAR93 .... 1:11.263

*Accelerated Exposure of Automotive Interior Trim Material Using Outdoor Under-Glass Controlled Sun-Tracking Temperature and Humidity Apparatus—SAE J2230 FEB93 .... 1:11.123

*Accelerated Exposure of Automotive Interior Trim Materials Using an Outdoor Under Glass Variable Angled Controlled Temperature Apparatus—SAE J2229 FEB93 .... 1:11.119

Lenses
Plastic Materials for Use in Optical Parts Such as Lenses and Reflectors of Motor Vehicle Lighting Devices—SAE J576 JUL91 .... 2:24.33

Numbering System
Automotive Plastic Parts Specification (A)—SAE J1345 FEB82 .... 1:11.148

Symbols
†Marking of Plastic Parts—SAE J1344 APR93 .... 1:11.131

Valves
*Materials for Plastic Check Valves for Vacuum Booster System—SAE J1875 JUN93 .... 2:25.40

PLUGS
Connectors and Plugs—SAE J856 .... 2:24.15

Drain and Filler
†Automotive Pipe, Filler, and Drain Plugs—SAE J531 JUN93 .... 2:22.241

†Automotive Straight Thread Filler and Drain Plugs—SAE J532 JUN93 .... 2:22.246

†Drain, Fill, and Level Plugs for Off-Road, Self-Propelled Work Machines—SAE J371 MAY93 .... 3:40.257

Spark
Preignition Rating of Spark Plugs (A)—SAE J549 JUN90 .... 2:23.96

SAE 17.6 Cubic Inch Spark Plug Rating Engine (A)—SAE J2203 JUN91 .... 2:23.98

Spark Plug Heat Rating Classifications—SAE J2162 JUN92 .... 2:23.97

Spark Plug Installation Sockets—SAE J548/2 JUN92 .... 2:23.95

Spark Plugs—SAE J548/1 JUN92 .... 2:23.83

POLYMERS
See: PLASTICS, URETHANE

POLYURETHANE
See: PLASTICS, URETHANE

PORTS
Connections for Fluid Power and General Use—Ports and Stud Ends with ISO 261 Threads and O-Ring Sealing Part 1: Port with O-Ring Seal in Truncated Housing—SAE J2244/1 DEC91 .... 2:22.290

POWDER METALS
See: METAL POWDERS

POWER CONSUMPTION
Radial Lip Seal Torque: Measurement Methods and Results—SAE J1971 JUN89 .... 3:29.151

Accessories
Information Relating to Duty Cycles and Average Power Requirements of Truck and Bus Engine Accessories—SAE J1343 OCT88 .... 3:36.57

Compressors
Test Method for Measuring Power Consumption of Air Conditioning and Brake Compressors for Trucks and Buses (A)—SAE J1340 APR90 .... 3:36.53

Engine Fans
Method for Determining Power Consumption of Engine Cooling Fan-Drive Systems (A)—SAE J1342 AUG89 .... 3:36.55

Test Method for Measuring Power Consumption of Truck and Bus Engine Cooling Fans (A)—SAE J1339 AUG89 .... 3:36.52

Pumps
Test Method for Measuring Power Consumption of Hydraulic Pumps for Trucks and Buses (A)—SAE J1341 APR90 .... 3:36.54

POWER SPECTRAL DENSITY
Sintered Powder Metal Parts: Ferrous—SAE J471d 1:8.01

POWER TAKE-OFF
†Auxiliary Power Take-Off Drives for Agricultural Tractors—SAE J717 APR93 .... 3:39.53

†Clearance Envelopes for Six-Bolt, Eight-Bolt, and Rear Truck Transmission Mounted Power Takeoffs (A)—SAE J772 DEC92 .... 3:29.33

†Drawbars—Agricultural Wheel Tractors—SAE J1548 FEB93 .... 3:39.90

†Flywheels for Industrial Engines Used with Industrial Power Take-Offs Equipped with Driving-Ring Type Overcenter Clutches and Engine-Mounted Marine Gears and Single Bearing Engine-Mounted Power Generators—SAE J620 MAY93 .... 3:40.256

†Front and Rear Power Take-Off for Agricultural Tractors (A)—SAE J1170 FEB93 .... 3:39.49

Industrial Power Takeoffs with Driving Ring-Type Overcenter Clutches—SAE J621 APR89 .... 3:40.243

Mounting Flanges and Power Take-Off Shafts for Small Engines—SAE J609a 2:26.26

†Openings for Six- and Eight-Bolt Truck Transmission Mounted Power Take-Offs—SAE J704 DEC92 .... 3:29.41

†Operating Requirements for Tractors and Power Take-Off Driven Implements (A)—SAE J721 FEB93 .... 3:39.52

†Personnel Protection for General Purpose Industrial Machines (A)—SAE J98 NOV92 .... 3:40.472

Power Take-Off Definitions and Terminology for Agricultural Tractors (A)—SAE J722 MAR91 .... 3:39.53

†Safety for Agricultural Tractors—SAE J208 JUN93 .... 3:39.69

---

* Technical Revision   † New Document   (D)—DODISS Adopted   (A)—ANSI Recognized

Subject Index

POWER TEST CODES
† Agricultural Tractor Test Code (OECD)—SAE **J2708** APR93 .... 3:39.05
Engine Power Test Code—Spark Ignition and Compression Ignition—Net Power Rating—SAE **J1349** JUN90 .... 2:26.16
Engine Power Test Code—Spark Ignition and Compression Ignition—Gross Power Rating—SAE **J1995** JUN90 .... 2:26.11
Passenger Car and Truck Automatic Transmission Test Code (A)—SAE **J651** JAN91 .... 3:29.28
Small Craft—Marine Propulsion Engine and Systems—Power Measurements and Declarations—SAE **J1228** NOV91 .... 3:41.12
Small Engine Power Rating Procedure—SAE **J1940** JUN89 .... 2:26.23
† Surface Vehicle Brake Manual—SAE **HS-24** JUL92 .... Available as a separate publication (See Related Technical Reports Section)

POWERTRAINS
See also: AXLES, CLUTCHES, DRIVESHAFTS, SPECIFIC TYPES OF ENGINES, TRANSMISSIONS
All-Wheel-Drive Drivetrain Schematic Symbol Standards—SAE **J2059** AUG92 .... 3:29.19
E/E Diagnostic Test Modes—SAE **J1979** DEC91 .... 2:23.30
* Introduction to All-Wheel Drive—SAE **HS-2300** DEC93 .... Available as a separate publication (See Related Technical Reports Section)
Powertrain Control Interface for Electronic Controls Used in Medium and Heavy Duty Diesel On-Highway Vehicle Applications (A)—SAE **J1922** DEC89 .... 3:36.177
Towed Vehicle Drivetrain Test Procedure—Passenger Cars, Vans, and Light Duty Trucks (A)—SAE **J1144** JUN91 .... 3:42.16

PREHEATERS
† Electric Engine Preheaters and Battery Warmers for Diesel Engines—SAE **J1310** JUN93 .... 3:40.500
Engine Preheaters (A)—SAE **J226** JAN86 .... 2:26.370
Selection and Application Guidelines for Diesel, Gasoline, and Propane Fired Liquid Cooled Engine Pre-Heaters—SAE **J1350** MAR88 .... 3:40.496

PRESSURE DAMPERS
Fuel Injection System Fuel Pressure Regulator and Pressure Damper—SAE **J1862** FEB90 .... 2:26.279

PRESSURE REGULATORS
Fuel Injection System Fuel Pressure Regulator and Pressure Damper—SAE **J1862** FEB90 .... 2:26.279
Pressure Relief for Cooling System (A)—SAE **J151** JUN91 .... 2:26.350
Radiator Caps and Filler Necks (A)—SAE **J164** JUN91 .... 2:26.373

PRESSURIZATION SYSTEMS
† Operator Enclosure Pressurization System Test Procedure—SAE **J1012** JUN93 .... 3:39.67

PREVENTIVE MAINTENANCE
† Principles of Engine Cooling Systems, Components and Maintenance—SAE **HS-40** JAN91 .... Available as a separate publication (See Related Technical Reports Section)

PRINTED CIRCUITS
See also: ELECTRIC CIRCUITS
Automotive Printed Circuits (A)—SAE **J771** APR86 .... 2:23.145

PRODUCT IDENTIFICATION NUMBER
Product Identification Numbering System of Off-Road Work Machines (A)—SAE **J1360** SEP87 .... 3:40.04

PROGRAMMABLE CONTROLLERS
* OEM/Vendor's Interface Specification for Vehicle Electronic Programming Station—SAE **J1924** DEC92 .... 2:23.522

PROPELLER SHAFTS
See: DRIVESHAFTS

PULLEYS
† Automotive Synchronous Belt Drives (A)—SAE **J1313** MAR93 .... 2:19.13
Automotive V-Belt Drives—SAE **J637** FEB89 .... 2:19.04
Automotive V-Ribbed Belt Drives and Test Methods—SAE **J1596** JUN89 .... 2:19.17
† SI (Metric) Synchronous Belts and Pulleys—SAE **J1278** MAR93 .... 2:19.09
V-Belts and Pulleys—SAE **J636** MAY92 .... 2:19.01
V-Ribbed Belts and Pulleys—SAE **J1459** AUG88 .... 2:19.15

PUMPS
Engine Water Pump Remanufacture Procedures and Acceptance Criteria—SAE **J1916** MAY89 .... 3:42.41
Information Relating to Duty Cycles and Average Power Requirements of Truck and Bus Engine Accessories—SAE **J1343** OCT88 .... 3:36.57
Method for Evaluating the Flow Properties of Pumpable Sealers—SAE **J2025** JUN89 .... 1:11.66
† Principles of Engine Cooling Systems, Components and Maintenance—SAE **HS-40** JAN91 .... Available as a separate publication (See Related Technical Reports Section)

Fuel Injection
* Diesel Engines—Fuel Injection Pump Testing—SAE **J1668** JUN93 .... 2:26.322
Diesel Fuel Injection Pump—Validation of Calibrating Nozzle Holder Assemblies (A)—SAE **J1549** APR88 .... 2:26.321
Diesel Fuel Injection—End Mounting Flanges for Fuel Injection Pumps (A)—SAE **J626** JUN88 .... 2:26.297
Diesel Injection Pump Testing—Part 2: Orifice Plate Flow Measurement (A)—SAE **J968/2** MAY91 .... 2:26.318
† Fuel Injection Equipment Nomenclature—SAE **J830** OCT92 .... 2:26.254
Tapers for Shaft Ends and Hubs for Fuel Injection Pumps (A)—SAE **J1419** FEB88 .... 2:26.245
Validation Testing of Electric Fuel Pumps for Gasoline Fuel Injection Systems (A)—SAE **J1537** JUN90 .... 2:26.291

Hydraulic
Hydraulic Power Pump Test Procedure (A)—SAE **J745** APR87 .... 3:40.111
Hydraulic Pump and Motor Mounting and Drive Dimensions (A)—SAE **J744** JUL88 .... 3:40.116
Test Method for Measuring Power Consumption of Hydraulic Pumps for Trucks and Buses (A)—SAE **J1341** APR90 .... 3:36.54

Seals
Engine Coolant Pump Seals—SAE **J780** JUN90 .... 2:26.415
Guide to the Application and Use of Engine Coolant Pump Face Seals (A)—SAE **J1245** JUN82 .... 2:26.417

QUALITY CONTROL
Brake Lining Quality Control Test Procedure (A)—SAE **J661** AUG87 .... 2:25.90
† Compressed Air for General Use—Part 1: Contaminants and Quality Classes—SAE **J1649/1** JUN93 .... 3:34.47
Design/Process Checklist for Vehicle Electronic Systems—SAE **J1938** OCT88 .... 2:23.578
General Qualification and Production Acceptance Criteria for Integrated Circuits in Automotive Applications—SAE **J1879** OCT88 .... 2:23.583
Manual on Shot Peening—SAE **HS-84** .... Available as a separate publication (See Related Technical Reports Section)
† Specific Gravity of Brake Lining—SAE **J380** FEB93 .... 2:25.89

Piston Rings
† Internal Combustion Engines—Piston Rings—Quality Requirements—SAE **J1996** OCT92 .... 2:26.74

---

* Technical Revision   † New Document   (D)—DODISS Adopted   (A)—ANSI Recognized

Subject Index

QUENCHING
General Characteristics and Heat Treatments of Steels—SAE **J412** JUN89 .... **1:2.05**

† High-Strength, Quenched, and Tempered Structural Steels—SAE **J368** MAR93 .... **1:1.164**

† Methods of Determining Hardenability of Steels—SAE **J406** JUN93 .... **1:1.23**

RADIAL FORCE
† Disc Wheel Radial Runout Low Point Marking—SAE **J2133** JUN93 .... **3:31.38**

Lip Force Measurement, Radial Lip Seals (A)—SAE **J1901** SEP90 .... **3:29.135**

RADIATION INSPECTION
Nondestructive Tests (A)—SAE **J358** FEB91 .... **1:3.67**

Penetrating Radiation Inspection (A)—SAE **J427** MAR91 **1:3.76**

RADIATORS
See also: **COOLING SYSTEMS, HEAT EXCHANGERS**

† Principles of Engine Cooling Systems, Components and Maintenance—SAE **HS-40** JAN91 .... Available as a separate publication *(See Related Technical Reports Section)*

Radiator Caps and Filler Necks (A)—SAE **J164** JUN91 . **2:26.373**

Terminology

† Radiator Nomenclature (A)—SAE **J631** APR93 .... **2:26.385**

RADIOGRAPHIC FILMS
Penetrating Radiation Inspection (A)—SAE **J427** MAR91 **1:3.76**

REAMERS
Numbering System for Reamers (A)—SAE **J2124** NOV90 **3:43.18**

REAR LAMPS
See: **BACKUP LAMPS, STOP LAMPS, TAIL LAMPS, TURN INDICATORS**

REAR VIEW MIRRORS
Passenger Car Rear Vision—SAE **J834a** **3:34.114**

Test Procedure for Determining Reflectivity of Rear View Mirrors—SAE **J964** JUN92 .... **3:34.107**

Vision Factors Considerations in Rear View Mirror Design—SAE **J985** OCT88 .... **3:34.110**

Motorcycles

Rear View Mirrors—Motorcycles—SAE **J268** MAY89 ... **3:37.06**

REAR WINDOWS
† Passenger Car Backlight Defogging System (A)—SAE **J953** APR93 .... **3:34.28**

Passenger Car Glazing—Electrical Circuits (A)—SAE **J216** DEC89 .... **3:34.104**

Stationary Glass Replacement—SAE **J1556** DEC85 ... **3:42.21**

RECREATIONAL VEHICLES
See also: **BICYCLES, MINIBIKES, MOPEDS, MOTORBOATS, MOTORCYCLES, SNOWMOBILES, TRAILERS**

Tire to Body Clearance Check for Recreational Vehicles—SAE **J1214** JAN92 .... **3:35.04**

Wheels—Recreational and Utility Trailer Test Procedure (A)—SAE **J1204** DEC89 .... **3:31.08**

RECYCLING
CFC-12 (R-12) Extraction Equipment for Mobile Automotive Air-Conditioning Systems—SAE **J2209** JUN92 .... **3:34.46**

Extraction and Recycle Equipment for Mobile Automotive Air-Conditioning Systems—SAE **J1990** MAR92 .... **3:34.43**

HFC-134a Recycling Equipment for Mobile Air-Conditioning Systems—SAE **J2210** DEC91 .... **3:34.58**

Recommended Service Procedure for the Containment of HFC-134a—SAE **J2211** DEC91 .... **3:34.57**

Standard of Purity for Recycled HFC-134a for Use in Mobile Air-Conditioning Systems—SAE **J2099** DEC91 .. **3:34.60**

Standard of Purity for Use in Mobile Air-Conditioning Systems—SAE **J1991** OCT89 .... **3:34.47**

REFLECTORS
Accelerated Exposure of Automotive Exterior Materials Using a Solar Fresnel-Reflective Apparatus—SAE **J1961** DEC88 .... **1:11.286**

Emergency Warning Device (Triangular Shape) (A)—SAE **J774** DEC89 .... **2:24.181**

Lighting and Marking of Agricultural Equipment on Highways—SAE **J137** JUN89 .... **3:39.60**

Plastic Materials for Use in Optical Parts Such as Lenses and Reflectors of Motor Vehicle Lighting Devices—SAE **J576** JUL91 .... **2:24.33**

Reflex Reflectors (D)—SAE **J594** MAY89 .... **2:24.177**

*Relfex Reflectors for use on Vehicles 2032 mm or More in Overall Width—SAE **J2041** JUN92 .... **2:24.178**

Retroreflective Materials for Vehicle Conspicuity (A)—SAE **J1967** MAR91 .... **3:34.112**

† SAE Ground Vehicle Lighting Standards Manual—SAE **HS-34** JAN93 .... Available as a separate publication *(See Related Technical Reports Section)*

Snowmobile and Snowmobile Cutter Lamps, Reflective Devices, and Associated Equipment (A)—SAE **J292** OCT84 .... **3:38.07**

REFRIGERANTS
*Alternate Refrigerant Consistency Criteria for use in Mobile Air Conditioning Systems—SAE **J658** JUN93 ... **3:34.63**

*Cautionary Statements for Handling HFC-134a During Mobile Air Conditioning Service—SAE **J1629** JUN93 ... **3:34.60**

CFC-12 (R-12) Extraction Equipment for Mobile Automotive Air-Conditioning Systems—SAE **J2209** JUN92 .... **3:34.46**

*Compatibility of Retrofit Refrigerants with Air Conditioning System Materials—SAE **J1662** JUN93 .... **3:34.69**

Extraction and Recycle Equipment for Mobile Automotive Air-Conditioning Systems—SAE **J1990** MAR92 .... **3:34.43**

*Fittings and Labels for Retrofit of CFC-12 (R12) Mobile Air Conditioning Systems to HFC-134a (R134a)—SAE **J1660** JUN93 .... **3:34.66**

HFC-134a (R-134a) Service Hose Fittings for Automotive Air-Conditioning Service Equipment—SAE **J2197** JUN92 **3:34.53**

HFC-134a Recycling Equipment for Mobile Air-Conditioning Systems—SAE **J2210** DEC91 .... **3:34.58**

*Procedure for Retrofitting CFC-12 (R12) Mobile Air Conditioning Systems to HFC-134a (R134a)—SAE **J1661** JUN93 .... **3:34.67**

*R134a Refrigerant Automotive Air Conditioning Hose—SAE **J2064** JUN93 .... **1:11.219**

*Rating Criteria for Electronic Refrigerant Leak Detectors—SAE **J1627** JUN93 .... **3:34.71**

Recommended Service Procedure for the Containment of R-12—SAE **J1989** OCT89 .... **3:34.42**

Recommended Service Procedure for the Containment of HFC-134a—SAE **J2211** DEC91 .... **3:34.57**

Safety and Containment of Refrigerant for Mechanical Vapor Compression Systems Used for Mobile Air-Conditioning Systems—SAE **J639** NOV91 .... **3:34.40**

*Selection Criteria for Retrofit Refrigerants to Replace R-12 in Mobile Air Conditioning Systems—SAE **J1657** JUN93 .... **3:34.61**

Service Hose for Automotive Air Conditioning—SAE **J2196** JUN92 .... **3:34.51**

Standard of Purity for Recycled HFC-134a for Use in Mobile Air-Conditioning Systems—SAE **J2099** DEC91 .. **3:34.60**

Standard of Purity for Use in Mobile Air-Conditioning Systems—SAE **J1991** OCT89 .... **3:34.47**

*Technican Procedure for Using Electronic Refrigerant Leak Detectors for Service of Mobile Air Conditioning Systems—SAE **J1628** JUN93 .... **3:34.72**

*Vehicle Testing Requirements for Replacement Refrigerants for CFC-12 (R12) Mobile Air Conditioning Systems—SAE **J659** JUN93 .... **3:34.64**

---

* Technical Revision   † New Document   (D)—DODISS Adopted   (A)—ANSI Recognized

Subject Index

REFRIGERATION
Tubing
† Refrigeration Tube Fittings—General Specifications—SAE **J513** JUN93 .... 2:22.33

REFUELING EMISSIONS
† Instrumentation and Techniques for Vehicle Refueling Emissions Measurement—SAE **J1045** MAY93 .... 1:13.137

REGULATIONS
* SAE Technical Standards Board Rules and Regulations—SAE **TSB 001** JUN93 .... 1:A.01

REINFORCED PLASTICS
See: COMPOSITE MATERIALS, FIBER REINFORCED PLASTICS

REMANUFACTURING
* Automotive Starter Drive Assembly Remanufacturing Procedures—SAE **J2241** MAR93 .... 3:42.46
* Automotive Starter Remanufactured Procedures—SAE **J2073** MAR93 .... 3:42.42
* Automotive Starter Solenoid Remanufacturing Procedures—SAE **J2242** MAR93 .... 3:42.47
Recommended Remanufacturing Procedures for Manual Transmission Clutch Assemblies—SAE **J1915** FEB90 .. 3:42.39
* Starter Armature Remanufacturing Procedures—SAE **J2240** MAR93 .... 3:42.45

REPAIR
See also: MAINTENANCE
Design Guidelines for Optimizing Automobile Collision Damage Resistance, Repairability, and Serviceability—SAE **J1555** NOV85 .... 3:42.19
Engine Water Pump Remanufacture Procedures and Acceptance Criteria—SAE **J1916** MAY89 .... 3:42.41
Flexible Bumper Repair—SAE **J1573** FEB87 .... 3:42.22
Identifying and Repairing High Strength Steel Vehicle Components—SAE **J1554** NOV85 .... 3:42.18
Inspection of Energy Absorbing Bumper Mounts (A)—SAE **J1571** MAY87 .... 3:42.25
Recommended Remanufacturing Procedures for Manual Transmission Clutch Assemblies—SAE **J1915** FEB90 .. 3:42.39
Stationary Glass Replacement—SAE **J1556** DEC85 .. 3:42.21
Unibody Weld Quality Testing (A)—SAE **J1827** MAY87 . 3:42.23
Uniform Reference and Dimensional Guidelines for Unibody Vehicles—SAE **J1828** MAY87 .... 3:42.28

Service Technicians
Assessing Technician Training (A)—SAE **J2018** JUL89 . 3:42.50
Developing Technician Training (A)—SAE **J2017** AUG89 . 3:42.48

RESISTANCE TESTING
American National Standard for Safety Glazing Materials for Glazing Motor Vehicles and Motor Vehicle Equipment Operating on Land Highways—Safety Code—ANSI **SAE Z26.1-1990** .... Available as a separate publication *(See Related Technical Reports Section)*

RESTRAINT SYSTEMS
See also: INFLATABLE RESTRAINTS, PASSIVE RESTRAINT SYSTEMS, SAFETY BELTS
Cab Sleeper Occupant Restraint System Test—SAE **J1948** MAR89 .... 3:34.267
Glossary of Automotive Inflatable Restraint Terms—SAE **J1538** APR88 .... 3:33.31
Guidelines for Evaluating Out-of-Position Vehicle Occupant Interactions with Deploying Airbags (A)—SAE **J1980** NOV90 .... 3:33.33
Identification of Automotive Air Bags—SAE **J1856** MAY89 .... 3:33.29
Occupant Restraint System Evaluation—Passenger Cars—SAE **J128** 3:33.12
† Operator Restraint System for Off-Road Work Machines—SAE **J386** JUN93 .... 3:40.491

Seat Belt Comfort, Fit, and Convenience (A)—SAE **J1834** JUN91 .... 3:34.267
† Vehicle Occupant Restraint Systems and Components Standards Manual—SAE **HS-13** MAY93 .... Available as a separate publication *(See Related Technical Reports Section)*

Child
Anchorage Provisions for Installation of Child Restraint Tether Straps in Rear Seating Positions—SAE **J1369** MAR92 .... 3:33.16
† Child Restraint Anchorages and Attachment Hardware (A)—SAE **J1368** JUN93 .... 3:33.14
* Guidelines for Evaluating Child Restraint System Interactions with Deploying Airbags—SAE **J2189** FEB93 . 3:33.20
Securing Child Restraint Systems in Motor Vehicle Rear Seats (A)—SAE **J1819** NOV90 .... 3:33.18

Head
Motor Vehicle Seating Manual—SAE **J782b** NOV80 ... 3:34.73

Passive
Instrumentation for Measuring Acoustic Impulses Within Vehicles (A)—SAE **J247** FEB87 .... 1:14.07
* The Air Bag Systems in Your Cars 'What the Public Needs to Know'—SAE **J2074** JUN93 .... 3:33.42

RETARDERS
Heavy Truck and Bus Retarder Downhill Performance Mapping Procedure (A)—SAE **J1489** JUN87 .... 3:36.124

Off-Road Vehicles
† Retardation Capability of Off-Highway Dumpers and Scrapers—SAE **J1430** FEB93 .... 3:40.298

RIDE
Automobile
Subjective Rating Scale for Evaluation of Noise and Ride Comfort Characteristics Related to Motor Vehicle Tires—SAE **J1060** 3:30.37

Off-Road Vehicles
† Measurement of Whole Body Vibration of the Seated Operator of Off-Highway Work Machines—SAE **J1013** AUG92 .... 3:39.92

Trucks and Buses
Measurement and Presentation of Truck Ride Vibrations (A)—SAE **J1490** JAN87 .... 3:34.252

RIMS
* Balance Weight and Rim Flange Design Specifications, Test Procedures, and Performance Recommendations—SAE **J1986** FEB93 .... 3:31.10
Dimensions—Wheels for Demountable Rims, Demountable Rims and Rim Spacers—Commercial Vehicles (A)—SAE **J851** JUN87 .... 3:36.98
Fastener Hardware for Wheels for Demountable Rims (A)—SAE **J1835** MAR90 .... 3:31.26
Labeling—Disc Wheel and Demountable Rims—Trucks (A)—SAE **J179** JUL84 .... 3:31.30
Off-Road Tire and Rim Classification—Construction Machines (A)—SAE **J751** APR86 .... 3:40.301
Off-Road Tire and Rim Classification—Forestry Machines (A)—SAE **J1440** APR86 .... 3:40.310
Off-Road Tire and Rim Selection and Application (A)—SAE **J1315** JAN91 .... 3:40.307
* Road Vehicles--Wheels for Commercial Vehicles and Multipurpose Passenger Vehicles--Fixing Nuts--Test Methods—SAE **J1965** FEB93 .... 3:36.107
Wheels/Rims—Trucks—Test Procedures and Performance Requirements (A)—SAE **J267** JAN91 .... 3:31.14

Terminology
Nomenclature—Wheels, Hubs, and Rims for Commercial Vehicles—SAE **J393** OCT91 .... 3:31.32
Wide Base Tire Rims and Wheels (A)—SAE **J876** JAN91 . 3:36.109

RINGS
See: O-RINGS, PISTON RINGS

---

* Technical Revision † New Document (D)—DODISS Adopted (A)—ANSI Recognized

Subject Index

RIPPERS
 Terminology
  Nomenclature—Rippers and Scarifiers (A)—SAE J733 JUN86 .... 3:40.45

RIVETS
 See also: FASTENERS
  Blind Rivets—Break Mandrel Type—SAE J1200 ...... 2:16.39
  Mechanical and Chemical Requirements for Nonthreaded Fasteners—SAE J430 .... 1:4.18
  Rivets and Riveting—SAE J492 .... 2:16.29
  Rivets for Brake Linings and Bolts for Brake Blocks—SAE J663b 2:25.94

ROAD ILLUMINATION DEVICES
 See: BACKUP LAMPS, CORNERING LAMPS, FOG LAMPS, HEADLAMPS, LAMPS, SPOT LAMPS, STOP LAMPS, TAIL LAMPS

ROAD LOAD
  Road Load Measurement and Dynamometer Simulation Using Coastdown Techniques—SAE J1263 JUN91 .... 2:26.470

ROAD TESTS
  Brake System Road Test Code—Passenger Car and Light-Duty Truck (A)—SAE J843 NOV90 .... 2:25.108
  *Electric Vehicle Acceleration, Gradeability, and Deceleration Test Procedure—SAE J1666 MAY93 ..... 3:28.12
  Fuel Economy Measurement Road Test Procedure—SAE J1082 JAN89 .... 2:26.459
  Fuel Economy Measurement—Road Test Procedure—Cold Start and Warm-Up Fuel Economy—SAE J1256 OCT88 .... 2:26.506
  *Vehicle Testing Requirements for Replacement Refrigerants for CFC-12 (R12) Mobile Air Conditioning Systems—SAE J659 JUN93 .... 3:34.64
 Corrosion
  Undervehicle Coupon Corrosion Tests—SAE J1293 JAN90 .... 1:3.48
 Fuel Consumption
  Joint TMC/SAE Fuel Consumption In-Service Test Procedure Type III (A)—SAE J1526 JUN87 .... 2:26.450

ROD ENDS
  Metric Spherical Rod Ends—SAE J1259 JUN89 ...... 2:17.16
  Performance Test Procedure—Ball Joints and Spherical Rod Ends (A)—SAE J1367 SEP81 .... 2:17.18
  Rod Ends and Clevis Pins—SAE J493 .... 2:16.28
  Spherical Rod Ends—SAE J1120 JUN89 .... 2:17.14

RODS
  Wrought Copper and Copper Alloys—SAE J463 SEP81 . 1:10.81
 Ferrous Materials
  Detection of Surface Imperfections in Ferrous Rods, Bars, Tubes, and Wires—SAE J349 FEB91 .... 1:3.63

ROLLERS
 See also: CONSTRUCTION AND INDUSTRIAL EQUIPMENT
  †Steering Frame Lock for Articulated Loaders and Tractors—SAE J276 NOV92 .... 3:40.353
 Brakes
  Braking Performance—Roller Compactors (A)—SAE J1472 JUN87 .... 3:40.288
  Braking Performance—Roller/Compactors (A)—SAE J1136 APR87 .... 3:40.300
 Noise
  Sound Measurement—Off-Road Self-Propelled Work Machines—Operator—Work Cycle (A)—SAE J1166 MAY90 .... 1:14.55
 Terminology
  Nomenclature—Rollers and Compactors (A)—SAE J1017 JAN86 .... 3:40.49

ROLLING
  Surface Rolling and Other Methods for Mechanical Prestressing of Metals—SAE J811 AUG81 .... 1:8.32

ROLLING RESISTANCE
 Tires
  Heavy Truck and Bus Retarder Downhill Performance Mapping Procedure (A)—SAE J1489 JUN87 .... 3:36.124
  Measurement of Passenger Car, Light Truck, and Highway Truck and Bus Tire Rolling Resistance—SAE J1270 MAR87 .... 3:30.21
  Rolling Resistance Measurement Procedure for Passenger Car, Light Truck, and Highway Truck and Bus Tires (A)—SAE J1269 MAR87 .... 3:30.18

ROLLOVER
  *Dolly Rollover Recommended Test Procedure—SAE J2114 APR93 .... 3:34.283
  †Roll-Over Protective Structures (ROPS) for Wheeled Agricultural Tractors (A)—SAE J2194 JUN93 .... 3:39.78

ROLLOVER PROTECTIVE STRUCTURES
 See: OPERATOR PROTECTIVE STRUCTURES

ROLLOVER THRESHOLDS
  *A Tilt-table Procedure for Measuring the Static Rollover Threshold for Heavy Trucks—SAE J2180 APR93 ..... 2:23.528

ROOFS
  Method for the Determination of Expansion and Water Absorption of Automotive Sealers (A)—SAE J1918 MAY88 .... 1:11.86
  Vehicle Roof Strength Test Procedure (A)—SAE J374 MAY91 .... 3:34.285

ROPS
 See: OPERATOR PROTECTIVE STRUCTURES

ROTORS
  Transmission Mounted Vehicle Speed Signal Rotor Specification—SAE J1377 JAN89 .... 2:21.17
  Turbocharger Gas Stand Test Code—SAE J1826 APR89 2:26.355

RUBBER
 See also: ELASTOMERS
  †Sponge and Expanded Cellular Rubber Products—SAE J18 JUL92 .... 1:11.56
  Standard Method for Determining Continuous Upper Temperature Resistance of Elastomers—SAE J2236 JUN92 .... 1:11.32
 Cups
  Motor Vehicle Brake Fluid (A)—SAE J1703 JUN91 .... 2:25.41
  Rubber Cups for Hydraulic Actuating Cylinders (A)—SAE J1601 NOV90 .... 2:25.175
 Floor Mats
  Automotive Rubber Mats (A)—SAE J80 JAN88 .... 1:11.185
 Foam
  Latex Foam Rubbers—SAE J17 JAN85 .... 1:11.53
 Hoses
  Automotive Air Conditioning Hose—SAE J51 MAY89 ... 1:11.249
  Power Steering Pressure Hose—High Volumetric Expansion Type (A)—SAE J188 OCT89 .... 1:11.241
  Power Steering Pressure Hose—Low Volumetric Expansion Type (A)—SAE J191 OCT89 .... 1:11.244
  Power Steering Pressure Hose—Wire Braid—SAE J190 1:11.248
  Power Steering Return Hose—Low Pressure (A)—SAE J189 DEC89 .... 1:11.243
 Latex
  Latex Dipped Goods and Coatings for Automotive Applications—SAE J19 JAN85 .... 1:11.63
  Latex Foam Rubbers—SAE J17 JAN85 .... 1:11.53
 Numbering System
  †Classification System for Rubber Materials—SAE J200 MAY93 .... 1:11.01
 O-Rings
  Rubber Rings for Automotive Applications—SAE J120a . 1:11.203
 Seals
  Brake Master Cylinder Reservoir Diaphragm Gasket—SAE J1605 MAR92 .... 2:25.22

---

* Technical Revision   † New Document   (D)—DODISS Adopted   (A)—ANSI Recognized

Subject Index

RUBBER (continued)
Seals (continued)
Low Water Tolerant Brake Fluids (A)—SAE **J1705** OCT88 .... 2:25.56

Motor Vehicle Brake Fluid (A)—SAE **J1703** JUN91 .... 2:25.41

Rubber Boots for Drum-Type Hydraulic Brake Wheel Cylinders (A)—SAE **J1604** OCT89 .... 2:25.24

Rubber Dust Boots for the Hydraulic Disc Brake Piston—SAE **J1570** SEP91 .... 2:25.181

Rubber Seals for Hydraulic Disc Brake Cylinders (A)—SAE **J1603** JUN90 .... 2:25.19

Seals—Evaluation of Elastohydrodynamic—SAE **J1002** JUN87 .... 3:29.137

SAFETY
See also: CRASHWORTHINESS, INJURIES, RESTRAINT SYSTEMS, SAFETY BELTS

Agricultural Equipment
Overhead Protection for Agricultural Tractors—Test Procedures and Performance Requirements—SAE **J167** APR92 .... 3:39.89

Rollover Protective Structures (ROPS) for Wheeled Agricultural Tractors (A)—SAE **J1194** MAY89 .... 3:39.71

Safety Alert Symbol for Agricultural, Construction and Industrial Equipment (A)—SAE **J284** JAN91 .... 3:39.68

† Safety for Agricultural Tractors—SAE **J208** JUN93 .... 3:39.69

† Steering Frame Lock for Articulated Loaders and Tractors—SAE **J276** NOV92 .... 3:40.353

Air Conditioning
*Cautionary Statements for Handling HFC-134a During Mobile Air Conditioning Service—SAE **J1629** JUN93 .... 3:34.60

Safety and Containment of Refrigerant for Mechanical Vapor Compression Systems Used for Mobile Air-Conditioning Systems—SAE **J639** NOV91 .... 3:34.40

*Selection Criteria for Retrofit Refrigerants to Replace R-12 in Mobile Air Conditioning Systems—SAE **J1657** JUN93 .... 3:34.61

Automobile
American National Standard for Safety Glazing Materials for Glazing Motor Vehicles and Motor Vehicle Equipment Operating on Land Highways—Safety Code—ANSI SAE **Z26.1-1990** .... Available as a separate publication (See Related Technical Reports Section)

Anchorage Provisions for Installation of Child Restraint Tether Straps in Rear Seating Positions—SAE **J1369** MAR92 .... 3:33.16

† Automotive Safety Glasses—SAE **J673** APR93 .... 3:34.100

Collision Deformation Classification—SAE **J224** MAR80 .... 3:34.235

Fixed Rigid Barrier Collision Tests—SAE **J850** NOV88 .... 3:34.281

Human Tolerance to Impact Conditions as Related to Motor Vehicle Design—SAE **J885** JUL86 .... 3:34.306

Instrumentation for Impact Test (A)—SAE **J211** OCT88 .... 3:34.270

Moving Rigid Barrier Collision Tests—SAE **J972** DEC88 .... 3:34.278

Passenger Car Door System Crush Test Procedure—SAE **J367** JUN80 .... 3:34.286

Safety Glazing Materials—Motor Vehicles (A)—SAE **J674** NOV90 .... 3:34.103

*The Air Bag Systems in Your Cars 'What the Public Needs to Know'—SAE **J2074** JUN93 .... 3:33.42

Construction and Industrial Equipment
On-Machine Alarm Test and Evaluation Procedure for Construction and General Purpose Industrial Machinery—SAE **J1446** MAY89 .... 1:14.50

Operator Precautions (A)—SAE **J153** MAY87 .... 3:40.470

† Personnel Protection for General Purpose Industrial Machines (A)—SAE **J98** NOV92 .... 3:40.472

Personnel Protection—Skid Steer Loaders (A)—SAE **J1388** JUN85 .... 3:40.474

Safety Alert Symbol for Agricultural, Construction and Industrial Equipment (A)—SAE **J284** JAN91 .... 3:39.68

† Steering Frame Lock for Articulated Loaders and Tractors—SAE **J276** NOV92 .... 3:40.353

Three Point Hitch (Type A) Backhoe Personnel Protection—SAE **J1518** JUN86 .... 3:40.353

Earthboring Machines
*Operating Precautions for Horizontal Earthboring Machines—SAE **J2023** JAN93 .... 3:40.20

Maintenance
Service Maintenance of SAE J1703 Brake Fluids in Motor Vehicle Brake Systems—SAE **J1707** NOV91 .... 2:25.53

Mowers
Industrial Flail Mowers and Power Rakes (A)—SAE **J1001** MAR87 .... 3:40.488

Industrial Rotary Mowers (A)—SAE **J232** DEC84 .... 3:40.477

Off-Road Vehicles
Access Systems for Off-Road Machines (D)—SAE **J185** JUN88 .... 3:40.383

Safety Signs (A)—SAE **J115** JAN87 .... 3:40.494

Unauthorized Starting or Movement of Machines (A)—SAE **J1083** JUL85 .... 3:40.96

Snowmobiles
Dynamic Cushioning Performance Criteria for Snowmobile Seats (A)—SAE **J89** JAN85 .... 3:38.13

† Recommendations for Children's Snowmobile—SAE **J1038** DEC92 .... 3:38.18

Terminology
Handbook of Motor Vehicle, Safety and Environmental Terminology—SAE **HS-215** .... Available as a separate publication (See Related Technical Reports Section)

Trucks and Buses
Collision Deformation Classification—SAE **J224** MAR80 .... 3:34.235

Fixed Rigid Barrier Collision Tests—SAE **J850** NOV88 .... 3:34.281

Rear Underride Guard Test Procedure (A)—SAE **J260** JUN90 .... 3:34.287

Safety Glazing Materials—Motor Vehicles (A)—SAE **J674** NOV90 .... 3:34.103

School Bus Stop Arm (A)—SAE **J1133** JUL89 .... 2:24.196

SAFETY BELTS
See also: RESTRAINT SYSTEMS

Human Tolerance to Impact Conditions as Related to Motor Vehicle Design—SAE **J885** JUL86 .... 3:34.306

Motor Vehicle Seat Belt Assembly Installations (A)—SAE **J800** APR86 .... 3:33.09

† Operator Restraint System for Off-Road Work Machines—SAE **J386** JUN93 .... 3:40.491

*SAE Automotive Textiles and Trim Standards Manual—SAE **HS-2700** JUN93 .... Available as a separate publication (See Related Technical Reports Section)

Seat Belt Comfort, Fit, and Convenience (A)—SAE **J1834** JUN91 .... 3:34.267

Seat Belt Hardware Performance Requirements—SAE **J141** 3:33.03

Seat Belt Hardware Test Procedure—SAE **J140a** 3:33.01

Securing Child Restraint Systems in Motor Vehicle Rear Seats (A)—SAE **J1819** NOV90 .... 3:33.18

† Vehicle Occupant Restraint Systems and Components Standards Manual—SAE **HS-13** MAY93 .... Available as a separate publication (See Related Technical Reports Section)

Anchorages
Anchorage Provisions for Installation of Child Restraint Tether Straps in Rear Seating Positions—SAE **J1369** MAR92 .... 3:33.16

Motor Vehicle Seat Belt Anchorages—Design Recommendations (A)—SAE **J383** APR86 .... 3:33.05

---

* Technical Revision   † New Document   (D)—DODISS Adopted   (A)—ANSI Recognized

Subject Index

SAFETY BELTS (continued)
Anchorages (continued)
Motor Vehicle Seat Belt Anchorages—Performance Requirements (A)—SAE **J385** APR86 .... 3:33.08

Motor Vehicle Seat Belt Anchorages—Test Procedure (A)—SAE **J384** APR86 .... 3:33.07

Webbing
Seat Belt Assembly Webbing Abrasion Performance Requirements (A)—SAE **J114** MAR86 .... 3:33.11

Seat Belt Assembly Webbing Abrasion Test Procedure (A)—SAE **J339** MAR86 .... 3:33.11

SAFETY CHAINS
Safety Chain of Full Trailers or Converter Dollies (A)—SAE **J697** MAY88 .... 3:36.98

Trailer Couplings, Hitches and Safety Chains—Automotive Type (A)—SAE **J684** MAY87 .... 3:35.01

SAFETY GLASS
See: GLASS, WINDOWS, WINDSHIELDS

SAFETY SIGNS
Safety Signs (A)—SAE **J115** JAN87 .... 3:40.494

SCARIFIERS
See also: CONSTRUCTION AND INDUSTRIAL EQUIPMENT
Terminology
Nomenclature—Rippers and Scarifiers (A)—SAE **J733** JUN86 .... 3:40.45

SCHOOL BUSES
† SAE Ground Vehicle Lighting Standards Manual—SAE **HS-34** JAN93 .... Available as a separate publication *(See Related Technical Reports Section)*

School Bus Stop Arm (A)—SAE **J1133** JUL89 .... 2:24.196

School Bus Warning Lamps (A)—SAE **J887** AUG87 .... 2:24.192

SCRAPERS
See also: CONSTRUCTION AND INDUSTRIAL EQUIPMENT, EARTHMOVING EQUIPMENT

† Capacity Rating—Elevating Scrapers—SAE **J957** ISO 6484- JUN93 .... 3:40.322

† Capacity Rating—Scraper, Open Bowl (A)—SAE **J741** JUN93 .... 3:40.319

Cutting Edge—Double Bevel Cross Sections (A)—SAE **J738** JUN86 .... 3:40.235

Hole Spacing for Scraper and Dozer Cutting Edges (A)—SAE **J737** AUG89 .... 3:40.231

Loading Ability Test Code—Scrapers (A)—SAE **J764** NOV85 .... 3:40.159

Brakes
Braking Performance—Rubber-Tired Construction Machines (A)—SAE **J1152** APR80 .... 3:40.289

Noise
Sound Measurement—Off-Road Self-Propelled Work Machines—Operator—Work Cycle (A)—SAE **J1166** MAY90 .... 1:14.55

Retarders
† Retardation Capability of Off-Highway Dumpers and Scrapers—SAE **J1430** FEB93 .... 3:40.298

Steering
Minimum Performance Criteria for Emergency Steering of Wheeled Earthmoving Construction Machines—SAE **J53** OCT84 .... 3:40.515

Terminology
Component Nomenclature—Scrapers (A)—SAE **J728** JUL90 .... 3:40.34

Identification Terminology of Earthmoving Machines (A)—SAE **J1057** SEP88 .... 3:40.05

Tire Guards
Tire Guards for Protection of Operator of Earthmoving Haulage Machines (A)—SAE **J321b** 3:40.352

SCREWS
See also: FASTENERS
Decarburization in Hardened and Tempered Threaded Fasteners (A)—SAE **J121** AUG83 .... 1:4.44

Mechanical and Material Requirements for Externally Threaded Fasteners (A)—SAE **J429** AUG83 .... 1:4.01

Mechanical and Material Requirements for Metric Externally Threaded Steel Fasteners (A)—SAE **J1199** SEP83 .... 1:4.06

Mechanical and Quality Requirements for Machine Screws—SAE **J82** JUN79 .... 1:4.14

Mechanical and Quality Requirements for Tapping Screws—SAE **J933** JUN79 .... 1:4.19

Metric Thread Rolling Screws—SAE **J1237** 1:4.29

Screw Threads (ANSI B1.1)—SAE **J475a** .... Available as a separate publication *(See Related Technical Reports Section)*

Steel Self-Drilling Tapping Screws—SAE **J78** JUN79 .... 1:4.21

Surface Discontinuities on Bolts, Screws, and Studs—SAE **J123c** 1:4.37

† Surface Discontinuities on General Application Bolts, Screws, and Studs—SAE **J1061** OCT92 .... 1:4.33

Test Methods for Metric Threaded Fasteners (A)—SAE **J1216** 1:4.10

Thread Rolling Screws—SAE **J81** JUN79 .... 1:4.26

SEALANTS
See also: ADHESIVES
Design Guide for Formed-In-Place Gaskets (A)—SAE **J1497** DEC88 .... 1:11.187

Electrocoat Compatibilities of Automotive Sealers—SAE **J1969** OCT88 .... 1:11.90

Method for Evaluating Material Separation in Automotive Sealers Under Pressure in Static Conditions—SAE **J1864** APR87 .... 1:11.92

Method for Evaluating the Adhesion Characteristics of Automotive Sealers—SAE **J1600** APR87 .... 1:11.82

Method for Evaluating the Flow Properties of Pumpable Sealers—SAE **J2025** JUN89 .... 1:11.66

Method for Evaluating the Paintable Characteristics of Automotive Sealers—SAE **J1800** APR87 .... 1:11.83

Method for the Determination of Expansion and Water Absorption of Automotive Sealers (A)—SAE **J1918** MAY88 .... 1:11.86

Method of Viscosity Test for Automotive Type Adhesives, Sealers, and Deadeners—SAE **J1524** NOV88 .... 1:11.78

Methods of Tests for Automotive-Type Sealers, Adhesives, and Deadeners—SAE **J243** 1:11.68

O.D. Coatings for Radial Lip-Type Shaft Seals—SAE **J1947** AUG89 .... 1:11.65

Overlap Shear Test for Automotive Type Sealant for Stationary Glass Bonding (A)—SAE **J1529** MAY86 .... 1:11.95

Overlap Shear Test for Sealant Adhesive Bonding of Automotive Glass Encapsulating Material to Body Opening—SAE **J1836** OCT88 .... 1:11.97

Synthetic Resin Plastic Sealers, Nondrying Type—SAE **J250** 1:11.65

* Thread Sealants—SAE **J1615** JUN93 .... 2:22.240

Urethane
Stationary Glass Replacement—SAE **J1556** DEC85 .... 3:42.21

SEALS
See also: GASKETS, O-RINGS, PISTON RINGS
Cast Iron Sealing Rings—SAE **J281** SEP80 .... 3:29.148

† Cast Iron Sealing Rings (Metric)—SAE **J1236** APR93 .... 3:29.156

† Fitting—O-Ring Face Seal—SAE **J1453** JUN93 .... 2:22.256

Front-Wheel-Drive Constant Velocity Joint Boot Seals—SAE **J2028** JUN92 .... 3:29.07

Guide to the Application and Use of Passenger Car Air-Conditioning Compressor Face Seals (A)—SAE **J1954** MAY90 .... 2:26.423

---

* Technical Revision   † New Document   (D)—DODISS Adopted   (A)—ANSI Recognized

SEALS (continued)
Rubber Rings for Automotive Applications—SAE J120a . 1:11.203
† Static and Reciprocating Elastomeric Trnamission Seals—SAE J654 MAR93 .... 3:29.145
† Test Method for Evaluating the Sealing Capability of Hose Connections with a PVT Test Facility—SAE J1610 APR93 .... 2:16.80

Elastohydrodynamic
Application Guide to Radial Lip Seals—SAE J946 OCT91 3:29.109
Seals—Evaluation of Elastohydrodynamic—SAE J1002 JUN87 .... 3:29.137

Hydraulic Cylinder
Cylinder Rod Wiper Seal Ingression Test (A)—SAE J1195 MAR86 .... 3:40.141
Hydraulic Cylinder Leakage Test—SAE J1336 JUN87 .. 3:40.346
Hydraulic Cylinder Rod Seal Endurance Test Procedure (A)—SAE J1374 MAY85 .... 3:40.143
Rubber Seals for Hydraulic Disc Brake Cylinders (A)—SAE J1603 JUN90 .... 2:25.19

Pump
Engine Coolant Pump Seals—SAE J780 JUN90 ...... 2:26.415
Guide to the Application and Use of Engine Coolant Pump Face Seals (A)—SAE J1245 JUN82 .... 2:26.417

Radial Lip
Application Guide to Radial Lip Seals—SAE J946 OCT91 3:29.109
Fluid Sealing Handbook—Radial Lip Seals—SAE HS-1417 .... Available as a separate publication (See Related Technical Reports Section)
Lip Force Measurement, Radial Lip Seals (A)—SAE J1901 SEP90 .... 3:29.135
O.D. Coatings for Radial Lip-Type Shaft Seals—SAE J1947 AUG89 .... 1:11.65
Radial Lip Seal Torque: Measurement Methods and Results—SAE J1971 JUN89 .... 3:29.151
Seals—Bond Test Fixture and Procedure (A)—SAE J1900 OCT90 .... 3:29.128
Seals—Terminology of Radial Lip (A)—SAE J111 JUN88 3:29.103
Seals—Testing of Radial Lip—SAE J110 DEC91 ...... 3:29.140

Rubber
Brake Master Cylinder Reservoir Diaphragm Gasket—SAE J1605 MAR92 .... 2:25.22
Motor Vehicle Brake Fluid (A)—SAE J1703 JUN91 .... 2:25.41
Rubber Seals for Hydraulic Disc Brake Cylinders (A)—SAE J1603 JUN90 .... 2:25.19

Transmission
† Static and Reciprocating Elastomeric Trnamission Seals—SAE J654 MAR93 .... 3:29.145

SEAT BELTS
See: SAFETY BELTS

SEATS
* SAE Automotive Textiles and Trim Standards Manual—SAE HS-2700 MAR93 .... Available as a separate publication (See Related Technical Reports Section)

Automobiles and Trucks
Accommodation Tool Reference Point—SAE J1516 MAR90 .... 3:34.247
Devices for Use in Defining and Measuring Vehicle Seating Accommodation—SAE J826 JUN92 ...... 3:34.90
Driver Hand Control Reach (A)—SAE J287 JUN88 .... 3:34.157
Driver Selected Seat Position (A)—SAE J1517 MAR90 . 3:34.250
Dynamic Flex Fatigue Test for Slab Polyurethane Foam—SAE J388 .... 3:34.98
Load Deflection Testing of Urethane Foams for Automotive Seating—SAE J815 MAY81 .... 3:34.100
Motor Vehicle Seating Manual—SAE J782b MAY80 ... 3:34.73
Motor Vehicle Seating Systems—SAE J879b 3:34.94
Securing Child Restraint Systems in Motor Vehicle Rear Seats (A)—SAE J1819 NOV90 .... 3:33.18
Truck Driver Shin-Knee Position for Clutch and Accelerator (A)—SAE J1521 MAR90 .... 3:34.263
Truck Driver Stomach Position (A)—SAE J1522 MAR90 3:34.265
Urethane for Automotive Seating—SAE J954 .... 3:34.99

Off-Road Vehicles
Determining Seat Index Point (A)—SAE J1163 JUN91 .. 3:40.391
Dynamic Durability Testing of Seat Cushions for Off-Road Work Machines (A)—SAE J1454 JAN86 .... 3:40.389
Force-Deflection Measurements of Cushioned Components of Seats for Off-Road Work Machines—SAE J1051 DEC88 .... 3:40.388
Operator's Seat Dimensions for Off-Road Self-Propelled Work Machines—SAE J899 DEC88 .... 3:40.395

Snowmobiles
Dynamic Cushioning Performance Criteria for Snowmobile Seats (A)—SAE J89 JAN85 .... 3:38.13

Vibration
Classification of Agricultural Wheeled Tractors for Vibration Tests of Operator Seats (A)—SAE J1386 JAN86 .... 3:40.401
Classification of Earthmoving Machines for Vibration Tests of Operator Seats (A)—SAE J1385 JUN83 ...... 3:40.398
† Vibration Performance Evaluation of Operator Seats (A)—SAE J1384 JUN93 .... 3:40.396

SECURITY SYSTEMS
E/E Data Link Security—SAE J2186 SEP91 .... 2:23.507

SENSORS
See also: ELECTRONIC EQUIPMENT
† Accelerator Pedal Position Sensor for Use with Electronic Controls in Medium- and Heavy-Duty Vehicle Applications—SAE J1843 APR93 .... 2:23.515
Chrysler Sensor and Control (CSC) Bus Multiplexing Network for Class 'A' Applications—SAE J2058 JUN90 2:23.422
* Class A Multiplexing Sensors—SAE J2057/3 JUN93 ... 2:23.413
Collision Detection Serial Data Communications Multiplex Bus—SAE J1567 AUG87 .... 2:23.481
* Equivalent Temperature—SAE J2234 JAN93 .... 3:34.25

Crash
Glossary of Automotive Inflatable Restraint Terms—SAE J1538 APR88 .... 3:33.31

Speed
Transmission Mounted Vehicle Speed Signal Rotor Specification—SAE J1377 JAN89 .... 2:21.17

SERVICE TECHNICIANS
Assessing Technician Training (A)—SAE J2018 JUL89 . 3:42.50
Developing Technician Training (A)—SAE J2017 AUG89 3:42.48

SERVICE VEHICLES

Lamps
360 Degree Warning Devices for Authorized Emergency, Maintenance, and Service Vehicles—SAE J845 MAR92 . 2:24.185
Flashing Warning Lamps for Authorized Emergency, Maintenance and Service Vehicles (A)—SAE J595 JAN90 .... 2:24.183
Gaseous Discharge Warning Lamp for Authorized Emergency, Maintenance, and Service Vehicles (A)—SAE J1318 APR86 .... 2:24.188
† SAE Ground Vehicle Lighting Standards Manual—SAE HS-34 JAN93 .... Available as a separate publication (See Related Technical Reports Section)

SERVICEABILITY
See also: MAINTENANCE, REPAIR

Construction and Industrial Equipment
Design Guidelines for Optimizing Automobile Collision Damage Resistance, Repairability, and Serviceability—SAE J1555 NOV85 .... 3:42.19

---
* Technical Revision   † New Document   (D)—DODISS Adopted   (A)—ANSI Recognized

Subject Index

SERVICEABILITY (continued)
Construction and Industrial Equipment (continued)
Engineering Design Serviceability Guidelines—Construction and Industrial Machinery—Maintainability Index—Off-Road Work Machines (A)—SAE **J817/2** MAR91 .... 3:40.522

Engineering Design Serviceability Guidelines—Construction and Industrial Machinery—Serviceability Definitions—Off-Road Work Machines (A)—SAE **J817/1** MAR91 .... 3:40.521

Identifying and Repairing High Strength Steel Vehicle Components—SAE **J1554** NOV85 .... 3:42.18

Stationary Glass Replacement—SAE **J1556** DEC85 ... 3:42.21

Off-Road Vehicles
Engineering Design Serviceability Guidelines—Construction and Industrial Machinery—Serviceability Definitions—Off-Road Work Machines (A)—SAE **J817/1** MAR91 .... 3:40.521

SHIELDING
† Personnel Protection for General Purpose Industrial Machines (A)—SAE **J98** NOV92 .... 3:40.472

† Recommendations for Children's Snowmobile—SAE **J1038** DEC92 .... 3:38.18

† Safety for Agricultural Tractors—SAE **J208** JUN93 .... 3:39.69

Shielding of Starter System Energization—SAE **J1493** OCT91 .... 3:40.266

SHOT
See: SHOT PEENING

SHOT PEENING
See also: GLASS BEADS

† Cast Shot and Grit Size Specifications for Peening and Cleaning—SAE **J444** MAY93 .... 1:8.11

Cast Steel Shot (A)—SAE **J827** MAR90 .... 1:8.09

† Cut Wire Shot—SAE **J441** JUN93 .... 1:8.05

Manual on Shot Peening—SAE **HS-84** .... Available as a separate publication *(See Related Technical Reports Section)*

Metallic Shot and Grit Mechanical Testing—SAE **J445** AUG84 .... 1:8.12

Procedures for Using Standard Shot Peening Test Strip (A)—SAE **J443** JAN84 .... 1:8.08

Size Classification and Characteristics of Ceramic Shot for Peening (A)—SAE **J1830** MAY87 .... 1:8.32

Size Classification and Characteristics of Glass Beads for Peening (A)—SAE **J1173** SEP88 .... 1:8.30

Specifications for Low Carbon Cast Steel Shot (A)—SAE **J2175** JUN91 .... 1:8.10

Test Strip, Holder and Gage for Shot Peening—SAE **J442** AUG79 .... 1:8.06

SHOVELS
Shovel Dipper, Clam Bucket, and Dragline Bucket Rating—SAE **J67** NOV84 .... 3:40.323

Crane
Nomenclature and Dimensions for Crane Shovels—SAE **J958** JUN89 .... 3:40.53

SIDE MARKER LAMPS
Clearance, Side Marker, and Identification Lamps—SAE **J592** JUN92 .... 2:24.170

Clearance, Sidemarker, and Identification Lamps for Use on Motor Vehicles 2032 mm or More in Overall Width (A)—SAE **J2042** JUN91 .... 2:24.172

† SAE Ground Vehicle Lighting Standards Manual—SAE **HS-34** JAN93 .... Available as a separate publication *(See Related Technical Reports Section)*

Side Turn Signal Lamps (A)—SAE **J914** NOV87 ...... 2:24.160

Snowmobile and Snowmobile Cutter Lamps, Reflective Devices, and Associated Equipment (A)—SAE **J292** OCT84 .... 3:38.07

SIGNAL PROCESSING
A Vehicle Network Protocol with a Fault Tolerant Multiplex Signal Bus—SAE **J1813** AUG87 .... 2:23.499

SIGNALING AND MARKING DEVICES
See: BACKUP LAMPS, CORNERING LAMPS, FOG LAMPS, HEADLAMPS, LAMPS, SIDE MARKER LAMPS, STOP LAMPS, TAIL LAMPS, TURN INDICATORS

SILENCERS
See also: EXHAUST SYSTEMS

Measurement Procedure for Determination of Silencer Effectiveness in Reducing Engine Intake or Exhaust Sound Level (A)—SAE **J1207** FEB87 .... 1:14.115

SIMULATION
* A Tilt-table Procedure for Measuring the Static Rollover Threshold for Heavy Trucks—SAE **J2180** APR93 ..... 2:23.528

Electrostatic Discharge Test for Vehicles—SAE **J1595** OCT88 .... 2:23.464

* Guidelines for Evaluating Child Restraint System Interactions with Deploying Airbags—SAE **J2189** FEB93 3:33.20

Brakes
Simulated Mountain Brake Performance Test Procedure—SAE **J1247** APR80 .... 2:25.123

Road Load
Road Load Measurement and Dynamometer Simulation Using Coastdown Techniques—SAE **J1263** JUN91 .... 2:26.470

Work Cycle
Acoustics—Measurement of Exterior Noise Emitted by Earthmoving Machinery—Dynamic Test Conditions (A)—SAE **J2104** JUN91 .... 1:14.81

Acoustics—Measurement of Noise Emitted by Earthmoving Machinery at the Operator's Position—Simulated Work Cycle Test Conditions (A)—SAE **J2105** JUN91 .... 1:14.87

SINTERED CARBIDES
Sintered Carbide Tools—SAE **J439a** 1:7.07

Sintered Tool Materials—SAE **J1072** 1:7.12

SINTERED METAL POWDERS
Sintered Powder Metal Parts: Ferrous—SAE **J471d** 1:8.01

SIRENS
See: WARNING SYSTEMS

SKIDDERS
See also: FORESTRY EQUIPMENT

Classification and Nomenclature Towing Winch for Skidders and Crawler Tractors—SAE **J1014** JAN85 .... 3:40.51

Specification Definitions—Clam Bunk Skidder (A)—SAE **J1824** JUN91 .... 3:40.81

Specification Definitions—Winches for Crawler Tractors and Skidders—SAE **J1158** JAN85 .... 3:40.79

Brakes
Braking Performance—Rubber-Tired Skidders (A)—SAE **J1178** JUN87 .... 3:40.97

Noise
Sound Measurement—Off-Road Self-Propelled Work Machines—Operator—Work Cycle (A)—SAE **J1166** MAY90 .... 1:14.55

Terminology
† Component Nomenclature—Articulated Log Skidder, Rubber-Tired (A)—SAE **J1109** JUN93 .... 3:40.62

Component Nomenclature—Skidder-Grapple (A)—SAE **J1111** MAR81 .... 3:40.64

Identification Terminology of Mobile Forestry Machines—SAE **J1209** JAN85 .... 3:40.25

Nomenclature—Clam Bunk Skidder (A)—SAE **J1353** APR87 .... 3:40.80

† Specification Definitions—Articulated Rubber-Tired Log Skidder (A)—SAE **J1110** JUN93 .... 3:40.73

---

* Technical Revision † New Document (D)—DODISS Adopted (A)—ANSI Recognized

Subject Index

SKIDDERS (continued)
Terminology (continued)
Specification Definitions—Skidder-Grapple (A)—SAE J1112 APR80 .... 3:40.75
Tires
Off-Road Tire and Rim Classification—Forestry Machines (A)—SAE J1440 APR86 .... 3:40.310

SLACK ADJUSTER
Automatic Slack Adjuster Performance Requirements—SAE J1513 APR90 .... 3:36.176
Automatic Slack Adjuster Test Procedure (A)—SAE J1462 MAY87 .... 3:36.175
Manual Slack Adjuster Performance Requirements (A)—SAE J1512 APR90 .... 3:36.176
Manual Slack Adjuster Test Procedure (A)—SAE J1461 APR91 .... 3:36.174

SLOPE OPERATION
Machine Slope Operation Test Code (A)—SAE J897 OCT85 .... 3:40.158

SLOW MOVING VEHICLES
Mounting Brackets and Socket for Warning Lamp and Slow-Moving Vehicle (SMV) Identification Emblem (A)—SAE J725 MAR91 .... 3:39.65
† Slow-Moving Vehicle Identification Emblem—SAE J943 JUN93 .... 3:39.62

SMALL ENGINES
See: DIESEL ENGINES, SPARK IGNITION ENGINES

SMOKE
Diesel Emission Production Audit Test Procedure (A)—SAE J1243 MAY88 .... 1:13.50
Diesel Engine Smoke Measurement—SAE J255a 1:13.56
Diesel Smoke Measurement Procedure (A)—SAE J35 SEP88 .... 1:13.64

SMOKE TUNNEL TESTING
Aerodynamic Flow Visualization Techniques and Procedures—SAE HS-1566 .... Available as a separate publication (See Related Technical Reports Section)

SMOKEMETERS
Diesel Engine Smoke Measurement—SAE J255a 1:13.56

SNOW BLOWERS
Noise
Bystander Sound Level Measurement Procedure for Small Engine Powered Equipment (A)—SAE J1175 MAR85 .... 1:14.97
Operator Ear Sound Level Measurement Procedure for Small Engine Powered Equipment (A)—SAE J1174 MAR85 .... 1:14.95

SNOWMOBILES
† Recommendations for Children's Snowmobile—SAE J1038 DEC92 .... 3:38.18
Snowmobile Drive Mechanisms—SAE J1279 DEC88 .... 3:38.18
Brakes
Brake System Test Procedure—Snowmobiles (A)—SAE J45 OCT84 .... 3:38.11
Service Brake System Performance Requirements—Snowmobiles—SAE J44 DEC91 .... 3:38.11
Snowmobile Brake Control Systems (A)—SAE J1282 OCT84 .... 3:38.09
† Surface Vehicle Brake Manual—SAE HS-24 JUL92 .... Available as a separate publication (See Related Technical Reports Section)
Control Systems
Snowmobile Brake Control Systems (A)—SAE J1282 OCT84 .... 3:38.09
Snowmobile Throttle Control Systems (A)—SAE J92 OCT84 .... 3:38.15

Speed Control Assurance for Snowmobiles (A)—SAE J1222 OCT84 .... 3:38.17
Electric Equipment
Maintenance of Design Voltage—Snowmobile Electrical Systems (A)—SAE J277 JUN90 .... 3:38.03
Fuel Tanks
Snowmobile Fuel Tanks (A)—SAE J288 NOV83 ...... 3:38.15
Hand Grips
Snowmobile Passenger Handgrips (A)—SAE J1062 OCT84 .... 3:38.09
Lamps
† SAE Ground Vehicle Lighting Standards Manual—SAE HS-34 JAN93 .... Available as a separate publication (See Related Technical Reports Section)
Snowmobile and Snowmobile Cutter Lamps, Reflective Devices, and Associated Equipment (A)—SAE J292 OCT84 .... 3:38.07
Snowmobile Headlamps (A)—SAE J280 JUN84 ...... 3:38.04
Snowmobile Stop Lamp (A)—SAE J278 OCT84 ...... 3:38.05
Snowmobile Tail Lamp (Rear Position Lamp) (A)—SAE J279 OCT84 .... 3:38.06
Noise
Exterior Sound Level for Snowmobiles (A)—SAE J192 MAR85 .... 1:14.36
Operational Sound Level Measurement Procedure for Snow Vehicles (A)—SAE J1161 MAR83 .... 1:14.39
Operator Ear Sound Level Measurement Procedure for Snow Vehicles (A)—SAE J1160 MAR83 .... 1:14.37
Seats
Dynamic Cushioning Performance Criteria for Snowmobile Seats (A)—SAE J89 JAN85 .... 3:38.13
Switches
Tests for Snowmobile Switching Devices and Components—SAE J68 DEC91 .... 3:38.08
Terminology
Snowmobile Definitions and Nomenclature—General—SAE J33 DEC91 .... 3:38.01
Vehicle Identification Number
Vehicle Identification Numbers—Snowmobiles (A)—SAE J64 APR88 .... 3:38.03

SOFTWARE
† Joint SAE/TMC Electronic Data Interchange Between Microcomputer Systems in Heavy-Duty Vehicle Applications—SAE J1587 AUG92 .... 2:23.531

SOLDERING
Automotive Metallurgical Joining—SAE J836a AUG70 .. 1:9.08
Welding, Brazing, and Soldering—Materials and Practices—SAE J1147 JUN83 .... 1:9.07
Copper Alloys
Wrought and Cast Copper Alloys—SAE J461 SEP81 .... 1:10.45

SOLDERS
Solders—SAE J473a 1:10.157

SOLENOIDS
Qualifications for Four-Way Subbase Mounted Air Valves for Automotive Manufacturing Applications—SAE J2051 JUN92 .... 3:43.21
Shielding of Starter System Energization—SAE J1493 OCT91 .... 3:40.266

SOUND
See also: ACOUSTICS, NOISE
Data Acquisition
Qualifying a Sound Data Acquisition System (D)—SAE J184 AUG87 .... 1:14.09
Earthmoving Equipment
† Sound Power Level Measurements of Earthmoving Machinery - Static and In-Place Dynamic Methods—SAE J1805 APR93 .... 1:14.90

---

* Technical Revision † New Document (D)—DODISS Adopted (A)—ANSI Recognized

Subject Index

SOUND (continued)
Measurement

Bystander Sound Level Measurement Procedure for Small Engine Powered Equipment (A)—SAE J1175 MAR85 .... 1:14.97

Laboratory Measurement of the Airborne Sound Barrier Performance of Automotive Materials and Assemblies (A)—SAE J1400 MAY90 .... 1:11.174

† Measurement of Exhaust Sound Levels of Stationary Motorcycles—SAE J1287 JUN93 .... 1:14.05

Measurement of Interior Sound Levels of Light Vehicles (A)—SAE J1477 JAN86 .... 1:14.19

Measurement of Light Vehicle Exhaust Sound Level Under Stationary Conditions—SAE J1169 MAR92 ..... 1:14.21

Measurement of Light Vehicle Stationary Exhaust System Sound Level Engine Speed Sweep Method—SAE J1492 MAR92 .... 1:14.24

Measurement of Noise Emitted by Accelerating Highway Vehicles—SAE J1470 MAR92 .... 1:14.32

Measurement Procedure for Determination of Silencer Effectiveness in Reducing Engine Intake or Exhaust Sound Level (A)—SAE J1207 FEB87 .... 1:14.115

Operator Ear Sound Level Measurement Procedure for Small Engine Powered Equipment (A)—SAE J1174 MAR85 .... 1:14.95

Shoreline Sound Level Measurement Procedure—SAE J1970 DEC91 .... 1:14.40

† Sound Measurement—Construction Site (A)—SAE J1075 JUN93 .... 1:14.99

Sound Measurement—Off-Road Self-Propelled Work Machines—Operator—Work Cycle (A)—SAE J1166 MAY90 .... 1:14.55

Sound Measurement—Trenching Machines—SAE J1262 OCT90 .... 1:14.109

† Sound Power Level Measurements of Earthmoving Machinery - Static and In-Place Dynamic Methods—SAE J1805 APR93 .... 1:14.90

Stationary Sound Level Measurement Procedure for Pleasure Motorboats—SAE J2005 DEC91 .... 1:14.41

Surface Vehicle Sound Measurement Manual—SAE HS-184 NOV89 .... Available as a separate publication (See Related Technical Reports Section)

SOUNDPROOFING

Acoustical and Thermal Materials Test Procedure (A)—SAE J1324 OCT89 .... 1:11.111

Laboratory Measurement of the Airborne Sound Barrier Performance of Automotive Materials and Assemblies (A)—SAE J1400 MAY90 .... 1:11.174

SPARK ARRESTERS

Multiposition Small Engine Exhaust System Fire Ignition Suppression (A)—SAE J335 SEP90 .... 2:26.351

Spark Arrester Test Carbon (A)—SAE J997 SEP90 .... 2:26.368

Spark Arrester Test Procedure for Large Size Engines (A)—SAE J342 JAN91 .... 2:26.360

Spark Arrester Test Procedure for Medium Size Engines (A)—SAE J350 JAN91 .... 2:26.365

SPARK IGNITION ENGINES

Engine Rotation and Cylinder Numbering (A)—SAE J824 JAN86 .... 2:26.28

Cold Weather Operation

Fuel Economy Measurement—Road Test Procedure—Cold Start and Warm-Up Fuel Economy—SAE J1256 OCT88 .... 2:26.506

† Low-Temperature Cranking Load Requirements of an Engine—SAE J1253 JUN93 .... 2:23.81

* SAE Cold Start and Driveability Procedure—SAE J1635 MAY93 .... 2:26.498

Cooling Systems

Engine Charge Air Cooler Nomenclature (A)—SAE J1148 JUN90 .... 2:26.392

Engine Cooling System Field Test (Air-to-Boil) (A)—SAE J819 MAR87 .... 3:40.157

† Glossary of Engine Cooling System Terms (A)—SAE J1004 APR93 .... 2:26.380

Guide to the Application and Use of Engine Coolant Pump Face Seals (A)—SAE J1245 JUN82 .... 2:26.417

† Principles of Engine Cooling Systems, Components and Maintenance—SAE HS-40 JAN91 .... Available as a separate publication (See Related Technical Reports Section)

Requirements for Engine Cooling System Filling, Deaeration and Drawdown Tests—SAE J1436 JUN89 . 2:26.379

Dipsticks

Engine and Transmission Dipstick Marking (A)—SAE J614b 2:26.233

Exhaust Emissions

Constant Volume Sampler System for Exhaust Emissions Measurement—SAE J1094 JUN92 .... 1:13.117

Instrumentation and Techniques for Exhaust Gas Emissions Measurement (A)—SAE J254 AUG84 ...... 1:13.104

* SAE Surface Vehicle Emissions Standards Manual—SAE HS-2400 FEB93 .... Available as a separate publication (See Related Technical Reports Section)

† Test Procedure for the Measurement of Gaseous Exhaust Emissions from Small Utility Engines (A)—SAE J1088 FEB93 .... 1:13.151

Fans

Engine Cooling Fan Structural Analysis (A)—SAE J1390 APR82 .... 2:26.400

Heavy Duty Non-Metallic Engine Cooling Fans—Material, Manufacturing and Test Considerations—SAE J1474 JAN85 .... 2:26.403

Flame Arresters

† Devices Providing Backfire Flame Control for Gasoline Engines in Marine Applications—SAE J1928 APR93 ... 3:41.44

Flywheels

Maximum Allowable Rotational Speed for Internal Combustion Engine Flywheels (A)—SAE J1456 JUN90 . 2:26.243

Fuel Consumption

Fuel Economy Measurement—Road Test Procedure—Cold Start and Warm-Up Fuel Economy—SAE J1256 OCT88 .... 2:26.506

Fuel Injection

Fuel Injection Nomenclature — Spark Ignition Engines (A)—SAE J1541 FEB88 .... 2:26.290

Gasoline Fuel Injector (A)—SAE J1832 NOV89 ...... 2:26.263

Fuels

† Alternative Automotive Fuels—SAE J1297 MAR93 .... 1:12.74

Identification Numbers

Engine and Transmission Identification Numbers—SAE J129 DEC81 .... 3:27.03

Ignition Systems

Ignition System Nomenclature and Terminology (A)—SAE J139 JUN90 .... 2:23.58

Lubricants

Engine Oil Performance and Engine Service Classification (Other Than "Energy-Conserving") (A)—SAE J183 JUN91 .... 1:12.01

† Engine Oil Tests (A)—SAE J304 JUN93 .... 1:12.19

† Engine Oil Viscosity Classification—SAE J300 MAR93 . 1:12.16

Fuel and Lubricant Tanks for Motorcycles—SAE J1241 . 3:37.11

† International Tests and Specifications for Automotive Engine Oils—SAE J2227 MAY93 .... 1:12.10

Lubricants for Two-Stroke-Cycle Gasoline Engines—SAE J1510 OCT88 .... 1:12.32

---

* Technical Revision   † New Document   (D)—DODISS Adopted   (A)—ANSI Recognized

SPARK IGNITION ENGINES (continued)

Lubricants (continued)
Standard Oil Filter Test Oil—SAE **J1260** APR89 ...... 2:26.350

The Automotive Lubricant Performance and Service Classification Maintenance Procedure (A)—SAE **J1146** JUN86 .... 1:12.87

† Two-Stroke-Cycle Gasoline Engine Lubricants Performance and Service Classification (A)—SAE **J2116** JUN93 .... 1:12.33

Noise
Engine Sound Level Measurement Procedure (A) (D)—SAE **J1074** FEB87 .... 1:14.114

Measurement Procedure for Determination of Silencer Effectiveness in Reducing Engine Intake or Exhaust Sound Level (A)—SAE **J1207** FEB87 .... 1:14.115

Octane Requirements
† Automotive Gasolines—SAE **J312** JAN93 .... 1:12.47

Performance
Certificates of Maximum Net Horsepower for Motor Trucks and Truck Tractors—SAE **J690** 3:36.65

Engine Power Test Code—Spark Ignition and Compression Ignition— Gross Power Rating—SAE **J1995** JUN90 .... 2:26.11

Engine Power Test Code—Spark Ignition and Compression Ignition—Net Power Rating—SAE **J1349** JUN90 .... 2:26.16

Information Relating to Duty Cycles and Average Power Requirements of Truck and Bus Engine Accessories—SAE **J1343** OCT88 .... 3:36.57

Procedure for Mapping Engine Performance—Spark Ignition and Compression Ignition Engines (A)—SAE **J1312** JAN90 .... 2:26.22

Preheaters
Engine Preheaters (A)—SAE **J226** JAN86 .... 2:26.370

Selection and Application Guidelines for Diesel, Gasoline, and Propane Fired Liquid Cooled Engine Pre-Heaters—SAE **J1350** MAR88 .... 3:40.496

Preignition Ratings
SAE 17.6 Cubic Inch Spark Plug Rating Engine (A)—SAE **J2203** JUN91 .... 2:23.98

Small
Bystander Sound Level Measurement Procedure for Small Engine Powered Equipment (A)—SAE **J1175** MAR85 .... 1:14.97

Mounting Flanges and Power Take-Off Shafts for Small Engines—SAE **J609a** 2:26.26

Multiposition Small Engine Exhaust System Fire Ignition Suppression (A)—SAE **J335** SEP90 .... 2:26.351

Operator Ear Sound Level Measurement Procedure for Small Engine Powered Equipment (A)—SAE **J1174** MAR85 .... 1:14.95

Procedure for Evaluating Transient Response of Small Engine Driven Generator Sets (A)—SAE **J1444** JUN91 . 2:26.23

Small Engine Power Rating Procedure—SAE **J1940** JUN89 .... 2:26.23

Small Spark Ignition Engine Test Code (A)—SAE **J607** AUG88 .... 2:26.22

Tapped and Flanged Exhaust Connections for Small Engines (A)—SAE **J624** SEP89 .... 3:40.255

† Test Procedure for the Measurement of Gaseous Exhaust Emissions from Small Utility Engines (A)—SAE **J1088** FEB93 .... 1:13.151

Spark Arresters
Multiposition Small Engine Exhaust System Fire Ignition Suppression (A)—SAE **J335** SEP90 .... 2:26.351

Spark Arrester Test Procedure for Medium Size Engines—SAE **J350** JAN91 .... 2:26.365

Terminology
† Emissions Terminology and Nomenclature—SAE **J1145** FEB93 .... 1:13.166

Engine Terminology and Nomenclature—General (A)—SAE **J604** JAN86 .... 2:26.01

Fuel Injection Nomenclature — Spark Ignition Engines (A)—SAE **J1541** FEB88 .... 2:26.290

Ignition System Nomenclature and Terminology (A)—SAE **J139** JUN90 .... 2:23.58

Thermal Efficiency
Heating Value of Fuels—SAE **J1498** MAY90 .... 1:12.82

Valves
Engine Poppet Valve Information Report—SAE **J775** JAN88 .... 1:1.165

SPARK PLUGS
Preignition Rating of Spark Plugs (A)—SAE **J549** JUN90 2:23.96

SAE 17.6 Cubic Inch Spark Plug Rating Engine (A)—SAE **J2203** JUN91 .... 2:23.98

Spark Plug Heat Rating Classifications—SAE **J2162** JUN92 .... 2:23.97

Spark Plug Installation Sockets—SAE **J548/2** JUN92 ... 2:23.95

Spark Plugs—SAE **J548/1** JUN92 .... 2:23.83

SPEED CONTROL
Maximum Allowable Rotational Speed for Internal Combustion Engine Flywheels (A)—SAE **J1456** JUN90 . 2:26.243

Automobiles
Automatic Vehicle Speed Control—Motor Vehicles—SAE **J195** DEC88 .... 2:21.16

Transmission Mounted Vehicle Speed Signal Rotor Specification—SAE **J1377** JAN89 .... 2:21.17

Snowmobiles
Speed Control Assurance for Snowmobiles (A)—SAE **J1222** OCT84 .... 3:38.17

SPEED SENSORS
Transmission Mounted Vehicle Speed Signal Rotor Specification—SAE **J1377** JAN89 .... 2:21.17

SPEEDOMETERS
See also: TACHOMETERS

Electric Speedometer Specification—On Road (A)—SAE **J1226** FEB83 .... 2:21.09

Factors Affecting Accuracy of Mechanically Driven Automotive Speedometer and Odometers—SAE **J862** JAN89 .... 2:21.01

Speedometer Test Procedure (A)—SAE **J1059** JUN84 . 2:21.08

Speedometers and Tachometers—Automotive—SAE **J678** DEC88 .... 2:21.03

SPLASH AND SPRAY PROTECTION
Rear Wheel Splash and Stone Throw Protection—SAE **J682** OCT84 .... 3:36.120

SPLINES
Involute Splines and Inspection Standard—ANSI **B92.1, 1a** .... Available as a separate publication (See Related Technical Reports Section)

Metric Involute Splines and Inspection Standard—ANSI **B92.2M** .... Available as a separate publication (See Related Technical Reports Section)

Parallel Side Splines for Soft Broached Holes in Fittings—SAE **J499a** 2:18.01

Part III: Inspection—Metric Module Involute Splines—ANSI **B92.2Ma** .... Available as a separate publication (See Related Technical Reports Section)

Serrated Shaft Ends (A)—SAE **J500** AUG89 .... 2:18.03

Shaft Ends—SAE **J501** 2:18.04

SPOT LAMPS
See also: LAMPS

Spot Lamps—SAE **J591** MAY89 .... 2:24.167

Subject Index

SPRINGS
See also: **SUSPENSION SYSTEMS**

Rated Suspension Spring Capacity—SAE **J274** JUN89 . 2:20.30

Belleville
Manual on Design and Manufacture of Coned Disk Springs (Belleville Springs) and Spring Washers—SAE **HS-1582** .... Available as a separate publication *(See Related Technical Reports Section)*

Coil
Helical Compression and Extension Spring Terminology (A)—SAE **J1121** JUL88 .... 2:20.17

† Internal Combustion Engines—Piston Rings Coil Spring Loaded Oil Control Rings—SAE **J2003** OCT92 .... 2:26.131

Conical
Conical Spring Washers—SAE **J773b** 2:16.19
Nut and Conical Spring Washer Assemblies—SAE **J238** 2:16.20

Helical
Helical Compression and Extension Spring Terminology (A)—SAE **J1121** JUL88 .... 2:20.17

Helical Springs: Specification Check Lists (A)—SAE **J1122** JUL88 .... 2:20.27

Manual on Design and Application of Helical and Spiral Springs—SAE **HS-J795** APR90 .... Available as a separate publication *(See Related Technical Reports Section)*

Leaf
Fatigue Testing Procedure for Suspension-Leaf Springs (A)—SAE **J1528** JUN90 .... 2:20.08

† Leaf Springs for Motor Vehicle Suspension—Made to Metric Units—SAE **J1123** NOV92 .... 2:20.01

† Leaf Springs for Motor Vehicle Suspension—Made to Customary U.S. Units—SAE **J510** NOV92 .... 2:20.09

Manual on Design and Application of Leaf Springs—SAE **HS-788** APR80 .... Available as a separate publication *(See Related Technical Reports Section)*

Pneumatic
Manual for Incorporating Pneumatic Springs in Vehicle Suspension Designs—SAE **HS-1576** .... Available as a separate publication *(See Related Technical Reports Section)*

Pneumatic Spring Terminology—SAE **J511** JUN89 .... 2:20.16

Seat
Motor Vehicle Seating Manual—SAE **J782b** JUN80 .... 3:34.73

Spiral
Manual on Design and Application of Helical and Spiral Springs—SAE **HS-J795** APR90 .... Available as a separate publication *(See Related Technical Reports Section)*

Steel
Hard Drawn Carbon Steel Valve Spring Quality Wire and Springs—SAE **J172** DEC88 .... 1:5.08

Hard Drawn Mechanical Spring Wire and Springs—SAE **J113** DEC88 .... 1:5.04

† Leaf Springs for Motor Vehicle Suspension—Made to Customary U.S. Units—SAE **J510** NOV92 .... 2:20.09

Music Steel Spring Wire and Springs—SAE **J178** DEC88 1:5.10

Oil Tempered Carbon Steel Spring Wire and Springs—SAE **J316** DEC88 .... 1:5.01

Oil Tempered Carbon Steel Valve Spring Quality Wire and Springs—SAE **J351** DEC88 .... 1:5.02

Oil Tempered Chromium-Vanadium Valve Spring Quality Wire and Springs—SAE **J132** DEC88 .... 1:5.03

Oil Tempered Chromium—Silicon Alloy Steel Wire and Springs—SAE **J157** DEC88 .... 1:5.07

Special Quality High Tensile, Hard Drawn Mechanical Spring Wire and Springs—SAE **J271** DEC88 .... 1:5.05

Stainless Steel 17-7 PH Spring Wire and Springs—SAE **J217** DEC88 .... 1:5.08

Stainless Steel, SAE 30302, Spring Wire and Springs—SAE **J230** DEC88 .... 1:5.09

Torsion Bar
Manual on Design and Manufacture of Torsion Bar Springs—SAE **HS-796** JUL90 .... Available as a separate publication *(See Related Technical Reports Section)*

STABILITY
See also: **HANDLING**

Cranes
Crane Load Stability Test Code (A)—SAE **J765** OCT90 . 3:40.210
Mobile Crane Stability Ratings (A)—SAE **J1289** APR81 . 3:40.223

Snowmobiles
† Recommendations for Children's Snowmobile—SAE **J1038** DEC92 .... 3:38.18

Terminology
Vehicle Dynamics Terminology—SAE **J670e** DEC76 ... 3:34.297

Trucks and Buses
*Braking, Stability, and Control Performance Test Procedures for Air-Brake- Equiped Truck Tractors—SAE **J1626** MAR93 .... 3:36.130

STAINING
Test Method for Determining Blocking Resistance and Associated Characteristics of Automotive Trim Materials—SAE **J912a** 1:11.165

Test Method for Measuring Wet Color Transfer Characteristics (A)—SAE **J1326** FEB85 .... 1:11.169

STARTERS
See also: **ACTUATORS**

*Automotive Starter Drive Assembly Remanufacturing Procedures—SAE **J2241** MAR93 .... 3:42.46

*Automotive Starter Remanufactured Procedures—SAE **J2073** MAR93 .... 3:42.42

*Automotive Starter Solenoid Remanufacturing Procedures—SAE **J2242** MAR93 .... 3:42.47

† Cranking Motor Application Considerations (A)—SAE **J1375** MAR93 .... 2:23.81

Electric Starting Motor Test Procedure (A)—SAE **J544** MAR88 .... 2:23.80

† Low-Temperature Cranking Load Requirements of an Engine—SAE **J1253** JUN93 .... 2:23.81

Shielding of Starter System Energization—SAE **J1493** OCT91 .... 3:40.266

*Starter Armature Remanufacturing Procedures—SAE **J2240** MAR93 .... 3:42.45

Starting Motor Mountings (A)—SAE **J542** JUN91 ...... 2:23.76

Starting Motor Pinions and Ring Gears (A)—SAE **J543** JUN88 .... 2:23.78

Unauthorized Starting or Movement of Machines (A)—SAE **J1083** JUL85 .... 3:40.96

Voltage Drop for Starting Motor Circuits—SAE **J541** FEB89 .... 2:23.74

Switches
Tests for Snowmobile Switching Devices and Components—SAE **J68** DEC91 .... 3:38.08

STARTING AIDS
† Electric Engine Preheaters and Battery Warmers for Diesel Engines—SAE **J1310** JUN93 .... 3:40.500

STEELS
See also: **CHEMICAL COMPOSITION, MECHANICAL PROPERTIES, PHYSICAL PROPERTIES**

† Cleanliness Rating of Steels by the Magnetic Particle Method—SAE **J421** MAY93 .... 1:3.57

Definitions for Macrostrain and Microstrain (A)—SAE **J932** AUG85 .... 1:3.81

---

* Technical Revision   † New Document   (D)—DODISS Adopted   (A)—ANSI Recognized

Subject Index

STEELS (continued)

Methods of Sampling Steel for Chemical Analysis (A)—SAE **J408** DEC83 .... 1:1.152

Microscopic Determination of Inclusions in Steels (A)—SAE **J422** DEC83 .... 1:3.59

† Selection and Use of Steels—SAE **J401** NOV92 ...... 1:1.06

Special Purpose Alloys ("Superalloys")—SAE **J467b** ... 1:10.128

Technical Report on Fatigue Properties—SAE **J1099** ... 1:3.81

Tensile Test Specimens (A)—SAE **J416** DEC83 ...... 1:3.01

Alloy

Carbon and Alloy Steels—SAE **J411** NOV89 .... 1:2.01

Chemical Compositions of SAE Alloy Steels (A)—SAE **J404** FEB91 .... 1:1.14

Estimated Mechanical Properties and Machinability of Steel Bars—SAE **J1397** MAY92 .... 1:2.26

Former SAE Standard and Former SAE EX-Steels—SAE **J1249** JAN89 .... 1:1.19

General Characteristics and Heat Treatments of Steels—SAE **J412** JUN89 .... 1:2.05

† Hardenability Bands for Carbon and Alloy H Steels (A)—SAE **J1268** JUN93 .... 1:1.48

High Strength Carbon and Alloy Die Drawn Steels—SAE **J935** JUN90 .... 1:2.04

† Leaf Springs for Motor Vehicle Suspension—Made to Metric Units—SAE **J1123** NOV92 .... 2:20.01

Oil Tempered Chromium-Vanadium Valve Spring Quality Wire and Springs—SAE **J132** DEC88 .... 1:5.03

Oil Tempered Chromium—Silicon Alloy Steel Wire and Springs—SAE **J157** DEC88 .... 1:5.07

Product Analysis—Permissible Variations from Specified Chemical Analysis of a Heat or Cast of Steel (A)—SAE **J409** DEC90 .... 1:1.153

Restricted Hardenability Bands for Selected Alloy Steels (A)—SAE **J1868** JUN90 .... 1:1.139

Bars

* Laboratory Measurement of the Composite Vibration Damping Properties of Materials on a Supporting Steel Bar—SAE **J1637** FEB93 .... 1:2.30

Carbon

Aging of Carbon Steel Sheet and Strip (A)—SAE **J763** APR91 .... 1:2.36

Brazed Double Wall Low Carbon Steel Tubing (A)—SAE **J527** JUN91 .... 2:22.209

Carbon and Alloy Steels—SAE **J411** NOV89 .... 1:2.01

* Categorization of Low Carbon Automotive Sheet Steel—SAE **J2096** JUL92 .... 1:3.25

Chemical Compositions of SAE Carbon Steels—SAE **J403** MAY92 .... 1:1.09

† Cut Wire Shot—SAE **J441** JUN93 .... 1:8.05

Estimated Mechanical Properties and Machinability of Steel Bars—SAE **J1397** MAY92 .... 1:2.26

Former SAE Standard and Former SAE EX-Steels—SAE **J1249** JAN89 .... 1:1.19

General Characteristics and Heat Treatments of Steels—SAE **J412** JUN89 .... 1:2.05

Glossary of Carbon Steel Sheet and Strip Terms—SAE **J940** DEC88 .... 1:2.34

Hard Drawn Carbon Steel Valve Spring Quality Wire and Springs—SAE **J172** DEC88 .... 1:5.08

Hard Drawn Mechanical Spring Wire and Springs—SAE **J113** DEC88 .... 1:5.04

† Hardenability Bands for Carbon and Alloy H Steels (A)—SAE **J1268** JUN93 .... 1:1.48

High Strength, Hot Rolled Steel Plates, Bars and Shapes—SAE **J1442** DEC88 .... 1:1.162

† Leaf Springs for Motor Vehicle Suspension—Made to Customary U.S. Units—SAE **J510** NOV92 .... 2:20.09

Methods of Measuring Decarburization (A)—SAE **J419** DEC83 .... 1:3.12

Oil Tempered Carbon Steel Spring Wire and Springs—SAE **J316** DEC88 .... 1:5.01

Oil Tempered Carbon Steel Valve Spring Quality Wire and Springs—SAE **J351** DEC88 .... 1:5.02

Product Analysis—Permissible Variations from Specified Chemical Analysis of a Heat or Cast of Steel (A)—SAE **J409** DEC90 .... 1:1.153

Properties of Galvanized Low Carbon Sheet Steels and Their Relation to Formability—SAE **J1852** SEP87 ..... 1:3.22

Properties of Low Carbon Sheet Steel and Their Relationship to Formability—SAE **J877** JUN84 .... 1:3.20

Seamless Low Carbon Steel Tubing Annealed for Bending and Flaring (A)—SAE **J524** JUN91 .... 2:22.206

Selecting and Specifying Hot and Cold Rolled Steel Sheet and Strip (A)—SAE **J126** JUN86 .... 1:3.27

Special Quality High Tensile, Hard Drawn Mechanical Spring Wire and Springs—SAE **J271** DEC88 .... 1:5.05

Split Type Bushings—Design and Application (A)—SAE **J835** DEC81 .... 2:26.201

Welded and Cold Drawn Low Carbon Steel Tubing Annealed for Bending and Flaring (A)—SAE **J525** JUN91 .... 2:22.207

Welded Flash Controlled Low Carbon Steel Tubing Normalized for Bending, Double Flaring, and Beading (A)—SAE **J356** JUN91 .... 2:22.213

Welded Low Carbon Steel Tubing (A)—SAE **J526** JUN91 .... 2:22.208

Cast

Automotive Steel Castings—SAE **J435c** 1:6.17

* Cast Steel Grit—SAE **J1993** MAR93 .... 1:8.13

Cast Steel Shot (A)—SAE **J827** MAR90 .... 1:8.09

Specifications for Low Carbon Cast Steel Shot (A)—SAE **J2175** JUN91 .... 1:8.10

Chromium-Silicon

Oil Tempered Chromium—Silicon Alloy Steel Wire and Springs—SAE **J157** DEC88 .... 1:5.07

Chromium-Vanadium

Oil Tempered Chromium-Vanadium Valve Spring Quality Wire and Springs—SAE **J132** DEC88 .... 1:5.03

Coated

Electroplating of Nickel and Chromium on Metal Parts—Automotive Ornamentation and Hardware—SAE **J207** FEB85 .... 1:10.150

Engine Poppet Valve Information Report—SAE **J775** JAN88 .... 1:1.165

Prevention of Corrosion of Motor Vehicle Body and Chassis Components—SAE **HS-J447** JUN81 .... Available as a separate publication (See Related Technical Reports Section)

Properties of Galvanized Low Carbon Sheet Steels and Their Relation to Formability—SAE **J1852** SEP87 ..... 1:3.22

Steel, High Strength, Hot Rolled Sheet and Strip, Cold Rolled Sheet, and Coated Sheet—SAE **J1392** JUN84 .. 1:1.159

Undervehicle Coupon Corrosion Tests—SAE **J1293** JAN90 .... 1:3.48

Formability

Properties of Galvanized Low Carbon Sheet Steels and Their Relation to Formability—SAE **J1852** SEP87 ..... 1:3.22

Properties of Low Carbon Sheet Steel and Their Relationship to Formability—SAE **J877** JUN84 .... 1:3.20

Glossary

Glossary of Carbon Steel Sheet and Strip Terms—SAE **J940** DEC88 .... 1:2.34

Grain Size

Grain Size Determination of Steels (A)—SAE **J418** DEC83 .... 1:3.10

Hardenability

Case Hardenability of Carburized Steels (A)—SAE **J1975** JUN91 .... 1:2.15

General Characteristics and Heat Treatments of Steels—SAE **J412** JUN89 .... 1:2.05

---

* Technical Revision   † New Document   (D)—DODISS Adopted   (A)—ANSI Recognized

Subject Index

STEELS (continued)

Hardenability (continued)

† Hardenability Bands for Carbon and Alloy H Steels (A)—SAE J1268 JUN93 ... 1:1.48

Mechanical Properties of Heat Treated Wrought Steels—SAE J413 JUN90 ... 1:2.24

† Methods of Determining Hardenability of Steels—SAE J406 JUN93 ... 1:1.23

Restricted Hardenability Bands for Selected Alloy Steels (A)—SAE J1868 JUN90 ... 1:1.139

Selection and Heat Treatment of Tool and Die Steels—SAE J437a ... 1:7.02

Heat Treatment

Case Hardenability of Carburized Steels (A)—SAE J1975 JUN91 ... 1:2.15

General Characteristics and Heat Treatments of Steels—SAE J412 JUN89 ... 1:2.05

† High-Strength, Quenched, and Tempered Structural Steels—SAE J368 MAR93 ... 1:1.164

Methods of Measuring Case Depth (A)—SAE J423 DEC83 ... 1:3.64

Selection and Heat Treatment of Tool and Die Steels—SAE J437a ... 1:7.02

High Strength

High Strength Carbon and Alloy Die Drawn Steels—SAE J935 JUN90 ... 1:2.04

High Strength, Hot Rolled Steel Plates, Bars and Shapes—SAE J1442 DEC88 ... 1:1.162

† High-Strength, Quenched, and Tempered Structural Steels—SAE J368 MAR93 ... 1:1.164

Identifying and Repairing High Strength Steel Vehicle Components—SAE J1554 NOV85 ... 3:42.18

Properties of Galvanized Low Carbon Sheet Steels and Their Relation to Formability—SAE J1852 SEP87 ... 1:3.22

Steel, High Strength, Hot Rolled Sheet and Strip, Cold Rolled Sheet, and Coated Sheet—SAE J1392 JUN84 ... 1:1.159

Hot Rolled

* Categorization of Low Carbon Automotive Sheet Steel—SAE J2096 JUL92 ... 1:3.25

Hot and Cold Rolled

High Strength, Hot Rolled Steel Plates, Bars and Shapes—SAE J1442 DEC88 ... 1:1.162

Properties of Low Carbon Sheet Steel and Their Relationship to Formability—SAE J877 JUN84 ... 1:3.20

Selecting and Specifying Hot and Cold Rolled Steel Sheet and Strip (A)—SAE J126 JUN86 ... 1:3.27

Sheet Steel Thickness and Profile (A)—SAE J1058 APR91 ... 1:3.19

Steel, High Strength, Hot Rolled Sheet and Strip, Cold Rolled Sheet, and Coated Sheet—SAE J1392 JUN84 ... 1:1.159

Surface Texture Measurement of Cold Rolled Sheet Steel (A)—SAE J911 JUN86 ... 1:9.01

Machinability

Estimated Mechanical Properties and Machinability of Steel Bars—SAE J1397 MAY92 ... 1:2.26

High Strength Carbon and Alloy Die Drawn Steels—SAE J935 JUN90 ... 1:2.04

Numbering System

Engine Poppet Valve Information Report—SAE J775 JAN88 ... 1:1.165

Potential Standard Steels—SAE J1081 DEC88 ... 1:1.18

SAE Numbering System for Wrought or Rolled Steel—SAE J402 DEC88 ... 1:1.08

Special Purpose Alloys ("Superalloys")—SAE J467b ... 1:10.128

Operator Protective Structures

Steel Products for Rollover Protective Structures (ROPS) and Falling Object Protective Structures (FOPS)—SAE J1119 DEC88 ... 3:40.371

Pipe

Metallic Air Brake System Tubing and Pipe (A)—SAE J1149 JUN91 ... 2:22.219

Piston Rings

† Internal Combustion Engines—Piston Rings—Half Keystone Rings (A)—SAE J2001 OCT92 ... 2:26.111

* Internal Combustion Engines—Piston Rings—Steel Rectangular Rings—SAE J2226 OCT92 ... 2:26.180

Plastic Deformation

Methods for Determining Plastic Deformation in Sheet Metal Stampings (A)—SAE J863 JUN86 ... 1:3.17

Sheet

* Categorization of Low Carbon Automotive Sheet Steel—SAE J2096 JUL92 ... 1:3.25

Sheet and Strip

Aging of Carbon Steel Sheet and Strip (A)—SAE J763 APR91 ... 1:2.36

Classification of Common Imperfections in Sheet Steel—SAE J810 MAR87 ... 1:3.30

Glossary of Carbon Steel Sheet and Strip Terms—SAE J940 DEC88 ... 1:2.34

Method for Determining Breakage Allowances for Sheet Steel (A)—SAE J424 JUN86 ... 1:3.66

Methods for Determining Plastic Deformation in Sheet Metal Stampings (A)—SAE J863 JUN86 ... 1:3.17

Properties of Galvanized Low Carbon Sheet Steels and Their Relation to Formability—SAE J1852 SEP87 ... 1:3.22

Properties of Low Carbon Sheet Steel and Their Relationship to Formability—SAE J877 JUN84 ... 1:3.20

Selecting and Specifying Hot and Cold Rolled Steel Sheet and Strip (A)—SAE J126 JUN86 ... 1:3.27

Sheet Steel Thickness and Profile (A)—SAE J1058 APR91 ... 1:3.19

Steel, High Strength, Hot Rolled Sheet and Strip, Cold Rolled Sheet, and Coated Sheet—SAE J1392 JUN84 ... 1:1.159

Surface Texture Measurement of Cold Rolled Sheet Steel (A)—SAE J911 JUN86 ... 1:9.01

Undervehicle Coupon Corrosion Tests—SAE J1293 JAN90 ... 1:3.48

Springs

† Leaf Springs for Motor Vehicle Suspension—Made to Customary U.S. Units—SAE J510 NOV92 ... 2:20.09

Springs and Wires

Hard Drawn Carbon Steel Valve Spring Quality Wire and Springs—SAE J172 DEC88 ... 1:5.08

Hard Drawn Mechanical Spring Wire and Springs—SAE J113 DEC88 ... 1:5.04

Manual on Design and Application of Helical and Spiral Springs—SAE HS-J795 APR90 ... Available as a separate publication (See Related Technical Reports Section)

Music Steel Spring Wire and Springs—SAE J178 DEC88 ... 1:5.10

Oil Tempered Carbon Steel Spring Wire and Springs—SAE J316 DEC88 ... 1:5.01

Oil Tempered Carbon Steel Valve Spring Quality Wire and Springs—SAE J351 DEC88 ... 1:5.02

Oil Tempered Chromium-Vanadium Valve Spring Quality Wire and Springs—SAE J132 DEC88 ... 1:5.03

Oil Tempered Chromium—Silicon Alloy Steel Wire and Springs—SAE J157 DEC88 ... 1:5.07

Special Quality High Tensile, Hard Drawn Mechanical Spring Wire and Springs—SAE J271 DEC88 ... 1:5.05

Stainless Steel 17-7 PH Spring Wire and Springs—SAE J217 DEC88 ... 1:5.08

Stainless Steel, SAE 30302, Spring Wire and Springs—SAE J230 DEC88 ... 1:5.09

Stainless

Chemical Compositions of SAE Wrought Stainless Steels—SAE J405 JAN89 ... 1:1.16

† Cut Wire Shot—SAE J441 JUN93 ... 1:8.05

---

* Technical Revision   † New Document   (D)—DODISS Adopted   (A)—ANSI Recognized

STEELS (continued)
Stainless (continued)
General Characteristics and Heat Treatments of Steels—SAE J412 JUN89 .... 1:2.05

Product Analysis—Permissible Variations from Specified Chemical Analysis of a Heat or Cast of Steel (A)—SAE J409 DEC90 .... 1:1.153

Stainless Steel 17-7 PH Spring Wire and Springs—SAE J217 DEC88 .... 1:5.08

Stainless Steel, SAE 30302, Spring Wire and Springs—SAE J230 DEC88 .... 1:5.09

Surface Hardness
*Cast Steel Grit—SAE J1993 MAR93 .... 1:8.13

Hardness Tests and Hardness Number Conversions—SAE J417 DEC83 .... 1:3.04

Methods of Measuring Case Depth (A)—SAE J423 DEC83 .... 1:3.64

†Surface Hardness Testing with Files—SAE J864 MAY93 .... 1:3.03

Tool & Die
Selection and Heat Treatment of Tool and Die Steels—SAE J437a 1:7.02

Tool and Die Steels—SAE J438b 1:7.01

Tubing
Brazed Double Wall Low Carbon Steel Tubing (A)—SAE J527 JUN91 .... 2:22.209

Diesel Engines—Steel Tubes for High Pressure Fuel Injection Pipes (Tubing)—SAE J1958 APR89 .... 2:26.260

Pressure Ratings for Hydraulic Tubing and Fittings (D)—SAE J1065 MAR92 .... 2:22.215

Seamless Low Carbon Steel Tubing Annealed for Bending and Flaring (A)—SAE J524 JUN91 .... 2:22.206

Welded and Cold Drawn Low Carbon Steel Tubing Annealed for Bending and Flaring (A)—SAE J525 JUN91 .... 2:22.207

Welded Flash Controlled Low Carbon Steel Tubing Normalized for Bending, Double Flaring, and Beading (A)—SAE J356 JUN91 .... 2:22.213

Welded Low Carbon Steel Tubing (A)—SAE J526 JUN91 .... 2:22.208

Welds
Unibody Weld Quality Testing (A)—SAE J1827 MAY87 .... 3:42.23

Wrought
Chemical Compositions of SAE Wrought Stainless Steels—SAE J405 JAN89 .... 1:1.16

General Characteristics and Heat Treatments of Steels—SAE J412 JUN89 .... 1:2.05

Mechanical Properties of Heat Treated Wrought Steels—SAE J413 JUN90 .... 1:2.24

Potential Standard Steels—SAE J1081 DEC88 .... 1:1.18

SAE Numbering System for Wrought or Rolled Steel—SAE J402 DEC88 .... 1:1.08

STEERING
See also: BALL STUDS
Performance Assurance of Remanufactured, Hydraulically-Operated Rack and Pinion Steering Gears (A)—SAE J1890 JUN88 .... 3:42.33

Construction and Industrial Equipment
Minimum Performance Criteria for Emergency Steering of Wheeled Earthmoving Construction Machines—SAE J53 OCT84 .... 3:40.515

Earthmoving Equipment
Minimum Performance Criteria for Emergency Steering of Wheeled Earthmoving Construction Machines—SAE J53 OCT84 .... 3:40.515

Locks
†Steering Frame Lock for Articulated Loaders and Tractors—SAE J276 NOV92 .... 3:40.353

Off-Road Vehicles
Steering for Off-Road, Rubber-Tired Machines (A)—SAE J1511 OCT90 .... 3:40.517

Power
*High-Temperature Power Steering Pressure Hose—SAE J2050 FEB93 .... 1:11.245

*High-Temperature Power Steering Return Hose—Low Pressure—SAE J2076 FEB93 .... 1:11.247

Power Steering Pressure Hose—High Volumetric Expansion Type (A)—SAE J188 OCT89 .... 1:11.241

Power Steering Pressure Hose—Low Volumetric Expansion Type (A)—SAE J191 OCT89 .... 1:11.244

Power Steering Pressure Hose—Wire Braid—SAE J190 .... 1:11.248

Power Steering Return Hose—Low Pressure (A)—SAE J189 DEC89 .... 1:11.243

Secondary Systems
Steering for Off-Road, Rubber-Tired Machines (A)—SAE J1511 OCT90 .... 3:40.517

Terminology
Vehicle Dynamics Terminology—SAE J670e OCT76 .... 3:34.297

Trucks and Buses
Information Relating to Duty Cycles and Average Power Requirements of Truck and Bus Engine Accessories—SAE J1343 OCT88 .... 3:36.57

Turning Ability and Off Tracking—Motor Vehicles (A)—SAE J695 DEC89 .... 3:36.78

STEERING WHEELS
Vision Obstruction
Describing and Measuring the Driver's Field of View—SAE J1050a 3:34.214

STOP LAMPS
See also: BACKUP LAMPS, TAIL LAMPS
*Central High Mounted Stop Lamp Standard for Use on Vehicles Less than 2032 mm Overall Width—SAE J1957 JUN93 .... 2:24.149

High Mounted Stop Lamps for Use on Vehicles 2032 mm or More in Overall Width—SAE J1432 OCT88 .... 2:24.147

†SAE Ground Vehicle Lighting Standards Manual—SAE HS-34 JAN93 .... Available as a separate publication (See Related Technical Reports Section)

Snowmobile and Snowmobile Cutter Lamps, Reflective Devices, and Associated Equipment (A)—SAE J292 OCT84 .... 3:38.07

Snowmobile Stop Lamp (A)—SAE J278 OCT84 .... 3:38.05

Stop Lamps for Use on Motor Vehicles 2032 mm or More in Overall Width (A)—SAE J1398 JUN91 .... 2:24.145

Stop Lamps for Use on Motor Vehicles Less Than 2032 mm in Overall Width (A)—SAE J586 DEC89 .... 2:24.143

Supplemental High Mounted Stop and Rear Turn Signal Lamps for Use on Vehicles Less Than 2032 mm in Overall Width (A)—SAE J186 DEC89 .... 2:24.155

Tail Lamps (Rear Position Lamps) for Use on Motor Vehicles Less Than 2032 mm in Overall Width—SAE J585 DEC91 .... 2:24.137

Switches
Mechanical Stop Lamp Switch (A)—SAE J249 JUN88 .... 2:24.150

Motorcycle Stop Lamp Switch—SAE J1167 JUN89 .... 3:37.07

Tests for Snowmobile Switching Devices and Components—SAE J68 DEC91 .... 3:38.08

STOPPING DISTANCES
Brake Performance—Rubber-Tired Earthmoving Machines (A)—SAE J1473 OCT90 .... 3:40.285

Brake System Road Test Code—Passenger Car and Light-Duty Truck (A)—SAE J843 NOV90 .... 2:25.108

Braking Performance—Roller Compactors (A)—SAE J1472 JUN87 .... 3:40.288

Braking Performance—Roller/Compactors (A)—SAE J1136 APR87 .... 3:40.300

Braking Performance—Rubber-Tired Construction Machines (A)—SAE J1152 APR80 .... 3:40.289

---
* Technical Revision   † New Document   (D)—DODISS Adopted   (A)—ANSI Recognized

Subject Index

STOPPING DISTANCES (continued)
Braking System Test Procedures and Braking Performance Criteria for Agricultural Tractors—SAE **J1041** OCT91 .... 3:39.29

*Braking, Stability, and Control Performance Test Procedures for Air-Brake- Equiped Truck Tractors—SAE **J1626** MAR93 .... 3:36.130

Minimum Performance Criteria for Braking Systems for Specialized Rubber-Tired, Self-Propelled Underground Mining Machines **(A)**—SAE **J1329** JUL89 .... 3:40.295

Stopping Distance Test Procedure **(A)**—SAE **J299** AUG87 .... 2:25.156

Wheel Slip Brake Control System Road Test Code—SAE **J46** JUN80 .... 2:25.158

STORAGE TANKS
CFC-12 (R-12) Extraction Equipment for Mobile Automotive Air-Conditioning Systems—SAE **J2209** JUN92 .... 3:34.46

Recommended Service Procedure for the Containment of R-12—SAE **J1989** OCT89 .... 3:34.42

STRESSES
Use of Terms Yield Strength and Yield Point **(A)**—SAE **J450** JUN91 .... 1:9.01

Cracking
Acoustic Emission Test Methods **(A)**—SAE **J1242** MAR91 .... 1:3.80

Chemical Stress Resistance of Polymers—SAE **J2016** JUN89 .... 1:11.154

Cranes
Cantilevered Boom Crane Structures—Method of Test **(A) (D)**—SAE **J1063** OCT80 .... 3:40.170

Rope Supported Lattice-Type Boom Crane Structures—Method of Test—SAE **J987** APR85 .... 3:40.161

Residual
Case Hardenability of Carburized Steels **(A)**—SAE **J1975** JUN91 .... 1:2.15

Manual on Shot Peening—SAE **HS-84** .... Available as a separate publication (See Related Technical Reports Section)

Residual Stress Measurements by X-Ray Diffraction—SAE **HS-J784a** .... Available as a separate publication (See Related Technical Reports Section)

Springs
Manual on Design and Application of Helical and Spiral Springs—SAE **HS-J795** APR90 .... Available as a separate publication (See Related Technical Reports Section)

Manual on Design and Application of Leaf Springs—SAE **HS-788** APR80 .... Available as a separate publication (See Related Technical Reports Section)

Manual on Design and Manufacture of Coned Disk Springs (Belleville Springs) and Spring Washers—SAE **HS-1582** .... Available as a separate publication (See Related Technical Reports Section)

STRUCTURAL ANALYSIS
Vehicle Roof Strength Test Procedure **(A)**—SAE **J374** MAY91 .... 3:34.285

Fans
Engine Cooling Fan Structural Analysis **(A)**—SAE **J1390** APR82 .... 2:26.400

Heavy Duty Non-Metallic Engine Cooling Fans—Material, Manufacturing and Test Considerations—SAE **J1474** JAN85 .... 2:26.403

STUDS
Connections for Fluid Power and General Use—Ports and Stud Ends with ISO 261 Threads and O-Ring Sealing Part 2: Heavy-Duty (S Series) Stud Ends—Dimensions, Design, Test Methods, and Requirements—SAE **J2244/2** DEC91 .... 2:22.293

*Connections for General use and Fluid Power-Ports and Stud Ends with ISO 725 Threads and O-Ring Sealing—Part 3: Light-Duty (L Series) Stud Ends—SAE **J1926/3** ISO 11 MAR93 .... 2:22.108

*Connections for General use and Fluid Power-Ports and Stud Ends with ISO 725 Trheads and O-Ring Sealing—Part 2: Heavy-Duty (S Series) Stud Ends—SAE **J1926/2** ISO 11 MAR93 .... 2:22.102

Decarburization in Hardened and Tempered Threaded Fasteners **(A)**—SAE **J121** AUG83 .... 1:4.44

Mechanical and Material Requirements for Externally Threaded Fasteners **(A)**—SAE **J429** AUG83 .... 1:4.01

Mechanical and Material Requirements for Metric Externally Threaded Steel Fasteners **(A)**—SAE **J1199** SEP83 .... 1:4.06

Surface Discontinuities on Bolts, Screws, and Studs—SAE **J123c** 1:4.37

†Surface Discontinuities on General Application Bolts, Screws, and Studs—SAE **J1061** OCT92 .... 1:4.33

SUBSTRATES
Metal to Metal Overlap Shear Strength Test for Automotive Type Adhesives **(A)**—SAE **J1523** JUN85 ... 1:11.78

Overlap Shear Test for Sealant Adhesive Bonding of Automotive Glass Encapsulating Material to Body Opening—SAE **J1836** OCT88 .... 1:11.97

Peel Adhesion Test for Glass to Elastomeric Material for Automotive Glass Encapsulation—SAE **J1907** OCT88 .. 1:11.99

Properties of Galvanized Low Carbon Sheet Steels and Their Relation to Formability—SAE **J1852** SEP87 ..... 1:3.22

Undervehicle Coupon Corrosion Tests—SAE **J1293** JAN90 .... 1:3.48

SULFUR
Exhaust Gases
Determination of Sulfur Compounds in Automotive Exhaust—SAE **J1280** JUN92 .... 1:13.72

SURFACE HARDNESS
See also: CASE HARDENING, HARDENABILITY

†Surface Hardness Testing with Files—SAE **J864** MAY93   1:3.03

Fasteners
Decarburization in Hardened and Tempered Threaded Fasteners **(A)**—SAE **J121** AUG83 .... 1:4.44

Mechanical and Material Requirements for Externally Threaded Fasteners **(A)**—SAE **J429** AUG83 .... 1:4.01

Mechanical and Material Requirements for Steel Nuts—SAE **J995** JUN79 .... 1:4.15

Sintered Carbide
Sintered Carbide Tools—SAE **J439a** 1:7.07

Steel
*Cast Steel Grit—SAE **J1993** MAR93 .... 1:8.13

Hardness Tests and Hardness Number Conversions—SAE **J417** DEC83 .... 1:3.04

Methods of Measuring Case Depth **(A)**—SAE **J423** DEC83 .... 1:3.64

SURFACE ROUGHNESS
Characterizing a Test Surface for Motorcycle Side Stand Retraction Performance Testing—SAE **J1846** JUN89 ... 3:37.17

Motorcycle Side Stand Retraction Performance Requirements—SAE **J1579** JUN89 .... 3:37.17

Motorcycle Side Stand Retraction Test Procedure—SAE **J1578** JUN89 .... 3:37.14

†Surface Discontinuities on General Application Bolts, Screws, and Studs—SAE **J1061** OCT92 .... 1:4.33

Surface Texture—SAE **J448a** 1:9.03

Surface Texture Control—SAE **J449a** 1:9.05

---

* Technical Revision   † New Document   (D)—DODISS Adopted   (A)—ANSI Recognized

Subject Index

SURFACE ROUGHNESS (continued)
  Surface Texture Measurement of Cold Rolled Sheet
  Steel (A)—SAE **J911** JUN86 .... 1:9.01

SUSPENSION SYSTEMS
  See also: BALL JOINTS, ROD ENDS, SPRINGS
  † Leaf Springs for Motor Vehicle Suspension—Made to
  Metric Units—SAE **J1123** NOV92 .... 2:20.01
  Manual for Incorporating Pneumatic Springs in Vehicle
  Suspension Designs—SAE **HS-1576** .... Available as a separate publication *(See Related Technical Reports Section)*
  Rated Suspension Spring Capacity—SAE **J274** JUN89 . 2:20.30
  Terminology
  Vehicle Dynamics Terminology—SAE **J670e** JUN76 ... 3:34.297

SWEEPERS
  *Self-Propelled Sweepers—SAE **J2130** APR93 .... 3:40.23

SWITCHES
  Door Courtesy Switch (A)—SAE **J2108** DEC91 .... 2:23.141
  Marine Electrical Switches—SAE **J1320** APR87 .... 3:41.28
  † SAE Ground Vehicle Lighting Standards Manual—SAE
  HS-34 JAN93 .... Available as a separate publication *(See Related Technical Reports Section)*
  Tests for Snowmobile Switching Devices and
  Components—SAE **J68** DEC91 .... 3:38.08
  Backup Lamps
  Backup Lamp Switch—SAE **J1076** MAR90 .... 2:24.175
  Blowers
  Electric Blower Motor Switch—SAE **J235** 2:23.189
  Hazard Warning Systems
  Hazard Warning Signal Switch (A) (D)—SAE **J910**
  OCT88 .... 2:24.184
  Headlamps
  Headlamp Beam Switching (A)—SAE **J564** MAR90 .... 2:24.08
  Headlamp Switch (A)—SAE **J253** DEC89 .... 2:24.176
  Semiautomatic Headlamp Beam Switching Devices (D)—
  SAE **J565** JUN89 .... 2:24.09
  Ignition
  Ignition Switch—SAE **J259** 2:23.141
  Stop Lamps
  Mechanical Stop Lamp Switch (A)—SAE **J249** JUN88 .. 2:24.150
  Motorcycle Stop Lamp Switch—SAE **J1167** JUN89 .... 3:37.07
  Turn Indicators
  Turn Signal Switch (D)—SAE **J589b** 2:24.159
  Windshield Washers
  Electric Windshield Washer Switch—SAE **J234** 2:23.188
  Windshield Wipers
  Electric Windshield Wiper Switch—SAE **J112a** 2:23.188

SYMBOLS
  All-Wheel-Drive Drivetrain Schematic Symbol
  Standards—SAE **J2059** AUG92 .... 3:29.19
  Dual Dimensioning (A)—SAE **J390** JUN82 .... 1:A.35
  *Introduction to All-Wheel Drive—SAE **HS-2300** JUN93 . Available as a separate publication *(See Related Technical Reports Section)*
  Recommended Practice for Bar-Coded Vehicle Emission
  Configuration Label (A)—SAE **J1892** MAY88 .... 3:42.34
  Safety Alert Symbol for Agricultural, Construction and
  Industrial Equipment (A)—SAE **J284** JAN91 .... 3:39.68
  Symbols and Color Codes for Maintenance Instructions,
  Container and Filler Identification (A)—SAE **J223** APR80 3:40.104
  Drives
  † Symbols for Hydrodynamic Drives (A)—SAE **J640**
  APR93 .... 3:29.01
  Lubricants
  † Automotive Lubricating Greases—SAE **J310** JUN93 ... 1:12.39

  Classification of Energy-Conserving Engine Oil for
  Passenger Cars, Vans, and Light-Duty Trucks—SAE
  J1423 FEB92 .... 1:12.29
  Motorboats
  Graphic Symbols for Boats (A)—SAE **J1449** FEB87 ... 3:41.31
  Operator Controls
  Graphical Symbols for Operator Controls and Displays on
  Off-Road Self-Propelled Work Machines—SAE **J1362**
  JUN92 .... 3:40.423
  Operator Controls and Displays on Motorcycles—SAE
  J107 APR80 .... 3:37.05
  Symbols for Motor Vehicle Controls, Indicators and
  Tell-Tales—SAE **J1048** MAR80 .... 3:34.288
  Plastics
  Automotive Plastic Parts Specification (A)—SAE **J1345**
  FEB82 .... 1:11.148
  † Marking of Plastic Parts—SAE **J1344** APR93 .... 1:11.131

SYNTHETIC RUBBER
  See: RUBBER

TACHOMETERS
  See also: ODOMETERS, SPEEDOMETERS
  Electric Tachometer Specification—SAE **J1399** JUN84 . 2:21.12
  Speedometers and Tachometers—Automotive—SAE
  J678 DEC88 .... 2:21.03

TAIL LAMPS
  See also: STOP LAMPS
  Backup Lamp Switch—SAE **J1076** MAR90 .... 2:24.175
  † Fog Tail Lamp (Rear Fog Light) Systems (A)—SAE
  J1319 JUN93 .... 2:24.141
  † SAE Ground Vehicle Lighting Standards Manual—SAE
  HS-34 JAN93 .... Available as a separate publication *(See Related Technical Reports Section)*
  Snowmobile and Snowmobile Cutter Lamps, Reflective
  Devices, and Associated Equipment (A)—SAE **J292**
  OCT84 .... 3:38.07
  Snowmobile Tail Lamp (Rear Position Lamp) (A)—SAE
  J279 OCT84 .... 3:38.06
  Supplemental High Mounted Stop and Rear Turn Signal
  Lamps for Use on Vehicles Less Than 2032 mm in
  Overall Width (A)—SAE **J186** DEC89 .... 2:24.155
  Tail Lamps (Rear Position Lamps) for Use on Vehicles
  2032 mm or More in Overall Width (A)—SAE **J2040**
  JUN91 .... 2:24.139
  Tail Lamps (Rear Position Lamps) for Use on Motor
  Vehicles Less Than 2032 mm in Overall Width—SAE
  J585 DEC91 .... 2:24.137

TAPS
  Numbering System for Standard Taps (A)—SAE **J2123**
  NOV90 .... 3:43.15

TECHNICAL REPORTS
  † Effective Dates of New or Revised Technical
  Reports (A)—SAE **J301** MAR93 .... 1:12.01
  Guidelines for Developing and Revising SAE
  Nomenclature and Definitions—SAE **J1115** 1:A.35
  Guidelines for Requests Received from Outside Sources
  for the ConAg Council to Originate or Review Technical
  Reports—SAE **J2129** NOV90 .... 3:40.03
  Preparation of SAE Technical Reports—SAE **TSB 002**
  JUN92 .... 1:A.05
  Rules for SAE Use of SI (Metric) Units—SAE **TSB 003**
  JUN92 .... 1:A.08
  *SAE Technical Standards Board Rules and
  Regulations—SAE **TSB 001** JUN93 .... 1:A.01
  Technical Committee Guidelines—SAE **TSB 004** JUN92 1:A.30

TEMPER DESIGNATIONS
  Aluminum Alloys
  Alloy and Temper Designation Systems for Aluminum—
  SAE **J993** JAN89 .... 1:10.02

---

  *Technical Revision   † New Document   (D)—DODISS Adopted   (A)—ANSI Recognized

Subject Index

TEMPER DESIGNATIONS (continued)
Aluminum Alloys (continued)
General Data on Wrought Aluminum Alloys (A)—SAE J454 FEB91 ... 1:10.26

General Information—Chemical Compositions, Mechanical and Physical Properties of SAE Aluminum Casting Alloys—SAE J452 JAN89 ... 1:10.06

Wrought Aluminum Applications Guidelines—SAE J1434 JAN89 ... 1:10.15

Magnesium Alloys
Magnesium Alloys—SAE J464 JAN89 ... 1:10.120

TEMPERING
† High-Strength, Quenched, and Tempered Structural Steels—SAE J368 MAR93 ... 1:1.164

TENSILE TESTS
Fasteners
Mechanical and Material Requirements for Externally Threaded Fasteners (A)—SAE J429 AUG83 ... 1:4.01

Mechanical and Material Requirements for Metric Externally Threaded Steel Fasteners (A)—SAE J1199 SEP83 ... 1:4.06

Test Methods for Metric Threaded Fasteners (A)—SAE J1216 ... 1:4.10

Thread Rolling Screws—SAE J81 JUN79 ... 1:4.26

Steel
Tensile Test Specimens (A)—SAE J416 DEC83 ... 1:3.01

TERMINALS
See: ELECTRIC TERMINALS

TERMINOLOGY
Guidelines for Developing and Revising SAE Nomenclature and Definitions—SAE J1115 ... 1:A.35

Ignition System Nomenclature and Terminology (A)—SAE J139 JUN90 ... 2:23.58

Preparation of SAE Technical Reports—SAE TSB 002 JUN92 ... 1:A.05

Use of Terms Yield Strength and Yield Point (A)—SAE J450 JUN91 ... 1:9.01

Vision Glossary—SAE J264 JUN88 ... 3:34.227

Acronyms
Recommended Practice for the Selection of Engineering Workstations—SAE J2026 NOV91 ... 3:43.01

Adaptive Products
Vehicle and Control Modifications for Drivers with Physical Disabilities Terminology—SAE J2094 JUN92 ... 3:34.196

Aerodynamics
Vehicle Aerodynamics Terminology (A)—SAE J1594 JUN87 ... 3:28.17

Agricultural Equipment
Power Take-Off Definitions and Terminology for Agricultural Tractors (A)—SAE J722 MAR91 ... 3:39.53

† Terminology for Agricultural Equipment—SAE J1150 OCT92 ... 3:39.01

Axles
Nomenclature and Terminology for Truck and Bus Drive Axles (A)—SAE J923 APR91 ... 3:29.72

Passenger Cars and Light Truck Axles (A)—SAE J2200 JAN91 ... 3:29.88

Backhoes
Identification Terminology of Earthmoving Machines (A)—SAE J1057 SEP88 ... 3:40.05

Nomenclature—Hydraulic Backhoes (A)—SAE J326 MAR86 ... 3:40.326

Specification Definitions—Hydraulic Backhoes (A)—SAE J49 APR80 ... 3:40.70

Belts
* Glossary Automatic Belt Tensioner—SAE J2198 JUL92 ... 2:19.19

Bicycles
A Dictionary of Terms for the Dynamics and Handling of Single Track Vehicles (Motorcycles, Mopeds, and Bicycles)—SAE J1451 FEB85 ... 3:37.01

Brakes
Antilock Brake System Review—SAE J2246 JUN92 ... 2:25.72

Automotive Brake Definitions and Nomenclature (A)—SAE J656 APR88 ... 2:25.70

† Surface Vehicle Brake Manual—SAE HS-24 JUL92 ... Available as a separate publication *(See Related Technical Reports Section)*

Bulldozers
Nomenclature and Specification Definitions—Dozers (A)—SAE J729 SEP86 ... 3:40.37

Catalysts
† Emissions Terminology and Nomenclature—SAE J1145 FEB93 ... 1:13.166

Clutches
Automotive Pull Type Clutch Terminology (A)—SAE J1479 APR91 ... 3:36.02

One-Way Clutches—Nomenclature and Terminology—SAE J1087 OCT90 ... 3:29.93

Compactors
Nomenclature—Rollers and Compactors (A)—SAE J1017 JAN86 ... 3:40.49

Computer Equipment
Recommended Practice for the Selection of Engineering Workstations—SAE J2026 NOV91 ... 3:43.01

Cooling Systems
Engine Charge Air Cooler Nomenclature (A)—SAE J1148 JUN90 ... 2:26.392

† Glossary of Engine Cooling System Terms (A)—SAE J1004 APR93 ... 2:26.380

Cranes
Nomenclature and Dimensions for Crane Shovels—SAE J958 JUN89 ... 3:40.53

Crawlers
Nomenclature—Crawler Tractor (A)—SAE J727 JAN86 ... 3:40.30

Drives
Hydrodynamic Drives Terminology (A)—SAE J641 JUL88 ... 3:29.02

Driveshafts
Universal Joints and Driveshafts—Nomenclature—Terminology—Application (A)—SAE J901 OCT90 ... 3:29.42

Dumpers
Component Nomenclature—Dumper Trailer (A)—SAE J734 JUL90 ... 3:40.47

Component Nomenclature—Dumpers (A)—SAE J1016 JUL90 ... 3:40.48

Identification Terminology of Earthmoving Machines (A)—SAE J1057 SEP88 ... 3:40.05

Dynamics
Vehicle Dynamics Terminology—SAE J670e SEP76 ... 3:34.297

Earthboring Machines
Classification, Nomenclature, and Specification Definitions for Horizontal Earthboring Machines (A)—SAE J2022 JUN91 ... 3:40.14

* Trenchless Equipment - (Horizontal Earthboring Machines) Operator Control Definitions—SAE J1611 JUN93 ... 3:40.21

Earthmoving Equipment
Identification Terminology of Earthmoving Machines (A)—SAE J1057 SEP88 ... 3:40.05

Electric Equipment
† Electrical/Electronic Systems Diagnostic Terms, Definitions, Abbreviations and Acronyms—SAE J1930 JUN93 ... 2:23.288

Nomenclature—Automotive Electrical Systems—SAE J831 ... 2:23.01

---

* Technical Revision   † New Document   (D)—DODISS Adopted   (A)—ANSI Recognized

Subject Index

TERMINOLOGY (continued)

Electronic Equipment
† Electrical/Electronic Systems Diagnostic Terms, Definitions, Abbreviations and Acronyms—SAE **J1930** JUN93 .... 2:23.288

Glossary of Automotive Electronic Terms—SAE **J1213** NOV82 .... 2:23.01

Glossary of Reliability Terminology Associated with Automotive Electronics—SAE **J1213/2** OCT88 .... 2:23.22

Glossary of Vehicle Networks for Multiplexing and Data Communications (A)—SAE **J1213/1** JUN91 .... 2:23.19

Engines
Engine Terminology and Nomenclature—General (A)—SAE **J604** JAN86 .... 2:26.01

† Glossary of Engine Cooling System Terms (A)—SAE **J1004** APR93 .... 2:26.380

Excavators
Identification Terminology of Earthmoving Machines (A)—SAE **J1057** SEP88 .... 3:40.05

Nomenclature and Dimensions for Hydraulic Excavators—SAE **J1193** NOV84 .... 3:40.59

Exhaust Emissions
† Emissions Terminology and Nomenclature—SAE **J1145** FEB93 .... 1:13.166

Handbook of Motor Vehicle, Safety and Environmental Terminology—SAE **HS-215** .... Available as a separate publication (See Related Technical Reports Section)

Feller Buncher
Component Nomenclature—Feller Buncher—SAE **J1254** JAN85 .... 3:40.63

Felling Head
Felling Head Terminology and Nomenclature—SAE **J1272** JAN85 .... 3:40.77

Fiberboard
Glossary of Fiberboard Terminology—SAE **J947** JAN85 . 1:11.156

Filters
Glossary of Terms Related to Fluid Filters and Filter Testing (A)—SAE **J1124** MAR87 .... 2:26.342

Foreign Equivalents
† Internal Combustion Engines—Piston Rings—Vocabulary—SAE **J1588** OCT92 .... 2:26.28

Forestry Equipment
Identification Terminology & Component Nomenclature—Knuckle Boom Log Loader (A)—SAE **J2055** JUN91 .... 3:40.67

Identification Terminology of Mobile Forestry Machines—SAE **J1209** JAN85 .... 3:40.25

Forks
Identification Terminology of Loaders/Tractors with Forks and Rough Terrain Forklifts—SAE **J1464** APR88 ...... 3:40.58

Forwarders
Nomenclature—Forwarder (A)—SAE **J1354** APR87 .... 3:40.88

Fuel Injection
† Fuel Injection Equipment Nomenclature—SAE **J830** OCT92 .... 2:26.254

Fuel Injection Nomenclature — Spark Ignition Engines (A)—SAE **J1541** FEB88 .... 2:26.290

Gears
† Planetary Gear(s)—Terminology—SAE **J646** APR93 ... 3:29.24

Governors
† Fuel Injection Equipment Nomenclature—SAE **J830** OCT92 .... 2:26.254

Graders
Component Nomenclature—Graders (A)—SAE **J870** JUL90 .... 3:40.36

Heat Treatment
Definitions of Heat Treating Terms—SAE **J415** JUN83 .. 1:2.37

Inflatable Restraints
Glossary of Automotive Inflatable Restraint Terms—SAE **J1538** APR88 .... 3:33.31

Lamps
Terminology—Motor Vehicle Lighting—SAE **J387** OCT88 2:24.01

Loaders
Component Nomenclature—Loader—SAE **J731** FEB85 . 3:40.39

Identification Terminology & Component Nomenclature—Knuckle Boom Log Loader (A)—SAE **J2055** JUN91 .... 3:40.67

Identification Terminology of Earthmoving Machines (A)—SAE **J1057** SEP88 .... 3:40.05

Identification Terminology of Loaders/Tractors with Forks and Rough Terrain Forklifts—SAE **J1464** APR88 ...... 3:40.58

Specification Definitions—Loaders—SAE **J732** JUN92 .. 3:40.65

Mopeds
A Dictionary of Terms for the Dynamics and Handling of Single Track Vehicles (Motorcycles, Mopeds, and Bicycles)—SAE **J1451** FEB85 .... 3:37.01

Motorcycles
A Dictionary of Terms for the Dynamics and Handling of Single Track Vehicles (Motorcycles, Mopeds, and Bicycles)—SAE **J1451** FEB85 .... 3:37.01

Definitions—Motorcycles—SAE **J213a** 3:37.01

*SAE Motorcycle Standards Manual—SAE **HS-2500** APR93 .... Available as a separate publication (See Related Technical Reports Section)

Mowers
Nomenclature—Industrial and Agricultural Mowers (A)—SAE **J990** MAR87 .... 3:40.475

Oil Coolers
Oil Cooler Nomenclature and Glossary (A)—SAE **J1244** AUG88 .... 2:26.405

Operator Controls
*Trenchless Equipment - (Horizontal Earthboring Machines) Operator Control Definitions—SAE **J1611** JUN93 .... 3:40.21

Pipelayers
Identification Terminology and Specification Definitions—Pipelayers and Side Booms, Tractor or Loader Mounted—SAE **J1295** JUN89 .... 3:40.55

Piston Rings
† Internal Combustion Engines—Piston Rings—Vocabulary—SAE **J1588** OCT92 .... 2:26.28

Radiators
† Radiator Nomenclature (A)—SAE **J631** APR93 .... 2:26.385

Rims
Nomenclature—Wheels, Hubs, and Rims for Commercial Vehicles—SAE **J393** OCT91 .... 3:31.32

Wide Base Tire Rims and Wheels (A)—SAE **J876** JAN91 3:36.109

Rippers
Nomenclature—Rippers and Scarifiers (A)—SAE **J733** JUN86 .... 3:40.45

Rollers
Nomenclature—Rollers and Compactors (A)—SAE **J1017** JAN86 .... 3:40.49

Safety
Handbook of Motor Vehicle, Safety and Environmental Terminology—SAE **HS-215** .... Available as a separate publication (See Related Technical Reports Section)

Scarifiers
Nomenclature—Rippers and Scarifiers (A)—SAE **J733** JUN86 .... 3:40.45

Scrapers
Component Nomenclature—Scrapers (A)—SAE **J728** JUL90 .... 3:40.34

Identification Terminology of Earthmoving Machines (A)—SAE **J1057** SEP88 .... 3:40.05

Skidders
† Component Nomenclature—Articulated Log Skidder, Rubber-Tired (A)—SAE **J1109** JUN93 .... 3:40.62

---

* Technical Revision   † New Document   (D)—DODISS Adopted   (A)—ANSI Recognized

Subject Index

TERMINOLOGY (continued)
Skidders (continued)
Component Nomenclature—Skidder-Grapple (A)—SAE J1111 MAR81 .... 3:40.64

Identification Terminology of Mobile Forestry Machines—SAE J1209 JAN85 .... 3:40.25

Nomenclature—Clam Bunk Skidder (A)—SAE J1353 APR87 .... 3:40.80

† Specification Definitions—Articulated Rubber-Tired Log Skidder (A)—SAE J1110 JUN93 .... 3:40.73

Specification Definitions—Skidder-Grapple (A)—SAE J1112 APR80 .... 3:40.75

Snowmobiles
Snowmobile Definitions and Nomenclature—General—SAE J33 DEC91 .... 3:38.01

Spark Ignition Engines
† Emissions Terminology and Nomenclature—SAE J1145 FEB93 .... 1:13.166

Springs
Helical Compression and Extension Spring Terminology (A)—SAE J1121 JUL88 .... 2:20.17

Pneumatic Spring Terminology—SAE J511 JUN89 .... 2:20.16

Steels
Glossary of Carbon Steel Sheet and Strip Terms—SAE J940 DEC88 .... 1:2.34

Tires
Military Tire Glossary (A)—SAE J2013 MAY91 .... 3:30.62

Towing
Towability Design Criteria and Equipment Use—Passenger Cars, Vans, and Light Duty Trucks (A)—SAE J1142 JUN91 .... 3:42.01

Tractors
Component Nomenclature—Construction Two- and Four-Wheel Tractors (A)—SAE J869 JUL90 .... 3:40.32

Identification Terminology of Loaders/Tractors with Forks and Rough Terrain Forklifts—SAE J1464 APR88 .... 3:40.58

Nomenclature—Industrial Tractors (Wheel) (A)—SAE J1092 JUN86 .... 3:40.31

Power Take-Off Definitions and Terminology for Agricultural Tractors (A)—SAE J722 MAR91 .... 3:39.53

Trailers
Nomenclature—Truck, Bus, Trailer (A)—SAE J687 JUN88 .... 3:36.01

Transmissions
Automatic Transmission Functions—Terminology (A)—SAE J649 JUL88 .... 3:29.27

† Automatic Transmission Hydraulic Control Systems—Terminology—SAE J648 APR93 .... 3:29.27

† Automotive Transmission Terminology—SAE J645 OCT92 .... 3:29.23

Trenching Equipment
† Classification, Nomenclature, and Specification Definitions for Trenching Machines—SAE J1382 JUN93 .... 3:40.41

Trucks and Buses
Nomenclature and Terminology for Truck and Bus Drive Axles (A)—SAE J923 APR91 .... 3:29.72

Nomenclature—Truck, Bus, Trailer (A)—SAE J687 JUN88 .... 3:36.01

Truck Identification Terminology (A)—SAE J1229 DEC81 .... 3:27.04

Turbochargers
Turbocharger Nomenclature and Terminology—SAE J922 JUL88 .... 2:26.04

Universal Joints
Universal Joints and Driveshafts—Nomenclature—Terminology—Application (A)—SAE J901 OCT90 .... 3:29.42

Wheels
Nomenclature—Wheels for Passenger Cars, Light Trucks, and Multipurpose Vehicles—SAE J1982 DEC91 .... 3:31.01

Nomenclature—Wheels, Hubs, and Rims for Commercial Vehicles—SAE J393 OCT91 .... 3:31.32

Wide Base Tire Rims and Wheels (A)—SAE J876 JAN91 .... 3:36.109

TEST BENCHES
* Diesel Engines—Fuel Injection Pump Testing—SAE J1668 JUN93 .... 2:26.322

TEXTILES
See: FABRICS, TRIM MATERIALS

THERMAL COMFORT
* Equivalent Temperature—SAE J2234 JAN93 .... 3:34.25

THERMOCOUPLES
Brake Disc and Drum Thermocouple Installation—SAE J79 .... 2:25.157

THORAX
Impact Tolerance
Human Mechanical Response Characteristics—SAE J1460 MAR85 .... 3:34.292

Human Tolerance to Impact Conditions as Related to Motor Vehicle Design—SAE J885 JUL86 .... 3:34.306

THREADS
† Automotive Tube Fittings—SAE J512 JUN93 .... 2:22.06

† Connections for General Use and Fluid Power-Ports and Stud Ends with ISO 725 Threads and O-ring Sealing—Part 1: Threaded Port with O-Ring Seal in Truncated Housing—SAE J1926/1 ISO 11 MAR93 .... 2:22.99

† Drain, Fill, and Level Plugs for Off-Road, Self-Propelled Work Machines—SAE J371 MAY93 .... 3:40.257

† Hydraulic Hose Fittings—SAE J516 JUN93 .... 2:22.114

† Hydraulic Tube Fittings—SAE J514 JUN93 .... 2:22.59

† Lubrication Fittings (A)—SAE J534 JUN93 .... 2:22.252

† Refrigeration Tube Fittings—General Specifications—SAE J513 JUN93 .... 2:22.33

Pipes
† Beaded Tube Hose Fittings—SAE J1231 JUN93 .... 2:22.158

Dryseal Pipe Threads—SAE J476a 2:15.01

* Thread Sealants—SAE J1615 JUN93 .... 2:22.240

Screws
Screw Threads (ANSI B1.1)—SAE J475a .... Available as a separate publication (See Related Technical Reports Section)

THROTTLES
† Marine Carburetors and Fuel Injection Throttle Bodies—SAE J1223 JUN93 .... 3:41.28

Marine Control Cable Connection—Engine Throttle Lever (A)—SAE J961 JUN91 .... 3:41.05

Snowmobile Throttle Control Systems (A)—SAE J92 OCT84 .... 3:38.15

TILLERS
See also: EARTHMOVING EQUIPMENT

Bystander Sound Level Measurement Procedure for Small Engine Powered Equipment (A)—SAE J1175 MAR85 .... 1:14.97

Operator Ear Sound Level Measurement Procedure for Small Engine Powered Equipment (A)—SAE J1174 MAR85 .... 1:14.95

TILT TABLE
* A Tilt-table Procedure for Measuring the Static Rollover Threshold for Heavy Trucks—SAE J2180 APR93 .... 2:23.528

TIN
See: TRANSAXLE IDENTIFICATION NUMBER (TIN), TRANSMISSION IDENTIFICATION NUMBER (TIN)

TIN ALLOYS
Bearing and Bushing Alloys—SAE J459 OCT91 .... 1:10.43

Bearing and Bushing Alloys—Chemical Composition of SAE Bearing and Bushing Alloys—SAE J460 OCT91 .... 1:10.41

TIRE CHAINS
See: CHAINS

---

* Technical Revision   † New Document   (D)—DODISS Adopted   (A)—ANSI Recognized

TIRE RIMS
See: RIMS

TIRES
See also: WHEELS

Agricultural Equipment
Agricultural Tractor Tire Dynamic Indices—SAE J1261 NOV84 .... 3:39.28
Agricultural Tractor Tire Loadings, Torque Factors, and Inflation Pressures (A)—SAE J709d 3:39.20
Liquid Ballast Table for Drive Tires of Agricultural Tractors (A)—SAE J884 MAR91 .... 3:39.25
Tire Selection Tables for Agricultural Tractors of Future Design (A)—SAE J711 MAR91 .... 3:39.26

Automobiles
Factors Affecting Accuracy of Mechanically Driven Automotive Speedometer and Odometers—SAE J862 JAN89 .... 2:21.01
† Laboratory Speed Test Procedure for Passenger Car Tires—SAE J1561 FEB93 .... 3:30.34
Laboratory Testing Machines and Procedures for Measuring the Steady State Force and Moment Properties of Passenger Car Tires—SAE J1107 FEB75 . 3:30.43
Laboratory Testing Machines for Measuring the Steady State Force and Moment Properties of Passenger Car Tires—SAE J1106 FEB75 .... 3:30.40
Measurement of Passenger Car, Light Truck, and Highway Truck and Bus Tire Rolling Resistance—SAE J1270 MAR87 .... 3:30.21
Passenger and Light Truck Tire Traction Device Profile Determination and Classification—SAE J1232 .... 3:36.115
Passenger Car and Light Truck Tire Dynamic Driving Traction in Snow (A)—SAE J1466 OCT85 .... 3:30.26
Passenger Car Tire Performance Requirements and Test Procedures—SAE J918c 3:30.01
Rolling Resistance Measurement Procedure for Passenger Car, Light Truck, and Highway Truck and Bus Tires (A)—SAE J1269 MAR87 .... 3:30.18
Subjective Rating Scale for Evaluation of Noise and Ride Comfort Characteristics Related to Motor Vehicle Tires—SAE J1060 .... 3:30.37
Test Procedure for Measuring Passenger Car Tire Revolutions per Mile—SAE J966 .... 3:30.36
Testing Machines for Measuring the Uniformity of Passenger Car and Light Truck Tires (A)—SAE J332 AUG81 .... 3:30.32
Wet or Dry Pavement Passenger Car Tire Peak and Locked Wheel Braking Traction—SAE J345a 3:30.24

Earthmoving Equipment
Off-Road Tire and Rim Selection and Application (A)—SAE J1315 JAN91 .... 3:40.307
Tire Guards for Protection of Operator of Earthmoving Haulage Machines (A)—SAE J321b 3:40.352
Ton Kilometer Per Hour Application (A)—SAE J1098 MAR91 .... 3:40.317
Ton-Kilometer Per Hour Test Procedure (A)—SAE J1015 MAR91 .... 3:40.314

Off-Road Vehicles
Off-Road Tire and Rim Classification—Construction Machines (A)—SAE J751 APR86 .... 3:40.301
Off-Road Tire and Rim Classification—Forestry Machines (A)—SAE J1440 APR86 .... 3:40.310
Off-Road Tire and Rim Selection and Application (A)—SAE J1315 JAN91 .... 3:40.307
Revolutions per Mile and Static Loaded Radius for Off-Road Tires—SAE J1544 JAN88 .... 3:40.307
Ton Kilometer Per Hour Application (A)—SAE J1098 MAR91 .... 3:40.317
Ton-Kilometer Per Hour Test Procedure (A)—SAE J1015 MAR91 .... 3:40.314

Terminology
Military Tire Glossary (A)—SAE J2013 MAY91 .... 3:30.62
Vehicle Dynamics Terminology—SAE J670e MAY76 .... 3:34.297

Trucks and Buses
† Disc Wheel Radial Runout Low Point Marking—SAE J2133 JUN93 .... 3:31.38
Heavy Truck and Bus Retarder Downhill Performance Mapping Procedure (A)—SAE J1489 JUN87 .... 3:36.124
Labeling—Disc Wheel and Demountable Rims—Trucks (A)—SAE J179 JUL84 .... 3:31.30
Measurement of Passenger Car, Light Truck, and Highway Truck and Bus Tire Rolling Resistance—SAE J1270 MAR87 .... 3:30.21
Passenger and Light Truck Tire Traction Device Profile Determination and Classification—SAE J1232 .... 3:36.115
Passenger Car and Light Truck Tire Dynamic Driving Traction in Snow (A)—SAE J1466 OCT85 .... 3:30.26
Pneumatic Tires for Military Tactical Wheeled Vehicles (A)—SAE J2014 APR91 .... 3:30.04
Rolling Resistance Measurement Procedure for Passenger Car, Light Truck, and Highway Truck and Bus Tires (A)—SAE J1269 MAR87 .... 3:30.18
Sound Level of Highway Truck Tires (A)—SAE J57 FEB87 .... 1:14.31
Test Procedures for Measuring Truck Tire Revolutions per Mile—SAE J1025 .... 3:30.36
Testing Machines for Measuring the Uniformity of Passenger Car and Light Truck Tires (A)—SAE J332 AUG81 .... 3:30.32
Tire Chain Clearance—Trucks, Buses (Except Suburban, Intercity, and Transit Buses), and Combinations of Vehicles—SAE J683 AUG85 .... 3:36.115
Truck Overall Widths Across Dual Tires—SAE J693 APR89 .... 3:36.111

Valves
Methods for Testing Snap-In Tubeless Tire Valves—SAE J1206 .... 3:30.38
Performance Requirements for Snap-In Tubeless Tire Valves—SAE J1205 .... 3:30.40

TOOLS
Numbering System for Reamers (A)—SAE J2124 NOV90 3:43.18
Numbering System for Standard Drills (A)—SAE J2122 NOV90 .... 3:43.11
Numbering System for Standard Taps (A)—SAE J2123 NOV90 .... 3:43.15
Selection and Heat Treatment of Tool and Die Steels—SAE J437a 1:7.02
Sintered Carbide Tools—SAE J439a 1:7.07
Sintered Tool Materials—SAE J1072 .... 1:7.12
Tool and Die Steels—SAE J438b 1:7.01

TORQUE
Radial Lip Seal Torque: Measurement Methods and Results—SAE J1971 JUN89 .... 3:29.151

Axles
Axle Efficiency Test Procedure (A)—SAE J1266 JUN90 . 3:29.85

Brakes
† Brake System Torque Balance Test Code Commercial Vehicles (A)—SAE J225 JUN93 .... 3:36.133
Brake Test Code—Inertia Dynamometer—SAE J667 ... 2:25.103

Fasteners
Torque-Tension Test Procedure for Steel Threaded Fasteners—SAE J174 .... 2:16.22

TORQUE CONVERTERS
Flywheels for Engine Mounted Torque Converters—SAE J927 NOV88 .... 3:40.254
Hydrodynamic Drive Test Code—SAE J643 JUN89 .... 3:29.05
Hydrodynamic Drives Terminology (A)—SAE J641 JUL88 3:29.02

TOWING
Automobiles
* Recovery Attachment Points for Passenger Cars, Vans, and Light Trucks—SAE J2069 JAN93 .... 3:42.21

Subject Index

TOWING (continued)
Automobiles and Trucks
Towability Design Criteria and Equipment Use—Passenger Cars, Vans, and Light Duty Trucks (A)—SAE J1142 JUN91 .... 3:42.01
Towed Vehicle Drivetrain Test Procedure—Passenger Cars, Vans, and Light Duty Trucks (A)—SAE **J1144** JUN91 .... 3:42.16
Towed Vehicle/Tow Equipment Attachment Test Procedure—Passenger Cars, Vans, and Light Duty Trucks (A)—SAE **J1143** JUN91 .... 3:42.12

Trailers
Brake System Road Test Code—Passenger Car and Light Duty Truck-Trailer Combinations (A)—SAE **J134** MAY85 .... 2:25.134
Service Brake System Performance Requirements—Passenger Car—Trailer Combinations—SAE **J135** MAY85 .... 2:25.143

Truck Trailers
Connection and Accessory Locations for Towing Multiple Trailers (A)—SAE **J849** NOV85 .... 3:36.93
Safety Chain of Full Trailers or Converter Dollies (A)—SAE **J697** MAY88 .... 3:36.98
Trailer Tow Bar Eye and Pintle Hook/Coupler Performance (A)—SAE **J847** NOV87 .... 3:36.95

TOXICITY
Refrigerants
*Selection Criteria for Retrofit Refrigerants to Replace R-12 in Mobile Air Conditioning Systems—SAE **J1657** JUN93 .... 3:34.61

TRACTION
Terminology
Vehicle Dynamics Terminology—SAE J670e JUN76 ... 3:34.297

Tire
Passenger and Light Truck Tire Traction Device Profile Determination and Classification—SAE **J1232** 3:36.115
Passenger Car and Light Truck Tire Dynamic Driving Traction in Snow (A)—SAE **J1466** OCT85 .... 3:30.26
Wet or Dry Pavement Passenger Car Tire Peak and Locked Wheel Braking Traction—SAE **J345a** 3:30.24

TRACTORS
See: AGRICULTURAL EQUIPMENT, CABS, CONSTRUCTION AND INDUSTRIAL EQUIPMENT, CRAWLERS

TRAILERS
See also: RECREATIONAL VEHICLES, TRUCK TRAILERS
Brakes
Brake System Road Test Code—Passenger Car and Light Duty Truck-Trailer Combinations (A)—SAE **J134** MAY85 .... 2:25.134
Service Brake System Performance Requirements—Passenger Car—Trailer Combinations—SAE **J135** MAY85 .... 2:25.143
† Surface Vehicle Brake Manual—SAE **HS-24** JUL92 .... Available as a separate publication (See Related Technical Reports Section)

Couplings
*Kingpin Wear Limits Commercial Trailers and Semi-Trailers—SAE **J2228** JUN93 .... 3:36.89
Trailer Couplings, Hitches and Safety Chains—Automotive Type (A)—SAE **J684** MAY87 .... 3:35.01

Electric Equipment
Five Conductor Electrical Connectors for Automotive Type Trailers (A)—SAE **J895** APR86 .... 2:23.171
Four- and Eight-Conductor Rectangular Electrical Connectors for Automotive Type Trailers—SAE **J1239** .. 2:23.173
† SAE Ground Vehicle Lighting Standards Manual—SAE **HS-34** JAN93 .... Available as a separate publication (See Related Technical Reports Section)

Parking Performance
Trailer Grade Parking Performance Test Procedure—SAE **J1452** OCT88 .... 2:25.155

Wheels
Wheels—Recreational and Utility Trailer Test Procedure (A)—SAE **J1204** DEC89 .... 3:31.08

Winches
Hand Winches—Boat Trailer Type—SAE **J1853** APR87 . 3:35.05

TRAINING
Assessing Technician Training (A)—SAE **J2018** JUL89 . 3:42.50
Developing Technician Training (A)—SAE **J2017** AUG89 3:42.48

TRANSAXLE IDENTIFICATION NUMBER (TIN)
Engine and Transmission Identification Numbers—SAE **J129** DEC81 .... 3:27.03

TRANSAXLES
Manual Transmission and Transaxle Efficiency and Parasitic Loss Measurement—SAE **J1540** APR92 ..... 3:29.61

TRANSDUCERS
See also: ELECTRONIC EQUIPMENT
Guide to Manifold Absolute Pressure Transducer Representative Test Method—SAE **J1346** JUN81 .... 2:23.510
Guide to Manifold Absolute Pressure Transducer Representative Specification—SAE **J1347** JUN81 ... 2:23.512
Manual Transmission and Transaxle Efficiency and Parasitic Loss Measurement—SAE **J1540** APR92 ..... 3:29.61
† Measurement of Whole Body Vibration of the Seated Operator of Off-Highway Work Machines—SAE **J1013** AUG92 .... 3:39.92
Radial Lip Seal Torque: Measurement Methods and Results—SAE **J1971** JUN89 .... 3:29.151

TRANSMISSION FLUIDS
See also: LUBRICANTS
Fluid for Passenger Car Type Automatic Transmissions—SAE **J311** APR86 .... 1:12.44
† Fuels and Lubricants Standards Manual—SAE **HS-23** FEB93 .... Available as a separate publication (See Related Technical Reports Section)
Powershift Transmission Fluid Classification—SAE **J1285** JAN85 .... 1:12.45
The Automotive Lubricant Performance and Service Classification Maintenance Procedure (A)—SAE **J1146** JUN86 .... 1:12.87
† Transmission Oil Cooler Hose—SAE **J1532** JUN93 .... 1:11.252

TRANSMISSION IDENTIFICATION NUMBER (TIN)
Engine and Transmission Identification Numbers—SAE **J129** DEC81 .... 3:27.03

TRANSMISSIONS
See also: DRIVES, GEARS, TORQUE CONVERTERS, TRANSMISSION FLUIDS
† Automatic Transmissions—Manual Control Sequence—SAE **J915** APR93 .... 3:29.32
† Clearance Envelopes for Six-Bolt, Eight-Bolt, and Rear Truck Transmission Mounted Power Takeoffs (A)—SAE **J772** DEC92 .... 3:29.33
Hydrodynamic Drive Test Code—SAE **J643** JUN89 .... 3:29.05
Hydrodynamic Drives Terminology (A)—SAE **J641** JUL88 3:29.02
† Openings for Six- and Eight-Bolt Truck Transmission Mounted Power Take-Offs—SAE **J704** DEC92 .... 3:29.41
Pull-Type Clutch—Transmission Installation Dimensions (A)—SAE **J1463** NOV84 .... 3:36.120
Recommended Remanufacturing Procedures for Manual Transmission Clutch Assemblies—SAE **J1915** FEB90 . 3:42.39
Transmission Mounted Vehicle Speed Signal Rotor Specification—SAE **J1377** JAN89 .... 2:21.17
Transmissions—Schematic Diagrams (A)—SAE **J647** APR90 .... 3:29.24

---
* Technical Revision † New Document (D)—DODISS Adopted (A)—ANSI Recognized

Subject Index

TRANSMISSIONS (continued)
Automatic
† Automatic Transmission Hydraulic Control Systems—Terminology—SAE **J648** APR93 .... 3:29.27
Fluid for Passenger Car Type Automatic Transmissions—SAE **J311** APR86 .... 1:12.44
*Park Standard for Automatic Transmissions—SAE **J2208** JUN93 .... 3:29.42
Passenger Car and Truck Automatic Transmission Test Code (A)—SAE **J651** JAN91 .... 3:29.28

Dipsticks
Engine and Transmission Dipstick Marking (A)—SAE **J614b** 2:26.233

Efficiency
Manual Transmission and Transaxle Efficiency and Parasitic Loss Measurement—SAE **J1540** APR92 ..... 3:29.61

Identification Numbers
Engine and Transmission Identification Numbers—SAE **J129** DEC81 .... 3:27.03

Lubricants
Axle and Manual Transmission Lubricant Viscosity Classification—SAE **J306** OCT91 .... 1:12.37
Axle and Manual Transmission Lubricants—SAE **J308** JUN89 .... 1:12.36
Fluid for Passenger Car Type Automatic Transmissions—SAE **J311** APR86 .... 1:12.44
Powershift Transmission Fluid Classification—SAE **J1285** JAN85 .... 1:12.45
The Automotive Lubricant Performance and Service Classification Maintenance Procedure (A)—SAE **J1146** JUN86 .... 1:12.87

Seals
Cast Iron Sealing Rings—SAE **J281** SEP80 .... 3:29.148
† Cast Iron Sealing Rings (Metric)—SAE **J1236** APR93 .. 3:29.156
Seals—Evaluation of Elastohydrodynamic—SAE **J1002** JUN87 .... 3:29.137
† Static and Reciprocating Elastomeric Trnamission Seals—SAE **J654** MAR93 .... 3:29.145

Terminology
Automatic Transmission Functions—Terminology (A)—SAE **J649** JUL88 .... 3:29.27
† Automatic Transmission Hydraulic Control Systems—Terminology—SAE **J648** APR93 .... 3:29.27
† Automotive Transmission Terminology—SAE **J645** OCT92 .... 3:29.23

TRENCHING EQUIPMENT
See also: **AGRICULTURAL EQUIPMENT, CONSTRUCTION AND INDUSTRIAL EQUIPMENT, OFF-ROAD VEHICLES**
Noise
Sound Measurement—Off-Road Self-Propelled Work Machines—Operator—Work Cycle (A)—SAE **J1166** MAY90 .... 1:14.55
Sound Measurement—Trenching Machines—SAE **J1262** OCT90 .... 1:14.109

Terminology
† Classification, Nomenclature, and Specification Definitions for Trenching Machines—SAE **J1382** JUN93   3:40.41

TRIM MATERIALS
See also: **FABRICS, FELTS, FIBERBOARD, PLASTICS, VINYL**
Abrasion Resistance Testing—Vehicle Exterior Graphics and Pin Striping—SAE **J1847** JUN89 .... 1:11.262
Accelerated Exposure of Automotive Interior Trim Components Using a Controlled Irradiance Water Cooled Xenon-Arc Apparatus—SAE **J1885** MAR92 .... 1:11.255
Anodized Aluminum Automotive Parts—SAE **J399** FEB85 1:10.38
*Decorative Anodizing Specification for Automotive Applications—SAE **J1974** JUN93 .... 1:10.39
Electroplate Requirements for Decorative Chromium Deposits on Zinc Base Materials Used for Exterior Ornamentation (A)—SAE **J1837** JUN91 .... 1:10.155
Electroplating of Nickel and Chromium on Metal Parts—Automotive Ornamentation and Hardware—SAE **J207** FEB85 .... 1:10.150
† Hot Odor Test for Insulation Materials—SAE **J1351** JUN93 .... 1:11.172
Hot Plate Method for Evaluating Heat Resistance and Thermal Insulation Properties of Materials—SAE **J1361** APR87 .... 1:11.171
Method of Testing Resistance to Crocking of Organic Trim Materials—SAE **J861** 1:11.114
Method of Testing Resistance to Scuffing of Trim Materials (A)—SAE **J365** FEB85 .... 1:11.163
Nonmetallic Trim Materials—Test Method for Determining the Staining Resistance to Hydrogen Sulfide Gas (A)—SAE **J322** DEC85 .... 1:11.116
*SAE Automotive Textiles and Trim Standards Manual—SAE **HS-2700** DEC93 .... Available as a separate publication *(See Related Technical Reports Section)*
Test Method for Determining Blocking Resistance and Associated Characteristics of Automotive Trim Materials—SAE **J912a** 1:11.165
Test Method for Determining Dimensional Stability of Automotive Textile Materials (A)—SAE **J883** FEB86 ... 1:11.114
Test Method for Determining Resistance to Abrasion of Automotive Bodycloth, Vinyl, and Leather, and the Snagging of Automotive Bodycloth (A)—SAE **J948** FEB86 .... 1:11.178
Test Method for Determining Window Fogging Resistance of Interior Trim Materials (A)—SAE **J275** FEB85 .... 1:11.117
Test Method for Measuring Weight of Organic Trim Materials (A)—SAE **J860** JAN85 .... 1:11.114
Test Method for Wicking of Automotive Fabrics and Fibrous Materials (A)—SAE **J913** FEB85 .... 1:11.163

Color
Instrumental Color Difference Measurement for Exterior Finishes, Textiles, and Colored Trim (A)—SAE **J1545** JUN86 .... 1:11.279
*Paint and Trim Code Location—SAE **J2235** JUN93 .... 3:33.44
Test Method for Determining Visual Color Match to Master Specimen for Fabrics, Vinyls, Coated Fiberboards, and Other Automotive Trim Materials (A)—SAE **J361** MAR85 .... 1:11.180

Flammability
Flammability of Automotive Interior Materials—Horizontal Test Method—SAE **J369** JUN89 .... 1:11.176

TRUCK TRACTORS
See: **CABS, TRUCKS AND BUSES**

TRUCK TRAILERS
See also: **TRAILERS**
Axles
Trailer Axle Alignment (A)—SAE **J875** JUN85 .... 3:36.112

Brakes
Air Brake Actuator Test Procedure—Truck-Tractor, Bus and Trailer (A)—SAE **J1469** JUN88 .... 3:36.157
Air Brake Gladhand Service (Control) and Emergency (Supply) Line Couplers—Trucks, Truck-Tractors, and Trailers—SAE **J318** SEP80 .... 3:36.139
Brake and Electrical Connection Locations—Truck-Tractor and Truck-Trailer (A)—SAE **J702** MAY85 . 3:36.114
Brake Force/Distribution Performance Guide—Commercial Vehicles (A)—SAE **J1854** JUN88 .... 3:36.170
† Surface Vehicle Brake Manual—SAE **HS-24** JUL92 .... Available as a separate publication *(See Related Technical Reports Section)*

Couplers
Air Brake Gladhand Service (Control) and Emergency (Supply) Line Couplers—Trucks, Truck-Tractors, and Trailers—SAE **J318** SEP80 .... 3:36.139

---
* Technical Revision   † New Document   (D)—DODISS Adopted   (A)—ANSI Recognized

Subject Index

TRUCK TRAILERS (continued)
Couplers (continued)
Connection and Accessory Locations for Towing Multiple Trailers (A)—SAE **J849** NOV85 .... 3:36.93

Fifth Wheel Kingpin Performance—Commercial Trailers and Semitrailers—SAE **J133** OCT87 .... 3:36.86

Fifth Wheel Kingpin, Heavy-Duty—Commercial Trailers and Semitrailers (A)—SAE **J848** JAN91 .... 3:36.92

*Kingpin Wear Limits Commercial Trailers and Semi-Trailers—SAE **J2228** JUN93 .... 3:36.89

Truck Tractor Semitrailer Interchange Coupling Dimensions (A)—SAE **J701** AUG84 .... 3:36.83

† Upper Coupler Kingpin—Commercial Trailers and Semitrailers (A)—SAE **J700** JUN93 .... 3:36.90

Electric Equipment
† Seven Conductor Electrical Connector for Truck-Trailer Jumper Cable—SAE **J560** JUN93 .... 2:23.175

Seven Conductor Jacketed Cable for Truck Trailer Connections—SAE **J1067** 2:23.178

Terminology
Nomenclature—Truck, Bus, Trailer (A)—SAE **J687** JUN88 .... 3:36.01

Towing
Connection and Accessory Locations for Towing Multiple Trailers (A)—SAE **J849** NOV85 .... 3:36.93

Safety Chain of Full Trailers or Converter Dollies (A)—SAE **J697** MAY88 .... 3:36.98

Trailer Tow Bar Eye and Pintle Hook/Coupler Performance (A)—SAE **J847** NOV87 .... 3:36.95

TRUCKS
Towing
*Recovery Attachment Points for Passenger Cars, Vans, and Light Trucks—SAE **J2069** JAN93 .... 3:42.21

TRUCKS AND BUSES
See also: TRUCK TRAILERS
Acceleration
Vehicle Acceleration Measurement—SAE **J1491** JUN90 .... 2:26.503

Accessories
Information Relating to Duty Cycles and Average Power Requirements of Truck and Bus Engine Accessories—SAE **J1343** OCT88 .... 3:36.57

Aerodynamics
Aerodynamic Flow Visualization Techniques and Procedures—SAE **HS-1566** .... Available as a separate publication (See Related Technical Reports Section)

SAE Wind Tunnel Test Procedure for Trucks and Buses (A)—SAE **J1252** JUL81 .... 3:36.61

Air Conditioning
Rating Air Conditioner Evaporator Air Delivery and Cooling Capacities (A)—SAE **J1487** MAY85 .... 3:36.58

Test Method for Measuring Power Consumption of Air Conditioning and Brake Compressors for Trucks and Buses (A)—SAE **J1340** APR90 .... 3:36.53

Axles
Nomenclature and Terminology for Truck and Bus Drive Axles (A)—SAE **J923** APR91 .... 3:29.72

Turning Ability and Off Tracking—Motor Vehicles (A)—SAE **J695** DEC89 .... 3:36.78

Brakes
Air Brake Actuator Test Procedure—Truck-Tractor, Bus and Trailer (A)—SAE **J1469** JUN88 .... 3:36.157

Air Brake Gladhand Service (Control) and Emergency (Supply) Line Couplers—Trucks, Truck-Tractors, and Trailers (A)—SAE **J318** SEP80 .... 3:36.139

Antilock Brake System Review—SAE **J2246** JUN92 .... 2:25.72

Automatic Slack Adjuster Test Procedure (A)—SAE **J1462** MAY87 .... 3:36.175

Automotive and Off-Highway Air Brake Reservoir Performance and Identification Requirements (A)—SAE **J10** OCT90 .... 3:36.132

Brake and Electrical Connection Locations—Truck-Tractor and Truck-Trailer (A)—SAE **J702** MAY85 .... 3:36.114

*Brake Performance and Wear Test Code Commercial Vehicle Inertia Dynamometer—SAE **J2115** JUN93 .... 2:25.106

Brake Power Rating Test Code—Commercial Vehicle Inertia Dynamometer (A)—SAE **J971** JUN91 .... 2:25.152

Brake Rating Horsepower Requirements—Commercial Vehicles—SAE **J257** MAR85 .... 3:36.137

Brake System Performance Requirements—Truck, Bus, and Combination of Vehicles (A)—SAE **J992** MAR85 .... 3:36.164

Brake System Rating Test Code—Commercial Vehicles—SAE **J880** MAR85 .... 3:36.147

Brake System Road Test Code—Passenger Car and Light-Duty Truck (A)—SAE **J843** NOV90 .... 2:25.108

Brake System Road Test Code—Passenger Car and Light Duty Truck-Trailer Combinations (A)—SAE **J134** MAY85 .... 2:25.134

Brake System Road Test Code—Truck, Bus, and Combination of Vehicles—SAE **J786a** 3:36.140

† Brake System Torque Balance Test Code Commercial Vehicles (A)—SAE **J225** JUN93 .... 3:36.133

*Braking, Stability, and Control Performance Test Procedures for Air-Brake-Equipped Truck Tractors—SAE **J1626** MAR93 .... 3:36.130

Hydraulic Master Cylinders for Motor Vehicle Brakes—Test Procedure (A)—SAE **J1153** JUN91 .... 2:25.29

Hydraulic Master Cylinders for Motor Vehicle Brakes—Performance Requirements (A)—SAE **J1154** JUN91 .... 2:25.34

Hydraulic Valves for Motor Vehicle Brake Systems—Performance Requirements—SAE **J1137** MAR85 .... 2:25.39

Hydraulic Wheel Cylinders for Automotive Drum Brakes (A)—SAE **J101** DEC89 .... 2:25.25

In-Service Brake Performance Test Procedure Passenger-Car and Light-Duty Truck—SAE **J201** MAY89 .... 2:25.85

† In-Service Brake Performance Test Procedure—Vehicles Over 4500 kg (10 000 lb)—SAE **J1250** NOV92 .... 2:25.86

Parking Brake Structural Integrity Test Procedure—Vehicles Over 10,000 Pounds (4500 KG) GVWR (A)—SAE **J1476** JUN87 .... 3:36.123

† Service Brake Structural Integrity Requirements—Vehicles Over 10 000 lb (4500 kg) GVWR (A)—SAE **J1404** JUN93 .... 3:36.171

† Service Brake Structural Integrity Test Procedure—Vehicles Over 4500 kg (10 000 lb) GVWR—SAE **J294** 3:36.138

Simulated Mountain Brake Performance Test Procedure—SAE **J1247** APR80 .... 2:25.123

Stopping Distance Test Procedure (A)—SAE **J299** AUG87 .... 2:25.156

† Surface Vehicle Brake Manual—SAE **HS-24** JUL92 .... Available as a separate publication (See Related Technical Reports Section)

Test Procedure for Air Reservoir Capacity—Highway Type Vehicles (A)—SAE **J1911** SEP90 .... 3:36.153

Truck and Bus Grade Parking Performance Test Procedure—SAE **J360** OCT88 .... 2:25.101

Wheel Slip Brake Control System Road Test Code—SAE **J46** JUN80 .... 2:25.158

Clutches
Automotive Pull Type Clutch Terminology (A)—SAE **J1479** APR91 .... 3:36.02

† Clutch Dimensions for Truck and Bus Applications (A)—SAE **J1806** NOV92 .... 3:36.119

Compressors
Test Method for Measuring Power Consumption of Air Conditioning and Brake Compressors for Trucks and Buses (A)—SAE **J1340** APR90 .... 3:36.53

---

* Technical Revision   † New Document   (D)—DODISS Adopted   (A)—ANSI Recognized

TRUCKS AND BUSES (continued)

Cornering
*Steady-State Circular Test Procedure for Trucks and Buses—SAE **J2181** JUN93 . . . . **3:36.68**

Crashworthiness
Collision Deformation Classification—SAE **J224** MAR80 **3:34.235**
Fixed Rigid Barrier Collision Tests—SAE **J850** NOV88 . **3:34.281**
Moving Rigid Barrier Collision Tests—SAE **J972** DEC88 **3:34.278**

Defrosters
Windshield Defrosting Systems Performance Guidelines—Trucks, Buses, and Multi-Purpose Vehicles (A)—SAE **J382** OCT84 . . . . **3:34.23**
Windshield Defrosting Systems Test Procedure—Trucks, Buses, and Multipurpose Vehicles (A) (D)—SAE **J381** JUN84 . . . . **3:34.21**

Dimensions
Average Vehicle Dimensions for Use in Designing Docking Facilities for Motor Vehicles (A)—SAE **J699** NOV85 . . . . **3:36.67**
Motor Truck CA Dimensions (A)—SAE **J691** SEP90 . . . **3:36.67**
†Motor Vehicle Dimensions (A)—SAE **J1100** JUN93 . . . . **3:34.118**
Motor Vehicle Fiducial Marks and Three Dimensional Reference System (A)—SAE **J182** NOV90 . . . . **3:34.229**
Truck Overall Widths Across Dual Tires—SAE **J693** APR89 . . . . **3:36.111**
Truck Tractor Semitrailer Interchange Coupling Dimensions (A)—SAE **J701** AUG84 . . . . **3:36.83**

Electronic Equipment
†Accelerator Pedal Position Sensor for Use with Electronic Controls in Medium- and Heavy-Duty Vehicle Applications—SAE **J1843** APR93 . . . . **2:23.515**
†Joint SAE/TMC Electronic Data Interchange Between Microcomputer Systems in Heavy-Duty Vehicle Applications—SAE **J1587** AUG92 . . . . **2:23.531**
Joint SAE/TMC Recommended Environmental Practices for Electronic Equipment Design (Heavy-Duty Trucks)—SAE **J1455** JAN88 . . . . **2:23.556**
*OEM/Vendor's Interface Specification for Vehicle Electronic Programming Station—SAE **J1924** DEC92 . . **2:23.522**
Serial Data Communications Between Microcomputer Systems in Heavy Duty Vehicle Applications (A)—SAE **J1708** OCT90 . . . . **2:23.553**

Engine Cooling Systems
On-Highway Truck Cooling Test Code (A)—SAE **J1393** JUN84 . . . . **2:26.388**

Exhaust Emissions
Instrumentation and Techniques for Exhaust Gas Emissions Measurement (A)—SAE **J254** AUG84 . . . . . . **1:13.104**

Fans
Method for Determining Power Consumption of Engine Cooling Fan-Drive Systems (A)—SAE **J1342** AUG89 . . . **3:36.55**

Fiducial Marks
†Motor Vehicle Dimensions (A)—SAE **J1100** JUN93 . . . . **3:34.118**
Motor Vehicle Fiducial Marks and Three Dimensional Reference System (A)—SAE **J182** NOV90 . . . . **3:34.229**

Flywheels
Flywheel Dimensions for Truck and Bus Applications (A)—SAE **J1857** MAY87 . . . . **2:26.237**

Fuel Consumption
Fuel Economy Measurement Test (Engineering Type) for Trucks and Buses (A)—SAE **J1376** JUL82 . . . . **3:36.31**
Joint RCCC/SAE Fuel Consumption Test Procedure (Short Term In-Service Vehicle) Type I (A)—SAE **J1264** OCT86 . . . . **3:36.06**
Joint TMC/SAE Fuel Consumption In-Service Test Procedure Type III (A)—SAE **J1526** JUN87 . . . . **2:26.450**
Joint TMC/SAE Fuel Consumption Test Procedure—Type II (A)—SAE **J1321** OCT86 . . . . **3:36.11**
Supportive Information Report for the Fuel Economy Measurement Test (Engineering Type) for Trucks and Buses (A)—SAE **J1420** AUG88 . . . . **3:36.44**

Fuel Tanks
Filler Pipes and Openings of Motor Vehicle Fuel Tanks (A)—SAE **J1140** FEB88 . . . . **2:26.433**
Fuel Tank Filler Conditions—Passenger Car, Multi-Purpose Passenger Vehicles, and Light Duty Trucks—SAE **J398** FEB88 . . . . **2:26.436**

Gradeability
Truck Ability Prediction Procedure (A)—SAE **J688** AUG87 . . . . **3:36.03**
Truck Ability Prediction Procedure—SAE **J688**—SAE **HS-82** . . . . Available as a separate publication (See Related Technical Reports Section)

Handling
*Steady-State Circular Test Procedure for Trucks and Buses—SAE **J2181** JUN93 . . . . **3:36.68**

Lamps
High Mounted Stop Lamps for Use on Vehicles 2032 mm or More in Overall Width—SAE **J1432** OCT88 . . . . **2:24.147**
†SAE Ground Vehicle Lighting Standards Manual—SAE **HS-34** JAN93 . . . . Available as a separate publication (See Related Technical Reports Section)

Lubricants
Classification of Energy-Conserving Engine Oil for Passenger Cars, Vans, and Light-Duty Trucks—SAE **J1423** FEB92 . . . . **1:12.29**

Noise
Exterior Sound Level for Heavy Trucks and Buses (A) (D)—SAE **J366** FEB87 . . . . **1:14.27**
Maximum Sound Level for Passenger Cars and Light Trucks (A)—SAE **J1030** FEB87 . . . . **1:14.17**
†Measurement of Exterior Sound Levels for Heavy Trucks Under Stationary Conditions—SAE **J1096** FEB93 . . . . . **1:14.29**
Measurement of Light Vehicle Exhaust Sound Level Under Stationary Conditions—SAE **J1169** MAR92 . . . . . **1:14.21**
Measurement of Light Vehicle Stationary Exhaust System Sound Level Engine Speed Sweep Method—SAE **J1492** MAR92 . . . . **1:14.24**
Measurement of Noise Emitted by Accelerating Highway Vehicles—SAE **J1470** MAR92 . . . . **1:14.32**
Sound Level for Passenger Cars and Light Trucks (D)—SAE **J986** OCT88 . . . . **1:14.14**
Sound Level for Truck Cab Interior—SAE **J336** OCT88 . **1:14.29**

Operator Controls
Automotive Adaptive Driver Controls, Manual (A)—SAE **J1903** AUG90 . . . . **3:34.198**
Design Criteria—Driver Hand Controls Location for Passenger Cars, Multi-Purpose Passenger Vehicles, and Trucks (10 000 GVW and Under)—SAE **J1138** **3:34.191**
Location and Operation of Instruments and Controls in Motor Truck Cabs (A)—SAE **J680** SEP88 . . . . **3:36.121**
Supplemental Information—Driver Hand Controls Location for Passenger Cars, Multi-Purpose Passenger Vehicles, and Trucks (10 000 GVW and Under)—SAE **J1139** **3:34.193**

Packaging
Accommodation Tool Reference Point—SAE **J1516** MAR90 . . . . **3:34.247**
Driver Selected Seat Position (A)—SAE **J1517** MAR90 . **3:34.250**
Truck Driver Shin-Knee Position for Clutch and Accelerator (A)—SAE **J1521** MAR90 . . . . **3:34.263**
Truck Driver Stomach Position (A)—SAE **J1522** MAR90 **3:34.265**

Performance
Certificates of Maximum Net Horsepower for Motor Trucks and Truck Tractors—SAE **J690** **3:36.65**

---

*Technical Revision   †New Document   (D)—DODISS Adopted   (A)—ANSI Recognized

Subject Index

TRUCKS AND BUSES (continued)

Performance (continued)

Truck Ability Prediction Procedure—SAE J688—SAE HS-82 .... Available as a separate publication (See Related Technical Reports Section)

Vehicle Acceleration Measurement—SAE J1491 JUN90   2:26.503

Power Take-Off

† Clearance Envelopes for Six-Bolt, Eight-Bolt, and Rear Truck Transmission Mounted Power Takeoffs (A)—SAE J772 DEC92 .... 3:29.33

† Openings for Six- and Eight-Bolt Truck Transmission Mounted Power Take-Offs—SAE J704 DEC92 .... 3:29.41

Pumps

Test Method for Measuring Power Consumption of Hydraulic Pumps for Trucks and Buses (A)—SAE J1341 APR90 .... 3:36.54

Restraint Systems

† Vehicle Occupant Restraint Systems and Components Standards Manual—SAE HS-13 MAY93 .... Available as a separate publication (See Related Technical Reports Section)

Retarders

Heavy Truck and Bus Retarder Downhill Performance Mapping Procedure (A)—SAE J1489 JUN87 .... 3:36.124

Ride

Measurement and Presentation of Truck Ride Vibrations (A)—SAE J1490 JAN87 .... 3:34.252

Rims

Nomenclature—Wheels, Hubs, and Rims for Commercial Vehicles—SAE J393 OCT91 .... 3:31.32

Wheels/Rims—Trucks—Test Procedures and Performance Requirements (A)—SAE J267 JAN91 .... 3:31.14

Rollover Threshold

* A Tilt-table Procedure for Measuring the Static Rollover Threshold for Heavy Trucks—SAE J2180 APR93 ..... 2:23.528

Safety

Rear Underride Guard Test Procedure (A)—SAE J260 JUN90 .... 3:34.287

Safety Glazing Materials—Motor Vehicles (A)—SAE J674 NOV90 .... 3:34.103

Splash and Spray Protection

Rear Wheel Splash and Stone Throw Protection—SAE J682 OCT84 .... 3:36.120

Terminology

Nomenclature and Terminology for Truck and Bus Drive Axles (A)—SAE J923 APR91 .... 3:29.72

Nomenclature—Truck, Bus, Trailer (A)—SAE J687 JUN88 .... 3:36.01

Nomenclature—Wheels, Hubs, and Rims for Commercial Vehicles—SAE J393 OCT91 .... 3:31.32

Tires

† Disc Wheel Radial Runout Low Point Marking—SAE J2133 JUN93 .... 3:31.38

Labeling—Disc Wheel and Demountable Rims—Trucks (A)—SAE J179 JUL84 .... 3:31.30

Measurement of Passenger Car, Light Truck, and Highway Truck and Bus Tire Rolling Resistance—SAE J1270 MAR87 .... 3:30.21

Passenger and Light Truck Tire Traction Device Profile Determination and Classification—SAE J1232 .... 3:36.115

Passenger Car and Light Truck Tire Dynamic Driving Traction in Snow (A)—SAE J1466 OCT85 .... 3:30.26

Pneumatic Tires for Military Tactical Wheeled Vehicles (A)—SAE J2014 APR91 .... 3:30.04

Rolling Resistance Measurement Procedure for Passenger Car, Light Truck, and Highway Truck and Bus Tires (A)—SAE J1269 MAR87 .... 3:30.18

Sound Level of Highway Truck Tires (A)—SAE J57 FEB87 .... 1:14.31

Test Procedures for Measuring Truck Tire Revolutions per Mile—SAE J1025 .... 3:30.36

Testing Machines for Measuring the Uniformity of Passenger Car and Light Truck Tires (A)—SAE J332 AUG81 .... 3:30.32

Tire Chain Clearance—Trucks, Buses (Except Suburban, Intercity, and Transit Buses), and Combinations of Vehicles—SAE J683 AUG85 .... 3:36.115

Truck Overall Widths Across Dual Tires—SAE J693 APR89 .... 3:36.111

Transmissions

Automatic Transmission Functions—Terminology (A)—SAE J649 JUL88 .... 3:29.27

† Automatic Transmissions—Manual Control Sequence—SAE J915 APR93 .... 3:29.32

† Automotive Transmission Terminology—SAE J645 OCT92 .... 3:29.23

Hydrodynamic Drive Test Code—SAE J643 JUN89 .... 3:29.05

Passenger Car and Truck Automatic Transmission Test Code (A)—SAE J651 JAN91 .... 3:29.28

Pull-Type Clutch—Transmission Installation Dimensions (A)—SAE J1463 NOV84 .... 3:36.120

Transmissions—Schematic Diagrams (A)—SAE J647 APR90 .... 3:29.24

Turning Behavior

Turning Ability and Off Tracking—Motor Vehicles (A)—SAE J695 DEC89 .... 3:36.78

Underride Guards

Rear Underride Guard Test Procedure (A)—SAE J260 JUN90 .... 3:34.287

Vehicle Identification Number

Truck and Truck Tractor Vehicle Identification Number Systems—SAE J1108 DEC81 .... 3:27.07

Truck Identification Terminology (A)—SAE J1229 DEC81   3:27.04

Truck Vehicle Identification Numbers—SAE J187 ..... 3:27.05

Vehicle Identification Numbers—SAE J853 DEC81 .... 3:27.03

Wheels

Dimensions—Wheels for Demountable Rims, Demountable Rims and Rim Spacers—Commercial Vehicles (A)—SAE J851 JUN87 .... 3:36.98

† Disc Wheel Radial Runout Low Point Marking—SAE J2133 JUN93 .... 3:31.38

Disc Wheel/Hub or Drum Interface Dimensions—Commercial Vehicles (A) (D)—SAE J694 SEP88 .... 3:36.100

Labeling—Disc Wheel and Demountable Rims—Trucks (A)—SAE J179 JUL84 .... 3:31.30

Nomenclature—Wheels, Hubs, and Rims for Commercial Vehicles (A)—SAE J393 OCT91 .... 3:31.32

Spoke Wheels and Hub Fatigue Test Procedures (A)—SAE J1095 JAN91 .... 3:31.20

Wheels/Rims—Trucks—Test Procedures and Performance Requirements (A)—SAE J267 JAN91 .... 3:31.14

Windshield Washers

Truck & Bus Multipurpose Windshield Washer System (A)—SAE J1944 MAY91 .... 3:34.10

Windshield Wipers

† Windshield Wiper Systems—Trucks, Buses, and Multipurpose Vehicles—SAE J198 JUN93 .... 3:34.12

Wiring Systems

Automobile, Truck, Truck-Tractor, Trailer, and Motor Coach Wiring—SAE J1292 OCT81 .... 2:23.142

Brake and Electrical Connection Locations—Truck-Tractor and Truck-Trailer (A)—SAE J702 MAY85 . 3:36.114

Five Conductor Electrical Connectors for Automotive Type Trailers (A)—SAE J895 APR86 .... 2:23.171

TRUNKS

Luggage Capacity

† Motor Vehicle Dimensions (A)—SAE J1100 JUN93 .... 3:34.118

TUBING

Flares for Tubing—SAE J533 JUN92 .... 2:22.203

---

* Technical Revision   † New Document   (D)—DODISS Adopted   (A)—ANSI Recognized

Subject Index

TUBING (continued)
Fluid Conductors and Connectors—SAE **HS-150** JUN92 ..... Available as a separate publication (See Related Technical Reports Section)

Formed Tube Ends for Hose Connections (A)—SAE J962 MAY86 ..... 2:22.202

*Seamless Copper-Nickel 90-10 Tubing—SAE **J1650** MAR93 ..... 2:22.211

Air Brakes
Metallic Air Brake System Tubing and Pipe (A)—SAE J1149 JUN91 ..... 2:22.219

Metric Nonmetallic Air Brake System Tubing (A)—SAE J1394 APR91 ..... 2:22.227

Nonmetallic Air Brake System Tubing (A)—SAE **J844** JUN90 ..... 2:22.221

Performance Requirements for SAE J844 Nonmetallic Tubing and Fitting Assemblies Used in Automotive Air Brake Systems (A)—SAE J1131 DEC87 ..... 2:22.231

Brakes
Automotive Hydraulic Brake System—Metric Tube Connections—SAE J1290 MAY89 ..... 2:25.165

Tubing—Motor Vehicle Brake System Hydraulic (A)—SAE J1047 JUN90 ..... 2:25.162

Connections
Connections for Fluid Power and General Use—Ports and Stud Ends with ISO 261 Threads and O-Ring Sealing Part 2: Heavy-Duty (S Series) Stud Ends—Dimensions, Design, Test Methods, and Requirements—SAE J2244/2 DEC91 ..... 2:22.293

Connections for Fluid Power and General Use—Ports and Stud Ends with ISO 261 Threads and O-Ring Sealing Part 1: Port with O-Ring Seal in Truncated Housing—SAE J2244/1 DEC91 ..... 2:22.290

†Hydraulic Flanged Tube, Pipe, and Hose Connections, Four-Bolt Split Flange Type—SAE J518 JUN93 ..... 2:22.190

*Metallic Tube Connections for Fluid Power and General Use-Test Methods for Threaded Hydraulic Fluid Power Connectors—SAE J1644 MAY93 ..... 2:22.285

SAE Quick Connector Specifications for Liquid Fuel Systems—SAE J2044 JUN92 ..... 2:26.442

Copper
Seamless Copper Tube (A)—SAE J528 JUN91 ..... 2:22.218

Copper Alloys
Wrought Copper and Copper Alloys—SAE J463 SEP81 ..... 1:10.81

Ferrous Materials
Detection of Surface Imperfections in Ferrous Rods, Bars, Tubes, and Wires—SAE J349 FEB91 ..... 1:3.63

Fittings
†Automotive Tube Fittings—SAE J512 JUN93 ..... 2:22.06

Coding Systems for Identification of Fluid Conductors and Connectors (D)—SAE J846 JUN89 ..... 2:22.01

†Fitting—O-Ring Face Seal—SAE J1453 JUN93 ..... 2:22.256

†Hydraulic Tube Fittings—SAE J514 JUN93 ..... 2:22.59

Pressure Ratings for Hydraulic Tubing and Fittings (D)—SAE J1065 MAR92 ..... 2:22.215

†Refrigeration Tube Fittings—General Specifications—SAE J513 JUN93 ..... 2:22.33

SAE Quick Connector Specifications for Liquid Fuel Systems—SAE J2044 JUN92 ..... 2:26.442

†Spherical and Flanged Sleeve (Compression) Tube Fittings—SAE J246 JUN93 ..... 2:22.21

Fuel Injection
Diesel Engines—Steel Tubes for High Pressure Fuel Injection Pipes (Tubing)—SAE J1958 APR89 ..... 2:26.260

Fuel Injection Pumps — High Pressure Pipes (Tubing) for Testing (A)—SAE J1418 DEC87 ..... 2:26.263

Fuel Systems
*Tube/Hose Assemblies—SAE J2045 OCT92 ..... 1:12.85

Nylon
Metric Nonmetallic Air Brake System Tubing (A)—SAE J1394 APR91 ..... 2:22.227

Steel
Brazed Double Wall Low Carbon Steel Tubing (A)—SAE J527 JUN91 ..... 2:22.209

Pressure Ratings for Hydraulic Tubing and Fittings (D)—SAE J1065 MAR92 ..... 2:22.215

Seamless Low Carbon Steel Tubing Annealed for Bending and Flaring (A)—SAE J524 JUN91 ..... 2:22.206

Welded and Cold Drawn Low Carbon Steel Tubing Annealed for Bending and Flaring (A)—SAE J525 JUN91 ..... 2:22.207

Welded Flash Controlled Low Carbon Steel Tubing Normalized for Bending, Double Flaring, and Beading (A)—SAE J356 JUN91 ..... 2:22.213

Welded Low Carbon Steel Tubing (A)—SAE J526 JUN91 ..... 2:22.208

Windshield Washers
Truck & Bus Multipurpose Windshield Washer System (A)—SAE J1944 MAY91 ..... 3:34.10

Windshield Washer Tubing (A)—SAE J1037 DEC87 ..... 1:11.254

TURBOCHARGERS
Spark Arrester Test Procedure for Large Size Engines (A)—SAE J342 JAN91 ..... 2:26.360

Turbocharger Gas Stand Test Code—SAE J1826 APR89 ..... 2:26.355

Terminology
Turbocharger Nomenclature and Terminology—SAE J922 JUL88 ..... 2:26.04

TURN INDICATORS
Combination Turn Signal Hazard Warning Signal Flashers (A)—SAE J2068 JAN90 ..... 2:24.165

Flasher Test (A)—SAE J823 JUN91 ..... 2:24.210

Front and Rear Turn Signal Lamps for Use on Motor Vehicles 2032 mm or More in Overall Width (A)—SAE J1395 JUN91 ..... 2:24.156

Motorcycle Turn Signal Lamps (A)—SAE J131 MAR83 ..... 2:24.136

†SAE Ground Vehicle Lighting Standards Manual—SAE **HS-34** JAN93 ..... Available as a separate publication (See Related Technical Reports Section)

Side Turn Signal Lamps (A)—SAE J914 NOV87 ..... 2:24.160

Supplemental High Mounted Stop and Rear Turn Signal Lamps for Use on Vehicles Less Than 2032 mm in Overall Width (A)—SAE J186 DEC89 ..... 2:24.155

Tail Lamps (Rear Position Lamps) for Use on Motor Vehicles Less Than 2032 mm in Overall Width—SAE J585 DEC91 ..... 2:24.137

†Turn Signal Flashers (A)—SAE J590 APR93 ..... 2:24.164

Turn Signal Lamps for Use on Motor Vehicles Less Than 2032 mm in Overall Width (A)—SAE J588 JUN91 ..... 2:24.153

Turn Signal Switch (D)—SAE J589b 2:24.159

TURNING BEHAVIOR
Trucks and Buses
Turning Ability and Off Tracking—Motor Vehicles (A)—SAE J695 DEC89 ..... 3:36.78

TWO STROKE CYCLE ENGINES
See also: SPARK IGNITION ENGINES

Engine Oil Performance and Engine Service Classification (Other Than "Energy-Conserving") (A)—SAE J183 JUN91 ..... 1:12.01

Engine Terminology and Nomenclature—General (A)—SAE J604 JAN86 ..... 2:26.01

Lubricants for Two-Stroke-Cycle Gasoline Engines—SAE J1510 OCT88 ..... 1:12.32

Two-Stroke-Cycle Engine Oil Miscibility/Fluidity Classification—SAE J1536 OCT88 ..... 1:12.33

†Two-Stroke-Cycle Gasoline Engine Lubricants Performance and Service Classification (A)—SAE J2116 JUN93 ..... 1:12.33

---

* Technical Revision † New Document (D)—DODISS Adopted (A)—ANSI Recognized

Subject Index

TWO STROKE CYCLE ENGINES (continued)
Performance
Engine Power Test Code—Spark Ignition and Compression Ignition— Gross Power Rating—SAE **J1995** JUN90 .... 2:26.11

Engine Power Test Code—Spark Ignition and Compression Ignition—Net Power Rating—SAE **J1349** JUN90 .... 2:26.16

ULTRASONIC TESTS
See also: **NONDESTRUCTIVE TESTS**

Detection of Surface Imperfections in Ferrous Rods, Bars, Tubes, and Wires—SAE **J349** FEB91 .... 1:3.63

Leakage Testing—SAE **J1267** DEC88 .... 1:3.73

Nondestructive Tests (A)—SAE **J358** FEB91 .... 1:3.67

Ultrasonic Inspection (A)—SAE **J428** MAR91 .... 1:3.79

UNDERCOATING
Corrosion Preventive Compound, Topside Vehicle Corrosion Protection—SAE **J1804** JUN89 .... 1:3.52

Corrosion Preventive Compound, Underbody Vehicle Corrosion Protection—SAE **J1959** JUN89 .... 1:3.55

Vibration Damping Materials and Underbody Coatings (A)—SAE **J671** APR82 .... 1:11.111

UNDERRIDE GUARDS
Rear Underride Guard Test Procedure (A)—SAE **J260** JUN90 .... 3:34.287

UNIVERSAL JOINTS
Universal Joints and Driveshafts—Nomenclature—Terminology—Application (A)—SAE **J901** OCT90 ..... 3:29.42

UPHOLSTERY
Fading
*Accelerated Exposure of Automotive Interior Trim Components Using a Controlled Irradiance Air-Cooled Xenon-Arc Apparatus—SAE **J2212** MAR93 .... 1:11.263

*Accelerated Exposure of Automotive Interior Trim Materials Using an Outdoor Under Glass Variable Angled Controlled Temperature Apparatus—SAE **J2229** FEB93 . 1:11.119

*Accelerated Exposure of Automotive Interior Trim Material Using Outdoor Under-Glass Controlled Sun-Tracking Temperature and Humidity Apparatus—SAE **J2230** FEB93 .... 1:11.123

URETHANE
Foams
Dynamic Flex Fatigue Test for Slab Polyurethane Foam—SAE **J388** 3:34.98

Load Deflection Testing of Urethane Foams for Automotive Seating—SAE **J815** MAY81 .... 3:34.100

Urethane for Automotive Seating—SAE **J954** 3:34.99

Sealers
Stationary Glass Replacement—SAE **J1556** DEC85 ... 3:42.21

V-BELTS
See: **BELTS**

VALVES
Qualifications for Four-Way Subbase Mounted Air Valves for Automotive Manufacturing Applications—SAE **J2051** JUN92 .... 3:43.21

Air Brakes
Air Brake Valves Test Procedure (A)—SAE **J1409** JUN88 .... 3:36.171

Air Brake Valve—Performance Requirements—SAE **J1410** OCT86 .... 3:36.173

Labeling Air Brake Valves with Their Performance (Input-Output) Characteristics (A)—SAE **J1860** AUG90 .... 3:36.164

Test Procedures for Determining Air Brake Valve Input-Output Characteristics—SAE **J1859** APR89 .... 3:36.162

Engine
Engine Poppet Valve Information Report—SAE **J775** JAN88 .... 1:1.165

Engine Terminology and Nomenclature—General (A)—SAE **J604** JAN86 .... 2:26.01

Hydraulic
Control Valve Test Procedure (A)—SAE **J747** MAY90 .. 3:40.130

Hydraulic Directional Control Valves, 3000 psi Maximum (A)—SAE **J748** MAR86 .... 3:40.120

†Hydraulic Valves for Motor Vehicle Brake Systems Test Procedure—SAE **J1118** JUN93 .... 2:25.35

Hydraulic Valves for Motor Vehicle Brake Systems—Performance Requirements—SAE **J1137** MAR85 ..... 2:25.39

Measuring and Reporting the Internal Leakage of a Hydraulic Fluid Power Valve (A)—SAE **J1235** MAR86 .. 3:40.123

Method of Measuring and Reporting the Pressure Differential-Flow Characteristics of a Hydraulic Fluid Power Valve (A)—SAE **J1117** MAR86 .... 3:40.120

Springs
Hard Drawn Carbon Steel Valve Spring Quality Wire and Springs—SAE **J172** DEC88 .... 1:5.08

Oil Tempered Carbon Steel Valve Spring Quality Wire and Springs—SAE **J351** DEC88 .... 1:5.02

Oil Tempered Chromium-Vanadium Valve Spring Quality Wire and Springs—SAE **J132** DEC88 .... 1:5.03

Tire
Methods for Testing Snap-In Tubeless Tire Valves—SAE **J1206** 3:30.38

Performance Requirements for Snap-In Tubeless Tire Valves—SAE **J1205** 3:30.40

VANS
See also: **Specific components or systems for Automobiles e.g. BRAKES, TIRES,**

Crashworthiness
Collision Deformation Classification—SAE **J224** MAR80 .. 3:34.235

Design Guidelines for Optimizing Automobile Collision Damage Resistance, Repairability, and Serviceability—SAE **J1555** NOV85 .... 3:42.19

Moving Rigid Barrier Collision Tests—SAE **J972** DEC88 .. 3:34.278

Dimensions
†Motor Vehicle Dimensions (A)—SAE **J1100** JUN93 ..... 3:34.118

Lubricants
Classification of Energy-Conserving Engine Oil for Passenger Cars, Vans, and Light-Duty Trucks—SAE **J1423** FEB92 .... 1:12.29

Towing
*Recovery Attachment Points for Passenger Cars, Vans, and Light Trucks—SAE **J2069** JAN93 .... 3:42.21

VEHICLE DYNAMICS
*Dolly Rollover Recommended Test Procedure—SAE **J2114** APR93 .... 3:34.283

*Steady-State Circular Test Procedure for Trucks and Buses—SAE **J2181** JUN93 .... 3:36.68

Vehicle Aerodynamics Terminology (A)—SAE **J1594** JUN87 .... 3:28.17

Vehicle Dynamics Terminology—SAE **J670e** JUN76 .... 3:34.297

VEHICLE ENCLOSURE
Measurement of Fuel Evaporative Emissions from Gasoline Powered Passenger Cars and Light Trucks Using the Enclosure Technique (A)—SAE **J171** APR91 . 1:13.95

VEHICLE IDENTIFICATION NUMBER (VIN)
See also: **ENGINE IDENTIFICATION NUMBER (EIN), TRANSAXLE IDENTIFICATION NUMBER (TIN), TRANSMISSION IDENTIFICATION NUMBER (TIN)**

Identification of Automotive Air Bags—SAE **J1856** MAY89 .... 3:33.29

Recommended Practice for Bar-Coded Vehicle Identification Number Label (A)—SAE **J1877** MAY88 ... 3:42.37

Vehicle Identification Number Systems—SAE **J272** DEC81 .... 3:27.01

Vehicle Identification Numbers—SAE **J853** DEC81 ... 3:27.03

World Manufacturer Identifier—SAE **J1044** DEC81 .... 3:27.05

Automobile
Passenger Car Identification Terminology—SAE **J218** DEC81 .... 3:27.02

---

* Technical Revision   † New Document   (D)—DODISS Adopted   (A)—ANSI Recognized

Subject Index

VEHICLE IDENTIFICATION NUMBER (VIN) (continued)
Automobile (continued)
Passenger Car Vehicle Identification Number System—
SAE **J273** DEC81 .... **3:27.02**
Snowmobile
Vehicle Identification Numbers—Snowmobiles (A)—SAE
J64 APR88 .... **3:38.03**
Trucks and Buses
Truck and Truck Tractor Vehicle Identification Number
Systems—SAE **J1108** DEC81 .... **3:27.07**
Truck Identification Terminology (A)—SAE **J1229** DEC81 **3:27.04**
Truck Vehicle Identification Numbers—SAE **J187** **3:27.05**

VEHICLE RANGE
*Electric Vehicle Energy Consumption and Range Test
Procedure—SAE **J1634** MAY93 .... **3:28.01**

VENTILATION
See also: AIR CONDITIONING
Performance Test for Air-Conditioned, Heated, and
Ventilated Off-Road Self-Propelled Work Machines (A)—
SAE **J1503** JUL86 .... **3:40.511**
*Personal Water craft Ventilation Systems—SAE **J2034**
JAN93 .... **3:41.49**
Test Procedure for Battery Flame Retardant Venting
Systems—SAE **J1495** MAR92 .... **2:23.72**

VIBRATION
L.E.D. Lighting Devices (A)—SAE **J1889** JUN88 ...... **2:24.19**
*Laboratory Measurement of the Composite Vibration
Damping Properties of Materials on a Supporting Steel
Bar—SAE **J1637** FEB93 .... **1:2.30**
Automobile
Recommended Environmental Practices for Electronic
Equipment Design—SAE **J1211** **2:23.228**
Damping
Vibration Damping Materials and Underbody
Coatings (A)—SAE **J671** APR82 .... **1:11.111**
Human Body
† Measurement of Whole Body Vibration of the Seated
Operator of Off-Highway Work Machines—SAE **J1013**
AUG92 .... **3:39.92**
Off-Road Vehicles
† Measurement of Whole Body Vibration of the Seated
Operator of Off-Highway Work Machines—SAE **J1013**
AUG92 .... **3:39.92**
Seats
Classification of Agricultural Wheeled Tractors for
Vibration Tests of Operator Seats (A)—SAE **J1386**
JAN86 .... **3:40.401**
Classification of Earthmoving Machines for Vibration
Tests of Operator Seats (A)—SAE **J1385** JUN83 ...... **3:40.398**
† Vibration Performance Evaluation of Operator
Seats (A)—SAE **J1384** JUN93 .... **3:40.396**
Terminology
Vehicle Dynamics Terminology—SAE **J670e** JUN76 ... **3:34.297**
Trucks and Buses
Joint SAE/TMC Recommended Environmental Practices
for Electronic Equipment Design (Heavy-Duty Trucks)—
SAE **J1455** JAN88 .... **2:23.556**
Measurement and Presentation of Truck Ride
Vibrations (A)—SAE **J1490** JAN87 .... **3:34.252**

VIN
See: VEHICLE IDENTIFICATION NUMBER (VIN)

VINYL
See also: PLASTICS, TRIM MATERIALS
Test Method for Determining Cold Cracking of Flexible
Plastic Materials—SAE **J323** **1:11.115**
Test Method for Determining Resistance to Abrasion of
Automotive Bodycloth, Vinyl, and Leather, and the
Snagging of Automotive Bodycloth (A)—SAE **J948**
FEB86 .... **1:11.178**

Test Method for Determining Visual Color Match to
Master Specimen for Fabrics, Vinyls, Coated
Fiberboards, and Other Automotive Trim Materials (A)—
SAE **J361** MAR85 .... **1:11.180**

VISCOSITY
Method of Viscosity Test for Automotive Type Adhesives,
Sealers, and Deadeners—SAE **J1524** NOV88 .... **1:11.78**
Methods of Tests for Automotive-Type Sealers,
Adhesives, and Deadeners—SAE **J243** **1:11.68**
Lubricants
Axle and Manual Transmission Lubricant Viscosity
Classification—SAE **J306** OCT91 .... **1:12.37**
Axle and Manual Transmission Lubricants—SAE **J308**
JUN89 .... **1:12.36**
† Engine Oil Viscosity Classification—SAE **J300** MAR93 .. **1:12.16**
† Fuels and Lubricants Standards Manual—SAE **HS-23**
FEB93 .... Available as a separate publication *(See Related Technical Reports Section)*
Lubricants for Two-Stroke-Cycle Gasoline Engines—SAE
J1510 OCT88 .... **1:12.32**
Lubricating Oil, Aircraft Piston Engine (Nondispersant
Mineral Oil) (A)—SAE **J1966** JUN91 .... **1:12.110**
Physical and Chemical Properties of Engine Oils (A)—
SAE **J357** JUN91 .... **1:12.24**
Plastics
Automotive Plastic Parts Specification (A)—SAE **J1345**
FEB82 .... **1:11.148**

VISION
Describing and Measuring the Driver's Field of View—
SAE **J1050a** **3:34.214**
*Headlamp Design Guidelines for Mature Drivers—SAE
J1606 MAR93 .... **3:34.225**
*Manual Controls for Mature Drivers—SAE **J2119** JUN93 **3:34.194**
Motor Vehicle Drivers' Eye Locations—SAE **J941** JUN92 **3:34.204**
Passenger Car Rear Vision—SAE **J834a** **3:34.114**
Photometric Guidelines for Instrument Panel Displays
That Accomodate Older Drivers—SAE **J2217** OCT91 .. **3:34.226**
Rear View Mirrors—Motorcycles—SAE **J268** MAY89 ... **3:37.06**
Vision Factors Considerations in Rear View Mirror
Design—SAE **J985** OCT88 .... **3:34.110**
Glossary
Vision Glossary—SAE **J264** JUN88 .... **3:34.227**

VOLATILITY
† Automotive Gasolines—SAE **J312** JAN93 .... **1:12.47**

VOLUMETRIC CAPACITY
Capacity Rating—Dumper Body and Trailer Body (A)—
SAE **J1363** JAN85 .... **3:40.321**
† Capacity Rating—Elevating Scrapers—SAE **J957** ISO
6484- JUN93 .... **3:40.322**
Capacity Rating—Loader Bucket—SAE **J742** FEB85 ... **3:40.324**
† Capacity Rating—Scraper, Open Bowl (A)—SAE **J741**
JUN93 .... **3:40.319**
† Excavator, Mini-Excavator, and Backhoe Hoe Bucket
Volumetrtic Rating—SAE **J296** JUN93 .... **3:40.329**
Shovel Dipper, Clam Bucket, and Dragline Bucket
Rating—SAE **J67** NOV84 .... **3:40.323**
Test Procedure for Air Reservoir Capacity—Highway
Type Vehicles (A)—SAE **J1911** SEP90 .... **3:36.153**

WARNING SYSTEMS
See also: STOP LAMPS, TURN INDICATORS
360 Degree Warning Devices for Authorized Emergency,
Maintenance, and Service Vehicles—SAE **J845** MAR92 . **2:24.185**
Alarm—Backup—Electric—Performance, Test, and
Application (A)—SAE **J994** MAR85 .... **1:14.45**
Combination Turn Signal Hazard Warning Signal
Flashers (A)—SAE **J2068** JAN90 .... **2:24.165**
Emergency Vehicle Sirens (A)—SAE **J1849** JUL89 .... **2:24.198**

---

* Technical Revision   † New Document   (D)—DODISS Adopted   (A)—ANSI Recognized

Subject Index

WARNING SYSTEMS (continued)
Emergency Warning Device (Triangular Shape) (A)—
SAE **J774** DEC89 .... 2:24.181
Flasher Test (A)—SAE **J823** JUN91 .... 2:24.210
† Flashing Warning Lamp for Agricultural Equipment—SAE
J974 JUN93 .... 3:39.59
Flashing Warning Lamp for Industrial Equipment (A)—
SAE **J96** MAR86 .... 3:40.261
Flashing Warning Lamps for Authorized Emergency,
Maintenance and Service Vehicles (A)—SAE **J595**
JAN90 .... 2:24.183
Gaseous Discharge Warning Lamp for Authorized
Emergency, Maintenance, and Service Vehicles (A)—
SAE **J1318** APR86 .... 2:24.188
Graphic Symbols for Boats (A)—SAE **J1449** FEB87 ... 3:41.31
Graphical Symbols for Operator Controls and Displays on
Off-Road Self-Propelled Work Machines—SAE **J1362**
JUN92 .... 3:40.423
Hazard Warning Signal Switch (A) (D)—SAE **J910**
OCT88 .... 2:24.184
Minimum Requirements for Wheel Slip Brake Control
System Malfunction Signals—SAE **J1230** OCT79 ..... 2:25.162
Mounting Brackets and Socket for Warning Lamp and
Slow-Moving Vehicle (SMV) Identification Emblem (A)—
SAE **J725** MAR91 .... 3:39.65
On-Machine Alarm Test and Evaluation Procedure for
Construction and General Purpose Industrial Machinery—
SAE **J1446** MAY89 .... 1:14.50
Safety Signs (A)—SAE **J115** JAN87 .... 3:40.494
School Bus Stop Arm (A)—SAE **J1133** JUL89 ...... 2:24.196
School Bus Warning Lamps (A)—SAE **J887** AUG87 ... 2:24.192
Supplemental High Mounted Stop and Rear Turn Signal
Lamps for Use on Vehicles Less Than 2032 mm in
Overall Width (A)—SAE **J186** DEC89 .... 2:24.155
Two-Block Warning and Limit Systems in Lifting Crane
Service—SAE **J1305** JUN87 .... 3:40.216
† Vehicular Hazard Warning Signal Flashers (A)—SAE
J945 JUN93 .... 2:24.164
Warning Lamp Alternating Flashers (A)—SAE **J1054**
OCT89 .... 2:24.166

WASHERS
Conical Spring Washers—SAE **J773b** 2:16.19
Manual on Design and Manufacture of Coned Disk
Springs (Belleville Springs) and Spring Washers—SAE
HS-1582 .... Available as a separate publication (See Related Technical Reports Section)
Mechanical and Material Requirements for Externally
Threaded Fasteners (A)—SAE **J429** AUG83 ..... 1:4.01
Nut and Conical Spring Washer Assemblies—SAE **J238** ... 2:16.20
Thrust Washers—Design and Application—SAE **J924**
JAN81 .... 2:26.206
Torque-Tension Test Procedure for Steel Threaded
Fasteners—SAE **J174** 2:16.22

WEAR
See also: ABRASION, FRICTION
Abrasive Wear—SAE **J965** AUG66 .... 1:9.22
*Kingpin Wear Limits Commercial Trailers and
Semi-Trailers—SAE **J2228** JUN93 .... 3:36.89
Brakes
Brake Test Code—Inertia Dynamometer—SAE **J667** ... 2:25.103
Fabrics
Accelerated Exposure of Automotive Interior Trim
Components Using a Controlled Irradiance Water Cooled
Xenon-Arc Apparatus—SAE **J1885** MAR92 ... 1:11.255
Dynamic Durability Testing of Seat Cushions for Off-Road
Work Machines (A)—SAE **J1454** JAN86 ..... 3:40.389
Method of Testing Resistance to Crocking of Organic
Trim Materials—SAE **J861** 1:11.114

Method of Testing Resistance to Scuffing of Trim
Materials (A)—SAE **J365** FEB85 .... 1:11.163
Test Method for Determining Dimensional Stability of
Automotive Textile Materials (A)—SAE **J883** FEB86 ... 1:11.114
Test Method for Determining Resistance to Abrasion of
Automotive Bodycloth, Vinyl, and Leather, and the
Snagging of Automotive Bodycloth (A)—SAE **J948**
FEB86 .... 1:11.178
Test Method for Determining Resistance to Abrasion;
Bearding; and Fiber Loss of Automotive Carpet
Materials (A)—SAE **J1530** JUN85 .... 1:11.107
Test Method for Measuring Weight of Organic Trim
Materials (A)—SAE **J860** JAN85 .... 1:11.114
Test Method for Wicking of Automotive Fabrics and
Fibrous Materials (A)—SAE **J913** FEB85 .... 1:11.163
Fiberboards
Fiberboard Crease Bending Test (A)—SAE **J119** FEB87 ... 1:11.166
Fiberboard Test Procedure (A)—SAE **J315** JAN85 .... 1:11.158

WELDING
Automotive Metallurgical Joining—SAE **J836a** JAN70 ... 1:9.08
Automotive Resistance Spot Welding Electrodes (AWS
D8.6-77)—SAE **HS-J1156** .... Available as a separate publication (See Related Technical Reports Section)
Methods of Tests for Automotive-Type Sealers,
Adhesives, and Deadeners—SAE **J243** 1:11.68
Specification for Automotive Frame Weld Quality—Arc
Welding (AWS D8.8-79)—SAE **HS-J1196** .... Available as a separate publication (See Related Technical Reports Section)
Specification for Automotive Weld Quality—Resistance
Spot Welding (AWS D8.7-78)—SAE **HS-J1188** ... Available as a separate publication (See Related Technical Reports Section)
Unibody Weld Quality Testing (A)—SAE **J1827** MAY87 ... 3:42.23
Welding, Brazing, and Soldering—Materials and
Practices—SAE **J1147** JUN83 .... 1:9.07
Aluminum Alloys
Wrought Aluminum Applications Guidelines—SAE **J1434**
JAN89 .... 1:10.15
Copper Alloys
Wrought and Cast Copper Alloys—SAE **J461** SEP81 ... 1:10.45

WHEEL CHOCKS
Wheel Chocks—SAE **J348** JUN90 .... 3:36.122

WHEEL RIMS
See: RIMS

WHEEL SLIP BRAKE CONTROL SYSTEMS
Minimum Requirements for Wheel Slip Brake Control
System Malfunction Signals—SAE **J1230** OCT79 ..... 2:25.162
Wheel Slip Brake Control System Road Test Code—SAE
J46 JUN80 .... 2:25.158

WHEELS
See also: RIMS
Agricultural and Industrial Equipment
† Industrial and Agricultural Disc Wheels (A)—SAE **J712**
APR93 .... 3:39.54
† Wheel Mounting Elements for Industrial and Agricultural
Disc Wheels (A)—SAE **J714** APR93 .... 3:39.57
Automobiles
Wheels—Impact Test Procedures—Road Vehicles (A)—
SAE **J175** JUN88 .... 3:31.07
Wheels—Passenger Cars—Performance Requirements
and Test Procedures—SAE **J328** MAR90 .... 3:31.06
Balance Weights
*Balance Weight and Rim Flange Design Specifications,
Test Procedures, and Performance Recommendations—
SAE **J1986** FEB93 .... 3:31.10

Subject Index

WHEELS (continued)

Bolts

Mechanical and Material Requirements for Wheel Bolts—SAE J1102 .... 1:4.18

Disc

† Disc Wheel Radial Runout Low Point Marking—SAE J2133 JUN93 .... 3:31.38

Disc Wheel/Hub or Drum Interface Dimensions—Commercial Vehicles (A) (D)—SAE J694 SEP88 ..... 3:36.100

† Industrial and Agricultural Disc Wheels (A)—SAE J712 APR93 .... 3:39.54

Labeling—Disc Wheel and Demountable Rims—Trucks (A)—SAE J179 JUL84 .... 3:31.30

† Wheel Mounting Elements for Industrial and Agricultural Disc Wheels (A)—SAE J714 APR93 .... 3:39.57

Wheels—Passenger Cars—Performance Requirements and Test Procedures (A)—SAE J328 MAR90 .... 3:31.06

Nuts

*Road Vehicles--Wheels for Commercial Vehicles and Multipurpose Passenger Vehicles--Fixing Nuts--Test Methods—SAE J1965 FEB93 .... 3:36.107

Off-Road Vehicles

Off-Road Tire and Rim Classification—Construction Machines (A)—SAE J751 APR86 .... 3:40.301

Off-Road Tire and Rim Classification—Forestry Machines (A)—SAE J1440 APR86 .... 3:40.310

Recreational Vehicles

Wheels—Recreational and Utility Trailer Test Procedure (A)—SAE J1204 DEC89 .... 3:31.08

Splash and Spray

Rear Wheel Splash and Stone Throw Protection—SAE J682 OCT84 .... 3:36.120

Terminology

Nomenclature—Wheels for Passenger Cars, Light Trucks, and Multipurpose Vehicles—SAE J1982 DEC91 ... 3:31.01

Nomenclature—Wheels, Hubs, and Rims for Commercial Vehicles—SAE J393 OCT91 .... 3:31.32

Wide Base Tire Rims and Wheels (A)—SAE J876 JAN91 ... 3:36.109

Trucks and Buses

Dimensions—Wheels for Demountable Rims, Demountable Rims and Rim Spacers—Commercial Vehicles (A)—SAE J851 JUN87 .... 3:36.98

† Disc Wheel Radial Runout Low Point Marking—SAE J2133 JUN93 .... 3:31.38

Disc Wheel/Hub or Drum Interface Dimensions—Commercial Vehicles (A) (D)—SAE J694 SEP88 ..... 3:36.100

Fastener Hardware for Wheels for Demountable Rims (A)—SAE J1835 MAR90 .... 3:31.26

Labeling—Disc Wheel and Demountable Rims—Trucks (A)—SAE J179 JUL84 .... 3:31.30

Nomenclature—Wheels, Hubs, and Rims for Commercial Vehicles—SAE J393 OCT91 .... 3:31.32

Rear Wheel Splash and Stone Throw Protection—SAE J682 OCT84 .... 3:36.120

Spoke Wheels and Hub Fatigue Test Procedures (A)—SAE J1095 JAN91 .... 3:31.20

Wheels/Rims—Trucks—Test Procedures and Performance Requirements (A)—SAE J267 JAN91 .... 3:31.14

WINCHES

Classification and Nomenclature Towing Winch for Skidders and Crawler Tractors—SAE J1014 JAN85 .... 3:40.51

Hand Winches—Boat Trailer Type—SAE J1853 APR87 ... 3:35.05

† Rating of Winches (A)—SAE J706 NOV90 .... 3:36.117

Specification Definitions—Winches for Crawler Tractors and Skidders—SAE J1158 JAN85 .... 3:40.79

WIND TUNNEL TESTING

See also: AERODYNAMICS

Aerodynamic Flow Visualization Techniques and Procedures—SAE HS-1566 .... Available as a separate publication (See Related Technical Reports Section)

Aerodynamic Testing of Road Vehicles—SAE J2071 MAR90 .... 3:28.18

*Aerodynamic Testing of Road Vehicles--Testing Methods and Procedures—SAE J2084 JAN93 .... 3:28.61

SAE Wind Tunnel Test Procedure for Trucks and Buses (A)—SAE J1252 JUL81 .... 3:36.61

WINDOWS

See also: GLASS, WINDSHIELDS

† Automotive Safety Glasses—SAE J673 APR93 .... 3:34.100

Stationary Glass Replacement—SAE J1556 DEC85 ... 3:42.21

Test Method for Determining Window Fogging Resistance of Interior Trim Materials (A)—SAE J275 FEB85 .... 1:11.117

WINDSHIELD WASHERS

Passenger Car Windshield Washer Systems (D)—SAE J942b 3:34.16

Switches

Electric Windshield Washer Switch—SAE J234 .... 2:23.188

Trucks and Buses

Truck & Bus Multipurpose Windshield Washer System (A)—SAE J1944 MAY91 .... 3:34.10

Tubing

Truck & Bus Multipurpose Windshield Washer System (A)—SAE J1944 MAY91 .... 3:34.10

Windshield Washer Tubing (A)—SAE J1037 DEC87 ... 1:11.254

WINDSHIELD WIPERS

Automobile

Passenger Car Windshield Wiper Systems—SAE J903c ... 3:34.07

Switches

Electric Windshield Wiper Switch—SAE J112a 2:23.188

Trucks and Buses

† Windshield Wiper Systems—Trucks, Buses, and Multipurpose Vehicles—SAE J198 JUN93 .... 3:34.12

WINDSHIELDS

See also: GLASS, WINDOWS

† Automotive Safety Glasses—SAE J673 APR93 .... 3:34.100

Light Transmittance of Automotive Windshields Safety Glazing Materials—SAE J1203 .... 3:34.102

Passenger Car Glazing Shade Bands (A)—SAE J100 MAR88 .... 3:34.104

Passenger Car Glazing—Electrical Circuits (A)—SAE J216 DEC89 .... 3:34.104

† Recommendations for Children's Snowmobile—SAE J1038 DEC92 .... 3:38.18

Stationary Glass Replacement—SAE J1556 DEC85 ... 3:42.21

Defrosters

† Passenger Car Windshield Defrosting Systems—SAE J902 APR93 .... 3:34.18

Windshield Defrosting Systems Performance Guidelines—Trucks, Buses, and Multi-Purpose Vehicles (A)—SAE J382 OCT84 .... 3:34.23

Windshield Defrosting Systems Test Procedure—Trucks, Buses, and Multipurpose Vehicles (A) (D)—SAE J381 JUN84 .... 3:34.21

WIRE

See also: WIRING SYSTEMS

Copper Alloys

Wrought Copper and Copper Alloys—SAE J463 SEP81 . 1:10.81

Ferrous Materials

Detection of Surface Imperfections in Ferrous Rods, Bars, Tubes, and Wires—SAE J349 FEB91 .... 1:3.63

---

* Technical Revision   † New Document   (D)—DODISS Adopted   (A)—ANSI Recognized

Subject Index

WIRE (continued)
Rope
Lifting Crane, Wire-Rope Strength Factors (A)—SAE J959 MAY91 .... 3:40.217

Steel
† Cut Wire Shot—SAE J441 JUN93 .... 1:8.05
Hard Drawn Carbon Steel Valve Spring Quality Wire and Springs—SAE J172 DEC88 .... 1:5.08
Hard Drawn Mechanical Spring Wire and Springs—SAE J113 DEC88 .... 1:5.04
Music Steel Spring Wire and Springs—SAE J178 DEC88 .... 1:5.10
Oil Tempered Carbon Steel Spring Wire and Springs—SAE J316 DEC88 .... 1:5.01
Oil Tempered Carbon Steel Valve Spring Quality Wire and Springs—SAE J351 DEC88 .... 1:5.02
Oil Tempered Chromium-Vanadium Valve Spring Quality Wire and Springs—SAE J132 DEC88 ... 1:5.03
Oil Tempered Chromium—Silicon Alloy Steel Wire and Springs—SAE J157 DEC88 .... 1:5.07
Special Quality High Tensile, Hard Drawn Mechanical Spring Wire and Springs—SAE J271 DEC88 .... 1:5.05
Stainless Steel 17-7 PH Spring Wire and Springs—SAE J217 DEC88 .... 1:5.08
Stainless Steel, SAE 30302, Spring Wire and Springs—SAE J230 DEC88 .... 1:5.09

WIRING SYSTEMS
Automobile, Truck, Truck-Tractor, Trailer, and Motor Coach Wiring—SAE J1292 OCT81 .... 2:23.142
Class A Application/Definition (A)—SAE J2057/1 JUN91 .... 2:23.407
*Class A Multiplexing Architecture Strategies—SAE J2057/4 JUN93 .... 2:23.415
*Class C Application Requirement Considerations—SAE J2056/1 JUN93 .... 2:23.366
Electrical Terminals—Blade Type—SAE J858a 2:23.179
Electrical Wiring Systems for Construction, Agricultural and Off-Road Machines—SAE J821 MAY85 .... 3:40.264
Five Conductor Electrical Connectors for Automotive Type Trailers (A)—SAE J895 APR86 .... 2:23.171
Four- and Eight-Conductor Rectangular Electrical Connectors for Automotive Type Trailers—SAE J1239 .. 2:23.173
Fusible Links (A)—SAE J156 APR86 .... 2:23.171
Low Tension Wiring and Cable Terminals and Splice Clips—SAE J163 .... 2:23.157
Marine Engine Wiring (A)—SAE J378 JAN88 .... 3:41.22
Seven Conductor Jacketed Cable for Truck Trailer Connections—SAE J1067 .... 2:23.178

WOODRUFF KEYS
See: KEYS

WORKSTATIONS
Recommended Practice for the Selection of Engineering Workstations—SAE J2026 NOV91 .... 3:43.01

WORLD MANUFACTURING SYSTEMS (WMI)
See also: VEHICLE IDENTIFICATION NUMBER (VIN)
Passenger Car Vehicle Identification Number System—SAE J273 DEC81 .... 3:27.02
World Manufacturer Identifier—SAE J1044 DEC81 ... 3:27.05

X-RAY DIFFRACTION
Residual Stress Measurements by X-Ray Diffraction—SAE HS-J784a .... Available as a separate publication (See Related Technical Reports Section)

ZINC ALLOYS
See also: CHEMICAL COMPOSITION, MECHANICAL PROPERTIES, PHYSICAL PROPERTIES
Zinc Alloy Ingot and Die Casting Compositions—SAE J468 DEC88 .... 1:10.139
Zinc Die Casting Alloys—SAE J469 JAN89 .... 1:10.139

Coated
Electroplating of Nickel and Chromium on Metal Parts—Automotive Ornamentation and Hardware—SAE J207 FEB85 .... 1:10.150

ZIRCONIUM OXIDE
Size Classification and Characteristics of Ceramic Shot for Peening (A)—SAE J1830 MAY87 .... 1:8.32
Size Classification and Characteristics of Ceramic Shot for Peening (A)—SAE J1830 MAY8 (A)—SAE J1830 MAY87 .... 1:8.32

---

* Technical Revision   † New Document   (D)—DODISS Adopted   (A)—ANSI Recognized

THREADS, FASTENERS, AND COMMON PARTS

15 Threads

SCREW THREADS—SAE J475a
SAE Standard

SAE Regular (NF Series) approved June 1911; SAE Fine (NEF Series) approved January 1915; SAE Special Pitch Series approved June 1926; Coarse (NC Series) approved January 1935; 8, 12, and 16 (National Series) approved January 1935. Last revised by Screw Threads Committee June 1964. Conforms in general to American Standard Unified Screw Threads, (including American National), ASA B1.1.

This report is available as ANSI B1.1 from American National Standards Institute, 1430 Broadway, New York, New York 10018.

DRYSEAL PIPE THREADS—SAE J476a
SAE Standard

Report of Miscellaneous Division approved March 1921 and last revised by Screw Threads Committee June 1961.
Values in Table 1 conform to those in Table 9, Limits on Crest and Root of proposed American Standard, Dryseal Pipe Threads, ASA B2.2

OUTLINE OF STANDARD

General Information
   Table 1—Limits on Crest and Root Truncation
   Table 2—Basic Dimensions (NPTF)
      Dryseal American Standard Taper Pipe Thread
   Table 3—Basic Dimensions (PTF—SAE Short External)
      Dryseal SAE Short External Taper Pipe Thread
   Table 4—Basic Dimensions (PTF—SAE Short Internal)
      Dryseal SAE Short Internal Taper Pipe Thread
   Table 5—Pipe Thread Limits (NPSF)
      Dryseal American Standard Fuel Internal Straight Pipe Thread
   Table 6—Pipe Thread Limit (NPSI)
      Dryseal American Intermediate Internal Straight Pipe Thread
Appendix A—Supplementary Thread Information
   Terminology
   Formulas
   Table A1—Blank Dimensions for External Pipe Thread (NPTF) and (PTF—SAE Short) Dryseal American Standard Taper Pipe Thread and Dryseal SAE Short External Taper Pipe Thread
   Table A2—Drilled Hole Specifications for Straight and Taper Pipe Threads
Appendix B—Chaser and Tap Information
   General Information
   Table B1—Tap and Chaser Teeth Crest and Root Flats
   Table B2—Taper Taps for Dryseal American Standard Fuel Internal Taper Pipe Thread (NPTF) Ground Thread Limits
   Table B3—Taper Taps for Dryseal SAE Short Internal Taper Pipe Thread, Ground Thread Limits
   Table B4—Straight Taps for Dryseal American Standard Fuel Internal Straight Pipe Thread (NPSF) Ground Thread Limits
   Table B5—Straight Taps for Dryseal American Intermediate Internal Straight Pipe Thread (NPSI) Ground Thread Limits
Appendix C—Dryseal Pipe Thread Gaging
   General Information
   Positional Gaging
   Step-Limit Gaging
   Turns Engagement Gaging
   Table C1—Basic Dimensions (L_1 and L_2) Basic-Notch Ring Gages
      Dryseal American Standard Taper Pipe Thread

   Table C2—Basic Dimensions (L_1) Basic-Notch Plug Gages
      Dryseal American Standard Taper Pipe Thread
   Table C3—Basic Dimensions (L_3) Basic-Notch Length Plug Gages
      Dryseal American Standard Taper Pipe Thread
   Table C4—Basic Dimensions (L_1) Step-Limit Thin-Ring Gages
      Dryseal American Standard Taper Pipe Thread
   Table C5—Basic Dimensions (L_2) Step-Limit Full-Ring Gages
      Dryseal American Standard Taper Pipe Thread
   Table C6—Basic Dimensions (L_1) Step-Limit Plug Gages
      Dryseal American Standard Taper Pipe Thread
   Table C7—Basic Dimensions (L_3) Length Step-Limit Plug Gages
      Dryseal American Standard Taper Pipe Thread
   Table C8—Basic Dimensions (L_1 Short) Step-Limit Thin-Ring Gages Dryseal SAE Short Taper Pipe Thread
   Table C9—Basic Dimensions (L_2 Short) Step-Limit Full-Ring Gages Dryseal SAE Short Taper Pipe Thread
   Table C10—Basic Dimensions (L_1 Short) Step-Limit Plug Gages
      Dryseal SAE Short Taper Pipe Thread and Dryseal American Standard Fuel Internal Straight Pipe Thread
   Table C11—Basic Dimensions (L_3 Short) Length Step Limit Plug Gages
      Dryseal SAE Short Taper Pipe Thread
   Table C12—Basic Dimensions (L_1) Step-Limit Plug Gages
      Dryseal American Intermediate Internal Straight Pipe Threads
Appendix D—Special and Fine Dryseal Pipe Threads
   General Information
   Dryseal Special Short Taper Pipe Thread
   Dryseal Special Extra Short Taper Pipe Thread
   Fine Thread Series
   Special Thread Series
   Formulae for Diameter and Length of Thread Designations
   Table D1—Basic Dimensions of Dryseal Taper Pipe Thread, Fine, F-PTF
   Table D2—Basic Dimensions of Dryseal Taper Pipe Thread, Special, SPL-PTF, for Thin Wall Nominal Size OD Tubing

GENERAL INFORMATION

Introduction—The Dryseal American Standard Taper Pipe Thread, the Dryseal American Fuel Internal Straight Pipe Thread and the Dryseal American Intermediate Internal Straight Pipe Thread covered by this standard conform with the American Standard ASA-B2.2. The Dryseal SAE-Short Taper Pipe Thread in this standard conforms with the Dryseal American Standard Taper Pipe Thread except for the length of thread, which is shortened for increased clearance and economy of material.

The significant feature of the Dryseal thread is controlled truncation at the crest and root to assure metal to metal contact coincident with or prior to flank contact. Contact at the crest and root prevents spiral leakage and insures pressure-tight joints without the use of a lubricant or sealer.

Lubricants, if not functionally objectionable, may be used to minimize the possibility of galling in assembly.

TRUNCATION—DRYSEAL AMERICAN STANDARD EXTERNAL AND INTERNAL PIPE THREADS
For Pressure-Tight Joints without Lubricant or Sealer

Thread Form—The angle between the flanks of the thread is 60 deg when measured on an axial plane and the line bisecting this angle is perpendicular to the axis of both the taper and straight threads.

Diametral taper of tapered threads is 0.75 in. ±0.06 in. per 12.00 in. of length.

Although the crests and roots of the Dryseal threads are theoretically flat, they may be rounded provided their contour is within the limits specified in Table 1.

Thread Series Symbols—The identification symbols which have been adopted for designating the various Dryseal Pipe Thread Series are as follows:
NPTF for Dryseal American Standard Taper Pipe Thread.
PTF—SAE for Dryseal SAE Short Taper Pipe Thread.
NPSF for Dryseal American Fuel Internal Straight Pipe Thread.
NPSI for Dryseal American Intermediate Internal Straight Pipe Thread.
Where: N stands for American Standard [formerly American (National) Standard].
P stands for Pipe
T stands for Taper
F stands for Fuel
S stands for Straight
I stands for Intermediate

Thread Designation—Dryseal pipe threads are designated by specifying in sequence the nominal size, number of threads per inch, form (Dryseal), and symbol of the thread series.
EXAMPLE: 1/8—27 DRYSEAL NPTF
1/8—27 DRYSEAL PTF—SAE SHORT
1/8—27 DRYSEAL NPSF
1/8—27 DRYSEAL NPSI

Straight Pipe Threads—An assembly with straight internal pipe threads and taper external pipe threads is frequently more advantageous than an all taper thread assembly, particularly in automotive and other allied industries where economy and rapid production are paramount considerations. Dryseal threads are not used on assemblies in which both components have straight pipe threads.

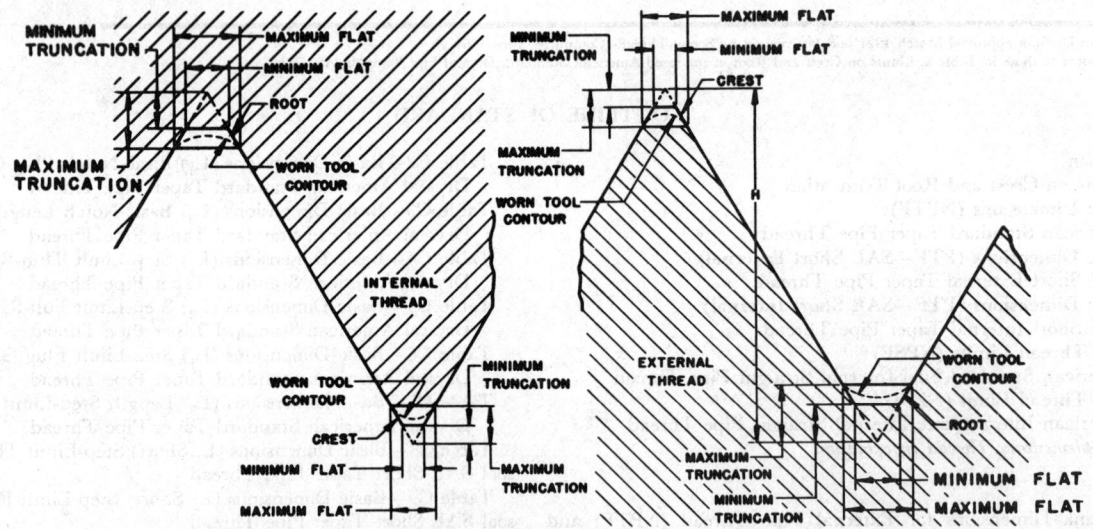

TABLE 1—LIMITS ON CREST AND ROOT TRUNCATION

Threads per in.		Depth of Sharp-V Thread, H	Truncation				Equivalent Width of Flat[a]			
			Min		Max		Min		Max	
		in.	Formula	in.	Formula	in.	Formula	in.	Formula	in.
27	Crest	0.03208	0.047p	0.0017	0.094p	0.0035	0.054p	0.0020	0.108p	0.0040
	Root		0.094p	0.0035	0.140p	0.0052	0.108p	0.0040	0.162p	0.0060
18	Crest	0.04811	0.047p	0.0026	0.078p	0.0043	0.054p	0.0030	0.090p	0.0050
	Root		0.078p	0.0043	0.109p	0.0061	0.090p	0.0050	0.126p	0.0070
14	Crest	0.06186	0.036p	0.0026	0.060p	0.0043	0.042p	0.0030	0.070p	0.0050
	Root		0.060p	0.0043	0.085p	0.0061	0.070p	0.0050	0.098p	0.0070
11-1/2	Crest	0.07531	0.040p	0.0035	0.060p	0.0052	0.046p	0.0040	0.069p	0.0060
	Root		0.060p	0.0052	0.090p	0.0078	0.069p	0.0060	0.103p	0.0090
8	Crest	0.10825	0.042p	0.0052	0.055p	0.0069	0.048p	0.0060	0.064p	0.0080[b]
	Root		0.055p	0.0069	0.076p	0.0095	0.064p	0.0080	0.088p	0.0110

[a] The major diameter of plug gages and minor diameter of ring gages used for gaging dryseal threads shall be truncated an amount sufficient to produce a flat width as shown in Appendix C, Tables C1-1 to C12-1 inclusive.

[b] There is reason to doubt the correctness of the 8 threads per in. flat widths on account of the volume of metal to be displaced.

DRYSEAL AMERICAN STANDARD TAPER PIPE THREAD (NPTF)

This series applies to both the external and internal threads of full length and is suitable for pipe joints in practically every type of service. These threads are generally conceded to be superior for strength and seal. Use of the tapered internal thread in hard or brittle materials having thin sections will minimize trouble from fracture. Dimensional data for (NPTF) threads is given in Table 2. See Appendix D for limitations of assembly of NPTF threads with other series Dryseal pipe threads.

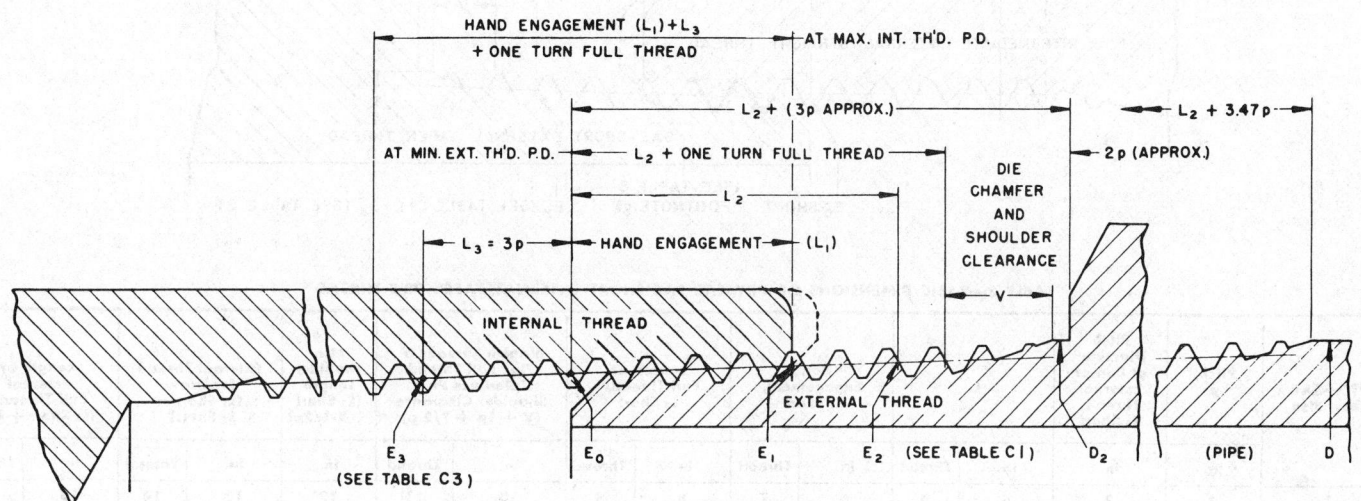

TABLE 2—BASIC DIMENSIONS OF DRYSEAL AMERICAN STANDARD TAPER PIPE THREAD[a]

NPTF Size	Pitch, p	Pitch Diameter at End of External Thread, E_0	Pitch Diameter at End of Internal Thread, E_1	Hand Engagement, L_1		Length of Full Thread,[b] L_2		Vanish Threads V Plus Full Thread Tolerance Plus Shoulder Clearance $(V + 1p + 1/2p)$		Shoulder Length $L_2 + (3p$ Approx)	External Thread for Draw $(L_2 - L_1)$		Length of Internal Full Thread,[c] $(L_1 + L_3)$		OD of Fitting, D_2	OD of Pipe, D
	in.	in.	in.	in.	Thread	in.	Thread	in.	Thread	in.	in.	Thread	in.	Thread	in.	in.
1	2	3	4	5	6	7	8	9	10	11	12	13	14	15	16	17
1/16 —27	0.03704	0.27118	0.28118	0.160	4.32	0.2611	7.05	0.1139	3.075	0.3750	0.1011	2.73	0.2711	7.32	0.315	0.3125
1/8 —27	0.03704	0.36351	0.37360	0.1615	4.36	0.2639	7.12	0.1112	3.072	0.3750	0.1024	2.76	0.2726	7.36	0.407	0.405
1-4 —18	0.05556	0.47739	0.49163	0.2278	4.10	0.4018	7.23	0.1607	2.892	0.5625	0.1740	3.13	0.3945	7.10	0.546	0.540
3/8 —18	0.05556	0.61201	0.62701	0.240	4.32	0.4078	7.34	0.1547	2.791	0.5625	0.1678	3.02	0.4067	7.32	0.681	0.675
1/2 —14	0.07143	0.75843	0.77843	0.320	4.48	0.5337	7.47	0.2163	3.028	0.7500	0.2137	2.99	0.5343	7.48	0.850	0.840
3/4 —14	0.07143	0.96768	0.98887	0.339	4.75	0.5457	7.64	0.2043	2.860	0.7500	0.2067	2.89	0.5533	7.75	1.060	1.050
1 —11-1/2	0.08696	1.21363	1.23863	0.400	4.60	0.6828	7.85	0.2547	2.929	0.9375	0.2828	3.25	0.6609	7.60	1.327	1.315
1-1/4 —11-1/2	0.08696	1.55713	1.58338	0.420	4.83	0.7068	8.13	0.2620	3.013	0.9688	0.2868	3.30	0.6809	7.83	1.672	1.660
1-1/2 —11-1/2	0.08696	1.79609	1.82234	0.420	4.83	0.7235	8.32	0.2765	3.180	1.0000	0.3035	3.49	0.6809	7.83	1.912	1.900
2 —11-1/2	0.08696	2.26902	2.29627	0.436	5.01	0.7565	8.70	0.2747	3.159	1.0312	0.3205	3.69	0.6969	8.01	2.387	2.375
2-1/2 —8	0.12500	2.71953	2.76216	0.682	5.46	1.1375	9.10	0.3781	3.025	1.5156	0.4555	3.64	1.0570	8.46	2.893	2.875
3 —8	0.12500	3.34062	3.38850	0.766	6.13	1.2000	9.60	0.3781	3.025	1.5781	0.4340	3.47	1.1410	9.13	3.518	3.500

[a] See general specifications preceding tables.
For gaging methods, gages, cut thread blanks, taps, drilled hole sizes, hole depths, and full thread lengths, see Appendixes A, B, and C.

[b] External thread tabulated full thread lengths include chamfers not exceeding one and one-half pitches (threads) length.

[c] Internal thread tabulated full thread lengths do not include countersink beyond the intersection of the pitch line and the chamfer cone (gaging reference point).

DRYSEAL SAE SHORT EXTERNAL TAPER PIPE THREAD (PTF—SAE SHORT EXTERNAL)

For assembly with Dryseal American Intermediate Internal Straight (Table 6) or Dryseal American Standard Taper (Table 2) Pipe Threads

External threads of this series conform in all respects with the NPTF threads except that the full thread length has been shortened by eliminating one thread from the small end. These threads are primarily intended for assembly with NPSI internal threads but may also be used with NPTF internal threads. They are not designed for and at extreme tolerance limits may not assemble with PTF—SAE Short or NPSF internal threads. Dimensional data for PTF—SAE Short External Threads is given in Table 3. See Appendix D for limitations of assembly of PTF—SAE Short external threads with other series Dryseal pipe threads.

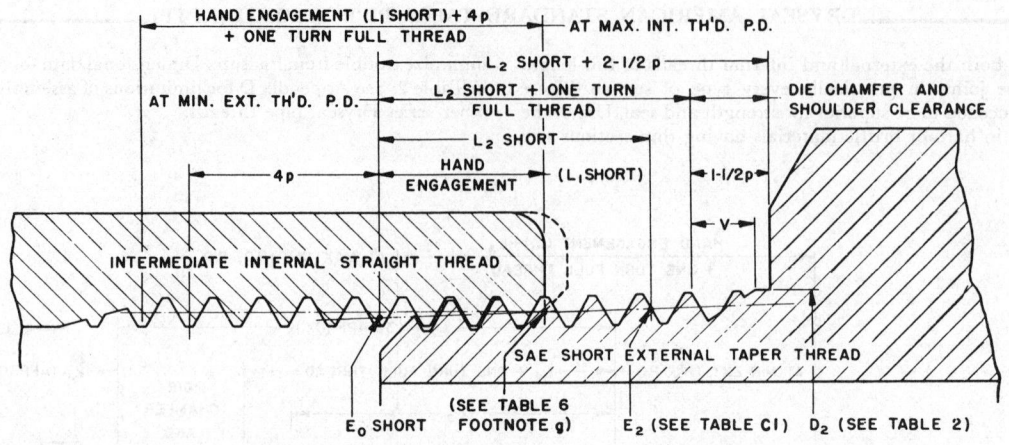

TABLE 3—BASIC DIMENSIONS OF DRYSEAL SAE SHORT EXTERNAL TAPER PIPE THREAD [a]

PTF—SAE Short Size	Pitch, p	Pitch Diameter at End of External Thread, E_0 Short	L_1		Hand Engagement, L_1 Short		Length of Full Thread,[b] L_2 Short		Vanish Threads V Plus Full Thread Tolerance Plus Shoulder Clearance $(V + 1p + 1/2\ p)$		Min Shoulder Length (L_2 Short + 2-1/2p)	External Thread for Draw (L_2 Short—L_1 Short)		Length of Internal Full Thread,[c] (L_1 Short + 4p)	
	in.	in.	in.	Thread	in.	Thread	in.	Thread	in.	Thread	in.	in.	Thread	in.	Thread
1	2	3	4	5	6	7	8	9	10	11	12	13	14	15	16
1/16—27	0.03704	0.27349	0.160	4.32	0.1230	3.32	0.2241	6.05	0.0926	2.50	0.3167	0.1011	2.73	0.2711	7.32
1/8 —27	0.03704	0.36582	0.1615	4.36	0.1244	3.36	0.2268	6.12	0.0926	2.50	0.3194	0.1024	2.76	0.2726	7.36
1/4 —18	0.05556	0.48086	0.2278	4.10	0.1722	3.10	0.3462	6.23	0.1389	2.50	0.4851	0.1740	3.13	0.3945	7.10
3/8 —18	0.05556	0.61548	0.240	4.32	0.1844	3.32	0.3522	6.34	0.1389	2.50	0.4911	0.1678	3.02	0.4067	7.32
1/2 —14	0.07143	0.76289	0.320	4.48	0.2486	3.48	0.4623	6.47	0.1786	2.50	0.6409	0.2137	2.99	0.5343	7.48
3/4 —14	0.07143	0.97214	0.339	4.75	0.2676	3.75	0.4743	6.64	0.1786	2.50	0.6528	0.2067	2.89	0.5533	7.75
1 —11-1/2	0.08696	1.21906	0.400	4.60	0.3130	3.60	0.5958	6.85	0.2174	2.50	0.8132	0.2828	3.25	0.6609	7.60
1-1/4 —11-1/2	0.08696	1.56256	0.420	4.83	0.3330	3.83	0.6198	7.13	0.2174	2.50	0.8372	0.2868	3.30	0.6809	7.83
1-1/2 —11-1/2	0.08696	1.80152	0.420	4.83	0.3330	3.83	0.6365	7.32	0.2174	2.50	0.8539	0.3035	3.49	0.6809	7.83
2 —11-1/2	0.08696	2.27445	0.436	5.01	0.3490	4.01	0.6695	7.70	0.2174	2.50	0.8869	0.3205	3.69	0.6969	8.01
2-1/2 —8	0.12500	2.72734	0.682	5.46	0.5570	4.46	1.0125	8.10	0.3125	2.50	1.3250	0.4555	3.64	1.0570	8.46
—8	0.12500	3.34844	0.766	6.13	0.6410	5.13	1.0750	8.60	0.3125	2.50	1.3875	0.4340	3.47	1.1410	9.13

[a] See general specifications preceding tables.
For gaging methods, gages, cut thread blanks, taps, drilled hole sizes, hole depths, and full thread lengths, see Appendixes A, B, and C.

[b] External thread tabulated full thread lengths include chamfers not exceeding one and one-half pitches (threads) length.

[c] Internal thread tabulated full thread lengths do not include countersink beyond the intersection of the pitch line and the chamfer cone (gaging reference point).

DRYSEAL SAE SHORT INTERNAL TAPER PIPE THREAD (PTF—SAE SHORT INTERNAL)
For assembly with American Standard External Taper Pipe Thread (Table 2)

Internal Threads of this series conform in all respects with the NPTF threads except that the full thread length has been shortened by eliminating on thread from the large end. These threads are primarily intended for assembly with NPTF external threads. They are not designed for and at extreme tolerance limits may not assemble with PTF—SAE Short external threads. Dimensional data for PTF—SAE Short Internal Threads is given in Table 4. See Appendix D for limitations of assembly of PTF—SAE Short internal threads with other series Dryseal pipe threads.

Trouble-free assemblies and pressure-tight joints without the use of lubricant or sealer can best be assured where both components are threaded with NPTF (full length) threads. This should be considered before specifying PTF—SAE Short External or Internal Thread.

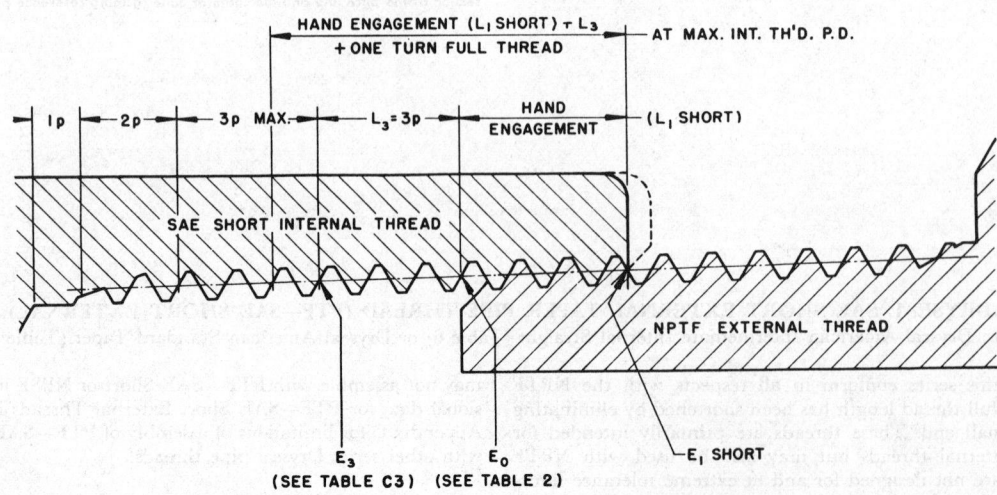

TABLE 4—BASIC DIMENSIONS OF DRYSEAL SAE SHORT INTERNAL TAPER PIPE THREAD [a]

PTF—SAE Short Size	Pitch p	Pitch Diameter at End of Internal Thread E₁ Short	L₁		Hand Engagement [b], L₁ Short		Length of Internal Full Thread [b] (L₁ Short + L₂)		Hole Depth for SAE Short Tap (Table B3)
	in.	in.	in.	Thread	in.	Thread	in.	Thread	in.
1	2	3	4	5	6	7	8	9	10
1/16—27	0.03704	0.27887	0.160	4.32	0.1230	3.32	0.2341	6.32	0.4564
1/8—27	0.03704	0.37129	0.1615	4.36	0.1244	3.36	0.2356	6.36	0.4578
1/4—18	0.05556	0.48815	0.2278	4.10	0.1722	3.10	0.3389	6.10	0.6722
3/8—18	0.05556	0.62354	0.240	4.32	0.1844	3.32	0.3511	6.32	0.6844
1/2—14	0.07143	0.77397	0.320	4.48	0.2486	3.48	0.4629	6.48	0.8915
3/4—14	0.07143	0.98441	0.339	4.75	0.2676	3.75	0.4819	6.75	0.9105
1—11-1/2	0.08696	1.23320	0.400	4.60	0.3130	3.60	0.5739	6.60	1.0956
1-1/4—11-1/2	0.08696	1.57795	0.420	4.83	0.3330	3.83	0.5939	6.83	1.1156
1-1/2—11-1/2	0.08696	1.81691	0.420	4.83	0.3330	3.83	0.5939	6.83	1.1156
2—11-1/2	0.08696	2.29084	0.436	5.01	0.3490	4.01	0.6099	7.01	1.1316
2-1/2—8	0.12500	2.75435	0.682	5.46	0.5570	4.46	0.9320	7.46	1.6820
3—8	0.12500	3.38069	0.766	6.13	0.6410	5.13	1.0160	8.13	1.7660

[a] See general specification preceding tables.
For gaging methods, gages, taps, drilled hole sizes, hole depths, and full thread lengths, see Appendixes A, B, and C.

[b] Internal thread tabulated full thread lengths do not include countersink beyond the intersection of the pitch line and the chamfer cone (gaging reference point).

DRYSEAL AMERICAN STANDARD FUEL INTERNAL STRAIGHT PIPE THREAD (NPSF)
For assembly with Dryseal American Standard External Taper Pipe Thread (Table 2)

Threads of this series are straight (cylindrical) instead of tapered. They are generally used in soft or ductile materials which will adjust at assembly to the taper of external threads but may also be used in hard or brittle materials where the section is heavy. These threads are primarily intended for assembly with full length NPTF external taper threads. Dimensional data for NPSF threads is given in Table 5. See Appendix D for limitations of assembly of NPSF internal threads with other series Dryseal pipe threads.

TABLE 5—DRYSEAL AMERICAN STANDARD FUEL INTERNAL STRAIGHT PIPE THREAD LIMITS [a]

NPSF Size	Pitch Diameter [b]		Minor Diameter [c]	Desired Min Length of Full Thread [g]	
	Max [d, e]	Min [e, f]	Min	in.	Thread
1	2	3	4	5	6
1/16—27	0.2803	0.2768	0.2482	0.31	8.44
1/8—27	0.3727	0.3692	0.3406	0.31	8.44
1/4—18	0.4904	0.4852	0.4422	0.47	8.44
3/8—18	0.6257	0.6205	0.5776	0.50	9.00
1/2—14	0.7767	0.7700	0.7133	0.66	9.19
3/4—14	0.9872	0.9805	0.9238	0.66	9.19
1—11-1/2	1.2365	1.2284	1.1600	0.78	8.98

Notes for Table 5

[a] See general specifications preceding tables.
For gaging methods, gages, taps, drilled hole sizes, hole depths, and full thread lengths, see Appendixes A, B, and C.
[b] The pitch diameter of the tapped hole as indicated by the taper plug gage is slightly larger than the values given due to the gage having to enter approximately 3/8 turn to engage first full thread.
[c] As the Dryseal American Standard pipe thread form is maintained, the major and minor diameters of the internal thread vary with the pitch diameter.
[d] Column 2 is the same as the E₁ pitch diameter of thread at large end of internal thread (Table 2) minus (small) 3/8 thread taper.
[e] Taps specified in Table B4 produce tapped holes to the above limits in cast iron, steel, and brass. In zinc and similar soft metals, they produce tapped holes approximately 0.001 smaller. Plug-gage turns engagement should be reduced accordingly.
[f] Column 3 is Column 2 reduced by 1-1/2 turns.
[g] Internal thread tabulated full thread lengths do not include countersink beyond the intersection of the pitch line and the chamfer cone (gaging reference point).

DRYSEAL AMERICAN INTERMEDIATE INTERNAL STRAIGHT PIPE THREAD (NPSI)
For assembly with Dryseal SAE Short External Taper (Table 3) or American Standard Taper Pipe Thread (Table 2)

Threads of this series are straight (cylindrical) instead of tapered. They are generally used in hard or brittle materials where the section is heavy and where there is little expansion at assembly with the external taper threads. These threads are primarily intended for assembly with PTF—SAE Short External Taper Threads, but will also assemble with full length NPTF External Taper Threads. Dimensional data for NPSI threads is given in Table 6. See Appendix D for limitations of assembly of NPSI internal threads with other series Dryseal pipe threads.

TABLE 6—DRYSEAL AMERICAN INTERMEDIATE INTERNAL STRAIGHT PIPE THREAD LIMITS [a]

NPSI Size	Pitch Diameter [b]		Minor Diameter [c]	Desired Min Length of Full Thread [g]	
	Max [d, e]	Min [e, f]	Min	in.	Thread
1	2	3	4	5	6
1/16—27	0.2826	0.2791	0.2505	0.31	8.44
1/8—27	0.3750	0.3715	0.3429	0.31	8.44
1/4—18	0.4938	0.4886	0.4457	0.47	8.44
3/8—18	0.6292	0.6240	0.5811	0.50	9.00
1/2—14	0.7812	0.7745	0.7180	0.66	9.19
3/4—14	0.9917	0.9850	0.9283	0.66	9.19
1—11-1/2	1.2420	1.2338	1.1655	0.78	8.98

Notes for Table 6

[a] See general specifications preceding tables.
For gaging methods, gages, taps, drilled hole sizes, hole depths, and full thread lengths see Appendixes A, B, and C.
[b] The pitch diameter of the tapped hole as indicated by the taper plug gage is slightly larger than the values given due to the gage having to enter approximately 3/8 turn to engage first full thread.
[c] As the Dryseal American Standard pipe thread form is maintained, the major and minor diameters of the internal thread vary with the pitch diameter.
[d] Column 2 is the E₁ pitch diameter of thread at large end of internal thread (Table 2) plus (large) 5/8 thread taper.
[e] Taps specified in Table B5 produce tapped holes to the above limits in cast iron, steel, and brass. In zinc and similar soft metals, they produce tapped holes approximately 0.001 smaller. Plug-gage turns engagement should be reduced accordingly.
[f] Column 3 is Column 2 reduced by 1-1/2 turns.
[g] Internal thread tabulated full thread lengths do not include countersink beyond the intersection of the pitch line and the chamfer cone (gaging reference point).

APPENDIX A—SUPPLEMENTARY THREAD INFORMATION

Terminology—For definitions of terms relating to size of parts, geometrical elements, or dimensions of threads see SAE Standards Screw Threads, Appendix A—Terminology.

DRYSEAL AMERICAN STANDARD AND SAE SHORT EXTERNAL TAPER PIPE THREAD BLANKS, CUT THREADS

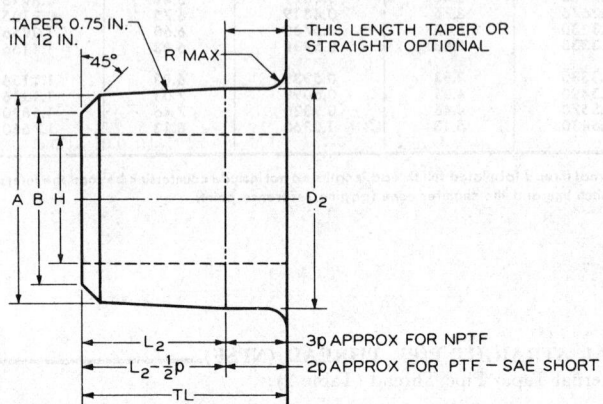

Formulas for Diameter and Length of Thread

Basic diameter and length of thread for different sizes given in Tables 2, 3, and 4, are based on the following formulas:

Basic pitch diameter of thread at small end of NPTF External Thread.

$$E_o = D - (0.05D + 1.1)p$$

Basic pitch diameter of thread at small end of PTF—SAE Short External Thread.

$$E_o \text{ Short} = D - (0.05D + 1.037)p$$

Basic pitch diameter of thread at large end of NPTF Internal Thread.

$$E_1 = E_o + (0.0625XL_1 \text{ Basic})$$

Basic pitch diameter of thread at large end of PTF—SAE Short Internal Thread.

$$E_1 \text{ Short} = E_o + (0.0625XL_1 \text{ Short})$$

Basic length of NPTF external full and effective length thread.

$$L_2 = (0.8D + 6.8)p$$

Basic length of PTF—SAE Short external full and effective length thread.

$$L_2 \text{Short} = (0.8D + 5.8)p$$

Basic length of NPTF internal full and effective length thread = L_1 Basic + L_3

Basic length of PTF—SAE Short internal full and effective length thread = L_1 Short + L_3

where D = outside diameter of pipe
P = pitch of thread in inches

NPSG (for oil and grease cup) is Dryseal American Standard Pipe Thread Form—use NPSF tap drill sizes.

The drilled hole sizes given above for Dryseal straight and taper internal pipe threads are the diameters produced by drills which are closest to the minimum minor diameters as shown in Table A2.

They represent the diameters of the holes which would be cut with a twist drill correctly ground when drilling a material without tearing or flow of metal. This is approximately the condition obtained when a correctly sharpened twist drill is cutting a hole in SAE 1112 or 1113 steel, or SAE 72 brass. When Dryseal taps are used, these holes produce an acceptable pipe thread with the required thread height.

When flat drills are used, the width of the cutting edge may have to be adjusted to produce a hole of the required diameter.

When hard metals and other similar materials are to be drilled and tapped, it may be found necessary to use a drill of slightly smaller diameter to produce a hole of a size that will make it possible for the tap to cut an acceptable pipe thread with the required thread height.

When soft metals and other similar materials are to be drilled and tapped, it may be found necessary to use a drill of slightly larger diameter to produce a hole of a size that will allow for a flow of the metal or material without loading the tap or tearing the material and make it possible for the tap to produce an acceptable pipe thread with the required thread height.

TABLE A1—DIMENSIONS OF DRYSEAL AMERICAN STANDARD EXTERNAL TAPER PIPE THREAD BLANKS (CUT THREADS)

Size	OD at Large End NPTF at L_2 Length D_2 PTF-SAE Short at $L_2 - 1/2$ p Length (Basic Thread One Turn Large with Max Truncation) +0.003 −0.000	OD at Small End, A		Chamfer Dia[b], B (Minor Dia[b] at Small End)	Min Length from Small End to Shoulder, TL		Corner Radius, R Max	Recommended Hole Size[a], H
		NPTF (Basic Thread Two Turns Large with Max Truncation) +0.003 −0.000	PTF-SAE Short (Basic Thread 2-1/2 Turns Large with Max Truncation)		NPTF L_2 +(3 p Approx)	PTF-SAE Short L_2 Short +(2-1/2 p Approx)		
1/16—27	0.315	0.301	0.302	0.23	0.38	0.3167	0.03	0.12
1/8 —27	0.407	0.393	0.394	0.32	0.38	0.3194	0.03	0.19
1/4 —18	0.546	0.523	0.525	0.42 +0.00	0.56	0.4851	0.06	0.28
3/8 —18	0.681	0.658	0.660	0.55	0.56	0.4911	0.06	0.41
1/2 —14	0.850	0.820	0.822	0.68 −0.02	0.75	0.6409	0.08	0.56
3/4 —14	1.060	1.029	1.031	0.89	0.75	0.6528	0.08	0.72
1 —11-1/2	1.327	1.289	1.292	1.12	0.94	0.8132	0.09	0.94
1-1/4 —11-1/2	1.672	1.633	1.636	1.46 +0.00	0.97	0.8372	0.09	1.25
1-1/2 —11-1/2	1.912	1.872	1.875	1.70	1.00	0.8539	0.09	1.47
2 —11-1/2	2.387	2.345	2.348	2.17 −0.03	1.03	0.8869	0.09	1.94
2-1/2 —8	2.893	2.829	2.833	2.59	1.52	1.3250	0.12	2.31
3 —8	3.518	3.450	3.454	3.21	1.58	1.3875	0.12	2.91

[a] The hole sizes recommended represent a desirable maximum, strength of wall being considered. However, as considerations other than wall strength frequently control the hole size in specific applications, the recommendations should not be construed as a requirement of this SAE Standard.

[b] External pipe threads shall be chamfered from a diameter (rounded to a two-place decimal) obtained by subtracting 0.016 in. for sizes below 1 in. and 0.025 in. for larger sizes from the minimum minor diameter at small end to produce a length of chamfered or partial thread equivalent to 1 to 1-1/2 times the pitch (rounded to a three-place decimal).

PIPE-THREAD DRILLED HOLE SIZES FOR DRYSEAL AMERICAN STANDARD INTERNAL PIPE THREAD

It should be understood that this table of drilled hole sizes is intended to help only the occasional user of drills in the application of this SAE Standard. When internal pipe threads are produced in larger quantities in a particular type of material and with specially designed machinery, it may be found to be more advantageous to use a drilled hole size not given in the table, even one requiring a nonstandard diameter drill size.

TABLE A2—PIPE-THREAD DRILLED HOLE SIZES

	Straight Pipe Thread						Taper Pipe Thread							Counter-sink
	Fuel (NPSF)		Intermediate (NPSI)		Desired Length of Full Thread	Hole Depth for Plug End Tap, Tables B4 and B5	NPTF (Not Reamed)				NPTF[d] (Taper Reamed)		Desired Length of Full Thread	Hole Depth for Standard Tap, Table B2	90 Deg × Dia[d]
							2 FF Thread[a]		4 FF Thread[a]						
Size	Minor Dia[c]	Drilled Hole Size	Minor Dia[c]	Drilled Hole Size			Minor Dia 2 Thread Small from Large End	Drilled Hole Size	Minor Dia 4 Thread Small from Large End	Drilled Hole Size	Desired Minor Dia at Smal. End	Drilled Hole Size			
	Min	+0.003 −0.001	Min	+0.003 −0.001	Min		Min	+0.003 −0.001	Min	+0.003 −0.001	Min	+0.003 −0.001	Min		
1/16—27	0.2482	0.2500	0.2505	0.2500	0.31	0.47	0.2480	0.2460	0.2434	0.2420	0.2356	0.2344	0.31	0.56	0.33
1/8 —27	0.3406	0.3437	0.3429	0.3437	0.31	0.47	0.3403	0.3390	0.3357	0.3320	0.3279	0.3281	0.31	0.56	0.42
1/4 —18	0.4422	0.4440	0.4457	0.4440	0.47	0.72	0.4417	0.4375	0.4348	0.4300	0.4241	0.4219	0.47	0.81	0.55
3/8 —18	0.5776	0.5781	0.5811	0.5781	0.50	0.72	0.5771	0.5781	0.5702	0.5700	0.5587	0.5625	0.50	0.81	0.69
1/2 —14	0.7133	0.7187	0.7180	0.7187	0.66	0.94	0.7127	0.7031	0.7038	0.6960	0.6873	0.6875	0.66	1.06	0.85
3/4 —14	0.9238	—	0.9283	—	0.66	0.94	0.9232	0.9219	0.9143	0.9062	0.8976	0.8906	0.66	1.06	1.06
1 —11-1/2	1.1600	—	1.1655	—	0.78	1.16	1.1593	1.1562	1.1484	1.1406	1.1290	1.1250	0.78	1.25	1.34
1-1/4 —11-1/2	—	—	—	—	—	—	1.5041	1.5000	1.4932	1.4844	1.4725	1.4687	0.81	1.31	1.58
1-1/2 —11-1/2	—	—	—	—	—	—	1.7430	1.7344	1.7321	1.7188	1.7115	1.7031	0.81	1.31	1.92
2 —11-1/2	—	—	—	—	—	—	2.2170	2.2187	2.2061	2.2031	2.1844	2.1875	0.81	1.31	2.39
2-1/2 —8	—	—	—	—	—	—	2.6488	2.6406	2.6336	2.6250	2.5983	2.5937	1.25	1.84	2.89
3 —8	—	—	—	—	—	—	3.2751	3.2656	3.2595	3.2500	3.2194	3.2187	1.34	1.91	3.52

Countersink tolerance column: +0.02 / −0.00 (sizes 1/16 through 1), +0.03 / −0.00 (sizes 1-1/4 and larger).

[a] NPTF (not reamed) drilled hole sizes are recommended for taper tapping without reaming. [NPTF (2 FF thread)] minimum minor diameter two threads small from large end and closest drilled hole sizes are recommended only for low pressure use. [NPTF (4 FF thread)] minimum minor diameter four threads small from large end and closest drilled hole sizes are recommended for all pressures. Thread lengths so produced are designated "Effective Thread" on drawings.

[b] NPTF (taper reamed) drilled hole sizes are recommended for taper reaming before tapping. They also are used without taper reaming by taper drilling or allowing the tap to act as a reamer. Thread lengths so produced are designated "Full or Complete Thread" on drawings.

[c] Minimum minor diameter for internal straight pipe threads is based upon minimum pitch diameter and minimum truncation and will vary with the pitch diameter.

[d] Internal pipe threads shall be countersunk 90 deg included angle to a diameter (rounded to a two- place decimal) obtained by adding 0.016 in. for sizes below 1 in. and 0.025 in. for larger sizes to the maximum major diameter at large end.

APPENDIX B—DRYSEAL PIPE THREAD TAPS AND CHASERS

General Information—While production taps will usually be purchased to specification, occasions may arise requiring adaptations of taps, dies, or chasers at hand.

American Standard Taper Taps, Dies, or Chasers (NPT) may be adapted for producing the Dryseal American Standard Taper Pipe Threads (NPTF) by truncating the outside diameter of taps and the inside diameter of dies or chasers the amount necessary to obtain flats shown in Table B1 for producing the limits on the product specified in Table 1. The pitch diameter of taps and dies or chasers so modified will remain standard. American Standard Coupling Straight Pipe Taps (NPSC) used for tapping the American Standard Coupling Straight Pipe Thread (NPSC), with the exception of one size only, may be adapted for tapping the Dryseal American Intermediate Straight Pipe Thread (NPSI) by truncating the outside diameter the amount necessary to obtain flats shown in Table B3 for producing the limits on the product shown in Table 1. The exception is the 1/4-18 size which has a minimum pitch diameter under that required. With the exception of one size only, taps designed to other standards cannot be adapted for tapping the Dryseal American Fuel Straight Pipe Thread (NPSF). The exception is the 1/8-27 American Standard Grease Fitting Tap (NPSG) which, if made in conformity with Tap Manufacturers' Standards of 1939 to 1941 issue, may be used without change for tapping of the 1/8-27 Dryseal American Fuel Straight Pipe Thread (NPSF).

Chamfer—2 to 3 threads.

Lead Tolerance—A maximum lead error of ±0.0005 in. in 1-in. of thread is permitted

Angle Tolerance—Error in half angle of ±30 min is permitted.

Taper Tolerance—A maximum taper error of ±1/32 in. per ft is permitted.

Marking—In addition to regular markings, Dryseal American Standard Taper Taps will be marked NPTF.

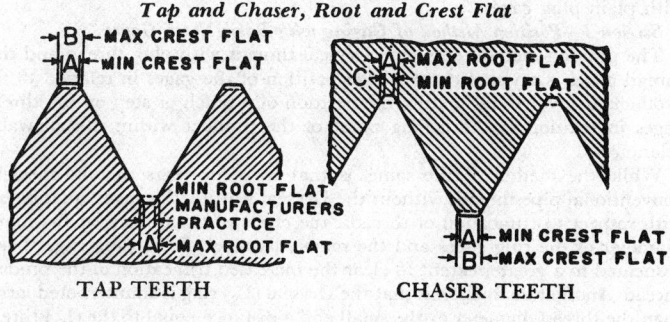

Tap and Chaser, Root and Crest Flat

TAP TEETH — CHASER TEETH

TABLE B1—WIDTH OF FLATS

Threads per in.	A	B	C
27	0.004	0.005	0.002
18	0.005	0.006	0.003
14	0.005	0.006	0.003
11-1/2	0.006	0.008	0.004
8	0.008	0.010	0.006[a]

[a] There is reason to doubt the correctness of the 8 threads per in. flat widths because of the volume of metal to be displaced.

TABLE B2—DRYSEAL TAPER PIPE TAPS FOR DRYSEAL AMERICAN STANDARD INTERNAL TAPER PIPE THREAD, GROUND THREAD LIMITS

PTF Size	Gage[a] Measure		Major Dia Flat		Minor Dia Flat[b]
	Min	Max	Min	Max	Max
1/16 —27	0.250	0.375	0.004	0.005	0.004
1/8 —27	0.250	0.375	0.004	0.005	0.004
1/4 —18	0.397	0.521	0.005	0.006	0.005
3/8 —18	0.392	0.516	0.005	0.006	0.005
1/2 —14	0.517	0.641	0.005	0.006	0.005
3/4 —14	0.503	0.627	0.005	0.006	0.005
1 —11-1/2	0.584	0.772	0.006	0.008	0.006
1-1/4 —11-1/2	0.592	0.780	0.006	0.008	0.006
1-1/2 —11-1/2	0.605	0.793	0.006	0.008	0.006
2 —11-1/2	0.573	0.761	0.006	0.008	0.006
2-1/2 —8[c]	0.831	1.019	0.008	0.010	0.008
3 —8[c]	0.831	1.019	0.008	0.010	0.008

[a] Distance small end of tap projects beyond face of American Standard Thin-Ring Gage.
[b] Minor diameter as specified or sharper.
[c] There is reason to doubt the correctness of the 8 threads per in. flat widths because of volume of metal to be displaced.

TABLE B3—DRYSEAL TAPER PIPE TAPS FOR DRYSEAL SAE SHORT INTERNAL TAPER PIPE THREAD, GROUND THREAD LIMITS

PTF—SAE Short Size	Gage[a] Measure		Major Dia Flat		Minor Dia Flat[b]
	Min (6p)	Max (7p)	Min	Max	Max
1/16—27	0.222	0.259	0.004	0.005	0.004
1/8 —27	0.222	0.259	0.004	0.005	0.004
1/4 —18	0.333	0.389	0.005	0.006	0.005
3/8 —18	0.333	0.389	0.005	0.006	0.005
1/2 —14	0.429	0.500	0.005	0.006	0.005
3/4 —14	0.429	0.500	0.005	0.006	0.005

[a] Distance small end of tap projects beyond face of American Standard Thin-Ring Gage.
[b] Minor diameter as specified or sharper.

Chamfer—$1\frac{1}{2}$ to 2 threads.
Marking—In addition to regular markings, Dryseal SAE Short Taper Taps will be marked PTF—SAE Short.
Tolerances—Same as for Dryseal American Standard Taper Taps.

TABLE B4—DRYSEAL STRAIGHT PIPE TAPS FOR DRYSEAL AMERICAN FUEL INTERNAL STRAIGHT PIPE THREAD, GROUND THREAD LIMITS

NPSF Size	Pitch Diameter			Major Dia Flat[a]		Minor Dia Flat[b]	Major Dia		
	Basic	Min	Max	Min	Max	Max	Basic	Min	Max
1/16—27	0.2812	0.2772	0.2777	0.004	0.005	0.004	0.3108	0.3008	0.3018
1/8 —27	0.3748	0.3696	0.3701	0.004	0.005	0.004	0.4044	0.3932	0.3942
1/4 —18	0.4899	0.4859	0.4864	0.005	0.006	0.005	0.5343	0.5239	0.5249
3/8 —18	0.6270	0.6213	0.6218	0.005	0.006	0.005	0.6714	0.6593	0.6603
1/2 —14	0.7784	0.7712	0.7717	0.005	0.006	0.005	0.8356	0.8230	0.8240
3/4 —14	0.9889	0.9817	0.9822	0.005	0.006	0.005	1.0460	1.0335	1.0345
1 —11-1/2	1.2386	1.2295	1.2305	0.006	0.008	0.006	1.3082	1.2933	1.2943

[a] For reference only. Major-diameter flats specified may be slightly larger or smaller with extreme combinations of pitch diameter, major diameter, and half angle.
[b] Minor diameter as specified or sharper.

Chamfer—Plug end 3 to 5 threads, intermediate end 2 to 3 threads, bottom end $1\frac{1}{2}$ to 2 threads.
Lead Tolerance—A maximum lead error of ±0.0005 in. in 1 in. of thread is permitted.
Angle Tolerance—Error in half angle of ±30 min permitted.
Marking—In addition to regular markings, Dryseal American Fuel Straight Pipe Taps will be marked NPSF.
Maximum Major Diameter equals minimum pitch diameter plus sharp-V thread height minus twice tool crest minimum truncation.
Minimum Major Diameter equals maximum major diameter minus tolerance.
Maximum Pitch Diameter equals the E_1 pitch diameter at large end of internal thread (Table 2) minus (small) $1\frac{1}{2}$ threads taper.
Minimum Pitch Diameter equals maximum diameter minus tolerance.

TABLE B5—DRYSEAL STRAIGHT PIPE TAPS FOR DRYSEAL AMERICAN INTERMEDIATE INTERNAL STRAIGHT PIPE THREAD, GROUND THREAD LIMITS

NPSI Size	Pitch Diameter			Major Dia Flat[a]		Minor Dia Flat[b]	Major Dia		
	Basic	Min	Max	Min	Max	Max	Basic	Min	Max
1/16—27	0.2812	0.2795	0.2800	0.004	0.005	0.004	0.3108	0.3031	0.3041
1/8 —27	0.3748	0.3719	0.3724	0.004	0.005	0.004	0.4044	0.3955	0.3965
1/4 —18	0.4899	0.4894	0.4899	0.005	0.006	0.005	0.5343	0.5274	0.5284
3/8 —18	0.6270	0.6248	0.6253	0.005	0.006	0.005	0.6714	0.6628	0.6638
1/2 —14	0.7784	0.7757	0.7762	0.005	0.006	0.005	0.8356	0.8275	0.8285
3/4 —14	0.9889	0.9862	0.9867	0.005	0.006	0.005	1.0460	1.0380	1.0390
1 —11-1/2	1.2386	1.2349	1.2359	0.006	0.008	0.006	1.3082	1.2987	1.2997

[a] For reference only. Major-diameter flats specified may be slightly larger or smaller with extreme combinations of pitch diameter, major diameter, and half angle.
[b] Minor diameter as specified or sharper.

Chamfer—Plug end 3 to 5 threads, intermediate end 2 to 3 threads, bottom end $1\frac{1}{2}$ to 2 threads.
Lead Tolerance—A maximum lead error of ±0.0005 in. in 1 in. of thread is permitted.
Angle Tolerance—Error in half angle of ±30 min permitted.
Marking—In addition to regular markings, American Intermediate Straight Pipe Taps will be marked NPSI.
Maximum Major Diameter equals minimum pitch diameter plus sharp-V thread height minus twice tool crest minimum truncation.
Minimum Major Diameter equals maximum major diameter minus tolerance.
Maximum Pitch Diameter equals the E_1 pitch diameter at large end of internal thread (Table 2) minus (small) $\frac{1}{2}$ thread taper.
Minimum Pitch Diameter equals maximum pitch diameter minus tolerance.

APPENDIX C—DRYSEAL PIPE-THREAD GAGING

General Information—There are three accepted methods of checking Dryseal pipe threads with threaded plug and ring gages. The methods separately described in the following sections are:

Section I—Position Method of Gaging with Basic Notch Gages.
Section II—Limit Method of Gaging with Step-Limit Gages.
Section III—Turns-Engagement Method of Gaging with Basic-Notch or Step-Limit Gages.

All methods of gaging external Dryseal threads involve the use of two ring thread gages, the (L_1) thin-ring thread gage for checking the pitch diameter over the hand engagement or (L_1) thread length and the (L_2) full-thread ring gage for checking the pitch diameter over the full thread length to insure adequate threads for wrench tightening.

All methods of gaging internal Dryseal threads involve the use of two plug thread gages, the (L_1) plug thread gage for checking the pitch diameter over the hand engagement or (L_1) thread length and the (L_3) plug thread gage for checking pitch diameter of the thread beyond the hand engagement length.

As indicated in the separate descriptions of the various gaging methods, coordination of the two ring thread gages for external threads and coordination of the two plug thread gages for internal threads control and check thread taper and length. The gages cannot be correlated, however, for external threads of minimum pitch diameter or internal threads of maximum pitch diameter unless the length of the threads is one thread longer than basic full thread length.

Working gages shall not be used where worn beyond the basic dimensions by more than $\frac{1}{2}$ turn (thread). Proper allowance shall be made for any variation from basic when using a gage.

The threads of tools and the threads of a percentage of the product or casts in the case of internal threads should be projected as a check on thread form and truncation. Although projection is strongly recommended, the truncation at major diameter of internal thread and minor diameter of external thread may be checked respectively with special plug and ring gages with thread angle reduced to clear the flank of the threads; and the truncation at minor diameter of internal taper thread and major diameter of external taper thread may be checked respectively with plain taper plug gages and plain taper ring gages. Internal straight thread truncation at minor diameter may be checked with plain plug gages.

Section I—Position Method of Gaging with Basic Notch Gages

The position method of gaging Dryseal threads with plug thread and ring thread gages is a visual check of the position of the gages in relation to the product. It involves estimating the position of a notch or step on the thread gages in relation to the gaging point of the product within the allowable tolerance.

While the method is the same as that used for years past in checking conventional pipe threads without the Dryseal feature, the gages are different with respect to truncation of threads, the crests of the threads at the minor diameter of the ring gages and the major diameter of the plug gages being truncated to a greater extent to clear the increased truncation of the product thread. Another distinction is that the Dryseal (L_2) ring is counterbored larger than the thread diameter at the small end a distance equal to the (L_1) thread length minus one pitch. Conventional rings and plugs, however, may be converted to Dryseal by grinding the crests to conform with the width of flats specified for Dryseal gages, and grinding a counterbore in the (L_2) ring gage.

The gages are turned or screwed hand-tight into or onto the threaded product, the position of the gage notch in relation to the product reference point being noted to determine whether the standoff exceeds the allowable tolerance. Allowance must be made for excessive chamfer at the small end of external threads and the large end of internal threads, the product reference point in the first instance being the beginning of the first thread on the chamfer, and in the second instance being the intersecton of the pitch diameter cone and the chamfer cone.

External Threads—Dryseal American Standard External Taper Pipe Threads (NPTF) are gaged with the NPTF (L_1) basic-notch Dryseal ring thread gages (Table C1) and the NPTF (L_2) basic-notch Dryseal ring thread gages (Table C1). Threads are within the allowable tolerance when the product reference point is flush with the gage reference point within a tolerance of plus (small) one turn, minus (large) one turn. As a check on taper, the (L_1) and (L_2) ring thread gages shall gage the same within $\frac{1}{2}$ turn.

Dryseal SAE Short External Taper Pipe Threads PTF—SAE Short, which are one thread shorter at the small end than standard full thread length, are gaged with the NPTF (L_1) basic-notch Dryseal ring thread gages (Table C1) and the NPTF (L_2) basic-notch Dryseal ring thread gages (Table C1). Threads are within the allowable tolerance when the product reference point is flush with the gage reference point within a tolerance of plus zero, minus (large) $1\frac{1}{2}$ turns. As a check on taper, the (L_1) and (L_2) ring thread gages shall gage the same within $\frac{1}{2}$ turn.

Internal Threads—Dryseal American Standard Internal Taper Pipe Threads (NPTF) are gaged with the NPTF (L_1) basic-notch Dryseal plug thread gages (Table C2) and the NPTF (L_3) basic-notch Dryseal plug thread gages (Table C3). Threads are within the allowable tolerance when the product reference point is flush with the gage notch within the following tolerances:

Plus (large) 1 turn, minus (small) 1 turn.

As a check on taper, the (L_1) and the (L_3) plug thread gages shall gage the same with relation to their respective notches within $\frac{1}{2}$ turn.

Dryseal SAE Short Internal Taper Pipe Threads PTF—SAE Short, which are one thread shorter at the large end than standard full thread length, are gaged with the NPTF (L_1) basic-notch Dryseal plug thread gages (Table C2) and the NPTF (L_3) basic-notch Dryseal plug thread gages (Table C3). Threads are within the allowable tolerance when the product reference point is flush with the gage notch within the following tolerances:

Plus (large) 0 turns, minus (small) $1\frac{1}{2}$ turns.

Dryseal American (National) Standard Fuel Internal Straight Pipe Threads (NPSF) are gaged with the NPTF (L_1) basic-notch Dryseal plug thread gages (Table C2) are within the allowable tolerance when the product reference point is flush with the gage notch within the following tolerances:

Plus (large) 0 turns, minus (small) $1\frac{1}{2}$ turns.

As depth gages without regard to gage notches, any of the (L_3) Dryseal plug thread gages may be used to check the full thread length of internal straight pipe threads.

Dryseal American (National) Standard Intermediate Internal Straight Pipe Threads (NPSI) are gaged with the NPTF (L_1) basic-notch Dryseal plug thread gages (Table C2). Threads are within the allowable tolerance when the product reference point is flush with the gage notch within the following tolerances:

Plus (large) 1 turn, minus (small) $\frac{1}{2}$ turn.

As depth gages without regard to gage notches, any of the (L_3) Dryseal plug thread gages may be used to check the full thread length of internal straight pipe threads.

Section II—Limit Method of Gaging with Step-Limit Gages

The limit-gage or step-limit method of checking threaded product with plug thread and ring thread gages is a visual check of the position of the gages in relation to the product. Plug and ring gages with maximum and minimum limit notches are provided for the different thread types: NPTF, PTF—SAE Short, NPSF, and NPSI. The location of the limit notches on the $\frac{1}{8}$- and $\frac{1}{4}$-in. plugs eliminates the necessity for gaging correction.

The gages are turned or screwed hand-tight into or onto the threaded product, the position of the product reference point in relation to the limit notches on the gage being noted. Allowance must be made for excessive chamfer at the small end of external threads and the large end of internal threads, the product reference point in the first instance being the beginning of the first thread on the chamfer, and in the second instance being the intersection of the pitch diameter cone and the chamfer cone.

External Threads—Dryseal American Standard External Taper Pipe Threads (NPTF) are gaged with the NPTF (L_1) step-limit Dryseal ring thread gages (Table C4) and the NPTF (L_2) step-limit Dryseal ring thread gages (Table C5). Threads are within the allowable tolerance when the product reference point is on or between the limit notches. As a check on taper, the (L_1) and the (L_2) Dryseal ring thread gages shall gage the same in relation to their respective notches within $\frac{1}{2}$ turn.

Dryseal SAE Short External Taper Pipe Threads PTF—SAE Short, which are one thread shorter at the small end than standard full thread length, are gaged with the PTF—SAE (L_1 short) step-limit Dryseal ring thread gages (Table C8) and the PTF—SAE (L_2 Short) step-limit ring thread gages (Table C9). Threads are within the allowable tolerance when the product reference point is on or between the limit notches. As a check on taper, the (L_1 Short) and the (L_2 Short) Dryseal ring thread gages shall gage the same with relation to their respective notches within $\frac{1}{2}$ turn.

Internal Threads—Dryseal American Standard Internal Taper Pipe Threads (NPTF) are gaged with the NPTF (L_1) step-limit Dryseal plug thread gages (Table C6) and the NPTF (L_3) step-limit Dryseal plug thread gages (Table C7). Threads are within the allowable tolerance when the product reference point is on or between the limit notches. As a check on taper, the (L_1) and (L_3) Dryseal plug thread gages shall gage the same with relation to their respective notches within $\frac{1}{2}$ turn.

Dryseal SAE Short Internal Taper Pipe Threads PTF—SAE Short, which are one thread shorter at the large end than standard full thread length, are gaged with the PTF—SAE (L_1 Short) step-limit Dryseal plug thread gages (Table C10) and the PTF—SAE (L_3 Short) step-limit Dryseal plug thread gages (Table C11). Threads are within the allowable tolerance when the product reference point is on or between the limit notches. As a check on taper, the (L_1 Short) and (L_3 Short) Dryseal plug thread gages shall gage the same with relation to their respective notches within $\frac{1}{2}$ turn.

Dryseal American Standard Fuel Internal Straight Pipe Threads (NPSF) are gaged with the NPSF (L_1 Short) step-limit Dryseal plug thread gages (Table C10). Threads are within the allowable tolerance when the product reference point is on or between the limit notches. As depth gages without regard to limit notches, any of the (L_3) Dryseal plug thread gages may be used to check the full thread length of internal straight pipe threads.

Dryseal American Standard Intermediate Internal Straight Pipe Threads (NPSI) are gaged with the NPSI (L_1) step-limit Dryseal plug thread gages (Table C12). Threads are within the allowable tolerance when the product reference point is on or between the limit notches. As depth gages without regard to limit notches, any of the (L_3) Dryseal plug thread gages may be used to check the full thread length of internal straight pipe threads.

Section III—Turns-Engagement Method of Gaging with Basic-Notch or Step-Limit Gages

The turns-engagement method of checking threaded product with plug thread and ring thread gages is a tactual check of the position of the gages in relation to the product. In checking by this method, either the basic-notch or the step-limit gages may be used. The gages are turned or screwed into or onto the threaded product and the turns to remove the gages are counted. This method compensates for gage chamfer and eliminates the variable of product chamfer.

Basic Turns Engagement of Ring Gages—The basic turns engagement of the (L_1) ring thread gages (Tables C1, C4, and C8) with Dryseal external taper pipe threads is the product of the (L_1) thread length of the ring gage used and the threads per inch, minus one turn to compensate for chamfer of the external threads and chamfer of the ring gages. (See accompanying tabulation of basic turns engagement.)

The basic turns engagement of the (L_2) ring thread gages (Tables C1, C5, and C9) with Dryseal external taper pipe threads is the product of the (L_2) thread length and the threads per inch, minus $1\frac{1}{4}$ turns to compensate for chamfer of the external threads and the chamfer and taper of the ring gages. (See accompanying tabulation of basic turns engagement.)

External Threads—Dryseal American Standard External Taper Pipe Threads (NPTF) by the turns-engagement method may be gaged with any combination of (L_1) and (L_2) Dryseal ring thread gages (Tables C1, C4, C5, C8, and C9). Nominal turns engagement equals basic turns engagement. The tolerance is plus (small) 1 turn, minus (large) 1 turn. As a check on taper, the difference in turns engagement with the (L_1) and the (L_2) Dryseal ring thread gages shall be within $\frac{1}{2}$ turn of the difference between the basic turns engagement of the ring gages.

Dryseal SAE Short External Taper Pipe Threads PTF—SAE Short, which are one thread shorter at the small end than standard full thread length, may be gaged by the turns-engagement method with any combination of (L_1) and (L_2) Dryseal ring thread gages (Tables C1, C4, C5, C8, and C9). Nominal turns engagement is one turn less than basic turns engagement. The tolerance is plus (small) 1 turn, minus (large) $\frac{1}{2}$ turn. As a check on taper, the difference in turns engagement with the (L_1) and the (L_2) Dryseal ring thread gages shall be within $\frac{1}{2}$ turn of the difference between the basic turns engagement of the ring gages.

Basic Turns Engagement of Plug Gages—The basic turns engagement of the (L_1) Dryseal plug thread gages (Tables C2, C6, C10, and C12) and Dryseal internal pipe threads is the product of the (L_1) thread length (Table 2) and the threads per inch, minus $\frac{1}{2}$ turn to compensate for chamfer on plug gages. (See accompanying tabulation of basic turns engagement.)

The basic turns engagement of the (L_3) Dryseal plug thread gages (Tables C3, C7, and C11) with Dryseal internal pipe threads is the (L_1) thread length (Table 2) plus three threads multiplied by the threads per inch, minus $\frac{3}{4}$ turn to compensate for chamfer and taper on plug gages. (See accompanying tabulation of basic turns engagement.)

Internal Threads—Dryseal American Standard Internal Taper Pipe Threads (NPTF) are gaged with any combination of (L_1) and (L_3) Dryseal plug thread gages (Tables C2, C3, C6, C7, C10, C11, and C12). The nominal turns engagement equals basic turns engagement. The tolerance is plus (large) 1 turn, minus (small) 1 turn. As a check on taper, the difference in turns engagement of the (L_1) and the (L_3) Dryseal plug thread gages shall not be less than $2\frac{1}{4}$ turns nor more than $3\frac{1}{4}$ turns.

Dryseal SAE Short Internal Taper Pipe Threads PTF—SAE Short, which are one thread shorter at the large end than standard full thread length, are gaged with any combination of (L_1) and (L_3) Dryseal plug thread gages (Tables C2, C3, C6, C7, C10, C11, and C12). The nominal turns engagement is one turn less than basic turns engagement. The tolerance is plus (large) 1 turn, minus (small) $\frac{1}{2}$ turn. As a check on taper, the difference in turns engagement of the (L_1) and the (L_3) Dryseal plug thread gages shall not be less than $2\frac{1}{4}$ turns nor more than $3\frac{1}{4}$ turns.

Dryseal American Standard Fuel Internal Straight Pipe Threads (NPSF) are gaged with any of the (L_1) Dryseal plug thread gages (Tables C2, C6, C10, and C12). The nominal turns engagement is one turn less than basic turns engagement. The tolerance is plus (large) 1 turn, minus (small) $\frac{1}{2}$ turn. As depth gages without regard to limit notches, any of the (L_3) Dryseal plug thread gages may be used to check the full thread length of internal straight pipe threads.

Dryseal American Standard Intermediate Internal Straight Pipe Threads (NPSI) are gaged with any of the (L_1) Dryseal plug thread gages (Tables C2, C6, C10, and C12). The nominal turns engagement equals basic turns engagement. The tolerance is plus (large) 1 turn, minus (small) $\frac{1}{2}$ turn. As depth gages without regard to limit notches, any of the (L_3) Dryseal plug thread gages may be used to check the full thread length of internal straight pipe threads.

BASIC TURNS ENGAGEMENT[a]

Size	L_1 Rings		All L_2 Rings	All L_1 Plugs	All L_3 Plugs
	Basic-Notch, Table C1	Step-Limit, Tables C4 and C8			
1/16 —27	3.32	3.32	5.80	3.82	6.57
1/8 —27	3.36	3.36	5.87	3.86	6.61
1/4 —18	3.10	3.10	5.98	3.60	6.35
3/8 —18	3.32	3.32	6.09	3.82	6.57
1/2 —14	3.48	3.48	6.22	3.98	6.73
3/4 —14	3.75	3.75	6.39	4.25	7.00
1 —11-1/2	3.60	3.60	6.60	4.10	6.85
1-1/4 —11-1/2	3.83	3.83	6.88	4.33	7.08
1-1/2 —11-1/2	3.83	3.83	7.07	4.33	7.08
2 —11-1/2	4.01	4.01	7.45	4.51	7.26
2-1/2 —8	4.46	4.46	7.85	4.96	7.71
3 —8	5.13	5.13	8.35	5.63	8.38

[a] Derivation of nominal turns engagement and tolerance for the different thread types, NPTF, PTF—SAE Short, NPSF, and NPSI, is explained in accompanying text.

DRYSEAL AMERICAN TAPER PIPE THREAD (L_1 AND L_2) RING GAGES

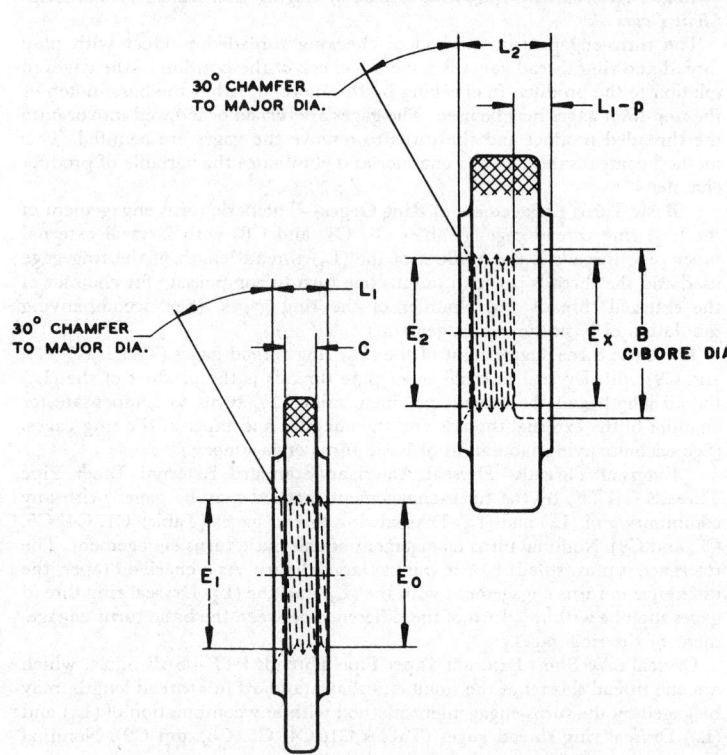

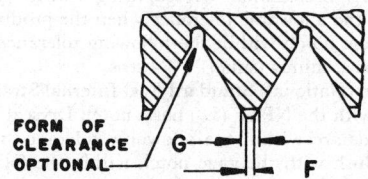

TABLE C1-1—THREAD FLATS

Threads per in.	F	G
27	0.0086	0.0107
18	0.0128	0.0160
14	0.0165	0.0206
11-1/2	0.0201	0.0251
8	0.0289	0.0361

Marking—In addition to the regular markings, Dryseal American Standard Taper Thread Thin-Ring Gages will be marked NPTF (L_1) and Full-Ring Gages will be marked NPTF (L_2) on the entering side of gage.

Thread Form—The threads in all particulars excepting truncation shall conform to American Standard Taper Pipe Thread practice. Crests of Threads at the minor diameter of ring gages and major diameter of plug gages shall be truncated 0.20p minimum to 0.25p maximum producing the minimum and maximum widths of flats specified in Table C1-1.

All other thread dimensions shall be within tolerances specified for the Dryseal American Standard Pipe Thread Working Plug Gages (ASA B2.2). Other gage details shall conform to American Gage Design Standards published in Commercial Standard CS8.

TABLE C1—BASIC DIMENSIONS OF DRYSEAL AMERICAN TAPER PIPE THREAD (L_1 AND L_2) BASIC-NOTCH RING GAGES

Size	(L_2) Basic-Notch Full-Ring Gages							(L_1) Basic-Notch Thin-Ring Gages				
	L_2	Pitch Dia, E_2	Minor Dia at Large End[a]	Pitch Dia at $L_1 - p$, E_x	Minor Dia at $L_1 - p$[a]	$L_1 - p$	B	L_1	Pitch Dia, E_1	Minor Dia at Large End[a]	Pitch Dia, E_0	Minor Dia at Small End[a]
1/16 —27	0.26113	0.28750	0.27024	0.27886	0.26160	0.12296	0.38	0.1600	0.28118	0.26392	0.27118	0.25392
1/8 —27	0.26385	0.38000	0.36274	0.37129	0.35403	0.12446	0.47	0.1615	0.37360	0.35634	0.36351	0.34625
1/4 —18	0.40178	0.50250	0.47661	0.48816	0.46227	0.17224	0.59	0.2278	0.49163	0.46574	0.47739	0.45150
3/8 —18	0.40778	0.63750	0.61161	0.62354	0.59765	0.18444	0.72	0.2400	0.62701	0.60112	0.61201	0.58612
1/2 —14	0.53371	0.79179	0.75850	0.77396	0.74067	0.24857	0.88	0.3200	0.77843	0.74514	0.75843	0.72514
3/4 —14	0.54571	1.00179	0.96850	0.98440	0.95111	0.26757	1.09	0.3390	0.98887	0.95558	0.96768	0.93439
1 —11-1/2	0.68278	1.25630	1.21577	1.23320	1.19267	0.31304	1.34	0.4000	1.23863	1.19810	1.21363	1.17310
1-1/4 —11-1/2	0.70678	1.60130	1.56077	1.57794	1.53741	0.33304	1.69	0.4200	1.58338	1.54285	1.55713	1.51660
1-1/2 —11-1/2	0.72348	1.84130	1.80077	1.81690	1.77637	0.33304	1.94	0.4200	1.82234	1.78181	1.79609	1.75556
2 —11-1/2	0.75652	2.31630	2.27577	2.29084	2.25031	0.34904	2.50	0.4360	2.29627	2.25574	2.26902	2.22849
2-1/2 —8	1.13750	2.79062	2.73237	2.75434	2.69609	0.55700	2.94	0.6820	2.76216	2.70391	2.71953	2.66128
3 —8	1.20000	3.41562	3.35737	3.38068	3.32243	0.54100	3.56	0.7660	3.38850	3.33025	3.34062	3.28237

[a] Minor diameter is based on crest minimum truncation of 0.20 p.

DRYSEAL AMERICAN TAPER PIPE THREAD (L_1) PLUG GAGES
Taper lock design, range $\tfrac{1}{8}$ to 3 in., inclusive

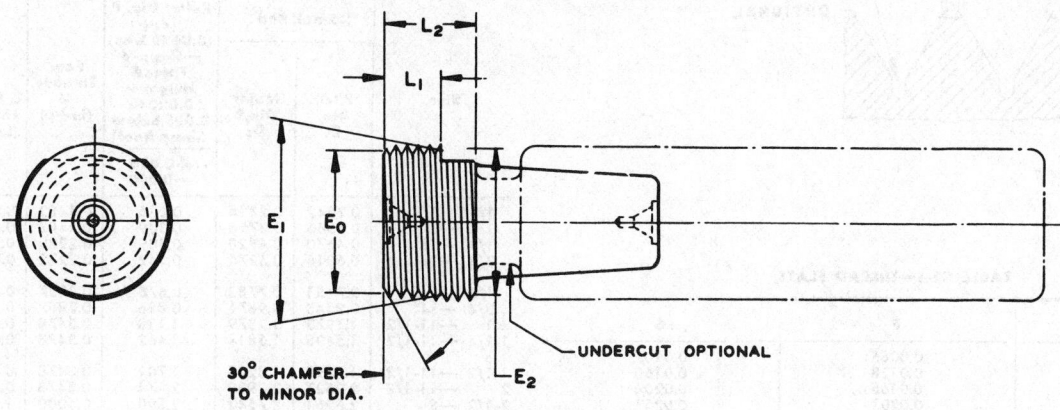

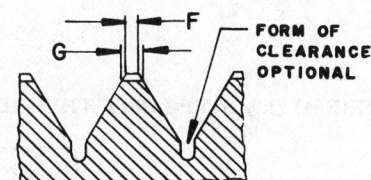

TABLE C2-1—THREAD FLATS

Threads per in.	F	G
27	0.0086	0.0107
18	0.0128	0.0160
14	0.0165	0.0206
11-1/2	0.0201	0.0251
8	0.0289	0.0361

TABLE C2—BASIC DIMENSIONS OF DRYSEAL AMERICAN TAPER PIPE THREAD (L_1) BASIC-NOTCH PLUG GAGES

Size	L_1	L_2	Small End		Gaging Notch		Large End	
			Pitch Dia, E_0	Major Dia[a]	Pitch Dia, E_1	Major Dia[a]	Pitch Dia, E_2	Major Dia[a]
1/16—27	0.1600	0.26113	0.27118	0.28844	0.28118	0.29844	0.28750	0.30476
1/8—27	0.1615	0.26385	0.36351	0.38077	0.37360	0.39086	0.38000	0.39726
1/4—18	0.2278	0.40178	0.47739	0.50328	0.49163	0.51752	0.50250	0.52839
3/8—18	0.2400	0.40778	0.61201	0.63790	0.62701	0.65290	0.63750	0.66339
1/2—14	0.3200	0.53371	0.75843	0.79170	0.77843	0.81170	0.79179	0.82506
3/4—14	0.3390	0.54571	0.96768	1.00095	0.98887	1.02214	1.00179	1.03506
1—11-1/2	0.4000	0.68278	1.21363	1.25416	1.23863	1.27916	1.25630	1.29683
1-1/4—11-1/2	0.4200	0.70678	1.55713	1.59766	1.58338	1.62391	1.60130	1.64183
1-1/2—11-1/2	0.4200	0.72348	1.79609	1.83662	1.82234	1.86287	1.84130	1.88183
2—11-1/2	0.4360	0.75652	2.26902	2.30955	2.29627	2.33680	2.31630	2.35683
2-1/2—8	0.6820	1.13750	2.71953	2.77778	2.76216	2.82041	2.79062	2.84887
3—8	0.7660	1.20000	3.34062	3.39887	3.38850	3.44675	3.41562	3.47387

[a] Major diameter is based upon crest minimum truncation of 0.20 p.

Marking—In addition to the regular markings, Dryseal American Standard Taper Pipe Thread (L_1) Plug Gages will be marked NPTF (L_1).

Thread Form—The threads in all particulars excepting truncation shall conform to American Standard Taper Pipe Thread practice. Crests of threads at the minor diameter of ring gages and major diameter of plug gages shall be truncated 0.20p minimum to 0.25p maximum producing the minimum and maximum widths of flats specified in Table C2-1.

All other thread dimensions shall be within tolerances specified for the Dryseal American Standard Pipe Thread Working Plug Gages (ASA B2.2). Other gage details shall conform to American Gage Design Standards published in Commercial Standard CS8.

DRYSEAL AMERICAN TAPER PIPE THREAD (L_3) LENGTH PLUG GAGES

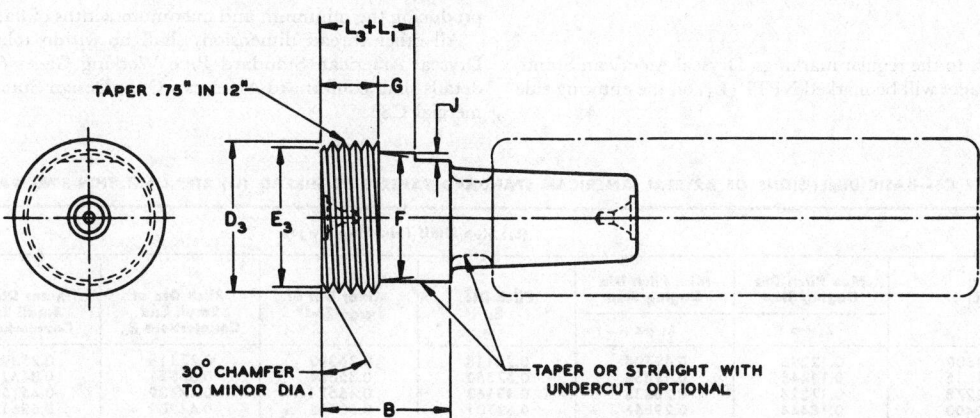

Marking—In addition to the regular markings, Dryseal American Standard Taper Pipe Thread (L_3) Plug Gages will be marked NPTF (L_3).

Thread Form—The threads in all particulars excepting truncation shall conform to American Standard Taper Pipe Thread practice. Crests of threads at major diameter shall be truncated 0.20p minimum to 0.25p maximum, producing the minimum and maximum widths of flat specified in Table C3-1.

All other thread dimensions shall be within tolerances specified for the Dryseal American Standard Pipe Thread Working Plug Gages (ASA B2.2). Other gage details shall conform to American Gage Design Standards published in Commercial Standard CS8.

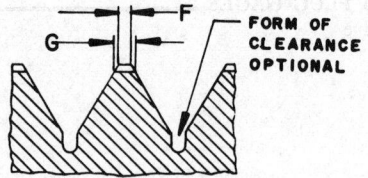

TABLE C3-1—THREAD FLATS

Threads per in.	F	G
27	0.0086	0.0107
18	0.0128	0.0160
14	0.0165	0.0206
11-1/2	0.0201	0.0251
8	0.0289	0.0361

TABLE C3—BASIC DIMENSIONS OF DRYSEAL AMERICAN TAPER PIPE THREAD (L_1) BASIC-NOTCH LENGTH PLUG GAGES

Size	Small End Pitch Dia, E_3	Small End Major Dia,[a] D_3	Relief Dia, F [E_3 + (0.0625 × 4p) — Sharp-V Thread Height — 0.020 to 0.025 below Sharp Root] +0.005 −0.000	Four Threads, G (L_1+p)	L_1 Plus 3 Threads (L_1+L_3)	Blank Length, B	Notch Depth, J +0.005 −0.000
1/16—27	0.2642	0.2815	0.216	0.1482	0.2711	0.38	0.030
1/8 —27	0.3566	0.3738	0.309	0.1482	0.2726	0.41	0.030
1/4 —18	0.4670	0.4928	0.409	0.2222	0.3945	0.50	0.030
3/8 —18	0.6016	0.6275	0.542	0.2222	0.4067	0.56	0.030
1/2 —14	0.7451	0.7783	0.676	0.2857	0.5343	0.69	0.040
3/4 —14	0.9543	0.9876	0.886	0.2857	0.5533	0.72	0.040
1 —11-1/2	1.1973	1.2379	1.118	0.3478	0.6609	0.88	0.050
1-1/4—11-1/2	1.5408	1.5814	1.462	0.3478	0.6809	0.88	0.050
1-1/2—11-1/2	1.7798	1.8203	1.701	0.3478	0.6809	0.88	0.050
2 —11-1/2	2.2527	2.2932	2.174	0.3478	0.6969	0.88	0.050
2-1/2 —8	2.6961	2.7543	2.590	0.5000	1.0570	1.50	0.050
3 —8	3.3172	3.3754	3.214	0.5000	1.1410	1.50	0.050

[a] Major diameter is based upon crest minimum truncation of 0.20 p.

DRYSEAL AMERICAN STANDARD TAPER PIPE THREAD (L_1) STEP-LIMIT THIN-RING GAGES

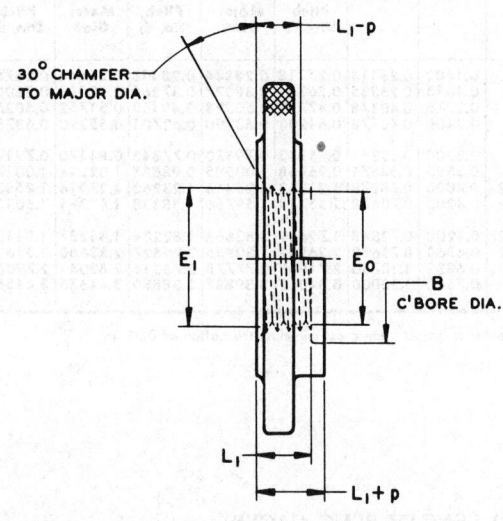

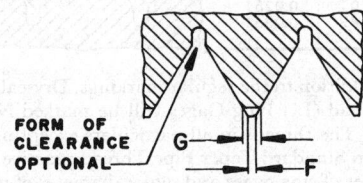

TABLE C4-1—THREAD FLATS

Threads per in.	F	G
27	0.0086	0.0107
18	0.0128	0.0160
14	0.0165	0.0206
11-1/2	0.0201	0.0251
8	0.0289	0.0361

Marking—In addition to the regular markings, Dryseal American Standard Taper Thread Ring Gages will be marked NPTF (L_1) on the entering side of gage.

Thread Form—The threads in all particulars excepting truncation shall conform to American Standard Taper Pipe Thread practice. Crests of threads at the minor diameter shall be truncated 0.20p minimum to 0.25p maximum, producing the minimum and maximum widths of flat specified in Table C4-1.

All other thread dimensions shall be within tolerances specified for the Dryseal American Standard Pipe Working Gages (ASA B2.2). Other gage details shall conform to American Gage Design Standard Published in Commercial CS8.

TABLE C4—BASIC DIMENSIONS OF DRYSEAL AMERICAN STANDARD TAPER PIPE THREAD (L_1) STEP-LIMIT THIN-RING GAGES

Size	(L_1) Step-Limit Thin-Ring Gages							
	L_1	Max Pitch Dia Gaging Step L_1−p	Min. Pitch Dia Gaging Step L_1+p	Pitch Dia, E_1	Minor Dia at Large End[a]	Pitch Dia at Small End Counterbore E_0	Minor Dia at Small End Counterbore[a]	B
1/16—27	0.1600	0.12296	0.19704	0.28118	0.26392	0.27118	0.25392	0.38
1/8 —27	0.1615	0.12446	0.19854	0.37360	0.35634	0.36351	0.34625	0.47
1/4 —18	0.2278	0.17224	0.28336	0.49163	0.46574	0.47739	0.45150	0.59
3/8 —18	0.2400	0.18444	0.29556	0.62701	0.60112	0.61201	0.58612	0.72
1/2 —14	0.3200	0.24857	0.39143	0.77843	0.74514	0.75843	0.72514	0.88
3/4 —14	0.3390	0.26757	0.41043	0.98887	0.95558	0.96768	0.93439	1.09
1 —11-1/2	0.4000	0.31304	0.48696	1.23863	1.19810	1.21363	1.17310	1.34
1-1/4—11-1/2	0.4200	0.33304	0.50696	1.58338	1.54285	1.55713	1.51660	1.69
1-1/2—11-1/2	0.4200	0.33304	0.50696	1.82234	1.78181	1.79609	1.75556	1.94
2 —11-1/2	0.4360	0.34904	0.52296	2.29627	2.25574	2.26902	2.22849	2.50
2-1/2 —8	0.6820	0.55700	0.80700	2.76216	2.70391	2.71953	2.66128	2.94
3 —8	0.7660	0.64100	0.89100	3.38850	3.33025	3.34062	3.28237	3.56

[a] Minor diameter is based on crest minimum truncation of 0.20 p.

DRYSEAL AMERICAN STANDARD TAPER PIPE THREAD (L_2) STEP-LIMIT FULL-RING GAGES

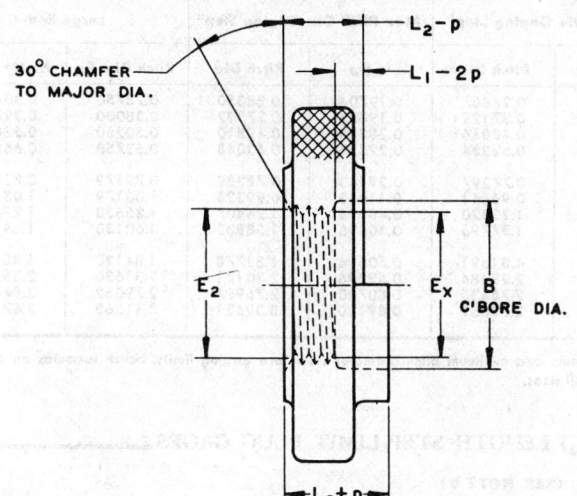

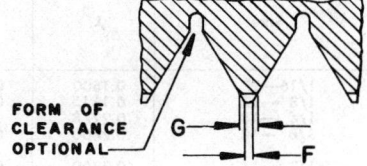

TABLE C5-1—THREAD FLATS

Threads per in.	F	G
27	0.0086	0.0107
18	0.0128	0.0160
14	0.0165	0.0206
11-1/2	0.0201	0.0251
8	0.0289	0.0361

Marking—In addition to the regular markings, Dryseal American Standard Taper Pipe Thread Ring Gages will be marked NPTF (L_2) on the entering side of gage.

Thread Form—The threads in all particulars excepting truncation shall conform to American Standard Taper Pipe Thread practice. Crests of threads at the minor diameter shall be truncated 0.20p minimum to 0.25p maximum, producing the minimum and maximum widths of flat specified in Table C5-1.

All other thread dimensions shall be within tolerances specified for the Dryseal American Standard Pipe Working Gages (ASA B2.2). Other gage details shall conform to American Gage Design Standard published in Commercial Standard CS8.

TABLE C5—BASIC DIMENSIONS OF DRYSEAL AMERICAN STANDARD TAPER PIPE THREAD (L_2) STEP-LIMIT FULL-RING GAGES

Size	(L_2) Step-Limit Full-Ring Gages								
	L_2	Max Pitch Dia Gaging Step L_1-p	Min Pitch Dia Gaging Step L_1+p	Pitch Dia, E_2	Minor Dia at Large End[a]	Pitch Dia at L_1 from Min Pitch Dia Gaging Step (E_x)	Minor Dia at Small End Counterbore[a]	L_1-2p	B
1/16—27	0.26113	0.22409	0.29817	0.28750	0.27024	0.27886	0.26160	0.08592	0.38
1/8—27	0.26385	0.22681	0.30089	0.38000	0.36274	0.37129	0.35403	0.08742	0.47
1/4—18	0.40178	0.34622	0.45734	0.50250	0.47661	0.48816	0.46227	0.11668	0.59
3/8—18	0.40778	0.35222	0.46334	0.63750	0.61161	0.62354	0.59765	0.12888	0.72
1/2—14	0.53371	0.46228	0.60514	0.79179	0.75850	0.77396	0.74067	0.17714	0.88
3/4—14	0.54571	0.47428	0.61714	1.00179	0.96850	0.98440	0.95111	0.19614	1.09
1—11-1/2	0.68278	0.59582	0.76974	1.25630	1.21577	1.23320	1.19267	0.22608	1.34
1-1/4—11-1/2	0.70678	0.61982	0.79374	1.60130	1.56077	1.57794	1.53741	0.24608	1.69
1-1/2—11-1/2	0.72348	0.63652	0.81044	1.84130	1.80077	1.81690	1.77637	0.24608	1.94
2—11-1/2	0.75652	0.66956	0.84348	2.31630	2.27577	2.29084	2.25031	0.26208	2.50
2-1/2—8	1.13750	1.01250	1.26250	2.79062	2.73237	2.75434	2.69609	0.43200	2.94
3—8	1.20000	1.07500	1.32500	3.41562	3.35737	3.38068	3.32243	0.51600	3.56

[a] Minor diameter is based on crest minimum truncation of 0.20 p.

DRYSEAL AMERICAN STANDARD TAPER PIPE THREAD (L_1) STEP-LIMIT PLUG GAGES
Taper lock design, range 1/8 to 3 in., inclusive

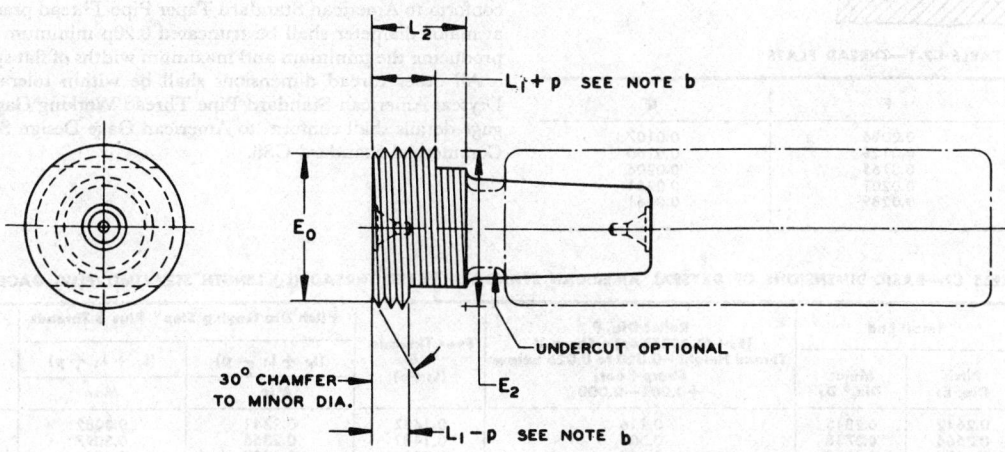

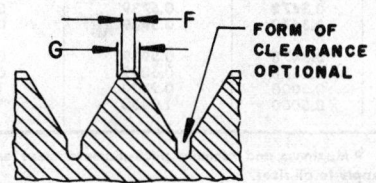

TABLE C6-1—THREAD FLATS

Threads per in.	F	G
27	0.0086	0.0107
18	0.0128	0.0160
14	0.0165	0.0206
11-1/2	0.0201	0.0251
8	0.0289	0.0361

Marking—In addition to the regular markings, Dryseal American Standard Taper Pipe Thread (L_1) Plug Gages will be marked NPTF (L_1).

Thread Form—The threads in all particulars excepting truncation shall conform to American Standard Taper Pipe Thread practice. Crests of threads at major diameter shall be truncated 0.20p minimum to 0.25p maximum, producing the minimum and maximum widths of flat specified in Table C6-1.

All other thread dimensions shall be within tolerances specified for the Dryseal American Standard Pipe Thread Working Plug Gages (ASA B2.2). Other gage details shall conform to American Gage Design Standards published in Commercial Standards CS8.

TABLE C6—BASIC DIMENSIONS OF DRYSEAL AMERICAN STANDARD TAPER PIPE THREAD (L₁) STEP-LIMIT PLUG GAGES

Size	L_1	L_2	Small End		Min Pitch Dia Gaging Step[b]		Max Pitch Dia Gaging Step[b]		Large End	
			Pitch Dia, E_0	Major Dia[a]	L_1-p	Pitch Dia	L_1+p	Pitch Dia	Pitch Dia, E_2	Major Dia[a]
1/16—27	0.1600	0.26113	0.27118	0.28844	0.12296	0.27887	0.19704	0.28350	0.28750	0.30476
1/8 —27	0.1615	0.26385	0.36351	0.38077	0.12446	0.37592	0.19854	0.37750	0.38000	0.39726
1/4 —18	0.2278	0.40178	0.47739	0.50328	0.17224	0.48816	0.28336	0.49510	0.50250	0.52839
3/8 —18	0.2400	0.40778	0.61201	0.63790	0.18444	0.62354	0.29556	0.63048	0.63750	0.66339
1/2 —14	0.3200	0.53371	0.75843	0.79172	0.24857	0.77397	0.39143	0.78289	0.79179	0.82508
3/4 —14	0.3390	0.54571	0.96768	1.00097	0.26757	0.98441	0.41043	0.99333	1.00179	1.03508
1 —11-1/2	0.4000	0.68278	1.21363	1.25416	0.31304	1.23320	0.48696	1.24407	1.25630	1.29683
1-1/4 —11-1/2	0.4200	0.70678	1.55713	1.59766	0.33304	1.57795	0.50696	1.58882	1.60130	1.64183
1-1/2 —11-1/2	0.4200	0.72346	1.79609	1.83662	0.33304	1.81691	0.50696	1.82778	1.84130	1.88183
2 —11-1/2	0.4360	0.75652	2.26902	2.30955	0.34904	2.29084	0.52296	2.30171	2.31630	2.35683
2-1/2 —8	0.6820	1.13750	2.71953	2.77778	0.55700	2.75435	0.80700	2.76997	2.79062	2.84887
3 —8	0.7660	1.20000	3.34062	3.39887	0.64100	3.38069	0.89100	3.39631	3.41562	3.47387

[a] Major diameter is based on crest minimum truncation of 0.20 p.
[b] Maximum and minimum pitch-diameter steps are gaging limits. Notch formulas on drawing apply to all sizes.

DRYSEAL AMERICAN STANDARD TAPER THREAD (L₃) LENGTH STEP-LIMIT PLUG GAGES

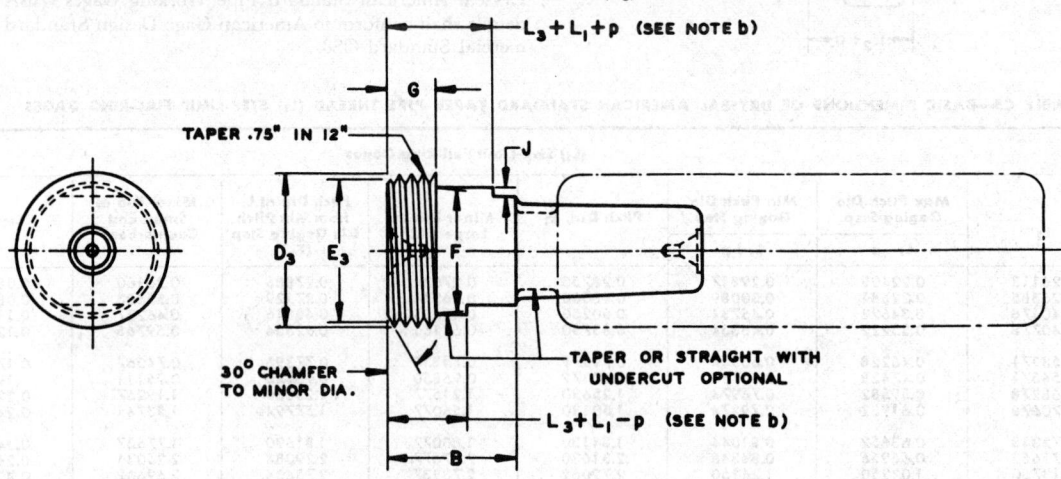

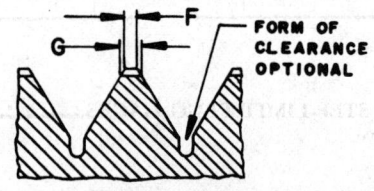

TABLE C7-1—THREAD FLATS

Threads per in.	F	G
27	0.0086	0.0107
18	0.0128	0.0160
14	0.0165	0.0206
11-1/2	0.0201	0.0251
8	0.0289	0.0361

Marking—In addition to the regular markings, Dryseal American Standard Taper Pipe Thread (L_3) Plug Gages will be marked PTF (L_3).

Thread Form—The threads in all particulars excepting truncation shall conform to American Standard Taper Pipe Thread practice. Crests of threads at major diameter shall be truncated 0.20p minimum and 0.25p maximum, producing the minimum and maximum widths of flat specified in Table C7-1.

All other thread dimensions shall be within tolerances specified for the Dryseal American Standard Pipe Thread Working Gages (ASA B2.2). Other gage details shall conform to American Gage Design Standards published in Commercial Standard CS8.

TABLE C7—BASIC DIMENSIONS OF DRYSEAL AMERICAN STANDARD TAPER THREAD (L₃) LENGTH STEP-LIMIT PLUG GAGES

Size	Small End		Relief Dia, F [$E_3+(0.0625 \times 4p)$—Sharp-V Thread Height—0.020 to 0.025 below Sharp Root] +0.005−0.000	Four Threads, G (L_3+p)	Pitch Dia Gaging Step[b] Plus 3 Threads		Blank Length B	Notch Depth, J +0.005 −0.000
	Pitch Dia, E_3	Major Dia,[a] D_3			(L_3+L_1-p) Min	(L_3+L_1+p) Max		
1/16—27	0.2642	0.2815	0.216	0.1482	0.2341	0.3082	0.38	0.030
1/8 —27	0.3566	0.3738	0.309	0.1482	0.2356	0.3097	0.41	0.030
1/4 —18	0.4670	0.4928	0.409	0.2222	0.3389	0.4500	0.50	0.030
3/8 —18	0.6016	0.6275	0.542	0.2222	0.3511	0.4622	0.56	0.030
1/2 —14	0.7451	0.7783	0.676	0.2857	0.4628	0.6057	0.69	0.040
3/4 —14	0.9543	0.9876	0.886	0.2857	0.4818	0.6247	0.72	0.040
1 —11-1/2	1.1973	1.2379	1.118	0.3478	0.5739	0.7478	0.88	0.050
1-1/4 —11-1/2	1.5408	1.5814	1.462	0.3478	0.5939	0.7678	0.88	0.050
1-1/2 —11-1/2	1.7798	1.8203	1.701	0.3478	0.5939	0.7678	0.88	0.050
2 —11-1/2	2.2527	2.2932	2.174	0.3478	0.6099	0.7838	0.88	0.050
2 1/2 —8	2.6961	2.7543	2.590	0.5000	0.9320	1.1820	1.50	0.050
3 —8	3.3172	3.3754	3.214	0.5000	1.0160	1.2660	1.50	0.050

[a] Major diameter is based upon crest minimum truncation of 0.20 p.
[b] Maximum and minimum pitch-diameter steps are gaging limits. Notch formulas on drawing apply to all sizes.

DRYSEAL SAE SHORT TAPER PIPE THREAD (L_1 SHORT) STEP-LIMIT THIN-RING GAGES

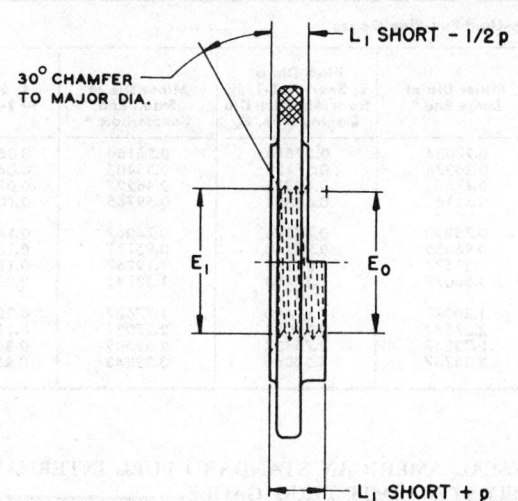

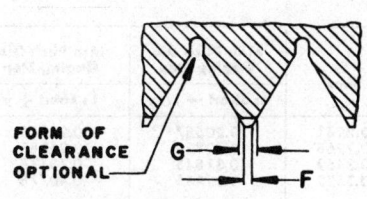

TABLE C8-1—THREAD FLATS

Threads per in.	F	G
27	0.0086	0.0107
18	0.0128	0.0160
14	0.0165	0.0206
11-1/2	0.0201	0.0251
8	0.0289	0.0361

TABLE C8—BASIC DIMENSIONS OF DRYSEAL SAE SHORT TAPER PIPE THREAD (L_1 SHORT) STEP-LIMIT THIN-RING GAGES

		(L_1 Short) Step-Limit Thin-Ring Gages					
Size	L_1 Short	Max Pitch Dia Gaging Step L_1 Short $- 1/2p$	Min Pitch Dia Gaging Step L_1 Short $+ p$	Pitch Dia, E_1	Minor Dia at Large End [a]	Pitch Dia at Min Pitch Dia Gaging Step, E_0	Minor Dia at Small End [a]
1/16—27	0.12296	0.10444	0.16000	0.28118	0.26392	0.27118	0.25392
1/8 —27	0.12446	0.10594	0.16150	0.37360	0.35634	0.36351	0.34625
1/4 —18	0.17224	0.14446	0.22780	0.49163	0.46574	0.47739	0.45150
3/8 —18	0.18444	0.15666	0.24000	0.62701	0.60112	0.61201	0.58712
1/2 —14	0.24857	0.21286	0.32000	0.77843	0.74514	0.75843	0.72514
3/4 —14	0.26757	0.23186	0.33900	0.98887	0.95558	0.96768	0.93439
1 —11-1/2	0.31304	0.26956	0.40000	1.23863	1.19810	1.21363	1.17310
1-1/4 —11-1/2	0.33304	0.28956	0.42000	1.58338	1.54285	1.55713	1.51660
1-1/2 —11-1/2	0.33304	0.28956	0.42000	1.82234	1.78181	1.79609	1.75556
2 —11-1/2	0.34904	0.30556	0.43600	2.29627	2.25574	2.26902	2.22849
2-1/2 —8	0.55700	0.49450	0.68200	2.76216	2.70391	2.71953	2.66128
3 —8	0.64100	0.57850	0.76600	3.38850	3.33025	3.34062	3.28237

[a] Minor diameter is based on crest minimum truncation of 0.20 p.

Marking—In addition to the regular markings, Dryseal SAE Short Taper Pipe Thread Ring Gages will be marked PTF—SAE Short (L_1 Short) on the entering side of gage.

Thread Form—The threads in all particulars excepting truncation shall conform to American Standard Taper Pipe Thread practice. Crests of threads at the minor diameter shall be truncated 0.20p minimum to 0.25p maximum, producing the minimum and maximum widths of flat specified in Table C8-1.

All other thread dimensions shall be within tolerances specified for the Dryseal American Standard Pipe Working Gages (ASA B2.2). Other gage details shall conform to American Gage Design Standard published in Commercial Standard CS8.

DRYSEAL SAE SHORT TAPER PIPE THREAD (L_2 SHORT) STEP-LIMIT FULL-RING GAGES

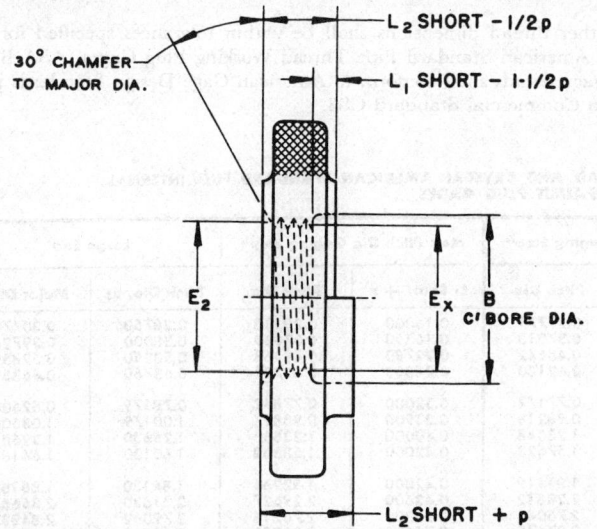

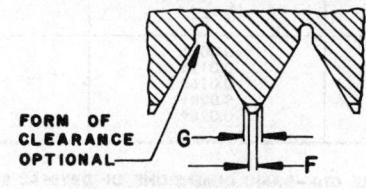

TABLE C9-1—THREAD FLATS

Threads per in.	F	G
27	0.0086	0.0107
18	0.0128	0.0160
14	0.0165	0.0206
11-1/2	0.0201	0.0251
8	0.0289	0.0361

Marking—In addition to the regular markings, Dryseal SAE Short Taper Pipe Thread Ring Gages will be marked PTF—SAE Short (L_2 Short) on the entering side of gage.

Thread Form—The threads in all particulars excepting truncation shall conform to American Standard Taper Pipe Thread practice. Crests of threads at the minor diameter shall be truncated 0.20p minimum to 0.25p maximum, producing the minimum and maximum widths of flat specified in Table C9-1.

All other thread dimensions shall be within tolerance specified for the Dryseal American Standard Pipe Working Gages (ASA B2.2). Other gage details shall conform to American Gage Design Standard published in Commercial Standard CS8.

TABLE C9—BASIC DIMENSIONS OF DRYSEAL SAE SHORT TAPER PIPE THREAD (L_2 SHORT) STEP-LIMIT FULL-RING GAGES

Size	L_2 Short	Max Pitch Dia Gaging Step L_2 Short $-1/2p$	Min Pitch Dia Gaging Step L_2 Short $+p$	Pitch Dia, E_2	Minor Dia at Large End [a]	Pitch Dia at L_2 Short $-1\text{-}1/2p$ from Min Pitch Dia Gaging Step, E_x	Minor Dia at Small End Counterbore [a]	L_1 Short $-1\text{-}1/2p$	B
1/16—27	0.2241	0.20557	0.26113	0.28750	0.27024	0.27886	0.26160	0.06740	0.38
1/8 —27	0.2268	0.20829	0.26385	0.38000	0.36274	0.37129	0.35403	0.06890	0.47
1/4 —18	0.3462	0.31845	0.40178	0.50250	0.47661	0.48816	0.46227	0.08891	0.59
3/8 —18	0.3522	0.32445	0.40778	0.63750	0.61161	0.62354	0.59765	0.10111	0.72
1/2 —14	0.4623	0.42657	0.53371	0.79179	0.75850	0.77396	0.74067	0.14143	0.88
3/4 —14	0.4743	0.43857	0.54571	1.00179	0.96850	0.98440	0.95111	0.16043	1.09
1 —11-1/2	0.5958	0.55235	0.68278	1.25630	1.21577	1.23320	1.19267	0.18260	1.34
1-1/4—11-1/2	0.6198	0.57635	0.70678	1.60130	1.56077	1.57794	1.53741	0.20260	1.69
1-1/2—11-1/2	0.6365	0.59305	0.72348	1.84130	1.80077	1.81690	1.77637	0.20260	1.94
2 —11-1/2	0.6695	0.62609	0.75652	2.31630	2.27577	2.29084	2.25031	0.21860	2.50
2-1/2— 8	1.0125	0.95000	1.13750	2.79062	2.73237	2.75434	2.69609	0.36950	2.94
3 — 8	1.0750	1.01250	1.20000	3.41562	3.35737	3.38068	3.32243	0.45350	3.56

[a] Minor diameter is based on crest minimum truncation of 0.20 p.

DRYSEAL SAE SHORT TAPER PIPE THREAD AND DRYSEAL AMERICAN STANDARD FUEL INTERNAL STRAIGHT PIPE THREAD (L_1 SHORT) STEP-LIMIT PLUG GAGES

Taper lock design, range 1/8 to 2 in., inclusive

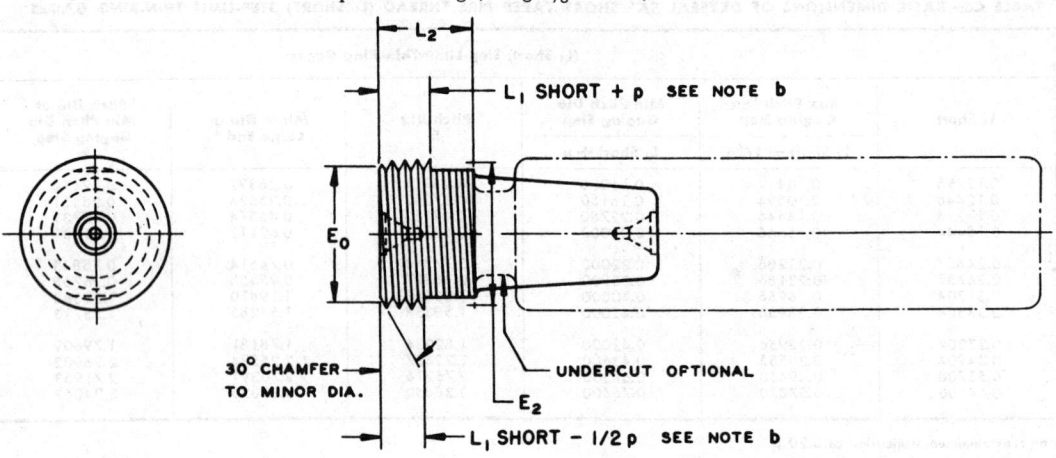

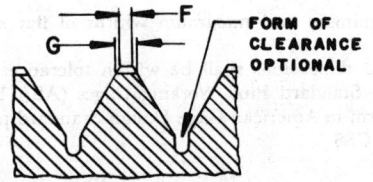

TABLE C10-1—THREAD FLATS

Threads per in.	F	G
27	0.0086	0.0107
18	0.0128	0.0160
14	0.0165	0.0206
11-1/2	0.0201	0.0251
8	0.0289	0.0361

Marking—In addition to the regular markings, Dryseal SAE Short Taper Pipe Thread L_1 Short Plug Gages will be marked PTF—SAE Short (L_1 Short). Dryseal American Standard Fuel Internal Straight Pipe Thread Taper Plug Gages will be marked NPSF (L_1 short).

Thread Form—The threads in all particulars excepting truncation shall conform to American Standard Taper Pipe thread practice. Crests of threads at major diameter shall be truncated 0.20p minimum to 0.25p maximum, producing the minimum and maximum widths of flat specified in Table C10-1.

All other thread dimensions shall be within tolerances specified for the Dryseal American Standard Pipe Thread Working Plug Gages (ASA B2.2). Other gage details shall conform to American Gage Design Standards published in Commercial Standard CS8.

TABLE C10—BASIC DIMENSIONS OF DRYSEAL SAE SHORT TAPER PIPE THREAD AND DRYSEAL AMERICAN STANDARD FUEL INTERNAL STRAIGHT PIPE THREAD (L_1 SHORT) STEP-LIMIT PLUG GAGES

Size	L_1 Short	L_2	Small End		Min Pitch Dia Gaging Step [b]		Max Pitch Dia Gaging Step [b]		Large End	
			Pitch Dia, E_0	Major Dia [a]	L_1 Short $-1/2p$	Pitch Dia	L_1 Short $+p$	Pitch Dia	Pitch Dia, E_2	Major Dia [a]
1/16—27	0.12296	0.26113	0.27118	0.28844	0.10444	0.27771	0.16000	0.28118	0.28750	0.30476
1/8 —27	0.12446	0.26385	0.36351	0.38077	0.10594	0.37013	0.16150	0.37360	0.38000	0.39726
1/4 —18	0.17224	0.40178	0.47739	0.50328	0.14446	0.48642	0.22780	0.49163	0.50250	0.52839
3/8 —18	0.18444	0.40778	0.61201	0.63790	0.15666	0.62180	0.24000	0.62701	0.63750	0.66339
1/2 —14	0.24857	0.53371	0.75843	0.79170	0.21286	0.77174	0.32000	0.77843	0.79179	0.82506
3/4 —14	0.26757	0.54571	0.96768	1.00095	0.23186	0.98218	0.33900	0.98887	1.00179	1.03506
1 —11-1/2	0.31304	0.68278	1.21363	1.25416	0.26956	1.23048	0.40000	1.23863	1.25630	1.29683
1-1/4—11-1/2	0.33304	0.70678	1.55713	1.59766	0.28956	1.57523	0.42000	1.58338	1.60130	1.64183
1-1/2—11-1/2 [c]	0.33304	0.72348	1.79609	1.83662	0.28956	1.81419	0.42000	1.82234	1.84130	1.88183
2 —11-1/2 [c]	0.34904	0.75652	2.26902	2.30955	0.30556	2.28812	0.43600	2.29627	2.31630	2.35683
2-1/2— 8 [c]	0.55700	1.13750	2.71953	2.77778	0.49450	2.75044	0.68200	2.76216	2.79062	2.84887
3 — 8 [c]	0.64100	1.20000	3.34062	3.39887	0.57850	3.37678	0.76600	3.38850	3.41562	3.47387

[a] Major diameter is based on crest minimum truncation of 0.20 p.
[b] Maximum and minimum pitch-diameter steps are gaging limits. Notch formulas on drawing apply to all sizes.
[c] For reference only above 1—11-1/2 NPSF.

DRYSEAL SAE SHORT TAPER PIPE THREAD (L_3 SHORT) LENGTH STEP-LIMIT PLUG GAGES

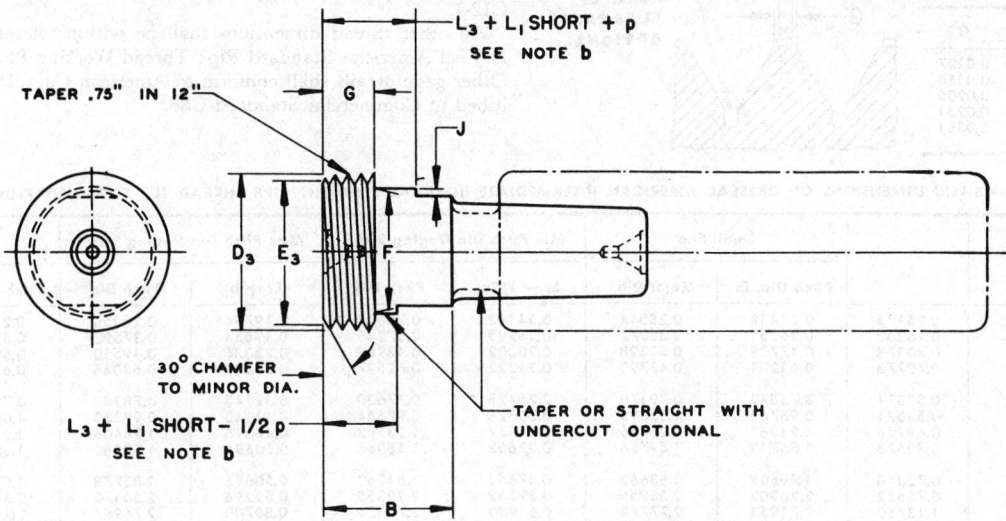

TABLE C11-1—THREAD FLATS

Threads per in.	F	G
27	0.0086	0.0107
18	0.0128	0.0160
14	0.0165	0.0206
11-1/2	0.0201	0.0251
8	0.0289	0.0361

Marking—In addition to the regular markings, Dryseal SAE Short Taper Pipe Thread (L_3) Plug Gages will be marked PTF—SAE Short (L_3 Short).

Thread Form—The threads in all particulars excepting truncation shall conform to American Standard Taper Pipe Thread practice. Crests of threads at major diameter shall be truncated 0.20p minimum to 0.25p maximum, producing the minimum and maximum widths of flat specified in Table C11-1.

All other thread dimensions shall be within tolerances specified for the Dryseal American Standard Pipe Thread Working Plug Gages (ASA B2.2). Other gage details shall conform to American Gage Design Standards published in Commercial Standard CS8.

TABLE C11—BASIC DIMENSIONS OF DRYSEAL SAE SHORT TAPER PIPE THREAD (L_3 SHORT) LENGTH STEP-LIMIT PLUG GAGES

Size	Small End Pitch Dia, E_3	Small End Major Dia,[a] D_3	Relief Dia, F [E_3 + (0.0625 × 4p) − Sharp-V Thread Height − 0.020 to 0.025 below Sharp Root] +0.005 −0.000	Four Threads, (G) ($L_2 + p$)	Pitch Dia Gaging Step[b] Plus 3 Threads ($L_2 + L_1$ Short − 1/2 p) Min	Pitch Dia Gaging Step[b] Plus 3 Threads ($L_2 + L_1$ Short + p) Max	Black Length, (B)	Notch Depth, (J) +0.005 −0.000
1/16—27	0.2642	0.2815	0.216	0.1482	0.2156	0.2711	0.38	0.030
1/8 —27	0.3566	0.3738	0.309	0.1482	0.2171	0.2726	0.41	0.030
1/4 —18	0.4670	0.4928	0.409	0.2222	0.3111	0.3945	0.50	0.030
3/8 —18	0.6016	0.6275	0.542	0.2222	0.3233	0.4067	0.56	0.030
1/2 —14	0.7451	0.7783	0.676	0.2857	0.4271	0.5343	0.69	0.040
3/4 —14	0.9543	0.9876	0.886	0.2857	0.4462	0.5533	0.72	0.040
1 —11-1/2	1.1973	1.2379	1.118	0.3478	0.5304	0.6609	0.88	0.050
1-1/4 —11-1/2	1.5408	1.5814	1.462	0.3478	0.5504	0.6809	0.88	0.050
1-1/2 —11-1/2	1.7798	1.8203	1.701	0.3478	0.5504	0.6809	0.88	0.050
2 —11-1/2	2.2527	2.2932	2.174	0.3478	0.5644	0.6969	0.88	0.050
2-1/2 —8	2.6961	2.7543	2.590	0.5000	0.8695	1.0570	1.50	0.050
3 —8	3.3172	3.3754	3.214	0.5000	0.9535	1.1410	1.50	0.050

[a] Major diameter is based upon crest minimum truncation of 0.20 p.
[b] Maximum and minimum pitch-diameter steps are gaging limits. Notch formulas on drawing apply to all sizes.

DRYSEAL AMERICAN INTERMEDIATE INTERNAL STRAIGHT PIPE THREAD (L_1) STEP-LIMIT PLUG GAGES

Taper lock design, range 1/8 to 1 in., inclusive

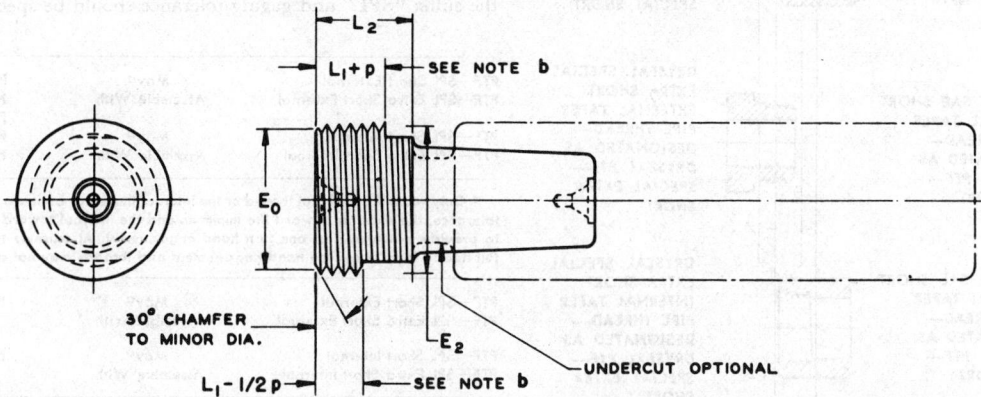

Marking—In addition to the regular markings, Dryseal American Intermediate Internal Straight Pipe Thread Taper Plug Gages will be marked NPSI (L_1).

Thread Form—The threads in all particulars excepting truncation shall conform to American Standard Taper Pipe Thread Practice. Crests of threads at major diameter shall be truncated 0.20p minimum to 0.25p maximum,

TABLE C12-1—THREAD FLATS

Threads per in.	F	G
27	0.0086	0.0107
18	0.0128	0.0160
14	0.0165	0.0206
11-1/2	0.0201	0.0251
8	0.0289	0.0361

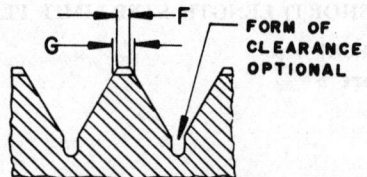

producing the minimum and maximum widths of flat specified in Table C12-1.

All other thread dimensions shall be within tolerances specified for the Dryseal American Standard Pipe Thread Working Plug Gages (ASA B2.2). Other gage details shall conform to American Gage Design Standards published in Commercial Standard CS8.

TABLE C12—BASIC DIMENSIONS OF DRYSEAL AMERICAN INTERMEDIATE INTERNAL STRAIGHT PIPE THREAD (L_1) STEP-LIMIT PLUG GAGES

Size	L_1	L_2	Small End		Min Pitch Dia Gaging Step[b]		Max Pitch Dia Gaging Step[b]		Large End	
			Pitch Dia, E_0	Major Dia[a]	$L_1 - 1/2p$	Pitch Dia	$L_1 + p$	Pitch Dia	Pitch Dia, E_2	Major Dia[a]
1/16—27	0.1600	0.26113	0.27118	0.28844	0.14148	0.28002	0.19704	0.28350	0.28750	0.30476
1/8 —27	0.1615	0.26385	0.36351	0.38077	0.14298	0.37245	0.19854	0.37592	0.38000	0.39726
1/4 —18	0.2278	0.40178	0.47739	0.50328	0.20002	0.48989	0.28336	0.49510	0.50250	0.52839
3/8 —18	0.2400	0.40778	0.61201	0.63790	0.21222	0.62527	0.29556	0.63048	0.63750	0.66339
1/2 —14	0.3200	0.53371	0.75843	0.79170	0.28428	0.77620	0.39143	0.78289	0.79179	0.82506
3/4 —14	0.3390	0.54571	0.96768	1.00095	0.30328	0.98664	0.41043	0.99333	1.00179	1.03506
1 —11-1/2	0.4000	0.68278	1.21363	1.25416	0.35652	1.23592	0.48696	1.24406	1.25630	1.29683
1-1/4 —11-1/2[c]	0.1200	0.70678	1.55713	1.59766	0.37652	1.58066	0.50696	1.58882	1.60130	1.64183
1-1/2 —11-1/2[c]	0.1200	0.72348	1.79609	1.83662	0.37652	1.81962	0.50696	1.82778	1.84130	1.88183
2 —11-1/2[c]	0.4360	0.75652	2.26902	2.30955	0.39252	2.29355	0.52296	2.30170	2.31630	2.35683
2-1/2 —8[c]	0.6820	1.13750	2.71953	2.77778	0.61950	2.75825	0.80700	2.76997	2.79062	2.84887
3 —8[c]	0.7660	1.20000	3.34062	3.39887	0.70350	3.38459	0.89100	3.39631	3.41562	3.47387

[a] Major diameter is based on crest minimum truncation of 0.20 p.
[b] Maximum and minimum pitch-diameter steps are gaging limits. Notch formulas on drawing apply to all sizes.
[c] For reference only

APPENDIX D—SPECIAL SHORT, SPECIAL EXTRA SHORT, FINE, AND SPECIAL DIAMETER PITCH COMBINATION DRYSEAL PIPE THREADS

General Information—The SAE Dryseal Pipe Thread Series are based on thread length. Full thread lengths and clearances for Dryseal Standard and SAE Short Series are shown in Tables 2, 3, and 4 of the standard and the differences between them are described in the text under the series headings. These full thread lengths and clearances should be used in design applications wherever possible.

Design limitations, economy of material, permanent installation or other limiting conditions may not permit the use of either of the full thread lengths and shoulder lengths in the preceding tables for the above thread series. To meet these conditions two special thread series have been established as shown in Fig. 1. The deviations from standard practice are described below.

Dryseal Special Short Taper Pipe Thread (PTF—SPL Short)—Threads of this series conform in all respects to the PTF—SAE Short threads except that the full thread length has been further shortened by eliminating one thread at the large end of external threads or eliminating one thread at the small end of internal threads. Gaging is the same as for PTF—SAE Short except the L_2 ring thread gage for external thread length and taper or the L_3 plug thread gage for internal thread length and taper cannot be used. Tolerance must be altered and co-ordinated as described in paragraph on Limitation of Assembly. The designation of this series thread is for example:

1/8—27 DRYSEAL PTF—SPL Short

Dryseal Special Extra Short Taper Pipe Thread (PTF—SPL Extra Short)—Threads of this series conform in all respects to the PTF—SAE Short threads except that the full thread length has been further shortened by eliminating two threads at the large end of external threads or eliminating two threads at the small end of internal threads. Gaging is the same as for PTF—SAE Short except the L_2 ring thread gage for external thread length and taper or the L_2 plug thread gage for internal thread length and taper cannot be used. Tolerance must be altered and co-ordinated as described in paragraph on Limitation of Assembly. The designation of this series thread is for example:

1/8—27 DRYSEAL PTF—SPL Extra Short

Limitation of Assembly—Standard combinations and applications of the various series Dryseal Pipe Threads are given in the preceding thread descriptions. However, where special combinations are used, additional considerations as outlined below must be observed. These should be designated with the suffix "SPL" and gaging tolerance should be specified.

PTF—SPL Short External	May[a]	PTF—SAE Short Internal
PTF—SPL Extra Short External	Assemble With	NPSF Internal
		PTF—SPL Short Internal
PTF—SPL Short Internal	May[a]	PTF—SPL Extra Short Internal
PTF—SPL Extra Short Internal	Assemble With	PTF—SAE Short External

[a] Only when the external thread or the internal thread or both are held closer than the standard tolerance, the external toward the minimum and the internal toward the maximum pitch diameter to provide a minimum of one turn hand engagement. At extreme tolerance limits the shortened full thread lengths reduce hand engagement and threads may not start

PTF—SPL Short External	May[a]	NPTF or NPSI Internal
PTF—SPL Extra Short External	Assemble With	
PTF—SPL Short Internal	May[a]	NPTF External
PTF—SPL Extra Short Internal	Assemble With	

[a] Only when both the internal thread and the external thread are held closer than the standard tolerance, the internal toward the minimum and the external toward the maximum pitch diameter to provide a minimum of two turns or wrench make up and sealing. At extreme tolerance limits the shortened full thread lengths reduce wrench make up and threads may not seal.

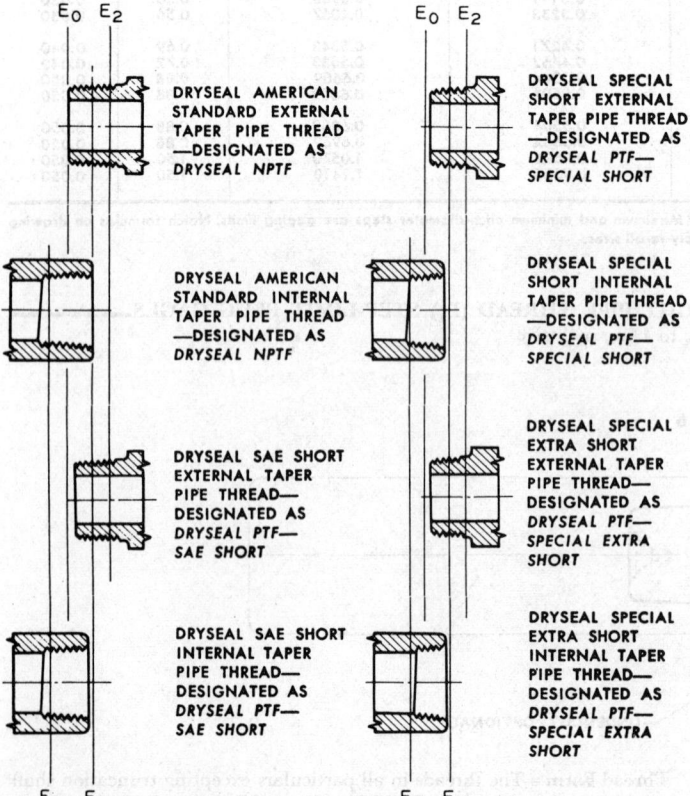

FIG. 1—THREAD LENGTH AND DESIGNATION

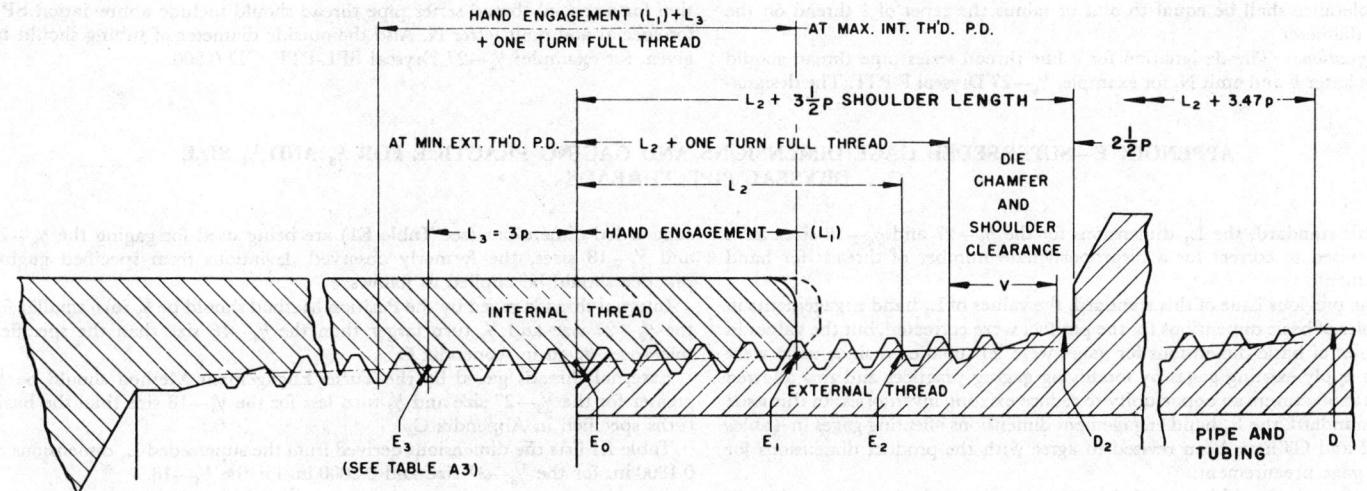

TABLE D1—BASIC DIMENSIONS OF DRYSEAL TAPER PIPE THREAD, FINE, F-PTF

F-PTF Size (Fine)	Pitch, p	Pitch Dia at Small End of External Thread, E_0	Pitch Dia at Large End of Internal Thread, E_1	Pitch Dia at Large End of External Thread, E_2	Pitch Dia at Small End of Internal Thread, E_3	Hand Engagement, L_1		Length of Full Thread,[a,b] Internal (L_1+L_3) and External (L_2)		Vanish Threads- V Plus Full Thread Tolerance Plus Shoulder Clearance, ($V+1p+1/2p$)		Shoulder Length, $L_2+3\text{-}1/2p$	Thread for Draw		OD of Fitting, D_2	OD of Pipe, D
	in.	in.	in.	in.	in.	in.	Thread	in.	Thread	in.	Thread	in.	in.	Thread	in.	in.
1	2	3	4	5	6	7	8	9	10	11	12	13	14	15	16	17
1/4—27	0.03704	0.49826	0.50807	0.51501	0.49132	0.157	4.23	0.268	7.23	0.1296	3.5	0.3975	0.1111	3.0	0.546	0.540
3/8—27	0.03704	0.63301	0.64307	0.65001	0.62607	0.161	4.34	0.272	7.34	0.1296	3.5	0.4015	0.1111	3.0	0.681	0.675
1/2—18	0.05556	0.77655	0.79205	0.80249	0.76613	0.248	4.47	0.415	7.47	0.1944	3.5	0.6096	0.1667	3.0	0.850	0.840
3/4—18	0.05556	0.98597	1.00210	1.01247	0.97555	0.258	4.64	0.424	7.64	0.1944	3.5	0.6189	0.1667	3.0	1.060	1.050
1 —14	0.07143	1.23173	1.25342	1.26679	1.21834	0.347	4.85	0.561	7.85	0.2500	3.5	0.8109	0.2143	3.0	1.327	1.315
1-1/4—14	0.07143	1.57550	1.59837	1.61181	1.56211	0.366	5.13	0.581	8.13	0.2500	3.5	0.8306	0.2143	3.0	1.672	1.660
1-1/2—14	0.07143	1.81464	1.83839	1.85176	1.80125	0.380	5.32	0.594	8.32	0.2500	3.5	0.8443	0.2143	3.0	1.912	1.900
2 —14	0.07143	2.28794	2.31338	2.32675	2.27455	0.407	5.70	0.621	8.70	0.2500	3.5	0.8714	0.2143	3.0	2.387	2.375

[a] External thread tabulated full thread lengths include chamfers not exceeding one and one-half pitches (threads) length.

[b] Internal thread tabulated full thread lengths do not include countersink beyond the intersection of the pitch line and the chamfer cone (gaging reference point).

TABLE D2—BASIC DIMENSIONS OF DRYSEAL TAPER PIPE THREAD, SPECIAL, SPL-PTF, FOR THIN WALL NOMINAL SIZE OD TUBING

Tubing Dia,[c] D	Threads per in.	Pitch, p	Pitch Dia at Small End of External Thread, E_0	Pitch Dia at Large End of Internal Thread, E_1	Pitch Dia at Large End of External Thread, E_2	Pitch Dia at Small End of Internal Thread, E_3	Hand Engagement, L_1		Length of Full Thread,[a,b] Internal (L_1+L_3) and External (L_2)		Thread for Draw	
in.		in.	in.	in.	in.	in.	in.	Thread	in.	Thread	in.	Thread
1	2	3	4	5	6	7	8	9	10	11	12	13
1/2	27	0.03704	0.45833	0.46806	0.47500	0.45139	0.1556	4.2	0.2667	7.2	0.1111	3.0
5/8	27	0.03704	0.58310	0.59306	0.60000	0.57616	0.1593	4.3	0.2704	7.3	0.1111	3.0
3/4	27	0.03704	0.70787	0.71806	0.72500	0.70093	0.1630	4.4	0.2741	7.4	0.1111	3.0
7/8	27	0.03704	0.83264	0.84306	0.85000	0.82570	0.1667	4.5	0.2778	7.5	0.1111	3.0
1	27	0.03704	0.95740	0.96805	0.97500	0.95046	0.1704	4.6	0.2815	7.6	0.1111	3.0

[a] External thread tabulated full thread lengths include chamfers not exceeding one and one-half pitches (threads) length.

[b] Internal thread tabulated full thread lengths do not include countersink beyond the intersection of the pitch line and the chamfer cone (gaging reference point).

[c] This denotes nominal outside diameter of tubing and should not be confused with nominal pipe diameter and thread designations.

Fine Thread Series—The need for finer pitches for nominal pipe sizes has brought into use applications of 27 threads per in. to ¼ and ⅜ in. pipe sizes. There may be other needs which require finer pitches for larger pipe sizes. It is recommended that the existing threads per in. be applied to the next size larger pipe size for a fine thread series such as shown in Table D1. This series applies to external and internal threads of full length and is suitable for applications where threads finer than NPTF are required.

Special Thread Series—Other applications of diameter-pitch combinations have also come into use where taper pipe threads are applied to nominal size thin wall tubing such as shown in Table D2. This series applies to external and internal threads of full length and is applicable to thin wall nominal outside diameter tubing. The pitch is uniform at 27 threads per in. Dimensions of other combinations of diameter and pitch, in addition to those listed in Table D2, may be developed by the use of formulae.

Formulae for Diameter and Length of Thread—Basic diameter and length of thread for sizes of Dryseal Taper Pipe Thread Fine (F-PTF), and Dryseal Taper Pipe Thread Special (SPL—PTF) given in Tables D1 and D2 are based on the following formulae:

D = outside diameter of pipe or tubing (in.)
p = pitch of thread (in.)
Diametral taper = 0.75 in. per 12.00 in. of length
Basic pitch diameter at small end of external thread
$E_0 = D - (0.05D + 1.1)p$
Basic pitch diameter at large end of internal thread
$E_1 = E_0 + 0.0625 L_1 = D - 0.8625 p$
Basic pitch diameter at large end of external thread
$E_2 = E_0 + 0.0625 L_2 = D - 0.675 p$
Basic pitch diameter at small end of internal thread
$E_3 = E_0 - 0.0625 L_3 = D - (0.05D + 1.2875) p$
Basic length of thread for hand engagement
$L_1 = (0.8D + 3.8) p$
Basic length of full and effective thread
$L_2 = (0.8D + 6.8) p$
Basic length of internal thread from end of hand engagement (E_0) to small end of internal thread (E_3)
$L_3 = 3p$

Tolerance shall be equal to plus or minus the taper of 1 thread on the diameter

Designations—The designation for a fine thread series pipe thread should include letter F and omit N, for example: ¼—27 Dryseal F-PTF. The designation for a special thread series pipe thread should include abbreviation SPL for special and omit letter N. Also the outside diameter of tubing should be given, for example: ½—27 Dryseal SPL-PTF, OD 0.500.

APPENDIX E—SUPERSEDED GAGE DIMENSIONS AND GAGING PRACTICE FOR ⅛ AND ¼ SIZE DRYSEAL PIPE THREADS

In this standard, the L_1 dimensions for the ⅛—27 and ¼—18 sizes have been revised to correct for a disproportionate number of threads for hand engagement.

In the previous issue of this standard, the values of L_1 hand engagements in the tables of basic dimensions for the product were corrected, but the values in the tables of basic dimensions for gages were left unaltered since users were able to apply existing gages by modifying gaging practices and this allowed gage manufacturers an opportunity to reduce existing inventories. In this issue of the standard, the L_1 hand engagement dimensions affecting gages in Tables C1, C2 and C3 have been revised to agree with the product dimensions for future gage procurement.

Therefore, it should be noted that where basic-notch thread gages having superseded dimensions (see Table E1) are being used for gaging the ⅛—27 and ¼—18 sizes, the formerly observed deviations from specified gaging practice should be applied as follows:

Internal threads gaged by the Position Method should be ½ turn smaller for the ⅛—27 size and ½ turn larger than the ¼—18 size than the specified tolerances given in Appendix C.

External threads gaged by the Turns Engagement Method should be ½ greater for the ⅛—27 size and ½ turn less for the ¼—18 size than the basic turns specified in Appendix C.

Table E1 lists the dimensions derived from the superseded L_1 dimensions of 0.1800 in. for the ⅛—27 size and 0.2000 in. for the ¼—18.

TABLE E1—BASIC DIMENSIONS OF SUPERSEDED BASIC-NOTCH GAGES

Size	L_2 Ring Gage			L_1 Plug and Ring Gages		L_1 Ring Gage	L_1 Plug Gage	L_3 Plug Gage
	Pitch Dia at L_1—p (E_1)	Minor Dia at L_1—p	L_1—p	L_1	Pitch Dia (E_1)	Minor Dia at Large End	Major Dia at Gaging Notch	3 Threads Plus L_1 (L_3+L_1)
1/8—27	0.37244	0.35518	0.14296	0.1800	0.37476	0.35750	0.39202	0.2911
1/4—18	0.48642	0.46053	0.14444	0.2000	0.48989	0.46400	0.51578	0.3667

16 Fasteners

SPRING NUTS
—SAE J891 JUN93
SAE Standard

Report of Fasteners Committee approved August 1964 and revised June 1970. Completely revised by the Iron and Steel Technical Committee SC8—Carbon and Alloy Steel Hardenability June 1993.

1. Scope—Included herein are complete general and dimensional specifications for metric and inch types of spring nuts recognized as SAE standard. These nuts are intended for general use where the engagement of a single thread on the mating screw is considered adequate for the application.

It should be noted that spring nuts having other configurations, dimensions, provisions for ground, etc., are available and manufacturers should be consulted.

2. References

2.1 Applicable Document—The following publication forms a part of this specification to the extent specified herein. The latest issue of SAE publications shall apply.

2.1.1 SAE PUBLICATION—Available from SAE, 400 Commonwealth Drive, Warrendale, PA 15096-0001.

SAE J478—Slotted and Recessed Head Screws (ANSI/ASME B18.6.5M & 18.6.7M)

3. General Specifications

3.1 Dimensional Tolerances—Dimensions and tolerances are given in both metric and inch units as designated. In many cases the metric units have been rounded to reflect metric modules rather than being true soft conversions of the corresponding inch dimensions. Tolerance on dimensions in Tables 1 to 9 and Figures 1 to 16 shall be ±0.25 mm (±0.010 in) unless otherwise specified.

3.2 Boss Detail—The detail of boss shall be such as to assemble readily and function satisfactorily with the specified screw and meet the performance requirements of this specification except as indicated otherwise.

Both the type "P" and "T" bosses are designed to function with all spaced threaded tapping screws in sizes 3.5 × 1.27 through 6.3 × 1.81 or (#6-20 through 1/4-14) with the exception of the type "T" boss of the 6.3 (1/4). The boss has been designed to perform satisfactorily with either a type AB or B tapping screw.

The sides of the Type P boss (Figure 1) shall be formed to provide an opening conforming to the helix of the mating thread. The opening shall be round and equal to, or slightly larger than, the minor diameter of the mating thread.

The prongs of the Type T boss (Figure 2) shall be formed to provide a circular opening conforming to the helix of the mating thread and, at the opening, the prongs shall be normal to the axis of the mating thread. The round portion of the opening shall be equal to, or slightly larger than, the minor diameter of the mating thread.

The size and formation of the helical opening (see Table 1) shall be such as to permit ready assembly of the specified screw or gage when inserted from the base of the boss at 90 degrees to the plane of the nut, or component thereof which contains the boss. For machine screw threads, basic GO thread plug gages shall be used to check assembly. For Type A and Type B pitch tapping screw threads, special gages conforming to the maximum limits of the screws may be used in place of the specified screws to check assembly.

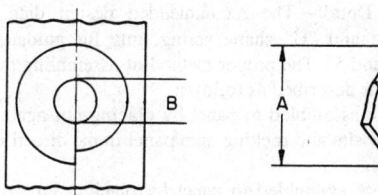

FIGURE 1—TYPE P BOSS

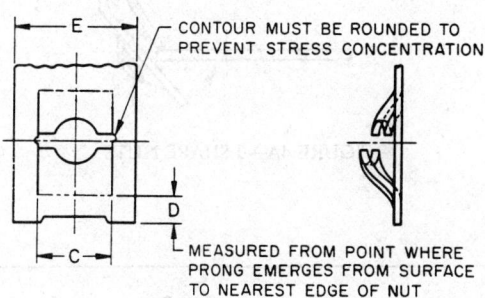

FIGURE 2—TYPE T BOSS

3.3 Retaining Extrusion Detail—The size and configuration of the extrusion in the lower leg of "J" shape and "U" shape spring nuts shall be such that nuts will meet the performance requirements of this specification. The size and relative location of the hole and extrusion to the boss shall be such that when nut is assembled onto a test panel having minimum hole size, located at maximum edge distance, the extrusion will snap into the hole and permit the specified screw of maximum size (or special threaded plug gage, Figure 3) to be assembled into the boss normal to the base of bass with interference at the extrusion or the sides of either hole. The screw or gage is to be entered into the boss until the head of the screw or shoulder on gage lightly contacts the bottom of the lower leg. The extrusion shall have a uniform shape and blend evenly from the specified height at point X into the upper surface of the lower leg at points Y and Y' as shown. The critical edges of the extrusion shall be free from burrs which would cause interference as spring nut is assembled to panel.

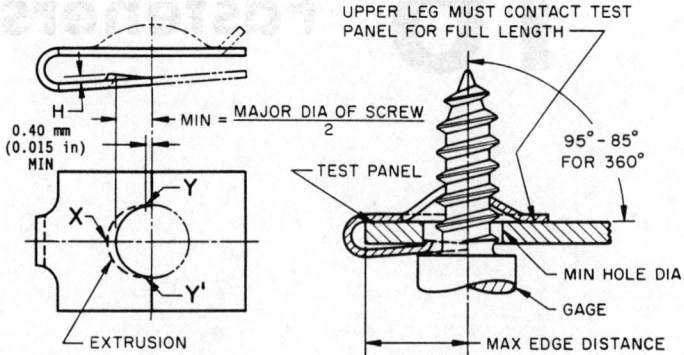

FIGURE 3—OPTIONAL EXTRUSION AND GAGING DETAIL

3.4 Material—Spring steel; SAE 1050 or higher carbon; suitably processed to meet the performance requirements of this specification.

3.5 Hardness—Hardness shall be as specified in Table 3.

3.6 Finish—Spring nuts are normally supplied with corrosion-resistant finish as specified by the purchaser. Nuts subjected to corrosion preventive treatment which might induce hydrogen embrittlement shall be baked or otherwise treated to obviate such embrittlement.

3.7 Workmanship—Spring nuts shall be free from cracks, burrs, splits, loose scale, or any other defects that might affect their serviceability.

3.8 Performance—Spring nuts shall perform in accordance with the requirements specified in Table 2 except as indicated otherwise.

3.9 Assembly Detail—The recommended design data pertaining to assembly of "J" shape and "U" shape spring nuts for guidance of users is presented in Tables 4 and 5. The proper method of assembling these corrosion-resistant nuts to panels is described as follows:

"J" shape nuts are assembled to panel by placing nut against the edge of the panel as shown opposite and rocking onto panel in the direction indicated by the arrow. See Figure 4A.

"U" shape nuts are assembled to panel by placing nut over edge of the panel as shown opposite and pushing onto the panel in the direction indicated by the arrow. See Figure 4B.

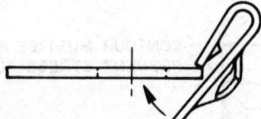

FIGURE 4A—J SHAPE NUTS

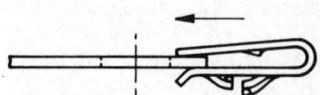

FIGURE 4B—U SHAPE NUTS

4. Tests and Test Fixtures for Evaluating Spring Nut Performance—Spring nuts shall be subjected to the following tests to determine conformance with the performance requirements specified in Table 2 except as indicated otherwise.

4.1 Test Plates and Screws for Tests—To assure uniformity of test results, the test plates and screws used for the tests shall conform to the following specifications.

Test plates shall have boundary dimensions and hole sizes as depicted in Figure 5. The thickness of test plates shall be equal to the mean of the specified panel range within a tolerance of ±0.03 mm (±0.001 in). The holes in test plates shall be located at the maximum edge distance specified for the particular spring nut within a tolerance of ±0.03 mm (±0.001 in). Test plates and panels shall have a minimum hardness of Rockwell C 50-54 (HV 515-580).

The screws used for test purposes shall conform to the specifications in SAE J478 (ANSI/ASME B18.6.5M and 18.6.7M) for the respective sizes and types. They shall be Hexagon Head style and a length which is compatible with the test fixture, with a 72 h salt spray corrosion resistant phosphate finish (ASTM B 117).

4.2 Torque Tests—Spring nut samples shall be assembled with a test screw onto a test plate and tightened to the recommended installation torque. For wide-range design spring nuts, this test shall be performed with a device capable of measuring the clamp load developed and, when assembly is tightened to the recommended installation torque, the clamp load obtained shall not be less than the value tabulated.

Upon disassembly, the boss shall return to a position that will accept reentry of the test screw.

The spring nut, when reassembled and tightened on the test plate, shall not strip the threads on the screw nor fail the nut boss at less than the ultimate torque specified.

4.3 Tensile Tests—When the spring nut on a test plate is assembled to suitable back-up plates at the recommended installation torque and pulled in a tensile testing machine, the spring nut shall meet the ultimate strengths specified.

The ultimate strength shall be considered reached when the boss or the thread on the screw is destroyed. In performing tensile test, care should be taken to assure there is no interference between the screw and the holes in the plates. A typical tensile test fixture is illustrated in Figure 6.

4.4 Preassembly and Retention Test—The "J" shape and "U" shape spring nuts shall preassemble onto test panels of thickness equal to the two extremes of the panel ranges specified, having minimum holes, located at the maximum edge distance. The extrusion in the lower leg shall snap into the hole and when nuts are so assembled, a pull force of 13.3 N (3 lb) minimum applied parallel to the upper leg in line with the axis of the nut shall be required to remove the nut from the panel.

TABLE 1—DETAIL OF BOSSES

Screw Thread Type	Screw Nominal Size mm	Type P Screw Nominal Size in	Type P Boss A Base Dia Ref mm	Type P Boss A Base Dia Ref in	Type P Boss B Hole Dia mm	Type P Boss B Hole Dia in	Type P Boss E Min Blank Width mm	Boss E Min Blank Width in
Machine	M3.5 × 0.6	6-32	6.6	0.26	2.6	0.104	8.6	0.340
Machine	M4 × 0.7	8-32	5.8	0.23	3.3	0.130	10.3	0.406
Machine	M5 × 0.8	10-24	7.1	0.28	3.6	0.143	12.7	0.500
Machine		1/4-20	9.4	0.37	4.9	0.193	14.3	0.562
Tapping	3.5	#6	6.4	0.25	2.7	0.105	8.6	0.340
Tapping	4.2	#8	7.1	0.28	3.1	0.123	10.3	0.406
Tapping	4.8	#10	7.1	0.28	3.6	0.142	12.7	0.500
Tapping	6.3	1/4 #14	9.6	0.38	4.9	0.193	14.3	0.562

TABLE 1—DETAIL OF BOSSES (CONTINUED)

Screw Thread Type	Screw Nominal Size mm	Screw Nominal Size in	Type T Boss C Width of Shear Basic mm	Type T Boss C Width of Shear Basic in	Type T Boss D[1] End of Slit to Edge Min mm	Type T Boss D[1] End of Slit to Edge Min in	Type T Boss E Min Blank Width mm	Type T Boss E Min Blank Width in
Machine	M3.5 × 0.6	6-32	4.0	0.157	1.3	0.050	8.0	0.312
Machine	M4 × 0.7	8-32	4.7	0.184	1.3	0.050	10.3	0.406
Machine	M5 × 0.8	10-24	5.3	0.210	1.3	0.050	9.5	0.375
Machine		1/4-20	6.9	0.270	2.3	0.090	12.7	0.500
Tapping	3.5	#6	4.0	0.157	1.3	0.050	8.0	0.312
Tapping	4.2	#8	4.7	0.184	1.5	0.060	10.3	0.406
Tapping	4.8	#10	5.3	0.210	2.0	0.080	12.7	0.500
Tapping	6.3	1/4 #14	6.9	0.270	2.3	0.090	14.3	0.562

NOTE 1—The tabulated values are applicable to standard spring nuts only. This factor shall be sufficient to meet the performance requirements for torque, tensile strength, and vibration as set forth in Table 2.

TABLE 2—PERFORMANCE REQUIREMENTS FOR SPRING NUTS

Screw Thread Type	Screw Nominal Size mm	Screw Nominal Size in	Recommended Installation Torque-Max N·m	Recommended Installation Torque-Max lb-in	Clamp-Load at Recommended Installation Torque-Min kN	Clamp-Load at Recommended Installation Torque-Min lb	Destructive Torque Min N·m	Destructive Torque Min lb-in
Machine	M3.5 × 0.6	6-32	0.6	6	0.36	80	1.0	8
Machine	M4 × 0.7	8-32	1.0	8	0.44	100	1.2	10
Machine	M5 × 0.8	10-24	1.6	14	0.62	140	2.0	17
Machine		1/4-20	4.0	35	1.47	330	5.0	45
Tapping	3.5	#6	1.4	12	1.06	240	2.0	17
Types	4.2	#8	2.2	20	1.78	400	2.8	25
AB & B	4.8	#10	4.0	35	2.45	550	4.9	44
	6.3	#14-1/4	7.0	60	3.34	750	9.0	80

TABLE 2—PERFORMANCE REQUIREMENTS FOR SPRING NUTS (CONTINUED)

Screw Thread Type	Screw Nominal Size mm	Screw Nominal Size in	Ultimate Tensile Strength Min Type P kN	Ultimate Tensile Strength Min Type P lb	Ultimate Tensile Strength Min Type T kN	Ultimate Tensile Strength Min Type T lb
Machine	M3.5 × 0.6	6-32	0.69	155	0.69	156
Machine	M4 × 0.7	8-32	0.89	200	0.84	189
Machine	M5 × 0.8	10-24	1.40	315	1.22	274
Machine		1/4-20	2.54	570	2.45	550
Tapping	3.5 × 1.3	#6-25	2.22	500	1.89	425
Types	4.2 × 1.4	#8-18	2.76	620	2.38	534
AB & B	4.8 × 1.6	#10-16	4.45	1000	2.99	672
	6.3 × 1.8	1/4-14	5.03	1130	5.15	1158

TABLE 3—FASTENER HARDNESS

Material Thickness mm	Material Thickness in	Rockwell Scale	Dial Reading	Conversion to Rockwell C Scale	Conversion to Vickers
up to 0.41	up to 0.016	15 N	80.4 to 85.5	40 to 50	390 to 515
0.43 to 0.61	0.017 to 0.024	30 N	59.5 to 68.5	40 to 50	390 to 515
0.64 to 0.99	0.025 to 0.039	45 N	43.1 to 55.0	40 to 50	390 to 515
1.02 and over	0.040 and over	C	40 to 50	40 to 50	390 to 515

16.04

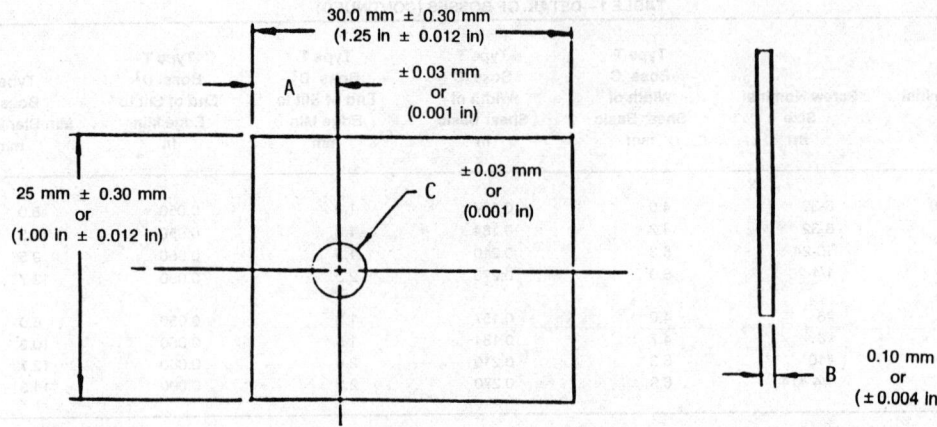

Machine and Tapping Screws Size mm	Machine and Tapping Screws Size in	A Edge to Center	B Plate Thickness	C Hole Dia mm	C Hole Dia in
3.5	#6	Equals Maximum	Equals Mean	6.35	0.250
4.2	#8	Edge Distance	of Panel Range	7.14	0.281
4.8	#10	Specified in	Specified in	7.92	0.312
6.3	#14-1/4	Assembly Data or on Part Drawing	Dimensional Tables	9.52	0.375

FIGURE 5—TEST PLATES

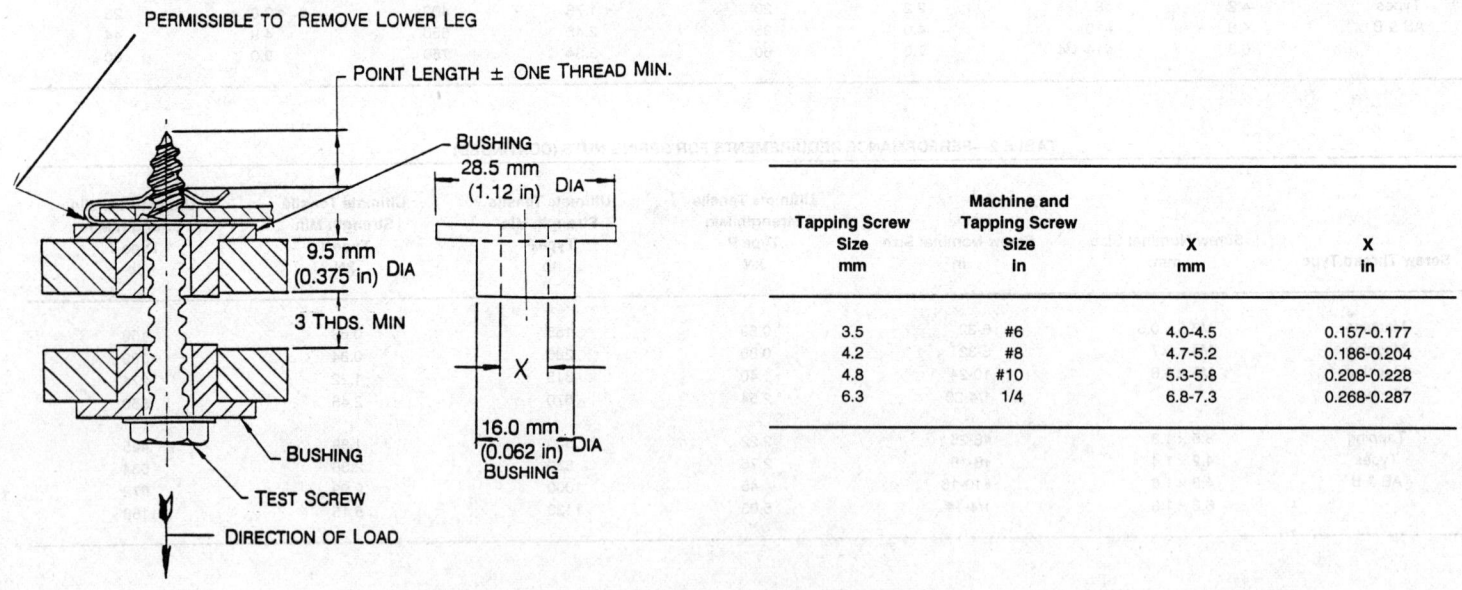

Tapping Screw Size mm	Machine and Tapping Screw Size in	X mm	X in
3.5	#6	4.0-4.5	0.157-0.177
4.2	#8	4.7-5.2	0.186-0.204
4.8	#10	5.3-5.8	0.208-0.228
6.3	1/4	6.8-7.3	0.268-0.287

FIGURE 6—TYPICAL TENSILE TEST FIXTURE

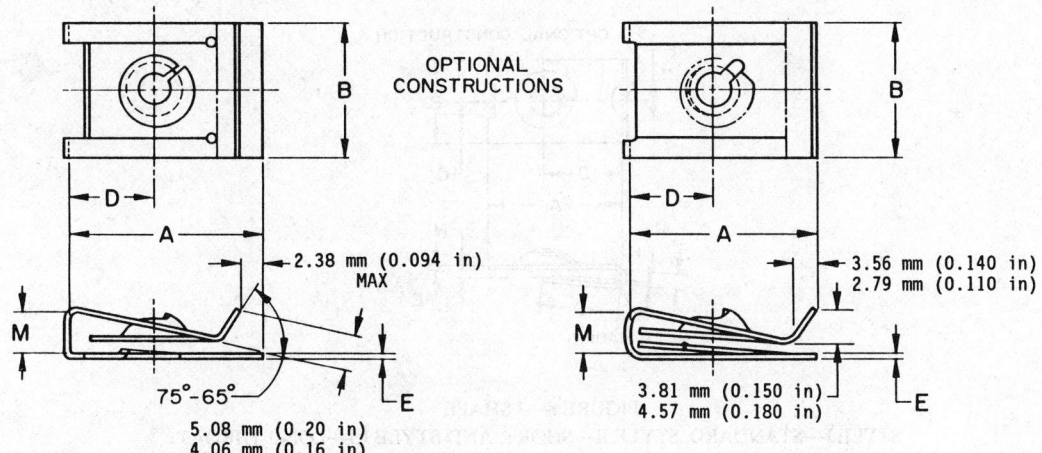

FIGURE 7—U SHAPE
SHORT THREAD STYLE AND LONG THROAT STYLE

TABLE 4—DIMENSIONS OF U SHAPE WIDE-RANGE DESIGN SPRING NUTS WITH OR WITHOUT RETAINING EXTRUSION (FIGURE 7)

Tapping Screw Size[1] mm	Tapping Screw Size[1] in	Panel Thickness Range mm	Panel Thickness Range in	Style	A Leg Length +0.38 -0.76 m	A Leg Length +0.015 -0.030 in	B Nut Width ±0.38	B Nut Width ±0.015	D Throat Depth mm	D Throat Depth in	E Stock Thickness mm	E Stock Thickness in	M Width at Fold mm	M Width at Fold in
3.5	#6	0.64	0.025	Short	19.56	0.770	13.58	0.535	8.64	0.34	0.46	0.018	3.81	0.150
3.5	#6	0.64	0.025	Short	19.56	0.770	13.58	0.535	9.14	0.36	0.46	0.018	3.81	0.150
3.5	#6	3.66	0.150	Long	25.4	1.000	13.58	0.535	14.48	0.57	0.68	0.027	4.04	0.190
3.5	#6	3.66	0.150	Long	25.4	1.000	13.58	0.535	15.24	0.60	0.68	0.027	4.04	0.190
3.5	#6	3.18	0.125	Short	19.56	0.770	13.58	0.535	8.64	0.34	0.46	0.018	6.60	0.260
3.5	#6	3.18	0.125	Short	19.56	0.770	13.58	0.535	9.14	0.36	0.46	0.018	6.60	0.260
3.5	#6	6.35	0.250	Long	25.4	1.000	13.58	0.535	14.48	0.57	0.68	0.027	7.11	0.280
3.5	#6	6.35	0.250	Long	25.4	1.000	13.58	0.535	15.24	0.60	0.68	0.027	7.11	0.280
4.2	#8	0.64	0.025	Short	19.56	0.770	13.58	0.535	8.64	0.34	0.58	0.023	3.81	0.150
4.2	#8	0.64	0.025	Short	19.56	0.770	13.58	0.535	9.14	0.36	0.58	0.023	3.81	0.150
4.2	#8	3.66	0.150	Long	25.4	1.000	13.58	0.535	14.48	0.57	0.76	0.030	4.04	0.190
4.2	#8	3.66	0.150	Long	25.4	1.000	13.58	0.535	15.24	0.60	0.76	0.030	4.04	0.190
4.2	#8	3.18	0.125	Short	19.56	0.770	13.58	0.535	8.64	0.34	0.58	0.023	6.60	0.260
4.2	#8	3.18	0.125	Short	19.56	0.770	13.58	0.535	9.14	0.36	0.58	0.023	6.60	0.260
4.2	#8	6.35	0.250	Long	25.4	1.000	13.58	0.535	14.48	0.57	0.76	0.030	7.11	0.280
4.2	#8	6.35	0.250	Long	25.4	1.000	13.58	0.535	15.24	0.60	0.76	0.030	7.11	0.280
4.8	#10	0.64	0.025	Short	19.56	0.770	16.12	0.635	8.64	0.34	0.71	0.028	3.81	0.150
4.8	#10	0.64	0.025	Short	19.56	0.770	16.12	0.635	9.14	0.36	0.71	0.028	3.81	0.150
4.8	#10	3.66	0.150	Long	25.4	1.000	16.12	0.635	14.48	0.57	0.88	0.035	4.04	0.190
4.8	#10	3.66	0.150	Long	25.4	1.000	16.12	0.635	15.24	0.60	0.88	0.035	4.04	0.190
4.8	#10	3.18	0.125	Short	19.56	0.770	16.12	0.635	8.64	0.34	0.71	0.028	6.60	0.260
4.8	#10	3.18	0.125	Short	19.56	0.770	16.12	0.635	9.14	0.36	0.71	0.028	6.60	0.260
4.8	#10	6.35	0.250	Long	25.4	1.000	16.12	0.635	14.48	0.57	0.88	0.035	7.11	0.280
4.8	#10	6.35	0.250	Long	25.4	1.000	16.12	0.635	15.24	0.60	0.88	0.035	7.11	0.280
6.3	1/4	0.64	0.025	Short	19.56	0.770	16.12	0.635	8.64	0.34	0.84	0.033	3.81	0.150
6.3	1/4	0.64	0.025	Short	19.56	0.770	16.12	0.635	9.14	0.36	0.84	0.033	3.81	0.150
6.3	1/4	3.66	0.150	Long	25.4	1.000	16.12	0.635	14.48	0.57	0.99	0.039	4.04	0.190
6.3	1/4	3.66	0.150	Long	25.4	1.000	16.12	0.635	15.24	0.60	0.99	0.039	4.04	0.190
6.3	1/4	3.18	0.125	Short	19.56	0.770	16.12	0.635	8.64	0.34	0.84	0.033	6.60	0.260
6.3	1/4	3.18	0.125	Short	19.56	0.770	16.12	0.635	9.14	0.36	0.84	0.033	6.60	0.260
6.3	1/4	6.35	0.250	Long	25.4	1.000	16.12	0.635	14.48	0.57	0.99	0.039	7.11	0.280
6.3	1/4	6.35	0.250	Long	25.4	1.000	16.12	0.635	15.24	0.60	0.99	0.039	7.11	0.280

[1] See Boss Detail under General Specifications for applicability of types and sizes. Type P spring nuts of similar proportions are also available in respective machine screw sizes 3.5 mm (#6), 4.2 mm (#8), 4.8 mm (#10), and 6.3 mm (1/4 in), and manufacturers should be consulted for dimensions.

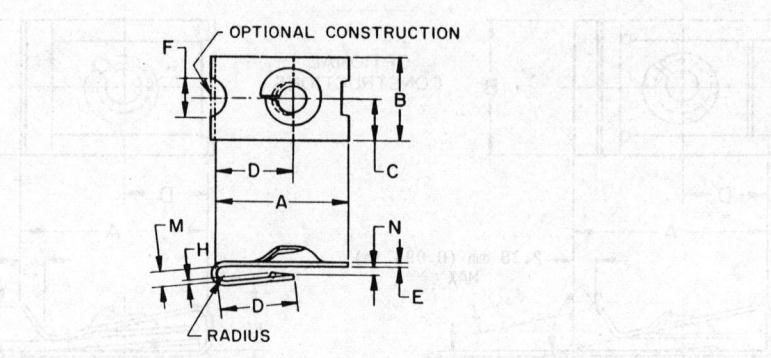

FIGURE 8—J SHAPE
STYLE I—STANDARD, STYLE II—SHORT, AND STYLE III—LONG THROAT

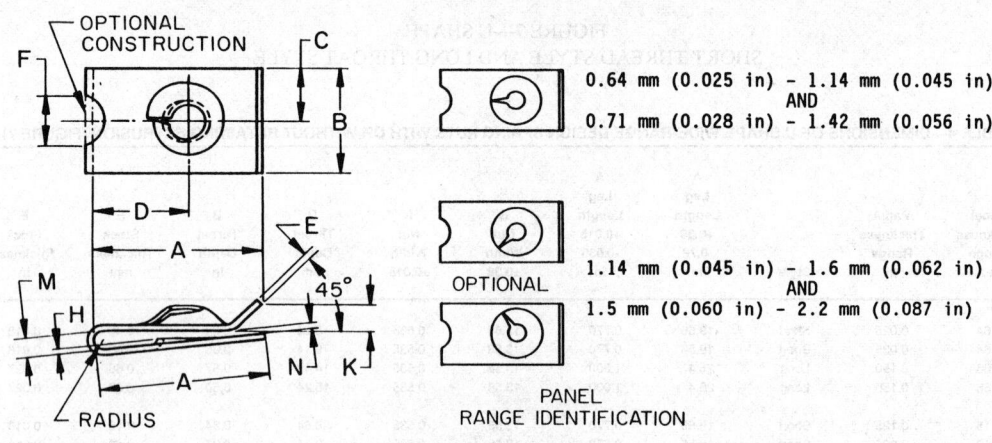

FIGURE 9—U SHAPE
STYLE—STANDARD, STYLE II—SHORT, AND STYLE III—LONG THROAT

TABLE 5—DIMENSIONS OF TYPE P, J SHAPE, AND U SHAPE REGULAR DESIGN SPRING NUTS (FIGURES 8 AND 9)

Tapping Screw Size[1] mm	Tapping Screw Size[1] in	Panel Thickness Range mm	Panel Thickness Range in	Style	A Leg Length J Shape mm ±0.5	A Leg Length J Shape in ±0.02	A Leg Length U Shape mm ±0.5	A Leg Length U Shape in ±0.02	B Nut Width mm	B Nut Width in	C Edge to Center mm	C Edge to Center in	D Throat Depth mm ±0.5	D Throat Depth in ±0.02
3.5	#6	0.64-1.14	0.025-0.045	I	13.5	0.53	15.0	0.59	8.6	0.34	4.3	0.17	8.4	0.33
3.5	#6	1.14-1.6	0.045-0.062	I	13.5	0.53	15.0	0.59	8.6	0.34	4.3	0.17	8.4	0.33
3.5	#6	0.64-1.14	0.025-0.045	II	11.9	0.47	13.5	0.53	8.6	0.34	4.3	0.17	6.9	0.27
3.5	#6	1.14-1.6	0.045-0.062	II	11.9	0.47	13.5	0.53	8.6	0.34	4.3	0.17	6.9	0.27
3.5	#6	0.64-1.14	0.025-0.045	III	17.8	0.70	19.3	0.76	8.6	0.34	4.3	0.17	13.2	0.52
3.5	#6	1.14-1.6	0.045-0.062	III	17.8	0.70	19.3	0.76	8.6	0.34	4.3	0.17	13.2	0.52
4.2	#8	0.64-1.14	0.025-0.045	I	15.5	0.61	17.0	0.67	10.2	0.40	5.1	0.20	9.9	0.39
4.2	#8	1.14-1.6	0.045-0.062	I	15.5	0.61	17.0	0.67	10.2	0.40	5.1	0.20	9.9	0.39
4.2	#8	0.64-1.14	0.025-0.045	II	13.0	0.51	14.5	0.57	10.2	0.40	5.1	0.20	7.1	0.26
4.2	#8	1.14-1.6	0.045-0.062	II	13.0	0.51	14.5	0.57	10.2	0.40	5.1	0.20	7.1	0.26
4.2	#8	0.64-1.14	0.025-0.045	III	20.6	0.81	22.4	0.88	10.2	0.40	5.1	0.20	14.7	0.58
4.2	#8	1.14-1.6	0.045-0.062	III	20.6	0.81	22.4	0.88	10.2	0.40	5.1	0.20	14.7	0.58
4.8	#10	0.64-1.14	0.025-0.045	I	16.3	0.64	17.8	0.70	11.4	0.50	5.7	0.25	9.9	0.39
4.8	#10	1.14-1.6	0.045-0.062	I	16.3	0.64	17.8	0.70	11.4	0.50	5.7	0.25	9.9	0.39
4.8	#10	0.64-1.14	0.025-0.045	II	15.5	0.61	16.3	0.64	11.4	0.50	5.7	0.25	8.4	0.33
4.8	#10	1.14-1.6	0.045-0.062	II	15.5	0.61	16.3	0.64	11.4	0.50	5.7	0.25	8.4	0.33
4.8	#10	0.64-1.14	0.025-0.045	III	21.1	0.83	22.6	0.87	11.4	0.50	5.7	0.25	14.7	0.58
4.8	#10	1.14-1.6	0.045-0.062	III	21.1	0.83	22.6	0.87	11.4	0.50	5.7	0.25	14.7	0.58
6.3	1/4	0.71-1.42	0.028-0.056	I	20.3	0.80	21.8	0.86	14.2	0.56	7.1	0.28	11.9	0.47
6.3	1/4	1.5-2.2	0.06-0.087	I	20.3	0.80	21.8	0.86	14.2	0.56	7.1	0.28	11.9	0.47
6.3	1/4	0.71-1.42	0.028-0.056	III	27.7	1.09	28.4	1.16	14.2	0.56	7.1	0.28	18.3	0.72
6.3	1/4	1.5-2.2	0.06-0.087	III	27.7	1.09	28.4	1.16	14.2	0.56	7.1	0.28	18.3	0.72

TABLE 5—DIMENSIONS OF TYPE P, J SHAPE, AND U SHAPE REGULAR DESIGN SPRING NUTS (FIGURES 8 AND 9) (CONTINUED)

Tapping Screw Size[1] mm	Tapping Screw Size[1] in	Style	E Stock Thickness mm Min	E Stock Thickness in Min	E Stock Thickness mm Max	E Stock Thickness in Max	F Hole Dia[2] mm	F Hole Dia[2] in	H Height mm +0.25 -0.00	H Height in +0.01 -0.00	K Tang Height mm ±0.38	K Tang Height in ±0.015	M Width at Fold mm	M Width at Fold in	N Gap Opening Min J Shape mm	N Gap Opening Min J Shape in	N Gap Opening Max U Shape mm	N Gap Opening Max U Shape in
3.5	#6	I	0.47	0.019	0.54	0.021	5.5	0.22	0.64	0.025	2.0	0.08	1.5	0.06	0.50	0.02	0.50	0.02
3.5	#6	I	0.47	0.019	0.54	0.021	5.5	0.22	0.64	0.025	2.0	0.08	2.0	0.08	1.0	0.04	1.0	0.04
3.5	#6	II	0.47	0.019	0.54	0.021	5.5	0.22	0.64	0.025	2.0	0.08	1.5	0.06	0.50	0.02	0.50	0.02
3.5	#6	II	0.47	0.019	0.54	0.021	5.5	0.22	0.64	0.025	2.0	0.08	2.0	0.08	1.0	0.04	1.0	0.04
3.5	#6	III	0.47	0.019	0.54	0.021	5.5	0.22	0.64	0.025	2.0	0.08	1.5	0.06	0.50	0.02	0.50	0.02
3.5	#6	III	0.47	0.019	0.54	0.021	5.5	0.22	0.64	0.025	2.0	0.08	2.0	0.08	1.0	0.04	1.0	0.04
4.2	#8	I	0.60	0.024	0.67	0.026	5.5	0.22	0.64	0.025	2.3	0.09	1.5	0.06	0.50	0.02	0.50	0.02
4.2	#8	I	0.60	0.024	0.67	0.026	5.5	0.22	0.64	0.025	2.3	0.09	2.0	0.08	1.0	0.04	1.0	0.04
4.2	#8	II	0.60	0.024	0.67	0.026	5.5	0.22	0.64	0.025	2.3	0.09	1.5	0.06	0.50	0.02	0.50	0.02
4.2	#8	II	0.60	0.024	0.67	0.026	5.5	0.22	0.64	0.025	2.3	0.09	2.0	0.08	1.0	0.04	1.0	0.04
4.2	#8	III	0.60	0.024	0.67	0.026	5.5	0.22	0.64	0.025	2.3	0.09	1.5	0.06	0.50	0.02	0.50	0.02
4.2	#8	III	0.60	0.024	0.67	0.026	5.5	0.22	0.64	0.025	2.3	0.09	2.0	0.08	1.0	0.04	1.0	0.04
4.8	#10	I	0.70	0.028	0.82	0.032	5.5	0.22	0.64	0.025	2.3	0.09	1.5	0.06	0.50	0.02	0.50	0.02
4.8	#10	I	0.70	0.028	0.82	0.032	5.5	0.22	0.64	0.025	2.3	0.09	2.0	0.08	1.0	0.04	1.0	0.04
4.8	#10	II	0.70	0.028	0.82	0.032	5.5	0.22	0.64	0.025	2.3	0.09	1.5	0.06	0.50	0.02	0.50	0.02
4.8	#10	II	0.70	0.028	0.82	0.032	5.5	0.22	0.64	0.025	2.3	0.09	2.0	0.08	1.0	0.04	1.0	0.04
4.8	#10	III	0.70	0.028	0.82	0.032	5.5	0.22	0.64	0.025	2.3	0.09	1.5	0.06	0.50	0.02	0.50	0.02
4.8	#10	III	0.70	0.028	0.82	0.032	5.5	0.22	0.64	0.025	2.3	0.09	2.0	0.08	1.0	0.04	1.0	0.04
6.3	1/4	I	0.70	0.028	0.82	0.032	5.5	0.22	0.64	0.025	2.8	0.11	2.0	0.08	1.0	0.04	1.0	0.04
6.3	1/4	I	0.70	0.028	0.82	0.032	5.5	0.22	0.64	0.025	2.8	0.11	2.5	0.10	1.2	0.05	1.2	0.05
6.3	1/4	III	0.70	0.028	0.82	0.032	5.5	0.22	0.64	0.025	2.8	0.11	2.0	0.08	1.0	0.04	1.0	0.04
6.3	1/4	III	0.70	0.028	0.82	0.032	5.5	0.22	0.64	0.025	2.8	0.11	2.5	0.10	1.2	0.05	1.2	0.05

[1] See Boss Detail under General Specifications for applicability of types and sizes. Type P spring nut of similar proportions are also available in respective machine screw sizes 3.5 mm (#6), 4.2 mm (#8), 4.8 mm (#10), and 6.3 mm (1/4 in), and manufacturers should be consulted for dimensions.
[2] Diameter of hole punched in blank before forming.

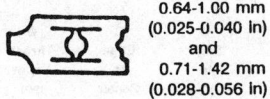
0.64-1.00 mm
(0.025-0.040 in)
and
0.71-1.42 mm
(0.028-0.056 in)

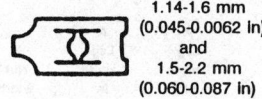
1.14-1.6 mm
(0.045-0.0062 in)
and
1.5-2.2 mm
(0.060-0.087 in)

PANEL RANGE IDENTIFICATION FOR J SHAPE NUTS

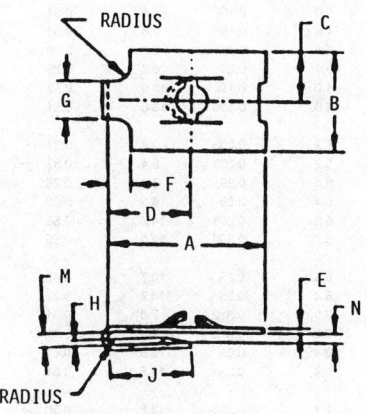

FIGURE 10A
STYLE I—STANDARD THROAT
AND STYLE III—LONG THROAT

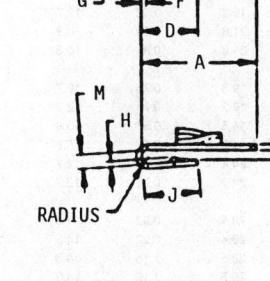

FIGURE 10B
STYLE II—SHORT THROAT

FIGURE 10—J SHAPE

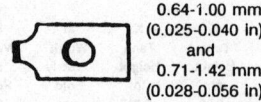
0.64-1.00 mm
(0.025-0.040 in)
and
0.71-1.42 mm
(0.028-0.056 in)

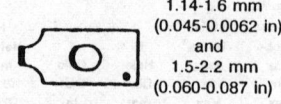
1.14-1.6 mm
(0.045-0.0062 in)
and
1.5-2.2 mm
(0.060-0.087 in)

PANEL RANGE IDENTIFICATION FOR U SHAPE NUTS

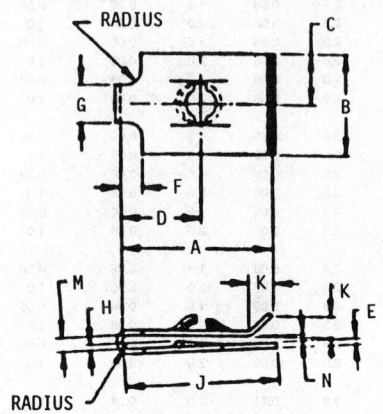

FIGURE 11A
STYLE I—STANDARD THROAT
AND STYLE III—LONG THROAT

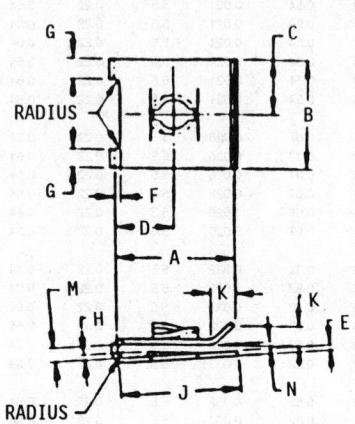

FIGURE 11B
STYLE II—SHORT THROAT

FIGURE 11—U SHAPE

TABLE 6—DIMENSIONS OF TYPE T, J SHAPE, AND U SHAPE REGULAR DESIGN SPRING NUTS (FIGURES 10A TO 10B)

Tapping Screw Size[1] mm	Tapping Screw Size[1] in	Panel Thickness range mm	Panel Thickness range in	Style	A Leg Length mm ±0.5	A Leg Length in ±0.02	B Nut Width mm	B Nut Width in	C Edge to Center mm	C Edge to Center in	D Throat Depth mm ±0.5	D Throat Depth in ±0.02	E Stock Thickness mm Min	E Stock Thickness mm Max	E Stock Thickness in Min	E Stock Thickness in Max
3.5	#6	0.64-1.14	0.025-0.045	I	16.3	0.64	7.9	0.312	4.0	0.156	8.6	0.34	0.027	0.023	0.69	0.58
3.5	#6	1.14-1.6	0.045-0.062	I	16.3	0.64	7.9	0.312	4.0	0.156	8.6	0.34	0.027	0.023	0.69	0.58
3.5	#6	0.64-1.14	0.025-0.045	II	12.2	0.48	12.7	0.50	6.4	0.25	6.7	0.26	0.027	0.023	0.69	0.58
3.5	#6	1.14-1.6	0.045-0.062	II	12.2	0.48	12.7	0.50	6.4	0.25	6.7	0.26	0.027	0.023	0.69	0.58
3.5	#6	0.64-1.14	0.025-0.045	III	20.1	0.81	7.9	0.312	4.0	0.156	13.0	0.51	0.027	0.023	0.69	0.58
3.5	#6	1.14-1.6	0.045-0.062	III	20.1	0.81	7.9	0.312	4.0	0.156	13.0	0.51	0.027	0.023	0.69	0.58
4.2	#8	0.64-1.14	0.025-0.045	I	17.3	0.68	10.3	0.406	5.2	0.203	9.4	0.37	0.03	0.026	0.76	0.66
4.2	#8	1.14-1.6	0.045-0.062	I	17.3	0.68	10.3	0.406	5.2	0.203	9.4	0.37	0.03	0.026	0.76	0.66
4.2	#8	0.64-1.14	0.025-0.045	II	13.2	0.52	12.7	0.50	6.4	0.25	6.6	0.26	0.03	0.026	0.76	0.66
4.2	#8	1.14-1.6	0.045-0.062	II	13.2	0.52	12.7	0.50	6.4	0.25	6.6	0.26	0.03	0.026	0.76	0.66
4.2	#8	0.64-1.14	0.025-0.045	III	21.8	0.86	10.3	0.406	5.2	0.203	14.0	0.55	0.03	0.026	0.76	0.66
4.2	#8	1.14-1.6	0.045-0.062	III	21.8	0.86	10.3	0.406	5.2	0.203	14.0	0.55	0.03	0.026	0.76	0.66
4.8	#10	0.64-1.14	0.025-0.045	I	19.3	0.76	12.7	0.50	6.4	0.25	10.7	0.42	0.033	0.029	0.84	0.74
4.8	#10	1.14-1.6	0.045-0.062	I	19.3	0.76	12.7	0.50	6.4	0.25	10.7	0.42	0.033	0.029	0.84	0.74
4.8	#10	0.64-1.14	0.025-0.045	II	14.5	0.57	15.9	0.625	7.9	0.312	7.8	0.31	0.033	0.029	0.84	0.74
4.8	#10	1.14-1.6	0.045-0.062	II	14.5	0.57	15.9	0.625	7.9	0.312	7.8	0.31	0.033	0.029	0.84	0.74
4.8	#10	0.64-1.14	0.025-0.045	III	24.4	0.96	12.7	0.50	6.4	0.25	15.5	0.61	0.033	0.029	0.84	0.74
4.8	#10	1.14-1.6	0.045-0.062	III	24.4	0.96	12.7	0.50	6.4	0.25	15.5	0.61	0.033	0.029	0.84	0.74
6.3	1/4	0.71-1.5	0.028-0.06	I	23.4	0.92	14.3	0.562	7.1	0.281	12.6	0.50	0.042	0.035	1.07	0.89
6.3	1/4	1.5-2.2	0.06-0.087	I	23.4	0.92	14.3	0.562	7.1	0.281	12.6	0.50	0.042	0.035	1.07	0.89
6.3	1/4	0.7-1.5	0.028-0.06	III	29.5	1.16	14.3	0.562	7.1	0.281	18.9	0.75	0.039	0.035	0.99	0.89
6.3	1/4	1.5-2.2	0.06-0.087	III	29.5	1.16	14.3	0.562	7.1	0.281	18.9	0.75	0.039	0.035	0.99	0.89

[1] See Boss Detail under General Specifications for applicability of types and sizes. Type T spring nuts of similar proportions are also available in respective machine screw sizes 3.5 mm (#6), 4.2 mm (#8), 4.8 mm (#10), and 6.3 mm (1/4 in), and manufacturers should be consulted for dimensions.

TABLE 6—DIMENSIONS OF TYPE T, J SHAPE, AND U SHAPE REGULAR DESIGN SPRING NUTS (FIGURES 10A TO 10B) (CONTINUED)

Tapping Screw Size[1] mm	Tapping Screw Size[1] in	Style	F Notch Depth mm ±0.5	F Notch Depth in ±0.02	G Width mm ±0.5	G Width in ±0.02	H Height mm ±0.13	H Height in ±0.005	J Leg Length mm ±0.5	J Leg Length in ±0.02	K Tang mm ±0.5	K Tang in ±0.02	N Gap Opening J Shape mm Min	N Gap Opening J Shape in Min	N Gap Opening U Shape mm Max	N Gap Opening U Shape in Max
3.5	#6	I	2.0	0.08	2.8	0.11	0.5	0.02	8.9	0.35	2.0	0.08	0.5	0.02	0.5	0.02
3.5	#6	I	2.0	0.08	2.8	0.11	0.5	0.02	8.9	0.35	2.0	0.08	1.0	0.04	1.0	0.04
3.5	#6	II	0.5	0.02	2.3	0.09	0.5	0.02	6.6	0.26	2.0	0.08	0.5	0.02	0.5	0.02
3.5	#6	II	0.5	0.02	2.3	0.09	0.5	0.02	6.6	0.26	2.0	0.08	1.0	0.04	1.0	0.04
3.5	#6	III	6.2	0.24	3.6	0.14	0.5	0.02	13.0	0.51	2.0	0.08	0.5	0.02	0.5	0.02
3.5	#6	III	6.2	0.24	3.6	0.14	0.5	0.02	13.0	0.51	2.0	0.08	1.0	0.04	1.0	0.04
4.2	#8	I	2.3	0.09	4.6	0.18	0.64	0.025	9.4	0.37	2.3	0.09	0.5	0.02	0.5	0.02
4.2	#8	I	2.3	0.09	4.6	0.18	0.64	0.025	9.4	0.37	2.3	0.09	1.0	0.04	1.0	0.04
4.2	#8	II	0.5	0.20	2.3	0.09	0.64	0.025	6.9	0.27	2.3	0.09	0.5	0.02	0.5	0.02
4.2	#8	II	0.5	0.20	2.3	0.09	0.64	0.025	6.9	0.27	2.3	0.09	1.0	0.04	1.0	0.04
4.2	#8	III	6.9	0.27	4.6	0.18	0.64	0.025	14.2	0.56	2.3	0.09	0.5	0.02	0.5	0.02
4.2	#8	III	6.9	0.27	4.6	0.18	0.64	0.025	14.2	0.56	2.3	0.09	1.0	0.04	1.0	0.04
4.8	#10	I	2.8	0.11	3.1	0.12	0.64	0.025	10.7	0.42	2.3	0.09	0.5	0.02	0.5	0.02
4.8	#10	I	2.8	0.11	3.1	0.12	0.64	0.025	10.7	0.42	2.3	0.09	1.0	0.04	1.0	0.04
4.8	#10	II	0.5	0.02	3.3	0.13	0.64	0.025	7.9	0.31	2.3	0.09	0.5	0.02	0.5	0.02
4.8	#10	II	0.5	0.02	3.3	0.13	0.64	0.025	7.9	0.31	2.3	0.09	1.0	0.04	1.0	0.04
4.8	#10	III	6.4	0.25	4.3	0.17	0.64	0.025	15.5	0.61	2.3	0.09	0.5	0.02	0.5	0.02
4.8	#10	III	6.4	0.25	4.3	0.17	0.64	0.025	15.5	0.61	2.3	0.09	1.0	0.04	1.0	0.04
6.3	1/4	I	1.5	0.06	4.6	0.18	0.76	0.03	12.5	0.49	2.8	0.11	1.0	0.04	1.0	0.04
6.3	1/4	I	1.5	0.06	4.6	0.18	0.76	0.03	12.5	0.49	2.8	0.11	1.2	0.05	1.2	0.05
6.3	1/4	III	7.9	0.31	6.4	0.25	0.76	0.03	19.0	0.75	2.8	0.11	1.0	0.04	1.0	0.04
6.3	1/4	III	7.9	0.31	6.4	0.25	0.76	0.03	19.0	0.75	2.8	0.11	1.2	0.05	1.2	0.05

[1] See Boss Detail under General Specifications for applicability of types and sizes. Type T spring nuts of similar proportions are also available in respective machine screw sizes 3.5 mm (#6), 4.2 mm (#8), 4.8 mm (#10), and 6.3 mm (1/4 in), and manufacturers should be consulted for dimensions.

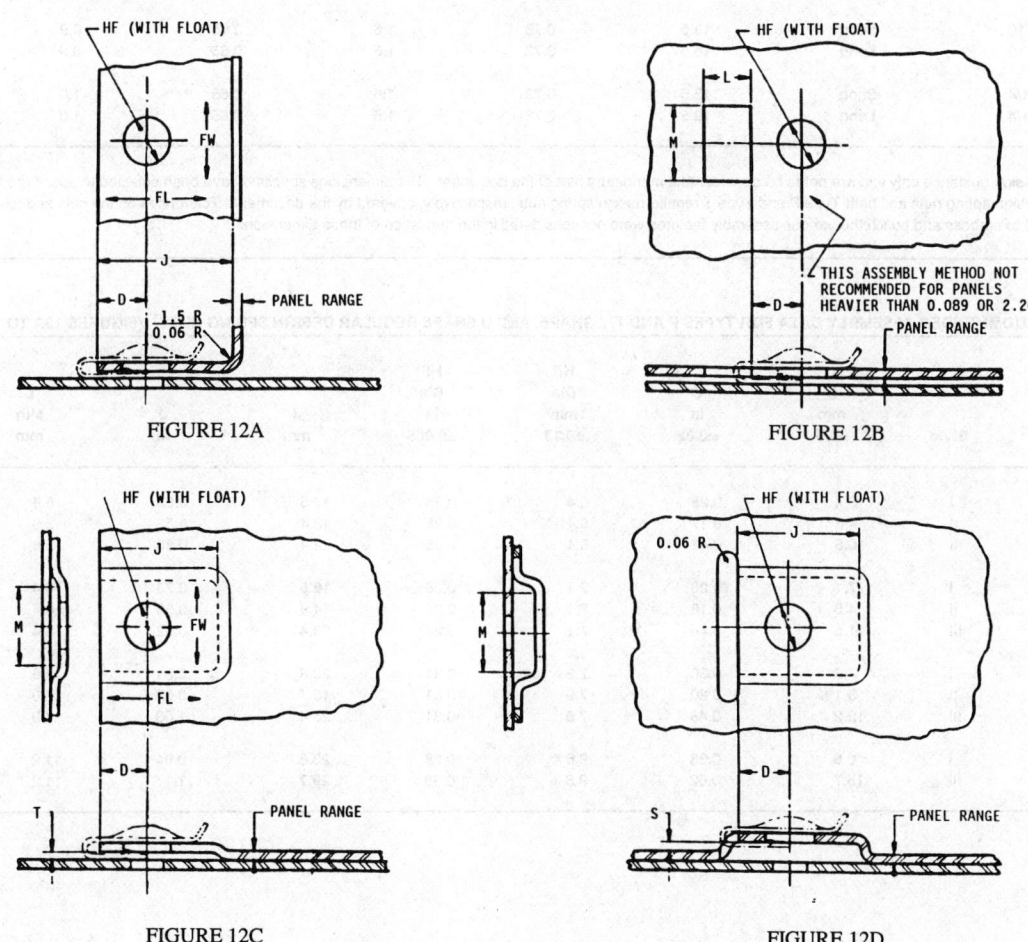

FIGURE 12A

FIGURE 12B

FIGURE 12C

FIGURE 12D

FIGURE 12—RECOMMENDED ASSEMBLY FOR WIDE-RANGE DESIGN SPRING NUTS

TABLE 7—RECOMMENDED ASSEMBLY DATA FOR WIDE-RANGE DESIGN SPRING NUTS[1] (FIGURES 12A TO 12D)

Tapping Screw Size mm	Tapping Screw Size in	Style	D mm ±0.5	D in ±0.02	HF Dia mm ±0.13	HF Dia in ±0.005	J mm	J in	L Min mm	L Min in
3.5	#6	Short	6.4	0.25	6.4	0.25	19.8	0.78	6.4	0.25
3.5	#6	Long	12.2	0.48	6.4	0.25	25.6	1.01	6.4	0.25
4.2	#8	Short	6.4	0.25	7.1	0.28	19.8	0.78	8.6	0.34
4.2	#8	Long	12.2	0.48	7.1	0.28	25.6	1.01	8.6	0.34
4.8	#10	Short	6.4	0.25	7.9	0.31	19.8	0.78	10.9	0.43
4.8	#10	Long	12.2	0.48	7.9	0.31	25.6	1.01	10.9	0.43
6.3	1/4	Short	6.4	0.25	9.6	0.38	19.8	0.78	12.7	0.50
6.3	1/4	Long	12.2	0.48	9.6	0.38	25.6	1.01	12.7	0.50

TABLE 7—RECOMMENDED ASSEMBLY DATA FOR WIDE-RANGE DESIGN SPRING NUTS[1] (FIGURES 12A TO 12D) (CONTINUED)

Tapping Screw Size mm	Tapping Screw Size in	Style	M Flat mm	M Flat in	S Min mm	S Min in	T Min mm	T Min in
3.5	#6	Short	16.0	0.63	1.6	0.65	0.6	0.027
3.5	#6	Long	16.0	0.63	1.6	0.65	0.6	0.027
4.2	#8	Short	16.0	0.63	1.6	0.65	0.8	0.03
4.2	#8	Long	16.0	0.63	1.6	0.65	0.8	0.03
4.8	#10	Short	18.5	0.73	1.6	0.65	0.9	0.035
4.8	#10	Long	18.5	0.73	1.6	0.65	0.9	0.035
6.3	1/4	Short	18.5	0.73	1.6	0.65	1.0	0.04
6.3	1/4	Long	18.5	0.73	1.6	0.65	1.0	0.04

[1] These data are intended for design guidance only and are not to be considered a mandatory part of the document. The dimensions specified have been selected to accommodate the optional constructions of wide-range design spring nuts and both Type P and Type T regular design spring nuts, respectively, covered by the document. Tolerances on the nuts and tolerances entailed in the manufacturing processes used to emboss and punch the various assembly features were not considered in the derivation of these dimensions.

TABLE 8—RECOMMENDED ASSEMBLY DATA FOR TYPES P AND T, J SHAPE, AND U SHAPE REGULAR DESIGN SPRING NUTS[1] (FIGURES 12A TO 12D)

Tapping Size mm	Tapping Size in	Style	D mm ±0.5	D in ±0.02	HF Dia mm ±0.13	HF Dia in ±0.005	J mm	J in	L Min mm	L Min in
3.5	#6	I	6.4	0.25	6.4	0.25	17.5	0.69	5.6	0.22
3.5	#6	II	4.3	0.17	6.4	0.25	13.4	0.53	5.6	0.22
3.5	#6	III	10.6	0.42	6.4	0.25	22.4	0.88	5.6	0.22
4.2	#8	I	7.1	0.23	7.1	0.28	18.5	0.73	6.4	0.25
4.2	#8	II	4.6	0.18	7.1	0.28	14.9	0.59	6.4	0.25
4.2	#8	III	11.6	0.46	7.1	0.28	23.4	0.92	6.4	0.25
4.8	#10	I	7.6	0.30	7.9	0.31	20.6	0.81	7.9	0.31
4.8	#10	II	5.1	0.20	7.9	0.31	15.7	0.62	7.9	0.31
4.8	#10	III	12.2	0.48	7.9	0.31	25.4	1.00	7.9	0.31
6.3	#14 or 1/4	I	0.6	0.38	9.6	0.38	23.8	0.94	11.2	0.44
6.3	#14 or 1/4	III	15.7	0.62	9.6	0.38	29.7	1.17	11.2	0.44

TABLE 8—RECOMMENDED ASSEMBLY DATA FOR TYPES P AND T, J SHAPE, AND U SHAPE REGULAR DESIGN SPRING NUTS[1] (FIGURES 12A TO 12D) (CONTINUED)

Tapping Size mm	Tapping Size in	Style	M Flat mm	M Flat in	S mm Min	S in Min	T mm Min	T in Min
3.5	#6	I	11.8	0.44	1.5	0.06	0.8	0.03
3.5	#6	II	14.9	0.59	1.5	0.06	0.8	0.03
3.5	#6	III	11.8	0.44	1.5	0.06	0.8	0.03
4.2	#8	I	12.7	0.50	1.5	0.06	0.8	0.033
4.2	#8	II	14.9	0.59	1.6	0.065	0.8	0.033
4.2	#8	III	12.7	0.50	1.6	0.065	0.8	0.033
4.8	#10	I	14.9	0.59	1.6	0.065	0.9	0.036
4.8	#10	II	18.2	0.72	1.6	0.065	0.9	0.036
4.8	#10	III	14.9	0.59	1.8	0.07	0.9	0.036
6.3	#14 or 1/4	I	16.8	0.66	2.2	0.085	1.1	0.042
6.3	#14 or 1/4	III	16.8	0.66	2.2	0.085	1.1	0.042

[1] These data are intended for design guidance only and are not to be considered a mandatory part of the document. The dimensions specified have been selected to accommodate the optional constructions of wide-range design spring nuts and both Type P and Type T regular design spring nuts, respectively, covered by the document. Tolerances on the nuts and tolerances entailed in the manufacturing processes used to emboss and punch the various assembly features were not considered in the derivation of these dimensions.

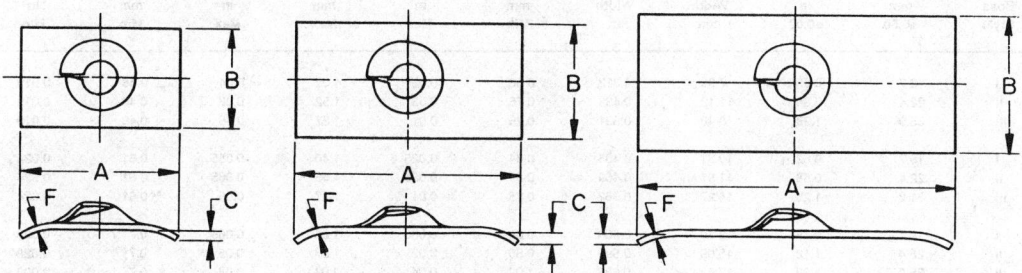

FIGURE 13—TYPE P SINGLE BOSS

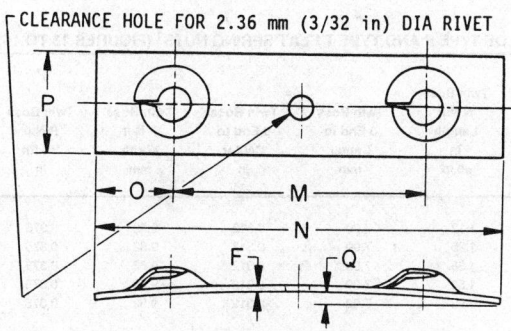

FIGURE 14—TYPE P TWIN BOSS

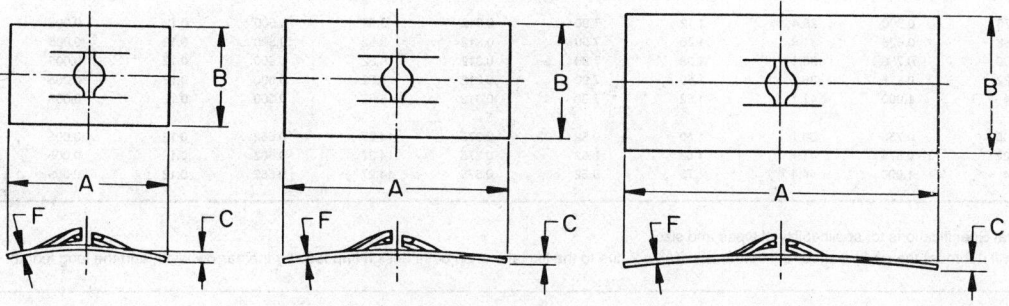

FIGURE 15—TYPE T SINGLE BOSS

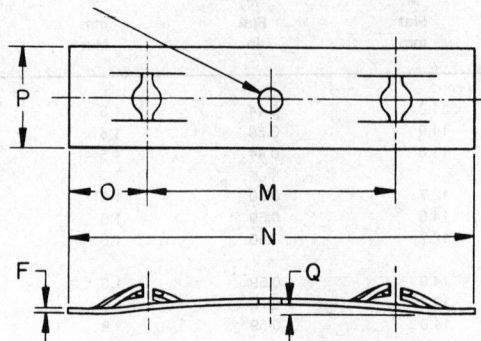

FIGURE 16—TYPE T TWIN BOSS

TABLE 9—DIMENSIONS OF TYPE P AND TYPE T FLAT SPRING NUTS[1] (FIGURES 13 TO 16)

Tapping Screw Size mm	Tapping Screw Size in	Single Boss Style	Single Boss A Nut Length mm ±0.50	Single Boss A Nut Length in ±0.02	Single Boss B Nut Width mm	Single Boss B Nut Width in	Single Boss C Arch Height mm Min	Single Boss C Arch Height in Min	Single Boss C Arch Height mm Max	Single Boss C Arch Height in Max	F Stock Thickness mm Min	F Stock Thickness in Min	F Stock Thickness mm Max	F Stock Thickness in Max
3.5	#6	I	12.7	0.50	7.90	0.312	0.50	0.02	1.02	0.04	0.48	0.019	0.66	0.026
3.5	#6	II	22.4	0.88	11.12	0.438	0.76	0.03	1.52	0.06	0.48	0.019	0.66	0.026
3.5	#6	III	35.0	1.38	13.48	0.531	0.25	0.01	1.27	0.05	0.48	0.019	0.66	0.026
4.2	#8	I	15.7	0.62	10.31	0.406	0.64	0.025	1.40	0.055	0.61	0.024	0.74	0.029
4.2	#8	II	22.4	0.88	11.91	0.469	0.89	0.035	1.65	0.065	0.61	0.024	0.74	0.029
4.2	#8	III	31.8	1.25	14.27	0.562	0.25	0.01	1.27	0.05	0.61	0.024	0.74	0.029
4.8	#10	I	19.1	0.75	12.70	0.500	0.89	0.035	1.65	0.065	0.71	0.028	0.84	0.033
4.8	#10	II	28.4	1.12	15.08	0.594	0.50	0.02	1.27	0.05	0.71	0.028	0.84	0.033
4.8	#10	III	35.0	1.38	17.48	0.688	1.02	0.04	2.03	0.08	0.71	0.028	0.84	0.033
6.3	#14 or 1/4	I	22.4	0.88	14.27	0.562	1.14	0.045	1.90	0.075	0.71	0.028	0.99	0.039
6.3	#14 or 1/4	II	26.9	1.06	15.88	0.625	0.50	0.02	1.27	0.05	0.71	0.028	0.99	0.039
6.3	#14 or 1/4	III	35.0	1.38	17.48	0.688	1.14	0.045	2.16	0.085	0.71	0.028	0.99	0.039

TABLE 9—DIMENSIONS OF TYPE P AND TYPE T FLAT SPRING NUTS[1] (FIGURES 13 TO 16) (CONTINUED)

Tapping Screw Size mm	Tapping Screw Size in	Twin Boss M Boss Center to Center mm	Twin Boss M Boss Center to Center in	Twin Boss N Nut Length mm ±0.50	Twin Boss N Nut Length in ±0.02	Twin Boss O End to Center mm	Twin Boss O End to Center in	Twin Boss P Nut Width mm	Twin Boss P Nut Width in	Twin Boss Q Arch Height mm Min	Twin Boss Q Arch Height in Min	Twin Boss Q Arch Height mm Max	Twin Boss Q Arch Height in Max
3.5	#6	12.70	0.500	28.4	1.12	7.90	0.312	9.52	0.375	0.12	0.005	1.14	0.045
3.5	#6	15.88	0.625	31.8	1.25	7.90	0.312	9.52	0.375	0.12	0.005	1.14	0.045
3.5	#6	19.05	0.750	35.1	1.38	7.90	0.312	9.52	0.375	0.12	0.005	1.14	0.045
3.5	#6	22.22	0.875	38.1	1.50	7.90	0.312	9.52	0.375	0.12	0.005	1.65	0.065
3.5	#6	25.4	1.000	41.1	1.62	7.90	0.312	9.52	0.375	0.12	0.005	1.65	0.065
4.2	#8	12.70	0.500	28.4	1.12	7.90	0.312	9.52	0.375	0.12	0.005	1.14	0.045
4.2	#8	15.88	0.625	31.8	1.25	7.90	0.312	9.52	0.375	0.12	0.005	1.14	0.045
4.2	#8	19.05	0.750	35.1	1.38	7.90	0.312	9.52	0.375	0.12	0.005	1.14	0.045
4.2	#8	22.22	0.875	38.1	1.50	7.90	0.312	9.52	0.375	0.12	0.005	1.65	0.065
4.2	#8	25.4	1.000	41.1	1.62	7.90	0.312	9.52	0.375	0.12	0.005	1.65	0.065
4.8	#10	12.70	0.500	28.4	1.12	7.90	0.312	9.52	0.500	0.12	0.005	1.14	0.045
4.8	#10	15.88	0.625	31.8	1.25	7.90	0.312	9.52	0.500	0.12	0.005	1.14	0.045
4.8	#10	19.05	0.750	35.1	1.38	7.90	0.312	9.52	0.500	0.12	0.005	1.14	0.045
4.8	#10	22.22	0.875	38.1	1.50	7.90	0.312	9.52	0.500	0.12	0.005	1.65	0.065
4.8	#10	25.4	1.000	41.1	1.62	7.90	0.312	9.52	0.500	0.12	0.005	1.65	0.065
6.3	#14 or 1/4	19.05	0.750	38.1	1.50	9.52	0.375	14.27	0.562	0.12	0.005	1.65	0.065
6.3	#14 or 1/4	22.22	0.875	41.1	1.62	9.52	0.375	14.27	0.562	0.12	0.005	1.65	0.065
6.3	#14 or 1/4	25.4	1.000	44.4	1.75	9.52	0.375	14.27	0.562	0.12	0.005	1.65	0.065

1. See Boss Detail under General Specifications for applicability of types and sizes.
2. The type P nuts in this style will not meet the performance requirements in Table 2 due to the limitations on boss design imposed by the narrow width and the long length.

PUSH-ON SPRING NUTS—SAE J892a

SAE Standard

Report of Fasteners Committee approved August 1964 and last revised June 1970.

GENERAL SPECIFICATIONS

Scope—It should be noted that push-on spring nuts having other configurations are available and manufacturers should be consulted.

Dimensional Tolerance—Tolerance on dimensions in the tables shall be plus and minus 0.010 in. unless otherwise specified. Tolerance on the thickness of material shall be plus and minus 0.001 in.

Boss—Size and formation of boss and other detail shall be such as to assemble readily and function satisfactorily with the specified stud.

Material—Spring steel suitably processed to meet the hardness requirements of this specification.

Hardness—Hardness shall be as follows:

Material Thickness	Rockwell Scale	Dial Reading	Conversion to Rockwell C Scale
Up to 0.016	15N	82.5–85.5	44–50
0.017–0.024	30N	63.0–68.5	44–50

Finish—Spring nuts are normally supplied with rustproof finish as specified by purchaser. Nuts subjected to corrosion preventative treatment which might induce hydrogen embrittlement shall be baked or otherwise treated to obviate such embrittlement.

Workmanship—Spring nuts shall be free from cracks, burrs, splits, loose scale, or any other defects which might affect their serviceability.

Application and Design—Where nut is to function only as a locking means, Style I is recommended. Where greater area of load distribution is a requirement, i.e.; nut is to function also as spanner washer, Style II is recommended. The Light Series are for use on plastic studs; the Medium Series are for use on soft metal studs, and the Heavy Series are for use on hardened metal or chromium plated studs.

Assembly Considerations—Since performance of push-on spring nuts is dependent upon the studs to which they are applied, it is essential that stud diameters and plating recommendations as set forth in Fig. 1 and Table 1 be adhered to as closely as possible. The actual stud length is determined by adding the thickness of mating panel or panels "T," through which the stud protrudes, to the factors tabulated under "C" (the minimum stud protrusion

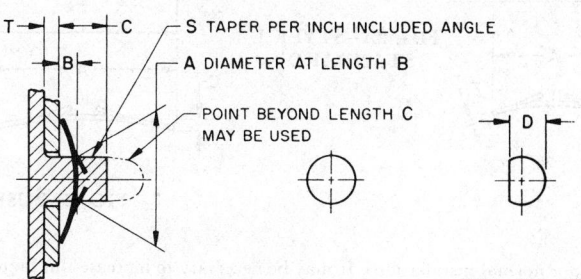

FIG. 1A—ROUND STUD   FIG. 1B—"D" SHAPED STUD

FIG. 1—RECOMMENDED STUD DESIGN

TABLE 1—DIMENSIONS OF RECOMMENDED STUDS (FIGS. 1A and 1B)

Nominal Stud Size	A[a] Stud Diameter Max	A[a] Stud Diameter Min	B Straight Length, Min	C Stud Length, Min	D Stud Width Max	D Stud Width Min	S Taper, Max
1/16	0.065	0.059	T + 0.24	T + 0.36	0.054	0.044	0.052
3/32	0.097	0.091	T + 0.24	T + 0.36	0.079	0.069	0.052
1/8	0.128	0.122	T + 0.24	T + 0.36	0.105	0.095	0.052
5/32	0.159	0.153	T + 0.24	T + 0.36	0.130	0.120	0.040
3/16	0.191	0.185	T + 0.24	T + 0.36	0.155	0.145	0.040
7/32	0.222	0.216	T + 0.24	T + 0.36	0.180	0.170	0.040
1/4	0.253	0.247	T + 0.24	T + 0.36	0.205	0.195	0.040
5/16	0.315	0.309	T + 0.24	T + 0.36	0.255	0.245	0.033
3/8	0.378	0.372	T + 0.24	T + 0.36	0.305	0.295	0.033

[a]Diameter limits include thickness of plating on studs. Chromium or nickel plating is permissible only on studs to be used with Heavy Series Nuts and then it is recommended that plating thickness along length of die cast studs be held to within 0.0015 in. wherever possible and that in no case should the plating thickness exceed 0.003 in.

TABLE 2—DIMENSIONS OF PUSH-ON SPRING NUTS (FIGS. 2A–2C)

Nominal Stud Size	Style	Series[a]	A Nut Length	B Nut Width	C Arch Height Max	C Arch Height Min	F Stock Thickness
1/16	I	Light	0.38	0.22	0.025	0.005	0.012
		Medium					0.014
		Heavy					0.017
	II	Light	0.56	0.34	0.040	0.010	0.012
		Medium					0.014
		Heavy					0.017
3/32	I	Light	0.45	0.23	0.040	0.010	0.012
		Medium					0.014
		Heavy					0.017
	II	Light	0.70	0.38	0.050	0.020	0.012
		Medium					0.014
		Heavy					0.017
1/8	I	Light	0.58	0.31	0.040	0.010	0.012
		Medium					0.014
		Heavy					0.017
	II	Light	0.45	0.50	0.080	0.050	0.012
		Medium					0.014
		Heavy					0.017
5/32	I	Light	0.56	0.38	0.040	0.010	0.012
		Medium					0.014
		Heavy					0.017
5/32	II	Light	0.88	0.56	0.075	0.045	0.012
		Medium					0.014
		Heavy					0.017
3/16	I	Light	0.62	0.38	0.060	0.030	0.012
		Medium					0.017
		Heavy					0.020
	II	Light	0.98	0.56	0.080	0.050	0.012
		Medium					0.017
		Heavy					0.020
7/32	I	Light	0.62	0.44	0.050	0.020	0.012
		Medium					0.017
1/4	I	Light	0.62	0.44	0.050	0.020	0.012
		Medium					0.017
		Heavy					0.020
	II	Light	0.98	0.62	0.095	0.065	0.012
		Medium					0.017
		Heavy					0.020
5/16	I	Light	0.69	0.50	0.060	0.030	0.014
		Medium					0.020
3/8	I	Light	0.75	0.56	0.060	0.030	0.014
		Medium					0.020

[a]See General Specifications, Application and Design.

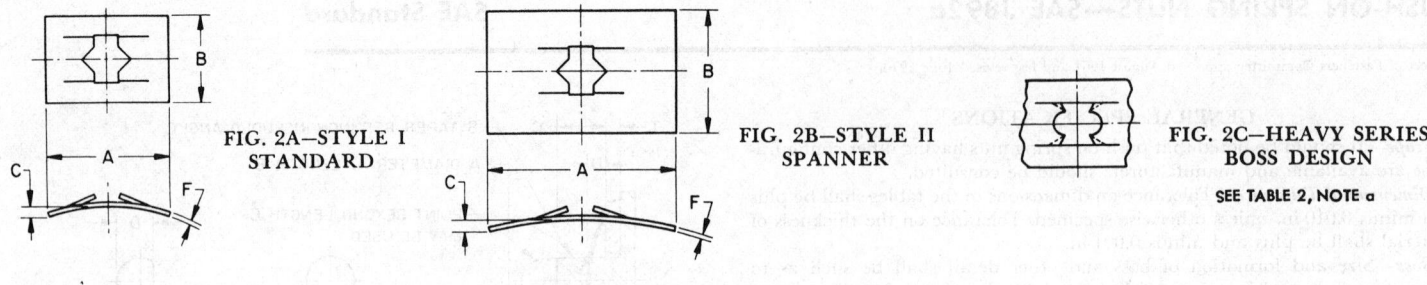

FIG. 2—PUSH-ON SPRING NUTS

required for normal installation). It may be necessary to increase this factor to provide adequate stud protrusion where uncompressed materials or mismatch of trim contours are encountered. It should be noted by users desiring to standardize on stud designs that the studs applicable to self-threading stamped nuts may be utilized for push-on spring nuts where economics justify and the additional stud protrusion is not objectionable.

Heavy Series Nuts are used on round studs only. All other nuts in this standard may be used on either round or "D" shaped studs. Nuts are used on round studs in applications where the assembly is permanent and on "D" shaped studs where disassembly is a consideration.

STEEL STAMPED NUTS OF ONE PITCH THREAD DESIGN—SAE J1053

SAE Standard

Report of Fasteners Committee approved August 1973.

1. General Specifications

1.1 Scope—Included herein are general, dimensional, and performance specifications for those types, styles, and sizes of stamped nuts of one pitch thread design recognized as SAE Standard. These nuts are intended for general use where the engagement of a single thread on the mating screw or unthreaded stud is considered adequate for the application.

It should be noted that stamped nuts having other sizes and configurations are available and manufacturers should be consulted.

1.2 Dimensional Tolerance—Tolerance on dimensions shown in the tables shall be ±0.010 in. unless otherwise specified.

1.3 Miscellaneous Dimensions—Taper on the sides of hexagon portions of nuts (angle between one side and the axis of nut) shall not exceed 1 deg, the maximum limit specified being the largest dimension. Tolerance on steel stock thickness shall be ±0.001 in. for thicknesses up to and including 0.028 in., and ±0.0015 in. for thicknesses exceeding 0.028 in.

1.4 Thread Embossments

1.4.1 Formed Thread Embossment—Detail of the thread engaging portion of formed thread type nuts shall be such as to permit nut to assemble readily with the specified screw and not strip or deform at the minimum torques shown in Table 2. The edges around the opening shall be spirally formed to conform to the helix of the mating thread and, as indicated on illustrations, the top or top and bottom corners on edges of holes shall be swaged to provide flats for bearing on flanks of the mating thread.

1.4.2 Self-Threading Embossment—The configuration of self-threading embossment may vary with manufacturer; however, the detail and formation of embossment shall be such as to enable the nuts to cut and/or form threads on cast or wire studs, conforming to the recommended stud designs contained in paragraph 3.2, at or below the maximum driving torques shown in Table 3.

1.5 Material—Nuts shall be fabricated from carbon spring steel suitably processed to meet the performance requirements of this standard.

1.6 Finish—Stamped nuts are normally supplied with finishes as specified by the purchaser. Nuts processed with supplemental finishes shall be suitably treated to obviate hydrogen embrittlement.

1.7 Workmanship—Stamped nuts shall be free from cracks, burrs, splits, loose scale, or any defects that might affect their serviceability.

2. Test Procedures and Performance Requirements

2.1 Formed Thread Embossment

2.1.1 Ultimate Torque Test—Insert hardened steel (Rockwell C53 min) unplated or uncoated test socket head cap screws of the respective size and 1.00 in. length, Class 3A thread, as-received with light coating of oil, into holes in the test fixture. The test fixture is to consist of a hardened steel (Rockwell C58-62) bar, 1.00 x 0.25 x 18.00 in. or equivalent, having 12 equally spaced test holes of the diameter given in Table 1 for respective size.

Hand assemble the test nuts to the test screws. In turn, hold each nut and tighten the test screw to the torque value shown in Table 2 for the respective size. The test shall be performed with a device capable of measuring the clamp load developed, and the load attained shall not be less than the minimum tension values specified in Table 2. After initial breakaway, the nuts must disassemble, by hand, from the test screws.

2.1.2 Embrittlement Test—Insert hardened steel (Rockwell C53 min) unplated or uncoated test socket head cap screws of the respective size and 1.00 in. length, Class 3A thread, as-received with light coating of oil, into holes in the test fixture described in paragraph 2.1.1.

Hand assemble new test nuts, from the same lot, to test screws. In turn, hold each nut and tighten test screw to the minimum torque value shown in

TABLE 1—TEST BAR HOLE SIZES, IN

Nominal Screw and Nut Size	Hole Diameter		Nominal Screw and Nut Size	Hole Diameter	
	Max	Min		Max	Min
6-32	0.149	0.144	1/4-20	0.262	0.257
8-32	0.178	0.173	5/16-18	0.328	0.323
10-24	0.204	0.199	3/8-16	0.391	0.386

TABLE 2—ULTIMATE TORQUE SPECIFICATIONS

Nominal Nut Size, in	Torque, lb-in	Tension, lb	Nominal Nut Size, in	Torque, lb-in	Tension, lb
	Min	Min		Min	Min
6-32	8	120	1/4-20	27	340
8-32	12	150	5/16-18	32	450
10-24	17	220	3/8-16	40	480

TABLE 3—TORQUE AND RELATED TENSION SPECIFICATIONS

Nominal Stud or Test Rod Dia, in	Test Rod Dia, in		Driving Torque, lb-in	Nut Flange Dia, in	Test Torque, lb-in	Tension, lb
	Max	Min	Max	Basic		Min
1/8	0.126	0.123	8	0.437	34	180
3/16	0.189	0.186	26	0.562	68	280
1/4	0.251	0.248	35	0.687	90	350

Table 2 for respective size. After 48 h in this state, the test nuts shall be examined. No cracks are permitted.

2.1.3 SCREW THREAD DAMAGE APPRAISAL TEST—Insert 12 nonheat treated, unplated or uncoated, steel screws of respective size and 1.00 in. length, into holes in the test fixture described in paragraph 2.1.1.

Hand assemble new test nuts, from the same lot, to test screws. In turn, hold each nut and tighten test screw to the torque value shown in Table 2 for the respective size. Remove nuts from screws and screws from test bar and examine threads on screws for visible damage. Continue test by assembling, with the fingers, untested nuts from same lot onto tested screws. The new nuts must pass over the area on the screw where the previously tested nut engaged the threads.

2.2 Self-Threading Embossment

2.2.1 STARTING EASE TEST—The test nut must start onto the chamfered end (0.030 x 45 deg) of an unplated or uncoated cold rolled steel (Rockwell 30T 78–81) rod of the diameter specified in Table 3, within one revolution of nut when applied with an appropriate socket affixed to a screwdriver handle.

2.2.2 ULTIMATE TORQUE TEST—Insert a test rod (see paragraph 2.2.1) into a suitable holding device exposing the chamfered end to a height equivalent to the nut height plus 0.125 in. or, for closed end nuts, equivalent to the wrenching height. Place an unplated or uncoated soft steel (Rockwell 30T 78–82) flat test plate on the exposed test rod. The test plate shall have a minimum thickness of 0.030 in., an inside diameter 0.031 in. larger than the diameter of test rod, and shall be at least 1.00 in. square. A new test plate shall be used for each torque test. The test rod and assembled plate must be retained in a suitable clamping device to prevent rotation of the rod and plate and tilting of the plate. The test shall be performed with a device capable of measuring the clamp load developed.

Assemble test nut on the test rod with a suitable torque indicating device. The maximum driving torque shall be recorded and this shall not exceed the maximum driving torque values shown in Table 3 for the respective size. At the torque test values specified in Table 3, the minimum tension values indicated shall be achieved.

2.2.3 EMBRITTLEMENT TEST—Assemble new test nut from the same lot to test rod using test torques shown in Table 3 for respective size. After 48 h, inspect the assembled nut for cracks. No cracks are permitted.

3. Design Criteria

3.1 Formed Thread Embossment—To insure proper starting of formed thread type stamped nuts, the length of the mating externally threaded component shall be such that it will protrude beyond the embossment in nut a minimum distance equivalent to two pitches (threads), exclusive of the length of any chamfer or point provision, under limit stack conditions. Recommended minimum protrusion lengths beyond panels with no allowance for pointing are presented in Fig. 1 and Table 4 for respective nut types.

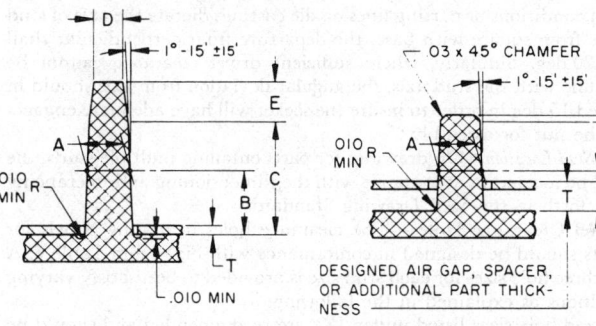

FIG. 2—DIE CAST STUD   FIG. 3—SHORT DIE CAST STUD

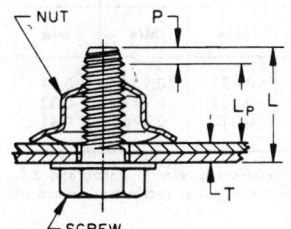   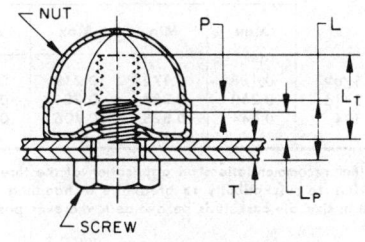

FIG. 1A—FACETED FLANGE TYPE NUTS   FIG. 1B—ACORN OR REGULAR TYPE NUTS

WHERE:
$L = L_P + T + P$
L = MINIMUM LENGTH OF SCREW OR STUD
L_P = MINIMUM PROTRUSION OF FULL FORM THREAD LENGTH BEYOND PANEL (SEE TABLE 4 FOR RESPECTIVE NUT TYPES)
L_T = MAXIMUM PROTRUSION OF MATING PART BEYOND PANEL ALLOWABLE FOR ACORN TYPE NUTS (SEE TABLE 4)
P = LENGTH OF POINT ON SCREW OR STUD
T = MAXIMUM THICKNESS OF PANEL OR PANELS TO BE ASSEMBLED, INCLUDING ALLOWANCE, IF NECESSARY, TO ACCOMMODATE MISMATCH OF SURFACES, ETC.

FIG. 1

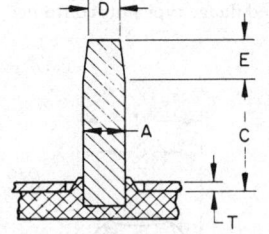   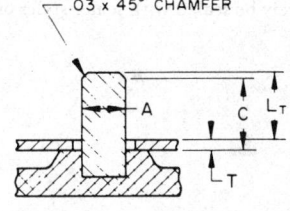

FIG. 4—WIRE OR ROD STUD   FIG. 5—SHORT WIRE OR ROD STUD

TABLE 4—PROTRUSION LENGTHS FOR FORMED THREAD TYPE STAMPED NUTS, IN

Nominal Thread Size	L_P		L_T
	Protrusion of Threaded Length on Mating Part Beyond Panel		Total Protrusion of Mating Part Beyond Panel
	Faceted Flange Type	Acorn or Regular Types	Acorn Types
	Min[a]	Min[b]	Max[c]
6-32	0.29	0.13	0.21
8-32	0.29	0.13	0.24
10-24	0.33	0.16	0.25
1/4-20	0.40	0.19	0.28
5/16-18	0.44	0.22	0.36
3/8-16	—	0.23	0.34

[a]Values shown are applicable to nuts shown in Table 7. For sealer styles, add height of uncompressed sealer.
[b]Values shown are applicable to nuts shown in Tables 8 and 9, respectively.
[c]Values shown apply to nuts shown in Table 8.

TABLE 5—RECOMMENDED DIMENSIONS OF STUDS FOR USE WITH SELF-THREADING TYPES OF STAMPED NUTS, IN

Nominal Stud Size	A[a]		B[b,c]		C[c]		D[a]	E[d]		L_T
	Stud Dia		Length		Length		Point Dia	Point Length		Stud Protrusion
			Faceted Flange Type	Acorn and Regular Types	Faceted Flange Type	Acorn and Regular Types				Acorn Type
	Max	Min	Min	Min	Min	Min	±0.005	Max	Min	Max
1/8[e]	0.128	0.122	T+0.24	T+0.08	T+0.45	T+0.12	0.073	0.175	0.145	0.26
3/16	0.191	0.185	T+0.24	T+0.11	T+0.45	T+0.16	0.130	0.190	0.150	0.36
1/4	0.253	0.247	T+0.27	T+0.11	T+0.45	T+0.16	0.180	0.260	0.220	0.38

[a]Maximum 0.003 in plating per side.
[b]Point on shank of die cast studs where A diameter must be within the specified limits.
[c]The T dimension in illustrations represents the distance from base of part to the bearing face of nut in the installed position. The factors to be added represent the minimum length required for normal installation of sealer styles of nuts. Where the factors specified would create an interference condition or otherwise be objectionable, it may be necessary to reduce the factor to that which is required for the respective nut size, type and style.
[d]On studs for acorn type nuts, it may be necessary to shorten point or apply the chamfer specified for short studs in order to keep protrusion of stud beyond panel within maximum permissible.
[e]Due to susceptibility to breakage in handling and processing, it is recommended use of 1/8 in size die cast studs be avoided wherever possible.

3.2 Self-Threading Embossment—To assure proper function and performance of self-threading types of stamped nuts and to provide flexibility for changing nut designs, it is essential that studs and clearance holes in mating panels be designed in conformance with the recommendations set forth in the following.

3.2.2 STUD DESIGN—Studs which are integral features of die cast components should comply as closely as possible with the recommendations presented in Figs. 2 and 3 and Table 5. Studs fabricated from wire or rod shall be in accord with recommendations shown in Figs. 4 and 5 and Table 5. Consideration should also be given to the recommendations for fillets, plating and alignment which follow:

3.2.2.1 *Fillets*—The fillet at the junction of stud with die casting base shall have as generous a radius as the design will permit, but not less than 0.010 in. Where the panel is to fit tight against the die casting, an annular relief groove should be provided in the die casting at base of stud to accommodate the fillet (see Fig. 2) and the fillet radius should be made larger wherever the design will permit.

3.2.2.2 *Angularity*—It is preferable that the axis of the stud be kept perpendicular to the base of the part or as nearly so as possible. However, where design conditions or parting lines on die castings dictate the axis of stud must deviate from square with base, the departure from perpendicular shall not exceed 20 deg. Similarly, where sufficient driver clearance cannot be provided in line with the stud axis, the angular deviation from axis should in no case exceed 15 deg in order to insure the socket will have adequate engagement with the nut for assembly.

3.2.2.3 *Stud Location*—On drawings for parts entailing multiple studs, the studs should be located in accordance with the dimensioning and tolerancing practices set forth in the SAE Drawing Standards.

3.3 Panel Clearance Holes—The clearance holes in mating panels for stamped nuts should be designed in conformance with Fig. 6 and Table 6. A selection of three hole sizes for each stud size is provided to best satisfy varying design conditions as explained in the following:

(a) Preferred hole sizes listed under "X" are recommended and should be used for all attachments requiring normal provisions for clearance and adjustment.

(b) Maximum clearance holes tabulated under "X_1" should be used only in applications where maximum adjustment capability is a requirement. These holes provide maximum clearance while assuring that the hole can effectively be sealed with sealer styles of the faceted flange type nuts contained herein.

(c) Minimum clearance holes shown under "X_2" may necessarily have to be used where the width of the part being fastened is at or approaches the minimum "Z" dimension. These holes provide adequate clearance for studs under limit stack conditions while insuring that the fastened part will cover the hole. It follows, therefore, that the "Z" dimension shall be the design criterion for the width of portions of parts adjacent to studs.

3.3.1 PANEL HOLE LOCATION—On drawings for panels, multiple holes shall be located in a manner which is compatible with that used to position studs on the mating part.

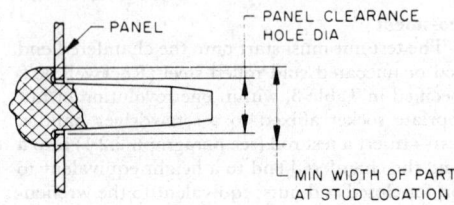

FIG. 6

TABLE 6—PANEL CLEARANCE HOLES, IN

Nominal Stud Dia	X		X_1		X_2		Z
	Clearance Hole Diameter[a]						Part Width
	Preferred		Maximum Clearance		Minimum Clearance		
	Max	Min	Max	Min	Max	Min	Min
1/8[b]	0.188	0.172	0.219	0.203	0.171	0.155	0.24
3/16	0.250	0.234	0.281	0.265	0.234	0.219	0.32
1/4	0.344	0.328	0.406	0.390	0.312	0.296	0.41

[a]For recommendations on application of the three choices offered, refer to paragraph 3.3.
[b]Due to susceptibility to breakage in handling and processing, it is recommended use of 1/8 in size die cast studs be avoided wherever possible.

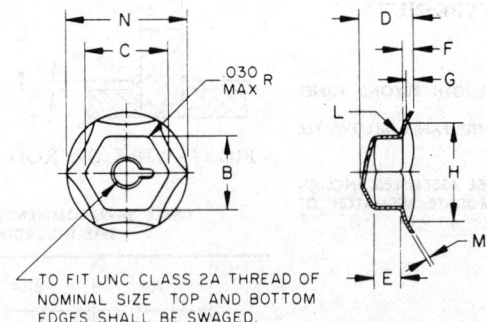

FIG. 7

TABLE 7—DIMENSIONS OF FORMED THREAD FACETED FLANGE TYPE STAMPED NUTS, IN[a]

Nominal Size[b] or Basic Thread Dia		Threads per in	B Hexagon Across Flats		C Hexagon Across Corners		D Overall Height		E Height of Flat	F Depth to Radius		G Dish Depth		H Dish Diameter		L Fillet Radius		M Stock Thickness	N Flange Diameter	
			Max	Min	Max	Min	Max	Min	Min	Max	Min	Max	Min	Max	Min	Max	Min	Basic	Max	Min
6	0.1380	32	0.312	0.306	0.360	0.348	0.218	0.198	0.067	0.017	0.007	—	—	0.401	0.387	0.033	0.027	0.013	0.442	0.432
8	0.1640	32	0.343	0.337	0.396	0.382	0.225	0.205	0.072	0.018	0.008	—	—	0.429	0.415	0.035	0.029	0.014	0.474	0.464
10	0.1900	24	0.375	0.369	0.433	0.418	0.246	0.226	0.076	0.019	0.009	—	—	0.457	0.443	0.037	0.031	0.018	0.505	0.495
1/4	0.2500	20	0.437	0.430	0.505	0.488	0.302	0.282	0.090	0.047	0.035	0.043	0.033	0.572	0.552	0.039	0.033	0.021	0.692	0.682
5/16	0.3125	18	0.500	0.492	0.577	0.557	0.330	0.310	0.100	0.057	0.045	0.050	0.040	0.674	0.650	0.041	0.035	0.023	0.817	0.807

[a]Sealer styles are also available, consult nut manufacturers.
[b]Where specifying nominal size in decimals, zeros preceding decimal and in fourth decimal place shall be omitted.

For recommended assembly data refer to Design Criteria in paragraph 3.
Additional requirements given in General Specifications shall apply.

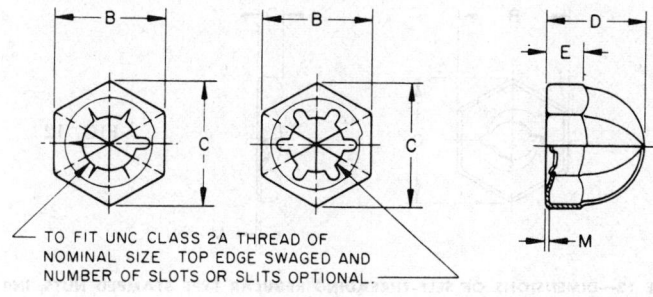

FIG. 8

TABLE 8—DIMENSIONS OF FORMED THREAD ACORN TYPE STAMPED NUTS, IN

Nominal Size[a] or Basic Thread Dia		Threads per in	B Hexagon Across Flats		C Hexagon Across Corners		D Overall Height ±0.010	E Height at Corner of Hexagon	M Stock Thickness
			Max	Min	Max	Min		Min	Basic
6	0.1380	32	0.312	0.306	0.360	0.348	0.261	0.084	0.013
8	0.1640	32	0.343	0.337	0.397	0.383	0.297	0.097	0.013
10	0.1900	24	0.375	0.368	0.433	0.418	0.324	0.110	0.017
1/4	0.2500	20	0.437	0.429	0.505	0.488	0.380	0.122	0.021
5/16	0.3125	18	0.562	0.553	0.650	0.627	0.484	0.157	0.024
3/8	0.3750	16	0.562	0.553	0.650	0.627	0.474	0.157	0.020

[a]Where specifying nominal size in decimals, zeros preceding decimal and in fourth decimal place shall be omitted.
For recommended assembly data, refer to Design Criteria in paragraph 3.
Additional requirements given in General Specifications shall apply.

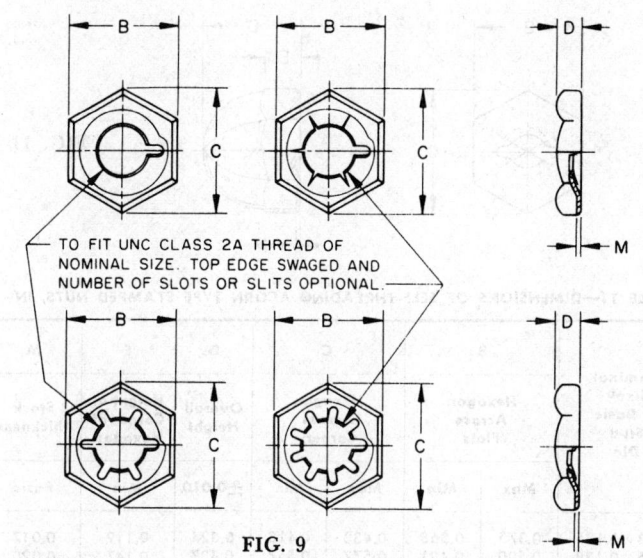

FIG. 9

TABLE 9—DIMENSIONS OF FORMED THREAD REGULAR TYPE STAMPED NUTS, IN

Nominal Size[a] or Basic Thread Dia		Threads per in	B Hexagon Across Flats		C Hexagon Across Corners		D Overall Height		M Stock Thickness
			Max	Min	Max	Min	Max	Min	Basic
6	0.1380	32	0.312	0.305	0.361	0.348	0.102	0.082	0.013
8	0.1640	32	0.343	0.336	0.397	0.383	0.109	0.089	0.013
10	0.1900	24	0.375	0.368	0.433	0.418	0.115	0.095	0.017
1/4	0.2500	20	0.437	0.429	0.505	0.488	0.133	0.113	0.021
5/16	0.3125	18	0.562	0.553	0.650	0.627	0.155	0.135	0.024
3/8	0.3750	16	0.625	0.615	0.722	0.697	0.166	0.146	0.027

[a]Where specifying nominal size in decimals, zeros preceding decimal and in fourth decimal place shall be omitted.
For recommended assembly data, refer to Design Criteria in paragraph 3.
Additional requirements given in General Specifications shall apply.

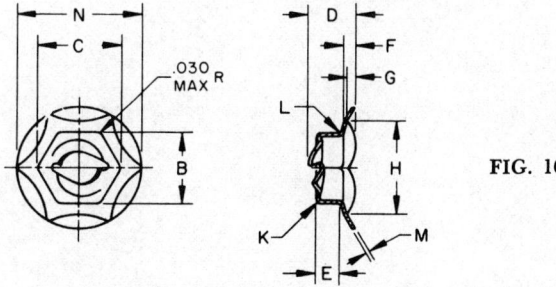

FIG. 10

TABLE 10—DIMENSIONS OF SELF-THREADING FACETED FLANGE TYPE STAMPED NUTS, IN[a]

Nominal Size[b] or Basic Stud Dia		B Hexagon Across Flats		C Hexagon Across Corners		D Overall Height		E Height of Flat	F Depth to Radius		G Dish Depth		H Dish Diameter		K Corner Radius		L Fillet Radius		M Stock Thickness	N Flange Diameter	
		Max	Min	Max	Min	Max	Min	Min	Max	Min	Max	Min	Max	Min	Max	Min	Max	Min	Basic	Max	Min
1/8	0.125	0.312	0.304	0.360	0.348	0.199	0.179	0.067	0.017	0.007	—	—	0.401	0.387	0.035	0.033	0.027	0.020	0.442	0.432
3/16	0.188	0.375	0.366	0.433	0.418	0.239	0.219	0.078	0.037	0.025	0.036	0.026	0.468	0.448	0.037	0.037	0.031	0.020	0.567	0.557
1/4	0.250	0.437	0.428	0.505	0.488	0.273	0.253	0.090	0.047	0.035	0.043	0.033	0.572	0.552	0.043	0.039	0.033	0.021	0.692	0.682

[a]Sealer styles are also available, consult nut manufacturers.
[b]Where specifying nominal size in decimals, zeros preceding decimal shall be omitted.
For recommended assembly data refer to Design Criteria in paragraph 3.
Additional requirements given in General Specifications shall apply.

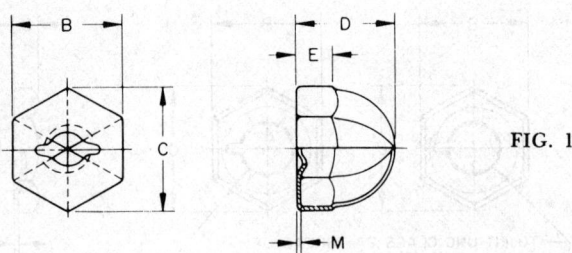

FIG. 11

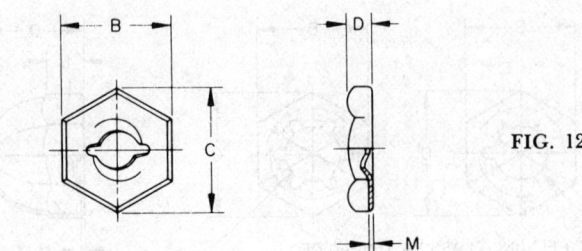

FIG. 12

TABLE 11—DIMENSIONS OF SELF-THREADING ACORN TYPE STAMPED NUTS, IN[a]

Nominal Size[b] or Basic Stud Dia		B Hexagon Across Flats		C Hexagon Across Corners		D Overall Height	E Height at Corner of Hexagon	M Stock Thickness
		Max	Min	Max	Min	±0.010	Min	Basic
1/8	0.125	0.375	0.368	0.433	0.418	0.324	0.119	0.017
3/16	0.188	0.500	0.491	0.577	0.557	0.437	0.147	0.020
1/4	0.250	0.562	0.553	0.650	0.628	0.484	0.166	0.024

[a]Sealer Styles are also available, consult nut manufacturers.
[b]Where specifying nominal size in decimals, zeros preceding decimal shall be omitted.
For recommended assembly data, refer to Design Criteria in paragraph 3.
Additional requirements given in General Specifications shall apply.

TABLE 12—DIMENSIONS OF SELF-THREADING REGULAR TYPE STAMPED NUTS, IN[a]

Nominal Size[b] or Basic Stud Dia		B Hexagon Across Flats		C Hexagon Across Corners		D Overall Height		M Stock Thickness
		Max	Min	Max	Min	Max	Min	Basic
1/8	0.125	0.312	0.306	0.360	0.348	0.110	0.090	0.017
3/16	0.188	0.500	0.492	0.577	0.557	0.139	0.119	0.019
1/4	0.250	0.500	0.492	0.577	0.557	0.150	0.130	0.026

[a]Sealer styles are also available, consult nut manufacturers.
[b]Where specifying nominal size in decimals, zeros preceding decimal shall be omitted.
For recommended assembly data refer to Design Criteria in paragraph 3.
Additional requirements given in General Specifications shall apply.

CONICAL SPRING WASHERS—SAE J773b — SAE Standard

Report of Fasteners Committee approved June 1961 and last revised February 1976.

1. Scope—This SAE Standard covers dimensional, material, and general specifications and methods of test for two types of general purpose conical spring washers, designated type L and type H, for use as loose washers over screws and bolts, and also for use as pre-assembled washers in screw and washer assemblies.

1.1 Both the type L and type H washers are available in three washer series (narrow, regular and wide), having varied proportions designed to fulfill specific application requirements for load distribution.

1.2 Where so specified by the user, washers shall be supplied with peripheral teeth.

1.3 All sizes and types of washers specified in this standard are not necessarily stock production items. Users should consult with manufacturers concerning availability.

2. Designation—Washers shall be specified or designated as shown in the following example:

Washer, Conical, 1/2, SAE Type L, Wide

3. Use and Application—Type L washers are intended for use with screws and bolts equivalent to SAE Grade 1 and 2. Type H washers are intended for use with SAE Grade 5 or equivalent bolts or screws (SAE J429).

3.1 The desired installed position of this washer is as near flat as possible. The flattening will occur at a load equal to approximately 27 500 bolt psi for the Type L washer and 60 000 bolt psi for the Type H washer. The spring return will vary due to the compromises in washer diameter, thickness, and tolerances, which have been made to maintain this standard in a commercial category (see Paragraph 7.1, Recovery Test).

3.2 The relatively high supporting load and spring return makes this washer very effective where bolt tension may be subject to loss due to such factors as compensating for wear, thermal expansion, or compression set.

3.3 When used to span over-size clearance holes, it is recommended that (1) if the full periphery is supported, at least 70% of the washer annular area be bearing or (2) if the periphery is partially supported, as over a slot, the slot should be no wider than 1 1/2 times the I.D. Narrow series should always be fully supported. Insufficient bearing will reduce spring return.

3.4 Washers with peripheral teeth are used for non-slip or positive electrical grounding purposes.

4. Dimensions—Dimensions of Type L and Type H conical spring washers are specified in Table 1.

4.1 Manufacturing Detail—Washers shall be symmetrical in shape. The radial section of the washer shall be flat to convex upward with flat preferred

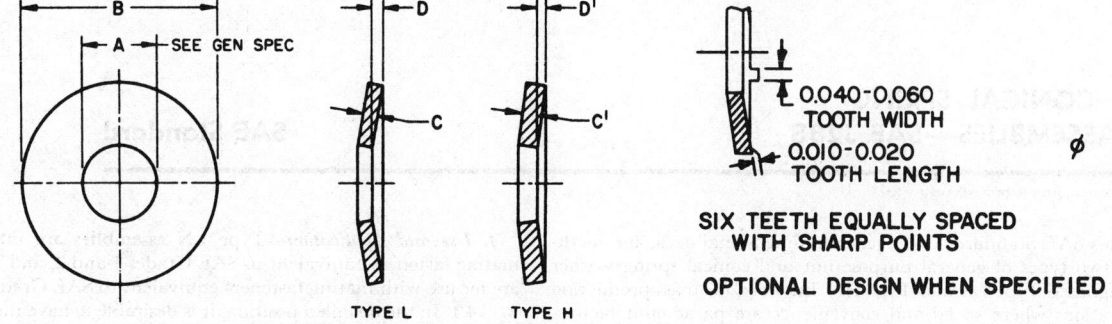

TABLE 1—DIMENSIONS OF CONICAL SPRING WASHERS, in

Nominal Screw or Bolt Size	Washer Series	A[a] ID		B OD		Type L				Type H			
						C Thickness			D Crown Height		C¹ Thickness		D¹ Crown Height	
		Min	Max	Max	Min	Nom	Max	Min	Min	Max	Nom	Max	Min	Min	Max
6	Narrow	0.151	0.156	0.320	0.307	0.025	0.029	0.023	0.010	0.016	0.035	0.040	0.033	0.015	0.025
	Regular			0.446	0.433	0.030	0.034	0.028	0.014	0.020	0.040	0.046	0.037	0.015	0.025
	Wide			0.570	0.557	0.030	0.034	0.028	0.021	0.031	0.040	0.046	0.037	0.019	0.029
8	Narrow	0.183	0.188	0.383	0.370	0.035	0.040	0.033	0.010	0.016	0.040	0.046	0.037	0.015	0.025
	Regular			0.508	0.495	0.035	0.040	0.033	0.020	0.030	0.045	0.050	0.042	0.016	0.026
	Wide			0.640	0.620	0.035	0.040	0.033	0.027	0.037	0.045	0.050	0.042	0.030	0.040
10	Narrow	0.203	0.208	0.446	0.433	0.035	0.040	0.033	0.010	0.016	0.050	0.056	0.047	0.015	0.025
	Regular			0.570	0.557	0.040	0.046	0.037	0.017	0.027	0.055	0.060	0.052	0.016	0.026
	Wide			0.765	0.743	0.040	0.046	0.037	0.026	0.036	0.055	0.060	0.052	0.024	0.034
12	Narrow	0.230	0.240	0.446	0.433	0.040	0.046	0.037	0.011	0.017	0.055	0.060	0.052	0.015	0.025
	Regular			0.640	0.620	0.040	0.046	0.037	0.023	0.033	0.055	0.060	0.052	0.016	0.026
	Wide			0.890	0.868	0.045	0.050	0.042	0.034	0.044	0.064	0.071	0.059	0.023	0.033
1/4	Narrow	0.271	0.281	0.515	0.495	0.045	0.050	0.042	0.014	0.024	0.064	0.071	0.059	0.015	0.025
	Regular			0.765	0.743	0.050	0.056	0.047	0.023	0.033	0.079	0.087	0.074	0.022	0.032
	Wide			1.015	0.993	0.055	0.060	0.052	0.030	0.040	0.079	0.087	0.074	0.029	0.039
5/16	Narrow	0.334	0.344	0.640	0.620	0.055	0.060	0.052	0.016	0.026	0.079	0.087	0.074	0.016	0.026
	Regular			0.890	0.868	0.064	0.071	0.059	0.031	0.041	0.095	0.103	0.090	0.019	0.029
	Wide			1.140	1.118	0.064	0.071	0.059	0.034	0.044	0.095	0.103	0.090	0.030	0.040
3/8	Narrow	0.396	0.406	0.765	0.743	0.071	0.079	0.066	0.015	0.025	0.095	0.103	0.090	0.015	0.025
	Regular			1.015	0.993	0.071	0.079	0.066	0.033	0.043	0.118	0.126	0.112	0.023	0.033
	Wide			1.265	1.243	0.079	0.087	0.074	0.037	0.047	0.118	0.126	0.112	0.035	0.045
7/16	Narrow	0.470	0.480	0.890	0.868	0.079	0.087	0.074	0.018	0.028	0.128	0.136	0.122	0.016	0.026
	Regular			1.140	1.118	0.095	0.103	0.090	0.031	0.041	0.128	0.136	0.122	0.028	0.038
	Wide			1.530	1.493	0.095	0.103	0.090	0.049	0.059	0.132	0.140	0.126	0.039	0.049
1/2	Narrow	0.530	0.540	1.015	0.993	0.100	0.108	0.094	0.021	0.031	0.142	0.150	0.136	0.020	0.030
	Regular			1.265	1.243	0.111	0.120	0.106	0.033	0.043	0.142	0.150	0.136	0.027	0.037
	Wide			1.780	1.743	0.111	0.120	0.106	0.052	0.062	0.152	0.160	0.146	0.042	0.052

[a] Not applicable to washers assembled with screw blanks. See General Specifications.

ϕ (see Fig. 1). Unless otherwise specified by the user, the direction of blanking the outside diameter should permit the sharper edge to be on the underside of the washer. Washers shall be free from sharp edges, burrs, cracks, checks, embrittlement, loose scale, and all other defects that might affect their serviceability.

4.2 Assembly Detail—The inside diameters of washers for pre-assembly on unthreaded screw blanks shall be optional, but shall be such that the washer will be retained on the screw after thread rolling, but shall not bind on the screw shank before and during tightening of the assembly.

5. Material and Hardness—Washers shall be made from SAE 1050 to 1065 carbon steel, fabricated and heat treated to a Rockwell hardness of 44-48 C scale (Rockwell hardness of C46-50 if austempering is used) or equivalent for loose washers, and 40-48 C scale or equivalent for pre-assembled washers, heat treated as an integral part of heat treated bolt or screw and washer assemblies. Washer hardness shall be checked by grinding or filing a flat spot on the top conical surface of the washer to rest on the anvil, with reading to be taken on the undisturbed inner face of the washer. If washer hardness, as obtained above, is not within specification, washers may be qualified by checking hardness on a cut-out section of the washer on which both sides have been ground. However, an excessive decarburized surface, especially on the lighter gage material, may be grounds for rejection if the performance of the washer is affected.

6. Finish—Electroplated washers or screw or bolt and washer assemblies shall be baked at 400°F as soon as practicable after plating, in order to relieve hydrogen or acid embrittlement. If washers so treated fail to meet the prescribed tests, the baking time and/or the temperature shall be increased, but not to approach annealing temperature.

7. Tests

ϕ **7.1 Recovery Test**—The washers shall retain at least one-third their original crown height after flattening between two hardened plates and release. (Note: Conical washers which have a higher angle of elevation than covered by this standard are not expected to have the same percentage of recovery.)

7.2 Embrittlement Test—As a constant quality control check, a minimum of 12 pieces shall be taken from each batch after plating or final finishing operations and subjected to a load test sufficient to flatten washers for a minimum period of 24 h. Upon examination after testing, washers shall not exhibit cracks or fractures.

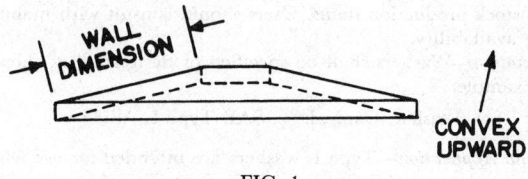

FIG. 1

NUT AND CONICAL SPRING WASHER ASSEMBLIES—SAE J238

SAE Standard

Report of Fasteners Committee approved August 1973.

1. Scope—This SAE Standard covers general, dimensional data, and methods of test for two types of general purpose nut and conical spring washer assemblies, designated Type LN and Type HN, intended for mass production and other operations where speed and convenience are paramount factors.

1.1 Both the Type LN and Type HN assemblies are available in three washer series (narrow, regular, and wide), having varied proportions designed to fulfill specific purposes of distributing the load over various areas, as shown in Table 1.

1.2 Where so specified by user, assemblies shall be supplied with toothed washers for nonslip or positive electrical grounding purposes. Toothed washers shall have six teeth, of proportions depicted in Fig. 1, equally spaced on the outer periphery. Teeth shall have sharp edges.

1.3 The inclusion of dimensional data in this standard is not intended to imply that all of the products described are stock production items. Users should consult with manufacturers concerning availability.

2. Designation—Nut and conical spring washer assemblies shall be specified or designated as shown in the following examples: $\frac{1}{4}$-20 nut and conical spring washer assembly, Type LN, wide; No. 10-24 nut and toothed conical spring washer assembly, Type HN, regular. (Unless otherwise specified, threads will be furnished as Class UNC 2B.)

3. Identification—Assemblies for No. 10 and $\frac{1}{4}$ in. nominal sizes are available in Types LN and HN. To identify the HN type in these sizes, parts should be finished in accordance with paragraph 7.

4. Use and Application—Type LN assemblies are intended for use with mating fasteners equivalent to SAE Grades 1 and 2, and Type HN assemblies are for use with mating fasteners equivalent to SAE Grade 5. (See SAE J429.)

4.1 In the installed position, it is desirable to have the washer compressed flat. Such flattening is designed to occur at a load in the bolt equivalent to approximately 27,500 psi for the Type LN assemblies and 60,000 psi for the Type HN assemblies.

4.2 The relatively high load supporting and spring return characteristics of the washer components make these assemblies very effective in applications where bolt tension may be subject to loss due to such factors as brinnelling, thermal set of parts, compression set of gaskets, etc.

5. Dimensions—All dimensions in this standard are in inches unless otherwise specified. Dimensions for both Type LN and Type HN assemblies are given in Table 1.

5.1 Nut Manufacturing Detail—The nut thickness specified in Table 1 is the overall distance, measured parallel to the axis of nut, from the top of nut to the surface which bears against top of washer. No transverse section through the nut between 25 and 75% of the actual nut thickness, as measured from the top of the nut, shall be less than the minimum width across flats. The maximum width across flats shall not be exceeded. Tops of nuts shall be flat. Corners on top and bottom of hexagon portion of nuts shall be chamfered to a diameter equal to the maximum width across flats within a tolerance of −15%. The length of chamfer at hexagon corners shall be 5–15% of the basic

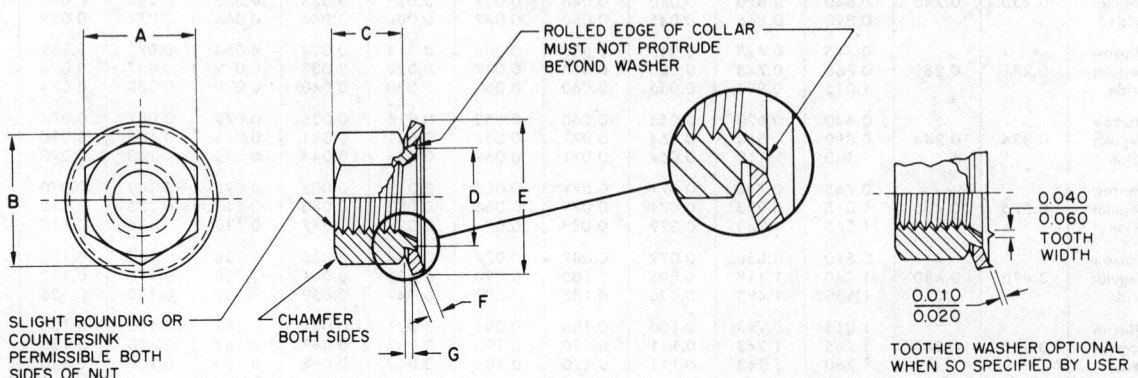

FIG. 1

TABLE 1—DIMENSIONS OF NUT AND CONICAL SPRING WASHER ASSEMBLIES

Nom Size	Basic Major Dia of Thread	Thds per in	Washer Series	Types LN and HN						Type LN								Type HN					
				Washer OD	Nut					Nut		Washer						Nut		Washer			
				E	A		B		D		C		F		G			C		F		G	
				±0.010	Max	Min	Min	Max	Min	Max	Min	Max	Min	Max	Min			Max	Min	Max	Min	Max	Min
No. 8	0.1540	32	Narrow Regular Wide	0.375 0.500 0.625	0.343	0.332	0.378	0.236	0.232	0.130	0.117	0.040 0.040 0.040	0.033 0.033 0.033	0.025 0.025 0.035	0.015 0.015 0.025			—	—	—	—	—	—
No. 10	0.1900	24	Narrow Regular Wide	0.438 0.562 0.750	0.375	0.365	0.413	0.274	0.270	0.130	0.117	0.040 0.040 0.046	0.033 0.033 0.037	0.025 0.025 0.030	0.015 0.015 0.020			0.207	0.187	0.043 0.051 0.056	0.037 0.042 0.047	0.025 0.025 0.030	0.015 0.015 0.020
1/4	0.2500	20	Narrow Regular Wide	0.625 0.750 1.000	0.437	0.428	0.488	0.332	0.328	0.193	0.178	0.051 0.056 0.065	0.042 0.047 0.055	0.025 0.025 0.030	0.015 0.015 0.025			0.226	0.212	0.065 0.079 0.087	0.055 0.066 0.074	0.025 0.025 0.030	0.015 0.015 0.020
5/16	0.3125	18	Narrow Regular Wide	0.750 1.000 1.125	0.500	0.489	0.557	0.405	0.400	—	—	—	—	—	—			0.273	0.258	0.079 0.103 0.103	0.066 0.090 0.090	0.025 0.030 0.032	0.015 0.020 0.022
3/8	0.3750	15	Narrow Regular Wide	1.000 1.125 1.250	0.562	0.551	0.628	0.470	0.465	—	—	—	—	—	—			0.337	0.320	0.103 0.120 0.120	0.090 0.106 0.106	0.025 0.032 0.035	0.015 0.022 0.025
7/16	0.4375	14	Narrow Regular Wide	1.125 1.250 1.500	0.687	0.675	0.768	0.550	0.545	—	—	—	—	—	—			0.385	0.365	0.126 0.136 0.136	0.112 0.122 0.122	0.027 0.036 0.036	0.017 0.026 0.026
1/2	0.5000	13	Narrow Regular Wide	1.250 1.500 1.750	0.750	0.736	0.840	0.610	0.605	—	—	—	—	—	—			0.448	0.427	0.140 0.150 0.150	0.126 0.136 0.136	0.027 0.035 0.035	0.017 0.025 0.025

thread diameter. The surface of chamfer may be slightly convex or rounded. A rounding or lack of fill at the junction of hexagon corners with chamfer shall be permissible provided the minimum width across corners is reached and maintained beyond a distance equal to 17.5% of the basic thread diameter from the chamfered faces.

5.1.1 TAPER OF SIDES OF HEX—Nut (angle between one side and the axis) shall not exceed 2 deg, the specified width across flats being the largest dimension.

5.2 Washer Manufacturing Detail—The washers shall be symmetrical in shape and shall be tumbled (except toothed washers) or otherwise processed to remove sharp edge at top inner periphery prior to assembly to nuts.

5.2.1 A diametral section through the washer shall show the surface element to be straight, subject to the following tolerances (see Fig. 2):

Wall Dimension	Tolerance (convex upward only), in
Up to 1/4	0.010
Over 1/4 to 1/2	0.015
Over 1/2	0.020

5.3 Assembly Detail—The size and shape of the hole in washers and the collar on the nuts shall be such that washers after assembly to nuts—by spinning, swaging, or staking of collar—will be firmly retained on the nuts and yet be free to rotate at a torque not to exceed 5 lb-in. The length of the collar on the nuts shall be such as to be wholly contained within the thickness of the washer after the assembly operation. No protusion of the collar beyond the washer in the retention area shall be permissible.

5.3.1 COLLAR CRACKS—Collar cracks may occur due to the application of pressure to the collar lip during assembly of the washer. Providing these cracks are limited to the contour of the collar, such cracks shall be permissible discontinuities and not considered cause for rejection of otherwise acceptable assemblies.

6. Material—Nut and washer components of assemblies shall be made from materials specified below:

6.1 Nuts shall be manufactured in accordance with SAE J995 (latest issue). Type LN shall be Grade 2 and Type HN shall be Grade 5.

6.2 Washers shall be made from SAE 1050 to 1065 carbon steel, fabricated and heat treated to a hardness of Rockwell C44-48 (or equivalent) and shall be capable of meeting the embrittlement tests set forth in paragraph 8.2. When the austempering process is used, washers shall be heat treated to a Rockwell C 46-50.

When heat treatment takes place after assembly of the washer and nut, a hardness range of Rockwell C 40-48 is permitted. Washer hardness shall be checked by grinding or filing a flat spot on the top side of the washer to rest on the anvil with the reading to be taken on the undisturbed inner face of the washer. If washer hardness, as thus obtained, is not within specification, washers may be qualified by checking hardness on a cutout section of the washer on which both sides have been ground flat and parallel. Excessive decarburization which adversely affects the performance of the washer may be grounds for rejection of the assembly.

7. Finish—Finish shall be as specified by purchaser. Where assemblies are to be used for electrical ground, cadmium or zinc plating is recommended. To identify the No. 10 and 1/4 in. nominal sizes Type HN when used for electrical grounding, surface treatment with yellow dichromate solution is recommended. Where electrical grounding is not a consideration, it is recommended that the No. 10 and 1/4 in. nominal sizes Type HN be phosphate coated.

7.1 Assemblies shall be free from hydrogen embrittlement or acid embrittlement. It is recommended that electroplated assemblies be baked at approximately 400°F for 3 h as soon as practicable after plating. If assemblies so treated fail to meet the test described in paragraph 8, the baking time and/or the baking temperature shall be increased.

8. Tests

8.1 Recovery Test—Conical washers shall not remain flat after deflection and release. The washers covered by this standard shall retain at least one-third the original minimum crown height after flattening between two hardened plates and release. (Note: Conical washers which have a higher angle of elevation than covered by this standard are not expected to have the same percentage of recovery.)

8.2 Embrittlement Test—As a constant quality control check, a minimum of 12 assemblies shall be taken from each batch after plating or final finishing operations and subjected to a load test sufficient to flatten washers for a minimum period of 24 h. Upon examination after testing, washers shall not exhibit any signs of cracks or fractures.

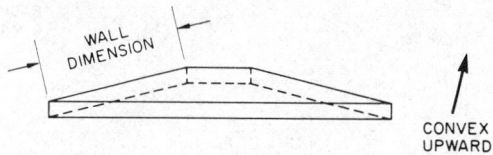

FIG. 2

TORQUE-TENSION TEST PROCEDURE FOR STEEL THREADED FASTENERS—SAE J174

SAE Recommended Practice

Report of Fasteners Committee approved June 1970. Editorial change April 1971.

Introduction—On some applications of threaded fasteners, it is desirable to control the amount of developed tension when a specified range of torque has been applied or the torque required to develop a specified range of tension. Accurate torque-tension relationships can be achieved only by uniquely defining and controlling the many related test parameters.

Scope—This test procedure is intended to provide a standard method for checking torque-tension relationships of nonprevailing torque type threaded steel fasteners 1/4 through 1 in. nominal diameters.

Test Material

Test Bolt—Bolts conforming to SAE J429, Grade 8, requirements shall be used to evaluate nuts. Threads shall gage to the same class of fit as the nut. Threads on all bolts shall be produced by rolling.

Bolts shall be free from burrs, loose scale, fins, and contamination. The finish shall be zinc phosphate and oil (dry to the touch), meeting a 72 hr salt spray life when tested in accordance with ASTM D 117.

Test Washer—Washer shall conform to the dimensional, metallurgical, mechanical, and finish requirements given in Table 1. Optionally, clipped washers or multihole plates or strips may be used providing they conform to the above requirements.

Test Nut—Nuts conforming to SAE J995, Grade 8, requirements shall be used to evaluate bolts. Threads shall gage to the same class of fit as the bolt.

Nuts shall be free from burrs, loose scale, and contamination. The finish shall be zinc phosphate and oil (dry to the touch), meeting a 72 hr salt spray life when tested in accordance with ASTM B 117.

NOTE: Lubricant shall neither be removed nor added to test material.

Test Equipment

Tension Measuring Device—The tension measuring device shall be capable of measuring the axial tension induced in the bolt as it is tightened. The device shall be accurate within ±5% of the test load.

Torque Measuring Device—The torque measuring device shall have an accuracy within ±1% of a given torque reading.

Test Socket—A hexagon socket is preferred, features shall be provided in the socket to prevent the socket from contacting either the test washer or the threaded end of the bolt.

Test Spacer (If Required)—The test spacer (used only for testing bolts) shall be placed under the nut. The spacer must be hardened to Rockwell C52 minimum and the faces shall be parallel to each other and perpendicular to the axis within 0.0005 in./in. The spacer hole diameter shall be equivalent to Table 1, dimension A, and minimum spacer wall thickness shall be equivalent to one-half the bolt diameter. A feature of preventing the nut and spacer from rotating shall be provided.

Test Method

Testing Bolt—The bolt, as received, shall be inserted in the tension measuring device with the test washer placed under the bolt head. The test nut, and spacer if required, shall be assembled onto the bolt by turning the bolt head until the bolt is seated against the hardened washer. The test shall be such that a minimum of two threads protrude through the nut.

The bolt shall then be continuously and uniformly tightened at a speed not to exceed 30 rpm, with a torque measuring device or equivalent means, until either the torque or the tension value, as required, is developed, at which time both torque and tension readings shall be recorded. NOTE: The nut must not have engaged incomplete bolt threads.

During all tests, the test washer shall be prevented from turning and contacting bolt shank. A new bolt, nut, and washer shall be used for each test.

Testing Nut—To test a nut, the nut and bolt exchange positions and the above procedure shall apply.

TABLE 1—TEST WASHERS

Nominal Fastener Size	Washer Dimensions				
	Inside Dia A[a]	Outside Dia B	Width D	Thickness C	
				Max	Min
1/4	0.281	0.750	0.656	0.080	0.073
5/16	0.344	0.875	0.776	0.080	0.073
3/8	0.406	1.000	0.892	0.080	0.073
7/16	0.469	1.125	1.018	0.080	0.073
1/2	0.531	1.312	1.152	0.121	0.114
9/16	0.625	1.500	1.274	0.121	0.114
5/8	0.688	1.625	1.422	0.121	0.114
11/16	0.750	1.687	1.500	0.121	0.114
3/4	0.812	1.750	1.678	0.160	0.153
7/8	0.969	1.875	1.916	0.160	0.153
1	1.025	2.000	2.184	0.160	0.153

[a] The washer ID is intended for use with finished hex bolts and all nuts. To accommodate other bolts with larger under head fillet radii, washer hole diameter shall be increased proportionally to allow bearing surface of bolt head to seat.

NOTES:
1. All dimensions are in inches.
2. Square washers are preferred. Use of round washers is acceptable during a transition period to exclusive use of square washers.
3. Material shall be carbon steel with a chemical composition of C, 0.48–0.60%; Mn, 0.60–1.50%; P, 0.035% max; and S, 0.045% max; quenched and tempered, with a surface hardness of Rockwell 15N 85–88, and a core hardness of Rockwell A73–78.
4. Washers shall be electrodeposited zinc plated to a coating thickness of 0.0002–0.0004 in. and shall be subjected to no additional surface treatment. As soon as practicable following plating, washers shall be baked for 1 hr at 375 ± 25 F. Plating thickness shall be checked in accordance with ASTM A 219 (Microscopic Test).
5. Washers shall be free from burrs and sharp edges.

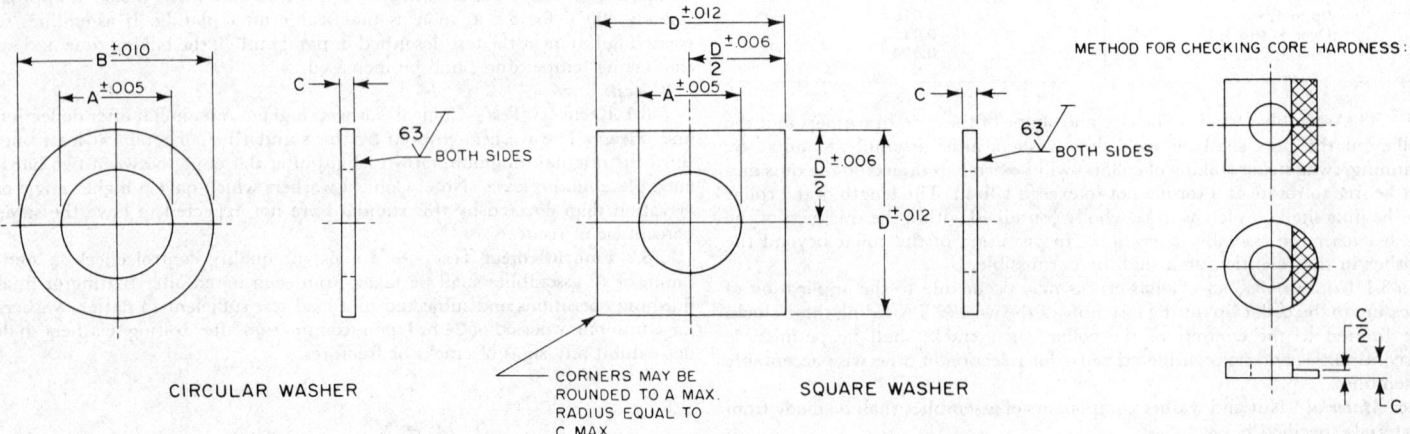

CIRCULAR WASHER   SQUARE WASHER (CORNERS MAY BE ROUNDED TO A MAX. RADIUS EQUAL TO C MAX.)   METHOD FOR CHECKING CORE HARDNESS

STRAIGHT PINS (SOLID)—SAE J495

SAE Standard

Report of Parts and Fittings Committee approved January 1957. Editorial change June 1964.

TABLE 1—DIMENSIONS OF STRAIGHT PINS, IN.[a]

Nominal	A, Pin Dia Max	A, Pin Dia Min	B Chamfer
0.062	0.0625	0.0605	0.015
0.094	0.0937	0.0917	0.015
0.109	0.1094	0.1074	0.015
0.125	0.1250	0.1230	0.015
0.156	0.1562	0.1542	0.015
0.188	0.1875	0.1855	0.015
0.219	0.2187	0.2167	0.015
0.250	0.2500	0.2480	0.015
0.312	0.3125	0.3095	0.030
0.375	0.3750	0.3720	0.030
0.438	0.4375	0.4345	0.030
0.500	0.500	0.4970	0.030

[a] These pins must be straight and free from burrs or any other defects that will affect their serviceability.

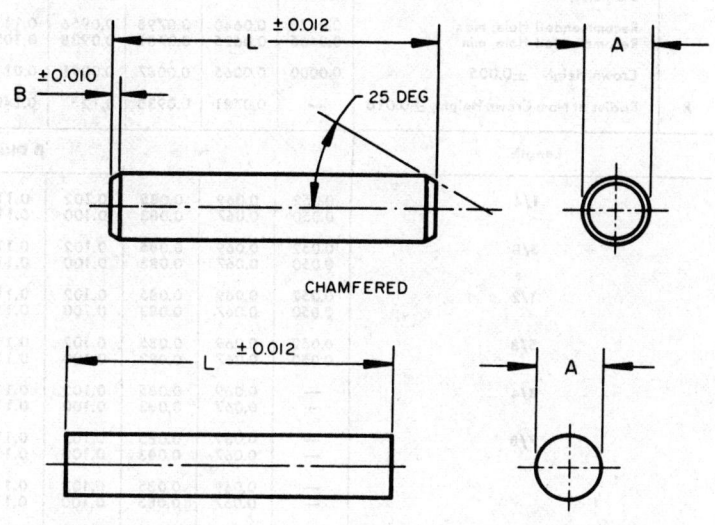

GROOVED STRAIGHT PINS—SAE J494

SAE Standard

Report of Parts and Fittings Committee approved May 1955 and last revised May 1959. Editorial change June 1962. Reaffirmed without change June 1964.

GENERAL DATA

Material—Cold drawn SAE 1112 or 1113 steel, alloy steel, stainless steel or copper alloy as specified by purchaser.

Finishes—Unless otherwise specified, steel pins shall have a flash plate of cadmium or zinc for protection of pins in transit or storage.

Defects—Grooved Straight Pins must be free from burrs and all other defects that might affect their use and serviceability.

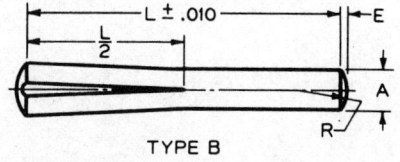
TYPE B

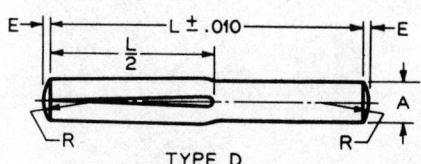
TYPE D

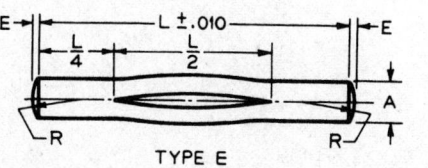
TYPE E

TABLE 1—TYPES B, D, AND E GROOVED STRAIGHT PINS, IN.

	Nominal Size	3/64	1/16	5/64	3/32	7/64	1/8	5/32	3/16	7/32	1/4	5/16	3/8	7/16	1/2
A	Diameter, max	0.0469	0.0625	0.0781	0.0938	0.1094	0.1250	0.1563	0.1875	0.2188	0.2500	0.3125	0.3750	0.4375	0.5000
	Diameter, min	0.0459	0.0615	0.0771	0.0928	0.1084	0.1230	0.1543	0.1855	0.2168	0.2480	0.3105	0.3730	0.4355	0.4980
	Recommended Hole, max	0.0478	0.0640	0.0798	0.0956	0.1113	0.1271	0.1587	0.1903	0.2219	0.2534	0.3166	0.3797	0.4428	0.5060
	Recommended Hole, min	0.0465	0.0625	0.0781	0.0938	0.1094	0.1250	0.1563	0.1875	0.2188	0.2500	0.3125	0.3750	0.4375	0.5000
	Crown Height, ±0.005	0.0000	0.0065	0.0087	0.0091	0.0110	0.0130	0.0170	0.0180	0.0220	0.0260	0.0340	0.0390	0.0470	0.0520
R	Radius at Nom Crown Height, ±0.010	—	0.0781	0.0938	0.125	0.1406	0.1562	0.1875	0.2500	0.2812	0.3125	0.3750	0.4688	0.5312	0.6250

Length	\	\	\	\	B Diameter, Max and Min Limits (Measured with Ring Gages)									
1/4	0.052	0.069	0.085	0.102	0.118	0.136	—	—	—	—	—	—	—	—
	0.050	0.067	0.083	0.100	0.116	0.132	—	—	—	—	—	—	—	—
3/8	0.052	0.069	0.085	0.102	0.118	0.136	0.168	0.200	—	—	—	—	—	—
	0.050	0.067	0.083	0.100	0.116	0.132	0.164	0.196	—	—	—	—	—	—
1/2	0.052	0.069	0.085	0.102	0.118	0.136	0.168	0.200	0.232	0.265	—	—	—	—
	0.050	0.067	0.083	0.100	0.116	0.132	0.164	0.196	0.228	0.261	—	—	—	—
5/8	0.052	0.069	0.085	0.102	0.118	0.136	0.168	0.200	0.232	0.265	0.331	—	—	—
	0.050	0.067	0.083	0.100	0.116	0.132	0.164	0.196	0.228	0.261	0.327	—	—	—
3/4	—	0.069	0.085	0.102	0.118	0.136	0.168	0.200	0.232	0.265	0.331	0.396	—	—
	—	0.067	0.083	0.100	0.116	0.132	0.164	0.196	0.228	0.261	0.327	0.392	—	—
7/8	—	0.069	0.085	0.102	0.118	0.136	0.168	0.200	0.232	0.265	0.331	0.396	0.461	—
	—	0.067	0.083	0.100	0.116	0.132	0.164	0.196	0.228	0.261	0.327	0.392	0.457	—
1	—	0.069	0.085	0.102	0.118	0.136	0.168	0.200	0.232	0.265	0.331	0.396	0.461	0.527
	—	0.067	0.083	0.100	0.116	0.132	0.164	0.196	0.228	0.261	0.327	0.392	0.457	0.523
1-1/4	—	—	—	0.102	0.118	0.136	0.168	0.200	0.232	0.265	0.331	0.396	0.461	0.527
	—	—	—	0.100	0.116	0.132	0.164	0.196	0.228	0.261	0.327	0.392	0.457	0.523
1-1/2	—	—	—	—	—	0.136	0.168	0.200	0.232	0.265	0.331	0.396	0.461	0.527
	—	—	—	—	—	0.132	0.164	0.196	0.228	0.261	0.327	0.392	0.457	0.523
1-3/4	—	—	—	—	—	—	0.167	0.200	0.232	0.265	0.331	0.396	0.461	0.527
	—	—	—	—	—	—	0.163	0.196	0.228	0.261	0.327	0.392	0.457	0.523
2	—	—	—	—	—	—	0.167	0.200	0.232	0.265	0.331	0.396	0.461	0.527
	—	—	—	—	—	—	0.163	0.196	0.228	0.261	0.327	0.392	0.457	0.523
2-1/4	—	—	—	—	—	—	—	0.199	0.232	0.265	0.331	0.396	0.461	0.527
	—	—	—	—	—	—	—	0.195	0.228	0.261	0.327	0.392	0.457	0.523
2-1/2	—	—	—	—	—	—	—	—	0.232	0.265	0.331	0.396	0.461	0.527
	—	—	—	—	—	—	—	—	0.228	0.261	0.327	0.392	0.457	0.523
2-3/4	—	—	—	—	—	—	—	—	0.231	0.264	0.331	0.396	0.461	0.527
	—	—	—	—	—	—	—	—	0.227	0.260	0.327	0.392	0.457	0.523
3	—	—	—	—	—	—	—	—	0.231	0.264	0.331	0.396	0.461	0.527
	—	—	—	—	—	—	—	—	0.227	0.260	0.327	0.392	0.457	0.523
3-1/4	—	—	—	—	—	—	—	—	—	0.264	0.330	0.395	0.461	0.527
	—	—	—	—	—	—	—	—	—	0.260	0.326	0.391	0.457	0.523
3-1/2	—	—	—	—	—	—	—	—	—	—	0.330	0.395	0.461	0.527
	—	—	—	—	—	—	—	—	—	—	0.326	0.391	0.457	0.523
3-3/4	—	—	—	—	—	—	—	—	—	—	—	0.395	0.460	0.527
	—	—	—	—	—	—	—	—	—	—	—	0.391	0.456	0.523
4	—	—	—	—	—	—	—	—	—	—	—	0.395	0.460	0.527
	—	—	—	—	—	—	—	—	—	—	—	0.391	0.456	0.523
4-1/4	—	—	—	—	—	—	—	—	—	—	—	0.395	0.460	0.526
	—	—	—	—	—	—	—	—	—	—	—	0.391	0.456	0.522
4-1/2	—	—	—	—	—	—	—	—	—	—	—	—	0.460	0.526
	—	—	—	—	—	—	—	—	—	—	—	—	0.456	0.522

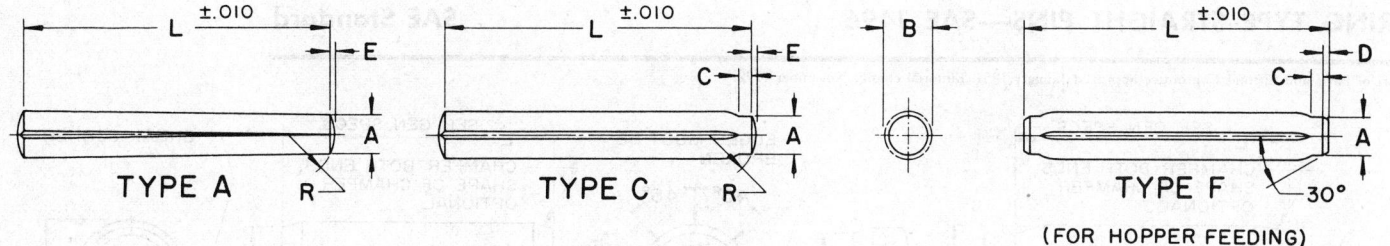

TABLE 2—TYPES A, C, AND F GROOVED STRAIGHT PINS, IN.

	Nominal Size	3/64	1/16	5/64	3/32	7/64	1/8	5/32	3/16	7/32	1/4	5/16	3/8	7/16	1/2
A	Diameter, max	0.0469	0.0625	0.0781	0.0938	0.1094	0.1250	0.1563	0.1875	0.2188	0.2500	0.3125	0.3750	0.4375	0.5000
	Diameter, min	0.0459	0.0615	0.0771	0.0928	0.1084	0.1230	0.1543	0.1855	0.2168	0.2480	0.3105	0.3730	0.4355	0.4980
	Recommended Hole, max	0.0478	0.0640	0.0798	0.0956	0.1113	0.1271	0.1587	0.1903	0.2219	0.2534	0.3166	0.3797	0.4428	0.5060
	Recommended Hole, min	0.0465	0.0625	0.0781	0.0938	0.1094	0.1250	0.1563	0.1875	0.2188	0.2500	0.3125	0.3750	0.4375	0.5000
E	Crown Height, ±0.005	0.0000	0.0065	0.0087	0.0091	0.0110	0.0130	0.0170	0.0180	0.0220	0.0260	0.0340	0.0390	0.0470	0.0520
R	Radius at Nom Crown Height, ±0.010	—	0.0781	0.0938	0.1250	0.1406	0.1562	0.1875	0.2500	0.2812	0.3125	0.3750	0.4688	0.5312	0.6250
C	Pilot Length		0.0312	0.0312	0.0312	0.0312	0.0312	0.0625	0.0625	0.0625	0.0625	0.0938	0.0938	0.0938	0.0938
D[a]	Chamfer Length (Type F Only)	—	0.0156	0.0156	0.0156	0.0156	0.0156	0.0312	0.0312	0.0312	0.0312	0.0469	0.0469	0.0469	0.0469
Length		**B Diameter, Max and Min Limits (Measured with Ring Gages)**													
1/8		0.052	—	—	—	—	—	—	—	—	—	—	—	—	—
		0.050	—	—	—	—	—	—	—	—	—	—	—	—	—
3/16		0.052	—	—	—	—	—	—	—	—	—	—	—	—	—
		0.050	—	—	—	—	—	—	—	—	—	—	—	—	—
1/4		0.052	0.069	0.085	0.102	0.118	0.136	—	—	—	—	—	—	—	—
		0.050	0.067	0.083	0.100	0.116	0.132	—	—	—	—	—	—	—	—
3/8		0.052	0.069	0.085	0.102	0.118	0.136	0.168	0.200	—	—	—	—	—	—
		0.050	0.067	0.083	0.100	0.116	0.132	0.164	0.196	—	—	—	—	—	—
1/2		0.052	0.069	0.085	0.102	0.118	0.136	0.168	0.200	0.232	0.265	—	—	—	—
		0.050	0.067	0.083	0.100	0.116	0.132	0.164	0.196	0.228	0.261	—	—	—	—
5/8		0.052	0.069	0.085	0.102	0.118	0.136	0.168	0.200	0.232	0.265	0.331	—	—	—
		0.050	0.067	0.083	0.100	0.116	0.132	0.164	0.196	0.228	0.261	0.327	—	—	—
3/4		—	0.069	0.085	0.102	0.117	0.136	0.168	0.200	0.232	0.265	0.331	0.396	—	—
		—	0.067	0.083	0.100	0.115	0.132	0.164	0.196	0.228	0.261	0.327	0.392	—	—
7/8		—	0.069	0.085	0.102	0.117	0.135	0.167	0.200	0.232	0.265	0.331	0.396	0.461	—
		—	0.067	0.083	0.100	0.115	0.131	0.163	0.196	0.228	0.261	0.327	0.392	0.457	—
1		—	0.069	0.085	0.102	0.116	0.135	0.167	0.200	0.232	0.265	0.331	0.396	0.461	0.527
		—	0.067	0.083	0.100	0.114	0.131	0.163	0.196	0.228	0.261	0.327	0.392	0.457	0.523
1-1/4		—	—	—	0.102	0.116	0.134	0.166	0.199	0.232	0.265	0.331	0.396	0.461	0.527
		—	—	—	0.100	0.114	0.130	0.162	0.195	0.228	0.261	0.327	0.392	0.457	0.523
1-1/2		—	—	—	—	—	0.134	0.166	0.199	0.231	0.264	0.331	0.396	0.461	0.527
		—	—	—	—	—	0.130	0.162	0.195	0.227	0.260	0.327	0.392	0.457	0.523
1-3/4		—	—	—	—	—	—	0.165	0.199	0.231	0.264	0.330	0.395	0.461	0.527
		—	—	—	—	—	—	0.161	0.195	0.227	0.260	0.326	0.391	0.457	0.523
2		—	—	—	—	—	—	0.165	0.198	0.231	0.264	0.330	0.395	0.460	0.527
		—	—	—	—	—	—	0.161	0.194	0.227	0.260	0.326	0.391	0.456	0.523
2-1/4		—	—	—	—	—	—	—	0.198	0.231	0.264	0.330	0.395	0.460	0.526
		—	—	—	—	—	—	—	0.194	0.227	0.260	0.326	0.391	0.456	0.522
2-1/2		—	—	—	—	—	—	—	—	0.230	0.263	0.329	0.395	0.460	0.526
		—	—	—	—	—	—	—	—	0.226	0.259	0.325	0.391	0.456	0.522
2-3/4		—	—	—	—	—	—	—	—	0.230	0.263	0.329	0.395	0.460	0.526
		—	—	—	—	—	—	—	—	0.226	0.259	0.325	0.391	0.456	0.522
3		—	—	—	—	—	—	—	—	0.229	0.262	0.329	0.394	0.459	0.525
		—	—	—	—	—	—	—	—	0.225	0.258	0.325	0.390	0.455	0.521
3-1/4		—	—	—	—	—	—	—	—	—	0.262	0.328	0.394	0.459	0.525
		—	—	—	—	—	—	—	—	—	0.258	0.324	0.390	0.455	0.521
3-1/2		—	—	—	—	—	—	—	—	—	—	0.328	0.393	0.458	0.524
		—	—	—	—	—	—	—	—	—	—	0.324	0.389	0.454	0.520
3-3/4		—	—	—	—	—	—	—	—	—	—	—	0.393	0.458	0.524
		—	—	—	—	—	—	—	—	—	—	—	0.389	0.454	0.520
4		—	—	—	—	—	—	—	—	—	—	—	0.392	0.457	0.523
		—	—	—	—	—	—	—	—	—	—	—	0.388	0.453	0.519
4-1/4		—	—	—	—	—	—	—	—	—	—	—	0.392	0.457	0.523
		—	—	—	—	—	—	—	—	—	—	—	0.388	0.453	0.519
4-1/2		—	—	—	—	—	—	—	—	—	—	—	—	0.456	0.522
		—	—	—	—	—	—	—	—	—	—	—	—	0.452	0.518

[a] On agreement between user and supplier a suitable radius may be substituted optionally for the chamfers on the ends of Type F pins for the 1/4 in. size and below.

SPRING TYPE STRAIGHT PINS—SAE J496 — SAE Standard

Report of Parts and Fittings Committee approved January 1957. Editorial change November 1972.

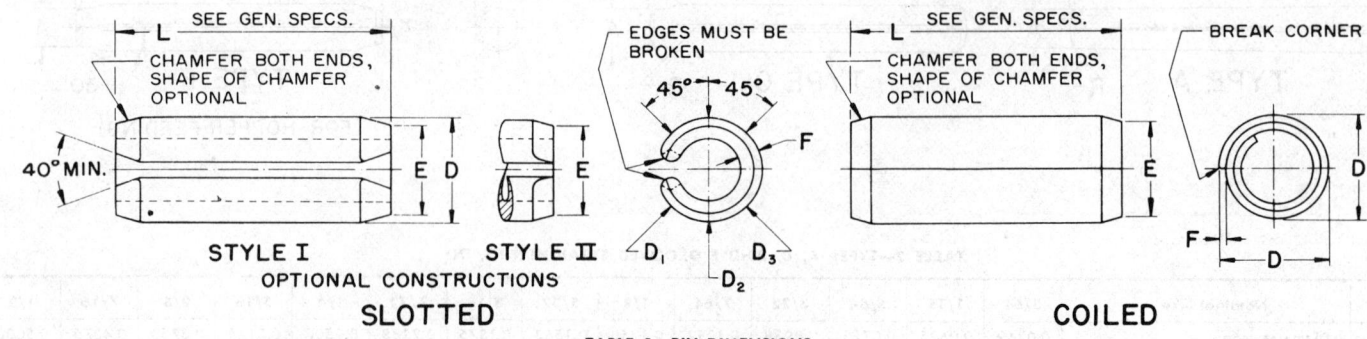

TABLE 1—PIN DIMENSIONS

Nominal Pin Size	D[a] Slotted Series A[b] and B[b]		D[a] Coiled Series A		D[a] Coiled Series B		D[a] Coiled Series C		E Slotted and Coiled Series A, B, and C	F Slotted Series A Nominal	F Slotted Series B Style I Nominal	F Slotted Series B Style II Nominal	F Coiled Series A Nominal	F Coiled Series B Nominal	F Coiled Series C Nominal	Double Shear Strength (lb)[e] Slotted and Coiled Series A Min	Double Shear Strength (lb)[e] Slotted and Coiled Series B Min	Double Shear Strength (lb)[e] Slotted and Coiled Series C Min	Recommended Hole Size Slotted Series A and B Max	Recommended Hole Size Slotted Series A and B Min	Recommended Hole Size Coiled Series A, B, and C Max	Recommended Hole Size Coiled Series A, B, and C Min
	Max	Min[c]	Max	Min	Max	Min	Max	Min	Max													
1/32	—	—	—	—	0.035	0.033	—	—	0.029[d]	—	—	—	0.003	—	—	—	75	—	—	—	0.0325	0.0310
3/64	—	—	—	—	0.052	0.049	—	—	0.045[d]	—	—	—	0.004	—	—	—	170	—	—	—	0.0485	0.0470
0.052	—	—	—	—	0.057	0.054	—	—	0.050[d]	—	—	—	0.004	—	—	—	230	—	—	—	0.0535	0.0520
1/16	0.069	0.066	0.070	0.066	0.071	0.067	0.072	0.067	0.059	0.012	—	0.006	0.007	0.005	0.003	425	300	160	0.065	0.062	0.065	0.061
5/64	0.086	0.083	0.086	0.082	0.087	0.083	0.088	0.083	0.075	0.018	—	0.008	0.007	0.005	0.003	650	480	260	0.081	0.078	0.081	0.077
3/32	0.103	0.099	0.103	0.098	0.104	0.099	0.105	0.099	0.091	0.022	0.012	0.012	0.010	0.007	0.005	1000	690	370	0.097	0.094	0.097	0.093
7/64	0.118	0.113	0.118	0.113	0.119	0.114	0.120	0.114	0.106	0.022	—	0.018	0.010	0.007	0.005	1410	940	510	0.112	0.109	0.112	0.108
1/8	0.135	0.131	0.136	0.130	0.137	0.131	0.138	0.131	0.122	0.028	0.012	0.018	0.014	0.010	0.007	1840	1000	660	0.129	0.125	0.129	0.124
9/64	0.149	0.145	0.151	0.145	0.152	0.146	0.153	0.146	0.136	0.028	—	0.022	0.014	0.010	0.007	2200	1550	830	0.141	0.140	0.144	0.139
5/32	0.167	0.162	0.168	0.161	0.170	0.163	0.171	0.163	0.152	0.032	0.018	0.022	0.017	0.011	0.007	2880	1750	1040	0.160	0.156	0.160	0.155
3/16	0.199	0.194	0.202	0.194	0.204	0.196	0.206	0.196	0.182	0.040	0.022	0.028	0.020	0.015	0.010	4140	2500	1500	0.192	0.187	0.192	0.185
7/32	0.232	0.226	0.235	0.226	0.238	0.229	0.240	0.229	0.214	0.048	0.028	0.032	0.024	0.017	0.011	5640	3760	2040	0.224	0.219	0.224	0.217
1/4	0.264	0.258	0.268	0.258	0.270	0.260	0.272	0.260	0.245	0.048	0.028	0.032	0.028	0.020	0.015	7360	4600	2660	0.256	0.250	0.256	0.248
5/16	0.328	0.321	0.340	0.327	0.341	0.327	0.342	0.327	0.306	0.062	—	0.040	0.032	0.024	0.017	11500	7670	4160	0.318	0.312	0.318	0.308
3/8	0.392	0.385	0.407	0.391	0.408	0.391	0.409	0.391	0.368	0.077	—	0.048	0.040	0.028	0.020	16580	11040	6000	0.382	0.375	0.382	0.368
7/16	0.456	0.448	0.475	0.457	0.476	0.457	0.478	0.457	0.430	0.077	—	0.048	0.047	0.036	0.024	20000	15020	8160	0.445	0.437	0.445	0.429
1/2	0.521	0.513	0.542	0.522	0.543	0.522	0.545	0.522	0.490	0.094	—	0.062	0.055	0.040	0.028	25800	19600	10640	0.510	0.500	0.510	0.490

[a] Maximum D shall be checked by a "GO" ring gage.
[b] Series designation applies to stock thickness, A being heaviest.
[c] Minimum D shall be the average of the D_1, D_2, and D_3 diameters.
[d] Series B coiled.
[e] Applies to pins made from SAE 1070 to 1095 steel and SAE 51410 or AISI 420 corrosion resistant steel. SAE 30302 stainless steel has a minimum shear strength equal to 85% of values shown for coiled pins.

GENERAL SPECIFICATIONS

Length and Availability—Table 3, Practical Length Increments and Ranges, indicates spring pin sizes. Information on availability of individual lengths in the various types, weights, and materials may be obtained from suppliers.

The tolerance on length for coiled type spring pins shall be ±0.010 in. for sizes up to and including 5/16 in.; and ±0.015 in. for sizes larger than 5/16 in.

The tolerance on length for slotted type spring pins shall be in accordance with the following tabulation:

Length Range	3/16 to 1 Incl	Over 1 to 2 Incl	Over 2 to 3 Incl	Over 3 to 4 Incl	Over 4 in.
Tolerance on Length	±0.015	±0.020	±0.025	±0.030	±0.035

Surface Treatment—Where corrosion preventive treatment applied to carbon steel spring pins is such that it might produce hydrogen embrittlement, the spring pins shall be baked or treated in such a manner as to obviate such embrittlement.

Material and Hardness—Hardness shall be tested in the following manner: For slotted pins the readings shall be taken near the center of a longitudinal flat ground on the pin at right angles to the slot. Coiled pins shall be ground or cut in half along the longitudinal axis and the hardness readings shall be taken on the inside surface of the outer half coil. Table 2 designates materials and the proper Rockwell scale to be used for the various wall thickness ranges.

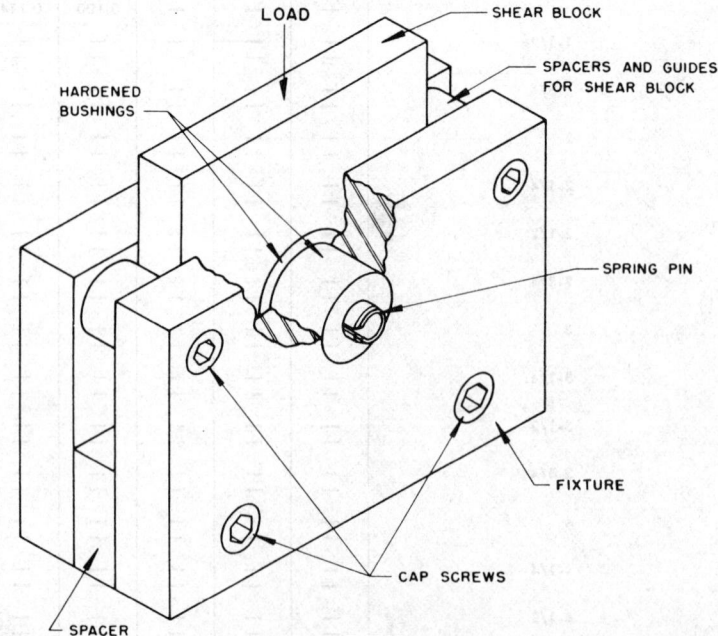

TYPICAL SPRING PIN SHEAR TEST FIXTURE

Defects—Spring pins shall be free from burrs, loose scale, seams, notches, sharp edges, or other defects which might affect their serviceability.

Performance—Spring pins shall withstand the minimum double shear loads specified in the dimensional tables when tested in accordance with the following procedure.

The shear test shall be performed in a fixture in which the pin support members and the member for applying the load shall have holes of a diameter conforming to the recommended hole size for the pins being tested and shall have a Rockwell hardness of C 58 or equivalent. The clearance between the supporting members and the loading member shall not exceed 0.005 in. and the shear plane shall be at least one diameter away from each end of the pin. Pins shall be located so that the slot is approximately at right angles to the line of application of the load. The speed of load application shall not exceed $\frac{1}{2}$ in. per min. Pins too short to be tested in double shear shall be tested by shearing two pins simultaneously in single shear. Spring pins which have been tested for shear strength shall show a ductile shear without longitudinal cracks.

TABLE 2—PIN HARDNESS

Wall Thickness Range	Over 0.001 to 0.010	Over 0.010 to 0.025	Over 0.025 to 0.050	Over 0.050 to 0.094
Rockwell Scale	Dph[a]-Tukon	15N	A	C
Pin Type / Material	Hardness Reading			
Slotted — SAE 1070–1095 Steel	458–562	83.6–87	73.6–78	46–53
Slotted — AISI 420 Corrosion Resistant Steel	413–545	82–86.6	72–77	43–52
Coiled — SAE 1070 Steel	393–515	80.4–85.5	70.4–75.9	40–50
Coiled — SAE 51410 or AISI 420 Corrosion Resistant Steel	393–515	80.4–85.5	70.4–75.9	40–50
Coiled — SAE 30302 Stainless Steel	—	80.4–85.5	70.4–75.9	40–50

[a] Diamond pyramidal hardness.

TABLE 3—PRACTICAL LENGTH INCREMENTS AND RANGES

Length	1/32[a]	3/64[a]	0.052[a]	1/16	5/64	3/32	7/64	1/8	9/64	5/32	3/16	7/32	1/4	5/16	3/8	7/16	1/2
1/8	●	●	●	●	●												
3/16	●	●	●	●	●	●											
1/4	●	●	●	●	●	●	●	●									
5/16	●	●	●	●	●	●	●	●	●	●							
3/8	●	●	●	●	●	●	●	●	●	●	●						
7/16	●	●	●	●	●	●	●	●	●	●	●	●					
1/2	●	●	●	●	●	●	●	●	●	●	●	●	●				
9/16	●	●	●	●	●	●	●	●	●	●	●	●	●				
5/8	●	●	●	●	●	●	●	●	●	●	●	●	●	●			
11/16	●	●	●	●	●	●	●	●	●	●	●	●	●	●			
3/4	●	●	●	●	●	●	●	●	●	●	●	●	●	●	●		
13/16	●	●	●	●	●	●	●	●	●	●	●	●	●	●	●		
7/8	●	●	●	●	●	●	●	●	●	●	●	●	●	●	●		
15/16	●	●	●	●	●	●	●	●	●	●	●	●	●	●	●	●	
1	●	●	●	●	●	●	●	●	●	●	●	●	●	●	●	●	●
1-1/8				●	●	●	●	●	●	●	●	●	●	●	●	●	●
1-1/4				●	●	●	●	●	●	●	●	●	●	●	●	●	●
1-3/8					●	●	●	●	●	●	●	●	●	●	●	●	●
1-1/2					●	●	●	●	●	●	●	●	●	●	●	●	●
1-5/8						●	●	●	●	●	●	●	●	●	●	●	●
1-3/4						●	●	●	●	●	●	●	●	●	●	●	●
1-7/8							●	●	●	●	●	●	●	●	●	●	●
2							●	●	●	●	●	●	●	●	●	●	●
2-1/4								●	●	●	●	●	●	●	●	●	●
2-1/2									●	●	●	●	●	●	●	●	●
2-3/4										●	●	●	●	●	●	●	●
3										●	●	●	●	●	●	●	●
3-1/4											●	●	●	●	●	●	●
3-1/2												●	●	●	●	●	●
3-3/4													●	●	●	●	●
4													●	●	●	●	●

[a] Coiled type only.

UNHARDENED GROUND DOWEL PINS—SAE J497 — SAE Standard

Report of Parts and Fittings Committee approved January 1957. Editorial change June 1964.

TABLE 1—DIMENSIONS OF UNHARDENED GROUND DOWEL PINS, IN.[a]

Nominal	Diameter, A		Chamfer, B
	Max	Min	
0.062	0.0600	0.0595	0.015
0.094	0.0912	0.0907	0.015
0.109	0.1068	0.1063	0.015
0.125	0.1223	0.1218	0.015
0.156	0.1535	0.1530	0.015
0.188	0.1847	0.1842	0.015
0.219	0.2159	0.2154	0.015
0.250	0.2470	0.2465	0.015
0.312	0.3094	0.3089	0.030
0.375	0.3717	0.3712	0.030
0.438	0.4341	0.4336	0.030
0.500	0.4964	0.4959	0.030
0.625	0.6211	0.6206	0.045
0.750	0.7458	0.7453	0.045
0.875	0.8705	0.8700	0.060
1.000	0.9952	0.9947	0.060

[a] Maximum diameters are graduated from 0.0005 on 0.062 in. pins to 0.0028 on 1.000 in. pins under the minimum commercial bar stock sizes.

ROD ENDS AND CLEVIS PINS—SAE J493 — SAE Standard

Report of Miscellaneous Division approved January 1915 and last revised by Parts and Fittings Committee May 1959. Editorial change June 1961.

GENERAL SPECIFICATIONS FOR CLEVIS PINS

Material—SAE 1010 or SAE 1111 steel or equivalent.

Heat Treatment—Clevis pins shall be supplied either soft or cyanide hardened as specified.

Defects—Clevis pins must be free from burrs, loose scale, sharp edges, and all other defects that might affect their serviceability.

Tolerances—General tolerances for all dimensions are ±0.010 unless otherwise specified.

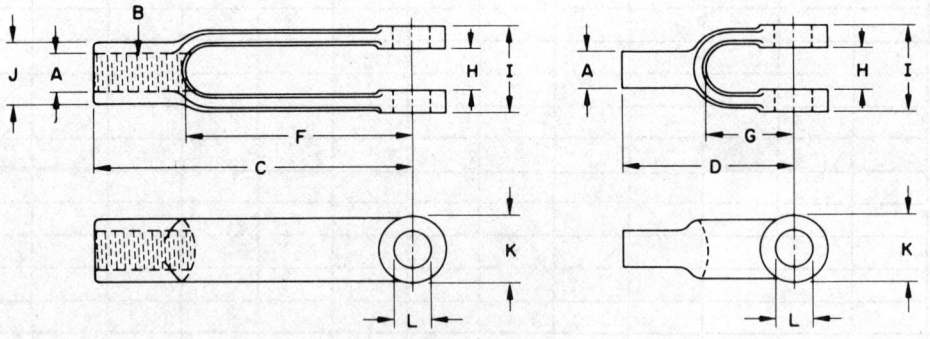

FIG. 1—ADJUSTABLE AND PLAIN YOKES FIG. 2—EYE

TABLE 1—ADJUSTABLE AND PLAIN YOKE AND EYE, FIGS. 1 AND 2

Series	A	Threads per in., B	C	D	E	F	G	H Fork ±0.010 Eye +0.000 −0.010	I ±0.010	J	K	L Nominal	L Tolerance Plus 0.001; Minus
Light	No. 10	32	1-9/16	1-1/4	1-1/4	1	7/16	3/16	7/16	5/16	3/8	3/16	0.001
	1/4	28	2	1-3/4	1-1/4	1-1/4	5/8	9/32	5/8	7/16	1/2	1/4	0.001
	5/16	24	2-1/4	2	1-3/8	1-7/16	3/4	11/32	3/4	1/2	19/32	5/16	0.001
	3/8	24	2-1/2	2-1/8	1-1/2	1-5/8	27/32	7/16	7/8	5/8	11/16	3/8	0.001
	7/16	20	2-7/8	2-1/4	1-5/8	1-7/8	1	1/2	1	23/32	13/16	7/16	0.001
	1/2	20	3	2-1/2	1-3/4	1-7/8	1-1/8	9/16	1-1/8	13/16	15/16	1/2	0.002
Heavy	1/2	20	4-3/16	2-1/2	1-3/4	3-1/16	1-1/8	9/16	1-1/8	13/16	15/16	1/2	0.002
	5/8	18	4-15/16	2-7/8	2	3-11/16	1-7/16	11/16	1-3/8	1-3/16	1-3/16	5/8	0.002
	3/4	16	6-1/16	3-5/8	2-3/8	4-9/16	1-11/16	13/16	1-1/2	1-1/4	1-7/16	3/4	0.002
	7/8	14	7-1/8	4	2-3/4	5-1/4	2	15/16	1-7/8	1-7/16	1-11/16	7/8	0.002
	1	14	8	4-1/2	3	6	2-1/2	1-1/16	2-1/8	1-5/8	1-15/16	1	0.002

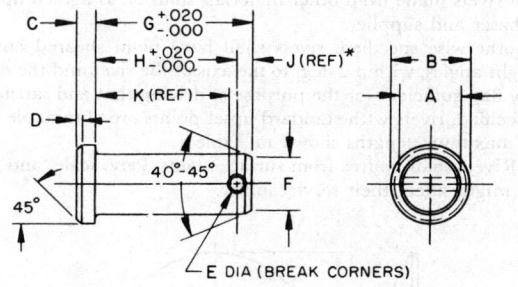

* THE "J" DIMENSION (DISTANCE FROM CENTERLINE OF HOLE TO END OF PIN) IS FOR CALCULATION ONLY. ON DETAIL DRAWINGS OF CLEVIS PINS, HOLE LOCATION WILL BE SHOWN AS THE DISTANCE FROM THE UNDERSIDE OF THE HEAD TO THE CENTERLINE OF THE HOLE

FIG. 3—PIN

TABLE 2—CLEVIS PINS, FIG. 3

Nominal Size	A Body Dia Max	A Body Dia Min	B Head Dia	C Head Height	D Head Chamfer	E Hole Dia Min	E Hole Dia Max	F Chamfer Dia Max	F Chamfer Dia Min	Length (H+J) Nom G[a]	H Under Head to Center of Hole	J Center of Hole to End Ref	K Under Head to Edge of Nom Hole Ref
3/16	0.186	0.181	0.312	0.062	0.016	0.073	0.088	0.152	0.147	0.578	0.484	0.094	0.445
1/4	0.248	0.243	0.375	0.094	0.031	0.073	0.088	0.214	0.209	0.766	0.672	0.094	0.633
5/16	0.311	0.306	0.438	0.094	0.031	0.104	0.119	0.265	0.259	0.938	0.812	0.125	0.757
3/8	0.373	0.368	0.500	0.125	0.031	0.104	0.119	0.327	0.321	1.062	0.938	0.125	0.883
7/16	0.436	0.431	0.562	0.156	0.047	0.104	0.119	0.390	0.384	1.188	1.062	0.125	1.008
1/2	0.496	0.491	0.625	0.156	0.047	0.136	0.151	0.439	0.431	1.359	1.203	0.156	1.133
5/8	0.621	0.616	0.812	0.203	0.062	0.136	0.151	0.564	0.556	1.609	1.453	0.156	1.383
3/4	0.746	0.741	0.938	0.250	0.078	0.167	0.182	0.678	0.668	1.906	1.719	0.188	1.633
7/8	0.871	0.866	1.031	0.312	0.094	0.167	0.182	0.803	0.793	2.156	1.969	0.188	1.883
1	0.996	0.991	1.188	0.344	0.109	0.167	0.182	0.928	0.918	2.406	2.219	0.188	2.133

[a] Tabulated lengths intended for use with standard clevises without spacers, where other lengths are required it is recommended that 1/16 in. increments be used.

RIVETS AND RIVETING—SAE J492

SAE Standard

Report of Parts and Fittings Division approved February 1928 and last revised by Fasteners Committee June 1961. Editorial change May 1968.

GENERAL SPECIFICATIONS FOR SMALL SOLID RIVETS

General—This small solid rivet standard covers the complete general and dimensional data for flat head, pan head, button head, truss head, countersunk head, copper's, tinner's and belt rivets. Design and assembly data are given in the Appendix—Rivet Selection and Design Considerations.

The inclusion of dimensional data in this standard is not intended to imply that all of the products described are stock production sizes. Consumers are requested to consult with manufacturers concerning stock production sizes.

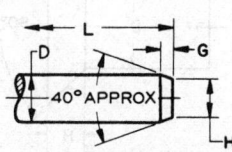

TABLE 1—DIMENSIONS OF STANDARD POINTS FOR SMALL SOLID RIVETS

Nominal Size or Basic Shank Dia	G Point Length Ref	H Point Dia Approx[a]	L Rivet Length Max
1/16 0.062	0.015	0.051	9/16
3/32 0.094	0.023	0.077	3/4
1/8 0.125	0.031	0.102	3/4
5/32 0.156	0.039	0.127	1
3/16 0.188	0.047	0.154	1
7/32 0.219	0.055	0.179	1-3/8
1/4 0.250	0.062	0.204	1-3/8
9/32 0.281	0.070	0.230	1-1/2
5/16 0.312	0.078	0.255	1-1/2
11/32 0.344	0.086	0.281	1-5/8
3/8 0.375	0.094	0.307	1-5/8
13/32 0.406	0.102	0.332	2
7/16 0.438	0.110	0.358	3

[a] No standard tolerances are contemplated.

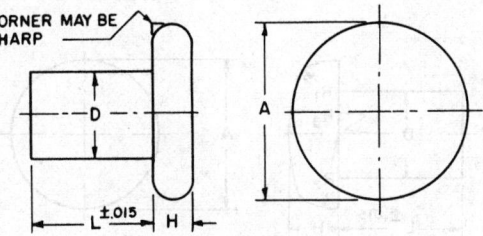

TABLE 2—FLAT HEAD RIVETS

Nominal Size or Basic Shank Dia	D Dia of Shank Max	D Dia of Shank Min	A Dia of Head Max	A Dia of Head Min	H Height of Head Max	H Height of Head Min
1/16 0.062	0.065	0.059	0.140	0.120	0.027	0.017
3/32 0.094	0.096	0.090	0.200	0.180	0.038	0.026
1/0 0.125	0.127	0.121	0.260	0.240	0.048	0.036
5/32 0.156	0.158	0.152	0.323	0.301	0.059	0.045
3/16 0.188	0.191	0.182	0.307	0.361	0.069	0.055
7/32 0.219	0.222	0.213	0.453	0.427	0.080	0.065
1/4 0.250	0.253	0.244	0.515	0.485	0.091	0.075
9/32 0.281	0.285	0.273	0.579	0.545	0.103	0.085
5/16 0.312	0.316	0.304	0.641	0.607	0.113	0.095
11/32 0.344	0.348	0.336	0.705	0.667	0.124	0.104
3/8 0.375	0.380	0.365	0.769	0.731	0.135	0.115
13/32 0.406	0.411	0.396	0.834	0.790	0.146	0.124
7/16 0.438	0.443	0.428	0.896	0.852	0.157	0.135

For dimensions and tolerances not specified above, see General Specifications.

Tolerances—The tolerances given on the dimensional tables are those for rivets made by the normal cold heading process. The tolerance for rivets made by the hot heading or forging process shall be as agreed upon between the purchaser and supplier.

Heads—Because the heads of these rivets are not machined or trimmed, the circumference may be slightly irregular and the edges may be rounded or flat.

Underhead Fillets—Rivets, other than countersunk type, shall be furnished with a definite fillet under the head but radius of fillet shall not exceed 10% of maximum shank diameter or 0.03 in., whichever is the smaller.

Material—Rivets shall be steel, copper, brass, aluminum, or other metals as specified by purchaser.

Suitable material for steel small solid rivets is covered by SAE Recommended Practice, Mechanical and Chemical Requirements for Non-threaded Fasteners—SAE J430.

Requirements of rivets made from other materials shall be as agreed upon between the purchaser and supplier.

Points—Unless otherwise specified, rivets shall have plain sheared ends. Ends shall be at right angles, within 2 deg, to the axis of the rivet and the end shall be reasonably flat, sufficient for the purpose of driving that end satisfactorily. When so specified, rivets with standard upset points are obtainable on lengths up to the maximum lengths shown in Table 1.

Workmanship—Rivets shall be free from surface seams, loose scale, and all other defects that might affect their serviceability.

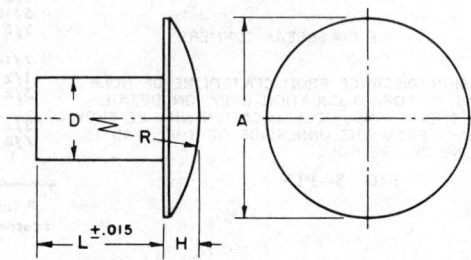

TABLE 5 — TRUSS HEAD OR WAGON BOX RIVETS

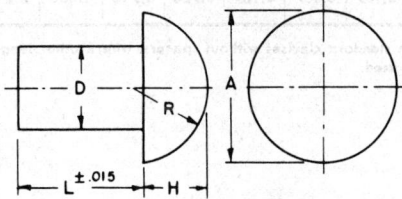

TABLE 3 — BUTTON HEAD RIVETS

Nominal Size or Basic Shank Dia	D Dia of Shank		A Dia of Head		H Height of Head		R Rad of Head
	Max	Min	Max	Min	Max	Min	Approx
1/16 0.062	0.065	0.059	0.122	0.102	0.052	0.042	0.055
3/32 0.094	0.096	0.090	0.182	0.162	0.077	0.065	0.084
1/8 0.125	0.127	0.121	0.235	0.215	0.100	0.088	0.111
5/32 0.156	0.158	0.152	0.290	0.268	0.124	0.110	0.138
3/16 0.188	0.191	0.182	0.348	0.322	0.147	0.133	0.166
7/32 0.219	0.222	0.213	0.405	0.379	0.172	0.158	0.195
1/4 0.250	0.253	0.244	0.460	0.430	0.196	0.180	0.221
9/32 0.281	0.285	0.273	0.518	0.484	0.220	0.202	0.249
5/16 0.312	0.316	0.304	0.572	0.538	0.243	0.225	0.276
11/32 0.344	0.348	0.336	0.630	0.592	0.267	0.247	0.304
3/8 0.375	0.380	0.365	0.684	0.646	0.291	0.271	0.332
13/32 0.406	0.411	0.396	0.743	0.699	0.316	0.294	0.358
7/16 0.438	0.443	0.428	0.798	0.754	0.339	0.317	0.387

For dimensions and tolerances not specified above, see General Specifications.

Nominal Size or Basic Shank Dia	D Dia of Shank		A Dia of Head		H Height of Head		R Radius of Head
	Max	Min	Max	Min	Max	Min	Approx
3/32 0.094	0.096	0.090	0.226	0.206	0.038	0.026	0.239
1/8 0.125	0.127	0.121	0.297	0.277	0.048	0.036	0.314
5/32 0.156	0.158	0.152	0.368	0.348	0.059	0.045	0.392
3/16 0.188	0.191	0.182	0.442	0.422	0.069	0.055	0.470
7/32 0.219	0.222	0.213	0.515	0.495	0.080	0.066	0.555
1/4 0.250	0.253	0.244	0.590	0.560	0.091	0.075	0.628
9/32 0.281	0.285	0.273	0.661	0.631	0.103	0.085	0.706
5/16 0.312	0.316	0.304	0.732	0.702	0.113	0.095	0.784
11/32 0.344	0.348	0.336	0.806	0.776	0.124	0.104	0.862
3/8 0.375	0.380	0.365	0.878	0.848	0.135	0.115	0.942
13/32 0.406	0.411	0.396	0.949	0.919	0.145	0.123	0.128
7/16 0.438	0.443	0.428	1.020	0.990	0.157	0.135	1.098

For dimensions and tolerances not specified above, see General Specifications.

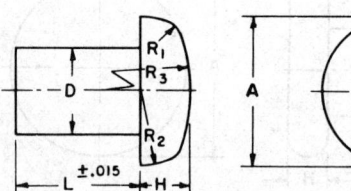

TABLE 4 — PAN HEAD RIVETS

Nominal Size or Basic Shank Dia	D Dia of Shank		A Dia of Head		H Height of Head		Radii of Head		
	Max	Min	Max	Min	Max	Min	R_1	R_2	R
							Approx		
1/16 0.062	0.065	0.059	0.118	0.098	0.040	0.030	0.019	0.052	0.217
3/32 0.094	0.096	0.090	0.173	0.153	0.060	0.048	0.030	0.080	0.326
1/8 0.125	0.127	0.121	0.225	0.205	0.078	0.066	0.039	0.106	0.429
5/32 0.156	0.158	0.152	0.279	0.257	0.096	0.082	0.049	0.133	0.535
3/16 0.188	0.191	0.182	0.334	0.308	0.114	0.100	0.059	0.159	0.641
7/32 0.219	0.222	0.213	0.391	0.365	0.133	0.119	0.069	0.186	0.754
1/4 0.250	0.253	0.244	0.444	0.414	0.151	0.135	0.079	0.213	0.858
9/32 0.281	0.285	0.273	0.499	0.465	0.170	0.152	0.088	0.239	0.963
5/16 0.312	0.316	0.304	0.552	0.518	0.187	0.169	0.098	0.266	1.070
11/32 0.344	0.348	0.336	0.608	0.570	0.206	0.186	0.108	0.292	1.176
3/8 0.375	0.380	0.365	0.663	0.625	0.225	0.205	0.118	0.319	1.286
13/32 0.406	0.411	0.396	0.719	0.675	0.243	0.221	0.127	0.345	1.392
7/16 0.438	0.443	0.428	0.772	0.728	0.261	0.239	0.137	0.372	1.500

For dimensions and tolerances not specified above, see General Specifications.

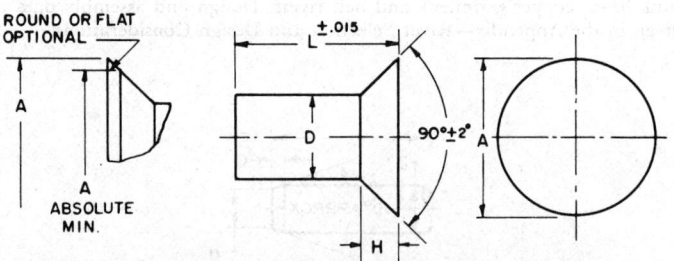

TABLE 6 — COUNTERSUNK HEAD RIVETS

Nominal Size or Basic Shank Dia	D Dia of Shank		A Dia of Head		H[a] Height of Head
	Max	Min	Max Sharp	Abs Min	
1/16 0.062	0.065	0.059	0.118	0.110	0.027
3/32 0.094	0.096	0.090	0.176	0.163	0.040
1/8 0.125	0.127	0.121	0.235	0.217	0.053
5/32 0.156	0.158	0.152	0.293	0.272	0.066
3/16 0.188	0.191	0.182	0.351	0.326	0.079
7/32 0.219	0.222	0.213	0.413	0.384	0.094
1/4 0.250	0.253	0.244	0.469	0.437	0.106
9/32 0.281	0.285	0.273	0.528	0.491	0.119
5/16 0.312	0.316	0.304	0.588	0.547	0.133
11/32 0.344	0.348	0.336	0.646	0.602	0.146
3/8 0.375	0.380	0.365	0.704	0.656	0.159
13/32 0.406	0.411	0.396	0.763	0.710	0.172
7/16 0.438	0.443	0.428	0.823	0.765	0.186

[a] Height of head, H, is given for construction purposes only. Variations in this dimension are controlled by the diameters A and D and the included angle of the head.

For dimensions and tolerances not specified above, see General Specifications.

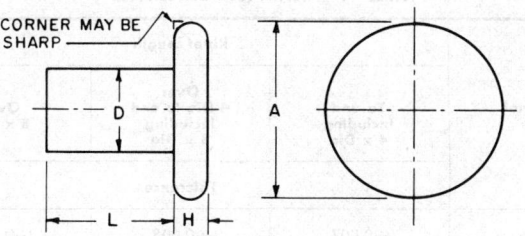

TABLE 7—TINNERS' RIVETS

Nominal Size[a]	D Dia of Shank		A Dia of Head		H Height of Head		Length	
	Max	Min	Max	Min	Max	Min	Max	Min
6 oz	0.081	0.075	0.213	0.193	0.028	0.016	0.135	0.115
8 oz	0.091	0.085	0.225	0.205	0.036	0.024	0.166	0.146
10 oz	0.097	0.091	0.250	0.230	0.037	0.025	0.182	0.162
12 oz	0.107	0.101	0.265	0.245	0.037	0.025	0.198	0.178
14 oz	0.111	0.105	0.275	0.255	0.038	0.026	0.198	0.178
1 lb	0.113	0.107	0.285	0.265	0.040	0.028	0.213	0.193
1-1/4 lb	0.122	0.116	0.295	0.275	0.045	0.033	0.229	0.209
1-1/2 lb	0.132	0.126	0.316	0.294	0.046	0.034	0.244	0.224
1-3/4 lb	0.136	0.130	0.331	0.309	0.049	0.035	0.260	0.240
2 lb	0.146	0.140	0.341	0.319	0.050	0.036	0.276	0.256
2-1/2 lb	0.150	0.144	0.311	0.289	0.069	0.055	0.291	0.271
3 lb	0.163	0.154	0.329	0.303	0.073	0.059	0.323	0.303
3-1/2 lb	0.168	0.159	0.348	0.322	0.074	0.060	0.338	0.318
4 lb	0.179	0.170	0.368	0.342	0.076	0.062	0.354	0.334
5 lb	0.190	0.181	0.388	0.362	0.084	0.070	0.385	0.365
6 lb	0.206	0.197	0.419	0.393	0.090	0.076	0.401	0.381
7 lb	0.223	0.214	0.431	0.405	0.094	0.080	0.416	0.396
8 lb	0.227	0.218	0.475	0.445	0.101	0.085	0.448	0.428
9 lb	0.241	0.232	0.490	0.460	0.103	0.087	0.463	0.443
10 lb	0.241	0.232	0.505	0.475	0.104	0.088	0.479	0.459
12 lb	0.263	0.251	0.532	0.498	0.108	0.090	0.510	0.490
14 lb	0.288	0.276	0.577	0.543	0.113	0.095	0.525	0.505
16 lb	0.304	0.292	0.597	0.563	0.128	0.110	0.541	0.521
18 lb	0.347	0.335	0.706	0.668	0.156	0.136	0.603	0.583

[a] Nominal size refers to the approximate weight of 1,000 rivets
For dimensions and tolerances not specified above, see General Specifications.

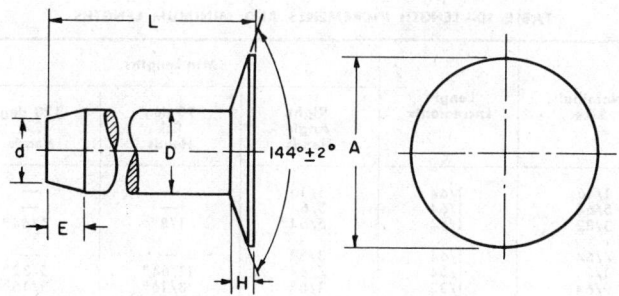

TABLE 8—COOPERS' RIVETS

Nominal Size[a]	D Dia of Shank		A Dia of Head		H Height of Head		Point		L Length	
							d Dia	E Length		
	Max	Min	Max	Min	Max	Min	Approx	Approx	Max	Min
1 lb	0.111	0.105	0.291	0.271	0.045	0.031	Not Pointed		0.249	0.219
1-1/4	0.122	0.116	0.324	0.302	0.050	0.036	Not Pointed		0.285	0.255
1-1/2	0.132	0.126	0.324	0.302	0.050	0.036	Not Pointed		0.285	0.255
1-3/4	0.136	0.130	0.324	0.302	0.052	0.034	Not Pointed		0.318	0.284
2	0.142	0.136	0.355	0.333	0.056	0.038	Not Pointed		0.322	0.288
3	0.158	0.152	0.386	0.364	0.058	0.040	0.123	0.062	0.387	0.353
4	0.168	0.159	0.388	0.362	0.058	0.040	0.130	0.062	0.418	0.388
5	0.183	0.174	0.419	0.393	0.063	0.045	0.144	0.062	0.454	0.420
6	0.206	0.197	0.482	0.456	0.073	0.051	0.160	0.094	0.498	0.457
7	0.223	0.214	0.513	0.487	0.076	0.054	0.175	0.094	0.561	0.523
8	0.241	0.232	0.546	0.516	0.081	0.059	0.182	0.094	0.597	0.559
9	0.248	0.239	0.578	0.548	0.085	0.063	0.197	0.094	0.601	0.563
10	0.253	0.244	0.578	0.548	0.085	0.063	0.197	0.094	0.632	0.594
12	0.263	0.251	0.580	0.546	0.086	0.060	0.214	0.094	0.633	0.575
14	0.275	0.263	0.611	0.577	0.091	0.065	0.223	0.094	0.670	0.612
16	0.285	0.273	0.611	0.577	0.089	0.063	0.223	0.094	0.699	0.641
18	0.285	0.273	0.642	0.608	0.108	0.082	0.230	0.125	0.749	0.691
20	0.316	0.304	0.705	0.671	0.128	0.102	0.250	0.125	0.769	0.711
3/8	0.380	0.365	0.800	0.762	0.136	0.106	0.312	0.125	0.840	0.778

[a] Nominal size refers to the approximate weight of 1,000 rivets.
For dimensions and tolerances not specified above, see General Specifications.

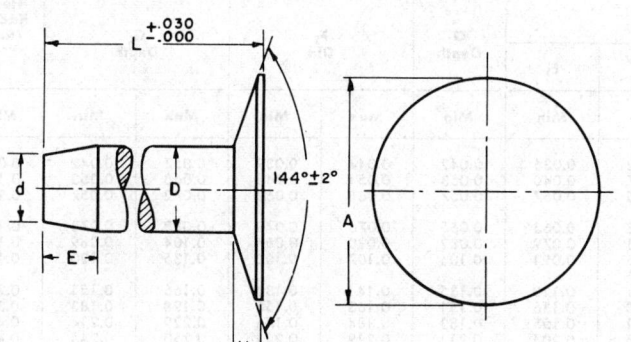

TABLE 9—BELT RIVETS

Nominal Size[a]	D Dia of Shank		A Dia of Head		H Height of Head		Point	
							d Dia	E Length
	Max	Min	Max	Min	Max	Min	Approx	Approx
14	0.085	0.079	0.260	0.240	0.042	0.030	0.065	0.078
13	0.097	0.091	0.322	0.302	0.051	0.039	0.073	0.078
12	0.111	0.105	0.353	0.333	0.054	0.040	0.083	0.078
11	0.122	0.116	0.383	0.363	0.059	0.045	0.097	0.078
10	0.136	0.130	0.417	0.395	0.065	0.047	0.109	0.094
9	0.150	0.144	0.448	0.426	0.069	0.051	0.122	0.094
8	0.167	0.161	0.481	0.455	0.072	0.054	0.135	0.094
7	0.183	0.174	0.513	0.487	0.075	0.056	0.151	0.125
6	0.206	0.197	0.606	0.580	0.090	0.068	0.165	0.125
5	0.223	0.214	0.700	0.674	0.105	0.083	0.185	0.125
4	0.241	0.232	0.921	0.893	0.138	0.116	0.204	0.141

[a] Nominal size refers to the Stubs iron wire gage number of the stock used in the shank of the rivet.
For dimensions and tolerances not specified above, see General Specifications.

GENERAL SPECIFICATIONS FOR TUBULAR RIVETS

General—This tubular rivet standard covers the complete general and dimensional data for oval head, truss head, flat head, 90- and 120-deg countersunk head semitubular rivets and oval head, truss head and countersunk head full tubular rivets. Design and assembly data are given in the Appendix—Rivet Selection and Design Considerations.

The inclusion of dimensional data in this standard is not intended to imply that all of the products described are stock production sizes. Consumers are requested to consult with manufacturers concerning stock production sizes.

Heads—The bearing surface of flat, oval, and truss head rivets shall be at right angles to the axis of the body within 2 deg. Heads of all tubular rivets shall not be eccentric with the shank beyond a tolerance of 3% of the maximum head diameter. Because the heads are not machined or trimmed, the circumference may be slightly irregular and the edges rounded or flat.

Underhead Fillets—Rivets, other than countersunk type, shall be furnished with a definite fillet under the head but radius of fillet shall not exceed 10% of maximum shank diameter.

Material—Tubular rivets shall be low carbon steel, or brass, standard with manufacturer; or stainless steel, aluminum, copper, or other metals as agreed upon between the purchaser and supplier.

Length—Length of rivets shall be measured as indicated in the illustrations for each head style. Tubular rivets are available in length increments specified in Table 10.

Tolerance on length of tubular rivets shall be as specified in Table 11.

Workmanship—Tubular rivet end irregularities shall not be such that usability of the rivet is impaired. Rivets shall be free from surface seams, splits, and all other defects that might affect their serviceability.

TABLE 10—LENGTH INCREMENTS AND MINIMUM LENGTHS

Nominal Size	Length Increments	Min Lengths		
		Right Angle Heads	90 deg Csk Heads	120 deg Csk Heads
1/16	1/64	1/16	—	—
5/64	1/64	5/64	—	—
3/32	1/64	5/64[a]	1/8[a]	7/64[a]
7/64	1/64	3/32	11/64[a]	5/32[a]
1/8	1/64	7/64[a]	3/16[a]	3/16[a]
9/64	1/32	1/8[a]	3/16[a]	3/16[a]
3/16	1/32	5/32[a]	1/4[a]	1/4[a]
7/32	1/16	3/16[a]	5/16[a]	9/32[a]
1/4	1/16	7/32[a]	11/32[a]	5/16[a]
5/16	1/16	1/4[a]		

[a] Hole depth to point of apex shall not exceed shank length for straight hole rivets of these lengths.

TABLE 11—TOLERANCES ON LENGTH

Nominal Size	Rivet Length		
	To and Including 4 x Dia	Over 4 Dia to and Including 8 x Dia	Over 8 x Dia
	Tolerance		
1/16	±0.007	±0.008	±0.010
5/64	±0.007	±0.008	±0.010
3/32	±0.007	±0.008	±0.010
7/64	±0.007	±0.008	±0.010
1/8	±0.007	±0.010	±0.015
9/64	±0.010	±0.012	±0.015
3/16	±0.010	±0.012	±0.015
7/32	±0.010	±0.015	±0.020
1/4	±0.010	±0.015	±0.020
5/16	±0.010	±0.015	±0.020

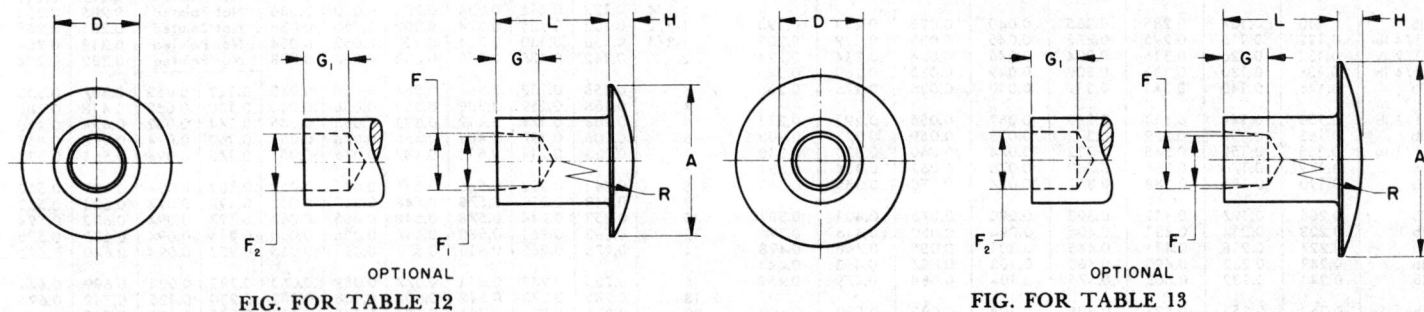

FIG. FOR TABLE 12   FIG. FOR TABLE 13

TABLE 12—DIMENSIONS OF OVAL HEAD SEMITUBULAR RIVETS

Nominal Size	D Dia of Shank		A Dia of Head		H Height of Head		Tapered Hole				Straight Hole				R Head Radius (Ref)
							Dia			G Depth	F₂ Dia		G₁ Depth		
							F		F₁						
	Max	Min	Max	Min	Max	Min	Max	Min	Min	Min	Max	Min	Max	Min	Min
1/16	0.061	0.058	0.114	0.104	0.020	0.014	0.046	0.042	0.036	0.042	0.044	0.039	0.057	0.042	0.084
5/64	0.075	0.072	0.133	0.123	0.023	0.017	0.053	0.047	0.040	0.053	0.051	0.045	0.068	0.053	0.101
3/32	0.089	0.085	0.152	0.142	0.026	0.020	0.069	0.065	0.057	0.057	0.068	0.062	0.072	0.057	0.120
7/64	0.099	0.095	0.192	0.182	0.032	0.026	0.076	0.072	0.063	0.065	0.076	0.070	0.088	0.073	0.158
1/8	0.123	0.118	0.223	0.213	0.038	0.030	0.095	0.091	0.079	0.082	0.090	0.084	0.104	0.089	0.183
9/64	0.146	0.141	0.239	0.229	0.045	0.035	0.112	0.106	0.091	0.104	0.107	0.100	0.135	0.120	0.182
3/16	0.188	0.182	0.318	0.306	0.065	0.055	0.145	0.139	0.120	0.135	0.141	0.134	0.166	0.151	0.232
7/32	0.217	0.210	0.444	0.430	0.075	0.061	0.166	0.158	0.136	0.151	0.163	0.155	0.198	0.183	0.381
1/4	0.252	0.244	0.507	0.493	0.085	0.071	0.191	0.181	0.155	0.183	0.184	0.176	0.229	0.214	0.439
5/16	0.310	0.302	0.570	0.554	0.100	0.086	0.235	0.225	0.201	0.214	0.219	0.211	0.260	0.245	0.473

For dimensions and tolerances not specified above, see General Specifications.

TABLE 13—DIMENSIONS OF TRUSS HEAD SEMITUBULAR RIVETS

Nominal Size	D Dia of Shank		A Dia of Head		H Height of Head		Tapered Hole				Straight Hole				R Head Radius (Ref)
							Dia			G Depth	F₂ Dia		G₁ Depth		
							F		F₁						
	Max	Min	Max	Min	Max	Min	Max	Min	Min	Min	Max	Min	Max	Min	Min
1/16	0.061	0.058	0.130	0.120	0.020	0.014	0.046	0.042	0.036	0.042	0.044	0.039	0.057	0.042	0.110
3/32	0.089	0.085	0.192	0.182	0.026	0.020	0.069	0.065	0.057	0.057	0.068	0.062	0.072	0.057	0.189
1/8	0.123	0.118	0.286	0.276	0.038	0.030	0.095	0.091	0.079	0.082	0.090	0.084	0.104	0.089	0.300
9/64	0.146	0.141	0.318	0.306	0.045	0.035	0.112	0.106	0.091	0.104	0.107	0.100	0.135	0.120	0.313
3/16	0.188	0.182	0.381	0.369	0.065	0.055	0.145	0.139	0.120	0.135	0.141	0.134	0.166	0.151	0.324

For dimensions and tolerances not specified above, see General Specifications.

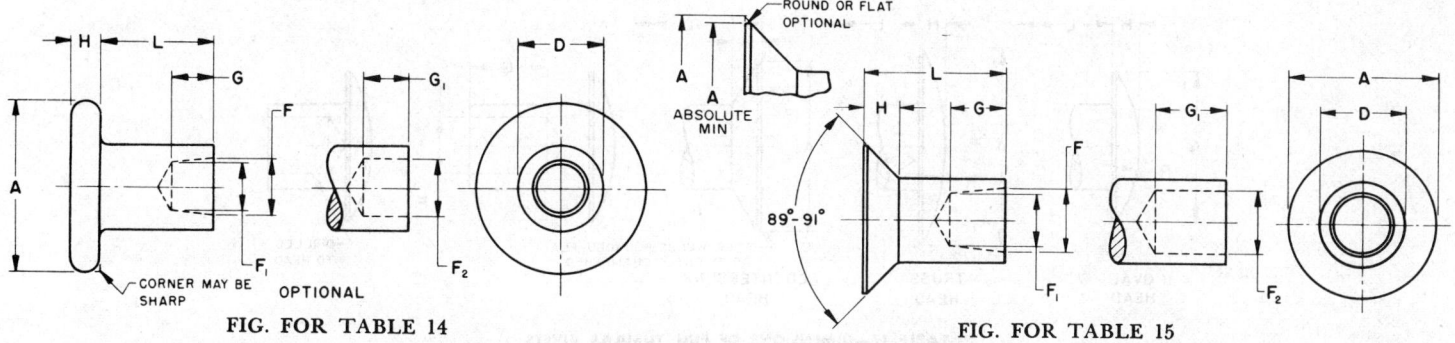

FIG. FOR TABLE 14   FIG. FOR TABLE 15

TABLE 14—DIMENSIONS OF FLAT HEAD SEMITUBULAR RIVETS

Nominal Size	D Dia of Shank		A Dia of Head		H Height of Head		Tapered Hole				Straight Hole			
							Dia			G Depth	F_2 Dia		G_1 Depth	
							F		F_1					
	Max	Min	Max	Min	Max	Min	Max	Min	Min	Min	Max	Min	Max	Min
1/16	0.061	0.058	0.114	0.104	0.027	0.023	0.046	0.042	0.036	0.042	0.044	0.039	0.057	0.042
3/32	0.089	0.085	0.161	0.151	0.034	0.028	0.069	0.065	0.057	0.057	0.068	0.062	0.072	0.057
1/8	0.123	0.118	0.223	0.213	0.041	0.034	0.095	0.091	0.079	0.082	0.090	0.084	0.104	0.089
9/64	0.146	0.141	0.317	0.307	0.052	0.042	0.112	0.106	0.091	0.104	0.107	0.100	0.135	0.120
3/16	0.188	0.182	0.381	0.369	0.067	0.057	0.145	0.139	0.120	0.135	0.141	0.134	0.166	0.151
1/4	0.252	0.244	0.507	0.493	0.090	0.076	0.191	0.181	0.155	0.183	0.184	0.176	0.229	0.214

For dimensions and tolerances not specified above, see General Specifications.

TABLE 15—DIMENSIONS OF 90 DEGREE COUNTERSUNK HEAD SEMITUBULAR RIVETS

Nominal Size	D Dia of Shank		A Dia of Head		H Height of Head	Tapered Hole				Straight Hole			
						Dia			G Depth	F_2 Dia		G_1 Depth	
						F		F_1					
	Max	Min	Max Sharp	Abs Min	Ref[a]	Max	Min	Min	Min	Max	Min	Max	Min
3/32	0.089	0.085	0.176	0.163	0.045	0.069	0.065	0.057	0.057	0.068	0.062	0.072	0.057
1/8	0.123	0.118	0.235	0.217	0.057	0.095	0.091	0.079	0.082	0.090	0.084	0.104	0.089
9/64	0.146	0.141	0.270	0.250	0.060	0.112	0.106	0.091	0.104	0.107	0.100	0.135	0.120
3/16	0.188	0.182	0.351	0.326	0.083	0.145	0.139	0.120	0.135	0.141	0.134	0.166	0.151
7/32	0.217	0.210	0.413	0.384	0.100	0.166	0.158	0.136	0.151	0.163	0.155	0.198	0.183
1/4	0.252	0.244	0.469	0.437	0.112	0.191	0.181	0.155	0.183	0.184	0.176	0.229	0.214

[a] Height of head, H, is given for reference purposes only. Variations in this dimension are controlled by diameters A and D and included angle of the head.

For dimensions and tolerances not specified above, see General Specifications.

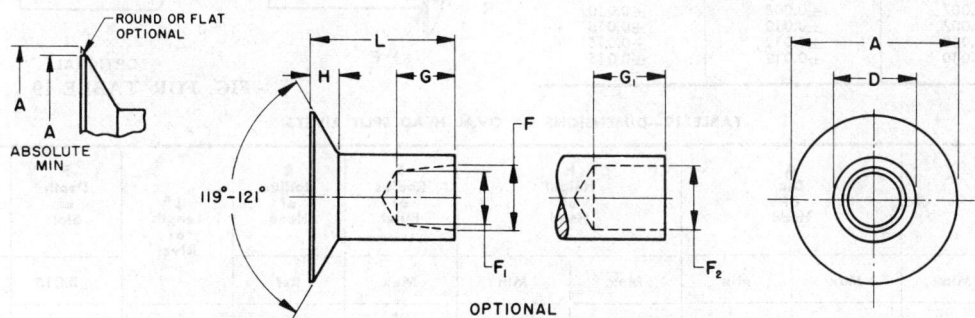

TABLE 16—DIMENSIONS OF 120 DEGREE COUNTERSUNK HEAD SEMITUBULAR RIVETS

Nominal Size	D Dia of Shank		A Dia of Head		H Height of Head	Tapered Hole				Straight Hole			
						Dia			G Depth	F_2 Dia		G_1 Depth	
						F		F_1					
	Max	Min	Max Sharp	Abs Min	Ref[a]	Max	Min	Min	Min	Max	Min	Max	Min
3/32	0.089	0.085	0.223	0.203	0.041	0.069	0.065	0.057	0.057	0.068	0.062	0.072	0.057
1/8	0.123	0.118	0.271	0.245	0.045	0.095	0.091	0.079	0.082	0.090	0.084	0.104	0.089
9/64	0.146	0.141	0.337	0.307	0.057	0.112	0.106	0.091	0.104	0.107	0.100	0.135	0.120
3/16	0.188	0.182	0.404	0.369	0.065	0.145	0.139	0.120	0.135	0.141	0.134	0.166	0.151
7/32	0.217	0.210	0.472	0.430	0.077	0.166	0.158	0.136	0.151	0.163	0.155	0.198	0.183
1/4	0.252	0.244	0.540	0.493	0.087	0.191	0.181	0.155	0.183	0.184	0.176	0.299	0.214

[a] Height of head, H, is given for reference purposes only. Variations in this dimension are controlled by diameters A and D and included angle of the head.

For dimensions and tolerances not specified above, see General Specifications.

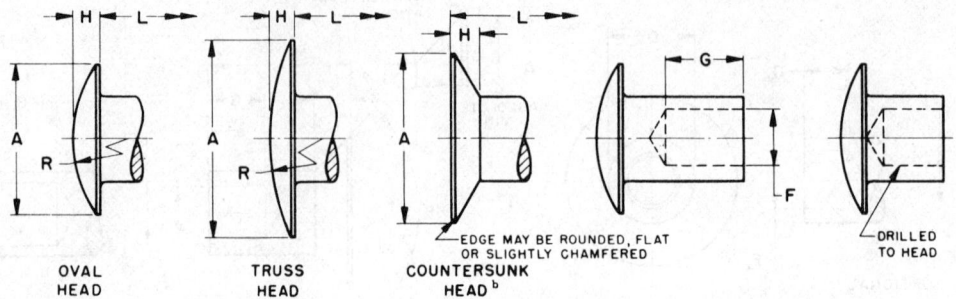

OVAL HEAD | TRUSS HEAD | COUNTERSUNK HEAD[b]

EDGE MAY BE ROUNDED, FLAT OR SLIGHTLY CHAMFERED

DRILLED TO HEAD

TABLE 17—DIMENSIONS OF FULL TUBULAR RIVETS

Head Style	Nominal Size	D Dia of Shank		A Dia of Head		H Height of Head		F Dia of Hole		G Depth of Hole	R Head Radius (Ref)
		Max	Min	Max	Min	Max	Min	Max	Min	Min[a]	Min
Oval	9/64	0.146	0.141	0.239	0.229	0.045	0.035	0.107	0.100	0.375	0.182
Truss	9/64	0.146	0.141	0.318	0.306	0.045	0.035	0.107	0.100	0.375	0.313
	3/16	0.188	0.182	0.381	0.369	0.065	0.055	0.141	0.134	0.375	0.324
Countersunk[b]	9/64	0.146	0.141	0.317	0.307	0.050	0.040	0.107	0.100	0.375	—
	3/16	0.188	0.182	0.381	0.369	0.060	0.048	0.141	0.134	0.375	—

[a] Full tubular rivets having length of 3/8 or shorter shall be drilled to head.
[b] Angle of head not specified since it is assumed this type of rivet would generally be used in soft materials and therefore form its own countersink.

For dimensions and tolerances not specified above, see General Specifications.

GENERAL SPECIFICATIONS FOR SPLIT RIVETS

General—This standard covers the complete general and dimensional data for oval head and countersunk head split rivets. Design and assembly data are given in the Appendix—Rivet Selection and Design Considerations.

Heads—The bearing surface of oval head split rivets shall be at right angles to the axis of the body within 2 deg. Because the heads are not machined or trimmed, the circumference may be slightly irregular and the edges may be rounded or flat.

Material—Split rivets shall be low carbon steel, brass, or other metals as agreed upon between the purchaser and supplier.

Workmanship—Rivets shall be free from surface seams, and all other defects that might affect their serviceability.

Length—Rivet length shall be measured as indicated in the illustrations for each head style. Tolerance on length shall be as specified in Table 18.

TABLE 18—TOLERANCES ON LENGTH

Nominal Size	Rivet Length		
	To and Including 4 x Dia	Over 4 to and Including 8 x Dia	Over 8 x Dia
	Tolerance		
3/32	±0.007	±0.008	±0.010
1/8	±0.007	±0.010	±0.015
9/64	±0.010	±0.012	±0.015
3/16	±0.010	±0.012	±0.015

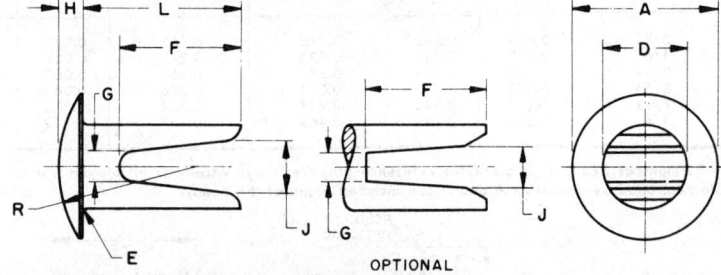

FIG. FOR TABLE 19

TABLE 19—DIMENSIONS OF OVAL HEAD SPLIT RIVETS

Nominal Size	D Dia of Shank		A Dia of Head		H Height of Head		E Radius of Fillet	R Radius of Head	L[a] Length of Rivet	F Depth of Slot ±0.015	Width of Slot	
											G ±0.005	J ±0.005
	Max	Min	Max	Min	Max	Min	Max	Ref				
3/32	0.092	0.086	0.151	0.141	0.027	0.019	0.010	0.130	3/16	0.156	0.030	0.037
									1/4	0.219	0.030	0.039
									5/16	0.250	0.030	0.039
									3/8 and over	0.312	0.030	0.039
1/8	0.122	0.114	0.223	0.213	0.037	0.027	0.014	0.210	3/16	0.156	0.040	0.047
									1/4	0.219	0.040	0.052
									5/16	0.266	0.040	0.057
									3/8 and over	0.312	0.040	0.057
9/64	0.152	0.144	0.317	0.307	0.049	0.037	0.018	0.304	3/16	0.156	0.050	0.060
									1/4	0.219	0.050	0.073
									5/16	0.281	0.050	0.078
									3/8	0.328	0.050	0.081
									7/16	0.344	0.050	0.083
									1/2 and over	0.391	0.052	0.077
3/16	0.195	0.185	0.350	0.338	0.064	0.050	0.022	0.300	1/4	0.219	0.065	0.120
									5/16	0.281	0.065	0.125
									3/8	0.312	0.065	0.127
									7/16	0.375	0.065	0.130
									1/2 and over	0.437	0.068	0.133

[a] Lengths over those tabulated shall be in increments of 1/16 in.

For dimensions and tolerances not specified above, see General Specifications.

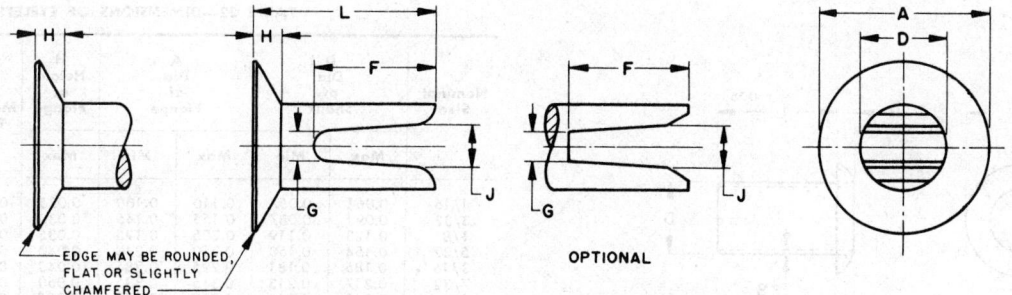

TABLE 20—DIMENSIONS OF COUNTERSUNK[a] HEAD SPLIT RIVETS

Nominal Size	D Dia of Shank		A Dia of Head		H Height of Head		L[b] Length of Rivet	F Depth of Slot ±0.016	Width of Slot	
									G ±0.005	J ±0.005
	Max	Min	Max	Min	Max	Min				
1/8	0.122	0.114	0.223	0.213	0.036	0.026	1/4	0.156	0.040	0.047
							5/16	0.219	0.040	0.052
							3/8	0.281	0.040	0.057
							7/16 and over	0.312	0.040	0.057
9/64	0.152	0.144	0.317	0.307	0.053	0.043	1/4	0.156	0.050	0.060
							5/16	0.219	0.050	0.073
							3/8	0.281	0.050	0.078
							7/16	0.328	0.050	0.081
							1/2	0.344	0.050	0.083
							9/16 and over	0.391	0.052	0.077
			0.380	0.370	0.062	0.052	1/4	0.156	0.050	0.060
							5/16	0.219	0.050	0.073
							3/8	0.281	0.050	0.078
							7/16	0.328	0.050	0.081
							1/2	0.344	0.050	0.083
							9/16 and over	0.391	0.052	0.077
3/16	0.195	0.185	0.443	0.431	0.061	0.051	5/16	0.219	0.065	0.120
							3/8	0.281	0.065	0.125
							7/16	0.312	0.065	0.127
							1/2	0.375	0.065	0.130
							9/16 and over	0.437	0.068	0.133

[a] Lengths over those tabulated shall be in increments of 1/16 in.
[b] Angle of head not specified since it is assumed this type of rivet would generally be used in soft materials and therefore form its own countersink.
For dimensions and tolerances not specified above, see General Specifications.

GENERAL SPECIFICATIONS FOR RIVET CAPS

General—This standard covers the complete general and dimensional data for rivet caps used with full tubular and split rivets where appearance is a consideration.

Materials—Rivet caps shall be brass or steel, standard with manufacturer.

Workmanship—Rivet caps shall be free from all defects that might affect their serviceability.

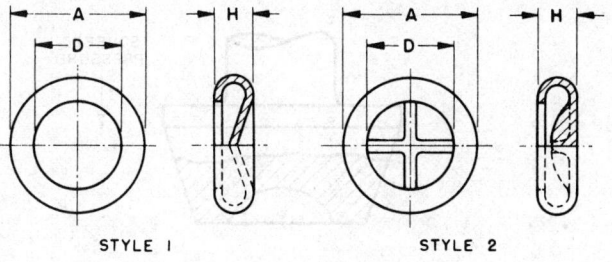

TABLE 21—DIMENSIONS OF RIVET CAPS

Style	D Dia of Hole		A Outside Dia		H Height	
	Max	Min	Max	Min	Max	Min
1[a]	0.233	0.203	0.288	0.258	0.098	0.068
	0.233	0.203	0.311	0.299	0.098	0.068
	0.233	0.203	0.358	0.346	0.098	0.068
2[b]	0.233	0.203	0.350	0.320	0.098	0.068
	0.281	0.251	0.442	0.412	0.129	0.099

[a] Style 1 rivet caps are designed for use with split rivets.
[b] Style 2 rivet caps are designed for use with full tubular rivets.

GENERAL SPECIFICATIONS FOR EYELETS

General—This standard covers the complete general and dimensional data for rolled flange eyelets. Design and assembly data are given in the Appendix —Rivet Selection and Design Considerations.

Flanges—Flanges of eyelets shall not be eccentric with the shank by more than 0.0075 in.

Material—Eyelets shall be brass, steel, or aluminum, standard with manufacturer.

Length—Length of eyelets shall be measured as indicated in the illustration. They are available in length increments of 1/32 in. between the limits specified in Table 22.

Workmanship—Eyelet end irregularities shall not be such that usability of the eyelet is impaired. Eyelets shall be free from surface seams, splits, and all other defects that might affect their serviceability.

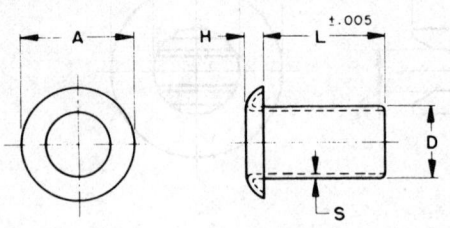

TABLE 22—DIMENSIONS OF EYELETS

Nominal Size	D Dia of Shank		A Dia of Flange		H Height of Flange	S[a] Material Thickness	L Available Lengths	
	Max	Min	Max	Min	Max		Max	Min
1/16	0.061	0.057	0.110	0.100	0.025	0.007	7/32	1/16
3/32	0.091	0.087	0.155	0.145	0.030	0.009	5/16	3/32
1/8	0.123	0.119	0.205	0.195	0.035	0.0095	11/32	3/32
5/32	0.154	0.150	0.250	0.240	0.040	0.010	11/32	3/32
3/16	0.185	0.181	0.295	0.285	0.045	0.0105	7/16	3/32
7/32	0.217	0.213	0.345	0.335	0.050	0.011	3/8	3/32
1/4	0.248	0.244	0.390	0.380	0.055	0.011	13/32	3/32

[a] Thicknesses tabulated are those from which eyelets are fabricated; therefore, thickness at shank may be slightly less than specified values.

APPENDIX—RIVET SELECTION AND DESIGN CONSIDERATIONS

General—This appendix is a guide intended to aid the user in the proper selection and application of rivets as a fastening means. It consists of general information on the advantages of riveting, various methods of riveting, selection of rivets and design considerations.

Advantages of Riveting—Riveting as a means of fastening is popular because of its simplicity, dependability, and low cost. Where the parts to be assembled do not normally need to be disassembled and, in the case of tubular, semitubular and split rivets, the tensile and fatigue strength of the joints made are not critical, riveting has many advantages. Some of the more outstanding of these are:

1. Metallic rivets are almost universally made by cold heading in high speed headers, and this makes a rivet a very economical fastener.
2. Investment in assembly equipment is low.
3. Maintenance costs of assembly equipment are low.
4. Rate of assembly is high and due to its simplicity, riveting lends itself to automation.
5. A minimum of skill is required to perform the operation.
6. Metallic or nonmetallic materials, or combinations thereof may be joined.
7. Rivets can be produced in a great variety of metals, ranging from low carbon steel to precious metals such as silver or gold.
8. Rivets may be used, not only as fasteners, but as functional components, such as pivots, electrical contacts, spacers, or supports.
9. Riveting normally requires no supplementary parts such as plain washers, lock washers, nuts, or safety wiring, nor are additional operations required such as assembly of nuts or locking devices as in the case of threaded fasteners.
10. Except for tubular, semitubular and split rivets, the rivet, when driven, usually fills the hole and prevents shifting of the parts joined.

Methods of Riveting—Riveting operations are performed by a number of methods, some of which are applicable only to particular types of rivets. The most commonly used methods are as follows:

Impact—This method employs a header die which strikes repetitive blows thus forming a head while the preformed end of the rivet is backed up with a tool called a buck or bucking bar. The header die may also be rotated while striking the repetitive blows. In machine riveting the buck is usually a part of the holding fixture. The method is applicable to solid rivets driven either hot or cold. Hot riveting is usually confined to large rivets used for structural purposes, while cold riveting is the method generally used for industrial applications on manufactured products. During the riveting operation the rivet material is displaced outward and downward into contact with the sides of the hole in which it is being assembled. The remainder of the material at rivet end forms the head. Upsetting of the shank can be controlled by using the proper impact force. See Fig. 1.

Squeeze—As its name implies, this method consists of applying steady pressure with a formed header die while the preformed end of rivet is backed up with a buck which may be made a part of the holding fixture. This method is applicable to solid rivets driven hot or cold. As in the case of impact riveting, the rivet material is displaced outward and downward into contact with the sides of the hole in which it is being assembled. The remainder of the material forms the head. See Fig. 2.

Clinch—This method of riveting involves forming the hollow end of tubular rivets and eyelets or prongs of split rivets back against the material being fastened and, depending on the shape and extent of the forming, is referred to as roll clinching, star or corrugated clinching, or scored clinching.

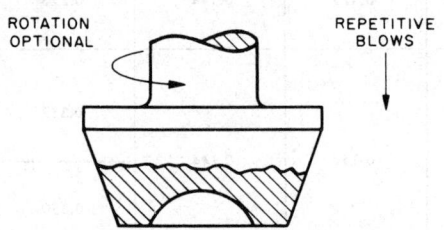

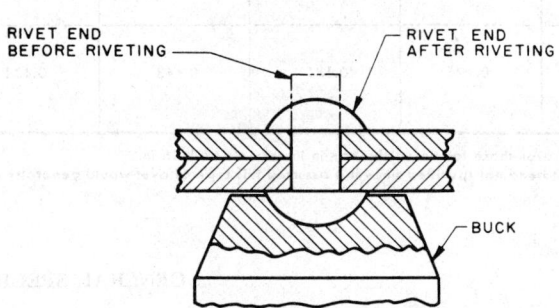

FIG. 1—IMPACT RIVETING

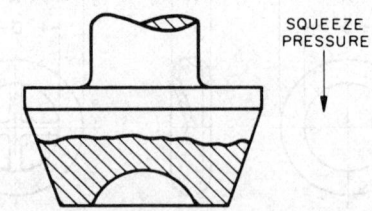

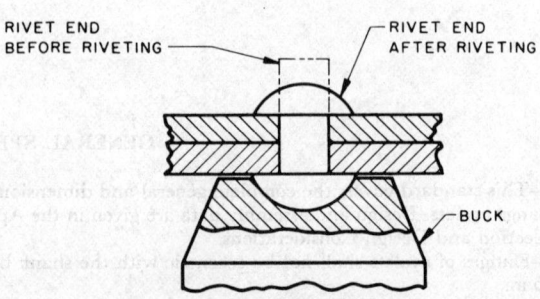

FIG. 2—SQUEEZE RIVETING

See Fig. 3. Roll clinching is accomplished by applying pressure with a formed header die, commonly called anvil, which turns or rolls the tubular shank or prongs of the rivet outward and over to bring it into contact with the part being assembled and is the method generally used to rivet semitubular and full tubular rivets and eyelets when used in metals or other hard materials. Star or corrugated clinching is accomplished by applying pressure with a formed header die which first splits or splays the tubular shank of the rivet and then turns or rolls the splayed portions outward and over to bring it into contact with the parts being assembled and is the method generally used to rivet full tubular rivets or eyelets when used in soft or resilient materials. When the splayed portions are actually turned back into the material being fastened, the method is often referred to as scored clinching. Where a finished appearance on both sides of the assembly is desirable, tubular and split rivets may be clinched into rivet caps designed for the purpose.

Shear—This method of riveting is accomplished by the use of a circular shear tool resembling a hollow punch. The method is applicable to solid rivets and the operating is performed cold. With the rivet properly bucked the tool having a hollow portion smaller than the rivet shank shears an annulus of material from the shank and with squeeze pressure upsets or displaces it into a flat annular head formed around the stub portion of the shank left by the hollow in the tool. The annular head is in contact with the part being assembled. The shearing action terminates flush with the top of the head thus leaving the head integral with the shank. See Fig. 4.

Staking—Staking consists of deforming the material of assembled rivets in such a way as to prevent their loosening or becoming disassembled under operating conditions. It does not include the forming of a head. It is done with a sharp tool at one or more points which forces the metal at these points tightly against the mating part. Where rivets are used in soft or thin materials and where light riveting is sufficient, the end of the rivet may be staked or slightly peened over the hole in a plain washer, commonly referred to as a riveting burr, to provide more bearing area on the staked side. See Fig. 5.

Rivet Selection—Requirement Considerations—With the wide variety of rivet types available, no fixed rule can be established to cover the selection of a type best suited for a given application. Generally, however, solid rivets are indicated for maximum strength while semitubular are preferred where cost is a prime factor and tensile or fatigue strength is not as critical. Full tubular rivets can, in some cases, be used with materials such as plastic, leather, canvas fabric, and wood in which the rivet under pressure pierces its own hole. The deep hole allows the slugs of pierced material to compress inside the rivet thereby exposing the required rivet material for clinching. Split rivets are also used extensively in soft materials such as those mentioned herein. The prongs pierce their own holes and are then clinched to effect the assembly. Split rivets may be used in the self-piercing and fastening of light gage metal as well. Semitubular rivets may also be used in the self-piercing and fastening of light metal wherever the appearance of the clinch is not important. Self-piercing riveting is economical and lends itself to high speed assembly operations.

Strength—A rivet is primarily strong only in shear. When set it is not stressed in tension. Thus, the designer must select the rivet size and material which will provide the necessary shear resistance needed in the application.

Diameter—The shear strength of a rivet is a direct function of the diameter so it is important to select a diameter which will provide the necessary shear strength.

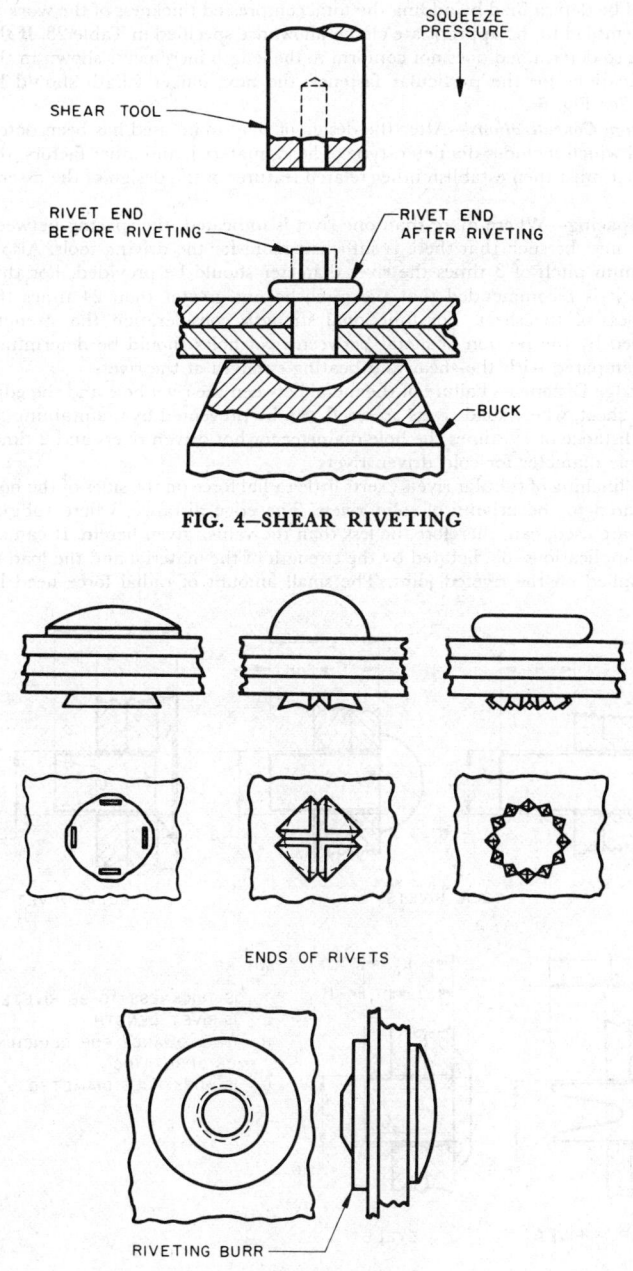

FIG. 4—SHEAR RIVETING

FIG. 5—STAKING

FIG. 3—CLINCH RIVETING

Head Design—The type of head specified will, of necessity, be dictated by the requirements of the application such as clearance, appearance, bearing area, and so forth. Round, truss, oval, flat, pan, and similar head styles with flat bearing surfaces provide good holding power at minimum cost. The use of flat head rivets where appearance is not a consideration minimizes tooling and production problems. Countersunk head rivets should be employed only where a flush surface is required since the countersinking or dimpling of parts to be fastened increases cost and production time on the assembly line.

Length—The length of rivet is affected by conditions such as the total compressed thickness of the members to be joined, the kind of rivet being used, the method of riveting being employed, the head style being formed, and the clearance hole into which the rivet is being assembled. The length of rivet required to provide optimum assembly conditions for a particular application can best be determined by experiment.

The following recommendations are often used to determine the length of various types of rivets for general applications and as a starting point in specific applications. The approximate length of solid rivets, when impact or squeeze riveted, required to form the head and fill the clearance space in the hole should be in excess of the thickness of the material to be riveted by an amount equal to approximately 0.75 to 1.00 times the rivet diameter for forming countersunk heads and from 1.3 to 1.7 times the rivet diameter for forming round or pan heads. See Fig. 6.

The approximate lengths for tubular rivets, split rivets, and eyelets should be determined by adding the total compressed thickness of the work to be assembled to the appropriate clinch allowance specified in Table 23. If the length so determined does not conform to the length increments shown in the specifications for the particular fastener, the next longer length should be used. See Fig. 6.

Design Considerations—After the design of rivet to be used has been determined which includes diameter, type of head, material, and other factors, the designer must then establish other related features of the design of the assembly.

Spacing—Where more than one rivet is indicated, the spacing between rivets must be such that there is sufficient room for the driving tools. Also a minimum pitch of 3 times the rivet diameter should be provided. For thin sheets it is recommended that the pitch be not greater than 24 times the thickness of the sheet. For functional strength consideration the strength afforded by the portion of metal between rivet holes should be determined and compared with the shear and bearing strength of the rivets.

Edge Distance—Failure of the metal between the rivet hole and the edge of the sheet, where solid rivets are used, can be prevented by maintaining an edge distance of $1\frac{1}{2}$ times the hole diameter for hot driven rivets and 2 times the hole diameter for cold driven rivets.

Clinching of tubular rivets exerts little radial force on the sides of the hole compared to the driving of solid rivets. The edge distance, where tubular rivets are used, can, therefore, be less than the values given herein. It can, in most applications, be dictated by the strength of the material and the load to be applied on the riveted joint. The small amount of radial force need be considered only where fastening very brittle materials such as ceramics and some plastics.

Accessibility—When using standard rivets, it is necessary to have both the preformed and driven head ends of the rivet accessible so that both the forming die and the buck may be properly used. Sufficient space should be provided to permit the use of power or manually operated rivet sets and bucks.

Hole Size—Holes should be held as close to the rivet shank diameter as possible and still permit easy and rapid assembly. Possible misalignment of holes must be considered in establishing hole sizes. Holes that are too large may result in buckling of the rivet shank or other detrimental effect when the rivet is being driven. The most suitable hole size for a given application can best be determined by experiment.

A general rule often applied to determine the hole size for solid rivets is to provide a clearance of from 0.003 to 0.008 over the maximum shank diameter of the rivet. The clearance can be increased to 0.015 where necessary for rivets $\frac{1}{4}$ in. and under and to 0.030 for rivets over $\frac{1}{4}$ in. in size. These increases, however, often result in poor riveting especially in applications requiring long grip lengths.

A general rule often applied to determine the preformed hole size for tubular rivets is to allow approximately 7% of the maximum shank diameter of the rivet for clearance. Recommended hole size values for these fasteners are given in Table 24. See Fig. 6.

Countersinking—Where the rivet heads on one or both sides of a riveted assembly must be flush with the surface, rivets with countersunk heads are used and the hole is countersunk to conform with the size and contour of the heads.

Dimpling—Dimpling involves the deformation of a sheet surface by pressure to form a countersunk recess on the one side and a corresponding projecting cone on the other. In the case of 2 or more sheets to be joined by riveting the dimpling is done on each sheet. When assembled the nesting of the dimples and projecting cones provides a large shear area and the rivet merely serves as a compression anchor to keep the dimples in contact. A relatively thin sheet may also be dimpled to match a countersunk recess in a thick sheet or manufactured part. Dimpling may be produced by a die set or by pressure exerted on the rivet head.

TABLE 23—RECOMMENDED CLINCH ALLOWANCES FOR TUBULAR AND SPLIT RIVETS AND EYELETS

Nominal Size	Clinch Allowances			
	Semitubular Rivets[a]	Full Tubular Rivets[b]	Split Rivets	Eyelets[a]
1/16	0.034	—	—	0.043
5/64	0.041	—	—	—
3/32	0.048	—	0.078	0.048
7/64	0.059	—	—	—
1/8	0.074	—	0.094	0.048
9/64	0.088	0.125	0.125	—
5/32	—	—	—	0.053
3/16	0.122	0.188	0.141	0.053
7/32	0.141	—	0.141	0.058
1/4	0.162	—	—	0.058
5/16	0.202	—	—	—

[a] For rolled clinch.
[b] For star or corrugated clinch, where roll clinch is desired, use semitubular values

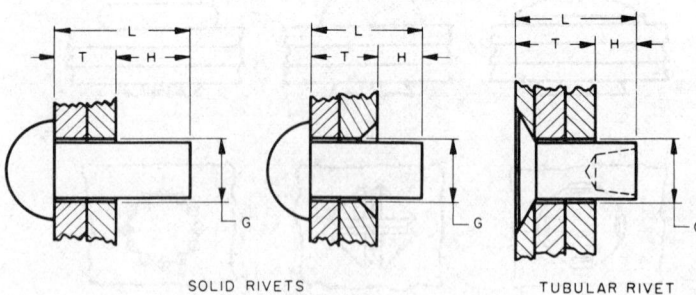

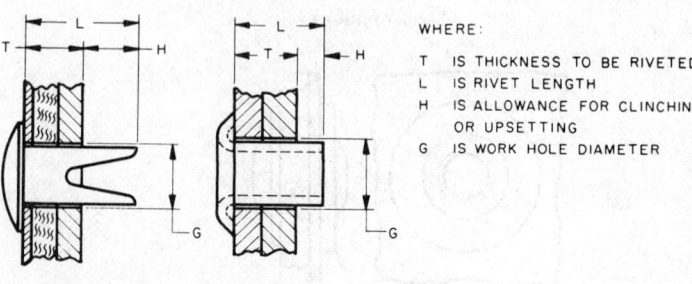

WHERE:
T IS THICKNESS TO BE RIVETED
L IS RIVET LENGTH
H IS ALLOWANCE FOR CLINCHING OR UPSETTING
G IS WORK HOLE DIAMETER

FIG. 6—TYPES OF RIVETS

TABLE 24—RECOMMENDED WORK HOLE DIAMETERS FOR TUBULAR AND SPLIT RIVETS AND EYELETS

Nominal Size	Tubular Rivets		Split Rivets		Eyelets	
	Hole Dia[a]	Drill Size	Hole Dia[a]	Drill Size	Hole Dia	Drill Size
1/16	0.064	No. 52	—	—	0.063	No. 52
5/64	0.081	No. 46	—	—	—	—
3/32	0.094	3/32	0.093	No. 42	0.093	No. 42
7/64	0.104	No. 37	—	—	—	—
1/8	0.129	No. 30	0.128	No. 30	0.125	1/8
9/64	0.152	No. 24	0.154	No. 23	—	—
5/32	—	—	—	—	0.156	5/32
3/16	0.196	No. 9	0.199	No. 8	0.188	No. 12
7/32	0.228	No. 1	—	—	0.219	7/32
1/4	0.261	G	—	—	0.250	1/4
5/16	0.328	21/64	—	—	—	—

[a] Applicable to full tubular and split rivets only where one of the parts to be assembled is prepunched or drilled.

BLIND RIVETS—BREAK MANDREL TYPE—SAE J1200   SAE Standard

Report of Fasteners Committee approved July 1977.

1. General Specifications

1.1 Scope—This standard establishes the dimensional, mechanical, and performance requirements of inch and metric break mandrel blind rivets suitable for use in joining the component parts of an assembly.

1.2 Definitions

1.2.1 BLIND RIVET—A blind rivet is a blind fastener which has a self-contained mechanical, chemical, or other feature which permits the formation of an upset on the blind end of the rivet and expansion of the rivet shank during rivet setting to join the component parts of an assembly.

1.2.2 BREAK MANDREL BLIND RIVET—Break mandrel blind rivets are pull mandrel-type blind rivets, where during the setting operation the mandrel is pulled into or against the rivet body and breaks at or near the junction of the mandrel shank and its upset end.

1.3 Designations—These rivets are designated by styles and grades as described below in addition to size, length, and finish.

1.3.1 RIVET STYLES—The two basic styles of break mandrel blind rivets are designated as protruding head and flush head. Protruding head rivets are available in two styles designated as regular head and large head. Flush head rivets are available in the 120 deg countersunk head.

1.3.2 RIVET GRADES—The material combination of break mandrel blind rivets are designated as grades, with each material combination representing a different combination of rivet body material and mandrel material as given in Table 1.

TABLE 1—GRADES OF BREAK MANDREL BLIND RIVETS

Grade Designation	Rivet Body Material	Mandrel Material
10	Aluminum Alloy 5050	Aluminum Alloy 7178 or 2024
11	Aluminum Alloy 5052	Aluminum Alloy 7178 or 2024
16	Aluminum Alloy 5154	Carbon Steel
18	Aluminum Alloy 5052	Carbon Steel
19	Aluminum Alloy 5056	Carbon Steel
20	Copper Alloy No. 110	Carbon Steel
30	Low Carbon Steel	Carbon Steel
40	Nickel-Copper Alloy (Monel)	Carbon Steel
50	Stainless Steel (300 Series)	Carbon Steel
51	Stainless Steel (300 Series)	Stainless Steel (300 Series)

1.4 Dimensions and Tolerances—The design of break mandrel type blind rivets shall be in accordance with the practice of the manufacturer providing the dimensions shown in Tables 6A, 6B, 7A, and 7B are maintained and rivets meet the mechanical and performance requirements of this standard. Tolerance on dimensions in tables, not designated otherwise, shall be ±0.010 in (0.25 mm).

1.5 Materials—Rivet bodies and mandrels shall be made of the material specified for the grade in Table 1. When the specific material analysis is not given, the analysis shall be selected by the manufacturer and shall be such to assure that rivets meet the mechanical and performance requirements specified under paragraph 2.

1.6 Finishes—Grade 30 rivet bodies are either zinc or cadmium plated with a minimum plating thickness of 0.00015 in (0.004 mm). Rivet bodies of all other grades are furnished plain (bare metal), unless otherwise specified. Because mandrels are discarded following rivet setting, mandrels of all materials may be furnished plain or with a protective coating at the option of the manufacturer, unless otherwise specified.

2. Mechanical and Performance Requirements

2.1 Shear Strength—Rivets, except those described in paragraph 2.2.1, shall have ultimate shear loads not less than the minimum ultimate shear loads specified for the applicable size and grade given in Tables 2A and 2B, when tested in accordance with paragraph 3.1.

2.2 Tensile Strength—Rivets, except those described in paragraph 2.2.1, shall have ultimate tensile loads not less than the minimum ultimate tensile loads specified for the applicable size and grade given in Tables 3A and 3B when tested in accordance with paragraph 3.2.

2.2.1 Grade 20 rivet is not subject to either shear or tensile testing. For all other grades, protruding head rivets with specified maximum grip lengths shorter than 1.0 times the nominal rivet diameter, and flush head rivets with specified maximum grip lengths shorter than 1.5 times the nominal rivet diameter shall not be subject to either shear or tensile testing.

2.3 Mandrel Break Load—While the rivet is being set, the axially applied load necessary to break the mandrel shall be within the limits specified for the applicable rivet size and grade in Tables 4A and 4B when tested in accordance with paragraph 3.3.

2.4 Mandrel Retention—The mandrel shall be retained within the rivet body such that a force in excess of 2 lb (8.9 N) is required to reduce the mandrel protrusion to its specified minimum.

TABLE 2A—ULTIMATE SHEAR LOADS OF BREAK MANDREL BLIND RIVETS

Nominal Rivet Size or Basic Shank Dia		Ultimate Shear Load[a] (Force) Min, lb				
		Grades 10, 11, & 18	Grades 16 & 19	Grade 30	Grade 40	Grades 50 & 51
3/32	0.0938	70	90	130	200	230
1/8	0.1250	120	170	260	350	420
5/32	0.1562	190	260	370	550	650
3/16	0.1875	260	380	540	800	950
1/4	0.2500	460	700	1000	1400	1700

[a] Grade 20 rivet is not subject to shear testing.

TABLE 2B—ULTIMATE SHEAR LOADS OF BREAK MANDREL BLIND RIVETS

Nominal Rivet Size mm	Ultimate Shear Load[a] (Force) Min, N				
	Grades 10, 11, & 18	Grades 16 & 19	Grade 30	Grade 40	Grades 50 & 51
2.4	310	400	580	890	1020
3.2	530	760	1160	1560	1870
4.0	850	1160	1650	2450	2890
4.8	1160	1690	2400	3560	4230
6.3	2050	3110	4450	6230	7560

[a] Grade 20 rivet is not subject to shear testing.

TABLE 3A—ULTIMATE TENSILE LOADS OF BREAK MANDREL BLIND RIVETS

Nominal Rivet Size or Basic Shank Dia		Ultimate Tensile Load[a] (Force) Min, lb				
		Grades 10, 11, & 18	Grades 16 & 19	Grade 30	Grade 40	Grades 50 & 51
3/32	0.0938	80	120	170	250	280
1/8	0.1250	150	220	310	450	530
5/32	0.1562	230	350	470	700	820
3/16	0.1875	320	500	680	1000	1200
1/4	0.2500	560	920	1240	1850	2100

[a] Grade 20 rivet is not subject to tensile testing.

TABLE 3B—ULTIMATE TENSILE LOADS OF BREAK MANDREL BLIND RIVETS

Nominal Rivet Size mm	Ultimate Tensile Load[a] (Force) Min, N				
	Grades 10, 11, & 18	Grades 16 & 19	Grade 30	Grade 40	Grades 50 & 51
2.4	360	530	760	1110	1250
3.2	670	980	1380	2000	2360
4.0	1020	1560	2090	3110	3650
4.8	1420	2220	3020	4450	5340
6.3	2490	4090	5520	8230	9340

[a] Grade 20 rivet is not subject to tensile testing.

TABLE 4A—MANDREL BREAK LOADS OF BREAK MANDREL BLIND RIVETS

Nominal Rivet Size or Basic Shank Dia		Limit	Mandrel Break Load[a] lb						
			Grades 10 & 11	Grades 16, 18, & 19	Grade 20	Grade 30	Grade 40	Grade 50	Grade 51
3/32	0.0938	Min	140	175	175	260	300	300	300
		Max	240	275	275	360	450	500	500
1/8	0.1250	Min	250	400	400	600	650	650	650
		Max	400	600	600	800	850	950	950
5/32	0.1562	Min	425	600	600	750	950	1150	1150
		Max	600	850	850	1000	1200	1450	1450
3/16	0.1875	Min	625	750	750	1150	1450	1400	1400
		Max	825	1050	1050	1450	1750	1900	1900
1/4	0.2500	Min	1100	1450	1450	1950	2500	3000	3000
		Max	1400	1850	1850	2350	2900	3600	3600

[a] Mandrel break load is defined as the load in pounds necessary to break the mandrel when setting break mandrel types of pull mandrel blind rivets.

TABLE 4B—MANDREL BREAK LOADS OF BREAK MANDREL BLIND RIVETS

Nominal Rivet Size or Basic Shank Dia mm	Limit	Mandrel Break Load[a] N						
		Grades 10 & 11	Grades 16, 18, & 19	Grade 20	Grade 30	Grade 40	Grade 50	Grade 51
2.4	Min	620	780	780	1160	1330	1330	1330
	Max	1070	1220	1220	1600	2000	2220	2220
3.2	Min	1110	1780	1780	2670	2890	2890	2890
	Max	1780	2670	2670	3560	3780	4230	4230
4.0	Min	1890	2670	2670	3340	4230	5120	5120
	Max	2670	3780	3780	4450	5340	6450	6450
4.8	Min	2780	3340	3340	5120	6450	6230	6230
	Max	3670	4670	4670	6450	7780	8450	8450
6.3	Min	4890	6450	6450	8670	11 100	13 300	13 300
	Max	6230	8230	8230	10 500	12 900	16 000	16 000

[a] Mandrel break load is defined as the load in Newtons necessary to break the mandrel when setting break mandrel types of pull mandrel blind rivets.

2.5 Blind Head Formation—The axially applied load necessary to upset the end of the rivet body, that is, form the blind side head, shall not exceed 80% of the actual mandrel break load, when tested in accordance with paragraph 3.3.

3. Test Methods

3.1 Shear Test—The test shall be comprised of loading a single lap joint assembled with one rivet so that the direction of applied load induces transverse shear against the rivet body. The test specimen shall be mounted in a tensile testing machine capable of applying load at a controllable rate. The grips shall be self-aligning and care shall be taken when mounting the specimen to assure that the load will be transmitted in a straight line through the test rivet.

The specimen shall be loaded at a speed of testing as determined with a free running cross head not less than 0.3 in (7.6 mm) nor greater than 0.5 in (13.0 mm)/min. Loading shall be continued until failure of the rivet occurs.

The maximum load in pounds or Newtons applied to the specimen coincident with or prior to rivet failure shall be recorded as the ultimate shear strength of the rivet.

The test specimen shall be comprised of two plates, of equal nominal thickness, axially aligned and assembled into a single lap joint with the test rivet, as shown in Fig. 1 and Tables 5A and 5B. The design of test plates may be modified to include holes for shear testing two or more rivets using the same plates. Such holes shall be located on the longitudinal centerline of the plate, and center distances between adjacent holes shall be at least 4 times the diameter of the larger test hole. Ends of plates may be drilled for pin-type mounting in testing machine. Plates shall be alloy steel, quenched and tempered to a hardness of Rockwell C46-50.

The test rivet shall be set with a setting tool standard for that type of rivet and in accordance with the setting procedures recommended by the rivet manufacturer.

TABLE 5A—DIMENSIONS OF TEST PLATES, IN (FIGS. 1, 2, AND 3)

Nominal Rivet Size or Basic Shank Dia		G Shear and Tensile Test Plate Hole Dia		G_1 Break Mandrel Test Restraining Plate Hole Dia	S End to Center Length	Shear and Tensile Test Plate Thickness	
						T Protruding Head Styles	T_1 Flush Head Styles
		Max	Min	Basic[a]	Min[b]	Min[c]	Min[d]
3/32	0.0938	0.100	0.098	0.067	0.375	0.047	0.056
1/8	0.1250	0.132	0.130	0.086	0.500	0.062	0.093
5/32	0.1562	0.164	0.162	0.105	0.625	0.078	0.117
3/16	0.1875	0.196	0.194	0.125	0.750	0.094	0.141
1/4	0.2500	0.260	0.258	0.161	1.000	0.125	0.188

[a] Values shown are equal to nominal mandrel diameter plus 0.010 in.
[b] Values shown are equal to 4 times basic shank diameter of rivet.
[c] Minimum values shown are equal to 0.50 times basic shank diameter of rivet. Maximum thickness shall not exceed 0.50 times maximum grip length specified for applicable rivet in Table 8A.
[d] Minimum values shown are equal to 0.75 times basic shank diameter of rivet. Maximum thickness shall not exceed 0.50 times maximum grip length specified for applicable rivet in Table 8A.
[e] The protrusion diameter of the mandrel (W diameter), including the point burr, shall be less than basic G_1 plate hole diameter.

TABLE 5B—DIMENSIONS OF TEST PLATES, MM (FIGS. 1, 2, AND 3)

Nominal Rivet Size or Basic Shank Dia	G Shear and Tensile Test Plate Hole Dia		G_1 Break Mandrel Test Restraining Plate Hole Dia	S End To Center Length	Shear and Tensile Test Plate Thickness	
					T Protruding Head Styles	T_2 Flush Head Styles
	Max	Min	Basic[a]	Min[b]	Min[c]	Min[d]
2.4	2.54	2.49	1.70	9.6	1.2	1.8
3.2	3.35	3.30	2.18	12.8	1.6	2.4
4.0	4.16	4.11	2.66	16.0	2.0	3.0
4.8	4.98	4.93	3.15	19.2	2.4	3.6
6.3	6.60	6.55	4.09	25.2	3.2	4.7

[a] Values shown are equal to nominal mandrel diameter plus 0.25 mm.
[b] Values shown are equal to 4 times basic shank diameter of rivet.
[c] Minimum values shown are equal to 0.50 times basic shank diameter of rivet. Maximum thickness shall not exceed 0.50 times maximum grip length specified for applicable rivet in Table 8B.
[d] Minimum values shown are equal to 0.75 times basic shank diameter of rivet. Maximum thickness shall not exceed 0.50 times maximum grip length specified for applicable rivet in Table 8B.
[e] The protrusion diameter of the mandrel (W diameter), including the point burr, shall be less than basic G_1 plate hole diameter.

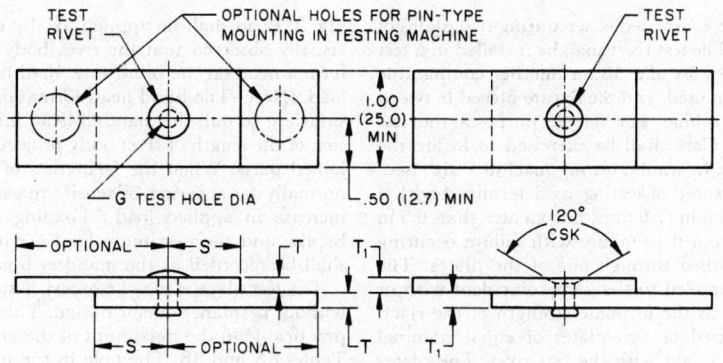

FIG. 1—TEST SPECIMENS FOR SHEAR TESTING BREAK MANDREL BLIND RIVETS

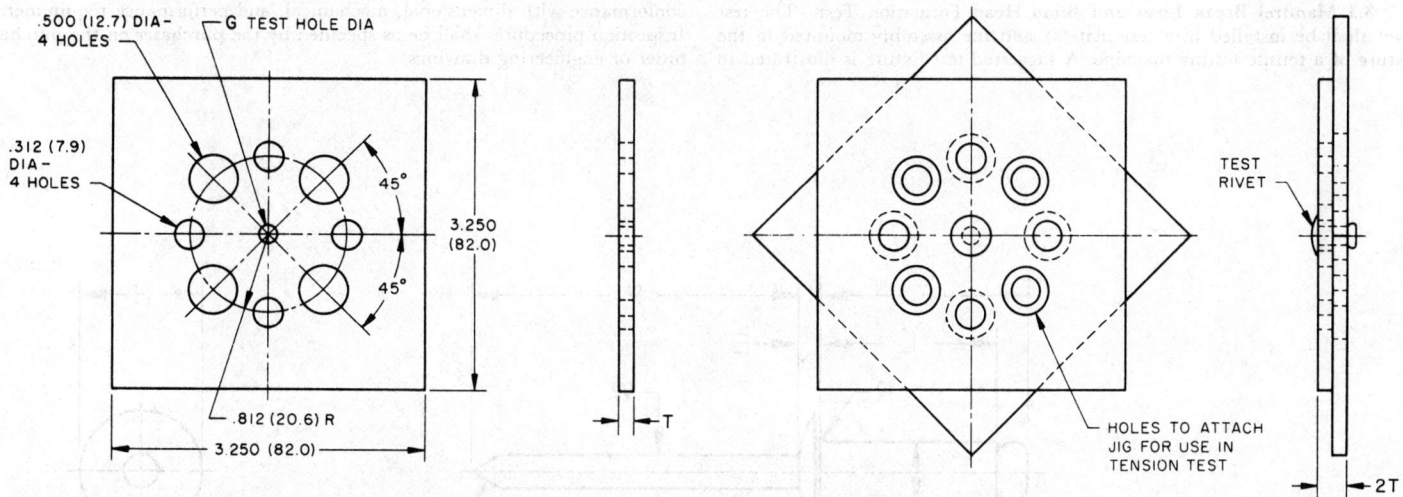

FIG. 2A—DETAIL OF PLATE USED FOR TENSION TESTS

FIG. 2B—ASSEMBLY OF TENSION TEST PLATES BEFORE ATTACHING TO JIG

FIG. 2—TEST FIXTURE FOR TENSILE TESTING BREAK MANDREL BLIND RIVETS

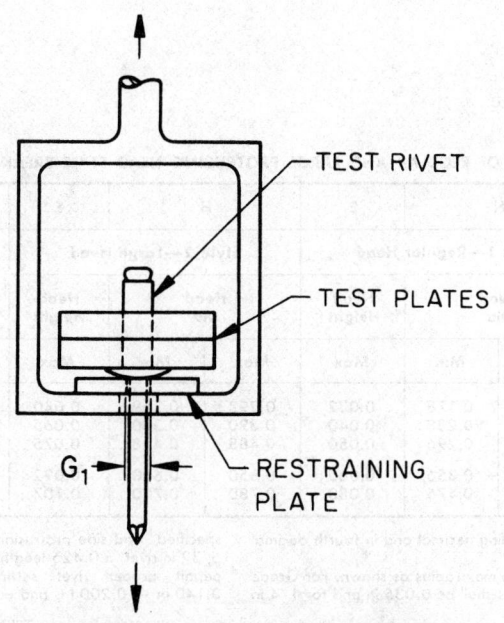

FIG. 3—TEST FIXTURE FOR TESTING MANDREL BREAK LOADS AND BLIND HEAD FORMATION

3.2 Tensile Test—The test shall be comprised of separating two plates of a joint assembled with one blind rivet. The test rivet shall be installed in a test fixture, as depicted in Fig. 2 and Tables 5A and 5B, or another comparable arrangement if an alternate test fixture is used, and the fixture placed between the compression heads of a testing machine. For referee purposes the test fixture shown in Fig. 2 shall be used. Care shall be exercised to locate the fixture at the center of the piston when hydraulic testing machines are used. Load shall be applied to the joint at a speed of testing, as determined with a free running cross head, not less than 0.3 in (7.6 mm) nor greater than 0.5 in (13.0 mm)/min. Loading shall be continued to failure with failure occuring when the rivet body fractures or is pulled through one of the plates. The maximum load in pounds (Newtons) applied to the joint coincident with or prior to rivet failure shall be recorded as the ultimate strength of the rivet.

The test specimen shall be comprised of two plates of equal nominal thickness, aligned and assembled into a joint with the test rivet. The plates shall be of alloy steel, quenched, and tempered to a hardness of Rockwell C46-50.

The test rivet shall be set with a setting tool which is standard for that type of rivet and in accordance with the setting procedures recommended by the rivet manufacturer.

3.3 Mandrel Break Load and Blind Head Formation Test—The test rivet shall be installed in a test plate(s), and the assembly mounted in the fixture of a tensile testing machine. A suggested test fixture is illustrated in Fig. 3. Load shall be applied axially to the mandrel. The load at which it is visually observed that the rivet body end is upset or otherwise deformed to form a head on the blind side, shall be recorded as the blind head formation load. (Note: The blind head formation load is a load applied to the mandrel sufficient to pull the mandrel head into the rivet body and initiate an expansion of the length of rivet body projecting beyond the blind side surface of the joined parts. When the formation of the blind side upset occurs there will normally be a period of tensile machine cross head travel with little or no increase in applied load.) Loading shall be continued until the mandrel breaks, and the maximum load occuring coincident with or prior to failure shall be recorded as the mandrel break load.

The test plate(s) may be of any material capable of supporting the test load without permanent deformation. Thickness of test plate(s) shall be as close as practicable to the maximum of the grip range of the test rivet as specified in Tables 8A and 8B. The hole in the test plate(s) shall conform to the recommended hole size given for the rivet size in Tables 8A and 8B.

The restraining plate shall be alloy steel, quenched, and tempered to a hardness of Rockwell C42-46. The hole in the plate shall conform to G diameter as specified in Tables 5A and 5B.

4. Inspection—Break mandrel blind rivets shall be inspected to determine conformance with dimensional, mechanical, and performance requirements. Inspection procedures shall be as specified by the purchaser on the purchase order or engineering drawings.

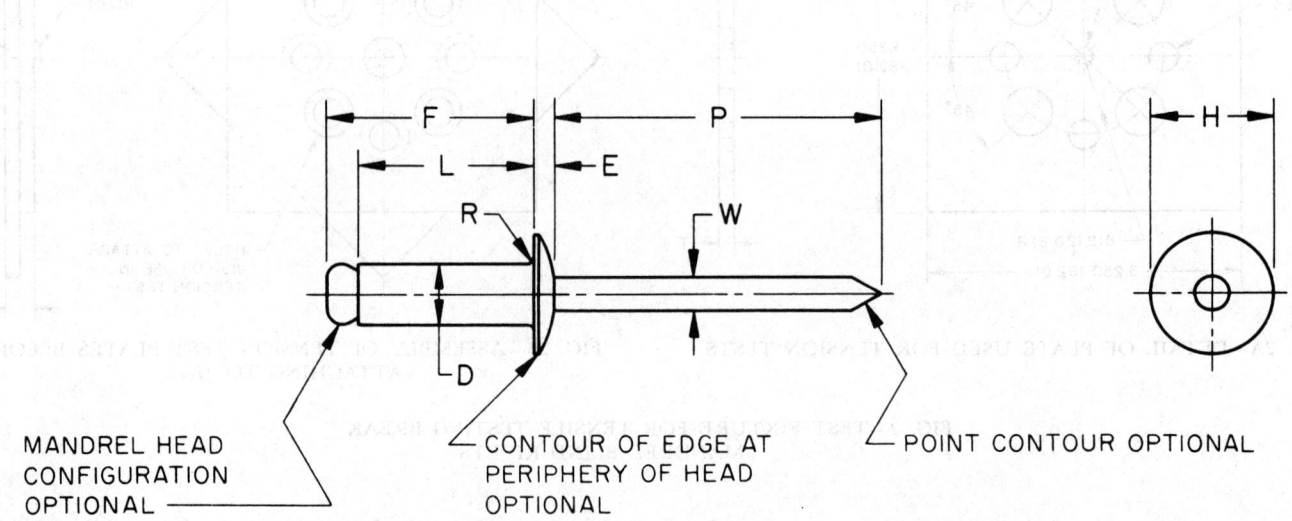

TABLE 6A—DIMENSIONS OF REGULAR AND LARGE PROTRUDING HEAD STYLE BREAK MANDREL BLIND RIVETS, IN

Nominal Rivet Size[a] or Basic Shank Dia		D Rivet Shank Dia		Style 1—Regular Head				Style 2—Large Head				Fillet Radius[b]	W Mandrel Dia	P Mandrel Protrusion	F Blind Side Protrusion[c]
				H Head Dia		E Head Height		H Head Dia		E Head Height					
		Max	Min	Max	Min	Max		Max	Min	Max		Max	Nom	Min	Max
3/32	0.0938	0.096	0.090	0.198	0.178	0.032		0.293	0.269	0.040		0.015	0.057	1.00	L + 0.100
1/8	0.1250	0.128	0.122	0.262	0.238	0.040		0.390	0.360	0.065		0.020	0.076	1.00	L + 0.120
5/32	0.1562	0.159	0.153	0.328	0.296	0.050		0.488	0.448	0.075		0.020	0.095	1.06	L + 0.140
3/16	0.1875	0.191	0.183	0.394	0.356	0.060		0.650	0.600	0.092		0.025	0.114	1.06	L + 0.160
1/4	0.2500	0.255	0.246	0.525	0.475	0.080		0.780	0.720	0.107		0.030	0.151	1.25	L + 0.180

[a] Where specifying nominal size in decimals, zeros preceding decimal and in fourth decimal place shall be omitted.

[b] The junction of head and shank shall have a fillet with a max radius as shown. For Grade 40, 50, and 51 rivets, the max fillet radius for 3/16 in rivets shall be 0.035 in and for 1/4 in rivets shall be 0.060 in.

[c] When computing the blind side protrusion (F), the max length of rivet (L), as given in Table 8A for the applicable grip shall be used. Minimum blind side clearance may be calculated by subtracting the actual grip (G), (that is, total thickness of the material to be joined), from the specified blind side protrusion (F). (Example: To join two plates, each 0.100 in thick, with a 5/32 in rivet, a 0.425 length rivet would be used. Minimum blind side clearance necessary to permit proper rivet setting would be L + 0.140 in − G, which is 0.425 in + 0.140 in − 0.200 in, and equals 0.365 in).

For application data see Table 8A.
Additional requirements given in General Specifications shall apply.

TABLE 6B—DIMENSIONS OF REGULAR AND LARGE PROTRUDING HEAD STYLE BREAK MANDREL BLIND RIVETS, MM

| Nominal Rivet Size or Basic Shank Dia | D Rivet Shank Dia | | Style 1—Regular Head | | | Style 2—Large Head | | | R Fillet Radius[a] | W Mandrel Dia | P Mandrel Protrusion | F Blind Side Protrusion[b] |
| | | | H Head Dia | | E Head Height | H Head Dia | | E Head Height | | | | |
	Max	Min	Max	Min	Max	Max	Min	Max	Max	Nom	Min	Max
2.4	2.44	2.29	5.03	4.52	0.81	7.44	6.83	1.02	0.4	1.45	25.0	L + 2.5
3.2	3.25	3.10	6.65	6.05	1.02	9.91	9.14	1.65	0.5	1.93	25.0	L + 3.0
4.0	4.04	3.89	8.33	7.52	1.27	12.40	11.38	1.90	0.5	2.41	27.0	L + 3.5
4.8	4.85	4.65	10.01	9.04	1.52	16.51	15.24	2.34	0.7	2.90	27.0	L + 4.0
6.3	6.48	6.25	13.33	12.07	2.03	19.81	18.29	2.72	0.8	3.84	31.0	L + 4.5

[a] The junction of head and shank shall have a fillet with a max radius as shown. For Grade 40, 50, and 51 rivets, the max fillet radius for 4.8 mm rivets shall be 0.9 mm and for 6.3 mm rivets shall be 1.5 mm.

[b] When computing the blind side protrusion (F), the max length of rivet (L) as given in Table 8B for the applicable grip shall be used. Minimum blind side clearance may be calculated by subtracting the actual grip (G), (that is, total thickness of the material to be joined), from the specified blind side protrusion (F). (Example: To join two plates, each 2.5 mm thick, with a 4.0 mm rivet, a 10.8 mm length rivet would be used. Minimum blind side clearance necessary to permit proper rivet setting would be L + 3.5 mm − G, which is 10.8 mm + 3.5 mm − 5.0 mm and equals 9.3 mm).

For application data see Table 8B.
Additional requirements given in General Specifications shall apply.

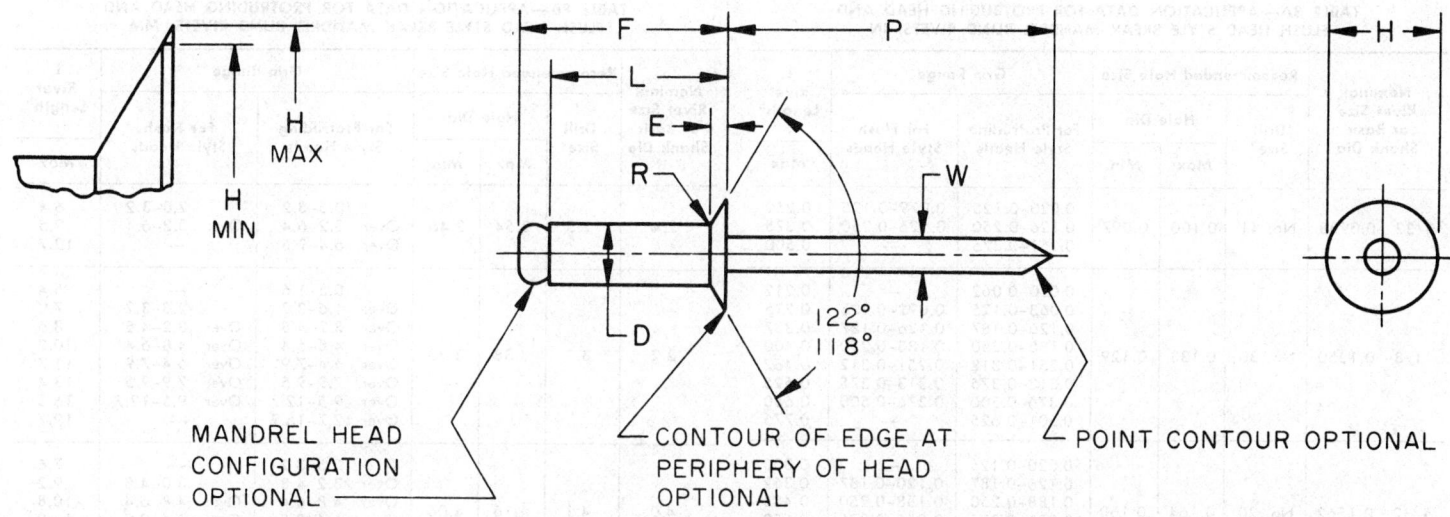

TABLE 7A—DIMENSIONS OF 120 DEG COUNTERSUNK FLUSH HEAD STYLE BREAK MANDREL BLIND RIVETS, IN

| Nominal Rivet Size[a] or Basic Shank Dia | | D Rivet Shank Dia | | H Head Dia[b] | | E Head Height[c] | R Fillet Radius | W Mandrel Dia | P Mandrel Protrusion | F Blind Side Protrusion[d] |
		Max	Min	Max	Min	Ref	Max	Nom	Min	Max
3/32	0.0938	0.096	0.090	0.187	0.161	0.027	0.020	0.057	1.00	L + 0.100
1/8	0.1250	0.128	0.122	0.233	0.207	0.031	0.025	0.076	1.00	L + 0.120
5/32	0.1562	0.159	0.153	0.294	0.268	0.040	0.030	0.095	1.06	L + 0.140
3/16	0.1875	0.191	0.183	0.361	0.335	0.050	0.035	0.114	1.06	L + 0.160

[a] Where specifying nominal size in decimals, zeros preceding decimal and in fourth decimal place shall be omitted.

[b] Max head diameter is calculated on nominal rivet diameter and nominal head angle extended to sharp corner. Min head diameter is absolute.

[c] Head height is given for reference purposes only. Variations in this dimension are controlled by the diameters (H) and (D) and the included angle of the head.

[d] When computing the blind side protrusion (F), the max length of rivet (L), as given in Table 8A for the applicable grip shall be used. Minimum blind side clearance may be calculated by subtracting the actual grip (G), (such as, total thickness of the material to be joined), from the specified blind side protrusion (F). (Example: To join two plates, each 0.187 in thick, with a 3/16 in rivet, a 0.575 length rivet would be used. Minimum blind side clearance necessary to permit proper rivet setting would be L + 0.160 in − G, which is 0.575 in + 0.160 in − 0.374 in which equals 0.361 in).

For application data see Table 8A.
Additional requirements given in General Specifications shall apply.

TABLE 7B—DIMENSIONS OF 120 DEG COUNTERSUNK FLUSH HEAD STYLE BREAK MANDREL BLIND RIVETS, MM

Nominal Rivet Size or Basic Shank Dia	D Rivet Shank Dia		H Head Dia[a]		E Head Height[b]	R Fillet Radius	W Mandrel Dia	P Mandrel Protrusion	F Blind Side Protrusion[c]
	Max	Min	Max	Min	Ref	Max	Nom	Min	Max
2.4	2.44	2.29	4.75	4.09	0.69	0.5	1.45	25.0	L + 2.5
3.2	3.25	3.10	5.92	5.26	0.79	0.7	1.93	25.0	L + 3.0
4.0	4.04	3.89	7.47	6.81	1.02	0.8	2.41	27.0	L + 3.5
4.8	4.85	4.65	9.17	8.51	1.27	0.9	2.90	27.0	L + 4.0

[a] Max head diameter is calculated on nominal rivet diameter and nominal head angle extended to sharp corner. Min head diameter is absolute.

[b] Head height is given for reference purposes only. Variations in this dimension are controlled by the diameters (H) and (D) and the included angle of the head.

[c] When computing the blind side protrusion (F), the max length of rivet (L), as given in Table 8B for the applicable grip shall be used. Minimum blind side clearance may be calculated by subtracting the actual grip (G), (that is, total thickness of the material to be joined), from the specified blind side protrusion (F). (Example: To join two plates, each 4.7 mm thick, with a 4.8 mm rivet, a 14.6 mm length rivet would be used. Minimum blind side clearance necessary to permit proper rivet setting would be L + 4.0 mm − G, which is 14.6 mm + 4.0 mm − 9.4 mm which equals 9.2 mm.)

For application data see Table 8B.

Additional requirements given in General Specifications shall apply.

TABLE 8A—APPLICATION DATA FOR PROTRUDING HEAD AND FLUSH HEAD STYLE BREAK MANDREL BLIND RIVETS, IN

Nominal Rivet Size or Basic Shank Dia	Recommended Hole Size			Grip Range		L Rivet Length[b]
	Drill Size[a]	Hole Dia		For Protruding Style Heads	For Flush Style Heads	Max
		Max	Min			
3/32 0.0938	No. 41	0.100	0.097	0.020–0.125	0.079–0.125	0.250
				0.126–0.250	0.126–0.250	0.375
				0.251–0.375		0.500
1/8 0.1250	No. 30	0.133	0.129	0.020–0.062	—	0.212
				0.063–0.125	0.092–0.125	0.275
				0.126–0.187	0.126–0.187	0.337
				0.188–0.250	0.188–0.250	0.400
				0.251–0.312	0.251–0.312	0.462
				0.313–0.375	0.313–0.375	0.525
				0.376–0.500	0.376–0.500	0.650
				0.501–0.625		0.775
5/32 0.1562	No. 20	0.164	0.160	0.020–0.125	—	0.300
				0.126–0.187	0.120–0.187	0.362
				0.188–0.250	0.188–0.250	0.425
				0.251–0.375	0.251–0.375	0.550
				0.376–0.500	0.376–0.500	0.675
				0.501–0.625		0.800
3/16 0.1875	No. 11	0.196	0.192	0.020–0.125	—	0.325
				0.126–0.187	0.151–0.187	0.387
				0.188–0.250	0.188–0.250	0.450
				0.251–0.375	0.251–0.375	0.575
				0.376–0.500	0.376–0.500	0.700
				0.501–0.625	0.501–0.625	0.825
				0.626–0.750	—	0.950
				0.751–0.875	—	1.075
				0.876–1.000	—	1.200
				1.001–1.125	—	1.325
1/4 0.2500	F	0.261	0.257	0.020–0.125	—	0.375
				0.126–0.250	—	0.500
				0.251–0.375	—	0.625
				0.376–0.500	—	0.750
				0.501–0.625	—	0.875
				0.626–0.750	—	1.000
				0.751–0.875	—	1.125
				0.876–1.000	—	1.250
				1.001–1.125	—	1.375
				1.126–1.250	—	1.500

[a] Recommended drill sizes are those which normally produce holes within the specified hole size limits.

[b] Where blind side clearances permit and it is economically feasible, rivets of the next longer length than those recommended for a given grip may be substituted to limit the number of different inventory items.

TABLE 8B—APPLICATION DATA FOR PROTRUDING HEAD AND FLUSH HEAD STYLE BREAK MANDREL BLIND RIVETS, MM

Nominal Rivet Size or Basic Shank Dia	Recommended Hole Size			Grip Range		L Rivet Length[b]
	Drill Size[a]	Hole Dia		For Protruding Style Heads	For Flush Style Heads	Max
		Max	Min			
2.4	2.5	2.54	2.46	Over 0.5–3.2	—	6.4
				Over 3.2–6.4	Over 2.0–3.2	9.5
				Over 6.4–9.5	Over 3.2–6.4	12.7
3.2	3.3	3.38	3.28	Over 0.5–1.6	—	5.4
				Over 1.6–3.2	—	7.0
				Over 3.2–4.8	Over 2.3–3.2	8.6
				Over 4.8–6.4	Over 3.2–4.8	10.2
				Over 6.4–7.9	Over 4.8–6.4	11.7
				Over 7.9–9.5	Over 6.4–7.9	13.4
				Over 9.5–12.7	Over 7.9–9.5	16.5
				Over 12.7–15.9	Over 9.5–12.7	19.7
4.0	4.1	4.16	4.06	Over 0.5–3.2	—	7.6
				Over 3.2–4.8	Over 3.0–4.8	9.2
				Over 4.8–6.4	Over 4.8–6.4	10.8
				Over 6.4–9.5	Over 6.4–9.5	14.0
				Over 9.5–12.7	Over 9.5–12.7	17.2
				Over 12.7–15.9		20.3
4.8	4.9	4.98	4.88	Over 0.5–3.2	—	8.3
				Over 3.2–4.8	Over 3.8–4.8	9.8
				Over 4.8–6.4	Over 4.8–6.4	11.5
				Over 6.4–9.5	Over 6.4–9.5	14.6
				Over 9.5–12.7	Over 9.5–12.7	17.8
				Over 12.7–15.9	Over 12.7–15.9	21.0
				Over 15.9–19.1		24.2
				Over 19.1–22.2		27.3
				Over 22.2–25.4		30.5
				Over 25.4–28.6		33.7
6.3	6.5	6.63	6.53	Over 0.5–3.2	—	9.5
				Over 3.2–6.4	—	12.7
				Over 6.4–9.5	—	15.9
				Over 9.5–12.7	—	19.1
				Over 12.7–15.9	—	22.2
				Over 15.9–19.1	—	25.4
				Over 19.1–22.2	—	28.6
				Over 22.2–25.4	—	31.8
				Over 25.4–28.6	—	34.9
				Over 28.6–31.8	—	38.1

[a] Recommended drill sizes are those which normally produce holes within the specified hole size limits.

[b] Where blind side clearances permit and it is economically feasible, rivets of the next longer length than those recommended for a given grip may be substituted to limit the number of different inventory items.

HOSE CLAMP SPECIFICATIONS—SAE J1508 JUN93   SAE Standard

Report of the Fasteners Committee approved June 1987. Completely revised by the Hose/Clamp Subcommittee of the SAE Coolant Hose Committee May 1991, and revised June 1993.

1. Scope—This SAE Standard covers twenty (20) types of clamps most commonly and suitably being used on OEM coolant, fuel, oil, vacuum, and emission systems.

1.1 Purpose—This document is compiled for the specific purpose of describing the basic characteristics and minimum performance requirements recommended by the manufacturers. No application recommendations are intended or implied.

2. References

2.1 Applicable Documents—The following publications form a part of this specification to the extent specified herein. The latest issue of SAE publications shall apply.

2.1.1 SAE PUBLICATIONS—Available from SAE, 400 Commonwealth Drive, Warrendale, PA 15096-0001.

SAE J178—Music Steel Wire and Spring
SAE J402—SAE Numbering System for Wrought or Rolled Steel
SAE J478—Slotted and Recessed Head Screws
SAE J1086—Metals and Alloys in the Unified Numbering System

2.1.2 ANSI AND IFI PUBLICATIONS—Available from ANSI, 11 West 42nd Street, New York, NY 10036-8002.

ANSI B1.1, 3M—Unified Inch Screw Thread
ANSI B1.3M—Screw Thread Gauging Systems for Dimensional Acceptability
IFI 112—High Performance Thread Rolling Screws

2.1.3 ASTM PUBLICATIONS—Available from ASTM, 1916 Race Street, Philadelphia, PA 19103-1187.

ASTM A 228—Standard Specification for Steel Wire, Music Spring Quality
ASTM B 117—Standard Method of Salt Spray (Fog) Testing

2.1.4 MILITARY PUBLICATIONS—Available from Commanding Officer, Naval Publications and Forms Center, 700 Robbins Avenue, Philadelphia, PA 19111.

MIL Std MS21044—Nut, Self-Locking, Hexagon, Regular Height, 250 °F, 125 KSI Ftu and 60 KSI Ftu
MIL Std MS21045—Nut, Self-Locking, Hexagon-Regular Height, 450 °F, 125 KSI Ftu
MIL Std MS39326—Clamp, Spring: Hose (Low Pressure) Type "E"

2.1.5 OTHER PUBLICATION

AISI—Material Standards

2.2 Definitions

2.2.1 FREE TORQUE—The torque value expressed in newton meters (pound inches) when the clamp is tightened four complete revolutions of the screw or nut, while in the free state. This value does not include any break-away effects due to staking or passage of the band ends beyond the screw head.

2.2.2 DURABILITY TORQUE—The maximum torque value applied to a clamp without evidence of deformation or excessive wear when tightened once over a steel mandrel.

2.2.3 INSTALLATION TORQUE—The recommended torque for installation of the clamp. This is generally expressed in terms of 50% to 75% of the rated "Durability Torque" for specific clamps. Installation Torque is sometimes referred to as Application Torque.

2.2.4 ULTIMATE TORQUE—The torque value at which the clamp develops deformation to a degree that it cannot be reused or no longer achieves its intended use.

3. Classification—For ease of handling the various clamp designs and modifications thereof; clamps have been grouped by their basic design and functional characteristics:

3.1 Group #1—Clamps which require torquing a screw or nut for installation.

3.1.1 "A" AND "AHH"—Dual body wires utilizing a machine screw with trunnion nut for the tightening mechanism. Screw position tangential to the diameter. See Figure 1 and Tables 1 and 1A.

3.1.2 "B" AND "D"—Flat band body stock utilizing a machine screw and square nut for the tightening mechanism. Screw position tangential to the diameter. See Figures 2, 3, and 3A and Tables 2 and 3.

3.1.3 "C"—Flat band body stock utilizing a bridge structure to position the machine screw and nut tightening mechanism perpendicular to the diameter. See Figure 4 and Tables 4 and 4A.

3.1.4 "F," "FEO," "FE," "HD," "I," "M," AND "MX"—A tangential worm drive screw engaging either pierced through slots or embossed threads. Those using pierced through slots are also available in extended band versions to protect soft hose compounds. See Figures 5 to 11 and Tables 5 to 14.

NOTE—"FE" means type "F," embossed slots; "FEO" means type "F," embossed slots with screw offset from centerline of the band.

3.1.5 "TB"—A fixed, tangential, T-bolt with a rotating locknut the turning of which draws both clamp ends together. Construction may employ either a floating bridge, tongue, or be of one piece (band) construction as standard. See Figures 12 and 13 and Table 15.

3.2 Group #2 (Types "E," "CTW," or "CTB")—Clamps which are either supplied in a locked, spring-loaded, full-open position or sprung open at installation and then released over the hose/fitting to create sealing due to the spring-like function.

3.2.1 "E"—Single round wire, heat-treated to spring temper. Ancillary specification MIL Std MS39326. See Figures 14 and 15 and Tables 16 and 17.

3.2.2 "CTB"—Flat band stock, heat-treated to spring temper. See Figure 16 and Table 18.

3.2.3 "CTW"—Dual rough pre-hardened spring wires, or wires heat-treated to spring temper. See Figure 17 and Tables 19 and 20.

3.3 Group #3—Hybrid clamps which require torquing of a screw, or nut, for installation but which also incorporate a means of storing energy for the spring-like function.

3.3.1 "SLA"—Basic type "A" clamp modified to incorporate a stack of spring washers for energy storage. See Figure 18.

3.3.2 "SLF"—Basic type "F" clamp modified to incorporate a stack of conical spring washers for energy storage. See Figure 20.

3.3.3 "SLHD"—Basic type "HD" clamp modified to incorporate a stack of conical spring washers for energy storage. See Figure 11 and Table 14.

3.3.4 "T"—Basic type "F" clamp utilizing a convoluted and heat-treated band for energy storage and a full, flanged inner shield. See Figures 20 and 21 and Table 21.

3.3.5 "SLTB"—Basic type "TB" with a coil spring for energy storage. See Figures 12 and 13 and Table 15.

4. General Requirements

4.1 Group #1—Clamps shall be supplied in the full open position. Those clamps using machine screws shall have the screws retained in the clamp by staking or other means agreeable to the user. Where so specified by the purchaser, types "B" and "D" clamps shall have provisions to retain the nut in base leg when axial pressure is applied to screw. All clamps shall close tight upon round steel mandrels of the sizes 4.1 indicated in the respective open and closed diameter charts. All clamps shall be free from burrs, seams, laps, loose scale, or any other defects that may affect their serviceability.

4.2 Group #2—Clamps type "E" and "CTW" shall be supplied in the free state, full-closed position. To assure that permanent deformation, resulting from opening the clamp at installation, does not occur—clamps shall be opened to a diameter no larger than that listed in column "A" (for each respective clamp type) and released to the free state at which point the clamps may not pass over a "NO-GO" size mandrel as listed in column "D," respectively. Clamps shall be free of burrs, heat-treat scale, and nicks that may affect their serviceability.

4.2.1 Type "CTB" clamps may be supplied in either the free-state (Table 18) or a locked, spring-loaded, full-opened position (Figure 16, b and c). The clamp shall be designed so as not to allow plastic deformation in the full-opened position. Clamps shall be free from burrs, seams, laps, loose scale, or any other defects that affect their performance.

4.3 Group #3—Clamps are governed by the General Requirements set forth for Group #1 clamps in 4.1.

16.46

TYPE "A" & "AHH"

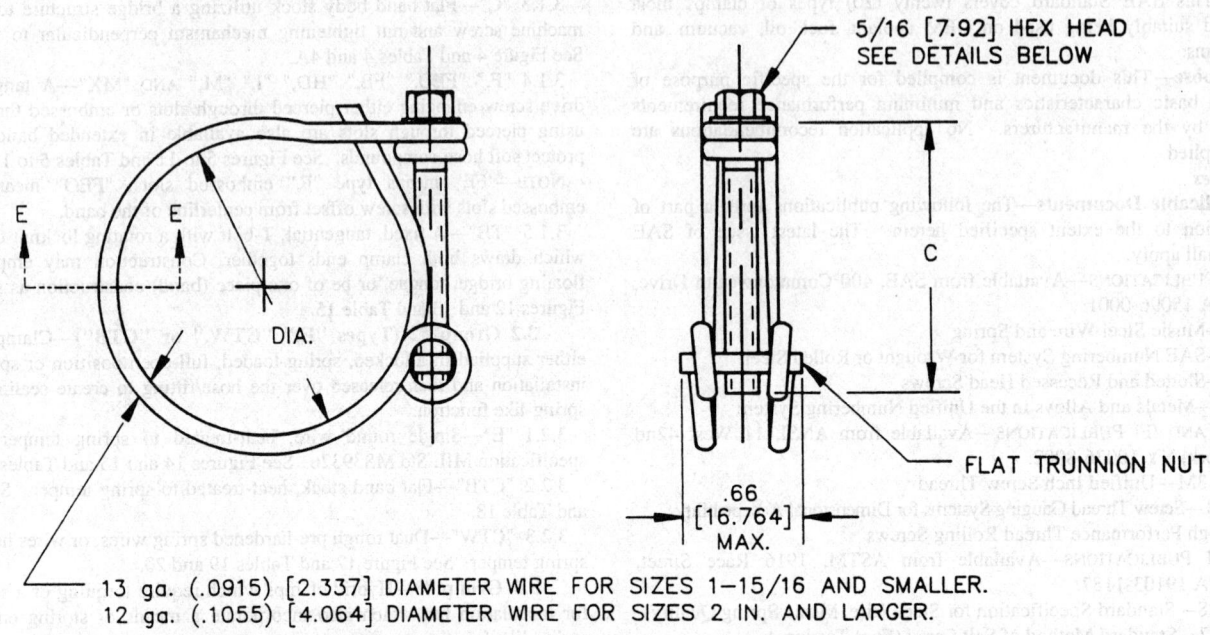

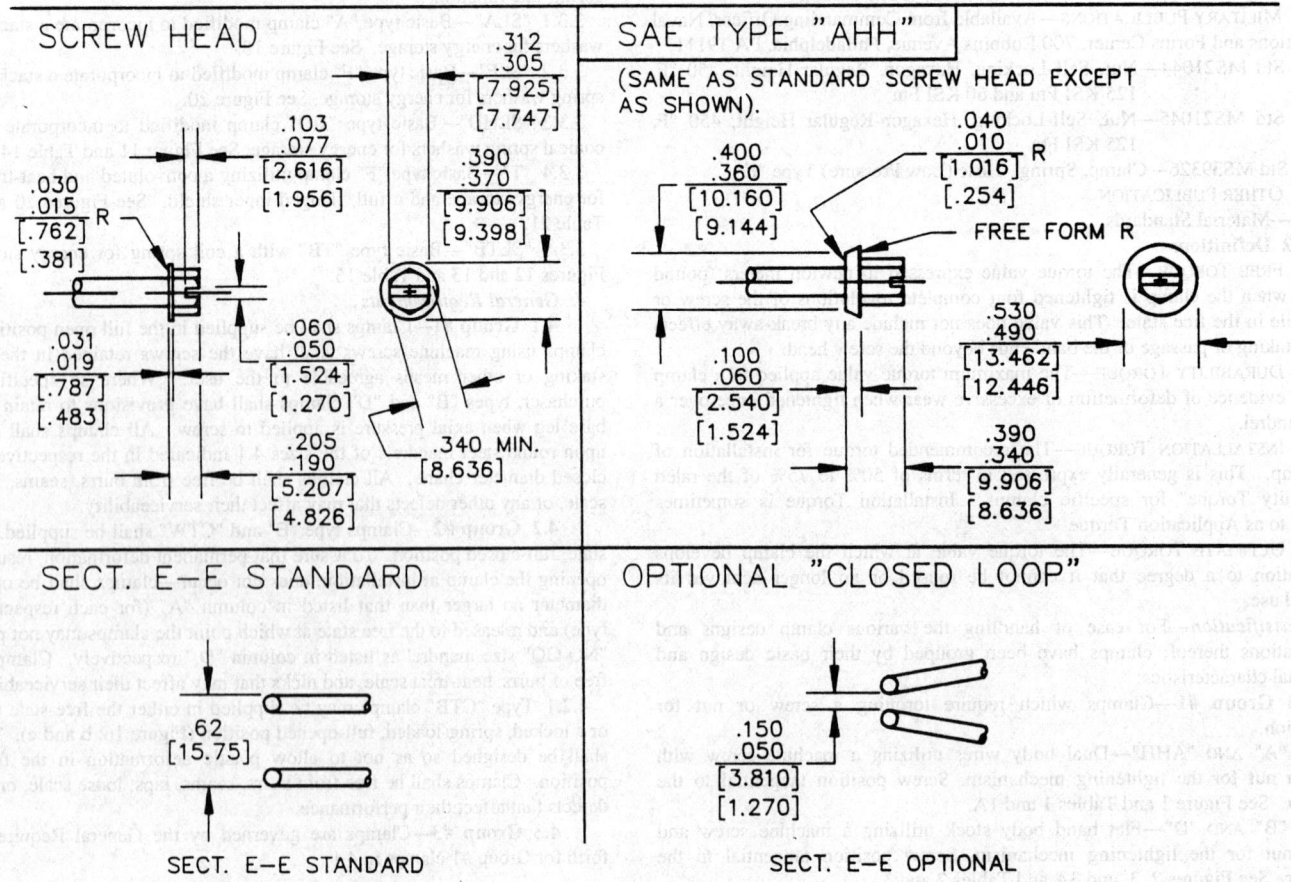

FIGURE 1—BASIC ENVELOPE DRAWING—INCH [METRIC—MM]

TABLE 1—TYPE "A," "AHH," AND "SLA"

SAE SIZE NO.	OPEN DIA. (MM)	CLOSED DIA. (MM)	ADJUST RANGE (MM)	SCREW LENGTH (MM)	SAE SIZE NO.	OPEN DIA. (MM)	CLOSED DIA. (MM)	ADJUST RANGE (MM)	SCREW LENGTH (MM)
16	12.70	11.18	1.52	21.59		52.07	45.97	6.10	37.59
18	14.22	12.19	2.03	21.59	66	52.32	47.75	4.57	37.59
20	15.75	13.97	1.78	21.59	68	53.85	49.28	4.57	37.59
22	17.53	14.73	2.79	21.59		53.85	48.51	5.33	37.59
24	19.05	16.26	2.79	31.24		54.61	49.28	5.33	37.59
26	20.57	17.53	3.05	31.24	70	55.63	50.04	5.59	37.59
28	22.35	19.05	3.30	31.24		55.88	50.29	5.59	37.59
	23.11	19.81	3.30	31.24		55.88	49.53	6.35	43.94
30	23.88	20.57	3.30	31.24	72	57.15	51.56	5.59	37.59
	24.64	21.34	3.30	31.24		56.39	49.96	6.35	43.94
	25.15	21.34	3.81	31.24		56.39	50.29	6.10	37.59
32	25.40	22.35	3.05	31.24		57.15	51.56	5.59	37.59
	26.16	23.11	3.05	31.24		57.15	50.80	6.35	43.94
34	26.92	23.88	3.05	31.24	74	58.67	53.85	4.83	37.59
	27.69	24.38	3.30	31.24		58.67	52.32	6.35	43.94
	28.20	24.38	3.81	31.24		60.20	54.61	5.59	43.94
36	28.45	24.13	4.32	31.24	76	60.45	55.63	4.83	37.59
	28.96	25.65	3.30	31.24		60.96	55.37	5.59	37.59
	29.21	25.91	3.30	31.24	78	61.98	57.15	4.83	37.59
38	30.23	26.92	3.30	31.24	80	63.50	57.91	5.59	37.59
	30.48	27.18	3.30	31.24	82	65.02	59.44	5.59	37.59
	30.99	27.69	3.30	31.24	84	66.55	61.21	5.33	37.59
40	31.75	27.69	4.06	31.24	86	68.33	62.74	5.59	37.59
	32.51	27.94	4.57	37.59	88	69.85	64.26	5.59	37.59
42	33.27	29.46	3.81	31.24	90	71.37	65.79	5.59	37.59
44	35.05	30.23	4.83	31.24		72.14	66.55	5.59	37.59
46	36.58	31.75	4.83	31.24	92	73.15	67.56	5.59	37.59
48	38.10	33.27	4.83	31.24	94	74.68	69.09	5.59	37.59
50	39.62	35.05	4.57	31.24	96	76.20	70.61	5.59	37.59
	39.62	34.29	5.33	37.59	98	77.72	72.14	5.59	37.59
52	41.15	36.58	4.57	31.24	100	79.25	73.91	5.33	37.59
	41.66	35.31	6.35	37.59	102	81.03	75.44	5.59	37.59
	42.42	36.25	6.10	43.94	104	82.55	76.96	5.59	37.59
	42.67	36.32	6.35	43.94		83.31	77.72	5.59	37.59
54	42.93	38.10	4.83	31.24	106	84.07	78.49	5.59	37.59
	42.93	37.08	5.84	37.59	108	85.85	80.26	5.59	37.59
	43.18	37.34	5.84	37.59	110	87.38	81.79	5.59	37.59
	43.43	37.01	6.35	43.94	112	88.90	82.55	6.35	43.94
	43.69	38.10	5.59	37.59	114	90.42	84.07	6.35	43.94
	44.20	38.10	6.10	43.94	116	91.95	85.85	6.10	43.94
56	44.45	39.62	4.83	37.59	118	93.73	87.38	6.35	43.94
	44.45	38.86	5.59	37.59	120	95.25	88.90	6.35	43.94
	44.70	38.35	6.35	43.94	122	96.77	90.42	6.35	43.94
	45.72	39.62	6.10	37.59	124	98.55	91.95	6.35	43.94
58	45.97	41.15	4.83	37.59	126	100.08	93.73	6.35	43.94
	46.74	41.91	4.83	37.59	128	101.60	95.25	6.35	43.94
	46.74	41.15	5.59	37.59	130	103.12	96.77	6.35	43.94
	46.99	41.15	5.84	37.59	132	104.65	98.55	6.10	43.94
60	47.75	42.93	4.83	37.59	134	106.43	100.08	6.35	43.94
	47.75	41.40	6.35	43.94	136	107.95	101.60	6.35	43.94
	48.01	41.61	6.35	43.94	138	109.47	103.12	6.35	43.94
	48.51	43.69	4.83	37.59	140	111.25	104.65	6.60	43.94
	48.51	42.93	5.59	37.59	142	112.78	106.43	6.35	43.94
62	49.28	44.45	4.83	37.59	144	114.30	107.95	6.35	43.94
	49.28	43.69	5.59	37.59	146	115.82	109.47	6.35	43.94
	49.28	42.93	6.35	43.94	148	117.35	111.25	6.10	43.94
	49.78	43.31	6.35	43.94	150	119.13	112.78	6.35	43.94
	49.78	43.94	5.84	37.59	152	120.65	114.30	6.35	43.94
	50.29	43.94	6.35	43.94	154	122.17	115.82	6.35	43.94
64	50.80	45.97	4.83	37.59	156	123.95	117.35	6.60	43.94
	50.80	45.21	5.59	37.59	158	125.48	119.13	6.35	43.94
	50.80	44.45	6.35	43.94	160	127.00	120.65	6.35	43.94
	51.56	45.47	6.10	43.94					

TABLE 1A—TYPE "A," "AHH," AND "SLA"

SAE SIZE NO.	OPEN DIA. (IN.)	CLOSED DIA. (IN.)	ADJUST RANGE (IN.)	SCREW LENGTH (IN.)	SAE SIZE NO.	OPEN DIA. (IN.)	CLOSED DIA. (IN.)	ADJUST RANGE (IN.)	SCREW LENGTH (IN.)
16	0.50	0.440	0.06	0.85		2.05	1.810	0.24	1.48
18	0.56	0.480	0.08	0.85	66	2.06	1.880	0.18	1.48
20	0.62	0.550	0.07	0.85	68	2.12	1.940	0.18	1.48
22	0.69	0.580	0.11	0.85		2.12	1.910	0.21	1.48
24	0.75	0.640	0.11	1.23		2.15	1.940	0.21	1.48
26	0.81	0.690	0.12	1.23	70	2.19	1.970	0.22	1.48
28	0.88	0.750	0.13	1.23		2.20	1.980	0.22	1.48
	0.91	0.780	0.13	1.23		2.20	1.950	0.25	1.73
30	0.94	0.810	0.13	1.23	72	2.22	1.967	0.25	1.73
	0.97	0.840	0.13	1.23		2.22	1.980	0.24	1.48
	0.99	0.840	0.15	1.23		2.25	2.030	0.22	1.48
32	1.00	0.880	0.12	1.23		2.25	2.030	0.22	1.48
	1.03	0.910	0.12	1.23		2.25	2.000	0.25	1.73
34	1.06	0.940	0.12	1.23	74	2.31	2.120	0.19	1.48
	1.09	0.960	0.13	1.23		2.31	2.060	0.25	1.73
	1.11	0.960	0.15	1.23		2.37	2.150	0.22	1.48
36	1.12	0.950	0.17	1.23	76	2.38	2.190	0.19	1.48
	1.14	1.010	0.13	1.23		2.40	2.180	0.22	1.48
	1.15	1.020	0.13	1.23	78	2.44	2.250	0.19	1.48
38	1.19	1.060	0.13	1.23	80	2.50	2.280	0.22	1.48
	1.20	1.070	0.13	1.23	82	2.56	2.340	0.22	1.48
	1.22	1.090	0.13	1.23	84	2.62	2.410	0.21	1.48
40	1.25	1.090	0.16	1.23	86	2.69	2.470	0.22	1.48
	1.28	1.100	0.18	1.48	88	2.75	2.530	0.22	1.48
42	1.31	1.160	0.15	1.23	90	2.81	2.590	0.22	1.48
44	1.38	1.190	0.19	1.23		2.84	2.620	0.22	1.48
46	1.44	1.250	0.19	1.23	92	2.88	2.660	0.22	1.48
48	1.50	1.310	0.19	1.23	94	2.94	2.720	0.22	1.48
50	1.56	1.380	0.18	1.23	96	3.00	2.780	0.22	1.48
	1.56	1.350	0.21	1.48	98	3.06	2.840	0.22	1.48
52	1.62	1.440	0.18	1.23	100	3.12	2.910	0.21	1.48
	1.64	1.390	0.25	1.48	102	3.19	2.970	0.22	1.48
	1.67	1.427	0.24	1.73	104	3.25	3.030	0.22	1.48
	1.68	1.430	0.25	1.73		3.28	3.060	0.22	1.48
54	1.69	1.500	0.19	1.23	106	3.31	3.090	0.22	1.48
	1.69	1.460	0.23	1.48	108	3.38	3.160	0.22	1.48
	1.70	1.470	0.23	1.48	110	3.44	3.220	0.22	1.48
	1.71	1.457	0.25	1.73	112	3.50	3.250	0.25	1.73
	1.72	1.500	0.22	1.48	114	3.56	3.310	0.25	1.73
	1.74	1.500	0.24	1.73	116	3.62	3.380	0.24	1.73
56	1.75	1.560	0.19	1.48	118	3.69	3.440	0.25	1.73
	1.75	1.530	0.22	1.48	120	3.75	3.500	0.25	1.73
	1.76	1.510	0.25	1.73	122	3.81	3.560	0.25	1.73
	1.80	1.560	0.24	1.48	124	3.88	3.620	0.26	1.73
58	1.81	1.620	0.19	1.48	126	3.94	3.690	0.25	1.73
	1.84	1.650	0.19	1.48	128	4.00	3.750	0.25	1.73
	1.84	1.620	0.22	1.48	130	4.06	3.810	0.25	1.73
	1.85	1.620	0.23	1.48	132	4.12	3.880	0.24	1.73
60	1.88	1.690	0.19	1.48	134	4.19	3.940	0.25	1.73
	1.88	1.630	0.25	1.73	136	4.25	4.000	0.25	1.73
	1.89	1.638	0.25	1.73	138	4.31	4.060	0.25	1.73
	1.91	1.720	0.19	1.48	140	4.38	4.120	0.26	1.73
	1.91	1.690	0.22	1.48	142	4.44	4.190	0.25	1.73
62	1.94	1.750	0.19	1.48	144	4.50	4.250	0.25	1.73
	1.94	1.720	0.22	1.48	146	4.56	4.310	0.25	1.73
	1.94	1.690	0.25	1.73	148	4.62	4.380	0.24	1.73
	1.96	1.705	0.25	1.73	150	4.69	4.440	0.25	1.73
	1.96	1.730	0.23	1.48	152	4.75	4.500	0.25	1.73
	1.98	1.730	0.25	1.73	154	4.81	4.560	0.25	1.73
64	2.00	1.810	0.19	1.48	156	4.88	4.620	0.26	1.73
	2.00	1.780	0.22	1.48	158	4.94	4.690	0.25	1.73
	2.00	1.750	0.25	1.73	160	5.00	4.750	0.25	1.73
	2.03	1.790	0.24	1.73					

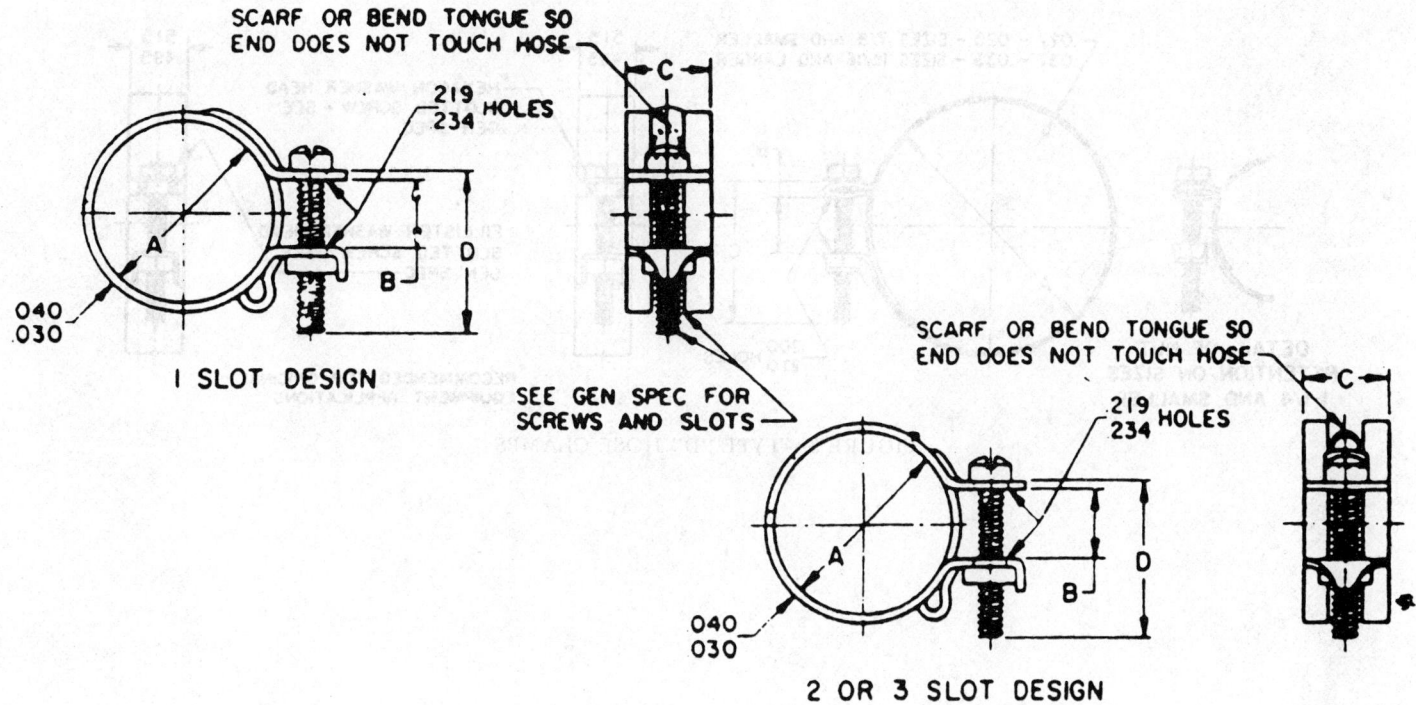

FIGURE 2—TYPE "B" HOSE CLAMPS

TABLE 2—DIMENSIONS OF TYPE "B" HOSE CLAMPS

SAE Size No.	A Dia			B[a] Gap	C Band Width ±0.01	D Screw Length Min	SAE Size No.	A Dia			B[a] Gap	C Band Width ±0.01	D Screw Length Min
	Nom	Open	Closed					Nom	Open	Closed			
18	0.50	0.58	0.44	0.38	0.50[b]	1.00	58	1.75	1.83	1.64	0.50	0.62[c]	1.12
20	0.56	0.64	0.48	0.38	0.50[b]	1.00	60	1.81	1.89	1.70	0.50	0.62[c]	1.12
22	0.62	0.70	0.55	0.38	0.50[b]	1.00	62	1.88	1.95	1.77	0.50	0.62[c]	1.12
24	0.69	0.77	0.61	0.38	0.50[b]	1.00	64	1.94	2.02	1.83	0.50	0.62[c]	1.12
26	0.75	0.83	0.67	0.38	0.50[b]	1.00	67	2.03	2.11	1.92	0.50	0.62[c]	1.12
28	0.81	0.89	0.73	0.38	0.50[b]	1.00							
30	0.88	0.95	0.80	0.38	0.50[b]	1.00	70	2.12	2.20	2.02	0.50	0.62[c]	1.12
32	0.94	1.02	0.86	0.38	0.50[b]	1.00	72	2.19	2.27	2.08	0.50	0.62[c]	1.12
35	1.03	1.11	0.95	0.38	0.50[b]	1.00	75	2.28	2.36	2.17	0.50	0.62[c]	1.12
36	1.06	1.14	0.98	0.38	0.50[b]	1.00	79	2.38	2.48	2.27	0.50	0.62[c]	1.25
38	1.12	1.20	1.02	0.38	0.50[b]	1.12	83	2.50	2.61	2.39	0.50	0.62[c]	1.25
40	1.19	1.27	1.08	0.50	0.50[b]	1.12	88	2.62	2.75	2.52	0.50	0.62[c]	1.25
42	1.25	1.33	1.14	0.50	0.62[c]	1.12	92	2.75	2.88	2.64	0.50	0.62[c]	1.25
44	1.31	1.39	1.20	0.50	0.62[c]	1.12	96	2.88	3.00	2.77	0.50	0.62[c]	1.25
46	1.38	1.45	1.27	0.50	0.62[c]	1.12	100	3.00	3.12	2.89	0.50	0.62	1.25
48	1.44	1.52	1.33	0.50	0.62[c]	1.12	104	3.12	3.25	3.02	0.50	0.62	1.25
50	1.50	1.58	1.39	0.50	0.62[c]	1.12	108	3.25	3.38	3.14	0.50	0.62	1.25
52	1.56	1.64	1.45	0.50	0.62[c]	1.12	112	3.38	3.50	3.27	0.50	0.62	1.25
54	1.62	1.70	1.52	0.50	0.62[c]	1.12	122	3.56	3.81	3.42	0.62	0.75	1.38
56	1.69	1.77	1.58	0.50	0.62[c]	1.12							

[a] Reference dimension. When gap is at value tabulated, clamp diameter shall approximate the nominal diameter.
[b] 0.62 in. width optional with user.
[c] 0.50 in. width optional with user.

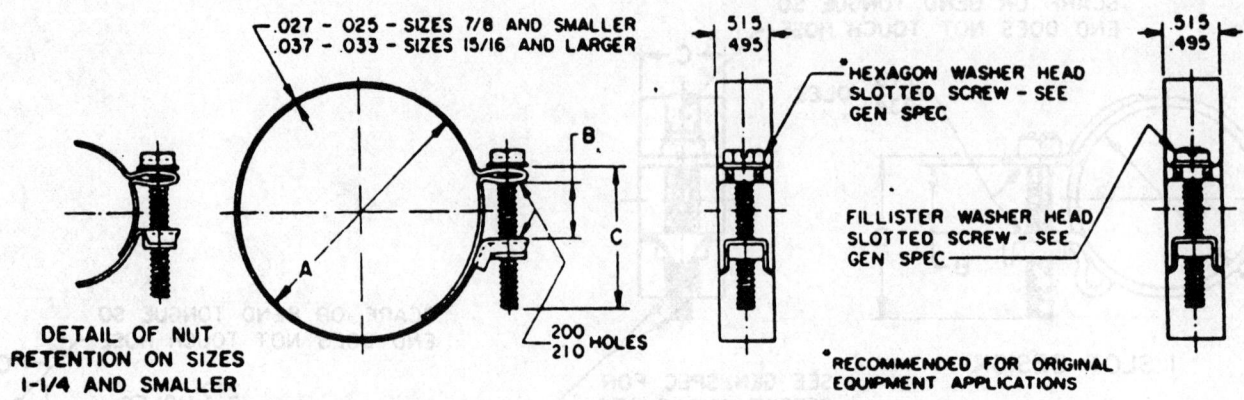

FIGURE 3—TYPE "D" HOSE CLAMPS

TABLE 3—DIMENSIONS OF TYPE "D" HOSE CLAMPS

SAE Size No.	A, Dia			B*, Gap	C, Screw Length Min	SAE Size No.	A, Dia			B*, Gap	C, Screw Length Min	SAE Size No.	A, Dia			B*, Gap	C, Screw Length Min
	Nom	Open	Closed				Nom	Open	Closed				Nom	Open	Closed		
23	0.62	0.72	0.53	0.38	1.12	83	2.50	2.59	2.34	0.62	1.38	143	4.38	4.47	4.16	0.75	1.50
25	0.69	0.78	0.59	0.38	1.12	85	2.56	2.66	2.41	0.62	1.38	145	4.44	4.53	4.22	0.75	1.50
27	0.75	0.84	0.66	0.38	1.12	87	2.62	2.72	2.47	0.62	1.38	147	4.50	4.59	4.28	0.75	1.50
29	0.81	0.91	0.72	0.38	1.12	89	2.69	2.78	2.53	0.62	1.38	149	4.56	4.66	4.34	0.75	1.50
31	0.88	0.97	0.78	0.38	1.12	91	2.75	2.84	2.59	0.62	1.38	151	4.62	4.72	4.41	0.75	1.50
33	0.94	1.03	0.84	0.38	1.12	93	2.81	2.91	2.66	0.62	1.38	153	4.69	4.78	4.47	0.75	1.50
35	1.00	1.09	0.91	0.38	1.12	95	2.88	2.97	2.72	0.62	1.38	155	4.75	4.84	4.53	0.75	1.50
37	1.06	1.16	0.97	0.38	1.12	97	2.94	3.03	2.78	0.62	1.38	157	4.81	4.91	4.59	0.75	1.50
39	1.12	1.22	1.03	0.38	1.12	99	3.00	3.09	2.84	0.62	1.38	159	4.88	4.97	4.66	0.75	1.50
41	1.19	1.28	1.06	0.50	1.25	101	3.06	3.16	2.91	0.62	1.38	161	4.94	5.03	4.72	0.75	1.50
43	1.25	1.34	1.12	0.50	1.25	103	3.12	3.22	2.97	0.62	1.38	163	5.00	5.09	4.78	0.75	1.50
45	1.31	1.41	1.19	0.50	1.25	105	3.19	3.28	3.03	0.62	1.38	165	5.06	5.16	4.84	0.75	1.50
47	1.38	1.47	1.25	0.50	1.25	107	3.25	3.34	3.09	0.62	1.38	167	5.12	5.22	4.91	0.75	1.50
49	1.44	1.53	1.31	0.50	1.25	109	3.31	3.41	3.16	0.62	1.38	169	5.19	5.28	4.97	0.75	1.50
51	1.50	1.59	1.38	0.50	1.25	111	3.38	3.47	3.22	0.62	1.38	171	5.25	5.34	5.03	0.75	1.50
53	1.56	1.66	1.44	0.50	1.25	113	3.44	3.53	3.28	0.62	1.38	173	5.31	5.41	5.09	0.75	1.50
55	1.62	1.72	1.50	0.50	1.25	115	3.50	3.59	3.34	0.62	1.38	175	5.38	5.47	5.16	0.75	1.50
57	1.69	1.78	1.56	0.50	1.25	117	3.56	3.66	3.34	0.75	1.50	177	5.44	5.53	5.22	0.75	1.50
59	1.75	1.84	1.62	0.50	1.25	119	3.62	3.72	3.41	0.75	1.50	179	5.50	5.59	5.28	0.75	1.50
61	1.81	1.91	1.69	0.50	1.25	121	3.69	3.78	3.47	0.75	1.50	181	5.56	5.66	5.34	0.75	1.50
63	1.88	1.97	1.75	0.50	1.25	123	3.75	3.84	3.53	0.75	1.50	183	5.62	5.72	5.41	0.75	1.50
65	1.94	2.03	1.81	0.50	1.25	125	3.81	3.91	3.59	0.75	1.50	185	5.69	5.78	5.47	0.75	1.50
67	2.00	2.09	1.88	0.50	1.25	127	3.88	3.97	3.66	0.75	1.50	187	5.75	5.84	5.53	0.75	1.50
69	2.06	2.16	1.94	0.50	1.25	129	3.94	4.03	3.72	0.75	1.50	189	5.81	5.91	5.59	0.75	1.50
71	2.12	2.22	2.00	0.50	1.25	131	4.00	4.09	3.78	0.75	1.50	191	5.88	5.97	5.66	0.75	1.50
73	2.19	2.28	2.06	0.50	1.25	133	4.06	4.16	3.84	0.75	1.50	193	5.94	6.03	5.72	0.75	1.50
75	2.25	2.34	2.12	0.50	1.25	135	4.12	4.22	3.91	0.75	1.50	195	6.00	6.09	5.78	0.75	1.50
77	2.31	2.41	2.19	0.50	1.25	137	4.19	4.28	3.97	0.75	1.50						
79	2.38	2.47	2.22	0.62	1.38	139	4.25	4.34	4.03	0.75	1.50						
81	2.44	2.53	2.28	0.62	1.38	141	4.31	4.41	4.09	0.75	1.50						

16.51

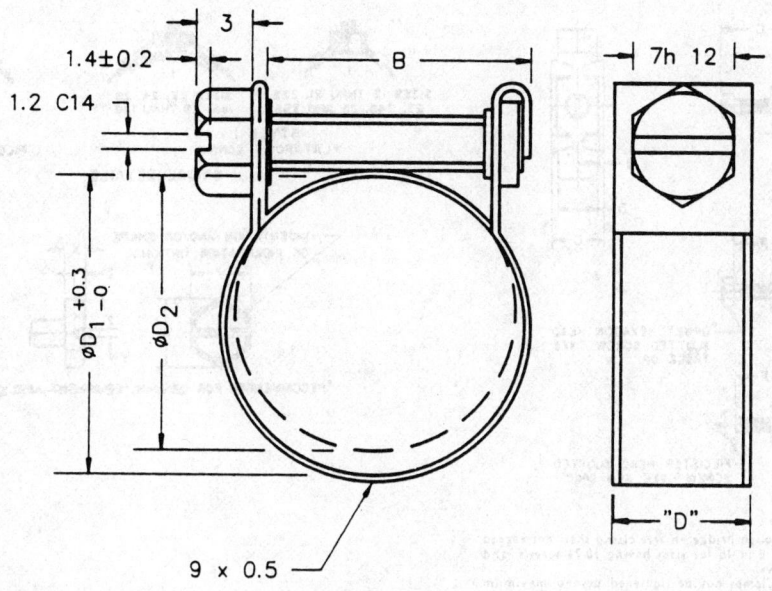

MANUFACTURERS DESIGNATION FOR SAE SIZE *	DIAMETER SUPPLIED IN INCHES MM	CLAMPING RANGE IN INCHES DECIMALS MM	B INCHES MM	D INCHES MM	RECOMMENDED TIGHTENING TORQUE (Nm)
8	.326 8.3	15/64–21/64 .234–.328 6.0–8.3	33/64 13.1	23/64 9.1	1.5
9	.366 9.3	9/32–3/8 .276–.375 7.0–9.5	33/64 13.1	23/64 9.1	1.5
10	.405 10.3	5/16–13/32 .315–.406 8.0–10.3	33/64 13.1	23/64 9.1	1.5
11	.444 11.3	23/64–29/64 .358–.453 9.0–11.5	33/64 13.1	23/64 9.1	1.5
12	.484 12.3	25/64–31/64 .394–.484 10.0–12.3	20/32 15.9	23/64 9.1	1.5
13	.523 13.3	7/16–17/32 .433–.531 11.0–13.5	20/32 15.9	23/64 9.1	1.5
14	.562 14.3	15/32–9/16 .479–.562 12.0–14.3	20/32 15.9	23/64 9.1	1.5
15	.602 15.3	33/64–39/64 .512–.609 13.0–15.5	20/32 15.9	23/64 9.1	1.5
16	.641 16.3	35/64–41/64 .551–.640 14.0–16.3	20/32 15.9	23/64 9.1	1.5
17	.681 17.3	19/32–11/16 .590–.685 15.0–17.5	20/32 15.9	23/64 9.1	1.5

t = BAND THICKNESS = [0.5 MM] 0.02"
* = IN THE ABSENCE OF AN APPROPRIATE SAE SIZE

FIGURE 3A—TYPE "D"

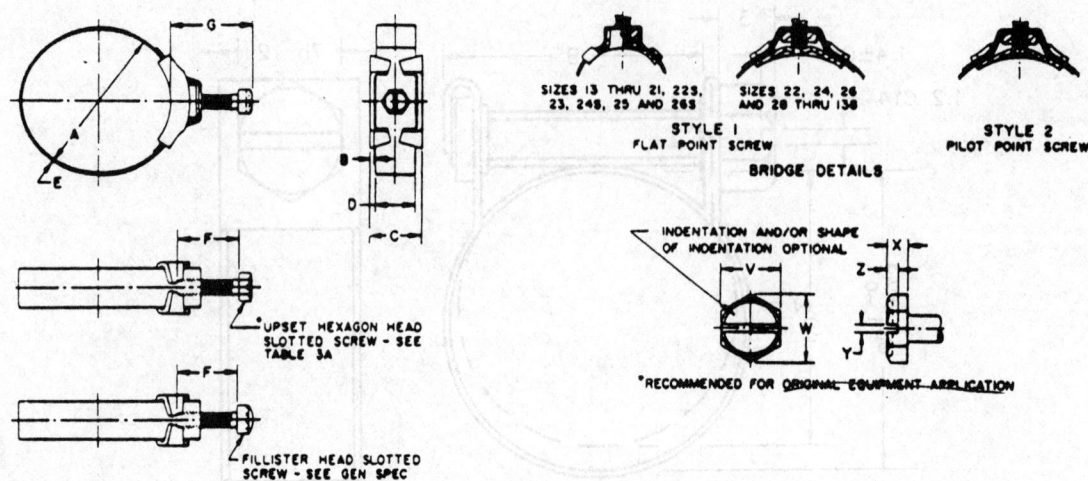

Torque required to draw band through bridge on free clamp shall not exceed 4 in.-lb for sizes having 6-32 screws, 8 in.-lb for sizes having 10-24 screws, and 10 in.-lb for sizes having 12-24 screws.

It is recommended that Type C Clamps not be tightened beyond maximum torques of 9 in.-lb for sizes having 6-32 screws, 22 in.-lb for sizes having 10-24 screws, and 30 in.-lb for sizes having 12-24 screws.

FIGURE 4—TYPE "C" HOSE CLAMPS

TABLE 4—DIMENSIONS OF TYPE "C" HOSE CLAMPS

SAE Size No.[a]	A Diameter Open	A Diameter Closed	B Bridge Stock Thickness ±0.002	C[b] Bridge Width Max	D Band Width ±0.010	E[c] Band Thickness ±0.001	F Screw Size and Length	G[b] Height Over Screw Max	SAE Size No.[a]	A Diameter Open	A Diameter Closed	B Bridge Stock Thickness ±0.002	C[b] Bridge Width Max	D Band Width ±0.010	E[c] Band Thickness ±0.001	F Screw Size and Length	G[b] Height Over Screw Max
13	0.40	0.34	0.035	0.41	0.281	0.010	6-32 x 0.50	0.64	46N	1.44	1.12	0.050	0.64	0.438	0.017	10-24 x 0.88	1.13
14	0.43	0.37	0.035	0.41	0.281	0.010	6-32 x 0.50	0.64	48	1.50	1.19	0.062	0.72	0.505	0.020	12-24 x 0.88	1.13
15	0.46	0.40	0.035	0.41	0.281	0.010	6-32 x 0.50	0.64	48N	1.50	1.19	0.050	0.64	0.438	0.017	10-24 x 0.88	1.13
16	0.50	0.37	0.035	0.41	0.281	0.010	6-32 x 0.50	0.64	50	1.56	1.25	0.062	0.72	0.505	0.020	12-24 x 0.88	1.13
17	0.53	0.40	0.035	0.41	0.281	0.010	6-32 x 0.50	0.64	52	1.62	1.31	0.062	0.72	0.505	0.020	12-24 x 0.88	1.13
18	0.56	0.43	0.035	0.41	0.281	0.010	6-32 x 0.50	0.64	54	1.69	1.38	0.062	0.72	0.505	0.020	12-24 x 0.88	1.13
19	0.59	0.46	0.035	0.41	0.281	0.010	6-32 x 0.50	0.64	56	1.75	1.44	0.062	0.72	0.505	0.020	12-24 x 0.88	1.13
20	0.62	0.50	0.035	0.41	0.281	0.010	6-32 x 0.50	0.64	58	1.81	1.50	0.062	0.72	0.505	0.020	12-24 x 0.88	1.13
21	0.65	0.53	0.035	0.41	0.281	0.010	6-32 x 0.50	0.64	60	1.88	1.56	0.062	0.72	0.505	0.020	12-24 x 0.88	1.13
22	0.69	0.38	0.050	0.64	0.438	0.017	10-24 x 0.88	1.13	62	1.94	1.62	0.062	0.72	0.505	0.020	12-24 x 0.88	1.13
22N	0.69	0.56	0.035	0.41	0.281	0.010	6-32 x 0.50	0.64	64	2.00	1.69	0.062	0.72	0.505	0.020	12-24 x 0.88	1.13
23	0.71	0.59	0.035	0.41	0.281	0.010	6-32 x 0.50	0.64	66	2.06	1.69	0.062	0.72	0.505	0.020	12-24 x 1.00	1.25
24	0.75	0.44	0.050	0.64	0.438	0.017	10-24 x 0.88	1.13	68	2.12	1.75	0.062	0.72	0.505	0.020	12-24 x 1.00	1.25
24N	0.75	0.62	0.035	0.41	0.281	0.010	6-32 x 0.50	0.64	70	2.19	1.81	0.062	0.72	0.505	0.020	12-24 x 1.00	1.25
25	0.78	0.66	0.035	0.41	0.281	0.010	6-32 x 0.50	0.64	72	2.25	1.88	0.062	0.72	0.505	0.020	12-24 x 1.00	1.25
26	0.81	0.50	0.050	0.64	0.438	0.017	10-24 x 0.88	1.13	74	2.31	1.94	0.062	0.72	0.505	0.020	12-24 x 1.00	1.25
26N	0.81	0.69	0.035	0.41	0.281	0.010	6-32 x 0.50	0.64	76	2.38	2.00	0.062	0.72	0.505	0.020	12-24 x 1.00	1.25
28	0.88	0.56	0.050	0.64	0.438	0.017	10-24 x 0.88	1.13	78	2.44	2.06	0.062	0.72	0.505	0.020	12-24 x 1.00	1.25
30	0.94	0.62	0.050	0.72	0.505	0.017	12-24 x 0.88	1.13	80	2.50	2.12	0.062	0.72	0.505	0.020	12-24 x 1.00	1.25
30N	0.94	0.62	0.050	0.64	0.438	0.017	10-24 x 0.88	1.13	82	2.56	2.19	0.062	0.72	0.505	0.020	12-24 x 1.00	1.25
32	1.00	0.69	0.050	0.72	0.505	0.017	12-24 x 0.88	1.13	84	2.62	2.25	0.062	0.72	0.505	0.020	12-24 x 1.00	1.25
32N	1.00	0.69	0.050	0.64	0.438	0.017	10-24 x 0.88	1.13	86	2.69	2.31	0.062	0.72	0.505	0.020	12-24 x 1.00	1.25
34	1.06	0.75	0.050	0.72	0.505	0.020	12-24 x 0.88	1.13	88	2.75	2.38	0.062	0.72	0.505	0.020	12-24 x 1.00	1.25
34N	1.06	0.75	0.050	0.64	0.438	0.017	10-24 x 0.88	1.13	90	2.81	2.44	0.062	0.72	0.505	0.020	12-24 x 1.00	1.25
36	1.12	0.81	0.050	0.72	0.505	0.020	12-24 x 0.88	1.13	92	2.88	2.50	0.062	0.72	0.505	0.020	12-24 x 1.00	1.25
36N	1.12	0.81	0.050	0.64	0.438	0.017	10-24 x 0.88	1.13	94	2.94	2.56	0.062	0.72	0.505	0.020	12-24 x 1.00	1.25
38	1.19	0.88	0.062	0.72	0.505	0.020	12-24 x 0.88	1.13	96	3.00	2.62	0.062	0.72	0.505	0.020	12-24 x 1.00	1.25
38N	1.19	0.88	0.050	0.64	0.438	0.017	10-24 x 0.88	1.13	100	3.12	2.75	0.062	0.72	0.505	0.020	12-24 x 1.00	1.25
40	1.25	0.94	0.062	0.72	0.505	0.020	12-24 x 0.88	1.13	104	3.25	2.88	0.062	0.72	0.505	0.020	12-24 x 1.00	1.25
40N	1.25	0.94	0.050	0.64	0.438	0.017	10-24 x 0.88	1.13	110	3.44	3.06	0.062	0.72	0.505	0.020	12-24 x 1.00	1.25
42	1.31	1.00	0.062	0.72	0.505	0.020	12-24 x 0.88	1.13	114	3.56	3.19	0.062	0.72	0.505	0.020	12-24 x 1.00	1.25
42N	1.31	1.00	0.050	0.64	0.438	0.017	10-24 x 0.88	1.13	118	3.69	3.31	0.062	0.72	0.505	0.020	12-24 x 1.00	1.25
44	1.38	1.06	0.062	0.72	0.505	0.020	12-24 x 0.88	1.13	138	4.31	3.94	0.062	0.72	0.505	0.020	12-24 x 1.00	1.25
44N	1.38	1.06	0.050	0.64	0.438	0.017	10-24 x 0.88	1.13									
46	1.44	1.12	0.062	0.72	0.505	0.020	12-24 x 0.88	1.13									

[a] The N suffix applied to SAE size numbers designates the smaller series clamp design where sizes overlap in two clamp designs.
[b] Reference dimension for clearance purposes only.
[c] For size numbers 30-138, clamps having 0.020 tabulated band thickness are also available with 0.018-0.016 and 0.027-0.025 band thickness where so specified by user.

16.53

TABLE 4A—DIMENSIONS OF HEXAGON SCREW HEADS

Screw Size	V Across Flats		W Across Corners	X Head Height		Y Slot Width		Z Slot Depth	
	Max	Min	Min	Max	Min	Max	Min	Max	Min
6	0.250	0.244	0.272	0.080	0.067	0.048	0.039	0.046	0.033
10	0.375	0.367	0.409	0.145	0.120	0.060	0.050	0.072	0.057
12	0.375	0.367	0.409	0.155	0.139	0.067	0.056	0.077	0.093

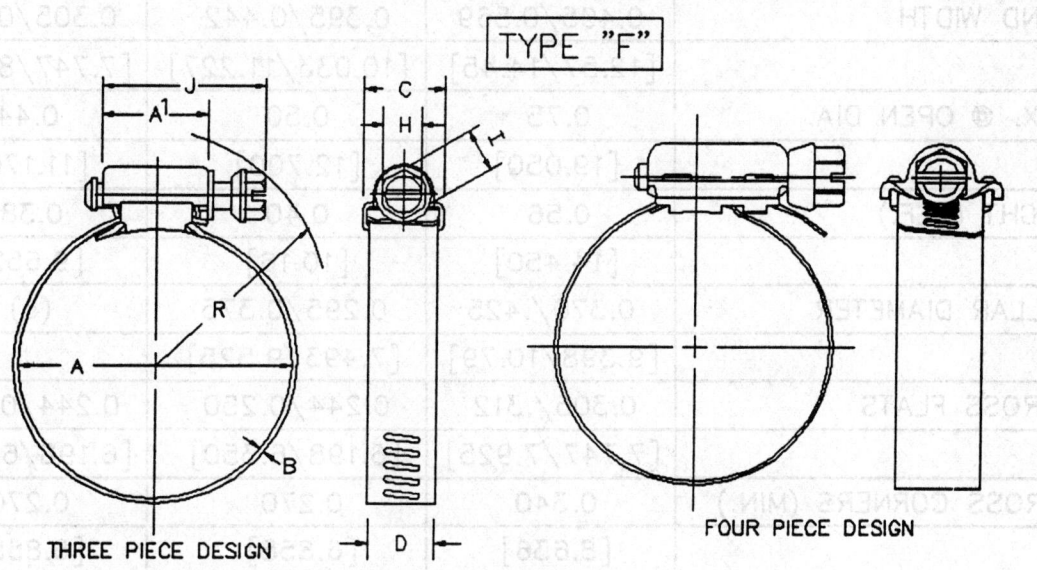

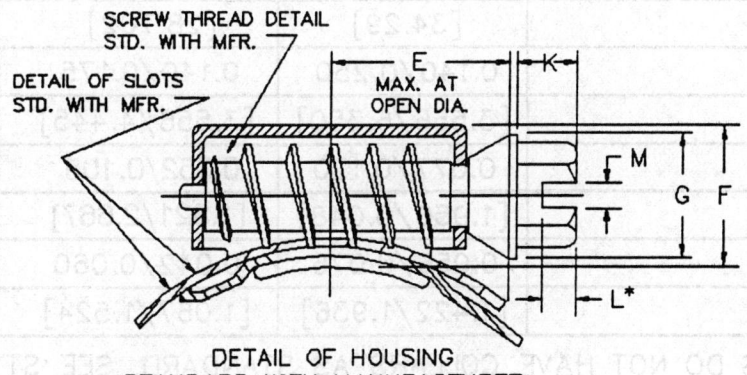

DETAIL OF HOUSING
STANDARD WITH MANUFACTURER
FOR DIMENSIONS RELATIVE TO THIS STYLE REFER TO TABLE 5
TYPE F, I, & M
FOR H.D. STYLE HOUSING DIMENSIONS REFER TO TABLE 14
NOTE: SCREW HOUSING DESIGN MAY VARY BY MANUFACTURER

(R) FIGURE 5—TYPE "F," "I," AND "M"

(R) TABLE 5—DIMENSIONS OF TYPE "F," "I," AND "M" CLAMP

DIMENSION	TYPE F	TYPE I	TYPE M
A[1] HSG LENGTH (REF.)	0.76	0.64	0.42
	[19.30]	[16.26]	[10.668]
B THICKNESS	0.021/0.031	0.019/0.030	0.019/0.026
	[0.533/0.787]	[0.483/0.762]	[0.483/0.660]
C HSG WIDTH (REF.)	0.81	0.53	0.60
	[20.570]	[13.462]	[15.240]
D BAND WIDTH	0.495/0.569	0.395/0.442	0.305/0.325
	[12.57/14.45]	[10.033/11.227]	[7.747/8.255]
E MAX. @ OPEN DIA.	0.75	0.50	0.44
	[19.050]	[12.700]	[11.176]
F HEIGHT (REF.)	0.56	0.40	0.38
	[14.450]	[10.16]	[9.652]
G COLLAR DIAMETER	0.370/.425	0.295/0.375	(a)
	[9.398/10.79]	[7.493/9.525]	
H ACROSS FLATS	0.305/.312	0.244/0.250	0.244/0.250
	[7.747/7.925]	[6.198/6.350]	[6.198/6.350]
I ACROSS CORNERS (MIN.)	0.340	0.270	0.270
	[8.636]	[6.858]	[6.858]
J LG. OF SCREW (MAX.)	1.35	1.13	0.80
	[34.29]	[28.702]	[20.32]
K HEX HEIGHT	0.140/0.250	0.140/0.175	0.150/0.185
	[3.556/6.350]	[3.556/4.445]	[3.810/4.699]
L SLOT DEPTH	0.077/0.120	0.052/0.105	0.052/0.105
	[1.956/3.048]	[1.321/2.667]	[1.321/2.667]
M SLOT WIDTH	0.056/0.076	0.042/0.060	0.042/0.060
	[1.422/1.936]	[1.067/1.524]	[1.067/1.524]

[a] TYPE M CLAMPS DO NOT HAVE COLLARS AS STANDARD. SEE STYLE 6.

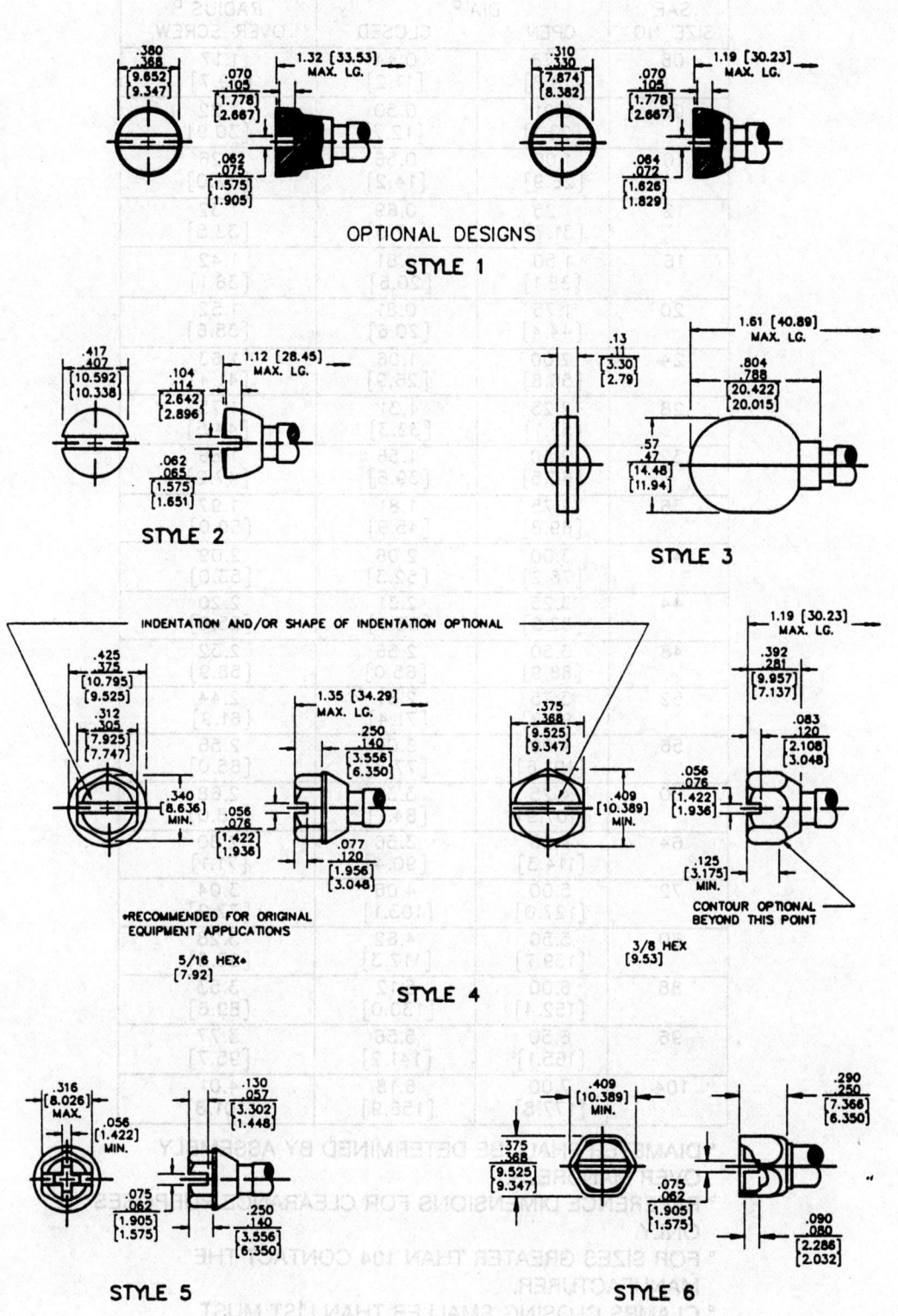

FIGURE 6—TYPE "F," "I," AND "M"

(R) TABLE 6—DIMENSIONS OF TYPE "F" HOSE CLAMPS

SAE SIZE NO.	A DIA [a] OPEN	CLOSED	R RADIUS [b] OVER SCREW
06	0.78 [19.8]	0.44 [11.2]	1.17 [29.7]
08	0.91 [23.1]	0.50 [12.7]	1.22 [30.9]
10	1.06 [26.9]	0.56 [14.2]	1.26 [32.0]
12	1.25 [31.7]	0.69 [17.5]	1.32 [33.5]
16	1.50 [38.1]	0.81 [20.6]	1.42 [36.1]
20	1.75 [44.4]	0.81 [20.6]	1.52 [38.6]
24	2.00 [50.8]	1.06 [26.9]	1.63 [41.4]
28	2.25 [57.1]	1.31 [33.3]	1.75 [44.5]
32	2.50 [63.5]	1.56 [39.6]	1.86 [47.2]
36	2.75 [69.8]	1.81 [45.9]	1.97 [50.0]
40	3.00 [76.2]	2.06 [52.3]	2.09 [53.0]
44	3.25 [82.5]	2.31 [58.6]	2.20 [55.8]
48	3.50 [88.9]	2.56 [65.0]	2.32 [58.9]
52	3.75 [95.2]	2.81 [71.4]	2.44 [61.9]
56	4.00 [101.6]	3.06 [77.7]	2.56 [65.0]
60	4.25 [107.9]	3.31 [84.1]	2.68 [68.0]
64	4.50 [114.3]	3.56 [90.4]	2.80 [71.1]
72	5.00 [127.0]	4.06 [103.1]	3.04 [77.2]
80	5.50 [139.7]	4.62 [117.3]	3.28 [83.3]
88	6.00 [152.4]	5.12 [130.0]	3.53 [89.6]
96	6.50 [165.1]	5.56 [141.2]	3.77 [95.7]
104	7.00 [177.8]	6.18 [156.9]	4.01 [101.8]

[a] DIAMETER SHALL BE DETERMINED BY ASSEMBLY OVER MANDRELS.
[b] REFERENCE DIMENSIONS FOR CLEARANCE PURPOSES ONLY.
[c] FOR SIZES GREATER THAN 104 CONTACT THE MANUFACTURER.
[d] CLAMPS CLOSING SMALLER THAN LIST MUST COMPLY WITH 8.1.

16.57

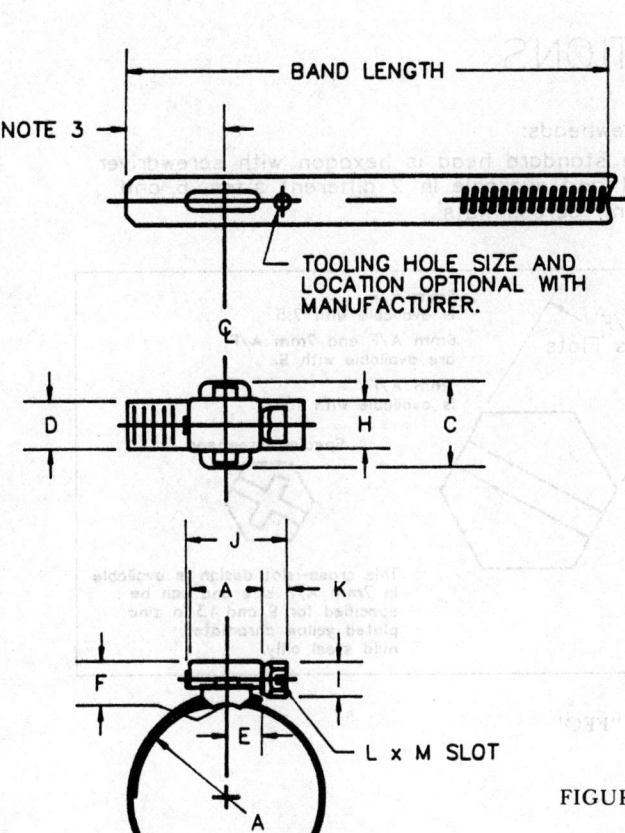

1. THREE SLOTS MAXIMUM, UNCOVERED BY LINER AT MAX. DIAMETER.
2. CLAMP SHAPE NEED NOT BE PERFECTLY ROUND AS LONG AS CLAMP WILL FREELY ACCEPT THE MAX. OPEN DIA. GAUGE.
3. BAND EXTENSION LENGTH OPTIONAL WITH MANUFACTURER FOR CONFORMANCE WITH NOTE 1.

FIGURE 7—TYPE "MX"

TABLE 7—DIMENSIONS OF TYPE "I" HOSE CLAMPS

SAE SIZE NO.	A DIA [a] OPEN	A DIA CLOSED	R RADIUS [b] OVER SCREW
06	0.78 [19.8]	0.44 [11.2]	1.00 [25.4]
08	0.91 [23.1]	0.50 [12.7]	1.03 [26.1]
10	1.06 [26.9]	0.56 [14.2]	1.09 [27.6]
12	1.25 [31.7]	0.69 [17.5]	1.12 [28.4]
16	1.50 [38.1]	0.81 [20.6]	1.25 [31.7]
20	1.75 [44.4]	0.81 [20.6]	1.38 [35.0]
24	2.00 [50.8]	1.06 [26.9]	1.50 [38.1]
28	2.25 [57.1]	1.31 [33.3]	1.62 [41.1]
32	2.50 [63.5]	1.56 [39.6]	1.75 [45.0]
36	2.75 [69.8]	1.81 [45.9]	1.87 [47.5]

[a] DIAMETER SHALL BE DETERMINED BY ASSEMBLY OVER MANDRELS.
[b] REFERENCE DIMENSIONS FOR CLEARANCE PURPOSES ONLY.
[c] LARGER SIZE CLAMPS AVAILABLE THROUGH MANUFACTURERS.

SPECIFICATIONS

Materials:
Both the 9mm and 13mm series are available in 5 different material types.

Material No.	
1	Zinc plated mild steel throughout. Can be yellow chromated for added corrosion protection.
2	Band and housing in stainless steel (430 SS) and zinc plated yellow chromated mild steel screw.
3	Stainless steel throughout. (430 SS)
4	Non magnetic stainless steel throughout. (304 SS)
5	High grade non magnetic stainless steel throughout. (316 SS)

Screwheads:
The standard head is hexagon with screwdriver slot and available in 2 different sizes, 6 and 7mm 'across flats'.

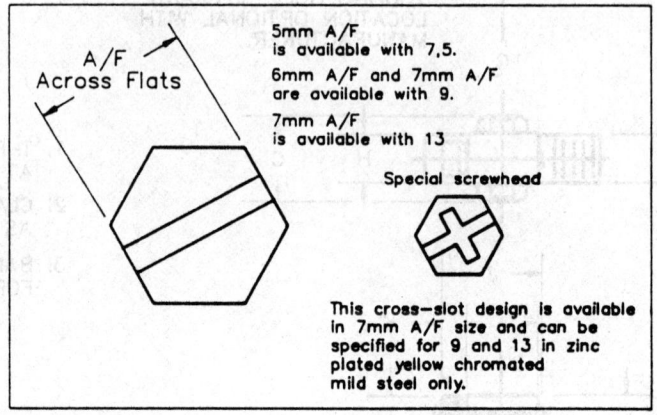

FIGURE 8—TYPE "FEO"

TABLE 8—DIMENSIONS OF TYPE "M" HOSE CLAMPS

SAE SIZE NO.	A DIA [a] OPEN	CLOSED	R RADIUS [b] OVER SCREW
04	0.62 [15.7]	0.25 [6.3]	0.77 [19.5]
06	0.78 [19.8]	0.44 [11.2]	0.91 [23.1]
08	0.91 [23.1]	0.50 [12.7]	0.96 [24.3]
10	1.06 [26.9]	0.56 [14.2]	1.03 [26.1]
12	1.25 [31.7]	0.69 [17.5]	1.09 [27.7]

[a] DIAMETER SHALL BE DETERMINED BY ASSEMBLY OVER MANDRELS.
[b] REFERENCE DIMENSIONS FOR CLEARANCE PURPOSES ONLY.
[c] LARGER SIZE CLAMPS AVAILABLE THROUGH MANUFACTURERS.

KEY:
- b – Bandwidth
- B – Housing width
- h – Housing height
- L – Housing + screw length
- s – Band thickness
- A/F – Screw head size

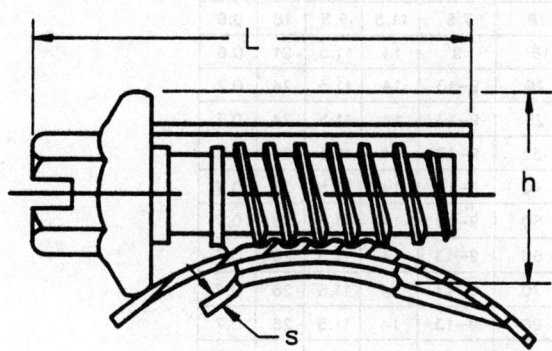

FIGURE 9—TYPE "FEO"

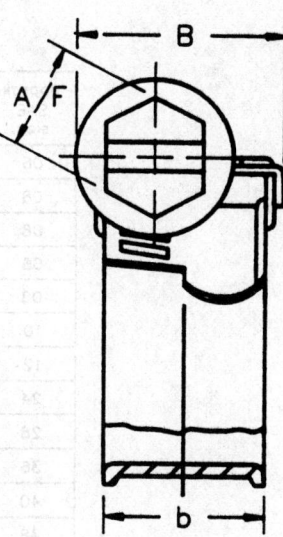

TABLE 9—TYPE "MX"

SAE CLAMP SIZE	OLD SAE REF.	CLAMP DIAMETER INCH		CLAMP DIAMETER METRIC	
		OPEN	CLOSE	OPEN	CLOSE
MX50		0.50	0.25	12.70	6.35
MX53		0.53	0.28	13.46	7.11
MX56		0.56	0.31	14.22	7.87
MX59		0.59	0.34	14.99	8.64
MX63	4	0.63	0.38	16.00	9.65
MX66		0.66	0.41	16.76	10.41
MX69		0.69	0.44	17.53	11.18
MX72		0.72	0.47	18.29	11.94
MX75		0.75	0.50	19.05	12.70
MX78	6	0.78	0.48	19.81	12.19
MX81		0.81	0.51	20.57	12.95
MX84		0.84	0.54	21.34	13.72
MX88		0.88	0.58	22.35	13.73
MX91	8	0.91	0.61	23.11	14.73
MX94		0.94	0.64	23.88	16.26
MX97		0.97	0.67	24.64	17.02
MX100		1.00	0.70	25.40	17.78
MX103		1.03	0.73	26.16	18.54
MX106	10	1.06	0.76	26.92	19.30
MX109		1.09	0.79	27.69	20.07
MX113		1.13	0.83	28.70	21.08
MX116		1.16	0.86	29.46	21.84
MX119		1.19	0.89	30.23	22.61
MX122		1.22	0.92	30.99	23.37
MX125	12	1.25	0.95	31.75	24.13

TABLE 10—TYPE "FEO"

Approx. SAE size	Clamping range in inches	Clamping range in mm	b	B	h	L	s
06	5/16"–1/2"	8 – 12	7.5	11.5	9.5	18	0.6
08	3/8"–5/8"	10 – 16	7.5	11.5	9.5	18	0.6
08	1/2"–3/4"	12 – 18	7.5	11.5	9.5	18	0.6
06	5/16"–5/8"	8 – 16	9	14	11.5	21	0.6
08	1/2"–3/4"	12 – 20	9–13	14	11.5	24	0.7
10	5/8"–1"	16 – 25	9–13	14	11.5	24	0.7
12	3/4"–1 1/4"	20 – 32	9–13	14	11.5	24	0.7
24	1"–1 5/8"	25 – 40	9–13	14	11.5	26	0.7
28	1 1/4"–2"	32 – 50	9–13	14	11.5	26	0.7
36	1 5/8"–2 3/8"	40 – 60	9–13	14	11.5	26	0.7
40	2"–2 3/4"	50 – 70	9–13	14	11.5	26	0.7
48	2 3/8"–3 1/8"	60 – 80	9–13	14	11.5	26	0.7
52	2 3/4"–3 1/2"	70 – 90	9–13	14	11.5	26	0.7
60	3 1/8"–4"	80 – 100	9–13	14	11.5	26	0.7
64	3 1/2"–4 3/8"	90 – 110	9–13	14	11.5	26	0.7
72	4"–4 3/4"	100 – 120	9–13	14	11.5	26	0.7
80	4 3/8"–5 1/8"	110 – 130	9–13	14	11.5	26	0.7
80	4 3/4"–5 1/2"	120 – 140	9–13	14	11.5	26	0.7
88	5 1/8"–5 7/8"	130 – 150	9–13	14	11.5	26	0.7
96	5 1/2"–6 1/4"	140 – 160∗	9–13	14	11.5	26	0.7

∗ Larger sizes available

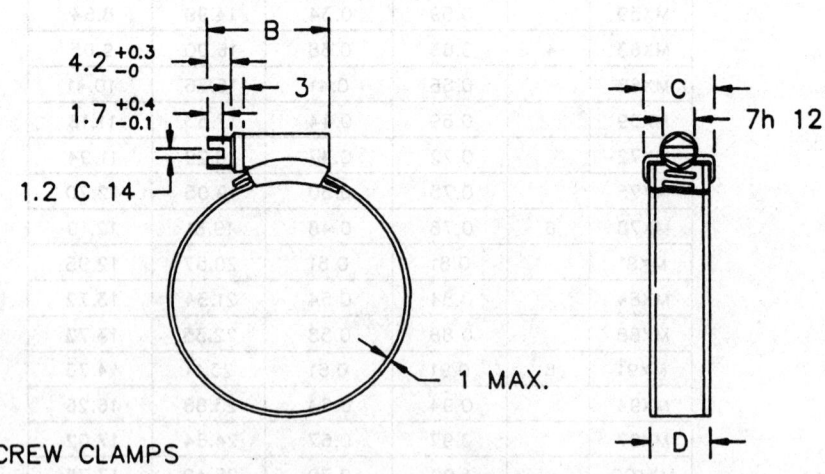

WORM DRIVE SCREW CLAMPS

FIGURE 10—TYPE "FE"

TABLE 11—TORQUE REQUIREMENTS FOR TYPE "FEO" CLAMPS

Ultimate Torque Ref. SAE	Ultimate Torque Clamp Range	Torque by Material No. #1	Torque by Material No. #2 through #5
Clamps with 9mm wide bands:			
6	8-16	22.2 2.5 Nm	26.6 3.0 Nm
8	12-20	35.4 4.0 Nm	39.8 4.5 Nm
64	90-110	35.4 4.0 Nm	39.8 4.5 Nm

Free torque for above 9mm clamps = 6.2 (0.7 Nm) max.

Clamps with 13mm wide bands:			
10	16-25	53.1 6.0 Nm	70.8 8.0 Nm
104	160-180	53.1 6.0 Nm	70.8 8.0 Nm
Except for: 36	40-60	62.0 7.0 Nm	N/A

Free torque for above 13mm clamps = 8.9 (1.0 Nm) max.

TABLE 12—TYPE "FE"

APPROX. SAE SIZE	CLAMPING RANGE IN		DIAMETER SUPPLIED IN		B INCHES MM	C INCHES MM	D INCHES MM	MINIMUM BREAKING TORQUE (Nm)
	INCHES	MM	INCHES	MM				
3	5/16–9/16	8–14	9/16	15	49/64 19.5	33/64 13	23/64 9	4.5
4	7/16–11/16	11–17	11/16	18				
6	1/2–13/16	13–20	13/16	21	27/32 21.5			
8	5/8–15/16	15–24	1	25				6.0
10	3/4–1 1/8	19–28	1 1/8	29	59/64 23.5			
12	7/8–1 1/4	22–32	1 5/16	33	1 25.5			
16	1 1/16–1 1/2	26–38	1 9/16	39				
20	1 1/4–1 3/4	32–44	1 3/4	45	1 5/32 29.5			7.0
24	1 1/2–2	38–50	2	51				
28	1 3/4–2 1/4	44–56	2 1/4	57				
32	2–2 9/16	50–65	2 5/8	66				
40	2 5/16–3	58–75	3	76		5/8 16	31/64 12.2	
44	2 11/16–3 3/8	68–85	3 3/8	86				
52	3–3 3/4	77–95	3 13/16	96	1 9/32 32.5			
64	3 7/16–4 7/16	87–112	4 7/16	113				
80	4 1/8–5 7/16	104–138	5 1/2	139				8.0
96	5 1/8–6 1/2	130–165	6 9/16	166				
104	5 7/8–7 1/8	150–180	7 1/8	181				
122	6 7/8–8 1/8	175–205	8 1/8	206				
138	7 7/8–9 1/8	200–231	9 1/8	232				
154	8 7/8–10 1/16	226–256	10 1/8	257				
170	9 7/8–11 1/8	251–282	11 1/8	283				
186	10 7/8–12 1/8	277–307	12 1/8	308				

t = BAND THICKNESS 0.04" (1 MM) MAX.

THE FREE TORQUE FOR A/M CLAMPS: (1.0 Nm) 8.9 MAX.

THE MINIMUM TORQUE ABOVE MUST BE TESTED ON A STEEL MANDREL, WITH THE MIN. DIAMETER SPECIFIED IN THE CLAMPING RANGE, I.E. 8, 11, 13 ETC, AS PER ABOVE.

TABLE 13—TORQUE REQUIREMENTS FOR TYPE "FE" CLAMPS

Clamp Range (mm)	Minimum Ultimate
8-14 to 13-20	39.8 in.lbs. (4.5 Nm)
15-24 to 26-38	53.1 in.lbs. (6.0 Nm)
32-44 to 50-65	62.0 in.lbs. (7.0 Nm)
58-75 to 277-307	70.8 in.lbs. (8.0 Nm)

16.63

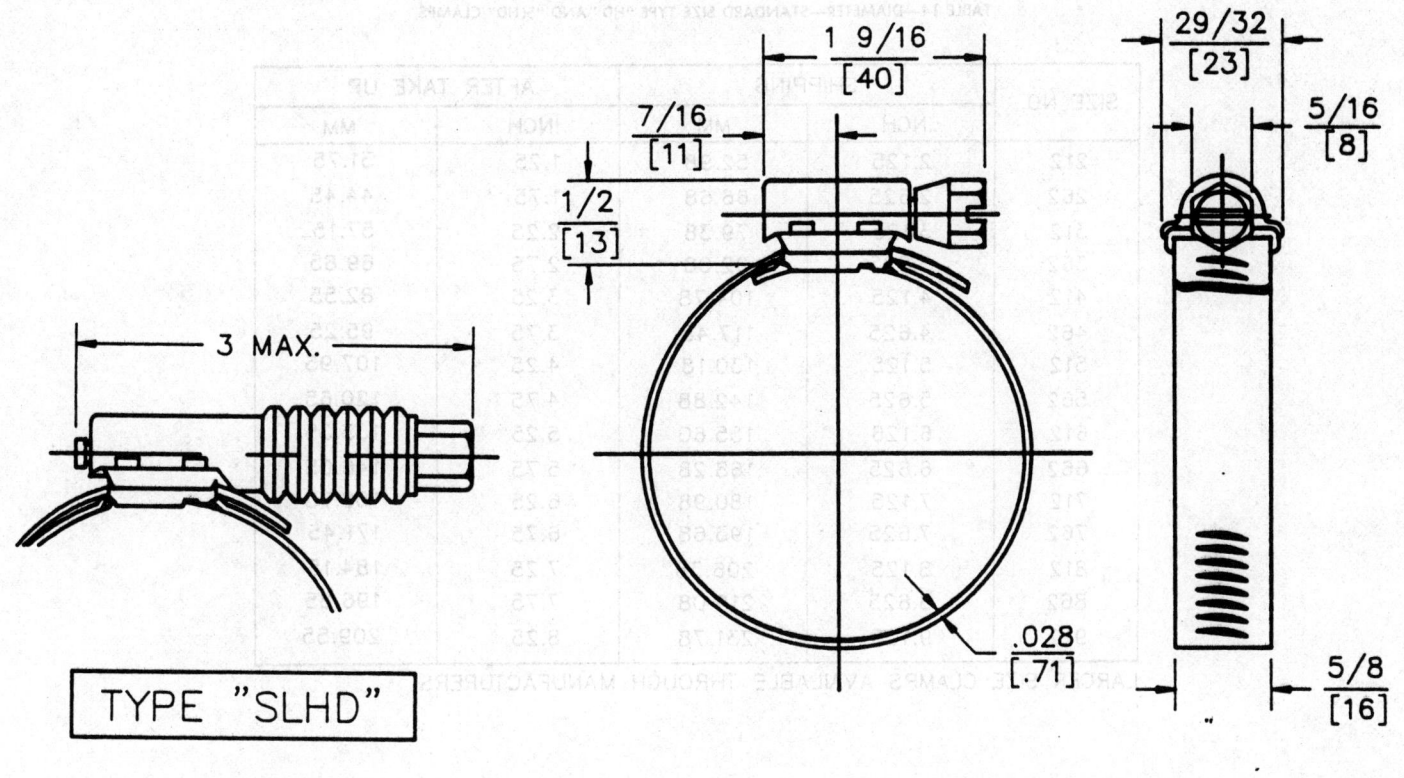

FIGURE 11—TYPE "HD" AND "SLHD"

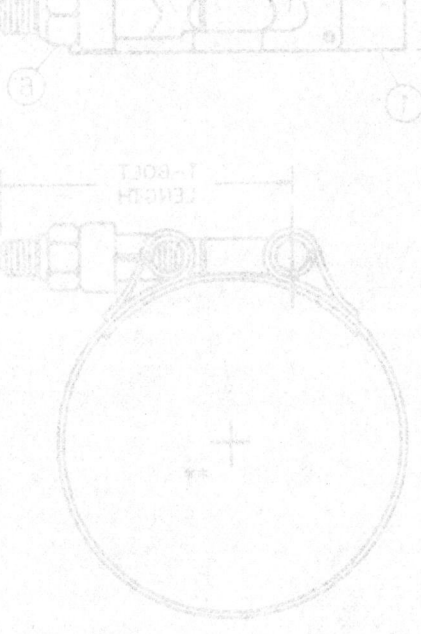

TABLE 14—DIAMETER—STANDARD SIZE TYPE "HD" AND "SLHD" CLAMPS

SIZE NO.	SHIPPING		AFTER TAKE UP	
	INCH	MM	INCH	MM
212	2.125	52.98	1.25	31.75
262	2.625	66.68	1.75	44.45
312	3.125	79.38	2.25	57.15
362	3.625	92.08	2.75	69.85
412	4.125	104.78	3.25	82.55
462	4.625	117.48	3.75	95.25
512	5.125	130.18	4.25	107.95
562	5.625	142.88	4.75	120.65
612	6.126	155.60	5.25	133.35
662	6.625	168.28	5.75	146.05
712	7.125	180.98	6.25	158.75
762	7.625	193.68	6.75	171.45
812	8.125	206.38	7.25	184.15
862	8.625	219.08	7.75	196.85
912	9.125	231.78	8.25	209.55

LARGER SIZE CLAMPS AVAILABLE THROUGH MANUFACTURERS.

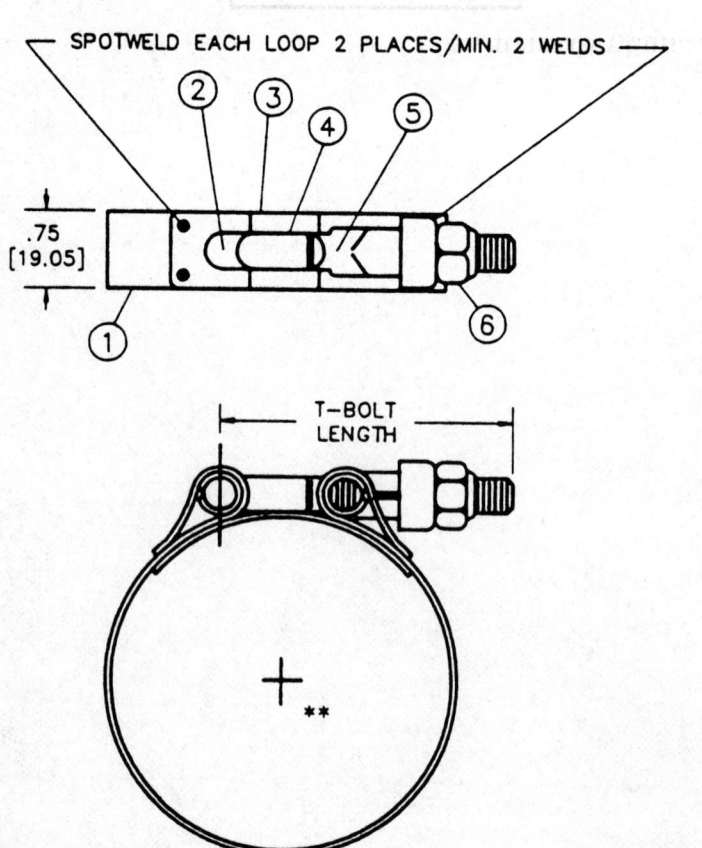

1. BAND
   AISI TYPE 201, 301, 302, 304 S.S
2. T-BOLT WRAPPER
3. TONGUE OR FLOATING BRIDGE, MFG'S OPTION
4. T-BOLT*
5. TRUNNION
6. NUT

* SEE 12.6.1 FOR DIMENSIONS
** FOR DIAMETER RANGE, SEE TABLE 15

NOTE: THESE ILLUSTRATIONS ARE FOR GRAPHIC PURPOSES ONLY. CONSTRUCTION MAY VARY ACCORDING TO MANUFACTURER.

FIGURE 12—TYPE "TB"

16.65

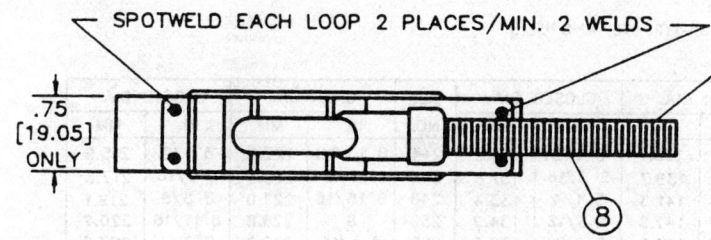

.75 [19.05] ONLY

SPOTWELD EACH LOOP 2 PLACES/MIN. 2 WELDS

.250-28UNF – 2A OR 3A, MFG'S OPTION

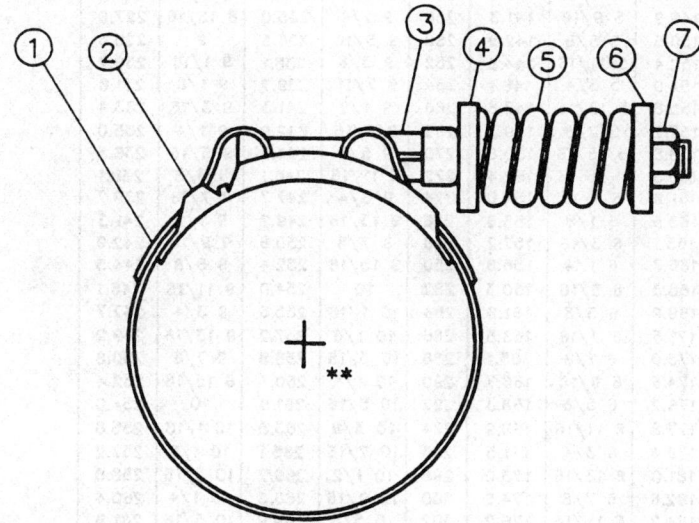

① BAND
AISI TYPE 201, 301, 302, OR 304 CRES
HALF HARD TEMPER

② FLOATING BRIDGE OR TONGUE, MFG'S OPTION
AISI TYPE 301, 302, OR 304 CRES
ANNEALED TEMPER

③ TRUNNION
C.Q. C.R.S CAD OR ZINC PLATED

④ WASHER
1.0 O.D X 0.281 I.D. x 0.109 THICK (ANSI-B 27.2)
STEEL - 0.20 CARBON MAX, COMM'L ZINC PLATED

⑤ COMPRESSION SPRING
0.187 DIA. MUSIC WIRE
SAE J178 (ASTM A228)
CAD OR ZINC PLATED

⑥ WASHER
1.0 O.D. x 0.443 I.D. x 0.084 THICK
STEEL - 0.20 CARBON MAX, COMM'L ZINC PLATED

⑦ T-NUT
0.250-28 UNF-2B
CARBON STL 12L 14, COMM'L ZINC PLATED

⑧ T-BOLT
0.250-28 UNF-2A X OR 3A PERMISSIBLE
C-1022-1038 STL, CAD, OR ZINC PLATED

** FOR DIAMETER RANGE, SEE TABLE 15

NOTE: THESE ILLUSTRATIONS ARE FOR GRAPHIC PURPOSES ONLY. CONSTRUCTION MAY VARY ACCORDING TO MANUFACTURER.

FIGURE 13—TYPE "SLTB"

TABLE 15—TYPE "TB" AND "SLTB"

SIZE NO.	OPEN DIA. IN.	OPEN DIA. MM	CLOSED DIA. IN.	CLOSED DIA. MM	SIZE NO.	OPEN DIA. IN.	OPEN DIA. MM	CLOSED DIA. IN.	CLOSED DIA. MM	SIZE NO.	OPEN DIA. IN.	OPEN DIA. MM	CLOSED DIA. IN.	CLOSED DIA. MM
28	2	50.8	1 3/4	44.5	136	5 7/16	138.1	5 1/8	130.2	244	8 13/16	223.8	8 1/2	215.9
30	2 1/16	52.4	1 13/16	46.0	138	5 1/2	139.7	5 3/16	131.8	246	8 7/8	225.4	8 9/16	217.5
32	2 3/16	55.6	1 7/8	47.6	140	5 9/16	141.3	5 1/4	133.4	248	8 15/16	227.0	8 5/8	219.1
34	2 1/4	57.2	1 15/16	49.2	142	5 5/8	142.9	5 5/16	134.9	250	9	228.6	8 11/16	220.7
36	2 5/16	58.7	2	50.8	144	5 11/16	144.5	5 3/8	136.5	252	9 1/16	230.2	8 3/4	222.3
38	2 3/8	60.3	2 1/16	52.4	146	5 3/4	146.1	5 7/16	138.1	254	9 1/8	231.8	8 13/16	223.8
40	2 7/16	61.9	2 1/8	54.0	148	5 13/16	147.6	5 1/2	139.7	256	9 3/16	233.4	8 7/8	225.4
42	2 1/2	63.5	2 3/16	55.6	150	5 7/8	149.2	5 9/16	141.3	258	9 1/4	235.0	8 15/16	227.0
44	2 9/16	65.1	2 1/4	57.2	152	5 15/16	150.8	5 5/8	142.9	260	9 5/16	236.5	9	228.6
46	2 5/8	66.7	2 5/16	58.7	154	6	152.4	5 11/16	144.5	262	9 3/8	238.1	9 1/16	230.2
48	2 11/16	68.3	2 3/8	60.3	156	6 1/16	154.0	5 3/4	146.1	264	9 7/16	239.7	9 1/8	231.8
50	2 3/4	69.9	2 7/16	61.9	158	6 1/8	155.8	5 13/16	147.6	266	9 1/2	241.3	9 3/16	233.4
52	2 13/16	71.4	2 1/2	63.5	160	6 3/16	157.2	5 7/8	149.2	268	9 9/16	242.9	9 1/4	235.0
54	2 7/8	73.0	2 9/16	65.1	162	6 1/4	158.8	5 15/16	150.8	270	9 5/8	244.5	9 5/16	236.5
56	2 15/16	74.6	2 5/8	66.7	164	6 5/16	160.3	6	152.4	272	9 11/16	246.1	9 3/8	238.1
58	3	76.2	2 11/16	68.3	166	6 3/8	161.9	6 1/16	154.0	274	9 3/4	247.7	9 7/16	239.7
60	3 1/16	77.8	2 3/4	69.9	168	6 7/16	163.5	6 1/8	155.6	276	9 13/16	249.2	9 1/2	241.3
62	3 1/8	79.4	2 13/16	71.4	170	6 1/2	165.1	6 3/16	157.2	278	9 7/8	250.8	9 9/16	242.9
64	3 3/16	81.0	2 7/8	73.0	172	6 9/16	166.7	6 1/4	158.8	280	9 15/16	252.4	9 5/8	244.5
66	3 1/4	82.6	2 15/16	74.6	174	6 5/8	168.3	6 5/16	160.3	282	10	254.0	9 11/16	246.1
68	3 5/16	84.1	3	76.2	176	6 11/16	169.9	6 3/8	161.9	284	10 1/16	255.6	9 3/4	247.7
70	3 3/8	85.7	3 1/16	77.8	178	6 3/4	171.5	6 7/16	163.5	286	10 1/8	257.2	9 13/16	249.2
72	3 7/16	87.3	3 1/8	79.4	180	6 13/16	173.0	6 1/2	165.1	288	10 3/16	258.8	9 7/8	250.8
74	3 1/2	88.9	3 3/16	81.0	182	6 7/8	174.6	6 9/16	166.7	290	10 1/4	260.4	9 15/16	252.4
76	3 9/16	90.5	3 1/4	82.6	184	6 15/16	176.2	6 5/8	168.3	292	10 5/16	261.9	10	254.0
78	3 5/8	92.1	3 5/16	84.1	186	7	177.8	6 11/16	169.9	294	10 3/8	263.5	10 1/16	255.6
80	3 11/16	93.7	3 3/8	85.7	188	7 1/16	179.4	6 3/4	171.5	296	10 7/16	265.1	10 1/8	257.2
82	3 3/4	95.3	3 7/16	87.3	190	7 1/8	181.0	6 13/16	173.0	298	10 1/2	266.7	10 3/16	258.8
84	3 13/16	96.8	3 1/2	88.9	192	7 3/16	182.6	6 7/8	174.6	300	10 9/16	268.3	10 1/4	260.4
86	3 7/8	98.4	3 9/16	90.5	194	7 1/4	184.2	6 15/16	176.2	302	10 5/8	269.9	10 5/16	261.9
88	3 15/16	100.0	3 5/8	92.1	196	7 5/16	185.7	7	177.8	304	10 11/16	271.5	10 3/8	263.5
90	4	101.6	3 11/16	93.7	198	7 3/8	187.3	7 1/16	179.4	306	10 3/4	273.1	10 7/16	265.1
92	4 1/16	103.2	3 3/4	95.3	200	7 7/16	188.9	7 1/8	181.0	308	10 13/16	274.6	10 1/2	266.7
94	4 1/8	104.8	3 13/16	96.8	202	7 1/2	190.5	7 3/16	182.6	310	10 7/8	276.2	10 9/16	268.3
96	4 3/16	106.4	3 7/8	98.4	204	7 9/16	192.1	7 1/4	184.2	312	10 15/16	277.8	10 5/8	269.9
98	4 1/4	108.0	3 15/16	100.0	206	7 5/8	193.7	7 5/16	185.7	314	11	279.4	10 11/16	271.5
100	4 5/16	109.5	4	101.6	208	7 11/16	195.3	7 3/8	187.3	316	11 1/16	281.0	10 3/4	273.1
102	4 3/8	111.1	4 1/16	103.2	210	7 3/4	196.9	7 7/16	188.9	318	11 1/8	282.6	10 13/16	274.6
104	4 7/16	112.7	4 1/8	104.8	212	7 13/16	198.4	7 1/2	190.5	320	11 3/16	284.2	10 7/8	276.2
106	4 1/2	114.3	4 3/16	106.4	214	7 7/8	200.0	7 9/16	192.1	322	11 1/4	285.8	10 15/16	277.8
108	4 9/16	115.9	4 1/4	108.0	216	7 15/16	201.6	7 5/8	193.7	324	11 5/16	287.3	11	279.4
110	4 5/8	117.5	4 5/16	109.5	218	8	203.2	7 11/16	195.3	326	11 3/8	288.9	11 1/16	281.0
112	4 11/16	119.1	4 3/8	111.1	220	8 1/16	204.8	7 3/4	196.9	328	11 7/16	290.5	11 1/8	282.6
114	4 3/4	120.7	4 7/16	112.7	222	8 1/8	206.4	7 13/16	198.4	330	11 1/2	292.1	11 3/16	284.2
116	4 13/16	122.2	4 1/2	114.3	224	8 3/16	208.0	7 7/8	200.0	332	11 9/16	293.7	11 1/4	285.8
118	4 7/8	123.8	4 9/16	115.9	226	8 1/4	209.6	7 15/16	201.6	334	11 5/8	295.3	11 5/16	287.3
120	4 15/16	125.4	4 5/8	117.5	228	8 5/16	211.1	8	203.2	336	11 11/16	296.9	11 3/8	288.9
122	5	127.0	4 11/16	119.1	230	8 3/8	212.7	8 1/16	204.8	338	11 3/4	298.5	11 7/16	290.5
124	5 1/16	128.6	4 3/4	120.7	232	8 7/16	214.3	8 1/8	206.4	340	11 13/16	300.0	11 1/2	292.1
126	5 1/8	130.2	4 13/16	122.2	234	8 1/2	215.9	8 3/16	208.0	342	11 7/8	301.6	11 9/16	293.7
128	5 3/16	131.8	4 7/8	123.8	236	8 9/16	217.5	8 1/4	209.6	344	11 15/16	303.2	11 5/8	295.3
130	5 1/4	133.4	4 15/16	125.4	238	8 5/8	219.1	8 5/16	211.1	346	12	304.8	11 11/16	296.9
132	5 5/16	134.9	5	127.0	240	8 11/16	220.7	8 3/8	212.7					
134	5 3/8	136.5	5 1/16	128.6	242	8 3/4	222.3	8 7/16	214.3					

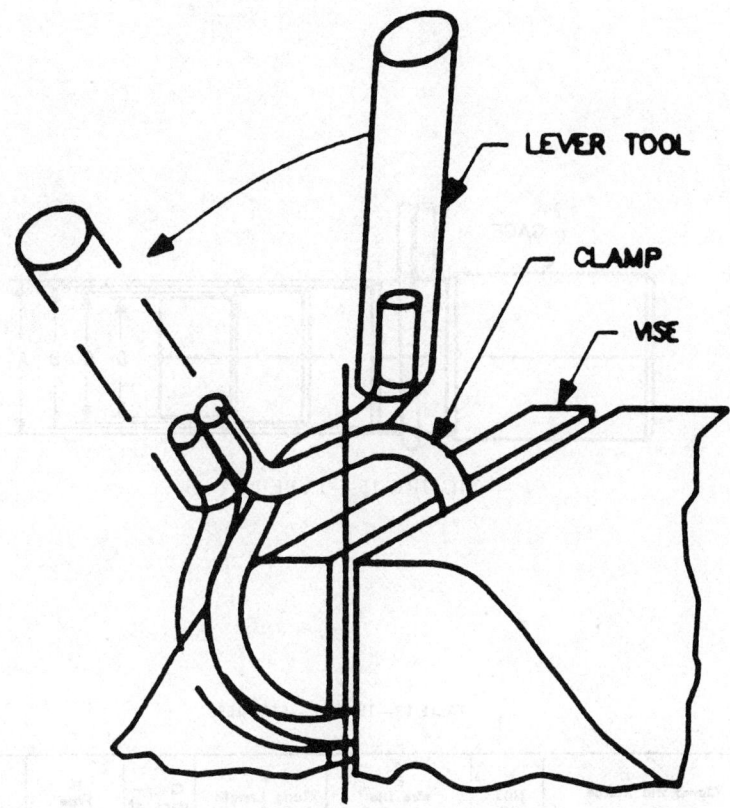

FIGURE 14—TYPE "E"

TABLE 16—TYPE "E"—CARBON

SAE Size No.	Clamp Dia Range[a]			D NOT GO Gage Dia	E Wire Dia[d]		F Length of Tang		G Clearance at Overlap	W Free Width Over Tangs	Y Overall Height	Z Gaging[e] Clearance
	A Max	B Nom	C Min		Max	Min	Max	Min	Max	Max	Ref	Max
6	0.380	0.375	0.370	0.350	0.083	0.081	0.38	0.34	0.015	0.88	1.06	0.005
7[b]	0.442	0.438	0.432	0.405	0.088	0.086	0.38	0.34	0.015	0.94	1.12	0.005
8[c]	0.510	0.500	0.490	0.462	0.093	0.091	0.38	0.34	0.025	1.00	1.19	0.005
9	0.573	0.562	0.551	0.520	0.108	0.106	0.38	0.34	0.025	1.06	1.38	0.006
10[b]	0.640	0.625	0.610	0.580	0.108	0.106	0.38	0.34	0.025	1.06	1.38	0.006
11[c]	0.703	0.688	0.671	0.635	0.113	0.111	0.38	0.34	0.025	1.12	1.50	0.006
12	0.770	0.750	0.730	0.690	0.113	0.111	0.38	0.34	0.031	1.19	1.50	0.008
13[b]	0.832	0.812	0.792	0.740	0.118	0.116	0.38	0.34	0.031	1.25	1.50	0.008
14[c]	0.900	0.875	0.850	0.800	0.123	0.121	0.38	0.34	0.031	1.25	1.62	0.008
15	0.968	0.938	0.906	0.855	0.123	0.121	0.38	0.34	0.062	1.25	1.69	0.008
16[b]	1.031	1.000	0.969	0.915	0.133	0.131	0.38	0.34	0.062	1.31	1.75	0.008
17[c]	1.090	1.062	1.034	0.960	0.143	0.141	0.41	0.34	0.062	1.50	1.88	0.010
18	1.150	1.125	1.100	1.030	0.153	0.151	0.41	0.34	0.062	1.62	2.00	0.010
19[b]	1.218	1.188	1.156	1.095	0.153	0.151	0.41	0.34	0.062	1.62	2.02	0.010
20[c]	1.280	1.250	1.219	1.145	0.153	0.151	0.41	0.34	0.062	1.75	2.00	0.010
21	1.344	1.312	1.281	1.210	0.163	0.161	0.41	0.34	0.062	1.75	2.31	0.010
22[b]	1.406	1.375	1.344	1.250	0.163	0.161	0.41	0.34	0.062	1.88	2.31	0.010
24	1.531	1.500	1.469	1.350	0.163	0.161	0.44	0.38	0.062	1.88	2.40	0.010
26	1.672	1.625	1.578	1.455	0.174	0.170	0.44	0.38	0.062	2.00	2.69	0.010
28	1.797	1.750	1.703	1.550	0.174	0.170	0.44	0.38	0.062	2.12	2.75	0.010
30	1.937	1.875	1.812	1.675	0.179	0.175	0.44	0.38	0.093	2.25	2.88	0.010
31	2.000	1.938	1.875	1.720	0.179	0.175	0.44	0.38	0.093	2.25	3.00	0.010
32	2.061	2.000	1.939	1.750	0.179	0.175	0.44	0.38	0.093	2.31	3.00	0.010
34	2.187	2.125	2.062	1.880	0.184	0.180	0.44	0.38	0.093	2.31	3.19	0.010
35	2.250	2.188	2.125	1.925	0.184	0.180	0.44	0.38	0.093	2.31	3.25	0.010
36	2.312	2.250	2.187	2.000	0.184	0.180	0.44	0.38	0.093	2.38	3.25	0.010
38	2.437	2.375	2.312	2.100	0.194	0.190	0.44	0.38	0.093	2.38	3.44	0.010
40	2.561	2.500	2.439	2.187	0.194	0.190	0.44	0.38	0.093	2.38	3.62	0.010
42	2.688	2.625	2.562	2.320	0.204	0.200	0.44	0.38	0.093	2.38	3.75	0.010

[a] To be used for corresponding gage diameter. Gage diameter tolerance +0.001, −0.000.
[b] These sizes shall be furnished with greenish hue. ⎤
[c] These sizes shall be furnished with reddish hue. ⎦ Optional when specified by purchaser.
[d] Wire diameters shown are before forming.
[e] Gage clearance per para. 12.2.1.2.

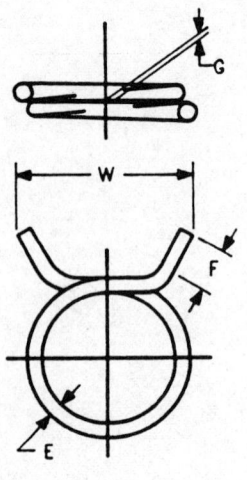
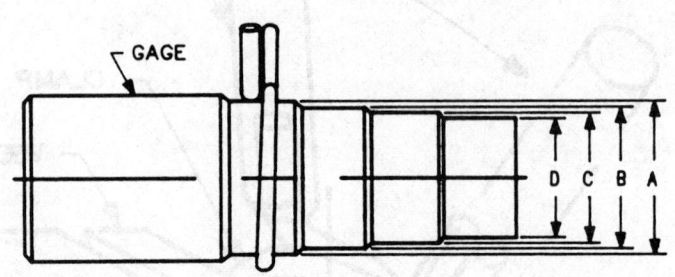
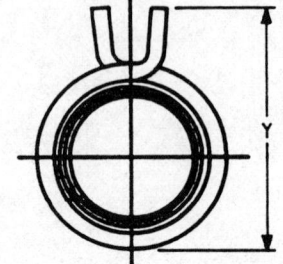

FIGURE 15—TYPE "E"

TABLE 17—TYPE "E"—STAINLESS

Size No.	Clamp Dia Range			D NOT GO Gage Dia	E Wire Dia		F Tang Length		G Clearance at Overlap	W Free Width	Y Height	Z Gage Clearance
	A Max	B Nom	C Min		Max	Min	Max	Min	Max	Max	Ref	Max
S-4	.253	.250	.247	.235	.039	.041	.38	.34	.015	.75	.68	.004
S-5	.315	.312	.309	.292	.052	.050	.38	.34	.015	.81	.68	.004
S-6	.380	.375	.370	.360	.067	.065	.38	.34	.015	.88	1.06	.004
S-7	.442	.438	.432	.415	.077	.075	.38	.34	.015	.94	1.12	.004
S-8	.510	.500	.490	.472	.083	.081	.38	.34	.025	1.00	1.19	.005
S-9	.573	.562	.551	.530	.093	.091	.38	.34	.025	1.06	1.38	.006
S-10	.640	.625	.610	.590	.107	.105	.38	.34	.025	1.06	1.38	.006
S-11	.703	.688	.671	.645	.107	.105	.38	.34	.025	1.12	1.50	.006
S-12	.770	.750	.730	.700	.107	.105	.38	.34	.031	1.18	1.50	.008
S-13	.832	.812	.792	.750	.113	.111	.38	.34	.031	1.25	1.50	.008
S-14	.900	.875	.850	.810	.121	.119	.38	.34	.031	1.25	1.62	.008
S-15	.968	.938	.906	.865	.121	.119	.38	.34	.062	1.25	1.69	.008
S-16	1.031	1.000	.969	.925	.121	.119	.38	.34	.062	1.31	1.75	.008
S-17	1.090	1.062	1.034	.970	.133	.131	.38	.34	.062	1.50	1.88	.010
S-18	1.150	1.125	1.100	1.040	.143	.131	.38	.34	.062	1.62	2.00	.010

16.69

$$F = \frac{F1 + F2 + F3}{3}$$

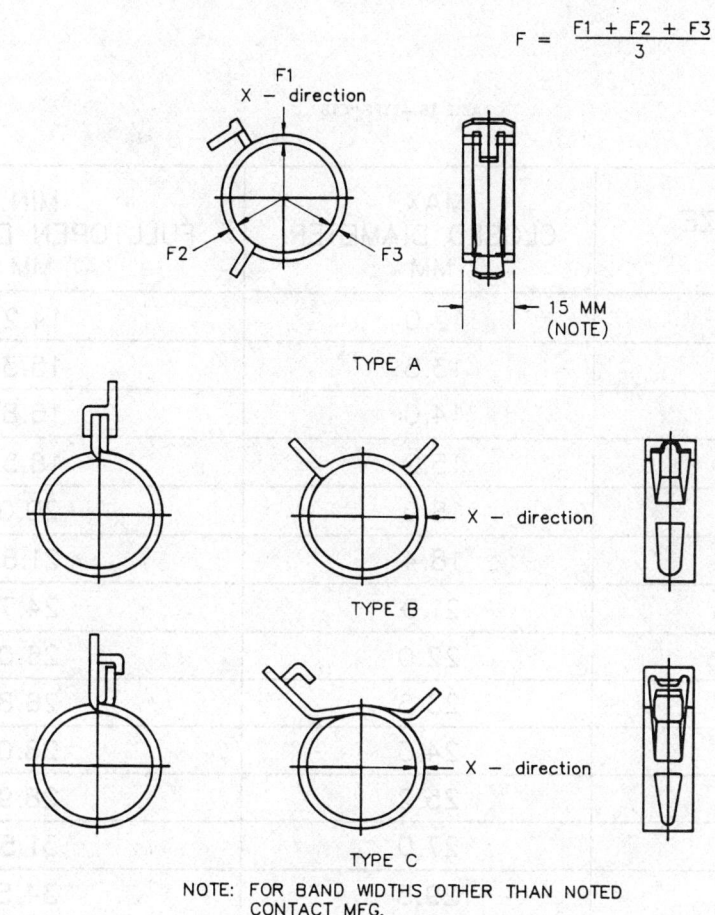

NOTE: FOR BAND WIDTHS OTHER THAN NOTED CONTACT MFG.

FIGURE 16—TYPE "CTB"

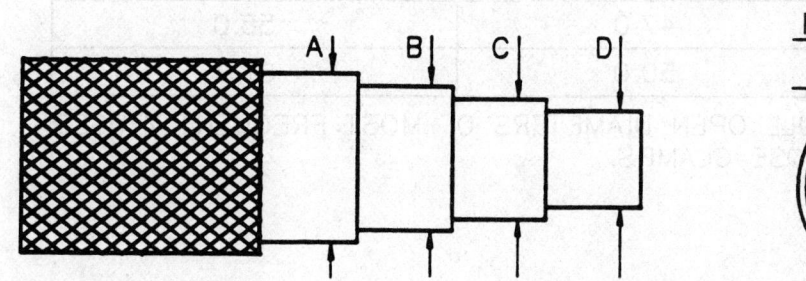 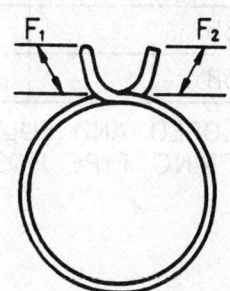 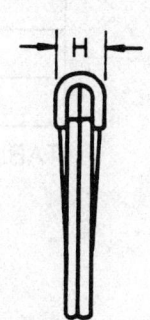

FIGURE 17—TYPE "CTW"

TABLE 18—TYPE "CTB"

NOMINAL SIZE CODE	MAX. CLOSED DIAMETER MM	MIN. FULL OPEN DIAMETER MM
13	12.0	14.2
14	13.5	15.3
15	14.0	16.8
17	15.2	18.5
19	18.0	20.0
20	18.4	21.6
23	21.0	24.7
24	22.0	26.0
25	23.5	26.8
26	24.3	28.0
27	25.2	28.9
29	27.0	31.5
32	29.5	34.5
35	31.5	38.0
38	34.5	41.5
40	35.5	42.5
42	37.5	44.5
44	38.5	46.5
47	41.5	50.0
50	43.5	53.0
51	44.0	54.0
53	46.0	55.0
55	47.0	58.0
58	50.0	61.0

TABLE 1: CLOSED AND FULL OPEN DIAMETERS OF MOST FREQUENTLY USED SPRING TYPE HOSE CLAMPS.

TABLE 19—TYPE "CTW"—METRIC

A	B	C	D	E	F₁	F₂	G	H
CLAMP DIAMETERS			NO	WIRE	REFERENCE DIM.		GAGE	REF.
MAX.	NOM.	MIN.	GO	SIZE	MAX.	MIN.	MAX.	DIM.
7.47	7.26	6.96	6.73	1.00	6.35	4.80	.105	6.35
7.80	7.60	7.30	7.10	1.00	6.35	4.80	.105	6.35
8.80	8.70	8.60	8.10	1.00	6.35	4.80	.105	6.35
9.65	9.50	9.40	8.90	1.00	6.35	4.80	.105	6.35
10.57	10.39	10.19	9.68	1.50	9.65	6.35	.153	7.10
11.25	11.13	11.00	10.28	1.50	9.65	6.35	.153	7.10
13.00	12.55	12.50	11.73	1.50	9.65	6.35	.153	7.10
14.10	13.73	13.36	12.36	1.50	9.65	6.35	.153	7.10
14.58	14.31	14.00	13.75	1.70	10.80	6.35	.153	8.25
15.93	15.60	15.11	14.10	1.70	10.80	6.35	.153	8.25
16.26	15.88	15.49	14.73	1.70	10.80	6.35	.153	8.25
16.81	16.41	15.93	14.88	1.70	10.80	6.35	.153	8.25
17.86	17.48	17.04	16.13	1.98	12.70	8.26	.203	9.14
18.69	18.19	17.70	16.51	1.98	12.70	8.26	.203	9.14
19.50	19.00	18.50	17.50	1.98	12.70	8.26	.203	9.14
20.62	20.19	19.61	18.25	1.98	12.70	8.26	.203	9.14
21.13	20.62	20.12	18.80	1.98	12.70	8.26	.203	9.14
22.75	22.13	21.50	20.25	2.19	13.97	9.53	.203	10.16
23.57	23.09	22.40	20.98	2.19	13.97	9.53	.203	10.16
24.59	23.83	23.01	21.72	2.19	13.97	9.53	.203	10.16
26.29	25.50	24.61	23.24	2.49	14.22	9.53	.254	11.43
27.68	26.97	26.26	24.38	2.49	14.22	9.53	.254	11.43
28.12	27.48	26.67	24.99	2.49	14.22	9.53	.254	11.43
29.21	28.58	27.94	26.16	2.49	14.22	9.53	.254	11.43
30.94	30.18	29.36	27.81	2.80	16.76	11.43	.254	12.19
32.00	31.29	30.38	28.37	2.80	16.76	11.43	.254	12.19
32.51	31.75	30.96	29.08	2.80	16.76	11.43	.254	12.19
34.14	33.32	32.54	30.73	2.80	16.76	11.43	.254	12.19
35.69	34.98	33.91	32.00	3.00	19.00	12.70	.254	13.72
36.40	35.59	34.59	32.49	3.00	19.00	12.70	.254	13.72
38.10	37.21	36.20	33.78	3.00	19.00	12.70	.254	13.72
38.89	38.10	37.31	34.29	3.20	19.00	12.70	.254	14.22
40.44	39.65	38.86	35.84	3.20	19.00	12.70	.254	14.22
42.98	41.28	40.08	37.47	3.20	19.00	12.70	.254	14.22
45.64	44.45	43.26	40.13	3.20	19.00	12.70	.254	14.22
49.20	47.63	46.02	43.69	3.20	19.00	12.70	.254	14.22
50.80	49.23	47.63	45.19	3.50	20.32	13.97	.254	14.99
52.35	50.80	49.25	46.48	3.50	20.32	13.97	.254	14.99
55.55	53.98	52.37	49.43	3.50	20.32	13.97	.254	14.99
57.15	55.55	53.98	50.17	3.50	20.32	13.97	.254	14.99
58.42	57.15	55.55	50.80	3.50	20.32	13.97	.254	14.99
71.00	69.85	60.00	63.00	3.80	21.60	13.97	.508	17.02
78.50	76.20	74.00	69.50	3.80	21.60	13.97	.508	17.02
85.00	82.55	80.00	75.00	4.00	21.60	13.97	.560	18.03
91.70	88.90	86.20	81.00	4.00	21.60	13.97	.560	18.03
98.20	95.25	92.30	87.00	4.00	21.60	13.97	.560	18.03
104.77	101.60	98.50	92.50	4.00	21.60	13.97	.560	18.03
111.40	107.95	104.50	98.00	4.20	21.60	13.97	.609	19.05
118.00	114.30	110.50	103.80	4.20	21.60	13.97	.609	19.05
124.68	120.65	116.50	109.35	4.20	21.60	13.97	.609	19.05
131.50	127.00	122.50	115.00	4.20	21.60	13.97	.609	19.05

NOTE: FOR EXPLANATION, SEE 12.5

TABLE 20—TYPE "CTW"—STANDARD

A	B	C	D	E	F₁	F₂	G	H
CLAMP DIAMETERS			NO	WIRE	REFERENCE DIM.		GAGE	REF.
MAX.	NOM.	MIN.	GO	SIZE	MAX.	MIN.	MAX.	DIM.
.294	.286	.274	.265	.039	.250	.190	.004	.250
.306	.301	.285	.280	.039	.250	.190	.004	.250
.345	.342	.339	.320	.039	.250	.190	.004	.250
.380	.375	.370	.350	.039	.250	.190	.004	.250
.416	.409	.401	.381	.059	.380	.250	.006	.280
.442	.438	.432	.405	.059	.380	.250	.006	.280
.510	.500	.490	.462	.059	.380	.250	.006	.280
.555	.539	.524	.484	.059	.380	.250	.006	.280
.573	.562	.551	.520	.070	.425	.250	.006	.325
.637	.614	.595	.555	.070	.425	.250	.006	.325
.640	.625	.610	.580	.070	.425	.250	.006	.325
.662	.646	.627	.586	.070	.425	.250	.006	.325
.703	.688	.671	.635	.078	.500	.325	.006	.325
.736	.716	.697	.650	.078	.500	.325	.008	.360
.770	.750	.730	.690	.078	.500	.325	.008	.360
.812	.795	.772	.720	.078	.500	.325	.008	.360
.832	.812	.792	.740	.078	.500	.325	.008	.360
.900	.875	.850	.800	.086	.550	.375	.008	.400
.928	.909	.882	.826	.086	.550	.375	.008	.400
.968	.938	.906	.855	.086	.550	.375	.008	.400
1.031	1.000	.969	.915	.098	.560	.375	.008	.450
1.090	1.062	1.034	.960	.098	.560	.375	.008	.450
1.107	1.082	1.050	.984	.098	.560	.375	.008	.450
1.150	1.125	1.100	1.030	.098	.560	.375	.008	.450
1.218	1.188	1.156	1.095	.110	.660	.450	.010	.480
1.260	1.232	1.196	1.117	.110	.660	.450	.010	.480
1.280	1.250	1.219	1.145	.110	.660	.450	.010	.480
1.344	1.312	1.281	1.210	.110	.660	.450	.010	.480
1.405	1.377	1.335	1.260	.118	.750	.500	.010	.540
1.433	1.401	1.362	1.279	.118	.750	.500	.010	.540
1.500	1.465	1.425	1.330	.118	.750	.500	.010	.540
1.531	1.500	1.469	1.350	.126	.750	.500	.010	.560
1.592	1.561	1.530	1.411	.126	.750	.500	.010	.560
1.692	1.625	1.578	1.475	.126	.750	.500	.010	.560
1.797	1.750	1.703	1.580	.126	.750	.500	.010	.560
1.937	1.875	1.812	1.720	.126	.750	.500	.010	.560
2.000	1.938	1.875	1.799	.137	.800	.550	.010	.590
2.061	2.000	1.939	1.830	.137	.800	.550	.010	.590
2.187	2.125	2.062	1.946	.137	.800	.550	.010	.590
2.250	2.187	2.125	1.975	.137	.800	.550	.010	.590
2.300	2.250	2.187	2.000	.137	.800	.550	.010	.590
2.795	2.750	2.638	2.480	.150	.850	.550	.020	.670
3.090	3.000	2.913	2.736	.150	.850	.550	.020	.670
3.346	3.250	3.150	2.953	.158	.850	.550	.022	.710
3.610	3.500	3.394	3.189	.158	.850	.550	.022	.710
3.866	3.750	3.634	3.425	.158	.850	.550	.022	.710
4.125	4.000	3.878	3.642	.158	.850	.550	.022	.710
4.386	4.250	4.114	3.858	.165	.850	.550	.024	.750
4.645	4.500	4.350	4.087	.165	.850	.550	.024	.750
4.909	4.750	4.587	4.305	.165	.850	.550	.024	.750
5.177	5.000	4.823	4.528	.165	.850	.550	.024	.750

NOTE: FOR EXPLANATION, SEE 12.5

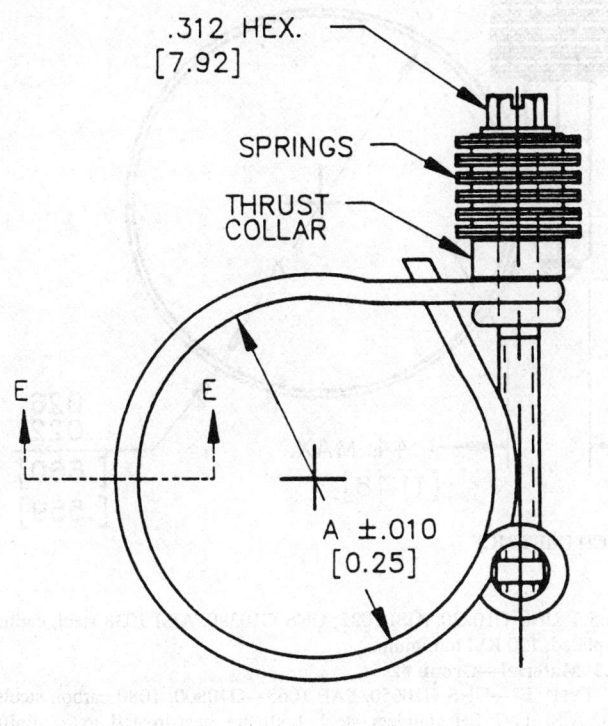
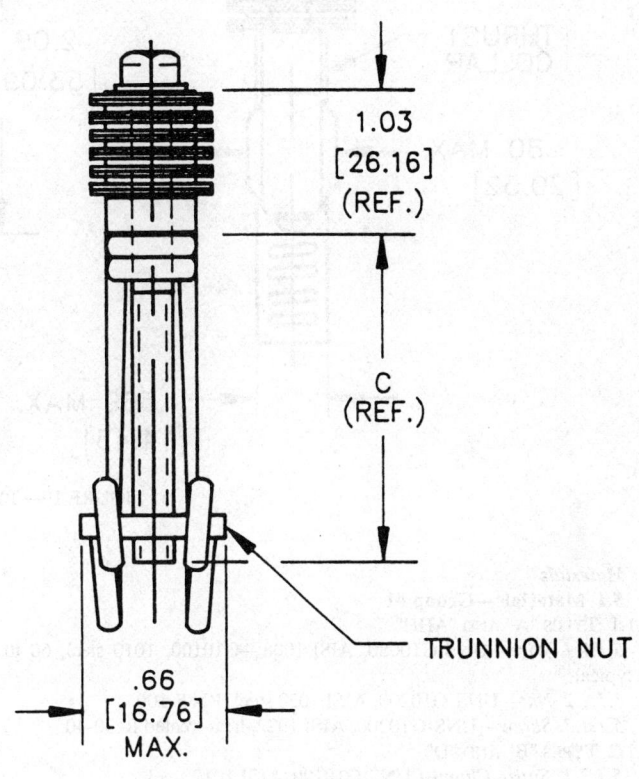
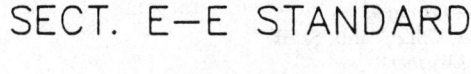
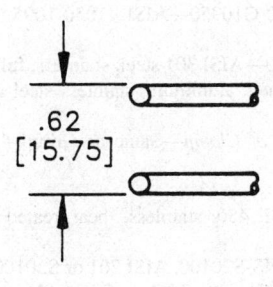
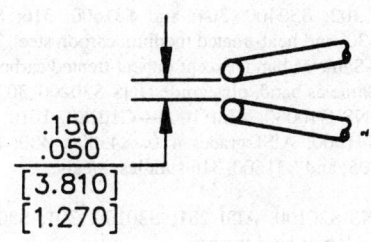

FIGURE 18—TYPE "SLA"—INCH [METRIC]

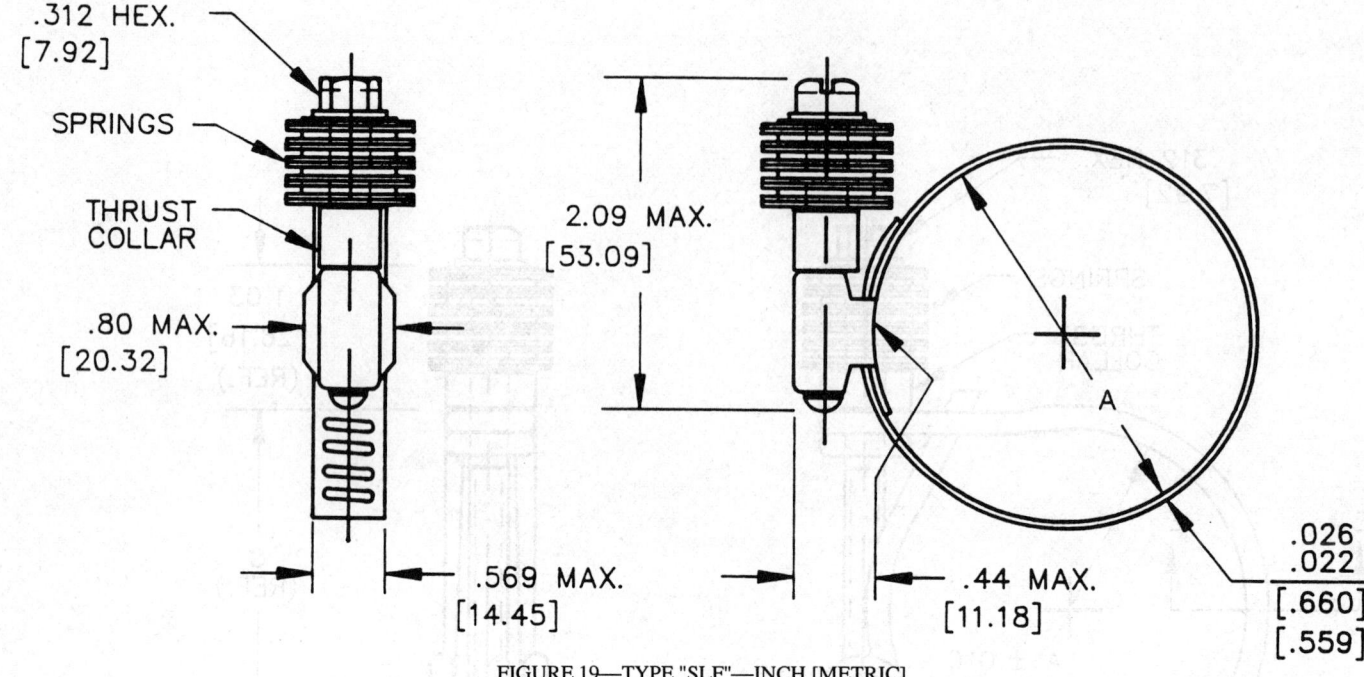

FIGURE 19—TYPE "SLF"—INCH [METRIC]

5. Materials
5.1 Materials—Group #1
5.1.1 TYPES "A" AND "AHH"

5.1.1.1 Wire—UNS-G10080, AISI 1008—G10100, 1010 steel, 60 to 80 KSI typical.

5.1.1.2 Nut—UNS-G10200, AISI 1020 steel, Rb85-100.

5.1.1.3 Screw—UNS-G10200, AISI 1022, heat-treated Rc30-40.

5.1.2 TYPES "B" AND "D"

5.1.2.1 Entire Clamp—UNS-G10100, AISI 1010 steel.

5.1.2.2 Entire Clamp—UNS-S30400, AISI 304 stainless (metric sizes per Figure 3A).

5.1.3 TYPE "C"

5.1.3.1 Band—UNS-G10100, AISI 1010 steel, except sizes #13 through #21, 22S, 23, 24S, 25, and 26S which are stainless steel grade.

5.1.3.2 Nut—Same as band (5.1.3.1) at manufacturer's option.

5.1.3.3 Screw—Same as band (5.1.3.1) at manufacturer's option.

5.1.3.4 Bridge—Same as band (5.1.3.1).

5.1.4 TYPES "F," "FEO," "FE," "HD," "I," "M," AND "MX"

5.1.4.1 Band—UNS-S20100, AISI Austenitic stainless grades 201; S30100, 301; S30200, 302; S30400, 304; and S31600, 316; S43000, AISI Ferritic stainless grade 430; and heat-treated medium carbon steel.

5.1.4.2 Housing—Same as band, except unheat-treated carbon steel.

5.1.4.3 Saddle—Same as band, plus grade UNS-S30200, 302 stainless.

5.1.4.4 Screw—UNS-G10060, AISI 1006—G10180, 1018; and G10211 10B21 carbon steels; S41000, AISI grades 410; S43000, 430; S30200, 302; S30400, 304; S30550, 305; and S31600, 316 stainless steels.

5.1.5 TYPE "TB"

5.1.5.1 Band—UNS-S20100, AISI 201; S30100, 301; S30200, 302; or S30400, 304 stainless steel; half hard temper.

5.1.5.2 Bridge—UNS-S30100, AISI 301; S30200, 302; S30400, 304; stainless steel, annealed, 1/4 hard, or 1/2 hard temper.

5.1.5.3 Trunnion—Low carbon steel cadmium plated or stainless steel (same grades as for "bridge").

5.1.5.4 Nut—UNS-G10200, AISI 1020—G10500, 1050 steel, cadmium or zinc plated.

5.1.5.5 T-Bolt

5.1.5.5.1 UNS-G40370, AISI 4037 alloy steel, heat-treated to 125 KSI minimum, cadmium or zinc plated.

5.1.5.5.2 UNS-S43100, AISI 431 stainless steel, heat-treated to 125 KSI minimum.

5.1.5.5.3 UNS-S66286, AISI A286 stainless steel, 130 KSI minimum.

5.1.5.5.4 UNS-S30200, AISI 302; or S30500, 305 stainless steel, 95 KSI minimum.

5.1.5.5.5 UNS-G10220, AISI 1022; UNS-G10380, AISI 1038 steel, cadmium or zinc plated, 120 KSI minimum.

5.2 Material—Group #2
5.2.1 TYPE "E"—UNS-G10650, SAE 1065—G10800, 1080 carbon steels; or S17700, AISI 17-7 PH stainless steel; both are heat-treated to a minimum Rockwell hardness of Rc50 to meet the performance and ductility requirements specified in Section 12.

5.2.2 TYPE "CTB"—Carbon steel or alloyed spring steels, heat-treated to Rc47-53 (mean of Rc50) to meet the performance requirements specified in Section 12.

5.2.3 TYPE "CTW"—Carbon or stainless steel as follows:

5.2.3.1 UNS-G10700, SAE 1070—G10850, 1085 carbon steel wire; pre-hardened to Rc50, then stress relieved after forming.

5.2.3.2 UNS-S17700—AISI 17-7PH stainless steel heat-treated to condition "C" by aging 1 h at 900 °F.

5.3 Material—Group #3
5.3.1 TYPES "SLA," "SLF," AND "SLHD"

5.3.1.1 Spring Washers

5.3.1.1.1 UNS-G10500-G10950—AISI 1050-1095 steel, heat-treated to Rc42-50.

5.3.1.1.2 UNS-G30100—AISI 301 steel, stainless, full hard.

5.3.1.2 Spacer—Steel, aluminum, stainless steel as supplied or furnished by manufacturer.

5.3.1.3 Remainder of Clamp—Same as Group #1—types "A," "F," and "HD," respectively.

5.3.2 TYPE "T"

5.3.2.1 Band—AISI 450 stainless, heat-treated Rc40-46. (No UNS number.)

5.3.2.2 Shield—UNS-S20100, AISI 201 or S30100, 301 stainless.

5.3.2.3 Balance of Clamp—Same as Group #1—type "F."

5.3.3 TYPE "SLTB"

5.3.3.1 Spring—Music wire per SAE J178 (ASTM A 228), diameter as recommended by manufacturer.

5.3.3.2 Washer—UNS-G10200, AISI 1020 steel.

5.3.3.3 Remainder of Clamp—Same as Group #1—type "TB."

5.4 General Materials Note
The materials listed in Section 5 describe those which are currently being used by the clamp industry. It serves only as a reference for the user and in no way implies that the current manufacturers are required to use the listed materials. As raw material prices move and new technologies emerge, the clamp manufacturers reserve the right to change the raw materials and processes used in their products so long as they can demonstrate that overall clamp performance has not been impaired.

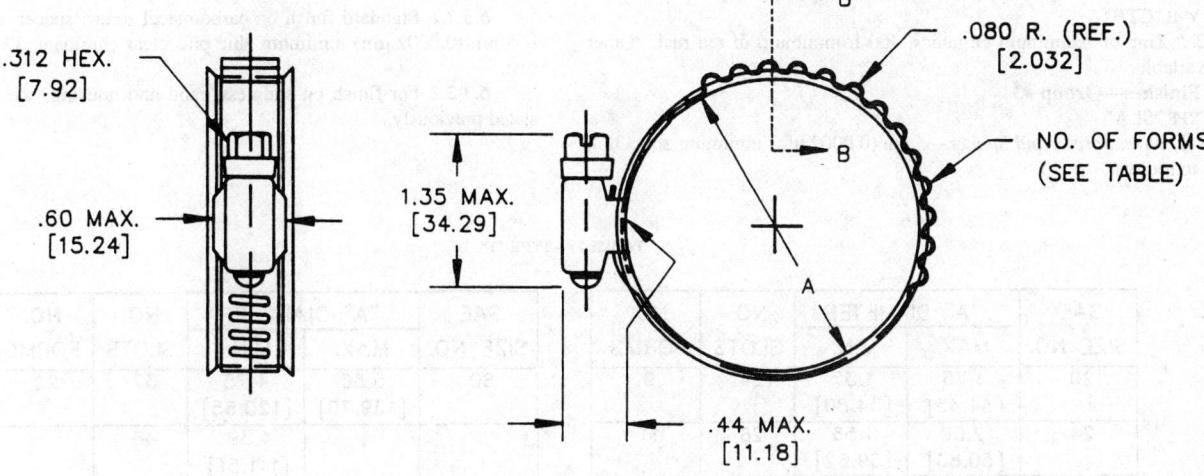

FIGURE 20—TYPE "T"

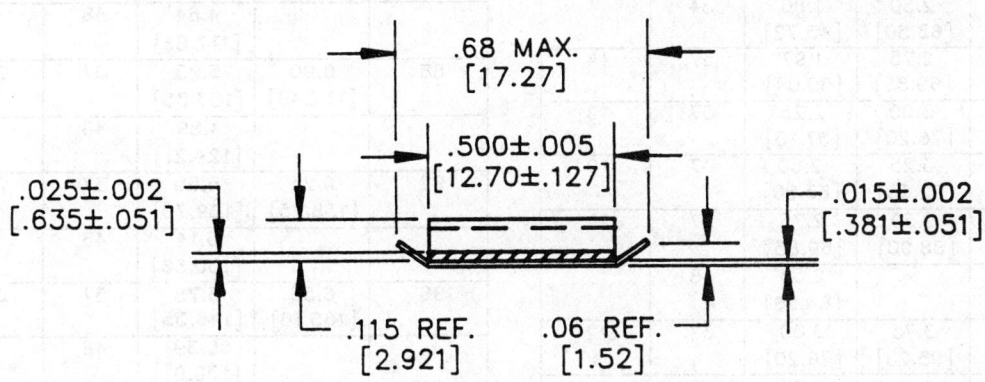

SECT. B–B
SCALE: 2:1

FIGURE 21—TYPE "T"—INCH [METRIC]

6. Finishes—General—Carbon steel components of clamps are normally supplied with rustproof finishes as specified by the purchaser. It is recommended that a reasonable latitude be allowed in the inspection of finishes on parts fabricated from precoated steel and the overlapping areas on clamps treated after assembly. All salt spray times (minimum hours) are per ASTM B 117.

NOTE—Magnitude of white corrosion and red rust permissible shall be determined between supplier and purchaser.

6.1 Finishes—Group #1
6.1.1 TYPES "A" AND "AHH"

6.1.1.1 Standard finish is 5 μm (0.0002 μin) minimum zinc on all external surface areas. Rated time to red rust is 32 h (minimum).

6.1.1.2 Optional—Zinc plus yellow chromate, 72 h minimum to white corrosion, 168 h minimum to red rust. Iridescence is acceptable.

6.1.1.3 Optional—Zinc Phos/electrodeposited black paint with oil sealer, 168 h minimum to red rust.

CAUTION—High lubricity lowers the clamp ultimate torque capacity.

6.1.1.4 Optional—Aluminum base coat/organic plus lube, silver grey, and black color, 400 h minimum to red rust.

6.1.1.5 Optional—Phosphate/zinc flake/organic, silver grey color, 240 h minimum to red rust.

6.1.2 TYPES "B" AND "D"

6.1.2.1 Standard finish is 5 μm (0.0002 μin) minimum zinc, 32 h minimum to red rust.

NOTE—For other finishes contact manufacturer.

6.1.3 TYPE "C"

6.1.3.1 Standard finish is 5 μm (0.0002 μin) minimum zinc, 32 h minimum to red rust.

NOTE—For other finishes contact manufacturer.

6.1.4 TYPES "F," "FEO," "FE," "HD," "I," "M," "MX," AND "TB"

6.1.4.1 Band/Housing/Saddle/Bridge/Trunnion—Generally these items are made of stainless steel and therefore are supplied as manufactured. Optional finishes vary with the manufacturer but generally include the following: Passivation, black oxide, and color chromating of zinc plated carbon steel parts.

6.1.4.2 Carbon Steel Screws/Nuts—Zinc plus chromate—thickness, chromate color and salt spray times vary, but are typically:

6.1.4.2.1 32 h minimum to red rust for 5 μm (0.0002 μin) minimum zinc plus clear chromate.

6.1.4.2.2 72 h minimum to red rust for 5 μm (0.0002 μin) minimum zinc plus yellow chromate.

6.1.4.3 Stainless steel screws do not receive plating. They can, however, be passivated, or black oxidized along with the clamp assembly.

6.1.5 TYPE "FE"

6.1.5.1 *Screw and Band*—Bright galvanized plus clear chromate.
6.1.5.2 *Housing*—Multiple coats of alkyd enamel paint optional.

6.2 Finishes—Group #2
6.2.1 TYPES "E" AND "CTW"

6.2.1.1 Zinc or Aluminum Organic, 400 h minimum to red rust.

6.2.1.2 Mechanical or electroplated zinc 5 μm (0.0002 μin) minimum plus chromate, 32 h minimum to red rust.
6.2.1.3 Stainless steel clamps are unfinished.
6.2.2 TYPE "CTB"
6.2.2.1 Zinc or Aluminum Organics, 400 h minimum to red rust. Other finishes available.
6.3 Finishes—Group #3
6.3.1 TYPE "SLA"
6.3.1.1 Wire, Screw, and Spacer—5 μm (0.0002 μin) minimum zinc, 32 h minimum to red rust.

6.3.1.2 Spring Washers—5 μm (0.0002 μin) minimum mechanical zinc plus clear chromate, 32 h minimum to red rust.
6.3.2 TYPE "SLF" AND "SLHD"
6.3.2.1 Standard finish on carbon steel screw, spacer, and spring washers is 5 μm (0.0002 μin) minimum zinc plus clear chromate, 32 h minimum to red rust.
6.3.2.2 For finish on stainless band and housing, see Type "F" or "HD" stated previously.

TABLE 21—TYPE "T"

SAE SIZE NO.	"A" DIAMETER MAX.	"A" DIAMETER MIN.	NO. SLOTS	NO. FORMS	SAE SIZE NO.	"A" DIAMETER MAX.	"A" DIAMETER MIN.	NO. SLOTS	NO. FORMS
20	1.75 [44.45]	1.37 [34.80]	24	9	80	5.50 [139.70]	4.75 [120.65]	37	22
							4.39 [111.51]	48	
24	2.00 [50.80]	1.56 [39.62]	26	11	84	5.75 [146.05]	5.00 [127.00]	37	22
28	2.25 [57.15]	1.63 [41.40]	32	12			4.64 [117.86]	48	
32	2.50 [63.50]	1.80 [45.72]	34	14	88	6.00 [152.40]	5.25 [133.35]	37	22
36	2.75 [69.85]	1.97 [50.04]	37	14			4.89 [124.21]	48	
40	3.00 [76.20]	2.25 [57.15]	37	19	92	6.25 [158.75]	5.50 [139.70]	37	22
44	3.25 [82.55]	2.50 [63.50]	37	19			5.14 [130.56]	48	
48	3.50 [88.90]	2.75 [69.85]	37	22	96	6.50 [165.10]	5.75 [146.05]	37	22
		2.53 [64.26]	48				5.39 [136.91]	48	
52	3.75 [95.25]	3.00 [76.20]	37	22	100	6.75 [171.45]	6.00 [152.40]	37	22
		2.64 [67.06]	48				5.64 [143.26]	48	
56	4.00 [101.60]	3.25 [82.55]	37	22	104	7.00 [177.80]	6.25 [158.75]	37	22
		2.89 [73.41]	48				5.89 [175.01]	48	
60	4.25 [107.95]	3.50 [88.90]	37	22	108	7.25 [184.15]	6.50 [165.10]	37	22
		3.14 [79.76]	48				6.14 [155.96]	48	
64	4.50 [114.30]	3.75 [95.25]	37	22	112	7.50 [190.50]	6.75 [171.45]	37	22
		3.39 [86.11]	48				6.39 [162.31]	48	
68	4.75 [120.65]	4.00 [101.60]	37	22	116	7.75 [196.85]	7.00 [177.80]	37	22
		3.64 [92.46]	48				6.64 [168.66]	48	
72	5.00 [127.00]	4.25 [107.95]	37	22	120	8.00 [203.20]	7.25 [184.15]	37	22
		3.89 [124.21]	48				6.89 [175.01]	48	
76	5.25 [133.35]	4.50 [114.30]	37	22					
		4.14 [105.16]	48						

NOTES:
1. SIZES LESS THAN NO. 20 ARE NOT AVAILABLE.
2. OTHER SIZES AVAILABLE THROUGH MANUFACTURERS — FOLLOW TYPE "F" FOR STANDARD SIZE INCREMENTS.

6.3.3 TYPE "T"

6.3.3.1 Screws—See Type "F" in 6.1.4.

6.3.3.2 Band—Finish on 450 stainless band (Custom Grade—No UNS designation) is as heat-treated with heat-tint color (typically copper-blue).

6.3.3.3 Housing and Shield—Plain.

6.3.4 TYPE "SLTB"

6.3.4.1 Bolt and Nut—See Group #1 Type "TB" in 5.1.5.

6.3.4.2 Stainless steel band, bridge, and trunnion are plain, as manufactured.

7. Threads

7.1 Types "A," "AHH," "B," "C," "D," and "SLA"—ANSI B1.1 Unified Inch Screw Threads Class 2A/2B, System 21 (B1.3M)—External and internal threads shall apply.

7.2 Type "F," "FEO," "HD," "I," "M," "MX," "SLF," "SLHD," and "T"—Modified buttress thread standard with manufacturer.

7.3 Type "TB," "SLTB"—ANSI B1.1 Unified Screw Threads, Class 3A/3B, System 21 (B1.3M)—External and internal threads shall apply.

8. Screws—Shall conform to the section on Machine Screws in SAE J478, except for special head and point details specified herein or unspecified detail specifically left to manufacturer's option.

8.1 Type "A," "AHH," "SLA"—Use 10-24 UNC hexagon washer head slotted or special high hex washer slotted machine screws per the illustrations in the document. Thread-forming screws conforming to IFI-112 Type TT may be used.

8.2 Type "B"—Use 10-24 hexagon washer head slotted, 10-24 Fillister head slotted, 10-24 Fillister washer head slotted, 12-24 Fillister head slotted or 12-24 round head cross recess screws, as specified.

8.3 Type "C"—Use 6-32 hexagon head slotted machine screws, and 10-24 or 12-24 upset hexagon head slotted machine screws, or 10-24 or 12-24 Fillister head slotted machine screws, with flat or pilot point, as specified.

8.4 Type "D"—Use 10-24 hexagon washer head slotted or Fillister washer head slotted machine screws.

8.5 Types "F," "HD," "I," "M," "MX," "SLF," "SLHD," and "T"—Use screws conforming to Styles 1, 2, 3, 4, 5, or 6 on Figure 6 and as noted in the tables. Unspecified details are standard with the individual manufacturer.

8.6 Type "FEO," "FE"—Use screws conforming to the manufacturer's specifications (see Figures 8 to 10 and Tables 10, 11, and 12).

8.7 Type "TB," "SLTB"—Use 10-32 UNF, 1/4-28 UNF, 5/16-24 UNF "T"-Bolts depending upon the clamp width and/or clamp open diameter.

9. Nuts

9.1 Types "A," "SLA," "B," "C," and "D"—Use square or rectangular nuts as indicative, of a size to suit clamp design, except Types "A" and "SLA" clamps, which use a flat trunnion nut standard with the manufacturers.

9.2 Type "TB"—Uses commercial quality and size, hex nuts with nylon locking feature conforming to MIL Std MS21044. All steel plated locknuts are also available, conforming to MIL Std MS21045.

9.3 Type "SLTB"—May require special manufactured nuts depending upon the method used for captivating the coil spring.

10. Identification

10.1 Type "A," "AHH," "SLA" clamps are not marked for size identification due to the limited available flat surface area.

10.2 Types "B" and "C" clamps shall be marked with SAE size number or fractional equivalent thereof.

10.3 Type "E" clamps may be marked for size as indicated on Figures 14 and 15 and Tables 16 and 17; some sizes can be color chromated/painted when specified by the purchaser.

10.4 Types "F," "FEO," "HD," "I," "M," "MX," "SLF," "SLHD," "T," "TB," and "SLTB" clamps shall be identified by size number stamped on the band. At manufacturer's option, manufacturer's name or trademarks may appear adjacent to size identifications and/or on the housing.

10.5 Type "CTB"—Nominal clamp sizes as being distinguished by a number or letter code, stamped on the clamp. See Table 18.

10.6 Type "CTW"—Same as Type "E."

11. Manufacturer's Notes

11.1 Type "B" clamps are normally manufactured with one slot in sizes up to and including No. 40; two slots in sizes No. 42 through No. 96 and three slots in sizes No. 100 and larger. Widths of slots and tongues shall not be greater than 40% of bandwidth and not less than 30% of bandwidth. Slots shall be centered in the bandwidth.

11.2 Type "CTW"—Working range is that difference between the "maximum" full open diameter (Dimension "A") and the "minimum" closed diameter (Dimension "C") of the clamp.

12. Clamp Performance, Acceptance Requirements, and Application Notes

12.1 Types "A," "AHH," "SLA"

12.1.1 CLAMP DIAMETERS—The "standard SAE" sizes and the nonstandard sizes currently available are listed in Table 1A. Additional sizes can be made available with the standard incremental open diameter size being 0.25 mm (0.010 in). The closed diameter, being a function of usable screw length, will be in accordance with Table 22.

12.1.2 TOLERANCE ON DIAMETERS

12.1.2.1 Open diameter manufacturing tolerance will be either +0.020/-0.000 (+0.51/-0.00) or +0.030/-0.000 (+0.76/-0.00) depending upon the manufacturer.

12.1.2.2 Closed Diameter—All clamps shall close tight on a steel mandrel of the sizes listed in Table 22 without any significant air gaps.

12.1.3 MINIMUM ULTIMATE TORQUE—When tested on a round steel mandrel of the open diameter less 0.06 (1.52), the clamp must withstand the following hand applied torques without failure.

12.1.3.1 For standard zinc plated, optional zinc chromate, aluminum/organic, and zinc/organic finishes—5.65 Nm (50 in-lb) minimum.

12.1.3.2 For phosphate/paint/oil type finish and for other oil bearing finishes—2.03 Nm (18 in-lb) minimum.

12.1.4 FREE TORQUE—For all type finishes, the free running torque measured near the clamps open diameter size shall not exceed 0.45 Nm (4.0 in-lb) in four revolutions of the screw. The torque value does not include any break-away effects due to screw staking method and/or optional finishes.

12.1.5 APPLICATION TORQUE

12.1.5.1 2.82 to 3.39 Nm (25 to 30 in-lb) for zinc and zinc chromated clamps.

12.1.5.2 1.13 to 1.36 Nm (10 to 12 in-lb) for phosphate/paint/oil type finishes.

12.2 Type "E" Clamps—Acceptability of type "E" clamps will be determined by the following tests and inspections:

12.2.1 EXPANSION AND PERMANENT SET—Expansion and permanent set of clamps shall be inspected by subjecting the clamps to the following tests and inspections in sequence:

12.2.1.1 Expand clamp to fit diameter "A" of gage.

NOTE—Care should be taken to avoid over-expansion during this operation.

12.2.1.2 Clamps shall be fitted respectively to gage diameter "B" and "C." When clamps are so fitted, a wire of "Z" diameter shall not pass between the gage and the clamp when inserted in a direction parallel to the axis of the gage.

12.2.1.3 In order to be sure that permanent set suffered by the material after assembly is within the prescribed limits for the best working range of the clamp and after being expanded to no greater than "A" diameter the clamp in the relaxed posture shall not fall off the "D" diameter gage.

12.2.2 BRITTLENESS (EMBRITTLEMENT)—Type "E" clamps subjected to corrosion preventive treatments which might produce hydrogen embrittlement shall be baked or otherwise treated to obviate such embrittlement and shall be capable of being expanded on a nominal diameter plug for a continuous 24 h period without signs of breaking or cracking.

12.2.3 DUCTILITY—Ductility of type "E" clamps shall be inspected by subjecting the clamps to the following tests:

12.2.3.1 The clamp shall be gripped in a vise in a manner such that the gripping edge of the vise will coincide with the clamp axis which bisects the angle between tangs as illustrated in Figures 14 and 15 and Tables 16 and 17. Clamp shall be expanded by moving the free tang as shown in Figures 14 and 15 and Tables 16 and 17 to a point where the free tang will position the stationary tang. There shall be no evidence of fracture during or after this test.

12.2.3.2 When clamp is expanded by movement of the free tang beyond the stationary tang to where the clamp fractures, the structure at the point of fracture shall show a fine grain and the clamp up to the instant of fracture shall deliver a tough springy reaction.

12.3 Type "F," "FEO," "FE," "HD," "I," "M," "MX," "SLF," "SLHD," and "T" Clamps

12.3.1 DURABILITY TORQUE—Screw threads and slots in the band shall show no evidence of deformation or excessive wear when clamps are tightened once on a steel mandrel to the applied screw torques in Table 23.

12.3.2 FREE TORQUE—With the band fully engaging the screw, the torque required to turn the screw four turns shall not exceed the values in Table 24.

12.3.3 INSTALLATION TORQUE—The suggested installation torque for a particular application must be established by the supplier and the user, given due consideration to the physical configurations, properties of the materials involved, and assembly tools to be used.

12.3.3.1 Installation Torque—Good practice indicates that the clamp types listed in Tables 23 and 24 be installed at 50% to 70% of their rated "Durability Torque."

12.4 Type "CTB"

12.4.1 WORKING RANGE—That difference between the "minimum" full open diameter and the "maximum" closed diameter of the clamp.

12.4.2 The open diameter shall be measured by means of a step gage with 0.10 mm (0.0039 in) increments.

12.4.3 The closed diameter shall be measured in the "X" direction as shown by Figure 16.

12.4.4 A minimum clamping force may be specified by the purchaser. Said minimum clamping force shall be determined on a three segment load cell, simultaneously measuring 3 forces at 120-degree intervals around the clamp's inside diameter, when the clamp is at its "nominal" size.

12.5 Type "CTW"

12.5.1 Clamps may not fall off the "NO-GO" size mandrel.

12.5.2 DIMENSIONAL CONTROL—The following tests are to be made using plug gages for each of the following four diameters:

12.5.2.1 "A" Diameter—The clamp when opened to its maximum limit must pass over a gage of the "A" diameter.

12.5.2.2 "B" and "C" Diameters—After being expanded to no greater than "A" diameter the clamps must be round within "G" (gauging dimension) when installed on "B" and "C" diameter gages.

12.5.2.3 "D" Diameter—In order to be sure that permanent set suffered by the material after assembly is within prescribed limits for the best working of the clamp and after being expanded to no greater than "A" diameter the clamp in the relaxed posture shall not fall off the "D" diameter gage.

12.6 Type "TB"

12.6.1 CLAMP SIZES—"T-Bolt" band clamps are available in three basic bolt sizes. See Table 25.

12.6.2 T-BOLT SECTION—Criterion based upon: temperature, tensile strength, and installation torque. See Table 26.

12.7 Type "SLTB"

12.7.1 CLAMP SIZES—The "spring loaded T-bolt" band clamps are only available with the 0.250-28UNF thread size.

12.7.2 MAXIMUM TORQUE—A maximum installation torque of 7.3 Nm (65 in-lb) is recommended for all clamp diameters.

12.7.3 TEMPERATURE—The type "SLTB" clamps are capable of 288 °C (550 °F) maximum service.

TABLE 22—CLAMP DIAMETERS MM (IN)

Open Diameter	Maximum Range	Screw Length
38.10 (1.50) & up	4.83 (0.19)	31.75 (1.25) nominal
39.62 (1.56) & up	5.59 (0.23)	38.10 (1.50) nominal
42.67 (1.69) & up	6.60 (0.26)	44.45 (1.75) nominal

TABLE 23—TORQUE NM (IN-LB) BY SCREW TYPE

Clamp Type	Carbon	410 ss	305 ss
"F"	5.65 (50)	6.78 (60)	6.78 (60)
"FEO"		See Figure 10	
"FE"		See Table 12	
"HD"	—	16.95 (150)	11.30 (100)
"I"	4.52 (40)	4.52 (40)	4.52 (40)
"M," "MX"	2.26 (20)	2.26 (20)	1.69 (15)
"SLF"	5.08 (45)	5.08 (45)	—
"SLHD"	—	14.12 (125)	—
"T"	8.47 (75)	8.47 (75)	8.47 (75)

TABLE 24—FREE TORQUE VALUES

Clamp Type	Max Free Torque Nm (in-lb)
"F"	0.45 (4.0)
"FEO"—9 mm	0.70 (6.2)
"FEO"—13 mm	1.01 (8.9)
"FE"	1.01 (8.9)
"HD," "SLHD"	0.68 (6.0)
"I"	0.45 (4.0)
"M," "MX"	0.45 (4.0)
"SLF"	0.45 (4.0)
"T"	0.45 (4.0)

TABLE 25—DIAMETRAL

Thread Size	T-Bolt Length	Minimum Clamp Dia.	Band Width Min/Max	Adjustment Plus/Minus	Band Thickness[1]
0.190-32	36.8 ± 1.52 (1.75) (0.06)	31.8 (1.25)	15.7 /25.4 (0.62)/(1.00)	1.5 / 3.0 (0.06)/(0.12)	0.508 to 0.635 (0.020 to 0.025)
0.190-32	57.2 ± 1.52 (2.25) (0.06)	76.2 (3.00)	15.7 /25.4 (0.62)/(1.00)	2.3 / 6.4 (0.09)/(0.25)	0.508 to 0.635 (0.020 to 0.025)
0.250-28	63.5 ± 69.9 (2.50) (2.75)	63.5 (2.50)	19.1 /38.1 (0.75)/(1.50)	2.0 / 5.1 (0.08)/(0.20)	0.635 (0.025)
0.250-28	88.9 ± 1.52 (3.50) (0.06)	177.8 (7.00)	19.1 /38.1 (0.75)/(1.50)	3.6 /12.2 (0.14)/(0.48)	0.635 (0.025)
0.312-24	88.9 ± 1.52 (3.50) (0.06)	76.2 (3.00)	22.4 /76.2 (0.88)/(3.00)	4.3 / 5.1 (0.17)/(0.20)	1.016 (0.040)
0.312-24	101.6 ± 1.52 (4.00) (0.06)	165.1 (6.50)	22.4 /76.2 (0.88)/(3.00)	9.9 /12.2 (0.39)/(0.48)	1.27 (0.050)

[1] Thickness Tolerances:

Thickness	Tolerance
0.020 to 0.025	±0.0015 (0.0381)
0.040	±0.002 (0.0508)
0.050	±0.003 (0.0767)

TABLE 26—T-BOLT SELECTION

T-Bolt Material	Rm. Temp. Maximum °C	Rm. Temp. Maximum °F	Tensile (KSI)	Maximum Installation Torque[1] Nm (in-lb) for Thread Sizes: 0.190-32	0.250-28	0.312-24
AISI 4037	288	550	125	5.6 (50)	7.9 (70)	22.6 (200)
AISI 431	288	550	125	5.6 (50)	7.9 (70)	22.6 (200)
AISI 302/305	427	800	95	4.5 (40)	6.8 (60)	16.9 (150)
AISI A286	427	800	130	5.6 (50)	7.9 (70)	22.6 (200)
AISI 1022-1038	232	450	120	5.0 (45)	7.3 (65)	19.6 (175)

[1] Maximum installation torque is that value recommended by the clamp manufacturer at which time the clamp shall achieve the intended purpose without destruction of the clamp or device it is applied to.

TYPE "F" CLAMPS FOR PLUMBING APPLICATIONS—SAE J1670 MAY93

SAE Standard

Report of the Hose/Clamp Subcommittee of the SAE Coolant Hose Committee approved May 1993.

1. **Scope**—This SAE Standard covers complete dimensional and general specifications for worm gear/worm drive hose clamps for general use in the plumbing industry.

   1.1 **Purpose**—To establish minimum functional guidelines for hose clamps intended for use in Plumbing application, herein referred to as Type "F" clamps.

2. **References**—There are no referenced publications specified herein.

3. **General Description**—Worm drive hose clamps for clamps with tangentially mounted buttress-like threaded screws, enclosed in a housing which is securely fastened to the band, which, in turn, is engaged with the screw. When the screw is rotated in a clockwise direction, the clamp becomes smaller and conversely a counterclockwise motion of the screw will eventually open the clamp.

4. **General Dimensions**—The following specifications tables and illustrations apply to Type "F" worm drive hose clamps.

   4.1 **Shipping Diameter**—Type "F" clamps will be supplied in an "A" Diameter, full open, still engaged. See Table 1.

   4.2 **Identification**—Clamps will be permanently marked with Country or Origin and/or manufacturer's identification.

   4.2.1 The SAE clamp size number shall be clearly marked on the band.

5. **Screws**—The screws shall conform to the specifications designated as follows:

   5.1 The screw head shall have an 8 mm (5/16 in) hex collar head screw as specified in Figure 1, Style 4, Slot optional.

   5.1.1 Screw threads shall be modified buttress external thread standard with manufacturer.

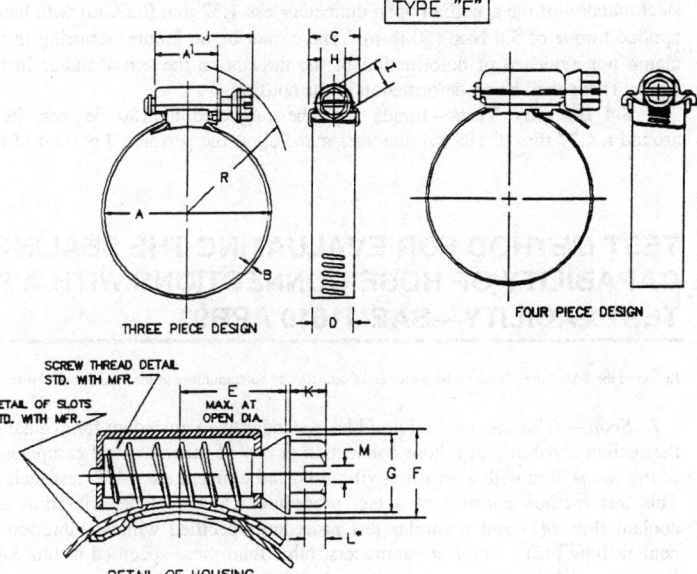

TABLE 1—DIMENSIONS OF TYPE F HOSE CLAMPS

SAE Size No.	A Dia.[1] Open mm	A Dia.[1] Open in	A Dia.[1] Closed mm	A Dia.[1] Closed in	R Radius[2] Over Screw mm	R Radius[2] Over Screw in
06	19.8	0.78	11.2	0.44	29.7	1.17
08	23.1	0.91	12.7	0.50	30.9	1.22
10	26.9	1.06	14.2	0.56	32.0	1.26
12	31.7	1.25	17.5	0.69	33.5	1.32
16	38.1	1.50	20.6	0.81	36.1	1.42
20	44.4	1.75	20.6	0.81	38.6	1.52
24	50.8	2.00	26.9	1.06	41.4	1.63
28	57.1	2.25	33.3	1.31	44.5	1.75
32	63.5	2.50	39.6	1.56	47.2	1.86
36	69.8	2.75	45.9	1.81	50.0	1.97
40	76.2	3.00	52.3	2.06	53.0	2.09
44	82.5	3.25	58.6	2.31	55.8	2.20
48	88.9	3.50	65.0	2.56	58.9	2.32
52	95.2	3.75	71.4	2.81	61.9	2.44
56	101.6	4.00	77.7	3.06	65.0	2.56
60	107.9	4.25	84.1	3.31	68.0	2.68
64	114.3	4.50	90.4	3.56	71.1	2.80
72	127.0	5.00	103.1	4.06	77.2	3.04
80	139.7	5.50	117.3	4.62	83.3	3.28
88	152.4	6.00	130.0	5.12	89.6	3.53
96	165.1	6.50	141.2	5.56	95.7	3.77
104	177.8	7.00	156.9	6.18	101.8	4.01

[1] Diameter shall be determined by assembly over mandrels.
[2] Reference dimensions for clearance purposes only.

NOTES:
For sizes greater than 104 contact the manufacturer.
Clamps closing smaller than list must comply with 8.1.

6. **Materials**—Screws, bands, and housings shall be fabricated from a minimum of 300 series stainless steel.

7. **Workmanship**—All clamps and components thereof shall be free of burrs, seams, loose scale, and other defects that might affect the performance.

DIMENSIONS OF CLAMPS

	Dimension	Type F mm	Type F in
A[1]	HSG Length (Ref.)	19.30	0.76
B	Thickness	0.533/0.787	0.021/0.031
C	HSG Width (Ref.)	20.570	0.81
D	Band Width	12.57/14.45	0.495/0.569
E	Max. at Open Dia.	19.050	0.75
F	Height (Ref.)	14.450	0.56
G	Collar Diameter	9.398/10.79	0.370/0.425
H	Across Flats	7.747/7.925	0.305/0.312
I	Across Corners (Min.)	8.636	0.340
J	Lg. of Screw (Max.)	34.29	1.35
K	Hex Height	3.556/6.350	0.140/0.250
L	Slot Depth (*Optional)	1.956/3.048	0.077/0.120
M	Slot Width	1.422/1.936	0.056/0.076

* Slot optional
[1] Reference dimension only

FIGURE 1—STAINLESS STEEL HOSE CLAMPS

8. **Test and Performance Requirements**—Clamp acceptability shall be determined by compliance with the following methods.

   8.1 **Clamping Diameter Range**—Clamps shall assemble over and close tight upon round mandrels equal to the corresponding open and closed diameters listed in Table 1. Diameters smaller than the diameters shown are permissible. For diameters greater than listed, contact the manufacturers.

   8.1.1 When tested for minimum and maximum open diameter, all threads must be fully engaged.

   8.2 **Free Running Torque**

8.2.1 FREE TORQUE—The torque value expressed in newton meters (pound inches) when the clamp is tightened four complete revolutions of the screw or nut, while in the free state. This value does not include any break-away effects due to staking or passage of the band ends beyond the screw head.

8.3 Durability Torque—Clamps shall be tightened once, over a round steel mandrel of the specified open diameters less 1.52 mm (0.06 in) with hand-applied torque of 5.6 N·m (50 lb-in) There shall be no failure occurring in the clamp nor evidence of deformation of the threads on the screw and/or in the band. There shall be no deformation of the housing.

8.4 Ductility Tests—Bands shall be subjected to 180 degrees, bend around a 4.77 mm (0.188 in) diameter mandrel, at the perforated portion of the band and then restraightened. The band shall at no time during or after the test exhibit cracks, breaking, or other indications of failure.

9. Installation Torque—The suggested installation torque for a particular application must be established by the user.

9.1 Manufacturer's recommended installation torque for all size TYPE "F" worm drive clamp is:

TYPE "F" = 3.44 N·m (30 lb-in) for all size and materials.

9.2 Assembly Tools—It is advised that the use of power tools to install worm drive clamps be of the stall torque type. Use of clutch type or impact type assembly tools is not recommended.

TEST METHOD FOR EVALUATING THE SEALING CAPABILITY OF HOSE CONNECTIONS WITH A PVT TEST FACILITY—SAE J1610 APR93

SAE Recommended Practice

Report of the SAE Hose/Clamp Performance and Compatibility Subcommittee of the SAE Coolant Hose Committee approved June 1992 and revised April 1993.

1. Scope—This test method provides a standardized procedure for evaluating the sealing capability of a hose connection or any of the individual components of the connection with a pressure, vibration, and temperature (PVT) test facility. This test method consists of a test procedure which includes vibration and coolant flow (#1) and a similar test procedure specified without vibration or coolant flow (#2). Any test parameters, other than those specified in this SAE Recommended Practice, are to be agreed to by the tester and the requestor.

2. References

2.1 Applicable Documents—The following publications form a part of this specification to the extent specified herein. The latest issue of SAE publications shall apply.

2.1.1 SAE PUBLICATIONS—Available from SAE, 400 Commonwealth Drive, Warrendale, PA 15096-0001.

SAE J962—Formed Tube Ends for Hose Connections
SAE J1508—Hose Clamps Specification

3. Materials and Equipment Required—Unless otherwise specified, the components for evaluating a hose connection should include:

3.1 Hose molded to consistent ID and OD and cut to a minimum length of 400 m (16 in).

3.2 Hose connectors per SAE J962.

3.3 Series Type F hose clamps per SAE J1508.

3.4 PVT test facility.

3.5 An ethylene glycol/water (50/50 volume) mixture with appropriate inhibitors.

4. Preparation of Test Components

4.1 All pertinent information of the components (ID, OD, type of materials, etc.) should be documented before assembly.

4.2 The sealing surfaces should be cleaned of all foreign materials for the test.

4.3 An ethylene glycol/water (50/50 volume) mixture should be used as an assembly aid for the hose and connector unless another lubricant or no lubricant is specified for the test.

4.4 The Type F clamps should be torqued to 3.4 N·m (30 in-lb) unless otherwise specified in the requestor's test requirements.

4.5 After assembly, each hose connection should be evaluated by pressure testing to 345 kPa (50 psi) for a minimum of 60 s unless otherwise specified in the requestor's test requirements.

5. Test Procedure #1—With Vibration and Coolant Flow

5.1 Raise the temperature of the chamber from ambient to 121 °C (250 °F) at an approximate rate of 2 to 4 °C (3 to 7 °F) per minute.

5.1.1 Heat the coolant to 113 °C ± 8 °C (235 °F ± 15 °F).

5.1.2 Flow coolant through hose assembly at an approximate rate of 20 L/min (5 gal/min).

5.1.3 Vibrate one end of the hose assembly at an amplitude of 12.5 mm (0.5 in) and a total displacement of 25 mm (1 in) at 0.5 Hz.

5.1.4 Pressure cycle the hose assembly from 69 kPa ± 14 kPa to 193 kPa ± 14 kPa (10 psi ± 2 psi to 28 psi ±2 psi) at a rate of 2 to 4 cycles per minute.

5.1.5 Estimated time—1 h.

5.2 Maintain the temperature of the assembly at 104 °C (220 °F) and continue the following test conditions for 1 h.

NOTE—The temperature of the hose assembly must be at a minimum of 104 °C (220 °F) for this 1 h period.

5.2.1 Continue flowing coolant through the hose assembly.

5.2.2 Continue vibrating the hose assembly.

5.2.3 Continue pressure cycling the hose assembly.

5.3 Reduce the chamber temperature to -40 °C (-40 °F) at an approximate rate of 2 to 4 °C (3 to 7 °F) per minute.

5.3.1 Coolant heaters should be off.

5.3.2 Flow of coolant should be off.

5.3.3 Vibration system should be off.

5.3.4 Pressure cycling should be off.

5.3.5 Estimated time—4.5 h.

5.4 Soak the hose assembly at a -40 °C (-40 °F) for 0.5 h.

(R) NOTE—The temperature of the hose assembly must be -32 °C (-25 °F) or lower for this 0.5 h period. The thermocouple should be placed between the hose at interface of tube and fitting and the tube before the clamp.

5.4.1 Coolant heaters should be off.

5.4.2 Flow of coolant should be off.

5.4.3 Vibration system should be off.

5.4.4 Pressure cycling should be off.

5.5 At -40 °C (-40 °F), after the 0.5 h soak, start the following test sequence:

5.5.1 Heat the coolant to 24 °C (75 °F).

5.5.2 Flow coolant through the hose assembly at an approximate rate of 20 L/min max (5 gal/min).

5.5.3 Vibrate the hose assembly at an amplitude of 12.5 mm (0.5 in) and a displacement of 25 mm (1.0 in) at a frequency of 0.5 Hz.

5.5.4 Pressure cycle the hose assembly from 69 kPa ± 14 kPa to 193 kPa ± 14 kPa (10 psi ± 2 psi to 28 psi ± 2 psi) at a rate of 2 to 4 cycles per minute.

5.5.5 Estimated time—0.5 h.

5.6 Raise the chamber temperature from -40 °C (-40 °F) to ambient and repeat the test cycle.

5.6.1 Estimated time to return to ambient—0.5 h.

5.6.2 One test cycle is approximately 8 h.

5.6.3 The end of the testing is 15 complete test cycles, approximately 120 h.

6. Test Procedure #2—Without Vibration or Coolant Flow—The test procedure for this requirement does not include vibration or coolant flow. This procedure is similar to Test Procedure #1 except for the steps specified for flow and vibration.

7. Report—The test report should document the pertinent details of the entire test including the actual test conditions at the time of a failure, if experienced. The waveform used in the pressure cycle is optional and should be stated in the test report. The criteria for a failure should be any visible coolant leakage from the assembly unless otherwise specified in the requestor's test requirements. The reported data should include the date, time, number of hours, number of PVT test cycles, chamber temperature, coolant temperature, coolant pressure, and any related comments.

17 Ball Studs and Joints

STEERING BALL STUDS AND SOCKET ASSEMBLIES—SAE J491 NOV87

SAE Recommended Practice

Report of the Parts and Fittings Division approved August 1922 and revised by the Steering and Suspension Ball Stud and Socket Committee November 1987.

1. General Specifications

1.1 Purpose—This SAE Recommended Practice has been established for the purpose of providing design criteria and suggested dimensional proportions which may be used for ball studs and ball stud socket assemblies as used on steering systems or control mechanisms of passenger vehicles, trucks and off-road equipment.

The recommended practice does not cover all applications. It is intended to provide assistance in obtaining functional satisfaction and interchangeability.

The inclusion of dimensional data in this report is not intended to imply that all the products described are stock production sizes. Consumers are requested to consult with manufacturers concerning stock production parts.

2. Terminology

2.1 Master Gage—A taper gage that serves as a standard or base, designed to specific dimensions within blueprint specifications.

2.2 Blueing—A nondrying light paste with a pigment or dye (such as "Prussian Blue") that colors a contacting surface. Blueing must be distributed evenly with minimum thickness on the taper surface of the master gage.

2.3 Blueing in a Master Gage—Blueing as applied to a master gage and gage forced onto a taper by hand pressure with a slight twisting motion followed by a rocking motion of the stud in the gage. (The gage should not rock.) Gage is removed for visual check of contact surface of taper.

3. Ball Studs

3.1 Selection of Size—Tensile and compressive forces and functional load requirements of the steering link must be considered in selecting the proper ball stud size.

The proper ball stud size for a specific application can be reasonably well estimated by considering the stud as a cantilever beam supported at the junction with the mating boss, and loaded radially through the ball end by a force, or forces, the magnitude and direction of which have been previously determined. The type of loading, ball stud and bearing material, heat treatment, and load requirements will influence the size of unit chosen for a specific application.

Design requirements which cannot be satisfied by the tabulated dimensions may be *adequately* fulfilled by deviating wherever necessary provided due consideration is given to the functional stresses.

3.2 Materials—Plain carbon and alloy steels are widely used for all stud fabrication. The principal requirements for either type of steel are: case hardenability to provide a wear resistant surface, good machinability or formability, and adequate core toughness to avoid brittle fracture under impact loading. For *plastic or rubber bearing* socket constructions, case hardening may not be required. A quench and tempered alloy steel may be the preferred material. Some of the more popular standard materials used for ball studs lie in the same category as SAE 1019, 1541, 4615, 8115, 8620, and 8640 steels.

3.3 Processing—Processing of ball studs is usually dictated by the size, volume of production and equipment available. Ball studs used on passenger car and light truck steering systems are usually cold formed. Larger ball studs of relatively low production volume such as used on heavy duty trucks and off-road equipment are often fabricated as machined parts. Hot upset forging methods have also been used. The forming or machining is followed by total or selective hardening. Depending on the application and method of fabrication, it may be desirable to control the surface finish.

3.4 Attachment—The ball stud locking taper is usually used in conjunction with screw thread and nut attachment to allow for repair or replacement. The taper (usually 1:8) is designed to attach into mating parts made of steel; however, by proper design the ball studs can be adapted to mate with other materials for which different tapers may be desirable.

The nut selection should be determined by design requirements.

The specified nut torque or seating force may draw the stud locking taper into the mating arm significantly depending on several variables such as amount of taper, size of taper, material, heat treat and surface. When using a slotted nut, the nut is to be tightened to specification and then tightened further if necessary to align a slot with the cotter hole.

The amount of stud taper draw-in should be determined experimentally for each specific application.

The boss thickness and hole gage diameter at the nut face of the boss must be so related to the stud gage diameter as to provide sufficient unexposed threads to allow for draw-in at specified nut torque (reference Fig. 5). To obtain the full stud cantilever strength, the face of the mounting boss should correspond nominally with the large end of the stud taper.

In addition to the screw thread and nut attachment method, locking

17.02

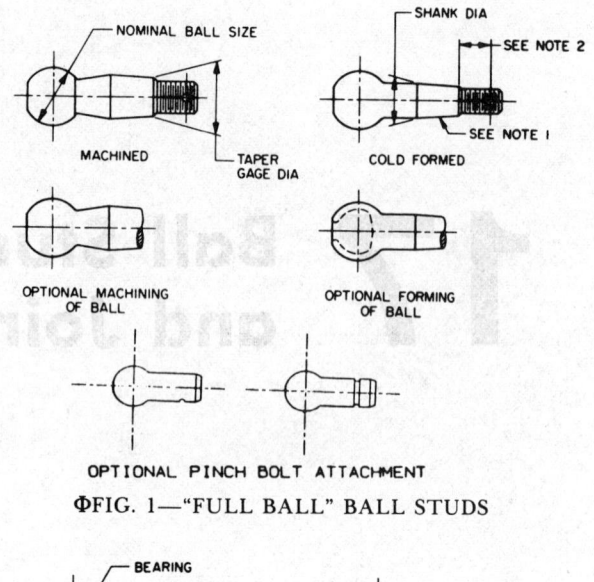

ΦFIG. 1—"FULL BALL" BALL STUDS

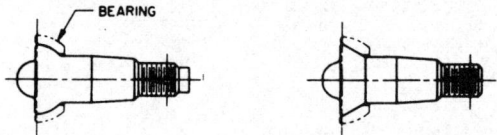

FIG. 2—"HALF BALL" BALL STUDS FOR SEPARATE SPHERICAL BEARINGS

NOTES:
1. Taper 1:8 on diameter unless other specified must show 60% minimum area of contact when blued in a master gage (See General Specifications).
2. Cotter pin hole diameter should conform to SAE J485. The hole location is determined by the attachment eye and nut size. The hole should be located to provide a minimum of 50% engagement of the cotter pin with nut slot after allowing for draw-in at specified nut torque. Cotter pin hole may be omitted and other than slotted nut used when application permits.

pinch bolt type attachment may be used. Permanent attachment means also may be used, such as riveting (upsetting), welding, spinning or staking. These methods may be used in conjunction with a locking taper or an interference press fit, in which case the stud is usually a straight shank end with a shoulder for locating purposes.

The permanently attached studs are used in conjunction with tubular ("horizontal type") sockets which can be assembled onto the stud (see Fig. 6). The screw thread and nut tapered studs are usually preassembled into ball stud sockets, but can also be used with horizontal type sockets.

4. Sockets

4.1 Selection of Size—The socket size is generally dependent upon the ball stud size necessary for a specific application; however, the tensile and compressive forces and functional load requirements of the steering link must be considered in selecting the proper socket size.

4.2 Materials—Steering link sockets are often made of SAE 1030, 1038, 1040, and 1541 steels; however, a number of standard and special steels may be used to provide the desired mechanical properties for each application. Tubular ("horizontal type") sockets may also be made from seamless or welded tubing, frequently in SAE 1010, 1020 and 1025 steels.

4.3 Processing and Attachment—Steering link sockets are usually machined forgings with threaded stems for attaching to mating components. The stems may have external or internal threads. For internal threads, the socket may be forged with integral clamp ears that are machined to provide for locking the threads. Instead of clamp ears, lock nuts at the end of the internal threaded stem may also be used. Sockets may be forged into both ends of an integral link. In applications of sockets with threaded stems for attachment to tubular intermediaries or turnbuckles, a right-hand thread is usually paired with a left-hand thread to provide fast assembly and adjustment.

4.4 Socket Types & Dust Seals—In addition to sockets shown in Fig. 4 and Fig. 5, various other constructions are available. Selection

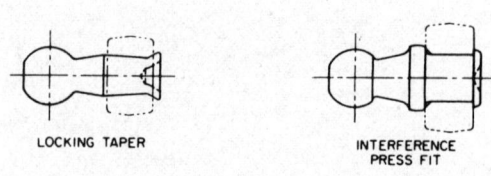

FIG. 3—PERMANENTLY ATTACHED STUDS—RIVETED, WELDED, SPUN, OR STAKED

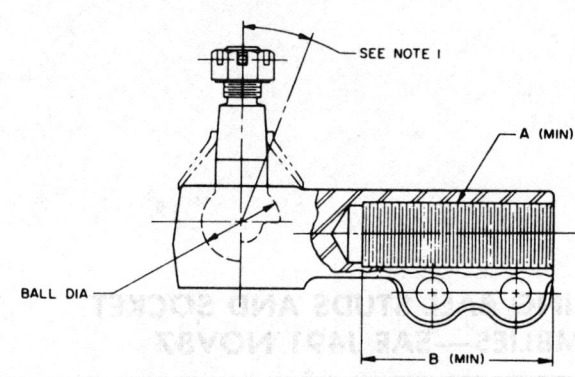

FIG. 4—TYPICAL BALL STUD SOCKET ASSEMBLY WITH INTERNAL THREADED STEM

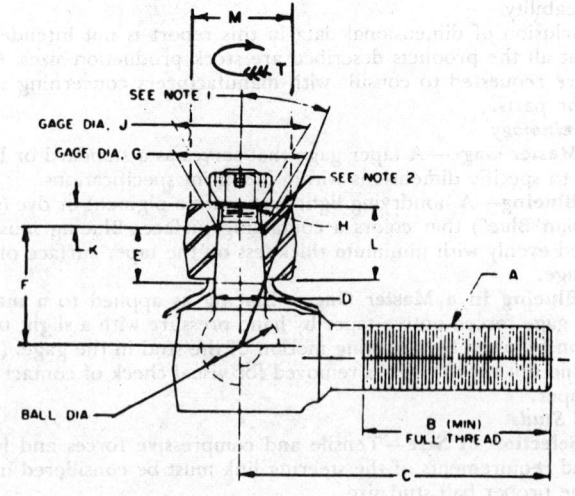

FIG. 5—TYPICAL BALL STUD SOCKET ASSEMBLY WITH EXTERNAL THREADED STEM

NOTES:
1. 0—Angular travel as required for a specific application.
2. Nut surface on boss must be sufficiently square with tapered hole to prevent excessive stress on threaded end.

TABLE 1A—STEERING-BALL STUD SOCKET ASSEMBLY DIMENSIONS[a] (DIMENSIONS IN UNITED STATES CUSTOMARY UNITS) SEE FIG. 5.

Nominal Ball Dia	Socket Thread		Socket Length C	Ball Stud					Attachment Arm				Stud Nut
	Size A	Length B		Shank Dia D	Taper Length E	Gage Dia Location F	Gage Dia G	Thread Size H	Gage Dia J	Draw-in Clearance K[b]	Thickness L	Boss Dia M[c]	Tightening Torque T[d]
in	in	in	in	in	in	in	in	in	in	in	in	in	lb-ft
5/8	1/2-20	1.74	2.62	0.469	0.41	1.06	0.418	3/8-24	0.402	0.128	0.50	1.25	15-30
3/4	9/16-18	1.94	2.94	0.547	0.45	1.20	0.490	7/16-20	0.473	0.136	0.56	1.38	30-45
7/8	11/16-18	2.06	3.12	0.625	0.52	1.36	0.560	1/2-20	0.543	0.136	0.62	1.50	35-55
1	11/16-18	2.21	3.38	0.703	0.72	1.62	0.613	9/16-18	0.590	0.184	0.88	1.75	55-80
1 1/8	7/8-18	2.75	3.88	0.781	0.84	1.88	0.675	5/8-18	0.652	0.184	1.00	1.88	80-110
1 1/4	1-16	3.19	4.25	0.875	0.94	2.03	0.758	5/8-18	0.731	0.216	1.12	12.00	80-110
1 1/2	1 1/8-12	3.75	4.88	1.031	1.12	2.44	0.890	3/4-16	0.863	0.216	1.31	2.25	100-140
1 3/4	1 1/4-12	—	5.75	1.250	1.28	2.81	1.074	7/8-14	1.043	0.248	1.50	2.75	120-170
2	—	—	—	1.350	1.50	3.22	1.166	1-12	1.131	0.280	1.75	3.00	140-220
2 1/4	—	—	—	1.510	1.78	3.72	1.285	1 1/8-12	1.250	0.280	2.00	3.25	180-270
2 1/2	—	—	—	1.700	2.06	4.34	1.441	1 1/4-12	1.406	0.280	2.25	3.50	230-320

[a] These dimensions may be varied as required for specific applications. See General Specification.
[b] Before tightening nut.
[c] The boss diameter or "hoop size" was determined by using the recommended nut tightening torque to determine "hoop size" at recommended thickness "L" with the stress level below the yield strength of medium carbon steel forgings.
[d] Ranges of tightening torque for 1:8 taper in medium carbon or alloy steel forgings. For other materials or tapers, these torque values must be adjusted. The torque values recommended are empirical values determined by combined experience of SAE Ball Stud and Tie Rod Socket Committee members.

TABLE 1B—STEERING-BALL STUD SOCKET ASSEMBLY DIMENSIONS (DIMENSIONS IN SI UNITS) SEE FIG. 5.

Nominal Ball Dia[a]	Socket Thread		Socket Length C	Ball Stud					Attachment Arm				Stud Nut
	Size A	Length B		Shank Dia D	Taper Length E	Gage Dia Location F	Gage Dia G	Thread Size H	Gage Dia J	Draw in Clearance K[b]N	Thickness L	Boss Dia M[c]	Tightening Torque T[d]
mm	mm	mm	mm	mm	mm	mm	mm	mm	mm	mm	mm	mm	N.m
16	M12×1.25	45	65	12	10.4	26	10.62	M10×1.25	10.21	3.3	13	32	24-40
20	M14×1.5	50	75	14	11.4	30	12.45	M12×1.25	12.01	3.5	14	35	40-60
22	M16×1.5	50	80	16	13.2	34	14.22	M12×1.25	13.79	3.5	16	38	40-60
25	M16×1.5	55	85	18	18.3	41	15.57	M14×1.5	14.99	4.7	22	44	75-110
28	M24×1.5	70	100	20	21.3	48	17.15	M16×1.5	16.56	4.7	25	48	110-150
32	M24×1.5	80	110	22	23.9	50	19.25	M16×1.5	18.57	5.5	28	50	110-150
38	M30×1.5	95	125	26	28.5	62	22.61	M20×1.5	21.92	5.5	33	58	140-190
44	M30×1.5	—	150	32	32.5	72	27.28	M24×1.5	26.49	6.3	38	70	180-230
50	—	—	—	34	38.1	82	29.62	M24×1.5	28.73	7.1	44	76	160-230
58	—	—	—	38	45.2	95	32.64	M30×1.5	31.75	7.1	50	82	240-370
64	—	—	—	44	52.3	110	36.60	M30×1.5	35.71	7.1	57	88	240-370

[a] These dimensions may be varied as required for specific applications. See General Specification.
[b] Before tightening nut.
[c] The boss diameter or "hoop size" was determined by using the recommended nut tightening torque to determine "hoop size" at recommended thickness "L" with the stress level below the yield strength of medium carbon steel forgings.
[d] Ranges of tightening torque for one unit per eight units of length taper in medium carbon or alloy steel forgings. For other materials or tapers, these torque values must be adjusted. The torque values recommended are empirical values determined by combined experience of SAE Ball Stud and Tie Rod Socket Committee members.

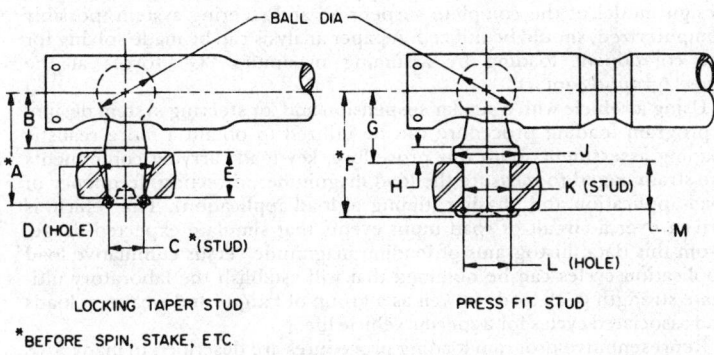

FIG. 6—TYPICAL TUBULAR ("HORIZONTAL TYPE") SOCKET ASSEMBLY FOR RIVETED, SPUN, STAKED, WELDED, OR THREADED TAPER STUD ATTACHMENT. (FOR THREADED TAPER STUDS, USE TABLE 1 DIMENSIONS.)

depends on application. The sockets may have partial spherical surfaces under the dust seal area. Dust seals may be of the sliding type or boot seal type (which are attached to the socket). For limited angularity sockets, as may be used for some truck applications, the dust seal may be of dense resilient material to preload the socket against the attachment boss so as to restrict rocking motion of the steering links about the link center lines.

When the ball stud is drawn into the taper hole (as listed under *Attachment* for ball studs, paragraph 3.4), the dust seal is preloaded to the proper amount to seal the socket against contamination. The socket and dust seal combination designed for the vehicle should not be modified in any way. The proper tools and extreme caution must be used when removing the socket for servicing attaching parts to prevent damaging the dust seal and ball stud threaded end.

4.5 Threads—Threads are Unified Class 2A of the size indicated in Table 1. Ends of threaded features should be chamfered approximately 0.02 in below the minor diameter; the length of chamfer to be 1/2 to 1 1/2 threads.

Lead error and other dimensional limits should be controlled. A reference for control which may be used is: American National Standard

ANSI B1.3M "Screw Thread Gaging Systems for Dimensional Acceptability."

In application of sockets with threaded stems, the length of engagement of stem threads with the attaching rod or sleeve depends on design and may vary, but should be approximately 2½ times the thread diameter. Under extremes of adjustment, where the application permits, engagement may be as low as 1½ times the thread diameter.

4.6 Lubrication—Lubrication fittings, if required, may be placed at any convenient location on the periphery or face of the socket provided the location does not create forging processing, assembly or functional complications.

4.7 Figures and Tables—The included Figures 4, 5, and 6 are intended to illustrate descriptions and table dimensions.

TABLE 2A—TYPICAL DIMENSIONS FOR TUBULAR SOCKET STUDS (DIMENSIONS IN UNITED STATES CUSTOMARY UNITS) SEE FIG. 6.

Nominal Ball Dia	Locking Taper Studs					Press Fit Studs						
	A	B	C	D	E	F	G	H	J	K	L	M
in	in	in	in	in	in	in	in	in	in	in	in	in
5/8	1.31	0.68	0.387	0.402	0.50	—	—	—	—	—	—	—
3/4	1.47	0.75	0.457	0.437	0.56	1.49	0.87	0.18	0.75	0.598	0.592	—
7/8	1.62	0.84	0.527	0.543	0.62	1.78	1.03	0.28	0.88	0.693	0.637	0.50
1	1.84	0.91	0.585	0.601	0.78	2.06	1.12	0.28	1.00	0.770	0.764	0.62
1 1/8	2.14	1.03	0.642	0.660	0.94	2.15	1.21	0.28	1.12	0.839	0.833	0.81
1 1/4	2.34	1.09	0.719	0.739	1.06	—	—	—	—	—	—	—
1 1/2	2.72	1.31	0.855	0.879	1.19	—	—	—	—	—	—	—
1 3/4	3.16	1.53	1.031	1.059	1.38	—	—	—	—	—	—	—
2	3.50	1.72	1.131	1.162	1.50	—	—	—	—	—	—	—
2 1/4	4.02	1.94	1.248	1.281	1.75	—	—	—	—	—	—	—
2 1/2	4.69	2.28	1.399	1.437	2.00	—	—	—	—	—	—	—

TABLE 2B—TYPICAL DIMENSIONS FOR TUBULAR SOCKET STUDS (DIMENSIONS IN SI UNITS) SEE FIG. 6.

Nominal Ball Dia	Locking Taper Studs					Press Fit Studs						
	A	B	C	D	E	F	G	H	J	K	L	M
mm	mm	mm	mm	mm	mm	mm	mm	mm	mm	mm	mm	mm
16	33	17	10.0	10.4	13	—	—	—	—	—	—	—
19	37	19	12.0	12.4	14	38.0	22.0	5	19	15.15	15.00	12.8
22	41	21	13.5	14.0	16	45.2	26.0	7	22	17.60	17.45	16.0
25	47	23	15.0	15.4	20	52.3	28.5	7	25	19.65	19.50	20.6
28	54	26	16.5	17.0	24	54.5	30.5	7	28	21.35	21.20	20.8
32	59	28	18.0	18.5	27	—	—	—	—	—	—	—
38	70	33	22.0	22.6	30	—	—	—	—	—	—	—
44	80	39	26.0	26.7	35	—	—	—	—	—	—	—
50	90	44	29.0	29.8	38	—	—	—	—	—	—	—
56	100	49	32.0	32.8	45	—	—	—	—	—	—	—
64	120	58	35.5	36.5	50	—	—	—	—	—	—	—

ϕ BALL STUD AND SOCKET ASSEMBLY—TEST PROCEDURES—SAE J193 FEB87

SAE Recommended Practice

Report of the Ball Stud and Tie Rod Socket Committee, approved August 1970 and completely revised by the Steering and Suspension Ball Stud and Socket Committee February 1987.

1. Scope—The test procedures describe a method to laboratory test suspension and steering system ball stud and/or socket assemblies for functional characteristics. This procedure is an extension of SAE J491b recommended practice on dimensional recommendations for ball studs towards a vehicle application. The tests are conducted either on ball studs individually or on complete integral assemblies representing the application.

2. Objective—To provide a uniform method of testing ball studs and ball stud and socket assemblies to ensure that the parts will meet functional requirements of the application.

3. Test Procedures—The test procedures for suspension and steering components with few exceptions can be similar because all ball stud and socket assemblies are subject to axial, lateral and longitudinal forces, differing only in the direction and magnitude of loading depending on the application.

The test procedures cover the following characteristics:
5.1 Ball Stud
5.1.1 Ball Stud Impact Strength
5.1.2 Ball Stud Yield
5.1.3 Ball Stud Tensile Load
5.2 Ball Stud and Socket
5.2.1 Ball Stud to Socket Rotating and Oscillating Torque
5.2.2 Ball Stud to Socket Axial End Movement
5.2.3 Ball Stud to Socket Cam-Out Strength
5.2.4 Ball Stud and Socket Assembly Fatigue and Heat Test
5.2.5 Ball Stud and Socket Pull-Out and Push-Out Strength
5.2.6 Ball Stud and Socket Angularity

4. Loading and Cycle Life—The loading used in the test procedures should be as representative as possible in magnitude and direction with loads encountered in the design application. Recommended cycle life is provided where applicable with each procedure.

To determine preliminary loading magnitude and direction, a layout design model of the complete suspension and steering system, possibly computerized, should be utilized. A paper analysis can be made solving for the component loading by assuming maximum "G" forces at the wheel/ground contact.

Using a vehicle with a similar suspension and/or steering system design, a program loading procedure can be utilized to obtain a more realistic loading assessment. With this procedure, key load carrying components are strain gaged to measure the load magnitudes, direction, frequency of load application and phasing (timing of load application). The vehicle is driven over a circuit of road input events that simulate expected usage. From this data, histograms of loading magnitude versus cumulative load application cycles can be obtained that will establish the laboratory ultimate strength peak loads as well as a group of fatigue and wear test loads and associated cycles for a specific vehicle life.

Representative program loading procedures are described in many SAE reports (SAE Report #660102, "Simulation of Field Loading in Fatigue Testing").

5. Objectives and Test Procedures
5.1 Ball Stud—Tests conducted on individual ball studs.
5.1.1 BALL STUD IMPACT STRENGTH
5.1.1.1 *Objective*—To determine the impact strength of the ball stud.
5.1.1.2 *Procedure*—The test is applicable to either suspension or steer-

ing system studs. Mount the stud in a rigid fixture as shown in Fig. 1. Lock stud in fixture by torquing the retaining nut to design specifications.

Apply an impact load to exceed the expected impact load in the vehicle application.

The stud must not fail by brittle fracture. Bending deflection must be 10 deg minimum.

Increase the impact load incrementally until a separation occurs to determine the load capability. Only one impact per stud is permissible.

5.1.2 BALL STUD YIELD

5.1.2.1 *Objective*—To determine at what load condition the ball stud will take a permanent set without fracture.

5.1.2.2 *Procedure*—The test is applicable to either suspension or steering system studs.

The test fixture is shown in Fig. 2. Grind a small flat on the head of the stud for accuracy of reading, to receive the dial indicator or other measuring device.

Install the ball stud in the fixture with the mating taper hole in such a manner that a load can be applied to the stud at right angles to the stud centerline and opposite the flat ground on the stud. Lock stud in fixture by torquing the retaining nut to design specification.

Preload the stud. Set dial indicator or other measuring device to zero. Take deflection and set readings in desired increments to permanent set range.

Stud yield load is equal to load required to permanently set stud without surface cracks or failure and should be used to select stud application.

5.1.3 BALL STUD TENSILE LOAD

5.1.3.1 *Objective*—To determine the tensile load capability of the ball stud.

5.1.3.2 *Procedure*—The test is applicable to either suspension or steering system studies depending on the intended application, but generally would be appropriate for a suspension ball stud where the predominant loading would be in a tensile direction.

Mount the stud in a load/deflection testing machine as shown in Fig. 3. *Caution:* Use care to prevent eccentric loading. A typical tensile load application rate is 5 mm/min. Record load and mode of fracture for each sample tested.

5.2 Ball Stud and Socket—Tests conducted on complete ball stud and socket assemblies.

5.2.1 BALL STUD TO SOCKET ROTATING AND OSCILLATING TORQUE

5.2.1.1 *Objective*—To ensure desired rotating and oscillating torque is obtained.

5.2.1.2 *Procedure*—The test is applicable to either suspension or steering system components. The assembly should be held in a manner to prevent addition of external clamping pressure which may affect torque readings.

FIG. 1—IMPACT TEST

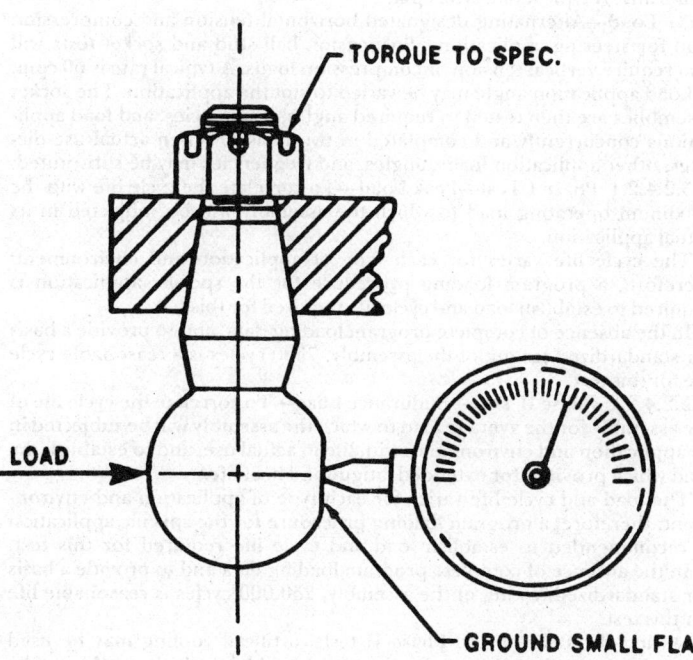

FIG. 2

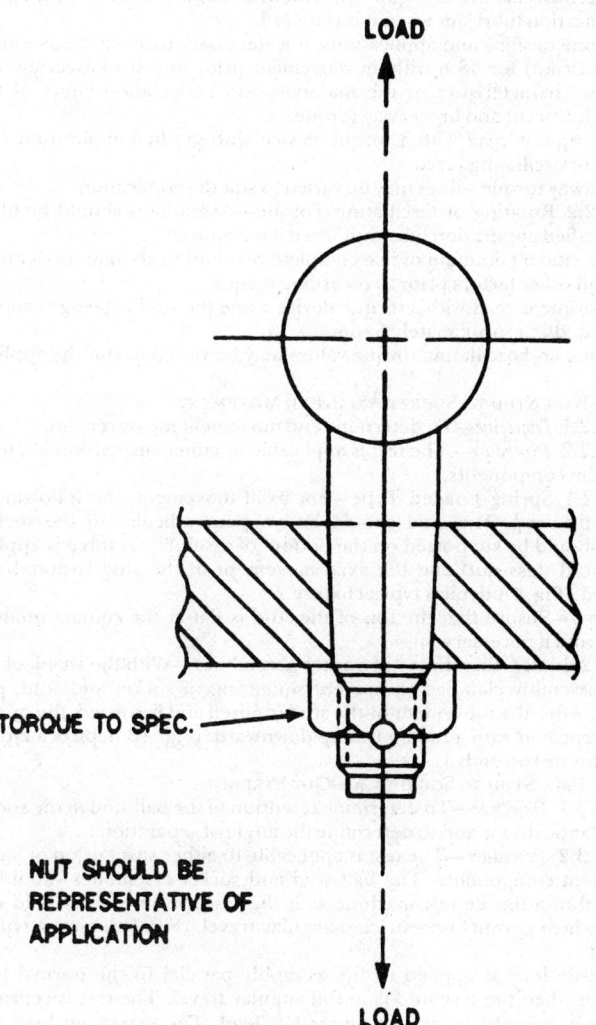

FIG. 3—TENSILE TEST

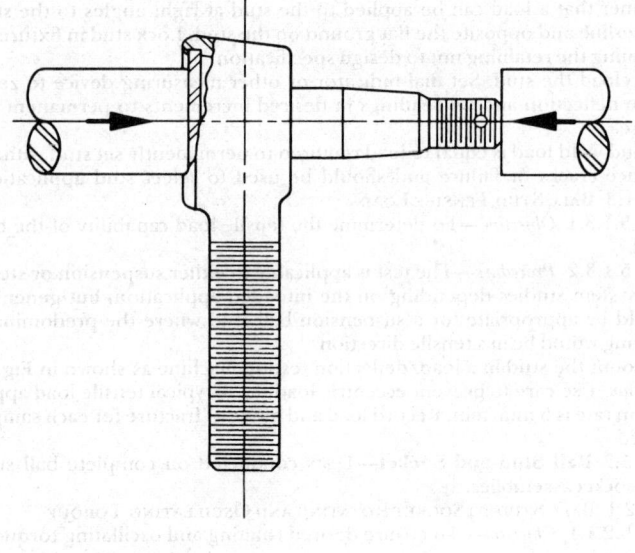

FIG. 4A

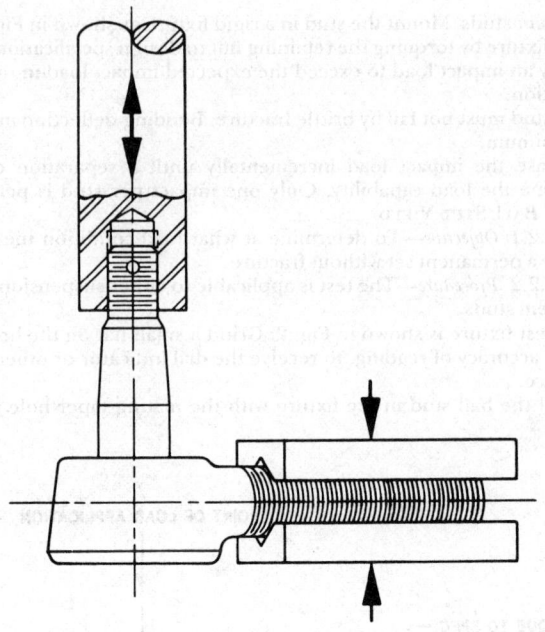

FIG. 4B

5.2.1.2.1 *Breakaway Torque*—Assemblies should be filled with specified application lubricant when it is required.

For some designs and applications, it is necessary to store the assembly (with lubricant) for 48 h without movement prior to test to ascertain the cold flow characteristics of the materials and congelation effect of the selected lubricant and breakaway torque.

The torque is read with a torque device with gradual application of a rotating or oscillating force.

Breakaway torque values may be varied to suit the application.

5.2.1.2.2 *Rotating or Oscillating Torque*—Assemblies should be filled with specified application lubricant when it is required.

Rotate stud a minimum of five complete revolutions to minimize congelation and other factors prior to recording torque.

The torque is read with a torque device while the stud is being revolved or oscillated at approximately 5 rpm.

Rotating and oscillating torque values may be varied to suit the application.

5.2.2 BALL STUD TO SOCKET AXIAL END MOVEMENT

5.2.2.1 *Objective*—To determine end movement measurement.

5.2.2.2 *Procedure*—The test is applicable to either suspension or steering system components.

5.2.2.2.1 *Spring Loaded Type*—For axial movement, the following is commonly used. The stud should be set perpendicular to the socket. Socket should be supported on the bottom of assembly. A force is applied to the stud (less nut) and the axial movement of the stud is noted and recorded. (Fig. 4A depicts typical fixture.)

NOTE—Ensure that the top of the stud is flat at the contact point of force (grind if necessary).

5.2.2.2.2 *All Other Types of Socket Assemblies*—With the shank of the socket assembly clamped to prevent squeezing of socket and stud, pull upward. After the movement of the stud is noted and recorded, the operation is repeated with a force pushing downward. (Fig. 4B depicts a typical fixture for tie rod ends.)

5.2.3 BALL STUD TO SOCKET CAM-OUT STRENGTH

5.2.3.1 *Objective*—To determine retention of the ball stud in the socket at angular positions and to determine the angle of separation.

5.2.3.2 *Procedure*—The test is applicable to either suspension or steering system components. The ball stud and socket assemblies should be mounted in a tensile test machine with the test specimen stud held in a fixture which permits unrestricted angular travel. (Fig. 5 depicts a typical fixture.)

A tensile load is applied to the assembly parallel to the normal load direction when the test stud is in full angular travel. The test is repeated with a new sample, using a compression load. The maximum load and angle induced prior to separating the stud from the socket is recorded.

5.2.4 BALL STUD AND SOCKET ASSEMBLY FATIGUE AND HEAT TEST

5.2.4.1 *Objective*—To determine fatigue and wear characteristics of ball stud and socket assemblies.

5.2.4.2 *Procedure*—This test is applicable to either suspension or steering system components. Use socket assemblies which have been tested according to paragraphs 5.2.1, 5.2.2 and 5.2.6 and found acceptable. Socket assemblies should be filled with recommended lubricant when required in the application. Socket assemblies should be installed, with seals when required, in a fixture by placing the taper shank in the mating tapered hole with the retaining nut torqued to design specification. For each type to be tested, the ball stud and socket assemblies may be modified to suit the test machine providing the modifications do not affect test results. Securely clamp the link in a manner to achieve the required motions. The following are typical motions which may be used. (Refer to Fig. 6 (A and B)).

(1) **Angular Oscillation**— ± 20 deg in a plan parallel to the link or 90 deg to axis of suspension control arm pivot centerline. A typical rate is 60 cpm.

(2) **Angular Rotation**— ± 40 deg measured about the ball stud shank centerline. A typical rate is 32 cpm.

(3) **Load**—Alternating designated horizontal tension and compression load for steering applications. Suspension ball stud and socket tests will also require vertical tension or compression loads. A typical rate is 60 cpm.

Load application angle may be varied to suit the application. The socket assemblies are then tested to required angles, frequencies, and load applications concurrently and completed in two phases. When actual use dictates, other application loads, angles, and frequencies may be substituted.

5.2.4.2.1 *Phase I Test—Peak Load*—To correlate the cycle life with the maximum operating load to which the assembly will be subjected in its actual application.

The cycle life varies for each type of application and environment; therefore, a program loading procedure for the specific application is required to establish load and cycle life required for this test.

In the absence of complete program loading data, and to provide a basis for standardized testing of the assembly, 7500 cycles is a reasonable cycle life for this test.

5.2.4.2.2 *Phase II Test—Endurance Load*—To correlate the cycle life of the assembly for the average load to which the assembly will be subjected in its application and environment, with life in actual use, and to establish the load which provides for extended fatigue and heat life.

The load and cycle life varies for each type of application and environment; therefore, a program loading procedure for the specific application is recommended to establish load and cycle life required for this test.

In the absence of complete program loading data and to provide a basis for standardized testing of the assembly, 250,000 cycles is reasonable life for this test.

During the Phase I and Phase II tests, artificial cooling may be used where deemed necessary to prevent heat build-up which would not be experienced in the application.

If the application of the ball stud and socket assembly includes environmental contamination, contaminants should be provided to correlate with these conditions in the test. This procedure will determine seal durability and effectiveness.

Typical test environments commonly used are:
(1) Ozone atmosphere
(2) Saline and dust 4.0 L of water, 50% saturated solution of common salt at 21-24°C, and 0.15 kg of SAE air cleaner test dust fine grade.
(3) Steam

Conditions to be examined at completion of test to determine adequacy for the application:
The rotating and oscillating torque condition (see Section 5.2.1).
The end movement (see Section 5.2.2).
The ball stud should be examined for any surface cracks, determined by a dye check or approved equivalent.
The ball stud should be examined for any local yielding, determined by a dimension check.
The internal components should be examined for damage.

5.2.5 BALL STUD AND SOCKET PULL-OUT AND PUSH-OUT STRENGTH

5.2.5.1 *Objective*—To determine the tension or compression loads which will separate the ball stud from the socket.

5.2.5.2 *Procedure*—This test is applicable to either suspension or steering system components. Assemble the ball stud and socket assembly in a load/deflection testing machine so that the load applied to the stud is perpendicular to the ball joint mounting surface.

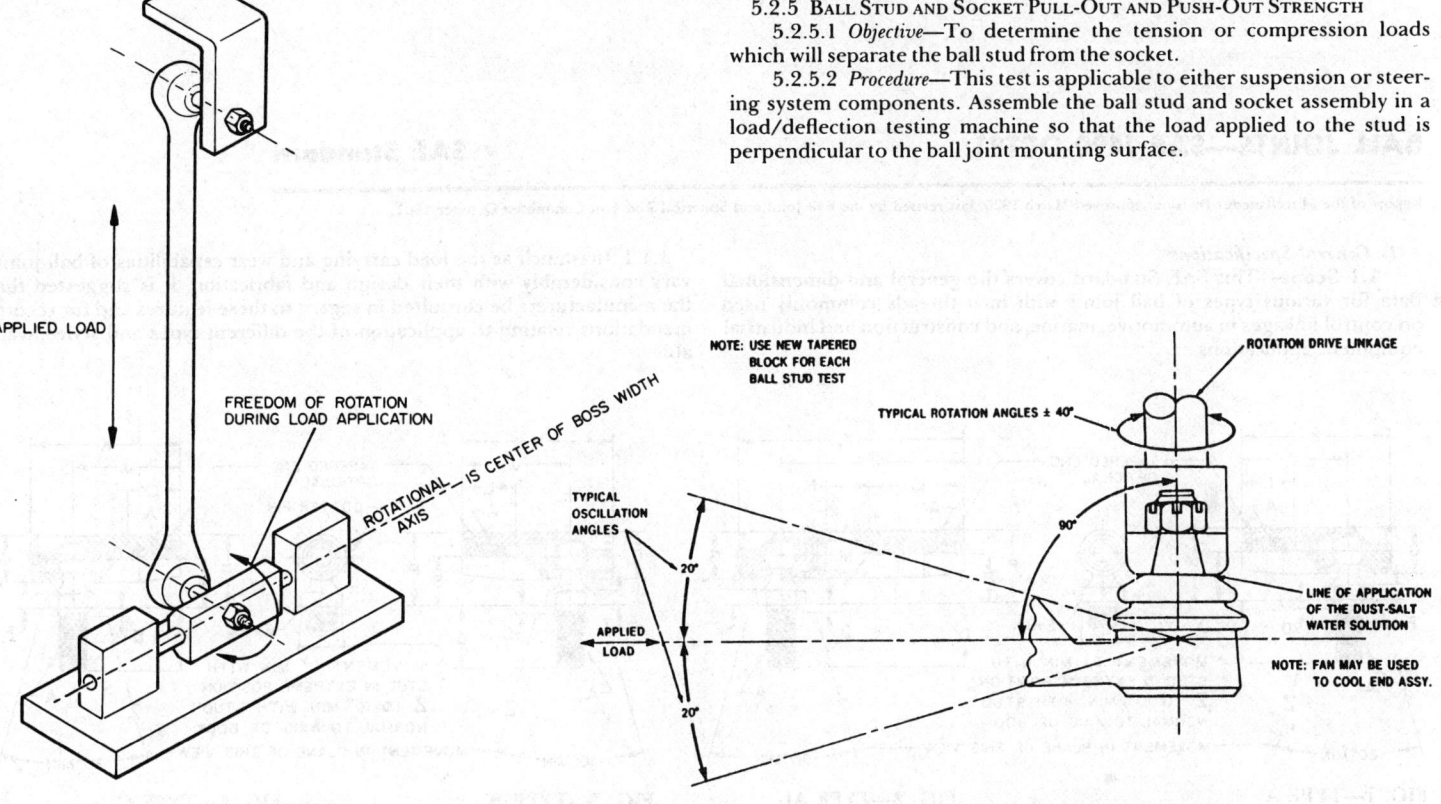

FIG. 5

FIG. 6A—TYPICAL STEERING LINKAGE BALL STUD & SOCKET ASSEMBLY

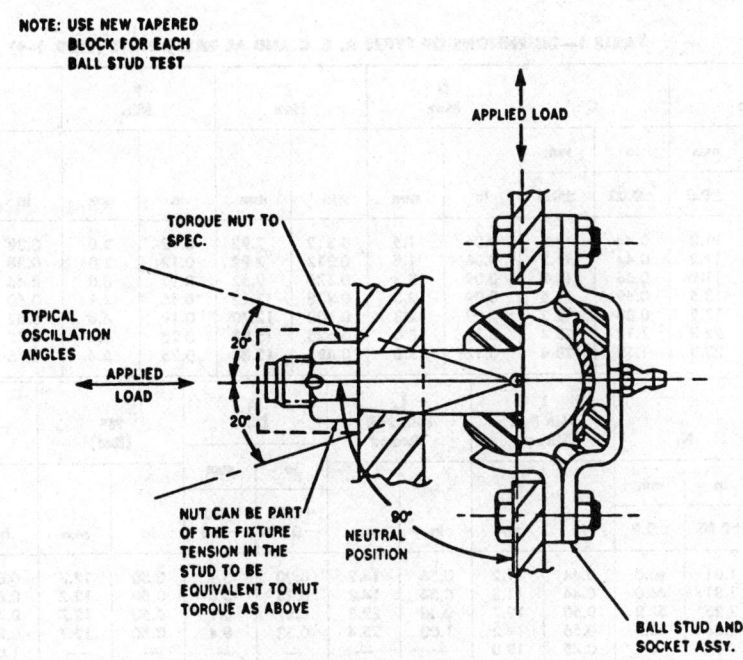

FIG. 6B—TYPICAL SUSPENSION BALL STUD & SOCKET ASSEMBLY

Measure and record the stud tensile load required to separate the stud from the socket assembly. A typical load application rate is 5 mm/min. Repeat the test with another sample using a compressive load.

The acceptance criteria for stud pull or push-out should be established from the calculated or measured loads based on the application.

5.2.6 BALL STUD AND SOCKET ANGULARITY

5.2.6.1 *Objective*—To determine if the socket throat is capable of providing the required stud angularity.

5.2.6.2 *Procedure*—This test is applicable to either suspension or steering system components. Assemble the ball stud and socket assembly into a rigid fixture simulating the assembly into a suspension arm or steering linkage.

Measure and record stud travel angularity both along the socket throat major axis and across the minor axis.

The acceptance criteria is based on the angularity requirements in the specific application adjusted to take into account the effect of dimensional variations and suspension/linkage compliance.

BALL JOINTS—SAE J490 OCT81 — SAE Standard

Report of the Miscellaneous Division, approved March 1920, last revised by the Ball Joint and Spherical Rod End Committee October 1981.

1. General Specifications

1.1 Scope—This SAE Standard covers the general and dimensional data for various types of ball joints with inch threads commonly used on control linkages in automotive, marine, and construction and industrial equipment applications.

1.1.1 Inasmuch as the load carrying and wear capabilities of ball joints vary considerably with their design and fabrication, it is suggested that the manufacturers be consulted in regard to these features and for recommendations relating to application of the different types and styles available.

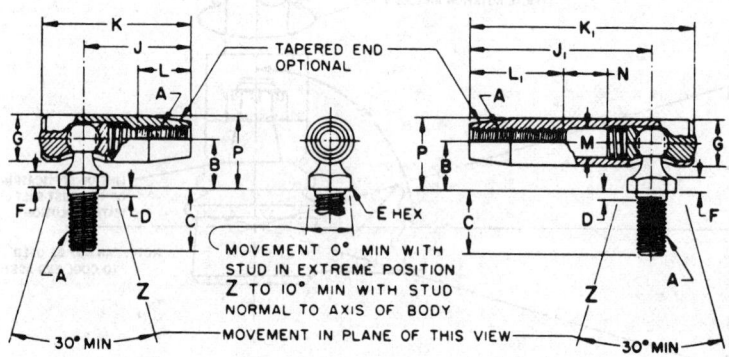

FIG. 1—TYPE A
CRIMPED END PLUG WITH SPRING CONSTRUCTION

FIG. 2—TYPE AL

FIG. 3—TYPE B
THREADED END PLUG WITH SPRING CONSTRUCTION

FIG. 4—TYPE C
THREADED END PLUG WITHOUT SPRING CONSTRUCTION

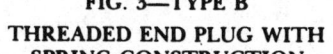

TABLE 1—DIMENSIONS OF TYPES A, B, C AND AL BALL JOINTS (FIGS. 1–4)

Nominal Ball Joint Size and Thread Diameter, A, in	Threads per in	B		C		D Max		E Hex		F Min		G Dia		J		J_1	
		in ±0.02	mm ±0.5	in ±0.02	mm ±0.5	in	mm	in	mm	in	mm	in	mm	in ±0.03	mm ±0.8	in ±0.03	mm ±0.8
No. 10 0.190	32	0.44	11.2	0.44	11.2	0.06	1.5	0.312	7.92	0.12	3.0	0.38	9.7	0.88	22.3	1.50	38.1
No. 12 0.216	32	0.44	11.2	0.44	11.2	0.06	1.5	0.312	7.92	0.12	3.0	0.38	9.7	0.88	22.3	1.50	38.1
1/4 0.250	28	0.47	11.9	0.56	14.2	0.09	2.3	0.375	9.52	0.12	3.0	0.44	11.2	0.97	24.6	1.81	46.0
5/16 0.3125	24	0.53	13.5	0.69	17.5	0.09	2.3	0.438	11.12	0.16	4.1	0.50	12.7	1.12	28.4	1.94	49.3
3/8 0.375	24	0.69	17.5	0.88	22.3	0.09	2.3	0.500	12.70	0.19	4.8	0.62	15.8	1.38	35.0	—	—
7/16 0.4375	20	0.88	22.3	1.12	28.4	0.12	3.0	0.625	15.88	0.25	6.4	0.75	19.0	1.94	49.3	—	—
1/2 0.500	20	0.88	22.3	1.12	28.4	0.12	3.0	0.625	15.88	0.25	6.4	0.75	19.0	1.94	49.3	—	—

Nominal Ball Joint Size and Thread Diameter, A, in	K		K_1		L Min Full Thread		L_1 Min Full Thread		M Dia		N[a] (Ref)		P[a] Max (Ref)		Stud Ball Diameter (Ref)[a]			
															Max		Min	
	in ±0.03	mm ±0.8	in ±0.03	mm ±0.8	in	mm	in	mm	in +0.01/−0.00	mm +0.3/−0.0	in	mm	in	mm	in	mm	in	mm
No. 10 0.190	1.25	31.8	1.81	46.0	0.44	11.2	0.56	14.2	0.20	5.1	0.50	12.7	0.65	16.5	0.255	6.48	0.250	6.35
No. 12 0.216	1.25	31.8	1.81	46.0	0.44	11.2	0.56	14.2	0.23	5.8	0.50	12.7	0.65	16.5	0.255	6.48	0.250	6.35
1/4 0.250	1.38	35.0	2.25	57.2	0.50	12.7	0.88	22.3	0.27	6.9	0.50	12.7	0.72	18.3	0.305	7.75	0.300	7.62
5/16 0.3125	1.56	39.6	2.38	60.5	0.56	14.2	1.00	25.4	0.33	8.4	0.50	12.7	0.81	20.6	0.350	8.89	0.345	8.76
3/8 0.375	1.94	49.3	—	—	0.75	19.0	—	—	—	—	—	—	1.03	26.2	0.424	10.77	0.419	10.64
7/16 0.4375	2.62	66.5	—	—	1.00	25.4	—	—	—	—	—	—	1.28	32.5	0.555	14.10	0.550	13.97
1/2 0.500	2.62	66.5	—	—	1.00	25.4	—	—	—	—	—	—	1.28	32.5	0.555	14.10	0.550	13.97

[a] These dimensions are given for design purposes only and are not intended for inspection.

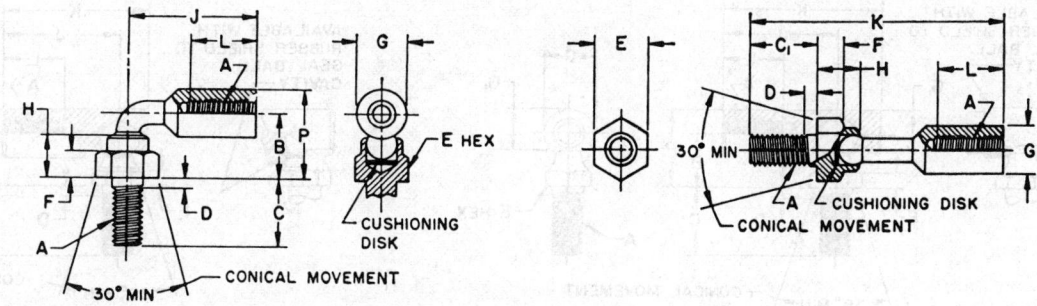

*TYPES D AND DS ARE NOT RECOMMENDED FOR APPLICATIONS INVOLVING TENSION OR SEVERE VIBRATION.

FIG. 5—TYPE D* FIG. 6—TYPE DS*

CUSHIONED TWO PIECE CONSTRUCTION

TABLE 2—DIMENSIONS OF TYPES D AND DS BALL JOINTS (FIGS. 5 AND 6)

Nominal Ball Joint Size and Thread Diameter, A, in	Thds per in	B in ±0.03	B mm ±0.8	C in ±0.02	C mm ±0.5	C_1 in ±0.02	C_1 mm ±0.5	D Max in	D Max mm	E Hex in	E Hex mm	F Min in	F Min mm	G Dia in	G Dia mm	H in ±0.03	H mm ±0.8	J in ±0.03	J mm ±0.8	K in ±0.03	K mm ±0.8	L Min Full Thread in	L Min Full Thread mm	P Max (Ref) in	P Max (Ref) mm
No. 10 0.190	32	0.53	13.5	0.44	11.2	0.56	14.2	0.06	1.5	0.375	9.52	0.19	4.8	0.28	7.1	0.33	8.4	1.03	26.2	2.03	51.6	0.50	12.7	0.70	17.8
No. 10 0.190	32	0.53	13.5	0.44	11.2	—	—	0.06	1.5	0.375	9.52	0.19	4.8	0.28	7.1	0.33	8.4	0.78	19.8	—	—	0.38	9.7	0.70	17.8
No. 12 0.216	24	0.53	13.5	0.56	14.2	0.56	14.2	0.06	1.5	0.375	9.52	0.19	4.8	0.28	7.1	0.33	8.4	1.03	26.2	2.03	51.6	0.50	12.7	0.70	17.8
No. 12 0.216	32	0.53	13.5	0.56	14.2	0.56	14.2	0.06	1.5	0.375	9.52	0.19	4.8	0.28	7.1	0.33	8.4	1.03	26.2	2.03	51.6	0.50	12.7	0.70	17.8
1/4 0.250	28	0.56	14.2	0.56	14.2	0.56	14.2	0.06	1.5	0.438	11.12	0.19	4.8	0.31	7.9	0.35	8.9	1.06	26.9	2.09	53.1	0.56	14.2	0.75	19.0
5/16 0.3125	24	0.69	17.5	0.69	17.5	0.69	17.5	0.09	2.3	0.562	14.28	0.28	7.1	0.44	11.2	0.45	11.4	1.31	33.3	2.63	66.8	0.69	17.5	0.94	23.9

[a] These dimensions are given for design purposes only and are not intended for inspection.

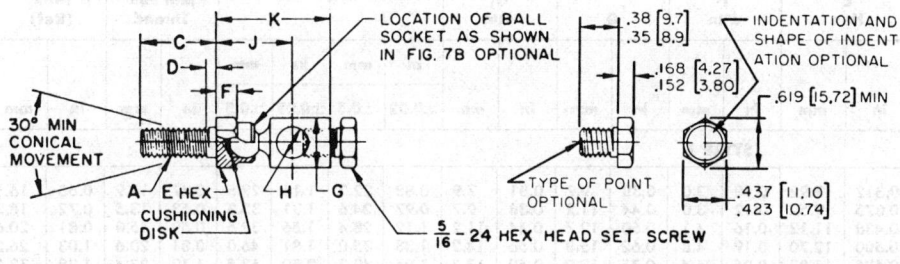

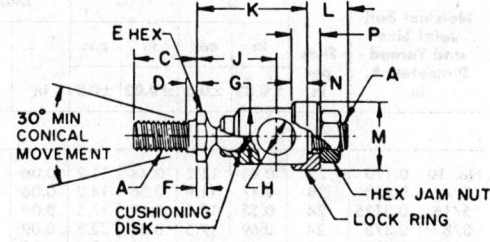

FIG. 7A—STYLE 1 FIG. 7B—STYLE 2

FIG. 7—TYPE DC

TABLE 3—DIMENSIONS OF TYPE DC BALL JOINTS (FIGS. 7A AND 7B)

Nominal Ball Joint Size and Thread Diameter, A, in	Thds per in	C in ±0.02	C mm ±0.5	D Max in	D Max mm	E Hex in	E Hex mm	F Min in	F Min mm	G Dia in	G Dia mm	H Dia in ±0.005	H Dia mm ±0.13	J in ±0.03	J mm ±0.8	K in ±0.03	K mm ±0.8	L in ±0.02	L mm ±0.5	M Dia in	M Dia mm	N in ±0.01	N mm ±0.3	P in ±0.005	P mm ±0.13
STYLE 1																									
No. 10 0.190	32	0.31	7.9	0.06	1.5	0.438	11.12	0.19	4.8	0.50	12.7	0.323	8.20	0.75	19.0	1.12	28.4	—	—	—	—	—	—	—	—
1/4 0.250	20	0.44	11.2	0.09	2.3	0.438	11.12	0.19	4.8	0.50	12.7	0.323	8.20	0.75	19.0	1.12	28.4	—	—	—	—	—	—	—	—
1/4 0.250	20	0.56	14.2	0.09	2.3	0.438	11.12	0.19	4.8	0.50	12.7	0.323	8.20	0.75	19.0	1.12	28.4	—	—	—	—	—	—	—	—
1/4 0.250	28	0.44	11.2	0.09	2.3	0.438	11.12	0.19	4.8	0.50	12.7	0.323	8.20	0.75	19.0	1.12	28.4	—	—	—	—	—	—	—	—
1/4 0.250	28	0.56	14.2	0.09	2.3	0.438	11.12	0.19	4.8	0.50	12.7	0.323	8.20	0.75	19.0	1.12	28.4	—	—	—	—	—	—	—	—
5/16 0.3125	24	0.62	15.8	0.09	2.3	0.438	11.12	0.19	4.8	0.50	12.7	0.323	8.20	0.75	19.0	1.12	28.4	—	—	—	—	—	—	—	—
5/16 0.3125	24	0.75	19.0	0.09	2.3	0.438	11.12	0.19	4.8	0.50	12.7	0.323	8.20	0.75	19.0	1.12	28.4	—	—	—	—	—	—	—	—
3/8 0.375	24	0.62	15.8	0.09	2.3	0.438	11.12	0.19	4.8	0.50	12.7	0.323	8.20	0.75	19.0	1.12	28.4	—	—	—	—	—	—	—	—
STYLE 2																									
No. 10 0.190	32	0.50	12.7	0.06	1.5	0.375	9.52	0.09	2.3	0.44	11.2	0.197	5.00	0.62	15.8	0.78	19.8	0.40	10.2	0.56	14.2	0.12	3.0	0.250	6.35
1/4 0.250	20	0.44	11.2	0.09	2.3	0.438	11.12	0.09	2.3	0.50	12.7	0.328	8.33	0.78	19.8	1.02	25.9	0.34	8.6	0.62	15.8	0.12	3.0	0.250	6.35
1/4 0.250	28	0.44	11.2	0.06	1.5	0.438	11.12	0.09	2.3	0.50	12.7	0.328	8.33	0.78	19.8	1.02	25.9	0.34	8.6	0.62	15.8	0.12	3.0	0.250	6.35
5/16 0.3125	24	0.62	15.8	0.09	2.3	0.438	11.12	0.11	2.8	0.56	14.2	0.380	9.65	0.75	19.0	1.03	26.2	0.53	13.5	0.75	19.0	0.19	4.8	0.344	8.74

17.10

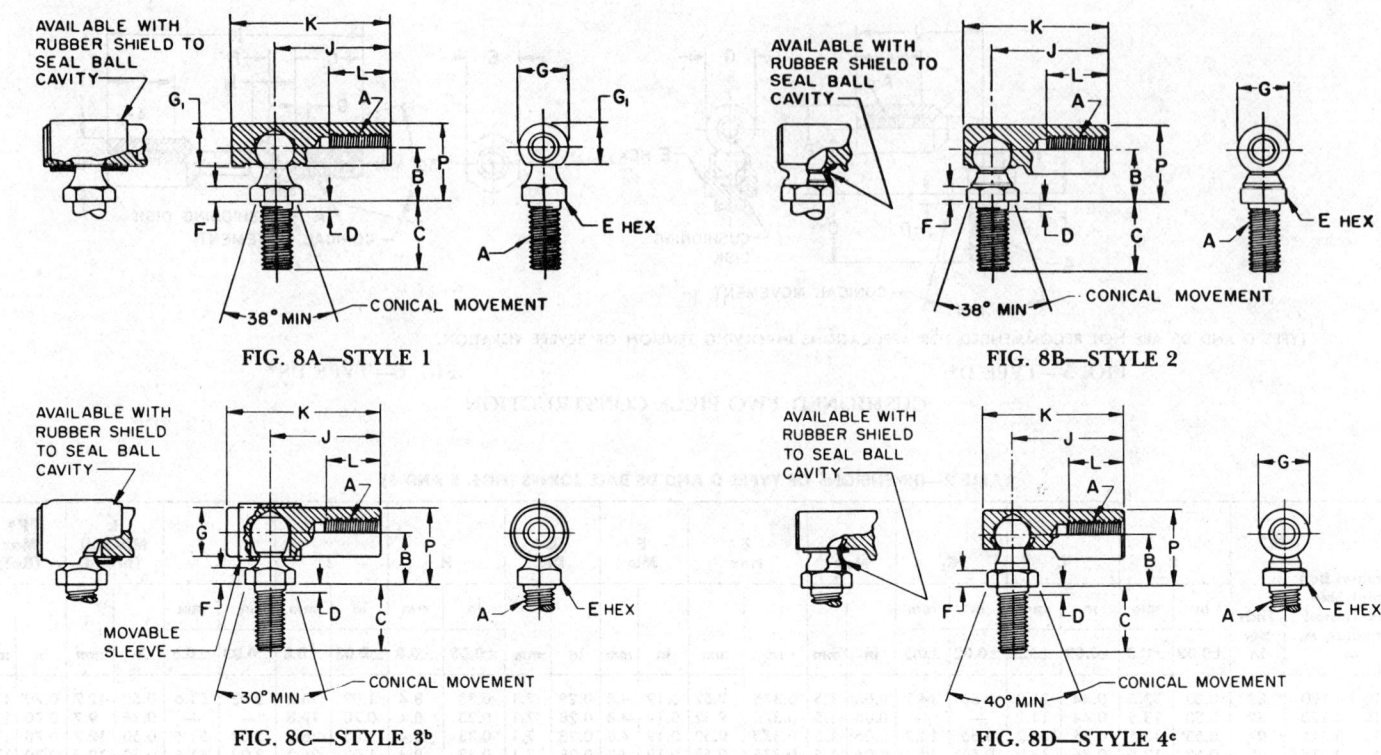

FIG. 8A—STYLE 1
FIG. 8B—STYLE 2
FIG. 8C—STYLE 3[b]
FIG. 8D—STYLE 4[c]

FIG. 8—TYPE G NONDETACHABLE CONSTRUCTION WITHOUT SPRING

TABLE 4—DIMENSIONS OF TYPE G BALL JOINTS (FIGS. 8A-8D)

Nominal Ball Joint Size and Thread Diameter, A, in		Thds per in	B		C		D Max		E Hex		F Min		G		G₁ Min		J		K		L Min Full Thread		P[a] Max (Ref)	
			in ±0.02	mm ±0.5	in ±0.02	mm ±0.5	in	mm	in	mm	in	mm	in	mm	in	mm	in ±0.02	mm ±0.5	in ±0.02	mm ±0.5	in	mm	in	mm
STYLE 1																								
No. 10	0.190	32	0.44	11.2	0.44	11.2	0.06	1.5	0.312	7.92	0.12	3.0	0.38	9.7	0.31	7.9	0.88	22.3	1.16	29.5	0.47	11.9	0.65	16.5
1/4	0.250	28	0.47	11.9	0.56	14.2	0.06	1.5	0.375	9.52	0.12	3.0	0.44	11.2	0.38	9.7	0.97	24.6	1.31	33.3	0.53	13.5	0.72	18.3
5/16	0.3125	24	0.53	13.5	0.69	17.5	0.09	2.3	0.438	11.12	0.16	4.1	0.50	12.7	0.44	11.2	1.12	28.4	1.56	39.6	0.59	15.0	0.81	20.6
3/8	0.375	24	0.69	17.5	0.88	22.3	0.09	2.3	0.500	12.70	0.19	4.8	0.62	15.8	0.56	14.2	1.38	35.0	1.81	46.0	0.81	20.6	1.03	26.2
7/16	0.4375	20	0.88	22.3	1.12	28.4	0.12	3.0	0.625	15.88	0.25	6.4	0.75	19.0	0.69	17.5	1.94	49.3	2.50	63.5	1.12	28.4	1.28	32.5
1/2	0.500	20	0.88	22.3	1.12	28.4	0.12	3.0	0.625	15.88	0.25	6.4	0.75	19.0	0.69	17.5	1.94	49.3	2.50	63.5	1.12	28.4	1.28	32.5
STYLE 2																								
No. 10	0.190	32	0.44	11.2	0.44	11.2	0.06	1.5	0.312	7.92	0.12	3.0	0.38	9.7	—	—	0.88	22.3	1.06	26.9	0.47	11.9	0.65	16.5
1/4	0.250	28	0.47	11.9	0.56	14.2	0.09	2.3	0.375	9.52	0.12	3.0	0.44	11.2	—	—	0.97	24.6	1.22	31.0	0.50	12.7	0.72	18.3
5/16	0.3125	24	0.53	13.5	0.69	17.5	0.09	2.3	0.438	11.12	0.16	4.1	0.50	12.7	—	—	1.12	28.4	1.41	35.8	0.56	14.2	0.81	20.6
3/8	0.375	24	0.69	17.5	0.88	22.3	0.09	2.3	0.500	12.70	0.19	4.8	0.62	15.8	—	—	1.38	35.0	1.69	42.9	0.75	19.0	1.03	26.2
7/16	0.4375	20	0.88	22.3	1.12	28.4	0.12	3.0	0.625	15.88	0.25	6.4	0.75	19.0	—	—	1.94	49.3	2.38	60.5	1.00	25.4	1.28	32.5
1/2	0.500	20	0.88	22.3	1.12	28.4	0.12	3.0	0.625	15.88	0.25	6.4	0.75	19.0	—	—	1.94	49.3	2.38	60.5	1.00	25.4	1.28	32.5
5/8	0.625	18	1.00	25.4	1.12	28.4	0.12	3.0	0.750	19.05	0.31	7.9	0.88	22.3	—	—	2.06	52.3	2.58	65.5	1.00	25.4	1.47	37.3
3/4	0.750	16	1.06	26.9	1.12	28.4	0.12	3.0	0.875	22.22	0.31	7.9	1.00	25.4	—	—	2.12	53.8	3.00	76.2	1.12	28.4	1.59	40.4
STYLE 3[b]																								
No. 10	0.190	32	0.44	11.2	0.44	11.2	0.06	1.5	0.312	7.92	0.12	3.0	0.38	9.7	—	—	0.88	22.3	1.16	29.5	0.47	11.9	0.65	16.5
1/4	0.250	28	0.47	11.9	0.56	14.2	0.06	1.5	0.375	9.52	0.12	3.0	0.44	11.2	—	—	0.97	24.6	1.31	33.3	0.53	13.5	0.72	18.3
5/16	0.3125	24	0.53	13.5	0.69	17.5	0.09	2.3	0.438	11.12	0.16	4.1	0.50	12.7	—	—	1.12	28.4	1.56	39.6	0.59	15.0	0.81	20.6
3/8	0.375	24	0.69	17.5	0.88	22.3	0.09	2.3	0.500	12.70	0.19	4.8	0.62	15.8	—	—	1.38	35.0	1.81	46.0	0.81	20.6	1.03	26.2
1/2	0.500	20	0.88	22.3	1.12	28.4	0.12	3.0	0.625	15.88	0.28	7.1	0.75	19.0	—	—	1.94	49.3	2.62	66.5	1.12	28.4	1.28	32.5
5/8	0.625	18	1.06	26.9	1.12	28.4	0.12	3.0	0.875	22.22	0.31	7.9	1.00	25.4	—	—	2.12	53.8	3.00	76.2	1.12	28.4	1.59	40.4
3/4	0.750	16	1.06	26.9	1.12	28.4	0.12	3.0	0.875	22.22	0.31	7.9	1.00	25.4	—	—	2.12	53.8	3.00	76.2	1.12	28.4	1.59	40.4
STYLE 4[c]																								
No. 10	0.190	32	0.47	11.9	0.44	11.2	0.06	1.5	0.375	9.52	0.12	3.0	0.44	11.2	—	—	0.97	24.6	1.22	31.0	0.44	11.2	0.72	18.3
1/4	0.250	28	0.47	11.9	0.56	14.2	0.09	2.3	0.375	9.52	0.12	3.0	0.44	11.2	—	—	0.97	24.6	1.22	31.0	0.50	12.7	0.72	18.3
5/16	0.3125	24	0.53	13.5	0.69	17.5	0.09	2.3	0.438	11.12	0.16	4.1	0.50	12.7	—	—	1.12	28.4	1.41	35.8	0.56	14.2	0.81	20.6
3/8	0.375	24	0.69	17.5	0.88	22.3	0.09	2.3	0.500	12.70	0.19	4.8	0.62	15.8	—	—	1.38	35.0	1.69	42.9	0.75	19.0	1.03	26.2
7/16	0.4375	20	0.88	22.3	1.12	28.4	0.12	3.0	0.625	15.88	0.25	6.4	0.75	19.0	—	—	1.94	49.3	2.38	60.5	1.00	25.4	1.28	32.5
1/2	0.500	20	0.88	22.3	1.12	28.4	0.12	3.0	0.625	15.88	0.25	6.4	0.75	19.0	—	—	1.94	49.3	2.38	60.5	1.00	25.4	1.28	32.5
5/8	0.625	18	1.00	25.4	1.12	28.4	0.12	3.0	0.750	19.05	0.31	7.9	0.88	22.3	—	—	2.06	52.3	2.58	65.5	1.00	25.4	1.47	37.3

[a] These dimensions are given for design purposes only and are not intended for inspection.
[b] Type G Style 3 ball joints are furnished with ball studs and ball cavities (ball stud only on 5/8 and 3/4 in sizes) hardened to assure longer wear.
[c] Type G, Style 4 ball joints in all sizes are furnished with both ball studs and ball sockets hardened to assure longer wear.

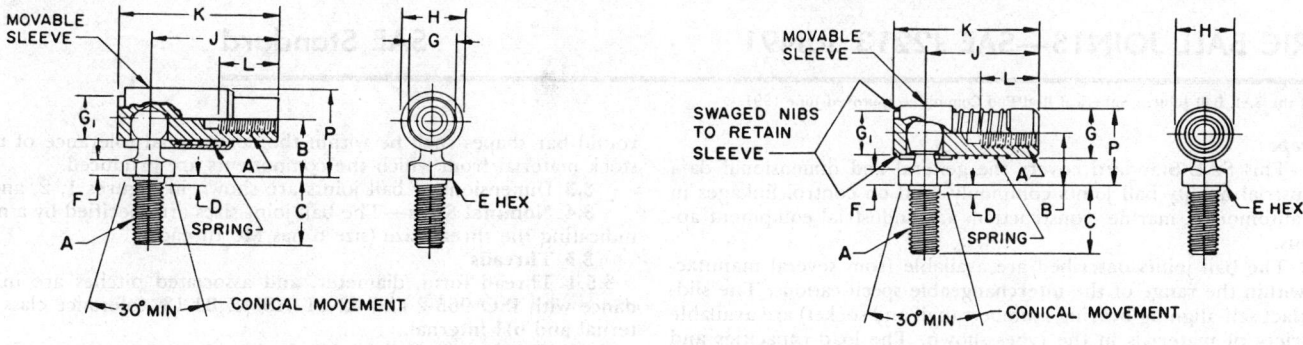

FIG. 9A—STYLE 1   FIG. 9B—STYLE 2

FIG. 9—TYPE S DETACHABLE CONSTRUCTION

TABLE 5—DIMENSIONS OF TYPE S BALL JOINTS (FIGS. 9A AND 9B)

Nominal Ball Joint Size and Thread Diameter, A, in		Thds per in	B		C		D Max		E Hex		F Min		G Dia		G_1 Dia		H Dia		J		K		L Min Full Thread		P[a] Max (Ref)	
			in ±0.02	mm ±0.5	in ±0.02	mm ±0.5	in	mm	in	mm	in	mm	in ±0.010	mm ±0.25	in ±0.010	mm ±0.25	in ±0.01	mm ±0.3	in ±0.02	mm ±0.5	in ±0.02	mm ±0.5	in	mm	in	mm
STYLE 1																											
No. 10	0.190	32	0.47	11.9	0.44	11.2	0.06	1.5	0.312	7.92	0.12	3.0	0.312	7.92	0.312	7.92	0.312	7.92	0.44	11.2	0.91	23.1	1.09	27.7	0.44	11.2	0.72	18.3
No. 12	0.216	32	0.47	11.9	0.44	11.2	0.06	1.5	0.312	7.92	0.12	3.0	0.312	7.92	0.312	7.92	0.312	7.92	0.44	11.2	0.91	23.1	1.09	27.7	0.44	11.2	0.72	18.3
1/4	0.250	28	0.47	11.9	0.56	14.2	0.06	1.5	0.312	7.92	0.12	3.0	0.312	7.92	0.312	7.92	0.44	11.2	0.91	23.1	1.09	27.7	0.50	12.7	0.72	18.3
5/16	0.3125	24	0.59	15.0	0.69	17.5	0.09	2.3	0.438	11.12	0.16	4.1	0.438	11.12	0.438	11.12	0.62	15.7	1.25	31.8	1.56	39.6	0.56	14.2	0.93	23.6
3/8	0.375	24	0.72	18.2	0.88	22.3	0.09	2.3	0.500	12.70	0.19	4.8	0.562	14.28	0.562	14.27	0.75	19.0	1.56	39.6	1.94	49.3	0.75	19.0	1.13	28.7
7/16	0.4375	20	0.97	24.6	1.12	28.4	0.12	3.0	0.625	15.88	0.25	6.4	0.750	12.70	0.750	19.05	1.00	25.4	2.03	51.6	2.53	64.3	1.00	25.4	1.50	38.1
1/2	0.500	20	0.97	24.6	1.12	28.4	0.12	3.0	0.625	15.88	0.25	6.4	0.750	12.70	0.750	19.05	1.00	25.4	2.03	51.6	2.53	64.3	1.00	25.4	1.50	38.1
5/8	0.625	18	1.12	28.4	1.12	28.4	0.12	3.0	0.750	19.05	0.31	7.9	0.875	22.22	0.875	22.22	1.12	28.4	2.31	58.7	2.88	73.2	1.00	25.4	1.71	43.4
STYLE 2																											
1/4	0.250	28	0.47	11.9	0.56	14.2	0.06	1.5	0.375	9.52	0.12	3.0	0.562	14.28	0.438	11.12	0.53	13.5	0.97	24.6	1.25	31.8	0.53	13.5	0.78	19.8
5/16	0.3125	24	0.53	13.5	0.69	17.5	0.09	2.3	0.438	11.12	0.16	4.1	0.625	15.88	0.500	12.70	0.59	15.0	1.12	28.4	1.45	36.8	0.59	15.0	0.87	22.1
3/8	0.375	24	0.69	17.5	0.88	22.3	0.09	2.3	0.500	12.70	0.19	4.8	0.750	19.05	0.625	15.88	0.75	19.0	1.38	35.0	1.75	44.4	0.81	20.6	1.09	27.7
1/2	0.500	20	0.88	22.3	1.12	28.4	0.12	3.0	0.625	15.88	0.28	7.1	0.938	23.82	0.750	19.05	0.89	22.6	1.94	49.3	2.38	60.5	1.12	28.4	1.39	35.3

[a] These dimensions are given for design purposes only and are not intended for inspection.

1.1.2 The inclusion of dimensional data in this standard is not intended to imply that all the products described are stock production sizes. Consumers are requested to consult with manufacturers concerning availability of stock production parts.

1.2 Dimensions and Tolerances—Except for nominal sizes and thread designations which are inch values only, dimensions and tolerances are given in both U. S. customary and SI units, as designated. Tabulated dimensions shall apply to the finished parts, plated or otherwise processed, as specified by the user. Limits on hexagon or round bar shapes shall be within the commercial tolerance of the bar stock material from which the components are produced.

1.3 Threads—Unified Standard Class 2A external threads and Class 2B internal threads shall apply to plain finish (unplated) parts. For externally threaded components with additive finish, the maximum diameters of Class 2A may be exceeded by the amount of the allowance; that is, the basic diameters (Class 2A maximum diameters plus the allowance) apply to an externally threaded part after plating. For internally threaded components with additive finish, the Class 2B diameters apply after plating. See ANSI B 1.1.

1.3.1 External threads shall be chamfered to a diameter 0.01 in (0.3 mm) less than the minor diameter to produce a length of chamfered or partial thread equivalent to ¾ to 1¼ times the pitch (rounded to a three-place decimal).

1.3.2 Internal threads shall be countersunk 90 deg included angle to a diameter 0.01 in (0.3 mm) greater than the major diameter of the thread (rounded to a two-place decimal).

1.4 Material

1.4.1 BALL JOINTS—Ball joints are normally made from low carbon free machining steel. The ball stud and mating plug components of Types A, AL, B, and C and the ball sockets on Type G, Styles 3 and 4, ball joints shall be case hardened unless otherwise specified. For special application, ball joints can be produced from alloy steel, corrosion resistant steel, brass, bronze, or other materials.

1.4.2 CUSHIONING DISCS—Cushioning discs shall be Neoprene, Buna N rubber, or equivalent material.

1.5 Finishes—Unless otherwise specified, carbon steel ball joints shall be furnished with cadmium or zinc protective finish and shall meet the requirements of 32 h salt spray test in accordance with ASTM B 117, Method of Salt Spray (Fog) Testing. At manufacturer's option, a subsequent chromate treatment may be used. Plated, hardened carbon steel components of ball joints (subject to hydrogen embrittlement) shall be baked or otherwise processed to obviate such embrittlement.

1.6 Lubrication—Unless otherwise specified by user, ball joints shall be supplied with ball sockets suitably lubricated in accordance with manufacturers practice.

1.7 Dust Covers—Where so specified by the user, Type G ball joints shall be supplied with an oil resistant rubber shield of such construction as to prevent dirt and dust from entering the ball cavity. However, shields for Style 3 are available in sizes ⅝ and ¾ only.

1.8 Workmanship—Ball joints must be free from burrs, loose scale, sharp edges, and any other defects which might affect their serviceability.

METRIC BALL JOINTS—SAE J2213 JUN91 — SAE Standard

Report of the SAE Ball Joint & Spherical Rod End Committee approved June 1991.

1. Scope

1.1 This SAE Standard covers the general and dimensional data for industrial quality ball joints commonly used on control linkages in metric automotive, marine, construction, and industrial equipment applications.

1.2 The ball joints described are available from several manufacturers within the range of the interchangeable specifications. The sliding contact self-aligning bearing members (ball and socket) are available in a variety of materials in the types shown. The load capacities and wear capabilities vary considerably with the design and fabrication. It is suggested that the manufacturers be consulted for recommendations for the type and design appropriate to particular applications.

1.3 The inclusion of dimensional data in the document is not intended to imply that all the products described are stock production sizes. Consumers are requested to consult with manufacturers concerning availability of stock production parts.

2. References

2.1 Applicable Documents—The following publications form a part of this specification to the extent specified herein. The latest issue of SAE Publications shall apply.

2.1.1 SAE PUBLICATIONS—Available from SAE, 400 Commonwealth Drive, Warrendale, PA 15096-0001.

SAE J490—Ball Joints

2.1.2 ANSI AND ISO PUBLICATIONS—Available from ANSI, 11 West 42nd Street, New York, NY 10036.

ANSI/ASME B1.13—Metric Screw Threads—M Profile

ISO 965-2—ISO general purpose metric screw threads—Tolerances—Part 2: Limits of sizes for general purpose bolt and nut threads—Medium quality

2.1.3 ASTM PUBLICATIONS—Available from ASTM, 1916 Race Street, Philadelphia, PA 19103.

ASTM B 117—Method of Salt Spray (Fog) Testing

3. Dimensions

3.1 All dimensions and tolerances are in millimeters. See SAE J490 for the ball joint specification with unified threads.

3.2 Tabulated dimensions shall apply to the finished parts, plated or otherwise processed, as specified by the user. Limits on hexagon of round bar shapes shall be within the commercial tolerance of the bar stock material from which the components are produced.

3.3 Dimensions of ball joints are shown in Figures 1, 2, and 3.

3.4 Nominal Sizes—The ball joint sizes are specified by a number indicating the thread size (size 6 has M6 threads).

3.5 Threads

3.5.1 Thread form, diameter, and associated pitches are in accordance with ISO 965-2 and ANSI/ASME B1.13, tolerance class 6g external and 6H internal.

3.5.2 The threads shall be right hand unless otherwise specified. The threads must be chamfered to insure a clean start according to good industrial practice.

4. Material and Finishes

4.1 Ball joints are normally made from low carbon free machining steel. The ball stud and the ball sockets on Type A Style 2 ball joints shall be case hardened unless otherwise specified. For special application, ball joints can be produced from alloy steel, corrosion resistant steel, brass, bronze, or other materials.

4.2 Retainer clips shall be made of high carbon steel. Springs shall be stainless steel.

4.3 Unless otherwise specified, carbon steel ball joints shall be furnished with a zinc protective finish with subsequent chromate treatment and shall meet the requirements of 72 h salt spray test in accordance with ASTM B 117.

4.3.1 Plated, hardened carbon steel components of ball joints (subject to hydrogen embrittlement) shall be baked or otherwise processed to obviate such embrittlement.

5. Lubrication

5.1 Unless otherwise specified by the user, ball joints shall be supplied with ball sockets suitably lubricated in accordance with the manufacturer's practice.

6. Dust Covers

6.1 Where so specified by the user, Type G Style 1 ball joints shall be supplied with an oil resistant rubber shield of such construction as to prevent dirt and dust from entering the ball cavity.

7. Workmanship

7.1 Ball joints must be free from burrs, loose scale, sharp edges, and any other defects which might affect their serviceability.

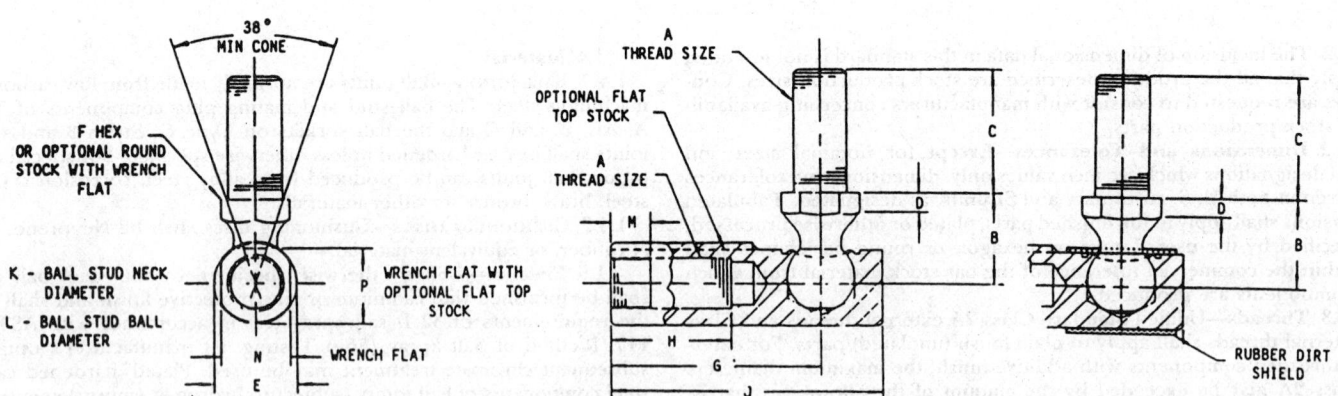

Size	A Thread Size	B ±0.5	C ±0.5	D MIN	D₁ MAX	E MAX	F MAX	G ±0.5	H MIN	J MAX	K MIN	L NOM	M ±0.5	N -0.25	P ±0.25
5	M5X0.8	9.0	11.0	2.4	3.0	10	8	22.0	12	30	4.25	7.60	6.5	8	8
6	M6X1	10.5	14.5	2.0	3.0	12	10	24.5	13	34	5.05	8.75	6.5	9	9
8	M8X1.25	12.5	17.5	2.4	3.5	13	13	28.5	15	40	6.10	10.25	7.0	11	11
10	M10X1.5	16.0	22.0	3.6	4.0	16	13	35.0	20	47	7.10	12.45	8.0	12	14
12	M12X1.75	20.5	28.5	4.4	4.5	20	16	49.0	28	64	9.45	15.60	9.5	16	16

FIGURE 1—DIMENSIONS OF TYPE A NONDETACHABLE BALL JOINT

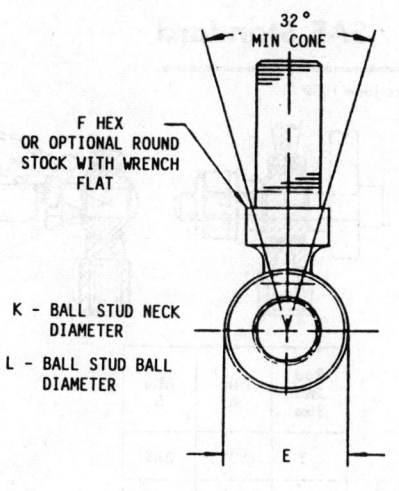
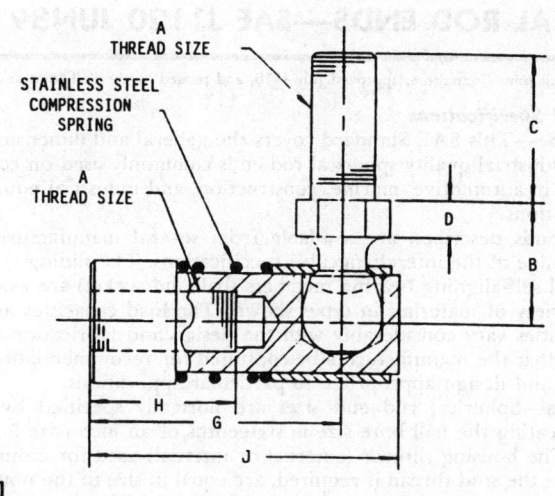

STYLE 1

Size	A Thread Size	B ±0.5	C ±0.5	D MIN	D₁ MAX	E MAX	F MAX	G ±0.5	H MIN	J MAX	K MIN	L NOM
6	M6X1	12.0	14.5	2.8	3.0	15	10	24.5	13	33	4.65	8.75
8	M8X1.25	13.5	17.5	3.6	3.5	17	13	28.5	15	37	5.60	10.25
10	M10X1.5	17.5	22.0	4.4	4.0	20	13	35.0	20	45	6.70	12.45
12	M12X1.75	22.0	28.5	6.8	4.5	24	16	49.0	28	61	8.90	15.60

FIGURE 2—DIMENSIONS OF TYPE B DETACHABLE BALL JOINTS

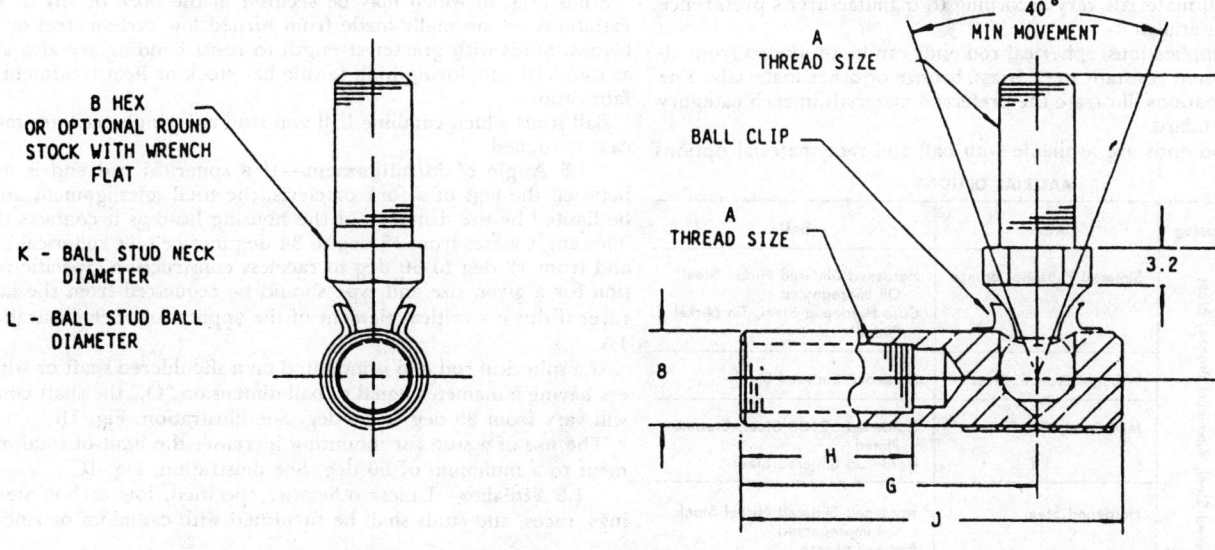

STYLE 2 (MP)

Size	A Thread Size	B ±0.5	C ±0.5	D MIN	D₁ MAX	F MAX	G ±0.5	H MIN	J MAX	K MIN	L ±0.15
5	M5X0.8	11.0	11.0	3.2	3	8	22.0	12	30	3.5	6.35
6	M6X1	11.0	14.5	3.2	3	8	24.5	13	33	3.5	6.35

FIGURE 3—DIMENSIONS OF TYPE B DETACHABLE BALL JOINTS—STYLE 2 (MP)

SPHERICAL ROD ENDS—SAE J1120 JUN89   SAE Standard

Report of the Ball Joint Committee, approved July 1975, and revised by the Ball Joint and Spherical Rod End Committee June 1989.

1. General Specifications

1.1 Scope—This SAE Standard covers the general and dimensional data for industrial quality spherical rod ends commonly used on control linkages in automotive, marine, construction, and industrial equipment applications.

The rod ends described are available from several manufacturers within the range of the interchangeable specifications. The sliding contact spherical self-aligning bearing members (ball and socket) are available in a variety of materials in types shown. The load capacities and wear capabilities vary considerably with the design and fabrication. It is suggested that the manufacturers be consulted for recommendations for the type and design appropriate to particular applications.

1.2 Sizes—Spherical rod end sizes are normally specified by a number indicating the ball bore size in sixteenths of an inch (size 5 = $5/16$ bore). The housing threads (external or internal) used for mounting, as well as the stud thread if required, are equal in size to the nominal ball bore. Sizes larger than those listed are available in both standard and special configurations.

1.3 Threads—Unified Standard fine thread series (UNF) Class 2A external threads and Class 2B internal threads shall apply to plain finish (unplated) parts. For externally threaded components with additive finish, the maximum diameters of Class 2A may be exceeded by the amount of the allowance: that is, the basic diameters (Class 2A maximum diameters plus the allowance) apply to an externally threaded part after plating. For internally threaded components with additive finish, the Class 2B diameters apply after plating. See SAE J475 (ANSI B1.1-1974).

Housing threads, left or right hand, may be specified as required. Standard studs are threaded right hand.

External and internal threads must be chamfered to insure a clean start according to good industrial practice. Roll formed internal and external threads are preferred.

1.4 Material—Spherical rod end housing members are normally made from low carbon steel turned, forged, headed, or press-stamped blanks.

Race and ball materials vary according to manufacturer's preference for bearing materials.

For special applications, spherical rod ends can be produced from alloy steel, corrosion resistant steel, brass, bronze or other materials. The charted combinations illustrate the preferred materials in each category available as standard.

Spherical rod ends are available with ball and race material options listed below:

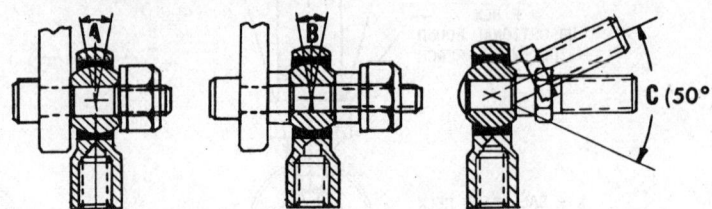

Rod End Size	Min A	Min B
3	10°	34°
4	14°	34°
5	12°	28°
6	10°	30°
7	14°	32°
8	10°	32°
10	14°	30°
12	14°	25°

FIG. 1—A—HOUSING STRIKES YOKE OR LEVER,
B—WASHER OR SHOULDERED SHAFT WITH DIA "O" STRIKES RACE ID
C—STUD STRIKES RACE ID

MATERIAL OPTIONS

Rod End	Housing	Race	Ball
Type A (Fig. 2)	Mild Steel, Alloy Steel, Stainless Steel, Hardened Steel, Aluminum Bronze, Brass	Sintered Phosphor Bronze	Hardened Sintered Nickel Steel, Oil Impregnated; Case Hardened Steel, Tin Nickel Plated
		Wrought Bronze, Brass	Hardened Sintered Steel
		Mild Steel, Cad Plated	Hardened 52100 Steel, Chrome Plated; Hardened Sintered Steel
		Hardened Steel	Hardened Sintered Nickel Steel, Oil Impregnated; Sintered Bronze, Oil Impregnated; Hardened 52100
Type B (Fig. 3)		Nylon Reinforced, Delrin, TFE Lined	Case Hardened Steel, Cad or Tin Nickel Plated; Hardened Sintered Nickel Steel, Oil Impregnated; Hardened 52100
Type C (Fig. 4)		None	Hardened 52100; Hardened Sintered Iron, Oil Impregnated; Case Hardened Steel, Tin Nickel Plated
Type D (Fig. 5)		None	Mild Steel—Case Hardened, Cad Plated

Studs (Fig. 6) which may be secured in the bore of any of the ball variations are normally made from turned low carbon steel or headed blanks. Studs with greater strength to resist bending are also available as standard, employing high tensile bar stock or heat treatment during fabrication.

Ball studs which combine ball and stud as a single part are mild steel case hardened.

1.5 Angle of Misalignment—If a spherical rod end is mounted between the legs of a fork or clevis, the total misalignment angle will be limited by the diameter of the housing head as it contacts the legs. This angle varies from 18 deg to 34 deg in race type spherical rod ends and from 12 deg to 30 deg in raceless construction. Specific information for a given size and type should be requested from the manufacturer if this is a critical element of the application. See illustration, Fig. 1A.

If a spherical rod end is mounted on a shouldered shaft or with washers having a diameter equal to ball dimension "O," the shaft cone angle will vary from 25 deg to 34 deg. See illustration, Fig. 1B.

The use of a stud for mounting increases the limit of total misalignment to a minimum of 50 deg. See illustration, Fig. 1C.

1.6 Finishes—Unless otherwise specified, low carbon steel housings, races, and studs shall be furnished with cadmium or zinc protec-

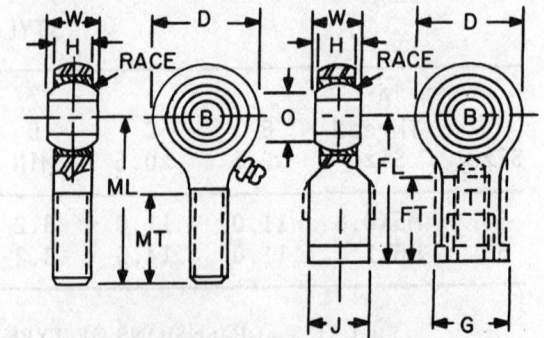

FIG. 2—TYPE A METALLIC RACE

tive finish and shall meet the requirements of 32 h Salt Spray Fog Testing in accordance with ASTM B 117. At manufacturer's option, a subsequent chromate treatment may be used. Black oxide treatment for studs may also be employed.

Hardened steel races shall be black oxide treated and oiled. Nonsintered balls and ball studs shall be plated according to manufacturer's preference for corrosion protection appropriate to their use as bearing elements.

1.7 Lubrication—Unless otherwise specified by the user, spherical rod ends shall be supplied with ball sockets suitably lubricated in accordance with manufacturer's practice, including vacuum impregnation of self-lubricating sintered bearing elements.

Grease fittings for supplemental lubrication are provided upon request for most types. Standard location is shown. Special locations at 12 o'clock and 3 o'clock positions are also available.

1.8 Workmanship—Industrial quality spherical rod ends must be free from burrs, loose scale, sharp edges, and any other defects.

1.9 Ball Bore Chamfer—Ball bores are chamfered at both faces to break the edge 0.005 in (0.13 mm) or up to a maximum of 0.03 in (0.8 mm) according to manufacturer's preference and method of fabrication. The user is cautioned against seating bolt heads against the ball face during mounting because bolt fillets under the head may distort or crack the ball. This is especially true of hex bolts and screws meeting ANSI B 18.2.1-1972 specifications. The use of a washer or other suitable alternate is recommended.

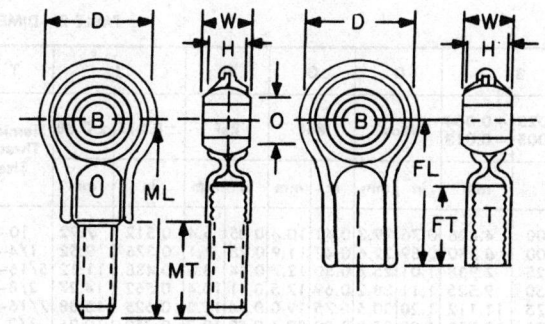

FIG. 5—TYPE D RACELESS, STAMPED HOUSING

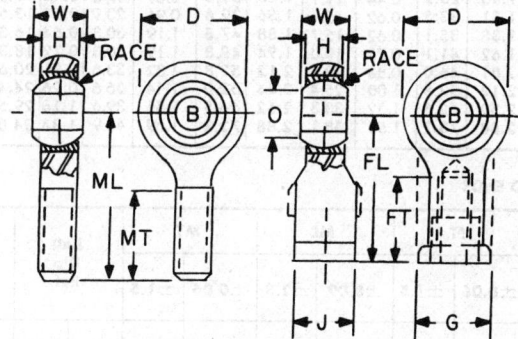

FIG. 3—TYPE B MOLDED RACE

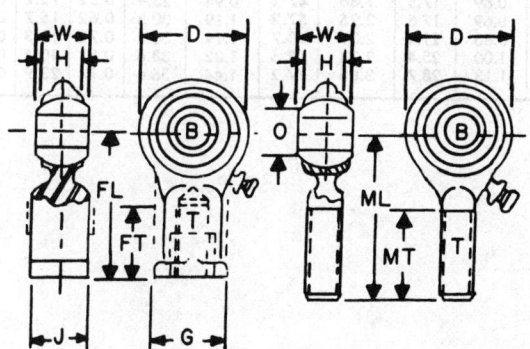

FIG. 4—TYPE C RACELESS

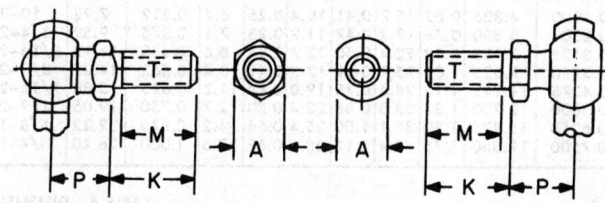

Rod End Size	A ±0.010		K ±0.03		M ±0.03		P ±0.04		T
	0.25		0.8		0.8		1.0		Nominal Thread Size
	in	mm	in	mm	in	mm	in	mm	
3	0.312	7.92	0.50	12.7	0.44	11.2	0.48	12.2	10-32
4	0.375	9.52	0.56	14.2	0.50	12.7	0.48	12.2	1/4-28
5	0.437	11.10	0.69	17.5	0.59	15.0	0.54	13.7	5/16-24
6	0.500	12.70	0.89	22.6	0.81	20.6	0.65	16.5	3/8-24
7	0.625	15.88	1.09	27.7	0.97	24.6	0.84	21.3	7/16-20
8	0.625	15.88	1.12	28.4	1.00	25.4	0.88	22.4	1/2-20
10	0.750	19.05	1.50	38.1	1.38	35.1	1.00	25.4	5/8-18
12	1.000	25.40	1.81	46.0	1.63	41.4	1.19	30.2	3/4-16

FIG. 6—STUD ASSEMBLED TO BALLS HAVING ANY STANDARD MATERIAL OPTIONS, ONE PIECE BALL STUD, LOW CARBON STEEL, CASE HARDENED

TABLE 1—DIMENSIONS FOR TYPE A ROD ENDS

Rod End Size	B +0.0025 −0.0005		D +0.064 −0.013		G Max		H Ref		J ±0.015		T Nominal Thread Size	W ±0.005		FL +0.06 −0.03		FT +1.5 −0.8		ML +0.06 −0.03		MT +1.5 −0.8		Ball Dia Ref		O Ref	
									±0.38			±0.13		±1.5		±0.06		±1.5		±0.06					
	in	mm	in	mm	in	mm	in	mm	in	mm		in	mm	in	mm	in	mm	in	mm	in	mm	in	mm	in	mm
3	0.1900	4.826	0.76	19.3	0.41	10.4	0.25	6.4	0.312	7.92	10-32	0.312	7.92	1.06	26.9	0.50	12.7	1.25	31.8	0.69	17.5	0.44	11.2	0.31	7.9
4	0.2500	6.350	0.89	22.6	0.47	11.9	0.28	7.1	0.375	9.52	1/4-28	0.375	9.52	1.31	33.3	0.69	17.5	1.56	39.6	0.94	23.9	0.51	13.0	0.35	8.9
5	0.3125	7.938	1.01	25.7	0.50	12.7	0.34	8.6	0.438	11.12	5/16-24	0.438	11.12	1.38	35.1	0.69	17.5	1.88	47.8	1.19	30.3	0.62	15.7	0.45	11.4
6	0.3750	9.525	1.11	28.2	0.69	17.5	0.41	10.4	0.562	14.27	3/8-24	0.500	12.70	1.62	41.1	0.88	22.4	1.94	49.3	1.19	30.3	0.72	18.3	0.52	13.2
7	0.4375	11.112	1.20	30.5	0.75	19.0	0.44	11.2	0.625	15.88	7/16-20	0.562	14.27	1.81	46.0	1.00	25.4	2.12	53.8	1.32	33.6	0.81	20.6	0.59	15.0
8	0.5000	12.700	1.39	35.3	0.88	22.4	0.50	12.7	0.750	19.05	1/2-20	0.625	15.88	2.12	53.8	1.13	28.7	2.44	62.0	1.44	36.6	0.94	23.9	0.70	17.8
10	0.6250	15.875	1.57	39.9	1.00	25.4	0.56	14.2	0.875	22.22	5/8-18	0.750	19.05	2.50	63.5	1.44	36.6	2.62	66.5	1.56	39.6	1.12	28.4	0.81	20.6
12	0.7500	19.050	1.82	46.2	1.12	28.4	0.69	17.5	1.000	25.40	3/4-16	0.875	22.22	2.88	73.2	1.69	42.9	2.88	73.2	1.69	42.9	1.32	33.5	1.02	25.9

TABLE 2—DIMENSIONS FOR TYPE B ROD ENDS

Rod End Size	B +0.0025 -0.0005		D +0.064 -0.013		G Max		H Ref		J ±0.015		T ±0.38		Nominal Thread Size	W ±0.005		FL ±0.13		FT +0.06 -0.03		ML +1.5 -0.8		MT +0.06 -0.03		Ball Dia Ref		O Ref	
	in	mm	in	mm	in	mm	in	mm	in	mm			in	mm	in	mm	in	mm	in	mm	in	mm	in	mm	in	mm
3	0.1900	4.826	0.76	19.3	0.41	10.4	0.25	6.4	0.312	7.92	10-32		0.312	7.92	1.06	26.9	0.50	12.7	1.25	31.8	0.69	17.5	0.44	11.2	0.31	7.9
4	0.2500	6.350	0.89	22.6	0.47	11.9	0.28	7.1	0.375	9.52	1/4-28		0.375	9.52	1.31	33.3	0.69	17.5	1.56	39.6	0.94	23.9	0.51	13.0	0.35	8.9
5	0.3125	7.938	1.01	25.7	0.50	12.7	0.34	8.6	0.438	11.12	5/16-24		0.438	11.12	1.38	35.1	0.69	17.5	1.88	47.8	1.19	30.3	0.62	15.7	0.45	11.4
6	0.3750	9.525	1.11	28.2	0.69	17.5	0.41	10.4	0.562	14.27	3/8-24		0.500	12.70	1.62	41.1	0.88	22.4	1.94	49.3	1.19	30.3	0.72	18.3	0.52	13.2
7	0.4375	11.112	1.20	30.5	0.75	19.0	0.44	11.2	0.625	15.88	7/16-20		0.562	14.27	1.81	46.0	1.00	25.4	2.12	53.8	1.32	33.6	0.81	20.6	0.59	15.0
8	0.5000	12.700	1.39	35.3	0.88	22.4	0.50	12.7	0.750	19.05	1/2-20		0.625	15.88	2.12	53.8	1.13	28.7	2.44	62.0	1.44	36.6	0.94	23.9	0.70	17.8
10	0.6250	15.875	1.51	38.4	1.00	25.4	0.56	14.2	0.875	22.22	5/8-18		0.750	19.05	2.50	63.5	1.44	36.6	2.62	66.5	1.56	39.6	1.12	28.4	0.81	20.6

TABLE 3—DIMENSIONS FOR TYPE A ROD ENDS

Rod End Size	B +0.0025 -0.0005		D +0.064 -0.013		G Max		H Ref		J ±0.015		T ±0.38		Nominal Thread Size	W ±0.005		FL ±0.13		FT +0.06 -0.03		ML +1.5 -0.8		MT +0.06 -0.03		Ball Dia Ref		O Ref	
	in	mm	in	mm	in	mm	in	mm	in	mm			in	mm	in	mm	in	mm	in	mm	in	mm	in	mm	in	mm
3	0.1900	4.826	0.62	15.7	0.41	10.4	0.25	6.4	0.312	7.92	10-32		0.312	7.92	1.06	26.9	0.44	11.1	1.25	31.8	0.69	17.5	0.45	11.4	0.35	8.9
4	0.2500	6.350	0.76	19.3	0.47	11.9	0.28	7.1	0.375	9.52	1/4-28		0.375	9.52	1.31	33.3	0.62	15.7	1.56	39.6	0.94	23.9	0.53	13.5	0.42	10.7
5	0.3125	7.938	0.88	22.4	0.50	12.7	0.34	8.6	0.438	11.12	5/16-24		0.438	11.12	1.38	35.1	0.62	15.7	1.88	47.8	1.19	30.3	0.64	16.3	0.49	12.4
6	0.3750	9.525	1.01	25.7	0.69	17.5	0.41	10.4	0.562	14.27	3/8-24		0.500	12.70	1.62	41.1	0.75	19.0	1.94	49.3	1.19	30.3	0.72	18.3	0.51	13.0
7	0.4375	11.112	1.12	28.4	0.75	19.0	0.44	11.2	0.625	15.88	7/16-20		0.562	14.27	1.81	46.0	0.88	22.2	2.12	53.8	1.32	33.6	0.81	20.6	0.58	14.7
8	0.5000	12.700	1.31	33.3	0.88	22.4	0.50	12.7	0.750	19.05	1/2-20		0.625	15.88	2.12	53.8	1.00	25.4	2.44	62.0	1.44	36.6	0.96	24.4	0.79	20.1
10	0.6250	15.875	1.50	38.1	1.00	25.4	0.56	14.2	0.875	22.22	5/8-18		0.750	19.05	2.50	63.5	1.32	33.3	2.62	66.5	1.56	39.6	1.16	29.5	0.92	23.4
12	0.7500	19.050	1.75	44.4	1.12	28.4	0.69	17.5	1.000	25.40	3/4-16		0.875	22.22	2.88	73.2	1.50	38.1	2.88	73.2	1.69	42.9	1.34	34.0	1.06	26.9

TABLE 4—DIMENSIONS FOR TYPE A ROD ENDS

Rod End Size	B +0.0025 -0.0005		D +0.064 -0.013		H Ref		T Nominal Thread Size	W ±0.005		FL ±0.13		FT ±1.5		ML ±2.3		MT ±1.5		Ball Dia Ref		O Ref	
	in	mm	in	mm	in	mm		in	mm	in	mm	in	mm	in	mm	in	mm	in	mm	in	mm
3	0.1900	4.826	0.78	18.8	0.25	6.4	10-32	0.312	7.92	1.06	26.9	0.50	12.7	1.50	38.1	0.69	17.5	0.44	11.2	0.31	7.9
4	0.2500	6.350	0.88	22.4	0.29	7.1	1/4-28	0.375	9.52	1.31	33.3	0.69	17.5	1.86	47.2	0.94	23.9	0.52	13.2	0.35	8.9
5	0.3125	7.938	1.05	26.7	0.31	7.9	5/16-24	0.438	11.12	1.38	35.1	0.69	17.5	2.25	57.2	1.19	30.3	0.62	15.7	0.45	11.4
6	0.3750	9.525	1.16	29.5	0.41	10.4	3/8-24	0.500	12.70	1.62	41.1	0.88	22.4	2.39	60.7	1.19	30.3	0.72	18.3	0.52	13.2
7	0.4375	11.112	1.37	34.8	0.44	11.2	7/16-20	0.562	14.27	1.91	48.5	1.00	25.4	2.74	69.6	1.32	33.6	0.81	20.6	0.59	15.0
8	0.5000	12.700	1.51	38.4	0.50	12.7	1/2-20	0.625	15.88	2.12	53.8	1.13	28.7	3.04	77.2	1.44	36.6	0.94	23.9	0.70	17.8

METRIC SPHERICAL ROD ENDS—SAE J1259 DEC89   SAE Standard

Report of the Ball Joint Spherical Rod End Committee, approved April 1980 and revised December 1989.

1. General Specifications

1.1 Scope—This SAE Standard covers the general and dimensional data for industrial quality spherical rod ends commonly used on control linkages in metric automotive, marine, construction, and industrial equipment applications.

The rod ends described are available from several manufacturers within the range of the interchangeable specifications. The sliding contact spherical self-aligning bearing members (ball and socket) are available in a variety of materials in the types shown. The load capacities and wear capabilities vary considerably with the design and fabrication. It is suggested that the manufacturers be consulted for recommendations for the type and design appropriate to particular applications.

1.2 Dimensions—All dimensions are in millimeters. See SAE J1120 for the U.S. Customary unit specification for spherical rod ends.

1.3 Sizes—The spherical rod end sizes are normally specified by a number indicating the ball bore in millimeters (size 5 = 5 mm). The housing threads (external or internal) used for mounting, as well as the stud thread if required, are equal in size to the nominal ball bore. Sizes larger than those listed are available in both standard and special configurations.

1.4 Threads—Thread form, diameter, and associated pitches are in accordance with ISO 965/II and ANSI B1.13, tolerance class 6g external and 6H internal.

The threads shall be right hand unless otherwise specified. The threads must be chamfered to insure a clean start according to good industrial practice.

1.5 Material—The spherical rod end housing members are normally made from low-carbon steel turned, forged, or headed.

The race and ball materials vary according to the manufacturer's preference for bearing materials.

For special applications, spherical rod ends can be produced from alloy steel, corrosion resistant steel, brass, bronze, or other materials. The charted combinations illustrate the preferred materials in each category available as standard.

The spherical rod ends are available with the ball and race material options listed below:

MATERIAL OPTIONS

Rod End	Housing	Race	Ball
Type A (Fig. 2)	Mild Steel, Alloy Steel, Stainless Steel, Hardened Steel, Aluminum Bronze, Brass	Sintered Phosphor Bronze	Hardened Sintered Nickel Steel, Oil Impregnated Case Hardened Steel, Tin Nickel Plated
		Wrought Bronze, Brass	Hardened Sintered Steel
		Mild Steel, Cad Plated	Hardened 52100 Steel, Chrome Plated Hardened Sintered Steel
		Hardened Steel	Hardened Sintered Nickel Steel, Oil Impregnated Sintered Bronze, Oil Impregnated Hardened 52100
Type B (Fig. 3)		Nylon Reinforced, Delrin,	Case Hardened Steel, Cad or Tin Nickel Plated
		TFE Lined	Hardened Sintered Nickel Steel, Oil Impregnated Hardened 52100
Type C (Fig. 4)		None	Hardened 52100 Hardened Sintered Iron, Oil Impregnated Case Hardened Steel, Tin Nickel Plated

The studs (Fig. 5), which may be secured in the bore of any of the ball variations, are normally made from turned low-carbon steel or headed blanks. The studs with greater strength to resist bending are also available by agreement between user and manufacturer.

The ball studs, which combine ball and stud as a single part, are mild steel case hardened.

1.6 Angle of Misalignment—If a spherical rod end is mounted between the legs of a fork or clevis, the total misalignment angle will be limited by that portion of the housing head that contacts the legs. This angle varies from 12-18 deg. Specific information for a given size and type should be requested from the manufacturer if this is a critical element of the application. See illustration, Fig. 1A.

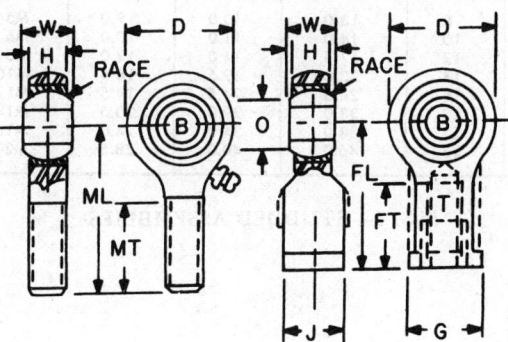

FIG. 2—TYPE A—METALLIC RACE

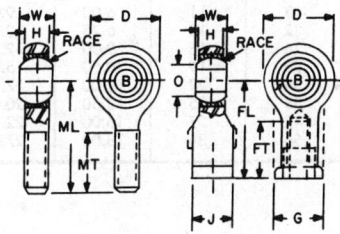

FIG. 3—TYPE B—MOLDED RACE

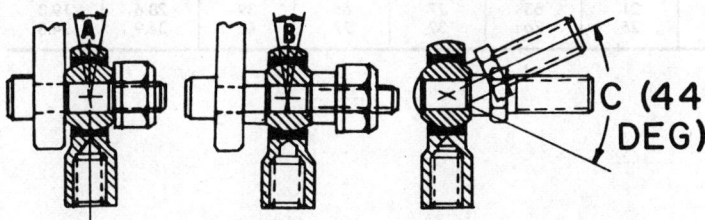

Rod End Size	Min A Deg	Min B Deg
5	13	24
6	12	24
8	14	26
10	14	26
12	14	26
14	18	30
16	17	30
20	17	28

FIG. 1—A—HOUSING STRIKES YOKE OR LEVER
B—WASHER OR SHOULDERED SHAFT WITH DIA "O" STRIKES RACE ID
C—STUD STRIKES RACE ID

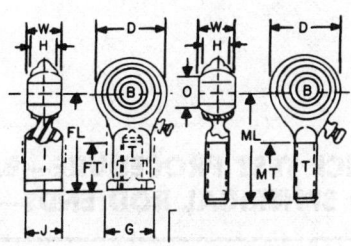

FIG. 4—TYPE C—RACELESS

If a spherical rod end is mounted on a shouldered shaft or with washers having a diameter equal to ball dimension "O", the shaft cone angle will vary from 24-30 deg. See illustration, Fig. 1B.

The use of a stud for mounting increases the limit of total misalignment to a minimum of 44 deg. See illustration, Fig. 1C.

1.7 Finishes—Unless otherwise specified, low-carbon steel housing, races, and studs shall be furnished with cadmium or zinc protective

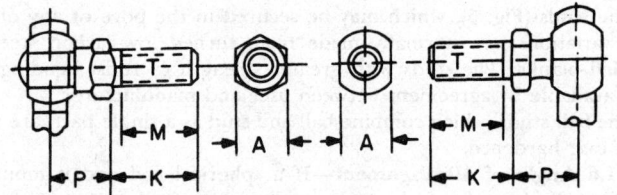

Rod End Size	A Ref	K ±0.25	M Min	P ±1	T Nominal Thread Size
5	8	13.0	10.0	9.0	M5 × 0.80
6	10	14.0	11.0	10.0	M6 × 1.00
8	12	17.5	14.0	12.0	M8 × 1.25
10	14	23.0	19.5	16.5	M10 × 1.50
12	16	28.5	24.5	19.5	M12 × 1.75
14	20	33.0	29.0	20.5	M14 × 2.00
16	22	38.0	34.0	24.0	M16 × 2.00
20	25	46.0	40.0	28.5	M20 × 1.50

FIG. 5—STUDDED ASSEMBLIES

finish and shall meet the requirements of 32 h Salt Spray (Fog) Testing in accordance with ASTM B 117. At manufacturer's option, a subsequent chromate treatment may be used. A black oxide treatment for studs may also be employed.

Nonsintered balls and ball studs shall be plated according to the manufacturer's preference for corrosion protection appropriate to their use as bearing elements.

1.8 Lubrication—Unless otherwise specified by the user, spherical rod ends shall be supplied with ball sockets suitably lubricated in accordance with the manufacturer's practice, including vacuum impregnation of self-lubrication sintered bearing elements.

The grease fittings for supplemental lubrication are provided upon request for most types. The standard location is shown. Special locations at 12 o'clock and 3 o'clock positions are also available.

1.9 Workmanship—Industrial quality spherical rod ends must be free from burrs, loose scale, sharp edges, and any other defects.

1.10 Ball Bore Chamfer—The ball bores are chamfered at both faces to break the edge 0.13 mm or up to a maximum of 0.8 mm according to the manufacturer's preference and the method of fabrication. The user is cautioned against seating bolt heads against the ball face during mounting because bolt fillets under the head may distort or crack the ball. The use of a washer or other suitable alternate is recommended.

φ TABLE 1—DIMENSIONS FOR ROD ENDS—TYPE A (FIG. 2), TYPE B (FIG. 3), AND TYPE C (FIG. 4)

Rod End Size	B +0.07 −0.00	D Max	G ±0.25	H ±0.15	J Ref	T Nominal Thread Size	W ±0.15	FL +1.5 −0.8	FT Min	ML +1.5 −0.8	MT Min	Ball Dia Ref	O Ref
5	5	18	11	6.00	9.0	M5 × 0.80	8	26	9	32	19	11.1	7.7
6	6	22	13	6.75	10.0	M6 × 1.00	9	29	11	35	21	12.7	8.9
8	8	26	16	9.00	12.5	M8 × 1.25	12	35	15	41	24	15.8	10.4
10	10	30	19	10.50	15.0	M10 × 1.50	14	42	19	47	28	19.1	12.9
12	12	34	22	12.00	17.5	M12 × 1.75	16	49	21	54	32	22.2	15.4
14	14	38	25	13.50	20.0	M14 × 2.00	19	56	24	59	35	25.4	16.8
16	16	42	27	15.00	22.0	M16 × 2.00	21	63	27	65	39	28.6	19.3
20	20	50	34	18.00	27.5	M20 × 1.50	25	76	32	77	46	34.9	24.3

PERFORMANCE TEST PROCEDURE—BALL JOINTS AND SPHERICAL ROD ENDS—SAE J1367 SEP81

SAE Standard

Report of the Ball Joint and Spherical Rod End Committee, approved September 1981.

1. Scope—The purpose of this test procedure is to provide a uniform method of testing commercial spherical rod end bearings to determine their performance characteristics under specific application situations. This procedure is an extension of the dimensional requirements for spherical rod end bearings as set forth in SAE J1120 and J1259. The loads, number of cycles, definition of failure, etc., are to be agreed to by the user and supplier. This procedure can also be used as the basis for testing ball joints covered by SAE J490.

2. Objective—To provide a testing format for spherical rod end bearings.

3. Performance Characteristics—The test procedure covers the following characteristics:

  3.1 Ball-to-race diametral clearance per paragraph 4.1.
  3.2 Ball-to-race axial clearance per paragraph 4.2.
  3.3 Ball-to-race rotational torque per paragraph 4.3.
  3.4 Radial static limit load per paragraph 4.4.

3.5 Axial static limit load per paragraph 4.5.
3.6 Fatigue and wear test per paragraph 4.6.

4. Test Objectives and Procedures

4.1 Ball-to-Race Diametral Clearance

4.1.1 OBJECTIVE—To determine the total diametral clearance of the ball within the race.

4.1.2 PROCEDURE—The following method shall be used to determine total diametral clearance. The ball shall be fastened to the fixture as shown in Fig. 1. A reversing force of 20 N is applied to the body and the total movement of the body is measured. The tightening torques shown in Fig. 1 shall be used for all tests.

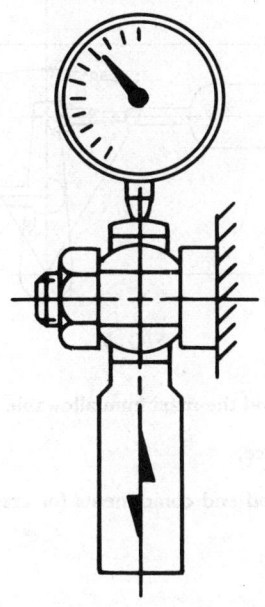

FIG. 1

Rod End Size		Nut Torque
SAE J1120	SAE J1259	N · m
3	5	0.9 ± 0.1
4	6	2.25 ± 0.25
5	8	3.4 ± 0.5
6	10	12 ± 4
7	12	25 ± 7
8	14	45 ± 7
10	16	100 ± 15
12	20	150 ± 20

4.2 Ball-to-Race Axial Clearance

4.2.1 OBJECTIVE—To determine the total axial clearance of the ball within the race.

4.2.2 PROCEDURE—The following method shall be used to determine total axial clearance. The body shall be clamped as shown in Fig. 2. A nut and bolt shall be assembled to the ball. A reversing force of 20 N is applied to the bolt, and the total axial movement of the bolt is measured.

4.3 Ball-to-Race Rotational Torque

4.3.1 OBJECTIVE—To ensure desired rotating torque is obtained.

4.3.2 PROCEDURE—The following method shall be used to determine the breakaway and rotational torque of the ball. The rod end shall be mounted as shown in Fig. 3.

4.3.2.1 Breakaway Torque—Rod ends which are designed to have zero clearance between the ball and race should have a control on breakaway torque. It may be desirable to condition the assembly for a specified period under certain conditions of temperature and humidity. After conditioning, torque shall be gradually applied through a torque-measuring device and the maximum torque determined.

4.3.2.2 No Load Rotational Torque—The ball should be rotated five revolutions prior to recording the torque. The torque is read with a torque-measuring device by the gradual application of a rotating force.

4.4 Radial Static Limit Load

4.4.1 OBJECTIVE—To check the operational radial static load and the radial static limit load capacity of the rod end. The operational radial static load is a load which can be applied a limited number of times to a rod end without impairing the operation.

4.4.2 PROCEDURE—The rod end shall be installed in a test fixture as shown in Fig. 3 using a loose nt on the shaft. A tensile load shall be applied at the rate of 1%/s.

4.4.2.1 Operational Radial Static Load—Apply the load to the required operational radial static load. Hold for 30 s. Release the load. Repeat for a total of five cycles. The rod end ball shall still be free to rotate.

4.4.2.2 Ultimate Radial Static Load—Apply the load to the required ultimate radial static load. Hold for 30 s. The rod end body and other components shall remain intact.

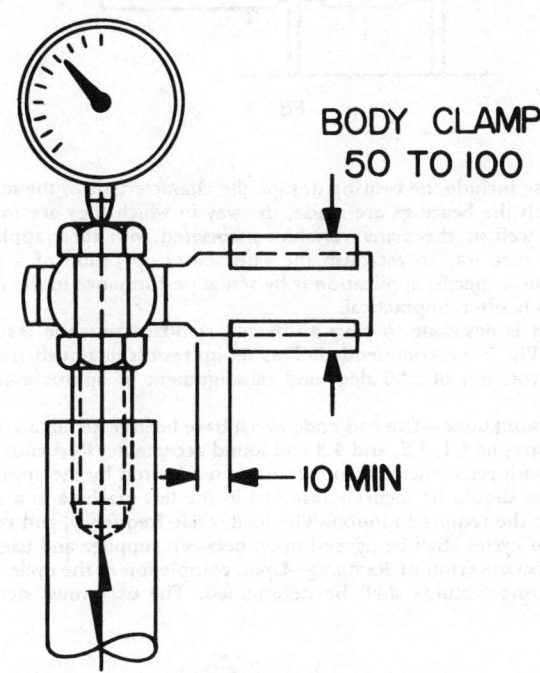

FIG. 2

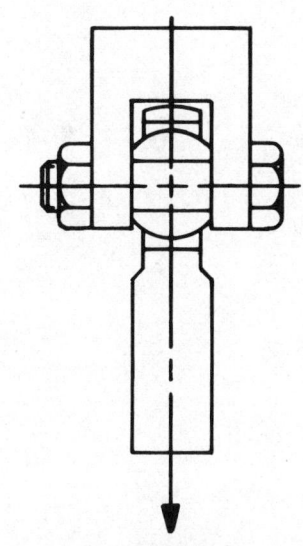

FIG. 3

4.5 Axial Static Limit Load

4.5.1 OBJECTIVE—To check the maximum axial load capacity of the rod end.

4.5.2 PROCEDURE—The rod end shall be mounted on a rigid fixture as shown in Fig. 4, so that only the rod end body is supported. The axial static load shall be applied to the ball at the rate of 1%/s and maintained on the bearing for 30 s. No push-out of the bearing components shall occur when the axial load is applied.

4.6 Fatigue and Wear Test

4.6.1 OBJECTIVE—To determine fatigue and wear characteristics.

4.6.2 DISCUSSION—Rod end performance is a function of many varia-

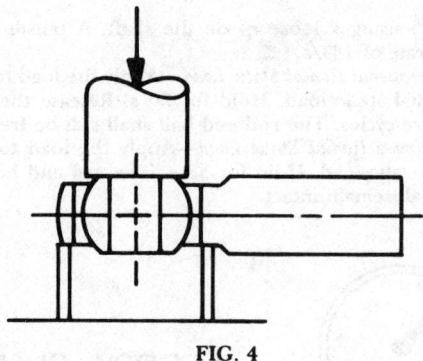

FIG. 4

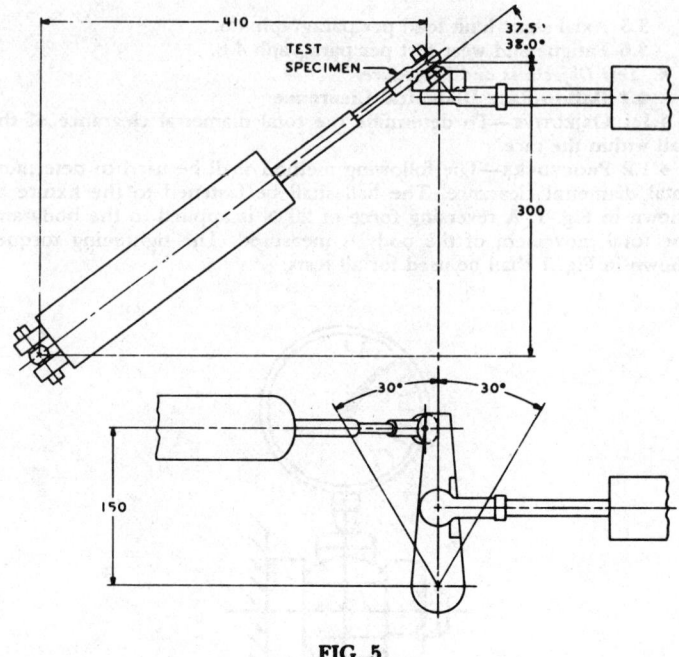

FIG. 5

bles. These include the bearing design, the characteristic of the materials from which the bearings are made, the way in which they are manufactured, as well as the many variables associated with their application. The only sure way to establish the satisfactory operation of a bearing selected for a specific application is by actual performance in the application. This is often impractical.

When it is desirable to have an overall standard test, the test set-up shown in Fig. 5 is recommended. This set-up results in a push/pull loading, ball rotation of ±30 deg, and misalignment of approximately ±5 deg.

4.6.3 PROCEDURE—Use rod ends which have been tested in accordance with paragraphs 4.1, 4.2, and 4.3 and found acceptable. Rod ends should be filled with recommended lubricant when required by the application. Assemblies should be securely fastened in the test machine in a manner to achieve the required motion. The load, cycle frequency, and required number of cycles shall be agreed upon between supplier and user.

4.6.4 EXAMINATION OF RESULTS—Upon completion of the cycle testing, the following features shall be determined. The user must determine the number of cycles and the maximum allowable wear for his particular application.

a) Diametral clearance,
b) Axial clearance,
c) Examination of rod end components for cracks and other signs of damage.

18 Splines

PARALLEL SIDE SPLINES FOR SOFT BROACHED HOLES IN FITTINGS—SAE J499a

SAE Standard

Report of Broaches Division approved January 1914, revised by Shaft Fittings Division March 1920, and reviewed January 1936. Last revised by ANSI B92 Committee—Involute Splines and Inspection—October 1975.

This Information Report along with SAE J500 and J501 is generally understood to be technically obsolete for the design of new applications. However, it is listed for those existing applications where it may be required. For the design of new applications, consult ANSI B92.1-1970—Involute Splines and Inspections Standard.

[The dimensions, given in inches, apply only to soft broached holes. The shaft dimensions depend upon the shape and material of the parts, their heat treatment, and methods of machining to give the required fit. The method and amount of "breaking" sharp corners and edges also depend upon the conditions and requirements of each application.

The formula for theoretical torque capacity (pressure on sides of spline) in inch-pounds per inch of bearing length (L) and at 1000 psi pressure is:

$$T = \text{Torque} = 1000 \times \text{No. of splines} \times \text{mean radius} \times h \times L$$

The tolerances allowed are for good construction and may be readily maintained by usual broaching methods. The tolerances selected for the large and small diameters will depend upon whether the fit between the mating parts, as finally made, is on the large or the small diameter. The other diameter, being designed for clearance may have a wider manufacturing tolerance. If the final fit between the parts is on only the sides of the spline, wider tolerances may be permitted on both the large and small diameters.]

Radii on corners of splines are not to exceed 0.015 in.

Splines shall not be more than 0.006 in per ft out of parallel with respect to the axis of the shaft.

No allowance is made for radii on corners or for clearance. Dimensions are intended to apply to only the soft broached hole. Allowance must be made for machining.

For values of D, W, d, h, and T for four-, six-, ten-, and sixteen-spline fittings, see Tables 2, 3, 4, and 5, respectively.

TABLE 1—W, h, AND d, IN TERMS OF LARGE DIAMETER, D

No. of Splines	W For All Fits	A Permanent Fit		B To Slide when Not under Load		C To Slide under Load	
		h	d	h	d	h	d
4	0.241[a]	0.075	0.850	0.125	0.750	—	—
6	0.250	0.050	0.900	0.075	0.850	0.100	0.800
10	0.156	0.045	0.910	0.070	0.860	0.095	0.810
16	0.098	0.045	0.910	0.070	0.860	0.095	0.810

[a] Four splines, for fits A and B only.

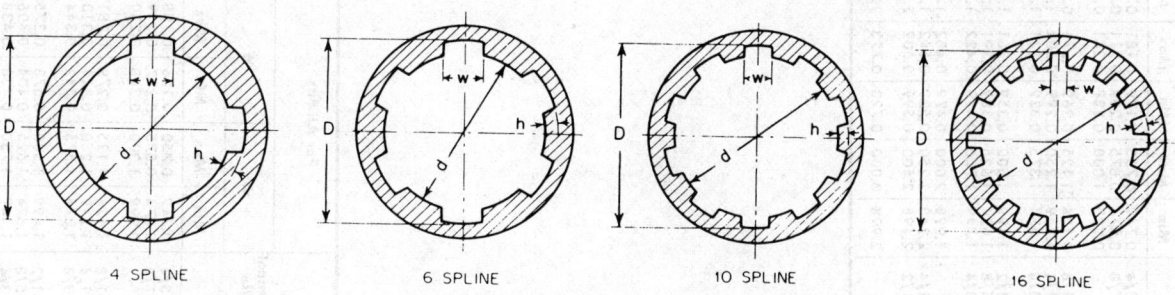

FIG. 1—DIMENSIONS FOR 4, 6, 10, AND 16 SPLINE FITTINGS (SEE TABLE 1)

TABLE 2—FOUR SPLINE FITTINGS

Nominal Dia	For All Fits D		W		4A, Permanent Fit d		h		T	4B, To Slide when Not under Load d		h		T
	Min	Max	Min	Max	Min	Max	Min	Max		Min	Max	Min	Max	
3/4	0.749	0.750	0.179	0.181	0.636	0.637	0.055	0.056	80	0.561	0.562	0.093	0.094	78
7/8	0.874	0.875	0.209	0.211	0.743	0.744	0.065	0.066	109	0.655	0.656	0.108	0.109	107
1	0.999	1.000	0.239	0.241	0.849	0.850	0.074	0.075	143	0.749	0.750	0.124	0.125	139
1-1/8	1.124	1.125	0.269	0.271	0.955	0.956	0.083	0.084	180	0.843	0.844	0.140	0.141	175
1-1/4	1.249	1.250	0.299	0.301	1.061	1.062	0.093	0.094	223	0.936	0.937	0.155	0.156	217
1-3/8	1.374	1.375	0.329	0.331	1.168	1.169	0.102	0.103	269	1.030	1.031	0.171	0.172	262
1-1/2	1.499	1.500	0.359	0.361	1.274	1.275	0.111	0.112	321	1.124	1.125	0.186	0.187	311
1-5/8	1.624	1.625	0.389	0.391	1.380	1.381	0.121	0.122	376	1.218	1.219	0.202	0.203	367
1-3/4	1.749	1.750	0.420	0.422	1.486	1.487	0.130	0.131	436	1.311	1.312	0.218	0.219	424
2	1.998	2.000	0.479	0.482	1.698	1.700	0.148	0.150	570	1.498	1.500	0.248	0.250	555
2-1/4	2.248	2.250	0.539	0.542	1.910	1.912	0.167	0.169	721	1.685	1.687	0.279	0.281	703
2-1/2	2.498	2.500	0.599	0.602	2.123	2.125	0.185	0.187	891	1.873	1.875	0.310	0.312	865
3	2.998	3.000	0.720	0.723	2.548	2.550	0.223	0.225	1283	2.248	2.250	0.373	0.375	1249

TABLE 3—SIX SPLINE FITTINGS

Nominal Dia	For All Fits D		W		6A, Permanent Fit d		h		T	6B, To Slide when Not under Load d		h		T	6C, To Slide when under Load d		c		T
	Min	Max	Min	Max	Min	Max	Min	Max		Min	Max	Min	Max		Min	Max	Min	Max	
3/4	0.749	0.750	0.186	0.188	0.674	0.675			80	0.637	0.638			117	0.599	0.600			152
7/8	0.874	0.875	0.217	0.219	0.787	0.788			109	0.744	0.745			159	0.699	0.700			207
1	0.999	1.000	0.248	0.250	0.899	0.900			143	0.850	0.851			208	0.799	0.800			270
1-1/8	1.124	1.125	0.279	0.281	1.012	1.013			180	0.955	0.956			263	0.899	0.900			342
1-1/4	1.249	1.250	0.311	0.313	1.124	1.125			223	1.062	1.063			325	0.999	1.000			421
1-3/8	1.374	1.375	0.342	0.344	1.237	1.238			269	1.168	1.169			393	1.099	1.100			510
1-1/2	1.499	1.500	0.373	0.375	1.349	1.350			321	1.274	1.275			468	1.199	1.200			608
1-5/8	1.624	1.625	0.404	0.406	1.462	1.463			376	1.380	1.381			550	1.299	1.300			713
1-3/4	1.749	1.750	0.436	0.438	1.574	1.575			436	1.487	1.488			637	1.399	1.400			827
2	1.998	2.000	0.497	0.500	1.798	1.800			570	1.698	1.700			833	1.598	1.600			1080
2-1/4	2.248	2.250	0.560	0.563	2.023	2.025			721	1.911	1.913			1052	1.798	1.800			1367
2-1/2	2.499	2.500	0.622	0.625	2.248	2.250			891	2.123	2.125			1300	1.998	2.000			1688
3	2.998	3.000	0.747	0.750	2.698	2.700			1283	2.548	2.550			1873	2.398	2.400			2430

TABLE 4—TEN SPLINE FITTINGS

Nominal Dia	For All Fits D		W		10A, Permanent Fit d		T	10B, To Slide when Not under Load d		T	10C, To Slide when under Load d		T
	Min	Max	Min	Max	Min	Max		Min	Max		Min	Max	
3/4	0.749	0.750	0.115	0.117	0.682	0.683	120	0.644	0.645	183	0.607	0.608	241
7/8	0.874	0.875	0.135	0.137	0.795	0.796	165	0.752	0.753	248	0.708	0.709	329
1	0.999	1.000	0.154	0.156	0.909	0.910	215	0.859	0.860	326	0.809	0.810	430
1-1/8	1.124	1.125	0.174	0.176	1.023	1.024	271	0.967	0.968	412	0.910	0.911	545
1-1/4	1.249	1.250	0.193	0.195	1.137	1.138	336	1.074	1.075	508	1.012	1.013	672
1-3/8	1.374	1.375	0.213	0.215	1.250	1.251	406	1.182	1.183	614	1.113	1.114	813
1-1/2	1.499	1.500	0.232	0.234	1.364	1.365	483	1.289	1.290	732	1.214	1.215	967
1-5/8	1.624	1.625	0.252	0.254	1.478	1.479	566	1.397	1.398	860	1.315	1.316	1135
1-3/4	1.749	1.750	0.271	0.273	1.592	1.593	658	1.504	1.505	997	1.417	1.418	1316
2	1.998	2.000	0.309	0.312	1.818	1.820	860	1.718	1.720	1302	1.618	1.620	1720
2-1/4	2.248	2.250	0.348	0.351	2.046	2.048	1088	1.933	1.935	1647	1.821	1.823	2176
2-1/2	2.498	2.500	0.387	0.390	2.273	2.275	1343	2.148	2.150	2034	2.023	2.025	2688
3	2.998	3.000	0.465	0.468	2.728	2.730	1934	2.578	2.580	2929	2.428	2.430	3869
3-1/2	3.497	3.500	0.543	0.546	3.182	3.185	2632	3.007	3.010	3987	2.832	2.835	5266
4	3.997	4.000	0.621	0.624	3.637	3.640	3438	3.437	3.440	5208	3.237	3.240	6878
4-1/2	4.497	4.500	0.699	0.702	4.092	4.095	4351	3.867	3.870	6591	3.645	3.645	8705
5	4.997	5.000	0.777	0.780	4.547	4.550	5371	4.297	4.300	8137	4.047	4.050	10746
5-1/2	5.497	5.500	0.855	0.858	5.002	5.005	6500	4.727	4.730	9846	4.452	4.455	13003
6	5.997	6.000	0.933	0.936	5.457	5.460	7735	5.157	5.160	11718	4.857	4.860	15475

TABLE 5—SIXTEEN SPLINE FITTINGS

Nominal Dia	For All Fits D		W		16A, Permanent Fit d		T	16B, To Slide when Not under Load d		T	16C, To Slide when under Load d		T
	Min	Max	Min	Max	Min	Max		Min	Max		Min	Max	
2	1.997	2.000	0.193	0.196	1.817	1.820	1375	1.717	1.720	2083	1.617	1.620	2751
2-1/2	2.497	2.500	0.242	0.245	2.273	2.275	2149	2.147	2.150	3255	2.022	2.025	4299
3	2.997	3.000	0.291	0.294	2.727	2.730	3094	2.577	2.580	4687	2.427	2.430	6190
3-1/2	3.497	3.500	0.340	0.343	3.182	3.185	4212	3.007	3.010	6378	2.832	2.835	8426
4	3.997	4.000	0.389	0.392	3.637	3.640	5501	3.437	3.440	8333	3.237	3.240	11005
4-1/2	4.497	4.500	0.438	0.441	4.092	4.095	6962	3.867	3.870	10546	3.642	3.645	13928
5	4.997	5.000	0.487	0.490	4.547	4.550	8595	4.297	4.300	13020	4.047	4.050	17195
5-1/2	5.497	5.500	0.536	0.539	5.002	5.005	10395	4.727	4.730	15754	4.452	4.455	20806
6	5.997	6.000	0.585	0.588	5.457	5.460	12377	5.157	5.160	18749	4.857	4.860	24760

SERRATED SHAFT ENDS—SAE J500 AUG89   SAE Recommended Practice

Report of Parts and Fittings Division approved 1922 and last revised by Parts and Fittings Committee June 1955. Revised by the Gears and Splines Committee August 1989.

(R) **1. Scope**—This SAE Recommended Practice is intended for service only. Use ANSI B 92.1 and 1a, Involute Splines and Inspection Standard.

(R) **2. Straight Shafts**—
N = Number of serrations.
b = Included angle of the space in the hole, and the tooth on the shaft.
The pitch diameter (PD) and hole are basic.
The pitch line is midway between the inner and outer sharp points.
The minimum hole with maximum shaft as measured across wires in Table 1, produce basic (no clearance) fit.
The wire diameter in Table 1 is the diameter that will bear on the pitch line.
Tolerance for diameter across wires = −0.001 on tooth thickness, for hole and shaft sizes from 1/8 to 1 3/4 in, and −0.0015 on sizes from 2 to 3 in, inclusive.
Tooth thickness on the shaft may be varied from the tolerance given, to secure desired fit.
Wc = Constant to be added to measurement across wires for shaft, subtracted for hole, for each 0.001-in increase in wire diameter used over wire size in Table 1.
When serrations are hobbed, the sides of teeth are involute. This departure from flat sides is slight and is ignored.

3. Formulas
3.1 Diameter over sharp points:
  (OD) = 1.0476479 PD for 36 serrations.
     = 1.0349592 PD for 48 serrations.
3.2 Diameter under sharp points:
  (RD) = 0.9523521 PD for 36 serrations.
     = 0.9650408 PD for 48 serrations.
3.3 Diameter of wire that will bear on pitch line of hole:
  (Wh) = 0.05309792 PD for 36 serrations.
     = 0.04133332 PD for 48 serrations.

Diameter of wire that will bear on pitch line of shaft:
  (Ws) = 0.06585005 PD for 36 serrations.
     = 0.0485955 PD for 48 serrations.
3.4 Measurement across wires:
  for hole = 0.9119441 PD for 36 serrations.
       = 0.9309375 PD for 48 serrations.
Measurement across wires:
  for shaft = 1.1113285 PD for 36 serrations.
        = 1.0823601 PD for 48 serrations.
3.5 Tolerance constant for measurement across wires per 0.001-in tooth

$$\text{thickness for hole} = 0.001 \frac{\sin 45}{\sin \left(45 - \frac{180}{N}\right)}$$

$$\text{for shaft} = 0.001 \frac{\sin \left(45 - \frac{180}{N}\right)}{\sin 45}$$

3.6 Wire diameter constant for use of other diameter wire than that which bears on pitch line, (Wc)

$$0.001 + \frac{0.001}{\sin \left(45 - \frac{180}{N}\right)} \text{ for hole.}$$

$$= 0.001 + \frac{0.001}{\cos 45} \text{ for shaft.}$$

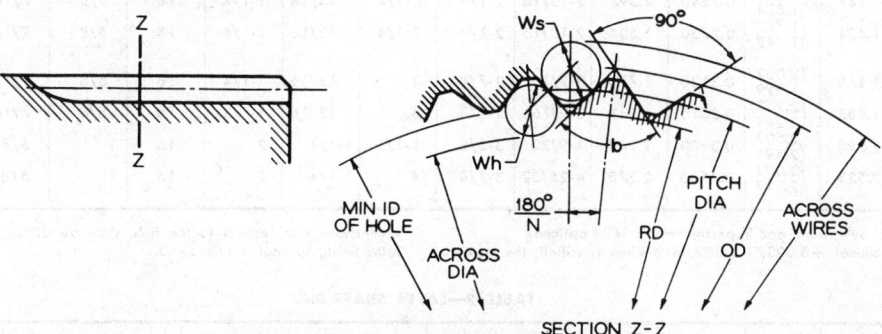

SECTION Z-Z

TABLE 1—DIMENSIONS OF HOLES AND SHAFTS, IN

Nominal Dia	Hole and Shaft			Theoretical Dia of Points		Hole						Shaft					
	Pitch Dia	N	b, Deg	OD	RD	Large Dia, min	Small Dia Max	Small Dia Min	Dia Across Wires Min	Dia Across Wires Max	Wire Size, Wh	Wc	Root Dia, max	Outside Dia Max	Outside Dia Min	Dia Across Wires Max	Dia Across Wires Min	Wire Size, Ws	Wc
1/8	0.122	36	80	0.1278	0.1162	0.125	0.118	0.117	0.1113	0.1124	0.0065	0.0026	0.116	0.124	0.123	0.1356	0.1347	0.0080	0.0024
3/16	0.182	36	80	0.1907	0.1733	0.187	0.176	0.175	0.1660	0.1671	0.0097	0.0026	0.174	0.186	0.185	0.2023	0.2014	0.0120	0.0024
1/4	0.243	36	80	0.2546	0.2314	0.250	0.235	0.234	0.2216	0.2227	0.0129	0.0026	0.233	0.249	0.248	0.2701	0.2692	0.0160	0.0024
5/16	0.303	36	80	0.3174	0.2886	0.312	0.293	0.292	0.2763	0.2774	0.0161	0.0026	0.291	0.311	0.310	0.3367	0.3358	0.0200	0.0024
3/8	0.363	36	80	0.3803	0.3457	0.375	0.352	0.351	0.3310	0.3321	0.0193	0.0026	0.350	0.374	0.373	0.4034	0.4025	0.0239	0.0024
1/2	0.485	36	80	0.5081	0.4619	0.500	0.469	0.468	0.4423	0.4434	0.0258	0.0026	0.467	0.499	0.498	0.5390	0.5381	0.0319	0.0024
5/8	0.605	36	80	0.6338	0.5762	0.625	0.584	0.583	0.5517	0.5528	0.0321	0.0026	0.582	0.624	0.623	0.6724	0.6715	0.0398	0.0024
3/4	0.733	48	82-1/2	0.7586	0.7074	0.750	0.716	0.714	0.6824	0.6835	0.0303	0.0025	0.713	0.749	0.747	0.7934	0.7925	0.0356	0.0024
7/8	0.855	48	82-1/2	0.8849	0.8251	0.875	0.835	0.833	0.7960	0.7971	0.0353	0.0025	0.832	0.874	0.872	0.9254	0.9245	0.0415	0.0024
1	0.977	48	82-1/2	1.0112	0.9428	1.000	0.954	0.952	0.9095	0.9106	0.0404	0.0025	0.951	0.999	0.997	1.0575	1.0566	0.0475	0.0024
1-1/8	1.098	48	82-1/2	1.1364	1.0596	1.125	1.071	1.069	1.0222	1.0233	0.0454	0.0025	1.068	1.124	1.122	1.1884	1.1875	0.0534	0.0024
1-1/4	1.220	48	82-1/2	1.2626	1.1773	1.250	1.190	1.188	1.1357	1.1368	0.0504	0.0025	1.187	1.249	1.247	1.3205	1.3196	0.0593	0.0024
1-3/8	1.342	48	82-1/2	1.3889	1.2951	1.375	1.309	1.307	1.2493	1.2504	0.0555	0.0025	1.306	1.374	1.372	1.4525	1.4516	0.0652	0.0024
1-1/2	1.464	48	82-1/2	1.5152	1.4128	1.500	1.428	1.426	1.3629	1.3640	0.0605	0.0025	1.425	1.499	1.497	1.5846	1.5837	0.0711	0.0024
1-3/4	1.708	48	82-1/2	1.7677	1.6483	1.750	1.666	1.664	1.5900	1.5911	0.0706	0.0025	1.663	1.749	1.747	1.8487	1.8478	0.0830	0.0024
2	1.952	48	82-1/2	2.0202	1.8838	2.000	1.904	1.902	1.8172	1.8188	0.0807	0.0025	1.901	1.999	1.997	2.1128	2.1114	0.0949	0.0024
2-1/4	2.196	48	82-1/2	2.2728	2.1192	2.250	2.142	2.140	2.0443	2.0459	0.0908	0.0025	2.139	2.249	2.247	2.3769	2.3755	0.1067	0.0024
2-1/2	2.440	48	82-1/2	2.5253	2.3547	2.500	2.380	2.378	2.2715	2.2731	0.1009	0.0025	2.377	2.499	2.497	2.6410	2.6396	0.1180	0.0024
2-3/4	2.684	48	82-1/2	2.7778	2.5902	2.750	2.618	2.616	2.4986	2.5002	0.1109	0.0025	2.615	2.749	2.747	2.9051	2.9037	0.1304	0.0024
3	2.928	48	82-1/2	3.0304	2.8256	3.000	2.856	2.854	2.7258	2.7274	0.1210	0.0025	2.853	2.999	2.997	3.1692	3.1678	0.1423	0.0024

SHAFT ENDS—SAE J501 — SAE Standard

Report of Broaches Division approved June 1914 and last revised by Parts and Fittings Committee May 1948.

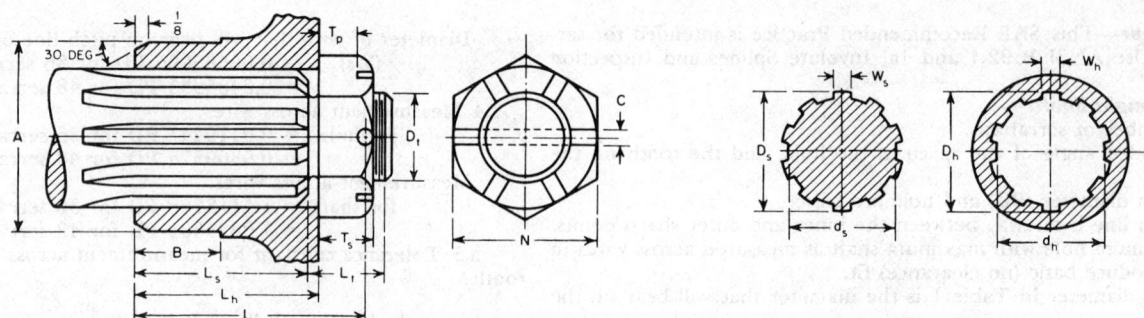

FIG. 1—PERMANENT FIT SPLINE SHAFT ENDS (SEE TABLE 1)

TABLE 1—PERMANENT FIT SPLINE SHAFT ENDS FOR UNIVERSAL JOINTS AND SIMILAR APPLICATIONS

Nominal Shaft Dia	10-Spline Shaft[a]			10-Spline Hole[a]			Hub Dimensions									C	A[b]	B
	D_s +0.000 −0.001	W_s +0.000 −0.0015	d_s +0.000 −0.010	D_h +0.000 −0.0015	W_h +0.000 −0.0015	d_h +0.000 −0.000	L_c	L_s	L_h	L_t	D_t	Threads per in.	T_s	T_p	N			
3/4	0.749	0.1170	0.632	0.751/0.749	0.1170	0.682	1-11/32	15/16	1	17/32	1/2	28	7/16	1/4	13/16	1/8	1-1/8	5/8
7/8	0.874	0.1370	0.745	0.876/0.874	0.1370	0.795	1-11/16	1-1/8	1-1/4	11/16	5/8	24	1/2	5/16	15/16	5/32	1-1/4	3/4
1	0.999	0.1560	0.859	1.001/0.999	0.1560	0.909	1-15/16	1-3/8	1-1/2	11/16	3/4	20	1/2	5/16	1-1/16	5/32	1-3/8	7/8
1-1/8	1.124	0.1760	0.973	1.126/1.124	0.1760	1.023	1-15/16	1-3/8	1-1/2	11/16	7/8	20	1/2	5/16	1-1/4	5/32	1-1/2	7/8
1-1/4	1.249	0.1950	1.087	1.251/1.249	0.1950	1.137	1-15/16	1-3/8	1-1/2	11/16	1	20	1/2	5/16	1-7/16	5/32	1-3/4	7/8
1-3/8	1.374	0.2150	1.200	1.376/1.374	0.2150	1.250	2-7/16	1-7/8	2	11/16	1	20	1/2	5/16	1-7/16	5/32	2	1
1-1/2	1.499	0.2340	1.304	1.501/1.499	0.2340	1.364	2-7/16	1-7/8	2	13/16	1-1/4	18	5/8	7/16	1-13/16	5/32	2-1/4	1
1-5/8	1.624	0.2540	1.347	1.627/1.624	0.2540	1.397	2-13/16	2-1/8	2-1/4	13/16	1-1/4	18	5/8	7/16	1-13/16	5/32	2-3/8	1-1/4
1-3/4	1.749	0.2730	1.454	1.752/1.749	0.2730	1.504	2-13/16	2-1/8	2-1/4	13/16	1-1/4	18	5/8	7/16	2-3/16	5/32	2-1/2	1-1/4
2	1.999	0.3120	1.668	2.002/1.999	0.3120	1.718	3-9/16	2-7/8	3	13/16	1-1/4	18	5/8	7/16	2-3/16	5/32	2-3/4	1-1/2
2-1/4	2.249	0.3510	1.883	2.252/2.249	0.3510	1.933	3-9/16	2-7/8	3	13/16	1-1/2	18	5/8	7/16	2-3/8	5/32	3	1-1/2
2-1/2	2.499	0.3900	2.098	2.502/2.499	0.3900	2.148	4-9/32	3-3/8	3-1/2	1-1/4	2	16	1	5/8	3-1/8	7/32	3-1/2	1-3/4
3	2.999	0.4680	2.528	3.002/2.999	0.4680	2.578	4-25/32	3-7/8	4	1-1/4	2	16	1	5/8	3-1/8	7/32	4	2

[a] SAE Standard, Involute Splines, Serrations, and Inspection—SAE J498 optional.
[b] Tolerance for ground finish, nominal +0.003, −0.002; and when specified, the maximum eccentricity with respect to the hole shall be 0.002 (indicator reading 0.004). Tolerance for lathe finish, nominal +1/32, −0.

TABLE 2—TAPER SHAFT END

Nominal Shaft Dia	D_s Shaft Dia		D_h Hole Dia		L_c	L_s	L_h	L_t	D_t	Threads per in.	T_p	T_p	Nut Width (Flats)	C	W		H +0.004 −0.000		Square Key		A[a]	B
	Max	Min	Max	Min											Max	Min	Max	Min	Max	Min		
1/4	0.250	0.249	0.248	0.247	9/16	5/16	3/8	5/16	#10	40	7/32	9/64	5/16	5/64	0.0625	0.0615	0.033		0.0635	0.0625	1/2	3/16
3/8	0.375	0.374	0.373	0.372	47/64	7/16	1/2	23/64	5/16	32	17/64	17/64	1/2	5/64	0.0937	0.0927	0.049		0.0947	0.0937	11/16	1/4
1/2	0.500	0.499	0.498	0.497	63/64	11/16	3/4	23/64	5/16	32	17/64	17/64	3/16	5/64	0.1250	0.1240	0.065		0.1260	0.1250	7/8	3/8
5/8	0.625	0.624	0.623	0.622	1-3/32	11/16	3/4	17/32	1/2	28	7/16	1/4	3/4	1/8	0.1562	0.1552	0.080		0.1572	0.1562	1-1/16	1/2
3/4	0.750	0.749	0.748	0.747	1-11/32	15/16	1	17/32	1/2	28	7/16	1/4	3/4	1/8	0.1875	0.1865	0.096		0.1885	0.1875	1-1/4	5/8
7/8	0.875	0.874	0.873	0.872	1-11/16	1-1/8	1-1/4	11/16	5/8	24	1/2	5/16	15/16	5/32	0.2500	0.2490	0.127		0.2510	0.2500	1-1/2	3/4
1	1.001	0.999	0.997	0.995	1-15/16	1-3/8	1-1/2	11/16	3/4	20	1/2	5/16	1-1/4	5/32	0.2500	0.2490	0.127		0.2510	0.2500	1-3/4	7/8
1-1/8	1.126	1.124	1.122	1.120	1-15/16	1-3/8	1-1/2	11/16	7/8	20	1/2	5/16	1-1/4	5/32	0.3125	0.3115	0.158		0.3135	0.3125	7/8	7/8
1-1/4	1.251	1.249	1.247	1.245	1-15/16	1-3/8	1-1/2	11/16	1	20	1/2	5/16	1-7/16	5/32	0.3125	0.3115	0.158		0.3135	0.3125	2-1/8	7/8
1-3/8	1.376	1.374	1.372	1.370	2-7/16	1-7/8	2	11/16	1	20	1/2	5/16	1-7/16	5/32	0.3750	0.3740	0.190		0.3760	0.3750	2	1
1-1/2	1.501	1.499	1.497	1.495	2-7/16	1-7/8	2	11/16	1	20	1/2	5/16	1-7/16	5/32	0.3750	0.3740	0.190		0.3760	0.3750	2-1/2	1
1-5/8	1.626	1.624	1.622	1.620	2-13/16	2-1/8	2-1/4	13/16	1-1/4	18	5/8	7/16	2-3/16	5/32	0.4375	0.4365	0.221		0.4385	0.4375	2-3/4	1-1/4
1-3/4	1.751	1.749	1.747	1.745	2-13/16	2-1/8	2-1/4	13/16	1-1/4	18	5/8	7/16	2-3/16	5/32	0.4375	0.4365	0.221		0.4385	0.4375	3	1-1/4
1-7/8	1.876	1.874	1.872	1.870	3-1/16	2-3/8	2-1/2	13/16	1-1/4	18	5/8	7/16	2-3/16	5/32	0.4375	0.4365	0.221		0.4385	0.4375	3-1/8	1-1/4
2	2.001	1.999	1.997	1.995	3-9/16	2-7/8	3	13/16	1-1/4	18	5/8	7/16	2-3/16	5/32	0.5000	0.4990	0.252		0.5010	0.5000	3-1/4	1-1/2
2-1/4	2.252	2.248	2.245	2.242	3-9/16	2-7/8	3	13/16	1-1/2	18	5/8	7/16	2-3/8	5/32	0.5625	0.5610	0.283		0.5640	0.5625	3-1/2	1-1/2
2-1/2	2.502	2.498	2.495	2.492	4-9/32	3-3/8	3-1/2	1-1/4	2	16	1	5/8	3-1/8	7/32	0.6250	0.6235	0.315		0.6265	0.6250	4	1-3/4
2-3/4	2.752	2.748	2.745	2.742	4-9/32	3-3/8	3-1/2	1-1/4	2	16	1	5/8	3-1/8	7/32	0.6875	0.6860	0.346		0.6890	0.6875	4-3/8	1-3/4
3	3.002	2.998	2.995	2.992	4-25/32	3-7/8	4	1-1/4	2	16	1	5/8	3-1/8	7/32	0.7500	0.7485	0.377		0.7515	0.7500	5	2
3-1/4	3.252	3.248	3.245	3.242	5-1/32	4-1/8	4-1/4	1-1/4	2	16	1	5/8	3-1/8	7/32	0.7500	0.7485	0.377		0.7515	0.7500	5	2-1/4
3-1/2	3.502	3.498	3.495	3.492	5-7/16	4-3/8	4-1/2	1-1/4	2-1/2	14	1	3/4	3-7/8	9/32	0.8750	0.8735	0.440		0.8765	0.8750	5-1/2	2-1/4
4	4.002	3.998	3.995	3.992	6-7/16	5-3/8	5-1/2	1-3/8	2-1/2	16	1-1/8	3/4	3-7/8	9/32	1.0000	0.9985	0.502		1.0015	1.0000	6-1/4	2-3/4

[a] Tolerance for ground finish, nominal +0.003, −0.002; and when specified, the maximum eccentricity with respect to the hole shall be 0.002 (indicator reading 0.004). Tolerance for lathe finish, nominal +1/32, −0.

(Continued on next page)

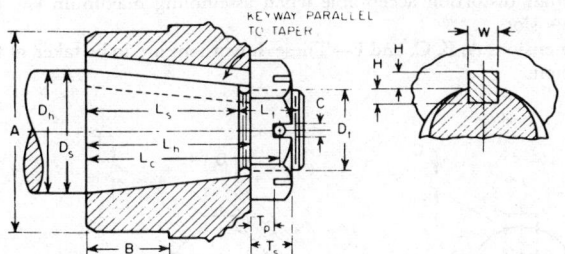

Taper per foot = 1.500 ± 0.002 in. Dimension H is measured normal to the key and at the large end of the taper.

C = cotter pin hole or slot. The centerline of the cotter pin hole shall be 90 deg from the position of the keyway, as shown in Fig. 2.

FIG. 2—TAPER SHAFT END (SEE TABLE 2)

WOODRUFF KEYS—SAE J502   SAE Standard

Report of Parts and Fittings Division approved February 1928 and last revised by Parts and Fittings Committee January 1956. Editorial change September 1972.

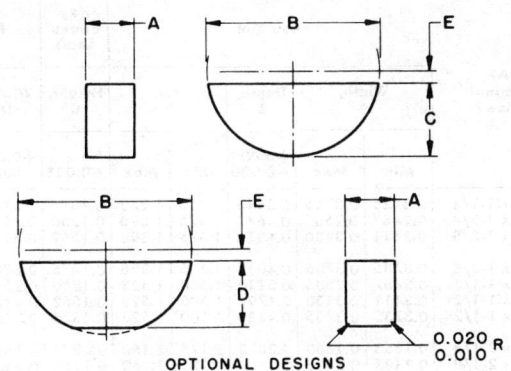

TABLE 1A—KEY DIMENSIONS

Part No.	SAE Nominal Size	Width, A +0.001 −0.000	Dia, B +0.000 −0.010	Height C +0.000 −0.005	Height D +0.000 −0.006	Height E Nominal	Key Area at Shear Line	Approximate Weight, Lb per 1000
201	1/16 x 1/4	0.0625	0.250	0.109	—	1/64	0.0145	0.6
206	1/16 x 5/16	0.0625	0.312	0.140	—	1/64	0.0184	0.7
207	3/32 x 5/16	0.0938	0.312	0.140	—	1/64	0.0264	0.9
211	1/16 x 3/8	0.0625	0.375	0.172	—	1/64	0.0225	0.9
212	3/32 x 3/8	0.0938	0.375	0.172	—	1/64	0.0328	1.3
213	1/8 x 3/8	0.1250	0.375	0.172	—	1/64	0.0420	1.5
1	1/16 x 1/2	0.0625	0.500	0.203	0.194	3/64	0.0296	1.3
2	3/32 x 1/2	0.0938	0.500	0.203	0.194	3/64	0.0434	1.9
3	1/8 x 1/2	0.1250	0.500	0.203	0.194	3/64	0.0512	2.5
4	3/32 x 5/8	0.0938	0.625	0.250	0.240	1/16	0.0523	3.0
5	1/8 x 5/8	0.1250	0.625	0.250	0.240	1/16	0.0716	3.9
6	5/32 x 5/8	0.1563	0.625	0.250	0.240	1/16	0.0871	4.9
61	3/16 x 5/8	0.1875	0.625	0.250	0.240	1/16	0.0105	5.8
7	1/8 x 3/4	0.1250	0.750	0.313	0.303	1/16	0.0884	6.1
8	5/32 x 3/4	0.1563	0.750	0.313	0.303	1/16	0.1086	7.5
9	3/16 x 3/4	0.1875	0.750	0.313	0.303	1/16	0.1279	9.0
91	1/4 x 3/4	0.2500	0.750	0.313	0.303	1/16	0.1623	12.0
10	5/32 x 7/8	0.1563	0.875	0.375	0.365	1/16	0.1294	11.0
11	3/16 x 7/8	0.1875	0.875	0.375	0.365	1/16	0.1531	13.0
12	7/32 x 7/8	0.2188	0.875	0.375	0.365	1/16	0.1813	14.9
A	1/4 x 7/8	0.2500	0.875	0.375	0.365	1/16	0.1976	17.0
13	3/16 x 1	0.1875	1.000	0.438	0.428	1/16	0.1781	17.0
14	7/32 x 1	0.2188	1.000	0.438	0.428	1/16	0.2100	20.1
15	1/4 x 1	0.2500	1.000	0.438	0.428	1/16	0.2320	23.0
B	5/16 x 1	0.3125	1.000	0.438	0.428	1/16	0.2811	29.0
16	3/16 x 1-1/8	0.1875	1.125	0.484	0.475	5/64	0.2007	22.0
17	7/32 x 1-1/8	0.2188	1.125	0.484	0.475	5/64	0.2320	25.0
18	1/4 x 1-1/8	0.2500	1.125	0.484	0.475	5/64	0.2622	29.0
C	5/16 x 1-1/8	0.3125	1.125	0.484	0.475	5/64	0.3193	36.0
19	3/16 x 1-1/4	0.1875	1.250	0.547	0.537	5/64	0.2284	27.1
20	7/32 x 1-1/4	0.2188	1.250	0.547	0.537	5/64	0.2608	31.8
21	1/4 x 1-1/4	0.2500	1.250	0.547	0.537	5/64	0.2955	36.0
D	5/16 x 1-1/4	0.3125	1.250	0.547	0.537	5/64	0.3621	45.0
E	3/8 x 1-1/4	0.3750	1.250	0.547	0.537	5/64	0.4243	54.0
22	1/4 x 1-3/8	0.2500	1.375	0.594	0.584	3/32	0.3259	43.0
23	5/16 x 1-3/8	0.3125	1.375	0.594	0.584	3/32	0.4003	54.0
F	3/8 x 1-3/8	0.3750	1.375	0.594	0.584	3/32	0.4705	65.0
24	1/4 x 1-1/2	0.2500	1.500	0.641	0.631	7/64	0.3562	50.0
25	5/16 x 1-1/2	0.3125	1.500	0.641	0.631	7/64	0.4384	63.0
G	3/8 x 1-1/2	0.3750	1.500	0.641	0.631	7/64	0.5166	75.0

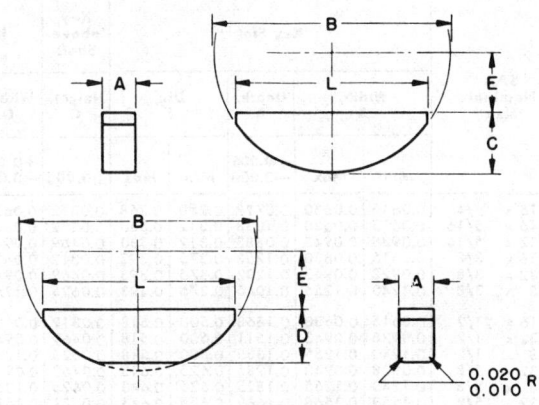

TABLE 1B—KEY DIMENSIONS

Part No.	SAE Nominal Size	Width, A +0.001 −0.000	Dia, B +0.000 −0.010	Height C +0.000 −0.005	Height D +0.000 −0.006	Height E Nominal	Length, L +0.000 −0.010	Key Area at Shear Line	Approximate Weight, Lb per 1000
126	3/16 x 2-1/8	0.1875	2.125	0.406	0.396	21/32	1.380	0.2578	23.4
127	1/4 x 2-1/8	0.2500	2.125	0.406	0.396	21/32	1.380	0.3437	31.2
128	5/16 x 2-1/8	0.3125	2.125	0.406	0.396	21/32	1.380	0.4296	39.3
129	3/8 x 2-1/8	0.3750	2.125	0.406	0.396	21/32	1.380	0.4833	47.2
26	3/16 x 2-1/8	0.1875	2.125	0.531	0.521	17/32	1.723	0.3222	36.3
27	1/4 x 2-1/8	0.2500	2.125	0.531	0.521	17/32	1.723	0.4178	48.2
28	5/16 x 2-1/8	0.3125	2.125	0.531	0.521	17/32	1.723	0.5062	60.1
29	3/8 x 2-1/8	0.3750	2.125	0.531	0.521	17/32	1.723	0.5868	72.3
Rx	1/4 x 2-3/4	0.2500	2.750	0.594	0.584	25/32	2.000	0.5000	64.8
Sx	5/16 x 2-3/4	0.3125	2.750	0.594	0.584	25/32	2.000	0.6286	80.8
Tx	3/8 x 2-3/4	0.3750	2.750	0.594	0.584	25/32	2.000	0.6943	96.6
Ux	7/16 x 2-3/4	0.4375	2.750	0.594	0.584	25/32	2.000	0.8253	112.9
Vx	1/2 x 2-3/4	0.5000	2.750	0.594	0.584	25/32	2.000	0.9094	129.3
R	1/4 x 2-3/4	0.2500	2.750	0.750	0.740	5/8	2.317	0.5718	91.6
S	5/16 x 2-3/4	0.3125	2.750	0.750	0.740	5/8	2.317	0.7071	114.2
T	3/8 x 2-3/4	0.3750	2.750	0.750	0.740	5/8	2.317	0.8319	136.6
U	7/16 x 2-3/4	0.4375	2.750	0.750	0.740	5/8	2.317	0.9499	159.2
V	1/2 x 2-3/4	0.5000	2.750	0.750	0.740	5/8	2.317	1.0606	191.8
30	3/8 x 3-1/2	0.3750	3.500	0.938	0.927	13/16	2.880	1.0781	216.0
31	7/16 x 3-1/2	0.4375	3.500	0.938	0.927	13/16	2.880	1.2371	252.0
32	1/2 x 3-1/2	0.5000	3.500	0.938	0.927	13/16	2.880	1.3905	288.0
33	9/16 x 3-1/2	0.5625	3.500	0.938	0.927	13/16	2.880	1.5368	325.0
34	5/8 x 3-1/2	0.6250	3.500	0.938	0.927	13/16	2.880	1.6755	359.0
35	11/16 x 3-1/2	0.6875	3.500	0.938	0.927	13/16	2.880	1.8062	399.0
36	3/4 x 3-1/2	0.7500	3.500	0.938	0.927	13/16	2.880	1.9281	435.0

Material—Keys are to be carbon steel or alloy heat treated steel as specified. Carbon steel keys are to be 0.30 carbon minimum, with hardness of Rockwell B 90 minimum. Alloy steel keys are to be SAE 2330 or 8630 steel, heat treated to a hardness of 40–50 RC; or other alloy steels having equal physical properties at the same hardness. Alloy heat treated keys are to be marked with depressions on the top to distinguish them from carbon steel keys.

Dimensions—Dimensions are to be as given in Tables 1A, 1B, and 2.

Width A (Shown in Illustrations Accompanying Tables)—Values shown were set with the maximum key slot width as that figure which will receive a key with the greatest amount of looseness consistent with assuring the key's sticking in the slot. Minimum key slot width is that figure permitting the largest shaft distortion acceptable when assembling maximum key in minimum key slot.

Dimensions A, B, C, and E—These dimensions are to be taken at the side intersection.

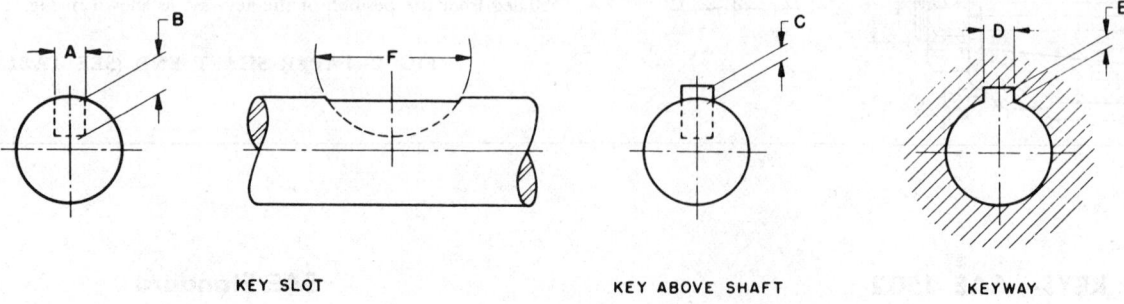

KEY SLOT   KEY ABOVE SHAFT   KEYWAY

TABLE 2—KEY SLOT AND KEYWAY DIMENSIONS

Part No.	SAE Nominal Size	Key Slot Width, A Min	Key Slot Width, A Max	Key Slot Depth, B +0.005 −0.000	Dia, F Min	Dia, F Max	Key above Shaft Height, C ±0.005	Keyway Width, D +0.002 −0.000	Keyway Depth, E +0.005 −0.000	Mfrs Part No.	SAE Nominal Size	Key Slot Width, A Min	Key Slot Width, A Max	Key Slot Depth, B +0.005 −0.000	Dia, F Min	Dia, F Max	Key above Shaft Height, C ±0.005	Keyway Width, D +0.002 −0.000	Keyway Depth, E +0.005 −0.000
201	1/16 x 1/4	0.0615	0.0630	0.0728	0.250	0.268	0.0312	0.0635	0.0372	E	3/8 x 1-1/4	0.3735	0.3755	0.3545	1.250	1.273	0.1875	0.3760	0.1935
206	1/16 x 5/16	0.0615	0.0630	0.1038	0.312	0.330	0.0312	0.0635	0.0372	22	1/4 x 1-3/8	0.2487	0.2505	0.4640	1.375	1.398	0.1250	0.2510	0.1310
207	3/32 x 5/16	0.0928	0.0943	0.0882	0.312	0.330	0.0469	0.0948	0.0529	23	5/16 x 1-3/8	0.3111	0.3130	0.4328	1.375	1.398	0.1562	0.3135	0.1622
211	1/16 x 3/8	0.0615	0.0630	0.1358	0.375	0.393	0.0312	0.0635	0.0372	F	3/8 x 1-3/8	0.3735	0.3755	0.4015	1.375	1.398	0.1875	0.3760	0.1935
212	3/32 x 3/8	0.0928	0.0943	0.1202	0.375	0.393	0.0469	0.0948	0.0529	24	1/4 x 1-1/2	0.2487	0.2505	0.5110	1.500	1.523	0.1250	0.2510	0.1310
213	1/8 x 3/8	0.1240	0.1255	0.1045	0.375	0.393	0.0625	0.1260	0.0685	25	5/16 x 1-1/2	0.3111	0.3130	0.4798	1.500	1.523	0.1562	0.3135	0.1622
1	1/16 x 1/2	0.0615	0.0630	0.1668	0.500	0.518	0.0312	0.0635	0.0372	G	3/8 x 1-1/2	0.3735	0.3755	0.4485	1.500	1.523	0.1875	0.3760	0.1935
2	3/32 x 1/2	0.0928	0.0943	0.1511	0.500	0.518	0.0469	0.0948	0.0529										
3	1/8 x 1/2	0.1240	0.1255	0.1355	0.500	0.518	0.0625	0.1260	0.0685	126	3/16 x 2-1/8	0.1863	0.1880	0.3073	2.125	2.160	0.0937	0.1885	0.0997
4	3/32 x 5/8	0.0928	0.0943	0.1981	0.625	0.643	0.0469	0.0948	0.0529	127	1/4 x 2-1/8	0.2487	0.2505	0.2760	2.125	2.160	0.1250	0.2510	0.1310
5	1/8 x 5/8	0.1240	0.1255	0.1825	0.625	0.643	0.0625	0.1260	0.0685	128	5/16 x 2-1/8	0.3111	0.3130	0.2448	2.125	2.160	0.1562	0.3135	0.1622
6	5/32 x 5/8	0.1553	0.1568	0.1669	0.625	0.643	0.0781	0.1573	0.0841	129	3/8 x 2-1/8	0.3735	0.3755	0.2135	2.125	2.160	0.1875	0.3760	0.1935
61	3/16 x 5/8	0.1863	0.1880	0.1513	0.625	0.643	0.0937	0.1885	0.0997	26	3/16 x 2-1/8	0.1863	0.1880	0.4323	2.125	2.160	0.0937	0.1885	0.0997
7	1/8 x 3/4	0.1240	0.1255	0.2455	0.750	0.768	0.0625	0.1260	0.0685	27	1/4 x 2-1/8	0.2487	0.2505	0.4010	2.125	2.160	0.1250	0.2510	0.1310
8	5/32 x 3/4	0.1553	0.1568	0.2299	0.750	0.768	0.0781	0.1573	0.0841	28	5/16 x 2-1/8	0.3111	0.3130	0.3698	2.125	2.160	0.1562	0.3135	0.1622
9	3/16 x 3/4	0.1863	0.1880	0.2143	0.750	0.768	0.0937	0.1885	0.0997	29	3/8 x 2-1/8	0.3735	0.3755	0.3385	2.125	2.160	0.1875	0.3760	0.1935
91	1/4 x 3/4	0.2487	0.2505	0.1830	0.750	0.768	0.1250	0.2510	0.1310	Rx	1/4 x 2-3/4	0.2487	0.2505	0.4640	2.750	2.785	0.1250	0.2510	0.1310
10	5/32 x 7/8	0.1553	0.1568	0.2919	0.875	0.895	0.0781	0.1573	0.0841	Sx	5/16 x 2-3/4	0.3111	0.3130	0.4328	2.750	2.785	0.1562	0.3135	0.1622
11	3/16 x 7/8	0.1863	0.1880	0.2763	0.875	0.895	0.0937	0.1885	0.0997	Tx	3/8 x 2-3/4	0.3735	0.3755	0.4015	2.750	2.785	0.1875	0.3760	0.1935
12	7/32 x 7/8	0.2175	0.2193	0.2607	0.875	0.895	0.1093	0.2198	0.1153	Ux	7/16 x 2-3/4	0.4360	0.4380	0.3703	2.750	2.785	0.2187	0.4385	0.2247
A	1/4 x 7/8	0.2487	0.2505	0.2450	0.875	0.895	0.1250	0.2510	0.1310	Vx	1/2 x 2-3/4	0.4985	0.5005	0.3390	2.750	2.785	0.2500	0.5010	0.2560
13	3/16 x 1	0.1863	0.1880	0.3393	1.000	1.020	0.0937	0.1885	0.0997	k	1/4 x 2-3/4	0.2487	0.2505	0.6200	2.750	2.785	0.1250	0.2510	0.1310
14	7/32 x 1	0.2175	0.2193	0.3237	1.000	1.020	0.1093	0.2198	0.1153	S	5/16 x 2-3/4	0.3111	0.3130	0.5888	2.750	2.785	0.1562	0.3135	0.1622
15	1/4 x 1	0.2487	0.2505	0.3080	1.000	1.020	0.1250	0.2510	0.1310	T	3/8 x 2-3/4	0.3735	0.3755	0.5575	2.750	2.785	0.1875	0.3760	0.1935
B	5/16 x 1	0.3111	0.3130	0.2768	1.000	1.020	0.1562	0.3135	0.1622	U	7/16 x 2-3/4	0.4360	0.4380	0.5263	2.750	2.785	0.2187	0.4385	0.2247
16	3/16 x 1-1/8	0.1863	0.1880	0.3853	1.125	1.145	0.0937	0.1885	0.0997	V	1/2 x 2-3/4	0.4985	0.5005	0.4950	2.750	2.785	0.2500	0.5010	0.2560
17	7/32 x 1-1/8	0.2175	0.2193	0.3697	1.125	1.145	0.1093	0.2198	0.1153										
18	1/4 x 1-1/8	0.2487	0.2505	0.3540	1.125	1.145	0.1250	0.2510	0.1310	30	3/8 x 3-1/2	0.3735	0.3755	0.7455	3.500	3.535	0.1875	0.3760	0.1935
C	5/16 x 1-1/8	0.3111	0.3130	0.3228	1.125	1.145	0.1562	0.3135	0.1622	31	7/16 x 3-1/2	0.4360	0.4380	0.7143	3.500	3.535	0.2187	0.4385	0.2247
19	3/16 x 1-1/4	0.1863	0.1880	0.4483	1.250	1.273	0.0937	0.1885	0.0997	32	1/2 x 3-1/2	0.4985	0.5005	0.6830	3.500	3.535	0.2500	0.5010	0.2560
										33	9/16 x 3-1/2	0.5610	0.5630	0.6518	3.500	3.535	0.2812	0.5635	0.2872
20	7/32 x 1-1/4	0.2175	0.2193	0.4327	1.250	1.273	0.1093	0.2198	0.1153	34	5/8 x 3-1/2	0.6235	0.6255	0.6205	3.500	3.535	0.3125	0.6260	0.3185
21	1/4 x 1-1/4	0.2487	0.2505	0.4170	1.250	1.273	0.1250	0.2510	0.1310	35	11/16 x 3-1/2	0.6860	0.6880	0.5893	3.500	3.535	0.3437	0.6885	0.3497
D	5/16 x 1-1/4	0.3111	0.3130	0.3858	1.250	1.273	0.1562	0.3135	0.1622	36	3/4 x 3-1/2	0.7485	0.7505	0.5580	3.500	3.535	0.3750	0.7510	0.3810

WOODRUFF KEY SLOTS AND KEYWAYS—SAE J503

SAE Information Report

Report of Parts and Fittings Division approved February 1928 and last revised by Parts and Fittings Committee January 1949. Editorial change May 1959.

Shaft Diameter, L—Decimal equivalents are given to four places in Table 1. All figures are calculated from this basic dimension. Any change in the shaft diameter from basic will necessarily change all other figures; and, in this case, should accurate dimensions be required, the formula given below should be used.

Versed Sine, G—The versed sines specified are determined from the following formula:

$$G = \frac{L}{2} - \sqrt{\frac{L^2 - A^2}{4}}$$

where A is the minimum width of the key.

Bottom of Key Slot to Opposite Side of Shaft, H—Obtain by subtracting the versed sine G, and depth of key slot, B, from the shaft diameter, L.

Top of Key to Opposite Side of Shaft, J—Obtain by subtracting the versed sine, G, from the shaft diameter, L, and then adding to this figure the height of key above shaft, C.

Bottom of Keyway to Opposite Side of Bore, K—Obtain by subtracting the versed sine, G, from the shaft diameter, L, and then adding to this figure the depth of keyway, E.

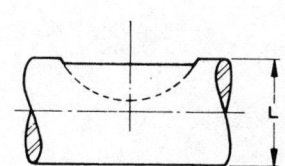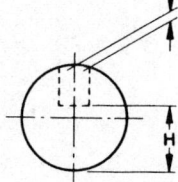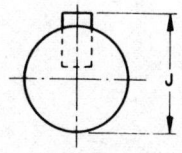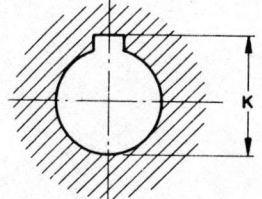

TABLE 1—VERSED SINE DIMENSION, G[a]

L Shaft Dia	Key Width													
	1/16	3/32	1/8	5/32	3/16	7/32	1/4	5/16	3/8	7/16	1/2	9/16	5/8	11/16	3/4
0.3125	0.0032	—	—	—	—	—	—	—	—	—	—	—	—	—
0.3437	0.0029	0.0065	—	—	—	—	—	—	—	—	—	—	—	—
0.3750	0.0026	0.0060	0.0107	—	—	—	—	—	—	—	—	—	—	—
0.4060	0.0024	0.0055	0.0099	—	—	—	—	—	—	—	—	—	—	—
0.4375	0.0022	0.0051	0.0091	—	—	—	—	—	—	—	—	—	—	—
0.4687	0.0021	0.0047	0.0085	0.0134	—	—	—	—	—	—	—	—	—	—
0.5000	0.0020	0.0044	0.0079	0.0125	—	—	—	—	—	—	—	—	—	—
0.5625	—	0.0039	0.0070	0.0111	0.0161	—	—	—	—	—	—	—	—	—
0.6250	—	0.0035	0.0063	0.0099	0.0144	0.0198	—	—	—	—	—	—	—	—
0.6875	—	0.0032	0.0057	0.0090	0.0130	0.0179	0.0235	—	—	—	—	—	—	—
0.7500	—	0.0029	0.0052	0.0082	0.0119	0.0163	0.0214	0.0341	—	—	—	—	—	—
0.8125	—	0.0027	0.0048	0.0076	0.0110	0.0150	0.0197	0.0312	—	—	—	—	—	—
0.8750	—	0.0025	0.0045	0.0070	0.0102	0.0139	0.0182	0.0288	—	—	—	—	—	—
0.9375	—	—	0.0042	0.0066	0.0095	0.0129	0.0170	0.0268	0.0391	—	—	—	—	—
1.0000	—	—	0.0039	0.0061	0.0089	0.0121	0.0159	0.0250	0.0365	—	—	—	—	—
1.0625	—	—	0.0037	0.0058	0.0083	0.0114	0.0149	0.0235	0.0342	—	—	—	—	—
1.1250	—	—	0.0035	0.0055	0.0079	0.0107	0.0141	0.0221	0.0322	0.0443	—	—	—	—
1.1875	—	—	0.0033	0.0052	0.0074	0.0102	0.0133	0.0209	0.0304	0.0418	—	—	—	—
1.2500	—	—	0.0031	0.0049	0.0071	0.0097	0.0126	0.0198	0.0288	0.0395	—	—	—	—
1.3750	—	—	—	0.0045	0.0064	0.0088	0.0115	0.0180	0.0261	0.0357	0.0471	—	—	—
1.5000	—	—	—	0.0041	0.0059	0.0080	0.0105	0.0165	0.0238	0.0326	0.0429	—	—	—
1.6250	—	—	—	0.0038	0.0054	0.0074	0.0097	0.0152	0.0219	0.0300	0.0394	0.0502	—	—
1.7500	—	—	—	—	0.0050	0.0069	0.0090	0.0141	0.0203	0.0278	0.0365	0.0464	—	—
1.8750	—	—	—	—	0.0047	0.0064	0.0084	0.0131	0.0189	0.0259	0.0340	0.0432	0.0536	—
2.0000	—	—	—	—	0.0044	0.0060	0.0078	0.0123	0.0177	0.0242	0.0318	0.0404	0.0501	—
2.1250	—	—	—	—	—	0.0056	0.0074	0.0116	0.0167	0.0228	0.0298	0.0379	0.0470	0.0572	0.0684
2.2500	—	—	—	—	—	—	0.0070	0.0109	0.0157	0.0215	0.0281	0.0357	0.0443	0.0538	0.0643
2.3750	—	—	—	—	—	—	—	0.0103	0.0149	0.0203	0.0266	0.0338	0.0419	0.0509	0.0608
2.5000	—	—	—	—	—	—	—	—	0.0141	0.0193	0.0253	0.0321	0.0397	0.0482	0.0576
2.6250	—	—	—	—	—	—	—	—	0.0135	0.0184	0.0240	0.0305	0.0377	0.0457	0.0547
2.7500	—	—	—	—	—	—	—	—	—	0.0175	0.0229	0.0291	0.0360	0.0437	0.0521
2.8750	—	—	—	—	—	—	—	—	—	0.0168	0.0219	0.0278	0.0344	0.0417	0.0498
3.0000	—	—	—	—	—	—	—	—	—	—	0.0210	0.0266	0.0329	0.0399	0.0476

[a] Listed for the different shaft sizes and keyway widths for reference in checking dimensions H, J, and K.

19 V-Belts

(R) V-BELTS AND PULLEYS—SAE J636 MAY92 — SAE Standard

Report of the Miscellaneous Division approved August 1915, completely revised July 1977, editorial change January 1978 and reaffirmed by the V-Belt Committee February 1987. Completely revised by the SAE Belt Drive Committee May 1992. Rationale statement available.

1. Scope—This specification covers standard dimensions, tolerances, and methods of measurement of V-belts and pulleys for automotive V-belt drives.

2. References—There are no referenced publications specified herein.

3. V-Belt Types—Automotive V-belts are produced in a variety of constructions in a basic trapezoidal shape. The inside circumference of the V-belt can be a plain straight line or corrugated by means of cogs or notches for the purpose of increasing the belt(s) flexibility for use with pulleys in the lower proposed diameter. Belts are to be dimensioned in such a way that they are functional in pulleys dimensioned as described in subsequent sections.

4. Pulleys—Pulleys are to conform to requirements of Figure 1 and Tables 1 and 2.

5. V-Belt Measurement—Belt length and SAE size are defined by using effective length and rideout as measured in standard pulleys. These are determined by use of a measuring fixture comprised of two pulleys of equal diameter, a method of applying force, and a means of measuring the center distance between the two pulleys. One of the two pulleys is fixed in position while the other is movable along a graduated scale. The fixture is shown schematically in Figure 2. Specifications for measuring pulley dimensions are given in Table 2 and Figure 3.

5.1 Length—To measure the length, the belt is placed on the measuring fixture at the force shown in Table 3, and rotated around the pulleys at least two revolutions of the belt to seat the belt properly in the pulley grooves and to divide the total force equally between the two strands of the belt. The midpoint of the center distance travel of the

TABLE 1A—V-BELT PULLEY DIMENSIONS, mm

SAE Size	Recommended[1] Min Effective Dia	A Groove Angle (deg) ±0.5	W Effective Groove Width	D Groove Depth Min	d Ball or Rod Dia (±0.013)	2K 2X Ball[4] Extension	2X[2]	S Groove[3] Spacing (±0.38)
6A	57	36	6.3	7	5.558	4.16	1.0	8.00
8A	57	36	8.0	9	7.142	5.63	1.3	10.49
10A	61	36	9.7	11	7.938	3.77	1.5	13.74
11A	70	36	11.2	13	9.525	5.88	1.8	15.01
13A	76	36	12.7	14	11.113	7.99	2.0	16.79
15A	76	34				6.42		
	Over 102	36	15.2	14	12.700	7.02	0	19.76
	Over 152	38				7.56		
17A	76	34				8.21		
	Over 102	36	16.8	15	14.288	8.82	0.5	21.36
	Over 152	38				9.38		
20A	89	34				11.77		
	Over 114	36	20.0	18	17.463	12.42	1.0	24.54
	Over 152	38				13.02		
23A	102	34				15.67		
	Over 152	36	23.1	21	20.638	16.33	1.5	27.71
	Over 203	38				16.94		

[1] Pulley effective diameters below those recommended should be used with caution, because power transmission and belt life may be reduced.
[2] 2X is to be subtracted from the effective diameter to obtain "pitch diameter" for speed ratio calculation.
[3] These values are intended for adjacent grooves of the same effective width (W). Choice of pulley manufacture or belt design parameter may justify variance from these values. The S dimension shall be the same on all multiple groove pulleys in a drive using matched belts.
[4] 2K dimensions are calculated in millimeters.

19.02

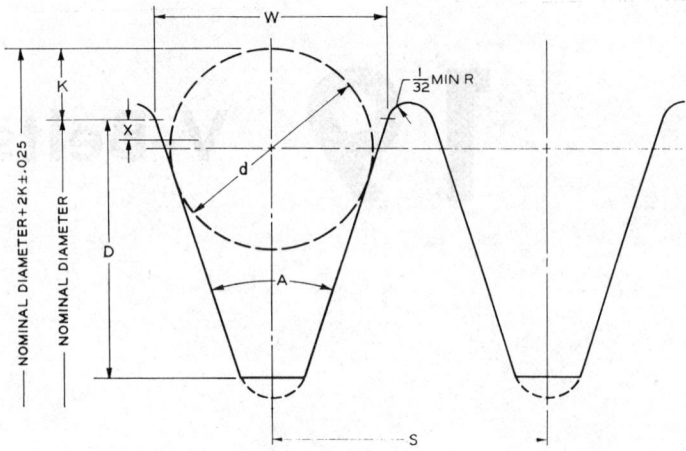

FIGURE 1—V-BELT PULLEY DIMENSIONS

NOTES: 1. The sides of the groove are to be 3.2 μm (125 μin) A. A. maximum.
2. Radial runout is not to exceed 0.38 mm (0.015 in) full indicator movement (FIM). Axial runout is not to exceed 0.38 mm (0.015 in) FIM. Runout in the two directions is measured separately with a ball mounted under spring pressure to follow the groove as the pulley is rotated. Diameter, load, and overhang conditions may require or permit variations in the above-specified runout limits.
3. Bottom corner radii optional, but if used, it shall be below the depth, D.
4. In pulleys for use with belts in multiple on common centers, the diameters over the ball gages are not to vary from groove to groove in the same pulley more than 0.05 mm/25 mm (0.002 in/in) of diameter, with top limit of 0.30 mm (0.012 in) for diameters 152 mm (6 in) and above.
5. Centerline of groove is to be 90 degrees ± 2 degrees with pulley axis.
6. The X dimension is radial. 2X is to be subtracted from the effective diameter to obtain "pitch diameter" for speed ratio calculations.

TABLE 1B—V-BELT PULLEY DIMENSIONS, in

SAE Size	Recommended[1] Min Effective Dia	A Groove Angle (deg) ±0.5	W Effective Groove Width	D Groove Depth Min	d Ball or Rod Dia (±0.0005)	2K 2X Ball Extension	2X[2]	S Groove[3] Spacing (±0.015)
0.250	2.25	36	0.248	0.276	0.2188	0.164	0.04	0.315
0.315	2.25	36	0.315	0.354	0.2812	0.222	0.05	0.413
0.380	2.40	36	0.380	0.433	0.3125	0.154	0.06	0.541
0.440	2.75	36	0.441	0.512	0.3750	0.231	0.07	0.591
0.500	3.00	36	0.500	0.551	0.4375	0.314	0.08	0.661
11/16 (0.600)	3.00	34	0.597	0.551	0.500	0.258	0.00	0.778
	Over 4.00	36				0.280		
	Over 6.00	38				0.302		
3/4 (0.660)	3.00	34	0.660	0.630	0.5625	0.328	0.02	0.841
	Over 4.00	36				0.352		
	Over 6.00	38				0.374		
7/8 (0.790)	3.50	34	0.785	0.709	0.6875	0.472	0.04	0.966
	Over 4.50	36				0.496		
	Over 6.00	38				0.520		
1 (0.910)	4.00	34	0.910	0.827	0.8125	0.616	0.06	1.091
	Over 6.00	36				0.642		
	Over 8.00	38				0.666		

[1] Pulley effective diameters below those recommended should be used with caution, because power transmission and belt life may be reduced.
[2] 2X is to be subtracted from the effective diameter to obtain "pitch diameter" for speed ratio calculation.
[3] These values are intended for adjacent grooves of the same effective width (W). Choice of pulley manufacture or belt design parameter may justify variance from these values. The S dimension shall be the same on all multiple groove pulleys in a drive using matched belts.

movable pulley defines the center distance and will be measured through one revolution of the belt minimum after the two seating revolutions. The belt effective length is equal to two times the center distance plus the effective pulley circumference. Standard belt center distance tolerances are shown in Table 4.

6.2 Rideout—The rideout standard and rideout tolerance are shown in Table 3. The rideout of a belt section is determined by measuring from a straight edge across the top of the belt to the rim of the measuring pulley, as shown in Figure 4.

6.3 Matched Belt Sets—For V-belts used in sets of two or more for a general application, the difference in center distance between the belts cannot exceed the values shown in Table 5.

7. Standard Lengths—Standard lengths up to and including 2032 mm (80 in) are to be 12.7 mm (1/2 in) increments. Standard lengths over 2032 mm (80 in) up to and including 2540 mm (100 in) are to be 25.4 mm (1 in) increments without fractions.

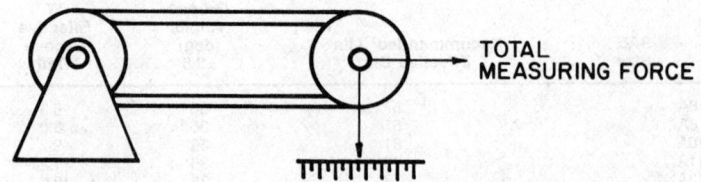

NOTE—The outside diameter and the effective diameter on the measuring pulley are one and the same.

FIGURE 2—DIAGRAM OF A FIXTURE FOR MEASURING V-BELTS

TABLE 2A—MEASURING PULLEY DIMENSIONS, mm

SAE Size	d_1 Effective Dia (±0.05)	Effective Pulley Circumference	A Groove Angle (deg) ±0.15	W Effective Groove Width	D Groove Depth Min	d Ball or Rod Dia (±0.013)	d_2 Dia[1] Over Balls or Rods (±0.05)
6A	97.03	304.8	36	6.3	7	5.558	101.18
8A	97.03	304.8	36	8.0	9	7.142	102.66
10A	97.03	304.8	36	9.7	11	7.938	100.80
11A	97.03	304.8	36	11.2	13	9.525	102.91
13A	97.03	304.8	36	12.7	14	11.113	105.02
15A	97.03	304.8	34	15.2	14	12.700	103.45
17A	97.03	304.8	34	16.8	16	14.288	105.24
20A	121.29	381.0	34	20.0	18	17.463	133.06
23A	121.29	381.0	34	23.1	21	20.638	136.96

[1] d_2 dimensions are calculated in millimeters.

TABLE 2B—MEASURING PULLEY DIMENSIONS, in

SAE Size US	d_1 Effective Dia (±0.002)	Effective Pulley Circumference	A Groove Angle (deg) ±0.15	W Effective Groove Width	D Groove Depth Min	d Ball or Rod Dia (±0.0005)	d_2 Dia Over Balls or Rods (±0.002)
0.250	3.820	12.000	36	0.248	0.276	0.2188	3.984
0.315	3.820	12.000	36	0.315	0.354	0.2812	4.042
0.380	3.820	12.000	36	0.380	0.433	0.3125	3.974
0.440	3.820	12.000	36	0.441	0.512	0.3750	4.051
0.500	3.820	12.000	36	0.500	0.551	0.4375	4.134
11/16 (0.600)	3.820	12.000	34	0.597	0.551	0.5000	4.078
3/4 (0.660)	3.820	12.000	34	0.660	0.630	0.5625	4.148
7/8 (0.790)	4.775	15.000	34	0.785	0.709	0.6875	5.247
1 (0.910)	4.775	15.000	34	0.910	0.827	0.8125	5.391

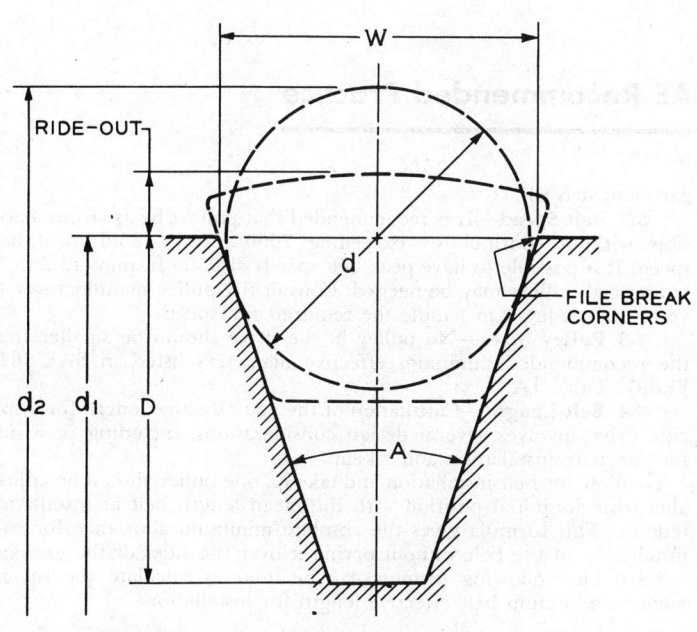

NOTE: The outside diameter and the effective diameter on the measuring pulley are one and the same.

FIGURE 3—MEASURING PULLEY DIMENSIONS

TABLE 3A—MEASURING CONDITIONS AND RIDEOUT, SI UNITS

SAE Size Metric	Total Measuring Force, N	Rideout[1] mm	Rideout[1] Tolerance mm
6A	222	0.8	±0.8
8A	222	0.8	±0.8
10A	267	1.5	±1.1
11A	267	1.0	±1.1
13A	267	1.5	±1.1
15A	267	2.3	±1.1
17A	356	2.3	±1.1
20A	445	2.3	±1.1
23A	534	2.3	±1.1

[1] The belt rideout, as measured along the circumference of the belt, must fall within the specified tolerance at all points with the exception of measurements at points of dimension variations inherent to the manufacturing process or product such as material splices, belt identifications, etc.

TABLE 3B—MEASURING CONDITIONS AND RIDEOUT, U.S. CUSTOMARY UNITS

SAE Size in	Total Measuring Force, lb	Rideout[1] in	Rideout[1] Tolerance in
0.250	50	0.031	±0.031
0.315	50	0.031	±0.031
0.380	60	0.060	±0.045
0.440	60	0.040	±0.045
0.500	60	0.060	±0.045
11/16 (0.600)	60	0.090	±0.045
3/4 (0.660)	80	0.090	±0.045
7/8 (0.790)	100	0.090	±0.045
1 (0.910)	120	0.090	±0.045

[1] The belt rideout, as measured along the circumference of the belt, must fall within the specified tolerance at all points with the exception of measurements at points of dimension variations inherent to the manufacturing process or product such as material splices, belt identifications, etc.

TABLE 4A—STANDARD BELT CENTER DISTANCE TOLERANCES, mm

Belt Length	Tolerance on Center Distance
1270 and less	±3.0
Over 1270-1524, incl.	±4.1
Over 1524-2032, incl.	±4.8
Over 2032-2540, incl.	±5.6

TABLE 4B—STANDARD BELT CENTER DISTANCE TOLERANCES, in

Belt Length	Tolerance on Center Distance
50 and less	±0.12
Over 50-60, incl.	±0.16
Over 60-80, incl.	±0.19
Over 80-100, incl.	±0.22

TABLE 5A—MAXIMUM CENTER DISTANCE DIFFERENCE FOR BELTS IN A SET, mm

SAE Size	
6A	0.8
8A	0.8
10A	1.0
11A	1.0
13A	1.0
15A	1.5
17A	1.5
20A	1.5
23A	1.5

TABLE 5B—MAXIMUM CENTER DISTANCE DIFFERENCE FOR BELTS IN A SET, in

SAE Size	
0.250	0.03
0.315	0.03
0.380	0.04
0.440	0.04
0.500	0.04
11/16 (0.600)	0.06
3/4 (0.660)	0.06
7/8 (0.790)	0.06
1 (0.910)	0.06

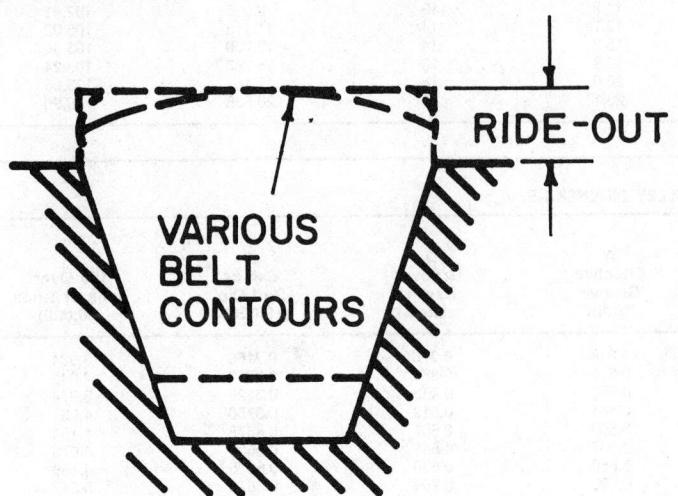

FIGURE 4—MEASURING BELT RIDE

φAUTOMOTIVE V-BELT DRIVES—SAE J637 FEB89

SAE Recommended Practice

Report of Engine Committee approved January 1954 and completely revised by V-Belt Committee February 1989.

1. Introduction—Selection and specification of belts have been major problems due to the lack of a recognized industry standard for classifying V-belts according to performance and quality level.

From the very beginning of the use of V-belts on automotive drives, the automotive manufacturers and the V-belt manufacturers have employed laboratory tests on the products for such purposes as product development, source approval, and quality verification. This standard is the result of the combined effort of the users and suppliers.

2. Scope—The following information is intended as a guide to be used for evaluating belt construction, source approval, and quality audit. This recommendation has been prepared from existing literature, including standards, specifications and data supplied by both producers and users.

These recommendations cover drive layout details and V-belt testing methods, including test layout, pulley diameters, torque loads, and guidance for interpreting test data. The application of these automotive V-belts is to power engine or vehicle accessories that are physically attached to the engine.

3. General Drive Layout Considerations

3.1 Power Transmission—When the engine is used to drive an external unit equipped with industrial type pulleys and belts, it is recommended that the power takeoff pulley on the engine be grooved according to the appropriate industrial standard. There are four such standards. Three of these standards are published by the RMA-MPTA (Rubber Manufacturers Association-Mechanical Power Transmission Association) and include Classical Multi-V-Belt (A, B, C, D and E belt sections), Narrow Multi-V-Belts (3, 5 and 8 V), and Single V-Belts (2, 3, 4 and 5 L). The fourth is published as an American Society of Agricultural Engineers standard, V-Belt Drives for Farm Machines.

The grooves in these four standards differ from each other in the reference dimensions. They are not interchangeable with SAE grooves which were standardized for engine accessory and other engine compartment drives.

3.2 Belt Speed—It is recommended that pulleys be as large as possible without continuously exceeding 7000 ft/min (35.6 m/s) belt speed. It is possible to have peak belt speeds of 8500 ft/min (43.2 m/s) but special pulleys may be needed. Consult the pulley manufacturer to verify the pulleys can handle the required rim speed.

3.3 Pulley Sizes—No pulley in the drive should be smaller than the recommended minimum effective diameters listed in SAE J636 FEB87, Table 1A.

3.4 Belt Length—Calculation of the belt effective length for a specific drive involves several design considerations, including provision for adequate installation and takeup.

To allow for belt installation and takeup, one pulley should be adjustable from its initial position with the mean length belt at installation tension. This formula gives the absolute minimum allowance for easy installation of the belt without prying it over the sides of the grooves.

3.4.1 The following formula can be used to calculate the recommended minimum belt effective length for installation:

$$\text{Min Belt EL} = (1.005)(L_1) + L_1 + C_I$$

Where:

L_1 = Effective belt length (addition of span lengths and effective arc lengths on the pulleys) around the drive with the tensioning pulley in the minimum position. The 1.005 factor provides for length change from slack to measuring tension.

L_2 = 2 × negative belt manufacturing center distance tolerance. (Table 4, SAE J636)

C_I = Length to account for belt worked into groove. Installation constant (C_I) found in Table 3.

3.4.2 Select a belt to be used that has a nominal effective length equal to or greater than the recommended minimum EL.

3.4.3 Calculate the maximum required effective length around the drive to provide for take-up:

TABLE 1A - TEST CONDITIONS[a]
PLAIN SECTION BELTS, (in)

SAE Belt Size	Standard Groove Width	Diameter Where Specified Groove Width Occurs (w/o Width Tol)		Driver Pulley Speed RPM ±2%	Load HP	Length Range	
		DR & DN Pulleys ±0.010	Tension Pulley ±0.010			Total	Preferred
0.250	0.248	4.750	2.250	4900	8.50 9.50 b	Under 40 40 - 50 Over 55	36 - 40 45 - 50 55 - 60
0.315	0.315	4.750	2.250	4900	9.25 10.25 11.25	Under 40 40 - 55 Over 55	36 - 40 45 - 50 55 - 60
0.380	0.380	4.750	2.500	4900	10.00 11.00 12.00	Under 40 40 - 55 Over 55	36 - 40 45 - 50 55 - 60
0.440	0.441	4.750	2.750	4900	10.75 11.75 12.75	Under 40 40 - 55 Over 55	36 - 40 45 - 50 55 - 60
0.500	0.500	5.000	3.000	4700	12.00 13.00 14.00	Under 40 40 - 55 Over 55	36 - 40 45 - 50 55 - 60
11/16 (0.600)	0.597	5.000	3.500	4700	13.00 14.50 15.00	Under 40 40 - 55 Over 55	36 - 40 45 - 50 55 - 60
3/4 (0.660)	0.660	5.000	3.625	4700	13.50 14.50 15.50	Under 40 40 - 55 Over 55	36 - 40 45 - 50 55 - 60
7/8 (0.790)	0.785	6.000	4.000	3900	b	b	b
1 (0.910)	0.910	7.000	4.625	3350	b	b	b

[a] = Groove details as given in SAE J636, Table 1A, Fig. 1
[b] = Values to be per agreement between user and manufacturer (insufficient usage for recommendations)

TABLE 1B - TEST CONDITIONS[a]
PLAIN SECTION BELTS, (mm)

SAE Belt Size	Standard Groove Width	Diameter Where Specified Groove Width Occurs (w/o Width Tol)		Driver Pulley Speed RPM ±2%	Load kW	Length Range	
		DR & DN Pulley ±0.25	Tension Pulley ±0.25			Total	Preferred
6A	6.3	120.5	57.0	4900	6.3 7.1 b	Under 1020 1020 - 1400 Over 1400	920 - 1020 1140 - 1270 1400 - 1520
8A	8.0	120.5	57.0	4900	6.9 7.6 8.4	Under 1020 1020 - 1400 Over 1400	920 - 1020 1140 - 1270 1400 - 1520
10A	9.7	120.5	63.5	4900	7.5 8.2 8.9	Under 1020 1020 - 1400 Over 1400	920 - 1020 1140 - 1270 1400 - 1520
11A	11.2	120.5	70.0	4900	8.0 8.8 9.5	Under 1020 1020 - 1400 Over 1400	920 - 1020 1140 - 1270 1400 - 1520
13A	12.7	127.0	76.0	4700	8.9 9.7 10.4	Under 1020 1020 - 1400 Over 1400	920 - 1020 1140 - 1270 1400 - 1520
15A	15.2	127.0	89.0	4700	9.7 10.4 11.2	Under 1020 1020 - 1400 Over 1400	920 - 1020 1140 - 1270 1400 - 1520
17A	16.8	127.0	92.0	4700	10.1 10.8 11.6	Under 1020 1020 - 1400 Over 1400	920 - 1020 1140 - 1270 1400 - 1520
20A	20.0	152.5	101.5	3900	b	b	b
23A	23.1	178.0	117.5	3350	b	b	b

[a] = Groove details as given in SAE J636, Table 1B, Fig. 1
[b] = Values to be per agreement between user and manufacturer (insufficient usage for recommendations)

TABLE 2A - TEST CONDITIONS[a]
COG, OR NOTCHED BELTS, (in)

SAE Belt Size	Standard Groove Width	Diameter Where Specified Groove Width Occurs (w/o Width Tol)		Driver Pulley Speed RPM ±2%	Load HP	Length Range	
		DR & DN Pulleys ±0.010	Tension Pulley ±0.010			Total	Preferred
0.250	0.248	b	b	b	b	b	b
0.315	0.315	b	b	b	b	b	b
0.380	0.380	4.750	2.250	4900	10.00	Under 40	36 - 40
					11.00	40 - 55	45 - 50
					12.00	Over 55	55 - 60
0.440	0.441	4.750	2.500	4900	10.75	Under 40	36 - 40
					11.75	40 - 55	45 - 50
					12.75	Over 55	55 - 60
0.500	0.500	5.000	2.750	4700	12.00	Under 40	36 - 40
					13.00	40 - 55	45 - 50
					14.00	Over 55	55 - 60
11/16 (0.600)	0.597	5.000	3.250	4700	13.00	Under 40	36 - 40
					14.50	40 - 55	45 - 50
					15.00	Over 55	55 - 60
3/4 (0.660)	0.660	5.000	3.375	4700	13.50	Under 40	36 - 40
					14.50	40 - 55	45 - 50
					15.50	Over 55	55 - 60
7/8 (0.790)	0.785	6.000	3.750	3900	b	b	b
1 (0.910)	0.910	7.000	4.375	3350	b	b	b

a = Groove details as given in SAE J636, Table 1A, Fig. 1
b = Values to be per agreement between user and manufacturer (insufficient usage for recommendations)

Maximum required belt path length = $(1.005) L_3 + L_4 + L_5 + L_6$

Where:
L_3 = Nominal Belt EL as defined in 3.4.2. The 1.005 factor accounts for elongation from measuring to installation tension.
L_4 = (0.01) (EL)—Allows 1% for tensile member growth and belt wear during service life.
L_5 = 2 X positive belt manufacturing center distance tolerance. (Table 4, SAE J636)
L_6 = Belt seating factor (0.38 in, 9.6 mm)

3.5 Pulley Misalignment—The recommended maximum misalignment between pulleys is 1/16 in/ft of span length (1.6 per 300 mm span length) or approximately 1/3 of 1 deg.

4. V-Belt Fatigue Test Method
The belt shall be mounted on a test layout as shown in Fig. 1 with pulley diameters and speeds as given in Tables 1 and 2. The horsepower (kilowatts) to be absorbed at the driven pulley shall be compatible with the tension pulley diameter and belt length as shown in Tables 1 and 2.

The driver pulley speed (rpm) shall be used in the torque load calculation and the torque load shall be kept constant without compensation for loss of driven pulley rpm resulting from belt slippage and creep.

$$\text{Torque, lb·in} = \frac{\text{Specified horsepower} \times 63025}{\text{Driver rpm}}$$

$$\text{Torque, N·m} = \frac{\text{Specified kilowatts} \times 9549}{\text{Driver rpm}}$$

Measurable parasitic loads due to bearing losses, lubricants, etc., shall be deducted from the specified horsepower (kilowatts) in the above calculation.

The tension shall be applied by weights equal in number of pounds to 10 times the number of units of the specified horsepower (in number of Newtons to 60 times the number of units of the specified kilowatts).

4.1 The test procedure shall be as follows:
4.1.1 Condition the belt by running 5 min under the prescribed test details but without the dynamometer load. Maintain a constant tension during this period by operating with the tension pulley center position unlocked.

4.1.2 Stop the machine, allow to stand for a minimum of 10 min and lock the tension pulley center position midway of the limits of travel during belt rotation.

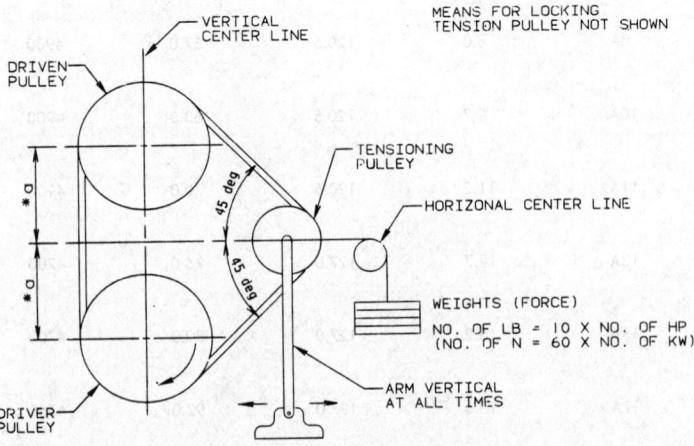

*Dimension a is adjusted for various length belts to maintain tension pulley midway vertically between driver and driven pulleys.

**45 deg is specified for initial test configuration and may change slightly with resets as test progresses.

FIG. 1 – V-BELT FATIGUE TEST

19.07

TABLE 2B - TEST CONDITIONS[a]
COG, OR NOTCHED BELTS, (mm)

SAE Belt Size	Standard Groove Width	Diameter Where Specified Groove Width Occurs (w/o Width Tol) DR & DN Pulley ±0.25	Tension Pulley ±0.25	Driver Pulley Speed RPM ±2%	Load kW	Length Range Total	Length Range Preferred
6A	6.3	b	b	b	b	b	b
8A	8.0	b	b	b	b	b	b
10A	9.7	120.5	57.0	4900	7.5 8.2 8.9	Under 1020 1020 - 1400 Over 1400	920 - 1020 1140 - 1270 1400 - 1520
11A	11.2	120.5	63.5	4900	8.0 8.8 9.5	Under 1020 1020 - 1400 Over 1400	920 - 1020 1140 - 1270 1400 - 1520
13A	12.7	127.0	70.0	4700	8.9 9.7 10.4	Under 1020 1020 - 1400 Over 1400	920 - 1020 1140 - 1270 1400 - 1520
15A	15.2	127.0	82.5	4700	9.7 10.4 11.2	Under 1020 1020 - 1400 Over 1400	920 - 1020 1140 - 1270 1400 - 1520
17A	16.8	127.0	85.5	4700	10.1 10.8 11.6	Under 1020 1020 - 1400 Over 1400	920 - 1020 1140 - 1270 1400 - 1520
20A	20.0	152.5	95.0	3900	b	b	b
23A	23.1	178.0	111.0	3350	b	b	b

[a] = Groove details as given in SAE J636, Table 1B, Fig. 1
[b] = Values to be per agreement between user and manufacturer (insufficient usage for recommendations)

TABLE 3 - CLEARANCE FACTOR FOR BELT INSTALLATION

SAE SIZE		C1			
ENGLISH	METRIC	ENGLISH		METRIC	
		SINGLE	MULTIPLE	SINGLE	MULTIPLE
0.250	6A	0.245	0.490	6.2	12.4
0.315	8A	0.240	0.480	6.1	12.2
0.380	10A	0.300	0.600	7.6	15.2
0.440	11A	0.300	0.600	7.6	15.2
0.500	13A	0.390	0.780	9.9	19.8
11/16	15A	0.350	0.700	8.9	17.8
3/4	17A	0.425	0.850	10.8	21.6
7/8	20A	0.485	0.970	12.3	24.6
1	23A	0.510	1.020	13.0	26.0

TABLE 4A - TEST CONDITIONS, (in)

Belt Type	Standard Groove Width	Dr & Dn Pulleys ±0.010	Tension Pulley ±0.010	Driver Speed rpm ±2%	Load hp	Length Range Total	Length Range Preferred
Plain Section	0.380	4.750	2.500	4900	11.0 12.5 13.5	Under 40 40-55 Over 55	36-40 45-50 55-60
Notched or Cog	0.380	4.750	2.250	4900	11.0 12.5 13.5	Under 40 40-55 Over 55	36-40 45-50 55-60

TABLE 4B - TEST CONDITIONS, (mm)

Belt Type	Standard Groove Width	Dr & Dn Pulleys ±0.25	Tension Pulley ±0.25	Driver Speed rpm ±2%	Load kW	Length Range Total	Length Range Preferred
Plain Section	9.7	120.5	63.5	4900	8.2 9.4 10.1	Under 1020 1020 - 1400 Over 1400	920-1020 1140-1270 1400-1520
Notched or Cog	9.7	120.5	57.0	4900	8.2 9.4 10.1	Under 1020 1020 - 1400 Over 1400	920-1020 1140-1270 1400-1520

4.1.3 Restart with the dynamometer load and run until the slip reaches 8% or until the belt will no longer transmit the load uniformly because of breakage or rough running.

4.1.4 Whenever the slip reaches 8%, stop the machine, allow to stand for a minimum of 20 min, unlock the tension pulley center, restore the initial tension, relock, and restart the machine.

4.1.5 Record the number of hours run and the number of resets (exclusive of the 5 min run-in).

4.1.6 The ambient temperature shall be 80-90°F (27-32°C). An increase in internal belt temperature will reduce belt life. Internal belt temperature is dependent upon ambient temperature as well as other test conditions.

5. Test Performance Guidelines

The test life which a belt must attain shall be according to agreement between user and manufacturer. However, typical curves of average test life versus belt length are shown in Fig. 2. The typical curves of Fig. 2 are constructed with the belt life varying as the 2.75 power of belt length for the test conditions given in Tables 1 and 2. The acceptable number of retensionings after the initial 5 min run-in shall be according to agreement between the manufacturer and user.

The belt manufacturer's test data on belts of a certain construction specification shall be considered valid for evaluation of all belts of the same construction specification regardless of the intended user. Belts shall be considered to be of the same construction specification when they are the same with respect to the manufacturer's cross section dimensions, material specifications, and method of manufacture.

In evaluating for part source approval and for production quality surveillance, test data for the entire length group containing a part in question shall be considered pertinent. The design of some test machines may not accommodate the shortest lengths shown in Fig. 2. In such cases, test data on some longer belt(s) of the same construction specification and within the length group 28-40 in (710-1020 mm) shall be used. Similarly, test data on belt(s) within the length group 56-68 in (1420-1730 mm) shall be used for lengths beyond 68 in (1730 mm).

Whether testing is performed for part source approval or for production quality surveillance, a realistic statistical guide to acceptability would be "not more than 10% of test lives shall be permitted to fall below 50% of the specified average life."

For part source approval, test data of the immediately preceding three month period shall be considered pertinent. When such data are not sufficient for the statistical evaluation, the manufacturer may have the option of submitting data for source approval on a "sample" of the part under consideration or on samples of the same length group and construction specification. Because the data would be limited to this situation, a guide to approval could be to permit no test results to be below 50% of the specified average life.

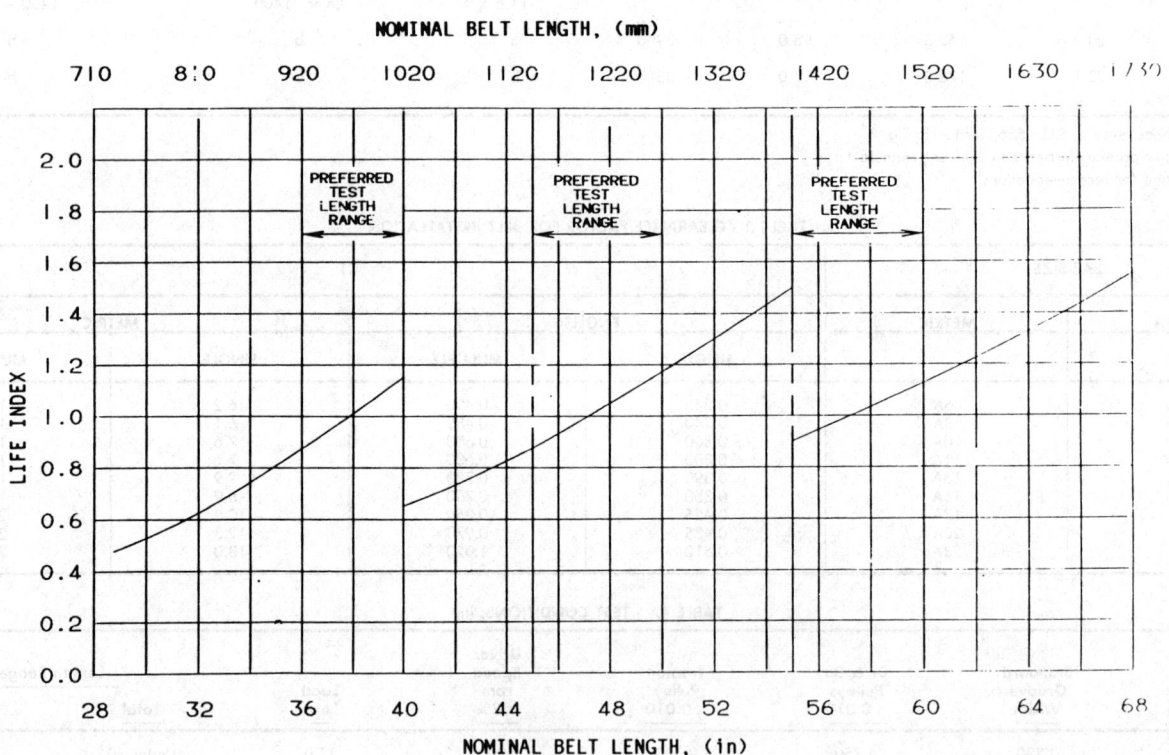

FIG. 2 – TYPICAL LIFE-LENGTH CURVES (FOR TEST CONDITIONS IN TABLES 1 AND 2)

APPENDIX A
STANDARDIZED LABORATORY TESTING OF NONSTANDARDIZED AUTOMOTIVE V-BELTS

The following information is supplementary to the test conditions for the nine standard SAE top width belts shown in Tables 1 and 2.

"High ride" belts have been tried in a number of different standard SAE top width sheave grooves. At least one of these "high ride" belts has had sufficient usage to be recognized and test conditions specified for it. This "high ride" belt is defined as a belt that has a nominal ride-out 0.105 in (2.7 mm) in an SAE 0.380 in (9.7 mm) pulley.

Considerations for the "high ride" belt in the standard 0.380 in (9.7 mm) groove are:

1. Improved belt life since the belt diameters are effectively larger, slightly increasing belt velocity and decreasing bending stress.
2. Decreased tension decay resulting from No. 1 above, and a slight increase in belt tensile width.
3. Change in speed ratio with a reduced driven speed when used on such accessories as alternator drives.
4. Questions on belt stability.

The test conditions shown in Table 4 are applicable to a "high ride" belt in SAE 0.380 in (9.7 mm) grooves.

SI (METRIC) SYNCHRONOUS BELTS AND PULLEYS—SAE J1278 MAR93

SAE Standard

Report of the V-Belt Committee approved October 1980, reaffirmed without change June 1986. Reaffirmed by the SAE Belt Drive Committee March 1993.

Foreword—This reaffirmed document has been changed only to reflect the new SAE Technical Standards Board format.

1. Scope
Synchronous belt drives consist of a toothed belt which mates with grooved pulleys to provide a precise speed ratio between the driver and driven pulleys. This SAE Standard covers the synchronous belt and pulley sections currently in use in automotive applications such as camshaft, distributor, and other underhood drives that may require synchronization. It also provides for future sections to be added as usage develops. Table 1 lists the sections currently in use.

TABLE 1—PULLEY GENERATING TOOL RACK FORM DIMENSIONS (mm)

Pulley Section	Diameter Range (No. of Grooves)	P_b Pitch ±0.003	±0.25 deg	h_g +0.05 -0.00	b_g +0.05 -0.00	r_b ±0.03	r_t ±0.03	2α
ST	10 and over	9.525	40	2.13	3.10	0.86	0.53	0.762
SU	14 thru 19	12.700	40	2.59	4.24	1.47	1.04	1.372
SU	over 19	12.700	40	2.59	4.24	1.47	1.42	1.372
STA	19 and over	9.525	40	2.13	3.10	0.86	0.71	1.372

2. References
There are no referenced publications specified herein.

3. Belt and Pulley Sections
Synchronous belt and pulley sections are defined primarily by pitch, which is the linear distance between the axes of two consecutive teeth when the belt is loaded to the prescribed measuring force. Figures 1 and 2 illustrate the location of the pitch line. A two- or three-letter designation is used to identify the standard sections. Two-letter designations identify a specific pitch, tooth form, and pitch line location. For example, an ST section has a pitch of 9.525 mm, while an SU section has a pitch of 12.700 mm. Since it is possible to have more than one section with the same pitch, a three-letter designation is used to identify a section that is a variation of a normal two-letter section. For example, an STA section has the same pitch as an ST section (9.525 mm) and the same tooth form as the ST section, but it utilizes the SU section pitch line location because of belt construction. As a result, STA section belts will not mesh properly with ST section pulleys even though the pitch and tooth forms are the same because it has a larger pitch line differential. Therefore, the STA section belt is a unique section and requires special pulleys designed for this particular section. Should another section with the ST (9.525 mm) pitch be standardized, it would have the section designation of STB.

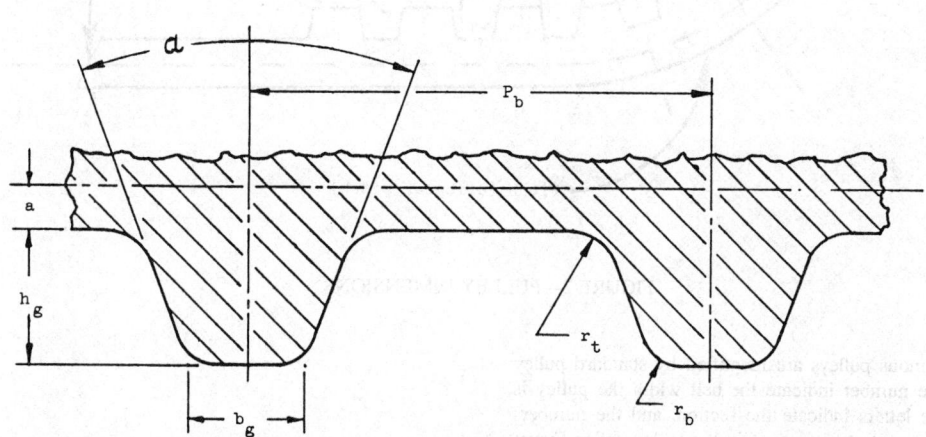

FIGURE 1—PULLEY GENERATING TOOL RACK FORM

4. Pulleys

4.1 Groove Profile—The groove profile is defined as the profile formed by the generating tool rack form described in Table 1 and Figure 1. The relationship of pitch diameter to outside diameter is illustrated in Figure 2.

4.2 Tolerances—Tolerances on pulleys shall conform to values shown in Table 2 and the accompanying footnotes.

TABLE 2—PULLEY TOLERANCES (mm)

Outside Diameter Range	Pitch to Pitch Tolerance Adjacent Grooves	Pitch to Pitch Tolerance Accumulative Over 90 deg
Up to 50, incl	±0.03	±0.09
Over 50 to 100, incl	±0.03	±0.11
Over 100 to 175, incl	±0.03	±0.13
Over 175 to 300, incl	±0.03	±0.15

Outside Diameter
Up to 50 mm, incl
For each additional 25 mm or portion thereof

Tolerance
+0.05 mm/-0.00 mm
+0.025 mm/-0.00 mm

Outside Diameter Runout[1]
Up to 75 mm, incl outside diameter
For each additional 25 mm or portion thereof

0.08 mm (max)
0.01 mm (max)

Axial Runout[1] **(Side Wobble)**
Up to 250 mm, incl outside diameter
For each additional 25 mm outside diameter over 25 mm add 0.01 mm

0.02 mm per 25 mm of diameter add 0.01 mm

Diametrical Taper
0.01 mm per 10 mm of face width

Groove Helix
0.01 mm per 10 mm of face width

---
[1] Full indicator movement

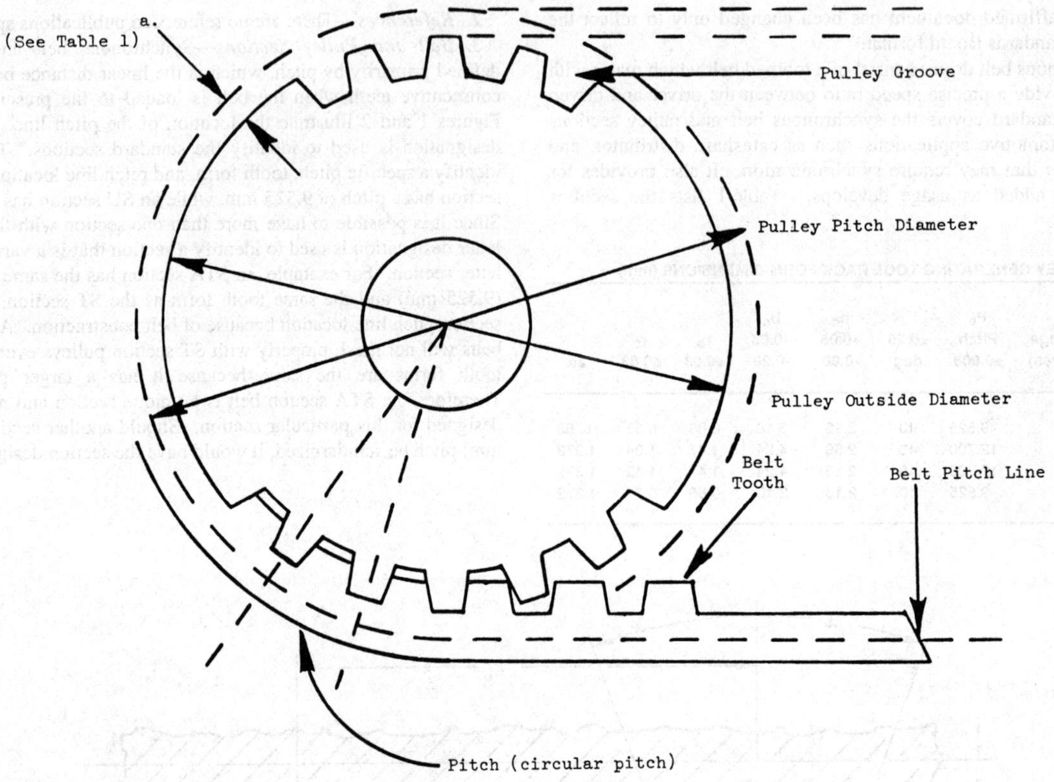

FIGURE 2—PULLEY DIMENSIONS

4.3 Designation—Synchronous pulleys are identified by standard pulley numbers. The first digits in the number indicate the belt width the pulley is designed to accommodate. The letters indicate the section, and the numbers following the letters indicate the number of grooves in the pulley. (See Figure 3.)

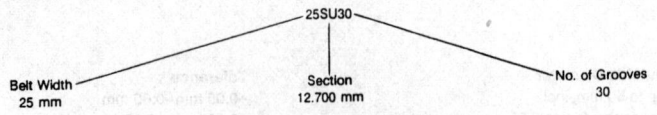

FIGURE 3—PULLEY DESIGNATION

5. Belts

5.1 Dimensions—Nominal dimensions of the synchronous belt sections are shown in Table 3 and Figure 4.

TABLE 3—NOMINAL BELT DIMENSIONS (mm)

Belt Section	Pitch	h_b	2β deg	h_t	b_t	r_{bb}	r_{bt}
ST	9.525	3.6	40	1.9	3.2	0.5	0.5
SU	12.700	4.1	40	2.3	4.4	1.0	1.0
STA	9.525	4.1	40	1.9	3.2	0.5	0.5

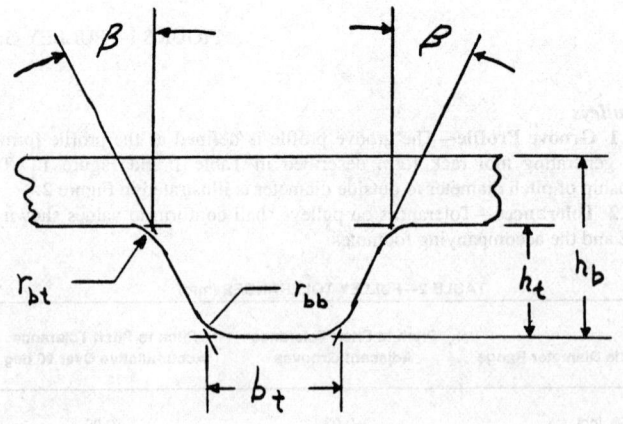

FIGURE 4—BELT SECTION

5.2 Tolerances—Tolerances on width shall conform to values shown in Table 4.

TABLE 4—BELT WIDTH TOLERANCES (mm)

Belt Width	Belt Length Range Up to 840, incl	Belt Length Range Over 840 to 1680, incl
Up to 40, incl	+0.6/-0.6	+0.6/-0.6
Over 40 to 50, incl	+0.8/-0.8	+1.0/-1.0

5.3 Designation—Belt sizes shall be identified by standard belt numbers. The first digits in the number indicate the belt width in millimeters. The letters indicate the belt section (pitch) designation, and the numbers following the letters indicate the pitch length in millimeters. For example, the number 25SU1500 indicates a belt 25 mm wide, SU section (pitch) of 12.7 mm, and a pitch length of 1500 mm. (See Figure 5.)

5.4 Length Determination—The pitch length of a synchronous belt shall be determined by placing the belt on a measuring fixture comprised of two pulleys of equal pitch diameter, a method of applying force and a means of measuring the center distance between the two pulleys. One of the two pulleys is in a fixed location while the other is movable along a graduated scale. The fixture is shown schematically in Figure 6. Measuring pulley dimensions and measuring force are specified in Tables 5 and 6 and Figure 7.

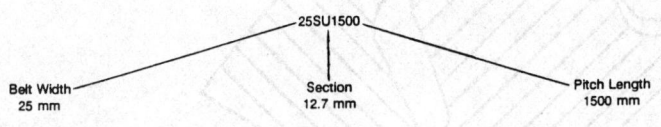

FIGURE 5—BELT DESIGNATION

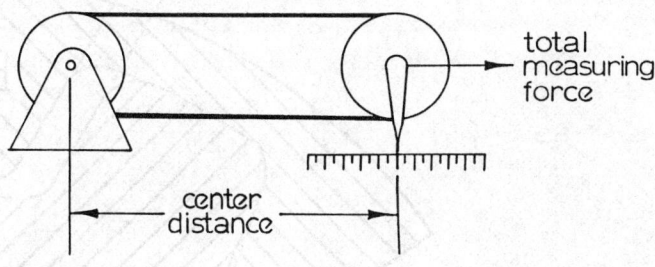

FIGURE 6—DIAGRAM OF FIXTURE FOR DETERMINING BELT PITCH LENGTH

TABLE 5—MEASURING PULLEY DIMENSIONS (mm)

Belt Section	No. of Grooves	Pitch Circumference	Outside Dia ±0.013	Outside Dia Runout FIM[1], max	Axial Runout (Side Wobble) FIM[1], max	Min Clearance[2]
ST	16	152.40	47.748	0.013	0.025	0.33
SU	20	254.00	79.479	0.013	0.025	0.38
STA	20	190.50	59.266	0.013	0.025	0.33

[1] Full indicator movement
[2] See Figure 7

TABLE 6—TOTAL MEASURING FORCE (N)

Belt Section	Belt Width (mm)															
	8	10	12	14	16	18	19	20	22	25	28	30	33	35	40	45	50
ST	55	75	100	125	145	165	175	185	210	240	275	295	330	355	410	470	530
SU	--	--	245	300	370	420	445	475	530	610	700	750	840	900	1050	1200	1350
STA	--	--	245	300	370	420	445	475	530	610	700	750	840	900	1050	1200	1350

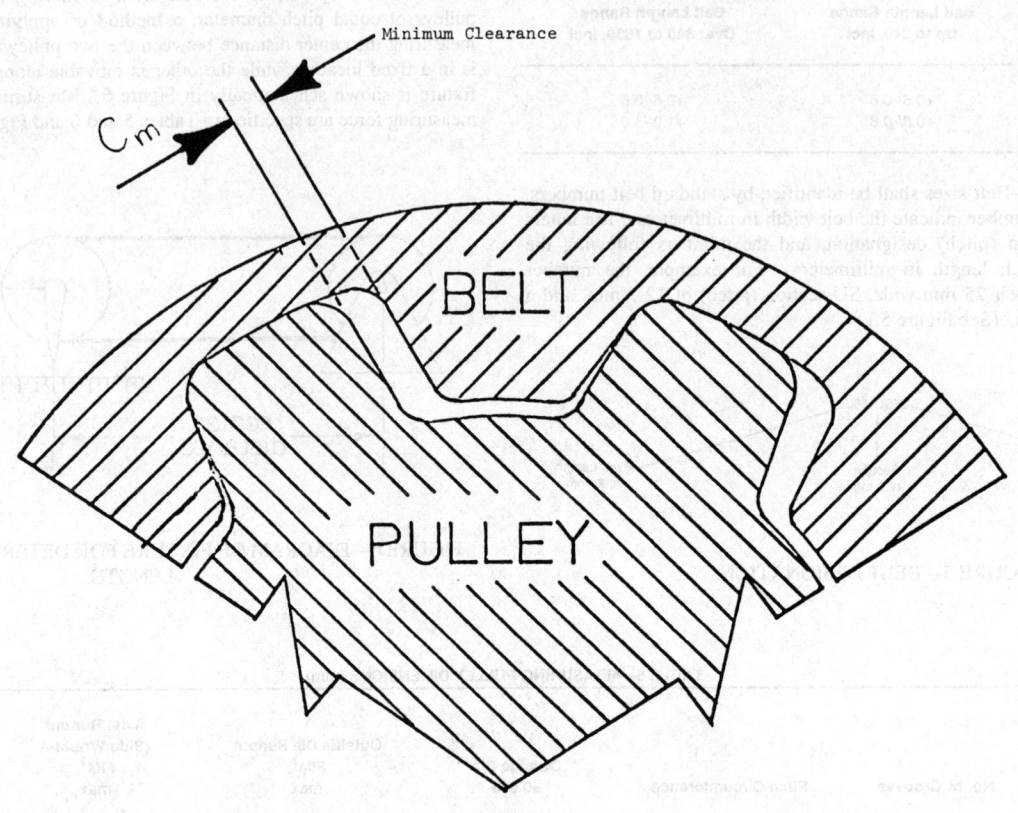

FIGURE 7—CLEARANCE BETWEEN MEASURING PULLEY AND BELT

In measuring the length of a synchronous belt, the belt should be rotated at least two revolutions of the belt in order to (a) seat the belt properly in the pulley grooves, (b) divide equally the total force between the two strands of the belt, and (c) determine the midpoint of the center distance travel of the movable pulley, which shall define the center distance. The pitch length shall be calculated by adding the pitch circumference of one of the measuring pulleys to twice the measured center distance between the two pulleys.

5.5 Length Tolerances—Belt length shall be within the limits specified in Table 7.

TABLE 7—BELT LENGTH TOLERANCES (mm)

Belt Length Range	Tolerance on Belt Pitch Length
Up to 400, incl	±0.46
Over 400 to 520, incl	±0.51
Over 520 to 770, incl	±0.61
Over 770 to 1020, incl	±0.66
Over 1020 to 1270, incl	±0.76
Over 1270 to 1525, incl	±0.81
Over 1525 to 1780, incl	±0.86
Over 1780 to 2040, incl	±0.91
Over 2040 to 2300, incl	±0.97
Over 2300 to 2560, incl	±1.02
Over 2560 to 3050, incl	±1.12

AUTOMOTIVE SYNCHRONOUS BELT DRIVES—SAE J1313 MAR93 SAE Standard

Report of the V-Belt Committee approved October 1980, reaffirmed without change June 1986. Reaffirmed by the SAE Belt Drive Committee March 1993.

Foreword—This reaffirmed document has been changed only to reflect the new SAE Technical Standards Board format.

1. *Scope*—The following information applies to automotive camshaft drives, distributor drives, or other underhood drives that may require synchronization. For other power transmission drives requiring synchronization, refer to Specifications for Drives using Synchronous Belts (MXL, XL, L, H, XH, and XXH belt sections) (IP 24/1978), published jointly by the Rubber Manufacturers Association (RMA), the Mechanical Power Transmission Association (MPTA), and the Rubber Association of Canada (RAC).

2. *References*

   2.1 *Applicable Documents*—The following publications form a part of this specification to the extent specified herein. The latest issue of SAE publications shall apply.

   2.1.1 SAE PUBLICATION—Available from SAE, 400 Commonwealth Drive, Warrendale, PA 15096-0001.

   SAE J1278—SI (Metric) Synchronous Belts and Pulleys

   2.1.2 OTHER PUBLICATION

   IP 24/1978—Published jointly by the Rubber Manufacturers Association (RMA), the Mechanical Power Transmission Association (MPTA), and the Rubber Association of Canada (RAC)

3. *Pulleys*

   3.1 *Minimum Pulley Diameters*—Minimum recommended pulley diameters are shown in Table 1.

TABLE 1—MINIMUM RECOMMENDED PULLEY DIAMETERS (mm)

Pulley Section	Pitch	Minimum Grooves	Minimum Pitch Diameter	Minimum Outside Diameter
ST	9.525	10	30.32	29.56
SU	12.700	14	56.60	55.23
STA	9.525	19	57.61	56.23

3.2 *Minimum Pulley Width*—The minimum pulley width between flanges (Figure 1) is determined by Equation 1:

$$1.5 \text{ (belt plus side tolerance)} + \text{nominal width} \quad (\text{Eq.1})$$

NOTE—Stack up tolerances should be handled between pulley manufacturer and the user.

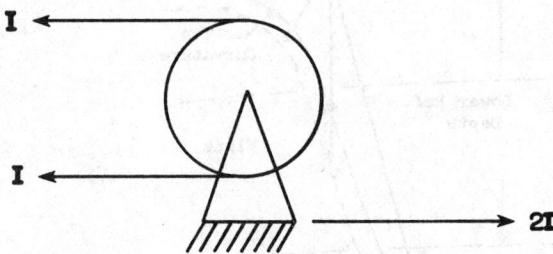

FIGURE 1—MINIMUM PULLEY WIDTH

3.3 *Pulley Finish*—A maximum surface finish of 2 μm Ra is normally satisfactory for standard drives. However, a maximum of 1 μm Ra finish is strongly recommended for crankshaft and other critical drive pulleys.

3.4 *Flanging*—Since a synchronous belt will have a tendency to ride to one side similar to a flat belt, it is necessary to contain it. Due to an inextensible tensile member, it is impossible to utilize a crown as is typical with flat belts. Therefore, flanges are used to guide the belt on the pulleys. The direction of track is controlled by the direction of rotation. (Any given belt will track opposite to its original track when the direction of rotation is reversed.) Since the direction of rotation is not usually furnished, and because of reversal applications, smaller driving pulleys are generally furnished with flanges on both sides.

3.5 *Flanged Pulleys*—Recommended flange dimensions are shown in Table 2 and Figure 2.

TABLE 2—FLANGE DIMENSIONS (mm)

Pulley Section	Minimum Flange Thickness	Minimum Flange Height
ST	1.3	1.6
SU	1.3	2.0
STA	1.3	2.4

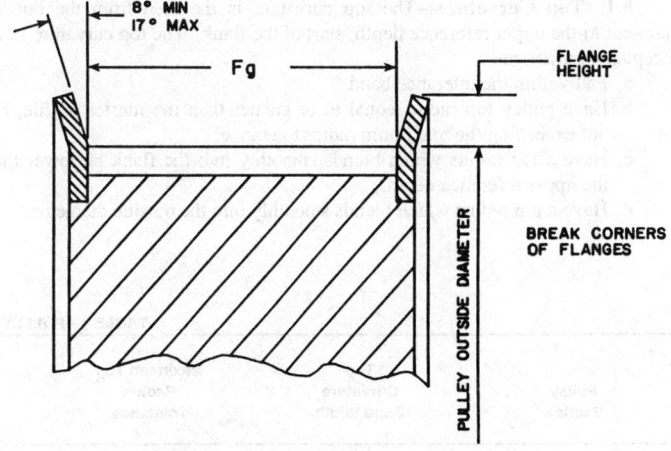

FIGURE 2—PULLEY FLANGES

3.6 *Selection of Flanged Pulleys*—On all two-pulley drives, the minimum flanging requirements are: two flanges on one pulley, or one flange on each pulley on opposite sides.

On drives where the center distance is more than eight times the diameter of the small pulley, both pulleys should be flanged on both sides. On vertical shaft drives, it is usually advisable to flange the bottom side of the larger pulley as well as both sides of the smaller pulley. This is a function of center distance, speed ratio, and belt width, and will vary with respective applications.

On multipoint drives, the minimum flanging requirements are two flanges on every other pulley, or one flange on every pulley alternating sides around the system.

4. *Recommended Use of Idlers*—The use of idlers should be restricted to those cases in which they are functionally necessary. The usual cases are:

   a. As a means of applying tension when pulley centers are not adjustable.
   b. To increase the number of teeth in mesh on the small pulley of relatively high ratio drives.

Idlers should be located on the slack side of the belt. For inside idlers, grooved pulleys are recommended up to 40 grooves. On larger diameters, flat uncrowned pulleys may be used. Outside idlers should be flat, uncrowned pulleys. Idler diameters should not be smaller than the smallest pulley diameter in the system.

Fixed idlers are recommended.

5. *Belts*

   5.1 *Maximum Belt Width*—Belt width should not exceed the small pulley diameter in order to avoid excessive belt side thrust.

6. *Installation Tension*—Installation tension varies considerably with respective users. This is a result of other factors involved in the drive, such as guards, clearance areas, etc., as well as individual belt manufacturers' recommendations. The formulae in Table 3 are offered for general guidance covering belt widths from 5 to 50 mm:

TABLE 3—INSTALLATION TENSION FORMULAE

Section	Installation Tension (N) Min		Installation Tension (N) Max
ST	5.5 b_S–17	≤ 1 ≤	7.6 b_S–24
SU	12 b_S–38	≤ 1 ≤	20 b_S–62
STA			

where nominal belt width = b_S in millimeters

7. Master Profile—The master profile is generated by the nominal pulley generating tool rack form (see Table 1 of SAE J1278) at a specific number of grooves and nominal pulley OD. Master profiles can be obtained from belt manufacturers.

8. Tolerances—Pulley groove tolerances are applied separately to the four general areas of the profile: top curvature, flank, bottom curvature, and depth.

8.1 Top Curvature—The top curvature is the area from the outside diameter to the upper reference depth, start of the flank. The top curvature of an acceptable pulley must:
 a. Fall within the tolerance band.
 b. Have pulley top radius equal to or greater than the master profile, but not exceeding the maximum radius tolerance.
 c. Have a top radius which blends smoothly into the flank no lower than the upper reference depth.
 d. Have a top radius which blends smoothly into the outside diameter.

8.2 Flank—The flank is the distance between the upper and the lower reference depths. The flank of an acceptable pulley must fall within the tolerance band and must be parallel to the master profile within 0.5 degrees.

8.3 Bottom Curvature—The bottom curvature is the area from the lower reference depth to the bottom of the groove profile. The bottom curvature of an acceptable pulley must fall within the tolerance band.

8.4 Depth—The depth of an acceptable pulley groove must fall within the tolerance band.

8.5 Lower Reference Depth—The lower reference depth is the point of tangency of the belt tooth bottom radius and the straight-sided belt tooth flank. This point has been selected because below it there is no contact between the belt tooth and pulley groove. It is measured radially from the pulley outside diameter.

8.6 Upper Reference Depth—The upper reference depth divides the profile into an area generated by the rack top radius and an area generated by the cutter flank. Hence, it determines the start of the involute portion of the groove profile. It is measured radially from the pulley outside diameter.

9. Procedure for Checking Pulley Grooves
 a. Check pulley grooves on a comparator against a master profile (see Table 4 and Figure 3).
 b. Line up the master profile with the outside diameter of the pulley.
 c. Determine if the subject profile falls within the tolerance bands in all four of the areas of the profile.
 d. Check to see that the top radius is equal to or greater than the master profile top radius, but does not exceed the band width tolerance and maximum radius tolerance.
 e. Check to see that all radii blend smoothly into the flank, groove bottom, and outside diameter.
 f. Check angle of the flank against the master profile flank.

TABLE 4—PULLEY GROOVE TOLERANCES (mm)

Pulley Section	Top Curvature Band Width	Maximum Top Radius Tolerance	Flank Band Width	Bottom Curvature Band Width	Depth Band Width	Upper Reference Depth
ST	0.04	±0.1/–0.0	0.05	0.05	0.05	0.5
SU	0.04	±0.1/–0.0	0.05	0.05	0.05	0.8
STA	0.04	±0.1/–0.0	0.05	0.05	0.05	0.5

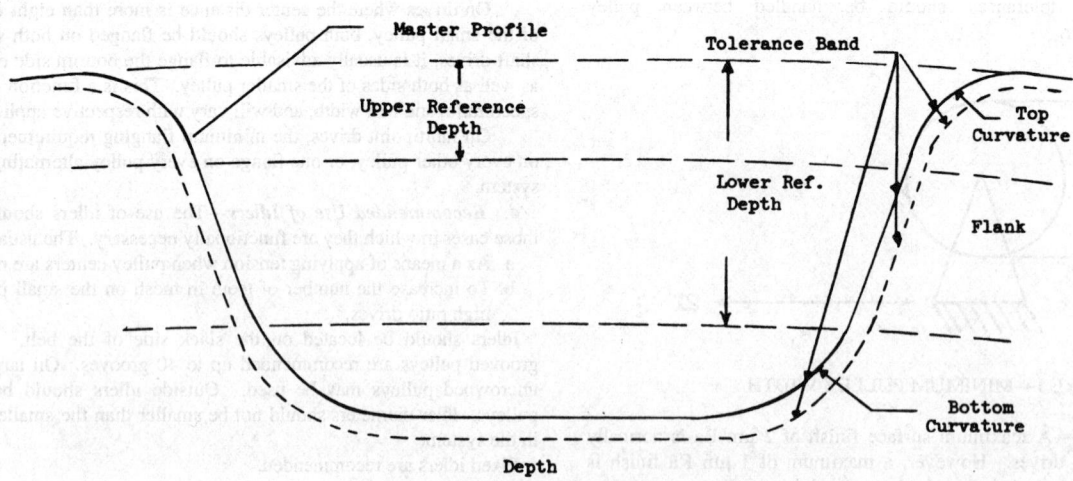

FIGURE 3—PULLEY GROOVE PROFILE

φV-RIBBED BELTS AND PULLEYS—SAE J1459 AUG88 — SAE Standard

Report of the V-Belt Committee, approved August 1984 and completely revised August 1988.

1. **Scope**—This standard covers dimensions, tolerances, and methods of measurement of V-ribbed belts and pulleys for use on automotive[1] drives.

2. **V-Ribbed Belts**—V-ribbed belts are produced in a variety of cross-sectional sizes which are given letter designations. It has been determined that the "PK" ("K") and "PL" ("L") section V-ribbed belts are applicable for automotive use. Because of different constructions and different methods of manufacture, the cross-sectional shape, dimensions, and included angle between the sidewalls of the belt may differ with different manufacturers. Belts are to be dimensioned in such a way that they are functional in pulleys dimensioned as described in a subsequent section.

3. **Pulleys For V-Ribbed Belts**—Pulleys are to conform to requirements of Fig. 1 and Tables 1A or 1B.

4. **V-Ribbed Belt Size**—The belt-cross-sectional size can be determined by the basic dimension S_g. The belt width for a cross section is determined by the number of ribs. The belt length is determined by measurement as described in a subsequent section. Belt size is designated by a standard series of alphanumeric characters. Belts measured on a metric length system are designated by the number of ribs followed by the belt-cross-sectional size ("PK" or "PL") and the effective length in millimeters. For example, 6PK1370 signifies a 6-rib "PK" section belt, with an effective length of 1370 mm. Belts measured on an inch-length system are designated by the effective length to the nearest tenth of an inch followed by the belt-cross-sectional size ("K" or "L") and the number of ribs. For example, 540K6 signifies a belt with an effective length of 54.0 in, a "K" cross section, and 6 ribs.

5. **V-Ribbed Pulley Size**—Pulley size is designated by the effective diameter, the groove-cross-sectional size, and the number of grooves. The effective diameter can be determined by measuring the diameter over the balls or rods and subtracting 2K. The groove-cross-sectional size can be determined by measuring the dimension S_g. The pulley width for a cross section is determined by the number of grooves.

6. **Measurement of V-Ribbed Belts**—The length of a V-ribbed belt is determined by use of a measuring fixture comprised of two pulleys of equal diameter, a method of applying force, and a means of measuring the center distance between the two pulleys. One of the two pulleys is fixed in position while the other is movable along a graduated scale. Both pulleys are allowed to rotate. The fixture is shown schematically

[1] For non-automotive drives, see Engineering Standard IP-26 published by: Rubber Manufacturers Association, Inc., 1901 Pennsylvania Avenue, Washington, DC 20006. Pulleys produced to Engineering Standard IP-26 will meet all of the requirements of this standard.

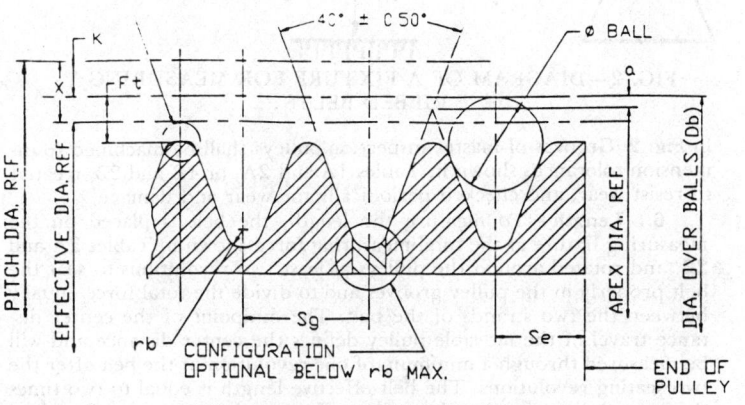
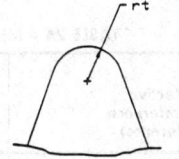

FIG. 1—STANDARD GROOVE DIMENSIONS

TABLE 1A – GROOVE DIMENSIONS – MILLIMETERS

Cross Section Size	Minimum Recommended Effective Diameter	S_g ±0.05	Se Min	rt min	rt max	rb max	Ft[b] max	2a	2k	2x[c]	ΦBALL ±0.010
PK	45	3.56	2.50	0.35	0.50	0.50	1.00	0.03	0.99	3.0	2.500
PL	75	4.70	3.30	0.40	0.70	0.40	1.75	0.82	2.36	1.54	3.500

[a] Summation of the deviations from Sg for any one belt groove set in a pulley shall not exceed ±0.30 mm.
[b] Ft is measured from the actual ride position of the ball or rod in the pulley.
[c] This number may vary with belt manufacturer.

TABLE 1B – GROOVE DIMENSIONS – INCHES

Cross Section Size	Minimum Recommended Effective Diameter	S_g ±0.002	Se Min	rt min	rt max	rb max	Ft[b] max	2a	2k	2x[c]	ΦBALL ±0.0004
K	1.8	0.140	0.100	0.014	0.020	0.020	0.039	0.001	0.039	0.118	0.0984
L	3.0	0.185	0.130	0.016	0.028	0.016	0.069	0.032	0.093	0.061	0.1378

[a] Summation of the deviations from Sg for any one belt groove set in a pulley shall not exceed ±0.012 in.
[b] Ft is measured from the actual ride position of the ball or rod in the pulley.
[c] This number may vary with belt manufacturer.

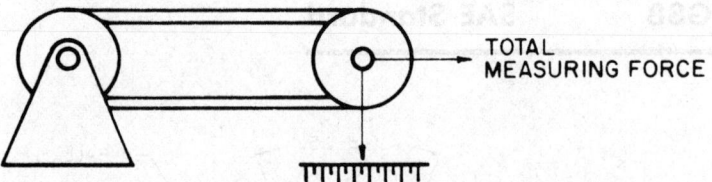

FIG. 2—DIAGRAM OF A FIXTURE FOR MEASURING V-RIBBED BELTS

in Fig. 2. Grooves of master inspection pulleys shall be machined to dimension tolerances shown in Tables 1A and 2A, or 1B and 2B, treated to resist wear, and checked periodically for wear and damage.

6.1 Length—To measure the length, the belt is placed on the measuring fixture at the total measuring force shown in Tables 2A and 2B, and rotated around the pulleys at least two revolutions to seat the belt properly in the pulley grooves and to divide the total force equally between the two strands of the belt. The midpoint of the center distance travel of the movable pulley defines the center distance and will be measured through a minimum of one revolution of the belt after the two seating revolutions. The belt effective length is equal to two times the center distance plus the pulley effective circumference. Standard belt center distance tolerances are shown in Tables 3A and 3B. For center distance tolerances less than standard, the belt manufacturer should be consulted.

7. Standard Lengths—Standard lengths up to and including 2000 mm (79 in) are to be in 10 mm (0.4 in) increments. Standard lengths over 2000 mm (79 in) up to and including 4000 mm (157 in) are to be in 25 mm (1.0 in) increments.

NOTES:
1. The sides of the groove are to be 3.2 µm (125 µin) A.A. maximum.
2. Radial and axial run-out is not to exceed 0.35 mm (0.014 in) full indicator movement (FIM). Run-out in the two directions is measured separately with a ball mounted under spring pressure to follow the groove as the pulley is rotated.
3. The diameters over the ball gauges are not to vary from groove to groove more than 0.25 mm (0.010 in) for any one belt groove set in a pulley.
4. Centerline of groove is to be 90.0 ± 0.5 deg with pulley axis.
5. The pitch diameter is used for calculation of speed ratio. The "x" dimension is radial. "2x" is to be added to the effective diameter to determine the pitch diameter.
6. The apex diameter may be helpful for layout of pulley grooves or production tooling. "2a" is subtracted from the diameter over the balls to determine the apex diameter.

TABLE 2A – MEASURING CONDITIONS – MILLIMETERS

Cross Section Size	Effective Diameter (reference)	Effective Circumference (reference)	d_B Ball or Rod Diameter ±0.010	Diameter Over Ball or Rods ±0.10	Total Measuring Force Per Rib (N)
PK	95.49	300	2.500	96.48	100
PL	159.15	500	3.500	161.51	200

TABLE 2B – MEASURING CONDITIONS – INCHES

Cross Section Size	Effective Diameter (reference)	Effective Circumference (reference)	d_g Ball or Rod Diameter ±0.0004	Diameter Over Ball or Rods ±0.004	Total Measuring Force Per Rib (lb)
K	3.759	11.8	0.0984	3.798	22
L	6.266	19.7	0.1378	6.359	45

TABLE 3A – STANDARD BELT CENTER DISTANCE TOLERANCES – MILLIMETERS

Belt Length	Tolerance on Center Distance
1250 and less	±3.2
Over 1250 – 1600, incl	±4.0
Over 1600 – 2000, incl	±4.8
Over 2000 – 2500, incl	±5.6
Over 2500 – 3150, incl	±6.4
Over 3150 – 4000, incl	±7.2

TABLE 3B – STANDARD BELT CENTER DISTANCE TOLERANCES – INCHES

Belt Length	Tolerance on Center Distance
49 and less	±0.13
Over 49 – 63, incl	±0.16
Over 63 – 79, incl	±0.19
Over 79 – 98, incl	±0.22
Over 98 – 124, incl	±0.25
Over 124 – 157, incl	±0.28

AUTOMOTIVE V-RIBBED BELT DRIVES AND TEST METHODS—SAE J1596 JUN89

SAE Recommended Practice

Report of the V-Belt Committee approved June 1989.

1. Scope—The following information covers engine accessory drive layout details and testing methods and includes test configurations, pulley diameters, power loads, and guidance for interpreting test data. This information has been prepared from existing literature, including standards and data supplied by both producers and users of V-ribbed belts.

When the engine is used to drive an external unit equipped with industrial type V-pulleys and belts, it is recommended that the power takeoff pulley on the engine be grooved according to the appropriate industrial standard. These standards are published by RMA-MPTA (Rubber Manufacturers Association – Mechanical Power Transmission Association); ISO (International Standards Organization) and ASAE (American Society of Agricultural Engineers). The grooves in those standards differ from each other in the reference dimensions, and they are not interchangeable with SAE grooves, which are standardized for engine accessory drives as covered in SAE J636 and J1459.

1.1 Purpose—This recommended practice is intended as a guide to be used for evaluating V-ribbed belt construction, source approval, or quality audit.

2. References
SAE J636, V-belts and Pulleys
SAE J1459, V-ribbed Belts and Pulleys

3. General Drive Layout Considerations

3.1 Belt Speed—It is recommended that the pulley diameters be as large as possible without exceeding 50 m/s belt speed.

3.2 Pulley Sizes—The inside pulleys should never be smaller than the tension pulley diameter specified in Tables 1 and 2 of this standard; however, belt life is directly related to pulley diameter. A more practical minimum pulley diameter, for increased belt life, would be 50 mm. (See paragraph 3.5 for backside pulley diameter recommendations.)

3.3 Belt Length—The calculation of the belt effective length (E.L.) for a specific drive involves several design considerations, including provision for adequate installation and take-up.

To allow for belt installation and take-up, one pulley should be adjustable from its initial position with the mean length belt at installation tension. This formula gives the minimum allowance for easy installation of the belt without prying or otherwise forcing it over the sides of the grooves.

3.3.1 The following formula can be used to calculate the recommended minimum belt E.L. for installation:

"K" Section: Inside pulleys
  Min. Belt E.L. = $1.003(L_1) + L_2 + .77 + 6.28(h)$

"K" Section: Flat backside pulley
  Min. Belt E.L. = $1.003(L_1) + L_2 + 6.28(h)$

"L" Section: Inside pulleys
  Min. Belt E.L. = $1.005(L_1) + L_2 + 1.15 + 6.28(h)$

"L" Section: Flat backside pulley
  Min. Belt E.L. = $1.005(L_1) + L_2 + 6.28(h)$

NOTE: The last pulley, over which the belt can be installed, should be considered and the appropriate belt installation formula used.

L_1 – The effective belt length (addition of span lengths and effective arc lengths on the pulleys) around the drive with the tensioning pulley in the minimum position. The 1.003 ("K" section) and 1.005 ("L" section) factors provide for length change from slack to measuring tension.

L_2 – The 2 X negative belt manufacturing center distance tolerance (Tables 3A and 3B, SAE J1459).

h – The flange height.

This formula also allows for full contact belt rib height for V-ribbed pulleys—0.77 ("K" section) and 1.15 ("L" section).

The metric formulas are the same with the following exception:
"K" section – 0.77 value is 20
"L" section – 1.15 value is 29

The belt length, flange height, and manufacturing tolerance are in inches in the English system and millimeters in the metric system.

3.3.2 Select a belt to be used that has a nominal effective length equal to or greater than the recommended minimum E.L.

3.3.3 Calculate the maximum required effective length path around the drive to provide for take-up:

The maximum required belt path length =

"K" section—$(1.0145)L_3 + L_4 + L_5$
"L" section—$(1.0190)L_3 + L_4 + L_5$

L_3 – The nominal effective belt length as defined in paragraph 3.3.2. The 1.0145 ("K" section) and 1.0190 ("L" section) factors account for elongation from measuring to installation tension and 1% for belt stretch and wear.

L_4 – "K" section – 0.12
   "L" section – 0.19
   Values for belt seating factor

L_5 – Positive manufacturing tolerance for belt length. (Tables 3A and 3B, SAE J1459.)

The formulas are in English units with belt length in inches. The metric units will be belt length in millimeters and the following seating factor changes:
"K" section – 0.12 value is 3.0
"L" section – 0.19 value is 4.8

3.4 Pulley Misalignment—The recommended maximum misalignment between pulleys is 0.58 mm per 100 mm (0.069 in/ft) of span length or approximately 0.33 deg.

3.5 Backside Pulleys—The application design with backside pulleys is acceptable provided their minimum diameter is 90 mm (3.54 in) for PK (K) section, and 140 mm (5.5 in) for PL (L) section.

4. V-Ribbed Belt Fatigue Test Method

4.1 Test Configuration—The belt shall be mounted on a test fixture layout as shown in Fig. 1 with the pulley diameters and speeds as given in Tables 1 and 2 and the installation clearance factor as shown in Table 3.

4.2 Test Load—The kilowatts (horsepower) to be absorbed at the driven pulley shall be as shown in Tables 1 and 2.

4.2.1 The driver pulley speed (rpm) shall be used in the torque load calculation, and the torque load shall be kept constant without compensation for loss of driven pulley rpm resulting from belt slippage.

$$\text{Torque, N} \cdot \text{m} = \frac{\text{Specified kilowatts} \times 9549}{\text{Driver rpm}}$$

$$\text{Torque, lbf} \cdot \text{in} = \frac{\text{Specified horsepower} \times 63025}{\text{Driver rpm}}$$

4.2.2 The test equipment shall be maintained to minimize parasitic loads due to bearing losses, lubricants, etc.

4.2.3 The belt tensioning force shall be equal in the number of Newtons to sixty times the number of specified kilowatt units (in number of pounds to ten times the number of horsepower units).

4.3 Temperature—The ambient temperature of the test fixture shall be controlled to 80°C ± 3 (175°F ± 5) within a suitable enclosure.

4.4 Test Procedure—The test procedure shall be as follows:

4.4.1 Set the required torque load.

4.4.2 Install the belt on the test fixture and apply the required belt tension. Condition the belt by running it for 5 min without the dynomometer load. Maintain a constant tension during this period by operating with the tension pulley center position unlocked. Stop the machine, allow it to stand for a minimum of 10 min and lock the tension pulley center position midway of the limits of travel during belt rotation.

4.4.3 Start the machine. Adjust the torque to the proper setting. Adjust the temperature to the proper setting.

4.4.4 Whenever the slip reaches 8%, stop the machine, allow it to stand for a minimum of 20 min, unlock the tension pulley center, restore the initial tension per 4.2.3 relock, and restart the machine.

4.4.5 (Exclusive of the 5 min run-in) - Record the number of hours run and the number of re-tensions.

5. Test Performance Guidelines

5.1 Test Life—The test life, which a belt must attain, shall be according to agreement between the manufacturer and user.

5.2 Acceptable Number of Re-tensions—The acceptable number of re-tensions after the initial 5 min run-in shall be according to agreement between the manufacturer and user.

5.3 Failure Criteria—A belt shall be considered failed when it no longer transmits the specified power because of breakage or it exceeds the number of agreed re-tensions. The belt manufacturer and user may agree on other acceptance/rejection criteria.

5.4 Length vs. Life Relationship—Generally speaking, the belt test life is a function of the belt length. The relationship shall be according to agreement between the belt manufacturer and user.

5.5 Definition of Construction Specifications—The belt manufacturer's test data on belts of a certain construction specification shall be considered valid for evaluation of all belts of the same construction specification when they are the same with respect to the manufacturer's cross section dimensions, material specifications, and method of manufacture.

5.6 Belt Test Lengths—The test belt range will be: 1020-1400 mm (40-55 in) with recommended length being 1200 mm (47.24 in). When evaluating for part source approval and for production quality surveillance, the test data from one belt length will be representative of all belt lengths of the same construction.

5.7 Statistical Relationships—Whether testing is performed for part source approval or for production surveillance, a realistic statistical guide to acceptability would be "not more than 10% of the test lives shall be permitted to fall below 50% of the specified average life."

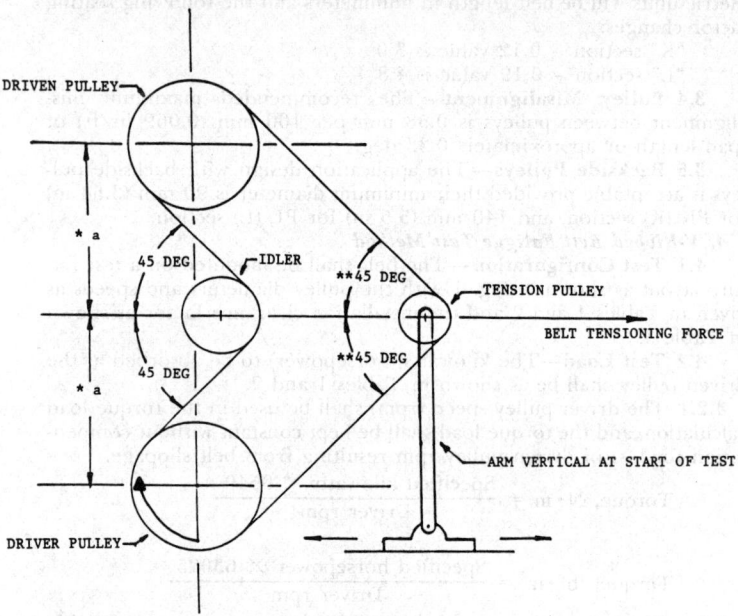

* Dimension a is adjusted for various length belts to maintain tension pulley midway vertically between driver and driven pulleys.

** 45 deg is specified for initial test configuration and may change slightly with resets as the test progresses.

FIG. 1—V-RIBBED BELT FATIGUE TEST

TABLE 1—TEST CONDITIONS - METRIC

SAE Belt Size	Effective Diameter			Driver Pulley Speed rpm ±2%	Load (KW)	Number of Ribs	Belt Length Range	
	Driver and Driven Pulleys ±0.25	Tension Pulley ±0.25	Idler Pulley ±0.25				Total	Preferred
PK	120.6 b	44.45 b	76.2 b	4900 b	10.4 b	3 b	1020-1400	1200
PL							1020-1400	1200

a = Groove details per SAE J1459
b = Not developed for Automotive belts

TABLE 2—TEST CONDITIONS - ENGLISH

SAE Belt Size	Effective Diameter			Driver Pulley Speed rpm ±2%	Load (HP)	Number of Ribs	Belt Length Range	
	Driver and Driven Pulleys ±0.010	Tension Pulley ±0.010	Idler Pulley ±0.010				Total	Preferred
K	4.750 b	1.750 b	3.000 b	4900 b	14 b	3 b	40.17-55.14	47.26
L							40.17-55.14	47.26

TABLE 3—CLEARANCE FACTOR FOR BELT INSTALLATION

SAE BELT SIZE		C VALUE	
Metric	English	mm	in
PK	K	3.07	0.12
PL	L	4.52	0.18

GLOSSARY—AUTOMATIC BELT TENSIONER—SAE J2198 JUL92

SAE Information Report

Report of the SAE Bolt Drive Committee approved July 1992.

1. **Scope**—This glossary was written to provide a consistent and uniform definition of terms used in describing the selection of an automatic belt tensioner as it applies to an automotive accessory drive system.
2. **References**—There are no referenced publications specified herein.
3. **System Related Tensioner Characteristics**

    3.1 **Tensioner Torque Requirement Curve (See Figure 1)**—The calculated torque curve required from a tensioning device which yields a constant belt tension throughout the tensioner range of travel in a system.

    3.2 **Positive/Negative Rate Geometry (See Figure 1)**—A positive/negative slope resulting from the linear approximation of the tensioner torque requirement curve.

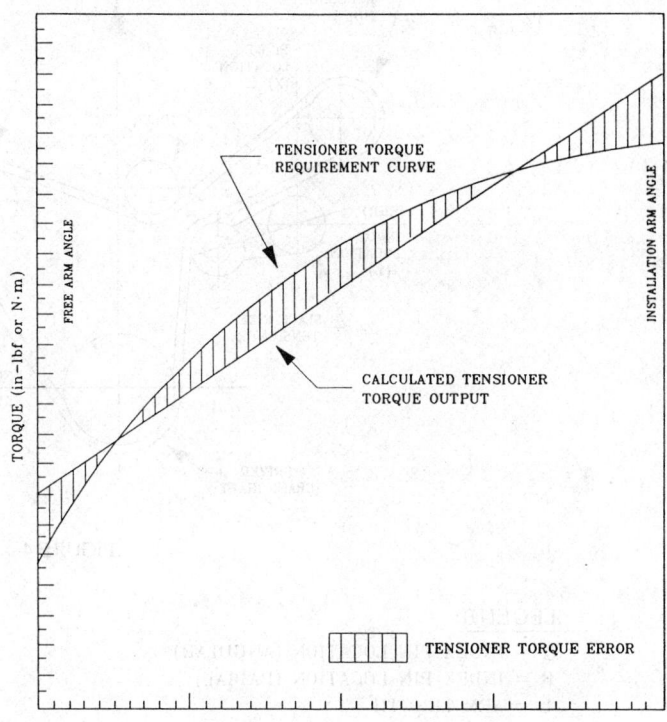

FIGURE 2—TENSIONER TORQUE ERROR CURVE

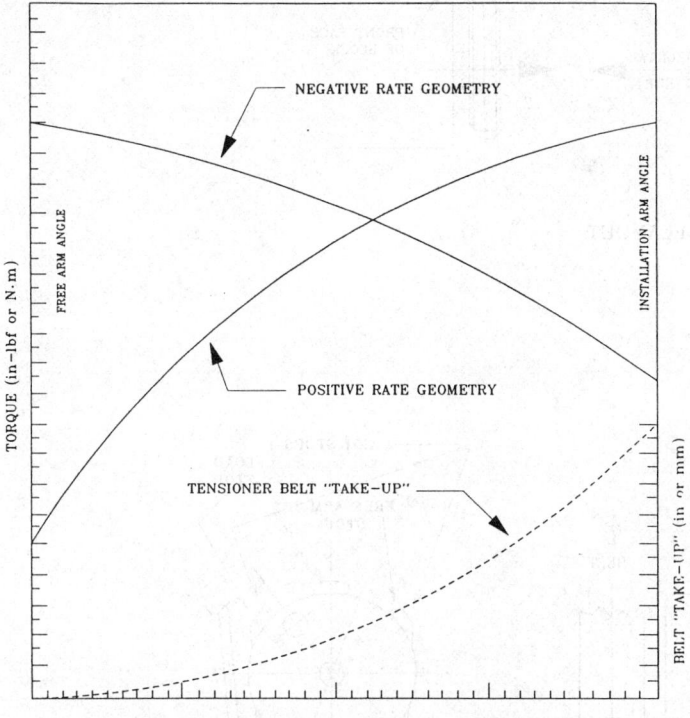

FIGURE 1—TENSIONER TORQUE REQUIREMENT CURVE

3.3 **Tension/Torque Error Curves (See Figures 2 and 3)**—The tension/torque curves representing calculated tensioner output as compared to the tensioner torque requirement curve.

3.4 **Tensioner Pivot Location (See Figure 4)**—The x,y coordinates of the tensioner pivot as compared to all other components in the system and the center of the driver pulley at (0,0).

3.5 **Tensioner Arm to Belt Bisector Angle (see ß, Figure 5)**—The angle between a line from the center of the tensioner pivot to the center of the tensioner pulley and the hub load vector at the tensioner pulley.

3.6 **Tensioner Belt "Take-Up" (See Figure 1)**—The amount of belt length change a tensioner will accommodate as a function of tensioner arm displacement within the tensioner operating range.

3.7 **Maximum Sustainable Tensioner Arm Amplitude**—Angular displacement of the tensioner arm in dynamic (running) system conditions.

3.8 **Slack Side Tensioner (See Figure 4)**—The location of the tensioner is between the driver pulley and the first load-carrying component in the direction of belt travel (routing idlers are not considered load-carrying components).

3.9 **Tight Side Tensioner**—The location of the tensioner is in any belt span other than the slack side of the driver pulley.

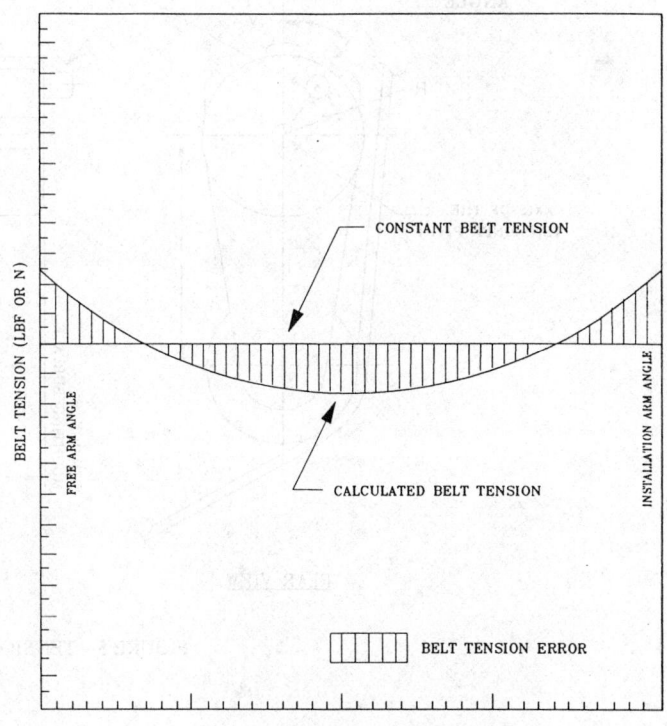

FIGURE 3—BELT TENSION ERROR CURVE

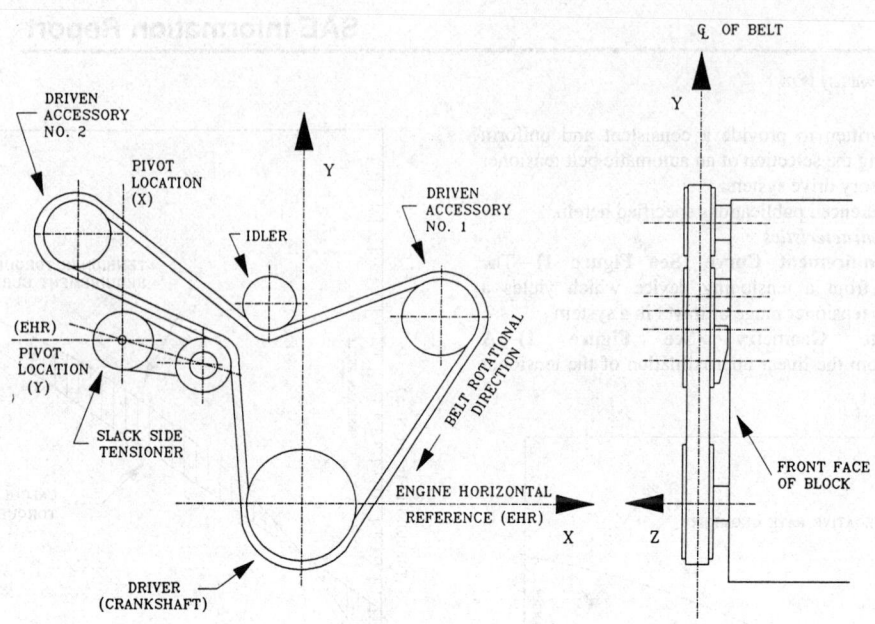

FIGURE 4—SYSTEM LAYOUT

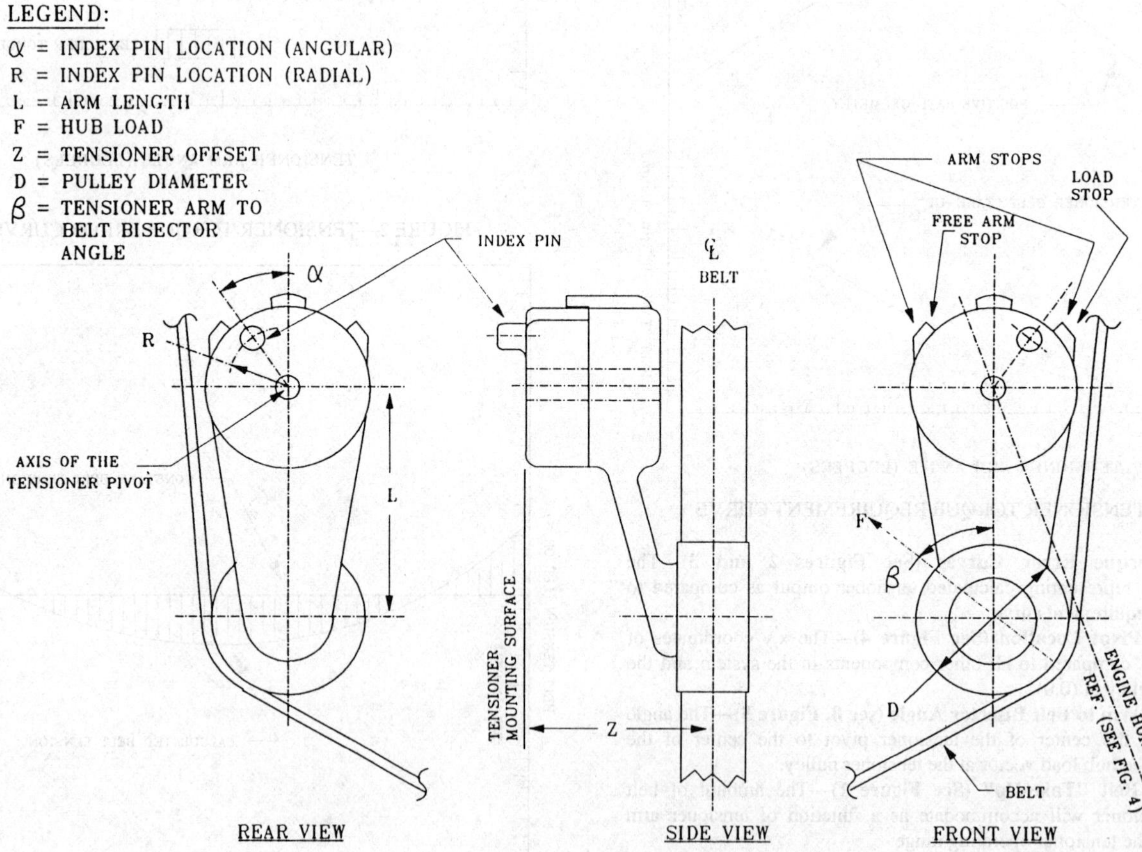

LEGEND:
α = INDEX PIN LOCATION (ANGULAR)
R = INDEX PIN LOCATION (RADIAL)
L = ARM LENGTH
F = HUB LOAD
Z = TENSIONER OFFSET
D = PULLEY DIAMETER
β = TENSIONER ARM TO BELT BISECTOR ANGLE

FIGURE 5—TENSIONER CHARACTERISTICS

4. Tensioner Specific Design Characteristics

4.1 Tensioner Structural Integrity (Load and Free Arm Stops, Index Pin) (See Figure 5)—The structural integrity of the physical tensioner arm stops and the tensioner index pin in consideration of assembly and service.

4.2 Tensioner Arm Angle (See Figures 4 and 6)—The angle of a line from the center of the tensioner pivot to the center of the tensioner pulley measured in system polar coordinates where 0° = positive x-axis.

4.3 Tensioner Nominal Arm Angle (See ß, Figure 6)—The tensioner arm angle which corresponds to the nominal length belt.

4.4 Tensioner Free Arm Angle (See α, Figure 6)—The tensioner arm angle which corresponds to the free state of the tensioner, i.e., without the belt installed.

4.5 Tensioner Installation Arm Angle (See γ, Figure 6)—The tensioner arm angle which corresponds to the maximum displacement of the tensioner arm to allow for belt installation.

4.6 Tensioner Arm Length (See L, Figure 5)—The distance between the center of the tensioner pivot and the center of the tensioner pulley.

4.7 Tensioner Pulley Diameter (See D, Figure 5)—Diameter of the pulley attached to the tensioner arm as defined by SAE standard for groove side pulleys. SAE standard for backside pulleys is TBD.

4.8 Tensioner Operating Range (See ϵ, Figure 6)—The difference between the tensioner arm angle at the shortest belt length and the tensioner arm angle at the longest belt length including stretch and wear.

4.9 Tensioner Arm/Pulley Inertia—The resistance to angular velocity changes about the pivot axis in the tensioner assembly.

4.10 Tensioner Torque Output Characteristics (See Figure 7)

4.10.1 TENSIONER TORQUE OUTPUT CURVE—The curve generated as a function of tensioner load output characteristics versus tensioner arm position as measured on a special fixture (hysteresis curve).

4.10.2 AVERAGE TORQUE OUTPUT—Calculated value from the tensioner torque output curve. The average of tensioner up and down stroke torque values according to Equation 1:

$$T_{average} = \frac{(T_{upstroke} + T_{downstroke})}{2} \qquad (Eq.1)$$

4.10.3 SPRING RATE—The rate of change in average torque output per degree of arm displacement.

4.10.4 TENSIONER TORQUE OUTPUT (AVERAGE) AT NOMINAL ARM ANGLE—The average of tensioner arm up and down stroke torque or force measurements at a tensioner arm position which corresponds to the mean belt length.

4.10.5 TENSIONER TORQUE OUTPUT TOLERANCE—The range of variation in tensioner average torque output.

4.10.6 SPRING SET—The change in tensioner torque output due to the initial stabilization of the various components in the tensioner assembly.

4.10.7 DAMPING (AT TENSIONER NOMINAL ARM ANGLE)—Friction existing in the tensioner. Procedure for calculating percent damping is shown in Equation 2:

$$\text{Percent Damping} = \frac{(T_{upstroke} - T_{average})}{T_{average}} \times 100 \qquad (Eq.2)$$

4.11 Tensioner Offset (See Figure 5)—The distance from tensioner mounting surface to theoretical centerline of belt.

4.12 Tensioner Arm/Pulley Angularity (Backside Pulleys at Tensioner Nominal Arm Angle) (See Figure 8)

4.12.1 TOE—The angle between the centerplane of the pulley and the plane of the tensioner mounting surface measured in a plane perpendicular to the hub load vector as a rotation about the hub load axis (right hand rule with the thumb pointing in the direction of the hub load = positive direction).

4.12.2 CAMBER—The angle between the centerplane of the pulley and the plane of the tensioner mounting surface measured in a plane passing through the hub load vector and pulley axis as a rotation about the axis perpendicular to the hub load vector (right hand rule with the thumb pointing in the direction of belt motion in a clockwise rotating system).

4.13 Bearing "Free Rock"—The total angular movement of the pulley due to internal bearing clearances as measured in a special fixture.

4.14 Index Pin Location (at Tensioner Free Arm Angle) (See Figure 5)—The angle between the tensioner arm centerline and a line from the axis of the tensioner pivot to the axis of the tensioner index pin and a distance R from the axis of the tensioner pivot.

4.15 Total Tensioner Arm Travel (See Figure 6)—The angular difference between the tensioner installation arm angle and the tensioner free arm angle.

4.16 Belt Installation Feature—The design feature used for rotating the tensioner from its free arm position to the maximum travel position for belt installation, i.e., pulley bolt head or tool lug.

LEGEND:
α = TENSIONER FREE ARM ANGLE
β = TENSIONER NOMINAL ARM ANGLE
γ = TENSIONER INSTALLATION ARM ANGLE
Δ = TOTAL TENSIONER ARM TRAVEL
ϵ = TENSIONER OPERATING RANGE
S = SHORTEST BELT POSITION
L = LONGEST BELT POSITION
EHR = ENGINE HORIZONTAL REFERENCE

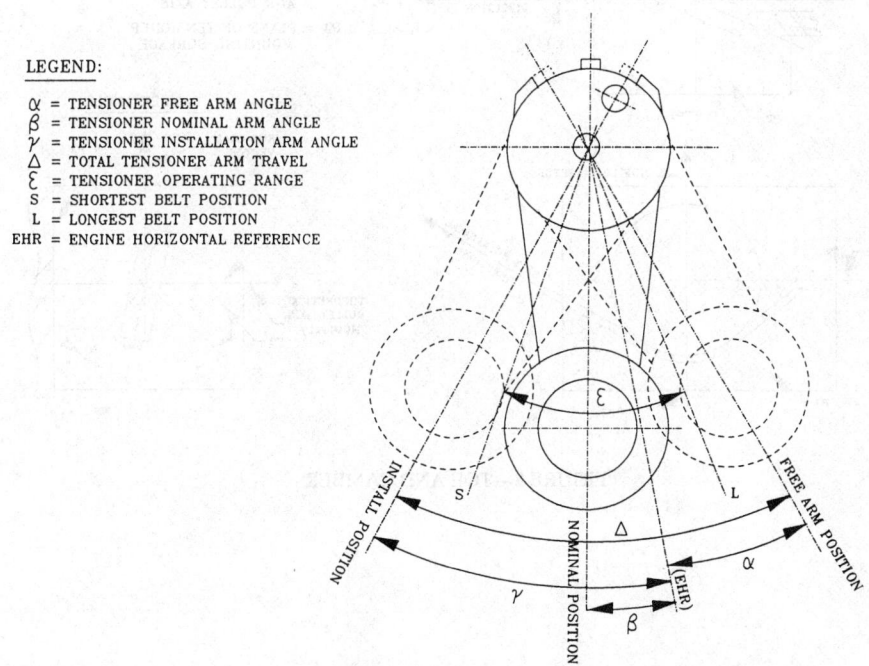

FIGURE 6—TENSIONER CHARACTERISTICS

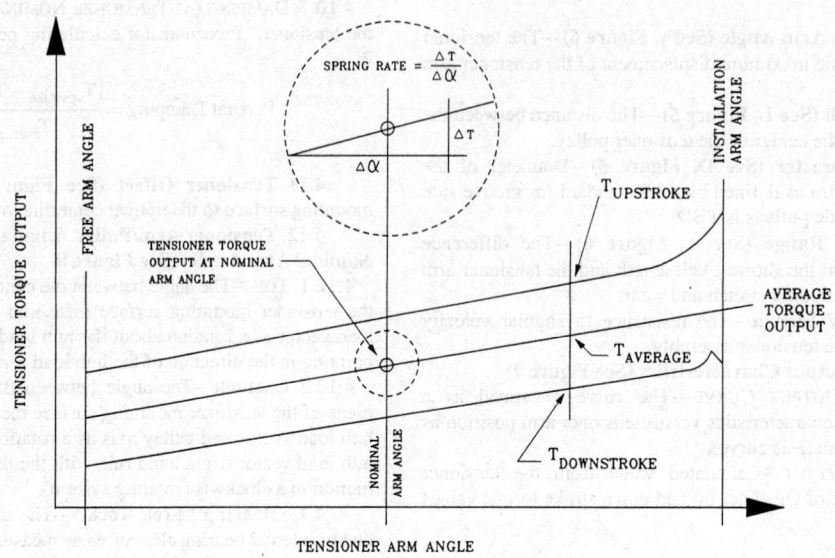

FIGURE 7—TENSIONER TORQUE OUTPUT

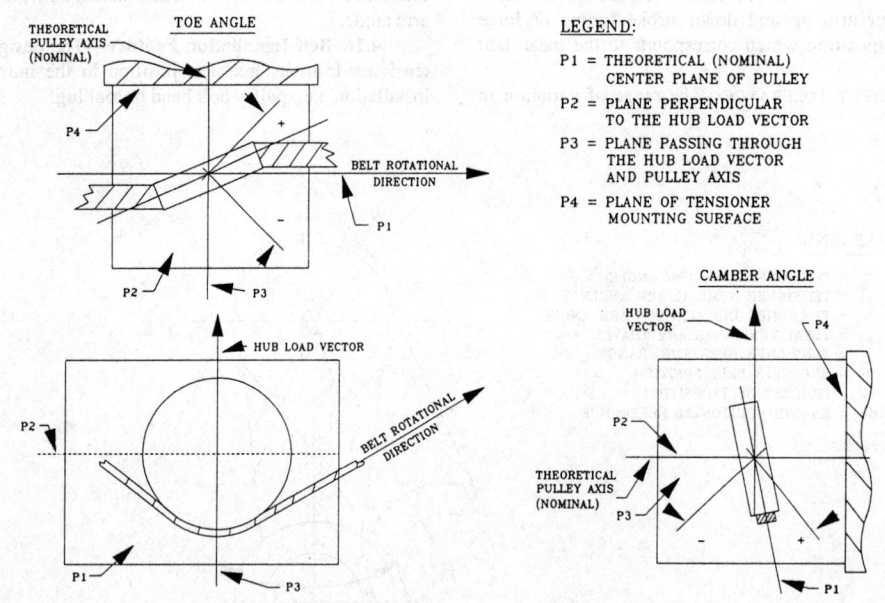

FIGURE 8—TOE AND CAMBER

LEGEND:
P1 = THEORETICAL (NOMINAL) CENTER PLANE OF PULLEY
P2 = PLANE PERPENDICULAR TO THE HUB LOAD VECTOR
P3 = PLANE PASSING THROUGH THE HUB LOAD VECTOR AND PULLEY AXIS
P4 = PLANE OF TENSIONER MOUNTING SURFACE

20 Springs

LEAF SPRINGS FOR MOTOR VEHICLE SUSPENSION—MADE TO METRIC UNITS—SAE J1123 NOV92

SAE Standard

Report of the Spring Committee, approved November 1975, second revision, Leaf Spring Subcommittee, May 1985. Reaffirmed by the SAE Leaf Spring Subcommittee of the SAE Spring Committee November 1992.

Foreword—This reaffirmed document has been changed only to reflect the new SAE Technical Standards Board format.

1. Scope

(NOTE—For leaf springs made to customary U.S. units, see SAE J510.)

This SAE Standard is limited to concise specifications promoting an adequate understanding between spring maker and spring user on all practical requirements in the finished spring. The basic concepts for the spring design and for many of the details have been fully dealt with in HS-J788.

2. References

2.1 Applicable Documents—The following publications form a part of this specification to the extent specified herein. The latest issue of SAE publications shall apply.

2.1.1 SAE PUBLICATIONS—Available from SAE, 400 Commonwealth Drive, Warrendale, PA 15096-0001.

SAE J419—Methods of Measuring Decarburization

SAE J510—Leaf Springs for Motor Vehicle Suspension—Made to Customary U.S. Units

HS-J788—Manual on Design and Application of Leaf Springs

3. Bar Sizes and Tolerances

Round edge flat spring steel has been adopted as the SAE standard.

The bars shall be of flat rolled steel having two flat surfaces and two rounded (convex) edges. They are subject to the tolerances shown in Table 1. These cross-section tolerances permit the two flat surfaces to be slightly concave. When that occurs, the radii of the arcs of the two concave surfaces shall be of approximately equal length.

The rounding of the convex edges shall be an arc with a radius of curvature that may vary from 65 to 85% of the thickness of the bar.

Bars shall be substantially straight and free from physical characteristics known as "kinks" or "twists" which render them unsatisfactory for spring manufacturing purposes.

Distortions due to a bar being bent about either major axis of section shall be measured with the bar against a flat checking surface so as to make contact with this surface near both bar ends. Gaps between the bar and the checking surface shall not exceed 4.0 mm/1 m of bar length out of contact with the checking surface when this bar length is greater than 1 m. Also, a gap between the bar and a straight edge 1 m long applied along any portion of the surface or edge of the bar, shall not exceed 4.0 mm.

It is recommended that all leaf spring bars which have been cold straightened be identified by the steel mill so that the spring manufacturer can use them selectively.

The bars which are generally provided in alloy steel shall be specified and rolled as in Table 2.

4. Surface Decarburization

Surface decarburization may reduce the fatigue durability of the springs; therefore, it is important that surface decarburization be at a minimum.

Hot rolled steel bars as received from the mills have some decarb, at least of the minimum Type 3 (see SAE J419), where more than 50% of the base carbon content remains at the surface (that is, some partial, but not more than 50% loss of carbon).

If decarb is of Type 2, where 50% or less of the base carbon content remains at the surface (that is, appreciable partial, but not total loss of carbon), the decarb normally does not exceed a depth of 0.25 mm for steels of thicknesses 5.00 to 12.50 mm, nor a depth of 0.50 mm for steels of thicknesses over 12.50 to 37.50 mm.

With sections over 25.00 mm in thickness, some of the hot rolled steel bars may have decarb of Type 1, in which virtually carbon-free ferrite (that is, total loss of carbon) exists for a measurable distance below the surface.

The depth of decarb varies from mill to mill, from rolling to rolling, and from bar to bar. The extent to which the depth and type of the decarb can be acceptable will be subject to agreement between the steel producer and the spring manufacturer.

The edges of the bars are somewhat higher in decarb than the flat surfaces; decarb on both the edges and the flat surfaces usually has greater depth with increased bar thickness.

After forging and non-atmospheric controlled heat treating, the spring leaves will have greater decarb. Scaling of the steel in this processing reduces the thickness of the leaf. While some of the surface decarb is removed with the scale, the final depth of decarb is usually greater than it was in the steel bars as received from the mills.

5. Definitions, Dimensions, and Tolerances

5.1 Leaf Spring—A spring of full elliptic, semi-elliptic, or quarter-elliptic shape with one or more leaves. The term "multi-leaf" has generally applied to springs of constant width and with stepped leaves, each of constant thickness except where leaf ends may be tapered in thickness. More recently, the term has been extended to include an assembly of stacked "single" leaves, each of which is characterized by tapering either in width or in thickness, or by a combination of both. Examples of multi-leaf springs are shown in Figures 1 to 6. Figure 7 shows a single leaf spring.

The leaves of a multi-leaf spring are usually held together with a center bolt and prevented from lateral shifting by alignment clips. Prior to assembly, the leaves are formed (cambered) and heat-treated by heating, quenching, and tempering to the required hardness. Quench dies or fixtures are used to maintain the required camber within tolerances.

5.2 Datum Line—Reference line used with many of the subsequently defined terms. In Figure 1 (where the springs are shown inverted as in a machine for load and rate checking), it is shown as the line X-X. On springs with eyes, the datum line passes through the centers of the eyes. On other springs, it passes through the points where the load is applied near the ends of the spring. These points must be indicated on the drawing. When load and rate are checked, the spring ends shall be free to move in the direction of the datum line.

5.3 Seat Angle Base Line (see Figure 1)—Reference line drawn through the terminal points of the active spring length at each eye, taken along the tension surface of the main leaf. On springs without eyes, the seat angle base line is coincident with the datum line.

5.4 Loaded Length—Distance between spring eye centers when the spring is deflected to the specified load position. On springs without eyes, it is the distance between the lines where load is applied under the specified conditions. Tolerance, ±3.0 mm.

5.5 Loaded Fixed End Length—Distance from the center of the fixed end eye to the projection on the datum line of the point where the centerline of the center bolt intersects the spring surface in contact with the spring seat. Tolerance, ±1.5 mm.

5.6 Straight Length—Distance between eye centers when the tension surface of the main leaf at the center bolt centerline is in the plane of the seat angle base line. The distance is measured parallel to the seat angle base line. Tolerance, ±3.0 mm.

5.7 Seat Length—Length of spring that is in actual engagement with the spring seat when installed on a vehicle at design height. It is always greater than the inactive length.

5.8 Inactive Length—Length of spring rendered inactive by the action of the U-bolts or clamping bolts.

5.9 Seat Angle (see Figure 1)—Angle between the tangent to the center of the spring seat and the seat angle base line. When the spring is viewed with the fixed end of the spring to the left as shown, and the load is applied to the shortest leaf from above, the seat angle may be specified as either positive (counterclockwise) or negative (clockwise), depending upon the angular direction in which the tangent to the center of the spring seat is disposed from the seat angle base line.

Consequently, with the spring in normal vehicle position so that the load is applied from below as shown in Figures 2, 4, 5, 6, and 7 and again with the fixed end of the spring to the left of the drawing, the seat angle is defined as positive when that tangent is disposed clockwise; and as negative when the tangent is disposed counterclockwise.

5.10 Finished Width—Width to which the spring leaves are ground or milled to give the edges a flat bearing surface. If the spring ends have a finished width, the required length of the finished edge must also be indicated. The usual tolerances for finished widths are as in Table 3.

5.11 Assembled Spring Width—Where more than one leaf constitutes a spring assembly, the overall width tolerance of the assembly within the spring seat length shall be as follows as in Table 4.

5.12 Stack Thickness—Aggregate of the nominal thicknesses of all leaves of the spring including any spacer plates which are part of the spring at the seat.

5.13 Leaf Ends—The leaf ends used most generally are:
   a. Square as sheared
   b. Trimmed to a shape
   c. Taper rolled
   d. Taper rolled; trimmed or forged to a shape or both

5.14 Surface Finish—Condition of the surface of the spring leaves after the steel has been heat treated and prior to coating.

5.14.1 "As Heat Treated" Finish—The surface of the spring leaves is in the condition as taken from the heat treating furnace where generally the leaves have a finish of oxide coating.

5.14.2 "Shot Peened" Finish—The tension surface of the spring leaves has been exposed to the shot peening operation where the oxide coating and scale are removed and a matte luster finish is produced.

5.14.3 "Ground or Polished Leaf Ends"—The bearing areas of leaves are ground or polished to produce a smooth surface for reduced friction. The distance or length to be ground or polished should be specified.

5.15 Protective Coating—Material added to surface of spring leaves or exposed areas of assembled springs. For additional information, see HS-J788.

5.16 Leaf Numbers (see Figure 1)—Leaves are designated by numbers, starting with the main leaf which is No. 1, the adjoining leaf is No. 2, and so on. If rebound leaves are used, the rebound leaf adjoining the main leaf is rebound leaf No. 1, the next one rebound leaf No. 2, and so on. (Rebound leaves are assembled adjacent to the side opposite the load bearing leaves.) Helper springs are considered as separate units.

5.17 Opening and Overall Height (see Figure 1)—Distance from the datum line to the point where the center bolt centerline intersects the surface of the spring that is in contact with the spring seat.

If the surface in contact with the seat is on the main leaf or a rebound leaf (as on underslung springs), this distance is called "opening."

If the surface in contact with the seat is on the shortest leaf (as on overslung springs), this distance is called "overall height."

TABLE 1—CROSS-SECTION TOLERANCES, mm

Width	Width Tolerance Minus 0.00	Tolerance in Thickness (±)[1] and in Flatness (-)[2] For Thickness 5.00-9.50	Tolerance in Thickness (±)[1] and in Flatness (-)[2] For Thickness 10.00-21.20	Tolerance in Thickness (±)[2] and in Flatness (-)[2] For Thickness 22.40-37.50	Maximum Difference in Thickness[3] For Thickness 5.00-9.50	Maximum Difference in Thickness[3] For Thickness 10.00-21.20	Maximum Difference in Thickness[3] For Thickness 22.40-37.50
40.0	+0.75	0.13	0.15	-	0.05	0.05	-
45.0	+0.75	0.13	0.15	-	0.05	0.05	-
50.0	+0.75	0.13	0.15	-	0.05	0.05	-
56.0	+0.75	0.13	0.15	-	0.05	0.05	-
63.0	+0.75	0.13	0.15	-	0.05	0.05	-
75.0	+1.15	0.15	0.20	0.30	0.08	0.10	0.15
90.0	+1.15	0.15	0.20	0.30	0.08	0.10	0.15
100.0	+1.15	0.15	0.20	0.30	0.08	0.10	0.15
125.0	+1.65	0.18	0.25	0.40	0.10	0.13	0.20
150.0	+2.30	-	0.30	0.50	-	0.15	0.25

[1] Thickness measurements shall be taken at the edge of the bar where the flat surfaces intersect the rounded edge.
[2] This tolerance represents the maximum amount by which the thickness at the center of the bar may be less than the thickness at the edges. Thickness at the center may never exceed the thickness at the edges.
[3] Maximum difference in thickness between the two edges of each bar.

TABLE 2—SPECIFIED WIDTHS AND THICKNESSES OF ALLOY STEEL BARS, mm

Widths	Widths	Thicknesses	Thicknesses	Thicknesses	Thicknesses	Thicknesses	Thicknesses
40.0	75.0	5.00	7.10	10.00	14.00	20.00	28.00
45.0	90.0	5.30	7.50	10.60	15.00	21.20	30.00
50.0	100.0	5.60	8.00	11.20	16.00	22.40	31.50
56.0	125.0	6.00	8.50	11.80	17.00	23.60	33.50
63.0	150.0	6.30	9.00	12.50	18.00	25.00	35.50
		6.70	9.50	13.20	19.00	26.50	37.50

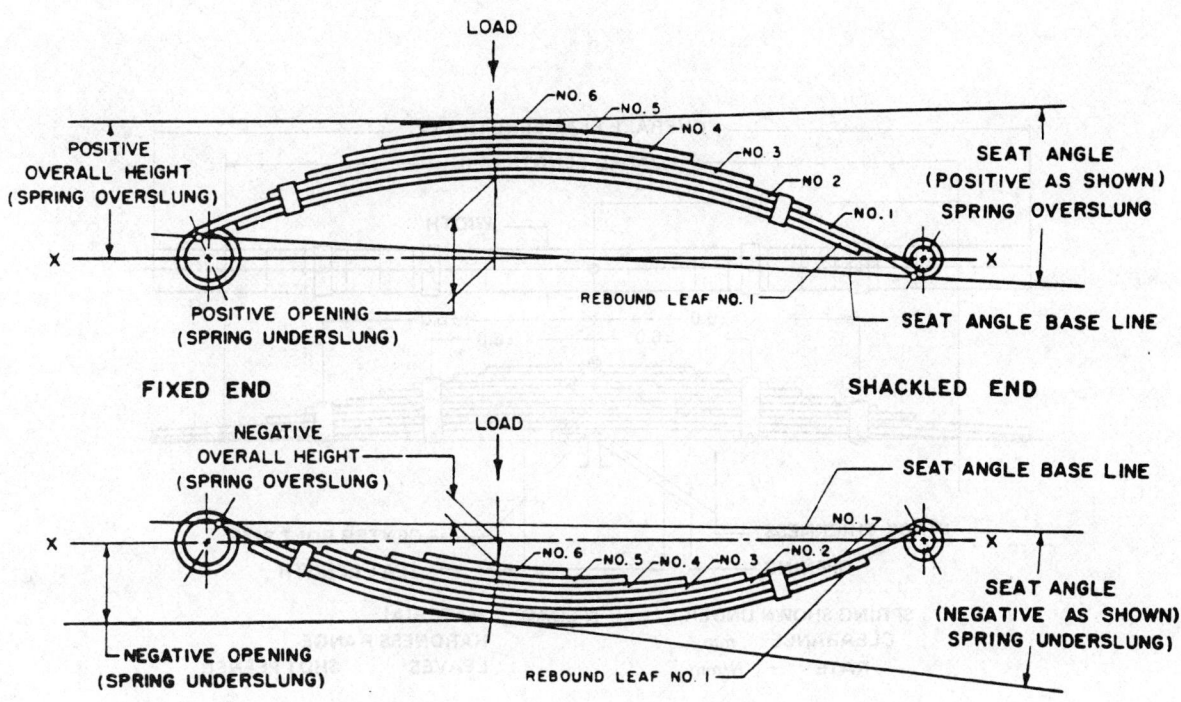

FIGURE 1—MEASUREMENT OF OPENING, OVERALL HEIGHT, AND SEAT ANGLE

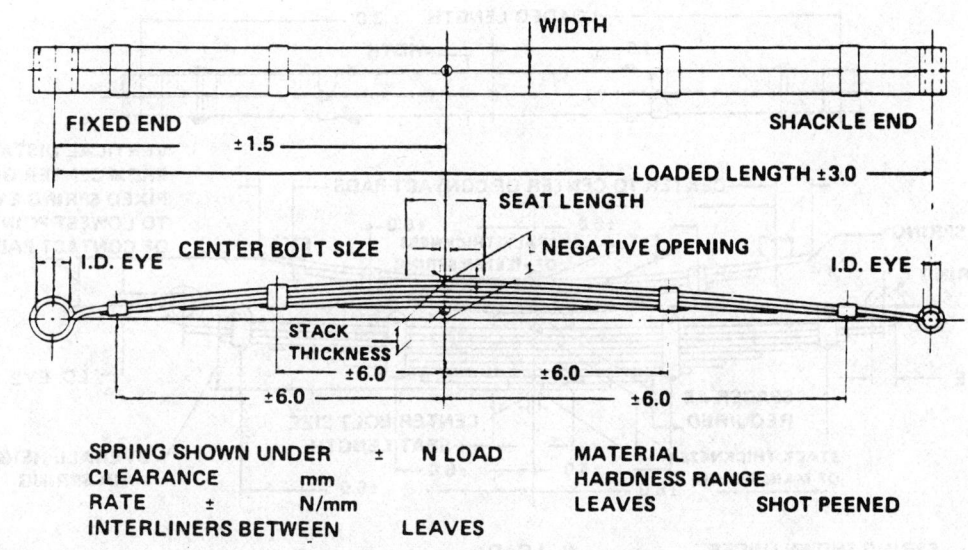

FIGURE 2—MINIMUM SPECIFICATION REQUIREMENTS FOR UNDERSLUNG SPRINGS WITH NEGATIVE OPENING

TABLE 3—TOLERANCES FOR FINISHED WIDTHS

Leaf Width Over	Leaf Width To and Including	Tolerance from Nominal Width +0.00
0	50	−0.25
50	63	−0.35
63	150	−0.50

TABLE 4—WIDTH TOLERANCE OF THE ASSEMBLY

Leaf Width Over	Leaf Width To and Including	Tolerance −0.000
0.0	63	+2.5
63	100	+3.0
100	125	+3.7
125	150	+4.4

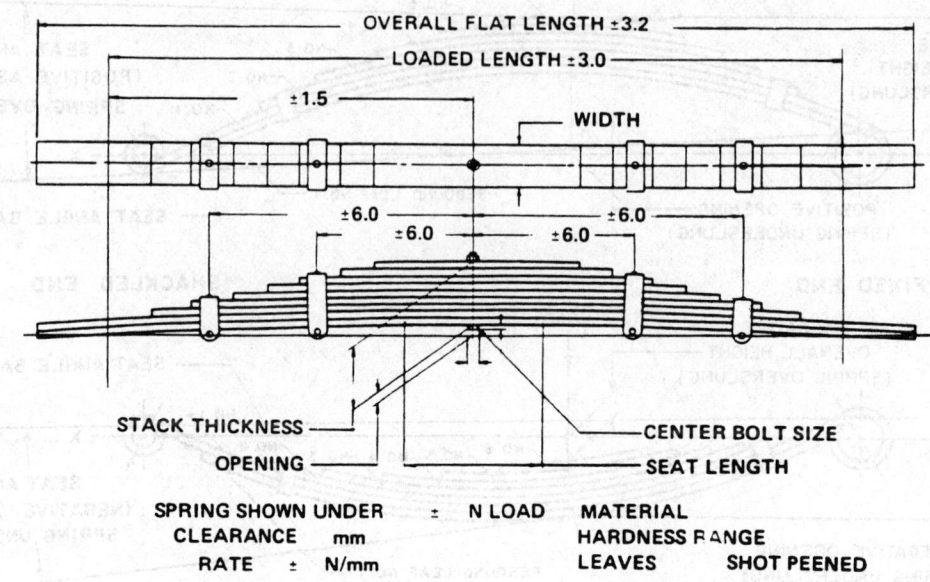

FIGURE 3—MINIMUM SPECIFICATION REQUIREMENTS FOR SPRINGS WITH PLAIN ENDS

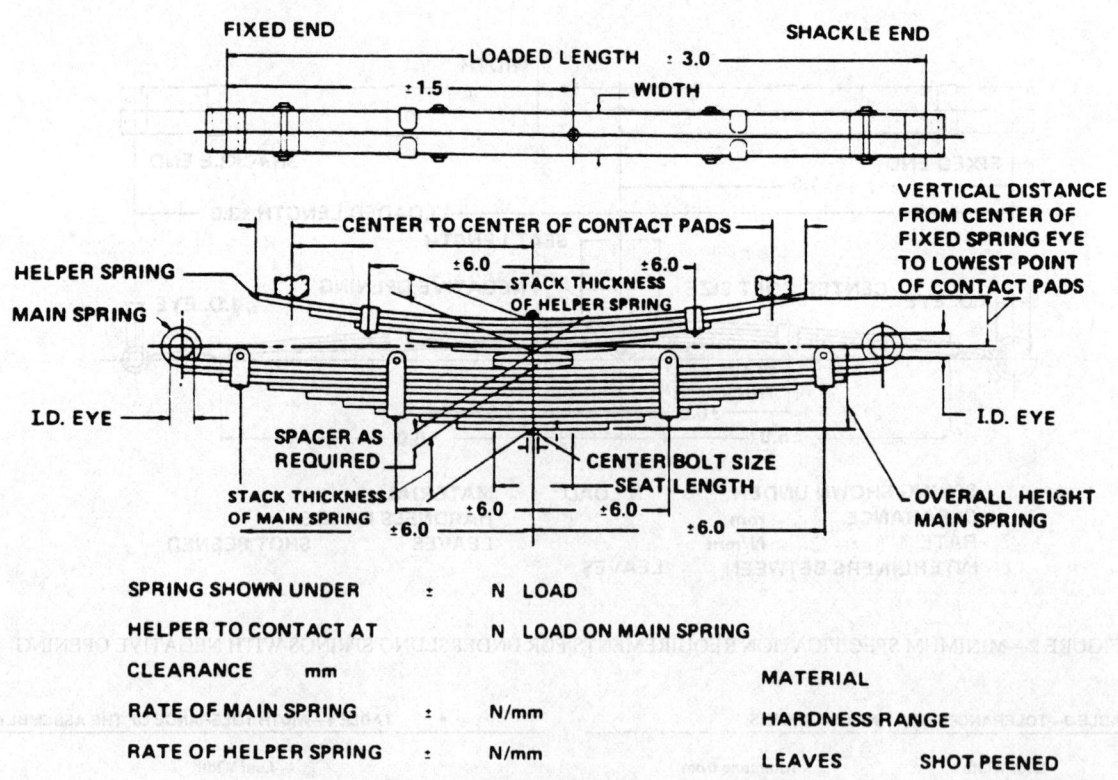

FIGURE 4—MINIMUM SPECIFICATION REQUIREMENTS FOR OVERSLUNG COMMERCIAL VEHICLE SPRINGS

"Opening" and "overall height" may be positive or negative (see Figure 1). They are specified dimensions not subject to a tolerance. See 5.19 on Load.

5.18 Clearance—Difference in opening, or overall height, between the design load position and the extreme position (of maximum stress) to which the spring can be deflected on the vehicle.

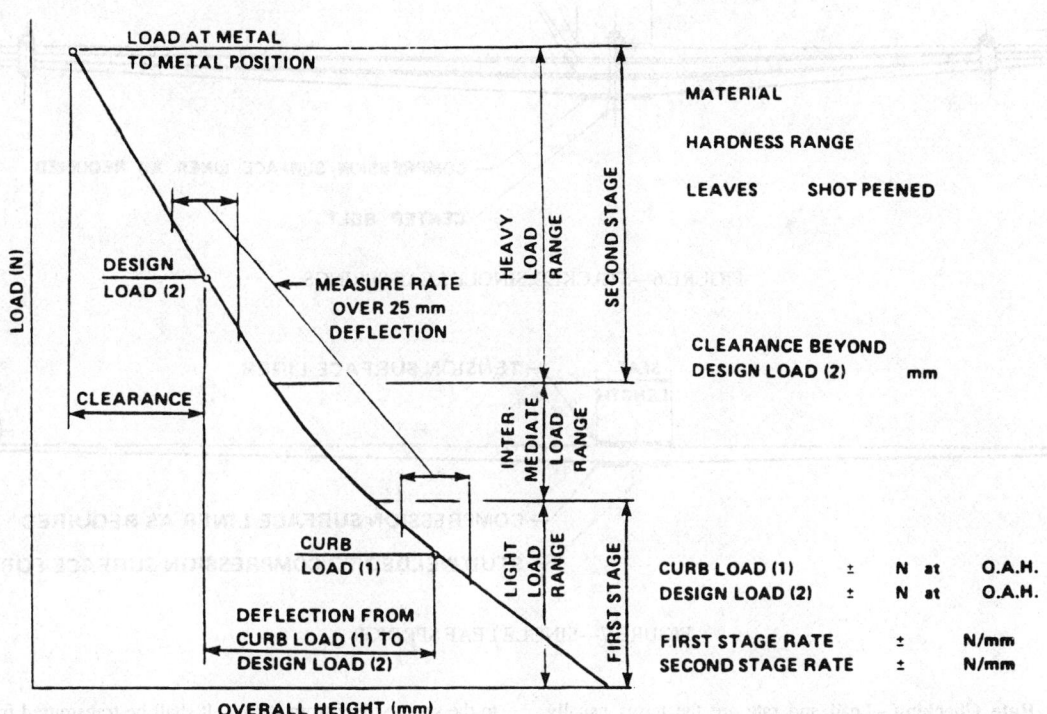

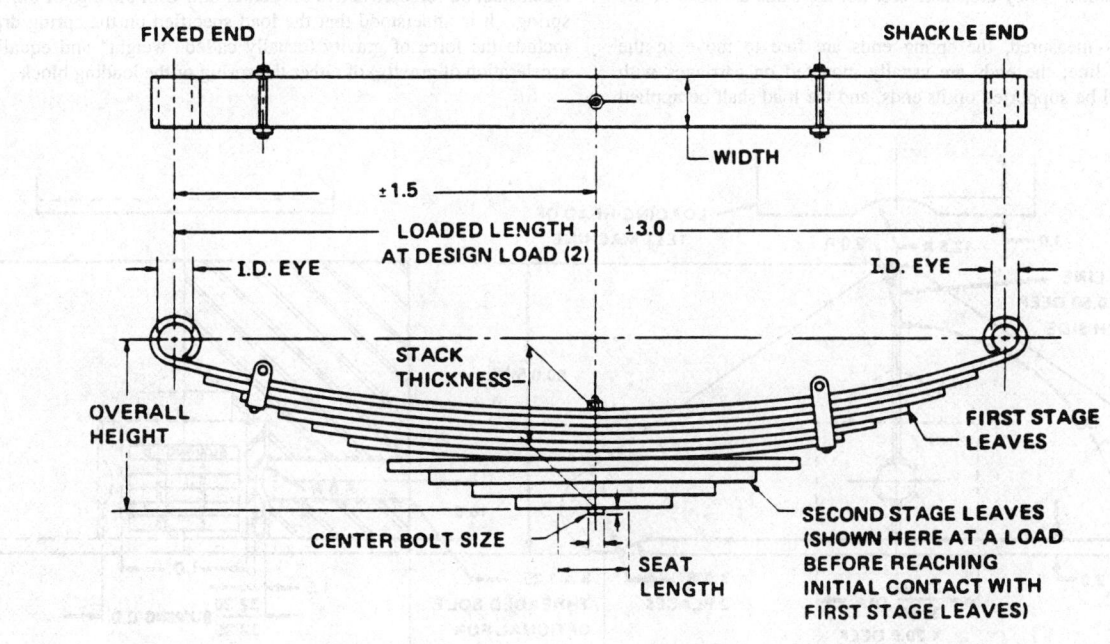

FIGURE 5—MINIMUM SPECIFICATION REQUIREMENTS FOR VARIABLE
RATE OR PROGRESSIVE RATE SPRINGS (OVERSLUNG TYPE SHOWN)

5.19 Load—The force exerted by the spring at the specified opening or overall height. The total tolerance on load at the specified overall height or opening is usually expressed as a load range (N) which is equivalent to a deflection (mm) at the nominal rate (N/mm). This deflection may be as small as 6.0 mm for a passenger car spring and as large as 13.0 mm for a heavy truck spring.

5.20 Rate—The change of load per unit of spring deflection (N/mm). For leaf springs, it is determined as one fiftieth (2%) of the difference between the loads measured 25 mm above and 25 mm below the specified position, unless otherwise specified (see Figure 5). The tolerance is usually held within ±5% on low rate springs and within ±7% on high rate springs.

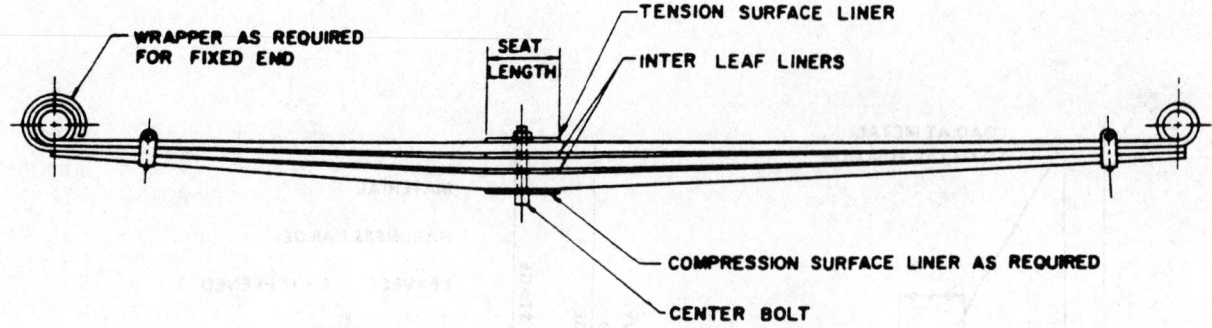

FIGURE 6—STACKED SINGLE LEAF SPRINGS

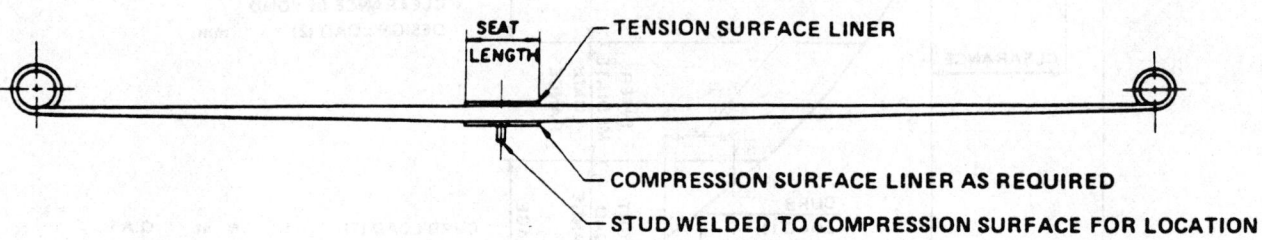

FIGURE 7—SINGLE LEAF SPRINGS

5.21 Load and Rate Checking—Load and rate are the terms usually employed to describe the basic characteristics of a leaf spring without center clamp and without shackles. They are, therefore, not the same as those of the installed spring.

When the load is measured, the spring ends are free to move in the direction of the datum line; the ends are usually mounted on carriages with rollers. The spring shall be supported on its ends, and the load shall be applied to the shortest leaf from above. It shall be transmitted from the testing machine head through a standard SAE loading block shown in Figure 8. The loading block shall be centered above the center bolt with the legs of the V resting on the spring. It is understood that the load specified on the spring drawing does not include the force of gravity (usually called "weight" and equaling mass times acceleration of gravity) of either the spring or the loading block.

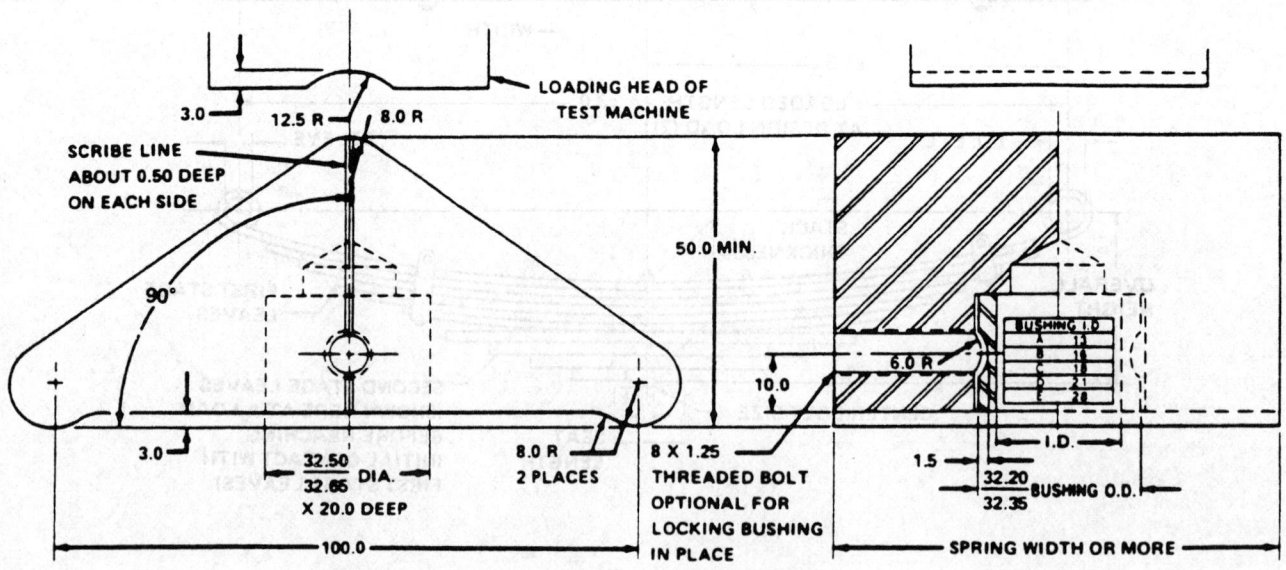

FIGURE 8—SPRING LOADING BLOCK

Just before the spring is checked for load or rate, it shall undergo a preloading operation. During the initial preloading by the spring maker, the spring shall be deflected at least to the position defined under 5.18 on Clearance. During any subsequent preloading, the spring shall be deflected only to and not beyond this "clearance position" in order to remove any temporary recovery from the set incurred during the initial preloading. After the spring has been preloaded, it shall be released to the free position before the load is applied for load and rate checking. For additional information on preloading, see HS-J788.

Load and rate shall be measured in terms of the forces exerted by the spring during compression of the spring (compression loads) and not during release of the spring (release loads). The compression load in any position shall be read only after the spring has been thoroughly rapped in that position with a plastic or soft metal hammer.

5.22 Specification Requirements—Minimum specification requirements are given in Figures 2 to 5.

5.23 Spring Eyes and Bushings—For some types of currently used spring eyes, spring ends, bushings, and shackle constructions, see HS-J788.

For eyes with specified inside diameter, the size and roundness of the eye should be checked by means of a round plug gage from which two opposite segments of 60 degrees have been removed. The gage shall have a taper on diameter per unit of length of 0.002:1 (see Figure 9). The gage shall be inserted into the eye three times from each side at angular positions differing by about 60 degrees. The eye is acceptable only if the gage reading on the side of the eye from which the gage is inserted is within the specified diametral limits at each of the six checks.

Also, the eye should be checked with a round plug, GO/NO GO gage, to determine if the eye is cone shaped or tapered. The GO diameter must pass completely through the eye and the NO GO diameter must not enter the eye from either side.

The total tolerance shall be 1% of the nominal diameter of the eye, except for large diameter eyes (40 mm or more), where bushing retention may require a smaller tolerance of 0.75% of the nominal eye diameter. For eye diameters of less than 25 mm, the minimum tolerance is 0.25 mm.

For a bushing where the ID may have been affected by pressing into the spring eye, it should be checked with a round plug gage. Total tolerance, 0.13 mm unless otherwise specified.

Eyes of the main leaf in the assembled spring, measured in the unloaded condition, shall be parallel to the surface at the spring seat, and square with a tangent to either edge of the main leaf at the spring seat, within ±1 degree.

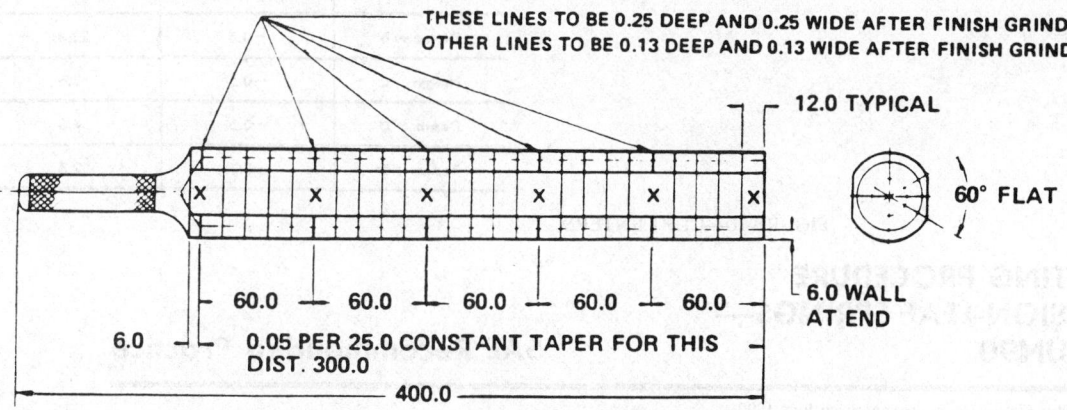

X-STAMP GAGE DIAMETERS AT THESE STATIONS

MATERIAL: STEEL - UNS G40270 (SAE 4027) OR EQUIVALENT

PROCESS: CARBURIZED AND HARDENED; CASE DEPTH 0.50 MIN.
SURFACE HARDNESS: HRC 58 MIN.

FIGURE 9—GAGE—LEAF SPRING EYE PLUG

5.24 Alignment Clips—Most surface vehicle leaf springs are fitted with clips of some form which serve primarily to prevent sidewise spread and vertical separation of the leaves.

Clips employed for passenger car springs show a great variety in design, but commercial vehicle springs are generally equipped with either bolt-type or clinch-type clips, see HS-J788. Dimensions must be chosen to suit the individual service requirement.

5.25 Center Bolt—The center bolt is required to hold the spring leaves together, and the center bolt head is used as a locating dowel during installation on the vehicle. For underslung springs, the head should be adjacent to the main leaf; for overslung springs, the head should be adjacent to the shortest leaf. The center bolt should not be depended upon to prevent the shifting of leaves due to driving and braking forces.

In most cases, center bolts are highly stressed in the handling of the springs and in service. Therefore, it is necessary to use bolts and nuts of high mechanical properties. See Table 5 for sizes.

5.26 Cup Center—Cup centers are often used in heavy-duty springs which may not safely depend on clamps and center bolts to prevent a shifting of the spring on the axle seat due to driving and braking forces.

When the main leaf is assembled adjacent to the axle seat as in underslung springs, the cup is hot forged in the main leaf only (away from the No. 2 leaf). When the shortest leaf is mounted above the axle seat as in overslung springs, all the leaves must be cupped toward the shortest leaf.

This method of cupping locks the main leaf to the axle seat. The horizontal forces which are applied to the main leaf will be resisted by the cup rather than the clamp and the center bolt.

There are several types of cup centers in general use, one of which is shown in Figure 10. The cup dimensions are listed according to center bolt diameter; however, the cup diameter should not exceed one-half the leaf width, and the cup depth should not exceed one-half the leaf thickness.

TABLE 5—RECOMMENDED CENTER BOLT AND NUT DIMENSIONS (mm)

Nominal Bolt Diameter	Threads Pitch	Threads Minimum Length	Bolt Head Size Diameter	Bolt Head Size Height	Nut Size Style 1 Width Across Flats (Max)	Nut Size Style 1 Width Across Corners (Max)	Nut Size Style 1 Thickness (Max)
8	1.25	25	12.0	6.0	13.0	15.01	6.6
10	1.5	25	15.0	7.0	15.0	17.32	9.0
12	1.75	30	17.0	8.0	18.0	20.78	10.7
16	2	35	20.0	10.0	24.0	27.71	14.5

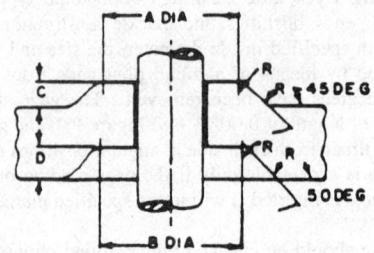

Dimension	Tolerance +0.0	For Use With Centerbolt Diameters	
		10, 12	16
Diameter A	−0.5	21.3	31.5
Diameter B	−0.5	22.4	33.0
Height C	−0.5	3.6	5.1
Depth D	−0.5	4.6	6.1
Radius R	−0.3	2.5	3.0

FIGURE 10—CUP CENTERS

FATIGUE TESTING PROCEDURE FOR SUSPENSION-LEAF SPRINGS— SAE J1528 JUN90

SAE Recommended Practice

Report of the Truck and Bus Chassis Committee approved June 1990.

1. Scope

1.1 Test Material—Only fully processed new springs which are representative of springs intended for the vehicle shall be used for the tests. No complete spring or separate leaf shall be used for more than one test.

2. Report Content

—To obtain uniform documentation, every report shall include detailed information on the following points, when applicable:

2.1 Geometry—Overall dimensions, location and dimensions of fracture sections including the location of the fracture initiation point shall be described in writing or by photographs.

2.2 Material and Manufacturing Process—The type of material, as well as essential steps in the manufacturing which may affect the test results, shall be specified. As examples, type of hardening, shot-peening under stress at given level, etc., can be mentioned.

The hardness shall be checked on critical surfaces, and the hardness distribution through the section shall be measured. Shot peen coverage and the microstructure shall be evaluated.

2.3 Fractography—Fracture surfaces shall be shown on photographs and the type of fracture discussed. The extension of fatigue fracture shall be measured and the crack starting points shall be examined.

2.4 Test Result Presentation—The individual test results may be clearly stated. Even noncritical events like cracks in secondary leaves must be recorded.

The results of the fatigue testing of a group of springs should be subjected to statistical analysis. The Weibull distribution is recommended. Also, it is recommended that minimum fatigue performance requirements be specified in terms of B10 life and population slope.

3. Vertical Loading Methods

3.1 General Directions

3.1.1 Equipment—The test machine shall be able to maintain the maximum and minimum specified force within ±2%. This can be accomplished by force control or control of deflection calibrated against a static force. In the latter case, dynamic and static spring rates must be considered.

3.1.2 Clamping—The spring shall be clamped at the center to simulate its installation in the vehicle. The clamping parts and assembly requirements must be specified by the vehicle manufacturer.

The clamping hardware shall be tightened to the torque values specified and verified throughout the test. The bolt torque should be measured and brought up to specification more frequently at the beginning of the test. Measurements are to be taken at 2000, 5000, and 10 000 cycles; then at 10 000 cycle intervals up to 50 000 cycles and every 50 000 cycles until completion of the tests are recommended.

3.1.3 Test Mountings—Unless otherwise specified, springs with eyes shall be free to move in the direction of the datum line; springs with slipper ends shall be tested on fixed mountings. The vehicle manufacturer shall specify the mountings on springs with other end configurations.

3.1.4 Rate of Testing—Springs shall be cycled at a rate of between 0.5 and 2 Hz. The cyclic rate shall be chosen so that surface temperature on the spring does not exceed 90°C. Fans may be used to provide cooling air.

3.2 Test Procedure

3.2.1 Static Load Test—The spring shall be loaded from zero up to the prescribed maximum deflection and back to zero. It is acceptable to apply the load in steps.

The force shall be measured at the center clamp. The vertical deflection of the spring center shall also be measured. The relation between the force and the deflection during a full cycle may be plotted in a diagram.

3.2.2 Fatigue Test—The spring shall be loaded from 1/2 g (g = design load) to maximum load experienced under actual vehicle conditions, typically 2g.

For validation, six springs for each design, shall be tested to failure or to the number of cycles specified.

Position limit switches shall be placed so that the test is terminated when the spring deflection has increased a prescribed distance (see 4.1).

In the case of a deflection controlled test, the spring rates shall be measured at uniform intervals during the test on at least one specimen in a test batch.

If the spring rate increased more than 5% during the test, the deflection should be corrected to keep the test peak forces constant. Regarding a corresponding decrease, see 4.1. If it is necessary to correct the deflection during the test as shown by the measurement of the spring rate, this shall be recorded and done in the same way, including the number of cycles, for all springs in the batch.

Measure the load at rated load position at 50 000 cycle intervals to determine load loss due to permanent set.

4. Fatigue Failure Criteria

4.1 Inability of Spring to Sustain Load—Normally, this is said to happen when deflection has increased 5 to 10% above the maximum total deflection at the test start or load loss at 50 000 cycles exceeds 5% of the load at test start.

4.2 Visible Crack in #1 Leaf or Visible Cracks in More Than Two Supporting Leaves.

LEAF SPRINGS FOR MOTOR VEHICLE SUSPENSION— MADE TO CUSTOMARY U.S. UNITS—SAE J510 NOV92

SAE Standard

Report of the Springs Division, approved August 1951, completely revised, Spring Committee, Leaf Spring Subcommittee, May 1985. Reaffirmed by the SAE Spring Committee November 1992.

Foreword—This reaffirmed document has been changed only to reflect the new SAE Technical Standards Board format.

1. Scope

NOTE—For leaf springs made to metric units, see SAE J1123.

This SAE Standard is limited to concise specifications promoting an adequate understanding between spring maker and spring user on all practical requirements in the finished spring. The basic concepts for the spring design and for many of the details have been fully addressed in HS-J788, SAE Information Report, Manual on Design and Application of Leaf Springs, which is available from SAE Headquarters.

2. References

2.1 Applicable Documents—The following publications form a part of this specification to the extent specified herein. The latest issue of SAE publications shall apply.

2.1.1 SAE PUBLICATIONS—Available from SAE, 400 Commonwealth Drive, Warrendale, PA 15096-0001.

SAE J419—Methods of Measuring Decarburization
SAE J1123—Leaf Springs for Motor Vehicle Suspension—Made to Metric Units

3. Bar Sizes and Tolerances

Round edge flat spring steel has been adopted as the SAE standard.

The bars shall be of flat rolled steel having two flat surfaces and two rounded (convex) edges. They are subject to the tolerances shown in Table 1. These cross-section tolerances permit the two flat surfaces to be slightly concave. When that occurs, the radii of the arcs of the two concave surfaces shall be of approximately equal length.

The rounding of the convex edges shall be an arc with a radius of curvature that may vary from 65 to 85% of the thickness of the bar.

Bars shall be substantially straight and free from physical characteristics known as kinks or twists which render them unsatisfactory for spring manufacturing purposes.

Distortions due to a bar being bent about either major axis of section shall be measured with the bar against a flat checking surface so as to make contact with this surface near both ends. Gaps between the bar and the checking surface shall not exceed 0.05 in/ft of bar length out of contact with the checking surface when this bar length is greater than 3 ft. Also, a gap between the bar and a straight edge, 3 ft long, applied along any portion of the surface or edge of the bar shall not exceed 0.15 in.

It is recommended that all leaf spring bars which have been cold straightened be identified by the steel mill so that the spring manufacturer can use them selectively.

Leaf spring bars are generally available in the following widths in inches: 1.75, 2.00, 2.25, 2.50, 3.00, 3.50, 4.00, 5.00, and 6.00.

Spring drawings shall specify steel of the following nominal thicknesses, in inches, to which all bars shall be rolled: 0.194, 0.204, 0.214, 0.225, 0.237, 0.249, 0.262, 0.276, 0.291, 0.307, 0.323, 0.341, 0.360, 0.380, 0.401, 0.423, 0.447, 0.473, 0.499, 0.527, 0.558, 0.590, 0.625, 0.662, 0.702, 0.744, 0.788, 0.836, 0.887, 0.941, 0.999, 1.061, 1.127, 1.197, 1.273, 1.354, and 1.440.

4. Surface Decarburization

Surface decarburization may reduce the fatigue durability of the springs; therefore, it is important that surface decarburization be at a minimum.

Hot-rolled steel bars as received from the mills have some decarb, at least of the minimum Type 3 (see SAE J419), where more than 50% of the base carbon content remains at the surface (i.e., some partial but not more than 50% loss of carbon).

If decarb is of Type 2, where 50% or less of the base carbon content remains at the surface (i.e., appreciable partial but not total loss of carbon), the decarb normally does not exceed a depth of 0.010 in for steels of thicknesses 0.194 to 0.499 in, nor a depth of 0.020 in for steels of thicknesses over 0.499 to 1.440 in.

With sections over 1.000 in thickness, some of the hot-rolled steel bars may have decarb of Type 1, in which virtually carbon-free ferrite (i.e., total loss of carbon) exists for a measurable distance below the surface.

The depth of decarb varies from mill to mill, from rolling to rolling, and from bar to bar. The extent to which the depth and type of the decarb can be acceptable will be subject to agreement between the steel producer and the spring manufacturer.

The edges of the bars are somewhat higher in decarb than the flat surfaces; decarb on both the edges and the flat surfaces usually has greater depth with increased bar thickness.

After forging and nonatmospheric controlled heat treating, the spring leaves will have greater decarb. Scaling of the steel in this processing reduces the thickness of the leaf. While some of the surface decarb is removed with the scale, the final depth of decarb is usually greater than it was in the steel bars as received from the mills.

5. Definitions, Dimensions, and Tolerances

5.1 Leaf Spring—A spring of full elliptic, semi-elliptic, or quarter-elliptic shape with one or more leaves. The term multi-leaf has generally applied to springs of constant width and with stepped leaves, each of constant thickness except where leaf ends may be tapered in thickness. More recently, the term has been extended to include an assembly of stacked single leaves, each of which is characterized by tapering either in width or in thickness or by a combination of both. Examples of multi-leaf springs are shown in Figures 1 to 6; Figure 7 shows a single leaf spring.

TABLE 1—CROSS-SECTION TOLERANCE, in

Nominal Width Over	Nominal Width To and Including	Tolerance in Width -0.00	For Thickness	Tolerance in Thickness[1] (±)	Tolerance in Flatness[2] (-)	Max Difference[3] in Thickness
0.00	2.50	+0.030	0.375 or under	0.005	0.005	0.002
			Over 0.375 to 0.875, incl	0.006	0.006	0.002
2.50	4.00	+0.045	0.375 or under	0.006	0.006	0.003
			Over 0.375 to 0.875, incl	0.008	0.008	0.004
			Over 0.875 to 1.500, incl	0.012	0.012	0.006
4.00	5.00	+0.065	0.375 or under	0.007	0.007	0.004
			Over 0.375 to 0.875, incl	0.010	0.010	0.005
			Over 0.875 to 1.500, incl	0.016	0.016	0.008
5.00	6.00	+0.090	Over 0.375 to 0.875, incl	0.012	0.012	0.006
			Over 0.875 to 1.500, incl	0.020	0.020	0.010

[1] Thickness measurements shall be taken at the edge of the bar where the flat surfaces intersect the rounded edge.
[2] This tolerance represents the maximum amount by which the thickness at the center of the bar may be less than the thickness of the edges. Thickness at the center may never exceed the thickness at the edges.
[3] Maximum difference in thickness between the two edges of each bar.

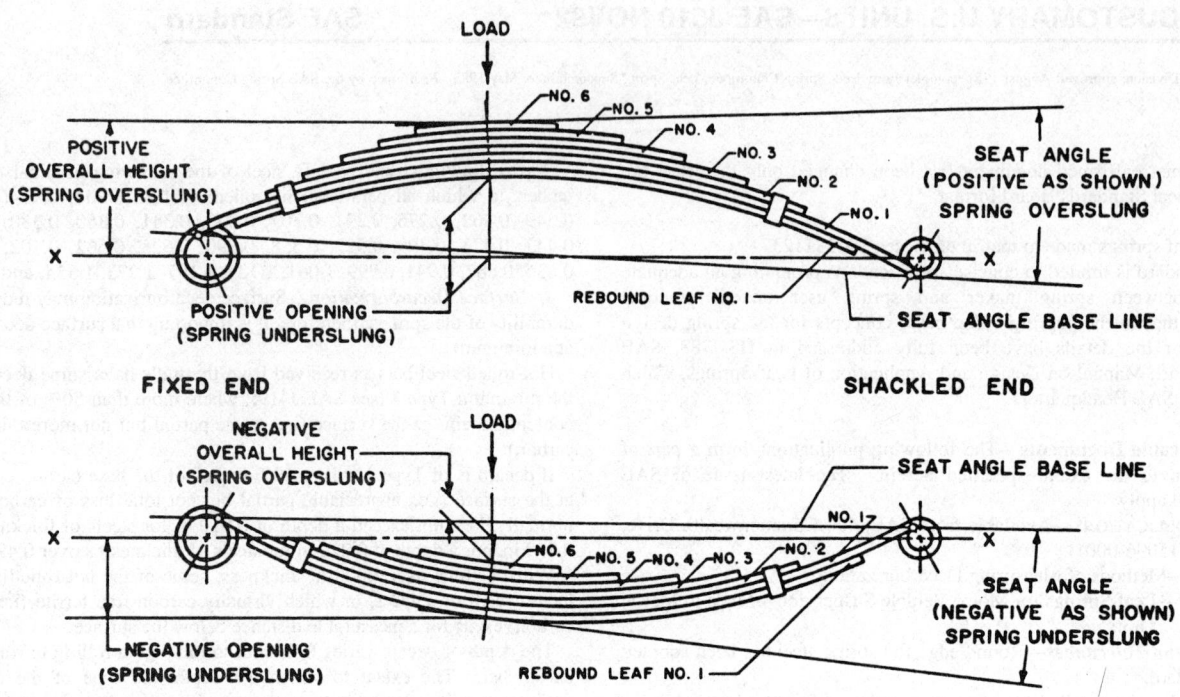

FIGURE 1—MEASUREMENT OF OPENING, OVERALL HEIGHT, AND SEAT ANGLE

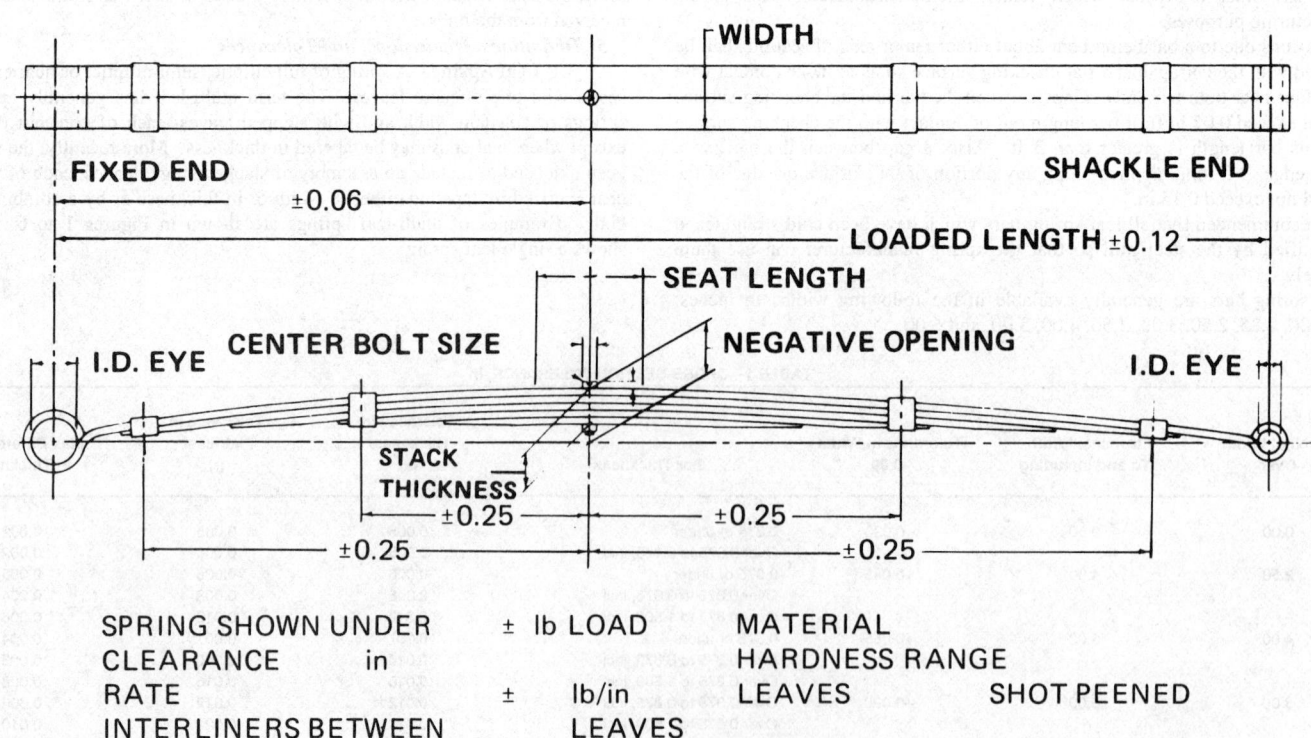

FIGURE 2—MINIMUM SPECIFICATION REQUIREMENTS FOR UNDERSLUNG SPRINGS WITH NEGATIVE OPENING

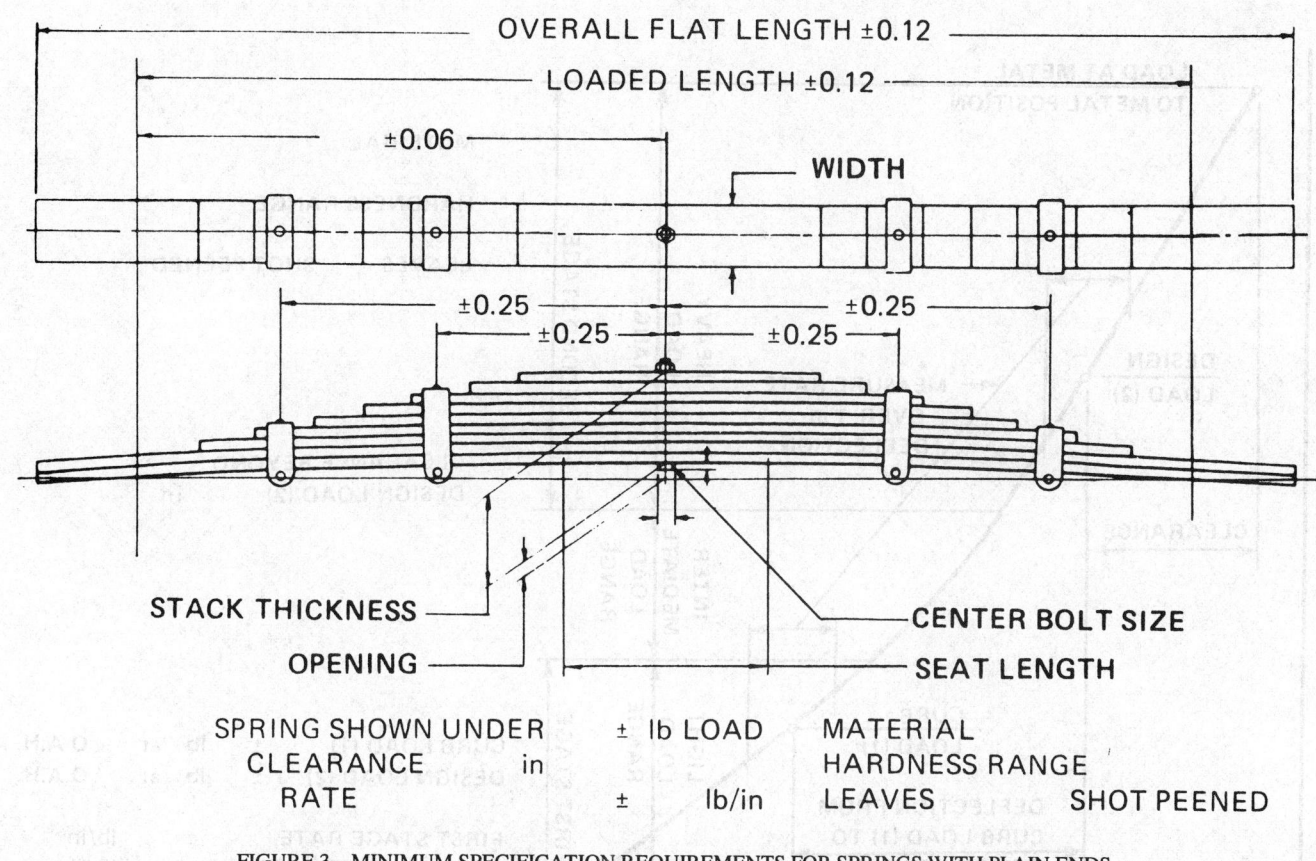

FIGURE 3—MINIMUM SPECIFICATION REQUIREMENTS FOR SPRINGS WITH PLAIN ENDS

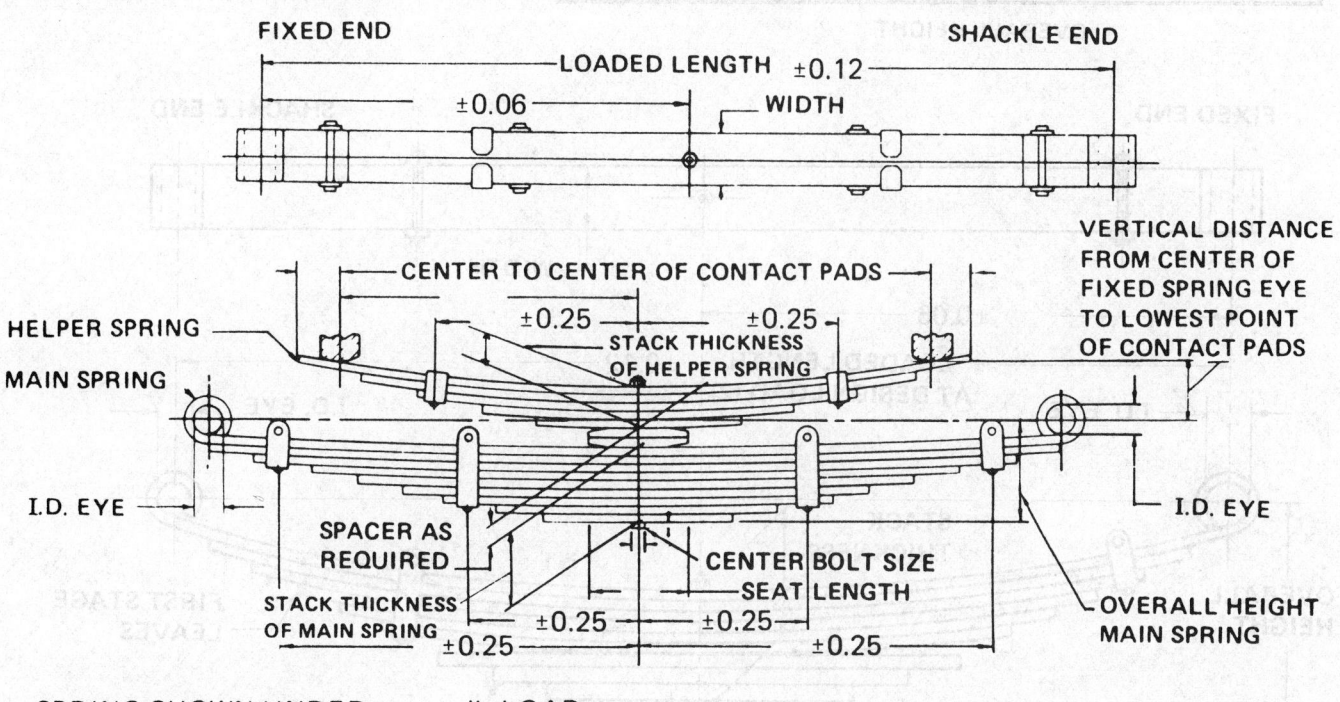

FIGURE 4—MINIMUM SPECIFICATION REQUIREMENTS FOR OVERSLUNG COMMERCIAL VEHICLE SPRINGS

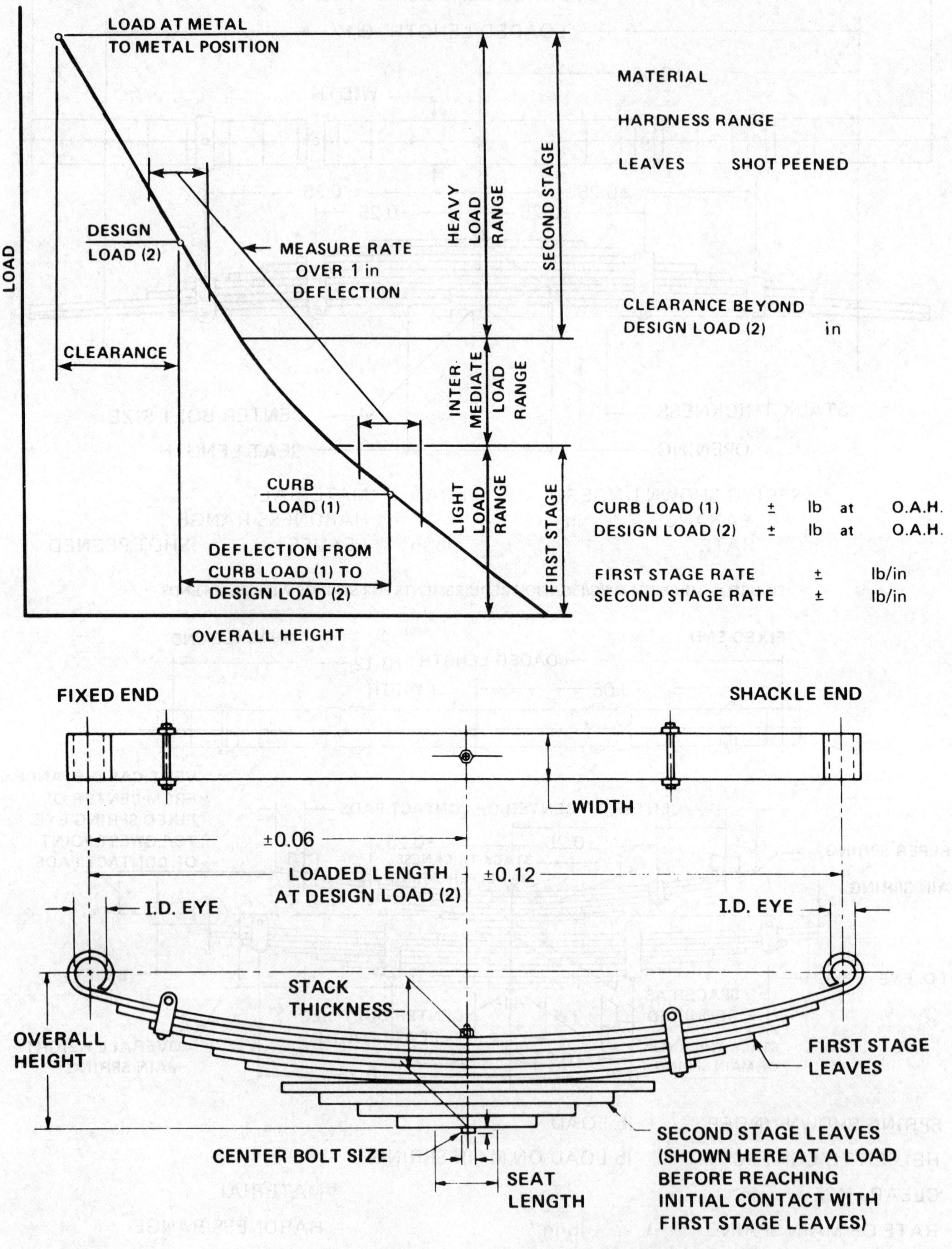

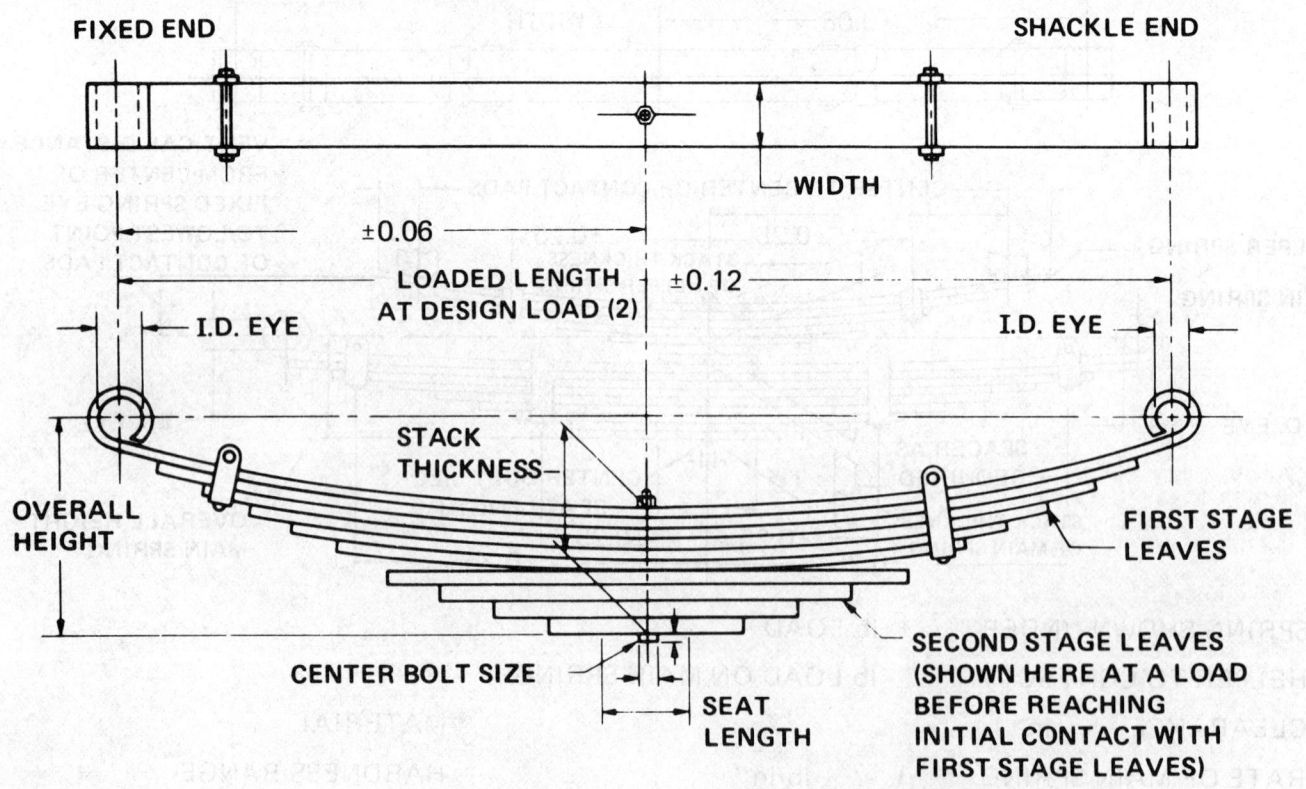

FIGURE 5—MINIMUM SPECIFICATION REQUIREMENTS FOR
VARIABLE RATE OR PROGRESSIVE RATE SPRINGS (OVERSLUNG TYPE SHOWN)

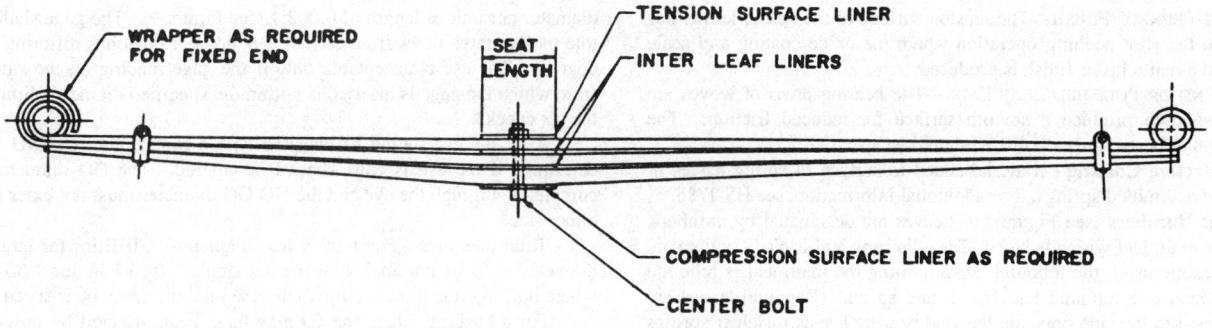

FIGURE 6—STACKED SINGLE LEAF SPRINGS

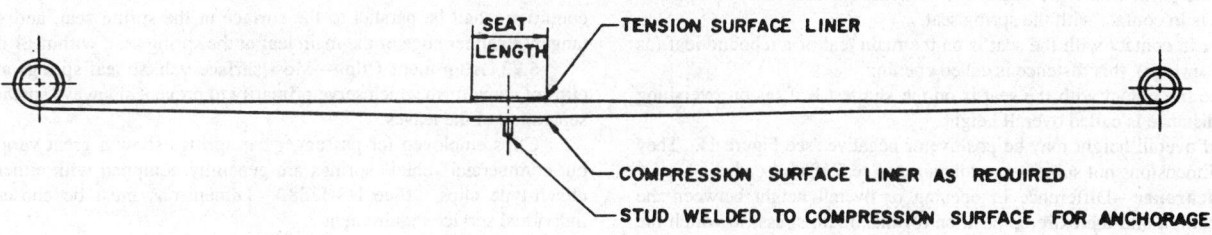

FIGURE 7—SINGLE LEAF SPRINGS

The leaves of a multi-leaf spring are usually held together with a center bolt and prevented from lateral shifting by alignment clips. Prior to assembly, the leaves are formed (cambered) and heat treated by heating, quenching, and tempering to the required hardness. Quench dies or fixtures are used to maintain the required camber within tolerances.

5.2 Datum Line—Reference line used with many of the subsequently defined terms. In Figure 1 (where the springs are shown inverted as in a machine for load and rate checking), it is shown as the line X-X. On springs with eyes, the datum line passes through the centers of the eyes. On other springs it passes through the points where the load is applied near the ends of the spring. These points must be indicated on the drawing. When load and rate are checked, the spring ends shall be free to move in the direction of the datum line.

5.3 Seat Angle Base Line (see Figure 1)—Reference line drawn through the terminal points of the active spring length at each eye, taken along the tension surface of the main leaf. On springs without eyes the seat angle base line is coincident with the datum line.

5.4 Loaded Length—Distance between spring eye centers when the spring is deflected to the specified load position. On springs without eyes, it is the distance between the lines where load is applied under the specified conditions. Tolerance, ±0.12 in.

5.5 Loaded Fixed End Length—Distance from the center of the fixed end eye to the projection on the datum line of the point where the centerline of the center bolt intersects the spring surface in contact with the spring seat. Tolerance, ±0.06 in.

5.6 Straight Length—Distance between eye centers when the tension surface of the main leaf at the center bolt centerline is in the plane of the seat angle base line. The distance is measured parallel to the seat angle base line. Tolerance, ±0.12 in.

5.7 Seat Length—Length of spring that is in actual engagement with the spring seat when installed on a vehicle at design height. It is always greater than the inactive length.

5.8 Inactive Length—Length of spring rendered inactive by the action of the U-bolts or clamping bolts.

5.9 Seat Angle (see Figure 1)—Angle between the tangent to the center of the spring seat and the seat angle base line. When the spring is viewed with the fixed end of the spring to the left, as shown, and the load is applied to the shortest leaf from above, the seat angle may be specified as either positive (counterclockwise) or negative (clockwise), depending upon the angular direction in which the tangent to the center of the spring seat is disposed from the seat angle base line.

Consequently, with the spring in normal vehicle position so that the load is applied from below as shown in Figures 2, 4, 5, 6, 7, and again with the fixed end of the spring to the left of the drawing, the seat angle is defined as positive when that tangent is disposed clockwise; and as negative when the tangent is disposed counterclockwise.

5.10 Finished Width—Width to which the spring leaves are ground or milled to give the edges a flat bearing surface. If the spring ends have a finished width, the required length of the finished edge must also be indicated. The usual tolerances for finished widths are as indicated in Table 2.

TABLE 2—TOLERANCES FOR FINISHED WIDTHS

Leaf Width Over	Leaf Width To and Including	Tolerance from Nominal Width +0.000
0.00	2.00	-0.010
2.00	2.50	-0.015
2.50		-0.020

5.11 Assembled Spring Width—Where more than one leaf constitutes a spring assembly, the overall width tolerance of the assembly within the spring seat length shall be as follows in Table 3:

TABLE 3—WIDTH TOLERANCE OF THE ASSEMBLY

Leaf Width Over	Leaf Width To and Including	Tolerance -0.000
0.00	2.50	+0.100
2.50	4.00	+0.120
4.00	5.00	+0.145
5.00	6.00	+0.175

5.12 Stack Thickness—Aggregate of the nominal thicknesses of all leaves of the spring including any liners and spacer plates which are part of the spring at the seat.

5.13 Leaf Ends—The leaf ends used most generally are: square as sheared; trimmed to a shape; taper rolled; and taper rolled, trimmed or forged to a shape, or both.

5.14 Surface Finish—Condition of the surface of the spring leaves after the steel has been heat treated and prior to coating.

5.14.1 "As Heat-Treated" Finish—The surface of the spring leaves is in the condition as taken from the heat treating furnace where generally the leaves have a finish of oxide coating.

5.14.2 "SHOT-PEENED" FINISH—The tension surface of the spring leaves has been exposed to the shot peening operation where the oxide coating and scale are removed and a matte luster finish is produced.

5.14.3 GROUND OR POLISHED LEAF ENDS—The bearing areas of leaves are ground or polished to produce a smooth surface for reduced friction. The distance or length to be ground or polished should be specified.

5.15 Protective Coating—Material added to surface of spring leaves or exposed areas of assembled springs. For additional information, see HS-J788.

5.16 Leaf Numbers (see Figure 1)—Leaves are designated by numbers, starting with the main leaf which is No. 1. The adjoining leaf is No. 2 and so on. If rebound leaves are used, the rebound leaf adjoining the main leaf is rebound leaf No. 1, the next one rebound leaf No. 2, and so on. (Rebound leaves are assembled adjacent to the side opposite the load bearing leaves.) Helper springs are considered as separate units.

5.16.1 OPENING AND OVERALL HEIGHT (see Figure 1)—Distance from the datum line to the point where the center bolt centerline intersects the surface of the spring that is in contact with the spring seat.

If the surface in contact with the seat is on the main leaf or a rebound leaf (as on underslung springs), this distance is called opening.

If the surface in contact with the seat is on the shortest leaf (as on overslung springs), this distance is called overall height.

Opening and overall height may be positive or negative (see Figure 1). They are specified dimensions not subject to a tolerance. See 5.18 on Load.

5.17 Clearance—Difference in opening or overall height between the design load position and the extreme position (of maximum stress) to which the spring can be deflected on the vehicle.

5.18 Load—The force exerted by the spring at the specified opening or overall height. The total tolerance on load at the specified overall height or opening is usually expressed as a load range (lb) which is equivalent to a deflection (in) at the nominal rate (lb/in). This deflection may be as small as 0.25 in for a passenger car spring and as large as 0.50 in for a heavy truck spring.

5.19 Rate—The change of load per unit of spring deflection (lb/in). For leaf springs it is determined as half of the difference between the loads measured 1 in above and 1 in below the specified load position, unless otherwise specified (see Figure 5). The tolerance is usually held within ±5% on low rate springs and within ±7% on high rate springs.

5.20 Load and Rate Checking—Load and rate are the terms usually employed to describe the basic characteristics of a leaf spring and, as specified on the spring drawing, refer to quantities measured on the spring without center clamp and without shackles. They are, therefore, not the same as those of the installed spring.

When the load is measured, the spring ends are free to move in the direction of the datum line; the ends are usually mounted on carriages with rollers.

The spring shall be supported on its ends, and the load shall be applied to the shortest leaf from above. It shall be transmitted from the testing machine head through a standard SAE loading block, shown in Figure 8. The loading block shall be centered over the center bolt with the legs of the V resting on the spring. It is understood that the load specified on the spring drawing does not include either the weight of the spring or the weight of the loading block.

Just before the spring is checked for load or rate, it shall undergo a preloading operation. During the initial preloading by the spring maker, the spring shall be deflected at least to the position defined under 5.17 on Clearance. During any subsequent preloading, the spring shall be deflected only to and not beyond this clearance position in order to remove any temporary recovery from the set incurred during the initial preloading. After the spring has been preloaded, it shall be released to the free position before the load is applied for load and rate checking. For additional information on preloading, see HS-J788.

Load and rate shall be measured in terms of the forces exerted by the spring during compression of the spring (compression loads) and not during release of the spring (release loads). The compression load in any position shall be read only after the spring has been thoroughly rapped in that position with a plastic or soft metal hammer.

5.21 Specification Requirements—Minimum specification requirements are given in Figures 2 to 5.

5.22 Spring Eyes and Bushings—For some types of currently used spring eyes, spring ends, bushings, and shackle constructions, see HS-J788.

For eyes with specified inside diameter, the size and roundness of the eye should be checked by means of a round plug gage from which two opposite segments of 60 degrees have been removed. The gage shall have a taper on diameter per unit of length of 0.002:1 (see Figure 9). The gage shall be inserted into the eye three times from each side at angular positions differing by about 60 degrees. The eye is acceptable only if the gage reading on the side of the eye from which the gage is inserted is within the specified diametral limits at each of the six checks.

Also, the eye should be checked with a round plug, GO/NO GO gage, to determine if the eye is cone shaped or tapered. The GO diameter must pass completely through the eye and the NO GO diameter must not enter the eye from either side.

Total tolerance—For 1 in or less diameter—0.010 in; for larger than 1 in diameter—1% of nominal diameter (example: 0.015 in for 1.50 in ID eye); where bushing retention is critical, the 1% tolerance may be reduced to 0.75%.

For a bushing where the ID may have been affected by pressing into the spring eye, it should be checked with a round plug gage. Total tolerance—0.005 in unless otherwise specified.

Eyes of the main leaf in the assembled spring, measured in the unloaded condition, shall be parallel to the surface at the spring seat, and square with a tangent to either edge of the main leaf at the spring seat, within ±1 degree.

5.23 Alignment Clips—Most surface vehicle leaf springs are fitted with clips of some form which serve primarily to prevent sideways spread and vertical separation of the leaves.

Clips employed for passenger car springs show a great variety in design, but commercial vehicle springs are generally equipped with either bolt-type or clinch-type clips. (See HS-J788.) Dimensions must be chosen to suit the individual service requirement.

5.24 Center Bolt—The center bolt is required to hold the spring leaves together, and the center bolt head is used as a locating dowel during installation on the vehicle. For underslung springs, the head should be adjacent to the main leaf; for overslung springs, the head should be adjacent to the shortest leaf. The center bolt should not be depended upon to prevent the shifting of leaves due to driving and braking forces.

In most cases, center bolts are highly stressed in the handling of the springs and in service. Therefore, it is necessary to use bolts and nuts of high mechanical properties. (See Table 4 for sizes.)

TABLE 4—RECOMMENDED CENTER BOLT AND NUT DIMENSIONS, IN

Bolt Diameter	Threads[1] Per Inch	Threads[1] Minimum Length	Head Size Nominal Diameter	Head Size Nominal Height	Nut Size Nominal Width Across Flats	Nut Size Nominal Thickness
5/16	24	1.00	1/2	1/4	1/2	17/64
3/8	24	1.00	9/16	5/16	9/16	21/64
7/16	20	1.25	5/8	3/8	11/16	3/8
1/2	20	1.25	3/4	7/16	3/4	7/16
5/8	18	1.50	7/8	9/16	15/16	35/64
3/4	16	1.75	1	5/8	1-1/8	41/64
7/8	14	2.00	1-1/8	11/16	1-5/16	3/4
1	12	2.25	1-5/16	25/32	1-1/2	55/64

[1] Threads are Unified Standard Fine, Classes 2A and 2B.

5.25 Cup Center—Cup centers are often used in heavy-duty springs which may not safely depend on clamps and center bolts to prevent a shifting of the spring on the axle seat due to driving and braking forces.

When the main leaf is assembled adjacent to the axle seat as in underslung springs, the cup is hot forged in the main leaf only (away from the No. 2 leaf); when the shortest leaf is mounted above the axle seat as in overslung springs, all the leaves must be cupped toward the shortest leaf.

This method of cupping locks the main leaf to the axle seat. The horizontal forces which are applied to the main leaf will be resisted by the cup rather than the clamp and the center bolt.

There are several types of cup centers in general use, one of which is shown in Figure 10. The cup dimensions are listed according to center bolt diameter; however, the cup diameter should not exceed one-half the leaf width, and the cup depth should not exceed one-half the leaf thickness.

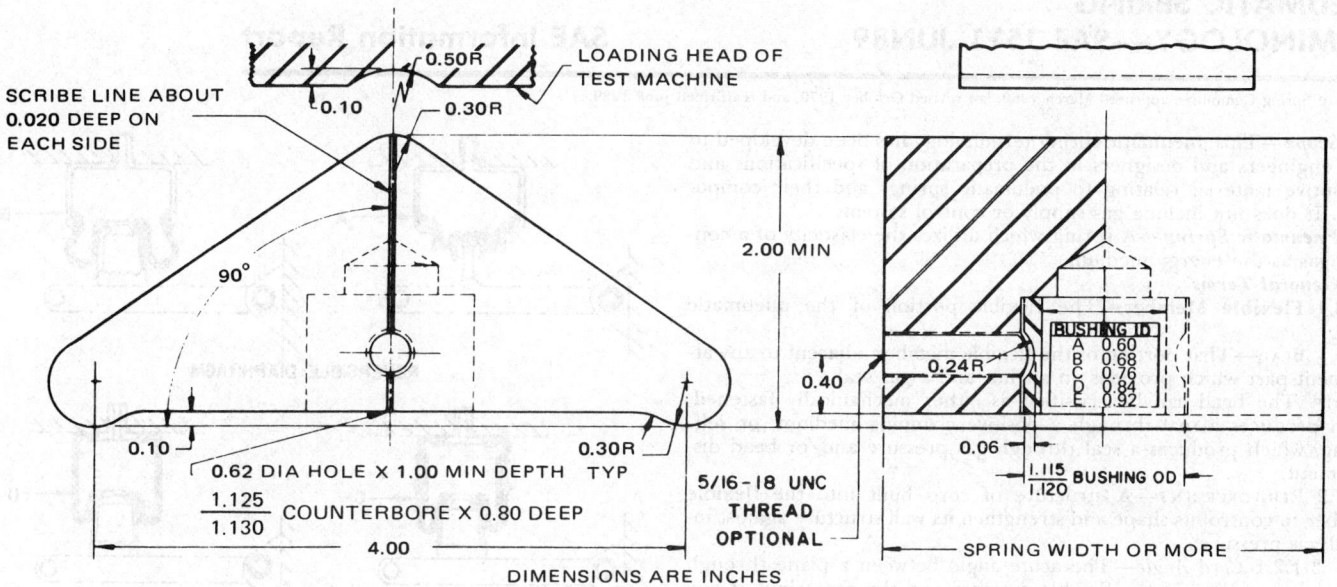

FIGURE 8—SPRING LOADING BLOCK

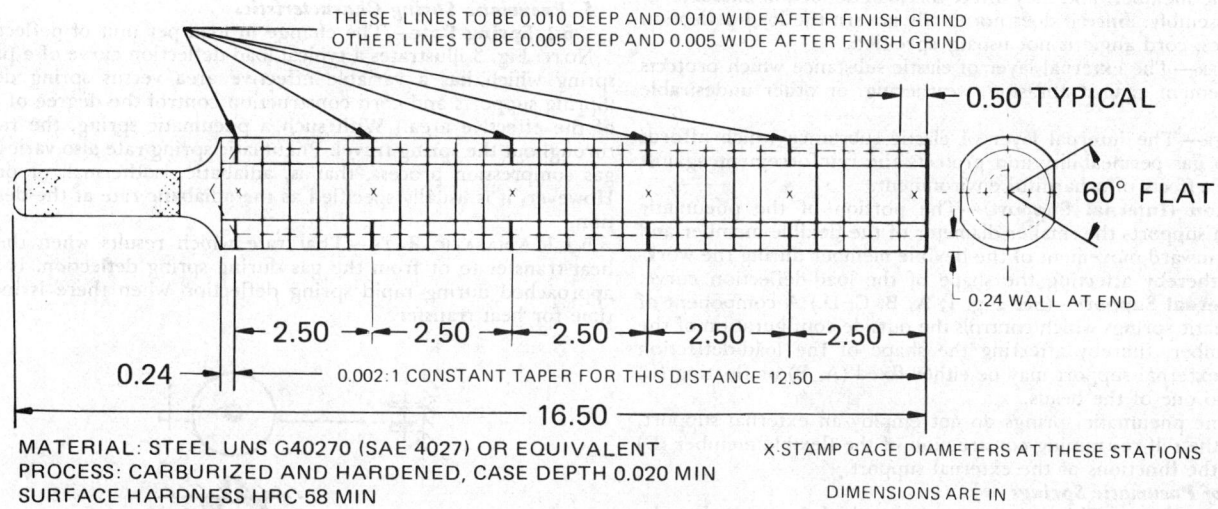

FIGURE 9—GAGE-LEAF SPRING EYE PLUG

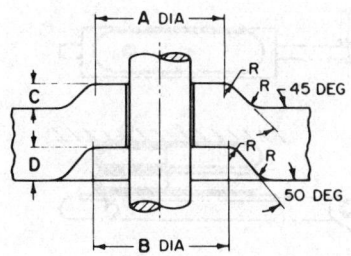

Dimension	Tolerance +0.00	Center Bolt Dia 5/16, 3/8	Center Bolt Dia 5/8, 3/4
Dia A	-0.02	0.84	1.24
Dia B	-0.02	0.88	1.30
Height C	-0.02	0.14	0.20
Depth D	-0.2	0.18	0.24
Radius R	-0.01	0.10	0.12

FIGURE 10—CUP CENTERS

PNEUMATIC SPRING TERMINOLOGY—SAE J511 JUN89

SAE Information Report

Report of Spring Committee approved March 1960, last revised October 1970, and reaffirmed June 1989.

1. Scope—This pneumatic spring terminology has been developed to assist engineers and designers in the preparation of specifications and descriptive material relating to pneumatic springs and their components. It does not include gas supply or control systems.

2. Pneumatic Spring—A spring which utilizes the elasticity of a confined gas as the energy medium.

3. General Terms

3.1 Flexible Member—The flexible portion of the pneumatic spring.

3.1.1 BEAD—That portion of the flexible member adjacent to any attachment part which provides an anchor and a gas seal.

NOTE: The bead can be classified as either mechanically fastened, which produces a seal through a positive clamping medium, or self-sealing, which produces a seal through gas pressure and/or bead displacement.

3.1.2 REINFORCEMENT—A structure of cord built into the flexible member to control its shape and strengthen its wall structure against internal gas pressure.

3.1.2.1 *Cord Angle*—The acute angle between a plane through the axial centerline of the flexible member and the centerline of any cord. This angle can pertain to the as-molded shape of the flexible member and will vary according to position of measurement and cross-sectional shape. It also can pertain to inflated shape and will vary according to position of measurement, cross-sectional shape, and inflation pressure.

NOTE: The cord angle is a determining factor of the inflated shape of the flexible member, and may affect the load-deflection characteristics of the assembly. Since it does not totally govern the load-deflection characteristics, cord angle is not usually specified.

3.1.3 COVER—The external layer of elastic substance which protects the reinforcement against abrasion, weathering, or other undesirable effects.

3.1.4 LINER—The internal layer of elastic substance which affords resistance to gas permeability and protects the reinforcement against aging or the effects of a harmful environment.

3.2 Piston (Internal Support)—The portion of the pneumatic spring which supports the smaller diameter of the flexible member and controls the inward movement of the flexible member during the working stroke, thereby affecting the shape of the load-deflection curve.

3.3 External Support—(See Fig. 1, A, B, C, D.) A component of some pneumatic springs which controls the outside configuration of the flexible member, thereby affecting the shape of the load-deflection curve. The external support may be either fixed (A, B) or floating (C) in relation to one of the beads.

NOTE: Some pneumatic springs do not employ an external support, but rely on the self-restraining construction of the flexible member (D) to perform the functions of the external support.

4. Types of Pneumatic Springs

4.1 Piston Type—This type uses a piston which is attached to the inner bead of a reversible flexible member. See Fig. 1.

4.1.1 REVERSIBLE DIAPHRAGM—In this type, the piston bead usually passes through the opposite bead of the flexible member. See Fig. 1, A and B.

4.1.2 REVERSIBLE SLEEVE—In this type, the piston bead travels within the flexible member and does not pass through the opposite bead. See Fig. 1, C and D.

4.2 Bellows Type—This type utilizes a nonreversible flexible member and relies upon its self-restraining characteristics to affect the load-deflection curve. See Fig. 2.

NOTE: The flexible member (round or oblong in section) may consist of one or more convolutions. A girdle ring is usually used between the convolutions of the round section multiconvolution bellows type pneumatic spring.

4.3 Piston and Cylinder Type—This type uses a piston and cylinder, but does not require a flexible member. A gas tight sliding seal is provided between the piston and cylinder.

4.4 Bladder Type—This type utilizes no integral reinforcement. It relies on being contained with a restrictive structure, such as a coil spring, for its support.

4.5 Hydropneumatic Type—This type contains both liquid and gas. Spring characteristics are provided by the confined gas, while damping may be provided by forcing the liquid through a restriction.

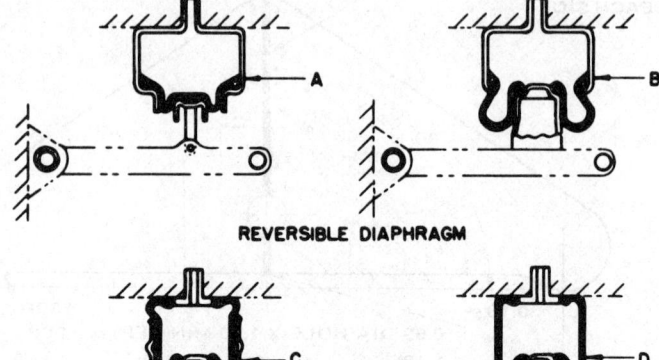

FIG. 1—PISTON TYPE PNEUMATIC SPRINGS

5. Pneumatic Spring Characteristics

5.1 Spring Rate—The change in load per unit of deflection.

NOTE: Fig. 3 illustrates a typical load-deflection curve of a pneumatic spring which has a variable effective area versus spring deflection. (Spring supports and cord construction control the degree of variation of the effective area.) With such a pneumatic spring, the rate varies throughout the spring travel. Pneumatic spring rate also varies with the gas compression process, that is, adiabatic, isothermal, or polytropic. However, it is usually specified as the adiabatic rate at the design position.

5.1.1 ADIABATIC RATE—That rate which results when there is no heat transfer to or from the gas during spring deflection. It is usually approached during rapid spring deflection when there is insufficient time for heat transfer.

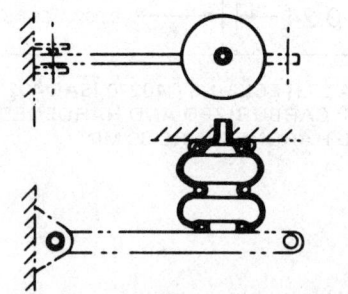

TWO CONVOLUTION CIRCULAR SECTION

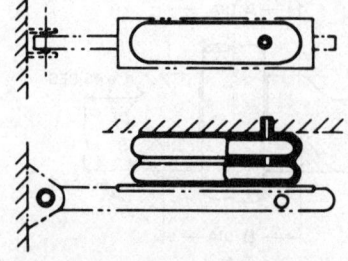

TWO CONVOLUTION OBLONG SECTION

FIG. 2—BELLOWS TYPE PNEUMATIC SPRINGS

5.1.2 ISOTHERMAL RATE—That rate which results when the spring deflects at a constant gas temperature. Isothermal rate is approached when the spring is deflected very slowly to allow time for the transfer of the heat.

5.1.3 POLYTROPIC RATE—That rate which results when there is limited heat transfer to or from the gas during spring deflection. Polytropic rate results during spring deflections which produce neither adiabatic nor isothermal rate.

5.2 **Working Volume**—The confined gas volume of the pneumatic spring. It is usually specified at design position.

5.3 **Design Position**—The selected position of the pneumatic spring which satisfies the vehicle requirements. It is usually specified by a dimension between reference points on the fixed and movable parts of the pneumatic spring.

5.4 **Total Spring Travel**—The distance between the extremes of the spring position measured at the spring axis. It is designated as the total of the compression and rebound deflections from the design position.

5.5 **Design Load**—The pneumatic spring load at design position.

5.6 **Design Pressure**—The internal gas pressure required to support the design load at the design position.

5.6.1 PRESSURE LIMITS—The minimum and maximum permissible pressures at the design position which provide satisfactory pneumatic spring operation.

5.7 **Effective Area**—A nominal area found by dividing the load of the pneumatic spring by its gas pressure at any given spring position.

6. *Color Coding to Identify Pneumatic Springs*—Pneumatic springs may be color coded for specific properties and operational environments by placing permanent color markings at least 0.25 inch in diameter on a visible section of the flexible member. The recommended guide for normal capabilities is shown in Table 1.

TABLE 1—IDENTIFICATION CODE

Color	Usage Characteristic	General Temperature Range, F
Yellow	Oil resistant	−20 to +150
Red	High temperature	−20 to +180
Green	Low temperature	−65 to +150
No color	General service	−20 to +150

Note: These color codes may be used in combination. Other temperature ranges and usage characteristics are available with special materials.

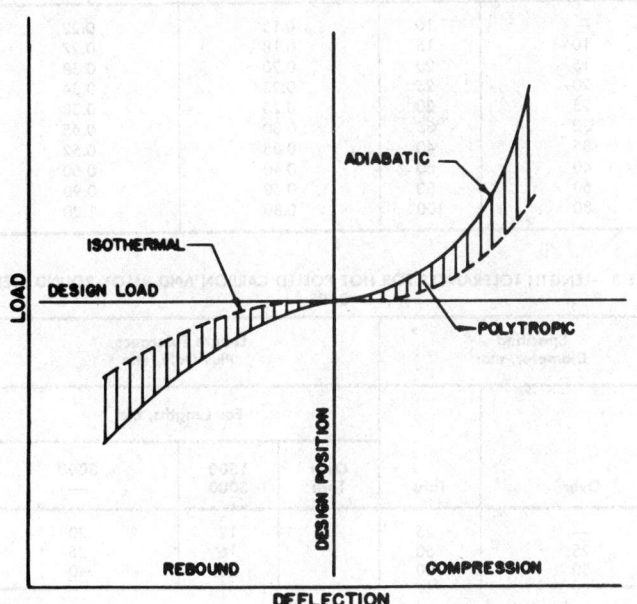

FIG. 3—PNEUMATIC SPRING LOAD-DEFLECTION CURVE

φHELICAL COMPRESSION AND EXTENSION SPRING TERMINOLOGY—SAE J1121 JUL88

SAE Recommended Practice

Report of Spring Committee approved November 1975 and completely revised July 1988.

1. *Scope*—The following recommended practice has been developed to assist engineers and designers in the preparation of specifications for the major types of helical compression and extension springs. It is restricted to a concise presentation of items which will promote an adequate understanding between spring manufacturer and spring user of the major practical requirements in the finished spring. Closer tolerances are obtainable where greater accuracy is required and the increased cost is justified.

For the basic concepts underlying the spring design and for many of the details, see the SAE Information Report MANUAL ON DESIGN AND APPLICATION OF HELICAL AND SPIRAL SPRINGS, SAE HS J795, which is available from SAE Headquarters in Warrendale, PA 15096. A uniform method for specifying design information is shown in the TYPICAL DESIGN CHECK LISTS FOR HELICAL SPRINGS, SAE J1122.

Two types of helical springs are considered:

Hot coiled compression springs for general automotive use as well as for motor vehicle suspensions.

Cold wound compression and extension springs for general automotive use.

This recommended practice uses SI (metric) units in accordance with the provisions of SAE J916 JUN82.

2. *Hot Coiled Springs*

2.1 **Materials and Heat Treatment**—Round spring steel bars are available in carbon and alloy analyses. The bars are generally used in the "as rolled" condition (either commercial hot rolled or precision hot rolled), but they may be centerless ground before coiling.

The heat treatment necessary to develop the required physical properties of the material may be accomplished by direct quench immediately after coiling, or by allowing the coiled spring to cool to a temperature below the critical, then reheating to the required temperature and quenching; the quench is followed by tempering to produce the specified hardness.

Table 1 lists available materials. Their hardenability limitations dictate maximum bar size. For tensile and torsional properties, see MANUAL, SAE HS J795, Chapter 2, Table 2.21.

TABLE 1—MATERIALS FOR HOT COILED COMPRESSION SPRINGS

Materials	Specification	Max. Bar* Dia., mm
Carbon Steels	SAE 1085	10
	SAE 1095	10
Carbon Boron Steel	SAE 15B62H	25
Alloy Steels	SAE 5150 H	10
	SAE 5160 H	20
	SAE 9260 H	10
	SAE 51B60H	30
	SAE 4161 H	60
	SAE 6150 H	10

*Based on a through hardened bar of 444 HB typical hardness ranges are 444 - 495 HB and 461 - 514 HB.

2.2 **Shot Peening**—Shot peening is used to increase the fatigue life of springs. It consists of subjecting the spring to a stream of metallic shot moving at high velocity. The peening action of the shot reduces the effect of surface defects and sets up beneficial stresses in a thin surface layer. It also results in cold working this layer. To be effective, the peening must reach the area of highest stress which for helical compression springs is the inside diameter of the coil.

The fatigue life of hot coiled springs is greatly impaired when the bar surface is afflicted by such flaws as impurities, cracks, seams, or decarburization, but it can be increased by the peening operation in the order of 4 to 1. Even the much better fatigue life attainable in hot-coiled springs with nearly perfect bar surface will be improved by peening in the order of more than 2 to 1. For further details see MANUAL, SAE HS J795, Chapters 1 and 4, also SHOT PEENING MANUAL, SAE HS 84 J808.

2.3 Presetting—Presetting (also called scragging, cold setting, or bulldozing) is an operation during the manufacturing process in which the spring is compressed beyond the yield point of the heat treated material. In preparation for this, the spring is coiled to a free length in excess of the designated free length. The yielding in the surface layers of the bar which occurs during presetting produces beneficial residual stresses, thus increasing the elastic limit and thereby reducing the chances of settling in subsequent service. The yielding causes the spring to take a permanent set, thus bringing it down to the designated free length. See also Preset Length, paragraph 2.6.3.

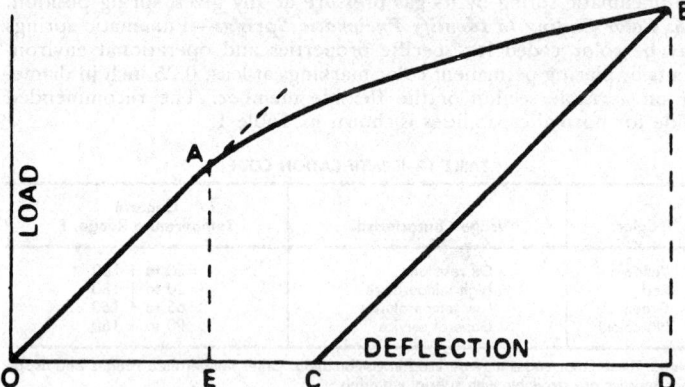

FIG. 1—TYPICAL LOAD-DEFLECTION DIAGRAM OF HELICAL SPRING DURING PRESETTING

2.3.1 WARM SETTING—In order to reduce the "sag" of "settling" of helical suspension springs which occurs when they are subjected to vehicle loading over time, it has become common practice to warm set the spring at an elevated temperature (usually about 200°C depending on the particular spring design). One theory holds that the major benefit of this operation results from an increase in the amount of strain hardening that occurs when the spring is stressed past the proportional limit (point "A" in Fig. 1). Increasing the temperature lowers the proportional limit to some stress lower than point "A", and therefore if the spring is still stressed to point "B", the amount of strain hardening that occurs is greater. This increase in strain hardening will reduce the dynamic or static settling (load loss) that occurs over the useful life of the spring.

A second theory is that a more effective beneficial residual stress pattern is set up over the bar cross section when a spring is warm set at elevated temperature.

It should be noted that a final (cold) presetting operation is still necessary.

In general, warm setting will decrease the load loss by more than 50%, depending on the working stress level.

2.4 Bar Diameter and Length—Round bars are hot rolled to any desired diameter between 9 and 100 mm. Table 2 shows the cross section tolerances for commercial hot rolled bars. Bars may be precision hot rolled with 50% of the tolerances in Table 2, or they may be centerless ground with 25% of the tolerances in Table 2.

Bars are commonly purchased in the exact length required to produce one spring. Tolerances for bar lengths are shown in Table 3.

2.5 Coil Diameter—The coil diameter can be expressed in terms of the mean coil diameter (D) which is used in the rate and stress formulae. However, coil diameter tolerances should be specified on either the inside diameter (ID) or the outside diameter (OD) of the coils, depending upon the importance of the respective dimensions to the user. Tolerances are shown in Table 4, based on coil diameter and spring length.

For motor vehicle suspension springs, it is customary to specify the ID in order to facilitate the coiling of a family of springs on a single arbor.

TABLE 2—CROSS SECTION TOLERANCES FOR HOT ROLLED CARBON AND ALLOY STEEL ROUND BARS

Specified Diameter, mm		Tolerance, Plus and Minus, mm	Out of Round, mm
Over	Thru		
--	10	0.15	0.22
10	15	0.18	0.27
15	20	0.20	0.30
20	25	0.23	0.34
25	30	0.25	0.38
30	35	0.30	0.45
35	40	0.35	0.52
40	60	0.40	0.60
60	80	0.60	0.90
80	100	0.80	1.20

TABLE 3—LENGTH TOLERANCES FOR HOT ROLLED CARBON AND ALLOY ROUND STEEL BARS

Specified Diameter, mm		Length Tolerance, Plus Only, mm	
		For Lengths, mm	
Over	Thru	Over Thru 1500 3000	3000 —
—	25	12	20
25	50	16	25
50	100	25	40

2.6 Spring Lengths—Spring lengths are to be measured after preloading (see Preload Length, paragraph 2.6.4) as the distance parallel to the spring axis between the end surfaces, or else between two reference points specified on the spring drawing.

2.6.1 FREE LENGTH—Free length is the length when no external load is applied. When load is specified, free length is used as a reference dimension only. When load is not specified, free length tolerance equals ±(1.5 mm + 4% of free-to-solid deflection).

2.6.2 SOLID LENGTH (SEE ALSO NUMBER OF COILS, PARAGRAPH 2.7)—Solid length is the length when the spring is compressed with an applied load sufficient to bring all coils in contact; for practical purposes, this applied load is taken to equal approximately 150% of the load beyond which no appreciable deflection takes place.

2.6.3 PRESET LENGTH—In the presetting operation (see Presetting, paragraph 2.3), the spring is usually compressed solid. However, if the stress at solid length is so high that the spring would be excessively distorted, the presetting operation may only be carried to a specified preset length. If more than one preset compression is desired, this must be specified on the drawing. See Also MANUAL, SAE HS J795, Chapters 1 and 4.

2.6.4 PRELOAD LENGTH—Preloading is the operation of deflecting the spring to the preload length in order to remove temporary recovery of free length before the spring is checked for load and rate.

If the spring was preset during the manufacturing process to the solid length, the preloading may also be carried to the solid length, but it may be restricted to a preload length slightly greater than the solid length, provided the maximum deflection during subsequent service will not go below the preload length.

If the spring was preset to a specified preset length greater than the solid length, the preloading should be restricted to a preload length greater than the preset length.

However, the preload length must not exceed the minimum spring length possible in the mechanism for which the spring is designed. In suspensions, this is called the "length at metal-to-metal position." The metal-to-metal contact will occur in the suspension mechanism when rubber bumpers are disregarded. The spring deflection from the specified loaded length to the metal-to-metal position is called "clearance."

2.6.5 LOADED LENGTH—Loaded length is the length while the load is being measured; it is a fixed dimension, with the tolerance applied to the load.

2.7 Number of Coils—Total number of coils (N_t) are counted tip to tip, active number of coils (N) are specified as the number of working coils at free length. With increasing load, N may progressively decrease due to the "bottoming out" effect. If no appreciable bottoming out occurs, the relationships between N and N_t are as shown in Table 5 which also gives the formulae for nominal solid length.

Since nominal solid length may be exceeded somewhat by actual solid length due to manufacturing variations, a frequent practice is to specify nominal solid length together with a maximum solid length, as shown in Table 6.

2.8 Spring Ends—Four types of ends are used (Fig. 2):

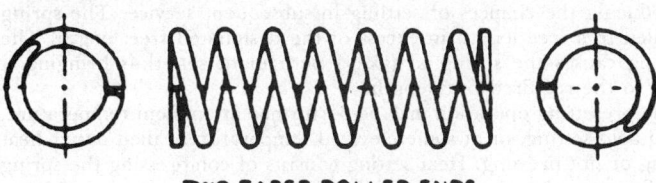

TWO TAPER ROLLED ENDS

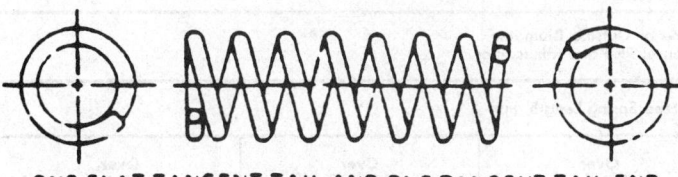

ONE FLAT TANGENT TAIL AND ONE TANGENT TAIL END

TWO PIGTAIL ENDS

FIG. 2—TYPICAL ENDS FOR HOT COILED COMPRESSION SPRINGS

1. A flat end formed from a tapered bar end. The bar end is usually tapered for a length equal to 2/3 coil and to a tip thickness of approximately 1/3 of the bar diameter. When the spring is coiled, the tip shall be in approximate contact with the adjacent coil and shall not protrude beyond the outside diameter by more than 20% of the bar diameter.

   When stipulated, the bearing surface of the spring end shall be ground perpendicular to the axis of the spring helix in order to produce a firm bearing. The actual ground bearing surface shall not be shorter than two-thirds of the mean coil circumference, nor narrower than half the width of the hot tapered surface of the bar. However, this grinding is usually not required if the tapering and coiling operations are performed adequately.

2. An untapered end coil formed substantially smaller than the central coils of the spring and in such a fashion as to have an outboard bearing surface perpendicular to the axis of the spring helix, the so-called "pigtail" end.

3. An untapered end coil formed as a helix having a pitch substantially equal to the bar diameter. To facilitate coiling, a straight end portion about 25 mm long is permitted to project tangent to the helix of this end construction, the so-called "tangent tail" end. The use of this type of end requires a spring seat formed at the same pitch of helix as that of the spring end.

4. An untapered end coil formed perpendicular to the axis of the spring helix for a circumference of at least 220 deg, the so-called "flat tangent tail" end. To facilitate coiling, a straight end portion about 25 mm long is permitted to project tangent to the outer circumference.

Springs can be specified to have any combination of the four types of ends. The combination of two tangent tail ends may involve a complex arrangement for indexing the spring seats, unless the design of every spring is adjusted to an identical number of total coils. Springs for general automotive use generally have two flat tapered ends. Spring ends and seats are usually so formed as to render approximately two-thirds to one coil inactive at each end.

2.9 Squareness of Ends—Unless otherwise specified, the tapered ends of any spring having an outside diameter to bar diameter ratio of 4 or more, and a free length to outside diameter of 4 or less, shall not deviate more than 3 deg from the perpendicular to the spring axis, as determined by standing the spring on its end and measuring the angular deviation of the outer helix from a perpendicular to the plate on which the spring is standing. In the case of a tangent tail end, the spring must stand on a seat with matching helical ramp. Tolerances for springs outside these limits are subject to special agreement.

2.10 Load—Load is the force in newtons (N) measured on the load testing machine required to deflect the spring to the specified loaded length. It is to be measured during compression of the spring (compression load) and not during release of the spring (release load), unless otherwise specified.

With loaded length fixed, the usual tolerance for motor vehicle suspension springs is expressed in terms of load equivalent to a deflection of ±5 mm at the nominal rate. Where the demand for greater accuracy warrants the cost of additional presetting or other operations, the load tolerance may be specified as low as ±1.50 mm at the nominal rate.

In the springs for general automotive use, the load tolerance (with loaded length fixed) typically equals ±(1.50 mm + 3% of free-to-solid deflection) × nominal rate. This tolerance is limited to springs where the free length does not exceed 900 mm, does not exceed six times the free-to-solid deflection, and is not less than 0.8 times the OD.

2.11 Rate—Rate is the change of load per unit length of spring deflection (N/mm).

In the springs for motor vehicle suspension, the rate is expressed in terms of the load increase per 25 mm deflection (N/25 mm). It is therefore determined as one-half the difference between the loads measured 25 mm above and 25 mm below the specified loaded length. Tolerance is ±3% with centerless ground or with precision rolled bars, and ±4% when commercial hot rolled bars are used.

In the springs for general automotive use, the rate is determined between 20 and 60% of the total deflection unless otherwise defined. Typical tolerance is ±5%. In non-critical applications, this may be increased to ±10%.

2.12 Direction of Coiling—For most applications, the direction of coiling is unimportant; however, right hand coiling is preferred because most spring manufacturers are so equipped. When direction of coiling is important, as in the case of concentrically nested springs, it must be specified for each component spring, maintaining opposite directions for adjacent springs. For tangent tail springs, the direction of coiling must conform with the installation conditions.

2.13 Uniformity of Pitch—The pitch of coils in a compression spring must be sufficiently uniform so that when the spring is compressed, unsupported laterally, to a length representing a deflection of 80% of the nominal free-to-solid deflection, none of the coils must be in contact with one another, excluding the inactive end coils. This requirement does not apply when the design of the spring calls for variable pitch, or when it is such that the spring cannot be compressed to solid length without lateral support.

2.14 Shaped and Variable Rate Coils—Many newer motor vehicle applications require specially shaped suspension coil springs, or springs with variable output characteristics. The coils which are specially shaped usually exhibit a partially conical or barrel form in order to satisfy restricted height, tire clearance, or other suspension requirements. In some cases, the ends of the spring may be offset in order to provide off center loading for suspension strut applications.

With regard to variable output characteristics, some springs are designed to provide a variable rate and corresponding frequency change, for improved height control, ride and handling. The variable rate characteristic is achieved by designing and producing the spring with very specific coil spacing such that active coil segments "bottom out" against a spring seat or against each other as the spring is deflected, thereby decreasing the effective number of active coils and increasing the rate. This effect is achieved with the greatest material and space efficiency if the bar is conically tapered over the length of the coils which bottom out. It should be pointed out, however, that special equipment is required to conically taper the bars. Also, it is important to note that coil-to-coil or coil-to-seat contact can cause undesirable noise.

2.15 Concentricity of Coils—At free length, the center of all coils must be concentric with the spring axis within 1.5 mm. This axis is the straight line connecting the centers of the end coils.

3. Cold Wound Springs

3.1 Material—Round wire sizes and tolerances may be found in the individual wire specifications, such as:

Music Wire	SAE J178
Carbon Steel Spring Wire - Oil Tempered	SAE J316
- Hard Drawn	SAE J113
- Special Quality High Tensile Hard Drawn	SAE J271
- Valve Spring Quality Oil Tempered	SAE J351
- Valve Spring Quality Hard Drawn	SAE J172
Chromium Vanadium Wire - Valve Spring Quality	SAE J132
Chromium Silicon Alloy Steel Wire	SAE J157
Stainless Steel Wire, SAE 30302	SAE J230
Stainless Steel Wire, 17-7 PH	SAE J217
Phosphor-Bronze Wire, SAE CA510	SAE J461
Beryllium-Copper Wire, SAE CA172	SAE J461
Brass Wire, SAE CA260	SAE J461
Silicon-Bronze Wire, SAE CA655	SAE J461

of presetting is most beneficial when design stresses are at or near the yield point, and settling prevents the spring from performing as required.

Presetting is an operation that is performed during the manufacturing of helical compression springs in which the spring is compressed beyond the yield point of the material. The yielding of the surface layers of the wire which occurs during the presetting produces beneficial residual stresses, thus increasing the elastic limit of the spring and thereby reducing the chances of settling in subsequent service. The spring is coiled to a free length in excess of the designated free length. The yielding causes the spring to take a permanent set, thus bringing it down to the required free length.

The presetting operation may be performed at ambient temperature, called cold setting, or at some elevated temperature, called either heat setting or hot pressing. Heat setting consists of compressing the spring on a fixture, subjecting the compressed spring to a temperature higher than the desired operating temperature for a time suitable to insure

TABLE 4—COIL DIAMETER TOLERANCES

	Inside or Outside Diameter Tolerance, Plus and Minus, mm				
	For Free Spring Length, mm				
For Specified or Computed Outside Diameter, mm	Up to 250	Over 250 thru 450	Over 450 thru 650	Over 650 thru 850	Over 850 thru 1050
75.0 thru 110.0	0.8	1.3	2.5	3.6	4.6
Over 110.0 thru 150.0	1.3	2.5	3.6	4.6	5.6
Over 150.0 thru 200.0	2.5	3.6	4.6	5.6	6.6
Over 200.0 thru 300.0	3.6	4.6	5.6	6.6	6.6

TABLE 5—FORMULAE FOR TOTAL COILS AND FOR NOMINAL SOLID LENGTH

End Configuration	Total Coils (N_t)	Nominal Solid Length (L_s)
Both ends taper rolled	N + 2	$1.01 d (N_t - 1) + 2t$
Both ends with tangent tail	N + 1.33	$1.01 d (N_t + 1)$
Both ends with pigtail	N + 1.50	$1.01 d (N_t - 1.25)$
Taper rolled plus tangent tail	N + 1.67	$1.01 d N_t + 1$
Taper rolled plus pigtail	N + 1.75	$1.01 d (N_t - 1) + t$
Tangent tail plus pigtail	N + 1.42	$1.01 d N_t$

where
d = bar diameter
t = tip thickness of taper rolled bar
1.01 = factor used to compensate for the cosine effect of the coil helix angle
The bracketed term in the solid length formula for springs with two pigtail ends may vary between (N_t − 0.90) and (N_t − 1.60), depending on the pigtail details.

TABLE 6—SPRING SOLID LENGTH TOLERANCES

Nominal Solid Length, mm		Maximum Deviation of Solid Length Above Nominal Solid Length, mm
Over	Thru	
—	175	1.5
175	250	2.5
250	325	3.0
325	400	4.0
400	475	4.8
475	550	5.5
550	625	6.5

3.2 Shot Peening—Shot peening is used to increase the fatigue life of springs. It consists of subjecting the spring to a stream of metallic shot moving at high velocity. The peening action of the shot reduces the effect of surface defects and sets up beneficial stresses in a thin surface layer. It also results in cold working this layer. To be effective, the peening must reach the area of highest stress which for helical compression and extension springs is the inside diameter of the coil.

Even when the wire surface is virtually flawless, the fatigue life of the cold-wound spring can be increased by peening in the order of more than 2 to 1. See MANUAL, SAE HS J795, Chapter 1, also SHOT PEENING MANUAL, SAE HS 84 J808.

3.3 Presetting—The need for presetting depends upon the design stresses, the application and its conditions and requirements. The use

complete penetration of the heat, and then cooling to room temperature before releasing.

Hot pressing consists of heating the spring in its free or relaxed position to some temperature for sufficient time to insure complete penetration; then, while the spring is at the temperature, it is compressed to some height below the installed or operating position and released.

3.4 Coil Diameter—Coil diameter tolerances can be specified on either the inside diameter (ID) or the outside diameter (OD) of the coils, depending upon the importance of the respective dimensions to the user. Tolerances are functions of the "Spring Index", which is the ratio of mean coil diameter (D) to wire diameter (d). They are to be considered as manufacturing tolerances and do not take into account the effects of changes in diameter due to applied loads. See Figs. 3 and 4.

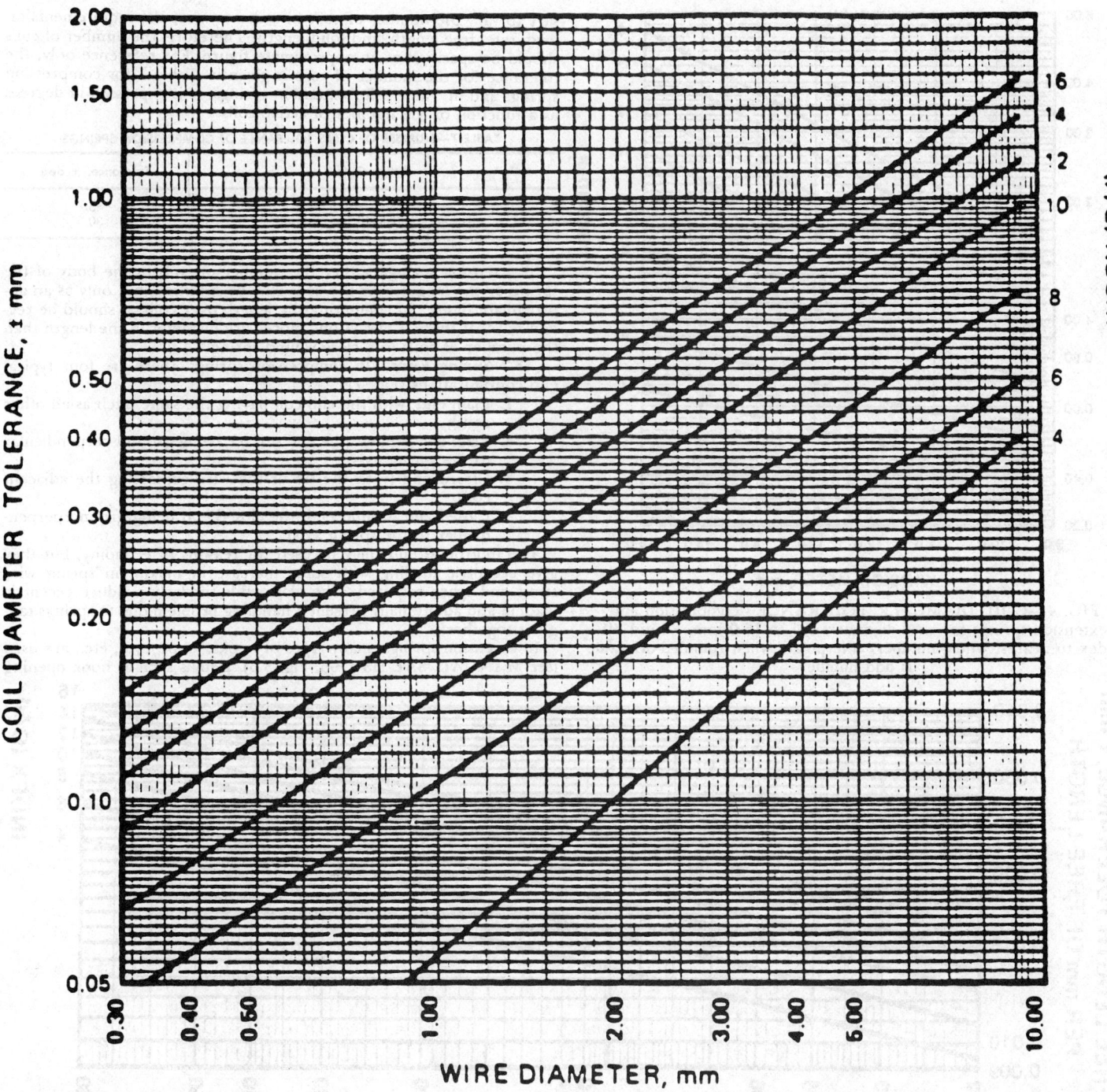

FIG. 3—COIL DIAMETER TOLERANCE - compression and extension springs for wire diameters 0.30 to 9.50 mm. Round off index to nearest whole number. Interpolate when the rounded-off value is an odd number. Use tolerance for 0.30 mm wire diameter when wire diameter is less than 0.30 mm.

3.5 Spring Lengths—Spring lengths of compression springs are overall dimensions measured parallel to the axis of the spring.

Spring lengths of extension springs are measured inside to inside of the hooks (overall length minus two wire diameters).

3.5.1 FREE LENGTH—Free length is the length under no load. When load is specified, free length is used as a reference dimension only. When load is not specified, free length is specified for control and inspection purposes by using Fig. 5 for compression springs and Fig. 6 for extension springs.

The tolerances in Fig. 3 are based on the number of active coils (N), the free length (L_o), and the spring index (D/d). With these parameters known, the N/L_o value is established on the abscissa, and the tolerance is found by multiplying the corresponding ordinate value by L_o. Round off the index to the nearest whole number and interpolate when this is an odd number. The tolerances shown in Fig. 5 are for springs with ends closed and ground. For springs with the ends closed but not ground, multiply by 1.7.

3.5.2 SOLID LENGTH (SEE ALSO NUMBER OF COILS, PARAGRAPH 3.6)—In compression springs, this is the length with all active coils closed, to be specified as a maximum dimension allowing the manufacturer any tolerance required by the variations in wire size, total coils, and the amount of grind at the ends; platings and coatings increase the wire diameter and must be considered.

For springs with ground ends, the maximum solid length is the total number of coils times the wire diameter; for springs with ends not ground, the solid length is the total number of coils plus one, times the wire diameter.

3.5.3 PRESET LENGTH—After the compression spring has been coiled to a free length in excess of the designated free length, it is compressed solid or to a specified preset length; this produces yielding, which re-

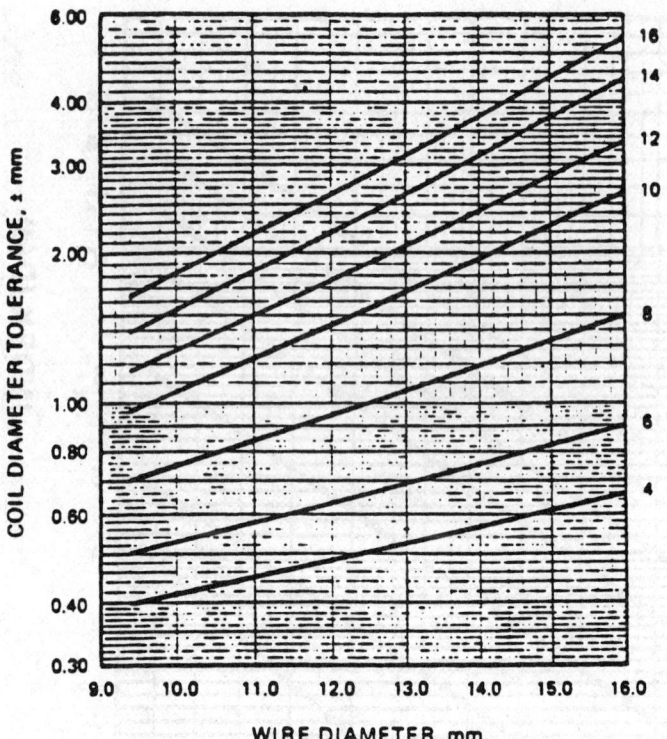

FIG. 4—COIL DIAMETER TOLERANCE - compression and extension springs for wire diameters 9.5 to 16.0 mm. Round off index to nearest whole number. Interpolate when rounded-off value is odd number.

sary to vary the number of coils in order to meet the requirements on load, rate, free length, and solid length. Therefore, the number of coils should be specified as an approximate figure. For reference only, the tolerance for the number of coils is given in Table 7 for compression springs and in Table 8 for extension springs. It is expressed in degrees as a function of the number of active coils.

TABLE 7—NUMBER OF COILS TOLERANCE OF COMPRESSION SPRINGS

Active Coils	Tolerance, ± deg
3 — 10	45
For each additional 10 coils, add	30

In extension springs, either the number of coils in the body of the spring or the length over the coils may be specified, but only as an approximate figure. In computing the length over coils, it should be recognized that there is always one more wire diameter in the length than the number of coils in a close-wound spring.

3.7 Spring Ends—In compression springs, there are four typical end configurations (Fig. 7):
1. Plain end (with the end coil having the same pitch as all other coils);
2. Plain end ground (the end surface being ground perpendicular to the spring axis);
3. Closed end (with the tip of the wire contacting the adjacent coil);
4. Closed and ground end (the closed end being ground perpendicular to the spring axis).

The unground ends may be used for reasons of economy, but they give eccentric loading with some increase in maximum spring wire stress and space required. The plain ends similarly produce eccentric loading and additionally present a handling problem due to springs tangling together.

In extension springs, many types of hooks, loops, eyes, etc., are used (see MANUAL, SAE HS J795, Fig. 3.3). Details such as hook opening

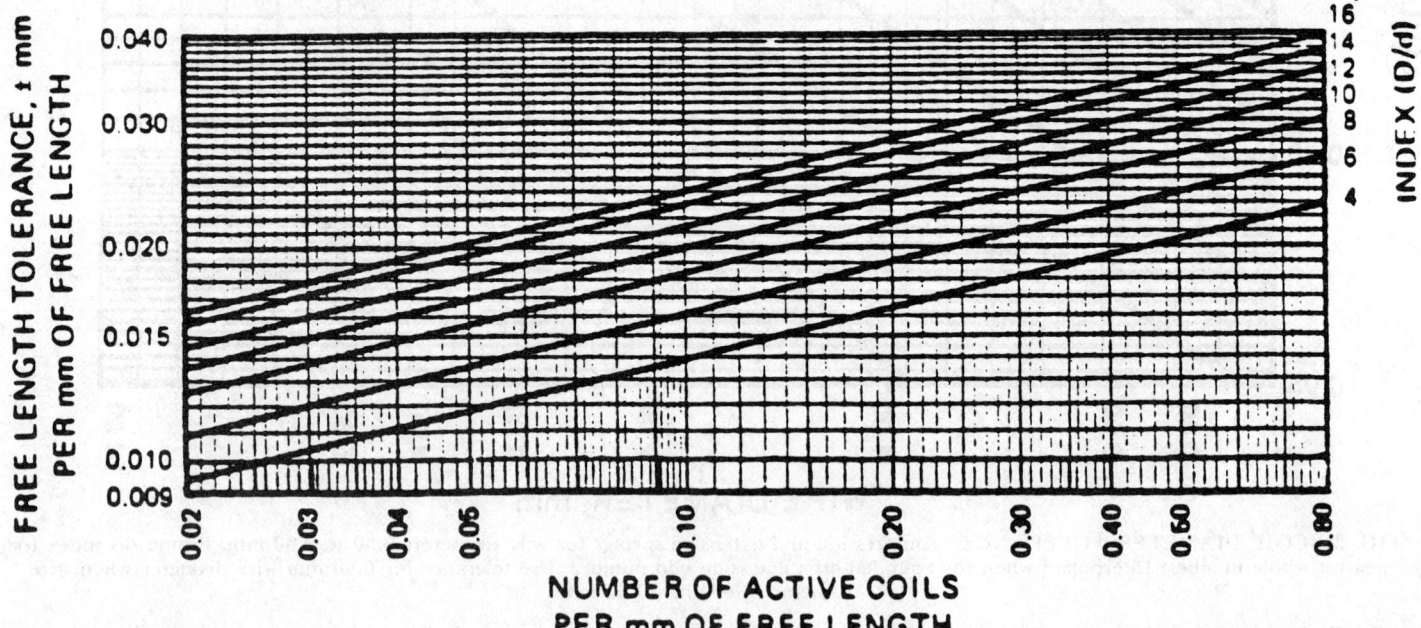

FIG. 5—FREE LENGTH TOLERANCE - compression springs. Round off index to nearest whole number. Interpolate when rounded-off value is odd number. These are tolerances for springs with ends closed and ground. For springs with ends closed but not ground, multiply by 1.7.

sults in bringing the spring to the designated free length. If more than one preset compression is desired, it must be specified on the drawing. See also MANUAL, SAE HS J795, Chapters 1 and 4.

3.5.4 LOADED LENGTH—This is the length while the load is being measured. It is a fixed reference dimension, with the tolerance applied to the load.

3.5.5 MAXIMUM EXTENDED LENGTH—Extension springs normally do not have a definite stop to their deflection, therefore the drawing specifications should include a statement of the maximum extended length which must be attained without encountering permanent set.

3.6 Number of Coils—In compression springs, it is often neces-

restraint of the loop within the body diameter should be specified on the drawing. The position of hooks relative to each other can be in line, at right angles, or at any other angular position as required. If this relative position is important, the spring drawing should emphasize the importance by a statement as well as by pictorial representation. Sharp bends in forming the end hooks should be avoided because they produce stress concentrations.

3.8 Squareness of Ends—In compression springs with closed and ground ends, the squareness of the ends, as measured in the unloaded position, is to be maintained within a limit of 3 deg with the axis of the spring.

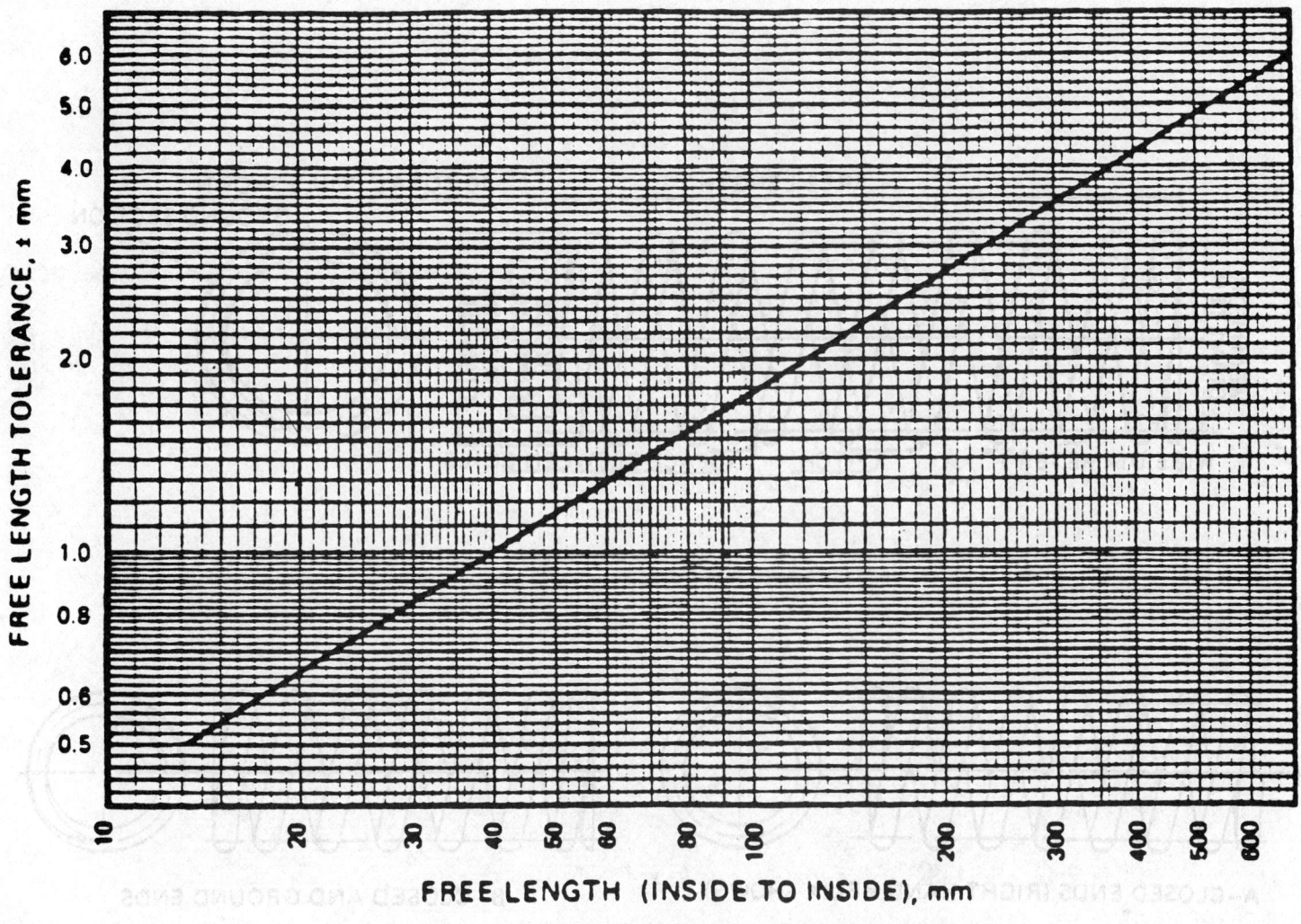

FIG. 6—FREE LENGTH TOLERANCE – EXTENSION SPRINGS

TABLE 8—NUMBER OF COILS TOLERANCE OF EXTENSION SPRINGS

Active Coils	Tolerance, ± deg	
	Close Wound	Open Wound
3	30	90
4 – 10	45	90
For each additional 10 coils, add	15	30

3.9 Load—Load is the force in newtons (N) measured on the load testing machine required to deflect the spring to the specified loaded length.

For compression springs, the load is to be measured during compression of the spring (compression load) unless otherwise specified. Tolerances are shown in Fig. 8 as functions of the nominal free length tolerance (Fig. 5) and the deflection from free length to loaded length. Round off the percent load tolerance values to the next larger whole number. Interpolate when this is an odd number and when it is between 8 and 25%.

For extension springs, the load is to be measured during extension of the spring. Tolerances are computed as the product of the appropriate tolerance factor from Fig. 9A and the appropriate multiplying factor from Fig. 9B.

Cold coiled extension springs may be wound with tension between the coils so that a load must be applied to separate them, the so-called initial tension in the spring.

3.10 Rate—Rate is the change of load per unit length of spring deflection (N/mm). The rate is to be determined between 20 and 60% of the total deflection. Tolerances depending on the number of active coils are given in Fig. 10.

3.11 Direction of Coiling—For most applications, the direction of coiling is unimportant; however, right hand coiling is preferred because most spring manufacturers are so equipped.

3.12 Uniformity of Pitch—The pitch of coils in a compression spring must be sufficiently uniform so that when the spring is compressed, unsupported laterally, to a length representing a deflection of 80% of the nominal free-to-solid deflection, none of the coils must be in contact with one another, excluding the inactive end coils. This requirement does not apply when the design of the spring calls for variable pitch, or when it is such that the spring cannot be compressed to solid length without lateral support.

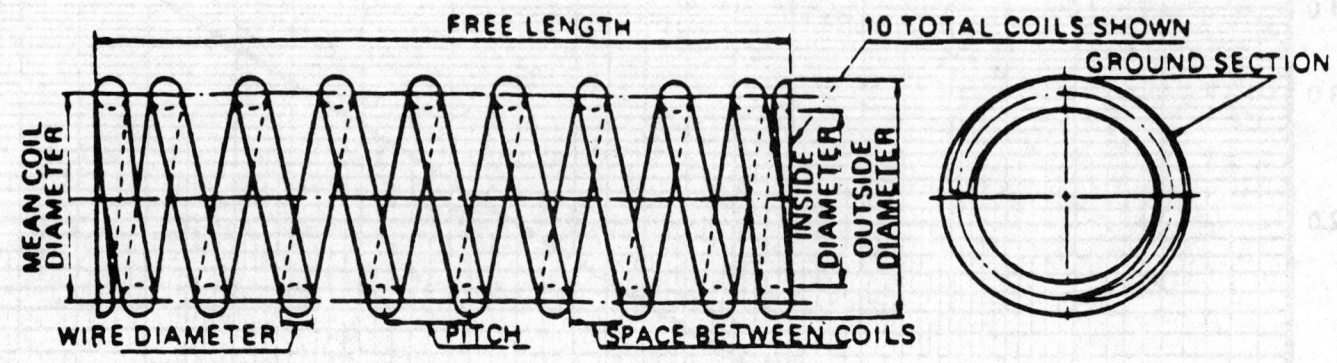

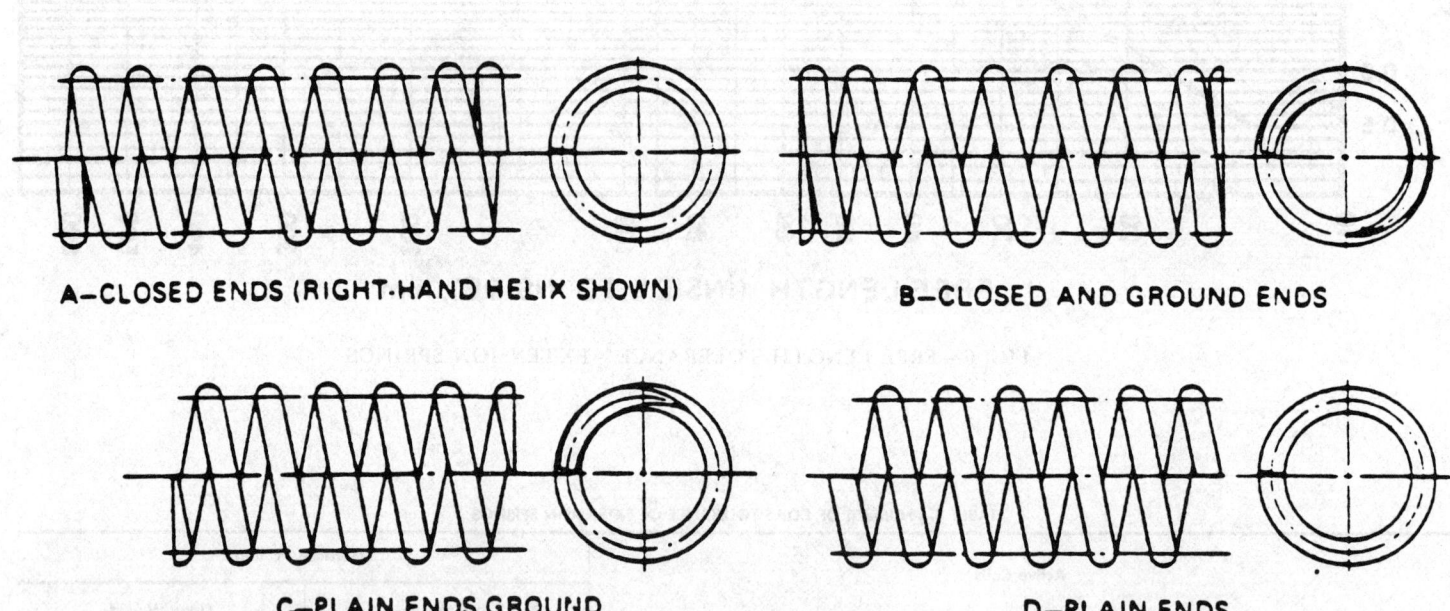

FIG. 7—TYPICAL ENDS OF HELICAL COMPRESSION SPRINGS

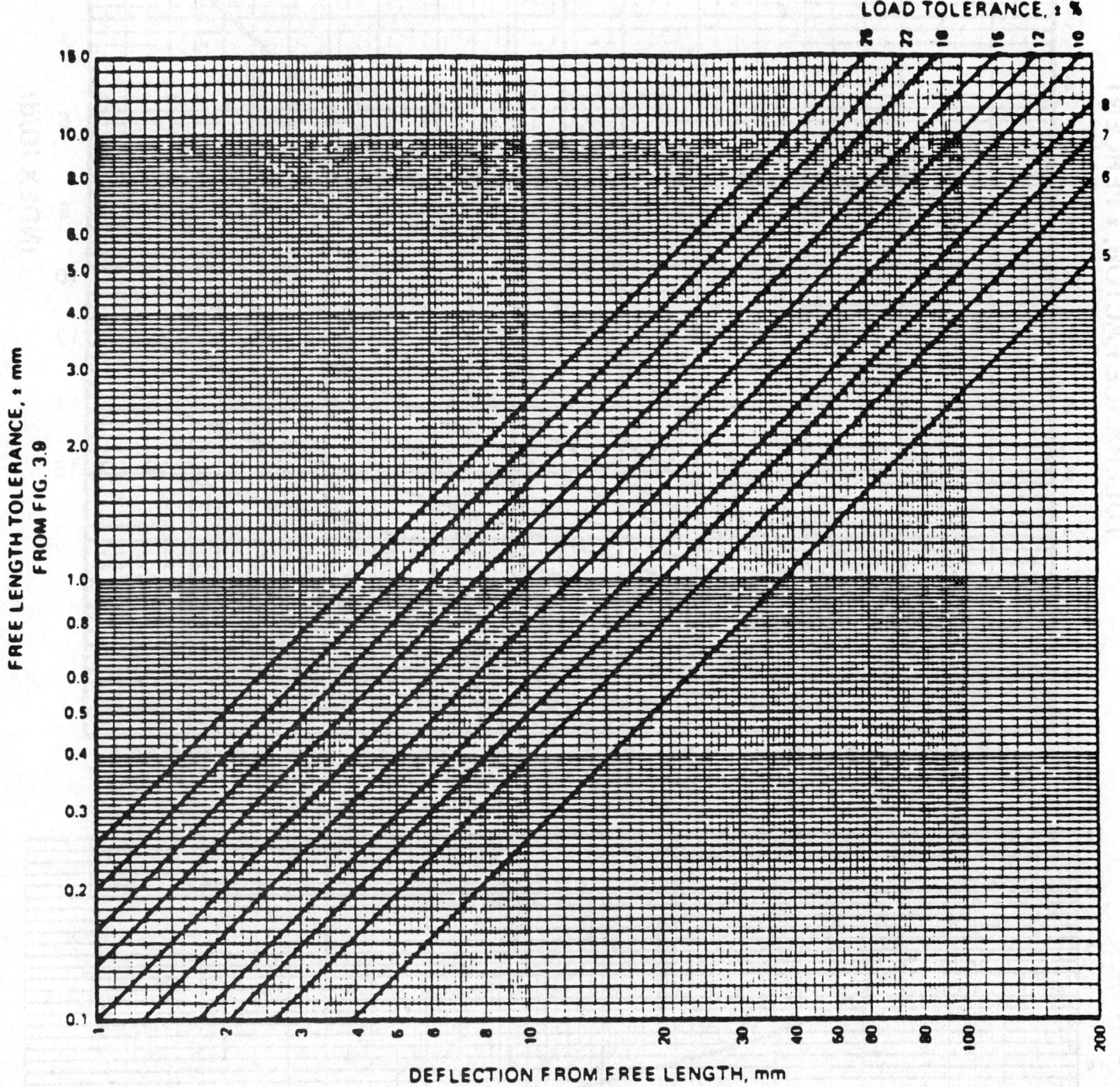

FIG. 8—LOAD TOLERANCE – compression springs, ±%. Enter chart from bottom with deflection from free length to loaded length and from left with free length tolerance of Fig. 4. Round off percent load tolerance values to next larger whole number. Interpolate when rounded-off value is odd and between 8 and 25%.

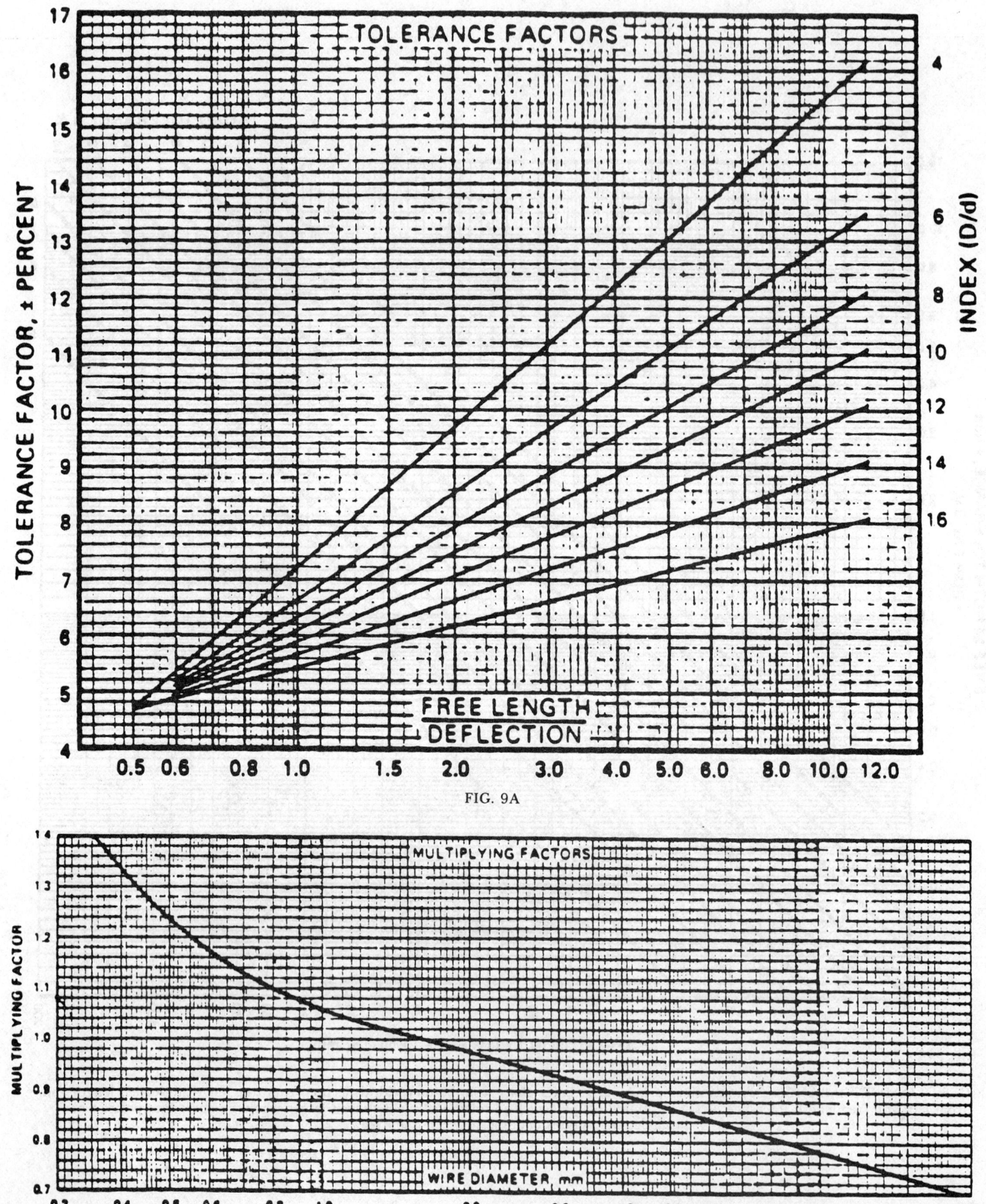

FIG. 9B—LOAD TOLERANCE – extension springs, ±%. To find load tolerance, multiply tolerance factor from Fig. 9A by multiplying factor from Fig. 9B.

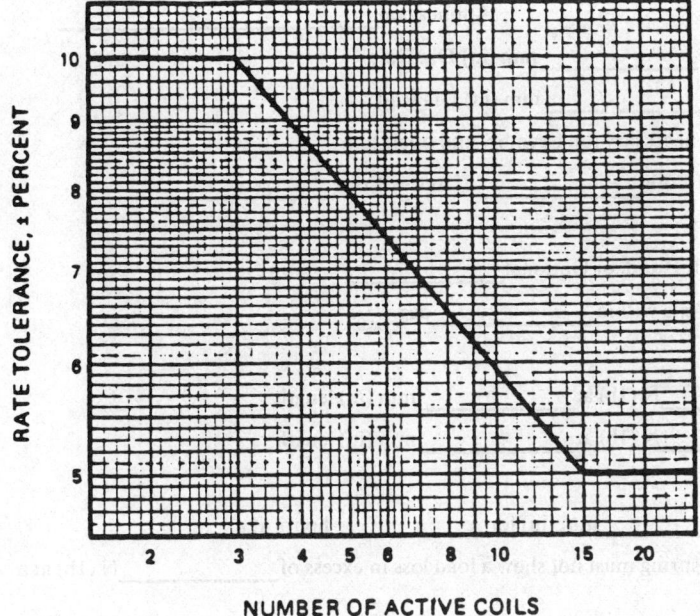

FIG. 10—RATE TOLERANCE – compression and extension springs. Rate and rate tolerance should be specified only when rate is functional. When rate specification is necessary, range of deflection over which it is to apply must be clearly identified. Deflection ranges for rate control should fall within 20–60% limit of total deflection because rate is likely to be variable outside this range.

φHELICAL SPRINGS: SPECIFICATION CHECK LISTS— SAE J1122 JUL88

SAE Recommended Practice

Report of Spring Committee approved February 1976 and completely revised July 1988.

1. Scope—This SAE Recommended Practice furnishes sample forms for helical compression, extension and torsion springs to provide a uniform method for specifying design information. It is not necessary to fill in all the data, but sufficient information must be supplied to fully describe the part and to satisfy the requirements of its application. For detailed information, see "Design and Application of Helical and Spiral Springs SAE HS J795 SEP82", also "Helical Compression and Extension Spring Terminology SAE J1121 NOV75".

Both these documents use SI (metric) Units in accordance with the provisions of SAE J916 MAY85, and so does SAE J1122. Here, however, the U.S. Customary Units (in, lb, psi) have been added in parentheses after each SI Unit for the convenience of the user who must furnish specifications on a project where all requirements are listed in non-metric terms.

TABLE 1—CONVERSION TABLE

To Convert from SI Unit to U.S. Customary Unit, Divide by the Factor
To Convert from U.S. Customary Unit to SI Unit, Multiply by the Factor

Quantity	SI Unit		Factor	U.S. Customary Unit	
Length	millimeter	mm	25.4 (Exactly)	inch	in
Area	square millimeter	mm^2	645.16 (Exactly)	square inch	in^2
Mass	kilogram	kg	0.453 592 4	pound-mass	lb$_m$
Force (or Load)	newton	N	4.448 222[a]	pound-force	lb$_f$
Moment	newton millimeter	N.mm	112.984 8	pound inch	lb$_f$.in
Linear Spring Rate	newton per millimeter	N/mm	0.175 126 8	pound per inch	lb$_f$/in
Torsional Spring Rate	newton mm per degree	N.mm/deg	112.984 8	pound inch per degree	lb$_f$.in/deg
Stress	megapascal	MPa	0.006 894 757[b]	pound per square inch	psi

[a] 4.448 222 = 0.453 592 37 · 9.806 650

where

9.806 650 = Acceleration of Gravity "g" adopted in 1901 by International Committee on Weights & Measures

[b] 0.006 894 757 = 4.448 2216 · 0.000 645 16

A - HELICAL COMPRESSION SPRINGS   Application_____

OD_____mm (in) to work inside_____mm (in) Dia Hole

ID_____mm (in) to work over _____mm (in) Dia Rod

Note: Specify only those diameters that are necessary for assembly and operation

Wire Dia _____mm (in)   Total Coils_____   Active Coils_____

Free Length_____mm (in) approx

Direction of Coil Winding: Right Hand, Left Hand, or Optional_____

Type of Ends_____Square with Axis within_____deg

Max Solid Length_____mm (in)

Load_____N (lb) ± _____N (lb) at _____mm (in) length

Load_____N (lb) ± _____N (lb) at _____mm (in) length

Rate_____N/mm (lb/in) Ref

After being compressed to a length of_____mm (in) for_____hours at a

temperature of_____°C (°F) the spring must not show a load loss in excess of_____N (lb) at a

length of_____mm (in)

Spring Index = Mean Coil Diameter/Wire Diameter (D/d)_____

Wahl Stress Correction Factor K_w_____

Stress at_____mm(in) length: _____MPa (psi) Corrected or
                                                     Uncorrected_____

Stress at_____mm(in) length: _____MPa (psi) Corrected or
                                                     Uncorrected_____

Material_____

Hardness or Tensile Strength_____

Surface Treatment/Finish_____

Identification_____

Remarks_____

B - HELICAL EXTENSION SPRINGS   Application_____

OD_____mm (in)   Wire Dia_____mm (in)

Number of Coils_____

Direction of Coil Winding: Right Hand, Left Hand, or Optional_____

Free Length Inside Hooks_____mm (in) approx

Type of Ends (use sketch if necessary)_____
_____

Load_____N (lb) ± _____N (lb) at _____mm (in) length

Load_____N (lb) ± _____N (lb) at _____mm (in) length

Initial Tension_____N (lb) Ref

Rate_____N/mm (lb/in) Ref

Stress at_____mm (in) length: _____MPa (psi) Corrected or Uncorrected_____

Stress at_____mm (in) length: _____MPa (psi) Corrected or Uncorrected_____

Max Extended Length without set_____mm (in)

Stress at Max Extended Length_____MPa (psi) Corrected or Uncorrected_____

Material_____

Hardness or Tensile Strength_____

Surface Treatment/Finish_____

Indentification_____

Remarks_____

C - HELICAL TORSION SPRINGS   Application _____

OD _____ mm (in)

ID _____ mm (in) to work over _____ mm (in) Dia Shaft

Note: Specify only those diameters that are necessary for assembly and operation

Wire Dia _____ mm (in)   Number of Coils _____

Max Free Length _____ mm (in)

Direction of Coil Winding: Right Hand or Left Hand _____

Type of Ends (use sketch if necessary) _____
_____

Moment _____ N mm (lb in) ± _____ N mm (lb in) at _____ deg between ends

Moment _____ N mm (lb in) ± _____ N mm (lb in) at _____ deg between ends

Rate _____ N mm/deg (lb in/deg) approx

Stress at _____ deg _____ MPa (psi)

Stress at _____ deg _____ MPa (psi)

Max Wound Position _____ deg

Stress at Max Wound Position _____ MPa (psi)

Material _____

Hardness or Tensile Strength _____

Surface Treatment/Finish _____

Indentification _____

Remarks _____

RATED SUSPENSION SPRING CAPACITY—SAE J274 JUN89

SAE Recommended Practice

Report of the Spring Committee, approved September 1972, completely revised June 1984, and reaffirmed June 1989.

1. Scope—The Rated Suspension Spring Capacity definition has been developed to assist engineers and designers in the preparation of specifications and descriptive material and values relating thereto.

2. Purpose—The following definition of Rated Suspension Spring Capacity is applicable to all types of suspensions designed for vehicles used predominantly on the highway. This capacity provides a basis for comparison of spring load carrying abilities in a particular suspension application. This definition is intended to clarify a commonly used term which has heretofore been used indiscriminately.

3. Definition—Rated Suspension Spring Capacity is a load rating assigned to each spring installation and vehicle application which will provide adequate spring durability and vehicle stability under all intended load conditions. The value of the load rating must equal or exceed that portion of the maximum allowable force of gravity (usually called 'weight' and equaling mass times acceleration of gravity) at the ground which relates directly to the spring. Therefore, the load rating is based on the total of sprung and unsprung forces of gravity (usually called 'sprung weight' and 'unsprung weight') of the loaded vehicle.

4. Related Terms

4.1 Spring—Includes all types of suspension springs (such as: leaf, coil, torsion bar, rubber, pneumatic, etc.).

4.2 Load Rating—Is expressed in the SI (metric) unit of load, the newton (1N = 1kg · 1 m/s^2), determined vertically with the vehicle on a horizontal plane. Here the acceleration in the equation is the 'acceleration of gravity' which, by International Agreement, is generally accepted as 9.806 650 m/s^2 on the surface of the earth. What is commonly called 'weight' is actually the force (or load) which requires an equal, but opposite force, to restrain the mass of a body against free fall. This 'force of gravity' (or 'gravitational pull') is proportional to the mass. Thus a body of 1 kg mass will 'weigh'

1 kg · 9.806 650 m/s^2 = 9.806 650 N.

4.3 Spring Installation—Any spring as used in a particular suspension.

4.4 Vehicle Application—The usage of the vehicle as intended by the vehicle manufacturer.

4.5 Adequate Spring Durability—The endurance life characteristics regarded as sufficient by the vehicle manufacturer to satisfy customer requirements.

4.6 Adequate Vehicle Stability—The ride and handling characteristics of the vehicle regarded by the vehicle manufacturer as sufficient for safe operation.

4.7 Intended Load Conditions—The various payloads and payload distribution applied to the vehicle within the prescribed limits of 'Gross Vehicle Weight (GVW)' or 'Vehicle Full Rated Load,' and component capacities as established by the vehicle manufacturer.

4.8 Sprung Weight and Unsprung Weight—Defined in paragraphs 4.1.1 and 4.1.4 of SAE VEHICLE DYNAMICS TERMINOLOGY-HS J670, as published by SAE in 1978.

4.9 Loaded Vehicle—A vehicle which satisfies the conditions described in paragraph 4.7.

4.10 Maximum Allowable Force of Gravity (Weight) at the Ground—The Vehicle Full Rated Load or GVW acting at the ground.

4.11 Related Directly to the Spring—The load at the ground, which is transmitted through the suspension components to the spring and includes that portion of the unsprung weight.

The Rated Suspension Spring Capacity does not indicate spring payload capability, but rather the total of payload and vehicle weight. The assignment of a Rated Suspension Spring Capacity value is the responsibility of the vehicle manufacturer.

21 Speedometers and Tachometers

FACTORS AFFECTING ACCURACY OF MECHANICALLY DRIVEN AUTOMOTIVE SPEEDOMETER AND ODOMETERS—SAE J862 JAN89

SAE Information Report

Report of the Speedometer and Tachometer Committee, approved June 1963, third revision June 1981, and reaffirmed January 1989.

1. Scope—This report is concerned with factors which affect accuracy of distance indication and speed indication of automotive type odometer speedometers. It is the intent to supply information regarding all items which affect the instrument.

2. Distance Indication—Distance traveled is indicated by a numbered set of wheels, called the odometer, normally viewed through a slot in the dial of the speedometer. The wheels incorporate gear teeth which engage a pinion interposed between each set of wheels. The odometer can then be said to be a set of gears with numerals on their outer surface. The odometer is driven by a system of reduction gearing within the speedometer instrument. This reduction gearing is, in turn, driven by the speedometer cable core. SAE J678 (JUN84) specifies that 1000 or 1001 revolutions of the speedometer cable core shall cause a 1 mile indication on the odometer (616 through 630 revolutions, depending on the odometer gear train drive, for 1 km) if driven from the transmission. In wheel driven speedometers, the nominal number of wheel revolutions per mile (kilometer) shall cause a 1 mile (km) indication on the odometer. Because of the positive gear driven mechanism, no slippage error occurs in the odometer.

3. Factors Affecting Odometer Accuracy

3.1 Overall Assembly in Vehicle—The ideal of achieving the exact nominal value of speedometer cable core revolutions in one unit of distance of vehicle travel can seldom be realized. This becomes apparent when consideration is given to the overall design problem.

3.1.1 The speedometer cable core is driven by a gear called the take off pinion gear which is driven by the worm drive gear connected to the transmission output shaft which, in turn, drives the wheels through the differential gears. The distance traveled is dependent on the number of tire revolutions in a mile (kilometer). By experimentation, a nominal figure of tire revolutions per unit of distance is determined for the vehicle. Knowing the differential ratio, it is possible to calculate the necessary ratio in the transmission and the take off pinion gear for the speedometer cable core to achieve 1000 or 1001 revolutions per mile or 616 through 630 revolutions per kilometer. In the case of an odometer/speedometer driven directly from a wheel, it is then necessary to calculate the proper gearing within the speedometer head itself to achieve nominal conditions.

3.1.2 The exact ratio frequently results in a fraction which must, of course, be rounded to a whole number of teeth for the take off pinion gear. This gear selection must be accurate enough to assure that the odometer records actual distances traveled within ± 4% at 20, 40, and 55 mph (32.2, 64.4, and 88.5 km/h).[1]

3.1.3 Because of the different axle ratios used, it is necessary in any one line of automobiles to have a variety of take off pinion gears with different numbers of teeth. The number of teeth in the worm drive gear is not readily subject to change, since the gear is assembled within the transmission and is usually uniform for any transmission model.

3.2 Tires and Load—Tires are elastic members subject to variations from nominal size caused by manufacturing tolerances, temperature, inflation pressure, wear, speed, and loading. A tire will change size due to aging, after it is placed on a rim and inflated. These size variations, plus differences in construction material, and in the type of tread on tires from the same or different manufacturers, can result in a different number of tire revolutions per unit of distance. It is obvious that these variations from the nominal originally selected can directly affect distance indication.

3.3 Speed—A tire may experience as much as a 3% change in revolutions per unit of distance from a 30 mph (48.3 km/h) speed to a 90 mph (144.8 km/h) speed due to a change in rolling radius by centrifugal force.

3.4 Analysis and Summary—Fig. 1 is a chart which demonstrates the magnitudes of error which might occur in odometer readings. The average individual effect will be less than the maximum indicated by the chart since some of the conditions tend to compensate for others. For instance, tire wear and aging growth are compensating factors. Tire wear has the effect of increasing odometer indication and the tire aging growth will decrease the indication. When reading the chart, however, it should be appreciated that the errors may be additive.

3.5 Corrective Measures

3.5.1 In the foregoing, it has been shown that there are factors present which cannot be economically reduced or controlled which will cause distance indication errors. Some of the factors, however, can, in some degree, be controlled by proper tire inflation and replacement of worn tires.

3.5.2 Inadvertent installation of an improper pinion gear for a particular axle ratio will, of course, result in considerable error in odometer reading. Such a condition is, however, easily remedied by installation of the correct take off pinion gear.

3.5.3 A vehicle operator, especially one who modified a standard vehicle, can determine his percentage of odometer error by driving an accurately known distance at approximately nominal operating conditions of speed, load, temperature, and proper tire inflation. The reading of the odometer shall be compared to the known distance traveled. A reading greater than the distance traveled indicates a plus error, conversely a reading less than the distance traveled indicates a minus error. For example, if the odometer indicates 5.2 miles (km) as compared to a nominal 5.0 miles (km) distance traveled, an error of 0.2 divided by 5.0 or + 4% exists. A grossly plus error may be compensated for by using a take off pinion gear with a greater number of teeth or a large minus error may be corrected with a take off pinion gear with less teeth.

4. Speed Indication—Speed indication in an automotive speedometer is commonly accomplished through the use of a principle known as eddy current drive. The speedometer cable core drives a magnet shaft of the speedometer to which a permanent magnet is affixed. This magnet is located inside an aluminum or copper speed cup. The speed cup is attached to the same spindle on which the speedometer pointer is affixed. Also affixed to this spindle is a hairspring. A force applied to the speed cup results in a controlled reaction of the speedometer pointer. As the magnet rotates inside the speed cup, a force proportional to the speed of rotation is developed, thus providing measurement of speed indicated on the dial (Fig. 2).

5. Speedometer Calibration—The speedometer is calibrated at room temperature 75°F (23.9°C) by the instrument manufacturer. See SAE J678 for the recommended calibration tolerances.

---

[1] See SAE J678 for specific state or local requirements.

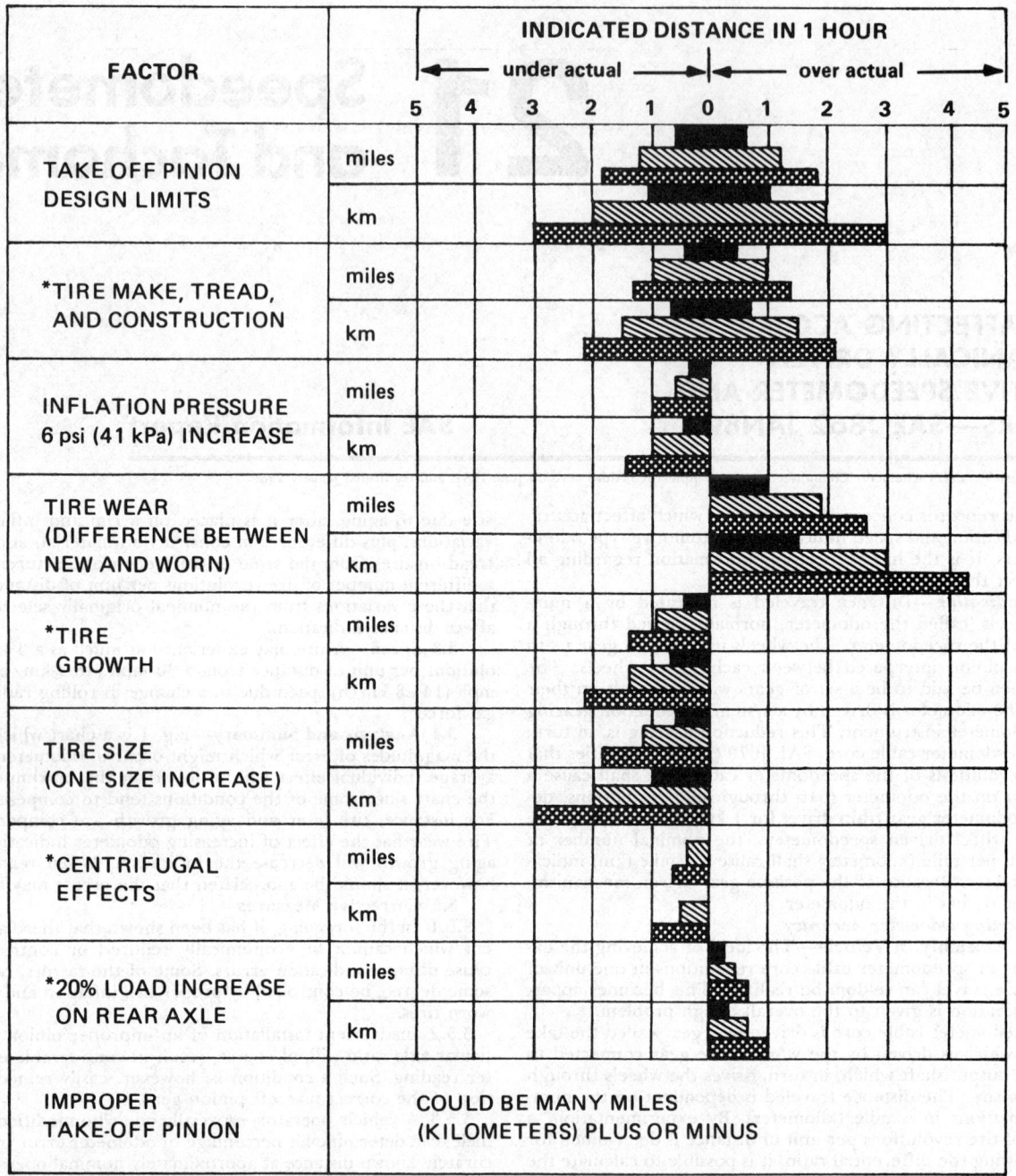

FIG. 1—FACTORS WHICH AFFECT ODOMETER READINGS
CROSS SECTION OF ALL U.S. MAKES

6. Factors Affecting Speedometer Accuracy

6.1 Drive Errors—Indication of speed is subject to the same errors as distance indication because the same speedometer cable core drives both the odometer and speed indicator. Some of the error may be compensated for by calibration of the speed indicator. Note should be taken that individual errors due to gearing, tire size, tire wear, tire pressure, speed, and load in the vehicle may be additive or subtractive to the speed indicator.

6.2 Temperature, Vibration, and Friction—The speed indication is affected by these factors. For information on the allowable variations due to these factors, see SAE J678 and SAE J1059.

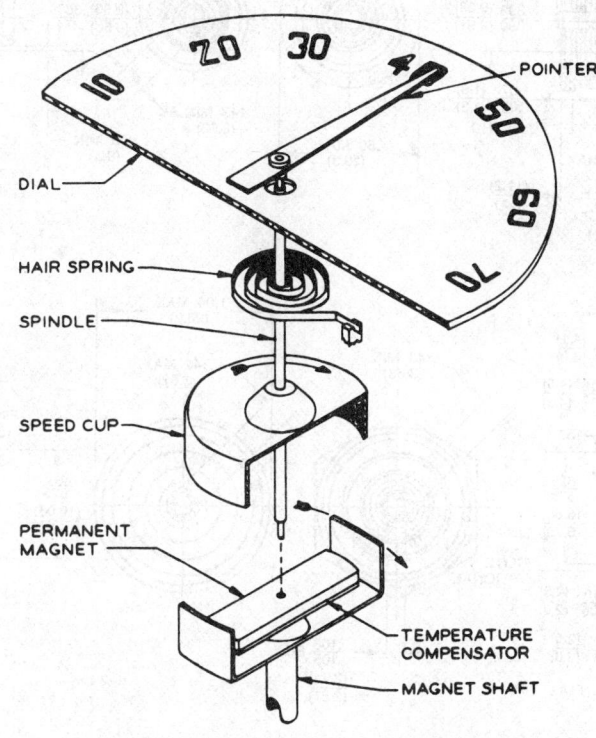

FIG. 2—EDDY CURRENT DRIVE

SPEEDOMETERS AND TACHOMETERS— AUTOMOTIVE—SAE J678 DEC88

SAE Recommended Practice

Report of the Parts and Fittings Division, approved January 1939, sixth revision, Speedometer and Tachometer Committee, June 1984, reaffirmed December 1988.

1. Scope—This SAE Recommended Practice applies to speedometers, odometers, and speedometer drives typical of passenger vehicles, buses, and trucks used for personal or commercial purposes. The method of determining wheel revolutions per unit distance (paragraph 2.1) and overall system design variation (paragraph 2.3.3) are applicable to passenger cars only. Comparable recommendations for trucks and buses are under development. The data of tachometers is applicable to vehicular use, as previously described, and also to stationary and marine engines and special vehicles.

2. Speedometer

2.1 Wheel Revolutions per Unit Distance[1]—The nominal number of vehicle wheel revolutions per mile (kilometer) is to be determined by the vehicle manufacturer and the information is to be used as a basis for design calculations of gearing and speedometer calibrations. Vehicle wheel revolutions shall be determined at 45 ± 2 mph (72.4± 3.2 km/h).[2] Tire inflation for measuring wheel revolutions is to be in accordance with the vehicle manufacturers' recommended pressure with tires at ambient test temperatures. This test is to be run immediately after a 5 mile (8 km) test run at 45 mph (72.4 km/h) to stabilize tire pressure, and with the vehicle at curb weight plus driver and one passenger [or 150 lb (68 kg)].

2.2 Types of Drive—The practice for cable-driven speedometers is to drive the system from either the transmission or the front wheel of the vehicle.

2.2.1 TRANSMISSION DRIVE—The design of a transmission drive for a speedometer requires that the vehicle manufacturer determine the nominal number of vehicle wheel revolutions per mile (kilometer). This information is then used as a basis for calculating the speedometer drive ratio. The number of teeth on the worm drive gear in the transmission and the number of teeth on the take-off pinion gear (drive ratio) are selected to provide a proper odometer drive and speed indication, as defined in paragraphs 2.3.1, 2.3.2, and 2.4.1. It is urged that those unfamiliar with conditions which affect vehicle wheel revolutions consult SAE J862 so that they may properly control conditions when determining the nominal value.

2.2.2 FRONT WHEEL DRIVE—This type of drive also requires that the nominal number of vehicle wheel revolutions be determined. The information is used to provide proper odometer drive gearing and for calibration purposes to achieve a proper speed indication.

2.2 Mileage Indication (Odometer)

2.3.1 INSTRUMENT DESIGN CONSIDERATION—Odometer tamper resistance should be considered when designing odometers for U. S. applications.

2.3.2 ALLOWABLE VARIATION WITHIN THE INSTRUMENT—The odometer shall indicate one mile for every 1000 or 1001 revolutions of the flexible shaft (one kilometer for every 616 thru 630 revolutions, depending on the odometer gear train drive) if driven from the transmission. The odometer of front wheel drive units shall indicate one mile (kilometer) when the flexible shaft is rotated a specified number of revolutions, as determined by the vehicle manufacturer.

---

[1] See also SAE J966.
[2] Direct metric conversions in parentheses.

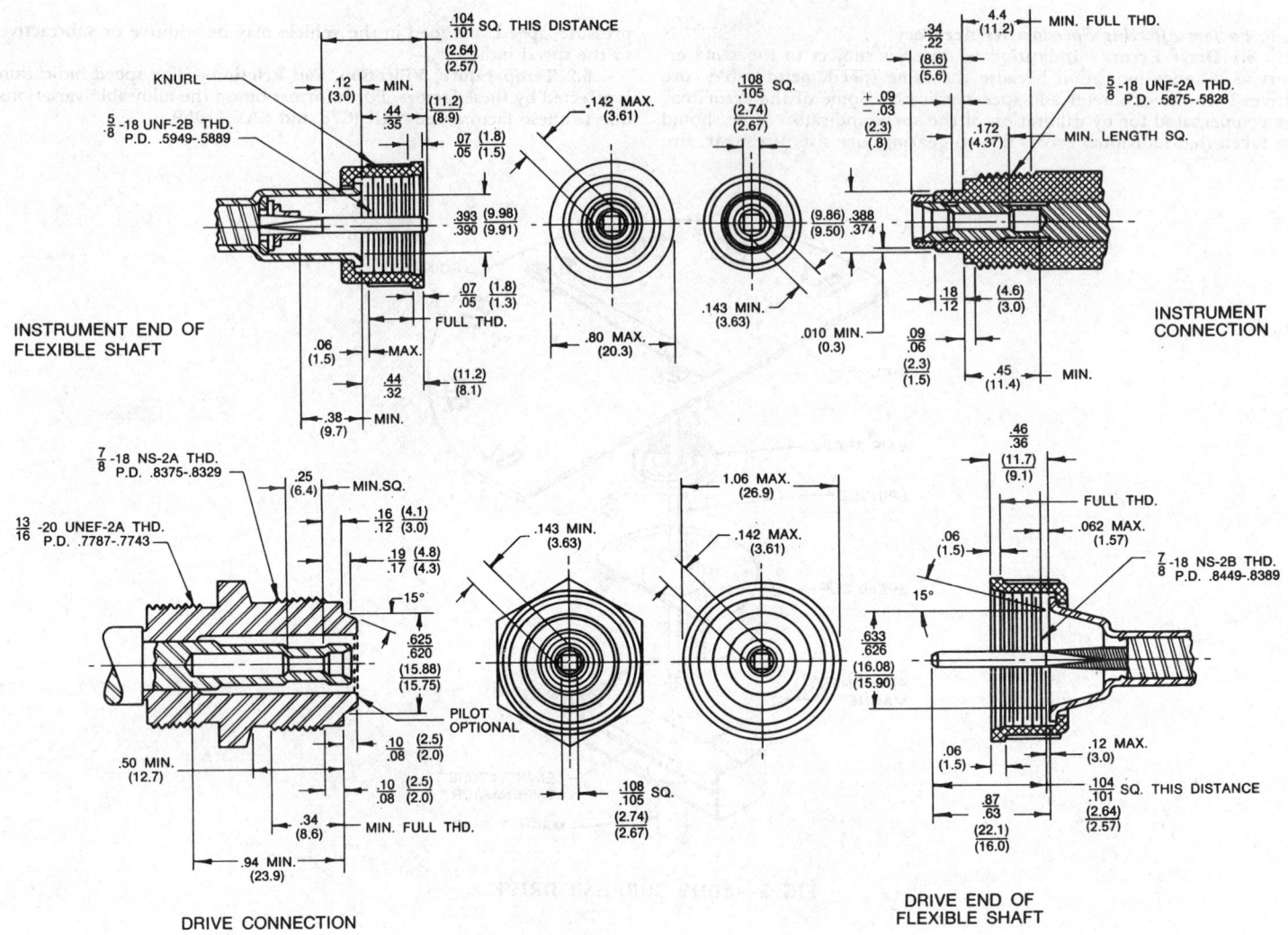

FIG. 1—LIGHT-DUTY DRIVE (FOR USE WITH EDDY CURRENT TYPE INSTRUMENTS)

2.3.3 OVERALL SYSTEM DESIGN VARIATION—The vehicle manufacturer shall specify odometer drive ratios that will produce one unit of distance indication within ±4% for each actual unit of distance travelled at 20, 40, and 55 mph (32.2, 64.4, and 88.5 km/h).[3] The design limits thus derived should not, however, be construed as absolute. Factors which cause variation from nominal wheel revolution per unit distance travelled under operating conditions are covered in SAE J862. It is recommended that SAE J862 be studied to determine probable effects on odometer accuracy under operating conditions.

2.4 Speed Indication

2.4.1 ALLOWABLE VARIATION WITHIN THE INSTRUMENT—Two philosophies exist in calibrating the instrument for overall vehicle system accuracy.

One: Calibrate the speedometer at the center of the graduations so that overall vehicle system accuracy will be within a specified amount, as established by the vehicle manufacturer. The intent is to have an instrument with no system bias. Generally ±4.0 mph (±6.4 km/h) is an acceptable limit for the vehicle system.

Two: Bias the speedometer calibration high so that overall vehicle system calibration will be higher than true vehicle speed. Generally +6.0, −0.0 mph (+9.6, −0.0 km/h) is an acceptable limit for the vehicle system.

The specific calibration of the instrument relative to vehicle system limits should be determined by the instrument manufacturer and vehicle manufacturer.[4] The speedometer head should be calibrated in approximately the same angular position that it will have when mounted in the vehicle.

Factors influencing vehicle system accuracy are described in Fig. 2 of SAE J862. In addition, the following factors also affect the speedometer head accuracy:

Temperature—The speed indication will vary with changes in temperature. Temperature affects the reaction between the speed cup and magnet—Refer to Fig. 3 in SAE J862. To counteract this effect, an element called a temperature compensator is incorporated in all speedometers. Due to variations in materials which cannot be perfectly controlled, the change in indicated speed from temperature compensation may be as much as ±4% between temperatures of 20° and 130°F (−7° to 55°C).

Vibration—Another condition which affects speed indication is vibration. At 55 mph (88.5 km/h) an effect of 0.5 to 1 mph (0.8 to 1.6 km/h) may be noted.

Friction—A minor error which affects speed indication is frictional lag. This is a condition in which the speedometer will not read exactly the same under acceleration and deceleration. Investigations indicate that the error from frictional lag is less than 1 mph (1.6 km/h) at 55 mph (88.5 km/h).

---

[3] Some states and local regulations require an accuracy of ±3.75% under specified test conditions for vehicles introduced into rental or leasing markets. Test conditions are specified in the National Bureau of Standards Handbook 44.

[4] The following regulatory requirements for speedometers should also be consulted:

Australia D Regulation	1019
Japan Article	46
United Nations	ECE R 39
European Economic Comm.	EEC 75/443

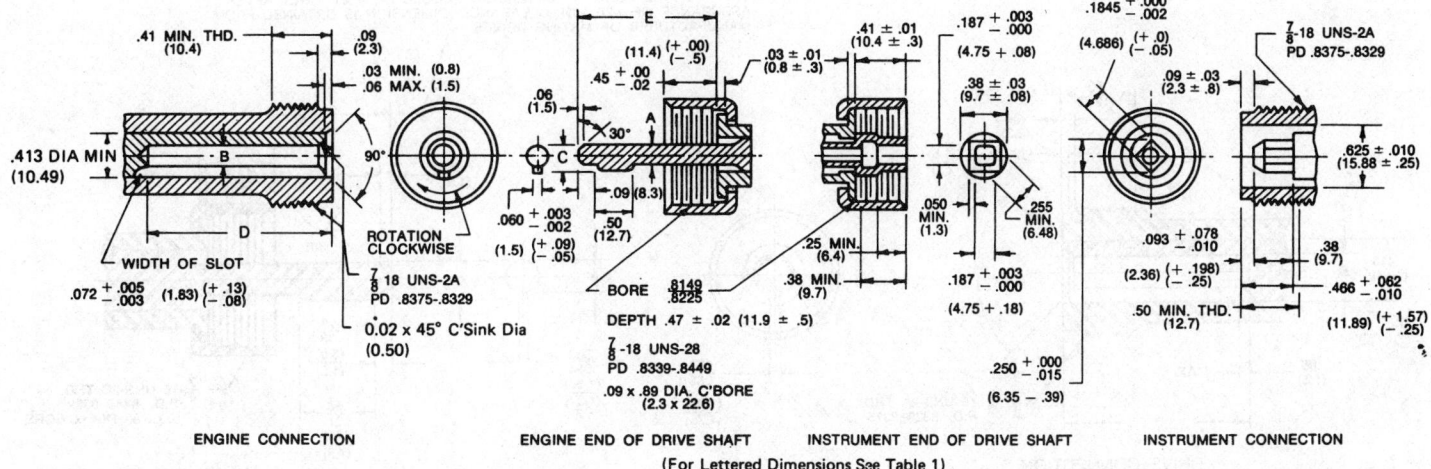

FIG. 2—SAE REGULAR DRIVE FOR SPEEDOMETERS AND TACHOMETERS (FOR LETTERED DIMENSIONS SEE TABLE 1)

2.5 Identification—Identification should appear on all speedometers. It should consist of a distinct marking of model number or part number on the instrument and the date of its manufacture. All kilometer speedometers should be identified by marking "Kilo" on the back of the speedometer case, unless identification of sufficient nature appears on the face dial or other suitable means of identification is applied, such as marking the speedometer head with a red dot. Drive ratio information shall appear on front-wheel drive instruments in an area readily visible when the instrument is removed.

3. Speedometer and Tachometer Drive—Flexible shafts for driving mechanical speedometers and tachometers shall consist of a flexible casing and a flexible cable capable of transmitting motion from a suitable takeoff to operate the instrument. Recommended takeoff and instrument fittings are shown in the following illustrations.

In routing of flexible shafts, bends of less than 6 in (150 mm) radius should be avoided. Figs. 1–5 and Table 1 give dimensions for SAE light, regular, square, and heavy-duty drive for speedometers and tachometers.

3.1 Miscellaneous Drive Ends of Flexible Shafts—Specific detail and dimensions to be determined between user and supplier.

3.1.1 Plug Type Lower Ferrule—This type of drive end (shown on Fig. 6) is widely used for speedometer drives in the automotive industry. No specific dimensional standard is recommended because dimensions vary depending on design of transmission with which it is used. The end of the cable has a standard 0.101–0.104 in (2.56–2.64 mm) square for passenger car service.

TABLE 1—DIMENSIONS FOR REGULAR AND HEAVY-DUTY DRIVE FOR SPEEDOMETER AND TACHOMETER
Ref. Figs. 2, 4, and 5

	SAE Size		
	5/32 in	3/16 in	13/64 in
A Shaft Dia.	0.152 (3.86 mm) +0.003 (+0.08 mm) −0.002 (−0.06 mm)	0.188 (4.78 mm) +0.000 (+0.00 mm) −0.005 (−0.13 mm)	0.203 (5.16 mm) ±0.003 (±0.08 mm)
B Drilled Hole	0.161 (4.09 mm) (No. 20 Drill)	0.191 (4.85 mm) (No. 11 Drill)	0.213 (5.41 mm) (No. 3 drill)
C Tang. ±0.005 (±0.12 mm)	0.200 (5.08 mm)	0.239 (6.07 mm)	0.245 (6.22 mm)
D Min.	Long Extension 1.26 (32 mm) Short Extension 0.94 (23.88 mm)		
E ±0.12 (±3.05 mm)	Long Extension 1.12 (28.45 mm) Short Extension 0.75 (19.05 mm)		

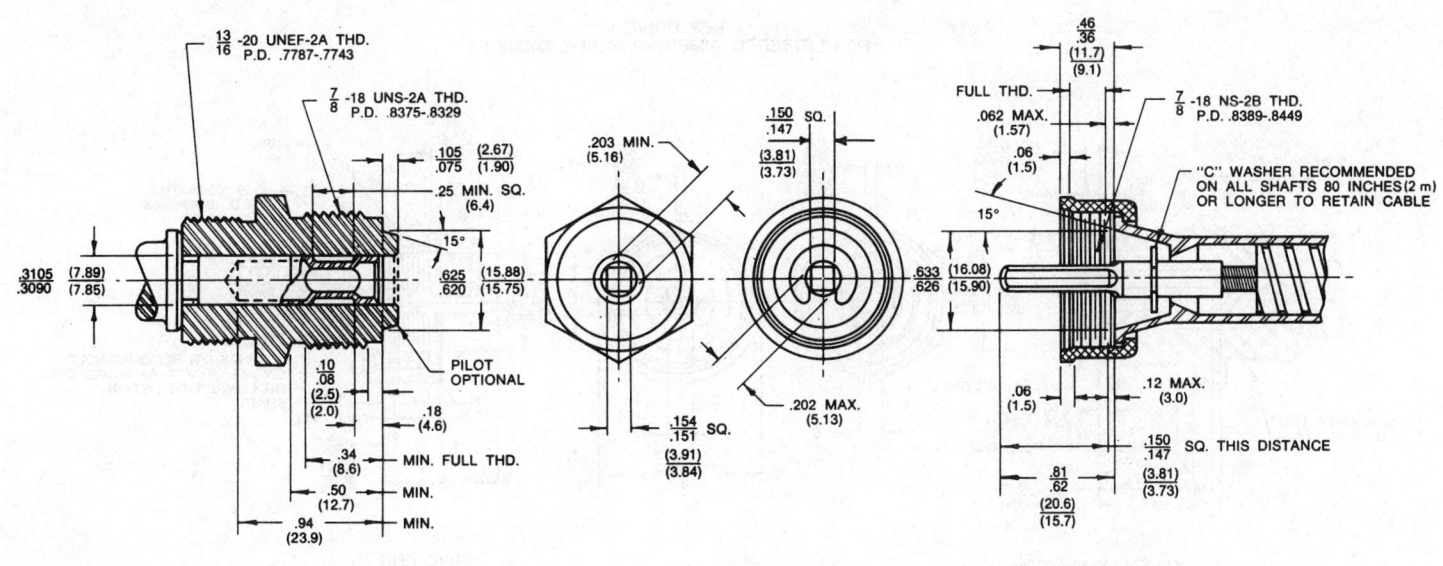

FIG. 3—SQUARE DRIVE

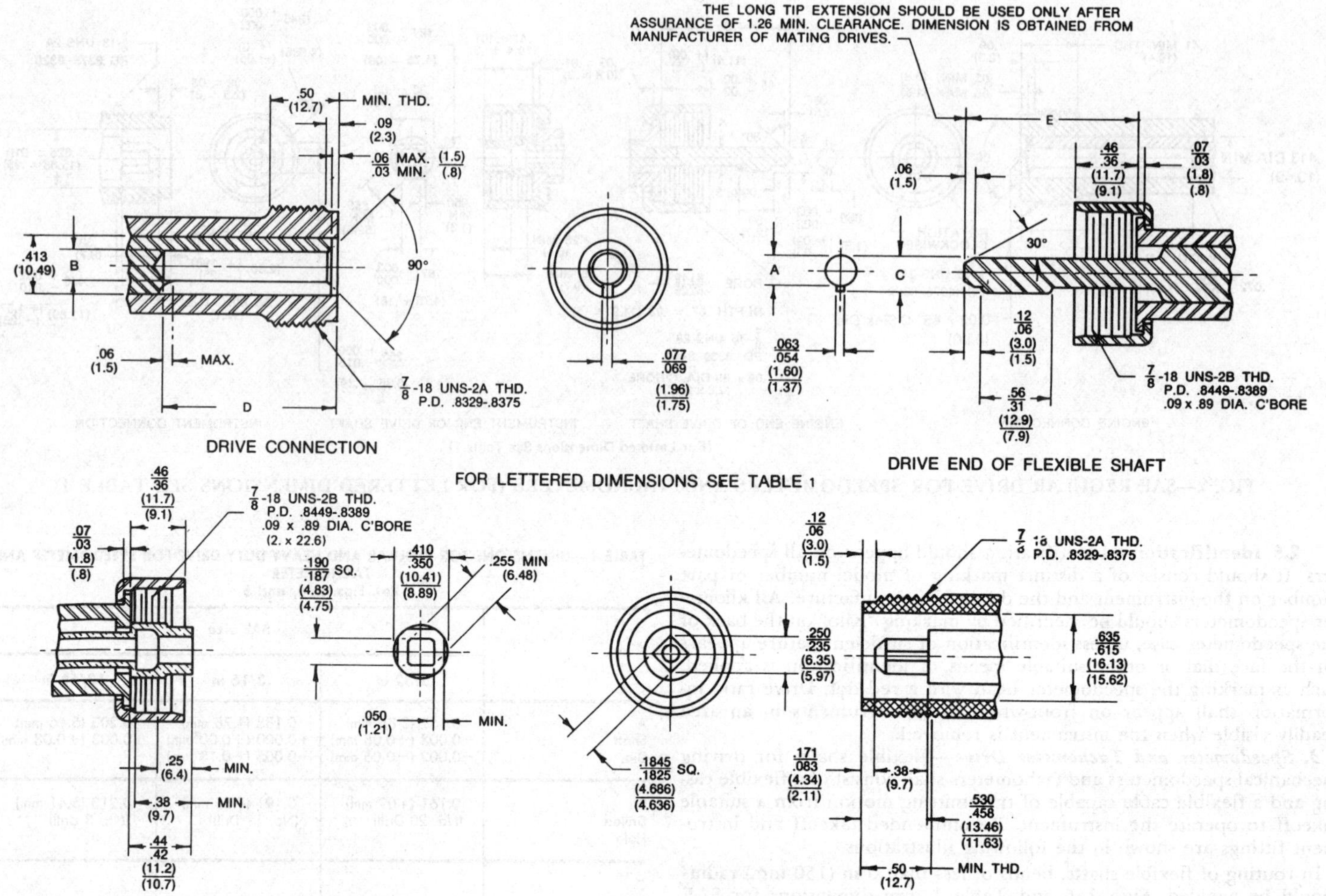

21.06

FIG. 4—HEAVY-DUTY DRIVE (FOR USE WITH CENTRIFUGAL TYPE INSTRUMENTS)

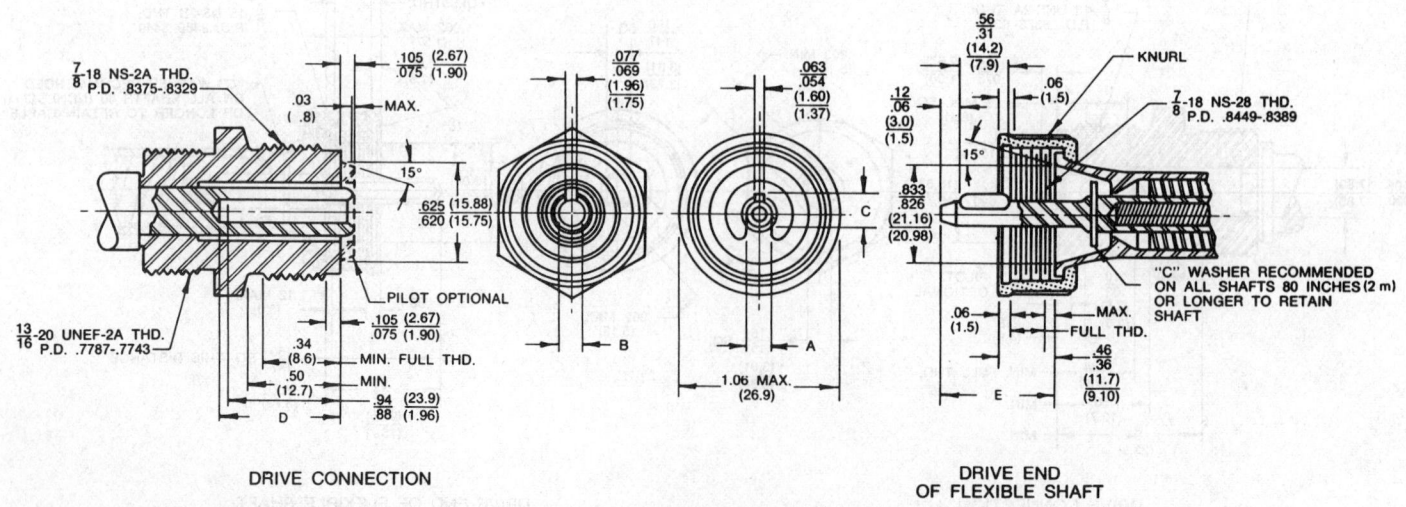

FIG. 5—HEAVY-DUTY DRIVE (FOR USE WITH EDDY CURRENT TYPE INSTRUMENTS AND KEY DRIVE)

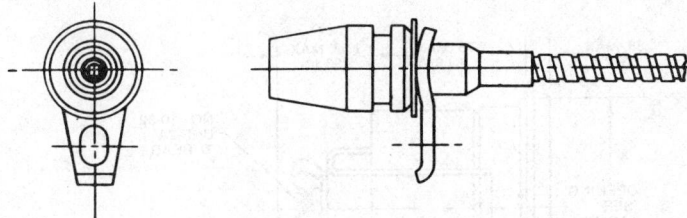

FIG. 6—PLUG TYPE LOWER FERRULE

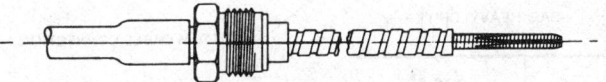

FIG. 7—FRONT-WHEEL DRIVE

3.1.2 FRONT-WHEEL DRIVE (SEE FIG. 7)—Some speedometer drives are taken from the vehicle front wheel. This involves a spindle machined to accept the flexible shaft drive end and providing a watertight joint. The cable terminates in a standard 0.101–0.104 in (2.56–2.64 mm) square for passenger car service. No specific dimensional standard is recommended because of the many variations in front wheel suspensions, steering mechanisms, and spindle designs which will affect the flexible shaft design.

3.1.3 QUICK CONNECT UPPER FERRULE—No specific dimensional standard is recommended because dimensions vary depending on the design of speedometer with which it is used. The end of the cable has a standard 0.101–0.104 in (2.56–2.64 mm) square for passenger car service.

4. Tachometers

4.1 Mechanical Eddy Current Tachometers

4.1.1 GENERAL—Illumination, waterproofness, and corrosion resistance are not within the scope of this recommended practice.

4.1.1.1 *Tachometer Drive Connections*—SAE regular, optional and heavy. Clockwise and counter clockwise rotation.

4.1.1.2 *Tachometer Dials*—Recommended dial ranges in rpm are: 0–2500, 0–4000, 0–6000, and 0–8000, for a minimun of 270 deg full deflection, clockwise or counter clockwise rotation. Light graduations, numerals, and pointers on a dark background and with graduations having outside numerals for highest accuracy in scale readings are also recommended.

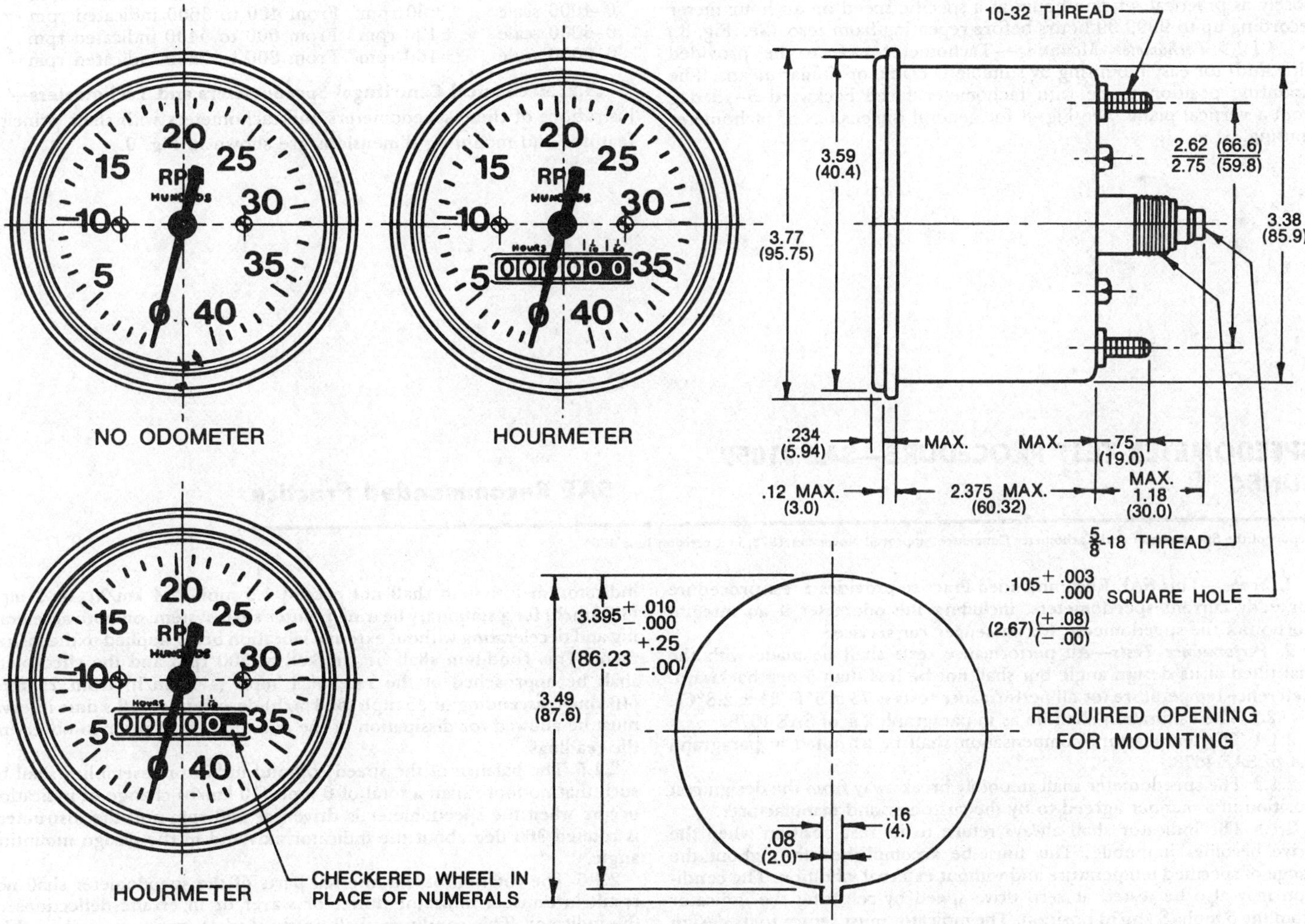

FIG. 8—MECHANICAL EDDY CURRENT TACHOMETER

21.08

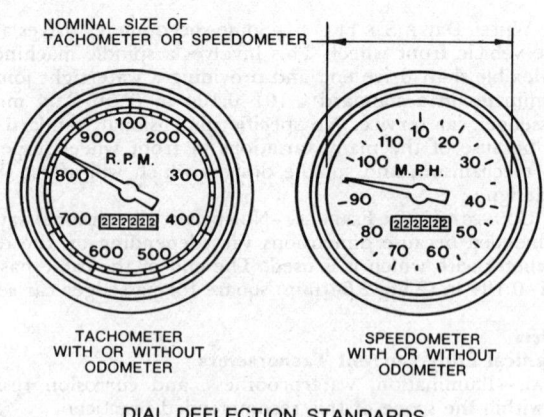
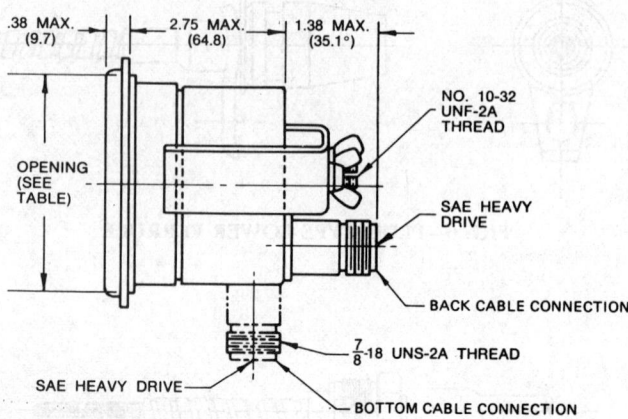

Tachometer or Speedometer Size, In.	Outlets		Size of Opening Required, Dia. +0.010 (+.25) −0.000 (−.000)
	Standard	Optional	
3	Back	Bottom	3.395 (86.23)
4	Bottom	Back	4.395 (111.63)
6 and over	Bottom	Back	5.250 (131.35)

FIG. 9—MECHANICAL CENTRIFUGAL TACHOMETERS AND SPEEDOMETERS

4.1.1.3 Tachometer Drive Ratio—Tachometer to indicate two times flexible shaft speed and to be driven 0.5 times engine speed.

4.1.1.4 Tachometer With Hour Meter—Hour meter to indicate, as closely as practical, engine hours at a specific speed on an hour meter recording up to 9999.99 hours before repeating from zero. (See Fig. 8.)

4.1.1.5 Tachometer Mounting—Tachometer case to be provided with studs for easy mounting by suitable U-clamp or similar means. The mounting position to be with tachometer faced backward 5-45 deg from a vertical plane. See Fig. 8 for general dimensions of tachometer housing.

4.1.1.6 Tachometer Calibration—Recommended calibration limits are as follows:

0–2500 scale	±50 rpm	From 250 to 2250 indicated rpm
0–4000 scale	±80 rpm	From 400 to 3600 indicated rpm
0–6000 scale	±120 rpm	From 600 to 5400 indicated rpm
0–8000 scale	±160 rpm	From 800 to 7200 indicated rpm

4.2 Mechanical Centrifugal Speedometers and Tachometers—Illustrations of these speedometers and tachometers with their principal features and mounting dimensions are shown in Fig. 9.

SPEEDOMETER TEST PROCEDURE—SAE J1059
JUN84
SAE Recommended Practice

Report of the Speedometer and Tachometer Committee, approved November 1973, first revision June 1984.

1. Scope—This SAE Recommended Practice provides a test procedure for eddy current speedometers, including the odometer if an integral portion of the speedometer, for passenger car service.

2. Performance Tests—All performance tests shall be made with the dial tilted at its design angle but shall not be less than 5 deg backward. Reference temperature for all performance tests is 75 ± 5°F (24 ± 2.8°C).

φ 2.1 The calibration shall be as in paragraph 2.4 of SAE J678.

2.1.1 The temperature compensation shall be as noted in paragraph 2.4 of SAE J678.

2.1.2 The speedometer shall smoothly break away from the design rest position in a manner agreed to by the customer and manufacturer.

2.1.3 The indicator shall always return to its rest position when the drive becomes immobile. This must be accomplished throughout the range of specified temperature and without external vibration. The condition may also be tested at zero drive speed by releasing the indicator from the 5 mph (8 km/h) position. The indicator must return to its design rest position.

2.1.4 The total backlash (hysteresis) in an instrument with a live bearing indicator shaft system shall not exceed 1.5 mph (2.4 km/h), or 3 mph φ (4.8 km/h) for a stationary bearing pointer shaft system, on both accelerat- φ ing and decelerating without external vibration being applied to the instrument. This condition shall be checked at 500 rpm and the checkpoint shall be approached at the rate of 1 mph (1.6 km/h) from 25 mph (40 km/h) ascending or 35 mph (56 km/h) descending. A 2 s time interval must be allowed for dissipation of the damping effect prior to observing the readings.

2.1.5 The balance of the speed cup and indicator assemblies shall be such that no more than a total of 6 mph (10 km/h) change of indication occurs when the speedometer is driven at 500 rpm and the instrument is rotated 360 deg about the indicator axis and in the design mounting angle.

2.1.6 The rotation of the internal parts of the speedometer shall not result in unusual indicator flutter or waver or in erratic deflections of the indicator. This condition shall be checked at random speeds and be observed from a 2 ft (0.6 m) distance perpendicular to and at 45 deg angle to the dial. During this test, the speedometer shall be driven by means which exclude excitement caused by or transmitted through the speedometer cable.

2.1.7 The speedometer shall be so damped that when being driven at 500 rpm and the indicator is physically displaced to 70 mph (110 km/h), and released, the indicator shall reverse direction not more than four times; and if it does not reverse direction, it shall return to the original reading within 1.5 s.

ϕ 2.1.8 The torque required to rotate the magnet shaft and odometer shall not exceed 0.00752 lbf·in (0.00085 N·m) for single odometer units
ϕ not more than 0.01328 lbf·in (0.00150 N·m) with total and trip odometers. During these tests, all odometer numerals shall be in operation. The test shall be conducted between a drive speed of 3 and 1000 rpm at a temperature of 75 ± 5°F (24 ± 2.8°C).

3. Vibration Tests

3.1 Test speedometers shall be vibrated for 3 h. For 1 h in each direction along three mutually perpendicular axis with a total excursion of 0.020 in (0.5 mm) and a frequency varying 16–50 c/s, the frequency shall be cycled from 16 to 50 to 16 over a 2 min period. The mounting to be at design angle but no less than 5 deg backward.

3.2 After completion of the vibration test, the performance deviation listed in Section 6 will be permitted.

4. Laboratory Endurance Tests

4.1 Endurance life tests for the speedometer shall be 50 000 miles (80 000 km) or a duration test for an equivalent number of driveshaft revolutions, with speed and temperature cycling as follows:

4.1.1 SPEED CYCLING—The speed shall be cycled from 167 rpm reverse to 1500 rpm forward to 167 rpm reverse every 2 min.

4.1.2 TEMPERATURE CYCLING—Elevate the test chamber to 120 ± 5°F (49 ± 2.8°C) each day for three consecutive days, 6 h per day, and speed cycle as per paragraph 4.1.1. The speedometers are operated at room temperature and 1500 rpm for the remainder of the 24 h period.

ϕ 4.1.3 Elevate the test chamber to 170–180°F (76–82°C) for 6 h one day each week. The speedometers are not operated during this heat cycle. The speedometers are then continued operating at room temperature and 1500 rpm for the remaining hours of the day.

4.1.4 Reduce the test chamber to 0 ± 5°F (−17.8 ± 2.8°C) for 6 h one day each week. During the first hour, the speedometer shall not be operated. During the next 5 h, the speedometer shall be operated according to the speed cycling test of paragraph 4.1.1, except that the maximum speed shall be 1000 rpm. The test sample is then to be operated at room temperature and 1500 rpm for the remainder of the day. Throughout the cold test and any subsequent testing, the speedometer shall not seize or exhibit an appreciable increase in noise level when tapped by a wooden drafting pencil held loosely in the fingers.

4.1.5 Test speedometers shall be run at room temperature and 1500 rpm constant speed for two days to complete the weekly cycle.

4.2 After completing the life test (paragraph 4.1), the performance shall be as specified by Section 6.

5. Vehicle Testing—Test speedometers shall be installed in test vehicles and subjected to a 25 000 mile (40 250 km) (or equivalent driveshaft revolutions) general endurance road test whereby a great variety of road conditions are encountered. After completing the vehicle test, the performance deviation permissible in Section 6 will be permitted.

6. Performance Checks after Vibration, Endurance, or Vehicle Tests—After completion of the endurance, vibration, or vehicle test, the performance of the instruments shall be checked against the readings taken during the initial performance check. Deviations are allowed as follows:

6.1 The calibration shall not deviate more than 3% of the test speed from the limits of paragraph 2.4 of SAE J678. ϕ

6.2 The temperature compensation shall be as stated in paragraph ϕ 2.4 of SAE J678.

6.3 The indicator must have a positive smooth breakaway movement from its design rest position, as agreed to by the manufacturer and customer.

6.4 The indicator must return to design rest position.

6.5 The indicator backlash (hysteresis) shall not exceed 2.5 mph (4.0 km/h) for live bearing units or 4.0 mph (6.5 km/h) for a stationary bearing unit.

6.6 The rotation of the internal parts of the speedometer shall not result in unusual indicator flutter or waver or in erratic deflections of the indicator. This condition shall be checked at random speeds and be observed from a 2 ft (0.6 m) distance perpendicular to and at a 45 deg angle to the dial. During this test, the speedometer shall be driven by means which exclude excitement caused by or transmitted through the speedometer cable.

6.7 The damping shall be within the original specification, except five reversals of direction shall be allowed.

6.8 The drive torque shall be no more than 0.00885 lbf·in (0.00100 ϕ N·m) for single odometer units or 0.01549 lbf·in (0.00175 N·m) for ϕ double odometer units. Tests shall be as described in paragraph 2.1.8. ϕ

6.9 The balance of the indicator assembly may change no more than 2 mph (3 km/h), as compared to the reading obtained as per paragraph ϕ 2.1.5.

ELECTRIC SPEEDOMETER SPECIFICATION—ON ROAD—SAE J1226 FEB83

SAE Recommended Practice

Report of the Speedometer and Tachometer Committee, approved February 1983.

1. Scope—This SAE Recommended Practice covers electric speedometer systems for general on-road (passenger car, multi-purpose passenger vehicle, truck, and bus) applications.

2. Purpose—To recommend design practices and test procedures for electric speedometers used in an on-road vehicle environment using the methods of determining wheel revolutions per unit of distance specified in SAE J678,[1] paragraph 2.1, and SAE J966.

3. Electric Speedometer System—A typical electric speedometer system consists of an indicating unit and sender unit with inter-connecting wiring. The indicating unit is made up of a speed indicator and distance indicator (odometer). In practice, the speed indicator is very similar to a conventional electric tachometer where the indication is proportional to the frequency of the input pulses. The distance indicator may utilize a counter such as described in SAE J862, paragraph 2, and may be driven by a stepping motor through a system of reduction gearing or by a solenoid. The stepping motor or solenoid in turn is driven by voltage pulses generated by an amplifier/divider circuit or from a switch in the sender. The divider circuit serves to reduce the frequency of the input pulses from the sender that are applied to the stepping motor or solenoid. The sender unit in most applications will be one of the following types: permanent magnet generator, magnetic switch, or magnetic sensor. Generally, they are either transmission or wheel mounted.

4. Factors Affecting Odometer System Accuracy

4.1 General—The overall accuracy of an electric speedometer distance indicator is subject to the same inaccuracies as those described in SAE J862. While the average effect shown in Fig. 1 of SAE J862 may not be directly applicable to trucks, the effect of being either over or under is the same.

4.2 Transmission Driven Senders—Referring to Fig. 1 of SAE J862, the factor "Take-Off Pinion Design Limits" is applicable to transmission driven senders only. Since the number of revolutions per unit distance for transmission driven senders is generally fixed at 1000 rpm at 60 mph (96.6 km/h), the accuracy of the odometer with respect to the rotation of the sender may be closely controlled by proper gearing or electrical division within the indicating unit itself.

4.3 Wheel Mounted Senders—Corrections for variations in rolling radius may be made by proper selection of the reduction gearing between the stepper motor and the odometer or of the sender excitation frequency. Where precise selection of the sender excitation frequency is possible,

---

[1] When reference is made to existing SAE specifications, the latest revision shall apply.

the overall odometer accuracy can be very high. However, rounding off of the reduction gearing and/or electrical division in the indicating unit can produce the same type of error that is obtained by rounding off any fractional number of teeth on the speedometer drive gear in the transmission.

5. *Distance Indication (Allowable System Variation)*

5.1 Overall Design Variation—The overall odometer accuracy shall be within −4% to +4% for each actual unit of distance of travel over the operating range of the instrument. The design limits should not, however, be construed as absolute under all operating conditions. Factors which cause variations from nominal wheel revolutions under operating conditions are covered in SAE J862. It is recommended that SAE J862 be studied to determine probable effects under service conditions.

5.1.1 ODOMETER INPUT—Inaccuracies contributed by the odometer and associated circuitry will be negligible with proper selection of the number of pulses per mile (kilometer) supplied to the unit. The actual number of sender pulses per mile (kilometer) can be negotiated between the user and the manufacturer.

5.2 Operating Range—The odometer shall meet the requirements of paragraphs 6.1 and 9.1.2 at any operating speed above 5 mph (8.05 km/h).

6. *Distance Indication (Allowable Instrument Variation)*

6.1 With nominal input frequency applied, the odometer shall indicate calculated mileage within ±0.3%.

6.2 For vehicles under 16 000 lb GVW, Federal Motor Vehicle Safety Standard 127 imposes certain requirements on the odometer and should be examined for the latest information.

7. *Factors Affecting Speed Indication*

7.1 Transmission Driven—The overall accuracy of speed indication is affected by the same errors as distance indication since the same sender drives both the odometer and speed indicator.

7.2 Wheel Driven—The speed indicator is calibrated for a nominal sender excitation frequency, therefore, the overall accuracy of speed indication is determined by the speed indicator calibration limits.

7.3 All Types—Speed indication may be affected by changes in ambient temperature and voltage.

8. *Speed Indication (Allowable Instrument and Sender Variation)*—The speed indicating unit shall be within the limits shown in Table 1 or Table 2 (consult speedometer vendor to determine proper table) when the sender is driven at the specified frequency at a temperature of 75 ± 5°F (24 ± 3°C) with nominal voltage applied. When analog displays are used, the spacing of the graduations on the speedometer dial may be non-linear to compensate for non-linearity in the system. It should be noted that variations in speedometer reading on the road may lie outside the limits of Table 1 due to the factors described in Fig. 1 of SAE J862. All calibration of speedometers during manufacture shall be made with the instrument in approximately the same angular position that it will have when mounted in the vehicle. See Environmental Conditions for allowable variation within the instrument due to changes in ambient temperatures and voltage.

TABLE 1—SPEED INDICATION LIMITS FOR SPEEDOMETER FULL SCALE = 85 mph (136.79 km/h) BIASED

Actual Speed	mph	20	40	55
Indicated Speed	mph	18.9—22.4	39.5—43.0	55—58.4
Actual Speed	km/h	30	70	90
Indicated Speed	km/h	28.1—33.8	69.3—75.0	90—95.6

TABLE 2—SPEED INDICATION LIMITS FOR SPEEDOMETER FULL SCALE = 85 mph (136.79 km/h) NOT BIASED

Actual Speed	mph	20	40	55
Indicated Speed	mph	18.3—21.7	38.3—41.7	53.3—56.7
Actual Speed	km/h	30	70	90
Indicated Speed	km/h	27.2—32.8	67.2—72.8	87.2—92.8

NOTE: The speedometer accuracies are a percentage of full scale. If for example a 60 mph full scale is used, the speedometer accuracy will be ± 1.2 mph. The total speedometer tolerance may be added to the actual speed to obtain the limits, as in Table 1, or may be split and added to and subtracted from the actual speed to obtain the limits, as in Table 2. Consult speedometer manufacturer to determine which method is used.

9. *Effects of Environmental Conditions*

9.1 Temperature (Allowable System Variation)

9.1.1 SPEED INDICATION—With nominal voltage applied, the speed indication shall not vary more than ±2% of full scale from the reading determined in paragraph 8 while the indicating unit is operating over the range of +20 to +130°F (−7 to +54°C) and the sender is operating over the range of −40 to +280°F (−40 to +138°C). No permanent damage shall result from operating the indicating unit in a range of −40 to +180°F (−40 to +82°C). Internal lighting, if any, shall not be operating during this test.

9.1.2 DISTANCE INDICATION—With nominal voltage applied, the distance indication shall not vary more than ±0.3% from a reading obtained at 75 ± 5°F (24 ± 3°C) while the instrument is operating over the range of +20 to +130°F (−7 to +54°C) and the sender is operating over the range of −40 to +280°F (−40 to +138°C). No permanent damage shall result from operating the indicating unit in a range of −40 to +180°F (−40 to +82°C). Internal lighting, if any, shall not be operating during this test.

9.2 Temperature Extremes (Sender Only)—It will be necessary to evaluate the specific application to specify the allowable temperature extremes.

9.3 Storage (Indicating Unit Only)—A 4 h exposure of the indicating unit to a temperature of −40 to +185°F (−40 to +85°C) shall result in no more than ±1% of full scale permanent calibration change from the reading determined in paragraph 8. The rate of temperature change during this test shall not exceed 3.6°F (2°C) per minute.

9.4 Voltage Variation (Indicating Unit)

9.4.1 SPEED INDICATION—The indication shall not change more than ±1% of full scale indication from the reading determined in paragraph 8 within the following voltage ranges:

12 Volt System	24 Volt System
12 to 16 VDC	24 to 32 VDC

Twelve and 24 V indicating units shall not change more than ±3% of full scale indication from the reading obtained in paragraph 8 at 11 and 22 V respectively.

9.4.2 DISTANCE INDICATION—At 75 ± 5°F (24 ± 3°C) the distance indication shall not vary more than ±0.3% when operating within the voltage ranges given in paragraph 9.4.1.

9.5 Abnormal Voltage Conditions

9.5.1 TRANSIENT VOLTAGE PROTECTION—The indicating unit shall be capable of withstanding supply voltage transients without permanent damage and shall remain within the calibration specification of paragraphs 6.1 and 8 at the conclusion of this test.

The instrument shall be connected and operated for a total of 1 h with a means provided to impress upon the nominal battery voltage a repetitive rectangular voltage pulse of plus and minus six times nominal battery voltage with a duration of 300 μs and 1% duty cycle with a current of no more than 1.0 A.

For applications with transient voltages having a magnitude, duration, or duty cycle exceeding the above requirements, contact the instrument manufacturer for recommendations.

9.5.2 OVERVOLTAGE AND REVERSE POLARITY—Provisions for protection against booster starts with double battery voltage and/or reversed polarity must be negotiated between the user and the manufacturer.

9.6 Moisture Resistance

9.6.1 HUMIDITY (INDICATING UNIT)—Indicating unit shall withstand exposure to 95% relative humidity at 100°F (38°C) for 48 h.

9.6.2 SALT SPRAY (SENDER UNIT)—Sender units shall be corrosion resistant and shall withstand a salt spray (fog) test of 48 h duration with 5% salt solution (Reference ASTM B117-73).

9.6.3 PERFORMANCE DEGRADATION—Allowable degradation during humidity and salt spray tests (paragraphs 9.6.1 and 9.6.2) is negotiable between the user and the manufacturer.

9.7 Vibration Test (Indicating Unit)—The indicating unit shall be capable of withstanding without mechanical or electrical failure, 3 h of vibration, 1 h along each of the three mutually perpendicular axes. One axis is to be parallel to the indicator shaft. The vibration test shall be run at a double amplitude (peak to peak) of 0.030 in (0.76 mm) with the frequency varying from 10–30–10 Hz at intervals of 1 min. After completion of test, the calibration shall remain within tolerances as specified in paragraphs 8 and 6.1.

9.8 Vibration Test (Sender Only)

9.8.1 TRANSMISSION MOUNTED—The sender shall be capable of withstanding 6 h of vibration without mechanical or electrical failure, 2 h along each of the three mutually perpendicular axes. One axis is to be perpendicular to the mounting plane. The vibration test shall be run at a double amplitude (peak to peak) of 0.020 in (0.51 mm) with the frequency varying from 10–120–10 Hz at intervals of 1 min.

9.8.2 Wheel Mounted Sender—The sender shall be capable of withstanding 6 h of vibration without mechanical or electrical failure, 2 h along each of the three mutually perpendicular axes. One axis is to be perpendicular to the mounting plane. The vibration shall be run at a double amplitude (peak to peak) of 0.040 in (1.02 mm) with frequency varying from 10–120–10 Hz at intervals of 1 min.

9.9 Shock Test (Indicating Unit Only)—The indicating unit shall be capable of withstanding without mechanical or electrical failure the following series of shocks and still maintain the calibration tolerances specified in paragraphs 8 and 6.1. The indicating unit shall be subjected to one shock in each direction along each of three mutually perpendicular axes. One axis is to be parallel to the indicator shaft. Each shock shall have an amplitude of 23–27 g, half sine of 9–13 milliseconds duration.

9.10 Shock Test (All Senders)—The sender shall be capable of withstanding, without mechanical or electrical failure, six shocks of 44–55 g, half sine of 9–13 milliseconds duration in each direction along each of three mutually perpendicular axes. One axis is to be perpendicular to the mounting plane.

10. Design Detail Recommendations (Indicating Unit Only)

10.1 When analog displays are used, the display shall be accomplished by a pointer or other indicator traversing in a clockwise or left to right direction as applicable to register increasing speed over a suitable scale on the indicating unit dial. Consult FMVSS 127 for any requirements concerning dial specifications.

10.2 Graduations shall be designed for the best practical legibility and accuracy of reading.

10.3 Unless otherwise specified, pointers and dial printing shall be white, dial background shall be low gloss black, and visible portions of the indicating unit should exhibit low reflectivity. The distance indicator shall have white numerals on a low gloss black background except for the tenths indicator which shall have black numerals on a white background.

10.4 The indicating unit case shall be provided with studs for mounting by suitable U-clamps or similar means.

10.5 Typical envelope, mounting studs and terminal designations are displayed in Figs. 1, 2, and 3.

11. Identification

11.1 Indicating Unit

11.1.1 To be legibly stamped on outside of case:
 a. Manufacturer's or user's part number.
 b. Manufacturer's or user's serial number and/or date of manufacture.

11.1.2 To be printed on dial and/or stamped on case: manufacturer's or user's name or trademark.

11.1.3 Electrical connections shall be clearly identified for proper wiring of instrument into circuit.

11.2 Sender—Sender identification is to be as agreed by the manufacturer and the user.

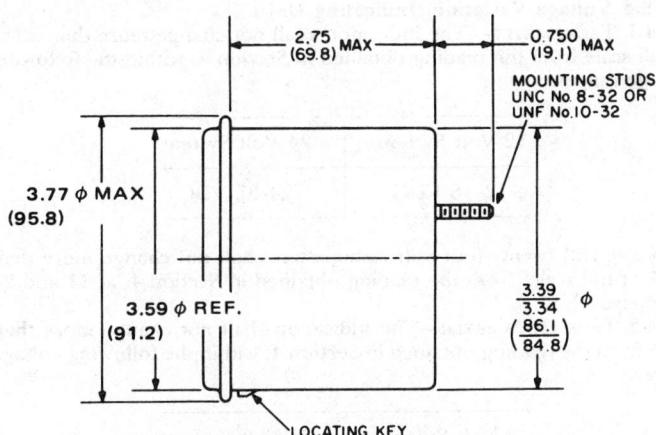

DIMENSIONS ARE INCHES (MILLIMETERS)

FIG. 1—ENVELOPE AND MOUNTING STUDS

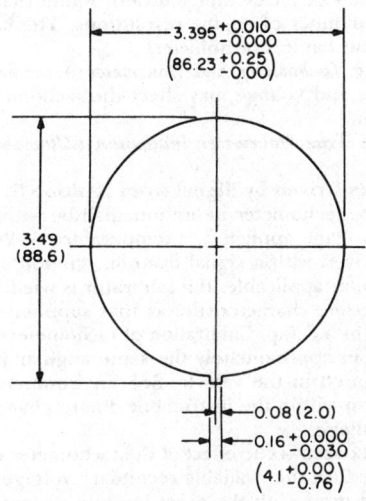

DIMENSIONS ARE INCHES (MILLIMETERS)

FIG. 2—MOUNTING CUTOUT DETAIL

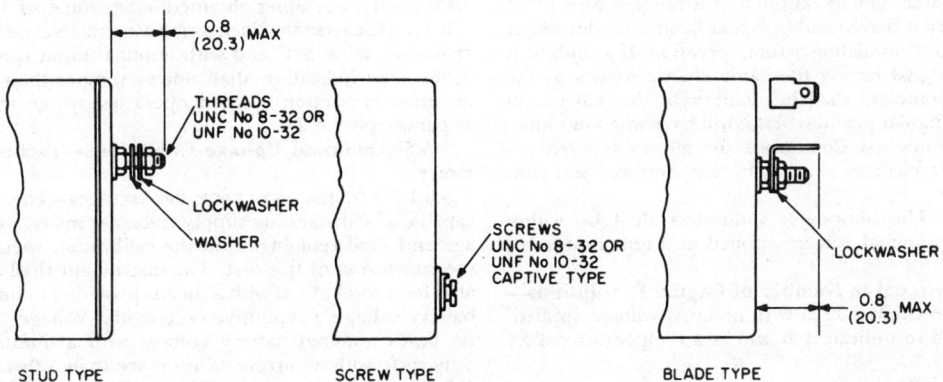

DIMENSIONS ARE INCHES (MILLIMETERS)

FIG. 3—TERMINALS

ELECTRIC TACHOMETER SPECIFICATION—SAE J1399 JUN84

SAE Recommended Practice

Report of the Speedometer and Tachometer Committee, Electric Subcommittee, approved September 1982. SAE J196 and J197 have been replaced by this report.

1. Scope—This SAE Recommended Practice establishes minimum requirements for electric tachometer systems with and without hourmeter or revolution counter, for general applications as follows:

Class 1—Passenger Car
Class 2—Bus and Truck
Class 3—Off-Road Vehicles

2. Electric Tachometer System—A typical electric tachometer system for engines using a Kettering ignition system or the newer electronic ignition systems, consists of an indicating unit that obtains a signal proportional to engine speed from the ignition system.

If the tachometer is intended for use on a diesel engine, a sender may be used to supply a signal proportional to engine speed. A signal may also be obtained from an a.c. tap on the alternator if the alternator is so equipped. If a sending unit is used, it will often be one of the following types: permanent magnet generator, magnetic switch, or magnetic sensor. The sender may be mounted on the engine outlet provided for mechanical tachometer cables, or it may be mounted so as to sense the number of teeth on the flywheel ring gear or some other location where a rotating element with teeth, slots, holes, or bosses may be sensed.

The indicating unit may contain an hourmeter. The hourmeter in an electric tachometer may be a true time indicator rather than an indication proportional to the number of engine revolutions. The latter indication is usually found in mechanical tachometers.

3. Factors Affecting Tachometer and Hourmeter Accuracy—Changes in ambient temperature and voltage may affect the tachometer and/or the hourmeter indication.

4. Tachometer and True Hourmeter Indication (Allowable System Variation)

4.1 Tachometers Driven by Signal from Ignition System or Alternator A.C. Tap—The tachometer indication shall be within ±2% of full scale with nominal voltage applied at a temperature of 24 ± 3°C when the tachometer is driven with a signal from an ignition system or from an alternator a.c. tap, as applicable. If a calibrator is used, it must supply a signal having the same characteristics as that supplied by an ignition system or an alternator a.c. tap. Calibration of tachometers shall be made with the instrument in approximately the same angular position that it will have when mounted in the vehicle. See Environmental Conditions for allowable variation within the instrument due to changes in ambient temperatures and voltage.

4.1.1 SECONDARY LOSSES—The effect of the tachometer on the ignition system should not reduce the available secondary voltage by more than 4%. Testing is to be done with the exact ignition system to be used in actual practice. The distributor is to be run with the coil input voltage held constant at 14.0 V and the coil secondary open circuited. The exact test procedure for measurements shall be established by the supplier and consumer.

4.2 Sender Driven Units—The tachometer indication shall be within ±2% of full scale with nominal voltage applied at a temperature of 24 ± 3°C, when the tachometer is driven with a signal from a sender either rotated or excited in a fashion simulating actual operation. If a calibrator is used, it must supply a signal having the same characteristics as the sender. Calibration of tachometers shall be made with the instrument in approximately the same angular position that it will have when mounted in the vehicle. See Environmental Conditions for allowable variation within the instrument due to changes in ambient temperatures and voltage.

4.3 True Hourmeter—The hourmeter indication shall be within ±2% of elapsed time with nominal voltage applied at a temperature of 24 ± 3°C.

4.4 Hourmeter Proportional to Number of Engine Revolutions—The time indication shall be within ±0.3% with nominal voltage applied, nominal input rpm required to indicate 1 h and at a temperature of 24 ± 3°C.

5. Effects of Environmental Conditions

5.1 Temperature (Allowable System Variation)

5.1.1 TACHOMETER INDICATION—With nominal voltage applied, the tachometer indication shall not vary more than ±2% of full scale from the reading determined in Section 4, while the indicating unit is operating over the range of −7 to +54°C. The sender (if required) is operating over the range of −40 to +138°C. No permanent damage shall result from operating the indicating unit in a range of −40 to +82°C.

5.1.2 TRUE HOURMETER—With nominal voltage applied, the time indication shall not vary more than ±1% from a reading obtained at 24 ± 3°C while the instrument is operating over the range of −7 to +54°C. No permanent damage shall result from operating the instrument in a range of −40 to +82°C.

5.1.3 HOURMETER PROPORTIONAL TO NUMBER OF ENGINE REVOLUTIONS—With nominal voltage applied and nominal input rpm required to indicate 1 h, the time indication shall not vary more than ±0.3% from a reading obtained in Section 4 while the instrument is operating over the range of −7 to +54°C. No permanent damage shall result from operating the instrument in a range of −40 to +82°C.

5.2 Temperature Extremes (Sender Only)—It will be necessary to evaluate the specific application to specify the allowable temperature extremes.

5.3 Storage Temperature (Indication Unit Only)

5.3.1 TACHOMETER—A 4 h exposure of the indicating unit to a temperature of −40 to +85°C shall result in no more than ±1% of full scale permanent calibration change from the reading obtained in Section 4. The rate of temperature change during this test shall not exceed 2°C/min.

5.3.2 HOURMETERS—A 4 h exposure of the indicating unit to a temperature of −40 to +85°C shall result in no more than ±1% permanent calibration change from the reading obtained in Section 4. The rate of temperature change during this test shall not exceed 2°C/min.

5.4 Voltage Variation (Indicating Unit)

5.4.1 TACHOMETER—The indication shall not change more than ±1% of full scale from the reading obtained in Section 4, within the following voltage ranges.

12 Volt System	24 Volt System
12–16 VDC	24–32 VDC

Twelve and twenty-four volt tachometers shall not change more than ±3% of full scale, from the reading obtained in Section 4, at 11 and 22 V respectively.

5.4.2 TRUE HOURMETER—The indication shall not change more than ±1% from the reading obtained in Section 4, within the following voltage ranges.

12 Volt System	24 Volt System
12–16 VDC	24–32 VDC

Twelve and twenty-four volt hourmeters shall not change more than ±3% from the reading obtained in Section 4, at 11 and 22 V respectively.

5.4.3 HOURMETER PROPORTIONAL TO NUMBER OF ENGINE REVOLUTIONS—At 24 ± 3°C and with nominal input rpm required to indicate 1 h, the time indication shall not vary more than ±0.3% from a reading obtained in Section 4 when operating within the voltage ranges given in paragraph 5.4.2.

5.5 Abnormal Voltage Conditions—Tachometer and True Hourmeter

5.5.1 TRANSIENT VOLTAGE PROTECTION—The indicating unit shall be capable of withstanding supply voltage transients without permanent damage and shall remain within the calibration specification of Section 4 at the conclusion of this test. The instrument shall be connected and operated for a total of 1 h with a means provided to impress upon the nominal battery voltage a repetitive rectangular voltage pulse of plus and minus six times nominal battery voltage with a duration of 300 μs and 1% duty cycle with a current of no more than 1.0 A.

5.5.2 OVERVOLTAGE AND REVERSE POLARITY—Provisions for protection against booster starts with double battery voltage and/or reversed polarity must be negotiated between the user and the manufacturer.

5.6 Moisture Resistance

5.6.1 HUMIDITY (INDICATING UNIT)—Indicating unit shall withstand exposure to 95% relative humidity at 38°C for 48 h.

5.6.2 SALT SPRAY (SENDER UNIT)—Sender units shall be corrosion resistant and shall withstand a salt spray (fog) test of 48 h duration with 5% salt solution (Reference ASTM B117-73).

5.7 Vibration Test (Indicating Unit)—The indicating unit shall be capable of withstanding without mechanical or electrical failure, 3 h of vibration, 1 h along each of the three mutually perpendicular axes. One of said axes is to be parallel to the indicator shaft. The vibration test shall be run at a double amplitude as specified in the table below with the frequency varying as shown in the table at intervals of 1 min. After completion of test, the calibration shall remain within tolerances as specified in Sections 4 and 5.

Class	Amplitude, DA (mm)	Frequency Range (Hz)	Max. Acceleration (g)
1	0.75	10–30–10	1.4 at 30 Hz
2	1.52	10–55–10	10 at 55 Hz
3	1.52	10–80–10	20 at 80 Hz

5.8 Vibration Test (Sender Only)

5.8.1 ENGINE MOUNTED—The sender shall be capable of withstanding 6 h of vibration without mechanical or electrical failure, 2 h along each of the three mutually perpendicular axes. One of said axes is to be parallel to the input shaft. The vibration test shall be run at a double amplitude of 0.50 mm with the frequency varying from 10–120–10 Hz at intervals of 1 min.

5.9 Shock Test (Indicating Unit Only)—The indicating unit shall be capable of withstanding without mechanical or electrical failure, the following series of shocks and still maintain the calibration tolerances specified in Sections 4 and 5. The indicating unit shall be subjected to an equal number of shocks in each direction along each of three mutually perpendicular axes. One of said axes is to be parallel to the indicator shaft. Each shock shall be half sine of 9–13 ms duration, the acceleration and total number of shocks per table below.

Class	Peak Acceleration (g)	Total Number
1	23–27	6
2	23–27	72
3	44–55	72

5.10 Shock Test (All Senders)—The sender shall be capable of withstanding, without mechanical or electrical failure, six shocks of 44–55

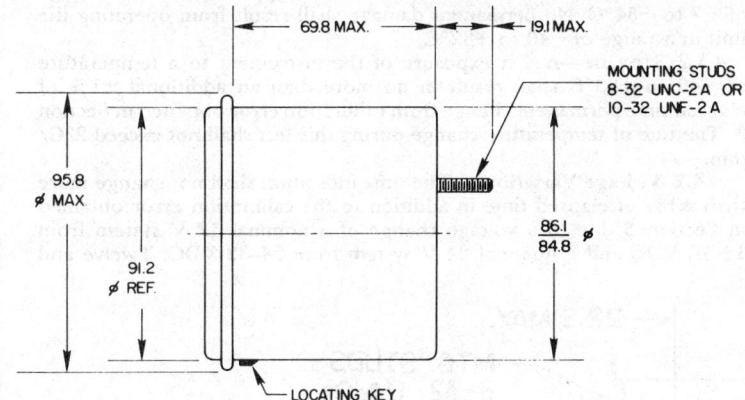

ALL DIMENSIONS ARE MILLIMETERS UNLESS OTHERWISE SPECIFIED

FIG. 1—ENVELOPE AND MOUNTING STUDS

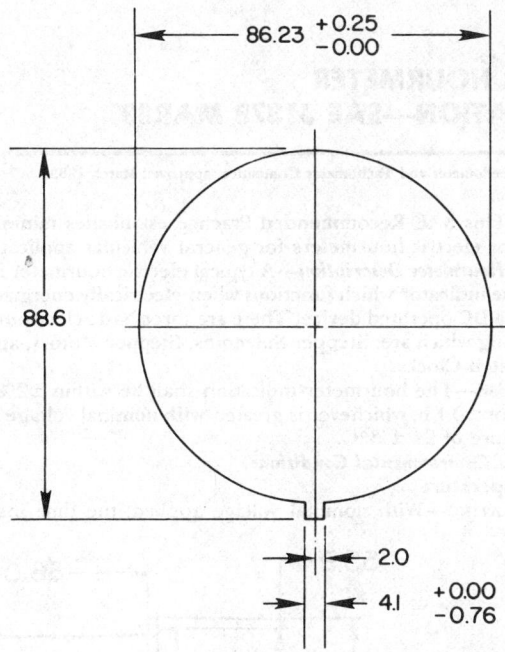

ALL DIMENSIONS ARE MILLIMETERS UNLESS OTHERWISE SPECIFIED

FIG. 2—MOUNTING CUTOUT DETAIL

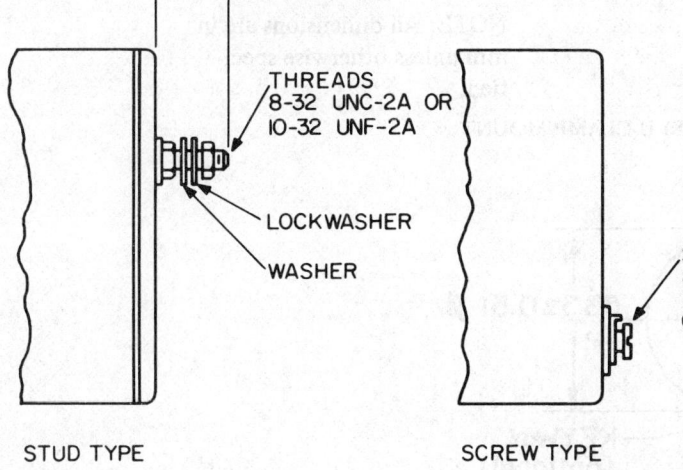
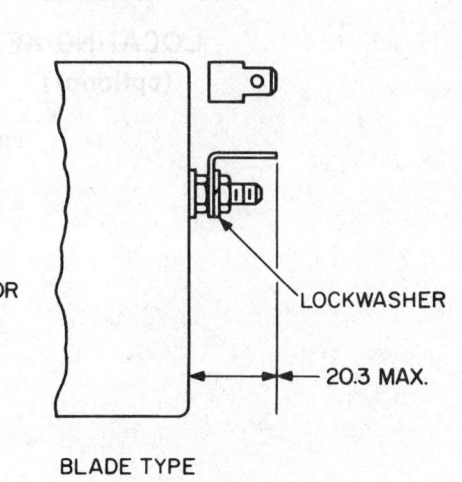

ALL DIMENSIONS ARE MILLIMETERS UNLESS OTHERWISE SPECIFIED

FIG. 3—TERMINALS

g, half sine of 9–13 ms duration in each direction along each of three mutually perpendicular axes. One of said axes is to be perpendicular to the mounting plane.

5.11 Design Detail Recommendations (Indicating Unit Only)

5.11.1 When analog displays are used, the display shall be accomplished by a pointer or other indicator traversing in a clockwise or left to right direction as applicable, to register increasing revolutions per minute over a suitable scale on the indicating unit dial.

5.11.2 Graduations shall be designed for the best practical legibility and accuracy of reading.

5.11.3 Unless otherwise specified: pointers and dial printing shall be white, dial background shall be low gloss black, and visible portions of the indicating unit should exhibit low reflectivity; the time or revolution indicator shall have white numerals on a low gloss black background except for the tenths indicator which shall have black numerals on a white background.

5.11.4 The indicating unit case shall be provided with studs for mounting by suitable U-clamps or similar means.

5.11.5 Typical envelope, mounting studs and terminal designations are displayed in Figs. 1–3.

5.12 Identification

5.12.1 INDICATING UNIT

5.12.1.1 To be legibly indicated on outside of case:
(a) Manufacturer's or user's part number.
(b) Manufacturer's or user's serial number and/or date of manufacture.

5.12.1.2 To be printed on dial and/or indicated on case: Manufacturer's or user's name or trademark.

5.12.1.3 Electrical connections shall be clearly identified for proper wiring of instrument into circuit.

5.12.2 SENDER—Sender identification is to be used as agreed between manufacturer and user.

ELECTRIC HOURMETER SPECIFICATION—SAE J1378 MAR83

SAE Recommended Practice

Report of the Speedometer and Tachometer Committee, approved March 1983.

1. Scope—This SAE Recommended Practice establishes minimum requirements for electric hourmeters for general vehicular applications.

2. Electric Hourmeter Description—A typical electric hourmeter is a true operating time indicator which functions when electrically energized. The hourmeter is a DC operated device. There are three basic electromechanical types among which are: Stepper Solenoids, Stepper Motors, and Electrically Operated Clocks.

3. Calibration—The hourmeter indication shall be within ±2% of the elapsed time or ±0.1 h, whichever is greater, with nominal voltage applied at a temperature of 24 ± 3°C.

4. Effects of Environmental Conditions

4.1 Temperature

4.1.1 OPERATING—With nominal voltage applied, the time indication shall not vary more than ±1% of elapsed time in addition to the calibration error obtained in Section 3, while the unit is operating over the range of −7 to +54°C. No permanent damage shall result from operating the unit in a range of −40 to +82°C.

4.1.2 STORAGE—A 4 h exposure of the instrument to a temperature of −40 to +85°C shall result in no more than an additional ±1% of elapsed time permanent change from calibration error obtained in Section 3. The rate of temperature change during this test shall not exceed 2°C/min.

4.2 Voltage Variations—The time indication shall not change more than ±1% of elapsed time in addition to the calibration error obtained in Section 3 due to a voltage change of a nominal 12 V system from 12–16 VDC and a nominal 24 V system from 24–32 VDC. Twelve and

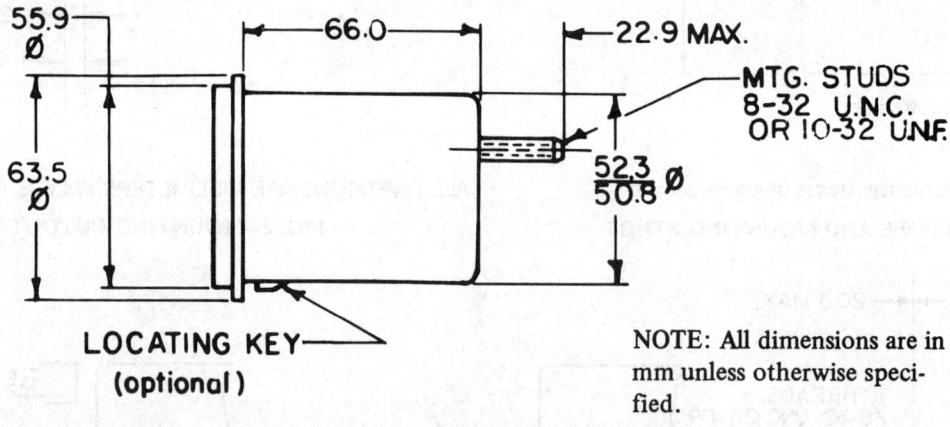

NOTE: All dimensions are in mm unless otherwise specified.

FIG. 1—(ENVELOPE) U-CLAMP MOUNT

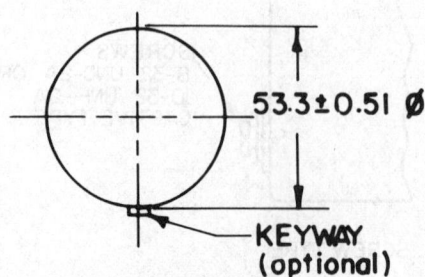

NOTE: All dimensions are in mm unless otherwise specified.

FIG. 1A—PANEL CUTOUT

24 V hourmeters shall not change more than ±3% from the reading obtained in Section 3 at 11 and 22 V respectively.

4.3 Abnormal Voltage Conditions

4.3.1 TRANSIENT PROTECTION—The instrument shall be capable of withstanding supply voltage transients without permanent damage and shall remain within the calibration specification of Section 3 at the conclusion of this test. The instrument shall be connected and operated for a total of 1 h with a means provided to impress upon the nominal battery voltage a repetitive rectangular voltage pulse of plus and minus six times nominal battery voltage with a duration of 300 μs and 1% duty cycle with a current of no more than 1.0 amp. For some applications which may have transient voltages having a magnitude, duration, or duty cycle exceeding the above requirements, contact the instrument manufacturer for recommendations. Further information on transients may be found in SAE J1113.

4.3.2 OVERVOLTAGE AND REVERSE POLARITY—Provisions for protection against booster starts with double battery voltage and/or reversed polarity must be negotiated between the user and the maufacturer.

4.4 Humidity—Instrument shall not have its function impaired due to exposure to 95% relative humidity at 38°C for 48 h.

4.5 Vibration Test—The electric hourmeter shall be capable of withstanding without mechanical or electrical failure 6 h of vibration, 2 h along each of the three mutually perpendicular axes, one axis to be perpendicular to mounting plane. The vibration tests shall be run at a double amplitude of 1.52 mm with the frequency varying from 10–80–10 Hz (20 g max) at intervals of 1 min. After completion of test, the calibration shall remain within tolerances as specified in Section 3.

4.6 Shock Test—The instrument shall be capable of withstanding without mechanical or electrical failure, the following series of shocks and still maintain the calibration tolerances specified in Section 3.

The unit shall be subjected to 12 shocks in each direction along each of the three mutually perpendicular axes (72 total shocks), one axis to be perpendicular to the mounting plane. Each shock shall have an amplitude of 44–55 g, half sine of 9–13 μs duration.

5. Design Detail Recommendations

5.1 Unless otherwise specified by user, dial printing shall be white, dial background shall be low gloss black, and visible portions of the instrument should exhibit low reflectivity; the indicating wheels or drums shall have white numerals on a low gloss black background, except for the tenths indicator, which shall have black numerals on a white background.

5.2 All exposed surfaces shall be corrosion resistant for limited exposure. (NOTE: If instruments are required for installations in extreme environments, contact manufacturer for recommendations.)

5.3 Instruments shall be moisture and dust resistant. (See NOTE—paragraph 5.2.)

5.4 The hourmeter case may be provided with studs for mounting by suitable U-clamps or similar means. Some hourmeters may have a flange for mounting.

5.5 Typical envelope, mounting studs, mounting flange, panel cutout, and terminal designations are displayed in Figs. 1–4.

5.6 Identification (Unless Otherwise Specified)

5.6.1 To be legibly indicated on outside of case:
  a. manufacturer's or user's part number
  b. manufacturer's or user's serial number and/or date of manufacture.

5.6.2 To be printed on dial and/or indicated on case: manufacturer's or user's name and/or trademark.

5.6.3 Electrical connections shall be clearly identified for proper wiring of instrument into circuit.

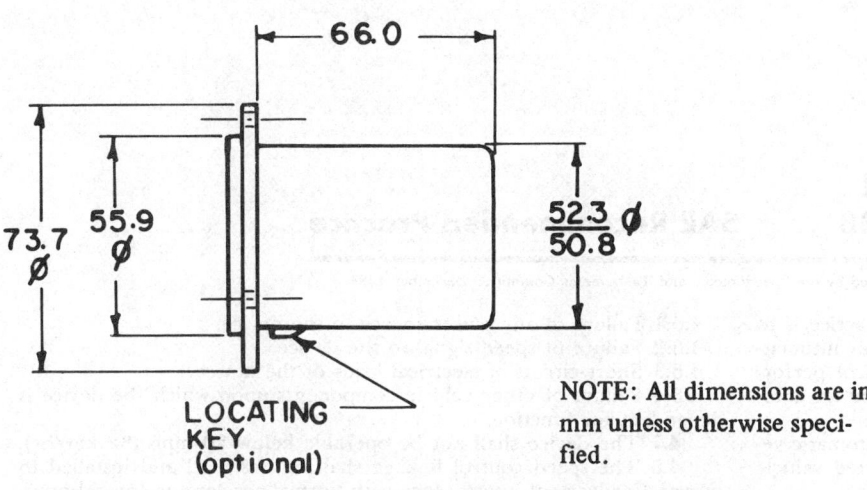

NOTE: All dimensions are in mm unless otherwise specified.

FIG. 2—(ENVELOPE) FLANGE MOUNT

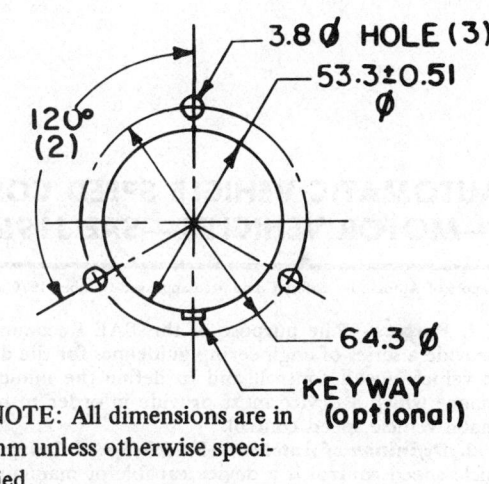

NOTE: All dimensions are in mm unless otherwise specified.

FIG. 2A—PANEL CUTOUT

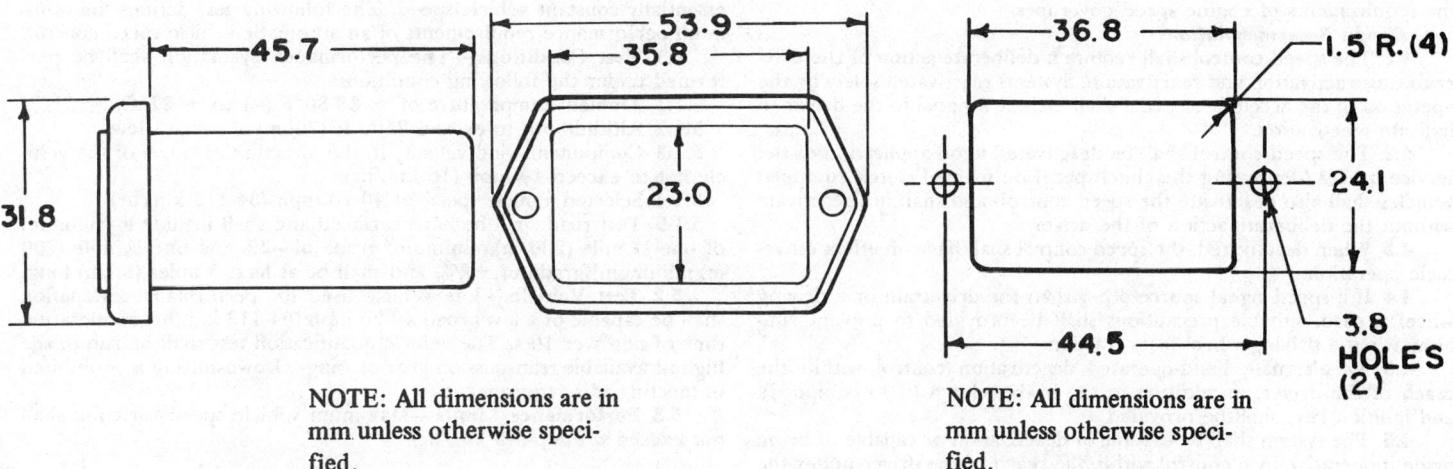

NOTE: All dimensions are in mm unless otherwise specified.

FIG. 3—(ENVELOPE) FLANGE MOUNT

NOTE: All dimensions are in mm unless otherwise specified.

FIG. 3A—PANEL CUTOUT

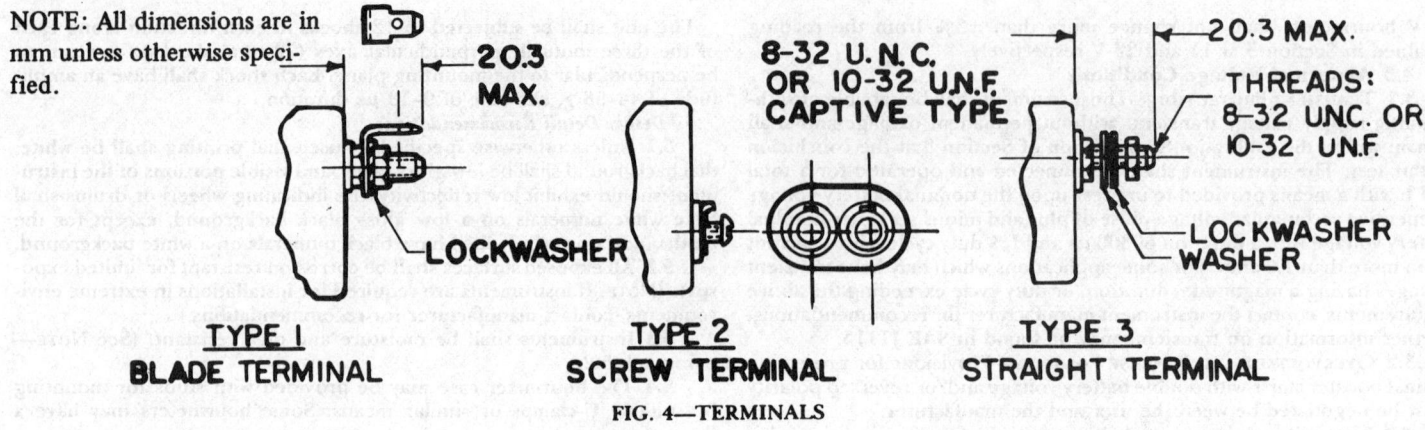

NOTE: All dimensions are in mm unless otherwise specified.

FIG. 4—TERMINALS

AUTOMATIC VEHICLE SPEED CONTROL —MOTOR VEHICLES—SAE J195 DEC88

SAE Recommended Practice

Report of Automotive Safety Committee approved October 1970 and reaffirmed by the Speedometer and Tachometer Committee December 1988.

1. Purpose—The purpose of this SAE Recommended Practice is to provide a series of engineering guidelines for the design of an automatic vehicle speed control, and to define the minimum control performance which a device must provide in order to be classified an automatic vehicle speed control.

2. Definition of Automatic Vehicle Speed Control—An automatic vehicle speed control is a device capable of maintaining selected vehicle speeds in the presence of changing road load conditions.

3. Scope—This SAE Recommended Practice is intended to apply only to the design of an automatic vehicle speed control. It is not intended to encourage or discourage the installation of automatic vehicle speed controls on any class of vehicles, nor is it intended to influence the requirements of engine speed governors.

4. Design Recommendations

4.1 The speed control shall require a deliberate action of the driver to cause activation and reactivation. Systems reactivated solely by the operation of the accelerator pedal shall include a signal to the driver to indicate reactivation.

4.2 The speed control shall be deactivated upon application of the service brakes (depressing the clutch pedal on manual clutch equipped vehicles shall also deactivate the speed control) and shall not reactivate without the deliberate action of the driver.

4.3 When deactivated, the speed control shall have no effect on vehicle operation.

4.4 If a speed signal source other than the drivetrain or a driving wheel is used, suitable precautions shall be provided to prevent runaway when a driving wheel loses traction.

4.5 An alternate hand-operated deactivation control within the reach of the driver, in addition to the brake, clutch (if so equipped), and ignition key, shall be provided.

4.6 The system shall be capable of deactivation or capable of being made inoperative by a control within the reach of the driver under the following conditions:

4.6.1 Failure of any power source to the device.

4.6.2 Failure of speed signal to the device.

4.6.3 Short circuit of electrical leads of the device.

4.6.4 Failure of other vehicle components upon which the device is dependent for function.

4.7 The device shall not be operable below 20 mph (32 km/hr).

4.8 The speed control linkage shall be designed and installed to prevent inadvertent interference with normal accelerator control operation under all operating and environmental conditions consistent with the environmental capabilities of the vehicle upon which it is installed.

5. Performance Requirements—The automatic vehicle speed control shall regulate the output power of the engine to provide a stable and essentially constant vehicle speed. The following test defines the minimum performance requirements of an automatic vehicle speed control.

5.1 *Test Conditions*—The performance evaluation shall be performed under the following conditions:

5.1.1 Ambient temperature of + 30-80°F (−1 to + 27°C).

5.1.2 Altitude not to exceed 2500 ft (760m) above sea level.

5.1.3 Component wind velocity in the direction of travel of the vehicle not to exceed 10 mph (16 km/hr).

5.1.4 Selected vehicle speed of 40-70 mph (64-113 km/hr).

5.1.5 Test road shall be hard surfaced and shall include a minimum of one 1/8 mile (200 m)(minimum) grade of −2% and one 1/8 mile (200 m)(minimum) grade of +2%, and shall be at least 5 miles (8 km) long.

5.2 *Test Vehicle*—The vehicle used for performance evaluation shall be capable of a level road 40-70 mph (64-113 km/hr) acceleration time of not over 16 s. The vehicle qualification test shall be run in the highest available transmission gear or range. Downshifting is prohibited in meeting this requirement.

5.3 *Performance Limits*—Maximum vehicle speed variation shall not exceed ±3 mph (5 km/hr).

TRANSMISSION MOUNTED VEHICLE SPEED SIGNAL ROTOR SPECIFICATION—SAE J1377 JAN89   SAE Recommended Practice

Developed by the Electric Speedometer and Tachometer Subcommittee, approved by the Speedometer and Tachometer Committee June 1984, and reaffirmed January 1989.

1. Scope—This SAE Recommended Practice covers the transmission output shaft mounted signal rotor and sensor mounting hole used for electronic speed sensing on Class 6, 7, and 8 highway vehicles.

2. Purpose—To standardize the number of discontinuities sensed (such as teeth or slots) per revolution of the transmission output shaft.

3. Electronic Speed Sensing—A gear, stamping or milled disc with discontinuities that are sensed by a sensor, usually a magnetic sensor consisting of a coil of wire–suitably terminated–wound around a cylindrical magnet and encased in a metal housing, may be used as a vehicle speed signal source.

The vehicle speed signal may then be used to activate electronic speedometers, cruise controls, vehicle speed limiting devices, and other components requiring vehicle speed information.

4. Transmission Signal Rotor—The signal rotor shall have 16 teeth or discontinuities, and shall turn at the same speed as the transmission output shaft. The eccentricity of the sensed portion of the signal rotor shall not exceed 0.25 mm total indicated runout. When magnetic sensors are used, the signal rotor, gear, disc, or stamping in the transmission shall have teeth, slots, holes, and other discontinuities and shall be a low carbon, low remanence ferrous or magnetic material, such as cast iron or steel. The discontinuities shall be at least 3 mm deep. It is desirable for the discontinuities and the metal separating them to be of approximately equal size. The discontinuity size or rotor face width shall be no smaller than 3 mm wide.

5. Sensor Mounting Hole—The hole for the magnetic sensor shall be a 3/4 in 16 UNF thread. The axis of the hole shall be perpendicular to the sensed surface of the transmission gear, disc, or signal rotor, and shall be positioned so as to center the sensor over the discontinuities being sensed. The external surface around the speed sensor mounting hole shall be flat and smooth for a diameter sufficient to accommodate a locknut. The requirement for sealing in transmission lubricant should be negotiated between the user and the transmission and sensor suppliers. SAE Hydraulic Tube fitting O-ring seals are suggested. See SAE J514 and J515. (Reference nominal tube size 1/2 in.)

22 Tubing, Hose and Fittings

(R) CODING SYSTEMS FOR IDENTIFICATION OF FLUID CONDUCTORS AND CONNECTORS—
SAE J846 JUN89
SAE Recommended Practice

Report of the Tube, Pipe, Hose, and Lubrication Fittings Committee, approved January, 1963, completely revised by the Fluid Conductors and Connectors Technical Committee, June 1989. This document is constantly under revision to include new codes/symbols.

NOTE:

It should be noted that the code numbers assigned to the applicable standards covered by the coding system could possibly change. Therefore, it is recommended that for the purpose of transmitting technical or engineering information relating to the various tube, pipe, and hose fittings, the applicable code numbers, standard number, and revision letter be specified for proper identification.

SCOPE:

This coding system is intended to provide a convenient means of identifying the various tube, pipe, and hose fittings and of transmitting technical or engineering information relating to them wherever drawings or other pictorial media may not be readily available. The code has been kept flexible to permit expansion to cover new fitting categories or styles and, if the need develops, the inclusion of materials. The system is also compatible with automatic data processing equipment.

It is not intended that this code should supersede established systems or means of identification. However, because the SAE code for automotive flare fittings shown in SAE J512 is also applicable to corresponding refrigeration fittings in SAE J513, both an SAE code and the existing code ANSI B70.1, Refrigeration Flare Fittings, are included throughout SAE J513. Therefore, it should be the prerogative of the user to apply that code which best satisfies his requirements.

GENERAL SPECIFICATIONS

Code

The code shall consist of two groups of numbers and one group of letters: the first group of numbers symbolizing the size identification, the second group of numbers symbolizing the fitting and hose identification, and the third group of letters symbolizing the material and assembly as delineated below.

1. Size Identification

The fitting size shall be identified by a series of dash numbers, each representing the size of the respective fitting ends. The size of the tube end shall precede the size of the pipe, hose, or other ends of the fitting. The dash shall be given in the sequence defined below:

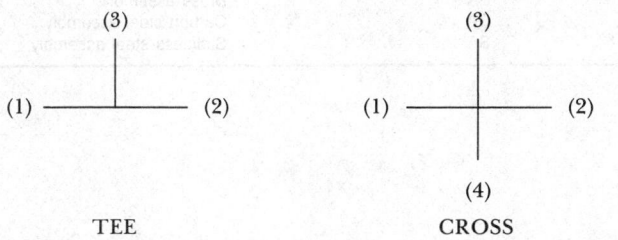

When special size combinations of tube-to-tube fitting ends are specified, the largest tube size shall precede the smaller tube size for unions and union elbows. For tees, the (1) shall be the larger tube size of (1) and (2). For crosses, the (1) shall be the largest tube size of (1) and (2) and (3) shall be equal to or larger than (4).

The dash size symbol, applicable to all tube ends and straight thread O-ring boss ends, shall consist of the number of sixteenth inch increments contained in the outside diameter of the tubing (nominal tube OD) they are designed to be used with, as listed below in Table 1.

TABLE 1—DASH NUMBERS FOR TUBE AND STRAIGHT THREAD ENDS

Nominal Tube OD, in	Dash Size Symbol	Nominal Tube OD, in	Dash Size Symbol	Nominal Tube OD, in	Dash Size Symbol
1/8	-2	7/16	-7	7/8	-14
3/16	-3	1/2	-8	1	-16
1/4	-4	9/16	-9	1-1/4	-20
5/16	-5	5/8	-10	1-1/2	-24
3/8	-6	3/4	-12	2	-32

The dash size symbol for pipe thread ends shall be the number of sixteenth inch increments contained in the nominal pipe thread size as listed below in Table 2.

TABLE 2—DASH NUMBERS FOR PIPE THREAD ENDS

Nominal Pipe Thread Size, in	Dash Size Symbol	Nominal Pipe Thread Size, in	Dash Size Symbol	Nominal Pipe Thread Size, in	Dash Size Symbol
1/16	-1	3/8	-6	1	-16
1/8	-2	1/2	-8	1-1/4	-20
1/4	-4	3/4	-12	1-1/2	-24
				2	-32

The dash size symbol for hose shall be the number of sixteenth inch increments contained in the inside diameter of the hose (nominal hose ID), except in the case of SAE 100R5 and 100R14 hose where it is equivalent to the number of sixteenth inch increments in the outside diameter of tubing having approximately the same inside diameter as the hose. See Table 3 for respective hose types.

The dash size symbol for 4-bolt split flange O-ring connections shall be the number of sixteenth inch increments contained in the nominal flange size as listed below in Table 4.

The dash size symbols for straight thread pipe plugs and filler

TABLE 3—DASH NUMBERS FOR 100R SERIES HOSES

Nominal Hose ID, in	Dash Size Symbols For Hose Type													
	SAE 100R1	SAE 100R2	SAE 100R3	SAE 100R4	SAE 100R5	SAE 100R6	SAE 100R7	SAE 100R8	SAE 100R9	SAE 100R10	SAE 100R11	SAE 100R12	SAE 100R13	SAE 100R14
1/8	—	—	—	—	—	—	—	—	—	—	—	—	—	-3
3/16	-3	-3	-3	—	-4	-3	-3	-3	—	-3	-3	—	—	-4
1/4	-4	-4	-4	—	-5	-4	-4	-4	—	-4	-4	—	—	-5
5/16	-5	-5	-5	—	-6	-5	-5	—	—	—	—	—	—	-6
3/8	-6	-6	-6	—	—	-6	-6	-6	-6	-6	-6	-6	—	-7
13/32	-6.5	—	—	—	-8	—	—	—	—	—	—	—	—	-8
1/2	-8	-8	-8	—	-10	-8	-8	-8	-8	-8	-8	-8	—	-10
5/8	-10	-10	-10	—	-12	-10	-10	-10	—	—	—	—	—	-12
3/4	-12	-12	-12	-12	—	-12	-12	-12	-12	-12	-12	-12	-12	-14
7/8	-14	-14	—	—	-16	—	—	—	—	—	—	—	—	-16
1	-16	-16	-16	-16	—	—	-16	-16	-16	-16	-16	-16	-16	-18
1-1/8	—	—	—	—	-20	—	—	—	—	—	—	—	—	-20
1-1/4	-20	-20	-20	-20	—	—	—	—	-20	-20	-20	-20	-20	—
1-3/8	—	—	—	—	-24	—	—	—	—	—	—	—	—	—
1-1/2	-24	-24	—	-24	—	—	—	—	-24	-24	-24	-24	-24	—
1-13/16	—	—	—	—	-32	—	—	—	—	—	—	—	—	—
2	-32	-32	—	-32	—	—	—	—	-32	-32	-32	-32	-32	—
2-3/8	—	—	—	—	-40	—	—	—	—	—	—	—	—	—
2-1/2	—	-40	—	-40	—	—	—	—	—	—	—	-40	—	—
3	—	—	—	-48	-48	—	—	—	—	—	—	—	—	—
3-1/2	—	—	—	-56	—	—	—	—	—	—	—	—	—	—
4	—	—	—	-64	—	—	—	—	—	—	—	—	—	—

TABLE 4—DASH NUMBERS FOR 4-BOLT SPLIT FLANGE CONNECTIONS

Nominal Flange Size, in	Dash Size Symbol	Nominal Flange Size, in	Dash Size Symbol	Nominal Flange Size, in	Dash Size Symbol
1/2	-8	1-1/2	-24	3-1/2	-56
3/4	-12	2	-32	4	-64
1	-16	2-1/2	-40	5	-80
1-1/4	-20	3	-48		

TABLE 5—DASH NUMBERS FOR STRAIGHT THREAD PLUGS

Nominal Straight Thread Size, in	Dash Size Symbol	Nominal Metric Thread Size, mm	Dash Size Symbol
5/16	-5	M10 x 1	-M10
3/8	-6	M14 x 1.25	-M14
1/2	-8	M18 x 1.5	-M18
5/8	-10		
3/4	-12		
7/8	-14		
1	-16		
1-1/4	-20		
1-1/2	-24		
1-3/4	-28		
2	-32		

and drain plugs shall be the number of sixteenth inch increments contained in the nominal straight thread size except in the case of metric thread sizes where the thread size shall be designated as the dash size listed in Table 5.

2. Fitting Identification

The fitting identification shall consist of a six-digit number made up of three groups of two digits each symbolizing in sequence: (a) the fitting type, (b) the fitting shape, and (c) the fitting connecting ends. (For convenient reference, the fitting identification codes applicable to fittings appearing in SAE J246, SAE J512, SAE J513, SAE J514, SAE J516, SAE J518, SAE J530, SAE J531, SAE J532, SAE J1453, and SAE J1926 are shown in brackets adjacent to the respective figure numbers.) The identification symbols for each of the three groups shall be as follows:

a. FITTING TYPE IDENTIFICATION—The two-digit symbols applicable to the various fitting types and styles shall be as tabulated below in Table 6A and 6B.

b. FITTING SHAPE IDENTIFICATION—The two-digit symbols applicable to the various shapes of fittings shall be as tabulated below in Table 7.

c. FITTING CONNECTING END IDENTIFICATION—The two-digit symbols applicable to the various threaded, hose, connecting ends or combinations thereof for the fittings covered shall be as tabulated below in Table 8.

3. Material and Assembly Identification

The material and assembly identification shall consist of two letters symbolizing the material and assembly of multiple tube fitting components supplied assembled rather than as separate pieces. For convenient reference, this code is applicable to SAE J246, SAE J512, and SAE J514. The identification symbols shall be as tabulated below:

Material Assembly Symbol	Material and Assembly
BA	Brass assembly
CA	Carbon steel assembly
SA	Stainless steel assembly

TABLE 6A—FITTING TYPE SYMBOLS
(Arranged Numerically)

FITTING TYPE SYMBOL	FITTING TYPE AND STYLES
01	45 deg flared, automotive-J512
02	
03	
04	Inverted flared, automotive-J512
05	
06	Tapered sleeve, automotive-J512
07	37 deg flared, hydraulic-J514
08	Flareless, hydraulic-J514
09	SAE O-ring boss, hydraulic-J1926
10	Flanged sleeve, nylon tube, automotive-J246
11	4-bolt split flange O-ring, hydraulic-J518
12	Spherical sleeve, copper tube, automotive-J246
13	Pipe, automotive-J530, J531
14	Pipe, hydraulic-J514
15	Male pipe, hose, permanently attached
16	Male pipe, hose, field attachable
17	Male pipe, hose, field attachable, segment clamp
18	SAE O-ring boss, hose, permanently attached
19	SAE O-ring boss, hose, field attachable
20	SAE O-ring boss, hose, field attachable, segment clamp
21	37 deg male flared, hose, permanently attached
22	37 deg male flared, hose, field attachable
23	37 deg male flared, hose, field attachable, segment clamp
24	37 deg female flared hose, permanently attached
25	37 deg female flared, hose, field attachable (2 piece)
26	37 deg female flared, hose, field attachable, segment clamp
27	37 deg female flared, hose, field attachable (3 piece)
28	45 deg male flared, hose, permanently attached
29	45 deg male flared, hose, field attachable
30	45 deg female flared, hose, permanently attached
31	45 deg female flared, hose, field attachable (2 piece)
32	45 deg female flared, hose, field attachable (3 piece)
33	Flareless, male, hose, permanently attached
34	Flareless, male, hose, field attachable
35	Flareless, male, hose, field attachable, segment clamp
36	Flareless, female, hose, permanently attached
37	Flareless, female, hose, field attachable
38	Flareless, female, hose, field attachable, segment clamp
39	Split flange, hose, standard pressure, permanently attached
40	Split flange, hose, standard pressure, field attachable
41	Split flange, hose, standard pressure, field attachable, segment clamp
42	Straight thread filler and drain plug-J532
43	Beaded tube hose end-J1231
45	Capillary, refrigeration-J513
49	Split flange, hose, high pressure, permanently attached
50	Split flange, hose, high pressure, field attachable
51	Split flange, hose, high pressure, field attachable, segment clamp
52	O-ring face seal, hydraulic-J1453
53	O-ring face seal, female, hose, permanently attached
54	O-ring face seal, female, hose, field attachable
55	O-ring face seal, male, hose, permanently attached
56	O-ring face seal, male, hose, field attachable
57	O-ring face seal, female, hose, field attachable, segment clamp
58	O-ring face seal, male, hose, field attachable, segment clamp
59	Female pipe, hose, permanently attached
60	Female pipe, hose, field attachable
61	Female pipe, hose, field attachable, segment clamp

TABLE 6B—FITTING TYPE SYMBOLS
(Arranged by Type)

FITTING TYPE OR STYLE	SYMBOL
Tube Fittings and Adapters - Automotive	
Pipe, automotive-J530, J531	13
45 deg flared, automotive-J512 (also refrigeration-J513)	01
Inverted flare, automotive-J512	04
Tapered sleeve, automotive-J512	06
Flanged sleeve, nylon tube, automotive-J246	10
Spherical sleeve, copper tube, automotive-J246	12
Straight thread filler and drain plug-J532	42
Capillary, refrigeration-J513	45
Tube Fittings and Adapters - Hydraulic	
Pipe, hydraulic-J514	14
37 deg flared, hydraulic-J514	07
Flareless, hydraulic-J514	08
SAE O-ring boss, hydraulic-J1926	09
4-bolt split flange O-ring, hydraulic-J518	11
O-ring face seal, hydraulic-J1453	52
Hose Fittings-J516	
Male pipe, hose, permanently attached	15
Male pipe, hose, field attachable	16
Male pipe, hose, field attachable, segment clamp	17
Female pipe, hose, permanently attached	59
Female pipe, hose, field attachable	60
Female pipe, hose, field attachable, segment clamp	61
45 deg male flared, hose, permanently attached	28
45 deg male flared, hose, field attachable	29
45 deg female flared, hose, permanently attached	30
45 deg female flared, hose, field attachable (2 piece)	31
45 deg female flared, hose, field attachable (3 piece)	32
Beaded tube hose end-J1231	43
37 deg male flared, hose, permanently attached	21
37 deg male flared, hose, field attachable	22
37 deg male flared, hose, field attachable, segment clamp	23
37 deg female flared, hose, permanently attached	24
37 deg female flared, hose, field attachable (2 piece)	25
37 deg female flared, hose, field attachable, segment clamp	26
37 deg female flared, hose, field attachable (3 piece)	27
Flareless, male, hose, permanently attached	33
Flareless, male, hose, field attachable	34
Flareless, male, hose, field attachable, segment clamp	35
Flareless, female, hose, permanently attached	36
Flareless, female, hose, field attachable	37
Flareless, female, hose, field attachable, segment clamp	38
SAE O-ring boss, hose, permanently attached	18
SAE O-ring boss, hose, field attachable	19
SAE O-ring boss, hose, field attachable, segment clamp	20
O-ring face seal, male, hose, permanently attached	55
O-ring face seal, male, hose, field attachable	56
O-ring face seal, male, hose, field attachable, segment clamp	58
O-ring face seal, female, hose, permanently attached	53
O-ring face seal, female, hose, field attachable	54
O-ring face seal, female, hose, field attachable, segment clamp	57
Split flange, standard pressure, hose, permanently attached	39
Split flange, standard pressure, hose, field attachable	40
Split flange, standard pressure, hose, field attachable, segment clamp	41
Split flange, high pressure, hose, permanently attached	49
Split flange, high pressure, hose, field attachable	50
Split flange, high pressure, hose, field attachable, segment clamp	51

TABLE 7—FITTING SHAPE SYMBOLS

Fitting Shape Symbol	Fitting Shape
01	Straight
02	90 deg elbow
03	45 deg elbow
04	Tee
05	Cross
06	Straight bulkhead union
07	90 deg bulkhead elbow union
08	45 deg bulkhead elbow union
09	Bulkhead tee
10	22-1/2 deg elbow
11	30 deg elbow
12	60 deg elbow
13	67-1/2 deg elbow
14	90 deg elbow short drop
15	90 deg elbow long drop
16	90 deg elbow extra long drop

TABLE 8—FITTING CONNECTING END SYMBOLS

Fitting Connecting End Symbol	Connecting Ends, Hose, and Combinations
01	Tube, all ends
02	Tube to external pipe
03	Tube to internal pipe
04	Tube to solder connection
05	Internal flare to external flare
06	Internal flare to external pipe
07	Internal flare union
08	Swivel flare union internal, all ends
09	Plug
10	Short nut, tube
11	Long nut, tube
12	CAP
13	Flare gasket
14	Flare seal bonnet
15	Tube sleeve
16	Reducing nut
17	Lock nut, small hex
18	Lock nut, large hex
19	Connector, large hex
20	Tube to SAE O-ring boss end
21	Tube to SAE straight swivel connection (female swivel)
22	Connector, long (tube to SAE O-ring boss end)
23	Reducer seat to tube
24	External pipe on run
25	External pipe on branch
26	Internal pipe on run
27	Internal pipe on branch
28	Straight thread on run
29	Straight thread on branch
30	Swivel straight pipe NPSM to external pipe
31	Swivel straight pipe NPSM to internal pipe
32	Tube to SAE straight swivel (female swivel) on run
33	Tube to SAE straight swivel (female swivel) on branch
34	Tube to internal pipe on run and external pipe on branch
35	Tube to external pipe on run and internal pipe on branch
36	Straight thread to SAE O-ring boss
37	External pipe to external pipe (nipple)
38	Internal pipe to internal pipe
39	Internal pipe to external pipe
40	Internal pipe inside external pipe, bushing
41	Seat insert, J512
42	100R1 hose, Type A
43	100R2 hose, Types A and B
44	100R3 hose
45	100R4 hose
46	100R5 hose
47	100R6 hose
48	100R7 hose
49	100R8 hose
50	100R9 hose, Type A
51	100R10 hose, Type A
52	100R11 hose
53	100R1 hose, Type AT
54	100R2 hose, Types AT and BT
55	
56	Internal pipe to SAE O-ring boss end
57	Swivel straight pipe NPSM to SAE O-ring boss end
58	Tube to bulkhead on run
59	Tube to bulkhead on branch
60	Beaded tube to male pipe-J1231
61	Split flange to flanged head, standard pressure series
62	Split flange to flanged head, high pressure series
63	Tube to pipe fusible-J513
64	Plug, pipe, fusible-J513
65	Tube to internal pipe, drum adapter-J513
66	Nut, short, refrigeration
67	Nut, long, refrigeration
68	Nut, short, reducing, refrigeration
69	Nut, long, reducing, refrigeration
70	Swivel straight pipe NPSM to external pipe on branch
71	Swivel straight pipe NPSM to internal pipe on branch
72	Tube to braze connection
73	
74	
75	100R9 hose, Type AT
76	100R10 hose, Type AT
77	100R12 hose
78	100R13 hose
79	100R14 hose, Type A
80	100R14 hose, Type B
81	SAE straight swivel (female swivel) to SAE O-ring boss end

Application of Code

The identification code shall be applied to the various fittings as depicted in the examples below.

1. Examples of Code Applied to Tube Fittings

The 45 deg flared tube connector for 1/8 in tube OD, shown in Fig. 1A and Table 2 of SAE J512, would be coded as follows:

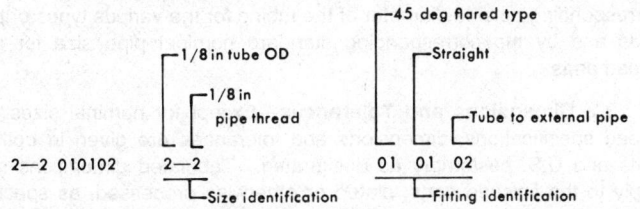

The hydraulic flareless tube tee for 3/8 in tube OD, shown in Fig. 22D and Table 7 of SAE J514, would be coded as follows:

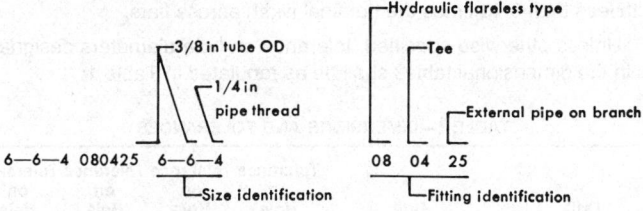

2. Examples of Code Applied to Hose Fittings

The male 45 deg flared type permanently attached style hose fitting for 1/4 in tube OD (-4 thread size) and 1/4 in ID SAE 100R1 hydraulic hose, shown in Fig. 8A and Table 5A of SAE J516, would be coded as follows:

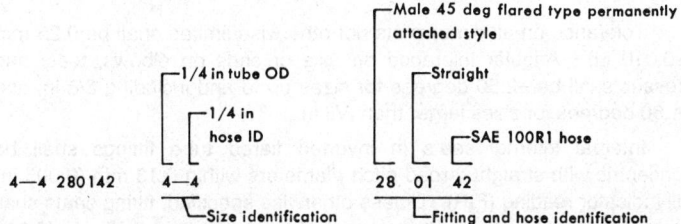

The 4-bolt split flange type field attachable screw style 45 deg angle hose fitting for 1/2 in flange size and 3/4 in ID SAE 100R2 hydraulic hose, shown in Fig. 17B and Table 9B of SAE J516, would be coded as follows:

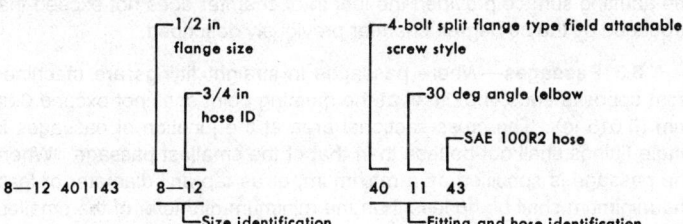

3. Example of Code Applied to 4-Bolt Split Flange Connections

The 4-bolt split flange connection for the 1-1/4 in flange size and split flange to flanged head, standard pressure series, shown in Fig. 1 and Table 1 of SAE J518, would be coded as follows:

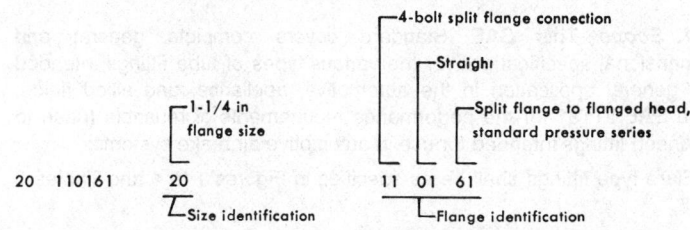

4. Example of Code Applied to Assembled Tapered Sleeve, Automotive, Union

The tapered sleeve, automotive union assembly for 1/4 in tube OD, the union shown in Fig. 12B and Table 8, the sleeve shown in Fig. 15 and Table 9, and the nut shown in Fig. 16 and Table 10 of SAE J512 would be coded as follows:

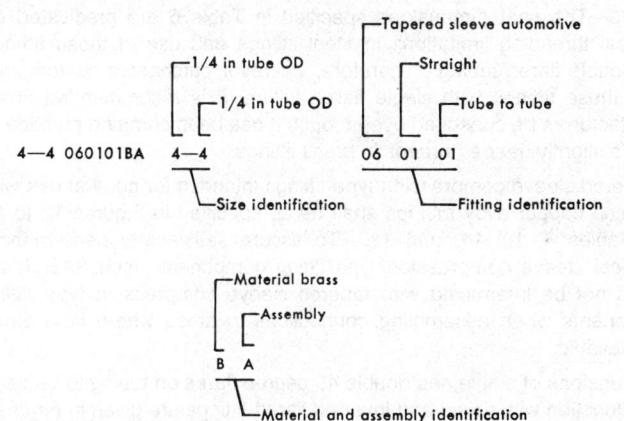

5. Example of Code Applied to Adapter Unions

The 90 deg adapter union in the 1/2 in size, shown in Fig. 25 and Table 10 of SAE J516, would be coded as follows:

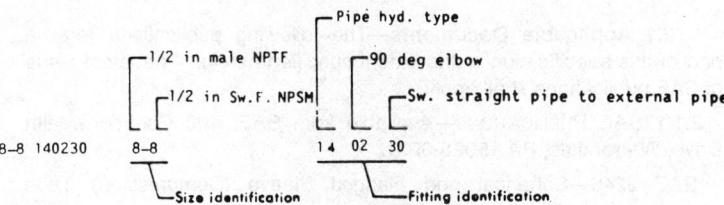

NOTE—For adapter unions shown in SAE J516 the order of thread designation shall be male pipe NPTF, female pipe NPTF, male O-ring and swivel female NPSM.

AUTOMOTIVE TUBE FITTINGS—SAE J512 JUN93

SAE Standard

Report of the Carburetor Fittings Division, approved June 1912, revised by the Fluid Conductors and Connectors Technical Committee November 1979, editorial change October 1980. Completely revised by the Fluid Conductors and Connectors Technical Committee June 1990. Revised by the SAE Fluid Conductors and Connectors Technical Committee SC1—Automotive and Hydraulic Tube and Fittings June 1992. Revised by the SAE Fluid Conductors and Connectors Technical Committee June 1993. Rationale statement available.

1. Scope—This SAE Standard covers complete general and dimensional specifications for the various types of tube fittings intended for general application in the automotive, appliance, and allied fields. See SAE J1131 for the performance requirements of reusable (push to connect) fittings intended for use in automotive air brake systems.

Flare type fittings shall be as specified in Figures 1 to 4 and Tables 3 to 5.

NOTE—For sizes 3/16 to 3/8 and 1/2 to 3/4 the flare type fittings depicted in Figures 1A to 3C are identical with the corresponding refrigeration tube fittings specified in SAE J513. Special size combination fittings 3/16 to 3/8 and 1/2 to 3/4 shall be as specified in SAE J513.

Inverted flared type fittings shall be as specified in Figures 5 to 11 and Tables 3, 6, 7, 8, and 9. Gages and gaging procedures pertaining to inverted flared tube fittings are given in Appendix A.

NOTE—The seat dimensions specified in Table 6 are predicated on practical threading limitations in steel fittings and use of these fittings with double flared tubing. Therefore, wherever purchasers contemplate using these fittings with single flared tubing, it is recommended fitting manufacturers be consulted even though it has been common practice to provide slightly deeper threads in brass fittings.

Tapered sleeve compression type fittings intended for general use with annealed copper alloy tubings shall be as specified in Figures 12 to 17 and Tables 3, 10, 11, and 12. To assure satisfactory performance, spherical sleeve compression type fitting components (see SAE J246) should not be intermixed with tapered sleeve compression type fitting components when assembling connections in areas where both types are available.

Dimensions of single and double 45 degree flares on tubing to be used in conjunction with flared and inverted flared fittings are given in Figure 2 and Table 3 of SAE J533.

The following general specifications supplement the dimensional data for all types of fittings contained in Tables 3 to 13 with respect to all unspecified detail.

2. References

2.1 Applicable Documents—The following publications form a part of this specification to the extent specified herein. The latest issue of SAE publications shall apply.

2.1.1 SAE PUBLICATIONS—Available from SAE, 400 Commonwealth Drive, Warrendale, PA 15096-0001.

SAE J246—Spherical and Flanged Sleeve (Compression) Tube Fittings
SAE J476—Dryseal Pipe Threads
SAE J513—Refrigeration Tube Fittings
SAE J533—Flares for Tubing
SAE J846—Coding Systems for Identification of Fluid Conductors and Connectors

2.1.2 ANSI PUBLICATION—Available from ANSI, 11 West 42nd Street, New York, NY 10036-8002.

ANSI B1.1—Screw Threads

2.1.3 ASTM PUBLICATION—Available from ASTM, 1916 Race Street, Philadelphia, PA 19103-1187.

ASTM B 117—Method of Salt Spray (Fog) Testing

3. General Specifications

3.1 Size Designations—Fitting sizes are designated by the corresponding outside diameter of the tubing for the various types of tube ends and by the corresponding standard nominal pipe size for pipe thread ends.

3.2 Dimensions and Tolerances—Except for nominal sizes and thread specifications, dimensions and tolerances are given in both SI units and U.S. customary as designated. Tabulated dimensions shall apply to the finished parts, plated or otherwise processed, as specified by the purchaser. Unless otherwise specified, the maximum and minimum across flats dimensions shall be within the commercial tolerance of bar or extruded stock from which the fittings are produced. The minimum across corners dimensions of hexagons shall be 1.092 times the nominal width across flats, but shall not result in a side flat width less than 0.43 times the nominal width across flats.

Unless otherwise specified, tolerance on hole diameters designated drill in the dimensional tables shall be as tabulated in Table 1:

TABLE 1—DIMENSIONS AND TOLERANCES

Drill Size Range mm	Drill Size Range in	Tolerance on Hole Diameter Plus mm	Tolerance on Hole Diameter Plus in	Tolerance on Hole Diameter Minus mm	Tolerance on Hole Diameter Minus in
0.343 thru 4.699	0.0135 thru 0.1850	0.08	0.003	0.05	0.002
4.762 thru 6.299	0.1875 thru 0.2480	0.10	0.004	0.05	0.002
6.350 thru 19.050	0.2500 thru 0.7500	0.15	0.006	0.08	0.003
19.251 thru 25.400	0.7579 thru 1.0000	0.18	0.007	0.10	0.004

Tolerance on all dimensions not otherwise limited shall be 0.25 mm (±0.010 in). Angular tolerance on axis of ends on elbows, tees, and crosses shall be ±2.50 degrees for sizes up to and including 3/8 in, and ±1.50 degrees for sizes larger than 3/8 in.

Integral internal seats in inverted flared tube fittings shall be concentric with straight thread pitch diameters within 0.13 mm (0.005 in) full indicator reading (FIR). Unless otherwise specified, fitting seats shall be concentric with straight thread pitch diameters within 0.25 mm (0.010 in) full indicator reading (FIR).

Where so illustrated and not otherwise specified, hexagon corners shall be chamfered 30 degrees ± 5 degrees to a diameter equal to the nominal width across flats, with a tolerance of -0.41 mm (-0.016 in); or 0.41 where design permits, corners may be chamfered to the diameter of the abutting surface provided the length of chamfer does not exceed that produced by the 30 degree chamfer previously described.

3.3 Passages—Where passages in straight fittings are machined from opposite ends, the offset at the meeting point shall not exceed 0.38 mm (0.015 in). The cross-sectional area at the junction of passages in angle fittings shall not be less than that of the smallest passage. Where the passage is specified as a maximum, or as tap drill diameter or less, the minimum shall be no less than the minimum diameter of the smallest passage in the fitting.

3.4 Wall Thickness—Unless otherwise designated, the wall thickness at any point on fittings shall not be less than the thickness established by the specified dimensions, tolerances, and eccentricities for inner and outer surfaces.

3.5 Contour—Details of contour shall be optional with manufacturer provided the tabulated dimensions are maintained and serviceability of the fitting is not impaired. Wrench flats on elbows and tees shall be optional. Where extruded or forged shapes are reduced to conserve material, the wall thickness, unless otherwise specified, shall not be less than the respective minimum values tabulated in Table 2:

TABLE 2—WALL THICKNESS

Nom Tube OD in	Wall Thickness Min Extruded Shape[a] mm	Wall Thickness Min Extruded Shape[a] in	Wall Thickness Min Forged Shape mm	Wall Thickness Min Forged Shape in
1/8	1.0	0.04	1.52	0.060
3/16	1.0	0.04	1.78	0.070
1/4	1.0	0.04	1.90	0.075
5/16	1.3	0.05	1.90	0.075
3/8	1.3	0.05	2.29	0.090
7/16	1.5	0.06	2.29	0.090
1/2	1.5	0.06	2.29	0.090
9/16	1.5	0.06	2.29	0.090
5/8	2.0	0.08	2.54	0.100
3/4	2.0	0.08	2.54	0.100
7/8	2.0	0.08	3.05	0.120
1	2.0	0.08	3.05	0.120

[a] Applies to reduction to one plane of shape only.

3.6 Straight Threads—Unified Standard Class 2A external threads and Class 2B internal threads with minor diameters, where specified, modified to Class 3B limits, shall apply to plain finish (unplated) fittings of all types. For externally threaded parts with additive finish, the maximum diameters of Class 2A may be exceeded by the amount of the allowance; that is, the basic diameters (Class 2A maximum diameters plus the allowance) shall apply to an externally threaded part after plating. For internally threaded parts with additive finish, the Class 2B diameters and modified minor diameters shall apply after plating.

The pitch diameter tolerance shall be the same as the corresponding diameter-pitch combination and the class of the Unified coarse and fine thread series or for special diameter-pitch combinations shall be based on diameter, pitch, and a length of engagement of 9 times the pitch. See ANSI B1.1.

Where external threads are produced by roll threading and the body is not undercut, the unthreaded portion of body adjacent to the shoulder may be reduced to the minimum pitch diameter.

External threads shall be chamfered to the diameter of abutting surfaces, or to the diameters specified, to produce a length of chamfered or partial thread equivalent to 3/4 to 1-1/4 times the pitch (rounded to a three-place decimal).

Internal threads shall be countersunk 90 degrees, included angle, to the diameters specified to the dimensional tables.

FLARED TYPE

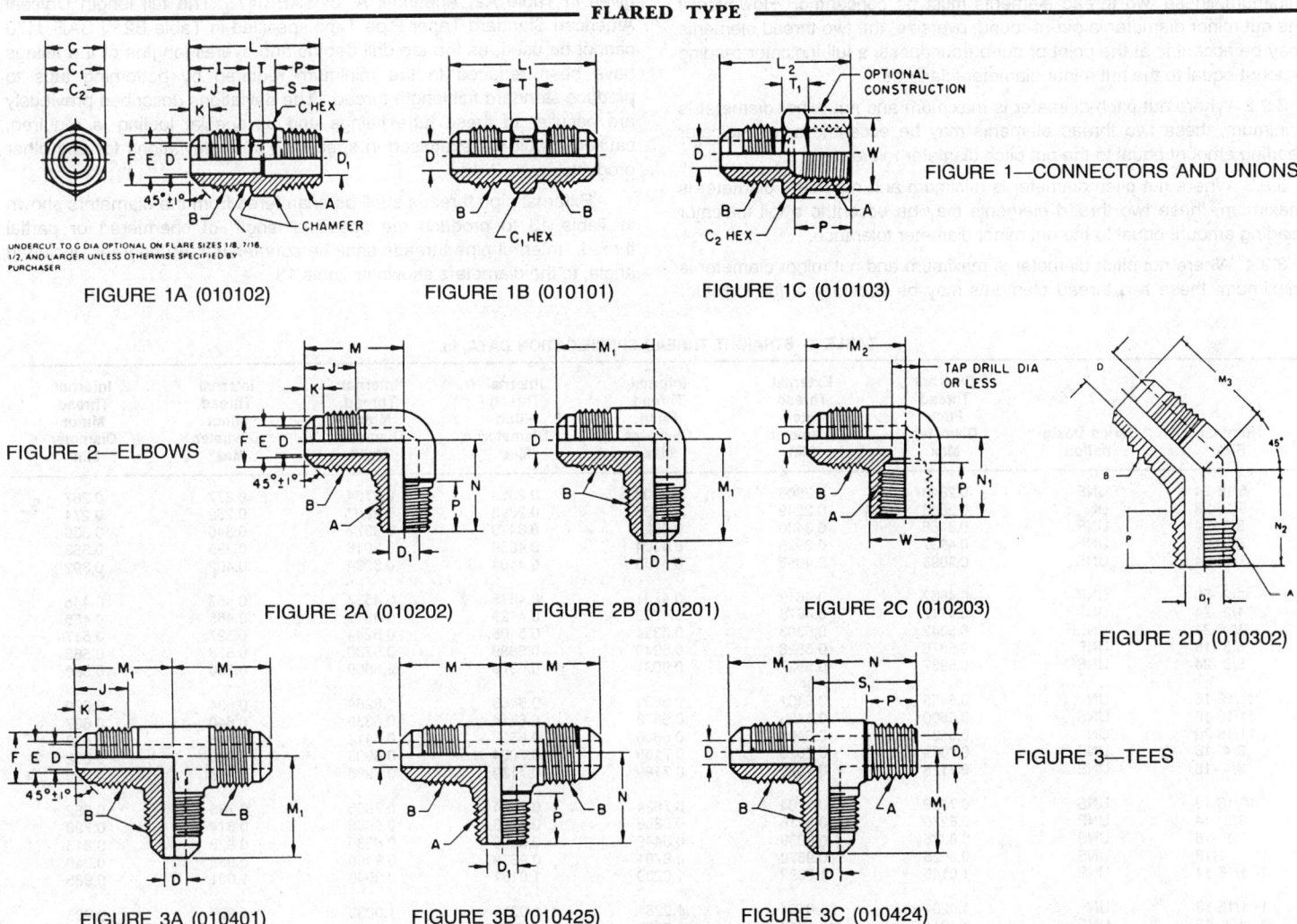

FIGURE 1A (010102) FIGURE 1B (010101) FIGURE 1C (010103)
FIGURE 1—CONNECTORS AND UNIONS

FIGURE 2—ELBOWS
FIGURE 2A (010202) FIGURE 2B (010201) FIGURE 2C (010203) FIGURE 2D (010302)

FIGURE 3A (010401) FIGURE 3B (010425) FIGURE 3C (010424)
FIGURE 3—TEES

NOTES: UNSPECIFIED DETAIL WITH RESPECT TO DIMENSIONS, TOLERANCES, CONTOURS, MATERIAL, WORKMANSHIP, ETC., MUST CONFORM TO GENERAL SPECIFICATIONS FOR AUTOMOTIVE TUBE FITTINGS. THE DIMENSIONAL DESIGNATIONS ON THE FIRST FIGURE IN EACH GROUP SHALL APPLY TO ALL OTHER FIGURES IN THAT GROUP EXCEPT AS SHOWN OTHERWISE. CODES SHOWN IN BRACKETS ADJACENT TO FIGURE NUMBERS REPRESENT RESPECTIVE FITTING IDENTIFICATION IN ACCORDANCE WITH SAE J846.

3.7 Thread Eccentricity Tolerances—The various thread elements of Class 2A external and Class 2B, modified, internal threads on tube fittings shall be concentric within the following limitations:

3.8 External Thread (Screw)

3.8.1 Where screw pitch diameter is maximum and screw major diameter is maximum, these two thread elements must be concentric. However, if the screw major diameter is out-of-round, undersize, these two thread elements may be eccentric at the point of out-of-roundness, a full indicator reading amount equal to the screw major diameter tolerance.

3.8.2 Where screw pitch diameter is minimum and screw major diameter is maximum, these two thread elements may be eccentric a full indicator reading amount equal to the screw pitch diameter tolerance.

3.8.3 Where screw pitch diameter is maximum and screw major diameter is minimum, these two thread elements may be eccentric a full indicator reading amount equal to the screw major diameter tolerance.

3.8.4 Where screw pitch diameter is minimum and screw major diameter is minimum, these two thread elements may be eccentric a full indicator reading amount equal to the sum of the screw pitch diameter tolerance and the screw major diameter tolerance.

3.9 Internal Thread (Nut)

3.9.1 Where nut pitch diameter is minimum and nut minor diameter is minimum, these two thread elements must be concentric. However, if the nut minor diameter is out-of-round, oversize, the two thread elements may be eccentric at the point of out-of-roundness, a full indicator reading amount equal to the nut minor diameter tolerance.

3.9.2 Where nut pitch diameter is maximum and nut minor diameter is minimum, these two thread elements may be eccentric a full indicator reading amount equal to the nut pitch diameter tolerance.

3.9.3 Where nut pitch diameter is minimum and nut minor diameter is maximum, these two thread elements may be eccentric a full indicator reading amount equal to the nut minor diameter tolerance.

3.9.4 Where nut pitch diameter is maximum and nut minor diameter is maximum, these two thread elements may be eccentric a full indicator reading amount equal to the sum of the nut pitch diameter tolerance and the nut minor diameter tolerance.

3.10 Pipe Threads—Pipe threads, unless there is specific authorization to the contrary, shall conform to the Dryseal American Standard Taper Pipe Thread (NPTF). At purchaser's option, the pipe thread on automotive tube fittings may be shortened in conformity with the SAE Short Dryseal Taper Pipe Thread (PTF-SAE Short). Specifications for pipe threads are given in detail in SAE J476.

The length of full form external thread shall not be shorter than L_2, plus one pitch (thread) for Dryseal NPTF and L_2, for Dryseal PTF-SAE Short, except that where thread is cut through into a relieved body or undercut on the fitting, the minimum full thread length may be reduced by one pitch (thread).

Where external pipe threads are produced by roll threading, the diameter of the unthreaded portion of shank adjacent to shoulder may be reduced to the E, basic diameter.

The tube fitting dimensions tabulated herein are based on length of the Dryseal American Standard Taper Pipe Thread (NPTF), it being the consensus of manufacturers and users that trouble-free assembly cannot be assured unless a full length is used. However, the tap drill depths and overall lengths specified in the tables for fittings with internal taper pipe threads are not consistent with the tap drill depths and overall thread lengths of the Dryseal American Standard Taper Pipe Threads (NPTF) given in Table A2, Appendix A, of SAE J476. The full length Dryseal American Standard Taper Pipe Taps specified in Table B2 of SAE J476 cannot be used, as the tap drill depths and overall lengths of the fittings have been reduced to the minimum required by bottoming taps to produce standard full length thread. The deviations described previously are peculiar to these tube fittings and as special tooling is required, caution should be exercised in specifying such deviations for any other products.

External pipe threads shall be chamfered from the diameters shown in Table 13 to produce the specified length of chamfered or partial thread. Internal pipe threads shall be countersunk 90 degrees, included angle, to the diameters shown in Table 13.

TABLE 3—STRAIGHT THREAD SPECIFICATION DATA, in

Nominal Size	Series Designation	External Thread Pitch Diameter Max	External Thread Pitch Diameter Min	Internal Thread Pitch Diameter Max	Internal Thread Pitch Diameter Min[a]	Internal Thread Minor Diameter Max[b]	Internal Thread Minor Diameter Max[c]	Internal Thread Minor Diameter Min[c]
5/16-24	UNF	0.2843	0.2806	0.2902	0.2854	0.2754	0.277	0.267
5/16-28	UN	0.2883	0.2849	0.2937	0.2893	0.2807	0.282	0.274
3/8 -24	UNF	0.3468	0.3430	0.3528	0.3479	0.3372	0.340	0.330
7/16-20	UNF	0.4037	0.3995	0.4104	0.4050	0.3916	0.395	0.383
7/16-24	UNS	0.4093	0.4055	0.4153	0.4104	0.3994	0.402	0.392
1/2 -20	UNF	0.4662	0.4619	0.4731	0.4675	0.4537	0.457	0.446
1/2 -24	UNS	0.4717	0.4678	0.4780	0.4729	0.4619	0.465	0.455
9/16-24	UNEF	0.5342	0.5303	0.5354	0.5405	0.5244	0.527	0.517
5/8 -18	UNF	0.5875	0.5828	0.5949	0.5889	0.5730	0.578	0.565
5/8 -24	UNEF	0.5967	0.5927	0.6031	0.5979	0.5869	0.590	0.580
11/16-16	UN	0.6455	0.6407	0.6531	0.6469	0.6284	0.634	0.620
11/16-18	UNS	0.6500	0.6455	0.6573	0.6514	0.6335	0.640	0.627
11/16-20	UN	0.6537	0.6494	0.6606	0.6550	0.6412	0.645	0.633
3/4 -16	UNF	0.7079	0.7029	0.7159	0.7094	0.6908	0.696	0.682
3/4 -18	UNS	0.7125	0.7079	0.7199	0.7139	0.6980	0.703	0.690
13/16-18	UNS	0.7750	0.7704	0.7824	0.7764	0.7605	0.765	0.752
7/8 -14	UNF	0.8270	0.8216	0.8356	0.8286	0.8068	0.814	0.798
7/8 -18	UNS	0.8375	0.8239	0.8449	0.8389	0.8230	0.828	0.815
1 -18	UNS	0.9625	0.9578	0.9701	0.9639	0.9480	0.953	0.940
1- 1/16-14	UNS	1.0145	1.0092	1.0230	1.0161	0.9940	1.001	0.985
1- 1/16-16	UN	1.0204	1.0154	1.0284	1.0219	1.0033	1.009	0.995
1- 1/4 -12	UNF	1.1941	1.1879	1.2039	1.1959	1.1698	1.178	1.160
1- 3/8 -12	UNF	1.3127	1.3190	1.3291	1.3209	1.2948	1.303	1.285

[a] These values are also the basic pitch diameter.
[b] Class 3B maximum minor diameter limits shall apply where so designated in respective dimensional tables.
[c] Class 2B minor diameter limits shall apply unless otherwise designated.

TABLE 4—DIMENSIONS OF CONNECTORS, UNIONS, ELBOWS, AND TEES
(Figures 1A to 3C)

Nom Tube OD in	B Dryseal Taper Thread NPTF[a] in	B Nom Thread Size in Class 2A Ext	C Nom in	C_1 Nom in	C_2 Nom in	D[e] Dia Drill mm	D[e] Dia Drill in
1/8	1/8	5/16 -24	7/16	5/16	9/16	1.98	0.078
3/16	1/8	3/8 -24	7/16	3/8	9/16	3.18	0.125
1/4	1/8	7/16 -20	7/16	7/16	9/16	4.78	0.188
5/16	1/8	1/2 -20	1/2	1/2	9/16	5.56	0.219
3/8	1/4	5/8 -18	5/8	5/8	11/16	7.14	0.281
7/16	1/4	11/16-16	11/16	11/16	11/16	7.92	0.312
1/2	3/8	3/4 -16	3/4	3/4	13/16	10.31	0.406
5/8	1/2	7/8 -14	7/8	7/8	1	12.70	0.500
3/4	1/2	1- 1/16-14	1- 1/16	1- 1/16	1- 1/16	15.88	0.625
7/8	3/4	1- 1/4 -12	1- 1/4	1- 1/4	1- 1/4	19.05	0.750
1	1	1- 3/8 -12	1- 3/8	1- 3/8	1- 1/2	22.22	0.875

Nom Tube OD in	D_1[e] Dia Drill mm	D_1[e] Dia Drill in	E Dia mm	E Dia in	F Dia mm	F Dia in	G[d] Dia mm +0.00 −0.25	G[d] Dia in +0.000 −0.010
1/8	5.56	0.219	2.77	0.109	5.94	0.234	6.35	0.250
3/16	5.56	0.219	3.96	0.156	7.54	0.297	—	—
1/4	5.56	0.219	5.56	0.219	8.74	0.344	—	—
5/16	5.56	0.219	6.35	0.250	10.31	0.406	—	—
3/8	7.92	0.312	7.92	0.312	13.49	0.531	—	—
7/16	7.92	0.312	8.74	0.344	14.63	0.578	15.14	0.596
1/2	10.31	0.406	11.13	0.438	16.28	0.641	16.74	0.659
5/8	14.27	0.562	13.49	0.531	19.05	0.750	19.56	0.770
3/4	14.27	0.562	18.26	0.719	23.83	0.938	24.33	0.958
7/8	19.05	0.750	20.24	0.797	28.58	1.125	28.65	1.128
1	23.82	0.938	23.83	0.938	31.75	1.250	31.83	1.253

Nom Tube OD in	I mm	I in	J[d] Full Thread Min mm	J[d] Full Thread Min in	K mm	K in	L[b] mm ±0.08	L[b] in ±0.03
1/8	9.7	0.38	7.9	0.31	3.0	0.12	23.4	0.92
3/16	11.2	0.44	9.7	0.38	3.0	0.12	25.4	1.00
1/4	12.7	0.50	10.4	0.41	4.1	0.16	26.9	1.06
5/16	14.2	0.56	11.9	0.47	4.8	0.19	29.5	1.16
3/8	15.7	0.62	13.7	0.54	5.6	0.22	36.6	1.44
7/16	17.5	0.69	14.2	0.56	6.4	0.25	38.9	1.53
1/2	19.0	0.75	16.8	0.66	6.4	0.25	41.1	1.62
5/8	22.4	0.88	19.3	0.76	7.1	0.28	50.8	2.00
3/4	25.4	1.00	22.9	0.90	7.1	0.28	55.6	2.19
7/8	28.4	1.12	24.1	0.95	9.7	0.38	60.2	2.37
1	28.4	1.12	24.6	0.97	9.7	0.38	66.5	2.62

TABLE 4—(Continued)

Nom Tube OD in	L_1 mm ±0.8	L_1 in ±0.03	L_2[b,c] mm ±0.8	L_2[b,c] in ±0.03	M mm ±0.8	M in ±0.03	M_1 mm ±0.8	M_1 in ±0.03
1/8	23.4	0.92	23.1	0.91	15.7	0.62	15.7	0.62
3/16	26.9	1.06	24.6	0.97	19.0	0.75	19.0	0.75
1/4	30.2	1.19	26.2	1.03	20.6	0.81	22.4	0.88
5/16	34.0	1.34	26.9	1.06	23.1	0.91	23.1	0.91
3/8	38.1	1.50	33.3	1.31	25.4	1.00	26.9	1.06
7/16	42.2	1.66	35.8	1.41	28.4	1.12	28.4	1.12
1/2	46.0	1.81	38.1	1.50	31.0	1.22	31.0	1.22
5/8	53.8	2.12	46.0	1.81	35.8	1.41	35.8	1.41
3/4	62.0	2.44	48.5	1.91	41.1	1.62	42.2	1.66
7/8	69.6	2.74	57.2	2.25	44.5	1.75	44.5	1.75
1	71.1	2.80	62.0	2.44	49.3	1.94	49.3	1.94

Nom Tube OD in	M_2 mm ±0.8	M_2 in ±0.03	M_3 mm ±0.8	M_3 in ±0.03	N[b] mm ±0.8	N[b] in ±0.03	N_1[b,c] mm ±0.8	N_1[b,c] in ±0.03
1/8	19.0	0.75	15.0	0.59	17.5	0.69	10.7	0.42
3/16	20.6	0.81	15.7	0.62	19.0	0.75	11.2	0.44
1/4	22.4	0.88	17.0	0.67	19.8	0.78	11.9	0.47
5/16	23.9	0.94	19.8	0.78	19.8	0.78	11.9	0.47
3/8	27.7	1.09	22.6	0.89	26.9	1.06	17.5	0.69
7/16	28.4	1.12	24.6	0.97	26.9	1.06	18.3	0.72
1/2	32.5	1.28	26.9	1.06	28.4	1.12	19.0	0.75
5/8	38.1	1.50	31.2	1.23	35.1	1.38	25.4	1.00
3/4	41.1	1.62	35.8	1.41	38.1	1.50	26.9	1.06
7/8	47.8	1.88	41.1	1.62	42.9	1.69	28.4	1.12
1	52.3	2.06	42.9	1.69	49.3	1.94	35.1	1.38

Nom Tube OD in	N_2[b] mm ±0.8	N_2[b] in ±0.03	P[b,c] Min mm	P[b,c] Min in	S[b,e] Max mm	S[b,e] Max in	S_1[b,e] Min mm	S_1[b,e] Min in
1/8	13.2	0.52	9.7	0.38	11.7	0.46	19.6	0.77
3/16	13.2	0.52	9.7	0.38	12.2	0.48	21.6	0.85
1/4	16.3	0.64	9.7	0.38	12.2	0.48	23.4	0.92
5/16	16.3	0.64	9.7	0.38	—	—	—	—
3/8	21.8	0.86	14.2	0.56	17.5	0.69	31.5	1.24
7/16	21.8	0.86	14.2	0.56	—	—	—	—
1/2	24.1	0.95	14.2	0.56	—	—	—	—
5/8	29.7	1.17	19.0	0.75	23.9	0.94	42.4	1.67
3/4	30.5	1.20	19.0	0.75	31.0	1.22	51.3	2.02
7/8	32.2	1.27	19.0	0.75	—	—	—	—
1	37.6	1.48	23.9	0.94	31.0	1.22	61.5	2.42

TABLE 4—(Continued)

Nom Tube OD in	T[f] Ref mm	T[f] Ref in	T[1] Min mm	T[1] Min in	W[g] Dia mm +0.0 −0.5	W[g] Dia in +0.00 −0.02
1/8	3.8	0.15	5.3	0.21	14.2	0.56
3/16	4.6	0.18	5.3	0.21	14.2	0.56
1/4	4.6	0.18	6.1	0.24	14.2	0.56
5/16	5.3	0.21	6.1	0.24	14.2	0.56
3/8	6.1	0.24	7.6	0.30	17.5	0.69
7/16	6.9	0.27	7.6	0.30	17.5	0.69
1/2	7.6	0.30	9.4	0.37	20.6	0.81
5/8	9.4	0.37	10.9	0.43	25.4	1.00
3/4	10.9	0.43	12.4	0.49	26.9	1.06
7/8	12.4	0.49	13.2	0.52	31.8	1.25
1	14.0	0.55	14.7	0.58	38.1	1.50

[a] Dryseal American Standard Taper Pipe Thread. See General Specifications.
[b] Where SAE Short Pipe Thread is authorized by purchaser, dimensions L, L_2, N, N_1, P, S, and S_1 are reduced in accordance with reduction of pipe thread length. See SAE J476 Tables 3 and 4.
[c] Tap drill depths given require use of bottoming taps to produce standard full thread length. For increased tap clearance, see Internal Taper Pipe Threads in the General Specifications.
[d] Where thread relief undercut is used, the last thread shall be chamfered 1/2 to 1 pitch long from G diameter and dimension 1 may be reduced by an amount equal to 1/2 pitch.
[e] At manufacturers' option, through passages in fittings shown in Figures 1A and 3C may conform with the smaller diameter specified or be counterbored to the larger diameter from the appropriate end for depths S or S_1, respectively.
[f] Minimum design thickness, not subject to inspection.
[g] Basic dimensions shown shall apply as minimum diameter or across flats for bosses. The −0.5 mm (−0.02 in) tolerance shall apply only to chamber diameter on full hexagon versions of fitting in Figure 1C.

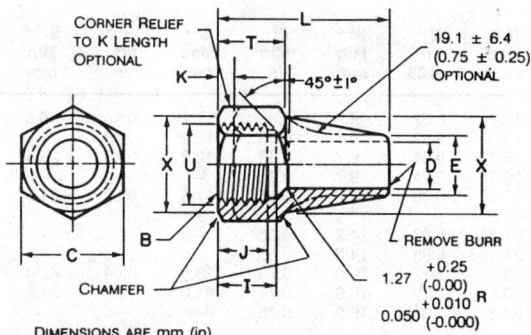

FIGURE 4—FLARED TYPE NUTS
(010110) SHORT NUT
(010111) LONG NUT

TABLE 5—DIMENSIONS OF NUTS (Figure 4)

Nom Tube OD in	B Nom Thread Size in Class 2B int	C Nom in	D Dia mm +0.13 −0.00	D Dia in +0.005 −0.000	E Dia mm	E Dia in	I mm	I in
1/8	5/16-24	3/8	3.30	0.130	6.4	0.25	5.6	0.22
3/16	3/8 -24	7/16	4.88	0.192	7.9	0.31	7.1	0.28
1/4	7/16-20	9/16	6.48	0.255	9.7	0.38	8.6	0.34
5/16	1/2 -20	5/8	8.05	0.317	11.2	0.44	9.7	0.38
3/8	5/8 -18	3/4	9.65	0.380	12.7	0.50	11.2	0.44
7/16	11/16-16	13/16	11.23	0.442	14.2	0.56	11.9	0.47
1/2	3/4 -16	7/8	12.83	0.505	15.7	0.62	13.5	0.53
5/8	7/8 -14	1- 1/16	16.00	0.630	19.1	0.75	16.8	0.66
3/4	1- 1/16-14	1- 1/4	19.18	0.755	22.4	0.88	19.8	0.78
7/8	1- 1/4 -12	1- 1/2	22.35	0.880	26.9	1.06	19.8	0.78
1	1- 3/8 -12	1- 5/8	25.53	1.005	30.2	1.19	20.6	0.81

Nom Tube OD in	J Full Thread Min mm	J Full Thread Min in	K mm	K in	L Long mm	L Long in	L Short mm	L Short in
1/8	4.1	0.16	2.3	0.09	19.1	0.75	12.7	0.50
3/16	5.6	0.22	2.3	0.09	20.6	0.81	15.7	0.62
1/4	7.1	0.28	3.0	0.12	23.9	0.94	19.1	0.75
5/16	7.9	0.31	3.0	0.12	28.4	1.12	22.4	0.88
3/8	9.7	0.38	3.0	0.12	33.3	1.31	25.4	1.00
7/16	10.4	0.41	3.0	0.12	38.1	1.50	26.9	1.06
1/2	11.9	0.47	3.0	0.12	41.1	1.62	28.4	1.12
5/8	15.0	0.59	3.0	0.12	47.8	1.88	33.3	1.31
3/4	17.5	0.69	3.0	0.12	55.6	2.19	38.1	1.50
7/8	17.5	0.69	3.0	0.12	58.7	2.31	41.1	1.62
1	17.5	0.69	3.0	0.12	63.5	2.50	46.0	1.81

Nom Tube OD in	T mm	T in	U Dia Min mm	U Dia Min in	U Dia Max mm	U Dia Max in	X Dia mm	X Dia in
1/8	5.6	0.22	8.4	0.33	8.9	0.35	9.7	0.38
3/16	7.1	0.28	9.9	0.39	10.4	0.41	11.2	0.44
1/4	9.7	0.38	11.4	0.45	11.9	0.47	14.2	0.56
5/16	11.2	0.44	13.0	0.51	13.5	0.53	15.7	0.62
3/8	12.7	0.50	16.3	0.64	17.0	0.67	19.1	0.75
7/16	14.2	0.56	18.0	0.71	18.8	0.74	20.6	0.81
1/2	15.7	0.62	19.6	0.77	20.3	0.80	22.4	0.88
5/8	19.1	0.75	22.9	0.90	23.6	0.93	26.9	1.06
3/4	22.4	0.88	27.4	1.08	28.2	1.11	31.8	1.25
7/8	25.4	1.00	32.3	1.27	33.0	1.30	38.1	1.50
1	25.4	1.00	35.3	1.39	36.1	1.42	41.1	1.62

3.11 Material and Manufacture—All types of automotive tube fittings shall be made from brass or steel as specified by the purchaser and, at manufacturer's option and in accordance with his process of manufacture, may be milled from the bar or forged. Brass shall be UNS C36000 (half-hard), UNS C34500, or UNS C35000 for bar or extruded stock and UNS C37700 for forgings.

(R) **3.12 Finish**—The external surfaces and threads of all carbon steel parts shall be plated or coated with a suitable material that passes a 72 h salt spray test in accordance with ASTM B 117. Any appearance of red rust during the 72 h salt spray test shall be considered failure, except for the following:

  a. All internal fluid passages.
  b. Edges such as hex points, serrations, and crests of threads where there may be mechanical deformation of the plating or coating typical of mass-produced parts or shipping effects.
  c. Areas where there is mechanical deformation of the plating or coating caused by crimping, flaring, bending, and other post-plate metal forming operations.
  d. Areas where the parts are suspended or affixed in the test chamber where condensate can accumulate.

NOTE—Cadmium plating is not preferred due to environmental reasons. Parts manufactured to this standard after January 1, 1997, shall not be cadmium plated. Internal fluid passages shall be protected from corrosion during storage. Changes in plating may affect assembly torques and require requalification, when applicable.

3.13 Workmanship—Workmanship shall conform to the best commercial practice to produce high-quality fittings. Fittings shall be free from all hanging burrs, loose scale, and slivers which might become dislodged in usage and all other defects which might affect their serviceability. All sealing surfaces must be smooth except that annular tool marks up to 2.5 μm (100 μin) maximum shall be permissible.

22.11

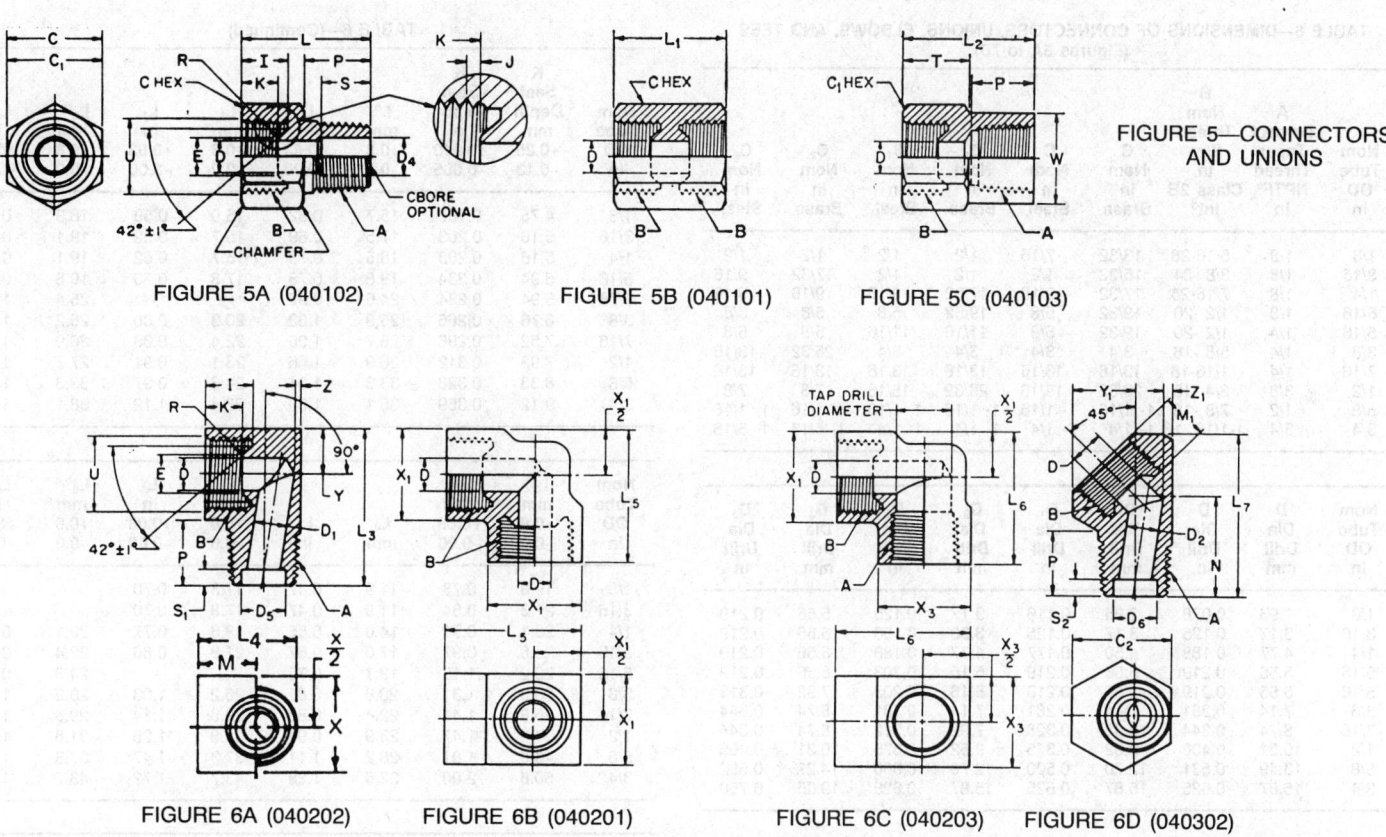

FIGURE 5A (040102) FIGURE 5B (040101) FIGURE 5C (040103)

FIGURE 5—CONNECTORS AND UNIONS

FIGURE 6A (040202) FIGURE 6B (040201) FIGURE 6C (040203) FIGURE 6D (040302)

FIGURE 6—ELBOWS

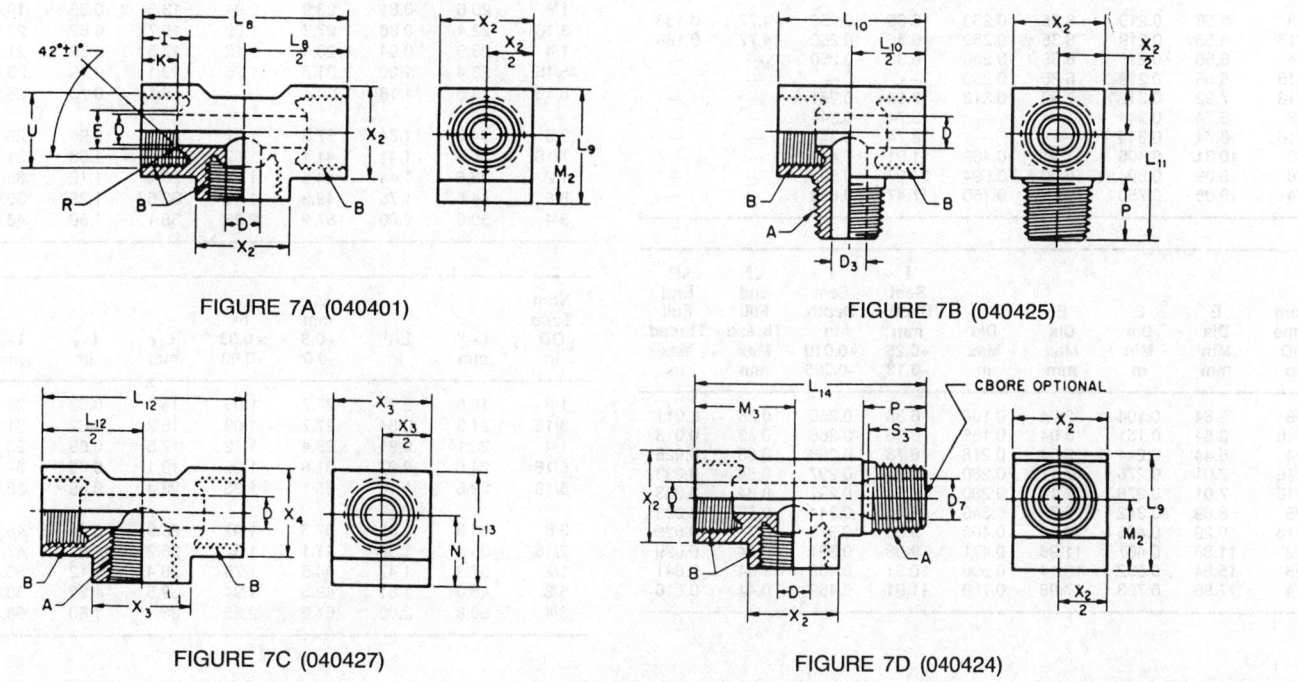

FIGURE 7A (040401) FIGURE 7B (040425)

FIGURE 7C (040427) FIGURE 7D (040424)

FIGURE 7—TEES

NOTES: UNSPECIFIED DETAIL WITH RESPECT TO DIMENSIONS, TOLERANCES, CONTOURS, MATERIAL, WORKMANSHIP, ETC., MUST CONFORM TO GENERAL SPECIFICATIONS FOR AUTOMOTIVE TUBE FITTINGS. THE DIMENSIONAL DESIGNATIONS ON THE FIRST FIGURE IN EACH GROUP SHALL APPLY TO ALL OTHER FIGURES IN THAT GROUP EXCEPT AS SHOWN OTHERWISE. CODES SHOWN IN BRACKETS ADJACENT TO FIGURE NUMBERS REPRESENT RESPECTIVE FITTING IDENTIFICATION IN ACCORDANCE WITH SAE J846.

TABLE 6—DIMENSIONS OF CONNECTORS, UNIONS, ELBOWS, AND TEES
(Figures 5A to 7D)

Nom Tube OD in	A Dryseal Taper Thread NPTF[a] in	B Nom Thread Size in Class 2B Int[d]	C Nom in Brass	C Nom in Steel	C_1 Nom in Brass	C_1 Nom in Steel	C_2 Nom in Brass	C_2 Nom in Steel
1/8	1/8	5/16-28	13/32	7/16	1/2	1/2	1/2	1/2
3/16	1/8	3/8 -24	15/32	1/2	1/2	1/2	17/32	9/16
1/4	1/8	7/16-24	17/32	9/16	17/32	9/16	9/16	9/16
5/16	1/8	1/2 -20	19/32	5/8	19/32	5/8	5/8	5/8
5/16	1/4	1/2 -20	19/32	5/8	11/16	11/16	5/8	5/8
3/8	1/4	5/8 -18	3/4	3/4	3/4	3/4	25/32	13/16
7/16	1/4	11/16-18	13/16	13/16	13/16	13/16	13/16	13/16
1/2	3/8	3/4 -18	29/32	15/16	29/32	15/16	7/8	7/8
5/8	1/2	7/8 -18	1-1/16	1-1/16	1-1/16	1-1/16	1-1/16	1-1/16
3/4	3/4	1-1/16 -16	1-1/4	1-1/4	1-1/4	1-1/4	1-5/16	1-5/16

Nom Tube OD in	D Dia Drill mm	D Dia Drill in	D_1 Dia Drill mm	D_1 Dia Drill in	D_2 Dia Drill mm	D_2 Dia Drill in	D_3 Dia Drill mm	D_3 Dia Drill in
1/8	1.98	0.078	2.95	0.116	3.17	0.125	5.56	0.219
3/16	3.17	0.125	3.17	0.125	3.96	0.156	5.56	0.219
1/4	4.77	0.188	4.50	0.177	4.77	0.188	5.56	0.219
5/16	5.56	0.219	5.56	0.219	5.16	0.203	5.56	0.219
5/16	5.56	0.219	5.56	0.219	5.16	0.203	7.92	0.312
3/8	7.14	0.281	7.14	0.281	7.14	0.281	8.74	0.344
7/16	8.74	0.344	8.33	0.328	7.92	0.312	8.74	0.344
1/2	10.31	0.406	9.52	0.375	9.52	0.375	10.31	0.406
5/8	13.49	0.531	12.70	0.500	12.70	0.500	14.27	0.562
3/4	15.87	0.625	15.87	0.625	15.87	0.625	19.05	0.750

Nom Tube OD in	D_4 Dia Drill mm	D_4 Dia Drill in	D_5 Dia Drill mm	D_5 Dia Drill in	D_6 Dia Drill mm	D_6 Dia Drill in	D_7 Dia Drill mm	D_7 Dia Drill in
1/8	5.56	0.219	6.35	0.250	6.35	0.250	4.77	0.188
3/16	5.56	0.219	6.35	0.250	6.35	0.250	4.77	0.188
1/4	5.56	0.219	6.35	0.250	6.35	0.250	—	—
5/16	5.56	0.219	6.35	0.250	—	—	—	—
5/16	7.92	0.312	7.92	0.312	7.14	0.281	—	—
3/8	8.74	0.344	—	—	8.74	0.344	—	—
7/16	8.74	0.344	—	—	8.74	0.344	—	—
1/2	10.31	0.406	11.91	0.469	11.91	0.469	—	—
5/8	15.09	0.594	15.09	0.594	14.27	0.562	—	—
3/4	19.05	0.750	19.05	0.750	17.47	0.688	—	—

Nom Tube OD in	E Dia Min mm	E Dia Min in	E Dia Max mm	E Dia Max in	I Seat Depth mm +0.25 -0.13	I Seat Depth in +0.010 -0.005	J[e] End Full Thread Max mm	J[e] End Full Thread Max in
1/8	2.64	0.104	2.74	0.108	6.35	0.250	0.28	0.011
3/16	3.84	0.151	3.94	0.155	6.76	0.266	0.33	0.013
1/4	5.44	0.214	5.54	0.218	6.76	0.266	0.66	0.026
5/16	7.01	0.276	7.11	0.280	7.54	0.297	0.84	0.033
5/16	7.01	0.276	7.11	0.280	7.54	0.297	0.84	0.033
3/8	8.69	0.342	8.79	0.346	8.74	0.344	0.58	0.023
7/16	10.29	0.405	10.39	0.409	9.53	0.375	0.74	0.029
1/2	11.86	0.467	11.96	0.471	9.93	0.391	0.74	0.029
5/8	15.04	0.592	15.14	0.596	10.31	0.406	1.04	0.041
3/4	17.86	0.703	17.98	0.708	11.91	0.469	0.41	0.016

TABLE 6—(Continued)

Nom Tube OD in	K Seat Depth mm +0.25 -0.13	K Seat Depth in +0.010 -0.005	L[b] mm +0.8 -0.0	L[b] in +0.03 -0.00	L_1 mm +0.8 -0.0	L_1 in +0.03 -0.00	L_2[b,c] mm +0.8 -0.0	L_2[b,c] in +0.03 -0.00
1/8	4.75	0.187	15.7	0.62	15.0	0.59	18.3	0.72
3/16	5.16	0.203	17.5	0.69	15.7	0.62	19.1	0.75
1/4	5.16	0.203	18.5	0.73	15.7	0.62	19.1	0.75
5/16	5.94	0.234	19.8	0.78	17.8	0.70	19.8	0.78
5/16	5.94	0.234	24.6	0.97	—	—	25.4	1.00
3/8	6.76	0.266	25.9	1.02	20.3	0.80	26.2	1.03
7/16	7.52	0.296	26.7	1.05	22.4	0.88	26.9	1.06
1/2	7.93	0.312	26.9	1.06	23.1	0.91	27.7	1.09
5/8	8.33	0.328	33.3	1.31	24.6	0.97	33.3	1.31
3/4	9.12	0.359	35.1	1.38	28.4	1.12	38.1	1.50

Nom Tube OD in	L_3[b] mm +0.8 -0.0	L_3[b] in +0.03 -0.00	L_4 mm	L_4 in	L_5 mm +0.8 -0.0	L_5 in +0.03 -0.00	L_6[b,c] mm +0.8 -0.0	L_6[b,c] in +0.03 -0.00
1/8	19.8	0.78	11.9	0.47	17.8	0.70	—	—
3/16	21.3	0.84	11.9	0.47	17.8	0.70	—	—
1/4	23.1	0.91	14.0	0.55	19.6	0.77	20.1	0.79
5/16	24.6	0.97	17.0	0.67	21.8	0.86	22.4	0.88
5/16	29.2	1.15	19.1	0.75	—	—	24.6	0.97
3/8	33.3	1.31	20.6	0.81	26.2	1.03	26.2	1.03
7/16	35.8	1.41	22.4	0.88	29.0	1.14	29.5	1.16
1/2	37.3	1.47	23.9	0.94	31.8	1.25	31.8	1.25
5/8	46.0	1.81	28.2	1.11	37.3	1.47	37.3	1.47
3/4	50.8	2.00	32.5	1.28	43.7	1.72	43.7	1.72

Nom Tube OD in	L_7[b] mm	L_7[b] in	L_8 mm +0.8 -0.0	L_8 in +0.03 -0.00	L_9 mm	L_9 in	L_{10} mm +0.8 -0.0	L_{10} in +0.03 -0.00
1/8	20.6	0.81	23.9	0.94	13.5	0.53	19.8	0.78
3/16	22.4	0.88	27.7	1.09	15.7	0.62	20.6	0.81
1/4	23.9	0.94	28.4	1.12	17.5	0.69	21.3	0.84
5/16	25.4	1.00	31.8	1.25	19.1	0.75	23.9	0.94
5/16	29.5	1.16	—	—	19.1	0.75	26.9	1.06
3/8	34.0	1.34	37.3	1.47	23.9	0.94	29.5	1.16
7/16	35.8	1.41	41.1	1.62	26.2	1.03	31.0	1.22
1/2	36.6	1.44	44.5	1.75	28.4	1.12	35.1	1.38
5/8	44.5	1.75	49.3	1.94	32.5	1.28	39.6	1.56
3/4	50.8	2.00	57.2	2.25	38.1	1.50	46.0	1.81

Nom Tube OD in	L_{11}[b] mm	L_{11}[b] in	L_{12} mm +0.8 -0.0	L_{12} in +0.03 -0.00	L_{13} mm	L_{13} in	L_{14}[b] mm	L_{14}[b] in
1/8	19.8	0.78	27.7	1.09	15.7	0.62	29.5	1.16
3/16	21.3	0.84	27.7	1.09	15.7	0.62	31.8	1.25
1/4	23.1	0.91	28.4	1.12	17.5	0.69	33.3	1.31
5/16	24.6	0.97	31.8	1.25	19.1	0.75	37.3	1.47
5/16	29.5	1.16	35.1	1.38	22.4	0.88	38.9	1.53
3/8	33.3	1.31	37.3	1.47	23.9	0.94	46.5	1.83
7/16	35.1	1.38	41.1	1.62	26.2	1.03	47.8	1.88
1/2	37.3	1.47	44.5	1.75	28.4	1.12	50.8	2.00
5/8	46.0	1.81	49.3	1.94	32.5	1.28	60.5	2.38
3/4	50.8	2.00	57.2	2.25	38.1	1.50	66.5	2.62

TABLE 6—(Continued)

Nom Tube OD in	M mm	M in	M_1 mm	M_1 in	M_2 mm	M_2 in	M_3 mm	M_3 in
1/8	6.9	0.27	5.6	0.22	8.4	0.33	11.9	0.47
3/16	6.9	0.27	6.4	0.25	9.9	0.39	13.5	0.53
1/4	8.4	0.33	6.9	0.27	10.7	0.42	14.2	0.56
5/16	11.9	0.47	8.6	0.34	11.4	0.45	15.7	0.62
5/16	11.4	0.45	5.8	0.23	11.4	0.45	15.7	0.62
3/8	13.5	0.53	9.7	0.38	14.2	0.56	19.1	0.75
7/16	15.0	0.59	10.4	0.41	15.7	0.62	20.6	0.81
1/2	15.0	0.59	9.6	0.38	17.0	0.67	22.4	0.88
5/8	17.0	0.67	11.4	0.45	19.1	0.75	24.6	0.97
3/4	19.1	0.75	12.7	0.50	22.4	0.88	28.4	1.12

(R) TABLE 6—(Continued)

Nom Tube OD in	N^b mm	N^b in	N_1 mm	N_1 in	$P^{b,c}$ Min mm	$P^{b,c}$ Min in	R Radius Max mm	R Radius Max in
1/8	13.2	0.52	9.9	0.39	9.7	0.38	0.25	0.010
3/16	14.0	0.55	9.9	0.39	9.7	0.38	0.25	0.010
1/4	14.7	0.58	10.7	0.42	9.7	0.38	0.91	0.036
5/16	14.2	0.56	11.4	0.45	9.7	0.38	0.91	0.036
5/16	21.1	0.83	12.7	0.50	14.2	0.56	0.91	0.036
3/8	21.3	0.84	14.2	0.56	14.2	0.56	0.91	0.036
7/16	21.8	0.86	15.7	0.62	14.2	0.56	0.91	0.036
1/2	23.1	0.91	17.0	0.67	14.2	0.56	0.91	0.036
5/8	27.7	1.09	19.1	0.75	19.1	0.75	0.91	0.036
3/4	31.0	1.22	22.4	0.88	19.1	0.75	0.91	0.036

Nom Tube OD in	S^b mm	S^b in	S_1^b mm	S_1^b in	S_2^b mm	S_2^b in	S_3^b mm	S_3^b in
1/8	4.8	0.19	6.35	0.250	2.36	0.093	9.7	0.38
3/16	5.8	0.23	2.36	0.093	2.36	0.093	9.7	0.38
1/4	7.1	0.28	2.36	0.093	2.36	0.093	—	—
5/16	—	—	2.36	0.093	—	—	—	—
5/16	12.7	0.50	12.70	0.500	4.32	0.170	—	—
3/8	10.4	0.41	—	—	2.36	0.093	—	—
7/16	—	—	—	—	2.36	0.093	—	—
1/2	—	—	3.96	0.156	2.36	0.093	—	—
5/8	13.5	0.53	5.56	0.219	4.77	0.188	—	—
3/4	13.5	0.53	7.95	0.312	6.35	0.250	—	—

Nom Tube OD in	T mm	T in	U Dia Min mm	U Dia Min in	U Dia Max mm	U Dia Max in	W Dia mm +0.00 -0.25	W Dia in +0.00 -0.01
1/8	8.6	0.34	8.1	0.32	8.6	0.34	12.7	0.50
3/16	9.7	0.38	9.9	0.39	10.4	0.41	12.7	0.50
1/4	9.7	0.38	11.4	0.45	11.9	0.47	13.5	0.53
5/16	10.4	0.41	13.0	0.51	13.5	0.53	15.0	0.59
5/16	11.2	0.44	13.0	0.51	13.5	0.53	17.5	0.69
3/8	11.9	0.47	16.3	0.64	17.0	0.67	19.1	0.75
7/16	12.7	0.50	17.8	0.70	18.5	0.73	20.6	0.81
1/2	13.5	0.53	19.6	0.77	20.3	0.80	23.1	0.91
5/8	15.7	0.62	22.6	0.89	23.4	0.92	26.9	1.06
3/4	17.5	0.69	27.4	1.08	28.2	1.11	31.8	1.25

TABLE 6—(Continued)

Nom Tube OD in	X mm	X in	X_1 mm	X_1 in	X_2 mm	X_2 in	X_3 mm	X_3 in
1/8	11.2	0.44	10.4	0.41	10.4	0.41	12.7	0.50
3/16	11.9	0.47	11.9	0.47	11.9	0.47	12.7	0.50
1/4	13.5	0.53	13.5	0.53	13.5	0.53	13.5	0.53
5/16	15.0	0.59	15.0	0.59	15.0	0.59	15.0	0.59
5/16	15.0	0.59	15.0	0.59	15.0	0.59	17.5	0.69
3/8	19.1	0.75	18.3	0.72	19.1	0.75	19.1	0.75
7/16	21.3	0.84	19.8	0.78	20.6	0.81	20.6	0.81
1/2	23.1	0.91	22.4	0.88	23.1	0.91	23.1	0.91
5/8	26.9	1.06	26.9	1.06	26.9	1.06	26.9	1.06
3/4	31.8	1.25	31.8	1.25	31.8	1.25	31.8	1.25

Nom Tube OD in	X_4 mm	X_4 in	Y mm +0.25 -0.00	Y in +0.010 -0.000	Y_1^f mm	Y_1^f in	Z Min mm	Z Min in	Z_1 Min mm	Z_1 Min in
1/8	11.9	0.47	5.21	0.205	7.6	0.30	0.8	0.03	2.0	0.08
3/16	11.9	0.47	5.84	0.230	8.4	0.33	0.8	0.03	1.5	0.06
1/4	13.5	0.53	6.60	0.260	9.1	0.36	0.8	0.03	1.3	0.05
5/16	15.0	0.59	7.37	0.290	10.7	0.42	1.3	0.05	2.3	0.09
5/16	15.0	0.59	7.37	0.290	7.9	0.31	—	—	1.5	0.06
3/8	19.1	0.75	9.40	0.370	12.7	0.50	0.8	0.03	1.3	0.05
7/16	20.6	0.81	10.54	0.415	13.5	0.53	1.3	0.05	0.8	0.03
1/2	23.1	0.91	11.43	0.450	13.5	0.53	0.8	0.03	1.5	0.06
5/8	26.9	1.06	13.34	0.525	16.3	0.64	1.3	0.05	1.5	0.06
3/4	31.8	1.25	15.75	0.620	19.8	0.78	1.3	0.05	2.0	0.08

[a] Dryseal American Standard Taper Pipe Thread. See General Specifications.
[b] Where SAE Short Pipe Thread is authorized by purchaser, dimensions L, L_2, L_3, L_5, L_7, L_{11}, L_{14}, N, P, S, S_1, S_2, and S_3 are reduced in accordance with reduction of pipe thread length. See SAE J476, Tables 3 and 4.
[c] Tap drill depths given require use of bottoming taps to produce standard full thread length. For increased tap clearance, see Internal Taper Pipe Threads in General Specifications.
[d] Class 3B minor diameter limits apply to copper alloy fittings and Class 2B minor diameter limits apply to steel fittings.
[e] End full thread is measured from face of cone seat to last full form thread of major diameter, see Appendix A. A minimum of 3/4 partial thread beyond the last full form thread is required.
[f] For steel parts, the Y_1 dimension shall be 8.9 mm (0.35 in) for 3/16 in tube size and 13.2 mm (0.52 in) for 3/8 in tube size.

3.14 Assembly Considerations—Where it is not objectionable from a function or production standpoint, the use of a compatible lubricant or sealant may be desirable in assembling Dryseal pipe threads on automotive tube fittings to minimize galling and effect a pressure-tight seal.

3.15 Wrenching Test—Steel nuts when assembled without tubing into mating brass fittings which are held securely by a suitable means, such as in a vise, shall be capable of being tightened by means of a standard open end wrench, having an opening as tabulated to the minimum torque values specified without failure (rounding) of the hexagon corner.

NOTE—The tabulated torque requirements should not in any case be misconstrued as being installation torques. They are intended solely for determining the adequacy of the hexagon corners. (See Table 8A.)

TABLE 6A—DIMENSIONS OF SPECIAL SIZE CONNECTORS (Figure 5A) AND ELBOWS (Figure 6A)

Nom Tube OD in	A Dryseal Taper Thread NPTF[a] in	B Nom Thread Size in Class 2B int[d]	C Nom in	D Dia Drill mm	D Dia Drill in	D_1 Dia Drill mm	D_1 Dia Drill in
1/4	1/4	7/16-24	9/16	4.78	0.188	4.78	0.188
3/8	1/8	5/8-18	3/4	7.14	0.281	5.56	0.219
3/8	3/8	5/8-18	3/4	7.14	0.281	7.92	0.312
1/2	1/4	3/4-18	29/32	10.31	0.406	7.14	0.281
1/2	1/2	3/4-18	29/32	10.31	0.406	10.31	0.406
5/8	3/8	7/8-18	1-1/16	13.49	0.531	11.10	0.437
3/4	1/2	1-1/16-16	1-1/4	15.82	0.625	13.49	0.531

Nom OD in	D_4 Dia Drill mm	D_4 Dia Drill in	D_5 Dia Drill mm	D_5 Dia Drill in	L^b +0.8 -0.0 mm	L^b +0.03 -0.00 in	L_3^b +0.8 -0.0 mm	L_3^b +0.03 -0.00 in
1/4	7.92	0.312	8.74	0.344	22.4	0.88	27.7	1.09
3/8	5.56	0.219	—	—	22.4	0.88	28.4	1.12
3/8	10.31	0.406	11.10	0.437	25.4	1.00	33.3	1.31
1/2	8.74	0.344	—	—	26.9	1.06	37.1	1.46
1/2	14.27	0.562	12.70	0.500	31.8	1.25	42.2	1.66
5/8	10.31	0.406	11.91	0.469	28.4	1.12	41.1	1.62
3/4	14.27	0.562	—	—	35.1	1.38	50.8	2.00

Nom Tube OD in	L_4 mm	L_4 in	M mm	M in	$P^{b,c}$ mm	$P^{b,c}$ in	$S^{b,e}$ mm	$S^{b,e}$ in
1/4	14.2	0.56	7.1	0.28	14.2	0.56	11.2	0.44
3/8	18.5	0.73	13.5	0.53	9.7	0.38	9.7	0.38
3/8	21.3	0.84	12.7	0.50	14.2	0.56	10.4	0.41
1/2	22.4	0.88	15.0	0.59	14.2	0.56	11.2	0.44
1/2	27.7	1.09	16.8	0.66	19.0	0.75	15.7	0.62
5/8	28.2	1.11	19.0	0.75	14.2	0.56	11.2	0.44
3/4	33.0	1.30	21.6	0.85	19.0	0.75	11.9	0.47

Nom Tube OD in	S_1^b mm	S_1^b in	X mm	X in	Y +0.25 -0.00 mm	Y +0.010 -0.000 in	Z Min mm	Z Min in
1/4	6.10	0.240	14.2	0.56	6.98	0.275	0.8	0.03
3/8	—	—	19.0	0.75	9.40	0.370	—	—
3/8	11.10	0.437	19.0	0.75	9.40	0.370	1.3	0.05
1/2	—	—	23.1	0.91	11.43	0.450	—	—
1/2	15.88	0.625	23.1	0.91	11.43	0.450	—	—
5/8	3.96	0.156	26.9	1.06	13.34	0.525	1.3	0.05
3/4	—	—	31.8	1.25	15.75	0.620	—	—

[a] Dryseal American Standard Taper Pipe Thread. See General Specifications.
[b] Where SAE Short Pipe Thread is authorized by purchaser, dimensions L, L_2, L_3, L_5, L_7, L_{11}, L_{14}, N, P, S, S_1, S_2, and S_3 are reduced in accordance with reduction of pipe thread length. See SAE J476, Tables 3 and 4.
[c] Tap drill depths given require use of bottoming taps to produce standard full thread length. For increased tap clearance, see Internal Taper Pipe Threads in General Specifications.
[d] Class 3B minor diameter limits apply to copper alloy fittings and Class 2B minor diameter limits apply to steel fittings.
[e] Measured from end containing the largest passage.

TABLE 7—DIMENSIONS OF PLUGS (Figure 8)

Nom Tube OD in	B Nom Thread Size in Class 2A Ext	C Hex Nom in	D Dia mm ±0.13	D Dia in ±0.005	E Dia mm +0.10 -0.00	E Dia in +0.004 -0.000	F Dia mm +0.00 -0.13	F Dia in +0.000 -0.005
3/16	3/8-24	3/8	4.77	0.188	5.92	0.233	7.37	0.290
1/4	7/16-24	7/16	4.77	0.188	7.52	0.296	9.02	0.355
5/16	1/2-20	1/2	6.35	0.250	9.09	0.358	10.59	0.417
3/8	5/8-18	5/8	7.92	0.312	11.33	0.446	12.89	0.507
7/16	11/16-18	11/16	9.52	0.375	12.95	0.510	14.73	0.580
1/2	3/4-18	3/4	11.10	0.437	14.50	0.571	16.26	0.640
5/8	7/8-18	7/8	14.27	0.562	17.68	0.696	19.58	0.771
3/4	1-1/16-16	1-1/16	16.66	0.656	21.97	0.865	23.80	0.937

Nom Tube OD in	J Full Thread Min mm	J Full Thread Min in	K mm	K in	L mm	L in
3/16	8.6	0.34	7.1	0.28	1.0	0.04	13.5	0.53
1/4	8.6	0.34	7.1	0.28	1.0	0.04	13.7	0.54
5/16	9.7	0.38	7.9	0.31	1.5	0.06	15.0	0.59
3/8	10.4	0.41	8.4	0.33	1.5	0.06	16.8	0.66
7/16	12.7	0.50	9.9	0.39	1.5	0.06	19.1	0.75
1/2	12.7	0.50	10.4	0.41	1.5	0.06	20.6	0.81
5/8	12.7	0.50	10.9	0.43	1.5	0.06	22.4	0.88
3/4	15.5	0.61	12.7	0.50	2.0	0.08	24.1	0.95

TABLE 8—DIMENSIONS OF NUTS (Figure 9)

Nom Tube OD in	B Nom Thread Size in Class 2A Ext	C^b mm +0.13 -0.00	C^b in +0.005 -0.000	D^c Dia mm +0.13 -0.00	D^c Dia in +0.005 -0.000	E Dia mm +0.10 -0.00	E Dia in +0.004 -0.000	G Dia mm +0.00 -0.13	G Dia in +0.000 -0.005
1/8	5/16-28	7.80	0.307	3.30	0.130	5.61	0.221	6.58	0.259
3/16	3/8-24	9.40	0.370	4.93	0.194	6.81	0.268	7.95	0.313
1/4	7/16-24	11.00	0.433	6.53	0.257	8.41	0.331	9.53	0.375
5/16	1/2-20	12.57	0.495	8.10	0.319	9.98	0.393	10.85	0.427
3/8	5/8-18	15.75	0.620	9.70	0.382	12.37	0.487	13.79	0.543
7/16	11/16-18	17.35	0.683	11.25	0.443	13.97	0.550	15.37	0.605
1/2	3/4-18	18.92	0.745	12.85	0.506	15.54	0.612	16.97	0.668
5/8	7/8-18	22.10	0.870	16.03	0.631	18.72	0.737	20.14	0.793
3/4	1-1/16-16	26.85	1.057	19.23	0.757	23.01	0.906	24.66	0.971

Nom Tube OD in	J Full Thread Min mm	J Full Thread Min in	L mm	L in	Q Min mm	Q Min in	V mm +0.00 -0.38	V in +0.000 -0.015
1/8	8.1	0.32	6.6	0.26	13.2	0.52	8.81	0.347	6.63	0.261
3/16	9.1	0.36	7.6	0.30	14.2	0.56	10.62	0.418	8.03	0.316
1/4	9.1	0.36	7.6	0.30	14.2	0.56	12.40	0.488	9.60	0.378
5/16	10.2	0.40	8.1	0.32	15.7	0.62	14.10	0.555	10.92	0.430
3/8	10.2	0.40	8.1	0.32	16.8	0.66	17.81	0.701	13.92	0.548
7/16	11.2	0.44	8.6	0.34	17.3	0.68	19.61	0.772	15.49	0.610
1/2	11.7	0.46	9.1	0.36	18.8	0.74	21.39	0.842	17.09	0.673
5/8	12.2	0.48	9.7	0.38	20.3	0.80	24.71	0.973	20.27	0.798
3/4	13.7	0.54	11.2	0.44	22.4	0.88	30.30	1.193	24.79	0.976

[a] On 3/16, 1/4, 5/16 and 3/8 in size nuts, the maximum hole diameter limits may be exceeded at the hexagon end and the hole slightly tapered to a depth not exceeding 2.5 mm (0.10 in) below the hexagon end. A slight flash at this point shall not be objectionable provided the tabulated hole diameter limits are maintained. A definite rounding at the edge of the hole and a slight rounding of hexagon corners shall also be permissible provided the nuts meet the Wrenching Test requirements in Table 8A.
[b] Tabulated values apply only to cold formed nuts and nuts made from steel bar stock. Limits for across flats on nuts made from brass bar stock shall conform with commercial tolerances on the bar stock.
[c] Hole diameter D and 90 degree seat shall be concentric with thread pitch diameter within 0.18 mm (0.007 in) full indicator reading (FIR).

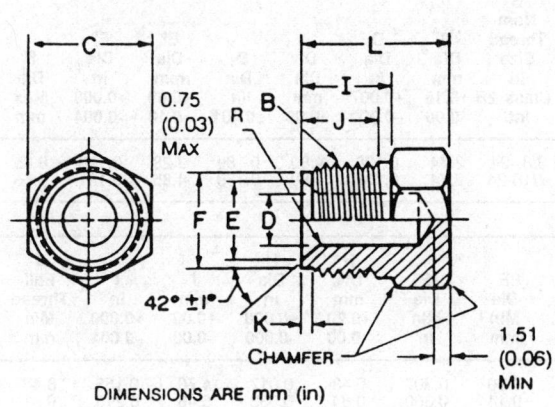

FIGURE 8—INVERTED FLARE PLUGS (040109)

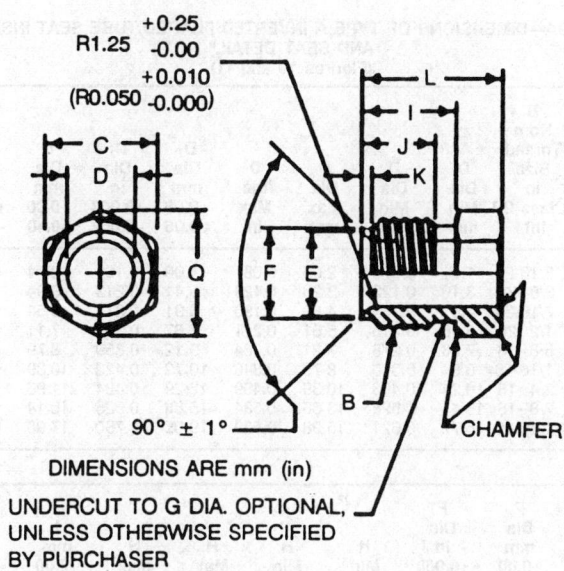

(R) FIGURE 9—INVERTED FLARE NUTS (040110)

TABLE 8A—WRENCHING TEST REQUIREMENTS

Nom Tube OD in	Torque Requirements for Steel Nuts N·m	Torque Requirements for Steel Nuts lb-in	Wrench Opening Max mm	Wrench Opening Max in
1/8	6.8	60	8.18	0.322
3/16	13.6	120	9.75	0.384
1/4	16.9	150	11.33	0.446
5/16	20.3	180	12.95	0.510
3/8	23.7	210	16.15	0.636
7/16	33.9	300	17.76	0.699
1/2	45.2	400	19.38	0.763
5/8	56.5	500	22.56	0.888
3/4	73.4	650	27.36	1.077

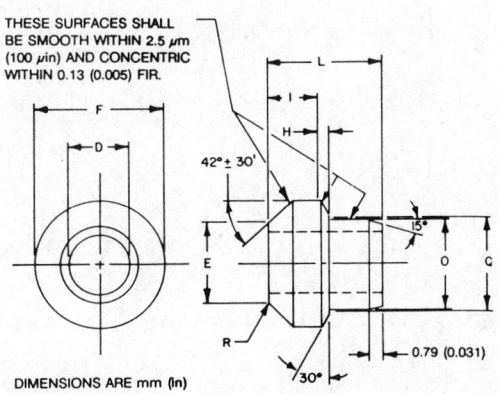

FIGURE 10—SEAT INSERT (040141)

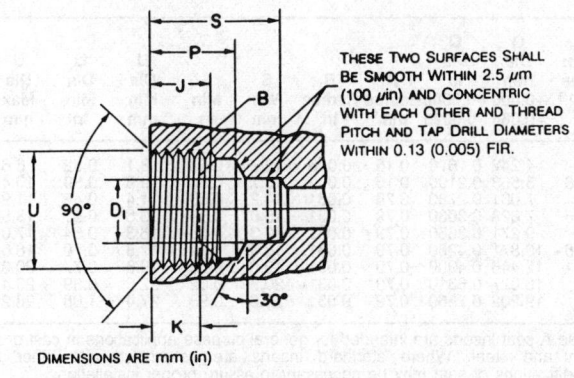

FIGURE 11—DETAIL OF SEAT ASSEMBLY

TABLE 9A—DIMENSIONS OF TYPE A INVERTED FLARED TUBE SEAT INSERTS AND SEAT DETAIL[a]
(Figures 10 and 11)

Nom Tube OD in	B Nom Thread Size in Class 2B Int.	D Dia Min mm	D Dia Min in	D Dia Max mm	D Dia Max in	D_1 Dia mm +0.08 -0.05	D_1 Dia in +0.003 -0.002	E Dia mm +0.00 -0.10	E Dia in +0.000 -0.004
1/8	5/16-28	1.91	0.075	2.06	0.081	4.09	0.161	2.74	0.108
3/16	3/8 -24	3.10	0.122	3.25	0.128	5.41	0.213	3.94	0.155
1/4	7/16-24	4.67	0.184	4.83	0.190	6.91	0.272	5.54	0.218
5/16	1/2 -20	5.46	0.215	5.61	0.221	7.67	0.302	7.11	0.280
3/8	5/8 -18	7.06	0.278	7.21	0.284	9.12	0.359	8.79	0.346
7/16	11/16-18	8.64	0.340	8.79	0.346	10.72	0.422	10.39	0.409
1/2	3/4 -18	10.24	0.403	10.39	0.409	12.29	0.484	11.96	0.471
5/8	7/8 -18	13.41	0.528	13.56	0.534	15.88	0.625	15.14	0.596
3/4	1-1/16-16	15.77	0.621	15.98	0.629	19.05	0.750	17.98	0.708

Nom Tube OD in	F Dia mm +0.00 -0.13	F Dia in +0.000 -0.005	H Min mm	H Min in	H Max mm	H Max in	I mm +0.00 -0.10	I in +0.000 -0.004
1/8	6.71	0.264	0.38	0.015	0.51	0.020	3.71	0.146
3/16	8.15	0.321	0.38	0.015	0.51	0.020	3.84	0.151
1/4	9.73	0.383	0.38	0.015	0.51	0.020	3.84	0.151
5/16	11.05	0.435	0.64	0.025	0.76	0.030	4.22	0.166
3/8	14.10	0.554	0.76	0.030	0.89	0.035	5.00	0.197
7/16	16.65	0.616	0.76	0.030	0.89	0.035	5.00	0.197
1/2	17.25	0.679	0.89	0.035	1.14	0.045	5.49	0.216
5/8	29.42	0.804	0.89	0.035	1.14	0.045	5.99	0.236
3/4	24.97	0.983	1.14	0.045	1.40	0.055	7.24	0.285

Nom Tube OD in	J Full Thread Min mm	J Full Thread Min in	K Ref mm	K Ref in	L mm	L in	O Dia mm +0.00 -0.08	O Dia in +0.000 -0.003	P mm ±0.13	P in ±0.005
1/8	5.44	0.214	4.75	0.187	7.1	0.28	4.09	0.161	8.26	0.325
3/16	5.72	0.225	5.16	0.203	7.6	0.30	5.41	0.213	8.84	0.348
1/4	5.72	0.225	5.16	0.203	8.1	0.32	6.91	0.272	8.84	0.348
5/16	6.55	0.258	5.94	0.234	8.4	0.33	7.67	0.302	9.96	0.392
3/8	7.34	0.289	6.76	0.266	9.6	0.38	9.12	0.359	11.56	0.455
7/16	8.13	0.320	7.52	0.296	9.9	0.39	10.72	0.422	12.29	0.484
1/2	8.53	0.336	7.93	0.312	10.2	0.40	12.29	0.484	13.18	0.519
5/8	8.92	0.351	8.33	0.328	11.7	0.46	15.88	0.625	14.10	0.555
3/4	10.52	0.414	9.12	0.359	13.7	0.54	19.05	0.750	16.13	0.635

Nom Tube OD in	Q Dia mm +0.000 -0.064	Q Dia in +0.0000 -0.0025	R Radius mm	R Radius in	S Min mm	S Min in	U Dia Min mm	U Dia Min in	U Dia Max mm	U Dia Max in
1/8	4.242	0.1670	0.15	0.006	12.7	0.50	8.1	0.32	8.6	0.34
3/16	5.563	0.2190	0.15	0.006	13.7	0.54	9.9	0.39	10.4	0.41
1/4	7.061	0.2780	0.79	0.031	14.2	0.56	11.4	0.45	11.9	0.47
5/16	7.823	0.3080	0.79	0.031	15.0	0.59	13.0	0.51	13.5	0.53
3/8	9.271	0.3650	0.79	0.031	17.3	0.68	16.3	0.64	17.0	0.67
7/16	10.871	0.4280	0.79	0.031	18.3	0.72	17.8	0.70	18.5	0.73
1/2	12.446	0.4900	0.79	0.031	19.0	0.75	19.6	0.77	20.3	0.80
5/8	16.027	0.6310	0.79	0.031	20.8	0.82	22.6	0.89	23.4	0.92
3/4	19.202	0.7560	0.79	0.031	23.6	0.93	27.4	1.08	28.2	1.11

[a] Type A seat inserts are intended for general purpose applications in cast or malleable iron and steel. Where standard inserts are assembled into other materials, modifications of seat may be necessary to assure proper installation.

TABLE 9B—DIMENSIONS OF TYPE B INVERTED FLARED TUBE SEAT INSERTS AND SEAT DETAIL[a]
(Figures 10 and 11)

Nom Tube OD in	B Nom Thread Size in Class 2B Int.	D mm +0.18 -0.00	D in +0.007 -0.000	D_1 Dia mm ±0.03	D_1 Dia in ±0.001	E[b] Dia mm +0.00 -0.10	E[b] Dia in +0.000 -0.004	F Dia Max mm	F Dia Max in
3/16	3/8 -24	2.74	0.108	4.80	0.189	3.28	0.129	8.05	0.317
1/4	7/16-24	2.74	0.108	4.80	0.189	4.85	0.191	9.65	0.380

Nom Tube OD in	F Dia Min mm	F Dia Min in	H Dia mm +0.20 -0.00	H Dia in +0.008 -0.000	I mm +0.00 -0.00	I in +0.000 -0.004	J Full Thread Min mm	J Full Thread Min in
3/16	7.80	0.307	0.43	0.017	4.70	0.185	8.48	0.334
1/4	9.14	0.360	0.84	0.033	5.46	0.215	9.91	0.390

Nom Tube OD in	K Max mm	K Max in	K Min mm	K Min in	L mm	L in	O Dia mm +0.00 -0.10	O Dia in +0.000 -0.004
3/16	6.20	0.244	5.69	0.224	8.6	0.34	4.85	0.191
1/4	6.60	0.260	5.84	0.230	9.9	0.39	4.85	0.191

Nom Tube OD in	P mm	P in	Q Dia mm +0.00 -0.05	Q Dia in +0.000 -0.002	R Radius mm ±0.03	R Radius in ±0.001	S Min mm	S Min in	U Dia Max mm	U Dia Max in
3/16	10.31	0.406	4.95	0.195	0.79	0.031	15.62	0.615	11.10	0.437
1/4	11.94	0.470	4.95	0.195	0.79	0.031	17.27	0.680	12.70	0.500

[a] Type B seat inserts are used extensively in cast or malleable iron and steel components of hydraulic brake systems. Where standard inserts are assembled into other materials, modifications of seat may be necessary to assure proper installation.
[b] On Type B inserts, E diameter defines center of R radius.

22.17

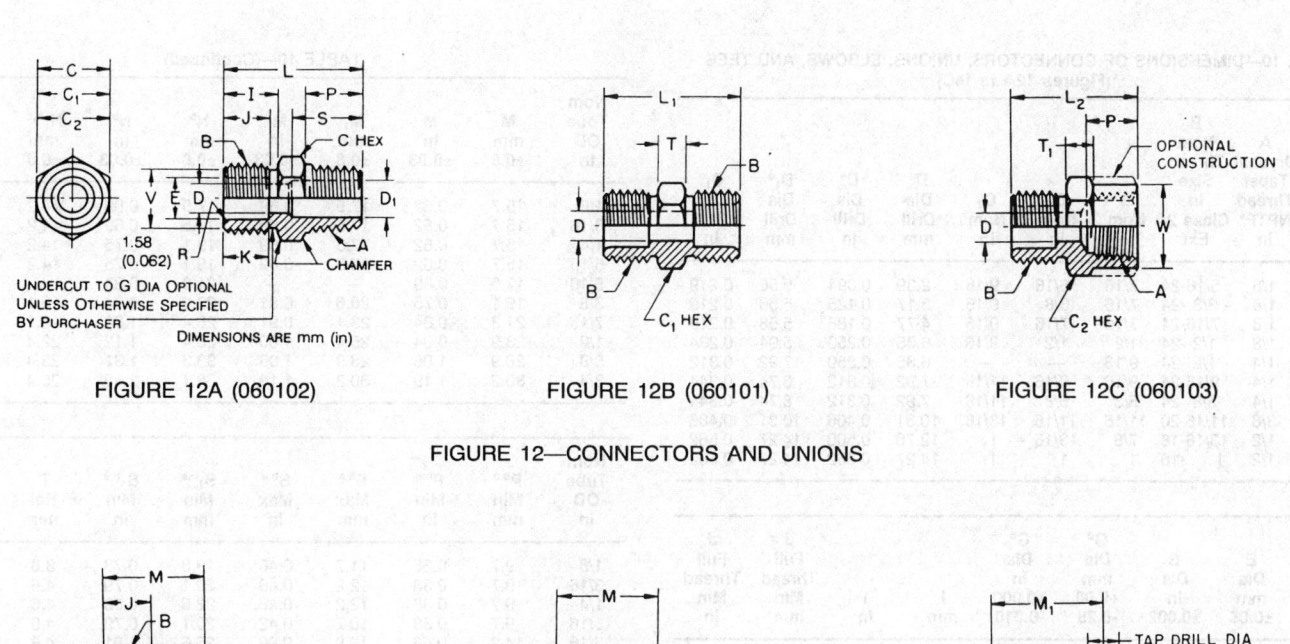

FIGURE 12A (060102)   FIGURE 12B (060101)   FIGURE 12C (060103)

FIGURE 12—CONNECTORS AND UNIONS

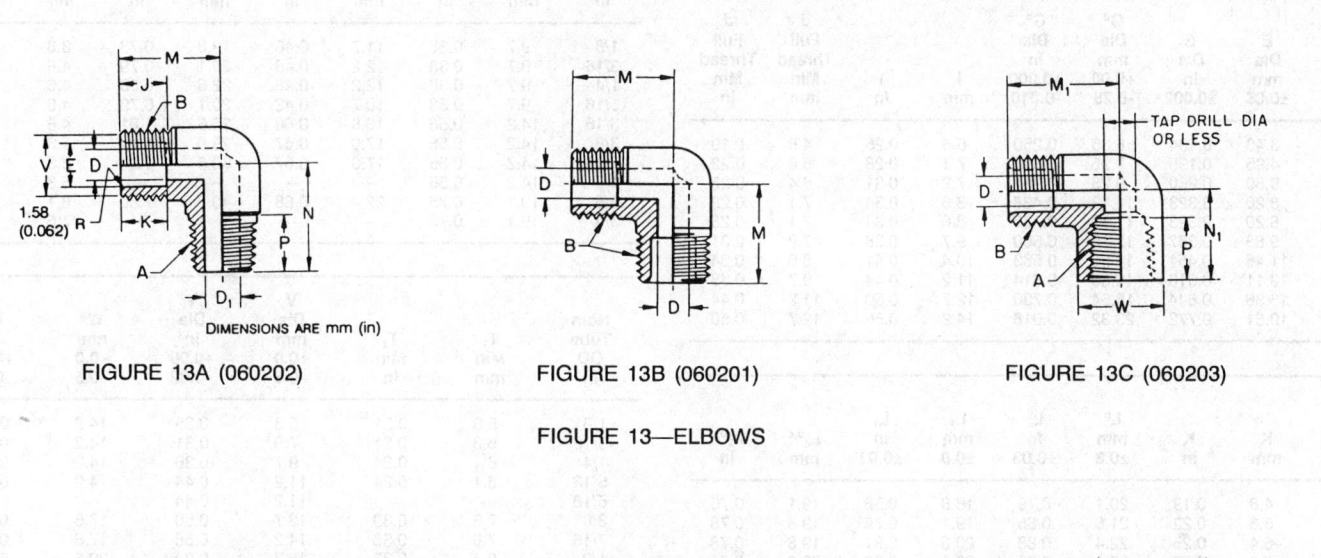

FIGURE 13A (060202)   FIGURE 13B (060201)   FIGURE 13C (060203)

FIGURE 13—ELBOWS

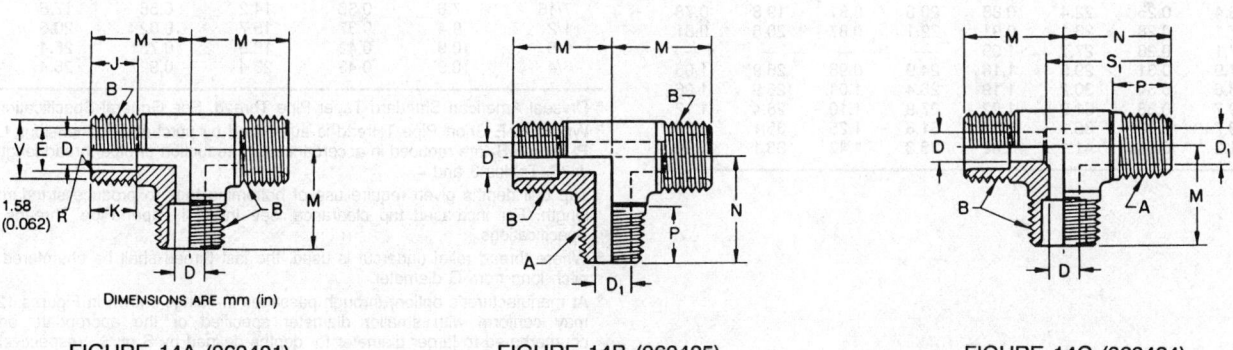

FIGURE 14A (060401)   FIGURE 14B (060425)   FIGURE 14C (060424)

FIGURE 14—TEES

NOTES: UNSPECIFIED DETAIL WITH RESPECT TO DIMENSIONS, TOLERANCES, CONTOURS, MATERIAL, WORKMANSHIP, ETC., MUST CONFORM TO GENERAL SPECIFICATIONS FOR AUTOMOTIVE TUBE FITTINGS. THE DIMENSIONAL DESIGNATIONS ON THE FIRST FIGURE IN EACH GROUP SHALL APPLY TO ALL OTHER FIGURES IN THAT GROUP EXCEPT AS SHOWN OTHERWISE. CODES SHOWN IN BRACKETS ADJACENT TO FIGURE NUMBERS REPRESENT RESPECTIVE FITTING IDENTIFICATION IN ACCORDANCE WITH SAE J846.

TABLE 10—DIMENSIONS OF CONNECTORS, UNIONS, ELBOWS, AND TEES
(Figures 12A to 14C)

Nom Tube OD in	A Dryseal Taper Thread NPTF[a] in	B Nom Thread Size in Class 2A Ext	C Nom in	C_1 Nom in	C_2 Nom in	D[e] Dia Drill mm	D[e] Dia Drill in	D_1[e] Dia Drill mm	D_1[e] Dia Drill in
1/8	1/8	5/16-24	7/16	5/16	9/16	2.39	0.094	5.56	0.219
3/16	1/8	3/8 -24	7/16	3/8	9/16	3.17	0.125	5.56	0.219
1/4	1/8	7/16-24	7/16	7/16	9/16	4.77	0.188	5.56	0.219
5/16	1/8	1/2 -24	1/2	1/2	9/16	6.35	0.250	5.94	0.234
5/16	1/4	1/2 -24	9/16	—	—	6.35	0.250	7.92	0.312
3/8	1/4	9/16-24	9/16	9/16	11/16	7.92	0.312	8.74	0.344
7/16	1/4	5/8 -24	5/8	5/8	11/16	7.92	0.312	8.74	0.344
1/2	3/8	11/16-20	11/16	11/16	13/16	10.31	0.406	10.31	0.406
5/8	1/2	13/16-18	7/8	13/16	1	12.70	0.500	14.27	0.562
3/4	1/2	1 -18	1	1	1	14.27	0.562	14.27	0.562

Nom Tube OD in	E Dia mm ±0.05	E Dia in ±0.002	G[d] Dia mm +0.00 -0.25	G[d] Dia in +0.000 -0.010	I mm	I in	J Full Thread Min mm	J Full Thread Min in
1/8	3.40	0.134	6.35	0.250	6.4	0.25	4.8	0.19
3/16	4.95	0.195	7.95	0.313	7.1	0.28	5.6	0.22
1/4	6.60	0.260	9.25	0.364	7.9	0.31	6.4	0.25
5/16	8.20	0.323	11.13	0.438	8.6	0.34	7.1	0.28
5/16	8.20	0.323	11.13	0.438	8.6	0.34	7.1	0.28
3/8	9.83	0.387	12.70	0.500	9.7	0.38	7.9	0.31
7/16	11.46	0.451	14.30	0.563	10.4	0.41	8.6	0.34
1/2	13.11	0.516	15.60	0.614	11.2	0.44	9.7	0.38
5/8	16.36	0.644	18.54	0.730	12.7	0.50	11.2	0.44
3/4	19.61	0.772	23.32	0.918	14.2	0.56	12.7	0.50

Nom Tube OD in	K mm	K in	L[b] mm ±0.8	L[b] in ±0.03	L_1 mm ±0.8	L_1 in ±0.03	L_2[b,c] mm	L_2[b,c] in
1/8	4.8	0.19	20.1	0.79	16.8	0.66	19.1	0.75
3/16	5.6	0.22	21.6	0.85	19.1	0.75	19.8	0.78
1/4	6.4	0.25	22.4	0.88	20.6	0.81	19.8	0.78
5/16	7.1	0.28	23.1	0.91	22.1	0.87	20.6	0.81
5/16	7.1	0.28	27.7	1.09	—	—	—	—
3/8	7.9	0.31	29.5	1.16	24.9	0.98	26.9	1.06
7/16	8.6	0.34	30.2	1.19	26.4	1.04	26.9	1.06
1/2	9.7	0.38	31.0	1.22	27.9	1.10	28.4	1.12
5/8	9.7	0.38	38.1	1.50	31.8	1.25	35.1	1.38
3/4	11.2	0.44	41.1	1.62	36.3	1.43	38.1	1.50

TABLE 10—(Continued)

Nom Tube OD in	M mm ±0.8	M in ±0.03	M_1 mm ±0.8	M_1 in ±0.03	N[b] mm ±0.8	N[b] in ±0.03	N_1[b,c] mm ±0.8	N_1[b,c] in ±0.03
1/8	15.7	0.62	17.5	0.69	17.5	0.69	14.2	0.56
3/16	15.7	0.62	17.5	0.69	17.5	0.69	14.2	0.56
1/4	15.7	0.62	17.5	0.69	19.1	0.75	14.2	0.56
5/16	15.7	0.62	17.5	0.69	19.1	0.75	14.2	0.56
5/16	17.5	0.69	—	—	21.3	0.84	—	—
3/8	19.1	0.75	20.6	0.81	23.9	0.94	19.1	0.75
7/16	21.3	0.84	23.1	0.91	25.4	1.00	19.1	0.75
1/2	23.9	0.94	25.4	1.00	28.4	1.12	22.4	0.88
5/8	26.9	1.06	26.9	1.06	33.3	1.31	25.4	1.00
3/4	30.2	1.19	30.2	1.19	38.1	1.50	25.4	1.00

Nom Tube OD in	P[b,c] Min mm	P[b,c] Min in	S[b,e] Max mm	S[b,e] Max in	S_1[b,e] Min mm	S_1[b,e] Min in	T[f] Ref mm	T[f] Ref in
1/8	9.7	0.38	11.7	0.46	19.8	0.78	3.8	0.15
3/16	9.7	0.38	12.2	0.48	20.1	0.79	4.6	0.18
1/4	9.7	0.38	12.2	0.48	22.6	0.89	4.6	0.18
5/16	9.7	0.38	10.7	0.42	20.1	0.79	4.6	0.18
5/16	14.2	0.56	16.8	0.66	25.6	1.01	4.6	0.18
3/8	14.2	0.56	17.0	0.67	29.0	1.14	5.3	0.21
7/16	14.2	0.56	17.0	0.67	30.5	1.20	5.3	0.21
1/2	14.2	0.56	—	—	—	—	5.3	0.21
5/8	19.1	0.75	22.4	0.88	40.6	1.60	6.1	0.24
3/4	19.1	0.75	—	—	—	—	7.6	0.30

Nom Tube OD in	T_1 Min mm	T_1 Min in	V Dia mm +0.0 -0.5	V Dia in +0.00 -0.02	W[g] mm +0.0 -0.5	W[g] in +0.00 -0.02
1/8	5.3	0.21	6.3	0.25	14.2	0.56
3/16	5.3	0.21	7.9	0.31	14.2	0.56
1/4	6.1	0.24	9.7	0.38	14.2	0.56
5/16	6.1	0.24	11.2	0.44	14.2	0.56
5/16	—	—	11.2	0.44	—	—
3/8	7.6	0.30	12.7	0.50	17.5	0.69
7/16	7.6	0.30	14.2	0.56	17.5	0.69
1/2	9.4	0.37	15.7	0.62	20.6	0.81
5/8	10.9	0.43	18.5	0.73	25.4	1.00
3/4	10.9	0.43	23.4	0.92	25.4	1.00

[a] Dryseal American Standard Taper Pipe Thread. See General Specifications.

[b] Where SAE Short Pipe THread is authorized by purchaser, dimensions L, L_2, N, N_1, P, S, and S_1 are reduced in accordance with reduction of pipe thread length. See SAE J476, Tables 3 and 4.

[c] Tap drill depths given require use of bottoming taps to produce standard full thread length. For increased tap clearance, see Internal Taper Pipe Threads in General Specifications.

[d] Where thread relief undercut is used, the last thread shall be chamfered to 1/2 to 1 pitch long from G diameter.

[e] At manufacturer's option, through passages in fittings shown in Figures 12A and 14C may conform with smaller diameter specified or the appropriate end may be counterbored to larger diameter for depths defined by S or S_1, respectively.

[f] Minimum design thickness, not subject to inspection.

[g] Basic dimensions shown shall apply as minimum diameter or across flats for bosses. The –0.5 mm (–0.02 in) tolerance shall apply only to chamfer diameter on full hexagon version of fitting in Figure 12C.

TABLE 11—DIMENSIONS OF TAPERED SLEEVES (Figure 15)

Nom Tube OD in	D Dia mm ±0.05	D Dia in ±0.002	E Dia mm ±0.08	E Dia in ±0.003	F Dia mm	F Dia in	L mm	L in
1/8	3.30	0.130	3.51	0.138	4.8	0.19	4.8	0.19
3/16	4.88	0.192	5.11	0.201	6.9	0.27	5.6	0.22
1/4	6.48	0.255	6.76	0.266	8.6	0.34	6.4	0.25
5/16	8.08	0.318	8.36	0.329	10.4	0.41	6.4	0.25
3/8	9.70	0.382	9.98	0.393	11.9	0.47	6.4	0.25
7/16	11.28	0.444	11.61	0.457	13.5	0.53	7.9	0.31
1/2	12.88	0.507	13.26	0.522	15.0	0.59	9.7	0.38
5/8	16.05	0.632	16.51	0.650	18.3	0.72	9.7	0.38
3/4	19.25	0.758	19.76	0.778	22.4	0.88	11.2	0.44

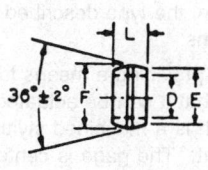

FIGURE 15—TAPERED SLEEVES (060115)

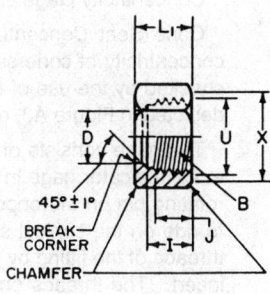

FIGURE 16—SHORT NUTS (060110)

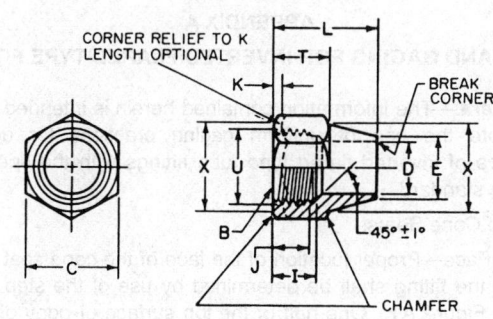

FIGURE 17—LONG NUTS (060111)

TABLE 12—DIMENSIONS OF SHORT AND LONG NUTS (Figures 16 and 17)

Nom Tube OD in	B Straight Thread Nom Size in Class 2B Int	C Hex in Nom	D Dia mm ±0.05	D Dia in ±0.002	E Dia mm	E Dia in	I mm +0.00 -0.25	I in +0.000 -0.010
1/8	5/16-24	3/8	3.30	0.130	6.4	0.25	6.4	0.25
3/16	3/8 -24	7/16	4.88	0.192	7.9	0.31	7.1	0.28
1/4	7/16-24	1/2	6.48	0.255	9.7	0.38	7.9	0.31
5/16	1/2 -24	9/16	8.08	0.318	11.2	0.44	7.9	0.31
3/8	9/16-24	5/8	9.70	0.382	12.7	0.50	8.6	0.34
7/16	5/8 -24	11/16	11.28	0.444	14.2	0.56	9.4	0.37
1/2	11/16-20	13/16	12.88	0.507	15.7	0.62	12.4	0.49
5/8	13/16-18	15/16	16.05	0.632	19.1	0.75	12.7	0.50
3/4	1 -18	1- 3/16	19.25	0.758	22.4	0.88	13.5	0.53

Nom Tube OD in	J Full Thread Min mm	J Full Thread Min in	K mm	K in	L mm	L in	L_1 mm	L_1 in
1/8	4.8	0.19	2.3	0.09	12.7	0.50	9.7	0.38
3/16	5.6	0.22	2.3	0.09	15.7	0.62	10.4	0.41
1/4	6.4	0.25	3.0	0.12	19.1	0.75	11.2	0.44
5/16	6.4	0.25	3.0	0.12	21.3	0.84	11.2	0.44
3/8	7.1	0.28	3.0	0.12	24.6	0.97	11.9	0.47
7/16	7.9	0.31	3.0	0.12	26.2	1.03	12.7	0.50
1/2	10.4	0.41	3.0	0.12	26.9	1.06	15.7	0.62
5/8	10.4	0.41	3.0	0.12	30.2	1.19	15.7	0.62
3/4	11.2	0.44	3.0	0.12	35.1	1.38	17.5	0.69

Nom Tube OD in	T mm	T in	U Dia Min mm	U Dia Min in	U Dia Max mm	U Dia Max in	X Dia mm	X Dia in
1/8	6.4	0.25	8.4	0.33	8.9	0.35	9.7	0.38
3/16	7.1	0.28	9.9	0.39	10.4	0.41	11.2	0.44
1/4	9.7	0.38	11.4	0.45	11.9	0.47	12.7	0.50
5/16	11.2	0.44	13.0	0.51	13.5	0.53	14.2	0.56
3/8	12.7	0.50	14.7	0.58	15.2	0.60	15.7	0.62
7/16	14.2	0.56	16.3	0.64	16.8	0.66	17.5	0.69
1/2	15.7	0.62	17.8	0.70	18.5	0.73	20.6	0.81
5/8	18.3	0.72	21.1	0.83	21.8	0.86	23.9	0.94
3/4	19.1	0.75	25.9	1.02	26.7	1.05	30.2	1.19

TABLE 13—DIMENSIONS FOR PIPE THREADS

Nominal Pipe Thread Size in	External Thread Chamfer Diameter[a] Max mm	External Thread Chamfer Diameter[a] Max in	External Thread Chamfer Diameter[a] Min mm	External Thread Chamfer Diameter[a] Min in	External Thread Length of Chamfered or Partial Thread Min mm	External Thread Length of Chamfered or Partial Thread Min in
1/8	8.1	0.32	7.6	0.30	0.94	0.037
1/4	10.7	0.42	10.2	0.40	1.42	0.056
3/8	14.0	0.55	13.5	0.53	1.42	0.056
1/2	17.3	0.68	16.8	0.66	1.80	0.071
3/4	22.6	0.89	22.1	0.87	1.80	0.071
1	28.4	1.12	27.7	1.09	2.21	0.087

Nominal Pipe Thread Size in	External Thread Length of Chamfered or Partial Thread Max mm	External Thread Length of Chamfered or Partial Thread Max in	Internal Thread Counter-sink Diameter[a] Min mm	Internal Thread Counter-sink Diameter[a] Min in	Internal Thread Counter-sink Diameter[a] Max mm	Internal Thread Counter-sink Diameter[a] Max in
1/8	1.40	0.055	10.7	0.42	11.2	0.44
1/4	2.13	0.084	14.0	0.55	14.5	0.57
3/8	2.13	0.084	17.5	0.69	18.0	0.71
1/2	2.72	0.107	21.6	0.85	22.1	0.87
3/4	2.72	0.107	26.9	1.06	27.4	1.08
1	3.30	0.130	34.0	1.34	34.8	1.37

[a] Tabulated diameters conform with Appendix A, SAE J476.

APPENDIX A

GAGES AND GAGING FOR INVERTED FLARED TYPE FITTINGS

A.1 General—The information contained herein is intended to provide and promote the use of uniform gaging practices for determining conformance of inverted flared type tube fittings with the specifications given in the standard.

Gaging of Cone Seats:

Depth of Face—Proper location of the face of the cone seat relative to the face of the fitting shall be determined by use of the step limit gage depicted in Figure A1. One-half of the top surface of body of this gage shall be machined and/or ground 0.38 mm (0.015 in) below the balance of surface to provide a low limit step. The gage pin shall have a minimum diameter equal to "E max" (see Table 4) and shall be of a length equivalent to the sum of the height of the high-limit side of gage body plus "K min" (see Table 4). The opposite faces of gage body and ends of gage pin shall be flat, smooth, parallel, and square with the axis of pin. The fit between gage pin and gage body shall be such that the pin alignment is maintained yet free to move, of its own weight, within the gage body. All gage components should be made from steel suitably hardened to assure adequate service life.

When this gage is placed on the face of fitting, in line with the axis of seat and with gage pin contacting face of seat, the top of gage pin must not protrude above the high limit nor be below the low limit top surfaces of the gage body for fitting to be acceptable.

Gaging of Internal Threads:

Size and Form—Conformance of internal threads with the dimensions specified shall be determined in accordance with standard thread gaging procedures.

Full Form Thread Depth—Suitability of the relationship between the last full form thread to the face of the cone seat shall be determined by use of the gage depicted in Figure A2. This gage shall have external threads, conforming in all respects with the corresponding GO thread plug gage for the respective thread size, of a minimum length equal to "K min" (see Table 4) plus one pitch (thread) and shall have the partial thread at starting end removed. Use of a thread relief undercut beyond a length equal to "K min" shall be permissible. The gage pin hole at starting end shall be suitably counterbored to clear the fitting cone seat to a depth equivalent to 1-1/2 pitches (threads). The length of the gage pin shall be equivalent to the overall length of the body of gage plus the difference between "J max" (see Table 4) and 1/2 pitch (thread). All other features of gage body, gage pin, fit, and material shall be as specified for the gage in Figure A1.

When this gage is assembled by hand into the fitting as far as the thread will permit, with pin contacting face of cone seat, the top of the gage pin must be flush with or protrude above the top surface of the gage body for fitting to be acceptable.

Concentricity Gage and Gaging:

Cone Seat Concentricity—Conformance with specified limitations on concentricity of cone seats with respect to thread pitch cylinder shall be checked by the use of functional gages of the type described herein and depicted in Figure A3, or equivalent means.

The gage consists of a body or frame providing a means for mounting a dial indicator gage in such a manner that it can be actuated through a rotating pin on the opposite end of which is a machined stylus designed to ride on the conical surface of the seat. The gage is centered on the threads of the fitting by means of an interchangeable male threaded gage insert. The threads on the insert shall conform in all respects with the corresponding GO thread plug gage for the respective thread size and the length of full form thread shall be equivalent to K min (see Table 3) plus one pitch (thread). The entering end of gage insert shall have a pilot of length equal to one pitch (thread) and diameter which will clear the minor diameter of the seat threads. The first thread beyond the pilot shall be chamfered or the partial portion shall be removed. Use of a thread relief undercut beyond a length equivalent to K min shall be permissible. The fit between the stylus extension pin and the hole through gage body and threaded gage insert shall be such that pin will move freely yet be retained in alignment. Fitting manufacturers may be consulted with regard to details of existing gages.

The fitting to be inspected shall be gaged in accordance with the following procedure:

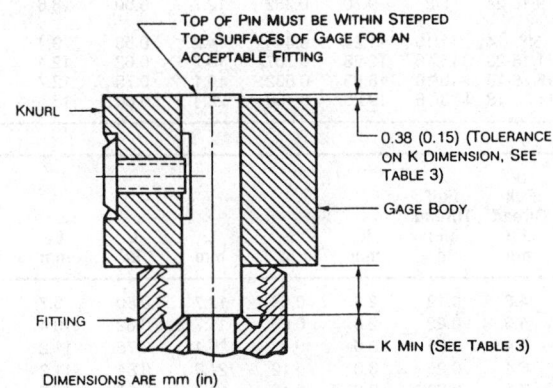

FIGURE A1—FLUSH PIN GAGE FOR CHECKING LOCATION OF FACE OF CONE SEAT

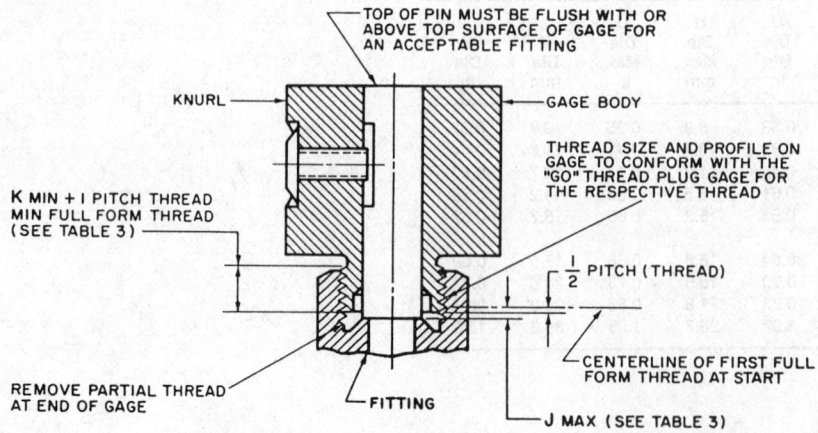

FIGURE A2—FLUSH PIN GAGE FOR CHECKING RELATION OF END OF FULL THREAD WITH FACE OF CONE SEAT

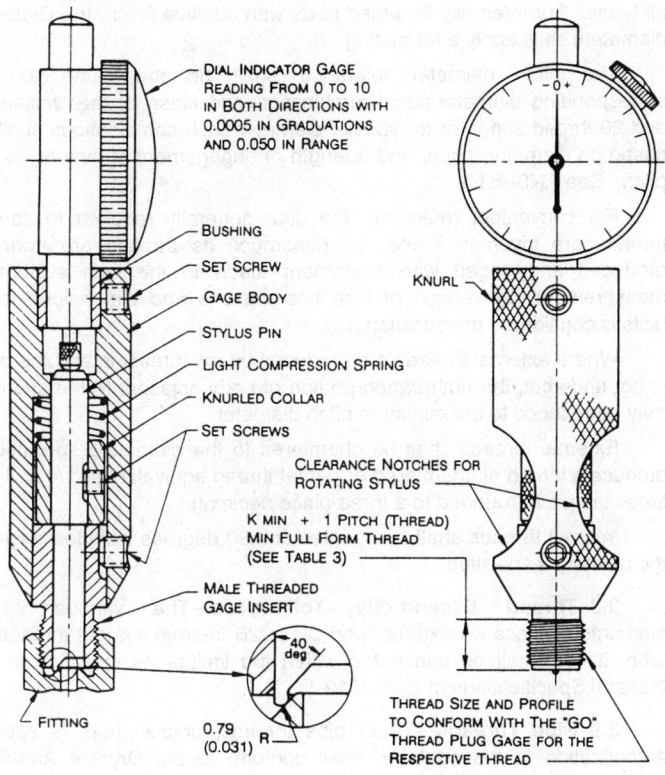

FIGURE A3—TYPICAL CONE SEAT CONCENTRICITY GAGE

1. The threaded end of gage shall be assembled by hand into the seat on the fitting as far as the thread will permit.
2. The indicator gage shall be zeroed and the stylus then rotated slowly through a complete 360 degree revolution by twisting the knurled collar through the finger notches in gage body.
3. The total runout indicated throughout the rotation of stylus shall be read and recorded. It shall not exceed 0.142 mm (0.0056 in) for the fitting to be acceptable. (The measured deviation is in a plane perpendicular to the plane of the cone seat. The concentricity variation, therefore, may be determined from Table A1, or by multiplying the dial indicator reading by the tangent of 42 degrees, or 0.90040.)

TABLE A1—CONCENTRICITY

Dial Reading mm 3 Pl.	Dial Reading in	Concentricity mm 4 Pl.	Concentricity in
0.013	0.0005	0.0114	0.00045
0.025	0.0010	0.0229	0.00090
0.038	0.0015	0.0343	0.00135
0.051	0.0020	0.0457	0.00180
0.064	0.0025	0.0572	0.00225
0.076	0.0030	0.0686	0.00270
0.089	0.0035	0.0800	0.00315
0.102	0.0040	0.0914	0.00360
0.114	0.0045	0.1029	0.00405
0.127	0.0050	0.1143	0.00450
0.140	0.0055	0.1257	0.00495
0.152	0.0060	0.1372	0.00540

SPHERICAL AND FLANGED SLEEVE (COMPRESSION) TUBE FITTINGS—SAE J246 JUN93

SAE Standard

Report of the Tube, Pipe, Hose, and Lubrication Fittings Committee approved May 1971. Revised July 1977, editorial change March 1981. Completely revised by the Fluid Conductors Technical Committee June 1990. Revised by the Fluid Conductors and Connectors Technical Committee SC1—Automotive and Hydraulic Tube and Fittings April 1991 and June 1992. Revised by the Fluid Conductors and Connectors Technical Committee June 1993. Rationale statement available.

1. Scope—This SAE Standard covers complete general and dimensional specifications for tube fittings of the spherical and flanged sleeve compression types for use in the piping of air brake systems on automotive vehicles. The spherical sleeve compression type Figures 1A to 5 and Tables 1 to 3 is intended for use with annealed copper alloy tubing per SAE J1149, Type 1. The flanged sleeve compression type Figures 6A to 11 and Tables 4 to 6 is intended for use with nylon tubing per SAE J844. It is not intended to restrict or preclude other designs of a tube fitting for use with SAE J844, air brake tubing. Performance requirements for SAE J844 are covered in SAE J1131. See SAE J1131 for the Performance Requirements of Reusable (Push to Connect) Fittings Intended for Use in Automotive Air Brake Systems.

CAUTION—To assure satisfactory performance, tapered sleeve compression type fitting components (SAE J512) should not be intermixed with the spherical or flanged sleeve components, nor should the spherical sleeve compression type components be intermixed with the flanged sleeve compression type components when assembling connection in areas where the three types are available.

2. References

2.1 Applicable Documents—The following publications form a part of this specification to the extent specified herein. The latest issue of SAE publications shall apply.

2.1.1 SAE PUBLICATIONS—Available from SAE, 400 Commonwealth Drive, Warrendale, PA 15096-0001.

SAE J246—Spherical and Flanged Sleeve (Compression) Tube Fittings

SAE J476—Dryseal Pipe Threads

SAE J512—Automotive Tube Fittings

SAE J844—Nonmetallic Air Brake System Tubing

SAE J846—Coding Systems for Identification of Fluid Conductors and Connectors

SAE J1131—Performance Requirements for SAE J844 Nonmetallic Tubing and Fitting Assemblies Used in Automotive Air Brake Systems

SAE J1149—Metallic Air Brake System Tubing and Pipe

2.1.2 ANSI PUBLICATION—Available from ANSI, 11 West 42nd Street, New York, NY 10036-8002.

ANSI/ASME B1.1-1989—Unified Inch Screw Thread (UN and UNR)

2.1.3 ASTM PUBLICATIONS—Available from ASTM, 1916 Race Street, Philadelphia, PA 19103-1187.

ASTM B 117—Method of Salt Spray (Fog) Testing

FMVSS 106—Brake Hoses

3. General Specifications—The following general specifications supplement the dimensional data contained in Tables 1 to 8 with respect to all unspecified details.

3.1 Identification—At manufacturer's option, or where so specified by the purchaser, the fittings, except sleeves, may be permanently and legibly marked air brake. The nut shall be permanently and legibly marked according to the current U.S. Department of Transportation FMVSS 106 Regulation (NHTSA). The location of such markings shall be optional with manufacturer.

3.2 Size Designations—Fitting sizes are designated by the corresponding nominal outside diameter of the tubing for the various sizes of tube ends and by the corresponding standard nominal pipe size for pipe thread ends.

3.3 Dimensions and Tolerances—Except for nominal size and thread specifications, dimensions and tolerances are given in both SI units and U.S. customary as designated. Tabulated dimensions shall apply to the finished parts. The maximum and minimum across flat dimensions shall be within the commercial tolerance of bar or extruded stock from which the fittings are produced. The minimum across corners dimensions of hexagons shall be 1.092 times the nominal width across flats, but shall not result in a side flat width less than 0.43 times the nominal width across flats.

Unless otherwise specified, tolerance on hole diameters designated "drill" in the dimensional tables shall be as tabulated in Table 9.

Tolerance on all dimensions not otherwise specified shall be ±0.25 mm (±0.010 in). Tube seat diameters E shall be concentric with straight thread pitch diameters within 0.25 mm (0.010 in) full indicator reading (FIR). Large seat diameters F shall be concentric with tube seat diameters E within 0.13 mm (0.005 in) full indicator reading (FIR). The surface of tube stop at base of seat shall be flat and perpendicular to the axis of thread. Angular tolerance on axis of ends of elbows and tees shall be ±2.50 degrees for sizes up to and including 3/8 in, and ±1.50 degrees for sizes larger than 3/8 in.

Where so illustrated, hexagon corners shall be chamfered 30 degrees ± 5 degrees to a diameter equal to the nominal width across flats, with a tolerance of -0.41 mm (-0.016 in); or, where design permits, corners may be chamfered to the diameter of the abutting surface, provided the length of chamfer does not exceed that produced by the 30 degree chamfer previously described.

3.4 Passages—Where passages in straight fittings are machined from opposite ends, the offset at the meeting point shall not exceed 0.38 mm (0.015 in). The cross-sectional area at the junction of passages in angle fittings shall not be less than that of the smallest passage. At manufacturer's option, all passages in a particular fitting may conform with the smallest diameter specified for that fitting. Where the passage is specified as tap drill diameter or less, the minimum shall be no less than the minimum diameter of the smallest passage in the fitting.

3.5 Wall Thickness—Unless otherwise designated, the wall thickness at any point on fittings shall not be less than the thickness established by the specified dimensions, tolerances, and eccentricities for inner and outer surfaces.

3.6 Contour—Details of contour shall be optional with manufacturer, provided the tabulated dimensions are maintained and serviceability of the fittings is not impaired. Wrench flats on elbows and tees shall be optional. Where extruded and forged shapes are reduced to conserve material, the wall thickness, unless otherwise specified, shall not be less than the respective minimum values tabulated in Table 10.

3.7 Straight Threads—Unified standard Class 2A external and Class 2B internal threads shall apply to plain finish (unplated) fittings of all types. For internally threaded parts with additive finish, the Class 2B diameters shall apply after plating.

The pitch diameter tolerance shall be the same as the corresponding diameter-pitch combination and class of the Unified 18 and 20 thread series or for special diameter-pitch combinations shall be based on diameter, pitch, and a length of engagement of nine times the pitch. See ANSI B1.1.

For convenient reference, the data generally required to specify threads are given in Table 7. (Inasmuch as threads are normally produced and gaged with equipment made to the inch system of measurement, conversion of size designations and dimensions to SI units is considered unnecessary.)

Where external threads are produced by roll threading and the body is not undercut, the unthreaded portion of body adjacent to the shoulder may be reduced to the minimum pitch diameter.

External threads shall be chamfered to the diameters specified to produce a length of chamfered or partial thread equivalent to 1/4 to 1-1/4 times the pitch (rounded to a three-place decimal).

Internal threads shall be countersunk 90 degrees included angle to the diameters specified.

3.8 Thread Eccentricity Tolerances—The various thread elements of Class 2A external and Class 2B internal straight threads on tube fittings shall be concentric within the limitations specified in the General Specifications of SAE J512.

3.9 Pipe Threads—Taper pipe threads, unless there is specific authorization to the contrary, shall conform to the Dryseal American Standard Taper Pipe Thread (NPTF). Specifications for pipe threads are given in detail in SAE J476.

The length of full form external thread shall not be shorter than L_2 plus one pitch (thread), except that where thread is cut through into a relieved body or undercut on the fitting, the minimum full threaded length may be reduced by one pitch (thread).

Where external pipe threads are produced by roll threading, the diameter of the unthreaded portion of shank adjacent to shoulder may be reduced to the E_2 basic pitch diameter.

The tube fitting dimensions tabulated herein are based on length of the Dryseal American Standard Taper Pipe Thread (NPTF), it being the consensus of manufacturers and users that trouble-free assembly cannot be assured unless a full length thread is used. However, the tap drill depths and overall lengths specified in the tables for fittings with internal taper pipe threads are not consistent with the tap drill depths and overall thread lengths of the Dryseal American Standard Taper Pipe Threads (NPTF) given in Table A2, Appendix A of SAE J476. The full length Dryseal American Standard Taper Pipe Taps specified in Table B2 of SAE J476 cannot be used, as the tap drill depths and overall lengths of the fittings have been reduced to the minimum required by bottoming taps to produce standard full thread length. The deviations described previously are peculiar to these tube fittings and as special tooling is required, caution should be exercised in specifying such deviations for any other products.

External pipe threads shall be chamfered from the diameters tabulated in Table 8 to produce the specified length of chamfered or partial thread. Internal pipe threads shall be countersunk 90 degrees included angle, to the diameters tabulated in Table 8.

3.10 Material and Manufacture—Fittings shall be made from brass and, at manufacturer's option and in accordance with his process of manufacture, may be milled from the bar and forged. Brass shall be UNS C36000 (half-hard), UNS C34500, or UNS C35000 for bar or extruded stock and UNS C37700 for forgings. Sleeves shown in Figure 4 shall be made from UNS C36000 brass and annealed to a maximum hardness of Rockwell F70. Sleeves shown in Figure 10 shall be made from UNS C36000 brass. Tube supports shown in Figure 9 shall be made from UNS C26000 or 300 series stainless steel, according to manufacturer's process standard. Nuts may be made from steel when so specified by the purchaser.

TABLE 1A—DIMENSIONS OF CONNECTORS, UNIONS, ELBOWS, AND TEES (FIGURES 1A TO 3C)

Nom Tube OD, in	A[a] Dryseal Taper Thread in	B, Nom Thread Size, in Class 2A Ext.	C Nom, in	C_1 Nom, in	C_2 Nom, in	D Dia Drill mm	D Dia Drill in	D_1 Dia Drill mm	D_1 Dia Drill in	E Dia mm +0.10 −0.00	E Dia in +0.004 −0.000	F Dia mm +0.10 −0.00	F Dia in +0.004 −0.000	G[b] Dia mm +0.00 −0.25	G[b] Dia in +0.000 −0.010
1/4	1/8	7/16–24	7/16	7/16	9/16	4.78	0.188	4.78	0.188	6.45	0.254	7.90	0.311	9.96	0.392
5/16	1/8	1/2–24	1/2	1/2	9/16	6.35	0.250	4.78	0.188	8.08	0.318	9.60	0.378	9.96	0.454
3/8	1/4	17/32–24	9/16	9/16	11/16	7.92	0.312	7.92	0.312	9.70	0.382	11.38	0.448	12.34	0.486
1/2	3/8	11/16–20	11/16	11/16	7/8	10.31	0.406	10.31	0.406	12.88	0.507	14.66	0.577	16.08	0.633
5/8	1/2	13/16–18	7/8	13/16	1-1/16	13.49	0.531	13.49	0.531	16.05	0.632	18.01	0.709	19.10	0.752
3/4	1/2	1–18	1	1	1-1/16	16.66	0.656	13.49	0.531	19.25	0.758	21.59	0.850	23.85	0.939
3/4	3/4	1–18	1-1/16	1-1/4	1-1/4	16.66	0.656	19.05	0.750	19.25	0.758	21.59	0.850	23.85	0.939
1	1	1-1/4–16	1-3/8	1-1/4	1-5/8	22.22	0.875	22.22	0.875	25.60	1.008	28.50	1.122	30.02	1.182

Nom Tube OD, in	I[c] mm	I[c] in	J Full Thread Min mm	J Full Thread Min in	K mm	K in	L mm ±0.8	L in ±0.03	L_1 mm ±0.8	L_1 in ±0.03	L_2 mm ±0.8	L_2 in ±0.03	M mm ±0.8	M in ±0.03	M_1 mm ±0.8	M_1 in ±0.03
1/4	8.4	0.33	6.9	0.27	6.4	0.25	22.9	0.90	21.6	0.85	21.6	0.85	16.0	0.63	17.8	0.70
5/16	8.6	0.34	7.1	0.28	7.1	0.28	23.9	0.94	22.9	0.90	21.8	0.86	17.0	0.67	17.8	0.70
3/8	11.2	0.44	9.7	0.38	7.9	0.31	31.0	1.22	27.9	1.10	30.2	1.19	20.3	0.80	22.9	0.90
1/2	13.5	0.53	11.2	0.44	11.2	0.44	34.0	1.34	33.3	1.31	32.5	1.28	23.9	0.94	26.4	1.04
5/8	15.0	0.59	12.7	0.50	11.2	0.44	40.4	1.59	36.3	1.43	38.4	1.51	27.9	1.10	30.5	1.20
3/4	16.8	0.66	13.5	0.53	14.2	0.56	42.9	1.69	40.6	1.60	40.1	1.58	30.5	1.20	31.2	1.23
3/4	16.8	0.66	13.5	0.53	14.2	0.56	43.7	1.72	—	—	41.4	1.63	32.5	1.28	34.5	1.36
1	18.3	0.72	15.0	0.59	19.8	0.78	50.8	2.00	45.2	1.78	48.5	1.91	36.6	1.44	40.4	1.59

Nom Tube OD, in	M_2 mm ±0.8	M_2 in ±0.03	M_3 mm ±0.8	M_3 in ±0.03	N mm ±0.8	N in ±0.03	N_1 mm ±0.8	N_1 in ±0.03	N_2 mm ±0.8	N_2 in ±0.03	P[a] min mm	P[a] min in	S Max mm	S Max in
1/4	12.7	0.50	16.0	0.63	17.0	0.67	13.7	0.54	16.3	0.64	9.7	0.38	—	—
5/16	13.7	0.54	17.0	0.67	17.8	0.70	14.5	0.57	16.3	0.64	9.7	0.38	11.4	0.45
3/8	18.3	0.72	20.3	0.80	23.6	0.93	19.8	0.78	21.8	0.86	14.2	0.56	—	—
1/2	21.6	0.85	23.9	0.94	25.4	1.00	21.1	0.83	24.1	0.95	14.2	0.56	—	—
5/8	23.9	0.94	27.9	1.10	31.8	1.25	27.4	1.08	29.7	1.17	19.1	0.75	—	—
3/4	28.4	1.12	31.8	1.25	34.0	1.34	29.0	1.14	30.5	1.20	19.1	0.75	20.3	0.80
3/4	30.5	1.20	—	—	34.0	1.34	29.0	1.14	30.5	1.20	19.1	0.75	21.8	0.86
1	32.3	1.27	36.6	1.44	42.2	1.66	36.6	1.44	37.6	1.48	23.9	0.94	—	—

Nom Tube OD, in	S_1 Min mm	S_1 Min in	T[d] Ref mm	T[d] Ref in	T_1 Min mm	T_1 Min in	V Dia Max mm	V Dia Max in	V Dia Min mm	V Dia Min in	W[e] Forged or Bar Min mm +0.0 −0.5	W[e] Forged or Bar Min in +0.00 −0.02
1/4	—	—	4.6	0.18	4.6	0.18	9.7	0.38	9.1	0.36	14.2	0.56
5/16	19.8	0.78	5.3	0.21	5.3	0.21	11.2	0.44	10.7	0.42	14.2	0.56
3/8	—	—	5.3	0.21	5.3	0.21	12.2	0.48	11.7	0.46	17.5	0.69
1/2	—	—	6.1	0.24	6.1	0.24	15.7	0.62	15.5	0.61	20.6	0.81
5/8	—	—	6.1	0.24	6.1	0.24	18.5	0.73	18.3	0.72	25.4	1.00
3/4	38.1	1.50	6.9	0.27	6.9	0.27	23.4	0.92	22.9	0.90	25.4	1.00
3/4	43.7	1.72	7.6	0.30	7.6	0.30	23.4	0.92	22.9	0.90	31.8	1.25
1	—	—	8.4	0.33	8.4	0.33	29.5	1.16	29.0	1.14	38.1	1.50

[a] Dryseal American Standard Taper Pipe Thread, except as noted in General Specifications.
[b] Where thread relief undercut is used, last thread shall be chamfered 1/2 to 1 pitch long from G diameter. Thread marks on surface of undercut shall be permissible.
[c] For elbows and tees, the L dimensions shown shall apply to turned or finished length. Where body is relieved beyond thread, length of turned boss L may be reduced to the minimum full thread length J.
[d] Minimum design thickness, not subject to inspection.
[e] Basic dimensions shown shall apply as min dia or across flats for bosses. The −0.5 mm (−0.02 in) tolerance shall apply only to chamfer on full hexagon version in Figure 1C.

TABLE 1B—DIMENSIONS OF SPECIAL SIZE FITTINGS,[b]
CONNECTORS, ELBOWS, AND TEES (FIGURES 1A, 1C, 2A, 2C, 2D, 3B, 3C, AND 3D)

Nom Tube OD, in	A^a Dryseal Taper Thread, in	A^a_1 Dryseal Taper Thread, in	B, Nom Thread Size, in Class 2A Ext.	C Nom, in	C_2 Nom, in	L mm ±0.8	L in ±0.03	L_2 mm ±0.8	L_2 in ±0.03	M mm ±0.8	M in ±0.03	M_1 mm ±0.8	M_1 in ±0.03
1/4	1/4		7/16–24	9/16	11/16	28.2	1.11	27.4	1.08	17.5	0.69	20.1	0.79
1/4	3/8	1/8	7/16–24	11/16	7/8	29.0	1.14	27.4	1.08	18.8	0.74	21.3	0.84
1/4	1/2		7/16–24	7/8	1-1/16	33.8	1.33	31.8	1.25	21.3	0.84	23.9	0.94
5/16	1/4		1/2–24	9/16	11/16	28.4	1.12	27.7	1.09	17.8	0.70	20.3	0.80
5/16	3/8	1/8	1/2–24	11/16	7/8	29.2	1.15	27.7	1.09	19.0	0.75	26.6	0.85
5/16	1/2		1/2–24	7/8	1-1/16	34.0	1.34	32.0	1.26	21.6	0.85	24.1	0.95
3/8	1/8		17/32–24	9/16	9/16	26.4	1.04	25.7	1.01	18.5	0.73	21.3	0.84
3/8	3/8	1/4	17/32–24	11/16	7/8	31.8	1.25	30.2	1.19	21.6	0.85	24.1	0.95
3/8	1/2		17/32–24	7/8	1-1/16	36.6	1.44	34.5	1.36	24.1	0.95	26.7	1.05
1/2	1/8		11/16–20	11/16	11/16	29.5	1.16	27.9	1.10	20.3	0.80	23.1	0.91
1/2	1/4		11/16–20	11/16	11/16	34.0	1.34	32.5	1.28	22.1	0.87	24.6	0.97
1/2	1/2	3/8	11/16–20	7/8	1-1/16	38.9	1.53	36.8	1.45	26.4	1.04	29.0	1.14
1/2	3/4		11/16–20	1-1/16	1-1/4	40.4	1.59	38.1	1.50	29.2	1.15	31.2	1.23
1/2	1		11/16–20	1-3/8	1-5/8	46.0	1.81	43.7	1.72	31.8	1.25	35.6	1.40
5/8	1/4		13/16–18	13/16	13/16	35.6	1.40	33.5	1.32	24.1	0.95	26.4	1.04
5/8	3/8	1/2	13/16–18	13/16	7/8	35.6	1.40	33.5	1.32	25.7	1.01	27.9	1.10
5/8	3/4		13/16–18	1-1/16	1-1/4	41.9	1.65	39.6	1.56	30.7	1.21	32.8	1.29
5/8	1		13/16–18	1-3/8	1-5/8	47.5	1.87	45.2	1.78	33.3	1.31	37.1	1.46
3/4	1/4		1–18	1	1	38.1	1.50	35.3	1.39	26.7	1.05	27.2	1.07
3/4	3/8		1–18	1	1	38.1	1.50	35.3	1.39	27.9	1.10	28.7	1.13
3/4	1		1–18	1-3/8	1-5/8	49.3	1.94	47.0	1.85	35.1	1.38	38.9	1.53
1	1/2	3/4	1-1/4–16	1-1/4	1-1/4	46.0	1.81	43.7	1.72	30.5	1.20	34.0	1.34
1	3/4		1-1/4–16	1-1/4	1-1/4	46.0	1.81	43.7	1.72	33.3	1.31	37.1	1.46

Nom Tube OD, in	M_2 mm ±0.8	M_2 in ±0.03	N mm ±0.8	N in ±0.03	N_1 mm ±0.8	N_1 in ±0.03	N_2 mm ±0.8	N_2 in ±0.03	N_3 mm ±0.8	N_3 in ±0.03	W_e mm +0.0 −0.5	W_e in +0.00 −0.02
1/4	15.5	0.61	22.4	0.88	18.3	0.72	21.8	0.86			17.5	0.69
1/4	16.5	0.65	22.1	0.87	18.3	0.72	24.1	0.95	23.9	0.94	20.6	0.81
1/4	17.3	0.68	26.9	1.06	23.1	0.91	29.7	1.17			25.4	1.00
5/16	15.7	0.62	23.1	0.91	19.0	0.75	21.8	0.86			17.5	0.69
5/16	16.8	0.66	23.1	0.91	19.0	0.75	24.1	0.95	23.9	0.94	20.6	0.81
5/16	17.5	0.69	27.7	1.09	23.9	0.94	29.7	1.17			25.4	1.00
3/8	18.3	0.72	19.0	0.75	15.2	0.60	17.3	0.68			14.2	0.56
3/8	19.3	0.76	23.4	0.92	19.8	0.78	24.1	0.95	25.4	1.00	20.6	0.81
3/8	20.1	0.79	28.2	1.11	24.6	0.97	29.7	1.17			25.4	1.00
1/2	21.6	0.85	20.8	0.82	16.5	0.65	19.6	0.77			14.2	0.56
1/2	21.6	0.85	25.4	1.00	21.1	0.83	24.1	0.95			17.5	0.69
1/2	22.4	0.88	30.2	1.19	25.9	1.02	29.7	1.17	31.8	1.25	25.4	1.00
1/2	27.2	1.07	30.0	1.18	25.7	1.01	30.5	1.20			31.8	1.25
1/2	27.4	1.08	35.1	1.38	30.7	1.21	37.6	1.48			38.1	1.50
5/8	23.9	0.94	26.9	1.06	22.6	0.89	24.8	0.98			17.5	0.69
5/8	23.9	0.94	26.9	1.06	22.6	0.89	24.8	0.98	34.0	1.34	20.6	0.81
5/8	28.7	1.13	31.8	1.25	27.4	1.08	30.5	1.20			31.8	1.25
5/8	29.0	1.14	36.6	1.44	32.3	1.27	37.6	1.48			38.1	1.50
3/4	28.4	1.12	29.2	1.15	24.1	0.95	25.7	1.01			17.5	0.69
3/4	28.4	1.12	29.2	1.15	24.1	0.95	25.7	1.01			20.6	0.81
3/4	30.7	1.21	38.9	1.53	33.8	1.33	37.6	1.48			38.1	1.50
1	32.3	1.27	37.3	1.47	31.8	1.25	32.8	1.29	37.3	1.47	25.4	1.00
1	32.3	1.27	37.3	1.47	31.8	1.25	32.8	1.29			31.8	1.25

(R) *3.11 Finish*—The external surfaces and threads of all carbon steel parts shall be plated or coated with a suitable material that passes a 72 h salt spray test in accordance with ASTM B 117. Any appearance of red rust during the 72 h salt spray test shall be considered failure, except for the following:

a. All internal fluid passages.

b. Edges such as hex points, serrations, and crests of threads where there may be mechanical deformation of the plating or coating typical of mass-produced parts or shipping effects.

22.25

c. Areas where there is mechanical deformation of the plating or coating caused by crimping, flaring, bending, and other post-plate metal forming operations.

d. Areas where the parts are suspended or affixed in the test chamber where condensate can accumulate.

NOTE—Cadmium plating is not preferred due to environmental reasons. Parts manufactured to this Standard after January 1, 1997, shall not be cadmium plated. Internal fluid passages shall be protected from corrosion during storage. Changes in plating may affect assembly torques and require requalification, when applicable.

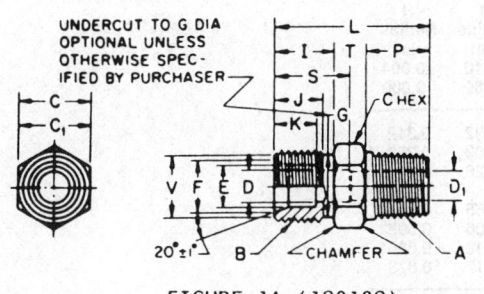
FIGURE 1A (120102)

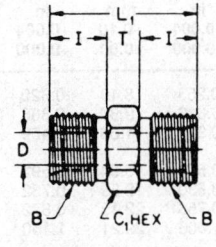
FIGURE 1B (120101)

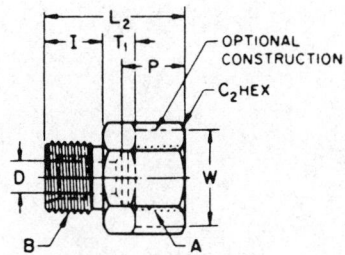
FIGURE 1C (120103)

FIGURE 1—CONNECTORS AND UNIONS

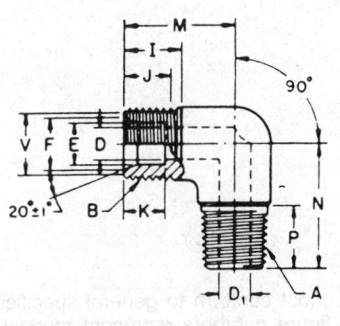
FIGURE 2A (120202)

FIGURE 2B (120201)

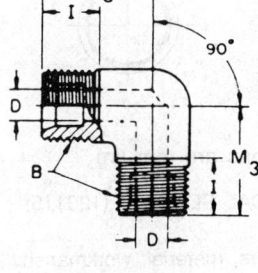
FIGURE 2C (120203)

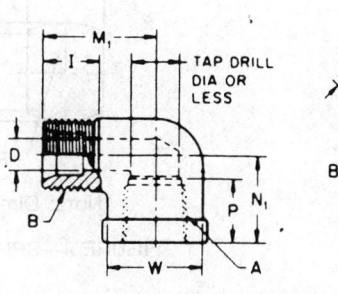
FIGURE 2D (120302)

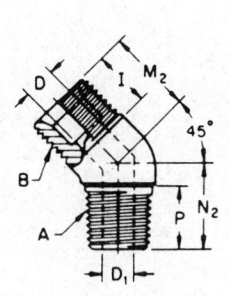

FIGURE 2—ELBOWS

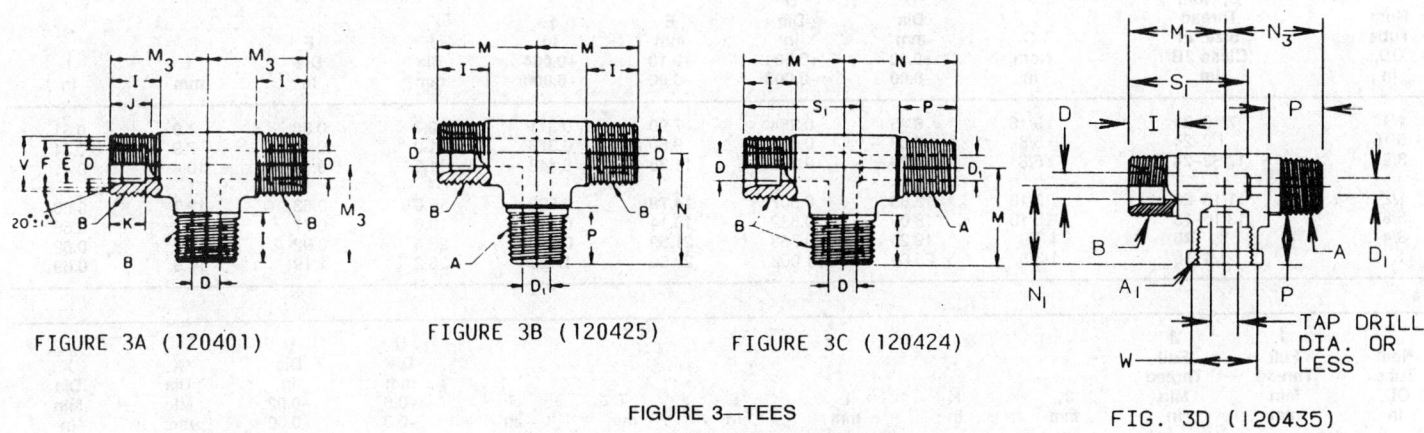

FIGURE 3A (120401)   FIGURE 3B (120425)   FIGURE 3C (120424)   FIG. 3D (120435)

FIGURE 3—TEES

NOTES: Unspecified detail with respect to dimensions, tolerances, contours, material, workmanship, etc., must conform to general specifications for spherical and flanged sleeve (compression) tube fittings. The dimensional designations on the first figure in each group shall apply to other figures in that group except as shown otherwise. Codes shown in brackets adjacent to figure numbers represent respective fitting identification in accordance with SAE J846.

FIGURES 1A THRU 3D

3.12 Workmanship—Workmanship shall conform to the best commercial practice to produce high-quality fittings. Fittings shall be free from all machining fluids, chips, hanging burrs, loose scale, and slivers which might become dislodged in usage and all other defects which might affect their serviceability. All seating surfaces for the spherical sleeve must be smooth except that annular tool marks up to 2.5 μm (100 μin) maximum shall be permissible.

3.13 Assembly Considerations—Where it is not objectionable from a function or production standpoint, the use of a compatible lubricant or sealant may be desirable in assembling Dryseal pipe threads on spherical and flanged sleeve (compression) tube fittings to minimize galling and effect a pressure-tight seal.

TABLE 2—DIMENSIONS OF SPHERICAL SLEEVES (FIGURE 4)

Nominal Tube OD, in	D Dia mm +0.10 -0.00	D Dia in +0.004 -0.000	F Dia mm +0.10 -0.00	F Dia in +0.004 -0.000	L mm ±0.13	L in ±0.005	R Radius mm +0.10 -0.00	R Radius in +0.004 -0.000
1/4	6.43	0.253	8.13	0.320	6.35	0.250	7.92	0.312
5/16	8.03	0.316	9.80	0.386	7.14	0.281	9.09	0.358
3/8	9.65	0.380	11.66	0.459	7.95	0.313	10.26	0.404
1/2	12.83	0.505	15.04	0.592	9.52	0.375	12.65	0.498
5/8	16.00	0.630	18.59	0.732	11.13	0.438	15.06	0.593
3/4	19.20	0.756	22.15	0.872	12.70	0.500	17.42	0.686
1	25.55	1.006	29.21	1.150	15.90	0.626	22.17	0.873

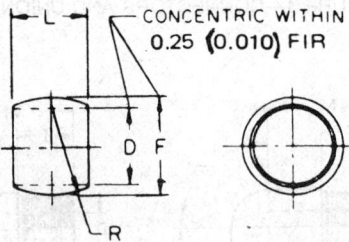

Note: Dimensions are mm (in)

FIGURE 4—SPHERICAL SLEEVES (120115)

Notes: Unspecified detail with respect to dimensions, tolerances, contours, material, workmanship, etc., must conform to general specifications for spherical and flanged sleeve (compression) tube fittings. Codes shown in brackets adjacent to figure numbers represent respective fitting identification in accordance with SAE J846.

TABLE 3—DIMENSIONS OF SPHERICAL SLEEVE NUTS (FIGURE 5)

Nom Tube OD, in	B, Nom Thread Size, in Class 2B Int	C Nom, in	D Dia mm +0.10 -0.00	D Dia in +0.004 -0.000	E Dia mm +0.10 -0.00	E Dia in +0.004 -0.000	F Dia mm	F Dia in	I mm	I in
1/4	7/16-24	9/16	6.45	0.254	7.90	0.311	9.7	0.38	7.9	0.31
5/16	1/2-24	5/8	8.08	0.318	9.60	0.378	11.2	0.44	7.9	0.31
3/8	17/32-24	5/8	9.70	0.382	11.38	0.448	11.9	0.47	10.4	0.41
1/2	11/16-20	13/16	12.88	0.507	14.78	0.582	16.0	0.63	12.7	0.50
5/8	13/16-18	15/16	16.05	0.632	18.14	0.714	19.6	0.77	14.2	0.56
3/4	1 -18	1-1/8	19.25	0.758	21.59	0.850	23.4	0.92	15.7	0.62
1	1-1/4-16	1-3/8	25.60	1.008	28.50	1.122	30.2	1.19	17.5	0.69

Nom Tube OD, in	J Full Thread Min mm	J Full Thread Min in	K mm	K in	L mm	L in	T mm	T in	U Dia mm +0.5 -0.0	U Dia in +0.02 -0.00	X Dia Min mm	X Dia Min in
1/4	6.4	0.25	3.3	0.13	19.0	0.75	9.7	0.38	11.4	0.45	13.5	0.53
5/16	6.4	0.25	3.3	0.13	22.4	0.88	12.7	0.50	13.0	0.51	15.0	0.59
3/8	8.9	0.35	3.3	0.13	28.7	1.13	13.5	0.53	13.7	0.54	15.5	0.61
1/2	10.7	0.42	3.3	0.13	31.8	1.25	16.0	0.63	17.8	0.70	19.8	0.78
5/8	12.2	0.48	4.8	0.19	35.1	1.38	18.3	0.72	21.1	0.83	23.1	0.91
3/4	13.7	0.54	4.8	0.19	39.6	1.56	19.0	0.75	25.9	1.02	27.7	1.09
1	15.2	0.60	4.8	0.19	42.9	1.69	23.9	0.94	32.3	1.27	34.0	1.34

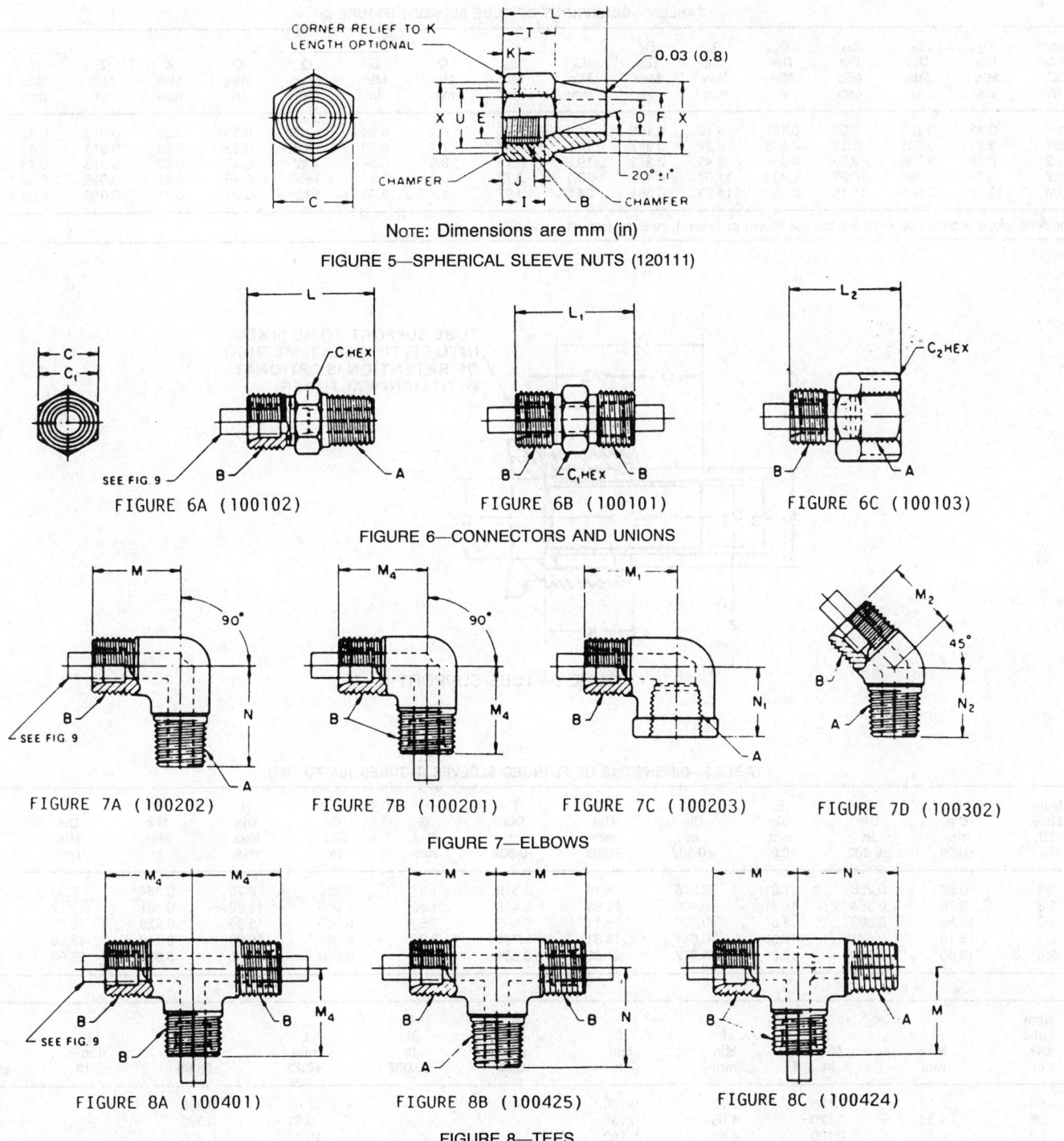

NOTE: Dimensions are mm (in)

FIGURE 5—SPHERICAL SLEEVE NUTS (120111)

FIGURE 6A (100102) FIGURE 6B (100101) FIGURE 6C (100103)

FIGURE 6—CONNECTORS AND UNIONS

FIGURE 7A (100202) FIGURE 7B (100201) FIGURE 7C (100203) FIGURE 7D (100302)

FIGURE 7—ELBOWS

FIGURE 8A (100401) FIGURE 8B (100425) FIGURE 8C (100424)

FIGURE 8—TEES

NOTES: Flanged sleeve type fittings having tube support features (Figures 6A to 8C) are intended only for use with nylon tubing per SAE J844 in conjunction with flanged sleeve in Figure 10 and flanged sleeve nut in Figure 11.
Unspecified detail with respect to dimensions, tolerances, contours, materials, workmanship, etc., must conform to general specifications for spherical and flanged sleeve (compression) tube fittings. All dimensions for Figures 6A to 8C shall correspond with respective Figures 1A to 3C and Table 1A and 1B except for addition of the tube support shown in Figure 9 and Table 4. Codes shown in brackets adjacent to figure numbers represent respective fitting identification in accordance with SAE J846.

FIGURE 5 TO 8C

TABLE 4—DIMENSIONS OF TUBE SUPPORT (FIGURE 9)

Nom Tube OD, in	D_2 Dia Min mm	D_2 Dia Min in	D_3 Dia Min mm	D_3 Dia Min in	D_3 Dia Max mm	D_3 Dia Max in	L_3 Min mm	L_3 Min in	O Min mm	O Min in	O Max mm	O Max in	Z Min mm	Z Min in	Z Max mm	Z Max in
1/4	3.12	0.123	4.09	0.161	4.19	0.165	11.4	0.45	6.1	0.24	7.6	0.30	0.30	0.012	0.43	0.017
3/8	5.21	0.205	6.17	0.243	6.27	0.247	14.5	0.57	7.4	0.29	8.9	0.35	0.33	0.013	0.43	0.017
1/2	8.28	0.326	9.35	0.368	9.45	0.372	19.0	0.75	8.6	0.34	10.2	0.40	0.33	0.013	0.48	0.019
5/8	9.75	0.384	10.97	0.432	11.07	0.436	20.1	0.79	10.2	0.40	11.7	0.46	0.41	0.016	0.56	0.022
3/4	12.73	0.501	14.15	0.557	14.25	0.561	24.6	0.97	11.7	0.46	13.2	0.52	0.41	0.016	0.66	0.026

[a] For dimensions depicted on Figure 9 but not shown in Table 4, refer to Table 1A and 1B.

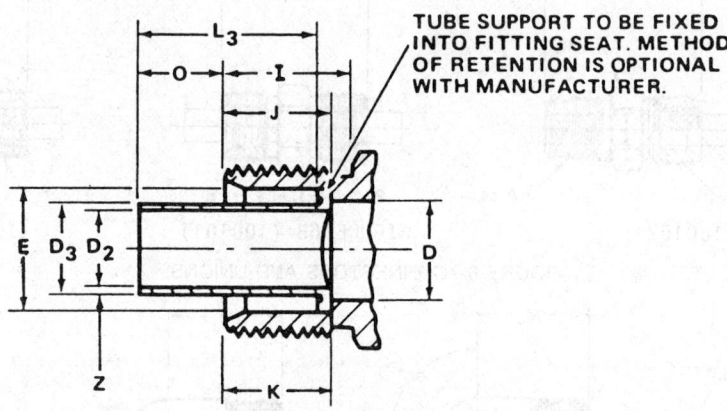

FIGURE 9—TUBE SUPPORT

TABLE 5—DIMENSIONS OF FLANGED SLEEVES (FIGURES 10A TO 10B)

Nom Tube OD, in	D Dia mm ±0.05	D Dia in ±0.002	E Dia mm ±0.05	E Dia in ±0.002	F Dia mm ±0.05	F Dia in ±0.002	G Ref mm	G Ref in	H Dia Max mm	H Dia Max in	H Dia Min mm	H Dia Min in
1/4	6.50	0.256	7.01	0.276	8.10	0.319	1.27	0.050	9.25	0.364	8.99	0.354
3/8	9.75	0.384	10.31	0.406	11.63	0.458	1.52	0.060	12.22	0.481	12.12	0.477
1/2	12.93	0.509	13.67	0.538	15.11	0.595	1.52	0.060	15.95	0.628	15.75	0.620
5/8	16.10	0.634	16.36	0.644	18.31	0.721	2.03	0.080	19.00	0.748	18.90	0.744
3/4	19.30	0.760	19.61	0.772	21.95	0.864	2.29	0.090	23.55	0.927	23.90	0.917

Nom Tube OD, in	I Max mm	I Max in	I Min mm	I Min in	K Dia mm ±0.05	K Dia in ±0.002	L mm ±0.13	L in ±0.005	M mm ±0.13	M in ±0.005
1/4	3.18	0.125	3.05	0.120	—	—	7.49	0.295	—	—
3/8	4.32	0.170	4.06	0.160	—	—	9.91	0.390	—	—
1/2	4.83	0.190	4.57	0.180	—	—	10.92	0.430	—	—
5/8	5.33	0.210	5.08	0.200	17.37	0.684	12.45	0.490	1.40	0.055
3/4	5.84	0.230	5.59	0.220	21.01	0.827	13.72	0.540	1.90	0.075

TABLE 6—DIMENSIONS OF FLANGED SLEEVE NUTS (FIGURE 11)

Nom Tube OD, in	B, Nom Thread Size, in Class 2B Int	C Hex Nom in	D Dia mm +0.10 −0.00	D Dia in +0.004 −0.000	E Dia mm +0.10 −0.00	E Dia in +0.004 −0.000	F Dia mm ±0.5	F Dia in ±0.02	I mm	I in
1/4	7/16–24	9/16	6.45	0.254	7.90	0.311	8.6	0.34	7.9	0.31
3/8	17/32–24	5/8	9.70	0.382	11.38	0.448	11.9	0.47	11.2	0.44
1/2	11/16–20	13/16	12.88	0.507	14.78	0.582	15.5	0.61	13.0	0.51
5/8	13/16–18	15/16	16.05	0.632	18.14	0.714	18.8	0.74	14.7	0.58
3/4	1–18	1-1/8	19.25	0.758	21.59	0.850	22.4	0.88	15.7	0.62

Continued on next page.

TABLE 6—DIMENSIONS OF FLANGED SLEEVE NUTS (FIGURE 11) (CONTINUED)

Nom Tube OD, in	J Full Thd Min mm	J Full Thd Min in	K mm	K in	L mm	L in	U Dia mm +0.5 -0.0	U Dia in +0.02 -0.00	X Dia Min mm	X Dia Min in
1/4	6.4	0.25	3.3	0.13	11.4	0.45	11.4	0.45	13.5	0.53
3/8	9.7	0.38	3.3	0.13	16.0	0.63	13.7	0.54	15.5	0.61
1/2	11.2	0.44	3.3	0.13	18.3	0.72	17.8	0.70	19.8	0.78
5/8	12.7	0.50	4.8	0.19	19.6	0.77	21.1	0.83	23.1	0.91
3/4	13.7	0.54	4.8	0.19	20.6	0.81	25.9	1.02	27.7	1.09

L
G
I
I
D
E
F
H

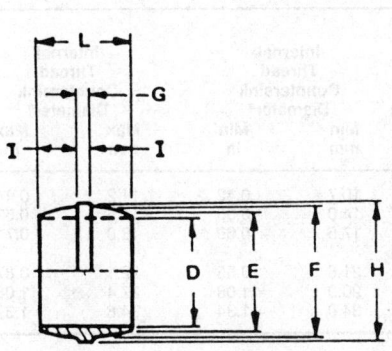

FIGURE 10A—FLANGED SLEEVE, 1/4, 3/8 AND 1/2 IN SIZES (100115)

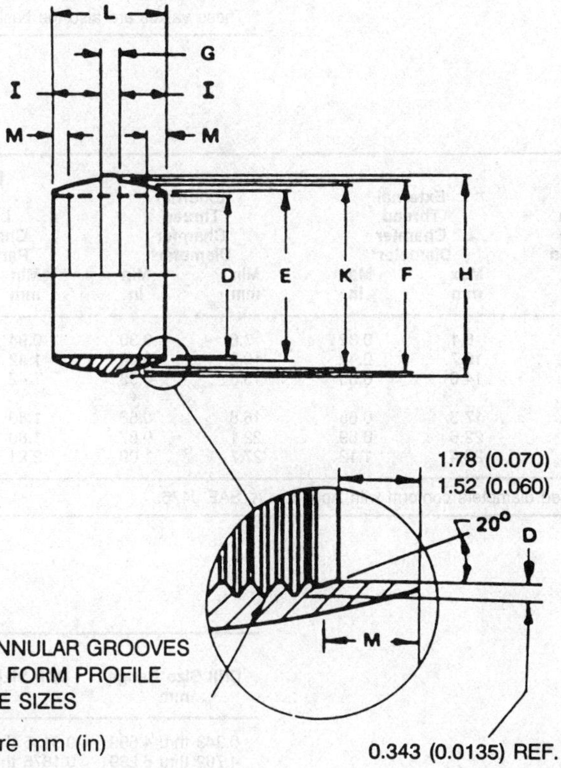

L
G
I
I
M
M
D
E
K
F
H
1.78 (0.070)
1.52 (0.060)
20°
D
M
0.64 (0.025) PITCH ANNULAR GROOVES
OF UNIFIED THREAD FORM PROFILE
OPTIONAL ON THESE SIZES
0.343 (0.0135) REF.

FIGURE 10A AND 10B

NOTE: Dimensions are mm (in)

FIGURE 10B—FLANGED SLEEVE, 5/8 AND 3/4 IN SIZES (100115)

NOTES: Unspecified detail with respect to dimensions, tolerances, contour, material, workmanship, etc., must conform to general specifications for spherical and flanged sleeve (compression) tube fittings. Codes shown in brackets adjacent to figure numbers represent respective fitting identification in accordance with SAE J846.

C
L
K
CORNER RELIEF TO K
LENGTH OPTIONAL
CHAMFER
0.38 (0.015) X 45° CORNER BREAK
X
U
E
T
D
F
20° ± 1°
CHAMFER
J
I
B

NOTE: Dimensions are mm (in)

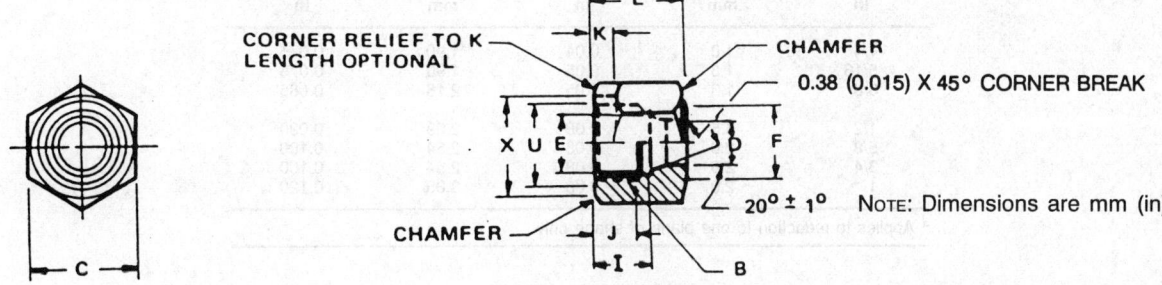

FIGURE 11—FLANGED SLEEVE NUTS (100110)

TABLE 7—STRAIGHT THREAD SPECIFICATION DATA, IN

Nominal Size	Series Designation	External Thread Pitch Diameter Max	External Thread Pitch Diameter Min	Internal Thread Pitch Diameter Max	Internal Thread Pitch Diameter Min[a]	Internal Thread Minor Diameter Max	Internal Thread Minor Diameter Min
7/16–24	UNS	0.4093	0.4055	0.4153	0.4104	0.402	0.392
1/2 –24	UNS	0.4717	0.4678	0.4780	0.4729	0.465	0.455
17/32–24	UNS	0.5030	0.4991	0.5092	0.5041	0.496	0.486
11/16–20	UN	0.6537	0.6494	0.6606	0.6550	0.645	0.633
13/16–18	UNS	0.7750	0.7704	0.7824	0.7764	0.765	0.752
1 –18	UNS	0.9625	0.9578	0.9701	0.9639	0.953	0.940
1-1/4 –16	UN	1.2079	1.2028	1.2160	1.2094	1.196	1.182

[a] These values are also the basic pitch diameter.

TABLE 8

Nom Pipe Thread Size, in	External Thread Chamfer Diameter[a] Max mm	External Thread Chamfer Diameter[a] Max in	External Thread Chamfer Diameter[a] Min mm	External Thread Chamfer Diameter[a] Min in	External Thread Length of Chamfered or Partial Thread Min mm	External Thread Length of Chamfered or Partial Thread Min in	External Thread Length of Chamfered or Partial Thread Max mm	External Thread Length of Chamfered or Partial Thread Max in	Internal Thread Countersink Diameter[a] Min mm	Internal Thread Countersink Diameter[a] Min in	Internal Thread Countersink Diameter[a] Max mm	Internal Thread Countersink Diameter[a] Max in
1/8	8.1	0.32	7.6	0.30	0.94	0.037	1.42	0.056	10.7	0.42	11.2	0.44
1/4	10.7	0.42	10.2	0.40	1.42	0.056	2.11	0.083	14.0	0.55	14.5	0.57
3/8	14.0	0.55	13.5	0.53	1.42	0.056	2.11	0.083	17.5	0.69	18.0	0.71
1/2	17.3	0.68	16.8	0.66	1.80	0.071	2.72	0.107	21.6	0.85	22.1	0.87
3/4	22.6	0.89	22.1	0.87	1.80	0.071	2.72	0.107	26.9	1.06	27.4	1.08
1	28.4	1.12	27.7	1.09	2.21	0.087	3.30	0.130	34.0	1.34	34.8	1.37

[a] Tabulated diameters conform with Appendix A, SAE J476.

TABLE 9

Drill Size Range mm	Drill Size Range in	Tolerance Plus mm	Tolerance Plus in	Tolerance Minus mm	Tolerance Minus in
0.343 thru 4.699	0.0135 thru 0.1850	0.08	0.003	0.05	0.002
4.762 thru 6.299	0.1875 thru 0.2480	0.10	0.004	0.05	0.002
6.350 thru 19.050	0.2500 thru 0.7500	0.15	0.006	0.08	0.003
19.251 thru 25.400	0.7579 thru 1.0000	0.18	0.007	0.10	0.004

TABLE 10

Nominal Tube OD, in	Wall Thickness, Min Extruded Shapes[a] mm	Wall Thickness, Min Extruded Shapes[a] in	Wall Thickness, Min Forged Shapes mm	Wall Thickness, Min Forged Shapes in
1/4	1.0	0.04	1.90	0.075
5/16	1.3	0.05	1.90	0.075
3/8	1.3	0.05	2.16	0.085
1/2	1.5	0.06	2.29	0.090
5/8	2.0	0.08	2.54	0.100
3/4	2.0	0.08	2.54	0.100
1	2.0	0.08	3.05	0.120

[a] Applies to reduction to one plane of shape only.

APPENDIX A
TABLES FOR CALCULATING DIMENSIONS ON SPECIAL SIZES

A.1 The tables in this Appendix present various factors to be used in determining the dimensions applicable to special size combination fittings not contained in SAE J246. (See Table 1A for calculated sizes.)

For any special size fitting, be it a connector, 45 or 90 degree elbow, tee, or cross, one end is always standard. Consider this end to be the largest on the fitting, it may then be used as the basis for establishing the stock size and the length (either overall or end to center) by deducting factors equivalent to the reduction in machining requirements from the appropriate standard lengths.

A.1.1 The factors applicable to the various end configurations and size reductions tabulated in Tables A1 to A9 were determined on the following basis:

a. Those pertaining to lengths were derived by maintaining the standard hexagon thickness for straight fittings and the standard centerline to machining start for shaped fittings.

b. Tables A1 and A2 were derived by subtracting the standard machining length required for the smaller end from that required for the larger end.

c. Table A3 factors are equal to one-half the difference in tube end thread diameters.

d. Table A4 factors are equal to one-half the difference in pipe end thread dimensions.

e. Table A5 factors are equal to one-half the difference in the tube end D drill diameter.

f. Table A6 factors are equal to one-half the difference in the pipe end W diameter or width.

A.1.2 Straight, Tube Size Reduced

L(L1, L2) special size = L(L1, L2) std size - factor from Table A2   (Eq.A1)

EXAMPLE—For a straight connector (Figures 1A and 1C) with 3/8 in tube and 3/8 in NPTF, the overall length would be determined as follows:

a. 34.0 mm (1.34 in) = L overall length for 1/2 tube to 3/8 NPTF from Table 7

b. 2.3 mm (0.09 in) = factor from Table A2 for 3/8 machining on 1/2 size fitting 31.8 mm (1.25 in) = overall length

The proper hex size will be the larger of the values for the tube size and external pipe size shown in Tables A7 and A8, respectively. The 3/8 in tube to 3/8 in NPTF fitting's hex size will (from Table A8) be 11/16 in.

A.1.3 Straight, Pipe Size Reduced

L(L1, L2) special size = L(L1, L2) std size - factor from Table A1   (Eq.A2)

A.1.4 Elbows with External Pipe Threads, Tube Size Reduced

M special size = M std size - factor from Table A2   (Eq.A3)

N special size = N std size - factor from Table A3   (Eq.A4)

EXAMPLE—For elbows and tees (Figures 2A, 3B, 3C) with 1/4 in tube and 1/4 in NPTF, the end to center length M would be derived as follows:

a. 20.3 mm (0.80 in) (M dimension for 3/8 tube to 1/4 NPTF from Table 1A) - 2.8 mm (0.11 in) (factor from Table A2 for 1/4 machined on 3/8 size fitting) = 17.5 mm (0.69 in) (end-to-center length)

The end to center length N would be derived as follows:

a. 23.6 mm (0.93 in) (N dimension for 3/8 tube to 1/4 NPTF from Table 1A) - 1.3 mm (0.05 in) (factor from Table A3 for 1/4 machined on 3/8 size fitting) = 22.3 mm (0.88 in) (end-to-center length)

The proper wrench pad width will be the larger value for the tube size and external pipe size shown in Tables A7 and A8, respectively. The wrench pad width for 1/4 in tube to 1/4 in NPTF fitting will (from Table A8) be 14.0 mm (0.55 in).

A.1.5 Elbows with External Pipe Thread, Pipe Size Reduced

M special size = M std size - factor from Table A4   (Eq.A5)

N special size = N std size - factor from Table A1   (Eq.A6)

A.1.6 Elbows, All Tube Ends

M_4 reduced end = M_4 std size - factor from Table A2   (Eq.A7)

M_4 end not reduced = M_4 std size - factor from Table A3   (Eq.A8)

A.1.7 Tees with External Pipe Thread, Tube Size Reduced—
Figure as two elbows. Use larger of the two figures obtained for the branch centerline-to-end dimension.

A.1.8 Tees with External Pipe Thread, Pipe Size Reduced—
Figure as two elbows. Use larger of the two figures obtained for the branch centerline-to-end dimension.

A.1.9 Tees, All Tube Ends—
Figure as two elbows. Use larger of the two figures obtained for the branch centerline-to-end dimension.

A.1.10 Elbow with Internal Threads, Tube Size Reduced

M_1 special size = M_1 std size - factor from Table A2   (Eq.A9)

N_1 special size = N_1 std size - factor from Table A5   (Eq.A10)

A.1.11 Elbows with Internal Pipe Threads, Pipe Size Reduced

M_1 special size = M_1 std size - factor from Table A6   (Eq.A11)

N_1 special size = N_1 std size - factor from Table A1   (Eq.A12)

A.1.12 45 Degree Elbow, Tube Size Reduced

M_2 special size = M_2 std size - factor from Table A2   (Eq.A13)

N_2 special size = N_2 std size   (Eq.A14)

A.1.13 45 Degree Elbow, Pipe Size Reduced

M_2 special size = M_2 std size   (Eq.A15)

N_2 special size = N_2 std size - factor from Table A1   (Eq.A16)

TABLE A1—LENGTH FACTORS STANDARD PIPE SIZE

Nom Pipe Size, in		1/4 mm	1/4 in	3/8 mm	3/8 in	1/2 mm	1/2 in	3/4 mm	3/4 in	1 mm	1 in
Reduced Pipe Size	1/8	4.6	0.18	4.6	0.18	9.7	0.38	9.7	0.38	14.2	0.56
	1/4	—	—	0.0	0.00	4.8	0.19	4.8	0.19	9.7	0.38
	3/8	—	—	—	—	4.8	0.19	4.8	0.19	9.7	0.38
	1/2	—	—	—	—	—	—	0.0	0.00	4.8	0.19
	3/4	—	—	—	—	—	—	—	—	4.8	0.19

TABLE A2—LENGTH FACTORS STANDARD MACHINING SIZE

Nom Tube OD, in		5/16 mm	5/16 in	3/8 mm	3/8 in	1/2 mm	1/2 in	5/8 mm	5/8 in	3/4 mm	3/4 in	1 mm	1 in
Reduced Tube Size	1/4	0.3	0.01	2.8	0.11	5.1	0.20	6.6	0.26	8.4	0.33	9.9	0.39
	5/16	—	—	2.5	0.10	4.8	0.19	6.4	0.25	8.1	0.32	9.7	0.38
	3/8	—	—	—	—	2.3	0.09	3.8	0.15	5.6	0.22	7.1	0.28
	1/2	—	—	—	—	—	—	1.5	0.06	3.3	0.13	4.8	0.19
	5/8	—	—	—	—	—	—	—	—	1.8	0.07	3.3	0.13
	3/4	—	—	—	—	—	—	—	—	—	—	1.5	0.06

TABLE A3—LENGTH FACTORS STANDARD MACHINING SIZE

Nom Tube OD, in		5/16 mm	5/16 in	3/8 mm	3/8 in	1/2 mm	1/2 in	5/8 mm	5/8 in	3/4 mm	3/4 in	1 mm	1 in
Reduced Tube Size	1/4	0.8	0.03	1.3	0.05	3.3	0.13	4.8	0.19	7.1	0.28	10.4	1.41
	5/16	—	—	0.5	0.02	2.3	0.09	4.1	0.16	6.4	0.25	9.7	0.38
	3/8	—	—	—	—	2.0	0.08	3.6	0.14	5.8	0.23	9.1	0.36
	1/2	—	—	—	—	—	—	1.5	0.06	4.1	0.16	7.1	0.28
	5/8	—	—	—	—	—	—	—	—	2.3	0.09	5.6	0.22
	3/4	—	—	—	—	—	—	—	—	—	—	3.3	0.13

TABLE A4—LENGTH FACTORS STANDARD PIPE SIZES

Nom Pipe Size, in		1/4 mm	1/4 in	3/8 mm	3/8 in	1/2 mm	1/2 in	3/4 mm	3/4 in	1 mm	1 in
Reduced Pipe Size	1/8	1.8	0.07	3.6	0.14	5.6	0.22	8.4	0.33	11.7	0.46
	1/4	—	—	1.8	0.07	3.8	0.15	7.1	0.28	9.9	0.39
	3/8	—	—	—	—	2.3	0.09	4.8	0.19	8.1	0.32
	1/2	—	—	—	—	—	—	2.8	0.11	6.1	0.24
	3/4	—	—	—	—	—	—	—	—	3.3	0.13

TABLE A5—LENGTH FACTORS STANDARD MACHINING SIZE

Nom Tube OD, in		5/16 mm	5/16 in	3/8 mm	3/8 in	1/2 mm	1/2 in	5/8 mm	5/8 in	3/4 mm	3/4 in	1 mm	1 in
Reduced Tube Size	1/4	0.8	0.03	1.5	0.06	2.8	0.11	4.3	0.17	5.8	0.23	8.6	0.34
	5/16	—	—	0.8	0.03	2.0	0.08	3.6	0.14	5.1	0.20	7.9	0.31
	3/8	—	—	—	—	1.3	0.05	2.8	0.11	4.3	0.17	7.1	0.28
	1/2	—	—	—	—	—	—	1.5	0.06	3.3	0.13	5.8	0.23
	5/8	—	—	—	—	—	—	—	—	1.5	0.06	4.3	0.17
	3/4	—	—	—	—	—	—	—	—	—	—	2.8	0.11

TABLE A6—LENGTH FACTORS STANDARD PIPE SIZE

Nom Pipe Size, in		1/4 mm	1/4 in	3/8 mm	3/8 in	1/2 mm	1/2 in	3/4 mm	3/4 in	1 mm	1 in
Reduced Pipe Size	1/8	1.5	0.06	3.3	0.13	5.6	0.22	8.9	0.35	11.9	0.47
	1/4	—	—	1.8	0.07	4.1	0.16	7.4	0.29	10.4	0.41
	3/8	—	—	—	—	2.5	0.10	5.6	0.22	8.9	0.35
	1/2	—	—	—	—	—	—	3.3	0.13	6.4	0.25
	3/4	—	—	—	—	—	—	—	—	3.3	0.13

TABLE A7—MINIMUM STOCK SIZE FOR TUBE ENDS

Nom Tube OD, in	Hexagon Width, Min in	Width over Wrench Pads, Min mm	Width over Wrench Pads, Min in
1/4	7/16	11.2	0.44
5/16	1/2	12.7	0.50
3/8	9/16	13.5	0.53
1/2	11/16	17.5	0.69
5/8	13/16	20.6	0.81
3/4	1	25.4	1.00
1	1-1/4	31.8	1.25

TABLE A8—MINIMUM STOCK SIZE FOR EXTERNAL PIPE ENDS

Nom Pipe Size, in	Hexagon Width, Min in	Width over Wrench Pads, Min mm	Width over Wrench Pads, Min in
1/8	7/16	10.4	0.41
1/4	9/16	14.0	0.55
3/8	11/16	17.5	0.69
1/2	7/8	21.6	0.85
3/4	1-1/16	26.9	1.06
1	1-3/8	33.8	1.33

TABLE A9—MINIMUM STOCK SIZE FOR INTERNAL PIPE ENDS

Nom Pipe Size, in	Hexagon Width, Min in	Width over Wrench Pads, Min mm	Width over Wrench Pads, Min in
1/8	9/16	14.2	0.56
1/4	11/16	17.5	0.69
3/8	7/8	20.6	0.81
1/2	1-1/16	25.4	1.00
3/4	1-1/4	31.8	1.25
1	1-5/8	38.1	1.50

REFRIGERATION TUBE FITTINGS
—GENERAL SPECIFICATIONS—SAE J513 JUN93

SAE Standard

Report of Parts and Fittings Division approved January 1936 and revised by Tube, Pipe, Hose, and Lubrication Fittings Committee December 1976. Editorial change October 1977. Completely revised by the Fluid Conductors and Connectors Technical Committee June 1990. Completely revised by the SAE Fluid Conductors and Connectors Technical Committee S1—Automotive and Hydraulic Tube and Fittings June 1992. Revised by the SAE Fluid Conductors and Connectors Technical Committee June 1993. Rationale statement available.

1. Scope—This SAE Standard covers complete general and dimensional specifications for refrigeration tube fittings of the flare type specified in Figures 1 to 42 and Tables 1 to 16. These fittings are intended for general use with flared annealed copper tubing in refrigeration applications.

Dimensions of single and double 45 degree flares on tubing to be used in conjunction with these fittings are given in Figure 2 and Table 1 of SAE J533.

The following general specifications supplement the dimensional data contained in Tables 1 to 16 with respect to all unspecified details.

2. References

2.1 Applicable Documents—The following publications form a part of this specification to the extent specified herein. The latest version of SAE publications shall apply.

2.1.1 SAE PUBLICATIONS—Available from SAE, 400 Commonwealth Drive, Warrendale, PA 15096-0001.

SAE J476—Dryseal Pipe Threads
SAE J512—Automotive Tube Fittings
SAE J533—Flares for Tubing
SAE J846—Coding Systems for Identification of Fluid Conductors and Connectors

2.1.2 ANSI PUBLICATIONS—Available from ANSI, 11 West 42nd Street, New York, NY 10036-8002.

ANSI B1.1—Screw Threads
ANSI B2.1—Pipe Threads
ANSI B70.1

2.1.3 ASTM PUBLICATION—Available from ASTM, 1916 Race Street, Philadelphia, PA 19103-1187.

ASTM B 117—Method of Salt Spray (Fog) Testing

3. Pressure Ratings and Service Limitations—Fittings covered by this document are satisfactory for operating pressures up to 3450 kPa (500 psi) and are suitable for use in systems conducting most fluorinated hydrocarbon refrigerants. Fitting manufacturers should be consulted for recommendations.

4. Size Designations—Fitting sizes throughout the dimensional tables are designated by the corresponding outside diameter of the tubing for flared type or solder type tube ends and by the corresponding standard nominal pipe size for pipe thread ends.

5. Dimensions and Tolerances—Except for nominal sizes and thread specifications, dimensions and tolerances are given in both SI and U.S. customary units as designated. Tabulated dimensions shall apply to the finished parts, plated or otherwise processed, as specified by the purchasers. Unless otherwise specified, the maximum and minimum across flat dimensions shall be within the commercial tolerance of bar or extruded stock from which the fittings are produced. The minimum across corners dimensions of hexagons shall be 1.092 times the nominal width across flats, but shall not result in a size flat width less than 0.43 times the nominal width across flats.

Unless otherwise specified, tolerance on hole diameters designated drill in the dimensional tables shall be as tabulated in Table 17.

Tolerance on all dimensions not otherwise limited shall be ±0.25 mm (±0.010 in). Fitting seats shall be concentric with the straight thread pitch diameters within 0.25 mm (0.010 in) full indicator reading (FIR). Angular tolerance on axis of ends on elbows, tees, and crosses shall be ±2.50 degrees for sizes up to and including 3/8 in, and ±1.50 degrees for sizes larger than 3/8 in.

Where so illustrated and not otherwise specified, hexagon corners shall be chamfered 30 degrees ± 5 degrees, to a diameter equal to the nominal width across flats, with a tolerance of -0.41 mm (-0.016 in); or where design permits, corners may be chamfered to the diameter of the abutting surface provided the length of chamfer does not exceed that produced by the 30 degree chamfer previously described.

6. Passages—Where passages in straight fittings are machined from opposite ends, the offset at the meeting point shall not exceed 0.38 mm (0.015 in). The cross-sectional area at the junction of passages in angle fittings shall not be less than that of the smallest passage. Where the passage is specified as a maximum or as tap drill diameter or less, the minimum shall be no less than the minimum diameter of the smallest passage in the fitting.

7. Wall Thickness—Unless otherwise designated, the wall thickness at any point on fittings shall not be less than the thickness established by the specified dimensions, tolerances, and eccentricities for inner and outer surfaces.

8. Contour—Details of contour shall be optional with the manufacturer, providing the tabulated dimensions are maintained and serviceability of the fittings is not impaired. Wrench flats on elbows and tees shall be optional. Where extruded or forged shapes are reduced to conserve material, the wall thickness, unless otherwise specified, shall not be less than the respective minimum values tabulated in Table 18.

9. Straight Threads—Unified Standard Class 2A external and Class 2B internal threads with minor diameters, where specified, modified to Class 3B limits shall apply to plain finish (unplated) fittings of all types. For externally threaded parts with additive finish, the maximum diameters of Class 2A may be exceeded by the amount of the allowance, that is, the basic diameters (Class 2A maximum diameters plus the allowance) shall apply after plating. For internally threaded parts with additive finish, the Class 2B diameters and modified minor diameters shall apply after plating.

The pitch diameter tolerance shall be the same as the corresponding diameter-pitch combination and class of the Unified fine thread series or for special diameter-pitch combinations shall be based on diameter, pitch, and a length of engagement of 9 times the pitch. See ANSI B1.1.

For convenient reference, the data generally required to specify threads are given in Table 1. (Inasmuch as threads are normally produced and gaged with equipment made to the inch system of measurement, conversion of size designations and dimensions to SI units is considered unnecessary.)

Where external threads are produced by roll threading and the body is not undercut, the unthreaded portion of body adjacent to the shoulder may be reduced to the minimum pitch diameter.

External threads shall be chamfered to the diameter of abutting surfaces, or to the diameters specified, to produce a length of chamfered or partial thread equivalent to 3/4 to 1-1/4 times the pitch (rounded to a three-place decimal).

TABLE 1—STRAIGHT THREAD SPECIFICATION DATA, in

Nominal Size	Series Designation	External Thread Pitch Dia Max	External Thread Pitch Dia Min	Internal Thread Pitch Dia Max	Internal Thread Pitch Dia Min[a]	Internal Thread Minor Dia[b] Max	Internal Thread Minor Dia[b] Min
5/16-24	UNF	0.2843	0.2806	0.2902	0.2854	0.2754	0.2670
3/8 -24	UNF	0.3468	0.3430	0.3528	0.3479	0.3372	0.3300
7/16-20	UNF	0.4037	0.3995	0.4104	0.4050	0.3916	0.3830
1/2 -20	UNF	0.4462	0.4619	0.4731	0.4675	0.4537	0.4460
5/8 -18	UNF	0.5875	0.5828	0.5949	0.5889	0.5730	0.5650
3/4 -16	UNF	0.7079	0.7029	0.7159	0.7094	0.6908	0.6820
7/8 -14	UNF	0.8270	0.8216	0.8356	0.8286	0.8068	0.7980
1-1/16-14	UNS	1.0145	1.0092	1.0230	1.0161	0.9940	0.9850

[a] These values are also the basic pitch diameter.
[b] Class 3B minor diameter limits.

TABLE 2

Nominal Pipe Thread Size in	External Thread Chamfer Diameter[a] Max mm	External Thread Chamfer Diameter[a] Max in	External Thread Chamfer Diameter[a] Min mm	External Thread Chamfer Diameter[a] Min in	External Thread Length of Chamfered or Partial Thread Min mm	External Thread Length of Chamfered or Partial Thread Min in
1/8	8.1	0.32	7.6	0.30	0.90	0.037
1/4	10.7	0.42	10.2	0.40	1.42	0.056
3/8	14.0	0.55	13.5	0.53	1.42	0.056
1/2	17.3	0.68	16.8	0.66	1.80	0.071
3/4	22.6	0.89	22.1	0.87	1.80	0.071

Nominal Pipe Thread Size in	External Thread Length of Chamfered or Partial Thread Max mm	External Thread Length of Chamfered or Partial Thread Max in	Internal Thread Countersink Diameter[a] Min mm	Internal Thread Countersink Diameter[a] Min in	Internal Thread Countersink Diameter[a] Max mm	Internal Thread Countersink Diameter[a] Max in
1/8	1.42	0.056	10.7	0.42	11.2	0.44
1/4	2.11	0.083	14.0	0.55	14.5	0.57
3/8	2.11	0.083	17.5	0.69	18.0	0.71
1/2	2.72	0.107	21.6	0.85	22.1	0.87
3/4	2.72	0.107	26.9	1.06	27.4	1.08

[a] Tabulated diameters conform with Appendix A, SAE J476.

Internal threads shall be countersunk 90 degrees included angle to the diameters specified in the dimensional tables.

Where external threads are produced by roll threading and the body is not undercut, the unthreaded portion of body adjacent to the shoulder may be reduced to the minimum pitch diameter.

External threads shall be chamfered from the diameter of abutting surface to produce a length of chamfered or partial thread equivalent to 3/4 to 1-1/4 times the pitch (rounded to a three-place decimal).

Internal threads shall be countersunk 90 degrees included angle to the diameters specified in the dimensional tables.

10. Thread Eccentricity Tolerances—The various thread elements of Class 2A external and Class 2B modified internal threads on tube fittings shall be concentric within the limitations specified in SAE J512.

11. Pipe Threads—Taper pipe threads, unless there is specific authorization to the contrary, shall conform to the Dryseal American Standard Taper Pipe Thread (NPTF). Specifications for pipe threads are given in detail in SAE J476.

The length of full form external thread shall not be shorter than L_2 plus one pitch (thread), except that where thread is cut through into a relieved body or undercut on the fitting, the minimum full thread length may be reduced by one pitch (thread).

Where external pipe threads are produced by roll threading, the diameter of the unthreaded portion of shank adjacent to shoulder may be reduced to the E_2 basic pitch diameter.

The tube fitting dimensions tabulated herein are based on length of the Dryseal American Standard Taper Pipe Thread (NPTF), it being the consensus of manufacturers and users that trouble-free assembly cannot be assured unless full length thread is used. However, the tap drill depths and overall lengths specified in the tables for fittings with internal taper pipe threads are not consistent with the tap drill depths and overall thread lengths of the Dryseal American Standard Taper Pipe Threads (NPTF) given in Table A2, Appendix A of SAE J476. The full length Dryseal American Standard Taper Pipe Taps specified in Table B2 of SAE J476 cannot be used, as the tap drill depths and overall lengths of the fittings have been reduced to the minimum required by bottoming taps to produce standard full thread length. The deviations described previously are peculiar to these tube fittings and as special tooling is required, caution should be exercised in specifying such deviations for any other products.

Straight pipe threads, where specified, shall conform to American Standard Straight Pipe Threads for Mechanical Joints (NPSM) in ANSI B2.1.

TABLE 3—DIMENSIONS OF CONNECTORS, UNIONS, ADAPTORS, ELBOWS, TEES, AND CROSSES
(Figures 1 to 20)

Nom Tube OD in	A Dryseal Pipe Thread NPTF[b]	B Straight Thread Nominal Size	C Hex in Nom	C_1 Hex in Nom	C_2 Hex in Nom	C_3 Hex in Nom	D[e] Drill mm	D[e] Drill in	D_1[e] Drill mm	D_1[e] Drill in
3/16	1/8	3/8 -24	7/16	3/8	9/16	1/2	3.18	0.125	5.56	0.219
1/4	1/8	7/16-20	7/16	7/16	9/16	5/8	4.78	0.188	5.56	0.219
5/16	1/8	1/2 -20	1/2	1/2	9/16	1- 1/16	5.56	0.219	5.56	0.219
3/8	1/4	5/8 -18	5/8	5/8	11/16	13/16	7.14	0.281	7.92	0.312
1/2	3/8	3/4 -16	3/4	3/4	13/16	15/16	10.31	0.406	10.31	0.406
5/8	1/2	7/8 -14	7/8	7/8	1	1- 1/16	12.70	0.500	14.27	0.562
3/4	1/2	1-1/16-14	1-1/16	1-1/16	1- 1/16	1- 5/16	15.88	0.625	14.27	0.562

Nom Tube OD in	D_2 Drill mm	D_2 Drill in	D_3[h] Dia mm ±0.025	D_3[h] Dia in ±0.0010	D_4 Drill mm	D_4 Drill in	D_5 Drill mm	D_5 Drill in	D_6 Tube ID mm	D_6 Tube ID in
3/16	3.96	0.156	4.864	0.1915	4.78	0.188	4.78	0.188	2.97	0.117
1/4	4.78	0.188	6.452	0.2540	5.56	0.219	6.35	0.250	4.57	0.180
5/16	6.35	0.250	8.039	0.3165	5.56	0.219	7.92	0.312	6.15	0.242
3/8	7.92	0.312	9.627	0.3790	8.74	0.344	9.52	0.375	7.75	0.305
1/2	11.13	0.438	12.802	0.5040	10.31	0.406	12.70	0.500	10.92	0.430
5/8	13.89	0.547	15.977	0.6290	14.27	0.562	15.88	0.625	14.10	0.555
3/4	17.48	0.688	19.152	0.7540	14.27	0.562	19.05	0.750	17.27	0.680

Nom Tube OD in	E Dia mm	E Dia in	F Dia mm	F Dia in	G[e] Dia mm +0.00 −0.25	G[e] Dia in +0.000 −0.010	I mm	I in	I_1 mm	I_1 in
3/16	3.96	0.156	7.54	0.297	—	—	11.2	0.44	7.1	0.28
1/4	5.56	0.219	8.74	0.344	—	—	12.7	0.50	8.6	0.34
5/16	6.35	0.250	10.31	0.406	—	—	14.2	0.56	9.7	0.38
3/8	7.92	0.312	13.49	0.531	—	—	15.7	0.62	11.2	0.44
1/2	11.13	0.438	16.28	0.641	16.74	0.659	19.0	0.75	13.5	0.53
5/8	13.49	0.531	19.05	0.750	19.56	0.770	22.4	0.88	16.8	0.66
3/4	18.26	0.719	23.83	0.938	24.33	0.958	25.4	1.00	19.8	0.78

External pipe threads shall be chamfered from the diameters tabulated in Table 2 to produce the specified length of chamfered or partial thread. Internal pipe threads shall be countersunk 90 degrees included angle to the diameters tabulated in Table 2.

12. Material and Manufacture—Fittings shall be made from SAE CA360 brass (half-hard), CA345, or CA350 brass bar or extruded shapes or from SAE CA377 brass forgings in accordance with the manufacturer's processes. Nuts may be made from SAE CA377 brass forging, or steel as specified by the purchaser. Seal bonnets and gaskets shall be made from copper conforming to SAE CA102, CA110, or CA122. As specified by purchaser, fusible metal alloys shall be supplied for temperature ranges 70 to 74, 95 to 104, 135 to 143 °C (158 to 165, 203 to 219, or 275 to 290 °F).

13. Finish—The external surfaces and threads of all carbon steel parts shall be plated or coated with a suitable material that passes a 72 h salt spray test in accordance with ASTM B 117. Any appearance of red rust during the 72 h salt spray test shall be considered failure, except for the following:

a. All internal fluid passages.

b. Edges such as hex points, serrations, and crests of threads where there may be mechanical deformation of the plating or coating typical of mass-produced parts or shipping effects.

c. Areas where there is mechanical deformation of the plating or coating caused by crimping, flaring, bending, and other post-plate metal forming operations.

d. Areas where the parts are suspended or affixed in the test chamber where condensate can accumulate.

NOTE—Cadmium plating is not preferred due to environmental reasons. Parts manufactured to this standard after January 1, 1997, shall not be cadmium plated. Internal fluid passages shall be protected from corrosion during storage. Changes in plating may affect assembly torques and require requalification, when applicable.

14. Workmanship—Workmanship shall conform to the best commercial practice to produce high-quality fittings. Fittings shall be free from all hanging burrs, loose scale, and slivers which might become dislodged in usage and all other defects which might affect their serviceability. All sealing surfaces must be smooth except that annular tool marks up to 2.5 µm (100 µin) maximum shall be permissible.

15. Assembly Considerations—Torque loads experienced during the assembly of the nut onto the flared ends of the fittings are variable due to the different metals used in the fittings and nuts, hardness of the tubing, manufacturing tolerances and conformance to the manufacturing recommendations. Therefore, it is recommended that flare nuts be assembled to the flared ends one quarter (1/4) turn past finger-tight. In the case of slight side loads which may exist during assembly, it may be necessary to "snug up" the mating parts with a wrench to achieve a finger-tight condition. The use of lubricants is not recommended.

16. Manufacturing Recommendations—Only circular tool marks concentric to the centerline of the sealing face will be permitted. Any such tool marks on the sealing face must be smooth within the limits of 2.5 µm (100 µin).

Flare fittings should be handled, shipped, and stored in a manner that protects the sealing surface from damage.

TABLE 3—(Continued)

Nom Tube OD in	J[c] Full Thread Min mm	J[c] Full Thread Min in	J₁ Full Thread Min mm	J₁ Full Thread Min in	K mm	K in	L mm ±0.8	L in ±0.03	L₁ mm ±0.8	L₁ in ±0.03
3/16	9.7	0.38	5.6	0.22	3.0	0.12	25.4	1.00	26.9	1.06
1/4	10.4	0.41	6.9	0.27	4.1	0.16	26.9	1.06	30.2	1.19
5/16	11.9	0.47	7.6	0.30	4.8	0.19	29.5	1.16	34.0	1.34
3/8	13.7	0.54	8.6	0.34	5.6	0.22	36.6	1.44	38.1	1.50
1/2	16.8	0.66	11.2	0.44	6.4	0.25	41.1	1.62	46.0	1.81
5/8	19.3	0.76	14.0	0.55	7.1	0.28	50.8	2.00	53.8	2.12
3/4	22.9	0.90	17.0	0.67	7.1	0.28	55.6	2.19	62.0	2.44

Nom Tube OD in	L₂ mm ±0.8	L₂ in ±0.03	L₃ mm ±0.8	L₃ in ±0.03	L₄ mm ±0.8	L₄ in ±0.03	L₅ mm ±0.8	L₅ in ±0.03	L₆ mm ±0.8	L₆ in ±0.03
3/16	24.6	0.97	23.9	0.94	23.9	0.94	20.6	0.81	22.4	0.88
1/4	26.2	1.03	25.4	1.00	26.9	1.06	23.1	0.91	25.4	1.00
5/16	26.9	1.06	27.7	1.09	28.4	1.12	23.9	0.94	26.9	1.06
3/8	33.3	1.31	30.2	1.19	33.3	1.31	32.5	1.28	31.8	1.25
1/2	38.1	1.50	36.6	1.44	39.6	1.56	35.1	1.33	36.6	1.44
5/8	46.0	1.81	44.4	1.75	46.0	1.81	42.2	1.66	42.9	1.69
3/4	48.5	1.91	52.3	2.06	52.3	2.06	47.8	1.88	50.8	2.00

Nom Tube OD in	L₇ Min mm	L₇ Min in	L₈ mm ±0.8	L₈ in ±0.03	M mm ±0.8	M in ±0.03	M₁ mm ±0.8	M₁ in ±0.03	M₂ mm ±0.8	M₂ in ±0.03
3/16	33.3	1.31	15.0	0.59	19.0	0.75	19.0	0.75	20.6	0.81
1/4	33.3	1.31	17.5	0.69	20.6	0.81	22.4	0.88	22.4	0.88
5/16	35.1	1.38	19.8	0.78	23.1	0.91	23.1	0.91	23.9	0.94
3/8	38.1	1.50	22.4	0.88	25.4	1.00	26.9	1.06	27.7	1.09
1/2	44.4	1.75	26.9	1.06	31.0	1.22	31.0	1.22	32.5	1.28
5/8	50.8	2.00	30.2	1.19	35.8	1.41	35.8	1.41	38.1	1.50
3/4	60.5	2.38	33.3	1.31	41.1	1.62	42.2	1.66	41.1	1.62

Nom Tube OD in	M₃ mm ±0.8	M₃ in ±0.03	M₄ mm ±0.8	M₄ in ±0.03	M₅ mm ±0.8	M₅ in ±0.03	M₆ mm ±0.8	M₆ in ±0.03	M₇ mm ±0.8	M₇ in ±0.03
3/16	—	—	—	—	18.3	0.72	15.0	0.59	15.7	0.62
1/4	23.9	0.94	19.8	0.78	20.6	0.81	15.7	0.62	17.0	0.67
5/16	—	—	—	—	23.1	0.91	16.8	0.66	19.8	0.78
3/8	29.5	1.16	24.6	0.97	26.2	1.03	18.3	0.72	22.6	0.89
1/2	34.0	1.34	28.4	1.12	31.0	1.22	21.3	0.84	26.9	1.06
5/8	—	—	—	—	35.8	1.41	26.2	1.03	31.2	1.23
3/4	—	—	—	—	41.1	1.62	31.8	1.25	35.8	1.41

Nom Tube OD in	N mm ±0.8	N in ±0.03	N₁ mm ±0.8	N₁ in ±0.03	N₂ mm ±0.8	N₂ in ±0.03	O mm	O in	P Min mm	P Min in
3/16	19.0	0.75	11.2	0.44	13.2	0.52	7.9	0.31	9.7	0.38
1/4	19.8	0.78	11.9	0.47	16.3	0.64	7.9	0.31	9.7	0.38
5/16	19.8	0.78	11.9	0.47	16.3	0.64	7.9	0.31	9.7	0.38
3/8	26.9	1.06	17.5	0.69	21.8	0.86	7.9	0.31	14.2	0.56
1/2	28.4	1.12	19.0	0.75	24.1	0.95	9.7	0.38	14.2	0.56
5/8	35.1	1.38	25.4	1.00	29.7	1.17	12.7	0.50	19.0	0.75
3/4	38.1	1.50	26.9	1.06	30.5	1.20	15.7	0.62	19.0	0.75

Nom Tube OD in	Q[d] mm	Q[d] in	S[e] Max mm	S[e] Max in	S₁ mm	S₁ in	S₂[e] Max mm	S₂[e] Max in	S₃[e] Min mm	S₃[e] Min in
3/16	—	—	12.2	0.48	7.9	0.31	10.4	0.41	21.6	0.85
1/4	—	—	12.2	0.48	7.9	0.31	—	—	23.4	0.92
5/16	—	—	—	—	7.9	0.31	10.7	0.42	—	—
3/8	17.5	0.69	17.5	0.69	7.9	0.31	11.2	0.44	31.5	1.24
1/2	—	—	—	—	9.7	0.38	13.7	0.54	—	—
5/8	23.9	0.94	23.9	0.94	12.7	0.50	17.5	0.69	42.4	1.67
3/4	—	—	31.0	1.22	15.7	0.62	21.3	0.84	51.3	2.02

22.37

TABLE 3—(Continued)

Nom Tube OD in	T[f] Ref mm	T[f] Ref in	T_1[f] Min mm	T_1[f] Min in	T_2[f] Ref mm	T_2[f] Ref in	U Dia Min mm	U Dia Min in	U Dia Max mm	U Dia Max in
3/16	4.6	0.18	5.3	0.21	3.8	0.15	9.9	0.39	10.4	0.41
1/4	4.6	0.18	6.1	0.24	4.6	0.18	11.4	0.45	11.9	0.47
5/16	5.3	0.21	6.1	0.24	5.3	0.21	13.0	0.51	13.5	0.53
3/8	6.1	0.24	7.6	0.30	6.1	0.24	16.3	0.64	17.0	0.67
1/2	7.6	0.30	9.4	0.37	7.6	0.30	19.6	0.77	20.3	0.80
5/8	9.4	0.37	10.9	0.43	7.6	0.30	22.9	0.90	23.6	0.93
3/4	10.9	0.43	12.4	0.49	7.6	0.30	27.4	1.08	28.2	1.11

Nom Tube OD in	W[g] Dia mm +0.00 -0.5	W[g] Dia in +0.00 -0.02	X[g] Dia mm +0.00 -0.5	X[g] Dia in +0.00 -0.02	Y Dia Min mm	Y Dia Min in	Z Min mm	Z Min in
3/16	14.2	0.56	12.7	0.50	7.1	0.28	1.5	0.06
1/4	14.2	0.56	15.7	0.62	8.6	0.34	1.3	0.05
5/16	14.2	0.56	17.5	0.69	10.2	0.40	1.5	0.06
3/8	17.5	0.69	20.6	0.81	12.2	0.48	1.5	0.06
1/2	20.6	0.81	23.9	0.94	15.2	0.60	2.0	0.08
5/8	25.4	1.00	26.9	1.06	18.8	0.74	2.5	0.10
3/4	26.9	1.06	33.3	1.31	21.8	0.86	2.5	0.10

[a] For reducing sizes of unions, internal flare to external flare adaptors, and 90 degree elbow unions, see Table 4; for reducing sizes of solder connectors and 90 degree solder elbows, see Table 5; for reducing sizes of connectors, internal pipe thread connectors, internal flare to external pipe adapters, 90 degree elbow, 45 degree elbow, and internal pipe thread 90 degree elbow, see Table 6; for reducing sizes of tees, see Tables 7 and 8.
[b] Dryseal American Standard Taper Pipe Thread.
[c] Where thread relief undercut is used, last thread shall be chamfered 1/2 to 1 pitch long from G diameter and dimension J may be reduced by an amount equal to 1/2 pitch.
[d] Available with three types of fusible alloys as specified in general specifications.
[e] At manufacturer's option through passages in fittings shown in Figures 1, 5, and 19 may conform with the smaller diameter specified or be counterbored to the larger diameter from the appropriate end for depths S, S_2 or S_3, respectively.
[f] Minimum design thickness, not subject to inspection.
[g] Basic dimensions shown shall apply as minimum diameter for bosses or across flats. The -0.5 mm (-0.02 in) tolerance shall apply only to chamfer diameters on full hexagon versions of fittings shown in Figures 4, 6 to 8.
[h] ID of solder cup shall not be out of round by more than 0.08 mm (0.0003 in).

TABLE 4—DIMENSIONS OF REDUCING UNIONS, REDUCING ADAPTERS, AND REDUCING ELBOW UNIONS (Figures 21 to 23)[a]

B[c] Nom Tube OD in	B_1[c] Nom Tube OD in	C in Nom	C_1 in Nom	D[d] Drill mm	D[d] Drill in	D_1[d] Drill mm	D_1[d] Drill in	D_2 Drill mm	D_2 Drill in	L mm ±0.8	L in ±0.03	L_1 mm ±0.8	L_1 in ±0.03
3/16	1/4	7/16	5/8	3.18	0.125	4.78	0.188	—	—	28.4	1.12	26.2	1.03
3/16	5/16	1/2	11/16	3.18	0.125	5.56	0.219	—	—	31.0	1.22	26.9	1.06
3/16	3/8	5/8	13/16	3.18	0.125	7.14	0.281	—	—	33.3	1.31	30.2	1.19
3/16	1/2	3/4	15/16	2.18	0.125	10.31	0.406	—	—	38.1	1.50	34.0	1.34
3/16	5/8	7/8	1-1/16	3.18	0.125	12.70	0.500	—	—	42.9	1.69	38.9	1.53
3/16	3/4	1-1/16	1-5/16	3.18	0.125	15.88	0.625	—	—	47.8	1.88	44.4	1.75

B[c] Nom Tube OD in	B_1[c] Nom Tube OD in	M mm ±0.8	M in ±0.03	M_1 mm ±0.8	M_1 in ±0.03	S[d] Max mm	S[d] Max in	T[e] Ref mm	T[e] Ref in	T_1 Min mm	T_1 Min in	X[f] Dia mm +0.0 -0.5	X[f] Dia in +0.00 -0.02
3/16	1/4	19.0	0.75	22.4	0.88	15.2	0.60	4.6	0.18	6.1	0.24	15.7	0.62
3/16	5/16	19.8	0.78	23.1	0.91	17.0	0.67	5.3	0.21	6.1	0.24	17.5	0.69
3/16	3/8	21.3	0.84	26.9	1.06	19.0	0.75	6.1	0.24	7.6	0.30	20.6	0.81
3/16	1/2	23.1	0.91	31.0	1.22	23.1	0.91	7.6	0.30	9.4	0.37	23.9	0.94
3/16	5/8	24.6	0.97	35.8	1.41	27.2	1.07	9.4	0.37	10.9	0.43	26.9	1.06
3/16	3/4	26.9	1.06	42.2	1.66	31.0	1.22	10.9	0.43	12.4	0.49	33.3	1.31

TABLE 4—(Continued)

B^c Nom Tube OD in	B_1^c Nom Tube OD in	C in Nom	C_1 in Nom	D^d Drill mm	D^d Drill in	D_1^d Drill mm	D_1^d Drill in	D_2 Drill mm	D_2 Drill in	L mm ±0.8	L in ±0.03	L_1 mm ±0.8	L_1 in ±0.03
1/4	3/16	7/16	1/2	4.78	0.188	3.18	0.125	—	—	28.4	1.12	24.6	0.97
1/4	5/16	1/2	11/16	4.78	0.188	5.56	0.219	—	—	32.5	1.28	28.4	1.12
1/4	3/8	5/8	13/16	4.78	0.188	7.14	0.281	—	—	35.1	1.38	31.0	1.22
1/4	1/2	3/4	15/16	4.78	0.188	10.31	0.406	—	—	39.6	1.56	35.1	1.38
1/4	5/8	7/8	1-1/16	4.78	0.188	12.70	0.500	—	—	44.4	1.75	39.6	1.56
1/4	3/4	1-1/16	1-5/16	4.78	0.188	15.88	0.625	—	—	49.3	1.94	42.9	1.69

B^c Nom Tube OD in	B_1^c Nom Tube OD in	M mm ±0.8	M in ±0.03	M_1 mm ±0.8	M_1 in ±0.03	S^d Max mm	S^d Max in	T^e Ref mm	T^e Ref in	T_1 Min mm	T_1 Min in	X^f Dia mm +0.0 −0.5	X^f Dia in +0.00 −0.02
1/4	3/16	22.4	0.88	19.0	0.75	15.2	0.60	4.6	0.18	5.3	0.21	12.7	0.50
1/4	5/16	21.3	0.84	23.1	0.91	17.0	0.67	5.3	0.21	6.1	0.24	17.5	0.69
1/4	3/8	23.1	0.91	26.9	1.06	19.0	0.75	6.1	0.24	7.6	0.30	20.6	0.81
1/4	1/2	24.6	0.97	31.0	1.22	23.1	0.91	7.6	0.30	9.4	0.37	23.9	0.94
1/4	5/8	26.2	1.03	35.8	1.41	27.2	1.07	9.4	0.37	10.9	0.43	26.2	1.06
1/4	3/4	28.4	1.12	42.2	1.66	31.0	1.22	10.9	0.43	12.4	0.49	33.3	1.31

B^c Nom Tube OD in	B_1^c Nom Tube OD in	C in Nom	C_1 in Nom	D^d Drill mm	D^d Drill in	D_1^d Drill mm	D_1^d Drill in	D_2 Drill mm	D_2 Drill in	L mm ±0.8	L in ±0.03	L_1 mm ±0.8	L_1 in ±0.03
5/16	3/16	1/2	1/2	5.56	0.219	3.18	0.125	4.78	0.188	31.0	1.22	25.4	1.00
5/16	1/4	1/2	5/8	5.56	0.219	4.78	0.188	—	—	32.5	1.28	27.7	1.09
5/16	3/8	5/8	13/16	5.56	0.219	7.14	0.281	—	—	36.6	1.44	31.8	1.25
5/16	1/2	3/4	15/16	5.56	0.219	10.31	0.406	—	—	41.1	1.62	35.8	1.41
5/16	5/8	7/8	1-1/16	5.56	0.219	12.70	0.500	—	—	46.0	1.81	40.4	1.59
5/16	3/4	1-1/16	1-5/16	5.56	0.219	15.88	0.625	—	—	50.8	2.00	46.0	1.81

B^c Nom Tube OD in	B_1^c Nom Tube OD in	M mm ±0.8	M in ±0.03	M_1 mm ±0.8	M_1 in ±0.03	S^d Max mm	S^d Max in	T^e Ref mm	T^e Ref in	T_1 Min mm	T_1 Min in	X^f Dia mm +0.0 −0.5	X^f Dia in +0.00 −0.02
5/16	3/16	23.1	0.91	19.8	0.78	17.0	0.67	5.3	0.21	5.3	0.21	12.7	0.50
5/16	1/4	23.1	0.91	21.3	0.84	17.0	0.67	5.3	0.21	6.1	0.24	15.7	0.62
5/16	3/8	24.6	0.97	26.9	1.06	19.0	0.75	6.1	0.24	7.6	0.30	20.6	0.81
5/16	1/2	26.2	1.03	31.0	1.22	23.1	0.91	7.6	0.30	9.4	0.37	23.9	0.94
5/16	5/8	27.7	1.09	35.8	1.41	27.2	1.07	9.4	0.37	10.9	0.43	26.9	1.06
5/16	3/4	30.2	1.19	42.2	1.66	31.0	1.22	10.9	0.43	12.4	0.49	33.3	1.31

B^c Nom Tube OD in	B_1^c Nom Tube OD in	C in Nom	C_1 in Nom	D^d Drill mm	D^d Drill in	D_1^d Drill mm	D_1^d Drill in	D_2 Drill mm	D_2 Drill in	L mm ±0.8	L in ±0.03	L_1 mm ±0.8	L_1 in ±0.03
3/8	3/16	5/8	5/8	7.14	0.281	3.18	0.125	4.78	0.188	33.3	1.31	26.2	1.03
3/8	1/4	5/8	5/8	7.14	0.281	4.78	0.188	6.35	0.250	35.1	1.38	28.4	1.12
3/8	5/16	5/8	11/16	7.14	0.281	5.56	0.219	—	—	36.6	1.44	30.2	1.19
3/8	1/2	3/4	15/16	7.14	0.281	10.31	0.406	—	—	42.9	1.69	36.6	1.44
3/8	5/8	7/8	1-1/16	7.14	0.281	12.70	0.500	—	—	47.8	1.88	41.1	1.62
3/8	3/4	1-1/16	1-5/16	7.14	0.281	15.89	0.625	—	—	52.3	2.06	46.7	1.84

B^c Nom Tube OD in	B_1^c Nom Tube OD in	M mm ±0.8	M in ±0.03	M_1 mm ±0.8	M_1 in ±0.03	S^d Max mm	S^d Max in	T^e Ref mm	T^e Ref in	T_1 Min mm	T_1 Min in	X^f Dia mm +0.0 −0.5	X^f Dia in +0.00 −0.02
3/8	3/16	26.9	1.06	21.3	0.84	19.0	0.75	6.1	0.24	6.1	0.24	15.7	0.62
3/8	1/4	26.9	1.06	23.1	0.91	19.0	0.75	6.1	0.24	6.1	0.24	15.7	0.62
3/8	5/16	26.9	1.06	24.6	0.97	19.0	0.75	6.1	0.24	6.1	0.24	17.5	0.69
3/8	1/2	27.7	1.09	31.0	1.22	23.1	0.91	7.6	0.30	9.4	0.37	23.9	0.94
3/8	5/8	29.5	1.16	35.8	1.41	27.2	1.07	9.4	0.37	10.9	0.43	26.9	1.06
3/8	3/4	31.8	1.25	42.2	1.66	31.0	1.22	10.9	0.43	12.4	0.49	33.3	1.31

TABLE 4—(Continued)

B^c Nom Tube OD in	B_1^c Nom Tube OD in	C in Nom	C_1 in Nom	D^d Drill mm	D^d Drill in	D_1^d Drill mm	D_1^d Drill in	D_2 Drill mm	D_2 Drill in	L mm ±0.8	L in ±0.03	L_1 mm ±0.8	L_1 in ±0.03
1/2	3/16	3/4	3/4	10.31	0.406	3.18	0.125	4.78	0.188	38.1	1.50	29.5	1.16
1/2	1/4	3/4	3/4	10.31	0.406	4.78	0.188	6.35	0.250	39.6	1.56	31.8	1.25
1/2	5/16	3/4	3/4	10.31	0.406	5.56	0.219	7.92	0.312	41.1	1.62	32.5	1.28
1/2	3/8	3/4	13/16	10.31	0.506	7.14	0.281	9.52	0.375	42.9	1.69	35.8	1.41
1/2	5/8	7/8	1-1/16	10.31	0.406	12.70	0.500	—	—	50.8	2.00	42.9	1.69
1/2	3/4	1-1/16	1-5/16	10.31	0.406	15.88	0.625	—	—	55.6	2.19	48.5	1.91

B^c Nom Tube OD in	B_1^c Nom Tube OD in	M mm ±0.8	M in ±0.03	M_1 mm ±0.8	M_1 in ±0.03	S^d Max mm	S^d Max in	T^e Ref mm	T^e Ref in	T_1 Min mm	T_1 Min in	X^f Dia mm +0.0 −0.5	X^f Dia in +0.00 −0.02
1/2	3/16	31.0	1.22	23.1	0.91	23.1	0.91	7.6	0.30	7.6	0.30	19.0	0.75
1/2	1/4	31.0	1.22	24.6	0.97	23.1	0.91	7.6	0.30	7.6	0.30	19.0	0.75
1/2	5/16	31.0	1.22	26.2	1.03	23.1	0.91	7.6	0.30	7.6	0.30	19.0	0.75
1/2	3/8	31.0	1.22	27.7	1.09	23.1	0.91	7.6	0.30	7.6	0.30	20.6	0.81
1/2	5/8	32.5	1.28	35.8	1.41	27.2	1.07	9.4	0.37	10.9	0.43	26.9	1.06
1/2	3/4	35.1	1.38	42.2	1.66	31.0	1.22	10.9	0.43	12.4	0.49	33.3	1.31

B^c Nom Tube OD in	B_1^c Nom Tube OD in	C in Nom	C_1 in Nom	D^d Drill mm	D^d Drill in	D_1^d Drill mm	D_1^d Drill in	D_2 Drill mm	D_2 Drill in	L mm ±0.8	L in ±0.03	L_1 mm ±0.8	L_1 in ±0.03
5/8	3/16	7/8	7/8	12.70	0.500	3.18	0.125	4.78	0.188	42.9	1.69	31.8	1.25
5/8	1/4	7/8	7/8	12.70	0.500	4.78	0.188	6.35	0.250	44.4	1.75	33.3	1.31
5/8	5/16	7/8	7/8	12.70	0.500	5.56	0.219	7.92	0.312	46.0	1.81	35.1	1.38
5/8	3/8	7/8	7/8	12.70	0.500	7.14	0.281	9.52	0.375	47.8	1.88	37.3	1.47
5/8	1/2	7/8	15/16	12.70	0.500	10.31	0.406	—	—	50.8	2.00	41.1	1.62
5/8	3/4	1-1/16	1-5/16	12.70	0.500	15.88	0.625	—	—	58.7	2.31	50.0	1.97

B^c Nom Tube OD in	B_1^c Nom Tube OD in	M mm ±0.8	M in ±0.03	M_1 mm ±0.8	M_1 in ±0.03	S^d Max mm	S^d Max in	T^e Ref mm	T^e Ref in	T_1 Min mm	T_1 Min in	X^f Dia mm +0.0 −0.5	X^f Dia in +0.00 −0.02
5/8	3/16	35.8	1.41	24.6	0.97	27.2	1.07	9.4	0.37	9.4	0.37	22.4	0.88
5/8	1/4	35.8	1.41	26.2	1.03	27.2	1.07	9.4	0.37	9.4	0.37	22.4	0.88
5/8	5/16	35.8	1.41	27.7	1.09	27.2	1.07	9.4	0.37	9.4	0.37	22.4	0.88
5/8	3/8	35.8	1.41	29.5	1.16	27.2	1.07	9.4	0.37	9.4	0.37	22.4	0.88
5/8	1/2	35.8	1.41	32.5	1.28	27.2	1.07	9.4	0.37	9.4	0.37	23.9	0.94
5/8	3/4	38.1	1.50	42.2	1.66	31.0	1.22	10.9	0.43	12.4	0.49	33.3	1.31

B^c Nom Tube OD in	B_1^c Nom Tube OD in	C in Nom	C_1 in Nom	D^d Drill mm	D^d Drill in	D_1^d Drill mm	D_1^d Drill in	D_2 Drill mm	D_2 Drill in	L mm ±0.8	L in ±0.03	L_1 mm ±0.8	L_1 in ±0.03
3/4	3/16	1-1/16	1-1/16	15.88	0.625	3.18	0.125	4.78	0.188	47.8	1.88	36.6	1.44
3/4	1/4	1-1/16	1-1/16	15.88	0.625	4.78	0.188	6.35	0.250	49.3	1.94	36.6	1.44
3/4	5/16	1-1/16	1-1/16	15.88	0.625	5.56	0.219	7.92	0.312	50.8	2.00	36.6	1.44
3/4	3/8	1-1/16	1-1/16	15.88	0.625	7.14	0.281	9.52	0.375	52.3	2.06	38.9	1.53
3/4	1/2	1-1/16	1-1/16	15.88	0.625	10.31	0.406	12.70	0.500	55.6	2.19	42.9	1.69
3/4	5/8	1-1/16	1-1/16	15.88	0.625	12.70	0.500	—	—	58.7	2.31	47.8	1.88

TABLE 4—(Continued)

B[c] Nom Tube OD in	B₁[c] Nom Tube OD in	M mm ±0.8	M in ±0.03	M₁ mm ±0.8	M₁ in ±0.03	S[d] Max mm	S[d] Max in	T[e] Ref mm	T[e] Ref in	T₁ Min mm	T₁ Min in	X[f] Dia mm +0.0 −0.5	X[f] Dia in +0.00 −0.02
3/4	3/16	42.2	1.66	26.9	1.06	31.0	1.22	10.9	0.43	10.9	0.43	26.9	1.06
3/4	1/4	42.2	1.66	28.4	1.12	31.0	1.22	10.9	0.43	10.9	0.43	26.9	1.06
3/4	5/16	42.2	1.66	30.2	1.19	31.0	1.22	10.9	0.43	10.9	0.43	26.9	1.06
3/4	3/8	42.2	1.66	31.8	1.25	31.0	1.22	10.9	0.43	10.9	0.43	26.9	1.06
3/4	1/2	42.2	1.66	35.1	1.38	31.0	1.22	10.9	0.43	10.9	0.43	26.9	1.06
3/4	5/8	42.2	1.66	38.1	1.50	31.0	1.22	10.9	0.43	10.9	0.43	26.9	1.06

[a] For flare dimensions shown on Figures 21 to 23 but not covered in Table 4, see corresponding dimensions for the specified Tube OD in Table 3.
[b] In these sizes the reducing unions and reducing elbows are the reverses of sizes already specified in Table.
[c] Where thread relief undercut is used last thread shall be chamfered 1/2 to 1 pitch long from G diameter and dimension J may be reduced by an amount equal to 1/2 pitch.
[d] At manufacturer's option through passages in fittings shown in Figure 21 may conform with the smaller diameter specified or be counterbored to the larger diameter from the appropriate end for depth S.
[e] Minimum design thickness, not subject to inspection.
[f] Basic dimensions shown shall apply as minimum for bosses. The −0.51 mm (−0.02 in) tolerance shall apply only to chamfer diameter on full hexagon version of fittings in Figure 22.

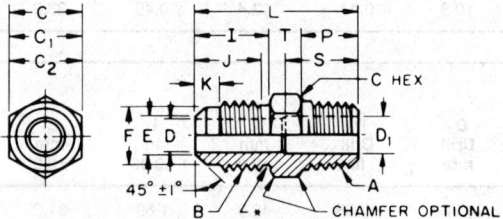

FIGURE 1—CONNECTOR (HALF UNION) (010102) (U1)

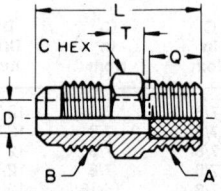

FIGURE 2—FUSIBLE CONNECTOR (HALF UNION) (010163) (FU)

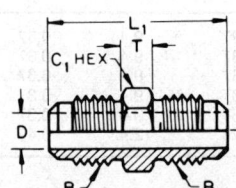

FIGURE 3—UNION (010101) (U2)

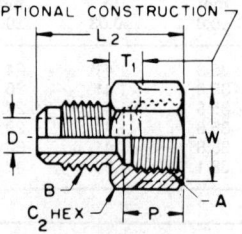

FIGURE 4—INTERNAL PIPE THREAD CONNECTOR (HALF UNION) (010103) (U3)

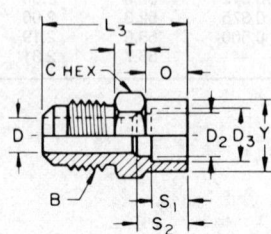

FIGURE 5—SOLDER CONNECTOR (HALF UNION) (010104) (US3)

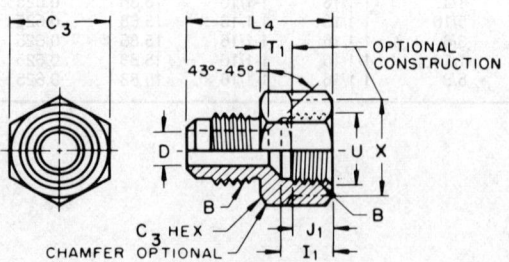

FIGURE 6—INTERNAL FLARE TO EXTERNAL FLARE ADAPTOR (010105) (UR3)

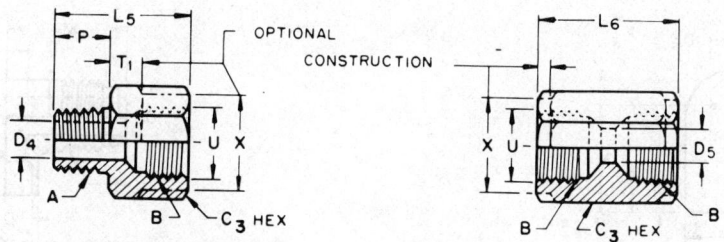

FIGURE 7—INTERNAL FLARE TO EXTERNAL (010107) (U4)

FIGURE 8—INTERNAL FLARE UNION (010108) (US4)

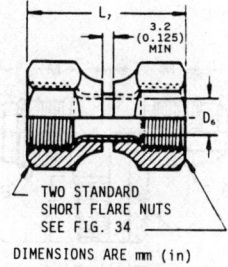

FIGURE 9—INTERNAL FLARE SWIVEL UNION PIPE ADAPTOR (010106) (U5)

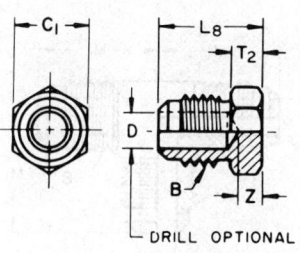

FIGURE 10—PLUG (010109) (P2)

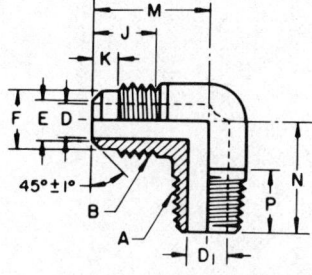

FIGURE 11—90 DEGREE ELBOW (010202) (E1)

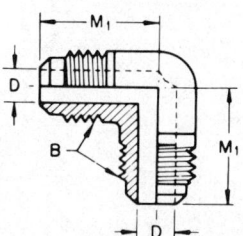

FIGURE 12—90 DEGREE ELBOW UNION (010201) (E2)

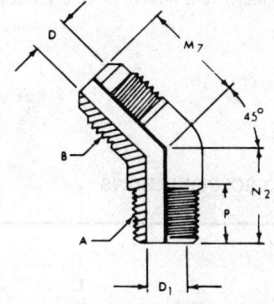

FIGURE 13—45 DEGREE ELBOW (010302) (E5)

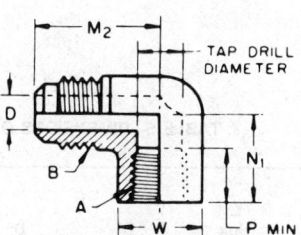

FIGURE 14—90 DEGREE INTERNAL PIPE THREAD ELBOW (010203) (E3)

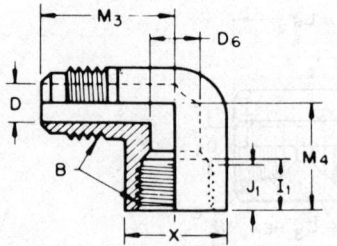

FIGURE 15—INTERNAL FLARE TO EXTERNAL FLARE 90 DEGREE ELBOW (010205) (E4)

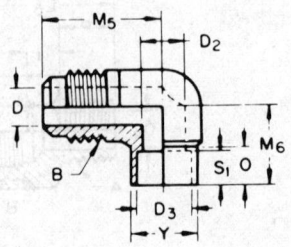

FIGURE 16—90 DEGREE SOLDER ELBOW (010425) (TI)

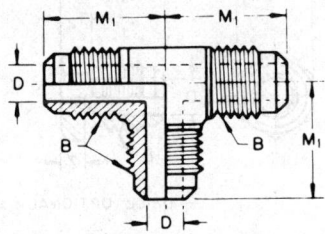

FIGURE 17—THREE-WAY TEE (010204) (ES)

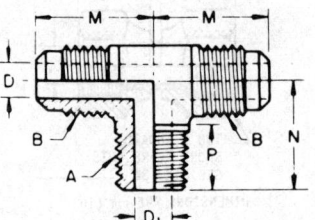

FIGURE 18—TWO-WAY TEE (010401) (T2)

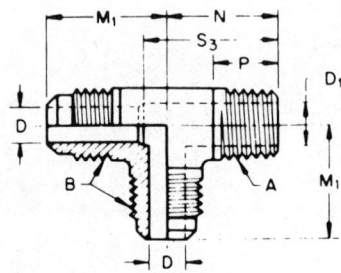

FIGURE 19—RIGHT ANGLE TWO-WAY TEE (010424) (T3)

NOTE—UNSPECIFIED DETAIL WITH RESPECT TO DIMENSIONS, TOLERANCES, CONTOUR, MATERIAL, WORKMANSHIP, ETC., MUST CONFORM TO GENERAL SPECIFICATIONS FOR REFRIGERATION TUBE FITTINGS. THE DIMENSIONAL DESIGNATIONS IN FIGURES 1, 6 AND 11 AND THE FIRST FIGURE IN EACH GROUP SHALL APPLY TO CORRESPONDING FEATURES OF OTHER FIGURES ON THIS PAGE UNLESS SHOWN OTHERWISE. THE ILLUSTRATIONS ON THIS PAGE APPLY TO TABLE 3. CODES SHOWN IN BRACKETS ADJACENT TO FIGURE NUMBERS REPRESENT RESPECTIVE FITTING IDENTIFICATION IN ACCORDANCE WITH SAE J846 (FIRST NUMBER) AND ANSI B70.1 (SECOND NUMBER).

TABLE 5—DIMENSIONS OF REDUCING SOLDER CONNECTORS AND REDUCING SOLDER ELBOWS
(Figures 24 and 25)[a]

B Nom Tube OD in	Solder Tube OD in	C Hex in Nom	D[c] Drill mm	D[c] Drill in	D_1[c] Drill mm	D_1[c] Drill in	D_2[e] Dia mm ±0.025	D_2[e] Dia in ±0.0010	L mm ±0.8	L in ±0.03	M mm ±0.8	M in ±0.03
3/16	1/8	3/8	3.18	0.125	2.39	0.094	3.277	0.1290	23.1	0.91	18.3	0.72
3/16	1/4	7/16	3.18	0.125	4.78	0.188	6.452	0.2540	23.9	0.94	18.3	0.72
3/16	5/16	7/16	3.18	0.125	6.35	0.250	8.039	0.3165	23.9	0.94	19.0	0.75
3/16	3/8	1/2	3.18	0.125	7.92	0.312	9.627	0.3790	23.9	0.94	19.8	0.78
3/16	1/2	5/8	3.18	0.125	11.13	0.438	12.802	0.5040	26.2	1.03	21.3	0.84
3/16	5/8	3/4	3.18	0.125	13.89	0.547	15.977	0.6290	30.2	1.19	23.1	0.91
3/16	3/4	7/8	3.18	0.125	17.48	0.688	19.152	0.7540	35.1	1.38	24.6	0.97
3/16	7/8	1	3.18	0.125	19.84	0.781	22.327	0.8790	39.6	1.56	26.9	1.06

TABLE 5—(Continued)

B Nom Tube OD in	Solder Tube OD in	M_1 mm ±0.8	M_1 in ±0.03	O mm	O in	S mm	S in	S_1[c] Max mm	S_1[c] Max in	T[d] Ref mm	T[d] Ref in	Y Dia Min mm	Y Dia Min in
3/16	1/8	15.0	0.59	7.9	0.31	7.9	0.31	13.2	0.52	3.8	0.15	5.6	0.22
3/16	1/4	15.0	0.59	7.9	0.31	7.9	0.31	10.4	0.41	4.6	0.18	8.6	0.34
3/16	5/16	15.7	0.62	7.9	0.31	7.9	0.31	10.4	0.41	4.6	0.18	10.2	0.40
3/16	3/8	16.8	0.66	7.9	0.31	7.9	0.31	10.4	0.41	4.6	0.18	12.2	0.48
3/16	1/2	19.8	0.78	9.7	0.38	9.7	0.38	12.4	0.49	5.3	0.21	15.2	0.60
3/16	5/8	24.6	0.97	12.7	0.50	12.7	0.50	16.0	0.63	6.1	0.24	18.8	0.74
3/16	3/4	29.5	1.16	15.7	0.62	15.7	0.62	19.8	0.78	7.6	0.30	21.8	0.86
3/16	7/8	35.1	1.38	19.0	0.75	19.0	0.75	23.9	0.94	9.4	0.37	24.9	0.98

B Nom Tube OD in	Solder Tube OD in	C Hex in Nom	D[c] Drill mm	D[c] Drill in	D_1[c] Drill mm	D_1[c] Drill in	D_2[e] Dia mm ±0.025	D_2[e] Dia in ±0.0010	L mm ±0.8	L in ±0.03	M mm ±0.8	M in ±0.03
1/4	1/8	7/16	4.78	0.188	2.39	0.094	3.277	0.1290	25.4	1.00	20.6	0.81
1/4	3/16	7/16	4.78	0.188	4.78	0.188	4.864	0.1915	25.4	1.00	20.6	0.81
1/4	5/16	7/16	4.78	0.188	6.35	0.250	8.039	0.3165	25.4	1.00	20.6	0.81
1/4	3/8	1/2	4.78	0.188	7.92	0.312	9.627	0.3790	25.4	1.00	21.3	0.84
1/4	1/2	5/8	4.78	0.188	11.13	0.438	12.802	0.5040	27.7	1.09	23.1	0.91
1/4	5/8	3/4	4.78	0.188	13.89	0.547	15.977	0.6290	31.8	1.25	24.6	0.97
1/4	3/4	7/8	4.78	0.188	17.48	0.688	19.152	0.7540	36.6	1.44	26.2	1.03
1/4	7/8	1	4.78	0.188	19.84	0.781	22.327	0.8790	41.1	1.62	28.4	1.12

B Nom Tube OD in	Solder Tube OD in	M_1 mm ±0.8	M_1 in ±0.03	O mm	O in	S mm	S in	S_1[c] Max mm	S_1[c] Max in	T[d] Ref mm	T[d] Ref in	Y Dia Min mm	Y Dia Min in
1/4	1/8	15.7	0.62	7.9	0.31	7.9	0.31	15.2	0.60	4.6	0.18	5.6	0.22
1/4	3/16	15.7	0.62	7.9	0.31	7.9	0.31	15.2	0.60	4.6	0.18	7.1	0.28
1/4	5/16	15.7	0.62	7.9	0.31	7.9	0.31	10.4	0.41	4.6	0.18	10.2	0.40
1/4	3/8	16.8	0.66	7.9	0.31	7.9	0.31	10.4	0.41	4.6	0.18	12.2	0.48
1/4	1/2	19.8	0.78	9.7	0.38	9.7	0.38	12.4	0.49	5.3	0.21	15.2	0.60
1/4	5/8	24.6	0.97	12.7	0.50	12.7	0.50	16.0	0.63	6.1	0.24	18.8	0.74
1/4	3/4	29.5	1.16	15.7	0.62	15.7	0.62	19.8	0.78	7.6	0.30	21.8	0.86
1/4	7/8	35.1	1.38	19.0	0.75	19.0	0.75	23.9	0.94	9.4	0.37	24.9	0.98

B Nom Tube OD in	Solder Tube OD in	C Hex in Nom	D[c] Drill mm	D[c] Drill in	D_1[c] Drill mm	D_1[c] Drill in	D_2[e] Dia mm ±0.025	D_2[e] Dia in ±0.0010	L mm ±0.8	L in ±0.03	M mm ±0.8	M in ±0.03
5/16	1/8	1/2	5.56	0.219	2.39	0.094	3.277	0.1290	27.7	1.09	23.1	0.91
5/16	3/16	1/2	5.56	0.219	3.96	0.156	4.864	0.1915	27.7	1.09	23.1	0.91
5/16	1/4	1/2	5.56	0.219	4.78	0.188	6.452	0.2540	27.7	1.09	23.1	0.91
5/16	3/8	1/2	5.56	0.219	7.92	0.312	9.627	0.3790	27.7	1.09	23.1	0.91
5/16	1/2	5/8	5.56	0.219	11.13	0.438	12.802	0.5040	29.5	1.16	24.6	0.97
5/16	5/8	3/4	5.56	0.219	13.89	0.547	15.977	0.6290	33.3	1.31	26.2	1.03
5/16	3/4	7/8	5.56	0.219	17.48	0.688	19.152	0.7540	38.1	1.50	27.7	1.09
5/16	7/8	1	5.56	0.219	19.84	0.781	22.327	0.8790	42.9	1.69	30.2	1.19

B Nom Tube OD in	Solder Tube OD in	M_1 mm ±0.8	M_1 in ±0.03	O mm	O in	S mm	S in	S_1[c] Max mm	S_1[c] Max in	T[d] Ref mm	T[d] Ref in	Y Dia Min mm	Y Dia Min in
5/16	1/8	16.8	0.66	7.9	0.31	7.9	0.31	17.0	0.67	5.3	0.21	5.6	0.22
5/16	3/16	16.8	0.66	7.9	0.31	7.9	0.31	17.0	0.67	5.3	0.21	7.1	0.28
5/16	1/4	16.8	0.66	7.9	0.31	7.9	0.31	17.0	0.67	5.3	0.21	8.6	0.34
5/16	3/8	16.8	0.66	7.9	0.31	7.9	0.31	10.7	0.42	5.3	0.21	12.2	0.48
5/16	1/2	19.8	0.78	9.7	0.38	9.7	0.38	12.4	0.49	5.3	0.21	15.2	0.60
5/16	5/8	24.6	0.97	12.7	0.50	12.7	0.50	16.0	0.63	6.1	0.24	18.8	0.74
5/16	3/4	29.5	1.16	15.7	0.62	15.7	0.62	19.8	0.78	7.6	0.30	21.8	0.86
5/16	7/8	35.1	1.38	19.0	0.75	19.0	0.75	23.9	0.94	9.4	0.37	24.9	0.98

TABLE 5—(Continued)

B Nom Tube OD in	Solder Tube OD in	C Hex in Nom	D[c] Drill mm	D[c] Drill in	D₁[c] Drill mm	D₁[c] Drill in	D₂[e] Dia mm ±0.025	D₂[e] Dia in ±0.0010	L mm ±0.8	L in ±0.03	M mm ±0.8	M in ±0.03
3/8	1/8	5/8	7.14	0.281	2.39	0.094	3.277	0.1290	30.2	1.19	26.2	1.03
3/8	3/16	5/8	7.14	0.281	3.96	0.156	4.864	0.1915	30.2	1.19	26.2	1.03
3/8	1/4	5/8	7.14	0.281	4.78	0.188	6.452	0.2540	30.2	1.19	26.2	1.03
3/8	5/16	5/8	7.14	0.281	6.35	0.250	8.039	0.3165	30.2	1.19	26.2	1.03
3/8	1/2	5/8	7.14	0.281	11.13	0.438	12.802	0.5040	31.8	1.25	26.2	1.03
3/8	5/8	3/4	7.14	0.281	13.89	0.547	15.977	0.6290	35.1	1.38	27.7	1.09
3/8	3/4	7/8	7.14	0.281	17.48	0.688	19.152	0.7540	39.6	1.56	29.5	1.16
3/8	7/8	1	7.14	0.281	19.84	0.781	22.327	0.8790	44.4	1.75	31.8	1.25

B Nom Tube OD in	Solder Tube OD in	M₁ mm ±0.8	M₁ in ±0.03	O mm	O in	S mm	S in	S₁[c] Max mm	S₁[c] Max in	T[d] Ref mm	T[d] Ref in	Y Dia Min mm	Y Dia Min in
3/8	1/8	18.3	0.72	7.9	0.31	7.9	0.31	19.0	0.75	6.1	0.24	5.6	0.22
3/8	3/16	18.3	0.72	7.9	0.31	7.9	0.31	19.0	0.75	6.1	0.24	7.1	0.28
3/8	1/4	18.3	0.72	7.9	0.31	7.9	0.31	19.0	0.75	6.1	0.24	8.6	0.34
3/8	5/16	18.3	0.72	7.9	0.31	7.9	0.31	19.0	0.75	6.1	0.24	10.2	0.40
3/8	1/2	19.8	0.78	9.7	0.38	9.7	0.38	13.0	0.51	6.1	0.24	15.2	0.60
3/8	5/8	24.6	0.97	12.7	0.50	12.7	0.50	16.0	0.63	6.1	0.24	18.8	0.74
3/8	3/4	29.5	1.16	15.7	0.62	15.7	0.62	19.8	0.78	7.6	0.30	21.8	0.86
3/8	7/8	31.8	1.25	19.0	0.75	19.0	0.75	23.9	0.94	9.4	0.37	24.9	0.98

B Nom Tube OD in	Solder Tube OD in	C Hex in Nom	D[c] Drill mm	D[c] Drill in	D₁[c] Drill mm	D₁[c] Drill in	D₂[e] Dia mm ±0.025	D₂[e] Dia in ±0.0010	L mm ±0.8	L in ±0.03	M mm ±0.8	M in ±0.03
1/2	1/8	3/4	10.31	0.406	2.39	0.094	3.277	0.1290	35.1	1.38	31.0	1.22
1/2	3/16	3/4	10.31	0.406	3.96	0.156	4.864	0.1915	35.1	1.38	31.0	1.22
1/2	1/4	3/4	10.31	0.406	4.78	0.188	6.452	0.2540	35.1	1.38	31.0	1.22
1/2	5/16	3/4	10.31	0.406	6.35	0.250	8.039	0.3165	35.1	1.38	31.0	1.22
1/2	3/8	3/4	10.31	0.406	7.92	0.312	9.627	0.3790	35.1	1.38	31.0	1.22
1/2	5/8	3/4	10.31	0.406	13.89	0.547	15.977	0.6290	39.6	1.56	31.0	1.22
1/2	3/4	7/8	10.31	0.406	17.48	0.688	19.152	0.7540	42.9	1.69	32.5	1.28
1/2	7/8	1	10.31	0.406	19.84	0.781	22.327	0.8790	47.8	1.88	35.1	1.38

B Nom Tube OD in	Solder Tube OD in	M₁ mm ±0.8	M₁ in ±0.03	O mm	O in	S mm	S in	S₁[c] Max mm	S₁[c] Max in	T[d] Ref mm	T[d] Ref in	Y Dia Min mm	Y Dia Min in
1/2	1/8	19.8	0.78	7.9	0.31	7.9	0.31	23.1	0.91	7.6	0.30	5.6	0.22
1/2	3/16	19.8	0.78	7.9	0.31	7.9	0.31	23.1	0.91	7.6	0.30	7.1	0.28
1/2	1/4	19.8	0.78	7.9	0.31	7.9	0.31	23.1	0.91	7.6	0.30	8.6	0.34
1/2	5/16	19.8	0.78	7.9	0.31	7.9	0.31	23.1	0.91	7.6	0.30	10.2	0.40
1/2	3/8	19.8	0.78	7.9	0.31	7.9	0.31	23.1	0.91	7.6	0.30	12.2	0.48
1/2	5/8	24.6	0.97	12.7	0.50	12.7	0.50	16.8	0.66	7.6	0.30	18.8	0.74
1/2	3/4	29.5	1.16	15.7	0.62	15.7	0.62	19.8	0.78	7.6	0.30	21.8	0.86
1/2	7/8	35.1	1.38	19.0	0.75	19.0	0.75	23.9	0.94	9.4	0.37	24.9	0.98

B Nom Tube OD in	Solder Tube OD in	C Hex in Nom	D[c] Drill mm	D[c] Drill in	D₁[c] Drill mm	D₁[c] Drill in	D₂[e] Dia mm ±0.025	D₂[e] Dia in ±0.0010	L mm ±0.8	L in ±0.03	M mm ±0.8	M in ±0.03
5/8	1/8	7/8	12.70	0.500	2.39	0.094	3.277	0.1290	39.6	1.56	35.8	1.41
5/8	3/16	7/8	12.70	0.500	3.96	0.156	4.864	0.1915	39.6	1.56	35.8	1.41
5/8	1/4	7/8	12.70	0.500	4.78	0.188	6.452	0.2540	39.6	1.56	35.8	1.41
5/8	5/16	7/8	12.70	0.500	6.35	0.250	8.039	0.3165	39.6	1.56	35.8	1.41
5/8	3/8	7/8	12.70	0.500	7.92	0.312	9.627	0.3790	39.6	1.56	35.8	1.41
5/8	1/2	7/8	12.70	0.500	11.13	0.438	12.802	0.5040	41.1	1.62	35.8	1.41
5/8	3/4	7/8	12.70	0.500	17.48	0.688	19.152	0.7540	47.8	1.88	35.8	1.41
5/8	7/8	1	12.70	0.500	19.84	0.781	22.327	0.8790	50.8	2.00	38.1	1.50

TABLE 5—(Continued)

B Nom Tube OD in	Solder Tube OD in	M_1 mm ±0.8	M_1 in ±0.03	O mm	O in	S mm	S in	S_1[c] Max mm	S_1[c] Max in	T[d] Ref mm	T[d] Ref in	Y Dia Min mm	Y Dia Min in
5/8	1/8	21.3	0.84	7.9	0.31	7.9	0.31	27.2	1.07	9.4	0.37	5.6	0.22
5/8	3/16	21.3	0.84	7.9	0.31	7.9	0.31	27.2	1.07	9.4	0.37	7.1	0.28
5/8	1/4	21.3	0.84	7.9	0.31	7.9	0.31	27.2	1.07	9.4	0.37	8.6	0.34
5/8	5/16	21.3	0.84	7.9	0.31	7.9	0.31	27.2	1.07	9.4	0.37	10.2	0.40
5/8	3/8	21.3	0.84	7.9	0.31	7.9	0.31	27.2	1.07	9.4	0.37	12.2	0.48
5/8	1/2	23.1	0.91	9.7	0.38	9.7	0.38	27.2	1.07	9.4	0.37	15.2	0.60
5/8	3/4	29.5	1.16	15.7	0.62	15.7	0.62	20.6	0.81	9.4	0.37	21.8	0.86
5/8	7/8	35.1	1.38	19.0	0.75	19.0	0.75	23.9	0.94	9.4	0.37	24.9	0.98

B Nom Tube OD in	Solder Tube OD in	C Hex in Nom	D[c] Drill mm	D[c] Drill in	D_1[c] Drill mm	D_1[c] Drill in	D_2[e] Dia mm ±0.025	D_2[e] Dia in ±0.0010	L mm ±0.8	L in ±0.03	M mm ±0.8	M in ±0.03
3/4	1/8	1-1/16	15.88	0.625	—	0.094	3.277	0.1290	44.4	1.75	41.1	1.62
3/4	3/16	1-1/16	15.88	0.625	—	0.156	4.864	0.1915	44.4	1.75	41.1	1.62
3/4	1/4	1-1/16	15.88	0.625	—	0.188	6.452	0.2540	44.4	1.75	41.1	1.62
3/4	5/16	1-1/16	15.88	0.625	—	0.250	8.039	0.3165	44.4	1.75	41.1	1.62
3/4	3/8	1-1/16	15.88	0.625	—	0.312	9.627	0.3790	44.4	1.75	41.1	1.62
3/4	1/2	1-1/16	15.88	0.625	—	0.438	12.802	0.5040	46.0	1.81	41.1	1.62
3/4	5/8	1-1/16	15.88	0.625	—	0.547	15.977	0.6290	49.3	1.94	41.1	1.62
3/4	3/4	1-1/16	15.88	0.625	—	0.781	22.327	0.8790	55.6	2.19	41.1	1.62

B Nom Tube OD in	Solder Tube OD in	M_1 mm ±0.8	M_1 in ±0.03	O mm	O in	S mm	S in	S_1[c] Max mm	S_1[c] Max in	T[d] Ref mm	T[d] Ref in	Y Dia Min mm	Y Dia Min in
3/4	1/8	23.9	0.94	7.9	0.31	7.9	0.31	31.0	1.22	10.9	0.43	5.6	0.22
3/4	3/16	23.9	0.94	7.9	0.31	7.9	0.31	31.0	1.22	10.9	0.43	7.1	0.28
3/4	1/4	23.9	0.94	7.9	0.31	7.9	0.31	31.0	1.22	10.9	0.43	8.6	0.34
3/4	5/16	23.9	0.94	7.9	0.31	7.9	0.31	31.0	1.22	10.9	0.43	10.2	0.40
3/4	3/8	23.9	0.94	7.9	0.31	7.9	0.31	31.0	1.22	10.9	0.43	12.2	0.48
3/4	1/2	25.4	1.00	9.7	0.38	9.7	0.38	31.0	1.22	10.9	0.43	15.2	0.60
3/4	5/8	28.4	1.12	12.7	0.50	12.7	0.50	31.0	1.22	10.9	0.43	18.8	0.74
3/4	7/8	35.1	1.38	19.0	0.75	19.0	0.75	24.6	0.97	10.9	0.43	24.9	0.98

[a] For flare dimensions shown on Figures 24 and 25 but not covered in Table 5, see corresponding dimensions for the specified Tube OD in Table 3.
[b] Where thread relief undercut is used, last thread shall be chamfered 1/2 to 1 pitch long from G diameter and dimension J may be reduced by an amount equal to 1/2 pitch.
[c] At manufacturer's option through passages in fittings shown in Figure 24 may conform with the smaller diameter specified or be counterbored to the larger diameter from appropriate end for depth S.
[d] Minimum design thickness, not subject to inspection.
[e] ID of solder cup shall not be out of round by more than 0.08 mm (0.003 in).

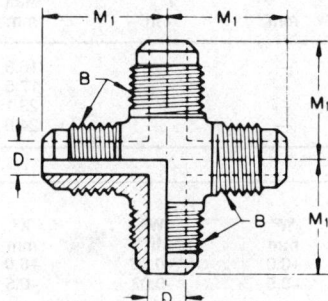

FIGURE 20—CROSS (010501) (C1)

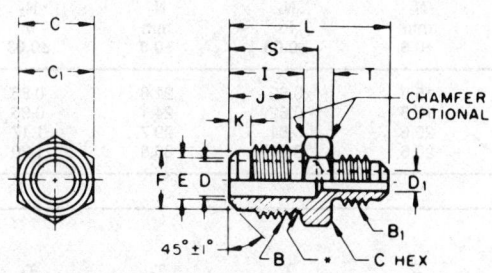

FIGURE 21—REDUCING UNION (010101) (UR2)

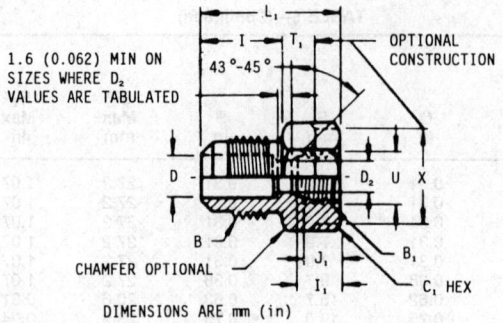

FIGURE 22—INTERNAL FLARE TO EXTERNAL FLARE REDUCING ADAPTOR
(010105) (UR3)

NOTE— UNSPECIFIED DETAIL WITH RESPECT TO DIMENSIONS, TOLERANCES, CONTOUR, MATERIAL, WORKMANSHIP, ETC., MUST CONFORM TO GENERAL SPECIFICATIONS FOR REFRIGERATION TUBE FITTINGS. THE ILLUSTRATIONS ON THIS PAGE APPLY TO TABLE 4. CODES SHOWN IN BRACKETS ADJACENT TO FIGURE NUMBERS REPRESENT RESPECTIVE FITTING IDENTIFICATION IN ACCORDANCE WITH SAE J846 (FIRST NUMBER) AND ANSI B70.1 (SECOND NUMBER).

TABLE 6—DIMENSIONS OF REDUCING CONNECTORS, REDUCING ADAPTORS, AND REDUCING ELBOWS (Figures 26 to 32)[a]

B Nom Tube OD in	A Dryseal Pipe Thread NPTF[b]	C Hex in Nom	C_1 Hex in Nom	C_2 Hex in Nom	D[d] Drill mm	D[d] Drill in	D_1[d] Drill mm	D_1[d] Drill in	D_2 Drill mm	D_2 Drill in	L mm ±0.8	L in ±0.03
3/16	1/4	9/16	11/16	9/16	3.18	0.125	7.92	0.312	4.78	0.188	30.2	1.19
3/16	3/8	11/16	13/16	11/16	3.18	0.125	10.31	0.406	4.78	0.188	31.8	1.22
3/16	1/2	7/8	1	7/8	3.18	0.125	14.27	0.562	4.78	0.188	38.1	1.44
3/16	3/4	1-1/16	1-1/4	1-1/16	3.18	0.125	19.05	0.750	4.78	0.188	41.1	1.50

B Nom Tube OD in	L_1 mm ±0.8	L_1 in ±0.03	L_2 mm ±0.8	L_2 in ±0.03	M mm ±0.8	M in ±0.03	M_1 mm ±0.8	M_1 in ±0.03	M_2 mm ±0.8	M_2 in ±0.03	N mm ±0.8	N in ±0.03
3/16	30.2	1.19	23.9	0.94	20.6	0.81	23.1	0.91	18.0	0.71	24.6	0.97
3/16	31.0	1.22	21.3	0.84	22.4	0.88	24.6	0.97	19.0	0.75	25.4	1.00
3/16	36.6	1.44	26.9	1.06	24.6	0.97	26.9	1.06	20.3	0.80	32.5	1.28
3/16	38.1	1.50	30.2	1.19	26.9	1.06	31.0	1.22	21.3	0.84	35.1	1.38

B Nom Tube OD in	N_1 mm ±0.8	N_1 in ±0.03	N_2 mm ±0.8	N_2 in ±0.03	P Min mm	P Min in	Q[e] mm	Q[e] in	S[d] Max mm	S[d] Max in
3/16	15.8	0.62	21.8	0.86	14.2	0.56	—	—	16.8	0.66
3/16	15.8	0.62	24.1	0.95	14.2	0.56	—	—	17.5	0.69
3/16	20.6	0.81	29.7	1.17	19.0	0.75	—	—	23.1	0.91
3/16	20.6	0.81	30.5	1.20	19.0	0.75	—	—	24.6	0.97

B Nom Tube OD in	T[f] Ref mm	T[f] Ref in	T_1 Min mm	T_1 Min in	T_2 Min mm	T_2 Min in	W[g] mm +0.0 −0.5	W[g] in +0.00 −0.02	X[g] mm +0.0 −0.5	X[g] in +0.00 −0.02
3/16	4.6	0.18	6.1	0.24	5.3	0.21	17.5	0.69	14.2	0.56
3/16	6.1	0.24	7.6	0.30	6.1	0.24	20.6	0.81	17.5	0.69
3/16	7.6	0.30	9.4	0.37	7.6	0.30	25.4	1.00	22.4	0.88
3/16	10.9	0.43	12.4	0.49	10.9	0.43	31.8	1.25	26.9	1.06

TABLE 6—(Continued)

B Nom Tube OD in	A Dryseal Pipe Thread NPTF[b]	C Hex in Nom	C₁ Hex in Nom	C₂ Hex in Nom	D[d] Drill mm	D[d] Drill in	D₁[d] Drill mm	D₁[d] Drill in	D₂ Drill mm	D₂ Drill in	L mm ±0.8	L in ±0.03
1/4	1/4	9/16	11/16	5/8	4.78	0.188	7.92	0.312	6.35	0.250	31.8	1.25
1/4	3/8	11/16	13/16	11/16	4.78	0.188	10.31	0.406	6.35	0.250	33.3	1.31
1/4	1/2	7/8	1	7/8	4.78	0.188	14.27	0.562	6.35	0.250	39.6	1.56
1/4	3/4	1-1/16	1-1/4	1-1/16	4.78	0.188	19.05	0.750	6.35	0.250	42.9	1.69

B Nom Tube OD in	L₁ mm ±0.8	L₁ in ±0.03	L₂ mm ±0.8	L₂ in ±0.03	M mm ±0.8	M in ±0.03	M₁ mm ±0.8	M₁ in ±0.03	M₂ mm ±0.8	M₂ in ±0.03	N mm ±0.8	N in ±0.03
1/4	31.8	1.25	26.2	1.03	23.1	0.91	24.6	0.97	19.0	0.75	23.9	0.94
1/4	32.5	1.28	23.9	0.94	23.9	0.94	26.2	1.03	20.1	0.79	26.2	1.03
1/4	38.1	1.50	26.9	1.06	26.2	1.03	28.4	1.12	21.6	0.85	32.5	1.28
1/4	39.6	1.56	30.2	1.19	28.4	1.12	32.5	1.28	22.6	0.89	35.1	1.38

B Nom Tube OD in	N₁ mm ±0.8	N₁ in ±0.03	N₂ mm ±0.8	N₂ in ±0.03	P Min mm	P Min in	Q[e] mm	Q[e] in	S[d] Max mm	S[d] Max in
1/4	16.8	0.66	21.8	0.86	14.2	0.56	16.8	0.66	16.8	0.66
1/4	16.8	0.66	24.1	0.95	14.2	0.56	17.5	0.69	17.5	0.69
1/4	21.3	0.84	29.7	1.17	19.0	0.75	—	—	23.1	0.91
1/4	21.3	0.84	30.5	1.20	19.0	0.75	—	—	24.6	0.97

B Nom Tube OD in	T[f] Ref mm	T[f] Ref in	T₁ Min mm	T₁ Min in	T₂ Min mm	T₂ Min in	W[g] mm +0.0 −0.5	W[g] in +0.00 −0.02	X[g] mm +0.0 −0.5	X[g] in +0.00 −0.02
1/4	4.6	0.18	6.1	0.24	6.1	0.24	17.5	0.69	15.7	0.62
1/4	6.1	0.24	7.6	0.30	6.1	0.24	20.6	0.81	17.5	0.69
1/4	7.6	0.30	9.4	0.37	7.6	0.30	25.4	1.00	22.4	0.88
1/4	10.9	0.43	12.4	0.49	10.9	0.43	31.8	1.25	26.9	1.06

B Nom Tube OD in	A Dryseal Pipe Thread NPTF[b]	C Hex in Nom	C₁ Hex in Nom	C₂ Hex in Nom	D[d] Drill mm	D[d] Drill in	D₁[d] Drill mm	D₁[d] Drill in	D₂ Drill mm	D₂ Drill in	L mm ±0.8	L in ±0.03
5/16	1/4	9/16	11/16	11/16	5.56	0.219	7.92	0.312	7.92	0.312	34.0	1.34
5/16	3/8	11/16	13/16	11/16	5.56	0.219	10.31	0.406	7.92	0.312	35.1	1.38
5/16	1/2	7/8	1	7/8	5.56	0.219	14.27	0.562	7.92	0.312	41.1	1.62
5/16	3/4	1-1/16	1-1/4	1-1/16	5.56	0.219	19.05	0.750	7.92	0.312	44.4	1.75

B Nom Tube OD in	L₁ mm ±0.8	L₁ in ±0.03	L₂ mm ±0.8	L₂ in ±0.03	M mm ±0.8	M in ±0.03	M₁ mm ±0.8	M₁ in ±0.03	M₂ mm ±0.8	M₂ in ±0.03	N mm ±0.8	N in ±0.03
5/16	32.5	1.28	27.7	1.09	24.6	0.97	26.2	1.03	20.6	0.81	23.9	0.94
5/16	33.3	1.31	25.4	1.00	25.4	1.00	27.7	1.09	21.6	0.85	26.2	1.03
5/16	38.9	1.53	28.4	1.12	27.7	1.09	30.2	1.19	23.1	0.91	32.5	1.28
5/16	40.4	1.52	30.2	1.19	30.2	1.19	34.0	1.34	24.1	0.95	35.1	1.38

B Nom Tube OD in	N₁ mm ±0.8	N₁ in ±0.03	N₂ mm ±0.8	N₂ in ±0.03	P Min mm	P Min in	Q[e] mm	Q[e] in	S[d] Max mm	S[d] Max in
5/16	16.8	0.66	21.8	0.86	14.2	0.56	—	—	17.0	0.67
5/16	16.8	0.66	24.1	0.95	14.2	0.56	—	—	17.5	0.69
5/16	21.3	0.84	29.7	1.17	19.0	0.75	—	—	23.1	0.91
5/16	21.3	0.84	30.5	1.20	19.0	0.75	—	—	24.6	0.97

TABLE 6—(Continued)

B Nom Tube OD in	T[f] Ref mm	T[f] Ref in	T_1 Min mm	T_1 Min in	T_2 Min mm	T_2 Min in	W[g] mm +0.0 −0.5	W[g] in +0.00 −0.02	X[g] mm +0.0 −0.5	X[g] in +0.00 −0.02
5/16	5.3	0.21	6.1	0.24	6.1	0.24	17.5	0.69	17.5	0.69
5/16	6.1	0.24	7.6	0.30	6.1	0.24	20.6	0.81	17.5	0.69
5/16	7.6	0.30	9.4	0.37	7.6	0.30	25.4	1.00	22.4	0.88
5/16	10.9	0.43	12.4	0.49	10.9	0.43	31.8	1.25	26.9	1.06

B Nom Tube OD in	A Dryseal Pipe Thread NPTF[b]	C Hex in Nom	C_1 Hex in Nom	C_2 Hex in Nom	D[d] Drill mm	D[d] Drill in	D_1[d] Drill mm	D_1[d] Drill in	D_2 Drill mm	D_2 Drill in	L mm ±0.8	L in ±0.03
3/8	1/8	5/8	5/8	13/16	7.14	0.281	5.56	0.219	—	—	31.8	1.25
3/8	3/8	11/16	13/16	13/16	7.14	0.281	10.31	0.406	9.52	0.375	36.6	1.44
3/8	1/2	7/8	1	7/8	7.14	0.281	14.27	0.562	9.52	0.375	42.9	1.69
3/8	3/4	1-1/16	1-1/4	1-1/16	7.14	0.281	19.05	0.750	9.52	0.375	46.0	1.81

B Nom Tube OD in	L_1 mm ±0.8	L_1 in ±0.03	L_2 mm ±0.8	L_2 in ±0.03	M mm ±0.8	M in ±0.03	M_1 mm ±0.8	M_1 in ±0.03	M_2 mm ±0.8	M_2 in ±0.03	N mm ±0.8	N in ±0.03
3/8	28.4	1.12	27.7	1.09	26.2	1.03	26.9	1.06	22.6	0.89	23.1	0.91
3/8	35.1	1.38	28.4	1.12	26.9	1.06	29.5	1.16	23.6	0.93	27.7	1.09
3/8	41.1	1.62	31.8	1.25	29.5	1.16	31.8	1.25	25.1	0.99	32.5	1.28
3/8	42.2	1.66	30.2	1.19	31.8	1.25	35.8	1.41	26.2	1.03	35.1	1.38

B Nom Tube OD in	N_1 mm ±0.8	N_1 in ±0.03	N_2 mm ±0.8	N_2 in ±0.03	P Min mm	P Min in	Q[e] mm	Q[e] in	S[d] Max mm	S[d] Max in
3/8	12.7	0.50	17.0	0.67	9.7	0.38	—	—	19.0	0.75
3/8	17.5	0.69	24.1	0.95	14.2	0.56	17.5	0.69	17.5	0.69
3/8	22.4	0.88	29.7	1.17	19.0	0.75	—	—	23.1	0.91
3/8	22.4	0.88	30.5	1.20	19.0	0.75	—	—	24.6	0.97

B Nom Tube OD in	T[f] Ref mm	T[f] Ref in	T_1 Min mm	T_1 Min in	T_2 Min mm	T_2 Min in	W[g] mm +0.0 −0.5	W[g] in +0.00 −0.02	X[g] mm +0.0 −0.5	X[g] in +0.00 −0.02
3/8	6.1	0.24	6.1	0.24	7.6	0.30	15.7	0.62	20.6	0.81
3/8	6.1	0.24	7.6	0.30	7.6	0.30	20.6	0.81	20.6	0.81
3/8	7.6	0.30	9.4	0.37	7.6	0.30	25.4	1.00	22.4	0.88
3/8	10.9	0.43	12.4	0.49	10.9	0.43	31.8	1.25	26.9	1.06

B Nom Tube OD in	A Dryseal Pipe Thread NPTF[b]	C Hex in Nom	C_1 Hex in Nom	C_2 Hex in Nom	D[d] Drill mm	D[d] Drill in	D_1[d] Drill mm	D_1[d] Drill in	D_2 Drill mm	D_2 Drill in	L mm ±0.8	L in ±0.03
1/2	1/8	3/4	3/4	15/16	10.31	0.406	5.56	0.219	—	—	36.6	1.44
1/2	1/4	3/4	3/4	15/16	10.31	0.406	7.92	0.312	—	—	41.1	1.62
1/2	1/2	7/8	1	15/16	10.31	0.406	14.27	0.562	12.70	0.500	46.0	1.81
1/2	3/4	1-1/16	1-1/4	1-1/16	10.31	0.406	19.05	0.750	12.70	0.500	49.3	1.94

B Nom Tube OD in	L_1 mm ±0.8	L_1 in ±0.03	L_2 mm ±0.8	L_2 in ±0.03	M mm ±0.8	M in ±0.03	M_1 mm ±0.8	M_1 in ±0.03	M_2 mm ±0.8	M_2 in ±0.03	N mm ±0.8	N in ±0.03
1/2	30.2	1.19	31.8	1.25	31.0	1.22	31.0	1.22	26.9	1.06	25.4	1.00
1/2	35.8	1.41	34.0	1.34	31.0	1.22	31.0	1.22	26.9	1.06	30.2	1.19
1/2	44.4	1.75	37.3	1.47	32.5	1.28	35.1	1.38	28.4	1.12	35.1	1.38
1/2	46.0	1.81	35.1	1.38	35.1	1.38	38.9	1.53	29.5	1.16	35.1	1.38

TABLE 6—(Continued)

B Nom Tube OD in	N₁ mm ±0.8	N₁ in ±0.03	N₂ mm ±0.8	N₂ in ±0.03	P Min mm	P Min in	Q[e] mm	Q[e] in	S[d] Max mm	S[d] Max in
1/2	14.2	0.56	19.3	0.76	9.7	0.38	—	—	23.1	0.91
1/2	19.0	0.75	24.1	0.95	14.2	0.56	—	—	23.1	0.91
1/2	23.9	0.94	29.7	1.17	19.0	0.75	—	—	23.1	0.91
1/2	23.9	0.94	30.5	1.20	19.0	0.75	—	—	24.6	0.97

B Nom Tube OD in	T[f] Ref mm	T[f] Ref in	T₁ Min mm	T₁ Min in	T₂ Min mm	T₂ Min in	W[g] mm +0.0 -0.5	W[g] in +0.00 -0.02	X[g] mm +0.0 -0.5	X[g] in +0.00 -0.02
1/2	7.6	0.30	7.6	0.30	9.4	0.37	19.0	0.75	23.9	0.94
1/2	7.6	0.30	7.6	0.30	9.4	0.37	19.0	0.75	23.9	0.94
1/2	7.6	0.30	9.4	0.37	9.4	0.37	25.4	1.00	23.9	0.94
1/2	10.9	0.43	12.4	0.49	10.9	0.43	31.8	1.25	26.9	1.06

B Nom Tube OD in	A Dryseal Pipe Thread NPTF[b]	C Hex in Nom	C₁ Hex in Nom	C₂ Hex in Nom	D[d] Drill mm	D[d] Drill in	D₁[d] Drill mm	D₁[d] Drill in	D₂ Drill mm	D₂ Drill in	L mm ±0.8	L in ±0.03
5/8	1/8	7/8	7/8	1-1/16	12.70	0.500	5.56	0.219	—	—	41.1	1.62
5/8	1/4	7/8	7/8	1-1/16	12.70	0.500	7.92	0.312	—	—	46.0	1.81
5/8	3/8	7/8	7/8	1-1/16	12.70	0.500	10.31	0.406	—	—	46.0	1.81
5/8	3/4	1-1/16	1-1/4	1-1/16	12.70	0.500	19.05	0.750	15.88	0.625	52.3	2.06

B Nom Tube OD in	L₁ mm ±0.8	L₁ in ±0.03	L₂ mm ±0.8	L₂ in ±0.03	M mm ±0.8	M in ±0.03	M₁ mm ±0.8	M₁ in ±0.03	M₂ mm ±0.8	M₂ in ±0.03	N mm ±0.8	N in ±0.03
5/8	32.5	1.28	35.8	1.41	35.8	1.41	35.8	1.41	31.2	1.23	26.9	1.06
5/8	38.1	1.50	38.9	1.53	35.8	1.41	35.8	1.41	31.2	1.23	31.8	1.25
5/8	40.4	1.59	39.6	1.56	35.8	1.41	35.8	1.41	31.2	1.23	31.8	1.25
5/8	48.5	1.91	42.2	1.66	36.6	1.44	42.2	1.66	32.3	1.27	38.1	1.50

B Nom Tube OD in	N₁ mm ±0.8	N₁ in ±0.03	N₂ mm ±0.8	N₂ in ±0.03	P Min mm	P Min in	Q[e] mm	Q[e] in	S[d] Max mm	S[d] Max in
5/8	15.7	0.62	20.1	0.79	9.7	0.38	—	—	27.2	1.07
5/8	20.6	0.81	24.9	0.98	14.2	0.56	—	—	27.2	1.07
5/8	20.6	0.81	24.9	0.98	14.2	0.56	—	—	27.2	1.07
5/8	25.4	1.00	30.5	1.20	19.0	0.75	—	—	24.6	0.97

B Nom Tube OD in	T[f] Ref mm	T[f] Ref in	T₁ Min mm	T₁ Min in	T₂ Min mm	T₂ Min in	W[g] mm +0.0 -0.5	W[g] in +0.00 -0.02	X[g] mm +0.0 -0.5	X[g] in +0.00 -0.02
5/8	9.4	0.37	9.4	0.37	10.9	0.43	22.4	0.88	26.9	1.06
5/8	9.4	0.37	9.4	0.37	10.9	0.43	22.4	0.88	26.9	1.06
5/8	9.4	0.37	9.4	0.37	10.9	0.43	22.4	0.88	26.9	1.06
5/8	10.9	0.43	12.4	0.49	10.9	0.43	31.8	1.25	26.9	1.06

B Nom Tube OD in	A Dryseal Pipe Thread NPTF[b]	C Hex in Nom	C₁ Hex in Nom	C₂ Hex in Nom	D[d] Drill mm	D[d] Drill in	D₁[d] Drill mm	D₁[d] Drill in	D₂ Drill mm	D₂ Drill in	L mm ±0.8	L in ±0.03
3/4	1/8	1-1/16	1-1/16	1-5/16	15.88	0.625	5.56	0.219	—	—	46.0	1.81
3/4	1/4	1-1/16	1-1/16	1-5/16	15.88	0.625	7.92	0.312	—	—	50.8	2.00
3/4	3/8	1-1/16	1-1/16	1-5/16	15.88	0.625	10.31	0.406	—	—	50.8	2.00
3/4	3/4	1-1/16	1-1/4	1-5/16	15.88	0.625	19.05	0.750	—	—	55.6	2.19

TABLE 6—(Continued)

B Nom Tube OD in	L_1 mm ±0.8	L_1 in ±0.03	L_2 mm ±0.8	L_2 in ±0.03	M mm ±0.8	M in ±0.03	M_1 mm ±0.8	M_1 in ±0.03	M_2 mm ±0.8	M_2 in ±0.03	N mm ±0.8	N in ±0.03
3/4	35.1	1.38	41.1	1.62	41.1	1.62	40.4	1.59	35.8	1.41	31.0	1.22
3/4	39.6	1.56	44.4	1.75	41.1	1.62	40.4	1.59	35.8	1.41	35.8	1.41
3/4	42.2	1.66	45.2	1.78	41.1	1.62	40.4	1.59	35.8	1.41	35.8	1.41
3/4	50.0	1.97	45.2	1.78	40.4	1.59	45.2	1.78	35.8	1.41	41.1	1.62

B Nom Tube OD in	N_1 mm ±0.8	N_1 in ±0.03	N_2 mm ±0.8	N_2 in ±0.03	P Min mm	P Min in	Q^e mm	Q^e in	S^d Max mm	S^d Max in
3/4	17.5	0.69	20.8	0.82	9.7	0.38	—	—	31.0	1.22
3/4	22.4	0.88	25.7	1.01	14.2	0.56	—	—	31.0	1.22
3/4	22.4	0.88	25.7	1.01	14.2	0.56	—	—	31.0	1.22
3/4	26.9	1.06	30.5	1.20	19.0	0.75	—	—	24.6	0.97

B Nom Tube OD in	T^f Ref mm	T^f Ref in	T_1 Min mm	T_1 Min in	T_2 Min mm	T_2 Min in	W^g mm +0.0 −0.5	W^g in +0.00 −0.02	X^g mm +0.0 −0.5	X^g in +0.00 −0.02
3/4	10.9	0.43	10.9	0.43	12.4	0.49	26.9	1.06	33.3	1.31
3/4	10.9	0.43	10.9	0.43	12.4	0.49	26.9	1.06	33.3	1.31
3/4	10.9	0.43	10.9	0.43	12.4	0.49	26.9	1.06	33.3	1.31
3/4	10.9	0.43	12.4	0.49	12.4	0.49	31.8	1.25	33.3	1.31

[a] For flare dimensions shown on Figures 26 to 32 but not given in Table 6, see corresponding dimensions for the specified Tube OD in Table 3.
[b] Dryseal American Standard Taper Pipe Thread.
[c] Where thread relief undercut is used, last thread shall be chamfered 1/2 to 1 pitch long from G diameter and dimension J may be reduced by an amount equal to 1/2 pitch.
[d] At manufacturer's option, through passages in fittings shown in Figure 26 may conform with the smaller diameter specified or be counterbored to the larger diameter from the appropriate end for depth S.
[e] Available with three types of fusible alloys as specified in General Specifications.
[f] Minimum design thickness, not subject to inspection.
[g] Basic dimensions shown shall apply as minimum for bosses. The −0.51 mm (−0.02 in) tolerance shall apply only to chamfer diameters on full hexagon versions of fittings shown in Figures 28 and 29.

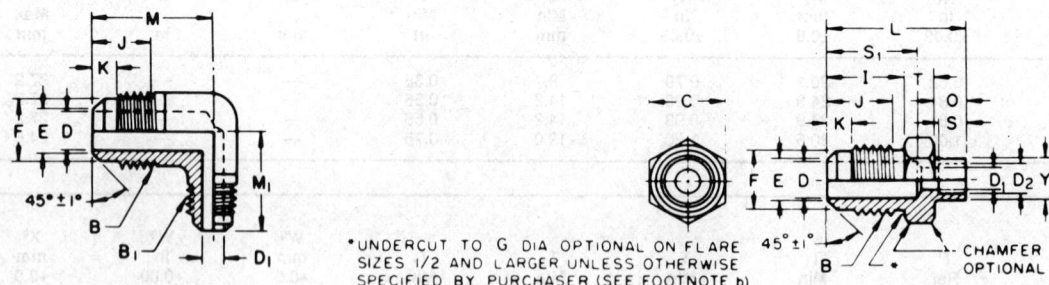

FIGURE 23—90 DEGREE REDUCING ELBOW UNION (010201) (ER2)

FIGURE 24—FLARE TO SOLDER REDUCING CONNECTOR (HALF UNION) (010104) (US3)

NOTE—UNSPECIFIED DETAIL WITH RESPECT TO DIMENSIONS, TOLERANCES, CONTOUR, MATERIAL, WORKMANSHIP, ETC., MUST CONFORM TO GENERAL SPECIFICATIONS FOR REFRIGERATION TUBE FITTINGS. THE ILLUSTRATIONS ON THIS PAGE APPLY TO TABLE 4. CODES SHOWN IN BRACKETS ADJACENT TO FIGURE NUMBERS REPRESENT RESPECTIVE FITTING IDENTIFICATION IN ACCORDANCE WITH SAE J846 (FIRST NUMBER) AND ANSI B70.1 (SECOND NUMBER).

TABLE 7—END TO CENTER DIMENSIONS OF FLARE TO FLARE ENDS ON REDUCING TEES[a]

B and B_1 Tube OD of Run in	End to Center ±0.8 mm ±0.03 in	B_2 Nominal Flare Sizes of Branch in 3/16 mm	B_2 Nominal Flare Sizes of Branch in 3/16 in	B_2 Nominal Flare Sizes of Branch in 1/4 mm	B_2 Nominal Flare Sizes of Branch in 1/4 in	B_2 Nominal Flare Sizes of Branch in 5/16 mm	B_2 Nominal Flare Sizes of Branch in 5/16 in
3/16	M or M_1	19.0	0.75	19.0	0.75	19.8	0.78
3/16	M_2	19.0	0.75	22.4	0.88	23.1	0.91
1/4	M or M_1	22.4	0.88	22.4	0.88	21.3	0.84
1/4	M_2	19.0	0.75	22.4	0.88	23.1	0.91
5/16	M or M_1	23.1	0.91	23.1	0.91	23.1	0.91
5/16	M_2	19.8	0.78	21.3	0.84	23.1	0.91
3/8	M or M_1	26.9	1.06	26.9	1.06	26.9	1.06
3/8	M_2	21.3	0.84	23.1	0.91	24.6	0.97
1/2	M or M_1	31.0	1.22	31.0	1.22	31.0	1.22
1/2	M_2	23.1	0.91	24.6	0.97	26.2	1.03
5/8	M or M_1	35.8	1.41	35.8	1.41	35.8	1.41
5/8	M_2	24.6	0.97	26.2	1.03	27.7	1.09
3/4	M or M_1	42.2	1.66	42.2	1.66	42.2	1.66
3/4	M_2	26.9	1.06	28.4	1.12	30.2	1.19

B and B_1 Tube OD of Run in	End to Center ±0.8 mm	B_2 Nominal Flare Sizes of Branch in 3/8 mm	B_2 Nominal Flare Sizes of Branch in 3/8 in	B_2 Nominal Flare Sizes of Branch in 1/2 mm	B_2 Nominal Flare Sizes of Branch in 1/2 in	B_2 Nominal Flare Sizes of Branch in 5/8 mm	B_2 Nominal Flare Sizes of Branch in 5/8 in	B_2 Nominal Flare Sizes of Branch in 3/4 mm	B_2 Nominal Flare Sizes of Branch in 3/4 in
3/16	M or M_1	21.3	0.84	23.1	0.91	24.6	0.97	26.9	1.06
3/16	M_2	26.9	1.06	31.0	1.22	35.8	1.41	42.2	1.66
1/4	M or M_1	23.1	0.91	24.6	0.97	26.2	1.03	28.4	1.12
1/4	M_2	26.9	1.06	31.0	1.22	35.8	1.41	42.2	1.66
5/16	M or M_1	24.6	0.97	26.2	1.03	27.7	1.09	30.2	1.19
5/16	M_2	26.9	1.06	31.0	1.22	35.8	1.41	42.2	1.66
3/8	M or M_1	26.9	1.06	27.7	1.09	29.5	1.16	31.8	1.25
3/8	M_2	26.9	1.06	31.0	1.22	35.8	1.41	42.2	1.66
1/2	M or M_1	31.0	1.22	31.0	1.22	32.5	1.28	35.1	1.38
1/2	M_2	27.7	1.09	31.0	1.22	35.8	1.41	42.2	1.66
5/8	M or M_1	35.8	1.41	35.8	1.41	35.8	1.41	38.1	1.50
5/8	M_2	29.5	1.16	32.5	1.28	35.8	1.41	42.2	1.66
3/4	M or M_1	42.2	1.66	42.2	1.66	42.2	1.66	42.2	1.66
3/4	M_2	31.8	1.25	35.1	1.38	38.1	1.50	42.2	1.66

[a] For flare and pipe thread dimensions shown on Figures 33 to 35, see corresponding dimensions for specified Tube OD and specified pipe thread size in Table 3. For passage diameters, see Tables 9 and 10.

TABLE 8—END TO CENTER DIMENSIONS OF FLARE TO PIPE ENDS ON REDUCING TEES[a]

B B_1 or B_2 Tube OD in	End to Center ±0.8 mm ±0.03 in	A_1 Dryseal Taper Thread NPTF[b] in 1/8 mm	A_1 Dryseal Taper Thread NPTF[b] in 1/8 in	A_1 Dryseal Taper Thread NPTF[b] in 1/4 mm	A_1 Dryseal Taper Thread NPTF[b] in 1/4 in	A_1 Dryseal Taper Thread NPTF[b] in 3/8 mm	A_1 Dryseal Taper Thread NPTF[b] in 3/8 in	A_1 Dryseal Taper Thread NPTF[b] in 1/2 mm	A_1 Dryseal Taper Thread NPTF[b] in 1/2 in	A_1 Dryseal Taper Thread NPTF[b] in 3/4 mm	A_1 Dryseal Taper Thread NPTF[b] in 3/4 in
3/16	M_3	19.0	0.75	20.6	0.81	22.4	0.88	24.6	0.97	26.9	1.06
3/16	N	19.0	0.75	24.6	0.97	25.4	1.00	32.5	1.28	35.1	1.38
1/4	M_3	20.6	0.81	23.1	0.91	23.9	0.94	26.2	1.03	28.4	1.12
1/4	N	19.8	0.78	23.9	0.94	26.2	1.03	32.5	1.28	35.1	1.38
5/16	M_3	23.1	0.91	24.6	0.97	25.4	1.00	27.7	1.09	30.2	1.19
5/16	N	19.8	0.78	23.9	0.94	26.2	1.03	32.5	1.28	35.1	1.38
3/8	M_3	26.2	1.03	25.4	1.00	26.9	1.06	29.5	1.16	31.8	1.25
3/8	N	23.1	0.91	26.9	1.06	27.7	1.09	32.5	1.28	35.1	1.38
1/2	M_3	31.0	1.22	31.0	1.22	31.0	1.22	32.5	1.28	35.1	1.38
1/2	N	25.4	1.00	30.2	1.19	28.4	1.12	35.1	1.38	35.1	1.38
5/8	M_3	35.8	1.41	35.8	1.41	35.8	1.41	35.8	1.41	36.6	1.44
5/8	N	26.9	1.06	31.8	1.25	31.8	1.25	35.1	1.38	38.1	1.50
3/4	M_3	41.1	1.62	41.1	1.62	41.1	1.62	41.1	1.62	40.4	1.59
3/4	N	31.0	1.22	35.8	1.41	35.8	1.41	38.1	1.50	41.1	1.62

[a] For flare and pipe thread dimensions shown on Figures 33 to 35, see corresponding dimensions for specified Tube OD and specified pipe thread size in Table 3. For passage diameters, see Tables 9 and 10.
[b] Dryseal American Standard Taper Pipe Thread.

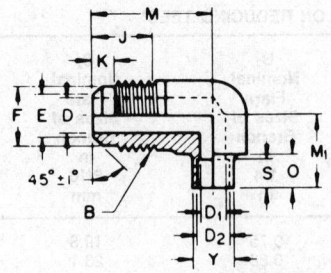

FIGURE 25—FLARE TO SOLDER 90 DEGREE REDUCING ELBOW (010204) (ES)

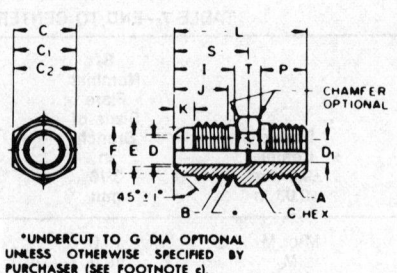

FIGURE 26—REDUCING CONNECTOR (HALF UNION) (010102) (U1)

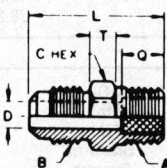

FIGURE 27—FUSIBLE REDUCING CONNECTOR (HALF UNION) (010163) (FU)

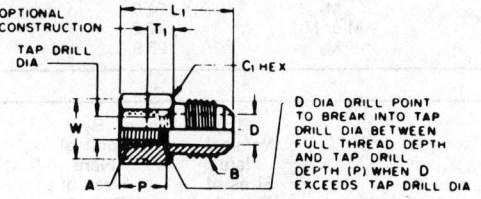

FIGURE 28—INTERNAL THREAD REDUCING CONNECTOR (HALF UNION) (010103) (U3)

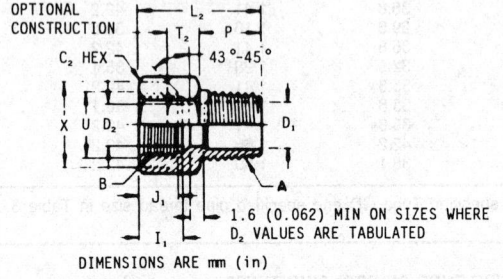

FIGURE 29—INTERNAL FLARE TO EXTERNAL PIPE REDUCING ADAPTOR (010106) (U5)

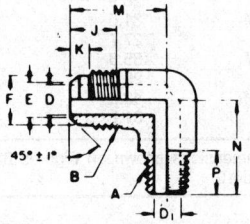

FIGURE 30—90 DEGREE REDUCING ELBOW (010202) (E1)

NOTE—UNSPECIFIED DETAIL WITH RESPECT TO DIMENSIONS, TOLERANCES, CONTOUR, MATERIAL, WORKMANSHIP, ETC., MUST CONFORM TO GENERAL SPECIFICATIONS FOR REFRIGERATION TUBE FITTINGS. THE DIMENSIONAL DESIGNATIONS IN FIGURES 1 AND 30 SHALL APPLY TO CORRESPONDING FEATURES OF OTHER FIGURES ON THIS PAGE UNLESS SHOWN OTHERWISE. THE ILLUSTRATIONS ON THIS PAGE APPLY TO TABLE 6. CODES SHOWN IN BRACKETS ADJACENT TO FIGURE NUMBERS REPRESENT RESPECTIVE FITTING IDENTIFICATION IN ACCORDANCE WITH SAE J846 (FIRST NUMBER) AND ANSI B70.1 (SECOND NUMBER).

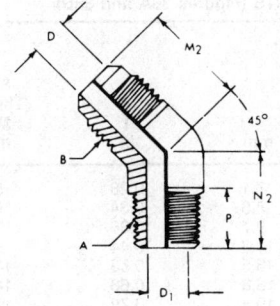

FIGURE 31—45 DEGREE REDUCING ELBOW (010302) (E5)

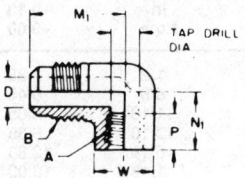

FIGURE 32—90 DEGREE INTERNAL PIPE THREAD REDUCING ELBOW (010203) (E3)

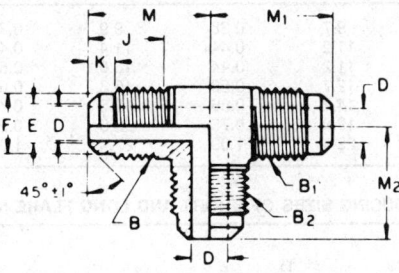

FIGURE 33—THREE-WAY REDUCING TEE (010401) (TR2)

Note—Unspecified detail with respect to dimensions, tolerances, contour, material, workmanship, etc., must conform to General Specifications for Refrigeration Tube Fittings. The illustrations on this page apply to Tables 7 to 10. Codes shown in brackets adjacent to figure numbers represent respective fitting identification in accordance with SAE J846 (first number) and ANSI B70.1 (second number).

TABLE 9—PASSAGE DIAMETERS THROUGH FLARE ENDS

Nom Tube OD in	D^a Dia Drill mm	D^a Dia Drill in
3/16	3.18	0.125
1/4	4.78	0.188
5/16	5.56	0.219
3/8	7.14	0.281
1/2	10.31	0.406
5/8	12.70	0.500
3/4	15.88	0.625

[a] At manufacturer's option, through passages in tees shown in Figures 33 to 35 having varying diameters at opposite ends may conform to the smaller diameters specified or be counterbored to the larger diameter from the appropriate end for a minimum depth equivalent to the maximum end to center length of that end, plus 1/2 the maximum passage through brand plus 0.3 mm (0.01 in).

TABLE 10—PASSAGE DIAMETERS THROUGH PIPE THREAD ENDS

Nom Pipe Size in	D^a Dia Drill mm	D^a Dia Drill in
1/8	5.56	0.219
1/4	7.92	0.312
3/8	10.31	0.406
1/2	14.27	0.562
3/4	19.05	0.750

[a] At manufacturer's option, through passages in tees shown in Figures 33 to 35 having varying diameters at opposite ends may conform to the smaller diameters specified or be counterbored to the larger diameter from the appropriate end for a minimum depth equivalent to the maximum end to center length of that end, plus 1/2 the maximum passage through branch plus 0.3 mm (0.01 in).

TABLE 11—DIMENSIONS FOR STANDARD SIZES OF SHORT AND LONG FLARE NUTS (Figures 36A and 36B)

Nom Tube OD in	B Straight Thread Nom Size in	C Hex in Nom	D Dia mm +0.13 -0.00	D Dia in +0.005 -0.000	E Dia Min mm	E Dia Min in	I mm	I in	J Full Thread Min mm	J Full Thread Min in
3/16	3/8 -24	1/2	4.88	0.192	10.4	0.41	7.1	0.28	5.6	0.22
1/4	7/16-20	5/8	6.48	0.255	11.9	0.47	8.6	0.34	6.9	0.27
5/16	1/2 -20	11/16	8.05	0.317	11.9	0.47	9.7	0.38	7.6	0.30
3/8	5/8 -18	13/16	9.65	0.380	15.0	0.59	11.2	0.44	8.6	0.34
1/2	3/4 -16	15/16	12.83	0.505	19.0	0.75	13.5	0.53	11.2	0.44
5/8	7/8 -14	1- 1/16	16.00	0.630	23.9	0.94	16.8	0.66	14.0	0.55
3/4	1-1/16-14	1- 5/16	19.18	0.755	28.4	1.12	19.8	0.78	17.0	0.67

Nom Tube OD in	L mm ±0.8	L in ±0.03	L_1 mm +2.3 -0.0	L_1 in +0.09 -0.00	T mm ±0.8	T in ±0.03	U Dia Min mm	U Dia Min in	U Dia Max mm	U Dia Max in	X mm +0.0 -0.8	X in +0.00 -0.03
3/16	22.4	0.88	13.5	0.53	9.7	0.38	9.9	0.39	10.4	0.41	12.7	0.50
1/4	23.9	0.94	15.0	0.59	11.2	0.44	11.4	0.45	11.9	0.47	15.7	0.62
5/16	23.9	0.94	15.7	0.62	11.2	0.44	13.0	0.51	13.5	0.53	17.5	0.69
3/8	26.9	1.06	17.5	0.69	12.7	0.50	16.3	0.64	17.0	0.67	20.6	0.81
1/2	30.2	1.19	20.6	0.81	14.2	0.56	19.6	0.77	20.3	0.80	23.9	0.94
5/8	36.6	1.44	23.9	0.94	19.0	0.75	22.9	0.90	23.6	0.93	26.9	1.06
3/4	44.5	1.75	28.4	1.12	25.4	1.00	27.4	1.08	28.2	1.11	33.3	1.31

TABLE 12—DIMENSIONS FOR REDUCING SIZES OF SHORT AND LONG FLARE NUTS (Figures 36A and 36B)

Nom Tube OD in	Nom Tube OD in	B Straight Thread Nom Size in	C Hex in Nom	D Dia mm +0.13 -0.00	D Dia in +0.005 -0.000	E Dia Min mm	E Dia Min in	I mm	I in	J Full Thread Min mm	J Full Thread Min in
1/4	3/16	7/16-20	5/8	4.88	0.192	10.4	0.41	8.6	0.34	6.9	0.27
5/16	3/16	1/2 -20	11/16	4.88	0.192	10.4	0.41	9.7	0.38	7.6	0.30
5/16	1/4	1/2 -20	11/16	6.48	0.255	11.9	0.47	9.7	0.38	7.6	0.30
3/8	1/4	5/8 -18	13/16	6.48	0.255	11.9	0.47	11.2	0.44	8.6	0.34
3/8	5/16	5/8 -18	13/16	8.05	0.317	11.9	0.47	11.2	0.44	8.6	0.34
1/2	3/8	3/4 -16	15/16	9.65	0.380	15.0	0.59	13.5	0.53	11.2	0.44
5/8	3/8	7/8 -14	1- 1/16	9.65	0.380	15.0	0.59	16.8	0.66	14.0	0.55
5/8	1/2	7/8 -14	1- 1/16	12.83	0.505	19.0	0.75	16.8	0.66	14.0	0.55
3/4	1/2	1-1/16-14	1- 5/16	12.83	0.505	19.0	0.75	19.8	0.78	17.0	0.67
3/4	5/8	1-1/16-14	1- 5/16	16.00	0.630	23.9	0.94	19.8	0.78	17.0	0.67

Nom Tube OD in	L mm ±0.8	L in ±0.03	L_1 mm +2.3 -0.0	L_1 in +0.09 -0.00	T mm ±0.8	T in ±0.03	U Dia Min mm	U Dia Min in	U Dia Max mm	U Dia Max in	X mm +0.8 -0.0	X in +0.03 -0.00
1/4	23.9	0.94	15.0	0.59	11.2	0.44	11.4	0.45	11.9	0.47	15.7	0.62
5/16	23.9	0.94	—	—	11.2	0.44	13.0	0.51	13.5	0.53	17.5	0.69
5/16	23.9	0.94	15.8	0.62	11.2	0.44	13.0	0.51	13.5	0.53	17.5	0.69
3/8	26.9	1.06	17.5	0.69	12.7	0.50	16.3	0.64	17.0	0.67	20.6	0.81
3/8	26.9	1.06	17.5	0.69	12.7	0.50	16.3	0.64	17.0	0.67	20.6	0.81
1/2	30.2	1.19	20.6	0.81	14.2	0.56	19.6	0.77	20.3	0.80	23.9	0.94
5/8	36.6	1.44	—	—	19.0	0.75	22.9	0.90	23.6	0.93	26.9	1.06
5/8	36.6	1.44	23.9	0.94	19.0	0.75	22.9	0.90	23.6	0.93	26.9	1.06
3/4	44.4	1.75	28.4	1.12	25.4	1.00	27.4	1.08	28.2	1.11	33.3	1.31
3/4	44.4	1.75	28.4	1.12	25.4	1.00	27.4	1.08	28.2	1.11	33.3	1.31

22.55

To determine correct end to center lengths on tees, each 90 degrees must be figured seperately as an elbow and the larger of the two branch lengths shall apply. See examples below.

Example: Find lengths for 5/8 × 1/2 × 3/8 in three-way tee.
1. From Table 7 obtain values for each 90 degrees separately.
2. Use larger of two M_2 dimensions as found.

Example: Find lengths for 5/8 × 3/8 × 1/2 in right angle two-way tee.
1. From Tables 7 and 8 obtain values for each 90 degrees separately.
2. Use the larger dimension M_2 or M_3 as found.

FROM TABLE 7

B Nominal Tube OD in	End to Center	B_2 3/8 mm	B_2 3/8 in
5/8	M	35.8	1.41
5/8	M_2	29.5	1.16

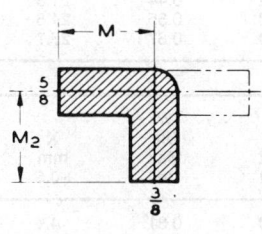

FROM TABLE 7

B Nominal Tube OD in	End to Center	B_2 1/2 mm	B_2 1/2 in
5/8	M	35.8	1.41
5/8	M_2	32.5	1.28

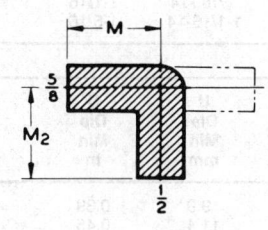

FROM TABLE 7

B_1 Nominal Tube OD in	End to Center	B_2 3/8 mm	B_2 3/8 in
1/2	M_1	31.0	1.22
1/2	M_2	27.7	1.09

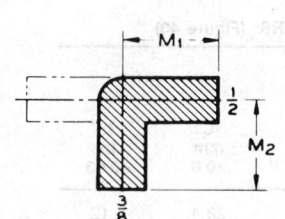

FROM TABLE 8

B_2 Nominal Tube OD in	End to Center	A 3/8 mm	A 3/8 in
1/2	M_3	31.0	1.22
1/2	N	28.4	1.22

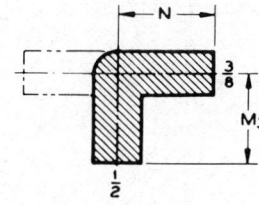

Result:

Dimension	mm	in
M	35.8	1.41
N	28.4	1.12
M_2	32.5	1.28

Result:

Dimension	mm	in
M	35.8	1.41
M_1	31.0	1.22
M_2	29.5	1.16

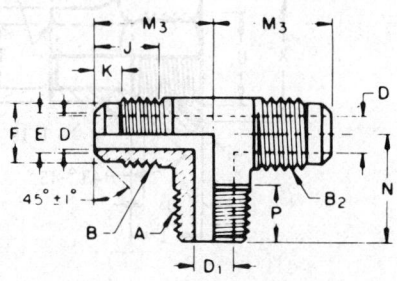

FIGURE 34—TWO-WAY REDUCING TEE (010425) (T1)

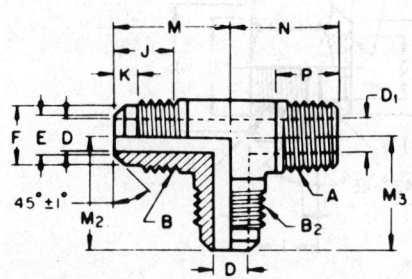

FIGURE 35—RIGHT ANGLE TWO-WAY REDUCING TEE (010424) (T3)

Note—Unspecified detail with respect to dimensions, tolerances, contour, material, workmanship, etc., must conform to General Specifications for Refrigeration Tube Fittings. Codes shown in brackets adjacent to figure numbers represent respective fitting identification in accordance with SAE J846 (first number) and ANSI B70.1 (second number).

22.56

TABLE 13—DIMENSIONS OF FLARE CAP, FLARE GASKET, AND FLARE SEAL BONNET (Figures 37 to 39)

Nom Tube OD in	B Straight Thread Nom Size Class 2B Int	C Hex in Nom	D Dia mm	D Dia in	I mm	I in	J Full Thread Min mm	J Full Thread Min in	L mm	L in	T Min mm	T Min in
3/16	3/8 -24	1/2	3.0	0.12	7.1	0.28	5.6	0.22	11.9	0.47	6.9	0.27
1/4	7/16-20	9/16	4.3	0.19	8.6	0.34	6.9	0.27	13.5	0.53	10.2	0.40
5/16	1/2 -20	5/8	5.6	0.22	9.7	0.38	7.6	0.30	15.7	0.62	10.2	0.40
3/8	5/8 -18	3/4	7.1	0.28	11.2	0.44	.8.6	0.34	17.5	0.69	11.7	0.46
1/2	3/4 -16	7/8	10.4	0.41	13.5	0.53	11.2	0.44	21.3	0.84	13.2	0.52
5/8	7/8 -14	1-1/16	12.7	0.50	16.8	0.66	14.0	0.55	24.6	0.97	15.7	0.62
3/4	1-1/16-14	1-5/16	15.7	0.62	19.8	0.78	17.0	0.67	27.7	1.09	18.8	0.74

Nom Tube OD in	U Dia Min mm	U Dia Min in	U Dia Max mm	U Dia Max in	V Dia mm ±0.8	V Dia in ±0.03	W Dia Max mm	W Dia Max in	X mm ±0.5	X in ±0.02	Y Min mm	Y Min in
3/16	9.9	0.39	10.4	0.41	3.0	0.12	7.9	0.31	4.1	0.16	4.1	0.16
1/4	11.4	0.45	11.9	0.47	4.8	0.19	9.1	0.36	5.6	0.22	5.6	0.22
5/16	13.0	0.51	13.5	0.53	5.6	0.22	10.7	0.42	7.1	0.28	7.1	0.28
3/8	16.3	0.64	17.0	0.67	7.1	0.28	14.0	0.55	9.1	0.36	8.6	0.34
1/2	19.6	0.77	20.3	0.80	10.4	0.41	16.8	0.66	11.9	0.47	10.4	0.41
5/8	22.9	0.90	23.6	0.93	12.7	0.50	19.6	0.77	15.5	0.61	11.2	0.44
3/4	27.4	1.08	28.2	1.11	15.7	0.62	24.1	0.95	18.3	0.72	14.2	0.56

TABLE 14—DIMENSIONS OF DRUM ADAPTERS[a] (Figure 40)

Nom Tube OD in	A Straight Internal Pipe Thread NPSM[b]	C Hex in Nom	D Drill mm	D Drill in	L mm ±0.8	L in ±0.03
1/4	1/2	1-1/8	4.78	0.188	28.4	1.12
1/4	3/4	1-1/4	4.78	0.188	28.4	1.12
3/8	3/4	1-1/4	71.4	0.281	31.8	1.25
1/2	3/4	1-1/4	10.31	0.406	35.1	1.38

[a] For flare and dimensions shown on Figure 40 but not specified in Table 14, see corresponding dimensions for the specified Tube OD in Table 3. Drum adaptor fittings are normally supplied with seal gasket for pipe thread end.
[b] American Standard Straight Pipe Thread for Mechanical Joints.
[c] Where thread relief undercut is used, last thread shall be chamfered 1/2 to 1 pitch long from G diameter and dimension J may be reduced by an amount equal to 1/2 pitch.

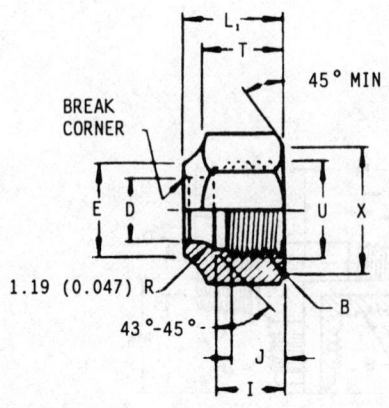

FIGURE 36A—SHORT NUTS (010166) (NS4) (010168) (NRS4)

DIMENSIONS ARE mm (in)

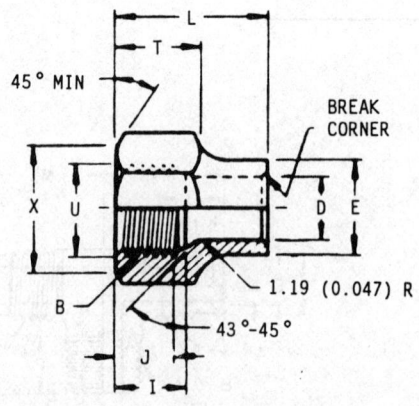

FIGURE 36B—LONG NUTS (010167) (N4) (010169) (NR4)

FIGURE 36—FLARE NUTS AND REDUCING FLARE NUTS

TABLE 15—DIMENSIONS OF FUSIBLE PIPE PLUGS[a] (Figure 41)

A Dryseal Pipe Thread NPTF[b]	C Hex in Nom	D Dia mm	D Dia in	L mm	L in	T mm	T in
1/8	7/16	5.6	0.219	14.2	0.56	4.8	0.19
1/4	9/16	6.4	0.250	19.0	0.75	4.8	0.19
3/8	3/4	9.5	0.375	19.8	0.78	5.6	0.22

[a] Plugs are available with three types of fusible alloys as specified in general specifications.
[b] Dryseal American Standard Taper Pipe Thread.

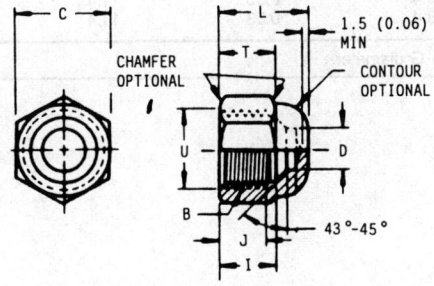

FIGURE 37—FLARE CAP (010112) (N5) (010113) (B2)

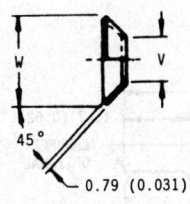

FIGURE 38—FLARE

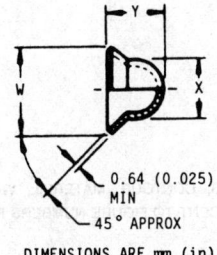

FIGURE 39—FLARE SEAL GASKET BONNET (010114) (B1)

TABLE 16—DIMENSIONS OF CAPILLARY TUBE CONNECTIONS (Figure 42)

Nom Tube OD in	B Straight Thread Nominal Size in	C Dia in Nom	D Dia mm ±0.13	D Dia in ±0.005	D Drill mm ±0.13	D Drill in ±0.005	D_1 Drill mm ±0.05	D_1 Drill in ±0.002
0.081	5/16-24	11/64	4.7	0.187	2.6	0.104	2.16	0.085
0.093	5/16-24	11/64	4.7	0.187	2.6	0.104	2.46	0.097

Nom Tube OD in	E Dia mm ±0.08	E Dia in ±0.003	G[a] Dia mm +0.00 −0.25	G[a] Dia in +0.000 −0.010	U Dia mm +0.5 −0.0	U Dia in +0.02 −0.00	U_1 Dia mm +0.5 −0.0	U_1 Dia in +0.02 −0.00
0.081	2.67	0.105	6.35	0.250	8.4	0.33	11.4	0.45
0.093	2.67	0.105	6.35	0.250	8.4	0.33	11.4	0.45

[a] Where thread relief undercut is used, the last thread shall be chamfered 1/2 to 1 pitch thread long from the G diameter.

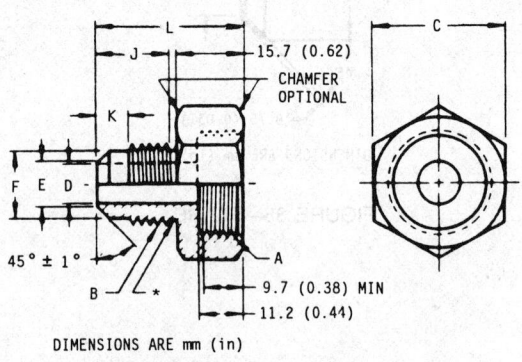

FIGURE 40—DRUM ADAPTOR (010165) (K)

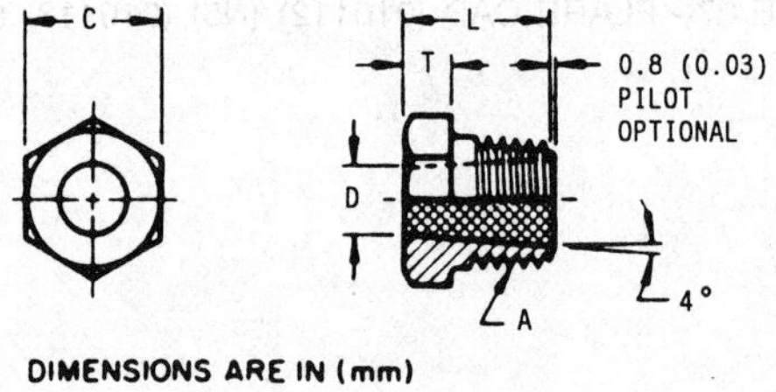

FIGURE 41—FUSIBLE PIPE PLUG (010164) (FP)

NOTE—UNSPECIFIED DETAIL WITH RESPECT TO DIMENSIONS, TOLERANCES, CONTOUR, MATERIAL, WORKMANSHIP, ETC., MUST CONFORM TO GENERAL SPECIFICATIONS FOR REFRIGERATION TUBE FITTINGS. CODES SHOWN IN BRACKETS ADJACENT TO FIGURE NUMBERS REPRESENT RESPECTIVE FITTING IDENTIFICATION IN ACCORDANCE WITH SAE J846 (FIRST NUMBER) AND ANSI B70.1 (SECOND NUMBER).

TABLE 17—DIMENSIONS AND TOLERANCES

Drill Size Range mm	Drill Size Range in	Tolerance Plus mm	Tolerance Plus in	Tolerance Minus mm	Tolerance Minus in
0.343 thru 4.699	0.0135 thru 0.1850	0.08	0.003	0.05	0.002
4.762 thru 6.299	0.1875 thru 0.2480	0.10	0.004	0.05	0.002
6.350 thru 19.050	0.2500 thru 0.7500	0.15	0.006	0.08	0.003
19.25 thru 25.400	0.7579 thru 1.0000	0.18	0.007	0.10	0.004

TABLE 18—WALL THICKNESS

Nominal Tube OD in	Wall Thickness Min Extruded Shapes[a] mm	Wall Thickness Min Extruded Shapes[a] in	Wall Thickness Min Forged Shapes mm	Wall Thickness Min Forged Shapes in
3/16	1.0	0.04	1.52	0.060
1/4	1.0	0.04	1.90	0.075
5/16	1.3	0.05	1.90	0.075
3/8	1.3	0.05	2.16	0.085
1/2	1.5	0.06	2.29	0.090
5/8	2.0	0.08	2.54	0.100
3/4	2.0	0.08	2.54	0.100

[a] Applies to reduction in one plane of shape only.

22.59

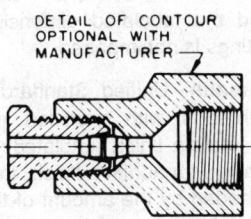

FIGURE 42A—CAPILLARY TUBE CONNECTION ASSEMBLY (450101BA) (CTN)

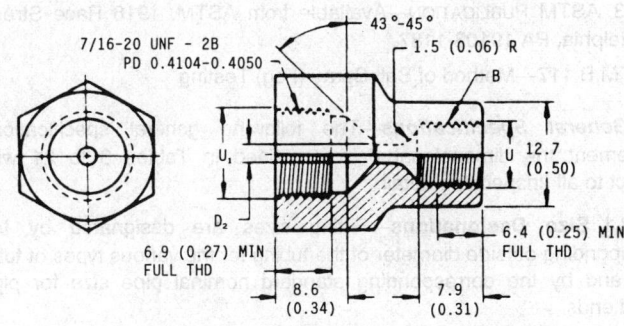

FIGURE 42B—CAPILLARY TUBE BODY (450101) (CTN)

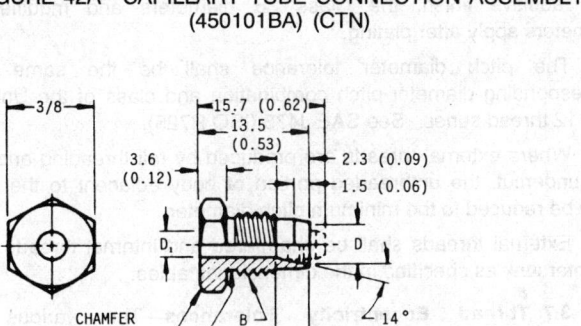

FIGURE 42C—CAPILLARY TUBE (450115) (CTN)

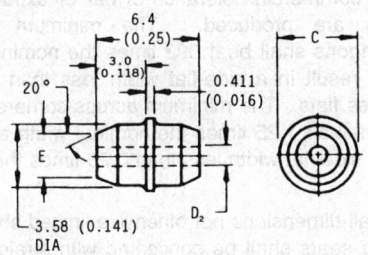

FIGURE 42D—CAPILLARY TUBE SLEEVE COMPRESSION SCREW (450110) (CTN)

FIGURE 42—CAPILLARY TUBE CONNECTION

NOTE—UNSPECIFIED DETAIL WITH RESPECT TO DIMENSIONS, TOLERANCES, CONTOUR, MATERIAL, WORKMANSHIP, ETC., MUST CONFORM TO GENERAL SPECIFICATIONS FOR REFRIGERATION TUBE FITTINGS. CODES SHOWN IN BRACKETS ADJACENT TO FIGURE NUMBERS REPRESENT RESPECTIVE FITTING IDENTIFICATION IN ACCORDANCE WITH SAE J846 (FIRST NUMBER) AND ANSI B70.1 (SECOND NUMBER).

HYDRAULIC TUBE FITTINGS—SAE J514 JUN93   SAE Standard

Report of the Construction and Industrial Machinery Technical Committee and the Tube, Pipe, Hose, and Lubrication Fittings Committee approved May 1950. Completely revised by the Fluid Conductors and Connectors Technical Committee May 1978, editorial change April 1980, and completely revised June 1990. Completely revised by the Fluid Conductors and Connectors Technical Committee SC1—Automotive and Hydraulic Tube and Fittings June 1991, revised June 1992, and revised June 1993. Rationale statement available.

1. Scope—This SAE Standard covers complete general and dimensional specifications for 37 degree flared and flareless types of hydraulic tube fittings and O-ring plugs. Also included are pipe fittings and adapter unions for use in conjunction with these tube fittings. These fittings are intended for general application in hydraulic systems on industrial equipment and commercial products.

These fittings are capable of providing leak-proof, full-flow connections in hydraulic systems operating at working pressures as specified in SAE J1065.

Since many factors influence the pressure at which a hydraulic system will or will not perform satisfactorily, the values shown in SAE J1065 should not be construed as a guaranteed minimum.

For any application, it is recommended that sufficient testing be conducted and reviewed by both the user and fitting manufacturer to assure that performance levels will be safe and satisfactory.

The standard is divided into six sections as follows:
Section 1—37 Degree Flare Tube Fittings
Section 2—Flareless Tube Fittings
Section 3—O-ring Plugs (for O-ring Ports see SAE J1926)
Section 4—Hydraulic Pipe Fittings (formerly SAE J926)
Section 5—Adapter Unions (formerly in SAE J516)
Section 6—Tables for Calculating Dimensions on Special Sizes

2. References

2.1 Applicable Documents—The following publications form a part of this specification to the extent specified herein. The latest issue of SAE publications shall apply.

2.1.1 SAE PUBLICATIONS—Available from SAE, 400 Commonwealth Drive, Warrendale, PA 15096-0001.

SAE J405—Chemical Compositions of SAE Wrought Stainless Steels
SAE J476—Dryseal Pipe Threads
SAE J512—Automotive Tube Fittings
SAE J533—Flares for Tubing
SAE J1065—Pressure Ratings for Hydraulic Tubing and Fittings

2.1.2 ANSI PUBLICATION—Available from ANSI, 11 West 42nd Street, New York, NY 10036-8002.

ANSI B1.20.3—Dryseal Pipe Threads (Inch)

2.1.3 ASTM PUBLICATION—Available from ASTM, 1916 Race Street, Philadelphia, PA 19103-1187.

ASTM B 117—Method of Salt Spray (Fog) Testing

3. General Specifications—The following general specifications supplement the dimensional data contained in Tables 3 to 21 with respect to all unspecified detail.

3.1 Size Designations—Fitting sizes are designated by the corresponding outside diameter of the tubing for the various types of tube ends and by the corresponding standard nominal pipe size for pipe thread ends.

See SAE J846 for proper coding and call-out.

3.2 Dimensions and Tolerances—Except for nominal sizes and thread specifications, dimensions and tolerances are given in both SI Units and U.S. Customary as designated. Tabulated dimensions shall apply to the finished parts, plated or otherwise processed, as specified by the purchasers. The maximum and minimum across-flat dimensions shall be within the commercial tolerance of bar or extruded stock from which the fittings are produced. The minimum across-corners dimensions of hexagons shall be 1.092 times the nominal width across flats, but shall not result in a side-flat width less than 0.43 times the nominal width across flats. The minimum across-corners dimensions of external squares shall be 1.25 times the nominal width across flats, but shall not result in a side-flat width less than 0.75 times the nominal width across the flats.

Tolerance on all dimensions not otherwise limited shall be ±0.4 mm (±0.016 in). Fitting seats shall be concentric with straight thread pitch diameters within 0.25 mm (0.010 in) full indicator movement (FIM).

Unless otherwise specified, tolerance on hole diameters designated drill in the dimensional tables shall be as tabulated in Table 1:

TABLE 1—DRILL TOLERANCES

Drill Size Range mm	Drill Size Range in	Tolerance, mm Plus	Tolerance, mm Minus	Tolerance, in Plus	Tolerance, in Minus
0.35– 6.25	0.0135–0.246	0.08	0.08	0.003	0.003
6.35–12.70	0.250–0.500	0.10	0.10	0.004	0.004
13.10–19.05	0.516–0.750	0.13	0.13	0.005	0.005
19.40–25.40	0.765–1.000	0.18	0.13	0.007	0.005
25.80–38.10	1.016–1.500	0.20	0.13	0.008	0.005
38.50	1.516 and over	0.25	0.13	0.010	0.005

Angular tolerance on axis of ends on elbows, tees, and crosses shall be ±2.50 degrees for 1/8 to 3/8 in tube fittings or 1/8 and 1/4 pipe fittings; ±1.50 degrees for 1/2 to 2 in O.D. tube fittings or 3/8 to 2 in pipe fittings.

Where so illustrated and not otherwise specified, hexagon corners shall be chamfered 15 to 30 degrees to a diameter equal to the width across flats, with a minus tolerance of 0.4 mm (0.016 in); or where design permits, corners may be chamfered to the diameter of the abutting surface providing the length of chamfer does not exceed that produced by the 30 degree chamfer previously described.

Alternatively, on connections other than SAE straight thread, a 5 degree chamfer starting at the undercut diameter behind the threads or outside diameter of the threads shall be allowed, providing the hex width at corners is not reduced below that produced by the 30 degree chamfer previously described.

3.3 Passages—Where passages in straight fittings are machined from opposite ends, the offset at the meeting point shall not exceed 0.4 mm (0.016 in). The cross-sectional area at the junction of passages in angle fittings shall not be less than that of the smallest passage.

3.4 Wall Thickness—Unless otherwise designated, the wall thickness at any point on fittings shall not be less than the thickness established by the specified dimensions, tolerances, and eccentricities for inner and outer surfaces.

3.5 Contour—Details of contour shall be optional with manufacturer provided the tabulated dimensions are maintained and serviceability of the fittings is not impaired.

3.6 Straight Threads—Unified Standard Class 2A external and Class 2B internal threads with modified minor diameters, where specified, shall apply to plain finish (unplated) fittings of all types. For externally threaded parts with additive finish, the maximum diameters of Class 2A may be exceeded by the amount of the allowance, that is, the basic diameters (Class 2A maximum diameters plus the allowance) apply to an externally threaded part after plating. For internally threaded parts with additive finish, the Class 2B diameters and modified minor diameters apply after plating.

The pitch diameter tolerance shall be the same as the corresponding diameter-pitch combination and class of the Unified fine and 12 thread series. See SAE J475 (ISO R725).

Where external threads are produced by roll threading and body is not undercut, the unthreaded portion of body adjacent to the shoulder may be reduced to the minimum pitch diameter.

External threads shall be chamfered and internal threads shall be countersunk as specified in the dimensional tables.

3.7 Thread Eccentricity Tolerances—The various thread elements of Class 2A external and Class 2B internal threads on tube fittings shall be concentric within the limitations specified under General Specifications in SAE J512.

3.8 Pipe Threads—Pipe threads, unless there is specific authorization to the contrary, shall conform to the Dryseal American Standard Taper Pipe Thread (NPTF). Specifications are given in detail in SAE J476 (ANSI B1.20.3).

The length of full form external thread shall not be shorter than L_2 plus one pitch (thread).

Where external pipe threads are produced by roll threading, the diameter of the unthreaded shank adjacent to shoulder may be reduced to the E_2 pitch diameter for brass fittings and to the root diameter on steel fittings.

External pipe threads shall be chamfered from the diameters tabulated below to produce the specified length of chamfer or partial thread. Internal pipe threads shall be countersunk 90 degrees, included angle, to the diameters tabulated in Table 2.

3.9 Material—Unless otherwise specified, fittings and ferrules shall be made from carbon steel. Flareless type ferrules in Figures 28 and 29 shall be made from SAE 1010, 1112, 1113, 1213, 12L14, or 1215 steel and cyanide hardened to a depth of 0.03 to 0.05 mm (0.0010 to 0.0019 in).

Stainless steel fittings shall be made from AISI Type 300 Series stainless steel of good quality.[1] Flareless type ferrules in Figures 28 and 29 shall be made from stainless steel of such hardness as to be capable of biting, fully annealed type 304 stainless steel tubing. Unless otherwise specified by the purchaser, stainless steel fittings shall be passivated. Carbon steel and stainless steel fittings fabricated from multiple components must be bonded together with materials having a melting point of not less than 996 °C (1825 °F).

Thirty-seven degree flared type and pipe type brass fittings shall be made from C36000 (CA360) one-half hard barstock or extruded shapes or C37700 (CA377) forgings.

---

[1] See SAE J405.

TABLE 2—PIPE THREAD CHAMFER DIAMETERS

Nominal Pipe Thread Size	External Thread Chamfer Dia Max mm	External Thread Chamfer Dia Max in	External Thread Chamfer Dia Min mm	External Thread Chamfer Dia Min in	External Thread Length of Chamfer or Partial Thread Min mm	External Thread Length of Chamfer or Partial Thread Min in	External Thread Length of Chamfer or Partial Thread Max mm	External Thread Length of Chamfer or Partial Thread Max in	Internal Thread Counter-sink Dia Min mm	Internal Thread Counter-sink Dia Min in	Internal Thread Counter-sink Dia Max mm	Internal Thread Counter-sink Dia Max in
1/8	8.1	0.32	7.6	0.30	0.94	0.037	1.40	0.055	10.7	0.42	11.2	0.44
1/4	10.7	0.42	10.2	0.40	1.42	0.056	2.13	0.084	14.0	0.55	14.5	0.57
3/8	14.0	0.55	13.5	0.53	1.42	0.056	2.13	0.084	17.5	0.69	18.0	0.71
1/2	17.3	0.68	16.8	0.66	1.80	0.071	2.72	0.107	21.6	0.85	22.1	0.87
3/4	22.6	0.89	22.1	0.87	1.80	0.071	2.72	0.107	26.9	1.06	27.4	1.08
1	28.4	1.12	27.7	1.09	2.21	0.087	3.30	0.130	34.0	1.34	34.8	1.37
1 1/4	37.1	1.46	36.3	1.43	2.21	0.087	3.30	0.130	42.7	1.68	43.4	1.71
1 1/2	43.2	1.70	42.4	1.67	2.21	0.087	3.30	0.130	48.8	1.92	49.5	1.95
2	55.1	2.17	54.4	2.14	2.21	0.087	3.30	0.130	60.7	2.39	61.5	2.42

Tabulated diameters conform with Appendix A of SAE J476.

(R) **3.10 Finish**—The external surfaces and threads of all carbon steel parts shall be plated or coated with a suitable material that passes a 72 h salt spray test in accordance with ASTM B 117. Any appearance of red rust during the 72 h salt spray test shall be considered failure, except for the following:

a. All internal fluid passages.
b. Edges such as hex points, serrations, and crests of threads where there may be mechanical deformation of the plating or coating typical of mass-produced parts or shipping effects.
c. Areas where there is mechanical deformation of the plating or coating caused by crimping, flaring, bending, and other post-plate metal forming operations.
d. Areas where the parts are suspended or affixed in the test chamber where condensate can accumulate.

NOTE—Cadmium plating is not preferred due to environmental reasons. Parts manufactured to this document after January 1, 1997, shall not be cadmium plated. Internal fluid passages shall be protected from corrosion during storage. Changes in plating may affect assembly torques and require requalification, when applicable.

3.11 Workmanship—Workmanship shall conform to the best commercial practice to produce high-quality fittings. Fittings shall be free from all hanging burrs, loose scale, and slivers which might become dislodged in usage and all other defects which might affect their serviceability. All sealing surfaces must be smooth except that annular tool marks up to 2.5 μm (100 μin) max A.A. shall be permissible.

3.12 Assembly Considerations—Use of a compatible lubricant is desirable in assembling dryseal pipe threads on hydraulic tube or pipe fittings to minimize galling and effect a pressure-tight seal.

The O-ring washer must be clinched to fitting with a tight slip fit to an interference fit. The slip fit shall be tight enough so that washer cannot be shaken loose to cause it to drop from its uppermost position by its own weight. The interference fit shall not require a locknut torque more than that indicated in Table 3 of SAE J1453. Position the washer farthest from the end of the fittings as shown in Figure 10A. Care must be taken not to clinch washer on the transition area between diameter Y and locknut thread which results in a loose washer when it is repositioned at assembly. Washer flatness allowance is given in Table 3 of SAE J1453. Any surface out of flatness must be uniform (not wavy) and concave with respect to the O-ring boss end of the fitting.

4. Section 1— 37 Degree Flare Tube Fittings—The 37 degree flared tube fittings shall be as shown in Figures 1 to 15 and Tables 3 to 7. Since the basic design of these fittings is derived form Air Force and Navy Standards for 37 degree flared fittings which meet a performance specification, any future changes should not be detrimental to the design or performance. Dimensions for double and single 37 degree flares on tubing to be used with these fittings are given in Figure 3 and Table 2 of SAE J533.

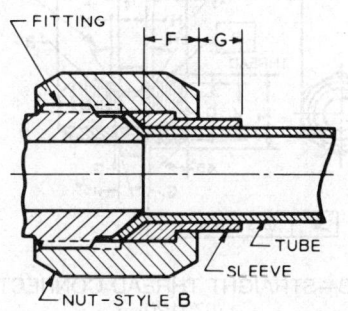

FIGURE 1A—THREE PIECE TUBE ASSEMBLY

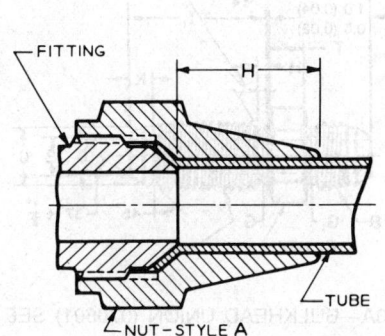

FIGURE 1B—TWO PIECE TUBE ASSEMBLY

FIGURE 1—DETAILS OF 37 DEGREE FLARED HYDRAULIC TUBE FITTING ASSEMBLIES

TABLE 3—DIMENSIONS OF 37 DEGREE FLARED HYDRAULIC TUBE FITTING ASSEMBLIES (FIGURE 1)

Nominal Tube OD	F (Ref) mm	F (Ref) in	G (Ref) mm	G (Ref) in	H (Ref) mm	H (Ref) in
1/8	4.8	0.19	3.0	0.12	12.2	0.48
3/16	6.4	0.25	3.0	0.12	14.2	0.56
1/4	4.8	0.19	4.3	0.17	15.0	0.59
5/16	7.6	0.30	2.8	0.11	16.8	0.66
3/8	7.1	0.28	4.8	0.19	17.5	0.69
1/2	7.9	0.31	4.8	0.19	20.6	0.81
5/8	9.7	0.38	6.9	0.27	23.9	0.94
3/4	9.1	0.36	6.4	0.25	26.2	1.03
7/8	9.7	0.38	7.9	0.31	30.2	1.19
1	8.6	0.34	10.4	0.41	33.3	1.31
1 1/4	8.6	0.34	9.9	0.39	38.1	1.50
1 1/2	12.7	0.50	14.0	0.55	38.9	1.53
2	14.0	0.55	13.5	0.53	44.4	1.75

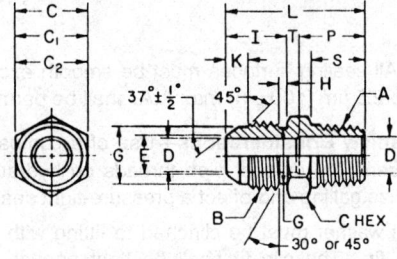

(R) FIGURE 2A—MALE CONNECTOR (070102) SEE NOTE J

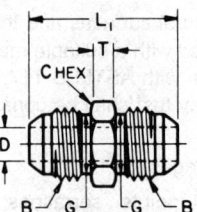

FIGURE 2B—UNION (070101)

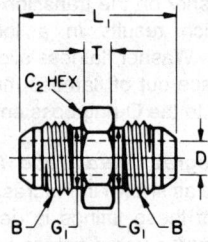

FIGURE 2C—LARGE HEX UNION* (070119)

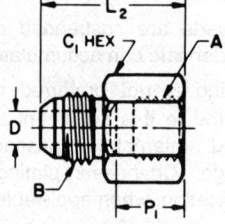

FIGURE 2D—FEMALE CONNECTOR (070103)

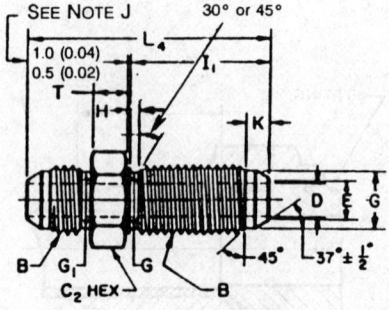

(R) FIGURE 3A—BULKHEAD UNION (070601) SEE NOTE M

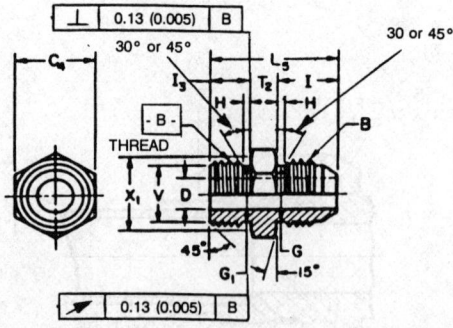

FIGURE 3B—STRAIGHT THREAD CONNECTOR SHORT* (070120)

22.63

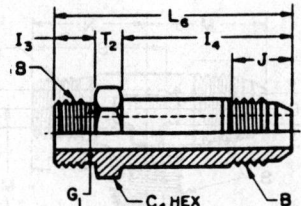

FIGURE 3C—STRAIGHT THREAD CONNECTOR LONG* (070122)

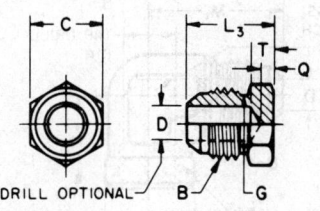

FIGURE 3D—PLUG (070109)

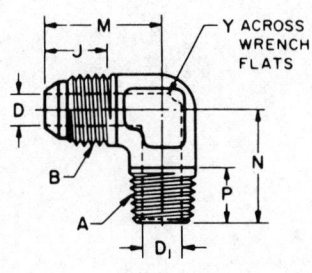

FIGURE 4A—90 DEGREE MALE ELBOW (070202)

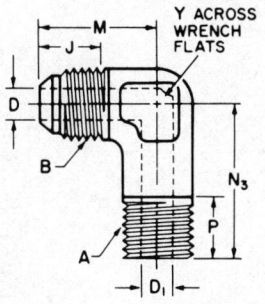

FIGURE 4B—90 DEGREE MALE LONG ELBOW (071502)

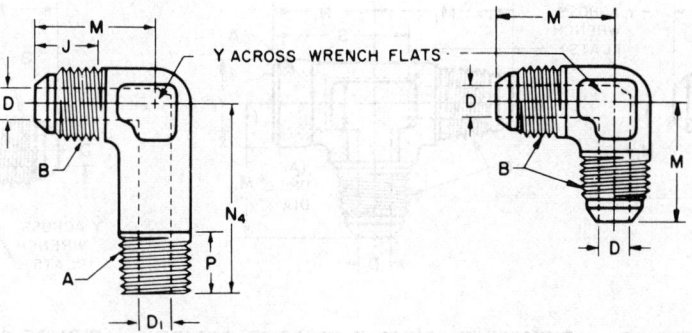

FIGURE 4C—90 DEGREE MALE EXTRA LONG ELBOW (071602)　　FIGURE 4D—90 DEGREE UNION ELBOW (070201)

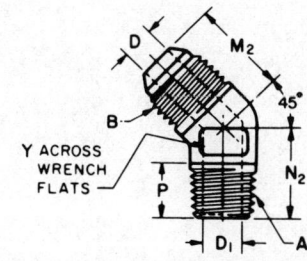

FIGURE 4E—45 DEGREE MALE ELBOW (070302)

22.64

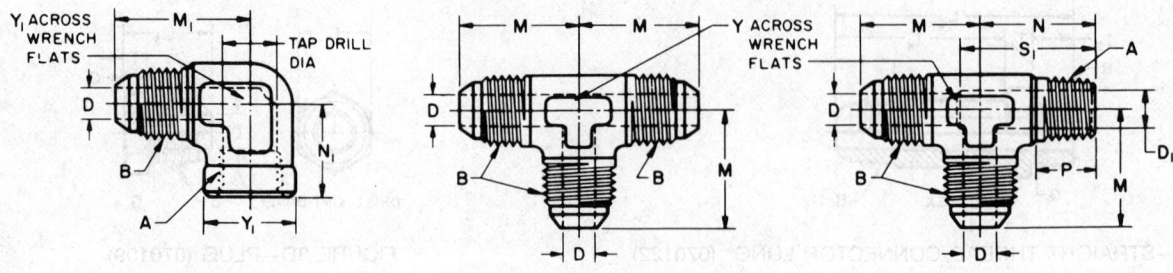

FIGURE 5A—90 DEGREE FEMALE ELBOW (070203)   FIGURE 5B—UNION TEE (070401)   FIGURE 5C—MALE RUN TEE (070424)

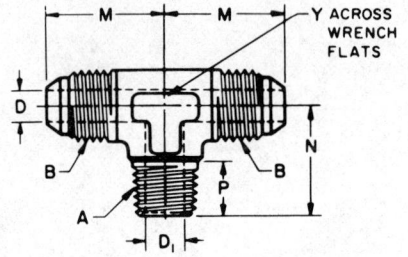

FIGURE 5D—MALE BRANCH TEE (070425)

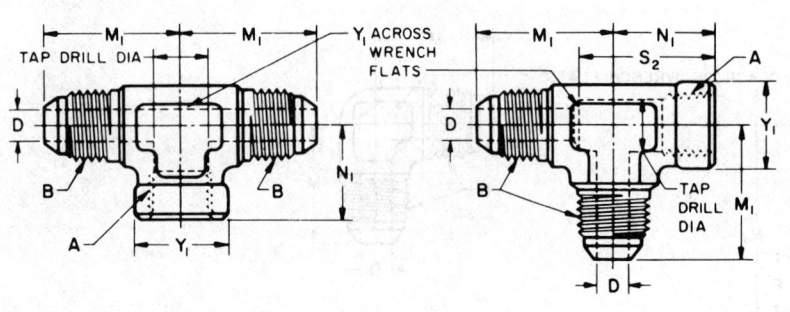 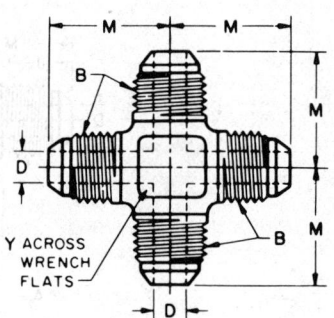

FIGURE 6A—FEMALE BRANCH TEE (070427)   FIGURE 6B—FEMALE RUN TEE (070426)   FIGURE 6C—CROSS (070501)

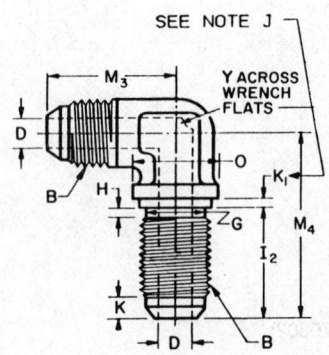 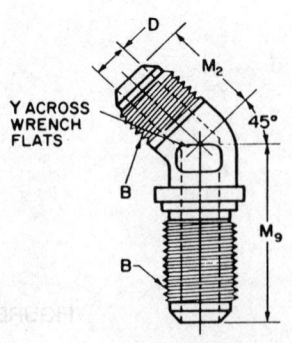

FIGURE 7A—90 DEGREE BULKHEAD ELBOW (070701)
SEE NOTE M

FIGURE 7B—45 DEGREE BULKHEAD ELBOW (070801)
SEE NOTE M

22.65

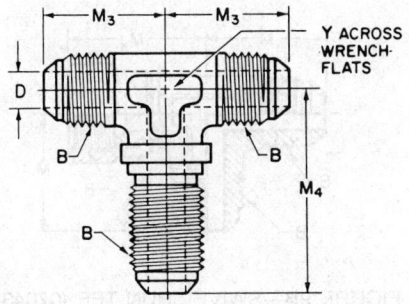

FIGURE 7C—BULKHEAD BRANCH TEE (070959) SEE NOTE M

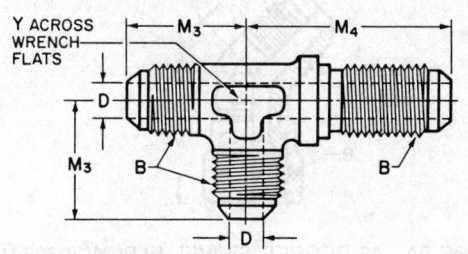

FIGURE 7D—BULKHEAD RUN TEE (070958) SEE NOTE M

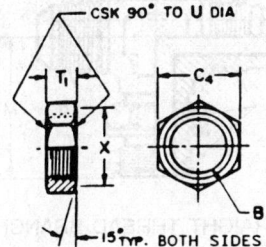

FIGURE 8A—STRAIGHT THREAD LOCKNUT* (070117)

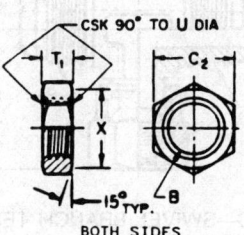

FIGURE 8B—BULKHEAD LOCKNUT (070118)

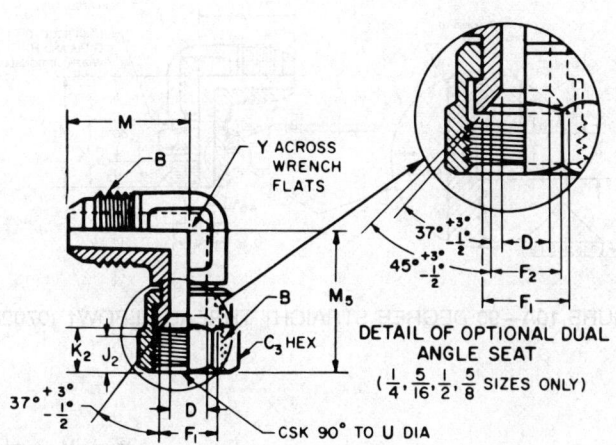

FIGURE 8C—90 DEGREE SWIVEL ELBOW (070221)

THE DESIGN AND METHOD OF ATTACHING THE SWIVEL NUT SHALL BE OPTIONAL WITH THE MANUFACTURER PROVIDING THE TABULATED DIMENSIONS ARE MAINTAINED AND THE NUT TURNS FREELY.

22.66

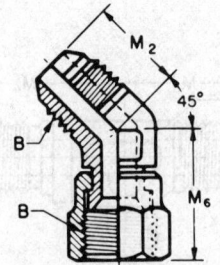

FIGURE 9A—45 DEGREE SWIVEL ELBOW (070321)

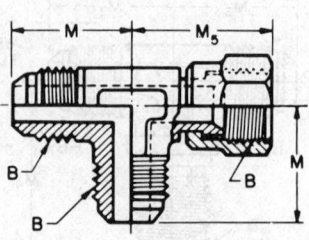

FIGURE 9B—SWIVEL RUN TEE (070432)

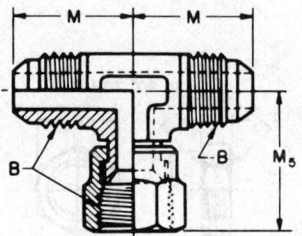

FIGURE 9C—SWIVEL BRANCH TEE (070433)

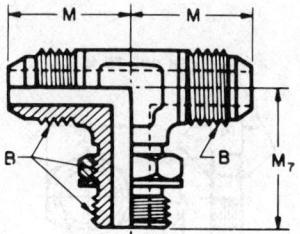

FIGURE 9D—STRAIGHT THREAD BRANCH TEE† (070429)

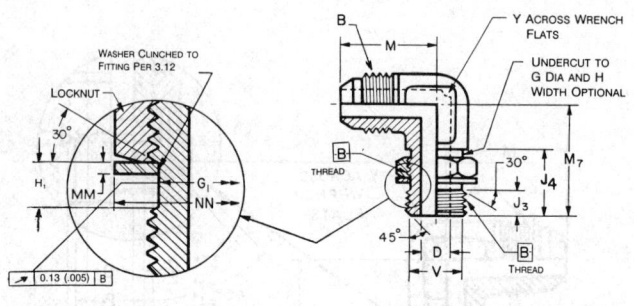

FIGURE 10A—90 DEGREE STRAIGHT THREAD ELBOW† (070220)

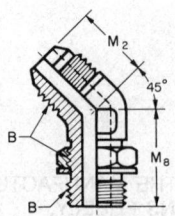

FIGURE 10B—45 DEGREE STRAIGHT THREAD ELBOW† (070320)

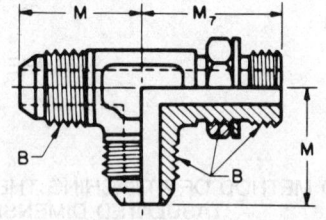

FIGURE 10C—STRAIGHT THREAD RUN TEE† (070428)

NOTES: UNSPECIFIED DETAIL WITH RESPECT TO DIMENSIONS, TOLERANCES, CONTOURS, MATERIAL, WORKMANSHIP, ETC., MUST CONFORM TO GENERAL SPECIFICATIONS FOR HYDRAULIC TUBE FITTINGS. THE DIMENSIONAL DESIGNATIONS FOR TUBE ENDS IN FIGURES 2A TO 10C, FOR SWIVEL ENDS IN FIGURES 8C TO 9C, FOR O-RING BOSS ENDS IN FIGURES 3B AND 3C, AND FOR ADJUSTABLE STRAIGHT THREAD ENDS IN FIGURES 9D TO 10C SHALL APPLY TO CORRESPONDING ENDS OF OTHER FIGURES ON THIS AND PRECEDING PAGE UNLESS SHOWN OTHERWISE. FIGURES 2A TO 10C ON THIS AND PRECEDING PAGE APPLY TO TABLE 4. CODES SHOWN IN BRACKETS ADJACENT TO FIGURE NUMBERS REPRESENT RESPECTIVE FITTING IDENTIFICATION IN ACCORDANCE WITH SAE J846.

*MODIFICATION OF 1/8-1 IN SIZES IN THESE TYPES OF FITTINGS FOR USE WITH MS 33649 (OR SUPERSEDED AND 10050) BOSSES IS SHOWN IN FIGURE 35 AND TABLE 15.

†IF DESIRED BY THE PURCHASER AND SO SPECIFIED, THESE FITTINGS MAY BE FURNISHED WITH LARGE HEXAGON LOCKNUT SHOWN IN FIGURE 8B.

(R) TABLE 4—DIMENSIONS OF ALL BODIES AND LOCKNUTS (FIGURES 2A TO 10C)

Nom Tube OD, in	A Dryseal Pipe Thread SAE J476 (ANSI B1.20.3)	B Thread Size, in SAE J475 (ISO R725) Class 2A Ext Class 2B Int g	C^h Hex Nom in	C_1^h Hex Nom in	C_2^h Hex Nom in	C_3^h Hex Min in	C_4^h Hex Nom in	D^n Drill mm	D^n Drill in	$D_{1b,n}$ Drill mm	$D_{1b,n}$ Drill in	E Dia mm ±0.08	E Dia in ±0.003	F_1 Dia mm ±0.25	F_1 Dia in ±0.010	F_2 Dia mm ±0.13	F_2 Dia in ±0.005
1/8	1/8-27	5/16-24	7/16	9/16	9/16	7/16	7/16	1.6	0.062	4.8	0.188	2.11	0.083	4.85	0.190	---	---
3/16	1/8-27	3/8-24	7/16	9/16	5/8	1/2	1/2	3.2	0.125	4.8	0.188	3.71	0.146	6.20	0.245	---	---
1/4	1/8-27	7/16-20	1/2	9/16	11/16	9/16	9/16	4.4	0.172	4.8	0.188	4.90	0.193	7.35	0.290	6.35	0.250
5/16	1/8-27	1/2-20	9/16	9/16	3/4	5/8	5/8	6.0	0.234	4.8	0.188	6.48	0.255	8.90	0.350	7.93	0.312
3/8	1/4-18	9/16-18	5/8	3/4	13/16	11/16	11/16	7.5	0.297	7.0	0.281	8.08	0.318	10.90	0.430	---	---
1/2	3/8-18	3/4-16	13/16	1	7/8	7/8	7/8	9.9	0.391	10.3	0.406	10.82	0.426	14.35	0.565	12.70	0.500
5/8	1/2-14	7/8-14	15/16	1-1/8	1-1/8	1	1	12.3	0.484	13.5	0.531	13.69	0.539	17.15	0.675	15.88	0.625
3/4	3/4-14	1-1/16-12	1-1/8	1-3/8	1-3/8	1-1/4	1-1/4	15.5	0.609	18.0	0.719	16.87	0.664	21.45	0.845	---	---
7/8	3/4-14	1-3/16-12	1-1/4	1-3/8	1-1/2	1-3/8	1-3/8	18.0	0.719	18.0	0.719	20.02	0.788	24.65	0.970	---	---
1	1-11-1/2	1-5/16-12	1-3/8	1-5/8	1-5/8	1-1/2	1-1/2	21.5	0.844	23.8	0.938	23.19	0.913	27.80	1.095	---	---
1-1/4	1-1/4-11-1/2	1-5/8-12	1-11/16	2	1-7/8	2	1-7/8	27.5	1.078	31.7	1.250	29.13	1.147	35.70	1.405	---	---
1-1/2	1-1/2-11-1/2	1-7/8-12	2	2-3/8	2-1/8	2-1/4	2-1/8	33.0	1.312	38.0	1.500	35.08	1.381	41.15	1.620	---	---
2	2-11-1/2	2-1/2-12	2-5/8	2-7/8	2-3/4	2-7/8	2-3/4	45.0	1.781	49.0	1.938	47.75	1.880	56.75	2.235	---	---

Nom Tube OD, in	G Dia mm +0.05 -0.25	G Dia in +0.002 -0.010	G_1^a Dia mm +0.05 -0.08	G_1^a Dia in +0.002 -0.003	H mm +0.8 -0.0	H in +0.030 -0.000	H_1 mm +0.3 -0.0	H_1 in +0.010 -0.000	I mm ±0.4	I in ±0.016	I_1 mm ±0.5	I_1 in ±0.02	I_2 mm ±0.5	I_2 in ±0.02	I_3 mm ±0.13	I_3 in ±0.005	I_4 mm ±0.5	I_4 in ±0.02	J^1 Full Thread Min mm	J^1 Full Thread Min in
1/8	6.35	0.250	6.35	0.250	1.6	0.063	3.2	0.125	11.4	0.448	28.2	1.11	23.4	0.92	7.54	0.297	29.7	1.17	11.00	0.433
3/16	7.92	0.312	7.95	0.313	1.6	0.063	3.3	0.131	12.2	0.479	28.2	1.11	23.4	0.92	7.54	0.297	31.8	1.25	11.81	0.464
1/4	9.25	0.364	9.25	0.364	1.9	0.075	4.0	0.156	14.0	0.550	30.5	1.20	25.9	1.02	9.14	0.360	35.3	1.39	13.59	0.535
5/16	10.82	0.426	10.85	0.427	1.9	0.075	4.0	0.156	14.0	0.550	30.5	1.20	25.9	1.02	9.14	0.360	36.8	1.45	13.59	0.535
3/8	12.22	0.481	12.24	0.482	2.1	0.083	4.0	0.156	14.1	0.556	32.5	1.28	27.7	1.09	9.93	0.391	39.6	1.56	13.74	0.541
1/2	16.74	0.659	16.76	0.660	2.4	0.094	4.8	0.187	16.7	0.657	36.6	1.44	31.8	1.25	11.13	0.438	47.8	1.88	16.31	0.642
5/8	19.61	0.772	19.63	0.773	2.7	0.107	5.6	0.219	19.3	0.758	40.1	1.58	35.3	1.39	12.70	0.500	53.1	2.09	18.87	0.743
3/4	23.95	0.943	24.00	0.945	3.2	0.125	5.9	0.234	21.9	0.864	44.4	1.75	39.6	1.56	15.09	0.594	63.5	2.50	21.56	0.849
7/8	27.13	1.068	27.18	1.070	3.2	0.125	5.9	0.234	22.6	0.890	44.4	1.75	39.6	1.56	15.09	0.594	68.3	2.69	22.22	0.875
1	30.30	1.193	30.35	1.195	3.2	0.125	5.9	0.234	23.1	0.911	44.4	1.75	39.6	1.56	15.09	0.594	72.1	2.84	22.76	0.896
1-1/4	38.25	1.506	38.28	1.507	3.2	0.125	5.9	0.234	24.3	0.958	45.7	1.80	40.9	1.61	15.09	0.594	88.1	3.47	23.95	0.943
1-1/2	44.58	1.755	44.60	1.756	3.2	0.125	5.9	0.234	27.5	1.083	46.0	1.81	41.1	1.62	15.09	0.594	98.6	3.88	27.13	1.068
2	60.45	2.380	60.48	2.381	3.2	0.125	5.9	0.234	33.9	1.333	53.1	2.09	48.5	1.91	15.09	0.594	122.9	4.84	33.48	1.318

Nom Tube OD, in	J_2 Full Thread Min mm	J_2 Full Thread Min in	J_3 Full Thread mm ±0.13	J_3 Full Thread in ±0.005	J_4 Min mm	J_4 Min in	K mm +0.4 -0.0	K in +0.016 -0.000	K_1^J mm ±0.5	K_1^J in ±0.02	K_2 mm +0.8 -0.4	K_2 in +0.030 -0.016	L mm ±0.5	L in ±0.02	L_1 mm ±0.5	L_1 in ±0.02	L_2 mm ±0.5	L_2 in ±0.02	L_3 mm ±0.5	L_3 in ±0.02	L_4 mm ±0.5	L_4 in ±0.02
1/8	6.38	0.251	5.94	0.234	15.0	0.59	4.5	0.177	2.4	0.094	7.9	0.312	28.2	1.11	29.7	1.17	28.4	1.12	17.8	0.70	47.5	1.87
3/16	7.16	0.282	5.94	0.234	15.0	0.59	4.5	0.177	2.4	0.094	8.3	0.328	29.0	1.14	31.2	1.23	28.7	1.13	18.5	0.73	48.3	1.90
1/4	7.62	0.300	6.73	0.265	18.5	0.73	4.9	0.193	2.4	0.094	8.7	0.344	31.0	1.22	34.8	1.37	30.2	1.19	20.3	0.80	52.6	2.07
5/16	8.41	0.331	6.73	0.265	18.5	0.73	4.9	0.193	2.4	0.094	9.5	0.375	31.0	1.22	34.8	1.37	29.7	1.17	20.3	0.80	52.6	2.07
3/8	8.46	0.333	7.92	0.312	19.1	0.75	5.0	0.198	2.4	0.094	9.5	0.375	36.3	1.43	35.8	1.41	35.6	1.40	21.3	0.84	55.4	2.18
1/2	9.52	0.375	8.74	0.344	21.9	0.86	6.4	0.253	3.2	0.125	10.7	0.422	38.9	1.53	41.1	1.62	39.6	1.56	23.9	0.94	62.0	2.44
5/8	11.71	0.461	9.93	0.391	25.4	1.00	6.8	0.266	3.2	0.125	12.7	0.500	48.0	1.89	47.8	1.88	48.0	1.89	27.9	1.10	69.6	2.74
3/4	11.91	0.469	11.91	0.469	29.0	1.14	8.0	0.315	3.2	0.125	14.3	0.562	52.3	2.06	54.9	2.16	52.3	2.06	32.5	1.28	78.5	3.09
7/8	13.11	0.516	11.91	0.469	29.0	1.14	8.0	0.315	3.2	0.125	14.7	0.578	53.1	2.09	56.1	2.21	52.3	2.06	33.3	1.31	79.2	3.12
1	14.30	0.563	11.91	0.469	29.0	1.14	8.0	0.315	3.2	0.125	15.1	0.594	58.4	2.30	57.2	2.25	59.7	2.35	33.8	1.33	79.8	3.14
1-1/4	14.30	0.563	11.91	0.469	29.0	1.14	9.3	0.367	3.2	0.125	15.9	0.625	62.2	2.45	61.7	2.43	63.2	2.49	36.8	1.45	84.1	3.31
1-1/2	16.79	0.661	11.91	0.469	29.0	1.14	9.6	0.378	3.2	0.125	18.6	0.734	68.1	2.68	69.8	2.75	66.5	2.62	41.9	1.65	89.4	3.52
2	21.44	0.844	11.91	0.469	29.0	1.14	11.7	0.461	3.2	0.125	23.8	0.938	79.0	3.11	86.4	3.40	75.4	2.97	52.1	2.05	106.7	4.20

TABLE 4—DIMENSIONS OF ALL BODIES AND LOCKNUTS (FIGURES 2A TO 10C) (CONTINUED)

Nom Tube OD, in	L_5 mm ±0.5	L_5 in ±0.02	L_6 mm ±0.5	L_6 in ±0.02	M mm ±0.8	M in ±0.03	M_1 mm ±0.8	M_1 in ±0.03	M_2 mm ±0.8	M_2 in ±0.03	M_3 mm ±0.8	M_3 in ±0.03	M_4 mm ±0.8	M_4 in ±0.03	M_5 mm ±1.5	M_5 in ±0.06	M_6 mm ±1.5	M_6 in ±0.06	M_7 mm ±0.8	M_7 in ±0.03	M_8 mm ±0.8	M_8 in ±0.03	M_9 mm ±0.8	M_9 in ±0.03
1/8	26.9	1.06	45.5	1.79	19.6	0.77	25.4	1.00	17.5	0.69	22.4	0.88	38.1	1.50	24.6	0.97	23.9	0.94	23.9	0.94	22.4	0.88	36.1	1.42
3/16	27.9	1.10	47.5	1.87	21.1	0.83	26.2	1.03	17.5	0.69	23.9	0.94	38.1	1.50	25.4	1.00	23.9	0.94	23.9	0.94	22.4	0.88	36.1	1.42
1/4	31.2	1.23	52.8	2.08	22.6	0.89	27.4	1.08	18.3	0.72	24.6	0.97	40.4	1.59	25.4	1.00	23.9	0.94	26.2	1.03	26.7	1.05	38.9	1.53
5/16	31.2	1.23	54.4	2.14	24.1	0.95	27.4	1.08	19.6	0.77	26.9	1.06	43.7	1.72	26.9	1.06	25.4	1.00	28.7	1.13	26.7	1.05	42.2	1.66
3/8	33.0	1.30	58.7	2.31	26.9	1.06	31.2	1.23	21.1	0.83	27.7	1.09	46.0	1.81	31.8	1.25	28.4	1.12	31.8	1.25	29.0	1.14	42.4	1.67
1/2	37.6	1.48	68.6	2.70	31.8	1.25	36.1	1.42	24.9	0.98	34.5	1.36	53.6	2.11	35.1	1.38	32.5	1.28	36.8	1.45	33.0	1.30	49.3	1.94
5/8	43.2	1.70	77.2	3.04	36.8	1.45	41.7	1.64	28.2	1.11	39.6	1.56	60.7	2.39	41.1	1.62	36.6	1.44	43.2	1.70	38.6	1.52	55.1	2.17
3/4	50.0	1.97	91.7	3.61	42.2	1.66	48.0	1.89	32.5	1.28	45.2	1.78	67.8	2.67	44.4	1.75	38.1	1.50	49.3	1.94	43.9	1.73	62.0	2.44
7/8	50.5	1.99	96.5	3.80	45.7	1.80	47.2	1.86	36.8	1.45	48.8	1.92	71.1	2.80	45.2	1.78	41.1	1.62	50.8	2.00	47.2	1.86	63.5	2.50
1	51.8	2.04	101.1	3.98	46.0	1.81	55.1	2.17	37.3	1.47	49.3	1.94	71.1	2.80	50.8	2.00	44.4	1.75	52.1	2.05	47.2	1.86	65.0	2.56
1 1/4	55.1	2.17	119.1	4.69	52.3	2.06	59.2	2.33	40.4	1.59	55.1	2.17	79.2	3.12	58.7	2.31	51.6	2.03	57.2	2.25	48.5	1.91	67.3	2.65
1 1/2	60.2	2.37	131.3	5.17	59.2	2.33	73.4	2.89	45.2	1.78	59.4	2.34	86.9	3.42	65.8	2.59	57.2	2.25	60.7	2.39	48.5	1.91	67.8	2.67
2	70.6	2.78	159.8	6.29	77.7	3.06	83.8	3.30	56.4	2.22	73.4	2.89	104.4	4.11	85.9	3.38	73.9	2.91	73.4	2.89	47.2	1.86	73.9	2.91

Nom Tube in	N mm ±0.8	N in ±0.03	N_1 mm ±0.8	N_1 in ±0.03	N_2 mm ±0.8	N_2 in ±0.03	N_3 mm ±0.8	N_3 in ±0.03	N_4 mm ±0.8	N_4 in ±0.03	O mm ±0.8	O in ±0.03	P Min mm	P Min in	P_1 Max mm	P_1 Max in	Q Min mm	Q Min in	S^b Max mm	S^b Max in	$S_1^{b,c}$ Min mm	$S_1^{b,c}$ Min in	S_2^c Min mm	S_2^c Min in
1/8	18.3	0.72	16.8	0.66	13.2	0.52	25.4	1.00	32.5	1.28	11.1	0.438	9.7	0.38	11.7	0.46	2.8	0.11	12.4	0.49	20.1	0.79	18.5	0.73
3/16	18.3	0.72	16.8	0.66	13.2	0.52	26.4	1.04	34.3	1.35	12.7	0.500	9.7	0.38	11.7	0.46	2.8	0.11	12.4	0.49	20.8	0.82	19.3	0.76
1/4	19.8	0.78	16.8	0.66	16.3	0.64	29.7	1.17	39.6	1.56	14.3	0.562	9.7	0.38	11.7	0.46	2.8	0.11	12.4	0.49	23.1	0.91	20.1	0.79
5/16	19.8	0.78	16.8	0.66	16.3	0.64	29.7	1.17	41.4	1.63	15.9	0.625	9.7	0.38	11.7	0.46	3.0	0.12	16.3	0.64	28.2	1.11	20.8	0.82
3/8	27.7	1.09	22.4	0.88	21.8	0.86	40.1	1.58	52.6	2.07	17.5	0.688	14.2	0.56	17.0	0.67	3.0	0.12	17.0	0.67	31.8	1.25	27.2	1.07
1/2	31.0	1.22	25.9	1.02	24.1	0.95	46.2	1.82	61.5	2.42	22.2	0.875	14.2	0.56	17.3	0.68	4.1	0.16	17.3	0.68	37.1	1.46	32.0	1.26
5/8	37.3	1.47	31.2	1.23	29.7	1.17	55.1	2.17	72.9	2.87	25.4	1.000	19.0	0.75	22.9	0.90	4.1	0.16	23.1	0.91	44.4	1.75	38.4	1.51
3/4	40.4	1.59	34.5	1.36	30.5	1.20	62.0	2.44	83.3	3.28	30.2	1.188	19.0	0.75	23.1	0.91	4.8	0.19	23.9	0.94	49.3	1.94	43.4	1.71
7/8	42.9	1.69	36.1	1.42	33.02	1.30	65.8	2.59	88.9	3.50	33.3	1.312	19.0	0.75	23.1	0.91	4.8	0.19	—	—	—	—	46.2	1.82
1	50.0	1.97	41.1	1.62	37.6	1.48	76.5	3.01	102.9	4.05	36.5	1.438	23.9	0.94	29.0	1.14	4.8	0.19	28.7	1.13	61.7	2.43	52.8	2.08
1 1/4	60.5	2.38	43.2	1.70	42.4	1.67	93.7	3.69	127.0	5.00	44.4	1.750	24.6	0.97	29.5	1.16	5.8	0.23	30.5	1.20	75.2	2.96	57.9	2.28
1 1/2	67.1	2.64	52.8	2.08	45.0	1.77	104.1	4.10	141.0	5.55	50.8	2.000	25.4	1.00	29.5	1.16	5.8	0.23	32.3	1.27	84.8	3.34	70.6	2.78
2	76.2	3.00	60.7	2.39	53.6	2.11	122.2	4.81	168.4	6.63	66.7	2.625	26.2	1.03	30.0	1.18	8.4	0.33	34.8	1.37	100.1	3.94	84.6	3.33

Nom Tube OD, in	T^d Ref mm	T^d Ref in	T_1 mm ±0.5	T_1 in ±0.02	T_2^d Ref mm	T_2^d Ref in	U Dia mm +0.4 −0.0	U Dia in +0.016 −0.000	V Dia mm ±0.25	V Dia in ±0.010	X Dia mm ±0.13	X Dia in ±0.005	X_1 Dia mm ±0.13	X_1 Dia in ±0.005
1/8	5.6	0.22	5.6	0.22	7.1	0.28	8.1	0.317	6.25	0.245	14.27	0.562	11.13	0.438
3/16	5.6	0.22	5.6	0.22	7.1	0.28	9.7	0.380	7.75	0.305	15.88	0.625	12.70	0.500
1/4	5.6	0.22	7.13	0.281	7.1	0.28	11.3	0.443	9.15	0.360	17.48	0.688	14.30	0.563
5/16	5.6	0.22	7.13	0.281	7.1	0.28	12.8	0.505	10.65	0.420	19.05	0.750	15.88	0.625
3/8	6.4	0.25	6.9	0.27	7.9	0.31	14.4	0.567	12.05	0.475	20.62	0.812	17.48	0.688
1/2	6.4	0.25	7.9	0.31	8.6	0.34	19.2	0.755	16.65	0.655	25.40	1.000	22.22	0.875
5/8	7.9	0.31	9.1	0.36	10.2	0.40	22.4	0.880	19.55	0.770	28.58	1.125	25.40	1.000
3/4	9.7	0.38	10.4	0.41	11.9	0.47	27.1	1.067	23.90	0.940	34.92	1.375	31.75	1.250
7/8	9.7	0.38	10.4	0.41	11.9	0.47	30.3	1.193	27.05	1.065	38.10	1.500	34.92	1.375
1	9.7	0.38	10.4	0.41	12.7	0.50	33.5	1.317	30.25	1.190	41.28	1.625	38.10	1.500
1 1/4	11.7	0.46	10.4	0.41	14.7	0.58	41.4	1.630	38.10	1.500	47.62	1.875	47.62	1.875
1 1/2	13.5	0.53	10.4	0.41	16.5	0.65	47.8	1.880	44.45	1.750	53.98	2.125	53.98	2.125
2	17.3	0.68	10.4	0.41	20.6	0.81	63.6	2.505	60.35	2.375	69.85	2.750	69.85	2.750

TABLE 4—DIMENSIONS OF ALL BODIES AND LOCKNUTS (FIGURES 2A TO 10C) (CONTINUED)

Nom Tube OD, in	Y Forging[f,k] mm +0.0	Y Forging[f] in +0.000	Y[f] Barstock Max mm	Y[f] Barstock Max in	Y_1[e,f] mm +0.0	Y_1[e,f] in +0.000	MM mm ±0.08	MM in ±0.003	NN Dia mm ±0.4	NN Dia in ±0.016	PP Dia mm ±0.25	PP Dia in ±0.010
1/8	11.1-0.8	0.438-0.030	—	—	14.3-0.8	0.562-0.030	0.76	0.030	12.8	0.504	6.25	0.245
3/16	11.1-0.8	0.438-0.030	—	—	14.3-0.8	0.562-0.030	0.76	0.030	14.6	0.575	7.75	0.305
1/4	11.1-0.8	0.438-0.030	14.3	0.562	14.3-0.8	0.562-0.030	0.89	0.035	16.5	0.650	9.15	0.360
5/16	14.3-0.8	0.562-0.030	15.9	0.625	14.3-0.8	0.562-0.030	0.89	0.035	18.3	0.722	10.65	0.420
3/8	14.3-0.8	0.562-0.030	20.6	0.812	19.0-0.8	0.750-0.030	0.89	0.035	20.2	0.794	12.05	0.475
1/2	19.0-0.8	0.750-0.030	22.2	0.875	22.2-0.8	0.875-0.030	1.04	0.041	25.7	1.010	16.65	0.655
5/8	22.2-0.8	0.875-0.030	28.6	1.125	27.0-0.8	1.062-0.030	1.27	0.050	29.3	1.155	19.55	0.770
3/4	27.0-1.0	1.062-0.040	34.9	1.375	33.3-1.0	1.312-0.040	1.27	0.050	36.7	1.444	23.90	0.940
7/8	33.3-1.0	1.312-0.040	38.1	1.500	33.3-1.0	1.312-0.040	1.27	0.050	40.4	1.589	27.05	1.065
1	33.3-1.0	1.312-0.040	41.3	1.625	41.3-1.0	1.625-0.040	1.27	0.050	44.0	1.732	30.25	1.190
1 1/4	41.3-1.0	1.625-0.040	54.0	2.125	47.6-1.0	1.875-0.040	1.27	0.050	55.0	2.165	38.10	1.500
1 1/2	47.6-1.0	1.875-0.040	57.2	2.250	65.1-1.0	2.562-0.040	1.27	0.050	62.3	2.454	44.45	1.750
2	63.5-1.0	2.500-0.040	82.6	3.250	71.4-1.0	2.812-0.040	1.27	0.050	80.3	3.160	—	—

[a] O-ring groove undercut must be smooth and free from tool marks.
[b] At manufacturer's option, through passages in Figures 2A and 5C may conform with the smaller diameter specified or the appropriate end may be counterbored to the larger diameter for depths S and S_1, respectively.
[c] Maximum depth shall be optional with manufacturer providing wall thickness is controlled in compliance with General Specifications.
[d] Minimum design thickness, not subject to inspection.
[e] The basic dimensions shown shall apply as minimum for boss diameters.
[f] For optional metric fitting flats see Appendix A.
[g] Unified class 2B thread shall apply to swivel nuts Figures 8C to 9C and with minor diameter modified to class 3B limits for locknuts Figures 8A and 8B.
[h] Across flat widths must fit standard wrench openings. See ANSI B18.2.2.
[j] Diameter of bulkhead pilot is the same as major thread diameter. Recommended pilot hole for bulkhead fittings is 0.4 mm (.016 in) over major thread diameter.
[l] J full thread minimum with thread runout. If undercut to G diameter and H width. Length of thread and undercut must not be less than l.
[m] Previously designed with the undercut shall remain as an option until five years from date of this revision.
[n] See Table 1 for tolerance.

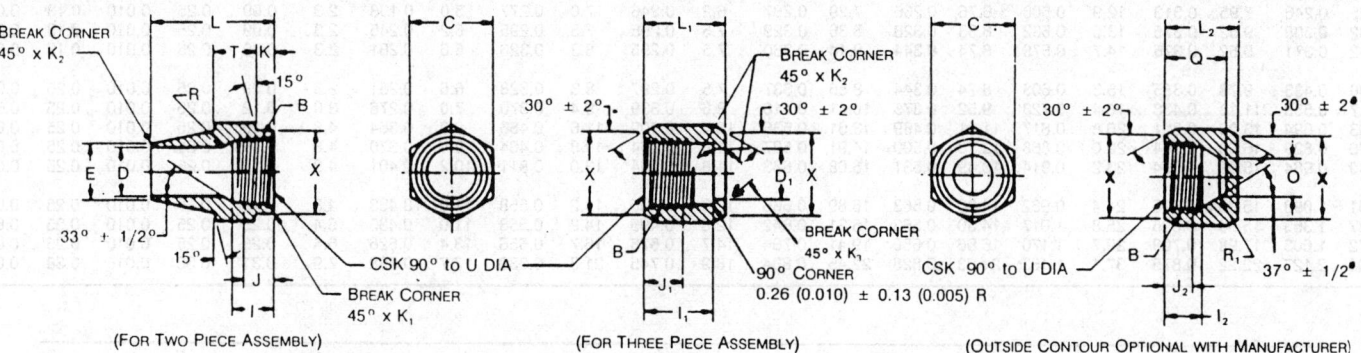

FIGURE 11—STYLE A NUT (070111)   (R) FIGURE 12—STYLE B NUT (070110)   FIGURE 13—CAP (070112)

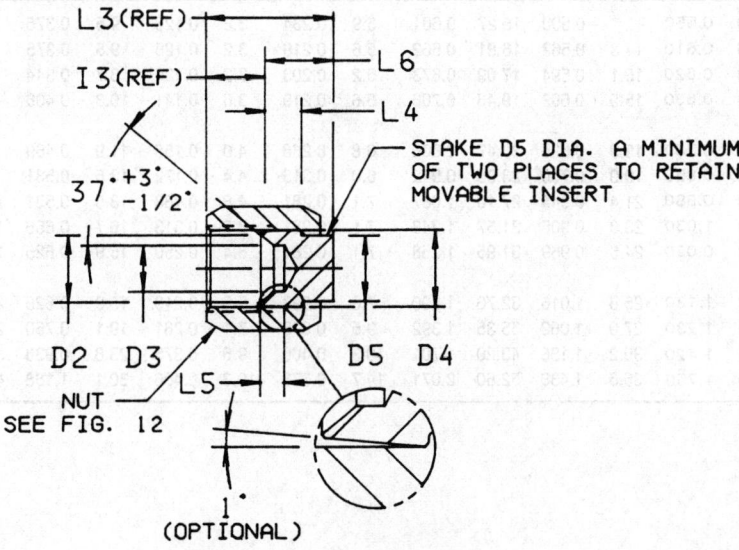

FIGURE 13A—CAP ASSEMBLY (MOVEABLE INSERT) (070112A)

TABLE 5—DIMENSIONS OF STYLE A AND B NUTS AND CAPS (FIGURES 11 TO 13A)

Nom Tube OD, in	B Thread Size, in SAE J475 (ISO R725) Class 2B Int	Thread[b] Minor Dia mm +0.13 -0.00	Thread[b] Minor Dia in +0.005 -0.000	C[a] Hex Min in	D Dia mm +0.08 -0.00	D Dia in +0.003 -0.000	D_1 Dia mm -0.00	D_1 Dia tol	D_1 Dia in	D_1 Dia tol -0.000	D_2 Dia mm +0.13 -0.00	D_2 Dia in +0.005 -0.000	D_3 Dia mm	D_3 Dia in	D_4 Dia mm +0.08 -0.00	D_4 Dia in +0.003 -0.000	D_5 Dia mm +0.13 -0.00	D_5 Dia in +0.005 -0.000
1/8	5/16-24	6.91	0.272	3/8	3.30	0.130	4.55	+0.10	0.179	+0.004	4.95	0.195	2.39	0.094	6.71	0.264	4.24	0.167
3/16	3/8-24	8.48	0.334	7/16	4.90	0.193	6.12	+0.10	0.241	+0.004	6.22	0.245	3.96	0.156	8.28	0.326	5.82	0.229
1/4	7/16-20	9.86	0.388	9/16	6.48	0.255	7.72	+0.10	0.304	+0.004	7.49	0.295	4.37	0.172	9.65	0.380	7.42	0.292
5/16	1/2-20	11.46	0.451	5/8	8.08	0.318	9.47	+0.10	0.373	+0.004	9.02	0.355	5.94	0.234	11.23	0.442	9.17	0.361
3/8	9/16-18	12.90	0.508	11/16	9.65	0.380	11.15	+0.10	0.439	+0.004	11.05	0.435	7.54	0.297	12.68	0.499	10.85	0.427
1/2	3/4-16	17.48	0.688	7/8	12.83	0.505	14.43	+0.13	0.568	+0.005	14.48	0.570	9.93	0.391	17.25	0.679	14.15	0.557
5/8	7/8-14	20.42	0.804	1	16.03	0.631	17.68	+0.13	0.696	+0.005	17.27	0.680	12.29	0.484	20.17	0.794	17.40	0.685
3/4	1 1/16-12	24.87	0.979	1 1/4	19.20	0.756	21.13	+0.13	0.832	+0.005	21.59	0.850	14.27	0.562	24.61	0.969	20.85	0.821
7/8	1 3/16-12	28.04	1.104	1 3/8	22.38	0.881	24.36	+0.13	0.959	+0.005	24.77	0.975	18.26	0.719	27.78	1.094	24.08	0.948
1	1 5/16-12	31.22	1.229	1 1/2	25.55	1.006	27.61	+0.13	1.087	+0.005	27.94	1.100	21.44	0.844	30.96	1.219	27.33	1.076
1 1/4	1 5/8-12	39.14	1.541	2	32.00	1.260	34.14	+0.15	1.344	+0.006	35.81	1.410	27.38	1.078	38.88	1.531	33.88	1.334
1 1/2	1 7/8-12	45.49	1.791	2 1/4	38.35	1.510	41.00	+0.15	1.614	+0.006	41.27	1.625	33.32	1.312	45.24	1.781	40.73	1.604
2	2 1/2-12	61.37	2.416	2 7/8	51.16	2.014	54.97	+0.15	2.164	+0.006	56.89	2.240	45.23	1.781	61.11	2.406	54.71	2.154

E Dia mm +0.13 -0.00	E Dia in +0.005 -0.000	I mm ±0.13	I in ±0.005	I_1 mm ±0.3	I_1 in ±0.010	I_2 mm ±0.13	I_2 in ±0.005	I_3 mm Ref	I_3 in Ref	J Full Thread Min mm	J Full Thread Min in	J_1 Full Thread Min mm	J_1 Full Thread Min in	J_2 Full Thread Min mm	J_2 Full Thread Min in	K mm ±0.5	K in ±0.02	K_1 mm ±0.13	K_1 in ±0.005	K_2 mm ±0.13	K_2 in ±0.005
4.65	0.183	7.16	0.282	11.7	0.460	5.94	0.234	5.74	0.226	5.5	0.215	6.2	0.246	4.2	0.167	2.3	0.09	0.25	0.010	0.13	0.005
6.25	0.246	7.95	0.313	12.9	0.506	6.76	0.266	7.29	0.287	6.2	0.246	7.0	0.277	5.0	0.198	2.3	0.09	0.25	0.010	0.13	0.005
7.82	0.308	9.52	0.375	13.5	0.532	8.33	0.328	8.36	0.329	7.5	0.295	7.5	0.295	6.2	0.245	2.3	0.09	0.25	0.010	0.13	0.005
9.42	0.371	9.52	0.375	14.7	0.579	8.74	0.344	9.14	0.360	7.5	0.295	8.3	0.326	6.6	0.261	2.3	0.09	0.25	0.010	0.13	0.005
11.00	0.433	9.78	0.385	15.3	0.603	8.74	0.344	8.56	0.337	7.5	0.297	8.3	0.328	6.6	0.261	2.3	0.09	0.25	0.010	0.25	0.010
14.17	0.558	11.13	0.438	18.4	0.723	9.52	0.375	10.41	0.410	8.6	0.339	9.4	0.370	7.0	0.276	3.0	0.12	0.25	0.010	0.25	0.010
17.63	0.694	13.23	0.521	20.8	0.817	11.91	0.469	13.61	0.536	10.4	0.409	11.6	0.456	9.0	0.354	4.8	0.19	0.25	0.010	0.25	0.010
21.06	0.829	15.09	0.594	22.0	0.868	12.70	0.500	14.91	0.587	11.8	0.464	11.8	0.464	9.4	0.370	4.8	0.19	0.25	0.010	0.25	0.010
24.49	0.964	15.09	0.594	23.2	0.914	13.49	0.531	16.08	0.633	11.8	0.464	13.0	0.511	10.2	0.401	4.8	0.19	0.25	0.010	0.25	0.010
27.91	1.099	15.88	0.625	24.4	0.962	14.30	0.563	16.89	0.665	12.6	0.495	14.2	0.558	11.0	0.433	4.8	0.19	0.25	0.010	0.25	0.010
34.37	1.353	15.88	0.625	25.8	1.017	14.30	0.563	16.31	0.642	12.6	0.495	14.2	0.558	11.0	0.433	6.4	0.25	0.25	0.010	0.25	0.010
40.72	1.603	17.98	0.708	29.7	1.170	16.66	0.656	19.41	0.764	14.7	0.578	16.7	0.656	13.4	0.526	6.4	0.25	0.25	0.010	0.25	0.010
54.03	2.127	22.22	0.875	37.1	1.462	21.03	0.828	22.45	0.884	18.9	0.745	21.3	0.839	17.7	0.698	7.9	0.31	0.38	0.015	0.38	0.015

K_3 mm Max	K_3 in Max	L mm ±0.5	L in ±0.02	L_1 mm ±0.5	L_1 in ±0.020	L_2 mm ±0.3	L_2 in ±0.01	L_3 mm Ref	L_3 in Ref	L_4 mm ±0.4	L_4 in ±0.016	L_5 mm ±0.4	L_5 in ±0.016	L_6 mm ±0.4	L_6 in ±0.016	O Dia mm ±0.13	O Dia in ±0.005	Q mm ±0.3	Q in ±0.010
0.5	0.020	21.3	0.84	14.0	0.550		0.500	15.27	0.601	5.9	0.234	3.2	0.125	9.5	0.375	1.57	0.062	11.1	0.438
0.5	0.020	23.9	0.94	15.5	0.610	113	0.562	16.81	0.662	5.6	0.219	3.2	0.125	9.5	0.375	3.18	0.125	12.7	0.500
0.5	0.020	25.4	1.00	15.8	0.620	15.1	0.594	17.09	0.673	5.2	0.203	3.2	0.125	8.7	0.344	4.37	0.172	13.5	0.531
0.5	0.020	26.9	1.06	17.3	0.680	15.5	0.609	19.46	0.766	5.6	0.219	3.6	0.141	10.3	0.406	5.94	0.234	13.9	0.547
0.5	0.020	27.7	1.09	18.5	0.730	15.9	0.625	20.47	0.806	6.8	0.266	4.0	0.156	11.9	0.469	7.54	0.297	14.3	0.562
0.5	0.020	32.5	1.28	21.6	0.850	19.0	0.750	23.93	0.942	8.0	0.313	4.4	0.172	13.5	0.531	9.93	0.391	15.9	0.625
1.0	0.040	37.6	1.48	24.9	0.980	21.4	0.844	27.10	1.067	7.1	0.281	4.8	0.188	13.5	0.531	12.29	0.484	18.3	0.719
1.0	0.040	42.2	1.66	26.2	1.030	23.0	0.906	31.57	1.243	7.1	0.281	8.0	0.313	16.7	0.656	15.47	0.609	19.8	0.781
1.0	0.040	46.0	1.81	27.7	0.090	24.6	0.969	31.95	1.258	7.1	0.281	6.4	0.250	15.9	0.625	18.26	0.719	21.4	0.844
1.0	0.40	49.3	1.94	28.7	1.130	25.8	1.016	32.76	1.290	7.6	0.297	5.6	0.219	15.9	0.625	21.44	0.844	21.8	0.859
1.0	0.40	55.6	2.19	31.2	1.230	27.0	1.062	35.35	1.392	9.5	0.375	7.1	0.281	19.1	0.750	27.38	1.078	23.0	0.906
1.0	0.40	58.7	2.31	36.1	1.420	30.2	1.188	43.20	1.701	10.3	0.406	9.5	0.375	23.8	0.938	33.32	1.312	26.2	1.031
1.0	0.40	69.8	2.75	44.5	1.750	36.5	1.438	52.60	2.071	14.7	0.578	10.3	0.406	30.1	1.188	45.24	1.781	32.5	1.281

TABLE 5—DIMENSIONS OF STYLE A AND B NUTS AND CAPS (FIGURES 11 TO 13A) (CONTINUED)

R Rad mm ±0.3	R Rad in ±0.01	R_1 Rad mm ±0.3	R_1 Rad in ±0.01	T mm ±0.5	T in ±0.02	U Dia mm +0.4 -0.0	U Dia in +0.016 -0.000	X mm ±0.3	X in ±0.01
0.8	0.03	0.8	0.03	6.4	0.25	8.05	0.317	9.1	0.36
0.8	0.03	0.8	0.03	6.9	0.27	9.65	0.380	10.7	0.42
0.8	0.03	0.8	0.03	8.4	0.33	11.25	0.443	13.7	0.54
0.8	0.03	0.8	0.03	8.4	0.33	12.83	0.505	15.2	0.60
1.3	0.05	1.5	0.06	8.6	0.34	14.40	0.567	17.0	0.67
1.5	0.06	1.5	0.06	11.4	0.45	19.18	0.755	21.8	0.86
1.5	0.06	1.5	0.06	13.2	0.52	22.35	0.880	24.9	0.98
2.0	0.08	1.5	0.06	16.3	0.64	27.10	1.067	31.5	1.24
2.3	0.09	1.5	0.06	17.5	0.69	30.30	1.193	34.5	1.36
2.3	0.09	1.5	0.06	18.5	0.73	33.45	1.317	37.6	1.48
2.3	0.09	1.5	0.06	18.5	0.73	41.40	1.630	50.3	1.98
2.8	0.11	1.5	0.06	21.1	0.83	47.75	1.880	56.9	2.24
2.8	0.11	1.5	0.06	23.4	0.92	63.63	2.505	72.6	2.86

a Across flat widths must fit standard wrench openings.
b Modified minor diameter.

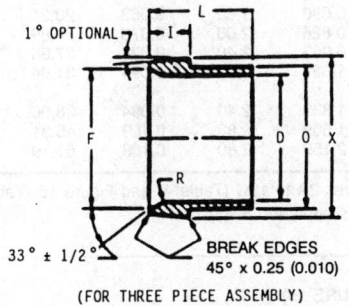

FIGURE 14—SLEEVE (070115)

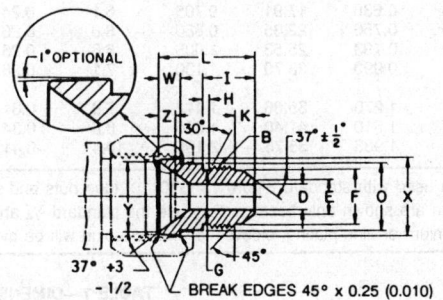

FIGURE 15—REDUCING ADAPTER (070123)

TABLE 6—DIMENSIONS FOR SLEEVES (FIGURE 14) FOR INCH TUBING

Nom Tube OD, in	D Dia mm +0.08 -0.00	D Dia in +0.003 -0.000	F Dia mm ±0.13	F Dia in ±0.005	I mm ±0.5	I in ±0.02	L mm ±0.5	L in ±0.02	O Dia mm +0.00 -0.08	O Dia in +0.000 -0.003	R Rad mm ±0.3	R Rad in ±0.01	X Dia mm +0.00 -0.08	X Dia in +0.000 -0.003
1/8	3.30	0.130	5.21	0.205	3.0	0.12	8.6	0.34	4.37	0.172	0.8	0.031	6.78	0.267
3/16	4.90	0.193	6.78	0.267	3.6	0.14	8.6	0.34	5.94	0.234	0.8	0.031	8.36	0.329
1/4	6.48	0.255	8.00	0.315	3.6	0.14	10.4	0.41	7.54	0.297	0.8	0.031	9.73	0.383
5/16	8.08	0.318	9.52	0.375	4.1	0.16	11.2	0.44	9.30	0.366	0.8	0.031	11.30	0.445
3/8	9.65	0.380	11.20	0.441	4.3	0.17	12.7	0.50	10.97	0.432	1.2	0.047	12.75	0.502
1/2	12.83	0.505	14.96	0.589	5.6	0.22	14.2	0.56	14.27	0.562	1.6	0.062	17.32	0.682
5/8	16.03	0.631	17.91	0.705	6.1	0.24	16.8	0.66	17.53	0.690	1.6	0.062	20.24	0.797
3/4	19.20	0.756	22.35	0.880	6.6	0.26	17.3	0.68	20.98	0.826	2.0	0.078	24.69	0.972
7/8	22.38	0.881	25.53	1.005	6.6	0.26	19.3	0.76	24.21	0.953	2.4	0.094	27.86	1.097
1	25.55	1.006	28.70	1.130	7.1	0.28	19.8	0.78	27.46	1.081	2.4	0.094	31.04	1.222
1 1/4	32.00	1.260	35.86	1.412	7.9	0.31	23.1	0.91	34.01	1.339	2.4	0.094	38.96	1.534
1 1/2	38.35	1.510	41.40	1.630	8.6	0.34	28.4	1.12	40.87	1.609	2.8	0.109	45.31	1.784
2	51.16	2.014	55.75	2.195	10.4	0.41	30.2	1.19	54.84	2.159	2.8	0.109	61.19	2.409

5. Section 2— Flareless Tube Fittings—The flareless tube fittings shall be as shown in Figures 16 to 32 and Tables 8 to 12. The basic design of these fittings is derived from existing military standards.

5.1 Assembly Instructions for Hydraulic Flareless Tube Fittings—These instructions apply to the assembly of hydraulic tube fittings of the flareless type given in Tables 9 to 12 and Figures 19 to 32.

The following instructions should be used to assure proper make-up of the fitting when assembled since the fitting depends on securing the ferrule to the tube by the cutting action of the ferrule into the tube.

5.1.1 Cut tube square and burr inside and outside corner (not excessive).

5.1.2 Assemble fitting by sliding nut over tubing with open end out. Slide ferrule on tubing with cutting edge out, the large head end should be inside of the nut. Lubricate the ferrule and the threads on the body and nut with oil or petrolatum. Insert tube into fitting.

5.1.3 Bottom the tube in the fitting, and tighten the nut until the ferrule just grips the tube. With a little experience, the mechanic can determine this point by feel. If the fittings are bench assembled, the gripping action can be determined by rotating the tube by hand as the nut is drawn down. When the tube can no longer be turned by hand, the ferrule has started to grip the tube.

TABLE 6A—DIMENSIONS FOR SLEEVES (FIGURE 14) FOR METRIC TUBING[c]

Nom Tube OD, mm	D Dia mm +0.08 -0.00	D Dia in +0.003 -0.000	F Dia mm ±0.13	F Dia in ±0.005	I mm ±0.5	I in ±0.02	L mm ±0.5	L in ±0.02	O Dia mm +0.00 -0.08	O Dia in +0.000 -0.003	R Rad mm ±0.30	R Rad in ±0.010	X Dia mm +0.00 -0.08	X Dia in +0.000 -0.003	in Size Body[a] and nut
6	6.13	0.241	8.00	0.315	3.6	0.14	10.4	0.41	7.54	0.297	0.80	0.031	9.73	0.383	1/4
8	8.13	0.320	9.52	0.375	4.1	0.16	11.2	0.44	9.30	0.366	0.80	0.031	11.30	0.445	5/16
10	10.13	0.399	11.20	0.441	4.3	0.17	12.7	0.50	10.97	0.432	1.20	0.047	12.75	0.502	3/8
12	12.13	0.478	14.96	0.589	5.6	0.22	14.2	0.56	14.27	0.562	1.60	0.062	17.32	0.682	1/2
16	16.15	0.636	17.91	0.705	6.1	0.24	16.8	0.66	17.53	0.690	1.60	0.062	20.24	0.797	5/8
19[b]	19.20	0.756	22.35	0.880	6.6	0.26	17.3	0.68	20.98	0.826	2.00	0.078	24.69	0.972	3/4
20	20.15	0.793	25.53	1.005	6.6	0.26	19.3	0.76	24.21	0.953	2.40	0.094	27.86	1.097	7/8
25	25.15	0.990	28.70	1.130	7.1	0.28	19.8	0.78	27.46	1.081	2.40	0.094	31.04	1.222	1
32	32.25	1.270	35.86	1.412	7.9	0.31	23.1	0.91	34.01	1.339	2.40	0.094	38.96	1.534	1 1/4
38[b]	38.35	1.510	41.40	1.630	8.6	0.34	28.4	1.12	40.87	1.609	2.80	0.109	45.31	1.784	1 1/2
50	50.36	1.983	55.75	2.195	10.4	0.41	30.2	1.19	54.84	2.159	2.80	0.109	61.19	2.409	2

[a] Metric sleeves are used with standard Figure 12 (070110) tube nuts and standard fitting bodies as shown in Figures 2A to 10C (Table 4) and Figure 15 (Table 7).
[b] 19 mm and 38 mm are shown only because they use the standard 3/4 and 1 1/2 size Figure 14 sleeve and there is apparent usage.
[c] In addition to cadmium or zinc plating, sleeves for metric tubing will be dyed light blue for identification.

TABLE 7—DIMENSIONS[a] OF REDUCING ADAPTER (FIGURE 15)

Tube Reduction, in	D Dia Ref mm	D Dia Ref in	L mm ±0.5	L in ±0.02	O Dia mm +0.00 -0.08	O Dia in +0.000 -0.003	W mm ±0.5	W in ±0.02	X Dia mm +0.00 -0.08	X Dia in +0.000 -0.003	Y Dia mm +0.13 -0.40	Y Dia in +0.005 -0.016	Z mm ±0.5	Z in ±0.02
3/8 × 1/4	4.4	0.172	24.6	0.97	10.97	0.432	4.3	0.17	12.75	0.502	11.20	0.441	—	—
1/2 × 1/4	4.4	0.172	25.4	1.00	14.27	0.562	5.6	0.22	17.32	0.682	14.96	0.589	—	—
1/2 × 3/8	7.5	0.297	25.4	1.00	14.27	0.562	5.6	0.22	17.32	0.682	14.96	0.589	—	—
5/8 × 1/4	4.4	0.172	26.2	1.03	17.53	0.690	5.8	0.23	20.24	0.797	17.91	0.705	—	—
5/8 × 3/8	7.5	0.297	26.2	1.03	17.53	0.690	5.8	0.23	20.24	0.797	17.91	0.705	—	—
3/4 × 1/4	4.4	0.172	27.7	1.09	20.98	0.826	6.9	0.27	24.69	0.972	22.35	0.880	10.4	0.41
3/4 × 3/8	7.5	0.297	27.7	1.09	20.98	0.826	6.9	0.27	24.69	0.972	22.35	0.880	—	—
3/4 × 1/2	10.0	0.391	30.2	1.19	20.98	0.826	6.9	0.27	24.69	0.972	22.35	0.880	—	—
1 × 3/4	15.5	0.609	37.3	1.47	27.46	1.081	7.1	0.28	31.04	1.222	28.70	1.130	—	—

[a] For dimensions shown on Figure 15 but not specified in above table, see corresponding dimensions for the specified outside diameter in Table 4.

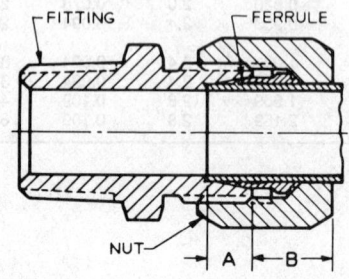

FIGURE 16A—ASSEMBLY WITH STYLE A FERRULE (FIGURE 28)

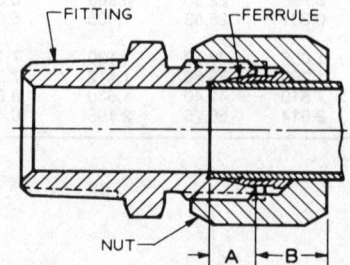

FIGURE 16B—ASSEMBLY WITH STYLE B FERRULE (FIGURE 29)

FIGURE 16—DETAILS OF FLARELESS HYDRAULIC TUBE FITTING ASSEMBLIES

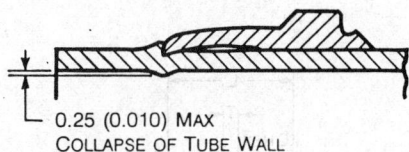

FIGURE 17—ENLARGED VIEW OF STYLE A FERRULE BITE

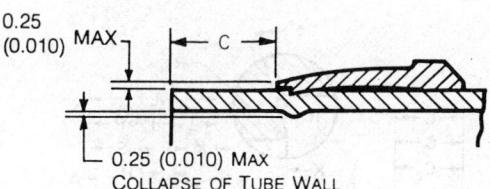

FIGURE 18—ENLARGED VIEW OF STYLE B FERRULE BITE

TABLE 8—DIMENSIONS OF FLARELESS HYDRAULIC TUBE FITTING ASSEMBLIES (FIGURES 16 AND 18)

Nominal Tube OD	A (Ref) mm	A (Ref) in	B (Ref) mm	B (Ref) in	C (Ref) mm	C (Ref) in
1/8	4.78	0.188	7.9	0.31	2.0	0.08
3/16	5.94	0.234	8.6	0.34	3.0	0.12
1/4	5.94	0.234	10.7	0.42	3.6	0.14
5/16	6.35	0.250	10.7	0.42	3.6	0.14
3/8	6.35	0.250	11.9	0.47	3.6	0.14
1/2	7.75	0.305	12.7	0.50	3.6	0.14
5/8	8.89	0.350	13.5	0.53	3.6	0.14
3/4	8.89	0.350	14.2	0.56	3.6	0.14
7/8	8.89	0.350	13.5	0.53	—	—
1	10.54	0.415	16.8	0.66	4.3	0.17
1 1/4	10.54	0.415	18.3	0.72	5.1	0.20
1 1/2	12.32	0.485	18.3	0.72	5.1	0.20
2	12.32	0.485	21.3	0.84	5.1	0.20

5.1.4 After the ferrule grips the tube, tighten the nut one full turn. This may vary slightly with different tubing materials, but for general practice, it is a good rule for the mechanic to follow.

5.1.5 The fittings can now be disassembled for inspection. The two styles of ferrules differ somewhat in inspection even though their principles of makeup and application are similar.

a. For the ferrule in Figure 28, the bite or cut into the tube can be readily seen since it is on the lead edge of the ferrule. The bite into the tube should show a definite groove where the ferrule cuts into the tube and peels the metal over the lead edge of the ferrule. See Figure 17 for further detail.

b. For the ferrule in Figure 29, the pilot at the end of the ferrule should be contacting or be within 0.038 mm (0.0015 in) of the tube for hard material or not more than 0.13 mm (0.005 in) on soft material. See Figure 18 for further detail. This is an indication that the cutting edge has performed its function and has taken a secure bite in the tube. The sleeve should be slightly sprung or arched.

For both styles of ferrules, the rounded or lead edge should show a good seat in the fitting, and the head or shoulder end should be collapsed tight against the tube. The ferrules should have no end movement; however, the ferrule may be rotated on the tube due to spring back of the material. The performance of the fitting is not affected if the ferrule rotates.

5.1.6 In production, it may be preferable to use a threaded presetting tool to preform the ferrule onto the tubing. The presetting tool is a counterpart of the fitting hardened to provide good wearing properties for repeated usage. When using the presetting tool, the assembly instructions are the same, as the presetting tool takes the place of the fitting. Care should be taken to keep the cam surface of the presetting tool free of defects since they would transfer themselves to the ferrule, which would result in improper seating when the fitting is installed.

5.1.7 In some installations, it may be necessary to use a mandrel to support the inside of the tube when setting the ferrule. This is only necessary when the tube wall is so thin or so soft that it will not resist the biting action of the ferrule without collapsing. The mandrel in this instance supports the tube and allows the ferrule to bite into the tube without deforming or collapsing. Because the use of a mandrel allows very little give in the tubing, the setting of the ferrule may be made with slightly fewer turns than described previously.

5.2 Reassembly Instructions for Flareless Fittings—After disassembly of the fitting joint, the flareless fitting can be reassembled by assembling the tube and ferrule into the socket of the fitting and threading the nut onto the fitting.

The operation of assembly up to the point at which the ferrule seats itself in the fitting can usually be accomplished by hand or with the use of a small wrench. If a wrench is required, only low torques are necessary to seat the ferrule.

When the ferrule is seated, an increase in the torque will be quite evident. When this point is reached, draw the nut up approximately 1/6 of a turn minimum, but not more than 1/3 of a turn, to complete the tightening operation.

NOTE—Instructions for assembling adjustable and swivel style hydraulic fittings in straight thread O-ring bosses are given immediately following the table of dimensions of O-ring boss plugs.

NOTE—Table 13 and Figure 33 have intentionally been deleted from this document. See SAE J1926, Specification for Straight Thread O-Ring Boss Port.

6. Section 3— O-Ring Plugs—Specifications for straight thread O-ring port into which connector and adjustable styles of hydraulic tube fittings assemble are given in SAE J1926. O-ring boss plugs shall be as shown in Figure 34 and Table 14. Modification of hexagon chamfers on standard fittings when used in MS 33649 (or superseded AND 10050) type O-ring bosses is covered in Figure 35 and Table 15. For specifications of O-rings used in conjunction with these fittings, see SAE J515. Assembly instruction for O-ring fittings are shown in Figures 36 and 37.

6.1 Modification of Hexagons on Standard Fittings to Accommodate MS 33649 (or Superseded AND 10050) Bosses— When 37 degree flared fittings shown in Figures 3B and 3C, flareless fittings shown in Figures 20B and 20C, or O-ring boss plugs shown in Figure 34, in sizes from 1/8 to 1 in inclusive, are to be used with MS 33649 (or superseded AND 10050) type straight thread O-ring bosses, the chamfer on the bearing face of the hexagon of these fittings shall be modified as shown in Figure 35.

6.2 Assembly Instructions for Adjustable Style Fittings in Straight Thread O-Ring Boss Ports—These instructions apply to the assembly of hydraulic fittings of the 37 degree flared type shown in Figures 9D, 10A, 10B, 10C, and 8A, and flareless type shown in Figures 26D, 27A, 27B, 27C, and 25A, and hydraulic O-rings, Figure 1 of SAE J515.

6.2.1 Lubricate the O-ring by coating with a light oil or petrolatum and install in the groove adjacent to the face of the metal back-up washer which is assembled at the extreme end of the groove as shown in Figure 36B.

22.74

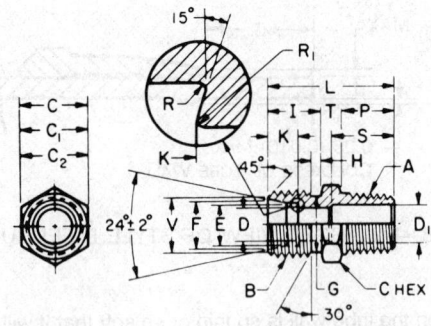

FIGURE 19A—MALE CONNECTOR (080102)

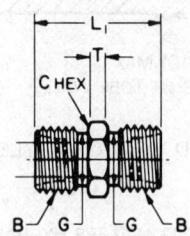

FIGURE 19B—UNION (080101)

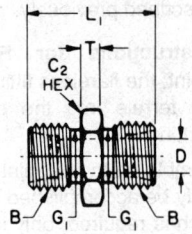

FIGURE 19C—LARGE HEX UNION* (080119)

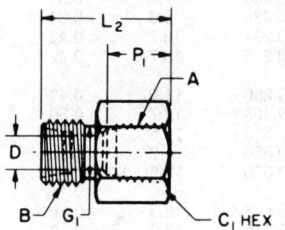

FIGURE 19D—FEMALE CONNECTOR (080103)

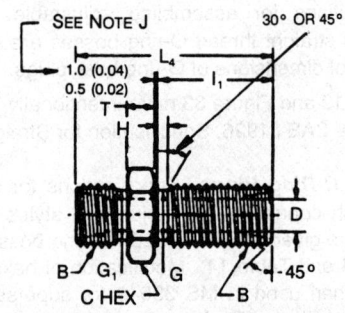

FIGURE 20A—BULKHEAD UNION (080601) SEE NOTE M

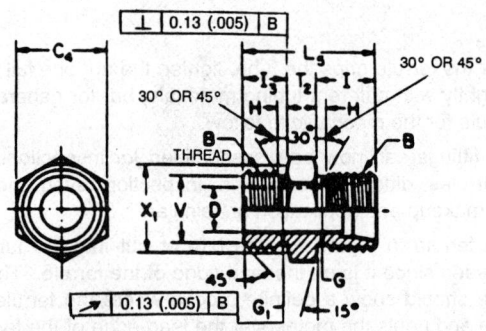

FIGURE 20B—STRAIGHT THREAD CONNECTOR SHORT* (080120)

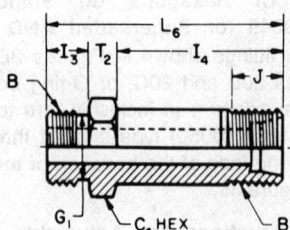

FIGURE 20C—STRAIGHT THREAD CONNECTOR LONG* (080122)

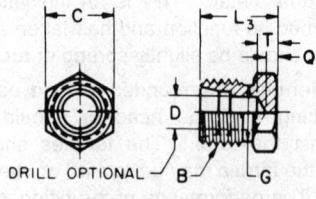

FIGURE 20D—PLUG (080109)

22.75

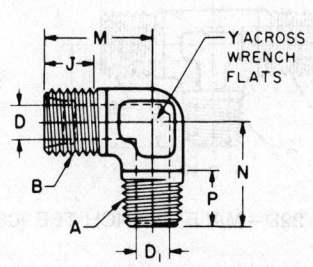

FIGURE 21A—90 DEGREE MALE ELBOW (080202)

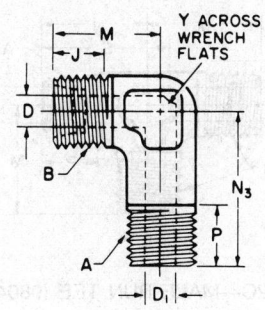

FIGURE 21B—90 DEGREE MALE LONG ELBOW (081502)

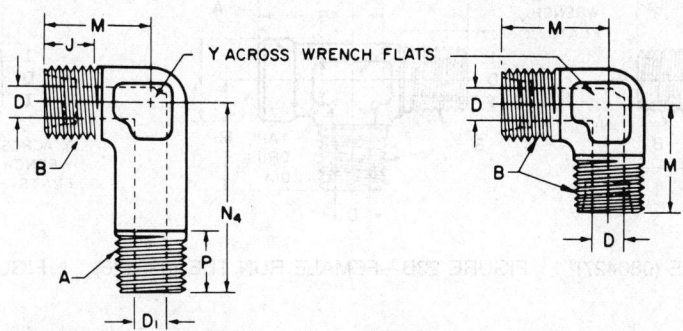

FIGURE 21C—90 DEGREE MALE EXTRA LONG ELBOW (081602)   FIGURE 21D—90 DEGREE UNION ELBOW (080201)

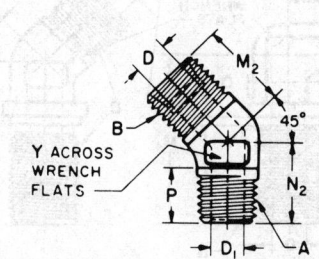

FIGURE 21E—45 DEGREE MALE ELBOW (080302)

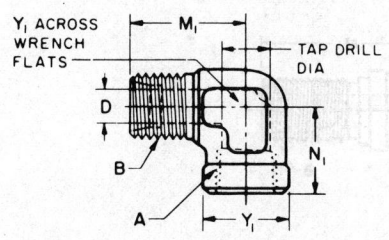

FIGURE 22A—90 DEGREE FEMALE ELBOW (080203)

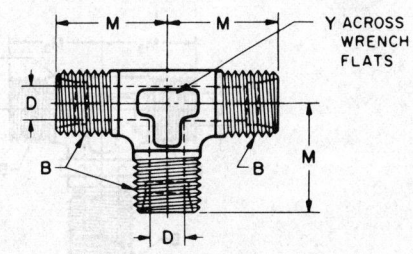

FIGURE 22B—UNION TEE (080401)

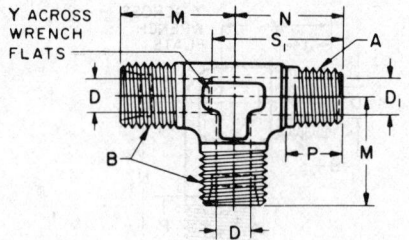

FIGURE 22C—MALE RUN TEE (080424)

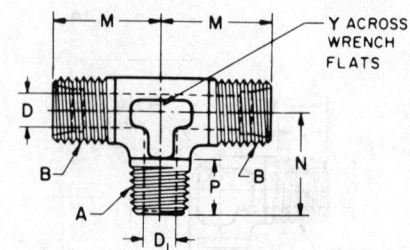

FIGURE 22D—MALE BRANCH TEE (080425)

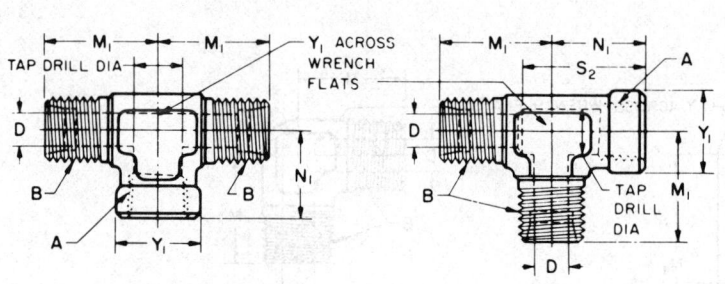

FIGURE 23A—FEMALE BRANCH TEE (080427)   FIGURE 23B—FEMALE RUN TEE (080426)   FIGURE 23C—CROSS (080501)

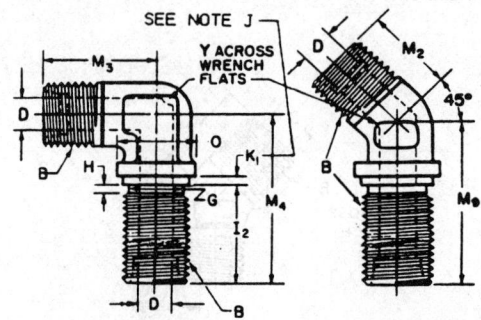

FIGURE 24A—90 DEGREE BULKHEAD ELBOW (080701) SEE NOTE M

FIGURE 24B—45 DEGREE BULKHEAD ELBOW (080801) SEE NOTE M

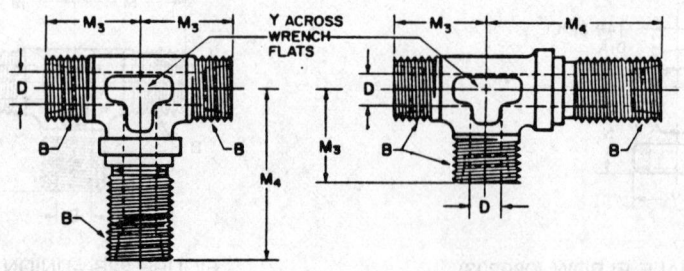

FIGURE 24C—BULKHEAD BRANCH TEE (080959) SEE NOTE M

FIGURE 24D—BULKHEAD RUN TEE (080958) SEE NOTE M

22.77

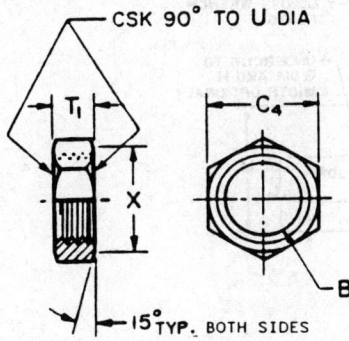

FIGURE 25A—STRAIGHT THREAD LOCKNUT* (080117)

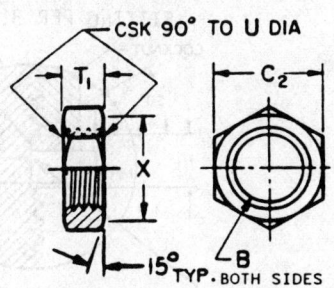

FIGURE 25B—BULKHEAD LOCKNUT (080118)

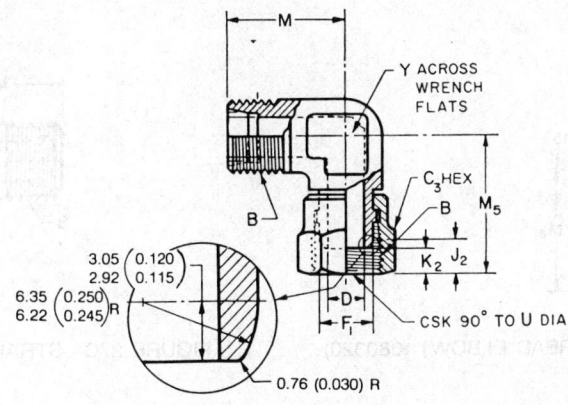

FIGURE 25C—90 DEGREE SWIVEL ELBOW (080221)

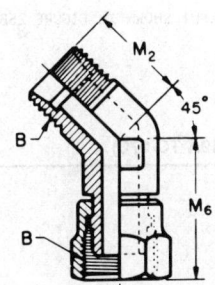

FIGURE 26A—45 DEGREE SWIVEL ELBOW (080321)

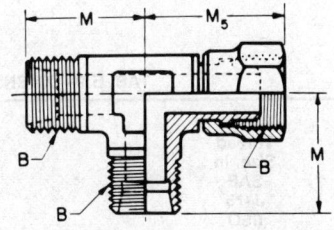

FIGURE 26B—SWIVEL RUN TEE (080432)

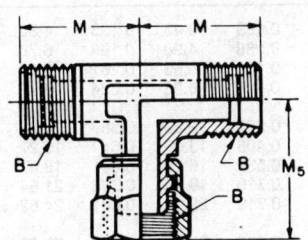

FIGURE 26C—SWIVEL BRANCH TEE (080433)

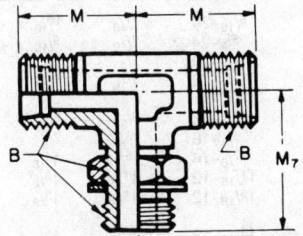

FIGURE 26D—STRAIGHT THREAD BRANCH TEE† (080429)

THE DESIGN AND METHOD OF ATTACHING THE SWIVEL NUT SHALL BE OPTIONAL WITH THE MANUFACTURER PROVIDING THE TABULATED DIMENSIONS ARE MAINTAINED AND THE NUT TURNS FREELY.

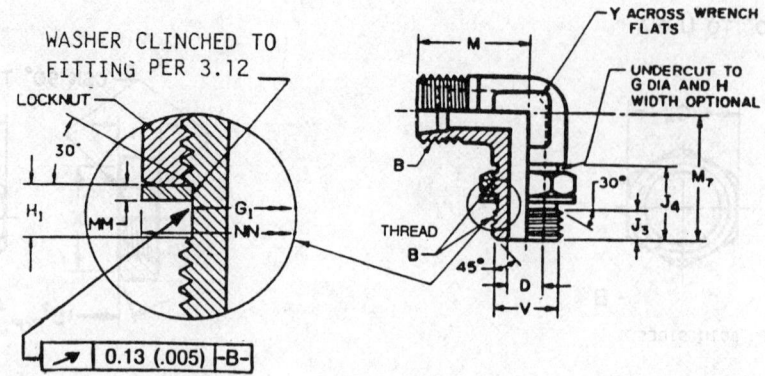

FIGURE 27A—90 DEGREE STRAIGHT THREAD ELBOW† (080220)

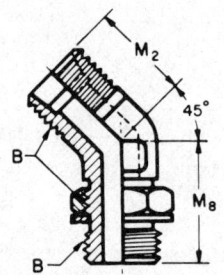

FIGURE 27B—45 DEGREE STRAIGHT THREAD ELBOW† (080320)

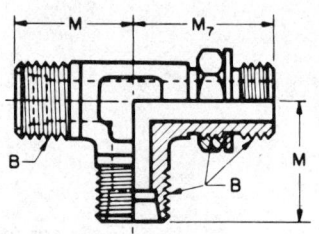

FIGURE 27C—STRAIGHT THREAD RUN TEE† (080428)

NOTES: UNSPECIFIED DETAIL WITH RESPECT TO DIMENSIONS, TOLERANCES, CONTOURS, MATERIAL, WORKMANSHIP, ETC., MUST CONFORM TO GENERAL SPECIFICATIONS FOR HYDRAULIC TUBE FITTINGS. THE DIMENSIONAL DESIGNATIONS FOR TUBE ENDS IN FIGURES 19A THRU 27C, FOR SWIVEL ENDS IN FIGURES 25C THRU 26C, FOR O-RING BOSS ENDS IN FIGURES 20B AND 20C, AND FOR ADJUSTABLE STRAIGHT THREAD ENDS IN FIGURES 26D THRU 27C SHALL APPLY TO CORRESPONDING ENDS OF OTHER FIGURES ON THIS AND PRECEDING PAGE UNLESS SHOWN OTHERWISE. FIGURES 19A THRU 27C ON THIS AND PRECEDING PAGE APPLY TO TABLE 9. CODES SHOWN IN BRACKETS ADJACENT TO FIGURE NUMBERS REPRESENT RESPECTIVE FITTING IDENTIFICATION IN ACCORDANCE WITH SAE J846.

*MODIFICATION OF 1/8–1 IN SIZES IN THESE TYPES OF FITTINGS FOR USE WITH MS 33649 (OR SUPERSEDED AND 10050) BOSSES IS SHOWN IN FIGURE 35 AND TABLE 15.

†IF DESIRED BY THE PURCHASER AND SO SPECIFIED, THESE FITTINGS MAY BE FURNISHED WITH LARGE HEXAGON LOCKNUT SHOWN IN FIGURE 25B.

TABLE 9—DIMENSIONS OF ALL BODIES AND LOCKNUTS (FIGURES 19A TO 27C)

Nom Tube OD, in	A Dryseal Pipe Thread, SAE J476 (ANSI B2.2)	B[f] Thread Size, in SAE J475 (ISO R725) Class 2A Ext Class 2B Int	C^g Hex Nom, in	C_1^g Hex Nom, in	C_2^g Hex Nom, in	C_3^g Hex Min, in	C_4^g Hex Nom, in	D^m Dia Drill mm	D^m Dia Drill in	$D_1^{b,m}$ Dia Drill mm	$D_1^{b,m}$ Dia Drill in	E Dia mm +0.10 −0.00	E Dia in +0.004 −0.000	F Dia mm +0.10 −0.00	F Dia in +0.004 −0.000	F_1 Dia mm +0.13 −0.00	F_1 Dia in +0.005 −0.000
1/8	1/8-27	5/16-24	7/16	9/16	9/16	7/16	7/16	2.4	0.093	4.8	0.188	3.43	0.135	4.80	0.189	4.32	0.170
3/16	1/8-27	3/8-24	7/16	9/16	5/8	1/2	1/2	3.2	0.125	4.8	0.188	4.98	0.196	6.78	0.267	6.35	0.250
1/4	1/8-27	7/16-20	1/2	9/16	11/16	9/16	9/16	5.2	0.203	4.8	0.188	6.63	0.261	8.10	0.319	8.13	0.320
5/16	1/8-27	1/2-20	9/16	9/16	3/4	5/8	5/8	6.0	0.234	4.8	0.188	8.23	0.324	9.70	0.382	9.65	0.380
3/8	1/4-18	9/16-18	5/8	3/4	13/16	11/16	11/16	7.0	0.281	7.0	0.281	9.80	0.386	11.20	0.441	11.30	0.445
1/2	3/8-16	3/4-16	13/16	7/8	1	7/8	7/8	10.7	0.422	10.3	0.406	13.06	0.514	15.27	0.601	14.55	0.573
5/8	1/2-14	7/8-14	15/16	1	1 1/8	1	1	12.7	0.500	13.5	0.531	16.28	0.641	18.47	0.727	18.16	0.715
3/4	3/4-14	1 1/16-12	1 1/8	1 3/8	1 3/8	1 1/4	1 1/4	16.6	0.656	18.0	0.719	19.46	0.766	21.64	0.852	21.21	0.835
7/8	3/4-14	13/16-12	1 1/4	1 3/8	1 1/2	1 3/8	1 3/8	18.0	0.719	18.0	0.719	22.63	0.891	24.82	0.977	24.10	0.949
1	1-11 1/2	1 5/16-12	1 3/8	1 5/8	1 5/8	1 1/2	1 1/2	22.2	0.875	23.8	0.938	25.81	1.016	27.99	1.102	26.92	1.060
1 1/4	1 1/4-11 1/2	1 5/8-12	1 11/16	2	1 7/8	2	1 7/8	27.8	1.093	31.7	1.250	32.26	1.270	34.42	1.355	33.66	1.325
1 1/2	1 1/2-11 1/2	1 7/8-12	2	2 3/8	2 1/4	2 1/4	2 1/8	34.1	1.344	38.0	1.500	38.61	1.520	40.74	1.604	40.39	1.590
2	2-11 1/2	2 1/2-12	2 5/8	2 7/8	2 3/4	2 7/8	2 3/4	46.0	1.813	49.0	1.938	51.36	2.022	53.54	2.108	53.21	2.095

(R) TABLE 9—DIMENSIONS OF ALL BODIES AND LOCKNUTS (FIGURES 19A TO 27C) (CONTINUED)

Nom Tube OD, In	G Dia mm +0.05 -0.30	G Dia In +0.002 -0.010	G_1[a] Dia mm +0.05 -0.08	G_1[a] Dia In +0.002 -0.003	H mm +0.8 -0.0	H In +0.030 -0.000	H_1 mm +0.3 -0.0	H_1 In +0.010 -0.000	I mm ±0.4	I In ±0.016	I_1 mm ±0.5	I_1 In ±0.02	I_2 mm ±0.5	I_2 In ±0.02	I_3 mm ±0.13	I_3 In ±0.005	I_4 mm ±0.5	I_4 In ±0.02	J[k] Full Thread Min mm	J[k] Full Thread Min In
1/8	6.35	0.250	6.35	0.250	1.6	0.063	3.2	0.125	9.5	0.375	25.9	1.02	21.1	0.83	7.54	0.297	27.4	1.08	9.1	0.360
3/16	7.95	0.313	7.95	0.313	1.6	0.063	3.3	0.131	10.7	0.422	26.9	1.06	22.4	0.88	7.54	0.297	30.2	1.19	10.3	0.407
1/4	9.25	0.364	9.25	0.364	1.9	0.075	4.0	0.156	11.5	0.453	28.4	1.12	23.9	0.94	9.14	0.360	33.8	1.33	11.1	0.438
5/16	10.85	0.427	10.85	0.427	1.9	0.075	4.0	0.156	11.5	0.453	28.4	1.12	23.9	0.94	9.14	0.360	35.3	1.39	11.1	0.438
3/8	12.24	0.482	12.24	0.482	2.1	0.083	4.0	0.156	11.9	0.469	29.7	1.17	24.9	0.98	9.93	0.391	37.6	1.48	11.5	0.454
1/2	16.76	0.660	16.76	0.660	2.4	0.094	4.7	0.187	14.3	0.562	33.3	1.31	28.4	1.12	11.13	0.438	45.7	1.80	13.9	0.547
5/8	19.63	0.773	19.63	0.773	2.7	0.107	5.6	0.219	15.9	0.625	36.8	1.45	32.3	1.27	12.70	0.500	51.6	2.03	15.5	0.610
3/4	24.00	0.945	24.00	0.945	3.2	0.125	5.9	0.234	17.5	0.688	39.6	1.56	35.1	1.38	15.09	0.594	61.5	2.42	17.1	0.673
7/8	27.18	1.070	27.18	1.070	3.2	0.125	5.9	0.234	17.5	0.688	39.6	1.56	35.1	1.38	15.09	0.594	66.3	2.61	17.1	0.673
1	30.35	1.195	30.35	1.195	3.2	0.125	5.9	0.234	17.5	0.688	39.6	1.56	35.1	1.38	15.09	0.594	70.4	2.77	17.1	0.673
1 1/4	38.28	1.507	38.28	1.507	3.2	0.125	5.9	0.234	17.5	0.688	39.6	1.56	35.1	1.38	15.09	0.594	86.1	3.39	17.1	0.673
1 1/2	44.60	1.756	44.60	1.756	3.2	0.125	5.9	0.234	17.5	0.688	39.6	1.56	35.1	1.38	15.09	0.594	96.8	3.81	17.1	0.673
2	60.48	2.381	60.48	2.381	3.2	0.125	5.9	0.234	17.5	0.688	45.0	1.77	40.1	1.58	15.09	0.594	121.4	4.78	17.1	0.673

Nom Tube OD, In	J_2 Full Thread Min mm	J_2 Full Thread Min In	J_3 Full Thread mm ±0.13	J_3 Full Thread In ±0.005	J_4 mm min	J_4 In min	K mm +0.40 -0.13	K In +0.016 -0.005	K_1[h] mm ±0.5	K_1[h] In ±0.02	K_2 mm +1.0 -0.5	K_2 In +0.04 -0.02	L mm ±0.5	L In ±0.02	L_1 mm ±0.5	L_1 In ±0.02	L_2 mm ±0.5	L_2 In ±0.02	L_3 mm ±0.5	L_3 In ±0.02
1/8	7.5	0.297	5.94	0.234	15.0	0.59	4.78	0.188	2.4	0.094	4.4	0.172	26.4	1.04	25.9	1.02	26.7	1.05	16.0	0.63
3/16	7.5	0.297	5.94	0.234	15.0	0.59	5.94	0.234	2.4	0.094	4.5	0.178	27.7	1.09	28.2	1.11	27.4	1.08	17.3	0.68
1/4	7.6	0.300	6.73	0.265	18.5	0.73	5.94	0.234	2.4	0.094	5.6	0.219	28.4	1.12	30.0	1.18	27.7	1.09	18.0	0.71
5/16	8.4	0.331	6.73	0.265	18.5	0.73	6.35	0.250	2.4	0.094	5.6	0.219	28.4	1.12	30.0	1.18	27.4	1.08	18.0	0.71
3/8	8.5	0.333	7.92	0.312	19.1	0.75	6.35	0.250	2.4	0.094	6.7	0.265	34.0	1.34	31.5	1.24	33.3	1.31	19.0	0.75
1/2	9.5	0.375	8.74	0.344	21.9	0.86	7.75	0.305	3.2	0.125	6.4	0.250	36.6	1.44	36.1	1.42	37.3	1.47	21.6	0.85
5/8	11.7	0.461	9.93	0.391	25.4	1.00	8.89	0.350	3.2	0.125	7.7	0.304	44.4	1.75	40.9	1.61	44.7	1.76	24.6	0.97
3/4	11.9	0.469	11.91	0.469	29.0	1.14	8.89	0.350	3.2	0.125	9.0	0.354	47.8	1.88	46.0	1.81	48.0	1.89	27.9	1.10
7/8	11.9	0.469	11.91	0.469	29.0	1.14	8.89	0.350	3.2	0.125	7.1	0.281	47.8	1.88	46.0	1.81	47.2	1.86	27.9	1.10
1	11.1	0.438	11.91	0.469	29.0	1.14	10.54	0.415	3.2	0.125	6.4	0.250	52.6	2.07	46.0	1.81	54.1	2.13	27.9	1.10
1 1/4	11.1	0.438	11.91	0.469	29.0	1.14	10.54	0.415	3.2	0.125	6.7	0.265	55.4	2.18	48.0	1.89	56.4	2.22	30.0	1.18
1 1/2	11.1	0.438	11.91	0.469	29.0	1.14	12.32	0.485	3.2	0.125	7.3	0.289	57.9	2.28	49.8	1.96	56.6	2.23	31.8	1.25
2	11.1	0.438	11.91	0.469	29.0	1.14	12.32	0.485	3.2	0.125	7.1	0.281	62.5	2.46	53.6	2.11	58.7	2.31	35.6	1.40

Nom Tube OD, in	L_4 mm ±0.5	L_4 in ±0.02	L_5 mm ±0.5	L_5 in ±0.02	L_6 mm ±0.5	L_6 in ±0.02	M mm ±0.8	M in ±0.03	M_1 mm ±0.8	M_1 in ±0.03	M_2 mm ±0.8	M_2 in ±0.03	M_3 mm ±0.8	M_3 in ±0.03	M_4 mm ±0.8	M_4 in ±0.03	M_5 mm ±1.5	M_5 in ±0.06	M_6 mm ±1	M_6 in ±0.06
1/8	43.4	1.71	25.1	0.99	43.2	1.70	19.6	0.77	21.1	0.83	16.3	0.64	20.3	0.80	34.5	1.36	24.9	0.98	24	0.95
3/16	45.7	1.80	26.4	1.04	46.0	1.81	21.1	0.83	21.1	0.83	16.3	0.64	23.9	0.94	36.6	1.44	25.9	1.02	24	0.95
1/4	48.0	1.89	28.7	1.13	51.3	2.02	22.6	0.89	22.6	0.89	17.8	0.70	24.1	0.95	38.6	1.52	26.7	1.05	24	0.98
5/16	48.0	1.89	28.7	1.13	52.8	2.08	24.1	0.95	24.1	0.95	19.1	0.75	26.7	1.05	42.2	1.66	29.0	1.14	26	1.05
3/8	50.3	1.98	30.7	1.21	56.6	2.23	26.7	1.05	26.7	1.05	21.1	0.83	27.4	1.08	43.2	1.70	32.5	1.28	30	1.20
1/2	56.4	2.22	35.1	1.38	66.5	2.62	31.8	1.25	31.2	1.23	24.9	0.98	33.8	1.33	50.0	1.97	37.3	1.47	33	1.33
5/8	63.0	2.48	39.9	1.57	75.7	2.98	36.1	1.42	36.1	1.42	27.4	1.08	38.6	1.52	57.7	2.27	40.9	1.61	37	1.48
3/4	69.1	2.72	45.5	1.79	89.7	3.53	40.1	1.58	40.1	1.58	32.3	1.27	41.7	1.64	63.0	2.48	45.0	1.77	40	1.61
7/8	69.1	2.72	45.5	1.79	94.5	3.72	42.2	1.66	41.1	1.62	34.0	1.34	43.2	1.70	64.8	2.55	45.7	1.80	40	1.58
1	69.1	2.72	46.2	1.82	99.3	3.91	43.9	1.73	43.9	1.73	34.5	1.36	43.9	1.73	66.3	2.61	47.2	1.86	43	1.70
1 1/4	71.1	2.80	48.3	1.90	117.1	4.61	48.0	1.89	52.8	2.08	36.8	1.45	51.3	2.02	73.4	2.89	52.8	2.08	45	1.80
1 1/2	72.9	2.87	50.0	1.97	129.5	5.10	51.3	2.02	65.5	2.58	38.6	1.52	55.9	2.20	80.5	3.17	55.9	2.20	48	1.92
2	82.0	3.23	54.1	2.13	158.2	6.23	62.2	2.45	67.1	2.64	46.5	1.83	60.7	2.39	95.8	3.77	64.0	2.52	52	2.05

(R) TABLE 9—DIMENSIONS OF ALL BODIES AND LOCKNUTS (FIGURES 19A TO 27C) (CONTINUED)

Nom Tube OD, in	M_7 mm ±0.8	M_7 in ±0.03	M_8 mm ±0.8	M_8 in ±0.03	M_9 mm ±0.8	M_9 in ±0.03	N mm ±0.8	N in ±0.03	N_1 mm ±0.8	N_1 in ±0.03	N_2 mm ±0.8	N_2 in ±0.03	N_3 mm ±0.8	N_3 in ±0.03	N_4 mm ±0.8	N_4 in ±0.03	O mm ±0.5	O in ±0.02	P Min	P Min
1/8	23.1	0.91	22.4	0.88	32.5	1.28	18.3	0.72	16.8	0.66	13.2	0.52	25.4	1.00	32.5	1.28	11.1	0.438	9.7	0.38
3/16	23.9	0.94	22.4	0.88	33.8	1.33	18.3	0.72	16.8	0.66	13.2	0.52	26.4	1.04	34.3	1.35	12.7	0.500	9.7	0.38
1/4	26.2	1.03	26.7	1.05	36.8	1.45	19.8	0.78	16.8	0.66	16.3	0.64	29.7	1.17	39.6	1.56	14.3	0.562	9.7	0.38
5/16	28.7	1.13	26.7	1.05	36.8	1.45	20.6	0.81	16.8	0.66	16.3	0.64	29.7	1.17	41.4	1.63	15.9	0.625	9.7	0.38
3/8	31.8	1.25	29.0	1.14	39.6	1.56	27.7	1.09	22.4	0.88	21.8	0.86	40.1	1.58	52.6	2.07	17.5	0.688	14.2	0.56
1/2	36.8	1.45	33.0	1.30	46.0	1.81	31.0	1.22	25.9	1.02	24.1	0.95	46.2	1.82	61.5	2.42	22.2	0.875	14.2	0.56
5/8	43.2	1.70	38.6	1.52	52.1	2.05	37.3	1.47	31.2	1.23	29.7	1.17	55.1	2.17	72.9	2.87	25.4	1.000	19.0	0.75
3/4	49.3	1.94	43.9	1.73	57.2	2.25	40.4	1.59	34.5	1.36	30.5	1.20	62.0	2.44	83.3	3.28	30.2	1.188	19.0	0.75
7/8	50.8	2.00	45.7	1.80	58.7	2.31	42.9	1.69	36.1	1.42	33.0	1.30	65.8	2.59	88.9	3.50	33.3	1.312	19.0	0.75
1	52.1	2.05	47.2	1.86	60.5	2.38	50.0	1.97	41.1	1.62	37.6	1.48	76.5	3.01	102.9	4.05	36.5	1.438	23.9	0.94
1 1/4	57.2	2.25	48.5	1.91	61.2	2.41	60.5	2.38	43.2	1.70	42.4	1.67	93.7	3.69	127.0	5.00	44.4	1.750	24.6	0.97
1 1/2	60.7	2.39	48.5	1.91	61.5	2.42	67.1	2.64	52.8	2.08	45.0	1.77	104.1	4.10	141.0	5.55	50.8	2.000	25.4	1.00
2	73.4	2.89	47.2	1.86	65.5	2.58	76.2	3.00	60.7	2.39	53.6	2.11	122.2	4.81	168.4	6.63	66.7	2.625	26.2	1.03

Nom Tube OD, in	P_1 Max mm	P_1 Max in	Q Min mm	Q Min in	R Rad Max mm	R Rad Max in	R_1 Rad Max mm	R_1 Rad Max in	S^b Max mm	S^b Max in	$S_1^{b,c}$ Min mm	$S_1^{b,c}$ Min in	S_2^c Min mm	S_2^c Min in	T^d Ref mm	T^d Ref in	T_1 mm ±0.5	T_1 in ±0.02	T_2^d Ref mm	T_2^d Ref in
1/8	11.7	0.46	2.8	0.11	0.3	0.010	0.1	0.005	12.4	0.49	20.6	0.81	19.0	0.75	5.6	0.22	5.6	0.22	7.1	0.28
3/16	11.7	0.46	2.8	0.11	0.4	0.016	0.1	0.005	12.4	0.49	20.8	0.82	19.3	0.76	5.6	0.22	5.6	0.22	7.1	0.28
1/4	11.7	0.46	2.8	0.11	0.4	0.016	0.1	0.005	14.0	0.55	27.2	1.07	20.3	0.80	5.6	0.22	7.1	0.28	7.1	0.28
5/16	11.7	0.46	3.0	0.12	0.4	0.016	0.3	0.010	14.0	0.55	28.2	1.11	20.8	0.82	5.6	0.22	7.1	0.28	7.1	0.28
3/8	17.0	0.67	3.0	0.12	0.4	0.016	0.3	0.010	—	—	—	—	26.9	1.06	6.4	0.25	6.9	0.27	7.9	0.31
1/2	17.3	0.68	4.0	0.16	0.4	0.016	0.3	0.010	17.0	0.67	38.1	1.50	32.3	1.27	6.4	0.25	7.9	0.31	8.6	0.34
5/8	22.9	0.90	4.0	0.16	0.4	0.016	0.3	0.010	23.1	0.91	44.7	1.76	38.6	1.52	7.9	0.31	9.1	0.36	10.2	0.40
3/4	23.1	0.91	4.8	0.19	0.4	0.016	0.3	0.010	23.9	0.94	49.5	1.95	43.9	1.73	9.7	0.38	10.4	0.41	11.9	0.47
7/8	23.1	0.91	4.8	0.19	0.4	0.016	0.3	0.010	—	—	—	—	46.2	1.82	9.7	0.38	10.4	0.41	11.9	0.47
1	29.0	1.14	4.8	0.19	0.4	0.016	0.3	0.010	28.7	1.13	62.2	2.45	53.3	2.10	9.7	0.38	10.4	0.41	12.7	0.50
1 1/4	29.5	1.16	5.8	0.23	0.6	0.025	0.4	0.016	30.5	1.20	75.7	2.98	58.2	2.29	11.7	0.46	10.4	0.41	14.7	0.58
1 1/2	29.5	1.16	5.8	0.23	0.6	0.025	0.4	0.016	32.3	1.27	85.3	3.36	71.1	2.80	13.5	0.53	10.4	0.41	16.5	0.65
2	30.8	1.18	8.4	0.33	0.6	0.025	0.4	0.016	34.8	1.37	100.3	3.95	84.8	3.34	17.3	0.68	10.4	0.41	20.6	0.81

Nom Tube OD, in	U Dia mm +0.4 −0.0	U Dia in +0.016 −0.000	V Dia mm ±0.25	V Dia in ±0.010	X Dia mm ±0.13	X Dia in ±0.005	X_1 Dia mm ±0.13	X_1 Dia in ±0.005	$Y^{n,o}$ mm +0.0	$Y^{n,o}$ in +0.000	Y_1^e mm +0.0	Y_1^e in +0.000	MM mm ±0.08	MM in ±0.003	NN Dia mm ±0.4	NN Dia in ±0.016
1/8	8.1	0.317	6.25	0.245	14.27	0.562	11.13	0.438	11.1-0.8	0.438-0.030	14.3-0.8	0.562-0.030	0.76	0.030	12.8	0.504
3/16	9.7	0.380	7.75	0.305	15.88	0.625	12.70	0.500	11.1-0.8	0.438-0.030	14.3-0.8	0.562-0.030	0.76	0.030	14.6	0.575
1/4	11.3	0.443	9.15	0.360	17.48	0.688	14.30	0.563	11.1-0.8	0.438-0.030	14.3-0.8	0.562-0.030	0.89	0.035	16.5	0.650
5/16	12.8	0.505	10.65	0.420	19.05	0.750	15.88	0.625	14.3-0.8	0.562-0.030	14.3-0.8	0.562-0.030	0.89	0.035	18.3	0.722
3/8	14.4	0.567	12.05	0.475	20.62	0.812	17.48	0.688	14.3-0.8	0.562-0.030	19.0-0.8	0.750-0.030	0.89	0.035	20.2	0.794
1/2	19.2	0.755	16.65	0.655	25.40	1.000	22.22	0.875	19.0-0.8	0.750-0.030	22.2-0.8	0.875-0.030	1.04	0.041	25.7	1.010
5/8	22.4	0.880	19.55	0.770	28.58	1.125	25.40	1.000	22.2-0.8	0.875-0.030	27.0-0.8	1.062-0.030	1.27	0.050	29.3	1.155
3/4	27.1	1.067	23.90	0.940	34.92	1.375	31.75	1.250	27.0-1.0	1.062-0.040	33.3-1.0	1.312-0.040	1.27	0.050	36.7	1.444
7/8	30.3	1.193	27.05	1.065	38.10	1.500	34.92	1.375	33.3-1.0	1.312-0.040	33.3-1.0	1.312-0.040	1.27	0.050	40.4	1.589
1	33.5	1.317	30.25	1.190	41.28	1.625	38.10	1.500	33.3-1.0	1.312-0.040	41.3-1.0	1.625-0.040	1.27	0.050	44.0	1.732
1 1/4	41.4	1.630	38.10	1.500	47.62	1.875	47.62	1.875	41.3-1.0	1.625-0.040	47.6-1.0	1.875-0.040	1.27	0.050	55.0	2.165
1 1/2	47.8	1.880	44.45	1.750	53.98	2.125	53.98	2.125	47.6-1.0	1.875-0.040	65.1-1.0	2.562-0.040	1.27	0.050	62.3	2.454
2	63.6	2.505	60.35	2.375	69.85	2.750	69.85	2.750	63.5-1.0	2.500-0.040	71.4-1.0	2.812-0.040	1.27	0.050	80.3	3.160

[a] O-ring groove undercut must be smooth and free from tool marks.
[b] At manufacturer's option, through passages in Figures 19A and 22C may conform with the smaller diameter specified or the appropriate end may be counterbored to the larger diameter for depths S and S_j, respectively.
[c] Maximum depth shall be optional with manufacturer providing wall thickness is controlled in compliance with General Specifications.
[d] Minimum design thickness, not subject to inspection.
[e] The basic dimensions shown shall apply as minimum for boss diameters.
[f] Unified class 2B thread shall apply to swivel nuts Figures 25C, 26A, 26B and 26C and with minor diameter modified to class 3B limits for locknuts Figures 25A and 25B.
[g] Across flat widths must fit standard wrench openings. See ANSI B18.2.2.
[h] Diameter of bulkhead pilot is the same as major thread diameter. Recommended pilot hole for bulkhead fittings is 0.4 mm (.016 in) over major thread diameter.
[k] J full thread minimum with thread runout. If undercut to G diameter and H width. Length of thread and undercut must not be less than I.
[l] Previously designed with the undercut shall remain as an option until five years from date of this revision.
[m] See Table 1 for tolerance.

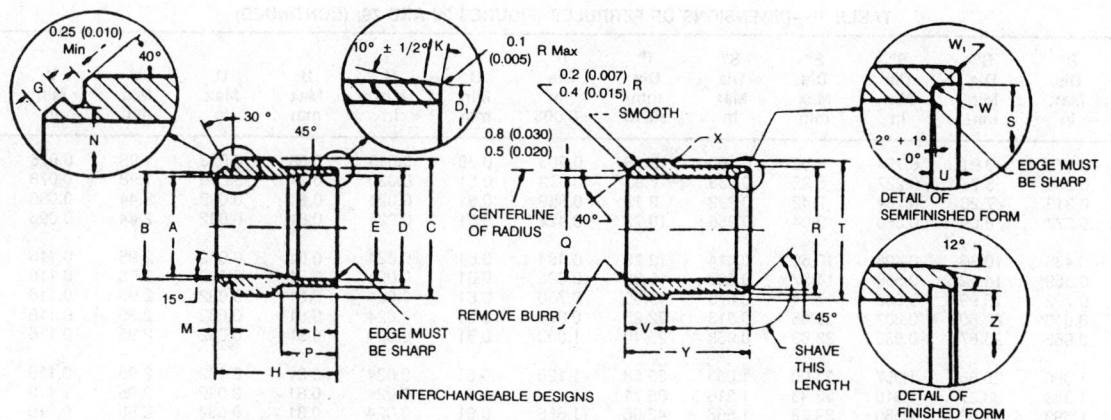

FIGURE 28—STYLE A FERRULE (080115A)   FIGURE 29—STYLE B FERRULE (080115B)

NOTE: FIGURES 28 AND 29 APPLY TO TABLE 10. CODES SHOWN IN BRACKETS ADJACENT TO FIGURE NUMBERS REPRESENT RESPECTIVE FITTING IDENTIFICATION IN ACCORDANCE WITH SAE J846.

TABLE 10—DIMENSIONS OF FERRULES (FIGURES 28 AND 29)

Nom Tube OD, in	A^a Dia mm +0.08 -0.000	A^a Dia in +0.003 -0.000	B Dia mm +0.00 -0.20	B Dia in +0.000 -0.008	C^a Dia mm +0.13 -0.00	C^a Dia in +0.005 -0.000	D^a Dia mm +0.00 -0.08	D^a Dia in +0.000 -0.003	D_1 Dia Ref mm	D_1 Dia Ref in	E^a Dia mm +0.08 -0.00	E^a Dia in +0.003 -0.000	G Ref mm	G Ref in	H mm ±0.08	H in ±0.003
1/8	3.30	0.130	3.81	0.150	6.02	0.237	4.52	0.178	4.3	0.171	3.56	0.140	0.69	0.027	7.32	0.288
3/16	4.90	0.193	5.38	0.212	7.80	0.307	6.32	0.249	6.1	0.242	5.21	0.205	0.69	0.027	8.36	0.329
1/4	6.48	0.255	6.98	0.275	9.32	0.367	7.92	0.312	7.7	0.303	6.81	0.268	0.69	0.027	9.22	0.363
5/16	8.08	0.318	8.59	0.338	10.92	0.430	9.52	0.375	9.3	0.366	8.38	0.330	0.69	0.027	9.32	0.367
3/8	9.65	0.380	10.16	0.400	12.50	0.492	11.18	0.440	10.9	0.431	9.98	0.393	0.74	0.029	9.98	0.393
1/2	12.83	0.505	13.59	0.535	16.84	0.663	14.91	0.587	14.7	0.577	13.23	0.521	1.12	0.044	10.90	0.429
5/8	16.03	0.631	16.79	0.661	19.81	0.780	18.11	0.713	17.9	0.703	16.43	0.647	1.07	0.042	11.23	0.442
3/4	19.20	0.756	19.96	0.786	23.50	0.925	21.29	0.838	21.0	0.828	19.61	0.772	1.27	0.050	12.06	0.475
7/8	22.38	0.881	23.14	0.911	26.42	1.040	24.46	0.963	24.2	0.953	22.78	0.897	1.32	0.052	12.06	0.475
1	25.55	1.006	26.31	1.036	30.15	1.187	27.64	1.088	27.4	1.078	25.96	1.022	1.37	0.054	12.06	0.475
1 1/4	32.00	1.260	32.74	1.289	36.73	1.446	32.06	1.341	33.8	1.331	32.38	1.275	1.57	0.062	12.06	0.475
1 1/2	38.35	1.510	39.09	1.539	43.03	1.694	40.39	1.590	40.1	1.580	38.71	1.524	1.57	0.062	12.06	0.475
2	51.16	2.014	51.79	2.039	56.13	2.210	53.19	2.094	52.9	2.084	51.46	2.026	1.78	0.070	12.93	0.509

Nom Tube OD, in	K mm +0.08 -0.05	K in +0.003 -0.002	L mm +0.40 -0.00	L in +0.016 -0.000	M mm +0.00 -0.15	M in +0.000 -0.006	N Dia mm +0.00 -0.13	N Dia in +0.000 -0.005	P Min mm	P Min in	P Max mm	P Max in	Q^b Dia mm +0.08 -0.00	Q^b Dia in +0.003 -0.000	R^b Dia Min mm	R^b Dia Min in
1/8	0.51	0.020	1.98	0.078	1.17	0.046	3.86	0.152	3.18	0.125	3.30	0.130	3.28	0.129	4.44	0.175
3/16	0.51	0.020	1.98	0.078	1.19	0.047	5.54	0.218	3.18	0.125	3.30	0.130	4.83	0.190	6.40	0.252
1/4	0.64	0.025	2.77	0.109	1.24	0.049	7.24	0.285	3.96	0.156	4.09	0.161	6.48	0.255	7.85	0.309
5/16	0.64	0.025	3.18	0.125	1.24	0.049	8.94	0.352	4.22	0.166	4.34	0.171	8.08	0.318	9.47	0.373
3/8	0.64	0.025	3.18	0.125	1.24	0.049	10.62	0.418	3.84	0.151	3.96	0.156	9.65	0.380	10.95	0.431
1/2	0.76	0.030	4.11	0.162	1.73	0.068	14.10	0.555	6.10	0.240	6.60	0.260	12.85	0.506	14.86	0.585
5/8	0.76	0.030	4.60	0.181	1.63	0.064	17.30	0.681	5.77	0.227	6.27	0.247	16.08	0.633	17.73	0.698
3/4	0.76	0.030	4.60	0.181	1.93	0.076	20.50	0.807	6.35	0.250	6.86	0.270	19.25	0.758	21.23	0.836
7/8	0.76	0.030	4.60	0.181	2.11	0.083	23.65	0.931	6.48	0.255	6.98	0.275	22.43	0.883	24.41	0.961
1	0.76	0.030	4.75	0.187	2.11	0.083	26.82	1.056	6.20	0.244	6.71	0.264	25.60	1.008	27.58	1.086
1 1/4	0.76	0.030	4.75	0.187	2.11	0.083	33.25	1.309	5.72	0.225	6.22	0.245	32.00	1.260	34.01	1.339
1 1/2	0.76	0.030	4.75	0.187	2.18	0.086	39.60	1.559	5.72	0.225	6.22	0.245	38.38	1.511	40.36	1.589
2	0.76	0.030	4.75	0.187	2.34	0.092	52.30	2.059	6.05	0.238	6.55	0.258	51.16	2.014	53.14	2.092

TABLE 10—DIMENSIONS OF FERRULES (FIGURES 28 AND 29) (CONTINUED)

Nom Tube OD, in	R[b] Dia Max mm	R[b] Dia Max in	S[b] Dia Min mm	S[b] Dia Min in	S[b] Dia Max mm	S[b] Dia Max in	T[b] Dia mm ±0.13	T[b] Dia in ±0.005	U Min mm	U Min in	U Max mm	U Max in	V Min mm	V Min in	V Max mm	V Max in
1/8	4.52	0.178	3.91	0.154	4.06	0.160	5.16	0.203	0.38	0.015	0.58	0.023	1.98	0.078	2.13	0.084
3/16	6.50	0.256	5.77	0.227	5.92	0.233	7.93	0.312	0.51	0.020	0.71	0.028	1.98	0.078	2.13	0.084
1/4	7.95	0.313	7.26	0.286	7.42	0.292	9.12	0.359	0.61	0.024	0.81	0.032	2.44	0.096	2.59	0.102
5/16	9.58	0.377	8.89	0.350	9.04	0.356	10.72	0.422	0.61	0.024	0.81	0.032	2.44	0.096	2.59	0.102
3/8	11.05	0.435	10.36	0.408	10.52	0.414	12.29	0.484	0.61	0.024	0.81	0.032	2.95	0.116	3.10	0.122
1/2	14.96	0.589	14.12	0.556	14.27	0.562	15.88	0.625	0.61	0.024	0.81	0.032	2.95	0.116	3.10	0.122
5/8	17.83	0.702	16.99	0.669	17.14	0.675	19.05	0.750	0.61	0.024	0.81	0.032	2.95	0.116	3.10	0.122
3/4	21.34	0.840	20.50	0.807	20.65	0.813	22.23	0.875	0.61	0.024	0.81	0.032	2.95	0.116	3.10	0.122
7/8	24.51	0.965	23.67	0.932	23.83	0.938	25.40	1.000	0.61	0.024	0.81	0.032	2.95	0.116	3.10	0.122
1	27.69	1.090	26.85	1.057	27.00	1.063	28.58	1.125	0.61	0.024	0.81	0.032	2.95	0.116	3.10	0.122
1 1/4	34.11	1.343	33.27	1.310	33.43	1.316	35.71	1.406	0.61	0.024	0.81	0.032	2.95	1.116	3.10	0.122
1 1/2	40.46	1.593	39.62	1.560	39.78	1.566	42.06	1.656	0.61	0.024	0.81	0.032	2.95	0.116	3.10	0.122
2	53.24	2.096	52.40	2.063	52.55	2.069	55.58	2.188	0.61	0.024	0.81	0.032	3.89	0.153	4.04	0.159

Nom Tube OD, in	W Rad Min mm	W Rad Min in	W Rad Max mm	W Rad Max in	W_1 Rad mm	W_1 Rad in	X Rad mm	X Rad in	Y Min mm	Y Min in	Y Max mm	Y Max in	Z Dia Min mm	Z Dia Min in	Z Dia Max mm	Z Dia Max in
1/8	0.08	0.003	0.15	0.006	0.2	0.007	0.3	0.010	6.98	0.275	7.14	0.281	3.76	0.148	3.96	0.156
3/16	0.08	0.003	0.15	0.006	0.2	0.007	0.5	0.020	6.98	0.275	7.14	0.281	5.61	0.221	5.82	0.229
1/4	0.08	0.003	0.15	0.006	0.2	0.007	0.5	0.020	8.46	0.333	8.61	0.339	7.11	0.280	7.32	0.288
5/16	0.08	0.003	0.15	0.006	0.2	0.007	0.5	0.020	8.46	0.333	8.61	0.339	8.74	0.344	8.94	0.352
3/8	0.08	0.003	0.15	0.006	0.2	0.007	0.5	0.020	9.45	0.372	9.60	0.378	10.21	0.402	10.41	0.410
1/2	0.20	0.008	0.36	0.014	0.3	0.010	0.5	0.020	9.45	0.372	9.60	0.378	13.97	0.550	14.17	0.558
5/8	0.20	0.008	0.36	0.014	0.3	0.010	0.5	0.020	10.46	0.412	10.62	0.418	16.84	0.663	17.04	0.671
3/4	0.20	0.008	0.36	0.014	0.3	0.010	0.5	0.020	10.46	0.412	10.62	0.418	20.35	0.801	20.55	0.809
7/8	0.20	0.008	0.36	0.014	0.3	0.010	0.5	0.020	10.46	0.412	10.62	0.418	23.52	0.926	23.72	0.934
1	0.20	0.008	0.36	0.014	0.3	0.010	0.5	0.020	10.46	0.412	10.62	0.418	26.70	1.051	26.90	1.059
1 1/4	0.20	0.008	0.36	0.014	0.3	0.010	0.5	0.020	10.46	0.412	10.62	0.418	33.12	1.304	33.32	1.312
1 1/2	0.20	0.008	0.36	0.014	0.3	0.010	0.5	0.020	10.46	0.412	10.62	0.418	39.47	1.554	39.67	1.562
2	0.20	0.008	0.36	0.014	0.3	0.010	0.5	0.020	11.43	0.450	11.58	0.456	52.25	2.057	52.45	2.065

[a] These diameters, A, C, D, and E must be concentric with 0.13 mm (0.005 in).
[b] These diameters, Q, R, S, and T must be concentric with 0.13 mm (0.005 in).

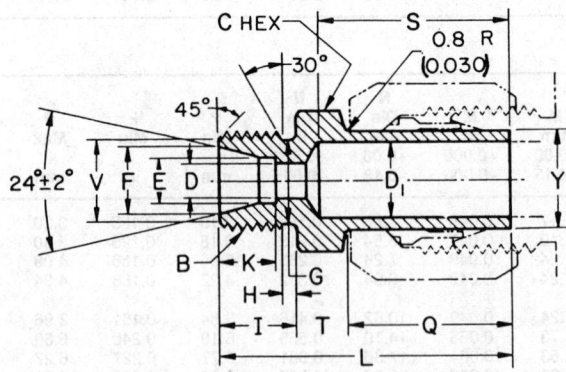

FIGURE 30—REDUCER (080123)

TABLE 11—DIMENSIONS[a] OF REDUCERS (FIGURE 30)

Tube Reduction in	B Thread Size, in SAE J475 (ISO R725) Class 2A Ext	C Hex Nom, in	D[c] Dia Drill mm	D[c] Dia Drill in	D₁[c] Dia Drill mm	D₁[c] Dia Drill in	L mm ±0.5	L in ±0.02	Q mm ±0.5	Q in ±0.02	S mm ±0.5	S in ±0.02	T[b] Ref mm	T[b] Ref in	Y Dia mm ±0.08	Y Dia in ±0.003
3/8 × 1/4	7/16-20	1/2	5.2	0.203	6.3	0.250	40.9	1.61	22.4	0.88	23.4	0.92	5.6	0.22	9.52	0.375
1/2 × 1/4	7/16-20	9/16	5.2	0.203	9.5	0.375	43.9	1.73	25.4	1.00	27.2	1.07	5.6	0.22	12.70	0.500
1/2 × 3/8	9/16-18	5/8	7.0	0.281	9.5	0.375	45.0	1.77	25.4	1.00	27.9	1.10	6.4	0.25	12.70	0.500
5/8 × 1/4	7/16-20	11/16	5.2	0.203	12.7	0.500	47.0	1.85	27.7	1.09	29.0	1.14	6.4	0.25	15.88	0.625
5/8 × 3/8	9/16-18	11/16	7.0	0.281	12.7	0.500	47.2	1.86	27.7	1.09	30.5	1.20	6.4	0.25	15.88	0.625
5/8 × 1/2	3/4-16	13/16	10.7	0.422	12.7	0.500	49.8	1.96	27.7	1.09	30.5	1.20	6.4	0.25	15.88	0.625
3/4 × 1/4	7/16-20	13/16	5.2	0.203	15.9	0.625	48.8	1.92	29.5	1.16	29.0	1.14	6.4	0.25	19.05	0.750
3/4 × 3/8	9/16-18	13/16	7.0	0.281	15.9	0.625	49.0	1.93	29.5	1.16	30.5	1.20	6.4	0.25	19.05	0.750
3/4 × 1/2	3/4-16	13/16	10.7	0.422	15.9	0.625	51.6	2.03	29.5	1.16	30.5	1.20	6.4	0.25	19.05	0.750
3/4 × 5/8	7/8-14	15/16	12.7	0.500	15.9	0.625	54.6	2.15	29.5	1.16	31.8	1.26	7.9	0.31	19.05	0.750
1 × 1/2	3/4-16	11/16	10.7	0.422	21.4	0.844	52.1	2.05	28.4	1.12	27.9	1.10	7.9	0.31	25.40	1.000
1 × 5/8	7/8-14	11/16	12.7	0.500	21.4	0.844	53.6	2.11	28.4	1.12	27.9	1.10	7.9	0.31	25.40	1.000
1 × 3/4	1 1/16-12	1 1/8	16.6	0.656	21.4	0.844	56.9	2.24	28.4	1.12	29.7	1.17	9.7	0.38	25.40	1.000
1 1/4 × 5/8	7/8-14	1 3/8	12.7	0.500	26.2	1.031	56.4	2.22	29.5	1.16	29.0	1.14	9.7	0.38	31.75	1.250
1 1/4 × 3/4	1 1/16-12	1 3/8	16.6	0.656	26.2	1.031	58.2	2.29	29.5	1.16	30.5	1.20	9.7	0.38	31.75	1.250
1 1/4 × 1	1 5/16-12	1 3/8	22.2	0.875	26.2	1.031	57.9	2.28	29.5	1.16	31.2	1.23	9.7	0.38	31.75	1.250

[a] For dimensions shown on Figure 30 but not specified in above table, see corresponding dimensions for the specified tube outside diameter in Table 9.
[b] Minimum design thickness, not subject to inspection.
[c] See Table 1 for tolerance.

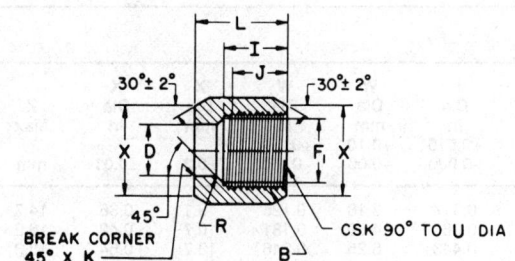

FIGURE 31—NUT (080110)

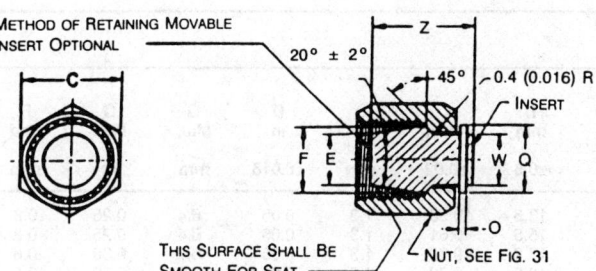

FIGURE 32—CAP ASSEMBLY (080112)

NOTE: Table 13 and Figure 33 have intentionally been deleted from this document. See SAE J1926, Specification for Straight Thread O-Ring Boss Port.

TABLE 12—DIMENSIONS OF NUT AND CAP ASSEMBLY (FIGURES 31 AND 32)

Nom Tube OD, in	B Thread Size, in SAE J475 (ISO R725) Class 2B Int	C Hex Nom, in	D Dia mm	D Dia tol mm -0.00	D Dia in	D Dia tol in -0.000	E Dia mm +0.13 -0.00	E Dia in +0.005 -0.000	F Dia mm ±0.5	F Dia in ±0.02	F_1 Dia mm ±0.13	F_1 Dia in ±0.005	I mm +0.40 -0.00	I in +0.016 -0.000	J Full Thread Min mm	J Full Thread Min in	K mm	K in
1/8	5/16-24	3/8	3.30	+0.10 -0.00	0.130	+0.004 -0.000	3.20	0.126	5.6	0.22	—	—	9.86	0.388	8.3	0.328	0.5	0.02
3/16	3/8-24	7/16	4.90	+0.10 -0.00	0.193	+0.004 -0.000	4.75	0.187	7.1	0.28	—	—	11.66	0.459	10.3	0.406	0.5	0.02
1/4	7/16-20	9/16	6.48	+0.10 -0.00	0.255	+0.004 -0.000	6.40	0.252	8.6	0.34	—	—	13.64	0.537	11.9	0.469	0.5	0.02
5/16	1/2-20	5/8	8.08	+0.10 -0.00	0.318	+0.004 -0.000	8.00	0.315	10.4	0.41	—	—	14.02	0.552	12.3	0.483	0.5	0.02
3/8	9/16-18	11/16	9.65	+0.10 -0.00	0.380	+0.004 -0.000	9.58	0.377	11.9	0.47	—	—	14.43	0.568	12.7	0.500	0.5	0.02
1/2	3/4-16	7/8	12.83	+0.13 -0.00	0.505	+0.005 -0.000	12.78	0.503	15.7	0.62	—	—	15.21	0.599	13.1	0.516	0.5	0.02
5/8	7/8-14	1	16.03	+0.13 -0.00	0.631	+0.005 -0.000	16.00	0.630	19.0	0.75	—	—	17.20	0.677	14.7	0.578	0.5	0.02
3/4	1 1/16-12	1 1/4	19.20	+0.13 -0.00	0.756	+0.005 -0.000	19.18	0.755	22.4	0.88	—	—	17.20	0.677	14.3	0.562	0.8	0.03
7/8	1 3/16-12	1 3/8	22.38	+0.13 -0.00	0.881	+0.005 -0.000	22.22	0.875	25.4	1.00	—	—	17.20	0.677	14.3	0.562	0.8	0.03
1	1 5/16-12	1 1/2	25.55	+0.13 -0.00	1.006	+0.005 -0.000	25.53	1.005	29.5	1.16	—	—	17.20	0.677	14.3	0.562	0.8	0.03
1 1/4	1 5/8-12	2	32.00	+0.15 -0.00	1.260	+0.006 -0.000	31.93	1.257	35.8	1.41	36.73	1.446	16.38	0.645	13.5	0.531	0.8	0.03
1 1/2	1 7/8-12	2 1/4	38.35	+0.15 -0.00	1.510	+0.006 -0.000	38.28	1.507	42.2	1.66	43.05	1.695	16.23	0.639	13.1	0.515	0.8	0.03
2	2 1/2-12	2 7/8	51.16	+0.15 -0.00	2.014	+0.006 -0.000	50.80	2.000	55.6	2.19	58.04	2.285	15.62	0.615	12.7	0.500	0.8	0.03

Nom Tube OD, in	L mm ±0.5	L in ±0.02	O mm ±0.4	O in ±0.016	Q Max mm	Q Max in	R Rad mm	R Rad in	U Dia mm +0.4 -0.0	U Dia in +0.016 -0.000	W Dia mm +0.10 -0.00	W Dia in +0.004 -0.000	X Dia mm ±0.3	X Dia in ±0.01	Z Max mm	Z Max in
1/8	13.5	0.53	1.3	0.05	6.4	0.25	0.8	0.031	8.1	0.317	3.18	0.125	9.1	0.36	14.7	0.58
3/16	15.5	0.61	1.3	0.05	6.4	0.25	0.8	0.031	9.7	0.380	4.60	0.181	10.7	0.42	16.2	0.64
1/4	17.8	0.70	1.3	0.05	9.7	0.38	0.8	0.031	11.3	0.443	6.25	0.246	13.7	0.54	18.3	0.72
5/16	18.3	0.72	1.3	0.05	12.7	0.50	0.8	0.031	12.8	0.505	7.87	0.310	15.2	0.60	18.8	0.74
3/8	19.0	0.75	1.3	0.05	12.7	0.50	0.8	0.031	14.4	0.567	9.45	0.372	17.0	0.67	19.5	0.77
1/2	21.3	0.84	1.5	0.06	15.7	0.62	1.2	0.047	19.2	0.755	12.65	0.498	21.8	0.86	23.6	0.93
5/8	23.4	0.92	1.5	0.06	19.0	0.75	1.2	0.047	22.4	0.880	15.88	0.625	24.9	0.98	23.9	0.94
3/4	24.6	0.97	1.5	0.06	22.4	0.88	1.2	0.047	27.1	1.067	19.05	0.750	31.5	1.24	23.6	0.93
7/8	25.4	1.00	1.5	0.06	25.4	1.00	1.2	0.047	30.3	1.193	22.22	0.875	34.5	1.36	24.4	0.96
1	26.7	1.05	2.3	0.09	31.8	1.25	1.6	0.062	33.5	1.317	25.40	1.000	37.6	1.48	27.4	1.08
1 1/4	26.7	1.05	3.0	0.12	38.1	1.50	1.6	0.062	41.4	1.630	31.75	1.250	50.3	1.98	30.2	1.19
1 1/2	26.2	1.03	4.1	0.16	44.4	1.75	1.6	0.062	47.8	1.880	38.10	1.500	56.9	2.24	32.7	1.29
2	28.4	1.12	4.1	0.16	57.2	2.25	1.6	0.062	63.6	2.505	50.80	2.000	72.6	2.86	32.3	1.27

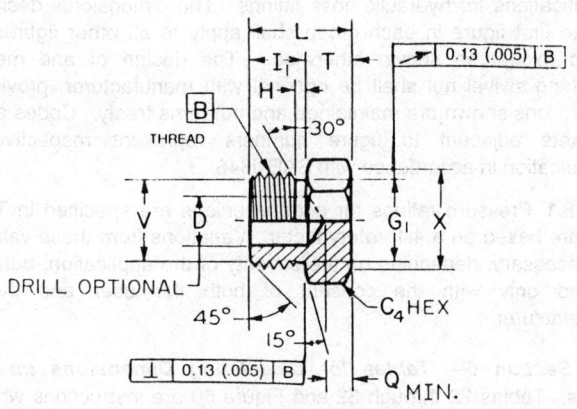

(R) FIGURE 34A—HEXAGON HEAD O-RING BOSS PLUG (090109A)

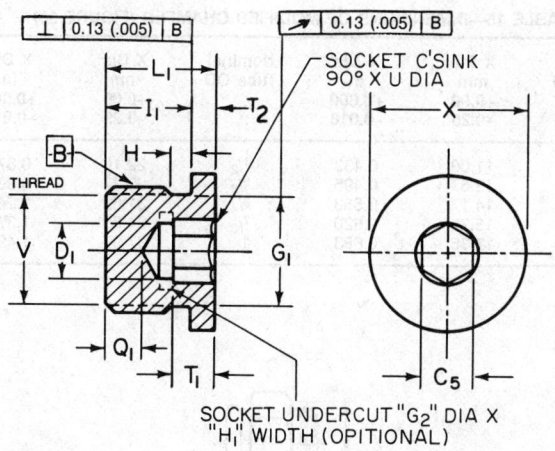

(R) FIGURE 34B—HEXAGON SOCKET O-RING BOSS PLUG (090109B)

(R) TABLE 14—DIMENSIONS[a] OF O-RING BOSS PLUGS (FIGURES 34A TO 34B)

Nom Tube OD, In	B Thread Size, In SAE J475 (ISO R725) Class 2A Ext, in	C_4 Hex Nom	C_5 Hex Socket mm +0.13 −0.00	C_5 Hex Socket in +0.005 −0.000	D^d Drill mm	D^d Drill in	D_1 Dia mm +0.13 −0.00	D_1 Dia in +0.005 −0.000	G_1^b Dia mm +0.05 −0.08	G_1^b Dia in +0.002 −0.003	G_2 Dia Optional mm +0.25 −0.00	G_2 Dia Optional in +0.010 −0.000	H mm +0.8 −0.0	H in +0.030 −0.000	H_1 Optional mm ±0.5	H_1 Optional in ±0.020
1/8	5/16-24	7/16	3.18	0.125	2.4	0.093	3.18	0.125	6.35	0.250	4.0	0.156	1.6	0.063	1.6	0.063
3/16	3/8-24	1/2	3.18	0.125	3.2	0.125	3.18	0.125	7.95	0.313	4.0	0.156	1.6	0.063	1.6	0.063
1/4	7/16-20	9/16	4.78	0.188	5.2	0.203	4.78	0.188	9.25	0.364	5.9	0.234	1.9	0.075	2.4	0.094
5/16	1/2-20	5/8	4.78	0.188	5.9	0.234	4.78	0.188	10.85	0.427	5.9	0.234	1.9	0.075	2.4	0.094
3/8	9/16-18	11/16	6.35	0.250	7.5	0.297	6.35	0.250	12.24	0.482	7.5	0.297	2.1	0.083	2.4	0.094
1/2	3/4-16	7/8	7.95	0.313	10.7	0.422	7.95	0.313	16.76	0.660	9.7	0.380	2.4	0.094	2.4	0.094
5/8	7/8-14	1	9.52	0.375	12.7	0.500	9.52	0.375	19.63	0.773	11.3	0.443	2.7	0.107	2.4	0.094
3/4	1 1/16-12	1 1/4	14.30	0.563	16.7	0.656	14.30	0.563	24.00	0.945	16.8	0.661	3.2	0.125	3.2	0.125
7/8	1 3/16-12	1 3/8	14.30	0.563	18.2	0.718	14.30	0.563	27.18	1.070	16.8	0.661	3.2	0.125	3.2	0.125
1	1 5/16-12	1 1/2	15.88	0.625	22.2	0.875	15.88	0.625	30.35	1.195	18.8	0.740	3.2	0.125	3.2	0.125
1 1/4	1 5/8-12	1 7/8	19.05	0.750	27.8	1.093	19.05	0.750	38.28	1.507	22.2	0.875	3.2	0.125	3.2	0.125
1 1/2	1 7/8-12	2 1/8	19.05	0.750	34.1	1.344	19.05	0.750	44.60	1.756	22.2	0.875	3.2	0.125	3.2	0.125
2	2 1/2-12	2 3/4	19.05	0.750	46.1	1.813	19.05	0.750	60.48	2.381	22.2	0.875	3.2	0.125	3.2	0.125

Nom Tube OD, in	I_1 mm ±0.13	I_1 in ±0.005	L mm ±0.5	L in ±0.02	L_1 Ref mm	L_1 Ref in	Q Min mm	Q Min in	Q_1 Min mm	Q_1 Min in	T^c Ref mm	T^c Ref in	T_1 Min Hexagon Depth mm	T_1 Min Hexagon Depth in	T_2 mm +0.00 −0.25	T_2 in +0.000 −0.010	V Dia mm ±0.13	V Dia in ±0.005	X Dia mm ±0.13	X Dia in ±0.005	U Dia mm ±0.3	U Dia in ±0.010
1/8	7.54	0.297	15.2	0.60	10.2	0.40	1.5	0.06	2.8	0.11	7.1	0.28	3.18	0.125	2.74	0.108	6.35	0.250	11.13	0.438	4.0	0.156
3/16	7.54	0.297	15.2	0.60	10.2	0.40	2.0	0.08	2.8	0.11	7.1	0.28	3.18	0.125	2.74	0.108	7.87	0.310	12.70	0.500	4.0	0.156
1/4	9.14	0.360	17.0	0.67	11.9	0.47	2.5	0.10	3.0	0.12	7.1	0.28	3.96	0.156	2.92	0.115	9.27	0.365	14.30	0.563	5.6	0.219
5/16	9.14	0.360	17.0	0.67	11.9	0.47	3.0	0.12	3.0	0.12	7.1	0.28	3.96	0.156	2.92	0.115	10.80	0.425	15.88	0.625	5.6	0.219
3/8	9.93	0.391	18.5	0.73	12.7	0.50	4.1	0.16	3.0	0.12	7.9	0.31	4.77	0.188	2.92	0.115	12.19	0.480	17.48	0.688	7.5	0.297
1/2	11.13	0.438	20.3	0.80	14.7	0.58	5.6	0.22	3.8	0.15	8.6	0.34	4.77	0.188	3.73	0.147	16.76	0.660	22.22	0.875	9.5	0.375
5/8	12.70	0.500	23.6	0.93	16.5	0.65	6.4	0.25	3.8	0.15	10.4	0.41	6.35	0.250	3.94	0.155	19.68	0.775	25.40	1.000	11.1	0.438
3/4	15.09	0.594	27.7	1.09	19.5	0.77	6.4	0.25	3.8	0.15	11.9	0.47	7.95	0.313	4.60	0.181	24.00	0.945	31.75	1.250	16.7	0.656
7/8	15.09	0.594	27.7	1.09	19.5	0.77	6.4	0.25	3.8	0.15	11.9	0.47	7.95	0.313	4.60	0.181	27.18	1.070	34.92	1.375	16.7	0.656
1	15.09	0.594	28.4	1.12	19.5	0.77	6.4	0.25	4.8	0.19	12.7	0.50	9.52	0.375	4.60	0.181	30.35	1.195	38.10	1.500	18.6	0.734
1 1/4	15.09	0.594	30.5	1.20	19.5	0.77	6.4	0.25	4.8	0.19	14.7	0.58	9.52	0.375	4.60	0.181	38.23	1.505	47.62	1.875	23.0	0.906
1 1/2	15.09	0.594	32.3	1.27	19.5	0.77	6.4	0.25	6.4	0.25	16.5	0.65	9.52	0.375	4.60	0.181	44.58	1.755	53.98	2.125	23.0	0.906
2	15.09	0.594	36.3	1.43	19.5	0.77	7.6	0.30	6.4	0.25	20.6	0.81	9.52	0.375	4.60	0.181	60.45	2.380	69.85	2.750	23.0	0.906

[a] Modification of 1/8 to 1 in sizes for use with MS 33649 (or superseded AND 10050) bosses is shown in Figure 35 and Table 15.
[b] O-ring groove undercut must be smooth and free of tool marks.
[c] Minimum design thickness, not subject to inspection.
[d] See Table 1 for tolerance.

TABLE 15—DIMENSIONS OF MODIFIED CHAMFER (FIGURE 35)

Nominal Tube OD	X Dia mm +0.00 −0.25	X Dia in +0.000 −0.010	Nominal Tube OD	X Dia mm +0.00 −0.25	X Dia in +0.000 −0.010
1/8	11.00	0.433	1/2	22.10	0.870
3/16	12.57	0.495	5/8	25.27	0.995
1/4	14.17	0.558	3/4	31.62	1.245
5/16	15.75	0.620	7/8	34.80	1.370
3/8	17.35	0.683	1	37.97	1.495

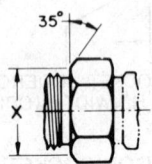

FIGURE 35—MODIFIED HEXAGON CHAMFER

6.2.2 Install the fitting into the SAE straight thread port (see SAE J1926) until the metal back-up washer contacts the face of the port as shown in Figure 36C.

6.2.3 Position the fitting by turning out (counterclockwise) up to a maximum of one turn. Holding the pad of the fitting with a wrench, tighten the locknut and washer against the face of the port as shown in Figure 36D.

6.3 Assembly Instructions for Swivel Style and O-Ring Boss Fittings in Straight Thread O-Ring Ports—These instructions apply to the assembly of hydraulic fittings of the 37 degree flared type shown in Figures 3B and 3C, flareless type shown in Figures 20B and 20C, and O-ring boss plugs shown in Figure 34.

6.3.1 Lubricate O-ring by coating with a light oil or petrolatum and install in the O-ring groove on the fitting.

6.3.2 Screw fitting into the straight thread port and tighten hexagon against the face of the port as shown in Figure 37C.

7. Section 4— *Hydraulic Pipe Fittings*—Hydraulic pipe fittings are shown in Figures 38 to 46 and Tables 16 to 19.

8. Section 5— *Adapter Unions*—Adapter union fittings are shown in Figures 47A to 49C and Tables 20 and 21.

NOTE—Unspecified detail with respect to dimensions, tolerances, contours, material workmanship, and so on, must conform to general specifications for hydraulic hose fittings. The dimensional designations on the first figure in each group shall apply to all other figures in that group except as shown otherwise. The design of and method of attaching swivel nut shall be optional with manufacturer, providing the dimensions shown are maintained and nut turns freely. Codes shown in brackets adjacent to figure numbers represent respective fitting identification in accordance with SAE J846.

8.1 Pressure ratings for adapter unions are specified in Table 20 and are based on a 4:1 safety factor. Variations from these values may be necessary, depending on the severity of the application, but shall be altered only with the consent of both the user and the fitting manufacturer.

9. Section 6— *Tables for Calculating Dimensions on Special Sizes*—Tables 22 through 32 and Figure 50 are instructions which may be used for determining the overall lengths, leg lengths, and stock sizes applicable to special size combination fittings not covered in the standard dimensional tabulations.

9.1 Tables for Calculating Dimensions on Special Sizes—Tables 22 to 32 present various factors to be used in determining the dimensions applicable to special size combination fittings not contained in SAE J514.

In Tables 23, 27, 28, and 29, no factors are given for extreme combination sizes because of differences in factors due to method of manufacture. These extreme conditions are rare and it is suggested a manufacturer be contacted for the proper dimension.

No factors are given for bulkhead or swivel ends as combinations are not generally specified for these fittings.

Tables 30 to 32 present the minimum stock size acceptable for the various machine ends.

For any nonstandard size fitting, be it a connector, 45 or 90 degree elbow, tee, or cross, one end is always standard, conforming to the SAE J514 tables of dimensions. Considering this end to be the largest on the fitting, it may then be used as a basis for establishing the stocksize and length (either over-all or end to center) for all other parts by deducting factors equivalent to the reduction in machining requirements from the appropriate standard lengths as shown in Figures 51A and 51B.

The factors applicable to the various end configurations and size reductions tabulated in the tables were determined on the following basis:

Those pertaining to lengths were derived by maintaining the standard hexagon thickness for straight fittings and the standard centerline to machining start for shaped fittings.

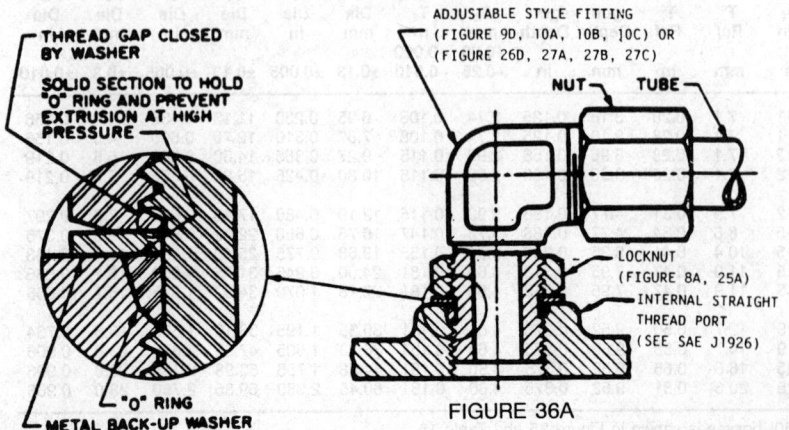
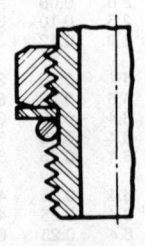
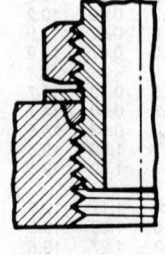
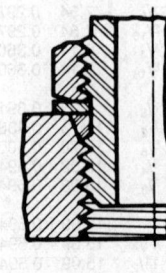

FIGURE 36A   FIGURE 36B   FIGURE 36C   FIGURE 36D

FIGURE 36—ADJUSTABLE STYLE FITTINGS

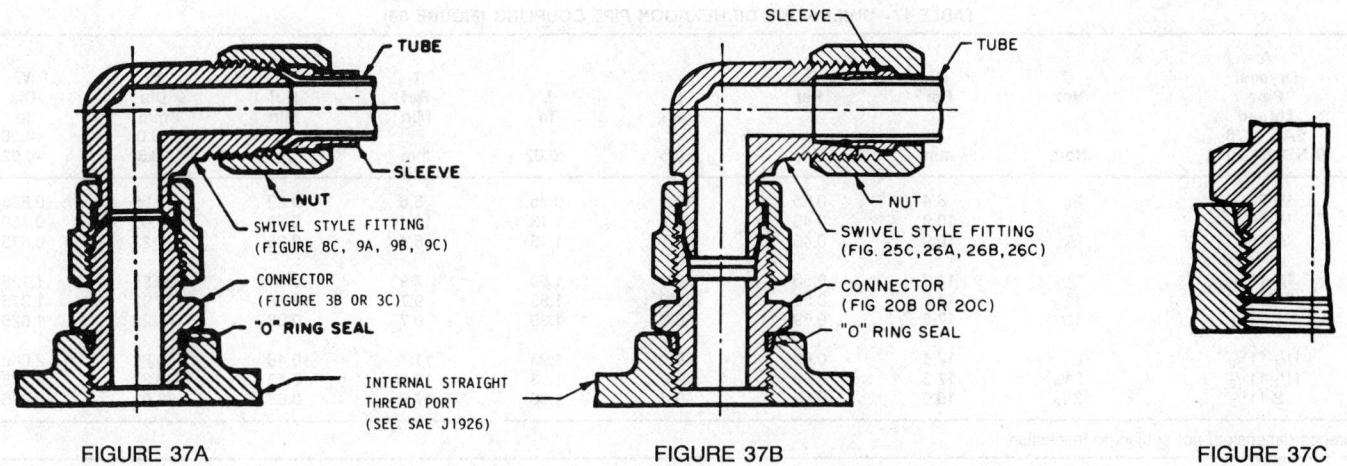

FIGURE 37A FIGURE 37B FIGURE 37C

FIGURE 37—SWIVEL STYLE FITTINGS

NOTES: UNSPECIFIED DETAIL WITH RESPECT TO DIMENSIONS, TOLERANCES, CONTOURS, MATERIAL AND WORKMANSHIP MUST CONFORM TO GENERAL SPECIFICATIONS FOR HYDRAULIC PIPE FITTINGS. CODES SHOWN IN BRACKETS ADJACENT TO FIGURE NUMBERS REPRESENT RESPECTIVE FITTING IDENTIFICATION IN ACCORDANCE WITH SAE J846.

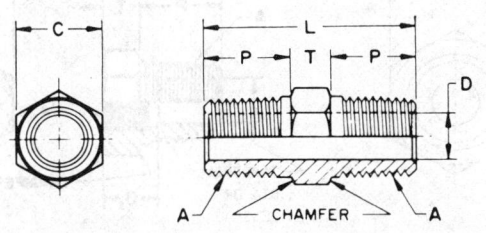

FIGURE 38—HEXAGON PIPE NIPPLE (140137)

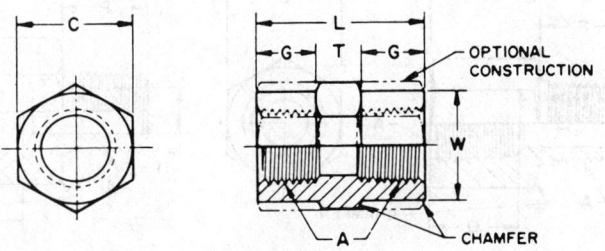

FIGURE 39—HEXAGON PIPE COUPLING (140138)

TABLE 16—DIMENSIONS OF HEXAGON PIPE NIPPLES (FIGURE 38)

A Dryseal Pipe Thread SAE J476 (ANSI B1.20.3)	C Hex Nom	D^b Drill mm	D^b Drill in	L mm ±0.5	L in ±0.02	P Min mm	P Min in	T_a Ref mm	T_a Ref in
1/8-27	7/16	4.8	0.188	26.9	1.06	9.7	0.38	5.6	0.22
1/4-18	5/8	7.0	0.281	36.8	1.45	14.2	0.56	6.4	0.25
3/8-18	3/4	10.3	0.406	36.8	1.45	14.2	0.56	6.4	0.25
1/2-14	7/8	13.5	0.531	48.0	1.89	19.0	0.75	7.9	0.31
3/4-14	1 1/8	18.0	0.719	49.8	1.96	19.0	0.75	9.7	0.38
1-11 1/2	1 3/8	23.8	0.938	59.4	2.34	23.9	0.94	9.7	0.38
1 1/4-11 1/2	1 3/4	31.7	1.250	63.0	2.48	24.6	0.97	11.7	0.46
1 1/2-11 1/2	2	38.0	1.500	66.3	2.61	25.4	1.00	13.5	0.53
2-11 1/2	2 1/2	49.0	1.938	71.6	2.82	26.2	1.03	17.3	0.68

[a] Minimum design thickness, not subject to inspection.
[b] See Table 1 for tolerance.

TABLE 17—DIMENSIONS OF HEXAGON PIPE COUPLING (FIGURE 39)

A Dryseal Pipe Thread SAE J476 (ANSI B1.20.3)	C Hex Nom	G_a Ref mm	G_a Ref in	L mm ±0.5	L in ±0.02	T Ref Min mm	T Ref Min in	W Dia mm +0.0 −0.5	W Dia in +0.00 −0.02
1/8-27	5/8	6.4	0.25	19.0	0.75	5.6	0.22	15.88	0.625
1/4-18	3/4	10.9	0.43	28.7	1.13	6.4	0.25	19.05	0.750
3/8-18	7/8	10.9	0.43	28.7	1.13	6.4	0.25	22.22	0.875
1/2-14	1 1/8	14.2	0.56	38.1	1.50	7.9	0.31	28.58	1.125
3/4-14	1 3/8	14.2	0.56	38.9	1.53	9.7	0.38	34.92	1.375
1-11 1/2	1 5/8	17.5	0.69	48.0	1.89	9.7	0.38	41.28	1.625
1 1/4-11 1/2	2	17.5	0.69	49.0	1.93	11.7	0.46	50.80	2.000
1 1/2-11 1/2	2 3/8	17.5	0.69	49.0	1.93	13.5	0.53	60.32	2.375
2-11 1/2	2 7/8	16.0	0.63	49.8	1.96	17.3	0.68	73.03	2.875

[a] Reference dimension, not subject to inspection.

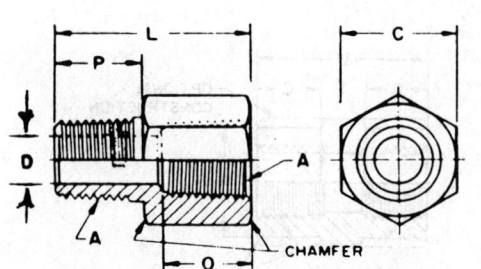

FIGURE 40—ADAPTER (140139)

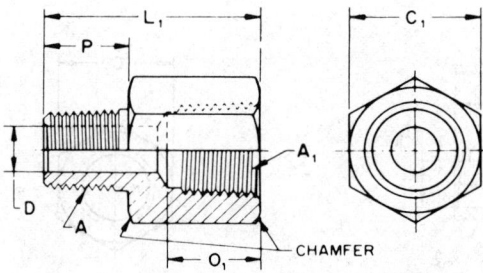

FIGURE 41—INCREASE ADAPTERS (140139)

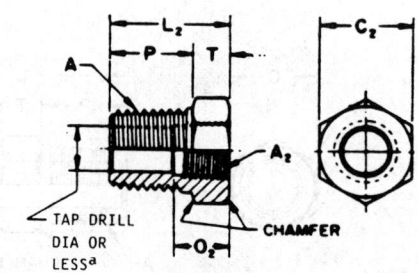

FIGURE 42—REDUCER BUSHING (140140)

TABLE 18—DIMENSIONS OF ADAPTERS, INCREASE ADAPTERS, AND REDUCER BUSHINGS (FIGURES 40, 41, AND 42)

Dryseal Pipe Thread SAE J476 (ANSI B1.20.3) A Adapter	Dryseal Pipe Thread SAE J476 (ANSI B1.20.3) A × A₁ Increase Adapter	Sryseal Pipe Thread SAE J476 (ANSI B1.20.3) A × A₂ Reducer Bushing	C Hex Nom	C_1 Hex Nom	C_2 Hex Nom	D^c Drill mm	D^c Drill in	L mm ±0.5	L in ±0.02	L_1 mm ±0.5	L_1 in ±0.02	L_2 mm ±0.5	L_2 in ±0.02	O Min mm	O Min in	O_1 Min mm	O_1 Min in	$O_2{}^a$ Min mm	$O_2{}^a$ Min in	P Min mm	P Min in	T^b Ref mm	T^b Ref in
1/8-27	1/8 × 1/4	—	5/8	3/4	—	4.8	0.188	26.4	1.04	30.7	1.21	—	—	9.7	0.38	14.2	0.56	—	—	9.7	0.38	—	—
1/4-18	1/4 × 3/8	1/4 × 1/8	3/4	7/8	5/8	7.0	0.281	35.3	1.39	36.6	1.44	21.6	0.85	14.2	0.56	14.7	0.58	9.7	0.38	14.2	0.56	6.4	0.25
3/8-18	3/8 × 1/2	3/8 × 1/4	7/8	1 1/8	3/4	10.3	0.406	36.6	1.44	42.7	1.68	21.6	0.85	14.7	0.58	19.0	0.75	14.2	0.56	14.2	0.56	6.4	0.25
1/2-14	1/2 × 3/4	1/2 × 3/8	1 1/8	1 3/8	7/8	13.5	0.531	47.5	1.87	49.0	1.93	27.9	1.10	19.0	0.75	19.6	0.77	14.7	0.58	19.0	0.75	7.9	0.31
3/4-14	3/4 × 1	3/4 × 1/2	1 3/8	1 5/8	1 1/8	18.0	0.719	49.0	1.93	55.4	2.18	29.7	1.17	19.6	0.77	23.9	0.94	19.0	0.75	19.0	0.75	9.7	0.38
1-11 1/2	1 × 1 1/4	1 × 3/4	1 5/8	2	1 3/8	23.8	0.938	60.2	2.37	62.5	2.46	34.5	1.36	23.9	0.94	23.9	0.94	19.6	0.77	23.9	0.94	9.7	0.38
1 1/4-11 1/2	1 1/4 × 1 1/2	1 1/4 × 1	2	2 3/8	1 3/4	31.7	1.250	63.2	2.49	63.5	2.50	37.3	1.47	23.9	0.94	23.9	0.94	23.9	0.94	24.6	0.97	11.7	0.46
1 1/2-11 1/2	1 1/2 × 2	1 1/2 × 1 1/4	2 3/8	2 7/8	2	38.0	1.500	64.3	2.53	66.8	2.63	39.9	1.57	23.9	0.94	24.6	0.97	23.9	0.94	25.4	1.00	13.5	0.53
2-11 1/2	—	2 × 1 1/2	2 7/8	—	2 1/2	49.0	1.938	67.6	2.66	—	—	44.5	1.75	24.6	0.97	—	—	23.9	0.94	26.2	1.03	17.3	0.68

[a] Beyond top drill depth O_2, hole may be reduced below tap drill diameter, but shall not be less than D diameter in corresponding external pipe size adapter.
[b] Minimum design thickness, not subject to inspection.
[c] See Table 1 for tolerance.

22.89

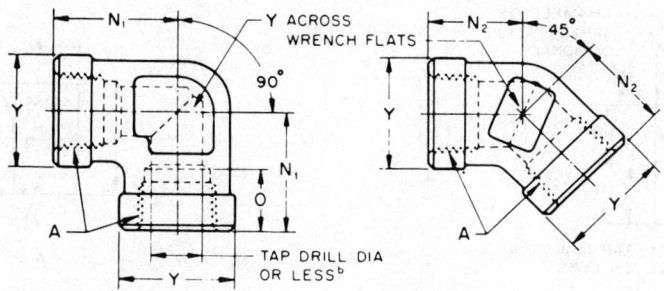

FIGURE 43A—90 DEGREE PIPE ELBOW (140238)　　　FIGURE 43B—45 DEGREE PIPE ELBOW (140338)

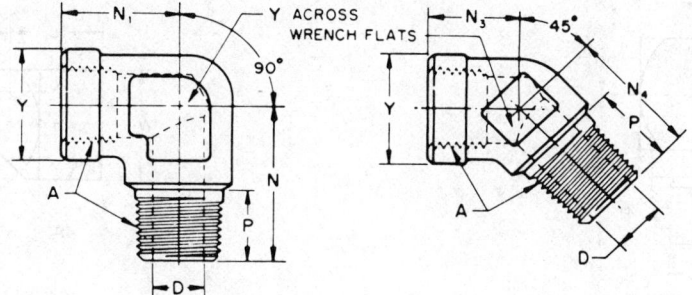

FIGURE 43C—90 DEGREE STREET ELBOW (140239)　　　FIGURE 43D—45 DEGREE STREET ELBOW (140339)

NOTE: FIGURES 43A TO 43D AND 46A TO 46C DEPICT FORGED CONSTRUCTION AND ARE OPTIONAL WITH SOLID OR FABRICATED BAR-STOCK CONSTRUCTION DEPICTED IN FIGURES 44A TO 45D.

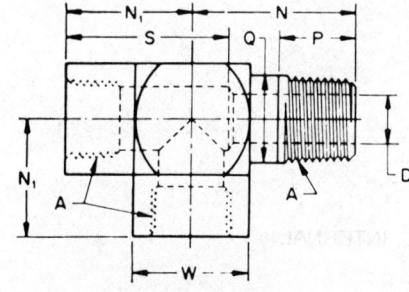　　　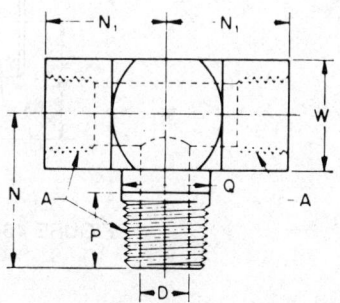

FIGURE 44A—PIPE TEE INTERNAL, EXTERNAL, INTERNAL (140424)　　　FIGURE 44B—PIPE TEE INTERNAL, INTERNAL, EXTERNAL (140425)

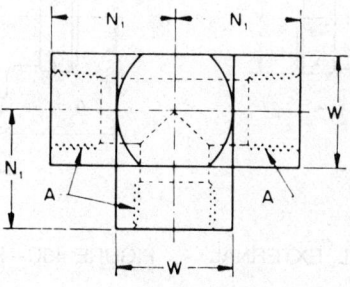

FIGURE 44C—PIPE TEE INTERNAL, INTERNAL, INTERNAL (140438)

22.90

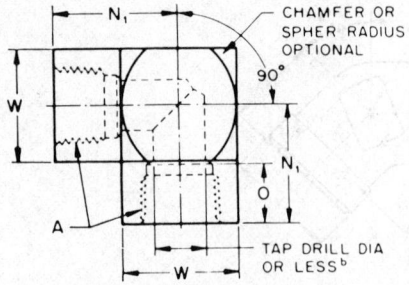

FIGURE 45A—90 DEGREE PIPE ELBOW (140238)

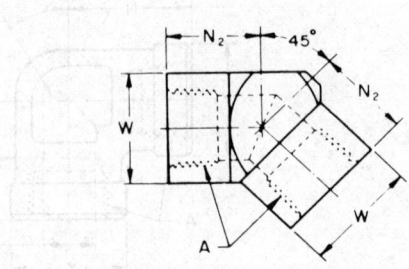

FIGURE 45B—45 DEGREE PIPE ELBOW (140338)

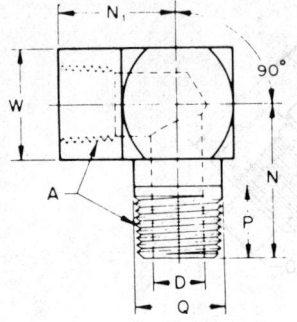

FIGURE 45C—90 DEGREE STREET ELBOW (140239)

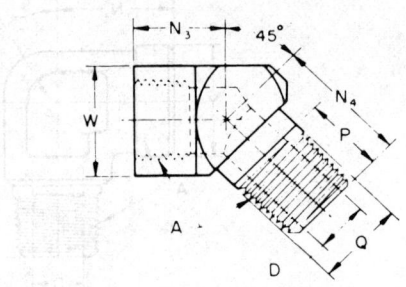

FIGURE 45D—45 DEGREE STREET ELBOW (140339)

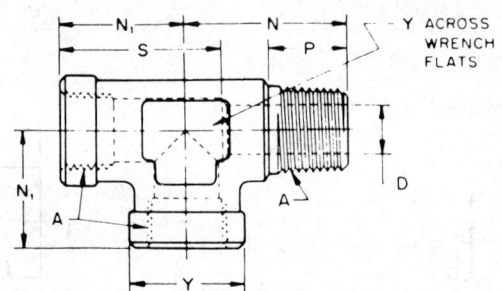

FIGURE 46A—PIPE TEE INTERNAL, EXTERNAL, INTERNAL (140424)

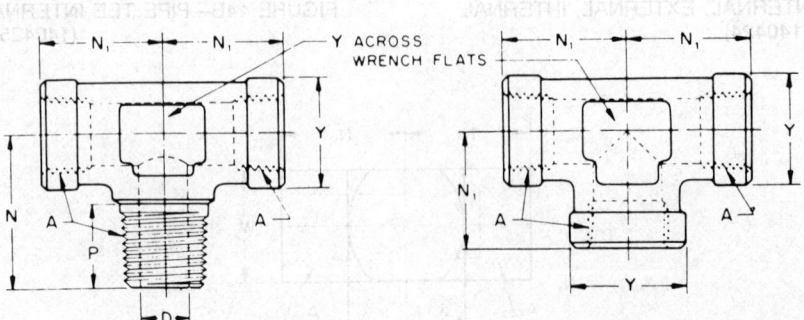

FIGURE 46B—PIPE TEE INTERNAL, INTERNAL, EXTERNAL (140425)

FIGURE 46C—PIPE TEE INTERNAL, INTERNAL, INTERNAL (140438)

NOTES: UNSPECIFIED DETAIL WITH RESPECT TO DIMENSIONS, TOLERANCES, CONTOURS, MATERIAL AND WORKMANSHIP MUST CONFORM TO GENERAL SPECIFICATIONS FOR HYDRAULIC PIPE FITTINGS. THE DIMENSIONAL DESIGNATIONS FOR TAP DRILL AND NOTES SHOWN ON FIGURES 6 AND 13 SHALL APPLY TO CORRESPONDING FEATURES OF FIGURES 7 TO 12 AND FIGURES 45 TO 46C UNLESS SHOWN OTHERWISE.

TABLE 19—DIMENSIONS OF FORGED AND BARSTOCK TYPES OF PIPE ELBOWS, STREET ELBOWS, AND PIPE TEES (FIGURES 43A TO 46C)

A Dryseal Pipe Thread SAE J476 ANSI B1.20.3)	D[a,d] Drill mm	D[a,d] Drill in	N mm ±0.8	N in ±0.03	N_1 mm ±0.8	N_1 in ±0.03	N_2 mm ±0.8	N_2 in ±0.03	N_3 mm ±0.8	N_3 in ±0.03	N_4 mm ±0.8	N_4 in ±0.03	O[a] Min mm	O[a] Min in	P Min mm	P Min in	Q Dia Min mm	Q Dia Min in	S[b] Min mm	S[b] Min in	W Square or Dia max mm	W Square or Dia max in	Y[c] mm +0.0	Y[c] in +0.000
1/8-27	4.8	0.188	19.8	0.78	16.8	0.66	12.7	0.50	11.9	0.47	18.3	0.72	9.6	0.38	9.6	0.38	11.2	0.44	23.9	0.94	15.7	0.62	14.3-0.8	0.562-0.030
1/4-18	7.0	0.281	27.7	1.09	22.4	0.88	17.5	0.69	15.7	0.62	26.7	1.05	14.2	0.56	14.2	0.56	14.2	0.56	29.0	1.14	19.0	0.75	19.0-0.8	0.750-0.030
3/8-18	10.3	0.406	31.0	1.22	25.9	1.02	19.0	0.75	18.3	0.72	26.9	1.06	14.7	0.58	14.2	0.56	17.3	0.68	33.8	1.33	22.3	0.88	22.2-0.8	0.875-0.030
1/2-14	13.5	0.531	37.3	1.47	31.2	1.23	23.9	0.94	23.1	0.91	34.0	1.34	19.0	0.75	19.0	0.75	22.4	0.88	41.1	1.62	28.4	1.12	27.0-1.0	1.062-0.040
3/4-14	18.0	0.719	40.4	1.59	34.5	1.36	25.4	1.00	24.6	0.97	35.1	1.38	19.6	0.77	19.0	0.75	26.9	1.06	47.2	1.86	35.0	1.38	33.3-1.0	1.312-0.040
1-11 1/2	23.8	0.938	50.0	1.97	41.1	1.62	30.2	1.19	28.4	1.12	43.7	1.72	23.8	0.94	23.8	0.94	35.1	1.38	56.6	2.23	41.1	1.62	41.3-1.0	1.625-0.040
1 1/4-11 1/2	31.7	1.250	60.5	2.38	43.2	1.70	36.6	1.44	41.4	1.63	45.7	1.80	23.8	0.94	24.6	0.97	42.9	1.69	62.7	2.47	50.8	2.00	47.6-1.0	1.875-0.040
1 1/2-11 1/2	38.0	1.500	67.1	2.64	52.8	2.08	37.1	1.46	42.9	1.69	52.3	2.06	23.8	0.94	25.4	1.00	48.3	1.90	75.4	2.97	60.5	2.38	65.1-1.0	2.562-0.040
2-11 1/2	49.0	1.938	76.2	3.00	60.7	2.39	40.4	1.59	55.6	2.19	54.6	2.15	24.6	0.97	26.2	1.03	60.5	2.38	89.4	3.52	73.2	2.88	71.4-1.0	2.812-0.040

[a] Beyond tap drill depth O, hole may be reduced to below tap drill diameter, but shall not be less than D drill for corresponding size. See Figures 43A and 45A.
[b] Maximum depth shall be optional with manufacturer provided that strength of fitting is not impaired.
[c] The basic dimension shown shall apply as minimum for boss diameter.
[d] See Table 1 for tolerance.

The factors shown in Tables 22, 24, 26, 28, and 29 were derived by subtracting the standard machining length required for the smaller end from that required for the larger standard end and rounding the result to a two-place decimal.

The factors given in Tables 23, 25, and 27 were derived by subtracting the standard machining length plus an allowance of 1-1/2 pitches (threads) for imperfect thread length required for the smaller end from the same value required for the larger end and rounding the result to a two-place decimal.

The minimum allowable stock size for the various types of ends are also tabulated for reference purposes.

TABLE 20—PRESSURE RATINGS OF ADAPTER UNIONS

A_1 Straight Pipe Thread (NPSM) d	Max Operating Pressure[a] MPa	Max Operating Pressure[a] psi
-2	34.5	5000
-4	34.5	5000
-6	27.6	4000
-8	24.1	3500
-12	15.5	2250
-16	13.8	2000
-20	11.2	1625
-24	8.6	1250
-32	7.8	1125

[a] Pressure ratings shown are the pressure ratings for the fitting and are not necessarily the pressure ratings applicable in an assembly with hydraulic hose. The applicable pressure rating for an assembly with hydraulic hose is the lower value specified for the hose and the fitting unless otherwise agreed to by the supplier and user.

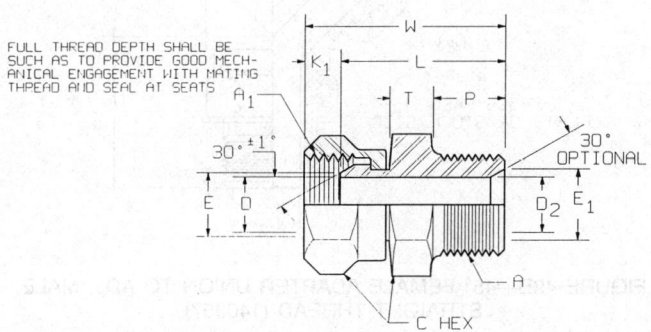

FIGURE 47A—FEMALE ADAPTER UNION TO MALE PIPE (140130)

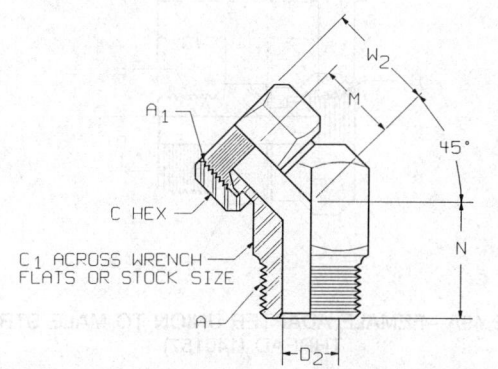

FIGURE 47B—45° FEMALE ADAPTER UNION TO MALE PIPE (140330)

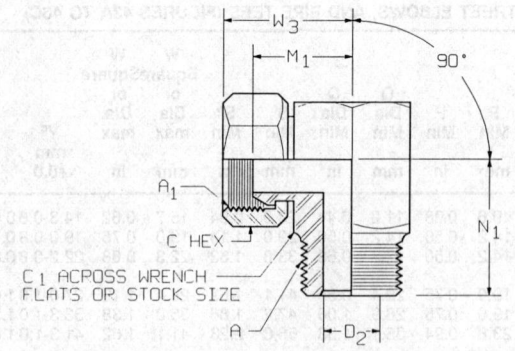

FIGURE 47C—90° FEMALE ADAPTER UNION TO MALE PIPE
(140230)

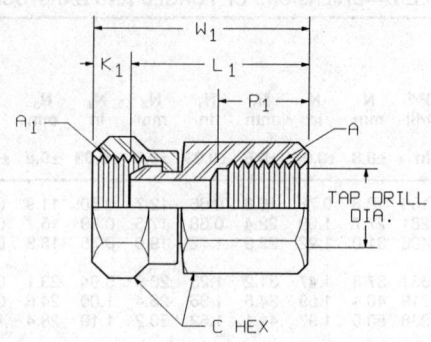

FIGURE 48A—FEMALE ADAPTER UNION TO FEMALE PIPE
(140131)

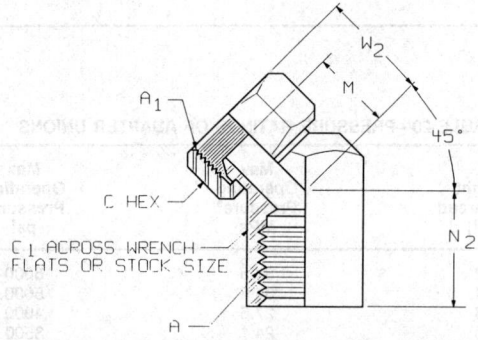

FIGURE 48B—45° FEMALE ADAPTER UNION TO FEMALE PIPE
(140331)

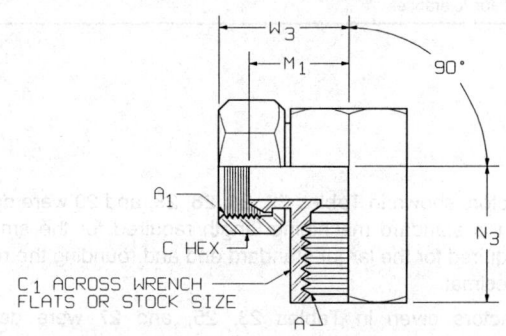

FIGURE 48C—90° FEMALE ADAPTER UNION TO FEMALE PIPE
(140231)

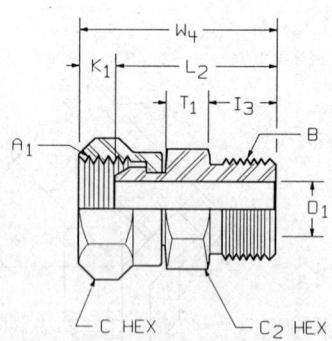

FIGURE 49A—FEMALE ADAPTER UNION TO MALE STRAIGHT THREAD (140157)

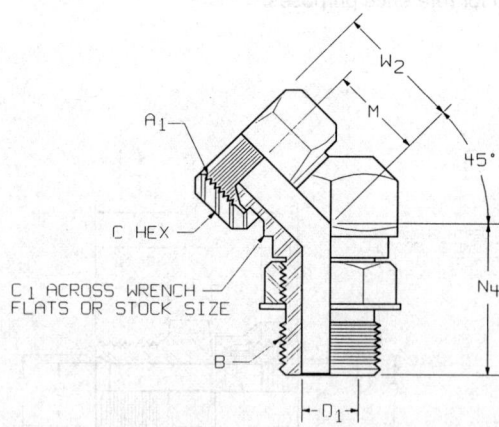

FIGURE 49B—45° FEMALE ADAPTER UNION TO ADJ. MALE STRAIGHT THREAD (140357)

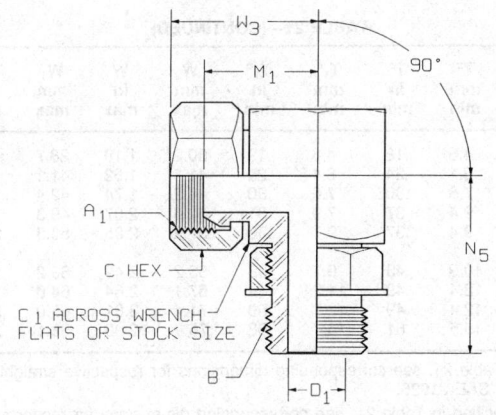

FIGURE 49C—90° FEMALE ADAPTER UNION TO ADJ. MALE STRAIGHT THREAD (140257)

TABLE 21—DIMENSIONS OF ADAPTER UNIONS (FIGURES 47A TO 49C)

Nominal Tube Size	Pipe Thread Dash Size	A Dryseal Pipe Thread (NPTF) SAE J476e (ANSI B1.20.3)	A₁ Straight Pipe Thread NPSMd	B Thread Size, in (ISO R725) SAE J475 Class 2A, Int Class 2B, Ext	C Hex Nom	C₁ Forging mm	C₁ Forging in	C₁ Bar Stock mm max	C₁ Bar Stock in max	C₂ Hex Nom	D Dia min mm	D Dia min in	D Dia max mm	D Dia max in	D₁ Min Dia mm	D₁ Min Dia in	D₂ Min Dia mm	D₂ Min Dia in
1/4	-2	1/8-27	1/8-27	7/16-20	9/16	11.1	.44	14.3	.56	9/16	2.8	.11	4.8	.19	2.8	.11	3.3	.13
3/8	-4	1/4-18	1/4-18	9/16-18	11/16	12.7	.50	20.6	.81	11/16	5.3	.21	5.8	.23	5.3	.21	5.6	.22
1/2	-6	3/8-18	3/8-18	3/4-16	7/8	19.1	.75	22.4	.88	7/8	8.1	.32	9.1	.36	8.4	.33	8.4	.33
5/8	-8	1/2-14	1/2-14	7/8-14	1	22.4	.88	28.7	1.13	1	11.2	.44	12.2	.48	11.4	.45	11.9	.47
3/4	-12	3/4-14	3/4-14	1 1/16-12	1 1/4	33.3	1.31	35.1	1.38	1 1/4	15.5	.61	16.8	.66	15.2	.60	16.5	.65
1	-16	1-11 1/2	1-11 1/2	1 5/16-12	1 1/2	33.3	1.31	41.4	1.63	1 1/2	19.8	.78	22.4	.88	21.1	.83	21.1	.83
1 1/4	-20	1 1/4-11 1/2	1 1/4-11 1/2	1 5/8-12	1 7/8	41.4	1.63	54.1	2.13	1 7/8	25.7	1.01	29.5	1.16	27.2	1.07	28.4	1.12
1 1/2	-24	1 1/2-11 1/2	1 1/2-11 1/2	1 7/8-12	2 1/8	47.8	1.88	57.2	2.25	2 1/8	31.8	1.25	35.0	1.38	33.0	1.30	34.5	1.36
2	-32	2-11 1/2	2-11 1/2	2 1/2-12	2 5/8	63.5	2.50	82.3	3.25	2 3/4	43.7	1.72	47.7	1.88	45.0	1.77	45.5	1.79

Nominal Tube Size	E Dia. mm ±0.25	E Dia. in ±.010	E₁ Dia. mm ±0.25	E₁ Dia. in ±.010	I₃ mm ±0.13	I₃ in ±.005	K₁ mm min	K₁ in min	L mm max	L in max	L₁ mm max	L₁ in max	L₂ mm max	L₂ in max	M mm max	M in max	M₁ mm max	M₁ in max	N mm max	N in max	N₁ mm max	N₁ in max	N₂ mm max	N₂ in max	N₃ mm max	N₃ in max	N₄ mm max	N₄ in max
1/4	5.9	.234	7.1	.281	9.1	.360	3.0	.12	25.4	1.00	23.9	.94	24.4	.96	17.8	.70	18.8	.74	18.5	.73	26.9	1.06	18.5	.73	19.1	.75	23.9	.94
3/8	7.5	.297	8.7	.344	9.9	.391	4.3	.17	35.3	1.39	35.3	1.38	31.3	1.23	20.8	.82	23.9	.94	25.4	1.00	33.3	1.31	25.4	1.00	25.4	1.00	30.5	1.20
1/2	10.7	.422	11.9	.469	11.1	.438	4.8	.19	37.3	1.47	35.3	1.39	32.8	1.29	24.1	.95	28.4	1.12	28.4	1.12	39.6	1.56	28.4	1.21	28.4	1.12	34.5	1.36
5/8	14.3	.562	15.9	.625	12.7	.500	6.6	.26	44.7	1.76	39.6	1.56	35.6	1.40	23.9	.94	28.2	1.11	36.6	1.44	47.5	1.87	36.6	1.44	34.8	1.37	40.1	1.58
3/4	19.1	.750	20.7	.813	15.1	.594	8.4	.33	46.7	1.84	45.0	1.77	42.2	1.66	28.7	1.13	35.3	1.39	39.6	1.56	52.3	2.06	39.6	1.56	39.6	1.56	45.5	1.79
1	24.6	.969	26.2	1.031	15.1	.594	8.4	.33	52.3	2.06	52.3	2.06	46.0	1.81	32.8	1.29	39.6	1.56	39.6	1.56	62.0	2.44	39.6	1.56	46.0	1.81	48.8	1.92
1 1/4	32.5	1.281	34.1	1.344	15.1	.594	10.2	.40	55.4	2.18	52.3	2.06	49.5	1.95	37.6	1.48	47.0	1.85	47.5	1.87	68.1	2.68	46.7	1.84	52.3	2.06	47.5	1.87
1 1/2	38.1	1.500	41.3	1.625	15.1	.594	10.2	.40	58.7	2.31	55.4	2.18	52.3	2.06	36.6	1.44	52.1	2.05	51.6	2.03	72.9	2.87	51.6	2.03	58.7	2.31	59.4	2.34
2	50.8	2.000	52.4	2.063	15.1	.594	10.2	.40	62.5	2.46	68.1	2.68	56.4	2.22	39.6	1.56	61.0	2.40	58.7	2.31	85.6	3.37	52.3	2.06	71.4	2.81	52.3	2.06

TABLE 21—(CONTINUED)

Nominal Tube Size	N_5 mm max	N_5 in max	P mm min	P in min	P_1 mm max	P_1 in max	T^c mm min	T^c in min	T_1^c mm min	T_1^c in min	W mm max	W in max	W_1 mm max	W_1 in max	W_2 mm max	W_2 in max	W_3 mm max	W_3 in max	W_4 mm max	W_4 in max
1/4	25.4	1.00	9.7	.38	11.7	.46	4.6	.18	4.8	.19	30.2	1.19	28.7	1.13	22.4	.88	23.1	.91	28.7	1.13
3/8	31.8	1.25	14.2	.56	17.0	.67	6.1	.24	6.4	.25	41.1	1.62	41.1	1.62	27.0	1.06	30.0	1.18	36.1	1.42
1/2	38.4	1.51	14.2	.56	17.3	.68	7.6	.30	7.6	.30	44.2	1.74	42.4	1.67	31.2	1.23	35.6	1.40	39.6	1.56
5/8	43.2	1.70	19.1	.75	22.9	.90	9.4	.37	7.9	.31	52.6	2.07	49.3	1.94	32.5	1.28	38.1	1.50	43.7	1.72
3/4	50.8	2.00	19.1	.75	23.1	.91	9.4	.37	9.7	.38	57.2	2.25	53.3	2.10	36.1	1.42	43.7	1.72	50.0	1.97
1	54.6	2.15	23.9	.94	29.0	1.14	10.9	.43	9.7	.38	63.2	2.49	63.2	2.49	38.4	1.51	49.0	1.93	55.9	2.20
1 1/4	64.3	2.53	24.6	.97	29.5	1.16	12.4	.49	11.2	.44	67.1	2.64	64.0	2.52	46.0	1.81	57.2	2.25	60.2	2.37
1 1/2	67.8	2.67	25.4	1.00	29.5	1.16	12.4	.49	12.7	.50	71.4	2.81	65.0	2.56	49.3	1.94	71.4	2.81	65.0	2.56
2	84.1	3.31	26.2	1.03	30.0	1.18	15.5	.61	15.7	.62	75.2	2.96	68.1	2.68	52.3	2.06	72.9	2.87	69.1	2.72

a For dimensions shown in Figure 49A but not specified in Table 21, see corresponding dimensions for respective straight thread size in Figure 3B and Table 4 of SAE J514. For dimensions of mating bosses, see Figure 1 and Table 1 of SAE J1926.
b For dimensions shown in Figure 49B and 49C, but not specified in Table 21, see corresponding dimensions for respective straight thread size in Figure 10A and Table 4 of SAE J514.
c Minimum design thickness, not subject to inspection.
d American Standard Straight Pipe Thread for Mechanical Joints.
e Dryseal American Standard Taper Pipe Thread.

TABLE 22—FACTORS FOR 37 DEGREE FLARED END ON STRAIGHT FITTINGS
(FIGURES 2A TO 2D, 3A TO 3C, AND 15)

Nominal Tube OD	3/16 mm	3/16 in	1/4 mm	1/4 in	5/16 mm	5/16 in	3/8 mm	3/8 in	1/2 mm	1/2 in	5/8 mm	5/8 in	3/4 mm	3/4 in	7/8 mm	7/8 in	1 mm	1 in	1 1/4 mm	1 1/4 in	1 1/2 mm	1 1/2 in	2 mm	2 in	Nominal Tube OD
1/8	0.8	0.03	2.5	0.10	2.5	0.10	2.8	0.11	5.3	0.21	7.9	0.31	10.7	0.42	11.2	0.44	11.7	0.46	13.0	0.51	16.3	0.64	22.6	0.89	1/8
3/16	—	—	1.8	0.07	1.8	0.07	2.0	0.08	4.6	0.18	7.1	0.28	9.9	0.39	10.4	0.41	10.9	0.43	12.2	0.48	15.2	0.60	21.6	0.85	3/16
1/4	—	—	—	—	0.0	0.00	0.3	0.01	2.8	0.11	5.3	0.21	7.9	0.31	8.6	0.34	9.1	0.36	10.4	0.41	13.5	0.53	19.8	0.78	1/4
5/16	—	—	—	—	—	—	0.3	0.01	2.8	0.11	5.3	0.21	7.9	0.31	8.6	0.34	9.1	0.36	10.4	0.41	13.5	0.53	19.8	0.78	5/16
3/8	—	—	—	—	—	—	—	—	2.5	0.10	5.1	0.20	7.9	0.31	8.4	0.33	9.1	0.36	10.2	0.40	13.5	0.53	19.8	0.78	3/8
1/2	—	—	—	—	—	—	—	—	—	—	2.5	0.10	5.3	0.21	5.8	0.23	6.4	0.25	7.6	0.30	10.9	0.43	17.3	0.68	1/2
5/8	—	—	—	—	—	—	—	—	—	—	—	—	2.8	0.11	3.3	0.13	3.8	0.15	5.1	0.20	8.4	0.33	14.7	0.58	5/8
3/4	—	—	—	—	—	—	—	—	—	—	—	—	0.8	0.03	1.3	0.05	2.3	0.09	5.6	0.22	11.9	0.47			3/4
7/8	—	—	—	—	—	—	—	—	—	—	—	—	—	—	0.5	0.02	1.8	0.07	4.8	0.19	11.2	0.44			7/8
1	—	—	—	—	—	—	—	—	—	—	—	—	—	—	—	—	1.3	0.05	4.3	0.17	10.7	0.42			1
1 1/4	—	—	—	—	—	—	—	—	—	—	—	—	—	—	—	—	—	—	3.3	0.13	9.7	0.38			1 1/4
1 1/2	—	—	—	—	—	—	—	—	—	—	—	—	—	—	—	—	—	—	—	—	6.4	0.25			1 1/2

(Reduced Machining Size rows shown above under Standard Machining Size columns)

9.2 Examples

9.2.1 For a 37 degree flared male connector (Figure 2A) with 1/2 tube OD and 3/4 NPTF, the overall length would be determined as follows:

    2.06 = L overall length for 3/4 tube to 3/4 NPTF from Table 2
   -0.21 = Factor from Table 22 for 1/2 machining on 3/4 size fitting
    1.85 = Overall length for the nonstandard male connector

Since the 3/4 NPTF is the larger machining, the hexagon width of 1-1/8 from Table 31 would apply.

9.2.2 For a 37 degree flared 90 degree male elbow (Figure 4A) with 1/2 OD tube and 3/4 NPTF, the end to center length M would be derived as follows:

    1.66 = M dimension for standard 3/4 37 degree end
   -0.24 = Factor from Table 23 for 1/2 machining on 3/4 size fitting
    1.42 = End to center length

Since the 3/4 NPTF is the standard end, the N end-to-center dimension would remain 1.59 as shown in Table 4. The wrench flat size would be as shown by the Y column in Table 4 for the 3/4 tube OD.

9.3 Tolerances

The following tolerances apply to nonstandard sizes:

Overall length of straight fittings = ±0.5 mm (±0.02 in)

Centerline to end on shaped fittings = ±1.5 mm (±0.06 in)

TABLE 23—FACTORS FOR 37 DEGREE FLARED END ON SHAPE FITTINGS (FIGURES 4A TO 4E, 5A TO 5D, 6A TO 6C, 7A TO 7D, 8C, 9A TO 9D, AND 10A TO 10C)

Reduced Machining Size	Nominal Tube OD	3/16 mm	3/16 in	1/4 mm	1/4 in	5/16 mm	5/16 in	3/8 mm	3/8 in	1/2 mm	1/2 in	5/8 mm	5/8 in	3/4 mm	3/4 in	7/8 mm	7/8 in	1 mm	1 in	1 1/4 mm	1 1/4 in	1 1/2 mm	1 1/2 in	2 mm	2 in	Nominal Tube OD
	1/8	0.8	0.03	2.8	0.11	2.8	0.11	3.3	0.13	—	—	—	—	—	—	—	—	—	—	—	—	—	—	—	—	1/8
	3/16	—	—	2.0	0.08	2.0	0.08	2.5	0.10	5.3	0.21	—	—	—	—	—	—	—	—	—	—	—	—	—	—	3/16
	1/4	—	—	—	—	0.0	0.00	0.3	0.01	3.3	0.13	6.1	0.24	—	—	—	—	—	—	—	—	—	—	—	—	1/4
	5/16	—	—	—	—	—	—	0.3	0.01	3.3	0.13	6.1	0.24	9.1	0.36	—	—	—	—	—	—	—	—	—	—	5/16
	3/8	—	—	—	—	—	—	—	—	2.8	0.11	5.8	0.23	8.9	0.35	9.7	0.38	—	—	—	—	—	—	—	—	3/8
	1/2	—	—	—	—	—	—	—	—	—	—	3.0	0.12	6.1	0.24	6.9	0.27	7.4	0.29	—	—	—	—	—	—	1/2
	5/8	—	—	—	—	—	—	—	—	—	—	—	—	3.0	0.12	3.8	0.15	4.3	0.17	5.6	0.22	—	—	—	—	5/8
	3/4	—	—	—	—	—	—	—	—	—	—	—	—	0.8	0.03	1.3	0.05	2.3	0.09	5.5	0.22	—	—	—	—	3/4
	7/8	—	—	—	—	—	—	—	—	—	—	—	—	—	—	0.5	0.02	1.8	0.07	4.8	0.19	11.2	0.44	—	—	7/8
	1	—	—	—	—	—	—	—	—	—	—	—	—	—	—	—	—	1.3	0.05	4.3	0.17	10.7	0.42	—	—	1
	1 1/4	—	—	—	—	—	—	—	—	—	—	—	—	—	—	—	—	—	—	3.3	0.13	9.7	0.38	—	—	1 1/4
	1 1/2	—	—	—	—	—	—	—	—	—	—	—	—	—	—	—	—	—	—	—	—	6.4	0.25	—	—	1 1/2

TABLE 24—FACTORS FOR FLARELESS STRAIGHT FITTINGS (FIGURES 19A TO 19D, 20A TO 20C, AND 30)

Reduced Machining Size	Nominal Tube OD	3/16 mm	3/16 in	1/4 mm	1/4 in	5/16 mm	5/16 in	3/8 mm	3/8 in	1/2 mm	1/2 in	5/8 mm	5/8 in	3/4 thru 2 mm	3/4 thru 2 in
	1/8	1.3	0.05	2.0	0.08	2.0	0.08	2.5	0.10	4.8	0.19	6.4	0.25	7.9	0.31
	3/16	—	—	0.8	0.03	0.8	0.03	1.3	0.05	3.6	0.14	5.1	0.20	6.6	0.26
	1/4	—	—	—	—	0.0	0.00	0.5	0.02	2.8	0.11	4.3	0.17	5.8	0.23
	5/16	—	—	—	—	—	—	0.5	0.02	2.8	0.11	4.3	0.17	5.8	0.23
	3/8	—	—	—	—	—	—	—	—	2.3	0.09	3.8	0.15	5.3	0.21
	1/2	—	—	—	—	—	—	—	—	—	—	1.5	0.06	3.0	0.12
	5/8	—	—	—	—	—	—	—	—	—	—	—	—	1.5	0.06
	3/4 to 1 1/2	—	—	—	—	—	—	—	—	—	—	—	—	0.0	0.00

TABLE 25—FACTORS FOR FLARELESS STRAIGHT FITTINGS (FIGURES 21A TO 21E, 22A TO 22D, 23A TO 23C, 24A TO 24D, 25C, 26A TO 26D, AND 27A TO 27C)

	Nominal Tube OD	\multicolumn{14}{c	}{Standard Machining Size}
		3/16 mm	3/16 in	1/4 mm	1/4 in	5/16 mm	5/16 in	3/8 mm	3/8 in	1/2 mm	1/2 in	5/8 mm	5/8 in	3/4 thru 2 mm	3/4 thru 2 in
Reduced Machining Size	1/8	1.3	0.05	2.5	0.10	2.5	0.10	3.0	0.12	5.8	0.23	7.6	0.30	9.7	0.38
	3/16	—	—	1.3	0.05	1.3	0.05	1.8	0.07	4.6	0.18	6.4	0.25	8.4	0.33
	1/4	—	—	—	—	0.0	0.00	0.5	0.02	3.3	0.13	5.1	0.20	7.1	0.28
	5/16	—	—	—	—	—	—	0.5	0.02	3.3	0.13	5.1	0.20	7.1	0.28
	3/8	—	—	—	—	—	—	—	—	2.8	0.11	4.6	0.18	6.6	0.26
	1/2	—	—	—	—	—	—	—	—	—	—	1.8	0.07	3.8	0.15
	5/8	—	—	—	—	—	—	—	—	—	—	—	—	2.0	0.08
	3/4 to 1 1/2	—	—	—	—	—	—	—	—	—	—	—	—	0.0	0.00

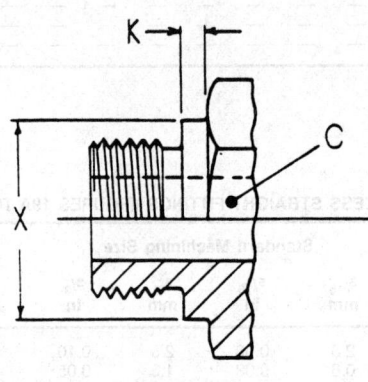

FIGURE 50

TABLE 26—FACTORS FOR NONADJUSTABLE STRAIGHT THREAD ENDS (FIGURES 3B, 3C, 20B, 20C)

	Nominal Tube OD	3/16 mm	3/16 in	1/4 mm	1/4 in	5/16 mm	5/16 in	3/8 mm	3/8 in	1/2 mm	1/2 in	5/8 mm	5/8 in	3/4 mm	3/4 in	7/8 mm	7/8 in	1 mm	1 in	1 1/4 mm	1 1/4 in	1 1/2 mm	1 1/2 in
Reduced Machining Size	1/8	0.0	0.00	-1.5	-0.06	+0.8	+0.03	0.0	0.00	-1.3	-0.05	-2.8	-0.11	-5.3	-0.21	-5.3	-0.21	-5.3	-0.21	-5.3	-0.21	-5.3	-0.21
	3/16	—	—	-1.5	-0.06	-1.5	-0.06	0.0	0.00	-1.3	-0.05	-2.8	-0.11	-5.3	-0.21	-5.3	-0.21	-5.3	-0.21	-5.3	-0.21	-5.3	-0.21
	1/4	—	—	—	—	0.0	0.00	-0.8	-0.03	+0.3	+0.01	-1.3	-0.05	-3.6	-0.14	-3.6	-0.14	-3.6	-0.14	-3.6	-0.14	-3.6	-0.14
	5/16	—	—	—	—	—	—	-0.8	-0.03	+0.3	+0.01	-1.3	-0.05	-3.6	-0.14	-3.6	-0.14	-3.6	-0.14	-3.6	-0.14	-3.6	-0.14
	3/8	—	—	—	—	—	—	—	—	-1.0	-0.04	+0.3	+0.01	-2.8	-0.11	-2.8	-0.11	-2.8	-0.11	-2.8	-0.11	-2.8	-0.11
	1/2	—	—	—	—	—	—	—	—	—	—	-1.5	-0.06	-0.8	-0.03	-0.8	-0.03	-0.8	-0.03	-0.8	-0.03	-0.8	-0.03
	5/8	—	—	—	—	—	—	—	—	—	—	—	—	-2.3	-0.09	+0.8	+0.03	+0.8	+0.03	+0.8	+0.03	+0.8	+0.03
	3/4	—	—	—	—	—	—	—	—	—	—	—	—	0.0	0.00	0.0	0.00	+3.3	+0.13	+3.3	+0.13	+3.3	+0.13
	7/8	—	—	—	—	—	—	—	—	—	—	—	—	—	—	0.0	0.00	+3.3	+0.13	+3.3	+0.13	+3.3	+0.13
	1	—	—	—	—	—	—	—	—	—	—	—	—	—	—	—	—	—	—	+4.1	+0.16	+4.1	+0.16
	1 1/4	—	—	—	—	—	—	—	—	—	—	—	—	—	—	—	—	—	—	—	—	+4.1	+0.16
	1 1/2	—	—	—	—	—	—	—	—	—	—	—	—	—	—	—	—	—	—	—	—	—	—

[a] Fittings involving hex sizes larger than "C" shall include a turned shoulder of diameter "X" and thickness "K" as shown in Figure 50. This turned shoulder permits the reduced straight thread end to seat in a standard port spotface.
For some combinations, additional length will be required to accommodate this feature. Therefore, all factors designated as plus in above table should be added to standard length dimensions to obtain applicable values.

	Nominal Tube OD	2 mm	2 in	Nominal Tube OD	X Dia mm ±0.3	X Dia in ±0.01	K mm ±0.5	K in ±0.02	C[a] Hex mm	C[a] Hex in
Reduced Machining Size	1/8	-5.3	-0.21	1/8	12.7	0.50	2.3	0.09	14.27	9/16
	3/16	-5.3	-0.21	3/16	14.2	0.56	2.3	0.09	15.88	5/8
	1/4	-3.6	-0.14	1/4	16.0	0.63	2.3	0.09	17.48	11/16
	5/16	-3.6	-0.14	5/16	17.5	0.69	2.3	0.09	19.05	3/4
	3/8	-2.8	-0.11	3/8	19.0	0.75	2.3	0.09	23.81	15/16
	1/2	-0.8	-0.03	1/2	23.9	0.94	3.3	0.13	25.40	1
	5/8	+0.8	+0.03	5/8	26.9	1.06	3.3	0.13	28.58	1 1/8
	3/4	+3.3	+0.13	3/4	33.3	1.31	3.3	0.13	31.92	1 3/8
	7/8	+3.3	+0.13	7/8	36.6	1.44	3.3	0.13	38.10	1 1/2
	1	+4.1	+0.16	1	39.6	1.56	4.1	0.16	41.28	1 5/8
	1 1/4	+4.1	+0.16	1 1/4	47.7	1.88	4.1	0.16	49.22	1 15/16
	1 1/2	+4.1	+0.16	1 1/2	54.1	2.13	4.1	0.16	55.58	2 3/16

TABLE 27—FACTORS FOR ADJUSTABLE STRAIGHT THREAD ENDS (FIGURES 9D, 10A TO 10C, 26D, AND 27A TO 27C)

	Nominal Tube OD	___ Standard Machining Size ___													
		3/16 mm	3/16 in	1/4 mm	1/4 in	5/16 mm	5/16 in	3/8 mm	3/8 in	1/2 mm	1/2 in	5/8 mm	5/8 in	3/4 thru 2 mm	3/4 thru 2 in
Reduced Machining Size	1/8	0.0	0.00	2.8	0.11	2.8	0.11	4.3	0.17	—	—	—	—	—	—
	3/16	—	—	2.8	0.11	2.8	0.11	4.3	0.17	7.6	0.30	—	—	—	—
	1/4	—	—	—	—	0.0	0.00	1.5	0.06	4.8	0.19	8.9	0.35	—	—
	5/16	—	—	—	—	—	—	1.5	0.06	4.8	0.19	8.9	0.35	—	—
	3/8	—	—	—	—	—	—	—	—	3.3	0.13	7.4	0.29	11.4	0.45
	1/2	—	—	—	—	—	—	—	—	—	—	4.1	0.16	8.1	0.32
	5/8	—	—	—	—	—	—	—	—	—	—	—	—	4.1	0.16
	3/4 to 1 1/2	—	—	—	—	—	—	—	—	—	—	—	—	0.0	0.00

TABLE 28—FACTORS FOR ALL MALE PIPE ENDS (FIGURES 2A, 4A TO 4C, 4E, 5C, 5D, 19A, 21A TO 21C, 21E, 22C AND 22D)

	Nominal Pipe Size	___ Standard Pipe Size ___													
		1/4 mm	1/4 in	3/8 mm	3/8 in	1/2 mm	1/2 in	3/4 mm	3/4 in	1 mm	1 in	1 1/4 mm	1 1/4 in	1 1/2 mm	1 1/2 in	2 mm	2 in
Reduced Pipe Size	1/8	4.8	0.19	4.8	0.19	9.7	0.38	9.7	0.38	—	—	—	—	—	—
	1/4	—	—	0.0	0.00	4.8	0.19	4.8	0.19	9.7	0.38	—	—	—	—
	3/8	—	—	—	—	4.8	0.19	4.8	0.19	9.7	0.38	10.4	0.41	—	—
	1/2	—	—	—	—	—	—	0.0	0.00	4.8	0.19	5.6	0.22	6.4	0.25
	3/4	—	—	—	—	—	—	—	—	4.8	0.19	5.6	0.22	6.4	0.25	7.1	0.28
	1	—	—	—	—	—	—	—	—	—	—	0.8	0.03	1.5	0.06	2.3	0.09
	1 1/4	—	—	—	—	—	—	—	—	—	—	—	—	0.8	0.03	1.5	0.06
	1 1/2	—	—	—	—	—	—	—	—	—	—	—	—	—	—	0.8	0.03

TABLE 29—FACTORS FOR ALL FEMALE PIPE ENDS (FIGURES 2D, 5A, 6A, 6B, 19D, 22A, 23A, AND 23B)

	Nominal Pipe Size	___ Standard Pipe Size ___													
		1/4 mm	1/4 in	3/8 mm	3/8 in	1/2 mm	1/2 in	3/4 mm	3/4 in	1 mm	1 in	1 1/4 mm	1 1/4 in	1 1/2 mm	1 1/2 in	2 mm	2 in
Reduced Pipe Size	1/8	5.3	0.21	5.6	0.22	11.2	0.44	11.4	0.45	—	—	—	—	—	—
	1/4	—	—	0.3	0.01	5.8	0.23	6.1	0.24	11.9	0.47	—	—	—	—
	3/8	—	—	—	—	5.6	0.22	5.8	0.23	11.7	0.46	12.2	0.48	—	—
	1/2	—	—	—	—	—	—	0.3	0.01	6.1	0.24	6.6	0.26	6.6	0.26
	3/4	—	—	—	—	—	—	—	—	5.8	0.23	6.4	0.25	6.4	0.25	6.9	0.27
	1	—	—	—	—	—	—	—	—	—	—	0.5	0.02	0.5	0.02	1.0	0.04
	1 1/4	—	—	—	—	—	—	—	—	—	—	—	—	0.0	0.00	0.5	0.02
	1 1/2	—	—	—	—	—	—	—	—	—	—	—	—	—	—	0.5	0.02

TABLE 30—MINIMUM STOCK SIZE FOR TUBE ENDS

Nominal Tube OD	Hexagon Width Minimum	Width Over Forged Pads Minimum
1/8	3/8	0.31
3/16	7/16	0.38
1/4	1/2	0.44
5/16	9/16	0.50
3/8	5/8	0.56
1/2	13/16	0.75
5/8	15/16	0.88
3/4	1 1/8	1.06
7/8	1 1/4	1.19
1	1 3/8	1.31
1 1/4	1 11/16	1.62
1 1/2	2	1.88
2	2 5/8	2.50

TABLE 31—MINIMUM STOCK SIZE FOR MALE PIPE ENDS

Nominal Pipe Size	Hexagon Width Minimum	Width Over Forged Pads Minimum
1/8	9/16	0.56
1/4	3/4	0.75
3/8	7/8	0.88
1/2	1 1/8	1.06
3/4	1 3/8	1.31
1	1 5/8	1.62
1 1/4	2	1.88
1 1/2	2 3/8	2.50
2	2 7/8	2.81

TABLE 32—MINIMUM STOCK SIZE FOR FEMALE PIPE ENDS

Nominal Pipe Size	Hexagon Width Minimum	Width Over Forged Pads Minimum
1/8	7/16	0.44
1/4	9/16	0.56
3/8	3/4	0.75
1/2	7/8	0.88
3/4	1 1/8	1.06
1	1 3/8	1.31
1 1/4	1 11/16	1.62
1 1/2	2	1.88
2	2 1/2	2.50

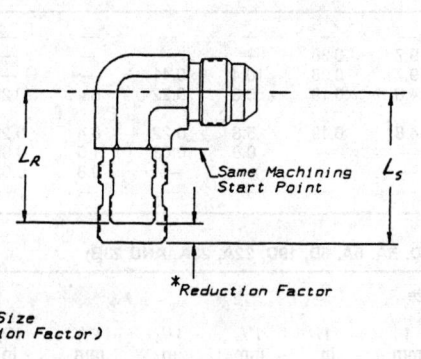

FIGURE 51A—JUMP SIZE CALCULATIONS—PORT SIDE

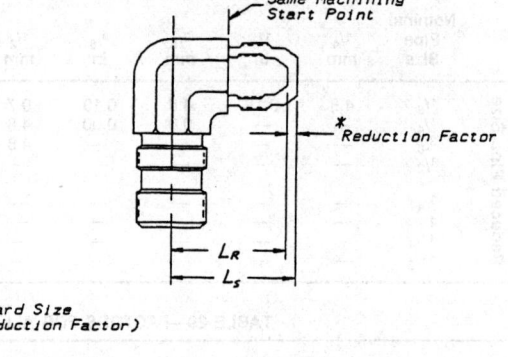

FIGURE 51B—JUMP SIZE CALCULATIONS—TUBE SIDE

(R) APPENDIX A
OPTIONAL METRIC FITTING FLAT DIMENSIONS

By agreement between user and supplier, the metric flat sizes in Table A1, shall be used in lieu of inch sizes.

TABLE A1—OPTIONAL METRIC FITTING FLAT DIMENSIONS

Size	Y Forging	Y Barstock Max	Y₁ Forging
−3	12.00/−0.8		14.00/−0.8
−4	12.00/−0.8	14	14.00/−0.8
−5	14.00/−0.8	17	14.00/−0.8
−6	14.00/−0.8	22	19.00/−0.8
−8	19.00/−0.8	22	22.00/−0.8
−10	22.00/−0.8	32	27.00/−1.0
−12	27.00/−1.0	36	32.00/−1.0
−14	32.00/−1.0	41	32.00/−1.0
−16	32.00/−1.0	41	41.00/−1.0
−20	41.00/−1.0	55	50.00/−1.0
−24	50.00/−1.0	60	65.00/−1.0
−32	65.00/−1.0	80	70.00/−1.0

(R) CONNECTIONS FOR GENERAL USE AND FLUID POWER—PORTS AND STUD ENDS WITH ISO 725 THREADS AND O-RING SEALING—PART 1: THREADED PORT WITH O-RING SEAL IN TRUNCATED HOUSING—SAE J1926/1 MAR93 / ISO 11926-1

SAE Standard

Report of the Fluid Conductors and Connectors Technical Committee approved August 1988. Completely revised by the SAE Fluid Conductors and Connectors Technical Committee SC1—Automotive and Hydraulic Tube and Fittings March 1993.

This document is technically equivalent to ISO 11926-1.

Foreword—SAE J1926 consists of the following parts, under the general title:

Connections for general use and fluid power—Ports and stud ends with ISO 725 threads and O-ring sealing:

— Part 1: Port With O-Ring Seal in Truncated Housing
— Part 2: Heavy-Duty (S Series) Stud Ends
— Part 3: Light-Duty (L Series) Stud Ends

These standards define performance requirements, dimensions, and designs for port and stud end connections for heavy-duty in Part 2 and light-duty in Part 3. Significant testing through 40 years of use has confirmed the performance requirements of these ports and stud ends. Stud ends in conformance with ISO 11926-2 are identical to those in conformance with SAE J1453, and stud ends in conformance with ISO 11926-3 are identical to those in conformance with SAE J514. Stud ends in conformance with SAE J1926-3 (ISO 11926-3) are used on fittings in ISO 8434-2.

In fluid power systems, power is transmitted and controlled through a fluid (liquid or gas) under pressure within an enclosed circuit. In general applications, a fluid may be conveyed under pressure. Components are connected through their threaded ports by fluid conductor fittings to tubes and pipes, or to hose fittings and hoses.

Ports are an integral part of fluid power components such as pumps, motors, valves, cylinders, etc.

1. Scope—This part of SAE J1926 specifies dimensions for fluid power and general use ports with inch threads to ISO 725 for use with adjustable and nonadjustable stud ends shown in SAE J1926-2 and SAE J1926-3.

Ports in accordance with this part of SAE J1926 may be used at working pressures up to 63 MPa for nonadjustable stud ends up to 40 MPa for adjustable stud ends. The permissible working pressure depends upon materials, design, working conditions, application, etc.

For threaded ports and stud ends specified in new designs for hydraulic fluid power applications, only SAE J2244 (ISO 6149) shall be used. Threaded ports and stud ends in accordance with ISO 1179, ISO 9974, and ISO 11926 shall not be used for new design in hydraulic fluid power applications.

Appendix A of this document is informative.

2. References

2.1 Applicable Documents—The following standards contain provisions which, through reference in this text, constitute provisions of this document. At the time of publication, the editions indicated were valid. All standards are subject to revision, and parties to agreements based on this document are encouraged to investigate the possibility of applying the most recent edition of the standards indicated as follows. Members of IEC and ISO maintain registers of currently valid International Standards.

2.1.1 SAE PUBLICATIONS—Available from SAE, 400 Commonwealth Drive, Warrendale, PA 15096-0001.

SAE J514—Hydraulic Tube Fittings

SAE J1453—Fitting—O-ring Face Seal

SAE J1644—Metallic Tube Connections for Fluid Power and General Use—Test Methods for Threaded Hydraulic Fluid Power Connectors

SAE J2244/1—Connections for Fluid Power and General Use—Parts and Stud Ends With ISO 261 Threads and O-ring Sealing—Part 1: Port With O-ring Seal in Truncated Housing

2.1.2 ISO PUBLICATIONS—Available from ANSI, 11 West 42nd Street, New York, NY 10036-8002.

ISO 263:1973—ISO inch screw threads—General plan and selection for screws, bolts and nuts—Diameter range 0,06 to 6 in

ISO 1101:1983—Technical drawings—Tolerancing of form, orientation, location and run-out—Generalities, definitions, symbols, indications on drawings

ISO 1302:1978—Technical drawings—Method of indicating surface texture on drawings

2.2 Related Publications—The following publications are provided for information purposes only and are not a required part of this document.

ISO 725:1978—ISO inch screw threads—Basic dimensions

ISO 1179-1:---[1]—Connections for general use and fluid power—Ports and stud ends with ISO 228-1 threads with elastomeric and metal-to-metal sealing—Part 1: Threaded port

ISO 1179-2:---[1]—Connections for general use and fluid power—Ports and stud ends with ISO 228-1 threads with elastomeric and metal-to-metal sealing—Part 2: Heavy duty (S series) and light duty (L series) stud ends with elastomeric sealing (type E)

ISO 1179-3:---[1]—Connections for general use and fluid power—Ports and stud ends with ISO 228-1 threads with elastomeric and metal-to-metal sealing—Part 3: Light duty (L series) stud end with sealing by O-ring with retaining ring (types G and H)

ISO 1179-4:---[1]—Connections for general use and fluid power—Ports and stud ends with ISO 228-1 threads with elastomeric and metal-to-metal sealing—Part 4: Stud end for general use only with metal-to-metal sealing (type B)

ISO 2306:1972—Drills for use prior to tapping screw threads

ISO 5598:1985—Fluid power systems and components—Vocabulary

ISO 6149-1:---[1]—Connections for fluid power and general use—Ports and stud ends with ISO 261 threads and O-ring sealing—Part 1: Port with O-ring seal in truncated housing

---

[1] To be published.

ISO 6149-2:---[1]—Connections for fluid power and general use—Ports and stud ends with ISO 261 threads and O-ring sealing—Part 2: Heavy duty (S series) stud ends—Dimensions, design, test methods and requirements

ISO 6149-3:---[1]—Connections for fluid power and general use—Ports and stud ends with ISO 261 threads and O-ring sealing—Part 3: Light duty (L series) stud ends—Dimensions, design, test methods and requirements

ISO 8434-2:---[1]—Metallic tube fittings for fluid power and general use—Part 2: 37° flared fittings

ISO 9974-1:---[1]—Connections for general use and fluid power—Ports and stud ends with ISO 261 threads with elastomeric and metal-to-metal sealing—Part 1: Threaded port

ISO 9974-2:---[1]—Connections for general use and fluid power—Ports and stud ends with ISO 261 threads with elastomeric and metal-to-metal sealing—Part 2: Stud end with elastomeric sealing (type E)

ISO 9974-3:---[1]—Connections for general use and fluid power—Ports and stud ends with ISO 261 threads with elastomeric and metal-to-metal sealing—Part 3: Stud end with metal-to-metal sealing (type B)

ISO 11926-1:---[1]—Connections for general use and fluid power—Ports and stud ends with ISO 261 threads and O-ring sealing—Part 1: Threaded port with O-ring seal in truncated housing

ISO 11926-2:---[1]—Connections for general use and fluid power—Ports and stud ends with ISO 261 threads and O-ring sealing—Part 2: Heavy duty (S series) stud ends

ISO 11926-3:---[1]—Connections for general use and fluid power—Ports and stud ends with ISO 261 threads and O-ring sealing—Part 3: Light duty (L series) stud ends

3. Definitions—For the purpose of this part of SAE J1926, the definitions given in ISO 5598 shall apply.

4. Port Size—The ports shall be specified by SAE J1926-1 and the thread size (without UNF or UN and 2B designation), separated by a colon, for example SAE J1926/1:1/2-20.

5. Dimensional Requirements—Ports shall conform to the dimensions in Figure 1 and Table 1.

6. Test Methods—Ports shall be tested along with stud ends per the test methods and requirements in SAE J1644.

7. Identification Statement—Use the following statement in test reports, catalogues, and sales literature when electing to comply with this part of SAE J1926:

Port conforms to SAE J1926/1, Connections for fluid power and general use—Ports and stud ends with ISO 725 threads and O-ring sealing—Part 1: Threaded port with O-ring seal in truncated housing.

8. Notes

8.1 Key Words—Fluid power, pipe fittings, standard connection, standard coupling, pipe joints, ports, stud ends, specifications, design, operating requirements, dimensions, designation, test methods, inch, straight thread, O-ring seal, high pressure

---
[1] To be published.

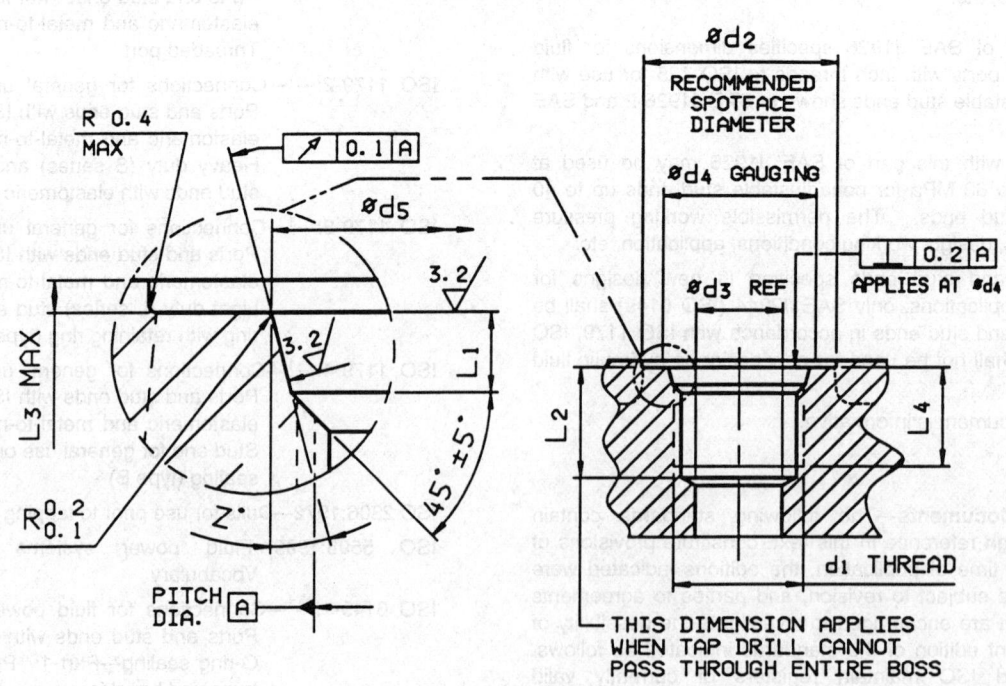

FIGURE 1—SAE J1926/1 PORT DETAIL

22.101

TABLE 1—SAE J1926/1 PORT DIMENSIONS

Dimensions in millimeters

Nominal Tube OD or Hose ID Inch Tubing Dash Size	Nominal Tube OD or Hose ID Inch Tubing mm	Nominal Tube OD or Hose ID Inch Tubing in	Nominal Tube OD or Hose ID Metric Tubing mm	d_1 Thread Size, in	d_2[1]	d_3[2] Ref	d_4 Min	d_5 ±0.05	L_1 ±0.2	L_2[3] Min	L_3[4] Max	L_4 Min Full Thread	Z° ±1°
-2	3.18	0.125	4	5/16-24 UNF-2B	17	1.6	11	9.15	2.1	12	1.6	10	12°
-3	4.76	0.188	5	3/8-24 UNF-2B	19	3.5	13	10.75	2.1	12	1.6	10	12°
-4	6.35	0.250	6	7/16-20 UNF-2B	21	4.5	15	12.45	2.6	14	1.6	11.5	12°
-5	7.94	0.312	8	1/2-20 UNF-2B	23	6	16	14.05	2.6	14	1.6	11.5	12°
-6	9.52	0.375	10	9/16-18 UNF-2B	25	7.5	18	15.70	2.7	15.5	1.6	12.7	12°
-8	12.70	0.500	12	3/4-16 UNF-2B	30	10	22	20.65	2.7	17.5	2.4	14.3	15°
-10	15.88	0.625	16	7/8-14 UNF-2B	34	12.5	26	24	2.7	20	2.4	16.7	15°
-12	19.05	0.750	20	1-1/16-12 UN-2B	41	16	32	29.2	3.5	23	2.4	19	15°
-14	22.22	0.875	22	1-3/16-12 UN-2B	45	18	35	32.4	3.5	23	2.4	19	15°
-16	25.40	1.000	25	1-5/16-12 UN-2B	49	21	38	35.55	3.5	23	3.2	19	15°
-20	31.75	1.250	30	1-5/8-12 UN-2B	58	27	48	43.55	3.5	23	3.2	19	15°
-24	38.10	1.500	38	1-7/8-12 UN-2B	65	33	54	49.9	3.5	23	3.2	19	15°
-32	50.80	2.000	50	2-1/2-12 UN-2B	88	45	70	65.75	3.5	23	3.2	19	15°

[1] If face of port is on a machined surface, dimensions d_2 and L_3 need not apply as long as R 0.2/0.1 is maintained to avoid damage to the O-ring during installation.

[2] Reference only, connecting hole application may require a different size.

[3] Tap drill depths given require use of a bottoming tap to produce the specified full thread lengths. Where standard taps are used, increase tap drill depths accordingly.

[4] Maximum recommended spotface depth to permit sufficient wrench grip for proper tightening of the fitting or locknut.

APPENDIX A—METRIC NUMBERS CONVERTED TO INCH DIMENSIONS

mm	in.
1.60	0.063
1.90	0.075
2.10	0.083
2.40	0.094
2.50	0.098
2.60	0.102
2.70	0.106
3.20	0.126
3.30	0.130
3.50	0.138
4.00	0.158
4.40	0.173
4.50	0.177
6.00	0.236
7.50	0.295
9.10	0.358
9.15	0.360
10.00	0.394
10.70	0.421
10.75	0.423
11.00	0.433
11.50	0.453
12.00	0.472
12.40	0.488
12.45	0.490

mm	in.
12.50	0.492
12.70	0.500
13.00	0.512
14.00	0.551
14.05	0.553
14.30	0.563
15.00	0.591
15.50	0.610
15.60	0.614
15.70	0.618
16.00	0.630
16.70	0.658
17.00	0.669
17.50	0.689
18.00	0.709
19.00	0.748
20.00	0.787
20.60	0.811
20.65	0.813
21.00	0.827
22.00	0.866
23.00	0.905
23.90	0.941
24.00	0.945
25.00	0.984

mm	in.
26.00	1.024
27.00	1.063
29.20	1.150
30.00	1.181
32.00	1.260
32.30	1.272
32.40	1.276
33.00	1.299
34.00	1.339
35.00	1.378
35.50	1.398
35.55	1.400
38.00	1.496
41.00	1.614
43.55	1.715
45.00	1.772
48.00	1.890
49.00	1.929
49.90	1.965
54.00	2.126
58.00	2.284
65.00	2.559
65.75	2.589
70.00	2.756
88.00	3.465

FIGURE A1—METRIC NUMBERS CONVERTED TO INCH DIMENSIONS

(R) CONNECTIONS FOR GENERAL USE AND FLUID POWER—PORTS AND STUD ENDS WITH ISO 725 THREADS AND O-RING SEALING—PART 2: HEAVY-DUTY (S SERIES) STUD ENDS—SAE J1926/2 MAR93 / ISO 11926-2

SAE Standard

Report of the Fluid Conductors and Connectors Technical Committee approved August 1988. Completely revised by the SAE Fluid Conductors and Connectors Technical Committee SC1—Automotive and Hydraulic Tube and Fittings March 1993.

This document is technically equivalent to ISO 11926-2.

Foreword—SAE J1926 consists of the following parts, under the general title:

Connections for general use and fluid power—Ports and stud ends with ISO 725 threads and O-ring sealing:

— Part 1: Port With O-Ring Seal in Truncated Housing
— Part 2: Heavy-Duty (S Series) Stud Ends
— Part 3: Light-Duty (L Series) Stud Ends

These standards define performance requirements, dimensions, and designs for port and stud end connections for heavy-duty in Part 2 and light-duty in Part 3. Significant testing through 40 years of use has confirmed the performance requirements of these ports and stud ends. Stud ends in conformance with ISO 11926-2 are identical to those in conformance with SAE J1453, and stud ends in conformance with ISO 11926-3 are identical to those in conformance with SAE J514. Stud ends in conformance with SAE J1926-3 (ISO 11926-3) are used on fittings in ISO 8434-2.

In fluid power systems, power is transmitted and controlled through a fluid (liquid or gas) under pressure within an enclosed circuit. In general applications, a fluid may be conveyed under pressure. Components are connected through their threaded ports by stud ends on fluid conductor fittings to tubes/pipes, or to hose fittings and hoses.

1. Scope—This part of SAE J1926 specifies dimensions, performance requirements, and test procedures for adjustable and nonadjustable heavy-duty (S series) stud ends with ISO 725 threads for use in fluid power and general applications and the O-rings used with them.

Stud ends in accordance with this part of SAE J1926 may be used at working pressures up to 63 MPa for nonadjustable stud ends and up to 40 MPa for adjustable stud ends. The permissible working pressure depends upon materials, design, working conditions, application, etc.

For threaded ports and stud ends specified in new designs for hydraulic fluid power applications, only SAE J2244 (ISO 6149) shall be used. Threaded ports and stud ends in accordance with ISO 1179, ISO 9974, and ISO 11926 shall not be used for new design in hydraulic fluid power applications.

Conformance to the dimensional information does not guarantee rated performance. Each manufacturer shall perform testing according to the specification contained in this document to ensure that components made to this document comply with the performance rating.

Appendices A and B of this document are normative; Appendix C of this document is informative.

2. References

2.1 Applicable Documents—The following standards contain provisions which, through reference in this text, constitute provisions of this document. At the time of publication, the editions indicated were valid. All standards are subject to revision, and parties to agreements based on this document are encouraged to investigate the possibility of applying the most recent edition of the standards indicated as follows. Members of IEC and ISO maintain registers of currently valid International Standards.

2.1.1 SAE PUBLICATIONS—Available from SAE, 400 Commonwealth Drive, Warrendale, PA 15096-0001.

SAE J515—Hydraulic O-ring

SAE J1644—Metallic Tube Connections for Fluid Power and General Use—Test Methods for Threaded Hydraulic Fluid Power Connectors

SAE J2244/2—Connections for Fluid Power and General Use—Ports and Stud Ends with ISO 261 Thread and O-ring Sealing—Part 2: Heavy-Duty (S Series) Stud Ends, Dimensions, Designs, Test Methods, and Requirements

2.1.2 ISO PUBLICATIONS—Available from ANSI, 11 West 42nd Street, New York, NY 10036-8002.

ISO 725:1978—ISO inch screw threads—Basic dimensions

ISO 1179-1:---[1]—Connections for general use and fluid power—Ports and stud ends with ISO 228-1 threads with elastomeric and metal-to-metal sealing—Part 1: Threaded port

ISO 1179-2:---[1]—Connections for general use and fluid power—Ports and stud ends with ISO 228-1 threads with elastomeric and metal-to-metal sealing—Part 2: Heavy duty (S series) and light duty (L series) stud ends with elastomeric sealing (type E)

ISO 1179-3:---[1]—Connections for general use and fluid power—Ports and stud ends with ISO 228-1 threads with elastomeric and metal-to-metal sealing—Part 3: Light duty (L series) stud end with sealing by O-ring with retaining ring (types G and H)

ISO 1179-4:---[1]—Connections for general use and fluid power—Ports and stud ends with ISO 228-1 threads with elastomeric and metal-to-metal sealing—Part 4: Stud end for general use only with metal-to-metal sealing (type B)

ISO 4759-1:1978—Tolerances for fasteners—Part 1: Bolts, screws and nuts with thread diameters between 1.6 (inclusive) and 150 mm (inclusive) and product grades A, B and C

ISO 5598:1985—Fluid power systems and components—Vocabulary

ISO 6149-1:---[1]—Connections for fluid power and general use—Ports and stud ends with ISO 261 threads and O-ring sealing—Part 1: Port with O-ring seal in truncated housing

ISO 6149-2:---[1]—Connections for fluid power and general use—Ports and stud ends with ISO 261 threads and O-ring sealing—Part 2: Heavy duty (S series) stud ends—Dimensions, design, test methods and requirements

ISO 6149-3:---[1]—Connections for fluid power and general use—Ports and stud ends with ISO 261 threads and O-ring sealing—Part 3: Light duty (L series) stud ends—Dimensions, design, test methods and requirements

---

[1] To be published.

ISO 8434-2:---[1]—Metallic tube fittings for fluid power and general use—Part 2: 37° Flared Fittings

ISO 9974-1:---[1]—Connections for general use and fluid power—Ports and stud ends with ISO 261 threads with elastomeric and metal-to-metal sealing—Part 1: Threaded port

ISO 11926-1:---[1]—Connections for general use and fluid power—Ports and stud ends with ISO 261 threads and O-ring sealing—Part 1: Threaded port with O-ring seal in truncated housing

ISO 11926-2:---[1]—Connections for general use and fluid power—Ports and stud ends with ISO 261 threads and O-ring sealing—Part 2: Heavy duty (S series) stud ends

ISO 11926-3:---[1]—Connections for general use and fluid power—Ports and stud ends with ISO 261 threads and O-ring sealing—Part 3: Light duty (L series) stud ends

2.2 Related Publications—The following publications are provided for information purposes only and are not a required part of this document.

2.2.1 ASTM Publications—Available from ASTM, 1916 Race Street, Philadelphia, PA 19103-1187.

ASTM B 117—Method of Salt Spray (Fog) Test

ASTM B 633—Standard Specifications for Electrodeposited Coatings of Zinc or Iron and Steel

2.2.2 ISO Publications—Available from ANSI, 11 West 42nd Street, New York, NY 10036-8002.

ISO 48:1979—Vulcanized rubbers—Determination of hardness (Hardness between 30 and 85 IRHD)

ISO 263:1973—ISO inch screw threads—General plan and selection for screws, bolts and nuts—Diameter range 0,06 to 6 in

ISO 1101:1983—Technical drawings—Tolerancing of form, orientation, location and run-out—Generalities, definitions, symbols, indications on drawings

ISO 1302:1978—Technical drawings—Method of indicating surface texture on drawings

ISO 3448:1975—Industrial liquid lubricants—ISO viscosity classification

ISO 3601-3:1987—Fluid systems—Sealing devices—O-rings—Part 3: Quality acceptance criteria

ISO 6803:1984—Rubber or plastic hoses and hose assemblies—Hydraulic pressure impulse test without flexing

ISO 9974-2:---[1]—Connections for general use and fluid power—Ports and stud ends with ISO 261 threads with elastomeric and metal-to-metal sealing—Part 2: Stud end with elastomeric sealing (type E)

ISO 9974-3:---[1]—Connections for general use and fluid power—Ports and stud ends with ISO 261 threads with elastomeric and metal-to-metal sealing—Part 3: Stud end with metal-to-metal sealing (type B)

3. Definitions—For the purpose of this part of SAE J1926, the definitions given in ISO 5598 and the following shall apply.

3.1 Adjustable Stud End—A stud end connector that allows for fitting orientation through final tightening of the locknut to complete the connection. This type of stud end is typically used on shaped fittings (e.g., tees, crosses, and elbows).

3.2 Nonadjustable Stud End—A stud end connector that does not require specific orientation before final tightening of the connection because it is only used on straight fittings.

4. Stud End Size Specifications—The stud ends shall be specified by SAE J1926/2 and the thread size, separated by a colon, for example, SAE J1926/2:1/2-20.

5. Requirements

5.1 Dimensions—Heavy-duty (S series) SAE J1926-2 stud ends shall conform to the dimensions in Figures 1A and 1B and Table 1. Hex tolerances across flats shall be according to ISO 4759-1 product grade C.

5.2 Working Pressure—Heavy-duty (S series) stud ends made of low-carbon steel shall be designed for use at the working pressures given in Table 2.

5.3 Performance—Heavy-duty (S series) stud ends made of low-carbon steel shall meet or exceed the burst and impulse pressures given in Table 2 when tested according to 5.5.

---
[1] To be published.

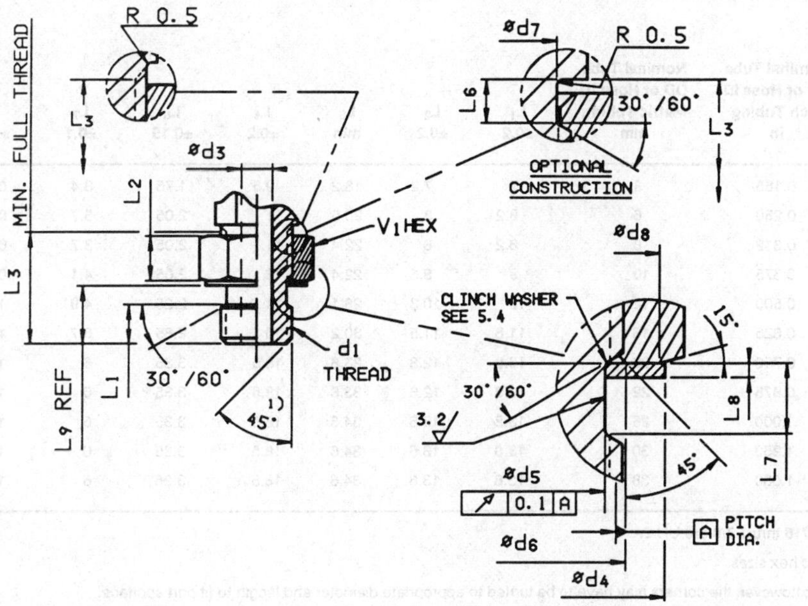

FIGURE 1A—ADJUSTABLE SAE J1926/2 HEAVY-DUTY (S SERIES) STUD END DETAIL

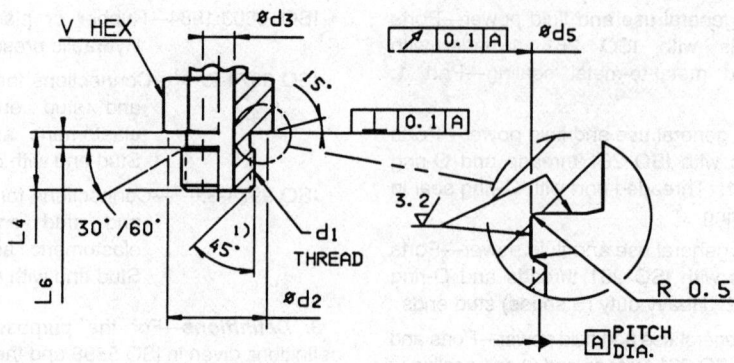

FIGURE 1B—NONADJUSTABLE SAE J1926/2 HEAVY-DUTY (S SERIES) STUD END DETAIL

TABLE 1—SAE J1926/2 HEAVY-DUTY (S SERIES) STUD END DIMENSIONS

Dimensions in millimeters

Nominal Tube OD or Hose ID Inch Tubing Dash Size	Nominal Tube OD or Hose ID Inch Tubing mm	Nominal Tube OD or Hose ID Inch Tubing in	Nominal Tube OD or Hose ID Metric Tubing mm	d_1[1] Thread Size in	ϕd_2 ±0.2	ϕd_3	ϕd_4 ±0.4	ϕd_5 +0.05 −0.08	ϕd_6 ±0.2	ϕd_7 +0 −0.3	ϕd_8 ±0.2
-3	4.76	0.188	5	3/8-24	12.5	3.2 +0.18/0	14.6	7.95	9.9	8	14.1
-4	6.35	0.250	6	7/16-20	14.1	4.5 +0.18/0	16.5	9.25	11.5	9.3	14.9
-5	7.94	0.312	8	1/2-20	15.7	6 +0.18/0	18.3	10.85	13	10.9	17.3
-6	9.52	0.375	10	9/16-18	17.3	7.5 +0.22/0	20.2	12.24	14.6	12.3	18.8
-8	12.70	0.500	12	3/4-16	22	10 +0.22/0	25.7	16.76	19.4	16.8	23.6
-10	15.88	0.625	16	7/8-14	25.2	12.5 +0.27/0	29.3	19.63	22.6	19.7	26.8
-12	19.05	0.750	20	1-1/16-12	37.5	15.5 +0.27/0	36.7	24	27.3	24	34.7
-14	22.22	0.875	22	1-3/16-12	34.7	18 +0.27/0	40.4	27.18	30.5	27.2	37.9
-16	25.40	1.000	25	1-5/16-12	37.9	21.5 +0.33/0	44	30.35	33.7	30.4	41.1
-20	31.75	1.250	30	1-5/8-12	47.4	27.5 +0.33/0	55	38.28	41.6	38.3	47.4
-24	38.10	1.500	38	1-7/8-12	53.8	33.5 +0.39/0	62.3	44.6	48	44.6	53.8

TABLE 1—SAE J1926/2 HEAVY-DUTY (S SERIES) STUD END DIMENSIONS (CONTINUED)

Dimensions in millimeters

Nominal Tube OD or Hose ID Inch Tubing Dash Size	Nominal Tube OD or Hose ID Inch Tubing mm	Nominal Tube OD or Hose ID Inch Tubing in	Nominal Tube OD or Hose ID Metric Tubing mm	L_1 ±0.2	L_2 ±0.2	L_3 min	L_4 ±0.2	L_6 ±0.15	L_7 ±0.1	L_8 ±0.08	L_9 ref	V[2,3] Hex	V_1[2] Hex
-3	4.76	0.188	5	7	7.2	18.2	9.5	1.75	3.4	0.8	9.6	12.70	14.29
-4	6.35	0.250	6	8.2	8	20.5	11	2.05	3.7	0.9	11	14.29	15.88
-5	7.94	0.312	8	8.2	8	22.4	11	2.05	3.7	0.9	11	15.88	17.46
-6	9.52	0.375	10	9	8.5	22.4	12	2.05	4.1	0.9	12.2	17.46	19.05
-8	12.70	0.500	12	10	10.3	26.1	14	2.25	4.9	1	13.9	22.22	23.81
-10	15.88	0.625	16	11.8	11.5	30.2	16	2.85	5.7	1.25	16.3	25.40	26.99
-12	19.05	0.750	20	13.8	12.8	33.8	18.5	3.35	6	1.25	18.6	31.75	34.93
-14	22.22	0.875	22	13.8	12.8	33.8	18.5	3.35	6	1.25	18.6	34.93	38.10
-16	25.40	1.000	25	13.8	13.6	34.6	18.5	3.35	6	1.25	18.6	38.10	41.28
-20	31.75	1.250	30	13.8	13.6	34.6	18.5	3.35	6	1.25	18.6	47.63	47.63
-24	38.10	1.500	38	13.8	13.6	34.6	18.5	3.35	6	1.25	18.6	53.98	53.98

[1] Sizes 3/8 thru 7/8 are UNF-2A, sizes 1-1/16 thru 1-7/8 are UN-2A.

[2] See Appendix A for recommended metric hex sizes.

[3] For jump sizes V hex size may be larger; however, the corners may have to be turned to appropriate diameter and length to fit port spotface.

TABLE 2—SAE J1926/2 HEAVY-DUTY (S SERIES) STUD END PRESSURES[1]

Units in megapascals[2]

Thread Size in	Stud End Styles Nonadjustable Working[1] Pressure	Stud End Styles Nonadjustable Test Pressure Burst	Stud End Styles Nonadjustable Test Pressure Impulse[3]	Stud End Styles Adjustable Working[1] Pressure	Stud End Styles Adjustable Test Pressure Burst	Stud End Styles Adjustable Test Pressure Impulse[3]
3/8-24 UNF-2A	63	252	83.8	40	160	53.2
7/16-20 UNF-2A	63	252	83.8	40	160	53.2
1/2-20 UNF-2A	63	252	83.8	40	160	53.2
9/16-18 UNF-2A	63	252	83.8	40	160	53.2
3/4-16 UNF-2A	63	252	83.8	40	160	53.2
7/8-14 UNF-2A	63	252	83.8	40	160	53.2
1-1/16-12 UN-2A	40	160	53.2	40	160	53.2
1-3/16-12 UN-2A	40	160	53.2	40	160	53.2
1-5/16-12 UN-2A	40	160	53.2	31.5	126	41.9
1-5/8-12 UN-2A	25	100	33.2	25	100	33.2
1-7/8-12 UN-2A	25	100	33.2	20	80	26.6

[1] These pressure ratings were established using fittings made of low-carbon steel and tested in accordance with 5.5.
[2] To convert from MPa to bar multiply by 10. To convert from MPa to psi multiply by 145.04.
[3] Cyclic endurance test pressure.

5.4 Adjustable Stud End Washer Fit and Flatness—The washer shall be clinched to the stud end with a tight slip fit to an interference fit. The slip fit shall be tight enough so that the washer cannot be shaken loose to cause it to drop from its uppermost position by its own weight. The locknut torque needed to move the washer at the maximum washer interference fit shall not exceed the torques given in Table 3.

Any washer surface that is out of flatness shall be uniform (i.e., not wavy) and concave with respect to the stud end and shall conform to the allowance given in Table 3.

TABLE 3—ADJUSTABLE STUD END WASHER TORQUE AND FLATNESS ALLOWANCE

Thread Size in	Maximum Nut Torque to Move Washer N·m[1]	Maximum Washer Flatness Allowance mm
3/8-24 UNF-2A	3	0.25
7/16-20 UNF-2A	4	0.25
1/2-20 UNF-2A	5	0.25
9/16-18 UNF-2A	7	0.25
3/4-16 UNF-2A	10	0.25
7/8-14 UNF-2A	12	0.25
1-1/16-12 UN-2A	15	0.40
1-3/16-12 UN-2A	18	0.40
1-5/16-12 UN-2A	20	0.40
1-5/8-12 UN-2A	25	0.50
1-7/8-12 UN-2A	30	0.50

[1] To convert from N·m to lbf·ft multiply by 0.737.

5.5 Test Methods—Stud ends shall be tested for burst and impulse per SAE J1644.

6. O-ring—O-rings used with heavy-duty (S series) SAE J1926/2 stud ends shall conform to the dimensions given in SAE J515.

7. Identification Statement—Use the following statement in test reports, catalogues, and sales literature when electing to comply with this part of SAE J1926:

"Heavy-duty (S series) stud end conforms to SAE J1926/2, Connections for general use and fluid power—Ports and stud ends with ISO 725 threads and O-ring sealing—Part 2: Heavy-duty (S series) stud ends."

8. Notes

8.1 Key Words—Fluid power, pipe fittings, standard connection, standard coupling, pipe joints, ports, stud ends, specifications, design, operating requirements, dimensions, designation, test methods, inch, straight thread, O-ring seal, high pressure

TABLE 4—SAE J1926/2 STUD END QUALIFICATION TORQUE

Thread Size in	Torque +10% -0 N·m[1]
3/8-24 UNF-2A	10
7/16-20 UNF-2A	20
1/2-20 UNF-2A	25
9/16-18 UNF-2A	35
3/4-16 UNF-2A	70
7/8-14 UNF-2A	100
1-1/16-12 UN-2A	170
1-3/16-12 UN-2A	215
1-5/16-12 UN-2A	270
1-5/8-12 UN-2A	285
1-7/8-12 UN-2A	370

[1] To convert from N·m to lbf·ft multiply by 0.737.

APPENDIX A
(NORMATIVE)

A.1 By agreement between user and supplier, these metric hex sizes, shown in Table A1, shall be used in lieu of inch hex.

TABLE A1—METRIC HEX SIZES

Thread Size in	Nonadjustable V Hex	Nonadjustable d_2 ±0.2 mm	Adjustable V_1 Hex	Adjustable d_8 ±0.2 mm
3/8-24 UNF-2A	12	11.8	14	13.8
7/16-20 UNF-2A	14	13.8	17	16.8
1/2-20 UNF-2A	17	16.8	17	16.8
9/16-18 UNF-2A	17	16.8	19	18.8
3/4-16 UNF-2A	22	21.8	24	23.8
7/8-14 UNF-2A	27	26.8	27	26.8
1-1/16-12 UN-2A	32	31.8	36	35.8
1-3/16-12 UN-2A	36	35.8	41	40.8
1-5/16-12 UN-2A	41	40.8	41	40.8
1-5/8-12 UN-2A	50	49.8	50	48.8
1-7/8-12 UN-2A	55	54.8	55	54.8

APPENDIX B
ASSEMBLY INSTRUCTIONS FOR ADJUSTABLE STYLE FITTINGS IN STRAIGHT THREAD O-RING PORT

B.1 Lubricate O-ring by coating with light oil or petrolatum and install in the groove adjacent to the face of the metal back-up washer which is assembled at the extreme end of the groove as shown in Figure B1.

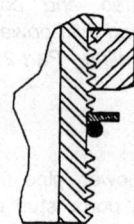

FIGURE B1—LOCKNUT BACKED OFF

B.2 Install the fitting into the SAE straight thread boss, Figure B2, until the metal back-up washer contacts the face of the boss as shown in Figure B2.

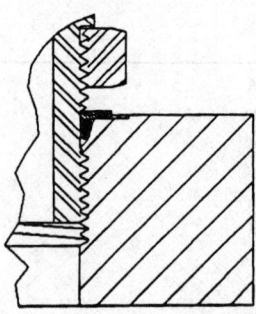

FIGURE B2—FITTING INSTALLED HAND TIGHT

B.3 Position the fitting by turning out (counterclockwise) up to a maximum of one turn (see Figure B3). Holding the pad of the fitting with a wrench, tighten the locknut and washer against the face as shown in Figure B4.

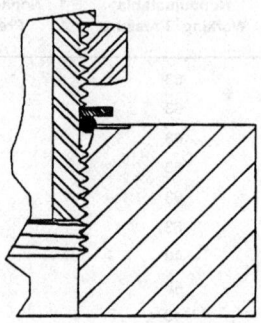

FIGURE B3—FITTING BACKED-OFF FOR ALIGNMENT (1 TURN MAXIMUM)

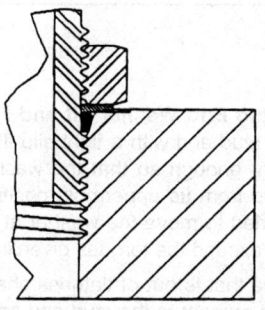

FIGURE B4—FITTING LOCKNUT TIGHTENED TO APPROPRIATE TORQUE

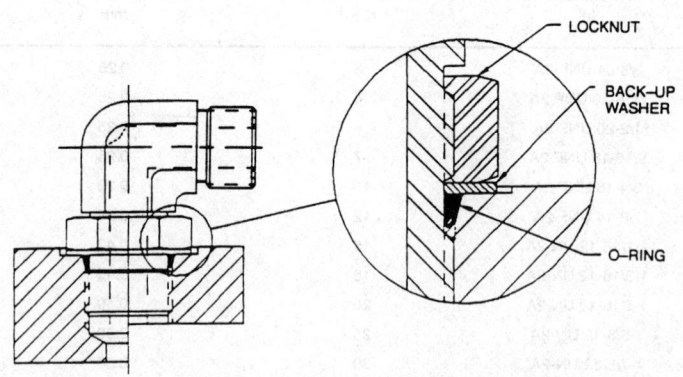

FIGURE B5—FINAL ASSEMBLY OF ADJUSTABLE STUD END

APPENDIX C
METRIC NUMBERS CONVERTED TO INCH DIMENSIONS

mm	in	mm	in	mm	in	mm	in	mm	in
0.025	0.001	5.70	0.224	14.00	0.551	21.80	0.858	34.90	1.374
0.05	0.002	6.00	0.236	14.10	0.555	22.00	0.866	34.93	1.375
0.08	0.003	7.00	0.276	14.11	0.555	22.20	0.874	35.80	1.409
0.10	0.004	7.20	0.283	14.29	0.563	22.22	0.875	36.70	1.445
0.13	0.005	7.50	0.295	14.30	0.563	22.40	0.882	37.50	1.476
0.15	0.006	7.95	0.313	14.40	0.567	22.60	0.890	37.70	1.484
0.18	0.007	8.00	0.315	14.60	0.575	23.60	0.929	37.90	1.492
0.20	0.008	8.20	0.323	14.90	0.587	23.80	0.937	38.10	1.500
0.22	0.009	8.50	0.335	15.50	0.610	23.81	0.937	38.28	1.507
0.25	0.010	9.00	0.354	15.70	0.618	24.00	0.945	38.30	1.508
0.27	0.011	9.25	0.364	15.88	0.625	25.20	0.992	40.40	1.591
0.30	0.012	9.30	0.366	15.90	0.626	25.40	1.000	40.80	1.606
0.33	0.013	9.50	0.374	16.00	0.630	25.70	1.012	41.10	1.618
0.39	0.015	9.60	0.378	16.30	0.642	26.10	1.028	41.28	1.625
0.40	0.016	9.70	0.382	16.50	0.650	26.80	1.055	41.30	1.626
0.80	0.031	9.90	0.390	16.76	0.660	26.99	1.063	41.40	1.630
0.90	0.035	10.00	0.394	16.80	0.661	27.00	1.063	41.60	1.638
1.00	0.039	10.30	0.405	17.30	0.681	27.10	1.067	44.00	1.732
1.25	0.049	10.85	0.427	17.46	0.687	27.18	1.070	44.60	1.756
1.60	0.063	10.90	0.429	17.50	0.689	27.20	1.071	47.40	1.866
1.75	0.069	11.00	0.433	18.00	0.709	27.30	1.075	47.60	1.874
1.90	0.075	11.30	0.445	18.20	0.716	27.50	1.083	47.63	1.875
2.05	0.081	11.50	0.453	18.30	0.720	29.30	1.153	47.80	1.882
2.10	0.083	11.80	0.465	18.50	0.728	30.20	1.189	48.00	1.890
2.25	0.089	12.00	0.472	18.60	0.732	30.30	1.193	48.80	1.921
2.40	0.094	12.20	0.480	18.80	0.740	30.35	1.195	49.80	1.961
2.70	0.106	12.24	0.482	19.00	0.748	30.40	1.197	53.80	2.118
2.85	0.112	12.30	0.484	19.05	0.750	30.50	1.201	53.98	2.125
3.20	0.126	12.50	0.492	19.20	0.756	31.75	1.250	54.00	2.126
3.35	0.132	12.70	0.500	19.40	0.764	31.80	1.252	54.80	2.157
3.40	0.134	12.80	0.504	19.63	0.773	33.50	1.319	55.00	2.165
3.70	0.146	13.00	0.512	19.70	0.776	33.70	1.327	62.30	2.453
4.10	0.161	13.60	0.535	20.20	0.795	33.80	1.331		
4.50	0.177	13.80	0.543	20.50	0.807	34.60	1.362		
4.90	0.193	13.90	0.547	21.50	0.846	34.70	1.366		

FIGURE C1—METRIC NUMBERS CONVERTED TO INCH DIMENSIONS

(R) CONNECTIONS FOR GENERAL USE AND FLUID POWER—PORTS AND STUD ENDS WITH ISO 725 THREADS AND O-RING SEALING—PART 3: LIGHT-DUTY (L SERIES) STUD ENDS—SAE J1926/3 MAR93 / ISO 11926-3

SAE Standard

Report of the Fluid Conductors and Connectors Technical Committee approved August 1988. Completely revised by the SAE Fluid Conductors and Connectors Technical Committee SC1—Automotive and Hydraulic Tube and Fittings March 1993.

This document is technically equivalent to ISO 11926-3.

Foreword—SAE J1926 consists of the following parts, under the general title:

Connections for general use and fluid power—Ports and stud ends with ISO 725 threads and O-ring sealing:

— Part 1: Port With O-Ring Seal in Truncated Housing

— Part 2: Heavy-Duty (S Series) Stud Ends

— Part 3: Light-Duty (L Series) Stud Ends

These standards define performance requirements, dimensions, and designs for port and stud end connections for heavy-duty in Part 2 and light-duty in Part 3. Significant testing through 40 years of use has confirmed the performance requirements of these ports and stud ends. Stud ends in conformance with ISO 11926-2 are identical to those in conformance with SAE J1453, and stud ends in conformance with ISO 11926-3 are identical to those in conformance with SAE J514. Stud ends in conformance with SAE J1926-3 (ISO 11926-3) are used on fittings in ISO 8434-2.

In fluid power systems, power is transmitted and controlled through a fluid (liquid or gas) under pressure within an enclosed circuit. In general applications, a fluid may be conveyed under pressure. Components are connected through their threaded ports by stud ends on fluid conductor fittings to tubes/pipes, or to hose fittings and hoses.

1. Scope—This part of SAE J1926 specifies dimensions, performance requirements, and test procedures for adjustable and nonadjustable light-duty (L series) stud ends with ISO 725 threads for use in fluid power and general applications and the O-rings used with them.

Stud ends in accordance with this part of SAE J1926 may be used at working pressures up to 40 MPa for nonadjustable stud ends and up to 31.5 MPa for adjustable stud ends. The permissible working pressure depends upon materials, design, working conditions, application, etc.

For threaded ports and stud ends specified in new designs for hydraulic fluid power applications, only SAE J2244 (ISO 6149) shall be used. Threaded ports and stud ends in accordance with ISO 1179, ISO 9974, and ISO 11926 shall not be used for new design in hydraulic fluid power applications.

Conformance to the dimensional information does not guarantee rated performance. Each manufacturer shall perform testing according to the specification contained in this document to ensure that components made to this document comply with the performance rating.

Appendices A and B of this document are normative; Appendix C of this document is informative.

2. References

2.1 Applicable Documents—The following standards contain provisions which, through reference in this text, constitute provisions of this document. At the time of publication, the editions indicated were valid. All standards are subject to revision, and parties to agreements based on this document are encouraged to investigate the possibility of applying the most recent edition of the standards indicated as follows. Members of IEC and ISO maintain registers of currently valid International Standards.

2.1.1 SAE PUBLICATIONS—Available from SAE, 400 Commonwealth Drive, Warrendale, PA 15096-0001.

SAE J515—Hydraulic O-ring

SAE J1644—Metallic Tube Connections for Fluid Power and General Use—Test Methods for Threaded Hydraulic Fluid Power Connectors

2.1.2 ISO PUBLICATIONS—Available from ANSI, 11 West 42nd Street, New York, NY 10036-8002.

ISO 48:1979—Vulcanized rubbers—Determination of hardness (Hardness between 30 and 85 IRHD)

ISO 263:1973—ISO inch screw threads—General plan and selection for screws, bolts, and nuts—Diameter range 9,06 to 6 in

ISO 725:1978—ISO inch screw threads—Basic dimensions

ISO 1101:1983—Technical drawings—Tolerancing of form, orientation, location, and run-out—Generalities, definitions, symbols, indications on drawings

ISO 1179-1:---[1]—Connections for general use and fluid power—Ports and stud ends with ISO 228-1 threads with elastomeric and metal-to-metal sealing—Part 1: Threaded port

ISO 1179-2:---[1]—Connections for general use and fluid power—Ports and stud ends with ISO 228-1 threads with elastomeric and metal-to-metal sealing—Part 2: Heavy duty (S series) and light duty (L series) stud ends with elastomeric sealing (type E)

ISO 1179-3:---[1]—Connections for general use and fluid power—Ports and stud ends with ISO 228-1 threads with elastomeric and metal-to-metal sealing—Part 3: Light duty (L series) stud end with sealing by O-ring with retaining ring (types G and H)

ISO 1179-4:---[1]—Connections for general use and fluid power—Ports and stud ends with ISO 228-1 threads with elastomeric and metal-to-metal sealing—Part 4: Stud end for general use only with metal-to-metal sealing (type B)

ISO 1302:1978—Technical drawings—Method of indicating surface texture on drawings

ISO 3448:1975—Industrial liquid lubricants—ISO viscosity classification

ISO 3601-3:1987—Fluid systems—Sealing devices—O-rings—Part 3: Quality acceptance criteria

ISO 4759-1:1978—Tolerances for fasteners—Part 1: Bolts, screws and nuts with thread diameters between 1.6 (inclusive) and 150 mm (inclusive) and product grades A, B and C

ISO 5598:1985—Fluid power systems and components—Vocabulary

ISO 6149-1:---[1]—Connections for fluid power and general use—Ports and stud ends with ISO 261 threads and O-ring sealing—Part 1: Port with O-ring seal in truncated housing

---

[1] To be published.

ISO 6149-2:---[1]—Connections for fluid power and general use—Ports and stud ends with ISO 261 threads and O-ring sealing—Part 2: Heavy duty (S series) stud ends—Dimensions, design, test methods and requirements

ISO 6149-3:---[1]—Connections for fluid power and general use—Ports and stud ends with ISO 261 threads and O-ring sealing—Part 3: Light duty (L series) stud ends—Dimensions, design, test methods and requirements

ISO 6803:1984—Rubber or plastic hoses and hose assemblies—Hydraulic pressure impulse test without flexing

ISO 8434-2:---[1]—Metallic tube fittings for fluid power and general use—Part 2: 37° Flared Fittings

ISO 9974-1:---[1]—Connections for general use and fluid power—Ports and stud ends with ISO 261 threads with elastomeric and metal-to-metal sealing—Part 1: Threaded port

ISO 9974-2:---[1]—Connections for general use and fluid power—Ports and stud ends with ISO 261 threads with elastomeric and metal-to-metal sealing—Part 2: Stud ends with elastomeric sealing (type E)

ISO 9974-3:---[1]—Connections for general use and fluid power—Ports and stud ends with ISO 261 threads with elastomeric and metal-to-metal sealing—Part 3: Stud ends with metal-to-metal sealing (type B)

ISO 11926-1:---[1]—Connections for general use and fluid power—Ports and stud ends with ISO 261 threads and O-ring sealing—Part 1: Threaded port with O-ring seal in truncated housing

ISO 11926-2:---[1]—Connections for general use and fluid power—Ports and stud ends with ISO 261 threads and O-ring sealing—Part 2: Heavy duty (S series) stud ends

ISO 11926-3:---[1]—Connections for general use and fluid power—Ports and stud ends with ISO 261 threads and O-ring sealing—Part 3: Light duty (L series) stud ends

2.1.3 ASTM PUBLICATIONS—Available from ASTM, 1916 Race Street, Philadelphia, PA 19103-1187.

ASTM B 117—Method of Salt Spray (Fog) Test

ASTM B 633—Standard Specifications for Electrodeposited Coatings of Zinc or Iron and Steel

3. Definitions—For the purpose of this part of SAE J1926, the definitions given in ISO 5598 and the following shall apply.

3.1 Adjustable Stud End—A stud end connector that allows for fitting orientation through final tightening of the locknut to complete the connection. This type of stud end is typically used on shaped fittings (e.g., tees, crosses, and elbows).

3.2 Nonadjustable Stud End—A stud end connector that does not require specific orientation before final tightening of the connection because it is only used on straight fittings.

4. Stud End Size Specifications—The stud ends shall be specified by SAE J1926/3 and the thread size, separated by a colon, for example, SAE J1926/3:1/2-20.

5. Requirements

5.1 Dimensions—Light-duty (L series) SAE J1926-3 stud ends shall conform to the dimensions in Figures 1A and 1B and Table 1. Hex tolerances across flats shall be according to ISO 4759-1 product grade C.

5.2 Working Pressure—Light-duty (L series) stud ends made of low-carbon steel shall be designed for use at the working pressures given in Table 2.

5.3 Performance—Light-duty (L series) stud ends made of low-carbon steel shall meet or exceed the burst and impulse pressures given in Table 2 when tested according to 5.5.

---
[1] To be published.

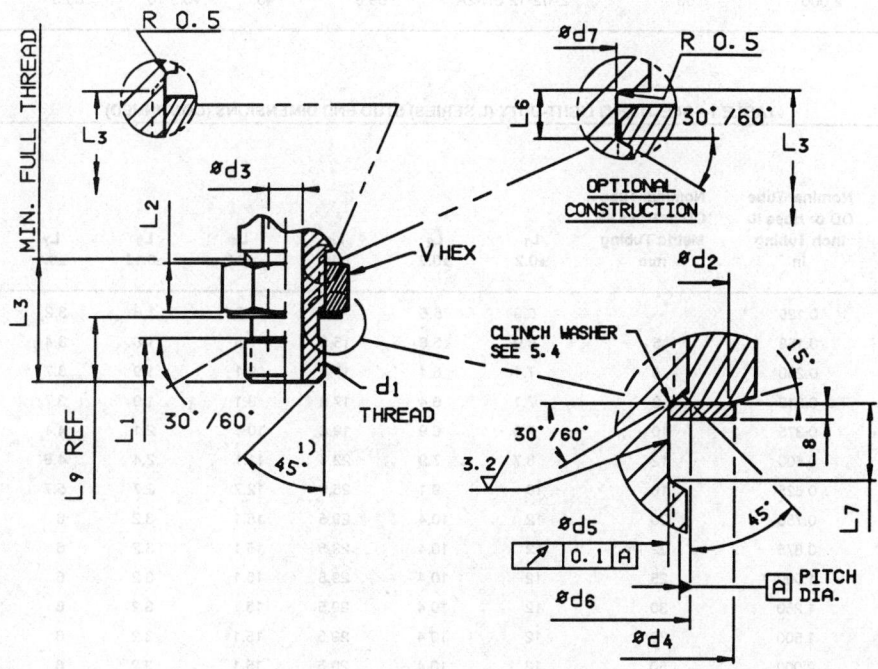

FIGURE 1A—ADJUSTABLE SAE J1926/3 LIGHT-DUTY (L SERIES) STUD END DETAIL

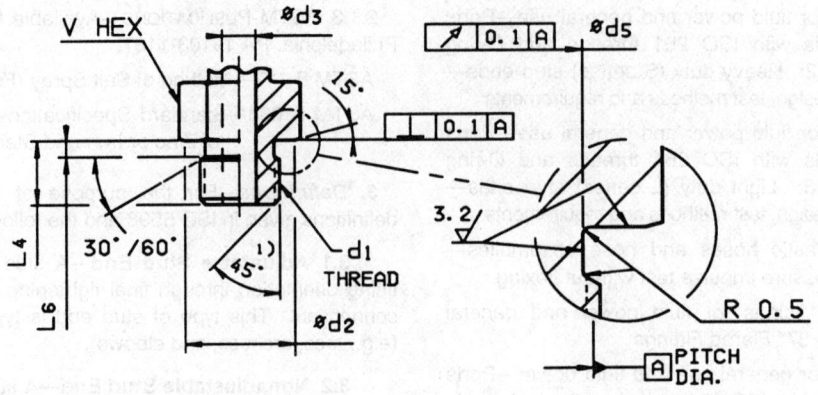

FIGURE 1B—NONADJUSTABLE SAE J1926/3 LIGHT-DUTY (L SERIES) STUD END DETAIL

TABLE 1—SAE J1926/3 LIGHT-DUTY (L SERIES) STUD END DIMENSIONS

Dimensions in millimeters

Nominal Tube OD or Hose ID Inch Tubing Dash Size	Nominal Tube OD or Hose ID Inch Tubing mm	Nominal Tube OD or Hose ID Inch Tubing in	Nominal Tube OD or Hose ID Metric Tubing mm	d_1 Thread Size in	ϕd_2 ±0.2	ϕd_3	ϕd_4 ±0.4	ϕd_5 +0.05 −0.08	ϕd_6 ±0.2	ϕd_7 +0 −0.3
-2	3.18	0.125	--	5/16-24 UNF-2A	10.9	1.6 +0.14/0	12.8	6.35	8.2	6.4
-3	4.76	0.188	5	3/8-24 UNF-2A	14.1	3.2 +0.18/0	14.6	7.95	9.9	8
-4	6.35	0.250	6	7/16-20 UNF-2A	15.7	4.5 +0.18/0	16.5	9.25	11.5	9.3
-5	7.94	0.312	8	1/2-20 UNF-2A	17.3	6 +0.22/0	18.3	10.85	13	10.9
-6	9.52	0.375	10	9/16-18 UNF-2A	18.8	7.5 +0.22/0	20.2	12.24	14.6	12.3
-8	12.70	0.500	12	3/4-16 UNF-2A	23.6	10 +0.22/0	25.7	16.76	19.4	16.8
-10	15.88	0.625	16	7/8-14 UNF-2A	26.8	12.5 +0.27/0	29.3	19.63	22.6	19.7
-12	19.05	0.750	20	1-1/16-12 UN-2A	34.7	15.5 +0.27/0	36.7	24	27.3	24
-14	22.22	0.875	22	1-3/16-12 UN-2A	37.9	18 +0.27/0	40.4	27.18	30.5	27.2
-16	25.40	1.000	25	1-5/16-12 UN-2A	41.1	21.5 +0.33/0	44	30.35	33.7	30.4
-20	31.75	1.250	30	1-5/8-12 UN-2A	47.4	27.5 +0.33/0	55	38.28	41.6	38.3
-24	38.10	1.500	38	1-7/8-12 UN-2A	53.8	33.5 +0.39/0	62.3	44.6	48	44.6
-32	50.80	2.000	50	2-1/2-12 UN-2A	69.8	45 +0.39/0	80.3	60.48	63.8	60.5

TABLE 1—SAE J1926/3 LIGHT-DUTY (L SERIES) STUD END DIMENSIONS (CONTINUED)

Dimensions in millimeters

Nominal Tube OD or Hose ID Inch Tubing Dash Size	Nominal Tube OD or Hose ID Inch Tubing mm	Nominal Tube OD or Hose ID Inch Tubing in	Nominal Tube OD or Hose ID Metric Tubing mm	L_1 ±0.2	L_2 ±0.2	L_3 min	L_4 ±0.2	L_6 ±0.15	L_7 ±0.1	L_8 ±0.08	L_9 ref	$V^{2,3}$ Hex
-2	3.18	0.125	--	5.9	5.6	15	7.5	1.6	3.2	0.8	8.3	11.11
-3	4.76	0.188	5	5.9	5.6	15.5	7.5	1.6	3.4	0.8	9.5	12.70
-4	6.35	0.250	6	7.1	6.4	17.8	9.1	1.9	3.7	0.9	10.9	14.29
-5	7.94	0.312	8	7.1	6.4	17.8	9.1	1.9	3.7	0.9	10.9	15.88
-6	9.52	0.375	10	7.9	6.9	19.6	10	2.1	4.1	0.9	12.1	17.46
-8	12.70	0.500	12	8.7	7.9	22.4	11.1	2.4	4.9	1	13.8	22.22
-10	15.88	0.625	16	10	9.1	25.9	12.7	2.7	5.7	1.25	16.1	25.40
-12	19.05	0.750	20	12	10.4	29.5	15.1	3.2	6	1.25	18.4	31.75
-14	22.22	0.875	22	12	10.4	29.5	15.1	3.2	6	1.25	18.4	34.93
-16	25.40	1.000	25	12	10.4	29.5	15.1	3.2	6	1.25	18.4	38.10
-20	31.75	1.250	30	12	10.4	29.5	15.1	3.2	6	1.25	18.4	47.63
-24	38.10	1.500	38	12	10.4	29.5	15.1	3.2	6	1.25	18.4	53.98
-32	50.80	2.000	50	12	10.4	29.5	15.1	3.2	6	1.25	18.4	69.85

[1] See Appendix A for recommended metric hex sizes.
[2] For jump sizes V hex size may be larger; however, the corners may have to be turned to appropriate diameter and length to fit port spotface.

TABLE 2—SAE J1926/3 LIGHT-DUTY (L SERIES) STUD END PRESSURES[1]

Units in megapascals[2]

Thread Size in	Stud End Styles Nonadjustable Working[1] Pressure	Stud End Styles Nonadjustable Test Pressure Burst	Stud End Styles Nonadjustable Test Pressure Impulse[3]	Stud End Styles Adjustable Working[1] Pressure	Stud End Styles Adjustable Test Pressure Burst	Stud End Styles Adjustable Test Pressure Impulse[3]
5/16-24 UNF-2A	31.5	126	41.9	31.5	126	41.9
3/8-24 UNF-2A	31.5	126	41.9	31.5	126	41.9
7/16-20 UNF-2A	31.5	126	41.9	31.5	126	41.9
1/2-20 UNF-2A	31.5	126	41.9	31.5	126	41.9
9/16-18 UNF-2A	31.5	126	41.9	25	100	33.3
3/4-16 UNF-2A	31.5	126	41.9	25	100	33.3
7/8-14 UNF-2A	25	100	33.3	20	80	26.6
1-1/16-12 UN-2A	25	100	33.3	20	80	26.6
1-3/16-12 UN-2A	20	80	26.6	16	64	21.3
1-5/16-12 UN-2A	20	80	26.6	16	64	21.3
1-5/8-12 UN-2A	16	64	21.3	12.5	50	16.6
1-7/8-12 UN-2A	16	64	21.3	12.5	50	16.6
2-1/2-12 UN-2A	12.5	50	16.6	10	40	13.3

[1] These pressure ratings were established using fittings made of low-carbon steel and tested in accordance with 5.5.
[2] To convert from MPa to bar multiply by 10. To convert from MPa to psi multiply by 145.04.
[3] Cyclic endurance test pressure.

5.4 Adjustable Stud End Washer Fit and Flatness—The washer shall be clinched to the stud end with a tight slip fit to an interference fit. The slip fit shall be tight enough so that the washer cannot be shaken loose to cause it to drop from its uppermost position by its own weight. The locknut torque needed to move the washer at the maximum washer interference fit shall not exceed the torques given in Table 3.

Any washer surface that is out of flatness shall be uniform (i.e., not wavy) and concave with respect to the stud end and shall conform to the allowance given in Table 3.

TABLE 3—ADJUSTABLE STUD END WASHER TORQUE AND FLATNESS ALLOWANCE

Thread Size in	Maximum Nut Torque to Move Washer N·m[1]	Maximum Washer Flatness Allowance mm
5/16-24 UNF-2A	2	0.25
3/8-24 UNF-2A	3	0.25
7/16-20 UNF-2A	4	0.25
1/2-20 UNF-2A	5	0.25
9/16-18 UNF-2A	7	0.25
3/4-16 UNF-2A	10	0.25
7/8-14 UNF-2A	12	0.25
1-1/16-12 UN-2A	15	0.40
1-3/16-12 UN-2A	18	0.40
1-5/16-12 UN-2A	20	0.40
1-5/8-12 UN-2A	25	0.50
1-7/8-12 UN-2A	30	0.50
2-1/2-12 UN-2A	40	0.50

[1] To convert from N·m to lbf·ft multiply by 0.737.

5.5 Test Methods—Stud ends shall be tested for burst and impulse per SAE J1644.

6. O-ring—O-rings used with light-duty (L series) SAE J1926/3 stud ends shall conform to the dimensions given in SAE J515.

7. Identification Statement—Use the following statement in test reports, catalogues, and sales literature when electing to comply with this part of SAE J1926:

"Light-duty (L series) stud end conforms to SAE J1926/3, Connections for general use and fluid power—Ports and stud ends with ISO 725 threads and O-ring sealing—Part 3: Light-duty (L series) stud ends."

8. Notes

8.1 Key Words—Fluid power, pipe fittings, standard connection, standard coupling, pipe joints, ports, stud ends, specifications, design, operating requirements, dimensions, designation, test methods, inch, straight thread, O-ring seal, high pressure

TABLE 4—SAE J1926/3 STUD END QUALIFICATION TORQUE

Thread Size in	Torque +10% -0 N·m[1]
5/16-24 UNF-2A	8
3/8-24 UNF-2A	10
7/16-20 UNF-2A	18
1/2-20 UNF-2A	25
9/16-18 UNF-2A	30
3/4-16 UNF-2A	50
7/8-14 UNF-2A	60
1-1/16-12 UN-2A	95
1-3/16-12 UN-2A	125
1-5/16-12 UN-2A	150
1-5/8-12 UN-2A	200
1-7/8-12 UN-2A	210
2-1/2-12 UN-2A	300

[1] To convert from N·m to lbf·ft multiply by 0.737.

APPENDIX A
(NORMATIVE)

A.1 By agreement between user and supplier, these metric hex sizes, shown in Table A1, shall be used in lieu of inch hex.

TABLE A1—METRIC HEX SIZES

Thread Size in	Nonadjustable & Adjustable V Hex, mm	Nonadjustable & Adjustable d_2 ±0.2 mm
5/16-24 UNF-2A	10	9.8
3/8-24 UNF-2A	12	11.8
7/16-20 UNF-2A	14	13.8
1/2-20 UNF-2A	17	16.8
9/16-18 UNF-2A	17	16.8
3/4-16 UNF-2A	22	21.8
7/8-14 UNF-2A	27	26.8
1-1/16-12 UN-2A	32	31.8
1-3/16-12 UN-2A	36	35.8
1-5/16-12 UN-2A	41	40.8
1-5/8-12 UN-2A	50	48.8
1-7/8-12 UN-2A	55	54.8
2-1/2-12 UN-2A	70	69.8

APPENDIX B
ASSEMBLY INSTRUCTIONS FOR ADJUSTABLE STYLE FITTINGS IN STRAIGHT THREAD O-RING PORT

B.1 Lubricate O-ring by coating with light oil or petrolatum and install in the groove adjacent to the face of the metal back-up washer which is assembled at the extreme end of the groove as shown in Figure B1.

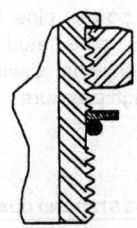

FIGURE B1—LOCKNUT BACKED OFF

B.2 Install the fitting into the SAE straight thread boss, Figure B2, until the metal back-up washer contacts the face of the boss as shown in Figure B2.

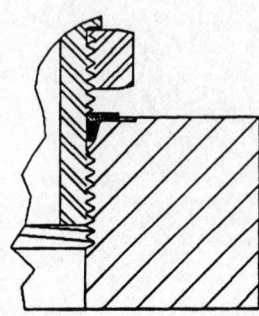

FIGURE B2—FITTING INSTALLED HAND TIGHT

B.3 Position the fitting by turning out (counterclockwise) up to a maximum of one turn (see Figure B3). Holding the pad of the fitting with a wrench, tighten the locknut and washer against the face as shown in Figure B4.

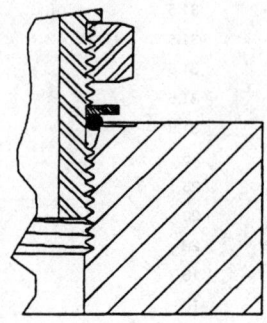

FIGURE B3—FITTING BACKED-OFF FOR ALIGNMENT (1 TURN MAXIMUM)

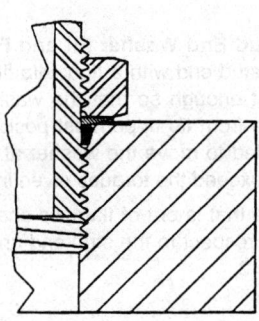

FIGURE B4—FITTING LOCKNUT TIGHTENED TO APPROPRIATE TORQUE

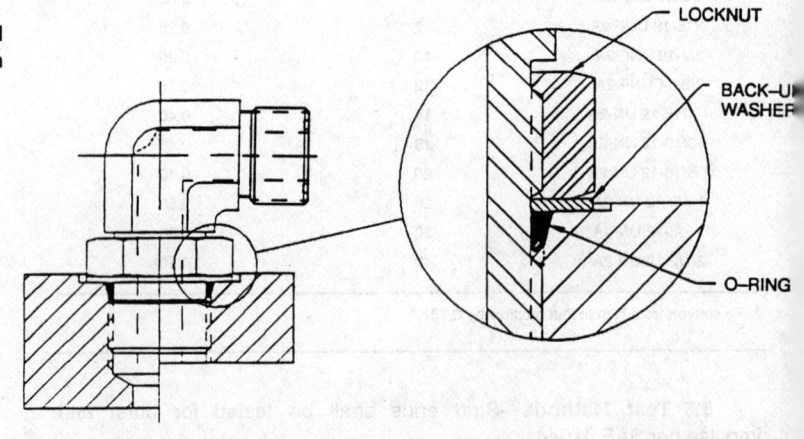

FIGURE B5—FINAL ASSEMBLY OF ADJUSTABLE STUD END

APPENDIX C
METRIC NUMBERS CONVERTED TO INCH DIMENSIONS

mm	in	mm	in	mm	in	mm	in	mm	in
0.025	0.001	5.70	0.224	12.80	0.504	22.00	0.866	38.30	1.508
0.04	0.002	5.90	0.232	13.00	0.512	22.40	0.882	40.40	1.591
0.05	0.002	6.00	0.236	13.80	0.543	22.60	0.890	40.80	1.606
0.08	0.003	6.35	0.250	14.00	0.551	23.60	0.929	41.00	1.614
0.10	0.004	6.40	0.252	14.10	0.555	23.80	0.937	41.10	1.618
0.13	0.005	6.90	0.272	14.40	0.567	24.00	0.945	41.30	1.626
0.14	0.006	7.10	0.279	14.30	0.563	25.70	1.012	41.40	1.630
0.15	0.006	7.50	0.295	14.60	0.575	25.90	1.020	41.60	1.638
0.18	0.007	7.90	0.311	15.00	0.591	26.80	1.055	44.00	1.732
0.20	0.008	7.95	0.313	15.10	0.594	27.00	1.063	44.60	1.756
0.22	0.009	8.00	0.315	15.50	0.610	27.10	1.067	45.00	1.772
0.25	0.010	8.20	0.323	15.70	0.618	27.18	1.070	47.40	1.866
0.27	0.011	8.30	0.327	15.80	0.622	27.20	1.071	47.60	1.874
0.30	0.012	8.70	0.342	15.90	0.626	27.30	1.075	47.80	1.882
0.33	0.013	9.10	0.358	16.10	0.634	27.50	1.083	48.00	1.890
0.39	0.015	9.25	0.364	16.50	0.650	29.30	1.153	48.80	1.921
0.40	0.016	9.30	0.366	16.76	0.660	29.50	1.161	49.80	1.961
0.50	0.020	9.50	0.374	16.80	0.661	30.30	1.193	50.00	1.968
0.80	0.031	9.70	0.382	17.00	0.669	30.35	1.195	53.80	2.118
0.90	0.035	9.80	0.386	17.30	0.681	30.40	1.197	54.00	2.126
1.00	0.039	9.90	0.390	17.50	0.689	30.50	1.201	54.80	2.157
1.25	0.049	10.00	0.394	17.80	0.701	31.80	1.252	55.00	2.165
1.60	0.063	10.40	0.409	18.00	0.709	32.00	1.260	59.80	2.354
1.90	0.075	10.85	0.427	18.30	0.720	33.00	1.299	60.48	2.381
2.10	0.083	10.90	0.429	18.40	0.724	33.50	1.319	60.50	2.382
2.40	0.094	11.10	0.437	18.80	0.740	33.70	1.327	62.30	2.453
2.70	0.106	11.30	0.445	19.00	0.748	33.80	1.331	63.60	2.504
3.20	0.126	11.50	0.453	19.20	0.756	34.70	1.366	63.80	2.512
3.40	0.134	11.80	0.466	19.40	0.764	34.90	1.374	69.80	2.748
3.70	0.146	12.00	0.472	19.60	0.772	35.80	1.409	70.00	2.756
4.10	0.161	12.10	0.476	19.70	0.776	36.00	1.417	80.00	3.161
4.40	0.173	12.24	0.482	19.63	0.773	36.70	1.445		
4.50	0.177	12.30	0.484	20.20	0.795	37.90	1.492		
4.90	0.193	12.50	0.492	21.50	0.846	38.10	1.500		
5.60	0.220	12.70	0.500	21.80	0.858	38.28	1.507		

FIGURE C1—METRIC NUMBERS CONVERTED TO INCH DIMENSIONS

HYDRAULIC HOSE FITTINGS—SAE J516 JUN93   SAE Standard

Report of the Construction and Industrial Machinery Technical Committee, Nonmetallic Materials Committee, and Tube, Pipe, Hose, and Lubrication Fittings Committee approved January 1952. Sixth revision, Hydraulic Hose and Hose Fittings Subcommittee, Fluid Conductors and Connectors Technical Committee November 1983. Completely revised by the Fluid Conductors and Connectors Technical Committee June 1990. Revised by the Fluid Conductors and Connectors Technical Committee S2—Hydraulic Hose and Hose Fittings June 1991, June 1992, and June 1993. Rationale statement available.

1. Scope—This SAE Standard provides general and dimensional specifications for the most common hose fittings used in conjunction with hydraulic hoses specified in SAE J517 and utilized in hydraulic systems on mobile and stationary equipment.

The general specifications contained in Sections 1 through 15 are applicable to all hydraulic hose fittings and supplement the detailed specifications for the 100R-series fittings contained in the later sections of this document.

This document shall be utilized as a procurement document only to the extent agreed upon by the manufacturer and user.

Refer to SAE J517 for specifications of hose and information on hose assemblies. SAE J1273 contains information on application factors affecting hose fittings, hose, and hose assemblies.

(R) THE RATED WORKING PRESSURE OF A HOSE ASSEMBLY COMPRISING SAE J516 FITTINGS AND SAE J517 HOSE SHALL NOT EXCEED THE LOWER OF THE TWO WORKING PRESSURE RATED VALUES.

The following are hose fitting types contained in this document:

a. Male dryseal pipe thread type hose fittings shall be as shown in Figures 1A to 1C for the respective styles, and in Tables 1A to 1M for the applicable hoses and sizes.

b. Male straight thread O-ring boss type hose fittings shall be as shown in Figures 2A and 2B for the respective styles, and in Tables 2A to 2G for the applicable hoses and sizes.

c. Male 37 degree flared type hose fittings shall be as shown in Figures 3A to 3C for the respective styles, and in Tables 3A to 3K for the applicable hoses and sizes.

d. Female 37 degree flared type hose fittings shall be as shown in Figures 4A to 7B for the respective styles and shapes, and in Tables 4A to 4M for the applicable hoses and sizes.

e. Male 45 degree flared type hose fittings shall be as shown in Figures 8A and 8B for the respective styles, and in Tables 5A to 5E for the applicable hoses and sizes.

f. Female 45 degree flared type hose fittings shall be as shown in Figures 9A to 12B for the respective styles and shapes, and in Tables 6A to 6E for the applicable hoses and sizes.

g. Male flareless type hose fittings shall be as shown in Figures 13A and 13B for the respective styles, and in Tables 7A and 7B for the applicable hoses and sizes.

h. Female flareless type hose fittings shall be as shown in Figures 14A to 14C for the respective styles, and in Tables 8A and 8B for the applicable hoses and sizes.

i. 4-bolt split flange type hose fittings shall be as shown in Figures 15A to 21C for the respective styles and shapes, and in Tables 9A to 9I2 for the applicable hoses and sizes.

j. Female O-ring face seal hose fittings shall be as shown in Figures 22A to 25B for the respective styles, and Tables 10A to 10L for the applicable hoses and sizes.

It is recommended that where step sizes or additional types of fittings are required, they be designed to conform with the specifications of this document insofar as they may apply. The following general specifications shall supplement the dimensional data contained in the tables with respect to all unspecified detail.

2. References

2.1 Applicable Documents—The following publications form a part of this specification to the extent specified herein. The latest issue of SAE publications shall apply.

2.1.1 SAE PUBLICATIONS—Available from SAE, 400 Commonwealth Drive, Warrendale, PA 15096-0001.

SAE J475—Screw Threads

SAE J476—Dryseal Pipe Threads

SAE J512—Automotive Tube Fittings

SAE J514—Hydraulic Tube Fittings

SAE J517—Hydraulic Hose

SAE J518—Hydraulic Flanged Tube, Pipe, and Hose Connections, 4-Bolt Split Flange Type

SAE J533—Flares for Tubing

SAE J846—Coding Systems for Identification of Fluid Conductors and Connectors

SAE J1273—Selection, Installation, and Maintenance of Hose and Hose Assemblies

SAE J1453—Fitting, O-Ring Face Seal

SAE J1926—Specification for Straight Thread O-Ring Boss Port

2.1.2 ANSI PUBLICATION—Available from ANSI, 11 West 42nd Street, New York, NY 10036-8002.

ANSI/ASME B1.20.1-1983—General Purpose Pipe Threads

2.1.3 ASTM PUBLICATION—Available from ASTM, 1916 Race Street, Philadelphia, PA 19103-1187.

ASTM B 117—Method of Salt Spray (Fog) Testing

3. Size Designations—The hose fitting size is generally designated by the fractional inch nominal hose inside diameter together with the nominal pipe or straight thread size or nominal split flange size for the respective fitting types. However, these sizes may also be designated by their dash sizes as follows:

3.1 The hose dash size is equivalent to the number of sixteenth inch increments in the hose inside diameter, except in the case of SAE 100R5 and 100R14 hose where it is equivalent to the number of sixteenth inch increments in the outside diameter of tubing having approximately the same inside diameter as the hose.

3.2 The pipe thread dash size is the number of sixteenth inch increments in the nominal pipe thread size.

3.3 The O-ring boss, 37 degree and 45 degree flared, and flareless type thread dash sizes correspond to the number of sixteenth inch increments in the outside diameter of the tubing with which they are designed to be used.

3.4 The 4-bolt split flange dash size is the number of sixteenth inch increments in the nominal flange size.

4. Fitting Identification—Permanently attached style fittings that are not assembled to hose by fitting manufacturers and all field attachable styles of hose fittings shall be permanently and legibly marked to identify the hose size and type on which they are designed to be used.

5. Dimensions and Tolerances—Tabulated dimensions shall apply to the finished parts, plated or otherwise processed, as specified by the purchaser. Dimensions over external contour of shell portion of fittings shown in the tables reflect the maximum envelope of products available. Dimensions W and Z are indicative of the minimum diameter hole through which the hose fitting will pass. Details of internal construction of the attaching portion of fittings are not specified and shall be optional with the manufacturer, providing the fittings, properly assembled onto the appropriate hose, will not be the point of failure when the assemblies are subjected to the various tests specified in SAE J517. In the case of field attachable styles of fittings, this requirement shall apply to a minimum of one reuse as well.

The maximum and minimum across flat dimensions shall be within the commercial tolerance of bar stock from which the fittings are produced. Formed or upset hexagon contours shall fit standard wrench size openings. The minimum across corners dimensions of external hexagons shall be 1.092 times the nominal width across flats, but shall not result in a side flat width of less than 0.43 times the nominal width across flats.

Tolerance on all dimensions not otherwise limited shall be ±0.016 in. Fitting seats shall be concentric with straight thread pitch diameters within 0.010 in full indicator reading (FIR).

6. Passages—The tabulated D dimensions reflect the minimum bore at any point through the fitting prior to assembly to the hose. The after assembly bore reduction will be a maximum of 10% starting at size 5/16[1] to 2 in. The 1/4 in and smaller D dimension is the minimum for after assembly bore. The reduction must be in the general shape of a venturi. Where passages in straight fittings are machined from opposite ends, the offset at the meeting point shall not exceed 0.015 in. On angle fittings, the cross-sectional area at the junction of fluid passages shall not be less than that of the smallest passage. This assembly passage definition does not apply to bent tubes.

7. Contour—Details of contour shall be optional with the manufacturer, providing the tabulated dimensions are maintained and serviceability of fittings is not impaired. The wrench clearance dimension Y, where specified, represents the width of the fitting hexagon T plus sufficient clearance in the shell portion of fitting adjacent to the hexagon to provide adequate space for application of a standard wrench to the hexagon without interfering with mating components during assembly.

8. Straight Threads—Unified Standard Class 2A external and Class 2B internal threads shall apply to plain finish (unplated) fittings having straight threads designated B. For externally threaded parts with additive finish, the maximum diameters of Class 2A may be exceeded by the amount of the allowance, that is, the basic diameters (Class 2A maximum diameters plus the allowance) shall apply to an externally threaded part after plating. For internally threaded parts with additive finish, the Class 2B diameters apply after plating.

The pitch diameter tolerance shall be the same as the corresponding diameter-pitch combination and class of the Unified fine and 12-thread series. See SAE J475.

Where external threads are produced by roll threading and the body is not undercut, the unthreaded portion adjacent to the shoulder may be reduced to the minimum pitch diameter.

External threads shall be chamfered and internal threads shall be countersunk as specified in the illustrations and dimensional tables.

9. Thread Eccentricity Tolerances—The various thread elements of Class 2A external and Class 2B internal threads on hose fittings shall be concentric within the limitations specified under General Specifications in SAE J512.

10. Pipe Threads—Taper pipe threads designated A in the illustrations and dimensional tables shall conform to the Dryseal American Standard Taper Pipe Thread (NPTF). Specifications are given in detail in SAE J476.

---

[1] Hose I.D.

The length of full form external thread shall not be shorter than L_2 plus one pitch (thread).

Where external pipe threads are produced by roll threading, the diameter of the unthreaded shank adjacent to shoulder may be reduced to the E_2 pitch diameter.

Straight internal pipe threads designated A1 in the illustrations and tables shall conform with free fitting American Standard Straight Pipe Threads for Mechanical Joints (NPSM) as specified in USA Standard, ANSI/ASME B1.20.1-1983 of latest issue.

External pipe threads shall be chamfered from the diameters tabulated in Table 1 to produce the specified length of chamfered or partial thread. Internal pipe threads shall be countersunk 90 degrees, included angle, to the diameters specified in Table 1:

TABLE 1—PIPE THREAD DIMENSIONS

Nominal Pipe Thread Size in	External Chamfer Dia[a] Max mm	External Chamfer Dia[a] Min mm	Thread Length of Chamfered or Partial Thread Max mm	Thread Length of Chamfered or Partial Thread Min mm	Internal Thread Countersink Dia[a] Max mm	Internal Thread Countersink Dia[a] Min mm
1/8	8.1	7.6	1.40	0.94	11.2	10.7
1/4	10.7	10.2	2.13	1.42	14.5	14.0
3/8	14.0	13.4	2.13	1.42	18.0	17.5
1/2	17.3	16.8	2.72	1.80	22.1	21.6
3/4	22.6	22.1	2.72	1.80	27.4	26.9
1	28.4	27.7	3.30	2.21	34.8	34.0
1-1/4	37.1	36.3	3.30	2.21	43.4	42.7
1-1/2	43.2	42.4	3.30	2.21	49.5	48.8
2	55.1	54.4	3.30	2.21	61.2	60.7

[a] Tabulated diameters conform with Appendix A, SAE J476.

11. Material—Fittings shall be made from materials of good quality, capable of withstanding the stresses resulting from hydraulic pressures equal to the minimum burst pressure of the applicable hose size and type to which they are assembled without failure.

(R) 12. Finish—The external surfaces and threads of all carbon steel parts shall be plated or coated with a suitable material that passes a 72 h salt spray test in accordance with ASTM B 117. Any appearance of red rust during the 72 h salt spray test shall be considered failure, except for the following:

a. All internal fluid passages.

b. Edges such as hex points, serrations, and crests of threads where there may be mechanical deformation of the plating or coating typical of mass-produced parts or shipping effects.

c. Areas where there is mechanical deformation of the plating or coating caused by crimping, flaring, bending, and other post-plate metal forming operations.

d. Areas where the parts are suspended or affixed in the test chamber where condensate can accumulate.

NOTE—Cadmium plating is not preferred due to environmental reasons. Parts manufactured to this standard after January 1, 1997, shall not be cadmium plated. Internal fluid passages shall be protected from corrosion during storage. Changes in plating may affect assembly torques and require requalification, when applicable.

13. Workmanship—Workmanship shall conform to the best commercial practice to produce high-quality fittings. Fittings shall be free from all hanging burrs, loose scales, and slivers which might become dislodged in usage, sharp edges, and all other defects that might affect their serviceability. All sealing surfaces must be smooth except that annular tool marks up to 100 μin maximum, unless specified otherwise, shall be permissible.

14. Assembly Considerations—Use of a compatible lubricant in assembling dryseal pipe threads on hose fittings may be desirable to minimize galling and effect a pressure-tight seal.

15. Pressure Ratings—Some hose pressure ratings in SAE J517 exceed the pressure ratings for the connection (threads, flanges, etc.) of certain types of hose fittings shown in SAE J516. SAE J518 and J1453 specify pressure ratings for 4-bolt split flange and O-ring face seal type hose fittings, respectively. For all other types of hose fitting connections consult the fitting manufacturer for pressure ratings. The pressure rating of a hose assembly comprising SAE J516 fittings and SAE J517 hose shall not exceed the lower of the two pressure rating values.

MALE DRYSEAL PIPE THREAD TYPE

NOTES: UNSPECIFIED DETAIL WITH RESPECT TO DIMENSIONS, TOLERANCES, CONTOURS, MATERIAL, WORKMANSHIP, AND SO ON, MUST CONFORM TO GENERAL SPECIFICATIONS FOR HYDRAULIC HOSE FITTINGS. THE DIMENSIONAL DESIGNATIONS ON THE FIRST FIGURE IN EACH GROUP SHALL APPLY TO ALL OTHER FIGURES IN THAT GROUP EXCEPT AS SHOWN OTHERWISE. CODES SHOWN IN BRACKETS ADJACENT TO FIGURE NUMBERS REPRESENT RESPECTIVE FITTING IDENTIFICATION, WITH XX SUBSTITUTED FOR THE HOSE TYPE CODE DEPICTED IN BRACKETS AT END OF RESPECTIVE TABLE TITLES, IN ACCORDANCE WITH SAE J846.

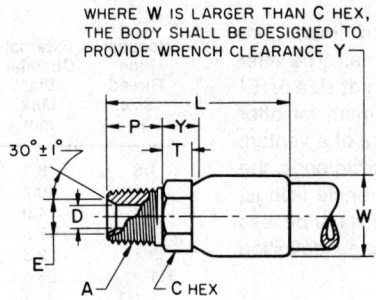

FIGURE 1A—PERMANENTLY ATTACHED STYLE (1501XX)

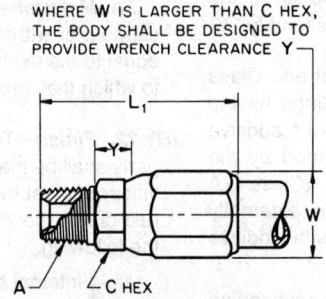

FIGURE 1B—FIELD ATTACHABLE SCREW STYLE (1601XX)

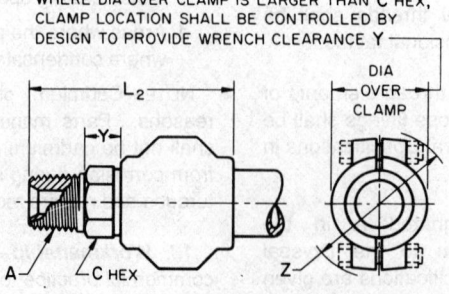

FIGURE 1C—FIELD ATTACHABLE SEGMENT CLAMP STYLE (1701XX)

FIGURE 1—MALE DRYSEAL PIPE THREAD TYPE HOSE FITTINGS

TABLE 1A—DIMENSIONS OF MALE DRYSEAL PIPE THREAD TYPE HOSE FITTINGS FOR USE ON SAE 100R1 HYDRAULIC HOSE (FIGURES 1A TO 1C) (CODES 42 AND 53)

Nominal SAE 100R1 Hose ID mm	Hose Dash Size	A Dryseal Taper Thread[a] NPTF	Pipe Thread Dash Size	C Min mm	D Dia Min mm	E Dia ±0.25 mm	L Max mm	L_1 Max mm	L_2 Max mm	P Min mm	T Min mm	W Max mm	Y Min mm	Z Max mm
4.8	-3	1/8	-2	11.1	2.3	7.14	54.9	54.9	—	9.7	4.8	19.0	9.4	—
4.8	-3	1/4	-4	14.3	2.3	8.74	57.2	60.5	—	14.2	5.6	19.0	10.4	—
6.4	-4	1/8	-2	11.1	2.8	7.14	58.7	56.9	—	9.7	4.8	22.1	9.4	—
6.4	-4	1/4	-4	14.3	2.8	8.74	63.2	61.5	—	14.2	5.6	22.1	10.4	—
7.9	-5	1/4	-4	14.3	5.1	8.74	63.2	62.7	—	14.2	5.6	23.9	10.4	—
7.9	-5	3/8	-6	17.5	5.1	11.91	63.2	62.7	—	14.2	6.4	23.9	12.7	—
9.5	-6	1/4	-4	14.3	5.3	8.74	67.6	65.0	—	14.2	5.6	27.7	10.4	—
9.5	-6	3/8	-6	17.5	6.6	11.91	67.6	65.0	—	14.2	6.4	27.7	12.7	—
9.5	-6	1/2	-8	22.2	6.6	15.88	70.4	70.6	—	19.0	7.9	27.7	14.2	—
10.3	-6.5	3/8	-6	17.5	7.1	11.91	67.6	65.0	—	14.2	6.4	29.5	12.7	—
10.3	-6.5	1/2	-8	22.2	7.1	15.88	70.4	71.4	—	19.0	7.9	29.5	14.2	—
12.7	-8	3/8	-6	17.5	8.1	11.91	67.6	73.9	—	14.2	6.4	31.2	12.7	—
12.7	-8	1/2	-8	22.2	9.7	15.88	72.4	78.5	78.5	19.0	7.9	31.2	14.2	63.5
15.9	-10	1/2	-8	22.2	12.2	15.88	80.0	89.7	—	19.0	7.9	34.8	14.2	—
15.9	-10	3/4	-12	27.0	12.7	20.65	80.0	89.7	—	19.0	9.7	34.8	15.5	—
19.0	-12	3/4	-12	27.0	15.5	20.65	91.7	89.7	86.4	19.0	9.7	40.4	15.5	76.2
25.4	-16	1	-16	34.9	19.8	26.19	97.5	107.2	100.8	23.9	9.7	51.3	15.5	91.2
31.8	-20	1-1/4	-20	42.9	25.7	34.14	109.5	119.4	111.3	24.6	11.2	62.5	23.9	101.6
38.1	-24	1-1/2	-24	50.8	31.8	41.28	116.6	127.0	118.9	25.4	12.7	69.6	23.9	114.3
50.8	-32	2	-32	61.9	44.2	52.40	131.1	139.7	139.7	26.2	17.5	88.1	28.4	130.0

[a] Dryseal American Standard Taper Pipe Thread.

TABLE 1B—DIMENSIONS OF MALE DRYSEAL PIPE THREAD TYPE HOSE FITTINGS FOR USE ON SAE 100R2 HYDRAULIC HOSE (FIGURES 1A TO 1C) (CODES 43 AND 54)

Nominal SAE 100R2 Hose ID mm	Hose Dash Size	A Dryseal Taper Thread[a] NPTF	Pipe Thread Dash Size	C Min mm	D Dia Min mm	E Dia ±0.25 mm	L Max mm	L_1 Max mm	L_2 Max mm	P Min mm	T Min mm	W Max mm	Y Min mm	Z Max mm
4.8	-3	1/8	-2	11.1	2.3	7.14	55.6	67.3	—	9.7	4.8	22.9	9.4	—
4.8	-3	1/4	-4	14.3	2.3	8.74	58.7	72.1	—	14.2	5.6	22.9	10.4	—
6.4	-4	1/8	-2	11.1	2.8	7.14	58.7	62.0	—	9.7	4.8	25.4	9.4	—
6.4	-4	1/4	-4	14.3	2.8	8.74	62.2	68.3	63.5	14.2	5.6	25.4	10.4	54.1
9.5	-6	1/4	-4	14.3	5.3	8.74	64.3	73.2	—	14.2	5.6	30.2	10.4	—
9.5	-6	3/8	-6	17.5	6.6	11.91	67.6	73.2	66.8	14.2	6.4	30.2	12.7	62.7
9.5	-6	1/2	-8	22.2	6.6	15.88	70.4	78.0	—	19.1	7.9	30.2	14.2	—
12.7	-8	3/8	-6	17.5	8.1	11.91	69.3	76.2	—	14.2	6.4	35.1	12.7	—
12.7	-8	1/2	-8	22.2	9.7	15.88	73.9	81.8	78.7	19.0	7.9	35.1	14.2	62.7
15.9	-10	1/2	-8	22.2	12.2	15.88	81.0	95.2	—	19.0	7.9	39.6	14.2	—
15.9	-10	3/4	-12	27.0	12.7	20.65	81.0	95.2	86.4	19.0	9.7	39.6	15.5	73.2
19.0	-12	3/4	-12	27.0	15.5	20.65	91.7	95.2	86.4	19.0	9.7	44.4	15.5	76.2
22.2	-14	1	-16	34.9	18.3	26.19	94.0	113.3	—	23.9	9.7	46.0	15.5	—
25.4	-16	1	-16	34.9	19.8	26.19	93.5	114.8	108.0	23.9	9.7	52.3	15.5	91.4
31.8	-20	1-1/4	-20	42.9	25.7	34.14	142.7	132.6	127.0	24.6	11.2	66.8	23.9	114.3
38.1	-24	1-1/2	-24	50.8	31.8	41.28	161.8	127.0	137.2	25.4	12.7	74.7	23.9	127.0
50.8	-32	2	-32	61.9	43.7	52.40	131.1	154.2	160.0	26.2	17.5	88.9	28.4	133.4

[a] Dryseal American Standard Taper Pipe Thread.

TABLE 1C—DIMENSIONS OF MALE DRYSEAL PIPE THREAD TYPE HOSE FITTINGS FOR USE ON SAE 100R3 HYDRAULIC HOSE (FIGURES 1A TO 1C) (CODE 44)

Nominal SAE 100R3 Hose ID mm	Hose Dash Size	A Dryseal Taper Thread[a] NPTF	Pipe Thread Dash Size	C Min mm	D Dia Min mm	E Dia ±0.25 mm	L Max mm	L_1 Max mm	L_2 Max mm	P Min mm	T Min mm	W Max mm	Y Min mm	Z Max mm
6.4	-4	1/8	-2	11.1	2.8	7.14	58.7	55.4	—	9.7	4.8	22.4	9.4	—
6.4	-4	1/4	-4	14.3	2.8	8.74	63.2	60.5	—	14.2	4.8	22.4	9.4	—
9.5	-6	1/4	-4	14.3	6.6	8.74	64.0	65.0	—	14.2	5.6	26.9	10.4	—
9.5	-6	3/8	-6	17.5	6.6	11.91	64.0	65.0	—	14.2	5.6	26.9	10.4	—
12.7	-8	3/8	-6	17.5	9.1	11.91	66.8	65.0	—	14.2	6.4	33.3	12.7	—
12.7	-8	1/2	-8	22.2	9.7	15.88	73.2	77.0	66.8	19.0	7.9	33.3	14.2	63.5
19.0	-12	3/4	-12	27.0	15.5	20.65	91.7	79.2	73.2	19.0	7.9	42.2	15.5	76.2
25.4	-16	1	-16	34.9	19.8	26.19	97.5	94.0	85.9	23.9	7.9	51.3	15.5	82.6
31.8	-20	1-1/4	-20	42.9	25.7	34.14	103.6	—	101.6	24.6	9.7	51.3	23.9	100.1

[a] Dryseal American Standard Taper Pipe Thread.

TABLE 1D—DIMENSIONS OF MALE DRYSEAL PIPE THREAD TYPE HOSE FITTINGS FOR USE ON SAE 100R4 HYDRAULIC HOSE (FIGURES 1A TO 1C) (CODE 45)

Nominal SAE 100R4 Hose ID mm	Hose Dash Size	A Dryseal Taper Thread[a] NPTF	Pipe Thread Dash Size	C Min mm	D Dia Min mm	E Dia ±0.25 mm	L Max mm	L_2 Max mm	P Min mm	T Min mm	W Max mm	Y Min mm	Z Max mm
19.0	-12	3/4	-12	27.0	15.5	20.65	66.8	91.7	19.0	7.9	43.9	15.5	76.2
25.4	-16	1	-16	35.0	19.8	26.19	73.2	97.5	23.9	7.9	51.1	15.5	82.6
31.8	-20	1-1/4	-20	42.9	25.7	34.14	77.0	104.6	24.6	9.7	58.7	23.9	100.1
38.1	-24	1-1/2	-24	50.8	31.8	41.28	90.4	114.3	25.4	12.7	66.0	23.9	110.2
50.8	-32	2	-32	61.9	43.7	52.40	101.6	124.0	26.2	14.2	80.8	28.4	125.5

[a] Dryseal American Standard Taper Pipe Thread.

TABLE 1E—DIMENSIONS OF MALE DRYSEAL PIPE THREAD TYPE HOSE FITTINGS FOR USE ON SAE 100R5 HYDRAULIC HOSE (FIGURES 1A TO 1B) (CODE 46)

Nominal SAE 100R5 Hose ID mm	Hose Dash Size	A Dryseal Taper Thread[a] NPTF	Pipe Thread Dash Size	C Min mm	D Dia Min mm	E Dia ±0.25 mm	L Max mm	L_1 Max mm	P Min mm	T Min mm	W Max mm	Y Min mm
4.8	-4	1/8	-2	11.1	2.3	7.14	46.2	45.7	9.7	4.8	20.1	9.4
4.8	-4	1/4	-4	14.3	2.3	8.74	50.8	52.3	14.2	4.8	20.1	10.4
6.4	-5	1/4	-4	14.3	2.8	8.74	51.8	52.3	14.2	4.8	22.1	10.4
7.9	-6	1/4	-4	14.3	5.1	8.74	52.6	55.6	14.2	4.8	23.9	10.4
10.3	-8	3/8	-6	17.5	7.1	11.91	54.9	64.5	14.2	6.4	27.4	12.7
12.7	-10	1/2	-8	22.2	9.7	15.88	61.0	74.7	19.0	7.9	33.3	14.2
15.9	-12	3/4	-12	27.0	12.7	20.65	62.0	83.8	19.0	9.7	38.6	15.5
22.2	-16	3/4	-12	27.0	18.3	20.65	68.6	77.7	19.0	9.7	43.9	15.5
22.2	-16	1	-16	34.9	18.3	26.19	73.4	81.0	23.9	9.7	43.9	15.5
28.6	-20	1-1/4	-20	42.9	23.9	34.14	77.2	88.1	24.6	11.2	51.3	23.9
34.9	-24	1-1/2	-24	50.8	29.5	41.28	85.9	90.2	25.4	12.7	58.7	23.9
46.0	-32	2	-32	61.9	40.9	52.40	96.3	104.1	26.2	17.5	73.4	28.4

[a] Dryseal American Standard Taper Pipe Thread.

NOTE: For SAE J1402 and DOT FMVSS 106 air brake applications, couplings used on 100R5 hose shall have a minimum assembled orifice no less than 66% of the nominal hose I.D.

TABLE 1F—DIMENSIONS OF MALE DRYSEAL PIPE THREAD TYPE HOSE FITTINGS FOR USE ON SAE 100R6 HYDRAULIC HOSE (FIGURES 1A TO 1B) (CODE 47)

Nominal SAE 100R6 Hose ID mm	Hose Dash Size	A Dryseal Taper Thread[a] NPTF	Pipe Thread Dash Size	C Min mm	D Dia Min mm	E Dia ±0.25 mm	L Max mm	L_1 Max mm	P Min mm	T Min mm	W Max mm	Y Min mm
4.8	-3	1/8	-2	11.1	2.3	7.14	58.7	29.2	9.7	3.0	16.0	9.4
4.8	-3	1/4	-4	14.3	2.3	8.74	63.5	35.3	14.2	4.3	16.0	10.4
6.4	-4	1/8	-2	11.1	2.8	7.14	58.7	31.8	9.7	3.0	19.0	9.4
6.4	-4	1/4	-4	14.3	2.8	8.74	63.5	36.8	14.2	4.3	19.0	10.4
6.4	-4	3/8	-6	17.5	2.8	11.91	64.3	—	14.2	6.4	19.0	12.7
7.9	-5	1/4	-4	14.3	5.1	8.74	65.0	36.8	14.2	3.0	20.6	10.4
7.9	-5	3/8	-6	17.5	5.1	11.91	68.1	38.6	14.2	6.4	20.6	12.7
9.5	-6	1/4	-4	14.3	5.3	8.74	68.1	39.1	14.2	3.0	25.4	10.4
9.5	-6	3/8	-6	17.5	6.6	11.91	74.7	39.1	14.2	6.4	25.4	12.7
12.7	-8	3/8	-6	17.5	9.1	11.91	74.7	41.4	14.2	6.4	28.4	12.7
12.7	-8	1/2	-8	22.2	9.7	15.88	79.0	47.8	19.0	7.9	28.4	14.2
15.9	-10	1/2	-8	22.2	12.2	15.88	81.3	—	19.0	7.9	26.9	14.2
15.9	-10	3/4	-12	27.0	12.7	20.65	89.4	—	19.0	9.7	26.9	15.5
19.0	-12	3/4	-12	27.0	14.5	20.65	101.6	79.5	19.0	7.9	36.8	15.5

[a] Dryseal American Standard Taper Pipe Thread.

TABLE 1G—DIMENSIONS OF MALE DRYSEAL PIPE THREAD TYPE HOSE FITTINGS FOR USE ON SAE 100R7 HYDRAULIC HOSE (FIGURES 1A TO 1B) (CODE 48)

Nominal SAE 100R7 Hose ID mm	Hose Dash Size	A Dryseal Taper Thread[a] NPTF	Pipe Thread Dash Size	C Min mm	D Dia Min mm	E Dia ±0.25 mm	L Max mm	L$_1$ Max mm	P Min mm	T Min mm	W Max mm	Y Min mm
4.8	-3	1/8	-2	11.1	2.3	7.14	58.7	50.8	9.7	4.8	19.0	9.4
4.8	-3	1/4	-4	14.3	2.3	8.74	63.5	52.6	14.2	5.6	19.0	10.4
6.4	-4	1/8	-2	11.1	2.8	7.14	58.7	52.6	9.7	4.8	20.6	9.4
6.4	-4	1/4	-4	14.3	2.8	8.74	63.5	57.2	14.2	5.6	20.6	10.4
7.9	-5	1/4	-4	14.3	5.1	8.74	64.3	60.5	14.2	5.6	22.4	10.4
7.9	-5	3/8	-6	17.5	5.1	11.9	65.0	60.5	14.2	5.8	22.4	12.7
9.5	-6	1/4	-4	14.3	5.3	8.74	68.1	66.8	14.2	5.6	25.7	10.4
9.5	-6	3/8	-6	17.5	6.6	11.9	68.1	66.8	14.2	5.8	25.7	12.7
9.5	-6	1/2	-8	22.2	6.6	15.9	74.7	71.6	19.0	7.9	25.7	14.2
12.7	-8	3/8	-6	17.5	8.1	11.9	74.7	74.4	14.2	5.8	31.2	12.7
12.7	-8	1/2	-8	22.2	9.1	15.9	79.0	79.5	19.0	7.9	31.2	14.2
15.9	-10	1/2	-8	22.2	12.2	15.9	81.3	—	19.0	7.9	34.8	14.2
19.0	-12	3/4	-12	27.0	14.5	20.7	89.4	88.9	19.0	8.4	40.4	15.5
25.4	-16	1	-16	34.9	19.8	26.2	101.6	101.6	23.9	9.1	49.5	15.5

[a] Dryseal American Standard Taper Pipe Thread.

TABLE 1H—DIMENSIONS OF MALE DRYSEAL PIPE THREAD TYPE HOSE FITTINGS FOR USE ON SAE 100R8 HYDRAULIC HOSE (FIGURE 1A) (CODE 49)

Nominal SAE 100R8 Hose ID mm	Hose Dash Size	A Dryseal Taper Thread[a] NPTF	Pipe Thread Dash Size	C Min mm	D Dia Min mm	E Dia ±0.25 mm	L Max mm	P Min mm	T Min mm	W Max mm	Y Min mm
4.8	-3	1/8	-2	11.1	2.3	7.14	58.7	9.7	4.8	19.0	9.4
4.8	-3	1/4	-4	14.3	2.3	8.74	63.5	14.2	5.6	19.0	10.4
6.4	-4	1/8	-2	11.1	2.8	7.14	58.7	9.7	4.8	20.6	9.4
6.4	-4	1/4	-4	14.3	2.8	8.74	63.5	14.2	5.6	20.6	10.4
9.5	-6	1/4	-4	14.3	5.3	8.74	68.1	14.2	5.6	25.7	10.4
9.5	-6	3/8	-6	17.5	6.6	11.91	68.1	14.2	5.8	25.7	12.7
9.5	-6	1/2	-8	22.2	6.6	15.88	74.7	19.0	7.9	25.7	14.2
12.7	-8	3/8	-6	17.5	8.1	11.91	74.7	14.2	5.8	31.2	12.7
12.7	-8	1/2	-8	22.2	9.7	15.88	79.0	19.0	7.9	31.2	14.2
15.9	-10	1/2	-8	22.2	12.2	15.88	81.3	19.0	7.9	36.8	14.2
19.0	-12	3/4	-12	27.0	14.5	20.65	89.4	19.0	8.4	40.4	15.5
25.4	-16	1	-16	34.9	19.8	26.19	101.6	23.9	9.1	49.5	15.5

[a] Dryseal American Standard Taper Pipe Thread.

TABLE 1I—DIMENSIONS OF MALE DRYSEAL PIPE THREAD TYPE HOSE FITTINGS FOR USE ON SAE 100R9 HYDRAULIC HOSE (FIGURES 1A TO 1B) (CODES 50 AND 75)

Nominal SAE 100R9 Hose ID mm	Hose Dash Size	A Dryseal Taper Thread[a] NPTF	Pipe Thread Dash Size	C Min mm	D Dia Min mm	E Dia ±0.25 mm	L Max mm	L$_1$ Max mm	P Min mm	T Min mm	W Max mm	Y Min mm
9.5	-6	3/8	-6	17.5	6.6	11.91	77.2	73.2	14.2	6.35	30.2	12.7
12.7	-8	1/2	-8	22.2	9.7	15.88	85.6	81.8	19.0	7.87	35.1	14.2
19.0	-12	3/4	-12	27.0	15.5	20.65	95.5	95.2	19.0	9.65	44.5	15.5
25.4	-16	1	-16	34.9	19.8	26.19	102.1	114.8	23.9	9.65	52.3	15.5
31.8	-20	1-1/4	-20	42.9	25.7	34.14	142.7	—	24.6	11.2	66.8	23.9
38.1	-24	1-1/2	-24	50.8	31.8	41.28	161.8	—	25.4	12.7	74.7	23.9
50.8	-32	2	-32	61.9	43.7	52.40	154.2	—	26.2	17.5	88.9	28.4

[a] Dryseal American Standard Taper Pipe Thread.

TABLE 1J—DIMENSIONS OF MALE DRYSEAL PIPE THREAD TYPE HOSE FITTINGS FOR USE ON SAE 100R10 HYDRAULIC HOSE (FIGURES 1A TO 1C) (CODES 51 AND 76)

Nominal SAE 100R10 Hose ID mm	Hose Dash Size	A Dryseal Taper Thread[a] NPTF	Pipe Thread Dash Size	C Min mm	D Dia Min mm	E Dia ±0.25 mm	L Max mm	L_1 Max mm	L_2 Max mm	P Min mm	T Min mm	W Max mm	Y Min mm	Z Max mm
6.4	-4	1/4	-4	14.3	2.8	8.74	75.7	76.2	63.5	14.2	5.6	28.7	10.4	54.1
9.5	-6	3/8	-6	17.5	5.6	11.91	83.3	82.6	66.8	14.2	6.4	28.7	12.7	62.7
12.7	-8	1/2	-8	22.2	8.6	15.88	101.1	101.6	78.7	19.0	7.9	31.8	14.2	62.7
19.0	-12	3/4	-12	27.0	12.7	20.65	112.5	114.3	86.4	19.0	9.7	42.9	15.5	79.2
25.4	-16	1	-16	34.9	17.3	26.19	131.8	133.4	103.6	23.9	9.7	50.8	15.5	107.2
31.8	-20	1-1/4	-20	42.9	22.1	34.14	153.7	152.4	116.6	24.6	11.2	62.2	17.8	126.0
38.1	-24	1-1/2	-24	50.8	27.7	41.28	166.4	171.4	139.4	25.4	12.7	76.2	19.3	133.1
50.8	-32	2	-32	63.5	39.6	52.40	195.6	203.2	169.7	26.2	17.5	81.8	28.4	166.6

[a] Dryseal American Standard Taper Pipe Thread.

TABLE 1K—DIMENSIONS OF MALE DRYSEAL PIPE THREAD TYPE HOSE FITTINGS FOR USE ON SAE 100R11 HYDRAULIC HOSE (FIGURE 1A) (CODE 52)

Nominal SAE 100R11 Hose ID mm	Hose Dash Size	A Dryseal Taper Thread[a] NPTF	Pipe Thread Dash Size	C Min mm	D Dia Min mm	E Dia ±0.25 mm	L Max mm	P Min mm	T Min mm	W Max mm	Y Min mm
25.4	-16	1	-16	34.9	18.5	26.19	133.9	23.9	9.7	55.9	15.5
31.8	-20	1-1/4	-20	42.9	24.1	34.14	147.3	24.6	11.2	66.8	18.5
38.1	-24	1-1/2	-24	50.8	30.5	41.28	170.7	25.4	12.7	74.7	20.1
50.8	-32	2	-32	61.9	41.9	52.40	195.6	26.2	17.5	88.9	21.8

[a] Dryseal American Standard Taper Pipe Thread.

(R) TABLE 1L—DIMENSIONS OF MALE DRYSEAL PIPE THREAD TYPE HOSE FITTINGS FOR USE ON SAE 100R12 HYDRAULIC HOSE (FIGURE 1A) (CODE 77)

Nominal SAE 100R12 Hose ID mm	Hose Dash Size	A Dryseal Taper Thread[a] NPTF	Pipe Thread Dash Size	C Min mm	D Dia Min mm	E Dia ±0.25 mm	L Max mm	P Min mm	T Min mm	W Max mm	Y Min mm
9.5	-6	3/8	-6	17.5	5.6	11.91	83.3	14.2	6.4	28.7	12.7
12.7	-8	1/2	-8	22.2	8.6	15.88	101.1	19.0	7.9	31.8	14.2
15.9	-10	3/4	-12	27.0	10.0	20.65	101.6	19.0	9.7	38.0	14.2
19.0	-12	3/4	-12	27.0	12.7	20.65	114.3	19.0	9.7	42.9	14.2
25.4	-16	1	-16	34.9	17.3	26.19	131.8	23.9	9.7	50.8	15.5
31.8	-20	1-1/4	-20	42.9	22.1	34.14	153.7	24.6	11.2	60.5	17.8
38.1	-24	1-1/2	-24	50.8	27.7	41.28	166.4	25.4	12.7	67.8	18.8
50.8	-32	2	-32	61.9	39.6	52.40	195.6	26.2	17.5	81.8	28.4

[a] Dryseal American Standard Taper Pipe Thread.

TABLE 1M—DIMENSIONS OF MALE DRYSEAL PIPE THREAD TYPE HOSE FITTINGS FOR USE ON SAE 100R13 HYDRAULIC HOSE (FIGURE 1A) (CODE 78)

Nominal SAE 100R13 Hose ID mm	Hose Dash Size	A Dryseal Taper Thread[a] NPTF	Pipe Thread Dash Size	C Min mm	D Dia Min mm	E Dia ±0.25 mm	L Max mm	P Min mm	T Min mm	W Max mm	Y Min mm
19.0	-12	3/4	-12	27.0	12.7	20.65	114.3	19.0	9.7	43.7	14.2
25.4	-16	1	-16	34.9	17.5	26.19	119.4	23.9	9.7	50.8	15.7
31.8	-20	1-1/4	-20	42.9	22.4	34.14	143.8	24.6	11.2	63.5	19.0
38.1	-24	1-1/2	-24	50.8	27.7	41.28	156.7	25.4	12.7	71.1	18.8
50.8	-32	2	-32	61.9	38.1	52.40	216.4	26.2	17.5	88.9	21.6

[a] Dryseal American Standard Taper Pipe Thread.

MALE STRAIGHT THREAD O-RING BOSS TYPE

NOTES: UNSPECIFIED DETAIL WITH RESPECT TO DIMENSIONS, TOLERANCES, CONTOURS, MATERIAL, WORKMANSHIP, AND SO ON, MUST CONFORM TO GENERAL SPECIFICATIONS FOR HYDRAULIC HOSE FITTINGS. THE DIMENSIONAL DESIGNATIONS ON THE FIRST FIGURE IN EACH GROUP SHALL APPLY TO ALL OTHER FIGURES IN THAT GROUP EXCEPT AS SHOWN OTHERWISE. CODES SHOWN IN BRACKETS ADJACENT TO FIGURE NUMBERS REPRESENT RESPECTIVE FITTING IDENTIFICATION, WITH XX SUBSITITUTED FOR THE HOSE TYPE CODE DEPICTED IN BRACKETS AT END OF RESPECTIVE TABLE TITLES, IN ACCORDANCE WITH SAE J846.

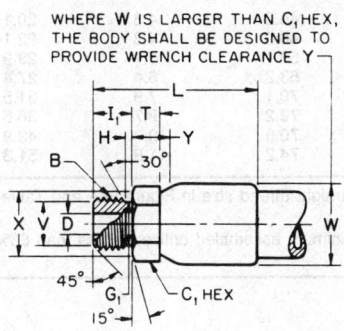
FIGURE 2A—PERMANENTLY ATTACHED STYLE (1801XX)

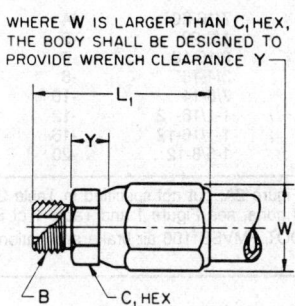
FIGURE 2B—FIELD ATTACHABLE SCREW STYLE (1901XX)

FIGURE 2—MALE STRAIGHT THREAD O-RING BOSS TYPE HOSE FITTINGS

TABLE 2A—DIMENSIONS OF MALE STRAIGHT THREAD O-RING BOSS TYPE HOSE FITTINGS FOR USE ON SAE 100R1 HYDRAULIC HOSE[a] (FIGURES 2A TO 2B) (CODES 42 AND 53)

Nominal SAE 100R1 Hose ID mm	Hose Dash Size	B Nominal Straight Thread Size	Thread Dash Size	C_1 Min mm	D Dia Min mm	L Max mm	L_1 Max mm	T Min mm	W Max mm	Y Min mm
7.9	-5	9/16-18	-6	17.5	5.1	62.7	62.7	7.9	23.9	10.4
9.5	-6	9/16-18	-6	17.5	6.6	64.5	64.5	7.9	27.7	10.4
9.5	-6	3/4-16	-8	22.2	6.6	68.3	68.3	8.6	27.7	12.7
10.3	-6.5	3/4-16	-8	22.2	6.6	68.3	68.3	8.6	29.5	12.7
12.7	-8	3/4-16	-8	22.2	9.7	70.1	76.2	8.6	31.2	14.2
12.7	-8	7/8-14	-10	25.4	9.7	72.6	78.7	7.9	31.2	14.2
15.9	-10	7/8-14	-10	25.4	12.2	77.7	86.9	7.9	34.8	14.2
15.9	-10	1-1/16-12	-12	31.8	12.7	81.3	90.9	9.7	34.8	17.5
19.0	-12	7/8-14	-10	25.4	12.2	81.3	90.9	7.9	40.4	14.2
19.0	-12	1-1/16-12	-12	31.8	14.5	81.3	90.9	9.7	40.4	17.5
19.0	-12	1-3/16-12	-14	34.9	15.5	84.1	93.7	9.7	40.4	17.5
25.4	-16	1-5/16-12	-16	38.1	19.8	94.5	107.7	12.7	51.3	19.0

[a] For dimensions shown in Figure 2A, but not specified in Table 2A, see corresponding dimensions for respective straight thread size in Figure 34A and Table 12 of SAE J514. For dimensions of mating boss ports, see Figure 1 and Table 1 of SAE J1926.

TABLE 2B—DIMENSIONS OF MALE STRAIGHT THREAD O-RING BOSS TYPE HOSE FITTINGS FOR USE ON SAE 100R2 HYDRAULIC HOSE[a] (FIGURE 2A) (CODES 43 AND 54)

Nominal SAE 100R2 Hose ID mm	Hose Dash Size	B Nominal Straight Thread Size	Thread Dash Size	C_1 Min mm	D Dia Min mm	L Max mm	T Min mm	W Max mm	Y Min mm
9.5	-6	9/16-18	-6	17.5	6.6	64.5	7.9	30.2	10.4
9.5	-6	3/4-16	-8	22.2	6.6	68.3	7.9	30.2	12.7
12.7	-8	3/4-16	-8	22.2	9.7	70.1	6.4	35.1	12.7
12.7	-8	7/8-14	-10	25.4	9.7	72.6	7.9	35.1	14.2
15.9	-10	7/8-14	-10	25.4	12.2	77.7	7.9	39.6	14.2
15.9	-10	1-1/16-12	-12	31.8	12.7	81.3	9.7	39.6	17.5
19.0	-12	7/8-14	-10	25.4	12.2	88.9	9.7	44.5	14.2
19.0	-12	1-1/16-12	-12	31.8	14.5	88.9	9.7	44.5	17.5
19.0	-12	1-3/16-12	-14	34.9	15.5	88.9	9.7	44.5	17.5
22.2	-14	1-3/16-12	-14	34.9	18.3	93.2	9.7	46.0	17.5
22.2	-14	1-5/16-12	-16	38.1	18.3	94.5	12.7	46.0	19.0
25.4	-16	1-5/16-12	-16	38.1	19.8	101.6	9.4	52.3	19.0
31.8	-20	1-5/8-12	-20	47.6	25.7	133.4	9.4	66.8	23.9

[a] For dimensions shown in Figure 2A, but not specified in Table 2B, see corresponding dimensions for respective straight thread size in Figure 34A and Table 12 of SAE J514. For dimensions of mating boss ports, see Figure 1 and Table 1 of SAE J1926.

TABLE 2C—DIMENSIONS OF MALE STRAIGHT THREAD O-RING BOSS TYPE HOSE FITTINGS FOR USE ON SAE 100R5 HYDRAULIC HOSE[a] (FIGURES 2A AND 2B) (CODE 46)

Nominal SAE 100R5 Hose ID mm	Hose Dash Size	B Nominal Straight Thread Size	Thread Dash Size	C_1 Min mm	D Dia Min mm	L Max mm	L_1 Max mm	T Min mm	W Max mm	Y Min mm
4.8	-4	7/16-20	-4	14.3	2.3	47.5	45.2	4.3	20.1	9.4
6.4	-5	1/2-20	-5	15.9	2.8	48.3	48.3	4.3	22.1	10.4
7.9	-6	9/16-18	-6	17.5	5.1	49.8	53.1	6.1	23.9	10.4
10.3	-8	3/4-16	-8	22.2	7.1	53.3	63.2	6.4	27.4	12.7
12.7	-10	7/8-14	-10	25.4	9.7	56.1	70.1	7.9	31.5	14.2
15.9	-12	1-1/16-12	-12	31.8	12.7	58.4	79.2	9.7	38.6	17.5
22.2	-16	1-1/16-12	-16	38.1	18.3	64.5	70.6	9.7	43.9	17.5
28.6	-20	1-5/8-12	-20	47.6	23.9	67.6	74.2	11.9	51.3	17.5

[a] For dimensions shown in Figure 2A, but not specified in Table 2C, see corresponding dimensions for respective straight thread size in Figure 34A and Table 12 of SAE J514. For dimensions of mating boss ports, see Figure 1 and Table 1 of SAE J1926.

NOTE: For SAE J1402 and DOT FMVSS 106 air brake applications, couplings used on 100R5 hose shall have a minimum assembled orifice no less than 66% of the nominal hose I.D.

TABLE 2D—DIMENSIONS OF MALE STRAIGHT THREAD O-RING BOSS TYPE HOSE FITTINGS FOR USE ON SAE 100R7 HYDRAULIC HOSE[a] (FIGURE 2A) (CODE 48)

Nominal SAE 100R7 Hose ID mm	Hose Dash Size	B Nominal Straight Thread Size	Thread Dash Size	C_1 Min mm	D Dia Min mm	L Max mm	T Min mm	W Max mm	Y Min mm
4.8	-3	7/16-20	-4	14.3	2.3	58.4	4.8	19.1	9.4
6.4	-4	7/16-20	-4	14.3	2.8	58.4	5.6	20.6	10.4
7.9	-5	9/16-18	-6	17.5	5.1	59.9	5.6	22.4	10.4
9.5	-6	9/16-18	-6	17.5	6.6	66.8	6.6	25.7	10.4
9.5	-6	3/4-16	-8	22.2	6.6	66.8	6.6	25.7	12.7
12.7	-8	3/4-16	-8	22.2	9.1	71.6	8.6	31.2	12.7
12.7	-8	7/8-14	-10	25.4	9.1	73.2	8.6	31.2	14.2
15.9	-10	7/8-14	-10	25.4	12.2	77.7	8.6	34.8	14.2
19.0	-12	1-1/16-12	-12	31.8	14.5	85.1	9.1	40.4	17.5
19.0	-12	1-3/16-12	-14	34.9	14.5	85.1	9.7	40.4	17.5
25.4	-16	1-5/16-12	-16	38.1	19.8	90.7	12.7	49.5	19.1

[a] For dimensions shown in Figure 2A, but not specified in Table 2D, see corresponding dimensions for respective straight thread size in Figure 34A and Table 12 of SAE J514. For dimensions of mating boss ports, see Figure 1 and Table 1 of SAE J1926.

TABLE 2E—DIMENSIONS OF MALE STRAIGHT THREAD O-RING BOSS TYPE HOSE FITTINGS FOR USE ON SAE 100R8 HYDRAULIC HOSE[a] (FIGURE 2A) (CODE 49)

Nominal SAE 100R8 Hose ID mm	Hose Dash Size	B Nominal Straight Thread Size	Thread Dash Size	C_1 Min mm	D Dia Min mm	L Max mm	T Min mm	W Max mm	Y Min mm
4.8	-3	7/16-20	-4	14.3	2.3	58.4	4.8	19.0	9.4
6.4	-4	7/16-20	-4	14.3	2.8	57.4	5.6	20.6	10.4
9.5	-6	9/16-18	-6	17.5	6.6	66.8	6.6	25.7	10.4
9.5	-6	3/4-16	-8	22.2	6.6	66.8	6.6	25.7	12.7
12.7	-8	3/4-16	-8	22.2	9.7	71.6	8.6	31.2	12.7
12.7	-8	7/8-14	-10	25.4	9.7	73.2	8.6	31.2	14.2
15.9	-10	7/8-14	-10	25.4	12.2	77.7	8.6	36.8	14.2
19.0	-12	1-1/16-12	-12	31.8	14.5	85.1	9.1	40.4	17.5
19.0	-12	1-3/16-12	-14	34.9	14.5	85.1	9.7	40.4	17.5
25.4	-16	1-5/16-12	-16	38.1	19.8	90.7	12.7	49.5	19.0

[a] For dimensions shown in Figure 2A, but not specified in Table 2E, see corresponding dimensions for respective straight thread size in Figure 34A and Table 12 of SAE J514. For dimensions of mating boss ports, see Figure 1 and Table 1 of SAE J1926.

TABLE 2F—DIMENSIONS OF MALE STRAIGHT THREAD O-RING BOSS TYPE HOSE FITTINGS FOR USE ON SAE 100R9 HYDRAULIC HOSE[a] (FIGURE 2A) (CODES 50 AND 75)

Nominal SAE 100R9 Hose ID mm	Hose Dash Size	B Nominal Straight Thread Size	Thread Dash Size	C_1 Min mm	D Dia Min mm	L Max mm	T Min mm	W Max mm	Y Min mm
9.5	-6	9/16-18	-6	17.5	6.6	70.4	7.9	30.2	10.4
12.7	-8	3/4-16	-8	22.2	9.7	70.6	8.6	35.1	12.7
19.0	-12	1-1/16-12	-12	31.8	14.5	90.9	9.7	44.5	17.5
25.4	-16	1-5/16-12	-16	38.1	19.8	101.6	12.7	52.3	19.0

[a] For dimensions shown in Figure 2A, but not specified in Table 2F, see corresponding dimensions for respective straight thread size in Figure 34A and Table 12 of SAE J514. For dimensions of mating boss ports, see Figure 1 and Table 1 of SAE J1926.

TABLE 2G—DIMENSIONS OF MALE STRAIGHT THREAD O-RING BOSS TYPE HOSE FITTINGS FOR USE ON SAE 100R10 HYDRAULIC HOSE[a] (FIGURES 2A AND 2B) (CODES 50 AND 76)

Nominal SAE 100R10 Hose ID mm	Hose Dash Size	B Nominal Straight Thread Size	Thread Dash Size	C Min mm	D Dia Min mm	L Max mm	L_1 Max mm	T Min mm	W Max mm	Y Min mm
6.4	-4	7/16-20	-4	14.3	2.8	70.6	63.5	5.6	28.7	10.4
9.5	-6	9/16-18	-6	17.5	5.6	79.5	63.5	6.4	28.7	12.7
12.7	-8	3/4-16	-8	22.2	8.6	92.2	69.8	7.9	31.8	12.7
19.0	-12	1-1/16-12	-12	31.8	12.7	107.4	104.1	9.7	42.9	13.0
25.4	-16	1-5/16-12	-16	38.1	17.3	120.1	120.6	9.7	50.8	14.5
31.8	-20	1-5/8-12	-20	47.6	22.1	131.1	133.4	11.2	62.2	23.9
38.1	-24	1-7/8-12	-24	54.0	27.7	142.2	146.0	12.7	76.2	23.9
50.8	-32	2-1/2-12	-32	69.8	39.6	163.1	165.1	17.6	81.8	28.4

[a] For dimensions shown in Figure 2A, but not specified in Table 2G, see corresponding dimensions for respective straight thread size in Figure 34A and Table 12 of SAE J514. For dimensions of mating boss ports, see Figure 1 and Table 1 of SAE J1926.

(R) TABLE 2H—DIMENSIONS OF MALE STRAIGHT THREAD O-RING BOSS TYPE HOSE FITTINGS FOR USE ON SAE 100R12 HYDRAULIC HOSE[a] (FIGURE 2A) (CODE 77)

Nominal SAE 100R12 Hose ID mm	Hose Dash Size	B Nominal Straight Thread Size	Thread Dash Size	C Min mm	D Dia Min mm	L Max mm	T Min mm	W Max mm	Y Min mm
9.5	-6	9/16-18	-6	17.5	5.6	79.5	6.4	28.7	12.7
12.7	-8	3/4-16	-8	22.2	8.6	92.2	7.9	31.8	12.7
15.9	-10	7/8-14	-10	25.4	10.0	99.0	8.0	38.0	13.0
19.0	-12	1-1/16-12	-12	31.8	12.7	107.4	9.7	42.9	13.0
25.4	-16	1-5/16-12	-16	38.1	17.3	120.1	9.7	50.8	14.5
31.8	-20	1-5/8-12	-20	47.6	22.1	131.1	11.0	60.5	23.9
38.1	-24	1-7/8-12	-24	54.0	27.7	142.2	12.7	67.8	23.9
50.8	-32	2-1/2-12	-32	69.8	39.6	163.1	17.5	81.8	28.4

[a] For dimensions shown in Figure 2A, but not specified in Table 2H, see corresponding dimensions for respective straight thread size in Figure 34A and Table 12 of SAE J514. For dimensions of mating boss ports, see Figure 1 and Table 1 of SAE J1926.

MALE 37-DEGREE FLARE TYPE

NOTES: UNSPECIFIED DETAIL WITH RESPECT TO DIMENSIONS, TOLERANCES, CONTOURS, MATERIAL, WORKMANSHIP, AND SO ON, MUST CONFORM TO GENERAL SPECIFICATIONS FOR HYDRAULIC HOSE FITTINGS. THE DIMENSIONAL DESIGNATIONS ON THE FIRST FIGURE IN EACH GROUP SHALL APPLY TO ALL OTHER FIGURES IN THAT GROUP EXCEPT AS SHOWN OTHERWISE. CODES SHOWN IN BRACKETS ADJACENT TO FIGURE NUMBERS REPRESENT RESPECTIVE FITTING IDENTIFICATION, WITH XX SUBSTITUTED FOR THE HOSE TYPE CODE DEPICTED IN BRACKETS AT END OF RESPECTIVE TABLE TITLES, IN ACCORDANCE WITH SAE J846.

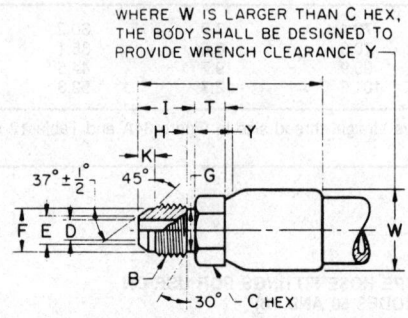

FIGURE 3A—PERMANENTLY ATTACHED STYLE (2101XX)

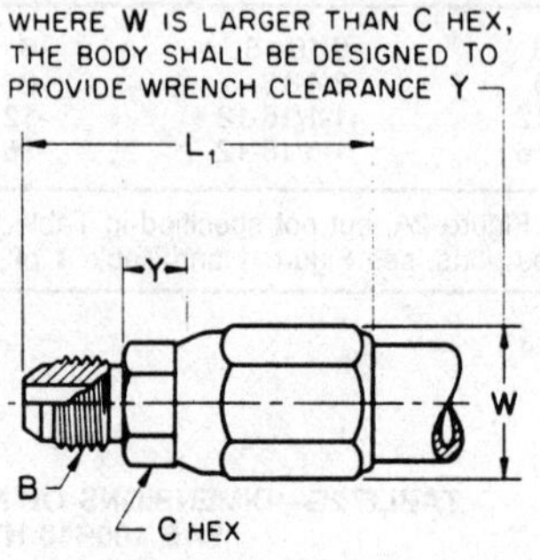

FIGURE 3B—FIELD ATTACHABLE SCREW STYLE (2201XX)

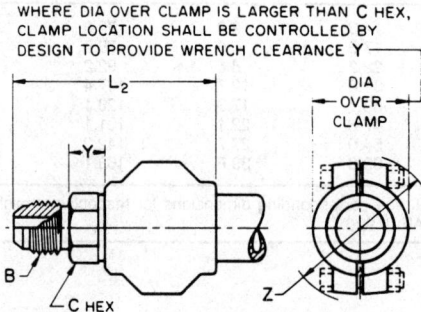

FIGURE 3C—FIELD ATTACHABLE SEGMENT CLAMP STYLE (2301XX)

FIGURE 3—MALE 37-DEGREE FLARED TYPE HOSE FITTINGS

TABLE 3A—DIMENSIONS OF MALE 37-DEGREE FLARED TYPE HOSE FITTINGS FOR USE ON SAE 100R1 HYDRAULIC HOSE[a] (FIGURES 3A TO 3C) (CODES 42 AND 53)

Nominal SAE 100R1 Hose ID mm	Hose Dash Size	B Nominal Straight Thread Size	Thread Dash Size	C Min mm	D Dia Min mm	L Max mm	L_1 Max mm	L_2 Max mm	T Min mm	W Max mm	Y Min mm	Z Max mm
4.8	-3	7/16-20	-4	11.1	2.3	59.2	59.2	—	5.6	19.0	9.4	—
6.4	-4	7/16-20	-4	11.1	2.8	63.0	61.2	—	5.6	22.1	9.4	—
6.4	-4	1/2-20	-5	12.7	2.8	63.0	61.2	—	5.6	22.1	10.4	—
6.4	-4	9/16-18	-6	14.3	2.8	63.2	61.2	—	5.6	22.1	10.4	—
7.9	-5	9/16-18	-6	14.3	5.1	63.2	62.0	—	5.6	23.9	10.4	—
9.5	-6	9/16-18	-6	14.3	6.6	64.0	65.0	—	5.6	27.7	10.4	—
9.5	-6	3/4-16	-8	19.0	6.6	66.8	67.6	—	6.4	27.7	12.7	—
10.3	-6.5	3/4-16	-8	19.0	6.6	67.8	67.6	—	6.4	29.5	12.7	—
12.7	-8	3/4-16	-8	19.0	9.7	68.6	78.2	—	6.4	31.2	12.7	—
12.7	-8	7/8-14	-10	22.2	9.7	72.4	80.8	80.8	6.4	31.2	14.2	63.5
15.9	-10	7/8-14	-10	22.2	12.2	84.3	86.9	—	6.4	34.8	14.2	—
15.9	-10	1-1/16-12	-12	27.0	12.7	81.3	90.9	—	7.9	34.8	15.5	—
15.9	-10	7/8-14	-10	22.2	12.2	94.5	90.9	—	7.9	40.4	14.2	—
19.0	-12	1-1/16-12	-12	27.0	14.5	94.5	90.9	87.6	7.9	40.4	15.5	76.2
19.0	-12	1-3/16-12	-14	27.0	15.5	95.2	93.2	89.9	9.7	40.4	15.5	76.2
25.4	-16	1-5/16-12	-16	33.3	19.8	96.8	106.4	95.2	9.7	51.3	17.5	92.2
31.8	-20	1-5/8-12	-20	41.3	25.7	112.8	120.1	112.0	11.7	62.5	21.3	101.6

[a] For dimensions shown in Figure 3A, but not specified in Table 3A, see corresponding dimensions for respective straight thread size in Figure 2A and Table 2 of SAE J514.

TABLE 3B—DIMENSIONS OF MALE 37-DEGREE FLARED TYPE HOSE FITTINGS FOR USE ON SAE 100R2
HYDRAULIC HOSE[a] (FIGURES 3A TO 3C) (CODES 43 AND 54)

Nominal SAE 100R2 Hose ID mm	Hose Dash Size	B Nominal Straight Thread Size	Thread Dash Size	C Min mm	D Dia Min mm	L Max mm	L_1 Max mm	L_2 Max mm	T Min mm	W Max mm	Y Min mm	Z Max mm
4.8	-3	7/16-20	-4	11.1	2.3	60.7	65.5	—	4.8	22.9	9.4	—
6.4	-4	7/16-20	-4	11.1	2.8	63.0	66.8	—	4.8	25.4	9.4	—
6.4	-4	1/2-20	-5	12.7	2.8	63.0	66.8	63.2	4.8	25.4	10.4	54.1
6.4	-4	9/16-18	-6	14.3	2.8	65.0	66.8	—	5.6	25.4	10.4	—
9.5	-6	9/16-18	-6	14.3	6.6	67.6	71.4	66.0	5.6	30.2	10.4	62.7
9.5	-6	3/4-16	-8	19.0	6.6	68.3	74.7	69.3	6.4	30.2	12.7	62.7
12.7	-8	3/4-16	-8	19.0	9.7	71.9	78.2	—	6.4	35.1	12.7	—
12.7	-8	7/8-14	-10	22.2	9.7	99.6	80.8	77.5	6.4	35.1	14.2	62.7
15.9	-10	7/8-14	-10	22.2	12.2	84.3	92.5	—	6.4	39.6	14.2	—
15.9	-10	1-1/16-12	-12	27.0	12.7	83.6	96.5	—	7.9	39.6	15.5	—
19.0	-12	7/8-14	-10	22.2	12.2	94.5	96.5	—	7.9	44.5	14.2	—
19.0	-12	1-1/16-12	-12	27.0	14.5	94.5	96.5	87.6	7.9	44.5	15.5	76.2
19.0	-12	1-3/16-12	-14	30.2	15.5	95.2	98.8	—	9.7	44.5	15.5	—
22.2	-14	1-3/16-12	-14	30.2	18.3	92.7	109.0	—	9.7	46.0	15.5	—
22.2	-14	1-5/16-12	-16	33.3	18.3	93.2	108.5	—	9.7	46.0	17.5	—
25.4	-16	1-5/16-12	-16	33.3	19.8	97.5	111.3	107.2	9.7	52.3	17.5	91.4
31.8	-20	1-5/8-12	-20	41.3	25.7	142.7	136.7	141.7	11.7	66.8	21.3	114.3

[a] For dimensions shown in Figure 3A, but not specified in Table 3B, see corresponding dimensions for respective straight thread size in Figure 2A and Table 2 of SAE J514.

TABLE 3C—DIMENSIONS OF MALE 37-DEGREE FLARED TYPE HOSE FITTINGS FOR USE ON SAE 100R3
HYDRAULIC HOSE[a] (FIGURES 3A AND 3B) (CODE 44)

Nominal SAE 100R3 Hose ID mm	Hose Dash Size	B Nominal Straight Thread Size	Thread Dash Size	C Min mm	D Dia Min mm	L Max mm	L_1 Max mm	T Min mm	W Max mm	Y Min mm
6.4	-4	7/16-20	-4	11.1	2.8	63.0	60.5	5.6	22.4	9.4
6.4	-4	1/2-20	-5	12.7	2.8	63.0	60.5	5.6	22.4	10.4
9.5	-6	9/16-18	-6	14.3	6.6	64.0	65.0	5.6	26.9	10.4
9.5	-6	3/4-16	-8	19.0	6.6	66.56	66.8	6.4	26.9	12.7
12.7	-8	3/4-16	-8	19.0	8.9	69.3	76.2	6.4	33.3	12.7
12.7	-8	7/8-14	-10	22.2	8.9	73.4	77.7	6.4	33.3	14.2
15.9	-10	7/8-14	-10	22.2	12.2	83.8	84.3	6.4	38.1	14.2
19.0	-12	7/8-14	-10	22.2	12.2	94.5	85.3	7.9	41.4	14.2
19.0	-12	1-1/16-12	-12	27.0	14.5	94.5	85.3	7.9	41.4	15.5
25.4	-16	1-5/16-12	-16	33.3	19.8	96.8	103.1	9.7	50.8	17.5

[a] For dimensions shown in Figure 3A, but not specified in Table 3C, see corresponding dimensions for respective straight thread size in Figure 2A and Table 2 of SAE J514.

TABLE 3D—DIMENSIONS OF MALE 37-DEGREE FLARED TYPE HOSE FITTINGS FOR USE ON SAE 100R4
HYDRAULIC HOSE[a] (FIGURE 3A) (CODE 45)

Nominal SAE 100R4 Hose ID mm	Hose Dash Size	B Nominal Straight Thread Size	Thread Dash Size	C Min mm	D Dia Min mm	L Max mm	T Min mm	W Max mm	Y Min mm
19.0	-12	1-1/16-12	-12	27.0	14.5	94.5	7.9	43.9	15.5
25.4	-16	1-5/16-12	-16	33.3	19.8	96.8	7.9	51.1	17.5
31.8	-20	1-5/8-12	-20	41.3	25.7	106.7	7.9	58.7	21.3

[a] For dimensions shown in Figure 3A, but not specified in Table 3D, see corresponding dimensions for respective straight thread size in Figure 2A and Table 2 of SAE J514.

TABLE 3E—DIMENSIONS OF MALE 37-DEGREE FLARED TYPE HOSE FITTINGS FOR USE ON SAE 100R5 HYDRAULIC HOSE[a] (FIGURES 3A AND 3B) (CODE 46)

Nominal SAE 100R5 Hose ID mm	Hose Dash Size	B Nominal Straight Thread Size	Thread Dash Size	C Min mm	D Dia Min mm	L Max mm	L_1 Max mm	T Min mm	W Max mm	Y Min mm
4.8	-4	7/16-20	-4	11.1	2.3	47.5	52.3	4.8	20.1	9.4
6.4	-5	1/2-20	-5	12.7	2.8	48.3	52.3	4.8	22.1	10.4
7.9	-6	9/16-18	-6	14.3	5.6	49.3	55.4	5.6	23.9	10.4
10.3	-8	3/4-16	-8	19.0	7.1	54.4	66.8	6.4	27.4	12.7
22.7	-10	7/8-14	-10	22.2	9.7	58.2	74.7	6.4	31.5	14.2
15.9	-12	1-1/16-12	-12	27.0	12.7	63.5	86.6	7.9	38.6	15.5
22.2	-16	1-5/16-12	-16	33.3	18.3	70.0	80.3	9.7	43.9	17.5
28.6	-20	1-5/8-12	-20	41.3	23.9	73.9	83.3	11.7	51.3	21.3
34.9	-24	1-7/8-12	-24	47.6	29.5	85.1	91.9	12.7	58.7	25.4
46.0	-32	2-1/2-12	-32	63.5	40.9	101.1	111.8	15.7	73.4	28.7

[a] For dimensions shown in Figure 3A, but not specified in Table 3E, see corresponding dimensions for respective straight thread size in Figure 2A and Table 2 of SAE J514.

NOTE: For SAE J1402 and DOT FMVSS 106 air brake applications, couplings used on 100R5 hose shall have a minimum assembled orifice no less than 66% of the nominal hose I.D.

TABLE 3F—DIMENSIONS OF MALE 37-DEGREE FLARED TYPE HOSE FITTINGS FOR USE ON SAE 100R6 HYDRAULIC HOSE[a] (FIGURE 3A) (CODE 47)

Nominal SAE 100R6 Hose ID mm	Hose Dash Size	B Nominal Straight Thread Size	Thread Dash Size	C Min mm	D Dia Min mm	L Max mm	T Min mm	W Max mm	Y Min mm
6.4	-4	7/16-20	-4	11.1	2.8	35.1	4.8	19.0	9.4
6.4	-4	1/2-20	-5	12.7	2.8	35.1	4.8	19.0	10.4
9.5	-6	9/16-18	-6	14.3	6.6	41.1	5.6	25.4	10.4
9.5	-6	3/4-16	-8	19.0	6.6	43.7	6.4	25.4	12.7
12.7	-8	3/4-16	-8	19.0	9.7	44.4	6.4	28.4	12.7
15.9	-10	7/8-14	-10	22.2	12.2	50.8	6.4	30.2	14.2
19.0	-12	1-1/16-12	-12	27.0	14.5	85.9	7.9	36.8	15.5

[a] For dimensions shown in Figure 3A, but not specified in Table 3F, see corresponding dimensions for respective straight thread size in Figure 2A and Table 2 of SAE J514.

TABLE 3G—DIMENSIONS OF MALE 37-DEGREE FLARED TYPE HOSE FITTINGS FOR USE ON SAE 100R7 HYDRAULIC HOSE[a] (FIGURE 3A) (CODE 48)

Nominal SAE 100R7 Hose ID mm	Hose Dash Size	B Nominal Straight Thread Size	Thread Dash Size	C Min mm	D Dia Min mm	L Max mm	T Min mm	W Max mm	Y Min mm
4.8	-3	7/16-20	-4	11.1	2.3	63.2	4.8	19.0	9.4
6.4	-4	7/16-20	-4	11.1	2.8	63.2	4.8	20.6	9.4
6.4	-4	1/2-20	-5	12.7	2.8	63.8	4.8	20.6	10.4
6.4	-4	9/16-18	-6	14.3	2.8	63.8	5.6	20.6	10.4
7.9	-5	9/16-18	-6	14.3	5.1	64.3	5.6	22.4	10.4
9.5	-6	9/16-18	-6	14.3	6.6	68.6	5.6	25.7	10.4
9.5	-6	3/4-16	-8	19.0	6.6	68.6	6.4	25.7	12.7
12.7	-8	3/4-16	-8	19.0	9.1	77.2	6.4	31.2	12.7
12.7	-8	7/8-14	-10	22.2	9.1	78.0	6.4	31.2	14.2
15.9	-10	7/8-14	-10	22.2	12.2	84.3	6.4	34.8	14.2
19.0	-12	1-1/16-12	-12	27.0	14.5	90.2	7.9	40.4	15.5
19.0	-12	1-3/16-12	-14	30.2	14.5	92.7	9.7	40.4	15.5
25.4	-16	1-5/16-12	-16	33.3	19.8	104.1	9.7	40.4	17.5

[a] For dimensions shown in Figure 3A, but not specified in Table 3G, see corresponding dimensions for respective straight thread size in Figure 2A and Table 2 of SAE J514.

22.127

TABLE 3H—DIMENSIONS OF MALE 37-DEGREE FLARED TYPE HOSE FITTINGS FOR USE ON SAE 100R8 HYDRAULIC HOSE[a] (FIGURE 3A) (CODE 49)

Nominal SAE 100R8 Hose ID mm	Hose Dash Size	B Nominal Straight Thread Size	Thread Dash Size	C Min mm	D Dia Min mm	L Max mm	T Min mm	W Max mm	Y Min mm
4.8	-3	7/16-20	-4	11.1	2.3	63.2	4.8	19.0	9.4
6.4	-4	7/16-20	-4	11.1	2.8	63.2	4.8	20.6	9.4
6.4	-4	1/2-20	-5	12.7	2.8	63.8	4.8	20.6	10.4
6.4	-4	9/16-18	-6	14.3	2.8	63.8	5.6	20.6	10.4
9.5	-6	9/16-18	-6	14.3	6.6	68.6	5.6	25.7	10.4
9.5	-6	3/4-16	-8	19.0	6.6	68.6	6.4	25.7	12.7
12.7	-8	3/4-16	-8	19.0	9.1	77.2	6.4	31.2	12.7
12.7	-8	7/8-14	-10	22.2	9.1	78.0	6.4	31.2	14.2
15.9	-10	7/8-14	-10	22.2	12.2	84.3	6.4	36.8	14.2
19.0	-12	1-1/16-12	-12	27.0	14.5	90.2	7.9	40.4	15.5
19.0	-12	1-3/16-12	-14	30.2	14.5	92.7	9.7	40.4	15.5
25.4	-16	1-5/16-12	-16	33.3	19.8	104.1	9.7	49.5	17.5

[a] For dimensions shown in Figure 3A, but not specified in Table 3H, see corresponding dimensions for respective straight thread size in Figure 2A and Table 2 of SAE J514.

TABLE 3I—DIMENSIONS OF MALE 37-DEGREE FLARED TYPE HOSE FITTINGS FOR USE ON SAE 100R9 HYDRAULIC HOSE[a] (FIGURES 3A AND 3B) (CODES 50 AND 75)

Nominal SAE 100R9 Hose ID mm	Hose Dash Size	B Nominal Straight Thread Size	Thread Dash Size	C Min mm	D Dia Min mm	L Max mm	L₁ Max mm	T Min mm	W Max mm	Y Min mm
9.5	-6	9/16-18	-6	14.3	6.6	74.4	71.4	5.6	30.2	10.4
12.7	-8	3/4-16	-8	19.0	9.7	76.2	78.2	6.4	35.1	12.7
19.0	-12	1-1/16-12	-12	27.0	14.5	98.3	98.8	7.9	44.5	15.5
25.4	-16	1-5/16-12	-16	33.3	19.8	101.3	111.3	9.7	52.3	17.5
31.8	-20	1-5/8-12	-20	41.3	25.7	142.7	—	11.7	66.8	21.3

[a] For dimensions shown in Figure 3A, but not specified in Table 3I, see corresponding dimensions for respective straight thread size in Figure 2A and Table 2 of SAE J514.

TABLE 3J—DIMENSIONS OF MALE 37-DEGREE HOSE FITTINGS FOR USE ON SAE 100R10 HYDRAULIC HOSE[a] (FIGURES 3A TO 3C) (CODES 51 AND 76)

Nominal SAE 100R10 Hose ID mm	Hose Dash Size	B Nominal Straight Thread Size	Thread Dash Size	C Min mm	D Dia Min mm	L Max mm	L₁ Max mm	L₂ Max mm	T Min mm	W Max mm	Y Min mm	Z Max mm
6.4	-4	7/16-20	-4	11.1	2.8	75.4	61.2	63.5	5.6	28.7	9.4	54.1
9.5	-6	9/16-18	-6	14.3	5.6	83.3	78.0	79.2	5.6	28.7	10.4	62.7
12.7	-8	3/4-16	-8	19.0	8.6	98.8	78.2	82.6	6.4	31.8	12.7	62.7
19.0	-12	1-1/16-12	-12	27.0	12.7	115.6	94.5	87.6	7.9	42.9	14.5	79.2
25.4	-16	1-5/16-12	-16	33.3	17.3	132.8	111.3	102.9	9.7	50.8	17.5	107.2
31.8	-20	1-5/8-12	-20	41.3	22.1	142.0	137.9	116.3	11.7	62.2	17.8	126.0
38.1	-24	1-7/8-12	-24	47.6	27.7	160.0	148.6	133.4	12.7	76.2	19.3	133.1
50.8	-32	2-1/2-12	-32	63.5	39.6	186.2	196.8	190.5	17.5	81.8	28.4	166.6

[a] For dimensions shown in Figure 3A, but not specified in Table 3J, see corresponding dimensions for respective straight thread size in Figure 2A and Table 2 of SAE J514.

(R) TABLE 3K—DIMENSIONS OF MALE 37-DEGREE FLARED TYPE HOSE FITTINGS FOR USE ON SAE 100R12 HYDRAULIC HOSE[a] (FIGURE 3A) (CODE 77)

Nominal SAE 100R12 Hose ID mm	Hose Dash Size	B Nominal Straight Thread Size	Thread Dash Size	C Min mm	D Dia Min mm	L Max mm	T Min mm	W Max mm	Y Min mm
9.5	-6	9/16-18	-6	14.3	5.6	83.3	5.6	28.7	10.4
12.7	-8	3/4-16	-8	19.0	8.6	98.8	6.4	31.8	12.7
15.9	-10	7/8-14	-10	22.2	10.0	108.0	7.9	38.0	14.5
19.0	-12	1-1/16-12	-12	27.0	12.7	117.3	7.9	42.9	14.5
25.4	-16	1-5/16-12	-16	33.3	17.3	132.8	9.7	50.8	15.7
31.8	-20	1-5/8-12	-20	41.3	22.1	143.5	11.2	60.5	17.5
38.1	-24	1-7/8-12	-24	47.6	27.7	160.0	12.7	67.8	18.8
50.8	-32	2-1/2-12	-32	63.5	39.6	186.2	15.7	81.8	22.4

[a] For dimensions shown in Figure 3A, but not specified in Table 3K, see corresponding dimensions for respective straight thread size in Figure 2A and Table 2 of SAE J514.

TABLE 3L—DIMENSIONS OF MALE 37-DEGREE FLARED TYPE HOSE FITTINGS FOR USE ON SAE 100R13 HYDRAULIC HOSE[a] (FIGURE 3A) (CODE 78)

Nominal SAE 100R13 Hose ID mm	Hose Dash Size	B Nominal Straight Thread Size	Thread Dash Size	C Min mm	D Dia Min mm	L Max mm	T Min mm	W Max mm	Y Min mm
19.0	-12	1-1/16-12	-12	27.0	12.7	117.3	7.9	43.7	15.7
25.4	-16	1-5/16-12	-16	33.3	17.5	127.0	9.7	50.8	15.7
31.8	-20	1-5/8-12	-20	41.3	22.4	143.5	11.2	63.5	17.5
38.1	-24	1-7/8-12	-24	47.6	27.7	159.0	12.7	71.1	18.8
50.8	-32	2-1/2-12	-32	63.5	38.1	224.3	15.7	88.9	22.4

[a] For dimensions shown in Figure 3A, but not specified in Table 3L, see corresponding dimensions for respective straight thread size in Figure 2A and Table 2 of SAE J514.

FEMALE 37-DEGREE FLARED TYPE

NOTES: UNSPECIFIED DETAIL WITH RESPECT TO DIMENSIONS, TOLERANCES, CONTOURS, MATERIAL, WORKMANSHIP, AND SO ON, MUST CONFORM TO GENERAL SPECIFICATIONS FOR HYDRAULIC HOSE FITTINGS. THE DIMENSIONAL DESIGNATIONS ON THE FIRST FIGURE IN EACH GROUP SHALL APPLY TO ALL OTHER FIGURES IN THAT GROUP EXCEPT AS SHOWN OTHERWISE. CODES SHOWN IN BRACKETS ADJACENT TO FIGURE NUMBERS REPRESENT RESPECTIVE FITTING IDENTIFICATION, WITH XX SUBSTITUTED FOR THE HOSE TYPE CODE DEPICTED IN BRACKETS AT END OF RESPECTIVE TABLE TITLES, IN ACCORDANCE WITH SAE J846.

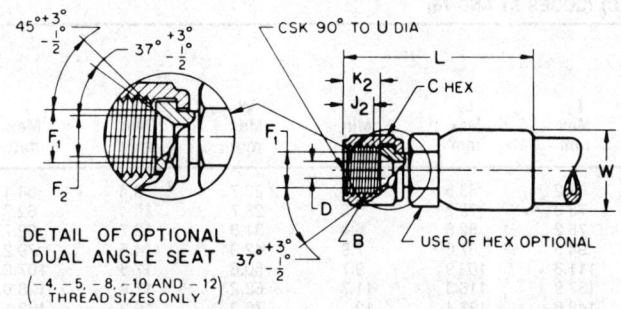

FIGURE 4A—PERMANENTLY ATTACHED STYLE (2401XX)

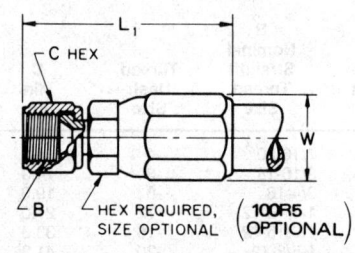

FIGURE 4B—FIELD ATTACHABLE SCREW STYLE (2501XX)

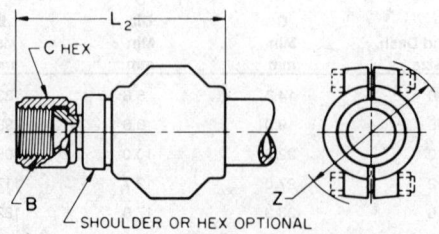

FIGURE 4C—FIELD ATTACHABLE SEGMENT CLAMP STYLE (2601XX)

FIGURE 4—FEMALE 37-DEGREE FLARED TYPE HOSE FITTINGS

22.129

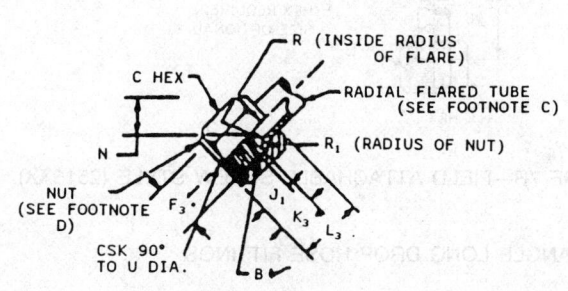

FIGURE 5A—OPTIONAL 37-DEGREE RADIAL FLARE SWIVEL END

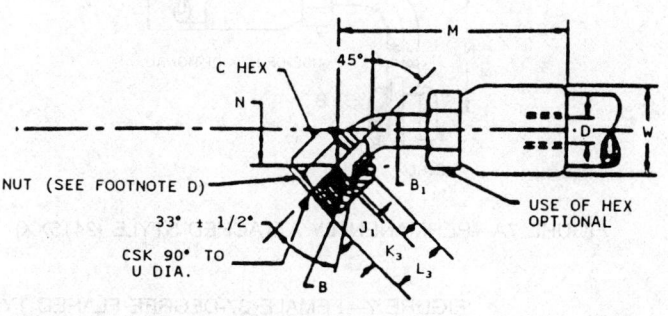

FIGURE 5B—PERMANENTLY ATTACHED STYLE WITH SLEEVE (2403XX)

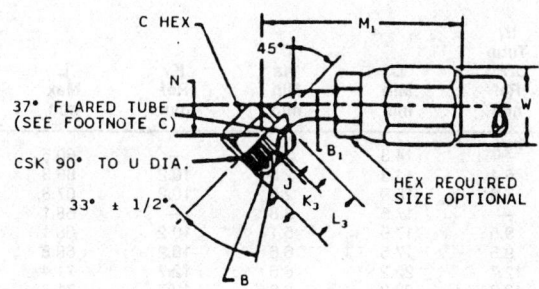

FIGURE 5C—FIELD ATTACHABLE SCREW STYLE (2503XX)

FIGURE 5—FEMALE 37-DEGREE FLARED TYPE 45-DEGREE ANGLE HOSE FITTINGS

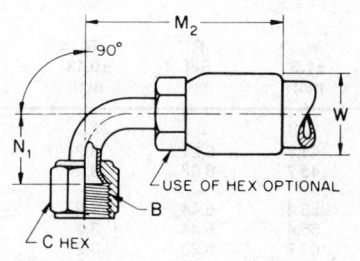

FIGURE 6A—PERMANENTLY ATTACHED STYLE (2414XX)

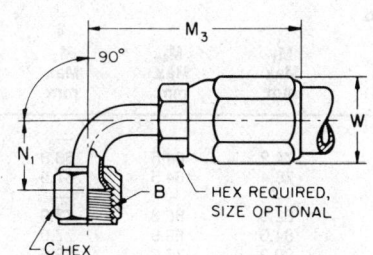

FIGURE 6B—FIELD ATTACHABLE SCREW STYLE (2514XX)

FIGURE 6—FEMALE 37-DEGREE FLARED TYPE 90-DEGREE ANGLE SHORT DROP HOSE FITTINGS

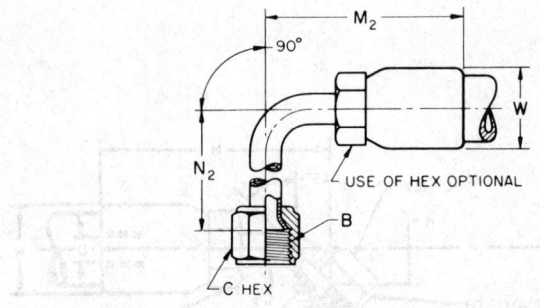

FIGURE 7A—PERMANENTLY ATTACHED STYLE (2415XX)

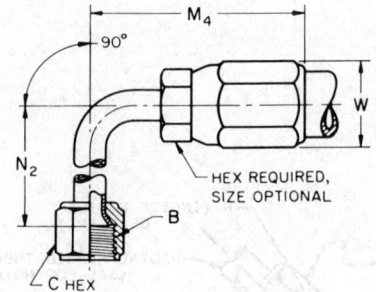

FIGURE 7B—FIELD ATTACHABLE SCREW STYLE (2515XX)

FIGURE 7—FEMALE 37-DEGREE FLARED TYPE 90-DEGREE ANGLE LONG DROP HOSE FITTINGS

TABLE 4A—DIMENSIONS OF FEMALE 37-DEGREE FLARED TYPE HOSE FITTINGS FOR USE ON SAE 100R1 HYDRAULIC HOSE[a] (FIGURES 4 TO 7) (CODES 42 AND 53)

Nominal SAE 100R1 Hose ID mm	Hose Dash Size	B Nominal Straight Thread Size[e]	Thread and Tube Dash Size	B_1 Tube OD[b,c] Ref mm	C Min mm	D Dia Min mm	K_3 Ref mm	L Max mm	L_1 Max mm	L_2 Max mm	L_3 ±0.3 mm	M Max mm
4.8	-3	7/16-20	-4	—	14.3	2.3	—	60.5	65.0	—	—	—
6.4	-4	7/16-20	-4	6.4	14.3	2.8	10.2	66.3	66.5	—	15.5	67.6
6.4	-4	1/2-20	-5	7.9	15.9	2.8	10.2	67.8	66.5	—	15.7	68.8
6.4	-4	9/16-18	-6	—	17.5	2.8	—	68.1	67.6	—	—	—
7.9	-5	9/16-18	-6	9.5	17.5	5.1	10.2	68.1	67.6	—	16.8	74.7
9.5	-6	9/16-18	-6	9.5	17.5	6.6	10.2	68.8	70.6	—	16.8	76.2
9.5	-6	3/4-16	-8	12.7	22.2	6.6	12.7	71.4	73.9	—	16.8	85.9
10.3	-6.5	3/4-16	-8	12.7	22.2	6.6	12.7	71.4	73.9	—	16.8	85.9
12.7	-8	3/4-16	-8	12.7	22.2	9.7	12.7	73.2	82.3	—	17.3	85.9
12.7	-8	7/8-14	-10	15.9	25.4	9.7	14.0	78.0	83.1	81.8	23.6	92.2
15.9	-10	7/8-14	-10	15.9	25.4	10.9	14.0	89.7	88.4	—	23.6	102.6
15.9	-10	1-1/16-12	-12	19.0	31.8	12.7	16.5	88.4	93.5	—	28.4	106.4
19.0	-12	7/8-14	-10	15.9	25.4	12.2	14.0	98.0	95.8	—	23.6	116.1
19.0	-12	1-1/16-12	-12	19.0	31.8	14.5	16.5	98.0	93.5	88.9	28.4	116.1
19.0	-12	1-3/16-12	-14	22.2	34.9	15.5	—	102.6	96.8	93.5	—	114.3
25.4	-16	1-5/16-12	-16	25.4	38.1	19.8	—	102.6	108.7	108.7	—	123.2
31.8	-20	1-5/8-12	-20	—	50.8	25.7	—	115.8	126.0	—	—	—

TABLE 4A—(CONTINUED)

Nominal SAE 100R1 Hose ID mm	Thread and Tube Dash Size	M_1 Max mm	M_2 Max mm	M_3 Max mm	M_4 Max mm	N ±1.5 mm	N_1 ±1.5 mm	N_2 ±1.5 mm	R Ref mm	R_1 ±0.13 mm	W Max mm	Z Max mm
4.8	-4	—	—	—	—	—	—	—	—	—	19.0	—
6.4	-4	74.2	64.5	66.8	66.8	8.4	17.3	45.7	5.21	3.2	22.1	—
6.4	-5	75.4	64.5	67.8	67.8	9.1	19.6	45.7	6.02	3.2	22.1	—
6.4	-6	—	—	—	—	—	—	—	—	—	22.1	—
7.9	-6	82.8	68.3	73.9	73.9	9.9	21.6	55.4	5.44	3.2	23.9	—
9.5	-6	84.6	69.9	77.0	77.0	9.9	21.6	55.4	5.44	3.2	27.7	—
9.5	-8	90.2	77.7	83.6	83.6	14.0	27.7	61.7	5.23	3.2	27.7	—
10.3	-8	90.2	77.7	83.6	83.6	14.0	27.7	61.7	5.23	3.2	29.5	—
12.7	-8	97.8	77.7	90.4	90.4	14.0	27.7	61.7	5.23	3.2	30.5	—
12.7	-10	104.6	82.6	96.0	96.0	16.0	31.2	65.3	7.77	4.8	30.5	63.5
15.9	-10	113.0	108.0	102.9	102.9	16.0	31.2	65.3	7.77	4.8	34.8	—
15.9	-12	115.1	96.5	107.2	107.2	19.8	46.2	94.7	6.60	3.8	34.8	—
19.0	-10	115.1	114.8	108.7	108.7	16.0	31.2	65.3	7.77	4.8	40.4	—
19.0	-12	115.1	114.8	108.7	108.7	19.8	46.2	94.7	6.60	3.8	40.4	76.2
19.0	-14	125.0	104.9	115.6	115.6	21.3	50.8	99.8	—	—	40.4	76.2
25.4	-16	146.8	122.9	132.1	132.1	22.6	54.4	110.0	—	—	51.3	91.2
31.8	-20	—	—	—	—	—	—	—	—	—	62.5	—

[a] For dimensions shown in Figures 4 to 7, but not specified in Table 4A, see corresponding dimensions for respective straight thread size in Figure 8C and Table 2 of SAE J514.
[b] Sleeves shown in Figure 5B shall conform with Figure 14 and Table 4 of SAE J514 for respective tube OD referenced above except on some sizes and shapes where sleeve length must be shortened.
[c] 37-degree flares shall conform with single or double 37-degree flares specified in Figure 3 and Table 2 of SAE J533 for respective tube OD referenced above. Dimension F3 of radial flares shall conform with dimension A in Table 2 of SAE J533.
[d] Nuts shown in Figure 5B shall conform with Figure 12 and Table 3 of SAE J514 for respective straight thread size.
[e] For dimensions shown in Figures 5 to 7, but not specified in Table 4A, see corresponding dimensions for respective straight thread size in Figures 11 and 13, and Table 3 of SAE J514.

TABLE 4B—DIMENSIONS OF FEMALE 37-DEGREE FLARED TYPE HOSE FITTINGS FOR USE ON SAE 100R2 HYDRAULIC HOSE[a] (FIGURES 4 TO 7) (CODES 43 AND 54)

Nominal SAE 100R2 Hose ID mm	Hose Dash Size	B Nominal Straight Thread Size	Thread and Tube Dash Size	B_1 Tube OD[b,c] Ref mm	C Min mm	D Dia Min mm	K_3 Ref mm	L Max mm	L_1 Max mm	L_2 Max mm	L_3 ±0.3 mm	M Max mm
4.8	-3	7/16-20	-4	—	14.3	2.3	—	62.0	67.8	—	—	—
6.4	-4	7/16-20	-4	6.4	14.3	2.8	10.2	66.3	73.2	—	15.5	71.1
6.4	-4	1/2-20	-5	7.9	15.9	2.8	10.2	67.8	74.7	66.0	15.7	73.7
6.4	-4	9/16-18	-6	—	17.5	2.8	—	68.1	74.7	—	—	—
9.5	-6	9/16-18	-6	9.5	17.5	6.6	10.2	69.1	80.3	70.9	16.8	79.2
9.5	-6	3/4-16	-8	12.7	22.2	6.6	12.7	74.4	80.3	71.4	16.8	88.1
12.7	-8	3/4-16	-8	12.7	22.2	9.7	12.7	76.7	84.6	—	16.8	93.5
12.7	-8	7/8-14	-10	15.9	25.4	9.7	14.0	78.5	85.9	84.3	23.6	102.1
15.9	-10	7/8-14	-10	15.9	25.4	10.9	14.0	89.7	94.7	—	23.6	108.5
15.9	-10	1-1/16-12	-12	19.0	31.8	12.7	16.5	88.4	103.6	—	28.4	114.8
19.0	-12	7/8-14	-10	15.9	25.4	12.2	14.0	98.0	103.6	—	23.6	117.3
19.0	-12	1-1/16-12	-12	19.0	31.8	14.5	16.5	98.0	103.6	94.7	28.4	117.3
19.0	-12	1-3/16-12	-14	22.2	34.9	15.5	—	102.6	105.2	—	—	120.9
22.2	-14	1-3/16-12	-14	22.2	34.9	18.3	—	99.1	114.3	—	—	114.3
22.2	-14	1-5/16-12	-16	25.4	38.1	18.3	—	100.1	115.3	—	—	120.7
25.4	-16	1-5/16-12	-16	25.4	38.1	19.8	—	106.4	115.3	114.0	—	140.2
31.8	-20	1-5/8-12	-20	—	50.8	25.7	—	166.6	138.9	133.4	—	—
38.1	-24	1-7/8-12	-24	—	57.2	31.8	—	170.7	148.6	146.0	—	—
50.8	-32	2-1/2-12	-32	—	73.0	43.7	—	177.8	165.9	171.7	—	—

TABLE 4B—(CONTINUED)

Nominal SAE 100R2 Hose ID mm	Thread and Tube Dash Size	M_1 Max mm	M_2 Max mm	M_3 Max mm	M_4 Max mm	N ±1.5 mm	N_1 ±1.5 mm	N_2 ±1.5 mm	R Ref mm	R_1 ±0.13 mm	W Max mm	Z Max mm
4.8	-4	—	—	—	—	—	—	—	—	—	22.9	—
6.4	-4	79.0	64.5	76.2	76.2	8.4	17.3	45.7	5.21	3.2	22.6	—
6.4	-5	77.2	66.5	71.1	71.1	9.1	19.6	45.7	6.02	3.2	25.4	54.1
6.4	-6	—	—	—	—	—	—	—	—	—	25.4	—
9.5	-6	86.9	71.9	84.3	84.3	9.9	21.6	55.4	5.44	3.2	30.2	62.7
9.5	-8	90.7	80.3	85.9	90.7	14.0	27.7	61.7	5.23	3.2	30.2	62.7
12.7	-8	100.3	85.6	92.7	92.7	14.0	27.7	61.7	5.23	3.2	31.5	—
12.7	-10	106.9	92.2	98.3	98.3	16.0	31.2	65.3	7.77	4.8	31.5	62.7
15.9	-10	113.0	108.0	105.2	105.2	16.0	31.2	65.3	7.77	4.8	39.6	—
15.9	-12	114.6	104.9	112.0	112.0	19.8	46.2	94.7	6.60	3.8	39.6	—
19.0	-10	122.2	114.8	112.0	120.9	16.0	31.2	65.3	7.77	4.8	44.4	—
19.0	-12	122.2	114.8	112.0	120.9	19.8	46.2	94.7	6.60	3.8	44.4	76.2
19.0	-14	129.5	108.0	120.1	120.1	21.3	50.8	99.8	—	—	44.4	—
22.2	-14	129.5	108.0	125.7	125.7	21.3	50.8	99.8	—	—	45.5	—
22.2	-16	146.8	114.3	132.1	132.1	22.6	54.4	110.0	—	—	45.5	—
25.4	-16	146.8	133.4	134.9	139.7	22.6	54.4	110.0	—	—	52.1	91.4
31.8	-20	—	—	—	—	—	—	—	—	—	66.8	114.3
38.1	-24	—	—	—	—	—	—	—	—	—	74.7	127.0
50.8	-32	—	—	—	—	—	—	—	—	—	88.9	133.4

[a] For dimensions shown in Figures 4 to 7, but not specified in Table 4B, see corresponding dimensions for respective straight thread size in Figure 8C and Table 2 of SAE J514.
[b] Sleeves shown in Figure 5B shall conform with Figure 14 and Table 4 of SAE J514 for respective tube OD referenced above except on some sizes and shapes where sleeve length must be shortened.
[c] 37-degree flares shall conform with single or double 37-degree flares specified in Figure 3 and Table 2 of SAE J533 for respective tube OD referenced above. Dimension F3 of radial flares shall conform with dimension A in Table 2 of SAE J533.
[d] Nuts shown in Figure 5B shall conform with Figure 12 and Table 3 of SAE J514 for respective straight thread size.
[e] For dimensions shown in Figures 5 to 7, but not specified in Table 4B, see corresponding dimensions for respective straight thread size in Figures 11 and 13, and Table 3 of SAE J514.

TABLE 4C—DIMENSIONS OF FEMALE 37-DEGREE FLARED TYPE HOSE FITTINGS FOR USE ON SAE 100R3 HYDRAULIC HOSE[a] (FIGURES 4 TO 7) (CODE 44)

Nominal SAE 100R3 Hose ID mm	Hose Dash Size	B Nominal Straight Thread Size	Thread Dash Size	C Min mm	D Dia Min mm	L Max mm	L_1 Max mm	L_2 Min mm	W Max mm	Z Min mm
6.4	-4	7/16-20	-4	14.3	2.8	66.3	63.5	—	22.4	—
6.4	-4	1/2-20	-5	15.9	2.8	67.8	63.5	—	22.4	—
9.5	-6	9/16-18	-6	17.5	6.6	70.6	70.4	—	26.9	—
9.5	-6	3/4-16	-8	22.2	6.6	70.9	69.8	—	26.9	—
12.7	-8	3/4-16	-8	22.2	9.7	74.4	78.7	—	33.3	—
12.7	-8	7/8-14	-10	25.4	9.7	77.2	82.6	—	33.3	—
15.9	-10	7/8-14	-10	25.4	10.9	89.7	83.8	—	38.1	—
19.0	-12	7/8-14	-10	25.4	12.2	98.0	85.9	—	41.4	—
19.0	-12	1-1/16-12	-12	31.8	14.5	98.0	85.9	—	42.2	—
25.4	-16	1-15/16-12	-16	38.1	19.8	102.6	95.2	93.5	50.8	82.6

[a] For dimensions shown in Figures 4 to 7, but not specified in Table 4C, see corresponding dimensions for respective straight thread size in Figure 8C and Table 2 of SAE J514.

TABLE 4D—DIMENSIONS OF FEMALE 37-DEGREE FLARED TYPE HOSE FITTINGS FOR USE ON SAE 100R4 HYDRAULIC HOSE[a] (FIGURES 4 TO 7) (CODE 45)

Nominal SAE 100R4 Hose ID mm	Hose Dash Size	B Nominal Straight Thread Size	Thread Dash Size	C Min mm	D Dia Min mm	L Max mm	W Max mm
19.0	-12	1-1/16-12	-12	31.8	14.5	98.0	43.9
25.4	-16	1-5/16-12	-16	38.1	19.8	102.6	51.1
31.8	-20	1-5/8-12	-20	50.8	25.7	106.7	58.7

[a] For dimensions shown in Figures 4 to 7, but not specified in Table 4D, see corresponding dimensions for respective straight thread size in Figure 8C and Table 2 of SAE J514.

TABLE 4E—DIMENSIONS OF FEMALE 37-DEGREE FLARED TYPE HOSE FITTINGS FOR USE ON SAE 100R5 HYDRAULIC HOSE[a] (FIGURES 4 TO 7) (CODE 46)

Nominal SAE 100R5 Hose ID mm	Hose Dash Size	B Nominal Straight Thread Size	Thread and Tube Dash Size	B_1 Tube OD[b,c] Ref mm	C Min mm	D Dia Min mm	K_3 Ref mm	L Max mm	L_1 Max mm	L_3 mm	M_1 Max mm	M_3 Max mm	M_4 Max mm	N mm	N_1 mm	N_3 Max mm	R Ref mm	R_1 mm	W Max mm
4.8	-4	7/16-20	-4	6.4	14.3	2.3	10.2	53.1	53.8	15.5	62.2	59.7	59.7	8.4	17.3	45.7	5.21	3.18	20.1
6.4	-5	1/2-20	-5	7.9	15.9	2.8	10.2	56.1	55.4	15.7	66.0	64.5	64.5	9.1	19.6	45.7	6.02	3.18	22.1
7.9	-6	9/16-18	-6	9.7	17.5	5.1	10.2	57.9	60.7	16.8	69.8	67.3	67.3	9.9	21.6	55.4	5.44	3.18	23.9
10.3	-8	3/4-16	-8	12.7	22.2	7.1	12.7	64.0	73.2	16.8	84.8	81.0	81.0	14.0	27.7	61.7	5.23	3.18	27.4
12.7	-10	7/8-14	-10	15.9	25.4	9.7	14.0	67.6	80.3	23.6	93.0	88.1	88.1	16.0	31.2	65.3	7.77	4.78	31.8
15.9	-12	1-1/16-12	-12	19.0	31.8	12.7	16.5	70.4	91.9	28.4	84.8	109.1	109.1	19.8	46.2	94.7	6.60	3.81	38.6
22.2	-16	1-5/16-12	-16	25.4	38.1	18.3	—	80.8	88.9	—	105.7	103.6	103.6	22.6	54.4	110.0	—	—	43.9
28.6	-20	1-5/8-12	-20	31.8	50.8	23.9	—	84.3	98.6	—	105.4	103.6	103.6	27.9	65.5	134.1	—	—	51.3
34.9	-24	1-7/8-12	-24	—	57.2	29.5	—	96.6	101.1	—	—	—	—	—	—	—	—	—	58.7
46.0	-32	2-1/2-12	-32	—	73.0	40.9	—	114.3	120.9	—	—	—	—	—	—	—	—	—	73.4

[a] For dimensions shown in Figures 4 to 7, but not specified in Table 4E, see corresponding dimensions for respective straight thread size in Figure 8C and Table 2 of SAE J514.
[b] Sleeves shown in Figure 5B shall conform with Figure 14 and Table 4 of SAE J514 for respective tube OD referenced above except on some sizes and shapes where sleeve length must be shortened.
[c] 37-degree flares shall conform with single or double 37-degree flares specified in Figure 3 and Table 2 of SAE J533 for respective tube OD referenced above. Dimension F3 of radial flares shall conform with dimension A in Table 2 of SAE J533.
[d] Nuts shown in Figure 5B shall conform with Figure 12 and Table 3 of SAE J514 for respective straight thread size.
[e] For dimensions shown in Figures 5 to 7, but not specified in Table 4E, see corresponding dimensions for respective straight thread size in Figures 11 and 13, and Table 3 of SAE J514.

NOTE: For SAE J1402 and DOT FMVSS 106 air brake applications, couplings used in 100R5 hose shall have a minimum assembled orifice no less than 66% of the nominal hose I.D.

TABLE 4F—DIMENSIONS OF FEMALE 37-DEGREE FLARED TYPE HOSE FITTINGS FOR USE ON SAE 100R6 HYDRAULIC HOSE[a] (FIGURES 4 TO 7) (CODE 47)

Nominal SAE 100R6 Hose ID mm	Hose Dash Size	B Nominal Straight Thread Size	Thread Dash Size	C Min mm	D Dia Min mm	L Max mm	L_1 Max mm	W Max mm
6.4	-4	7/16-20	-4	14.3	2.8	38.1	39.6	19.0
6.4	-4	1/2-20	-5	15.9	2.8	38.1	39.6	19.0
7.9	-5	1/2-20	-5	15.9	5.1	40.1	42.9	22.4
7.9	-5	9/16-18	-6	17.5	5.1	40.1	37.1	22.4
9.5	-6	9/16-18	-6	17.5	6.6	45.2	46.7	25.4
9.5	-6	3/4-16	-8	22.2	6.6	46.0	46.7	25.4
12.7	-8	3/4-16	-8	22.2	9.7	46.0	47.8	28.4
12.7	-8	7/8-14	-10	25.4	9.7	48.5	—	28.4
15.9	-10	7/8-14	-10	25.4	12.2	57.4	58.9	30.2
19.0	-12	1-1/16-12	-12	31.8	14.5	85.9	83.8	36.8

[a] For dimensions shown in Figures 4 to 7, but not specified in Table 4F, see corresponding dimensions for respective straight thread size in Figure 8C and Table 2 of SAE J514.

TABLE 4G—DIMENSIONS OF FEMALE 37-DEGREE FLARED TYPE HOSE FITTINGS FOR USE ON SAE 100R7 HYDRAULIC HOSE[a] (FIGURES 4 TO 7) (CODES 48)

Nominal SAE 100R7 Hose ID mm	Hose Dash Size	B Nominal Straight Thread Size	Thread and Tube Dash Size	B_1 Tube OD Ref mm	C Min mm	D Dia Min mm	L Max mm	L_1 Max mm	M Max mm	M_2 Max mm	N ±1.5 mm	N_1 ±1.5 mm	N_2 ±1.5 mm	W Max mm
4.8	-3	7/16-20	-4	—	14.3	2.3	66.3	51.3	76.2	72.4	8.4	17.3	45.7	19.0
6.4	-4	7/16-20	-4	6.14	14.3	2.8	66.3	56.4	87.6	82.0	8.4	17.3	45.7	20.6
6.4	-4	1/2-20	-5	7.9	15.9	2.8	66.5	56.4	88.9	83.3	9.1	19.6	45.7	20.6
6.4	-4	9/16-18	-6	—	17.5	2.8	67.1	56.4	—	—	—	—	—	20.6
7.9	-5	9/16-18	-6	9.5	17.5	5.1	67.8	64.3	90.2	84.1	9.9	21.6	55.4	22.4
9.5	-6	9/16-18	-6	9.5	17.5	6.6	74.2	70.4	94.7	88.9	9.9	21.6	55.4	25.7
9.5	-6	3/4-16	-8	12.7	22.2	6.6	76.2	70.4	103.6	96.8	14.0	27.7	61.7	25.7
12.7	-8	3/4-16	-8	12.7	22.2	9.1	80.8	79.2	107.2	100.1	14.0	27.7	61.7	31.2
12.7	-8	7/8-14	-10	15.9	25.4	9.1	83.6	79.2	115.8	106.7	16.0	31.2	65.3	31.2
15.9	-10	7/8-14	-10	15.9	25.4	10.9	89.7	—	118.4	111.8	16.0	31.2	65.3	34.8
19.0	-12	1-1/16-12	-12	19.0	31.8	14.5	91.7	88.4	120.9	117.1	19.8	46.2	94.7	40.4
19.0	-12	1-3/16-12	-14	22.2	34.9	14.5	92.2	82.8	—	—	—	—	—	40.4
25.4	-16	1-5/16-12	-16	25.4	38.1	19.8	105.7	94.5	140.2	133.4	22.6	54.4	110.0	49.5

[a] For dimensions shown in Figures 7 to 7, but not specified in Table 4G, see corresponding dimensions for respective straight thread size in Figure 8C and Table 2 of SAE J514.
[b] Sleeves shown in Figure 5B shall conform with Figure 14 and Table 4 of SAE J514 for respective tube OD referenced above except on some sizes and shapes where sleeve length must be shortened.
[c] Flares shall conform with single or double 37-degree flares specified in Figure 3 and Table 2 of SAE J533 for respective tube OD referenced above.
[d] Nuts shown in Figure 5B shall conform with Figure 12 and Table 3 of SAE J514 for respective straight thread size.
[e] For dimensions shown in Figures 5 to 7, but not specified in Table 4G, see corresponding dimensions for respective straight thread size in Figures 11 and 13, and Table 3 of SAE J514.

TABLE 4H—DIMENSIONS OF FEMALE 37-DEGREE FLARED TYPE HOSE FITTINGS FOR USE ON SAE 100R8 HYDRAULIC HOSE[a] (FIGURES 4 TO 7) (CODE 49)

Nominal SAE 100R8 Hose ID mm	Hose Dash Size	B Nominal Straight Thread Size	Thread and Tube Dash Size	B_1 Tube OD Ref mm	C Min mm	D Dia mm	L Max mm	L_1 Max mm	M Min mm	M_1 Max mm	M_2 Min mm	M_3 Max mm	M_4 Max mm	N ±1.5 mm	N_1 ±1.5 mm	N_2 ±1.5 mm	W Max mm
4.8	-3	7/16-20	-4	—	14.3	2.3	66.3	—	76.2	—	72.4	—	—	8.4	17.3	45.7	19.0
6.4	-4	7/16-20	-4	6.4	14.3	2.8	66.3	—	87.6	—	82.0	—	—	8.4	17.3	45.7	20.6
6.4	-4	1/2-20	-5	7.9	15.9	2.8	66.5	—	88.9	—	83.3	—	—	9.1	19.6	45.7	20.6
6.4	-4	9/16-18	-6	—	17.5	2.8	67.1	—	—	—	—	—	—	—	—	—	20.6
9.5	-6	9/16-18	-6	9.5	17.5	6.6	72.9	—	94.7	—	88.9	—	—	9.9	21.6	55.4	25.7
9.5	-6	3/4-16	-8	12.7	22.2	6.6	75.9	—	103.6	—	96.8	—	—	14.0	27.7	61.7	25.7
12.7	-8	3/4-16	-8	12.7	22.2	9.7	80.8	—	107.2	—	100.1	—	—	14.0	27.7	61.7	31.2
12.7	-8	7/8-14	-10	15.9	25.4	9.7	83.6	—	115.8	—	106.7	—	—	16.0	31.2	65.3	31.2
15.9	-10	7/8-14	-10	15.9	25.4	10.9	89.7	—	118.4	—	111.8	—	—	16.0	31.2	65.3	36.8
19.0	-12	1-1/16-12	-12	19.0	31.8	14.5	91.7	—	120.9	—	117.1	—	—	19.8	46.2	94.7	40.4
19.0	-12	1-3/16-12	-14	22.2	34.9	14.5	92.2	—	—	—	—	—	—	—	—	—	40.4
25.4	-16	1-5/16-12	-16	25.4	38.1	19.8	105.7	—	140.2	—	133.4	—	—	22.6	54.4	110.0	49.5

[a] For dimensions shown in Figures 4 to 7, but not specified in Table 4H, see corresponding dimensions for respective straight thread size in Figure 8C and Table 2 of SAE J514.
[b] Sleeves shown in Figure 5B shall conform with Figure 14 and Table 4 of SAE J514 for respective tube OD referenced above except on some sizes and shapes where sleeve length must be shortened.
[c] Flares shall conform with single or double 37-degree flares specified in Figure 3 and Table 2 of SAE J533 for respective tube OD referenced above.
[d] Nuts shown in Figure 5B shall conform with Figure 12 and Table 3 of SAE J514 for respective straight thread size.
[e] For dimensions shown in Figures 5 to 7, but not specified in Table 4H, see corresponding dimensions for respective straight thread size in Figures 11 and 13, and Table 3 of SAE J514.

TABLE 4I—DIMENSIONS OF FEMALE 37-DEGREE FLARED TYPE HOSE FITTINGS FOR USE ON SAE 100R9 HYDRAULIC HOSE[a] (FIGURES 4 TO 7) (CODES 50 AND 75)

Nominal SAE 100R9 Hose ID mm	Hose Dash Size	B Nominal Straight Thread Size	Thread and Tube Dash Size	B_1 Tube OD[b,c] Ref mm	C Min mm	D Dia Min mm	L Max mm	L_1 Max mm	M Max mm	M_1 Max mm	M_2 Max mm	M_3 Min mm	N ±3.0 mm	N_1 ±3.0 mm	W Max mm
9.5	-6	9/16-18	-6	9.5	17.5	6.6	80.0	80.3	79.2	86.9	71.9	84.3	9.9	21.6	30.2
12.7	-8	3/4-16	-8	12.7	22.2	9.7	85.9	84.6	93.5	100.3	91.4	92.7	14.0	27.7	31.5
19.0	-12	1-1/16-12	-12	19.0	31.8	14.5	98.0	103.6	117.3	122.2	114.8	112.0	19.8	46.2	44.5
25.4	-16	1-5/16-12	-16	25.4	38.1	19.8	109.0	115.3	140.2	146.8	133.4	134.9	22.6	54.4	52.1
31.8	-20	1-5/8-12	-20	—	50.8	25.7	159.0	—	—	—	—	—	—	—	66.8
38.1	-24	1-7/8-12	-24	—	57.2	31.8	170.7	—	—	—	—	—	—	—	74.7
50.8	-32	2-1/2-12	-32	—	73.0	43.7	177.8	—	—	—	—	—	—	—	88.9

[a] For dimensions shown in Figures 4 to 7, but not specified in Table 4I, see corresponding dimensions for respective straight thread size in Figure 8C and Table 2 of SAE J514.
[b] Sleeves shown in Figure 5B shall conform with Figure 14 and Table 4 of SAE J514 for respective tube OD referenced above except on some sizes and shapes where sleeve length must be shortened.
[c] Flares shall conform with single or double 37-degree flares specified in Figure 3 and Table 2 of SAE J533 for respective tube OD referenced above.
[d] Nuts shown in Figure 5B shall conform with Figure 12 and Table 3 of SAE J514 for respective straight thread size.
[e] For dimensions shown in Figures 5 to 7, but not specified in Table 4I, see corresponding dimensions for respective straight thread size in Figures 11 and 13, and Table 3 of SAE J514.

TABLE 4J—DIMENSIONS OF FEMALE 37-DEGREE FLARED TYPE HOSE FITTINGS FOR USE ON SAE 100R10 HYDRAULIC HOSE[a] (FIGURES 4 TO 7) (CODES 51 AND 76)

Nominal SAE 100R10 Hose ID mm	Hose Dash Size	B Nominal Straight Thread Size	Thread and Tube Dash Size	B_1 Tube OD[b,c] Ref mm	C Min mm	D Dia Min mm	L Max mm	L_1 Max mm	L_2 Max mm	M Max mm
6.4	-4	7/16-20	-4	6.4	14.3	2.8	66.3	66.5	67.3	73.7
9.5	-6	9/16-18	-6	9.5	17.5	5.6	81.3	83.8	71.1	87.4
12.7	-8	3/4-16	-8	12.7	22.2	8.6	98.0	82.3	82.6	103.6
19.0	-12	1-1/16-12	-12	19.0	31.8	12.7	114.0	98.0	95.2	130.6
25.4	-16	1-5/16-12	-16	25.4	38.1	17.3	135.6	113.8	109.0	142.7
31.8	-20	1-5/8-12	-20	31.8	50.8	22.1	158.2	143.3	121.2	169.2
38.1	-24	1-7/8-12	-24	38.1	57.2	27.7	173.5	171.4	136.9	189.7
50.8	-32	2-1/2-12	-32	50.8	73.0	39.6	207.8	203.2	178.6	254.0

TABLE 4J—DIMENSIONS OF FEMALE 37-DEGREE FLARED TYPE HOSE FITTINGS FOR USE ON SAE 100R10 HYDRAULIC HOSE[a] (FIGURES 4 TO 7) (CODES 51 AND 76)—(CONTINUED)

Nominal SAE 100R10 Hose ID mm	Thread and Tube Dash Size	M_1 Max mm	M_2 Max mm	M_3 Max mm	M_4 Max mm	N mm	N_1 mm	N_2 mm	W Max mm	Z Max mm
6.4	-4	74.2	69.6	66.8	66.8	8.4	17.3	45.7	28.7	54.1
9.5	-6	84.6	82.8	77.0	77.0	9.9	21.6	55.4	28.7	62.7
12.7	-8	97.8	99.6	90.4	90.4	14.0	27.7	61.7	31.8	62.7
19.0	-12	115.1	134.4	113.0	113.0	19.8	46.2	94.7	42.9	79.2
25.4	-16	146.8	160.0	139.7	139.7	22.6	54.4	110.0	50.8	107.2
31.8	-20	188.5	172.2	171.4	171.4	43.2	86.4	134.1	62.2	126.0
38.1	-24	212.9	195.1	191.0	191.0	50.8	100.3	157.2	76.2	133.1
50.8	-32	285.8	285.8	285.8	285.8	69.8	139.7	222.2	81.8	166.6

[a] For dimensions shown in Figures 4 to 7, but not specified in Table 4J, see corresponding dimensions for respective straight thread size in Figure 8C and Table 2 of SAE J514.
[b] Sleeves shown in Figure 5B shall conform with Figure 14 and Table 4 of SAE J514 for respective tube OD referenced above except on some sizes and shapes where sleeve length must be shortened.
[c] Flares shall conform with single or double 37-degree flares specified in Figure 3 and Table 2 of SAE J533 for respective tube OD referenced above.
[d] Nuts shown in Figure 5B shall conform with Figure 12 and Table 3 of SAE J514 for respective straight thread size.
[e] For dimensions shown in Figures 5 to 7, but not specified in Table 4J, see corresponding dimensions for respective straight thread size in Figure 11C and Table 3 of SAE J514.

TABLE 4K—DIMENSIONS OF FEMALE 37-DEGREE FLARED TYPE HOSE FITTINGS FOR USE ON SAE 100R11 HYDRAULIC HOSE[a] (FIGURE 4A) (CODE 52)

Nominal SAE 100R11 Hose ID mm	Hose Dash Size	B Nominal Straight Thread Size	Thread and Tube Dash Size	C Nom mm	D Dia Min mm	L Max mm	W Max mm
19.0	-12	1-1/16-12	-12	31.8	13.5	115.8	46.2
25.4	-16	1-5/16-12	-16	38.1	18.5	140.7	55.9
31.8	-20	1-5/8-12	-20	50.8	24.1	153.7	66.8
38.1	-24	1-7/8-12	-24	57.2	30.5	171.2	74.7
50.8	-32	2-1/2-12	-32	73.0	41.9	207.0	88.9

[a] For dimensions shown in Figures 4 to 7, but not specified in Table 4K, see corresponding dimensions for respective straight thread size in Figure 8C and Table 2 of SAE J514.

(R) TABLE 4L—DIMENSIONS OF FEMALE 37-DEGREE FLARED TYPE HOSE FITTINGS FOR USE ON SAE 100R12 HYDRAULIC HOSE[a] (FIGURES 4A, 5A, 5B, AND 6A) (CODE 77)

Nominal SAE 100R12 Hose ID mm	Hose Dash Size	B Nominal Straight Thread Size	Thread and Tube Dash Size	C Nom mm	D Dia Min mm	L Max mm	M Max mm	M_2 Max mm	N mm	N_1 mm	W Max mm
9.5	-6	9/16-18	-6	17.5	5.6	81.3	87.4	82.8	9.9	21.6	28.7
12.7	-8	3/4-16	-8	22.2	8.6	98.0	109.5	99.6	14.0	27.7	31.8
15.9	-10	7/8-14	-10	25.4	10.0	101.6	—	—	—	—	38.0
19.0	-12	1-1/16-12	-12	31.8	12.7	114.0	136.4	136.7	30.5	61.0	42.9
25.4	-16	1-5/16-12	-16	38.1	17.3	135.6	159.0	160.0	38.9	76.2	50.8
31.8	-20	1-5/8-12	-20	50.8	22.1	158.5	207.0	173.5	43.2	88.9	60.5
38.1	-24	1-7/8-12	-24	57.2	27.7	173.5	232.9	209.6	50.8	101.6	67.8
50.8	-32	2-1/2-12	-32	73.0	39.6	207.8	274.8	248.4	69.8	139.7	81.8

[a] For dimensions shown in Figures 4 to 7, but not specified in Table 4L, see corresponding dimensions for respective straight thread size in Figure 8C and Table 2 of SAE J514.

TABLE 4M—DIMENSIONS OF FEMALE 37-DEGREE FLARED TYPE HOSE FITTINGS FOR USE ON SAE 100R13 HYDRAULIC HOSE[a] (FIGURES 4 AND 6) (CODE 78)

Nominal SAE 100R13 Hose ID mm	Hose Dash Size	B Nominal Straight Thread Size	Thread and Tube Dash Size	C Min mm	D Dia Min mm	L Max mm	M Max mm	M_2 Max mm	N mm	N_1 mm	W Max mm
19.0	-12	1-1/16-12	-12	31.8	12.7	113.5	171.2	136.7	30.5	61.0	43.7
25.4	-16	1-5/16-12	-16	38.1	17.5	126.2	153.7	153.9	38.9	76.2	50.8
31.8	-20	1-5/8-12	-20	50.8	22.4	158.5	190.5	173.5	43.2	88.9	63.5
38.1	-24	1-7/8-12	-24	57.2	27.7	160.0	232.9	209.6	50.8	101.6	71.1
50.8	-32	2-1/2-12	-32	73.0	38.1	198.6	274.8	248.4	69.8	139.7	88.9

[a] For dimensions shown in Figures 4 to 7, but not specified in Table 4M, see corresponding dimensions for respective straight thread size in Figure 8C and Table 2 of SAE J514.

MALE 45-DEGREE FLARED TYPE

NOTES: UNSPECIFIED DETAIL WITH RESPECT TO DIMENSIONS, TOLERANCES, CONTOURS, MATERIAL, WORKMANSHIP, AND SO ON, MUST CONFORM TO GENERAL SPECIFICATIONS FOR HYDRAULIC HOSE FITTINGS. THE DIMENSIONAL DESIGNATIONS ON THE FIRST FIGURE IN EACH GROUP SHALL APPLY TO ALL OTHER FIGURES IN THAT GROUP EXCEPT AS SHOWN OTHERWISE. CODES SHOWN IN BRACKETS ADJACENT TO FIGURE NUMBERS REPRESENT RESPECTIVE FITTING IDENTIFICATION, WITH XX SUBSITITUTED FOR THE HOSE TYPE CODE DEPICTED IN BRACKETS AT END OF RESPECTIVE TABLE TITLES, IN ACCORDANCE WITH SAE J846.

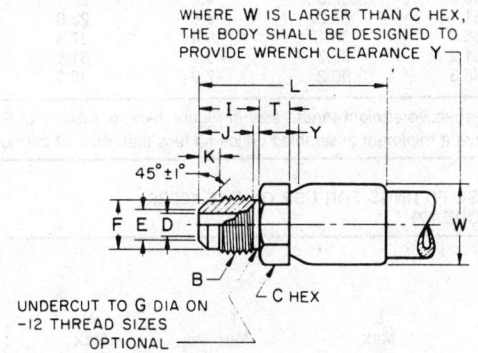

FIGURE 8A—PERMANENTLY ATTACHED STYLE (2801XX)

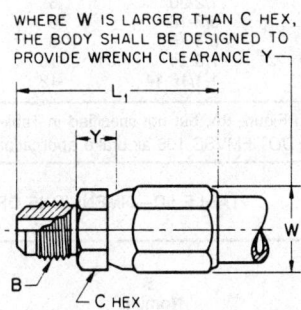

FIGURE 8B—FIELD ATTACHABLE SCREW STYLE (2901XX)

FIGURE 8—MALE 45-DEGREE FLARED TYPE HOSE FITTINGS

TABLE 5A—DIMENSIONS OF MALE 45-DEGREE FLARED TYPE HOSE FITTINGS FOR USE ON SAE 100R1 HYDRAULIC HOSE[a] (FIGURES 8A AND 8B) (CODES 42 AND 53)

Nominal SAE 100R1 Hose ID mm	Hose Dash Size	B Nominal Straight Thread Size	Thread Dash Size	C Min mm	D Dia Min mm	L Max mm	L_1 Max mm	T Max mm	W Max mm	Y Min mm
6.4	-4	7/16-20	-4	11.1	2.8	60.2	60.2	5.6	22.1	9.4
6.4	-4	1/2-20	-5	12.7	2.8	60.2	61.5	5.6	22.1	10.4
7.9	-5	9/16-18	-6	15.9	5.1	62.0	64.0	5.6	23.9	10.4
9.5	-6	9/16-18	-6	15.9	6.6	63.5	66.8	5.6	27.7	10.4
9.5	-6	3/4-16	-8	19.0	6.6	72.4	66.8	6.4	27.7	12.7
10.3	-6.5	3/4-16	-8	19.0	6.6	72.4	69.8	6.4	29.5	12.7
12.7	-8	3/4-16	-8	19.0	9.7	72.4	78.2	6.4	31.2	12.7
12.7	-8	7/8-14	-10	22.2	9.7	75.2	81.3	6.4	31.2	14.2
15.9	-10	7/8-14	-10	22.2	12.7	86.9	90.4	6.4	34.8	14.2
15.9	-10	1-1/16-12	-12	27.0	12.7	81.3	91.4	7.9	34.8	15.5
19.0	-12	7/8-14	-10	22.2	12.2	81.3	91.4	7.9	40.4	14.2
19.0	-12	1-1/16-12	-12	27.0	14.5	81.3	91.4	7.9	40.4	15.5

[a] For dimensions shown in Figure 8A, but not specified in Table 5A, see corresponding dimensions for respective straight thread size in Figure 1A and Table 2 of SAE J512.

TABLE 5B—DIMENSIONS OF MALE 45-DEGREE FLARED TYPE HOSE FITTINGS FOR USE ON SAE 100R3 HYDRAULIC HOSE[a] (FIGURE 8A) (CODE 44)

Nominal SAE 100R3 Hose ID mm	Hose Dash Size	B Nominal Straight Thread Size	Thread Dash Size	C Min mm	D Dia Min mm	L Max mm	T Min mm	W Max mm	Y Min mm
6.4	-4	7/16-20	-4	11.1	2.8	47.8	5.6	22.4	9.4
6.4	-4	1/2-20	-5	12.7	2.8	47.8	5.6	22.4	10.4
9.5	-6	9/16-18	-6	15.9	6.6	58.7	5.6	26.9	10.4
9.5	-6	3/4-16	-8	19.0	6.6	60.5	6.4	26.9	12.7
12.7	-8	3/4-16	-8	19.0	9.7	63.5	6.4	33.3	12.7
12.7	-8	7/8-14	-10	22.2	9.7	66.8	6.4	33.3	14.2
15.9	-10	7/8-14	-10	22.2	12.2	86.9	6.4	36.8	14.2
19.0	-12	7/8-14	-10	22.2	12.2	71.4	7.9	42.2	14.2
19.0	-12	1-1/16-12	-12	27.0	14.5	71.4	7.9	42.2	15.5

[a] For dimensions shown in Figure 8A, but not specified in Table 5B, see corresponding dimensions for respective straight thread size in Figure 1A and Table 2 of SAE J512.

TABLE 5C—DIMENSIONS OF MALE 45-DEGREE FLARED TYPE HOSE FITTINGS FOR USE ON SAE 100R5 HYDRAULIC HOSE[a] (FIGURES 5A TO 8B) (CODE 46)

Nominal SAE 100R5 Hose ID mm	Hose Dash Size	B Nominal Straight Thread Size	Thread Dash Size	C Min mm	D Dia Min mm	L Max mm	L_1 Max mm	T Min mm	W Max mm	Y Min mm
4.8	-4	7/16-20	-4	11.1	2.3	46.2	50.8	4.8	20.1	9.4
6.4	-5	1/2-20	-5	12.7	2.8	48.5	52.3	4.8	22.1	10.4
7.9	-6	5/8-18	-6	15.9	5.1	51.1	57.2	5.6	23.9	10.4
10.3	-8	3/4-16	-8	19.0	7.1	56.6	69.1	6.4	27.4	12.7
12.7	-10	7/8-14	-10	22.2	9.7	61.2	77.7	6.4	31.5	14.2
15.9	-12	1-1/16-14	-12	27.0	12.7	65.3	90.2	7.9	38.6	15.5

[a] For dimensions shown in Figure 8A, but not specified in Table 5C, see corresponding dimensions for respective straight thread size in Figure 1A and Table 2 of SAE J512.
NOTE: For SAE J1402 and DOT FMVSS 106 air brake applications, couplings used in 100R5 hose shall have a minimum assembled orifice no less than 66% of the nominal hose I.D.

TABLE 5D—DIMENSIONS OF MALE 45-DEGREE FLARED TYPE HOSE FITTINGS FOR USE ON SAE 100R6 HYDRAULIC HOSE[a] (FIGURE 8A) (CODE 47)

Nominal SAE 100R6 Hose ID mm	Hose Dash Size	B Nominal Straight Thread Size	Thread Dash Size	C Min mm	D Dia Min mm	L Max mm	T Min mm	W Max mm	Y Min mm
4.8	-3	3/8-24	-3	9.5	2.3	31.8	3.8	16.0	8.9
4.8	-3	7/16-20	-4	11.1	2.3	31.8	3.8	16.0	8.9
6.4	-4	3/8-24	-3	9.5	2.8	34.3	3.8	19.0	8.9
6.4	-4	7/16-20	-4	11.1	2.8	35.1	3.8	19.0	8.9
6.4	-4	1/2-20	-5	12.7	2.8	35.1	3.8	19.0	8.9
7.9	-5	1/2-20	-5	12.7	5.1	35.6	3.8	22.4	8.9
7.9	-5	5/8-18	-6	15.9	5.1	37.3	4.8	22.4	10.2
9.5	-6	5/8-18	-6	15.9	6.6	40.6	4.8	25.4	10.2
9.5	-6	3/4-16	-8	19.0	6.6	43.7	4.8	25.4	10.9
12.7	-8	3/4-16	-8	19.0	9.7	45.2	4.8	28.4	10.9
15.9	-10	7/8-14	-10	22.2	12.2	50.8	6.4	30.2	12.7
19.0	-12	1-1/16-14	-12	27.0	14.5	88.9	7.9	36.8	15.5

[a] For dimensions shown in Figure 8A, but not specified in Table 5D, see corresponding dimensions for respective straight thread size in Figure 1A and Table 2 of SAE J512.

TABLE 5E—DIMENSIONS OF MALE 45-DEGREE FLARED TYPE HOSE FITTINGS FOR USE ON SAE 100R7 HYDRAULIC HOSE[a] (FIGURE 8A) (CODE 48)

Nominal SAE 100R7 Hose ID mm	Hose Dash Size	B Nominal Straight Thread Size	Thread Dash Size	C Min mm	D Dia Min mm	L Max mm	T Min mm	W Max mm	Y Min mm
4.8	-3	7/16-20	-4	11.1	2.3	61.7	5.6	19.0	9.4
6.4	-4	7/16-20	-4	11.1	2.8	61.7	5.6	20.6	9.4
6.4	-4	1/2-20	-5	12.7	2.8	63.5	5.6	20.6	10.4
7.9	-5	5/8-18	-6	15.9	5.1	66.0	5.6	22.4	10.4
9.5	-6	5/8-18	-6	15.9	6.6	69.6	5.6	25.7	10.4
9.5	-6	3/4-16	-8	19.0	6.6	71.1	6.4	25.7	12.7
12.7	-8	3/4-16	-8	19.0	9.1	79.2	6.4	31.2	12.7
12.7	-8	7/8-14	-10	22.2	9.4	79.2	6.4	31.2	14.2
15.9	-10	7/8-14	-10	22.2	12.2	86.9	6.4	34.8	14.2
19.0	-12	1-1/16-14	-12	27.0	14.5	93.2	7.9	40.4	15.5

[a] For dimensions shown in Figure 8A, but not specified in Table 5E, see corresponding dimensions for respective straight thread size in Figure 1A and Table 2 of SAE J512.

FEMALE 45-DEGREE FLARED TYPE

NOTES: UNSPECIFIED DETAIL WITH RESPECT TO DIMENSIONS, TOLERANCES, CONTOURS, MATERIAL, WORKMANSHIP, AND SO ON, MUST CONFORM TO GENERAL SPECIFICATIONS FOR HYDRAULIC HOSE FITTINGS. THE DIMENSIONAL DESIGNATIONS ON THE FIRST FIGURE IN EACH GROUP SHALL APPLY TO ALL OTHER FIGURES IN THAT GROUP EXCEPT AS SHOWN OTHERWISE. CODES SHOWN IN BRACKETS ADJACENT TO FIGURE NUMBERS REPRESENT RESPECTIVE FITTING IDENTIFICATION, WITH XX SUBSITITUTED FOR THE HOSE TYPE CODE DEPICTED IN BRACKETS AT END OF RESPECTIVE TABLE TITLES, IN ACCORDANCE WITH SAE J846.

SWIVEL ENDS ON ANGLE FITTINGS MAY HAVE A MACHINED FLARE SEAT AND SWIVEL NUT AS SHOWN IN FIGURE 9A, A RADIAL FLARED TUBE AND MATING NUT SEAT AS SHOWN IN FIGURE 10A, OR A 45-DEGREE FLARED TUBE AND MATING NUT SEAT AS SHOWN IN FIGURE 10B.

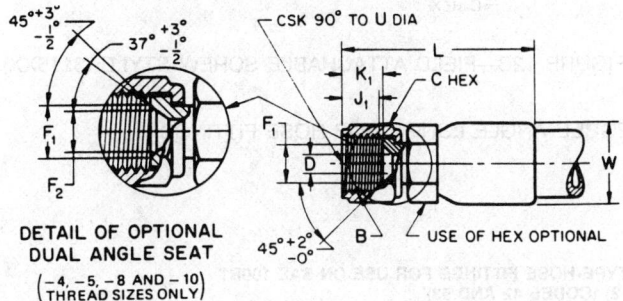

FIGURE 9A—PERMANENTLY ATTACHED STYLE (3001XX)

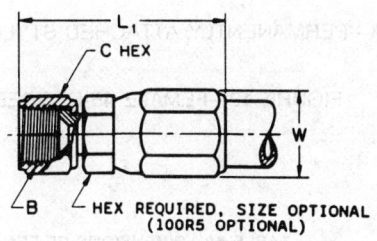

FIGURE 9B—FIELD ATTACHABLE SCREW STYLE (3101XX)

FIGURE 9—FEMALE 45-DEGREE FLARED TYPE HOSE FITTINGS

SWIVEL ENDS ON ANGLE FITTINGS MAY HAVE A MACHINED FLARE SEAT AND SWIVEL NUT AS SHOWN IN FIG. 9A, OR A 45-DEG FLARED TUBE AND MATING NUT SEAT AS SHOWN IN FIG. 10A.

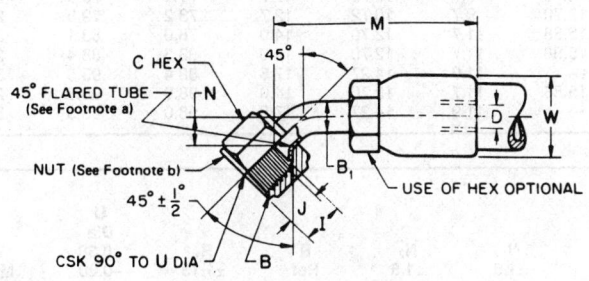

FIGURE 10A—PERMANENTLY ATTACHED STYLE (3003XX)

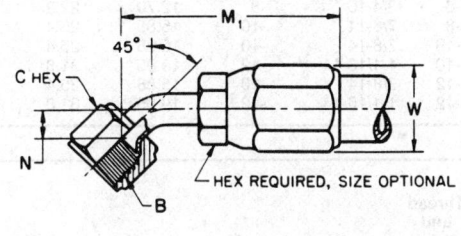

FIGURE 10B—FIELD ATTACHABLE SCREW STYLE (3103XX)

FIGURE 10—FEMALE 45-DEGREE FLARED TYPE 45-DEGREE ANGLE HOSE FITTINGS

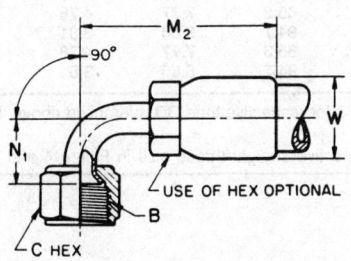

FIGURE 11A—PERMANENTLY ATTACHED STYLE (3014XX)

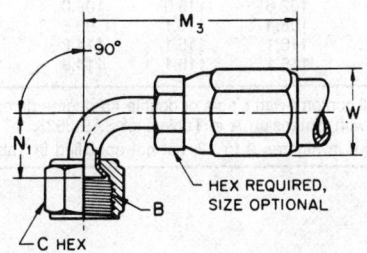

FIGURE 11B—FIELD ATTACHABLE SCREW STYLE (3114XX)

FIGURE 11—FEMALE 45-DEGREE FLARED TYPE 90-DEGREE ANGLE SHORT DROP HOSE FITTINGS

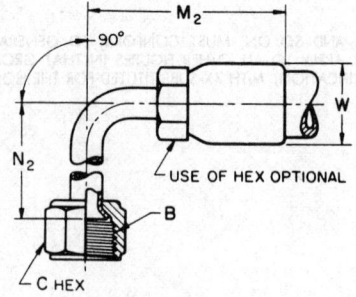

FIGURE 12A—PERMANENTLY ATTACHED STYLE (3015XX)

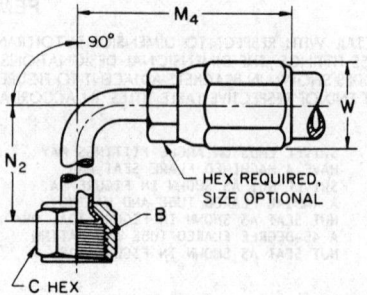

FIGURE 12B—FIELD ATTACHABLE SCREW STYLE (3115XX)

FIGURE 12—FEMALE 45-DEGREE FLARED TYPE 90-DEGREE ANGLE LONG DROP HOSE FITTINGS

TABLE 6A—DIMENSIONS OF FEMALE 45-DEGREE FLARED TYPE HOSE FITTINGS FOR USE ON SAE 100R1 HYDRAULIC HOSE (FIGURES 9 TO 12) (CODES 42 AND 53)

Nominal SAE 100R1 Hose ID mm	Hose Dash Size	Nominal Straight Thread Size	Thread and Tube Dash Size	B_1 Tube OD[a] Ref mm	C mm	D Min mm	F_1 Dia ±0.25 mm	F_2 Dia ±0.13 mm	J_1 Full Thread Min mm	K_1 +0.38 -0.00 mm	K_4 Ref mm	L Max mm	L_1 Max mm	L_4 ±0.03 mm
6.4	-4	7/16-20	-4	6.35	14.3	2.8	7.49	6.35	7.6	8.74	10.2	66.3	66.5	15.5
6.4	-4	1/2-20	-5	7.92	15.9	2.8	9.07	7.92	8.4	9.52	10.2	67.8	66.5	15.7
7.9	-5	5/8-18	-6	9.52	19.0	5.1	11.94	—	9.1	10.72	9.7	65.3	67.6	18.5
9.5	-6	5/8-18	-6	9.52	19.0	6.6	11.94	—	9.1	10.72	9.7	68.3	70.6	18.5
9.5	-6	3/4-16	-8	12.70	22.2	6.6	14.53	12.70	9.7	10.72	12.7	71.4	72.4	20.8
10.3	-6.5	3/4-16	-8	12.70	22.2	6.6	14.53	12.70	9.7	10.72	12.7	71.4	73.9	20.8
12.7	-8	3/4-16	-8	12.70	22.2	9.7	14.53	12.70	9.7	10.72	12.7	73.2	79.5	20.8
12.7	-8	7/8-14	-10	15.88	25.4	9.7	17.27	15.88	11.7	12.70	14.0	78.0	83.1	23.6
15.9	-10	7/8-14	-10	15.88	25.4	10.9	17.27	15.88	11.7	12.70	14.0	83.3	88.4	23.6
15.9	-10	1-1/16-14	-12	19.05	31.8	12.7	21.59	—	11.9	14.27	17.5	88.4	93.5	33.0
19.0	-12	7/8-14	-10	15.88	25.4	12.2	17.27	15.88	11.7	12.70	14.0	98.0	93.5	23.6
19.0	-12	1-1/16-14	-12	19.05	31.8	14.5	21.59	—	11.9	14.27	17.5	98.0	93.5	33.0

Nominal SAE 100R1 Hose ID mm	Thread and Tube Dash Size	M Max mm	M_1 Max mm	M_2 Max mm	M_3 Max mm	M_4 Max mm	N ±1.5 mm	N_1 ±1.5 mm	N_2 ±1.5 mm	R Ref mm	R_1 ±1.13 mm	U Dia +0.38 -0.00 mm	W Max mm
6.4	-4	67.6	74.2	64.5	66.8	66.8	8.4	17.3	45.7	5.21	3.18	11.25	22.1
6.4	-5	68.8	75.4	64.5	67.8	67.8	9.1	19.6	45.7	6.02	3.18	12.83	22.1
7.9	-6	74.7	82.8	68.3	73.9	73.9	9.9	21.6	55.4	5.44	3.18	16.00	23.9
9.5	-6	76.2	84.6	69.8	77.0	77.0	9.9	21.6	55.4	5.44	3.18	16.00	27.7
9.5	-8	85.9	90.2	77.7	83.6	83.6	14.0	27.7	61.7	5.23	3.18	19.18	27.7
10.3	-8	85.9	90.2	77.7	83.6	83.6	14.0	27.7	61.7	5.23	3.18	19.18	29.5
12.7	-8	85.9	97.8	77.7	90.4	90.4	14.0	27.7	61.7	5.23	3.18	19.18	31.2
12.7	-10	92.2	104.6	82.6	96.0	96.0	16.0	31.2	65.3	7.77	4.78	22.35	31.2
15.9	-10	102.6	113.0	108.0	102.9	102.9	16.0	31.2	65.3	7.77	4.78	22.35	34.8
15.9	-12	106.4	115.1	96.5	107.2	107.0	19.8	46.2	94.7	6.60	3.81	27.10	34.8
19.0	-10	116.1	115.1	114.8	108.7	108.7	16.0	31.2	65.3	7.77	4.78	22.35	40.4
19.0	-12	116.1	115.1	114.8	108.7	108.7	19.8	46.2	94.7	6.60	3.81	27.10	40.4

[a] 45-degree flares shall conform with single or double 45-degree flares specified in Figure 2 and Table 1 of SAE J533 for respective tube OD referenced above. Dimension F3 of radial flares shall conform with dimension A in Table 1 of SAE J533.

[b] For dimensions shown in Figures 9 to 12, but not specified in Table 6A, see corresponding dimensions for respective straight thread size in Figure 4 and Table 3 of SAE J512.

TABLE 6B—DIMENSIONS OF FEMALE 45-DEGREE HOSE FITTINGS FOR USE ON SAE 100R3 HYDRAULIC HOSE (FIGURES 9 TO 12) (CODE 44)

Nominal SAE 100R3 Hose ID mm	Hose Dash Size	B Nominal Straight Thread Size	Thread Dash Size	C Min mm	D Dia Min mm	F_1 Dia ±0.25 mm	F_2 Dia ±0.13 mm	J_1 Full Thread Min mm	K_1 +0.38 -0.00 mm	L Max mm	U Dia +0.38 -0.00 mm	W Max mm
6.4	-4	7/16-20	-4	14.3	2.8	7.49	6.35	7.6	8.74	52.3	11.25	22.4
6.4	-4	1/2-20	-5	15.9	2.8	9.07	7.92	8.4	9.52	52.3	12.83	22.4
9.5	-6	5/8-18	-6	19.0	6.6	11.94	—	9.7	10.72	61.2	16.00	26.9
9.5	-6	3/4-16	-8	22.2	6.6	14.53	12.70	9.7	10.72	66.8	19.18	26.9
12.7	-8	3/4-16	-8	22.2	9.7	14.53	12.70	9.7	10.72	68.6	19.18	33.3
12.7	-8	7/8-14	-10	25.4	9.7	17.27	15.88	11.7	12.70	73.2	22.35	33.3
15.9	-10	7/8-14	-10	25.4	10.9	17.27	15.88	11.7	12.70	74.9	22.35	38.1
19.0	-12	7/8-14	-10	25.4	12.2	17.27	15.88	11.7	12.70	77.0	22.35	42.2
19.0	-12	1-1/16-14	-12	31.8	14.5	21.59	—	11.9	14.27	77.0	27.10	42.2

[a] 45-degree flares shall conform with single or double 45-degree flares specified in Figure 2 and Table 1 of SAE J533 for respective tube OD referenced above. Dimension F3 of radial flares shall conform with dimension A in Table 1 of SAE J533.
[b] For dimensions shown in Figures 9 to 12, but not specified in Table 6B, see corresponding dimensions for respective straight thread size in Figure 4 and Table 3 of SAE J512.

TABLE 6C—DIMENSIONS OF FEMALE 45-DEGREE FLARED TYPE HOSE FITTINGS FOR USE ON SAE 100R5 HYDRAULIC HOSE (FIGURES 9 TO 12) (CODE 46)

Nominal SAE 100R5 Hose ID mm	Hose Dash Size	B Nominal Straight Thread Size	Thread and Tube Dash Size	B_1 Tube OD[a] Ref mm	C Min mm	D Dia Min mm	F_1 Dia ±0.25 mm	F_2 Dia ±0.13 mm	J_1 Full Thread Min mm	K_1 +0.38 -0.00 mm	K_4 Ref mm	L Max mm	L_1 Max mm	L_4 ±0.3 mm
4.8	-4	7/16-20	-4	6.35	14.3	2.3	7.49	6.35	7.6	8.74	10.2	53.1	53.8	15.5
6.4	-5	1/2-20	-5	7.92	15.9	2.8	9.07	7.92	8.4	9.52	10.2	56.1	55.4	15.7
7.9	-6	5/8-18	-6	9.52	19.0	5.1	11.94	—	9.7	10.72	9.7	57.9	60.5	18.5
10.3	-8	3/4-16	-8	12.70	22.2	7.1	14.53	12.70	9.7	10.72	12.7	64.0	73.2	20.8
12.7	-10	7/8-14	-10	15.88	25.4	9.7	17.27	15.88	11.7	12.70	14.0	67.6	80.3	23.6
15.9	-12	1-1/16-14	-12	19.05	31.8	12.7	21.59	—	11.9	14.27	17.5	70.4	90.9	33.0

NOTE: For SAE J1402 and DOT FMVSS 106 air brake applications, couplings used on 100R5 hose shall have a minimum assembled orifice no less than 66% of the nominal hose I.D.

TABLE 6C—(CONTINUED)

Nominal SAE 100R5 Hose ID mm	Thread Dash Size	M_1 Max mm	M_3 Max mm	M_4 Max mm	N ±0.15 mm	N_1 ±0.15 mm	N_2 ±0.15 mm	R Ref mm	R_1 ±0.13 mm	U Dia +0.38 -0.15 mm	W Max mm
4.8	-4	62.2	59.7	74.9	8.4	17.3	45.7	5.21	3.18	11.25	20.1
6.4	-5	66.0	64.5	64.5	9.1	19.6	48.0	6.02	3.18	12.83	22.1
7.9	-6	69.0	68.3	68.3	9.9	21.6	55.4	5.44	3.18	16.00	23.9
10.3	-8	84.8	81.0	81.0	14.0	25.4	61.7	5.23	3.18	19.18	27.4
12.7	-10	93.0	88.1	88.1	16.0	31.2	65.3	7.77	4.78	22.35	31.5
15.9	-12	110.2	109.0	109.0	19.8	46.2	94.7	6.60	3.81	27.10	38.6

[a] 45-degree flares shall conform with single or double 45-degree flares specified in Figure 2 and Table 1 of SAE J533 for respective tube OD referenced above. Dimension F3 of radial flares shall conform with dimension A in Table 1 of SAE J533.
[b] For dimensions shown in Figures 9 to 12, but not specified in Table 6C, see corresponding dimensions for respective straight thread size in Figure 4 and Table 3 of SAE J512.

TABLE 6D—DIMENSIONS OF FEMALE 45-DEGREE HOSE FITTINGS FOR USE ON SAE 100R6 HYDRAULIC HOSE (FIGURES 9 TO 12) (CODE 47)

Nominal SAE 100R6 Hose ID mm	Hose Dash Size	B Nominal Straight Thread Size	Thread Dash Size	C Min mm	D Dia Min mm	F_1 Dia ±0.013 mm	F_2 Dia ±0.013 mm	J_1 Full Thread Min mm	K_1 +0.38 -0.00 mm	L Max mm	L_1 Max mm	U Dia +0.38 -0.00 mm	W Max mm
4.8	-3	7/16-20	-4	14.3	2.3	7.49	6.35	7.6	8.74	36.1	—	11.25	16.0
6.4	-4	7/16-20	-4	14.3	2.8	7.49	6.35	7.6	8.74	38.1	39.6	11.25	19.0
6.4	-4	1/2-20	-5	15.9	2.8	9.07	7.92	8.4	9.52	38.1	—	12.83	19.0
7.9	-5	1/2-20	-5	15.9	2.8	9.07	7.92	8.4	9.52	40.1	42.9	12.83	22.4
7.9	-5	5/8-18	-6	19.0	5.1	11.94	—	9.7	10.72	40.6	—	16.00	22.4
9.5	-6	5/8-18	-6	19.0	6.6	11.94	—	9.7	10.72	45.2	46.7	16.00	25.4
12.7	-8	3/4-16	-8	22.2	9.7	14.53	12.70	9.7	10.72	46.7	47.8	19.18	28.4
15.9	-10	7/8-14	-10	25.4	12.2	17.27	15.88	11.7	12.70	56.6	59.9	22.35	30.2
19.0	-12	1-1/16-14	-12	31.8	14.5	21.59	—	11.9	14.27	90.4	88.1	27.10	36.8

[a] 45-degree flares shall conform with single or double 45-degree flares specified in Figure 2 and Table 1 of SAE J533 for respective tube OD referenced above. Dimension F3 of radial flares shall conform with dimension A in Table 1 of SAE J533.
[b] For dimensions shown in Figures 9 to 12, but not specified in Table 6D, see corresponding dimensions for respective straight thread size in Figure 4 and Table 3 of SAE J512.

TABLE 6E—DIMENSIONS OF FEMALE 45-DEGREE HOSE FITTINGS FOR USE ON SAE 100R7 HYDRAULIC HOSE[a] (FIGURES 9 TO 12) (CODE 48)

Nominal SAE 100R7 Hose ID mm	Hose Dash Size	B Nominal Straight Thread Size	Thread and Tube Dash Size	B_1 Tube OD[a] Ref mm	C Min mm	D Dia Min mm	F_1 Dia mm	F_2 Dia mm	J_1 Full Thread Min mm	K_1 mm	L Max mm	M Max mm	M_2 Max mm	N mm	N_1 mm	N_2 mm	U Dia mm	W Max mm
6.4	-4	7/16-20	-4	6.35	14.3	2.8	7.49	6.35	7.6	8.74	66.	87.6	82.0	8.4	17.3	45.7	11.25	20.6
6.4	-4	1/2-20	-5	7.92	15.9	2.8	9.07	7.92	8.4	9.52	66.5	88.9	83.3	9.1	19.6	45.7	12.83	20.6
7.9	-5	5/8-18	-6	9.52	19.0	5.1	11.94	—	9.1	10.72	68.6	90.2	84.1	9.9	21.6	55.4	16.00	22.4
9.5	-6	5/8-18	-6	9.52	19.0	6.6	11.94	—	9.1	10.72	74.2	94.7	88.9	9.9	21.6	55.4	16.00	25.7
9.5	-6	3/4-16	-8	12.70	22.2	6.6	14.53	12.70	9.7	10.72	76.2	103.6	96.8	14.0	27.7	61.7	19.18	25.7
12.7	-8	3/4-16	-8	12.70	22.2	9.1	14.53	12.70	9.7	10.72	80.8	107.2	100.1	14.0	27.7	61.7	19.18	31.2
12.7	-8	7/8-14	-10	15.88	25.4	9.1	17.27	15.88	11.7	12.70	83.6	115.8	106.7	16.	31.2	65.3	22.35	31.2
15.9	-10	7/8-14	-10	15.88	25.4	10.9	17.27	15.88	11.7	12.70	86.4	118.4	111.8	16.0	31.2	65.3	22.35	34.8
19.0	-12	1-1/16-14	-12	19.05	31.8	14.5	21.59	—	11.9	14.27	91.7	120.9	117.1	19.8	46.2	94.7	27.10	40.5

[a] Flares shall conform with single or double 45-degree flares specified in Figure 2 and Table 1 of SAE J533 for respective tube OD referenced above.
[b] For dimensions shown in Figures 10 to 12, but not specified in Table 6E, see corresponding dimensions for respective straight thread size in Figure 4 and Table 3 of SAE J512.

MALE FLARELESS TYPE

NOTES: UNSPECIFIED DETAIL WITH RESPECT TO DIMENSIONS, TOLERANCES, CONTOURS, MATERIAL, WORKMANSHIP, AND SO ON, MUST CONFORM TO GENERAL SPECIFICATIONS FOR HYDRAULIC HOSE FITTINGS. THE DIMENSIONAL DESIGNATIONS ON THE FIRST FIGURE IN EACH GROUP SHALL APPLY TO ALL OTHER FIGURES IN THAT GROUP EXCEPT AS SHOWN OTHERWISE. CODES SHOWN IN BRACKETS ADJACENT TO FIGURE NUMBERS REPRESENT RESPECTIVE FITTING IDENTIFICATION, WITH XX SUBSTITUTED FOR THE HOSE TYPE CODE DEPICTED IN BRACKETS AT END OF RESPECTIVE TABLE TITLES, IN ACCORDANCE WITH SAE J846.

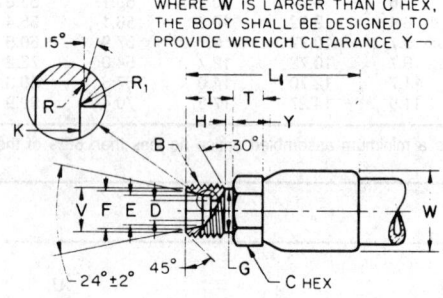

FIGURE 13A—PERMANENTLY ATTACHED STYLE (3301XX)

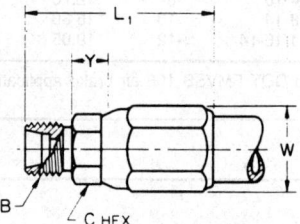

FIGURE 13B—FIELD ATTACHABLE SCREW STYLE (3401XX)

FIGURE 13—MALE FLARELESS TYPE HOSE FITTINGS

TABLE 7A—DIMENSIONS OF MALE FLARELESS TYPE HOSE FITTINGS FOR USE ON SAE 100R1 HYDRAULIC HOSE[a] (FIGURE 13) (CODES 42 AND 53)

Nominal SAE 100R1 Hose ID mm	Hose Dash Size	B Nominal Straight Thread Size	Thread Dash Size	C Min mm	D Dia Min mm	L Max mm	L_1 Max mm	T Min mm	W Max mm	Y Min mm
6.4	-4	7/16-20	-4	11.1	2.8	60.5	57.9	5.6	22.1	9.4
7.9	-5	9/16-18	-6	14.3	5.1	59.9	59.9	5.6	23.9	10.4
9.5	-6	9/16-18	-6	14.3	6.6	61.7	61.5	5.6	27.7	10.4
9.5	-6	3/4-16	-8	19.0	6.6	64.5	64.5	5.6	27.7	12.7
10.3	-6.5	3/4-16	-8	19.0	6.6	64.5	64.5	6.4	29.5	12.7
12.7	-8	3/4-16	-8	19.0	9.7	66.3	73.9	6.4	31.2	12.7
12.7	-8	7/8-14	-10	22.2	9.7	67.8	75.4	6.4	31.2	14.2
15.9	-10	7/8-14	-10	22.2	12.2	77.2	82.3	6.4	34.8	14.2
15.9	-10	1-1/16-12	-12	31.8	12.7	77.0	88.1	7.9	34.8	15.5
19.0	-12	7/8-14	-10	22.2	12.2	90.2	88.1	7.9	40.4	14.2
19.0	-12	1-1/16-12	-12	31.8	14.5	90.2	88.1	7.9	40.4	15.5
25.4	-16	1-5/16-12	-16	38.1	19.8	91.2	101.6	9.7	51.3	17.5

[a] For dimensions shown in Figure 13A, but not specified in Table 7, see corresponding dimensions for respective straight thread size in Figure 19A and Table 7 of SAE J514.

22.141

TABLE 7B—DIMENSIONS OF MALE FLARELESS TYPE HOSE FITTINGS FOR USE ON SAE 100R2 HYDRAULIC HOSE[a] (FIGURE 13) (CODES 43 AND 54)

Nominal SAE 100R2 Hose ID mm	Hose Dash Size	B Nominal Straight Thread Size	Thread Dash Size	C Min mm	D Dia Min mm	L Max mm	L_1 Max mm	T Min mm	W Max mm	Y Min mm
6.4	-4	7/16-20	-4	11.1	2.8	60.5	64.3	4.8	25.4	9.4
9.5	-6	9/16-18	-6	14.3	6.6	61.7	69.1	5.6	30.2	10.4
9.5	-6	3/4-16	-8	19.0	6.6	64.5	72.4	6.4	30.2	12.7
12.7	-8	3/4-16	-8	19.0	9.7	66.3	75.2	6.4	35.1	12.7
12.7	-8	7/8-14	-10	22.2	9.7	67.8	76.7	6.4	35.1	14.2
15.9	-10	7/8-14	-10	22.2	12.2	77.2	84.6	6.4	39.6	14.2
15.9	-10	1-1/16-12	-12	31.8	12.7	77.0	92.2	7.9	39.6	15.5
19.0	-12	7/8-14	-10	22.2	12.2	90.2	92.2	7.9	44.4	14.2
19.0	-12	1-1/16-12	-12	31.8	14.5	90.2	92.2	7.9	44.4	15.5
25.4	-16	1-5/16-12	-16	38.1	19.8	91.2	105.4	9.7	52.3	17.5

[a] For dimensions shown in Figure 13A, but not specified in Table 7B, see corresponding dimensions for respective straight thread size in Figure 19A and Table 7 of SAE J514.

FEMALE FLARELESS TYPE

NOTES: UNSPECIFIED DETAIL WITH RESPECT TO DIMENSIONS, TOLERANCES, CONTOURS, MATERIAL, WORKMANSHIP, AND SO ON, MUST CONFORM TO GENERAL SPECIFICATIONS FOR HYDRAULIC HOSE FITTINGS. THE DIMENSIONAL DESIGNATIONS ON THE FIRST FIGURE IN EACH GROUP SHALL APPLY TO ALL OTHER FIGURES IN THAT GROUP EXCEPT AS SHOWN OTHERWISE. CODES SHOWN IN BRACKETS ADJACENT TO FIGURE NUMBERS REPRESENT RESPECTIVE FITTING IDENTIFICATION, WITH XX SUBSTITUTED FOR THE HOSE TYPE CODE DEPICTED IN BRACKETS AT END OF RESPECTIVE TABLE TITLES, IN ACCORDANCE WITH SAE J846.

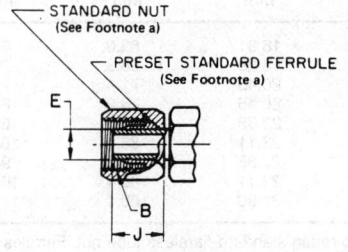

FIGURE 14A—OPTIONAL PRESET FERRULE END CONSTRUCTION

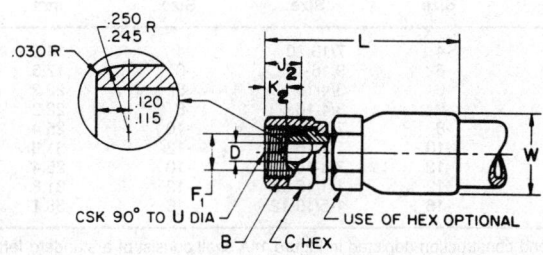

FIGURE 14B—PERMANENTLY ATTACHED STYLE (3601XX)

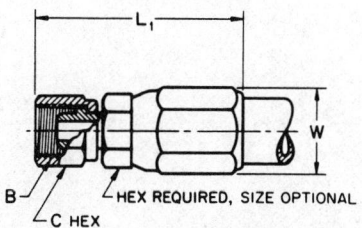

FIGURE 14C—FIELD ATTACHABLE SCREW STYLE (3701XX)

FIGURE 14—FEMALE FLARELESS TYPE HOSE FITTINGS

TABLE 8A—DIMENSIONS OF FEMALE FLARELESS TYPE HOSE FITTINGS FOR USE ON SAE 100R1 HYDRAULIC HOSE[b] (FIGURE 14) (CODES 42 AND 53)

Nominal SAE 100R1 Hose ID mm	Hose Dash Size	B Nominal Straight Thread Size	Thread Dash Size	C Min mm	D Dia Min mm	E Dia mm	J Ref mm	L Max mm	L₁ Max mm	W Max mm
6.4	-4	7/16-20	-4	14.3	2.8	6.35	16.61	61.5	66.5	22.1
7.9	-5	9/16-18	-6	17.5	5.1	9.52	18.29	62.3	67.6	23.9
9.5	-6	9/16-18	-6	17.5	6.6	9.52	18.29	68.3	70.6	27.7
9.5	-6	3/4-16	-8	22.2	6.6	12.70	20.45	71.4	73.9	27.7
10.3	-6.5	3/4-16	-8	22.2	6.6	12.70	20.45	71.4	73.9	29.5
12.7	-8	3/4-16	-8	22.2	9.7	12.70	20.45	73.2	82.3	31.2
12.7	-8	7/8-14	-10	25.4	9.7	15.88	22.35	78.0	83.1	31.2
15.9	-10	7/8-14	-10	25.4	12.2	15.88	22.35	89.4	87.6	34.8
15.9	-10	1-1/16-12	-12	31.8	12.7	19.05	23.11	88.4	92.2	34.8
19.0	-12	7/8-14	-10	25.4	12.2	15.88	22.35	88.4	92.2	40.4
19.0	-12	1-1/16-12	-12	31.8	14.5	19.05	23.11	88.4	92.2	40.4
25.4	-16	1-5/16-12	-16	38.1	19.8	25.40	27.30	100.1	108.7	51.3

[a] Optional end construction depicted in Figure 14A shall consist of a standard ferrule preset into a tubular nipple to retain standard flareless tube nut. Ferrules shall conform to Figures 28 or 29 and Table 8 of SAE J514 for nominal tube OD corresponding to E diameter above; and nuts shall conform to Figure 31 and Table 10 for corresponding straight thread size.
[b] For dimensions shown in Figure 14B, but not specified in Table 8A, see corresponding dimensions for respective straight thread size in Figure 25C and Table 7 of SAE J514.

TABLE 8B—DIMENSIONS OF FEMALE FLARELESS TYPE HOSE FITTINGS FOR USE ON SAE 100R2 HYDRAULIC HOSE[b] (FIGURE 14) (CODES 43 AND 54)

Nominal SAE 100R2 Hose ID mm	Hose Dash Size	B Nominal Straight Thread Size	Thread Dash Size	C Min mm	D Dia Min mm	E Dia mm	J Ref mm	L Max mm	L₁ Max mm	W Max mm
6.4	-4	7/16-20	-4	14.3	2.8	6.35	16.61	63.0	67.8	25.4
9.5	-6	9/16-18	-6	17.5	6.6	9.52	18.29	68.3	76.2	30.2
9.5	-6	3/4-16	-8	22.2	6.6	12.70	20.45	71.4	79.2	30.2
12.7	-8	3/4-16	-8	22.2	9.7	12.70	20.45	73.2	84.6	35.1
12.7	-8	7/8-14	-10	25.4	9.7	15.88	22.35	78.0	85.9	35.1
15.9	-10	1-1/16-12	-12	31.8	12.7	19.05	23.11	88.4	103.6	39.6
19.0	-12	7/8-14	-10	25.4	12.2	15.88	22.35	88.4	92.2	44.4
19.0	-12	1-1/16-12	-12	31.8	14.5	19.05	23.11	88.4	103.6	44.4
25.4	-16	1-5/16-12	-16	38.1	19.8	25.40	27.30	100.1	115.3	52.3

[a] Optional end construction depicted in Figure 14A shall consist of a standard ferrule preset into a tubular nipple to retain standard flareless tube nut. Ferrules shall conform to Figures 28 or 29 and Table 8 of SAE J514 for nominal tube OD corresponding to E diameter above; and nuts shall conform to Figure 31 and Table 10 for corresponding straight thread size.
[b] For dimensions shown in Figure 14B, but not specified in Table 8B, see corresponding dimensions for respective straight thread size in Figure 25C and Table 7 of SAE J514.

22.143

4-BOLT SPLIT FLANGE TYPE

NOTES: UNSPECIFIED DETAIL WITH RESPECT TO DIMENSIONS, TOLERANCES, CONTOURS, MATERIAL, WORKMANSHIP, AND SO ON, MUST CONFORM TO GENERAL SPECIFICATIONS FOR HYDRAULIC HOSE FITTINGS. THE DIMENSIONAL DESIGNATIONS ON THE FIRST FIGURE IN EACH GROUP SHALL APPLY TO ALL OTHER FIGURES IN THAT GROUP EXCEPT AS SHOWN OTHERWISE. CODES SHOWN IN BRACKETS ADJACENT TO FIGURE NUMBERS REPRESENT RESPECTIVE FITTING IDENTIFICATION, WITH XX SUBSTITUTED FOR THE HOSE TYPE CODE DEPICTED IN BRACKETS AT END OF RESPECTIVE TABLE TITLES, IN ACCORDANCE WITH SAE J846.

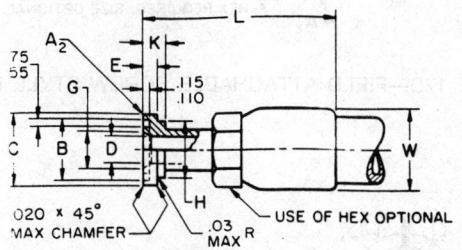

FIGURE 15A—PERMANENT ATTACHED STYLE (3901XX)

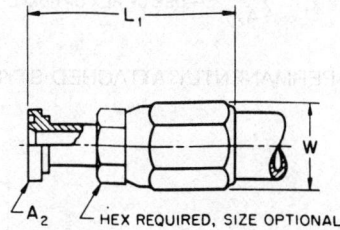

FIGURE 15B—FIELD ATTACHABLE SCREW STYLE (4001XX)

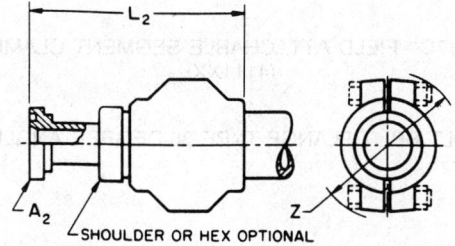

FIGURE 15C—FIELD ATTACHABLE SEGMENT CLAMP STYLE
(4101XX)

FIGURE 15—4-BOLT SPLIT FLANGE TYPE HOSE FITTINGS

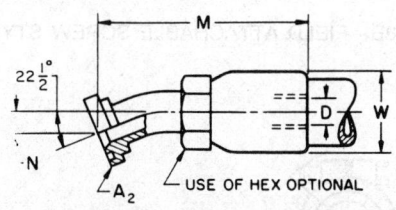

FIGURE 16A—PERMANENTLY ATTACHED STYLE (3910XX)

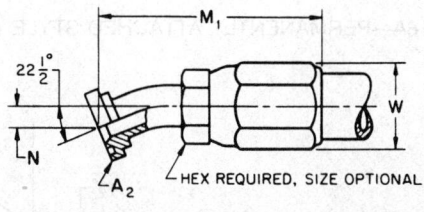

FIGURE 16B—FIELD ATTACHABLE SCREW STYLE (4010XX)

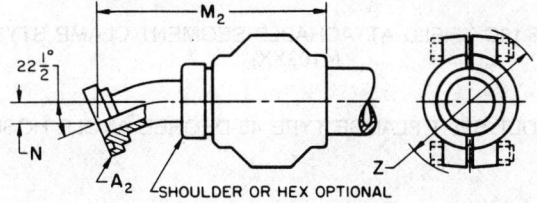

FIGURE 16C—FIELD ATTACHABLE SEGMENT CLAMP STYLE
(4110XX)

FIGURE 16—4-BOLT SPLIT FLANGE TYPE 22½-DEGREE ANGLE HOSE FITTINGS

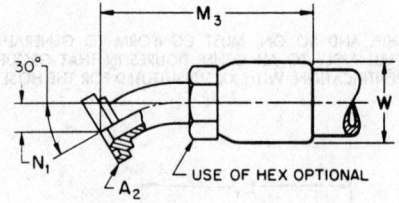

FIGURE 17A—PERMANENTLY ATTACHED STYLE (3911XX)

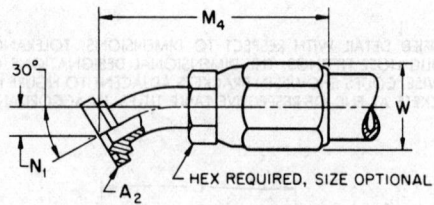

FIGURE 17B—FIELD ATTACHABLE SCREW STYLE (4011XX)

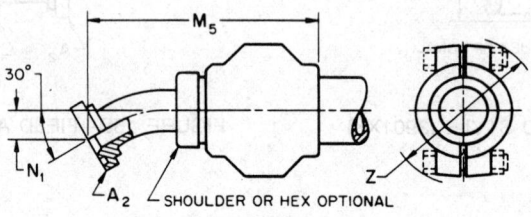

FIGURE 17C—FIELD ATTACHABLE SEGMENT CLAMP STYLE (4111XX)

FIGURE 17—4-BOLT SPLIT FLANGE TYPE 30-DEGREE ANGLE HOSE FITTINGS

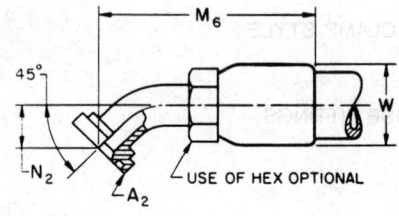

FIGURE 18A—PERMANENTLY ATTACHED STYLE (3903XX)

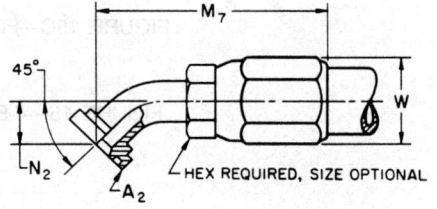

FIGURE 18B—FIELD ATTACHABLE SCREW STYLE (4003XX)

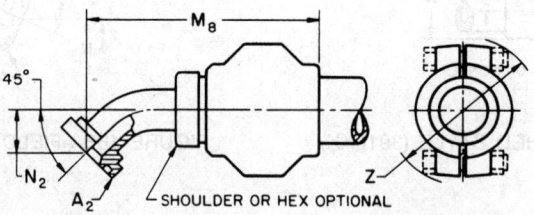

FIGURE 18C—FIELD ATTACHABLE SEGMENT CLAMP STYLE (4103XX)

FIGURE 18—4-BOLT SPLIT FLANGE TYPE 45-DEGREE ANGLE HOSE FITTINGS

22.145

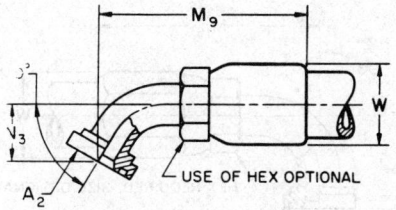

FIGURE 19A—PERMANENTLY ATTACHED STYLE (3912XX)

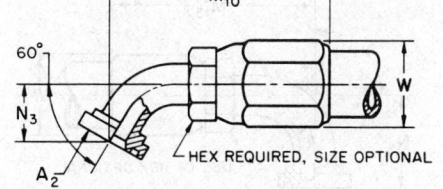

FIGURE 19B—FIELD ATTACHABLE SCREW STYLE (4012XX)

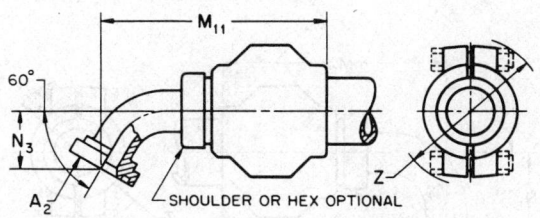

FIGURE 19C—FIELD ATTACHABLE SEGMENT CLAMP STYLE (4112XX)

FIGURE 19—4-BOLT SPLIT FLANGE TYPE 60-DEGREE ANGLE HOSE FITTINGS

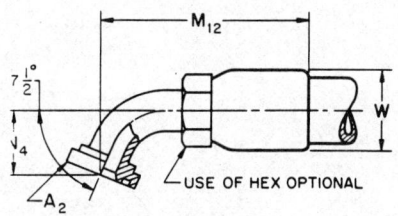

FIGURE 20A—PERMANENTLY ATTACHED STYLE (3913XX)

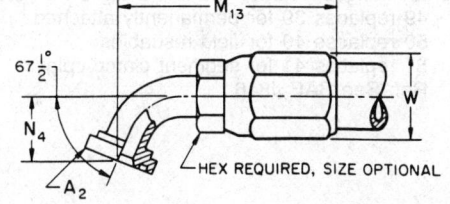

FIGURE 20B—FIELD ATTACHABLE SCREW STYLE (4013XX)

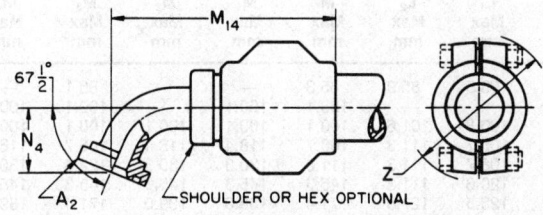

FIGURE 20C—FIELD ATTACHABLE SEGMENT CLAMP STYLE (4113XX)

FIGURE 20—4-BOLT SPLIT FLANGE TYPE 67½-DEGREE ANGLE HOSE FITTINGS

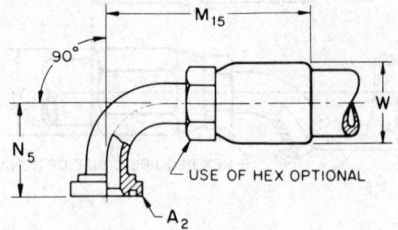

FIGURE 21A—PERMANENTLY ATTACHED STYLE (3902XX)

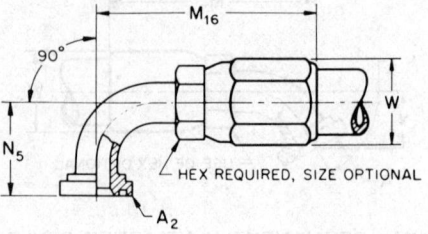

FIGURE 21B—FIELD ATTACHABLE SCREW STYLE (4002XX)

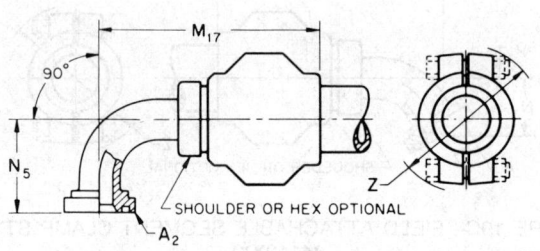

FIGURE 21C—FIELD ATTACHABLE SEGMENT CLAMP STYLE (4102XX)

FIGURE 21—4-BOLT SPLIT FLANGE TYPE 90-DEGREE ANGLE HOSE FITTINGS

NOTE: The Code Identification numbers shown in Figures 15A through 21C are Standard Pressure (Code 61) Split Flanges. For High Pressure (Code 62) flanges, replace the first 2 digits with the numbers shown below:
49 replaces 39 for permanently attached
50 replaces 40 for field resuables
51 replaces 41 for segment clamp cplgs.
Ref: See SAE J846

TABLE 9A—DIMENSIONS OF CODE 61 4-BOLT SPLIT FLANGE TYPE HOSE FITTINGS FOR USE ON SAE 100R1 HYDRAULIC HOSE[a] (FIGURES 15A TO 21C) (CODES 42 AND 53)

Nominal SAE 100R1 Hose ID mm	Hose Dash Size	A_2 Nominal Flange Size mm	Flange Dash Size	D Dia Min mm	L Max mm	L_1 Max mm	L_2 Max mm	M Max mm	M_1 Max mm	M_2 Max mm	M_3 Max mm	M_4 Max mm	M_5 Max mm	M_6 Max mm	M_7 Max mm	M_8 Max mm	M_9 Max mm
12.7	-8	12.7	-8	9.7	76.2	93.5	85.9	85.3	—	—	86.1	—	85.3	86.1	99.3	86.1	85.3
15.9	-10	19.0	-12	12.7	88.9	—	—	100.1	100.1	—	100.1	100.1	—	103.1	—	—	97.5
19.0	-12	19.0	-12	14.5	88.9	100.8	101.6	100.1	100.1	100.1	100.1	100.1	100.1	103.1	118.9	103.1	97.5
25.4	-16	25.4	-16	19.8	88.9	108.7	111.3	106.7	118.4	118.4	106.7	118.4	118.4	110.2	140.0	122.2	111.3
31.8	-20	31.8	-20	25.7	106.2	108.7	111.3	111.8	130.3	130.3	122.4	130.3	130.3	127.0	138.2	138.2	119.1
38.1	-24	38.1	-24	31.8	114.3	120.6	111.3	145.3	145.3	145.3	145.3	145.3	145.3	151.6	151.6	151.6	155.7
50.8	-32	50.8	-32	43.7	136.7	128.5	127.0	171.4	189.0	189.0	171.4	189.0	189.0	193.0	196.8	196.8	204.0

Nominal SAE 100R1 Hose ID mm	Flange Dash Size	M_{10} Max mm	M_{11} Max mm	M_{12} Max mm	M_{13} Max mm	M_{14} Max mm	M_{15} Max mm	M_{16} Max mm	M_{17} Max mm	N mm	N_1 mm	N_2 mm	N_3 mm	N_4 mm	N_5 mm	W Max mm	Z Max mm
12.7	-8	—	85.3	85.3	—	85.3	83.8	90.4	83.8	9.7	12.7	19.8	26.7	31.8	41.4	31.2	63.5
15.9	-12	103.9	—	97.5	103.9	—	93.7	100.1	—	—	9.7	13.5	21.3	29.5	33.3	54.1	34.8	—
19.0	-12	103.9	103.9	97.5	103.9	103.9	93.7	108.7	100.1	11.2	14.7	25.4	35.8	40.6	54.1	40.4	76.2
25.4	-16	124.0	124.0	111.3	124.0	124.0	108.0	132.1	120.6	11.2	15.7	26.9	38.1	44.4	60.5	51.3	91.2
31.8	-20	144.5	144.5	119.1	144.5	144.5	122.2	135.6	135.6	12.7	18.3	29.2	42.2	48.3	66.5	62.5	101.6
38.1	-24	161.0	161.0	155.7	161.0	161.0	155.7	152.4	152.4	16.0	22.4	35.8	50.8	50.8	79.2	69.5	114.3
50.8	-32	209.6	158.8	204.0	209.6	209.6	193.8	203.2	203.2	22.4	31.8	50.8	73.2	82.6	114.3	88.1	130.0

[a] For dimensions of flanged head shown in Figure 15A, but not specified in Table 9A, see corresponding dimensions for respective nominal flange size in Figure 3 and Table 1A of SAE J518.

TABLE 9B—DIMENSIONS OF CODE 61 4-BOLT SPLIT FLANGE TYPE HOSE FITTINGS FOR USE ON SAE 100R2 HYDRAULIC HOSE[a] (FIGURES 15A TO 21C) (CODES 43 AND 54)

Nominal SAE 100R2 Hose ID mm	Hose Dash Size	A_2 Nominal Flange Size mm	Flange Dash Size	D Dia Min mm	L Max mm	L_1 Max mm	L_2 Max mm	M Max mm	M_1 Max mm	M_2 Max mm	M_3 Max mm	M_4 Max mm	M_5 Max mm	M_6 Max mm	M_7 Max mm	M_8 Max mm	M_9 Max mm
12.7	-8	12.7	-8	9.7	91.7	96.0	88.9	88.9	96.0	87.9	88.9	—	90.4	91.9	101.6	91.2	91.9
12.7	-8	19.0	-12	9.7	82.6	98.6	88.9	90.4	—	90.4	90.4	—	91.9	91.9	—	92.7	91.9
15.9	-10	19.0	-12	12.7	70.6	—	—	111.3	—	—	113.8	—	—	116.1	—	—	116.8
19.0	-12	19.0	-12	14.5	96.8	108.0	93.0	101.6	112.8	102.6	104.6	115.3	105.2	106.7	118.9	108.7	108.0
19.0	-12	25.4	-16	14.5	86.6	108.0	93.0	101.6	116.8	106.7	104.6	119.4	109.2	106.7	121.9	111.8	108.0
22.2	-14	25.4	-16	18.3	82.6	—	—	—	—	—	—	—	—	—	—	—	—
25.4	-16	25.4	-16	19.8	108.7	120.4	102.4	111.8	123.4	117.9	115.3	121.9	121.9	119.9	138.4	125.5	120.6
25.4	-16	31.8	-20	19.8	110.2	120.4	109.7	113.3	123.4	120.9	117.3	125.0	125.0	120.6	138.4	128.0	121.4
31.8	-20	31.8	-20	25.7	136.7	135.9	130.3	153.9	142.7	142.7	158.8	154.4	148.3	166.6	164.3	151.9	171.4
31.8	-20	38.1	-24	25.7	136.7	135.9	130.3	153.9	—	—	144.3	158.8	—	148.3	166.6	—	151.9	171.4
38.1	-24	38.1	-24	31.8	155.4	142.5	134.9	177.3	165.6	165.6	180.3	172.7	172.7	186.2	177.8	177.8	187.5
38.1	-24	50.8	-32	31.8	155.4	142.5	134.9	177.3	—	165.6	180.3	—	172.7	186.2	—	177.8	187.5
50.8	-32	50.8	-32	43.7	138.2	161.8	162.6	185.9	206.2	165.6	193.0	213.4	221.0	198.1	218.4	226.1	210.3

Nominal SAE 100R2 Hose ID mm	Flange Dash Size	M_{10} Max mm	M_{11} Max mm	M_{12} Max mm	M_{13} Max mm	M_{14} Max mm	M_{15} Max mm	M_{16} Max mm	M_{17} Max mm	N mm	N_1 mm	N_2 mm	N_3 mm	N_4 mm	N_5 mm	W Max mm	Z Max mm
12.7	-8	93.0	91.9	89.4	93.0	90.4	85.9	92.7	88.9	9.7	12.7	19.8	26.7	31.8	41.4	35.1	66.6
12.7	-12	—	92.7	90.4	—	91.2	85.9	—	88.9	9.7	13.5	21.3	29.5	33.3	42.9	35.1	66.6
15.9	-12	—	—	115.8	—	—	113.0	—	—	9.7	13.5	21.3	29.5	33.3	54.1	39.6	—
19.0	-12	118.9	108.7	106.4	118.1	108.0	103.1	111.8	101.6	11.2	14.7	25.4	35.8	40.6	54.1	44.4	79.2
19.0	-16	121.9	111.8	106.4	120.6	110.5	103.1	114.3	104.1	11.2	14.7	25.4	35.8	40.6	54.1	44.4	79.2
22.2	-16	—	—	—	—	—	—	—	—	—	—	—	—	—	46.0	—
25.4	-16	132.1	129.3	122.9	129.6	128.6	117.3	130.6	124.5	11.2	15.7	26.9	38.1	44.4	60.5	52.3	91.4
25.4	-20	132.1	130.0	122.9	129.6	129.3	117.3	130.6	127.0	11.9	16.8	27.7	39.6	46.0	60.5	52.3	91.4
31.8	-20	154.9	154.9	171.4	152.4	152.4	165.1	160.8	152.4	12.7	18.3	29.2	42.2	48.3	66.5	66.8	114.3
31.8	-24	—	154.9	171.4	—	153.2	165.1	160.8	152.4	13.5	19.0	31.0	44.4	50.8	68.3	66.8	114.3
38.1	-24	179.3	179.3	185.9	176.8	176.8	185.4	178.1	175.3	16.0	22.4	35.8	50.8	50.8	79.2	74.7	127.0
38.1	-32	—	179.3	185.9	—	176.8	185.4	—	175.3	16.0	22.4	35.8	50.8	57.2	79.2	74.7	127.0
50.8	-32	230.6	238.3	207.8	228.1	235.7	198.1	218.9	226.1	22.4	31.8	50.8	73.2	82.6	114.3	88.9	149.4

[a] For dimensions of flanged head shown in Figure 15A, but not specified in Table 9B, see corresponding dimensions for respective nominal flange size in Figure 3 and Table 1A of SAE J518.

TABLE 9C—DIMENSIONS OF CODE 61 4-BOLT SPLIT FLANGE TYPE HOSE FITTINGS FOR USE ON SAE 100R4 HYDRAULIC HOSE[a] (FIGURES 15A, 15C, 16A, 16C, 17A, 18A, 18C, 19A, 19C, 20A, 20C, 21A AND 21C) (CODE 45)

Nominal SAE 100R4 Hose ID mm	Hose Dash Size	A_2 Nominal Flange Size mm	Flange Dash Size	D_1 Dia Min mm	L Min mm	L_2 Min mm	M Max mm	M_2 Max mm	M_3 Max mm	M_5 Max mm	M_6 Max mm	M_8 Max mm	M_9 Max mm
19.0	-12	19.0	-12	14.5	82.3	76.2	84.1	86.6	86.6	88.9	89.7	92.2	90.4
25.4	-16	25.4	-16	19.8	89.7	82.6	93.7	104.9	96.8	108.0	103.1	112.0	103.1
31.8	-20	31.8	-20	25.7	111.3	96.8	101.6	118.9	104.9	124.0	114.3	131.8	115.8
38.1	-24	38.1	-24	31.8	127.5	104.9	131.8	131.8	138.2	139.7	142.7	146.0	143.8
50.8	-32	50.8	-32	42.9	136.7	122.4	163.6	177.8	169.9	182.6	178.6	190.5	190.2
63.5	-40	63.5	-40	57.2	127.0	125.5	139.7	152.4	148.3	162.1	165.9	178.6	176.3
76.2	-48	76.2	-48	68.3	155.4	139.7	184.2	173.0	194.6	185.7	215.9	204.7	228.6

Nominal SAE 100R4 Hose ID mm	Flange Dash Size	M_{11} Max mm	M_{12} Max mm	M_{14} Max mm	M_{15} Max mm	M_{17} Max mm	N ±3.0 mm	N_1 ±3.0 mm	N_2 ±3.0 mm	N_3 ±3.0 mm	N_4 ±3.0 mm	N_5 ±3.0 mm	W Max mm	Z Max mm
19.0	-12	93.5	89.7	92.2	86.6	88.9	11.2	14.7	25.4	35.8	40.6	54.1	43.9	76.2
25.4	-16	114.3	102.4	112.8	100.1	111.3	11.2	15.7	26.9	38.1	44.4	60.5	51.1	82.6
31.8	-20	138.2	115.1	131.8	120.6	129.3	12.7	18.3	29.2	42.2	48.3	66.5	58.7	100.1
38.1	-24	155.7	144.3	152.4	134.1	146.0	16.0	22.4	35.8	50.8	50.8	79.2	66.0	110.2
50.8	-32	203.2	189.0	203.2	190.5	196.8	22.4	31.8	50.8	50.8	82.6	114.3	80.8	125.5
63.5	-40	189.0	179.3	192.0	177.8	190.5	15.7	23.9	43.7	66.5	79.2	117.3	99.8	139.7
76.2	-48	217.4	231.6	223.8	230.9	221.0	18.3	28.4	51.6	79.2	94.5	139.7	113.3	157.2

[a] For dimensions of flanged head shown on Figure 15A, but not specified in Table 9C, see corresponding dimensions for respective nominal flange size in Figure 3 and Table 1A of SAE J518.

TABLE 9D—DIMENSIONS OF CODE 61 4-BOLT SPLIT FLANGE TYPE HOSE FITTINGS FOR USE ON SAE 100R5 HYDRAULIC HOSE[a] (FIGURES 15B, 16B, 17B, 18B, 19B AND 21B) (CODE 46)

Nominal SAE 100R5 Hose ID mm	Hose Dash Size	A_2 Nominal Flange Size mm	Flange Dash Size	D Dia Min mm	L_1 Max mm	M_1 Max mm	M_4 Max mm	M_7 Min mm	M_{10} Max mm	M_{16} Min mm	N ±3.0 mm	N_1 ±3.0 mm	N_2 ±3.0 mm	N_3 ±3.0 mm	N_5 ±3.0 mm	W Max mm
10.3	-8	12.7	-8	7.1	71.6	90.2	—	88.9	—	74.7	12.7	—	25.4	—	41.1	27.4
15.9	-12	19.0	-12	12.7	81.8	101.3	103.6	103.4	106.9	96.5	12.7	17.5	25.4	41.1	53.8	38.6
22.2	-16	25.4	-16	18.3	71.9	91.7	85.9	98.0	99.6	91.7	12.7	12.7	28.4	41.7	60.5	43.9
28.6	-20	31.8	-20	23.9	92.5	98.6	113.6	105.7	108.7	103.6	12.7	24.4	28.4	41.7	63.5	51.3
34.9	-24	38.1	-24	29.5	103.4	103.6	102.1	112.8	121.7	113.8	12.7	14.7	28.7	50.8	69.8	58.7
46.0	-32	50.8	-32	40.9	126.5	117.9	120.4	133.1	—	138.4	12.7	16.5	31.8	—	82.6	73.4

[a] For dimensions of flanged head shown in Figure 15A, but not specified in Table 9D, see corresponding dimensions for respective nominal flange size in Figure 3 and Table 1A of SAE J518.

NOTE: For SAE J1402 and DOT FMVSS 106 air brake applications, couplings used on 100R5 hose shall have a minimum assembled orifice no less than 66% of the nominal hose I.D.

TABLE 9E—DIMENSIONS OF CODE 61 4-BOLT SPLIT FLANGE TYPE HOSE FITTINGS FOR USE ON SAE 100R9 HYDRAULIC HOSE[a] (FIGURES 15A TO 21C) (CODES 50 AND 75)

Nominal SAE 100R9 Hose ID mm	Hose Dash Size	A_2 Nominal Flange Size mm	Flange Dash Size	D_1 Dia Min mm	L Max mm	L_1 Max mm	M Max mm	M_1 Max mm	M_3 Max mm	M_4 Max mm	M_6 Max mm	M_7 Max mm	M_9 Max mm
12.7	-8	12.7	-8	9.7	98.0	96.0	102.1	96.0	104.4	—	104.4	101.6	104.4
19.0	-12	19.0	-12	14.5	111.3	108.0	120.6	112.8	122.7	115.3	127.0	118.9	127.0
25.4	-16	25.4	-16	19.8	123.2	120.4	132.6	123.4	134.6	121.9	137.4	138.4	137.7
31.8	-20	31.8	-20	25.7	142.2	—	164.1	—	169.2	—	166.6	—	171.4
38.1	-24	38.1	-24	31.8	155.4	—	178.1	—	180.3	—	186.2	—	187.5
50.8	-32	50.8	-32	43.7	189.7	—	210.3	—	216.4	—	216.4	—	227.1

Nominal SAE 100R9 Hose ID mm	Flange Dash Size	M_{10} Max mm	M_{12} Max mm	M_{13} Max mm	M_{15} Max mm	M_{16} Max mm	N ±4.8 mm	N_1 ±4.8 mm	N_2 ±4.8 mm	N_3 ±4.8 mm	N_4 ±4.8 mm	N_5 ±4.8 mm
12.7	-8	93.0	103.6	93.0	97.3	92.7	9.7	13.5	24.6	28.4	34.8	41.9
19.0	-12	118.9	120.4	118.1	127.0	111.8	12.4	17.0	26.7	39.6	42.7	57.2
25.4	-16	132.1	136.1	129.5	131.8	130.6	13.7	16.0	29.2	42.4	47.2	63.8
31.8	-20	—	159.5	—	165.1	—	14.0	21.3	30.5	43.7	50.5	69.1
38.1	-24	—	185.9	—	185.4	—	18.3	22.9	37.3	55.6	63.5	85.9
50.8	-32	—	225.8	—	214.1	—	23.9	34.0	53.3	79.2	88.4	123.7

[a] For dimensions of flanged head shown on Figure 15A, but not specified in Table 9E, see corresponding dimensions for respective nominal flange size in Figure 3 and Table 1A of SAE J518.

TABLE 9F1—DIMENSIONS OF CODE 61 4-BOLT SPLIT FLANGE TYPE HOSE FITTINGS FOR USE ON SAE 100R10 HYDRAULIC HOSE[a] (FIGURES 15A TO 21C) (CODES 51 AND 76)

Nominal SAE 100R10 Hose ID mm	Hose Dash Size	A_2 Nominal Flange Size mm	Flange Dash Size	D Dia Min mm	L Max mm	L_2 Max mm	M Max mm	M_2 Max mm	M_3 Max mm	M_5 Max mm	M_6 Max mm	M_8 Max mm
19.0	-12	19.0	-12	12.7	133.4	133.4	133.4	133.4	134.1	134.1	141.5	141.5
25.4	-16	25.4	-16	17.3	144.0	144.0	166.6	166.6	165.1	165.1	162.6	162.6
31.8	-20	31.8	-20	22.1	166.4	166.4	183.9	183.9	182.4	182.4	188.0	188.0
38.1	-24	38.1	-24	27.7	177.5	177.5	190.2	190.2	183.9	183.9	209.3	209.3
50.8	-32	50.8	-32	39.6	207.5	207.5	241.8	241.8	227.6	227.6	269.2	269.2

Continued on next page.

TABLE 9F1—(CONTINUED)

Nominal SAE 100R10 Hose ID mm	Flange Dash Size	M_9 Max mm	M_{11} Max mm	M_{12} Max mm	M_{14} Max mm	M_{15} Max mm	M_{17} Max mm	N ±3.3 mm	N_1 ±3.3 mm	N_2 ±3.3 mm	N_3 ±3.3 mm	N_4 ±3.3 mm	N_5 ±3.3 mm	W Max mm	Z Max mm
19.0	-12	139.2	139.2	138.2	138.2	127.0	127.0	12.2	16.8	26.7	38.4	42.9	55.6	42.9	79.2
25.4	-16	179.8	179.8	174.0	174.0	160.0	160.0	12.7	16.8	27.9	40.9	47.2	62.0	50.8	107.2
31.8	-20	200.4	200.4	194.8	194.8	199.1	199.1	14.5	19.3	31.8	44.7	50.8	68.1	62.2	126.0
38.1	-24	202.7	202.7	195.6	195.6	200.4	200.4	17.0	23.1	36.6	52.8	61.0	80.8	76.2	133.1
50.8	-32	259.1	259.1	231.9	231.9	259.1	259.1	23.9	33.3	52.8	76.2	86.4	114.3	81.8	166.6

[a] For dimensions of flanged head shown in Figure 15A, but not specified in Table 9F2, see corresponding dimensions for respective nominal flange size in Figure 3 and Table 1A of SAE J518.

TABLE 9F2—DIMENSIONS OF CODE 61 4-BOLT SPLIT FLANGE TYPE HOSE FITTINGS FOR USE ON SAE 100R10 HYDRAULIC HOSE[a] (FIGURES 15 TO 21) (CODES 51 AND 76)

Nominal SAE 100R10 Hose ID mm	Hose Dash Size	A_2 Nominal Flange Size mm	Flange Dash Size	D Dia Min mm	L Max mm	L_2 Max mm	M_6 Max mm	M_8 Max mm	M_{15} Max mm	M_{17} Max mm	N_2 ±3.3 mm	N_5 ±3.3 mm	W Max mm	Z Max mm
19.0	-12	19.0	-12	12.7	133.4	133.4	141.5	141.5	134.4	134.4	25.4	54.1	42.9	79.2
25.4	-16	25.4	-16	17.3	149.9	149.9	164.3	164.3	160.0	160.0	30.2	68.6	50.8	107.2
31.8	-20	31.8	-20	22.1	166.4	166.4	191.5	191.5	181.4	181.4	35.8	76.5	62.2	126.0
38.1	-24	38.1	-24	27.7	193.0	193.0	217.7	217.7	200.4	200.4	45.0	91.2	76.2	133.1
50.8	-32	50.8	-32	39.6	219.7	219.7	281.4	281.4	259.1	259.1	63.8	131.6	81.8	166.6

[a] For dimensions of flanged head shown in Figure 15A, but not specified in Table 9F2, see corresponding dimensions for respective nominal flange size in Figure 3 and Table 1B of SAE J518.

TABLE 9G—DIMENSIONS OF CODE 61 4-BOLT SPLIT FLANGE TYPE HOSE FITTINGS FOR USE ON SAE 100R11 HYDRAULIC HOSE (FIGURES 15 TO 21) (CODES 52 AND 76)

Nominal SAE 100R11 Hose ID mm	Hose Dash Size	A_2 Nominal Flange Size mm	Flange Dash Size	D Dia Min mm	L Max mm	M Max mm	M_3 Max mm	M_6 Max mm	M_9 Max mm	M_{12} Max mm	M_{15} Max mm	N ±3.3 mm	N_1 ±3.3 mm	N_2 ±3.3 mm	N_3 ±4.8 mm	N_4 ±4.8 mm	N_5 ±4.8 mm	W Max mm
25.4	-16	25.4	-16	18.5	162.1	165.1	168.1	171.4	173.0	171.4	168.1	12.7	17.8	28.7	41.4	47.0	63.5	55.9
31.8	-20	31.8	-20	24.1	74.2	179.1	185.9	182.4	187.5	184.7	184.7	12.7	19.0	30.2	44.4	50.8	69.1	67.3
38.1	-24	38.1	-24	30.5	196.8	200.7	207.8	212.1	216.7	215.9	212.3	16.5	23.6	38.1	53.3	58.7	82.6	74.7
50.8	-32	50.8	-32	41.9	238.8	254.0	264.2	270.5	275.3	276.4	269.2	22.9	33.0	53.3	76.2	85.9	119.1	88.9

[a] For dimensions of flanged head shown in Figure 15A, but not specified in Table 9G, see corresponding dimensions for respective nominal flange size in Figure 3 and Table 1A of SAE J518.

(R) TABLE 9H1—DIMENSIONS OF CODE 61 4-BOLT SPLIT FLANGE TYPE HOSE FITTINGS FOR USE ON SAE 100R12 HYDRAULIC HOSE[a] (FIGURES 15A, 16A, 17A, 18A, 19A, 20A, AND 21A) (CODE 77)

Nominal SAE 100R12 Hose ID mm	Hose Dash Size	A_2 Nominal Flange Size mm	Flange Dash Size	D Dia Min mm	L Max mm	M Max mm	M_3 Max mm	M_6 Max mm	M_9 Max mm	M_{12} Max mm	M_{15} Max mm	N ±3.3 mm	N_1 ±4.1 mm	N_2 ±6.4 mm	N_3 ±6.4 mm	N_4 ±6.4 mm	N_5 ±8.9 mm
12.7	-8	12.7	-8	8.9	108.2	110.0	110.0	110.5	110.5	110.5	110.5	9.7	12.7	19.8	26.7	31.8	41.4
15.9	-10	19.0	-12	10.0	120.7	120.7	120.7	127.0	139.2	138.2	127.0	12.2	16.8	26.7	38.4	42.9	55.9
19.0	-12	19.0	-12	12.7	133.4	133.4	133.4	141.5	139.2	138.2	127.0	12.2	16.8	26.7	38.4	42.9	55.9
25.4	-16	25.4	-16	17.3	144.0	166.6	165.1	162.6	179.8	174.0	160.0	12.7	16.8	27.9	40.6	47.0	62.0
31.8	-20	31.8	-20	22.0	166.4	183.9	182.4	188.0	200.4	194.8	199.1	14.5	19.3	31.8	44.7	50.8	68.1
38.1	-24	38.1	-24	27.7	177.5	190.2	183.9	209.3	202.7	195.6	200.4	17.0	23.1	36.6	52.8	61.0	79.2
50.8	-32	50.8	-32	38.0	207.5	241.8	227.6	269.2	259.1	231.9	259.1	23.9	33.3	52.8	76.2	86.4	118.0

[a] For dimensions of flanged head shown in Figure 15A, but not specified in Table 9H1, see corresponding dimensions for respective nominal flange size in Figure 3 and Table 1A of SAE J518.

(R) TABLE 9H2—DIMENSIONS OF CODE 62 4-BOLT SPLIT FLANGE TYPE HOSE FITTINGS FOR USE ON SAE 100R12 HYDRAULIC HOSE[a] (FIGURES 15A, 16A, 17A, 18A, 19A, 20A, AND 21A) (CODE 77)

Nominal SAE 100R12 Hose ID mm	Hose Dash Size	A_2 Nominal Flange Size mm	Flange Dash Size	D Dia Min mm	L Max mm	M Max mm	M_3 Max mm	M_6 Max mm	M_9 Max mm	M_{12} Max mm	M_{15} Max mm	N ±3.3 mm	N_1 ±4.1 mm	N_2 ±6.4 mm	N_3 ±6.4 mm	N_4 ±6.4 mm	N_5 ±8.9 mm
15.9	-10	19.0	-12	10.0	137.9	134.9	142.5	145.3	144.8	143.8	136.7	12.7	16.3	28.7	37.6	42.9	57.9
19.0	-12	19.0	-12	12.7	137.9	134.9	142.5	145.3	144.8	143.8	136.7	12.7	16.3	28.7	37.6	42.9	57.9
25.4	-16	25.4	-16	17.3	149.9	167.6	165.9	164.3	179.8	174.0	160.0	12.7	18.3	30.0	40.6	47.0	66.3
31.8	-20	31.8	-20	22.0	166.4	205.0	198.1	197.1	202.4	197.1	190.8	14.0	20.3	32.3	46.0	52.3	71.4
38.1	-24	38.1	-24	27.7	193.0	231.9	230.9	220.7	240.3	218.4	209.3	17.8	25.4	41.1	56.9	63.5	86.6
50.8	-32	50.8	-32	38.0	219.7	294.9	290.6	281.4	279.4	269.2	259.1	25.4	35.8	57.2	82.6	86.4	122.9

[a] For dimensions of flanged head shown in Figure 15A, but not specified in Table 9H2, see corresponding dimensions for respective nominal flange size in Figure 3 and Table 1A of SAE J518.

TABLE 9I2—DIMENSIONS OF CODE 61 4-BOLT SPLIT FLANGE TYPE HOSE FITTINGS FOR USE ON SAE 100R13 HYDRAULIC HOSE[a] (FIGURES 15A TO 21A) (CODE 78)

Nominal SAE 100R13 Hose ID mm	Hose Dash Size	A_2 Nominal Flange Size mm	Flange Dash Size	D Dia Min mm	L Max mm	M Max mm	M_3 Max mm	M_6 Max mm	M_9 Max mm	M_{12} Max mm	M_{15} Max mm	N ±3.3 mm	N_1 ±4.1 mm	N_2 ±6.4 mm	N_3 ±6.4 mm	N_4 ±6.4 mm	N_5 ±8.9 mm	W Max mm
19.0	-12	19.0	-12	12.7	137.9	134.9	142.5	145.3	144.8	143.8	136.7	12.7	16.3	28.7	37.6	42.9	57.9	43.7
25.4	-16	25.4	-16	17.5	148.6	167.6	165.9	160.3	166.6	174.0	156.7	12.7	18.3	30.0	40.6	47.0	66.3	50.8
31.8	-20	31.8	-20	22.4	163.3	205.0	198.1	197.1	202.4	197.1	190.8	14.0	20.3	32.3	46.0	52.3	71.4	63.5
38.1	-24	38.1	-24	27.7	191.8	231.9	230.9	220.7	240.3	218.4	209.3	17.8	25.4	41.1	56.9	63.5	86.6	71.1
50.8	-32	50.8	-32	38.1	234.2	302.5	298.2	288.8	300.5	290.6	280.9	25.4	35.8	57.2	82.6	86.4	122.9	88.9

[a] For dimensions of flanged head shown on Figure 15A, but not specified in Table 9I2, see corresponding dimensions for respective nominal flange size in Figure 3 and Table 1B of SAE J518.

22.151

FEMALE O-RING FACE SEAL TYPE

NOTES: UNSPECIFIED DETAIL WITH RESPECT TO DIMENSIONS, TOLERANCES, CONTOURS, MATERIAL, WORKMANSHIP, AND SO ON, MUST CONFORM TO GENERAL SPECIFICATIONS FOR HYDRAULIC HOSE FITTINGS. THE DIMENSIONAL DESIGNATIONS ON THE FIRST FIGURE IN EACH GROUP SHALL APPLY TO ALL OTHER FIGURES IN THAT GROUP EXCEPT AS SHOWN OTHERWISE. CODES SHOWN IN BRACKETS ADJACENT TO FIGURE NUMBERS REPRESENT RESPECTIVE FITTING IDENTIFICATION, WITH XX SUBSTITUTED FOR THE HOSE TYPE CODE DEPICTED IN BRACKETS AT END OF RESPECTIVE TABLE TITLES, IN ACCORDANCE WITH SAE J846.

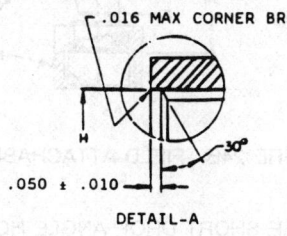
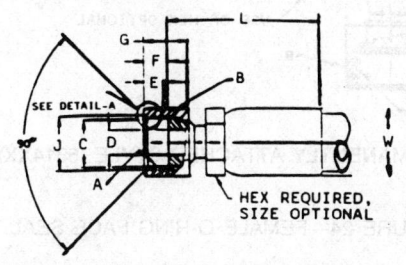

FIGURE 22A—PERMANENTLY ATTACHED STYLE (5301XX)

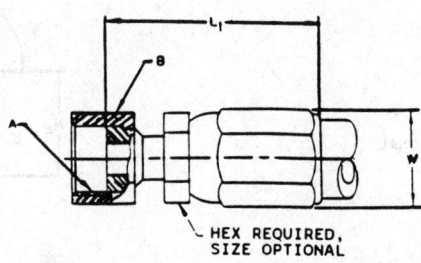

FIGURE 22B—FIELD ATTACHABLE SCREW STYLE (5401XX)

FIGURE 22—FEMALE O-RING FACE SEAL TYPE HOSE FITTINGS

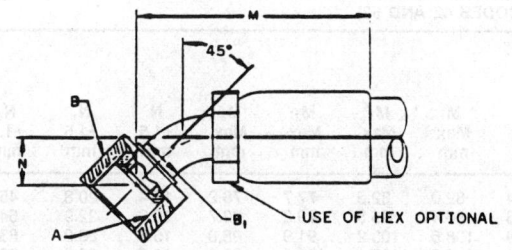
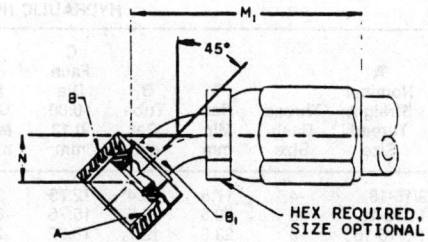

FIGURE 23A—PERMANENTLY ATTACHED STYLE (5303XX)   FIGURE 23B—FIELD ATTACHABLE SCREW STYLE (5403XX)

FIGURE 23—FEMALE O-RING FACE SEAL TYPE 45-DEGREE ANGLE HOSE FITTINGS

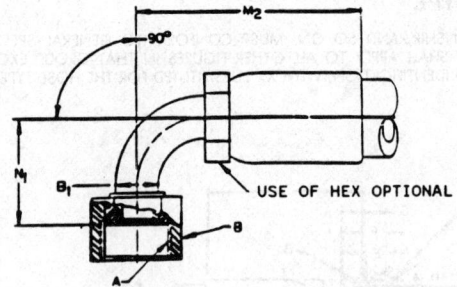

FIGURE 24A—PERMANENTLY ATTACHED STYLE (5314XX)

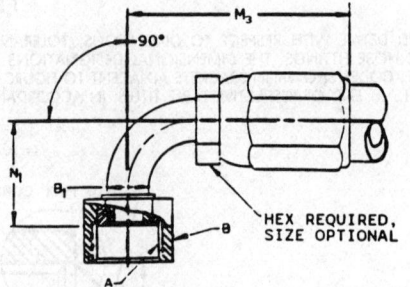

FIGURE 24B—FIELD ATTACHABLE SCREW STYLE (5414XX)

FIGURE 24—FEMALE O-RING FACE SEAL TYPE 90-DEGREE SHORT DROP ANGLE HOSE FITTING

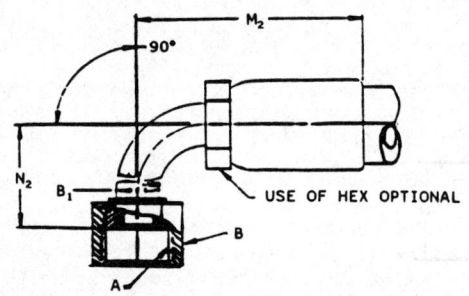

FIGURE 25A—PERMANENTLY ATTACHED STYLE (5315XX)

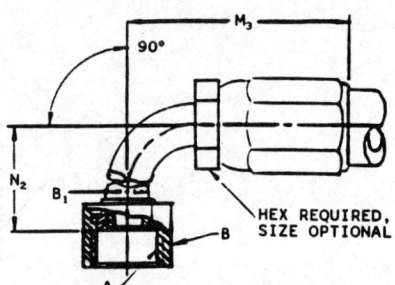

FIGURE 25B—FIELD ATTACHABLE SCREW STYLE (5415XX)

FIGURE 25—FEMALE O-RING FACE SEAL TYPE 90-DEGREE LONG DROP HOSE

TABLE 10A—DIMENSIONS OF FEMALE O-RING SEAL TYPE HOSE FITTINGS FOR USE ON SAE 100R1 HYDRAULIC HOSE[a] (FIGURES 22 TO 25) (CODES 42 AND 53)

Nominal SAE 100R1 Hose ID mm	Hose Dash Size	A Nominal Straight Thread Size	Thread Dash Size	B Hex Min mm	B_1 Tube Ref mm	C Face Dia +0.00 -0.13 mm	D Dia Min mm	F Ref mm	L Max mm	L_1 Max mm	M Max mm	M_1 Max mm	M_2 Max mm	M_3 Max mm	N ±1.5 mm	N_1 ±1.5 mm	N_2 ±1.5 mm	W Max mm
6.4	-4	9/16-18	-4	17.5	6.4[b]	12.75	2.8	8.1	73.7	73.4	82.0	82.6	77.7	76.2	10.4	20.8	45.7	22.1
9.5	-6	11/16-16	-6	20.6	9.5	15.75	6.6	9.7	82.6	83.6	86.4	95.8	79.2	92.7	10.9	22.9	54.1	27.7
12.7	-8	1-3/16-16	-8	23.8	12.7	18.92	7.6	10.9	88.9	85.9	108.5	105.2	91.9	98.0	15.0	29.2	63.8	31.2
15.9	-10	1-14	-10	28.6	15.9	23.44	10.9	13.5	95.5	92.5	111.3	118.1	97.3	102.9	16.5	32.3	70.1	34.8
19.0	-12	1-3/16-12	-12	34.9	19.0	27.86	14.5	14.5	106.7	101.3	121.9	118.1	119.1	115.6	21.1	47.8	96.0	40.4
25.4	-16	1-7/16-12	-16	41.3	25.4	34.21	19.8	14.7	109.5	117.9	134.9	142.7	133.4	141.2	23.9	56.1	114.3	51.3
31.8	-20	1-11/16-12	-20	47.6	31.8	40.56	25.7	14.7	119.1	134.1	162.3	177.3	162.1	177.0	25.4	63.8	129.3	62.5
38.1	-24	2-12	-24	57.2	38.1	48.51	31.8	14.7	132.1	124.2	182.9	136.7	184.4	189.0	27.2	68.6	140.7	69.6

[a] For dimensions shown in Figures 22 to 25, but not specified in Table 10A to 10L, see corresponding dimensions for respective straight thread size in Figure 4 of SAE J1453.
[b] 5/16 tube is acceptable for this size.

TABLE 10B—DIMENSIONS OF FEMALE O-RING FACE SEAL TYPE HOSE FITTINGS FOR USE ON SAE 100R2 HYDRAULIC HOSE[a] (FIGURES 22 TO 25) (CODES 42 AND 53)

Nominal SAE 100R2 Hose ID mm	Hose Dash Size	A Nominal Straight Thread Size	Thread Dash Size	B Hex Min mm	B_1 Tube Ref mm	C Face Dia +0.00 -0.13 mm	D Dia Min mm	F Ref mm	L Max mm	L_1 Max mm	M Max mm	M_1 Max mm	M_2 Max mm	M_3 Max mm	N ±1.5 mm	N_1 ±1.5 mm	N_2 ±1.5 mm	W Max mm
6.4	-4	9/16-18	-4	17.5	6.4[b]	12.75	2.8	8.1	73.7	73.4	82.0	82.6	77.7	76.2	10.4	20.8	45.7	25.4
9.5	-6	11/16-16	-6	20.6	9.5	15.75	6.6	9.7	82.6	83.6	86.4	95.8	79.2	92.7	10.9	22.9	54.1	30.2
12.7	-8	3/16-16	-8	23.8	12.7	18.92	9.7	10.9	88.9	87.6	108.5	105.2	91.9	98.0	15.0	29.2	63.8	35.1
15.9	-10	1-14	-10	28.6	15.9	23.44	10.9	13.5	95.5	92.5	111.3	118.1	97.3	102.9	16.5	32.3	70.1	39.6
19.0	-12	1-3/16-12	-12	34.9	19.0	27.86	14.5	14.5	106.7	104.1	121.9	118.1	119.1	115.6	21.1	47.8	96.0	44.4
25.4	-16	1-7/16-12	-16	41.3	25.4	34.21	19.8	14.7	109.5	121.2	134.9	142.7	133.4	141.2	23.9	56.1	114.3	52.3
31.8	-20	1-11/16-12	-20	47.6	31.8	40.56	25.7	14.7	119.1	136.4	162.3	117.3	162.1	177.0	25.4	63.8	129.3	66.8
38.1	-24	2-12	-24	57.2	38.1	48.51	31.8	14.7	132.1	136.4	182.9	147.6	184.4	189.0	27.2	68.6	140.7	74.7

[a] For dimensions shown in Figures 22 to 25, but not specified in Tables 10A to 10L, see corresponding dimensions for respective straight thread size in Figure 4 of SAE J1453.
[b] 5/16 tube is acceptable for this size.

TABLE 10C—DIMENSIONS OF FEMALE O-RING FACE SEAL TYPE HOSE FITTINGS FOR USE ON SAE 100R3 HYDRAULIC HOSE[a] (FIGURES 22 AND 25) (CODE 44)

Nominal SAE 100R3 Hose ID mm	Hose Dash Size	A Nominal Straight Thread Size	Thread Dash Size	B Hex Min mm	B_1 Tube Ref mm	C Face Dia +0.00 -0.13 mm	D Dia Min mm	F Ref mm	L Max mm	M Max mm	M_2 Max mm	N ±1.5 mm	N_1 ±1.5 mm	N_2 ±1.5 mm	W Max mm
6.4	-4	9/16-18	-4	17.5	6.4[b]	12.75	2.8	8.1	73.7	82.0	77.7	10.4	20.8	45.7	25.4
9.5	-6	11/16-18	-6	20.6	9.5	15.75	6.6	9.7	82.6	86.4	79.2	10.9	22.9	54.1	30.2
12.7	-8	13/16-16	-8	23.8	12.7	18.92	9.7	10.9	88.9	108.5	91.9	15.0	29.2	63.8	35.1
15.9	-10	1-14	-10	28.6	15.9	23.44	10.9	13.5	95.5	111.3	97.3	16.5	32.3	70.1	39.6
19.0	-12	1-3/16-12	-12	34.9	19.0	27.86	14.5	14.5	106.7	121.9	119.1	21.1	47.8	96.0	44.4
25.4	-16	1-7/16-12	-16	41.3	25.4	34.21	19.8	14.7	109.5	142.7	133.4	23.9	56.1	114.3	52.3

[a] For dimensions shown in Figures 22 to 25, but not specified in Tables 10A to 10L, see corresponding dimensions for respective straight thread size in Figure 4 of SAE J1453.
[b] 5/16 tube is acceptable for this size.

TABLE 10D—DIMENSIONS OF FEMALE O-RING FACE SEAL TYPE HOSE FITTINGS FOR USE ON SAE 100R4 HYDRAULIC HOSE[a] (FIGURES 22 TO 25) (CODE 45)

Nominal SAE 100R4 Hose ID mm	Hose Dash Size	A Nominal Straight Thread Size	Thread Dash Size	B Hex Min mm	B_1 Tube Ref mm	C Face Dia +0.00 -0.13 mm	D Dia Min mm	F Ref mm	L Max mm	M Max mm	M_2 Max mm	N ±1.5 mm	N_1 ±1.5 mm	N_2 ±1.5 mm	W Max mm
19.0	-12	1-3/16-12	-12	34.9	19.0	27.86	14.5	14.5	106.7	121.9	119.1	21.1	47.8	96.0	43.9
25.4	-16	1-7/16-12	-16	41.3	25.4	34.21	19.8	14.7	109.5	134.9	133.4	23.9	56.1	114.3	51.1
31.8	-20	1-11/16-12	-20	47.6	31.8	40.56	25.7	14.7	119.1	162.3	162.1	25.4	63.8	129.3	58.7

[a] For dimensions shown in Figures 22 to 25, but not specified in Tables 10A to 10L, see corresponding dimensions for respective straight thread size in Figure 4 of SAE J1453.

TABLE 10E—DIMENSIONS OF FEMALE O-RING FACE SEAL TYPE HOSE FITTINGS FOR USE ON SAE 100R5 HYDRAULIC HOSE[a] (FIGURES 22 TO 25) (CODE 46)

Nominal SAE 100R5 Hose ID mm	Hose Dash Size	A Nominal Straight Thread Size	Thread Dash Size	B Hex Min mm	B_1 Tube Ref mm	C Face Dia +0.00 -0.13 mm	D Dia Min mm	F Ref mm	L_1 Max mm	M_1 Max mm	M_3 Max mm	N ±1.5 mm	N_1 ±1.5 mm	N_2 ±1.5 mm	W Max mm
4.8	-4	9/16-18	-4	17.5	6.4[b]	12.75	2.3	8.1	63.2	66.3	59.9	10.4	20.8	45.7	20.3
7.9	-6	11/16-16	-6	20.6	9.5	15.75	5.1	9.7	73.9	73.4	70.4	10.9	22.9	54.1	23.9
10.3	-8	13/16-16	-8	23.8	12.7	18.92	7.1	10.9	85.6	94.0	86.9	15.0	29.2	63.8	27.7
12.7	-10	1-14	-10	28.6	15.9	23.44	9.7	13.5	91.7	96.8	90.9	16.5	32.3	70.1	33.0
15.9	-12	1-3/16-12	-12	34.9	19.0	27.86	12.7	14.5	100.6	115.8	113.3	21.1	47.8	96.0	40.4
22.2	-16	1-7/16-12	-16	41.3	25.4	34.21	18.3	14.7	94.7	115.8	113.3	23.9	56.1	114.3	47.8
28.6	-20	1-11/16-12	-20	47.6	31.8	40.56	23.9	14.7	98.6	138.4	138.2	25.4	63.8	129.3	55.1
34.9	-24	2-12	-24	57.2	38.1	48.51	29.5	14.7	104.1	176.3	159.8	27.2	68.6	140.7	66.0

[a] For dimensions shown in Figures 22 to 25, but not specified in Tables 10A to 10L, see corresponding dimensions for respective straight thread size in Figure 4 of SAE J1453.
[b] 5/16 tube is acceptable for this size.

NOTE: For SAE J1402 and DOT FMVSS 106 air brake applications, couplings used on 100R5 hose shall have a minimum assembled orifice no less than 66% of the nominal hose I.D.

TABLE 10F—DIMENSIONS OF FEMALE O-RING FACE SEAL TYPE HOSE FITTINGS FOR USE ON SAE 100R6 HYDRAULIC HOSE[a] (FIGURES 22 TO 25) (CODE 47)

Nominal SAE 100R6 Hose ID mm	Hose Dash Size	A Nominal Straight Thread Size	Thread Dash Size	B Hex Min mm	B_1 Tube Ref mm	C Face Dia +0.00 -0.13 mm	D Dia Min mm	F Ref mm	L Max mm	W Max mm
6.4	-4	9/16-18	-4	17.5	6.4[b]	12.75	2.8	8.1	—	25.4
9.5	-6	11/16-18	-6	20.6	9.5	15.75	6.6	9.7	—	30.2
12.7	-8	13/16-16	-8	23.8	12.7	18.92	9.7	10.9	—	35.1
15.9	-10	1-14	-10	28.6	15.9	23.44	12.2	13.5	—	39.6

[a] For dimensions shown in Figures 22 to 25, but not specified in Tables 10A to 10L, see corresponding dimensions for respective straight thread size in Figure 4 of SAE J1453.
[b] 5/16 tube is acceptable for this size.

TABLE 10G—DIMENSIONS OF FEMALE O-RING FACE SEAL TYPE HOSE FITTINGS FOR USE ON SAE 100R7 HYDRAULIC HOSE[a] (FIGURES 22 TO 25) (CODE 48)

Nominal SAE 100R7 Hose ID mm	Hose Dash Size	A Nominal Straight Thread Size	Thread Dash Size	B Hex Min mm	B_1 Tube Ref mm	C Face Dia +0.00 -0.13 mm	D Dia Min mm	F Ref mm	L Max mm	M Max mm	M_2 Max mm	N ±1.5 mm	N_1 ±1.5 mm	N_2 ±1.5 mm	W Max mm
6.4	-4	9/16-18	-4	17.5	6.4[b]	12.75	2.8	8.1	72.9	89.9	87.1	10.4	20.8	45.7	21.6
9.5	-6	11/16-16	-6	20.6	9.5	15.75	6.6	9.7	82.6	96.0	88.9	10.9	22.9	54.1	25.7
12.7	-8	13/16-16	-8	23.8	12.7	18.92	9.7	10.9	87.1	108.5	100.1	15.0	29.2	63.8	29.5
15.9	-10	1-14	-10	28.6	15.9	23.44	10.9	13.5	94.5	113.8	114.3	16.5	32.3	70.1	39.6
19.0	-12	1-3/16-12	-12	34.9	19.0	27.86	14.5	14.5	95.5	127.0	117.3	21.1	47.8	96.0	40.4
25.4	-16	1-7/16-12	-16	41.3	25.4	34.21	19.8	14.7	113.3	146.3	146.3	23.9	56.1	114.3	51.3

[a] For dimensions shown in Figures 22 to 25, but not specified in Tables 10A to 10L, see corresponding dimensions for respective straight thread size in Figure 4 of SAE J1453.
[b] 5/16 tube is acceptable for this size.

TABLE 10H—DIMENSIONS OF FEMALE O-RING FACE SEAL TYPE HOSE FITTINGS FOR USE ON SAE 100R8 HYDRAULIC HOSE[a] (FIGURES 22 TO 25) (CODE 49)

Nominal SAE 100R8 Hose ID mm	Hose Dash Size	A Nominal Straight Thread Size	Thread Dash Size	B Hex Min mm	B_1 Tube Ref mm	C Face Dia +0.00 -0.13 mm	D Dia Min mm	F Ref mm	L Max mm	M Max mm	M_2 Max mm	N ±1.5 mm	N_1 ±1.5 mm	N_2 ±1.5 mm	W Max mm
6.4	-4	9/16-18	-4	17.5	6.4[b]	12.75	2.8	8.1	72.9	89.9	87.1	10.4	20.8	45.7	21.6
9.5	-6	11/16-16	-6	20.6	9.5	15.75	6.6	9.7	74.4	96.0	88.9	10.9	22.9	54.1	25.7
12.7	-8	13/16-16	-8	23.8	12.7	18.92	9.7	10.9	87.1	108.5	100.1	15.0	29.2	63.8	29.5
15.9	-10	1-14	-10	28.6	15.9	23.44	10.9	13.5	94.5	—	—	16.5	32.3	70.1	39.6
19.0	-12	1-3/16-12	-12	34.9	19.0	27.86	14.5	14.5	95.5	123.7	117.3	21.1	47.8	96.0	40.4
25.4	-16	1-7/16-12	-16	41.3	25.4	34.21	19.8	14.7	113.3	146.3	146.3	23.9	56.1	114.3	51.3

[a] For dimensions shown in Figures 22 to 25, but not specified in Tables 10A to 10L, see corresponding dimensions for respective straight thread size in Figure 4 of SAE J1453.
[b] 5/16 tube is acceptable for this size.

TABLE 10I—DIMENSIONS OF FEMALE O-RING FACE SEAL TYPE HOSE FITTINGS FOR USE ON SAE 100R9 HYDRAULIC HOSE[a] (FIGURES 22 TO 25) (CODES 50 AND 75)

Nominal SAE 100R9 Hose ID mm	Hose Dash Size	A Nominal Straight Thread Size	Thread Dash Size	B Hex Min mm	B_1 Tube Ref mm	C Face Dia +0.00 -0.13 mm	D Dia Min mm	F Ref mm	L Max mm	L_1 Max mm	M Max mm	M_1 Max mm	M_2 Max mm	M_3 Max mm	N ±1.5 mm	N_1 ±1.5 mm	N_2 ±1.5 mm	W Max mm
9.5	-6	11/16-18	-6	20.6	9.5	15.75	6.6	9.7	74.2	83.6	80.8	95.8	86.4	77.0	10.9	22.9	54.1	27.7
12.7	-8	13/16-16	-8	23.8	12.7	18.92	9.7	10.9	84.6	87.6	96.0	105.2	91.4	90.4	15.0	29.2	63.8	31.2
19.0	-12	1-3/16-12	-12	34.9	19.0	27.86	14.5	14.5	96.5	104.1	111.5	118.1	114.8	108.7	21.1	47.8	96.0	40.4
25.4	-16	1-7/16-12	-16	41.3	25.4	34.21	19.8	14.7	107.2	121.4	128.3	142.7	133.4	132.1	23.9	56.1	114.3	51.3
31.8	-20	1-11/16-12	-20	47.6	31.8	40.56	25.7	14.7	116.8	136.4	152.4	177.3	162.1	144.3	25.4	63.8	129.3	62.5

[a] For dimensions shown in Figures 22 to 25, but not specified in Tables 10A to 10L, see corresponding dimensions for respective straight thread size in Figure 4 of SAE J1453.

TABLE 10J—DIMENSIONS OF FEMALE O-RING FACE SEAL TYPE HOSE FITTINGS FOR USE ON SAE 100R10 HYDRAULIC HOSE[a] (FIGURES 22 TO 25) (CODES 51 AND 76)

Nominal SAE 100R10 Hose ID mm	Hose Dash Size	A Nominal Straight Thread Size	Thread Dash Size	B Hex Min mm	B_1 Tube Ref mm	C Face Dia +0.00 -0.13 mm	D Dia Min mm	F Ref mm	L Max mm	M Max mm	M_2 Max mm	N ±1.5 mm	N_1 ±1.5 mm	N_2 ±1.5 mm	W Max mm
9.5[b]	-6	11/16-18	-6	20.6	9.5	15.75	5.8	9.7	86.9	85.3	81.8	10.9	22.9	54.1	32.5
12.7[b]	-8	13/16-16	-8	23.8	12.7	18.92	8.9	10.9	101.6	104.9	98.6	15.0	29.2	63.8	41.7
19.0	-12	1-3/16-12	-12	34.9	19.0	27.86	14.0	14.5	128.0	153.9	151.1	21.1	47.8	96.0	50.8
25.4	-16	1-7/16-12	-16	41.3	25.4	34.21	19.8	14.7	145.8	206.0	179.1	23.9	56.1	114.3	59.7
31.8	-20	1-11/16-12	-20	47.6	31.8	40.56	25.7	14.7	162.6	207.0	206.8	25.4	63.8	129.3	69.8
38.1	-24	2-12	-24	57.2	38.1	48.51	31.8	14.7	174.2	228.1	236.2	27.2	68.6	140.7	84.3

[a] For dimensions shown in Figures 22 to 25, but not specified in Tables 10A to 10L, see corresponding dimensions for respective straight thread size in Figure 4 of SAE J1453.
[b] Hose pressure rating exceeds O-ring face seal fitting rating in noted sizes.

TABLE 10K—DIMENSIONS OF FEMALE O-RING FACE SEAL TYPE HOSE FITTINGS FOR USE ON SAE 100R11 HYDRAULIC HOSE[a] (FIGURES 22 TO 25) (CODE 52)

Nominal SAE 100R11 Hose ID mm	Hose Dash Size	A Nominal Straight Thread Size	Thread Dash Size	B Hex Min mm	B_1 Tube Ref mm	C Face Dia +0.00 -0.13 mm	D Dia Min mm	F Ref mm	L Max mm	M Max mm	M_2 Max mm	N ±1.5 mm	N_1 ±1.5 mm	N_2 ±1.5 mm	W Max mm
19.0[b]	-12	1-3/16-12	-12	34.9	19.0	27.86	13.7	14.5	128.0	153.9	151.1	21.1	47.8	96.0	43.9
25.4	-16	1-7/16-12	-16	41.3	25.4	34.21	19.8	14.7	145.8	206.0	179.1	23.9	56.1	114.3	55.9
31.8	-20	1-11/16-12	-20	47.6	31.8	40.56	25.7	14.7	162.6	207.0	206.8	25.4	63.8	129.3	58.7
38.1	-24	2-12	-24	57.2	38.1	48.51	31.8	14.7	174.2	228.1	236.2	27.2	68.6	140.7	74.7

[a] For dimensions shown in Figures 22 to 25, but not specified in Tables 10A to 10L, see corresponding dimensions for respective straight thread size in Figure 4 of SAE J1453.
[b] Hose pressure rating exceeds O-ring face seal fitting rating in noted sizes.

TABLE 10L—DIMENSIONS OF FEMALE O-RING FACE SEAL TYPE HOSE FITTINGS FOR USE ON SAE 100R12 HYDRAULIC HOSE[a] (FIGURES 22 TO 25) (CODE 77)

Nominal SAE 100R12 Hose ID mm	Hose Dash Size	A Nominal Straight Thread Size	Thread Dash Size	B Hex Min mm	B_1 Tube Ref mm	C Face Dia +0.00 -0.13 mm	D Dia Min mm	F Ref mm	L Max mm	M Max mm	M_2 Max mm	N ±1.5 mm	N_1 ±1.5 mm	N_2 ±1.5 mm	W Max mm
9.5	-6	11/16-16	-6	20.6	9.5	15.75	6.6	9.7	76.2	95.2	86.4	10.9	22.9	54.1	32.5
12.7	-8	13/16-16	-8	23.8	12.7	18.92	8.1	10.9	91.4	111.8	104.6	15.0	29.2	63.8	41.7
19.0	-12	1-3/16-12	-12	34.9	19.0	27.86	13.7	14.5	114.3	127.0	127.0	21.1	47.8	96.0	50.8
25.4	-16	1-7/16-12	-16	41.3	25.4	34.21	19.8	14.7	127.0	147.1	152.4	23.9	56.1	114.3	59.7
31.8	-20	1-11/16-12	-20	47.6	31.8	40.56	25.7	14.7	139.7	175.8	177.8	25.4	63.8	129.3	69.8
38.1	-24	2-12	-24	57.2	38.1	48.51	31.8	14.7	152.4	185.2	203.2	27.2	68.6	140.7	84.3

[a] For dimensions shown in Figures 22 to 25, but not specified in Tables 10A to 10L, see corresponding dimensions for respective straight thread size in Figure 4 of SAE J1453.

HOSE AND HOSE ASSEMBLIES FOR MARINE APPLICATIONS—SAE J1942 JUN93

SAE Standard

Report of the Fluid Conductors and Connectors Technical Committee approved March 1989. Completely revised by the SAE Fluid Conductors and Connectors Technical Committee SC2—Hydraulic Hose and Hose Fittings, May 1992, and revised June 1993. Rationale statements available.

1. Scope—SAE J1942, developed through the cooperative efforts of the U.S. Coast Guard and SAE, became effective August 28, 1991[1], as the official document for nonmetallic flexible hose assemblies for marine use.

This SAE Standard covers specific requirements for several styles of hose and/or hose assemblies in systems on board commercial vessels inspected and certificated by the U.S. Coast Guard. It is intended that this document establish hose constructions and performance levels that are essential to safe operations in the marine environment. Refer to SAE J1273 for selection, installation, and maintenance of hose and hose assemblies.

(R) A Hose Assemblies Listing of the products which can be used for the applications described in Table 1 is available from SAE as SAE J1942/1.

2. References

2.1 Applicable Documents—The following publications form a part of this specification to the extent specified herein. The latest issue of SAE publications shall apply.

2.1.1 SAE PUBLICATIONS—Available from SAE, 400 Commonwealth Drive, Warrendale, PA 15096-0001.

SAE J343—Test and Procedures for SAE 100 R Series Hydraulic Hose and Hose Assemblies

SAE J517—Hydraulic Hose

SAE J1273—Selection, Installation, and Maintenance of Hose and Hose Assemblies

SAE J1475—Hydraulic Hose Fittings for Marine Applications

SAE J1942/1—SAE Marine Hose Assemblies List

2.1.2 ASTM PUBLICATIONS—Available from ASTM, 1916 Race Street, Philadelphia, PA 19103-1187.

ASTM D 1141-52—Specification for Substitute Ocean Water

2.1.3 MSHA PUBLICATIONS

MSHA 30 CFR 18.65—Conservation of Power and Water Resources—Subchapter B—Regulations Under the Federal Power Act

3. Hose Application/Construction—Hose construction and performance shall conform to Table 1.

4. Fittings—Fittings shall conform to SAE J1475 where applicable; only hose and fitting combinations that have been tested and passed the requirements of this document as hose assemblies are acceptable. Push-on fittings, quick disconnect couplings, and fittings with a single worm-gear clamp or a single band around the hose, are unacceptable.

5. Qualification Tests—For qualification to this document, hose and/or assemblies made therefrom shall conform to the tests and requirements specified in Table 1 for each hose application.

Testing shall conform to SAE J343 except as noted.

(R) Manufacturers may have their hose assemblies listed in SAE J1942/1 by submitting a statement to SAE certifying that all the applicable requirements in SAE J1942 are met for their specific hose assemblies. The certification shall also include such information as the make and model of the hose assembly, maximum allowable working pressure for the particular application(s), whether a fire sleeve is required, etc., in order that the hose assembly will be accurately listed in SAE J1942/1.

---

[1] Ref: Federal Register/Vol. 56, No. 145/Monday, July 29, 1991/Rules and Regulations

---

TABLE 1—HOSE APPLICATION/CONSTRUCTION

Code	Application	Maximum Service Pressure	Hose Reinforcement/ Construction	Requirements	Notes
HF	All Services	1 3	Plies or braids of steel wire with or without textile[2]	Sections 6, 7, 8, 9. 10, 11 (SAE J517 may be substituted for Section 9)	Acceptable for Codes H, VW NVW, and F applications [3]
H	Fluid Power (Hydraulic Systems)	1	Plies or braids of steel wire or textile[2]	Sections 6, 7, 9, 10, 11 (SAE J517 may be substituted for Section 9)	Acceptable for Codes H, VW, and NVW applications
F	Lube Oil and Fuel Systems	1	Plies or braids of steel wire with or without textile[2]	Sections 6, 7, 8, 9 (impulse per 9E not required) 10, 11	Acceptable for Codes F, VW, and NVW applications
VW	Vital and Nonvital Fresh and Salt Water	1	Plies or braids of steel wire or textile[2]	Sections 6, 7, 10, 11	Acceptable for Codes VW, and NVW applications
NVW	Nonvital Water and Pneumatic	0.34 MPa	Optional	Sections 6, 7, 10, 11	Acceptable for Code NVW application only

[1] As rated by SAE J517 or as rated by manufacturer.

[2] Wire helix construction may be used on suction and return lines in conjunction with a textile reinforcement.

[3] Maximum service pressure for lube oil and fuel systems applications (Code F) may be less than maximum service pressure for other systems applications, e.g., Code H. Refer to manufacturer's catalog and Hose Assemblies List, SAE J1942/1.

6. Immersion-Burst Test—One 450 mm assembly, uncapped, shall be completely immersed in a nonpressurized, closed container filled with synthetic sea water conforming to ASTM D 1141-52 for 48 h ± 1 h at 70 °C ± 1 °C. The assembly shall then be removed and held for 48 h ± 1 h in air at room temperature 21 °C ± 2.5 °C. Following this aging, the assembly shall be subjected to the burst test specified in SAE J343.

Burst shall not occur at a pressure less than four times the rated operating pressure. Within 1/2 h following the burst test, the hose shall be cut apart and the reinforcement examined for signs of corrosion and/or deterioration. The wire of wire-reinforced hose shall not show red rust.

7. Flame Resistance—The hose cover shall pass the MSHA (Mine Safety Health Administration) flame resistance requirements of 30 CFR 18.65. In lieu of testing the hose cover, it may be protected by a fire sleeve of suitable material that conforms to the flame resistance criteria of 30 CFR 18.65.

8. 2-1/2 Min Fire Test—For hose 51 mm inside diameter and smaller, three assemblies are to be consecutively tested for fire resistance. For hoses larger than 51 mm inside diameter, one hose assembly shall be tested. Free hose length measured between the fittings shall be 400 to 600 mm. At least one end fitting shall be positioned to be engulfed in the flame. The hose shall be positioned 230 mm above the top edge of an open pan the size of 215 × 355 × 13 mm. Sufficient heptane shall be added to the pan to provide for a 2-1/2 min burn.

Thermocouples shall be mounted so as to sense the flame temperature in the same plane and elevation as the hose assembly. The assembly shall be pressurized with water to the maximum operating pressure and maintained during the burning portion of the test. Following ignition of the heptane, timing shall begin and the temperature shall be monitored. The temperature shall reach a minimum of 650 °C but shall not exceed 730 °C. (If 650 °C is not reached, the test must be repeated with a new sample. If 730 °C is exceeded, results may be discarded and the test repeated.)

At the end of 2-1/2 min of fire exposure, the flame shall be extinguished and pressure relieved. Water from a 915 mm shall flow through the assembly. Failure to achieve free flow shall constitute failure. With free flow established, the assembly shall be pressurized to the maximum operating pressure for 30 s. Leakage during the fire exposure or subsequent pressure test shall constitute failure. Reference Figures 1 and 2 for fire test set-up and test chamber. (NOTE—Hose assemblies may use protective fire sleeves.)

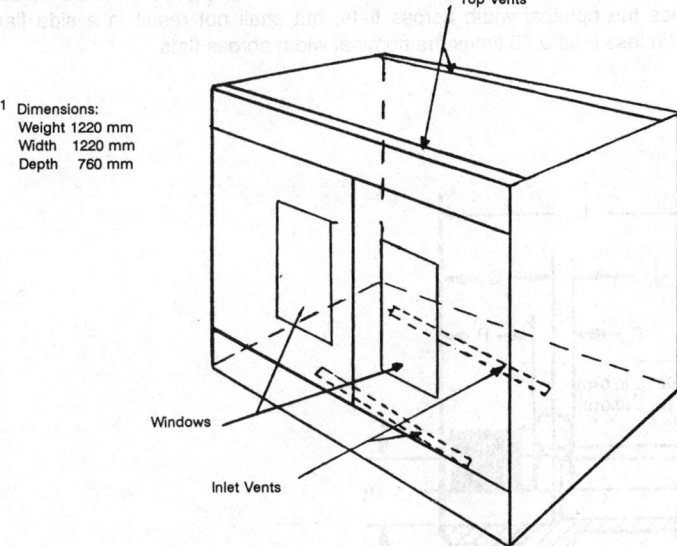

[1] Dimensions may vary slightly provided that all other test parameters in Section 8 are met.

FIGURE 1—FIRE TEST CABINET

9. Non-SAE J517 Hose/Hose Assemblies—Products not conforming to SAE J517 shall meet the following:

a. Proof test per SAE J343, no leakage allowed.
b. Change in length per SAE J343, not to exceed +2 and −4% (does not apply to wire helix type hose).
c. Burst test per SAE J343, minimum burst shall be at least four times the rated operating pressure.
d. Cold bend test per SAE J343, no cracking or leakage allowed. Testing to be done at the manufacturer's minimum recommended temperature.
e. Impulse test per SAE J343. Conduct test at 125% of operating pressure for 200 000 cycles at a fluid temperature of 100 °C. No leakage or other malfunction is allowed. (Impulse not required on wire helix type hose.)

10. Inspection Tests—The tests required in Table 1 shall be repeated at least every 3 years except for the 2-1/2 min fire test (Section 8) which does not need to be rerun unless there is a change in the construction or material of the listed hose.

All test results shall be maintained on file, for review, for a period of 6 years; except that test results of the 2-1/2 min fire test shall be maintained for at least 5 years after termination of hose production.

Test reports may be requested for inspection, at any time, by the U.S. Coast Guard.

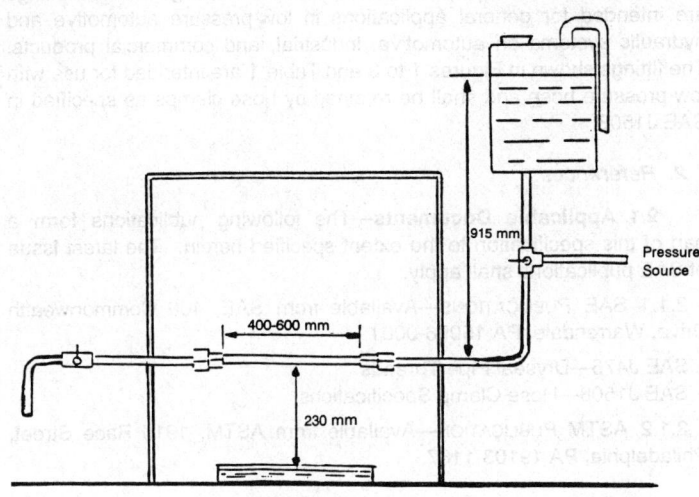

FIGURE 2—TEST SET-UP

11. Marking—Hose meeting SAE J517 is to be identified in accordance with SAE J517. Non-SAE J517 hose shall contain, as a minimum, the following information: maximum operating pressure, manufacturer's name and part number, and hose size. In addition, as an option to expedite U.S. Coast Guard inspection, hoses may be marked with the propeller symbol " ⚜ " as shown in Figure 3, followed by the appropriate alphabetical code from Table 1. Examples of hose markings, with the optional propeller symbol, are shown in Figure 4:

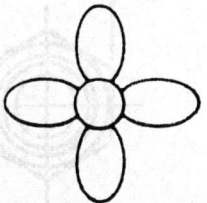

FIGURE 3—PROPELLER SYMBOL HOSE MARKING

SAE J517 Hose—Good for all Service Applications
ABC Co. MT12 3/4" I.D. SAE 100R2AT 15.5 MPa W.P.
FEB89 ⚜ HF

Non-SAE J517 Hose—Good for Fuel & Lube Oil Service
ABC Co. JK08 1/2" I.D. 3.4 MPa W.P.
FEB89 ⚜ F

FIGURE 4—EXAMPLES OF HOSE MARKINGS WITH THE OPTIONAL PROPELLER SYMBOL

(R) The manufacturer's catalog and the Hose Assemblies List, SAE J1942/1, shall be consulted prior to final selection and installation of all marine hose assemblies to ensure that they are adequate for the intended service application.

BEADED TUBE HOSE FITTINGS—SAE J1231 JUN93   SAE Standard

Report of the Fluid Conductors and Connectors Technical Committee approved May 1978 and revised October 1988. Completely revised by the SAE Fluid Conductors and Connectors Technical Committee SC2—Hydraulic Hose and Hose Fittings June 1992. Revised by the Fluid Conductors and Connectors Technical Committee June 1993. Rationale statement available.

(R) **1. Scope**—This SAE Standard covers complete general and dimensional specifications for beaded tube hose fittings. These fittings are intended for general applications in low-pressure automotive and hydraulic systems on automotive, industrial, and commercial products. The fittings shown in Figures 1 to 3 and Table 1 are intended for use with low-pressure hose and shall be retained by hose clamps as specified in SAE J1508.

2. References

2.1 Applicable Documents—The following publications form a part of this specification to the extent specified herein. The latest issue of SAE publications shall apply.

2.1.1 SAE PUBLICATIONS—Available from SAE, 400 Commonwealth Drive, Warrendale, PA 15096-0001.

SAE J476—Dryseal Pipe Threads

(R) SAE J1508—Hose Clamp Specifications

2.1.2 ASTM PUBLICATION—Available from ASTM, 1916 Race Street, Philadelphia, PA 19103-1187.

ASTM B 117—Method of Salt Spray (Fog) Testing

3. General Specifications—The following general specifications supplement the dimensional data for all types of fittings contained in Tables 1 and 2 with respect to all unspecified detail.

3.1 Size Designations—Fitting sizes are designated by the corresponding inside diameter of the hose I.D. for the various sizes of formed tube ends and by the corresponding standard nominal pipe size for pipe thread ends.

3.2 Dimensions and Tolerances—Except for nominal sizes and thread specifications, dimensions and tolerances are given in both U.S. Customary and SI units as designated. Tabulated dimensions shall apply to the finished parts, plated or otherwise processed, as specified by the purchaser. Unless otherwise specified, the maximum and minimum across flats dimensions shall be within the commercial tolerance of bar or extruded stock from which the fittings are produced. The minimum across corners dimensions of hexagons shall be 1.092 times the nominal width across flats, but shall not result in a side flat width less than 0.43 times the nominal width across flats.

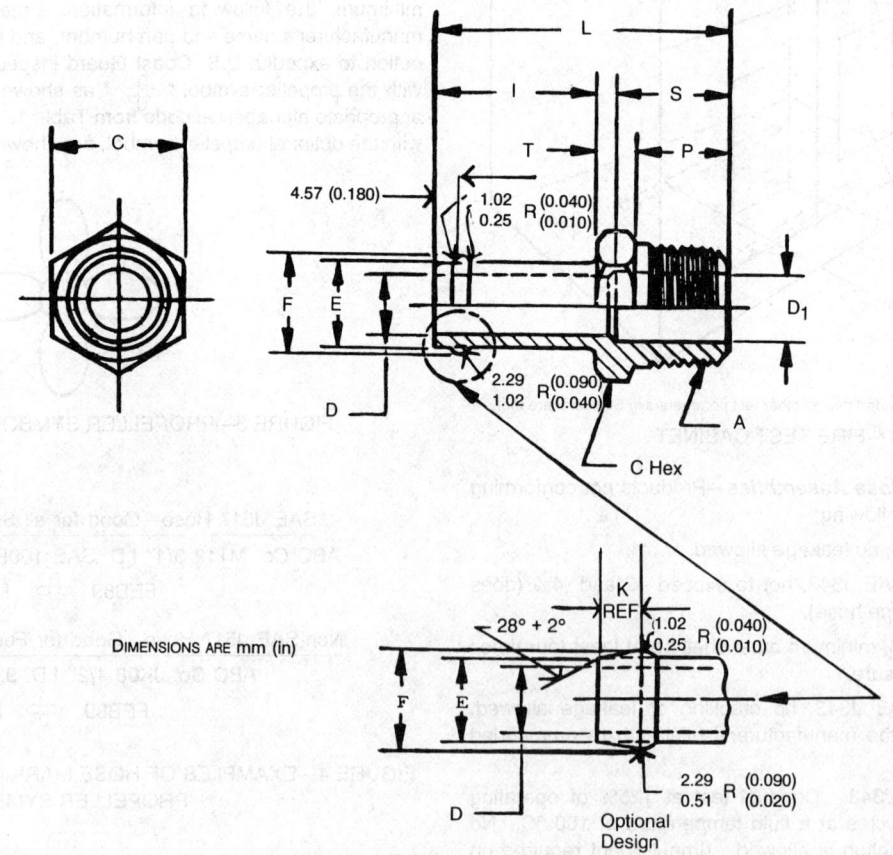

FIGURE 1—CONNECTOR (430160)

TABLE 1—DIMENSIONS OF HOSE PUSH-ON CONNECTOR, 90 DEGREE ELBOW AND 45 DEGREE ELBOW (FIGURES 1 TO 3)

Nominal Hose I.D. in	A Dryseal Taper Thread NPTF[1] in	C Nominal in	D[3] Dia Drill mm	D[3] Dia Drill in	D_1[3] Dia Drill mm	D_1[3] Dia Drill in	E Dia mm ±0.13	E Dia in ±0.005	F Dia mm ±0.13	F Dia in ±0.005	I mm	I in	K[5] Ref mm	K[5] Ref in	L[2] mm ±0.8	L[2] in ±0.03
3/16	1/8	7/16	3.18	0.125	5.56	0.219	4.78	0.188	5.59	0.220	28.19	1.110	2.03	0.080	42.4	1.67
1/4	1/8	7/16	4.78	0.188	5.56	0.219	6.35	0.250	7.37	0.290	28.70	1.130	2.54	0.100	42.9	1.69
1/4	1/4	9/16	4.78	0.188	8.74	0.344	6.35	0.250	7.37	0.290	28.70	1.130	2.54	0.100	48.3	1.90
5/16	1/8	7/16	6.35	0.250	5.56	0.219	7.92	0.312	9.14	0.360	28.70	1.130	2.54	0.100	42.9	1.69
5/16	1/4	9/16	6.35	0.250	8.74	0.344	7.92	0.312	9.14	0.360	28.70	1.130	2.54	0.100	48.3	1.90
3/8	1/8	1/2	7.54	0.297	5.56	0.219	9.52	0.375	10.92	0.430	28.70	1.130	2.54	0.100	42.9	1.69
3/8	1/4	9/16	7.54	0.297	8.74	0.344	9.52	0.375	10.92	0.430	28.70	1.130	2.54	0.100	48.3	1.90
3/8	3/8	11/16	7.54	0.297	10.31	0.406	9.52	0.375	10.92	0.430	28.70	1.130	2.54	0.100	49.0	1.93
1/2	1/4	5/8	10.31	0.406	8.74	0.344	12.70	0.500	14.22	0.560	29.46	1.160	3.05	0.120	49.0	1.93
1/2	3/8	11/16	10.31	0.406	10.31	0.406	12.70	0.500	14.22	0.560	29.46	1.160	3.05	0.120	49.8	1.96
1/2	1/2	7/8	10.31	0.406	14.27	0.562	12.70	0.500	14.22	0.560	29.46	1.160	3.05	0.120	54.6	2.15
1/2	3/4	1-1/16	10.31	0.406	19.05	0.750	12.70	0.500	14.22	0.560	29.46	1.160	3.05	0.120	56.1	2.21
5/8	3/8	3/4	12.70	0.500	10.31	0.406	15.88	0.625	17.53	0.690	29.46	1.160	3.05	0.120	49.8	1.96
5/8	1/2	7/8	12.70	0.500	14.27	0.562	15.88	0.625	17.53	0.690	29.46	1.160	3.05	0.120	54.6	2.15
5/8	3/4	1-1/16	12.70	0.500	19.05	0.750	15.88	0.625	17.53	0.690	29.46	1.160	3.05	0.120	56.1	2.21
3/4	3/8	7/8	15.88	0.625	10.31	0.406	19.05	0.750	20.83	0.820	29.46	1.160	3.05	0.120	49.8	1.96
3/4	1/2	7/8	15.88	0.625	14.27	0.562	19.05	0.750	20.83	0.820	29.46	1.160	3.05	0.120	54.6	2.15
3/4	3/4	1-1/16	15.88	0.625	19.05	0.750	19.05	0.750	20.83	0.820	29.46	1.160	3.05	0.120	56.1	2.21
1	1/2	1-1/8	21.44	0.844	14.27	0.562	25.40	1.000	26.92	1.060	29.46	1.160	3.05	0.120	54.6	2.15
1	3/4	1-1/8	21.44	0.844	19.05	0.750	25.40	1.000	26.92	1.060	29.46	1.160	3.05	0.120	56.1	2.21
1	1	1-3/8	21.44	0.844	23.80	0.937	25.40	1.000	26.92	1.060	29.46	1.160	3.05	0.120	61.7	2.43

TABLE 1—DIMENSIONS OF HOSE PUSH-ON CONNECTOR, 90 DEGREE ELBOW AND 45 DEGREE ELBOW (FIGURES 1 TO 3) (CONTINUED)

Nominal Hose I.D. in	M mm ±0.8	M in ±0.03	M_1 mm ±0.8	M_1 in ±0.03	N[2] mm ±0.8	N[2] in ±0.03	N_1[2] mm ±0.8	N_1[2] in ±0.03	P[2] min mm	P[2] min in	S[2,3] max mm	S[2,3] max in	T[4] Ref mm	T[4] Ref in
3/16	35.8	1.41	34.8	1.37	18.0	0.71	17.0	0.67	9.7	0.38	11.7	0.46	4.6	0.18
1/4	36.3	1.43	34.8	1.37	18.8	0.74	17.0	0.67	9.7	0.38	11.2	0.44	4.6	0.18
1/4	38.1	1.50	37.3	1.47	15.4	0.92	20.1	0.79	14.2	0.56	15.5	0.61	5.3	0.21
5/16	36.3	1.43	34.0	1.34	19.8	0.78	18.0	0.71	9.7	0.38	30.7	1.21	4.6	0.18
5/16	38.1	1.50	35.3	1.39	24.4	0.96	21.3	0.84	14.2	0.56	16.3	0.64	5.3	0.21
3/8	36.3	1.43	33.3	1.31	20.6	0.81	19.8	0.78	9.7	0.38	30.7	1.21	4.6	0.18
3/8	38.1	1.50	35.3	1.39	25.1	0.99	23.1	0.91	14.2	0.56	15.2	0.60	5.3	0.21
3/8	39.9	1.57	38.1	1.50	25.1	0.99	21.3	0.84	14.2	0.56	15.2	0.60	6.1	0.24
1/2	38.9	1.53	34.5	1.36	26.2	1.03	25.1	0.99	14.2	0.56	32.0	1.26	5.3	0.21
1/2	40.6	1.60	36.3	1.43	26.2	1.03	16.5	0.65	14.2	0.56	—	—	6.1	0.24
1/2	42.7	1.68	39.6	1.56	31.0	1.22	29.2	1.15	19.0	0.75	21.3	0.84	6.1	0.24
1/2	45.5	1.79	41.7	1.64	31.0	1.22	27.7	1.09	19.0	0.75	21.1	0.85	7.6	0.30
5/8	40.6	1.60	35.1	1.38	27.7	1.09	26.2	1.03	14.2	0.56	35.6	1.40	6.1	0.24
5/8	42.7	1.68	37.8	1.49	32.5	1.28	32.0	1.26	19.0	0.75	22.4	0.88	6.1	0.24
5/8	45.5	1.79	40.1	1.58	32.5	1.28	30.2	1.19	19.0	0.75	22.1	0.87	7.6	0.30
3/4	40.6	1.60	33.0	1.30	29.2	1.15	28.7	1.13	14.2	0.56	34.0	1.34	6.1	0.24
3/4	42.7	1.68	35.8	1.41	34.0	1.34	33.8	1.33	19.0	0.75	32.5	1.28	6.1	0.24
3/4	45.5	1.79	37.8	1.49	34.0	1.34	32.5	1.28	19.0	0.75	22.9	0.90	7.6	0.30
1	42.7	1.68	32.0	1.26	37.3	1.47	40.4	1.59	19.0	0.75	31.0	1.22	6.1	0.24
1	45.5	1.79	33.8	1.33	37.3	1.47	38.9	1.53	19.0	0.75	33.8	1.33	7.6	0.30
1	46.2	1.82	39.4	1.55	42.2	1.66	39.6	1.56	23.9	0.94	29.0	1.14	8.4	0.33

[1] Dryseal American Standard Pipe Thread. See General Specifications.
[2] Where SAE Short Pipe Thread is authorized by purchaser, dimensions L, N, N_1, and S are reduced in accordance with the reduction of pipe thread length. See SAE J476 Tables 3 and 4.
[3] At manufacturer's option through passages in fitting shown in Figure 1 may conform with the smaller diameter specified or be counterbored to the larger diameter for the appropriate end for depth S.
[4] Minimum design thickness, not subject to inspection.
[5] For reference purposes only, not intended for inspection.

Unless otherwise specified, tolerance on hole diameters designated drill in the dimensional tables shall be as tabulated in Table 3.

Tolerances on all dimensions not otherwise limited shall be ±0.25 mm (±0.010 in). Angular tolerance on axis of ends on elbows shall be ±2.50 degrees for sizes up to and including 3/8 in and ±1.50 degrees for sizes larger than 3/8 in. The F diameter shall be concentric with the E diameter within 0.25 mm (0.010 in) full indicator reading (FIR).

Where so illustrated and not otherwise specified, hexagon corners shall be chamfered 30 degrees ± 5 degrees to a diameter equal to the nominal width across flats, with a tolerance of -0.41 mm (-0.016 in); or where design permits, corners may be chamfered to the diameter of the abutting surface provided the length of chamfer does not exceed that produced by the 30 degree chamfer previously described.

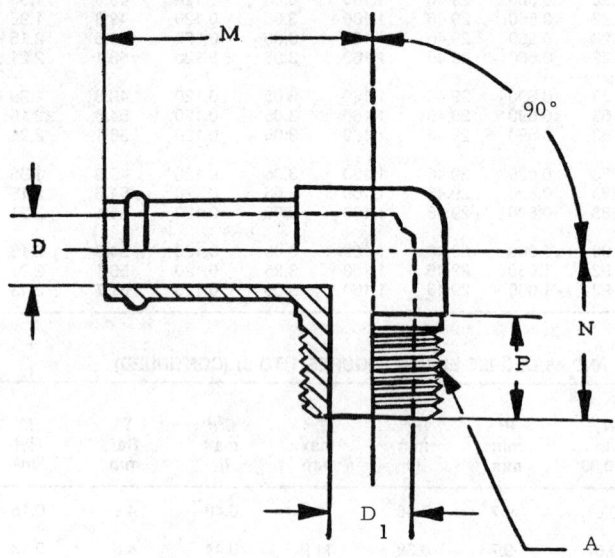

FIGURE 2—90° ELBOW (430260)

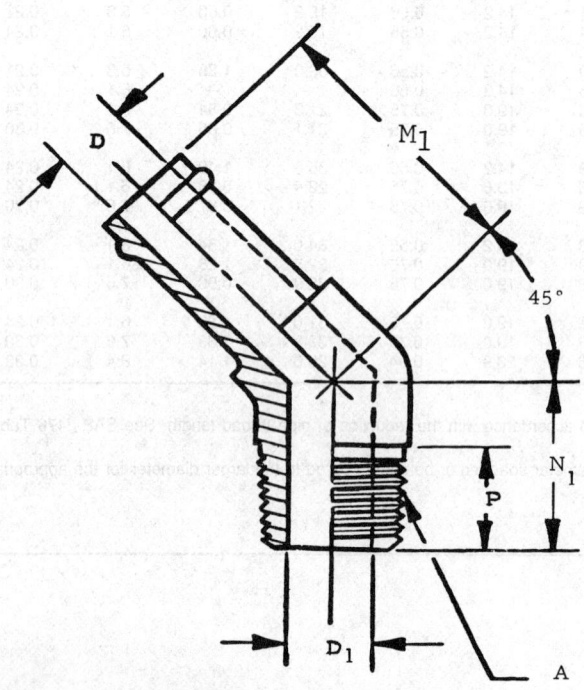

FIGURE 3—45° ELBOW (430360)

3.3 Passages—Where passages in straight fittings are machined from opposite ends, the offset at the meeting point shall not exceed 0.38 mm (0.015 in). The cross-sectional area at the junction of passages in angle fittings shall not be less than that of the smallest passage. Where the passage is specified as a maximum, the minimum shall be no less than the minimum diameter of the smallest passage in the fitting.

3.4 Wall Thickness—Unless otherwise designated, the wall thickness at any point on fittings shall not be less than the thickness established by the specified dimensions, tolerances, and eccentricities for inner and outer surfaces.

3.5 Contour—Details of contour shall be optional with manufacturer provided the tabulated dimensions are maintained and serviceability of the fittings is not impaired. Wrench flats on elbows shall be optional. Where extruded or forged shapes are reduced to conserve material, the wall thickness, unless otherwise specified, shall not be less than the respective minimum values tabulated in Table 4.

3.6 Pipe Threads—Pipe threads, unless there is specific authorization to the contrary, shall conform to the Dryseal American Standard Taper Pipe Thread (NPTF). At purchaser's option, the pipe thread on beaded tube hose fittings may be shortened in conformity with the SAE short Dryseal Taper Pipe Thread (PTF-SAE Short). Specifications for pipe threads are given in detail in SAE J476.

The length of full form external thread shall not be shorter than L, plus one pitch (thread) for Dryseal NPTF and L, for Dryseal PTF-SAE Short, except that where thread is cut through into a relieved body or undercut on the fitting, the minimum full thread length may be reduced by one pitch (thread).

Where external pipe threads are produced by roll threading, the diameter of the unthreaded portion of shank adjacent to shoulder may be reduced to the E, basic diameter.

The tube fitting dimensions tabulated herein are based on length of the Dryseal American Standard Taper Pipe Thread (NPTF), it being the consensus of manufacturers and users that trouble-free assembly cannot be assured unless a full length is used.

External pipe threads shall be chamfered from the diameters shown in Table 2 to produce the specified length of chamfered or partial thread.

3.7 Material and Manufacture—Beaded tube hose fittings shall be made from brass, steel, or multiple component braze design, as specified by the purchaser and, at manufacturer's option and in accordance with his process of manufacture, may be milled from the bar or forged. Brass shall be SAE UNS C36000 (half-hard), SAE UNS C34500, or SAE UNS C35000 for bar or extruded stock and SAE UNS C37700 for forgings.

(R)   **3.8 Finish**—The external surfaces and threads of all carbon steel parts shall be plated or coated with a suitable material that passes a 72 h salt spray test in accordance with ASTM B 117. Any appearance of red rust during the 72 h salt spray test shall be considered failure, except for the following:

a. All internal fluid passages.

b. Edges such as hex points, serrations, and crests of threads where there may be mechanical deformation of the plating or coating typical of mass-produced parts or shipping effects.

c. Areas where there is mechanical deformation of the plating or coating caused by crimping, flaring, bending, and other post-plate metal forming operations.

d. Areas where the parts are suspended or affixed in the test chamber where condensate can accumulate.

NOTE—Cadmium plating is not preferred due to environmental reasons. Parts manufactured to this standard after January 1, 1977, shall not be cadmium plated. Internal fluid passages shall be protected from corrosion during storage. Changes in plating may affect assembly torques and require requalification, when applicable.

3.9 Workmanship—Workmanship shall conform to the best commercial practice to produce high-quality fittings. Fittings shall be free from all hanging burrs, loose scale, and slivers which might become dislodged in usage and all other defects which might affect their serviceability. All sealing surfaces must be smooth except that annular tool marks up to 2.5 m (100 in) maximum shall be permissible. Fittings manufactured from multiple component braze design shall not leak under 862 kPa (125 psi) or vacuum test of 84 kPa (25 in of Hg).

3.10 Assembly Considerations—Where it is not objectionable from a function or production standpoint, the use of compatible lubricant or sealant may be desirable in assembling Dryseal pipe threads on beaded tube hose fittings to minimize galling and effect a pressure-tight seal.

TABLE 2—DIAMETERS FOR CHAMFER OF EXTERNAL PIPE THREADS

Nominal Pipe Thread Size, in	External Thread Chamfer Dia[1] max mm	External Thread Chamfer Dia[1] max in	External Thread Chamfer Dia[1] min mm	External Thread Chamfer Dia[1] min in	External Thread Length of Chamfered or Partial Thread min mm	External Thread Length of Chamfered or Partial Thread min in	External Thread Length of Chamfered or Partial Thread max mm	External Thread Length of Chamfered or Partial Thread max in
1/8	8.1	0.32	7.6	0.30	0.94	0.037	1.40	0.055
1/4	10.7	0.42	10.2	0.40	1.42	0.056	2.13	0.084
3/8	14.0	0.55	13.5	0.53	1.42	0.056	2.13	0.084
1/2	17.3	0.68	16.8	0.66	1.80	0.071	2.72	0.107
3/4	22.6	0.89	22.1	0.87	1.80	0.071	2.72	0.107
1	28.4	1.12	27.7	1.09	2.21	0.087	3.30	0.130

[1] Tabulated diameters conform with Appendix A, SAE J476.

TABLE 3—TOLERANCE ON HOLE DIAMETERS DESIGNATED DRILL

Drill Size Range mm	Drill Size Range in	Tolerance on Hole Diameter Plus mm	Tolerance on Hole Diameter Plus in	Tolerance on Hole Diameter Minus mm	Tolerance on Hole Diameter Minus in
0.343- 4.699	0.0135-0.1850	0.08	0.003	0.05	0.002
4.762- 6.299	0.1875-0.2480	0.10	0.004	0.05	0.002
6.350-19.050	0.2500-0.7500	0.15	0.006	0.08	0.003
19.251-25.400	0.7579-1.0000	0.18	0.007	0.10	0.004

TABLE 4—WALL THICKNESS

Nominal Pipe Thread Size, in	Wall Thickness, min Extruded Shape[1] mm	Wall Thickness min Extruded Shape[1] in	Wall Thickness min Forged Shape mm	Wall Thickness min Forged Shape in
1/8	1.0	0.004	1.52	0.060
1/4	1.3	0.005	1.90	0.075
3/8	1.5	0.06	2.29	0.090
1/2	2.0	0.08	2.54	0.100
3/4	2.0	0.08	2.54	0.100
1	2.0	0.08	3.05	0.120

[1] Applies to reduction to one plane of shape only.

HYDRAULIC HOSE—SAE J517 JUN93   SAE Standard

Report of the Construction and Industrial Technical Committee, Nonmetallic Materials Committee, and Tube, Pipe, Hose, and Lubrication Fittings Committee, approved January 1952. Completely revised by the Fluid Conductors and Connectors Technical Committee May 1989. Completely revised by the Fluid Conductors and Connectors Technical Committee SC2—Hydraulic Hose and Hose Fittings, April 1991, and revised June 1993. Rationale statements available.

1. Scope—This SAE Standard provides general, dimensional, and performance specifications for the most common hoses used in hydraulic systems on mobile and stationary equipment.

The general specifications contained in Sections 1 through 12 are applicable to all hydraulic hoses and supplement the detailed specifications for the 100R-series hoses contained in the later sections of this document. (See Table 1.)

This document shall be utilized as a procurement document only to the extent as agreed upon by the manufacturer and user.

(R) The rated working pressure of a hose assembly comprising SAE J517 hose and hose fitting end connections per SAE J516, J518, J1453, etc., shall not exceed the lower of the respective SAE working pressure rated values.

2. References

2.1 Applicable Documents—The following publications form a part of this specification to the extent specified herein. The latest issue of SAE publications shall apply.

2.1.1 SAE PUBLICATIONS—Available from SAE, 400 Commonwealth Drive, Warrendale, PA 15096-0001.

SAE J343—Tests and Procedures for SAE 100R Series Hydraulic Hose and Hose Assemblies

SAE J516—Hydraulic Hose Fittings

SAE J1273—Selection, Installation, and Maintenance of Hose and Hose Assemblies

SAE J1401—Road Vehicle—Hydraulic Brake Hose Assemblies for Use with Nonpetroleum-Base Hydraulic Fluids

2.1.2 ASTM PUBLICATION—Available from ASTM, 1916 Race Street, Philadelphia, PA 19103-1187.

ASTM D 792—Test Method for Specific Gravity (Relative Density) and Density of Plastics by Displacement

2.1.3 FEDERAL STANDARD

Federal Standard 595A

2.1.4 MILITARY PUBLICATIONS—Available from Naval Publications and Forms Center, 700 Robbins Avenue, Philadelphia, PA 19111.

MIL-H-83282

MIL-H-8446
MIL-L-7808

2.2 Related Publications—The following documents contain provisions which, through reference in this text, constitute provisions of this document. At the time of publication, the editions indicated were valid. All standards are subject to revision, and parties to agreements based on this document are encouraged to investigate the possibility of applying the most recent edition of the document indicated as follows. Members of IEC and ISO maintain registers of currently valid International Standards.

2.2.1 SAE PUBLICATIONS—Available from SAE, 400 Commonwealth Drive, Warrendale, PA 15096-0001.

SAE J343 APR91—Test and Procedures for SAE 100R Series Hydraulic Hose and Hose Assemblies

SAE J517 APR91—Hydraulic Hose

SAE J1176 MAY86—External Leakage Classifications for Hydraulic Systems

SAE J1273 MAY86—Selection, Installation, and Maintenance of Hose and Hose Assemblies

SAE J1401 JUN90—Color Assignments to Hose Manufacturers - Appendix B in SAE J1401 JUN85

SAE J1927 OCT88—Cumulative Damage Analysis for Hydraulic Hose Assemblies

SAE J1942 MAR89—Hose and Hose Assemblies for Marine Applications

2.2.2 ISO PUBLICATIONS—Available from ANSI, 11 West 42nd Street, New York, NY 10036-8002.

ISO 1402-1984—Rubber and plastics hoses and hose assemblies—Hydrostatic testing

ISO 1436-1978—Rubber products—Hoses and hose assemblies—Wire reinforced hydraulic type

ISO 1817-1985—Rubber vulcanized—Determination of the effect of liquids

ISO 4671-1984—Rubber and plastic hoses and hose assemblies—Methods of measurement of dimensions

ISO 6803-1984—Rubber or plastic hoses and hose assemblies—Hydraulic pressure impulse test without flexing

ISO 6945-1983—Rubber hoses—Determination of abrasion resistance of the outer cover

3. Age Control—Age control of rubber hose is a method for designating a period of time, following vulcanization, during which it is reasonable to expect that this product retains full capabilities for rendering the intended service, provided it has been stored as prescribed in SAE J1273.

Hose and hose assemblies are affected by exposure to ozone, oxygen, heat, sunlight, rain, and other similar environmental factors. Storage of bulk hose and hose assemblies should be in such a manner that exposure to these environmental factors is controlled as much as possible. (See SAE J1273 for selection, installation, and maintenance of hose and hose assemblies.)

Hose and hose assemblies should be stored, handled, shipped, and used in such a manner as to facilitate first-in first-out usages based on manufacturing date on hose or hose assembly.

Hose, in bulk form or in hose assemblies passing visual inspection and proof test, shall be acceptable for use up to and including 40 quarters (10 years) from the date of manufacture to the time received by the user. Shelf life of thermoplastic and polytetrafluoroethylene hose is considered to be unlimited.

4. Fittings—Hydraulic hose is rarely used without fittings. Hose with fittings attached are commonly referred to as hose assemblies. The general and dimensional standards for hydraulic hose fittings are contained in SAE J516.

5. Hose Assemblies—Hose assemblies may be fabricated by the manufacturer, an agent for or customer of the manufacturer, or by the user. Fabrication of permanently attached fittings to hydraulic hose requires specialized assembly equipment. Field attachable fittings (screw style and segment clamp style) can usually be assembled without specialized equipment although many manufacturers provide equipment to assist in this operation.

SAE J517 hose from one manufacturer is usually not compatible with SAE J516 fittings supplied by another manufacturer. It is the responsibility of the fabricator to consult the manufacturer's written assembly instructions or the manufacturers directly before intermixing hose and fittings from two manufacturers. Similarly, assembly equipment from one manufacturer is usually not interchangeable with that of another manufacturer. It is the responsibility of the fabricator to consult the manufacturer's written instructions or the manufacturer directly for proper assembly equipment. Always follow the manufacturer's instructions for proper preparation and fabrication of hose assemblies.

6. Application Factors—Hydraulic hose assemblies have a finite life and factors which will reduce life include:

a. Flexing the hose to less than the specified minimum bend radius

b. Twisting, pulling, kinking, crushing, or abrading the hose

c. Operating above maximum and below minimum temperature

d. Exposing the hose to surge pressures above the maximum operating pressure

e. Intermixing hose, fittings, or assembly equipment not recommended by the manufacturer or not following the manufacturer's instructions for fabricating hose assemblies

Surge pressures, noted in item d, are rapid and transient rises in pressure. Surge pressures will not be indicated on many common pressure gauges and can best be identified on electronic measuring instruments with a high-frequency response.

Refer to SAE J1273 for additional information on application factors.

The test requirements for each 100R-series hose is detailed in this document. Specific procedures for conducting each test are contained in SAE J343. The specified tests provide a baseline for performance capability of a hose assembly under controlled laboratory conditions. It is recognized that hydraulic systems will seldom duplicate these test parameters precisely. The hydraulic designer must consider the system demands on the hose assembly and correlate those demands with the specified requirements, with particular concern for the frequency and amplitude of pressure fluctuations. Rapid pressure cycling with a return to zero accelerates fatigue failures.

7. Size Designations—Hose sizes are normally designated by the nominal hose inside diameter expressed in fractions of inches, or by a dash number which, except for SAE 100R5 and 100R14, represents the number of sixteenth inch increments in the hose inside diameter. For these exceptions, the dash number represents the number of sixteenth inch increments in the outside diameter of tubing having approximately the same inside diameter as that of the hose. See dimensional tables for the respective hoses.

8. Hose Identification—Except for hose with a wire braided exterior, the entire length of hose shall be legibly marked with one or more stripes parallel to the longitudinal axis. Marking shall include, but is not limited to, the SAE hose specification number, including type designation where applicable, the hose dash size number, the fractional (in) nominal hose inside diameter, and the date of manufacture, repeated with the first letter of each repeat not more than 762 mm (30 in) from the first letter of that preceding.

TABLE 1—SUMMARY OF SAE J517 100R-SERIES HOSE MAXIMUM OPERATING PRESSURE $\frac{MPa}{PSI}$

Nominal Hose I.D. Size, in	100R1	100R2	100R3	100R4	100R5	100R6	100R7	100R8	100R9	100R10	100R11	100R12	100R13	100R14
1/8														10.3/1500
3/16	20.7/3000	34.5/5000	10.3/1500		20.7/3000	3.4/500	20.7/3000	34.5/5000		68.9/10000	86.2/12500			10.3/1500
1/4	19.0/2750	34.5/5000	8.6/1250		20.7/3000	2.8/400	19.0/2750	34.5/5000		60.3/8750	77.6/11250			10.3/1500
5/16	17.2/2500	29.3/4250	8.3/1200		15.5/2250	2.8/400	17.2/2500							10.3/1500
3/8	15.5/2250	27.6/4000	7.8/1125			2.8/400	15.5/2250	27.6/4000	31.0/4500	51.7/7500	68.9/10000	27.6/4000		10.3/1500
13/32	15.5/2250				13.8/2000									6.9/1000
1/2	13.8/2000	24.1/3500	6.9/1000		12.1/1750	2.8/400	13.8/2000	24.1/3500	27.6/4000	43.1/6250	51.7/7500	27.6/4000		5.5/800
5/8	10.3/1500	19.0/2750	6.0/875		10.3/1500	2.4/350	10.3/1500	19.0/2750				27.6/4000		5.5/800
3/4	8.6/1250	15.5/2250	5.2/750	2.1/300		2.1/300	8.6/1250	15.5/2250	20.7/3000	34.5/5000	43.1/6250	27.6/4000	34.5/5000	5.5/800
7/8	7.8/1125	13.8/2000			5.5/800									5.5/800
1	6.9/1000	13.8/2000	3.9/565	1.7/250			6.9/1000	13.8/2000	20.7/3000	27.6/4000	34.5/5000	27.6/4000	34.5/5000	5.5/800
1-1/8					4.3/625									4.1/600
1-1/4	4.3/625	11.2/1625	2.6/375	1.4/200					17.2/2500	20.7/3000	24.1/3500	20.7/3000	34.5/5000	
1-3/8					3.4/500									
1-1/2	3.4/500	8.6/1250		1.0/150					13.8/2000	17.2/2500	20.7/3000	17.2/2500	34.5/5000	
1-13/16					2.4/350									
2	2.6/375	7.8/1125		0.7/100					13.8/2000	17.2/2500	20.7/3000	17.2/2500	34.5/5000	
2-3/8					2.4/350									
2-1/2		6.9/1000		0.4/62								17.2/2500		
3				0.4/56	1.4/200									
4				0.2/35										

NOTE: Minimum burst of 100R hoses is at least 4 times operating pressure

(R) Electrically nonconductive 100R7 and 100R8 thermoplastic hoses shall have an orange-colored cover. (Reference color, chip #22510, Federal Standard 595A.) Also, in addition to the information required previously, the word nonconductive or electrically nonconductive shall appear in each marking repeat.

The date of manufacture may be expressed as month, day, and year (2/19/88), month and year (2/88), or quarter and year (1Q88) at the option of the manufacturer.

Date of manufacture is optional with the manufacturer on SAE 100R7 and 100R8.

SAE J517 hoses are referenced by listing, in sequence, the 100R number (100R1, 100R7, etc.), the hose type letters (A, AT, B, or BT) where applicable, and the hose dash size number (-4, -16, -24, etc.).

Examples: 100R2AT-8   1/2 in I.D., 2 Wire, Type AT
           100R4-32   2 in I.D., Suction Hose
           100R14B-16 7/8 in I.D., Elect. Cond. PTFE Tube

For hose with a wire braided exterior, information shall be incorporated on a tag or tape applied to each coil or length of bulk hose. Additionally, except for 100R14, a colored yarn shall be incorporated into the wall of the hose, identifying the manufacturer. The color shall be as designated by the Rubber Manufacturers Association. (See SAE J1401, Appendix B.)

9. Orientation of Offset Elbow Ends in a Hose Assembly—For double elbow and assemblies, it is imperative that the method of description and measurement provide the desired displacement rather than its mirror image. To achieve this, either end may be selected as the reference point, provided angle displacement is determined appropriately (clockwise or counterclockwise) for the reference selected.

As shown in Figure 1, with the centerline of the near end as a base reference, angular displacement is measured counterclockwise to the centerline of the far end.

As shown in Figure 2, with the centerline of the far end as a base reference, angular displacement is measured clockwise to the centerline of the near end.

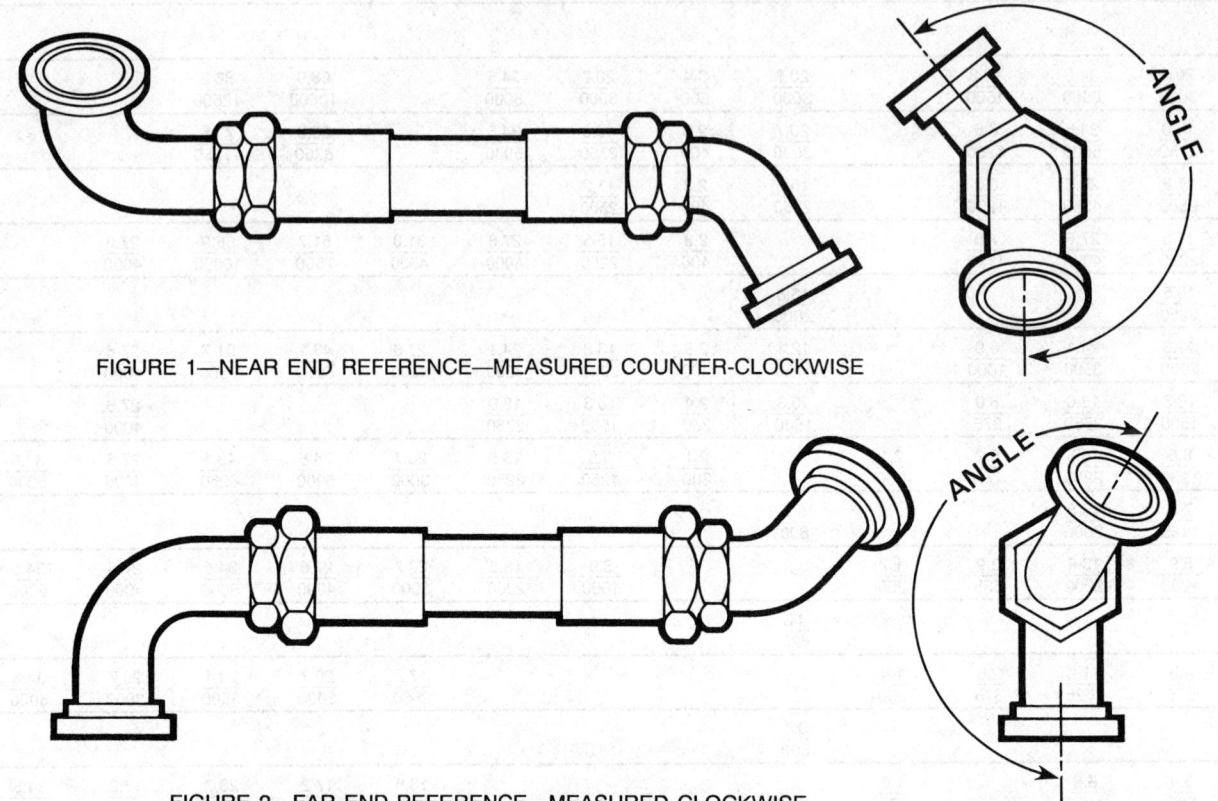

FIGURE 1—NEAR END REFERENCE—MEASURED COUNTER-CLOCKWISE

FIGURE 2—FAR END REFERENCE—MEASURED CLOCKWISE

Displacement angle may have any value up to 360 degrees. Please note that making the angle determination in the wrong direction will result in an unacceptable part.

Unless otherwise specified, a tolerance of ±3 degrees is acceptable for assembly lengths up to 610 mm (24 in) inclusive, and ±5 degrees for assembly lengths over 610 mm (24 in).

10. Assembly Length—Unless otherwise specified, assembly length shall be the overall length measured from the extreme end of one fitting to the extreme end of the other, except for O-ring face seal type fittings which shall be measured from the sealing face. Where elbow fittings are used, measurement shall be to the centerline of the sealing surface of the elbow end. (The sealing surface of female flared elbow fittings shall be the centerline of the outer end of the cone seat.) See Figure 3.

Method of measurement should be specified. Tolerances on assembly length shall be as in Table 2.

TABLE 2—TOLERANCES ON ASSEMBLY LENGTH

Length	Tolerance (Plus or Minus) mm[1]	Tolerance (Plus or Minus) in
Up to 305 mm (12 in) incl	3	0.13
Over 305 mm (12 in) thru 457 mm (18 in) incl	5	0.19
Over 457 mm (18 in) thru 914 mm (36 in) incl	7	0.25
Over 914 mm (36 in)	1%[2]	1%[2]

[1] mm values are rounded approximations from base inch tolerances and should not be used as a base to convert back into inches.
[2] Measured to nearest whole millimeter (or tenth of an inch).

11. Tests—Unless otherwise agreed upon between the manufacturer and purchaser, tests for evaluating conformance of product with specifications shall be on the basis of Qualification Tests and Inspection Tests set forth in this document. Tests may be conducted by the

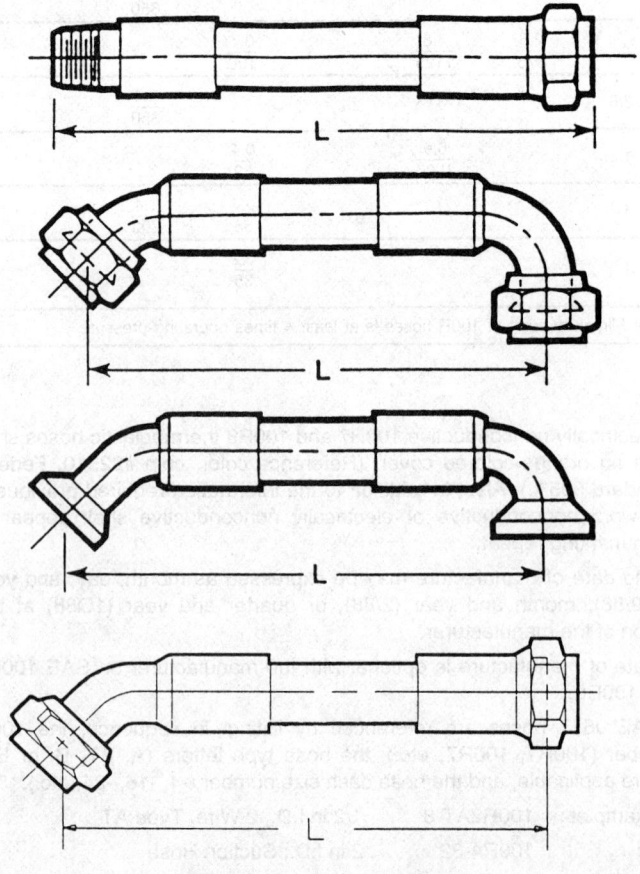

FIGURE 3—O-RING FACE SEAL

manufacturer, the purchaser, or both, as decreed by the purchaser. The tests, sampling, and criteria applicable to both test classifications are given in the detailed specifications and tables for each hose. All tests shall be conducted in accordance with the procedures in SAE J343.

12. Retests and Rejection—In the event of failure of one or more samples to meet any of the tests specified, the product shall be resampled and retested for the test or tests in which it failed. Twice the number of samples designated under the initial test procedure shall be selected from the lot in question for such retests, and failure of any of the retested samples shall be cause for rejection of the entire lot.

13. Steel Wire Reinforced, Rubber Covered Hydraulic Hose (SAE 100R1)—This section covers hose for use with petroleum- and water-base hydraulic fluids within a temperature range of -40 to +100 °C (-40 to +212 °F). Operating temperatures in excess of +100 °C (+212 °F) may materially reduce the life of the hose. Maximum operating pressure, minimum bend radius, and other performance data are specified in Table 3.

It should be noted that the detailed specifications which follow shall be supplemented by the general specifications given at the beginning of this document.

13.1 Dimensions—Dimensions and tolerances applicable to this hose are given in Table 3. The inside diameter of hose shall be concentric with outside diameter of hose and the outer surface of the reinforcement within the limits in Table 4.

13.2 Hose Construction

13.2.1 TYPE A—This hose shall consist of an inner tube of oil-resistant synthetic rubber, a single wire braid reinforcement, and an oil- and weather-resistant synthetic rubber cover. A ply or braid of suitable material may be used over the inner tube and/or over the wire reinforcement to anchor the synthetic rubber to the wire.

13.2.2 TYPE AT—This hose shall be of the same construction as Type A, except having a cover designed to assemble with fittings which do not require removal of the cover or a portion thereof.

13.3 Qualification Tests—For qualification to this section, hose and/or hose assemblies made therefrom shall conform to the following tests and requirements:

13.3.1 DIMENSIONAL CHECK TEST (ALL SAMPLES)—Shall conform to dimensions in Table 3 and these detailed specifications.

13.3.2 PROOF TEST (ALL SAMPLES)—Shall not leak at the proof pressure.

13.3.3 CHANGE IN LENGTH TEST (ONE SAMPLE)—Shall not exceed +2% to -4% change when pressurized to operating pressure.

13.3.4 BURST TEST (ONE 460 MM (18.00 IN) FREE HOSE LENGTH ASSEMBLY)—Shall not leak or fail below the minimum burst pressure.

13.3.5 LEAKAGE TEST (TWO 300 MM (12.00 IN) FREE HOSE LENGTH ASSEMBLIES)—Shall not leak or fail.

13.3.6 COLD BEND TEST (ONE ASSEMBLY)—Shall exhibit no cover cracks or leakage. Exposure shall be at -40 °C (-40 °F).

TABLE 3—DIMENSIONS AND SPECIFICATIONS FOR SAE 100R1 HOSE

Nominal SAE 100R1 Hose ID Size, in	Hose Dash Size	Hose ID Basic mm	Hose ID Basic in	Hose ID Tolerance Plus mm	Hose ID Tolerance Plus in	Hose ID Tolerance Minus mm	Hose ID Tolerance Minus in	Reinforcement Dia Max mm	Reinforcement Dia Max in	Reinforcement Dia Min mm	Reinforcement Dia Min in	Hose OD Type A Max mm	Hose OD Type A Max in	Hose OD Type A Min mm	Hose OD Type A Min in
3/16	-3	4.8	0.188	0.6	0.023	0.2	0.008	10.1	0.398	8.9	0.352	13.5	0.531	11.9	0.469
1/4	-4	6.4	0.250	0.6	0.023	0.2	0.008	11.7	0.461	10.6	0.416	16.7	0.656	15.1	0.594
5/16	-5	7.9	0.312	0.6	0.023	0.2	0.008	13.3	0.523	12.1	0.477	18.3	0.719	16.7	0.656
3/8	-6	9.5	0.375	0.6	0.023	0.2	0.008	15.7	0.617	14.5	0.571	20.6	0.812	19.0	0.750
13/32	-6.5	10.3	0.406	0.8	0.031	0.4	0.015	16.4	0.648	15.3	0.602	21.4	0.844	19.8	0.781
1/2	-8	12.7	0.500	0.8	0.031	0.4	0.015	19.0	0.750	17.5	0.688	23.8	0.938	22.2	0.875
5/8	-10	15.9	0.625	0.8	0.031	0.4	0.015	22.2	0.875	20.6	0.812	27.0	1.062	25.4	1.000
3/4	-12	19.0	0.750	0.8	0.031	0.4	0.015	26.2	1.031	24.6	0.969	31.0	1.219	29.4	1.156
7/8	-14	22.2	0.875	0.8	0.031	0.4	0.015	29.4	1.156	27.8	1.094	34.1	1.344	32.5	1.281
1	-16	25.4	1.000	1.0	0.040	0.4	0.015	34.1	1.344	32.5	1.281	39.3	1.547	36.9	1.453
1-1/4	-20	31.8	1.250	1.2	0.047	0.4	0.015	41.7	1.641	39.3	1.547	47.6	1.875	44.4	1.750
1-1/2	-24	38.1	1.500	1.2	0.047	0.4	0.015	48.0	1.891	45.6	1.797	54.0	2.125	50.8	2.000
2	-32	50.8	2.000	1.2	0.047	0.4	0.015	61.9	2.438	58.7	2.312	68.3	2.688	65.1	2.562

Nominal SAE 100R1 Hose ID Size, in	Hose OD Type AT Max mm	Hose OD Type AT Max in	Cover Thickness[1] Type AT Max mm	Cover Thickness[1] Type AT Max in	Cover Thickness[1] Type AT Min mm	Cover Thickness[1] Type AT Min in	Min Burst Pressure MPa	Min Burst Pressure psi	Proof Pressure MPa	Proof Pressure psi	Max Operating Pressure MPa	Max Operating Pressure psi	Min Bend Radius[2] mm	Min Bend Radius[2] in
3/16	12.5	0.494	1.52	0.060	0.76	0.030	82.7	12000	41.4	6000	20.7	3000	89	3.50
1/4	14.1	0.557	1.52	0.060	0.76	0.030	75.8	11000	37.9	5500	19.0	2750	102	4.00
5/16	15.7	0.619	1.52	0.060	0.76	0.030	68.9	10000	34.5	5000	17.2	2500	114	4.50
3/8	18.1	0.713	1.52	0.060	0.76	0.030	62.0	9000	31.0	4500	15.5	2250	127	5.00
13/32	18.9	0.744	1.52	0.060	0.76	0.030	62.0	9000	31.0	4500	15.5	2250	140	5.50
1/2	21.5	0.846	1.52	0.060	0.76	0.030	55.2	8000	27.6	4000	13.8	2000	178	7.00
5/8	24.7	0.971	1.52	0.060	0.76	0.030	41.4	6000	20.7	3000	10.3	1500	203	8.00
3/4	28.6	1.127	1.52	0.060	0.76	0.030	34.5	5000	17.2	2500	8.6	1250	241	9.50
7/8	31.8	1.252	1.52	0.060	0.76	0.030	31.0	4500	15.5	2250	7.8	1125	279	11.00
1	36.6	1.440	1.52	0.060	0.76	0.030	27.6	4000	13.8	2000	6.9	1000	305	12.00
1-1/4	44.8	1.766	2.03	0.080	1.02	0.040	17.2	2500	8.6	1250	4.3	625	419	16.50
1-1/12	52.0	2.047	2.54	0.100	1.27	0.050	13.8	2000	6.9	1000	3.4	500	508	20.00
2	65.9	2.594	2.54	0.100	1.27	0.050	10.3	1500	5.2	750	2.6	375	635	25.00

[1] Cover thickness shall be measured by means of a dial indicator depth gage having a rounded foot placed parallel to the hose, bridging a groove obtained by stripping a 12.5 to 25.4 mm (0.50 to 1.00 in) width of cover from the hose. A mandrel should be placed in the hose bore to insure freedom from misalignment.
[2] Bend radius measured at inside of bend.

TABLE 4—HOSE CONCENTRICITY (100R1)

Nominal Hose ID, in	Concentricity, FIR ID to OD mm	Concentricity, FIR ID to OD in	Concentricity, FIR ID to Reinforcement mm	Concentricity, FIR ID to Reinforcement in
Up to 1/4, incl	0.8	0.030	0.4	0.017
Over 1/4 to 7/8, incl	1.0	0.040	0.6	0.024
Over 7/8	1.3	0.050	0.8	0.031

13.3.7 OIL RESISTANCE TEST—After 70 h immersion at 100 °C (212 °F) in ASTM No. 3 oil, the volume change of hose inner tube and cover specimens shall be between 0% and +100%.

13.3.8 OZONE RESISTANCE TEST (TWO SAMPLES)—Specimens shall be subjected to an atmosphere comprised of air and ozone with an ozone partial pressure of 50 mPa (50 parts ozone per 100 million parts of air at standard atmospheric conditions) at an ambient temperature of 40 °C (104 °F). After 70 h exposure, specimens shall not show evidence of cracking or deterioration when viewed with seven-power magnification while still in a stressed condition.

13.3.9 IMPULSE TEST (FOUR UNAGED ASSEMBLIES)—Hose assemblies, when tested at 125% of operating pressure for hose sizes 25.4 mm (1 in) nominal ID and smaller and 100% of operating pressure for hose sizes 31.8 mm (1-1/4 in) nominal ID and larger, with 100 °C (212 °F) circulating petroleum-base test fluid, shall withstand a minimum of 150 000 cycles without leakage or other malfunction.

13.3.10 VISUAL EXAMINATION (ALL SAMPLES)

13.4 Inspection Tests—Inspection tests listed as follows shall be performed on two samples representing each lot of 150 to 3000 m (500 to 10 000 ft) of bulk hose or 100 to 10 000 assemblies. Lots of less than 150 m (500 ft) of hose or 100 assemblies need not be subjected to these tests if a lot has been tested and met the requirements within the previous 12-month period. Requirements shall be same as for corresponding Qualification Tests:

a. Dimensional Check Test (see 13.3.1)
b. Proof Test (see 13.3.2)
c. Change in Length Test (see 13.3.3)
d. Burst Test (see 13.3.4)

In addition, all hose and/or hose assemblies made therefrom shall be subjected to visual examination.

14. High-Pressure, Steel Wire Reinforced, Rubber Covered Hydraulic Hose (SAE 100R2)—This section covers hose for use with petroleum- and water-base hydraulic fluids within a temperature range of -40 to +100 °C (-40 to +212 °F). Operating temperatures in excess of +100 °C (+212 °F) may materially reduce the life of the hose. Maximum operating pressure, minimum bend radius, and other performance data are specified in Table 5.

TABLE 5—DIMENSIONS AND SPECIFICATIONS FOR SAE 100R2 HOSE

Nominal SAE 100R2 Hose ID Size, in	Hose Dash Size	Hose ID Basic mm	Hose ID Basic in	Hose ID Tolerance Plus mm	Hose ID Tolerance Plus in	Hose ID Tolerance Minus mm	Hose ID Tolerance Minus in	Reinforcement Dia Max mm	Reinforcement Dia Max in	Reinforcement Dia Min mm	Reinforcement Dia Min in	Hose OD Types A and B Max mm	Hose OD Types A and B Max in	Hose OD Types A and B Min mm	Hose OD Types A and B Min in
3/16	-3	4.8	0.188	0.6	0.023	0.2	0.008	11.7	0.461	10.6	0.416	16.7	0.656	15.1	0.594
1/4	-4	6.4	0.250	0.6	0.023	0.2	0.008	13.3	0.523	12.1	0.477	18.3	0.719	16.7	0.656
5/16	-5	7.9	0.312	0.6	0.023	0.2	0.008	14.9	0.586	13.7	0.539	19.8	0.781	18.3	0.719
3/8	-6	9.5	0.375	0.6	0.023	0.2	0.008	17.3	0.681	16.1	0.633	22.2	0.875	20.6	0.812
1/2	-8	12.7	0.500	0.8	0.031	0.4	0.015	20.6	0.812	19.0	0.750	25.4	1.000	23.8	0.938
5/8	-10	15.9	0.625	0.8	0.031	0.4	0.015	23.8	0.938	22.2	0.875	28.6	1.125	27.0	1.062
3/4	-12	19.0	0.750	0.8	0.031	0.4	0.015	27.8	1.094	26.2	1.031	32.5	1.281	31.0	1.219
7/8	-14	22.2	0.875	0.8	0.031	0.4	0.015	31.0	1.219	29.4	1.156	35.7	1.406	34.1	1.344
1	-16	25.4	1.000	1.0	0.040	0.4	0.015	35.7	1.406	34.1	1.344	40.9	1.609	38.5	1.516
1-1/4	-20	31.8	1.250	1.2	0.047	0.4	0.015	45.6	1.797	43.2	1.703	52.4	2.062	49.2	1.938
1-1/2	-24	38.1	1.500	1.2	0.047	0.4	0.015	52.0	2.047	49.6	1.953	58.7	2.312	55.6	2.188
2	-32	50.8	2.000	1.2	0.047	0.4	0.015	64.7	2.547	62.3	2.453	71.4	2.812	68.3	2.688
2-1/2	-40	63.5	2.500	1.6	0.062	0.4	0.015	77.8	3.062	74.6	2.937	84.1	3.312	80.9	3.187

Nominal SAE 100R2 Hose ID Size, in	Hose OD Types AT and BT Max mm	Hose OD Types AT and BT Max in	Cover Thickness[1] Types AT and BT Max mm	Cover Thickness[1] Types AT and BT Max in	Cover Thickness[1] Types AT and BT Min mm	Cover Thickness[1] Types AT and BT Min in	Min Burst Pressure MPa	Min Burst Pressure psi	Proof Pressure MPa	Proof Pressure psi	Max Operating Pressure MPa	Max Operating Pressure psi	Min Bend Radius[2] mm	Min Bend Radius[2] in
3/16	14.1	0.557	1.52	0.060	0.79	0.031	137.9	20000	68.9	10000	34.5	5000	89	3.50
1/4	15.7	0.619	1.52	0.060	0.79	0.031	137.9	20000	68.9	10000	34.5	5000	102	4.00
5/16	17.3	0.682	1.52	0.060	0.79	0.031	117.2	17000	58.6	8500	29.3	4250	114	4.50
3/8	19.7	0.777	1.52	0.060	0.79	0.031	110.3	16000	55.2	8000	27.6	4000	127	5.00
1/2	23.1	0.908	1.52	0.060	0.79	0.031	96.5	14000	48.3	7000	24.1	3500	178	7.00
5/8	26.3	1.034	1.52	0.060	0.79	0.031	75.8	11000	37.9	5500	19.0	2750	203	8.00
3/4	30.2	1.190	1.52	0.060	0.79	0.031	62.0	9000	31.0	4500	15.5	2250	241	9.50
7/8	33.4	1.315	1.52	0.060	0.79	0.031	55.2	8000	27.6	4000	13.8	2000	279	11.00
1	38.9	1.531	2.16	0.085	1.07	0.042	55.2	8000	27.6	4000	13.8	2000	305	12.00
1-1/4	49.6	1.953	2.54	0.100	1.27	0.050	44.8	6500	22.4	3250	11.2	1625	419	16.50
1-1/2	56.0	2.203	2.54	0.100	1.27	0.050	34.5	5000	17.2	2500	8.6	1250	508	20.00
2	68.6	2.703	2.54	0.100	1.27	0.050	31.0	4500	15.5	2250	7.8	1125	635	25.00
2-1/2	—	—	—	—	—	—	27.6	4000	13.8	2000	6.9	1000	762	30.00

[1] Cover thickness shall be measured by means of a dial indicator depth gage having a rounded foot placed parallel to the hose, bridging a groove obtained by stripping a 12.5 to 25.4 mm (0.50 to 1.00 in) width of cover from the hose. A mandrel should be placed in the hose bore to insure freedom from misalignment.
[2] Bend radius measured at inside of bend.

It should be noted that the detailed specifications which follow shall be supplemented by the general specifications given at the beginning of this document.

14.1 Dimensions—Dimensions and tolerances applicable to this hose are given in Table 5. The inside diameter of hose shall be concentric with outside diameter of hose and the outer surface of the reinforcement within the limits in Table 6.

TABLE 6—HOSE CONCENTRICITY (100R2)

Nominal Hose ID, in	Concentricity, FIR ID to OD mm	Concentricity, FIR ID to OD in	Concentricity, FIR ID to Reinforcement mm	Concentricity, FIR ID to Reinforcement in
Up to 1/4, incl	0.8	0.030	0.5	0.021
Over 1/4 to 7/8, incl	1.0	0.040	0.7	0.028
Over 7/8	1.3	0.050	0.9	0.035

14.2 Hose Construction—The hose shall consist of an inner tube of oil-resistant synthetic rubber, steel wire reinforcement according to hose type detailed in 14.2.1 through 14.2.4 and an oil- and weather-resistant synthetic rubber cover. A ply or braid of suitable material may be used over the inner tube and/or over the wire reinforcement to anchor the synthetic rubber to the wire.

14.2.1 TYPE A—This hose shall have two braids of wire reinforcement.

14.2.2 TYPE B—This hose shall have two spiral plies and one braid of wire reinforcement.

14.2.3 TYPE AT—This hose shall be of the same construction as Type A, except having a cover designed to assemble with fittings which do not require removal of the cover or a portion thereof.

14.2.4 TYPE BT—This hose shall be of the same construction as Type B, except having a cover designed to assemble with fittings which do not require removal of the cover or a portion thereof.

14.3 Qualification Tests—For qualification to this section, hose and/or hose assemblies made therefrom shall conform to the following tests and requirements:

14.3.1 DIMENSIONAL CHECK TEST (ALL SAMPLES)—Shall conform to dimensions in Table 5 and these detailed specifications.

14.3.2 PROOF TEST (ALL SAMPLES)—Shall not leak at the proof pressure.

14.3.3 CHANGE IN LENGTH TEST (ONE SAMPLE)—Shall not exceed +2% to -4% change when pressurized to operating pressure.

14.3.4 BURST TEST (ONE 460 MM (18.00 IN) FREE HOSE LENGTH ASSEMBLY)—Shall not leak or fail below the minimum burst pressure.

14.3.5 LEAKAGE TEST (TWO 300 MM (12.00 IN) FREE HOSE LENGTH ASSEMBLIES)—Shall not leak or fail.

14.3.6 COLD BEND TEST (ONE ASSEMBLY)—Shall exhibit no cover cracks or leakage. Exposure shall be at -40 °C (-40 °F).

14.3.7 OIL RESISTANCE TEST—After 70 h immersion at 100 °C (212 °F) in ASTM No. 3 oil, the volume change of hose inner tube and cover specimens shall be between 0% and +100%.

14.3.8 OZONE RESISTANCE TEST (TWO SAMPLES)—Specimens shall be subjected to an atmosphere comprised of air and ozone with an ozone partial pressure of 50 mPa (50 parts ozone per 100 million parts of air at standard atmospheric conditions) at an ambient temperature of 40 °C (104 °F). After 70 h exposure, specimens shall not show evidence of cracking or deterioration when viewed with seven-power magnification while still in a stressed condition.

14.3.9 IMPULSE TEST (FOUR UNAGED ASSEMBLIES)—Hose assemblies, when tested at 133% of operating pressure with 100 °C (212 °F) circulation petroleum-base test fluid, shall withstand a minimum of 200 000 cycles without leakage or other malfunction.

14.3.10 VISUAL EXAMINATION (ALL SAMPLES)

14.4 Inspection Tests—Inspection tests listed (a through d), shall be performed on two samples representing each lot of 150 to 300 m (500 to 10 000 ft) of bulk hose or 100 to 10 000 assemblies. Lots of less than 150 m (500 ft) of hose or 100 assemblies need not be subjected to these tests if a lot has been tested and met the requirements within the previous 12-month period. Requirements shall be same as for corresponding Qualification Tests:

a. Dimensional Check Test (see 14.3.1)

b. Proof Test (see 14.3.2)

c. Change in Length Test (see 14.3.3)

d. Burst Test (see 14.3.4)

In addition, all hose and/or hose assemblies made therefrom shall be subjected to visual examination.

15. Double Fiber Braid (Nonmetallic), Rubber Covered Hydraulic Hose (SAE 100R3)

—This section covers hose for use with petroleum- and water-base hydraulic fluids within a temperature range of -40 to +100 °C (-40 to +212 °F). Operating temperatures in excess of +100 °C (+212 °F) may materially reduce the life of the hose. Maximum operating pressure, minimum bend radius, and other performance data are specified in Table 7.

It should be noted that the detailed specifications which follow shall be supplemented by the general specifications given at the beginning of this document.

15.1 Dimensions—Dimensions and tolerances applicable to this hose are given in Table 7. The inside diameter and outside diameter of the hose shall be concentric within the limits in Table 8.

15.2 Hose Construction—The hose shall consist of an inner tube of oil-resistant synthetic rubber, two braids of suitable textile yarn, and an oil- and weather-resistant synthetic rubber cover.

15.3 Qualification Tests—For qualification to this section, hose and/or hose assemblies made therefrom shall conform to the following tests and requirements:

15.3.1 DIMENSIONAL CHECK TEST (ALL SAMPLES)—Shall conform to dimensions in Table 7 and these detailed specifications.

15.3.2 PROOF TEST (ALL SAMPLES)—Shall not leak at the proof pressure.

15.3.3 CHANGE IN LENGTH TEST (ONE SAMPLE)—Shall not exceed +2 to -4% change when pressurized to operating pressure.

15.3.4 BURST TEST (ONE 460 MM (18.00 IN) FREE HOSE LENGTH ASSEMBLY)—Shall not leak or fail below the minimum burst pressure.

15.3.5 LEAKAGE TEST (TWO 300 MM (12.00 IN) FREE HOSE LENGTH ASSEMBLIES)—Shall not leak or fail.

15.3.6 COLD BEND TEST (ONE ASSEMBLY)—Shall exhibit no cover cracks or leakage. Exposure shall be at -40 °C (-40 °F).

15.3.7 OIL RESISTANCE TEST—After 70 h immersion at 100 °C (212 °F) in ASTM No. 3 oil, the volume change of hose inner tube and cover specimens shall be between 0% and +100%.

15.3.8 OZONE RESISTANCE TEST (TWO SAMPLES)—Specimens shall be subjected to an atmosphere comprised of air and ozone with an ozone partial pressure of 50 mPa (50 parts ozone per 100 million parts of air at standard atmospheric conditions) at an ambient temperature of 40 °C (104 °F). After 70 h exposure, specimens shall not show evidence of cracking or deterioration when viewed with seven-power magnification while still in a stressed condition.

15.3.9 IMPULSE TEST (FOUR UNAGED ASSEMBLIES)—Hose assemblies, when tested at 133% of operating pressure, with 100 °C (212 °F) circulating petroleum-base test fluid, shall withstand a minimum of 200 000 cycles without leakage or other malfunction.

15.3.10 VISUAL EXAMINATIONS (ALL SAMPLES)

TABLE 7—DIMENSIONS AND SPECIFICATIONS FOR SAE 100R3 HOSE

Nominal SAE 100R3 Hose ID Size, in	Hose Dash Size	Hose ID Basic mm	Hose ID Basic in	Hose ID Tolerance Plus mm	Hose ID Tolerance Plus in	Hose ID Tolerance Minus mm	Hose ID Tolerance Minus in	Hose OD Max mm	Hose OD Max in	Hose OD Min mm	Hose OD Min in
3/16	−3	4.8	0.188	0.6	0.025	0.3	0.010	13.5	0.531	11.9	0.469
1/4	−4	6.4	0.250	0.6	0.025	0.3	0.010	15.1	0.594	13.5	0.531
5/16	−5	7.9	0.312	0.6	0.025	0.3	0.010	18.3	0.719	16.7	0.656
3/8	−6	9.5	0.375	0.6	0.025	0.3	0.010	19.8	0.781	18.3	0.719
1/2	−8	12.7	0.500	0.8	0.030	0.3	0.010	24.6	0.969	23.0	0.906
5/8	−10	15.9	0.625	0.8	0.030	0.3	0.010	27.8	1.094	26.2	1.031
3/4	−12	19.0	0.750	0.8	0.030	0.3	0.010	32.5	1.281	31.0	1.219
1	−16	25.4	1.000	0.8	0.030	0.3	0.010	39.3	1.547	36.9	1.453
1-1/4	−20	31.8	1.250	1.1	0.045	0.4	0.015	46.0	1.812	42.9	1.688

Nominal SAE 100R3 Hose ID Size, in	Min Burst Pressure MPa	Min Burst Pressure psi	Proof Pressure MPa	Proof Pressure psi	Max Operating Pressure MPa	Max Operating Pressure psi	Min Bend Radius[1] mm	Min Bend Radius[1] in
3/16	41.4	6000	20.7	3000	10.3	1500	76	3.00
1/4	34.5	5000	17.2	2500	8.6	1250	76	3.00
5/16	33.1	4800	16.5	2400	8.3	1200	102	4.00
3/8	31.0	4500	15.5	2250	7.8	1125	102	4.00
1/2	27.6	4000	13.7	2000	6.9	1000	127	5.00
5/8	24.1	3500	12.1	1750	6.0	875	140	5.50
3/4	20.7	3000	10.3	1500	5.2	750	152	6.00
1	15.5	2250	7.8	1125	3.9	565	203	8.00
1-1/4	10.3	1500	5.2	750	2.6	375	254	10.00

[1] Bend radius measured at inside of bend.

TABLE 8—HOSE CONCENTRICITY (100R3)

Nominal Hose ID, in	Concentricity, FIR mm	Concentricity, FIR in
Up to 1/4, incl	0.8	0.030
Over 1/4 to 3/4, incl	1.0	0.040
Over 3/4	1.3	0.050

15.4 Inspection Tests—Inspection tests listed as follows shall be performed on two samples representing each lot of 150 to 3000 m (500 to 10 000 ft) of bulk hose or 100 to 10 000 assemblies. Lots of less than 150 m (500 ft) of hose or 100 assemblies need not be subjected to these tests if a lot has been tested and met the requirements within the previous 12-month period. Requirements shall be same as for corresponding Qualification Tests:

 a. Dimensional Check Test (see 15.3.1)
 b. Proof Test (see 15.3.2)
 c. Change in Length Test (see 15.3.3)
 d. Burst Test (see 15.3.4)

In addition, all hose and/or hose assemblies made therefrom shall be subjected to visual examination.

16. Wire Inserted Hydraulic Suction Hose (SAE 100R4)—This section covers hose for use in low pressure and vacuum applications with petroleum- and water-base hydraulic fluids within a temperature range of -40 to +100 °C (-40 to +212 °F). Operating temperatures in excess of 100 °C (212 °F) may materially reduce the life of the hose. Maximum operating pressure, minimum bend radius, and other performance data are specified in Table 9.

It should be noted that the detailed specifications which follow shall be supplemented by the general specifications given at the beginning of this document.

16.1 Dimensions—Dimensions and tolerances applicable to this hose are given in Table 9.

16.2 Hose Construction—The hose shall consist of an inner tube of oil-resistant synthetic rubber, a reinforcement consisting of a ply or plies of woven or braided textile fibers with a suitable spiral of body wire, and an oil- and weather-resistant synthetic rubber cover.

16.3 Qualification Tests—For qualification to this section, hose and/or hose assemblies made therefrom shall conform to the following tests and requirements:

16.3.1 Dimensional Check Test (All Samples)—Shall conform to dimensions in Table 9.

16.3.2 Proof Test (All Samples)—Shall not leak at the proof pressure.

16.3.3 Burst Test (One 460 mm (18.00 in) Free Hose Length Assembly)—Shall not leak or fail below the minimum burst pressure.

16.3.4 Cold Bend Test (One Assembly)—Shall exhibit no cover cracks or leakage. Exposure shall be at -40 °C (-40 °F).

16.3.5 Oil Resistance Test—After 70 h immersion at 100 °C (212 °F) in ASTM No. 3 oil, the volume change of hose inner tube and cover specimens shall be between 0% and +100%.

16.3.6 Ozone Resistance Test (Two Samples)—Specimens shall be subjected to an atmosphere comprised of air and ozone with an ozone partial pressure of 50 mPa (50 parts ozone per 100 million parts of air at standard atmospheric conditions) at an ambient temperature of 40 °C (104 °F). After 70 h exposure, specimens shall not show evidence of cracking or deterioration when viewed with seven-power magnification while still in a stressed condition.

16.3.7 Resistance to Vacuum Test (One Sample)—After exposure for 5 min at absolute pressure of 17 kPa (25 in Hg), there shall be no evidence of hose blistering or collapse.

16.3.8 Visual Examination (All Samples)

TABLE 9—DIMENSIONS AND SPECIFICATIONS FOR SAE 100R4 HOSE

Nominal SAE 100R4 Hose ID Size, in	Hose Dash Size	Hose ID Basic mm	Hose ID Basic in	Hose ID Tolerance Plus mm	Hose ID Tolerance Plus in	Hose ID Tolerance Minus mm	Hose ID Tolerance Minus in	Hose OD Max mm	Hose OD Max in	Min Burst Pressure MPa	Min Burst Pressure psi	Proof Pressure MPa	Proof Pressure psi	Max Operating Pressure MPa	Max Operating Pressure psi	Min Bend Radius[1] mm	Min Bend Radius[1] in
3/4	-12	19.0	0.750	0.8	0.031	0.8	0.031	34.9	1.375	8.3	1200	4.1	600	2.1	300	127	5.00
1	-16	25.4	1.000	0.8	0.031	0.8	0.031	41.3	1.625	6.9	1000	3.4	500	1.7	250	152	6.00
1-1/4	-20	31.8	1.250	1.2	0.047	1.2	0.047	50.8	2.000	5.5	800	2.8	400	1.4	200	203	8.00
1-1/2	-24	38.1	1.500	1.2	0.047	1.2	0.047	57.2	2.250	4.1	600	2.1	300	1.0	150	254	10.00
2	-32	50.8	2.000	1.6	0.062	1.6	0.062	69.9	2.750	2.8	400	1.4	200	0.7	100	305	12.00
2-1/2	-40	63.5	2.500	1.6	0.062	1.6	0.062	82.6	3.250	1.7	250	0.9	125	0.4	62	356	14.00
3	-48	76.2	3.000	1.6	0.062	1.6	0.062	95.3	3.750	1.5	225	0.8	112	0.4	56	457	18.00
3-1/2	-56	88.9	3.500	1.6	0.062	1.6	0.062	107.9	4.250	1.2	180	0.6	90	0.3	45	533	21.00
4	-64	101.6	4.000	1.6	0.062	1.6	0.062	120.7	4.750	1.0	140	0.5	70	0.2	35	610	24.00

[1] Bend radius measured at inside of bend.

16.4 Inspection Tests—Inspection tests listed below shall be performed on two samples representing each lot of 150 to 3000 m (500 to 10 000 ft) of bulk hose or 100 to 10 000 assemblies. Lots of less than 150 m (500 ft) of hose or 100 assemblies need not be subjected to these tests if a lot has been tested and met the requirements within the previous 12-month period. Requirements shall be same as for corresponding Qualification Tests:

  a. Dimensional Check Test (see 16.3.1)
  b. Proof Test (see 16.3.2)
  c. Vacuum Test (see 16.3.3)
  d. Burst Test (see 16.3.4)

In addition, all hose and/or hose assemblies made therefrom shall be subjected to visual examination.

17. Single Wire Braid, Textile Covered Hydraulic Hose (SAE 100R5)—This section covers hose for use with petroleum- and water-base hydraulic fluids within a temperature range of -40 to +100 °C (-40 to +212 °F). Operating temperatures in excess of +100 °C (+212 °F) may materially reduce the life of the hose. Maximum operating pressure, minimum bend radius, and other performance data are specified in Table 10.

It should be noted that the detailed specifications which follow shall be supplemented by the general specifications given at the beginning of this document.

17.1 Dimensions—Dimensions and tolerances applicable to this hose are given in Table 10. The inside diameter and outside diameter of the hose shall be concentric within the limits in Table 11.

17.2 Hose Construction—The hose shall consist of an inner tube of oil-resistant synthetic rubber and two textile braids separated by a high tensile steel wire braid. All braids are to be impregnated with an oil- and mildew-resistant synthetic rubber compound.

17.3 Qualification Tests—For qualification to this specification, hose and/or hose assemblies made therefrom shall conform to the following tests and requirements:

17.3.1 DIMENSIONAL CHECK TEST (ALL SAMPLES)—Shall conform to dimensions in Table 10 and these detailed specifications.

17.3.2 PROOF TEST (ALL SAMPLES)—Shall not leak at the proof pressure.

17.3.3 CHANGE IN LENGTH TEST (ONE SAMPLE)—Shall not exceed +2% to -4% change when pressurized to operating pressure.

17.3.4 BURST TEST (ONE 460 MM (18.00 IN) FREE HOSE LENGTH ASSEMBLY)—Shall not leak or fail below the minimum burst pressure.

17.3.5 LEAKAGE TEST (TWO 300 MM (12.00 IN) FREE HOSE LENGTH ASSEMBLIES)—Shall not leak or fail.

17.3.6 COLD BEND TEST (ONE ASSEMBLY)—Shall exhibit no cover leakage. Exposure shall be at -40 °C (-40 °F).

17.3.7 OIL RESISTANCE TEST—After 70 h immersion at 100 °C (212 °F) in ASTM No. 3 oil, the volume change of hose inner tube and cover specimens shall be between 0% and +100%.

TABLE 10—DIMENSIONS AND SPECIFICATIONS FOR SAE 100R5 HOSE

Nominal SAE 100R5 Hose ID Size, in	Hose Dash Size	Hose ID Basic mm	Hose ID Basic in	Hose ID Tolerance Plus mm	Hose ID Tolerance Plus in	Hose ID Tolerance Minus mm	Hose ID Tolerance Minus in	Hose OD Max mm	Hose OD Max in	Hose OD Min mm	Hose OD Min in	Min Burst Pressure MPa	Min Burst Pressure psi	Proof Pressure MPa	Proof Pressure psi	Max Operating Pressure MPa	Max Operating Pressure psi	Min Bend Radius[1] mm	Min Bend Radius[1] in
3/16	-4	4.8	0.188	0.7	0.026	0.0	0.000	13.7	0.539	12.7	0.500	82.7	12000	41.4	6000	20.7	3000	76	3.00
1/4	-5	6.4	0.250	0.8	0.031	0.0	0.000	15.3	0.601	14.3	0.562	82.7	12000	41.4	6000	20.7	3000	86	3.38
5/16	-6	7.9	0.312	0.8	0.031	0.0	0.000	17.6	0.695	16.7	0.656	62.0	9000	31.0	4500	15.5	2250	102	4.00
13/32	-8	10.3	0.406	0.8	0.031	0.0	0.000	20.0	0.789	18.9	0.743	55.2	8000	27.6	4000	13.8	2000	117	4.62
1/2	-10	12.7	0.500	1.0	0.039	0.0	0.000	24.0	0.945	22.8	0.899	48.3	7000	24.1	3500	12.1	1750	140	5.50
5/8	-12	15.9	0.625	1.1	0.042	0.0	0.000	28.0	1.101	26.8	1.055	41.4	6000	20.7	3000	10.3	1500	165	6.50
7/8	-16	22.2	0.875	1.1	0.042	0.0	0.000	32.2	1.266	30.6	1.203	22.1	3200	11.0	1600	5.5	800	187	7.38
1-1/8	-20	28.6	1.125	1.2	0.047	0.0	0.000	38.9	1.531	37.3	1.469	17.2	2500	8.6	1250	4.3	625	229	9.00
1-3/8	-24	34.9	1.375	1.2	0.047	0.0	0.000	45.2	1.781	43.7	1.719	13.8	2000	6.9	1000	3.4	500	267	10.50
1-13/16	-32	46.0	1.812	1.2	0.047	0.0	0.000	57.6	2.266	55.2	2.172	9.7	1400	4.8	700	2.4	350	337	13.25
2-3/8	-40	60.3	2.375	1.6	0.062	0.0	0.000	74.2	2.922	71.8	2.828	9.7	1400	4.8	700	2.4	350	610	24.0
3	-48	76.2	3.000	1.6	0.062	0.0	0.000	91.7	3.609	89.3	3.515	5.5	800	2.7	400	1.4	200	838	33.0

[1] Bend radius measured at inside of bend.

TABLE 11—HOSE CONCENTRICITY (100R5)

Nominal Hose ID, in	Concentricity, FIR mm	Concentricity, FIR in
Up to 13/32, incl	0.6	0.020
Over 13/32	0.8	0.030

TABLE 13—HOSE CONCENTRICITY (100R6)

Nominal Hose ID, in	Concentricity, FIR mm	Concentricity, FIR in
Up to 1/4, incl	0.8	0.030
Over 1/4	1.0	0.040

17.3.8 IMPULSE TEST (FOUR UNAGED ASSEMBLIES)—Hose assemblies, when tested at 125% of operating pressure for hose sizes 22.2 mm (7/8 in) nominal ID and smaller and 100% of operating pressure for hose sizes 28.6 mm (1-1/8 in) nominal ID and larger, with 93 °C (200 °F) circulating petroleum-base test fluid, shall withstand a minimum of 150 000 cycles for hose sizes 22.2 mm (7/8 in) and smaller and a minimum of 100 000 cycles for hose sizes 28.6 mm (1-1/8 in) nominal ID and larger, without leakage or other malfunction. Hose sizes 1-1/8 in nominal ID and larger shall be tested straight.

17.3.9 VISUAL EXAMINATION (ALL SAMPLES)

17.4 Inspection Tests—Inspection tests listed as follows shall be performed on two samples representing each lot of 150 to 3000 m (500 to 10 000 ft) of bulk hose or 100 to 10 000 assemblies. Lots of less than 150 m (500 ft) of hose or 100 assemblies need not be subjected to these tests if a lot has been tested and met the requirements within the previous 12-month period. Requirements shall be same as for corresponding Qualification Tests:

    a. Dimensional Check Test (see 17.3.1)

    b. Proof Test (see 17.3.2)

    c. Change in Length Test (see 17.3.3)

    d. Burst Test (see 17.3.4)

In addition, all hose and/or hose assemblies made therefrom shall be subjected to visual examination.

18. Single Fiber Braid (Nonmetallic), Rubber Covered Hydraulic Hose (SAE 100R6)—This section covers hose for use with petroleum- and water-base hydraulic fluids within a temperature range of -40 to +100 °C (-40 to 212 °F). Operating temperatures in excess of +100 °C (+212 °F) may materially reduce the life of the hose. Maximum operating pressure, minimum bend radius, and other performance data are specified in Table 12.

It should be noted that the detailed specifications which follow shall be supplemented by the general specifications given at the beginning of this document.

18.1 Dimensions—Dimensions and tolerances applicable to this hose are given in Table 12. The inside diameter and outside diameter of the hose shall be concentric within the limits in Table 13.

18.2 Hose Construction—The hose shall consist of an inner tube of oil-resistant synthetic rubber, and one braided ply of suitable textile yarn, and an oil- and weather-resistant synthetic rubber cover.

18.3 Qualification Tests—For qualification to this section, hose and/or hose assemblies made therefrom shall conform to the following tests and requirements:

18.3.1 DIMENSIONAL CHECK TEST (ALL SAMPLES)—Shall conform to dimensions in Table 12 and these detailed specifications.

18.3.2 PROOF TEST (ALL SAMPLES)—Shall not leak at the proof pressure.

18.3.3 CHANGE IN LENGTH TEST (ONE SAMPLE)—Shall not exceed +2% to -4% change when pressurized to operating pressure.

18.3.4 BURST TEST (ONE 460 MM (18.00 IN) FREE HOSE LENGTH ASSEMBLY)—Shall not leak or fail below the minimum burst pressure.

18.3.5 LEAKAGE TEST (TWO 300 MM (12.00 IN) FREE HOSE LENGTH ASSEMBLIES—Shall not leak or fail.

18.3.6 COLD BEND TEST (ONE ASSEMBLY)—Shall exhibit no cover cracks or leakage. Exposure shall be at -40 °C (-40 °F).

18.3.7 OIL RESISTANCE TEST—After 70 h immersion at 100 °C (212 °F) in ASTM No. 3 oil, the volume change of hose inner tube and cover specimens shall be between 0% and +100%.

18.3.8 OZONE RESISTANCE TEST (TWO SAMPLES)—Specimens shall be subjected to an atmosphere comprised of air and ozone with an ozone partial pressure of 50 mPa (50 parts ozone per 100 million parts of air at standard atmospheric conditions) at an ambient temperature of 40 °C (104 °F). After 70 h exposure, specimens shall not show evidence of cracking or deterioration when viewed with seven-power magnification while still in a stressed condition.

18.3.9 VISUAL EXAMINATION (ALL SAMPLES)

18.4 Inspection Tests—Inspection tests listed as follows shall be performed on two samples representing each lot of 150 to 3000 m (500 to 10 000 ft) of bulk hose or 100 to 10 000 assemblies. Lots of less than 150 m (500 ft) of hose or 100 assemblies need not be subjected to these tests if a lot has been tested and met the requirements within the previous 12-month period. Requirements shall be same as for corresponding Qualification Tests:

    a. Dimensional Check Test (see 18.3.1)

    b. Proof Test (see 18.3.2)

    c. Change in Length Test (see 18.3.3)

    d. Burst Test (see 18.3.4)

In addition, all hose and/or hose assemblies made therefrom shall be subjected to visual examination.

TABLE 12—DIMENSIONS AND SPECIFICATIONS FOR SAE 100R6 HOSE

Nominal SAE 100R6 Hose ID Size, in	Hose Dash Size	Hose ID Basic mm	Hose ID Basic in	Hose ID Tolerance Plus mm	Hose ID Tolerance Plus in	Hose ID Tolerance Minus mm	Hose ID Tolerance Minus in	Hose OD Max mm	Hose OD Max in	Hose OD Min mm	Hose OD Min in	Min Burst Pressure MPa	Min Burst Pressure psi	Proof Pressure MPa	Proof Pressure psi	Max Operating Pressure MPa	Max Operating Pressure psi	Min Bend Radius[1] mm	Min Bend Radius[1] in
3/16	-3	4.8	0.188	0.6	0.025	0.3	0.010	11.9	0.469	10.3	0.406	13.8	2000	6.9	1000	3.4	500	51	2.00
1/4	-4	6.4	0.250	0.6	0.025	0.3	0.010	13.5	0.531	11.9	0.469	11.0	1600	5.5	800	2.8	400	64	2.50
5/16	-5	7.9	0.312	0.6	0.025	0.3	0.010	15.1	0.594	13.5	0.531	11.0	1600	5.5	800	2.8	400	76	3.00
3/8	-6	9.5	0.375	0.6	0.025	0.3	0.010	16.7	0.656	15.1	0.594	11.0	1600	5.5	800	2.8	400	76	3.00
1/2	-8	12.7	0.500	0.8	0.030	0.3	0.010	20.6	0.812	19.0	0.750	11.0	1600	5.5	800	2.8	400	102	4.00
5/8	-10	15.9	0.625	0.8	0.030	0.3	0.010	23.8	0.938	22.2	0.875	9.7	1400	4.8	700	2.4	350	127	5.00
3/4	-12	19.0	0.750	0.8	0.030	0.3	0.010	25.4	1.000	27.8	1.093	8.3	1200	4.1	600	2.1	300	152	6.00

[1] Bend radius measured at inside of bend.

19. Thermoplastic Hydraulic Hose (SAE 100R7)—This section covers thermoplastic hose for use with petroleum-and water-base, and synthetic hydraulic fluids within a temperature range of -40 to +93 °C (-40 to +200 °F). Operating temperatures in excess of +93 °C (+200 °F) may materially reduce the life of the hose. Maximum operating pressure, minimum bend radius, and other performance data are specified in Table 14.

(R) Electrically nonconductive 100R7 hose is available for use in applications where there is potential of contact with high voltage sources.

(R) To be classified "nonconductive," a hose must pass the Electrical Conductivity Test described in 19.4.10.

(R) Nonconductive hose is identified by its orange cover and lay line.

It should be noted that the detailed specifications which follow shall be supplemented by the general specifications given at the beginning of this document.

19.1 Dimensions—Dimensions and tolerances applicable to this hose are given in Table 14. The inside diameter and outside diameter of the hose shall be concentric within the limits in Table 15.

TABLE 14—DIMENSIONS AND SPECIFICATIONS FOR SAE 100R7 HOSE

Nominal SAE 100R7 Hose ID Size, in	Hose Dash Size	Hose ID Basic mm	Hose ID Basic in	Hose ID Tolerance Plus mm	Hose ID Tolerance Plus in	Hose ID Tolerance Minus mm	Hose ID Tolerance Minus in	Max Hose OD mm	Max Hose OD in	Min Burst Pressure MPa	Min Burst Pressure psi	Proof Pressure MPa	Proof Pressure psi	Max Operating Pressure MPa	Max Operating Pressure psi	Min Bend Radius[1] mm	Min Bend Radius[1] in
3/16	-3	4.8	0.188	0.6	0.023	0.2	0.008	11.4	0.450	82.7	12000	41.1	6000	20.7	3000	89	3.50
1/4	-4	6.4	0.250	0.6	0.023	0.2	0.008	13.7	0.538	75.8	11000	37.9	5500	19.0	2750	102	4.00
5/16	-5	7.9	0.312	0.6	0.023	0.2	0.008	15.6	0.615	68.9	10000	34.5	5000	17.2	2500	114	4.50
3/8	-6	9.5	0.375	0.8	0.031	0.2	0.008	18.4	0.725	62.0	9000	31.0	4500	15.5	2250	127	5.00
1/2	-8	12.7	0.500	0.8	0.031	0.4	0.015	22.5	0.885	55.2	8000	27.6	4000	13.8	2000	178	7.00
5/8	-10	15.9	0.625	0.8	0.031	0.4	0.015	25.8	1.015	41.4	6000	20.7	3000	10.3	1500	203	8.00
3/4	-12	19.0	0.750	0.8	0.031	0.4	0.015	28.6	1.125	34.5	5000	17.2	2500	8.6	1250	241	9.50
1	-16	25.4	1.000	1.0	0.040	0.4	0.015	36.7	1.445	27.6	4000	13.8	2000	6.9	1000	305	12.00

[1] Bend radius measured at inside of bend.

TABLE 15—HOSE CONCENTRICITY (100R7)

Nominal Hose ID, in	Concentricity, FIR mm	Concentricity, FIR in
Up to 1/4, incl	0.8	0.030
Over 1/4 to 3/4, incl	1.0	0.040
Over 3/4	1.3	0.050

19.2 Hose Construction—The hose shall consist of a thermoplastic inner tube resistant to hydraulic fluids with suitable synthetic fiber reinforcement and a hydraulic fluid- and weather-resistant thermoplastic cover.

(R) Nonconductive 100R7 hose must be identified with an orange cover and appropriate lay line. (Reference color, orange chip #22510, Federal Standard 595A.)

19.3 Fitting Compatibility—Fittings for thermoplastic hose may not necessarily be interchangeable. Therefore, it is recommended that fittings and hose be properly matched. Fitting and/or hose manufacturers should be consulted for recommendations.

19.4 Qualification Tests—For qualification to this section, hose and/or hose assemblies made therefrom shall conform to the following tests and requirements:

19.4.1 DIMENSIONAL CHECK TEST (ALL SAMPLES)—Shall conform to dimensions in Table 14 and these detailed specifications.

19.4.2 PROOF TEST (ALL SAMPLES)—Shall not leak at the proof pressure.

19.4.3 CHANGE IN LENGTH TEST (ONE SAMPLE)—Shall not exceed ±3% change when pressurized to operating pressure.

19.4.4 BURST TEST (ONE 460 MM (18.00 IN) FREE HOSE LENGTH ASSEMBLY)—Shall not leak or fail at the minimum burst pressure.

19.4.5 LEAKAGE TEST (TWO 300 MM (12.00 IN) FREE HOSE LENGTH ASSEMBLIES)—Shall not leak or fail.

19.4.6 COLD BEND TEST (ONE ASSEMBLY)—Shall exhibit no cracks or leakage. Exposure shall be at -40 °C (-40 °F).

19.4.7 OIL RESISTANCE TEST—After 70 h immersion at 100 °C (212 °F) in ASTM No. 3 oil, the volume change of the hose inner tube and cover specimens shall be between -15% and +35%.

19.4.8 OZONE RESISTANCE TEST (TWO SAMPLES)—Specimens shall be subjected to an atmosphere comprised of air and ozone with an ozone partial pressure of 50 mPa (50 parts ozone per 100 million parts of air at standard atmospheric conditions) at an ambient temperature of 40 °C (104 °F). After 70 h exposure, specimens shall not show evidence of cracking or deterioration when viewed with seven-power magnification while still in a stressed condition.

19.4.9 IMPULSE TEST (FOUR UNAGED ASSEMBLIES)—Hose assemblies, when tested at 125% of operating pressure, with 93 °C (200 °F) circulating petroleum-base test fluid, shall withstand a minimum of 150 000 cycles without leakage or other malfunction.

19.4.10 ELECTRICAL CONDUCTIVITY TEST—The maximum leakage shall not exceed 50 µA when subjected to 75 kV/305 mm (75 kV/ft) for 5 min. (This test shall not be applicable to hose with pinpricked outer cover.)

19.4.11 VISUAL EXAMINATION (ALL SAMPLES)

19.5 Inspection Tests—Inspection tests listed as follows shall be performed on two samples representing each lot of 150 to 3000 m (500 to 10 000 ft) of bulk hose or 100 to 10 000 assemblies. Lots of less than 150 m (500 ft) of hose or 100 assemblies need not be subjected to these tests if a lot has been tested and met the requirements within the previous 12-month period. Requirements shall be same as for corresponding Qualification Tests:

a. Dimensional Check Test (see 19.4.1)
b. Proof Test (see 19.4.2)
c. Change in Length Test (see 19.4.3)
d. Burst Test (see 19.4.4)

In addition, all hose and/or hose assemblies made therefrom shall be subjected to visual examination.

20. High-Pressure Thermoplastic Hydraulic Hose (SAE 100R8)—This section covers thermoplastic hose for use with petroleum-, water-base, and synthetic hydraulic fluids within a temperature range of -40 to +93 °C (-40 to +200 °F). Operating temperatures in excess of +93 °C (+200 °F) may materially reduce the life of the hose. Maximum operating pressure, minimum bend radius, and other performance data are specified in Table 16.

(R) Electrically nonconductive 100R8 hose is available for use in applications where there is potential of contact with high voltage sources.

(R) To be classified "nonconductive," a hose must pass the Electrical Conductivity Test described in 20.4.10.

(R) Nonconductive hose is identified by its orange cover and lay line.

TABLE 16—DIMENSIONS AND SPECIFICATIONS FOR SAE 100R8 HOSE

Nominal SAE 100R8 Hose ID Size, in	Hose Dash Size	Hose ID Basic mm	Hose ID Basic in	Hose ID Tolerance Plus mm	Hose ID Tolerance Plus in	Hose ID Tolerance Minus mm	Hose ID Tolerance Minus in	Max Hose OD mm	Max Hose OD in	Min Burst Pressure MPa	Min Burst Pressure psi	Proof Pressure MPa	Proof Pressure psi	Max Operating Pressure MPa	Max Operating Pressure psi	Min Bend Radius[1] mm	Min Bend Radius[1] in
3/16	-3	4.8	0.188	0.6	0.023	0.2	0.008	14.6	0.575	137.9	20000	68.9	10000	34.5	5000	89	3.50
1/4	-4	6.4	0.250	0.6	0.023	0.2	0.008	16.8	0.660	137.9	20000	68.9	10000	34.5	5000	102	4.00
3/8	-6	9.5	0.375	0.8	0.031	0.2	0.008	20.3	0.800	110.3	16000	55.2	8000	27.6	4000	127	5.00
1/2	-8	12.7	0.500	0.8	0.031	0.4	0.015	24.6	0.970	96.5	14000	48.3	7000	24.1	3500	178	7.00
5/8	-10	15.9	0.625	0.8	0.031	0.4	0.015	29.8	1.175	75.8	11000	37.9	5500	19.0	2750	203	8.00
3/4	-12	19.0	0.750	0.8	0.031	0.4	0.015	33.0	1.300	62.0	9000	31.0	4500	15.5	2250	241	9.50
1	-16	25.4	1.000	1.0	0.040	0.4	0.015	38.6	1.520	55.2	8000	27.6	4000	13.8	2000	305	12.00

[1] Bend radius measured at inside of bend.

20.1 Dimensions—Dimensions and tolerances applicable to this hose are given in Table 16. The inside diameter and outside diameter of the hose shall be concentric within the limits in Table 17.

TABLE 17—HOSE CONCENTRICITY (100R8)

Nominal Hose ID, in	Concentricity, FIR mm	Concentricity, FIR in
Up to 1/4, incl	0.8	0.030
Over 1/4 to 3/4, incl	1.0	0.040
Over 3/4	1.3	0.050

20.2 Hose Construction—The hose shall consist of a thermoplastic inner tube resistant to hydraulic fluids with suitable synthetic fiber reinforcement and a hydraulic fluid- and weather-resistant thermoplastic cover.

(R) Nonconductive 100R8 hose must be identified with an orange cover and appropriate lay line. (Reference color, orange chip #22510, Federal Standard 595A.)

20.3 Fitting Compatibility—Fittings for thermoplastic hose may not necessarily be interchangeable. Therefore, it is recommended that fittings and hose be properly matched. Fitting and/or hose manufacturers should be consulted for recommendations.

20.4 Qualification Tests—For qualification to this section, hose and/or hose assemblies made therefrom shall conform to the following tests and requirements:

20.4.1 DIMENSIONAL CHECK TEST (ALL SAMPLES)—Shall conform to dimensions in Table 16 and these detailed specifications.

20.4.2 PROOF TEST (ALL SAMPLES)—Shall not leak at the proof pressure.

20.4.3 CHANGE IN LENGTH TEST (ONE SAMPLE)—Shall not exceed ±3% change when pressurized to operating pressure.

20.4.4 BURST TEST (ONE 460 MM (18.00 IN) FREE HOSE LENGTH ASSEMBLY)—Shall not leak or fail at the minimum burst pressure.

20.4.5 LEAKAGE TEST (TWO 300 MM (12.00 IN) FREE HOSE LENGTH ASSEMBLIES)—Shall not leak or fail.

20.4.6 COLD BEND TEST (ONE ASSEMBLY)—Shall exhibit no cracks or leakage. Exposure shall be at -40 °C (-40 °F).

20.4.7 OIL RESISTANCE TEST—After 70 h immersion at 100 °C (212 °F) in ASTM No. 3 oil, the volume change of hose inner tube and cover specimens shall be between -15% and +35%.

20.4.8 OZONE RESISTANCE TEST (TWO SAMPLES)—Specimens shall be subjected to an atmosphere comprised of air and ozone with an ozone partial pressure of 50 mPa (50 parts ozone per 100 million parts of air at standard atmospheric conditions) at an ambient temperature of 40 °C (104 °F). After 70 h exposure, specimens shall not show evidence of cracking or deterioration when viewed with seven-power magnification while still in a stressed condition.

20.4.9 IMPULSE TEST (FOUR UNAGED ASSEMBLIES)—Hose assemblies, when tested at 133% of operating pressure, with 93 °C (200 °F) circulating petroleum-base test fluid, shall withstand a minimum of 200 000 cycles without leakage or other malfunction.

20.4.10 ELECTRICAL CONDUCTIVITY TEST—The maximum leakage shall not exceed 50 µA when subjected to 246 kV/m (75 kV/ft) for 5 min. (This test shall not be applicable to hose with pinpricked outer cover.)

20.4.11 VISUAL EXAMINATION (ALL SAMPLES)

20.5 Inspection Tests—Inspection tests listed as follows shall be performed on two samples representing each lot of 150 to 3000 m (500 to 10 000 ft) of bulk hose or 100 to 10 000 assemblies. Lots of less than 150 m (500 ft) of hose or 100 assemblies need not be subjected to these tests if a lot has been tested and met the requirements within the previous 12-month period. Requirements shall be same as for corresponding Qualification Tests:

a. Dimensional Check Test (see 20.4.1)

b. Proof Test (see 20.4.2)

c. Change in Length Test (see 20.4.3)

d. Burst Test (see 20.4.4)

In addition, all hose and/or hose assemblies made therefrom shall be subjected to visual examination.

21. High-Pressure, Four-Spiral Steel Wire Reinforced, Rubber Covered Hydraulic Hose (SAE 100R9)—This section covers hose for use with petroleum- and water-base fluids within a temperature range of -40 to +100 °C (-40 to +212 °F). Operating temperatures in excess of +100 °C (+212 °F) may materially reduce the life of the hose. Maximum operating pressure, minimum bend radius, and other performance data are specified in Table 18.

It should be noted that the detailed specifications which follow shall be supplemented by the general specifications given at the beginning of this document.

21.1 Dimensions—Dimensions and tolerances applicable to this hose are given in Table 18. The inside diameter of hose shall be concentric with outside diameter of hose and the outer surfaces of the reinforcement within the limits in Table 19.

21.2 Construction

TABLE 18—DIMENSIONS AND SPECIFICATIONS FOR SAE 100R9 HOSE

Nominal SAE 100R9 Hose ID Size, in	Hose Dash Size	Hose ID Basic mm	Hose ID Basic in	Hose ID Tolerance Plus mm	Hose ID Tolerance Plus in	Hose ID Tolerance Minus mm	Hose ID Tolerance Minus in	Reinforcement Dia Max mm	Reinforcement Dia Max in	Reinforcement Dia Min mm	Reinforcement Dia Min in	Hose OD Type A Max mm	Hose OD Type A Max in	Hose OD Type A Min mm	Hose OD Type A Min in
3/8	-6	9.5	0.375	0.6	0.023	0.2	0.008	18.0	0.710	16.9	0.664	22.2	0.875	20.6	0.812
1/2	-8	12.7	0.500	0.8	0.031	0.4	0.015	21.0	0.828	19.4	0.766	25.4	1.000	23.8	0.938
3/4	-12	19.0	0.750	0.8	0.031	0.4	0.015	28.2	1.109	26.6	1.047	32.2	1.266	30.6	1.203
1	-16	25.4	1.000	1.0	0.040	0.4	0.015	36.1	1.422	34.5	1.360	40.9	1.609	38.5	1.515
1-1/4	-20	31.8	1.250	1.2	0.047	0.4	0.015	45.6	1.797	43.3	1.703	52.4	2.062	49.2	1.938
1-1/2	-24	38.1	1.500	1.2	0.047	0.4	0.015	52.0	2.047	49.6	1.953	58.7	2.312	55.6	2.188
2	-32	50.8	2.000	1.2	0.047	0.4	0.015	66.2	2.608	63.9	2.515	73.0	2.875	69.9	2.750

TABLE 18—DIMENSIONS AND SPECIFICATIONS FOR SAE 100R9 HOSE (CONTINUED)

Nominal SAE 100R9 Hose ID Size, in	Hose OD Type AT Max mm	Hose OD Type AT Max in	Cover Thickness[1] Type AT Max mm	Cover Thickness[1] Type AT Max in	Cover Thickness[1] Type AT Min mm	Cover Thickness[1] Type AT Min in	Min Burst Pressure MPa	Min Burst Pressure psi	Proof Pressure MPa	Proof Pressure psi	Max Operating Pressure MPa	Max Operating Pressure psi	Min Bend Radius[2] mm	Min Bend Radius[2] in
3/8	21.1	0.831	1.6	0.062	0.8	0.031	124.1	18000	62.0	9000	31.0	4500	127	5.00
1/2	24.3	0.958	2.0	0.078	1.0	0.039	110.3	16000	55.2	8000	27.6	4000	178	7.00
3/4	31.9	1.255	2.0	0.078	1.0	0.039	82.7	12000	41.4	6000	20.7	3000	241	9.50
1	40.5	1.594	2.4	0.094	1.1	0.042	82.7	12000	41.4	6000	20.7	3000	305	12.00
1-1/4	50.7	1.997	2.8	0.109	1.3	0.050	68.9	10000	34.5	5000	17.2	2500	419	16.50
1-1/2	—	—	—	—	—	—	55.2	8000	27.6	4000	13.8	2000	508	20.00
2	—	—	—	—	—	—	55.2	8000	27.6	4000	13.8	2000	660	26.00

[1] Cover thickness shall be measured by means of a dial indicator depth gage having a rounded foot placed parallel to the hose, bridging a groove obtained by stripping a 12.5 to 25.4 mm (0.50 to 1.00 in) width of cover from the hose. A mandrel should be placed in the hose bore to insure freedom from misalignment.
[2] Bend radius measured at inside of bend.

TABLE 19—HOSE CONCENTRICITY (100R9)

Nominal Hose ID, in	Concentricity, FIR ID to OD mm	Concentricity, FIR ID to OD in	Concentricity, FIR ID to Reinforcement mm	Concentricity, FIR ID to Reinforcement in
Up to 1/4, incl	0.8	0.030	0.5	0.021
Over 1/4 to 7/8, incl	1.0	0.040	0.7	0.028
Over 7/8	1.3	0.050	0.9	0.035

21.2.1 Type A—This hose shall consist of an inner tube of oil-resistant synthetic rubber, four-spiral plies of wire wrapped in alternating directions, and an oil- and weather-resistant synthetic rubber cover. A ply or braid of suitable material may be used over the inner tube and/or over the wire reinforcement to anchor the synthetic rubber to the wire.

21.2.2 Type AT—This hose shall be of the same construction as Type A, except having a cover designed to assemble with fittings which do not require removal of the cover or a portion thereof.

21.3 Qualification Tests—For qualification to this section, hose and/or hose assemblies made therefrom shall conform to the following tests and requirements:

21.3.1 Dimensional Check Test (All Samples)—Shall conform to dimensions in Table 18 and these detailed specifications.

21.3.2 Proof Test (All Samples)—Shall not leak at the proof pressure.

21.3.3 Change in Length Test (One Sample)—Shall not exceed +2%, -4% change when pressurized to operating pressure.

21.3.4 Burst Test (One 460 mm (18.00 in) Free Hose Length Assembly)—Shall not leak or fail below the minimum burst pressure.

21.3.5 Leakage Test (Two 300 mm (12.00 in) Free Hose Length Assemblies)—Shall not leak or fail.

21.3.6 Cold Bend Test (One Assembly)—Shall exhibit no cover cracks or leakage. Exposure shall be at -40 °C (-40 °F).

21.3.7 Oil Resistance Test—After 70 h immersion at 100 °C (212 °F) in ASTM No. 3 oil, the volume change of hose inner tube and cover specimens shall be between 0% and +100%.

21.3.8 Ozone Resistance Test (Two Samples)—Specimens shall be subjected to an atmosphere comprised of air and ozone with an ozone partial pressure of 50 mPa (50 parts ozone per 100 million parts of air at standard atmospheric conditions) at an ambient temperature of 40 °C (104 °F). After 70 h exposure, specimens shall not show evidence of cracking or deterioration when viewed with seven-power magnification while still in a stressed condition.

21.3.9 Impulse Test (Four Unaged Assemblies)—Hose assemblies, when tested at 133% of operating pressure with 100 °C (212 °F) circulating petroleum-base test fluid, shall withstand a minimum of 200 000 cycles for sizes 9.5 and 12.7 mm (3/8 and 1/2 in), and 300 000 cycles for all other sizes without leakage or other malfunction.

21.3.10 Visual Examination (All Samples)

21.4 Inspection Tests—Inspection tests listed as follows shall be performed on two samples representing each lot of 150 to 3000 m (500 to 10 000 ft) of bulk hose or 100 to 10 000 assemblies. Lots of less than 150 m (500 ft) of hose or 100 assemblies need not be subjected to these tests if a lot has been tested and met the requirements within the previous 12-month period. Requirements shall be same as for corresponding Qualification Tests:

a. Dimensional Check Test (see 21.3.1)

b. Proof Test (see 21.3.2)

c. Change in Length Test (see 21.3.3)

d. Burst Test (see 21.3.4)

In addition, all hose and/or hose assemblies made therefrom shall be subjected to visual examination.

22. Heavy-Duty, Four-Spiral Steel Wire Reinforced, Rubber Covered Hydraulic Hose (SAE 100R10)—This section covers hose for use with petroleum- and water-base fluids within a temperature range of -40 to +100 °C (-40 to +212 °F). Operating temperatures in excess of +100 °C (+212 °F) may materially reduce the life of the hose. Maximum operating pressure, minimum bend radius, and other performance data are specified in Table 20.

It should be noted that the detailed specifications which follow shall be supplemented by the general specifications given at the beginning of this document.

TABLE 20—DIMENSIONS AND SPECIFICATIONS FOR SAE 100R10 HOSE

Nominal SAE 100R10 Hose ID Size, in	Hose Dash Size	Hose ID Basic mm	Hose ID Basic in	Hose ID Tolerance Plus mm	Hose ID Tolerance Plus in	Hose ID Tolerance Minus mm	Hose ID Tolerance Minus in	Reinforcement Dia Max mm	Reinforcement Dia Max in	Reinforcement Dia Min mm	Reinforcement Dia Min in	Hose OD Type A Max mm	Hose OD Type A Max in	Hose OD Type A Min mm	Hose OD Type A Min in	Hose OD Type AT Max mm	Hose OD Type AT Max in
3/16	-3	4.8	0.188	0.6	0.023	0.2	0.008	15.9	0.625	14.3	0.563	19.8	0.781	18.3	0.719	—	—
1/4	-4	6.4	0.250	0.6	0.023	0.2	0.008	17.4	0.687	15.8	0.625	21.4	0.844	19.8	0.781	—	—
3/8	-6	9.5	0.375	0.6	0.023	0.2	0.008	20.6	0.812	19.0	0.750	24.6	0.969	23.0	0.906	—	—
1/2	-8	12.7	0.500	1.0	0.039	0.2	0.008	24.6	0.969	23.0	0.907	28.6	1.125	27.0	1.062	—	—
3/4	-12	19.0	0.750	1.2	0.047	0.0	0.000	32.5	1.281	30.9	1.219	37.3	1.469	35.7	1.406	36.8	1.450
1	-16	25.4	1.000	1.6	0.063	0.0	0.000	40.5	1.594	38.9	1.532	45.6	1.797	43.3	1.703	45.5	1.790
1-1/4	-20	31.8	1.250	1.6	0.063	0.0	0.000	47.2	1.859	44.8	1.765	52.4	2.062	49.2	1.938	52.3	2.060
1-1/2	-24	38.1	1.500	1.6	0.063	0.0	0.000	53.6	2.109	51.1	2.015	58.7	2.312	55.6	2.188	58.7	2.310
2	-32	50.8	2.000	1.8	0.070	0.0	0.000	67.1	2.640	64.6	2.546	72.2	2.844	69.1	2.719	72.1	2.840

TABLE 20—DIMENSIONS AND SPECIFICATIONS FOR SAE 100R10 HOSE (CONTINUED)

Nominal SAE 100R10 Hose ID Size, in	Cover Thickness[1] Type AT Max mm	Cover Thickness[1] Type AT Max in	Cover Thickness[1] Type AT Min mm	Cover Thickness[1] Type AT Min in	Min Burst Pressure MPa	Min Burst Pressure psi	Proof Pressure MPa	Proof Pressure psi	Max Operating Pressure MPa	Max Operating Pressure psi	Min Bend Radius[2] mm	Min Bend Radius[2] in
3/16	—	—	—	—	275.8	40000	137.9	20000	68.9	10000	102	4.00
1/4	—	—	—	—	241.3	35000	120.6	17500	60.3	8750	127	5.00
3/8	—	—	—	—	206.8	30000	103.4	15000	51.7	7500	152	6.00
1/2	—	—	—	—	172.4	25000	86.2	12500	43.1	6250	203	8.00
3/4	2.0	0.078	1.0	0.039	137.9	20000	68.9	10000	34.5	5000	279	11.00
1	2.4	0.094	1.2	0.047	110.3	16000	55.2	8000	27.6	4000	356	14.00
1-1/4	2.8	0.109	1.4	0.054	82.7	12000	41.4	6000	20.7	3000	457	18.00
1-1/2	2.8	0.109	1.4	0.054	68.9	10000	34.5	5000	17.2	2500	559	22.00
2	2.8	0.109	1.4	0.054	68.9	10000	34.5	5000	17.2	2500	711	28.00

[1] Cover thickness shall be measured by means of a dial indicator depth gage having a rounded foot placed parallel to the hose, bridging a groove obtained by stripping a 12.5 to 25.4 mm (0.50 to 1.00 in) width of cover from the hose. A mandrel should be placed in the hose bore to insure freedom from misalignment.
[2] Bend radius measured at inside of bend.

22.1 Dimensions—Dimensions and tolerances applicable to this hose are given in Table 20. The inside diameter of hose shall be concentric with outside diameter of hose and the outer surface of the reinforcement within the limits in Table 21.

TABLE 21—HOSE CONCENTRICITY (100R10)

Nominal Hose ID, in	Concentricity, FIR ID to OD mm	Concentricity, FIR ID to OD in	Concentricity, FIR ID to Reinforcement mm	Concentricity, FIR ID to Reinforcement in
Up to 1/4, incl	0.8	0.030	0.5	0.021
Over 1/4 to 7/8, incl	1.0	0.040	0.7	0.028
Over 7/8	1.3	0.050	0.9	0.035

22.2 Construction

22.2.1 TYPE A—This hose shall consist of an inner tube of oil-resistant synthetic rubber, four-spiral plies of heavy wire wrapped in alternating directions, and an oil- and weather-resistant synthetic rubber cover. A ply or braid of suitable material may be used over the inner tube and/or over the wire reinforcement to anchor the synthetic rubber to the wire.

22.2.2 TYPE AT—This hose shall be of the same construction as Type A, except having a cover designed to assemble with fittings which do not require removal of the cover or a portion thereof.

22.3 Qualification Tests—For qualification to this section, hose and/or hose assemblies made therefrom shall conform to the following tests and requirements:

22.3.1 DIMENSIONAL CHECK TEST (ALL SAMPLES)—Shall conform to dimensions in Table 20 and these detailed specifications.

22.3.2 PROOF TEST (ALL SAMPLES)—Shall not leak at the proof pressure.

22.3.3 CHANGE IN LENGTH TEST (ONE SAMPLE)—Shall not exceed +2%, -4% when pressurized to operating pressure.

22.3.4 BURST TEST (ONE 460 MM (18.00 IN) FREE HOSE LENGTH ASSEMBLY)—Shall not leak or fail below the minimum burst pressure.

22.3.5 LEAKAGE TEST (TWO 300 MM (12.00 IN) FREE HOSE LENGTH ASSEMBLIES)—Shall not leak or fail.

22.3.6 COLD BEND TEST (ONE ASSEMBLY)—Shall exhibit no cover cracks or leakage. Exposure shall be at -40 °C (-40 °F). On sizes larger than 19 mm (3/4 in), tube and cover samples may be substituted for the hose bending test.

22.3.7 OIL RESISTANCE TEST—After 70 h immersion at 100 °C (212 °F) in ASTM No. 3 oil, the volume change of hose inner tube and cover specimens shall be between 0% and +100%.

22.3.8 OZONE RESISTANCE TEST (TWO SAMPLES)—Specimens shall be subjected to an atmosphere comprised of air and ozone with an ozone partial pressure of 50 mPa (50 parts ozone per 100 million parts of air at standard atmospheric conditions) at an ambient temperature of 40 °C (104 °F). After 70 h exposure, specimens shall not show evidence of cracking or deterioration when viewed with seven-power magnification while still in a stressed condition.

22.3.9 IMPULSE TEST (FOUR UNAGED ASSEMBLIES)—Hose assemblies, when tested at 133% of operating pressure with 100 °C (212 °F) circulating petroleum-base test fluid, shall withstand a minimum of 400 000 cycles without leakage or other malfunctions.

22.3.10 VISUAL EXAMINATION (ALL SAMPLES)

22.4 Inspection Tests—Inspection tests listed as follows shall be performed on two samples representing each lot of 150 to 3000 m (500 to 10 000 ft) of bulk hose or 100 to 10 000 assemblies. Lots of less than 150 m (500 ft) of hose or 100 assemblies need not be subjected to these tests if a lot has been tested and met the requirements within the previous 12-month period. Requirements shall be same as for corresponding Qualification Tests:

 a. Dimensional Check Test (see 22.3.1)

 b. Proof Test (see 22.3.2)

 c. Change in Length Test (see 22.3.3)

 d. Burst Test (see 22.3.4)

In addition, all hose and/or hose assemblies made therefrom shall be subjected to visual examination.

23. Heavy-Duty Six-Spiral Steel Wire Reinforced, Rubber Covered Hydraulic Hose (SAE 100R11)—This section covers hose for use with petroleum- and water-base fluids within a temperature range of -40 to +100 °C (-40 to 212 °F). Operating temperatures in excess of +100 °C (+212 °F) may materially reduce the life of the hose. Maximum operating pressure, minimum bend radius, and other performance data are specified in Table 22.

It should be noted that the detailed specifications which follow shall be supplemented by the general specifications given at the beginning of this document.

TABLE 22—DIMENSIONS AND SPECIFICATIONS FOR SAE 100R11 HOSE

Nominal SAE 100R11 Hose ID Size, in	Hose Dash Size	Hose ID Basic mm	Hose ID Basic in	Hose ID Tolerance Plus mm	Hose ID Tolerance Plus in	Hose ID Tolerance Minus mm	Hose ID Tolerance Minus in	Reinforcement Dia Max mm	Reinforcement Dia Max in	Reinforcement Dia Min mm	Reinforcement Dia Min in	Hose OD Max mm	Hose OD Max in	Hose OD Min mm	Hose OD Min in
3/16	-3	4.8	0.188	0.6	0.023	0.2	0.008	19.1	0.750	17.5	0.688	23.0	0.906	21.4	0.844
1/4	-4	6.4	0.250	0.6	0.023	0.2	0.008	20.6	0.812	19.1	0.750	24.6	0.969	23.0	0.906
3/8	-6	9.5	0.375	0.6	0.023	0.2	0.008	23.8	0.938	22.2	0.875	27.8	1.094	26.2	1.031
1/2	-8	12.7	0.500	1.0	0.039	0.2	0.008	27.8	1.094	26.2	1.031	31.8	1.250	30.2	1.188
3/4	-12	19.0	0.750	1.2	0.047	0.0	0.000	35.7	1.406	34.1	1.344	40.5	1.594	38.9	1.531
1	-16	25.4	1.000	1.6	0.062	0.0	0.000	44.0	1.734	41.7	1.641	49.6	1.953	47.2	1.859
1-1/4	-20	31.8	1.250	1.6	0.062	0.0	0.000	50.4	1.984	48.0	1.891	56.4	2.219	53.2	2.094
1-1/2	-24	38.1	1.500	1.6	0.062	0.0	0.000	56.7	2.234	54.4	2.140	62.7	2.469	59.5	2.344
2	-32	50.8	2.000	1.8	0.070	0.0	0.000	71.0	2.797	68.6	2.703	77.0	3.031	73.8	2.906
2-1/2	-40	63.5	2.500	1.8	0.070	0.0	0.000	86.3	3.399	83.5	3.289	92.9	3.656	89.7	3.531

Nominal SAE 100R11 Hose ID Size, in	Min Burst Pressure MPa	Min Burst Pressure psi	Proof Pressure MPa	Proof Pressure psi	Max Operating Pressure MPa	Max Operating Pressure psi	Min Bend Radius[1] mm	Min Bend Radius[1] in
3/16	344.7	50000	172.4	25000	86.2	12500	102	4.00
1/4	310.3	45000	155.1	22500	77.6	11250	127	5.00
3/8	275.8	40000	137.9	20000	68.9	10000	152	6.00
1/2	206.8	30000	103.4	15000	51.7	7500	203	8.00
3/4	172.4	25000	86.2	12500	43.1	6250	279	11.00
1	137.9	20000	68.9	10000	34.5	5000	356	14.00
1-1/4	96.5	14000	48.3	7000	24.1	3500	457	18.00
1-1/2	82.7	12000	41.4	6000	20.7	3000	559	22.00
2	82.7	12000	41.4	6000	20.7	3000	711	28.00
2-1/2	69.0	10000	34.5	5000	17.2	2500	914	36.00

[1] Bend radius measured at inside of bend.

23.1 Dimensions—Dimensions and tolerances applicable to this hose are given in Table 22. The inside diameter of hose shall be concentric with outside diameter of hose and the outer surface of the reinforcement within the limits in Table 23.

23.2 Construction—This hose shall consist of an inner tube of oil-resistant synthetic rubber, six-spiral plies of wire wrapped in alternating directions, and an oil- and weather-resistant synthetic rubber cover. A ply or braid of suitable material may be used over the inner tube and/or over the wire reinforcement to anchor the synthetic rubber to the wire.

23.3 Qualification Tests—For qualification to this section, hose and/or hose assemblies made therefrom shall conform to the following tests and requirements:

23.3.1 DIMENSIONAL CHECK TEST (ALL SAMPLES)—Shall conform to dimensions in Table 22 and these detailed specifications.

23.3.2 PROOF TEST (ALL SAMPLES)—Shall not leak at the proof pressure.

TABLE 23—HOSE CONCENTRICITY (100R11)

Nominal Hose ID, in	Concentricity, FIR ID to OD mm	Concentricity, FIR ID to OD in	Concentricity, FIR ID to Reinforcement mm	Concentricity, FIR ID to Reinforcement in
Up to 1/4, incl	0.8	0.030	0.5	0.021
Over 1/4 to 7/8, incl	1.0	0.040	0.7	0.028
Over 7/8	1.3	0.050	0.9	0.035

23.3.3 CHANGE IN LENGTH TEST (ONE SAMPLE)—Shall not exceed +2%, -4% change when pressurized to operating pressure.

23.3.4 BURST TEST (ONE 460 MM (18.00 IN) FREE HOSE LENGTH ASSEMBLY)—Shall not leak or fail below the minimum burst pressure.

23.3.5 LEAKAGE TEST (TWO 300 MM (12.00 IN) FREE HOSE LENGTH ASSEMBLIES)—Shall not leak or fail.

23.3.6 COLD BEND TEST (ONE ASSEMBLY)—Shall exhibit no cover cracks or leakage. Exposure shall be at -40 °C (-40 °F). On sizes larger than 19 mm (3/4 in), tube and cover samples may be substituted for the hose bending test.

23.3.7 OIL RESISTANCE TEST—After 70 h immersion at 100 °C (212 °F) in ASTM No. 3 oil, the volume change of hose inner tube and cover specimens shall be between 0% and +100 %.

23.3.8 OZONE RESISTANCE TEST (TWO SAMPLES)—Specimens shall be subjected to an atmosphere comprised of air and ozone with an ozone partial pressure of 50 mPa (50 parts ozone per 100 million parts of air at standard atmospheric conditions) at an ambient temperature of 40 °C (104 °F). After 70 h exposure, specimens shall not show evidence of cracking or deterioration when viewed with seven-power magnification while still in a stressed condition.

23.3.9 IMPULSE TEST (FOUR UNAGED ASSEMBLIES)—Hose assemblies, when tested at 133% of operating pressure with 100 °C (212 °F) circulating petroleum-base test fluid, shall withstand a minimum of 400 000 cycles without leakage or other malfunctions.

23.3.10 VISUAL EXAMINATION (ALL SAMPLES)

23.4 Inspection Tests—Inspection tests listed as follows shall be performed on two samples representing each lot of 150 to 3000 m (500 to 10 000 ft) of bulk hose or 100 to 10 000 assemblies. Lots of less than 150 m (500 ft) of hose or 100 assemblies need not be subjected to these tests if a lot has been tested and met the requirements within the previous 12-month period. Requirements shall be same as for corresponding Qualification Tests:

a. Dimensional Check Test (see 23.3.1)
b. Proof Test (see 23.3.2)
c. Change in Length Test (see 23.3.3)
d. Burst Test (see 23.3.4)

In addition, all hose and/or hose assemblies made therefrom shall be subjected to visual examination.

24. Heavy-Duty, High Impulse, Four-Spiral Wire Reinforced, Rubber Covered Hydraulic Hose (SAE 100R12)—This section covers hose for use with petroleum- and water-base fluids within a temperature range of -40 to +121 °C (-40 to +250 °F). Operating temperatures in excess of +121 °C (+250 °F) may materially reduce the life of the hose. Maximum operating pressure, minimum bend radius, and other performance data are specified in Table 24.

It should be noted that the detailed specifications which follow shall be supplemented by the general specifications given a the beginning of this document.

TABLE 24—DIMENSIONS AND SPECIFICATIONS FOR SAE 100R12 HOSE

Nominal SAE 100R12 Hose ID Size, in	Hose Dash Size	Hose ID Basic mm	Hose ID Basic in	Hose ID Tolerance Plus mm	Hose ID Tolerance Plus in	Hose ID Tolerance Minus mm	Hose ID Tolerance Minus in	Reinforcement OD Max mm	Reinforcement OD Max in	Reinforcement OD Min mm	Reinforcement OD Min in	Hose OD Max mm	Hose OD Max in	Hose OD Min mm	Hose OD Min in
3/8	-6	9.5	0.375	0.6	0.023	0.2	0.008	17.8	0.700	16.6	0.654	21.0	0.828	19.5	0.766
1/2	-8	12.7	0.500	0.8	0.031	0.4	0.015	21.5	0.846	19.9	0.784	24.6	0.966	23.0	0.904
5/8	-10	15.9	0.625	0.8	0.031	0.4	0.015	25.4	0.999	23.8	0.937	28.2	1.111	26.6	1.049
3/4	-12	19.0	0.750	0.8	0.031	0.4	0.015	28.4	1.120	26.9	1.059	31.5	1.241	29.9	1.179
1	-16	25.4	1.000	1.0	0.040	0.4	0.015	35.7	1.405	34.1	1.344	39.2	1.542	36.8	1.448
1-1/4	-20	31.8	1.250	1.2	0.047	0.4	0.015	45.1	1.777	42.7	1.683	48.6	1.912	45.4	1.788
1-1/2	-24	38.1	1.500	1.2	0.047	0.4	0.015	51.6	2.032	49.2	1.938	55.0	2.167	51.9	2.043
2	-32	50.8	2.000	1.2	0.047	0.4	0.015	64.8	2.553	62.5	2.459	68.3	2.688	65.1	2.564

Nominal SAE 100R12 Hose ID Size, in	Min Burst Pressure MPa	Min Burst Pressure psi	Proof Pressure MPa	Proof Pressure psi	Max Operating Pressure MPa	Max Operating Pressure psi	Min Bend Radius[1] mm	Min Bend Radius[1] in
3/8	110.3	16000	55.2	8000	27.6	4000	127	5.00
1/2	110.3	16000	55.2	8000	27.6	4000	178	7.00
5/8	110.3	16000	55.2	8000	27.6	4000	203	8.00
3/4	110.3	16000	55.2	8000	27.6	4000	241	9.50
1	110.3	16000	55.2	8000	27.6	4000	305	12.00
1-1/4	82.7	12000	41.4	6000	20.7	3000	419	16.50
1-1/2	69.0	10000	34.5	5000	17.2	2500	508	20.00
2	69.0	10000	34.5	5000	17.2	2500	635	25.00

[1] Bend radius measured at inside of bend.

24.1 Dimensions—Dimensions and tolerances applicable to this hose are given in Table 24. The inside diameter of hose shall be concentric with outside diameter of hose and the outer surface of the reinforcement within the limits of Table 25.

TABLE 25—HOSE CONCENTRICITY (100R12)

Nominal Hose ID, in	Concentricity, FIR ID to OD mm	Concentricity, FIR ID to OD in	Concentricity, FIR ID to Reinforcement mm	Concentricity, FIR ID to Reinforcement in
Up to 3/4, incl	1.0	0.040	0.7	0.028
Over 3/4	1.3	0.050	0.9	0.035

24.2 Hose Construction—This hose shall consist of an inner tube of oil-resistant synthetic rubber, four-spiral plies of wire wrapped in alternating directions, and an oil- and weather-resistant synthetic rubber cover. A ply or braid of suitable material may be used over or within the inner tube and/or over the wire reinforcement to anchor the synthetic rubber to the wire.

24.3 Qualification Tests—For qualification to this section, hose and/or hose assemblies made therefrom shall conform to the following tests and requirements:

24.3.1 DIMENSIONAL CHECK TEST (ALL SAMPLES)—Shall conform to dimensions in Table 24 and these detailed specifications.

24.3.2 PROOF TEST (ALL SAMPLES)—Shall not leak at the proof pressure.

24.3.3 CHANGE IN LENGTH TEST (ONE SAMPLE)—Shall not exceed +2% change when pressurized to operating pressure.

24.3.4 BURST TEST (ONE 460 MM (18.00 IN) FREE HOSE LENGTH ASSEMBLY)—Shall not leak or fail below the minimum burst pressure.

24.3.5 LEAKAGE TEST (TWO 300 MM (12.00 IN) FREE HOSE LENGTH ASSEMBLY)—Shall not leak or fail.

24.3.6 COLD BEND TEST (ONE ASSEMBLY)—Shall exhibit no cover cracks or leakage. Exposure shall be at -40 °C (-40 °F).

24.3.7 OIL RESISTANCE TEST—After 70 h immersion at 121 °C (250 °F) in ASTM No. 3 oil, the volume change of the hose inner tube and cover specimens shall be between 0% and 100% and cover specimens shall be between 0% and 125%.

24.3.8 OZONE RESISTANCE TEST (TWO SAMPLES)—Specimens shall be subjected to an atmosphere comprised of air and ozone with an ozone partial pressure of 50 mPa (50 parts ozone per 100 million parts of air at standard atmospheric conditions) at an ambient temperature of 40 °C (104 °F). After 70 h exposure, specimens shall not show evidence of cracking or deterioration when viewed with seven-power magnification while still in a stressed condition.

24.3.9 IMPULSE TEST (FOUR UNAGED ASSEMBLIES)—Hose assemblies, when tested at 133% of the maximum operating pressure per Table 24 with 121 °C (250 °F) circulating petroleum-base fluid shall withstand a minimum of 500 000 cycles without leakage or other malfunction.

24.3.10 VISUAL EXAMINATION (ALL SAMPLES)

24.4 Inspection Tests—Inspection tests listed as follows shall be performed on two samples representing each lot of 150 to 3000 m (500 to 10 000 ft) of bulk hose or 100 to 10 000 assemblies. Lots of less than 150 m (500 ft) of hose or 100 assemblies need not be subjected to these tests if a lot has been tested and met the requirements within the previous 12-month period. Requirements shall be the same as for the corresponding Qualification Tests:

a. Dimensional Check Test (see 24.3.1)
b. Proof Test (see 24.3.2)
c. Change in Length Test (see 24.3.3)
d. Burst Test (see 24.3.4)

In addition, all hose and/or hose assemblies made therefrom shall be subjected to visual examination.

25. Heavy-Duty, High Impulse, Multiple Spiral Wire Reinforced, Rubber Covered Hydraulic Hose (SAE 100R13)—This section covers hose for use with petroleum- and water-base fluids within a temperature range of -40 to +121 °C (-40 to +250 °F). Operating temperatures in excess of +121 °C (+250 °F) may materially reduce the life of the hose. Maximum operating pressure, minimum bend radius and other performance data are specified in Table 26.

It should be noted that the detailed specifications which follow shall be supplemented by the general specifications given at the beginning of this document.

TABLE 26—DIMENSIONS AND SPECIFICATIONS FOR SAE 100R13 HOSE

Nominal SAE 100R13 Hose ID Size, in	Hose Dash Size	Hose ID Basic mm	Hose ID Basic in	Hose ID Tolerance Plus mm	Hose ID Tolerance Plus in	Hose ID Tolerance Minus mm	Hose ID Tolerance Minus in	Reinforcement OD Max mm	Reinforcement OD Max in	Reinforcement OD Min mm	Reinforcement OD Min in	Hose OD Max mm	Hose OD Max in	Hose OD Min mm	Hose OD Min in
3/4	-12	19.0	0.750	0.8	0.031	0.4	0.015	29.8	1.174	28.2	1.112	33.2	1.306	31.0	1.220
1	-16	25.4	1.000	1.0	0.040	0.4	0.015	36.4	1.435	34.9	1.373	39.8	1.567	37.6	1.481
1-1/4	-20	31.8	1.250	1.2	0.047	0.4	0.015	48.0	1.889	45.6	1.795	51.3	2.021	48.3	1.903
1-1/2	-24	38.1	1.500	1.2	0.047	0.4	0.015	55.5	2.183	53.1	2.089	58.8	2.315	55.8	2.197
2	-32	50.8	2.000	1.2	0.047	0.4	0.015	69.3	2.727	66.9	2.633	72.7	2.862	69.5	2.738

Nominal SAE 100R13 Hose ID Size, in	Min Burst Pressure MPa	Min Burst Pressure psi	Proof Pressure MPa	Proof Pressure psi	Max Operating Pressure MPa	Max Operating Pressure psi	Min Bend Radius[1] mm	Min Bend Radius[1] in
3/4	137.9	20000	69.0	10000	34.5	5000	241	9.50
1	137.9	20000	69.0	10000	34.5	5000	305	12.00
1-1/4	137.9	20000	69.0	10000	34.5	5000	419	16.50
1-1/2	137.9	20000	69.0	10000	34.5	5000	508	20.00
2	137.9	20000	69.0	10000	34.5	5000	635	25.00

[1] Bend radius measured at inside of bend.

25.1 Dimensions—Dimensions and tolerances applicable to this hose are given in Table 26. The inside diameter shall be concentric with the outside diameter of the hose and the outer surface of the wire reinforcement within the limits in Table 27.

TABLE 27—HOSE CONCENTRICITY (100R13)

Nominal Hose ID, in	Concentricity, FIR ID to OD mm	Concentricity, FIR ID to OD in	Concentricity, FIR ID to Reinforcement mm	Concentricity, FIR ID to Reinforcement in
3/4 in	1.0	0.040	0.7	0.028
Over 3/4 in	1.3	0.050	0.9	0.035

25.2 Construction—This hose shall consist of an inner tube of oil-resistant synthetic rubber, multiple-spiral plies of heavy wire wrapped in alternating directions and an oil- and weather-resistant synthetic rubber cover. A ply or braid of suitable material may be used over or within the inner tube and/or over the wire reinforcement to anchor the synthetic rubber to the wire.

25.3 Qualification Tests—For qualification to this section, hose and/or hose assemblies made therefrom shall conform to the following tests and requirements:

25.3.1 DIMENSIONAL CHECK TEST (ALL SAMPLES)—Shall conform to the dimensions in Table 26 and these detailed specifications.

25.3.2 PROOF TEST (ALL SAMPLES)—Shall not leak at proof pressure.

25.3.3 CHANGE IN LENGTH TEST (ONE SAMPLE)—Shall not exceed ±2% change when pressurized to operating pressure.

25.3.4 BURST TEST (ONE 460 MM (18.00 IN) FREE HOSE LENGTH ASSEMBLY)—Shall not leak or fail below minimum burst pressure.

25.3.5 LEAKAGE TEST (TWO 300 MM (12.00 IN) FREE HOSE LENGTH ASSEMBLIES)—Shall not leak or fail.

25.3.6 COLD BEND TEST (ONE ASSEMBLY)—Shall exhibit no cover cracks or leakage. Exposure shall be at -40 °C (-40 °F). On sizes larger than 19 mm (3/4 in), tube and cover samples may be substituted for the hose bending test.

25.3.7 OIL RESISTANCE TEST—After 70 h immersion at 121 °C (250 °F) in ASTM No. 3 oil, the volume change of the hose inner tube shall be between 0% and +100%, and of the hose cover between 0% and +125%.

25.3.8 OZONE RESISTANCE TEST (TWO SAMPLES)—After 70 h exposure in an atmosphere comprised of 50 parts ozone per 100 million parts of air at an ambient temperature of 38 °C (100 °F), specimens shall show no evidence of cracks or deterioration when viewed with seven-power magnification while still in a stressed condition.

25.3.9 IMPULSE TEST (FOUR UNAGED SAMPLES)—Hose assemblies when tested at 120% of maximum operating pressure per Table 26 with 121 °C (250 °F) circulating petroleum-base fluid shall withstand a minimum of 500 000 cycles without leakage or other malfunction.

25.3.10 VISUAL EXAMINATION (ALL SAMPLES)

25.4 **Inspection Tests**—Inspection tests listed as follows shall be performed on two samples representing each lot of 150 to 3000 m (500 to 10 000 ft) of bulk hose or 100 to 10 000 hose assemblies. Lots of less than 150 m (500 ft) of hose or 100 assemblies need not be subjected to these tests if a lot has been tested and met the requirements within the previous 12-month period. Requirements shall be the same as for the corresponding Qualification Tests:

26. PTFE Lined Hydraulic Hose (SAE 100R14)

This section covers hose for use with petroleum-, synthetic-, and water-base hydraulic fluids within a temperature range of -54 to +204 °C (-65 to +400 °F). Operating temperatures in excess of 204 °C (400 °F) may reduce the life of the hose drastically. Maximum operating pressure, minimum bend radius, and other performance data are specified in Table 28. This does not exclude utilization of the hose in a multiplicity of applications for which PTFE is suitable. For such applications consult the manufacturer.

It should be noted that the detailed specifications which follow shall be supplemented by the general specifications given in the beginning of this document.

TABLE 28—DIMENSIONS AND SPECIFICATIONS FOR SAE 100R14 HOSE

Nominal SAE 100R14 Hose ID Size, in	Hose Dash Size	Hose ID Basic mm	Hose ID Basic in	Hose ID Tolerance Plus mm	Hose ID Tolerance Plus in	Hose ID Tolerance Minus mm	Hose ID Tolerance Minus in	Hose OD Max mm	Hose OD Max in	Hose OD Min mm	Hose OD Min in	Min Burst Pressure MPa	Min Burst Pressure psi	Proof Pressure MPa	Proof Pressure psi	Max Operating Pressure MPa	Max Operating Pressure psi	Min Bend Radius[1] mm	Min Bend Radius[1] in
1/8	3	3.2	0.125	0.6	0.025	0.4	0.015	6.8	0.268	5.3	0.208	82.7	12000	41.4	6000	10.3	1500	38	1.50
3/16	4	4.8	0.188	0.4	0.017	0.4	0.015	8.2	0.324	7.1	0.278	68.9	10000	34.5	5000	10.3	1500	51	2.00
1/4	5	6.4	0.250	0.5	0.018	0.4	0.015	10.1	0.397	8.9	0.351	62.0	9000	31.0	4500	10.3	1500	76	3.00
5/16	6	7.9	0.312	0.5	0.018	0.4	0.015	11.6	0.458	10.4	0.408	55.2	8000	27.6	4000	10.3	1500	102	4.00
3/8	7	9.5	0.375	0.5	0.020	0.4	0.015	13.4	0.526	12.2	0.480	48.3	7000	24.1	3500	10.3	1500	127	5.00
13/32	8	10.3	0.406	0.6	0.025	0.4	0.015	14.3	0.562	12.9	0.510	41.4	6000	20.7	3000	6.9	1000	133	5.25
1/2	10	12.7	0.500	0.6	0.025	0.4	0.015	16.8	0.663	15.3	0.603	41.4	6000	20.7	3000	5.5	800	165	6.50
5/8	12	15.9	0.625	0.6	0.025	0.6	0.025	20.1	0.793	18.6	0.733	34.5	5000	17.2	2500	5.5	800	197	7.75
3/4	14	19.0	0.750	0.6	0.025	0.6	0.025	23.3	0.917	21.3	0.840	27.6	4000	13.8	2000	5.5	800	229	9.00
7/8	16	22.2	0.875	0.8	0.030	0.8	0.030	26.9	1.061	24.6	0.970	24.1	3500	12.1	1750	5.5	800	229	9.00
1	18	25.4	1.000	0.8	0.030	0.8	0.030	29.8	1.175	27.8	1.095	24.1	3500	12.1	1750	5.5	800	305	12.00
1-1/8	20	28.6	1.125	0.8	0.030	0.8	0.030	33.5	1.320	31.9	1.258	17.2	2500	8.6	1250	4.1	600	406	16.00

[1] Bend radius measured at inside of bend.

Nominal Hose ID in	Tube Dimensions Gauge mm	Tube Dimensions Gauge in	Tube Dimensions Plus mm	Tube Dimensions Plus in	Tube Dimensions Minus mm	Tube Dimensions Minus in	Concentricity, FIR mm	Concentricity, FIR in
1/8 thru 1/2	0.8	0.030	0.1	0.005	0.1	0.005	0.3	0.010
5/8	0.8	0.030	0.3	0.010	0.1	0.005	0.3	0.010
3/4 thru 1	0.9	0.035	0.3	0.010	0.1	0.005	0.3	0.010
1-1/8	1.0	0.040	0.3	0.010	0.1	0.005	0.3	0.010

26.1 Dimensions—Dimensions and tolerances applicable to this hose are given in Table 28.

26.2 Hose Construction

26.2.1 TYPE A—This hose shall consist of an inner tube of polytetrafluoroethylene (PTFE), reinforced with a single braid of 303XX series stainless steel.

26.2.2 TYPE B—This hose shall be of the same construction as Type A but shall have the additional feature of an electrically conductive inner surface so as to preclude buildup of an electrostatic charge.

26.3 Fitting Compatibility—Fittings for PTFE lined hose may not be interchangeable. Therefore, it is recommended that fittings and hose be properly matched. Fitting and/or hose manufacturers should be consulted for recommendations.

26.4 Qualification Tests—For qualification to this section, hose and/or hose assemblies made therefrom shall conform to the following tests and requirements:

26.4.1 DIMENSIONAL CHECK TEST (ALL SAMPLES)—Shall conform to the dimensions in Table 28 and these detailed specifications.

26.4.2 PROOF TEST (ALL SAMPLES)—Shall not leak at proof pressure.

26.4.3 CHANGE IN LENGTH TEST (ONE SAMPLE)—Shall not exceed +2% to -4% change when pressurized to max operating pressure.

26.4.4 BURST TEST (ONE 460 MM (18 IN) FREE HOSE LENGTH ASSEMBLY)—Shall not leak or fail below the minimum burst pressure.

26.4.5 LEAKAGE TEST (TWO 300 MM (912 IN) FREE HOSE LENGTH ASSEMBLIES)—Shall not leak or fail.

26.4.6 COLD BEND TEST (ONE ASSEMBLY)—Shall exhibit no leakage. Exposure shall be at -54 °C (-65 °F).

26.4.7 IMPULSE TEST (FOUR UNAGED ASSEMBLIES)—Hose assemblies, when tested at 125% of operating pressure for hose sizes 22.2 mm (7/8 in) nominal ID and smaller and 100% of operating pressure for hose sizes 25.4 mm (1 in) nominal ID and larger, with 204 °C (400 °F) synthetic-base test fluid, shall withstand a minimum of 150 000 cycles without leakage or other malfunction. Test fluid should conform to MIL-H-83282, MIL-H-8446, or MIL-L-7808.

26.4.8 SPECIFIC GRAVITY (TUBE SAMPLE)—Shall be within 2.120 and 2.210 at 25 °C ± 1 °C (77 °F ± 2 °F) using ASTM D 792, Method A-1. Two drops of wetting agent shall be added to the water.

26.4.9 CONDUCTIVITY (ONE SAMPLE 330 MM (13 IN) CUT LENGTH)—Type B hose only. Size 8 and smaller shall be capable of conducting a direct current of 6 μA or greater, and size 10 and larger, 12 μA or greater, with 1000 V DC applied between electrodes (see J343 Figure 2).

26.4.10 VISUAL EXAMINATION (ALL SAMPLES)

26.5 Inspection Tests—Inspection tests listed as follows shall be performed on two samples representing each lot of 150 to 3000 m (500 to 10 000 ft) of bulk hose or 100 to 10 000 assemblies. Lots of less than 150 m (500 ft) of hose or 100 assemblies need not be subjected to these tests if a lot has been tested and met the requirements within the previous 12-month period. Requirements shall be the same as for corresponding Qualification Tests:

 a. Dimensional Check Test (see 26.4.1)
 b. Proof Test (see 26.4.2)
 c. Change in Length Test (see 26.4.3)
 d. Burst Test (see 26.4.4)
 e. Specific Gravity (see 26.4.8)

In addition, all hose and/or hose assemblies made therefrom shall be subjected to visual examination.

(R) 27. Compact High-Pressure One and Two Steel Wire Reinforced Rubber Cover Hydraulic Hose (SAE 100R16)

27.1 Foreword—Compact wire braid hydraulic hoses have been increasingly used over the recent years and it is considered that their status in the hydraulic hose field merits the development of an SAE Standard.

Hoses of this type are smaller in diameter than two braid hoses with similar performance characteristics specified in SAE 100R2, which gives them the ability to operate at smaller bend radii. They are also lighter in weight and their compactness offers advantages when minimal space is available in installations. This can be manufactured as either a one braid or two braid design, the hose must be identified as to the number of braids.

27.2 Scope—This SAE Standard specifies requirements for a compact high-pressure, steel wire reinforced one or two braid, rubber covered hydraulic hose (SAE 100RXX). This section covers hose of internal diameter from 6.3 mm to 33 mm, for use with common hydraulic fluids such as mineral oils, soluble oils, oil and water emulsions, aqueous glycol solution and water, at temperatures ranging from -40 °C to 100 °C.

The hose is not suitable for use with castor oil-based and ester-based fluids.

27.3 High-Pressure, Steel Wire Reinforced, Rubber Covered Hydraulic Hose (SAE 100R16)—This section covers hose for use with petroleum- and water-base hydraulic fluids within a temperature range of -40 to +100 °C. Operating temperatures in excess of +100 °C may materially reduce the life of the hose. Maximum operating pressure, minimum bend radius, and other performance data are specified in Table 29.

It should be noted that the detailed specifications which follow shall be supplemented by the general specifications given at the beginning of this document.

TABLE 29—DIMENSIONS AND SPECIFICATIONS FOR SAE 100R16 HOSE

Nominal SAE 100R16 Hose ID Size mm	Hose Dash Size mm	Hose ID Tolerance Plus mm	Hose ID Tolerance Minus mm	Reinforcement Dia Max mm	Hose OD Max mm	Cover Thickness[1] Max mm	Cover Thickness[1] Min mm	Min Burst Pressure MPa	Proof Pressure MPa	Max Operating Pressure MPa	Min Bend Radius[2] mm
6.4	-6	0.6	0.2	12.32	14.48	1.52	0.76	137.9	69.0	34.5	50.8
7.9	-8	0.6	0.2	13.33	15.75	1.52	0.76	117.2	58.6	29.3	57.2
9.5	-10	0.6	0.2	15.88	18.80	1.52	0.76	110.3	55.2	27.6	63.5
12.7	-13	0.8	0.4	19.05	21.97	1.52	0.76	96.5	48.3	24.1	88.9
15.9	-16	0.8	0.4	22.48	25.40	1.52	0.76	75.8	37.9	19.0	101.6
19.0	-19	0.8	0.4	26.29	28.96	1.52	0.76	62.0	31.0	15.5	120.6
25.4	-25	1.0	0.4	34.04	36.58	1.52	0.76	55.2	27.6	13.8	152.4
31.8	-32	1.2	0.4	41.91	44.32	2.03	1.02	44.8	22.4	11.2	209.5

[1] Cover thickness shall be measured by means of a dial gage having a rounded foot placed parallel to the hose, bridging a groove obtained by stripping a 12.5 to 25.4 mm width of cover from the hose. A mandrel should be placed in the hose bore to insure freedom from misalignment.

[2] Bend radius measured at the inside of bend.

27.4 Dimensions—Dimensions and tolerances applicable to this hose are given in Table 29. The inside diameter of hose shall be concentric with outside diameter of hose and the outer surface of the reinforcement within the limits of Table 30.

27.5 Hose Construction—The hose shall consist of an inner tube of oil-resistant synthetic rubber, steel wire reinforcement according to hose design (one or two braids) and an oil- and weather-resistant synthetic rubber cover. A ply or braid of suitable material may be used over the inner tube and/or over the wire reinforcement to anchor the synthetic rubber to the wire.

27.6 Qualification Tests—For qualification to this section, hose and/or hose assemblies made therefrom shall conform to the following tests and requirements:

27.6.1 DIMENSIONAL CHECK TEST (ALL SAMPLES)—Shall conform to dimensions in Table 30 and these detailed specifications.

27.6.2 PROOF TEST (ALL SAMPLES)—Shall not leak at the proof pressure.

27.6.3 CHANGE IN LENGTH TEST (ONE SAMPLE)—Shall not exceed +2% to -4% change when pressurized to operating pressure.

27.6.4 BURST TEST (ONE 460 MM FREE HOSE LENGTH ASSEMBLY)—Shall not leak or fail below the minimum burst pressure.

27.6.5 LEAKAGE TEST (TWO 300 MM FREE HOSE LENGTH ASSEMBLIES)—Shall not leak or fail.

27.6.6 COLD BEND TEST (ONE ASSEMBLY)—Shall exhibit no cover cracks or leakage. Exposure shall be at -40 °C.

TABLE 30—HOSE CONCENTRICITY (100R16)

Nominal Hose ID mm	Concentricity, FIR ID to OD mm	Concentricity, FIR ID to Reinforcement mm
Up to 6.4, incl	0.8	0.5
Over 6.4 to 22.2, incl	1.0	0.7
Over 22.2	1.3	0.9

27.6.7 OIL RESISTANCE TEST—After 70 h immersion at 100 °C in ASTM No. 3 oil, the volume change of hose inner tube and cover specimens shall be between 0% and +100%.

27.6.8 OZONE RESISTANCE TEST (TWO SAMPLES)—Specimens shall be subjected to an atmosphere comprised of air and ozone with an ozone partial pressure of 50 mPa (50 parts ozone per 100 million parts of air at standard atmospheric conditions) at an ambient temperature of 40 °C. After 70 h exposure, specimens shall not show evidence of cracking or deterioration when viewed with seven-power magnification while still in a stressed condition.

27.6.9 IMPULSE TEST (FOUR UNAGED ASSEMBLIES)—Hose assemblies, when tested at 133% of operating pressure with 100 °C circulating petroleum-base test fluid, shall withstand a minimum of 200 000 cycles without leakage or other malfunction.

27.6.10 VISUAL EXAMINATION (ALL SAMPLES)

27.7 Inspection Tests—Inspection tests listed as follows shall be performed on two samples representing each lot of 150 to 300 m (500 to 10 000 ft) of bulk hose or 100 to 10 000 assemblies. Lots of less than 150 m (500 ft) of hose or 100 assemblies need not be subjected to those tests if a lot has been tested and met the requirements within the previous 12-month period. Requirements shall be same as for corresponding Qualifications Tests:

    a. Dimensional Check (see 27.6.1)
    b. Proof Test (see 27.6.2)
    c. Change in Length Test (see 27.6.3)
    d. Burst Test (see 27.6.4)

In addition, all hose and/or hose assemblies made therefrom shall be subjected to visual examination.

CUMULATIVE DAMAGE ANALYSIS FOR HYDRAULIC HOSE ASSEMBLIES— SAE J1927 OCT88

SAE Information Report

Report of the Materials, Processes, and Parts Division approved October 1988.

1. Introduction

This report is intended to provide the hydraulic system analyst with a procedure which will assist in the selection and use of high-pressure wire reinforced hydraulic hose assemblies. Many construction, agricultural, industrial or commercial equipment systems utilize hydraulic hose assemblies that are subjected to irregular cyclic pressure variations (cannot be approximated by a constant amplitude pressure cycle). This SAE Information Report relates damage done by pressure cycles with the pressure-life performance curve for the hose assembly being evaluated, using a linear damage rule to predict fatigue life similar to that used for predicting metal fatigue life. More detailed information on the subject may be found in SAE Paper No. 880713, SAE Test Program on Cumulative Damage for Hydraulic Hose Assemblies. The accuracy of cumulative damage calculations is directly related to proper measurement of the service pressure history and pressure-life performance for the hose assembly being evaluated. Final selection of a hose assembly must also consider installation and maintenance as noted in SAE J1273.

2. Background

In the current SAE J517 for hydraulic hose, each style and size hose is assigned a maximum operating pressure rating to assure the user reasonable service life in a wide variety of applications. (See Appendix A for further explanation of nomenclature used in this document.) This rating is based on assessment of many factors, including repeated pressure cycling under controlled laboratory conditions at a pressure equal to or greater than the assigned maximum operating pressure. This standard test procedure, as detailed in SAE J343 minimizes variables so as to provide a baseline for performance capability. SAE J517 takes note that actual pressure cycling in a hydraulic system will seldom duplicate those test parameters precisely. It provides basic hose construction details; SAE J343 establishes standard test procedures - which may be utilized to develop data at pressures other than the established test value for any hose assembly. SAE J1273 provides a guide for selection, installation, and maintenance for hose and hose assemblies.

Many hydraulic systems will be subjected to variable amplitude pressure cycles, and in some of these the highest surge peaks will occur only a few times. Strict interpretation of SAE J517 indicates the use of hose with a maximum operating pressure equal to or greater than the highest peak - even if that peak should occur only once in the life of the system. If this were done, the hose would have more reinforcement, larger outside diameter, less flexibility, and higher cost than might be needed. It should not be necessary to use a hose that has over a million cycle capability at a pressure which is likely to occur only a few times in the application. What is needed, then, is a design verification procedure that assures adequate fatigue life for applications with variable amplitude pressure usage where the majority of the peaks are between 100 and 200% of rated pressure.

2.1 Cumulative damage of hydraulic hose is in many ways comparable to cumulative damage of metal components. The design verification of metal components is done in two ways depending on whether loading is of constant amplitude or variable amplitude:

    a) Size the part cross section so the constant amplitude (maximum) load will not produce a stress at the highest stressed area in excess of the material "endurance limit stress." This is a stress that will not cause failure in a million cycles of loading.

    b) Do a cumulative damage analysis on the variable amplitude load or strain history to predict fatigue life as is normally done to evaluate prototype machines in the ground vehicle industry.

2.2 In a similar manner, hydraulic hose design verification should be done in two ways:

    a) Select a hose assembly so the constant amplitude (maximum) pressure during service is less than the SAE J517 rated pressure.

    b) Do a cumulative damage analysis with the variable amplitude pressure history to determine if the fatigue life will meet the design life of the product. The block diagram in Fig. 1 illustrates how both the pressure history and the P-N curve information are the essential inputs for this procedure.

For both metal parts and hydraulic hose assemblies, the first method is appropriate when the load (pressure) is known to be a large number of cyclic applications of nearly constant amplitude. However, if the load (pressure) involves infrequent cycling or includes only a few large peaks, then the first method may lead to designs that are heavy, bulky, inflexible, and more costly than need be. For variable amplitude loads or pressure applications, it is logical to do a cumulative damage analysis to judge the adequacy of the design.

Other factors, such as internal temperature, ambient temperature, and ozone exposure, for all intents and purposes, have not been considered in this cumulative damage analysis procedure. Long-term exposure to extreme limits or high levels of these elements could affect the overall hose assembly life.

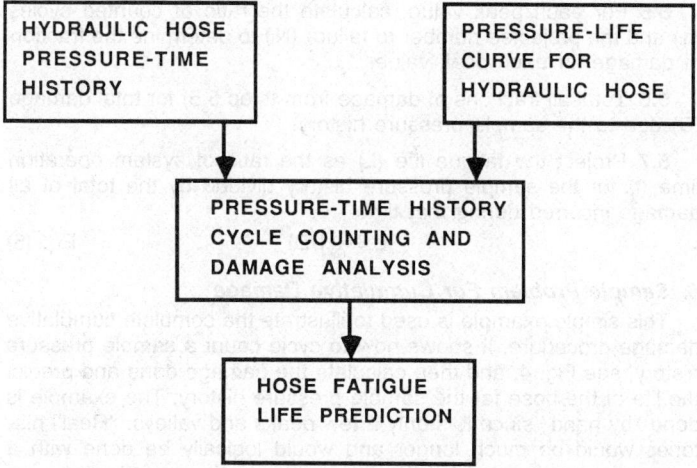

FIGURE 1—BLOCK DIAGRAM FOR HOSE FATIGUE ANALYSIS

3. Use Of P-N Curves In Design

A cumulative damage analysis is needed for a hydraulic system subjected to variable amplitude pressure cycles if system pressures in excess of rated pressure are to be in the design consideration. An approach for this analysis is to use the Reference P-N curve as defined in the previously mentioned reference paper and shown in Fig. 2. Verification that the actual hose assemblies have a P-N curve in excess of the Reference P-N curve is essential. The equation for the Reference P-N curve in Fig. 2 is:

$$P_a = P_b(N)^s \qquad \text{Eq. (1)}$$

where:

P_a = Zero-to-Max amplitude of pressure cycle
N = Cycles to failure at pressure amplitude P_a
P_b = Burst pressure (one cycle life)
s = Slope of curve on log-log plot

By using two known points on this line:
- 1 cycle at 400% (Burst Point)
- 200 000 cycles at 133% (Impulse Point)

the values of P_b and s can be found:

$$P_b = 400 \text{ and } s = -0.0902$$

and:

$$P_a = 400(N)^{-0.0902} \qquad \text{Eq. (2)}$$

Rearranged to a more useful form to solve for N cycles at any pressure P_a this becomes:

$$N = \left[\frac{400}{P_a}\right]^{11.086} \qquad \text{Eq. (3)}$$

If a different impulse point (for example, 500 000 cycles at 133%) were used, this would result in a different slope for the P-N curve. The Hydraulic Design Analyst has the option of using other Reference P-N curves in the analysis as long as there are data to demonstrate the actual P-N curve is in excess of the Reference P-N curve. In addition, the analyst can increase statistical confidence by requiring a low percentage failure P-N curve (for example, a B_{10} Curve) for test data for the hose assembly being considered to be in excess of the Reference P-N curve used in the analysis. (This is explained in greater detail in SAE Paper No. 880713.)

In order to use this prediction procedure, a sample pressure history has to be available. The cycle counting procedure for a pressure history is explained in more detail in the next section. The fundamental theory for damage accumulation uses zero-to-max pressure cycles since the P-N curve is determined from zero-to-max test pressure cycles. This is slightly different than for stress-life or strain-life curves for metal where there are fully reversed loads (zero mean stress). It is appropriate to resolve pressure cycles to zero-to-max cycles (mean pressure equals one-half of the maximum) since any possible negative pressure must be very small compared to the "high pressures" generally used in wire reinforced hydraulic hose assemblies. The damage is accumulated in a linear manner as is done for metal fatigue:

$$D = \sum_{i=1}^{j} \frac{n_i}{N_i} \qquad \text{Eq. (4)}$$

where:

D = Damage done (D = 1 assumes failure)
n_i = Number of cycles in the history at amplitude i
N_i = Number of cycles to cause failure at amplitude i
j = Number of different amplitudes in the history

This equation and the P-N curve illustrate that when subjected to repeated constant amplitude pressure cycles above a given level, a hose assembly will eventually fail. The higher this constant amplitude pressure, the shorter the life. If the pressure is large enough, then failure can occur in one cycle. This is a burst test and is represented by the one cycle pressure value at the left end of the P-N curve and is analogous to the ultimate strength of metal parts. Damage due to a variable amplitude pressure history (which has been resolved to zero-to-max cycles) is then simply the sum of the n_i/N_i for all the different amplitudes in the history. When this summation equals one, failure is predicted.

If a typical pressure history of a given length of time (t) (time factor is machine operation time, not calendar time) is analyzed, Eq. (4) will give a decimal fraction for the damage done. Then the fatigue life prediction is:

$$\text{Life}(L) = t(1/D) \qquad \text{Eq. (5)}$$

For example, if D = 0.01, one one-hundredth of damage has been done in t units of time and the total expected life would be one hundred times t units of time for the total life.

The procedure will work with either constant or variable amplitude pressure histories. Fig. 1 illustrates how the P-N curve and the pressure history are the two necessary inputs for the procedure. If the user of the procedure has test data to demonstrate that the actual P-N curve is in excess of the reference P-N curve, then life predictions will be conservative for the sample pressure history that is used in the analysis.

4. Pressure History Cycle Counting

Typical hydraulic hose pressure histories are variable amplitude and are almost totally positive pressure. Depending on the hydraulic component that the hose may be connected to, it is sometimes possible to have a partial vacuum in a hose. In the worst possible case, this could never be more than one atmosphere, which is small compared to the maximum that "high pressure" hoses normally experience, but should be avoided in system design as the impact on useful life is far greater than a numerical relationship would suggest. For cycle counting, pressure cycle histories will be considered to be all positive pressures. As indicated in the previous section, the cycle counting procedure needs to be designed to resolve the pressure history into zero-to-maximum pressure cycles since this is the type

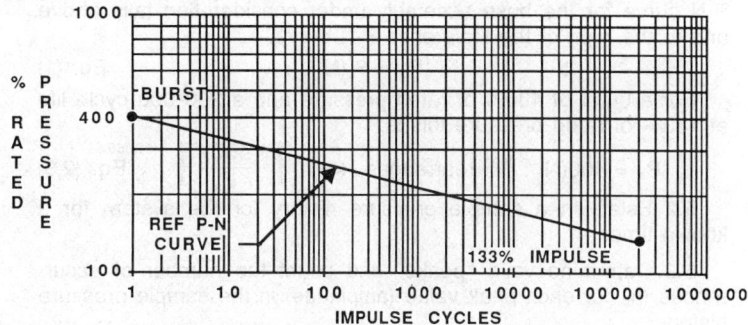

FIGURE 2—HYDRAULIC HOSE PRESSURE-LIFE PERFORMANCE CURVE (SECTION 3)

of pressure cycle used to determine the P-N curves.

A cycle counting procedure is discussed in detail in SAE Paper No. 880713 and suggests the following three steps:
a) Tabulate the pressure peaks from the history and assume each is followed by a minimum pressure value of zero.
b) A pressure peak is defined as a maximum value that is preceded and followed by pressure minimums of a specified lower magnitude (threshold). A pressure maximum that does not meet this requirement is not counted.
c) Threshold must be selected based on engineering judgment of the analyst.

As a result of the above steps, all cycles that are counted will be zero-to-max cycles and can be used directly in the damage Eq. (4). The threshold value needs to be at least 35-50% of the hose rated pressure to avoid counting cycles that are small pressure undulations and not pressure cycles that cause significant fatigue damage. Engineering judgment must be used to select the threshold to make an appropriate cycle count. This cycle counting procedure assumes the pressure minimums following significant maximums are zero. This was done to keep the procedure simple, practical, and conservative but not ultraconservative when only a few large pressure cycles are in the history. If, however, the service history does have a large number of cycles at the largest value, this evaluation procedure will still give the appropriate cycle count and life prediction. It will work with constant amplitude histories just as well as with variable amplitude histories.

An important part of the cycle counting is the ability to determine when a maximum should be kept as a peak. Fig. 3 shows the four possible cases that can occur. Only case one is considered a valid peak and kept for the cycle counting. From Fig. 3:

Case I
$R_1 > T$ and $R_2 > T$
Count P_1

Case II
$R_1 > T$ and $R_2 \leq T$
Discard P_1 and V_2
Keep V_1 as Valley
Consider next Peak and Valley

Case III
$R_1 \leq T$ and $R_2 > T$
Discard P_1 and V_1
Keep V_2 as Valley
Consider next Peak and Valley

Case IV
$R_1 \leq T$ and $R_2 \leq T$
Discard P_1 and highest Valley
Keep lower Valley
Consider next Peak and Valley

5. Summary Of Procedure To Predict Hose Assembly Life:

5.1 The analyst must have data to demonstrate that the actual P-N curve for the hose assembly under consideration falls above and to the right of the Reference P-N curve.

$$P_a = P_b(N)^s \qquad \text{Eq. (1)}$$

For a burst of 400% of rated pressure and a 200 000 cycle life at 133% of rated pressure this is:

$$P_a = 400(N)^{-0.0902} \text{ rearranged } N = \left[\frac{400}{P_a}\right]^{11.086} \qquad \text{Eq. (2,3)}$$

5.2 Establish a sample pressure history for the system for a known time.

5.3 Determine valid "peaks" and count the number of occurrences (n_i) for each peak value (amplitude) in the sample pressure history.

5.4 For each peak value, calculate cycles to failure (N_i) from the Reference P-N curve (step 5.1).

5.5 For each peak value, calculate the ratio of counted cycles (n_i) and the projected number to failure (N_i) to determine the fraction of damage for each peak value.

5.6 Total all fractions of damage from (step 5.5) for total damage (D) due to the sample pressure history.

5.7 Project the fatigue life (L) as the ratio of system operation time (t) for the sample pressure history divided by the total of all damage incurred during that time.

$$L = t(1/D) \qquad \text{Eq. (5)}$$

6. Sample Problem For Cumulative Damage

This simple example is used to illustrate the complete cumulative damage procedure. It shows how to cycle count a sample pressure history, see Fig. 4, and then calculate the damage done and predict the life of the hose for the sample pressure history. The example is done "by hand" since it is only a few peaks and valleys. "Real" histories would be much longer and would logically be done with a computer program that has been programmed to consider the various special cases of cycle counting and choice of threshold.

Tabulate original maximum-minimum sequence as potential peaks and valleys.

V-P-V-P-V-......P-V

Note history is assumed to start and end with a valley of zero.

a	b	c	d	e	f	g	h	i	j	k	l
0	120	40	140	120	150	50	160	100	110	30	180

m	n	o	p	q	r	s	t	u	v	w	x	y
70	130	110	170	40	80	20	110	90	110	40	150	0

Follow procedure in the previous section and Fig. 3 to determine which potential peaks will be counted.

For this example assume threshold is 35%.

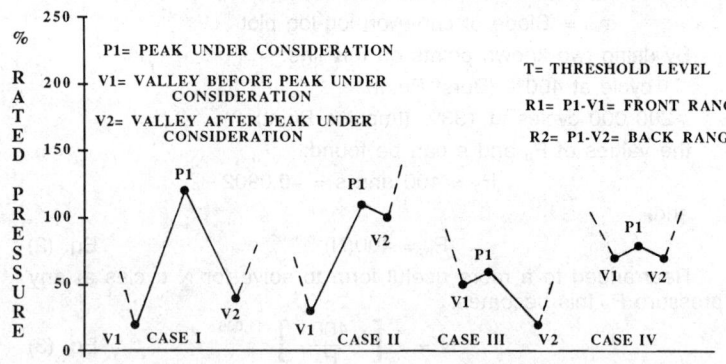

FIGURE 3—PEAK-VALLEY COMBINATIONS (SECTION 4)

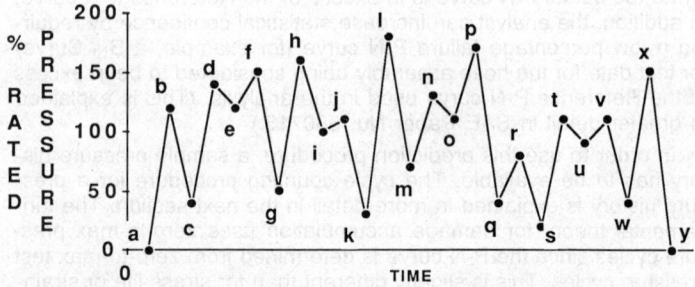

FIGURE 4—PRESSURE-TIME HISTORY (SECTION 6)

Counted peaks which result are:

b	f	h	l	p	r	t	x
120	150	160	180	170	80	110	150

Then for a hose assembly that has a P-N curve in excess of Reference P-N curve, use Eq. (3) to calculate N_i for each P_a. Table 1 shows the tabulated results for the sample pressure history.

Using Eq. (4)

$$\Sigma n_i/N_i = 29.8 \times 10^{-5}$$

Assume pressure history is for 0.5 h. Then from Eq. (5)

$$\text{Life}(L) = \frac{0.5}{29.8 \times 10^{-5}} = 1679 \text{ h}$$

APPENDIX A

1. Nomenclature
To facilitate understanding of the concept and life calculation procedure used in this document and in related documents, the following nomenclature is used.

2. Pressure History
The time oriented variations of internal pressure in a hydraulic system (hose assembly). This may be tabulated by listing a sequence of relative maximums and minimums from recorded pressure versus time data. Significant maximums and minimums are called peaks and valleys. (See section on cycle counting.) A peak is defined as a maximum both preceded and followed by a minimum less than the peak by a specified amount or threshold (differential pressure). A valley is defined as the smallest minimum between significant peaks. Note that it is possible for peaks to be lower than valleys in cases where they are not adjacent. Likewise, valleys could be greater than non-adjacent peaks.

3. Threshold (Differential Pressure)
The magnitude of pressure difference (differential pressure) between a maximum and adjacent minimum in a pressure history that is considered significant by the hydraulic design analyst. This threshold (differential pressure) must be chosen by the analyst and is usually at least 35% of the hose rated pressure. If both the differential pressure before and after a maximum are equal to or greater than the threshold, then that maximum is defined to be a significant peak in the pressure history. (See Fig. 3.)

4. Sample Pressure History
A representative recording for a given length of time of the pressure history for a hydraulic hose. Generally, a sequence of peaks and valleys for a given length of time, requiring a choice of threshold to disregard insignificant pressure variations.

5. Constant Amplitude Pressure History
A pressure history where all the peaks are of similar magnitude with the valleys near zero.

6. Variable Amplitude Pressure History
A pressure history where the peaks and valleys are irregular.

7. Rated Pressure
The reference pressure or "nominal design pressure" for a hose assembly from which other pressures are based.

8. Operating Pressure
(Used in SAE J517.) Same as rated pressure.

9. Maximum Operating Pressure
(Used in SAE J517.) Same as rated pressure.

10. Minimum Burst Pressure
Defined as 400% of rated pressure.

11. Burst Pressure
The actual pressure at which a hose assembly fails when subjected to slowly increasing hydrostatic pressure (see SAE J343).

12. Surge Pressure
(Used in SAE J517 and J1273.) A rapid and transient rise in pressure in a pressure history.

13. Impulse Test Pressure
A laboratory test pressure level to which a hose assembly is repeatedly subjected with near zero valleys between peaks. (See SAE J343.) Generally given in percent of rated pressure and is usually greater than 100%.

14. Impulse Life
The number of cycles to failure for a hose assembly when subjected to cyclic testing for a given impulse test pressure.

15. Fatigue Life
The predicted time to failure (length of operational service) for a hose assembly in a working hydraulic system, based on a sample pressure history and cumulative damage calculations.

16. Cumulative Damage Calculation
A procedure to calculate fatigue life of a hose assembly by relating pressure cycles and fatigue damage for a sample pressure history with a given P-N curve.

17. Pressure-Life (P-N) Curve
The relationship between impulse test pressure, P, and impulse life, N, for a given type and size of hose. Can be plotted as a line on a log-log chart of percent of rated pressure and cycles to failure.

18. Reference P-N Curve
A straight line relation on a P-N plot connecting one cycle at 400% (minimum burst) with the impulse test point (for example, 200 000 cycles at 133%). In this form (percent of rated pressure versus life), hoses with different rated pressure all plot with the same Reference P-N curve. P-N curves can also be plotted with pressure units rather than percent of rated pressure. In this form, a "family" of parallel P-N curves result for different rated pressures.

19. 10% P-N Curve-B_{10} Curve
A P-N curve for test data where 10% failure of a population of hoses would occur. Can be used to increase the statistical confidence of life prediction (see SAE paper 880713).

TABLE 1—TABULATED DATA

PRESSURE AS A % OF RATED PRESSURE	CALCULATED CYCLES TO CAUSE FAILURE @ PRESSURE i	COUNTED CYCLES IN HISTORY @ PRESSURE i	DEGREE OF DAMAGE EFFECTED BY n_i
P_a	N_i	n_i	n_i/N_i
80	56×10^6	1	0.178×10^{-7}
110	16.4×10^5	1	6.09×10^{-7}
120	62.6×10^4	1	16.0×10^{-7}
150	52.8×10^3	2	3.79×10^{-5}
160	25.8×10^3	1	3.87×10^{-5}
170	13.2×10^3	1	7.59×10^{-5}
180	6.99×10^3	1	14.3×10^{-5}

(R) SELECTION, INSTALLATION, AND MAINTENANCE OF HOSE AND HOSE ASSEMBLIES—
SAE J1273 NOV91 — **SAE Recommended Practice**

Report of the Fluid Conductors and Connectors Technical Committee, approved September 1979 and reaffirmed May 1986. Completely revised by the SAE Fluid Conductors and Connectors Technical Committee SC2—Hydraulic Hose and Hose Fittings November 1991. Rationale statement available.

1. Scope

Hose (also includes hose assemblies) has a finite life and there are a number of factors which will reduce its life.

This SAE Recommended Practice is intended as a guide to assist system designers and/or users in the selection, installation, and maintenance of hose. The designers and users must make a systematic review of each application and then select, install, and maintain the hose to fulfill the requirements of the application. The following are general guidelines and are not necessarily a complete list.

WARNING—IMPROPER SELECTION, INSTALLATION, OR MAINTENANCE MAY RESULT IN PREMATURE FAILURES, BODILY INJURY, OR PROPERTY DAMAGE

2. References

2.1 Applicable Documents

The following publications form a part of this specification to the extent specified herein. The latest issue of SAE publications shall apply.

2.1.1 SAE Publications—Available from SAE, 400 Commonwealth Drive, Warrendale, PA 15096-0001.

J516—Hydraulic Hose Fittings
J517—Hydraulic Hose

3. Selection

The following is a list of factors which must be considered before final hose selection can be made.

3.1 Pressure

After determining the system pressure, hose selection must be made so that the recommended maximum operating pressure is equal to or greater than the system pressure. Surge pressures higher than the maximum operating pressure will shorten hose life and must be taken into account by the hydraulic designer.

3.2 Suction

Hoses used for suction applications must be selected to insure the hose will withstand the negative pressure of the system.

3.3 Temperature

Care must be taken to insure that fluid and ambient temperatures, both static and transient, do not exceed the limitations of the hose. Special care must be taken when routing near hot manifolds.

3.4 Fluid Compatibility

Hose selection must assure compatibility of the hose tube, cover, and fittings with the fluid used. Additional caution must be observed in hose selection for gaseous applications.

3.5 Size

Transmission of power by means of pressurized fluid varies with pressure and rate of flow. The size of the components must be adequate to keep pressure losses to a minimum and avoid damage to the hose due to heat generation or excessive turbulence.

3.6 Routing

Attention must be given to optimum routing to minimize inherent problems.

3.7 Environment

Care must be taken to insure that the hose and fittings are either compatible with or protected from the environment to which they are exposed. Environmental conditions such as ultraviolet light, ozone, salt water, chemicals, and air pollutants can cause degradation and premature failure, and, therefore, must be considered.

3.8 Mechanical Loads

External forces can significantly reduce hose life. Mechanical loads which must be considered include excessive flexing, twist, kinking, tensile or side loads, bend radius, and vibration. Use of swivel-type fittings or adapters may be required to insure no twist is put into the hose. Unusual applications may require special testing prior to hose selection.

3.9 Abrasion

While a hose is designed with a reasonable level of abrasion resistance, care must be taken to protect the hose from excessive abrasion which can result in erosion, snagging, and cutting of the hose cover. Exposure of the reinforcement will significantly accelerate hose failure.

3.10 Proper End Fitting

Care must be taken to ensure that proper compatibility exists between the hose and coupling selected based on the manufacturer's recommendations substantiated by testing to industry standards such as SAE J517. End fitting components from one manufacturer are usually not compatible with end fitting components supplied by another manufacturer (i.e., using a hose fitting nipple from one manufacturer with a hose socket from another manufacturer). It is the responsibility of the fabricator to consult the manufacturer's written instructions or the manufacturer directly for proper end fitting componentry.

3.11 Length

When establishing proper hose length, motion absorption, hose length changes due to pressure, as well as hose and machine tolerances must be considered.

3.12 Specifications and Standards

When selecting hose, government, industry, and manufacturers' specifications and recommendations must be reviewed as applicable.

3.13 Hose Cleanliness

Hose components vary in cleanliness levels. Care must be taken to insure that the assemblies selected have an adequate level of cleanliness for the application.

3.14 Electrical Conductivity

Certain applications require that hose be nonconductive to prevent electrical current flow. Other applications require the hose to be sufficiently conductive to drain off static electricity. Hose and fittings must be chosen with these needs in mind.

4. Installation

After selection of proper hose, the following factors must be considered by the installer.

4.1 Pre-Installation Inspection

Prior to installation, a careful examination of the hose must be performed. All components must be checked for correct style, size, and length. In addition, the hose must be examined for cleanliness, I.D. obstructions, blisters, loose cover, or any other visible defects.

4.2 Follow Manufacturers' Assembly Instructions

Hose assemblies may be fabricated by the manufacturer, an agent for or customer of the manufacturer, or by the user. Fabrication of permanently attached fittings to hydraulic hose requires specialized assembly equipment. Field-attachable fittings (screw style and segment clamp style) can usually be assembled without specialized equipment although many manufacturers provide equipment to assist in this operation.

SAE J517 hose from one manufacturer is usually not compatible with SAE J516 fittings supplied by another manufacturer. It is the responsibility of the fabricator to consult the manufacturer's written assembly instructions or the manufacturers directly before intermixing hose and fittings from two manufacturers. Similarly, assembly equipment from one manufacturer is usually not interchangeable with that of another manufacturer. It is the responsibility of the fabricator to consult the manufacturer's written instructions or the manufacturer directly for proper assembly equipment. Always follow the manufacturer's instructions for proper preparation and fabrication of hose assemblies.

4.3 Minimum Bend Radius
Installation at less than minimum bend radius may significantly reduce hose life. Particular attention must be given to preclude sharp bending at the hose/fitting juncture.

4.4 Twist Angle and Orientation
Hose installations must be such that relative motion of machine components produces bending of the hose rather than twisting.

4.5 Securement
In many applications, it may be necessary to restrain, protect, or guide the hose to protect it from damage by unnecessary flexing, pressure surges, and contact with other mechanical components. Care must be taken to insure such restraints do not introduce additional stress or wear points.

4.6 Proper Connection of Ports
Proper physical installation of the hose requires a correctly installed port connection while insuring that no twist or torque is put into the hose.

4.7 Avoid External Damage
Proper installation is not complete without insuring that tensile loads, side loads, kinking, flattening, potential abrasion, thread damage, or damage to sealing surfaces are corrected or eliminated.

4.8 System Check Out
After completing the installation, all air entrapment must be eliminated and the system pressurized to the maximum system pressure and checked for proper function and freedom from leaks.

NOTE—Avoid potential hazardous areas while testing.

5. Maintenance

Even with proper selection and installation, hose life may be significantly reduced without a continuing maintenance program. Frequency should be determined by the severity of the application and risk potential. A maintenance program should include the following as a minimum.

5.1 Hose Storage
Hose products in storage can be affected adversely by temperature, humidity, ozone, sunlight, oils, solvents, corrosive liquids and fumes, insects, rodents, and radioactive materials. Storage areas should be relatively cool and dark and free of dust, dirt, dampness, and mildew.

5.2 Visual Inspection
Any of the following conditions requires replacement of the hose:
a. Leaks at fitting or in hose (leaking fluid is a fire hazard)
b. Damaged, cut, or abraded cover (any reinforcement exposed)
c. Kinked, crushed, flattened, or twisted hose
d. Hard, stiff, heat cracked, or charred hose
e. Blistered, soft, degraded, or loose cover
f. Cracked, damaged, or badly corroded fittings
g. Fitting slippage on hose

5.3 Visual Inspection
The following items must be tightened, repaired, or replaced as required:
a. Leaking port conditions
b. Clamps, guards, shields
c. Remove excessive dirt buildup
d. System fluid level, fluid type, and any air entrapment

5.4 Functional Test
Operate the system at maximum operating pressure and check for possible malfunctions and freedom from leaks.

NOTE—Avoid potential hazardous areas while testing.

5.5 Replacement Intervals
Specific replacement intervals must be considered based on previous service life, government or industry recommendations, or when failures could result in unacceptable down time, damage, or injury risk.

TEST AND PROCEDURES FOR SAE 100R SERIES HYDRAULIC HOSE AND HOSE ASSEMBLIES—SAE J343 JUN93 SAE Standard

Report of the Tube, Pipe, Hose, and Lubrication Fittings Committee approved June 1968. Revised by the Fluid Conductors and Connectors Technical Committee May 1989, and completely revised May 1990. Revised by the Fluid Conductors and Connectors Technical Committee SC2—Hydraulic Hose and Hose Fittings, April 1991 and June 1993. Rationale statements available.

1. Scope
This SAE Standard is intended to establish uniform methods for the testing and performance evaluation of the SAE 100R series of hydraulic hose and hose assemblies. The specific tests and performance criteria applicable to each variety of hose and/or assemblies made therefrom are set forth in the respective specifications of SAE J517.

2. References

2.1 Applicable Documents
The following publications form a part of this specification to the extent specified herein. The latest issue of SAE publications shall apply.

2.1.1 SAE PUBLICATION—Available from SAE, 400 Commonwealth Drive, Warrendale, PA 15096-0001.

SAE J517—Hydraulic Hose

2.1.2 ASTM PUBLICATIONS—Available from ASTM, 1916 Race Street, Philadelphia, PA 19103-1187.

ASTM D 380—Standard Methods of Testing Rubber Hose

ASTM D 518—Test Method for Rubber Deterioration—Surface Cracking

ASTM D 622—Methods of Testing Rubber Hose for Automotive Air and Vacuum Brake System

ASTM D 1149—Test Method for Rubber Deterioration—Surface Ozone Cracking in a Chamber (Flat Specimens)

3. Test Procedures
The test procedures described in the current issue of ASTM D 380 shall be followed. However, in cases of conflict between the ASTM specifications and those described as follows, the latter shall take precedence.

4. Standard Tests

4.1 Dimensional Check Test
The hose shall conform to all dimensions tabulated in the applicable specification.

Reinforcement diameter and finished outside diameter (R) measurements shall be made by calculation from measurement of the outside circumference. Use of a flexible tape graduated to read the diameter directly shall be acceptable.

Inside diameter measurements shall be made by means of suitable expanding ball or telescoping gages.

Concentricity shall be measured both over the reinforcement and the finished outside diameter using either a dial indicator gage or a micrometer. The foot of the measuring instrument contacting the inside of the hose shall be rounded to conform to the curvature of the hose. The readings shall be taken at 90 degree intervals around the hose and acceptability based on the total variation between high and low readings.

Inside and outside diameter measurements shall be made at a minimum distance of 25.4 mm (1 in) and concentricity measurements at a minimum distance of 12.7 mm (0.50 in), back from the ends of the hose.

4.2 Proof Test—Hose and/or hose assemblies shall be hydrostatically tested to the specified proof pressure for a period of not less than 30 s nor more than 60 s. There shall be no indication of failure or leakage.

4.3 Change in Length Test—Measurements for the determination of elongation or contraction shall be conducted on a previously untested, unaged hose assembly having at least 300 mm (12 in) length of free hose between hose couplings. The hose assembly shall be attached to the pressure source and pressurized to the specified pressure for a period of 30 s, after which time the pressure shall be released. After allowing the hose to restabilize for a period of 30 s following pressure release, reference marks 250 mm (10 in) apart shall be accurately placed on the hose outer cover, midway between the hose couplings. The hose shall then be repressurized to the specified pressure for a period of 30 s, after which time, while the hose is pressurized, the distance between the reference marks shall be measured. This length shall be the final length.

All hose measurements shall be made while the hose is in a straight position without any external end load.

The change in length shall be computed using the following equation:

$$\% \text{ Change} = \frac{(\text{Final length} - \text{Original length})\, 100}{\text{Original length}}$$

$(-\%)$ Change = contraction
$(+\%)$ Change = elongation (Eq.1)

4.4 Burst Test—Hose and/or hose assemblies on which the end fittings have been attached not over 30 d shall be subjected to a hydrostatic pressure increased at a constant rate so as to attain the specified minimum burst pressure within a period of not less than 15 s nor more than 30 s. There shall be no leakage, hose burst, or indication of failure below the specified minimum burst pressure.

4.5 Cold Bend Test—Hose and/or hose assemblies shall be subjected to the specified temperature for 24 h in a straight position. After this time and while still at the specified temperature, the sample shall be evenly and uniformly bent over a mandrel having a diameter equal to twice the minimum specified bend radius. Bending shall be accomplished within a period of not less than 8 s nor more than 12 s.

Hoses of less than 25.4 mm (1 in) nominal inside diameter shall be bent 180 degrees over the mandrel and hoses of 25.4 mm (1 in) nominal inside diameter and larger shall be bent 90 degrees over the mandrel.

After bending, the sample shall be allowed to warm to room temperature, then visually examined for cover cracks and subjected to the proof test. There shall be no cover cracks or leakage. (In lieu of the bending test, hoses of 25.4 mm (1 in) nominal inside diameter and over may be considered acceptable if samples of tube and cover pass the Low Temperature Test on Tube and Cover of ASTM D 380.)

4.6 Oil Resistance Test—After 70 h immersion in ASTM No. 3 oil at the designated temperature, the volume change of specimens taken from the hose inner tube and cover shall be within the specified limits.

4.7 Ozone Resistance Test—Hydraulic hose shall be tested for resistance of the cover compound to ozone in accordance with the latest issue of ASTM D 380, except that the mandrel shall be a diameter twice the minimum bend radius specified in the individual hose standard, and the cover shall be examined at the completion of the test under 7X magnification.

4.8 Impulse Test—Impulse testing shall be conducted with unaged hose assemblies and, where the individual standard requires, also with aged assemblies.

The test assemblies shall be impulsed on suitable equipment with the hose bent to the minimum bend radius. Hoses of less than 25.4 mm (1 in) nominal inside diameter shall be bent either 90 or 180 degrees and hoses of 25.4 mm (1 in) nominal inside diameter and over shall be bent 90 degrees. Individual standards may designate impulsing in a straight position on specific sizes.

The test assembly free length of hose measured between couplings shall be computed using the following equations:

$$90 \text{ degree bend free length} = \frac{\pi\,(\text{Min bend radius})}{2} + 2\,(\text{hose OD}) \quad (\text{Eq.2})$$

$$180 \text{ degree bend free length} = \pi\,(\text{Min bend radius}) + 2\,(\text{hose OD}) \quad (\text{Eq.3})$$

Straight free length = 356 to 457 mm (14 to 18 in) (Eq.4)

Where aged samples are required, refer to the individual standards.

The test fluid shall be circulated through the assemblies at the specified temperature with a tolerance of ±3 °C (±5 °F). The impulse rate shall be 30 to 75 cpm at the specified pressure. Circulation of the test fluid shall be at a rate which will maintain uniform bore temperature. Cooling or heating of the test chamber shall not be permitted, except when individual standards require testing with synthetic base test fluids at a temperature higher than 150 °C (302 °F). When such higher temperatures are required, the impulse test fluid need not be circulated if both the fluid and the assemblies are externally heated in the test chamber, at the specified temperature with a tolerance of ±5 °C (±9 °F).

The impulse pressure curve must fall entirely within the shaded area of Figure 1 and should conform as closely as possible to the curve as shown. Unless failure occurs first, the impulse test shall continue for the specified number of cycles.

It is recommended the test fluid be changed frequently to prevent breakdown.

4.9 Leakage Test—Unaged hose assemblies on which the end fittings have been attached not over 30 d shall be subjected to a hydrostatic pressure equal to 70% of the specified minimum burst pressure for a period of 5 to 5.5 min and then reduced to zero after which the 70% of minimum burst pressure shall be reapplied for another 5 min. There shall be no leakage or evidence of failure. This test is to be considered a destructive test and sample shall be destroyed.

4.10 Visual Examination of Product—All bulk hose shall be visually inspected to see that the hose identification has been properly applied and all assemblies shall be inspected to see that the correct fittings are properly installed.

4.11 Electrical Conductivity Test (for thermoplastic hose only)—Hose assemblies having a free length of 152 mm ± 13 mm (6 in ± 0.5 in) without fluid and capped to prevent entry of moisture shall be exposed to a minimum of 85% relative humidity at 24 °C ± 3 °C (75 °F ± 5 °F) for a period of 168 h. Surface moisture shall be removed prior to testing.

Conditioned assemblies shall have one end fitting attached to the lead from a source of 60 Hz sinusoidal, 37.5 kV (rms) electricity. This lead shall be suspended by dry fabric strings so that the hose hangs free, at least 600 mm (2 ft) from any extraneous objects. The lower end of the hose shall be connected to ground through a 1000 to 1 000 000 Ω resistor, keeping the resistor near the end of the hose. A suitable AC voltmeter shall be connected across the resistor, using a fully shielded cable with the shielding well grounded. Thirty-seven and one-half kV shall be applied to the specimen for 5 min and a current reading taken. This current shall not exceed the value specified.

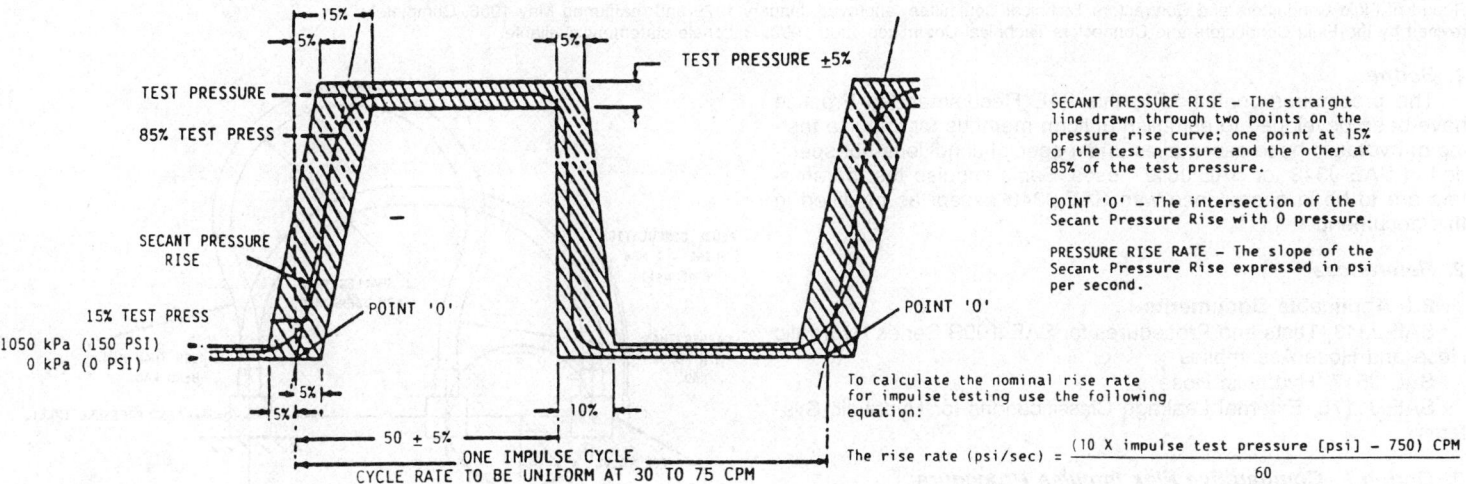

FIGURE 1—IMPULSE PRESSURE CURVE

4.12 Electrical Conductivity Test (PTFE hose only)—Test specimen shall be a 330 mm (13 in) cut length of hose with fitting attached to one end and the reinforcing braid flared away from the PTFE tube on the opposite end to prevent contact with the free end of the tube. The inner surface of the tube shall be cleaned, first with naphtha dry cleaning fluid or Stoddard solvent, and then with isopropyl alcohol to remove surface contamination, followed by thorough drying at room temperature.

Relative humidity shall be kept below 70% and room temperature between 16 °C (60 °F) and 32 °C (90 °F).

The specimen shall be mounted in a vertical position as shown in Figure 2. The adapter at the base is simply a convenient means of assuring proper electrical contact if a swivel female fitting is chosen, and may be omitted if a male fitting is used. In either case, the electrode must be insulated from ground.

A mercury or salt water solution electrode shall be provided at the upper end as shown, by inserting a nonmetallic plug with an O-ring seal to a distance of 76 mm (3 in) from the end of the tubing, thus providing a test length of 254 mm (10 in). Mercury or salt water solution shall then be added to a level 25 mm (1 in) above the plug. Any suitable conductor to this electrode may be used, including a threaded end attached to the plug if so desired. Concentration of salt water, if used, shall be 450 g NaCl per liter of H_2O.

1000 volts DC shall be applied between the upper electrode and the lower electrode (adapter or male fitting hex). The current shall be measured with an instrument with a sensitivity of at least 1 µA (1×10^{-6} A).

4.13 Resistance to Vacuum Test—The hose shall not blister nor show any other indication of failure when subjected to the specified vacuum for a period of 5 min. Where practicable, one end of the hose shall be equipped with a transparent cap and electric light to permit visual examination for failure. Where the length or size of the hose precludes visual examination, failure shall be determined by inability to pass through the hose a ball or cylinder 6.4 mm (0.250 in) less in diameter than the bore of hoses of 12.7 mm (1/2 in) nominal inside diameter and larger. For hoses under 12.7 mm (1/2 in) nominal inside diameter, a ball or cylinder 3.2 mm (0.125 in) smaller in diameter than the bore shall be used.

4.14 Cubical Expansion Test—Cubical expansion tests shall be run in accordance with the current issue of ASTM D 380.

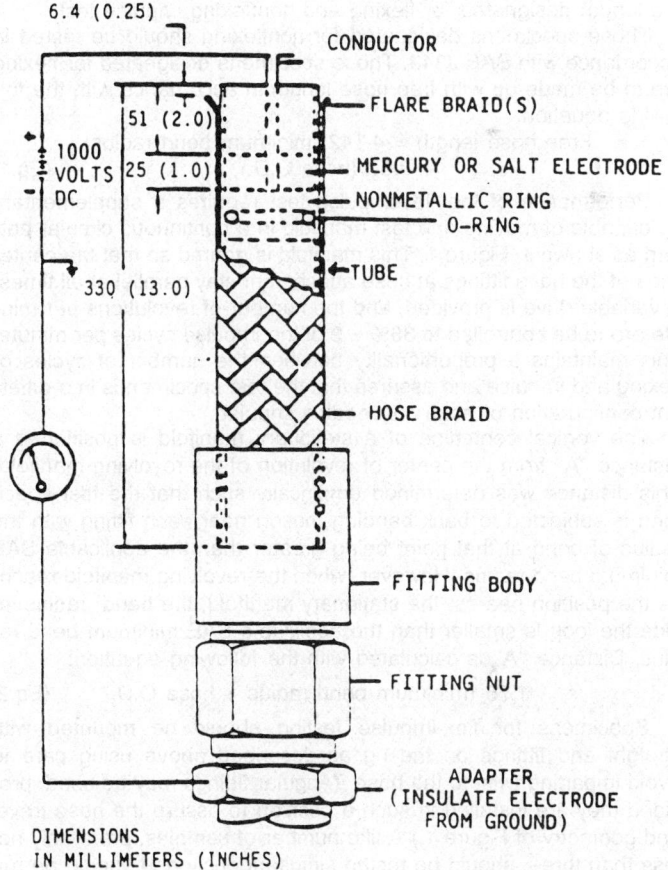

FIGURE 2—CONDUCTIVITY TEST DIAGRAM

(R) OPTIONAL IMPULSE TEST PROCEDURES FOR HYDRAULIC HOSE ASSEMBLIES—SAE J1405 JUN90

SAE Recommended Practice

Report of Fluid Conductors and Contractors Technical Committee, approved January 1979 and reaffirmed May 1986. Completely revised by the Fluid Conductors and Connectors Technical Committee June 1990. Rationale statement available.

1. Scope

The procedures contained in this SAE Recommended Practice have been developed to establish uniform methods for impulse testing of hydraulic hose assemblies under special conditions not specified in SAE J343 for SAE J517 hoses. Basic impulse test parameters are to be in accordance with SAE J343 except as modified in this document.

2. References

2.1 Applicable Documents

SAE J343, Tests and Procedures for SAE 100R Series Hydraulic Hose and Hose Assemblies
SAE J517, Hydraulic Hose
SAE J1176, External Leakage Classifications for Hydraulic Systems

3. Option I—Comparative Flex Impulse Procedure

3.1 Purpose

To generate comparative impulse test data, with and without flexing. This test procedure minimizes impulse test variables to provide comparative data between flexing and nonflexing to determine the effect on the ultimate life of hose. This test is not a requirement for SAE J517.

3.2 Test Procedure

For optimum validity of comparison, test specimens should be cut from a continuous length of hose with alternate samples along the length designated for flexing and nonflexing impulse test.

Those specimens designated for nonflexing should be tested in accordance with SAE J343. Those specimens designated for flexing are to be made up with free hose length in accordance with the following equation:

$$\text{Free hose length} = 4.142 \text{ (minimum bend radius)} + 3.57 \text{ (hose O.D.)} \quad \text{(Eq.1)}$$

Performance of the flex-impulse test requires a supplementary rig capable of moving one test manifold in a continuous circular pattern as shown in Figure 1. This manifold is geared so that the center lines of the hose fittings at hose attachment stay parallel at all times. A variable drive is provided, and the number of revolutions per minute are to be controlled to 36% ± 2 of the impulse cycles per minute. This maintains a proportionality between the number of cycles of flexing and impulse and assures that the test specimen is in a different configuration of each succeeding impulse.

The vertical centerline of a stationary manifold is positioned a distance "A" from the center of revolution of the revolving manifold. This distance was determined empirically such that the test specimen is subjected to back bending motion near each fitting with the radius of bend at that point being greater than the applicable SAE minimum bend radius. However, when the revolving manifold reaches the position nearest the stationary manifold, the bend[1] radius inside the loop is smaller than the applicable SAE minimum bend radius. Distance "A" is calculated with the following equation:

$$1.75 \text{ (minimum bend radius + hose O.D.} \quad \text{(Eq.2)}$$

Specimens for flex-impulse testing should be mounted with straight end fittings on the rig as described above using care to avoid imparting twist to the hose. (Angular fittings may be used, provided they are installed in such a position to assure the hose travel and geometry of Figure 1.) A like number of samples, preferably not less than three, should be tested simultaneously and should be run to failure.

[1] Violation of the minimum bend radius for this test does not imply that such violation is recommended in applications.

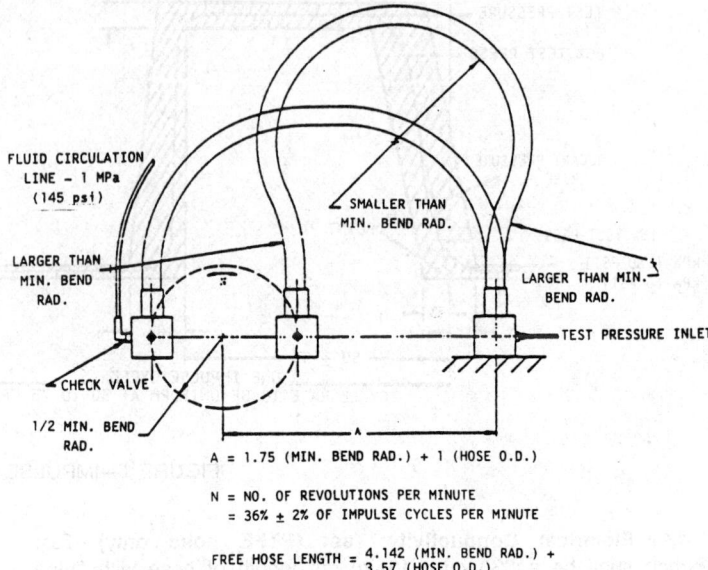

A = 1.75 (MIN. BEND RAD.) + 1 (HOSE O.D.)

N = NO. OF REVOLUTIONS PER MINUTE
  = 36% ± 2% OF IMPULSE CYCLES PER MINUTE

FREE HOSE LENGTH = 4.142 (MIN. BEND RAD.) + 3.57 (HOSE O.D.)

FIGURE 1—FLEX IMPULSE TEST HOSE GEOMETRY

To accelerate completion of the test for comparative purposes, a pressure based on actual burst values of the hose is recommended, with flexing and nonflexing specimens to be tested at the same pressure. Suggested procedure is to first determine the average burst strength for the test length of hose and from this calculate the impulse test pressure as 35% of average burst. If this test procedure does not produce failures within the desired range, a higher or lower percentage may be used.

4. Option II—Flex Impulse Test

4.1 Purpose

To establish requirements for impulse testing with the addition of flexing. This is a specialized test which is not a requirement of SAE J517, nor is it specified in SAE J343. It is intended to provide a standard method to flex-impulse hose assemblies when flexing is deemed necessary.

4.2 Test Procedure

Four unaged hose assemblies for flexing are to be made up with free hose length in accordance with the following equation:

$$\text{Free hose length} = 4.142 \text{ (minimum bend radius)} + 3.57 \text{ (hose O.D.)} \quad \text{(Eq.3)}$$

Performance of the flex-impulse test requires a supplementary rig capable of moving one test manifold in a continuous circular pattern as shown in Figure 1. This manifold is geared so that the center lines of the hose fittings at hose attachment stay parallel at all times. A variable drive is provided, and the number of revolutions per minute are to be controlled to 36% ± 2 of the impulse cycles per minute. This maintains a proportionality between the number of cycles of flexing and impulse and assures that the test specimen is in a different configuration on each succeeding impulse.

The vertical centerline of a stationary manifold is positioned a distance "A" from the center of revolution of the revolving manifold. This distance was determined empirically such that the test specimen is subjected to back bending motion near each fitting with the radius of bend at that point being greater than the applicable SAE minimum bend radius. However, when the revolving manifold reaches the position nearest the stationary manifold, the bend radius inside the loop is smaller than the applicable SAE minimum bend radius.[2] Distance "A" is calculated with the following equation:

$$1.75 \text{ (minimum bend radius)} + \text{hose O.D.} \quad \text{(Eq.4)}$$

Specimens for flex-impulse testing should be mounted with straight end fittings on the rig as described above using care to avoid imparting twist to the hose. (Angular fittings may be used provided they are installed in such a position to assure the hose travel and geometry of Figure 1.)

4.3 Test Requirements

The hose assemblies shall be tested at the impulse pressures, temperatures and minimum bend radii, for the minimum number of impulse cycles, as specified in SAE J517 for 100R series hoses. Other test parameters, as agreed upon by the supplier and/or user may be used.

5. Option III—Cool Down Leakage Test

[2] (See footnote 1)

5.1 Purpose

To establish requirements for performing a cold start leakage test to be used in conjunction with both flexing or nonflexing impulse tests.

5.2 Test Procedure

The impulse test unit shall be shut down at 40% ± 10 and 90% ± 10 of the required number of impulse cycles and allowed to cool until the test oil and hose assemblies reach a temperature of 30°C ± 3 (85°F ± 5). Accelerated cool down procedures, i.e. fans, heat exchangers, etc., may be used to speed the cooling process. Check test assemblies to assure they are clean and dry. With oil heater turned off, resume the test and observe and note leakage for 1000 impulse cycles. The acceptable rate of leakage shall be as agreed upon by the supplier and/or user. (See SAE J1176 for leakage classes.)

After completing the 1000 impulse cycles, turn on oil heater and continue the impulse test.

If leakage is noted during the cool down cycle, notation shall also be made as to whether or not a seal-off was effected as the temperature came back up. Results are applicable only to the specific hose construction and size, hose fitting design and size, and fitting assembly technique.

(R) TESTS AND PROCEDURES FOR HIGH TEMPERATURE TRANSMISSION OIL HOSE, ENGINE LUBRICATING OIL HOSE, AND HOSE ASSEMBLIES—SAE J1019 JUN90

SAE Standard

Report of Tube, Pipe, Hose, and Lubrication Fittings Committee approved April 1973 and reaffirmed by the Fluid Conductors and Connectors Technical Committee, May 1986. Completely revised by the Fluid Conductors and Connectors Technical Committee June 1990.

1. Scope

This SAE Standard is intended to establish uniform methods for testing and evaluation of hose and hose assemblies for use in high temperature transmission oil systems and high temperature lubricating oil systems using petroleum base oils within a temperature range of −40° to 150°C (−40° to 302°F) and a maximum working pressure of 1.5 MPa (217 psi). Hose construction, dimensions, identification, and hose fitting configurations shall be agreed upon by the supplier and user.

2. References

2.1 Applicable Documents

ASTM D 380
ASTM D 518
ASTM D 622
ASTM D 1149

3. Performance Tests

3.1 Preconditioning

The test hose or hose assemblies shall be conditioned at room temperature a minimum of 24 h prior to testing.

3.2 Qualification Tests

For qualification to this document, hose and hose assemblies made therefrom shall conform to the following tests and requirements:

3.2.1 PROOF TEST—Hose and hose assemblies shall be hydrostatically tested to 3 MPa (435 psi) for a period of not less than 30 s nor more than 60 s. There shall be no indication of failure or leakage.

3.2.2 LEAKAGE TEST—Two previously untested unaged hose assemblies having 300 mm ± 3 (12 in ± 1/8) length of free hose between fittings, on which the hose fittings have been attached for not over 30 days, shall be subjected to a hydrostatic pressure of 4.2 MPa (609 psi) for a period of 5 to 5.5 min and then reduced to zero after which 4.2 MPa (609 psi) shall be reapplied for another 5 to 5.5 min. There shall be no leakage or other evidence of failure. This shall be considered a destructive test and the samples shall be destroyed.

3.2.3 CHANGE IN LENGTH TEST—Measurements for the determination of elongation or contraction shall be conducted on two previously untested, unaged hose assemblies having at least 300 mm (12 in) length of free hose between hose fittings. The hose assemblies shall be attached to the pressure source and pressurized to 1.5 MPa (217 psi) for a period of 30 s, after which time the pressure shall be released. After allowing the hose to restabilize for a period of 30 s following pressure release, reference marks 254 mm (10 in) apart shall be accurately placed upon the hose outer cover, midway between the hose fittings. This length shall be the "original length".

The hose assemblies shall then be repressurized to 1.5 MPa (217 psi) for a period of 30 s, after which time, while the hose is pressurized, the distance between the reference marks shall be measured. This length shall be the "final length". Change in length shall be computed using the following equation:

$$\text{Percent change} = \frac{(\text{Final Length} - \text{Original Length}) \times 100}{\text{Original Length}} \quad \text{(Eq. 1)}$$

(Minus percent) Change = Contraction

(Plus percent) Change = Elongation

The percent change shall be within the agreed limits.

3.2.4 BURST TEST—Two unaged hose assemblies having at least 300 mm (12 in) length of free hose between fittings, on which the hose fittings have been attached for not over 30 days, shall be subjected to a hydrostatic pressure increasing at a constant rate so as to attain 6 MPa (870 psi) within a period of not less than 15 s nor more than 30 s. There shall be no leakage, hose burst, or other indication of failure below 6 MPa (870 psi). This shall be considered a destructive test and the sample shall be destroyed.

3.2.5 Tensile Test—Two hose assemblies having at least 300 mm (12 in) length of free hose between fittings shall be tested. The rate of separation of the head of the testing machine shall be 0.42 mm/s ± 0.04 (1 in/min ± 0.1). The minimum force required to separate the hose from a hose fitting shall be 1 kN (225 lbs). This shall be considered a destructive test and the samples shall be destroyed.

3.2.6 Cold Flexibility Test—Two hose assemblies shall be subjected to −40°C ± 2 (−40°F ± 3.6) for 24 h in a straight position. After this time and while still at the specified temperature, each hose assembly shall be evenly and uniformly bent over a mandrel equal to 8 times the nominal hose outside diameter (or twice the minimum bend radius, if specified). Bending shall be accomplished within a period of not less than 8 s and no more than 12 s. Hoses of less than 25 mm (1 in) nominal inside diameter shall be bent 180 degrees over the mandrel and hoses of 25 mm (1 in) nominal inside diameter and larger shall be bent 90 degrees over the mandrel.

After flexing, each sample shall be allowed to warm to room temperature, then visually examined for cover cracks and subjected to the Proof Test (3.2.1). There shall be no leakage, or cracks on the cover. In lieu of the flexing test, hoses of 25 mm (1 in) nominal inside diameter and larger shall be considered acceptable if samples of tube and cover pass the Low Temperature Test on Tube and Cover, Section 25 of ASTM D 380 of latest revision.

3.2.7 Temperature Cycling Test—Four hose assemblies having not less than 300 mm (12 in) nor more than 1000 mm (39 in) length of free hose between hose fittings shall be tested. Two shall be filled with transmission Type F fluid and two with oil conforming to MIL-L-2104, and lightly capped. The hose assemblies shall be placed in a circulating air oven at 150°C ± 2 (302°F ± 3.6) for 2 h, removed from the air oven, and allowed to stabilize to room temperature for 1 h minimum; then placed in a cold box at −40°C ± 2 (−40°F ± 3.6) for 2 h, removed from the cold box, and allowed to stabilize to room temperature for 1 h minimum. Ten such cycles shall be conducted, after which the hose assemblies shall be stabilized to room temperature for 1 h minimum and subjected to the Proof Test (3.2.1). There shall be no leakage or failure. This shall be considered a destructive test and the samples shall be destroyed.

3.2.8 Ozone Test—Two samples of the cover compound shall be tested in accordance with the latest issue of ASTM D 622, procedure 9, and ASTM D 1149. Where space limitations prohibit use of a hose, specimen cover stock tested in accordance with ASTM D 518, procedure B, may be substituted. After 70 h exposure in an atmosphere comprised of air and ozone wth an ozone partial pressure of 50 mPa (50 parts ozone per 100 million parts of air at standard atmospheric conditions) at ambient temperature of 40°C (104°F), specimens shall show no evidence of cracks or deterioration when viewed with 7 power magnification while still in a stressed condition. This shall be considered a destructive test and the samples shall be destroyed.

3.2.9 High Temperature Circulation Test—A minimum of two hose assemblies having at least 355 mm (14 in) length of free hose between hose fittings shall be mounted on a circulating oil test unit in a straight configuration. The ambient temperature shall be 93°C ± 11 (200°F ± 20) and the oil temperature 150°C ± 2 (302°F ± 3.6) between inlet and outlet. Oil conforming to MIL-L-2104 shall be circulated through the hose assemblies at a pressure between 0.35 and 0.69 MPa (50 to 100 psi). Entrained air in the oil must be kept to a minimum and caution must be exercised so that the hose assemblies do not come into contact with the heating elements and are located to permit good air circulation. The test fluid shall be changed every 375 h ± 25. Tests are to be run continuously except for oil change and addition or removal of samples. All shutdown time is to be recorded. After 750 h ± 5, the test assemblies shall be removed, the oil drained, and allowed to cool for a minimum of 4 h. The samples shall then be bent around a mandrel having a diameter of 12 times the inside diameter of the hose. The time required to bend the hose around the mandrel shall be between 8 and 12 s. Rubber covered hose shall be examined visually for cover cracks. No cracks are permitted. The assemblies shall then be subjected to Proof Test (3.2.1), with the hoses in the straight position. There shall be no failure or leakage through the hose or at hose fitting juncture. All tests are to be completed within 24 h of removal of samples from the circulating oil test unit. This shall be considered a destructive test and the samples shall be destroyed.

4. Inspection Test

Inspection tests and lot sizes for inspection shall be negotiated between user and seller.

HYDRAULIC FLANGED TUBE, PIPE, AND HOSE CONNECTIONS, FOUR-BOLT SPLIT FLANGE TYPE—SAE J518 JUN93

SAE Standard

Report of Construction and Industrial Machinery Technical Committee and Tube, Pipe, Hose, and Lubrication Fittings Committee approved February 1952, revised by Tube, Pipe, Hose, and Lubrication Fittings Commitee May 1972, and reaffirmed by Fluid Conductors and Connectors Technical Committee December 1987. Completely revised by the SAE Fluid Conductors and Connectors Technical Committee SC2—Hydraulic Hose and Hose Fittings April 1991, and revised June 1993. Rationale statement available.

1. Scope—This SAE Standard covers complete general and dimensional specifications for the flanged heads and split flange clamp halves applicable to four-bolt split flange type tube, pipe, and hose connections with appropriate references to the O-ring seals and attaching components used in their assembly. (See Figures 1 and 2.) Also included are recommended port dimensions and port design considerations.

The flanged heads specified are incorporated into fittings having suitable means for attachment to tubes, pipes, or hoses to provide connection ends. These connections are intended for application in hydraulic systems, on industrial and commercial products, where it is desired to avoid the use of threaded connections.

(R) THE RATED WORKING PRESSURE OF A HOSE ASSEMBLY COMPRISING SAE J518 HOSE CONNECTIONS AND SAE J517 HOSE SHALL NOT EXCEED THE LOWER OF THE TWO WORKING PRESSURE RATED VALUES.

Flanged heads shall be as specified in Figure 3 and Table 1. Split flange clamp halves shall be as specified in Figure 4 and Table 1. Port dimensions and spacing shall be as specified in Figure 5 and Table 2.

O-ring seals, having nominal dimensions as indicated in Table 1, are used in conjunction with these connections. They shall conform to the seals specified in SAE J120, Table on Dimensions and Tolerances.

Bolts for use with these connections shall be of the sizes and lengths indicated in Table 1. They shall be of SAE Grade 5 material or better as specified in SAE J429. Socket head cap screws of SAE Grade 5 material or better are acceptable.

Lock washers, if used, shall be in accordance with the light spring lock washers specified in SAE J489, Dimensions of Light, Medium, Heavy, Extra Heavy, and Hi Collar Spring Lock Washers, and of sizes applicable to the corresponding bolts.

The following general specifications supplement the dimensional data contained in Table 1 with respect to all unspecified detail.

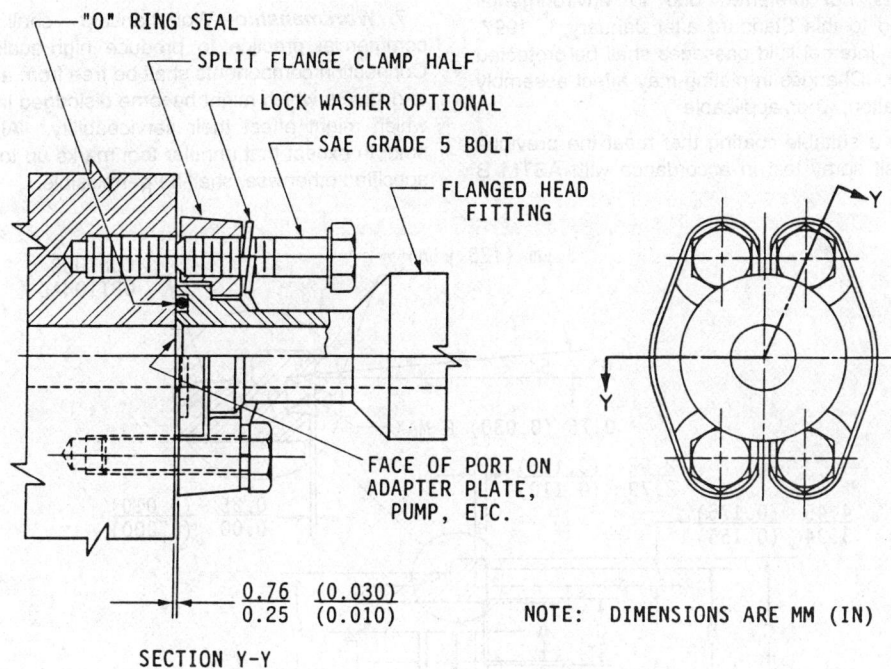

FIGURE 1—ASSEMBLED SPLIT FLANGED CONNECTION

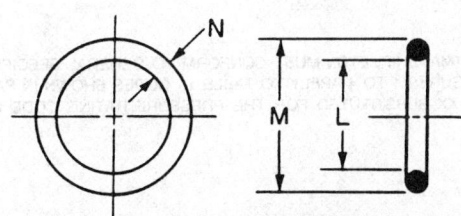

FIGURE 2—O-RING SEAL

2. References

2.1 Applicable Documents—The following publications form a part of this specification to the extent specified herein. The latest issue of SAE publications shall apply.

2.1.1 SAE PUBLICATIONS—Available from SAE, 400 Commonwealth Drive, Warrendale, PA 15096-0001.

SAE J120—Rubber Rings for Automotive Applications
SAE J429—Mechanical and Material Requirements for Externally Threaded Fasteners
SAE J489—Lock Washers
SAE J517—Hydraulic Hose
SAE J846—Coding Systems for Identification of Fluid Conductors and Connectors

2.1.2 ASTM PUBLICATION—Available from ASTM, 1916 Race Street, Philadelphia, PA 19103.

ASTM B 117—Method of Salt Spray (Fog) Testing

3. Size Designation—Four-bolt split flange connection sizes are designated by the nominal flange size which corresponds to the maximum inside diameter of the hole through the flanged head.

4. Dimensions and Tolerances—Tabulated dimensions and tolerances shall apply to the finished parts, plated or otherwise processed, as specified by the purchaser. Tolerances on all dimensions for flanged heads, split flange clamp halves, and ports not otherwise limited shall be ±0.4 mm (0.016 in).

5. Material—Flanged heads shall be made of steel. Split flange clamp halves shall be made from a material with the properties in Table 3.

(R) **6. Finish**—The external surfaces and threads of all carbon steel parts shall be plated or coated with a suitable material that passes a 72 h salt spray test in accordance with ASTM B 117. Any appearance of red rust during the 72 h salt spray test shall be considered failure, except for the following:

a. All internal fluid passages.

b. Edges such as hex points, serrations, and crests of threads where there may be mechanical deformation of the plating or coating typical of mass-produced parts or shipping effects.

c. Areas where there is mechanical deformation of the plating or coating caused by crimping, flaring, bending, and other post-plate metal forming operations.

d. Areas where the parts are suspended or affixed in the test chamber where condensate can accumulate.

NOTE—Cadmium plating is not preferred due to environmental reasons. Parts manufactured to this Standard after January 1, 1997, shall not be cadmium plated. Internal fluid passages shall be protected from corrosion during storage. Changes in plating may affect assembly torques and require requalification, when applicable.

Bolts shall be finished with a suitable coating that meet the previous requirements after a 16 h salt spray test in accordance with ASTM B 117. Lock washers may have a plain (natural) finish or a suitable coating.

7. Workmanship—Workmanship shall conform to the best commercial practice to produce high-quality connection components. Connection components shall be free from all hanging burrs, loose scale, and slivers which might become dislodged in usage and all other defects which might affect their serviceability. All sealing surfaces must be smooth except that annular tool marks up to 3 μm (100 μin) max, unless specified otherwise, shall be permissible.

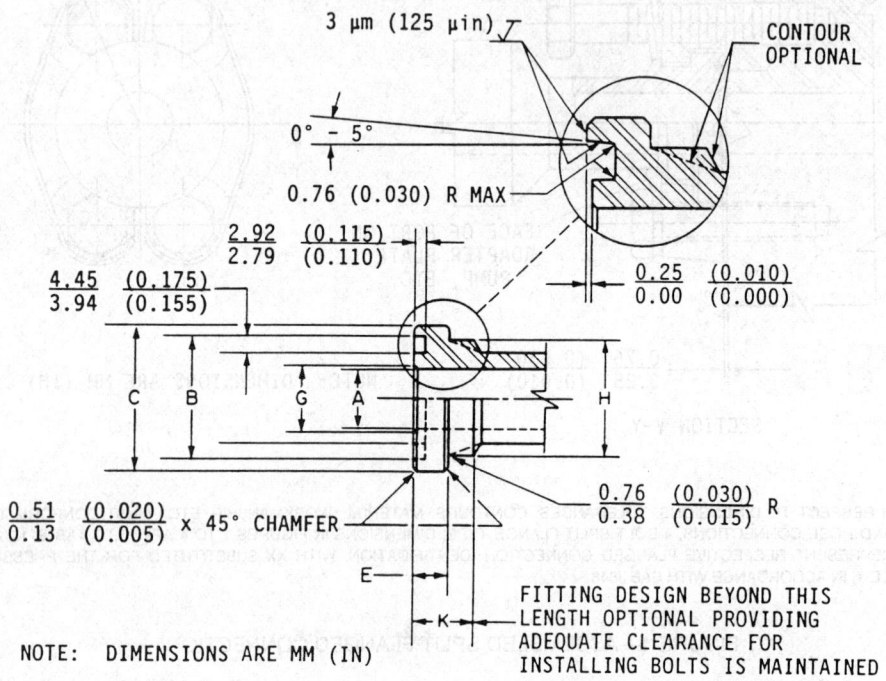

NOTE: DIMENSIONS ARE MM (IN)

FIGURE 3—FLANGED HEAD

NOTES: UNSPECIFIED DETAIL WITH RESPECT TO DIMENSIONS, TOLERANCES, CONTOURS, MATERIAL, WORKMANSHIP, ETC., MUST CONFORM TO GENERAL SPECIFICATIONS OF HYDRAULIC FLANGED TUBE, PIPE, AND HOSE CONNECTIONS, 4-BOLT SPLIT FLANGE TYPE, DIMENSIONS IN FIGURES 1 TO 4 APPLY TO TABLE 1. CODES SHOWN IN PARENTHESES ADJACENT TO FIGURE NUMBERS REPRESENT RESPECTIVE FLANGED CONNECTION IDENTIFICATION, WITH XX SUBSTITUTED FOR THE PRESSURE RATING CODE DEPICTED IN RESPECTIVE SUBHEADINGS OF TABLE 1, IN ACCORDANCE WITH SAE J846.

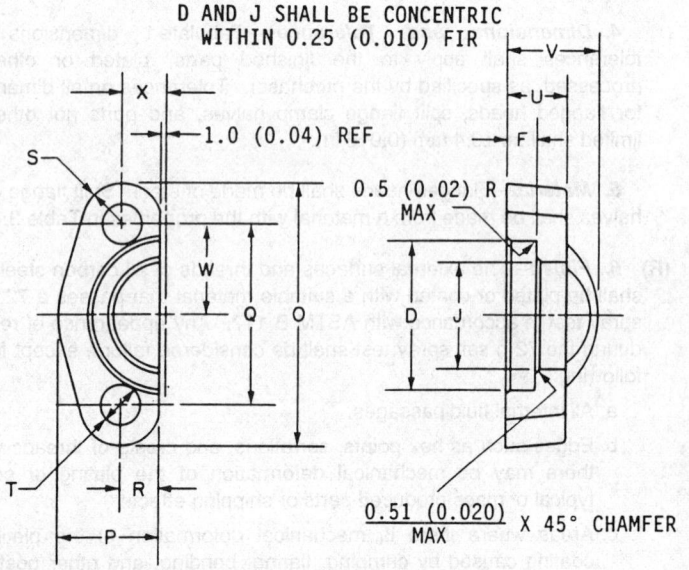

NOTE: DIMENSIONS ARE MM (IN)

FIGURE 4—SPLIT FLANGE CLAMP HALF (1101XX)

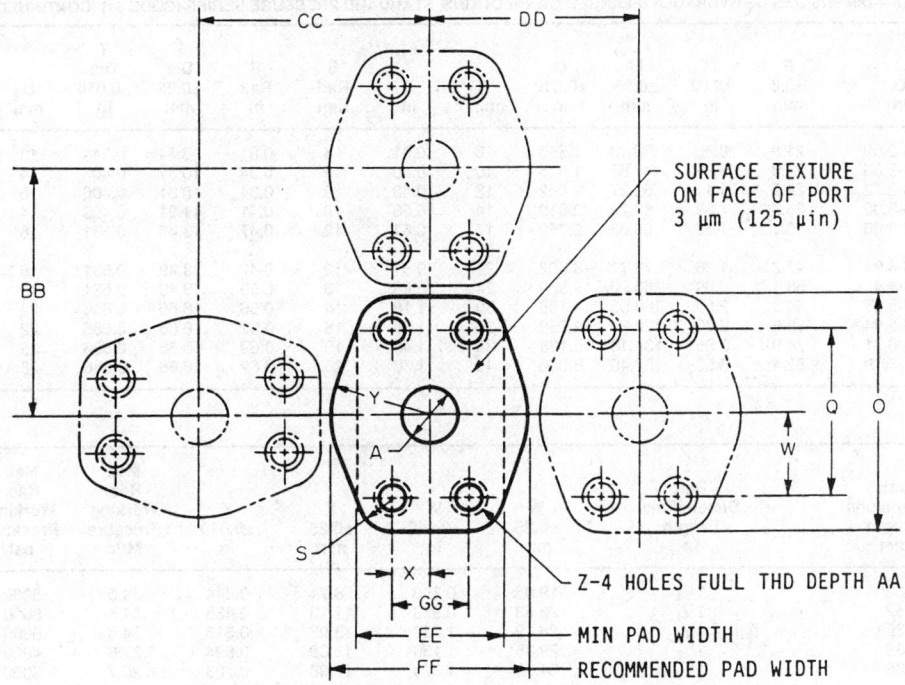

FIGURE 5—PORT DIMENSIONS FOR HYDRAULIC FLANGED, TUBE, PIPE, AND HOSE CONNECTIONS, FOUR-BOLT SPLIT FLANGE TYPE

TABLE 1A—DIMENSIONS OF HYDRAULIC FLANGED CONNECTIONS, STANDARD PRESSURE SERIES (CODE 61)

Nominal Flange Size, in	Flange Dash Size	A Dia Max mm	A Dia Max in	B Dia mm	B Dia in	C Dia ±0.25 mm	C Dia ±0.010 in	D Dia ±0.25 mm	D Dia ±0.010 in	E ±0.13 mm	E ±0.005 in	F ±0.13 mm	F ±0.005 in
1/2	-8	13	0.50	25.53-25.40	1.005-1.000	30.18	1.188	30.96	1.219	6.73	0.265	6.22	0.245
3/4	-12	19	0.75	31.88-31.75	1.255-1.250	38.10	1.500	38.89	1.531	6.73	0.265	6.22	0.245
1	-16	25	1.00	39.75-39.62	1.565-1.560	44.45	1.750	45.24	1.781	8.00	0.315	7.49	0.295
1 1/4	-20	32	1.25	44.58-44.45	1.755-1.750	50.80	2.000	51.59	2.031	8.00	0.315	7.49	0.295
1 1/2	-24	38	1.50	53.98-53.72	2.125-2.115	60.33	2.375	61.09	2.406	8.00	0.315	7.49	0.295
2	-32	51	2.00	63.50-63.25	2.500-2.490	71.42	2.812	72.24	2.844	9.53	0.375	9.02	0.355
2 1/2	-40	64	2.50	76.33-76.07	3.005-2.995	84.12	3.312	84.94	3.344	9.53	0.375	9.02	0.355
3	-48	76	3.00	92.08-91.82	3.625-3.615	101.60	4.000	102.39	4.031	9.53	0.375	9.02	0.355
3 1/2	-56	89	3.50	104.52-104.01	4.115-4.095	114.30	4.500	115.09	4.531	11.23	0.422	10.72	0.422
4	-64	102	4.00	117.22-116.71	4.615-4.595	127.00	5.000	127.79	5.031	11.23	0.442	10.72	0.422
5	-80	127	5.00	142.62-142.11	5.615-5.595	152.40	6.000	153.19	6.031	11.23	0.442	10.72	0.422

Nominal Flange Size, in	G Dia Max mm	G Dia Max in	H Dia Max mm	H Dia Max in	J Dia ±0.25 mm	J Dia ±0.010 in	K Ref mm	K Ref in	L ID Ref mm	L ID Ref in	M OD Ref mm	M OD Ref in	N Dia Ref mm	N Dia Ref in	O-Ring Size No.
1/2	14	0.56	24	0.94	24.26	0.955	13	0.50	18.64	0.734	25.70	1.012	3.53	0.139	210
3/4	21	0.81	32	1.25	32.13	1.265	14	0.56	24.99	0.984	32.05	1.262	3.53	0.139	214
1	27	1.06	38	1.50	38.48	1.515	14	0.56	32.92	1.296	39.98	1.574	3.53	0.139	219
1 1/4	33	1.31	43	1.70	43.69	1.720	14	0.56	37.69	1.484	44.75	1.762	3.53	0.139	222
1 1/2	40	1.56	50	1.98	50.80	2.000	16	0.62	47.22	1.859	54.28	2.137	3.53	0.139	225
2	52	2.06	62	2.45	62.74	2.470	16	0.62	56.74	2.234	63.80	2.512	3.53	0.139	228
2 1/2	65	2.56	74	2.92	74.93	2.950	18	0.69	69.44	2.734	76.50	3.012	3.53	0.139	232
3	78	3.06	90	3.55	90.93	3.580	19	0.75	85.32	3.359	92.38	3.637	3.53	0.139	237
3 1/2	90	3.56	102	4.00	102.36	4.030	22	0.88	98.02	3.859	105.08	4.137	3.53	0.139	241
4	103	4.06	114	4.50	115.06	4.530	25	1.00	110.72	4.359	117.78	4.637	3.53	0.139	245
5	129	5.06	140	5.50	140.46	5.530	28	1.12	136.12	5.359	143.18	5.637	3.53	0.139	253

TABLE 1A—DIMENSIONS OF HYDRAULIC FLANGED CONNECTIONS, STANDARD PRESSURE SERIES (CODE 61) (CONTINUED)

Nominal Flange Size, in	O mm	O in	P ±0.8 mm	P ±0.03 in	Q ±0.25 mm	Q ±0.010 in	R mm	R in	S Rad mm	S Rad in	T Dia ±0.25 mm	T Dia ±0.010 in	U mm	U in	V mm	V in
1/2	54.9-53.1	2.16-2.09	21.8	0.86	38.10	1.500	8	0.31	8	0.31	8.74	0.344	13	0.50	19	0.75
3/4	65.8-64.3	2.59-2.53	24.9	0.98	47.63	1.875	10	0.40	9	0.34	10.31	0.406	14	0.56	22	0.88
1	70.6-69.1	2.78-2.72	28.2	1.11	52.37	2.062	12	0.48	9	0.34	10.31	0.406	16	0.62	24	0.94
1 1/4	80.3-78.5	3.16-3.09	35.3	1.39	58.72	2.312	14	0.56	10	0.41	11.91	0.469	14	0.56	22	0.88
1 1/2	94.5-93.0	3.72-3.66	40.1	1.58	69.85	2.750	17	0.67	12	0.47	13.49	0.531	16	0.62	25	1.00
2	103.1-100.1	4.06-3.94	47.2	1.86	77.77	3.062	21	0.81	12	0.47	13.49	0.531	16	0.62	26	1.03
2 1/2	115.8-112.8	4.56-4.44	53.1	2.09	88.90	3.500	24	0.96	13	0.50	13.49	0.531	19	0.75	38	1.50
3	136.7-133.4	5.38-5.25	64.3	2.53	106.38	4.188	30	1.18	14	0.56	16.66	0.656	22	0.88	41	1.62
3 1/2	153.9-150.9	6.06-5.94	68.6	2.70	120.65	4.750	34	1.34	16	0.62	16.66	0.656	22	0.88	28	1.12
4	163.6-160.3	6.44-6.31	74.9	2.95	130.18	5.125	38	1.49	16	0.62	16.66	0.656	25	1.00	35	1.38
5	185.7-182.6	7.31-7.19	89.4	3.52	152.40	6.000	45	1.78	16	0.62	16.66	0.656	28	1.12	41	1.62

Nominal Flange Size, in	Bolt Dimensions Thread	Bolt Dimensions Length mm	Bolt Dimensions Length in	W ±0.25 mm	W ±0.010 in	X ±0.25 mm	X ±0.010 in	Max Rec. Working Pressure MPa	Max Rec. Working Pressure psi	Rec. Bolt Torque Range N·m	Rec. Bolt Torque Range lb·in
1/2	5/16-18	32	1 1/4	19.05	0.750	8.74	0.344	34.5	5000	20-25	175-225
3/4	3/8-16	32	1 1/4	23.83	0.938	11.13	0.438	34.5	5000	28-40	250-350
1	3/8-16	32	1 1/4	26.19	1.031	13.08	0.515	34.5	5000	37-48	325-425
1 1/4	7/16-14	38	1 1/2	29.36	1.156	15.09	0.594	27.6	4000	48-62	425-550
1 1/2	1/2-13	38	1 1/2	34.93	1.375	17.86	0.703	20.7	3000	62-79	550-700
2	1/2-13	38	1 1/2	38.89	1.531	21.44	0.844	20.7	3000	73-90	650-800
2 1/2	1/2-13	44	1 3/4	44.45	1.750	25.40	1.000	17.2	2500	107-124	950-1100
3	5/8-11	44	1 3/4	53.19	2.094	30.96	1.219	13.8	2000	186-203	1650-1800
3 1/2	5/8-11	51	2	60.33	2.375	34.93	1.375	3.4	500	158-181	1400-1600
4	5/8-11	51	2	65.07	2.562	38.89	1.531	3.4	500	158-181	1400-1600
5	5/8-11	57	2 1/4	76.20	3.000	46.02	1.812	3.4	500	158-181	1400-1600

TABLE 1B—DIMENSIONS OF HYDRAULIC FLANGED CONNECTIONS, HIGH PRESSURE SERIES (CODE 62)

Nominal Flange Size, in	Flange Dash Size	A Dia Max mm	A Dia Max in	B Dia mm	B Dia in	C Dia ±0.25 mm	C Dia ±0.010 in	D Dia ±0.25 mm	D Dia ±0.010 in	E ±0.13 mm	E ±0.005 in	F ±0.13 mm	F ±0.005 in
1/2	-8	13	0.50	25.53-25.40	1.005-1.000	31.75	1.250	32.54	1.281	7.75	0.305	7.24	0.285
3/4	-12	19	0.75	31.88-31.75	1.255-1.250	41.28	1.625	42.06	1.656	8.76	0.345	8.26	0.325
1	-16	25	1.00	39.75-39.62	1.565-1.560	47.63	1.875	48.41	1.906	9.53	0.375	9.02	0.355
1 1/4	-20	32	1.25	44.58-44.45	1.755-1.750	53.98	2.125	54.76	2.156	10.29	0.405	9.78	0.385
1 1/2	-24	38	1.50	53.98-53.72	2.125-2.115	63.50	2.500	64.29	2.531	12.57	0.495	12.07	0.475
2	-32	51	2.00	63.50-63.25	2.500-2.490	79.38	3.125	80.16	3.156	12.57	0.495	12.07	0.475

Nominal Flange Size, in	G Dia Max mm	G Dia Max in	H Dia Max mm	H Dia Max in	J Dia ±0.25 mm	J Dia ±0.010 in	K Ref mm	K Ref in	L ID Ref mm	L ID Ref in	M OD Ref mm	M OD Ref in	N Dia Ref mm	N Dia Ref in	O-Ring Size No.
1/2	14	0.56	24	0.94	24.64	0.970	14	0.56	18.64	0.734	25.70	1.012	3.53	0.139	210
3/4	21	0.81	32	1.25	32.51	1.280	18	0.69	24.99	0.984	32.05	1.262	3.53	0.139	214
1	27	1.06	38	1.50	38.86	1.530	21	0.81	32.92	1.296	39.98	1.574	3.53	0.139	219
1 1/4	33	1.31	44	1.72	44.45	1.750	25	1.00	37.69	1.484	44.75	1.762	3.53	0.139	222
1 1/2	40	1.56	51	2.00	51.56	2.030	30	1.19	47.22	1.859	54.28	2.137	3.53	0.139	225
2	52	2.06	67	2.62	67.56	2.660	38	1.50	56.74	2.234	63.80	2.512	3.53	0.139	228

Nominal Flange Size, in	O mm	O in	P ±0.8 mm	P ±0.03 in	Q ±0.25 mm	Q ±0.010 in	R mm	R in	S Rad mm	S Rad in	T Dia ±0.25 mm	T Dia ±0.010 in	U mm	U in	V mm	V in
1/2	57.2-55.6	2.25-2.19	22.6	0.89	40.49	1.594	8	0.32	8	0.31	8.74	0.344	16	0.62	22	0.88
3/4	72.1-70.6	2.84-2.78	29.0	1.14	50.80	2.000	11	0.43	10	0.41	10.31	0.406	19	0.75	28	1.12
1	81.8-80.3	3.22-3.16	33.8	1.33	57.15	2.250	13	0.51	12	0.47	11.91	0.469	24	0.94	33	1.31
1 1/4	96.0-94.5	3.78-3.72	37.6	1.48	66.68	2.625	15	0.59	14	0.56	13.49	0.531	27	1.06	38	1.50
1 1/2	114.3-111.3	4.50-4.38	46.5	1.83	79.38	3.125	17	0.68	17	0.66	16.66	0.656	30	1.19	43	1.69
2	134.9-131.8	5.31-5.19	55.9	2.20	96.82	3.812	21	0.84	18	0.72	19.84	0.781	37	1.44	52	2.06

TABLE 1B—DIMENSIONS OF HYDRAULIC FLANGED CONNECTIONS, HIGH PRESSURE SERIES (CODE 62) (CONTINUED)

Nominal Flange Size, in	Bolt Dimensions Thread	Bolt Dimensions Length mm	Bolt Dimensions Length in	W ±0.25 mm	W ±0.010 in	X ±0.25 mm	X ±0.010 in	Max Rec. Working Pressure MPa	Max Rec. Working Pressure psi	Rec. Bolt Torque Range N·m	Rec. Bolt Torque Range lb·in
1/2	5/16-18	32	1 1/4	20.24	0.797	9.12	0.359	41.4	6000	20-25	175-225
3/4	3/8-16	38	1 1/2	25.40	1.000	11.91	0.469	41.4	6000	34-45	300-400
1	7/16-14	44	1 3/4	28.58	1.125	13.89	0.547	41.4	6000	56-68	500-600
1 1/4	1/2-13	44	1 3/4	33.32	1.312	15.88	0.625	41.4	6000	85-102	750-900
1 1/2	5/8-11	57	2 1/4	39.67	1.562	18.26	0.719	41.4	6000	158-181	1400-1600
2	3/4-10	70	2 3/4	48.41	1.906	22.23	0.875	41.4	6000	271-294	2400-2600

TABLE 2A—PORT DIMENSIONS FOR BOLTED FLANGE CONNECTIONS, STANDARD PRESSURE SERIES

Nominal Flange Size, in	Flange Dash Size	A Dia +0.0/-1.5 mm	A Dia +0.00/-0.06 in	O mm	O in	FF mm	FF in	Q ±0.25 mm	Q ±0.010 in	GG ±0.25 mm	GG ±0.010 in	S Rad mm	S Rad in	W mm	W in
1/2	-8	12.7	0.50	54	2.12	46	1.81	38.10	1.500	17.48	0.688	8	0.31	19	0.75
3/4	-12	19.1	0.75	65	2.56	52	2.06	47.63	1.875	22.23	0.875	9	0.34	24	0.94
1	-16	25.4	1.00	70	2.75	59	2.31	52.37	2.062	26.19	1.031	9	0.34	26	1.03
1 1/4	-20	31.8	1.25	79	3.12	73	2.88	58.72	2.312	30.18	1.188	10	0.41	29	1.16
1 1/2	-24	38.1	1.50	94	3.69	83	3.25	69.85	2.750	35.71	1.406	12	0.47	35	1.38
2	-32	50.8	2.00	102	4.00	97	3.81	77.77	3.062	42.88	1.688	12	0.47	39	1.53
2 1/2	-40	63.5	2.50	114	4.50	109	4.28	88.90	3.500	50.80	2.000	13	0.50	44	1.75
3	-48	76.2	3.00	135	5.31	131	5.16	106.38	4.188	61.93	2.438	14	0.56	53	2.09
3 1/2	-56	88.9	3.50	152	6.00	140	5.50	120.65	4.750	69.85	2.750	16	0.62	60	2.38
4	-64	101.6	4.00	162	6.38	152	6.00	130.18	5.125	77.77	3.062	16	0.62	65	2.56
5	-80	127.0	5.00	184	7.25	181	7.12	152.40	6.000	92.08	3.625	16	0.62	76	3.00

Nominal Flange Size, in	X mm	X in	Y Rad mm	Y Rad in	Z Thread UNC-2B	AA Min mm	AA Min in	BB[1] Min mm	BB[1] Min in	CC[1] Min mm	CC[1] Min in	DD[1] Min mm	DD[1] Min in	EE Min mm	EE Min in
1/2	9	0.34	23	0.91	5/16-18	24	0.94	56	2.22	52	2.06	49	1.91	33	1.31
3/4	11	0.44	26	1.03	3/8-16	22	0.88	68	2.66	61	2.41	55	2.16	41	1.62
1	13	0.52	29	1.16	3/8-16	22	0.88	72	2.84	67	2.62	61	2.41	48	1.88
1 1/4	15	0.59	37	1.44	7/16-14	28	1.12	82	3.22	78	3.09	75	2.97	54	2.12
1 1/2	18	0.70	41	1.62	1/2-13	27	1.06	96	3.78	90	3.56	85	3.34	64	2.50
2	21	0.84	49	1.91	1/2-13	27	1.06	104	4.09	102	4.00	99	3.91	76	3.00
2 1/2	25	1.00	54	2.14	1/2-13	30	1.19	117	4.59	114	4.50	111	4.38	89	3.50
3	31	1.22	66	2.58	5/8-11	30	1.19	137	5.41	136	5.34	133	5.25	106	4.19
3 1/2	35	1.38	70	2.75	5/8-11	33	1.31	155	6.09	148	5.84	142	5.59	119	4.69
4	39	1.53	76	3.00	5/8-11	30	1.19	164	6.47	160	6.28	155	6.09	132	5.19
5	46	1.81	90	3.56	5/8-11	33	1.31	186	7.34	185	7.28	183	7.22	151	6.19

[1] Dimensions BB, CC, and DD provide 1.5 mm (0.06 in) clearance between flanges, dimensionally on the high limit, when the same size flanges are used on adjacent ports. These dimensions do not apply when more than one size of flanges are used on adjacent ports.

TABLE 2B—PORT DIMENSIONS FOR BOLTED FLANGE CONNECTIONS, HIGH PRESSURE SERIES

Nominal Flange Size, in	Flange Dash Size	A Dia +0.0 -1.5 mm	A Dia +0.00 -0.06 in	O mm	O in	FF mm	FF in	Q ±0.25 mm	Q ±0.010 in	GG ±0.25 mm	GG ±0.010 in	S Rad mm	S Rad in	W mm	W in
1/2	-8	12.7	0.50	56	2.22	48	1.88	40.49	1.594	18.24	0.718	8	0.31	20	0.80
3/4	-12	19.1	0.75	71	2.81	60	2.38	50.80	2.000	23.80	0.937	10	0.41	25	1.00
1	-16	25.4	1.00	81	3.19	70	2.75	57.15	2.250	27.76	1.093	12	0.47	28	1.12
1 1/4	-20	31.8	1.25	95	3.75	78	3.06	66.68	2.625	31.75	1.250	14	0.56	33	1.31
1 1/2	-24	38.1	1.50	113	4.44	95	3.75	79.38	3.125	36.50	1.437	17	0.66	40	1.56
2	-32	50.8	2.00	133	5.25	114	4.50	96.82	3.812	44.45	1.750	18	0.72	49	1.91

Nominal Flange Size, in	X mm	X in	Y Rad mm	Y Rad in	Z Thread UNC-2B	AA Min mm	AA Min in	BB[1] Min mm	BB[1] Min in	CC[1] Min mm	CC[1] Min in	DD[1] Min mm	DD[1] Min in	EE Min mm	EE Min in
1/2	9	0.36	24	0.94	5/16-18	21	0.81	59	2.34	56	2.22	53	2.09	38	1.50
3/4	12	0.47	30	1.19	3/8-16	24	0.94	75	2.94	70	2.75	66	2.59	48	1.88
1	14	0.55	35	1.38	7/16-14	27	1.06	84	3.31	80	3.16	75	2.97	54	2.12
1 1/4	16	0.62	39	1.53	1/2-13	25	1.00	99	3.88	90	3.56	83	3.25	60	2.38
1 1/2	18	0.72	48	1.88	5/8-11	35	1.38	116	4.56	108	4.25	101	3.97	70	2.75
2	22	0.88	57	2.25	3/4-10	38	1.50	137	5.38	128	5.03	120	4.72	86	3.38

[1] Dimensions BB, CC, and DD provide 1.5 mm (0.06 in) clearance between flanges, dimensionally on the high limit, when the same size flanges are used on adjacent ports. These dimensions do not apply when more than one size of flanges are used on adjacent ports.

TABLE 3—MATERIAL PROPERTIES

Standard series—1/2 in (-8) size	Minimum yield, 221 MPa (32 000 psi)
	Minimum elongation, 3%
All other sizes	Minimum yield, 414 MPa (60 000 psi)
	Minimum elongation, 3%
High pressure series—all sizes	Minimum yield, 331 MPa (48 000 psi)
	Minimum elongation, 3%

(R) CLIP FASTENER FITTING
—SAE J1467 JUN93

SAE Standard

Report of the Fluid Conductors and Connectors Technical Committee approved December 1988 and completely revised June 1993. Rationale statement available.

1. Scope—This SAE Standard covers material and dimensional requirements of steel clip fastener fittings. These fittings are intended for use in hydraulic systems on industrial equipment primarily in mining applications.

2. References

2.1 Applicable Document—The following publication forms a part of this specification to the extent specified herein.

2.1.1 ASTM PUBLICATION—Available from ASTM, 1916 Race Street, Philadelphia, PA 19103-1187.

ASTM B 117—Method of Salt Spray (Fog) Testing

3. Size Designation—Fitting sizes are designated by the corresponding nominal inside diameter of hose. See Table 1.

TABLE 1—NOMINAL SIZE DESIGNATIONS

Nominal SAE Dash Size	Nominal Hose I.D. mm
-4	6.35
-6	9.52
-8	12.70
-12	19.05
-16	25.40
-20	31.75
-24	38.10
-32	50.80

4. Material and Manufacture

4.1 Material

4.1.1 MALE AND FEMALE—The material used in the manufacture of the male and female components shall be steel such as AISI 12L14, 1137, 1141, or other free cutting steels having a minimum yield strength of 193 MPa and a minimum tensile strength of 345 MPa. See Figures 1 and 2.

4.1.2 O-RING—The standard clip fastener O-ring shall be manufactured from an elastomeric material that is compatible with the fluid being conveyed. Suitable materials include nitrile (NBR) rubber or viton having a minimum Shore `A' hardness of 80 durometer. See Figure 3.

4.1.3 BACK-UP RING—The clip fastener back-up ring shall be manufactured from a material that is compatible with the fluid being conveyed. Suitable materials include acetal homopolymers, polyamide, or Teflon (PTFE). See Figure 3.

4.1.4 STAPLES—The clip fastener staple (clip) shall be manufactured from corrosion-resistant steel or spring steel. Contour and details of staple are optional with manufacturer providing that interchangeability of the male and female is not affected. See Figure 4.

4.2 Finish—The external surfaces and threads of all carbon steel parts shall be plated or coated with a suitable material that passes a 72 h salt spray test in accordance with ASTM B 117. Any appearance of red rust during the 72 h salt spray test shall be considered failure, except for the following:

a. All internal fluid passages.

b. Edges such as hex points, serrations, and crests of threads where there may be mechanical deformation of the plating or coating typical of mass-produced parts or shipping effects.

c. Areas where there is mechanical deformation of the plating or coating caused by crimping, flaring, bending, and other post-plate metal forming operations.

d. Areas where the parts are suspended or affixed in the test chamber where condensate can accumulate.

NOTE—Cadmium plating is not preferred due to environmental reasons. Parts manufactured to this document after January 1, 1977, shall not be cadmium plated. Internal fluid passages shall be protected from corrosion during storage. Changes in plating may affect assembly torques and require requalification, when applicable.

4.3 Workmanship—Workmanship shall conform to the best commercial practice to produce high-quality fittings. Fittings must be free from visual contaminants, all hanging burrs, loose scale, and slivers which might be dislodged in usage, and any other defects that might affect the function of the parts.

4.4 Construction—Fittings may be made by forging, cold heading, or machined from bar stock. Carbon steel fittings fabricated from multiple components may be bonded together by copper brazing, silver brazing, welding, or other suitable processes.

4.5 Dimensions—The dimensions for the components shown in Figure 5 shall be in accordance with Figures 1 through 4.

NOTE—The alternate methods of fabricating the female may be used providing the envelope dimensions are not affected to the extent that interchangeability becomes a problem. See Figure 5.

5. Protection—Sealing surfaces and threads (both internal and external) shall be protected by the manufacturer from nicks, scratches, or any damage that is detrimental to their function.

22.198

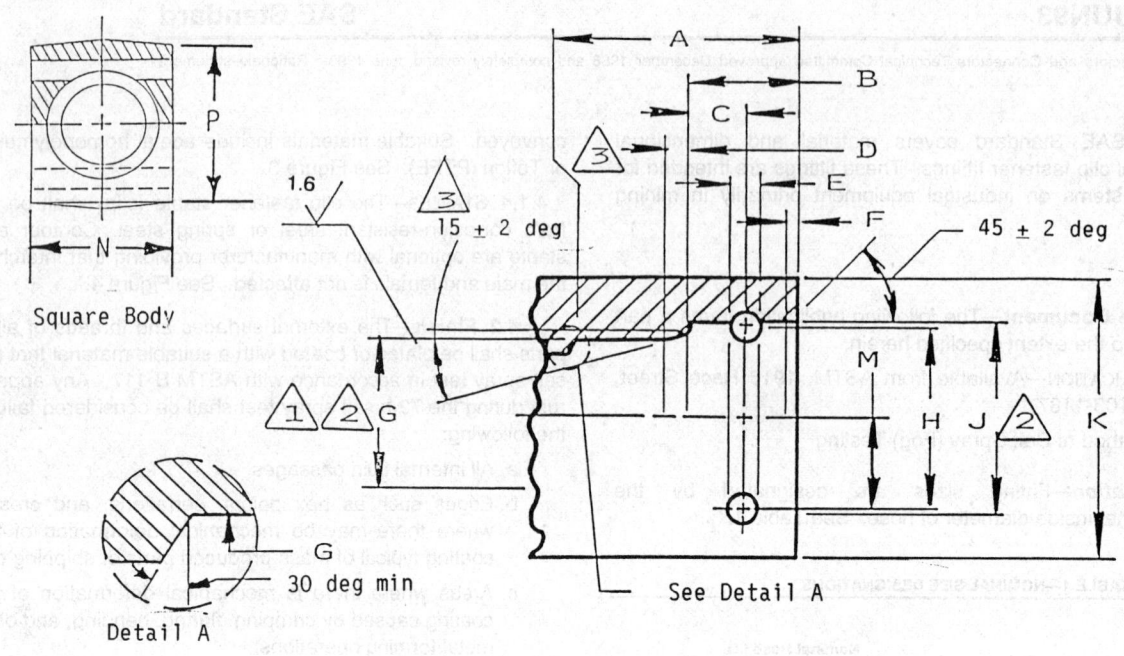

Nominal SAE Dash Size	A min mm	B ±0.5 mm	C mm	D ±0.10 mm	E mm	F max mm	G ±0.025 mm	H mm	J ±0.08 mm	K dia mm	M ±0.13 mm	N ±0.5 mm	P ±0.5 mm
-4	27.43	15.5	7.7 / 7.0	2.0	6.15 / 5.89	1.1	10.03	13.3 / 12.8	15.16	27.2 / 23.0	6.50	21.0	22.3
-6	27.43	15.5	7.7 / 7.0	2.0	6.15 / 5.89	1.1	14.02	18.3 / 17.8	20.17	32.0 / 30.0	9.02	26.0	30.0
-8	27.43	15.5	7.7 / 7.0	2.0	6.15 / 5.89	1.1	18.03	22.3 / 21.8	24.16	36.6 / 34.9	11.0	30.0	35.0
-12	27.43	15.5	7.7 / 7.0	2.0	6.15 / 5.89	1.1	24.03	27.3 / 26.8	29.16	45.2 / 41.0	13.51	35.0	41.0
-16	32.51	20.5	9.7 / 9.0	2.0	8.69 / 8.30	1.1	31.01	36.3 / 35.7	39.12	55.1 / 52.4	18.0	48.0	53.0
-20	32.51	20.5	9.7 / 9.0	2.0	8.69 / 8.30	1.1	38.02	43.4 / 42.8	46.15	65.3 / 60.0	21.50	55.0	60.0
-24	34.54	20.5	11.7 / 11.0	2.0	9.19 / 8.30	1.1	47.02	52.4 / 51.6	55.22	73.0 / 70.0	26.0	--	--
-32	34.54	20.5	11.7 / 11.0	2.0	9.19 / 8.30	1.1	56.00	61.4 / 60.6	64.21	82.6 / 80.0	30.50	--	--

4   The female body can be manufactured as a swivel type where the design and method of attachment shall be optional with the manufacturer (see Figure 5).

△3   Optional Design: $1.0 \times 30° \pm 2°$ (1.5)

△2   These diameters must be concentric within 0.05 TIR.

△1   "G" diameter to run to the full depth of "A" dimension.

FIGURE 1—FEMALE CLIP FASTENER BODY

22.199

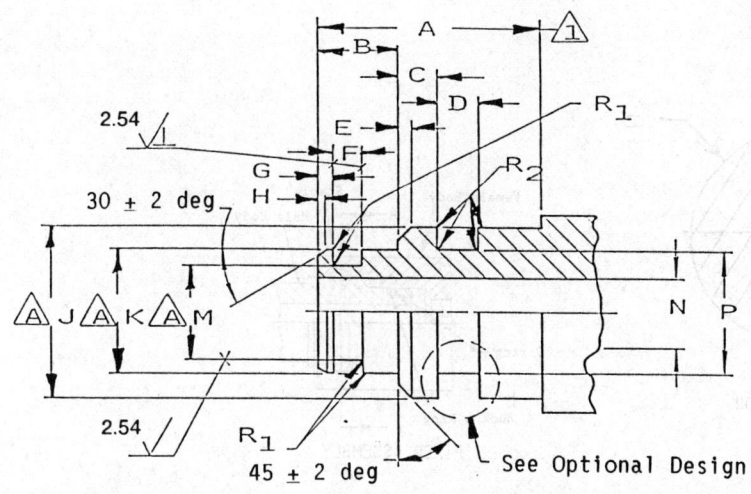
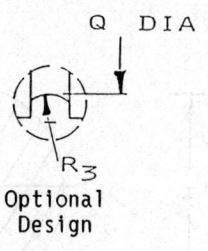

ALL DIAMETERS WITHIN THIS LENGTH MUST BE CONCENTRIC WITHIN 0.13 TIR. EXCEPT DIAMETERS ⚠ MUST BE CONCENTRIC WITHIN 0.005 TIR.

Nominal SAE Dash Size	A min mm	B mm	C ±0.1 mm	D ±0.1 mm	E ±0.2 mm	F ±0.010 mm	G mm	H ±0.25 mm	J ±0.08 mm	K +0.05 -0.08 mm	M mm	N mm	P ±0.1 mm	Q ±0.1 mm	R_1 max mm	R_2 max mm	R_3 ref mm
-4	27.74	11.25 10.74	4.9	5.1	1.5	3.1	3.1 2.6	1.0	14.90	9.90	6.86 6.73	4.7 2.8	8.40	7.42	0.5	0.7	3.4
-6	27.74	11.25 10.74	4.9	5.1	1.5	3.1	3.1 2.6	1.0	19.90	13.90	10.85 10.72	8.2 6.6	13.40	12.40	0.5	0.7	3.4
-8	27.74	11.25 10.74	4.9	5.1	1.5	3.6	3.1 2.6	1.0	23.90	17.90	14.05 13.92	11.4 9.7	17.40	16.40	0.5	0.7	3.4
-12	27.74	11.25 10.74	4.9	5.1	1.5	3.6	3.1 2.6	1.0	28.90	23.90	20.04 19.91	17.0 15.4	22.40	21.39	0.5	0.7	3.4
-16	32.90	11.25 10.74	5.9	7.1	1.5	3.6	3.1 2.6	1.0	38.80	30.90	27.05 26.92	23.0 19.8	29.90	28.78	0.5	0.7	5.3
-20	32.90	11.25 10.74	5.9	7.1	1.5	3.6	3.1 2.6	1.0	45.90	37.90	34.04 33.91	30.0 24.8	36.90	35.79	0.5	0.7	5.3
-24	36.30	13.28 12.78	7.9	7.1	1.5	5.1	3.1 2.6	1.0	54.90	46.90	42.00 41.90	35.8 30.0	45.90	44.78	0.6	0.7	5.3
-32	36.30	13.28 12.78	7.9	7.1	1.5	5.1	3.1 2.6	1.0	63.90	55.90	51.00 50.90	45.0 40.0	54.90	53.77	0.6	0.7	5.3

FIGURE 2—MALE CLIP FASTENER BODY

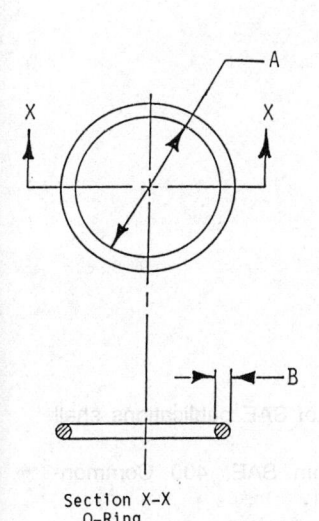
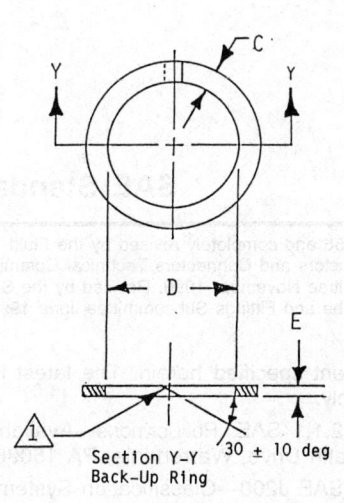

Nominal SAE Dash Size	A Dia mm	B Dia mm	C mm	D Dia mm	E mm
-4	6.16 5.84	2.08 1.92	1.65 1.55	6.96 6.86	.90 .70
-6	10.20 9.80	2.08 1.92	1.65 1.55	10.95 10.85	.90 .70
-8	13.25 12.75	2.59 2.41	2.06 1.96	14.15 14.05	.90 .70
-12	19.40 18.60	2.59 2.41	2.06 1.96	20.14 20.04	.90 .70
-16	25.40 24.60	2.59 2.41	2.06 1.96	27.15 27.05	.90 .70
-20	33.40 32.60	2.59 2.41	2.06 1.96	34.14 34.04	.90 .70
-24	40.50 39.50	3.10 2.90	2.57 2.46	42.11 42.01	1.85 1.40
-32	50.50 49.50	3.10 2.90	2.57 2.46	51.10 51.00	1.85 1.40

2. All surfaces must be smooth and free from irregularities.

⚠ Cut must be clean and sharp.

FIGURE 3—O-RING SEAL AND BACK-UP RING

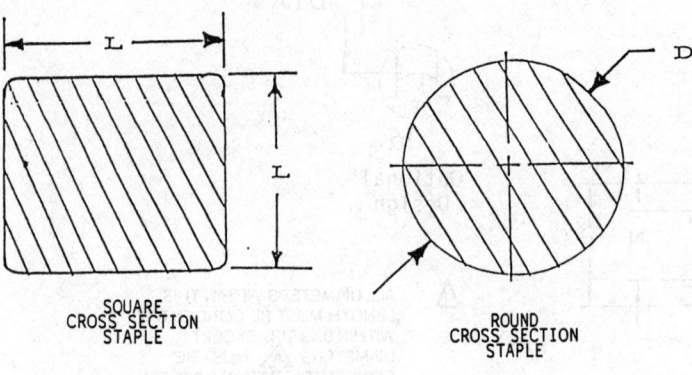

Nominal SAE Dash Size	Cross Section Type Square L Length mm	Cross Section Type Round D Diameter mm
-4	4.07 3.93	4.78 4.72
-6	4.07 3.93	4.78 4.72
-8	4.07 3.93	4.78 4.72
-12	4.07 3.93	4.78 4.72
-16	6.08 5.92	6.40 6.35
-20	6.08 5.92	6.40 6.35
-24	6.08 5.92	6.40 6.35
-32	6.08 5.92	6.40 6.35

FIGURE 4—STAPLE CROSS SECTIONS

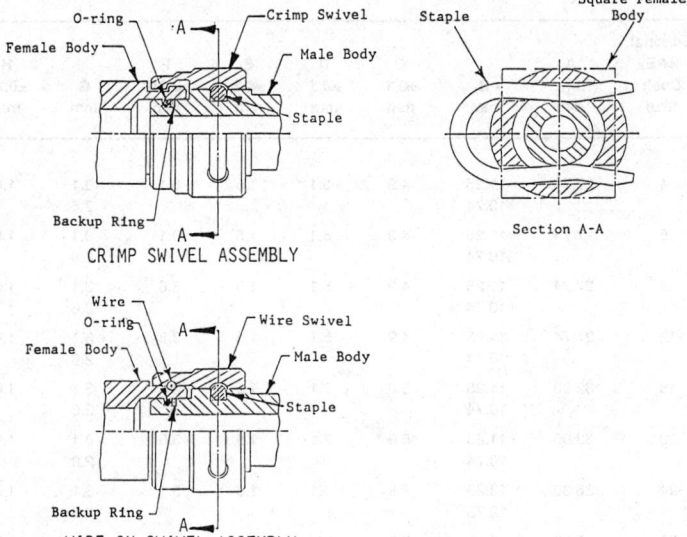

FIGURE 5—TYPICAL ASSEMBLIES OF SWIVEL AND FIXED CONNECTIONS

HYDRAULIC O-RING
—SAE J515 JUN92

SAE Standard

Report of the Tube, Pipe, Hose, and Lubrication Fittings Committee approved January 1956 and completely revised by the Fluid Conductors and Connectors Technical Committee June 1989. Revised by the Fluid Conductors and Connectors Technical Committee June 1990. Completely revised by the Fluid Conductors and Connectors Technical Committee November 1990. Revised by the SAE Fluid Conductors and Connectors Technical Committee S1—Automotive and Hydraulic Tube and Fittings Subcommittee June 1992. Rationale statement available.

1. Scope
This SAE Standard outlines material and dimensional requirements for O-rings used with SAE J1926 port and SAE J1453 fittings for hydraulic applications.

2. References

2.1 Applicable Documents
The following publications form a part of this specification to the extent specified herein. The latest issue of SAE publications shall apply.

2.1.1 SAE PUBLICATIONS—Available from SAE, 400 Commonwealth Drive, Warrendale, PA 15096-0001.

SAE J200—Classification System for Rubber Materials for Automotive Applications

SAE J1453—Fitting O-Ring Face Seal

SAE J1926—Specification for Straight Thread O-Ring Boss Port

2.1.2 ASTM Publications—Available from ASTM, 1916 Race Street, Philadelphia, PA 19103.

ASTM D 1329—Evaluating Rubber Property—Retraction at Lower Temperatures (TR Test)

ASTM D 2000—Rubber Products in Automotive Applications

3. Type 1
Petroleum base and nonflammable water base hydraulic fluids (see Tables 1 and 2).

3.1 General Service
High-pressure applications of pneumatics, water base hydraulic fluids, lubricating oils, hydraulic oils, and gasoline.

(R) ### 3.2 Temperature Range
−35 to +120 °C (−30 to +250 °F)

3.3 Shore Hardness
90 pts ± 5 pts

3.4 Elongation
100% min

(R) ### 3.5 Tensile
10.3 MPa min (1500 psi min)

3.6 Compound
Nitrile (Buna N) to ASTM D 2000 or SAE J200 4CH915B14E015E035

3.7 Lubrication
When assembling Type 1 O-rings with O-ring style fittings, the O-ring shall be coated with the fluid used or petrolatum, before assembly to ease installation.

4. Type 2
Nonflammable phosphate ester base hydraulic fluids (see Tables 1 and 2).

4.1 General Service
High-pressure application of nonflammable hydraulic fluids of the phosphate ester base type.

(R) ### 4.2 Temperature Range
−40 to 100 °C (−40 to 212 °F)

4.3 Shore Hardness A
80 pts ± 5 pts

4.4 Elongation
150% min

(R) ### 4.5 Tensile
10.3 MPa min (1500 psi min)

4.6 Compound
EPDM to ASTM D 2000 or SAE J200 3BA815A14B13F17

4.7 Lubrication
When assembling Type 2 O-rings with O-ring style fittings, lubricate the O-ring with the fluid used in the system. Do not use a petroleum-based lubricant.

5. Type 3
High temperature for hydraulic fluids (see Tables 1 and 2).

5.1 General Service
High-pressure, high-temperature applications of pneumatic, water base hydraulic fluids, lubricating oils, hydraulic oils, and fuels.

(R) ### 5.2 Temperature Range
−40 to 205 °C (−40 to 400 °F)

5.3 Shore Hardness A
90 pts ± 5 pts

5.4 Elongation
100% min

(R) ### 5.5 Tensile
10.0 MPa min (1450 psi min)

5.6 Compound
Fluorocarbon to ASTM D 2000 or SAE J200 M7HK910A1-11B38EF31E088Z1 or 7HK914A1-11B38EF31E088Z1 Z1 = TR10 temperature −15 °C (5 °F) or lower (similar to MIL-R-83248 Type II, Class 2).

5.7 Lubrication
When assembling Type 3 O-rings with O-ring style fittings, the O-ring shall be coated with fluid used or petrolatum before assembly to ease installation.

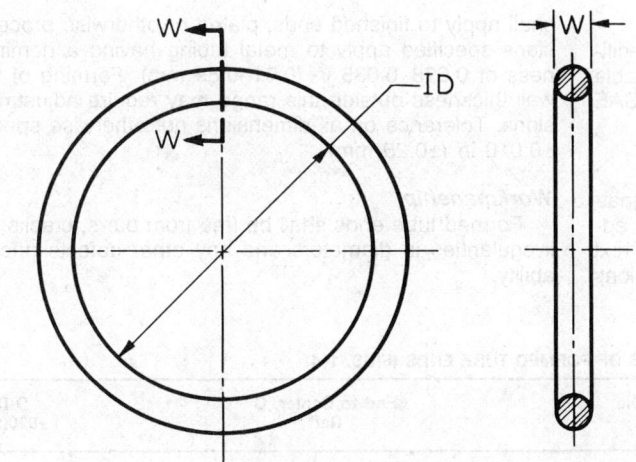
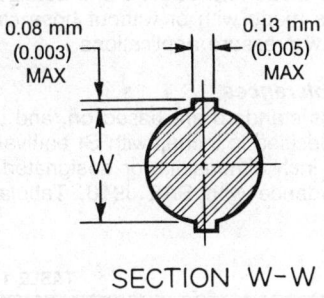

(R) FIGURE 1—O-RING SEAL

(R) TABLE 1—DIMENSIONS FOR O-RINGS FOR USE WITH SAE J1926 PORT
(DIMENSIONS IN MILLIMETERS AND INCHES—FIGURE 1)

Nominal Tubing Outer Diameter (mm)	Nominal Tubing Outer Diameter (in)	O-Ring Size Type 1	O-Ring Size Type 2	I.D. (mm)	I.D. (in)	W Width (mm)	W Width (in)
3.18	0.125	2-1	2-2	6.07 ± 0.13	0.239 ± 0.005	1.63 ± 0.08	0.064 ± 0.003
4.75	0.187	3-1	3-2	7.65 ± 0.13	0.301 ± 0.005	1.63 ± 0.08	0.064 ± 0.003
6.35	0.250	4-1	4-2	8.92 ± 0.13	0.351 ± 0.005	1.83 ± 0.08	0.072 ± 0.003
7.92	0.312	5-1	5-2	10.52 ± 0.13	0.414 ± 0.005	1.83 ± 0.08	0.072 ± 0.003
9.53	0.375	6-1	6-2	11.89 ± 0.13	0.468 ± 0.005	1.98 ± 0.08	0.078 ± 0.003
12.70	0.500	8-1	8-2	16.36 ± 0.13	0.644 ± 0.005	2.21 ± 0.08	0.087 ± 0.003
15.88	0.625	10-1	10-2	19.18 ± 0.13	0.755 ± 0.005	2.46 ± 0.08	0.097 ± 0.003
19.05	0.750	12-1	12-2	23.47 ± 0.15	0.924 ± 0.006	2.95 ± 0.10	0.116 ± 0.004
22.23	0.875	14-1	14-2	26.62 ± 0.15	1.048 ± 0.006	2.95 ± 0.10	0.116 ± 0.004
25.40	1.000	16-1	16-2	29.74 ± 0.15	1.171 ± 0.006	2.95 ± 0.10	0.116 ± 0.004
31.75	1.250	20-1	20-2	37.47 ± 0.25	1.475 ± 0.010	3.00 ± 0.10	0.118 ± 0.004
38.10	1.500	24-1	24-2	43.69 ± 0.25	1.720 ± 0.010	3.00 ± 0.10	0.118 ± 0.004
50.80	2.000	32-1	32-2	59.36 ± 0.25	2.337 ± 0.010	3.00 ± 0.10	0.118 ± 0.004

TABLE 2—DIMENSIONS FOR FACE SEAL O-RINGS FOR USE WITH SAE J1453 FITTINGS
(DIMENSIONS IN MILLIMETERS AND INCHES—FIGURE 1)

Nominal Tubing Outer Diameter (mm)	Nominal Tubing Outer Diameter (in)	SAE Dash Size	O-Ring Size No. per SAE J120	I.D. (mm)	I.D. (in)	O.D. Ref. (mm)	O.D. Ref. (in)	W Width (±0.08) (mm)	W Width (±0.003) (in)
6.35	0.250	-4	011	7.65 ± 0.13	0.301 ± 0.005	11.20	0.441	1.78	0.070
9.53	0.375	-6	012	9.25 ± 0.13	0.364 ± 0.005	12.80	0.504	1.78	0.070
12.70	0.500	-8	014	12.42 ± 0.13	0.489 ± 0.005	15.98	0.629	1.78	0.070
15.88	0.625	-10	016	15.60 ± 0.13	0.614 ± 0.005	19.15	0.754	1.78	0.070
19.05	0.750	-12	018	18.77 ± 0.13	0.739 ± 0.005	22.33	0.879	1.78	0.070
25.40	1.000	-16	021	23.52 ± 0.15	0.926 ± 0.006	27.08	1.066	1.78	0.070
31.75	1.250	-20	025	29.87 ± 0.15	1.176 ± 0.006	33.43	1.316	1.78	0.070
38.10	1.500	-24	029	37.82 ± 0.25	1.489 ± 0.010	41.38	1.629	1.78	0.070

FORMED TUBE ENDS FOR HOSE CONNECTIONS—SAE J962 MAY86

SAE Standard

Report of Tube, Pipe, Hose and Lubrication Fittings Committee approved June 1966, reaffirmed by the Fluid Conductors and Connectors Technical Committee, May 1986.

GENERAL SPECIFICATIONS

Scope

This SAE Standard covers the dimensional and general specifications applicable to those formed tube end configurations suitable for hose connections made with or without hose clamps (see SAE J536) in relatively low pressure applications.

Dimensions and Tolerances

Dimensions in this standard are based on, and unless designated otherwise, are specified in inches with SI equivalents shown adjacent to respective inch dimensions or designated mm in the text and tables in accordance with SAE J916. Tabulated dimensions shall apply to finished ends, plated or otherwise processed. Dimensions specified apply to metal tubing having a nominal wall thickness of 0.028–0.035 in (0.71–0.89 mm). Forming of tube having a wall thickness outside this range may require adjustment of dimensions. Tolerance on all dimensions not otherwise specified shall be ±0.010 in (±0.25 mm).

Workmanship

Formed tube ends shall be free from burrs, cracks, sharp edges, irregularities in diameters and any other defects affecting serviceability.

TABLE 1—DIMENSIONS OF FORMED TUBE ENDS (FIGS. 1–4)

Nominal Tube OD	Outside Dia, A		End Dia, B		End to Center, C Ref[a]		D Dia ±020(0.51)	
in	in	mm	in	mm	in	mm	in	mm
3/16	0.220	5.59	0.180	4.57	0.080	2.03	0.220	5.59
1/4	0.290	7.37	0.240	6.10	0.100	2.54	0.290	7.37
5/16	0.360	9.14	0.310	7.87	0.100	2.54	0.360	9.14
3/8	0.430	10.92	0.380	9.65	0.100	2.54	0.430	10.92
7/16	0.490	12.45	0.440	11.18	0.100	2.54	0.490	12.45
1/2	0.560	14.22	0.500	12.70	0.120	3.05	0.560	14.22
9/16	0.620	15.75	0.560	14.22	0.120	3.05	0.620	15.75
5/8	0.690	17.53	0.630	16.00	0.120	3.05	0.690	17.53

[a] For reference purposes only, not intended for inspection.

22.203

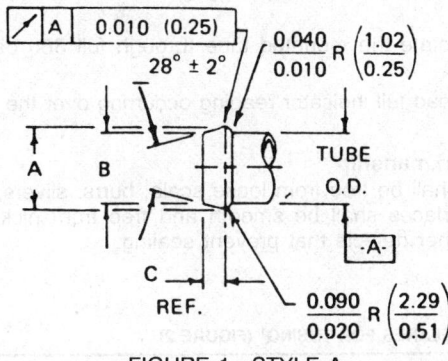

FIGURE 1—STYLE A

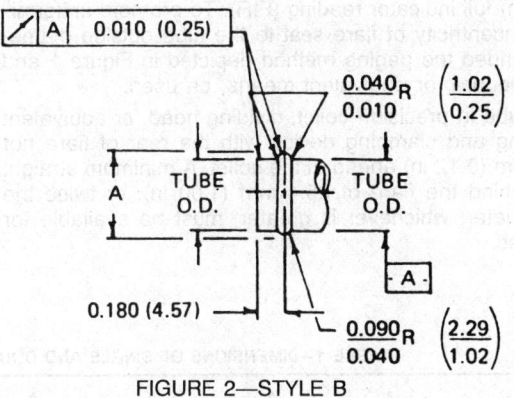

FIGURE 2—STYLE B

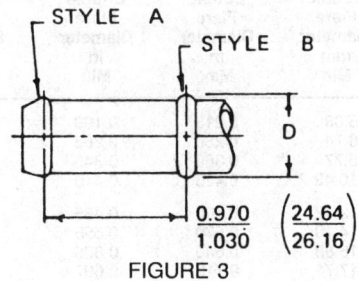

FIGURE 3

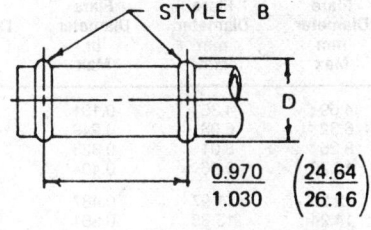

FIGURE 4

(R) FLARES FOR TUBING—SAE J533 JUN92   SAE Standard

Report of Parts and Fittings Technical Committee approved February 1947 and last revised by Tube, Pipe, Hose, and Lubrication Fittings Committee January 1972. Completely revised by the SAE Fluid Conductors and Connectors Technical Committee S1—Automotive and Hydraulic Tube and Fittings June 1992. Rationale statement available.

1. Scope

This SAE Standard covers specifications for 37 degree and 45 degree single and double flares for tube ends intended for use with 37 degree flared tube fittings and 45 degree flared or inverted flared tube fittings, respectively.

2. References

2.1 Applicable Documents

The following publications form a part of this specification to the extent specified herein. The latest issue of SAE publications shall apply.

2.1.1 SAE PUBLICATIONS—Available from SAE, 400 Commonwealth Drive, Warrendale, PA 15096-0001.

SAE J514—Hydraulic Tube Fittings
SAE TSB 003—Rules for Use of SI (Metric) Units

3. General Specifications

3.1 Dimensions

Dimensions in this Standard are based on and, unless designated otherwise, are specified in metric units, with inch equivalents shown in parentheses located adjacent to respective metric dimensions on illustrations or designated "in" in text and tables, in accordance with SAE TSB 003.

Single and double 45 degree flares shall conform to the dimensions specified in Figure 2 and Table 1.

Single and double 37 degree flares shall conform to the dimensions specified in Figure 3 and Table 2.

The following general specifications supplement the dimensional data with respect to unspecified detail and apply to both 37 degree and 45 degree flares for tubing.

3.2 Burring Prior to Flaring

To assure producing satisfactory flares, it may be necessary to perform burring operations on the tube end prior to flaring. Smoothly breaking the inside corner before single flaring ferrous, and some nonferrous, tubing is normally required to eliminate the cutoff burr which might otherwise create leakage paths across a substantial portion of the flare. Smoothly breaking the outside corner prior to single flaring, or both outside and inside corners prior to double flaring, shall be permissible on any tube material to minimize splitting.

Inasmuch as the specified dimensions shall prevail, whether or not the corners are broken, the quality of the finished flare shall be the only criterion applied to the burring operation.

3.3 Concentricity

Flare seat shall be concentric with tube outside diameter within

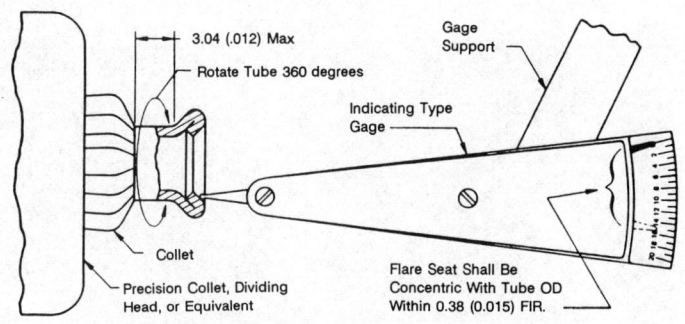

FIGURE 1—TYPICAL FLARE CONCENTRICITY GAGE

0.38 mm (0.015 in) full indicator reading (FIR). To promote uniformity in checking concentricity of flare seat to the tube outside diameter, it is recommended the gaging method depicted in Figure 1 and the following procedure, or equivalent means, be used.

3.3.1 Mount tube in precision collet, dividing head, or equivalent rotational centering and clamping device with the rear of flare not more than 3.04 mm (0.12 in) ahead of the collet. A minimum straight length of tube behind the flare of 25.4 mm (1.00 in), or twice the tube outside diameter, whichever is greater, must be available for mounting purposes.

3.3.2 Place stylus of indicator gage on the coined portion of flare seat.

3.3.3 Rotate the mounted tube through full 360 degree revolution.

3.3.4 Read full indicator reading occurring over the 360 degrees of rotation.

3.4 Workmanship

Flares shall be free from loose scale, burrs, slivers, and cracks. Seating surfaces shall be smooth and free from nicks, pit marks, and any other defects that prevent sealing.

TABLE 1—DIMENSIONS OF SINGLE AND DOUBLE 45-DEGREE FLARES FOR TUBING[1] (FIGURE 2)

Nominal Tube OD mm	Nominal Tube OD in	A Single Flare Diameter mm Max	A Single Flare Diameter mm Min	A Single Flare Diameter in Max	A Single Flare Diameter in Min	A_1 Double Flare Diameter mm Max	A_1 Double Flare Diameter mm Min	A_1 Double Flare Diameter in Max	A_1 Double Flare Diameter in Min	B Single Flare Radius mm ±0.25	B Single Flare Radius in ±0.01
3.18	1/8	4.59	4.35	0.181	0.171	5.41	5.03	0.213	0.198	0.51	0.02
4.76	3/16	6.32	6.08	0.249	0.239	7.11	6.74	0.280	0.265	0.51	0.02
6.35	1/4	8.25	8.01	0.325	0.315	9.14	8.77	0.360	0.345	0.51	0.02
7.94	5/16	10.26	9.86	0.404	0.388	10.79	10.42	0.425	0.410	0.51	0.02
9.52	3/8	12.36	11.97	0.487	0.471	12.70	12.32	0.500	0.485	0.51	0.02
11.11	7/16	14.24	13.85	0.561	0.545	14.47	14.10	0.570	0.555	0.51	0.02
12.70	1/2	15.82	15.42	0.623	0.607	16.25	15.88	0.640	0.625	0.51	0.02
14.29	9/16	17.17	16.77	0.676	0.660	18.08	17.71	0.712	0.697	0.51	0.02
15.88	5/8	18.99	18.60	0.748	0.732	19.60	19.23	0.772	0.757	0.51	0.02
19.05	3/4	23.26	22.86	0.916	0.900	23.16	22.79	0.912	0.897	0.51	0.02
22.22	7/8	26.44	26.04	1.041	1.025	—	—	—	—	0.51	0.02
25.40	1	29.38	28.99	1.157	1.141	—	—	—	—	0.51	0.02

Nominal Tube OD mm	Nominal Tube OD in	B_1 Double Flare Radius mm ±0.25	B_1 Double Flare Radius in ±0.01	C Double Flare Coined Seat Length mm Min	C Double Flare Coined Seat Length in Min	D^2 Single Flare Wall Thickness mm Max	D^2 Single Flare Wall Thickness in Max	$D_1{}^2$ Double Flare Wall Thickness mm Max	$D_1{}^2$ Double Flare Wall Thickness in Max
3.18	1/8	1.02	0.04	1.02	0.040	0.88	0.035	0.63	0.025
4.76	3/16	1.02	0.04	1.02	0.040	0.88	0.035	0.71	0.028
6.35	1/4	1.02	0.04	1.02	0.040	1.24	0.049	0.83	0.035
7.94	5/16	1.02	0.04	1.57	0.062	1.24	0.049	0.88	0.035
9.52	3/8	1.02	0.04	1.57	0.062	1.65	0.065	1.24	0.049
11.11	7/16	1.02	0.04	1.57	0.062	1.65	0.065	1.24	0.049
12.70	1/2	1.02	0.04	1.57	0.062	2.10	0.083	1.24	0.049
14.29	9/16	1.02	0.04	1.57	0.062	2.10	0.083	1.24	0.049
15.88	5/8	1.02	0.04	1.57	0.062	2.41	0.095	1.24	0.049
19.05	3/4	1.02	0.04	1.57	0.062	2.76	0.109	1.24	0.049
22.22	7/8	—	—	—	—	2.76	0.109	—	—
25.40	1	—	—	—	—	3.04	0.120	—	—

[1] It is not the intent of this document to define the appropriateness of fittings to be used in conjunction with the flares specified. Considerations such as the effects of wall thickness on working pressures, length of thread engagements, etc., shall be the responsibility of the user. See SAE J514.

[2] Recommended maximum nominal wall thickness of tubing normally considered suitable for flaring to the above specifications.

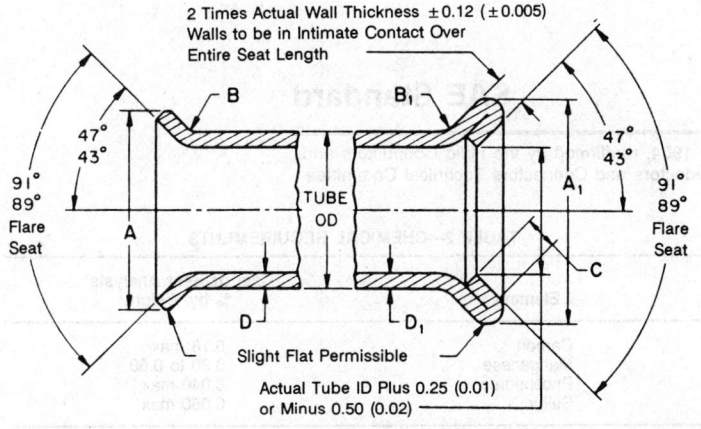

FIGURE 2—SINGLE AND DOUBLE 45 DEGREE FLARES FOR TUBING

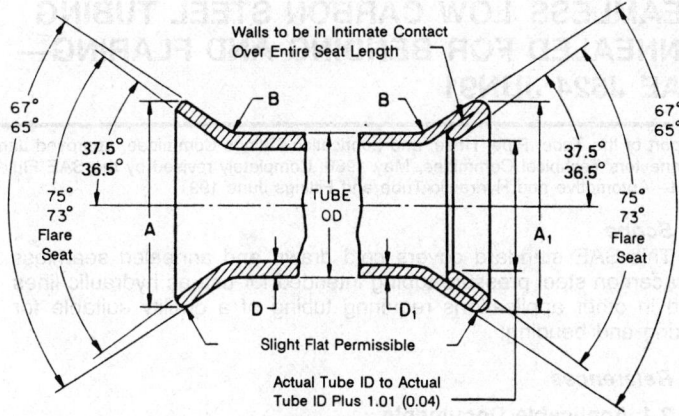

FIGURE 3—SINGLE AND DOUBLE 37 DEGREE FLARES FOR TUBING

TABLE 2—DIMENSIONS OF SINGLE AND DOUBLE 37-DEGREE FLARES FOR TUBING[1] (FIGURE 3)

Nominal Tube OD mm	Nominal Tube OD in	A Single Flare Diameter mm Max	A Single Flare Diameter mm Min	A Single Flare Diameter in Max	A Single Flare Diameter in Min	A_1 Double Flare Diameter mm Max	A_1 Double Flare Diameter mm Min	A_1 Double Flare Diameter in Max	A_1 Double Flare Diameter in Min
3.18	1/8	5.08	4.58	0.200	0.180	5.08	4.58	0.200	0.180
4.76	3/16	7.11	6.61	0.280	0.260	7.11	6.61	0.280	0.260
6.35	1/4	9.14	8.64	0.360	0.340	9.14	8.64	0.360	0.340
7.94	5/16	10.92	10.16	0.430	0.400	10.92	10.16	0.430	0.400
9.52	3/8	12.44	11.69	0.490	0.460	12.44	11.69	0.490	0.460
12.70	1/2	16.76	16.01	0.660	0.630	16.76	16.01	0.660	0.630
15.88	5/8	20.06	19.31	0.790	0.760	20.06	19.31	0.790	0.760
19.05	3/4	24.13	23.37	0.950	0.920	24.13	23.37	0.950	0.920
22.22	7/8	27.17	26.42	1.070	1.040	27.17	26.42	1.070	1.040
25.40	1	30.48	29.72	1.200	1.170	30.48	29.72	1.200	1.170
28.58	1-1/8	35.05	34.29	1.380	1.350	35.05	34.29	1.380	1.350
31.75	1-1/4	38.35	37.60	1.510	1.480	38.35	37.60	1.510	1.480
38.10	1-1/2	43.94	43.18	1.730	1.700	43.94	43.18	1.730	1.700
44.45	1-3/4	53.59	52.84	2.110	2.080	53.59	52.84	2.110	2.080
50.80	2	59.94	59.19	2.360	2.330	59.94	59.19	2.360	2.330

Nominal Tube OD mm	Nominal Tube OD in	B Radius mm ±0.5	B Radius in ±0.02	$D^{[2]}$ Single Flare Wall Thickness mm Max	$D^{[2]}$ Single Flare Wall Thickness in Max	$D_1^{[2]}$ Double Flare Wall Thickness mm Max	$D_1^{[2]}$ Double Flare Wall Thickness in Max
3.18	1/8	0.8	0.03	0.88	0.035	0.63	0.025
4.76	3/16	0.8	0.03	0.88	0.035	0.71	0.028
6.35	1/4	0.8	0.03	1.65	0.065	0.88	0.035
7.94	5/16	0.8	0.03	1.65	0.065	0.88	0.035
9.52	3/8	1.0	0.04	1.65	0.065	1.24	0.049
12.70	1/2	1.5	0.06	2.10	0.083	1.24	0.049
15.88	5/8	1.5	0.06	2.41	0.095	1.24	0.049
19.05	3/4	2.0	0.08	2.76	0.109	1.24	0.049
22.22	7/8	2.0	0.08	2.76	0.109	1.65	0.065
25.40	1	2.3	0.09	3.04	0.120	1.65	0.065
28.58	1-1/8	2.3	0.09	3.04	0.120	1.65	0.065
31.75	1-1/4	2.3	0.09	3.04	0.120	1.65	0.065
38.10	1-1/2	2.8	0.11	3.04	0.120	1.65	0.065
44.45	1-3/4	2.8	0.11	3.04	0.120	1.65	0.065
50.80	2	2.8	0.11	3.40	0.134	1.65	0.065

[1] It is not the intent of this document to define the appropriateness of fittings to be used in conjunction with the flares specified. Considerations such as the effects of wall thickness on working pressures, length of thread engagements, etc., shall be the responsibility of the user.

[2] Recommended maximum nominal wall thickness of tubing normally considered suitable for flaring to the above specifications.

SEAMLESS LOW CARBON STEEL TUBING ANNEALED FOR BENDING AND FLARING— SAE J524 JUN91

SAE Standard

Report of the Tube, Pipe, Hose, and Lubrication Fittings Committee, approved January 1954, reaffirmed by the Fluid Conductors and Connectors Technical Committee, May 1986. Completely revised by the SAE Fluid Conductors and Connectors Technical Committee SC1—Automotive and Hydraulic Tube and Fittings June 1991.

1. Scope

This SAE standard covers cold drawn and annealed seamless low carbon steel pressure tubing intended for use as hydraulic lines and in other applications requiring tubing of a quality suitable for flaring and bending.

2. References

2.1 Applicable Documents

The following publications form a part of this specification to the extent specified herein. The latest issue of SAE publications shall apply.

2.1.1 SAE PUBLICATIONS—Available from SAE, 300 Commonwealth Drive, Warrendale, PA 15096-0001.

SAE J409—Product Analysis—Permissible Variations from Specified Chemical Analysis of a Heat or Cast of Steel

SAE J514—Hydraulic Tube Fittings

3. Manufacture

The tubing shall be cold drawn to size and after forming shall be annealed in such a manner as to produce a finished product which will meet all requirements of this document.

4. Dimensions and Tolerances

The tolerances applicable to tubing outside diameter are shown in Table 1. The wall thickness shall not vary more than ±10% for tubing having 12.7 mm (0.50 in) or larger, nominal inside diameter nor more than ±15% for tubing having a smaller nominal inside diameter. Tubing outside diameter and wall thickness shall be as specified by purchaser.

TABLE 1—TUBING OUTSIDE DIAMETER TOLERANCES

Nominal Tubing OD[1, 2]		OD Tolerance ±	
mm	in	mm	in
Up to 25.4	Up to 1.00	0.10	0.004
Over 25.4 to 38.1 inclusive	Over 1.00 to 1.50 inclusive	0.15	0.006
Over 38.1 to 50.8 inclusive	Over 1.50 to 2.00 inclusive	0.20	0.008
Over 50.8 to 88.9 inclusive	Over 2.00 to 3.50 inclusive	0.25	0.010

[1] The actual outside diameter shall be the average of the maximum and minimum outside diameters as determined at any one cross section through the tubing.

[2] Refer to SAE J514 for nominal tubing outside diameters to be used in conjunction with standard hydraulic tube fittings.

5. Quality

Lengths of finished tubing shall be reasonably straight and have smooth ends free from burrs. Tubing shall be free from scale and injurious defects and have a workmanlike finish. Surface imperfections, such as handling marks, die marks, or shallow pits, shall not be considered injurious defects provided the imperfections are within the tolerances specified for diameter and wall thickness. The removal of such surface imperfections is not required.

The inside of tubing shall be clean and free from any contamination that cannot be removed readily by cleaning agents normally used in manufacturing.

6. Material

Tubing shall be made from low carbon steel conforming to the following chemical composition as shown in Table 2.

7. Mechanical Properties

The finished tubing shall have mechanical properties as tabulated in Table 3.

TABLE 2—CHEMICAL REQUIREMENTS

Element	Cast or Heat Analysis[1] % by Weight
Carbon	0.18 max
Manganese	0.30 to 0.60
Phosphorus	0.040 max
Sulfur	0.050 max

[1] Check analysis tolerance shall be as specified in SAE J409, Table 1.

TABLE 3—MECHANICAL PROPERTIES

Yield Strength, min	170 MPa (25 000 psi)
Ultimate Strength, min	310 MPa (45 000 psi)
Elongation in 50 mm (2 in), min	35%[1]
Hardness (Rockwell B), max	65[2]

[1] For tubing having nominal outside diameter of 9.5 mm (0.375 in) or less, and/or wall thicknesses of 0.9 mm (0.035 in) or less, a minimum elongation of 25% is permissible.

[2] The hardness test shall not be required on tubing with a nominal wall thickness of less than 1.65 mm (0.065 in). Such tubing shall meet all other mechanical properties and performance requirements.

8. Performance Requirements

The finished tubing shall satisfactorily meet the following performance tests. Test specimens shall be taken from tubing which has not been subjected to cold working after the anneal of the finished sized tubing.

8.1 Flattening Test

A section approximately 75 mm (3 in) in length, cut from the finished tubing, shall not crack or show any flaws when flattened between parallel plates to a distance equal to three times the wall thickness of the section under test. Superficial ruptures resulting from minor surface imperfections shall not be considered cause for rejection.

8.2 Expansion Test

A test specimen shall be taken from every shipment or every 460 m (1500 ft), whichever is smaller, of finished tubing and subjected to expansion over a hardened tapered plug having a slope of 0.1:1.0 until the outside diameter has been expanded 25% without evidence of cracking or flaws.

8.3 Pressure Proof Test

Unless otherwise specified, tubing supplied under this document shall have been tested hydrostatically, with no evidence of failure, at a pressure which will subject the material to a hoop (circumferential) stress of 140 MPa (20 000 psi). Test pressures shall be as determined by Barlow's formula for thin hollow cylinders under pressure.

$$P = \frac{2TS}{D} \qquad \text{(Eq.1)}$$

where:

D = outside diameter of tubing, mm (in)

P = hydrostatic pressure, MPa (psi)

S = allowable unit stress of material = 140 MPa (20 000 psi)

T = minimum wall thickness of tubing, mm (in)

No tube shall be tested beyond a hydrostatic pressure of 35 MPa (5000 psi), unless so specified.

8.4 Nondestructive Electric Test

In lieu of the hydrostatic test, when mutually agreed upon by the purchaser and manufacturer, all tubing shall be tested by passing it through an electric eddy current tester which is capable of detecting defects that would prevent the tubing from passing the hydrostatic pressure proof test.

9. Corrosion Protection

The inside and outside of the finished tubing shall be protected against corrosion during shipment and normal storage. If a corrosion preventive compound is applied, it shall be such that after normal storage periods it can readily be removed by cleaning agents normally used in manufacturing.

(R) WELDED AND COLD DRAWN LOW CARBON STEEL TUBING ANNEALED FOR BENDING AND FLARING—SAE J525 JUN91

SAE Standard

Report of the Pipe, Hose, and Lubrication Fittings Committee, approved April 1958, reaffirmed by the Fluid Conductors and Connectors Technical Committee, May 1986. Completely revised by the SAE Fluid Conductors and Connectors Technical Committee SC1—Automotive and Hydraulic Tube and Fittings June 1991.

1. Scope

This SAE Standard covers cold worked and annealed electric resistance welded single wall low carbon steel pressure tubing intended for use as hydraulic lines and in other applications requiring tubing of a quality suitable for flaring and bending.

2. References

2.1 Applicable Documents

The following publications form a part of this specification to the extent specified herein. The latest issue of SAE publications shall apply.

2.1.1 SAE Publications—Available from SAE, 400 Commonwealth Drive, Warrendale, PA 15096-0001.

SAE J409—Product Analysis—Permissible Variations from Specified Chemical Analysis of a Heat or Cast of Steel.

SAE J514—Hydraulic Tube Fittings

2.1.2 ASTM Publications—Available from ASTM, 1916 Race Street, Philadelphia, PA 19103.

ASTM A 370—Methods and Definitions for Mechanical Testing of Steel Products.

3. Manufacture

The tubing shall be made from a single strip of steel shaped into a tubular form, the edges of which are joined and sealed by electric resistance welding. After forming and welding, the tubing shall be normalized and subjected to a cold working operation that shall result in a 15% minimum reduction in cross-sectional area, of which at least 8% shall consist of a reduction in wall thickness. Subsequent to cold working, the tubing shall be annealed in such a manner as to produce a finished product which will meet all requirements of this document. Tubing that has been pickled to remove scale shall be suitably treated to eliminate any embrittlement induced by the pickling process.

4. Dimensions and Tolerances

The tolerances applicable to tubing outside diameter, inside diameter, and wall thickness are shown in Table 1. Tubing shall be subject to any two of the tolerances specified, as designated by the purchaser.

5. Quality

Lengths of finished tubing shall be reasonably straight and have smooth ends free from burrs. Tubing shall be free from scale and injurious defects and have a workmanlike finish. Surface imperfections such as handling marks, die marks, or shallow pits shall not be considered injurious defects provided the imperfections are within the tolerances specified for diameter and wall thickness. The removal of such surface imperfections is not required. There shall be no dimensional indications of the presence of the weld.

The inside of tubing shall be clean and free from any contamination that cannot be readily removed by cleaning agents normally used in manufacturing.

6. Material

Tubing shall be made from low carbon steel conforming to the following chemical composition in Table 2.

TABLE 1—TUBING OUTSIDE DIAMETER AND WALL THICKNESS TOLERANCE

Nominal Tubing OD[1,2] mm	Nominal Tubing OD[1,2] in	Tolerance OD ±mm	Tolerance OD ±in	Tolerance ID ±mm	Tolerance ID ±in	Tolerance ± Wall Thickness (%)
Up to 9.5	Up to 0.38	0.05	0.002	0.05	0.002	15
Over 9.5 to 15.9 inclusive	Over 0.38 to 0.63 inclusive	0.06	0.0025	0.06	0.0025	10
Over 15.9 to 50.8 inclusive	Over 0.63 to 2.00 inclusive	0.08	0.003	0.08	0.003	10
Over 50.8 to 63.5 inclusive	Over 2.00 to 2.50 inclusive	0.10	0.004	0.10	0.004	10
Over 63.5 to 76.2 inclusive	Over 2.50 to 3.00 inclusive	0.13	0.005	0.13	0.005	10
Over 76.2 to 101.6 inclusive	Over 3.00 to 4.00 inclusive	0.15	0.006	0.15	0.006	10

[1] The actual outside diameter shall be the average of the maximum and minimum outside diameters as determined at any one cross section through the tubing.
[2] Refer to SAE J514 for nominal tubing outside diameters to be used in conjunction with standard hydraulic tube fittings.

TABLE 2—CHEMICAL REQUIREMENTS

Element	Cast or Heat Analysis[1] % by Weight
Carbon	0.18 max
Manganese	0.30-0.60
Phosphorus	0.040 max
Sulfur	0.050 max

[1] Check analysis tolerance shall be as specified in SAE J409, Table 3.

7. Mechanical Properties

The finished tubing shall have mechanical properties as tabulated below:

TABLE 3—MECHANICAL PROPERTIES

Yield Strength, min	170 MPa (25 000 psi)
Ultimate Strength, min	310 MPa (45 000 psi)
Elongation in 50 mm (2 in), min	35%[1]
Hardness (Rockwell B scale), max	65[2]

[1] For tubing having nominal outside diameter of 9.5 mm (0.375 in) or less, and/or wall thicknesses of 0.9 mm (0.035 in) or less, a minimum elongation of 25% is permissible.
[2] The hardness test shall not be required on tubing with a nominal wall thickness of less than 1.65 mm (0.065 in). Such tubing shall meet all other mechanical properties and performance requirements.

8. Performance Requirements

The finished tubing shall satisfactorily meet the following performance tests. Test specimens shall be taken from tubing which has not been subjected to cold working after the anneal of the finished sized tubing.

8.1 Flattening Test

A section approximately 75 mm (3 in) in length, cut from the finished tubing, shall not crack or show any flaws when flattened between parallel plates to a distance equal to three times the wall thickness of the section under test. Superficial ruptures resulting from minor surface imperfections shall not be considered cause for rejection.

8.2 Reverse Flattening Test

A test specimen shall be taken from every shipment or every 460 m (1500 ft), whichever is smaller, of finished tubing and split longitudinally 90 degrees on each side of the weld. The section containing the weld shall be opened and flattened with the weld at the point of maximum bend. There shall be no evidence of cracks or lack of penetration or overlaps resulting from flash removal in the weld.

Refer to ASTM A 370, paragraph T5(B), reverse flattening test.

8.3 Expansion Test

A test specimen shall be taken from every shipment or every 460 m (1500 ft), whichever is smaller, of finished tubing and subjected to expansion over a hardened tapered plug having a slope of 0.1:1.0 until the outside diameter has been expanded 25% without evidence of cracking or flaws.

8.4 Pressure Proof Test

Unless otherwise specified, tubing supplied under this document shall have been tested hydrostatically, with no evidence of failure, at a pressure which will subject the material to a hoop (circumferential) stress of 140 MPa (20 000 psi). Test pressures shall be as determined by Barlow's formula for thin hollow cylinders under pressure:

$$P = \frac{2TS}{D} \qquad (Eq.1)$$

where:

D = outside diameter of tubing, mm (in)
P = hydrostatic pressure, MPa (psi)
S = allowable unit stress of material = 140 MPa (20 000 psi)
T = minimum wall thickness of tubing, mm (in)

No tube shall be tested beyond a hydrostatic pressure of 35 MPa (5000 psi) unless so specified.

8.5 Nondestructive Electric Test

In lieu of the hydrostatic test, where mutually agreed upon by the purchaser and manufacturer, all tubing shall be tested by passing it through an electric eddy current tester which is capable of detecting defects that would prevent the tubing from passing the hydrostatic pressure proof test.

9. Corrosion Protection

The inside and outside of the finished tubing shall be protected against corrosion during shipment and normal storage. If a corrosion preventive compound is applied, it shall be such that after normal storage periods it can readily be removed by cleaning agents normally used in manufacturing.

WELDED LOW CARBON STEEL TUBING —SAE J526 JUN91

SAE Standard

Report of the Tube, Pipe, Hose, and Lubrication Fittings Committee, approved January 1952, reaffirmed by the Fluid Conductors and Connectors Technical Committee, May 1986. Completely revised by the Fluid Conductors and Connectors Technical Committee June 1990. Revised by the SAE Fluid Conductors and Connectors Technical Committee SC1—Automotive and Hydraulic Tube and Fittings June 1991.

1. Scope

This SAE Standard covers welded single wall low carbon steel tubing intended for general automotive applications and other similar uses.

2. References

(R) 2.1 Applicable Documents

The following publications form a part of this specification to the intent specified herein. The latest issue of SAE publications should apply.

2.11 SAE PUBLICATIONS—Available from SAE, 400 Commonwealth Drive, Warrendale, PA 15096-0001.

SAE J533—Flares for Tubing

3. Manufacture

The tubing shall be made from a single strip of steel shaped into a tubular form, the edges of which are joined and sealed by a suitable butt welding process. After welding, the bead shall be removed from the outside to provide a smooth round surface and the tubing shall be processed in such a manner as to produce a finished product which will meet all requirements of this document.

(R) 4. Dimensions and Tolerances

The standard nominal diameters and the applicable dimensions and tolerances are shown in Table 1.

5. Quality

Finished tubing shall be clean, smooth, and round, both inside and outside; and shall be free from scale and injurious defects. A slight weld bead and splatter on the inside surface shall be permissible but must be held to the minimum consistent with good welding practice. Surface imperfections such as handling marks, die marks, or shallow pits shall not be considered injurious defects provided such imperfections are within the tolerances specified for diameter and wall thickness.

The inside of tubing shall be clean and free from any contamination which will impair the processing or serviceability of the tubing.

6. Material

Tubing shall be made from low carbon steel, such as UNS G10080 or UNS G10100.

(R) 7. Mechanical Properties

The finished tubing shall have mechanical properties as tabulated in Table 2.

8. Performance Requirements

The finished tubing shall satisfactorily meet the following performance tests. As designated therein, test specimens having minimum lengths equivalent to two times the tubing outside diameter or 50 mm (2 in), whichever is greater, shall be taken from tubing which has not been subjected to cold working after the final processing of the finished size tubing.

8.1 Flaring Test

A test specimen having squared and deburred ends shall withstand being double flared at one end to the requirements of SAE

TABLE 1—TUBING DIMENSIONS AND TOLERANCES[1]

Dash Size	Nominal Tubing OD		Outside Diameter[2] Basic		Outside Diameter[2] Tolerance		Wall Thickness[3] Basic		Wall Thickness[3] Basic	
	mm	in	mm	in	±mm	±in	mm	in	mm	in
-2	3.18	1/8	3.18	0.125	0.002	0.05	0.025	0.64	0.005	0.13
-3	4.76	3/16	4.78	0.188	0.003	0.08	0.028	0.71	0.005	0.13
-4	6.35	1/4	6.35	0.250	0.003	0.08	0.028	0.71	0.003	0.08
-5	7.94	5/16	7.92	0.312	0.003	0.08	0.028	0.71	0.003	0.08
-6	9.53	3/8	9.53	0.375	0.003	0.08	0.028	0.71	0.003	0.08
-7	11.11	7/16	11.13	0.438	0.004	0.10	0.030	0.76	0.003	0.08
-8	12.70	1/2	12.70	0.500	0.004	0.10	0.030	0.76	0.003	0.08
-8	12.70	1/2	12.70	0.500	0.004	0.10	0.035	0.89	0.0035	0.09
-9	14.29	9/16	14.27	0.562	0.004	0.10	0.030	0.76	0.003	0.08
-10	15.88	5/8	15.88	0.625	0.004	0.10	0.035	0.89	0.0035	0.09

[1] Other sizes may be specified by agreement between the supplier and the user.
[2] The actual outside diameter shall be the average of the maximum and minimum outside diameters as determined at any one cross section through the tubing.
[3] The tolerances listed represent the maximum permissible deviation at any point.

J533 without evidence of splitting or flaws. The test specimen shall be held firmly and squarely in the die, and the punch, while being forced down gradually, shall be guided parallel to the axis of the tubing.

8.2 Hardness Test
The hardness test shall not be required, it being recognized that hardness will be satisfactory if the tubing meets all other mechanical properties and performance requirements set forth in this document.

8.3 Bending Test
The finished tubing shall withstand bending on a centerline radius equal to three times the tubing outside diameter without undue reduction of area or flattening where proper bending fixtures are used.

(R) 8.4 Pressure Proof Test
Unless otherwise specified, the finished tubing shall withstand a hydrostatic proof test, with no evidence of failure, at a pressure[1] which will subject the material to a hoop (circumferential) stress of 140 MPa (20 000 psi). Test pressures shall be as determined by Barlow's formula for thin hollow cylinders under pressure.

$$P = \frac{2TS}{D} \quad \text{(Eq.1)}$$

[1] No tube shall be tested beyond a hydrostatic pressure of 35 MPa (5000 psi) unless so specified.

TABLE 2—MECHANICAL REQUIREMENTS

Yield Strength, min (0.2% offset)	170 MPa (25 000 psi)
Tensile Strength, min	290 MPa (42 000 psi)
Elongation in 50 mm (2 in)	14-40%
Hardness (Rockwell 30 T scale), max	65

where:

D = outside diameter of tubing, mm (in)
P = hydrostatic pressure, MPa (psi)
S = allowable unit stress of material = 140 MPa (20 000 psi)
T = minimum wall thickness of tubing, mm (in)

8.5 Nondestructive Electric Test
In lieu of the hydrostatic test, where mutually agreed upon by the purchaser and manufacturer, all tubing shall be tested by passing it through an electric eddy current tester which is capable of detecting defects that would prevent the tubing from passing the hydrostatic pressure proof test.

9. Corrosion Protection
The inside and outside of the finished tubing shall be protected against corrosion during shipment and normal storage. If a corrosion preventive compound is applied, it shall be such that after normal storage periods it can readily be removed by cleaning agents normally used in manufacturing.

BRAZED DOUBLE WALL LOW CARBON STEEL TUBING—SAE J527 JUN91

SAE Standard

Report of the Tube, Pipe, Hose, and Lubrication Fittings Committee, approved January 1952, fourth revision by the Fluid Conductors and Connectors Technical Committee January 1983. Completely revised by the Fluid Conductors and Connectors Technical Committee, June 1990. Revised by the SAE Fluid Conductors and Connectors Technical Committee SC1—Automotive and Hydraulic Tube and Fittings June 1991.

1. Scope
This SAE Standard covers brazed double wall low carbon steel tubing intended for general automotive applications and other similar uses.

(R) 2. References

2.1 Applicable Documents
The following publications form a part of this specification to the extent specified herein. The latest issue of SAE Publications shall apply.

2.1.1 SAE Publications—Available from SAE, 400 Commonwealth Drive, Warrendale, PA 15096-0001.

SAE J533—Flares for Tubing

3. Manufacture
The tubing shall be made from a single or double strip of steel shaped into the form of a double wall tubing, the seams of which are secured and sealed by copper brazing in a controlled atmosphere. The braze shall be uniform with no evidence of a bead on either the inside or outside of the tubing. The tubing shall be processed in such a manner as to produce a finished product which will meet all requirements of this document.

4. Dimensions and Tolerances
The standard nominal diameters and the applicable dimensions and tolerances are shown in Table 1.

(R) TABLE 1—TUBING DIMENSIONS AND TOLERANCES[1]

Dash Size	Nominal Tubing OD		Outside Diameter[1] Basic		Outside Diameter[1] Tolerance		Dash Size	Wall Thickness Basic		Wall Thickness Tolerance[2]	
	mm	in	mm	in	±mm	±in		mm	in	±mm	±in
–2	3.18	0.125	3.18	0.125	0.05	0.002	–2	0.64	0.025	0.13	0.005
–3	4.76	0.188	4.78	0.188	0.08	0.003	–3	0.71	0.028	0.08	0.003
–4	6.35	0.250	6.35	0.250	0.08	0.003	–4	0.71	0.028	0.08	0.003
–5	7.94	0.312	7.92	0.312	0.08	0.003	–5	0.71	0.028	0.08	0.003
–6	9.53	0.375	9.53	0.375	0.08	0.003	–6	0.71	0.028	0.08	0.003
–7	11.11	0.438	11.13	0.438	0.10	0.004	–7	0.76	0.030	0.08	0.003
–8	12.70	0.500	12.70	0.500	0.10	0.004	–8	0.89	0.035	0.09	0.0035
–9	14.29	0.562	14.27	0.562	0.10	0.004	–9	0.89	0.035	0.09	0.0035
–10	15.88	0.625	15.88	0.625	0.10	0.004	–10	0.89	0.035	0.09	0.0035

[1] The actual outside diameter shall be the average of the maximum and minimum outside diameters as determined at any one cross section through the tubing.

[2] The tolerances listed represent the maximum cc permissible deviation at any point.

5. Quality

Finished tubing shall be clean, smooth, and round, both inside and outside; and shall be free from scale and injurious defects. Surface imperfections such as handling marks, die marks, or shallow pits shall not be considered injurious defects provided such imperfections are within the tolerances specified for diameter and wall thickness.

The inside of tubing shall be clean and free from any contamination which will impair the processing or serviceability of the tubing.

6. Material

Tubing shall be made from low carbon steel, such as UNS G10080 or UNS G10100.

7. Mechanical Properties

The finished tubing shall have mechanical properties as tabulated in Table 2:

(R) TABLE 2—MECHANICAL PROPERTIES

Yield Strength, min (0.2% offset)	170 MPa (25 000 psi)
Tensile Strength, min	290 MPa (42 000 psi)
Elongation in 50 mm (2 in)	14 to 40%
Hardness (Rockwell 30 T scale), max	65

8. Performance Requirements

The finished tubing shall satisfactorily meet the following performance tests. Test specimens shall be taken from tubing which has not been subjected to cold working after the final processing of the finished sized tubing.

8.1 Flaring Test

A test section cut from the finished tubing, having squared and deburred ends, shall withstand being double flared at one end to the dimensions shown in SAE J533. The test section shall be held firmly and squarely in the die and the punch, while being forced down, shall be guided parallel to the axis of the tubing. The flare shall exhibit no evidence of splitting or flaws except that a separation of the outer lap joint with area A (Figure 1) shall be permissible providing it does not exceed 0.12 in (3.1 mm) in length and is confined to the outer thickness only. Seam separation shall not be permissible in the following areas:

a. AREA B—The flare seat, defined as the surface within the 90 degrees included angle. Conical surface shall be smooth and free from cracks or other irregularities which could cause leaks after assembly.
b. AREA C—The surface beyond the length of the double thickness created by the flare.

8.2 Bending Test

The finished tubing shall withstand bending on a centerline radius equal to three times the tubing outside diameter without undue reduction of area or flattening where proper bending fixtures are used.

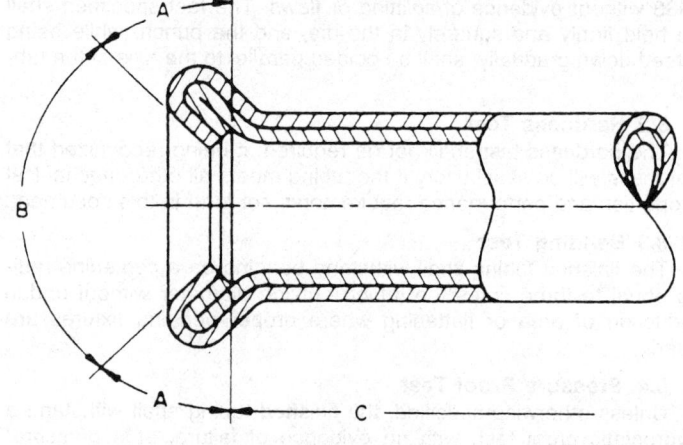

FIGURE 1

(R) 8.3 Pressure Proof Test

Unless otherwise specified, the finished tubing shall withstand a hydrostatic proof test, with no evidence of failure, at a pressure which will subject the material to a hoop (circumferential) stress of 140 MPa (20 000 psi). Test pressures shall be determined by Barlow's formula for thin hollow cylinders under pressure:

$$P = \frac{2TS}{D} \qquad (Eq.1)$$

where:

D = outside diameter of tubing, mm (in)
P = hydrostatic pressure, MPa (psi)
S = allowable unit stress of material = 140 MPa (20 000 psi)
T = minimum wall thickness of tubing, mm (in)

No tube shall be tested beyond a hydrostatic pressure of 35 MPa (5000 psi) unless so specified.

8.4 Nondestructive Electric Test

In lieu of the hydrostatic test, where mutually agreed upon by the purchaser and manufacturer, all tubing shall be tested by passing it through an electric eddy current tester which is capable of detecting defects that would prevent the tubing from passing the hydrostatic pressure proof test.

9. Corrosion Protection

The inside and outside of the finished tubing shall be protected against corrosion during shipment and normal storage. If a corrosion preventive compound is applied, it shall be such that after normal storage periods it can readily be removed by cleaning agents normally used in manufacturing.

SEAMLESS COPPER-NICKEL 90-10 TUBING—SAE J1650 MAR93   SAE Standard

Report of the SAE Fluid Conductors and Connectors Technical Committee S1—Automotive and Hydraulic Tube and Fittings approved March 1993.

1. Scope
This SAE Standard covers seamless copper-nickel tubing for use in hydraulic brake pressure conductors, general automotive applications, and other similar uses.

2. References

2.1 Applicable Document
The following publication forms a part of this specification to the extent specified herein. The latest issue of SAE publications shall apply.

2.1.1 SAE Publication
Available from SAE, 400 Commonwealth Drive, Warrendale, PA 15096-0001.

SAE J533—Flares for Tubing

3. Manufacture
This tubing shall be made from 90-10 copper-nickel and cold drawn to size. It shall then be annealed in such a manner as to produce a finished product which will meet all of the requirements in this document.

4. Dimensions and Tolerances
The standard nominal diameters and the applicable dimensions and tolerances are shown in Table 1.

TABLE 1—TUBING DIMENSIONS AND TOLERANCES[1]

Dash Size	Nominal Tubing OD mm	Nominal Tubing OD in	Outside Diameter[1] Basic	Outside Diameter[1] Tolerance ± mm	Outside Diameter[1] Tolerance ± in	Wall Thickness Basic mm	Wall Thickness Basic in	Wall Thickness Tolerance[2] ± mm	Wall Thickness Tolerance[2] ± in
-2	3.18	0.125	0.125	0.05	0.002	0.64	0.025	0.13	0.005
-3	4.76	0.188	0.188	0.08	0.003	0.71	0.028	0.08	0.003
-4	6.35	0.250	0.250	0.08	0.003	0.71	0.028	0.08	0.003
-5	7.94	0.312	0.312	0.08	0.003	0.71	0.028	0.08	0.003
-6	9.52	0.375	0.375	0.08	0.003	0.71	0.028	0.08	0.003
-7	11.11	0.438	0.438	0.10	0.004	0.76	0.030	0.08	0.003
-8	12.70	0.500	0.500	0.10	0.004	0.89	0.035	0.09	0.0035
-9	14.29	0.562	0.562	0.10	0.004	0.89	0.035	0.09	0.0035
-10	15.88	0.625	0.625	0.10	0.004	0.89	0.035	0.09	0.0035

[1] The actual outside diameter shall be the average of the maximum and minimum outside diameters as determined at any one cross section through the tubing.
[2] The tolerances listed represent the maximum permissible deviation at any point.

5. Quality
Finished tubing shall be clean, smooth, and round, both inside and outside, and shall be free from scale and injurious defects. Surface imperfections such as handling marks, die marks, or shallow pits shall not be considered injurious defects provided such defects are within the tolerances specified for diameter and wall thickness. The inside of the tubing shall be clean and free from any contamination which will impair the processing or serviceability of the tubing.

6. Material
Tubing shall be made from copper-nickel alloy UNS C70600.

7. Mechanical Properties
The finished tubing shall have mechanical properties as tabulated in Table 2:

TABLE 2—MECHANICAL PROPERTIES

Properties	Values
Yield Strength, min (0.2% offset)	110 MPa (16 000 psi)
Tensile Strength, min	290 MPa (42 000 psi)
Elongation in 2 in (50 mm)	14-40%
Hardness (Rockwell 30 T scale), max	65

8. Performance Requirements
The finished tubing shall satisfactorily meet the following performance requirements. Test specimens shall be taken from tubing which has not been subjected to cold working after the final processing of the finished sized tubing.

8.1 Flaring Test
A test section cut from the finished tubing, having squared and deburred ends, shall withstand being double flared at one end to the dimensions shown in SAE J533. The test section shall be held firmly and squarely in the die and the punch, while being forced down, shall be guided parallel to the axis of the tubing. The flare shall exhibit no evidence of splitting or flaws in area A (Figure 1).

   a. Area B—The flare seat, defined as the surface within the 90-degree included angle. Conical surface shall be smooth and free from cracks or other irregularities which could cause leaks after assembly.
   b. Area C—The surface beyond the length of the double thickness created by the flare.

8.2 Bending Test
The finished tubing shall withstand bending on a centerline radius equal to three times the tubing outside diameter without undue reduction of area or flattening where proper bending fixtures are used.

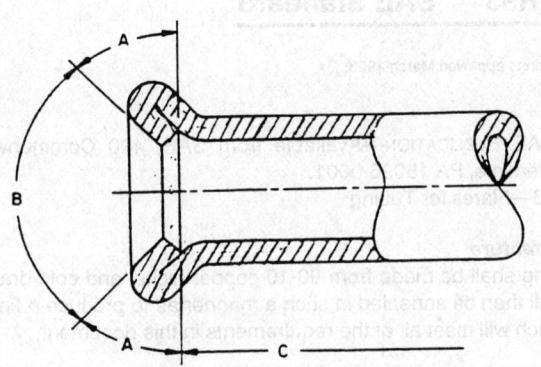

FIGURE 1—INVERTED FLARE

8.3 Pressure Proof Test

Unless otherwise specified, the finished tubing shall withstand a hydrostatic proof test with no evidence of failure, at a pressure which will subject the material to a yield stress of 110 MPa (16 000 psi). Test pressures shall be determined by Barlow's formula for thin hollow cylinders under tension.

$$P = \frac{2TS}{D} \qquad \text{(Eq.1)}$$

where:
- D = outside diameter of tubing, mm (in)
- P = hydrostatic pressure, MPa (psi)
- S = allowable unit stress of material = 110 MPa (16 000 psi)
- T = minimum wall thickness of tubing, mm (in)

No tube shall be tested beyond a hydrostatic pressure of 35 MPa (5000 psi) unless so specified.

8.4 Nondestructive Electric Test

In lieu of the hydrostatic test, where mutually agreed upon by the purchaser and manufacturer, all tubing shall be tested by passing it through an electric eddy current tester which is capable of detecting defects that would prevent the tubing from passing the hydrostatic proof test.

9. Corrosion Protection

The inside and outside of the finished tubing shall be protected against corrosion during shipment and normal storage. If a corrosion preventive compound is applied, it shall be such that after normal storage periods it can readily be removed by cleaning agents normally used in manufacturing.

(R) WELDED FLASH CONTROLLED LOW CARBON STEEL TUBING NORMALIZED FOR BENDING, DOUBLE FLARING, AND BEADING—SAE J356 JUN91

SAE Standard

Report of the Tube, Pipe, Hose, and Lubrication Fittings Committee, approved July 1968, reaffirmed by the Fluid Conductors and Connectors Technical Committee, May 1986. Completely revised by the SAE Fluid Conductors and Connectors Technical Committee SC1—Automotive and Hydraulic Tube and Fittings June 1991.

1. Scope

This SAE Standard covers normalized electric resistance welded flash controlled single-wall, low-carbon steel pressure tubing intended for use as pressure lines and in other applications requiring tubing of a quality suitable for bending, double flaring, beading, and brazing.

2. References

2.1 Applicable Documents

The following publications form a part of this specification to the extent specified herein. The latest issue of SAE publications shall apply.

2.1.1 SAE Publications—Available from SAE, 400 Commonwealth Drive, Warrendale, PA 15096-0001.

SAE J409—Product Analysis—Permissible Variations from Specified Chemical Analysis of a Heat or Cast of Steel

SAE J514—Hydraulic Tube Fittings

SAE J533—Flares for Tubing

2.1.2 ASTM Publications—Available from ASTM, 1916 Race Street, Philadelphia, PA 19103

ASTM A 370—Methods and Definitions for Mechanical Testing of Steel Products

3. Manufacture

The tubing shall be made from a single strip of steel shaped into a tubular form, the edges of which are joined and sealed by electric resistance welding. After forming and welding, the outside flash shall be removed to provide a smooth surface. The inside flash shall be of uniform contour, free from saw-tooth peaks, and controlled in height by seam welding techniques or by cutting, but not by hammering or rolling. The inside flash height shall conform to the following as in Table 1.

TABLE 1—INSIDE FLASH HEIGHT

Nominal Wall Thickness		Maximum Flash Height[1]			
		Thru 25.4 mm OD (1.000 in)		Over 25.4 mm OD (1.000 in)	
mm	in	mm	in	mm	in
0.90	Thru 0.035	0.13	0.005	0.25	0.010
0.90 thru 1.65	Over 0.035 thru 0.065	0.20	0.008	0.25	0.010
1.65	Over 0.065	0.25	0.010	0.25	0.010

[1] For tubes having an ID greater than 8 mm (0.312 in), the height of the inside weld flash shall be measured with a ball micrometer having a 3.96 mm ± 0.41 (0.156 in ± 0.016) radius on the anvil or ball point. For tubes having an ID 8 mm (0.312 in) or less, screw thread micrometers shall be used. The height of the flash shall be the difference between the thickness of the tubing wall at the point of maximum height of the flash and the average of the wall thickness measured at points adjacent to both sides of the flash.

The tubing shall be normalized to produce a finished product which will meet all requirements of this document.

4. Dimensions and Tolerances

The tolerances applicable to tubing outside diameter are shown in Table 2. The tolerances applicable to tubing wall thickness are shown in Table 3. Particular attention shall be given to areas adjacent to the weld to insure against thin spots and/or sharp indentations.

TABLE 2—TUBING OUTSIDE DIAMETER TOLERANCE

Nominal Tubing OD[1,2]		Tube OD Tolerance	
mm	in	± mm	± in
9.50	Thru 0.375	0.06	0.0025
9.50-15.88	Over 0.375-0.625	0.08	0.003
15.88-28.57	Over 0.625-1.125	0.09	0.0035
28.57-50.80	Over 1.125-2.000	0.13	0.005
50.80-63.50	Over 2.000-2.500	0.15	0.006
63.50-76.20	Over 2.500-3.000	0.20	0.008
76.20-88.90	Over 3.000-3.500	0.23	0.009
88.90-101.60	Over 3.500-4.000	0.25	0.010

[1] OD measurements shall be taken at least 50 mm (2.0 in) from the end of the tubing.
[2] Refer to SAE J514 for nominal tubing OD to be used in conjunction with standard hydraulic tube fittings and SAE J533 for recommended max nominal wall thickness for double flaring.

TABLE 3—TUBING WALL THICKNESS TOLERANCES,[1] IN

Nominal Wall Thickness[2]	Nominal Tubing Outside Diameter Thru 1.000 Plus[3]/Minus	Nominal Tubing Outside Diameter Over 1.000 thru 2.000 Plus[3]/Minus	Nominal Tubing Outside Diameter Over 2.000 thru 4.000 Plus[3]/Minus
0.028	0.002/0.003	0.003/0.003	0.004/0.003
0.035	0.002/0.004	0.003/0.004	0.004/0.004
0.049	0.002/0.005	0.003/0.005	0.004/0.005
0.065	0.004/0.006	0.005/0.008	0.006/0.008
0.083	0.004/0.006	0.006/0.008	0.007/0.008
0.095	0.004/0.006	0.006/0.010	0.007/0.010
0.109	0.004/0.006	0.008/0.010	0.009/0.010
0.120	0.004/0.008	0.008/0.010	0.009/0.010
0.134	0.004/0.008	0.008/0.010	0.009/0.010
0.148	— —	0.008/0.011	0.009/0.011
0.165	— —	0.008/0.011	0.009/0.011
0.180	— —	0.008/0.011	0.009/0.011
0.203	— —	0.008/0.012	0.009/0.012
0.220	— —	0.008/0.012	0.009/0.012
0.238	— —	0.013/0.018	0.014/0.018
0.259	— —	0.013/0.020	0.014/0.020

[1] Millimeter conversions of the inch tolerances are:

mm	in	mm	in	mm	in
0.05	0.002	0.18	0.007	0.30	0.012
0.08	0.003	0.20	0.008	0.33	0.013
0.10	0.004	0.23	0.009	0.36	0.014
0.13	0.005	0.25	0.010	0.46	0.018
0.15	0.006	0.28	0.011	0.51	0.020

[2] For intermediate wall thicknesses, the tolerance for the next heavier wall thickness shall apply.
[3] Plus tolerances include allowance for crown on flat rolled steel.

5. Quality

Lengths of finished tubing shall be reasonably straight and have smooth ends free from burrs. Finished tubing shall be free from scale and injurious imperfections and shall have a workmanlike finish. Outside surface imperfections such as handling marks, straightening marks, light die marks, or shallow pits shall not be considered injurious, provided the imperfections are not detrimental to the function of the tubing. The removal of such surface imperfections shall not be required.

The inside surface shall be free of weld splatter, pits, and all other injurious imperfections detrimental to the function of the tubing.

6. Material

Tubing shall be made from low carbon hot or cold rolled steel conforming to the chemical composition shown in Table 4. If rimmed steel is used, it shall be single strand. The steel shall be made by the open hearth basic oxygen, or electric furnace process. A ladle analysis of each heat shall be made to determine the percentages of the elements specified. The chemical composition thus determined shall be reported to the purchaser, or his representative, if requested, and shall conform to the requirements specified. If a check analysis is required, the tolerances shall be as specified in SAE J409, Table 3.

TABLE 4—CHEMICAL REQUIREMENTS

Element	Cast or Heat Analysis, Wgt %
Carbon	0.18 max
Manganese	0.30 thru 0.60
Phosphorus	0.04 max
Sulfur	0.05 max

7. Mechanical Properties

The finished tubing shall have mechanical properties as tabulated in Table 5.

TABLE 5—MECHANICAL PROPERTIES

Yield Strength, min	170 MPa (25 000 psi)
Ultimate Strength, min	310 MPa (45 000 psi)
Elongation in 50 mm (2 in), min	35%[1]
Hardness (Rockwell B), max	65[2]

[1] For tubing having nominal outside diameter of 9.5 mm (0.375 in) or less, and/or wall thicknesses of 0.9 mm (0.035 in) or less, a minimum elongation of 25% is permissible.

[2] The hardness test shall not be required on tubing with a nominal wall thickness of less than 1.65 mm (0.065 in). Such tubing shall meet all other mechanical properties and performance requirements.

8. Performance Requirements

The finished tubing shall satisfactorily meet the following performance tests. As designated therein, test specimens having minimum length equivalent to two times the tubing outside diameter or 50 mm (2 in), whichever is greater, shall be taken from finished tubing, as manufactured. All tests shall be conducted at room temperature.

8.1 Flattening Test

A test specimen shall be taken from every shipment or every 460 m (1500 ft), whichever is smaller, of finished tubing and flattened between parallel plates to a distance equal to three times the actual wall thickness of the specimen under test without any cracking or flaws. The weld shall be placed at 90 degrees from the direction of applied force. Superficial ruptures resulting from minor surface imperfections shall not be considered cause for rejection.

8.2 Reverse Flattening Test

A test specimen shall be taken from every shipment or every 460 m (1500 ft), whichever is smaller, of finished tubing and split longitudinally 90 degrees on each side of the weld. The section containing the weld shall be opened and flattened with the weld at the point of maximum bend. There shall be no evidence of cracks or metal flaking, or lack of weld penetration or overlaps resulting from flash control or flash removal in the weld.

8.3 Expansion Test

A test specimen shall be taken from every shipment or every 460 m (1500 ft), whichever is smaller, of finished tubing and subjected to expansion over a hardened tapered plug having a slope of 0.1:1.0 until the outside diameter has been expanded 25% without evidence of cracking or flaws.

The tubing shall be capable of being double flared as shown in SAE J533 without evidence of cracking or flaws. Refer to footnote 2 of Table 1 for tubing OD and wall thickness subject to this capability requirement. Double flaring tests shall not be required.

8.4 Hardness Test

One hardness test shall be made on a specimen from each production lot of tubing. The hardness test shall be made on the inside surface of the specimen. The hardness test shall not be required on tubing with a nominal wall thickness less than 1.7 mm (0.065 in). Such tubing shall meet all other mechanical properties and performance requirements.

8.5 Tensile Test

One tension test, in accordance with ASTM A 370, shall be made on a specimen from each production lot of tubing. If the percentage of elongation of the test specimen is less than that specified and/or any part of the fracture is more than 19 mm (0.75 in) from the center of the gage length, as indicated by scribe marks on the specimen before testing, a retest shall be allowed.

8.6 Pressure Proof Test

Unless otherwise specified, the finished tubing shall withstand a hydrostatic proof test, with no evidence of failure, at an actual pressure of 35 MPa (5000 psi) or at a hoop (circumferential) fiber stress of 140 MPa (20 000 psi), whichever is less. Test pressures shall be determined by the following formula:

$$P = \frac{2TS}{D} \quad \text{(Eq.1)}$$

where:

P = hydrostatic test pressure, MPa (psi) (35 MPa (5000 psi) max)

T = allowable minimum wall thickness of tubing, mm (in)

S = 140 MPa (20 000 psi) allowable fiber stress (80% of min yield strength)

D = nominal outside diameter of tubing, mm (in)

8.7 Nondestructive Electric Test

In lieu of the hydrostatic test, when mutually agreed upon by the purchaser and manufacturer, all tubing shall be tested by passing it through an electric eddy current tester which is capable of detecting defects that would prevent the tubing from passing the hydrostatic pressure proof test.

8.8 Test Specimens

Test specimens for mechanical tests shall be smooth on the ends and free from flaws. If any test specimen exhibits burrs, flaws, or defective machining, before testing, it may be discarded and another specimen may be selected.

8.9 Test Certificate

A certificate of compliance to the performance requirements shall be furnished to the purchaser by the producer if requested in the purchase agreement.

9. Cleanliness

The inside and outside surfaces of the finished tubing shall be commercially bright, clean and free from grease, oxide scale, carbon deposits and any other contamination that cannot be readily removed by cleaning agents normally used in manufacturing plants.

10. Corrosion Protection

The inside and outside surfaces of the finished tubing shall be protected against corrosion during shipment and normal storage. If a corrosion preventive compound is applied, it shall be such that after normal storage periods, it can be readily removed by cleaning agents normally used in manufacturing plants.

(R) PRESSURE RATINGS FOR HYDRAULIC TUBING AND FITTINGS—SAE J1065 MAR92

SAE Information Report

Report of Tube, Pipe, Hose and Lubrication Fittings Committee approved January 1974. Completely revised by the SAE Fluid Conductors and Connectors Technical Committee S1—Automotive and Hydraulic Tube and Fittings March 1992.

1. Scope

This SAE Information Report is intended to provide design guidance in the selection of steel tubing and related tube fittings for general hydraulic system applications. The information presented herein is based on tubing products which conform to SAE J524, SAE J525, and SAE J356, and is subject to due consideration being given to the following limitations:

1.1 Since many factors influence the pressure at which a hydraulic system will or will not perform satisfactorily, this report should not be used as a "standard" nor "specification," and the values shown herein should not be construed as "guaranteed" minimum.

1.1.1 Within the fluid power industry, many criteria are used for determining the pressure capability of tubing. Consideration is given to specified minimum yield or fiber stress factors, to calculated yield or burst pressures, and to yield or burst pressures determined by actual test. Also, varying design factors are applied, commensurate with the total system conditions. Thus, it is impractical to set down specific allowable working pressures that will satisfy all design criteria. It is considered desirable, however, to provide guidelines on the subject such as are published in this report.

1.1.2 Factors such as the thinning of tube walls due to forming operations, shock loads, and vibration characteristics of the system must also be considered.

2. References

2.1 Applicable Documents

The following publications form a part of this specification to the extent specified herein. The latest issue of SAE publications shall apply.

2.1.1 SAE PUBLICATIONS—Available from SAE, 400 Commonwealth Drive, Warrendale, PA 15096-0001.

SAE J356—Welded Flash Controlled Low Carbon Steel Tubing Normalized for Bending, Double Flaring, and Beading
SAE J514—Hydraulic Tube Fittings
SAE J524—Seamless Low Carbon Steel Tubing Annealed for Bending and Flaring
SAE J525—Welded and Cold Drawn Low Carbon Steel Tubing Annealed for Bending and Flaring
SAE J533—Flares for Tubing

3. Hydraulic Tubing

Three normally acceptable reference working pressures (psi) for each combination of diameter and wall thickness of commonly used tubing calculated using three generally accepted formulae and reflecting two popular design stress factors are presented in Figures 1 and 2. The designer, therefore, may select the desired tubing on the basis of the value which best suits the intended application and satisfies any preference he may have regarding formulation.

3.1 Formulae

The formulae from which the three values tabulated vertically opposite each tube size were derived are, respectively:

3.1.1 The Barlow formula:

$$P = \frac{2ST}{D} \quad \text{(Eq. 1)}$$

3.1.2 The Boardman formula

$$P = \frac{2ST}{D - 0.8T} \quad \text{(Eq. 2)}$$

3.1.3 The Lamé formula

$$P = S \left(\frac{D^2 - d^2}{D^2 + d^2} \right) \quad \text{(Eq. 3)}$$

where:

D = nominal outside diameter of tubing, mm
d = nominal inside diameter of tubing, mm
P = hydrostatic working pressure, MPa
S = allowable fiber stress of material, MPa
T = nominal wall thickness of tubing, mm

3.2 The values shown in Figure 1, reflecting a design factor of approximately 4:1, are based on an allowable fiber stress of 86 MPa (12,500 psi) which is equivalent to 50% of the minimum yield point and approximately 28% of the minimum ultimate strength of the tubing.

3.3 The values shown in Figure 2, reflecting a design factor of approximately 3:1, are based on an allowable fiber stress of 117 MPa (17,000 psi) which is equivalent to 68% of the minimum yield point and approximately 38% of the minimum ultimate strength of the tubing.

4. Hydraulic Tube Fittings

When properly assembled in conjunction with appropriate tubing selected from respective tables (flared where applicable, in accordance with SAE J533), the hydraulic tube fittings specified in SAE J514 are capable of providing leak-proof full-flow connections in hydraulic systems operating at working pressures designated in the following:

4.1 At a design factor of approximately 4:1, these fittings should be suitable for use with tubings selected from Figure 1 which have a reference working pressure of 20.5 MPa (3000 psi) or less.

4.2 At a design factor of approximately 3:1, these fittings should be suitable for use with tubings selected from Figure 2 which have a reference working pressure of 34.5 MPa (5000 psi) or less.

4.3 These fittings are also capable of higher working pressures in hydraulic systems wherein conditions will permit the use of smaller tube sizes, increased wall thickness (37 degree flared tube fittings are limited in this regard) and/or reduction of the design factor. Prior to such use, however, it is recommended that sufficient testing be conducted and reviewed by both the user and fittings manufacturers to assure performance levels will be satisfactory.

Nominal Tube OD mm (in)	See Note(a)	Nominal Tube Wall Thickness, mm/(in)											
		0.71 (0.028)	0.89 (0.035)	1.24 (0.049)	1.65 (0.065)	2.11 (0.083)	2.41 (0.095)	2.77 (0.109)	3.05 (0.120)	3.40 (0.134)	3.76 (0.148)	3.96 (0.156)	4.78 (0.188)
3.18 (0.125)	1	38.5	48.5										
	2	47.0	62.0										
	3	46.0	58.5										
4.77 (0.188)	1	26.0	32.0										
	2	29.5	38.0										
	3	29.5	37.5										
6.35 (0.250)	1	19.4	24.0	34.0	45.0								
	2	21.5	27.0	40.0	56.5								
	3	21.5	27.0	39.5	54.0								
7.92 (0.312)	1	15.5	19.4	27.0	36.0								
	2	16.6	21.5	31.0	43.0								
	3	16.8	21.5	31.0	42.5								
9.53 (0.375)	1	12.8	16.2	22.5	30.0	38.5	44.0						
	2	13.8	17.2	25.0	35.0	46.0	55.0						
	3	13.8	17.6	25.0	34.5	45.0	52.5						
12.70 (0.500)	1		12.0	17.0	22.5	28.5	33.0	37.5	41.5				
	2		12.8	18.2	25.0	33.0	38.5	45.5	51.5				
	3		12.8	18.6	25.0	33.0	38.5	44.5	49.5				
15.88 (0.625)	1		9.7	13.5	18.0	23.0	26.0	30.0	33.0				
	2		10.0	14.5	19.6	25.5	30.0	35.0	39.0				
	3		10.4	14.5	19.6	26.0	30.0	35.0	38.5				
19.05 (0.750)	1		7.9	11.4	14.8	19.0	21.5	25.0	27.5				
	2		8.3	11.8	16.2	21.0	24.0	28.5	31.5				
	3		8.3	12.0	16.2	21.0	24.5	28.5	31.5				
22.23 (0.875)	1		6.9	9.7	12.8	16.2	18.6	21.5	23.5				
	2		7.2	10.0	13.5	17.6	20.5	24.0	26.5				
	3		7.2	10.4	13.8	18.0	20.5	24.0	27.0				
25.40 (1.000)	1		6.0	8.3	11.0	14.0	16.2	18.6	20.5	23.0	25.5		
	2		6.2	8.6	11.8	15.2	17.6	20.5	23.0	26.0	29.0		
	3		6.2	9.0	12.0	15.5	18.0	20.5	23.0	26.0	29.0		
28.58 (1.125)	1			7.6	10.0	12.8	14.5	16.6	18.2	20.5	23.0		
	2			7.9	10.4	13.5	15.5	18.2	20.0	23.0	25.5		
	3			7.9	10.6	13.8	15.8	18.2	20.5	23.0	25.5		
31.75 (1.250)	1			6.9	9.0	11.4	13.0	15.2	16.6	18.6	20.5	21.5	26.0
	2			6.9	9.3	12.0	13.8	16.2	18.0	20.5	22.5	24.0	29.5
	3			6.9	9.3	12.0	14.2	16.2	18.2	20.5	23.0	24.0	29.5
38.10 (1.500)	1				7.6	9.7	11.0	12.4	13.8	15.5	16.8	18.0	21.5
	2				7.6	10.0	11.4	13.5	14.8	16.6	18.6	19.6	24.0
	3				7.9	10.0	11.8	13.5	14.8	16.8	18.6	19.6	24.0
44.45 (1.750)	1				6.4	8.3	9.3	10.6	11.8	13.2	14.5	15.5	18.6
	2				6.6	8.6	9.7	11.4	12.4	14.2	15.5	16.6	20.5
	3				6.6	8.6	10.0	11.4	12.8	14.2	15.8	16.6	20.5
50.80 (2.000)	1				5.5	7.2	8.3	9.3	10.4	11.4	12.8	13.5	16.2
	2				5.9	7.2	8.6	9.7	11.0	12.0	13.5	14.5	17.6
	3				5.9	7.6	8.6	10.0	11.0	12.4	13.8	14.5	17.6
57.15 (2.250)	1				4.8	6.2	7.2	8.3	9.3	10.4	11.4	12.0	14.5
	2				5.2	6.6	7.6	8.6	9.7	10.6	12.0	12.8	15.5
	3				5.2	6.6	7.6	8.6	9.7	11.0	12.0	12.8	15.5

Notes:
Wall Thickness Having Values Shown To Right Of Bold Line Are Not Normally Considered Suitable For 37 Deg. Single Flaring To SAE J533.

Pressures For Tube Sizes Not Shown May Be Calculated Using The Formulae Given in Paragraph 3.1.

(a) Pressure values listed opposite numbers 1, 2, and 3 for each tube OD were derived from the Barlow, Boardman, and Lame formulas, respectively, with 117 MPa allowable stress factor.

(b) 1 MPa = 145 psi

FIGURE 1—REFERENCE WORKING PRESSURES AT APPROXIMATELY 4:1 DESIGN FACTOR, MPa (see note b)

Nominal Tube OD mm (in)	See Note(a)	Nominal Tube Wall Thickness, mm/(in)											
		0.71 (0.028)	0.89 (0.035)	1.24 (0.049)	1.65 (0.065)	2.11 (0.083)	2.41 (0.095)	2.77 (0.109)	3.05 (0.120)	3.40 (0.134)	3.76 (0.148)	3.96 (0.156)	4.78 (0.188)
3.18 (0.125)	1	52.5	65.5										
	2	64.0	84.5										
	3	62.5	79.5										
4.77 (0.188)	1	35.0	44.0										
	2	39.5	51.5										
	3	40.0	51.0										
6.35 (0.250)	1	26.0	33.0	46.0	61.0								
	2	29.0	37.0	54.5	77.0								
	3	29.0	37.0	54.0	73.5								
7.92 (0.312)	1	21.0	26.0	37.0	48.5								
	2	23.0	29.0	42.0	58.5								
	3	23.0	29.0	42.0	57.5								
9.53 (0.375)	1	17.6	21.5	30.5	40.5	52.0	59.5						
	2	18.6	24.0	34.0	47.0	63.0	74.5						
	3	19.0	24.0	34.5	47.0	61.5	71.5						
12.70 (0.500)	1		16.6	23.0	30.5	39.0	44.5	51.0	56.0				
	2		17.2	25.0	34.0	45.0	52.5	62.0	69.5				
	3		17.6	25.0	34.0	45.0	52.0	60.5	67.0				
15.88 (0.625)	1		13.2	18.2	24.5	31.0	35.5	41.0	45.0				
	2		13.8	19.6	26.5	35.0	40.5	47.5	53.0				
	3		13.8	20.0	27.0	35.0	40.5	47.5	53.0				
19.05 (0.750)	1		11.0	15.2	20.5	26.0	29.5	34.0	37.5				
	2		11.4	16.2	21.5	28.5	33.0	38.5	43.0				
	3		11.4	16.2	22.0	28.5	33.5	38.5	43.0				
22.23 (0.875)	1		9.3	13.2	17.6	22.5	25.5	29.5	32.0				
	2		9.7	13.8	18.6	24.0	28.0	32.5	36.0				
	3		9.7	13.8	18.6	24.0	28.5	33.0	36.0				
25.40 (1.000)	1		8.3	11.4	15.2	19.4	22.5	25.5	28.5	31.5	35.0		
	2		8.3	12.0	16.2	20.5	24.0	28.0	31.0	35.5	39.5		
	3		8.6	12.0	16.2	21.0	24.5	28.5	31.5	35.5	39.5		
28.58 (1.125)	1			10.4	13.5	17.2	19.6	23.0	25.0	28.0	30.5		
	2			10.6	14.2	18.2	21.5	24.5	27.0	31.0	34.5		
	3			10.6	14.5	18.6	21.5	25.0	27.5	31.0	35.0		
31.75 (1.250)	1			9.3	12.0	15.5	18.0	20.5	22.5	25.0	28.0	29.5	35.0
	2			9.7	12.8	16.6	19.0	22.0	24.5	27.5	30.5	32.5	40.0
	3			9.7	12.8	16.6	19.4	22.0	24.5	27.5	31.0	33.0	40.5
38.10 (1.500)	1				10.0	13.2	14.8	16.8	18.6	21.0	23.0	24.5	29.5
	2				10.6	13.5	15.5	18.0	20.0	22.5	25.0	26.6	33.0
	3				10.6	13.8	15.8	18.2	20.5	23.0	25.5	27.0	33.0
44.45 (1.750)	1				8.6	11.0	12.8	14.5	16.2	18.0	20.0	21.0	25.0
	2				9.0	11.8	13.5	15.5	16.8	19.0	21.5	22.5	27.5
	3				9.0	11.8	13.5	15.5	17.2	19.4	21.5	23.0	28.0
50.80 (2.000)	1				7.6	9.7	11.0	12.8	14.2	15.8	17.2	18.2	22.0
	2				7.9	10.0	11.8	13.5	14.8	16.6	18.2	19.6	24.0
	3				7.9	10.0	11.8	13.5	14.8	16.8	18.6	19.6	24.0
57.15 (2.250)	1				6.7	8.6	10.0	11.4	12.4	13.8	15.5	16.2	19.6
	2				6.9	9.0	10.4	11.8	13.2	14.8	16.2	17.2	21.0
	3				6.9	9.0	10.4	12.00	13.2	14.8	16.6	17.2	21.0

Notes:

Wall Thickness Having Values Shown To Right Of Bold Line Are Not Normally Considered Suitable For 37 Deg. Single Flaring To SAE J533.

Pressures For Tube Sizes Not Shown May Be Calculated Using The Formulae Given in Paragraph 3.1.

(a) Pressure values listed opposite numbers 1, 2, and 3 for each tube OD were derived from the Barlow, Boardman, and Lame formulas, respectively, with 117 MPa allowable stress factor.

(b) 1 MPa = 145 psi

FIGURE 2—REFERENCE WORKING PRESSURES AT APPROXIMATELY 3:1 DESIGN FACTOR, MPa (see note b)

(R) SEAMLESS COPPER TUBE—SAE J528 JUN91 — SAE Standard

Report of the Tube, Pipe, Hose, and Lubrication Fittings Committee, approved January 1953, reaffirmed by the Fluid Conductors and Connectors Technical Committee, May 1986. Completely revised by the SAE Fluid Conductors and Connectors Technical Committee SC1—Automotive and Hydraulic Tube and Fittings June 1991.

1. Scope

This SAE Standard covers minimum requirements for soft (061) annealed seamless copper tube intended for automotive and general purposes. (Comparable specification is ASTM B 75. Other copper tube is covered in SAE J463.)

2. References

2.1 Applicable Documents

The following publications form a part of this specification to the extent specified herein. The latest issue of SAE publications shall apply.

2.1.1 SAE PUBLICATIONS—Available from SAE, 400 Commonwealth Drive, Warrendale, PA 15096-0001.

SAE J463—Wrought Copper and Copper Alloys

2.1.2 ASTM PUBLICATIONS—Available from ASTM, 1916 Race Street, Philadelphia, PA 19103.

ASTM B 75—Specification for Seamless Copper Tube

3. Manufacture

The tube shall be cold drawn to size and after forming shall be annealed in such a manner as to produce a finished product which will meet all requirements of this document.

4. Dimensions and Tolerances

Tube furnished to this standard shall conform to the dimensional tolerances shown in Table 1 for the size of tube specified by the purchaser. (Standard nominal sizes are listed.)

5. Quality

The finished tube shall be clean, smooth, and round, free from internal and external mechanical imperfections, and shall have a bright appearance.

6. Material

Unless otherwise specified by purchaser, tube shall be made from any one of the materials listed in Table 2. (UNS C12200 is most commonly used.) Average grain size of the tube shall be 0.040 mm, minimum.

7. Mechanical Properties

Tube shall conform to Table 3:

8. Expansion Test

Samples of tube (selected from sections which have not been subjected to cold working after anneal of the finished sized tube) shall be cut square and deburred. These shall be expanded on a hardened and ground tapered steel pin having an included angle of 60 degrees until the outside diameter is increased 40%. Care should be taken to keep the axes of the pin and the tube in line during the expansion operation. The test may be made in a die to restrict the expansion to 40%. The expanded tube shall show no cracking or rupture visible to the unaided eye.

9. Hydrostatic Test

Unless otherwise specified, tube shall show no evidence of weakness or defects when subjected to an internal hydrostatic pressure sufficient to subject the material to a hoop (circumferential) fiber stress of 40 MPa (6000 psi) determined by the following formula for thin, hollow cylinders under pressure. The tube need not be tested at a hydrostatic pressure of over 7 MPa (1000 psi) unless so specified.

$$P = \frac{2St}{D - 0.8t} \quad \text{(Eq.1)}$$

where:

P = hydrostatic pressure, MPa (psi)
t = minimum thickness of tube wall, mm (in)
D = basic outside diameter of tube, mm (in)
S = allowable stress of the material = 40 MPa (6000 psi)

10. Embrittlement Test

The tube is expected to pass the following test although the actual performance of the test is not required under this specification unless specifically stipulated by the purchaser:

a. Heat the cleaned or degreased specimens for 20 min minimum at a temperature of 850 °C ± 25 (1562 °F ± 45) in a furnace in which the atmosphere is at least 10% of hydrogen by volume. Then quench the specimens immediately and rapidly in water or in the same atmosphere with minimum contact with air.
b. Polish and etch if desired, cross-sectional test specimens taken transverse to, and bounded by, an original surface of the material. Examine the prepared surface microscopically under illumination at a magnification of 75 to 200 diameters inclusive. Specimens shall show no passing or open grain structure characteristic of embrittlement.

TABLE 1—TUBING DIMENSIONS AND TOLERANCES

Nominal Tubing OD		Outside Diameter[1] Basic		Outside Diameter[2] Tolerance		Wall Thickness Basic		Wall Thickness[2] Tolerance	
mm	in	mm	in	± mm	± in	mm	in	± mm	± in
3.18	1/8	3.18	0.125	0.05	0.0020	0.76	0.030	0.08	0.0030
4.76	3/16	4.78	0.188	0.05	0.0020	0.76	0.030	0.063	0.0025
6.35	1/4	6.35	0.250	0.05	0.0020	0.76	0.030	0.063	0.0025
7.94	5/16	7.92	0.312	0.05	0.0020	0.81	0.032	0.063	0.0025
9.53	3/8	9.53	0.375	0.05	0.0020	0.81	0.032	0.063	0.0025
12.70	1/2	12.70	0.500	0.05	0.0020	0.81	0.032	0.063	0.0025
15.88	5/8	15.88	0.625	0.05	0.0020	0.89	0.035	0.063	0.0025
19.05	3/4	19.05	0.750	0.063	0.0025	0.89	0.035	0.063	0.0025

[1] The actual outside diameter shall be the average of the maximum and minimum outside diameters as described at any one cross section through the tubing.
[2] The tolerances listed represent the maximum permissible deviation at any point.

TABLE 2—CHEMICAL COMPOSITION, WEIGHT %

SAE Alloy No.[1]	UNS No.[2]	Similar ASTM Copper No.[3]	Copper, min	Phosphorus	Arsenic
CA102	C10200	102 (was OF)	99.95	—	—
CA120	C12000	120 (was DLP)	99.90	0.004-0.012	—
CA122	C12200	122 (was DHP)	99.90	0.015-0.040	—
—	—	142 (was DPA)	99.40	0.015-0.040	0.15-0.50

[1] SAE J463.
[2] Unified Numbering System.
[3] ASTM B 75.

TABLE 3—MECHANICAL PROPERTIES

Ultimate Strength (Tensile), min	205 MPa (30 000 psi)
Yield Strength (Tensile), min[1]	62.0 MPa (9 000 psi)

[1] At 0.5% extension under load.

(R) METALLIC AIR BRAKE SYSTEM TUBING AND PIPE—SAE J1149 JUN91

SAE Standard

This standard was formerly designated SAE J844 approved June 1963, completely revised July 1976, and reaffirmed by the Fluid Conductors and Connectors Technical Committee December 1987. Completely revised by the SAE Fluid Conductors and Connectors Technical Committee SC1—Automotive and Hydraulic Tube and Fittings June 1991.

1. Scope

This SAE Standard covers minimum requirements for two types of metallic tubing and pipe as used in automotive air brake systems. It includes material and performance specifications, corrosion precautions, and installation recommendations. Copper tubing is designated Type 1, and galvanized steel pipe Type 2.

2. References

2.1 Applicable Documents

The following publications form a part of this specification to the extent specified herein. The latest issue of SAE publications shall apply.

2.1.1 SAE PUBLICATIONS—Available from SAE, 400 Commonwealth Drive, Warrendale, PA 15096-0001.

SAE J463—Wrought Copper and Copper Alloys
SAE J476—Dryseal Pipe Threads

2.1.2 ASTM PUBLICATIONS—Available from ASTM, 1916 Race Street, Philadelphia, PA 19103.

ASTM A 120—Specification for Black and Hot-Dipped Zinc-Coated (Galvanized) Welded and Seamless Steel Pipe for Ordinary Uses

ASTM A 370—Methods and Definitions for Mechanical Testing of Steel Products

ASTM E 8—Methods of Tension Testing of Metallic Materials

ASTM E 62—Method of Test for Antimony in Copper and Copper Base Alloys

ASTM E 79—Methods for Estimating the Average Grain Size of Wrought Copper and Copper Base Alloys

3. Corrosion Precautions

In the design and selection of air brake system components, adequate provision shall be made to control corrosion due to galvanic coupling of widely dissimilar metals and alloys when such materials used for tubing, pipe, fittings, and attaching or supporting parts are in intimate contact with each other. Also, adequate provision shall be made to protect the tubing, pipe, and fittings from oxygen concentration cell type of corrosion. Where soft nonmetallic cushions are used to prevent metal-to-metal contact between supporting components and the tubing, pipe, and fittings, the cushioning material shall be such that it will not absorb and retain significant amounts of water.

4. Installation Recommendations

The tubing or pipe installed in air brake systems shall be supported in such a manner as to minimize fatigue conditions. Metal-to-metal contact should be avoided by the use of soft nonmetallic cushions at points of support to control chafing and fretting. Tubing or pipe shall be protected against road hazards either by installation in a protected location or by providing adequate shielding at vulnerable areas. Protective loom, where used, shall be both water and acid resistant.

5. Specifications

5.1 Type 1—Copper Tubing

This material specification covers the minimum requirements for seamless annealed copper tubing that shall be used for automotive air brake lines.

5.1.1 MANUFACTURE—The tubing shall be seamless cold drawn to size and bright annealed as a final operation in such a manner as to produce a finished product which will meet all requirements of

TABLE 1—DIMENSIONS AND TOLERANCES OF AIR BRAKE TUBING

Nominal Tubing OD (in)	Outside Diameter[1] Specified		Outside Diameter[1] Tolerance±		Wall Thickness (min)	
	mm	in	mm	in	mm	in
1/4	6.35	0.250	0.05	0.002	0.75	0.0295
5/16	7.92	0.312	0.05	0.002	0.75	0.0295
3/8	9.53	0.375	0.05	0.002	0.75	0.0295
7/16	11.10	0.437	0.05	0.002	1.160	0.0455
1/2	12.70	0.500	0.05	0.002	1.160	0.0455
5/8	15.88	0.625	0.05	0.002	1.160	0.0455
3/4	19.05	0.750	0.06	0.0025	1.160	0.0455
1	25.40	1.000	0.06	0.0025	1.160	0.0455

[1] The actual outside diameter shall be the average of the maximum and minimum outside diameters as determined at any one cross section through the tubing.

5.1.2 DIMENSIONS AND TOLERANCES—The finished tubing shall conform to the dimensions and tolerances shown in Table 1, for the nominal diameter specified by the purchaser.

5.1.3 QUALITY—The finished tubing shall be clean, smooth, and round, free from internal and external mechanical imperfections, corrosion, scale, seams, and cracks.

5.1.4 MATERIAL—The tubing shall be made from phosphorized, low residual phosphorus copper conforming to SAE J463, UNS C12200 which has the chemical composition as in Table 2:

TABLE 2—CHEMICAL REQUIREMENTS

Element	Ladle Analysis % by Weight
Copper	99.90 min
Phosphorus	0.015-0.040

5.1.5 MECHANICAL PROPERTIES—The finished tubing shall have mechanical properties as tabulated in Table 3:

TABLE 3—MECHANICAL PROPERTIES, COPPER TUBING

Yield Strength MPa (psi) min[1]	Tensile Strength MPa (psi) min	Elongation in 50 mm (2 in), % min Tubing OD	
		19 mm ($^3/_4$ in) and smaller	Over 19 mm ($^3/_4$ in)
62 (9000)	210 (30 000)	30	40

[1] At 0.5% extension under load.

5.1.6 GRAIN SIZE—The tubing shall be furnished in either of two temper conditions with grain size as tabulated in Table 4:

TABLE 4—GRAIN SIZE

Temper	Grain Size, mm
Light Annealed	0.015-0.040
Soft Annealed	0.040 min

5.1.7 PERFORMANCE REQUIREMENTS—The finished tubing shall satisfactorily meet the following performance tests. Test specimens shall be taken from tubing which has not been subjected to cold working after the anneal of the finished sized tubing.

5.1.8 FLARING TEST—A test section cut from the finished tubing, having squared and deburred ends, shall withstand being flared at one end over a polished tapered mandrel of 60 degrees included angle until the actual average outside diameter is increased 40% without evidence of splitting or flaws. The axis of the mandrel and axis of the tubing shall be kept parallel during the flaring process and the test may be made in a die to restrict the expansion to 40%.

5.1.9 PRESSURE PROOF TEST—Unless otherwise specified, tubing supplied under this document shall withstand, with no evidence of failure, a hydrostatic proof test at a pressure equivalent to a hoop (circumferential) stress of 62 MPa (9000 psi). The test pressures shall be as determined from Barlow's formula for thin hollow cylinders under pressure:

$$P = \frac{2TS}{D} \quad \text{(Eq.1)}$$

where:

D = outside diameter of tubing, mm (in)
P = hydrostatic pressure, MPa (psi)
S = allowable unit stress of material = 62 MPa (9000 psi)
T = minimum wall thickness of tubing, mm (in)

The test pressure at a yield strength of 62 MPa (9000 psi) for the minimum wall thicknesses allowed are given in Table 5.

TABLE 5—HYDROSTATIC TEST PRESSURES FOR AIR BRAKE TUBING

Nominal Tubing OD, in	Hydrostatic Test Pressure MPa	Hydrostatic Test Pressure psi	Nominal Tubing OD, in	Hydrostatic Test Pressure MPa	Hydrostatic Test Pressure psi
$^1/_4$	14.50	2100	$^1/_2$	11.00	1600
$^5/_{16}$	11.70	1700	$^5/_8$	8.95	1300
$^3/_8$	9.65	1400	$^3/_4$	6.90	1000
$^7/_{16}$	12.40	1800	1	5.50	800

5.1.10 AIR PRESSURE TEST—Each length of finished tubing shall be tested at the maximum operating air pressure, as specified by the purchaser. The tubing shall show no leakage at the test pressure. An electric eddy current test may be substituted for the air pressure test, providing the rejection limits are such that the hydrostatic and air pressure requirements can be guaranteed.

5.1.11 IDENTIFICATION—Tubing shall be permanently and legibly marked at intervals not greater than 381 mm (15 in) with the words Air Brake.

5.1.12 METHODS OF TEST—All tests to determine conformance with the foregoing specifications shall be conducted in accordance with the following ASTM Standards:

5.1.12.1 *Chemical Analysis*—See ASTM E 62.

5.1.12.2 *Grain Size*—See ASTM E 79.

5.1.12.3 *Tensile*—See ASTM E 8.

5.2 Type 2—Galvanized Steel Pipe

This material specification covers the minimum requirements for pipe that shall be used in automotive air brake lines.

5.2.1 SPECIFICATIONS—Welded or seamless steel pipe shall be Schedule 40, Zinc Coated (galvanized by the hot dip process), and manufactured in accordance with ASTM A 120.

5.2.2 DIMENSIONS AND TOLERANCES—The finished pipe shall conform to the dimensions and tolerances listed for the several nominal sizes in Tables 6A and 6B.

5.2.3 PIPE THREADS—Both ends of lengths of pipe shall be threaded after coating, unless there is specific authorization to the contrary, to conform to Dryseal American Standard Taper Thread (NPTF). Specifications for pipe threads are given in detail in SAE J476.

5.2.4 MECHANICAL PROPERTIES—The steel in the finished pipe, including the weld, shall have mechanical properties as tabulated in Table 7.

5.2.5 PRESSURE PROOF TESTS PER TEST METHOD ASTM A 370, SUPPLEMENT II

5.2.5.1 *Hydrostatic Test*—Unless otherwise specified, each length of pipe shall be tested at the mill to a hydrostatic pressure of 4850 kPa (700 psi). For nominal sizes over 25.4 mm (1 in) dia see ASTM A 120.

5.2.5.2 *Nondestructive Electric Test*—In lieu of the hydrostatic test, if mutually agreeable to purchaser and the manufacturer, each pipe may be tested by passing it through an electric eddy current tester which is capable of detecting defects 1.57 mm (0.062 in) in length and one-half the wall thickness, or defects of any length completely penetrating the wall. Such tests shall be made on the welded seam and the adjacent metal affected thereby.

5.2.5.3 *Corrosion Protection*—The inside and outside surfaces of the pipe shall be coated with zinc by the hot dip process. The coating shall weigh at least 610 g/m^2 (2.0 oz/ft^2) of total surface. Tests to determine whether product meets this requirement shall be conducted in accordance with ASTM A 120.

5.2.5.4 *Bending*—Pipe shall be used for essentially straight runs; however, generous curves having a radius in excess of 20 times the outside diameter shall be permitted. In no case shall heat be used to facilitate bending of pipe.

TABLE 6A—DIMENSIONS AND TOLERANCES OF PIPE FOR AIR BRAKE USE, MM

Nominal Pipe Size	Outside Diameter Specified mm	Outside Diameter Tolerance min	Outside Diameter Tolerance max	Inside Diameter (Ref)	Wall Thickness Specified	Wall Thickness min	Threads per in	Nominal Weight Plain Ends kg/m±5%
1/8	10.29	9.50	10.67	6.83	1.73	1.52	27	0.36
1/4	13.72	12.93	14.10	9.25	2.24	1.96	18	0.63
3/8	17.14	16.36	17.53	12.53	2.31	2.03	18	0.85
1/2	21.34	20.55	21.72	15.80	2.77	2.41	14	1.27
3/4	26.67	25.88	27.05	20.93	2.87	2.51	14	1.68
1	33.40	32.61	33.78	26.64	3.38	2.95	11.5	2.50

TABLE 6B—DIMENSIONS AND TOLERANCES OF PIPE FOR AIR BRAKE USE, IN

Nominal Pipe Size	Outside Diameter Specified	Outside Diameter Tolerance Plus	Outside Diameter Tolerance Minus	Inside Diameter (Ref)	Wall Thickness Specified	Wall Thickness min	Threads per in	Weight Per ft,[1] lb ±5%
1/8	0.405	0.016	0.031	0.269	0.068	0.060	27	0.24
1/4	0.540	0.016	0.031	0.364	0.088	0.077	18	0.42
3/8	0.675	0.016	0.031	0.493	0.091	0.080	18	0.57
1/2	0.840	0.016	0.031	0.622	0.109	0.095	14	0.85
3/4	1.050	0.016	0.031	0.824	0.113	0.099	14	1.13
1	1.315	0.016	0.031	1.049	0.133	0.116	11.5	1.68

[1] Nominal Weight Plain Ends.

TABLE 7—MECHANICAL PROPERTIES, GALVANIZED STEEL PIPE

Yield Strength, MPa (psi), min	Elongation in 50 mm (2 in), %
170 (25 000)	14-40

R) NONMETALLIC AIR BRAKE SYSTEM TUBING—SAE J844 JUN90

SAE Standard

Report of the Tube, Pipe, Hose, and Lubrication Fittings Committee, approved June 1963 and completely revised by the Fluid Conductors and Connectors Technical Committee October 1988. Completely revised by the Fluid Conductors and Connectors Technical Committee June 1990.

1. Scope[1]

This SAE Standard covers the minimum requirements for nonmetallic tubing as manufactured for use in air brake systems. Non-reinforced products are designated type A and reinforced products type B. It is not intended to cover tubing for any portion of the system which operates below −40°C (−40°F), above +93°C (+200°F), above a maximum working gage pressure of 150 psi (1030 kPa), or in an area subject to attack by battery acid. This tubing is intended for use in the brake system for connections which maintain a basically fixed relationship between components during vehicle operation. Coiled tube assemblies required for those installations where flexing occurs are covered by this document and SAE J1131 to the extent of setting minimum requirements on the essentially straight tube and tube fitting connections which are used in the construction of such assemblies.[2]

2. References

2.1 Applicable Documents

SAE J246 Spherical and Flanged Sleeve (Compression) Tube Fittings

SAE J1131 Performance Requirements for SAE J844 Nonmetallic Tubing and Fitting Assemblies Used in Automotive Air Brake Systems

SAE J1149 Metallic Air Brake System Tubing and Pipe

3. Installation and Assembly Recommendations

3.1 End Fittings

End fittings are to be assembled to the tubing in accordance with the fitting manufacturer's recommendations. The fitting may be of the design shown in SAE J246, or any other design suitable for use with nonmetallic air brake tubing. Performance test requirements for nonmetallic air brake assemblies are covered in SAE J1131.

3.2 Noncoiled Tubing

Noncoiled tubing should not be used in flexing applications such as frame to axle.

3.3 Support and Routing

When installed in a vehicle this tubing shall be routed and supported so as to:
a. Eliminate chafing, abrasion, kinking, or other mechanical damage
b. Minimize fatigue conditions
c. Be protected against road hazards by installation in a protected location or by providing adequate shielding at vulnerable areas
d. Not to be exposed to temperatures, internal or external, over

---

[1] See SAE J1149 for Metallic Air Brake System Tubing and Pipe.
[2] Federal regulations covering designed requirements and accepted applications for coiled tube assemblies are set forth in 49CFR393.45.

+93°C (+200°F) or below −40°C (−40°F)
  e. Not to be exposed to attack by battery acid
  f. Avoid excessive sag

4. Identification

Air brake tubing shall be labeled in a contrasting color with the legend repeated every 15 in (380 mm) or less along the entire length of tubing in legible block capital letters.

The following minimum information, in the order listed, is required. Additional information and/or another lay line may be added, if necessary.
  a. Airbrake
  b. SAE J844
  c. Type, A or B
  d. Nominal, tubing O.D. in fractions of in - 1/4, 3/8, 1/2, etc. (6.4, 9.5, 12.7 mm)
  e. Tubing manufacturer's identification

5. Manufacture

The tubing shall be manufactured to comply with the requirements outlined in this document.

6. Construction

Type A tubing shall consist of a single wall extrusion of 100% virgin nylon (polyamide) containing additives which provide heat and light resistance. Type B tubing shall consist of a core extrusion of 100% virgin nylon (polyamide) containing additives which provide heat resistance. This core shall be reinforced with polyester braid or equivalent, and covered with a protective jacket of 100% virgin nylon (polyamide) containing additives which provide heat and light resistance. The protective covering shall be bonded to the core through the interstices of the braid. The inner core and outer jacket shall be of contrasting colors.

7. Dimensions and Tolerances

The tubing shall conform to dimensions shown in Table 1 under all conditions of moisture. Conformance with this requirement shall be determined on samples which have been subjected to 110°C (230°F)[3] for 4 h[4] in a circulating air oven, and on separate samples which have been immersed in boiling water for 2 h. Dimensional tests shall be made after samples have been returned to room temperature for 1/2 to 3 h.

8. Mechanical Properties

The tubing shall conform to the mechanical properties shown in Table 2, when tested according to the methods outlined in this document.

---

[3] All test temperatures specified may vary by ±5°F (±3°C).
[4] All times are minimum unless otherwise specified.

9. Performance Requirements

The tubing shall satisfactorily meet the following performance tests (see footnotes 3, 4, 5, 6, 7, and 8).

9.1 Leak Test[6]

The tubing manufacturer shall subject each continuous length of tubing to test at a gage pressure of 200 psi (1380 kPa) with an appropriate gas for a period of time sufficient to determine the presence of any leaks. Defective sections shall be cut off and scrapped. The remaining tubing shall be recoupled at the points where defective sections were removed and again subjected to the 200 psi (1380 kPa) pressure test. The procedure shall be repeated until all sections of tubing designated for distribution to users have successfully withstood the test.

9.2 Moisture Absorption[5]

Expose sample of tubing for 24 h in a circulating air oven at 110°C (230°F). Remove from oven, weigh immediately and expose for 100 h at 100% relative humidity and 24°(75°F). Within 5 min from humidity conditioning, wipe surface moisture from both the interior and exterior surfaces of the tubing and re-weigh. Moisture absorption shall not exceed 2% by weight.

9.3 Ultraviolet Resistance[5]

Place sample of tubing on a turntable 17 in (430 mm) in diameter, rotating at 33 rpm ± 3, with a RS-4* sunlamp or equivalent centrally located 9 in (230 mm) above the table. Expose for 1200 h using a new bulb that has been seasoned for 50 h prior to test. Do not permit temperature of tubing to exceed 49°C (120°F) during the test (a fan cooling unit may be utilized). Immediately following this exposure, subject the tubing to the impact test shown in Figure 1. Subject tubing to room temperature burst test as specified in 9.10. Tubing shall withstand no less than 80% of the burst pressure shown in Table 2.

*RS-4 sunlamp is manufactured by General Electric Company[7]
Cuyahoga Lamp Plant
  Nela Park
  Noble Road
  Cleveland, OH 44112

*RS-4 sunlamp is available from George W. Gates Co., Inc.
  P.O. Box 216
  Hempsted Turnpike and Lucille Ave.
  Franklin Square
  Long Island, NY 11010

---

[5] A Qualification Test.
[6] Normally an Inspection Test conducted on each lot of tubing, and where a lot is defined as "the output of one production shift of one size and color of tubing."
[7] The manufacturer and distributor of the sunlamp is listed due to the fact that at the present time this is the only known supplier.

TABLE 1—DIMENSIONS AND TOLERANCES

Type of Tubing	Nominal Tubing OD	Outside Diameter max in	Outside Diameter max mm	Outside Diameter min in	Outside Diameter min mm	Inside Diameter Basic in	Inside Diameter Basic mm	Wall Thickness Basic in	Wall Thickness Basic mm	Wall Thickness Tolerances in	Wall Thickness Tolerances mm
A	1/8	0.128	3.25	0.122	3.10	0.079	2.01	0.023	0.58	±0.003	±0.08
A	1/4	0.253	6.43	0.247	6.27	0.170	4.32	0.040	1.02	±0.003	±0.08
A	5/16	0.316	8.03	0.308	7.82	0.232	5.89	0.040	1.02	±0.004	±0.10
B	3/8	0.379	9.63	0.371	9.42	0.251	6.38	0.062	1.57	±0.004	±0.10
B	1/2	0.505	12.83	0.495	12.57	0.376	9.55	0.062	1.57	±0.004	±0.10
B	5/8	0.630	16.00	0.620	15.75	0.441	11.20	0.092	2.34	±0.005	±0.13
B	3/4	0.755	19.18	0.745	18.92	0.566	14.38	0.092	2.34	±0.005	±0.13

TABLE 2—MECHANICAL PROPERTIES

Type of Tubing	Nominal Tubing OD	Minimum Burst Pressure at 24°C (75°F)a psi	Minimum Burst Pressure at 24°C (75°F)a kPa	Minimum Bend Radius in	Minimum Bend Radius mm	Maximum Stiffness lbf	Maximum Stiffness N
A	1/8	1000	6900	0.37	9.4	1	4.4
A	1/4	1200	8300	1.00	25.4	2	8.9
A	5/16	1000	6900	1.25	31.8	6	27.0
B	3/8	1400	9700	1.50	38.1	8	36.0
B	1/2	950	6600	2.00	50.8	20	89.0
B	5/8	900	6200	2.50	63.5	50	222.0
B	3/4	800	5500	3.00	76.2	80	356.0

aWith moisture content of tubing 0.06% maximum.

*RS-4 sunlamp is a 100 W, 3010 lm mercury arc lamp with an outer glass jacket which eliminates wavelengths below 285 nanometers.

9.4 Cold Temperature Flexibility[5]
Expose sample of tubing for 24 h in a circulating air oven at 110°C (230°F). Remove from oven and within 30 min expose for 4 h at −40°C (−40°F). Also expose a mandrel at −40°C (−40°F) having a diameter equal to 12 times the nominal diameter of the tubing. (In order to obtain uniform temperatures, the tubing and mandrel may be supported by a nonmetallic surface during the entire period of test.) Immediately following this exposure, bend tubing 180 degrees over the mandrel accomplishing the bending motion within a period of 4 to 8 s. The tubing shall show no evidence of fracture.

9.5 Heat Aging[5]
Three separate heat aging tests shall be conducted; each phase shall be run on separate tubing samples. Subject tubing to room temperature burst test as specified in 9.10. Tubing shall withstand 80% of the burst pressure shown in Table 2.
  a. Phase 1—Bend samples of tubing 180 degrees around a mandrel having a diameter equivalent to twice the minimum bend radius specified in Table 2. While in this position, expose tubing and mandrel for 72 h in an air circulating oven at 110°(230°F). Remove from oven and permit tubing to return to 24°C (75°F) while still on the mandrel. Within 30 min after stabilization at 24°C (75°F), return the tubing to a straight position in a minimum of 4 s, then rebend (against the set) 180 degrees around the mandrel, accomplishing the bending motion within a period of 4 to 8 s.
  b. Phase 2—Expose samples of tubing for 72 h in a circulating air oven at 110°C (230°F). Remove from oven and permit tubing to return to 24°C (75°F). Within 30 min after stabilization at 24°C (75°F), subject tubing to the impact test shown in Figure 1.
  c. Phase 3—Immerse samples of tubing in boiling water for 2 h. Remove from water and permit to return to 24°C (75°F). Within 30 min after stabilization at 24°C (75°F), subject tubing to the impact test shown in Figure 1.

9.6 Resistance to Zinc Chloride[5]
Bend tubing to the minimum bend radius shown in Table 2. While in this position, immerse in a 50% (by weight) aqueous solution of zinc chloride for 200 h at 75°F (24°C). Remove from solution. Tubing shall show no evidence of cracking on the outside diameter.

NOTE: Fresh, anhydrous zinc chloride should be used to make up a concentration of 50% (by weight) aqueous solution (specific gravity of 1.576 or a Baume rating of 53 degrees at 60°F (15.6°C)).

9.7 Resistance to Methyl Alcohol[5]
Bend tubing to the minimum bend radius shown in Table 2. While in this position, immerse in 95% methyl alcohol for 200 h at 24°(75°F). Remove from solution. Tubing shall show no evidence of cracking.

9.8 Stiffness[5]
Use samples 11 in (280 mm) long. Insert a rod of suitable size into the tubing to maintain a straight position within ±0.125 in (3.2 mm). Expose tubing and rod for 24 h in a circulating air oven at 110°C (230°F). Remove from oven and permit tubing and rod to return to 24°C (75°F). Within 30 min after stabilization at 24°C (75°F), remove rod and subject tubing to stiffness test shown in Figure 2. Tubing shall require no more force than specified in Table 2 to deflect 2 in (51 mm).

9.9 Boiling Water Stabilization and Burst Test[5]
Immerse tubing in boiling water for 2 h. Remove from water and subject to the room temperature burst test as specified in 9.10. Tubing shall withstand no less than 80% of the burst pressure shown in Table 2.

9.10 Room Temperature Burst Test[6]
Tubing shall be stabilized for 0.5 to 3.0 h at 24°C (75°F) and tested by increasing pressure at a constant rate to reach the specified minimum burst pressure in Table 2 within a time period of 3 to 15 s. Tubing that bursts below the pressure specified in Table 2 shall be rejected.

9.11 Cold Temperature Impact[6]
Condition tubing by exposing one half the samples for 24 h at 110°C (230°F) in a circulating air oven, and one half the samples in boiling water for 2 h; then expose all the samples to −40°C (−40°F) for 4 h. Also, expose impact test apparatus, shown in Figure 1, to −40°C (−40°F). While tubing and apparatus are at this cold temperature (approximately −40°F), subject tubing to impact as specified. The tubing shall show no evidence of cracks. After impact testing, permit tubing to return to 24°C (75°F). Within 30 min after stabilization at 24°C (75°F), subject tubing to room temperature burst test as specified in 9.10. Tubing shall withstand at least 80% of the burst pressure shown in Table 2. Sample size shall be 10 specimens per lot. In the event of any failures, a second sample from the same lot consisting of 20 specimens shall be tested. If another failure occurs, the lot shall be rejected.

9.12 Adhesion Test[6]
9.12.1 This test applies only to the reinforced products, Type B.

9.12.2 CONDITION—This test shall be conducted at 24°C (75°F) ambient temperature.

NOMINAL TUBE O.D.	HOLE DIA D in	HOLE DIA D mm
1/8	0.156	(3.96)
1/4	0.281	(7.14)
5/16	0.343	(8.71)
3/8	0.406	(10.31)
1/2	0.531	(13.49)
5/8	0.656	(16.66)
3/4	0.800	(20.32)

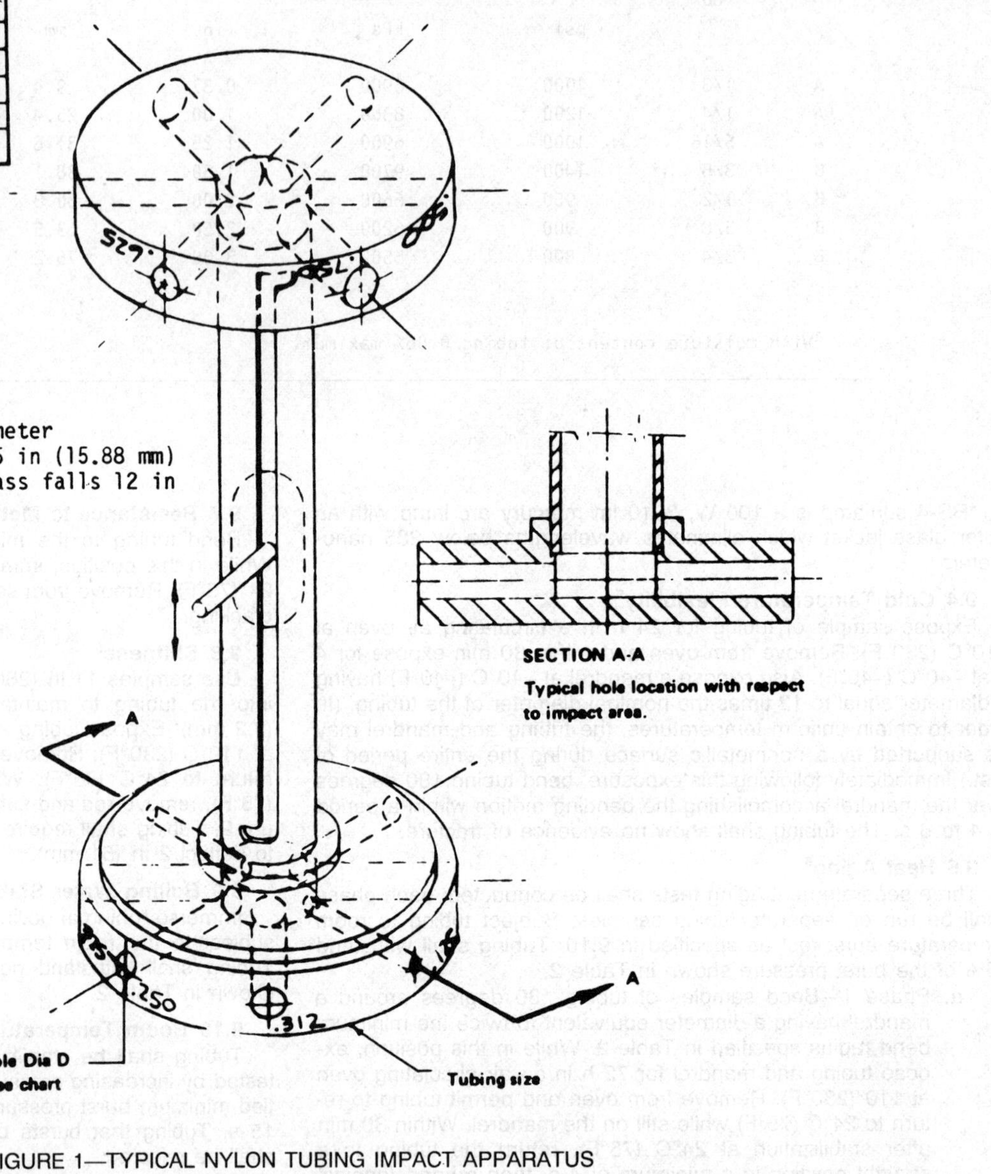

NOTE: Impact apparatus may be drilled to accept any combination of tube sizes listed in chart

1 lb (0.454 kg) mass, with a diameter of 1.25 in (31.75 mm) and a 0.625 in (15.88 mm) spherical radius on both ends. Mass falls 12 in (304.8 mm)

SECTION A-A
Typical hole location with respect to impact area.

FIGURE 1—TYPICAL NYLON TUBING IMPACT APPARATUS

9.12.3 PROCEDURE AND REQUIREMENTS—Cut a strip of tubing into a 0.25 in (6.0 mm) wide helical coil equal in length to five times the circumference of the tubing. Bend the helical coil in reverse of coiling so as to expose the braid gap between the outer jacket and core tube section. Start by working a sharp knife blade into the braid gap to initiate separation, and then attempt to separate the outer jacket from the core tube at the braid interstices. The bonded surface (excluding the braided area) between the outer jacket and core section shall be inseparable for the entire test sample length.

9.13 Heat Aging Adhesion Test[5]

9.13.1 PROCEDURE—Subject samples to Phase 1 of the heat aging test procedure per 9.5.

9.13.2 REQUIREMENTS—After completion of the Phase 1 procedure, the tubing shall meet the requirements of 9.12.

9.14 Collapse Resistance Test Procedure[5]

9.14.1 GENERAL—All tests are to be conducted at room temperature 93°C (75°F)[3] unless otherwise specified.

9.14.2 PREPARATION OF TEST SAMPLES—Three samples shall be prepared for testing. The free tube length of the samples shall be as follows:

3.14 X (min kink radius) + 10 X (tube O.D.) +
2 X (length of supporting pin)

9.14.3 TEST PROCEDURE—Place a reference mark at the middle of each sample and measure the cross-section diameter (Minor Diameter [unbent]) at this point and record.

NOTE: See Figures 3 and 4 for location of minor diameters.

9.14.4 Carefully install the samples on a bend test fixture (as shown in Figure 5) in a 180 degree bend condition. The tube shall be bent in the direction of the natural curvature of the tube. Samples prepared per 9.14.2 shall be bent to a radius equal to the minimum kink radius called out in Table 3.

9.14.5 Age samples on test fixture at 93°C (200°F)[3] for 24 h[4].

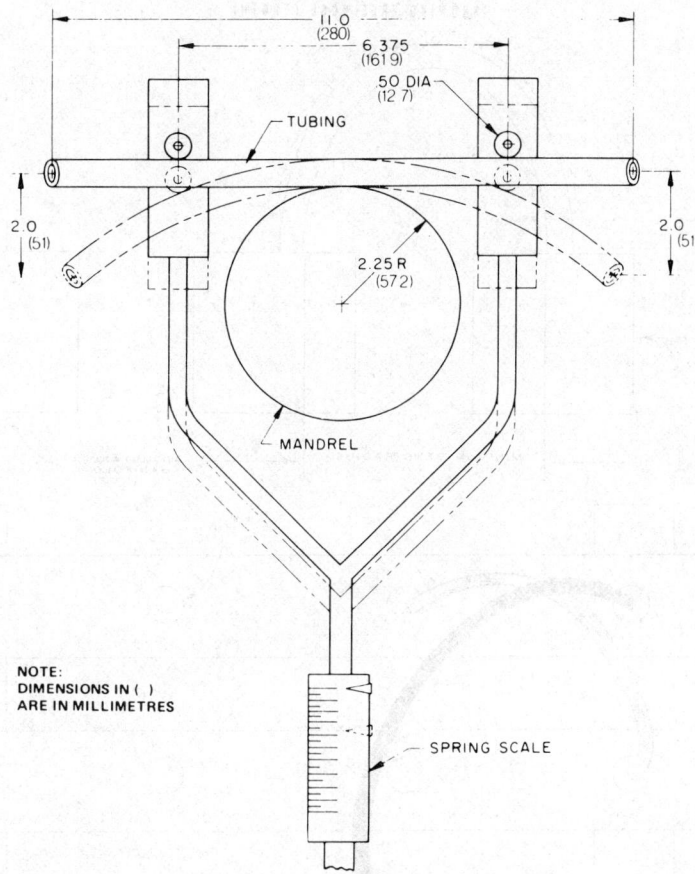

FIGURE 2—STIFFNESS TEST APPARATUS

Allow the samples to cool to room temperature. While the samples are on the test fixture, measure the minor diameter [bent]. Collapse of greater than 20% is considered a failure (see Eq.1).

$$\text{Percent Collapse} = \frac{\text{Minor O.D. [unbent]} - \text{Minor O.D. [bent]}}{\text{Minor O.D. [unbent]}} \times 100 \quad \text{(Eq.1)}$$

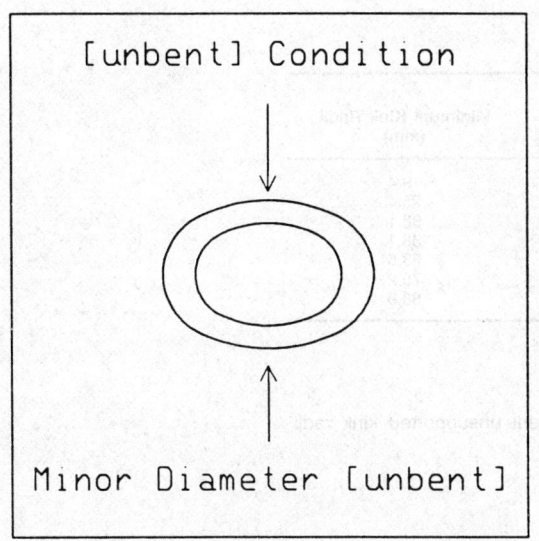

FIGURE 3—MINOR DIAMETER (UNBENT)

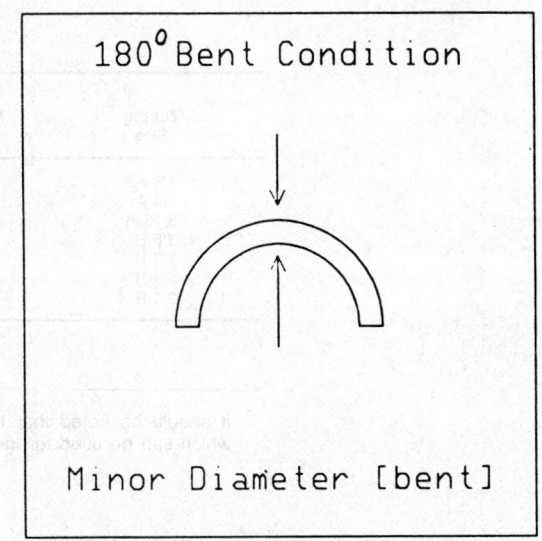

FIGURE 4—MINOR DIAMETER (BENT)

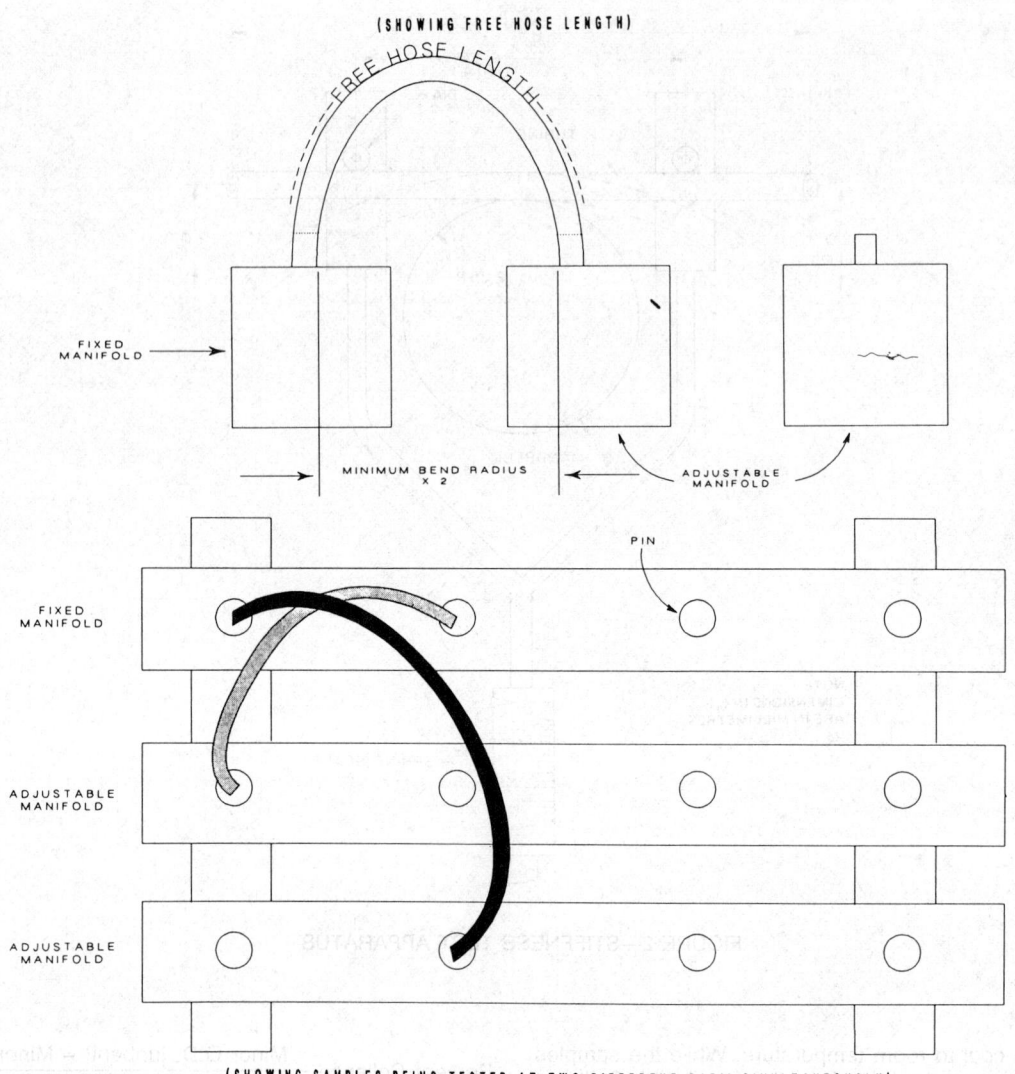

FIGURE 5—BEND TEST FIXTURE

TABLE 3[8]

Tubing Size	Minimum Kink Radii (in)	Minimum Kink Radii (mm)
1/8 A	0.37	9.4
1/4 A	1.00	25.4
5/16 A	1.50	38.1
3/8 B	1.50	38.1
1/2 B	2.50	63.5
5/8 B	3.00	76.2
3/4 B	3.50	88.9

[8] It should be noted that these values represent unsupported kink radii which can be used for installation purposes.

(R) METRIC NONMETALLIC AIR BRAKE SYSTEM TUBING—SAE J1394 APR91

SAE Standard

22.227

Report of the Fluid Conductors and Connectors Technical Committee, approved April 1983 and revised May 1989. Completely revised by the SAE Fluid Conductors and Connectors Technical Committee S4—Air Brake Tubing and Tube Fittings Subcommittee April 1991.

1. Scope[1]

This SAE Standard covers the minimum requirements for metric sizes of nonmetallic tubing as manufactured for use in air brake systems. Nonreinforced products are designated type A and reinforced products type B. It is not intended to cover tubing for any portion of the system that operates below –40 °C (–40 °F), above +93 °C (+200 °F), above a maximum working gage pressure of 1.0 MPa (150 psi), or in an area subject to attack by battery acid. This tubing is intended for use in the brake system for connections that maintain a basically fixed relationship between components during vehicle operation. Coiled tube assemblies required for those installations where flexing occurs are covered by this standard and SAE J1131 to the extent of setting minimum requirements on the essentially straight tube and tube fitting connections, which are used in the construction of such assemblies.[2]

2. References

2.1 Applicable Documents

The following publications form a part of this specification to the extent specified herein. The latest issue of SAE publications shall apply.

2.1.1 SAE PUBLICATIONS—Available from SAE, 400 Commonwealth Drive, Warrendale, PA 15096-0001.

SAE J246—Spherical and Flanged Sleeve (Compression) Tube Fittings

SAE J844—Nonmetallic Air Brake System Tubing

SAE J1131—Performance Requirements for SAE J844d Nonmetallic Tubing and Fittings Assemblies Used in Automotive Air Brake Systems

SAE J1149—Metallic Air Brake System Tubing and Pipe

2.1.2 GOVERNMENT PUBLICATIONS—Available from Superintendent of Documents, U.S. Government Printing Office, Washington, DC 20402

49CFR393.45

3. Installation and Assembly Recommendations

3.1 End Fittings

End fittings are to be assembled to the tubing in accordance with the fitting manufacturer's recommendations. The fitting may be of the design shown in the proposed metric version of SAE J246, or any other design suitable for use with metric size nonmetallic air brake tubing. Performance test requirements for nonmetallic air brake assemblies are covered in SAE J1131.

3.2 Noncoiled Tubing

Noncoiled tubing should not be used in flexing applications such as frame to axle.

3.3 Support and Routing

When installed in a vehicle, this tubing shall be routed and supported so as to:
  a. Eliminate chafing, abrasion, kinking, or other mechanical damage.
  b. Minimize fatigue conditions.
  c. Be protected against road hazards by installation in a protected location or by providing adequate shielding at vulnerable areas.
  d. Not be exposed to temperatures, internal or external, over +93 °C (+200 °F) or below –40 °C (–40 °F).
  e. Not be exposed to attack by battery acid.
  f. Avoid excessive sag.

4. Identification

Air brake tubing shall be labeled in contrasting color with the legend repeated every 380 mm (15 in) or less along the entire length of tubing in legible block capital letters.

The following minimum information, in the order listed, is required. Additional information and/or another lay line may be added, if necessary.
  a. Metric airbrake
  b. SAE J1394
  c. Type A or B
  d. Nominal tubing OD in mm—6, 8, 10, 12 or 16
  e. Tubing manufacturer's identification

5. Manufacture

The tubing shall be manufactured to comply with the requirements outlined in this document.

6. Construction

Type A tubing shall consist of a single wall extrusion of 100% virgin nylon (polyamide) containing additives that provide heat and light resistance. Type B tubing shall consist of a core extrusion of 100% virgin nylon (polyamide) containing additives that provide heat resistance. This core shall be reinforced with polyester braid or equivalent and covered with a protective jacket of 100% virgin nylon (polyamide) containing additives that provide heat and light resistance. The protective covering shall be bonded to the core through the interstices of the braid. The inner core and outer jacket shall be of contrasting colors.

7. Dimensions and Tolerances

The tubing shall conform to dimensions shown in Table 1 under all conditions of moisture. Conformance with this requirement shall be determined on samples that have been subjected to 110 °C (230 °F)[3] for 4 h[4] in a circulating air oven, and on separate samples that have been immersed in boiling water for 2 h. Dimensional tests shall be made after samples have been returned to room temperature for 0.5 to 3.0 h.

8. Mechanical Properties

The tubing shall conform to the mechanical properties shown in Table 2, when tested according to the method outlined in this document.

9. Performance Requirements

The tubing shall satisfactorily meet the following performance tests (see Footnotes 3, 4, 5, and 6).

9.1 Leak Test[5]

The tubing manufacturer shall subject each continuous length of tubing to test at a gage pressure of 1.4 MPa (200 psi) with an appropriate gas for a period of time sufficient to determine the pres-

---

[1] The metric values contained herein are to be regarded as standard; the in-lb values in parentheses may only be approximate. See SAE J844 for nonmetallic air brake system tubing (inch-dimensioned) and SAE J1149 for metallic air brake system tubing and pipe.

[2] Federal regulations covering designed requirements and accepted applications for coiled tube assemblies are set forth in 49CFR393.45.

[3] All test temperatures specified may vary by ±3 °C (±5 °F).

[4] All times are minimum unless otherwise specified.

[5] An inspection test conducted on each lot of tubing and where a lot is defined as "the output of one production shift of one size and color of tubing."

TABLE 1—DIMENSIONS AND TOLERANCES

Tubing Type	Tubing Size OD mm	Tubing Size OD in	Tubing Size ID mm	Tubing Size ID in	Minimum Wall Thickness mm	Minimum Wall Thickness in	OD Tolerances mm	OD Tolerances in	ID Tolerances mm	ID Tolerances in
A	6.0	0.236	4.0	0.157	0.9	0.035	±0.1	±0.004	±0.1	±0.004
A	8.0	0.315	6.0	0.236	0.9	0.035	±0.1	±0.004	±0.1	±0.004
B	10.0	0.393	7.0	0.275	1.35	0.053	±0.15	±0.006	±0.15	±0.006
B	12.0	0.472	9.0	0.354	1.35	0.053	±0.15	±0.006	±0.15	±0.006
B	16.0	0.629	12.0	0.472	1.8	0.071	±0.15	±0.006	±0.15	±0.006

TABLE 2—MECHANICAL PROPERTIES

Tubing Type	Nominal Tubing OD mm	Minimum Burst Pressure at 24 °C (75 °F)[1] MPa	Minimum Burst Pressure at 24 °C (75 °F)[1] psi	Test Bend Radius[2] mm	Test Bend Radius[2] in	Maximum Stiffness N	Maximum Stiffness lbf
A	6	7.6	1100	20	0.75	9	2
A	8	6.2	900	32	1.25	27	6
B	10	8.2	1200	38	1.50	36	8
B	12	6.9	1000	45	1.75	90	20
B	16	6.0	875	70	2.75	225	50

[1] With moisture content of tubing 0.06% max.
[2] For test purposes only.

ence of any leaks. Defective sections shall be cut off and scrapped. The remaining tubing shall be recoupled at the points where defective sections were removed and again subjected to the 1.4 MPa (200 psi) pressure test. The procedure shall be repeated until all sections of tubing designated for distribution to users have successfully withstood the test.

9.2 Moisture Absorption[6]

Expose sample of tubing for 24 h in a circulating air oven at 110 °C (230 °F). Remove from oven, weigh immediately, and expose for 100 h at 100% relative humidity and 24 °C (75 °F). Within 5 min from humidity conditioning, wipe surface moisture from both the interior and exterior surfaces of the tubing and reweigh. Moisture absorption shall not exceed 2% by weight.

9.3 Ultraviolet Resistance[7]

Place sample of tubing on a turntable 430 mm (17 in) in diameter, rotating at 33 rpm ± 3, with a RS-4* sunlamp or equivalent centrally located 230 mm (9 in) above the table. Expose for 1200 h using a new bulb that has been seasoned for 50 h prior to test. Do not permit temperature of tubing to exceed 50 °C (122 °F) during the test (a fan cooling unit may be utilized). Immediately following this exposure, subject the tubing to the impact test shown in Figure 1. Subject tubing to room temperature burst test as specified in 9.10. Tubing shall withstand no less than 80% of the burst pressure shown in Table 2.

*RS-4—sunlamp is manufactured by
General Electric Company[8]
Cuyahoga Lamp Plant
Nela Park
Noble Road
Cleveland, OH 44112

---

[6] A qualification test.
[7] A qualification test.
[8] The manufacturer and distributor of the sunlamp are listed due to the fact that at the present time they are the only known suppliers.

*RS-4 sunlamp is available from
George W. Gates Co. Inc.
P.O. Box 216
Hempsted Turnpike and Lucille Ave.
Franklin Square
Long Island, NY 11010

9.4 Cold Temperature Flexibility[9]

Expose sample of tubing for 24 h in a circulating air oven at 110 °C (230 °F). Remove from oven and within 30 min expose for 4 h at −40 °C (−40 °F). Also expose a mandrel at −40 °C (−40 °F) having a diameter equal to twelve times the nominal diameter of the tubing. (In order to obtain uniform temperatures, the tubing and mandrel may be supported by a nonmetallic surface during the entire period of test.) Immediately following this exposure, bend tubing 180 degrees over the mandrel, accomplishing the bending motion within a period of 4 to 8 s. The tubing shall show no evidence of fracture.

9.5 Heat Aging[10]

Three separate heat aging tests shall be conducted; each phase shall be run on separate tubing samples. Subject tubing to room temperature burst test as specified in 9.10. Tubing shall withstand 80% of the burst pressure shown in Table 2.

9.5.1 Phase 1—Bend samples of tubing 180 deg around a mandrel having a diameter equivalent to twice the bend radius specified in Table 2. While in this position, expose tubing and mandrel for 72 h in an air circulating oven at 110 °C (230 °F). Remove from oven and permit tubing to return to 24 °C (75 °F) while still on the mandrel. Within 30 min after stabilization at 24 °C (75 °F), return the tubing to a straight position in a minimum of 4 s, then rebend (against the set) 180 deg around the mandrel, accomplishing the bending motion within a period of 4 to 8 s.

---

[9] A qualification test.
[10] A qualification test.

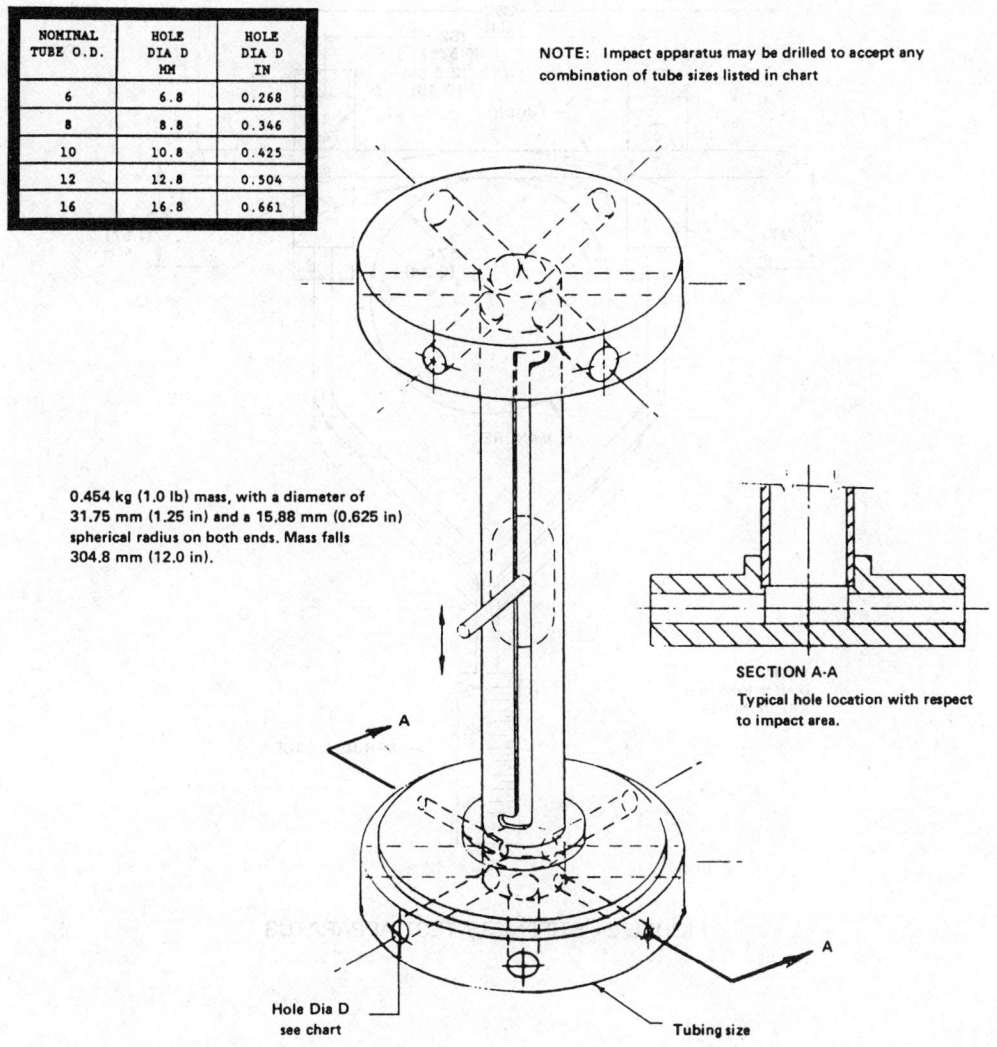

FIGURE 1—TYPICAL NYLON TUBING IMPACT APPARATUS

9.5.2 PHASE 2—Expose samples of tubing for 72 h in a circulating air oven at 110 °C (230 °F). Remove from oven and permit tubing to return to 24 °C (75 °F). Within 30 min after stabilization at 24 °C (75 °F), subject tubing to the impact test shown in Figure 1.

9.5.3 PHASE 3—Immerse samples of tubing in boiling water for 2 h. Remove from water and permit to return to 24 °C (75 °F). Within 30 min after stabilization at 24 °C (75 °F), subject tubing to the impact test shown in Figure 1.

9.6 Resistance to Zinc Chloride[10]
Bend tubing to the bend radius shown in Table 2. While in this position, immerse in a 50% (by weight) aqueous solution of zinc chloride for 200 h at 24 °C (75 °F). Remove from solution. Tubing shall show no evidence of cracking on the outside diameter.

NOTE—Fresh, anhydrous zinc chloride should be used to make a concentration of 50% (by weight) aqueous solution (specific gravity of 1.576 or a Baume rating of 53 degrees at 16 °C [61 °F]).

9.7 Resistance to Methyl Alcohol[10]
Bend tubing to the bend radius shown in Table 2. While in this position, immerse in 95% methyl alcohol for 200 h at 24 °C (75 °F). Remove from solution. Tubing shall show no evidence of cracking.

9.8 Stiffness[11]
Use samples 280 mm (11 in) long. Insert a rod of suitable size into the tubing to maintain a straight position within ±3 mm (0.120 in). Expose tubing and rod for 24 h in a circulating air oven at 110 °C (230 °F). Remove from oven and permit tubing and rod to return to 24 °C (75 °F). Within 30 min after stabilization at 24 °C (75 °F), remove rod and subject tubing to the stiffness test shown in Fig. 2. Tubing shall require no more force than specified in Table 2 to deflect 50 mm (1.97 in).

9.9 Boiling Water Stabilization and Burst Test[11]
Immerse tubing in boiling water for 2 h. Remove from water and subject to the room temperature burst test as specified in 8.10. Tubing shall withstand no less than 80% of the burst pressure shown in Table 2.

9.10 Room Temperature Burst Test[12]
Tubing shall be stabilized 0.5 to 3.0 h at 24 °C (75 °F) and tested

---
[11] A qualification test.
[12] An inspection test conducted on each lot of tubing and where a lot is defined as "the output of one production shift of one size and color of tubing."

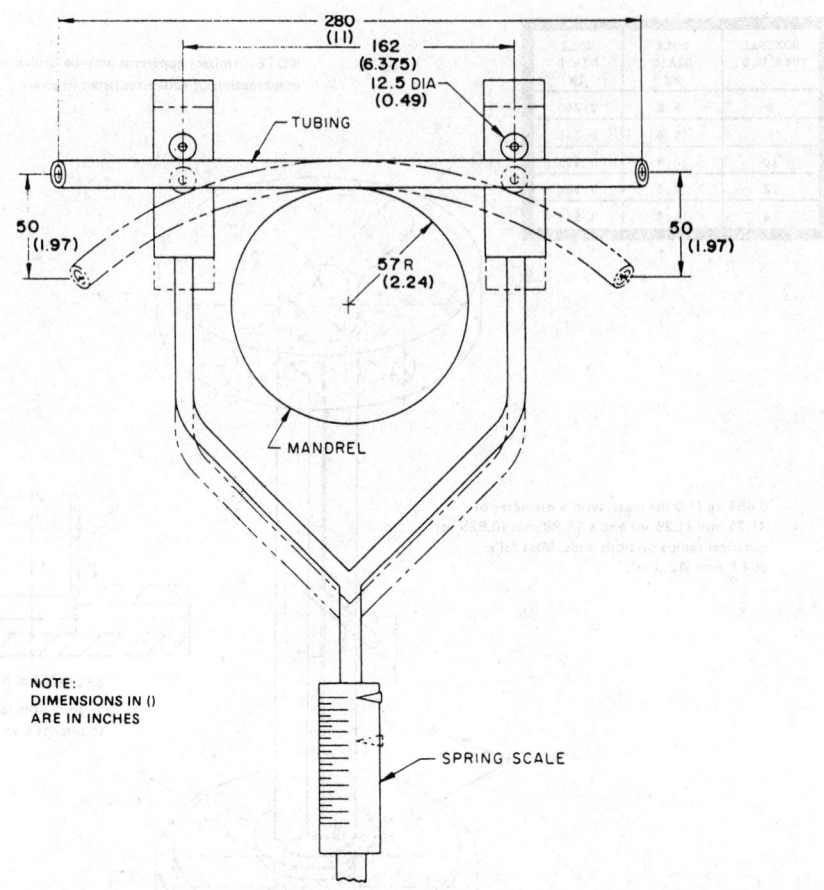

FIGURE 2—STIFFNESS TEST APPARATUS

by increasing pressure at a constant rate to reach the specified minimum burst pressure in Table 2 within a time period of 3 to 15 s. Tubing that bursts below the pressure specified in Table 2 shall be rejected.

9.11 Cold Temperature Impact[12]

Condition tubing by exposing one half of the samples for 24 h at 110 °C (230 °F) in a circulating air oven and one half of the samples in boiling water for 2 h; then expose all the samples to −40 °C (−40 °F) for 4 h. Also, expose impact test apparatus, shown in Figure 1, to −40 °C (−40 °F). While tubing and apparatus are at this cold temperature (approximately −40 °C), subject tubing to impact as specified. The tubing shall show no evidence of cracks. After impact testing, permit tubing to return to 24 °C (75 °F). Within 30 min after stabilization at 24 °C (75 °F), subject tubing to room temperature burst test as specified in 9.10. Tubing shall withstand at least 80% of the burst pressure shown in Table 2. Sample size shall be 10 specimens per lot. In the event of any failures, a second sample from the same lot consisting of 20 specimens shall be tested. If another failure occurs, the lot shall be rejected.

9.12 Adhesion Test[12]

9.12.1 This test applies only to the reinforced products, Type B.

9.12.2 Condition—This test shall be conducted at 24 °C (75 °F) ambient temperature.

9.12.3 Procedure and Requirements—Cut a strip of tubing into a 6.0 mm (0.25 in) wide helical coil equal in length to five times the circumference of the tubing. Bend the helical coil in reverse of coiling so as to expose the braid gap between the outer jacket and the core tube section. Start by working a sharp knife blade into the braid gap to initiate separation, and then attempt to separate the outer jacket from the core tube at the braid interstices. The bonded surface (excluding the braided area) between the outer jacket and core section shall be inseparable for the entire test sample length.

9.13 Heat Aging Adhesion Test[13]

9.13.1 Procedure—Subject samples to Phase 1 of the heat aging test procedure per 9.5.

9.13.2 Requirements—After completion of the Phase 1 procedure, the tubing shall meet the requirements of 9.12.

---

[13] A qualification test.

PERFORMANCE REQUIREMENTS FOR SAE J844 NONMETALLIC TUBING AND FITTING ASSEMBLIES USED IN AUTOMOTIVE AIR BRAKE SYSTEMS— SAE J1131 DEC87

SAE Standard

Report of the Tube, Pipe, Hose, and Lubrication Fittings Committee approved January 1976 and reaffirmed by the Fluid Conductors and Connectors Technical Committee December 1987.

1. Scope

This standard is intended to establish uniform methods of testing SAE J844 tubing and fitting assemblies as used in automotive air brake systems.

This standard also establishes minimum qualifications for tensile and pressure capabilities, vibrational durability under cyclic temperatures, serviceability, and fitting compatibility requirements. The specific tests and performance criteria applicable to the tubing are set forth in SAE J844.

> NOTE: The test values contained in this performance standard are for test purposes only. For environmental and usage limitations see SAE J844. Fittings—A type of fitting for use with SAE J844 nonmetallic tubing is included in SAE J246b; however, it is not intended to restrict or preclude the use of other designs of fittings that comply with this standard.

2. Tension Tests

2.1 Description

Both hot and cold tensile tests shall be conducted with different unaged assemblies (fittings attached within 30 days of test date). Tests consist of subjecting the assembly to increasing tension load in a suitable testing machine until the specified force values or elongation percentages are obtained.

2.2 Apparatus

A tension testing machine with suitable indicating device shall be used for the tension test. The fixtures for holding the test specimens shall be arranged so that the tubing and fittings have a straight center line corresponding to the direction of the machine pull. The lower part of the fixture shall be equipped with a container of sufficient dimensions to submerge the required length of tubing in water. A means of heating the water to boiling shall be provided.

2.3 Test Specimens

Obtain tubing specimens from current production stock and cut to a length sufficient to obtain 6 in ± 0.25 in (152 mm) of tubing between end fittings after assembly. Assemble fittings to the tubing using the manufacturer's recommendations.

2.4 Procedure

2.4.1 HOT PULL—Place the test specimen in the tensile machine with the lower fitting and 4 in $^{+0.25}_{-0.0}$ in (102 mm) of tubing submerged below the surface of the water such that the outside diameter is exposed to the water. Bring the water to a boil and continue boiling for 5 min $^{+0.5}_{-0.0}$ min. Apply load at a rate of pull of 1 in (25 mm) per min.

2.4.2 CONDITIONED PULL TEST—Soak test specimen in air at –40 ± 5°F (–40 ± 3°C) for 30 min $^{+0.5}_{-0.0}$ min, normalize at room temperature, and submerge in boiling water for 15 min. Repeat for a total of four complete cycles. Allow the test specimen to normalize at room temperature for 30 min. Conduct the tensile test within 30 min after the normalizing period while at ambient temperature of 75 ± 5°F (24 ± 3°C). Apply load at a rate of pull of 1 in (25 mm) per min.

2.5 Requirements

The test specimen shall elongate 50%, that is, 6 in (152 mm) increased to 9 in (229 mm), or shall withstand the load shown in the following table, without causing separation from the fitting.

3. Vibration Test

3.1 Description

This test is designed to evaluate the effects of vibration under varying ambient temperatures on a tubing and fitting assembly. Leakage rate is used to gage acceptability.

Nominal Tubing OD in	Nominal Tubing OD mm[a]	Tensile Load lb	Tensile Load N
1/8	3.2	15	67
1/4	6.4	50	222
5/16	7.9	75	334
3/8	9.5	150	667
1/2	12.7	200	890
5/8	15.9	325	1446
3/4	19.0	350	1557

[a] For Reference Only.

3.2 Apparatus

Equipment capable of vibrating one end of the test specimen at 600 cpm through 0.5 in (12.7 mm) displacement in a plane perpendicular to the tube while the other end is held rigid. The distance between the static and vibrating heads is to be such that when the assembly is displaced 0.5 in, no parallel pull to the longitudinal axis of the assembly will occur. The equipment must be capable of controlling the ambient air temperature between –40 ± 5°F (–40 ± 3°C) and 220 ± 5°F (104 ± 3°C) and of applying 120 ± 10 psig (827 ± 69 kPa) dry air to the test lines. A mass flow meter capable of determining air leakage shall be provided.

3.3 Test Specimens

Cut tubing specimens to a length sufficient to obtain 18 in (457 mm) between fittings after assembly. Assemble identical fittings to the tubing using the manufacturer's recommendation. Fitting attaching nuts are not permitted to be retightened during the test.

3.4 Procedure

Allowing 0.5 in (12.7 mm) slack, mount the lines straight in the vibrating machine. Oscillate one end of the lines at 600 ± 20 cycles per min through a total stroke of 0.5 in (12.7 mm) for a total of 1 000 000 $^{+50\,000}_{-0.0}$ cycles, while maintaining an internal pressure of 120 ± 10 psig (827 ± 69 kPa) using dry air. Starting at 220 ± 5°F (104 ± 3°C), vary the ambient air temperature from 220 ± 5°F (104 ± 3°C) to –40 ± 5°F (–40 ± 3°C) at 250 000 vibration cycle intervals (approximately 7 h intervals). Using a mass flow meter observe for fitting leakage during and after the test. Check nut tightness after completing the test.

3.5 Requirements

The test specimen is considered a failure if leakage exceeds 50 cm³/min at –40 ± 5°F (–40 ± 3°C) or 25 cm³/min at 70 ± 5°F (21 ± 3°C). The fitting is considered a failure if the attaching nut becomes loose. This is defined as follows:

1. Record the initial tightening torque.
2. At the conclusion of the test, attempt to tighten the nut further by applying 20% of the initial tightening torque in the tightening direction. Do not apply a higher torque and do not apply any torque or force in the loosening direction.
3. If the nut moves at all under the 20% torque, it shall be defined as a loose nut and failure of the test. Record the movement in degrees to reach 20%.

4. Proof and Burst Pressure Test

4.1 Description

This test is intended to evaluate fitting retention at proof pressure (50% of minimum burst) and at minimum burst pressure as listed in the latest issue of SAE J844.

4.2 Apparatus

The test apparatus consists of a suitable source of hydraulic pressure and the necessary gages and piping.

4.3 Test Specimen
Cut tubing specimens to obtain 12 in (305 mm) between fittings after assembly. Assemble fittings to the tubing using the manufacturer's recommendations.

4.4 Procedure
Plug one end of the test specimen and mount in the apparatus with the end unrestrained. Apply proof pressure at room temperature, 75 ± 5°F (24 ± 3°C) to the test specimen and hold for 30 s. Increase pressure at a constant rate so as to reach the specified minimum burst pressure within a time period of 3–15 s.

4.5 Requirements
Fittings shall not separate from the tubing nor shall the assembly visibly leak at less than specified minimum burst pressure.

5. Serviceability Test

5.1 Description
This test is intended to evaluate the effects of repeated assembly and disassembly of a tubing and fitting assembly. Leakage rate is used to gage acceptability.

5.2 Apparatus
The test apparatus consists of a suitable source of pneumatic pressure and the necessary gages and piping. A mass flow meter capable of determining air leakage shall be provided.

5.3 Test Specimens
Cut tubing specimens to obtain 12 in (305 mm) between fittings after assembly. Assemble fittings to the tubing using manufacturer's recommendations.

5.4 Procedure
The tubing and fitting connections shall be disassembled and reassembled for a minimum of five times. After the fifth reassembly, pressurize the test specimens with air to 120 ± 10 psig (827 ± 69 kPa) at room temperature, 75 ± 5°F (24 ± 3°C), and check for leakage.

5.5 Requirements
Leakage rate must not exceed 25 cm^3/min.

6. Fitting Compatability Test

6.1 Description
This test is intended to evaluate the effects of high and low temperatures on fitting performance.

6.2 Apparatus
The test apparatus consists of a suitable source of hydraulic pressure, 450 ± 10 psig (3103 ± 69 kPa), and necessary gages and piping in environmental test chambers at 200 ± 5°F (93 ± 3°C) and –40 ± 5°F (–40 ±3°C).

6.3 Test Specimens
Cut tubing specimens to obtain 12 in (305 mm) between fittings after assembly. Assemble fittings to the tubing using manufacturer's recommendations.

6.4 Procedure
Fill test specimens with hydraulic fluid and subject to 200 ± 5°F (93 ± 3°C), and atmospheric pressure for 24 h $^{+1}_{-0.0}$ h then apply internal pressure of 450 ± 10 psig (3103 ± 69 kPa) for 5 min while maintaining temperature of 200 ± 5°F (93 ± 3°C). Reduce to atmospheric pressure and permit test specimens to return to room temperature; then subject test specimens to –40 ± 5°F (–40 ± 3°C) and atmospheric pressure for 24 h. Apply internal pressure of 450 ± 10 psig (3103 ± 69 kPa) for 5 min $^{+0.5}_{-0.0}$ min while maintaining temperature of –40 ± 5°F (–40 ± 3°C).

6.5 Requirements
Tubing shall not rupture or disconnect from the fittings.

AUTOMOTIVE PIPE FITTINGS—SAE J530 JUN93 SAE Standard

Report of the Parts and Fittings Committee, approved February 1948, revised by the Tube, Pipe, Hose, and Lubrication Fittings Committee December 1973, editorial change January 1981. Revised by the SAE Fluid Conductors and Connectors Technical Committee SC1—Automotive and Hydraulic Tube and Fittings June 1992. Rationale statement available. Revised by the SAE Fluid Conductors and Connectors Technical Committee SC1—Automotive and Hydraulic Tube and Fittings June 1993. Rationale statement available.

1. Scope
This SAE Standard includes complete general and dimensional specifications for those types of pipe fittings commonly used in the automotive and other mass production industries where the use of lubricants or sealers is objectionable. The automotive pipe fittings shown in Figures 1 to 17 and Tables 1 to 6 are intended for general automotive and similar applications involving low or medium pressures or in conjunction with automotive tube fittings in piping systems.

2. References

2.1 Applicable Documents
The following publications form a part of this specification to the extent specified herein. The latest issue of SAE publications shall apply.

2.1.1 SAE PUBLICATIONS—Available from SAE, 400 Commonwealth Drive, Warrendale, PA 15096-0001.

SAE J476—Dryseal Pipe Threads

SAE J846—Coding Systems for Identification of Fluid Conductors and Connectors

2.1.2 ASTM PUBLICATION—Available from ASTM, 1916 Race Street, Philadelphia, PA 19103-1187.

ASTM B 117—Method of Salt Spray (Fog) Testing

3. General Specifications

3.1 Dimensions and Tolerances
Except for nominal sizes and thread specifications, dimensions and tolerances are given in both SI and U.S. customary units as designated. Tabulated dimensions shall apply to the finished fittings, plated or otherwise processed, as specified by the purchaser. Unless otherwise specified, maximum and minimum across flats dimensions shall be within the commercial tolerance of bar or extruded stock from which the fittings are produced. The minimum across corner dimensions of external hexagons shall be 1.092 times the nominal width across flats, but shall not result in a side flat width less than 0.43 times the nominal width across flats. The minimum across corner dimensions of external squares shall be 1.25 times the nominal width across flats, but shall not result in a side flat width less than 0.75 times the nominal width across flats. Unless otherwise specified, tolerance on hole diameters designated drill in the dimensional tables shall be as tabulated in Table 7.

Tolerance on all dimensions not otherwise limited shall be ±0.25 mm (±0.010 in). Angular tolerance on axis of ends on elbows and tees shall be ±2.50 degrees for sizes up to and including 3/8 in, and ±1.50 degrees for sizes larger than 3/8 in.

3.2 Wall Thickness
Unless otherwise designated, the wall thickness at any point on fittings shall not be less than the thickness established by the specified dimensions, tolerances, and eccentricities for inner and outer surfaces.

AUTOMOTIVE PIPE FITTINGS

NOTES: UNSPECIFIED DETAIL WITH RESPECT TO DIMENSIONS, TOLERANCES, CONTOURS, MATERIAL, WORKMANSHIP, ETC., MUST CONFORM TO GENERAL SPECIFICATIONS FOR AUTOMOTIVE PIPE FITTINGS. CODES SHOWN IN BRACKETS ADJACENT TO FIGURE NUMBERS REPRESENT RESPECTIVE FITTING IDENTIFICATION IN ACCORDANCE WITH SAE J846 (FEBRUARY, 1979).

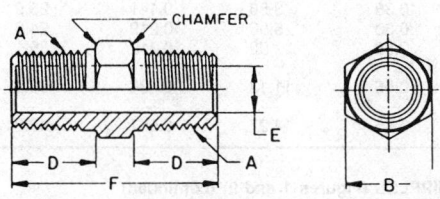

FIGURE 1—HEXAGON NIPPLE (130137)

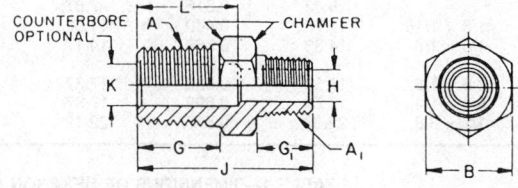

FIGURE 2—HEXAGON REDUCER NIPPLE (130137)

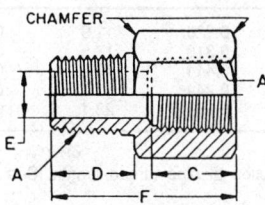

FIGURE 3—ADAPTER (130139)

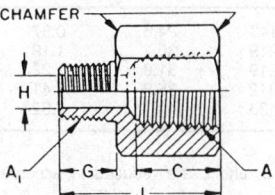

FIGURE 4—REDUCER ADAPTER (130139)

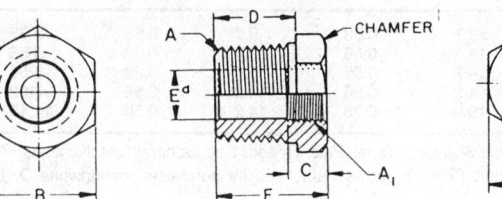

FIGURE 5—REDUCER BUSHING (130140)

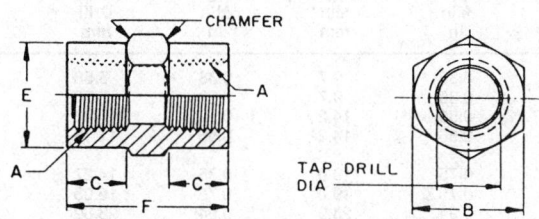

FIGURE 6—COUPLING (130138)

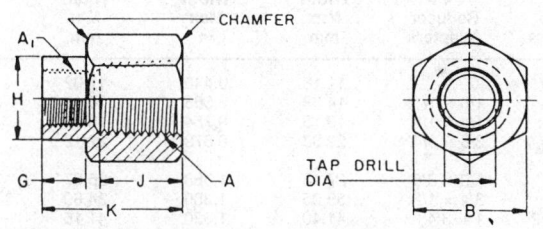

FIGURE 7—REDUCER COUPLING (130138)

3.3 Contour—Details of contour shall be optional with the manufacturer provided the tabulated dimensions are maintained and serviceability of the fittings is not impaired. Wrench flats on elbows and tees shall be optional. Where extruded or forged shapes are reduced to conserve material, the wall thickness, unless otherwise specified, shall not be less than the respective minimum values tabulated in Table 8.

3.4 Passages—Where passages in straight fittings are machined from opposite ends, the offset at the meeting point shall not exceed 0.38 mm (0.015 in). The cross-sectional area at the junction of passages in angle fittings shall not be less than that of the smaller passage.

3.5 Pipe Threads—The pipe threads, unless there is specific authorization to the contrary, shall conform with the Dryseal American Standard Taper Pipe Thread (NPTF). At purchaser's option, the pipe thread may be shortened in conformity with the SAE Short Dryseal Taper Pipe Thread (PTF-SAE Short). Specifications for pipe threads are given in detail in SAE J476 (June, 1961). The pipe fitting dimensions tabulated herein are based on length of the Dryseal American Standard Taper Pipe Thread (NPTF), it being the consensus of manufacturers and users that trouble-free assembly and pressure-tight joints without lubricant or sealer cannot be assured unless a full-length thread is used.

However, the tap drill depths and the overall lengths specified in the tables for fittings with internal taper pipe threads are not consistent with the tap drill depths and the overall thread lengths of the Dryseal American Standard Taper Pipe Threads (NPTF) specified in Table A2, Appendix A of SAE J476. The full-length Dryseal American Standard Taper Pipe Taps specified in Table B2, Appendix B of SAE J476 cannot be used as the tap drill depths, and overall lengths of the fittings have been reduced to the minimum required by bottoming taps to produce standard full thread length. The deviations described herein are peculiar to automotive pipe fittings. As special tooling is required, caution should be exercised in specifying the deviations for any other products. External pipe threads shall be chamfered from the diameters tabulated in Table 9 to produce the specified length of chamfered or partial thread. Internal pipe threads shall be countersunk 90 degrees included angle, to the diameters shown in Table 9.

TABLE 1—DIMENSIONS OF HEXAGON NIPPLES AND REDUCER NIPPLES (Figures 1 and 2)

Dryseal Taper Thread NPTF[a] in A Hexagon Nipples	Dryseal Taper Thread NPTF[a] in A × A₁ Hexagon Reducer Nipples	All Nipples B Hexagon Width Max mm	All Nipples B Hexagon Width Max in	All Nipples B Hexagon Width Min mm	All Nipples B Hexagon Width Min in	Nipples D Shoulder Length[b] mm	Nipples D Shoulder Length[b] in	Nipples E Drill Dia mm	Nipples E Drill Dia in	Nipples F Overall Length[b] mm	Nipples F Overall Length[b] in
1/16-27	—	8.03	0.316	7.87	0.310	9.7	0.38	3.58	0.141	23.9	0.94
1/8-27	1/8 × 1/16	11.18	0.440	11.02	0.434	9.7	0.38	5.56	0.219	24.6	0.97
1/4-18	1/4 × 1/8	14.38	0.566	14.17	0.558	14.2	0.56	7.92	0.312	35.1	1.38
3/8-18	3/8 × 1/8	17.58	0.692	17.37	0.684	14.2	0.56	11.13	0.438	35.8	1.41
—	3/8 × 1/4	17.58	0.692	17.37	0.684	—	—	—	—	—	—
1/2-14	1/2 × 3/8	22.33	0.879	22.12	0.871	19.0	0.75	14.27	0.562	46.0	1.81

TABLE 1—DIMENSIONS OF HEXAGON NIPPLES AND REDUCER NIPPLES (Figures 1 and 2) (continued)

Dryseal Taper Thread NPTF[a] in A × A₁ Hexagon Reducer Nipples	Reducer Nipples G Shoulder Length[b] Min mm	Reducer Nipples G Shoulder Length[b] Min in	Reducer Nipples G₁ Shoulder Length[b] Min mm	Reducer Nipples G₁ Shoulder Length[b] Min in	Reducer Nipples H Drill Dia[c] mm	Reducer Nipples H Drill Dia[c] in	Reducer Nipples J Overall Length[b] mm	Reducer Nipples J Overall Length[b] in	Reducer Nipples Counterbore K Max Dia[c] mm	Reducer Nipples Counterbore K Max Dia[c] in	Reducer Nipples Counterbore L Max Depth[b,c] mm	Reducer Nipples Counterbore L Max Depth[b,c] in
1/8 × 1/16	9.7	0.38	9.7	0.38	3.58	1.141	24.6	0.97	5.66	0.223	11.9	0.47
1/4 × 1/8	14.2	0.56	9.7	0.38	5.56	0.219	30.2	1.19	8.08	0.310	17.5	0.69
3/8 × 1/8	14.2	0.56	9.7	0.38	5.56	0.219	31.0	1.22	11.28	0.444	17.5	0.69
3/8 × 1/4	14.2	0.56	14.2	0.56	7.92	0.312	35.8	1.41	11.28	0.444	17.5	0.69
1/2 × 3/8	19.0	0.75	14.2	0.56	11.13	0.438	41.1	1.62	14.43	0.568	23.1	0.91

[a] Dryseal American Standard Taper Pipe Thread. See General Specifications.
[b] Where SAE Short Pipe Thread is authorized by purchaser, dimensions D, F, G, G₁, J, and L are reduced in accordance with reduction of pipe thread length. See General Specifications.
[c] At manufacturers option, through passages may conform with the smaller diameter specified or be counterbored to the larger diameter for the depth specified.

TABLE 2—DIMENSIONS OF ADAPTERS AND REDUCER ADAPTERS (Figures 3 and 4)

Dryseal Taper Thread NPTF[a] in A Adapters	Dryseal Taper Thread NPTF[a] in A × A₁ Reducer Adapters	All Adapters B Hexagon Width Max mm	All Adapters B Hexagon Width Max in	All Adapters B Hexagon Width Min mm	All Adapters B Hexagon Width Min in	All Adapters C Tap Drill Depth[b,c] Min mm	All Adapters C Tap Drill Depth[b,c] Min in	Adapters D Shoulder Length[b] Min mm	Adapters D Shoulder Length[b] Min in	Adapters E Dia Drill mm	Adapters E Dia Drill in
1/16-27	—	11.18	0.440	11.02	0.434	9.7	0.38	9.7	0.38	3.58	0.141
1/8-27	1/8 × 1/16	14.38	0.566	14.17	0.558	9.7	0.38	9.7	0.38	5.56	0.219
1/4-18	1/4 × 1/8	19.15	0.754	18.95	0.746	14.2	0.56	14.2	0.56	7.92	0.312
3/8-18	3/8 × 1/4	22.33	0.879	22.12	0.871	14.2	0.56	14.2	0.56	11.13	0.438
1/2-14	1/2 × 3/8	27.13	1.068	26.87	1.058	19.0	0.75	19.0	0.75	14.27	0.562
3/4-14	3/4 × 1/2	35.05	1.380	34.80	1.370	19.0	0.75	19.0	0.75	19.05	0.750
1-11-1/2	1 × 3/4	41.40	1.630	41.15	1.620	23.9	0.94	23.9	0.94	23.82	0.938

TABLE 2—DIMENSIONS OF ADAPTERS AND REDUCER ADAPTERS (Figures 3 and 4) (continued)

Dryseal Taper Thread NPTF[a] in A Adapters	Adapters Dryseal Taper Thread NPTF[a] in A × A₁ Reducer Adapters	Adapters F Overall Length[b] mm	Adapters F Overall Length[b] in	Reducer Adapters G Shoulder Length[b] Min mm	Reducer Adapters G Shoulder Length[b] Min in	Reducer Adapters H Dia Drill mm	Reducer Adapters H Dia Drill in	Adapters H Overall Length[b] mm	Adapters H Overall Length[b] in	Adapters J Overall Length[b] mm	Adapters J Overall Length[b] in
1/16-27	—	21.3	0.84	—	—	3.58	0.141	21.3	0.84		
1/8-27	1/8 × 1/16	22.4	0.88	9.7	0.38	5.56	0.219	26.9	1.06		
1/4-18	1/4 × 1/8	31.8	1.25	9.7	0.38	7.92	0.312	31.8	1.25		
3/8-18	3/8 × 1/4	31.8	1.25	14.2	0.56	11.13	0.438	37.3	1.47		
1/2-14	1/2 × 3/8	42.2	1.66	14.2	0.56	14.27	0.562	42.9	1.69		
3/4-14	3/4 × 1/2	42.9	1.69	19.0	0.75	19.05	0.750	47.8	1.88		
1-11-1/2	1 × 3/4	52.3	2.06	19.0	0.75						

[a] Dryseal American Standard Taper Pipe Thread. See General Specifications.
[b] Where SAE Short Pipe Thread is authorized by purchaser, dimensions C, F, G, and J are reduced in accordance with reduction of pipe thread length. See General Specifications.
[c] Tap drill depths given require use of bottoming taps to produce standard full thread lengths. See General Specifications.

AUTOMOTIVE PIPE FITTINGS—CAST TYPE

NOTES: UNSPECIFIED DETAIL WITH RESPECT TO DIMENSIONS, TOLERANCES, CONTOURS, MATERIAL, WORKMANSHIP, ETC., MUST CONFORM TO GENERAL SPECIFICATIONS FOR AUTOMOTIVE PIPE FITTINGS. THE DIMENSIONAL DESIGNATIONS ON THE FIRST FIGURE IN EACH GROUP SHALL APPLY TO ALL OTHER FIGURES IN THAT GROUP EXCEPT AS SHOWN OTHERWISE. CODES SHOWN IN BRACKETS ADJACENT TO FIGURE NUMBERS REPRESENT RESPECTIVE FITTING IDENTIFICATION IN ACCORDANCE, WITH SAE J846 (FEBRUARY, 1979).

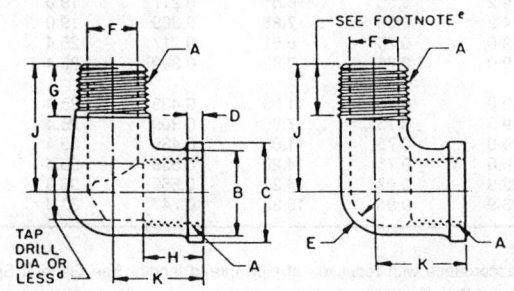

FIGURE 8—90 DEGREE STREET ELBOWS (130239)

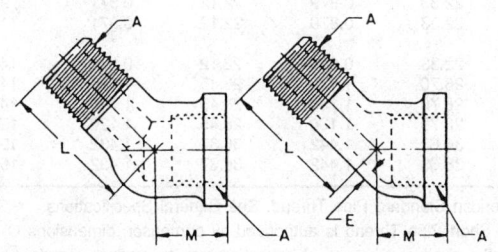

FIGURE 9—45 DEGREE STREET ELBOWS (130339)

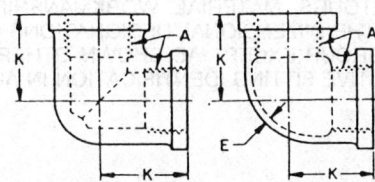

FIGURE 10—90 DEGREE PIPE ELBOWS (130238)

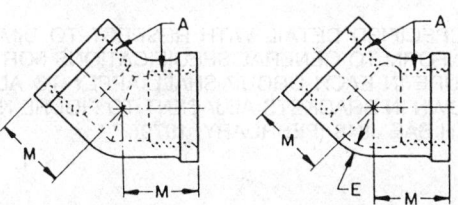

FIGURE 11—45 DEGREE PIPE ELBOWS (130338)

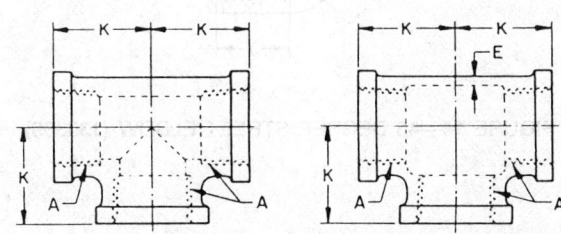

FIGURE 12A—INTERNAL, INTERNAL, INTERNAL TEES (130438)

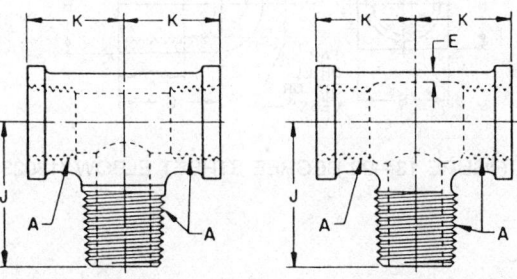

FIGURE 12B—INTERNAL, INTERNAL, EXTERNAL TEES (130425)

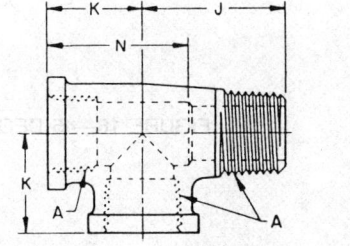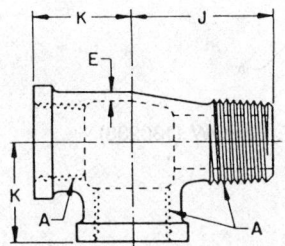

FIGURE 12C—INTERNAL, EXTERNAL, INTERNAL TEES (130424)

TABLE 3—DIMENSIONS OF REDUCER BUSHINGS (Figure 5)

Dryseal Taper Thread NPTF[a], in A × A₁	B Hexagon Width Max mm	B Hexagon Width Max in	B Hexagon Width Min mm	B Hexagon Width Min in	C Tap Drill Depth[b,c] Min mm	C Tap Drill Depth[b,c] Min in	D Shoulder Length[b] Min mm	D Shoulder Length[b] Min in	E Hole Dia[d] Min mm	E Hole Dia[d] Min in	F Overall Length[b] mm	F Overall Length[b] in
1/8 × 1/16	11.18	0.440	11.02	0.434	9.7	0.38	9.7	0.38	3.53	0.139	14.2	0.56
1/4 × 1/8	14.38	0.566	14.17	0.558	9.7	0.38	14.2	0.56	5.51	0.217	19.0	0.75
3/8 × 1/8	17.58	0.692	17.37	0.684	9.7	0.38	14.2	0.56	5.51	0.217	19.0	0.75
3/8 × 1/4	19.15	0.754	18.95	0.746	14.2	0.56	14.2	0.56	7.85	0.309	19.0	0.75
1/2 × 1/8	22.33	0.879	22.12	0.871	9.6	0.38	19.0	0.75	5.51	0.217	25.4	1.00
1/2 × 1/4	22.33	0.879	22.12	0.871	14.2	0.56	19.0	0.75	7.85	0.309	25.4	1.00
1/2 × 3/8	22.33	0.879	22.12	0.871	14.2	0.56	19.0	0.75	11.05	0.435	25.4	1.00
3/4 × 1/4	28.70	1.130	28.45	1.120	14.2	0.56	19.0	0.75	7.85	0.309	25.4	1.00
3/4 × 3/8	28.70	1.130	28.45	1.120	14.2	0.56	19.0	0.75	11.05	0.435	25.4	1.00
3/4 × 1/2	28.70	1.130	28.45	1.120	19.0	0.75	19.0	0.75	14.20	0.559	25.4	1.00
1 × 1/2	36.63	1.442	36.37	1.432	19.0	0.75	23.9	0.94	14.20	0.559	33.3	1.31
1 × 3/4	36.63	1.442	36.37	1.432	19.0	0.75	23.9	0.94	18.98	0.747	33.3	1.31

[a] Dryseal American Standard Pipe Thread. See General Specifications.
[b] Where SAE Short Pipe Thread is authorized by purchaser, dimensions C, D, and F are reduced in accordance with reduction of pipe thread length. See General Specifications.
[c] Tap drill depths given require use of bottoming taps to produce standard full thread lengths. See General Specifications.
[d] At manufacturer's option, hole may conform to tap drill diameter or may be reduced beyond tap drill depth C, but in no case shall it be smaller than E diameter specified.

AUTOMOTIVE PIPE FITTINGS—EXTRUDED OR BAR STOCK TYPE

NOTES: UNSPECIFIED DETAIL WITH RESPECT TO DIMENSIONS, TOLERANCES, CONTOURS, MATERIAL, WORKMANSHIP, ETC., MUST CONFORM TO GENERAL SPECIFICATIONS FOR AUTOMOTIVE PIPE FITTINGS. THE DIMENSIONAL DESIGNATIONS ON THE FIRST FIGURE IN EACH GROUP SHALL APPLY TO ALL OTHER FIGURES IN THAT GROUP EXCEPT AS SHOWN OTHERWISE. CODES SHOWN IN BRACKETS ADJACENT TO FIGURE NUMBERS REPRESENT RESPECTIVE FITTING IDENTIFICATION IN ACCORDANCE WITH SAE J846 (FEBRUARY, 1979).

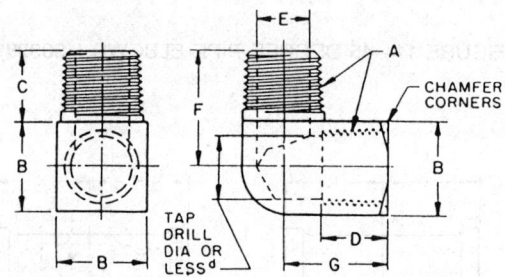

FIGURE 13—90 DEGREE STREET ELBOW (130239)

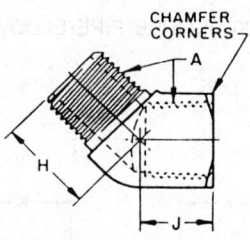

FIGURE 14—45 DEGREE STREET ELBOW (130339)

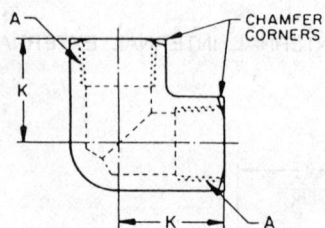

FIGURE 15—90 DEGREE PIPE ELBOW (130238)

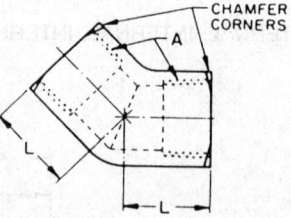

FIGURE 16—45 DEGREE, PIPE ELBOW (130338)

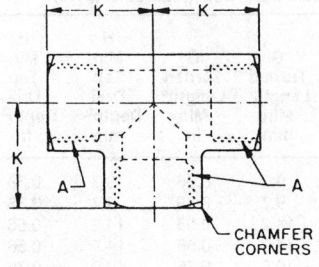

FIGURE 17A—INTERNAL, INTERNAL, EXTERNAL (130438)

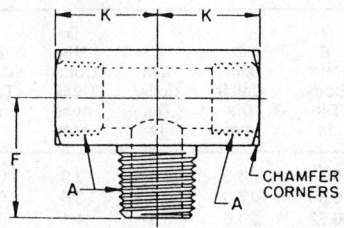

FIGURE 17B—INTERNAL, INTERNAL, INTERNAL (130425)

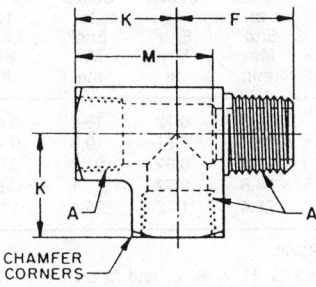

FIGURE 17C—INTERNAL, EXTERNAL, INTERNAL (130424)

FIGURE 17—TEES

TABLE 4—DIMENSIONS OF COUPLINGS AND REDUCER COUPLINGS (Figures 6 and 7)

Dryseal Taper Thread NPTF[a], in A Coupling	Dryseal All Taper Thread NPTF[a], in $A \times A_1$ Reducer Coupling	All Couplings B Hexagon Width Max mm	All Couplings B Hexagon Width Max in	All Couplings B Hexagon Width Min mm	Couplings B Hexagon Width Min in	Couplings C Shoulder Length[b] mm	Couplings C Shoulder Length[b] in	Couplings E Min Body Dia mm	Couplings E Min Body Dia in	Couplings F Overall Length[b] mm	Couplings F Overall Length[b] in
1/16-27	—	11.18	0.440	11.02	0.434	7.1	0.28	11.2	0.44	19.0	0.75
1/8-27	1/8 × 1/16	14.38	0.566	14.17	0.558	6.9	0.27	14.2	0.56	19.0	0.75
1/4-18	1/4 × 1/8	19.15	0.754	18.95	0.746	11.2	0.44	19.0	0.75	28.4	1.12
3/8-18	3/8 × 1/8	22.33	0.879	22.12	0.871	10.7	0.42	22.4	0.88	28.4	1.1
—	3/8 × 1/4	22.33	0.879	22.12	0.871	—	—	—	—	—	—
1/2-4	1/2 × 1/8	27.13	1.068	26.87	1.058	15.0	0.59	26.9	1.06	38.1	1.50
—	1/2 × 1/4	27.13	1.068	26.87	1.058	—	—	—	—	—	—
—	1/2 × 3/8	27.13	1.068	26.87	1.058	—	—	—	—	—	—

TABLE 4—DIMENSIONS OF COUPLINGS AND REDUCER COUPLINGS (Figures 6 and 7) (continued)

Dryseal Taper Thread NPTF[a], in A Coupling	Dryseal Taper Thread NPTF[a], in $A \times A_1$ Reducer Coupling	Reducer Couplings G Shoulder Length[b] mm	Reducer Couplings G Shoulder Length[b] in	Reducer Couplings H Min Body Dia mm	Reducer Couplings H Min Body Dia in	Reducer Couplings J Min Tap Drill Depth[b,c] mm	Reducer Couplings J Min Tap Drill Depth[b,c] in	Reducer Couplings K Overall Length[b] mm	Reducer Couplings K Overall Length[b] in
1/16-27	—	—	—	—	—	—	—	—	—
1/8-27	1/8 × 1/16	7.9	0.31	12.7	0.50	9.7	0.38	19.8	0.78
1/4-18	1/4 × 1/8	7.9	0.31	14.2	0.56	14.2	0.56	24.6	0.97
3/8-18	3/8 × 1/8	6.4	0.25	14.2	0.56	14.2	0.56	23.9	0.94
—	3/8 × 1/4	11.9	0.47	19.0	0.75	14.2	0.56	29.5	1.16
1/2-4	1/2 × 1/8	6.4	0.25	14.2	0.56	19.0	0.75	30.2	1.19
—	1/2 × 1/4	8.6	0.34	19.0	0.75	19.0	0.75	32.5	1.28
—	1/2 × 3/8	11.2	0.44	22.4	0.88	19.0	0.75	35.1	1.38

[a] Dryseal American Standard Taper Pipe Thread. See General Specifications.
[b] Where SAE Short Pipe Thread is authorized by purchaser, dimensions C, F, G, J, and K are reduced in accordance with reduction of pipe thread length. See General Specifications.
[c] Tap drill depths given require use of bottoming taps to produce standard full thread length. See General Specifications.

TABLE 5—DIMENSIONS OF CAST TYPE STREET ELBOWS, PIPE ELBOWS, AND PIPE TEES (Figures 8 to 12)

A Dryseal Taper Thread NPTF[a], in	B Min Body Dia mm	B Min Body Dia in	C Min Collar Dia mm	C Min Collar Dia in	D Min Collar Thickness mm	D Min Collar Thickness in	E Min Wall Thickness mm	E Min Wall Thickness in	F Drill Dia[f] mm	F Drill Dia[f] in	G Turned Length[b] Min mm	G Turned Length[b] Min in	H Min Tap Drill Depth[b,c] mm	H Min Tap Drill Depth[b,c] in	J Center to End Max mm	J Center to End Max in
1/16-27	11.2	0.44	13.5	0.53	3.0	0.12	2.0	0.08	3.58	0.141	9.7	0.38	9.7	0.38	21.3	0.84
1/8-27	14.2	0.56	17.0	0.67	3.6	0.14	2.0	0.08	5.56	0.219	9.7	0.38	9.7	0.38	24.1	0.95
1/4-18	19.3	0.72	20.6	0.81	4.1	0.16	2.0	0.08	7.92	0.312	14.2	0.56	14.2	0.56	29.2	1.15
3/8-18	22.4	0.88	25.4	1.00	4.3	0.17	2.3	0.09	11.13	0.438	14.2	0.56	14.2	0.56	33.0	1.30
1/2-14	26.2	1.03	29.7	1.17	4.8	0.19	2.3	0.09	14.27	0.562	19.0	0.75	19.0	0.75	39.6	1.56

TABLE 5—DIMENSIONS OF CAST TYPE STREET ELBOWS, PIPE ELBOWS, AND PIPE TEES (Figures 8 to 12) (continued)

A Dryseal Taper Thread NPTF[a], in	J Center to End[b] Min mm	J Center to End[b] Min in	K Center to End[b] Max mm	K Center to End[b] Max in	K Center to End[b] Min mm	K Center to End[b] Min in	L Center to End[b] Max mm	L Center to End[b] Max in	L Center to End[b] Min mm	L Center to End[b] Min in	M Center to End[b] Max mm	M Center to End[b] Max in	M Center to End[b] Min mm	M Center to End[b] Min in	N Drill Depth mm	N Drill Depth in
1/16-27	19.8	0.78	13.5	0.53	11.9	0.47	18.3	0.72	16.8	0.66	11.2	0.44	9.7	0.38	16.8	0.66
1/8-27	22.6	0.89	14.7	0.58	13.2	0.52	20.6	0.81	19.0	0.75	11.4	0.45	9.9	0.39	19.0	0.75
1/4-18	27.2	1.07	19.3	0.76	17.3	0.68	23.4	0.92	21.3	0.84	15.2	0.60	13.2	0.52	24.6	0.97
3/8-18	30.5	1.20	22.4	0.88	19.8	0.78	24.6	0.97	22.1	0.87	17.0	0.67	14.5	0.57	—	—
1/2-14	36.6	1.44	27.4	1.08	24.4	0.96	28.4	1.12	25.4	1.00	21.3	0.84	18.3	0.72	—	—

[a] Dryseal American Standard Taper Pipe Thread. See General Specifications.
[b] Where SAE Short Pipe Thread is authorized by purchaser, dimensions G, H, J, K, L, and M are reduced in accordance with reduction of pipe thread length. See General Specifications.
[c] Tap drill depths given require use of bottoming taps to produce standard full thread length. See General Specifications.
[d] Hole diameters may be reduced beyond tap drill depth H, but shall not be less than F specified for corresponding size. (See Figure 8.)
[e] Minimum pipe thread length where body is relieved or undercut shall not be shorter than L_2 plus one turn (thread) full thread. Thread length may be reduced one pitch (thread) if thread is cut through into relief or undercut. See SAE J476 and Figure 8.
[f] 1/16, 1/8, and 1/4 in size cast fittings are generally produced from solid castings and have drilled passage holes, 3/8 and 1/2 in size cast fittings are generally produced with cored passage holes and may have internal minimum full thread length of 9.1 and 10.9 mm (0.36 and 0.43 in), respectively.

TABLE 6—DIMENSIONS OF EXTRUDED AND FORGED TYPE STREET ELBOWS, PIPE ELBOWS, AND PIPE TEES (Figures 13 to 17)

A Dryseal Taper Thread NPTF[a], in	B Body Size mm	B Body Size in	C Turned Length[b] Min mm	C Turned Length[b] Min in	D Min Tap Drill Depth[b,c] mm	D Min Tap Drill Depth[b,c] in	E Drill Dia[d] mm	E Drill Dia[d] in	F Center to End[b] ±0.8 mm	F Center to End[b] ±0.03 in	G Center to End[b] ±0.8 mm	G Center to End[b] ±0.03 in
1/16-27	11.11	7/16	9.7	0.38	9.7	0.38	3.58	0.141	5.0	0.59	11.4	0.45
1/8-27	14.29	9/16	9.7	0.38	9.7	0.38	5.56	0.219	16.8	0.66	12.2	0.48
1/4-18	17.46	11/16	14.2	0.56	14.2	0.56	7.92	0.312	23.1	0.91	18.3	0.72
3/8-18	20.64	13/16	14.2	0.56	14.2	0.56	11.13	0.438	24.6	0.97	19.8	0.78
1/2-14	25.40	1	19.0	0.75	19.0	0.75	14.27	0.562	31.8	1.25	26.2	1.03
3/4-14	31.75	1-1/4	19.0	0.75	19.0	0.75	19.05	0.750	35.1	1.38	28.4	1.12
1-11-1/2	38.10	1-1/2	23.9	0.94	23.9	0.94	23.82	0.938	42.9	1.69	35.8	1.41

TABLE 6—DIMENSIONS OF EXTRUDED AND FORGED TYPE STREET ELBOWS, PIPE ELBOWS, AND PIPE TEES (Figures 13 to 17) (continued)

A Dryseal Taper Thread NPTF[a], in	H Center to End[b] ±0.8 mm	H Center to End[b] ±0.03 in	J Center to End[b] ±0.8 mm	J Center to End[b] ±0.03 in	K Center to End[b] ±0.8 mm	K Center to End[b] ±0.03 in	L Center to End[b] ±0.8 mm	L Center to End[b] ±0.03 in	M Drill Depth mm	M Drill Depth in
1/16-27	11.9	0.47	9.7	0.38	12.7	0.50	11.2	0.44	16.8	0.66
1/8-27	12.7	0.50	9.7	0.38	14.0	0.55	11.4	0.45	19.0	0.75
1/4-18	18.3	0.72	14.2	0.56	19.8	0.78	16.8	0.66	26.2	1.03
3/8-18	19.8	0.78	14.2	0.56	21.3	0.84	17.5	0.69	29.7	1.17
1/2-14	25.4	1.00	19.0	0.75	27.7	1.09	23.1	0.91	37.6	1.48
3/4-14	26.9	1.06	19.0	0.75	29.5	1.16	23.9	0.94	42.2	1.66
1-11-1/2	34.0	1.34	23.9	0.94	38.6	1.52	30.2	1.19	53.8	2.12

[a] Dryseal American Standard Taper Pipe Thread. See General Specifications.
[b] Where SAE Short Pipe Thread is authorized by purchaser, dimensions C, D, F, G, H, J, K, and L are reduced in accordance with reduction of pipe thread length. See General Specifications.
[c] Tap drill depths given require use of bottoming taps to produce standard full thread length. See General Specifications.
[d] Hole diameter may be reduced beyond tap drill depth D but shall not be less than E specified for corresponding size. (See Figure 13.)

3.6 Material and Manufacture—Pipe fittings may be made from cast iron, malleable iron, steel, stainless steel, brass, or aluminum alloy as specified by the purchaser, by casting, forging, milling from the bar, or upsetting from a grade of material free from defects which will affect their serviceability. However, all varieties and sizes of pipe fittings may not be currently available in the aforementioned materials. Nipples, adapters, bushings, and couplings are generally available in brass and steel. Cast elbows and tees are generally available in malleable iron for sizes 1/4 in and over and in brass. Extruded and forged elbows and tees are generally available in brass and steel.

(R) **3.7 Finish**—The external surfaces and threads of all carbon steel parts shall be plated or coated with a suitable material that passes a 72 h salt spray test in accordance with ASTM B 117. Any appearance of red rust during the 72 h salt spray test shall be considered failure, except for the following:

a. All internal fluid passages.

b. Edges such as hex points, serrations, and crests of threads where there may be mechanical deformation of the plating or coating typical of mass-produced parts or shipping effects.

c. Areas where there is mechanical deformation of the plating or coating caused by crimping, flaring, bending, and other post-plate metal forming operations.

d. Areas where the parts are suspended or affixed in the test chamber where condensate can accumulate.

NOTE—Cadmium plating is not preferred due to environmental reasons. Parts manufactured to this standard after January 1, 1997, shall not be cadmium plated. Internal fluid passages shall be protected from corrosion during storage. Changes in plating may affect assembly torques and require requalification, when applicable.

3.8 Workmanship—Workmanship shall conform to the best commercial practice to produce high-quality fittings. Fittings shall be free from all hanging burrs, loose scale, and slivers which might become dislodged in usage and all other defects which might affect serviceability.

TABLE 7—DRILL TOLERANCES

Drill Size Range mm	Drill Size Range in	Tolerance on Hole Diameter Plus mm	Tolerance on Hole Diameter Plus in	Tolerance on Hole Diameter Minus mm	Tolerance on Hole Diameter Minus in
0.343 thru 4.699	0.0135 thru 0.1850	0.08	0.003	0.05	0.002
4.762 thru 6.299	0.1875 thru 0.2480	0.10	0.004	0.05	0.002
6.350 thru 19.050	0.2500 thru 0.7500	0.15	0.006	0.08	0.003
19.25 thru 25.400	0.7579 thru 1.0000	0.18	0.007	0.10	0.004

TABLE 8—MINIMUM WALL THICKNESS

Nominal Pipe Thread Size, in	Wall Thickness Min[a] mm	Wall Thickness Min[a] in
1/16	1.0	0.04
1/8	1.0	0.04
1/4	1.3	0.05
3/8	1.5	0.06
1/2	2.0	0.08
3/4	2.0	0.08
1	2.0	0.08

[a] Applies to reduction to one plane only on extruded shapes.

TABLE 9—PIPE THREAD CHAMFER DIAMETERS

Nominal Pipe Thread Size in	External Thread Chamfer Diameter[a] Max mm	External Thread Chamfer Diameter[a] Max in	External Thread Chamfer Diameter[a] Min mm	External Thread Chamfer Diameter[a] Min in	External Thread Length of Chamfered or Partial Min mm	External Thread Length of Chamfered or Partial Min in	External Thread Length of Chamfered or Partial Max mm	External Thread Length of Chamfered or Partial Max in	Internal Thread Countersink Diameter[a] Min mm	Internal Thread Countersink Diameter[a] Min in	Internal Thread Countersink Diameter[a] Max mm	Internal Thread Countersink Diameter[a] Max in
1/16	5.8	0.23	5.3	0.21	0.94	0.037	1.42	0.056	8.4	0.33	8.9	0.35
1/8	8.1	0.32	7.6	0.30	0.94	0.037	1.42	0.056	10.7	0.42	11.2	0.44
1/4	10.7	0.42	10.2	0.40	1.42	0.056	2.11	0.083	14.0	0.55	14.5	0.57
3/8	14.0	0.55	13.5	0.53	1.42	0.056	2.11	0.083	17.5	0.69	18.0	0.71
1/2	17.3	0.68	16.8	0.66	1.80	0.071	2.72	0.107	21.6	0.85	22.1	0.87
3/4	22.6	0.89	22.1	0.87	1.80	0.071	2.72	0.107	26.9	1.06	27.4	1.08
1	28.4	1.12	27.7	1.09	2.21	0.087	3.30	0.130	34.0	1.34	34.8	1.37

[a] Tabulated diameters conform with Appendix A, SAE J476 (June, 1961).

THREAD SEALANTS—SAE J1615 JUN93

SAE Recommended Practice

Report of the SAE Fluid Conductors and Connectors Technical Committee SC1—Automotive and Hydraulic Tube and Fittings, approved June 1993.

1. Scope—Male pipe threads, including male dryseal pipe threads, when made into assemblies or installed into ports, will generally leak if not covered with a sealant.

This SAE Recommended Practice is intended as a guide to assist designers and/or users in the selection and application of various types of thread sealants. The designers and users must make a systematic review of each type and application and then select the sealant to fulfill the requirements of the application. The following are general guidelines and are not necessarily a complete list.

2. References—There are no referenced publications specified herein.

3. Types of Sealant

3.1 PTFE tape applied as joints are assembled.

3.2 Pre-applied paste.

3.3 Paste applied as joints are assembled.

4. Application of PTFE Tape

4.1 Inspect threads to be sure they are not damaged nor contain slivers, burrs, dirt, or other contaminants.

4.2 Looking from the leading end of the male thread, wrap the tape clockwise circumferentially around the thread. Overlap each spiral wrap of tape approximately 1/2 the width of the tape so that no more than two plies are applied. Be careful to leave the first 1/2 to 1-1/2 threads bare. Each wrap should be wound so that the tape is tight on the threads.

4.3 When assembling, each taped threaded end should be put together two full turns past finger tight on sizes up to 1/2 in male pipe thread. On larger sizes, each threaded end should be put together 1-1/2 to 2-1/2 full turns past finger tight. (Caution—During assembly, shredding of the tape can occur with consequent contamination of the system.)

4.4 Each taped threaded end must be able to pass the test requirements detailed in 7.1 and 8.1. PTFE tape is not recommended for repositioning or reassembly.

5. Application of Pre-Applied Paste

5.1 Inspect threads to be sure they are not damaged nor contain slivers, burrs, dirt, or other contaminants.

5.2 Apply paste evenly, without air pockets, around the circumference of the threaded area, leaving the first 1/2 to 1-1/2 threads unpasted and then extending to completely cover a minimum of the next three threads.

5.3 For recommended weights of sealant, if that method is used or specified, follow manufacturer's recommendation or submit parts to the acceptance test detailed in Section 7 or 8.

5.4 See manufacturer's recommendation for drying times and temperatures. Before use of a threaded part in an assembly, coatings should be firm without being tacky.

5.5 When assembling, each prepasted end should be put together two full turns past finger tight on sizes up to 1/2 in male pipe thread. On larger sizes, each threaded end should be put together 1-1/2 to 2-1/2 full turns past finger tight.

5.6 Each prepasted end must be able to pass the test requirements detailed in Section 7 or 8.

6. Application of Paste at Time of Assembly

6.1 Inspect threads to be sure they are not damaged nor contain slivers, burrs, dirt, or other contaminants.

6.2 Apply paste evenly around the circumference over the first four or five male pipe threads, being careful to avoid air pockets.

6.3 When assembling, each pasted end should be put together two full turns past finger tight on sizes up to 1/2 in male pipe thread. On larger sizes, each threaded end should be put together 1-1/2 to 2-1/2 full turns past finger tight.

6.4 Each pasted end must be able to pass the test requirements detailed in Section 7 or 8.

7. Functional Tests for Pneumatic Applications (When Using Brass or Aluminum Fittings)

7.1 **Leakage After Initial Installation**—Male pipe threads, sealed and assembled into a female pipe thread in accordance with this document, shall not leak when subjected to 0.8 MPa (120 psig) air.

7.2 **Leakage After Reuse**—After 24 h have elapsed, remove the samples used in 7.1 and reassemble them two full turns past finger tight. Subject them to 0.8 MPa (120 psig) air. To pass, no leakage is allowed.

7.3 Repeat 7.2 three additional times, each at 24 h intervals. To pass, no leakage is allowed.

8. Functional Tests for Pneumatic Applications (When Using Steel Fittings)

8.1 **Leakage After Initial Installation**—Male pipe threads, sealed and assembled into a female pipe thread in accordance with this document, shall not leak when subjected to air at the working pressure of the hose or tubing used in the assembly.

8.2 **Leakage After Reuse**—After 24 h have elapsed, remove the samples used in 8.1 and reassemble them two full turns past finger tight. Subject them to air at the working pressure of the hose or tubing used in the assembly. To pass, no leakage is allowed.

8.3 Repeat 8.2 three additional times, each at 24 h intervals. To pass, no leakage is allowed.

9. Assembly Recommendations

9.1 To minimize the possibility of a leaking threaded joint after assembling male to female pipe threads, neither end should be backed out (loosened) once the assembly has been made.

9.2 If positioning of a shaped part like an elbow or a tee must be accomplished, thread the shaped part in approximately 1 to 1-1/2 turns past finger tight and tighten further to the desired position.

AUTOMOTIVE PIPE, FILLER, AND DRAIN PLUGS—SAE J531 JUN93

SAE Standard

Report of the Parts and Fittings Committee, approved February 1948, revised by the Tube, Pipe, Hose, and Lubrication Fittings Committee December 1973, editorial change January 1981. Reaffirmed by the SAE Fluid Conductors and Connectors Technical Committee S1—Automotive and Hydraulic Tube and Fittings February 1992, and revised June 1993. Rationale statement available.

1. Scope
This SAE Standard includes complete general and dimensional specifications for those types of pipe, filler, and drain plugs (shown in Figures 1 to 6 and Tables 1 to 4) commonly used in automotive and related industrial applications.

2. References

2.1 Applicable Documents—The following publications form a part of this specification to the extent specified herein. The latest issue of SAE publications shall apply.

2.1.1 SAE PUBLICATIONS—Available from SAE, 400 Commonwealth Drive, Warrendale PA 15096-0001.

SAE J476—Dryseal Pipe Threads

SAE J846—Coding Systems for Identification of Fluid Conductors and Connectors

2.1.2 ASTM PUBLICATION—Available from ASTM, 1916 Race Street, Philadelphia, PA 19103-1187.

ASTM B 117—Method of Salt Spray (Fog) Testing

3. General Specifications

3.1 Dimensions and Tolerances—Except for nominal sizes and thread specifications, dimensions and tolerances are given in both SI and U.S. customary units as designated. Tabulated dimensions shall apply to the finished plugs, plated, hardened, or otherwise processed, as specified by the purchaser. The minimum across corner dimensions of external hexagons shall be 1.092 times the nominal width across flats. The minimum across corner dimensions of external squares shall be 1.25 times the nominal width across flats, but shall not result in a side flat width less than 0.75 times the nominal width across flats. At maximum material condition, the radii at corners of hexagon and square sockets in broached and upset plugs shall not exceed 0.13 mm (0.005 in). Tolerance on dimensions not otherwise limited shall be ±0.25 mm (±0.010 in).

3.2 Pipe Threads—The pipe threads on automotive pipe plugs, unless there is specific authorization to the contrary, shall conform with the Dryseal American Standard Taper Pipe Thread (NPTF) and be gaged accordingly. The automotive pipe plug dimensions are based on the length of the NPTF thread and are intended for assembly with all types of Dryseal taper and straight internal threads. It is the consensus of manufacturers and users that trouble-free assembly and pressure-tight joints without lubricant or sealer cannot be assured unless a full-length thread is used. The pipe threads on automotive filler and drain plugs, unless there is specific authorization to the contrary, shall conform with the Dryseal SAE Short Taper Pipe Thread (PTF-SAE Short) and be gaged accordingly. The automotive filler and drain plug dimensions are based on the length of the (PTF-SAE Short) thread and are primarily intended for assembly with Dryseal American Standard Taper (NPTF) or Dryseal American Standard Intermediate Straight (NPSI) internal pipe threads in installations where it is desirable to limit the entry of the small end of the plug. Limitations on other applications of this thread are explained in SAE J476.

SQUARE AND HEXAGON HEAD

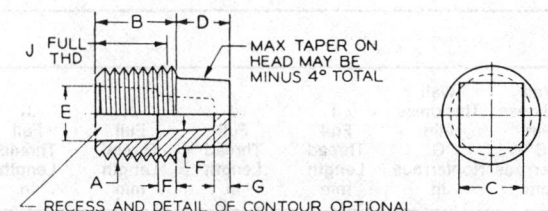

FIGURE 1A—SQUARE INSIDE HEAD PIPE PLUGS (130109A)

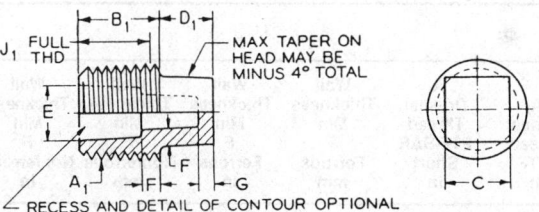

FIGURE 1B—SQUARE INSIDE HEAD FILLER AND DRAIN PLUGS[a] (130109B)

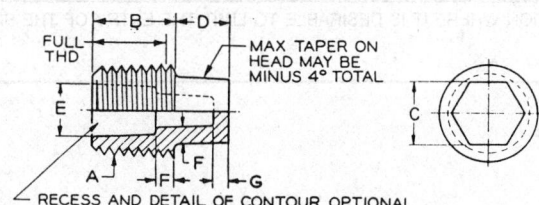

FIGURE 1C—HEXAGON INSIDE HEAD PIPE PLUGS (130109C)

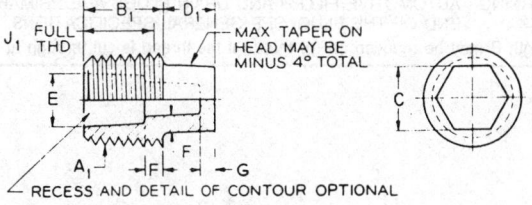

FIGURE 1D—HEXAGON INSIDE HEAD FILLER AND DRAIN PLUGS[a] (130109D)

FIGURE 1—SQUARE AND HEXAGON INSIDE HEAD PLUGS

CODES SHOWN IN BRACKETS ADJACENT TO FIGURE NUMBERS REPRESENT RESPECTIVE FITTING IDENTIFICATION IN ACCORDANCE TO SAE J846.

TABLE 1—CHAMFER DIMENSIONS

Nominal Dryseal Pipe Thread Size in	Chamfer Dia at Small End of Plugs of All Types[a] Max mm	Chamfer Dia at Small End of Plugs of All Types[a] Max in	Chamfer Dia at Small End of Plugs of All Types[a] Min mm	Chamfer Dia at Small End of Plugs of All Types[a] Min in	Chamfer Dia at Large End of Countersunk Headless Plugs Max mm	Chamfer Dia at Large End of Countersunk Headless Plugs Max in	Chamfer Dia at Large End of Countersunk Headless Plugs Min mm	Chamfer Dia at Large End of Countersunk Headless Plugs Min in	Length of Chamfer or Partial Thread Max mm	Length of Chamfer or Partial Thread Max in	Length of Chamfer or Partial Thread Min mm	Length of Chamfer or Partial Thread Min in
1/16	5.8	0.23	5.3	0.21	6.4	0.25	5.8	0.23	1.42	0.056	0.94	0.037
1/8	8.1	0.32	7.6	0.30	8.6	0.34	8.1	0.32	1.42	0.056	0.94	0.037
1/4	10.7	0.42	10.2	0.40	11.4	0.45	10.9	0.43	2.11	0.083	1.42	0.056
3/8	14.0	0.55	13.5	0.53	14.7	0.58	14.2	0.56	2.11	0.083	1.42	0.056
1/2	17.3	0.68	16.8	0.66	18.3	0.72	17.8	0.70	2.72	0.107	1.80	0.071
3/4	22.6	0.89	22.1	0.87	23.6	0.93	23.1	0.91	2.72	0.107	1.80	0.071
1	28.4	1.12	27.7	1.09	29.7	1.17	29.0	1.14	3.30	0.130	2.21	0.087

[a] Tabulated diameters conform with Appendix A, SAE J476.

TABLE 2—DIMENSIONS OF SQUARE AND HEXAGON INSIDE HEAD PIPE, FILLER, AND DRAIN PLUGS (FIGURES 1A TO 1D)[a]

A Dryseal Thread NPTF, in	A_1 Dryseal Thread PTF-SAE Short, in	B Body Length[b] mm	B Body Length[b] in	B_1 Body Length[b] mm	B_1 Body Length[b] in	C Head Width mm	C Head Width in	D Head Height, Square Head mm	D Head Height, Square Head in	D_1 Head Height, Hex Inside Head mm	D_1 Head Height, Hex Inside Head in	E Recess Dia, Max Ferrous mm	E Recess Dia, Max Ferrous in	E Recess Dia, Max Nonferrous mm	E Recess Dia, Max Nonferrous in
1/16-27	1/16-27	8.38	0.330	7.37	0.290	5.44	0.214	4.52	0.178	4.14	0.163	—	—	—	—
1/16-27	1/16-27	8.89	0.350	7.87	0.310	5.61	0.221	4.90	0.193	4.52	0.178	—	—	—	—
1/8-27	1/8-27	8.38	0.330	7.37	0.290	7.01	0.276	6.10	0.240	5.72	0.225	—	—	—	—
1/8-27	1/8-27	8.89	0.350	7.87	0.310	7.19	0.283	6.48	0.255	6.10	0.240	—	—	—	—
1/4-18	1/4-18	12.57	0.495	11.30	0.445	9.40	0.370	7.11	0.280	6.60	0.260	—	—	—	—
1/4-18	1/4-18	13.34	0.525	12.06	0.475	9.58	0.377	7.62	0.300	7.11	0.280	—	—	—	—
3/8-18	3/8-18	12.57	0.495	11.30	0.445	10.87	0.428	7.87	0.310	7.24	0.285	7.9	0.31	9.1	0.36
3/8-18	3/8-18	13.34	0.525	12.06	0.475	11.18	0.440	8.51	0.335	7.87	0.310				
1/2-14	1/2-14	16.76	0.660	14.99	0.590	14.05	0.553	9.65	0.380	8.89	0.350	9.7	0.38	13.5	0.53
1/2-14	1/2-14	17.78	0.700	16.00	0.630	14.35	0.565	10.41	0.410	9.65	0.380				
3/4-14	3/4-14	17.02	0.670	15.24	0.600	15.62	0.615	11.18	0.440	10.41	0.410	14.2	0.56	18.3	0.72
3/4-14	3/4-14	18.03	0.710	16.26	0.640	15.93	0.627	11.94	0.470	11.18	0.440				
1-11-1/2	1-11-1/2	21.08	0.830	19.05	0.750	20.40	0.803	12.70	0.500	11.68	0.460	19.0	0.75	23.6	0.93
1-11-1/2	1-11-1/2	22.10	0.870	20.07	0.790	20.70	0.815	13.72	0.540	12.70	0.500				

A Dryseal Thread NPTF, in	A_1 Dryseal Thread PTF-SAE Short, in	Wall Thickness Min F Ferrous mm	Wall Thickness Min F Ferrous in	Wall Thickness Min F Nonferrous mm	Wall Thickness Min F Nonferrous in	Wall Thickness Min G Ferrous mm	Wall Thickness Min G Ferrous in	Wall Thickness Min G Nonferrous mm	Wall Thickness Min G Nonferrous in	J Full Thread Length mm	J Full Thread Length in	J_1 Full Thread Length mm	J_1 Full Thread Length in
1/16-27	1/16-27	—	—	—	—	—	—	—	—	7.6	0.30	6.6	0.26
1/8-27	1/8-27	—	—	—	—	—	—	—	—	7.6	0.30	6.9	0.27
1/4-18	1/4-18	—	—	—	—	—	—	—	—	11.7	0.46	10.4	0.41
3/8-18	3/8-18	3.3	0.13	2.8	0.11	3.3	0.13	2.0	0.08	11.7	0.46	10.4	0.41
1/2-14	1/2-14	4.1	0.16	3.0	0.12	4.1	0.16	2.3	0.09	15.5	0.61	13.5	0.53
3/4-14	3/4-14	4.6	0.18	3.3	0.13	4.6	0.18	2.5	0.10	15.7	0.62	14.0	0.55
1-11-1/2	1-11-1/2	5.1	0.20	3.6	0.14	5.1	0.20	2.8	0.11	19.6	0.77	17.5	0.69

[a] WARNING—AUTOMOTIVE FILLER AND DRAIN PLUGS ARE PRIMARILY INTENDED FOR INSTALLATION WHERE IT IS DESIRABLE TO LIMIT THE ENTRY OF THE SMALL END OF THE PLUG. SEE GENERAL SPECIFICATIONS.
[b] Length B may be reduced on (p) thread if the thread is cut through at head corners.

HEXAGON OUTSIDE HEAD

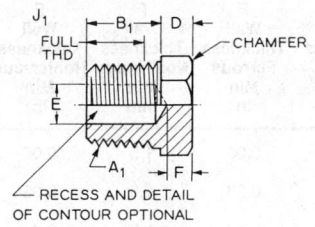
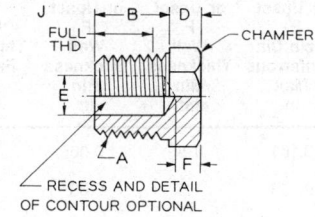
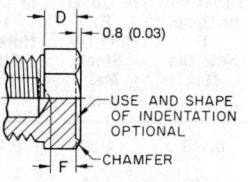
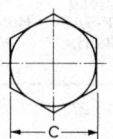

FIGURE 2A—HEXAGON OUTSIDE HEAD PIPE PLUGS (130109E)

FIGURE 2B—HEXAGON OUTSIDE HEAD FILLER AND DRAIN PIPE PLUGS[a] (130109F)

CODES SHOWN IN BRACKETS ADJACENT TO FIGURE NUMBERS REPRESENT RESPECTIVE FITTING IDENTIFICATION IN ACCORDANCE TO SAE J846.

TABLE 3—DIMENSIONS OF HEXAGON OUTSIDE HEAD PIPE, FILLER, AND DRAIN PLUGS[a] (FIGURES 2A AND 2B)

A Dryseal Thread NPTF, in	A_1 Dryseal Thread PTF-SAE Short, in	B Shoulder Length mm	B Shoulder Length in	B_1 Shoulder Length mm	B_1 Shoulder Length in	C Hex (Nom) in	D Head Height mm	D Head Height in	E Recess Dia, Max mm	E Recess Dia, Max in	F Wall Thickness Min mm	F Wall Thickness Min in	J Full Thread mm	J Full Thread in	J_1 Full Thread mm	J_1 Full Thread in
1/16-27	1/16-27	9.4	0.37	8.1	0.32	5/16	3.84	0.151	2.5	0.10	2.3	0.09	7.6	0.30	6.6	0.26
1/16-27	1/16-27	—	—	—	—	—	4.11	0.162	—	—	—	—	—	—	—	—
1/8-27	1/8-27	9.4	0.37	8.1	0.32	7/16	4.60	0.181	4.1	0.16	3.0	0.12	7.6	0.30	6.9	0.27
1/8-27	1/8-27	—	—	—	—	—	4.93	0.194	—	—	—	—	—	—	—	—
1/4-18	1/4-18	14.2	0.56	12.4	0.49	9/16	4.60	0.181	6.4	0.25	3.0	0.12	11.7	0.46	10.4	0.41
1/4-18	1/4-18	—	—	—	—	—	4.93	0.194	—	—	—	—	—	—	—	—
3/8-18	3/8-18	14.2	0.56	12.4	0.49	11/16	5.38	0.212	9.7	0.38	4.1	0.16	11.7	0.46	10.4	0.41
3/8-18	3/8-18	—	—	—	—	—	5.77	0.227	—	—	—	—	—	—	—	—
1/2-14	1/2-14	19.0	0.75	16.3	0.64	7/8	5.38	0.212	12.7	0.50	4.1	0.16	15.5	0.61	13.5	0.53
1/2-14	1/2-14	—	—	—	—	—	5.77	0.227	—	—	—	—	—	—	—	—
3/4-14	3/4-14	19.0	0.75	16.5	0.65	1-1/16	7.72	0.304	17.5	0.69	4.8	0.19	15.7	0.62	14.0	0.55
3/4-14	3/4-14	—	—	—	—	—	8.20	0.323	—	—	—	—	—	—	—	—
1-11-1/2	1-11-1/2	23.9	0.94	20.6	0.81	1-5/16	7.72	0.304	22.4	0.88	4.8	0.19	19.6	0.77	17.5	0.69
1-11-1/2	1-11-1/2	—	—	—	—	—	8.20	0.323	—	—	—	—	—	—	—	—

[a] WARNING—AUTOMOTIVE FILLER AND DRAIN PLUGS ARE PRIMARILY INTENDED FOR INSTALLATION WHERE IT IS DESIRABLE TO LIMIT THE ENTRY OF THE SMALL END OF THE PLUG. SEE GENERAL SPECIFICATIONS.
[b] Length B may be reduced one (p) thread if thread is cut through at head corners.

TABLE 4—DIMENSIONS OF SQUARE AND HEXAGON COUNTERSUNK HEADLESS PIPE PLUGS AND HEADLESS FILLER AND DRAIN PLUGS[a] (FIGURES 3 TO 6)

A Dryseal Thread NPTF, in	A_1 Dryseal Thread PTF-SAE Short, in	B Body Length[b] mm	B Body Length[b] in	B_1 Body Length[b] mm	B_1 Body Length[b] in	C Socket Depth, Min mm	C Socket Depth, Min in	C_1 Socket Depth, Min mm	C_1 Socket Depth, Min in	Broached or Upset D Socket Width[c] mm	Broached or Upset D Socket Width[c] in	Cast D_1 Socket Width[c] mm	Cast D_1 Socket Width[c] in	Broached or Upset D_1[c] Socket Width in	Cast D_1 Socket Width mm
1/16-27	1/16-27	7.37	0.290	6.35	0.250	3.0	0.12	2.3	0.09	3.30	0.130	3.3	0.13	0.156	3.96
1/16-27	1/16-27	7.87	0.310	6.86	0.270	—	—	—	—	3.43	0.135	3.6	0.14	0.161	4.09
1/8-27	1/8-27	7.37	0.290	6.60	0.260	3.0	0.12	2.3	0.09	4.88	0.192	4.8	0.19	0.188	4.78
1/8-27	1/8-27	7.87	0.310	7.11	0.280	—	—	—	—	5.00	0.197	5.3	0.21	0.193	4.90
1/4-18	1/4-18	11.30	0.445	10.03	0.395	4.8	0.19	4.1	0.16	6.48	0.255	6.6	0.26	0.250	6.35
1/4-18	1/4-18	12.06	0.475	10.80	0.425	—	—	—	—	6.60	0.260	7.1	0.28	0.255	6.48
3/8-18	3/8-18	11.30	0.445	10.03	0.395	4.8	0.19	4.1	0.16	8.10	0.319	8.1	0.32	0.313	7.95
3/8-18	3/8-18	12.06	0.475	10.80	0.425	—	—	—	—	8.23	0.324	8.9	0.35	0.318	8.08
1/2-14	1/2-14	14.99	0.590	13.21	0.520	6.4	0.25	4.8	0.19	9.70	0.382	9.7	0.38	0.375	9.52
1/2-14	1/2-14	16.00	0.630	14.22	0.560	—	—	—	—	9.83	0.387	10.4	0.41	0.380	9.65
3/4-14	3/4-14	15.24	0.600	13.46	0.530	7.9	0.31	4.8	0.19	12.90	0.508	13.0	0.51	0.563	14.30
3/4-14	3/4-14	16.26	0.640	14.48	0.570	—	—	—	—	13.03	0.513	13.7	0.54	0.568	14.43
1-11-1/2	1-11-1/2	19.05	0.750	17.02	0.670	9.7	0.38	6.4	0.25	12.90	0.508	13.0	0.51	0.625	15.88
1-11-1/2	1-11-1/2	20.07	0.790	18.03	0.710	—	—	—	—	13.03	0.513	13.7	0.54	0.630	16.00

TABLE 4—DIMENSIONS OF SQUARE AND HEXAGON COUNTERSUNK HEADLESS PIPE PLUGS AND HEADLESS FILLER AND DRAIN PLUGS[a] (FIGURES 3 TO 6)
(CONTINUED)

A Dryseal Thread NPTF, in	A_1 Dryseal Thread PTF-SAE Short, in	Broached or Upset E Hole Dia Max mm	Broached or Upset E Hole Dia Max in	Broached or Upset E_1 Hole Dia Steel Max mm	Broached or Upset E_1 Hole Dia Steel Max in	Broached or Upset E_1 Hole Dia Nonferrous Max mm	Broached or Upset E_1 Hole Dia Nonferrous Max in	Broached or Upset F Wall Thickness Min mm	Broached or Upset F Wall Thickness Min in	Cast F Wall Thickness Ferrous Min mm	Cast F Wall Thickness Ferrous Min in	Cast F Wall Thickness Nonferrous Min mm	Cast F Wall Thickness Nonferrous Min in
1/16-27	1/16-27	3.63	0.143	4.09	0.161	4.09	0.161	1.5	0.06	1.5	0.06	1.5	0.06
1/8-27	1/8-27	5.31	0.209	4.90	0.193	4.90	0.193	1.5	0.06	2.0	0.08	1.5	0.06
1/4-18	1/4-18	7.06	0.278	6.48	0.255	6.63	0.261	1.5	0.06	2.5	0.10	1.8	0.07
3/8-18	3/8-18	8.76	0.345	8.08	0.318	8.20	0.323	1.5	0.06	3.3	0.13	2.0	0.08
1/2-14	1/2-14	10.41	0.410	9.68	0.381	9.80	0.386	2.3	0.09	4.1	0.16	2.3	0.09
3/4-14	3/4-14	14.05	0.553	14.48	0.570	14.48	0.570	2.3	0.09	4.6	0.18	2.5	0.10
1-11-1/2	1-11-1/2	14.05	0.553	16.08	0.633	16.08	0.633	3.0	0.12	5.1	0.20	2.8	0.11

[a] WARNING—AUTOMOTIVE FILLER AND DRAIN PLUGS ARE PRIMARILY INTENDED FOR INSTALLATION WHERE IT IS DESIRABLE TO LIMIT THE ENTRY OF THE SMALL END OF THE PLUG. SEE GENERAL SPECIFICATIONS.
[b] Thread must be full or complete thread for length B and B_1.
[c] Tabulated limits shall be maintained for a distance equal to one-half the specified socket depth. Width at top and bottom portions of the socket may slightly exceed maximum.

SQUARE AND HEXAGON COUNTERSUNK HEADLESS

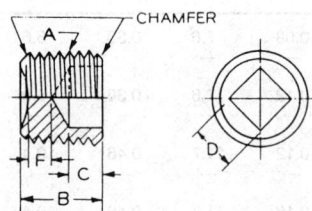 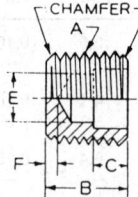 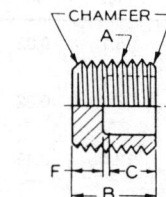

FIGURE 3A—UPSET (130109G)   FIGURE 3B—BROACHED (130109H)   FIGURE 3C—E CAST (130109J)

FIGURE 3—SQUARE COUNTERSUNK HEADLESS PIPE PLUGS
(NPTF)

External pipe threads shall be chamfered or rounded from the diameters tabulated in Table 1 to produce a length of chamfered or partial thread as specified. The threads on countersunk headless types of plugs shall be chamfered on both ends to the dimensions shown.

Related specifications covering blank sizes, dies, chasers, and gages are shown in SAE J476.

3.3 Material and Manufacture—Plugs may be made from low carbon steel, cast iron, malleable iron, brass, bronze, or aluminum alloy as specified by purchaser, by casting, milling from the bar, or upsetting from a grade of material free of defects which will affect their serviceability.

(R) **3.4 Finish**—The external surfaces and threads of all carbon steel parts shall be plated or coated with a suitable material that passes a 72 h salt spray test in accordance with ASTM B 117. Any appearance of red rust during the 72 h salt spray test shall be considered failure, except for the following:

a. All internal fluid passages.

b. Edges such as hex points, serrations, and crests of threads where there may be mechanical deformation of the plating or coating typical of mass-produced parts or shipping effects.

c. Areas where there is mechanical deformation of the plating or coating caused by crimping, flaring, bending, and other post-plate metal forming operations.

d. Areas where the parts are suspended or affixed in the test chamber where condensate can accumulate.

NOTE—Cadmium plating is not preferred due to environmental reasons. Parts manufactured to this standard after January 1, 1997, shall not be cadmium plated. Internal fluid passages shall be protected from corrosion during storage. Changes in plating may affect assembly torques and require requalification, when applicable.

3.5 Workmanship—Workmanship shall conform to the best commercial practice to produce high-quality parts. Plugs shall be free from all hanging burrs, loose scale, and slivers which might become dislodged in usage and all other defects which might affect their serviceability.

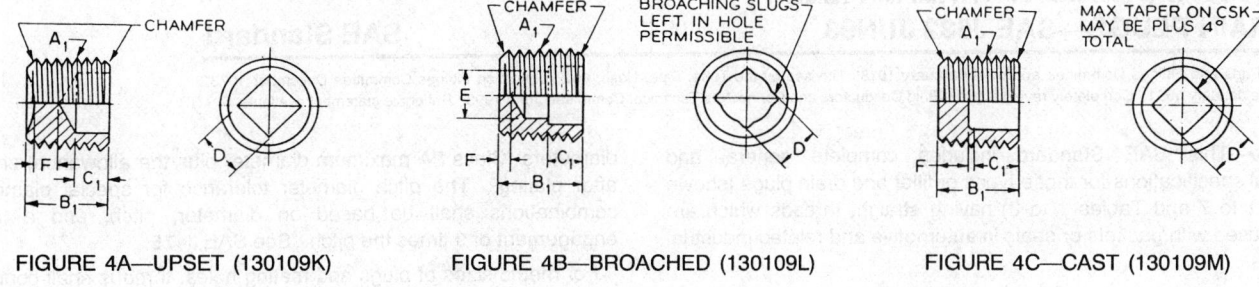

FIGURE 4A—UPSET (130109K)   FIGURE 4B—BROACHED (130109L)   FIGURE 4C—CAST (130109M)

FIGURE 4—SQUARE COUNTERSUNK HEADLESS FILLER AND DRAIN PLUGS (PTF)[a]

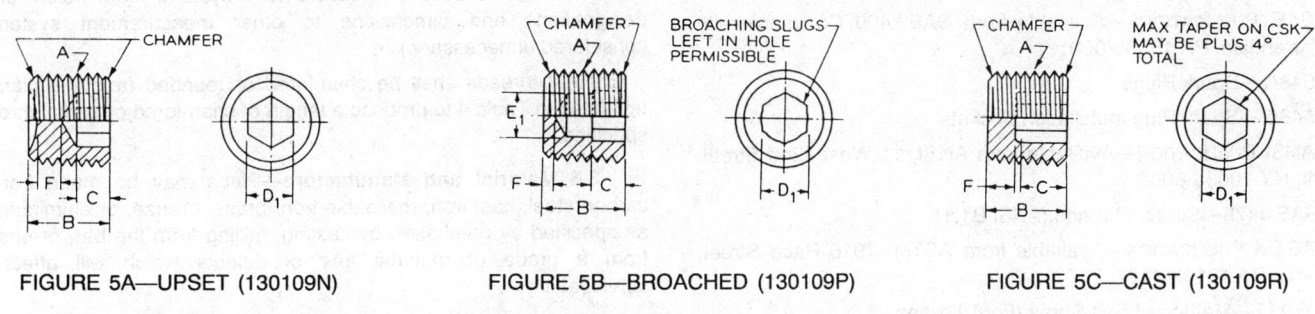

FIGURE 5A—UPSET (130109N)   FIGURE 5B—BROACHED (130109P)   FIGURE 5C—CAST (130109R)

FIGURE 5—HEXAGON COUNTERSUNK HEADLESS PIPE PLUGS (NPTF)

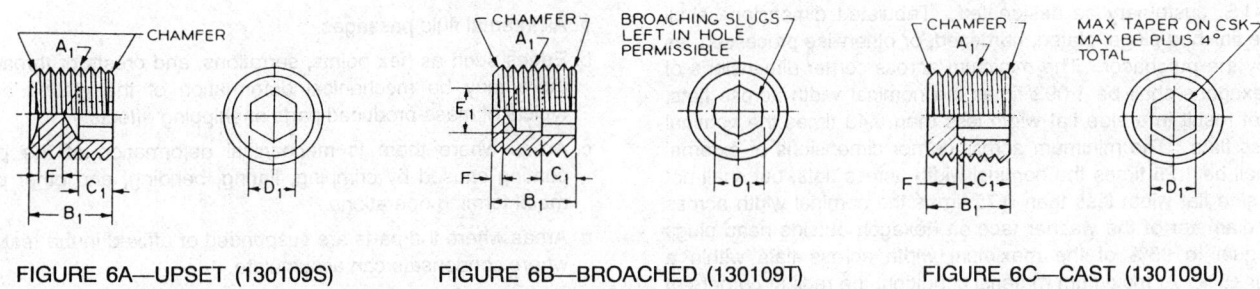

FIGURE 6A—UPSET (130109S)   FIGURE 6B—BROACHED (130109T)   FIGURE 6C—CAST (130109U)

FIGURE 6—HEXAGON COUNTERSUNK HEADLESS FILLER AND DRAIN PLUGS (PTF)[a]

CODES SHOWN IN BRACKETS ADJACENT TO FIGURE NUMBERS REPRESENT RESPECTIVE FITTING IDENTIFICATION IN ACCORDANCE TO SAE J846.

(R) AUTOMOTIVE STRAIGHT THREAD FILLER AND DRAIN PLUGS—SAE J532 JUN93

Report of the Parts and Fittings Committee approved February 1918. Revised by the Tube, Pipe, Hose, and Lubrication Fittings Committee December 1973, editorial change January 1981. Completely revised by the Fluid Conductors and Connectors Technical Committee June 1993. Rationale statement available.

1. Scope—This SAE Standard includes complete general and dimensional specifications for those types of filler and drain plugs (shown in Figures 1 to 7 and Tables 1 to 3) having straight threads which are commonly used with gaskets or seals in automotive and related industrial applications.

2. References

2.1 Applicable Documents—The following publications form a part of this specification to the extent specified herein. The latest issue of SAE publications shall apply.

2.1.1 SAE PUBLICATIONS—Available from SAE, 400 Commonwealth Drive, Warrendale, PA 15096-0001.

SAE J548/1—Spark Plugs

SAE J548/2—Spark Plug Installation Sockets

2.1.2 ANSI PUBLICATION—Available from ANSI, 11 West 42nd Street, New York, NY 10036-8002.

ANSI/SAE J475—Screw Threads (ANSI B1.1)

2.1.3 ASTM PUBLICATION—Available from ASTM, 1916 Race Street, Philadelphia, PA 19103-1187.

ASTM B 117—Method of Salt Spray (Fog) Testing

3. General Specifications

3.1 Dimensions and Tolerances—Except for nominal sizes and thread specifications, dimensions and tolerances are given in both SI units and U.S. customary as designated. Tabulated dimensions shall apply to the finished plugs, plated, hardened, or otherwise processed, as specified by the purchaser. The minimum across corner dimensions of external hexagons shall be 1.092 times the nominal width across flats, but shall not result in a side flat width less than 0.43 times the nominal width across flats. The minimum across corner dimensions of external squares shall be 1.25 times the nominal width across flats, but shall not result in a side flat width less than 0.75 times the nominal width across flats. The diameter of the washer face on hexagon outside head plugs shall be equal to 95% of the maximum width across flats within a tolerance of ±5%. At maximum material condition, the radii at corners of hexagon and square sockets in broached or upset plugs shall not exceed 0.13 mm (0.005 in).

Tolerance on dimensions not otherwise limited shall be ±0.25 mm (±0.010 in).

3.2 Straight Threads—Unified standard Class 2A external and Class 2B internal threads shall apply to inch sizes of plain finish (unplated) plugs and holes into which they assemble. For externally threaded parts with additive finish, the maximum diameters of Class 2A may be exceeded by the amount of the allowance, that is, the basic diameters (Class 2A maximum diameter plus the allowance) shall apply after plating. The pitch diameter tolerance for special diameter-pitch combinations shall be based on diameter, pitch, and a length of engagement of 9 times the pitch. See SAE J475.

For metric sizes of plugs and mating holes, threads shall conform with SAE J548.

For convenient reference, the data generally required to specify threads is given in Table 1 for both the plugs and mating holes. (Inasmuch as threads are normally produced and gaged with equipment made to the respective measurement system, conversion of size designations and dimensions to other measurement systems is considered unnecessary.)

External threads shall be chamfered or rounded from the diameters tabulated in Table 4 to produce a length of chamfered or partial thread as specified.

3.3 Material and Manufacture—Plugs may be made from low carbon steel, cast iron, malleable iron, brass, bronze, or aluminum alloy as specified by purchaser, by casting, milling from the bar, or upsetting from a grade of material free of defects which will affect their serviceability.

3.4 Finish—The external surfaces and threads of all carbon steel parts shall be plated or coated with a suitable material that passes a 72 h salt spray test in accordance with ASTM B 117. Any appearance of red rust during the 72 h salt spray test shall be considered failure, except for the following:

a. All internal fluid passages.

b. Edges such as hex points, serrations, and crests of threads where there may be mechanical deformation of the plating or coating typical of mass-produced parts or shipping effects.

c. Areas where there is mechanical deformation of the plating or coating caused by crimping, flaring, bending, and other post-plate metal forming operations.

d. Areas where the parts are suspended or affixed in the test chamber where condensate can accumulate.

NOTE—Cadmium plating is not preferred due to environmental reasons. Parts manufactured to this Standard after January 1, 1997, shall not be cadmium plated. Internal fluid passages shall be protected from corrosion during storage. Changes in plating may affect assembly torques and require requalification, when applicable.

3.5 Workmanship—Workmanship shall conform to the best commercial practice to produce high-quality parts. Plugs shall be free from all hanging burrs, loose scale, and slivers which might become dislodged in usage and all other defects which might affect their serviceability.

TABLE 1—STRAIGHT THREAD SIZES (EXTERNAL AND INTERNAL) (FIGURES 1 TO 7)

Nom Size, in	Series Designation	External Thread Pitch Diameter Max mm	External Thread Pitch Diameter Max in	External Thread Pitch Diameter Min mm	External Thread Pitch Diameter Min in	Internal Thread Pitch Diameter Max mm	Internal Thread Pitch Diameter Max in	Internal Thread Pitch Diameter Min mm	Internal Thread Pitch Diameter Min in	Internal Thread Minor Diameter Max mm	Internal Thread Minor Diameter Max in	Internal Thread Minor Diameter Min mm	Internal Thread Minor Diameter Min in
5/16-24	UNF		0.2843		0.2806		0.2902		0.2854		0.277		0.267
3/8-24	UNF		0.3468		0.3430		0.3528		0.3479		0.340		0.330
1/2-20	UNF		0.4662		0.4619		0.4731		0.4675		0.457		0.446
5/8-18	UNF		0.5875		0.5828		0.5949		0.5889		0.578		0.565
3/4-16	UNF		0.7079		0.7029		0.7159		0.7094		0.696		0.682
7/8-18	UNS		0.8375		0.8329		0.8449		0.8389		0.828		0.815
1-18	UNS		0.9625		0.9578		0.9701		0.9639		0.953		0.940
1-1/4-18	UNEF		1.2124		1.2075		1.2202		1.2139		1.203		1.190
1-1/2-18	UNEF		1.4624		1.4574		1.4704		1.4639		1.452		1.440
1-3/4-16	UN		1.7078		1.7025		1.7163		1.7094		1.696		1.682
2-16	UN		1.9578		1.9524		1.9664		1.9594		1.946		1.932
Metric Thread Sizes													
10 × 1	—	9.335		9.238		9.446		9.350		8.954		8.844	
14 × 1.25	—	13.155		13.048		13.297		13.188		12.962		12.499	
18 × 1.5	—	16.980		16.853		17.153		17.026		16.426		16.266	

SQUARE AND HEXAGON HEAD

CODES SHOWN IN BRACKETS ADJACENT TO FIGURE NUMBERS REPRESENT RESPECTIVE FITTING IDENTIFICATION IN ACCORDANCE WITH SAE J846.

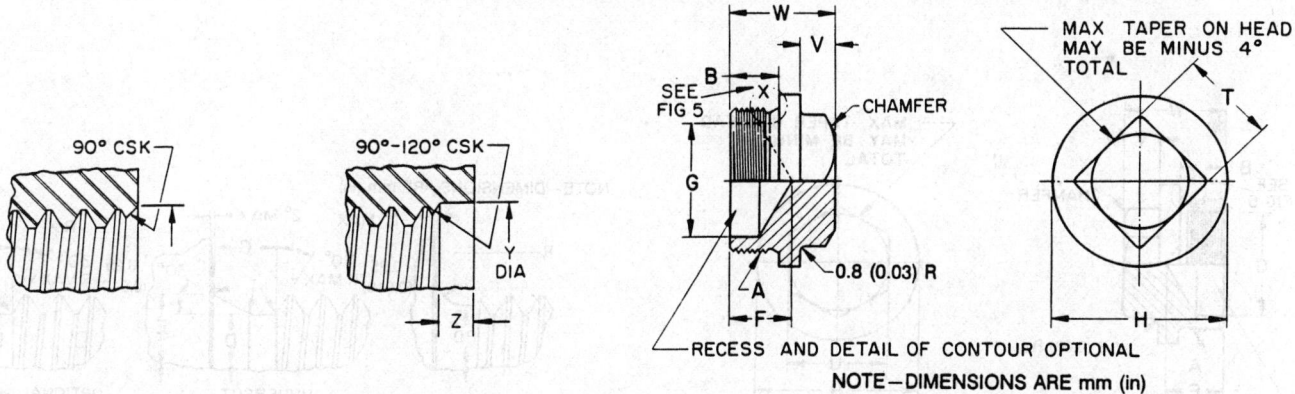

FIGURE 1—RECOMMENDED HOLE DATA

FIGURE 3—SQUARE HEAD PLUG (420109B)

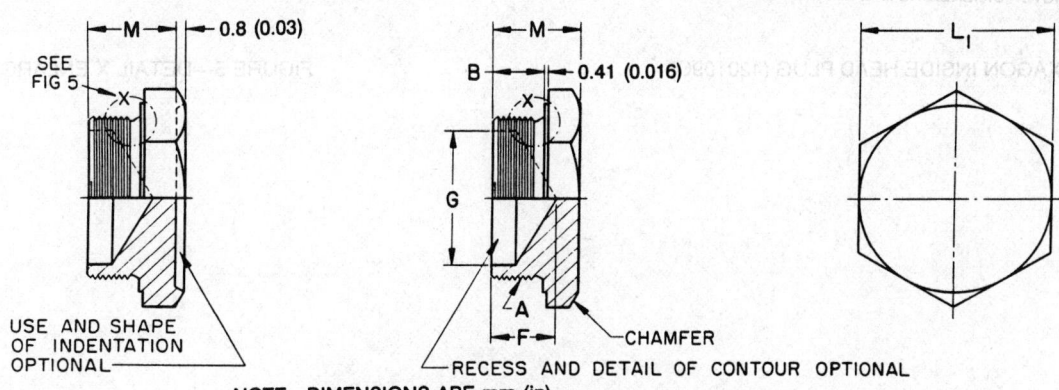

FIGURE 2—HEXAGON OUTSIDE HEAD PLUG (420109A)

TABLE 2A—DIMENSIONS OF HEXAGON OUTSIDE HEAD AND SQUARE, AND HEXAGON INSIDE HEAD FILLER AND DRAIN PLUGS (FIGURES 2 TO 5)

Nom Size, in	B Shoulder Length mm	B Shoulder Length in	C Relief Width mm	C Relief Width in	D Relief Dia mm	D Relief Dia in	E Pilot Dia mm	E Pilot Dia in	F Recess Depth mm	F Recess Depth in	G Recess Dia, Max mm	G Recess Dia, Max in	H Flange Dia mm	H Flange Dia in	R Fillet Radius[1] Approx mm	R Fillet Radius[1] Approx in
5/16	7.9	0.31	2.3	0.09	6.40 / 6.22	0.252 / 0.245	8.33 / 8.15	0.328 / 0.321	—	—	—	—	14.2	0.56	1.07	0.042
3/8	7.9	0.31	2.3	0.09	7.98 / 7.80	0.314 / 0.307	9.93 / 9.75	0.391 / 0.384	—	—	—	—	15.7	0.62	1.07	0.042
1/2	8.6	0.34	2.3	0.09	10.87 / 10.69	0.428 / 0.421	13.11 / 12.93	0.516 / 0.509	10.4	0.41	6.4	0.25	19.0	0.75	1.27	0.050
5/8	9.7	0.38	3.0	0.12	13.84 / 13.64	0.545 / 0.537	16.28 / 16.08	0.641 / 0.633	11.9	0.47	9.7	0.38	22.4	0.88	1.42	0.056
3/4	9.7	0.38	3.0	0.12	16.76 / 16.54	0.660 / 0.651	19.46 / 19.23	0.766 / 0.757	11.9	0.47	12.7	0.50	25.4	1.00	1.57	0.062
7/8	10.4	0.41	3.0	0.12	20.14 / 19.94	0.793 / 0.785	22.63 / 22.43	0.891 / 0.883	12.7	0.50	14.2	0.56	28.4	1.12	1.42	0.056
1	11.2	0.44	3.0	0.12	23.32 / 23.11	0.918 / 0.910	25.81 / 25.60	0.016 / 1.008	14.2	0.56	17.5	0.69	31.8	1.25	1.42	0.056
1-1/4	11.9	0.47	3.0	0.12	29.64 / 29.44	1.167 / 1.159	32.16 / 31.95	1.266 / 1.258	15.0	0.59	23.9	0.94	38.1	1.50	1.42	0.056
1-1/2	12.7	0.50	3.0	0.12	35.99 / 35.79	1.417 / 1.409	38.51 / 38.30	1.516 / 1.508	17.5	0.69	28.4	1.12	44.4	1.75	1.42	0.056
1-3/4	14.2	0.56	3.0	0.12	42.09 / 41.86	1.657 / 1.648	44.86 / 44.63	1.766 / 1.757	19.0	0.75	35.1	1.38	50.8	2.00	1.57	0.062
2	14.2	0.56	3.0	0.12	48.44 / 48.21	1.907 / 1.898	51.21 / 50.98	2.016 / 2.007	19.0	0.75	41.1	1.62	57.2	2.25	1.57	0.062
Metric Thread Sizes																
10	7.9	0.31	2.3	0.09	8.53 / 8.36	0.336 / 0.329	10.41 / 10.24	0.410 / 0.403	—	—	—	—	15.7	0.62	1.00	0.039
14	8.6	0.34	2.3	0.09	12.19 / 12.01	0.480 / 0.473	14.40 / 14.22	0.567 / 0.560	10.4	0.41	7.9	0.31	20.6	0.81	1.25	0.049
18	9.7	0.38	3.0	0.12	15.82 / 15.60	0.623 / 0.614	18.39 / 18.16	0.724 / 0.715	11.9	0.47	11.2	0.44	23.9	0.94	1.50	0.059

[1] See detail X in Figure 5.

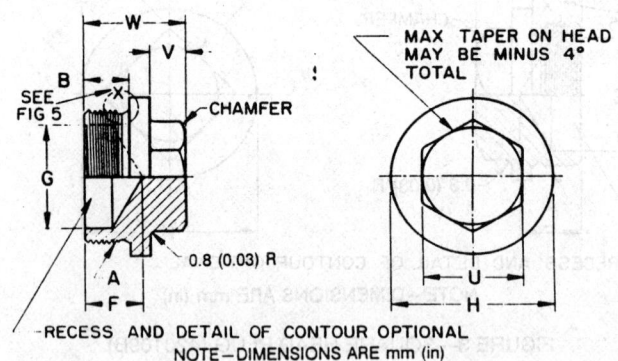

FIGURE 4—HEXAGON INSIDE HEAD PLUG (420109C)

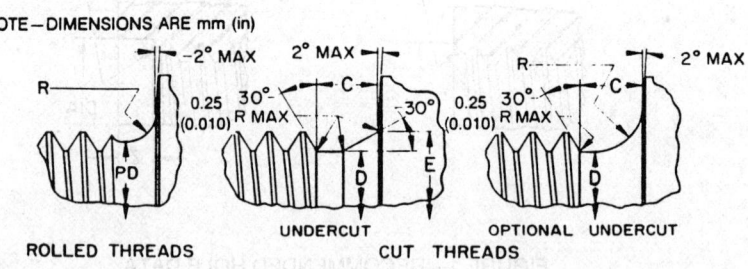

FIGURE 5—DETAIL X ENLARGED

TABLE 2B—DIMENSIONS OF HEXAGON OUTSIDE HEAD AND SQUARE, AND HEXAGON INSIDE HEAD FILLER AND DRAIN PLUGS (FIGURES 2 TO 5)

Nom Size, in	R₁ Fillet Radius[1] Approx mm	R₁ Fillet Radius[1] Approx in	Hex Outside Head L Hex Width mm	Hex Outside Head L Hex Width in	Hex Outside Head M Overall Length mm	Hex Outside Head M Overall Length in	Square and Hexagon Inside Head T Square Size mm	Square and Hexagon Inside Head T Square Size in	Square and Hexagon Inside Head U Hex Size mm	Square and Hexagon Inside Head U Hex Size in	Square and Hexagon Inside Head V Square and Hex Height mm	Square and Hexagon Inside Head V Square and Hex Height in	Square and Hexagon Inside Head W Overall Length mm	Square and Hexagon Inside Head W Overall Length in
5/16	1.57	0.062	14.35 / 14.00	0.565 / 0.551	12.7	0.50	5.61 / 5.44	0.221 / 0.214	7.19 / 7.01	0.283 / 0.276	5.6	0.22	16.8	0.66
3/8	1.57	0.062	15.93 / 15.54	0.627 / 0.612	12.7	0.50	7.19 / 7.01	0.283 / 0.276	8.00 / 7.72	0.315 / 0.304	6.4	0.25	17.5	0.69
1/2	1.90	0.075	19.10 / 18.72	0.752 / 0.737	13.5	0.53	9.58 / 9.40	0.377 / 0.370	11.18 / 10.87	0.440 / 0.428	6.4	0.25	18.3	0.72
5/8	2.11	0.083	22.28 / 21.84	0.877 / 0.860	15.7	0.62	11.18 / 10.87	0.440 / 0.428	12.75 / 12.42	0.502 / 0.489	7.1	0.28	19.8	0.78
3/4	2.39	0.094	25.45 / 24.97	1.002 / 0.983	15.7	0.62	12.75 / 12.45	0.502 / 0.490	14.35 / 14.00	0.565 / 0.551	7.1	0.28	20.6	0.81
7/8	2.11	0.083	28.63 / 28.09	1.127 / 1.106	16.8	0.66	14.35 / 14.05	0.565 / 0.553	15.93 / 15.54	0.627 / 0.612	7.9	0.31	22.4	0.88
1	2.11	0.083	31.80 / 31.24	1.252 / 1.230	19.0	0.75	15.93 / 15.62	0.627 / 0.615	20.70 / 20.27	0.815 / 0.798	7.9	0.31	23.1	0.91
1-1/4	2.11	0.083	38.15 / 37.52	1.502 / 1.477	19.8	0.78	20.70 / 20.40	0.815 / 0.803	25.45 / 24.97	1.002 / 0.983	8.6	0.34	25.4	1.00
1-1/2	2.11	0.083	44.50 / 43.82	1.752 / 1.725	22.4	0.88	23.88 / 23.57	0.940 / 0.928	28.63 / 28.09	1.127 / 1.106	9.7	0.38	28.4	1.12
1-3/4	2.39	0.094	50.85 / 50.14	2.002 / 1.974	23.9	0.94	28.63 / 28.32	1.127 / 1.115	31.80 / 31.24	1.252 / 1.230	11.2	0.44	31.8	1.25
2	2.39	0.094	57.20 / 56.69	2.252 / 2.232	23.9	0.94	33.40 / 33.07	1.315 / 1.302	38.15 / 37.52	1.502 / 1.477	11.2	0.44	33.3	1.31
Metric Thread Sizes														
10	1.50	0.059	15.93 / 15.54	0.627 / 0.612	12.7	0.50	7.19 / 7.01	0.283 / 0.276	8.00 / 7.72	0.315 / 0.304	6.4	0.25	17.5	0.69
14	1.88	0.074	20.68 / 20.29	0.814 / 0.799	13.5	0.53	9.58 / 9.40	0.377 / 0.370	11.18 / 10.87	0.440 / 0.428	6.4	0.25	18.3	0.72
18	2.25	0.089	23.88 / 23.39	0.940 / 0.921	15.7	0.62	12.75 / 12.45	0.502 / 0.490	14.35 / 14.00	0.565 / 0.551	7.1	0.28	20.6	0.81

[1] See detail X in Figure 5.

SQUARE AND HEXAGON SOCKET HEAD

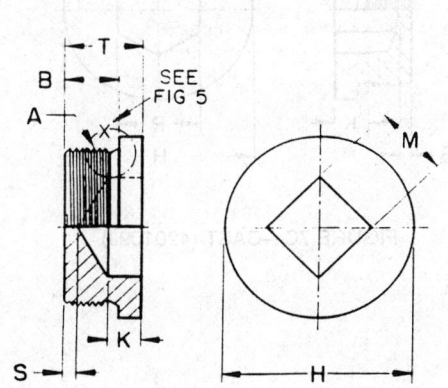

FIGURE 6A—UPSET (420109D)

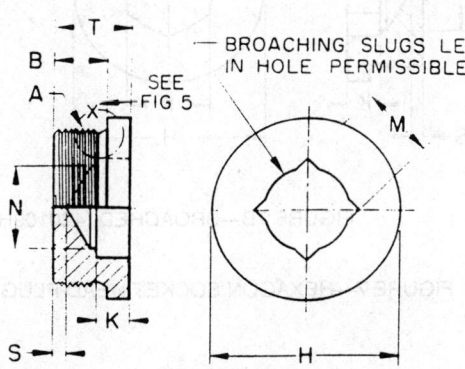

FIGURE 6B—BROACHED (420109E)

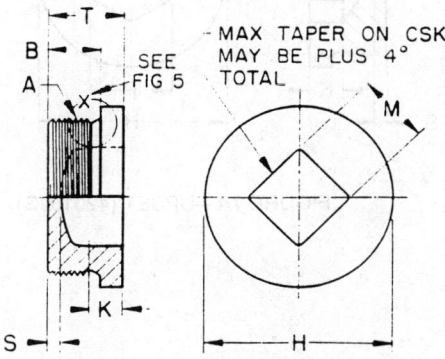

FIGURE 6C—CAST (420109F)

FIGURE 6—SQUARE SOCKET HEAD PLUGS

TABLE 3A—DIMENSIONS OF SQUARE AND HEXAGON SOCKET HEAD FILLER AND DRAIN PLUGS (FIGURES 5 TO 7)

Nom Size, in	B Shoulder Length mm	B Shoulder Length in	C Relief Width mm	C Relief Width in	D Relief Dia mm	D Relief Dia in	E Pilot Dia mm	E Pilot Dia in	H Flange Dia mm	H Flange Dia in	K Socket Depth, Min mm	K Socket Depth, Min in	S Wall Thickness, Min mm	S Wall Thickness, Min in	T Overall Length mm	T Overall Length in
5/16	7.9	0.31	2.3	0.09	6.40 / 6.22	0.252 / 0.245	8.33 / 815	0.328 / 0.321	14.2	0.56	3.0	0.12	3.0	0.12	11.2	0.44
3/8	7.9	0.31	2.3	0.09	7.98 / 7.80	0.314 / 0.307	9.93 / 9.75	0.391 / 0.384	15.7	0.62	3.0	0.12	3.0	0.12	11.2	0.44
1/2	8.6	0.34	2.3	0.09	10.87 / 10.69	0.428 / 0.421	13.11 / 12.93	0.516 / 0.509	19.0	0.75	4.8	0.19	3.0	0.12	11.9	0.47
5/8	9.7	0.38	3.0	0.12	13.84 / 13.64	0.545 / 0.537	16.28 / 16.08	0.641 / 0.633	22.4	0.88	4.8	0.19	3.0	0.12	12.7	0.50
3/4	9.7	0.38	3.0	0.12	16.76 / 16.54	0.660 / 0.651	19.46 / 19.23	0.766 / 0.757	25.4	1.00	6.4	0.25	3.0	0.12	13.2	0.52
7/8	10.4	0.41	3.0	0.12	20.14 / 19.94	0.793 / 0.785	22.63 / 22.43	0.891 / 0.883	28.4	1.12	6.4	0.25	3.0	0.12	14.2	0.56
1	11.2	0.44	3.0	0.12	23.32 / 23.11	0.918 / 0.910	25.81 / 25.60	1.016 / 1.008	31.8	1.25	6.4	0.25	3.0	0.12	15.0	0.59
1-1/4	11.9	0.47	3.0	0.12	29.64 / 29.44	1.167 / 1.159	32.16 / 31.95	1.266 / 1.258	38.1	1.50	7.9	0.31	3.0	0.12	16.8	0.66
1-1/2	12.7	0.50	3.0	0.12	35.99 / 35.79	1.417 / 1.409	38.51 / 38.30	1.516 / 1.508	44.4	1.75	9.7	0.38	3.0	0.12	19.0	0.75
1-3/4	14.2	0.56	3.0	0.12	42.09 / 41.86	1.657 / 1.648	44.86 / 44.63	1.766 / 1.757	50.8	2.00	9.7	0.38	3.0	0.12	20.6	0.81
2	14.2	0.56	3.0	0.12	48.44 / 48.21	1.907 / 1.898	51.21 / 50.98	2.016 / 2.007	57.2	2.25	9.7	0.38	3.0	0.12	22.4	0.88
Metric Thread Sizes																
10	7.9	0.31	2.3	0.09	8.53 / 8.36	0.336 / 0.329	10.41 / 10.24	0.410 / 0.403	15.7	0.62	3.0	0.12	3.0	0.12	11.2	0.44
14	8.6	0.34	2.3	0.09	12.19 / 12.01	0.480 / 0.473	14.40 / 14.22	0.567 / 0.560	20.6	0.81	4.8	0.19	3.0	0.12	11.9	0.47
18	9.7	0.38	3.0	0.12	15.82 / 15.60	0.623 / 0.614	18.39 / 18.16	0.724 / 0.715	23.9	0.94	6.4	0.25	3.0	0.12	13.5	0.53

[1] See detail X in Figure 5.

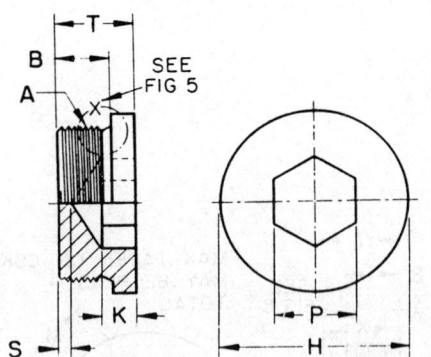
FIGURE 7A—UPSET (420109G)

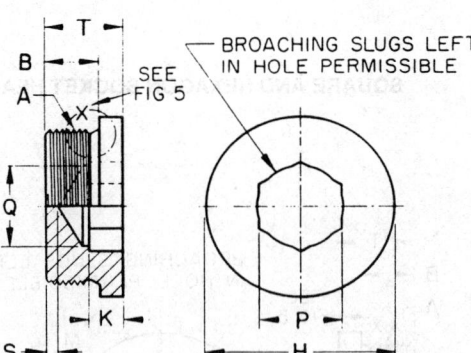
FIGURE 7B—BROACHED (420109H)

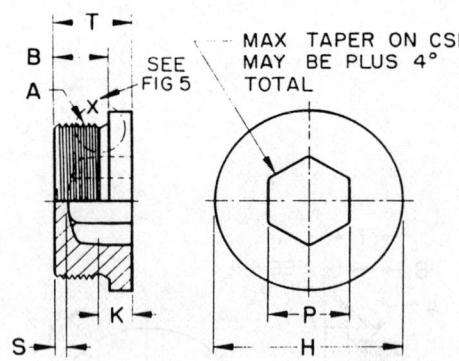
FIGURE 7C—CAST (420109J)

FIGURE 7—HEXAGON SOCKET HEAD PLUGS

22.251

TABLE 3B—DIMENSIONS OF SQUARE AND HEXAGON SOCKET HEAD FILLER AND DRAIN PLUGS (FIGURES 5 TO 7)

Nom Size, in	R Fillet Radius[1] Approx mm	R Fillet Radius[1] Approx in	R_4 Fillet Radius[1] Approx mm	R_4 Fillet Radius[1] Approx in	Square Socket Broached or Upset M^1 Socket Width mm	Square Socket Broached or Upset M^1 Socket Width in	Square Socket Broached or Upset N Hole Dia Max mm	Square Socket Broached or Upset N Hole Dia Max in	Square Socket Cast M Socket Width mm	Square Socket Cast M Socket Width in	Hexagon Socket Broached or Upset P^1 Socket Width mm	Hexagon Socket Broached or Upset P^1 Socket Width in	Hexagon Socket Broached or Upset Q Hole Dia Max mm	Hexagon Socket Broached or Upset Q Hole Dia Max in	Hexagon Socket Cast P Socket Width mm	Hexagon Socket Cast P Socket Width in
5/16	1.07	0.042	1.57	0.062	3.30 3.43	0.130 0.135	3.63	0.143	3.3 3.6	0.13 0.14	3.96 4.09	0.156 0.161	4.09	0.161	3.96 4.09	0.156 0.161
3/8	1.07	0.042	1.57	0.062	4.88 5.00	0.192 0.197	5.31	0.209	4.8 5.3	0.19 0.21	4.78 4.90	0.188 0.193	4.90	0.193	4.78 4.90	0.188 0.193
1/2	1.27	0.500	1.90	0.075	6.48 6.60	0.255 0.260	7.06	0.278	6.6 7.1	0.26 0.28	6.35 6.48	0.250 0.255	6.48	0.255	6.35 6.48	0.250 0.255
5/8	1.42	0.056	2.11	0.083	8.10 8.23	0.319 0.324	8.76	0.345	8.1 8.9	0.32 0.35	7.95 8.08	0.313 0.318	8.08	0.318	7.95 8.08	0.313 0.318
3/4	1.57	0.062	2.39	0.094	9.70 9.83	0.382 0.387	10.41	0.410	9.7 10.4	0.38 0.41	9.53 9.65	0.375 0.380	9.68	0.381	9.53 9.65	0.375 0.380
7/8	1.42	0.056	2.11	0.083	10.49 10.62	0.413 0.418	11.25	0.443	10.4 11.2	0.41 0.44	12.70 12.83	0.500 0.505	12.85	0.506	12.70 12.83	0.500 0.505
1	1.42	0.056	2.11	0.083	10.49 10.62	0.413 0.418	11.25	0.443	10.4 11.2	0.41 0.44	14.30 14.43	0.563 0.568	14.48	0.570	14.30 14.43	0.563 0.568
1-1/4	1.42	0.056	2.11	0.083	12.90 13.03	0.508 0.513	14.05	0.553	13.0 13.7	0.51 0.54	15.88 16.00	0.625 0.630	16.08	0.633	15.88 16.00	0.625 0.630
1-1/2	1.42	0.056	2.11	0.083	16.05 16.18	0.632 0.637	17.25	0.679	16.0 16.8	0.63 0.66	19.05 19.18	0.750 0.755	19.23	0.757	19.05 19.30	0.750 0.760
1-3/4	1.57	0.062	2.39	0.094	19.28 19.41	0.759 0.764	20.83	0.820	19.3 20.3	0.76 0.80	22.23 22.35	0.875 0.880	22.43	0.883	22.23 22.48	0.875 0.885
2	1.57	0.062	2.39	0.094	22.45 22.58	0.884 0.889	24.00	0.945	22.6 23.6	0.89 0.93	28.58 28.70	1.125 1.130	28.80	1.134	28.58 28.83	1.125 1.135
Metric Thread Sizes																
10	1.00	0.039	1.50	0.059	4.88 5.00	0.192 0.197	5.31	0.209	4.8 5.3	0.19 0.21	4.78 4.90	0.188 0.193	4.90	0.193	4.78 4.90	0.188 0.193
14	1.25	0.049	1.88	0.074	6.48 6.60	0.255 0.260	7.06	0.278	6.6 7.1	0.26 0.28	6.35 6.48	0.250 0.255	6.48	0.255	6.35 6.48	0.250 0.255
18	1.50	0.059	2.25	0.089	9.70 9.83	0.382 0.387	10.41	0.410	9.7 10.4	0.38 0.41	9.53 9.65	0.375 0.380	9.68	0.381	9.53 9.65	0.375 0.380

[1] See detail X in Figure 5.

TABLE 4 (FIGURE 1)

Nom Size, in	Series Designation	External Thread Chamfer Dia Max mm	External Thread Chamfer Dia Max in	External Thread Chamfer Dia Min mm	External Thread Chamfer Dia Min in	External Thread Length of Chamfer or Partial Thread Max mm	External Thread Length of Chamfer or Partial Thread Max in	External Thread Length of Chamfer or Partial Thread Min mm	External Thread Length of Chamfer or Partial Thread Min in	Internal Thread Y CSK or C'Bore Dia Basic mm	Internal Thread Y CSK or C'Bore Dia Basic in	Internal Thread Z C'Bore Depth Basic mm	Internal Thread Z C'Bore Depth Basic in
5/16-24	UNF	6.4	0.25	6.1	0.24	1.32	0.052	0.79	0.031	8.6	0.34	1.07	0.042
3/8-24	UNF	7.9	0.31	7.6	0.30	1.32	0.052	0.79	0.031	10.2	0.40	1.07	0.042
1/2-20	UNF	10.9	0.43	10.7	0.42	1.57	0.062	0.97	0.038	13.5	0.53	1.27	0.050
5/8-18	UNF	13.7	0.54	13.2	0.52	1.75	0.069	1.07	0.042	16.8	0.66	1.42	0.056
3/4-16	UNF	16.8	0.66	16.3	0.64	1.98	0.078	1.19	0.047	19.8	0.78	1.57	0.062
7/8-18	UNS	20.1	0.79	19.6	0.77	1.75	0.069	1.07	0.042	23.1	0.91	1.42	0.056
1-18	UNS	23.4	0.92	22.9	0.90	1.75	0.069	1.07	0.042	26.2	1.03	1.42	0.056
1-1/4-18	UNEF	29.7	1.17	29.2	1.15	1.75	0.069	1.07	0.042	32.5	1.28	1.42	0.056
1-1/2-18	UNEF	36.1	1.42	35.6	1.40	1.75	0.069	1.07	0.042	38.9	1.53	1.42	0.056
1-3/4-16	UN	42.2	1.66	41.7	1.64	1.98	0.078	1.19	0.047	45.2	1.78	1.57	0.062
2-16	UN	48.5	1.91	48.0	1.89	1.98	0.078	1.19	0.047	51.6	2.03	1.57	0.062
Metric Thread Sizes													
10 × 1	—	8.4	0.33	8.1	0.32	1.25	0.049	0.75	0.030	10.8	0.42	1.00	0.039
14 × 1.25	—	12.2	0.48	11.9	0.47	1.56	0.061	0.94	0.037	14.8	0.58	1.25	0.049
18 × 1.5	—	15.7	0.62	15.2	0.60	1.88	0.074	1.12	0.044	18.8	0.74	1.50	0.059

LUBRICATION FITTINGS—SAE J534 JUN93

SAE Standard

Report of the Parts and Fittings Committee approved January 1949. Revised by the Tube, Pipe, Hose, and Lubrication Fittings Committee October 1973. Editorial change January 1981. Revised by the SAE Fluid Conductors and Connectors Technical Committee June 1993. Rationale statement available.

1. Scope—This SAE Standard covers complete general and dimensional specifications for the various types of lubrication fittings and related threaded components intended for general application in the automotive and allied fields.

(R) **2. References**

2.1 Applicable Documents—The following publications form a part of this specification to the extent specified herein. The latest issue of SAE publications shall apply.

2.1.1 SAE PUBLICATION—Available from SAE, 400 Commonwealth Drive, Warrendale, PA 15096-0001.

SAE J476a—Dryseal Pipe Threads

2.1.2 ASTM PUBLICATION—Available from ASTM, 1916 Race Street, Philadelphia, PA 19103-1187.

ASTM B 117—Method of Salt Spray (Fog) Testing

3. General Specifications

3.1 Designations—Lubrication fittings are designated by the type and size of the threaded ends and the configuration of the fitting (Figures 1a, 1b, 1c).

3.2 Dimensions and Tolerances—Except for nominal sizes and thread designations, dimensions and tolerances are given in both SI units and U.S. customary, as designated in Table 1. Tabulated dimensions shall apply to the finished parts, plated or otherwise processed, as specified by the purchaser. Tolerance on all dimensions not otherwise limited shall be ±0.3 mm. The maximum and minimum across flats dimensions shall be within the commercial tolerance of bar or extruded stock from which the fittings are produced. The minimum across corners dimensions of hexagons shall be 1.092 times the nominal width across flats, but shall not result in a side flat width less than 0.43 times the nominal width across flats.

3.3 Check Valve—All the standard hydraulic lubrication fittings contained herein are supplied with ball check valves. Fittings without valves are not recommended by the lubrication fitting industry.

3.4 Contour—Details of contour shall be optional with the manufacturer, provided the tabulated dimensions are maintained and serviceability of the fittings is not impaired.

3.5 Pipe Threads—The pipe threads on fittings, unless there is specific authorization to the contrary, shall conform with the specifications given in detail in SAE J476 for the designated thread series, except that external thread crests may have greater maximum truncation due to manufacturing practices. Experience has shown that the crest of the threads on lubrication fittings, intended for use with grease, does not have to conform to Dryseal American Standard Form to function satisfactorily. The deviations from standard Dryseal practice are peculiar to lubrication fittings and as special considerations are involved, it is not advisable to use them in any other application of pipe thread practice.

External pipe threads shall be chamfered from a diameter (rounded to a two-place decimal) obtained by subtracting 0.41 mm from the minimum minor diameter at the small end, with a minus tolerance on the diameter of 0.5 mm, to produce a length of chamfered or partial thread equivalent to 1 to 1-1/2 times pitch (rounded to a three-place decimal). See Appendix A of SAE J476.

Internal pipe threads shall be countersunk 90 degrees included angle to a diameter (rounded to a two-place decimal) obtained by adding 0.41 mm to the maximum major diameter at the large end with a plus tolerance on the diameter of 0.5 mm. See Appendix A of SAE J476.

Recommended assembly considerations for the various combinations of Dryseal pipe threads are given under the respective standard thread series and the paragraph headed Limitation of Assembly, Appendix D, in SAE J476.

3.6 1/4-28 Taper Thread—External taper threads designated SAE-LT shall be Unified Standard Form 1/4-28 with 19.0 mm ± 1.5 mm, diametral taper per 304.8 mm of length. The pitch diameter measured at start of thread on small end shall be 5.733 to 3.649 mm.

Threads shall be chamfered 0.91 to 1.37 mm long from a diameter of 5.1 mm with a tolerance of -0.5 mm.

It is recommended that SAE-LT taper threads be assembled into 1/4-28 UNF, Class 3B, straight threaded holes having a modified maximum minor diameter of 5.466 mm to insure 75% minimum thread height.

3.7 Special Thread Forming Threads—The 1/4-28 special taper thread forming thread and the 1/8-27 pipe special thread forming thread, where specified, shall conform to the dimensions specified in Figure 4 and Table 2. Fittings employing these threads may be driven or spun into unthreaded holes of diameters recommended and they are generally either marked or colored to provide ready identification.

3.8 Material and Manufacture—Unless otherwise specified, fittings shall be made from steel standard with the manufacturer. At the manufacturer's option, caps for water pump fittings may be made from brass, steel, or aluminum.

The greasing end of fittings shall be hardened. They shall have a case depth of 0.13 to 0.23 mm and minimum hardness of 83 on the Rockwell 15N scale. The threaded end on special thread forming fittings shall also be hardened.

(R) **3.9 Finish**—The external surfaces and threads of all carbon steel parts shall be plated or coated with a suitable material that passes a 72 h salt spray test in accordance with ASTM B 117. Any appearance of red rust during the 72 h salt spray test shall be considered failure, except for the following:

a. All internal fluid passages.
b. Edges such as hex points, serrations, and crests of threads where there may be mechanical deformation of the plating or coating typical of mass-produced parts or shipping effects.
c. Areas where there is mechanical deformation of the plating or coating caused by crimping, flaring, bending, and other post-plate metal forming operations.
d. Areas where the parts are suspended or affixed in the test chamber where condensate can accumulate.

NOTE—Cadmium plating is not preferred due to environmental reasons. Parts manufactured to this standard after January 1, 1997, shall not be cadmium plated. Internal fluid passages shall be protected from corrosion during storage. Changes in plating may affect assembly torques and require requalification, when applicable.

3.10 Related Fittings—Figures 5, 6, 7a, and 7b designate special adapters used in conjunction with lubrication fittings.

3.11 Workmanship—Fittings shall be free from burrs, loose scale, sharp edges, and all other defects that might affect their serviceability.

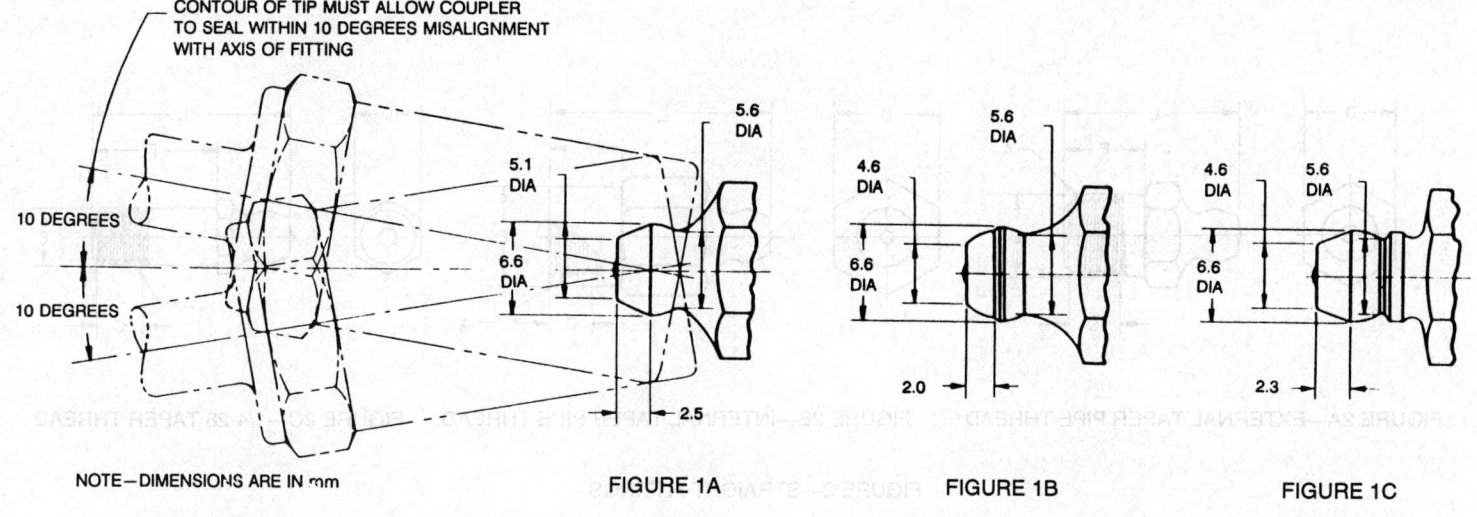

FIGURE 1A   FIGURE 1B   FIGURE 1C

FIGURE 1—OPTIONAL TIPS FOR LUBRICATION FITTINGS

TABLE 1—DIMENSIONS OF STRAIGHT AND ELBOW FITTINGS (FIGURES 2 AND 3)

Type	A Thread	B Angle, ±3 deg	C Effective Thread Length, min mm	D Hex Width Across Flats, Nom, in	E Shank Dia mm	F Shank Length mm ±0.8	L Overall Length mm ±1.0	M Overall Height mm ±1.0
Straight Fittings	1/8-27 Dryseal-PTF special extra short	---	4.6	7/16	10.2	7.1	16.8	---
	1/8-27 Dryseal-PTF special short	---	5.6	7/16	10.2	19.3	32.0	---
	1/8-27 Dryseal-PTF special short	---	5.6	7/16	10.2	32.5	44.7	---
	1/8-27 Dryseal-PTF special short	---	5.6	7/16	10.2	55.4	66.5	---
	1/8 pipe special thread forming	---	3.6	7/16	10.2	6.1	15.7	---
	1/8-27 Dryseal-NPTF internal thread	---	7.1	1/2	12.2	8.1	25.4	---
	1/4-28 taper thread (SAE-LT)	---	2.5	5/16	6.6	4.6	13.7	---
	1/4-28 taper thread (SAE-LT)	---	5.1	5/16	6.6	8.6	17.3	---
	1/4-28 taper thread (SAE-LT)	---	5.1	5/16	6.6	15.7	23.9	---
	1/4-28 special taper thread forming	---	2.5	5/16	6.50	5.1	14.0	---
Elbow Fittings	1/8-27 Dryseal-PTF special short	30	5.6	7/16	10.2	7.6	22.9	14.2
	1/8-27 Dryseal-PTF special short	30	5.6	7/16	10.2	32.0	53.3	14.2
	1/8 pipe special thread forming	30	3.6	7/16	10.2	5.1	21.8	14.2
	1/8-27 Dryseal-PTF special short	45	5.6	7/16	10.2	7.6	21.8	16.3
	1/4-28 taper thread (SAE-LT)	45	2.5	3/8	6.6	5.1	20.8	14.7
	1/4-28 taper thread (SAE-LT)	45	5.1	3/8	6.6	7.6	23.9	14.7
	1/4-28 special taper thread forming	45	2.5	3/8	6.50	4.8	20.3	14.7
	1/8-27 Dryseal-PTF special short	65	5.6	7/16	10.2	7.6	21.8	18.3
	1/8-27 Dryseal-PTF special short	65	5.6	7/16	10.2	14.2	30.0	18.3
	1/8 pipe special thread forming	65	3.6	7/16	10.2	5.1	19.8	18.3
	1/8-27 Dryseal-PTF special short	90	5.6	7/16	10.2	7.6	21.3	18.3
	1/8-27 Dryseal-PTF special short	90	5.6	7/16	10.2	32.0	46.2	18.3
	1/8 pipe special thread forming	90	3.6	7/16	10.2	5.1	19.3	18.3
	1/4-28 taper thread (SAE-LT)	90	2.5	3/8	6.6	5.1	19.3	16.8
	1/4-28 special taper thread forming	90	2.5	3/8	6.50	4.8	19.3	16.8
	1/8-27 Dryseal-PTF special short	105	5.6	7/16	10.2	7.6	26.9	19.3

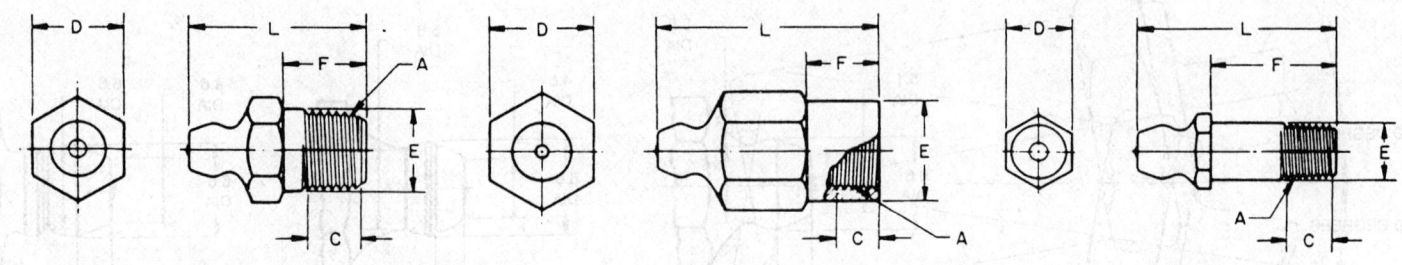

FIGURE 2A—EXTERNAL TAPER PIPE THREAD   FIGURE 2B—INTERNAL TAPER PIPE THREAD   FIGURE 2C—1/4-28 TAPER THREAD

FIGURE 2—STRAIGHT FITTINGS

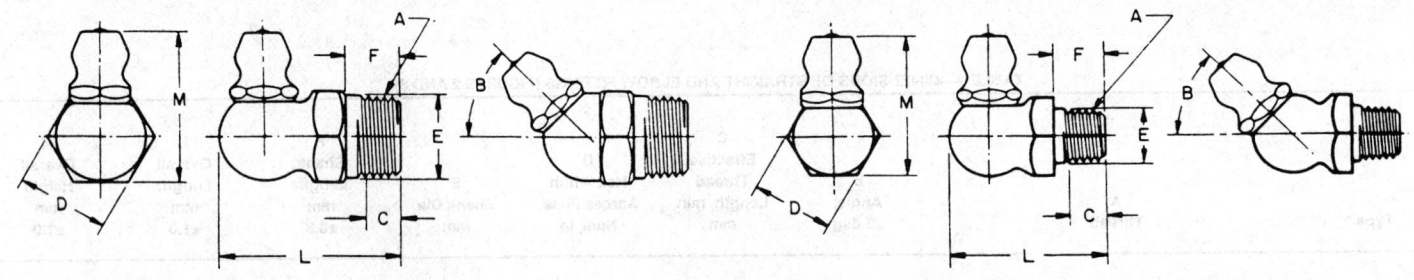

FIGURE 3A—TAPER PIPE THREAD   FIGURE 3B—1/4-28 TAPER THREAD

FIGURE 3—ELBOW FITTINGS

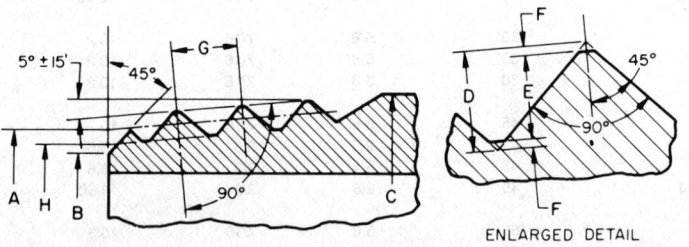

FIGURE 4—SPECIAL THREAD FORMING THREADS

TABLE 2—DIMENSIONS OF SPECIAL THREAD FORMING THREADS (FIGURE 4)

Nominal Thread Size	A Pitch Dia at Small End mm	B Chamfer Dia mm	C Shank Dia mm	D Height of Sharp V Thread mm	E Height of Truncated Thread mm	F Height of Truncated at Crest and Root mm	G Pitch mm	H Root Dia at Small End mm	Recommended Hole Dia[1] mm
1/4-28	5.654	5.1	6.58	0.452	0.427	0.069	0.907	5.28	5.97
Spl Taper	5.476	4.6	6.43		0.315	0.013		5.11	5.84
1/8-27	9.070	8.4	10.24	0.470	0.445	0.074	0.940	8.69	9.65
Spl Pipe	8.892	7.9	10.06		0.323	0.013		8.51	9.47

22.255

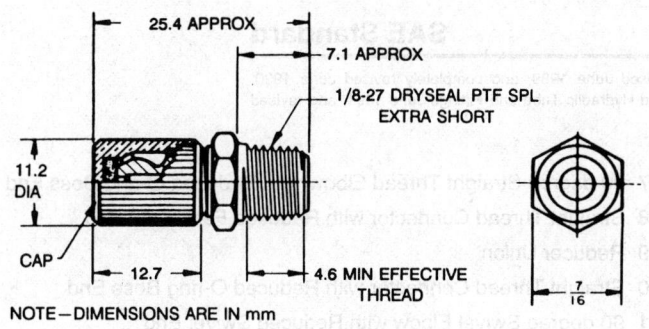

FIGURE 5—WATER PUMP FITTING

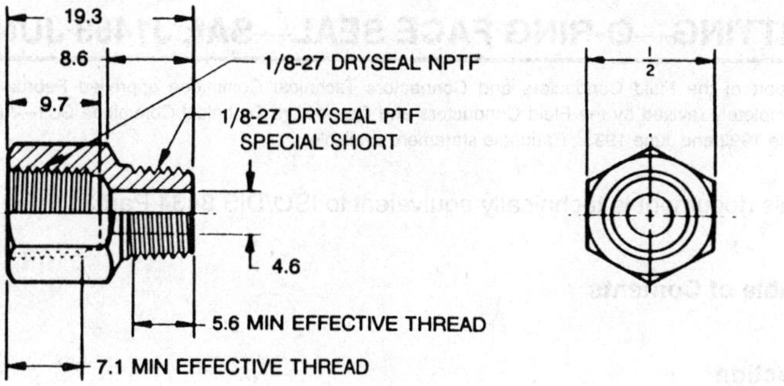

FIGURE 6—EXTENSION

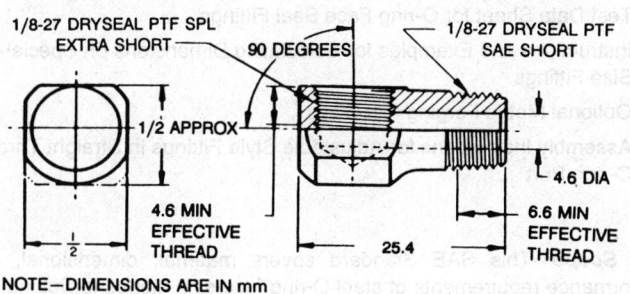

FIGURE 7A

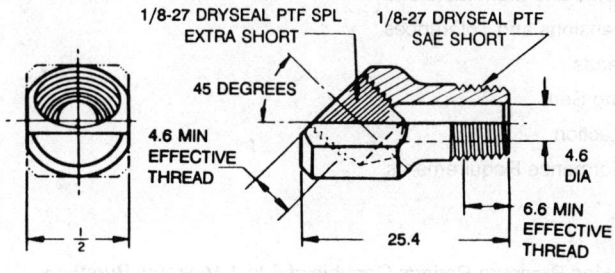

FIGURE 7B

FIGURE 7—ELBOW ADAPTERS

NOTE—UNSPECIFIED DETAIL WITH RESPECT TO DIMENSIONS, TOLERANCES, CONTOURS, MATERIAL, WORKMANSHIP, ETC., MUST CONFORM TO GENERAL SPECIFICATIONS FOR LUBRICATION FITTINGS.

FITTING—O-RING FACE SEAL—SAE J1453 JUN93

SAE Standard

Report of the Fluid Conductors and Connectors Technical Committee approved February 1987, revised June 1989, and completely revised June 1990. Completely revised by the Fluid Conductors and Connectors Technical Committee SC1—Automotive and Hydraulic Tube and Fittings June 1991, and revised June 1992 and June 1993. Rationale statement available.

This document is technically equivalent to ISO/DIS 8434 Part 3.

Table of Contents

Section
1. Scope
2. References
3. Size Designation
4. Material and Manufacture
5. Dimensions and Tolerances
6. Threads
7. Fitting Seat
8. Protection
9. Performance Requirements

Table
1. Working Pressure Ratings Capable of 4 to 1 Minimum Burst
2. Drill Tolerances
3. Maximum Torque to Move Washer
4. Working, Proof, and Minimum Burst Pressures
5. Qualification Test Torque Requirements
6. Fitting Dimensions (Figures 5 to 31)
7. Factors for Face Seal End on Shaped Fittings (Figure 36)
8. Factors for Adjustable O-ring Boss End on Shaped Fittings (Figure 37)
9. Factors for Face Seal End on Straight Fittings (Figures 38 to 39)
10. Dimensions for Straight O-ring Boss Jump Fittings (Figure 40)
11. Factors for Swivel End on Shaped Fittings (Figure 41)

Figure
1. Typical Example Tube and/or Hose Connection to Adapter
1a. Example of Tube and/or Hose Connection to Swivel Elbow
2. Hex Tolerances
3. Face Seal Dimensions
4. Nut
5 to 25 Standard Fittings
26 to 31 Straight Thread Connectors
32 to 33 Standard Sleeve and Plug Assembly
34 Reducing and Expanding Sleeves
35 Reducer Assembly
35a Reducer Assembly
36 90 degree Reducer Elbows
37 90 degree Straight Thread Elbow with Reduced O-ring Boss End
38 Straight Thread Connector with Reduced Face Seal End
39 Reducer Union
40 Straight Thread Connector with Reduced O-ring Boss End
41 90 degree Swivel Elbow with Reduced Swivel End

Appendix
A Surface Acceptance Criteria for Face Seal Fittings
B Test Data Sheet for O-ring Face Seal Fittings
C Instructions and Examples for Calculating Dimensions on Special Size Fittings
D Optional Metric Forging Flat Sizes
E Assembly Instructions for Adjustable Style Fittings in Straight Thread O-ring Port

(R) **1. Scope**—This SAE Standard covers material, dimensional, and performance requirements of steel O-ring face seal fittings for tubing and the O-ring face seal interface and nut portion of hose stem assemblies for nominal tube or hose diameters 6.35 mm (0.250 in) through 38.1 mm (1.500 in). These fittings are intended for general application and hydraulic systems on industrial equipment and commercial products, where elastomeric seals are acceptable to overcome leakage and variations in assembly procedures. These fittings are capable of providing leak-proof full flow connections in hydraulic systems operating from 95 kPa (28 in Hg) vacuum to working pressures shown in Table 1. Since many factors influence the pressure at which a hydraulic system will or will not perform satisfactorily, these values should not be construed as guaranteed minimums. For any application, it is recommended that sufficient testing be conducted and reviewed by both the user and manufacturer to assure that required performance levels will be safe. See Figure 1.

The rated working pressure of a hose assembly comprising SAE J1453 hose stem connections and SAE J517 hose shall not exceed the lower of the two working pressure rated values.

TABLE 1—WORKING PRESSURE RATINGS CAPABLE OF 4 TO 1 MINIMUM BURST

Nom SAE Dash Size	Nom Tube OD mm	Nom Tube OD in	Straight Fittings MPa	Straight Fittings psi	Adjustable Style Fittings MPa	Adjustable Style Fittings psi
-4	6.35	0.250	41.3	6000	41.3	6000
-6	9.52	0.375	41.3	6000	41.3	6000
-8	12.70	0.500	41.3	6000	41.3	6000
-10	15.88	0.625	41.3	6000	41.3	6000
-12	19.05	0.750	41.3	6000	41.3	6000
-16	25.40	1.000	41.3	6000	34.5	5000
-20	31.75	1.250	27.5	4000	27.5	4000
-24	38.10	1.500	27.5	4000	20.7	3000

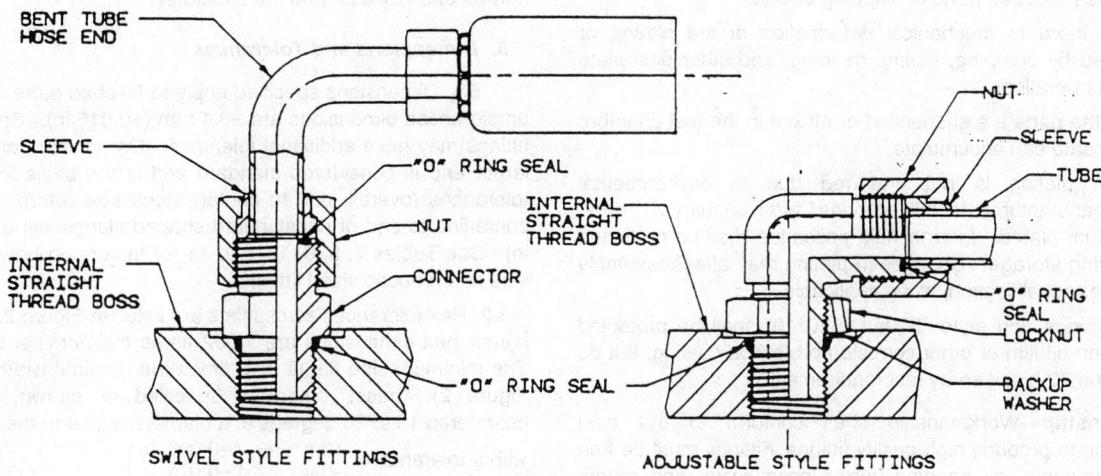

FIGURE 1—TYPICAL EXAMPLE OF TUBE AND/OR HOSE CONNECTION TO ADAPTER

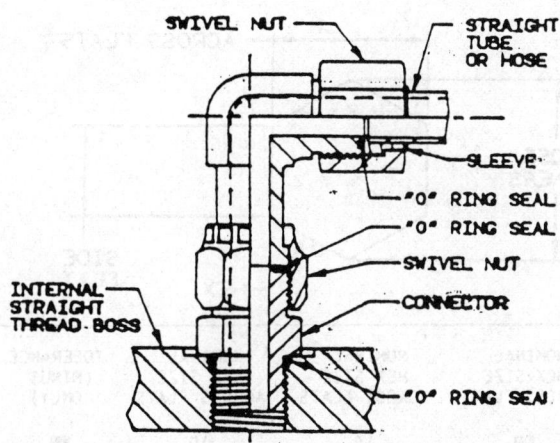

FIGURE 1a—EXAMPLE OF TUBE AND/OR HOSE CONNECTION TO SWIVEL ELBOW

2. References

2.1 Applicable Documents—The following publications form a part of this specification to the extent specified herein. The latest issue of SAE publications shall apply.

2.1.1 SAE Publications—Available from SAE, 400 Commonwealth Drive, Warrendale, PA 15096-0001.

SAE J515—Hydraulic "O" Rings

SAE J343—Tests and Procedures for SAE 100R Series Hydraulic Hose and Hose Assemblies

SAE J846—Coding Systems for Identification of Fluid Conductors and Connectors

2.1.2 ASTM Publication—Available from ASTM, 1916 Race Street, Philadephia, PA 19103-1187.

ASTM B 117—Method of Salt Spray (Fog) Testing

2.1.3 ANSI Publications—Available from ANSI, 11 West 42nd Street, New York, NY 10036-8002.

ANSI B1.1—Screw Thread

ANSI B46.1—Surface Texture

3. Size Designation

3.1 Fitting sizes are designated by the nominal outside diameter of the tubing or nominal inside diameter of hose. (See SAE J846.)

4. Material and Manufacture

4.1 Material—Fittings shall be made from manufacturer's standard steel that will fulfill the performance requirements in Section 9.

(R) 4.2 Finish—The external surfaces and threads of all carbon steel parts shall be plated or coated with a suitable material that passes a 72 h salt spray test in accordance with ASTM B 117. Any appearance of red rust during the 72 h salt spray test shall be considered failure, except for the following:

a. All internal fluid passages.

b. Edges such as hex points, serrations, and crests of threads where there may be mechanical deformation of the plating or coating typical of mass-produced parts or shipping effects.

c. Areas where there is mechanical deformation of the plating or coating caused by crimping, flaring, bending, and other post-plate metal forming operations.

d. Areas where the parts are suspended or affixed in the test chamber where condensate can accumulate.

NOTE—Cadmium plating is not preferred due to environmental reasons. Parts manufactured to this standard after January 1, 1997, shall not be cadmium plated. Internal fluid passages shall be protected from corrosion during storage. Changes in plating may affect assembly torques and require requalification, when applicable.

Braze-on type fittings and style "B" nut (520110) shall be protected from corrosion by an oil film or other corrosion protection coating, but do not need to meet the 72 h salt spray test requirement.

4.3 Workmanship—Workmanship shall conform to the best commercial practice to produce high-quality fittings. Fittings must be free from visual contaminants, all hanging burrs, loose scale, and slivers which might be dislodged in usage, and any other defects that might affect the function of the parts.

4.4 Construction—Fittings may be made by forging, cold heading, or machined from bar stock. Carbon steel fittings fabricated from multiple components must be bonded together with materials having a melting point of not less than 996 °C (1825 °F).

4.5 Fitting bodies and tube nuts must be permanently marked with individual supplier's trademark or code identifier, unless otherwise agreed upon by user and manufacturer.

5. Dimensions and Tolerances

5.1 Dimensions specified apply to finished parts. Tolerances on all untoleranced dimensions are ±0.4 mm (±0.016 in). Special nonstandard fittings may have additional tolerance. On nonstandard size fittings, the larger end is considered standard and is the basis for dimensions and tolerances (overall, end to center, stock size, etc.). The tolerance on centerline to end of nonstandard shaped fittings will be ±1.5 mm (±0.06 in). See Tables 7, 8, 9, 10, and 11 for factors on how to calculate jump size and reduced sized fittings.

5.2 Hex tolerances across flats are listed in Figure 2. Minimum across corner hex dimensions are 1.092 times the nominal width across flats. The minimum side flat is 0.43 times the nominal width across flats (see Figure 2). Unless otherwise specified or shown, hex corners are chamfered 15 to 30 degrees to a diameter equal to the width across flats, with a tolerance of $^{+0.0}_{-0.4}$ mm $\left(^{+0.000}_{-0.016} \text{ in}\right)$.

Alternatively, on connections other than SAE straight thread, a 5 degree chamfer starting at the undercut diameter behind the threads or outside diameter of the threads shall be allowed, providing the hex width at corners is not reduced below that produced by the 30 degree chamfer previously described.

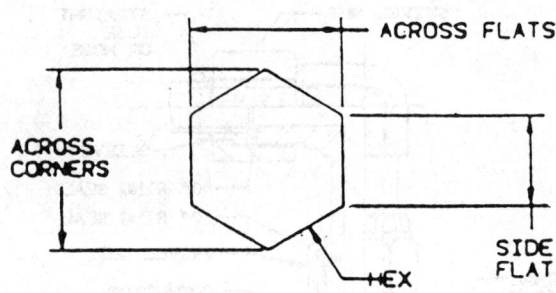

NOMINAL HEX SIZE ACROSS FLATS mm OVER	NOMINAL HEX SIZE ACROSS FLATS mm INCL	NOMINAL HEX SIZE ACROSS FLATS in OVER	NOMINAL HEX SIZE ACROSS FLATS in INCL	TOLERANCE (MINUS ONLY) mm	TOLERANCE (MINUS ONLY) in
—	19.05	—	0.750	0.3	0.012
19.05	25.40	0.750	1.000	0.4	0.016
25.40	34.92	1.000	1.375	0.5	0.020
34.92	AND UP	1.375	AND UP	0.8	0.031

FIGURE 2—HEX TOLERANCES

(R) 5.3 The across flat dimensions of elbows, tees, bulkhead, and swivel fittings shown as "JJ forging" or "JJ barstock" in Table 6 are intended to be for nominal inch wrench sizes with a minus tolerance only. The basic forging size may be increased up to the maximum size shown for barstock, but the size selected must be a nominal inch wrench size across flats with minus tolerance only.

NOTE—For optional metric stock sizes, see Appendix D.

5.4 Where passages in straight fittings are machined from opposite ends, the offset at the meeting point shall not exceed 0.4 mm (0.016 in). No cross-sectional area at a junction of passages shall be less than that of the smallest passage.

5.5 Angular tolerance on axis of ends on elbows, tees, and crosses is ±2.5 degrees for sizes up to 9.52 mm (0.375 in) and ±1.5 degrees for all larger sizes.

5.6 Details of contour are optional with the manufacturer, providing the tabulated dimensions are maintained. Wrench flats on elbows and tees must meet the tabulated dimensions. Abrupt reduction of a section must be avoided. Junctions of small external sections and adjoining relatively heavy sections must be blended by means of ample fillets.

5.7 Tolerances on hole diameters designated as drill diameter in other dimensional tables shall be as shown in Table 2.

TABLE 2—DRILL TOLERANCES

Drill Size Range mm	Drill Size Range in	Tolerance Plus mm	Tolerance Minus mm	Tolerance Plus in	Tolerance Minus in
0.35 - 6.25	0.0135 - 0.246	0.08	0.08	0.003	0.003
6.35 - 12.70	0.250 - 0.500	0.10	0.10	0.004	0.004
13.10 - 19.05	0.516 - 0.750	0.13	0.13	0.005	0.005
19.40 - 25.40	0.765 - 1.000	0.18	0.13	0.007	0.005
25.80 - 38.10	1.016 - 1.500	0.20	0.13	0.008	0.005
38.50 - and up	1.516 - and up	0.25	0.13	0.010	0.005

5.8 The O-ring washer must be clinched to fitting with a tight slip fit to an interference fit. The slip fit shall be tight enough so that washer cannot be shaken loose to cause it to drop from its uppermost position by its own weight. The interference fit shall not require a locknut torque more than that indicated in Table 3. Position the washer farthest from the end of the fittings as shown in Figure 9. Care must be taken not to clinch washer on the transition area between diameter Y and locknut thread which results in a loose washer when it is repositioned at assembly. Washer flatness allowance is given in Table 3. Any surface out of flatness must be uniform (not wavy) and concave with respect to the O-ring boss end of the fitting. Maximum torque to move washer is shown in Table 3.

6. Threads

6.1 Straight threads must be class 2A or 2B in accordance with ANSI B1.1 screw thread specification, except for internal thread minor diameter (see Figure 4, Note C). External class 2A threads which are plated or coated may exceed 2A diameters but shall not exceed maximum 3A diameters. Internal threads of all classes must be within specified limits after plating or coating.

6.2 When external threads are produced by thread rolling and the body is not undercut, the unthreaded area adjacent to the shoulder may be reduced to the minimum pitch diameter.

7. Fitting Seat

7.1 The face seal end dimensions shall conform to Figure 3. Surface finish of O-ring groove, and mating surfaces, shall be as shown in Figure 3 in accordance with ANSI B46.1.

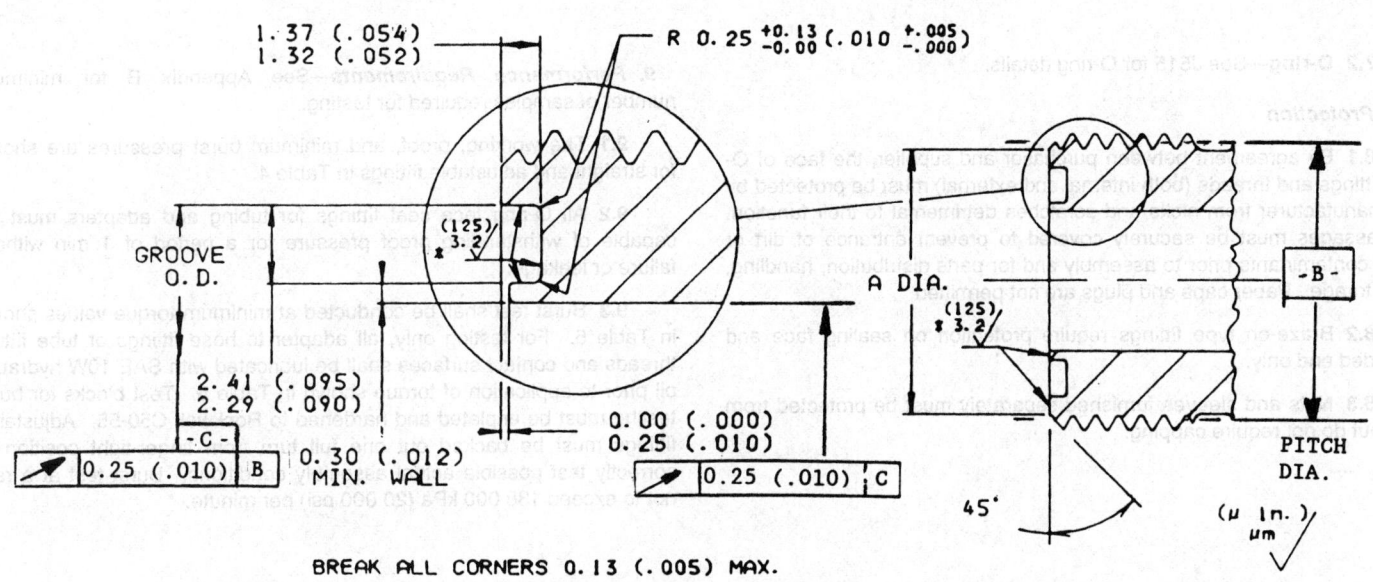

Nom SAE Dash Size	Nom Tube Dia mm	Nom Tube Dia in	Groove OD mm	Groove OD in	A Chamfer Dia mm	A Chamfer Dia in	O-Ring Seal SAE J515 Seal Size (90 Durometer, Shore A)
-4	6.35	0.250	11.00 ± 0.08	0.433 ± 0.003	12.15 ± 0.15	0.478 ± 0.005	-011
-6	9.52	0.375	12.60 ± 0.08	0.496 ± 0.003	15.10 ± 0.25	0.594 ± 0.010	-012
-8	12.70	0.500	15.77 ± 0.08	0.621 ± 0.003	18.25 ± 0.25	0.719 ± 0.010	-014
-10	15.88	0.625	19.00 ± 0.08	0.748 ± 0.003	22.60 ± 0.40	0.890 ± 0.015	-016
-12	19.05	0.750	22.17 ± 0.10	0.873 ± 0.004	27.00 ± 0.40	1.063 ± 0.015	-018
-16	25.40	1.000	26.87 ± 0.10	1.058 ± 0.004	33.35 ± 0.40	1.313 ± 0.015	-021
-20	31.75	1.250	33.25 ± 0.13	1.309 ± 0.005	39.75 ± 0.40	1.564 ± 0.015	-025
-24	38.10	1.500	41.17 ± 0.13	1.621 ± 0.005	47.65 ± 0.40	1.876 ± 0.015	-029

(R) FIGURE 3—FACE SEAL DIMENSIONS

TABLE 3—MAXIMUM TORQUE TO MOVE WASHER AND WASHER FLATNESS

Nom Tube OD	Nom Tube OD	Nom SAE Dash Size	Thd Size Inch	Nom Hex Size	Nom Hex Size	Maximum Torque To Move Washer	Maximum Torque To Move Washer	Washer Flatness Allowance	Washer Flatness Allowance
mm	in			mm	in	N·m	lb·in	mm	in
3.18	0.125	-2	5/16-24	12.70	0.500	1	9	0.25	0.010
4.76	0.187	-3	3/8-24	14.29	0.562	3	26	0.25	0.010
6.35	0.250	-4	7/16-20	15.88	0.625	4	35	0.25	0.010
7.94	0.312	-5	1/2-20	17.46	0.687	5	44	0.25	0.010
9.52	0.375	-6	9/16-18	19.05	0.750	7	62	0.25	0.010
12.70	0.500	-8	3/4-16	23.81	0.937	10	88	0.25	0.010
15.88	0.625	-10	7/8-14	26.99	1.062	12	106	0.25	0.010
19.05	0.750	-12	1-1/16-12	34.92	1.375	15	133	0.40	0.016
22.22	0.875	-14	1-3/16-12	38.10	1.500	18	159	0.40	0.016
25.40	1.000	-16	1-5/16-12	41.28	1.625	20	177	0.40	0.016
31.75	1.250	-20	1-5/8-12	47.62	1.875	25	221	0.50	0.020
38.10	1.500	-24	1-7/8-12	53.98	2.125	30	265	0.50	0.020
50.80	2.000	-32	2-1/2-12	69.85	2.750	40	353	0.50	0.020

7.2 O-ring—See J515 for O-ring details.

8. Protection

8.1 By agreement between purchaser and supplier, the face of O-ring fittings and threads (both internal and external) must be protected by the manufacturer from nicks and scratches detrimental to their function. All passages must be securely covered to prevent entrance of dirt or other contaminants prior to assembly and for parts distribution, handling, and storage. Paper caps and plugs are not permitted.

8.2 Braze-on type fittings require protection on sealing face and threaded end only.

8.3 Nuts and sleeves furnished separately must be protected from rust but do not require capping.

9. Performance Requirements

—See Appendix B for minimum number of samples required for testing.

9.1 The working, proof, and minimum burst pressures are shown for straight and adjustable fittings in Table 4.

9.2 All O-ring face seal fittings for tubing and adapters must be capable of withstanding proof pressure for a period of 1 min without failure or leakage.

9.3 Burst test shall be conducted at minimum torque values shown in Table 5. For testing only, all adapter to hose fittings or tube fitting threads and contact surfaces shall be lubricated with SAE 10W hydraulic oil prior to application of torque shown in Table 5. Test blocks for burst testing must be unplated and hardened to Rockwell C50-55. Adjustable fittings must be backed out one full turn from finger-tight position to correctly test possible actual assembly conditions. Burst test at a rate not to exceed 138 000 kPa (20 000 psi) per minute.

TABLE 4—WORKING, PROOF, AND MINIMUM BURST PRESSURES

Nom SAE Dash Size	Straight Fittings Working MPa	Straight Fittings Working psi	Straight Fittings Proof MPa	Straight Fittings Proof psi	Straight Fittings Min Burst MPa	Straight Fittings Min Burst psi	Adjustable Fittings Working MPa	Adjustable Fittings Working psi	Adjustable Fittings Proof MPa	Adjustable Fittings Proof psi	Adjustable Fittings Min Burst MPa	Adjustable Fittings Min Burst psi
-4	41.3	6000	82.5	12 000	165.0	24 000	41.3	6000	82.5	12 000	165.0	24 000
-6	41.3	6000	82.5	12 000	165.0	24 000	41.3	6000	82.5	12 000	165.0	24 000
-8	41.3	6000	82.5	12 000	165.0	24 000	41.3	6000	82.5	12 000	165.0	24 000
-10	41.3	6000	82.5	12 000	165.0	24 000	41.3	6000	82.5	12 000	165.0	24 000
-12	41.3	6000	82.5	12 000	165.0	24 000	41.3	6000	82.5	12 000	165.0	24 000
-16	41.3	6000	82.5	12 000	165.0	24 000	34.5	5000	69.0	10 000	138.0	20 000
-20	27.5	4000	55.0	8000	110.0	16 000	27.5	4000	55.0	8000	110.0	16 000
-24	27.5	4000	55.0	8000	110.0	16 000	20.7	3000	41.3	6000	82.5	12 000

TABLE 5—QUALIFICATION TEST TORQUE REQUIREMENTS

NOM SAE DASH SIZE	NOMINAL TUBE OD (mm)	NOMINAL TUBE OD (in)	O-RING FACE SEAL END THREAD SIZE (in)	O-RING FACE SEAL END SWIVEL NUT TORQUE (N·m)	O-RING FACE SEAL END SWIVEL NUT TORQUE (lb-ft)	OVER TORQUE (N·m)	OVER TORQUE (lb-ft)	SAE O-RING BOSS END THREAD SIZE (in)	SAE O-RING BOSS END STRAIGHT FITTING OR LOCKNUT TORQUE (N·m)	SAE O-RING BOSS END STRAIGHT FITTING OR LOCKNUT TORQUE (lb-ft)
-3	4.76	0.188	a	a	a	a	a	3/8-24	11-13	8-10
-4	6.35	0.250	9/16-18	14-16	10-12	32	24	7/16-20	20-22	14-16
-5	7.94	0.312	a	a	a	a	a	1/2-20	24-27	18-20
-6	9.52	0.375	11/16-16	24-27	18-20	54	40	9/16-18	33-35	24-26
-8	12.70	0.500	13/16-16	43-47	32-35	81	60	3/4-16	68-78	50-60
-10	15.88	0.625	1-14	60-68	46-50	136	100	7/8-14	98-110	72-80
-12	19.05	0.750	1-3/16-12	90-95	65-70	180	140	1-1/16-12	170-183	125-135
-14	22.22	0.875	1-3/16-12	90-95	65-70	180	140	1-3/16-12	215-245	160-180
-16	25.40	1.000	1-7/16-12	125-135	92-100	270	200	1-5/16-12	270-300	200-220
-20	31.75	1.250	1-11/16-12	170-190	125-140	380	280	1-5/8-12	285-380	210-280
-24	38.10	1.500	2-12	200-225	150-165	450	330	1-7/8-12	370-490	270-360

a O-ring face seal type end not defined for this tube size.

9.4 All tube fittings and adapters must pass a cyclic endurance test for one million cycles at 133% of corresponding working pressures for straight and adjustable fittings per Table 4. Cycle test to be conducted at minimum torque values shown in Table 5. For testing only, all adapter to hose fittings or tube fitting threads and contact surfaces shall be lubricated with SAE 10W hydraulic oil prior to application of torque shown in Table 5. Cycle rate shall be uniform at 30 to 75 cpm and shall conform to magnitude and frequency to the wave pattern shown in Figure 1 of SAE J343.

9.5 Components that require brazing to assemble and all style "B" 520110 nuts must be processed through a copper braze cycle of 996 to 1150 °C, with the cycle time sufficiently long to permit parts to reach above temperature and then be cooled in a protective atmosphere to prevent scaling. After annealing and before burst, cyclic endurance, or torque testing, the nuts must be plated per 4.2.

9.6 Fitting shall be capable of withstanding 95 MPa (28 in Hg) vacuum without leakage for 5 min.

(R) 9.7 The O-ring face seal interface and nut portion of hose and tube stem assemblies must meet proof, burst, and impulse requirements of the hose to which they are attached.

9.8 The standard face seal O-ring used for test shall be nitrile (NBR) rubber with a durometer "A" hardness of 90. (See Figure 3.)

9.9 Fitting swivel nuts shall be capable of withstanding the over torque qualification test with no indication of failure. For testing only, fitting threads and contact surfaces shall be lubricated with SAE 10W hydraulic oil prior to application of over torque specified in Table 5. For torque testing, use an unplated steel mandrel hardened to Rockwell C40-45. Fittings shall be restrained during test and the wrench shall be located at the threaded end of the nut hex.

Definition of failure after torque testing.

a. The nut cannot be removed by hand after breakaway.
b. The nut cannot swivel freely by hand.
c. The nut will not retract to its original position by hand.
d. Any visible cracks or severe deformation that would render nut unusable.

9.10 Parts which pass burst test or over torque test must not be tested further, used, or returned to stock.

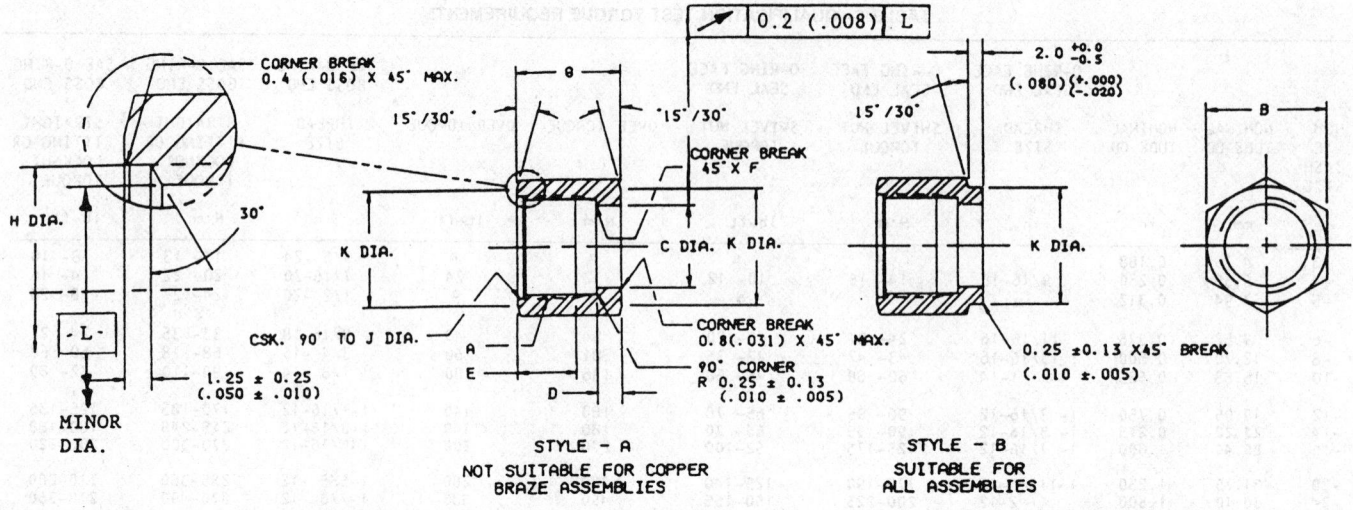

NOM TUBE OD	NOM TUBE OD	A THREAD SIZE	THREAD MINOR DIA[c]	THREAD MINOR DIA[c]	THREAD MINOR DIA[c]	THREAD MINOR DIA[c]	B[a] HEX	B[a] HEX	C DIA	C DIA	D LG	D LG
mm	in	in	mm min	mm max	in min	in max	mm ±0.08	in ±0.003	mm ±0.25	in ±0.010		
6.35	0.250	9/16-18	12.90	13.08	0.508	0.515	17.46	0.687	10.46	0.412	2.50	0.098
9.52	0.375	11/16-16	15.90	16.10	0.626	0.634	20.64	0.812	13.51	0.532	3.00	0.118
12.70	0.500	13/16-16	19.08	19.28	0.751	0.759	23.81	0.938	16.56	0.652	5.00	0.197
15.88	0.625	1-14	23.60	23.83	0.929	0.938	28.58	1.125	21.06	0.829	4.00	0.157
19.05	0.750	1-3/16-12	28.02	28.32	1.103	1.115	34.92	1.375	24.13	0.950	5.00	0.197
[b]22.22	0.875	1-3/16-12	28.02	28.32	1.103	1.115	34.92	1.375	25.04	0.986	5.00	0.197
25.40	1.000	1-7/16-12	34.37	34.67	1.353	1.365	41.28	1.625	29.06	1.144	6.00	0.236
31.75	1.250	1-11/16-12	40.72	41.02	1.603	1.615	47.62	1.875	35.94	1.415	6.00	0.236
38.10	1.500	2-12	48.67	48.97	1.916	1.928	57.15	2.250	43.89	1.728	6.00	0.236

NOM TUBE OD	NOM TUBE OD	E MIN FULL THREAD	E MIN FULL THREAD	F CHAMFER	F CHAMFER	G LG	G LG	H DIA	H DIA	J DIA	J DIA	K DIA	K DIA
mm	in	mm min	in min	mm ±0.13	in ±0.005	mm ±0.25	in ±0.010	mm ±0.25	in ±0.010	mm +0.4 -0.0	in +0.016 -0.000	mm ±0.3	in ±0.01
6.35	0.250	9.0	0.35	0.13	0.005	14.70	0.580	—	—	14.4	0.567	17.0	0.67
9.52	0.375	10.3	0.41	0.13	0.005	17.00	0.670	—	—	17.6	0.693	20.3	0.80
12.70	0.500	11.9	0.47	0.13	0.005	21.00	0.827	—	—	20.8	0.819	23.5	0.92
15.88	0.625	14.5	0.57	0.25	0.010	23.50	0.925	25.91	1.020	—	—	28.2	1.11
19.05	0.750	15.5	0.61	0.25	0.010	26.00	1.024	30.68	1.208	—	—	34.6	1.36
[b]22.22	0.875	15.5	0.61	0.25	0.010	30.86	1.215	30.68	1.208	—	—	34.6	1.36
25.40	1.000	16.0	0.63	0.38	0.015	27.80	1.095	37.03	1.458	—	—	41.0	1.61
31.75	1.250	16.0	0.63	0.38	0.015	27.80	1.095	43.38	1.708	—	—	47.2	1.86
38.10	1.500	16.0	0.63	0.38	0.015	27.80	1.095	51.31	2.020	—	—	56.9	2.24

[a] For hex tolerance, see 5.2 and Figure 2.
[b] For use with expanding sleeve 19.05 to 22.22, see Figure 34.
[c] Modified minor dia (not the minor dia listed in ANSI B1.1).

FIGURE 4—NUT (520110)

22.263

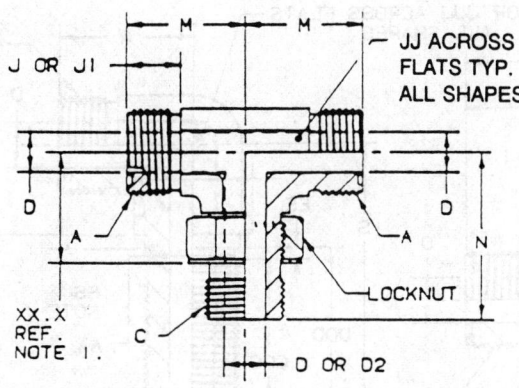

(R) FIGURE 5—STRAIGHT THREAD BRANCH TEE (520429)

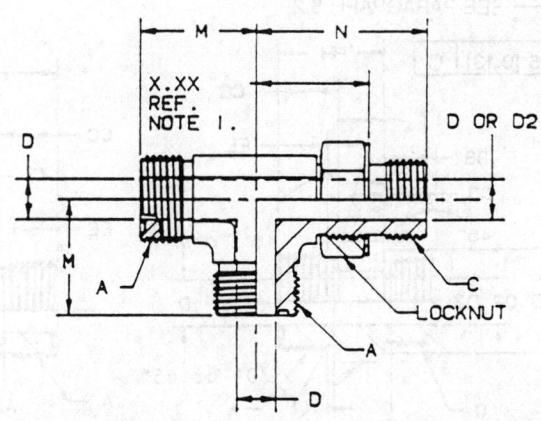

FIGURE 6—STRAIGHT THREAD RUN TEE (520428)

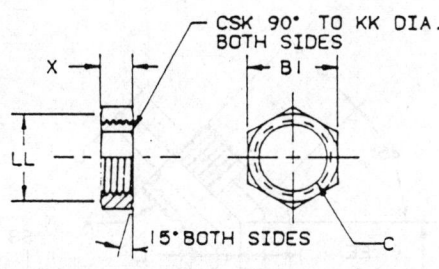

FIGURE 7—STRAIGHT THREAD LOCKNUT (520117)

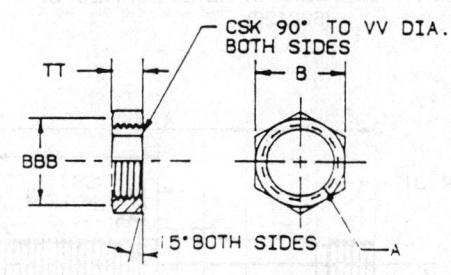

FIGURE 8—BULKHEAD LOCKNUT (520118)

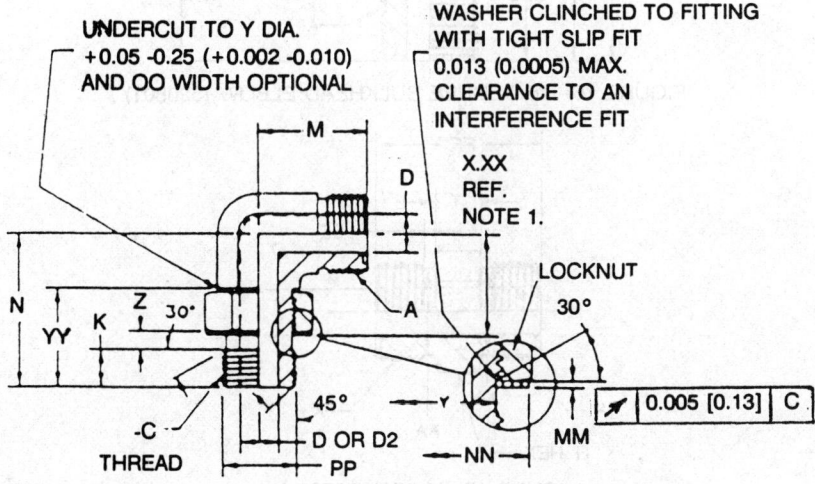

(R) FIGURE 9—90 DEGREE STRAIGHT THREAD ELBOW (520220)

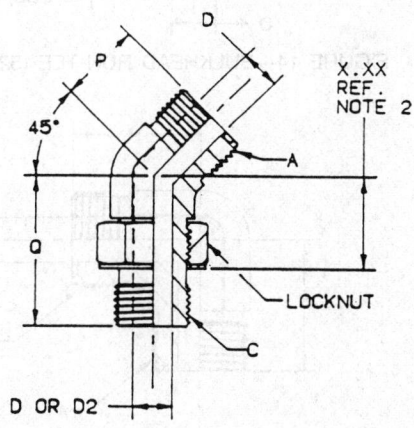

FIGURE 10—45 DEGREE STRAIGHT THREAD ELBOW (520320)

NOTES:
1. Ref dim. calculated by formula ref = N−(K+Z−MM).
2. Ref dim. calculated by formula ref = Q−(K+Z−MM).
For O-ring groove dimensions and thread chamfer diameter, see Figure 3.

FIGURES 5 TO 25—FITTINGS

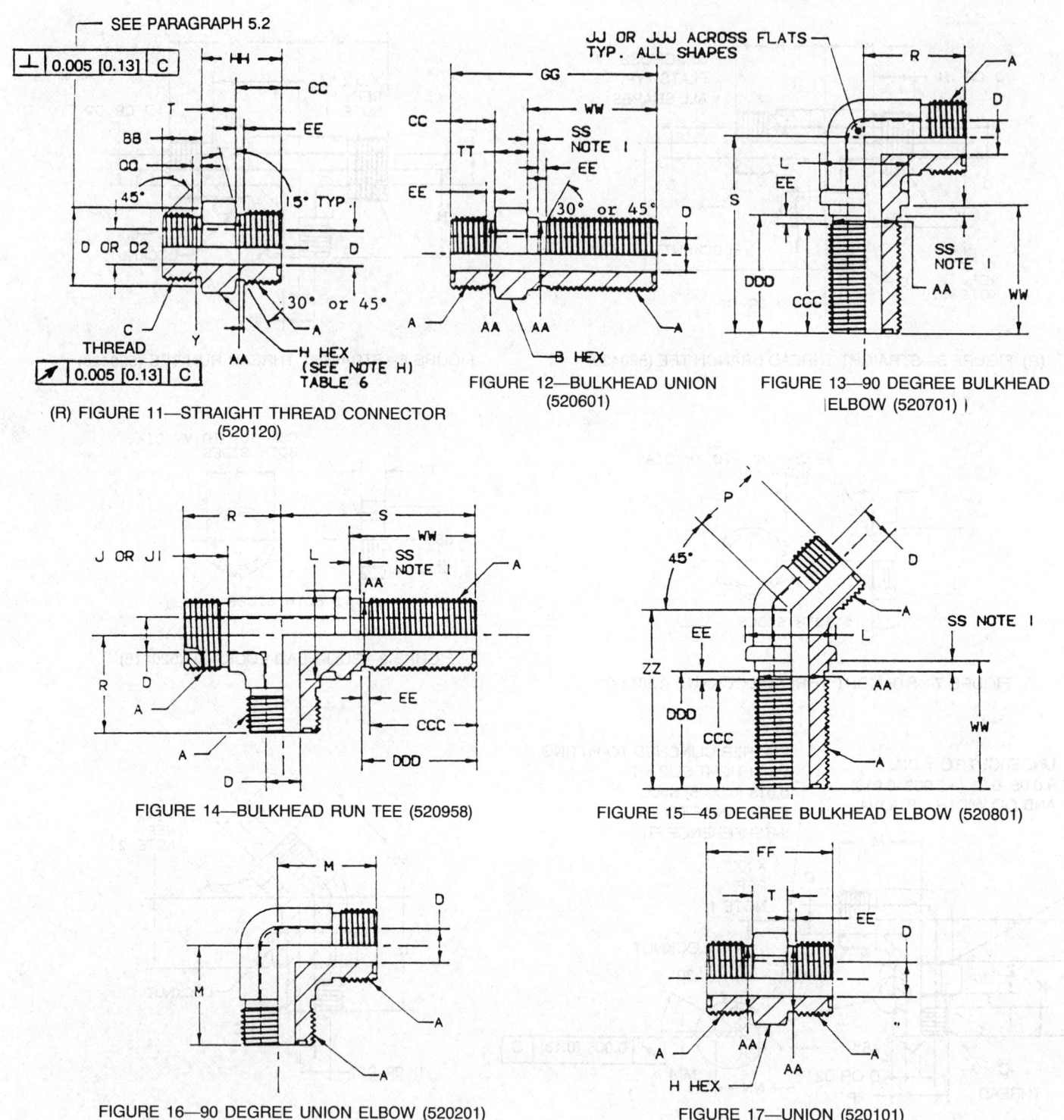

NOTE 1: Recommended clearance hole for bulkhead fittings is 0.4 (0.015) over major thread diameter. Diameter of SS is same as major thread diameter.
For O-ring groove dimensions and thread chamfer diameter, see Figure 3.

FIGURES 5 TO 25—(CONTINUED)

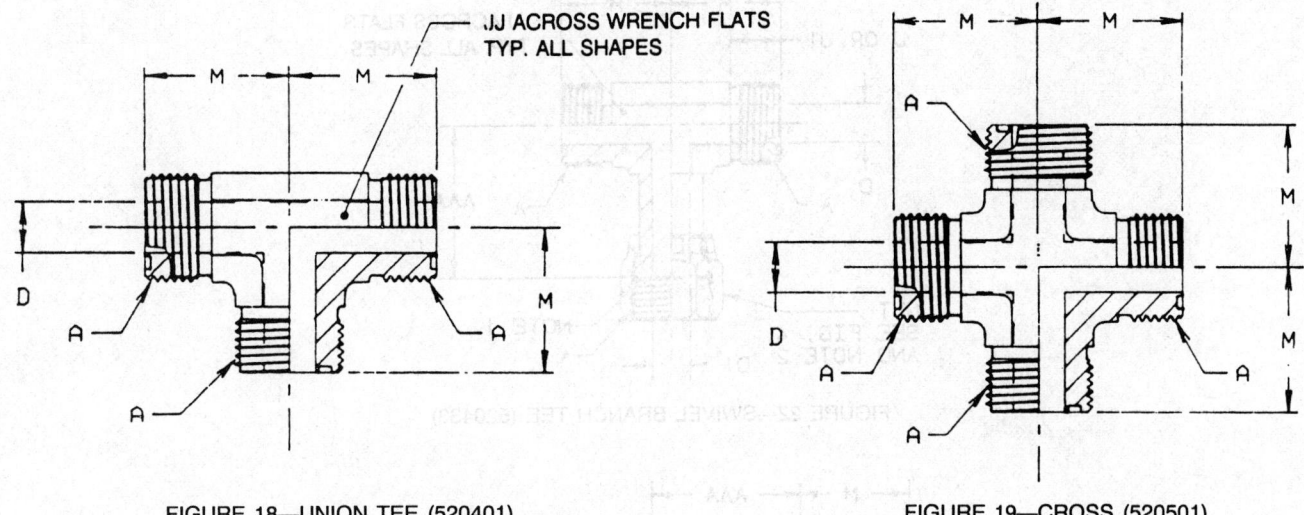

FIGURE 18—UNION TEE (520401)

FIGURE 19—CROSS (520501)

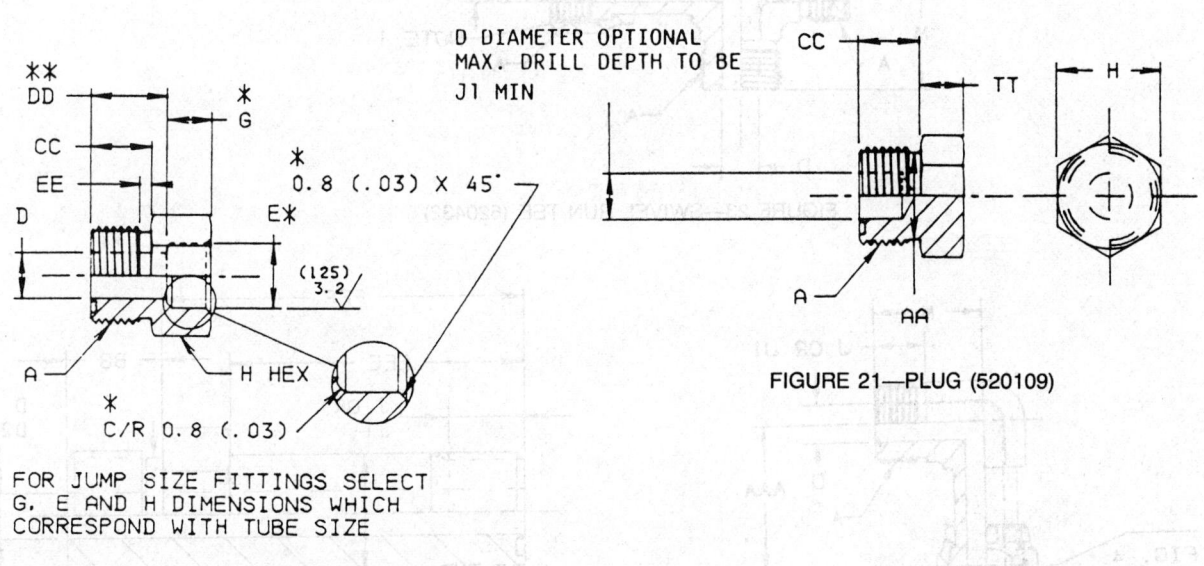

FIGURE 20—MALE CONNECTOR (520104)

FIGURE 21—PLUG (520109)

*Dimensions are for silver brazing. Other dimensions may apply for other joining methods.
**DD dimension remains constant for jump size fittings.
For O-ring groove dimensions and thread chamfer diameter, see Figure 3.

FIGURES 5 TO 25—(CONTINUED)

22.266

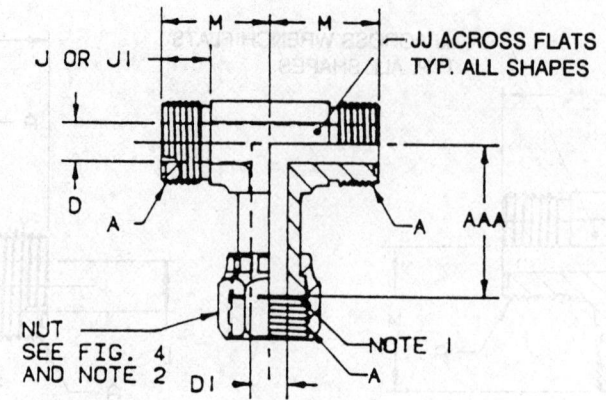

FIGURE 22—SWIVEL BRANCH TEE (520433)

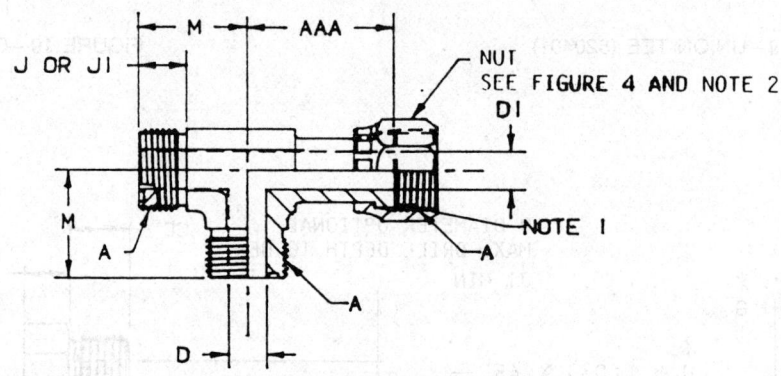

FIGURE 23—SWIVEL RUN TEE (520432)

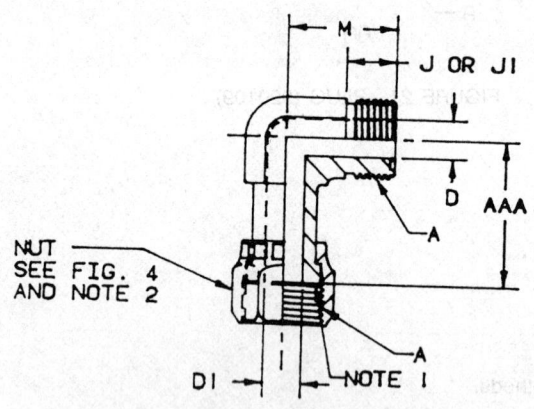

FIGURE 24—90 DEGREE SWIVEL ELBOW (520221)

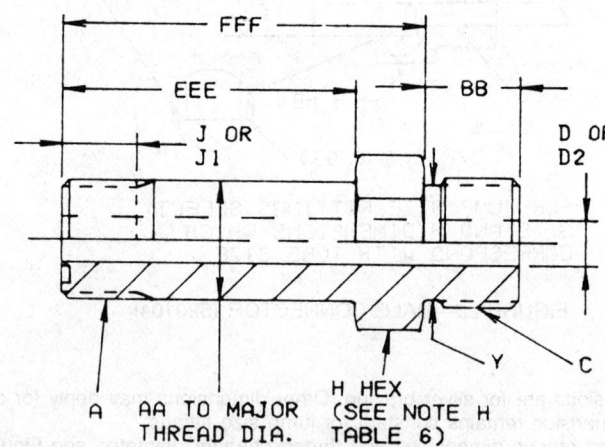

FIGURE 25—STRAIGHT THREAD CONNECTOR, LONG (520122)

Notes:
1. Shoulder face must be flush or exposed when nut is fully retracted. See Figure 32 for sleeve shoulder dimensions.
2. The design and method of attaching the swivel nut shall be optional with the manufacturer, providing the tabulated dimensions are maintained, the nut turns freely and meets the performance requirements in Section 9.

FIGURES 5 TO 25—(CONTINUED)

22.267

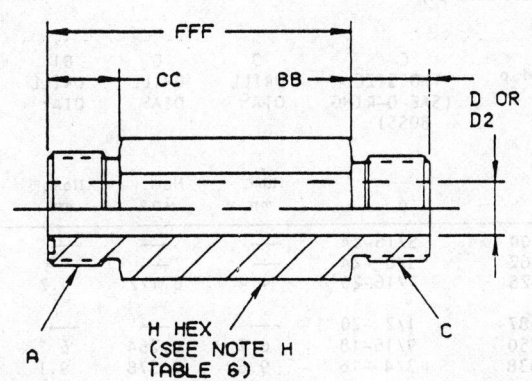

FIGURE 26—STRAIGHT THREAD CONNECTOR, LONG HEX (520122)

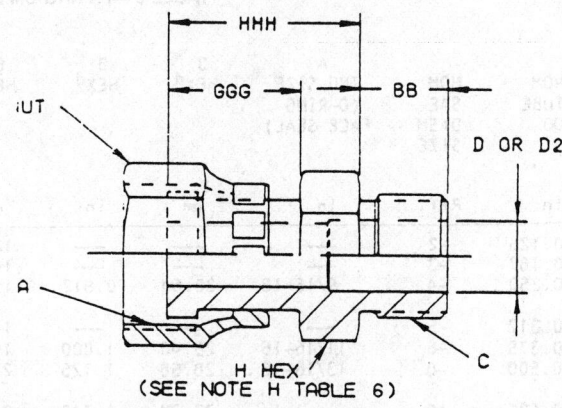

FIGURE 27—STRAIGHT THREAD SWIVEL CONNECTOR (520181)

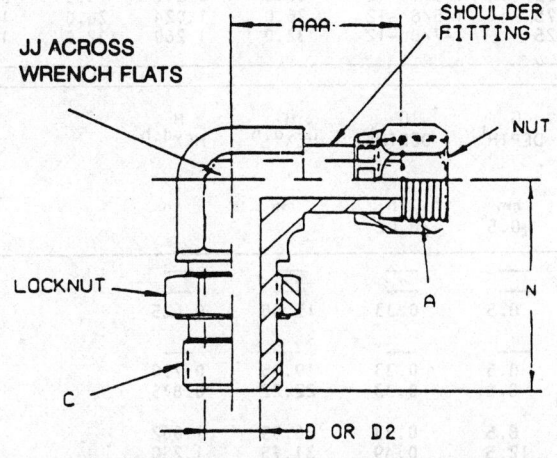

(R) FIGURE 28—90 DEGREE STRAIGHT THREAD SWIVEL ELBOW CONNECTOR (520281)

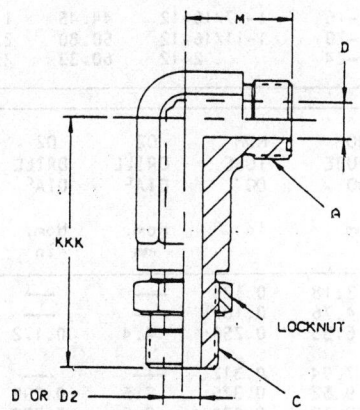

FIGURE 29—90 DEGREE STRAIGHT THREAD ELBOW, LONG (521520)

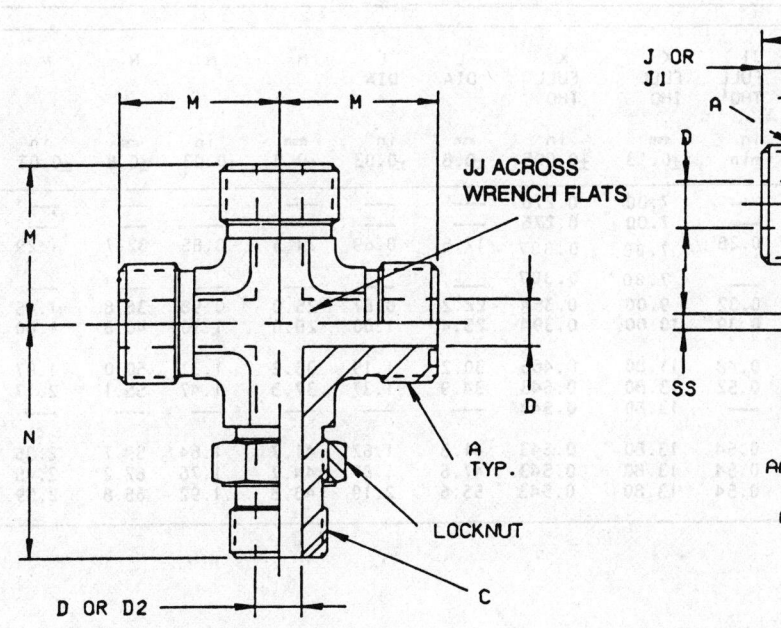

(R) FIGURE 30—STRAIGHT THREAD CROSS (520520)

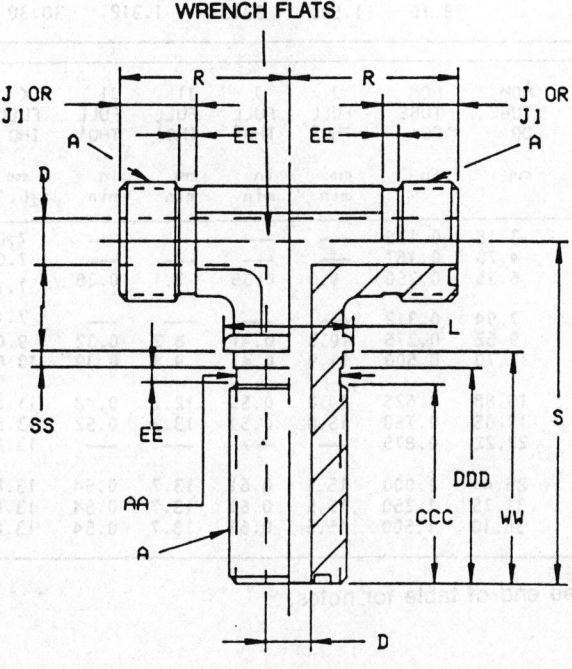

(R) FIGURE 31—BULKHEAD BRANCH TEE (520959)

TABLE 6*—FITTING DIMENSIONS (FIGURES 5 TO 31)[n]

NOM TUBE OD	NOM TUBE OD	NOM SAE DASH SIZE	A THD SIZE (O-RING FACE SEAL)	B HEX[d]	B HEX[d]	B1 HEX[d,p]	B1 HEX[d,p]	C THD SIZE (SAE O-RING BOSS)	D DRILL DIA[c]	D DRILL DIA[c]	D1 DRILL DIA[c]	D1 DRILL DIA[c]
mm	in	Ref.	in	mm	in	mm	in	in	Nom. mm	Nom. in	Nom. mm	Nom. in
3.18	0.125	-2	—	—	—	12.70	0.500	5/16-24	—	—	—	—
4.76	0.187	-3	—	—	—	14.29	0.562	3/8-24	—	—	—	—
6.35	0.250	-4	9/16-18	20.64	0.812	15.88	0.625	7/16-20	4.4	0.172	4.2	0.165
7.94	0.312	-5	—	—	—	17.46	0.687	1/2-20	—	—	—	—
9.52	0.375	-6	11/16-16	25.40	1.000	19.05	0.750	9/16-18	6.7	0.264	6.7	0.264
12.70	0.500	-8	13/16-16	28.58	1.125	23.81	0.938	3/4-16	9.6	0.378	9.1	0.358
15.88	0.625	-10	1-14	33.34	1.312	26.99	1.062	7/8-14	12.3	0.484	11.5	0.453
19.05	0.750	-12	1-3/16-12	38.10	1.500	34.92	1.375	1-1/16-12	15.5	0.609	13.9	0.547
22.22	0.875	-14	—	—	—	38.10	1.500	1-3/16-12	—	—	—	—
25.40	1.000	-16	1-7/16-12	44.45	1.750	41.28	1.625	1-5/16-12	20.6	0.811	19.9	0.783
31.75	1.250	-20	1-11/16-12	50.80	2.000	47.62	1.875	1-5/8-12	26.0	1.024	26.0	1.024
38.10	1.500	-24	2-12	60.33	2.375	53.98	2.125	1-7/8-12	32.0	1.260	32.0	1.260

NOM TUBE OD	NOM TUBE OD	D2 DRILL DIA[c]	D2 DRILL DIA[c]	E DIA[j]	E DIA[j]	G DEPTH[k]	G DEPTH[k]	H HEX[d,h]	H HEX[d,h]
mm	in	Nom. mm	Nom. in	mm ±0.05	in ±0.002	mm ±0.5	in ±0.02	mm	in
3.18	0.125	—	—	—	—	—	—	—	—
4.76	0.187	—	—	—	—	—	—	—	—
6.35	0.250	4.4	0.172	6.50	0.256	8.5	0.33	15.88	0.625
7.94	0.312	—	—	—	—	—	—	—	—
9.52	0.375	7.5	0.295	9.68	0.381	8.5	0.33	19.05	0.750
12.70	0.500	9.9	0.390	12.85	0.506	8.5	0.33	22.22	0.875
15.88	0.625	12.3	0.484	16.03	0.631	8.5	0.33	26.99	1.062
19.05	0.750	15.5	0.609	19.23	0.757	12.5	0.49	31.75	1.250
22.22	0.875	—	—	—	—	—	—	—	—
25.40	1.000	21.4	0.843	25.58	1.007	14.0	0.55	38.10	1.500
31.75	1.250	27.4	1.079	31.95	1.258	14.0	0.55	44.45	1.750
38.10	1.500	33.3	1.312	38.30	1.508	14.0	0.55	53.98	2.125

NOM TUBE OD	NOM TUBE OD	J FULL THD[e]	J FULL THD[e]	J1 FULL THD[f]	J1 FULL THD[f]	K FULL THD	K FULL THD	L DIA	L DIA	M	M	N	N
mm	in	mm min	in min	mm min	in min	mm ±0.13	in ±0.005	mm ±0.8	in ±0.03	mm ±0.8	in ±0.03	mm ±0.8	in ±0.03
3.18	0.125	—	—	—	—	7.00	0.276	—	—	—	—	—	—
4.76	0.187	—	—	—	—	7.00	0.276	—	—	—	—	—	—
6.35	0.250	9.0	0.35	7.1	0.28	7.80	0.307	17.5	0.69	21.5	0.85	32.7	1.29
7.94	0.312	—	—	—	—	7.80	0.307	—	—	—	—	—	—
9.52	0.375	10.3	0.41	8.2	0.32	9.00	0.354	22.2	0.87	25.0	0.98	36.8	1.45
12.70	0.500	11.9	0.47	9.8	0.39	10.00	0.394	25.4	1.00	28.0	1.10	40.6	1.60
15.88	0.625	14.0	0.55	12.2	0.48	11.80	0.465	30.2	1.19	33.3	1.31	50.0	1.97
19.05	0.750	15.0	0.59	13.2	0.52	13.80	0.543	34.9	1.37	37.3	1.47	55.1	2.17
22.22	0.875	—	—	—	—	13.80	0.543	—	—	—	—	—	—
25.40	1.000	15.5	0.61	13.7	0.54	13.80	0.543	41.3	1.62	41.7	1.64	59.7	2.35
31.75	1.250	15.5	0.61	13.7	0.54	13.80	0.543	47.6	1.87	44.7	1.76	62.2	2.45
38.10	1.500	15.5	0.61	13.7	0.54	13.80	0.543	55.6	2.19	48.8	1.92	65.8	2.59

*See end of table for notes.

(R) TABLE 6*—(CONTINUED)

NOM TUBE OD	NOM TUBE OD	P	P	Q	Q	R	R	S	S	T[a]	T[a]	X[i]	X[i]
mm	in	mm	in	mm	in	mm	in	mm	in	mm Ref.	in Ref.	mm	in
		±0.8	±0.03	±0.8	±0.03	±0.8	±0.03	±0.8	±0.03			±0.5	±0.02
3.18	0.125	---	---	---	---	---	---	---	---	---	---	7.2	---
4.76	0.187	---	---	---	---	---	---	---	---	---	---	7.2	---
6.35	0.250	16.0	0.63	30.0	1.18	22.5	0.89	47.0	1.85	7.9	0.31	8.0	0.31
7.94	0.312	---	---	---	---	---	---	---	---	---	---	8.0	0.31
9.52	0.375	18.8	0.74	33.0	1.30	26.0	1.02	52.0	2.05	8.7	0.34	8.5	0.33
12.70	0.500	20.3	0.80	36.3	1.43	29.0	1.14	55.5	2.18	9.8	0.39	10.3	0.41
15.88	0.625	23.4	0.92	44.7	1.76	34.5	1.36	63.0	2.48	11.6	0.46	11.5	0.45
19.05	0.750	25.9	1.02	50.0	1.97	38.5	1.52	67.2	2.65	13.1	0.52	12.8	0.50
22.22	0.875	---	---	---	---	---	---	---	---	---	---	12.8	0.50
25.40	1.000	30.0	1.18	52.3	2.06	42.5	1.67	71.2	2.80	14.3	0.56	13.6	0.54
31.75	1.250	32.0	1.26	53.6	2.11	45.5	1.79	75.5	2.97	16.3	0.64	13.6	0.54
38.10	1.500	36.8	1.45	53.6	2.11	49.5	1.95	79.5	3.13	18.1	0.71	13.6	0.54

NOM TUBE OD	NOM TUBE OD	Y[b] DIA	Y[b] DIA	Z	Z	AA	AA	BB	BB	CC	CC	DD	DD
mm	in	mm	in	mm	in	mm	in	mm	in	mm	in	mm	in
		+0.15 −0.08	+0.002 −0.003	+0.3 −0.0	+0.012 −0.000	±0.15	±0.006	±0.13	±0.005	±0.4	±0.016	±0.5	±0.02
3.18	0.125	6.35	0.250	3.2	0.126	—	—	9.50	0.374	—	—	—	—
4.76	0.187	7.95	0.313	3.3	0.130	—	—	9.50	0.374	—	—	—	—
6.35	0.250	9.25	0.364	4.0	0.156	2.14	0.478	11.00	0.433	9.8	0.386	13.5	0.53
7.94	0.312	10.85	0.427	4.0	0.156	—	—	11.00	0.433	—	—	—	—
9.52	0.375	12.24	0.482	4.0	0.156	15.09	0.594	12.00	0.472	11.2	0.441	14.6	0.57
12.70	0.500	16.76	0.660	4.8	0.187	18.26	0.719	14.00	0.551	12.8	0.504	16.2	0.64
15.88	0.625	19.63	0.773	5.6	0.219	22.76	0.896	16.00	0.630	15.5	0.610	18.9	0.74
19.05	0.750	24.00	0.945	5.9	0.234	27.15	1.069	18.50	0.728	17.0	0.670	21.2	0.83
22.22	0.875	27.18	1.070	5.9	0.234	—	—	18.50	0.728	—	—	—	—
25.40	1.000	30.35	1.195	5.9	0.234	33.50	1.319	18.50	0.728	17.5	0.689	24.6	0.97
31.75	1.250	38.28	1.507	5.9	0.234	39.85	1.569	18.50	0.728	17.5	0.689	24.6	0.97
38.10	1.500	44.60	1.756	5.9	0.234	47.78	1.881	18.50	0.728	17.5	0.689	24.6	0.97

NOM TUBE OD	NOM TUBE OD	EE	EE	FF	FF	GG	GG	HH	HH	JJ FORGING[g]	JJ FORGING[g]
mm	in	mm	in	mm	in	mm	in	mm	in	mm	in
		+0.4 −0.0	+0.016 −0.000	±0.5	±0.02	±0.5	±0.02	±0.5	±0.02	+0.00	+0.00
3.18	0.125	---	---	---	---	---	---	---	---		
4.76	0.187	---	---	---	---	---	---	---	---		
6.35	0.250	2.1	0.083	27.5	1.08	48.2	1.90	17.7	0.70	14.29−0.80	0.562−0.030
7.94	0.312	---	---	---	---	---	---	---	---		
9.52	0.375	2.4	0.094	31.1	1.22	53.2	2.09	19.9	0.78	19.05−0.80	0.750−0.030
12.70	0.500	2.4	0.094	35.4	1.39	58.3	2.30	22.6	0.89	19.05−0.80	0.750−0.030
15.88	0.625	2.7	0.106	42.6	1.68	66.6	2.62	27.1	1.07	26.99−1.00	1.062−0.040
19.05	0.750	3.2	0.125	47.1	1.86	69.2	2.72	30.1	1.19	30.16−1.00	1.188−0.040
22.22	0.875	---	---	---	---	---	---	---	---		
25.40	1.000	3.2	0.125	49.3	1.94	70.2	2.76	31.8	1.25	36.51−1.00	1.438−0.040
31.75	1.250	3.2	0.125	51.3	2.02	70.2	2.76	33.8	1.33	41.28−1.00	1.625−0.040
38.10	1.500	3.2	0.125	53.1	2.09	70.2	2.76	35.6	1.40	47.62−1.00	1.875−0.040

(R) TABLE 6*—(CONTINUED)

NOM TUBE OD	NOM TUBE OD	JJ BARSTOCK[g]	JJ BARSTOCK[g]	KK DIA	KK DIA	LL DIA	LL DIA	MM	mm	NN DIA	NN DIA
mm	in	mm max	in max	mm +0.4 -0.0	in +0.016 -0.000	mm ±0.25	in ±0.010	mm ±0.08	in ±0.003	mm ±0.4	in ±0.016
3.18	0.125	—	—	8.1	0.317	11.13	0.438	0.76	0.030	12.8	0.504
4.76	0.187	—	—	9.7	0.380	12.70	0.500	0.76	0.030	14.6	0.575
6.35	0.250	19.05	0.750	11.3	0.445	15.88	0.625	0.89	0.035	16.5	0.650
7.94	0.312	—	—	12.8	0.505	15.88	0.625	0.89	0.035	18.3	0.722
9.52	0.375	25.40	1.000	14.4	0.567	19.05	0.750	0.89	0.035	20.2	0.794
12.70	0.500	30.16	1.188	19.2	0.756	23.81	0.937	1.04	0.041	25.7	1.010
15.88	0.625	33.34	1.312	22.4	0.882	26.99	1.062	1.27	0.050	29.3	1.155
19.05	0.750	38.10	1.500	27.1	1.067	34.92	1.375	1.27	0.050	36.7	1.444
22.22	0.875	—	—	30.3	1.193	34.92	1.375	1.27	0.050	40.4	1.589
25.40	1.000	47.62	1.875	33.5	1.319	41.28	1.625	1.27	0.050	44.0	1.732
31.75	1.250	57.15	2.250	41.4	1.630	47.62	1.875	1.27	0.050	55.0	2.165
38.10	1.500	63.50	2.500	47.8	1.882	53.98	2.125	1.27	0.050	62.3	2.454

NOM TUBE OD	NOM TUBE OD	PP DIA	PP DIA	QQ	QQ	SS	SS	TT	TT	VV	VV	WW	WW
mm	in	mm +.25 -.25	in +.010 -.010	mm +0.4 -0.0	in +0.016 -0.000	mm ±0.5	in ±0.02	mm ±0.5	in ±0.02	mm +0.4 -0.0	in +0.016 -0.000	mm ±0.5	in ±0.02
3.18	0.125	6.25	.245	—	—	1.5	0.06	—	—	—	—	—	—
4.76	0.187	7.75	.305	—	—	1.5	0.06	—	—	—	—	—	—
6.35	0.250	9.15	.360	1.9	0.075	1.5	0.06	6.9	0.27	14.4	0.567	31.5	1.24
7.94	0.312	10.65	.420	—	—	1.5	0.06	—	—	—	—	—	—
9.52	0.375	12.05	.475	2.1	0.083	1.5	0.06	8.0	0.31	17.6	0.693	34.0	1.34
12.70	0.500	16.65	.655	2.4	0.094	1.5	0.06	9.0	0.35	20.8	0.819	36.5	1.44
15.88	0.625	19.55	.770	2.7	0.106	1.5	0.06	10.4	0.41	25.5	1.004	40.6	1.60
19.05	0.750	23.90	.940	3.2	0.125	1.5	0.06	10.4	0.41	30.3	1.193	41.7	1.64
22.22	0.875	27.05	1.065	—	—	1.5	0.06	—	—	—	—	—	—
25.40	1.000	30.25	1.190	3.2	0.125	1.5	0.06	10.4	0.41	36.6	1.441	42.2	1.66
31.75	1.250	38.10	1.500	3.2	0.125	1.5	0.06	10.4	0.41	43.0	1.693	42.2	1.66
38.10	1.500	44.45	1.750	3.2	0.125	1.5	0.06	10.4	0.41	50.9	2.004	42.2	1.66

NOM TUBE OD	NOM TUBE OD	YY	YY	ZZ	ZZ	AAA	AAA	BBB DIA	BBB DIA	REF CCC FULL THD[m]	REF CCC FULL THD[m]	DDD FULL THD[l]	DDD FULL THD[l]
mm	in	mm min	in min	mm ±0.8	in ±0.03	mm ±1.5	in ±0.06	mm ±0.25	in ±0.010	mm min	in min	mm min	in min
3.18	0.125	18.2	0.72	—	—	—	—	—	—	—	—	—	—
4.76	0.187	18.8	0.72	—	—	—	—	—	—	—	—	—	—
6.35	0.250	20.5	0.81	44.0	1.73	26.4	1.04	20.64	0.812	27.9	1.10	29.0	1.14
7.94	0.312	22.4	0.88	—	—	—	—	—	—	—	—	—	—
9.52	0.375	22.4	0.88	48.5	1.91	29.2	1.15	25.40	1.000	29.1	1.15	31.5	1.24
12.70	0.500	26.1	1.03	51.0	2.01	37.9	1.49	28.58	1.125	31.1	1.22	33.5	1.32
15.88	0.625	30.2	1.19	56.6	2.23	41.2	1.62	33.34	1.312	34.9	1.37	37.6	1.48
19.05	0.750	33.8	1.33	60.7	2.39	46.3	1.82	38.10	1.500	35.5	1.40	38.7	1.52
22.22	0.875	33.8	1.33	—	—	—	—	—	—	—	—	—	—
25.40	1.000	34.6	1.36	65.2	2.57	53.3	2.10	44.45	1.750	36.0	1.42	39.2	1.54
31.75	1.250	34.6	1.36	67.0	2.64	58.2	2.29	50.80	2.000	36.0	1.42	39.2	1.54
38.10	1.500	34.6	1.36	67.0	2.64	61.2	2.41	60.33	2.375	36.0	1.42	39.2	1.54

TABLE 6*—(CONTINUED)

NOM TUBE OD	NOM TUBE OD	EEE	EEE	FFF	FFF	GGG	GGG	HHH	HHH	KKK	KKK
mm	in	mm ±0.5	in ±0.02	mm ±0.5	in ±0.02	mm ±1.5	in ±0.06	mm +1.5 -0.0	in +0.06 -0.00	mm ±0.8	in ±0.03
3.18	0.125	---	---	---	---	---	---	---	---	---	---
4.76	0.187	---	---	---	---	---	---	---	---	---	---
6.35	0.250	33.8	1.33	41.7	1.64	15.0	0.59	26.2	1.03	56.6	2.23
7.94	0.312	---	---	---	---	---	---	---	---	---	---
9.52	0.375	37.0	1.46	45.7	1.80	17.3	0.68	28.2	1.11	66.3	2.61
12.70	0.500	44.5	1.75	53.8	2.12	21.3	0.84	35.3	1.39	74.9	2.95
15.88	0.625	52.6	2.07	63.5	2.50	23.9	0.94	37.8	1.49	89.2	3.51
19.05	0.750	64.0	2.52	77.0	3.03	26.2	1.03	41.2	1.62	100.8	3.97
22.22	0.875	---	---	---	---	---	---	---	---	---	---
25.40	1.000	72.9	2.87	86.6	3.41	28.2	1.11	49.0	1.93	114.6	4.51
31.75	1.250	86.6	3.41	102.4	4.03	28.2	1.11	49.0	1.93	126.5	4.98
38.10	1.500	97.0	3.82	115.1	4.53	28.2	1.11	49.0	1.93	139.2	5.48

[a] Nominal design thickness, not subject to inspection.
[b] O-ring groove has a surface texture of 3.2 micrometer.
[c] See Table 2 for tolerance.
[d] For hex tolerance, see 5.2 and Figure 2.
[e] J full THD min with thread runout.
[f] J1 full THD min with undercut to AA dia. Length of THD and undercut width must not be less than CC minimum.
[g] See 5.3.
[h] For 31.75 mm (1.250 in) NOM tube OD, use 47.62 mm (1.875 in) hex for (520120) connector.
[i] When purchased, assembled as part of a fitting, only minimum thickness applies to inspection.
[j] Actual bore size depends on joining process. (Dimensions given are for silver braze.)
[k] Actual length of engagement depends on joining process. (Dimensions given are for silver braze.)
[l] DDD full THD min with thread runout.
[m] CCC full THD min with undercut and AA dia. Length of THD and undercut width must not be less than DDD minimum.
[n] See Tables 7, 8, 9, 10, or 11 for "Jump Fitting" length factors.
[p] Standard fittings not defined for 3.18 mm, 4.76 mm, 7.94 mm, and 22.22 mm. All fittings are jump sizes, therefore, hex size will vary. See Tables 7 or 8.

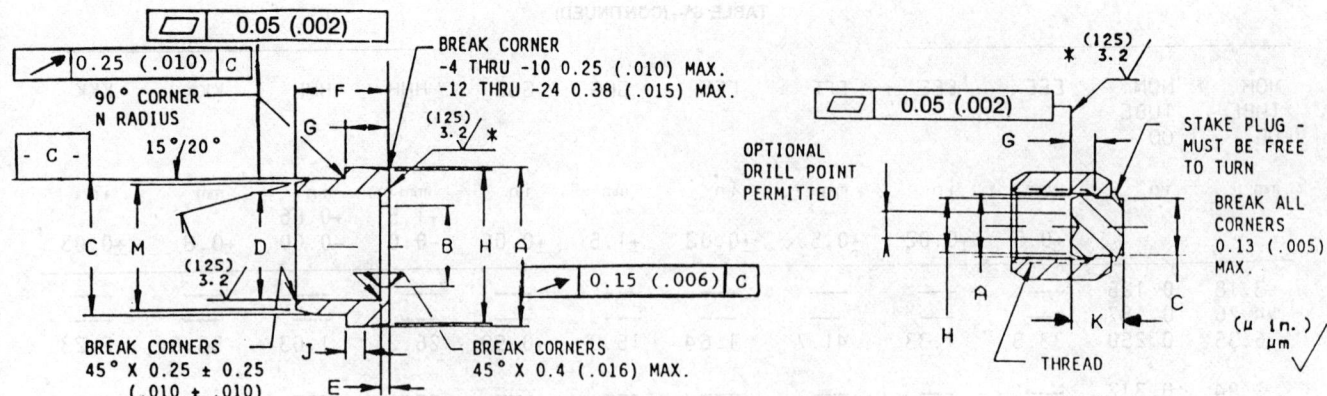

FIGURE 32—STANDARD SLEEVE (520115)

FIGURE 33—CAP ASSEMBLY (520112)

Nom Tube OD mm	Nom Tube OD in	Metric Tube[d] mm	A Dia mm +0.00 -0.13	A Dia in +0.000 -0.005	B Drill Dia[a] Nom mm	B Drill Dia[a] Nom in	C Dia mm ±0.08	C Dia in ±0.003	D Dia[b] mm ±0.05	D Dia[b] in ±0.002	E mm ±0.13	E in ±0.005
6.35	0.250	6	12.75	0.502	4.4	0.172	10.21	0.402	6.50	0.256	1.00	0.040
9.52	0.375	10	15.75	0.620	6.7	0.264	13.26	0.522	9.68	0.381	1.00	0.040
12.70	0.500	12	18.92	0.745	9.6	0.378	16.31	0.642	12.85	0.506	1.00	0.040
15.88	0.625	16	23.44	0.923	12.3	0.484	20.75	0.817	16.03	0.631	1.50	0.060
19.05	0.750	19	27.86	1.097	15.5	0.609	23.77	0.936	19.23	0.757	1.50	0.060
25.40	1.000	25	34.21	1.347	20.6	0.811	28.70	1.130	25.58	1.007	1.50	0.060
31.75	1.250	32	40.56	1.597	26.0	1.024	35.59	1.401	31.95	1.257	1.50	0.060
38.10	1.500	38	48.51	1.910	32.0	1.260	43.54	1.714	38.30	1.507	1.50	0.060

Nom Tube OD mm	Nom Tube OD in	F[c,d] mm ±0.3	F[c,d] in ±0.012	G mm ±0.13	G in ±0.005	H Dia mm ±0.13	H Dia in ±0.005	J mm ±0.5	J in ±0.02	K Dia mm ±0.5	K Dia in ±0.02	L mm max	L in max	M Dia mm ±0.13	M Dia in ±0.005	N Corner Radius mm +0.13 -0.00	N Corner Radius in +0.005 -0.000
6.35	0.250	9.5	0.374	4.00	0.157	---	---	---	---	8.5	0.34	4.0	0.157	8.51	0.335	0.13	0.005
9.52	0.375	9.5	0.374	4.50	0.177	---	---	---	---	9.5	0.37	5.0	0.197	11.81	0.465	0.13	0.005
12.70	0.500	9.5	0.374	5.00	0.197	---	---	---	---	12.0	0.47	8.0	0.315	15.11	0.595	0.13	0.005
15.88	0.625	10.5	0.413	6.00	0.236	22.61	0.890	1.3	0.05	12.0	0.47	10.0	0.394	19.18	0.755	0.25	0.010
19.05	0.750	14.0	0.551	6.50	0.256	27.00	1.063	1.3	0.05	13.5	0.53	10.0	0.394	22.10	0.870	0.25	0.010
25.40	1.000	15.5	0.610	7.00	0.276	33.35	1.313	1.3	0.05	15.0	0.59	15.0	0.591	28.07	1.105	0.50	0.020
31.75	1.250	15.5	0.610	7.00	0.276	39.73	1.564	1.3	0.05	15.0	0.59	15.0	0.591	34.04	1.340	0.50	0.020
38.10	1.500	15.5	0.610	7.00	0.276	47.65	1.876	1.3	0.05	15.0	0.59	15.0	0.591	41.91	1.650	0.50	0.020

[a] See Table 2 for tolerance. When purchased assembled, hole diameter may be reduced to tube ID.
[b] Actual bore size depends on joining process. (Dimensions given are for silver braze of inch tube.)
[c] Actual length of sleeve depends on joining process. (Dimensions given are for silver braze of inch tube.)
[d] Sleeve may be used to adapt to metric tube.

(R) FIGURES 32 TO 33—STANDARD SLEEVE AND CAP ASSEMBLY

*NOTE: See Appendix A for details on surface finish requirements.

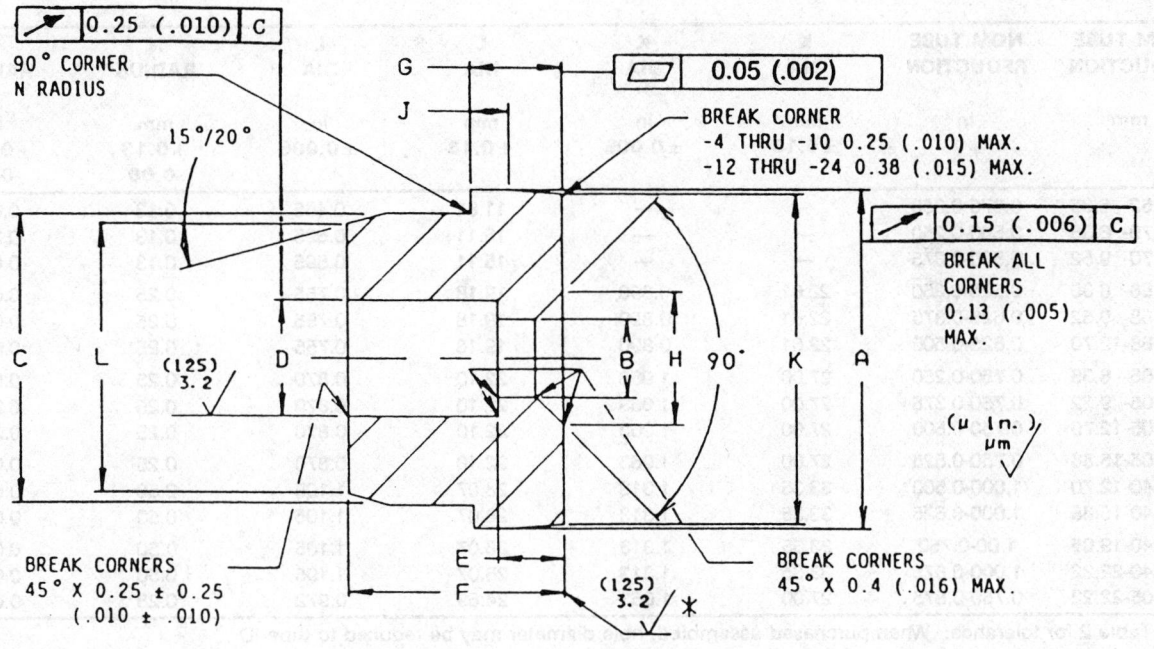

(R) FIGURE 34—REDUCING AND EXPANDING SLEEVES

NOM TUBE REDUCTION mm	NOM TUBE REDUCTION in	A DIA mm +0.00 -0.13	A DIA in +0.000 -0.005	B DRILL DIA[a] nom mm	B DRILL DIA[a] nom in	C DIA mm ±0.08	C DIA in ±0.003	D DIA[b] mm ±0.05	D DIA[b] in ±0.002	E mm ±0.13	E in ±0.005
9.52- 6.35	0.375-0.250	15.75	0.620	4.4	0.172	13.26	0.522	6.50	0.256	2.00	0.079
12.70- 6.35	0.500-0.250	18.92	0.745	4.4	0.172	16.31	0.642	6.50	0.256	3.50	0.138
12.70- 9.52	0.500-0.375	18.92	0.745	6.7	0.264	16.31	0.642	9.68	0.381	3.50	0.138
15.88- 6.35	0.625-0.250	23.44	0.923	4.4	0.172	20.75	0.817	6.50	0.256	5.00	0.197
15.88- 9.52	0.625-0.375	23.44	0.923	6.7	0.264	20.75	0.817	9.68	0.381	5.00	0.197
15.88-12.70	0.625-0.500	23.44	0.923	9.6	0.378	20.75	0.817	12.85	0.506	5.00	0.197
19.05- 6.35	0.750-0.250	27.86	1.097	4.4	0.172	23.77	0.936	6.50	0.256	6.00	0.236
19.05- 9.52	0.750-0.375	27.86	1.097	6.7	0.264	23.77	0.936	9.68	0.381	6.00	0.236
19.05-12.70	0.750-0.500	27.86	1.097	9.6	0.378	23.77	0.936	12.85	0.506	6.00	0.236
19.05-15.88	0.750-0.625	27.86	1.097	12.3	0.484	23.77	0.936	16.03	0.631	6.00	0.236
25.40-12.70	1.000-0.500	34.21	1.347	9.6	0.378	28.70	1.130	12.85	0.506	7.00	0.276
25.40-15.88	1.000-0.625	34.21	1.347	12.3	0.484	28.70	1.130	16.03	0.631	7.00	0.276
25.40-19.05	1.000-0.750	34.21	1.347	15.5	0.609	28.70	1.130	19.23	0.757	4.50	0.177
25.40-22.22	1.000-0.875	34.21	1.347	18.0	0.709	28.70	1.130	22.40	0.882	3.00	0.118
[d]19.05-22.22	0.750-0.875	27.86	1.097	15.5	0.609	24.69	0.972	22.40	0.882	1.50	0.060

*Note: See Appendix A for details on surface finish requirements.

FIGURE 34—REDUCING SLEEVE (520115)

NOM TUBE REDUCTION	NOM TUBE REDUCTION	K DIA	K DIA	L DIA	L DIA	N RADIUS	N RADIUS
mm	in	mm ±0.13	in ±0.005	mm ±0.13	in ±0.005	mm +0.13 -0.00	in +0.005 -0.00
9.52- 6.35	0.375-0.250	—	—	11.81	0.465	0.13	0.005
12.70- 6.35	0.500-0.250	—	—	15.11	0.595	0.13	0.005
12.70- 9.52	0.500-0.375	—	—	15.11	0.595	0.13	0.005
15.88- 6.35	0.625-0.250	22.61	0.890	19.18	0.755	0.25	0.010
15.88- 9.52	0.625-0.375	22.61	0.890	19.18	0.755	0.25	0.010
15.88-12.70	0.625-0.500	22.61	0.890	19.18	0.755	0.25	0.010
19.05- 6.35	0.750-0.250	27.00	1.063	22.10	0.870	0.25	0.010
19.05- 9.52	0.750-0.375	27.00	1.063	22.10	0.870	0.25	0.010
19.05-12.70	0.750-0.500	27.00	1.063	22.10	0.870	0.25	0.010
19.05-15.88	0.750-0.625	27.00	1.063	22.10	0.870	0.25	0.010
25.40-12.70	1.000-0.500	33.35	1.313	28.07	1.105	0.50	0.020
25.40-15.88	1.000-0.625	33.35	1.313	28.07	1.105	0.50	0.020
25.40-19.05	1.00-0.750	33.35	1.313	28.07	1.105	0.50	0.020
25.40-22.22	1.000-0.875	33.35	1.313	28.07	1.105	0.50	0.020
[d]19.05-22.22	0.750-0.875	27.00	1.033	24.69	0.972	0.25	0.010

[a] See Table 2 for tolerance. When purchased assembled, hole diameter may be reduced to tube ID.
[b] Actual bore size depends on joining process. (Dimensions given are for silver braze of inch tube.)
[c] Actual length of sleeve depends on joining process. (Dimensions given are for silver braze of inch tube.)
[d] For use with special nut for expanding 19.05 fitting to 22.22 tubing. See 22.22 nut Figure 4.

NOM TUBE REDUCTION	NOM TUBE REDUCTION	F[c]	F[c]	G	G	H	H	J	J
mm	in	mm ±0.3	in ±0.012	mm ±0.13	in ±0.005	mm +0.0 -0.5	in +0.00 -0.02	mm ±0.5	in ±0.02
9.52- 6.35	0.375-0.250	10.5	0.413	4.50	0.177	6.7	0.26	—	—
12.70- 6.35	0.500-0.250	12.0	0.472	5.00	0.197	9.6	0.38	—	—
12.70- 9.52	0.500-0.375	12.0	0.472	5.00	0.197	9.6	0.38	—	—
15.88- 6.35	0.625-0.250	13.5	0.531	6.00	0.236	12.3	0.48	1.3	0.05
15.88- 9.52	0.625-0.375	13.5	0.531	6.00	0.236	12.3	0.48	1.3	0.05
15.88-12.70	0.625-0.500	13.5	0.531	6.00	0.236	12.3	0.48	1.3	0.05
19.05- 6.35	0.750-0.250	14.5	0.571	6.50	0.256	15.5	0.61	1.3	0.05
19.05- 9.52	0.750-0.375	14.5	0.571	6.50	0.256	15.5	0.61	1.3	0.05
19.05-12.70	0.750-0.500	14.5	0.571	6.50	0.256	15.5	0.61	1.3	0.05
19.05-15.88	0.750-0.625	14.5	0.571	6.50	0.256	15.5	0.61	1.3	0.05
25.40-12.70	1.000-0.500	15.5	0.609	7.00	0.276	20.6	0.81	1.3	0.05
25.40-15.88	1.000-0.625	15.5	0.609	7.00	0.276	20.6	0.81	1.3	0.05
25.40-19.05	1.000-0.750	17.0	0.669	7.00	0.276	20.6	0.81	1.3	0.05
25.40-22.22	1.000-0.875	17.0	0.669	7.00	0.276	20.6	0.81	1.3	0.05
[d]19.05-22.22	0.750-0.875	16.5	0.650	11.50	0.453	—	—	1.3	0.05

(R) FIGURE 34—(CONTINUED)

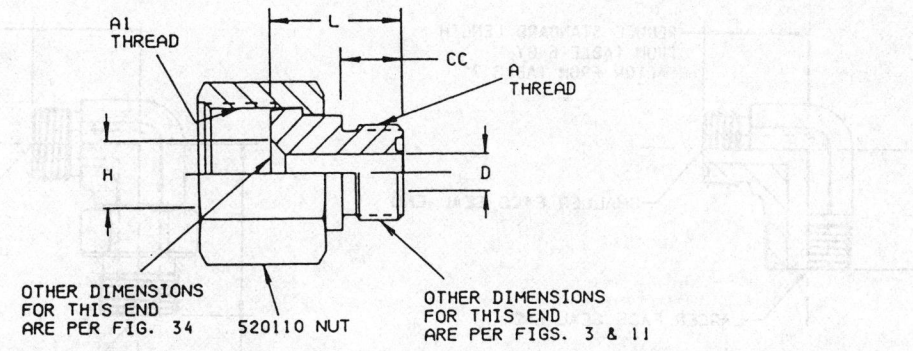

Nominal Tube Reduction MM	Nominal Tube Reduction IN	A Thread	A1 Thread	D(a) Dia. MM +0.0 -0.5	D(a) Dia. IN +0.00 -0.02	H Dia. MM +0.5 -0.5	H Dia. IN +0.20 -0.20	L MM	L IN	CC MM +0.4 -0.4	CC IN +0.016 -0.016
12.70- 6.35	0.500-0.250	9/16-18	13/16-16	4.4	0.172	9.6	0.378	21.8	0.86	9.8	0.386
15.88- 6.35	0.625-0.250	9/16-18	1-14	4.4	0.172	12.3	0.484	22.8	0.90	9.8	0.386
15.88- 9.52	0.625-0.375	11/16-16	1-14	6.7	0.264	12.3	0.484	24.1	0.95	11.2	0.441
15.88-12.70	0.625-0.500	13/16-16	1-14	9.6	0.378	12.3	0.484	25.7	1.01	12.8	0.504
19.05- 6.35	0.750-0.250	9/16-18	1-3/16-12	4.4	0.172	15.5	0.609	24.8	0.98	9.8	0.386
19.05- 9.52	0.750-0.375	11/16-18	1-3/16-12	6.7	0.264	15.5	0.609	26.2	1.03	11.2	0.441
19.05-12.70	0.750-0.500	13/16-16	1-3/16-12	9.6	0.378	15.5	0.609	27.7	1.09	12.8	0.504
25.40-12.70	1.000-0.500	13/16-16	1-7/16-12	9.6	0.378	20.6	0.811	29.2	1.15	12.8	0.504
25.40-15.88	1.000-0.625	1-14	1-7/16-12	12.3	0.484	20.6	0.811	32.0	1.26	15.5	0.610
31.75-19.05	1.250-0.750	1-3/16-12	1-11/16-12	15.5	0.609	26.0	1.024	33.5	1.32	17.0	0.670
38.10-25.40	1.500-1.000	1-7/16-12	2-12	20.6	0.811	32.0	1.260	34.0	1.34	17.5	0.689
38.10-31.75	1.500-1.250	1-11/16-12	2-12	26.0	1.024	32.0	1.260	34.0	1.34	17.5	0.689

(a) For tolerances see Table 2

FIGURE 35—REDUCER ASSEMBLY (520123)

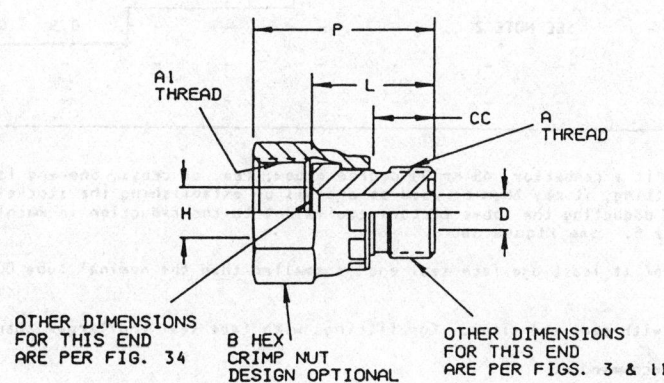

Nominal Tube Reduction MM	Nominal Tube Reduction IN	A Thread	A1 Thread	B(b) Hex (MIN) MM	B(b) Hex (MIN) IN	D(a) Dia. MM +0.0 -0.5	D(a) Dia. IN +0.000 -0.020	H Dia. MM +1.15 -1.15	H Dia. IN +0.045 -0.045	P MM +0.5 -0.5	P IN +0.020 -0.020	L MM +0.4 -0.4	L IN +0.016 -0.016	CC MM	CC IN
9.52- 6.35	0.375-0.250	9/16-18	11/16-16	20.64	0.812	4.4	0.172	6.7	0.264	28.2	1.11	19.56	0.770	9.8	0.386
12.70- 9.52	0.500-0.375	11/16-16	13/16-16	23.81	0.938	6.7	0.264	9.6	0.378	33.4	1.31	22.35	0.880	11.2	0.441
19.05-15.88	0.750-0.625	1-14	1-3/16-12	34.92	1.375	12.3	0.484	15.5	0.609	44.0	1.73	29.46	1.160	15.5	0.610
25.40-19.05	1.000-0.750	1-3/16-12	1-7/16-12	41.28	1.625	15.5	0.609	20.6	0.811	47.8	1.88	33.02	1.300	17.0	0.670
31.75-25.40	1.250-1.000	1-7/16-12	1-11/16-12	47.62	1.875	20.6	0.811	26.0	1.024	52.1	2.05	38.46	1.514	17.5	0.689

(a) For drill tolerances see Table 2.
(b) For hex tolerances see Figure 2.

FIGURE 35A—REDUCER ASSEMBLY (520123)

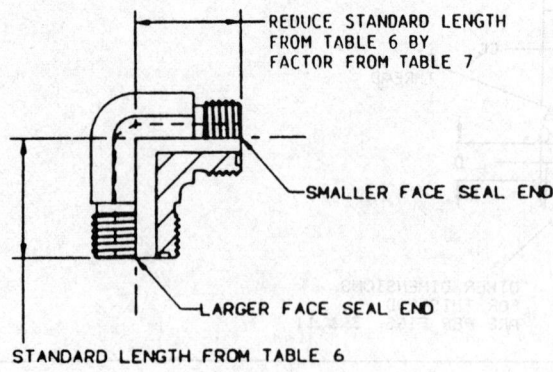
90° UNION ELBOW (520201)

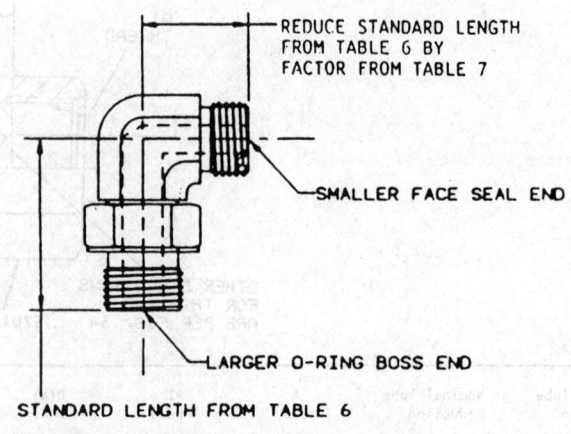
90° STRAIGHT THREAD ELBOW (520220)

FIGURE 36—90 DEGREE REDUCER ELBOWS

TABLE 7—FACTORS FOR FACE SEAL END ON SHAPED FITTINGS[1] (FIGURE 36)

			STANDARD MACHINING SIZE (LARGEST END OF FITTING)												
DASH SIZE			-6	-6	-8	-8	-10	-10	-12	-12	-16	-16	-20	-20	-24	-24
	NOMINAL TUBE OD mm	NOMINAL TUBE OD in	mm	in	mm	in	mm	in	mm	in	mm	in	mm	in	mm	in
			9.52	0.375	12.70	0.500	15.88	0.625	19.05	0.750	25.40	1.000	31.75	1.250	38.10	1.500
-4	6.35	0.250	1.5	0.06	3.3	0.13	5.6	0.22	7.1	0.28	-	-	-	-	-	-
-6	9.52	0.375	-	-	1.5	0.06	4.1	0.16	5.6	0.22	6.1	0.24	-	-	SEE NOTE 3	
-8	12.70	0.500	-	-	-	-	2.5	0.10	3.8	0.15	4.3	0.17	4.3	0.17	-	-
-10	15.88	0.625	-	-	-	-	-	-	1.5	0.06	2.0	0.08	2.0	0.08	2.0	0.08
-12	19.05	0.750	-	-	SEE NOTE 2		-	-	-	-	0.5	0.02	0.5	0.02	0.5	0.02
-16	25.40	1.000	-	-	-	-	-	-	-	-	-	-	0.0	0.00	0.0	0.00
-20	31.75	1.250	-	-	-	-	-	-	-	-	-	-	-	-	0.0	0.00

NOTES:

For any nonstandard size fitting, be it a connector, 45 or 90 degree elbow, tee, or cross, one end is always standard. Considering this end to be the largest on the fitting, it may then be used as a basis of establishing the stocksize and length (either overall or end to center) for all other ends by deducting the above factors equivalent to the reduction in machining requirements from the appropriate standard lengths in Table 6. See Figure 36.

1. To be used when nominal tube OD of at least one face seal end is smaller than the nominal tube OD or adjustable O-ring boss of the other end.

2. No factor required for fittings with same end sizes. For fittings with face seal end larger than other end, use Table 8.

3. Multiple-jump fittings are not recommended.

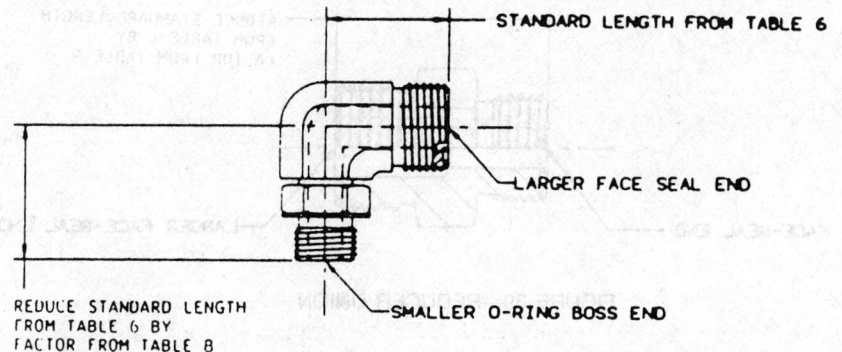

FIGURE 37—90 DEGREE STRAIGHT THREAD ELBOW WITH REDUCED O-RING BOSS END

TABLE 8—FACTORS FOR ADJUSTABLE O-RING BOSS END ON SHAPED FITTINGS[1] (FIGURE 37)

DASH SIZE	NOMINAL TUBE OD mm	NOMINAL TUBE OD in	-4 mm 6.35	-4 in 0.250	-6 mm 9.52	-6 in 0.375	-8 mm 12.70	-8 in 0.500	-10 mm 15.88	-10 in 0.625	-12 mm 19.05	-12 in 0.750	-16 mm 25.40–50.80	-16 in 1.000–2.000	-20 mm 31.75	-20 in 1.250	-24 mm 38.10	-24 in 1.500
-2	3.18	0.125	2.5	0.10	4.8	0.19	-	-	-	-	-	-	-	-	-	-	-	-
-3	4.76	0.188	2.5	0.10	4.8	0.19	8.6	0.34	-	-	SEE NOTE 3		-	-	-	-	-	-
-4	6.35	0.250	-	-	2.0	0.08	6.1	0.24	10.4	0.41	-	-			-	-	-	-
-5	7.94	0.312	-	-	2.0	0.08	5.8	0.23	10.4	0.41	-	-			-	-	-	-
-6	9.52	0.375	-	-	-	-	4.1	0.16	8.4	0.33	12.4	0.49			-	-	-	-
-8	12.70	0.500					-	-	4.3	0.17	8.4	0.33	9.4	0.37	-	-	-	-
-10	15.88	0.625		SEE NOTE 2			-	-	-	-	4.1	0.16	4.8	0.19	4.8	0.19	-	-
-12	19.05	0.750											0.7	0.03	0.7	0.03	0.7	0.03
-16	25.40	1.000																
-20	31.75	1.250																

NOTES:

For any nonstandard size fitting, 45 or 90 degree elbow, tee, or cross, one end is always standard. Considering this end to be the largest on the fitting, it may then be used as a basis of establishing the stocksize and length (either overall or end to center) for all other ends by deducting the above factors from the standard drop length for standard size fittings in Table 6. See Figure 37.

1. To be used when nominal tube OD for the adjustable O-ring boss end is smaller than the nominal tube OD of the face seal end.
2. No factor required for fittings with same end sizes. For fittings with face seal end smaller than O-ring boss end, see Table 7.
3. Multiple-jump fittings are not recommended.

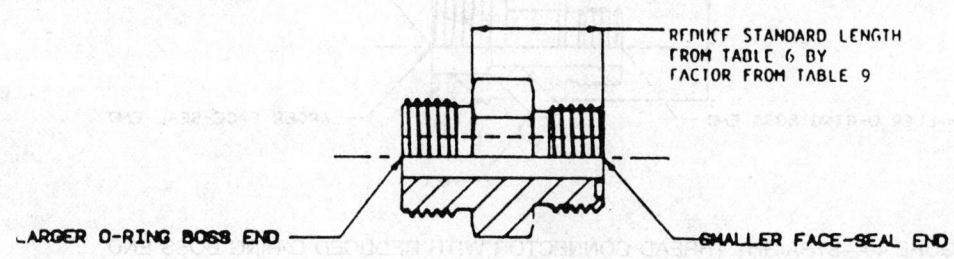

FIGURE 38—STRAIGHT THREAD CONNECTOR WITH REDUCED FACE SEAL END

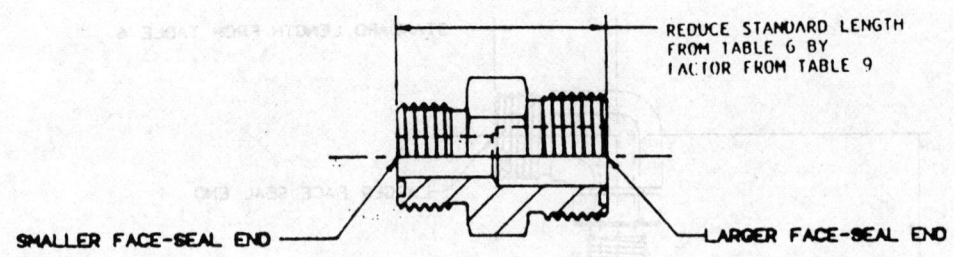

FIGURE 39—REDUCER UNION

TABLE 9—FACTORS FOR FACE SEAL END ON STRAIGHT FITTINGS[1] (FIGURES 38 TO 39)

			\multicolumn{14}{c	}{STANDARD MACHINING SIZE (LARGEST END OF FITTING)}
			-6	-6	-8	-8	-10	-10	-12	-12	-16	-16	-20	-20	-24	-24
DASH SIZE	NOMINAL TUBE OD mm	NOMINAL TUBE OD in	mm	in	mm	in	mm	in	mm	in	mm	in	mm	in	mm	in
			9.52	0.375	12.70	0.500	15.88	0.625	19.05	0.750	25.40	1.000	31.75	1.250	38.10	1.500
-4	6.35	0.250	1.4	0.05	3.0	0.12	5.7	0.22	7.2	0.28	7.7	0.30	7.7	0.30	7.7	0.30
-6	9.52	0.375	-	-	1.5	0.06	4.3	0.17	5.8	0.23	6.3	0.25	6.3	0.25	6.3	0.25
-8	12.70	0.500	-	-	-	-	2.7	0.11	4.2	0.17	4.7	0.19	4.7	0.19	4.7	0.19
-10	15.88	0.625	-	-	-	-	-	-	1.5	0.06	2.0	0.08	2.0	0.08	2.0	0.08
-12	19.05	0.750	-	-	SEE NOTE 2		-	-	-	-	0.5	0.02	0.5	0.02	0.5	0.02
-16	25.40	1.000	-	-	-	-	-	-	-	-	-	-	0.0	0.00	0.0	0.00
-20	31.75	1.250	-	-	-	-	-	-	-	-	-	-	-	-	0.0	0.00
MIN HEX			17.46	0.688	22.22	0.875	25.40	1.000	31.75	1.250	38.10	1.500	47.62	1.875	53.98	2.125

(REDUCED MACHINING SIZE on left axis)

NOTES:
For any nonstandard size fitting, one end, the largest, is always standard. It may then be used as a basis of establishing the stocksize and length (either overall or end to center) for all other ends by deducting the above factors equivalent to the reduction in machining requirements from the appropriate standard length in Table 6. See Figures 38 and 39.

1. To be used when nominal tube OD of one face seal end or O-ring boss end is larger than the nominal tube OD of the other end.

2. No factor required for fittings with same end sizes. For fittings with face seal end larger than O-ring boss end, see Table 10.

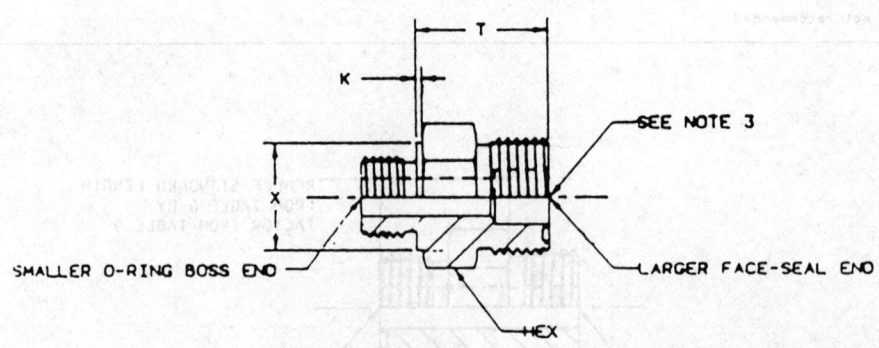

FIGURE 40—STRAIGHT THREAD CONNECTOR WITH REDUCED O-RING BOSS END

TABLE 10—DIMENSIONS FOR STRAIGHT O-RING BOSS JUMP FITTINGS[1] (FIGURE 40)

DASH SIZE	NOMINAL TUBE OD mm	NOMINAL TUBE OD in	FACE SEAL END SIZE, T													
			-4 mm 6.35	-4 in 0.250	-6 mm 9.52	-6 in 0.375	-8 mm 12.70	-8 in 0.500	-10 mm 15.88	-10 in 0.625	-12 mm 19.05	-12 in 0.750	-16 mm 25.4	-16 in 1.000	-20 mm 31.75	-20 in 1.250
-2	3.18	0.125	21.0	0.83	23.1	0.91	25.5	1.00	29.7	1.17	33.0	1.30	34.3	1.35	36.3	1.43
-3	4.76	0.188	17.7	0.70	23.1	0.91	25.5	1.00	29.7	1.17	33.0	1.30	34.3	1.35	36.3	1.43
-4	6.35	0.250	-	-	23.1	0.91	25.5	1.00	29.7	1.17	33.0	1.30	34.3	1.35	36.3	1.43
-5	7.94	0.312	-	-	19.8	0.78	25.5	1.00	29.7	1.17	33.0	1.30	34.3	1.35	36.3	1.43
-6	9.52	0.375	-	-	-	-	25.5	1.00	29.7	1.17	33.0	1.30	34.3	1.35	36.3	1.43
-8	12.70	0.500	-	-	-	-	-	-	31.2	1.23	34.5	1.36	35.8	1.41	37.8	1.49
-10	15.88	0.625	-	-	-	-	-	-	-	-	34.5	1.36	35.8	1.41	37.8	1.49
-12	19.05	0.750	-	-	SEE NOTE 2		-	-	-	-	-	-	35.8	1.41	37.8	1.49
-14	22.22	0.875	-	-	-	-	-	-	-	-	-	-	31.0	1.22	37.8	1.49
-16	25.40	1.000	-	-	-	-	-	-	-	-	-	-	-	-	39.3	1.55
-20	31.75	1.250	-	-	-	-	-	-	-	-	-	-	-	-	-	-
HEX			15.86	0.625	19.05	0.750	22.22	0.875	26.99	1.062	31.75	1.250	38.10	1.500	44.45	1.750

T (SEE FIGURE 33)

NOTES:

1. To be used when nominal tube OD of O-ring boss end is smaller than nominal tube OD of the face seal end.
2. No shoulder required for these sizes. For fittings with face seal end smaller than O-ring boss end, use Table 9.

TABLE 10 (CONTINUED)

DASH SIZE	NOMINAL TUBE OD mm	NOMINAL TUBE OD in	FACE SEAL END SIZE		SEE FIGURE 40			
			-24 mm 38.10	-24 in 1.500	K mm ±0.5	K in ±0.02	X DIA mm ±0.3	X DIA in ±0.01
-2	3.18	0.125	38.1	1.50	4.1	0.16	12.7	0.50
-3	4.76	0.188	38.1	1.50	4.1	0.16	14.2	0.56
-4	6.35	0.250	38.1	1.50	4.1	0.16	16.0	0.63
-5	7.94	0.312	38.1	1.50	4.1	0.16	17.5	0.69
-6	9.52	0.375	38.1	1.50	4.1	0.16	19.0	0.75
-8	12.70	0.500	39.6	1.56	5.6	0.22	23.9	0.94
-10	15.88	0.625	39.6	1.56	5.6	0.22	26.9	1.06
-12	19.05	0.750	39.6	1.56	5.6	0.22	33.3	1.31
-14	22.22	0.875	39.6	1.56	5.6	0.22	36.6	1.44
-16	25.40	1.000	41.1	1.62	7.1	0.28	39.6	1.56
-20	31.75	1.250	41.1	1.62	7.1	0.28	47.7	1.88
HEX			53.98	2.125	-	-	-	-

NOTES:

1. To be used when nominal tube OD of O-ring boss end is smaller than nominal tube OD of the face seal end.
2. No shoulder required for these sizes. For fittings with face seal end smaller than O-ring boss end, use Table 9.
3. Optional drill permitted on this end to size and depth illustrated in Figure 21 plug.

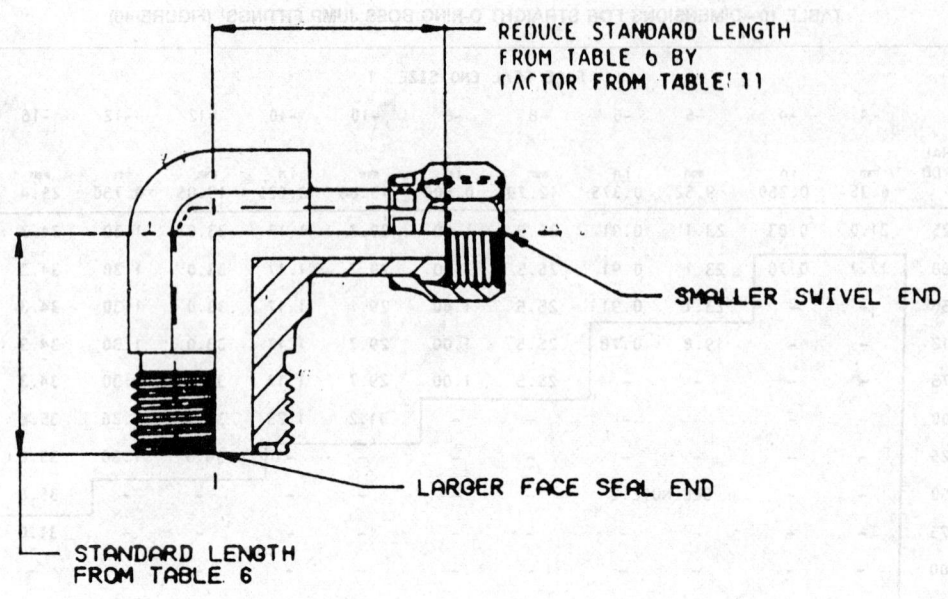

FIGURE 41—90 DEGREE SWIVEL ELBOW WITH REDUCED SWIVEL END

TABLE 11—FACTORS FOR SWIVEL END ON SHAPED FITTINGS[1] (FIGURE 41)

DASH SIZE	NOMINAL TUBE OD mm	NOMINAL TUBE OD in	STANDARD MACHINING SIZE (LARGEST END OF FITTING)													
			-6	-6	-8	-8	-10	-10	-12	-12	-16	-16	-20	-20	-24	-24
			mm	in	mm	in	mm	in	mm	in	mm	in	mm	in	mm	in
			9.52	0.375	12.70	0.500	15.88	0.625	19.05	0.750	25.40	1.000	31.75	1.250	38.10	1.500
-4	6.35	0.250	2.3	0.09	6.4	0.25	8.6	0.34	11.1	0.44	13.2	0.52	13.2	0.52	13.2	0.52
-6	9.52	0.375	-	-	4.1	0.16	6.6	0.26	8.9	0.35	10.7	0.42	10.7	0.42	10.7	0.42
-8	12.70	0.500	-	-	-	-	2.5	0.10	5.1	0.20	5.8	0.27	6.8	0.27	6.8	0.27
-10	15.88	0.625	-	-	-	-	-	-	2.5	0.10	4.3	0.17	4.3	0.17	4.3	0.17
-12	19.05	0.750	-	-	-	-	SEE NOTE 2		-	-	1.8	0.07	1.8	0.07	1.8	0.07
-16	25.4	1.000	-	-	-	-	-	-	-	-	-	-	0.0	0.00	0.0	0.00
-20	31.75	1.250	-	-	-	-	-	-	-	-	-	-	0.0	0.00	0.0	0.00

REDUCED MACHINING SIZE (left side of table)

SEE NOTE 3 (at -16 / -4 row)

NOTES:

For any nonstandard size fitting, be it a connector, 45 or 90 degree elbow, tee, or cross, one end is always standard. Considering this end to be the largest on the fitting, it may then be used as a basis of establishing the stocksize and length (either overall or end to center) for all other ends by deducting the above factors equivalent to the reduction in machining requirements from the appropriate standard lengths in Table 6 of SAE J1453.

1. To be used when nominal tube OD of at least one shoulder end is smaller than the nominal tube OD of the standard end.

2. No factor required for fittings with the same end sizes. For fittings with shoulder end larger than other end, use Table 7 or 8 of SAE J1453.

3. Multiple jump fittings are not recommended.

APPENDIX A
SURFACE ACCEPTANCE CRITERIA FOR FACE SEAL FITTINGS

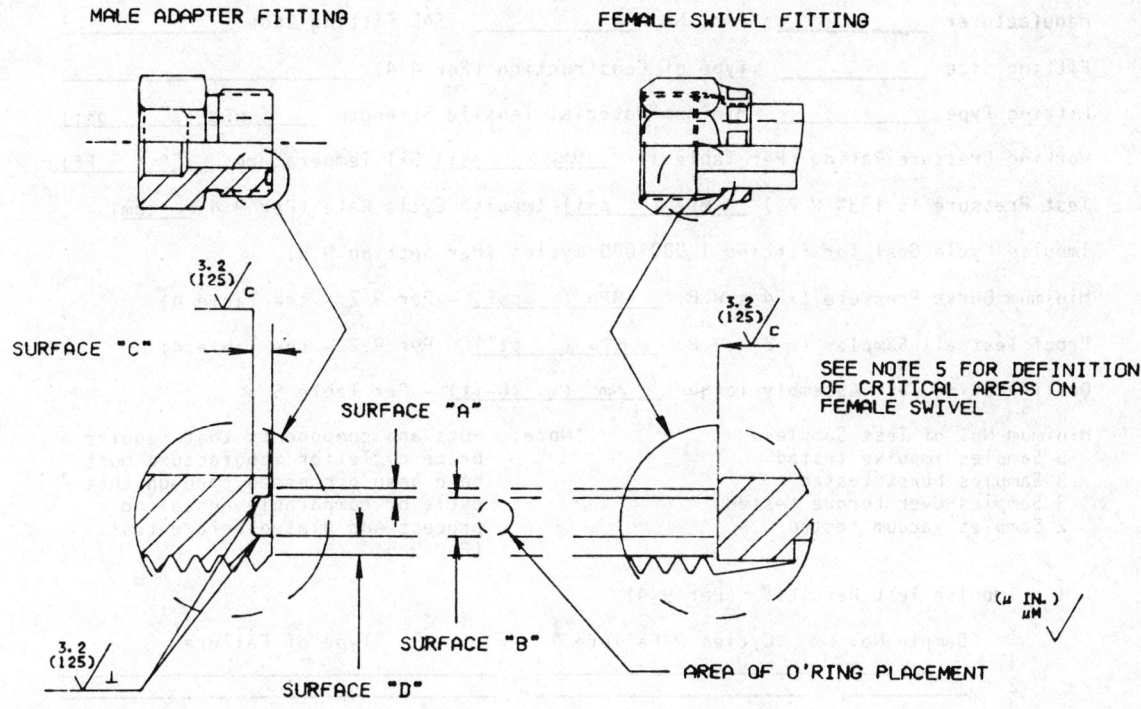

THE ENLARGED DETAIL ILLUSTRATIONS ABOVE ARE ALIGNED TO REPRESENT THE APPROXIMATE POSITION OF THE MALE FITTING WITH THE FEMALE SWIVEL FITTING WHEN CONNECTED. SURFACES DESIGNATED APPLY TO BOTH MALE AND FEMALE SURFACES.

NOTES:
1. Surface Roughness—3.2 µm (125 µin) maximum arithmetical average on surfaces specified in Figure A1.
2. Raised extrusions are not permitted on sealing surfaces "B" and "C" above.
3. Annular (circumferential) tool marks up to 3.2 µm (125 µin) Ra maximum are acceptable. Scratches with a width greater than 0.13 mm (.005 in) running perpendicular, radial, or spiral to the fitting I.D. on surface "B" are not acceptable. Surface mars with no depth or height are acceptable.
4. On surfaces "A" and "D," surface imperfections are allowed providing they do not inhibit assembly of fittings.
5. For clarification of the female swivel fitting which does not have the three distinct surfaces present, Table A1 defines this surface location. The surface I.D. and O.D. dimensions represent the total female face area contacted by the O-ring in each male fitting O-ring groove.

FIGURE A1—SURFACE ACCEPTANCE CRITERIA FOR FACE SEAL FITTINGS

TABLE A1—DIMENSIONS FOR SURFACE ACCEPTANCE CRITERIA

NOM TUBE OD	NOM TUBE OD	NOM DASH SIZE	MALE FACE SEAL FITTING THREAD SIZE	SURFACE "B" MINIMUM OD	SURFACE "B" MINIMUM OD	SURFACE "B" MAXIMUM ID	SURFACE "B" MAXIMUM ID
mm	in	Ref.	Ref.	mm	in	mm	in
6.35	0.250	-4	9/16-18	11.08	0.436	6.10	0.240
9.52	0.375	-6	11/16-16	12.68	0.499	7.70	0.303
12.70	0.500	-8	13/16-16	15.85	0.624	10.87	0.428
15.88	0.625	-10	1-14	19.08	0.751	14.10	0.555
19.05	0.750	-12	1-3/16-12	22.27	0.877	17.25	0.679
25.40	1.000	-16	1-7/16-12	26.97	1.062	21.95	0.864
31.75	1.250	-20	1-11/16-12	33.38	1.314	28.30	1.114
38.10	1.500	-24	2-12	41.30	1.626	36.22	1.426

APPENDIX B
TEST DATA SHEET FOR O-RING FACE SEAL FITTINGS

Manufacturer _____ Part No. _____ SAE Fitting Code _____

Fitting Size _____ Type of Construction (Per 4.4) _____

Fitting Type _____ Minimum Material Tensile Strength _____ MPa (_____ psi)

Working Pressure Rating (Per Table 1) _____ MPa (_____ psi) Oil Temperature _____ C° (_____ F°)

Test Pressure (= 133% W.P.) _____ MPa (_____ psi) Impulse Cycle Rate (Per 9.4) _____ cpm

Impulse Cycle Goal for Fitting 1 000 000 Cycles (Per Section 9.4)

Minimum Burst Pressure (= 4 × W.P. _____ MPa (_____ psi) - Per 9.2 - see Table 4)

Proof Test all Samples (= 2 × W.P. _____ MPa (_____ psi) - Per 9.2 - see Table 4)

Qualification Test Assembly Torque _____ Nm (_____ lb-ft) - Per Table 5

Minimum No. of Test Samples:
  6 Samples impulse tested
  3 Samples burst tested
  3 Samples over torque tested
  2 Samples vacuum tested

*Note: Nuts and components that require a braze cycle for manufacture must have been processed through that cycle or comparable annealing process and plated before test. (Per 9.5)

I. Impulse Test Results: (Per 9.4)

    Sample No.   Cycles @ Failure   Type of Failure
  1. _____ _____ _____
  2. _____ _____ _____
  3. _____ _____ _____
  4. _____ _____ _____
  5. _____ _____ _____
  6. _____ _____ _____

II. Burst Test Results: (Per 9.3 - see Tables 4 and 5)

    Sample No.   *Nut Hardness   Pressure @ Failure   Type of Failure
  1. _____ _____ RB/RC _____ MPa (_____ psi) _____
  2. _____ _____ RB/RC _____ MPa (_____ psi) _____
  3. _____ _____ RB/RC _____ MPa (_____ psi) _____

III. Over Torque Test Results: (Per 9.9 - see Table 5)

    Nut Type   *Nut Hardness   Torque @ Failure   Type of Failure
  1. _____ _____ RB/RC _____ Nm (_____ lb-ft) _____
  2. _____ _____ RB/RC _____ Nm (_____ lb-ft) _____
  3. _____ _____ RB/RC _____ Nm (_____ lb-ft) _____

IV. Vacuum Test Results: (Per 9.6)

    Sample No.   Temperature   Test Pressure   Type of Failure
  1. _____ _____ C° (_____ F°) _____ In Hg _____
  2. _____ _____ C° (_____ F°) _____ In Hg _____

V. Conclusions: _____

VI. Dimensions: List Any Exception _____

NOTE: The above test should be conducted on each of the following types and sizes of fittings to ensure overall performance of all configurations: Figure 4 (520110), Figure 9 (520220), Figure 11 (520120), Figure 24 (520221), and Figure 32 (520115).

APPENDIX C
INSTRUCTIONS AND EXAMPLES FOR CALCULATING DIMENSIONS ON SPECIAL SIZE FITTINGS

Tables 7, 8, 9, 10, and 11 present various factors to be used in determining the dimensions applicable to special size combination fittings not contained in SAE J1453.

Tables 7 and 8 for Note 3 conditions have no factors for extreme conditions due to methods of manufacture. These extreme conditions are rare and it is suggested a manufacturer be contacted for the proper dimensions.

For any nonstandard size fitting (a fitting where the tube ends are not the same size, e.g., -6, -6, -6) be it connector, 45 or 90 degree elbow, tee, or cross, one end is always standard conforming to SAE J1453 table of dimensions. Considering this to be the largest on the fitting, it may then be used as a basis for establishing the stock size and length (either overall or end to center) for all other ports by deducting factors equivalent to the reduction in machining requirements from the appropriate standard lengths.

The factors applicable to the various end configurations and size reductions tabulated in the tables were determined on the following basis:

a. Length dimensions were derived by maintaining the standard hexagon thickness for straight fittings, and the standard centerline to machining start for shaped fittings.

b. Factors given in Tables 7 and 8 were derived by subtracting the standard machining length plus an allowance of 1-1/2 pitches (threads) for imperfect thread length required for the smaller end from the same value required for the larger end and rounding the result to a two-place decimal.

c. Factors given in Tables 9, 10, and 11 were derived by subtracting the standard machining length required for the smaller end from that required for the larger standard end and rounding the result to a two-place decimal.

NOTE—See Figures C1a and C1b.

Example: Straight Thread Connector Short, Figure 11 (520120)

3 Situations—

(1) Even Sizes, -8 ORFS and -8 O-ring end sizes. Read data from Table 6.

(2) ORFS end > O-ring end—Use with Figure 40.

Example: -10 ORFS and -8 O-ring

Read "T" in Table 10 for -10 = 31.2 mm (1.23 in) and Read Hex from Table 6 = 26.99 mm (1.062 in).

Also note a shoulder is required. Read K and X dimensions from Table 10, K = 5.6 mm (0.22 in) and X = 23.9 mm (0.94 in).

All remaining dimensions are from Table 6.

(3) O-ring end > ORFS end—Use with Figure 38.

Example: -10 O-ring and -8 ORFS

Read dimensions given for -10 ORFS in Table 6 even though the ORFS end is -8, -10 is the largest end.

Read HH from Table 6 for -10 = 27.1 mm (1.07 in).

Since the fitting is a -8 ORFS, using Table 9 subtract 2.7 mm (0.11 in) from 27.1 mm (1.07 in) and HH = 24.4 mm (0.96 in).

All remaining dimensions come from Table 6.

Example: 45 Degree Straight Thread, Figure 10 (520320)

3 Situations—

(1) Even Sizes, -8 ORFS and -8 O-ring end sizes. Read data from Table 6.

(2) ORFS end > O-ring end—Use with Figure 37.

Example: -10 ORFS and -8 O-ring

Read Q = 44.7 mm (1.76 in) from Table 6 for -10 ORFS. From Table 8 subtract 4.3 mm (0.17 in) from the 44.7 mm (1.76 in) length, thus the drop length Q = 40.4 mm (1.59 in).

All other dimensions come from Table 6.

(3) O-ring end > ORFS end—Use with Figure 36.

Example: -16 O-ring and -8 ORFS

The O-ring end is considered the largest end of the fitting and Q for -16 = 52.3 mm (2.06 in).

Reference dimension under washer = Q - (K + Z - MM)

- in millimeters: = 52.3 - (13.80 + 5.9 - 1.27)
  = 33.87 mm
- in inches: = 2.06 - (0.543 + 0.234 - 0.05)
  = 1.33 in

P dimension must be adjusted since it is a -8 end. Read P for -16 = 30.0 mm (1.18 in) and from Table 7 read the deduction length of 4.3 mm (0.17 in).

- in millimeters: P = 30.0 - 4.3 = 25.7 mm
- in inches: P = 1.18 - 0.17 = 1.01 in

All other dimensions come from Table 6.

22.284

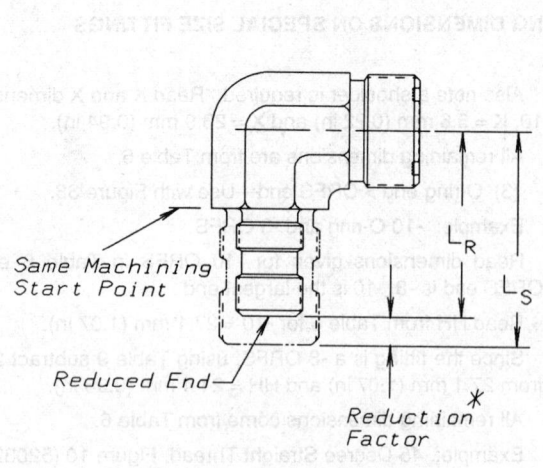
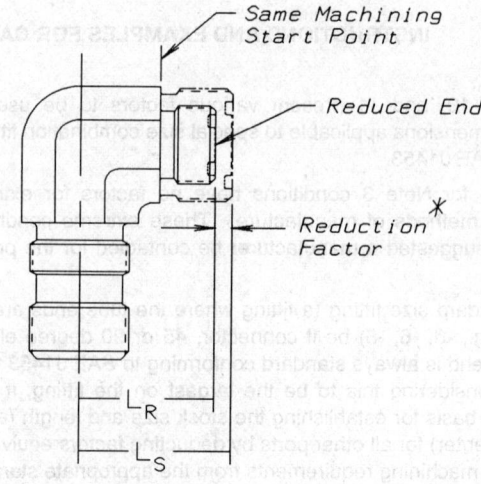

L_S – Standard Size
L_R = (L_S – Reduction Factor)

L_S – Standard Size
L_R = (L_S – Reduction Factor)

* Reduction Factor Equals:

$$\left(\begin{array}{c}\text{MIN. PERF. THRD. Length} + 1\ 1/2 \\ \text{Pitches of Standard Size.}\end{array}\right) - \left(\begin{array}{c}\text{MIN. PERF. THRD. Length} + 1\ 1/2 \\ \text{Pitches of Smaller Size.}\end{array}\right)$$

FIGURE C1A—PORT SIDE FIGURE C1B—TUBE SIDE

FIGURE C1—JUMP SIZE CALCULATIONS

(R) APPENDIX D
OPTIONAL METRIC FORGING FLAT SIZES

By agreement between user and supplier, these metric flat sizes shall be used in lieu of inch sizes.

Tube Size	Optional Metric Forging Sizes mm/mm
–4	14.00/13.20
–6	17.00/16.20
–8	19.00/18.20
–10	24.00/23.20
–12	27.00/26.00
–16	36.00/35.00
–20	41.00/40.00
–24	50.00/49.00

(R) APPENDIX E
ASSEMBLY INSTRUCTIONS FOR ADJUSTABLE STYLE FITTINGS IN STRAIGHT THREAD O-RING PORT

1. Lubricate O-ring by coating with system fluid or compatible lubricant and install in the groove adjacent to the face of the metal back-up washer which is assembled at the extreme end of the thread undercut as shown in Figure E2.

2. Install the fitting into the SAE straight thread port until the metal back-up washer contacts the face of the port as shown in Figure E3.

3. Position the fitting by turning out (counterclockwise) up to a maximum of one turn as shown in Figure E4. Holding the pad of the fitting with a wrench, tighten the locknut and washer against the face as shown in Figure E5; torque the nut to the manufacturer's recommendation.

FIGURE E1—FINAL ASSEMBLY OF ADJUSTABLE STUD END

FIGURE E2—LOCKNUT BACKED OFF

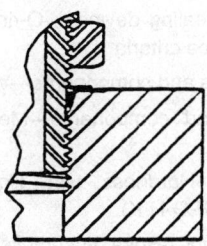
FIGURE E3—FITTING INSTALLED HAND TIGHT

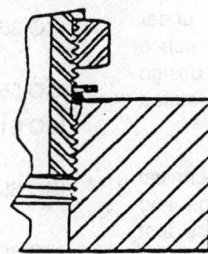
FIGURE E4—FITTING BACKED-OFF FROM ALIGNMENT (1 TURN MAXIMUM)

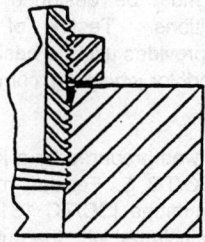
FIGURE E5—FITTING LOCKNUT TIGHTENED TO APPROPRIATE TORQUE

METALLIC TUBE CONNECTIONS FOR FLUID POWER AND GENERAL USE—TEST METHODS FOR THREADED HYDRAULIC FLUID POWER CONNECTORS—SAE J1644 MAY93   SAE Standard

Report of the Fluid Conductors and Connectors Technical Committee SC1—Automotive and Hydraulic Tube and Fittings, approved May 1993. Rationale statement available.

This document is technically equivalent to ISO 8434-5.

TABLE OF CONTENTS

1. Scope
2. References
2.1 Applicable Documents
2.2 Definitions
3. General Requirements
3.1 Test Samples
3.2 Test Temperatures
3.3 Test Report
4. Test Procedures
4.1 Proof Pressure Test
4.2 Failure Pressure Test (Burst)
4.3 Cyclic Endurance Test
4.4 Vacuum Test
4.5 Overtorque Test
5. Marking
6. Key Words
APPENDIX A   Normative—Test Data Form
APPENDIX B   Informative—Applicable Tests for Various Fitting and Stud End Standards

Foreword—In fluid power systems, power is transmitted and controlled through a fluid (liquid or gas) under pressure within an enclosed circuit. Components must be designed to meet these requirements under varying conditions. Testing of components to meet performance requirements provides users a basis of assurance for determining design application and for checking component compliance with their stated requirements.

ISO 8434-5 was prepared by a joint working group between Technical Committee ISO/TC 5, Ferrous metal pipes and metallic fittings and Technical Committee ISO/TC 131, Fluid Power Systems. These test methods were included with the publication of several ISO standards, but because of its generic application to ports, studs, and tube connections for both metric and inch connections, it was agreed a separate standard should be established. Additional tests are being considered for inclusion by the ISO Joint Working Group.

1. Scope—This SAE Standard specifies uniform methods for the testing and performance evaluation of threaded metallic tube connections and stud ends of ports for hydraulic fluid power.

1.1 Application—**Tests outlined within this document are independent of each other and document the method to follow for each test; see the appropriate component SAE or ISO standard for the test(s) requirements and the performance criteria.**

Appendix A of SAE J1644 is normative; Appendix B is informative.

2. References

2.1 Applicable Documents—The following standards contain provisions which, through reference in this text, constitute provisions of this document. At the time of publication, the editions indicated were valid. All standards are subject to revision, and parties to agreements based on this document are encouraged to investigate the possibility of applying the most recent edition of the standards indicated as follows. Members of IEC and ISO maintain registers of currently valid International Standards.

2.1.1 SAE Publications—Available from SAE, 400 Commonwealth Drive, Warrendale, PA 15096-0001.

SAE J514—Hydraulic Tube Fittings

SAE J1453—Fitting—O-ring Face Seal

SAE J1926/2—Connections for Fluid Power and General Use—Ports and Stud Ends With ISO 725 Threads and O-ring Sealing Part 2: Heavy-Duty (S Series) Stud Ends—Dimensions, Design, Test Methods, and Requirements

SAE J1926/3—Connections for Fluid Power and General Use—Ports and Stud Ends With ISO 725 Threads and O-ring Sealing Part 3: Light-Duty (L Series) Stud Ends—Dimensions, Design, Test Methods, and Requirements

SAE J2244/2—Connections for Fluid Power and General Use—Ports and Stud Ends With ISO 261 Threads and O-ring Sealing Part 2: Heavy-Duty (S Series) Stud Ends—Dimensions, Design, Test Methods, and Requirements

2.1.2 ISO Publications—Available from ANSI, 11 West 42nd Street, New York, NY 10036-8002.

ISO 48:1979—Vulcanized rubbers—Determination of hardness (Hardness between 30 and 85 IRHD)

ISO 261—ISO general purpose metric screw threads—General plan

ISO 725—ISO inch screw threads—Basic dimensions

ISO 1179—Pipe connections, threaded to ISO 228/1, for plain end steel and other metal tubes in industrial applications

ISO 3448:1975—Industrial liquid lubricants—ISO viscosity classification—first edition

ISO 3601-3:1987—Fluid systems—Sealing devices—O-rings—Part 3: Quality acceptance criteria

ISO 5598:1985—Fluid power systems and components—Vocabulary

ISO 6149—Fluid power systems and components—Metric ports—Dimensions and design

ISO 6508:1986—Metallic materials—Hardness test—Rockwell test (scales A-B-C-D-E-F-G-H-K)

ISO 6605:1986—Hydraulic fluid power—Hose assemblies—Methods of test

ISO 8434-5—Metallic tube connections for fluid power and general use—Part 5: Test methods for threaded hydraulic fluid power connectors

ISO 9974—Connections for general use and fluid power—Port and stud ends with ISO 261 threads with elastomeric or metal-to-metal sealing

2.2 Definitions—For the purpose of this document, the definitions given in ISO 5598 shall apply.

3. General Requirements

3.1 Test Samples—All components tested shall be tested in the final form (as the customer receives the part) including annealing as required for brazing.

3.2 Test Temperatures—Test temperature (ambient and fluid) shall be 15 to 35 °C unless otherwise specified in the controlling ISO standard.

3.3 Test Report—Test results and test conditions shall be reported on the test data form in Appendix A.

FIGURE 1—WARNING SYMBOL

Warning— Some of the tests described in this document are considered hazardous; it is therefore essential that, in conducting these tests, all appropriate safety precautions are strictly adhered to. Attention is drawn to the danger of burst, fine jets (which can penetrate the skin), and energy release of expanding gases. To reduce the hazard to energy release, bleed air out of test specimens prior to pressure testing. Tests shall be set-up and performed by properly trained personnel.

4. Test Procedures

4.1 Proof Pressure Test

4.1.1 Principle—Three samples shall be tested to confirm that the specified connection(s) meet(s) or exceed(s) a ratio of 2:1 between the proof and working pressure for 60 s minimum at proof pressure without any visual sign of leakage.

4.1.2 Materials

4.1.2.1 Test Block—Test blocks shall be unplated and hardened to 45 to 55 HRC per ISO 6508. The distance between the centerlines of test ports shall be a minimum of 1.5 × the port diameter. The distance between the port centerline and the edge of the test block shall be a minimum of 1 × the port diameter.

4.1.2.2 Test Seals (If Applicable)—Unless otherwise specified, seals shall be nitrile (NBR) rubber with a hardness of 85, +10/-0 IHRD when measured per ISO 48. Seals shall conform to their respective dimensional requirements, and O-rings shall meet or exceed the quality requirement grade N in ISO 3601-3, if applicable.

4.1.3 PROCEDURES

4.1.3.1 Thread Lubrication—For testing only, threads and contact surfaces shall be lubricated with hydraulic oil with a viscosity of VG 32 per ISO 3448 prior to application of torque.

4.1.3.2 Torque—Tube connections and stud ends shall be tested at the required minimum torques of the respective connector standard, if specified. Otherwise test at the minimum torque values supplied by the manufacturer. Adjustable stud torques shall be applied after being backed out one full turn from finger-tight position to correctly test the worst possible actual assembly conditions.

4.1.3.3 Pressure Rise Rate—The rate of pressure rise shall be constant and chosen to reach the final pressure between 30 and 60 s.

4.1.4 REUSE OF COMPONENTS—Parts which pass this test may be used for other tests or in production.

4.2 Failure Pressure Test (Burst)

4.2.1 PRINCIPLE—Three samples shall be tested to confirm that the specified connection(s) shall be capable of withstanding the minimum of four times the working pressure, without failure.

4.2.2 MATERIALS—Use the same materials as in 4.1.2.

4.2.3 PROCEDURES—Use the same procedures as in 4.1.3.

4.2.4 REUSE OF COMPONENTS—Parts which pass this test **shall not** be tested further, used, or returned to stock.

4.3 Cyclic Endurance Test

4.3.1 PRINCIPLE—Six samples, when tested at their respective impulse pressure, shall pass a cyclic endurance test for 1 000 000 cycles without leakage or component failure.

4.3.2 MATERIALS—Use the same materials as in 4.1.2.

4.3.3 PROCEDURES

4.3.3.1 Thread Lubrication—Apply lubricant per 4.1.3.1.

4.3.3.2 Torques—Apply torque per 4.1.3.2.

4.3.3.3 Cycle and Pressure Rise Rate—Cycle rate shall be uniform at 1 Hz ± 0.25 Hz and shall conform to the wave pattern shown in ISO 6605, except the pressure rise rate nominal slope shall be calculated using Equation 1:

$$R = 10P - k \quad (Eq.1)$$

where:

R = Rate of rise (MPa/s)

P = Nominal square wave test pressure in megapascals

k = 5 MPa

4.3.4 REUSE OF COMPONENTS—Parts which pass this test **shall not** be tested further, used, or returned to stock.

4.4 Vacuum Test

4.4.1 PRINCIPLE—Two samples shall be capable of withstanding a vacuum of 6.5 kPa (0.065 bar) absolute pressure for 5 min without leakage.

4.4.2 MATERIALS—Use the same materials as per 4.1.2.

4.4.3 PROCEDURES

4.3.3.1 Thread Lubrication—Apply lubricant per 4.1.3.1.

4.3.3.2 Torques—Apply torque as per 4.1.3.2.

4.4.4 REUSE OF COMPONENTS—Parts which pass this test may be used for other tests or in production.

4.5 Overtorque Test

4.5.1 PRINCIPLE—Six samples shall be capable of withstanding the overtorque qualification test when tested to the overtorque values shown in their respective standards.

4.5.2 MATERIALS—An unplated thread steel mandrel hardened to 40 to 45 HRC per ISO 6508 shall be used.

4.5.3 PROCEDURES

4.5.3.1 Thread Lubrication—Apply lubricant per 4.1.3.1.

4.5.3.2 Wrenching Requirements—Connections shall be restrained during the test and the wrench shall be located at the threaded end of the nut hex.

4.5.4 REUSE OF COMPONENTS—Parts which pass this test **shall not** be tested further, used, or returned to stock.

5. Marking

Identification statement—(Reference to this document)

Use the following statement in test reports, catalogues, and sales literature when electing to comply with this standard. "Test methods for metallic tube connectors conform to SAE J1644, Metallic Tube Connections for Fluid Power and General Use—Test Methods for Threaded Hydraulic Fluid Power Connectors."

6. Key Words—Fluid power, pipe fittings, ports, stud ends, specifications, design, operating requirements, dimensions, designation, test methods, metric, straight thread, O-ring seal, high pressure, proof, burst, test, fittings, vacuum, overtorque, procedures, rise rate.

APPENDIX A
NORMATIVE—TEST DATA FORM

A.1 See Figure A1.

SAE J1644 TEST DATA FORM

Specifications for connection being tested: ISO standard _____ Material type _____
Manufacturer _____ Test Facility _____
Stud end type _____ Size _____

Proof test results: (Three samples minimum tested)   Test pressure _____ MPa

Sample no.	Hardness	Torque	Pressure @ failure	Type of failure
1.	____ HRB	____ N·m	____ MPa	____
2.	____ HRB	____ N·m	____ MPa	____
3.	____ HRB	____ N·m	____ MPa	____

Over–pressure test results: (Three samples minimum tested) Test pressure _____ MPa

Sample no.	Hardness	Torque	Pressure @ failure	Type of failure
1.	____ HRB	____ N·m	____ MPa	____
2.	____ HRB	____ N·m	____ MPa	____
3.	____ HRB	____ N·m	____ MPa	____

Cycle endurance test results: (Six samples minimum tested) Test pressure _____ MPa

Sample no.	Hardness	Torque	Cycles @ failure	Type of failure
1.	____ HRB	____ N·m	____	____
2.	____ HRB	____ N·m	____	____
3.	____ HRB	____ N·m	____	____
4.	____ HRB	____ N·m	____	____
5.	____ HRB	____ N·m	____	____
6.	____ HRB	____ N·m	____	____

Vacuum test results: (Two samples minimum tested) Test pressure _____ kPa absolute

Sample no.	Hardness	Torque	Pressure absolute	Type of failure
1.	____ HRB	____ N·m	____ kPa	____
2.	____ HRB	____ N·m	____ kPa	____

Overtorque test results: (Six samples minimum tested)   Test torque _____ N·m

Sample no.	Nut hardness	Torque at failure		Type of failure
1.	____ HRB	____	N·m	____
2.	____ HRB	____	N·m	____
3.	____ HRB	____	N·m	____
4.	____ HRB	____	N·m	____
5.	____ HRB	____	N·m	____
6.	____ HRB	____	N·m	____

Conclusions: Pass/fail with reason for failure. _____

Dimensions: List any exceptions: _____

Name (printed/typed) and signature of person certifying this report:
_____   Date: _____

FIGURE A1—TEST DATA FORM

APPENDIX B
INFORMATIVE—APPLICABLE TESTS FOR VARIOUS FITTING AND STUD END STANDARDS

B.1 See Figure B1.

Fitting and stud end standards	Tests per SAE J1644 (ISO 8434–5)				
	Proof pressure	Failure pressure (Burst)	Cyclic endurance	Vacuum	Overtorque
Fittings –					
ISO 8434–1 24° compression fittings	✔	✔			
SAE J514 (ISO 8434–2) 37° flared fittings	✔	✔			
SAE J1453 (ISO 8434–3) O–ring face seal fittings	✔	✔	✔	✔	✔
ISO 8434–4 24° cone connectors with O–ring weld–on nipples	✔	✔			
Stud ends –					
SAE J2244/2 (ISO 6149–2)[1)] ISO 261 metric threads for heavy duty (S series) stud ends with O–ring sealing	✔	✔	✔		
ISO 6149–3[1)] ISO 261 metric threads for light duty (L series) stud ends with O–ring sealing	✔	✔	✔		
ISO 1179–2 ISO 228–1 Whitworth threads for heavy and light duty stud ends with elastomeric sealing	✔	✔	✔		
ISO 1179–3 ISO 228–1 Whitworth threads for light duty stud ends with sealing by O–ring with retaining ring	✔	✔	✔		
ISO 1179–4 ISO 228–1 Whitworth threads for stud ends with metal–to–metal sealing	✔	✔			
SAE J1926/2 (ISO 11926–2) ISO 725 inch threads for heavy duty (S series) stud ends with O–ring sealing	✔	✔	✔		
SAE J1926/3 (ISO 11926–3) ISO 725 inch threads for light duty (L series) stud ends with O–ring sealing	✔	✔	✔		
ISO 9974–2 ISO 261 metric threads for stud ends with elastomeric sealing (type E)	✔	✔	✔		
ISO 9974–3 ISO 261 metric threads for stud ends with metal–to–metal sealing (type B)	✔	✔	✔		
[1)] Preferred for new design					

FIGURE B1—APPLICABLE TESTS FOR VARIOUS FITTING AND STUD END STANDARDS

CONNECTIONS FOR FLUID POWER AND GENERAL USE—PORTS AND STUD ENDS WITH ISO 261 THREADS AND O-RING SEALING PART 1: PORT WITH O-RING SEAL IN TRUNCATED HOUSING—SAE J2244/1 DEC91

SAE Standard

Report of the SAE Fluid Conductors and Connectors Technical Committee SC1—Automotive and Hydraulic Tube and Fitting approved December 1991. Rationale statement available.

This document is technically equivalent to ISO 6149-1 except as noted in the Foreword.

Foreword

SAE J2244/ISO 6149 Parts 1 and 2 were prepared by SAE FCC-TC-SCI, Automotive and Hydraulic Tube and Fitting Subcommittee and ISO/TC 131 Fluid Power Systems. SAE J2244/ISO 6149 consists of the following parts under the general title: Connections for Fluid Power and General Use—Ports and Stud ends with ISO 261 threads and O-ring sealing:

Part 1: Port with O-Ring Seal in Truncated Housing

Part 2: Heavy-duty (S Series) Stud Ends—Dimensions, Design, Test Methods, and Requirements

The two parts of SAE J2244 constitute a revision of ISO 6149:1980. This revision defines performance requirements, dimensions, and designs for port and heavy-duty (S series) stud ends. Significant testing was conducted to confirm the performance requirements of stud ends made from carbon steel. ISO 6149-2 applies to fittings detailed in ISO 8434 parts 1, 3, and 4.

SAE J2244 Parts 1 and 2 are technically equivalent to ISO 6149 parts 1 and 2, respectively. Parts produced to either standard will interchange with parts produced to the other standard. **Two main differences exist between the SAE standards and the ISO standards: size M30 × 2 is included in the SAE standard but not in the ISO standard and the tube ODs have been shown in the SAE standard for the port sizes.** The SAE subcommittee chose not to include ISO 6149-3, a light-duty stud end, within SAE J2244 to minimize part proliferation.

In fluid power systems, power is transmitted and controlled through a fluid (liquid or gas) under pressure within an enclosed circuit. In general applications, a fluid may be conveyed under pressure. Components are connected through their threaded ports by fluid conductor fittings to tubes and pipes, or to hose fittings and hoses.

Ports are an integral part of fluid power components, such as pumps, motors, valves, cylinders, etc.

1. Scope

This part of SAE J2244 specifies dimensions for fluid power metric ports for use with adjustable and nonadjustable stud ends shown in SAE J2244/2.

Ports in accordance with this part of SAE J2244 may be used at working pressures up to 63 MPa for nonadjustable stud ends and 40 MPa for adjustable stud ends. The permissible working pressure depends upon materials, design, working conditions, application, etc.

For threaded ports and stud ends specified in new designs in hydraulic fluid power applications, only SAE J2244 shall be used. Threaded ports and stud ends in accordance with ISO 1179, ISO 9974, and SAE J1926 (ISO 11926) shall not be used for new designs in hydraulic fluid power applications.

2. References

2.1 Applicable Documents

The following standards contain provisions which, through reference in this text, constitute provisions of this Standard. At the time of publication, the editions indicated were valid. All standards are subject to revision, and parties to agreements based on this Standard are encouraged to investigate the possibility of applying the most recent edition of the standards indicated as follows. Members of IEC and ISO maintain registers of currently valid International Standards.

2.1.1 ISO Publications—Available from ANSI, 11 West 42nd Street, New York, NY 10036-8002.

ISO 261:1973—ISO general purpose metric screw threads—General plan

ISO 1179-1:- [1]—Connections for fluid power and general use—Ports and stud ends with ISO 228-1 threads with elastomeric and metal-to-metal sealing—Part 1: Threaded port

ISO 1179-2:- [1]—Connections for fluid power and general use—Ports and stud ends with ISO 228-1 threads with elastomeric and metal-to-metal sealing—Part 2: Heavy duty (S series) and light duty (L series) stud ends with elastomeric sealing (type E)

ISO 1179-3:- [1]—Connections for fluid power and general use—Ports and stud ends with ISO 228-1 threads with elastomeric and metal-to-metal sealing—Part 3: Light duty (L series) stud ends with sealing by O-ring with retaining ring (types G and H)

---
[1] To be published

22.291

ISO 1179-4:- [1]—Connections for fluid power and general use—Ports and stud ends with ISO 228-1 threads with elastomeric and metal-to-metal sealing—Part 4: Stud end for general use only with metal-to-metal sealing (type B)

ISO 1302:1978—Technical drawings—Method of indicating surface texture on drawings

ISO 2306:1972—Drills for use prior to tapping screw threads

ISO 5598:1985—Fluid power systems and components—Vocabulary

ISO 6149-1:- [1]—Connections for fluid power and general use—Ports and stud ends with ISO 261 threads and O-ring sealing—Part 1: Port with O-ring seal in truncated housing

ISO 6149-2:- [1]—Connections for fluid power and general use—Ports and stud ends with ISO 261 threads and O-ring sealing—Part 2: Heavy duty (S series) stud ends—Dimensions, design, test methods and requirements

ISO 6149-3:- [1]—Connections for fluid power and general use—Ports and stud ends with ISO 261 threads and O-ring sealing—Part 3: Light duty (L series) stud ends—Dimensions, design, test methods and requirements

ISO 7789:- [1]—Hydraulic fluid power—Two, three- and four-port screw-in cartridge valve cavities

ISO 9974-1:- [1]—Connections for fluid power and general use—Ports and stud ends with ISO 261 threads and elastomeric sealing ring and metal-to-metal sealing—Part 1: Threaded port

ISO 9974-2:- [1]—Connections for fluid power and general use—Ports and stud ends with ISO 261 threads and elastomeric sealing ring and metal-to-metal sealing—Part 2: Stud end with elastomeric sealing (type E)

ISO 9974-3:- [1]—Connections for fluid power and general use—Ports and stud ends with ISO 261 threads and elastomeric sealing ring and metal-to-metal sealing—Part 3: Stud end with metal-to-metal sealing (type S)

ISO 11926-1:- [1]—Connections for fluid power and general use—Ports and stud ends with ISO 725 threads and O-ring sealing—Part 1: Threaded port

ISO 11926-2:- [1]—Connections for fluid power and general use—Ports and stud ends with ISO 725 threads and O-ring sealing—Part 2: Heavy duty (S series) stud end

ISO 11926-3:- [1]—Connections for fluid power and general use—Ports and stud ends with ISO 725 threads and O-ring sealing—Part 3: Light duty (L series) stud end

2.2 U.S. References Identical to ISO References

2.2.1 ANSI Publications—Available from ANSI, 11 West 42nd Street, New York, NY 10036-8002.

ANSI/ASME B1.13M—83—Metric Screw Threads—M Profile

2.2.2 SAE Publications—Available from SAE, 400 Commonwealth Drive, Warrendale, PA 15096-0001.

SAE J343 APR91—Test and Procedures for SAE 100R Series Hydraulic Hose and Hose Assemblies

SAE J1926 AUG88—Specifications for Straight Thread O-ring Boss Port

SAE J2244/2 DEC91—Connections for Fluid Power and General Use—Ports and Stud Ends with ISO 261 Threads and O-ring Sealing—Part 2:

---

[1] To be published

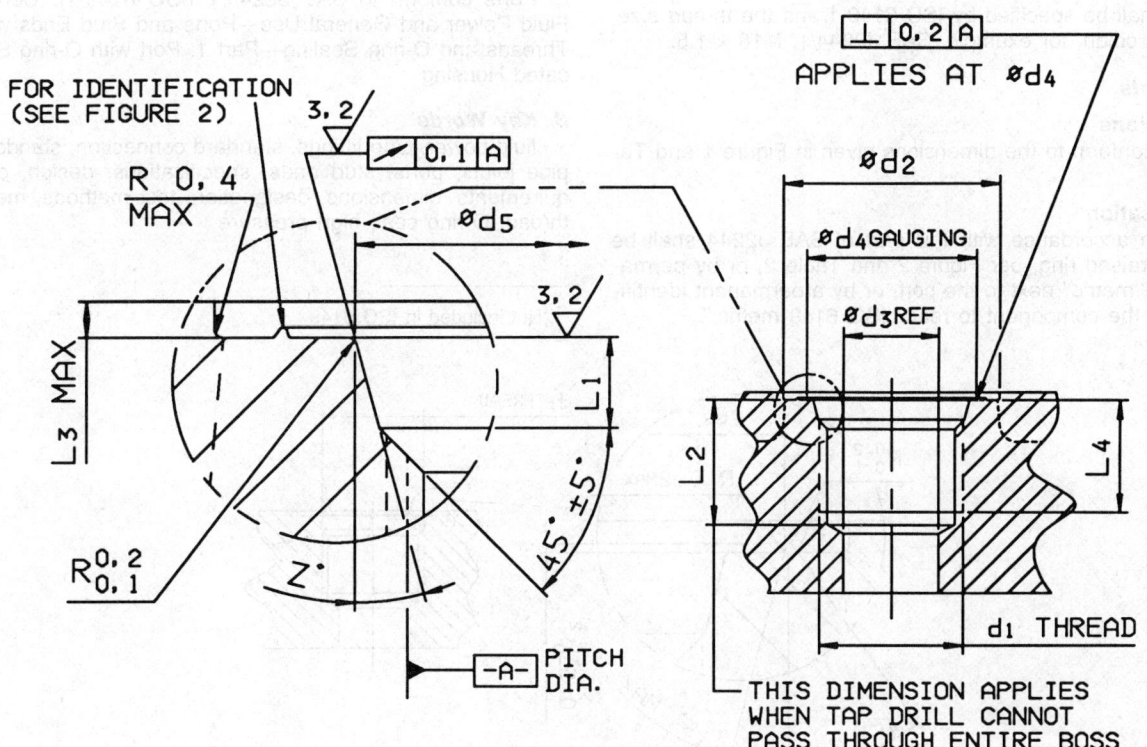

FIGURE 1—SAE J2244/1 PORT DETAIL

TABLE 1—SAE J2244/1 PORT DIMENSIONS

Dimensions in Millimeters

Tube OD	Inch Nominal Tube Dash Size	Inch Nominal Tube OD min	Inch Nominal Tube OD in	Thread Size d_1[1]	d_2 Large [2] min	d_2 Small [3] min	d_3[4] Ref	d_4	d_5 +0.1 0	L_1 +0.4 0	L_2[5] Min	L_3 Max	L_4 Min Full thread	Z° ±1°
4	-2	3.18	0.125	M8 × 1	17	14	3	12.5	9.1	1.6	11.5	1	10	12°
5	-3	4.76	0.188	M10 × 1	20	16	4.5	14.5	11.1	1.6	11.5	1	10	12°
6	-4	6.35	0.250	M12 × 1.5	23	19	6	17.5	13.8	2.4	14	1.5	11.5	15°
8	-5	7.94	0.312	M14 × 1.5 [6]	25	21	7.5	19.5	15.8	2.4	14	1.5	11.5	15°
10	-6	9.52	0.375	M16 × 1.5	28	24	9	22.5	17.8	2.4	15.5	1.5	13	15°
12	-8	12.7	0.500	M18 × 1.5	30	26	11	24.5	19.8	2.4	17	2	14.5	15°
16	-10	15.88	0.625	M22 × 1.5	34	29	14	27.5	23.8	2.4	18	2	15.5	15°
20	-12	19.05	0.750	M27 × 2	40	34	18	32.5	29.4	3.1	22	2	19	15°
22	-14	22.22	0.875	M30 × 2 [7]	43	38	18	36.5	32.4	3.1	22	2	19	15°
25	-16	25.4	1.000	M33 × 2	49	43	23	41.5	35.4	3.1	22	2.5	19	15°
30	-20	31.75	1.250	M42 × 2	60	52	30	50.5	44.4	3.1	22.5	2.5	19.5	15°
38	-24	38.10	1.500	M48 × 2	66	57	36	55.5	50.4	3.1	25	2.5	22	15°
50	-32	50.80	2.000	M60 × 2	76	67	44	65.5	62.4	3.1	27.5	2.5	24.5	15°
				M20 × 1.5 [8]	32	27		25.5	21.8	2.4		2	14.5	15°

[1] Per ISO 261 tolerance class 6H. Tap drill per ISO 2306 class 6H.
[2] Spotface diameter with identification ridge.
[3] Spotface diameter without identification ridge.
[4] Reference only, connecting hole application may require a different size.
[5] Tap drill depths given require use of a bottoming tap to produce the specified full thread lengths. Where standard taps are used increase tap drill depths accordingly.
[6] Preferred for diagnostic port applications.
[7] Not included in ISO 6149.
[8] For plug for cartridge valve cavity only. (See ISO 7789.)

Heavy duty (S Series) Stud Ends—Dimensions, Design, Test Methods and Requirements

3. Definitions
For the purposes of this part of SAE J2244, the definitions given in ISO 5598 shall apply.

4. Port Size
The ports shall be specified by ISO 6149-1 and the thread size, separated by a colon, for example, SAE J2244/1: M18 × 1.5.

5. Requirements

5.1 Dimensions
Ports shall conform to the dimensions given in Figure 1 and Table 1.

5.2 Identification
Each port in accordance with this part of SAE J2244 shall be identified by a raised ring, per Figure 2 and Table 2, or by permanently marking "metric" next to the port, or by a permanent identification label on the component to read "ISO 6149 metric."

6. Test Methods
Ports shall be tested along with stud ends per the test methods and requirements in SAE J2244/2.

7. Identification Statement
Use the following statement, except for M30 × 2,[2] in test reports, catalogues, and sales literature when electing to comply with this part of SAE J2244 (ISO 6149-2):

Ports conform to SAE J2244/1 (ISO 6149-1), Connections for Fluid Power and General Use—Ports and Stud Ends with ISO 261 Threads and O-ring Sealing—Part 1: Port with O-ring Seal in Truncated Housing

8. Key Words
fluid power, pipe fittings, standard connection, standard coupling, pipe joints, ports, stud ends, specifications, design, operating requirements, dimensions, designation, test methods, metric, straight thread, O-ring seal, high pressure

[2] Not included in ISO 6149

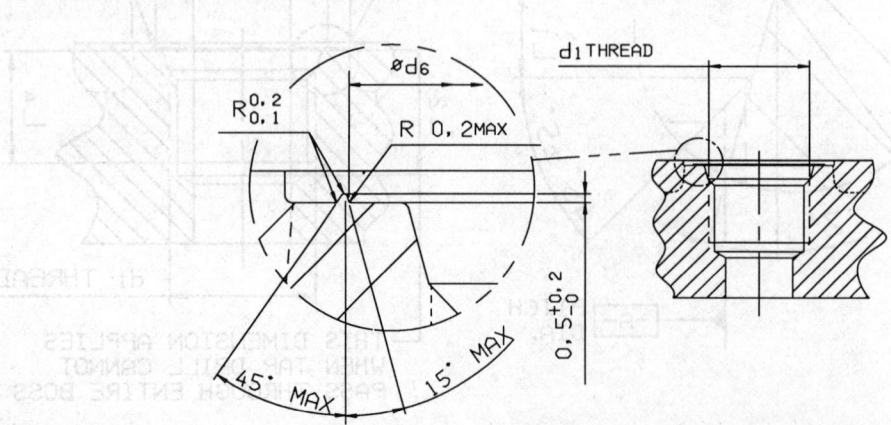

FIGURE 2—SAE J2244/1 PORT IDENTIFICATION DETAIL (SEE 5.2)

TABLE 2—SAE J2244/1 PORT IDENTIFICATION DIMENSIONS
Dimensions in Millimeters

Thread Size d_1	d_6 +0.5 / 0
M8 × 1	14
M10 × 1	16
M12 × 1.5	19
M14 × 1.5	21
M16 × 1.5	24
M18 × 1.5	26
M22 × 1.5	29
M27 × 2	34
M30 × 2[1]	38
M33 × 2	43
M42 × 2	52
M48 × 2	57
M60 × 2	67
M20 × 1.5[2]	27

[1] Not included in ISO 6149.
[2] For plug for cartridge valve cavity only. (See ISO 7789.)

CONNECTIONS FOR FLUID POWER AND GENERAL USE— PORTS AND STUD ENDS WITH ISO 261 THREADS AND O-RING SEALING PART 2: HEAVY-DUTY (S SERIES) STUD ENDS— DIMENSIONS, DESIGN, TEST METHODS, AND REQUIREMENTS —SAE J2244/2 DEC91

SAE Standard

Report of the SAE Fluid Conductors and Connectors Technical Committee SC1—Automotive and Hydraulic Tube and Fitting approved December 1991. Rationale statement available.

This document is technically equivalent to ISO 6149-2, except as noted in the foreword.

Foreword

SAE J2244 parts 1 and 2 were prepared by SAE FCCTC-SC1, Automotive and Hydraulic Tube and Fitting Subcommittee and ISO/TC 131, Fluid power systems. SAE J2244 consists of the following parts under the general title: Connections for Fluid Power and General Use—Ports and Stud Ends with ISO 261 Threads and O-ring Sealing—

Part 1: Port with O-ring Seal in Truncated Housing

Part 2: Heavy-duty (S series) Stud Ends—Dimensions, Design, Test Methods, and Requirements

The two parts of SAE J2244 constitute a revision of ISO 6149:1980. This revision defines performance requirements, dimensions, and designs for port and heavy-duty (S series) stud ends. Significant testing was conducted to confirm the performance requirements of stud ends made from carbon steel. ISO 6149-2 applies to fittings detailed in ISO 8434 parts 1, 3, and 4.

SAE J2244 Parts 1 and 2 are technically equivalent to ISO 6149 parts 1 and 2, respectively. Parts produced to either standard will interchange with parts produced to the other standard. **Two main differences exist between the SAE standards and the ISO standards: size M30 × 2 is included in the SAE standard but not in the ISO standard and the tube ODs have been shown in the SAE standard for the port sizes.** The SAE subcommittee chose not to include ISO 6149-3, a light-duty stud end, within SAE J2244 to minimize part proliferation.

Appendix A of this standard is normative.

In fluid power systems, power is transmitted and controlled through a fluid (liquid or gas) under pressure within an enclosed circuit. In general applications, a fluid may be conveyed under pressure. Components are connected through their threaded ports by stud ends on fluid conductor fittings to tubes and pipes, or to hose fittings and hoses.

1. Scope

This part of SAE J2244 specifies dimensions, performance requirements, and test procedures for metric adjustable and nonadjustable heavy-duty (S series) stud ends and O-rings.

Stud ends in accordance with this part of SAE J2244 may be used at working pressures up to 63 MPa for nonadjustable stud ends and 40 MPa for adjustable stud ends. The permissible working pressure depends upon materials, design, working conditions, application, etc.

For threaded ports and stud ends for use in new designs in hydraulic fluid power applications, only SAE J2244 shall be used. Threaded ports and stud ends in accordance with ISO 1179, ISO 9974, and SAE J1926 (ISO 11926) shall not be used for new designs in hydraulic fluid power applications.

Conformance to the dimensional information in this standard does not guarantee rated performance. Each manufacturer shall perform testing according to the specification contained in this standard to ensure that components made to this standard comply with the performance ratings.

2. References

2.1 Applicable Documents

The following standards contain provisions which, through reference in this text, constitute provisions of this Standard. At the time of publication, the editions indicated were valid. All standards are subject to revision, and parties to agreements based on this Standard are encouraged to investigate the possibility of applying the most recent edition of the standards indicated as follows. Members of IEC and ISO maintain registers of currently valid International Standards.

2.1.1 ISO Publications—Available from ANSI, 11 West 42nd Street, New York, NY 10036-8002.

ISO 48:1979—Vulcanized rubbers—Determination of hardness (Hardness between 30 and 85 IRHD)

ISO 261:1973—ISO general purpose metric screw threads—General plan

ISO 1179-1:- [1]—Connections for fluid power and general use—Ports and stud ends with ISO 228-1 threads with elastomeric and metal-to-metal sealing—Part 1: Threaded port

ISO 1179-2:- [1]—Connections for fluid power and general use—Ports and stud ends with ISO 228-1 threads with elastomeric and metal-to-metal sealing—Part 2: Heavy duty (S series) and light duty (L series) stud ends with elastomeric sealing (type E)

ISO 1179-3:- [1]—Connections for fluid power and general use—Ports and stud ends with ISO 228-1 threads with elastomeric and metal-to-metal sealing—Part 3: Light duty (L series) stud ends with sealing by O-ring with retaining ring (types G and H)

ISO 1179-4:- [1]—Connections for fluid power and general use—Ports and stud ends with ISO 228-1 threads with elastomeric and metal-to-metal sealing—Part 4: Stud end for general use only with metal-to-metal sealing (type B)

ISO 1302:1978—Technical drawings—Method of indicating surface texture on drawings

ISO 3448:1975—Industrial liquid lubricants—ISO viscosity classification

ISO 3601-3:1987—Fluid systems—Sealing devices—O-rings—Part 3: Quality acceptance criteria

ISO 4759-1:1978—Tolerances for fasteners—Part 1: Bolts, screws and nuts with thread diameters between 1.6 (inclusive) and 150 mm (inclusive) and product grades A, B and C

ISO 5598:1985—Fluid power systems and components—Vocabulary

ISO 6149-1:- [1]—Connections for fluid power and general use—Ports and stud ends with ISO 261 threads and O-ring sealing—Part 1: Port with O-ring seal in truncated housing

ISO 6149-2:- [1]—Connections for fluid power and genral use—Ports and stud ends with ISO 261 threads and O-ring sealing—Part 2: Heavy duty (S series) stud ends—Dimensions, design, test methods and requirements

ISO 6149-3:- [1]—Connections for fluid power and general use—Ports and stud ends with ISO 261 threads and O-ring sealing—Part 3: Light duty (L series) stud ends—Dimensions, design, test methods and requirements

ISO 6803:1984—Rubber or plastic hoses and hose assemblies—Hydraulic pressure impulse test without flexing

ISO 7789:- [1]—Hydraulic fluid power—Two-, three- and four-port screw-in cartridge valve cavities

ISO 9974-1:- [1]—Connections for fluid power and general use—Ports and stud ends with ISO 261 threads and elastomeric sealing ring and metal-to-metal sealing—Part 1: Threaded port

ISO 9974-2:- [1]—Connections for fluid power and general use—Ports and stud ends with ISO 261 threads and elastomeric sealing ring and metal-to-metal sealing—Part 2: Stud end with elastomeric sealing (type E)

ISO 9974-3:- [1]—Connections for fluid power and general use—Ports and stud ends with ISO 261 threads and elastomeric sealing ring and metal-to-metal sealing—Part 3: Stud end with metal-to-metal sealing (type B)

ISO 11926-1:- [1]—Connections for fluid power and general use—Ports and stud ends with ISO 725 threads and O-ring sealing—Part 1: Threaded port

ISO 11926-2:- [1]—Connections for fluid power and general use—Ports and stud ends with ISO 725 threads and O-ring sealing—Part 2: Heavy duty (S series) stud end

ISO 11926-3:- [1]—Connections for fluid power and general use—Ports and stud ends with ISO 725 threads and O-ring sealing—Part 3: Light duty (L series) stud end

2.2 U.S. References Identical to ISO References

2.2.1 ANSI Publications—Available from ANSI, 11 West 42nd Street, New York, NY 10036-8002.

ANSI/ASME B1.13M—83, Metric Screw Threads—M Profile

2.2.2 SAE Publications—Available from SAE, 400 Commonwealth Drive, Warrendale, PA 15096-0001.

SAE J343 APR91—Test and Procedures for SAE 100R Series Hydraulic Hose and Hose Assemblies

SAE J1926 AUG88—Specifications for Straight Thread O-ring Boss Port

SAE J2244/1/ISO 6149-1:DEC91—Connections for Fluid Power and General Use—Ports and Stud Ends With ISO 261 Threads and O-ring Sealing—Part 1: Port With O-ring Seal in Truncated Housing

3. Definitions

For the purposes of this part of SAE J2244, the definitions given in ISO 5598 and the following definitions shall apply:

3.1 Adjustable Stud End

A stud end that allows for orientation before final tightening of the connection.

3.2 Nonadjustable Stud End

A stud end that does not allow for orientation before final tightening of the connection.

4. Stud End Size

The stud ends shall be specified by SAE J2244/2 and the thread size, separated by a colon, for example, SAE J2244/2:M18 X 1.5.

5. Requirements

5.1 Dimensions

Heavy-duty (S series) stud ends shall conform to the dimensions given in Figures 1A and 1B and Table 1. Hex tolerances across flats shall be according to ISO 4759-1, product grade C.

5.2 Working Pressure

Heavy-duty (S series) stud ends made of carbon steel shall be designed for use at the working pressures given in Table 2.

5.3 Performance

Heavy-duty (S series) stud ends made of carbon steel shall meet or exceed the burst and impulse pressures given in Table 2, when tested according to Section 7.

---

[1] To be published

22.295

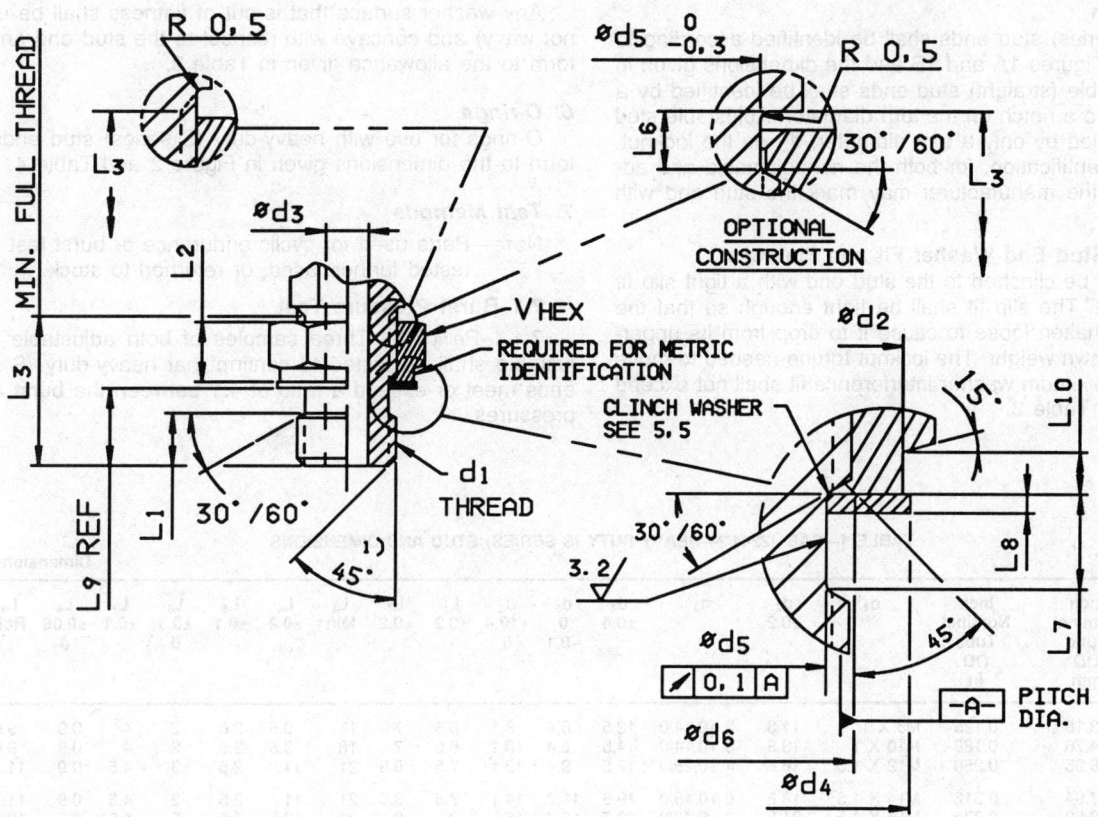

1) Chamfer to minor diameter of threads

FIGURE 1A—ADJUSTABLE SAE J2244/2 HEAVY-DUTY (S SERIES) STUD END DETAIL

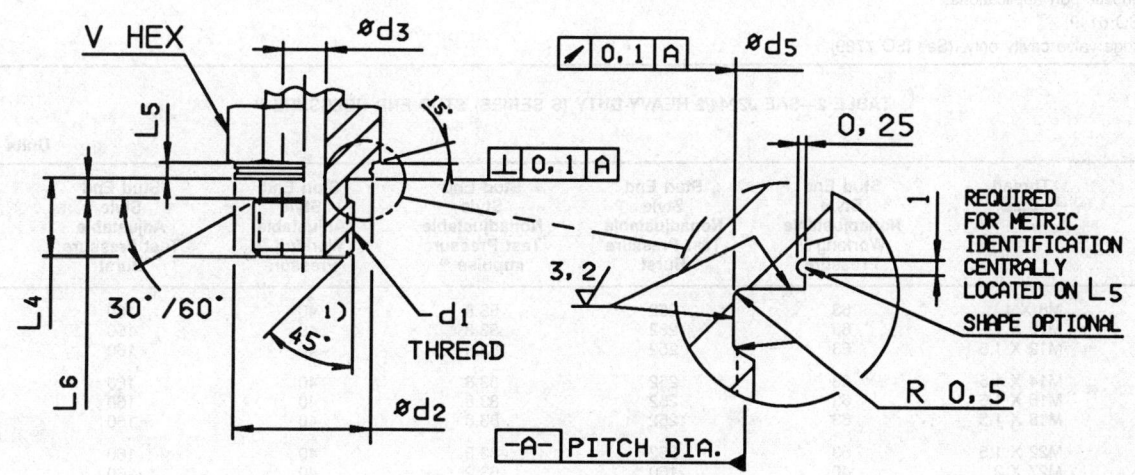

1) Chamfer to minor diameter of threads

FIGURE 1B—NONADJUSTABLE SAE J2244/2 HEAVY-DUTY (S SERIES) STUD END DETAIL

5.4 Identification

Heavy-duty (S series) stud ends shall be identified according to the detail shown in Figures 1A and 1B and the dimensions given in Table 1. Nonadjustable (straight) stud ends shall be identified by a turn diameter, d_2, and a notch on the turn diameter. Adjustable stud ends shall be identified by only a turn diameter, d_2, on the locknut. In addition to this identification, for both the nonadjustable and adjustable stud ends, the manufacturer may mark the stud end with the word "metric."

5.5 Adjustable Stud End Washer Fit and Flatness

The washer shall be clinched to the stud end with a tight slip fit to an interference fit. The slip fit shall be tight enough so that the washer cannot be shaken loose to cause it to drop from its uppermost position by its own weight. The locknut torque needed to move the washer at the maximum washer interference fit shall not exceed the torques given in Table 3.

Any washer surface that is out of flatness shall be uniform (i.e., not wavy) and concave with respect to the stud end and shall conform to the allowance given in Table 3.

6. O-rings

O-rings for use with heavy-duty (S series) stud ends shall conform to the dimensions given in Figure 2 and Table 4.

7. Test Methods

NOTE—Parts used for cyclic endurance or burst test shall not be tested further, used, or returned to stock.

7.1 Burst Pressure Test

7.1.1 PRINCIPLE—Three samples of both adjustable and nonadjustable shall be tested to confirm that heavy-duty (S series) stud ends meet or exceed a ratio of 4:1 between the burst and working pressures.

TABLE 1—SAE J2244/2—HEAVY-DUTY (S SERIES) STUD AND DIMENSIONS

Dimensions in Millimeters

Tube OD	Inch Nominal Tube Dash Size	Inch Nominal Tube OD mm	Inch Nominal Tube OD in	d_1[1]	d_2 ±0.2	d_3	d_4 ±0.4	d_5 0 -0.1	d_6 +10.4 0	L_1 ±0.2	L_2 ±0.2	L_3 Min	L_4 ±0.2	L_5 ±0.1	L_6 ±0.3 0	L_7 ±0.1	L_8 ±0.08 0	L_9 Ref.	L_{10} ±0.1	V Hex
4	-2	3.18	0.125	M8 X 1	11.8	2 +0.14/0	12.5	6.4	8.1	6.5	7	18	9.5	2.5	2	4	0.9	9.6	1.5	12
5	-3	4.76	0.188	M10 X 1	13.8	3 +0.14/0	14.5	8.4	10.1	6.5	7	18	9.5	2.5	2	4	0.9	9.6	1.5	14
6	-6	6.35	0.250	M12 X 1.5	16.8	4 +0.18/0	17.5	9.7	12.1	7.5	8.5	21	11	2.5	3	4.5	0.9	11.1	2	17
8	-5	7.94	0.312	M14 X 1.5[2]	18.8	6 +0.18/0	19.5	11.7	14.1	7.5	8.5	21	11	2.5	3	4.5	0.9	11.1	2	19
10	-6	9.52	0.375	M16 X 1.5	21.8	7 +0.22/0	22.5	13.7	16.1	9	9	23	12.5	2.5	3	4.5	0.9	12.6	2	22
12	-8	12.7	0.500	M18 X 1.5	23.8	9 +0.22/0	24.5	15.7	18.1	10.5	10.5	26	14	2.5	3	4.5	0.9	14.1	2.5	24
16	-10	15.88	0.625	M22 X 1.5	26.8	12 +0.27/0	27.5	19.7	22.1	11	11	27.5	15	2.5	3	5	1.25	14.8	2.5	27
20	-12	19.05	0.750	M27 X 2	31.8	15 +0.27/0	32.5	24	27.1	13.5	13.5	33.5	18.5	2.5	4	6	1.25	18.3	2.5	32
22	-14	22.22	0.875	M30 X 2[3]	35.8	18 +0.33/0	36.5	27	30.1	13.5	13.5	33.5	18.5	2.5	4	6	1.25	18.3	2.5	36
25	-16	25.4	1.000	M33 X 2	40.8	20 +0.33/0	41.5	30	33.1	13.5	13.5	33.5	18.5	3	4	6	1.25	18.8	3	41
30	-20	31.75	1.250	M42 X 2	49.8	26 +0.33/0	50.5	39	42.1	14	14	34.5	19	3	4	6	1.25	18.8	3	50
38	-24	38.10	1.500	M48 X 2	54.8	32 +0.39/0	55.5	45	48.1	16.5	15	38	21.5	3	4	6	1.25	21.3	3	55
50	-32	50.80	2.000	M60 X 2	64.8	40 +0.39/0	65.5	57	60.1	19	17	42.5	24	3	4	6	1.25	23.8	3	65
				M20 X 1.5[4]	26.8		17.7			14	2.5	3				2.5				

[1] Thread Class 6 g per ISO 261.
[2] Preferred for diagnostic port applications.
[3] Not included in ISO 6149.
[4] For plug for cartridge valve cavity only. (See ISO 7789).

TABLE 2—SAE J2244/2 HEAVY-DUTY (S SERIES) STUD END PRESSURE [1]

Units in Megapascals [2]

Tube OD	Thread Size	Stud End Style Nonadjustable Working [1] Pressure	Stud End Style Nonadjustable Test Pressure Burst	Stud End Style Nonadjustable Test Pressure Impulse [3]	Stud End Style Adjustable Working [1] Pressure	Stud End Style Adjustable Test Pressure Burst	Stud End Style Adjustable Test Pressure Impulse [3]
4	M8 X 1	63	252	83.8	40	160	53.2
5	M10 X 1	63	252	83.8	40	160	53.2
6	M12 X 1.5	63	252	83.8	40	160	53.2
8	M14 X 1.5	63	252	83.8	40	160	53.2
10	M16 X 1.5	63	252	83.8	40	160	53.2
12	M18 X 1.5	63	252	83.8	40	160	53.2
16	M22 X 1.5	63	252	83.8	40	160	53.2
20	M27 X 2	40	160	53.2	40	160	53.2
22	M30 X 2	40	160	53.2	40	160	53.2
25	M33 X 2	40	160	53.2	31.5	126	41.9
30	M42 X 2	25	100	33.2	25	100	33.2
38	M48 X 2	25	100	33.2	20	80	26.6
50	M60 X 2	25	100	33.2	16	64	21.3
For plug for cartridge valve cavity only (See ISO 7789)							
	M20 X 1.5	40	160	53.2	—	—	—

[1] These pressures were established using fittings made of carbon steel when tested in accordance with Section 7.
[2] To convert from MPa to bar multiply by 10. (10 bar/MPa)
[3] Cyclic endurance test pressure.

TABLE 3—ADJUSTABLE STUD END WASHER TORQUE AND FLATNESS ALLOWANCE

Thread Size	Maximum Nut Torque to Move Washer N·m	Maximum Washer Flatness Allowance mm
M8 X 1	1	0.25
M10 X 1	3	0.25
M12 X 1.5	4	0.25
M14 X 1.5	5	0.25
M16 X 1.5	7	0.25
M18 X 1.5	10	0.25
M22 X 1.5	12	0.25
M27 X 2	15	0.40
M30 X 2	18	0.40
M33 X 2	20	0.40
M42 X 2	25	0.50
M48 X 2	30	0.50
M60 X 2	40	0.50

7.1.2 Materials

7.1.2.1 *Test Block and Stud Ends*—Test blocks shall be unplated and hardened to 45-55 HRC. Stud ends shall be made from carbon steel and plated.

7.1.2.2 *Test O-rings*—Unless otherwise specified, O-rings shall be made from nitrile (NBR) rubber with a hardness of 85 +10/–0 IRHD when measured per ISO 48. O-rings shall conform to the dimensions given in Table 4 and shall meet or exceed the quality requirement grade N in ISO 3601-3.

7.1.3 Procedures

7.1.3.1 *Thread Lubrication*—For testing only, threads and contact surfaces shall be lubricated with hydraulic oil with a viscosity of VG 32 per ISO 3448 prior to the application of torque.

7.1.3.2 *Stud End Torque*—Stud ends shall be tested after application of the torques given in Table 5. Adjustable stud locknut torques shall be applied after the stud end has been backed out one full turn from finger tight position, to correctly test the worst possible actual assembly conditions.

7.1.3.3 *Pressure Rise Rate*—During the burst test, the rate of pressure rise shall not exceed 138 MPa per minute.

7.1.4 Test Report
Test results and conditions shall be reported on the test data form in Appendix A.

7.2 Cyclic Endurance (Impulse) Test

7.2.1 Principle
Six samples of both adjustable and nonadjustable stud ends, when tested at their respective impulse pressures, shall pass a cyclic endurance test of 1 000 000 cycles.

7.2.2 Materials
Use the same materials as per 7.1.2.

7.2.3 Procedures

7.2.3.1 *Thread Lubrication*—Apply lubricant per 7.1.3.1.

7.2.3.2 *Stud End Torques*—Apply torque per 7.1.3.2.

7.2.3.3 *Cycle and Pressure Rise Rate*—Cycle rate shall be uniform at 0.5 to 1.3 Hz and shall conform to the wave pattern shown in SAE J343 (ISO 6803).

7.2.4 Test Report
Test results and conditions shall be reported on the test data form in Appendix A.

8. Identification Statement

Use the following statement, except for M30 x 2,[2] in test reports, catalogues, and sales literature when electing to comply with this part of SAE J2244 (ISO 6149-2):

Heavy-duty (S series) stud end conforms to SAE J2244/2 (ISO 6149-2), Connections for Fluid Power and General Use—Ports and Stud Ends with ISO 261 Threads and O-ring Sealing—Part 2: Heavy-duty (S series) Stud Ends—Dimensions, Design, Test Methods, and Requirements

---

[2] Not included in ISO 6149

9. Key Words

fluid power, pipe fittings, standard connection, standard coupling, pipe joints, ports, stud ends, specifications, design, operating requirements, dimensions, designation, test methods, metric, straight thread, O-ring seal, high pressure

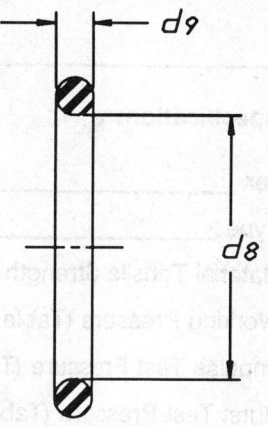

FIGURE 2—O-RING DETAIL

TABLE 4—SAE J2244/2 STUD END O-RING DIMENSIONS

Dimensions in Millimeters

Thread Size	Inside Diameter d_8	Inside Diameter d_8 tol. ±	Cross Section Diameter d_9	Cross Section Diameter d_9 tol ±
M8 X 1	6.1	0.20	1.6	0.08
M10 X 1	8.1	0.20	1.6	0.08
M12 X 1.5	9.3	0.20	2.2	0.08
M14 X 1.5	11.3	0.20	2.2	0.08
M16 X 1.5	13.3	0.20	2.2	0.08
M18 X 1.5	15.3	0.20	2.2	0.08
M22 X 1.5	19.3	0.22	2.2	0.08
M27 X 2	23.6	0.24	2.9	0.09
M30 X 2 [1]	26.6	0.26	2.9	0.09
M33 X 2	29.6	0.29	2.9	0.09
M42 X 2	38.6	0.37	2.9	0.09
M48 X 2	44.6	0.43	2.9	0.09
M60 X 2	56.6	0.51	2.9	0.09
M20 X 1.5 [2]	17.3	0.22	2.2	0.08

[2] For plug for cartridge valve cavity only, (See ISO 7789.)
[1] Not included in ISO 6149.

TABLE 5—TORQUE REQUIREMENTS FOR STUD END QUALIFICATION TEST

Thread Size	Torque +10% N·m –0
M8 X 1	10
M10 X 1	20
M12 X 1.5	35
M14 X 1.5	45
M16 X 1.5	55
M18 X 1.5	70
M22 X 1.5	100
M27 X 2	170
M30 X 2	215
M33 X 2	310
M42 X 2	330
M48 X 2	420
M60 X 2	500
M20 X 1.5 [1]	80

[1] For plug for cartridge valve cavity only. (See ISO 7789.)

APPENDIX A
(Normative)

A.1 See Figure A1.

SAE J2244/2 PORT AND STUD END TEST DATA FORM

_____

Stud end specification:

Manufacturer _____ Test Facility _____

Stud End Type _____ Size _____

Minimum Material Tensile Strength _____ MPa

Stud End Working Pressure (Table 2) _____ MPa

Stud End Impulse Test Pressure (Table 2) _____ MPa

Stud End Burst Test Pressure (Table 2) _____ MPa

Qualification Test Assembly Torque (Table 3) _____ N·m

Burst Test Results: (Three samples minimum burst tested)

Sample No.	Pressure @ Failure	Torque	Hardness	Type of Failure
1. _____	_____ MPa	_____ N·m	_____	_____
2. _____	_____ MPa	_____ N·m	_____	_____
3. _____	_____ MPa	_____ N·m	_____	_____

Cyclic endurance test results: (Six samples minimum Impulse tested)

Sample No.	Cycles @ Failure	Torque	Hardness	Type of Failure
1. _____	_____	_____ N·m	_____	_____
2. _____	_____	_____ N·m	_____	_____
3. _____	_____	_____ N·m	_____	_____
4. _____	_____	_____ N·m	_____	_____
5. _____	_____	_____ N·m	_____	_____
6. _____	_____	_____ N·m	_____	_____

Conclusions: Pass/fail and why- _____

_____

Dimensions/ List any exception: _____

_____

_____

Name (printed/typed) and signature of person certifying report:

_____ Date: _____

FIGURE A1—SAE J2244/2 PORT AND STUD END TEST DATA FORM

ELECTRICAL/ ELECTRONIC EQUIPMENT AND LIGHTING

23 Electrical/Electronic Equipment

NOMENCLATURE—AUTOMOTIVE ELECTRICAL SYSTEMS—SAE J831

SAE Standard

Report of Electrical Equipment Committee approved June 1962.

Purpose and Scope—The purpose of this standard is to define terms relating to automotive electrical systems to facilitate clear understanding as to their meaning and promote uniformity of nomenclature in engineering discussions, technical papers, and specifications.

Generator, Electric DC, Mechanical Rectification—An electric d-c generator with mechanical rectification is a device which transforms mechanical power into direct current electrical power (d-c) by means of a field structure for excitation, a generating winding, and a segmented commutator for rectification, all of which are contained in an integral package.

Generator, Electric DC, Diode Rectification—An electric d-c generator with diode rectification is a device which transforms mechanical power into direct current electrical power (d-c) by means of a field structure for excitation, a generating winding, and a network of diodes for rectification, all of which are contained in an integral package.

Generator, Electric AC—An electric a-c generator or alternator is a device which transforms mechanical power into alternating current electrical power (a-c) by means of a field structure for excitation and a generating winding. The frequency of the alternating current is determined by the speed of the device.

Output at Idle—Generator "output at idle" on vehicles is the current from the charging system which is available at engine idle speed. The current may be used to supply the connected electrical loads, to charge the battery, or may be divided between the electrical loads and the battery in an infinite number of combinations.

GLOSSARY OF AUTOMOTIVE ELECTRONIC TERMS—SAE J1213 NOV82

SAE Information Report

Report of the Electronic Systems Committee approved June 1978 and completely revised November 1982. Formerly HS J1213 NOV82.

Introduction—This Glossary has been compiled to serve for reference in an effort to assist communications between the automotive engineer and the electronics engineer.

Scope—This Glossary confines its content to the specific field of electronic systems and subsystems as they pertain to the automotive engineer.

Source of Terminology—A letter or letters in parentheses on the right hand margin of each term indicate the source, from a subcommittee of the SAE Electronic Systems Committee, of the particular term. They are:
- (D)—Display Subcommittee
- (DG)—Diagnostics Subcommittee
- (J)—First issue of SAE J1213, as prepared by the Design Guides Subcommittee (1978)
- (M)—Microprocessor Subcommittee
- (T)—Transducer Subcommittee

Additionally, a parenthetical expression may appear at the end of the text for a particular term. These indicate a further specific source for the term and are:
- (ASTM)—American Society for Testing and Materials, "Compilation of ASTM Standard Definitions," Fourth Edition, 1979, Philadelphia.
- (GRAF)—"Modern Dictionary of Electronics," Rudolf F. Graf, Howard W. Sams & Co., Inc., Indianapolis, 1977.
- (IEEE)—Institute of Electrical and Electronics Engineers, Standard Dic-

tionary of Electrical and Electronic Terms, ANSI/IEEE 100-1977, Wiley-Interscience, New York.

(ISA)—Instrument Society of America, 'Electrical Transducer Nomenclature and Terminology," (ISA S37.1-1982), Research Triangle Park, North Carolina.

(ISO)—ISO 4092, Road Vehicles—Diagnostic systems for motor vehicles—Vocabulary (Available from ANSI, New York) 1978.

Reference—For additional or expanded descriptions and definitions of SI units of measure, see NBS Special Publication 330, "The International System of Units (SI)," U.S. Department of Commerce, National Bureau of Standards, U.S. Government Printing Office, Washington, DC, 1981.

GLOSSARY OF AUTOMOTIVE ELECTRONIC TERMS

ABSORPTION FACTOR (A) (D)
The absorption factor is the ratio of the light absorbed to the incident light.

AC PLASMA PANEL (D)
A gas discharge display using AC coupled electrodes. Typical units utilize the inherent memory properties and are mainly used for graphic applications.

ACCELERATION ERROR (T)
The maximum difference, at any measurand value within the specified range, between output readings taken with and without the application of specified constant acceleration along specified axes. (ISA)

ACCELERATION, TRANSDUCER (T)
An acceleration perpendicular to the sensitive axis of the transducer.

ACCEPTANCE TEST (GENERAL) (T)
(A) A test to demonstrate the degree of compliance of a device with purchaser's requirements. (B) A conformance test demonstrates the quality of the units of a consignment, without implication of contractual relations between buyer and seller. (IEEE)

ACCESS TIME (M)
The time interval between the request for information and the instant this information is available.

ACCUMULATOR (M)
In computing, one of the operands for arithmetic and logic operations is commonly held in the accumulator, and the result of the operation becomes the new stored data.

ACCURACY (M)
The degree of freedom from error; that is, the degree of conformity to truth or to a rule. Accuracy is contrasted with precision. For example, four-place numerals are less precise than six-place numerals, nevertheless a properly computed four-place numeral might be more accurate than an improperly computed six-place numeral.

(T)
The ratio of the error to the full-scale output or the ratio of the error to the output, as specified, expressed in percent. Note 1: Accuracy may be expressed in terms of units of measurand, or as within percent of full scale output. Note 2: Use of the term accuracy should be limited to generalized descriptions of characteristics. It should not be used in specifications. The term error is preferred in specifications and other specific descriptions of transducer performance. (ISA)

ACTIVE DISPLAY (D)
An active display is one that emits light. Typical examples are gas discharge and incandescent.

ACTIVE ELEMENT (J)
A component capable of producing power gain such as a transistor, tunnel diode, thyristor, etc. Also active device, active component.

ACTIVE FILTER (J)
A device employing passive network elements and amplifiers used for transmitting or rejecting signals in certain frequency ranges or for controlling the relative output of signals in certain frequency ranges or for controlling the relative output of signals as a function of frequency. (GRAF)

ACTIVE TRANSDUCER (T)
A transducer whose output waves are dependent upon sources of power, apart from that supplied by any of the actuating waves, which power is controlled by one or more of the waves. Note: The definition of active transducer is a restriction of the more general active network; that is, one in which there is an impressed driving force. (IEEE)

ACTUATOR (T)
A transducer whose output is a force or torque and usually involves motion.

ADDER (M)
Switching circuit that combines binary bits to generate the sum and carry of these bits.

ADDRESS (M)
An expression, usually numerical, which designates a specific location in a storage or memory device.

ADDRESS FORMAT (M)
1. The arrangement of the address parts of an instruction. The expression "plus-one" is frequently used to indicate that one of the addresses specifies the location of the next instruction to be executed, such as one-plus-one, two-plus-one, three-plus-one, four-plus-one.
2. The arrangement of the parts of a single address such as those required for identifying channel, module, track, etc., in a disc system.

ADDRESS REGISTER (M)
A register in which an address is stored.

ALGORITHM (M)
A prescribed sequence of well-defined rules or operations for the solution of a problem in a specified number of steps.

ALPHANUMERIC CODE (M)
A code whose code set consists of letters, digits, and associated special characters.

ALPHANUMERIC DISPLAY (D)
An alphanumeric display is one which can present numeric, common alphabetic characters and sometimes other special symbols. Dot matrix and segmented (with more than 7 segments) displays are typical examples.

ALU (M)
Arithmetic logic unit, a computational subsystem which performs the mathematical operations of a digital system.

AMBIENT CONDITIONS (J)
The conditions (pressure, temperature, etc.) of the surrounding medium.

AMBIENT PRESSURE ERROR (T)
The maximum change in output, at any measurand value within the specified range, when the ambient pressure is changed between specified values. (ISA)

AMERICAN WIRE GAUGE (AWG) (J)
System of numerical designations for wire size based on specified ranges of circular mil area. American wire gauge starts with 4/0 (0000) at the largest size going to 3/0 (000), 2/0 (00), 1/0 (0), 1, 2, and up to 40 and beyond for the smallest sizes.

AMPERE (A) (J)
The standard unit for measuring the strength of an electric current. The rate of flow of a charge in a conductor or conducting medium of one coulomb per second.

AMPLIFIER (J)
A device, circuit, or component which produces as an output an enlarged reproduction of the essential features of its input.

AMPLITUDE MODULATION (AM) (J)
Modulation in which the amplitude of a wave is the characteristic subject to variation. (GRAF)

ANALOG (J)
Of or pertaining to the general class of devices or circuits in which the output varies as a continuous function of the input.

ANALOG COMPUTER (J)
A computer which represents numerical quantities as electrical and physical variables and manipulates these variables in accomplishing solutions to mathematical problems.

ANALOG OUTPUT (T)
Transducer output which is a continuous function of the measurand except as modified by the resolution of the transducer. (ISA)

ANALOG REPRESENTATION (M)
A representation that does not have discrete values but is continuously variable.

AND GATE (J)
A combinational logic element such that the output channel is in its one state, if and only if, each input channel is in its one state.

ANGSTRÖM (D)
The angström is a commonly used unit of length for light wavelength measurements equal to 10^{-10} meter. Nanometer is the preferred SI unit.

ANNUNCIATOR
An arrangement of indicators, tripped by relays, for indicating which of a number of circuits has operated a sound signalling device.

ANODE (J)
The positive pole (+) in batteries, galvanic cells, or plating apparatus. In diodes, the positive lead.

(M)
A more positive lead when conducting in the forward direction.

ANTI-REFLECTION COATING (D)
A coating applied to the front surface of a display device or optical element to reduce the reflection of incident light. This improves readability in high ambient light conditions.

APOSTILB (ASB) (D)
A unit of luminance equal to -1 candela/m^2 (1 unit). A surface of 1 m^2 all points of which are 1 m from a 1 candela source has a luminance of 1

apostilb.

APPARENT CANDLEPOWER (D)
The apparent candlepower (luminous intensity) of an extended source measured at a specific distance is the candlepower of a point source that would produce the same illumination at the same distance.

ARITHMETIC SHIFT (M)
1. A shift that does not affect the sign position.
2. A shift that is equivalent to the multiplication of a number by a positive or negative integral power of the radix.

ARRAY LOGIC (M)
A logic network whose configuration is a rectangular array of intersections of its input-output leads, with elements connected at some of these intersections. The network usually functions as an encoder or decoder.

AS-NEW (DG)
Having performance capabilities equal to level of performance of units when offered for sale after manufacture.

ASSEMBLE (M)
To prepare a machine language program from a symbolic language program by substituting absolute operation codes for symbolic operation codes and absolute or relocatable addresses for symbolic addresses.

ASSEMBLER (M)
A computer program that assembles.

ASYMMETRICAL LIGHT DISTRIBUTION (D)
An asymmetrical light distribution is one in which the emitted or reflected light distribution is not the same for conjugate viewing angles.

ASYNCHRONOUS DEVICE (M)
A device in which the speed of operation is not related to any frequency in the system to which it is connected.

ATTITUDE ERROR (T)
The error due to the orientation of the transducer relative to the direction in which gravity acts upon the transducer. (ISA)

AVALANCHE BREAKDOWN (J)
In a semi-conductor diode, a nondestructive breakdown caused by the cumulative multiplication of carriers through field-induced impact ionization. (GRAF)

AVALANCHE DIODE (J)
Also called breakdown diode. A silicon diode that has a high ratio of reverse-to-forward resistance until avalanche breakdown occurs. After breakdown, the voltage drop across the diode is essentially constant and is independent of the current. Used for voltage regulating and voltage limiting. Originally called Zener diode before it was found that the Zener effect had no significant role in the operation of diodes of this type.

BANDWIDTH (J)
The range within the limits of a band. The least frequency interval of a wave form. The range of frequencies of a device, within which its performance, with respect to some characteristic, conforms to a specified standard.

BARRIER LAYER (J)
See Depletion Layer.

BASE (J)
(Transistor) A region that lies between an emitter and a collector of a transistor and into which minority carriers are injected. (IEEE)

BASE RESISTANCE (J)
Resistance in series with the base lead in the common T equivalent circuit of a transistor. (GRAF)

BATTERY (J)
A DC voltage source which converts chemical, nuclear, thermal or solar energy into electrical energy. (GRAF)

BATTERY-BACKED (M)
(Not defined) A technique of using a standby power source of the battery variety to supply power to a system or part of a system for maintaining essential data, etc., during the loss of main power.

"BEST STRAIGHT LINE" (T)
A line midway between the two parallel straight lines closest together and enclosing all output vs. measurand values on a calibration curve. (ISA)

BIAS (J)
To influence or dispose to one direction, as for example, with a direct voltage or with a spring. (IEEE)

(T)
A constant or systematic error as opposed to a random error. It manifests itself as a persistent positive or negative deviation of the method average from the accepted reference value. (ASTM)

BIDIRECTIONAL DIODE-THYRISTOR (J)
A two terminal thyristor having substantially the same switching behavior in the first and third quadrants of the principal voltage-current characteristic. (IEEE)

BINARY (J)
A characteristic or property involving a selection, choice, or condition in which there are but two possible alternatives.

BINARY CODED DECIMAL (BCD) (M)
A binary numbering system for coding decimal numbers in groups of 4 bits. The binary value of these 4-bit groups ranges from 0000 to 1001, and codes the decimal digits "0" to "9." To count to 9 takes 4 bits; to count to 99 takes two groups of 4 bits; to count to 999 takes three groups of 4 bits, etc.

BIPOLAR (J)
Having to do with a device in which both majority and minority carriers are present. In connection with IC's, the term describes a specific type of construction; bipolar and MOS are the two most common types of IC construction. (GRAF)

BIT (BINARY DIGIT) (J)
The smallest element of information in binary language. A contraction of binary digit. These characters in system (computer) language signify "on" and "off" (1 and 0). Word length, memory capacity, etc., can be expressed in number of "bits."

BLACK BODY (D)
A body which has an absorption factor of 1 for all wavelengths of interest is a black body. The radiation from a black body is a function solely of its temperature and is useful as a standard for light measurements.

BLEEDER RESISTOR (J)
A resistor used to draw a fixed current. Also used to discharge a filter capacitor after the circuit is de-energized. (GRAF)

BLOCK (M)
1. A set of things, such as words, characters, or digits handled as a unit.
2. A collection of continuous records recorded as a unit. Blocks are separated by block gaps and each block may contain one or more records.

BLOCK DIAGRAM (M)
A diagram of a system, instrument, or computer in which the principal parts are represented by suitable associated geometrical figures to show both the basic functions and the functional relationships among the parts.

B-MULTIPLIER (J)
See Darlington Amplifier.

BOMB TEST (T)
A form of leak test in which enclosures are immersed in a fluid which is then pressurized for the purpose of driving it through possible leak passages and thus into the internal cavities where its presence will usually cause some form of electrical disturbance. (ASTM)

BOOLEAN ALGEBRA (J)
The algebra of logic named for mathematician George Boole using alphabetic symbols to stand for logical variables and "zero" and "one" to represent states. AND, OR, NOT are the three basic logic operations in this algebra. NAND and NOR are combinations of the three basic operations.

BOOTSTRAP (M)
A technique or device designed to bring itself into a desired state by means of its own action; e.g., a machine routine whose first few instructions are sufficient to bring the rest of itself into the computer from an input device.

BORROW (M)
An arithmetically negative carry.

BRANCHING (M)
A method of selecting, on the basis of results, the next operation to execute while the program is in progress. (See conditional jump.)

BREAK POINT (T)
1. The point in the pressure-time curve that is preceded by a pressure drop of exactly 2 psi (13.8 kPa) within 15 min and succeeded by a drop of not less than 2 psi (13.8 kPa) in 15 min.
2. The junction of two confluent straight-line segments of a plotted curve. Note: In the asymptotic approximation of a log-gain versus log-frequency relation in a Bode diagram, the value of the abscissa is called the corner frequency. (IEEE)
3. (a) Pertaining to a type of instruction, instruction digit, or other condition used to interrupt or stop a computer at a particular place in a routine when manually requested. (b) A place in a routine where such an interruption occurs or can be made to occur. (IEEE)

BREAKDOWN VOLTAGE (J)
See Dielectric Strength.

(T)
The voltage at which a disruptive discharge takes place through or over the surface of the insulation. (IEEE)

BREAKDOWN VOLTAGE RATING(S) (T)
The DC or sinusoidal AC voltage which can be applied across specified insulated portions of a transducer without causing arcing or conduction above a specified current value across the insulating material. (ISA)

BRIDGE RESISTANCE (T)
See Input Impedance and Output Impedance.

BRIGHTNESS (LUMINANCE) (D)

Brightness or luminance is the luminous intensity per projected area normal to the line of observation.

BRIGHTNESS RATIO (D)
Brightness ratio is the ratio of the brightnesses of any two surfaces. When the two surfaces are adjacent, the brightness ratio is called the brightness contrast.

BUFFER (M)
An isolating circuit used to avoid reaction of a driven circuit on the corresponding driver circuit. Also, a storage device used to compensate for a difference in the rate of flow of information or the time of occurrence of events when transmitting information from one device to another.

BULBS (DG)
Integral incandescent lamps exclusive of socket, lens, reflector, etc., that might be included in a "lamp assembly".

BURST PRESSURE RATING (S) (T)
The pressure which may be applied to the sensing element or the case (as specified) of a transducer without rupture of either the sensing element or transducer case as specified. Note: (1) minimum number of applications and time duration of each application must be specified, (2) in the case of transducers intended to measure a property of a pressurized fluid, burst pressure is applied to the portion subjected to the fluid. (ISA)

BUS (M)
As applied to computer technology, one or more conductors used as a path over which information is transmitted.

BYTE (M)
An IBM developed term used to indicate a specific number of consecutive bits treated as a single entity. A byte is most often considered to consist of eight bits which as a unit can represent one character or two numerals.

CALIBRATION (T)
A test during which known values of measurand are applied to the device and corresponding output readings are recorded under specified conditions.

CALIBRATION CURVE (T)
A graphical representation of the calibration record. (ISA)

CALIBRATION TRACEABILITY (T)
The relation of a transducer calibration, through a specified step-by-step process, to an instrument or group of instruments calibrated by the National Bureau of Standards. (ISA)

CALIBRATION UNCERTAINTY (T)
The maximum calculated error in the output values, shown in a calibration record, due to causes not attributable to the transducer. (ISA)

CALL (M)
To transfer control to a specified closed subroutine.

CANDELA (cd) (D)
The candela is the unit of luminous intensity. It is defined as 1/60 of the intensity of 1 square cm of a blackbody radiator at the temperature of solidification of platinum (2046K). The candela replaces the candle as the standard unit.

CANDLEPOWER (D)
See luminous intensity.

CAPACITANCE (C) (J)
In a system of conductors and dielectrics, that property which permits the storage of electrically separated charges when potential differences exist between the conductors. Its value is expressed as the ratio of a quantity of electricity to a potential difference (Q/V). (IEEE)

(M)
In second sentence, "electricity" should read "charge."

CAPACITOR (J)
(Condenser) A device consisting of two electrodes separated by a dielectric, which may be air, for introducing capacitance into an electric circuit. (IEEE)

CARRIER (J)
An AC voltage having a frequency suitably high to be modulated by electrical signals.

(M)
Not defined with respect to semi-conductors, but used in definition of "base," "bipolar," etc.

CARRY (M)
1. One or more digits, produced in connection with an arithmetic operation on one digit place of two or more numerals in positional notation, that are forwarded to another digit place for processing there.
2. The number represented by the digit or digits in definition 1 above.
3. Most commonly, a digit as defined in definition 1 above that arises when the sum or product of two or more digits equals or exceeds the radix of the number representation system.
4. Less commonly, a borrow.
5. To forward a carry.
6. The command directing that a carry be forwarded.

CARRY LOOK-AHEAD (M)
A type of adder in which the inputs to several stages are examined and the proper carries are produced simultaneously, rather than propagated through a sequence of operations.

CASCADE (J)
An arrangement of two or more similar circuits or amplifying stages in which the output of one provides the input of the next. (GRAF)

CASE PRESSURE (T)
See burst pressure rating, proof pressure or reference pressure.

CATHODE (J)
A general name for any negative electrode. (GRAF)

CATHODE-RAY TUBE (CRT) (J)
An electron-beam tube in which the beam can be focused to a small cross section on a luminescent screen and varied in position and intensity to produce a visible pattern. (IEEE)

CENTRAL PROCESSOR UNIT (CPU) (M)
The section of a computer that contains the arithmetic, logic and control circuits. In some systems, it may also include the memory unit and the operator's console. Also called main frame in larger systems. It performs arithmetic operations, controls instruction processing, and provides timing signals and other housekeeping operations.

CHANNEL (M)
A path along which signals can be sent or received; e.g., data channel, output channel.

CHARACTER (M)
Any of a set of symbols used to express information, singularly or in combination. Characters are not necessarily limited to the numerals 0 through 9 and the standard alphabet but can also include punctuation marks, typewriter symbols, or any other single symbol used for expressing information.

CHARACTERISTIC CURVE (D)
A characteristic curve is a curve expressing a relation between two variable properties of a device; e.g., luminance and volts.

CHECK BIT (M)
A binary check digit; e.g., a parity bit (used for validation of data).

CHIP (J)
1. A single substrate on which all the active and passive elements of an electronic circuit have been fabricated. A chip is not ready for use until it is packaged and provided with terminals for connection to the outside world. (Not true, however, for hybrid circuits such as IAR and TFI). Also called a die. (GRAF)
2. A leadless discrete component such as a resistor or capacitor intended for surface mounting to printed circuit boards or film hybrid substrates.

CHIP SETS (J)
A term describing the microprocessor chip in addition to RAM's, ROM's and interface I/O devices. Chip sets, mounted on a board are also referred to as the CPU portion of the microcomputer.

CHROMATICITY (D)
Chromaticity is the expression of dominant wavelength and purity.

CHROMATICITY DIAGRAM (D)
A chromaticity diagram is a two dimensional plot used to define a given color by the proportions of the three primary colors which when mixed match the given color.

CLOCK (J)
1. A device that generates periodic signals used for synchronization. (IEEE)
2. A device that measures and displays time.

CLOSED LOOP (M)
A control system implementation in which information on output parameter is used to improve the system accuracy and/or response.

CODE (M)
1. A set of unambiguous rules specifying the way in which data may be represented; e.g., the set of correspondence in the standard code for information interchange. Synonymous with coding schemes.
2. In data processing, to represent data or a computer program in a symbolic form that can be accepted by a data processor.

COIL (J)
See Inductor.

COLLECTOR (TRANSISTOR) (J)
A region through which primary flow of charge carriers leaves the base. (IEEE)

COLOR (D)
Color is a subjective visual response determined mainly by the wavelength of light emitted or reflected by a body.

COLOR TEMPERATURE (D)
Color temperature is the temperature of a black body whose emitted light is a match for the given color.

COLORANT (D)

A colorant is any substance such as a dye, pigment, or paint used to produce color in an object.

COMBINATORIAL LOGIC SYSTEM (M)
Digital system not utilizing memory elements.

COMMON COLLECTOR AMPLIFIER (J)
A transistor amplifier in which the collector element is common to both the input and output circuit. Also known as an emitter-follower and a grounded-collector amplifier. (GRAF)

COMMON MODE REJECTION (J)
A measure of how well a differential amplifier ignores a signal which appears simultaneously and is in phase at both input terminals. Also called in-phase rejection. (GRAF)

COMPENSATION (T)
A modifying or supplementary action (also, the effect of such action) intended to improve performance with respect to some specified characteristic. (IEEE)
Provision of a supplemental device, circuit, or special materials to counteract known sources of systematic error. (ISA)

COMPILE (M)
(Computing) The process of converting a computer program from another programming language into machine-recognizable codes.

COMPILER (M)
A program that compiles.

COMPLEMENT NOTATION (M)
A system of notation where positive binary numbers are identical to positive numbers in sign and magnitude notation, but where negative numbers are the exact complement of the magnitude of the corresponding positive value.

COMPLEMENTARY MOS (CMOS) (J)
Pertaining to N- and P-channel enhancement-mode devices fabricated compatibly on a silicon chip and connected into push-pull complementary digital circuits. These circuits offer low quiescent power dissipation and potentially high speeds, but they are more complex than circuits in which only one channel type is used. (GRAF)

COMPUTER (J)
Any device capable of accepting information, applying prescribed processes to the information and supplying the results of the process.

(DG)
An electronic device that can perform substantial computation on data including numerous arithmetic or logic operations without intervention by a human operator. A true computer must possess at least the following five characteristics: provision for data input, a control unit, a storage or memory capability, an arithmetic logic section, and a provision for discharging its output in usable form.

CONDITIONAL JUMP (M)
A jump that occurs if specified criteria are met. (See branching.)

CONDUCTIVITY (J)
The ability to transmit heat or electricity. Electrical conductivity is expressed in terms of the current per unit of applied voltage. The reciprocal of resistivity.

CONNECTING DEVICES (DG)
On or off vehicle devices for the necessary mechanical and/or electrical connections between the motor vehicle and the diagnostic equipment. (ISO)

CONSTANT CURRENT SOURCE (J)
A regulated source which acts to keep its output current constant in spite of changes in load, line or temperature while the output voltage changes by whatever amount is necessary to maintain the constant output current. (GRAF)

CONTAMINANT (T)
1. A foreign material present on or in the contact surface. (ASTM)
2. An impurity or foreign substance present in or on a material which affects one or more properties of the material.

CONTINUOUS RATING (J)
The rating applicable to specified operation for a specific uninterrupted length of time. (ISA)

CONTRAST RATIO (CONTRAST) (D)
Contrast ratio is the quotient of the luminous intensity of an activated area of a display and the luminous intensity of the same point in the unactivated state. Contrast ratio (relative contrast) is also used as the quotient between the active and background area luminous intensities of a display. A complete definition requires that the ambient light conditions be specified for a given contrast ratio value.

CONTROL HIERARCHY (M)
Design development used in complex systems to ensure an order of priority to several controls coming from more than one source.

CONTROLLER (M)
Digital subsystem responsible for implementing "how" a system is to function. Not to be confused with "timing," as timing tells the system "when" to perform its function.

COUNTER (M)
A digital circuit which counts input pulses and will give an output pulse after receiving a predetermined number of input pulses.

CREEP (J)
A change in output occurring over a specific time period while the input and all environmental conditions are held constant.

(T)
A change in output occurring over a specific time period while the measurand and all environmental conditions are held constant.

CRITICAL DAMPING (J)
The value of damping which provides the most rapid transient response without overshoot. Operation between underdamping and overdamping. (GRAF)

CROSS-AXIS SENSITIVITY, CROSS SENSITIVITY (T)
See Transverse Sensitivity.

CROSSTALK (M)
Interference which appears in a given channel but has its origin in another channel.

CURRENT DENSITY (J)
The amount of electric current passing through a given cross-sectional area of a conductor. (GRAF)

CURVE OF LIGHT DISTRIBUTION (D)
A curve of light distribution is a curve showing the variation of luminous intensity of a display with angle of observation.

CYCLE (M)
1. An interval of space or time in which one set of events of phenomena is completed.
2. Any set of operations that is repeated regularly in the same sequence. The operations may be subject to variations on each repetition.

DAMPING (J)
The transitory decay of the amplitude of a free oscillation of a system, associated with energy loss from the system. (IEEE)

(T)
Dissipation of energy in a vibrating or oscillating system.

DAMPING CONSTANT (T)
The component of applied force which is 90 deg out of phase with the deformation, divided by the velocity of deformation. (ASTM)

DAMPING RATIO (J)
The ratio of the degree of actual damping to the degree of damping required for critical damping. (GRAF)

DARLINGTON AMPLIFIER (J)
A transistor circuit which, in its original form, consists of two transistors in which the collectors are tied together and the emitter of the first transistor is directly coupled to the base of the second transistor. Therefore, the emitter current of the first transistor equals the base current of the second transistor. This connection of two transistors can be regarded as a compound transistor with three terminals. (GRAF)

DARLINGTON PAIR (J)
See Darlington amplifier.

D'ARSONVAL CURRENT (J)
A high-frequency, low voltage current of comparatively high amperage. (GRAF)

DATA (M)
1. A representation of facts, concepts, or instructions in a formalized manner suitable for communication, interpretation, or processing by humans or automatic means.
2. Any representations such as characters or analog quantities to which meaning is or might be assigned.

DATA BUS (M)
Most microprocessors communicate externally through the use of a data bus. Most are bi-directional; e.g., capable of transferring data to and from the CPU, storage and peripheral devices.

DATA PROCESSING (M)
The execution of a systematic sequence of operations performed upon data. Synonymous with information processing.

DATA PROCESSOR (M)
A device capable of performing data processing, including desk calculators, punched card machines, and computers. Synonymous with processor.

DC PLASMA PANEL (D)
A gas discharge panel using electrodes in direct contact with the gas. Typical units are for numeric and alphanumeric applications.

DEAD BAND (T)
The range through which the measured signal can be varied without initiating response.

DEBUG (M)

An instruction, program or action designed in microprocessor software to search for, correct, and/or eliminate sources of errors in programming routines. There are many types of "bugs" or glitches that can be located by single step testers, specifically designed programs, or operational procedures.

DECIBEL (J)

One-tenth of a bel, the number of decibels denoting the ratio of the two of power being ten times the logarithm to the base 10 of this ratio. Note: the abbreviation dB is commonly used for the term decibel. With P1 and P2 designating two amounts of power and n the number of decibels denoting their ratio:

$$n = 10 \log 10 \ (P1/P2) \text{ decibel}$$

When the conditions are such that ratios of currents or ratios of voltages (or analogous quantities in other fields) are the square roots of the corresponding power ratios, the number of decibels by which they differ is expressed by the following equations:

$$n = 20 \log 10 \ (I1/I2) \text{ decibel}$$
$$n = 20 \log 10 \ (V1/V2) \text{ decibel}$$

Where I1/I2 and V1/V2 are the given current and voltage ratios, respectively. By extension, these relations between numbers of decibels and ratios of currents or voltages are sometimes applied where these ratios are not the square roots of the corresponding power ratios; to avoid confusion, such usage should be accompanied by a specific statement of this application, such extensions of the term described should preferably be avoided. (IEEE)

DECIMAL (M)

1. Pertaining to a characteristic or property involving a selection, choice, or condition in which there are ten possibilities.
2. Pertaining to the number representation system with a radix of ten.

DECIMAL DIGIT (M)

In decimal notation, one of the characters 0 through 9.

DECODER (M)

A conversion circuit that accepts digital input information—in the case of a memory address decoder, a binary address information—that appears as a small number of lines and selects and activates one line of a large number of output lines.

DELAY (J)

1. The amount of time by which an event is retarded.
2. The amount of time by which a signal is delayed.

Note: It may be expressed in time (milliseconds, microseconds, etc.) or in number of characters (pulse times, word times, major cycles, minor cycles, etc.). (IEEE)

DELAY LINE (J)

(Electronic Computers)

1. Originally, a device utilizing wave propagation for producing a time delay of a signal.
2. Commonly, any real or artificial transmission line or equivalent device designed to introduce delay. (IEEE)

DEPLETION LAYER (J)

In a semi-conductor, the region in which the mobile carrier charge density is insufficient to neutralize the net fixed charge density of donors and acceptors. (GRAF)

DETECTOR (T)

1. FM Broadcast Receiver—A) A device to effect the process of detection; B) A mixer in a superheterodyne receiver. Note: In definition B), the device is often referred to as a first detector and the device is not used for detection as defined above.
2. Electromagnetic Energy—A device for the indication of the presence of electromagnetic fields. Note: In combination with an instrument, a detector may be employed for the determination of the complex field amplitudes. See also auxiliary device to an instrument.
3. Nuclear Power Generating Stations—Any device for converting radiation flux to a signal suitable for observation and measurement.

DEVICE CONTROL CHARACTER (M)

A control character intended for the control of ancillary devices associated with a data processing or telecommunication system, usually for switching devices "on" or "off."

DIAGNOSIS (DG)

The careful investigation of symptoms and facts leading to the determination of the cause of a problem.

DIAGNOSTIC (M)

Pertaining to the detection and isolation of a malfunction or mistake.

DIAGNOSTICIAN (DG)

One who is trained and experienced in the diagnosis of the modifying adjectives. (As in vehicle diagnostician, or radio diagnostician.)

DIAGNOSTIC CENTER (DG)

A term coined in the mid 1960's to describe "state of health" vehicle inspection stations. Usually involves the measurement of ten or more component characteristics or outputs and comparing these measured values against new part specifications. The comparison process is sometimes performed semi-automatically.

DIAGNOSTIC EQUIPMENT (DG)

Equipment that can identify specific faults or provide specific information as to what vehicle components should be adjusted, repaired, or replaced. (Example: A vacuum gage is test equipment, not diagnostic equipment.)

DIAGNOSTIC SENSOR (DG)

Any sensing device used partly or exclusively for diagnostic purposes, to provide information on conditions of a motor vehicle or parts of it. (ISO)

Built-in Diagnostic Sensor: A sensor for diagnostic purposes, which is part of the permanent equipment of a motor vehicle. (ISO)

Plug-in Diagnostic Sensor: A sensor for diagnostic purposes, which is part of the off-vehicle diagnostic equipment, to be fitted prior to diagnosis to a receiving device already provided on the motor vehicle for this purpose. (ISO)

Clip-on Diagnostic Sensor: A sensor for diagnostic purposes, which is part of the off-vehicle diagnostic equipment, to be attached to the motor vehicle prior to diagnosis. No special on-vehicle equipment is necessary. (ISO)

Reference Cylinder Sensor: Sensor for diagnostic purposes, the signal of which corresponds to the beginning of the reference cylinder ignition pulse in the case of a spark ignition engine, or to the beginning of the reference cylinder injection in the case of a compression ignition engine. (ISO)

DIAGNOSTIC TESTER (DG)

Self-contained autonomous off-vehicle equipment permitting motor vehicle diagnosis. (ISO)

DIELECTRIC (T)

A medium in which it is possible to maintain an electric field with little or no supply of energy from outside sources.

DIELECTRIC CONSTANT (J)

The property that determines the electrostatic energy stored per unit volume for unit potential gradient. Note: This numerical value usually is given relative to a vacuum. (IEEE)

DIELECTRIC STRENGTH (J)

(Electric Strength) (Breakdown Strength) The potential gradient at which electric failure or breakdown occurs. To obtain the true dielectric strength, the actual maximum gradient must be considered or the test piece and electrodes must be designed so that uniform gradient is obtained. The value obtained for the dielectric strength in practical tests will usually depend on the thickness of the material and on the method and conditions of test. (IEEE)

DIFFUSE REFLECTION (D)

Diffuse reflection is that in which the light is reflected in all directions.

DIFFUSE-SPECULAR (D)

Diffuse specular surfaces are those which are essentially diffuse but contain an outer layer of glazed material which reflects specularly. A typical example is porcelain enamel.

DIFFUSE TRANSMISSION (D)

Diffuse transmission is that in which the transmitted light is emitted in all directions.

DIFFUSE TRANSMISSION FACTOR (D)

The diffuse transmission factor is the ratio of the diffuse transmitted light to the incident light.

DIFFUSING SURFACE (D)

A diffusing surface is one which breaks up (diffuses) the incident light and distributes it angularly approximately in accordance with Lambert's Law. Rough plaster is a typical example of such a surface.

DIGIT (M)

A symbol that represents one of the non-negative integers smaller than the radix. For example, in decimal notation, a digit is one of the characters from 0 to 9. Synonymous with numeric character.

DIGITAL COMPUTER (J)

A computer that processes information in numerical form. Electronic digital computers generally use binary or decimal notation and process information by repeated high speed use of the fundamental arithmetic processes of addition, subtraction, multiplication and division.

DIGITAL OUTPUT (T)

Transducer output that represents the magnitude of the measurand in the form of a series of discrete quantities coded in a system of notation. (ISA)

DIGITAL-TO-ANALOG (D/A) CONVERTER (J)

A device which transforms digital data into analog data by translating digital magnitude to equivalent voltage level.

DIGITIZE (M)

To use numeric characters to express or represent data; e.g., to obtain from an analog representation of a physical quantity, a digital representation of the quantity.

DIMMING RANGE (D)
Dimming range is a dimensionless number giving the ratio of maximum to minimum luminous intensity of an active display.

DIODE (J)
(Electronic Tube) A two electrode electron tube containing an anode and a cathode. (Semi-conductor) A semi-conductor device having two terminals and exhibiting a non-linear voltage-current characteristic; in more restricted usage, a semi-conductor device that has the asymmetrical voltage-current characteristic exemplified by a single P-N junction. (IEEE)

DIODE TRANSISTOR LOGIC (DTL) (J)
A logic circuit that uses diodes at the input to perform the electronic logic function that activates the circuit transistor output. In monolithic circuits, the DTL diodes are a positive level logic and function or a negative level or function. The output transistor acts as an inverter to result in the circuit becoming a positive nand or a negative nor function. (GRAF)

DIP (J)
Abbreviation for Dual In-line Package. (GRAF)

DIPOLE ANTENNA (J)
Any one of a class of antennas producing the radiation pattern approximating that of an elementary electronic dipole. Note: Common usage considers a dipole to be a metal radiating structure that supports a line current distribution similar to that of a thin straight wire, a half wavelength long, so energized that the current has two modes, one at each of the far ends. (IEEE)

DIRECT ACCESS (M)
1. Pertaining to the process of obtaining data from, or placing data into storage where the time required for such access is independent of the location of the data most recently obtained or placed in storage.
2. Pertaining to a storage device in which the access time is effectively independent of the location of the data.
3. Synonymous with random access.

DIRECT ADDRESSING (M)
Method of programming that has the address pointing to the location of data or the instruction that is to be used.

DIRECTIVITY (T)
The solid angle or the angle in a specified plane over which sound or radiant energy incident on a transducer is measured within specified tolerances and a specified band of measurand frequencies. (ISA)

DIRECT MEMORY ACCESS CHANNEL (DMA) (M)
A method of input-output for a system that uses a small processor whose sole task is that of controlling input-output. With DMA, data are moved into or out of the system without program intervention.

DITHER (T)
(Control Circuits) A useful oscillation of small amplitude introduced to overcome the effects of friction, hysteresis or clogging. (IEEE)

DITHERING (T)
The application of intermittent or oscillatory forces just sufficient to minimize static friction within the transducer.

DOMINANT WAVELENGTH (D)
The wavelength of radiant energy of a single frequency that matches the color of the light under test when combined in suitable proportion with the radiant energy of the reference standard.

DOT MATRIX DISPLAY (D)
A form of display organized in an x-y matrix of light emitting dots. Such a display may be used for numeric, alphanumeric, or graphic information.

DOUBLE PRECISION (M)
Pertaining to the use of two computer words to represent a number.

DRAIN (D) (J)
In a field effect transistor, the element that corresponds to the collector of a transistor. (GRAF)

DRIFT (J)
An undesired change in output over a period of time, which change is not a function of the input.

(T)
A change in output over a period of time, which change is not a function of the measurand. (ISA)

DUMP (M)
1. To copy the contents of all or part of a storage, usually from an internal storage into an external storage.
2. A process as in definition 1 above.
3. The data resulting from the process as in definition 1 above.

DUTY CYCLE (J)
The ratio of the time "on" of a device or system divided by the total cycle time (i.e., "on" plus time "off"). For a device that normally runs intermittently rather than continuously; the amount of time a device operates as opposed to its idle time.

DYNAMIC STORAGE ELEMENTS (M)
Storage elements which contain storage cells that must be refreshed at appropriate intervals to prevent the loss of information content.

DYNAMOMETER (DG)
An energy absorbing device designed to allow measured dynamic operation of a vehicle's drive train while the vehicle remains stationary.

ECL CIRCUITS (EMITTER-COUPLED LOGIC) (M)
Nonsaturated bipolar logic in which the emitters of the input logic transistors are coupled to the emitter of a reference transistor.

EDDY CURRENTS (T)
(General) Those currents that exist as a result of voltages induced in the body of a conducting mass by a variation of magnetic flux. Note: The variation of magnetic flux is the result of a varying magnetic field or of a relative motion of the mass with respect to the magnetic field. (IEEE)

EDGE TRIGGERING (M)
Activation of a circuit at the edge of the pulse as it begins its change. Circuits then trigger at the edge of the input pulse rather than sensing a level change.

EDIT (M)
To modify the form or format of data; e.g., to insert or delete characters such as page numbers or decimal points.

EEROM (M)
Electrical eraseable read only memory. The contents of this type of memory may be electronically erased and new information programmed into the device. Also known as EAROM; or electrically alterable ROM.

ELECTRIC (DG)
Any device or system operating on the direct flow of electrons or the low frequency (below 1,000 Hz) alternating flow of electrons.

ELECTROCHROMIC DISPLAY (D)
A passive display whose operating principle is the electric field control of a chemical reaction. One of the chemical species is transparent, the other is colored, allowing electric field control of light transmission or reflection.

ELECTROLUMINESCENCE (J)
Luminescence resulting from a high-frequency discharge through a gas or from application of an alternating current to a layer of phosphor. (GRAF)

ELECTROLUMINESCENT DISPLAY (D)
An active display whose operating principle is the emission of light from a phosphor which is excited by energy from an electric field. AC and DC operated types are common.

ELECTROMAGNETIC COMPATABILITY (EMC) (J)
The ability of electronic communications equipment, subsystems and systems to operate in their intended environments without suffering or causing unacceptable degradation of performance as a result of unintentional electromagnetic radiation or response. (GRAF)

(T)
The capability of electronic equipment or systems to be operated in the intended operational electromagnetic environment at designed levels of efficiency.

ELECTROMAGNETIC INTERFERENCE (EMI) (J)
Electromagnetic phenomena which, either directly or indirectly, can contribute to the degradation in performance of an electronic receiver or system. (GRAF)

(T)
Impairment of a wanted electromagnetic signal by an electromagnetic disturbance. See also electromagnetic compatibility.

ELECTROMAGNETIC WAVES (J)
The radiant energy produced by the oscillation of an electric charge. (GRAF)

ELECTROMOTIVE FORCE (emf) (J)
The force which may cause current to flow when there is a difference of potential between two points. (GRAF)

ELECTRON (J)
One of the natural elementary constituents of matter. It carries a negative electric charge of one electronic unit. (GRAF)

ELECTRONIC (DG)
Any device or system in which electrons flow through a vacuum, gas, or semiconductor. Normally operative in frequency range above 1,000 Hz.

ELECTRONIC DISPLAY (D)
An electronic display is a device which presents information to a viewer by control of light, either emitted, reflected, or transmitted, in response to an electrical driving signal.

ELECTROPHORETIC DISPLAY (D)
A passive display whose operating principle is the movement of particles in a fluid under the influence of an electric field.

ELECTROSTATIC STORAGE (M)
A storage device that stores data as electrostatically charged areas on a dielectric surface.

EMISSIVITY (D)
The ratio of light emitted from the source in question to the light emitted from a black body at the same temperature.

EMITTER (J)
A region from which charge carriers that are minority carriers in the base are injected into the base. (IEEE)

EMULATE (M)
To imitate one system with another such that the imitating system accepts the same data, executes the same programs, and achieves the same results as the imitated system.

ENCODE (M)
To apply a set of unambiguous rules specifying the way in which data may be represented such that a subsequent decoding is possible. Synonymous with code.

END-AROUND CARRY (M)
A carry generated in the most significant digit place and sent directly to the least significant place.

ENTRY POINT (M)
In a routine, any place to which control can be passed.

EQUIVALENCE (M)
A logic operator having the property that if P is a statement, Q is a statement, R is a statement, ... then the equivalence of P, Q, R, ... is true if and only if all statements are true or all statements are false.

ERASE (M)
To obliterate information from a storage medium; e.g., to clear, to overwrite.

ERG (D)
The unit of energy in the CGS system is the erg. One erg = 10^{-7} joules.

ERROR (M)
Any discrepancy between a computed, observed or measured quantity and the true, specified or theoretically correct value or condition.

(T)
The algebraic difference between the indicated value and the true value of the measurand. (ISA)

ERROR BAND (T)
The band of maximum deviations of output values from a specified reference line or curve due to those causes attributable to the device.

EXCLUSIVE OR (J)
A logic operator having the property that if P is a statement and Q is a statement, then P exclusive OR Q is true if either but not both statements are true, false if both are true or both are false. (IEEE)

(M)
A modified form of the OR function which is true when one and only one input is true.

EXECUTE (M)
That portion of a computer cycle during which a selected control word or instruction is accomplished.

EXPONENT (M)
In a floating point representation, the numeral or a group of numerals representing a number, that indicates the power to which the base is raised.

EYE RECEPTORS (D)
Eye receptors, rods and cones, are the specialized structures in the eye which respond to visible light.

FAILURE MODE (DG)
A particular manner of malfunction.

FAILURE MODE ANALYSIS (DG)
The critical review of the design and function of an item to identify and describe all primary ways of malfunctioning and the effect of each on system performance.

FAILURE (TOTAL) (T)(DG)
The termination of the ability of an item to perform its required function.

FALL TIME (D)
Fall time is the time required for the specified optical property to fall from 90 to 10 percent of its maximum excursion after a step function decrease in the driving signal level.

The time required for the leading edge of a pulse to fall from 90 to 10 percent of its initial value. It is proportional to the time constant and is a measure of the steepness of the waveform. Also, the measured length of the time required for an output voltage of a digital circuit to change once the change has started.

The length of time for the output of a transducer to fall from a large specified percentage of its initial value to a small specified percentage of its initial value as a result of a step change of measurand.

FALL TIME DELAY (D)
Fall time delay is the time required for the specified optical or electrical property to fall from its steady state level to 90 percent of that level after a step function decrease in signal level.

FEEDBACK (J)
The recycling of a portion of the output to the input of a system. Systems employing feedback are called closed-loop systems.

FEEDBACK AMPLIFIER (J)
An amplifier that uses a passive network to return a portion of the output signal to modify the performance of the amplifier. (GRAF)

FERRITES (J)
Chemical compounds of iron oxide and other metallic oxides combined with ceramic material. They have ferromagnetic properties but are poor conductors of electricity. Hence, they are useful where ordinary ferromagnetic materials (which are good electrical conductors) would cause too great a loss of electrical energy. (GRAF)

FERROMAGNETIC MATERIAL (J)
Material whose relative permeability is greater than unity and depends upon the magnetizing force. A ferromagnetic material usually has relatively high values of relative permeability and exhibits hysteresis. (IEEE)

FET (FIELD EFFECT TRANSISTOR) (J)
A semi-conductor device in which the resistance between the source and drain terminals depends on a field produced by a voltage applied to the gate terminal. (GRAF)

FETCH (M)
That portion of a computer cycle during which the next instruction is retrieved from memory.

FIELD (M)
In a record, a specified area used for a particular category of data; e.g., a set of bit locations in a computer word used to express the address of the operand.

FILTER (D)
A device placed between a display and the observer for the purpose of modifying the optical characteristics of the display. Filters are generally used to improve contrast ratio and sometimes to change color. Examples include neutral density, colored, and polarizing types.

(J)
A selective network which may contain resistors, inductors, capacitors or active elements which offers comparatively little opposition to certain frequencies or to direct current, while blocking or attenuating other frequencies. (GRAF)

FIXED-POINT BINARY NUMBER (M)
A binary number represented by a sign bit and one or more number bits with a binary point fixed somewhere between two neighboring bits.

FLAG (M)
1. Any of various types of indicators used for identification; e.g., a wordmark.
2. A character that signals the occurrence of some condition, such as the end of a word, its sign, its parity, etc.
3. Synonymous with mark, sentinel, tag.

FLAT PACK (J)
A flat, rectangular integrated circuit or hybrid-circuit package with coplanar leads. (GRAF)

FLIP-FLOP (M)
(Storage Element) A circuit having two stable states and the capability of changing from one state to another with the application of a control signal and remaining in that state after removal of signals.

FLOATING-POINT BINARY NUMBER (M)
A binary number expressed in exponential notation. That is, a part of the binary word represents the mantissa and is a part of the exponent.

FLOW CHART (M)
A graphical representation for the definition, analysis, or solution of a problem, in which symbols are used to represent operations, data, flow, equipment, etc.

FLUX (J)
(Magnetic) The sum of all the lines of force in a magnetic field crossing a unit area per unit time.

FLUX DENSITY (J)
Flux per unit area perpendicular to the direction of the flux. (GRAF)

FONT (D)
A font is the set of symbols which a display can present. The particular form of certain characters which may be displayed in alternative forms must be given to define a font. For example, b or 6 are two ways to display a six on a seven segment display.

FOOTCANDLE (fc) (D)
The footcandle is the unit of illumination given by the illumination on a surface one square foot in area on which there is a flux of one lumen. A surface all points of which are one foot from a one candela source has an illumination of one footcandle.

FOOTLAMBERT (fl) (D)
The footlambert is a unit of brightness (luminance) equal to the uniform brightness of a perfectly diffusing surface emitting or reflecting light at the rate of one lumen per square foot. The average brightness of a reflecting

surface is the product of the illumination in footcandles times the reflection factor of the surface. A one square foot surface all points of which are one foot from a one candela source has a luminance of one footlambert.

FORMAT (M)
The arrangement of data.

FORTRAN (M)
(FORmula TRANslating system) A programming language primarily used to express computer programs by arithmetic formulas.

FORWARD VOLTAGE (Vf) (J)
The voltage across a semi-conductor diode associated with the flow of forward current. The P-region is at a positive potential with respect to the N-region.

FRAME TIME (D)
In a multiplexed display, the frame time is the time for one complete scan of the display.

FREQUENCY MODULATED OUTPUT fm (FREQUENCY MODULATION) (J)
A scheme for modulating a carrier frequency in which the amplitude remains constant but the carrier frequency is displaced in frequency proportionally to the amplitude of the modulating signal. A frequency modulation broadcast system is practically immune to atmospheric and man-made interference.

(T)
An output in the form of frequency deviations from a center frequency, where the deviation is a function of the input.

FREQUENCY OUTPUT (T)
An output in the form of frequency which varies as a function of the applied measurand (e.g., angular speed or flow rate). (ISA)

FREQUENCY, NATURAL (T)
The frequency of free (not forced) oscillations of the sensing element of a fully assembled transducer.

FREQUENCY, RESONANT (T)
The measurand frequency at which a transducer responds with maximum output amplitude.

FREQUENCY RESPONSE (J)
A measure of how the gain or loss of a circuit device or system varies with the frequencies applied to it. Also, the portion of the frequency spectrum which can be sensed by a device within specified limits of error.

(T)
The change with frequency of the output/measurand amplitude ratio (and of the phase difference between output and measurand), for a sinusoidally varying measurand applied to a transducer within a stated range of measurand frequencies. (ISA)

FRICTION ERROR (T)
The maximum change in output at any measurand value within the specified range, before and after minimizing friction within the transducer by dithering. (ISA)

FRICTION-FREE ERROR BAND (T)
The error band applicable at room conditions and with frictions within the transducer minimized by dithering.

FULL-ADDER (M)
A logic circuit like the half-adder, but with a provision for a carry-in from a preceding addition.

FULL SCALE (T)
See Range.

FULL-SCALE OUTPUT (T)
The algebraic difference between the end points for bi-directional sensors, the difference between zero and the larger end point.

FUNCTION (M)
A specific purpose of an entity, or its characteristic action.

FUSIBLE LINK MEMORY (M)
A read only memory in which the program or data is inserted after the chip is packaged. Once programmed, the contents of this memory cannot be erased.

GAGE FACTOR (T)
A measure of the ratio of the relative change of resistance to the relative change in length of a resistive strain transducer (strain gage). (ISA)

GAGE PRESSURE (T)
A term used to indicate the difference between the absolute pressure and atmospheric pressure. This term is seldom used in a vacuum technology. (ASTM)

GAIN (J)
Any increase in power when a signal is transmitted from one point to another. Usually expressed in decibels. (GRAF)

GAS DISCHARGE DISPLAY (D)
An active display whose operating principle is the emission of light by a rare gas mixture ionized by an electric field. AC and DC types are common.

GATE (J)
1. A device or element that, depending upon one or more specified inputs, has the ability to permit or inhibit the passage of a signal.
2. (Electronic Computers) A) A device having one output channel and one or more input channels, such that the output channel state is completely determined by the contemporaneous input channel states, except during switching transients. B) A combinational logic element having at least one input channel. C) An AND gate. D) An OR gate.
3. In a field effect transistor, the electrode that is analogous to the base of a transistor or the grid of a vacuum tube. (GRAF)

GCS (J)
Abbreviation for gate controlled switch. (GRAF)

GENERAL-PURPOSE COMPUTER (M)
A computer that is designed to handle a wide variety of problems.

GLITCH (M)
A transient signal or operation, usually in error.

GRAPHIC DISPLAY (D)
A graphic display is one which can display general graphic information such as histograms or graphs.

HALF-ADDER (M)
A logic circuit capable of adding two binary numbers with no provision for a carry-in from a preceding addition.

HALL EFFECT (J)
The development of a transverse electric potential gradient in a current carrying conductor or semi-conductor upon the application of a magnetic field.

(T)
The development of a transverse electric potential gradient or voltage in a current carrying conductor or semi-conductor upon the application of a magnetic field.

HARDWARE (J)
1. Mechanical, magnetic, electrical or electronic devices; physical equipment (contrasted with software).
2. Particular circuits of functions built into a system. (IEEE)

(M)
Physical equipment, as opposed to the computer program or method of use; e.g., mechanical, magnetic, electrical, or electronic devices.

HARMONIC DISTORTION (J)
The production of harmonic frequencies at the output by the nonlinearity of a system when a sinusoidal input is applied. (GRAF)

HEAT SINK (J)
A mounting base, usually metallic, that dissipates, carries away, or radiates into the surrounding atmosphere the heat generated within a semi-conductor device. (GRAF)

HELIUM BOMBING (T)
A method of testing for leaks in which hermetically sealed units containing an internal volume are subjected to a helium pressure prior to being bell jar tested. If leaks are present in the sealed unit, the helium pressure will drive some helium into the internal volume and this may be subsequently detected during bell jar testing.

HENRY
The henry is the inductance of a closed circuit in which an electromotive force of 1 volt is produced when the electric current in the circuit varies uniformly at the rate of one ampere per second.

HERTZ (J)
The unit of frequency, one cycle per second. (IEEE)

HIGH-THRESHOLD LOGIC (HTL) (J)
Logic with a high noise margin, used primarily in industrial applications, it closely resembles DTL, except that in HTL a reverse-biased emitter junction is used as a threshold element operating as a Zener diode. A typical noise margin is 6 volts with a 15-volt supply. (GRAF)

HOLE (J)
In the electronic valence structure of a semi-conductor, a mobile vacancy which acts like a positive electronic charge with a positive mass. (GRAF)

HOLE CONDUCTION (J)
The apparent movement of a hole to the more negative terminal in a semi-conductor. Since the hole is positive, this movement is equivalent to a flow of positive charges in that direction.

HYBRID CIRCUIT (J)
A circuit which combines the thin-film (or thick film) and semi-conductor technologies. Generally, the passive components are made by thin-film techniques, and the active components by semi-conductor techniques. (GRAF)

HYSTERESIS (J)
The difference between the response of a unit or system to an increasing and a decreasing signal. Hysteretical behavior is characterized by inability to retrace exactly on the reverse swing a particular locus of input/output conditions. (GRAF)

(T)

In a transducer, the maximum difference in output, at any measurand value within the specified range, when the value is approached first with increasing and then with decreasing measurand. (ISA)

I²L (M)
A semi-conductor technology which can be used for both analog and digital functions on the same chip. I²L is an acronym for current (I) Injected Logic.

ILLUMINATION (ILLUMINANCE) (D)
Illumination is the density of luminous flux on a surface. It is equal to the flux divided by the area when the surface is uniformly illuminated.

IMMEDIATE ADDRESS MODE (M)
Pertaining to an instruction in which an address part contains the value of an operand rather than its address.

IMPEDANCE (Z) (J)
The total opposition offered by a component or circuit to the flow of alternating or varying current. Impedance is expressed in ohms and is similar to the actual resistance in a direct current circuit. Impedance may be computed as $Z = E/I$, where E is the applied AC voltage and I is the resulting alternating current flow in the circuit.

INCANDESCENT DISPLAY (D)
An active display whose operating principle is the emission of light from a heated filament in vacuum or an inert gas.

INCIPIENT FAILURE (DG)
A failure condition which is just commencing or is in an initial stage and is beginning to show or reveal itself to a suitable sensor and detector.

INDEXED ADDRESS (M)
An address that is modified by the content of an index register prior to or during the execution of a computer instruction.

INDEXING (M)
In computers, a method of address modification that is implemented by means of index registers.

INDEX REGISTER (M)
A register whose content may be added to or subtracted from the operand address prior to or during the execution of a computer instruction.

INDICATOR (DG)
A device that makes information available about a measured characteristic but does not store the information nor initiate a responsive or corrective action.

INDIRECT ADDRESSING (M)
Programming method that has the initial address being the storage location of a word that contains another address. This indirect address is then used to obtain the data to be operated upon.

INDUCTANCE (J)
The property of an electric circuit by which a varying current in it produces a varying magnetic field that induces voltage in the same circuit or in a nearby circuit—measured in henrys.

INDUCTIVE (T)
Converting a change of measurand into a change of the self-inductance of a single coil.

INDUCTOR (J)
A device consisting of one or more associated windings, with or without a magnetic core, for introducing inductance into an electric circuit. (IEEE)

INFRARED (D)
The infrared is that portion of the light spectrum with wavelength greater than the visible.

INPUT IMPEDANCE (J)
The impedance a transducer presents to a source. The effective impedance seen looking into the input terminals of an amplifer, circuit details, signal level, and frequency. (GRAF)

INPUT/OUTPUT DEVICES (I/O) (M)
Computer hardware through which data are entered into or transmitted from a digital system or by which data are recorded for immediate or future use.

INSPECTION (DG)
To measure an item or characteristic of an item and compare the observed value(s) with pre-established standards.

INSTALL (DG)
To fix an item in its appointed location and make all integral connections necessary for the item's function.

INSTRUCTION (M)
A statement that specifies an operation and the values or locations of its operands.

INSTRUCTION COUNTER (M)
A counter that indicates the location of the next computer instruction to be interpreted.

INSTRUCTION REGISTER (M)
A register that stores an instruction for execution.

INSTRUMENT RESPONSE (T)
The behavior of the instrument output as a function of the measured signal, both with respect to time. (ASTM)

INSTRUMENT STANDARD (T)
A standard calibrated for use with a particular instrument; usually not suitable for use with an instrument of a different type.

INSULATION RESISTANCE (S) (T)
The resistance measured between specified insulated portions of a transducer when a specified DC voltage is applied at room conditions unless otherwise stated.

INSULATOR (J)
A high resistance device that supports or separates conductors to prevent a flow of current between them or to other objects. (GRAF)

INTEGRATED CIRCUIT (J)
A combination of interconnected circuit elements inseparably associated on or within a continuous substrate. Note: To further define the nature of an integrated circuit, additional modifiers may be prefixed. Examples are: (1) Dielectric-isolated monolithic integrated circuit. (2) Beam lead monolithic integrated circuit. (3) Silicon-chip tantalum thin-film hybrid integrated circuit. (IEEE)

INTERFACE (M)
A shared boundary. An interface might be a hardware component to link two devices or it might be a portion of storage or registers accessed by two or more computer programs.

INTERLEAVE (OR INTERLACE) (M)
To assign successive storage location numbers to physically separated memory storage locations. This serves to reduce access time.

INTERMITTENT RATING (T)
The rating applicable to specified operation over a specified number of time intervals of specified duration; the length of time between these time intervals must also be specified. (ISA)

INTERPOLATION (M)
A mathematical process by which intermediate values are computed between two known data points.

INTERRUPT (M)
To stop a process in such a way that it can be resumed.

INTERVAL OF CALIBRATION (T)
The elapsed time permitted between calibrations as required by the pertinent specifications, or when not specified, as determined under procedures in this method. (ASTM)

INVERSE VOLTAGE (J)
The effective voltage across a rectifier during the half-cycle when current does not flow. (GRAF)

ION IMPLANTATION (J)
A method of semi-conductor doping in which impurities that have been ionized and accelerated to a high velocity penetrate the semi-conductor surface and become deposited in the interior. (GRAF)

JERK (T)
The time rate of change of acceleration. Expressed in ft/s^3, cm/s^3, gm/s.

JFET (J)
Abbreviation for junction field effect transistor.

JOULE (D)
The joule is the unit of energy in the SI system.

JUMP (J)
(Electronic Computation)
1. To (conditionally or unconditionally) cause the next instruction to be obtained from a storage location specified by an address part of the current instruction when otherwise it would be specified by some convention.
2. An instruction that specifies a jump. (IEEE)

(M)
A departure from the normal sequence of executing instructions in a computer.

JUMP CONDITIONS (J)
Conditions defined in a transition table that determine the proper sequence for performing a jump.

KEEP ALIVE MEMORY (M)
A memory device which will lose information if power is removed but is normally supplied continuous power (via battery backed when main power is lost) so that the information is retained.

LABEL (M)
One or more characters used to identify a statement or an item of data in a computer program.

LAMBERT (D)
The lambert is a unit of luminance equal to 1 candela/cm² which is equal to 1 stilb. A one centimeter square surface, all points of which are one centimeter from a one candela source, has a luminance of one lambert.

LAMBERT'S LAW (D)
A surface where the light emitted in a given direction has an intensity that

varies as the cosine of the angle between the emitted ray and the normal to the surface is said to be perfectly diffusing and obey Lambert's Law. Since projected area also varies as the cosine of the angle, the luminous intensity per projected area or luminance is constant for all angles.

LAMBERTIAN SURFACE (D)
A lambertian surface is one which obeys Lambert's Law. White bond paper is a reasonable approximation to such a surface.

LANGUAGE (M)
A set of representations, conventions, and rules used to convey information.

LARGE-SCALE INTEGRATION (LSI) (J)
1. The simultaneous achievement of large area circuit chips and high density of component packaging for the express purpose of cost reduction by maximization of the number of system interconnections made at the chip level.
2. Monolithic digital IC's with a typical complexity of 100 or more gates or gate-equivalent circuits. The number of gates per chip used to define LSI depends on the manufacturer. The term sometimes describes hybrid IC's built with a number of MSI or LSI chips. (GRAF)

LATCH (J)
A feedback loop used in a symmetrical digital circuit (such as a flip-flop) to retain a state. (GRAF)

LCD DISPLAY (D)
A liquid crystal display is a passive display whose operating principle is the change in light transmission or polarization in a liquid crystal under the influence of an electric field.

LEAD FRAME (J)
1. A metal frame that holds the leads of a plastic encapsulated package (DIP) in place before encapsulation and is cut away after encapsulation. (GRAF)
2. A stamped or etched strip of metal that incorporates the connecting leads of a device or circuit and retains their respective position through the fabrication process. The indexing and retaining portions are cut away leaving only the leads prior to electrical test.

LEAK DETECTOR (T)
A device for detecting, locating and/or measuring leakage.

LEAKAGE RATE (T)
The maximum rate at which a liquid or gas is permitted or determined to leak through a barrier. (ISA)

LEAST-SQUARES LINE (T)
The straight line for which the sum of the squares of the residuals (deviations) is minimized. (ISA)

LED DISPLAY (D)
An active display whose operating principle is the emission of light due to current flow in a semi-conductor diode.

LEGIBILITY (D)
See Readability.

LIFE, OPERATING (T)
The specified minimum length of time over which the specified continuous and intermittent rating of a transducer applies without change in transducer performance beyond specified tolerances. (ISA)

LIFE, STORAGE (T)
The specified minimum length of time over which a transducer can be exposed to specified storage conditions without changing its performance beyond specified tolerances. (ISA)

LIFETIME (D)
1. A characteristic expressing the useful operating life of a display. The nature of the endpoint, such as final luminance, and the operating conditions should be specified for a complete definition.
2. The time that a minority charge carrier can exist in a semiconducting material before it recombines with an opposite charge.

LIGHT-EMITTING DIODE (J)
A PN junction that emits light when biased in the forward direction.

LIGHTS (DG)
Devices that emit radiant energy of wavelength perceptible by the average human eye—400 to 700 nanometers.

LIMITS (DG)
The set of values that describe the established boundaries of acceptance. The same characteristic may have several sets of limits depending upon the time or basis of establishing acceptance (e.g., new part limits, overhaul limits, and safety limits).

LINEARITY (J)
The relationship between two quantities when a change in a second quantity is directly proportionate to a change in the first quantity. Also, deviation from a straight-line response to an input signal. (GRAF)

(T)
The closeness with which a curve of a function approximates a specified time. (IEEE)

LINEARITY, END POINT (T)
Linearity referred to the end-point line.

LINEARITY, INDEPENDENT (T)
Linearity referred to the "best straight line."

LINEARITY, LEAST SQUARES (T)
Linearity referred to the least-squares line.

LINEARITY, TERMINAL (T)
Linearity referred to the terminal line.

LOAD (T)
1. The force in weight units applied to a body.
2. The weight of the contents of the container or transportation device.
3. A qualitative term denoting the contents of a container.
4. The element or circuit driven by the output of a device or circuit.

LOAD IMPEDANCE (T)
The impedance presented in the output terminals of a transducer by the associated external circuitry.

LOADING ERROR (T)
An error due to the effect of the load impedance on the transducer output.

LOGIC (J)
A mathematical approach to the solution of complex situations by the use of symbols to define basic concepts. In computers and information-processing networks, the systematic method that governs the operations performed on the information, usually with each step influencing the one that follows. (GRAF)

LOGIC SHIFT (M)
A shift that affects all positions.

LOOK-UP (M)
A technique to utilize data in a table; to execute that technique.

LOOP (M)
A sequence of instructions that is executed repeatedly until a terminal condition prevails.

LUMEN (lm) (D)
The lumen is the unit of luminous flux. It is equal to the flux through a unit solid angle from a point source of one candela. The flux on a unit surface, all points of which are at unit distance from a one candela source, is one lumen.

LUMINANCE (D)
Luminance is the preferred term for brightness.

LUMINOUS EFFICACY (K) (D)
Luminous efficacy is the ratio of the luminous flux emitted by a source to the radiant power emitted by the source and has units of lumens per watt. The ratio is calculated using a standard spectral response curve for the average human observer.

LUMINOUS EFFICIENCY (D)
Luminous efficiency is the ratio, expressed as a percent, of the actual luminous efficacy of a source to the maximum possible luminous efficacy of 673 lumen per watt for a monochromatic source operating at 555 nm. Luminous efficiency is also used, loosely, for the ratio of luminous flux emitted by the source to its input power in watts. When used in this manner, the units are lumens per watt instead of a percent. Therefore, it is used as a relative figure of merit for the power consumption of various displays. Typical values for current active displays are in the range of .05 to .5 lumen per watt.

LUMINOUS ENERGY (QUANTITY OF LIGHT) (D)
Luminous energy is the product of the luminous flux and the time maintained. It is expressed in lumen-seconds.

LUMINOUS FLUX (D)
Luminous flux is the amount of light energy emitted per unit time. The unit is the lumen.

LUMINOUS INTENSITY (D)
Luminous intensity is the luminous flux emitted by a source per unit solid angle. The unit is the candela. Practical units for electronic displays are the milli and micro candela.

LUX (lx) (D)
The lux is the SI unit of illumination and is equal to one lumen per square meter.

MACHINE CODE (M)
An operations code that a machine is designed to recognize.

MACHINE LANGUAGE (M)
A language that is used directly by a machine.

MACROINSTRUCTION (M)
An instruction in a source language that is equivalent to a specified sequence of machine instructions.

MACROPROGRAMMING (M)
Programming with macroinstructions.

MAGNETIC PARTICLE DISPLAY (D)
A passive display whose operating principle is the orientation of perma-

nently magnetized particles under the influence of an applied magnetic field.

MAGNETO RESISTIVE EFFECT (J)
The change in the resistance of a conductor or semi-conductor due to the application of a magnetic field.

MAIN FRAME (M)
The basic processing portion of a computing system, generally containing some basic storage, the arithmetic logic unit, and a group of registers.

MAJOR LOOP COMPUTATIONS (M)
Refers to those tasks which are performed at the lowest repetition rate. There are an integer number of minor loops per major loop.

MAJORITY CARRIER (M)
Refers to the predominant carrier in N-type semi-conductors, because there are more electrons than holes. Likewise, holes are the majority carrier in P-types, since they outnumber the electrons.

MASK (M)
1. A pattern of characters that is used to control the retention or elimination of portions of another pattern of characters.
2. A filter.

MATRIX (M)
1. In mathematics, a two-dimensional rectangular array of quantities. Matrices are manipulated in accordance with the rules of matrix algebra.
2. In computers, a logic network in the form of array of input leads and output leads with logic elements connected at some of their intersections.
3. By extension, an array of any number of dimensions.

MEASURAND (T)
A physical quantity, property or condition which is measured.

MEMORY (J)
(Electronic Computation) See Storage.

METALIZATION (J)
The deposition of a thin-film pattern of a conductive material onto a substrate to provide interconnection of electronic components or to provide conductive pads for interconnections. (GRAF)

MICROCOMPUTER (J)
A complete system capable of performing minicomputer functions, through a much lower power range. It is a combination of the chip sets; interface I/O along with the auxiliary circuits, power supply and control console.

MICROINSTRUCTION (M)
1. A small, single, short, add, shift or delete type of command.
2. A bit pattern that is stored in a microprogram memory word and specifies the operation of the individual LSI computing elements and related subunits, such as main memory and input/output interfaces.

MICRON (J)
A unit of length equal to 10^{-6} meter.

MICROPROCESSOR (J)
The digital processor on a chip which performs arithmetic logic and control logic. It is the basic building block of a microcomputer system.

(M)
The term applied to computer processor that is contained on one or a small number of chips.

MICROPROGRAMMING (M)
Control technique used to implement the stored program control function. Typically the technique is to use a preprogrammed read-only memory chip to contain several control sequences which normally occur together.

MINORITY CARRIER (J)
The less predominant carrier in a semi-conductor. Electrons are the minority carriers in P-type semi-conductors since there are fewer electrons than holes. Holes are the minority carriers in N-types since they are outnumbered by electrons. (GRAF)

MINOR LOOP COMPUTATIONS (M)
Refers to those tasks which must be performed at the highest repetition rate.

MNEMONIC SYMBOL (M)
A symbol chosen to assist the human memory; e.g., an abbreviation such as "MPY" for "multiply."

MONITOR (DG)
Something that reminds or warns. A device that measures a characteristic either constantly or on demand, compares the measured value against a standard or pre-established limit, and initiates a signaling action.

MONOLITHIC (J)
An integrated circuit which is built on a single slice of silicon substrate.

MNOS (J)
Abbreviation for metal-nitride-oxide semi-conductor. (GRAF)

MOS (J)
Abbreviation for metal oxide semi-conductor.

MOS TRANSISTOR (METAL-OXIDE SEMI-CONDUCTOR TRANSISTOR) (M)
An active semi-conductor device in which a conducting channel is induced in the region between two electrodes by a voltage applied to an insulated electrode on the surface of the region.

MOUNTING ERROR (T)
The error resulting from mechanical deformation of the transducer caused by mounting the transducer and making all measurand and electrical connections. (ISA)

MULTIPLEX (M)
To interleave or simultaneously transmit two or more messages on a single channel.

(D)
A multiplexed display is one organized to use fewer drivers than the number of emitting points by time sharing them along one or both axes of the display. A given display element then only operates for a fraction of the time per scan of the complete display but due to the persistence of vision of the eye, appears to be lit steadily at an average luminous intensity.

MULTIPLEXING (J)
The process of combining several measurements for transmission over the same signal path. There are two widely used methods of multiplexing: time division and frequency division. Time division utilizes the principle of time sharing among measurement channels. Frequency division utilizes the principle of frequency sharing among information channels where the data from each channel are used to modulate sinusoidal signals called subcarriers so that the resultant signal representing each channel contains only frequencies in a restricted narrow frequency range. Multiplex radio transmission, for instance, is the simultaneous transmission of two signals over a common carrier wave. (GRAF)

"NAND" AND "NOR" OPERATIONS (M)
The grouping of an AND gate followed by an inverter is called a NOT AND or NAND gate. If all the inputs have a value of 1, the output is 0, and if any of the inputs have a value of 0, the output will be 1. (Note that this is just the opposite of an AND gate.)

NAND GATE (J)
A combination of a NOT function and an AND function in a binary circuit that has two or more inputs and one output. (GRAF)

NEGATIVE FEEDBACK (DEGENERATION) (J)
A process by which a part of the output signal of an amplifying circuit is fed back to the input. (GRAF)

NEGATIVE LOGIC (M)
Logic in which the more-negative voltage represents the "1" state. The less-negative voltage represents the "0" state.

NIT (D)
The NIT is the SI unit of luminance and is equal to one candela per square meter.
(Obsolete—See also Candela, Footlambert, and Lumen.)

NMOS (N-TYPE MOS) (J)
MOS devices made on P-type silicon substrates where the active carriers are electrons flowing between N-type source and drain contacts.

NOISE (J)
Unwanted disturbances superimposed on a useful signal that tend to obscure its information content.

NOISE IMMUNITY (M)
A measure of the insensitivity of a logic circuit to triggering or reaction to spurious or undesirable electrical signals or noise, largely determined by the signal swing of the logic. Noise can occur in either of two directions, positive or negative.

NON-VOLATILE MEMORY (NVM) (J)
Electronic memory which is not lost during power off conditions.

(M)
A memory with the ability to maintain information without power applied. Usually used in reference to ROM or EAROM devices.

NON-VOLATILE RAM (NVRAM) (M)
A read-write RAM device that can preserve, through the use of non-volatile cells, data content without power. It has the speed of a RAM with non-volatility. (Not defined.)

NOR GATE (J)
An OR gate followed by an inverter to form a binary circuit in which the output is logic zero if any of the inputs is one, and vice versa. (GRAF)

NPN TRANSISTOR (J)
A transistor with a P-type base and N-type collector and emitter. (GRAF)

N-TYPE MATERIAL (J)
A crystal of pure semi-conductor material to which has been added an impurity so that electrons serve as the majority charge carriers. (GRAF)

NULL (J)
A condition (typically a condition of balance) which results in a minimum absolute value of output. Often specified as the calibration point when the least error can be tolerated by the associated control system.

(T)

A condition, such as a balance, which results in a minimum absolute value of output.

NULL BALANCE (T)
(Automatic Null-Balancing Electric Instruments) The condition that exists in the circuits of an instrument when the difference between an opposing electrical quantity within the instrument and the measured signal does not exceed the dead band. Note: The value of the opposing electrical quantity produced within the instrument is related to the position of the end device. See also control system, feedback, measurement system.

NUMERIC DISPLAY (D)
A numeric display is one which is capable of displaying only the numerals 0 to 9.

OBJECT CODE (M)
Output from a compiler or assembler which is itself executable machine code or is suitable for processing to produce executable machine code.

OBJECT LANGUAGE (M)
The language to which a statement is translated.

OFF-BOARD (DG)
Equipment or tools that are not a normal part of the content of the delivered new vehicle. Usually refers to special purpose shop or garage equipment that is attached to the vehicle when it is to be used.

OFF-VEHICLE EQUIPMENT (DG)
All off-vehicle devices for diagnostic purposes. (ISO)

OHM (J)
The unit of resistance. One ohm is the value of resistance through which a potential of one volt will maintain a current of one ampere. (GRAF)

ON-BOARD (DG)
Tools and/or equipment available as part of the delivered new vehicle.

ON-VEHICLE EQUIPMENT (DG)
All on-vehicle devices for diagnostic purposes. (ISO)

OPERAND (M)
That which is operated upon. An operand is usually identified by an address part of an instruction.

OPERATING TEMPERATURE (D)
Operating temperature is the ambient temperature at which a display device operates. The actual display temperature will be greater depending upon power dissipated. The operating temperature range is the range where practical display operation is possible. A complete definition requires the specification of parameters such as efficiency, life, or luminance which may be functions of temperature.

OPERATING TEMPERATURE RANGE (T)
The range of ambient temperatures, given by their extremes, within which a device is intended to operate; (i.e.) within this range of ambient temperature, all tolerances specified for temperature error, temperature error band, temperature gradient error, thermal zero shift and thermal sensitivity shift are applicable.

OPERATION (M)
1. A defined action; namely, the act of obtaining a result from one or more operands in accordance with a rule that completely specifies the result for any permissible combination of operands.
2. The set of such acts specified by such a rule, or the rule itself.
3. The act specified by a single computer instruction.
4. A program step undertaken or executed by a computer; e.g., addition, multiplication, extraction, comparison, shift, transfer. The operations are usually specified by the operator part of an instruction.
5. The event or specific action performed by a logic element.

OPERATIONAL AMPLIFIER (J)
An amplifier that performs various mathematical operations. Also called OP-AMP. (GRAF)

OPERATION CODE (M)
A code that represents specific operations. Synonymous with instruction code.

OR (M)
A logic operator having the property that if P and Q are logic quantities, then the quantity "P OR Q" assumes values as defined by the following table:

P	Q	P OR Q
0	0	0
0	1	1
1	0	1
1	1	1

The OR operator is represented in both electrical and FORTRAN terminology by a "plus;" i.e., P plus Q.

OR GATE (J)
A multiple-input gate circuit whose output is energized when any one or more of the inputs is in a prescribed state. Used in digital logic.

(M)
1. An electric circuit that implements the OR operator.
2. That specific gate which implements the OR operator.

OSCILLATOR (J)
An electronic device which generates alternating current power at a frequency determined by the values of certain constants in its circuits. (GRAF)

OUTPUT IMPEDANCE (T)
The impedance across the output terminals of a transducer presented by the transducer to the associated external circuitry. (ISA)

OUTPUT NOISE (T)
The unwanted RMS, peak or peak-to-peak (as specified) AC component of a transducer's DC output in the absence of variations. In the measurand and the environment.

OUTPUT REGULATION (T)
The change in output due to a change in excitation.

OVERHAUL (DG)
To dismantle, inspect, repair/replace parts as required, reassemble, and test a vehicle or system to a level of functional performance satisfactory for continued use, but not necessarily equal to new system performance levels.

OVERLOAD (T)
The maximum magnitude of measurand that can be applied to a transducer without causing a change in performance beyond specified tolerance. (ISA)

OVERRANGE (T)
See Overload.

OVERSHOOT (T)
The amount of output measured beyond the final steady output value, in response to a step change in the measurand. (ISA)

OVERVOLTAGE (GENERAL) (T)
A voltage above the normal rated voltage or the maximum operating voltage of a device or circuit. A direct test overvoltage is a voltage above the peak of the line alternating voltage. See also insulation testing.

PACK (M)
To compress data in a storage medium by taking advantage of known characteristics of the data in such a way that the original data can be recovered; e.g., to compress data in a storage medium by making use of bit or byte locations that would otherwise go unused.

PARALLEL OPERATION (M)
The organization of data manipulating within circuitry wherein all the digits of a word are transmitted simultaneously on separate lines in order to speed up operation.

PARALLEL PROCESSING (J)
Pertaining to the simultaneous execution of two or more sequences of instructions by a computer having multiple arithmetic or logic units. (IEEE)

PARAMETER (M)
A variable that is given a constant value for a specific purpose or process.

PARITY BIT (M)
A check bit appended to an array of binary digits to make the sum of all the binary digits, including the check bit, always odd or always even.

PARITY CHECK (M)
The technique of adding one bit to a digital word to make the total number of binary ones or zeros either always even or always odd. This type of checking will indicate a single error in data but will not indicate the location of the error.

PARTIAL FAILURE (DG)
Failures resulting from deviation in characteristics beyond specified limits but not such as to cause complete lack of the required function.

PASSIVE DISPLAY (D)
A passive display is one which modulates the transmission or reflection of external light. Typical examples are liquid crystal and electrochromic.

PASSIVE TRANSDUCER (T)
A transducer that has no source of power other than the input signal(s), and whose output signal-power cannot exceed that of the input. Note: The definition of a passive transducer is a restriction of the more general passive network; that is, one containing no impressed driving forces. (IEEE)

PERFECT DIFFUSION (D)
Perfect diffusion is that in which light is scattered uniformly in all directions by the diffusing medium.

PERFORMANCE (DG)
Degree of effectiveness of operation.

PERIPHERAL EQUIPMENT (M)
Units which work in conjunction with a computer but are not a part of it.

PERMEABILITY (J)
The measure of how much better a given material is than air as a path for magnetic lines or force. It is equal to the magnetic induction (B) in gausses, divided by the magnetizing force (H) in oersteds. (GRAF)

PHOT (ph) (T)

The phot is the CGS unit of illuminance and is equal to one candela per square centimeter.

PHOTOCELL (PHOTOELECTRIC CELL) (J)

1. A solid-state photosensitive electron device in which use is made of the variation of the current-voltage characteristic as a function of incident radiation.

2. A device exhibiting photovoltaic or photoconductive effects. (IEEE)

PHOTOMETER (T)

An instrument used for the measurement of photometric quantities. Photometers may be used for luminance or luminous intensity measurements. Spectral and time response are important characteristics.

PHOTOMETRIC (T)

Photometric units are quantities referring to light related to the human eye and hence have meaning only in the visible spectrum. The candela and the lumen are the basic photometric units.

PHOTOPIC CURVE (T)

The photopic curve is the curve giving relative eye response as a function of light wavelength for photopic vision. A standard response curve is used to allow calibration of electronic instruments for photometric quantities.

PHOTOPIC EYE RESPONSE (LIGHT ADAPTATION) (D)

Photopic vision is that dependent on the response of the cones in the eye and occurs at light intensities above approximately 3 nit.

PIEZOELECTRIC (J)

The property of certain crystals which: (1) Produce a voltage when subjected to a mechanical stress, (2) Undergo mechanical stress when subject to a voltage. (GRAF)

PLA (PROGRAMMABLE LOGIC ARRAY) (M)

An integrated circuit that employs ROM matrices to combine sum and product terms of logic networks.

PLASMA (J)

A gas made up of charged particles. Note: Usually plasmas are neutral, but not necessarily so, as, for example, the space charge in an electron tube. (IEEE)

PMOS (P-TYPE MOS) (J)

MOS devices made on an N-type silicon substrate where the active carriers are holes flowing between P-type source and drain controls.

PNP TRANSISTOR (J)

A transistor consisting of two P-type regions separated by an N-type region. (GRAF)

PNPN DIODE (J)

A semi-conductor device which may be regarded as a two transistor structure with two separate emitters feeding a common collector. (GRAF)

POLARIZER (T)

A material used to generate polarized light from a non-polarized source. A polarizer may select linear or circularly polarized light. Typical uses include liquid crystal displays and contrast enhancement filters for high ambient applications.

POSITIVE FEEDBACK (REGENERATION) (J)

The process by which the amplification is increased by having part of the power in the output returned to the input in order to reinforce the input power. (GRAF)

POSITIVE LOGIC (M)

Logic in which the more positive voltage represents the "1", state; the less positive voltage represents the "0" state.

POTENTIAL (J)

The difference in voltage between two points of a circuit. Frequently, one point is assumed to be ground which has zero potential. (GRAF)

POTTING (J)

An embedding process for parts that are assembled in a container or can into which the insulating material is poured, with the container remaining an integral part as the outer surface of the finished unit. (GRAF)

PRECISION (T)

See Repeatability and Stability.

PRIMARY LUMINOUS STANDARD (D)

A primary luminous standard is one by which the unit of light is established and from which the values of other standards are derived. A satisfactory primary standard must be reproducible from specifications.

PRIORITY INTERRUPT (M)

Designation given to method of providing some interrupt commands to have precedence over others, thus giving one condition of operation priority over another.

PROBLEM ORIENTED LANGUAGE (M)

A programming language designed for the convenient expression of a given class of problems.

PROCESSOR (M)

1. In hardware, a data processor.

2. In software, a computer program that includes the compiling, assembling, translating, and related functions, for a specific programming language, COBOL processor or FORTRAN processor.

PROGRAM (M)

1. A series of actions proposed in order to achieve a certain result.

2. Loosely, a routine.

3. To design, write, and test a program as in definition 1 above.

4. Loosely, to write a routine.

PROM (J)

An acronym for programmable read only memory. An electronic memory which may be permanent (non-volatile) or semi-permanent (erasable electronically or with ultra-violet light) and therefore able to be programmed one or more times.

PROPAGATION DELAY (M)

The time required for a change in logic level to be transmitted through an element or a chain of elements.

P-TYPE MATERIAL (J)

A semi-conductor material that has been doped with an excess of acceptor impurity atoms, so that free holes are produced in the material. (GRAF)

PURITY (D)

Purity is the relative brightness of the spectrum color (dominant wavelength) to the white continuous background spectrum in a color mixture.

PUSHDOWN LIST (M)

A list that is constructed and maintained so that the item to be retrieved is the most recently stored item in the list; i.e., last in, first out.

PUSHDOWN STACK (M)

A register which implements a pushdown list.

PUSHUP LIST (M)

A list that is constructed and maintained so that the next item to be retrieved and removed is the oldest item still in the list; i.e., first in, first out (FIFO).

RADIO FREQUENCY INTERFERENCE (RFI) (J)

Radio frequency energy of sufficient magnitude to have an influence on the operation of other electronic equipment. (GRAF)

RADIOMETRIC (T)

Radiometric quantities are properties of light based on purely physical measurement without relation to the eye response. Standard units are the watt and joule.

RAM (J)

An acronym for random access memory. A memory that has stored information immediately available when addressed regardless of the previous memory address location. As the memory words can be selected in any order, there is equal access time to all.

RANDOM ERROR (D)

The chance variation encountered in all experimental work despite the closest possible control of variables. It is characterized by the random occurrence of both positive and negative deviations from the mean value for the method, the algebraic average of which will approach zero in a long series of measurements. (ASTM)

RANGE (T)

The measurand values, over which a transducer is intended to measure, specified by their upper and lower limits.

REACTION TIME (T)

The interval between the beginning of a stimulus and the beginning of the response of an observer. (IEEE)

READABILITY (D)

Readability is a subjective measure of the ease with which a given display can be read under given conditions. Character size and format, brightness, color and contrast ratio are important determining parameters.

REAL TIME (M)

1. Pertaining to the actual time during which a physical process transpires.

2. Pertaining to the performance of a computation during the actual time that the related physical process transpires, in order that results of the computation can be used in guiding the physical process.

REBUILD (DG)

To completely dismantle, inspect, repair/replace parts as required, reassemble, and test a vehicle or system to level of functional performance equal to as-new levels.

RECOVERY TIME (T)

The time interval, after a specified event (e.g., overload, excitation transients, output shortcircuiting) after which a transducer again performs within its specified tolerances. (ISA)

RECTIFIER (J)

A device which, by virtue of its asymmetrical conduction characteristic, converts an alternating current into a unidirectional current. (GRAF)

REDUNDANCY (M)

1. The employment of several devices, each performing the same function in order to improve the reliability of a particular function.

2. In the transmission of information, that fraction of the gross information content of a message which can be eliminated without loss of essential information.

REFERENCE PRESSURE (T)
The pressure relative to which a differential-pressure device measures pressure.

REFERENCE PRESSURE ERROR (T)
The error resulting from variations of a differential-pressure transducer's reference pressure within the applicable reference pressure range. (ISA)

REFERENCE JUNCTION (T)
That junction of a thermocouple which is held at a known temperature.

REFERENCE VOLTAGE (ANALOG COMPUTERS) (T)
A voltage used as a standard of reference, usually the nominal full scale of the computer.

REFLECTANCE (REFLECTION FACTOR) (D)
Reflectance is the ratio of the light reflected to the incident light.

REFLECTANCE, SPECTRAL (D)
Spectral reflectance is the reflectance as a function of the wavelength of the incident light.

REFLECTION FACTOR (DIFFUSE) (D)
The diffuse reflection factor is the ratio of the diffuse reflected light to the incident light.

REFLECTION FACTOR (REGULAR) (D)
The regular reflection factor is the ratio of the regularly reflected light to the incident light.

REFRESH (M)
1. Method which restores charge on capacitance which deteriorates because of leakage.
2. Method which restores the state of elements in a memory array that are depleted by read out.

REFRESH RATE (D)
Refresh rate is the frequency at which a multiplexed display repeats its information. Rates of greater than 60 Hz are typically employed to prevent perceptible flicker.

REGISTER (M)
Temporary storage for digital data.

REGULAR TRANSMISSION FACTOR (D)
The regular transmission factor is the ratio of the regularly transmitted light to the incident light.

REGULATED POWER SUPPLY (J)
A unit which maintains a constant output voltage or current for changes in line voltage, output load, ambient temperature or time. (GRAF)

REGULATION (OVERALL POWER SUPPLIES) (J)
The maximum amount that the output will change as a result of the specified change in line voltage, output load, temperature or time. Note: Line regulation, load regulation, stability, and temperature coefficient are defined and usually specified separately. (IEEE)

RELATIVE ADDRESS (M)
Some identifier used to indicate the position of a memory location in a computer routine relative to the base address as opposed to the memory location's absolute address.

RELAY (J)
An electric device that is designed to interpret input conditions in a prescribed manner and after specified conditions are met, to respond to cause contact operation or similar abrupt change in associated electric control circuits. Notes: (1) Inputs are usually electric, but may be mechanical, thermal, or other quantities. Limit switches and similar simple devices are not relays. (2) A relay may consist of several units, each responsive to specified inputs, the combination providing the desired performance characteristic. (IEEE)

RELIABILITY (T)
1. The ability of an item to perform a required function under stated conditions for a stated period of time. Note: The term reliability is also used as a reliability characteristic denoting a probability of success or a success ratio.
2. The probability that a device will function without failure over a specified time period or amount of usage. Note: 1) Definition 2 is most commonly used in engineering applications. In any case where confusion may arise, specify the definition being used. 2) The probability that the system will perform its function over the specified time should be equal to or greater than the reliability.

RELUCTIVE PATH (T)
Converting a change of measurand into an AC voltage change by a change in the reluctance path between two or more coils or separated portions of one coil when AC excitation is applied to the coil(s). (ISA)

REPAIR (DG)
To restore a malfunctioning item to serviceable use by maintenance action.

REPEATABILITY (T)
The ability of a device (transducer) to reproduce output readings when the same measurand value is applied to it consecutively, under the same conditions, and in the same direction.

RESISTIVITY (J)
The measure of the resistance of a material to electric current either through its volume or on a surface. (GRAF)

RESISTOR (J)
A device the primary purpose of which is to introduce resistance into an electric circuit. Note: Resistor as used in electric circuits for purposes of operation, protection, or control, commonly consists of an aggregation of units. Resistors as commonly supplied consist of wire, metal ribbon, cast metal or carbon compounds supported by or imbedded in an insulation medium. The insulating medium may enclose and support the resistance material as in the case of porcelain-tube type or the insulation may be provided only at the points of support as in the case of heavy-duty ribbon or cast iron grids mounted in metal frames. (IEEE)

RESISTOR-CAPACITOR-TRANSISTOR LOGIC (RCTL) (J)
A logic circuit design that employs a resistor and a speed-up capacitor in parallel for each input of the gate. A transistor's base is connected to one end of the RC network. A positive voltage on the RC input will energize the transistor and turn it on so that the output voltage is nearly zero volts. This circuit is a positive NOR or negative NAND when NPN transistors are used in the circuit. (GRAF)

RESISTOR-TRANSISTOR LOGIC (RTL) (J)
A form of logic that has a resistor as the input component that is coupled to the base of an NPN transistor. As in RCTL, the transistor is an inverting element that produces the positive NOR gate or the negative NAND gate function. (GRAF)

RESIST PLATING (J)
Any metallic plating which, when deposited on a conductive area, prevents the areas underneath from being plated. (GRAF)

RESOLUTION (T)
1. The act of deriving from a sound, scene or other form of intelligence, a series of discrete elements wherefrom the original may subsequently be synthesized.
2. The degree to which nearly equal values of a quantity can be discriminated.
3. The fineness of detail in a reproduced spatial pattern.
4. The degree to which a system or a device distinguishes fineness of detail in a spatial pattern.

RESONANCE (T)
1. The reinforced vibration of a body exposed to the vibration at the frequency of another body.
2. An oscillation of large amplitude in a mechanical or electrical system caused by a relatively small periodic stimulus of the same or nearly the same period as the natural oscillation period of the system.

RESONANT FREQUENCY (J)
The frequency at which a given system or object will respond with maximum amplitude when driven by an external sinusoidal force of constant amplitude. (GRAF)

RESPONSE TIME (D)
Response time is the time for the optical property (luminance, reflectance, etc.) of a display to respond to a step function change in the driving signal.

(T)
The length of time required for the output of a transducer to change to a specified percentage of its final value as a result of a step change of measurand.

RIPPLE (T)
Periodic modulations in the output of a transducer having a DC output. (IEEE)

RISE TIME (D)
Rise time is the time required for the specified optical property to rise from 10 to 90 percent of its maximum excursion after a step function increase in the driving signal level.

(J)
The time required for the leading edge of a pulse to rise from 10-90% of its final value. It is proportionate to the time constant and is a measure of the steepness of the wavefront. Also, the measured length of time required for an output voltage of a digital circuit to change once the change has started. (GRAF)

(T)
The length of time for the output of a transducer to rise from a small specified percentage of its final value to a large specified percentage of its final value as a result of a step-change of measurand. (ISA)

RISE TIME DELAY (D)

Rise time delay is the time required for the specified optical or electrical property to rise from its steady state level to 10 percent of its maximum excursion after a step function increase in the driving signal.

ROM (J)

An acronym for read only memory. A memory which permits the reading of a predetermined pattern of zeros and ones. This predetermined information is stored in the ROM at the time of its manufacture. A ROM is analogous to a dictionary where a certain address results in predetermined information output.

ROUTINE (M)

A set of computer instructions arranged in a correct sequence and used to direct a computer in performing one or more desired operations.

SAMPLE-AND-HOLD CIRCUIT (M)

A circuit that performs the operation of looking at a voltage level during a short time period and accurately storing that voltage level for a much longer time period.

SATURATION (J)

A circuit condition whereby an increase in the driving or input signal no longer produces a change in the output. (GRAF)

SATURATION VOLTAGE (J)

Generally, the voltage excursion at which a circuit self-limits (i.e., is unable to respond to excitation in a proportional manner). (GRAF)

SCHOTTKY BARRIER (J)

A simple metal to semi-conductor interface that exhibits a nonlinear impedance. (GRAF)

SCOTOPIC EYE RESPONSE (DARK ADAPTATION) (D)

Scotopic vision is that dependent on the rods in the eye and occurs at light intensities equal and below approximately 3×10^{-5} nit. Color vision is lost when only rods are active.

SCRATCHPAD MEMORY (M)

A small local memory utilized to facilitate local data handling on a temporary basis. RAMs are usually used in microcomputer applications.

SECONDARY STANDARD (D)

A secondary standard is one calibrated by comparison with a primary standard.

SEGMENT DISPLAY (D)

A display organized with characters made up of segments. Typical configurations are seven segments for numeric use and fourteen or sixteen segments for alphanumeric use.

SELF-GENERATING (T)

Providing an output signal without applied excitation. Examples are piezoelectric, electromagnetic and thermoelectric transducers.

SELF-HEATING (T)

Internal heating resulting from electrical energy dissipated within the transducer.

SEMI-CONDUCTOR (J)

An electronic conductor, with resistivity in the range between metals and insulators, in which the electric-charge-carrier concentration increases with increasing temperature over some temperature range. Note: Certain semi-conductors possess two types of carriers; namely, negative electrons and positive holes.

SEMI-CONDUCTOR CONTROLLED RECTIFIER (SCR) (J)

An alternate name used for the reverse-blocking triode-thrysistor. Note: The name of the actual semi-conductor material (selenium, silicon, etc.) may be substituted in place of the word semi-conductor in the name of the components. (IEEE)

SENSING ELEMENT (T)

That part of the transducer which responds directly to the measurand. (ISA)

SENSITIVITY (J)

Measure of the ability of a device or circuit to react to a change in some input. Also, the minimum or required level of an input necessary to obtain rated output.

(T)

The ratio of the change in transducer output to a change in the value of the measurand.

SENSITIVITY SHIFT (T)

A change in the slope of the calibration curve due to a change in sensitivity. (ISA)

SENSOR (T)

A device directly responsive to the value of the measured quantity. A transducer which converts a parameter (at a test point) to a form suitable for measurement (by the test equipment).

See also diagnostic sensor.

SEQUENCING (M)

Control method used to cause a set of steps to occur in a particular order.

SEQUENTIAL LOGIC SYSTEMS (M)

Digital system utilizing memory elements.

SERIAL ACCUMULATOR (M)

A register which receives data bits in serial or sequence and temporarily holds the data for future use.

SERIAL OPERATION (M)

The organization of data manipulation within circuitry wherein the digits of a word are transmitted one at a time along a single line. The serial mode of operation is slower than parallel operation but utilizes less complex circuitry.

SERIAL-PARALLEL (J)

Pertaining to processing that includes both serial and parallel processing, such as one that handles decimal digits serially but handles the bits that comprise a digit in parallel. (IEEE)

SERIAL TRANSMISSION (DATA TRANSMISSION, TELECOMMUNICATION) (J)

Used to identify a system wherein the bits of a character occur serially in time. Implies only a single transmission channel. Also called serial by bit. (IEEE)

SERVO (T)

A transducer type in which the output of the transduction element is amplified and fed back so as to balance the forces applied to the sensing element or its displacements. The output is a function of the feedback signal. (ISA)

SET-UP TIME (M)

The minimum amount of time that data must be present at an input to ensure data acceptance when the device is clocked.

SHIFT (M)

A movement of data to the right or left.

SHIFT REGISTER (J)

1. A logic network consisting of a series of memory cells such that a binary code can be caused to shift into the register by serial input to only the first cell.

2. A register in which the stored data can be moved to the right or left.

SIGNAL (J)

1. A visual, audible or other indication used to convey information.

2. The intelligence, message or effect to be conveyed over a communication system.

3. A signal wave; the physical embodiment of a message.

4. (Computing Systems) The event or phenomenon that conveys data from one point to another.

5. (Control, Industrial Control) Information about a variable that can be transmitted in a system. (IEEE)

SIGNAL GENERATOR (J)

A shielded source of voltage or power, the output level and frequency of which are calibrated, and usually variable over a range. Note: The output of known waveform is normally subject to one or more forms of calibrated modulation. (IEEE)

SIGNAL-TO-NOISE-RATIO (J)

The ratio of the value of the signal to that of the noise. Notes: (A) This ratio is usually in terms of peak values in the case of impulse noise and in terms of the root-mean-square values in the case of the random noise. (B) Where there is a possibility of ambiguity, suitable definitions of the signal and noise should be associated with the terms; as, for example: peak-signal to peak-noise ratio; root-mean-square signal to root-mean-square noise ratio; peak-to-peak signal to peak-to-peak noise ratio, etc. (C) This ratio may be often expressed in decibels. (D) This ratio may be a function of the bandwidth of the transmission system. (IEEE)

SIGN AND MAGNITUDE NOTATION (M)

A system of notation where numbers are represented by a sign bit and one or more number bits.

SILICON-ON-SAPPHIRE (SOS) (J)

Pertaining to the technology in which monocrystalline silicon films are epitaxially deposited onto a single-crystal sapphire substrate to form a structure for the fabrication of dielectrically isolated elements. (GRAF)

SILO MEMORY (M)

Reads out stored data in a first-in/first-out mode. Also known as FIFO.

SIMULATE (M)

1. To represent certain features of the behavior of a physical or abstract system by the behavior of another system.

2. To represent the functioning of a device, system or computer program by another; e.g., to represent the functioning of one computer by another, to represent the behavior of a physical system by the execution of a computer program, to represent a biological system by a mathematical model.

SIMULATOR (M)

A device, system or computer program that represents certain features of the behavior of a physical or abstract system.

SKIP (M)

To ignore one or more instructions in a sequence of instructions.

SOFTWARE (J)
1. Computer programs, routines, programming languages and systems.
2. The collection of related utility, assembly and other programs that are desirable for properly presenting a given machine to a user.
3. Detailed procedures to be followed whether expressed as programs for a computer or as procedures for an operator or other person.
4. Documents, including hardware manuals and drawings, computer-program listings and diagrams, etc.

(M)
5. Items such as those in 1, 2, 3, and 4, as contrasted with hardware. (IEEE)

(M)
A set of computer programs, procedures, and possibly associated documentation concerned with the operation of a data processing system; e.g., compilers, library routines, manuals, circuit diagrams.

SOLID-STATE (J)
Pertaining to circuits and components using semi-conductors. (See solid-state device.) (GRAF)

SOLID-STATE DEVICE (J)
Any element that can control current without moving parts, heated filaments, or vacuum gaps. All semi-conductors are solid-state devices, although not all solid-state devices are semi-conductors (e.g., transformers). (GRAF)

SOLID-STATE RELAY (J)
A relay constructed exclusively of solid-state components.

SOUND PRESSURE, P(ML^{-1}T^{-2}): Pa (T)
A fluctuating pressure superimposed on the static atmospheric pressure. Its magnitude can be expressed in several ways such as instantaneous sound pressure or peak sound pressure, but the unqualified term means root-mean-square sound pressure. (ASTM)

SOUND PRESSURE LEVEL (T)
The sound pressure level, in decibels, is 20 times the logarithm to the base 10 of the ratio of the sound pressure to the reference pressure. The reference pressure must be explicitly stated. Unless otherwise explicitly stated, it is to be understood that the sound pressure is the effective root-mean-square sound pressure. (IEEE)

SOURCE(S) (OR SOURCE ELECTRODE) (J)
In a field effect transistor, the electrode that is analogous to the emitter of a transistor or the cathode of a vacuum tube. (GRAF)

SOURCE IMPEDANCE (J)
The impedance which a source of energy presents to the input terminal of a device. (GRAF)

SOURCE LANGUAGE (M)
The language from which a statement is translated.

SOURCE PROGRAM (M)
A computer program written in a source language.

SPAN (T)
The algebraic difference between the limits of the range.

SPECIFICATION (DG)
A description of an item or system that defines in detail its functional performance capabilities; and the installation, environmental, and operational requirements or limitations.

SPECTRAL RESPONSE CURVE (D)
A spectral response curve is one giving the relative response of a system (such as a photodetector) as a function of the measuring wavelength.

SPECTRUM (D)
A spectrum is a curve giving the relative light output of a source as a function of the wavelength.

SPECULAR (REGULAR) REFLECTION (D)
Specular reflection is that in which the angle of incidence equals the angle of reflection.

STABILITY (J)
The ability of a component or device to maintain its nominal operating characteristics after being subjected to changes in temperature, environment, current, and time. (GRAF)

(T)
The ability of a transducer to retain its performance characteristics for a relatively long period of time.

STANDARD, INSTRUMENT (T)
A standard calibrated for use with a particular instrument; usually not suitable for use with an instrument of a different type.

STATE (M)
The condition of an input or output of a circuit as to whether it is a logic "1" or a logic "0." The state of a circuit (gate or flip-flop) refers to its output. A flip-flop is said to be in the "1" state when its 0 output is "1." A gate is in the "1" state when its output is "1."

STATIC CALIBRATION (T)
A calibration performed under room conditions and in the absence of any vibration, shock, or acceleration (unless one of these is the measurand). (ISA)

STATIC STORAGE ELEMENTS (M)
Storage elements which contain storage cells that retain their information as long as power is applied unless the information is altered by external excitation.

STEADY-STATE (J)
A condition in which circuit values remain essentially constant, occurring after all initial transients or fluctuating conditions have settled down. (GRAF)

STILB (sb) (D)
The stilb is the CGS unit of luminance equal to one candela per square centimeter.

STORAGE (ELECTRONIC COMPUTATION) (J)
1. The act of storing information.
2. Any device in which information can be stored, sometimes called a memory device.
3. In a computer, a section used primarily for storing information. Such a section is sometimes called a memory or store (British). Notes: (A) The physical means of storing information may be electrostatic, ferroelectric, magnetic, acoustic, optical, chemical, electronic, electric, mechanical, etc., in nature. (B) Pertaining to a device in which data can be entered, in which it can be held, and from which it can be retrieved at a later time. (IEEE)

STORAGE TEMPERATURE RANGE (D)
The storage temperature range is the extreme of temperature to which a non-operating device may be subjected for long periods (greater than 500 hours) without damage.

STORED PROGRAM (M)
A set of instructions in memory specifying the operation to be performed.

STRAIN GAGE (T)
Converting a change of measurand into a change or resistance due to strain. (ISA)

SUBROUTINE (M)
A subroutine is a series of computer instructions to perform a specific task for many other routines. It is distinguishable from a main routine in that it requires as one of its parameters a location specifying where to return to the main program after its function has been accomplished.

SUBSTRATE (J)(T)
1. The supporting material on or in which the parts of a device such as an integrated circuit are attached or made.
2. The physical material upon which an electronic circuit is fabricated. Used primarily for mechanical support but may serve a useful electrical or thermal function.

SYMMETRICAL LIGHT DISTRIBUTION (D)
A symmetrical light distribution is one in which the curve of emitted light distribution as a function of viewing angle is substantially the same for conjugate angles.

SYMPTOM (DG)
Any perceptible change in a mechanism, material, structure or system which indicates any of the modes or kinds of failure.

SYNCHRONOUS CIRCUIT (M)
A circuit in which all ordinary operations are controlled by spaced signals from a master clock.

SYSTEM (M)
1. An assembly of methods, procedures or techniques, united by regulated interaction to form an organized whole.
2. An organized collection of men, machines, and methods required to accomplish a set of specific functions.

TABLE DRIVEN (M)
A technique of using table look-ups to implement a control or processing sequence.

TABLE LOOK-UP (M)
A procedure for obtaining the function value corresponding to an argument from a table of function values.

TEMPERATURE ERROR (T)
The maximum change in output, at any measurand value within the specified range, when the transducer temperature is changed from room temperature to specified temperature extremes.

TEMPERATURE ERROR BAND (T)
The error band applicable over stated environmental temperature limits.

TEMPERATURE GRADIENT ERROR (T)
The transient deviation in output of a transducer at a given measurand value when the ambient temperature or the measurand fluid temperature changes at a specified rate between specified magnitudes. (ISA)

TEMPORARY STORAGE (M)
In programming, storage locations reserved for intermediate results.

TEST (DG)
A procedure or action taken to measure, under real or simulated conditions, parameters of a system or component.

TEST EQUIPMENT (DG)
Devices used in the performance of a test. Test equipment is not necessarily diagnostic equipment.

THERMAL RESISTOR (J)
An electronic device which makes use of the change in resistivity of a semi-conductor with changes in temperature. (GRAF)

THERMAL RUNAWAY (J)
A condition in which the dissipation in a transistor or other device increases so rapidly with higher temperature that the temperature keeps on rising. (GRAF)

THERMISTOR (J)
A solid-state semi-conducting device, the electrical resistance of which varies with the temperature. Its temperature coefficient of resistance is high, nonlinear and negative. (GRAF)

(M)
A specific resistor whose temperature coefficient of resistance is unusually high; it is also nonlinear and negative. It is often made by sintering mixtures of oxide powders of various metals and is thus a solid-state semi-conductor material in many types of shapes as disks, flakes, rods, etc., to which contact wires are attached; the resistance of the unit varies with temperature changes, and thus acts as a sensor for temperature change.

THERMOCOUPLE (J)
Also called thermal junction. A device for measuring temperature where two electrical conductors of dissimilar metals are joined at the point of heat application and a resulting voltage difference, directly proportional to the temperature, is developed across the free ends and is measured potentiometrically. (GRAF)

(T)
A pair of dissimilar conductors so joined at two points that an electromotive force is developed by the thermoelectric effects when the junctions are at different temperatures. (IEEE)

THERMOELECTRIC (emf) (T)
The algebraic sum of the potential differences associated with 1) maintaining the junctions of two dissimilar conductors at different temperatures and 2) keeping a conductor in a thermal gradient. (ASTM)

THRESHOLD (D)
The minimum driving signal level at which a perceptible change in a specified optical property takes place. A multiplexed display requires that a threshold exist.

THICK-FILM (J)
Pertaining to a film pattern usually made by applying conductive and insulating materials to a ceramic substrate by a silk-screen process. Thick films can be used to form conductors, resistors and capacitors. (GRAF)

THIN-FILM (J)
A film of conductive or insulating material, usually deposited by sputtering or evaporation, that may be made in a pattern to form electronic components and conductors on a substrate or used as insulation between successive layers of component. (GRAF)

THYRISTOR (J)
A bistable semi-conductor device comprising three or more junctions that can be switched from the off state to the on state or vice versa, such switching occurring within at least one quadrant of the principal voltage current characteristic. (IEEE)

TIME CONSTANT (T)
The length of time required for the output of a transducer to rise to 63% of its final value as a result of a step change of measurand. (ISA)

TIME DELAY (T)
The time interval between the manifestation of a signal at one point and the manifestation or detection of the same signal at another point. (IEEE)

TRANSDUCER (J)
A device by means of which energy can flow from one or more transmission systems or media to one or more other transmission systems or media. Note: The energy transmitted by these systems or media may be of any form (for example, it may be electric, mechanical or acoustical), and it may be of the same form or different forms in the various input and output systems or media. (IEEE)

(DG)
A device which converts energy from one form or power level to another. Often used to interface unit under test with test or diagnostic equipment.

TRANSDUCTION ELEMENT (T)
The electrical portion of a transducer in which the output originates.

TRANSFER (M)
Same as Jump.

TRANSFER FUNCTION (T)
A mathematical, graphical or tabular statement of the influence which a system or element has on a signal or action compared at input and output terminals.

TRANSFORMER (J)
A device consisting of a winding with tap or taps, or two or more coupled windings with or without a magnetic core for introducing mutual coupling between electric circuits. (IEEE)

TRANSIENT (J)
1. A phenomenon caused in a system by a sudden change in conditions and which persists for a relatively short time after the change.
2. A temporary increase or decrease of the voltage or current. Transients may take the form of spikes or surges.

TRANSIENT SUPPRESSION NETWORKS (T)
Capacitors, resistors or inductors so placed as to control the discharge of stored-after-transfer function energy banks. They are commonly used to suppress transients caused by switching.

TRANSISTOR (J)
An active semi-conductor device with three or more terminals. (IEEE)

TRANSISTOR-TRANSISTOR LOGIC (ABBREVIATED TTL OR T²L) (J)
Also called multi-emitter transistor logic. A logic circuit design similar to DTL, with the diode inputs replaced by a multiple emitter transistor. In a four-input DTL gate, there are four diodes at the input. A four-input TTL gate will have four emitters of a single transistor as the input element. TTL gates using NPN transistors are positive-level NAND gates or negative-level NOR gates. (GRAF)

TRANSLATE (M)
To transform statements from one language to another without significantly changing the meaning.

TRANSMISSION FACTOR (D)
The transmission factor is the ratio of the transmitted light to the incident light.

TRANSVERSE SENSITIVITY (T)
The sensitivity of a transducer to transverse acceleration or other transverse measurand.

TRIAC (J)
A five-layer NPNPN device that is equivalent to two SCRs connected in antiparallel with a common gate. It provides switching action for either polarity of applied voltage and can be controlled in either polarity from the single gate electrode. (GRAF)

TRUTH TABLE (M)
A chart that tabulates and summarizes all the combinations of possible states of the inputs and outputs of a circuit. It tabulates what will happen at the output for a given input combination.

2's COMPLEMENT NOTATION (M)
A system of notation where positive binary numbers are identical to positive numbers in sign and magnitude notation, but where 1 must be added to 1's complement notation to obtain negative numbers.

ULTRAVIOLET (D)
The ultraviolet is that portion of the light spectrum with wavelengths shorter than the visible (less than 3900 angström).

UNIJUNCTION TRANSISTOR (J)
A three terminal semi-conductor device exhibiting stable open-circuit, negative resistance characteristics. (GRAF)

VACUUM FLUORESCENT DISPLAY (D)
An active display whose operating principle is the emission of light from a phosphor excited by electrons emitted from a filament in vacuum.

VAR (J)
Abbreviation for volt ampere reactive. The unit of reactive power, as opposed to real power in watts. One VAR is equal to one reactive volt-ampere. (GRAF)

VARIABLE (M)
A quantity that can assume any of a given set of values.

VARISTOR (J)
A two-electrode semi-conductor device with a voltage-dependent nonlinear resitance that drops markedly as the applied voltage is increased. (GRAF)

VIBRATION ERROR (T)
The maximum change in output, at any measurand value within the specified range, when vibration levels of specified amplitude and range of frequencies are applied to the transducer along specified axes. (ISA)

VIEWING ANGLE (D)
The angle between the viewer's line of sight and the normal to the display surface is the viewing angle. The included viewing angle is the sum of the viewing angle to each side of the normal over which the display is usable. The angle should be specified at a defined level of an optical property such as luminance, luminous intensity, or contrast ratio. For asymmetrical light distributions, the viewing angle plane should also be specified.

VISIBLE (D)
The visible is that portion of the light spectrum to which the human eye is sensitive, typically 390 to 770 nm.

VOLATILE MEMORY (J)
An electronic memory (RAM) which temporarily stores data that is lost when the power is turned off.

VOLATILE STORAGE (M)
A storage device in which stored data are lost when the applied power is removed.

VOLT (J)
The unit of voltage or potential difference in SI units. The volt is the voltage between two points of a conducting wire carrying a constant current of one ampere, when the power dissipated between these points is one watt. (IEEE)

WARM-UP PERIOD (T)
The period of time, starting with the application of excitation to the transducer, required to assure that the transducer will perform within all specified tolerances. (ISA)

WAVEGUIDE (J)
1. Broadly, a system of material boundaries capable of guiding electromagnetic waves.
2. More specifically, a transmission line comprising a hollow conducting tube within which electromagnetic waves may be propagated or a solid dielectric or dielectric filled conductor for the same purpose.
3. A system of material boundaries or structures for guiding transverse-electromagnetic mode, often and originally a hollow metal pipe for guiding electromagnetic waves. (IEEE)

WATT (J)
A unit of the electric power required to do work at the rate of one joule per second. It is the power expended when one ampere of direct current flows through a resistance of one ohm. (GRAF)

(D)
The watt is the SI unit of power equal to one joule per second.

WORD (M)
A character string or a bit string considered as an entity.

WORKING STANDARD (D)
A working standard is any standardized device for daily use in comparison or measurement. Such a standard is called traceable if its calibration is related to that of a standard maintained by a central government standards laboratory—the National Bureau of Standards (NBS) in the U.S.

WORKING STORAGE (M)
Working storage may be included in various microcontroller systems as an option. Typical uses are as a small buffer storage or as storage for intermediate program data. The working storage is viewed by the processor in much the same manner as is the interface vector. The storage is often organized as two pages each containing 128 bytes. A byte is accessed by storing its address in a specific register.

WRITE ENABLE (M)
Also called read/write or R/W. The control signal to a storage element or a memory that activates the write mode or operation. Conversely when not in the write mode, the read mode is active.

WRITE TIME (M)
The time that the appropriate level must be maintained on the write enable line and that data must be present to guarantee successful writing of data in the memory.

ZENER DIODE (J)
A two layer device that, above a certain reverse voltage (the Zener value), has a sudden rise in current. If forward-biased, the diode is an ordinary rectifier, but when reversed-biased, the diode exhibits a typical knee, or sharp break, in its current-voltage graph. The voltage across the device remains essentially constant for any further increase of reverse current, up to the allowable dissipation rating. The Zener diode is a good voltage regulator, over voltage protector, voltage reference, level shifter, etc. True Zener breakdown occurs at less than six volts. (See also Avalanche Diode.)

ZENER EFFECT (J)
A reverse current breakdown due to the presence of a high electrical field at the junction of a semi-conductor or insulator. (GRAF)

ZERO SHIFT (T)
A change in the zero-measurand output over a specified period of time and at room conditions. (ISA)

(R) GLOSSARY OF VEHICLE NETWORKS FOR MULTIPLEXING AND DATA COMMUNICATIONS— SAE J1213/1 JUN91 — SAE Information Report

Report of the Electrical and Electronics Systems Technical Committee approved April 1988. Completely revised by the SAE Vehicle Network for Multiplexing and Data Communications Standards Committee June 1991.

1. Scope—This SAE Information Report provides definition for terms (words and phrases) which are generally used within the SAE in describing network and data communication issues. In many cases, these definitions are different from those of the same or similar terms found in nonautomotive organizations, such as the Institute of Electrical and Electronic Engineers (IEEE). The Vehicle Networks for Multiplexing and Data Communications committee has found it useful to collect these specific terms and definitions into this document so documents related to the multiplexing and data communications issues will not need an extensive definitions section.

This document is intended to be the central reference for terms and definitions related to multiplexing and data communications and as such is intended to apply equally to Passenger Car, Truck and Bus, and Construction and Agriculture organizations within SAE. An attempt to use common terms across these applications has been made, so that these organizations can all utilize this glossary.

As terms are introduced for individual SAE documents, these terms will be considered for incorporation into this glossary. Any SAE document, of course, has the ability to define terms specific to that document which may be different from the definitions herein. However, authors are encouraged to use these terms as defined herein to reduce confusion.

Many of the terms in this revision of the document have been contributed by the related International Standards Organization (ISO) from their document ISO/TC22/SC3/WG1 N310E, dated November 1987.

This document shall be periodically reviewed for new terms to be included and possible modifications to the existing definitions. The addition of new terms is probable as the organization further documents Class A and Class C issues.

2. References

2.1 Applicable Documents—The following publications form a part of this specification to the extent specified herein.

2.1.1 ISO PUBLICATIONS—Available from ANSI, 11 West 42nd Street, New York, NY 10036

ISO/TC22/SC3/WG1 N310E—Glossary

3. Terms and Definitions

Acknowledgement—A type of response which is used to indicate whether a message has been received properly. Acknowledgements can be positive, indicating the message was received, or negative, indicating the message was not received.

Application-To-Application Delay Time—Delay time in a communication network, starting at the request for transmission and terminating at the presentation of the transferred information to the respective application(s).

Arbitration—Making a decision relative to a controversy existing at the inputs of the nodes involved.

Arbitration-Based Protocol—A form of contention-based protocol where contention is evaluated bit-by-bit. Arbitration-based protocols typically require that the physical length of the signal bus be sufficiently short such that the propagation time for signal transmission is significantly less than one information bit time.

Arbitration Field—Bits within the message frame attributed to each message for controlling the arbitration.

Availability—The decimal fraction of the time during which a system is capable of performing its required functions. Availability is determined from reliability (Mean Time Between Failures) and repair time (Mean Time to Restore) by the relationship A = MTBF / (MTBF + MTTR). To avoid ambiguity, it is possible to define different levels or modes of operation and to determine an availability for each level or mode.

Baseband—The band of frequencies occupied by unmodulated signals. The ratio of the information bandwidth (upper limit minus the lower limit of the frequency band) to the center frequency is typically larger than unity.

Baseband Communications—A communications method in which the transmitted signal is in its unmodulated form and not changed by modulation.

Baud Rate—Transfer rate, measured in bauds which is signal transitions per second.

Bit Rate—Bits per time during transmission, independent of bit representation.

Bit-by-Bit Contention—A contention-based arbitration, whereby the contention created by simultaneous access of multiple nodes on the network is resolved bit-by-bit. For example, bits may be represented as dominant and passive on the physical layer, dominant bits overriding passive ones in case of contention. For this example, the message with a dominant bit in the arbitration field survive without destruction, all others discontinue transmission. This procedure is repeated through all bits of the arbitration field.

Bridge—A node used to connect two networks that use similar protocols, as differentiated from a gateway.

Broadband Communications—A communications method in which the transmitted signal is the original signal modulated onto a carrier. The form of modulation can be amplitude, frequency, or phase.

Broadcast Communications—The transmission of information to more than one receiver, as differentiated from node-to-node communications.

Bus—Topology of a communication network, where all nodes are reached by links, which allow transmission in both directions.

Carrier—A wave suitable for modulation by an information bearing signal to be transmitted over a communications medium. For a waveform to be considered to include a carrier, it shall have a separate carrier, as in amplitude and frequency modulation, rather than an inherent carrier, as in pulse width modulation.

Carrier Sense—The ability of a receiver to sense if another node is transmitting (providing a listen before talking capability) or to sense if any node is transmitting (adding a listening while talking capability). These capabilities permit the design of contention based protocols.

Carrier Sense Multiple Access (CSMA)—An access method in which a node on a multiple node signal bus waits for an idle bus before attempting to transmit.

Carrier Sense Multiple Access With Collision Detection (CSMA/CD)—A type of CSMA method whereby when a node which is attempting to transmit detects that another node is also attempting to transmit, it delays for a time determined by a predefined contention algorithm.

Centralized Control—An organization of a control system whereby a central control element exercises control over the remainder of the system (see also Master/Slave).

Class A System—A multiplex system whereby vehicle wiring is reduced by the transmission and reception of multiple signals over the same signal bus between nodes that would have been accomplished by individual wires in a conventionally wired vehicle. The nodes used to accomplish multiplexed body wiring typically did not exist in the same or similar form in a totally conventional wired vehicle.

Class B System—A multiplex system whereby data (e.g. parametric data values) is transferred between nodes to eliminate redundant sensors and other system elements. The nodes in this form of a multiplex system typically already existed as stand-alone modules in a conventionally wired vehicle.

Class C System—A multiplex system whereby high data rate signals typically associated with real time control systems, such as engine controls and anti-lock brakes, are sent over the signal bus to facilitate distributed control and to further reduce vehicle wiring.

Closed System—A system, consisting of nodes interconnected by a common communications medium (signal bus), which does not permit the easy addition of modules developed by another manufacturer and temporary connection to other networks, as in an open system.

Coaxial Cable—A communications medium (signal bus) with two concentric conductors separated by dielectric material(s) resulting in low losses and a relatively constant specific impedance over a wide range of frequencies.

Collision Avoidance Protocol—A protocol in which a node waits a fixed period of time after the end of the last message or after detecting that another node is also attempting to transmit before trying to transmit or to transmit again. Each node in a network is assigned a different fixed period of time.

Communication Control Instruction—Instruction which controls the process of message transfer in a communication system.

Communication Integrity—Feature of a communication system, that the information is transferred uncorrupted and arrives at its destination(s) without modification.

Contention—A state of the bus in which two or more transmitters are turned on simultaneously to conflicting states.

Contention-Based Protocol—A protocol or organization of communications data where the nodes providing the data seek the use of the communications medium (signal bus) and the bus is awarded based upon the application of a predefined algorithm (typically a priority structure) to the information the nodes provide.

Data Consistency—A feature of communications in some multiplex wiring systems whereby it is determined and ensured that all required recipients of a message have received the message accurately before acting upon it simultaneously. This feature is desirable in, for example, ensuring that all four lamps are turned on at once or that all four brakes are energized simultaneously.

Distributed Control—An organization of a control system whereby control logic elements are physically located in several different places, as differentiated from a centralized organization of control.

Driven Line Length—Total line length driven by any one node.

Driver—A solid state device used to transfer electrical power to the next stage, which may be another driver, an electrical load (power driver), a wire or cable (line driver), a display (display driver), etc.

Dynamic Priority—Priority which may be altered during system operation.

Encoding—A method of how to represent information bits in data processing or communication systems.

Error—The inability of a system to perform properly caused by faults.

Error Message—Special message within a communication network informing all nodes about an error.

Failure—The inability of a system to perform properly caused by faults.

Fault—The loss of proper function of a component (hardware or software) and can be either permanent or intermittent.

Fault Tolerance—Ability of a system to survive a certain number of failures while performing its required functions, but possibly with some degraded characteristics.

Fiber Optics—A communications medium (signal bus) consisting of either individual fibers or an assemblage of transparent glass or plastic fiber(s) bundled essentially together parallel to one another. This fiber or bundle of fibers has the ability to transmit light along its axis by a process of total internal reflection.

Fiber Optics Receiver—An assembly which accomplishes the receive function in fiber optics communications, typically consisting of a photodetector (either a photodiode or a photo-transistor) and a preamplifier.

Fiber Optics Transmitter—A unit which accomplishes the driver function in fiber optics communications, typically consisting of a light emitting diode (LED) and an LED drive circuit. In contrast with the preamplifier of a fiber optics receiver, the LED drive circuit is not required to be, and typically is not, packaged with the LED.

Fixed Priority—Priority preassigned before the start of system operation.

Flexibility—Ability of the system to function with nodes manufactured by various suppliers.

Flexible System—A system, consisting of nodes interconnected by a common communications medium (signal bus) according to established standards, to which nodes manufactured by another supplier can be added.

Frequency Division Multiplex Protocol—A protocol where the meaning of a bit of information on the signal bus utilizes the principle of frequency sharing among information channels where the data from each channel are used to modulate sinusoidal signals called sub-carriers so that the resultant signal representing each channel contains only frequencies in a restricted narrow frequency range. Multiplex radio transmission, for instance, is the simultaneous transmission of two signals over a common carrier wave.

Functional Addressing—Labeling of messages based on their operation code or data content.

Functional Assignment of Wires—Functional specification for each wire on the physical layer and labeling it with a name.

Gateway—A node used to connect networks that use different protocols, as differentiated from a bridge. A gateway acts as a protocol converter.

Global Error—Error in a communication network, which is similarly detected in all nodes.

Ground Bus—The portion of the wiring serving all multiplex system nodes which provides ground potential and a return path for the current drawn by the node.

Initialization—Parameterization and eventual configuration of a system during start-up.

Internode Distance—Signal line length between any two nodes.

Length of Communication Medium—Maximum distance between any two nodes.

Line Driver—A solid state device (driver) used to transfer electrical energy to a wire or cable communications medium (signal bus), performing the transmit portion of the transceive function.

Line Receiver—A solid state device used to receive electrically transmitted signals from a wire or cable communications medium (signal bus), performing the receive portion of the transceive function.

Link—A relatively simple communications system which is capable of supporting communications between exactly two nodes, typically located at the physical ends of the system.

Listening-Mode—On-mode, receive only; i.e., transmission is not allowed.

Local Error—Error in a communication network, which is detected only in some of the nodes.

Master/Slave—A type of system whereby one node (a module) acts as a master or central unit and controls the actions of the other nodes, designated as slaves or remote units.

Message Administration—The portion of a communication protocol specifying how to handle and to buffer entire messages and respective control bits, e.g. what messages shall be transmitted when, or whether messages shall be received and how are they presented to the application.

Message Frame—A portion of a communication protocol within the message transfer, specifying the arrangement and meaning of bits or bit fields in the sequence of transfer across the transmission medium.

Message Latency—The time required by a system to access the medium so as to begin the delivery of information. Message latency is measured from the time that a node is ready to send specific information to the time of the start of the transmission of this information which will ultimately be successful. Thus, the total time required to successfully send a desired message will be the sum of the message latency and the message transmission time.

Message Rate—Number of completed message transmissions per unit of time.

Message Transfer—The portion of the protocol dealing with the organization, meaning, and timing associated with the bits of data. Message transfer deals with what bits must be sent and when they must be sent in accomplishing the transmission of a message.

Module—A subassembly with intelligence which typically accepts inputs from switches and/or sensors and provides outputs to actuators, lamps, and/or displays. In a multiplex system, a module is one type of node.

Monitoring—During transmission of a message, the actual physical signal on the transmission line is fed back into the transmitting node in order to be compared with the transmitted reference signal. Monitoring may be used for error detection, allowing to safely detect all global errors in a communication network.

Multi-Master—A system partitioned into several modules, where more than one module may temporarily control the communication network, sending information to or requesting information from other modules.

Multiple Receiver—Ability to address more than one node with a single message.

Multiplex—To interleave or simultaneously transmit two or more messages on a single channel.

Multiplex Bus—The wiring serving all multiplex system nodes and including the signal, power, and ground buses.

Multiplexing—The process of combining several messages for transmission over the same signal path. There are two widely used methods of multiplexing: time division and frequency division. (See the separate definitions of each.)

Network—A system capable of supporting communications by three or more nodes.

Network Access Scheme—Method used to award the communication network to one of the nodes for the transmission of a message.

Node—Any subassembly of a multiplex system which communicates on the signal bus. In addition to modules, nodes may include other devices which contain the intelligence necessary to support these communications. A node includes a transceiver.

Node-to-Node Communications—The transmission of information to a single receiver, as differentiated from broadcast communications.

Non-Return to Zero (NRZ)—A data bit format where the voltage or current value (typically voltage) determines the data bit value (typically one or zero).

Object Layer—The portion of a communication protocol specifying how to handle and to buffer entire messages and their respective control bits, and how to interface the actual application.

Off-Mode—System behavior when switched off.

On-Mode—System behavior in full operation.

Open System—A system, consisting of nodes interconnected by a common communications medium (signal bus) according to established standards, which will support temporary connections to manufacturing networks, diagnostics, and other local area networks.

Physical Addressing—Labeling of messages for the physical address location of their source and/or destination(s). This is independent of its geographic location, connector pin, and/or wire identification assignments.

Physical Layer—The properties of the communications medium (signal bus) which can be determined by electrical measurements, such as voltages, currents, impedances, rise times, etc.

Power Bus—The portion of the wiring serving all multiplex system nodes which provides the electrical power to the nodes, providing the electrical energy used by the node and its associated electrical loads.

Power Driver—A solid state device (driver) capable of turning on and off electrical loads requiring electrical power significantly in excess of semiconductor logic levels. A power driver can drive actuators, lamps, motors, etc. A power driver provides the output function typically provided by a switch in a conventional automotive wiring system, but does not provide its input function.

Priority—Attribute of a message controlling its ranking during arbitration. A high priority increases the probability that a message wins the arbitration process.

Protocol—A formal set of conventions or rules for the exchange of information between nodes, including the procedures for establishing and controlling transmissions on the multiplex signal bus (message administration) and the organization, meaning, and timing associated with the bits of data (message transfer).

Pulse Width Modulation—A data bit format where the width of a pulse of constant voltage or current determines the value (typically one or zero) of the data transmitted.

Receiver—A device that converts electrical or optical signals used for transmission back to information or data signals.

Recovery From Error Time—Time delay between detection of an error and restart of regular operation. This may include a re-initialization and reconfiguration.

Redundancy—Added features in a communication and/or data processing system which are not essential for the specified operation but which allow the detection of errors or continuation of operation in case of defects.

Reply Time of Acknowledge—The time between the end of the source-originated portion of the message and the start of the destination-originated portion of the same message.

Response—A message or portion of a message initiated by a receiving node as a result of a message transmitted by a different node. A response can be an acknowledgement or response data, and it can be appended to the original message (immediate response) or a unique message (separate response).

Response Data—A response to a message which provides the data (information) requested in the message.

Ring Topology—A bus topology with the ends of the bus (line or group of lines) tied together.

Self-Blocking—The assignment process of multiple tasks to a single execution instance is called self-blocking, if it may get stuck without a chance for recovery.

Signal Bus—The wire(s) in the portion of the wiring serving all multiplex system nodes which are dedicated to communications between the nodes.

Sleep-Mode—Node behavior on a low power consumption standby, waiting to be switched on by a message. This is distinct from an Off-Mode where there is no power consumption, disconnected from the power supply.

Star Topology—Nodes connected by links to a central unit which acts as a central processor or switching point.

Switch—A mechanically operated device for making, breaking, or changing the connections in an electrical circuit. In a conventional automotive wiring system, a switch serves as both the input device and the output device which provides electrical energy to the load. In a multiplex system, the input device is designated as a switch, and the output device is designated as a driver.

Synchronization—Procedure to ensure a desired timing for interrelated actions and/or processes.

System Architecture—The organization of a multiplex system including, but not necessarily limited to, the location and ranking of logic or decision making elements, and the types and methods of communications between these elements.

System Elasticity—The capability to easily add or delete nodes and functions, permitting the multiplex wiring system to be easily expanded or contracted as required. Reprogramming of units not added or deleted should be minimized in an elastic system. This expansion and contraction may be due to model to model variations, year to year changes, or the desire for new features and accessories.

Time Division Multiplex Protocol—A protocol where the meaning of a piece of information on the signal bus is determined by its relationship (first, second, third, etc.) to the start of the message or bit stream. In a time division multiplex protocol, data can be interleaved on a bit-by-bit, byte-by-byte, or block-by-block basis.

Token— The symbol of authority passed between nodes in a token passing protocol. Possession of this symbol identifies the node currently in control of the medium.

Token-Passing Protocol—A protocol where a node which has communicated passes the control of the bus, including the right to communicate to another node, at the end of the message via a token.

Topology—The configuration of the interconnected elements of a system.

Total Transmission Latency—The time required to transfer a message from the transmitting node, measured from the moment it is prepared to send the message, until it is correctly received by the targeted receiver. It may include a retry strategy delay if the initial exchange is not successful.

Transceiver—An electrical circuit which both transmits (line driver portion) and receives (line receiver portion).

Transfer Delay Time—Time delay between request for and completion of transmission, which is the addition of transmission latency and transmission times.

Transfer Layer—The portion of a communication protocol specifying sequential arrangement and meaning of bits or bit fields within a message. Eventually the transfer layer may also cover the arbitration scheme, error detection, and handling and strategies for fault confinement.

Transfer Rate—Information bits per unit of time during transmission, equivalent to bit rate.

Transmission Time—Time duration for the transmission of a message, depending on message length and transfer rate.

Transmitter—A device that converts information or data signals to electrical or optical signals so that these signals can be sent over a communications medium (signal bus).

Twisted Pair—A cable composed of two insulated conductors twisted about one another.

Variable Pulse Width (VPW) Modulation (VPWM)—A method of using both bus state and pulse width to encode bit information. This encoding technique is used to reduce the number of bus transitions for a given bit rate. One embodiment would define a One ("1") as a dominant short pulse or a passive long pulse while a Zero ("0") would be defined as a long dominant pulse or a short passive pulse. Since a message is comprised of random 1's and 0's, general byte or message times cannot be predicted in advance.

GLOSSARY OF RELIABILITY TERMINOLOGY ASSOCIATED WITH AUTOMOTIVE ELECTRONICS—SAE J1213/2 OCT88

SAE Information Report

Report of the Electrical/Electronic Systems Technical Committee approved October 1988.

1. Introduction—This glossary has been compiled to assist, by serving as a reference, in the communication between the automotive electronics engineer and the reliability engineer.

2. Scope—This compilation of terms, acronyms and symbols was drawn from usage which should be familiar to those working in automotive electronics reliability. Terms are included which are used to describe how items, materials and systems are evaluated for reliability, how they fail, how failures are modeled and how failures are prevented. Terms are also included from the disciplines of designing for reliability, testing and failure analysis as well as the general disciplines of Quality and Reliability Engineering. This glossary is intended to augment SAE J1213, Glossary of Automotive Electronic Terms.

3. References—A listing of applicable military and other organizational reference documents, from which many of these terms and definitions were drawn, is provided as a source of alternate or related definitions.

MILITARY

MIL-STD-105D	Sampling Procedures and Tables for Inspection by Attributes
MIL-STD-202E	Test Methods for Electronic and Electrical Component Parts
MIL-HDBK-217D	Reliability Prediction of Electronic Equipment
MIL-STD-280A	Definitions of Item Levels, Item Exchangeability, Models, and Related Terms
MIL-STD-414	Sampling Procedures and Tables for Inspection by Variables for Percent Defective
MIL-STD-756B	Reliability Models and Prediction
MIL-STD-781C	Reliability Design Qualification and Production Acceptance Tests - Experimental
MIL-STD-790B	Reliability Assurance Program for Electronic Parts Specification
MIL-STD-810C	Environmental Test Methods
MIL-STD-883B	Test Methods and Procedures for Microelectronics
MIL-STD-1313	Microelectronics Terms and Definitions
MIL-Q-9858A	Quality Program Requirements
AR-92	Quality Program Requirements
MIL-S-195001	Semiconductor Devices, General Specification for
MIC M38510D	Microcircuits, General Specifications for
MIL-STD-470	Maintainability Program Requirements for Systems and Equipments
MIL-STD-471A	Maintainability Verification/Demonstration/Evaluation
MIL-HDBK-472	Maintainability Prediction
MIL-STD-891	Contractor Parts Control and Standardization Program
MIL-STD-701	Preferred and Guidance List of Semiconductor Devices
MIL-STD-198A	Selection and Use of Capacitors
MIL-STD-199B	Selection and Use of Resistors
MIL-STD-1562	List of Standard Microcircuits
MIL-STD-976	Certification Requirements for JAN Microcircuits

EIA

Reliability Bulletin No. 1	A General Guide for Technical Reporting of Electronic Reliability Measurement
Reliability Bulletin No. 4A	Reliability Qualifications
Reliability Bulletin No. 5	Equipment Reliability Specification Guideline
Reliability Bulletin No. 10	Selection and Validation of Low Population and/or State of the Art Parts
Reliability Bulletin No. 9	Failure Mode and Effects Analyses
Reliability Bulletin No. 8	Equipment Burn-In
Engineering Bulletin No. 17	User Guidelines for Quality and Reliability Assurance of LSI Components
Engineering Bulletin No. 11	User Guidelines for Microelectronic Reliability Estimation
JEDEC Standard No. 22	Test Methods and Procedures for Solid State Devices Used in Transportation/Automotive Applications

OTHER

Glossary and Tables for Statistical Quality Control	–ASQC 1973
PROCUREMENT QUALITY CONTROL 2nd Edition	–ASQC
Quality Systems Terminology	– ANSI/ASQC A3-1978
Terms, Symbols and Definitions for Acceptance Sampling	– ANSI/ASQC A2-1978
Definitions, Symbols, Formulas and Tables for Control Charts	– ANSI/ASQC A1-1978
How to Speak Fluent Quality	– National Semiconductor Co.
Reliability Design Handbook RDH 376	– Reliability Analysis Center

Analysis Techniques for Mechanical Reliability WPS-1 1987 Desk Manual	–	Reliability Analysis Center – Microelectronic Manufacturing and Testing
Glossary of Automotive Electronic Terms	–	SAE HS J1213 1982
Automotive Electronics Reliability Handbook	–	SAE AE-9 1987
The American Heritage Dictionary	–	2nd College Edn. 1982 Houghton Mifflin Co.

4. Reliability Glossary

ACCELERATED LIFE TEST

A life test under test conditions that are more severe than usual operating conditions. It is necessary that a relationship between test severity and the probability distribution of life be ascertainable.

ACCELERATION FACTOR

(1) The factor by which the failure rate can be increased by an increased environmental stress.

(2) The ratio between the times necessary to obtain the same portion of failure in two equal samples under two different sets of stress conditions, involving the same failure modes and mechanisms.

ACCEPT/REJECT TEST

A test, the result of which will be the action to accept or reject something, for example, an hypothesis or a batch of incoming material.

ACCEPTABLE QUALITY LEVEL (AQL)

The maximum percent defective which can be considered satisfactory as a process average, or the percent defect whose probability of rejection is designated by α.

ACCEPTANCE NUMBER

The largest number of defects that can occur in an acceptance sampling plan and still have the lot accepted.

ACCEPTANCE SAMPLING PLAN

An accept/reject test whose purpose is to accept or reject a lot of items or material.

ACCESSIBILITY

A measure of the relative ease of admission to the various areas of an item.

ACHIEVED RELIABILITY

The reliability demonstrated at a given point in time under specified conditions of use and environment.

ACTIVATION ENERGY

(1) The energy level at which a specific microelectronic failure mechanism becomes active (in electron-volts).

(2) The slope of the time-temperature regression line in the Arrhenius equation (in electron-volts).

ACTIVE ELEMENT

A part that converts or controls energy, for example, transistor, diode, electron tube, relay.

ACTIVE ELEMENT GROUP

An active element and its associated supporting (passive) parts, for example, an amplifier circuit, a relay circuit, a pump and its plumbing and fittings.

AGING

The effect whereby the probability density function of strength is changed (strength is reduced) with time.

ALLOCATION

The process of assigning reliability requirements to individual units to attain the desired system reliability.

ALPHA PARTICLE INDUCED SOFT ERRORS

Integrated circuit memory transient errors due to emission of alpha particles during radioactive decay of uranium or thorium contamination in the IC packaging material.

AMBIENT

Used to denote surrounding, encompassing, or local conditions. Usually applied to environments, for example, ambient temperature, ambient pressure.

APPORTIONMENT

Synonym of Allocation.

ARITHMETIC MEAN

The arithmetic mean of n numbers is the sum of the n numbers, divided by n.

ARRHENIUS MODEL

A mathematical representation of the dependence of failure rate on absolute temperature and activation energy. The model assumes that degradation of a performance parameter is linear with time with the failure rate a function of temperature stress. The temperature dependence is taken to be the exponential function:

$$\theta_1 = \theta_2 \exp[(E/k)(1/T_2 - 1/T_1)]$$

θ_1 = mean time to failure at T_1
θ_2 = mean time to failure at T_2
T = junction temperature in K
E = activation energy in eV
k = Boltzmann's constant ($8.617 \cdot 10^5$ eV/K)

ARRHENIUS ACCELERATION FACTOR

The acceleration factor F is the factor by which the time to fail can be reduced by increased temperature.

$$F = \theta_1/\theta_2 = \exp(E/k)(1/T_2 - 1/T_1)$$

ASSESSMENT

(1) A critical appraisal, including qualitative judgments about an item, such as importance of analysis results, design criticality and failure effect.

(2) The use of test data and/or operational service data to form estimates of population parameters and to evaluate the precision of these estimates.

ATTRIBUTE

A term used to designate a method of measurement whereby units are examined by noting the presence (or absence) of some characteristic or attribute in each of the units in the group under consideration and by counting how many units do (or do not) possess it. Inspection by attributes can be of two kinds - either the unit of product is classified simply as defective or nondefective, or the number of defects in the unit of product is counted, with respect to a given requirement or set of requirements.

ATTRIBUTE TESTING

Testing to evaluate whether or not an item possesses a specified attribute.

AUTOMATIC TEST EQUIPMENT (ATE)

Test equipment that contains provisions for automatically performing a series of pre-programmed tests.

AVAILABILITY (OPERATIONAL READINESS)

The probability that at any point in time the system is either operating satisfactorily or ready to be placed in operation on demand when used under stated conditions.

AVERAGE

A general term. It often means arithmetic mean, but can refer to s-expected value, median, mode, or some other measure of the general location of the data values.

AVERAGE OUTGOING QUALITY (AOQ)

The average quality of outgoing product after 100% inspection of rejected lots, with replacement by good units of all defective units found in inspection.

AVERAGE OUTGOING QUALITY LIMIT (AOQL)

The maximum average outgoing quality (AOQ) for a sampling plan.

BAKE-OUT

To subject an unsealed item to an elevated temperature to drive out moisture and unwanted gases prior to other process or sealing.

BATHTUB CURVE

A plot of failure rate of an item (whether repairable or not) vs. time. The failure rate initially decreases, then stays reasonably constant, then begins to rise rather rapidly. It has the shape of a bathtub. Not all items have this behavior.

BIAS

(1) The difference between the s-expected value of an estimator and the value of the true parameter.

(2) Applied voltage.

BINOMIAL DISTRIBUTION

The probability of r, or fewer successes in n independent trials, given a probability of success p in a single trial, is given by the cumulative binomial distribution

$$\Pr(x \leq r) = F(r; p, n) = \sum_{x=0}^{r} \binom{n}{x} p^x (1-p)^{n-x}$$

BINOMIAL FUNCTION

The probability of exactly x successes in n independent trials, given a probability of success p in a single trial, is given by the binomial probability function:

$$f(x; p, n) = \binom{n}{x} p^x (1-p)^{n-x}, \quad x = 0, 1, 2, \ldots n, \quad 0 \leq p \leq 1$$

BOND
(1) An interconnection which performs a permanent electrical and/or mechanical function.
(2) To join with adhesives.

BOND LIFT OFF
The failure mode whereby the bonded lead separates the surface to which it was attached.

BOND STRENGTH
In wire bonding, the pull force at rupture of the bond interface.

BREADBOARD MODEL
A preliminary assembly of parts to test the feasibility of an item or principle without regard to eventual design or form. Usually refers to a small collection of electronic parts.

BURN-IN
The initial operation of an item to stabilize its characteristics, and to minimize infant mortality in the field.

CAPABILITY
(1) A measure of the ability of an item to achieve mission objectives given the conditions during the mission.
(2) The spread of performance of a process in a state of statistical control; the amount of variation from common causes identified after all special causes of variation have been eliminated.

C CHART
Control chart for number of nonconformities observed in some specified inspection. The units should be alike in size and in the apparent likelihood of the existence of the nonconformity, in order that the area of opportunity for nonconformity be constant from unit to unit.

CENTRAL LINE
The line on a control chart that represents the average or median value of the items being plotted. It is shown as a solid line.

CHECKOUT
Tests or observations on an item to determine its condition or status.

COEFFICIENT OF VARIATION
The standard deviation divided by the mean, multiplied by 100 and expressed as a percentage.

COMPLEXITY LEVEL
A measure of the number of active elements required to perform a specific system function.

COMPONENT
A self-contained combination of parts, subassemblies, or assemblies which perform a distinctive function in the overall operation of an equipment. Often used interchangeably with (electronic) part.

CONFIDENCE
A specialized statistical term referring to the reliance to be placed in an assertion about the value of a parameter of a probability distribution.

CONFIDENCE COEFFICIENT
(1) A measure of assurance that a statement based upon statistical data is correct.
(2) The probability that an unknown parameter lies within a stated interval or is greater or less than some stated value.

CONFIDENCE INTERVAL
The interval within which it is asserted that the parameter of a probability distribution lies.

CONFIDENCE LEVEL
Equals $1-\alpha$ where α = the risk (%).

CONFIDENCE LIMIT
A bound of a confidence interval.

CONSISTENCY
A statistical term relating to the behavior of an estimator as the sample size becomes very large. An estimator is consistent if it converges to the population value as the sample size becomes large.

CONSTANT FAILURE RATE
(1) A term characterizing the instantaneous failure rate in the middle, or "useful life" period of the Bathtub Curve model of item life.
(2) A term characterizing the hazard rate, h(t), of an item having an exponential reliability function.

CONTAMINATION
A general term used to describe an unwanted material that adversely affects the physical or electrical characteristics of an item.

CONTINUOUS SAMPLING PLAN
In acceptance sampling, a plan intended for application to a continuous flow of individual units of product that involves acceptance or rejection on a unit-by-unit inspection and sampling. Continuous sampling plans are usually characterized by requiring that each period of 100% inspection be continued until a specified number of consecutively inspected units are found clear of defects.

CORRECTIVE ACTION
A documented design, material or process change to correct the true cause of a failure. Part replacement with a like item does not constitute appropriate corrective action. Rather, the action should make it impossible for that failure to happen again.

CORRELATION
A form of statistical dependence between two variables. Unless stated otherwise, linear correlation is implied.

CORRELATION COEFFICIENT
A number between -1 and $+1$, which indicates the degree of linear relationship between two sets of numbers. Coefficients of -1 and $+1$ represent perfect linear agreement between two variables, while a coefficient of zero implies none.

CORRODE
To dissolve or wear away gradually, especially by chemical action.

CORROSION
The deterioration of a substance (usually a metal) because of a reaction with its environment, or with a corroding agent.

COSMETIC DEFECT
A variation from the conventional appearance of an item such as a slight change in its color, not necessarily detrimental to service performance.

COST-EFFECTIVENESS
A measure of the value received for the resources expended.

CRACK
Evidence of a full or partial break without separation of parts.

CRAZE
A network of fine cracks in the surface.

CREEP
(1) Elongation or fracture resulting from loads sustained over relatively long time at high temperatures.
(2) The dimensional change with time of a material under load.

CRITICALITY
A relative measure of the consequences of a failure.

CRITICALITY ANALYSIS
A procedure by which each potential failure mode is evaluated and ranked according to the combined influences of severity and probability of occurrence.

CUMULATIVE DISTRIBUTION FUNCTION (CDF)
The probability that the random variable takes on any value less than or equal to a value x, that is,
$$Cdf(x) = pr(X \leq x)$$
The unreliability function with regard to failures
$$F(t) = Pr(T \leq t)$$

CUT SET
In a Fault Tree, any basic event or combination of basic events whose occurrence will cause the top event to occur.

DEBUGGING
The period of "shakedown operation" of a finished equipment performed prior to placing it in use. During this period, defective parts and workmanship errors are corrected under test conditions that closely simulate field operation.

DECAP
To de-encapsulate. To remove the cover or plastic encapsulant on an item.

DECREASING FAILURE RATE
(1) A term characterizing the instantaneous failure rate in the first or "infant mortality" period of the Bathtub Curve model or product life.
(2) A term characterizing the hazard rate h(t) of an item having a Weibull reliability function with slope B < 1.

DEFECT
A deviation of an item from some ideal state. The ideal state is usually given in a formal specification.

DEFECT, CRITICAL
A defect that could result in hazardous or unsafe conditions for individuals using, maintaining or depending on the item.

DEFECTIVE
A unit of product which contains one or more defects.

DEGRADATION
A gradual deterioration in performance as a function of time.

DELTA LIMITS
The difference between initial and final readings usually associated with the difference between the zero time readings on a life test and

the final readings. Determine how much parameters shift during the test.

DEMONSTRATED

That which has been proven by the use of concrete evidence gathered under specified conditions.

DEPENDABILITY

A measure of the item operating condition at one or more points during the mission, including the effects of Reliability, Maintainability and Survivability, given the item condition(s) at the start of the mission. It may be stated as the probability that an item will (1) enter or occupy any one of its required operational modes during a specified mission and (2) perform the functions associated with those operational modes.

DERATING

The intentional reduction of stress/strength ratio in the application of an item, usually for the purpose of reducing the occurrence of stress-related failures.

DESIGN ADEQUACY

The probability that the system will satisfy effectiveness requirements, given that the system design satisfied the design specification.

DEVICE

Any subdivision of a system; synonym for ITEM.

DISCRIMINATION RATIO

A measure of the distance between two specific points of the operating characteristic curve which are used to define the acceptance sampling plan.

DISSOCIATION

The breakdown of a substance into two or more constituents.

DISTRIBUTION

Generally short for Cumulative Distribution Function.

DOWNTIME

The total time during which the system is not in condition to perform its intended function.

DURABILITY

The probability that an item will operate as specified under stated conditions without a wearout failure; a special case of reliability.

DUTY CYCLE

The ratio of the time "on" of a device or system divided by the total cycle time. For a device that normally runs intermittently rather than continuously, the amount of time a device operates as opposed to its idle time.

EARLY FAILURE PERIOD

That period of life, after assembly, in which failures occur at an initially high rate because of the presence of defective parts and workmanship defects.

EDX SPECTROMETER

Generally used with a scanning electron microscope (SEM) to provide elemental analysis of X-rays generated on the region being hit by the electron beam.

EFFECTIVENESS

The capability of the system or device to perform its function.

EFFICIENCY

A statistical term relating to the dispersion in values of an ESTIMATOR. It is between zero and one.

ELECTROMAGNETIC COMPATABILITY (EMC)

The capability of electronic equipment to function in the intended electromagnetic environment at designed levels of efficiency.

ELECTROMIGRATION

Dendritic or filamentary growth of a metal (for example, silver) in an electric field.

EMISSION SPECTROGRAPH

An instrument which identifies the presence of elements by burning the sample in an air plasma and analyzing the resultant optical spectrum.

ENGINEERING, HUMAN

The science of studying the man-machine relation in order to minimize the effects of human error and fatigue and thereby provide a more reliable operating system.

ENGINEERING, RELIABILITY

The science of including those factors in the basic design which will assure the required degree of reliability.

ENVIRONMENT

The aggregate of all external conditions and influences affecting the life and development of the product.

ELECTRICAL OVERSTRESS (EOS)

The electrical stressing of electronic components beyond specifications.

ELECTROSTATIC DISCHARGE (ESD)

The transfer of electric charge between bodies at different electrostatic potentials caused by direct contact or by an electrostatic field.

ESTIMATOR

A statistic, which is derived from a sample, used to infer a value of a parameter of an assumed distribution model.

EXPECTED VALUE

The mean or average, defined as: If x is a random variable and F(x) is its CDF

$$E(x) = \int x dF(x),$$ where the integration is over all x.

For continuous variables with a pdf, this reduces to

$$E(x) = \int x \, pdf(x) \, dx.$$

For discrete random variables with a pmf, this reduces to

$$E(x) = \Sigma x_n \, p(x_n)$$ where the sum is over all n.

EXPONENTIAL DISTRIBUTION

The probability density function

$$f(t) = \lambda \exp(-\lambda t)$$

where λ, the failure rate is constant.

EXPONENTIAL MODEL

In reliability engineering, a model based on the assumption that times t between successive failures are described by the exponential distribution.

EVALUATION

A broad term used to encompass prediction, measurement and demonstration.

EXTREME VALUE DISTRIBUTION

The asymptotic distribution of the smallest extreme from a statistical distribution; used to model capacitor breakdown voltage, time to failure of corrosion, etc. The pdf is $p(x) = \exp(x) \exp[-\exp(x)]$.

EYRING MODEL

An accelerated life test model in which failure rate is related to temperature. Given by $\lambda = T\exp(A-B/T)$.

FAILURE

The termination of the ability of an item to perform its required function.

FAILURE ANALYSIS

The identification of the failure mode, the failure mechanism and the cause. Often includes physical dissection.

FAILURE, CATASTROPHIC

A sudden change in the operating characteristics of an item resulting in a complete loss of useful performance.

FAILURE, DEGRADATION

A failure that occurs as a result of a gradual or partial change in the operating characteristics of an item.

FAILURE EFFECT

The consequences a failure mode has on the operation, function or status of an item.

FAILURE, INCIPIENT

A degradation failure which is just beginning to exist.

FAILURE, INDUCED

A failure caused by a physical condition external to the failed item.

FAILURE, INFANT

A failure that occurs during the early life of an item.

FAILURE, INHERENT

A failure basically caused by a physical condition or phenomenon internal to the failed item.

FAILURE, INITIAL

The first failure to occur in use.

FAILURE, LATENT

A malfunction that occurs as a result of a previous exposure to a condition that did not result in an immediately detectable failure.

FAILURE MECHANISM

The mechanical, chemical or other process that results in a failure.

FAILURE MODE

The effect or manner by which a failure is observed. Generally describes the way the failure occurs.

FAILURE MODES AND EFFECTS ANALYSIS (FMEA)

A systematic, organized procedure for evaluating potential failures in an operating system.

FAILURE MODES, EFFECTS AND CRITICALITY ANALYSIS (FMECA)

An analysis of possible modes of failure, their causes, effects, their criticalities and expected frequencies of occurrence.

FAILURE, NONRELEVANT

A failure not applicable to the computation or reliability.

FAILURE, PRIMARY

A failure whose occurrence is not caused by other failures.

FAILURE, RANDOM

A failure whose occurrence is not predictable in an absolute sense but is predictable in a probabilistic sense.

FAILURE RATE

(1) The conditional probability that an item will fail just after time t, given the item has not failed up to time t.
(2) The number of failures of an item per unit measure of life (cycles, time, miles, events, etc.) as applicable for the item.

FAILURE, RELEVANT

A failure attributable to a deficiency of design, manufacture or materials of the failed device, applicable to the computation of reliability.

FAILURE, SECONDARY

A failure caused directly or indirectly by the failure of another item.

FAILURE, WEAROUT

A failure whose time of occurrence is governed by rapidly increasing failure rate.

FATIGUE

Cracking or fracture from cyclic loads.

FAULT

An attribute which adversely affects the reliability of a device.

FAULT TREE ANALYSIS (FTA)

A method of reliability analysis in which a logical block diagram is used to indicate contributing lower level events.

FIT

A contraction of Failure unIT, having a value of failures per 10^9 component-hours.

FORCED DEFECT

A failure induced by stress testing.

FOREIGN MATERIAL

The presence of an object or material which comes from some source external to the part or system.

FREEDOM, DEGREE OF

The number of observations that are free to vary at random, regardless of the restrictions imposed by the mathematics describing the statistic.

FUNCTIONAL FAILURE

A failure whereby a device does not perform its intended function when the inputs or controls are correct.

GAMMA DISTRIBUTION

An important distribution in statistical queuing theory. Given by:

$$pdf(x) = \frac{\lambda(\lambda t)^{k-1} e^{-\lambda t}}{\Gamma(k)}, \text{ where } \Gamma(k) = \int_0^\infty u^{k-1} e^{-u} du$$

$$\lambda, \beta, t > 0$$

GAUSSIAN DISTRIBUTION (See NORMAL DISTRIBUTION)

A 2-parameter distribution with

$$pdf(x) = \frac{1}{\sigma\sqrt{2\pi}} \exp -\left[\frac{1}{2}\left(\frac{x-u}{\sigma}\right)^2\right]$$

GEOMETRIC MEAN

The geometric mean of n numbers is the nth root of their product.

GLASS TRANSITION TEMPERATURE

The temperature at which an amorphous polymer changes from a hard and relatively brittle condition to a viscous or rubbery condition.

GO, NO-GO

The result of a test of an attribute. It is either good or bad.

GOODNESS OF FIT

A statistical term that quantifies how likely a sample was to have come from a given probability distribution.

HAZARD RATE h(t)

(1) At a particular time, the rate of change of the number of items that have failed divided by the number of items surviving.
(2) Represents the probability that an item still functioning at time t will fail in the interval (t, t+Δt), where Δt is an infinitesimal time increment. Hazard rate is synonymous with conditional failure rate or instantaneous failure rate.

$$h(t) = \lim_{t \to 0} \frac{R(t) - R(t+\Delta t)}{t \, R(t)}$$

$$h(t) = f(t)/R(t)$$

HERMETICITY

The effectiveness of the seal of microelectronic and semiconductor devices with designed internal cavities.

HOMOGENEOUS

Of the same or similar nature. Uniform in structure or composition.

HUMAN FACTORS

A body of scientific facts about human characteristics. The term covers all biomedical and psychosocial considerations. It includes but is not limited to principles and applications in the areas of human engineering, personnel selection, training, life support, job performance aids and performance evaluation.

HYPOTHESIS, NULL

An hypothesis that there is no difference between some characteristics of the parent populations of several different samples, that is, that the samples come from similar populations. A conjecture about the true state of nature, that if true, will only rarely be rejected as the outcome of an experiment or measurement.

INCREASING FAILURE RATE

(1) A term characterizing the instantaneous failure rate in the third or "wearout" period of the Bathtub Curve model of product life.
(2) A term characterizing the hazard rate h(t) of an item having a Normal reliability function, for instance.

INFANT MORTALITY

Premature catastrophic failures occurring at a much greater rate than during the useful life period prior to the onset of substantial wearout.

INSPECTION

The examination and testing of supplies and services (including, when appropriate, raw materials, components and intermediate assemblies) to determine whether they conform to specified requirements.

INSPECTION BY ATTRIBUTES

Inspection whereby either the unit of product or characteristic thereof is classified simply as defective or nondefective, or the number of defects in the unit or product is counted with respect to a given requirement.

INSPECTION BY VARIABLES

Inspection wherein certain quality characteristics of a sample are evaluated with respect to a continuous numerical scale and expressed as precise points along this scale. Variable inspections record the degree of conformance of the unit with specified requirements for the quality characteristics involved.

INSPECTION LEVEL

An indication of the relative size of the sample to the size of the lot.

INSPECTION LOT

A collection of units of product bearing identification and treated as a unique entity from which a sample is to be taken and inspected to determine conformance with the acceptability criteria.

ITEM

An all-inclusive term, to include assemblies, subassemblies accessories, parts, equipment and services, applied to what is being discussed.

LIFE TEST

A test, usually of several items, made for the purpose of estimating some characteristic(s) of the probability distribution of life.

LOG NORMAL DISTRIBUTION

The model of a random variable whose logarithm follows the Normal function with parameters μ and σ. It is a life model for a process whose value results from the multiplication of many small errors. Its pdf-

$$f(x:\mu,\sigma) = 1/(\sigma x \sqrt{2\pi}) \exp[-1/2\sigma^2 (\ln x - \mu)^2]$$

$$x > 0$$
$$\sigma > 0$$
$$-\infty < \mu < \infty$$

LONGEVITY

Length of useful life of a product to its ultimate wearout requiring complete rehabilitation. This is a term generally applied in the definition of a safe, useful life for an equipment or system under the conditions of storage and use to which it will be exposed during its lifetime.

LOT

A group of units from a particular device type submitted each time for inspection and/or testing is called a lot.

LOT QUALITY

The true fraction defective in a lot.

LOT REJECT RATE (LRR)
The lot reject rate is the percentage of lots rejected from the lots evaluated.

LOT TOLERANCE PERCENT DEFECTIVE (LTPD)
The percent defective which is to be accepted a minimum or arbitrary fraction of the time, or that percent defective whose probability of rejection is designated by β.

MAINTAINABILITY
A characteristic of design and installation which is expressed as the probability that an item will be retained in or restored to a specified condition within a given period of time, when the maintenance is performed in accordance with prescribed procedures and resources.

MAINTENANCE, PREVENTIVE
The maintenance performed in an attempt to retain an item in a specified condition by providing systematic inspection, detection and prevention of incipient failure.

MARGIN TESTING
Testing in which item environments such as line voltage or temperature are changed to reversibly worsen the performance. Its purpose is to find how much margin is left in the item for its degradation.

MEAN
(1) The expected value of a random variable.
(2) The first moment of a probability distribution about its origin. As specifically defined and modified, for example, the arithmetic mean (sums), the geometric mean (products), the harmonic mean (reciprocals), logarithmic mean, etc.

MEAN LIFE (Θ)
The arithmetic average of lifetimes of all items considered.

$$\Theta = \int_0^T R(t)dt = \int_0^T t\,pdf(t)dt$$

where $R(t)$ = the s-reliability of the item
T = the interval over which the mean life is desired, usually the useful life

MEAN-LIFE-BETWEEN-FAILURES
This concept is the same as Mean Life except that it is for repaired items and is the mean up-time of the item. The formula is the same as for Mean Life except that $R(t)$ is interpreted as the distribution of up-times.

MEAN-TIME-BETWEEN-FAILURES (MTBF)
For a particular interval, the total functioning life of a population of an item divided by the total number of failures within the population during the measurement interval. The definition holds for time, cycles, miles, events or other measure of life units. A basic measure of reliability of repairable items.

MEAN-TIME-BETWEEN-MAINTENANCE (MTBM)
The mean of the distribution of the time intervals between maintenance actions (either preventive, corrective or both).

MEAN-TIME-TO-FAILURE (MTTF)
For nonrepaired items, the mean life.

MEAN-TIME-TO-REPAIR (MTTR)
The total corrective maintenance time divided by the total number of corrective maintenance actions during a given period of time.

$$MTTR = \int_0^T G(t)dt$$

where $G(t)$ = Cdf of repair time
T = Maximum allowed repair time

MEDIAN
The median of a distribution of one random variable X of the discrete or continuous type is a value of x such that $Pr(X \leq x) = 1/2$ and $Pr(X \geq x) = 1/2$, the middle value.

MISSION
The objective or task, together with the purpose, which clearly indicates the action to be taken.

MISSION RELIABILITY
The probability of success of an item to perform its required function for the duration of its intended mission.

MISSION PROFILE
The mission profile describes the events and conditions, including times and time spans, associated with a specific operational usage of an item. It is one segment of the operational cycle.

MODE
The mode of a distribution of one random variable X of discrete or continuous type is a value of x that maximizes the pdf f(x).

MODEL
A mathematical representation of a process. In Reliability there are two primary modeling concepts.
(1) A statistical function describing a life characteristic.
(2) A description of the reliability connectivity of the parts of a system.

MODEL, PARALLEL
A representation of the connection of the parts in a system such that the failure of all parts so connected is required for failure of that section of the system.

MODEL, SERIES
A representation of the connection of parts of a system such that failure of any part so connected will cause failure of that section of the system.

MODULE
An item which is packaged and is part of the next higher level of assembly.

NORMAL DISTRIBUTION (See GAUSSIAN)
The most prominant continuous distribution in statistics, frequently referred to as the Gaussian or bell-shaped distribution. Its density function is

$$f(x;\mu,\sigma) = \frac{1}{\sigma\sqrt{2\pi}} \exp\left[-\frac{(x-\mu)^2}{2\sigma^2}\right], \quad -\infty < x < \infty, \; -\infty < \mu < \infty, \; \sigma > 0$$

with mean μ and variance σ^2. The theoretical justification for the normal distribution lies in the central-limit theorem, which shows that under very broad conditions the distribution of the average of n independent observations from any distribution approaches a normal distribution as n becomes large.

NORMAL VARIABLE
A random variable that is normally distributed. In situations where the random variable represents the total effect of many "small" independent causes, each with mutually independent errors, the central limit theorem leads to the prospect the variable will be normally distributed.

OPERATING CHARACTERISTIC (OC CURVE)
A curve showing the relation between the probability of acceptance and either lot quality or process quality, whichever is applicable.

OPERATIONAL READINESS
The probability that, at any point in time, the system is either operating satisfactorily or ready to be placed in operation on demand when used under stated conditions, including stated allowable warning time. Thus, total calendar time is the basis for computation of operational readiness.

OVERCOAT
A thin film of insulating material over micro-circuit elements to provide mechanical protection or prevention of contamination.

OVERSTRESS
A condition wherein the severity levels of operation are more than usual or more than the specification.

PART
(1) An item that will not be disassembled for maintenance.
(2) The least subdivision of a system.
(3) An item which cannot ordinarily be disassembled without being destroyed.

PARTS PER MILLION (PPM)
Describing fractional defective, PPM is obtained by multiplying percent defective by 10 000; for example, 0.01% = 100 ppm.

PASSIVE ELEMENT
An element that is not active, that is, does not control energy; for example, a resistor, capacitor or an inductor.

PERCENTAGE DEFECTIVE
That proportion of a lot which is defective. This is the figure of merit in the population domain which characterizes quality control measurements and differentiates it from reliability.

PHYSIOCHEMICAL INSTABILITY
Change from an initial material bulk property, such as strength, resiliency, volume, composition, etc., as a result of age, pressure, temperature, etc.

PIN HOLE
A microscopic hole through an insulating (glass) layer. A defect.

POPULATION
The totality of the set of items, units, measurements, etc., real or conceptual, that is under consideration.

PROBABILITY
(1) Classical: If an event can occur in N equally likely and different ways, and if n of these ways have an attribute A, then the probability of the occurrence of A, denoted Pr(A) is defined as n/N.
(2) Frequency: If an experiment is conducted N times, and outcome A occurs n times, then the limit of n/N as N becomes large, is defined as the probability of A, denoted as Pr(A).
(3) Subjective: The probability Pr(A) is a measure of the degree of belief one holds in a specified proposition A.

PROBABILITY DENSITY FUNCTION
A continuous f(x) is a pdf if $f(x) \geq 0$, $-\infty < x < \infty$ and

$$\int_{-\infty}^{\infty} f(x)dx = 1$$

PROBABILITY MASS FUNCTION
A discrete f(x) is a pmf if f(x) = 0 for all x, except for a finite countable set of values of x for which f(x) > 0, and
$\sum_x f(x) = 1$, for all x such that f(x) > 0

PROBABILITY FUNCTION
The probability function is defined in terms of its pdf or pmf. If there is a pdf, F(x) is continuous and is defined by

$$F(x) = \int_{-\infty}^{x} f(x)dx$$

If there is a pmf, F(x) is discrete and is defined by

$$F(x) = \sum_x f(x)$$

PROBABILITY DISTRIBUTION
A mathematical function with specific properties which describes the probability that a random variable will take on a value or a set of values. If the random variable is continuous and well-behaved enough, there will be a pdf. If the random variable is discrete, there will be a pmf.

PROBABILITY PAPER
Graph paper constructed so that cumulative distribution curves plot as straight lines. Paper is available for normal, log-normal, Weibull, and several other distributions.

PROCESS AVERAGE (PA)
The total number of units rejected over an extended period of time divided by the total number of units produced over the same period of time.

PULL TEST
A test to determine the bond strength of a lead to an interconnecting surface, usually perpendicular to the surface, by pulling to failure.

PURPLE PLAGUE
One of several gold-aluminum compounds formed when bonding gold to aluminum and activated by exposure to moisture and high temperature, resulting in brittle, time-based bond failure.

QUALIFICATION
The entire process by which products are obtained from manufacturers or distributors, examined and tested, and then identified on a Qualified Products List.

QUALITY
(1) The composite of all characteristics or attributes, including performance, of an item.
(2) A measure of the degree to which an item conforms to applicable specification and workmanship standards.
(3) A property which refers to the tendency of an item to be made to specific specifications or the customer's express needs, or both. See current publications by Juran, Deming, Crosby, et al.

QUALITY ASSURANCE
A system of activities whose purpose it is to provide assurance that the overall quality control job is, in fact, being done effectively.

QUALITY CHARACTERISTICS
Those properties of an item or process which can be measured, reviewed or observed and which are identified in the drawings, specifications or constructual requirements. Reliability becomes a quality characteristic when so defined.

QUALITY CONTROL (QC)
The overall system of activities whose purpose is to provide a quality of product or service which meets the needs of users; also, the use of such a system.

RANDOMNESS
The occurrence of an event in accordance with the laws of chance.

RANDOM EVENT
The occurrence of an event affected by chance alone. For example, heads or tails on a flipped coin occurs at random.

RANDOM SAMPLE
As commonly used in acceptance sampling theory, the process of selecting sample units in such a manner that all units under consideration have the same probability of being selected.

RANDOM VARIABLE (r.v.)
A function defined on a sample space, or a transformation which associates a real number with each point in a sample space.

REDUNDANCY
The existence of more than one means for accomplishing a given function.

REDUNDANCY, ACTIVE
A type of redundancy where all items in the group are operating simultaneously.

REDUNDANCY, STANDBY
A type of redundancy where the alternative means of performing the function is inoperative until needed and is switched on upon failure of the primary means of performing the function.

REGRESSION ANALYSIS
A mathematical means to fit an assumed model to data containing errors by minimizing the sum of squared deviations from the fit.

RELIABILITY
The probability that a device will function without failure over a specified time period or amount of usage at stated conditions.

RELIABILITY GROWTH
The increase in reliability as a result of the effort, the resource commitment to improve design, purchasing, production, and inspection procedures.

RELIABILITY, INHERENT
The potential reliability of an item present in its design.

RELIABILITY, INTRINSIC
The probability that a device will perform its specified function, determined on the basis of a statistical analysis of the failure rates and other characteristics of the parts and components which comprise the device.

RELIABILITY, PREDICTED
The process of quantitatively assessing whether a proposed or actual equipment design will meet a specified reliability requirement.

RELIABILITY WITH REPAIR
The reliability that can be achieved when preventive maintenance is allowed.

REPAIRABILITY
The probability that a failed system will be restored to operable condition in a specified active repair time.

RISK
The probability of rendering a wrong decision based on inadequate data or analysis. The probability of an undesired outcome.

RISK, CONSUMER'S (β)
For a given sampling plan, the probability of acceptance for a designated numerical value of relatively poor submitted quality.

RISK, PRODUCER'S (α)
For a given sampling plan, the probability of rejection for a designated numerical value of relatively good submitted quality.

RISK PRIORITY NUMBER
In an FMEA, the product of the Occurrence Ranking, the Severity Ranking and the Detection Ranking.

SAFETY
The conservation of human life and its effectiveness, and the prevention of damage to items, consistent with mission requirements.

SAMPLE
A random selection of units from a lot, usually made for the purpose of evaluating the characteristics of the lot.

SCANNING ELECTRON MICROSCOPE (SEM)
An instrument which provides a visual image of the surface of an item. It scans an electron beam over the surface of a sample held in a vacuum and measures any of several resultant particle counts or energies. Provides depth of field and resolution significantly exceeding light

counts microscopy and may be used at magnifications exceeding 50 000 times.

SCREENING
The process of performing 100% inspection, or exposure to stress, on product lots and removing the defective units from the lots.

SCREENING TEST
A test or combination of tests intended to remove unsatisfactory items or those likely to exhibit early failures.

SERVICEABILITY
A measure of the degree to which servicing of an item will be accomplished within a given time under specified conditions.

SERVICING
The replenishment of consumables needed to keep an item in operating condition, but not including any other preventive maintenance or any corrective maintenance.

SEVERITY LEVEL
A general term implying the degree to which an environment will cause damage or shorten life, or both.

SHEAR TEST
Test of the shear strength of various attachments, for example, die attach, wire bond, wire weld, contact weld, etc. by application of force in the plane.

SHORT CIRCUIT
An abnormal connection of relatively low impedance, whether made intentionally or accidentally between two points of different electric potential.

SIGNIFICANCE
Results that show deviations between an hypothesis and the observations used as a test of the hypothesis greater than can be explained by random variation or chance alone, are called statistically significant.

SIGNIFICANCE LEVEL
The probability of rejecting the null hypothesis when it is actually true.

SNEAK CIRCUIT
An unexpected path or logic flow within a system which, under certain conditions, can initiate an undesired function or inhibit a desired function.

SOFT ERROR
Temporary memory content error due to intrusion of an alpha particle, for instance.

STANDARD DEVIATION
The square root of the variance.

STATISTIC
A value calculated from a sample which is used to estimate some characteristic of a population.

STATISTICAL CONTROL
Control of a process by statistical methods. A process is said to be in a state of statistical control if the variations among the sampling results from it can be attributed to a stable pattern of chance causes.

STATISTICAL MODEL
A probability distribution as a representation of time to failure.

STEP STRESS TEST
A test consisting of several stress levels applied sequentially for periods of equal duration to a sample. During each period, a stated stress level is applied, and the stress level is increased from one step to the next.

STORAGE LIFE (SHELF LIFE)
The length of time an item can be stored under specified conditions and still meet specified requirements.

STRESS
A general and ambiguous term used as an extension of its meaning in mechanics as that which could cause failure. It does not distinguish between those things which cause permanent damage (deterioration) and those things which do not.

STRESS RELIEF
A design means to minimize the effects of stress, for example, a cable clamp or a conformal coat.

STRESS, COMPONENT
The stresses on component parts during testing, assembly or use which affect the failure rate and hence the reliability of the parts. Voltage, power temperature and thermal environmental stress are included.

SUBASSEMBLY
A replaceable combination of parts which is an element of an assembly.

SUBSYSTEM
A major subdivision of a system which performs a specified function in the overall operation of a system.

SURVIVABILITY
The measure of the degree to which an item will withstand hostile man-made environment and not suffer abortive impairment of its ability to accomplish its designated mission.

SURVIVOR FUNCTION Sf(t)
The Reliability distribution function. The probability that an item will survive to time t.
$$Sf(t) = Pr(T>t) \quad t > 0, T \text{ a r.v.}$$

SUSPENDED ITEM
An item removed from test prior to failure.

SYSTEM
A combination of complete operation equipments, assemblies, components, parts or accessories interconnected to perform a specific operational function.

SYSTEM EFFECTIVENESS
A measure of the degree to which an item can be expected to achieve a set of specific mission requirements and which may be expressed as a function of availability, dependability and capability.

TEMPERATURE CYCLE
A stress test where the temperature of the medium (usually air) surrounding the test items is varied in a predetermined manner over the temperature range in such a way that the internal item temperature is kept at a fixed minimal increment from the medium temperature.

TEST TO FAILURE
The practice of inducing increased electrical and mechanical stresses in order to determine the maximum capability of a device so that conservative use in subsequent applications will, thereby, increase its life through the derating determined by these tests.

THERMAL ENDURANCE
The time at a selected temperature for a material or system of materials to deteriorate to some predetermined level of electrical, mechanical or chemical performance under prescribed conditions of test.

THERMAL FATIGUE
The failure of materials subjected to alternating heating and cooling.

THERMAL SHOCK
A stress test in which the temperature of the medium surrounding the test items is varied as rapidly as possible in order to create large, cyclic temperature gradients in the test items.

TIME, ACTIVE
That time during which an item is operational.

TIME, DOWN
That element of time during which the item is not in condition to perform its intended function.

TIME, MISSION
That element of uptime during which the item is performing its designated mission.

TIME, UP
That element of active time during which an item is either alert, reacting or performing a mission.

VARIABLE
In testing, the characteristic under examination which can have many values.

VARIANCE
The average of the squares of the deviations of individual values from their average. It is a measure of dispersion of a random variable. The second moment about the mean of a pdf given by

$$\text{var} = \int_{-\infty}^{\infty} (x-\mu)^2 f(x)dx$$

WARRANTY
A written guarantee of the performance of a product in which the maker for a specific period of time or other variable will be responsible for the repair or replacement of defective items.

WEAR
The mechanical removal of surface material by adhesion or abrasion.

WEAROUT
The process of attrition which results in an increase of hazard rate with increasing age (cycles, miles, events or time) as applicable for the item.

WEIBULL DISTRIBUTION
A general distribution, which is suitable for describing the life characteristic of a large group of problems. The general expression for the Weibull cumulative distribution function is defined, for $F(t=o)=0$

as

$$F(t) = 1 - e^{-\left(\frac{t}{\Theta}\right)^\beta}$$

β = Weibull Slope - The shape parameter of the distribution and equal to the slope of the line drawn through the failure data plotted on Weibull probability paper.

Θ = Characteristic Life - The scale parameter of the distribution and always equal to the life at 63.2% cumulative failure.

X CHART
Control chart for averages X of values in a subgroup.

X-RAY SPECTROMETER
Spectrographic analysis to characterize elements which are present.

YIELD
Elongation or fracture resulting from a single application of load in a relatively short period of time.

5. Acronyms, Abbreviations and Symbols

AGREE	Advisory Group on Reliability of Electronic Equipment
AOQ	Average Outgoing Quality
AOQL	Average Outgoing Quality Limit
AQL	Acceptance Quality Level
ATE	Automatic Test Equipment
BITE	Built-In Test Equipment
Cdf	Cumulative Distribution Function
CFR	Constant Failure Rate
COO	Cost of Ownership
C_p	Process Potential Index
C_{pk}	Process Performance for Two-Sided Spec Limits
DECAP	De-encapsulation
DFR	Declining Failure Rate
EDX	Energy Dispersive X-Ray
EDS	Energy Dispersive Spectrometer
EMC	Electromagnetic Compatability
EMI	Electromagnetic Interference
EOM	Ease of Maintenance
EOS	Electrical Overstress
ESD	Electrostatic Discharge
ESS	Environmental Stress Screening
FAR	Failure Analysis Request/Report
FIT	Failure Unit
FMA	Failure Modes Analysis
FMEA	Failure Modes and Effects Analysis
FMECA	Failure Modes and Effects Criticality Analysis
FRACAS	Failure Reporting, Analysis and Corrective Action System
FTA	Fault Tree Analysis
HAST	Highly Accelerated Stress Test
h(t)	Hazard Function
IFR	Increasing Failure Rate
JAN	Joint Army Navy
LCC	Life Cycle Cost
LSC	Logistic Support Cost
LTPD	Lot Tolerance Percent Defective
MRB	Material Review Board
MTBF	Mean Time Between Failure
MTBM	Mean Time Between Maintenance
MTTF	Mean Time to Failure
MTTR	Mean Time to Repair
NPF	No Problem Found
NTF	No Trouble Found
OC	Operating Characteristic (Curve)
ORLA	Optimum Repair Level Analysis
PA	Product Assurance / Process Average
PDA	Process Defect Average
pdf	Probability Density Function
pmf	Probability Mass Function
PPL	Preferred Parts List
PPM	Parts Per Million
PRST	Probability Ratio Sequential Test
QA	Quality Assurance
QC	Quality Control
QPL	Qualified Products List
R	Reliability
RAC	Reliability Analysis Center
RPM	Reliability Planning and Management
r.v.	Random Variable
s	Sample Standard Deviation
s-	Statistical (Prefix)
SCA	Sneak Circuit Analysis
SEM	Scanning Electron Microscope
Sf(t)	Survivor Function
SPC	Statistical Process Control
TNI	Trouble Not Identified
z(t)	Hazard Function
α	Producer's Risk
β	Weibull Slope, Consumer's Risk
λ	Failure Rate
μ	Mean
Θ	Mean Life
σ	Standard Deviation

E/E DIAGNOSTIC TEST MODES
—SAE J1979 DEC91

SAE Recommended Practice

Report of the SAE Vehicle E/E Systems Diagnostic Standards Committee approved December 1991. Rationale statement available.

Foreword—This SAE Recommended Practice describes the implementation of the Diagnostic Test Modes necessary to meet California On-Board Diagnostic (OBD II) requirements for emission related test data. This document is one of several prepared by task forces of the SAE E/E Diagnostics Committee in order to satisfy the proposed regulations. The development of these documents has been coordinated so that they are compatible with each other and with the legislation. Other documents necessary in addition to this document are:

SAE J1930—E/E Systems Diagnostic Terms, Definitions, Abbreviations, and Acronyms
SAE J1962—Diagnostic Connector
SAE J1978—OBD II Scan Tool
SAE J2012—Recommended Format and Messages for Diagnostic Trouble Codes

In addition, the diagnostic data communication link to be utilized with these recommended practices is specified by the regulation to be as specified in one of the following documents:

SAE J1850—Class B Data Communication Network Interface
ISO 9141—Road vehicles—Diagnostic systems—CARB requirements for interchange of digital information

TABLE OF CONTENTS

1. Scope
2. References
2.1 Applicable Documents
2.1.1 SAE Publications
2.1.2 ISO Publications
2.1.3 California ARB Publications
2.2 Definitions
2.2.1 Absolute Throttle Position Sensor
2.2.2 Bank
2.2.3 Base Fuel Schedule
2.2.4 Calculated Load Value
2.2.5 Continuous Monitoring
2.2.6 Fuel Trim
3. Technical Requirements
3.1 Diagnostic Test Mode General Conditions
3.1.1 Multiple Responses to a Single Data Request
3.1.2 Response Time
3.1.3 Minimum Time Between Requests from Scan Tool
3.1.4 Data Not Available

3.1.5 Maximum Values
3.2 Diagnostic Message Format
3.2.1 Addressing Method
3.2.2 Maximum Message Length
3.2.3 Diagnostic Message Format
3.2.4 Header Bytes
3.2.5 Data Bytes
3.2.6 Non-Data Bytes Included in Diagnostic Messages With J1850
3.2.7 Non-Data Bytes Included in Diagnostic Messages With ISO 9141 CARB
3.2.8 Bit Position Convention
3.3 Allowance for Expansion and Enhanced Diagnostic Test Modes
4. Test Modes
4.1 Mode $01—Request Current Powertrain Diagnostic Data
4.1.1 Functional Description
4.1.2 Message Data Bytes
4.2 Mode $02—Request Powertrain Freeze Frame Data
4.2.1 Functional Description
4.2.2 Message Data Bytes
4.3 PIDs for Modes $01 and $02
4.4 Mode $03—Request Emission-Related Powertrain Diagnostic Trouble Codes
4.4.1 Functional Description
4.4.2 Message Data Bytes
4.4.3 Powertrain Diagnostic Trouble Code Example
4.5 Mode $04—Clear/Reset Emission-Related Diagnostic Information
4.5.1 Functional Description
4.5.2 Message Data Bytes
4.6 Mode $05—Request Oxygen Sensor Monitoring Test Results
4.6.1 Functional Description
4.6.2 Message Data Bytes

1. Scope—This SAE Recommended Practice defines diagnostic test modes, and request and response messages, necessary to be supported by vehicle manufacturers and test tools to meet the requirements of the California OBD II regulations, which pertain to vehicle emission-related data only. These messages are intended to be used by any service tool capable of performing California OBD II mandated diagnostics.

Diagnostic Test Modes included in this document are:
  a. Mode $01—Request Current Powertrain Diagnostic Data
     Analog inputs and outputs
     Digital inputs and outputs
     System status information
     Calculated values
  b. Mode $02—Request Powertrain Freeze Frame Data
     Analog inputs and outputs
     Digital inputs and outputs
     System status information
     Calculated values
  c. Mode $03—Request Emission-Related Powertrain Diagnostic Trouble Codes
  d. Mode $04—Clear/Reset Emission-Related Diagnostic Information
  e. Mode $05—Request Oxygen Sensor Monitoring Test Results

For each test mode, this document includes:
  a. Functional descriptions of test mode
  b. Request and response message formats

2. References

2.1 Applicable Documents—The following publications form a part of this specification to the extent specified herein. The latest issue of SAE publications shall apply.

2.1.1 SAE PUBLICATIONS—Available from SAE, 400 Commonwealth Drive, Warrendale, PA 15096-0001.
  SAE J1850—Class B Data Communication Network Interface
  SAE J1930—E/E Systems Diagnostic Terms, Definitions, Abbreviations, and Acronyms
  SAE J1962—Diagnostic Connector
  SAE J1978—OBD II Scan Tool
  SAE J2012—Recommended Format and Messages for Diagnostic Trouble Codes
  SAE J2186—Diagnostic Data Link Security
  SAE J2190—Enhanced E/E Diagnostic Test Modes

2.1.2 ISO PUBLICATIONS—Available from ANSI, 11 West 42nd Street, New York, NY 10036-8002.
  ISO 9141 CARB—Road vehicles—Diagnostic systems—CARB requirements for interchange of digital information

2.1.3 CALIFORNIA ARB PUBLICATIONS
  Mail out #91-27 Title 13, California Code of Regulations, Section 1968.1 Malfunction and Diagnostic System Requirements—1994 and Subsequent Model-Year Passenger Cars, Light-Duty Trucks, and Medium-Duty Vehicles With Feedback Fuel Control Systems

2.2 Definitions—Most terms for components and systems contained in this document are included in SAE J1930. This section includes additional definitions of terms not included in SAE J1930.

2.2.1 ABSOLUTE THROTTLE POSITION SENSOR—This value is intended to represent the throttle opening. For systems where the output is proportional to the input voltage, this value is the percent of maximum input signal. For systems where the output is inversely proportional to the input voltage, this value is 100% minus the percent of maximum input signal. Throttle position at idle will usually indicate greater than 0%, and throttle position at wide open throttle will usually indicate less than 100%.

2.2.2 BANK—The group of cylinders which feed an oxygen sensor. Bank 1 contains the Number 1 cylinder.

2.2.3 BASE FUEL SCHEDULE—The fuel calibration schedule programmed into the Powertrain Control Module or PROM when manufactured or when updated by some off-board source, prior to any learned on-board correction.

2.2.4 CALCULATED LOAD VALUE—An indication of the current airflow divided by peak airflow, where peak airflow is corrected for altitude, if available. Mass airflow and barometric pressure sensors are not required for this calculation. This definition provides a unitless number that is not engine specific, and provides the service technician with an indication of the percent engine capacity that is being used (with wide open throttle as 100%).

$$CLV = \frac{\text{Current airflow}}{\text{Peak airflow (@ sea level)}} \times \frac{\text{Atmospheric pressure (@ sea level)}}{\text{Barometric pressure}} \times 100\%$$

(Eq. 1)

2.2.5 CONTINUOUS MONITORING—Sampling at a rate no less than two samples per second.

2.2.6 FUEL TRIM—Feedback adjustments to the base fuel schedule. Short-term fuel trim refers to dynamic or instantaneous adjustments. Long-term fuel trim refers to much more gradual adjustments to the fuel calibration schedule than short-term trim adjustments. These long-term adjustments compensate for vehicle differences and gradual changes that occur over time.

3. Technical Requirements

3.1 Diagnostic Test Mode General Conditions—These guidelines are necessary to ensure proper operation of both the test equipment and the vehicle during diagnostic procedures. Test equipment, when using messages defined in this document, should not affect normal operation of the emission control system.

3.1.1 MULTIPLE RESPONSES TO A SINGLE DATA REQUEST—The messages contained in this document are functional messages, which means the off-board test equipment will request data without knowledge of which module on the vehicle will respond. In some vehicles, multiple modules may respond with the information requested. In addition, a single module may send multiple responses to a single request. Any test device requesting information must, therefore, have provisions for receiving multiple responses.

3.1.2 RESPONSE TIME—For SAE J1850 network interfaces, the on-board systems should respond to a request within 100 ms of a request or a previous response. With multiple responses possible from a single request, this allows as much time as is necessary for all modules to access the data link and transmit their response(s). If there is no response within this time period, the tool can either assume no response will be received, or if a response has already been received, that no more responses will be received.

For ISO 9141 CARB interfaces, response time requirements are specified in the ISO 9141 CARB document.

3.1.3 MINIMUM TIME BETWEEN REQUESTS FROM SCAN TOOL—For SAE J1850 network interfaces, a tool should always wait for a response from the previous request, or "no response" timeout before sending another request. In no case should a request be sent less than 100 ms after the previous request.

For ISO 9141 CARB interfaces, required times between requests are specified in the ISO 9141 CARB document.

3.1.4 DATA NOT AVAILABLE—There will be no reject message for a request for data if the data value is not supported by the module.

3.1.5 MAXIMUM VALUES—If the data value exceeds the maximum value possible to be sent, the on-board system should send the maximum value possible ($FF or $FFFF). The tool should display the maximum value or an indication of data too high. This is not normally critical for real time diagnostics, but in the case of a misfire at 260 km/h with resulting freeze frame data stored, this will be very valuable diagnostic information.

3.2 Diagnostic Message Format

3.2.1 ADDRESSING METHOD—Functional addressing will be used for all generic Diagnostic Test Mode messages because the test tool does not know which system on the vehicle has the information that is needed.

3.2.2 MAXIMUM MESSAGE LENGTH—SAE J1850 defines required message elements and maximum message lengths that effectively limit the number of bytes that can be defined by this document to 12 bytes.

3.2.3 DIAGNOSTIC MESSAGE FORMAT—To conform to the SAE J1850 limitation on message length, diagnostic messages specified in this document begin with a three byte header, have a maximum of 7 data bytes, require ERR (error detection byte), and allow RSP (in-frame response byte), as shown in Figure 1:

3.2.4 HEADER BYTES—The first three bytes of all diagnostic messages are the header bytes. The value of the first header byte is dependent on the bit rate of the data link and the type of message, as shown in 3.2.3. The second byte has a value that depends on the type of message, either a request or a response. The third header byte is the physical address of the device sending the message.

Device address $F1 should be used for an OBD II Scan Tool, or any other tool that does not have a special reason to use another address. Other service tools should use addresses in the range from $F0 to $FD. The response to all request messages in this document will be independent of the address of the test equipment requesting the information.

Vehicle manufacturers should not use the J1979 header bytes for any purpose other than diagnostic messages. When they are used, they must conform to this specification.

3.2.5 DATA BYTES—The maximum number of data bytes available to be specified in this document is 7. The first data byte following the header is the test mode, and the remaining 6 bytes vary depending on the specific test mode. Each unique diagnostic message defined in this document is a fixed length, although not all messages are the same length. For modes $01 and $02, message length is determined by Parameter Identification (PID). For Mode $05, message length is determined by Test ID. For other modes, the message length is determined by the mode. This enables the tools to check for proper message length, and to recognize the end of the message without waiting for possible additional data bytes.

3.2.6 NON-DATA BYTES INCLUDED IN DIAGNOSTIC MESSAGES WITH SAE J1850

Error Byte (ERR) is defined as optional in SAE J1850, but is required in all request and response messages defined in this document. All diagnostic messages will use a Cyclic Redundancy Check (CRC) as the error detection byte.

In-frame response (RSP) is defined as optional in SAE J1850. For messages defined in this document, the RSP byte is required in all request and response messages at 41.6 Kbps, and is not allowed for messages at 10.4 Kbps.

SAE J1850 defines additional message elements that may be included in Diagnostic Messages. Use of these message elements is beyond the scope of this document, but need to be considered when defining total diagnostic messages.

3.2.7 NON-DATA BYTES INCLUDED IN DIAGNOSTIC MESSAGES WITH ISO 9141 CARB—Messages will include a checksum, defined in ISO 9141 CARB, after the data bytes instead of the CRC used with SAE J1850.

There is no provision for an in-frame response in ISO 9141 CARB.

3.2.8 BIT POSITION CONVENTION—Some data byte values in this document include descriptions that are based on bit positions within the byte. The convention used in this document is that the Most Significant Bit (MSB) is referred to as "bit 7," and the Least Significant Bit (LSB) is referred to as "bit 0," as shown in Figure 2:

3.3 Allowance for Expansion and Enhanced Diagnostic Test Modes—This document allows for the addition of Diagnostic Test Modes both as industry standards and manufacturer specific modes. Enhanced Diagnostic Test Modes will be defined in a separate SAE document, J2190. That document will reserve functional test modes $00 through $0F to be defined in J1979 if needed to accommodate future legislated requirements.

4. Test Modes

4.1 Mode $01—Request Current Powertrain Diagnostic Data

4.1.1 FUNCTIONAL DESCRIPTION—The purpose of this mode is to allow access to current emission related data values, including analog inputs and outputs, digital inputs and outputs, and system status information. The request for information includes a Parameter Identification (PID) value that indicates to the on-board system the specific information requested. PID definitions, scaling information, and display formats are included in this document.

The on-board module will respond to this message by transmitting the requested data value last determined by the system. All data values returned for sensor readings will be actual readings, not default or substitute values used by the system because of a fault with that sensor.

Not all PIDs are applicable or supported by all systems. PID $00 is a bit-encoded PID that indicates, for each module, which PIDs that module supports. PID $00 must be supported by all modules that respond to a Mode $01 request as defined in this document, because diagnostic tools that conform to SAE J1978 use the presence of a response by the vehicle to this request to determine which protocol is supported for OBD II communications.

4.1.2 MESSAGE DATA BYTES—(See Figure 3.)

4.2 Mode $02—Request Powertrain Freeze Frame Data

4.2.1 FUNCTIONAL DESCRIPTION—The purpose of this mode is to allow access to emission related data values which were stored during the freeze frame required by OBD II. This mode allows expansion to meet manufacturer specific requirements not necessarily related to the required freeze frame, and not necessarily containing the same data values as the required freeze frame. The request for information includes a Parameter Identification (PID) value that indicates to the on-board system the specific information requested. PID definitions, scaling information,

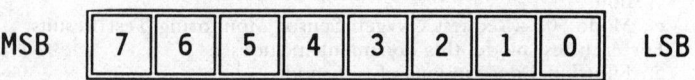

FIGURE 2—BIT POSITION WITHIN A DATA BYTE

Header Bytes (Hex)			Data Bytes								
Priority /Type	Target Address	Source Address	#1	#2	#3	#4	#5	#6	#7	ERR	RSP
Diagnostic Request at 10.4 Kbps (J1850 and ISO 9141 CARB)											
68	6A	Fx	Maximum 7 data bytes							Yes	No
Diagnostic Response at 10.4 Kbps (J1850 and ISO 9141 CARB)											
48	6B	addr	Maximum 7 data bytes							Yes	No
Diagnostic Request at 41.6 Kbps (J1850)											
61	6A	Fx	Maximum 7 data bytes							Yes	Yes
Diagnostic Response at 41.6 Kbps (J1850)											
41	6B	addr	Maximum 7 data bytes							Yes	Yes

FIGURE 1—DIAGNOSTIC MESSAGE FORMAT

	Data Bytes (Hex)						
	#1	#2	#3	#4	#5	#6	#7
Request Current Powertrain Diagnostic Data							
Request Powertrain Diagnostic Data	01	PID					
Report Current Powertrain Diagnostic Data							
Report Powertrain Diagnostic Data	41	PID	data A	data B (opt)	data C (opt)	data D (opt)	

FIGURE 3—MESSAGE DATA BYTES

and display formats for the required freeze frame are included in this document.

The on-board module will respond to this message by transmitting the requested data value stored by the system at the time of the first detected system fault. All data values returned for sensor readings will be actual readings, not default or substitute values used by the system because of a fault with that sensor.

Not all PIDs are applicable or supported by all systems. PID $00 is a bit-encoded PID that indicates, for each module, which PIDs that module supports. Therefore, PID $00 must be supported by all modules that respond to a Mode $02 request as defined in this document.

The frame number byte will indicate $00 for the OBD II mandated freeze frame data. Manufacturers may optionally save additional freeze frames and use this mode to obtain that data by specifying the freeze frame number in the request. If a manufacturer uses these additional freeze frames, they will be stored under conditions defined by the manufacturer, and contain data specified by the manufacturer.

4.2.2 MESSAGE DATA BYTES—(See Figure 4.)

	Data Bytes (Hex)						
	#1	#2	#3	#4	#5	#6	#7
Request Powertrain Freeze Frame Data							
Request Powertrain Freeze Frame Data	02	PID	frame no.				
Report Powertrain Freeze Frame Data							
Report Powertrain Freeze Frame Data	42	PID	frame no.	data A	data B (opt)	data C (opt)	data D (opt)

FIGURE 4—MESSAGE DATA BYTES

4.3 PIDs for Modes $01 and $02—(See Figure 5.)

4.4 Mode $03—Request Emission-Related Powertrain Diagnostic Trouble Codes

4.4.1 FUNCTIONAL DESCRIPTION—The purpose of this mode is to enable the off-board test device to obtain stored emission-related powertrain trouble codes. This should be a two-step process for the test equipment.

a. Step 1—Send a Mode $01, PID $01 request to get the number of stored emission-related powertrain trouble codes from all modules that have this available. Each on-board module that has stored codes will respond with a message that includes the number of stored codes which that module can report. If a module capable of storing powertrain codes does not have stored codes, then that module shall respond with a message indicating zero codes are stored.

b. Step 2—Send a Mode $03 request for all stored emission-related powertrain codes. Each module that has codes stored will respond with one or more messages, each containing up to 3 codes. If no codes are stored in the module, then the module will not respond to this request.

If additional trouble codes are set between the time that the number of codes are reported by a module, and the stored codes are reported by a module, then the number of codes reported could exceed the number expected by the tool. In this case, the tool should repeat this cycle until the number of codes reported equals the number expected based on the Mode 1 response.

Diagnostic trouble codes are transmitted in two bytes of information for each code. The first two bits (high order) of the first byte for each code will be zeroes to indicate a powertrain code (refer to SAE J2012 for additional interpretation of this structure). The second two bits will indicate the first digit of the diagnostic code (0 through 3). The second nibble of the first byte and the entire second byte are the next three digits of the actual code reported as Binary Coded Decimal (BCD). A powertrain trouble code transmitted as $0143 should be displayed as P0143. (See Figure 6.)

If less than three trouble codes are reported, the response messages

Modes *		PID (Hex)	Description	Min ($00) or ($0000)	Max ($FF) or ($FFFF)	SI (Metric) Scaling/bit and display	English scaling/bit and display
$01	$02						
X	X	00	PIDs supported ($01 - $20): Module responds with a message that contains 4 bytes of bit-encoded information, each bit indicating support or non-support of a PID where: 0 = PID not supported by this module 1 = PID supported by this module			Byte bit PID Data A 7 $01 Data A 6 $02 . . . Data B 7 $09 . . . Data D 0 $20	

X		01	Data A - Number of emission-related powertrain trouble codes and MIL status:
			bits 0-6:
			  Number of codes stored in this module
			bit 7:
			  0 = MIL not commanded ON by this module
			  1 = MIL commanded ON by this module
			
			Data B and Data C - Each bit indicates support or non-support of an on-board diagnostic evaluation:
			  Data B covers continuous monitoring tests
			  Data C covers tests run at least once per trip
			where:
			  0 = test not supported by this module
			  1 = test supported by this module
			
			Data D - Each bit indicates status of on-board diagnostic evaluation for this module, corresponding to tests included in Data C:
			  0 = test complete, or not applicable
			  1 = test not complete
						Data B:
						bit Evaluation supported
						 0 Misfire monitoring
						 1 Fuel system monitoring
						 2 Comprehensive component monitoring
						 3 unused
						 4 unused
						 5 unused
						 6 unused
						 7 unused
						
						Data C and Data D:
						bit Evaluation supported / status
						 0 Catalyst
						 1 Heated catalyst
						 2 Evaporative purge system
						 3 Secondary air system
						 4 A/C system refrigerant
						 5 Oxygen sensor
						 6 Oxygen sensor heater
						 7 EGR system	

FIGURE 5—PIDS FOR MODES $01 AND $02

Modes * $01	$02	PID (Hex)	Description	Min ($00) or ($0000)		Max ($FF) or ($FFFF)		SI (Metric) Scaling/bit and display	English scaling/bit and display
	X	02	Trouble code that caused CARB required freeze frame data storage (2 byte value - $0000 indicates no freeze frame data)	00	00	09	99	Pxxxx	

X	X	03	Data A: Fuel system status for bank 1
			Data B: Fuel system status for bank 2 ($00 if not used)
			
			For each data byte, no more than one bit at a time can be set to a 1 to indicate the status of that bank, where:
			  bit 0 = Open loop - has not yet satisfied conditions to go closed loop
			  bit 1 = Closed loop - using oxygen sensor(s) as feedback for fuel control
			  bit 2 = Open loop due to driving conditions (power enrichment, deceleration enleanment)
			  bit 3 = Open loop due to detected system fault
			  bit 4 = Closed loop, but fault with at least one oxygen sensor - may be using single oxygen sensor for fuel control
			  bits 5-7 = reserved						

X	X	04	Calculated load value	0%		100%		100/255% xxx.x%	
X	X	05	Engine coolant temperature	-40°C		215°C		1°C with -40°C offset xxx°C	xxx°F
X	X	06	Short term fuel trim - Bank 1 (use if only 1 fuel trim value)	-100.00% (lean)		99.22% (rich)		100/128% (0% at 128) xxx.x%	
X	X	07	Long term fuel trim - Bank 1	"		"		"	
X	X	08	Short term fuel trim - Bank 2	"		"		"	
X	X	09	Long term fuel trim - Bank 2	"		"		"	

FIGURE 5 – PIDS FOR MODES $01 AND $02 (CONTINUED)

used to report diagnostic trouble codes should be padded with $00 to fill 7 data bytes. This maintains the required fixed message length for all messages.

4.4.2 MESSAGE DATA BYTES—(See Figure 7.)

4.4.3 POWERTRAIN DIAGNOSTIC TROUBLE CODE EXAMPLE (ASSUME 10.4 Kbps)—(See Figure 8.)

4.5 Mode $04—Clear/Reset Emission-Related Diagnostic Information

4.5.1 FUNCTIONAL DESCRIPTION—The purpose of this mode is to provide a means for the external test device to command on-board modules to clear all emission-related diagnostic information. This includes:

    Clear number of diagnostic trouble codes (Mode $01, PID $01)
    Clear diagnostic trouble codes (Mode $03)
    Clear trouble code for freeze frame data (Mode $01, PID $02)
    Clear freeze frame data (Mode $02)
    Clear oxygen sensor test data (Mode $05)
    Reset status of system monitoring tests (Mode $01, PID $01)

Other manufacturer specific "clearing/resetting" actions may also occur in response to this request.

4.5.2 MESSAGE DATA BYTES—(See Figure 9.)

4.6 Mode $05—Request Oxygen Sensor Monitoring Test Results

4.6.1 FUNCTIONAL DESCRIPTION—The purpose of this mode is to allow access to the on-board oxygen sensor monitoring test results as required in OBD II regulations. Use of this mode is optional, depending on the method used by the vehicle manufacturer to comply with the requirement for oxygen sensor monitoring.

The request for test results includes a Test ID value that indicates the information requested. Test value definitions, scaling information, and display formats are included in this document.

Many methods may be used by different manufacturers to comply with this requirement. If data values are to be reported using these messages that are different from those predefined in this document, ranges of test values have been assigned that can be used that have standard units of measure. The tool can convert these values and display them in the proper units.

The on-board module will respond to this message by transmitting the requested test data last determined by the system.

The operation of this diagnostic mode in the on-board module is different from Mode $01. Mode $01 reports data value(s) that are stored internally at a single, or multiple contiguous, locations in memory. Mode $05 can report data values that are stored in non-contiguous memory locations. Test results will be stored in RAM, and test limits, if the value is a calculated value, would normally be stored in ROM. Therefore, the on-board software has additional requirements to respond to this request than it does for Mode $01 requests.

Not all test values are applicable or supported by all systems.

4.6.2 MESSAGE DATA BYTES—(See Figures 10, 11, and 12.)

Modes *		PID (Hex)	Description	Min ($00) or ($0000)	Max ($FF) or ($FFFF)	SI (Metric) Scaling/bit and display	English scaling/bit and display
$01	$02						
X	X	0A	Fuel pressure (gage)	0 kPaG	765 kPaG	3 kPaG xxx kPaG	xx.x psig
X	X	0B	Intake manifold absolute pressure	0 kPaA	255 kPaA	1 kPaA xxx kPaA	xx.x in. Hg
X	X	0C	Engine RPM (2 byte value - high byte/low byte)	0 r/min	16,383.75 r/min	1/4 r/min xxxxx r/min	
X	X	0D	Vehicle speed	0 km/h	255 km/h	1 km/h xxx km/h	xxx MPH
X		0E	Ignition timing advance for #1 cylinder (not including mechanical advance)	-64°	+63.5°	1/2° with 0° at 128 xx.x°	
X		0F	Intake air temperature	-40°C	215°C	1°C with -40°C offset xxx°C	xxx°F
X		10	Air flow rate from MAF sensor (2 byte value - high byte/low byte)	0 gm/sec	655.35 gm/sec	.01 gm/sec xxx.xx gm/sec	xxxx.x lb/min
X		11	Absolute throttle position sensor	0%	100%	100/255% xxx.x%	

X		12	Commanded secondary air status (if supported, one, and only one bit at a time can be set to a 1)
			    bit 0 1 = upstream of first catalytic converter
			    bit 1 1 = downstream of first catalytic converter inlet
			    bit 2 1 = atmosphere / off
			    bits 3 - 7 = reserved				

FIGURE 5 – PIDS FOR MODES $01 AND $02 (CONTINUED)

Modes *		PID (Hex)	Description	Min ($00) or ($0000)	Max ($FF) or ($FFFF)	SI (Metric) Scaling/bit and display	English scaling/bit and display
$01	$02						
X		13	Location of oxygen sensors, where sensor 1 is closest to the engine. Each bit indicates the presence or absence of an oxygen sensor at the following location: bit Sensor location Alternative sensor location 0 Bank 1 - Sensor 1 Bank 1 - Sensor 1 1 Bank 1 - Sensor 2 Bank 1 - Sensor 2 2 Bank 1 - Sensor 3 Bank 2 - Sensor 1 3 Bank 1 - Sensor 4 Bank 2 - Sensor 2 4 Bank 2 - Sensor 1 Bank 3 - Sensor 1 5 Bank 2 - Sensor 2 Bank 3 - Sensor 2 6 Bank 2 - Sensor 3 Bank 4 - Sensor 1 7 Bank 2 - Sensor 4 Bank 4 - Sensor 2 where: 1 = sensor present at that location 0 = sensor not present at that location				
X		14 15 16 17 18 19 1A 1B	Bank 1 - Sensor 1 Bank 1 - Sensor 2 Bank 1 - Sensor 3 Bank 1 - Sensor 4 Bank 2 - Sensor 1 Bank 2 - Sensor 2 Bank 2 - Sensor 3 Bank 2 - Sensor 4 for each sensor: Data A - Oxygen sensor output voltage Data B - short term fuel trim associated with this sensor ($FF if this sensor is not used in the calculation)	 0 volt -100.00% (lean)	 1.275 volt 99.22% (rich)	 .005 volt x.xxx volt 100/128% (0% at 128) xxx.x%	This scaling assumes a nominal 1 volt full scale oxygen sensor; any sensor with a different full scale value should be normalized to provide nominal full scale at $C8 (200 decimal).
		1C-1F	Unused - reserved for future expansion				

FIGURE 5—PIDS FOR MODES $01 AND $02 (CONTINUED)

Modes *		PID (Hex)	Description	Min ($00) or ($0000)	Max ($FF) or ($FFFF)	SI (Metric) Scaling/bit and display	English scaling/bit and display
$01	$02						
X		20	PIDs supported ($21 - $40):			Byte bit PID Data A 7 $21 Data A 6 $22 . . . Data B 7 $29 . . . Data D 0 $40	
		21-3F	Reserved - to be specified in J2190, if needed				
X		40	PIDs supported ($41 - $60):				
X		41-FF	Reserved for future expansion				

* NOTE: An "X" in the column under Mode $01 or $02 indicates that this value is included in OBD II regulations to be supported for this mode. Refer to the latest OBD II regulations to determine if each value is required to be supported on a given vehicle, or only required if available.

FIGURE 5 — PIDS FOR MODES $01 AND $02 (CONTINUED)

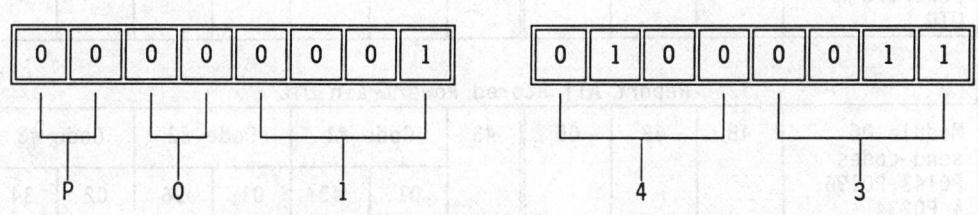

FIGURE 6 — DIAGNOSTIC TROUBLE CODE ENCODING EXAMPLE

	Data Bytes (Hex)						
	#1	#2	#3	#4	#5	#6	#7
Request number of codes from all modules							
Request number of Powertrain DTC	01	01					
Report number of codes (each module)							
Report number of stored powertrain DTC	41	01	# DTC & MIL	Eval. Supp. #1	Eval. Supp. #2	Eval. Status	
Request codes from all modules							
Request powertrain DTC	03						
Report codes (each module)							
Report powertrain DTC	43	Code #1		Code #2 or 00 00		Code #3 or 00 00	

NOTE — Refer to SAE J2012 for encoding method for trouble codes.

FIGURE 7 — MESSAGE DATA BYTES

	Header Bytes (Hex)			Data Bytes (Hex)						
	P/T	Tgt	Src	#1	#2	#3	#4	#5	#6	#7
Request Powertrain DTC										
Request number of Powertrain DTC	68	6A	F1	01	01					
Report Number of Powertrain DTC										
Module 06 has 6 stored DTC	48	6B	06	41	01	06	00	00	00	
Module C3 has 1 stored DTC	48	6B	C3	41	01	01	00	00	00	
Module 2B has 0 stored DTC	48	6B	2B	41	01	00	00	00	00	
Module 3E has 2 stored DTC and MIL ON	48	6B	3E	41	01	82	00	00	00	
Request All Stored Powertrain DTC										
Request powertrain DTC	68	6A	F1	03						
Report All Stored Powertrain DTC										
Module 06 send codes P0143,P0196, & P0234	48	6B	06	43	Code #1		Code #2		Code #3	
					01	43	01	96	02	34
Module C3 send code P0443	48	6B	C3	43	Code #1					
					04	43	00	00	00	00
Module 06 send codes P0357,P0531, & P0661	48	6B	06	43	Code #4		Code #5		Code #6	
					03	57	05	31	06	61
Module 3E send codes P0112 & P0445	48	6B	3E	43	Code #1		Code #2			
					01	12	04	45	00	00

FIGURE 8 — POWERTRAIN DIAGNOSTIC TROUBLE CODE EXAMPLE (ASSUME 10.4 Kbps)

	Data Bytes (Hex)						
	#1	#2	#3	#4	#5	#6	#7
Request to Clear/Reset Emission-Related Diagnostic Information							
Clear Powertrain DTC	04						
Report when Emission-Related Diagnostic Information is Reset							
Powertrain DTC cleared	44						

FIGURE 9 — MESSAGE DATA BYTES

	Data Bytes (Hex)						
	#1	#2	#3	#4	#5	#6	#7
Request Oxygen Sensor Test Results							
Request Oxygen Sensor Test Results	05	Test ID	O2S #				
Report Oxygen Sensor Test Results							
Report Oxygen Sensor Test Results	45	Test ID	O2S #	test value	min limit (opt)	max limit (opt)	

FIGURE 10 — MESSAGE DATA BYTES

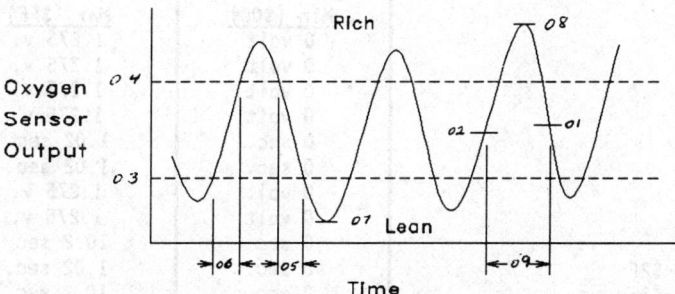

FIGURE 11 — TEST ID VALUE EXAMPLE

Data Byte	Description	Min ($00)	Max ($FF)	Scaling/bit
2	Which Test ID: $01 - Rich to lean sensor threshold voltage (constant) $02 - Lean to rich sensor threshold voltage (constant) $03 - Low sensor voltage for switch time calculation (constant) $04 - High sensor voltage for switch time calculation (constant) $05 - Rich to lean sensor switch time (calculated) $06 - Lean to rich sensor switch time (calculated) $07 - Minimum sensor voltage for test cycle (calculated) $08 - Maximum sensor voltage for test cycle (calculated) $09 - Time between sensor transitions (calculated) $0A-$1F - reserved $20-$2F - values with units of time less than 1.02 seconds $30-$3F - values with units of time less than 10.2 seconds $40-$4F - values with units of voltage less than 1.275 volts $50-$5F - values with units of voltage less than 12.75 volts $60-$6F - values with units of Hertz less than 25.5 Hz $70-$7F - values with units of counts less than 255 counts $80-$FF - manufacturer specific values / units			

3	Oxygen sensor location (one, and only one bit can be set to a 1):
	bit Sensor location Alternative sensor location
	 0 Bank 1 - Sensor 1 Bank 1 - Sensor 1
	 1 Bank 1 - Sensor 2 Bank 1 - Sensor 2
	 2 Bank 1 - Sensor 3 Bank 2 - Sensor 1
	 3 Bank 1 - Sensor 4 Bank 2 - Sensor 2
	 4 Bank 2 - Sensor 1 Bank 3 - Sensor 1
	 5 Bank 2 - Sensor 2 Bank 3 - Sensor 2
	 6 Bank 2 - Sensor 3 Bank 4 - Sensor 1
	 7 Bank 2 - Sensor 4 Bank 4 - Sensor 2
	where:
	   1 = sensor present at that location
	   0 = sensor not present at that location			

4	Test value: Test value $01	0 volt	1.275 v.	.005 v.
	Test value $02	0 volt	1.275 v.	.005 v.
	Test value $03	0 volt	1.275 v.	.005 v.
	Test value $04	0 volt	1.275 v.	.005 v.
	Test value $05	0 sec.	1.02 sec.	.004 sec.
	Test value $06	0 sec.	1.02 sec.	.004 sec.
	Test value $07	0 volt	1.275 v.	.005 v.
	Test value $08	0 volt	1.275 v.	.005 v.
	Test value $09	0 sec.	10.2 sec.	.04 sec.
	Test value $20-$2F	0 sec.	1.02 sec.	.004 sec.
	Test value $30-$3F	0 sec.	10.2 sec.	.04 sec.
	Test value $40-$4F	0 volt	1.275 v.	.005 volt
	Test value $50-$5F	0 volt	12.75 v.	.05 volt
	Test value $60-$6F	0 Hz	25.5 Hz	.1 Hz
	Test value $70-$7F	0 counts	255 counts	1 count
5	Minimum test limit (only for calculated test result)	see data byte #4	see data byte #4	see data byte #4
6	Maximum test limit (only for calculated test result)	see data byte #4	see data byte #4	see data byte #4

NOTE—Current oxygen sensors are nominally 1 V full scale. If an oxygen sensor is used with a different nominal output, the output voltage should be normalized to 1 V. Full scale should be reported as $C8 (decimal 200), which allows for reporting an overvoltage condition.

FIGURE 12—MESSAGE DATA BYTE DESCRIPTION

ENHANCED E/E DIAGNOSTIC TEST MODES—SAE J2190 JUN93

SAE Recommended Practice

Report of the SAE Vehicle E/E System Diagnostics Standards Committee approved June 1993.

TABLE OF CONTENTS

1.	Scope
2.	References
2.1.1	SAE Publications
2.1.2	ISO Publications
2.1.3	California ARB Documents
2.1.4	Federal EPA Documents
2.2	Definitions
3.	Technical Requirements
3.1	Test Mode Values
3.2	Physical Addressing
3.3	Miscellaneous Requirements
3.3.1	Message Response Time
3.3.2	Automatic Return to Normal Operation
3.3.3	Diagnostic Message Length
3.3.4	Message Response
4.	Test Modes
4.1	Modes $00 through $0F - Physically Addressed SAE J1979 Messages
4.2	Mode $10 - Initiate Diagnostic Operation
4.2.1	Functional Description
4.2.2	Message Data Bytes
4.3	Mode $11 - Request Module Reset
4.3.1	Functional Description
4.3.2	Message Data Bytes
4.4	Mode $12 - Request Diagnostic Freeze Frame Data
4.4.1	Functional Description
4.4.2	Message Data Bytes
4.5	Mode $13 - Request Diagnostic Trouble Code Information
4.5.1	Functional Description
4.5.2	Message Data Bytes
4.6	Mode $14 - Clear Diagnostic Information
4.6.1	Functional Description
4.6.2	Message Data Bytes
4.7	Mode $17 - Request Status of Diagnostic Trouble Codes
4.7.1	Functional Description
4.7.2	Message Data Bytes
4.8	Mode $18 - Request Diagnostic Trouble Codes by Status
4.8.1	Functional Description
4.8.2	Message Data Bytes
4.9	Mode $20 - Return to Normal Operation
4.9.1	Functional Description
4.9.2	Message Data Bytes
4.10	Modes $21 to $23 - Request Diagnostic Data
4.10.1	Functional Description
4.10.2	Mode $21 Message Data Bytes
4.10.3	Mode $22 Message Data Bytes
4.10.4	Mode $23 Message Data Bytes
4.11	Mode $24 - Request Scaling and Offset / PID
4.11.1	Functional Description
4.11.2	Message Data Bytes
4.11.3	Scaling Byte 1
4.11.4	Scaling Byte 2
4.11.5	Scaling Bytes 3 to n
4.12	Mode $25 - Stop Transmitting Requested Data
4.12.1	Functional Description
4.12.2	Message Data Bytes
4.13	Mode $26 - Specify Data Rates
4.13.1	Functional Description
4.13.2	Message Data Bytes
4.14	Mode $27 - Security Access Mode
4.14.1	Functional Description
4.14.2	Message Data Bytes
4.15	Mode $28 - Disable Normal Message Transmission
4.15.1	Functional Description
4.15.2	Message Data Bytes
4.16	Mode $29 - Enable Normal Message Transmission
4.16.1	Functional Description
4.16.2	Message Data Bytes
4.17	Mode $2A - Request Diagnostic Data Packet(s)
4.17.1	Functional Description
4.17.2	Message Data Bytes
4.18	Mode $2B - Dynamically Define Data Packet by Single Byte Offsets
4.18.1	Functional Description
4.18.2	Message Data Bytes
4.18.3	Message Example
4.19	Mode $2C - Dynamically Define Diagnostic Data Packet
4.19.1	Functional Description
4.19.2	Message Data Bytes
4.19.3	Message Example
4.20	Mode $2F - Input/Output Control by PID
4.20.1	Functional Description
4.20.2	Message Data Bytes
4.21	Mode $30 - Input/Output Control by Data Value ID
4.21.1	Functional Description
4.21.2	Message Data Bytes
4.22	Modes $31 to $33 - Perform Diagnostic Routine by Test Number
4.22.1	Functional Description
4.22.2	Message Data Bytes
4.23	Modes $34 to $37 - Data Transfer
4.23.1	Functional Description
4.23.2	Message Data Bytes
4.24	Modes $38 to $3A - Perform Diagnostic Routine at a Specified Address 48
4.24.1	Functional Description
4.24.2	Message Data Bytes
4.25	Mode $3B - Write Data Block
4.25.1	Functional Description
4.25.2	Message Data Bytes
4.26	Mode $3C - Read Data Block
4.26.1	Functional Description
4.26.2	Message Data Bytes
4.27	Mode $3F - Test Device Present (no operation performed)
4.27.1	Functional Description
4.27.2	Message Data Bytes
4.28	Mode $7F - General Response Message
4.28.1	Functional Description
4.28.2	Message Data Bytes
4.28.3	Message Examples
4.28.4	Response Codes
Appendix A - Diagnostic Test Mode Assignments	

1. Scope—This SAE Recommended Practice describes the implementation of Enhanced Diagnostic Test Modes, which are intended to supplement the legislated Diagnostic Test Modes defined in SAE J1979. Modes are defined for access to emission related test data beyond what is included in SAE J1979, and for non-emission related data. This document describes the data byte values for diagnostic messages transmitted between diagnostic test equipment, either on-vehicle or off-vehicle, and vehicle electronic control modules. No distinction is made between test modes for emission related and non-emission related diagnostics. These messages can be used with a diagnostic serial data link such as described in SAE J1850 or ISO 9141-2.

For each test mode, this document includes a functional description of the test mode, request and report message data byte content, and an example if useful for clarification.

2. References

2.1 Applicable Documents—The following publications form a part of this specification to the extent specified herein. The latest issue of SAE publications shall apply.

2.1.1 SAE PUBLICATIONS—Available from SAE, 400 Commonwealth Drive, Warrendale, PA 15096-0001.

SAE J1850—Class B Data Communication Network Interface

SAE J1930—E/E Systems Diagnostic Terms, Definitions, Abbreviations and Acronyms

SAE J1962—Diagnostic Connector

SAE J1978—OBD II Scan Tool
SAE J1979—E/E Diagnostic Test Modes
SAE J2012—Recommended Format and Messages for Diagnostic Trouble Codes
SAE J2178—Class B Data Communication Network Messages
SAE J2186—E/E Data Link Security

2.1.2 ISO PUBLICATIONS—Available from ANSI, 1 West 42nd Street, New York, NY 10036-8002.

ISO 9141-2—Road vehicles—Diagnostic systems—CARB requirements for interchange of digital information

2.1.3 CALIFORNIA ARB DOCUMENTS

Mail out #93-27—Title 13, California Code of Regulations, Section 1968.1 Malfunction and Diagnostic System May 21, 1993Requirements—1994 and Subsequent Model-Year Passenger Cars, Light-Duty Trucks, and Medium-Duty Vehicles and Engines

2.1.4 FEDERAL EPA DOCUMENTS

40 CFR Part 86—Control of Air Pollution From New Motor Vehicles and New Motor Vehicle Engines; Federal Register Regulations Requiring On-board Diagnostics

2.2 Definitions

2.2.1 DATA BYTES—Bytes between header bytes and error detection byte.

2.2.2 DIAGNOSTIC TEST MODE—See SAE J1930.

2.2.3 OFFSET—A number used to refer to a data value by specifying its relative position in a list of data values.

2.2.4 PID—PARAMETER IDENTIFICATION NUMBER—A unique identifier used to refer to a specific data value within a module.

2.2.5 SEED / KEY—See SAE J2186

2.2.6 DPID—DATA PACKET IDENTIFICATION NUMBER—An identifier used to refer to a **set** of data values within a module.

2.2.7 BYTE—A group of eight bits of data, bits 0 through bit 7, where bit 7 is the most significant bit and bit 0 is the least significant bit.

2.2.8 $—Prefix defining a hexadecimal number

2.2.9 NIBBLE—Four bits of data. A byte may be split into two nibbles.

2.2.10 EPA—ENVIRONMENTAL PROTECTION AGENCY (FEDERAL AGENCY)

2.2.11 CARB—CALIFORNIA AIR RESOURCES BOARD (STATE AGENCY)

2.2.12 OBD II—Second generation of On-Board Diagnostic regulations required by the California Air Resources Board.

3. Technical Requirements

3.1 Test Mode Values

Figure 1 indicates the assignment of test mode values, and indicates the type of message and where they are defined, either in SAE J1979, this document, or by the manufacturer.

Test mode values $00 - $0F and $40 - $4F are reserved to be defined in SAE J1979, which currently only includes functionally addressed messages. Usage of shaded test mode values will be defined in this document. Appendix A indicates the test mode values currently defined by both SAE J1979 and SAE J2190.

There is a one-to-one correspondence between request messages and response messages, with bit 6 of the test mode value indicating the message type.

Test Mode Value	Message type (bit 6)	Where defined
00 - 0F	Request (bit 6 = 0)	SAE J1979
10 - 1F		
20 - 2F		SAE J2190
30 - 3F		
40 - 4F	Response to Modes $00-$3F (bit 6 = 1)	SAE J1979
50 - 5F		
60 - 6F		SAE J2190
70 - 7F		
80 - 8F	Request (bit 6 = 0)	Reserved for future expansion as needed
90 - 9F		
A0 - AF		Defined by vehicle manufacturer
B0 - BF		
C0 - CF	Response to Modes $80-$BF (bit 6 = 1)	Reserved for future expansion as needed
D0 - DF		
E0 - EF		Defined by vehicle manufacturer
F0 - FF		

FIGURE 1—TEST MODE VALUES

3.2 Physical Addressing

Physical addressing is used for all diagnostic test mode messages defined in this document. Typically when using SAE J1850, this type of addressing requires that the target address of the device for which the message is intended be included in the header of the message. In ISO 9141, the addressing is typically done during initialization. Only that device being addressed will respond to the request. Each device will need to be assigned a unique address to be used for communication purposes.

3.3 Miscellaneous Requirements

3.3.1 MESSAGE RESPONSE TIME—The vehicle controllers should respond to a diagnostic request within 100 ms of a request. If there is no response within this time period, the tool can assume no response will be received.

3.3.2 AUTOMATIC RETURN TO NORMAL OPERATION—During a diagnostic procedure, the on-board controllers will often be put into an abnormal mode of operation to aid in diagnostics. Examples are to report diagnostic data periodically, to disable normal message transmission, or substitute an input or output parameter. In practice, the test procedure should return all on-board controllers to a normal mode of operation at the end of the procedure.

If a test device is disconnected from the vehicle before the on-board controllers have returned to normal operation, the controllers should automatically detect that the tool is disconnected and return to a normal mode of operation. This should be accomplished by the on-board controller looking for a diagnostic message from the test device, which should be apparent from the message header. If a diagnostic message is not received for a 5 s period of time, the on-board controller should automatically return to normal operation.

Some diagnostic procedures may require more than 5 s without a required diagnostic message. For these cases, this document specifies a "test tool present" message to be transmitted by the test device at least once every 5 s if there is no other message. On-board controllers should consider this as a diagnostic message and continue in the current mode of operation.

On-board controllers should also return to normal operation at power up or after a controller reset.

3.3.3 DIAGNOSTIC MESSAGE LENGTH—This document only defines the data bytes to be used in a diagnostic message. The actual number of data bytes in a message, referred to as "n" in the message data byte descriptions in this document, depends on the requirements of the test mode and the maximum number of bytes available in a diagnostic message. The maximum number of bytes available depends on the maximum allowable message length and the number of bytes used as a message header, error detection byte(s), or for other protocol specific purposes.

For example, the SAE J1850 data link effectively limits the total number of bytes in a message to 12. If the message strategy includes a 3 byte header, 1 byte CRC, and 1 byte in-message response, then 7 data bytes are available for SAE J2190 functionality. If the message strategy includes a 3 byte header and 1 byte CRC, then 8 data bytes are available.

If other data links are used as the diagnostic data link, such as ISO 9141, there may be other message length limitations and, therefore, different limits on the number of data bytes available for SAE J2190 functionality.

3.3.4 MESSAGE RESPONSE—All diagnostic request messages, except "Test Device Present (Mode $3F)," should receive a response from the target module. A response is optional for the "Test Device Present" message. There are two types of response messages, specific and general. Specific response messages are uniquely paired with each request message, and repeat enough data bytes of the request message to identify the specific request. The "General Response Message (Mode $7F)" has a single test mode value. The response message to be used for a request depends on the type of the request and whether or not the module can perform the request.

When the request is for data and the module can perform the request (positive response), only the specific form of the response can be used. The response will either be a single message, multiple messages with additional data, or periodic multiple messages with updated data values.

If the request is for action, such as a request for entering a mode of operation or command to perform a function, and the module can perform the request (positive response), either the specific response message or general response message can be used. Those requests for which either response type is a valid response have both response types indicated in this document.

If the request for either data or action cannot be performed by the module (negative response), only the general response message can be used.

Figure 2 summarizes the valid response types for different types of requests.

	Positive response	Negative response
Request for data	Specific response only	General response only
Request for action	Specific or general response allowed	General response only

FIGURE 2—VALID RESPONSE TYPES

With the General Response Message a response code is returned. Response codes can indicate a negative or positive response. Standard response codes are included in the section with the definition of the General Response Message.

4. Test Modes

4.1 Modes $00 through $0F—Physically Addressed SAE J1979 Messages
Modes $00 through $0F are reserved to be defined in SAE J1979. The data byte content of the response to an SAE J1979 request when the device is addressed physically should be identical to the data byte content of the response when the device is addressed functionally. The only device that will respond to this request is the device to which the request is directed.

The response to one of these requests will be a Mode $40 through $4F message.

The same PID list as specified in SAE J1979 will apply when the device is physically addressed using the same test mode value. Although SAE J1979 reserves 256 PID values to be defined in that document, SAE J1979 specifies that PID values $21 to $3F are reserved to be defined in this document. Possible values which may be considered to be included are:
- Number of header bytes and format
- Use of CRC versus checksum for error detection
- Use of in-message response
- Maximum number of data bytes allowed (based on header and in-message response)
- Use of manufacturer specific pins in the SAE J1962 connector

4.2 Mode $10—Initiate Diagnostic Operation

4.2.1 FUNCTIONAL DESCRIPTION—The purpose of this mode is to inform devices on the serial data link that a diagnostic tool is ready to start diagnostic procedures using the data link. The on-board devices may need to alter their normal operation in order for the diagnostic procedures to be effective. One example is that the system may need to reduce the amount of data being sent on the data link to make more time available for the diagnostic messages. Another possibility is that the system may need to degrade its normal operation in order to be able to process the diagnostic requests. Systems that rely on data from modules being diagnosed may not receive data from those modules as frequently as normal. Use of this mode could inform the receiving modules that diagnostic procedures are in progress and they should not set diagnostic trouble codes based on not receiving data at the normal frequency.

An optional "level of diagnostics" byte may be used to indicate how much the normal operation must be altered in order to accommodate the extra diagnostic procedures. These levels and their corresponding amount of change must be predefined in the module by the module manufacturer. The level of diagnostics can be changed by sending a subsequent Mode $10 command.

4.2.2 MESSAGE DATA BYTES—(See Figure 3.)

	Data Bytes (Hex)			
	#1	#2	#3 to #n-1	#n
Request from Tool to Vehicle				
Request diagnostic operation	10	Level (opt.)		
Positive Response from Vehicle to Tool (either of the following is a valid response)				
Confirm diagnostic operation	50	Level (opt.)		
General response	7F	Optional data bytes may be included (see Mode $7F description).		Resp. Code - 00

FIGURE 3—MESSAGE DATA BYTES FOR MODE $10

4.3 Mode $11—Request Module Reset

4.3.1 FUNCTIONAL DESCRIPTION—This mode requests the module to effectively perform a module power on reset. The response message may be sent either before or after the module is reset. The response message may also be sent by modules whenever they perform a power on reset, whether requested by this test mode or not.

An optional data byte may be included in the request to indicate different levels of reset. Examples are:
First time ever connected to vehicle
First time after battery disconnect
After a full normal power down / power up cycle
After minor power supply interruption

4.3.2 MESSAGE DATA BYTES—(See Figure 4.)

	Data Bytes (Hex)			
	#1	#2	#3 to #n-1	#n
Request from Tool to Vehicle				
Request module reset	11	Level (opt.)		
Positive Response from Vehicle to Tool (either of the following is a valid response)				
Confirm module reset	51	Level (opt.)		
General response	7F	Optional data bytes may be included (see Mode $7F description).		Resp. Code - 00

FIGURE 4—MESSAGE DATA BYTES FOR MODE $11

4.4 Mode $12—Request Diagnostic Freeze Frame Data

4.4.1 FUNCTIONAL DESCRIPTION—The purpose of this mode is to allow access to data values which were stored during freeze frame conditions specified by the vehicle manufacturer. Data content, data format, and method of retrieval are specified by the vehicle manufacturer. Typical uses for this mode are to report data stored upon detection of a system malfunction. Multiple frames of data may be stored before and/or after the malfunction is detected. The request for information includes a frame number followed by an optional indication of the data requested.

If the optional data byte is not used, or is $00, then all data for the requested freeze frame will be reported. If the optional byte is used, then the vehicle manufacturer will define the different methods used to specify which data is requested.

The on-board module will respond to this message by transmitting the requested data.

4.4.2 MESSAGE DATA BYTES—(See Figure 5.)

	Data Bytes (Hex)			
	#1	#2	#3	#4 to #n
Request from Tool to Vehicle				
Request diagnostic freeze frame data	12	Frame No.	Method to request data (optional) examples are: Request all data (not included or 00) Request by offset (1 byte) Request by PID (2 bytes) Request by memory address (3 bytes) Request by Data Packet ID Request starting address for data DTC which caused data to be stored Other manufacturer specific method	
Positive Response from Vehicle to Tool				
Report diagnostic freeze frame data	52	Frame No.	Data byte	Additional optional data bytes that are in response to the request may be added to fill the message up to the maximum number of available data bytes.

FIGURE 5—MESSAGE DATA BYTES FOR MODE $12

4.5 Mode $13—Request Diagnostic Trouble Code Information

4.5.1 FUNCTIONAL DESCRIPTION—The purpose of this mode is to enable a diagnostic test tool to obtain stored emission and non-emission related diagnostic trouble code (DTC) information.

This mode includes the option to request DTC information by function, where function is either powertrain, body, chassis, or undefined. These are the functions used in SAE J2012 to group trouble codes. The first two bits of the first nibble for requests by function are encoded using the same convention as when codes are reported:

00 - powertrain

01 - chassis
10 - body
11 - undefined

For requests by function using this test mode, the second two bits of the first nibble are 00, and the second nibble is all zeroes. This translates to the following function groups for DTC requests:

$00 - powertrain codes
$40 - chassis codes
$80 - body codes
$C0 - undefined

In addition, this document defines $FF as the function group to request DTC information for all functions.

This test mode can be used to either request diagnostic trouble codes, or request the number of diagnostic trouble codes. There are two ways this mode can request DTCs.

All codes can be requested by including only the test mode value in the request.

Codes can be requested by function group (Powertrain, Chassis, Body, Undefined, or All) by sending the function group followed by $00 as data bytes 2 and 3 ($00 00, $40 00, $80 00, $C0 00, or $FF 00). Requesting codes for all function groups by including $FF 00 yields the same response as not including function group in the request, but may be desired by a manufacturer for consistency with other messages.

There are also two ways this mode can request the number of DTCs.

Number of codes can be requested by including the function group as data byte 2, and not including data byte 3 ($FF as data byte 2 must be supported as a minimum to request the total number of DTC).

Number of codes can be requested by function group (Powertrain, Chassis, Body, Undefined, or All) by sending the function group followed by $FF as data bytes 2 and 3 ($00 FF, $40 FF, $80 FF, $C0 FF, or $FF FF). Requesting number of codes for all function groups by including $FF FF yields the same response as including only $FF as data byte 2 in the request, but may be desired by a manufacturer for consistency with other messages.

The response to a Mode $13 request for DTC will be one or more Mode $53 messages. If no codes are stored in the module, then the module will respond with one of the following:

Mode $53 with no additional data bytes
Mode $53 padded with $00 00 to fill response

Diagnostic trouble codes are transmitted as two bytes. The first two bits (high order) of the first byte for each code will be zeroes to indicate a powertrain code (refer to SAE J2012 - "Recommended Format and Messages for Diagnostic Trouble Codes" for additional interpretation of this structure). The second two bits will indicate the first digit of the diagnostic code (0 through 3). The second nibble of the first byte and the entire second byte are the next three digits of the actual code reported as Binary Coded Decimal (BCD). A powertrain trouble code transmitted as $0143 should be interpreted as P0143, as shown in Figure 6.

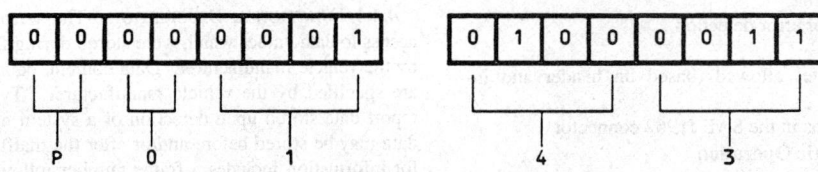

FIGURE 6—DIAGNOSTIC TROUBLE CODE EXAMPLE

4.5.2 MESSAGE DATA BYTES—(See Figure 7.)

	Data Bytes (Hex)			
	#1	#2	#3	#4 to #n
Request from Tool to Vehicle (either of the following messages can be used to request DTC)				
Request all DTC	13			
Request DTC by function group where group is: 00 - powertrain 40 - chassis 80 - body C0 - undefined FF - all	13	Group	00	
Request from Tool to Vehicle (either of the following messages can be used to request number of DTC)				
Request number of DTC by function where group is: 00 - powertrain 40 - chassis 80 - body C0 - undefined FF - all	13	Group		
FF in byte 3 is an optional byte used to maintain a fixed length request	13	Group	FF	
Positive Response from Vehicle to Tool (Multiple messages may be required to report the DTCs)				
Report stored DTC - response to request for DTC	53	Diagnostic trouble code or 00 00 (optional)		Additional optional diagnostic trouble codes may be added to fill the message up to the maximum number of available data bytes.
Report # of stored DTC - response to request for number of DTC	53	Number of codes		

NOTE—Refer to SAE J2012 for encoding method for trouble codes.

FIGURE 7—MESSAGE DATA BYTES FOR MODE $13

4.6 Mode $14—Clear Diagnostic Information

4.6.1 FUNCTIONAL DESCRIPTION—The purpose of this mode is to provide a means for the external test device to command on-board modules to clear all diagnostic information. This information includes primarily diagnostic trouble codes, but can also include freeze frame data or other on-board test results that may be stored as a result of the trouble code being set. This extra information is device dependent.

There are three ways this mode can be used. If only the test mode value is included in the request, then all diagnostic information stored in the module is to be cleared.

Diagnostic information can optionally be cleared by function (Powertrain, Chassis, or Body) by sending P0000, C0000, or B0000 as data bytes 2 and 3. These must be encoded using the same convention as when codes are reported (see description for Mode $13). The first two bits of the first nibble are:

00 - powertrain
01 - chassis
10 - body
11 - undefined

This translates to the following values for data bytes #2 and #3:

$00 00 - powertrain codes
$40 00 - chassis codes
$80 00 - body codes
$C0 00 - undefined

Diagnostic information can also optionally be cleared for a single trouble code by including the trouble code to clear as optional bytes in the request. The information stored with an individual trouble code would need to be clearly identified within the module and cleared at the same time.

Caution when clearing by single trouble code: This capability should not be provided and used unless consideration is given to the possible consequences of this option. If a single fault caused multiple trouble codes to be set, and a service technician found a defective component, repaired it, and cleared a single trouble code associated with that fault, then the device would still contain trouble codes that were set based on that fault. The next technician to check codes would read those codes, and if he did not know the vehicle history, he may try to repair the vehicle based only on the remaining trouble codes, which could waste time and result in replacement of good components.

Allowing use of this feature to clear emission related trouble codes should be considered very carefully to guard against improper diagnosis. Difficult interpretation can result if the system allowed clearing a single emission related trouble code. CARB required information could report that trouble codes are stored, but they would not be reported when an SAE J1979 Mode $04 message is sent to request those codes. Freeze frame data could be stored for a given trouble code, but that code would not be reported as stored.

4.6.2 MESSAGE DATA BYTES—(See Figure 8.)

	Data Bytes (Hex)			
	#1	#2	#3	#4 to #n-1	#n
Request from Tool to Vehicle					
Clear stored diagnostic information	14	Data bytes 2 and 3 are optional if not used - clear all information if 00 00 - clear all powertrain information if 40 00 - clear all chassis information if 80 00 - clear all body information if C0 00 - clear all Cx xx DTC information if FF 00 - clear all information any other value - trouble code to clear			
Positive Response from Vehicle to Tool (either of the following is a valid response)					
Diagnostic information cleared	54	repeat data bytes 2 and 3 of request, if used			
General response	7F	Optional data bytes may be included (see Mode $7F description).			Resp. Code - 00

FIGURE 8—MESSAGE DATA BYTES FOR MODE $14

4.7 Mode $17—Request Status of Diagnostic Trouble Codes

4.7.1 FUNCTIONAL DESCRIPTION—The purpose of this mode is to provide a means to determine the status of diagnostic trouble codes.

This mode is nearly identical in operation to Mode $13, except that only a single DTC, followed by the status of that DTC, is reported in a single response. This mode also allows a request for the status of a single DTC. The discussion of function groups and the method of reporting DTCs is identical to the description in Mode $13.

This test mode can be used to either request status of diagnostic trouble codes, or request the number of diagnostic trouble codes with a status bit set. There are three ways this mode can request DTCs.

All codes can be requested by only including the test mode value in the request.

Codes can be requested by function group (Powertrain, Chassis, Body, Undefined, or All) by sending the function group followed by $00 as data bytes 2 and 3 ($00 00, $40 00, $80 00, $C0 00, or $FF 00). Requesting codes for all function groups by including $FF 00 yields the same response as not including function group in the request, but may be desired by a manufacturer for consistency with other messages.

Status of a single trouble code can be requested by including the trouble code as data bytes 2 and 3 in the request.

There are two ways this mode can request the number of DTCs with a status bit set.

Number of codes can be requested by including the function group as data byte 2, and not including data byte 3 ($FF as data byte 2 must be supported as a minimum to request the total number of DTC).

Number of codes can be requested by function group (Powertrain, Chassis, Body, Undefined, or All) by sending the function group followed by $FF as data bytes 2 and 3 ($00 FF, $40 FF, $80 FF, $C0 FF, or $FF FF). Requesting number of codes for all function groups by including $FF FF yields the same response as including only $FF as data byte 2 in the request, but may be desired by a manufacturer for consistency with other messages.

Multiple Mode $57 response messages may be reported due to a single Mode $17 request message, depending on the type of request and the number of diagnostic trouble codes stored in the module. Each response message will report the status of a single DTC. If no codes are stored in the module, then the module will respond with one of the following:

Mode $57 with no additional data bytes
Mode $57 with $00 00 for DTC and $00 for status

4.7.2 MESSAGE DATA BYTES—(See Figure 9.)

	Data Bytes (Hex)			
	#1	#2	#3	#4
Request from Tool to Vehicle (any of the following messages can be used to request DTC)				
Request status of all DTC	17			

Request status of DTC by function group where group:
  00 - powertrain
  40 - chassis
  80 - body
  C0 - undefined
  FF - all
	17	Group	00	

Request status of specific DTC	17	Diagnostic Trouble Code		
Request from Tool to Vehicle (either of the following messages can be used to request number of DTC)				

Request number of DTC with a status bit set by function where group is:
  00 - powertrain
  40 - chassis
  80 - body
  C0 - undefined
  FF - all
	17	Group		

FF in byte 3 is an optional byte used to maintain a fixed length request	17	Group	FF	
Positive Response from Vehicle to Tool (Multiple messages may be required to report the DTCs)				
Report status of stored DTC	57	Diagnostic trouble code or 00 00 (optional)		Status of DTC or 00 (optional if DTC not reported): same definition as Data Byte #2 for Mode $18
Report # of stored DTC with any status bit set	57	Number of codes		

FIGURE 9—MESSAGE DATA BYTES FOR MODE $17

4.8 Mode $18—Request Diagnostic Trouble Codes by Status

4.8.1 FUNCTIONAL DESCRIPTION—This mode is used to retrieve diagnostic trouble codes based on the status by which they have been stored. There are various conditions under which codes are generated, for example, under normal customer driving conditions versus "on demand" codes which are generated by tests performed by the service technician. Another difference is that some codes illuminate a warning lamp while other codes do not. Some codes are retained in memory for at least 40 warmup cycles while some codes are retained only for one ignition/power up cycle.

This test mode is similar in operation to Mode $17, except that an additional data byte is inserted as data byte #2 to indicate the specific status bits of interest. Data bytes #3 and #4 in Mode $18 requests have the same meaning as data bytes #2 and #3 in Mode $17 requests.

There are many philosophies among the various manufacturers for storing codes. Bit definitions have been assigned to the more commonly accepted methods. Codes stored under that definition can be requested by setting the appropriate bit(s) to "1" to retrieve the set of desired codes.

Table 1 shows the interpretation of each bit of the status byte. This definition is used for both Modes $17 and $18. Multiple bits can be set to "1". When requesting codes using Mode $18, codes with any of the requested status bits will be reported.

Multiple Mode $58 response messages may be reported to a single request, depending on the number of diagnostic trouble codes stored in the module. Each response message will report up to three DTCs for which at least one of the requested status bits is set. If no codes are stored in the module that meet the requested status, then the module will respond with one of the following:

Mode $58 with Status Requested and no additional data bytes
Mode $58 with $00 00 as data bytes 3 and 4
Mode $58 with $00 00 repeated to fill response

TABLE 1—INTERPRETATION OF EACH BIT OF THE STATUS BYTE

bit	status
7	Warning lamp illuminated for this code
6	Warning lamp pending for this code, not illuminated but malfunction was detected
5	Warning lamp was previously illuminated for this code, malfunction not currently detected, code not yet erased
4	Stored trouble code
3	Manufacturer specific status
2	Manufacturer specific status
1	Current code - present at time of request
0	Maturing/intermittent code - insufficient data to consider as a malfunction

4.8.2 MESSAGE DATA BYTES—(See Figure 10.)

	Data Bytes (Hex)				
	#1	#2	#3	#4	#5 to #n
Request from Tool to Vehicle (either of the following messages can be used to request DTC)					
Request all DTC with status	18	Status			
Request all DTC with status and function group where group is: 00 - powertrain 40 - chassis 80 - body C0 - undefined FF - all	18	Status	Group	00	
Request from Tool to Vehicle (either of the following messages can be used to request number of DTC)					
Request number of DTC with any specific status bits set by function group 00 - powertrain 40 - chassis 80 - body C0 - undefined FF - all	18	Status	Group		
FF in byte 4 is an optional byte used to maintain a fixed length request	18	Status	Group	FF	
Positive Response from Vehicle to Tool					
Report DTC by status - response to request with test mode and status	58	Status req'd	Diagnostic trouble code or 00 00 (optional)	Additional optional diagnostic trouble codes may be added to fill the message up to the maximum number of available data bytes.	
Report DTC by status - response to request with test mode, status, and data byte #3	58	Status req'd	number of DTC with req'd status		

FIGURE 10—MESSAGE DATA BYTES FOR MODE $18

4.9 Mode $20—Return to Normal Operation

4.9.1 FUNCTIONAL DESCRIPTION—The on-board device will return to the normal mode of operation when this message is received. All normal algorithms and normal communications will be resumed. All active diagnostic modes will be terminated without sending additional request or response messages. If the module had been unlocked using security access, then the module should also be locked.

4.9.2 MESSAGE DATA BYTES—(See Figure 11.)

	Data Bytes (Hex)		
	#1	#2 to #n-1	#n
Request from Tool to Vehicle			
Request return to normal operation	20		
Positive Response from Vehicle to Tool (either of the following is a valid response)			
Confirm return to normal operation	60		
General response	7F	Optional data bytes may be included (see Mode $7F description).	Resp. Code - 00

FIGURE 11—MESSAGE DATA BYTES FOR MODE $20

4.10 Modes $21 to $23—Request Diagnostic Data

4.10.1 FUNCTIONAL DESCRIPTION—The purpose of these modes is to request data values, such as analog inputs and outputs, digital inputs and outputs, freeze frame data, calculated values, bit mapped fault code data, and system status information. The request for information can be by one of three methods:

Mode $21 - Offset (1 byte)
Mode $22 - Parameter Identification (PID) value (2 bytes)
Mode $23 - Memory Address (3 bytes)

System designers have the flexibility to maintain a table including PID numbers and data values, without having consecutive PID numbers. In this case, the tool can request either by PID number or offset. A request by offset is generally an easier method for the software to retrieve data because it does not need to search for the PID in the table. Mode $24 allows use of the PID number one time to get the offset, with successive requests by offset for better system performance.

Figure 12 is an example of a table that could be included in the system module to utilize all of these modes. The offset is the row number of the table and is not a table entry. The location of the information within this table can be calculated by multiplying the offset times the number of bytes in each row. The location in memory can then be calculated by adding the location within the table to the base location of the table in memory.

Offset	PID1	PID2	ADDR1	ADDR2	ADDR3	SCALING1	SCALING2
00	00	01	00	1B	FE		
01	00	0A	00	2C	12		
02	01	FE	00	BA	1F		
--	--	--	--	--	--	--	--
--	--	--	--	--	--	--	--
xx	04	2C	00	65	8A		

FIGURE 12—OFFSET/PID/ADDRESS TABLE EXAMPLE

A Mode $21 request by offset would calculate the location of the offset information in the table, find the address and scaling for that value, and return the appropriate number of bytes of information starting at that address.

A Mode $22 request by PID would search through the table for the requested PID, find the address and scaling for that value, and return the appropriate number of bytes of information starting at that address.

A Mode $23 request by address would ignore this table and report data starting at the address specified in the request.

4.10.2 MODE $21 MESSAGE DATA BYTES—(See Figure 13.)

		Data Bytes (Hex)					
		#1	#2	#3	#4	#5	#6 to #n-1	#n
		Request from Tool to Vehicle						
Request data by offset		21	offset	Value for Byte 3 (optional): not included - 1 response 00 - stop sending data 01 - send 1 response 02 - repeat at slow rate 03 - repeat at medium rate 04 - repeat at fast rate 05 to FF - manufacturer defined conditional response, where the condition is pre-defined in the module - multiple conditions may be defined Value for Byte 4 (optional if Byte 3 supported): 00 - send data until requested to stop nn - maximum number of responses to send				
		Vehicle Response to Request for Data (Multiple response messages will be sent if data rate requested periodic data reporting)						
Report data by offset		61	offset	data byte	Additional optional data bytes that are in response to the request may be added to fill the message up to the maximum number of available data bytes.			
		Vehicle Response to Request to stop sending data (either of the following is a valid response)						
Report stop sending data by offset		61	offset					
General response		7F	Optional data bytes may be included (see Mode $7F description).					Resp. Code - 00

FIGURE 13—MESSAGE DATA BYTES FOR MODE $21

4.10.3 MODE $22 MESSAGE DATA BYTES—(See Figure 14.)

		Data Bytes (Hex)					
		#1	#2	#3	#4	#5	#6 to #n-1	#n
		Request from Tool to Vehicle						
Request data by PID		22	PID (high byte)	PID (low byte)	Value for Byte 4 (optional): same definition as Data Byte #3 for Mode $21 Value for Byte 5 (optional if Byte 4 supported): 00 - send data until requested to stop nn - max. number of responses to send			
		Vehicle Response to Request for Data (Multiple response messages will be sent if data rate requested periodic data reporting)						
Report data by PID		62	PID (high byte)	PID (low byte)	data byte	Additional optional data bytes that are in response to the request may be added to fill the message up to the maximum number of available data bytes.		
		Vehicle Response to Request to stop sending data (either of the following is a valid response)						
Report stop sending data by PID		62	PID (high byte)	PID (low byte)				
General response		7F	Optional data bytes may be included (see Mode $7F description).					Resp. Code - 00

FIGURE 14—MESSAGE DATA BYTES FOR MODE $22

4.10.4 MODE $23 MESSAGE DATA BYTES—(See Figure 15.)

		Data Bytes (Hex)					
		#1	#2	#3	#4	#5	#6 to #n-1	#n
		Request from Tool to Vehicle						
Request data by memory address		23	addr (high byte)	addr (mid byte)	addr (low byte)	Value for Byte 5 (optional): same definition as Data Byte #3 for Mode $21 Value for Byte 6 (optional if Byte 5 supported): 00 - send data until requested to stop nn - max. number of responses to send		
		Vehicle Response to Request for Data (Multiple response messages will be sent if data rate requested periodic data reporting)						
Report data by memory address		63	addr (mid byte)	addr (low byte)	data byte	Additional optional data bytes that are in response to the request may be added to fill the message up to the maximum number of available data bytes.		
		Vehicle Response to Request to stop sending data (either of the following is a valid response)						
Report stop sending data by memory address		63	addr (mid byte)	addr (low byte)				
General response		7F	Optional data bytes may be included (see Mode $7F description).					Resp. Code - 00

FIGURE 15—MESSAGE DATA BYTES FOR MODE $23

4.11 Mode $24—Request Scaling and Offset / PID

4.11.1 FUNCTIONAL DESCRIPTION—The purpose of Mode $24 is to request either scaling and PID information by specifying offset, or request scaling and offset information by specifying PID. $FF should be included in the request for the data byte value not being specified. If both offset and Parameter ID are specified, and the parameter ID is not correct for the offset specified, unexpected data values can be returned. Operation of this mode is similar to either Mode $21 or $22, depending on which value was specified in the request. The response would include offset, PID, and scaling for the requested entry. Any data byte values not supported by the module would report $FF.

Mode $24 can also be used to determine the PID table by incrementing the offset in repeated Mode $24 requests.

4.11.2 MESSAGE DATA BYTES—(See Figure 16.)

		Data Bytes (Hex)			
		#1	#2	#3	#4	#5 to #n
		Request from Tool to Vehicle				
Request scaling and offset		24	offset ($FF if N/A)	PID (high byte - $FF if N/A)	PID (low byte - $FF if N/A)	
		Positive Response from Vehicle to Tool				
Report scaling and offset		64	offset	PID (high byte - $FF if N/A)	PID (low byte - $FF if N/A)	Scaling bytes may be added to fill the message up to the maximum number of available data bytes.

FIGURE 16—MESSAGE DATA BYTES FOR MODE $24

4.11.3 SCALING BYTE 1
High nibble (bits 7 to 4)

data type
- 0000 - Unsigned numeric (1 to 4 bytes)
- 0001 - Signed numeric (1 to 4 bytes)
- 0010 - Bit mapped, reported without mask
- 0011 - Bit mapped, reported with mask
- 0100 - BCD (1 to 4 bytes)
- 0101 - State encoded
- 0110 - ASCII
- 0111 - Signed floating point (ANSI/IEEE Std 754-1985)
- 1000 - Packet
- 1001 - Formula
- 1010 thru 1111 - Reserved

Low nibble (bits 3 to 0)
number of bytes of value
- 0000 - Reserved
- 0001 - 1 byte
- 0010 - 2 bytes
- 0011 - 3 bytes
- 0100 - 4 bytes
- 0101 thru 1111 - Reserved - may be greater than 4 for non-numeric data (ASCII or bit-mapped)

The high order nibble of the first byte defines the type of information encoding which is used to represent the parameter. The low order nibble can be read directly to determine the number of bytes used to represent the parameter.

4.11.3.1 Unsigned Numeric—This encoding uses a common binary weighting scheme to represent a value by means of discrete incremental steps. One byte affords 256 steps; two bytes yields 65 536 steps, etc. For example, vehicle speed may be encoded with a single unsigned numeric parameter ranging from 0 to 255 km/h. This unsigned numeric value may be up to 4 bytes in length.

4.11.3.2 Signed Numeric—This encoding uses a two's complement binary weighting scheme to represent a value by means of discrete incremental steps. One byte affords 256 steps; two bytes yields 65,536 steps, etc. A value can be encoded with a single unsigned numeric parameter ranging from -127 to 128. This signed numeric value may be up to 4 bytes in length.

4.11.3.3 Bit Mapped—Bit mapped encoding uses individual bits or small groups of bits to represent status. For every bit which represents status, a corresponding mask bit is required as part of the parameter definition. The mask indicates the validity of the bit for particular applications. A bit mapped parameter may contain 2 bytes; one representing status and one containing the validity mask, or 4 bytes; the first two representing status and the second two containing validity masks. Reference J2178, Part 2 for a discussion of bit mapped data values.

4.11.3.4 BCD—Conventional BCD encoding is used to represent two numeric digits per byte. The upper nibble is used to represent the most significant digit (0 - 9), and the lower nibble the least significant digit (0 - 9). Up to 8 characters (4 bytes) may comprise a single parameter.

4.11.3.5 State Encoded—This encoding uses a common binary weighting scheme to represent up to 256 distinct states. An example is a parameter which represents the status of the ignition switch. Codes "00," "01," "02," and "03" may indicate ignition off, locked, run, and start, respectively. The representation is always limited to 1 byte.

4.11.3.6 ASCII—Conventional ASCII encoding is used to represent up to 128 standard characters (MSB = logic 0). An additional 128 custom characters may be represented with MSB = logic 1. Up to four ASCII characters may comprise a single parameter.

4.11.3.7 Signed Floating Point—Floating point encoding is used for data that needs to be represented in floating point or scientific notation. Standard IEEE formats shall be used.

4.11.3.8 Packet—Packets contain multiple data values, usually related, each with unique scaling. Scaling information is not included for the individual values.

4.11.3.9 Formula—Different formulas are currently being discussed to be included. The formula type and constant values used in those formulas will be specified in scaling bytes, or reported as a different test mode.

4.11.4 SCALING BYTE 2—This byte is only relevant for signed and unsigned numbers and for bit mapped parameters. For other parameter types, this field is "padded" with $00. When used with signed and unsigned numbers, this byte specifies scaling by indicating the binary point location. The binary point may be moved 127 places to the right ($FF) or 128 places to the left ($00). The binary point is initially assumed to be to the right of the LSB without this information.

When this byte is used for bit mapped parameters that are reported without a mask, this byte indicates which bits of the PID are supported for the current application. When bit mapped data is reported with a mask, this byte is not used.

When the data type is "formula," this byte is used to indicate the formula type, where:
- $00 - $7F are reserved for SAE to define
- $80 - $FF are reserved for the manufacturer to define

4.11.5 SCALING BYTES 3 TO N—Optional bytes to be included if multiple bytes of bit mapped data are reported. This mode is more limiting in the number of masks that can be reported than the modes that allow retrieval of the data.

4.12 Mode $25—Stop Transmitting Requested Data

4.12.1 FUNCTIONAL DESCRIPTION—The purpose of Mode $25 is to stop all data transmission that was started by any test mode that can request repetitive data. If only an individual data request is to be stopped, then that value can be stopped by sending a request message for the mode that requested the data with a "0" for the data rate. In practice, if a message is already in an outgoing message queue ready to be sent, this would probably not remove that message from the queue, but it would prevent additional messages from being added to the queue. Response is either Mode $65 or general response.

4.12.2 MESSAGE DATA BYTES—(See Figure 17.)

	Data Bytes (Hex)		
	#1	#2 to #n-1	#n
Request from Tool to Vehicle			
Request all data stop	25		
Positive Response from Vehicle to Tool (either of the following is a valid response)			
Confirm all data stopped	65		
General response	7F	Optional data bytes may be included (see Mode $7F description).	Resp. Code - 00

FIGURE 17—MESSAGE DATA BYTES FOR MODE $25

4.13 Mode $26—Specify Data Rates

4.13.1 FUNCTIONAL DESCRIPTION—The purpose of Mode $26 is to specify the setting of data rates for data to be transmitted by request messages that can request repetitive data. Those modes include "Data Rate" as a data byte in the request message. Modules will normally have default rates associated with slow, mid, and fast. The slow rate should be about one or two samples per second, the mid rate should be normal handheld scan tool rates of about four to ten samples per second, and the fast rate 20 or more samples per second. Default rates are defined within each module and should be based on the frequency of change typical for the value reported, uses for that data, and the capability of the processor.

4.13.2 MESSAGE DATA BYTES—(See Figure 18.)

	Data Bytes (Hex)					
	#1	#2	#3	#4	#5 to #n-1	#n
Request from Tool to Vehicle						
Request data rate values	26	Slow Rate * - 40 msec per bit	Mid Rate * - 2 msec per bit	Fast Rate * - .1 msec per bit		
Positive Response from Vehicle to Tool (either of the following is a valid response)						
Confirm data rate values set	66					
General response	7F	Optional data bytes may be included (see Mode $7F description).				Resp. Code - 00

* NOTE: A value of $00 should cause the data to be reported as fast as possible. A value of $FF will cause no change in the current data rate.

FIGURE 18—MESSAGE DATA BYTES FOR MODE $26

4.14 Mode $27—Security Access Mode

4.14.1 FUNCTIONAL DESCRIPTION—The primary purpose of this mode is to restrict unauthorized intrusion into the on-board controller. Improper programs could potentially damage the electronics or other vehicle components or risk the vehicle's compliance to emission or safety standards. This mode is intended to be used to implement the data link security measures defined in SAE J2186.

The external device will request the controller to "Unlock" itself by sending Request #1. The controller will respond by sending a "Seed" using Response #1. The external device will respond by returning a "Key" number back to the controller using Request #2. The controller would compare this "Key" to one internally stored. If the two numbers agree, then the controller will enable ("Unlock") the external device's access to specific test modes and indicate that with Mode $67, Response #2. If upon two attempts of a Request #2 where the two keys do not compare, then the controller will insert a 10 s time delay before allowing further attempts. This time delay will also be required before responding to a Mode $27 request #1 for each controller power-on.

If a device supports security, but is already unlocked when a Request #1 is received, that device should respond with a Response #1 message with a seed of $00 00. A test device could use this method to determine if a device is locked by checking for a non-zero seed.

The security system will not prevent normal diagnostic or vehicle communications between external devices and the on-board controller. Proper "Unlocking" of the controller is a prerequisite to the external device's ability to perform some of the more critical functions such as reading specific memory locations within the controller, downloading information to specific locations, or downloading routines for execution by the controller. In other words, the only access to the controller permitted while in a "locked" mode is through the product specific software. This permits the product specific software to protect itself from unauthorized intrusion.

Devices that provide security should support reject messages if a secure mode is requested while the device is locked. The reject message to be returned is a general response message Mode $7F, with a response code $33 to indicate the product is secured.

Some devices could support multiple levels of security, either for different functions controlled by the device, or to allow different capabilities to be exercised. These additional levels of security can be accessed by using requests #3 and #4, etc. The second data byte of the request for seed should always be an odd number, and the second data byte of the message to send the key should be the next even number.

4.14.2 MESSAGE DATA BYTES—(See Figure 19.)

	Data Bytes (Hex)					
	#1	#2	#3	#4	#5 to #n-1	#n
Request #1 from Tool to Vehicle (data byte #2 can be an odd number greater than $01 if additional levels of security are supported)						
Request for "Seed"	27	01 (03, etc.)				
Positive Response #1 from Vehicle to Tool						
Return "Seed"	67	01 (03, etc.)	Seed (00 00 if module is not locked)	Optional data bytes may be specified by the vehicle manufacturer		
Request #2 from Tool to Vehicle (data byte #2 can be an even number one greater than data byte #2 of request #1 if additional levels of security are supported)						
Send "Key"	27	02 (04, etc.)	Key			
Positive Response #2 from Vehicle to Tool						
Security access	67	02 (04, etc.)	Optional data bytes may be specified by the vehicle manufacturer			
General response	7F	Optional data bytes may be included (see Mode $7F description).				Resp. Code - 00

FIGURE 19—MESSAGE DATA BYTES FOR MODE $27

4.15 Mode $28—Disable Normal Message Transmission

4.15.1 FUNCTIONAL DESCRIPTION—The purpose of this mode is to inhibit the on-board device from transmitting normal operating data on the link, while still performing other functions normally. The device will continue to operate in whatever diagnostic mode it was operating in prior to the Mode $28 command.

If an unsafe or undesirable vehicle operating condition would result from the lack of normal messages, then this mode could cause all nonessential messages to be inhibited. The optional manufacturer specific "level" data byte could be used to indicate which normal mode messages to disable. Defining which messages to disable is determined by the system designer to ensure safe vehicle operation. When using this test mode, some provision must be made to allow for other devices that rely on information from the silenced device so that they do not set diagnostic trouble codes due to the lack of required information.

One use for this test mode is to reduce message traffic on the data link. This would make more time available for diagnostic messages. Another use for this mode is to allow the test equipment to emulate the remote device for diagnostic purposes. In this scenario, the test device would send a Mode $28 request to the device to be emulated. The test device would then respond to all normal communication request messages directed to that device, most likely with data intended to cause a known response by a system that uses the information in the response. The test device can then observe the actions of those systems.

4.15.2 MESSAGE DATA BYTES—(See Figure 20.)

	Data Bytes (Hex)			
	#1	#2	#3 to #n-1	#n
Request from Tool to Vehicle				
Request disable normal message transmission	28	Level (opt.)		
Positive Response from Vehicle to Tool (either of the following is a valid response)				
Confirm normal message transmission disabled	68	Level (opt.)		
General response	7F	Optional data bytes may be included (see Mode $7F description).		Resp. Code - 00

FIGURE 20—MESSAGE DATA BYTES FOR MODE $28

4.16 Mode $29—Enable Normal Message Transmission

4.16.1 FUNCTIONAL DESCRIPTION—The purpose of this mode is to cause an on-board device to resume normal communications after previously disabling these messages by a Mode $28 command. The device will continue to operate in whatever diagnostic modes were set by previous messages, such as executing on-board routines.

4.16.2 MESSAGE DATA BYTES—(See Figure 21.)

	Data Bytes (Hex)		
	#1	#2 to #n-1	#n
Request from Tool to Vehicle			
Request enable normal message transmission	29		
Positive Response from Vehicle to Tool (either of the following is a valid response)			
Confirm normal message transmission enabled	69		
General response	7F	Optional data bytes may be included (see Mode $7F description).	Resp. Code - 00

FIGURE 21—MESSAGE DATA BYTES FOR MODE $29

4.17 Mode $2A—Request Diagnostic Data Packet(s)

4.17.1 FUNCTIONAL DESCRIPTION—The purpose of this mode is to request diagnostic data packets that contain data values, such as analog inputs and outputs, digital inputs and outputs, freeze frame data, calculated values, bit mapped fault code data, and system status information. This Mode differs from Modes $21 through $23 in that multiple data packets can be requested, each containing multiple data values. These data packets can either be predefined in the on-board control module or dynamically defined using Mode $2B or $2C.

The on-board module will respond to this message by transmitting multiple response messages specific to that device. The data packet ID Bytes in the request allow multiple Mode $2A data packets to be requested. Data packets should be defined which include data that is commonly used together, or changes together.

Uses for these different data packets include data such as wheel speeds for an ABS system. Data can be returned quickly with a minimal length request and response. Another use for these different data packets is to return values that do not change, such as VIN and option content. These values need to be known one time only during testing. Other data bytes may contain present values of analog and discrete device I/O, device software flags and status words, and failure codes.

4.17.2 MESSAGE DATA BYTES—(See Figure 22.)

	Data Bytes (Hex)				
	#1	#2	#3	#4 to #n-1	#n
Request from Tool to Vehicle					
Request diagnostic data packets	2A	Data rate *	DPID #	Additional optional data packet IDs may be requested to fill the message up to the maximum number of available data bytes.	
Positive Response from Vehicle to Tool (Multiple response messages will be sent if multiple DPIDs requested, or if data rate requested periodic data reporting)					
Report of diagnostic data packet - repeated for each requested data packet	6A	DPID #	Data Byte	Additional optional data bytes that are in response to the request may be added to fill the message up to the maximum number of available data bytes.	
Vehicle Response to Request to stop sending data (either multiple $6A messages or a single $7F message is a valid response)					
Report stop sending diagnostic data packets	6A	DPID# of DPID req.	(any additional DPID #s in this message could appear to be data)		
Repeat report stop sending diagnostic data packets for each DPID # requested	6A	DPID # of DPID req.			
General response	7F	Optional data bytes may be included (see Mode $7F description).			Resp. code - 00

* same data rate definition as Data Byte #3 for Mode $21

FIGURE 22—MESSAGE DATA BYTES FOR MODE $2A

4.18 Mode $2B—Dynamically Define Data Packet by Single Byte Offsets

4.18.1 FUNCTIONAL DESCRIPTION—This test mode can be used to dynamically define data packets that can subsequently be requested using a Mode $2A request. This mode is a special case alternative to using multiple Mode $2C messages, because a single request message can be used to completely define the contents of a data packet. All other methods to specify the data values to be included in data packets using Mode $2C would also require multiple messages, therefore, special modes to build data packets by PID or memory address are not included.

If the parameter for any offset value is longer than 1 byte, then offset data bytes in the request should be filled with $FF for each additional byte in the data value requested (see message example). This forces the data value byte locations in the response message to mirror the offset value byte locations in the request message.

4.18.2 MESSAGE DATA BYTES—(See Figure 23.)

4.18.3 MESSAGE EXAMPLE—Figure 24 is an example of messages required to define a data packet by offset and then request that data packet. The data packet can be defined with a single message and requested with another message.

	Data Bytes (Hex)				
	#1	#2	#3	#4 to #n-1	#n
Request from Tool to Vehicle					
Define diagnostic data packet	2B	DPID #	Offset	Additional optional data offset values may be added to fill the message up to the maximum number of available data bytes.	
Positive Response from Vehicle to Tool (either of the following is a valid response)					
Confirm diagnostic data packet defined	6B	DPID #			
General response	7F	Optional data bytes may be included (see Mode $7F description).			Resp. Code - 00

FIGURE 23—MESSAGE DATA BYTES FOR MODE $2B

	Data Bytes (Hex)							
	#1	#2	#3	#4	#5	#6	#7	#8
Define Diagnostic Data Packet #22 consisting of: Value at offset $30 (1 byte value) Value at offset $12 (2 byte value) Value at offset $26 (2 byte value) Value at offset $05 (1 byte value)								
Define diagnostic data packet #22	2B	22	30	12	FF	26	FF	05
Positive Response from Vehicle to Tool (either of the following is a valid response)								
Confirm data packet defined #22	6B	22						
General response	7F	2B	22	Resp. Code - 00				
Request from Tool to Vehicle								
Request data packet #22	2A	Data rate	22					
Positive Response from Vehicle to Tool								
Report data packet #22	6A	22	1 byte value from offset $30	2 byte value from offset $12		2 byte value from offset $26		1 byte value from offset $05

FIGURE 24—DATA PACKET EXAMPLE

4.19 Mode $2C—Dynamically Define Diagnostic Data Packet

4.19.1 FUNCTIONAL DESCRIPTION—This test mode can be used to dynamically define data packets that can subsequently be requested using a Mode $2A request. This single test mode is an alternative to multiple modes to define data packets by offset, PID, and memory address.

This mode allows data packets to be defined in different ways, and also allows a single data packet to include data specified in different ways. Multiple messages must be sent to define a single data packet. Each message constructs a portion of the data packet, and includes an indication of how the data is being requested, the starting byte of the data packet, and the number of bytes to be specified for the data packet.

Data packets can be shortened by sending a data packet definition with the starting byte indicating the first data byte not to be included, and 0 as the length. This should be interpreted as 0 bytes at the starting location, which means that data byte is not in the response message.

Data packet definitions can be cleared by sending a message with 1 as the starting data byte and 0 as the number of bytes. Requesting this data packet would result in a response which would include the mode value, data packet ID, and no data.

Data packet definitions can be removed by sending a message with 0 as the starting data byte and 0 as the number of bytes. This would free memory space, if needed.

An optional feature of this mode is to request the present definition for the data packet. This is accomplished by sending a Mode $2B with the data packet ID as the only additional data byte. The result is multiple $6B responses, each including data bytes that resemble the $2B messages that created the data packet.

4.19.2 MESSAGE DATA BYTES—(See Figure 25.)

		Data Bytes (Hex)			
	#1	#2	#3	#4 to #n-1	#n
Request from Tool to Vehicle					

Define diagnostic data packet	2C	DPID #	Data byte #3
			bits 7,6
			  if 00 - define by offset (1 byte)
			  if 01 - define by PID (2 bytes)
			  if 10 - define by memory address (3 bytes)
			  if 11 - manufacturer defined
			bits 5,4,3
			  starting byte for data, where 001 is the first byte after the DPID #
			bits 2,1,0
			  number of data bytes for this parameter
				Data bytes starting at byte #4 contain either the offset, PID, memory address, or manufacturer defined value for the data to be included in the packet	

Positive Response from Vehicle to Tool (either of the following is a valid response)					
Confirm diagnostic data packet defined	6C	DPID #	Request data byte #3		
General response	7F	Optional data bytes may be included (see Mode $7F description).			Resp. Code - 00

FIGURE 25—MESSAGE DATA BYTES FOR MODE $2C

4.19.3 MESSAGE EXAMPLE—Figure 26 shows the messages required to define a data packet and then request that data packet. Data packet #26 is to be defined as:

    Value for PID $01 04 (1 byte data value)
    Value at offset $32 (2 byte data value)
    2 bytes of data at memory address $24 81 70

This definition requires three sets of request / report messages, and the data packet request requires one set of messages.

4.20 Mode $2F—Input/Output Control by PID

4.20.1 FUNCTIONAL DESCRIPTION—This capability allows the tester to verify proper operation of external input and output components and circuitry by isolation techniques. Real world sensor inputs can be temporarily bypassed, and direct control of output devices can be achieved. Since substitution may cause the control module to operate in a manner which is unsuitable or unsafe for "on-the-road" operation, precautions must be taken to ensure safe operation. The substituted value is used only for the duration of the diagnostic procedure, and when the module is returned to normal operation, or control of the data value is returned to the vehicle, then the substituted value reverts back to the normal value determined by the control system.

Parameter IDs (PIDs) are assigned by the manufacturer for those input, output, and intermediate (calculated) values that are allowed to be substituted. The PID assignments can be the same PID values that are used with a Mode $22 request for data by PID.

4.20.2 MESSAGE DATA BYTES—(See Figure 27.)

	Data Bytes (Hex)						
	#1	#2	#3	#4	#5	#6	#7
Request from Tool to Vehicle							
PID $01 04 is 1 byte value to be included at data byte 3 (1 byte after DPID #)	2C	26	49 ---------- 01 001 001 1 1	01	04		
Positive Response from Vehicle to Tool (either of the following is a valid response)							
PID $01 04 at data byte 3	6C	26	49				
General response	7F	2C	26	49	Resp. Code - 00		
Request from Tool to Vehicle							
Offset $32 is 2 byte value to be included at data byte 4 (2 bytes after DPID #)	2C	26	12 ---------- 00 010 010 2 2	32			
Positive Response from Vehicle to Tool (either of the following is a valid response)							
Offset $32 starting at data byte 4	6C	26	12				
General response	7F	2C	26	12	Resp. Code - 00		
Request from Tool to Vehicle							
2 bytes starting at address $24 81 70 to be included at data bytes 6 and 7	2C	26	A2 ---------- 10 100 010 4 2	24	81	70	
Positive Response from Vehicle to Tool (either of the following is a valid response)							
Addresses $24 81 70 and $24 81 71 at data bytes 6 and 7	6C	26	A2				
General response	7F	2C	26	A2	Resp. Code - 00		
After the data packet is defined, the tool can request the data packet from the vehicle							
Request data packet #26	2A	26					
Positive Response from Vehicle to Tool							
Report data packet #26	6A	26	1 byte value for PID $01 04	2 byte value from offset $32		2 bytes at memory address $24 81 70	

FIGURE 26—DATA PACKET DEFINITION EXAMPLE

4.21 Mode $30—Input/Output Control by Data Value ID

4.21.1 FUNCTIONAL DESCRIPTION—This mode is identical in purpose and function to Mode $2F, except that the parameter to be controlled is specified by a 1 byte Data Value ID instead of a 2 byte PID.

Data value IDs are assigned by the manufacturer for those input, output, and intermediate (calculated) values that are allowed to be substituted.

4.21.2 MESSAGE DATA BYTES—(See Figure 28.)

FIGURE 27—MESSAGE DATA BYTES FOR MODE $2F

	Data Bytes (Hex)					
	#1	#2	#3	#4	#5 to #n-1	#n
Request from Tool to Vehicle						
Request Input/Output control by PID	2F	PID (high byte)	PID (low byte)	Data Byte #4 (optional) if not specified - control of identified data value is disabled if specified - value is substituted for identified data value in the module Data Bytes after #4 may be used if needed to specify substitute value		
Positive Response from Vehicle to Tool (either of the following is a valid response)						
Confirm Input/Output control by PID	6F	PID (high byte)	PID (low byte)	Req. data byte 4 (opt)		
General response	7F	Optional data bytes may be included (see Mode $7F description).				Resp. Code - 00

FIGURE 28—MESSAGE DATA BYTES FOR MODE $30

	Data Bytes (Hex)				
	#1	#2	#3	#4 to #n-1	#n
Request from Tool to Vehicle					
Request Input/Output control by data value ID	30	Data value ID	Data Byte #3 (optional) if not specified - control of identified data value is disabled if specified - value is substituted for identified data value in the module Data Bytes after #3 may be used if needed to specify substitute value		
Positive Response from Vehicle to Tool (either of the following is a valid response)					
Confirm Input/Output control by data value ID	70	Data value ID	Req. data byte 3 (opt)		
General response	7F	Optional data bytes may be included (see Mode $7F description).			Resp. Code - 00

4.22 Modes $31 to $33—Perform Diagnostic Routine by Test Number

4.22.1 FUNCTIONAL DESCRIPTION—The purpose of these modes is to execute diagnostic tests and obtain test results. The diagnostic routines are manufacturer defined and executed in the control module by referencing the test number. Each module may support up to 255 distinct diagnostic routines. Definition of the routines is defined by the module manufacturer.

Reported results may consist of any set of return information, such as measured test values or fault codes.

Three sets of messages are used to perform tests resident in an on-board control module. These messages perform three functions:

Mode $31 - Enter/start Diagnostic Routine by Test Number
Mode $32 - Exit/stop Diagnostic Routine by Test Number
Mode $33 - Request Diagnostic Routine Results by Test Number

The tests are known by the module and identified by a test number. They are usually permanently stored in the module, but may have been downloaded by Modes $34 and $36.

These test numbers could either be tests that run instead of normal operating code, or could be routines that are enabled in this mode and execute with the normal operating code. In the first case, normal system operation for the controller being tested is not possible. In the second case, multiple diagnostic routines can be enabled that run while all other parts of the system are functioning normally.

Any combination of the messages can be used, depending on the implementation in the module. Some examples are:

Tester sends a Mode $31 message to start a test. The module reports that the test has started with a Mode $71 and runs the test until a Mode $32 is sent to stop the test. The module exits the test and informs the tester with a Mode $72 that the test has stopped. The tester then requests test results with a Mode $33, and the module reports the results with a Mode $73 message.

Tester sends a Mode $31 message to start a test. The Module confirms the test has started, exits when the test is done and reports results using Mode $73.

Tester sends a Mode $31 message to start a test. The Module starts the test and periodically reports results using Mode $73 until a Mode $32 is sent to stop the test.

The on-board controller starts a test automatically, stops the test automatically and reports results using a Mode $73 message automatically. This requires no request messages from the tester.

These examples demonstrate both maximum and minimum use of the available message. Test equipment needs to know what tests are available, which messages are required, and how to interpret the results.

The general purpose response, Mode $7F, may also be used instead of a Mode $71 or Mode $72 message to acknowledge entry or exit from a diagnostic routine.

4.22.2 MESSAGE DATA BYTES—(See Figure 29.)

	Data Bytes (Hex)				
	#1	#2	#3	#4 to #n-1	#n
Request from Tool to Vehicle					
Request entry by test number	31	Test #	Additional optional data bytes may be added to the request to fill the message up to the maximum number of available data bytes.		
Positive Response from Vehicle to Tool (either of the following is a valid response)					
Report entry by test number	71	Test #	Additional optional data bytes that are in response to the request may be added to fill the message up to the maximum number of available data bytes.		
General response	7F	Optional data bytes may be included (see Mode $7F description).			Resp. code - 00
Request from Tool to Vehicle					
Request exit by test number	32	Test #	Additional optional data bytes may be added to the request to fill the message up to the maximum number of available data bytes.		
Positive Response from Vehicle to Tool (either of the following is a valid response)					
Report exit by test number	72	Test #	Additional optional data bytes that are in response to the request may be added to fill the message up to the maximum number of available data bytes.		
General response	7F	Optional data bytes may be included (see Mode $7F description).			Resp. code - 00
Request from Tool to Vehicle					
Request results by test number	33	Test #	Additional optional data bytes may be added to the request to fill the message up to the maximum number of available data bytes.		
Positive Response from Vehicle to Tool					
Report results by test number	73	Test #	Additional optional data bytes that are in response to the request may be added to fill the message up to the maximum number of available data bytes.		

Note: Mode $33 reports the results of a test that was run on request from a tool. This differs from Mode $05, which reports the results of the latest on-board monitoring test that was run to meet OBD regulations.

FIGURE 29—MESSAGE DATA BYTES FOR MODES $31 THROUGH $33

4.23 Modes $34 to $37—Data Transfer

4.23.1 FUNCTIONAL DESCRIPTION—These modes allow the transfer of data either from a tester to an on-board module or from an on-board module to a tester. Modes defined are:

Mode $34 - Request Download - tool to module
Mode $35 - Request Upload - module to tool

Mode $36 - Data Transfer
Mode $37 - Request Data Transfer Exit

Download is defined as data transfer from the tool to the controller in the vehicle. Upload is defined as data transfer from the controller in the vehicle to the tool.

4.23.2 MESSAGE DATA BYTES—(See Figure 30.)

		Data Bytes (Hex)	
	#1	#2 to #n-1	#n
Request from Tool to Vehicle			
Request download	34	Messages used to request download are defined by the vehicle/module manufacturer. This Mode value is reserved for a download request message.	
Positive Response from Vehicle to Tool (either of the following is a valid response)			
Report download status	74	Messages used to request download are defined by the vehicle/module manufacturer. This Mode value is reserved for the response to a Mode $34.	
General response	7F	Optional data bytes may be included (see Mode $7F description).	Resp. code - 00
Request from Tool to Vehicle			
Request upload	35	Messages used to request upload are defined by the vehicle/module manufacturer. This Mode value is reserved for an upload request message.	
Positive Response from Vehicle to Tool (either of the following is a valid response)			
Report upload status	75	Messages used to request download are defined by the vehicle/module manufacturer. This Mode value is reserved for the response to a Mode $35.	
General response	7F	Optional data bytes may be included (see Mode $7F description).	Resp. code - 00
Transfer Data			
Data transfer	36	Messages used to perform the data transfer are defined by the vehicle/module manufacturer. This Mode value is reserved for the data transfer request message.	
Data transfer	76	Messages used to perform the data transfer are defined by the vehicle/module manufacturer. This Mode value is reserved for the response to a Mode $36.	
Request from Tool to Vehicle			
Request exit	37	Messages used to exit data transfer are defined by the vehicle/module manufacturer. This Mode value is reserved for the exit data transfer request message.	
Positive Response from Vehicle to Tool (either of the following is a valid response)			
Report exit	77	Messages used to exit data transfer are defined by the vehicle/module manufacturer. This Mode value is reserved for the response to a Mode $37.	
General response	7F	Optional data bytes may be included (see Mode $7F description).	Resp. code - 00

FIGURE 30—MESSAGE DATA BYTES FOR MODES $34 THROUGH $37

4.24 Modes $38 to $3A—Perform Diagnostic Routine at a Specified Address

4.24.1 FUNCTIONAL DESCRIPTION—These test modes are used to execute code resident in the on-board controller at the specified address. This executable code may be permanently stored in the controller or may have been downloaded using Modes $34 and $36.

These three sets of messages are used to perform tests resident in an on-board control module. These messages perform three functions:

Mode $38 - Enter Diagnostic Routine by Address
Mode $39 - Exit Diagnostic Routine by Address
Mode $3A - Request Diagnostic Routine Results

4.24.2 MESSAGE DATA BYTES—(See Figure 31.)

	Data Bytes (Hex)						
	#1	#2	#3	#4	#5	#6 to #n-1	#n
Request from Tool to Vehicle							
Request test entry by starting address	38	Starting Address			(opt)	Additional optional data bytes may be added to the request to fill the message up to the maximum number of available data bytes.	
		high byte	mid byte	low byte			
Positive Response from Vehicle to Tool (either of the following is a valid response)							
Report entry	78	Starting Address			(opt)		
		high byte	mid byte	low byte			
General response	7F	Optional data bytes may be included (see Mode $7F description).					Resp. Code - 00
Request from Tool to Vehicle							
Request test exit by starting address	39	Starting Address			(opt)	Additional optional data bytes may be added to the request to fill the message up to the maximum number of available data bytes.	
		high byte	mid byte	low byte			
Positive Response from Vehicle to Tool (either of the following is a valid response)							
Report exit by starting address	79	Starting Address			Exit Status (opt)	Additional optional data bytes that are in response to the request may be added to fill the message up to the maximum number of available data bytes.	
		high byte	mid byte	low byte			
General response	7F	Optional data bytes may be included (see Mode $7F description).					Resp. code - 00
Request/Report Diagnostic Routine Results							
Request results	3A	(opt.)				Additional optional data bytes may be added to the request to fill the message up to the maximum number of available data bytes.	
Positive Response from Vehicle to Tool							
Report results	7A	Additional optional data bytes that are in response to the request may be added to fill the message up to the maximum number of available data bytes.					

FIGURE 31—MESSAGE DATA BYTES FOR MODES $38 THROUGH $3A

4.25 Mode $3B—Write Data Block

4.25.1 FUNCTIONAL DESCRIPTION—The purpose of this mode is to provide a means for the external test device to change the contents of a data block. The data block numbers and associated memory locations need to be known by the on-board device. This mode does not allow off-board test equipment to change any memory locations other than for those data blocks predefined in the on-board device.

The number of data bytes included in the request message for each block number depends on the system design and intent of the usage for that block.

Possible uses for this mode are:
Clear non-volatile memory
Reset learned values in a single table
Set option content
Set Vehicle Identification Number (VIN)
Change calibration values

This mode can be used to clear or reset tables stored in non-volatile memory. A data block number could be defined as memory locations that need to be either cleared or reset to predefined initial values, such as learned values in a table. Sending the data block number with no data bytes could cause the on-board system to clear/reset all learned values associated with that data block number to known nominal values.

Another use is to allow the VIN to be input. A data block number could be defined to expect a portion of the VIN. Multiple messages would be required to send the entire 17-character VIN. Whatever data values were included in the request message would be stored in the memory locations known by the on-board system for VIN.

In some systems, this mode could be used for programming calibrations or vehicle specific information in the assembly plant or service without using secured methods for reprogramming. If desired that those values not be changed after production, the capability to modify those data blocks could be disabled after the initial change.

4.25.2 MESSAGE DATA BYTES—(See Figure 32.)

	Data Bytes (Hex)			
	#1	#2	#3 to #n-1	#n
Request from Tool to Vehicle				
Request to write data block	3B	Block number	Additional optional data value bytes to be written to memory may be added to fill the message up to the maximum number of available data bytes.	
Positive Response from Vehicle to Tool (either of the following is a valid response)				
Confirm data block written	7B	Block number		
General response	7F	Optional data bytes may be included (see Mode $7F description).		Resp. code - 00

FIGURE 32—MESSAGE DATA BYTES FOR MODE $3B

4.26 Mode $3C—Read Data Block

4.26.1 FUNCTIONAL DESCRIPTION—The purpose of this mode is to provide a means for the external test device to read the contents of a data block. The data block numbers and associated memory locations need to be known by the on-board device. This mode does not allow off-board test equipment to read any memory locations other than for those data blocks predefined in the on-board device.

The number of data bytes included in the Response for each block number depends on the system design and intent of the usage for that block.

Possible uses for this mode are to read:
  Learned values in a single table
  Option content
  Vehicle Identification Number (VIN)
  Calibration values

4.26.2 MESSAGE DATA BYTES—(See Figure 33.)

	Data Bytes (Hex)		
	#1	#2	#3 to #n
Request from Tool to Vehicle			
Request to read data block	3C	Block number	
Positive Response from Vehicle to Tool			
Report contents of data block	7C	Block number	Additional optional data bytes that are in response to the request may be added to fill the message up to the maximum number of available data bytes.

FIGURE 33—MESSAGE DATA BYTES FOR MODE $3C

4.27 Mode $3F—Test Device Present (no operation performed)

4.27.1 FUNCTIONAL DESCRIPTION—This mode can be used to indicate a tool or test device is present. This message may be required, in the absence of other diagnostic messages, to prevent modules from automatically returning to normal operation. The presence of this message will indicate that the system should remain in a diagnostic mode of operation.

This message will normally be sent to all modules in the vehicle. Because this message is not a specific request for a controller to do something, there is no required response message to the test tool present message. However, if the message strategy includes a response to this, a general response message should be used. If the vehicle contains multiple modules, then multiple responses will be sent, using available data link time.

4.27.2 MESSAGE DATA BYTES—(See Figure 34.)

	Data Bytes (Hex)		
	#1	#2 to #n-1	#n
Request from Tool to Vehicle			
Test tool present	3F		
Positive Response from Vehicle to Tool (optional)			
General response	7F	Optional data bytes may be included (see Mode $7F description).	Resp. Code - 00

FIGURE 34—MESSAGE DATA BYTES FOR MODE $3F

4.28 Mode $7F—General Response Message

4.28.1 FUNCTIONAL DESCRIPTION—Physical requests for information must be responded to by the intended receiver of the request. This message is intended to be a generic response to any request message when data does not need to be returned to the test device. This is an alternative to a specific response when the request is for an action to be performed by the controller. This is also a reject message (negative response) for requests for data or for an action to be performed. See the "Message Response" section for a more detailed discussion about the use of this message.

The response may be an acknowledge that the message was received, or may include a code indicating the reason for rejection if a valid response is not possible. This response assumes that a valid message has been received, and is not intended to inform the sender of an invalid message or to report system failures. Valid response codes are included in this section. Response codes from $00 to $7F are reserved to be defined in this document, and codes from $80 to $FF are reserved to be defined by the vehicle manufacturer.

The format for the general response message is a $7F for the first data byte after the header, optional data bytes that repeat data bytes of the request, and the response code. If a fixed message length is required and there are not enough data bytes in the request, the message should be padded with $00 to fill the remaining data bytes.

4.28.2 MESSAGE DATA BYTES—(See Figure 35.)

	Data Bytes (Hex)				
	#1	#2 to #n-1 (optional)			#n
Response from Vehicle to Tool					
General response	7F	Optional data bytes that repeat data bytes in the request message, in order to uniquely identify the request message, or in order to maintain a common message length, may be inserted between the Mode Value and the Response Code. The last data byte of this message is always the response code.			Resp. Code
		Request Mode value (opt)	Data byte #2 of request (opt)	Data byte #3 of request (opt)	etc.

FIGURE 35—MESSAGE DATA BYTES FOR MODE $7F

4.28.3 MESSAGE EXAMPLES—Figure 36 shows an example of a Mode $3B request containing 8 data bytes. Four valid general response messages are shown, with others possible to be defined by the vehicle manufacturer. Note that the response code is always the last data byte of the response message.

	Data Bytes (Hex)							
	#1	#2	#3	#4	#5	#6	#7	#8
Example Request from Tool to Vehicle								
Request to write block of memory	3B	Block number	Data value #1	Data value #2	Data value #3	Data value #4	Data value #5	Data value #6
Example Responses from Vehicle to Tool (any of the following is a valid response)								
Simplest form of general response	7F	Resp. Code						
Include test mode value to identify test mode requested	7F	3B	Resp. Code					
Include test mode value and block number requested	7F	3B	Block number	Resp. Code				
Include as much of request message as will fit in response	7F	3B	Block number	Data value #1	Data value #2	Data value #3	Data value #4	Resp. Code

FIGURE 36—GENERAL RESPONSE EXAMPLES

4.28.4 RESPONSE CODES—Figure 37 includes response codes and descriptions to be used with the general response messages.

Resp Code	Response Code Description
00	AFFIRMATIVE RESPONSE This response code is used to indicate affirmative response to a request message. It indicates that the requested action will be taken or is already complete.
10	GENERAL REJECT This response code is used to indicate a negative response to a request message. No reason or explanation for the rejection is given.
11	MODE NOT SUPPORTED This response code is used to indicate negative response to a request message. It indicates that the requested action will not be taken because the ECU does not support the requested mode.
12	SUB-FUNCTION NOT SUPPORTED or INVALID FORMAT This response code is used to indicate negative response to a request message. It indicates that the requested action will not be taken because the ECU does not support the arguments of the requested mode or the format of the argument bytes do not match the prescribed format for the specified mode.
21	BUSY -- REPEAT REQUEST This response code indicates that the ECU is temporarily too busy to perform the requested operation. In this circumstance repetition of the request will eventually result in an affirmative response. This code may be returned while an ECU is in the process of clearing stored codes or fetching information. Whenever completion of the requested operation will exceed the maximum response time limit, it is appropriate to use the BUSY -- REPEAT REQUEST message to lengthen the acceptable response period.
22	CONDITIONS NOT CORRECT or REQUEST SEQUENCE ERROR This response code is used to indicate negative response to a request message. It indicates that the requested action will not be taken because the ECU prerequisite conditions are not met. This request may elicit an affirmative response at another time. This code may occur when sequence sensitive requests are issued in the wrong order.
23	ROUTINE NOT COMPLETE This response code is used to indicate that the message was properly received and the routine is in process, but not yet completed.
31	REQUEST OUT OF RANGE This response code is used to indicate negative response to a request message. It indicates that the requested action will not be taken because the ECU detects the request message contains a data byte which attempts to substitute a value beyond its range of authority (ex: attempting to substitute a data byte of 111 when the data is only defined to 100).
33	SECURITY ACCESS DENIED This response code is used to indicate negative response to a request message. It indicates that the requested action will not be taken because the ECU's security strategy has not been satisfied by the tester.
34	SECURITY ACCESS ALLOWED This response code is used to indicate positive response to a request message. It indicates that the ECU's security strategy has been satisfied by the tester.
35	INVALID KEY This response code is used to indicate negative response to a request message. It indicates that security access has not been given because the ECU's security key was not matched by the tester (this counts as an attempt to gain security access).
36	EXCEED NUMBER OF ATTEMPTS This response code is used to indicate negative response to a request message. It indicates that the requested action will not be taken because the Tester has unsuccessfully attempted to gain security access more times than the ECU's security strategy will allow.
37	REQUIRED TIME DELAY NOT EXPIRED This response code is used to indicate negative response to a request message. It indicates that the requested action will not be taken because the tester's latest attempt to gain security access was initiated before the ECU's required timeout period had elapsed.
40	DOWNLOAD NOT ACCEPTED This response code indicates that an attempt to download to an ECU's memory cannot be accomplished due to some fault condition.
41	IMPROPER DOWNLOAD TYPE This response code indicates that an attempt to download to an ECU's memory cannot be accomplished because the module does not support the type of download being attempted.
42	CAN'T DOWNLOAD TO SPECIFIED ADDRESS This response code indicates that an attempt to download to an ECU's memory cannot be accomplished because the module does not recognize the target address for the download as being available.
43	CAN'T DOWNLOAD NUMBER OF BYTES REQUESTED This response code indicates that an attempt to download to an ECU's memory cannot be accomplished because the module does not recognize the number of bytes requested for the download as being available.
44	READY FOR DOWNLOAD This response code indicates that a download to an ECU's memory can be accomplished.
50	UPLOAD NOT ACCEPTED This response code indicates that an attempt to upload from an ECU's memory cannot be accomplished due to some fault condition.
51	IMPROPER UPLOAD TYPE This response code indicates that an attempt to upload from an ECU's memory cannot be accomplished because the module does not support the type of upload being attempted.
52	CAN'T UPLOAD FROM SPECIFIED ADDRESS This response code indicates that an attempt to upload from an ECU's memory cannot be accomplished because the module does not recognize the target address for the upload as being available.
53	CAN'T UPLOAD NUMBER OF BYTES REQUESTED This response code indicates that an attempt to upload from an ECU's memory cannot be accomplished because the module does not recognize the number of bytes requested for the upload as being available.
54	READY FOR UPLOAD This response code indicates that an upload from an ECU's memory can be accomplished.
61	NORMAL EXIT WITH RESULTS AVAILABLE This response code is used to indicate response to a request message. It indicates that the requested exit action will be taken, or is already complete, and that additional information is available in the form of data, stored codes, or PIDs.

Continued

FIGURE 37—GENERAL RESPONSE CODES

62	NORMAL EXIT WITHOUT RESULTS AVAILABLE
	This response code is used to indicate response to a request message. It indicates that the requested exit action will be taken, or is already complete, but no additional information is available in the form of data, stored codes, or PIDs.
63	ABNORMAL EXIT WITH RESULTS
	This response code is used to indicate response to an exit request message. It indicates that the requested exit action will be taken, or is already complete. It is also indicates that the requested execution did not occur in the expected manner and that additional information is available in the form of data, stored codes, or PIDs.
64	ABNORMAL EXIT WITHOUT RESULTS
	This response code is used to indicate response to a exit request message. It indicates that the requested exit action will be taken, or is already complete. It is also indicates that the requested execution did not occur in the expected manner, but additional information is not available.
71	TRANSFER SUSPENDED
	This response code indicates that a block transfer operation was halted due to some fault, but will be completed later.
72	TRANSFER ABORTED
	This response code indicates that a block transfer operation was halted due to some fault, and will not be completed later.
73	BLOCK TRANSFER COMPLETE / NEXT BLOCK
	This response code indicates that a block transfer operation has been completed successfully.

Continued

FIGURE 37—GENERAL RESPONSE CODES (CONTINUED)

74	ILLEGAL ADDRESS IN BLOCK TRANSFER
	This response code indicates that the starting address included in the block transfer request message is either out of range, protected, the wrong type of memory for receiving data, or cannot be written to for some reason.
75	ILLEGAL BYTE COUNT IN BLOCK TRANSFER
	This response code indicates that the number of data bytes included in the block transfer request message is either more than the block transfer can accommodate, requires more memory than is available at the requested starting address, or cannot be handled by the block transfer software.
76	ILLEGAL BLOCK TRANSFER TYPE
	This response code is used to indicate that the block transfer type included in the request for block transfer is not valid for this application.
77	BLOCK TRANSFER DATA CHECKSUM ERROR
	This response code is used to indicate that the block transfer data checksum calculated for the block transfer message does not agree with the expected value.
78	BLOCK TRANSFER MESSAGE CORRECTLY RECEIVED
	This response code is used to indicate that the block transfer message was received correctly, and that any parameters included in the message were valid, but the action to be performed may not be completed yet. This response code can be used to indicate that the message was properly received and does not need to be retransmitted, but the receiving device is not yet ready to receive another block transfer.
79	INCORRECT BYTE COUNT DURING BLOCK TRANSFER
	This response code indicates that the number of bytes that was expected to be sent was not the same as the number of bytes received.
80 to FF	MANUFACTURER SPECIFIC CODES
	These response codes are reserved to be specified by the vehicle manufacturer.

FIGURE 37—GENERAL RESPONSE CODES (CONTINUED)

APPENDIX A
DIAGNOSTIC TEST MODE ASSIGNMENTS

A.1 See Table A1.

TABLE A1—DIAGNOSTIC TEST MODE ASSIGNMENTS

Mode	Message Name
00	
01	Request Current Powertrain Diagnostic Data
02	Request Powertrain Freeze Frame Data
03	Request Powertrain Diagnostic Trouble Codes
04	Request to Clear/Reset Diagnostic Trouble Codes
05	Request Oxygen Sensor Monitoring Test Results
06	Request On-Board Monitoring Test Results
07	Request Pending Powertrain Trouble Codes
08	
09	
0A	
0B	
0C	
0D	
0E	
0F	
10	Initiate Diagnostic Operation
11	Request Module Reset
12	Request Diagnostic Freeze Frame Data
13	Request Diagnostic Trouble Code Information
14	Clear Diagnostic Information
15	
16	
17	Request Status of Diagnostic Trouble Codes
18	Request Diagnostic Trouble Codes by Status
19	
1A	
1B	
1C	
1D	
1E	
1F	
20	Return to Normal Operation
21	Request Diagnostic Data by Offset
22	Request Diagnostic Data by PID
23	Request Diagnostic Data by Memory Address
24	Request Scaling and Offset/PID
25	Request to Stop Transmitting Data
26	Specify Setting of Data Rates
27	Data Link Security Access
28	Disable Normal Message Transmission
29	Enable Normal Message Transmission
2A	Request Diagnostic Data Packet(s)
2B	Define Diagnostic Data Packet by Offset
2C	Define Diagnostic Data Packet
2D	
2E	
2F	Input/Output Control by PID
30	Input/Output Control by Data Value ID
31	Request Start Diagnostic Routine by Test No.
32	Request Stop Diagnostic Routine by Test No.
33	Request Diagnostic Routine Results by Test No.
34	Request Download
35	Request Upload
36	Transfer Data

23.57

Mode	Message Name	Mode	Message Name
37	Request Stop Data Transfer	5C	
38	Request Start Diagnostic Routine by Address	5D	
39	Request Stop Diagnostic Routine by Address	5E	
3A	Request Diagnostic Routine Results by Address	5F	
3B	Write Data Block		
3C	Read Data Block	60	Confirm Normal Operation
3D		61	Report Diagnostic Data by Offset
3E		62	Report Diagnostic Data by PID
3F	Test Device Present - No Operation Performed	63	Report Diagnostic Data by Memory Address
40		64	Report Parameter Offset and Scaling/Mask
41	Report Current Powertrain Diagnostic Data	65	Confirm Data Stopped
42	Report Powertrain Freeze Frame Data	66	Confirm Data Rates Set
43	Report Powertrain Diagnostic Trouble Codes	67	Confirm Security Access
44	Report Diagnostic Trouble Codes Cleared	68	Confirm Normal Message Transmission Disabled
45	Report Oxygen Sensor Test Results	69	Confirm Normal Message Transmission Enabled
46	Report On-Board Moinitoring Test Results	6A	Report Diagnostic Data Packet(s)
47	Report Pending Powertrain Trouble Codes	6B	Confirm Data Packet Defined by Offset
48		6C	Confirm Data Packet Defined
49		6D	
4A		6E	
4B		6F	Confirm Input/Output Control by PID
4C			
4D		70	Confirm Input/Output Control by Data Value ID
4E		71	Confirm Diagnostic Routine Started by Test No.
4F		72	Confirm Diagnostic Routine Stopped by Test No.
		73	Report Diagnostic Routine Results by Test No.
50	Confirm Diagnostic Operation	74	Report Download Status
51	Confirm Module Reset	75	Report Upload Status
52	Report Diagnostic Freeze Frame Data	76	Data Transfer
53	Report Diagnostic Trouble Code Information	77	Confirm Data Transfer Stopped
54	Confirm Diagnostic Information Cleared	78	Confirm Diagnostic Routine Started by Address
55		79	Confirm Diagnostic Routine Stopped by Address
56		7A	Report Diagnostic Routine Results by Address
57	Report Status of Diagnostic Trouble Codes	7B	Confirm Data Block Written
58	Report Diagnostic Trouble Codes by Status	7C	Report Contents of Data Block
59		7D	
5A		7E	
5B		7F	General Response Message (accept/reject)

(R) IGNITION SYSTEM NOMENCLATURE AND TERMINOLOGY—SAE J139 JUN90

SAE Recommended Practice

Report of Electrical Equipment Committee approved January 1970. Completely revised by the Ignition Systems Standards Committee June 1990. Rationale statement available.

1. Scope—To provide standard terminology and definitions with regard to ignition systems for spark-ignited internal combustion engines.

2. References

2.1 Applicable Documents

SAE J973, Ignition System Measurement Procedure

3. Types of Ignition Systems

3.1 Electronic—A system in which the coil current is controlled by semiconductors. The semiconductors can be controlled by mechanical breaker points or other means.

3.2 Breakerless—A system like 3.1 except the semiconductors are controlled by means other than mechanical breaker points.

3.3 Distributorless (Static Distribution System)—A system which omits the conventional spark voltage distributor.

3.4 Magneto—A system which utilizes a permanent magnet on a rotating part of the engine to generate energy. It may be conventional (3.5.1), electronic (3.1), breakerless (3.2), or distributorless (3.3).

3.5 Inductive—A system which stores primary energy in a coil (inductor). (The energy can be discharged into a coil by mechanical breaker points, semiconductors or other means.)

3.5.1 CONVENTIONAL (KETTERING)—A system which consists of a coil, spark voltage distributor, battery and mechanical breaker points to control coil current.

3.6 Capacitor Discharge (C.D.)—A system which stores primary energy in a capacitor.

NOTE: As an example, an ignition system could be an electronic, breakerless, distributorless system.

4. Parameters

4.1 Available Secondary (Spark) Voltage—The minimum voltage at the spark plug terminal with the terminal open-circuited and insulated from ground. Voltage to be measured under specified conditions.

4.2 Required Secondary (Spark) Voltage—The maximum voltage required at the spark plug terminal to breakdown the spark plug gap. Voltage to be measured under specified conditions.

Voltage should be measured under full load (wide-open throttle) and a variety of part-load conditions, transients and cold.

4.3 Ignition Voltage Reserve—The difference between the available and required secondary (spark) voltages.

An adequate reserve is necessary for the ignition system to tolerate moisture, corona of the ignition cable, partially fouled spark plugs, etc.

4.4 Open-Circuit Coil Secondary Voltage—The voltage measured at the coil output terminal with secondary cable disconnected.

4.5 Loaded Secondary Voltage—The voltage measured at the spark plug terminal with the secondary cable disconnected from the spark plug and a noninductive (1% M ohm ± 1, 10 watt 0.0005%/V maximum voltage coefficient, dielectric strength that exceeds the system voltage) load resistor connected to the cable spark plug terminal.

4.5.1 SECONDARY VOLTAGE AT PRIMARY CURRENT SWITCH ON—Voltage which is induced in secondary winding due to rate of change of primary current at switch on.

4.6 Supply Voltage—The direct current (DC) voltage at the input terminals of the ignition system, under specified conditions.

4.7 Peak Coil Primary Voltage—The peak of the first half-cycle of the voltage at the coil primary terminals after discharge of the ignition.

4.8 Arc Voltage—The instantaneous voltage observed across the spark-gap during arcing.

4.9 Spark Current—The instantaneous current observed passing through the spark-gap electrodes during arcing.

4.10 Spark Energy—The energy dissipated between the spark-gap electrodes as determined by the integral of the product of spark voltage and spark current during current flow.

4.10.1 SPARK ENERGY—Optional method (see SAE J973).

4.11 Spark Duration—The length of time a spark is established across a spark-gap (in the spark-gap) under specified conditions.

4.12 Rise-Time—The time required (microseconds) for the secondary available voltage to rise from 10 to 90% of the peak voltage under specified conditions.

4.12.1 Rise time gradient 10 to 90% KV divided by the rise time in microseconds.

4.13 Minimum Operating Specified Speed (Cut-in)—The minimum engine speed at which the ignition system distributes a specified spark voltage, conditions of test to be specified.

4.14 Average Supply Current—The DC input current to an ignition system, under specified conditions.

4.15 Peak Coil Current—The peak current flowing through the coil primary winding, under specified conditions.

4.16 Coil Interruption Current—The peak current flowing through the coil primary winding at the time of interruption.

4.17 Timing Lag—The interval between the timing event and occurrence of a 12 KV spark under specified conditions. (Usually expressed in engine degrees per 1000 engine RPM.)

4.18 Energizing Interval (Dwell Time or Dwell Angle)—The interval during which the capacitor (CD ignition) is being charged or the coil current (inductive ignition) is flowing.

4.19 Ignition Coil—A transformer with an air or magnetic core used to step-up a low primary voltage to a high secondary voltage.

4.19.1 SINGLE ENDED IGNITION COIL—An ignition coil with a single output secondary winding.

4.19.2 DOUBLE ENDED IGNITION COIL—An ignition coil with one secondary winding that has a high tension terminal at each end of the winding.

4.19.3 GROUND SPARK—A spark which takes place simultaneously at the exhaust stroke of another cylinder when a spark occurs at the compression stroke of a cylinder.

4.20 Distributor—A device which distributes the spark voltage to the various spark plugs.

4.21 External Primary (Ballast) Resistor—A resistor, if used, that is connected in series with the coil primary circuit.

4.22 Ignition Cables–Metallic—A high voltage cable with metallic conductors. This cable routes the high voltage from the coil to distributor and distributor to spark plugs.

4.23 Ignition Cables–Nonmetallic—A high voltage cable similar to ignition cables–metallic (4.22) except with nonmetallic conductors such as carbon fibres.

4.24 Ignition Trigger Device—The device used to initiate the discharge of the energy stored in the ignition system.

4.25 Stored Energy—Theoretically, the amount of energy stored in the storage element (capacitor or coil) of the ignition system. This value does not take into consideration inefficiencies or losses in the system.

4.25.1 INDUCTIVE SYSTEM:

$$W = 1/2\, Li^2 \qquad (Eq.1)$$

Where:

W = Energy (joules) stored in coil inductive field
L = Coil primary inductive (henrias)
i = Coil primary interruption current (amperes)

4.25.2 CAPACITOR DISCHARGE SYSTEM:

$$W = 1/2\, CV^2 \qquad (Eq.2)$$

Where:

W = Energy (joules) stored in the storage capacitor
C = Storage capacitance (farads)
V = Voltage (volts) across the storage capacitor at the moment discharge begins

(R) IGNITION SYSTEM MEASUREMENTS PROCEDURE—SAE J973 JUN93

SAE Recommended Practice

Report of the Electrical Equipment Committee approved October 1966 and revised January 1973. Revised by the Ignition Systems Standards Committee August 1991. Rationale statement available. Completely revised by the SAE Ignition Systems Standards Committee June 1993. Rationale statement available.

1. Scope—This SAE Recommended Practice is intended to provide any technical person or group interested in ignition system design and/or evaluation with the specific equipment, conditions, and methods which will produce test results definitive and reproducible for his own work and yet sufficiently standardized to be acceptable to other groups working on battery ignition systems for automotive engines.

2. References

2.1 Applicable Document—The following publication forms a part of this specification to the extent specified herein. The latest issue of SAE publications shall apply.

2.1.1 SAE PUBLICATION—Available from SAE, 400 Commonwealth Drive, Warrendale, PA 15096-0001.

SAE AIR84—Ignition Peak Voltage Measurements

3. DC Source—The source of DC voltage to be used in ignition system measurements shall be a variable DC power supply having a 10 to 90% transient recovery time of not more than 50 µs over the load range encountered in use. It must have no more than 10 mV variation in average voltage from no load to full ignition system load and no more than 50 mV peak-to-peak ripple over the same load range. This power supply shall be shunted by a suitably tapped automotive-type lead acid battery and be positioned immediately adjacent to the test area so that the source impedance of a vehicle is simulated as closely as possible.

4. Ignition System Definition—The ignition system as defined for the tests tabulated in this report shall consist of:

a. A coil. This can be the conventional induction coil or an air or magnetic core transformer.

b. A coil external resistor or resistors if the coil being tested requires an external resistor.

c. A distributor. This is defined as any device which incorporates a timing mechanism, a spark advance mechanism or mechanisms, and a spark distribution mechanism, all of which have a proper angular interrelationship to themselves and, through a mechanical drive, to the engine.

d. High voltage, metal conductor ignition cables: coil to distributor—455 mm (18 in) long, distributor to spark gap—610 mm (24 in) long. Metal conductor cables are specified to eliminate the varying effects of the different kinds of cable with high impedance conductors. Resistance per foot, as well as inductance of spark plug cables built to suppress radiation, can be quite different from manufacturer to manufacturer.

NOTE—Some ignition systems may not function properly with metallic secondary cables due to EMI and may require low resistance inductance cables.

e. Any auxiliary switching means implicit with the system being tested such as a transistorized control unit.

The preceding devices shall be interconnected as the manufacturer recommends or similar to the conventional system illustrated in Figure 1.

5. System Load—The load connected to the ignition system shall be a multigap spark gap test stand, each gap being individually variable, the number of gaps used being the same as the number of towers on the distributor cap. Using an 8-cylinder distributor as an example, seven gaps will be set to fire at a nominal 12 kV, the remaining gap will be opened to the point where it never can fire. Attached to the nonfiring gap, by not less than 305 mm (1 ft) of secondary ignition cable, will be a high quality (dissipation factor of 3% or less), high voltage, 50 pico farad capacitor (this can be a section of shielded ignition cable) to simulate the capacitance of the cables and spark plugs as normally encountered on a vehicle, and at suitable times a low voltage coefficient (0.0005%/V max), noninductive approximately 10 W, 1.0 MΩ resistor. The resistor simulates lead or carbon fouled spark plugs.

For certain tests, as designated in Section 6, the capacitive and resistive loads will be directly connected to the coil high voltage tower with the coil not firing.

6. Measurements to be Made

6.1 Group A

6.1.1 AVAILABLE VOLTAGE AT SPARK PLUG—This measurement is fundamental to spark ignition. Comparing available voltage to voltage required to fire spark plugs (in a given engine) determines the adequacy of the ignition system. (See Figure 2A.)

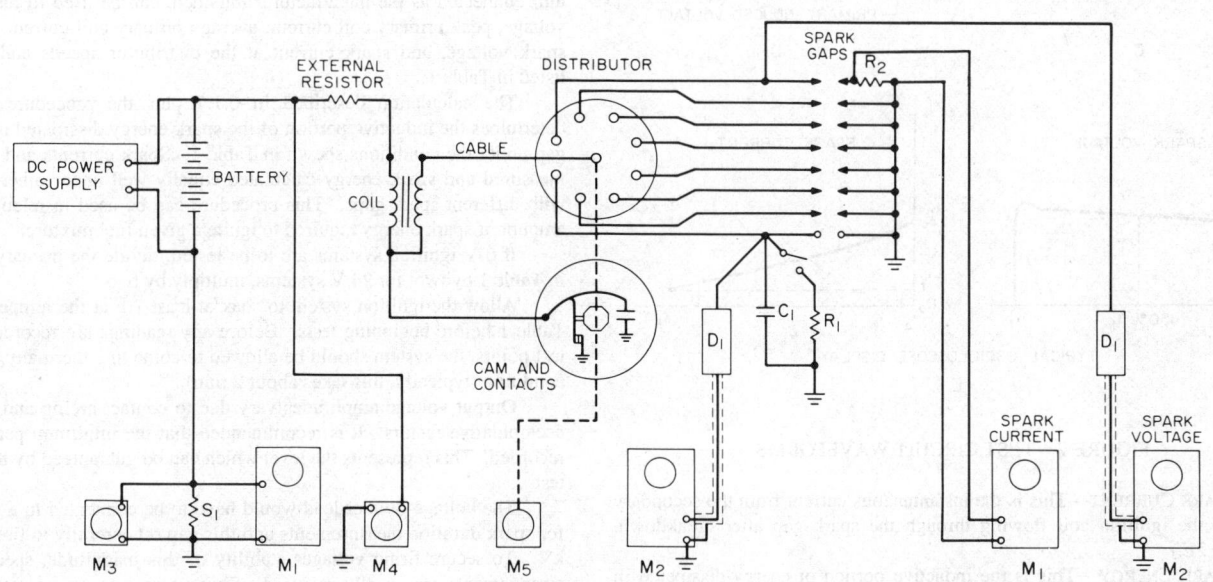

M_1 = CURRENT MEASURING OSCILLOSCOPE
M_2 = VOLTAGE MEASURING OSCILLOSCOPE
M_3 = DC AMMETER
M_4 = DC VOLTMETER
M_5 = TACHOMETER
C_1 = SHUNT CAPACITY ($C_1 + D_1$ = 50 MMF)
S_1 = 0.1 OHM METER SHUNT
D_1 = VOLTAGE DIVIDER
R_1 = 1.0 MEGOHM, 10 WATT
R_2 = SECONDARY CURRENT SENSING RESISTOR 100 OHM, 1/2 WATT - CARBON ±1%

FIGURE 1—TEST CIRCUIT ARRANGEMENT FOR GROUP A TESTS

6.1.2 PEAK COIL PRIMARY CURRENT—This measurement indicates energy into the coil ($E = 1/2 Li^2$) and must be controlled to insure adequate distributor contact life. (See Figure 2B.)

NOTE—Contacts are only used in distributors with mechanical switching.

6.1.3 AVERAGE COIL PRIMARY CURRENT—This measurement determines the average current draw of the system with respect to the DC source (alternator, generator, battery, etc.).

6.1.4 SPARK DURATION—Within limits, this measurement is indicative of the igniting capability of a spark under marginal fuel conditions. It also is an indication of the amount of erosion which will occur on spark plug electrodes due to electrical means. Because of the complexity of both of these areas, however, experience is required to use this information effectively. (See Figure 2C.)

6.1.5 SPARK VOLTAGE—This is the instantaneous voltage observed across the spark gap halfway through the discharge. (See Figure 2E.)

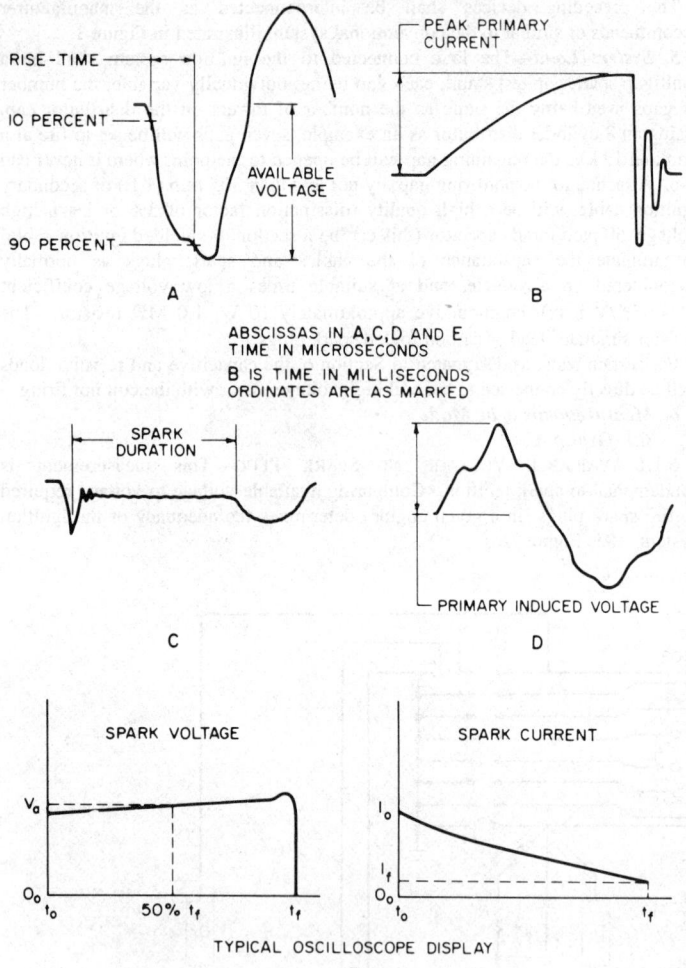

FIGURE 2—TEST CIRCUIT WAVEFORMS

6.1.6 SPARK CURRENT—This is the instantaneous current from the secondary winding of the ignition coil flowing through the spark gap after breakdown. (See Figure 2E.)

6.1.7 SPARK ENERGY—This is the inductive portion of energy dissipated in the spark after breakdown. It is calculated as shown:

$$E_{spark} = \frac{V_a(t_f - t_o)(i_f + i_o)}{2} \quad (Eq.1)$$

where:

t_o and i_o = initial values of time and current of the spark after breakdown
t_f and i_f = final values of time and current of the spark after breakdown
V_a = spark voltage at $(t_f - t_o)/2$

6.2 Group B

6.2.1 COIL SECONDARY VOLTAGE RISETIME—This measurement is an indication of the ability of an ignition system to fire shunted (fouled) spark plugs. The shorter the risetime, the less system energy is lost across the fouled shunt and the more voltage is available to fire the plug. (See Figure 2A.)

6.2.2 COIL PRIMARY INDUCED VOLTAGE—This measurement is useful with respect to distributor contact life on conventional ignition systems and is a measure of the stress on a semiconductor power switch in inductive energy storage ignition systems. (See Figure 2D.) This measurement is not applicable to capacitor discharge ignition systems.

7. Test Equipment

7.1 A voltage divider and oscilloscope for measuring high voltage as defined in SAE AIR84 should be used to measure available voltage, risetime, and spark duration.

7.2 An oscilloscope with a maximum risetime of 0.035 µs and with a minimum band pass of 10MC (ref. Tektronix 535A with a type L plug-in unit) with its input connected across a noninductive meter shunt which is in series with the coil primary for peak coil primary current measurements. The sensing resistor shall not have a resistance greater than 0.1 Ω. The oscilloscope must have a minimum deflection sensitivity of 50 mV/cm.

7.3 A good quality DC ammeter of the permanent magnet-moving coil type should be used for average coil primary current measurements. The meter range selected should easily allow reading resolutions of at least 0.1 A.

7.4 The same oscilloscope required in 7.2 should be used to measure primary induced voltage.

7.5 A good quality DC voltmeter with an input resistance of at least 1000 Ω V and with sufficient resolution to easily indicate differences of 0.1 V. To achieve this resolution the full scale deflection should be appropriate to the voltage rating of the ignition system being tested.

7.6 A distributor drive stand and attached tachometer which will have:

a. An eccentricity between the mounting fixture and drive of 0.076 mm (0.003 in) maximum.
b. A continuously variable speed adjustment with a total speed variation between 15 and 3500 rpm possible.
c. Speed stability within 5% at any given speed.
d. A tachometer accurate within 3% of indicated speed and independent of the electrical portion of the ignition system.

8. Procedures

8.1 Group A Tests—The conventional circuit arrangement as shown in Figure 1 with instrumentation in place, or modified with an auxiliary switching unit connected as the manufacturer intended, can be used to measure available voltage, peak primary coil current, average primary coil current, spark duration, spark voltage, and spark current at the distributor speeds and input voltages listed in Table 1.

The calculation described in 6.1.7 plus the procedure described here determines the inductive portion of the spark energy dissipated in a 12 kV spark gap under the conditions shown in Table 1. Spark currents and voltages can be measured and spark energy calculated equally well under other conditions and with different spark gaps. This procedure can be used in relating the effective amount of spark energy required to ignite a given fuel mixture.

If 6 V ignition systems are to be tested, divide the primary voltages listed in Table 1 by two; for 24 V systems, multiply by two.

Allow the ignition system to soak at least 1 h at the temperatures listed in Table 1 before beginning tests. Before any readings are recorded at any of the test points, the system should be allowed to come to a thermally stable operating condition (typically, this takes about 2 min).

Output voltage amplitudes vary due to contact arcing and other small but accumulative factors. It is recommended that the minimum peak amplitude be recorded. This represents the level which can be guaranteed by the system under test.

The voltage divider lead would have to be connected to a firing spark gap for spark duration measurements and this gap set carefully to fire at 12 kV ± 1/2 kV. To secure firing voltages stability of this magnitude, special gaps and/or arrangements are usually required. Firing across a surface may help stability. Firing a gap under pressure using a dry inert gas and spherical electrodes also helps.

When environmental equipment is used to control ambient test temperatures, care must be taken that wire and/or cable lengths and, consequently, impedances do not affect test results.

During simulated starting tests, the system shall be operated under conditions simulating vehicle application: that is, if primary resistor in series with coil is normally bypassed during vehicle cranking, resistor should be bypassed for this portion of bench tests.

TABLE 1—TEST CONDITIONS FOR GROUP A TESTS

Distributor rpm	Primary Volts	Environment Temperature °C	Environment Temperature °F	Operating Condition
20	5.0	−29 ± 1	−20 ± 2	Cold Starting
30	5.0	−29 ± 1	−20 ± 2	Cold Starting
40	5.0	−29 ± 1	−20 ± 2	Cold Starting
50	11.0	27 ± 3	80 ± 5	Hot Starting
60	11.0	27 ± 3	80 ± 5	Hot Starting
70	11.0	27 ± 3	80 ± 5	Hot Starting
250	14.0	27 ± 3	80 ± 5	Running
500	14.0	27 ± 3	80 ± 5	Running
750	14.0	27 ± 3	80 ± 5	Running
1000	14.0	27 ± 3	80 ± 5	Running
1250	14.0	27 ± 3	80 ± 5	Running
1500	14.0	27 ± 3	80 ± 5	Running
1750	14.0	27 ± 3	80 ± 5	Running
2000	14.0	27 ± 3	80 ± 5	Running
2250	14.0	27 ± 3	80 ± 5	Running
2500	14.0	27 ± 3	80 ± 5	Running
2750	14.0	27 ± 3	80 ± 5	Running
3000	14.0	27 ± 3	80 ± 5	Running

8.2 Group B Tests—The circuit arrangement shown in Figure 3 is appropriate to measure the coil's primary induced voltage and secondary voltage. When the 1.0 MΩ resistor is connected, it is also appropriate to measure the risetime of the secondary voltage. The distributor and spark gaps are dispensed within these tests, as the waveform irregularities they introduce add nothing to the results and make stabilized patterns on the oscilloscope difficult to achieve. Oscillograph M_1 is used to measure primary induced voltage in this case. These measurements should be made at an ambient temperature of 27 °C ± 3 °C (80 °F ± 5 °F), a distributor speed of 1000 rpm, and a primary voltage of 14 V. Primary induced voltage test results are usually more meaningful if compared to secondary voltage values measured simultaneously. A satisfactory ratio of secondary voltage to primary induced voltage should be established by each group making these tests if they wish to insure that neither contacts nor semiconductors are overstressed.

Because risetime is measured between 10 and 90% of the peak voltage amplitude, it is usually easier to photograph the oscillograph waveform than to attempt to read this figure directly. Most manufacturers of oscilloscopes furnish compatible cameras for this purpose.

8.3 Ignition Coil Energy Measurement (Zener Technique)

8.3.1 The measurement procedure measures the energy dissipated in a Zener string connected on the secondary side. The measured energy is an indication of:

a. Energy stored in the ignition coil at the time of interruption of the primary current and
b. The efficiency of the ignition coil in transferring that energy to the zener string. The energy delivered to the spark plug gap in an engine will be different.

The Zener Technique is used to provide a more reproducible measurement, eliminating the variability and RF noise of the spark gap.

8.3.2 TEST PROCEDURE

8.3.2.1 Place the ignition coil to be tested in the circuit shown in Figure 4.

8.3.2.2 Stabilize the coil at 25 °C ± 5 °C or at another temperature consistent with usage or manufacturers recommendations.

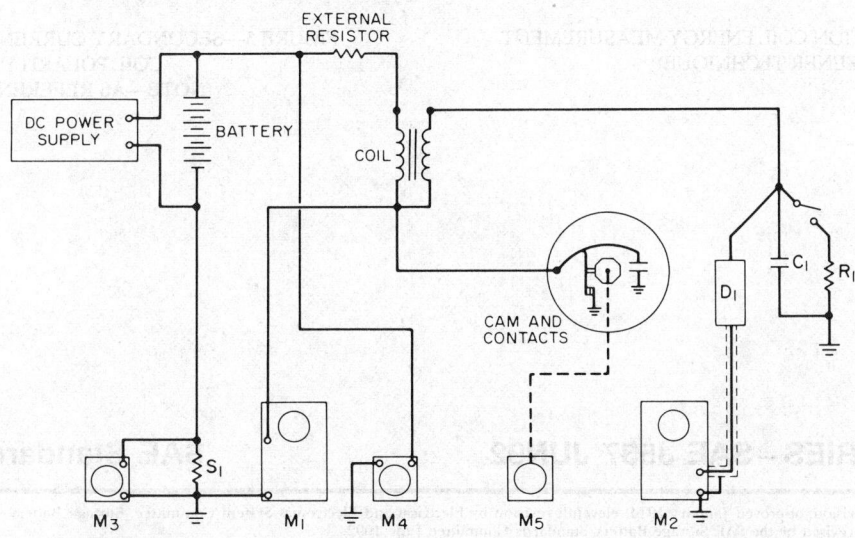

M_1 = PRIMARY VOLTAGE MEASURING OSCILLOSCOPE
M_2 = SECONDARY VOLTAGE MEASURING OSCILLOSCOPE
M_3 = DC AMMETER
M_4 = DC VOLTMETER
M_5 = TACHOMETER
C_1 = SHUNT CAPACITY ($C_1 + D_1$ = 50 MMF)
S_1 = 0.1 OHM METER SHUNT
D_1 = VOLTAGE DIVIDER
R_1 = 1.0 MEGOHM, 10 WATT

FIGURE 3—TEST CIRCUIT ARRANGEMENT FOR GROUP B TESTS

8.3.2.3 Adjust the input (power supply, ignition module, oscillator) so that the primary break current is at the manufacturers recommended coil interruption current.

8.3.2.4 Energy delivered to the Zener string by the ignition coil can most accurately be obtained by using a sampling (or digital) oscilloscope and the following procedure:

a. Multiply Zener voltage and Zener current waveforms to get power waveform.
b. Energy is the area under the power waveform over the duration of energy delivery.

When a sampling scope is not available, the approximate formula is described in 8.3.2.5.

8.3.2.5 Read and record the secondary output characteristics listed as follows (as shown in Figure 5):

V_z = Zener voltage in volts
I_o = Initial current in milliamps
I_f = Final current in milliamps
I_k = Current corresponding to knee in milliamps
T = Duration in milliseconds during which energy delivery to the Zener string is complete
T_k = Time at predominating knee in milliseconds or default of 0.5 ms if no predominating knee present
I_a = Average secondary current in milliamperes
I_p = Peak secondary current in milliamperes

I_o, I_k, I_f are converted from the voltage at the "low voltage probe" with the Equation 2:

$$I = V/R = V/100 \text{ or } I = 10 \text{ mA/V} \quad \text{(Eq.2)}$$

8.3.2.6 Calculate the spark energy in millijoules by Equation 3:

$$\text{Energy} = V_z * I_a * T \quad \text{(Eq.3)}$$

where:

$$I_a = 1/2\,(T_k I_p/T + I_K)$$

NOTE—Secondary current can be measured using an alternate method (rather than measuring the voltage across a 100 Ω resistor) as long as the method used is at least as accurate as the method shown.

The Zener string technique shown in Figure 4 can present a problem to the user if the polarity of the coil secondary is not determined before the measurements are made. An incorrect polarity can destroy the 5 kV -2 A blocking diode. This will create a condition where the DC voltage across the Zener string will not be restored to the baseline before the next current waveform begins, resulting in erroneous readings.

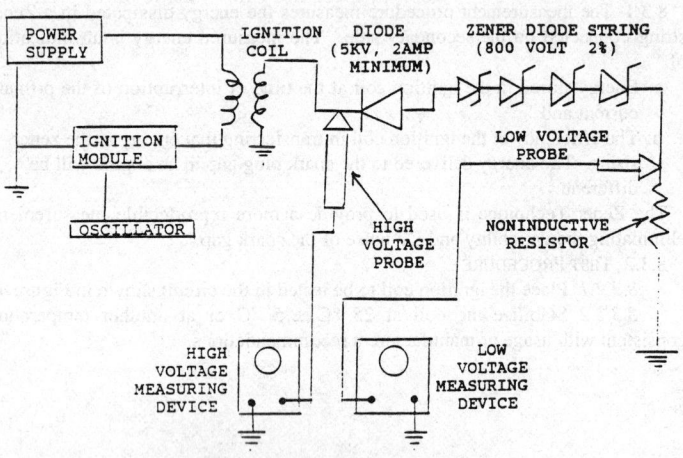

FIGURE 4—IGNITION COIL ENERGY MEASUREMENT
(ZENER TECHNIQUE)

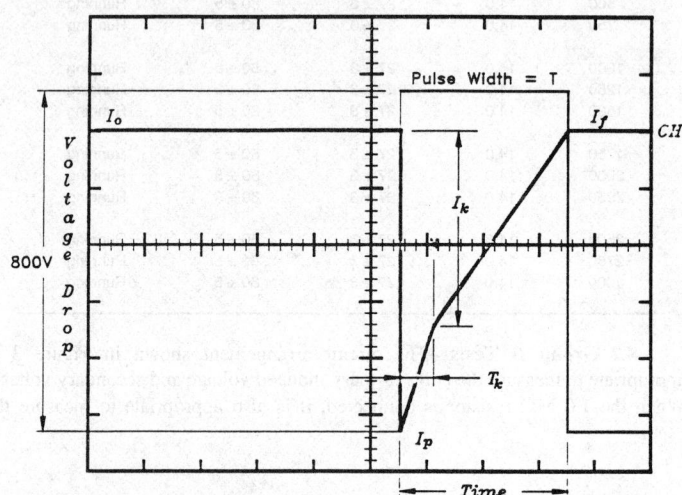

FIGURE 5—SECONDARY CURRENT WAVEFORM WITH CORRECT
COIL POLARITY CIRCUIT
(NOTE—AS REFERENCED TO FIGURE 4)

(R) STORAGE BATTERIES—SAE J537 JUN92 — SAE Standard

Report of the Electrical Equipment Division, approved January 1914, eleventh revision by Electrical and Electronics System Committee, Storage Battery Subcommittee June 1986. Completely revised by the SAE Storage Battery Standards Committee June 1992.

1. Scope—This SAE Standard serves as a guide for testing procedures and as a publication providing a record of current production batteries, their ratings, and container description. Any battery with planned significant usage may be submitted for inclusion in the Tables of use. (Reference Section 5 and Chart A1 of the Appendix.)

1.1 The ratings submitted are to be based on procedures described in this document. The ratings submitted must be of a level that when any subsequent significant sample is tested in accordance with this document, that at least 90% of the batteries shall meet the ratings. The choice of 90% compliance recognizes that batteries consist of many plates and require chemical-electrical formation procedures and small variations in test conditions and procedures can affect the performance of individual batteries.

1.2 The ratings and container descriptions listed in Chart A1 of the Appendix are provided as a reference of current battery usage. It is recommended that potential suppliers and customers establish specification and compliance requirements prior to any sample evaluation since individual customer requirements may vary from this document.

1.3 Applications—This document applies to lead-acid types of storage batteries used in motor vehicles, motorboats, tractors, and starting, lighting, and ignition (SLI) applications which use regulated charging systems.

2. References

2.1 Applicable Documents—The following publications form a part of this specification to the extent specified herein. The latest issue of SAE publications shall apply.

2.1.1 SAE PUBLICATIONS—Available from SAE, 400 Commonwealth Drive, Warrendale, PA 15096-0001.

SAE J240 JUN91—Life Test for Automotive Storage Batteries
SAE J1495 MAR92—Test Procedure for Battery Flame Retardant Venting Systems
SAE J2185 NOV91—Life Test for Heavy Duty Storage Batteries

3. Electrical Testing Procedure—Individual battery performance values are to be determined by the procedures outlined under Sampling, Conditioning, and Sequence of Tests. Battery classifications, dimensions, and ratings are given in the tables.

DANGER OF EXPLODING BATTERIES

Batteries contain sulfuric acid and they produce explosive mixtures of hydrogen and oxygen. Because self-discharge action generates hydrogen gas even when the battery is not in operation, make sure batteries are stored and worked on in a well ventilated area. ALWAYS wear safety goggles and a face shield when working on or near batteries. When working with batteries:

a. Always wear proper eye, face, and hand protection.
b. Keep all sparks, flames, and cigarettes away from the battery.
c. Do not remove or damage vent caps.
d. Cover vent caps with a damp cloth.
e. Make sure work area is well ventilated.
f. Never lean over battery while boosting, testing, or charging.

3.1 Sampling—Compliance determination samples shall be selected from normal production. Batteries tested should be new, unused, and no older than 60 days from date of manufacture.

3.2 Battery Conditioning and Charging

3.2.1 CHARGING TEMPERATURES—Charging must not be started if electrolyte temperature is below 16 °C (60 °F) and during charge the temperature must be maintained between 16 °C (60 °F) and 43 °C (110 °F).

3.2.2 ELECTROLYTE STRENGTH—Batteries shall be tested with the electrolyte as supplied by the manufacturer.

3.2.2.1 *Corrections to Specific Gravity*—Electrolyte specific gravity readings are to be corrected to a standard temperature of 27 °C (80 °F). Specific gravity decreases as liquid temperature increases, and vice versa. Test measurements made at other temperatures (T) must be corrected to the standard temperature using the following equations:

Specific gravity at 27 °C = measured value + 0.0007 (T−27) (Eq. 1)

where:
T is in °C

Specific gravity at 80 °F = measured value + 0.0004 (T−80) (Eq. 2)

where:
T is in °F

Electrolyte temperature shall be measured above the plates in an intermediate cell.

3.2.2.2 *Corrections to Voltage*—When constant current charging, the on charge terminal voltage is to be corrected to a standard temperature of 27 °C (80 °F). It must be reduced with increased temperature and vice versa, using the following formulae:

Terminal voltage at 27 °C = reading + (no. cells × 0.0063 [T−27]) (Eq. 3)

where:
T is in °C

Terminal voltage at 80 °F = reading + (no. cells × 0.0035 [T−80]) (Eq. 4)

where:
T is in °F

Electrolyte temperature shall be measured above the plates in an intermediate cell. Corrections do not need to be made to open circuit voltage readings or to voltages measured during discharge.

3.2.3 CONSTANT CURRENT CHARGING

3.2.3.1 *Batteries With Access To Electrolyte*—Charging at appropriate constant current must be continued until the criteria establishing full charge are achieved.

The constant current is to be set at a whole ampere value which lies between 1/2 and 3/4 of 1% of the cranking performance rating (in amperes) (unless the ratio of cold cranking rating [amperes] divided by reserve capacity rating [minutes] is less than 5.0—then, the constant current is to be set at a whole number ampere value between 3/4 and 1% of the cranking performance rating).

Batteries which are similar in design and which have been discharged to approximately the same extent may be series connected for recharge up to the voltage capacity of the charger unit.

Two basic criteria may be used to recognize when the battery is fully charged:

a. When the temperature corrected specific gravity of the electrolyte is constant within ±0.002 over three successive hourly intervals, or...
b. When the temperature corrected on charge terminal voltage at the constant current does not change by more than 0.008 V per cell per hour over three successive hourly intervals.

3.2.3.2 *Batteries Without Access to Electrolyte*—In batteries which operate with excess liquid electrolyte, provision to measure specific gravities and temperatures is to be made by carefully cutting holes through the cover in locations recommended by the manufacturer. These holes must be capable of being closed to retain integrity of the venting system. Adjustments to electrolyte strength or volume in such batteries are not permitted during preparation or testing.

3.2.4 CONSTANT VOLTAGE CHARGING—Constant voltage charging is authorized, but not recommended unless the correct applied voltage for the battery is known. Applied voltages will need to be selected between 2.40 V/cell and 2.75 V/cell as battery chemistries and designs differ. When the correct applied voltage is not known, testing laboratories should request information on voltage and charging time from the battery manufacturer.

For constant voltage charging, the charger output must be capable of at least 25 A per battery when parallel connected units are charged.

In conjunction with constant voltage charging, a mixing charge at a constant current rate, and for a fixed time period and in accordance with recommendations of the battery manufacturer, is allowed to promote electrolyte mixing and to ensure complete recharge.

3.2.5 VALVE REGULATED BATTERIES—No provision for directly reading electrolyte specific gravities or temperature can be made on batteries using electrolyte in gel form or absorbed in separators. Charging of valve regulated batteries should be carried out in accordance with the recommendation of the manufacturer.

3.2.6 DRY CHARGED OR SIMILAR BATTERIES WHICH NEED ELECTROLYTE TO ACTIVATE—If an activation test is required, refer to customer specifications; otherwise, fill according to the battery manufacturer's instructions, charge and condition according to 3.2.1 through 3.2.4, and then test as any other filled and charged battery according to Section 3.3.

3.3 Sequence of Tests

3.3.1 Charge battery according to methods given under Conditioning and Charging (3.2), and repeat this before each discharge.

3.3.2 Perform tests according to the sequence in Table 1.

3.4 New batteries may require extra conditioning, not afforded by test event 2, in determining their true reserve capacity, therefore, the highest reserve capacity test value obtained for each battery in test events 3, 6, or 8 shall be used as the reserve capacity performance of that battery. Statistical use of these data must be agreed upon by customer and supplier.

3.5 Reserve Capacity Test—Fully charge the battery according to 3.2. Allow it to stand at room temperature for 4 to 96 h.

During the stand period, regulate battery temperature so that electrolyte temperature, measured above the plates in an intermediate cell, is stabilized at 27 °C ± 3 °C (80 °F ± 5 °F) before the start of the discharge.

Discharge the battery at 25 A ± 0.1 A. During discharge, using any convenient method, maintain electrolyte temperature within the range 24 °C (75 °F) to 32 °C (90 °F). Results will not be considered valid if electrolyte temperature moves outside this range before the end of the discharge.

End the discharge when the voltage across the battery terminals has fallen to the equivalent of 1.75 V/cell, noting the discharge duration in minutes and the electrolyte temperature at the cut-off point.

Correct the discharge duration for final temperature different from 27 °C (80 °F) using the formulae which follow and record the corrected time as the Reserve Capacity achieved.

$$M_c = M_r [1 - 0.009(T_{final} - 27)] \quad \text{(Eq. 5)}$$

where:
M_c = minutes corrected to 27 °C (80 °F)
M_r = minutes actually run
T_{final} = temperature of electrolyte above the plates in an intermediate cell at end of discharge, °C
0.009 = temperature correction factor

For T_{final} in °F use:

$$M_c = M_r [1 - 0.005(T_{final} - 80)] \quad \text{(Eq. 6)}$$

3.6 Rechargeability and Charge Rate Acceptance—This test determines the battery capability to accept charge at low temperature when fully discharged, and to determine the rate at which the battery would accept the charge from a voltage regulated charging system which has adequate current capacity.

The charging equipment used in this test should be capable of a minimum current output in amps equivalent to 50% of the rated reserve capacity value in minutes.

3.6.1 Using the same battery discharged through reserve capacity test (3.5) or specifically discharged to 1.75 V/cell at the reserve capacity rate per 3.5, place the battery in a cold chamber until electrolyte temperature above the plates in an intermediate cell has stabilized at 0 °C ± 1 °C (32 °F ± 2 °F).

3.6.2 With the battery in a cold chamber, at 0 °C ± 1 °C (32 °F ± 2 °F) ambient, charge it at a constant potential equivalent to 2.4 V/cell. Measure the current input value after 10 min of charging.

3.6.3 Continue to charge for a total duration of 120 min. Discontinue

TABLE 1—SEQUENCE OF TESTS

	Standard Test Sequence	Minimum Required to Conduct any Optional Tests (Event 9)
1. Dry Charge Battery Activation (if Required)—Paragraph 3.2.6	X	
2. Preconditioning—Paragraph 3.2	X	X
3. Reserve Capacity Test—Paragraph 3.5	X	X
4. Charge Rate Acceptance Test—Paragraph 3.6	X	
5. Cold Cranking Test at -18 °C (0 °F)—Paragraph 3.7	X	X
6. Reserve Capacity Test	X	*1
7. Cold Cranking Test at -29 °C (-20 °F)	X	*2
8. Reserve Capacity Test	*3	*1
9. Optional tests		
a. J240, paragraph 3.8.1		
b. J1495, paragraph 3.8.2		
c. J2185, paragraph 3.8.3		
d. Vibration, paragraph 3.8.4		
e. Gassing Rate Characteristic, paragraph 3.8.5		

[1] Test events 6 and 8 are not required if the Reserve Capacity rating is met in event 3.
[2] Optional. Test event 6 required if this test is run.
[3] Test event 8 is not required if the Reserve Capacity rating is met in events 3 or 6.

the charging, remove the battery from cold chamber and raise the battery temperature until the electrolyte temperature above the plates in an intermediate cell has stabilized at 27 °C ± 3 °C (80 °F ± 5 °F).

3.6.4 Discharge the battery at 25 A ± 0.1 A. During discharge, using any convenient method, maintain electrolyte temperature within the range 24 °C (75 °F) to 32 °C (90 °F). Results will not be considered valid if electrolyte temperature moves outside this range before the end of the discharge.

End the discharge when the voltage across the battery terminals has fallen to the equivalent of 1.75 V/cell, noting the discharge duration in minutes and the electrolyte temperature at the cut-off point.

Correct the discharge duration for final temperature different from 27 °C (80 °F) using the formulae from 3.5.

3.6.5 ACCEPTANCE CRITERIA FOR THESE TESTS

3.6.5.1 *Rechargeability*—Current input after 10 min of charging (3.6.2) shall be at least 3% of the battery cold cranking rate.

3.6.5.2 *Charge Rate Acceptance*—The percent ratio of the discharge time in minutes as obtained after 120 min recharge (3.6.3) to the original reserve capacity value (3.5) shall be at least 50%.

3.7 Cold Cranking Test—The following test is a measure of the cranking capability of a battery at the rating temperature.[1]

Fully charge the battery according to 3.2. Allow it to stand at room temperature for 8 to 96 h.

Place the battery in an ambient held at the rating temperature (typically 16 h) until the electrolyte above the plates of an intermediate cell has stabilized at the rating temperature ±0.5 °C (±1 °F).

With the battery in an ambient at the rating temperature,[1] discharge the battery at the rating current shown in Chart A1 of the Appendix. The rating current shall be held constant ±1 A throughout the discharge. Measure battery terminal voltage under load at the end of 30 s.

The acceptance criterion for this test is that the battery terminal voltage at 30 s shall be equivalent to 1.2 V/cell or greater.

3.8 Optional Durability Tests

3.8.1 SAE J240 JUN91—Life Test for Automotive Storage Batteries.

3.8.2 SAE J1495 MAR92—Test Procedure for Battery Flame Retardant Venting Systems.

3.8.3 SAE J2185 NOV91—Life Test for Heavy Duty Storage Batteries.

3.8.4 VIBRATION TEST—This vibration test is to determine the ability of a battery to withstand G-levels similar to those developed in on-road applications of passenger car and light truck vehicles without suffering mechanical damage, loss of capacity, or loss of electrolyte.

3.8.4.1 *Equipment*—LAB vibration machines ARV-30 X 40-400 or similar type (even number of counter rotating shafts, vertical vibration component only); or U. S. Army Ordnance Vibration Machine as shown on Drawing No. D7070340.

3.8.4.2 *Procedure*

3.8.4.2.1 One or more fully charged batteries at 27 °C ± 3 °C (80 °F ± 5 °F) shall be placed on the vibration machines recommended for this test (see 4.2). The batteries shall be symmetrically balanced on the LAB vibrator. On the Ordnance vibrator, each battery shall be oriented symmetrically along the table centerline parallel to the shaft. On the Ordnance vibrator, the battery plates shall be oriented parallel to the axis of the rotating shaft of the machine.

3.8.4.2.2 The batteries shall be firmly held down by a top holddown frame bearing on the top four edges of the battery or by an optional top bar or bottom holddown. The holddown nuts should be torqued about 4.5 Nm (40 in-lb) for two bolt top holddowns. The holddown torque shall be slightly less for top hold-downs with more than two bolts. Optional bottom holddowns can normally be tightened up firmly without concern for excess torque. In no instance shall warpage of the battery case exceed 1.3 mm (0.05 in) per side due to excess holddown force.

3.8.4.2.3 The electrolyte shall be at the level recommended by the manufacturer. During vibration there shall be no electrolyte loss.

3.8.4.2.4 Batteries shall be vibrated for 4 h at a total acceleration of 3.5 G ± 0.2 G (32 Hz ± 2 Hz). Each 4 h of vibration shall represent one unit of vibration.

3.8.4.2.5 The total G-level shall be calculated from the vertical and two horizontal components by Equation 7. On the Ordnance table, the vertical component shall be determined near the high G level end of the battery (upward rotation side of shaft). On the LAB table, the vertical component shall be taken as the average of both ends of the battery. Vertical G-level measurements "near battery ends" will be made on the vibration table on an extension of the battery centerline as close as possible to the battery's end wall for the particular measuring device used. On both LAB and Ordnance tables, the x-and y- horizontal components shall be taken as the average of the four ends of the two opposing table edges. Horizontal G-level measurements (8) shall be made 76 mm (3 in) from the corners on the vibration table's sides. G-level may be determined by the use of an accelerometer or independent excursion and frequency measurements as agreed between the supplier and customer.

$$\text{G-Force} = 1.3 \times (\text{frequency})^2 \times \sqrt{[(\text{Vertical excursion})^2 + (\text{Horizontal X-excursion})^2 + (\text{Horizontal Y-excursion})^2]} \quad \text{(Eq. 7)}$$

Units: Frequency—hertz, excursions—mm

All measurements shall be made with batteries in place on the vibration machine. (See Figure A1, Appendix.)

3.8.4.2.6 After each unit of vibration, immediately discharge the battery at 27 °C ± 3 °C (80 °F ± 5 °F) at its specified -18 °C (0 °F) cold cranking rate on Chart A1 of the Appendix. The 30 s voltage must meet the 1.2 V/cell minimum requirement. If the 30 s voltage requirement is met, recharge according to methods given under Conditioning and repeat steps 3.8.4.2.4 and 3.8.4.2.5 until failure or a specified number of units has been successfully achieved.

3.8.4.2.7 The battery will be rated at a number of units it can survive and meet the 30 s 1.2 V/cell requirement. As an option to this test procedure, two or more units of 4 h may be run continuously without intervening performance tests. If a failure is recognized, at the end of a multiple unit sequence, only the units completed prior to the failed multiple unit shall count as being successfully completed.

3.8.5 BCI RECOMMENDED STORAGE BATTERY SPECIFICATIONS, STARTING,

---

[1] Rating temperature for purpose of this test is either -18 °C or -29 °C (0 °F or -20 °F) as shown in Chart A1 of the Appendix.

LIGHTING, AND IGNITION TYPES: **Gassing Rate Characteristic.** See Appendix.

4. Specifications

4.1 Battery Description—The location and polarity of the terminal posts and position of handles, when used, shall be as shown in Chart A2 of the Appendix.

4.2 Type Designations and Markings—Type letters, numbers, or symbols, which shall enable the user to determine ratings from the manufacturer's catalogs, shall be stamped or molded on the case or cell connectors or on a self-adhesive, electrolyte resistant label permanently attached to the top, end, or side of the battery.

4.3 Terminal Posts—Polarity shall be plainly marked as follows: The positive terminal shall be identified by **Pos, P** or + on the terminal or on the cover near the terminal. See Figures A2, A3, A4, and A4a in the Appendix.

5. Battery Identification and Classification Tables—The Charts in the Appendix describe batteries fitted as original equipment in applications defined in Section 1. A listing does not imply that a battery is immediately or generally available on a non-exclusive use basis, or that its use by the vehicle builder continues without change in specification. The Storage Battery Subcommittee will act on requests for entries to the Charts (or deletions) following the procedure subsequently described.

5.1 Any member may request listing of a new battery in the Charts. Requests must be accompanied by all the specification details necessary to complete the entry.

5.2 New batteries will be allocated an Assembly Figure Number and an SAE identification number.

5.2.1 The Assembly Figure Number will be the number of the plan in Chart A2 of the Appendix which corresponds to the cell, terminal and hold-down position of the new battery, with a suffix added to indicate hold-down style as detailed in Section 6. If no appropriate plan is shown in Chart A2, a new figure taking the next sequential number should be proposed for incorporation.

5.2.2 The SAE Identification Number will consist of the Battery Council International (BCI) Group Size Number followed by the -18 °C (0 °F) Cranking Performance Ampere Rating. When there may not yet be a BCI Group Size Number appropriate to a new battery, a temporary identification may be allocated. It will consist of the symbols TFG followed by the Assembly Figure Number followed by the -18 °C (0 °F) Cranking Performance Ampere Rating.

5.2.3 If a new battery has a combination of BCI Group Size Number and -18 °C (0 °F) Cranking Performance Ampere Rating which is identical to a battery already listed, or proposed, it will be distinguished from the earlier entry by adding to the Identification Number a slant line followed by a suffix 1, 2, etc.

5.3 Batteries will first be listed in Chart A1(Current) for a five year period. The entry will then be transferred to Chart A1a (Non-Current).

5.3.1 A member may request Subcommittee approval for the continuance of the entry in Chart A1 because the battery remains in use as an original equipment fitment. An approval extension will remain in force for three years without need for a further request.

5.3.2 A member may request Subcommittee approval for the prior removal or transfer of an entry.

5.3.3 For information for vehicle users, entries transferred to Chart A1a (Non-Current) will remain listed for ten clear years, when the Chairman will propose deletion.

5.4 Because of the possible need for frequent revision, Charts A1, A1a, and A2 will be considered to be supplements to this document.

6. Battery Container Design for Bottom Hold-Down

6.1 Batteries which have either ledges or recesses for the hold-down shall be designated as shown in hold-down Designs 2, 2A, 3 or 4 (Figures A4, A5, A7, and A8 in the Appendix). (Unless specified, all dimensions are ±0.3 mm (±0.01 in), all angles ±1 degree, all radii ±0.8 mm (±0.03 in).)

6.2 Batteries which have ledges or recesses in the sides for the hold-down shall be designated as shown in hold-down Designs 2, 2A or 4 (Figures A5, A6, and A7 in the Appendix). Those with Designs 2 and 2a will have the letter J and with Design 4 will have the letter K added after the assembly figure.

6.3 Batteries which have ledges on the ends for the hold-down shall be of the design shown in hold-down Design 3 (Figure A7, Appendix). This design is used on Figures 17 and 18 in Chart A2.

APPENDIX A

A.1 Gassing Rate Characteristic—Reprinted with permission of Battery Council International.

Charging at a constant voltage, a fully charged battery will accept a current which is characteristic of its design and construction. Unused current is dissipated in electrolyzing water from the electrolyte into hydrogen and oxygen gases, which escape. A measurement of gas evolution rate or the current accepted at a charging voltage typical of a vehicle electrical system when related to the reservoir of electrolyte above the plates provides a basis for comparing battery designs in respect to their ability to withstand service water losses. Both measures are useful, but there is no generally applicable factor correlating them. Both gas emission rates and currents are small values which demand accurate measurement and avoidance of leakage losses.

A.1.1 Procedure for Steady-State Charging Current Measurement

1. Complete Cranking Performance and Reserve Capacity tests, then fully charge the battery as described in "Pretest Conditioning and Charging Procedure."
2. Place the battery in an oven or water circulating bath held at a temperature of 51.7 °C ± 1.1 °C (125 °F ± 2 °F).
3. Apply a 14.1 V ± 0.1 V charging voltage to a 12 V battery (7.05 V ± 0.05 V for 6 V batteries) for 16 to 18 h.
4. After this period of conditioning, commence regulation of temperature to 51.7 °C ± 0.6 °C (125 °F ± 1 °F) and of charging voltage measured across battery terminals to 14.1 V ± 0.01 V (7.05 V ± 0.005 V for a 6 V battery).
5. Continue charging under these conditions for 2 h and then monitor charging current at 15 min intervals.
6. When charging current has stabilized so that variations over three successive readings are 2% or less than the average of the three successive readings, record that average as the Steady-State Charging Current in amperes.

A.1.2 Procedure For Gassing Rate Measurement

1. Complete Cranking Performance and Reserve Capacity tests, then fully charge the battery as described in "Pretest Conditioning and Charging Procedure."
2. Verify the absence of intercell and perimeter cover-case leaks by applying an air pressure of 1 lb/in^2 to each cell individually. This pressure must be maintained for 15 s after disconnecting the source of pressurization.
3. Place the battery in an oven or water circulating bath held at a temperature of 51.7 °C ± 1.1 °C (125 °F ± 2 °F) and prepare it for gas collection and measurement through water displacement out of inverted burettes. Gas must be collected separately from at least three cells.
   Collector burette volumes (in milliliters) must equal at least 300 times the Steady-State Charging Current but need not be more than 100 mL. Connecting tubes should be glass or metal as far as practical to reduce hydrogen escape by permeation.
   Cells set up for gas collection must have normal vents sealed off. Cells not set up for gas collection require provision for venting.
4. Follow steps 3, 4, 5 and 6 to bring the battery into Steady-State Charging Current Condition with the evolved gases bubbling through the water used for displacement.
5. Start collection of gases and continue until volume collected in any one burette reaches 100 mL, but not longer than 30 min.
6. Record individual volumes of gases collected, the times of collection, the room temperature and the barometric pressure.
7. Using data appropriate only to the cell showing the greatest gas volume, calculate the volume of gas corrected to S.T.P. (Standard Temperature and Pressure) and divide by time of collection to determine the characteristic Gassing Rate as milliliters per minute.

SAE No.	Assembly Figure No.	Electrical Test Values		Reserve Capacity min	Maximum Overall Dimensions					
		Cold Cranking			Length		Width		Height	
		at -18°C (0°F) A See Note b	at -29°C (-20°F) A See Note b		mm	in	mm	in	mm	in
6 Volt Batteries										
1-475	2	475	380	159	231	9.13	181	7.13	231	9.13
1-545	2	545	460	185	231	9.13	181	7.13	231	9.13
2-520	2	520	410	192	263	10.38	181	7.13	231	9.13
2-560	2	560	480	220	263	10.38	181	7.13	238	9.38
2-650	2	650	545	245	263	10.38	181	7.13	238	9.38
2-775	2	775	610	295	263	10.38	181	7.13	238	9.38
2E-595	5	595	510	210	492	19.38	104	4.13	238	9.38
3EH-830	5	830	675	340	492	19.38	110	4.34	248	9.77
4EH-880	5	880	700	420	492	19.38	127	5.00	248	9.77
2N-495	1	495	420	170	254	10.00	141	5.57	228	9.00
4-700	2	700	570	275	333	13.13	181	7.13	238	9.3S
4-720	2	720	590	280	333	13.13	181	7.13	238	9.38
4-860	2	860	750	380	330	13.03	178	7.04	241	9.S2
5D-800	2	800	675	340	349	13.75	181	7.13	238	9.39
7D-900	2	900	650	430	428	16.88	193	7.63	276	10.88
7DS-900	2	900	650	430	405	15.94	193	7.63	276	10.88
12 Volt Batteries										
3EE-290	9	290	230	85	490	19.32	110	4.35	225	8.88
3ET425	9-C	425	340	120	490	19.32	110	4.35	249	9.82
20H-235	10-C	235	170	45	198	7.82	173	6.82	238	9.38
21-325	10-J	325	250	68	208	8.19	173	6.81	222	8.77
22F-260	11-F-J	260	190	50	241	9.50	173	6.82	214	8.46
22F-305	11-F-J	305	210	75	241	9.50	172	6.79	207	8.17
22NF-245	11-F	245	185	52	239	9.44	139	5.50	226	8.91
22R 290	11	290	215	72	228	8.99	173	6.84	227	8.97
22R-290/1	11-L-J	290	210	65	227 a	8.90 a	174	6.86	214	8.44
22R-350	11-L-J	350	270	88	227 a	8.90 a	174	6.86	214	8.44
24-255	10	255	190	60	200	10.25	173	0.82	225	8.88
24-255	10	255	190	60	260	10.25	173	6.82	225	8.88
24-285	10-J-C	285	220	75	200	10.25	173	0.82	225	8.88
24-285	10-J-C	285	220	75	260	10.25	173	6.82	225	8.88
24-305	10	305	210	75	200	10.25	173	6.82	225	8.88
24-305	10	305	210	75	260	10.25	173	6.82	225	8.88
24-305/1	10	305	230	68	200	10.25	171	6.75	225	8.88
24-305/1	10	305	230	68	260	10.25	171	6.75	225	8.88
24-375	10-C	375	300	86	200	10.25	171	6.75	225	8.88
24-375	10-C	375	300	86	260	10.25	171	6.75	225	8.88
24-385/1	10-J	385	280	95	200	10.25	172	6.79	220	8.07
24-385/1	10-J	385	280	95	260	10.25	172	6.79	220	8.67
24-385/2	10-J	385	305	110	200	10.25	172	6.79	222	8.70
24-385/2	10-J	385	305	110	260	10.25	172	6.79	222	8.76
24-410	10-J	410	310	110	200	10.25	173	6.82	220	8.90
24-410	10-J	410	310	110	260	10.25	173	6.82	226	8.90
24-440	10-J	440	350	102	260	10.25	173	6.81	222	8.77
24H-305	10	305	280	98	200	10.25	173	6.82	238	9.38
24H-365	10	365	280	98	260	10.25	173	6.82	238	9.38
24R-350	11	350	280	99	261	10.30	173	6.84	227	8.97
24R-350	11	350	280	99	201	10.30	173	6.84	227	8.97
24R-380	10-J	380	290	113	200	10.25	173	6.82	247	9.7S
24R-440	11-L-J	440	320	120	260 a	10.35 a	174	6.86	227	8.94
24R-455	11-J	455	340	135	261	10.27	173	6.81	228	8.97
24T-380	10-J	380	290	113	260	10.25	173	6.82	247	9.75
25-430	10-J	430	270	100	222	8.77	170	6.67	224	8.82
27-300	10-J	300	280	110	300	12.04	173	6.82	225	8.88
27-360	10-J	360	280	110	306	12.04	173	6.82	225	8.88
27-440	10-C	440	350	102	304	12.00	171	6.75	222	8.7S
27-440	10-C	440	350	102	304	12.00	171	6.75	222	8.75
27-500	10	500	400	140	300	12.00	173	6.81	222	8.75
27-500	10	500	400	140	306	12.06	173	6.81	222	8.75
27-620	10	620	496	162	305	12.00	173	6.81	223	8.75
27H-435	10	435	340	125	297	11.72	173	6.82	238	9.38
27H-435	10	435	340	125	297	11.72	173	6.82	238	9.38
27HF-425	11-F	425	320	136	317	12.50	173	6.82	232	9.15
27HF-435	11-F	435	340	125	317	12.50	173	6.82	232	9.15
27R-430	11	430	320	125	305	12.01	173	6.81	227	8.95
27R-455	11	455	355	136	305	12.01	173	6.81	232	9.15
29H-600	10	600	455	170	333	13.10	180	7.12	248	9.75

CHART A1 — BATTERY CLASSIFICATIONS, RATINGS, AND DIMENSIONS

SAE No.	Assembly Figure No.	Electrical Test Values		Reserve Capacity min	Maximum Overall Dimensions					
		Cold Cranking			Length		Width		Height	
		at -18°C (0°F) A See Note b	at -29°C (-20°F) A See Note b		mm	in	mm	in	mm	in
29H-625	10	625	470	170	333	13.10	180	7.12	248	9.75
29NF-290	11-F	290	235	80	330	13.00	141	5.56	228	9.00
30H-460	10	460	330	158	342	13.50	173	6.82	238	9.38
30H-580	10	580	480	175	342	13.50	172	6.81	233	9.21
31-475	18	475	375	130	333	13.13	173	6.80	239	9.41
31-550	18	550	410	130	333	13.13	173	6.80	239	9.41
31-580	18	580	450	180	333	13.13	173	6.80	239	9.41
31-600	18	600	455	170	333	13.10	180	7.12	249	9.79
31-625	18	625	490	160	333	13.13	173	6.80	239	9.41
31-625	18	625	470	170	333	13.10	180	7.12	249	9.79
31-625	18	625	470	180	333	13.13	173	6.80	239	9.41
31-750	18	750	560	160	333	13.13	173	6.80	239	9.41
31-900	18	900	675	160	333	13.13	173	6.80	239	9.41
32N-350	11	350	280	115	361	14.25	139	5.50	226	8.91
34-500	10-J	500	400	110	260	10.25	173	6.81	200	7.88
34-600	10-J	600	450	120	260	10.25	173	6.81	200	7.88
34-650	10-J	650	485	125	260	10.25	173	6.81	200	7.88
34-685	10-J	685	510	125	260	10.25	173	6.81	200	7.88
34R-575	11	575	NA	110	260	10.25	173	6.81	200	7.88
35-390	11	390	NA	NA	230	9.06	175	6.88	225	8.88
50-400	10	400	300	85	323	12.72	134	6.00	254	10.00
50-600	10	600	450	108	323	12.72	134	6.00	254	10.00
51-405	10	405	NA	70	238	9.38	129	5.06	223	8.81
51R-405	11	405	NA	70	238	9.38	129	5.06	223	8.81
52-310	19-K	310	220	60	186	7.34	154	6.04	212	8.36
53-210	14	210	155	40	331	13.07	121	4.79	211	8.32
55-380	19-K	380	275	75	218	8.60	154	6.04	212	8.36
56-450	19-K	450	330	90	254	10.02	154	6.04	212	8.36
56-505	19-K	505	379	86	253	9.96	153	6.02	211	8.31
58-540	21-K	540	405	100	255	10.06	183	7.19	177	6.94
58R-540	23-K	540	405	75	255	10.06	183	7.19	177	6.94
58R-540	23-K	540	405	100	255	10.06	183	7.19	177	6.94
60-360	12	360	280	110	331	13.07	159	6.27	225	8.88
61-310	20-K	310	220	60	192	7.57	160	6.30	225	8.86
62-380	20-K	380	275	75	225	8.87	160	6.30	225	8.86
63-450	20-K	450	330	90	258	10.14	160	6.30	225	8.86
64-475	20-K	475	355	120	296	11.64	160	6.30	225	8.86
64-535	20-K	535	400	120	296	11.06	162	6.38	225	8.88
65-650	21-K	650	485	130	306	12.06	190	7.50	192	7.56
65-850	21-K	850	640	165	306	12.06	190	7.50	192	7.56
70-260	17-J	260	200	58	208	8.19	179	7.05	196	7.70
70-260	17-J	400	300	80	208	8.19	179	7.05	196	7.70
72-275/1	17-J	275	210	60	231	9.10	184	7.27	222	8.77
74-335	17-J	335	270	98	260	10.25	184	7.27	222	8.77
74-410	17-J	410	310	110	260	10.25	184	7.27	222	8.77
74-455	17-J	455	360	140	260	10.25	184	7.27	222	8.77
75-500	17-J	500	400	90	230	9.10	179	7.05	196	7.70
75-525	17-J	525	390	90	230	9.06	179	7.06	196	7.56
75-600	17-J	600	450	90	230	9.06	179	7.06	196	7.69
75-630	17-J	630	470	90	230	9.06	179	7.06	196	7.69
75-690	17-J	690	515	90	230	9.06	179	7.06	196	7.69
76-1075	17-J	1075	800	175	334	13.12	179	7.06	216	8.50
76-970	17-J	970	725	150	333	13.13	179	7.05	216	8.50
76-750	17-J	750	600	175	333	13.13	179	7.05	216	8.50
78-540	17-J	540	405	115	260	10.25	179	7.06	196	7.69
78-600	17-J	600	450	115	260	10.25	179	7.06	196	7.69
78-690	17-J	690	515	115	260	10.25	179	7.06	196	7.69
78-730	17-J	730	545	115	260	10.25	179	7.06	196	7.69
78-770	17-J	770	575	115	260	10.25	179	7.06	196	7.69
85-550	11	550	440	90	230	9.06	173	6.81	203	8.00
86-550	10	550	440	90	230	9.06	173	6.81	203	8.00
4D-640	8	640	450	285	539 c	21.25 c	222	8.75	276	10.88
4D-800	8	800	640	310	539 c	21.25 c	222	8.75	276	10.88
8D-900	8	900	650	430	539 c	21.25 c	282	11.13	276	10.88
U1-160	10	160	110	23	198	7.80	133	5.25	187	7.38
U1-200	10	200	150	32	198	7.80	133	5.25	187	7.38

a Add 1.3 mm (0.5 in) for Lifting Ledges.
b For batteries having double insulation, deduct 15% from the rating values for cold cranking.
c Dimensions over handles.

CHART A1 — BATTERY CLASSIFICATIONS, RATINGS, AND DIMENSIONS (CONTINUED)

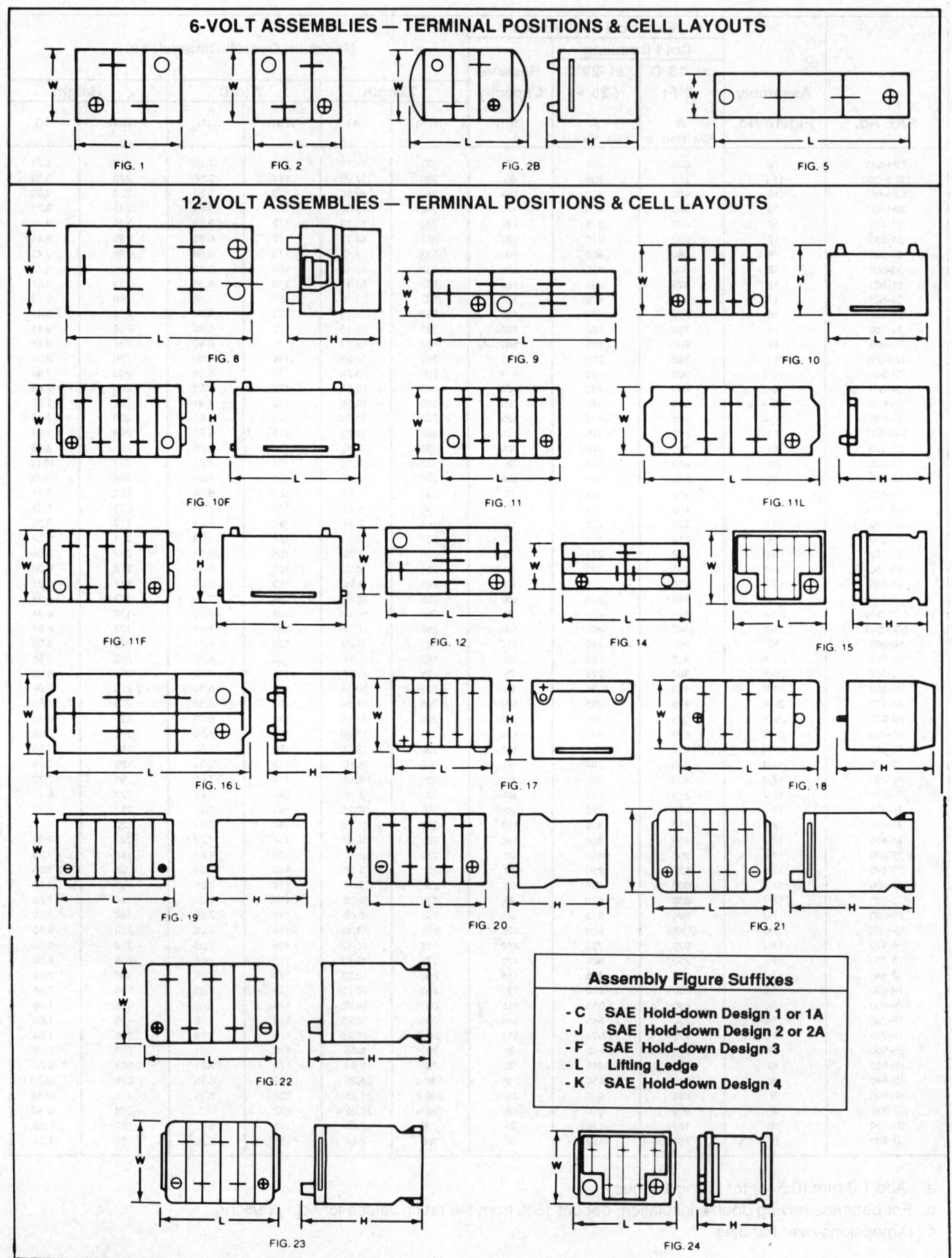

CHART A2—IDENTIFICATION SELECTION CHART

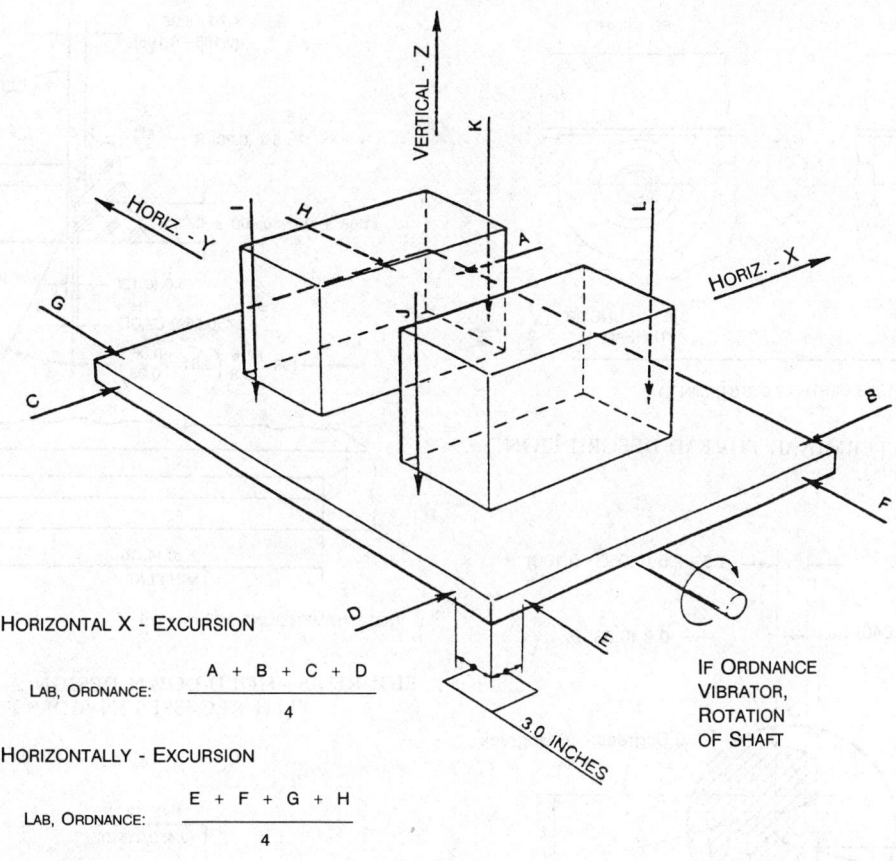

FIGURE A1 — EXCURSION MEASUREMENTS FOR G-LEVEL CALCULATIONS

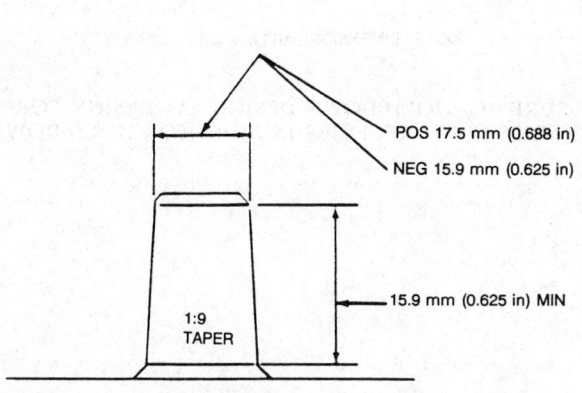

FIGURE A2 — TERMINAL POST DIMENSIONS

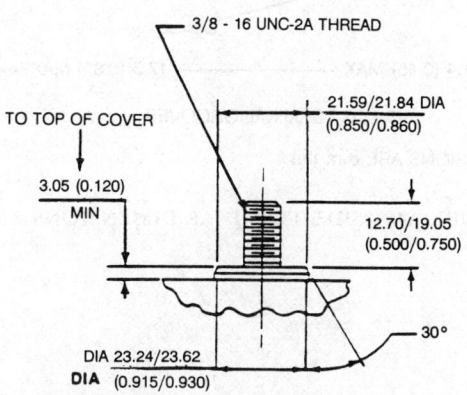

NOTE - DIMENSIONS ARE mm (in)

CAUTION - STUD LENGTH, CABLE EYELET THICKNESS AND TERMINAL NUT MUST BE COMPATIBLE TO INSURE RELIABLE CONNECTIONS. CONSULT BATTERY SUPPLIER FOR SPECIFIC STUD LENGTH.

FIGURE A3 — STUD TERMINAL DIMENSIONS

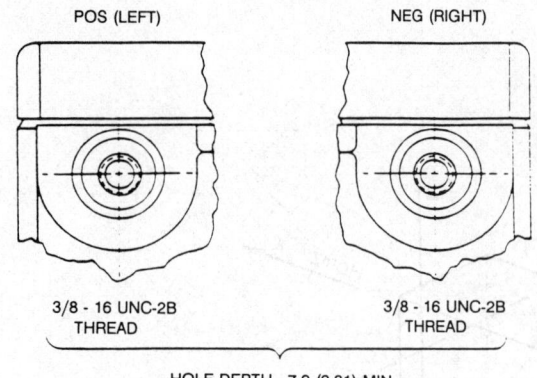

FIGURE A4 – SIDE TERMINAL THREAD DESCRIPTION

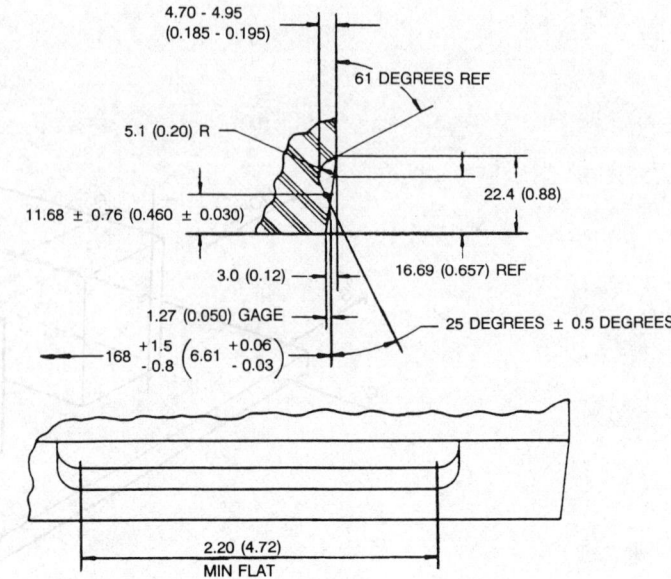

FIGURE A5 – HOLD-DOWN DESIGN 2 – DESIGN FOR BATTERIES WITH RECESSES IN SIDES FOR HOLD-DOWN

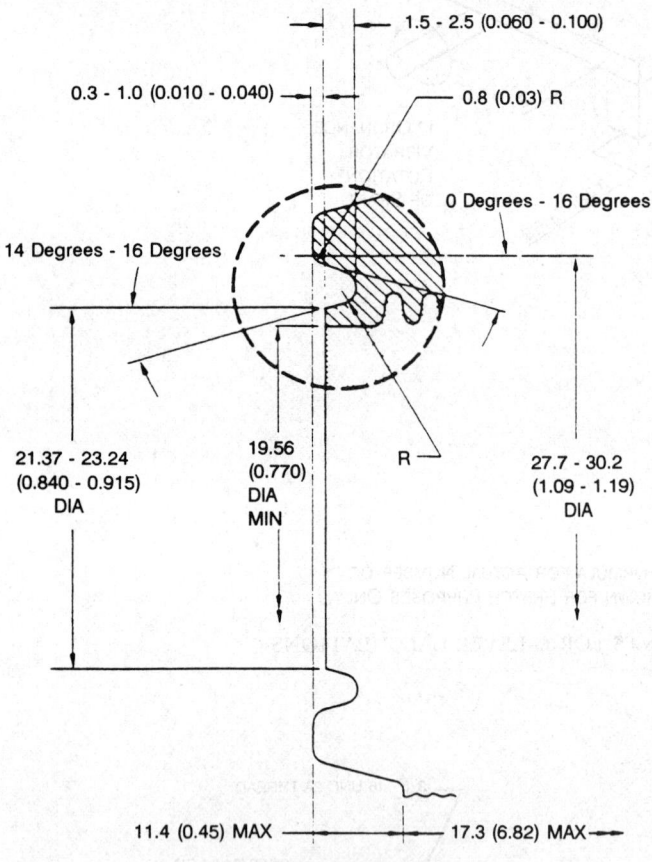

FIGURE A4a – SIDE TERMINAL DIMENSIONS

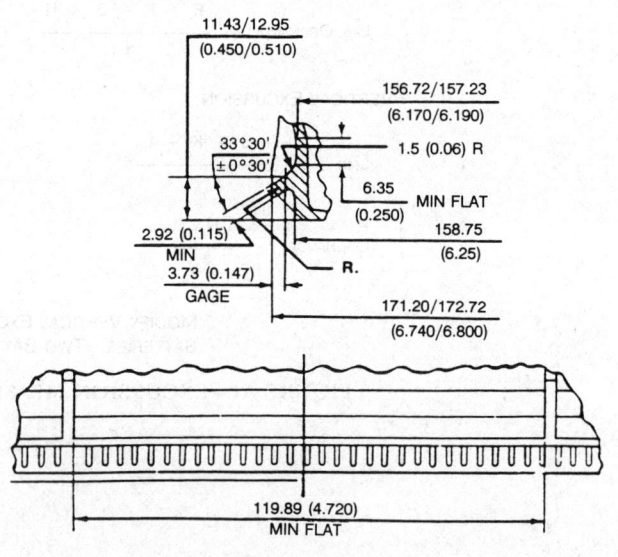

FIGURE A6 – HOLD-DOWN DESIGN 2A – DESIGN FOR BATTERIES WITH RECESSES IN SIDES FOR HOLD-DOWN

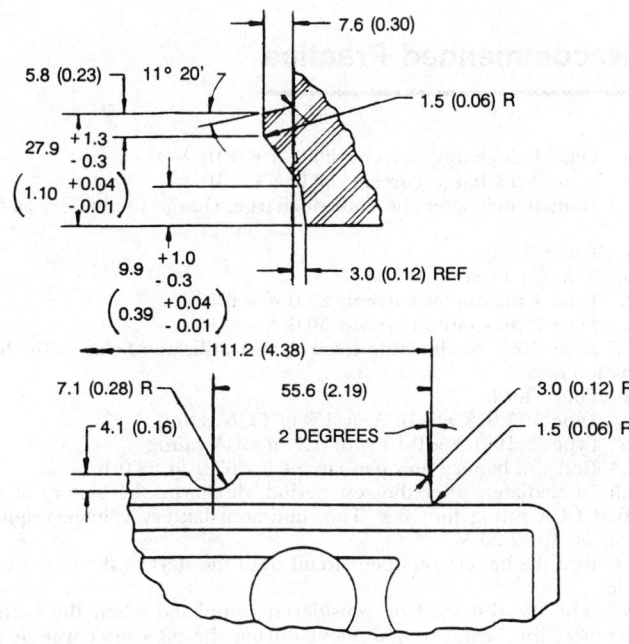

FIGURE A7—HOLD-DOWN DESIGN 3—DESIGN FOR BATTERIES WITH LEDGES ON ENDS FOR HOLD-DOWN

FIGURE A8—HOLD-DOWN DESIGN 4—DESIGN FOR BATTERIES WITH RECESSES IN SIDES FOR HOLD-DOWN

LIFE TEST FOR AUTOMOTIVE STORAGE BATTERIES—SAE J240 JUN93

SAE Standard

Report of the Electrical Equipment Committee approved May 1971, second revision by the Storage Battery Subcommittee June 1982. Completely revised by the SAE Storage Battery Standards Committee July 1991. Revised by the SAE Storage Battery Standards Committee June 1993. Rationale statement available.

(R) **1. Scope**—This SAE Standard applies to 12 V, automotive storage batteries of 180 min or less reserve capacity. This life test simulates automotive service when the battery operates in a voltage regulated charging system. It subjects the battery to charge and discharge cycles comparable to those encountered in automotive service. Other performance and dimensional information is contained in the latest issue of SAE J537.

This document is intended as a guide toward standard practice, but may be subject to change to keep pace with experience and technical advances.

2. Reference

2.1 Applicable Document—The following publication forms a part of this specification to the extent specified herein. The latest issue of SAE publications shall apply.

2.1.1 SAE PUBLICATION—Available from SAE, 400 Commonwealth Drive, Warrendale, PA 15096-0001.

SAE J537—Storage Batteries

3. Testing Procedure

3.1 Cycle life testing shall begin within sixty days of the final nondestructive test as shown in 3.3 of SAE J537 (Table 1).

3.2 The battery is tested in a water bath maintained at 41 °C ± 3 °C (105 °F ± 5 °F).

3.3 Water level of the bath specified in 3.2 is to be maintained at a height equal to or greater than 75% of the overall height of the battery container or within 1/2 in of the metal bushing of side terminal batteries.

3.4 The test cycle is performed as follows:

Discharge 4 min ± 1 s at 25 A ± 0.1 A.
Charge:
a. Maximum voltage (at battery cable terminals): 14.8 V ± 0.03 V
b. Maximum rate: 25 A ± 0.1 A
c. Time: 10 min ± 3 s

3.5 Battery is continuously cycled for 100 (+10, -0) h (example: Monday noon until 4:00 P.M. Friday). A switching delay of not more than 10 s is permitted from termination of charge to start of discharge and termination of discharge to start of charge.

3.6 The battery is given a 60 to 72 h stand on open circuit in the 41 °C (105 °F) water bath.

3.7 With the battery at the temperature obtained in 3.6, discharge at a rate equal to its -18 °C (0 °F) cold cranking rate in amperes (see SAE J537) to 1.20 V per cell, or a minimum discharge time of 30 s, whichever occurs first.

3.8 Replace battery on the life test without a separate recharge. Start on the "charge" portion of the cycle.

3.9 The life test shall be considered complete when the battery fails to maintain 1.2 V per cell for a minimum of 30 s on the manual discharge (3.7) for two consecutive 100 to 110 h test periods.

3.10 Water should be added to the electrolyte as required during the cycling portion of the test except to those batteries described as maintenance free.

4. Notes— The only change to this document was to upgrade it to a Standard.

LIFE TEST FOR HEAVY-DUTY STORAGE BATTERIES—SAE J2185 NOV91 — SAE Recommended Practice

Report of the SAE Storage Battery Standards Committee approved November 1991.

1. Scope—This SAE Recommended Practice applies to 12 V storage batteries which operate in a voltage regulated charging system. It simulates heavy-duty applications by subjecting the battery to deeper discharge and charge cycles than those encountered in starting a vehicle. The deeper discharge and charge cycles in service may come from a combination of the following conditions:
  a. Frequent occurrences of total electrical load exceeding the alternator output.
  b. Frequent occurrences of battery system supplying the electrical loads when the engine is not operating.
  c. Frequent occurrences of prolonged vehicle storage combined with high vehicle key-off loads.

Batteries will be classified into two types for this life test. Type 1 applies to batteries with reserve capacity of 200 min or less. Type 2 applies to batteries with reserve capacity greater than 200 min. ("C" value for Type 1 equals 25.0; "C" value for Type 2 equals 50.0.)

2. References—There are no referenced publications specified herein.

3. Testing Procedure

3.1 Place the fully charged test battery in a water bath maintained at 50 °C ± 1.7 °C (122 °F ± 3 °F) throughout the duration of the life test.

3.2 The test is conducted on a weekly cycle as follows:

3.2.1 Alternate charging and discharging 26 times, starting with charge portion first.

3.2.1.1 *Charge*
a. Time: 2.5 h
b. Voltage: 14.80 V ± 0.05 V
c. Type 1 maximum current: 25.0 A ± 0.10 A
d. Type 2 maximum current: 50.0 A ± 0.10 A

3.2.1.2 *Discharge*
a. Time: 1.0 h
b. Type 1 discharge current: 25.0 A ± 0.10 A
c. Type 2 discharge current: 50.0 A ± 0.10 A

3.2.2 Immediately after the 26th discharge, charge the battery as follows:
a. Time: 2.5 h
b. Voltage: 14.80 V ± 0.05 V
c. Type 1 maximum current: 25.0 A ± 0.10 A
d. Type 2 maximum current: 50.0 A ± 0.10 A

3.2.3 After 3.2.2, to eliminate electrolyte stratification, charge the battery as follows:
a. Time: 4.0 h
b. Type 1: 5.0 A ± 0.10 A or 1% of CCA rating
c. Type 2: 10.0 A ± 0.10 A or 1% of CCA rating

3.2.4 Rest the battery on open circuit for 57.5 to 68.0 h.

3.2.5 Immediately after the rest period, discharge the battery at the specified CCA rating for 50 s. The minimum battery voltage requirement at 50 s is 7.20 V.

3.2.6 Rest the battery on open circuit until the start of the next weekly cycle.

3.3 The life test shall be considered completed when the battery fails to meet the 7.20 V requirement during the 50 s discharge at the specified CCA rating. The life shall be defined as the number of cycles that have successfully passed the weekly 50 s discharge requirements. Cycles that occur during the week of failure should not be counted in the total. The cycle life may also be shown as total AH delivered by multiplying the number of cycles by the discharge current corresponding to the battery type classification.

3.4 Water shall not be added to the electrolyte during the test except to those batteries described as requiring periodic maintenance and water additions.

(R) TEST PROCEDURE FOR BATTERY FLAME RETARDANT VENTING SYSTEMS— SAE J1495 MAR92 — SAE Standard

Report of the Electrical and Electronics Systems Technical Committee approved May 1986. Completely revised by the Storage Battery Standards Committee March 1992.

1. Scope—This SAE Standard details procedures for testing lead-acid SLI (starting, lighting, and ignition), Heavy-Duty, EV (electric vehicle) and RV (recreational vehicle) batteries to determine the effectiveness of the battery venting system to retard the propagation of an externally ignited flame of battery gas into the interior of the battery where an explosive mixture is usually present.

NOTE—At this time 1992, there is no known comparable ISO Standard.

2. References—There are no referenced publications specified herein.

3. Safety Precautions and Procedures

WARNING—Testing of a battery venting system can result in an explosion. Extreme caution must be exercised to avoid personal injury. Absolutely no testing should be permitted where the prescribed safety precautions and procedures are not followed or exceeded.

3.1 All test apparatus, except the charging source, must be fully contained in an externally vented explosion test chamber, for example, Figure 1.

3.2 The battery charging source must be located outside the explosion test chamber convenient to the control of the testing personnel. The charging circuit must have two emergency disconnect switches located (1) readily accessible to the testing personnel and (2) at a remote position at least 3 m (10 ft) from the explosion test chamber. These disconnect switches are intended for emergency use only, since their use may damage some types of chargers.

3.3 A suitable test area should be designated, for example, 3 m² (32 ft²) or more. Signs restricting unauthorized persons from this area should be posted and observed while any electrical circuit in the explosion test chamber is or could be energized.

3.4 During testing, entry to the area in which the explosion test chamber is located should be clearly marked to restrict all persons not fully familiar with all safety requirements and not wearing full protection from the hazard to be encountered.

3.5 Smoking, open flames, unprotected lights, or other spark sources must not be permitted in the area during testing.

3.6 Full face protection devices must be worn by all persons within the restricted area.

3.7 The battery spark source circuit must have an emergency disconnect switch to the testing personnel.

3.8 The exhaust fan of the explosion test chamber should be operated during the entire spark test procedure. On completion of any test sequence, charging and sparking circuits into the explosion test chamber must have been interrupted for at least 5 min with the exhaust fan operating before anyone is permitted access to the chamber. This time interval allows any hydrogen to be purged from the chamber and to preclude the possibility of a delayed explosion occurring due to a sustained

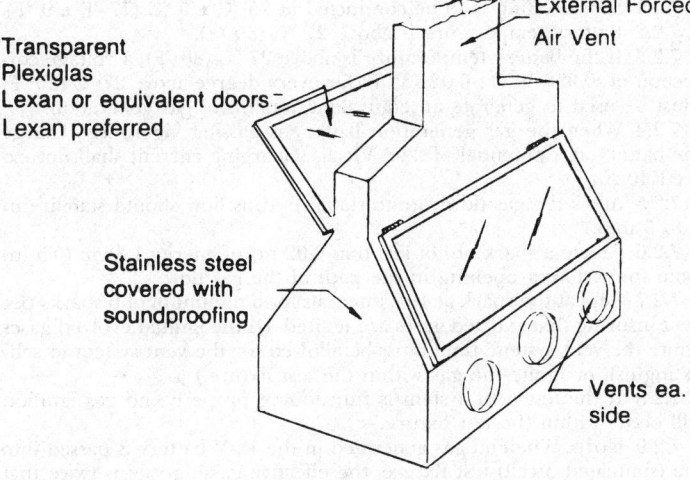

FIGURE 1 — TEST CHAMBER

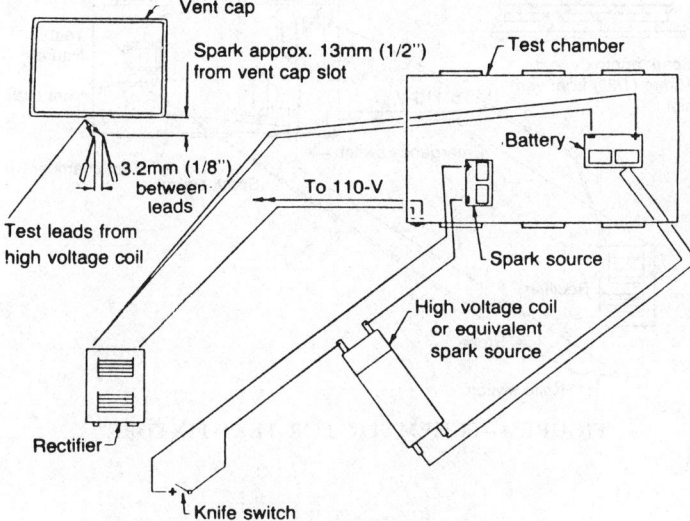

FIGURE 2A — SCHEMATIC FOR TEST ON BATTERY

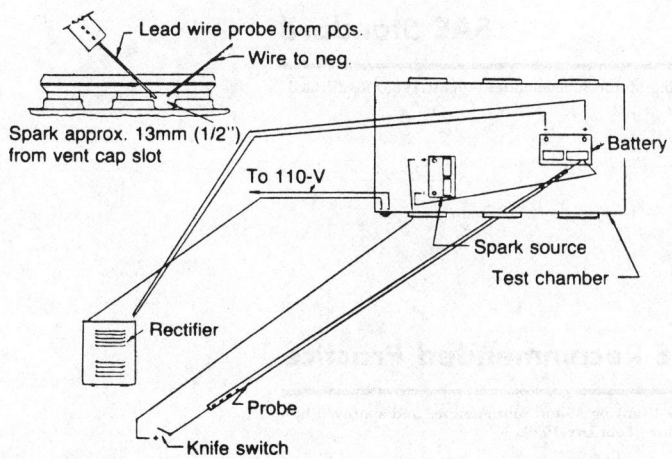

FIGURE 2B — SCHEMATIC FOR TEST ON BATTERY

"hidden" flame.

WARNING — HYDROGEN GAS CAN BURN WITHOUT VISIBLE FLAME..

4. Equipment Required for Spark Test

4.1 Where the spark test is to be conducted on a battery

4.1.1 An explosion chamber, for example, Figure 1, with an explosion-proof fan of adequate size to produce approximately one chamber volume change per minute, vented directly to the exterior of the building.

4.1.2 A battery charging source capable of constant voltage or current control, with at least 40 A output at 17.5 V.

4.1.3 A fully-charged 12-V battery to serve as an ignition source. This battery should be equipped with a functional flame retardant venting system.

4.1.4 Battery on which the test is to be performed.

4.1.5 Wiring and fixture equivalent to those shown in Figures 2A or 2B.

4.2 Where the spark test is to be conducted on a test fixture

4.2.1 An explosion chamber, for example, Figure 1, with an explosion-proof exhaust fan of adequate size to produce approximately one chamber volume change per minute, vented directly to the exterior of the building.

4.2.2 A charging source capable of constant voltage or current control, to be at least a 40 A output at 17.5 V.

4.2.3 A fully charged 12-V battery to serve as an ignition source. This battery should be equipped with a functional flame retardant venting system.

4.2.4 A second fully charged battery to serve as a gas mixture source. This battery must be vented only through the test fixture or functional flame retardant venting system.

4.2.5 A test fixture, for example, Figure 3.

4.2.6 Tubing and fittings equivalent to those shown in Figure 4.

5. Equipment Arrangement and Spark Test Preparation

5.1 Where the spark test is to be conducted on a battery

5.1.1 Arrange test apparatus as shown in Figure 2A or alternate Figure 2B.

5.1.2 Before spark testing, the test battery system should be checked for gas leakage at any place other than the vent opening, for example, with a soap solution.

5.1.3 Prior to the start of the test, the spark source battery must be fully charged.

5.1.4 Prior to the start of the test, the battery to which the test vent is affixed must be fully charged and gassing vigorously.

5.1.5 Prior to the start of the test, the test vent must be preconditioned as in 6.1.

5.2 Where the spark test is to be conducted on the test fixture

5.2.1 Arrange the test apparatus as shown in Figure 4. Note that the gas inlet to the test fixture must be well below the water level as shown in Figure 3 to prevent ignited gases from entering the gas generating battery.

5.2.2 Fill the test fixture with water to a level 3 mm (1/8 in) below the underside of the top. Place the hold-down frame over a 1 mil thickness of polyethylene film cut as shown in Figure 3. Place the frame, with film in place, over the four studs so that the film covers the open area between the fixture and the frame. Tighten the frame down finger tight with wing nuts to insure a gas-tight seal around the gasket. Fit the vent system to be tested into the fixture.

5.2.3 Before spark testing, the whole system should be checked for gas leakage at any place other than the vent opening, for example, with a soap solution.

5.2.4 Prior to the start of the test, the spark source battery must be fully charged.

5.2.5 Prior to the start of the test, the gas generating battery must be fully charged and gassing vigorously.

5.2.6 Prior to the start of the test, the test vent must be preconditioned as in 6.1.

6. Preconditioning the Vent System

6.1 Standard Preconditioning — Preconditioning shall consist of subjecting the battery to which the venting system is attached to an overcharge of 2 to 4 A for a period of 16 to 24 h. This will put the vent system in an acid moistened state as is typical of its in-service condition. The spark test shall be conducted within 1 h of completion of the preconditioning charge.

6.2 Hot and Cold Cycle Preconditioning (Optional)

6.2.1 Place the vent system in a cold box at -29 °C (-20 °F) for 16 h.

6.2.2 Remove from cold box and place it in an oven at 71 °C (160 °F) for 8 h.

6.2.3 Repeat this sequence for a total of 3 full cycles.

6.2.4 At the completion of the third full cycle, condition the vent system as in 6.1.

7. Spark Test Procedure

7.1 When the spark test is to be conducted on a battery

7.1.1 Spark testing is to be conducted at 25 °C ± 5 °C (77 °F ± 9 °F).

7.1.2 Battery temperature should be 27 °C (80 °F).

7.1.3 If the battery temperature is above 27 °C (80 °F), a voltage correction of -0.038 V/°C (-0.021 V/°F) for every degree above 27 °C (80 °F) must be used to guarantee an equivalent amount of gas generation.

7.1.4 When the test battery is gassing vigorously, charge the battery at a potential of 2.92 V/cell. (Charging current shall not exceed 40 A.)

7.1.5 Allow the gas flow rate to stabilize. (Gas flow should stabilize in 1 to 5 min.)

7.1.6 Create a spark of not less that 0.02 mJ of energy 1.3 cm (0.5 in) from the test vent opening in the path of the gas flow.

7.1.7 Repeat the spark at 10 s intervals for a minimum of 6 sparks per vent insuring that evolved gases are ignited. (If the ignited evolved gases ignite the battery, time must be allowed for the battery to self-extinguish or ignite the gas within the battery.)

7.1.8 If the vent system is functioning properly, no gas ignition will occur within the test battery.

7.2 Where the spark test is to be conducted on the test fixture

7.2.1 Spark testing is to be conducted at 25 °C ± 5 °C (77 °F ± 9 °F).

7.2.2 Battery temperature is above 27 °C (80 °F).

7.2.3 If the battery temperature is above 27 °C (80 °F), a voltage correction of -0.038 V/°C (-0.021 V/°F) for every degree above 27 °C (80 °F) must be used to generate an equivalent amount of gas generation.

7.2.4 When the gas generation battery is gassing vigorously, charge the battery at a potential of 2.92 V/cell. (Charging current shall not exceed 40 A.)

7.2.5 Allow the gas flow rate to stabilize. (Gas flow should stabilize in 1 to 5 min.)

7.2.6 Create a spark of not less that 0.02 mJ of energy 1.3 cm (0.5 in) from the test vent opening in the path of the gas flow.

7.2.7 Repeat the spark at 10 s intervals for a minimum of 6 sparks per vent insuring that evolved gases are ignited. (If the ignited evolved gases ignite the vent system, time must be allowed for the vent system to self-extinguish or ignite the gas within the test fixture.)

7.2.8 If the test vent system is functioning properly, no gas ignition will occur within the test fixture.

7.2.9 NOTE: When all gas generated in the 12-V battery is passed into the (simulated 3 cell) test fixture, the effective gassing rate is twice that indicated by the charger current reading, for either single or gang vents.

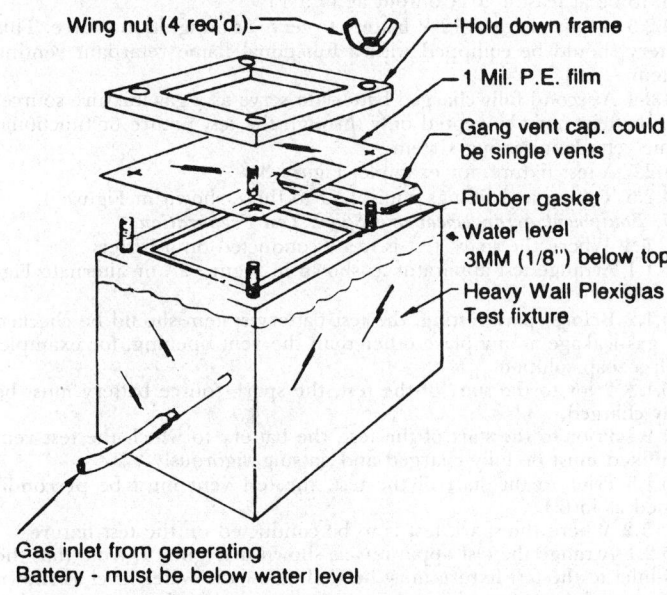

FIGURE 3—TEST FIXTURE

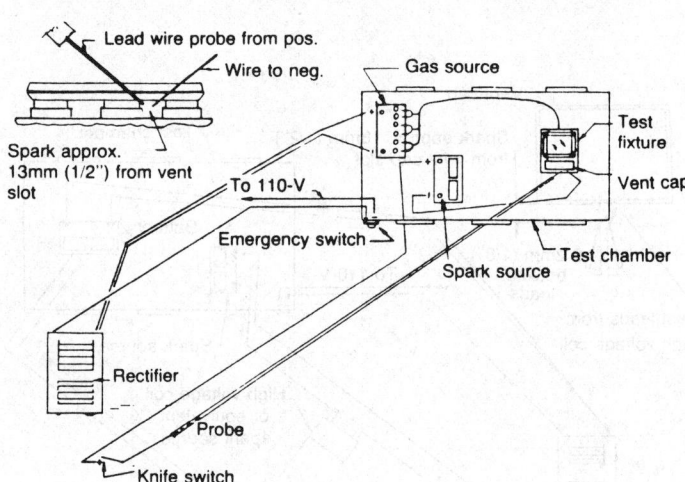

FIGURE 4—SCHEMATIC FOR TEXT FIXTURE

GROUNDING OF STORAGE BATTERIES— SAE J538 JUL89

SAE Standard

Report of the Electrical Equipment Committee, approved December 1955, second revision by Cranking Motor Subcommittee August 1983. Reaffirmed by the Cranking Motor Standards Committee July 1989.

The negative side of the storage battery shall be securely and adequately grounded so that the voltage drop to the starting motor is held within the limits specified in the current edition of SAE J541.

VOLTAGE DROP FOR STARTING MOTOR CIRCUITS—SAE J541 FEB89

SAE Recommended Practice

Report of the Electrical Equipment Division, approved January 1932, second revision prepared by the Cranking Motor Subcommittee and approved by the Electrical Equipment Committee July 1983. Reaffirmed by the Cranking Motor Standards Committee February 1989.

The starting motor circuits in motor vehicles, excluding motors, relays, and solenoids, shall be designed so that the difference between the voltage at the storage battery terminals and the starting motor terminals including connections shall not exceed those shown in Table 1. The voltage drop per hundred amps ($v_d/100_a$) is defined with a normal circuit temperature of 68°F (20°C). Since there is no finite division between "light and medium duty" and "heavy duty" applications, it is important that the engineer exercise judgment to make sure that the voltage drop selected does not have an adverse effect on the overall performance.

TABLE 1—MAXIMUM VOLTAGE DROP

System Voltage	(V_d/100$_a$)	Use
6V	0.12V	Light and Medium Duty
12V	0.200V	Light and Medium Duty
24V	0.400V	Light and Medium Duty
12V	0.10V	Heavy Duty
24V—32	0.17V	Heavy Duty

VOLTAGES FOR DIESEL ELECTRICAL SYSTEMS— SAE J539 MAR87

SAE Recommended Practice

Report of the Electrical Equipment Division approved January 1939, revised September 1976 and reaffirmed by the Electrical and Electronic Systems Technical Committee March 1987.

This SAE Recommended Practice is intended to apply to lamps, batteries, heaters, radios, and similar equipment for operation with mobile or automotive diesel engines. Twenty-four V systems have long been used for heavy duty services because 24 V permit operating 12 V systems in series-parallel. Thirty-two V systems have been used for marine, railroad-car lighting, and other uses.

Generators, storage batteries, starting motors, lighting, and auxiliary electrical equipment shall be for nominal system ratings of 12, 24, or 32 V as determined by the power requirements of the application. It is recommended that no intermediate voltages be considered except that a 30 V system may be used when cranking requirements permit and no lighting or auxiliary electrical equipment is involved.

The combination of a 24 V starting motor and two 12 V batteries connected in series for cranking is considered practical where it can be adapted to the installation. The batteries are reconnected in parallel for charging from a 12 V generator/alternator and for operating lights and other auxiliary equipment, or charged separately and used individually for lights and other electrical equipment.

VEHICLE SYSTEM VOLTAGE—INITIAL RECOMMENDATIONS —SAE J2232 JUN92

SAE Information Report

Report of the Dual/Higher Voltage Vehicle Electrical Systems Committee approved June 1992. Rationale statement available.

Foreword—It is the opinion of the committee that a set of guidelines and standards related to higher voltage levels will ensure the safety, shorten the introduction time, and lower the cost of such systems. A primary area of concern has been the potential for an increase in electrical shock hazard. Of secondary importance has been the need for some standardization of higher voltage levels. The committee feels that settling these broad issues now will allow development efforts to be more focused, thus shortening the total development time for higher voltage systems. The committee's review of the available technical literature and the industry expertise of our members leads to the recommendations discussed as follows.

1. Scope—This SAE Information Report is a summary of the initial recommendations of the SAE committee on Dual/Higher Voltage Vehicle Electrical Systems regarding the application of higher voltages in vehicle systems. This document does not attempt to address the technical merits of specific voltages or electrical system architectures.

2. References

2.1 Applicable Documents—The following publications form a part of this specification to the extent specified herein.

"Effects of Current Passing Through the Human Body, Part 1: General Aspects," International Electrotechnical Commission Report IEC-479 Part 1, Second edition, 1984.

"Effects of Current Passing Through the Human Body, Part 2: Special Aspects," International Electrotechnical Commission Report IEC-479 Part 2, Second edition, 1987.

R. J. Sandel and J. V. Hellmann, "Activities of the SAE Committee on Dual/Higher Voltage Vehicle Electrical Systems," presented at the 1991 SAE Future Transportation Technology Conference, Portland, Oregon.

3. Technical Requirements

3.1 Protection From Contact—Protection against direct contact to electrical circuits shall not be necessary if the possible contact voltages do not exceed the permissible levels of 65 VDC, including periodic ripple, or 50 VAC, RMS. The application of voltages above these levels shall not be discouraged, but additional protection against direct contact shall be necessary.

A study of previous work in relation to human tolerance to electrical shock was conducted and the various information sources were examined and thoroughly discussed by the members of the committee. Based on this study, an understanding of present vehicle wiring practices, and the safety record of today's 12 and 24 V systems, the 65 VDC and 50 VAC levels are seen as appropriate from a practical standpoint. Particularly useful in the formation of this recommendation has been the International Electrotechnical Commission's (IEC) Publication 479, parts 1 and 2.

It is recognized that with a system of present architecture utilizing a higher system voltage, say 50 VDC, transient levels will exceed 65 V. The committee has deferred consideration of the shock hazard of non-periodic transient (less than 300 ms) voltages to a future time. Also, the committee has not addressed the protection against contact necessary for voltages above the recommended limits. However, other industries demonstrate daily that such contact can be appropriately prevented.

3.2 Standard Storage Voltages—Standard storage battery voltages shall be defined as those normally associated with 3, 6, 12, and 24 cell lead-acid storage batteries (6, 12, 24, and 50 VDC nominal, respectively). The application of battery systems other than lead-acid shall not be discouraged. Additionally, the number of distinct generation, supply, or storage voltages that may be present in a vehicle shall not be limited.

The lead-acid battery system is currently the only system used commercially in starting, lighting, and ignition (SLI) applications on internal combustion engine automobiles. The committee feels that lead-acid will continue to be the SLI system of choice for a significant period of time. Standard voltages higher than 50 V may be defined in the future but will require a yet undefined degree of contact protection as their charge voltage will likely exceed 65 V.

As vehicle electrical requirements continue to evolve, battery systems other than lead-acid may become a viable alternative and should be able to meet the same nominal storage voltages as lead-acid. We feel that this option should not be limited. Similarly, limiting the voltage levels in magnitude or count will only restrict system options that may otherwise be appropriate solutions.

4. Notes

4.1 Key Words—Human Shock Tolerance, Personnel Protection, Shock Tolerance, Standard Voltage, System Voltage, Admissible Contact Voltage, High Voltage, Electrical System

(R) STARTING MOTOR MOUNTINGS— SAE J542 JUN91

SAE Recommended Practice

Report of the Electrical Equipment Division approved August 1917 and completely revised by the Electrical and Electronic Systems Technical Committee April 1987. Completely revised by the SAE Cranking Motor Standards Committee June 1991.

1. Scope—The purpose of this SAE Recommended Practice is to provide standardized dimensions for mounting starting motors. (See Figures 1 through 4.)

It is recommended that a full register diameter having a minimum depth of 2.54 mm (0.100 in) be provided in the flywheel housing to insure proper control of gear center distance and clearance between pitch diameters. The clearance between the starting motor pilot diameter and the register diameter in the flywheel housing should be 0.03 mm (0.001 in) minimum to 0.25 mm (0.010 in) maximum.

Text noted with an asterisk in Figures 1, 2, and 3, should not exceed root radius of pinion in order to provide clearance for the flywheel.

The face of the starting motor mounting flange should be relieved at its junction with the pilot diameter to avoid mounting interference with flywheel housing.

For backlash allowance between the pinion and ring gear refer to SAE J543.

Dimensional units—millimeter (inch).

2. References

2.1 Applicable Documents—The following publications form a part of this specification to the extent specified herein. The latest issue of SAE publications shall apply.

2.1.1 SAE PUBLICATIONS—Available from SAE, 400 Commonwealth Drive, Warrendale, PA 15096-0001.

SAE J543—Starting Motor Pinions and Ring Gears

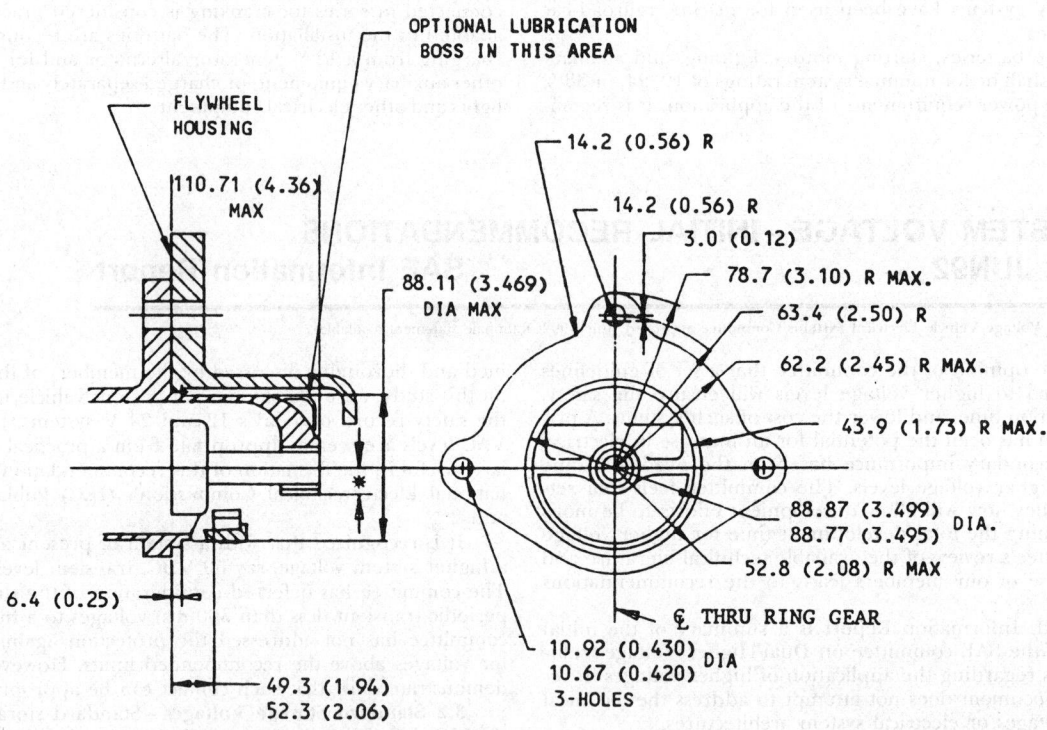

FIGURE 1—TYPE NO. 1 MOUNTING FLANGE MEDIUM DUTY

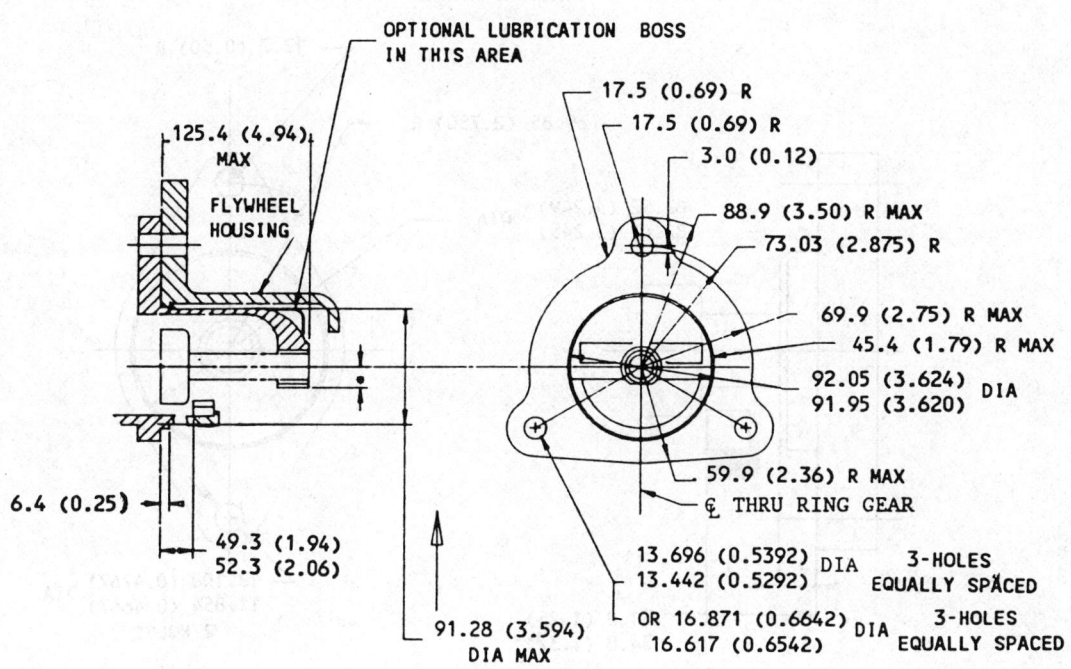

FIGURE 2—TYPE NO. 3 MOUNTING FLANGE HEAVY DUTY

NOTE: Type No. 2 mounting flange is the same as No. 3 except the mounting holes are 10.92 (0.430)—10.67 (0.420) dia.

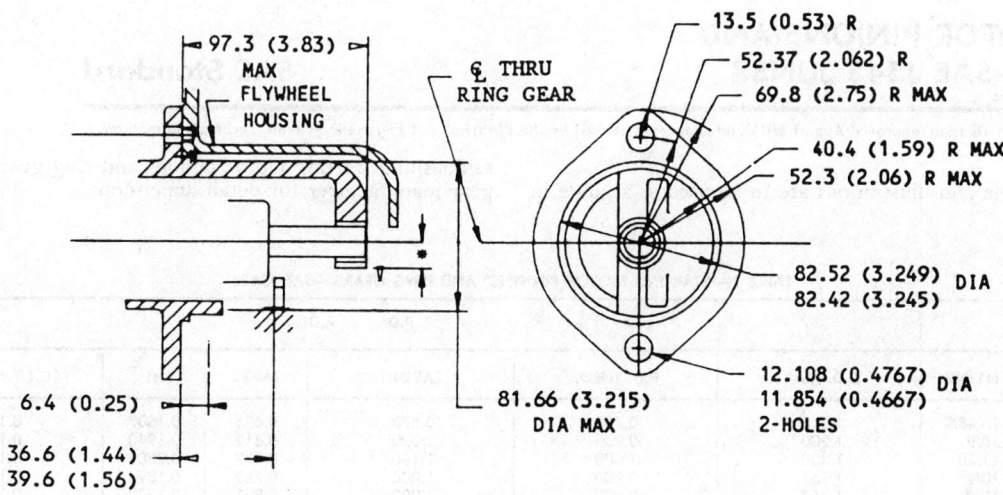

FIGURE 3—TYPE NO. 4 MOUNTING FLANGE LIGHT DUTY

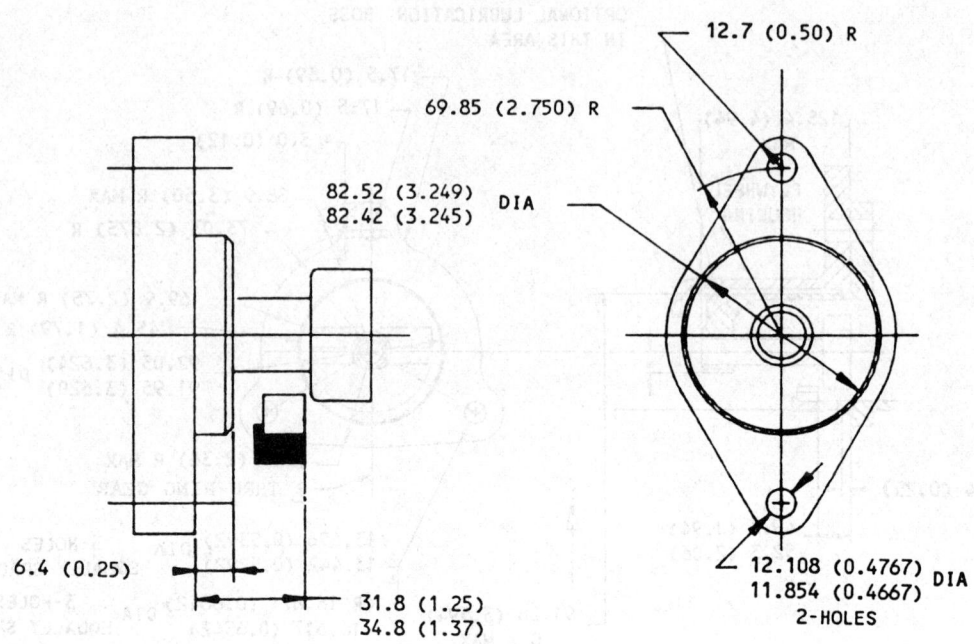

FIGURE 4—TYPE NO. 5 MOUNTING FLANGE LIGHT DUTY

φSTARTING MOTOR PINIONS AND RING GEARS—SAE J543 JUN88

SAE Standard

Report of Electrical Equipment Division, approved August 1917 and completely revised by the Electrical and Electronic Systems Technical Committee June 1988.

1. The following table and illustrations are to be used as a guide in establishing starting motor pinions and ring gear designs. Consult the gear manufacturer for detail dimensions.

TABLE 1—STARTING MOTOR PINIONS[a] AND RING GEARS—SAE J543c

P1/P2[b]	P.A.	N1/N2	O.D. MAX.	P.D. THEO.	P.D. LAYOUT	R.D. MAX.	HT	C.T.T.[e] MAX.	W2 MAX.
12/14	12	10.48/9	1.016	0.750	0.873	0.695	0.1607	0.161	0.416
10	14-1/2	10/9	1.200	0.900	1.000	0.812	0.1940	0.194	0.498
10	14-1/2	11/10	1.300	1.000	1.100	0.839	0.2305	0.194	0.498
10/12	20	10/9	1.167	0.900	1.000	0.789	0.1890	0.188	0.484
10/12	20	10/9	1.167	0.900	1.000	0.821	0.1730	0.198	0.495
10/12	20	11/10	1.267	1.000	1.100	0.889	0.1890	0.187	0.485
10/12	20	12/11	1.367	1.100	1.200	0.989	0.1890	0.187	0.486
10/12	20	13/12	1.467	1.200	1.300	1.089	0.1890	0.187	0.488
8/10	20	10/9	1.450	1.125	1.250	1.000	0.2250	0.245	0.615
8/10	20	11/10	1.575	1.250	1.375	1.125	0.2250	0.245	0.617
8/10	20	12/11	1.700	1.375	1.500	1.250	0.2250	0.245	0.618
8/10	20	13/12	1.825	1.500	1.625	1.375	0.2250	0.245	0.619
8/10	20	13/13	1.825[d]	1.625	1.625	1.393	0.2160	0.196	0.577
8/10	20	14/13	1.950	1.625	1.750	1.525	0.2125	0.244	0.622
6/8	20	11/11	2.083[d]	1.833	1.833	1.521	0.2810	0.262	0.765
6/8	20	12/11	2.240	1.833	2.000	1.688	0.2760	0.317	0.814

[a]Dimensions are for maximum metal conditions. Tolerances will result in increased clearances.
[b]The two diametral pitch gear data are based on Fellows stub tooth system.
[c]If larger root diameter is desired, consult gear manufacturers.
[d]Standard Blank. All others oversize blanks.
[e]Circular Tooth Thickness.

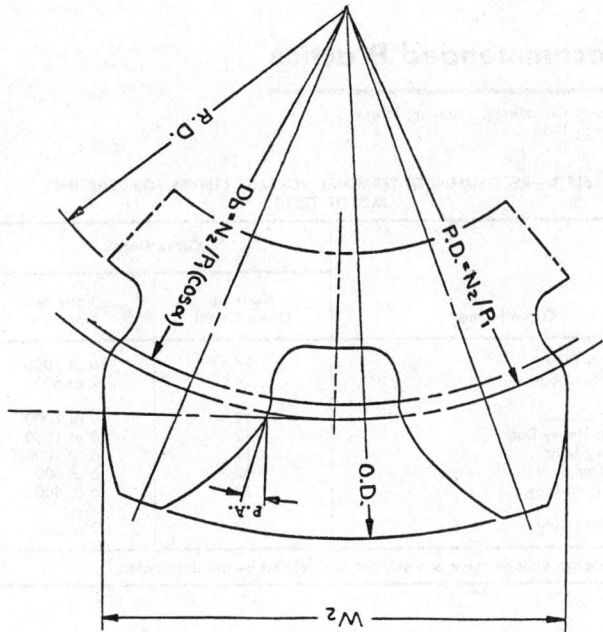

FIG. 1

2. Calculations for Spur Gears

Outside Diameter (O.D.) = $N1/P1 + 2/P2$
Theoretical Pitch Diameter (P.D.) = $N2/P1$
Layout Pitch Diameter = $N1/P1$
Base Circle Diameter (Db) = P.D. × COS P.A.

Where:

N1 - Blank Size
N2 - Number of Teeth
P1/P2 - Diametral Pitch for Fellows stub tooth. Use P1 to determine No. of Teeth, Pitch Diameter and Tooth Thickness. Use P2 to determine Addendum, Dedendum and Tooth Depth.
P.A. - Pressure Angle

3. Ring Gear and Pinion Installation—Backlash is necessary for free meshing and running of the pinion with the ring gear. Backlash may be obtained by increasing the center distance as in Fig. 2 or by reducing the tooth thickness.

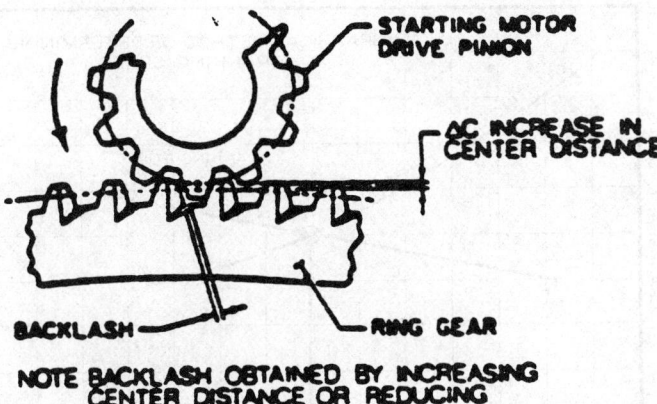

FIG. 2

4. Center Distance—The formula for calculating center distance (C.D.) is:

$$C.D. = \frac{\text{No. Ring Gear Teeth (Blank)}^a + \text{No. Pinion Teeth (Blank)}^a}{2 \times \text{Diametral Pitch}^b} + \Delta C^c$$

where:

a = the number of teeth is equal to the number used to determine blank size. A blank is a disk or cylinder of such size as to relate to a standard gear of standard addendum, dedendum and given number of teeth. To increase tooth strength and improve cranking ratio, many pinion gears are cut on an oversize blank (example: 10 teeth on 11 tooth blank). In this example, 11 would be used for the number of pinion teeth in calculating center distance.

b = for fractional diametral pitch (example: 8/10 pitch), use the numerator (8 in this example) for center distance calculation.

c = ΔC is the increase in center distance to obtain backlash. See Fig. 2. If backlash is obtained by reducing tooth thickness, omit ΔC from the C.D. formula.

Center Distance for starter drives, for 12, 12/14 and finer pitch, is held to theoretical values, with a 0.010 in tolerance. Backlash is designed into the ring gear by thinning the teeth and should be 0.010-0.030 inch. Spread of center for 10/12 and courser pitch should be 0.020 to 0.40 in to produce backlash of 0.015-0.035 inch.

5. Ring Gear Design—Ring gears of 10/12 pitch and finer are normally not chamfered. Gears coarser than 10/12 pitch should be chamfered in accordance with Fig. 3.

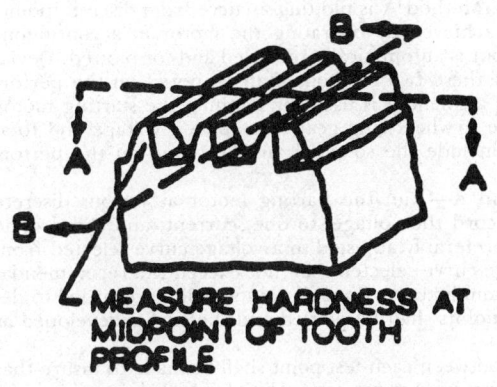

MEASURE HARDNESS AT MIDPOINT OF TOOTH PROFILE

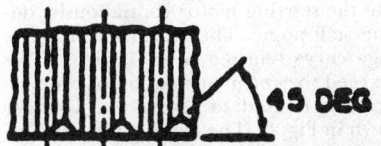

45 DEG

CHAMFER EDGE WHEN REQUIRED ± 1/64 OF CENTERLINE OF TOOTH

VIEW A-A

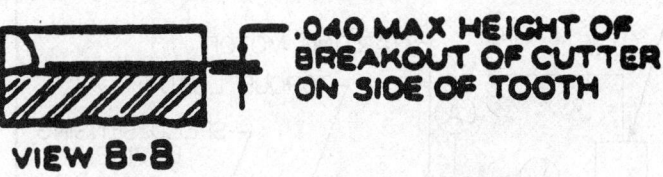

.040 MAX HEIGHT OF BREAKOUT OF CUTTER ON SIDE OF TOOTH

VIEW B-B

FIG. 3

6. Ring Gear Hardness—Hardness range for typical ring gears after assembly is:

8/10 pitch and coarser Rockwell C45-52
10/12 pitch and finer Rockwell C48-55

ELECTRIC STARTING MOTOR TEST PROCEDURE—SAE J544 MAR88

SAE Recommended Practice

Report of the Electrical Equipment Division, approved January 1945, third revision by the Electrical Equipment Committee, Cranking Motor Subcommittee, May 1982, and reaffirmed by the Electrical and Electronic Systems Technical Committee March 1988.

1. Introduction—Prior to 1981, SAE J544 addressed both starting motor and generator performance curves. Review of this technical report for improvement indicated that a recommended test procedure for development of the performance curves was needed and that starting motors and generators should be addressed in separate SAE technical reports. The generator performance curve information is now contained in SAE J56. The SAE J544 identity has been retained for testing the output performance and plotting the performance curves of starting motors.

2. Purpose—This SAE Recommended Practice provides a standard procedure for testing the output performance and plotting the performance curve of electric starting motors, and a graphical method of determining engine cranking speed.

3. Testing Procedure—The motor shall be mounted in a test stand as shown in Fig. 1. For larger starting motors, the torque may be measured directly at the motor axis with a special test end frame because of torque limitation of test equipment. The torque measurement may be recorded and should be identified on the performance curve as either a frame reaction at the motor axis or at the torque loading point shown in Fig. 1.

NOTE: If the latter test method is used, the effect of inertia should be taken into account.

Performance curves are established by running the starting motor in one of two methods. Method A is plotting a curve from discrete points while Method B is achieved by operating the motor in a continuous mode while the output is automatically recorded and/or plotted. Deviations from either of these two methods shall be noted on the performance curve. When a solenoid is used for meshing the starting motor pinion gear with the flywheel ring gear, the applied voltage and total current draw shall include the solenoid and be noted on the performance curve.

3.1 Test Method A—Run the starting motor at various discrete torque loads and record the voltage, torque, current, and speed. The terminal voltage is preferably adjusted to a voltage curve selected from Table 1. The voltage curve selected shall not exceed the recommendation of the motor manufacturer. Enough points shall be recorded to develop a curve. The points shall be plotted and the curves developed as shown in Fig. 2.

Cooling intervals between each test point shall be made to insure that the effects of temperature changes are negligible. Ambient test temperature shall be noted on the performance curve.

3.2 Test Method B—Operate the starting motor continuously, decreasing the torque load from the stall point. The terminal voltage is preferably maintained to a voltage curve selected from Table 1. The voltage curve selected shall not exceed the recommendation of the motor manufacturer. Plotting equipment is used to record the voltage, torque, current, and speed as shown in Fig. 2. The loading rate is to be such that the effect of temperature change is negligible. Ambient test temperature shall be noted on the performance curve.

TABLE 1—RECOMMENDED TERMINAL VOLTAGE CURVES FOR STARTING MOTOR TESTS

Curve Name	Curve Data[a]	
	Volts at Open Circuit	Volts at Amperes
24V Heavy Duty	24	16 at 1000
24V Standard Duty	24	12 at 600
12V High Output	12	10 at 1000
12V Extra Heavy Duty	12	8 at 1000
12V Heavy Duty	12	6 at 1000
12V Medium Duty	12	6 at 600
12V Standard Duty	12	6 at 400
6V Standard Duty	6	2 at 800

[a]The terminal voltage curve is a straight line defined by the data points.

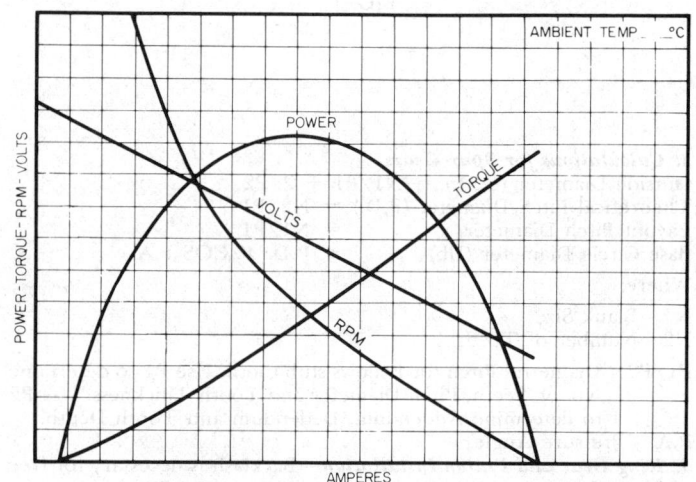

FIG. 2—STANDARD FORM FOR STARTING MOTOR CHARACTERISTIC CURVES

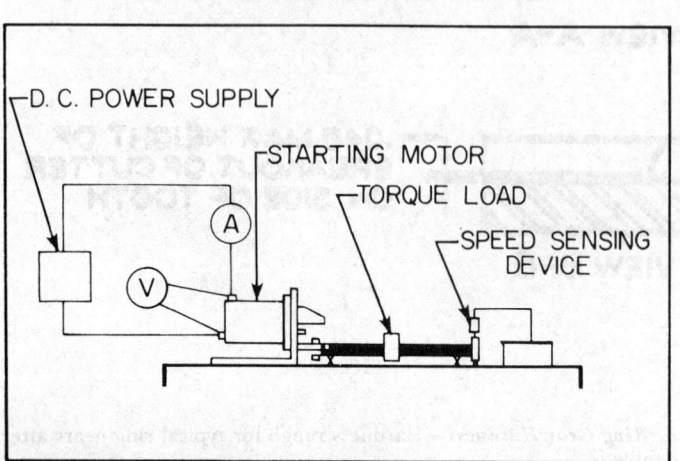

FIG. 1—TYPICAL TEST SET-UP FOR STARTING MOTORS

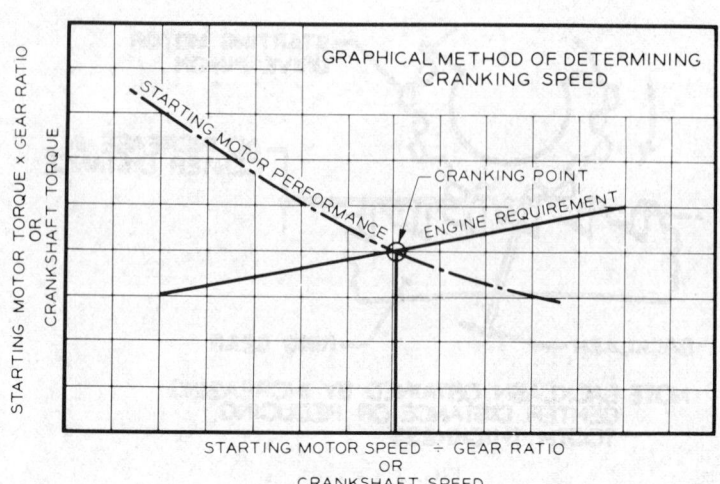

FIG. 3—METHOD OF DETERMINING CRANKING SPEEDS

4. Instrumentation—The voltmeter and the ammeter shall have an accuracy of ±1% of the full scale reading (full scale value not to exceed the maximum performance reading by more than 50%). The speed sensing device and torque measuring device shall have an accuracy of ±1% of the actual reading.

5. Graphical Method of Determining Engine Cranking Speed—With the starting motor performance curve generated by the above procedure and a cranking demand curve developed for a particular engine (Reference: Low Temperature Cranking Load Requirements of an Engine—SAE J1253 MAR86), the cranking speed can be determined by using the graphical method shown in Fig. 3.

The starting motor performance curve may need to be corrected for temperature and terminal voltages expected on the actual application. The motor performance corrections can be obtained from the motor manufacturer.

LOW-TEMPERATURE CRANKING LOAD REQUIREMENTS OF AN ENGINE—SAE J1253 JUN93 SAE Recommended Practice

Report of Cranking Motor Subcommittee approved February 1979 and revised by the Electrical and Electronic Systems Technical Committee March 1986. Revised by the Cranking Motor Standards Committee June 1993.

1. Scope—The electrical cranking system components, which include the battery, cables, and cranking motor, must be carefully selected to provide the necessary speed to start an engine under the most severe climatic conditions for which the system is intended. Engine cranking loads increase with cold temperatures, therefore, the initial selection of these components needs to consider low-temperature engine torque requirements. To insure an adequate electrical cranking system is obtained, it is important that proper test procedures are used for obtaining the cranking load requirements of the engine.

2. References

2.1 Applicable Documents—The following publications form a part of this specification to the extent specified herein. The latest issue of SAE publications shall apply.

2.1.1 SAE PUBLICATIONS—Available from SAE, 400 Commonwealth Drive, Warrendale, PA 15096-0001.

SAE J300—Engine Oil Viscosity Classification
SAE J544—Electric Starting Motor Test Procedure

3. Procedure—The following test procedure is recommended for obtaining low-temperature cranking torque requirements:

3.1 Engine Preparation

3.1.1 The engine to be tested should be equipped with all accessories that provide parasitic loads, such as power steering pump, automatic transmission, etc.

3.1.2 The engine, if new, should be run in to stabilize friction loads—equivalent to 1500 miles (2400 kilometers) or 18 h at 2400 engine rpm.

3.1.3 The engine is winterized with anti-freeze solution for the temperature at which the test will be run.

3.1.4 The engine oil selected for the low temperature test should be representative of the high limit viscosity for the SAE grade recommended by the engine manufacturer for the operating temperature range (refer to SAE J300). Sufficient oil of the same viscosity should be obtained for the complete test program so variations in test results can be minimized.

3.1.5 Fuel dilution of the engine oil will reduce its viscosity, therefore, to avoid this possibility, the cranking test is run without fuel in the carburetor, or with fuel system cut off.

3.1.6 To prepare the engine for test, the engine is warmed up and oil drained hot. This procedure should be repeated two times to assure complete change of oil when oil grade change is made. The oil filter is changed for the final fill. When the same grade of oil is used for other test temperatures and/or additional test days, the engine warm up procedure is repeated and only one drain is required.

3.1.7 Install a thermocouple in the center of the greatest mass of oil so soak temperatures can be monitored.

3.1.8 Equip engine with necessary instrumentation to provide cranking speed, battery voltage, cranking motor voltage, and current data. (The cranking speed can be determined from oscillographic current or voltage traces by calculation of the time span between the current or voltage peaks caused by the cylinder compression loads. The mean cranking speed is obtained over two consecutive revolutions. The mean torque is obtained by measuring the mean cranking current over the same period and calculated as described in 3.2.)

3.1.9 Prior to starting the cold soak period, warm up the engine for approximately 5 to 10 min to circulate oil, run carburetor bowl dry and disable ignition or cut off fuel system, and adjust throttle plate to the idle position.

3.1.10 The engine with the calibrated motor is soaked at the test temperature for a period of 16 to 24 h, which can be monitored by the oil thermocouple.

3.2 Cranking Motor Preparation—The cranking motor is used to measure the engine cranking torque. To minimize performance variances, a new cranking motor should be "run in" until the motor performance becomes stabilized prior to calibration which is determining the speed, torque, and current under load using a standard SAE terminal voltage curve (Reference Table 1 of SAE J544) unless otherwise specified.

After completion of the cranking load tests, a recalibration curve should be run to verify initial performance.

NOTE—Since torque is proportional to cranking motor current, determination of engine torque can be calculated by obtaining the cranking motor running torque corresponding to the cranking motor current measured at the test temperature from the performance characteristics of the calibrated cranking motor and multiplying this value by the proper flywheel ring gear to cranking motor pinion gear ratio.

3.3 Cranking Load Tests

3.3.1 A sufficient number of cranking tests should be run to obtain a curve of average torque versus average engine speed over an approximate range of 30 to 120 engine rpm for gasoline engines, 50 to 150 rpm for direct injection diesel engines, and 120 to 220 rpm for small indirect injection diesel engines.

3.3.2 To obtain the range of speeds required to plot the torque curve, various battery capacities or a regulated DC power supply that can simulate desired battery conditions are used to supply the appropriate cranking motor terminal voltages. The batteries are not required to be cold soaked but should be maintained at full charge.

3.3.3 The cranking time for each test should be approximately 10 s with readings between 5 to 10 s used as the plotting points. Allow a minimum of 30 min additional soak time before performing the next cranking test.

3.3.4 Using the test data, calibrated cranking motor performance characteristics and engine ring gear to cranking motor pinion gear ratio, calculate the engine torque requirements for each test speed and plot an engine torque requirement curve as shown in Figure 3 of SAE J544.

3.3.5 It should be noted that since gear efficiencies have been neglected, the torque measured is not true engine torque but that as seen by the cranking motor. However, it provides suitable design and application data for determining cranking motor requirements.

3.3.6 Once the engine torque requirement curve has been determined and the speed required to start the engine is known, cranking motor performance requirements for the engine application can be determined.

CRANKING MOTOR APPLICATION CONSIDERATIONS—SAE J1375 MAR93 SAE Recommended Practice

Report of the Electrical Equipment Committee, Cranking Motor Subcommittee, approved March 1982. Revised by the SAE Cranking Motor Standards Committee March 1993.

1. Scope—This SAE Recommended Practice identifies some basic and general conditions that should be considered when making electrical cranking motor applications.

2. References

2.1 Applicable Documents—The following publications form a part of this specification to the extent specified herein. The latest issue of SAE publications shall apply.

2.1.1 SAE PUBLICATIONS—Available from SAE, 400 Commonwealth Drive, Warrendale, PA 15096-0001.

SAE J541—Voltage Drop for Starting Motor Circuits
SAE J543—Starting Motor Pinions and Ring Gears
SAE J1253—Low Temperature Cranking Load Requirements of an Engine

3. Application Conditions for Consideration

(R) 3.1 Components for cranking motor system shall be selected according to the current edition of SAE J1253. A vehicle "owner's manual" recommended cranking cycle shall be limited to 15 s "on" followed by 2 min "off" for automotive gasoline engines. On diesel applications, allow a maximum of 30 s cranking time followed by 2 min "off."

3.2 Pinion and ring gear data and center distances shall be compatible with the current edition of SAE J543.

3.3 Maximum voltage drop in the battery cables shall not exceed limits shown in the current edition of SAE J541.

3.4 Wire size versus length for the control circuit of magnetic switching devices shall be selected to insure proper design function.

3.5 It is recommended that solenoid-actuated positive shift starting motors be mounted with the solenoid location at least 15 degrees above the horizontal center line.

3.6 Engine rotation and mounting arrangements shall be known for fitting the cranking motor. System voltage shall be specified, and not exceeded.

3.7 Heat shields shall be provided as required to insure protection from damage to insulating materials and function of magnetic devices.

3.8 All relevant environmental conditions including wet or dry transmission clutch construction shall be considered.

ELECTRICAL GENERATING SYSTEM (ALTERNATOR TYPE) PERFORMANCE CURVE AND TEST PROCEDURE—SAE J56 JUN83

SAE Recommended Practice

Report of the Electrical Equipment Committee, approved September 1978, first revision by the Electrical Generating Systems Subcommittee June 1983.

1. *Purpose*—The purpose of this SAE Recommended Practice is to provide a standard test procedure for the development of alternator output performance, and a standard form for plotting performance curves.

2. *Instrumentation*

2.1 Voltmeter and ammeter should have an accuracy of ±0.5% of full scale.

2.2 Speed sensing device should have an accuracy of ±1%.

3. *Test Procedure*

3.1 Alternator test stand with suitable instrumentation is shown in Fig. 1.

3.2 The test conditions require that the alternator output voltage ϕ be maintained at the applicable test voltage listed below, ±1%, with the field current to be supplied by the alternator.

ϕ Voltage Levels					
Basic System Battery Voltage	6.0	12.0	24.0	30.0	32.0
Test Voltage	7.0	14.0	28.0	35.0	37.5

3.3 Output performance curve data is obtained by driving the alternator at room ambient, 24 ± 3°C (75.0 ± 5°F) at various discrete speeds, and adjusting the load to maintain constant output test voltage. Temperature stabilization must be attained at each discrete speed.

3.4 When the alternator is normally used with a built in or external solid state type regulator having inherent voltage drop, the performance test should be run with the regulator adjusted or re-connected for *full field current* condition, or an equivalent voltage drop should be connected in the field circuit if the test is performed without a regulator.

3.5 Deviations from the above procedure should be noted on the performance curve.

4. *Curve*—The output performance curve is made by plotting output current (amps) as the ordinate versus alternator speed (rpm) as the abscissa. (See Fig. 2.)

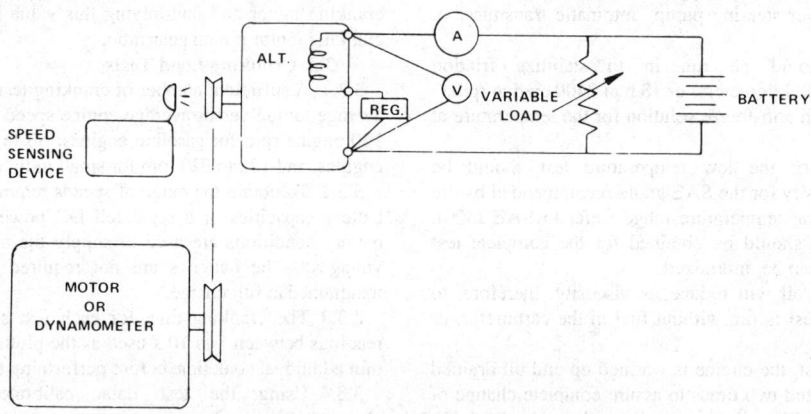

FIG. 1—TYPICAL TEST SETUP FOR OUTPUT PERFORMANCE

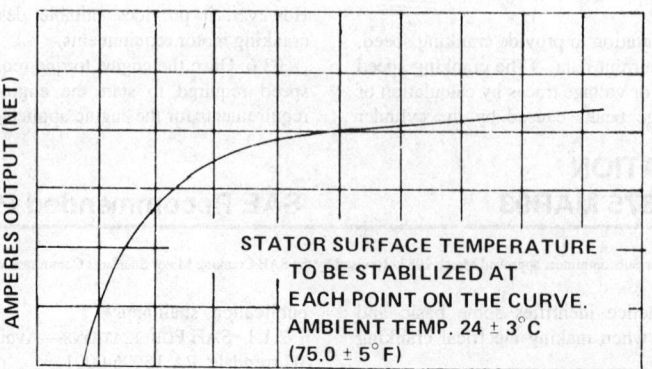

FIG. 2—STANDARD FORM FOR PERFORMANCE CURVE

(R) SPARK PLUGS—SAE J548/1 JUN92   SAE Standard

Report of Miscellaneous Division approved January 1915, Aircraft Division approved March 1918, Motorcycle Division approved August 1919, Electrical Equipment Division approved May 1930, and revised by the Electrical and Electronic Systems Technical Committee August 1987. Completely revised by the SAE Spark Plug Task Force June 1992. Rationale statement available.

1. Scope and Purpose—This SAE Standard applies only to spark plugs used for ground vehicles and stationary engines.

This document is intended to serve as a guide to dimensions common to the majority of current production spark plugs and future applications. It is not the intent of this document to prohibit the manufacture of spark plugs having dimensions differing from those presented. Many applications exist which require specialized or nonstandard spark plugs. It is recommended that this document be used in spark plug design and engine applications wherever possible. Whenever design situations arise that prevent the use of one of these standard spark plugs, a spark plug manufacturer should be contacted for guidance.

Figures 1 to 13 and Tables 1 to 6 show typical configurations of unshielded and shielded spark plug designs, their dimensional characteristics, installation, threaded hole, and spark plug thread sizes.

1.1 Thread Gages—In order to keep the wear on threading tools within permissible limits, the threads on the spark plug GO (ring) gage shall be truncated to the maximum minor diameter of the spark plug, and the tapped hole GO (plug) gage to the minimum major diameter of the tapped hole. The plain plug gage for checking the minor diameter of the tapped hole shall be the minimum specified.

2. References—There are no referenced publications specified herein.

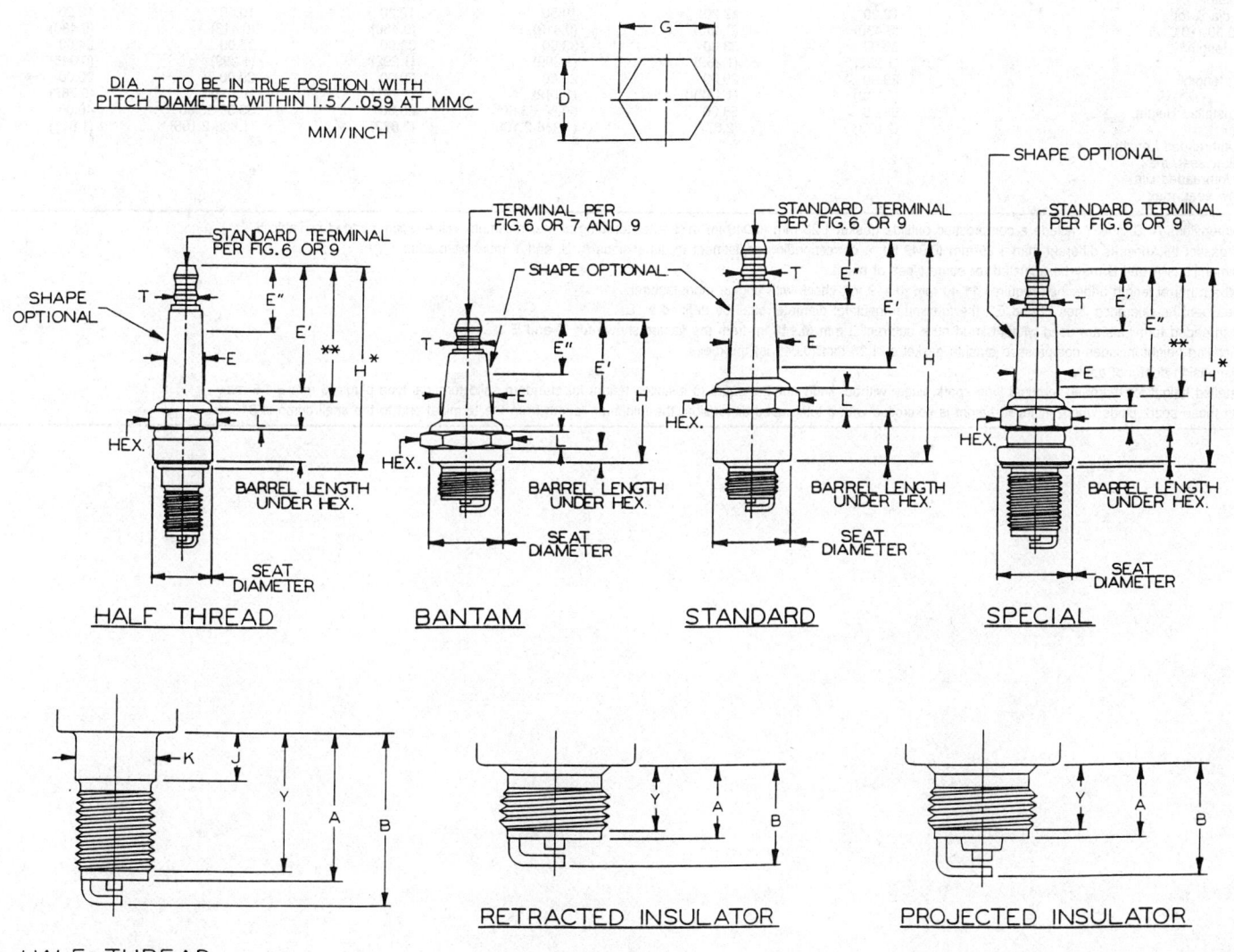

FIGURE 1—FLAT SEAT SPARK PLUGS (SEE TABLE 1)

TABLE 1A—FLAT SEAT SPARK PLUGS (FIGURE 1) mm (in)
M14 × 1.25 (PREFERRED)

Dimension	Short Reach	Normal Reach	Special Normal Reach	Long Reach	Special Long Reach	Bantam
A, Plug Reach, tol.[1] ±0.25 (±0.010)	9.53 (0.375)	12.70 (0.500)	12.70 (0.500)	19.00 (0.748)	19.00 (0.748)	9.53 (0.375)
B, Plug Depth, max[1,2] (retracted insulator)	13.60 (0.535)	17.80 (0.701)	17.80 (0.701)	24.40 (0.960)		13.60 (0.535)
B, Plug Depth, max[1,2] (projected insulator)	17.20 (0.677)	21.00 (0.827)	21.00 (0.827)	27.00 (1.063)	27.00 (1.063)	16.00 (0.630)
Y, Threaded Length & tol[1] ±0.30 (±0.012)	9.02 (0.355)	11.70 (0.461)	11.70 (0.461)	18.00 (0.709)	18.00 (0.709)	9.02 (0.355)
Hex Size:						
D, across flats	20.40-20.80 (0.803-0.819)	20.40-20.80 (0.803-0.819)	15.73-15.95 (0.619-0.628)	20.40-20.80 (0.803-0.819)	15.73-15.95 (0.619-0.628)	18.80-19.00 (0.740-0.748)
G, across corners, min	23.00 (0.906)	23.00 (0.906)	17.50 (0.689)	23.00 (0.906)	17.50 (0.689)	21.00 (0.827)
L, hex length, min	4.00 (0.157)	4.00 (0.157)	4.00 (0.157)	4.00 (0.157)	4.00 (0.157)	3.00 (0.118)
Barrel Length Under hex, min	11.40[3] (0.449)	10.00 (0.394)	8.75 (0.344)	10.00 (0.394)	8.75 (0.344)	3.00 (0.118)
Seat Dia., max	20.80 (0.819)	20.80 (0.819)	19.00 (0.748)	20.80 (0.819)	19.00 (0.748)	19.00 (0.748)
Insulator						
E, dia & tol[4] ± 0.30 (±0.012)	12.20 (0.480)	12.20 (0.480)	10.50 (0.413)	12.20 (0.480)	10.50 (0.413)	12.20 (0.480)
E', length[4]	33.00 (1.299)	33.00 (1.299)	33.00 (1.299)	33.00 (1.299)	33.00 (1.299)	24.00 (0.945)
E'', length[4]	29.00 (1.142)	29.00 (1.142)	29.00 (1.142)	29.00 (1.142)	29.00 (1.142)	20.00 (0.787)
H, Installed Height, max	68.00 (2.677)	68.00 (2.677)	50.50-53.50[5] (1.988-2.106)	68.00 (2.677)	50.50-53.50[5] (1.988-2.106)	46.00 (1.811)
J, Unthreaded Length below seat, max	6	6	6	6	6	6
K, Unthreaded Dia below seat, max	6	6	6	6	6	6

[1] Dimensions A, B, and Y include a compressed outside gasket 1.25 mm (0.049 in) thick after torquing at the maximum value recommended in Table 5.
If gasket thickness is different than 1.25 mm (0.049 in), a corresponding adjustment to dimensions A, B, and Y must be made.

[2] Dimensions A and B may be adjusted for some types of plugs.

[3] Where barrel length under hex requires 11.40 mm (0.449 in), check with engine manufacturer.

[4] Between the reference lines E' and E'', the maximum insulator diameter shall be defined by E.
If threaded terminals are used without knurl nuts, subtract 3 mm (0.118 in) from the values shown for E' and E''.

[5] Installed height includes compressed outside gasket of 1.25 mm (0.049 in) thickness.

[6] Dimension does not apply.

*Installed height for threaded terminal type spark plugs (without knurl nuts) is equal to installed height for standard solid terminal type plugs -3 mm ± 1.5 mm.

**On those spark plugs where installed height is controlled with a toleranced dimension, the minimum length from the terminal end to the shell crimp is 34 mm.

TABLE 1B — FLAT SEAT SPARK PLUGS (FIGURE 1) mm (in)

Dimension	M12 × 1.25 Normal Reach	M12 × 1.25 Long Reach	Normal Reach	M10 × 1.0 Long Reach	M10 × 1.0 Half Thread
A, Plug Reach, tol[1] ±0.25 (±0.010)	12.70 (0.500)	19.00 (0.748)	12.70 (0.500)	19.00 (0.748)	19.00 (0.748)
B, Plug Depth, max[1,2] (retracted insulator)	15.80 (0.622)	22.35 (0.880)	15.80 (0.622)	22.35 (0.880)	
B, Plug Depth, max[1,2] (projected insulator)	19.00 (0.748)	27.00— (1.063)	19.00 (0.748)	27.00 (1.063)	27.00 (1.063)
Y, Threaded Length, tol[1] ±0.30 (±0.012)	11.70 (0.461)	18.00 (0.709)	11.70 (0.461)	18.00 (0.709)	18.00 (0.709)
Max Size:					
D, across flats	15.73-15.95 (0.619-0.628)	15.73-15.95 (0.619-0.628)	15.73-15.95 (0.619-0.628)	15.73-15.95 (0.619-0.628)	15.73-15.95 (0.619-0.628)
G, across corners, min	17.50 (0.689)	17.50 (0.689)	17.50 (0.689)	17.50 (0.689)	17.50 (0.689)
L, hex length, min	4.00 (0.157)	4.00 (0.157)	4.00 (0.157)	4.00 (0.157)	4.00 (0.157)
Barrel Length Under hex, min	6.00 (0.236)	6.00 (0.236)	6.00 (0.236)	6.00 (0.236)	6.00 (0.236)
Seat dia, max	17.50 (0.689)	17.50 (0.689)	15.95 (0.628)	15.95 (0.628)	15.95 (0.628)
Insulator					
E, dia & tol[3] ±0.30 (±0.012)	10.50 (0.413)	10.50 (0.413)	10.50 (0.413)	10.50 (0.413)	10.50 (0.413)
E', length[3]	33.00 (1.299)	33.00 (1.299)	33.00 (1.299)	33.00 (1.299)	33.00 (1.299)
E'', length[3]	29.00 (1.142)	29.00 (1.142)	29.00 (1.142)	29.00 (1.142)	29.00 (1.142)
H, Installed Height, max	50.50-53.50[4,5] (1.988-2.106)	50.50-53.50[4,5] (1.988-2.106)	50.50-53.50[4,5] (1.988-2.106)	50.50-53.50[4,5] (1.988-2.106)	50.50-53.50[4,5] (1.988-2.106)
J, Unthreaded Length below seat, max	[6]	[6]	[6]	[6]	6.30 (0.248)
K, Unthreaded Dia below seat, max	[6]	[6]	[6]	[6]	11.00 (0.433)

[1] Dimensions A, B, and Y include a compressed outside gasket 1.25 mm (0.049 in) thick after torquing at the maximum value recommended in Table 5.
If gasket thickness is different than 1.25 mm (0.049 in), a corresponding adjustment to dimensions A, B, and Y must be made.

[2] Dimensions A and B may be adjusted for some types of plugs.

[3] Between the reference lines E' and E'', the maximum insulator diameter shall be defined by E.
If threaded terminals are used without knurl nuts, subtract 3 mm (0.118 in) from the values shown for E' and E''.

[4] Installed height includes compressed outside gasket of 1.25 mm (0.049 in) thickness.

[5] Installed height of 61 mm max, allowed for plugs of low thermal value (IMEP) and long reach as well as for normal reach plugs.

[6] Dimension does not apply.

*Installed height for threaded terminal type spark plugs (without knurl nuts) is equal to installed height for standard solid terminal type plugs - 3 mm ± 1.5 mm.

**On those spark plugs where installed height is controlled with a toleranced dimension, the minimum length from the terminal end to the shell crimp is 34 mm.

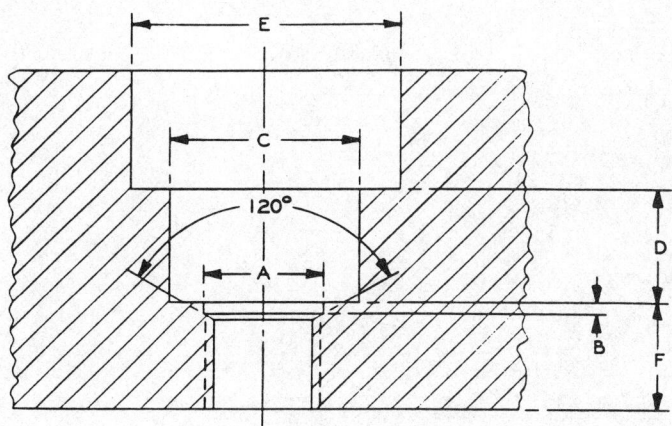

FIGURE 2 — HEAD COUNTERBORE — FLAT SEAT (SEE TABLE 2)

TABLE 2—HEAD COUNTERBORE—FLAT SEAT (FIGURE 2) mm (in)

M Thread Size and Reach	A Tolerance	B Max	C Min	D Max	E Min	F[2] Min
M14 × 1.25 Standard:						
Short reach	14.25-14.50 (0.561-0.571)	1.70 (0.067)	22.00 (0.866)	10.20 (0.402)	31.00 (1.220)	8.46 (0.333)
Normal reach	14.25-14.50 (0.561-0.571)	1.70 (0.067)	22.00 (0.866)	9.00 (0.354)	31.00 (1.220)	11.63 (0.458)
Long reach	14.25-14.50 (0.561-0.571)	1.70 (0.067)	22.00 (0.866)	9.00 (0.354)	31.00 (1.220)	18.00 (0.709)
Special long reach	14.25-14.50 (0.561-0.571)	1.70 (0.067)	21.20 (0.835)	3.00 (0.118)	24.00 (0.945)	18.00 (0.709)
M12 × 1.25 Bantam	14.25-14.50 (0.561-0.571)	[1]	[1]	[1]	[1]	8.46 (0.333)
M12 × 1.25:						
Normal reach	12.24-12.50 (0.482-0.492)	1.70 (0.067)	19.00 (0.748)	3.00[3] (0.118)	29.00 (1.142)	11.63 (0.458)
Long reach	12.24-12.50 (0.482-0.492)	1.70 (0.067)	19.00 (0.748)	3.00[3] (0.118)	29.00 (1.142)	18.00 (0.709)
M10 × 1.00:						
Normal reach	10.37-10.50 (0.408-0.413)	1.00 (0.039)	17.02 (0.670)	5.00 (0.197)	25.00 (0.984)	11.63 (0.458)
Long reach	10.37-10.50 (0.408-0.413)	1.00 (0.039)	17.02 (0.670)	5.00 (0.197)	25.00 (0.984)	18.00 (0.709)
Half thread	12.00 max (0.472)	5.80 min (0.228)	17.00 (0.669)	5.00 (0.197)	25.00 (0.984)	18.00 (0.709)

[1] Dimension does not apply.
[2] This F dimension shall be of sufficient length to ensure that the end of the spark plug thread does not project into the combustion chamber at any point when the gasket is tightened to maximum compression.
[3] Dimension of 3.00 mm encouraged for new designs but not required. Dimension of 5.00 mm allowed.

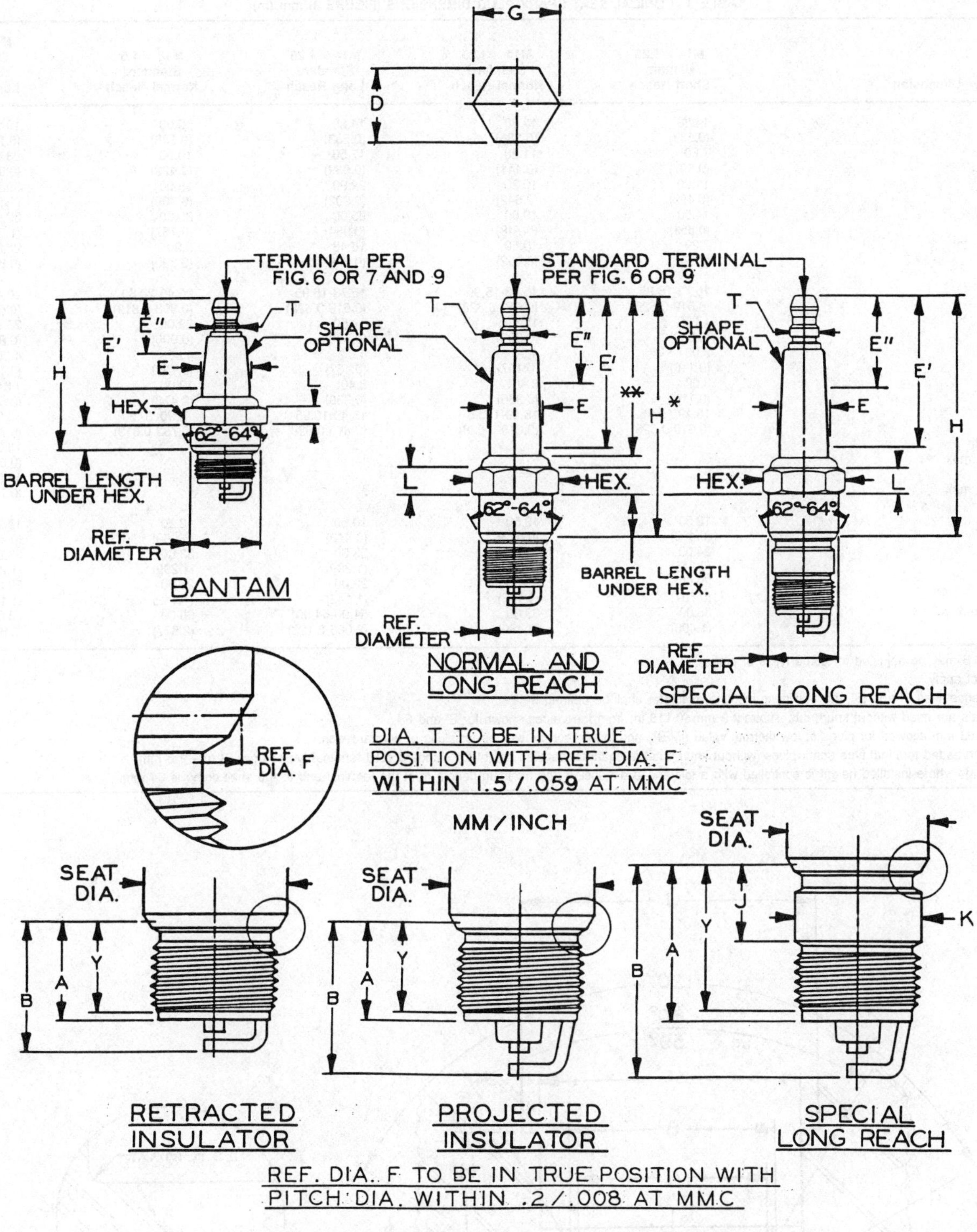

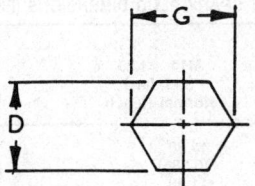

FIGURE 3—CONICAL SEAT SPARK PLUGS (SEE TABLE 3)

TABLE 3—CONICAL SEAT SPARK PLUG DIMENSIONS (FIGURE 3) mm (in)

Dimension	M14 × 1.25 Bantam Short Reach	M14 × 1.25 Standard Normal Reach	M14 × 1.25 Standard Long Reach	M18 × 1.5 Standard Normal Reach	M18 × 1.5 Standard Special Long Reach
F, Reference dia	14.80 (0.583)	14.80 (0.583)	14.80 (0.583)	19.00 (0.748)	19.00 (0.748)
A, Plug Reach, tol[1] ± 0.025 (±0.010)	7.80 (0.307)	11.20 (0.441)	17.50 (0.689)	10.90 (0.429)	23.83 (0.938)
B, Plug Depth, max[1] (retracted, ins.)	11.90 (0.469)	16.30 (0.642)	22.90 (0.902)	18.00 (0.709)	30.23 (1.190)
B, Plug Depth, max[1] (projected ins)	14.00 (0.551)	19.00 (0.748)	25.00 (0.984)	20.00 (0.787)	32.00 (1.260)
Y, Threaded Length, tol ±0.30 (±0.012)	7.29 (0.287)	10.20 (0.402)	16.48 (0.649)	9.90 (0.390)	22.80 (0.898)
Hex Size:					
D, across flats	15.73-15.95 (0.619-0.628)	15.73-15.95 (0.619-0.628)	15.73-15.95 (0.619-0.628)	20.40-20.80 (0.803-0.819)	20.40-20.80 (0.803-0.819)
G, across corners, min	17.50 (0.689)	17.50 (0.689)	17.50 (0.689)	23.00 (0.906)	23.00 (0.906)
L, hex length, min	3.00 (0.118)	4.00 (0.157)	4.00 (0.157)	4.00 (0.157)	4.00 (0.157)
Barrel Length under hex, min	3.00 (0.118)	8.40 (0.330)	8.40 (0.330)	12.00 (0.472)	14.48 (0.570)
Seat dia	15.49-15.95 (0.610-0.628)	15.49-15.95 (0.610-0.628)	15.49-15.95 (0.610-0.628)	19.90-20.80 (0.783-0.819)	19.90-20.80 (0.783-0.819)
J, Unthreaded Length below conical seat, max	[2]	[2]	[2]	[2]	13.84 (0.545)
K, Unthreaded dia below conical seat, max	[2]	[2]	[2]	[2]	17.93 (0.706)
Insulator:					
E, dia & tol[3] ±0.30 (±0.012)	10.50 (0.413)	10.50 (0.413)	10.50 (0.413)	12.20 (0.480)	12.20 (0.480)
E', length[3]	24.00 (0.945)	33.00 (1.299)	33.00 (1.299)	33.00 (1.299)	33.00 (1.299)
E', length[3]	20.00 (0.787)	29.00 (1.142)	29.00 (1.142)	29.00 (1.142)	29.00 (1.142)
H, Installed Height max	38.00 (1.496)	63.00 (2.480)	51.00-54.00[4] (2.008-2.126)	68.00 (2.677)	63.00 (2.677)

[1] Dimensions A and B may be adjusted for some types of plugs.
[2] Dimension does not apply.
[3] Between the reference lines E' and E'', the maximum insulator diameter shall be defined by E.
   If threaded terminals are used without knurl nuts, subtract 3 mm (0.118 in) from the values shown for E' and E''.
[4] Installed height of 63 mm allowed for plugs of low thermal value (IMEP) and long reach as well as for plugs of normal reach.
*Installed height for threaded terminal type spark plugs (without knurl nuts) is equal to installed height for standard solid terminal type plugs - 3 m ± 1.5 mm.
**On those spark plugs where installed height is controlled with a toleranced dimension, the minimum length from the terminal end to the shell crimp is 34 mm.

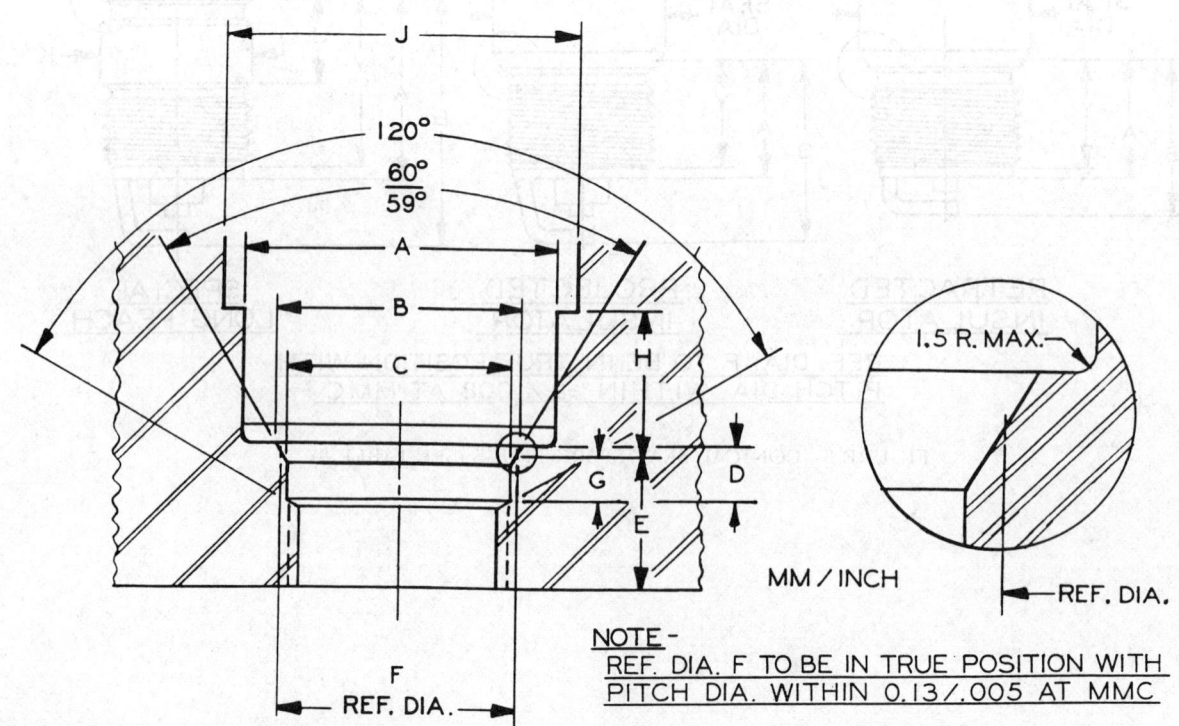

FIGURE 4—HEAD COUNTERBORE—FLAT & CONICAL SEAT (SEE TABLE 4)

TABLE 4 – HEAD COUNTERBORE – CONICAL SEAT (FIGURE 4) mm (in)

Thread Size and Reach	A Min	B Tolerance	C Tolerance	D Tolerance	E[3] Min	F Nom	G[1] Max	H[1] Max	J Min
M14, Short	17.50 (0.689)	15.10-15.40 (0.594-0.606)	14.25-14.50 (0.561-0.571)	2.00-2.30 (0.079-0.091)	7.90 (0.311)	14.80 (0.583)	[2]	2.00 (0.079)	25.00 (0.984)
M14, Normal	17.50 (0.689)	15.10-15.40 (0.594-0.606)	14.25-14.50 (0.561-0.571)	2.00-2.30 (0.079-0.091)	11.25 (0.443)	14.80 (0.583)	[2]	5.50 (0.217)	25.00 (0.984)
M14, Long	17.50 (0.689)	15.10-15.40 (0.594-0.606)	14.25-14.50 (0.561-0.571)	2.00-2.30 (0.079-0.091)	17.55 (0.441)	14.80 (0.583)	[2]	5.50 (0.217)	25.00 (0.984)
M18, Normal	22.00 (0.866)	19.56-19.81 (0.770-0.780)	18.24-18.39 (0.718-0.724)	2.00-2.30 (0.079-0.091)	11.20 (0.441)	19.00 (0.748)	[2]	9.00 (0.354)	31.00 (1.220)
M18, Special long	22.35 (0.880)	19.56-19.81 (0.770-0.780)	18.29-18.42 (0.720-0.725)	2.00-2.30 (0.079-0.091)	[2]	19.00 (0.748)	24.16 (0.951)	11.18 (0.440)	28.45 (1.120)
M14, Combination seat[4]	21.20 (0.835)	15.10-15.40 (0.594-0.606)	14.25-14.40 (0.561-0.567)	2.00-2.30 (0.079-0.091)	17.55 (0.691)	14.80 (0.583)	2.15 (0.085)	3.00 (0.118)	24.00 (0.945)

[1] Lengths are to the reference diameter. See Figure 4 as applicable.
[2] Dimension not applicable.
[3] This dimension shall be of sufficient length to ensure that the end of the spark plug thread does not project into the combustion chamber at any point when the spark plug is installed at the maximum recommended torque.
[4] Dimensions apply only for spark plugs having 16 mm (0.630 in) shell hexagon diameter.

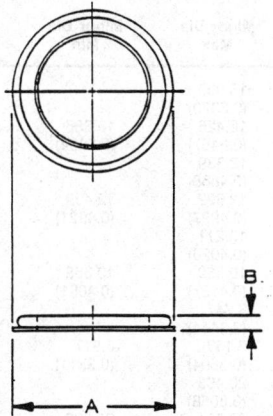 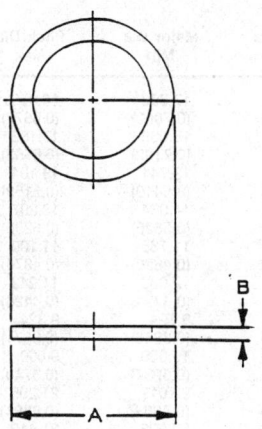

FIGURE 5 – OUTSIDE SHELL GASKET (SEE TABLE 5)

TABLE 5 – INSTALLATION TORQUE AND COMPRESSED GASKET DIMENSIONS (SEE FIGURE 5) mm (in)

Plug Size	Cast Iron Heads Installation Torque[2] N·m (lb-ft)	Cast Iron Heads Compressed Gasket Thickness mm (in) B dim.	Cast Iron Heads Gasket OD Max mm (in) A dim.	Aluminum Heads Installation Torque[2] N·m (lb-ft)	Aluminum Heads Compressed Gasket Thickness mm (in) B dim.	Aluminum Heads Gasket OD Max mm (in) A dim.
M14, Folded Steel	35-40 (26-30)	1.14-1.45 (0.045-0.057)	20.83 (0.820)	20-30 (15-22)	1.14-1.50 (0.045-0.059)	20.83 (0.820)
M18, Folded Steel	43-52 (32-38)	1.22-1.45 (0.048-0.057)	25.15 (0.990)	38-46 (28-34)	1.09-1.40 (0.043-0.055)	25.15 (0.990)
Solid copper	43-52 (32-38)	1.96 (0.077)	24.59 (0.968)	38-46 (28-34)	2.01 (0.079)	24.51 (0.965)
M12, Folded Steel	15-25 (11-18)	1.00-1.45 (0.040-0.057)	16.51 (0.650)	15-25 (11-18)	1.00-1.45 (0.040-0.057)	16.51 (0.650)
M10, Folded Steel	10-15 (7-11)	1.00-1.45 (0.040-0.057)	14.73 (0.580)	10-15 (7-11)	1.00-1.37 (0.040-0.054)	14.73 (0.580)
M14, Conical Seat (gasketless)	9-20 (7-15) service[1]			9-20 (7-15) service[1]		
M18, Conical Seat (gasketless)	20-27 (15-20) service[1]			20-27 (15-20) service[1]		
7/8-18, Folded Steel	67-74 (50-55)	1.68-1.88 (0.066-0.074)	27.94 (1.100)			

[1] Consult with engine manufacturer for original installation in engines. Without torque wrench, 1/16 turn after finger tight.
[2] The installation torque recommendations apply to new spark plugs without lubricant on the threads. If threads are lubricated, the torque value shall be reduced by approximately 1/3 to avoid overstressing.

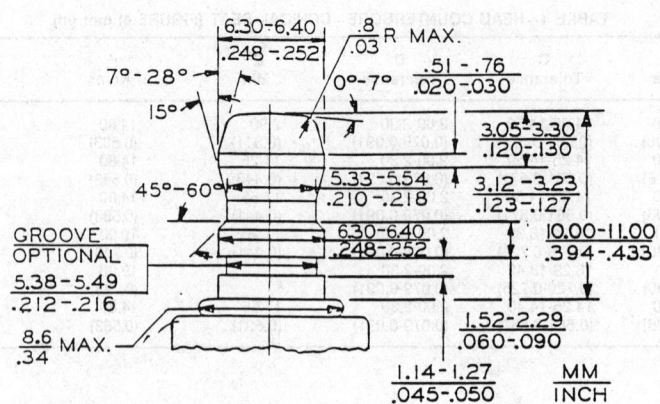

FIGURE 6 – SOLID POST TYPE TERMINAL

TABLE 6 – FINISHED SPARK PLUG AND TAPPED HOLE THREAD SIZES mm (in)

Size		Major Dia Max	Major Dia Min	Pitch Dia Max	Pitch Dia Min	Minor Dia Max	Minor Dia Min	Go Thread Ring Gage Minor Dia	Go Thread Plug Gage Major Dia	Plain Plug Gage Minor Dia Min
M18 × 1.5	Plug	17.933 (0.7060)	17.803 (0.7009)	16.959 (0.6677)	16.853 (0.6635)	16.053 (0.6320)		16.053 (0.6320)		
	Hole		18.039 (0.7102)	17.153 (0.6753)	17.026 (0.6703)	16.426 (0.6467)	16.266 (0.6404)		18.039 (0.7102)	16.266 (0.6404)
M14 × 1.25	Plug	13.868 (0.5460)	13.741 (0.5410)	13.104 (0.5159)	12.997 (0.5117)	12.339 (0.4858)		12.339 (0.4858)		
	Hole		14.034 (0.5525)	13.297 (0.5235)	13.188 (0.5192)	12.692 (0.4997)	12.499 (0.4921)		14.034 (0.5525)	12.499 (0.4921)
M12 × 1.25	Plug	11.862 (0.4670)	11.735 (0.4620)	11.100 (0.4370)	10.998 (0.4330)	10.211 (0.4020)		10.211 (0.4020)		
	Hole		12.000 (0.4724)	11.242 (0.4426)	11.188 (0.4405)	10.559 (0.4157)	10.366 (0.4081)		11.935 (0.4699)	10.366 (0.4081)
M10 × 1.0	Plug	9.974 (0.3927)	9.794 (0.3856)	9.324 (0.3671)	9.212 (0.3627)	8.747 (0.3444)		8.747 (0.3444)		
	Hole		10.000 (0.3937)	9.500 (0.3740)	9.350 (0.3681)	9.153 (0.3604)	8.917 (0.3511)		10.000 (0.3937)	8.917 (0.3511)
7/8-18	Plug	22.200 (0.8740)	22.017 (0.8668)	21.295 (0.8384)	21.191 (0.8343)	20.493 (0.8068)		20.493 (0.8068)		
	Hole		22.225 (0.8750)	21.412 (0.8430)	21.308 (0.8389)	20.851 (0.8209)	20.698 (0.8149)		22.225 (0.8750)	20.698 (0.8149)
M4 × 0.7	Term	3.944 (0.1553)	3.804 (0.1498)	3.489 (0.1374)	3.399 (0.1338)	3.085 (0.1215)		3.085 (0.1215)		
	Nut		4.000 (0.1575)	3.663 (0.1442)	3.545 (0.1396)	3.422 (0.1347)	3.242 (0.1276)		4.000 (0.1575)	

NOTE – Preferred M14 and M18 for new applications.

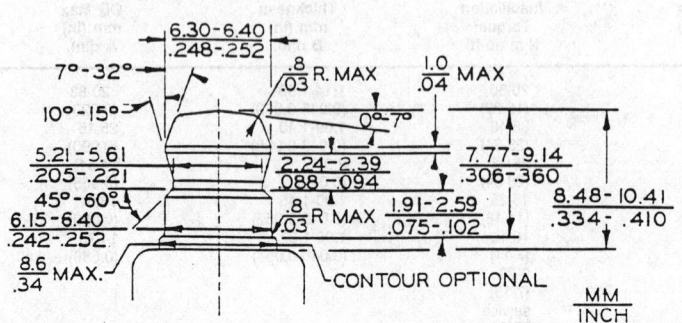

FIGURE 7 – ALTERNATE BANTAM SPARK PLUG SOLID POST TYPE TERMINAL

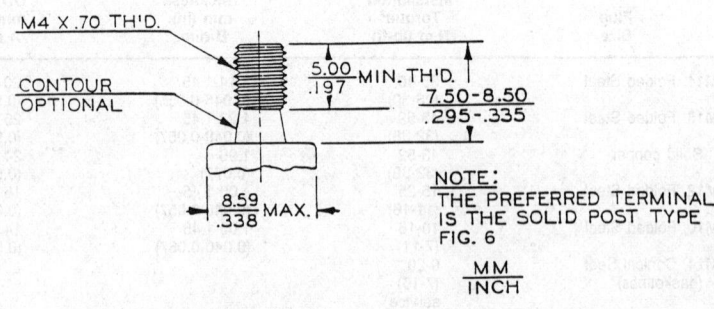

FIGURE 8 – THREADED POST TERMINAL (SEE TABLE 6) (6e OR 7e CLASS)

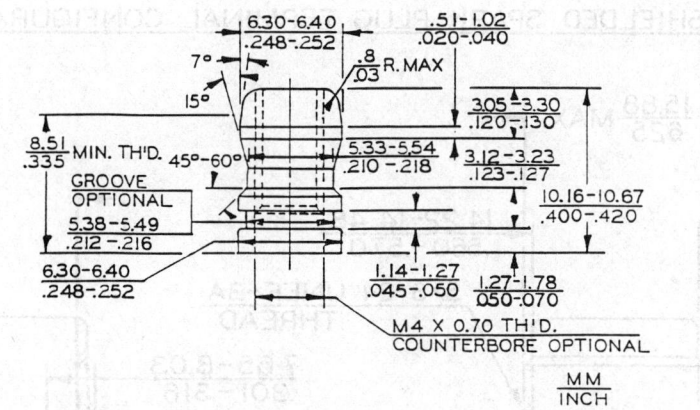

[1]Same External Dimensions as for Solid Post Terminal
FIGURE 9 — NUT TYPE FOR THREADED POST TERMINAL[1] (SEE TABLE 6) (6H CLASS)

SHIELDED SPARK PLUG TERMINAL CONFIGURATION

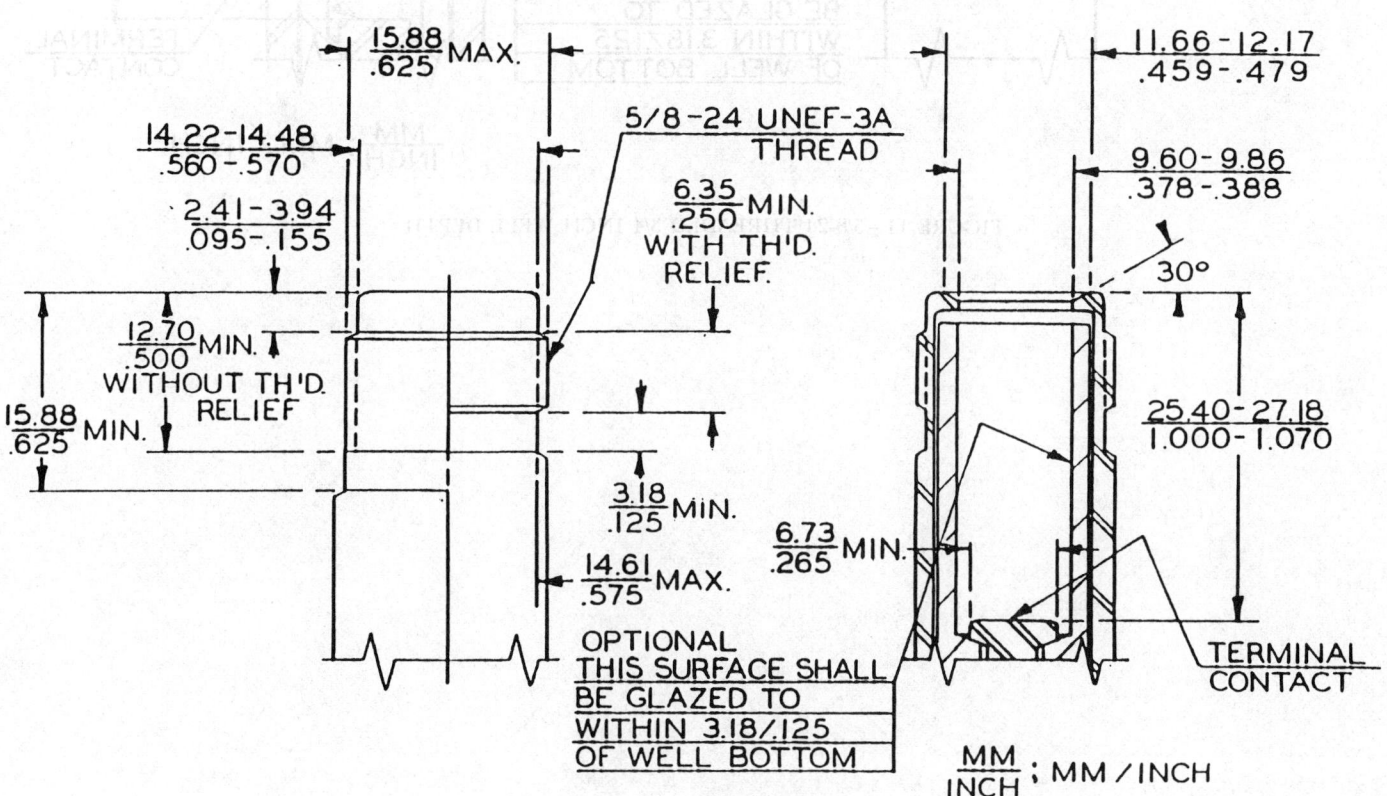

FIGURE 10 — 5/8-24 THREADED 1 INCH WELL DEPTH

SHIELDED SPARK PLUG TERMINAL CONFIGURATION

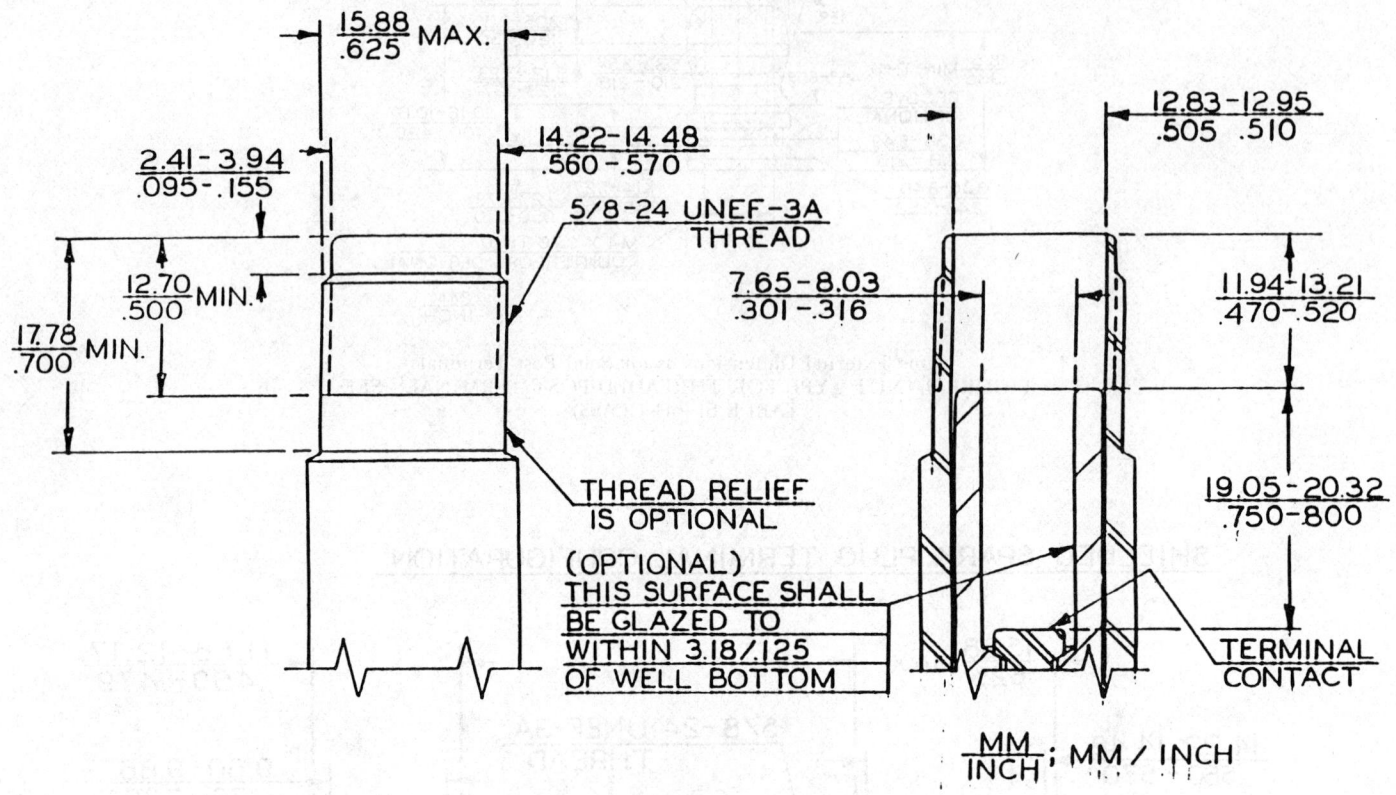

FIGURE 11 – 5/8-24 THREADED 3/4 INCH WELL DEPTH

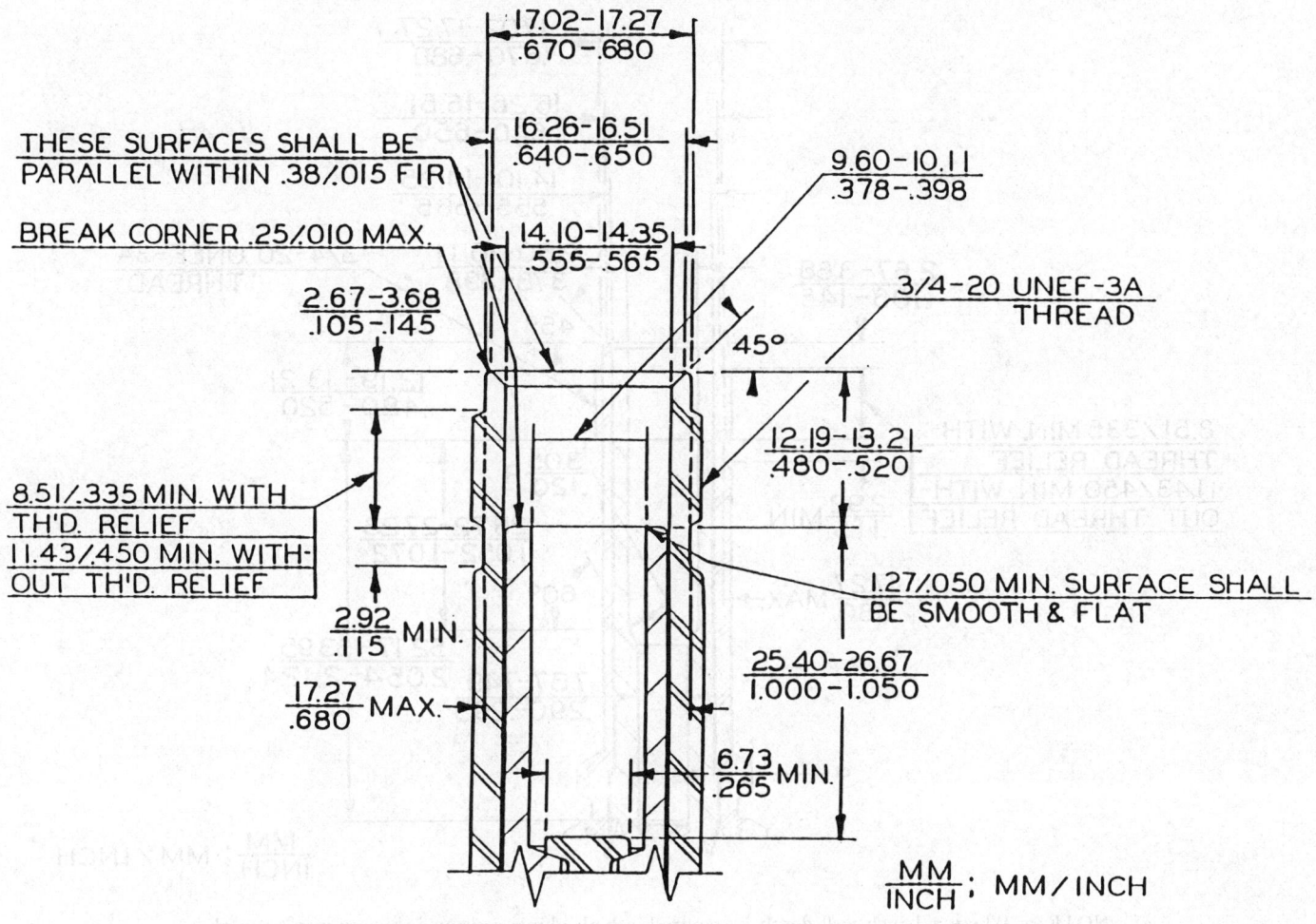

FIGURE 12 – 3/4-20 THREADED 1 INCH WELL DEPTH

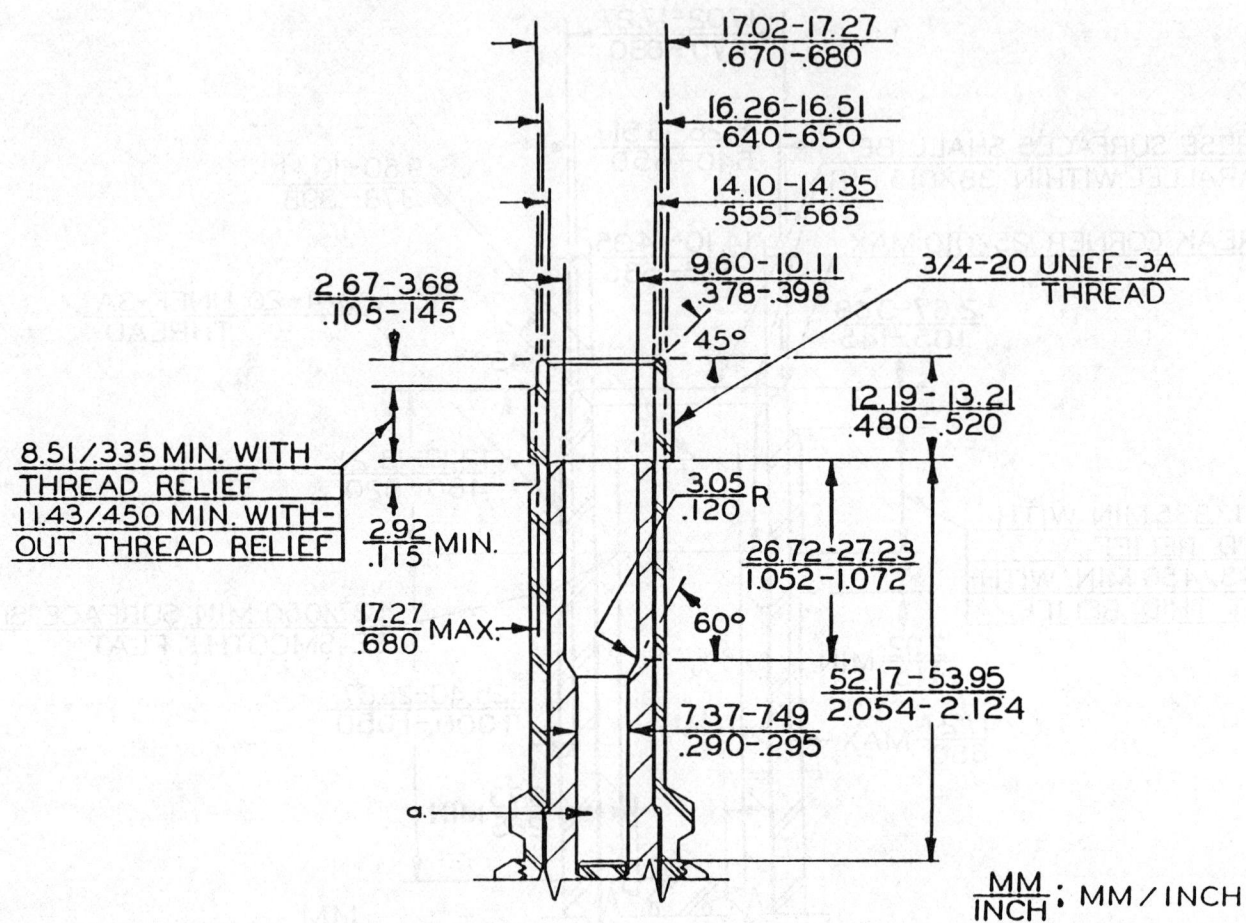

NOTE: a. Where a 1 inch well depth is required, a high ohmic resistor or spacer may be used.
FIGURE 13 – 3/4-20 THREADED 2-1/8 INCH WELL DEPTH

SPARK PLUG INSTALLATION SOCKETS—
SAE J548/2 JUN92

SAE Standard

Report of the SAE Spark Plug Task Force approved June 1992. Rationale statement available.

1. Scope—This SAE Standard applies to spark plug installation sockets of the long length type which are to be used for installing spark plugs of the most commonly used sizes for the North American market.

2. References

2.1 Applicable Documents—The following publications form a part of this specification to the extent specified herein.

2.1.1 SAE PUBLICATIONS—Available from SAE, 400 Commonwealth Drive, Warrendale, PA 15096-0001.

SAE J105—Hex Bolts

2.1.2 ANSI PUBLICATIONS—Available from ANSI, 11 West 42nd Street, New York, NY 10036-8002.

ANSI B 107.1—Socket Wrenches, Hand (in series)

ANSI/ASME B 107.5M—Socket Wrenches, Hand (metric series)

3. The spark plug installation socket shall be of the long length type and the conventional single hexagon design. A replaceable oil and heat resisting retaining bushing or other equally suitable device shall be installed inside the socket to prevent damage to the spark plug insulator and for aligning and retaining the spark plug for easy installation and removal. This bushing should be pliable and compressible to minimize side loading to the spark plug insulator. The socket shall be similar to Figure 1 except for the following:

a. The retaining bushing is not shown.
b. The drive end may be reduced in diameter for marking purposes and, at the option of the manufacturer, a male hexagon drive may be provided on the drive end, in addition to the female drive specified herein. It is recommended the socket be provided with a 3/8 in female square drive. Unless otherwise specified, the socket should comply with the standards set forth in ANSI Standards B 107.1 and B 107.5M. The sockets shall conform to Table 1.

4. Recommended Optional Design for 15.73/15.95 Hex Size Spark Plugs—The following option is recommended to maximize clearance between the side wall of the socket and the spark plug insulator. This is to reduce the probability of damaging the spark plug insulator by side loading. A suitable material and wall thickness must be used to provide sufficient wall strength for service requirements. As stated in the standard socket description previously, a bushing or other suitable device should be used to retain and centralize the spark plug in the socket. However, the device should be designed to allow the insulator to move off center without receiving a significant side loading. See Figure 2 and Table 2.

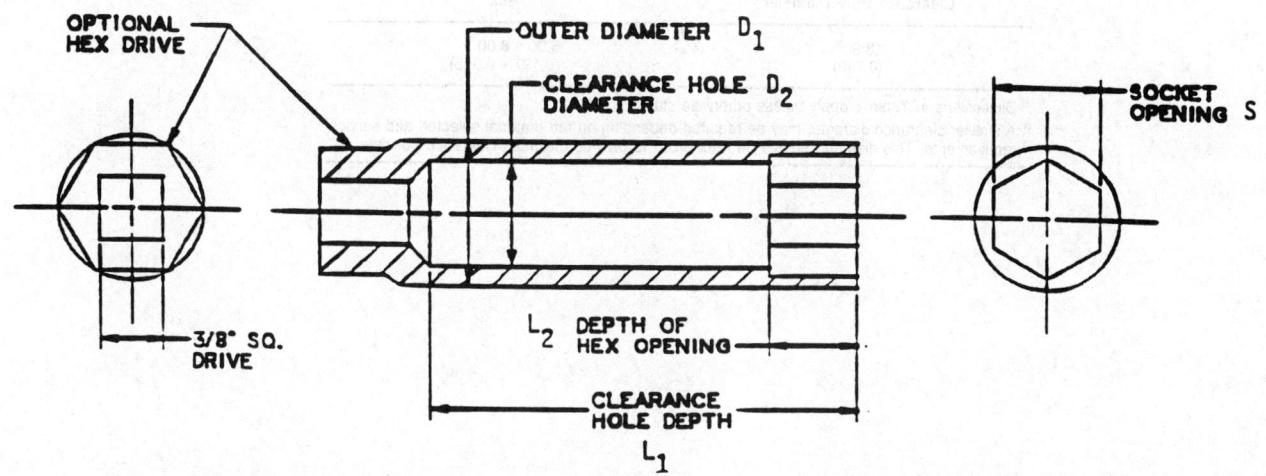

FIGURE 1—SPARK PLUG INSTALLATION DESIGN

TABLE 1—SOCKET DIMENSIONS

Spark Plug Hex Size Across Flats	S Minimum Socket Opening[1]	S Maximum Socket Opening[1]	L_2 Depth of Hex Opening	D_2 Minimum Clearance Hole Diameter[2]	L_1 Minimum Clearance Hole Depth[2]	D_1 Maximum Outer Diameter
15.73 - 15.95 (0.619 - 0.628)	16.05 (0.632)	16.18 (0.637)	9.50 - 13.00 (0.374 - 0.512)	15.90 (0.626)	63.00[3] (2.480)	22.00 (0.866)
18.80 - 19.00 (0.740 - 0.748)	19.06 (0.750)	19.36 (0.762)	9.50 - 13.00 (0.374 - 0.512)	18.00 (0.744)	56.00 (2.205)	25.50 (1.004)
20.40 - 20.80 (0.803 - 0.819)	20.86 (0.821)	21.16 (0.833)	9.50 - 13.00 (0.374 - 0.512)	20.00 (0.811)	67.00 (2.638)	27.50 (1.083)

[1] These openings vary from SAE J105 wrench openings to (a) accommodate both ISO and SAE spark plug maximum hex sizes and to (b) improve the socket hex interface with the spark plug hex to reduce the probability of the socket being tipped and contacting the insulator, causing damage.

[2] The proposed minimum clearance hole diameters and depths are based upon clearance to existing spark plug designs. This provides optimum fit conditions to prevent damage to the spark plug without subjecting socket manufacturers to undue restrictions.

[3] A minimum clearance hold depth of 63.00 mm was obtained using maximum permissible spark plug standard dimensions. A survey of major North American spark plug suppliers revealed that a minimum clearance hold depth of 57.4 mm would provide sufficient clearance to accommodate all of the spark plug designs. To maximize clearance between the socket end and other components, it is acceptable that the minimum clearance hole depth be 57.4 mm.

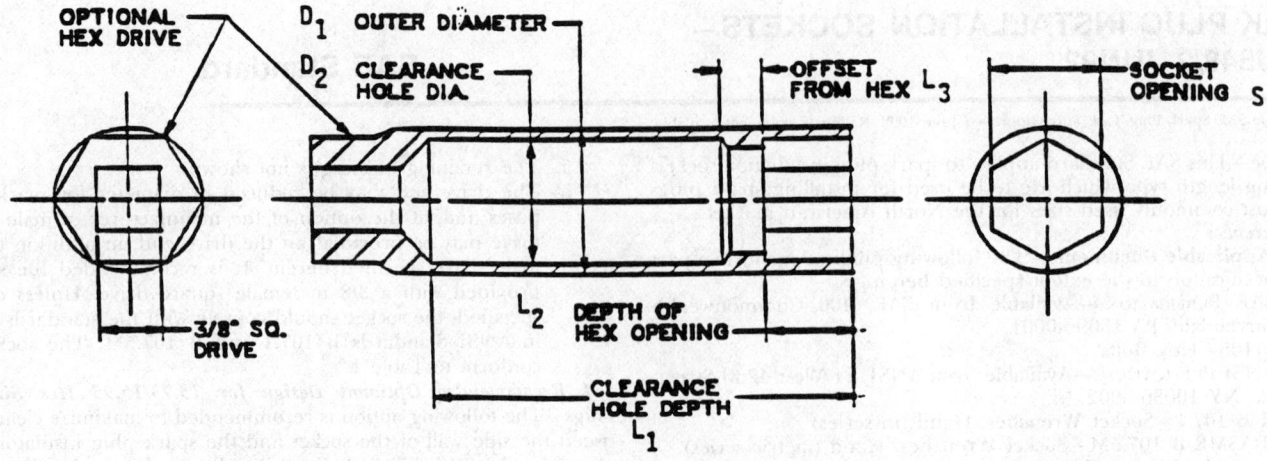

FIGURE 2—OPTIONAL SPARK PLUG INSTALLATION DESIGN

TABLE 2—OPTIONAL SOCKET DIMENSIONS[1]

D_2 Recommended Minimum Clearance Hole Diameter	L_3 Off Set From Hex
18.80[2] (0.740)	5.00 - 8.00 (0.197 - 0.315)

[1] Dimensions in Table 1 apply unless otherwise stated.
[2] A smaller clearance diameter may be required depending on the material selected and service requirements. This diameter should be maximized to provide the most clearance possible.

(R) PREIGNITION RATING OF SPARK PLUGS—SAE J549 JUN90

SAE Recommended Practice

Report of the Electrical Equipment Committee, approved December 1947, second revision by the Spark Plug 17.6 Rating Engine Panel June 1982. Completely revised by the Ignition Standards Committee June 1990.

1. Scope—This SAE Recommended Practice describes the equipment and procedures used in obtaining preignition ratings of spark plugs.

1.1 The spark plug preignition ratings obtained with the equipment and procedure specified herein are useful for comparative purposes and are not to be considered as absolute values since different numerical values may be obtained in different laboratories.

2. References

2.1 Applicable Documents—SAE HS840, Manual for the SAE 17.6 Cubic Inch Spark Plug Rating Engine, including Maintenance and Overhaul

3. Equipment—SAE 17.6 engine[1] with the cylinder barrel having knurled and chemically treated surface and compression piston rings chromium plated.

4. Speed—The nominal speed is to be 2700 rpm, but is not to be over 2765 rpm when firing, nor below 2670 rpm when motoring.

5. Compression Ratio—5.6:1.

6. Spark Advance—30 degrees Before Top Dead Center (BTDC) for nonaviation plugs, 40 degrees BTDC for aviation plugs or nonaviation plugs that cannot be rated at 30 degrees BTDC.

7. Ignition Source—Magneto or approved alternate.

8. Spark Plug Installation—The thread in the spark plug hole opening should conform in size and length to the standards established by SAE for the rating engine.

8.1 SAE recommended torque values should be used when installing plugs in the engine.

8.1.1 Reducer bushings or adaptors should not be used.

9. Fuel—98% - one degree Benzene, 2% - Specification MIL-L-

---

[1] See SAE HS840, Manual for the SAE 17.6 Cubic Inch Spark Plug Rating Engine, including Maintenance and Overhaul currently—1989 under rev.

6082D Grade 1100 SAE60 NONADDITIVE aviation oil, with 3 cc/gal (0.8 ml/L) T.E.L. added.

10. Fuel Injection Timing—The fuel injection pump port shall begin to close 60 degrees ± 5 of crankshaft angle After Top Dead Center (ATDC) on the intake stroke.

11. Fuel Circulation Rate—1/2 gal/min ± 1/4 (2 L/min ± 1).

12. Fuel Injection Pump—The gallery pressure of the fuel injection pump is to be 15 psi ± 2 (100 kPa ± 10).

13. Fuel Pressures-Injection—750 psi (5170 kPa) minimum.

14. Mixture Strength—The mixture strength is that which gives maximum thermal plug temperature.

15. Inlet Air Temperature—225°F ± 5 (107°C ± 3).

16. Inlet Air Humidity—75 g ± 25 of moisture/lb (0.453 kg) of dry air.

17. Coolant—The coolant should be water plus 1 g/gal (3 L) of an inhibitor. The total dissolved and suspended solids should not exceed 120 ppm.

18. Jacket Inlet Temperature
  a. With pressure cooling control: 225°F ± 5 (107°C ±3)
  b. With insert head engine: 190°F ± 2 (88°C ± 1)

19. Coolant Flow—5 gal/min ± 1/2 (20 L/min ± 2).

20. Crankcase Oil—Oil is to be nonadditive SAE 120 aviation oil.

21. Oil Pressure
  a. In main bearings, 95 psi ± 5 (650 kPa ± 40)
  b. In valve gear, 15 psi (100 kPa) minimum at operating temperature

22. Oil Temperature—190°F ± 10 (88°C ± 5).

23. Oil Quantity—Oil level is maintained at the center of the oil level sight glass.

24. Operating Conditions—The plug rating is that Indicated Mean Effective Pressure (IMEP) value obtained on the engine at a point when the supercharge pressure is 1 in Hg (3.37 kPa) below the preignition point.

24.1 Preignition Point—The following steps are recommended to attain the preignition point.

24.1.1 The supercharge pressure is increased in 4 in Hg (13.5 kPa) increments until preignition occurs as indicated by a rapid rise in thermal plug temperature. At each setting, the mixture strength is adjusted such that a maximum thermal plug temperature is obtained and held for 3 min.

24.1.2 When preignition occurs, the fuel supply is instantly cut off and the supercharge pressure is decreased 2 in Hg (6.7 kPa) at which point the fuel is turned on and again adjusted for maximum thermal plug temperature. This condition should be held for 3 min or until preignition again occurs.

24.1.3 If preignition occurs after Step 24.1.2, the supercharge pressure should be reduced by 1 in Hg (3.37 kPa) again adjusting for optimum thermal temperature until stable engine operation for 3 min is obtained or preignition occurs. If preignition occurs, refer to Step 24.1.5.

24.1.4 If, after Step 24.1.2 stable engine operation is obtained, the supercharge pressure should be increased by 1 in Hg (3.37 kPa), again adjusting for optimum thermal plug temperature until stable engine operation for 3 min is obtained or preignition occurs. If preignition occurs, refer to Step 24.1.5.

24.1.5 Friction torque should be measured at supercharge pressure 1 in Hg (3.37 kPa) below the preignition point (or previous stabilized setting prior to preignition), and within 30 s after the engine ceases to fire.

24.1.6 Rating data may be verified using a plug that has a rating point at least 50 IMEP above the plugs that have been rated.

25. Calculation of IMEP

$$\text{Indicated HP} = \text{Friction HP} + \text{Brake HP} \quad (Eq.1)$$

$$IHP = \frac{2700}{5252} T_F + \frac{2700}{5252} T_B$$

$$IHP = 0.51 (T_F + T_B) = \frac{\text{Plan}}{33\ 000}$$

$$0.51 (T_F + T_B) - (0.04)(0.01)P = IMEP$$

$$IMEP = 8.65 (T_F + T_B)$$

T_F — Friction Torque

T_B — Brake Torque

IMEP — Indicated Mean Effective Pressure

SPARK PLUG HEAT RATING CLASSIFICATIONS —SAE J2162 JUN92

SAE Standard

Report of the SAE Spark Plug Task Force of the SAE Ignition Standards Committee approved June 1992. Rationale statement available.

Foreword—This SAE Standard is based on U. S. Federal Standard 143, dated August 10, 1962, entitled "Spark Plug, Heat Rating of, Method of Classifying."

1. Scope—This SAE Standard details a uniform method for classifying heat ratings of unshielded spark plugs.

2. References

2.1 Applicable Documents—The following publications form a part of this specification to the extent specified herein. The latest issue of SAE publications shall apply.

2.1.1 SAE PUBLICATIONS—Available from SAE, 400 Commonwealth Drive, Warrendale, PA 15096-0001.

SAE J549—Preignition Rating of Spark Plugs

2.1.2 U. S. FEDERAL STANDARDS—Available from Superintendent of Documents, U. S. Government Printing Office, Washington, DC 20402.

Federal Standard 143—Spark Plug, Heat Rating of, Method of Classifying

2.2 Definitions—The following term is defined as follows for use in reference to this document.

2.2.1 IMEP—Indicated Mean Effective Pressure (refer to SAE J549 for formula to calculate IMEP).

3. Classifications—Table 1 shall constitute the complete list of heat rating descriptions:

TABLE 1—CLASSIFICATIONS

Heat Rating	IMEP (kPa)	IMEP (psi)
Hot	621-827	90-120
Medium Hot	827-1379	120-200
Medium	1379-2069	200-300
Medium Cold	2069-2413	300-350
Cold	2413-2758+	350-400+

SAE 17.6 CUBIC INCH SPARK PLUG RATING ENGINE—SAE J2203 JUN91

SAE Standard

Report of the SAE Ignition Systems Standards Committee approved June 1991.

Foreword—This manual was originally prepared under the auspices of the SAE Ignition Research Committee by the Spark Plug Rating Engine Standardization Panel of the Aircraft Piston Engine Ignition Subcommittee. In 1974, the Spark Plug Rating Engine Standardization Panel was placed under the jurisdiction of the SAE Electrical Equipment Committee.

This manual defines the standard engine to be used in determining spark plug preignition ratings. The engine is known as the SAE 17.6 Cubic Inch[1] Spark Plug Rating Engine. The background of its design, development, and applications is contained in SAE publication SP-243.

In addition to describing the engine, this manual deals with maintenance and overhaul instructions for the engine. Appendices providing engine manufacturing tolerances, replacement limits, and engine bill of materials are included. The manual also includes the procedure for rating spark plugs.

The 17.6 engine has been used for many years in the spark plug industry to classify spark plugs by their preignition rating. Correlation of these ratings among the various test agencies has been accomplished with limited success primarily due to engine variations. This correlation difficulty prompted the Aircraft Piston Engine Ignition Subcommittee of the SAE Ignition Research Committee to investigate methods of standardizing and improving this engine. The Ethyl Corporation (which originated and owns some of the patterns for the 17.6 engine) consented to the incorporation of improvements in the engine by SAE, with the provision that the Ethyl Corporation patterns not be changed.

The Spark Plug Rating Engine Standardization Panel, which was established to standardize and improve this engine, consists of persons who are closely associated with the use or manufacture of the engine. The sum of their individual experiences and the many special projects conducted by the panel have been gathered into this manual.

Conformance with the engine description and rating procedure included in this manual and the diligent following of the Maintenance, Overhaul, and Operation instructions will result in more uniform spark plug rating data from each engine and a closer rating correlation between engines.

This manual will be revised periodically to reflect engine improvements that have been developed and thoroughly evaluated. Comments, advice, or recommendations concerning the manual or the engine that it defines will be welcomed by this panel and should be sent to SAE Headquarters for consideration.

An engine of this type may be obtained from the Laboratory Equipment Corporation (Labeco), Mooresville, Indiana, and all part numbers herein mentioned are those of Labeco, unless otherwise specifically stated.

This edition of the manual includes only the 5750 series engine since it is the only type that has been manufactured in the last few years. The older type 5000 series was covered thoroughly in a previous edition of the manual (publication date, July 1964). The 5750 series engine (Figures 1a and 1b) differs from the 5000 series in that it incorporates a Lanchester-type of balancing system consisting of two counter-rotating, chain-driven, counterbalancing shafts, rotating at crankshaft speed, to dampen the unbalanced portion of the connecting rod and piston assembly.

TABLE OF CONTENTS

Foreword
1. *Scope*
2. *References*
   2.1 Applicable Documents
   2.2 Abbreviations
3. *Cylinder Assembly*
4. *Crankcase Assembly*
5. *Air Induction System*
6. *Ignition System*
   6.1 Magneto Ignition System
   6.2 Alternate Ignition Systems
7. *Fuel System*
8. *Cooling System*
9. *Lubrication System*
   9.1 Oil Filter
   9.2 Alternate Oil Filter
10. *Exhaust System*
11. *Crankcase Breather System*
    11.1 Standard System
12. *Air Supply System*
13. *Maintenance and Overhaul Procedure*
    13.1 General
    13.2 Detailed Disassembly of 5750 Engine
14. *Engine Run-In Schedule*
15. *Operating Instructions*
    15.1 Operating Instructions Conditions
    15.2 Step-by-Step Procedure
    15.3 Plug Rating

APPENDICES
Appendix A—Manufacturing Tolerances and Replacement Limits
Appendix B—Standard Spark Plug Inserts
Appendix C—Spark Plug Installation Torque
Appendix D—Bill of Material for 5750 Engine Assembly, SAE Spark Plug Rating w/Insert Type Head

1. *Scope*—This SAE Standard defines the standard engine to be used in determining spark plug preignition ratings. The engine is known as the SAE 17.6 Cubic Inch Spark Plug Rating Engine.

2. *References*

2.1 Applicable Documents—The following publications form a part of this specification to the extent specified herein. The latest issue of SAE publications shall apply.

2.1.1 SAE PUBLICATIONS—Available from SAE, 400 Commonwealth Drive, Warrendale, PA 15096-0001.

SAE J973—Ignition System Measurement Procedure
SAE SP-243—Proceedings of the 28th Automotive Technology Development Contractors Coordination Meeting, 27
AS 840—Manual, July 1964

2.2 Abbreviations

abc	after bottom center
abs.	absolute
assy.	assembly
atc	after top center
bbc	before bottom center
bdc	bottom dead center
bp	boiling point
brg.	bearing
brkt	bracket
btc	before top center
cap	capscrew
°C	degrees Centigrade
C.B.	counter balance
cc	cubic centimeters
cyl	cylinder
cm	centimeter
deg.	degrees
Dia.	diameter
etc.	and so forth
°F	degrees Fahrenheit
gal	gallons
HD	head
hex	hexagon
h	hours
Hg	mercury
/	per
ID	inside diameter
IMEP	indicated mean effective pressure
in	inches
K.O.	knock out
lb-ft	pounds-feet

---

[1] With the advent of the metric system, the metric notation should be 288.6 cc for the ending displacement. However, since the term "17.6" is quite familiar in the industry, it will be retained in that form.

23.99

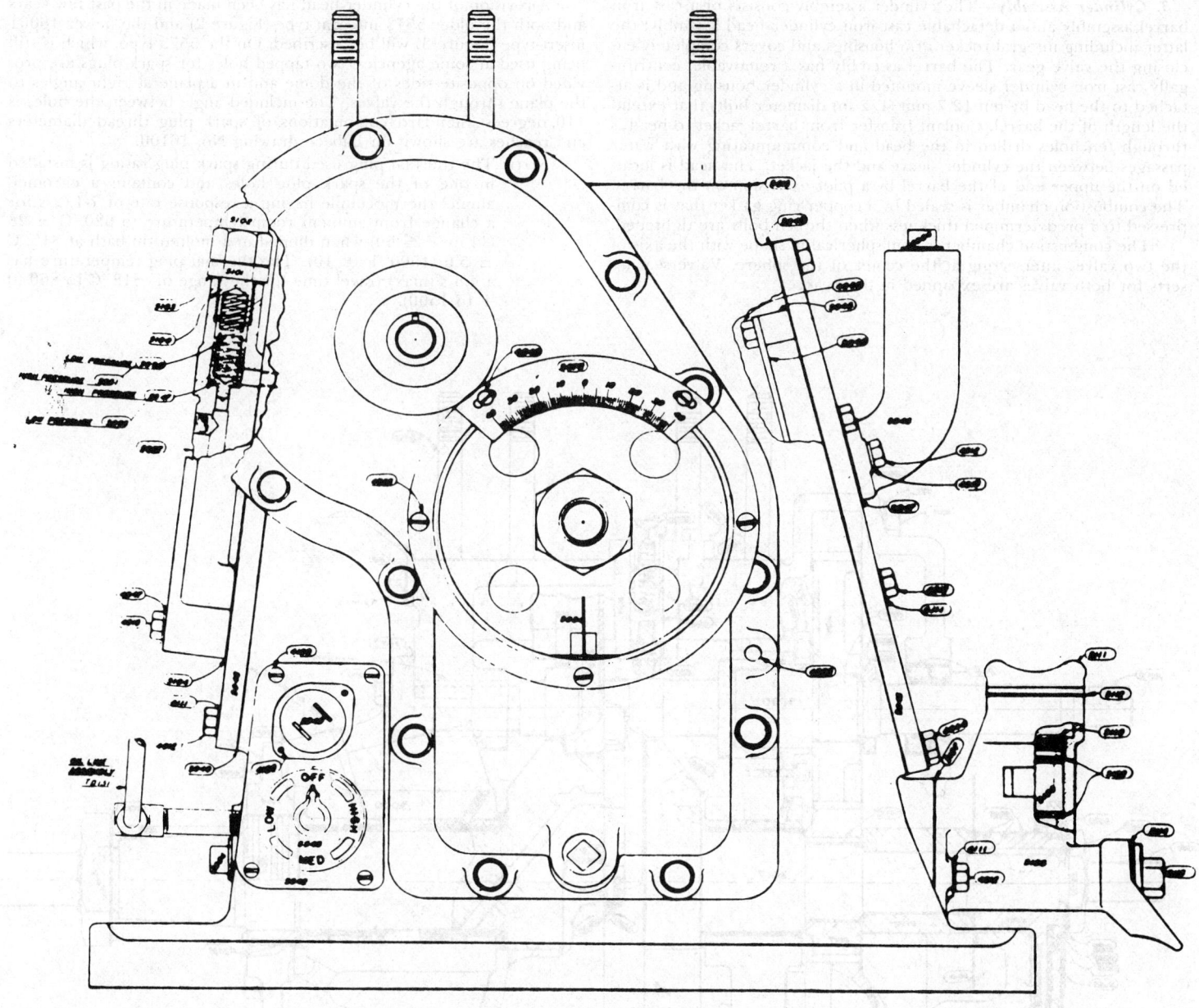

FIGURE 1A—THE 5750 ENGINE

NOTE: Shown for illustrative purposes only. Detailed drawings may be obtained from Laboratory Equipment Corp. Mooresville, Indiana

m	meter	psi	pounds per square inch
mm	millimeter	qt	quart
Mach.	machine	rd.	round
Mfg.	manufacturer	rpm	revolutions per minute
min	minimum or minute	SAE	Society of Automotive Engineers, Inc.
misc	miscellaneous	s	seconds
mnt.	mounting	soc	socket
No.	number	spkt.	sprocket
NPT	National pipe thread	std	standard
O.A.L.	overall length	tdc	top dead center
OD	outside diameter	V	volts
oz	ounces	W	watts
P.F.	press fit	X	by

3. Cylinder Assembly—The cylinder assembly consists of a cast iron barrel assembly and a detachable cast iron cylinder head assembly; the latter including integral rocker arm housings and covers completely enclosing the valve gear. The barrel assembly has a removable, centrifugally cast iron cylinder sleeve mounted in a cylinder housing and is attached to the head by ten 12.7 mm (1/2 in) diameter bolts that extend the length of the barrel. Coolant transfer from barrel jacket to head is through ten holes drilled in the head and communicating with water passages between the cylinder sleeve and the jacket. The head is located on the upper end of the barrel by a pilot extension on the barrel. The combustion chamber is sealed by a copper ring gasket that is compressed to a predetermined thickness when the ten bolts are tightened.

The combustion chamber is hemispherical in shape with the axis of the two valves intersecting at the center of the sphere. Valve seat inserts for both valves are expanded in the head.

A revision of the cylinder head has been made in the past few years and both the older 5573 integral type (Figure 2) and the newer 16001 insert-type (Figure 3) will be described. On the 5573 type, which is still being used at some agencies, two tapped holes for spark plugs are provided on opposite sides of the dome and in a plane at right angles to the plane through the valves. The included angle between the holes is 110 degrees. Standard combinations of spark plug thread diameters and reaches are shown on Labeco drawing No. 16100.

NOTE: The thermal plug used during spark plug rating is installed in one of the spark plug holes and contains a chromel-alumel thermocouple having a response rate of 7-1/2 s for a change from ambient room temperature to 620 °C ± 28 (1150 °F ± 50) when dipped in a molten tin bath at 815 °C ± 5.6 (1500 °F ± 10). The thermal plug temperature has a 4.5 s (max) travel time for the range of −18 °C to 860 (0 °F to 1500).

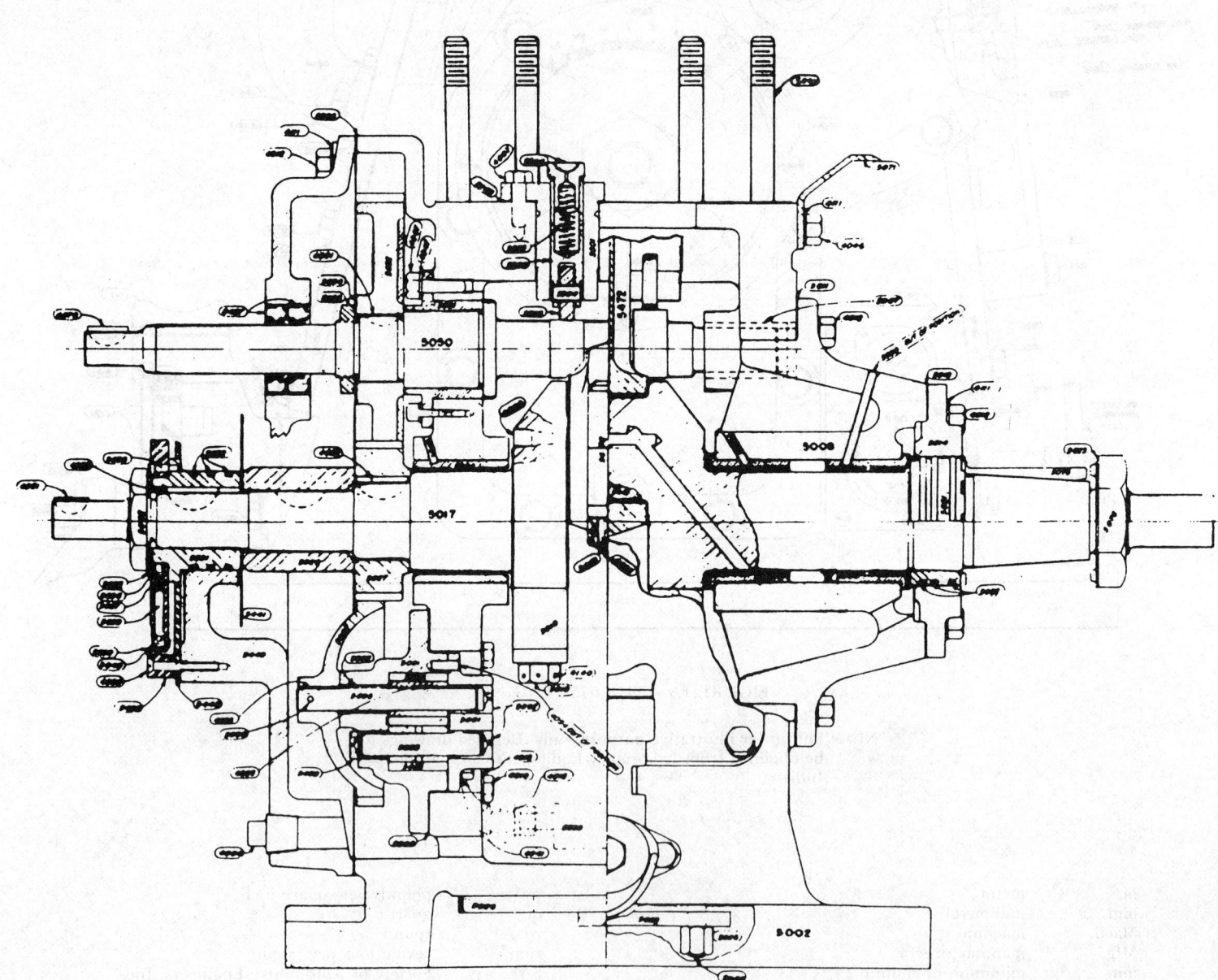

FIGURE 1B—THE 5750 ENGINE

NOTE: Shown for illustrative purposes only. Detailed drawings may be obtained from Laboratory Equipment Corp., Mooresville, Indiana

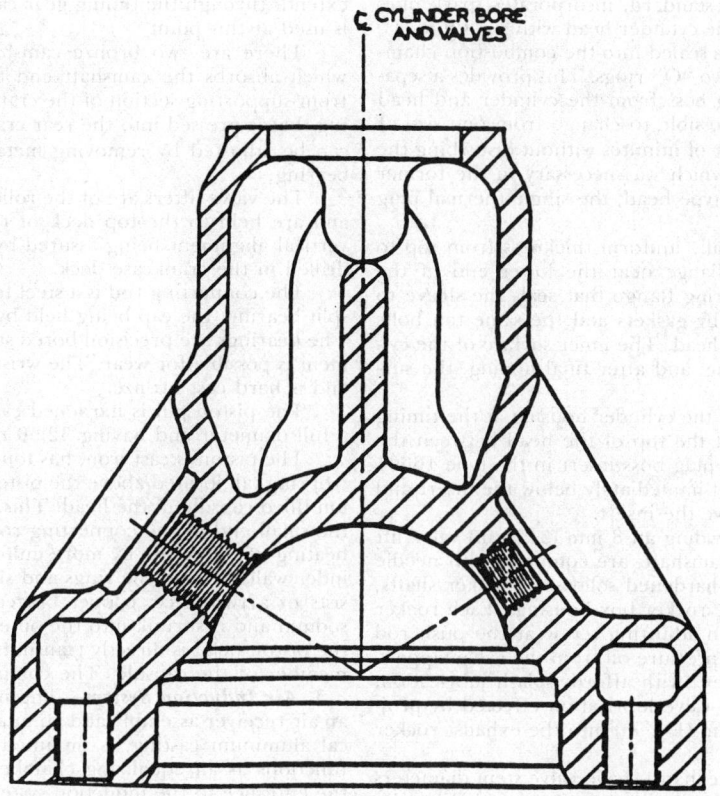

FIGURE 2—INTEGRAL TYPE HEAD (PART # 3573)

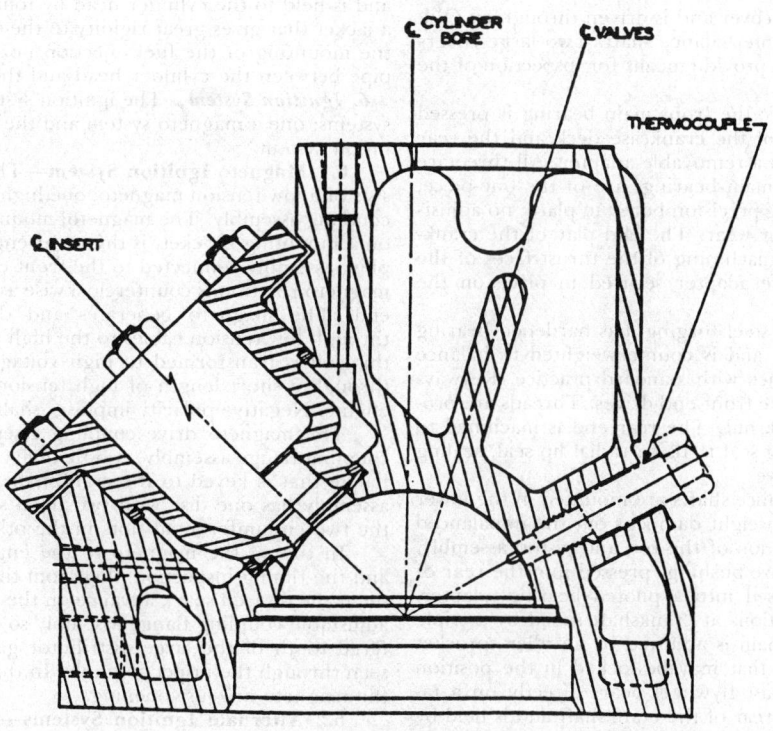

FIGURE 3—INSERT TYPE HEAD (PART # 16001)

The 16001 head, which is now standard, incorporates spark plug boss inserts that are mounted into the cylinder head with six 8 mm (1/16 in) studs. This spark plug insert is sealed into the combustion chamber and from the atmosphere with two "O" rings. This provides a separate water jacket for the spark plug boss from the cylinder and head water jacket. This insert makes it possible to change from one size of spark plug to another size in a matter of minutes without disturbing the cylinder, cylinder head, or piston, which was necessary in the former 5573 head design. With the 16001 type head, the single thermal plug remains installed at all times.

The cylinder sleeve is of generally uniform thickness from top to bottom, except for a small outer flange near the lower end of the sleeve. This flange engages a steel ring flange that seals the sleeve to the cylinder housing and the head by gaskets and the same ten bolts that hold the barrel assembly to the head. The inner surface of the cylinder is knurled before finish honing; and after final honing, the surface is Parco-Lubrize treated.

Coolant enters the lower end of the cylinder housing at the timing gear end and leaves the assembly at the top of the head between the rocker boxes. Coolant for the spark plug boss insert in the type 16001 cylinder head enters the insert jacket immediately below the insert and leaves the insert jacket directly above the insert.

The cast steel rocker arms, providing an 8 mm (5/16 in) valve lift for a 6.33 mm (1/4 in) lift of the camshaft, are equipped with needle bearings operating on floating case-hardened solid steel rocker shafts, secured by cover plates bolted to the rocker box housings. Each rocker has a roller at the valve end and an adjusting screw at the push rod end. The valve gear is lubricated by pressure oil from the valve tappets, through a hole in the adjusting screw, with affords splash lubrication, supplemented by additional exhaust valve lubrication effected by projecting the push rod housing 12.7 mm (1/2 in) into the exhaust rocker box.

The valves, one intake and one exhaust, have valve stem diameters and lengths considerably greater than those generally provided for the valve head diameters used. Each valve is operated by two valve springs that provide satisfactory operation up to and including 3200 rpm.

4. Crankcase Assembly—The gear end is considered the front end of the crankcase and the flywheel end, the rear of the crankcase. The crankcase consists of an extremely rigid iron casting with drilled oil passages allowing pressure lubrication to all bearing surfaces. The crankcase from the main bearing to the base houses the two counter-rotating, chain-driven counterbalance shafts. The timing gear case cover encloses the timing gears, the chain drive for the counterbalance shafts and the chain tension idler sprocket. An oil pump is mounted on the outside of the timing gear case cover and is driven through an Oldham coupling by the left-hand counterbalance shaft. Two large covers bolted to the sides of the crankcase provide means for inspection of the crankcase interior.

There are three main bearings; the front main bearing is pressed into the front supporting section of the crankcase deck and the rear two main bearings are pressed into a removable adapter. All three are locked in place by taper pins. All main bearings are of the one-piece, steel-backed silver grid type and are precision bored in place; no adjustment is provided to compensate for wear. The end play of the crankshaft is controlled by dimensional machining of the thrust faces of the two inner main bearings, with the adapter secured in place on the crankcase with the proper gasket.

The crankshaft is a very rigid steel forging, has hardened bearing journals to insure minimum wear, and is counterweighted to balance the centrifugal weight in accordance with standard practice. Keyways are provided for flywheel and all the front end drives. Threads are provided for the crankshaft front lock nut. The rear end is machined to use a radial lip seal. The front ring seal is also a radial lip seal, sealing against the timing disc spacer sleeve.

The lead weighted counterbalance shafts are mounted in the lower part of the case. Their unbalance weight dampens out the unbalanced forces generated by the upper portion of the rod and piston assembly. These shafts are mounted on bronze bushings pressed into the rear of the case and front bushings pressed into a piloted bearing adapter. They are driven in opposite directions at crankshaft speed by a triple chain drive. The tension of the chain is adjusted by an idler sprocket mounted on an eccentric bushing that may be locked in the position giving the desired tension. The cast flywheel bears directly on a tapered and hardened section at the rear of the crankshaft and is held by key and lock nut.

The camshaft, driven through helical gears, is carburized steel with case-hardened bearing journals and cams. The front of the camshaft extends through the timing gear case for an auxiliary drive. An oil seal is used at this point.

There are two bronze camshaft bearings: (a) the front bearing, which absorbs the camshaft end thrust, is bolted by a flange to the front supporting section of the crankcase; (b) the rear bearing is a bushing that is pressed into the rear crankcase supporting section. End play can be adjusted by removing metal from the inner face of the front bearing.

The valve lifters are of the roller-type. The guides are iron castings and are held to the top deck of the crankcase by capscrews, positive vertical alignment being assured by shoulders that fit in piloting holes drilled in the crankcase deck.

The connecting rod is a steel forging that has a precision shell-type split bearing, the cap being held by two bolts of generous proportions. The bearings are precision bored steel backed silver grid and no adjustment is possible for wear. The wrist pin bushing is a press fit in the rod and is hard cast bronze.

The piston pin is hardened carburized steel, is solid, and employs a full diameter and having 32.50 mm (1-9/32 in) spherical radii.

The piston is cast iron, has four compression rings and one oil control ring (all located above the piston pin boss), and incorporates a sodium filled capsule in the head. This capsule, cooled by an oil spray from the small end of the connecting rod, is used to prevent localized overheating of the piston by more uniformly dissipating the heat to the cylinder wall through the rings and skirt and to the oil. The capsule consists of a two-piece, copper brazed chamber that totally encloses the sodium and is shrunk into the outer casting. Pressure on the middle of the piston head is directly transmitted to the piston bosses by the inner member of the capsule. The compression ratio of the engine is 5.6:1.

5. Air Induction System—The induction system basically consists of an air receiver assembly and an intake pipe. The air receiver, a cylindrical aluminum casting, is mounted at the top of the intake pipe and functions as an equalizing chamber to provide a constant pressure at the entrance to the induction system. It contains a standpipe whose inside diameter, 22.2 mm (7/8 in), is equal to the pipe passage diameter and an air filter consisting of four layers of bronze screen (two of 110 mesh, and two of 22 mesh) to prevent pipe scale and the like from entering the cylinder. The air enters the receiver tangentially and is drawn off at the standpipe entrance near the top of the receiver. Two thermocouples are located in the air receiver; one is connected to a controller to maintain the air inlet temperature at 107.2 °C ± 2.8 (225 °F ± 5) (see Section 11), and the other is used for indicating the temperature.

The intake pipe is an iron casting in the form of a 90 degree bend and is held to the cylinder head by four studs. Surrounding the pipe is a jacket that gives great rigidity to the section. Provisions are made for the mounting of the fuel injection nozzle on either side of the intake pipe between the cylinder head and the air receiver.

6. Ignition System—The ignition system may consist of two alternate systems; one a magneto system and the other a condenser discharge ignition system.

6.1 Magneto Ignition System—The magneto ignition system consists of a low tension magneto, one high tension coil, and magneto drive coupling assembly. The magneto, mounted independently of the engine on a mounting bracket, is driven at engine speed through a drive coupling assembly connected to the front extension of the crankshaft. The magneto rotation is counterclockwise as viewed from the magneto drive end. The magneto generates and distributes low voltage current through low tension cables to the high tension coil. The low voltage by this coil is transformed to high voltage by this coil and is conducted through a short length of high tension cable to the spark plug in the engine. Negative polarity impulses shall be delivered to the spark plug.

The magneto drive coupling assembly consists of one adjustable coupling flange assembly, two flexible couplings, and a driving coupling flange that is keyed to the crankshaft. The adjustable coupling flange assembly has one disc with two fixed screws that can be positioned in the two circumferential slots in the other disc.

In timing the magneto to the engine, remove the breaker cover and the timing inspection plug from the magneto. With the crankshaft set at the desired spark advance on the compression stroke, position the adjustable coupling flange assembly so that the white dot on the chamfered tooth of the large distributor gear lines up with the pointer as seen through the inspection hole. In this position, the breaker points of the magneto are just opening.

6.2 Alternate Ignition Systems—Any commercially available automotive ignition system such as a capacitive discharge system, electronic breakerless system, or an inductive breaker type system will be satisfactory as long as it fulfills the following system specifications:

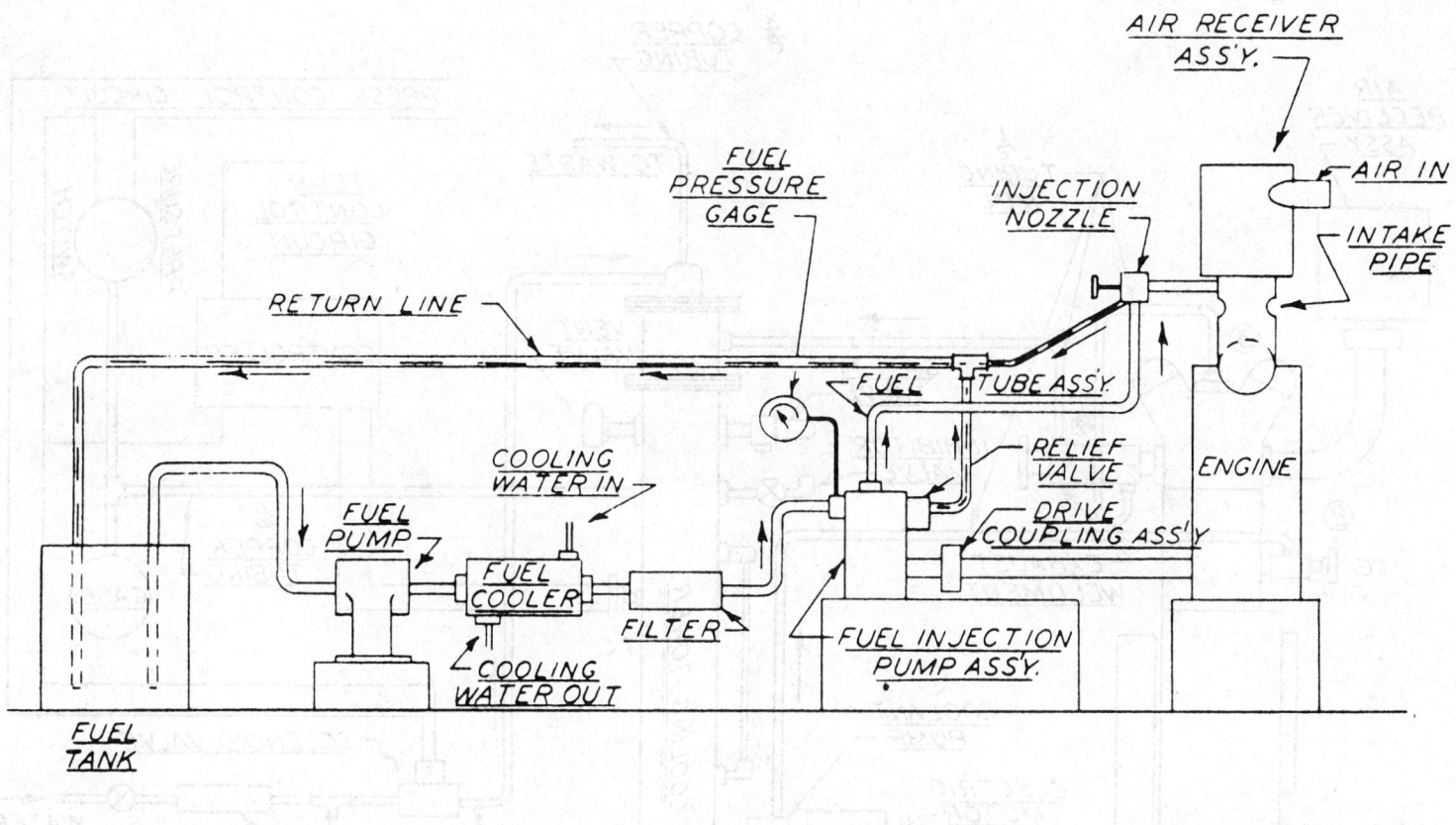

FIGURE 4—SUGGESTED FUEL SYSTEM

Open Circuit Voltage: 24 KV (minimum)
Rise Time: 50 μs (maximum)
Arc Duration: 60 μs (minimum)
Polarity: Negative

All measurements are to be taken with the spark plug firing at the following conditions:

Spark plug gap at 0.635 mm (0.025 in)
Engine running as follows: 2700 rpm
    100 inches (2540 mm) Hg-Supercharge
Spark Timing-30 degree B.T.D.C.

Voltage measurements are to be made in accordance with SAE J973.

7. Fuel System—The fuel system consists of a fuel supply pump, filter, fuel cooler, fuel injection pump assembly, injection nozzle, and fuel tank as shown in Figure 4.

The gear type, positive displacement, fuel supply pump, is driven at 600 rpm and has a capacity of 2.0 L/min ± 1.0 (1/2 gal/min ± 1/4) at this speed.

The filter is a multiple disc edge type with 0.038 mm (0.0015 in) spacing. To reduce difficulty during engine operation due to fuel contamination, it is suggested that the fuel be filtered through a 2 μm filter before delivery to the fuel system.

The fuel injection pump assembly, a single cylinder American Bosch type APEIE-70P 300/3 is mounted on the same mounting bracket that supports the magneto. The pump has a variable delivery rate and is driven at half engine speed through a drive coupling assembly (similar to that used for the magneto) connected to the front extension of the engine camshaft. The pump outlet connection contains a spring loaded relief valve to maintain a pressure in the pump gallery of 100 kPa ± 15 (15 psi ± 2) to reduce vapor locking. A water cooling element is installed in the pump gallery through which cold water is circulated to maintain the fuel temperature within the desired range of 16 °C to 32 °C (60 °F to 90 °F).

A portion of the fuel, determined by the injection pump control rod setting, is passed to the injection nozzle and the balance is returned to the fuel tank. The injection pump lubricant, SAE 30 oil or castor oil, should be changed at least every 50 h. The timing of the injection pump is accomplished by setting the engine flywheel at 60 degree atc on the intake stroke, and coupling the injection pump to the engine camshaft with the scribed line on the tapered shaft of the pump aligned with the "R" line on the pump endplate, for clockwise rotation of the pump as viewed from the drive end. When aligned, the bypass port of the pump is closed and fuel delivery to the nozzle begins.

The injection nozzle is mounted on the upper end of the intake pipe and sprays fuel directly accross the passage at right angles to the air flow direction.

8. Cooling System—With those agencies that still utilize the older integral type cylinder head wherein the coolant temperature must be maintained at 130 °C (265 °F), a pressure type cooling system is used. A detailed description of a suggested type when using the integral head is well documented in the AS840 Manual published in July 1964.

Where the spark plug insert type cylinder head configuration is utilized, the above pressurized system may be used. However, experience has shown that since coolant temperatures required on this type head are only 88 °C (190 °F), a system operated at atmospheric pressure is the more desirable. Figure 5 illustrates a suggested cooling system of this type. It consists basically of a coolant pump, heat exchanger, and expansion tank with auxiliary plumbing to effect coolant distribution to both the spark plug insert and the combustion chamber jacket. The coolant pump may be of the centrifugal type with enough capacity to circulate coolant at a rate of approximately 19 L/min (5 gal/min) under operating conditions.

The heat exchanger may be a commercial unit which has a rating of approximately 95 000 Btu/h (6650 gr.-cal./s) to heat up tap water entering at room temperature to the required 88 °C (190 °F).

Adequate temperature controllers are to be placed at the inlet points to both the cylinder head jacket and the spark plug insert such that 88 °C (190 °F) inlet temperatures are maintained.

Distilled or treated water is used for the coolant to prevent formation of mineral deposits in the cooling system. Since this system operates at atmospheric pressure, the expansion tank should be elevated to a position such that the coolant level is above the highest point in the engine. Make-up coolant may be added as required.

9. Lubrication System—The schematic layout of the complete lubrication system for the 5750 engine is shown in Figure 6. The oil pump is mounted externally on the front timing gear case cover and driven

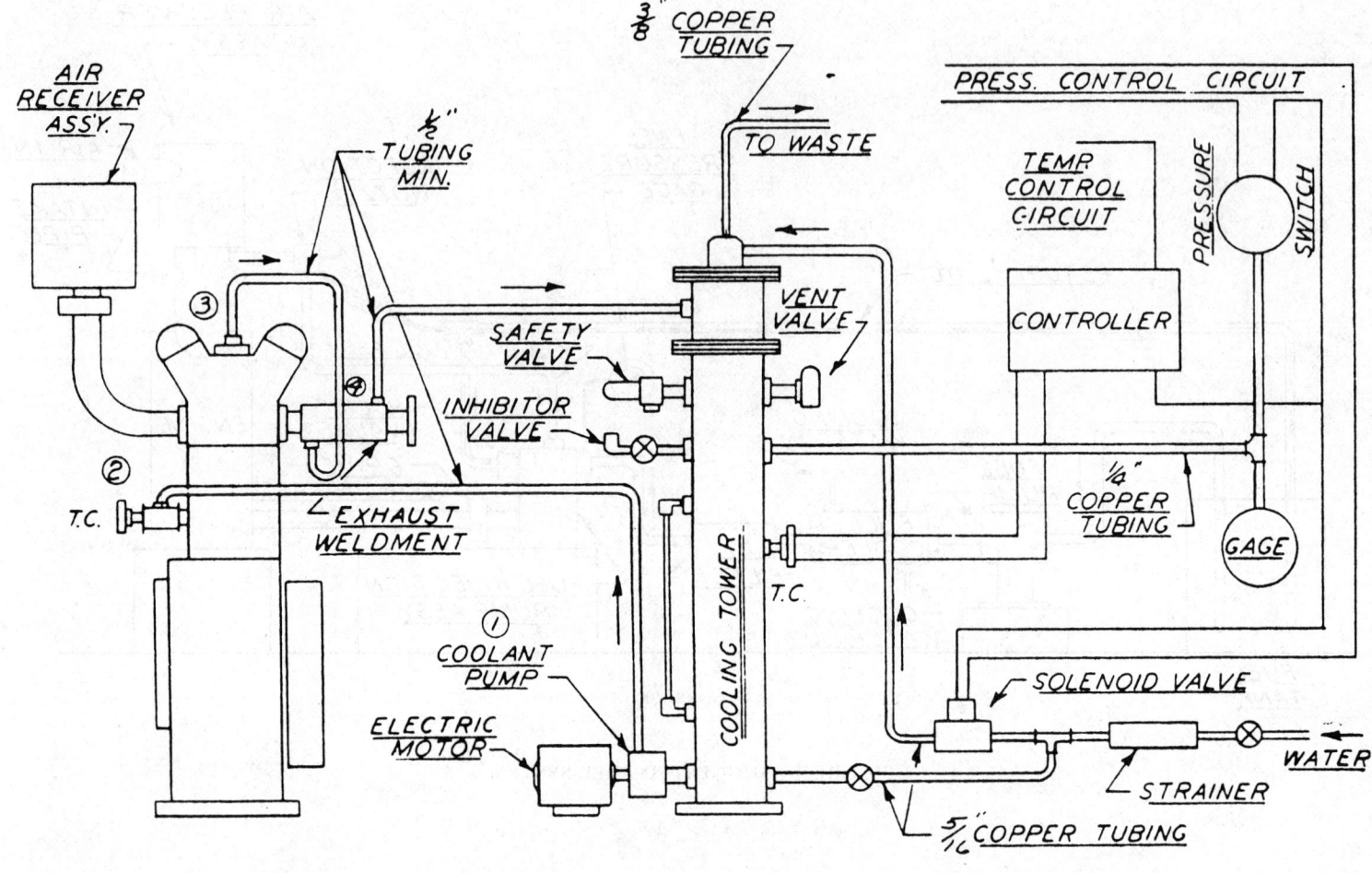

FIGURE 5—SUGGESTED COOLING SYSTEM

through an Oldham coupling by the left-hand counterbalance shaft. All bearings are pressure lubricated through a single pressure relief valve set to 690 kPa ± 35 (100 psi ± 5). The timing gears and the counterbalance drive chain are lubricated by splash from a metered hole in the right-hand counterbalance drive shaft sprocket. The valves and rocker arms are lubricated by bleed-off oil that comes through the camshaft to the valve lifters and through the pushrod to the rocker arms. The pressure to the valve gear is 70 to 105 kPa (10 to 15 psi) depending on the clearance in the camshaft bearings and between the valve lifter and valve lifter guide.

The oil cooler is fabricated with steel tubing and is identical in detail to the coolant heat exchanger (Figure 4) with the exception of length.

9.1 The oil filter is a heavy-duty edge type design 0.1 mm (0.0035 in) spacing between discs to withstand high pressures and so made that the disc edges may be cleaned without disassembly. Three taps are provided for oil drain, oil entrance, and oil exit, respectively.

9.2 Alternate Oil Filter—An aircraft-style oil filter in conjunction with a remote filter head assembly may be used as an alternate to the disc-type originally supplied with the rating engine. The filter should have a burst pressure test value of 2760 kPa (400 psi) minimum and should not have an internal bypass system. The filter head should have a bypass system with a visual and electrical indicator to indicate when filter is being bypassed. These filters are a spin-on type and should be replaced at regular intervals. It is recommended these filters have a 45 µm or less particulate restriction.

Four tubular cartridge-type oil heaters are located on the front of the crankcase and extend into the base of the crankcase. Three crankcase heating capacities (285, 570, and 1140W) are available.

The lubricating oil is nonadditive aviation SAE 120 type. Twenty cc of DAG Dispersion NO. 2404 to each 2.2 L (gallon) of oil may be used to reduce varnish formation on moving engine parts. The crankcase oil level indicator is incorporated in a casting that is bolted to the side of the crankcase, the oil level being maintained halfway up the sight glass with the engine at rest. (Crankcase capacity 2.2 L (2.5 qt)—entire system, approximately 5.7 L (6 qts). The oil sump should be drained, completely cleaned, and refilled every 50 h.

10. Exhaust System—The exhaust pipe weldment, held to the cylinder head by four studs, consists of a 25.4 cm (10 in) long steel tube with steel flanges brazed to each end. The tube is jacketed approximately two-thirds of its length by another steel tube, the assembly to serve as a coolant heat exchanger.

Although the primary purpose of this coolant heat exchanger is to aid in maintaining cylinder jacket coolant inlet temperature when the engine is being operated at low boost and hence low power, it also is very effective in: (a) avoiding corrosion and cracking of the exhaust pipe weldment; (b) removing exhaust heat from the immediate vicinity of the engine for the comfort of the operator; (c) in avoiding seizure of the nuts and studs holding it to the cylinder head.

A recommended system, which has given a minimum of trouble and has been widely used, is one in which the exhaust gasses are cooled by a water spray, the resultant mixture passing through a section of rubber covered steam hose to the exhaust pipe. Such a system has the double advantage of cooling the exhaust pipe for operator comfort and preventing exhaust pipe leaks occurring from cracking of seams or welds due to excessive temperatures. The water spray nozzle is welded to a 90 degree elbow, which is in turn welded to a flange, the unit being held to the exhaust weldment by stainless steel bolts and nuts. The elbow faces in a downward direction away from the operator, the water-exhaust mixture going into a 10.2 cm (4 in) pipe that is led outside the building into a 10.2 cm (4 in) tee. Two pieces of pipe are screwed into this tee; one extending vertically above the building parapet on which a muffler may be placed for quieting purposes, the other extending vertically downward to an exhaust sump. The water drain in the sump is held at a level approximately 12.7 cm (5 in) above the bottom end of this lower section and serves both as a water seal and as a back pressure relief valve in the event the upper vertical stack becomes plugged.

A simple back pressure alarm may be made by inserting a wire into each leg of a U-tube that contains a solution of water and salt, one wire immersed in and the second wire located above the electrolyte. A simple electrical circuit is made by connecting a bell, a 6 V power supply, and the U-tube "switch" in series. When the back pressure increases to raise the electrolyte sufficiently to contact the second wire, the circuit is completed to ring the bell.

The remainder of the exhaust system for the engine may be left to the discretion of the test laboratory provided a few precautions are taken. In general, it is recommended that precautions be taken to avoid any resonant effect that may cause alternating high and low back pressures. Such a resonant effect may be easily overcome by any of a variety of damping methods, such as elbows, surge chambers, and so on. Back pressure in the system should be limited to avoid difficulty in cylinder exhaust scavenging due to valve overlap resulting in abnormal cylinder head temperatures. Any possible water trap in the exhaust system should be avoided. It is recommended to slant the exhaust down from the weldment to prevent collection of moisture when the engine is not in operation. This pipe may be water jacketed for additional heat removal from the test area.

Provisions should be made so that the exhaust pipe may readily be disconnected and plugged in the event that the engine is not to be run for a prolonged period to avoid exhaust pipe weldment corrosion from exhaust acids.

11. Crankcase Breather System

11.1 Standard System—The 5750 engine has a casting attached to the left side cover plate. This casting is tapped for 12.7 mm (1/2 in) pipe. An elbow may be inserted here and a short length of pipe extended vertically or to an exhaust system. A baffle on the inside of the plate prevents splash leakage.

12. Air Supply System—The air supply system consists of a compressor, an air/water separator tank, a float-operated valve, a water circulatory pump, a water level alarm, a normally closed solenoid valve, an air pressure regulator, two air heaters, an auto transformer, an air temperature controller, and miscellaneous electrical equipment.

The compressor used is of the centrifugal displacement type of pump (Nash MD574) and consists of a round, multiblade rotor that revolves freely in an elliptical casting partially filled with water. The rotor blades are curved and project radially from the hub and form, with the side shrouds, a series of pockets around the periphery. The rotor revolves at a speed high enough to throw the liquid out from the center by centrifugal force, resulting in a solid ring of liquid revolving in a casing at the same speed as the rotor, but following the elliptical shape of the casing. As the liquid follows the casing and withdraws from the rotor, the air is pulled in through two inlet ports located around the hub of the rotor and connected with the pump inlet. As the liquid is forced

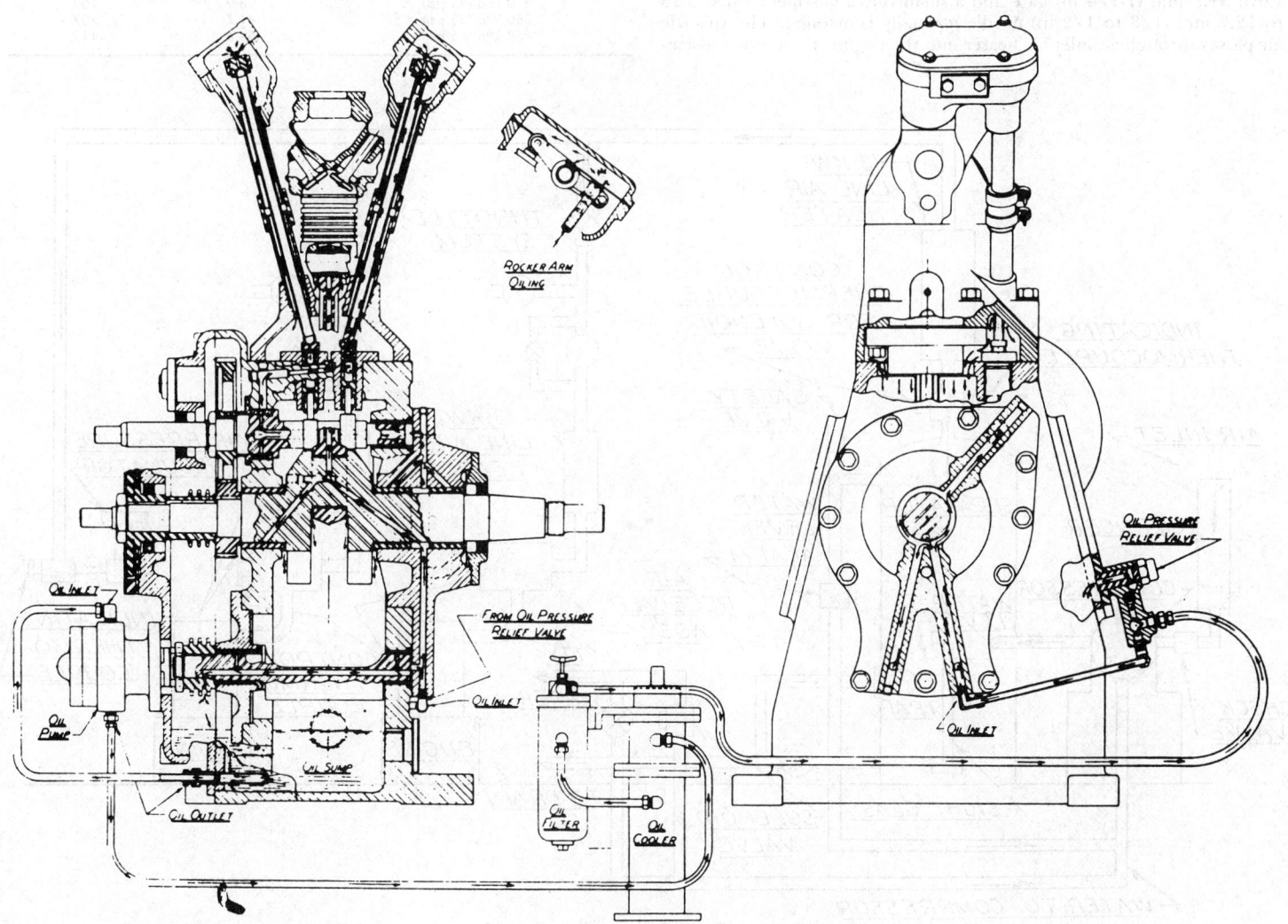

FIGURE 6—#5750 ENGINE LUBRICATION SYSTEM

back into the rotor chamber by the casing, the air trapped in the chamber is compressed and forced out through two discharge ports located around the hub of the rotor and connected to the pump outlet. The water supplied to the pump takes up the heat of compression, the surplus water being discharged with the air.

The air/water separator consists of a tank that acts as a centrifugal separator by removing the sealing water from the air. As the mixture of air and water enters the separator tangentially, the water falls to the bottom and is dumped by a float-operated discharge valve located about one-third of the way up the tank. Vertical baffles rise several inches above the water level and prevent the water in the base from spinning in a vortex and climbing the sides of the separator. Air is drawn off through a delivery pipe that projects some 3 in into the dome to prevent swirling water on the dome surface from creeping into the discharge air. The interior of the separator is galvanized as it is subjected to rather severe corrosive conditions due to being violently scrubbed with air-saturated water. Couplings are welded to the tank for drain, water level sight glass, thermocouples, and pressure taps.

The air delivered from the separator is in a saturated condition and may be cooled below its dew point and deposit water in the lines if the surrounding temperature conditions are suitable. In order to prevent such deposition of moisture, the air is discharged from the separator into a 3 kw line air heater. Current is supplied to an automatically-controlled heater to raise the temperature of the air a sufficient amount so that it will remain above the tank temperature to the next air heater located adjacent to the engine. Constant pressure is held at any predetermined value in the system by an air pressure regulator of the differential pressure diaphragm type, which bleeds off any excess air not used by the engine.

The air pressure delivered to the engine is controlled by a large valve 31.8 mm (1-1/4 in) gate and a small fine adjustment valve 3.18 to 12.7 mm (1/8 to 1/2 in) needle manually controlled. The throttle air passes through an inlet air heater into the engine air receiver assembly. With the equipment in this sequence, the expansion of the air at the throttle valves occurs before the heat is applied and regulation of manifold pressure and temperature is simplified. The inlet air heater consists of an enclosed 3 kw electric unit connected to the air receiver assembly by a flexible tube, preferably metallic, as rubber hose is likely to char. The inlet air heat is automatically controlled to 107.2 °C ± 2.8 (225 °F ± 5) by suitable temperature control connected to a thermocouple located in the air receiver assembly. The across-the-line load of the heater is carried by a suitable normally open contactor, the hold-down coil current being supplied by the controller.

The schematic layout of the air supply system is shown in Figure 7. As may be seen, the air is delivered to the compressor through a silencer and a check valve and fed together with the sealing water into the separator. The silencer is used to lower the noise level, the check valve to prevent the water from being blown back through the compressor when it is shut down. From the separator, the air goes through the line air heater to the engine throttle valves, through the inlet air heater into the engine. All the piping is 31.8 mm (1-1/4 in) galvanized and lagging is recommended for all long runs.

To control the moisture content of the supercharging air at 75 grains ± 25 (434 gms) of water per pound of dry air, the air pressure and air temperature in the tank must be held to essentially constant values. The temperature for various pressures to maintain this moisture content are as in Table 1:

TABLE 1—TEMPERATURE FOR VARIOUS PRESSURES TO MAINTAIN MOISTURE CONTENT

System Pressure	Separator Tank Air Temperatures	Separator Tank Air Temperatures
	C	F
310 kPa (45 psi)	38.9	102
380 kPa (55 psi)	41.7	107
468 kPa (65 psi)	44.4	112

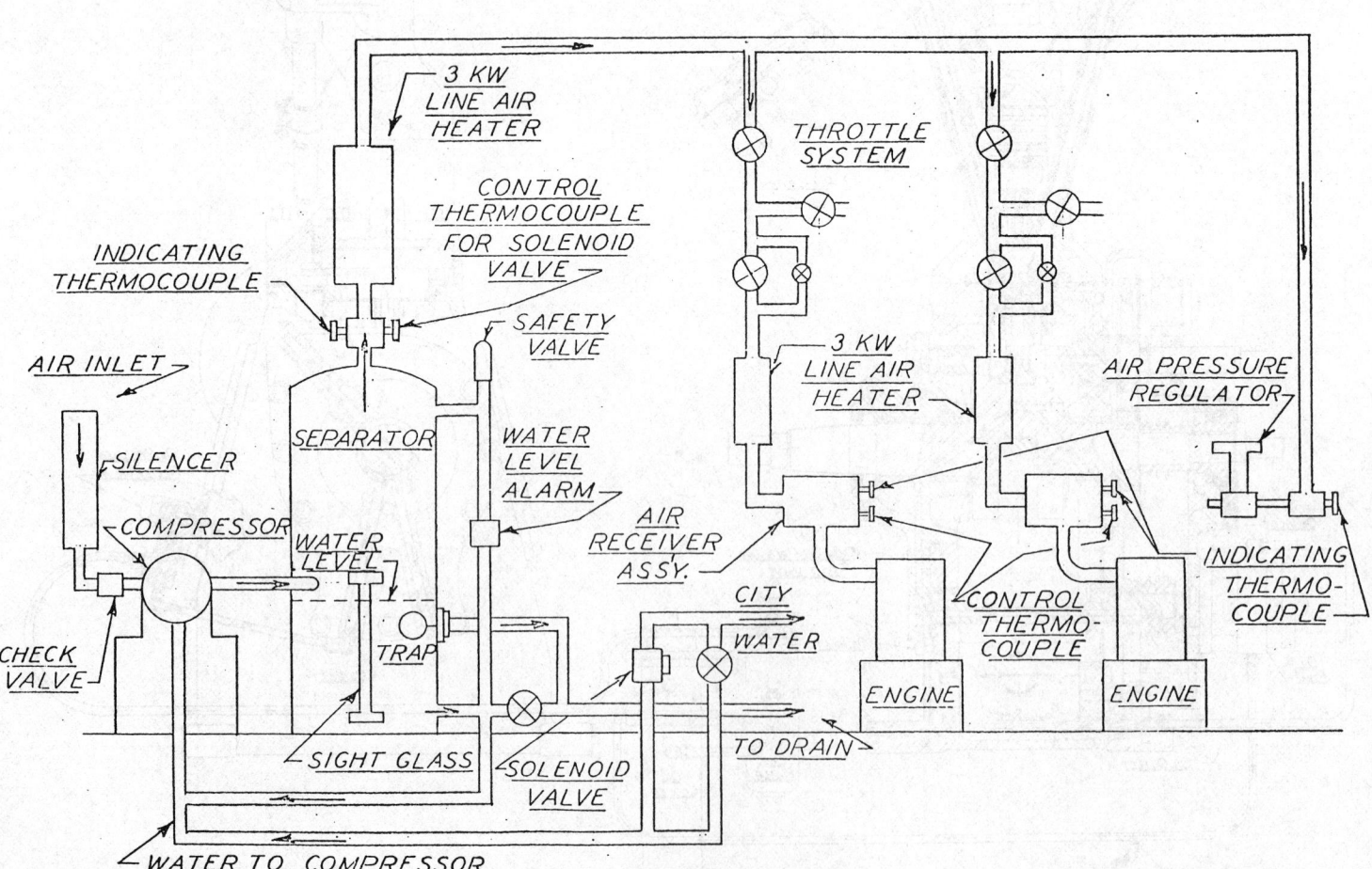

FIGURE 7—AIR SUPPLY SYSTEM

These temperatures are controlled by automatic regulation of the amount of water being admitted to the inlet of the water circulating pump, which in turn supplies the water under pressure to the compressor. This circulating pump must be started and pressure developed before the compressor is started. The sealing water is now in circulation through the system, it warms up due to heat of compression in the compressor. When the discharge air temperature reaches the specified separator tank air temperature, a normally closed solenoid valve is opened and cold water enters the system. The water supply must be at least 69 kPa (10 psi) higher than the compressed air pressure. The excess water goes out through the water float in the separator to the drain. When the discharge air temperature drops below the specified value, the solenoid closes and the water is again warmed by the compressor. A thermocouple mounted in the exit air line leading from the separator actuates the solenoid valve.

Two safety devices are incorporated in the installation. Mounted on the side of the separator is a water level alarm that rings a bell if the water rises in the tank and warns the operator that the water float valve has stuck. This can be connected to the compressor and/or the engine to automatically shut it off if desired.

To prevent the line air heater from burning up in the event the compressor fails, the line to the holding coil in the contactor for the heater is wired to the load side of the starter for the compressor motor. In this way, either seizure of the compressor or momentary power failure throws out the starter switch that has thermal overload protection and cuts the power to the heater.

> WARNING: Extreme caution should be exercised in completely shutting off the throttle valves to the engine if the compressor is allowed to run while the engine is shut down for any length of time. If the throttle valve is just barely cracked to atmospheric pressure, the expansion of the humidified air results in the water falling out of the air in the pipe downstream from the valve and entering the air receiver assembly. Although this water may not be of sufficient depth to flow over the standpipe in the receiver and into the intake pipe with the engine stopped, there may readily be enough lying in the bottom of the receiver to climb its walls in a vortex as the airflow through the receiver increases due to its tangential air entry. As the engine is of such small displacement, sufficient water may collect in the inlet air heater or throttle valve assembly to wreck the engine if allowed to enter the cylinder. If there is the slightest doubt that water has collected in the system, first, drain the receiver by the plug provided at the bottom of the casting; second, remove the spark plug and motor the engine with gradually increasing air velocity into the intake pipe. Such a procedure will safely remove all water and bent connecting rods, caved-in pistons and broken cylinders will be avoided.

13. Maintenance and Overhaul Procedure

13.1 General—It is strongly recommended that inspection of engine components be avoided unless there are obvious signs of trouble. Frequent teardown of the engine not only is unnecessary and time-consuming, but greatly increases the possibility of damage to the engine parts through careless handling.

With proper attention, the crankcase should run 5000 h before teardown inspection and overhaul are required. The need of an overhaul or replacement of any engine part or assembly in most instances is quite evident. A deep rumbling type of knock usually denotes main bearing failure; a high-pitched rattle, a loose wrist pin, and a high-pitched howl or whine indicates timing gear trouble. Oil seepage at the camshaft or crankshaft extensions through the crankcase denotes oil seal failures. Excessive clatter in the rocker boxes indicates either wear of the rocker roller pin, wear of the rocker arm thrust washer, valve spring interference, or excessive tappet clearance. Loss of oil pressure may denote wear in the pump, loosening of the pump body from the crankcase, a plugged inlet line, a relief valve stuck open, or bearing failure. Runaway coolant temperatures may mean either vapor locking in the coolant pump, seizure of the pump, or failure of the driving motor. Missing may be caused by spark plug failure, ignition cable failure, magneto trouble, injection pump plunger sticking, vapor locking of the fuel in the injection pump, fuel supply pump failure or perhaps by simply being too lean. Continued experience with the engine will make the operator familiar with the general noise level of the unit and more able to diagnose any symptoms accurately. In the event of any sign of distress, the fault should be found and repaired immediately, not allowed to continue until major damage has been done to the engine.

The need for having a valve job or reringing is not usually as evident and will be covered in some detail. Under normal routine operation, valve reconditioning periods of 150 h are sufficiently conservative, but engine performance is still the best indication for the need of an overhaul as service under conditions of severe preignition at high IMEP may bring the time period under 100 h. It is desirable to check the compression pressure periodically. Compression pressure should be approximately 790 kPa (115 psi) at 900 rpm. At any fixed set of engine conditions, there is a definite boost-IMEP relationship that is a straight line function as shown in Figure 8 and should be used to determine when the engine is in good condition. At high power levels, plotted points will fall below this curve if valves or rings are bad. A positive valve check may be made by removing the intake pipe and exhaust pipe weldment, turning the engine flywheel by hand until the piston is at bdc on the compression stroke, pouring gasoline into the valve ports covering the valve heads, and bringing the piston to tdc on the compression stroke by turning the flywheel with the hands. If valves or seats are in bad condition, the leakage of air past the valve through the gasoline is readily visible.

The most positive check of oil pumping is inspection of the piston head by means of a light (a medical diagnostic type is best) inserted through a spark plug hole. If the cylinder bore looks scuffed or scored and the piston head flooded with oil, it can readily be assured that the rings are also scuffed and possibly either stuck or broken. If the cylinder bore looks good and more than a film of oil is present on the piston head after shutting down the engine from 2700 rpm, reringing is generally indicated providing the rings are not new or have not just been cleaned. If the rings are either new or have been recently removed from the piston for cleaning, additional running is necessary to establish a good seal between the ring faces and the cylinder bore. Usually this can be accomplished by operating the engine at 2.415 MPa (350 psi) IMEP for 2 to 3 hrs.

The most common reason for high oil consumption is excessive ring side and end clearance. The compression rings have a minimum of 0.1 mm (0.004 in) of chrome plate and can readily accept 0.05 mm (0.002 in) average wear on the face without possible danger of wearing through the plate. Thus, an end clearance increase of 0.3 mm (0.012 in) could be tolerated. The limits of the end and side clearances are listed in Appendix A. Rings that show any signs of scuffing should be replaced. If one ring requires replacement, all rings should be replaced.

13.2 Detailed Disassembly of 5750 Engine

13.2.1 REMOVAL OF CYLINDER ASSEMBLY
 a. Disconnect all accessories.
 b. Remove intake pipe and exhaust pipe weldment.
 c. Remove rocker box covers.
 d. Fasten a wire clip, Part No. 5700, to each push rod and its respective rocker arm to prevent the push rods from falling out as the cylinder is lifted.
 e. Remove the six nuts holding the assembly to the crankcase.
 f. Bring the engine to bdc.
 g. Lift the assembly from the crankcase, being sure not to allow the piston to fall against the crankcase.
 h. Remove the push rods.

13.2.2 REMOVAL OF THE PISTON
 a. Repeat 13.2.1.
 b. Push out full floating piston pin and remove piston.
 c. Remove the piston rings being careful not to spread the rings more than necessary for removal. A perfect circle ring expander may be used.

13.2.3 REMOVAL OF CYLINDER HEAD ASSEMBLY
 a. Repeat 13.2.1.
 b. Loosen the clamps on the push rod housing hoses and push the hoses down onto the lower push rod housing.
 c. Remove the ten bolts holding the head assembly to the cylinder.
 d. Remove the head from the cylinder. These will pull apart easily once the gasket seals are broken loose.

13.2.4 REMOVAL OF THE CYLINDER SLEEVE.[2]
 a. Repeat 13.2.3.
 b. Remove the 9.5 mm (3/8 in) socket heat cap. Screws on the lower face of the sleeve flange.
 c. Remove the sleeve from its housing.

13.2.5 REMOVAL OF THE VALVE GEAR
 a. Repeat 13.2.3.
 b. Remove the rocker shaft cover.
 c. Push out the full floating rocker shaft with the fingers.

---

[2] Not to be done unless replacement is necessary.

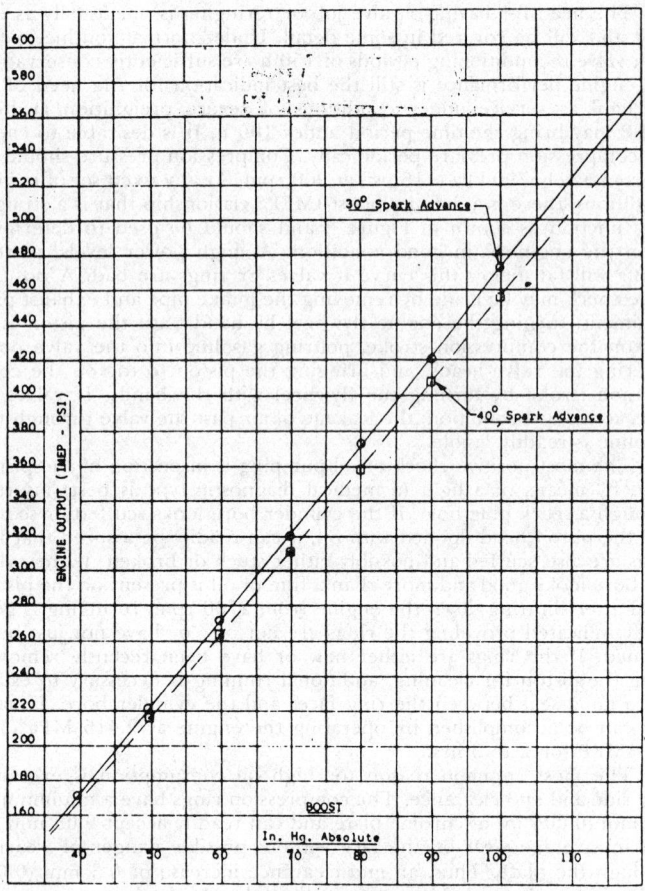

FIGURE 8—ENGINE BOOST VERSUS OUTPUT POWER CURVES AT MAXIMUM THERMAL PLUG TEMPERATURE

  d. Lift out the rocker arms and thrust washers.
  e. Compress the valve springs and remove the valve spring retaining keys. Use compressing tool, Part No. 5254.
  f. Remove valve springs, retainers, spacer, and valves.
13.2.6 Removal of the Ignition Timing Disc
  a. Remove the ignition timing disc quadrant support.
  b. Bend back the ear of the lock washer that anchors the nut on the front end of the crankshaft.
  c. Remove the front crankshaft nut and lock washer.
  d. Remove the disc by pulling with the fingers.
13.2.7 Removal of the Timing Disc Space Sleeve from the Crankshaft (if present)
  a. Remove with puller, Part No. 5702.
13.2.8 Removal of the Oil Pump
  a. Remove four 9.5 mm (3/8 in) nuts and washer.
  b. Remove oil pump and Oldham coupling.
13.2.9 Removal of Timing Gear Case Cover
  a. Repeat 13.2.7.
  b. Remove the cap screws holding the casting to crankcase.
  c. Remove the cover. Use cap screws in the two tapped holes provided and jack the cover loose.
13.2.10 Removal of the Counterbalance Drive Chain
  a. Remove socket head locking screw from idler sprocket bushing bolt.
  b. Remove idler sprocket bushing bolt.
  c. Remove idler sprocket and bushing.
  d. Remove drive chain.
13.2.11 Removal of the Camshaft Assembly
  a. Repeat 13.2.1.
  b. Repeat 13.2.9.
  c. Remove the valve lifter guides from the crankcase top deck.
  d. Remove the valve lifters through the crankcase.
  e. Remove the cap screws holding the front shaft bearing to the crankcase.
  f. Remove the camshaft together with its driving gear and front camshaft bearing.
13.2.12 Removal of the Camshaft[3]
  a. Repeat Steps a, b, c, and d of 13.2.11.
  b. Bend back the ear of the lock washer that anchors the nut on the front of the camshaft.
  c. Remove the camshaft nut and lock washer. Use socket wrench, Part No. 5703.
  d. Repeat Steps e and f of 13.2.11.
  e. Remove the timing gear from the camshaft. Use an arbor press to press the camshaft out of gear, taking care not to foul the front bearing.
13.2.13 Removal of the Crankshaft Sprocket and Timing Gear
  a. Repeat 13.2.9.
  b. Remove crankshaft sprocket with suitable puller.
  c. Remove timing gear. Use the puller, Part No. 5704.
13.2.14 Removal of the Flywheel
  a. Bend back the ear of the lock washer that anchors the flywheel nut.
  b. Remove the flywheel nut and lock washer. Use the socket Part No. 5705.
  c. Thread on the collar, Part No. 5706, over the crankshaft threads.
  d. Remove the flywheel. Use a suitable puller and do use a chain fall or get help to lift it off the shaft.
13.2.15 Removal of the Crankshaft Rear Oil Retainer
  a. Remove the flywheel key.
  b. Remove six cap screws and washers.
  c. Remove crankshaft oil retainer. Use 9.5 mm (3/8 in) cap screws in the two holes tapped in the flange.
  d. Remove gasket.
13.2.16 Removal of the Connecting Rod
  a. Repeat 13.2.2.
  b. Remove the crankcase side cover assembly (breather side).
  c. Remove the cotter keys in the connecting rod bolts and the nuts.
  d. Remove the connecting rod cap. Use a composition hammer and tap the cap lightly, first on one side and then on the other.
  e. Remove the connecting rod.
13.2.17 Removal of the Crankshaft Rear Bearing Adapter
  a. Repeat 13.2.14.
  b. Repeat 13.2.15.
13.2.18 Removal of the Counterbalance Assembly
  a. Remove six cap screws and washers.
  b. Remove assembly. Note locating dowel on top flange.
13.2.19 Removal of the Counterbalance Shafts
  a. Mount the counterbalance assembly in a soft jaw vise, using the flats on the counterbalance shafts.
  b. Bend back ear of the lock washer on right-hand shaft.
  c. Remove both nuts with suitable wrench.
  d. Remove sprockets and bearing adapter.

13.3 Detailed Inspection and Assembly of 5750 Engine
13.3.1 Counterbalance Assembly
  a. Inspect the counterbalance shaft bearing journals for galling and wear. See Appendix A for dimensions.
  b. Inspect front and rear counterbalance shaft bushings for wear. See Appendix A for dimensions. Replace if required. Bushings are a push fit.
  c. Inspect sprockets for excessive wear and replace, if required.
  d. Insert counterbalance shafts in bearing adapter. Clamp in a soft jaw vise. (The shafts are interchangeable.)
  e. Replace sprockets. Dowel pin hole in adapter indicates top. Use vertical keyway on left-hand shaft and horizontal keyway on right-hand shaft. This is to align the chain oiling hole in the sprocket and the shaft.
  f. Replace nuts using a lock washer on the right-hand shaft only. The nut on the left-hand shaft has a slot for an Oldham coupling.
  g. Install assembly in crankcase. No gasket used.
13.3.2 Crankshaft
  a. Inspect the crankshaft for galling and for wear of the bearing journals. See Appendix A for dimensions.

---
[3] Only to be done if obviously damaged.

b. Inspect main bearings for any sign of failure and wear. See Appendix A for dimensions.
c. Insert crankshaft through the front main bearing, being careful not to nick bearing with threads or shoulders of crankshaft.
d. Clean mating surfaces of crankcase and rear main bearing adapter, removing any nicks or burrs.
e. Install gasket dry or with soft soap.
f. Install rear main bearing adapter, being careful not to nick bearings on shoulders of crankshaft. Alignment is assured by the piloted shoulder on the adapter.

13.3.3 CRANKSHAFT REAR OIL RETAINER
a. Clean mating surfaces of crankcase, rear bearing adapter and rear oil retainer. Inspect oil seal and replace if necessary.
b. Install gasket.
c. Install retainer on adapter.

13.3.4 CRANKSHAFT TIMING GEAR
a. Inspect the gear teeth, bore and faces, and remove any nicks and burrs.
b. Insert the Woodruff drive key in the crankshaft.
c. Align the gear on the shaft with the side outward having an "X" on one tooth. Tap gently with a composition hammer to start.
d. Press the gear on the crankshaft. Push it on using the front crankshaft lock nut and the tool, Part No. 5708.
e. Install crankshaft chain sprocket.

13.3.5 CAMSHAFT
a. Inspect cams and bearing journals for signs of galling or wear and replace if necessary.
b. Inspect camshaft bearings. See Appendix A for clearance.
c. Install the camshaft and its front bearing.
d. Check end play. See Appendix A for clearance.

13.3.6 CAMSHAFT DRIVE GEAR
a. Install Woodruff key in the camshaft.
b. Reinstall gear using original keyway. Use arbor press.
c. Install camshaft assembly with the "X" marks on the gears mating.
d. Install the camshaft gear lock nut and lock washer.
e. Bend a shoulder of the lock washer over a flat on the camshaft nut.

When new timing gears are to be installed, the assembly procedure is as follows: Install the crankshaft timing gear on the crankshaft so that the puller holes on the front face of the gear face outward. Normally select the center keyway of the camshaft gear for the initial timing check. The front of the camshaft gear may be identified by the 9.5 mm (3/8 in) hub extending from the web to the face of the gear; the rear has a 3.2 mm (1/8 in) hub extending to the face of the gear. Install the Woodruff key in the camshaft and press the gear on to the camshaft, using the center keyway in the gear and the front of the gear extending outward.

Set the crank angle and flywheel at 28 degrees atc on the flywheel indicator. Using the intake valve lifter and intake lobe of the camshaft for the initial setting, install the camshaft assembly mating with the crankshaft gear in such a position that the intake valve lifter is raised approximately 1.0 mm (0.040 in) on the ramp of the camshaft in the direction of engine rotation on the opening side of the intake cam lobe, with the camshaft bolted in the running position.

Then, using an indicator on the intake valve lifter, turn the flywheel toward tc or to a point before top center where there is no movement on the indicator needle. Then turn the flywheel in the direction of rotation and observe when you get 1.0 mm (0.040 in) lift on the intake valve lifter from the indicator and check flywheel degrees to see if the timing is within the limits listed in this section.

If the valve lifter rise does not fall within the limits, it will be necessary to shift the camshaft and gear assembly one or more teeth in the proper direction to bring the timing within the limits. If this operation does not bring the timing within the limits, then remove the camshaft gear and reinstall it on the camshaft, using one of the other two keyways of the gear; reassemble camshaft and gear assembly in the crankcase and repeat preceding procedure. Since this is a cut-and-try procedure, it may be necessary to try all three keyways of the cam gear before the timing will follow the timing data.

After the engine is timed correctly, set the engine on tdc of the firing stroke and make suitable markings on the teeth of the timing gears and camshaft gear keyway.

The camshaft timing, with the engine completely assembled and valve clearance set at 1.26 mm (0.050 in) is as follows (Figure 9):

Intake valve opens at 28 degree atc ± 5
Intake valve closes at 22 degree abc ± 5
Exhaust valve opens at 23 degree bbc ± 5
Exhaust valve closes at 1 degree btc ± 5

The valve clearance then must be reset to 0.46 mm (0.018 in) before running the engine.

NOTE: When checking the timing, some thought must be given that wear on the rocker arm rollers and pins, valve lifter rollers, hubs, and pins will cause some lag in the timing characteristics; so for a true check, all above parts should be within the recommended clearances. A 10 degree tolerance is allowed for a used cam before replacement is required.

13.3.7 VALVE LIFTER ASSEMBLIES AND GUIDES
a. Inspect parts for galling or wear. Replace if necessary. See Appendix A for clearance.
b. Install assembly and guide as a unit holding fingers under valve lifter during installation to prevent dropping into crankcase.

13.3.8 COUNTERBALANCE DRIVE CHAIN
a. Place crankshaft at top dead center.
b. Install counterbalance drive chain over crankshaft sprocket, under left-hand counterbalance drive shaft sprocket and over right-hand counterbalance drive shaft sprocket. Arrows on counterbalance shafts should point down.
c. Install idler sprocket bushing into idler sprocket.
d. Insert bolt. Tighten chain by turning eccentric bushing until it has 6.35 mm (1/4 in) deflection measured midway between the idler and crankshaft sprockets.
e. Lock idler sprocket bolt to idler sprocket bushing with Allen set screw. With chain tight, the arrows on the counterbalance shafts may not be exactly parallel.

13.3.9 TIMING GEAR CASE COVER
a. Inspect mating surfaces of crankcase and cover.
b. Inspect oil seals and replace if necessary.
c. Install gasket and timing gear case cover.

13.3.10 INSTALL TIMING DISC SPACER SLEEVE

13.3.11 OIL PUMP
a. Inspect pump for bushing wear and back lash. See Appendix A for clearances.
b. Insert Oldham coupling.
c. Inspect mating faces of pump and timing gear cover and install gasket.
d. Mount oil pump.

13.3.12 FLYWHEEL
a. Lift the flywheel and slip it on the crankshaft over the flywheel key with dummy nut, Part No. 5706, on threads. Remove dummy nut after the flywheel is in place.
b. Install the flywheel nut and lock washer. Tighten until the flywheel is well driven onto the crankshaft taper. Bend one side of the washer over the nut as an anchor.

13.3.13 CONNECTING ROD
a. Inspect the big end bearing and replace if scored or cracked. See Appendix A for dimensions. Avoid scratching the bearing during measurement. Use a snap gage, not inside calipers.
b. Inspect the wrist pin bushing. See Appendix A for fit.
c. Install by lowering upper end together with its bearing onto crankshaft journal and raising the cap with its bearing into place.
d. Draw up the connecting rod belts, using a torque wrench and 61.0 to 67.8 N·m (45 to 50 lb-ft) torque. Use cotter pins to secure the nuts to the bolts.

13.3.14 CYLINDER HEAD
a. Remove spark plug insert and spark plug insert gaskets.
b. Remove combustion chamber deposits.
c. Inspect the valve seat inserts for looseness. This may be determined by inserting a close fitting pilot in the valve guide and measuring seat concentricity before and after tapping the insert on its back shoulder. Replace inserts if found loose.
d. Inspect the valve guides. Replace if scuffed, if bell-mouthed more than 0.6 mm (0.0025) or if I.D. of the intake and exhaust is greater than 10.9 mm (0.438 in). If new guides are needed, do not hammer them in. Do press them in with the tool, Part No. 5709, and finish ream to 10.9 mm (0.4365 in) for both the intake and exhaust.
e. Install a close fitting pilot in the valve guide and measure the runout of the inserts with a suitable indicator. If the runout exceeds 0.5 mm (0.002 in) or if the insert is pitted, burned or corroded, refacing is necessary. Using a suitable grinder and 45 degree stones with true faces, grind the insert, first using a coarse stone and finishing with a fine stone, removing as little metal as

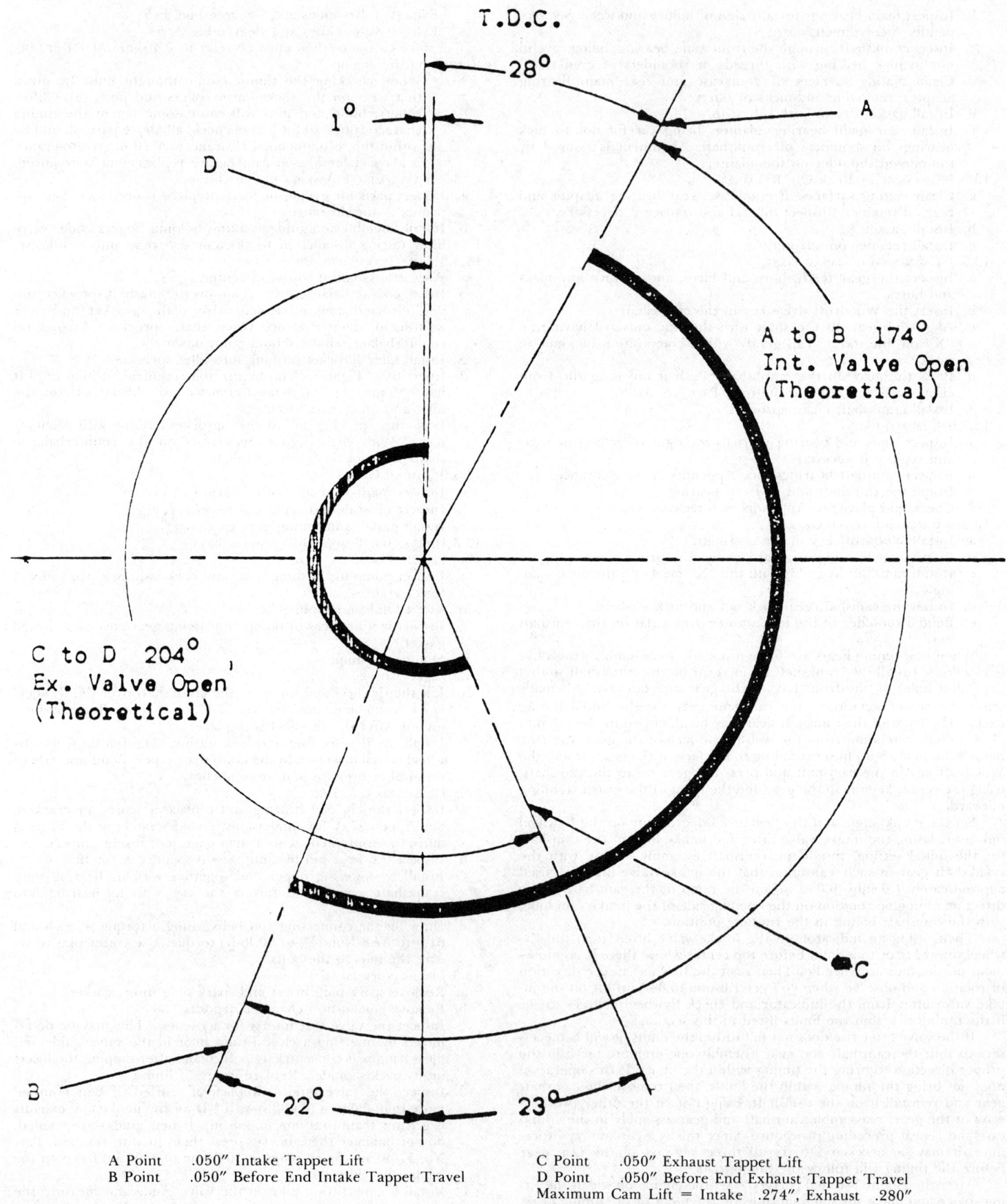

FIGURE 9—VALVE TIMING DIAGRAM 1.26 MM (0.050 IN) (VALVE CLEARANCE)
(COURTESY OF LABORATORY EQUIPMENT CORP.)

necessary. Runout of the ground face should not exceed 0.025 mm (0.001 in). See Figures 10 to 13 for refacing limits and method of checking. Care should be exercised that intake valves are not installed in exhaust seats. To assist in identification, a dimple has been machined in the intake valve heads, and a bump has been left on the exhaust valve heads. The previously mentioned drawings illustrate these identifying markings.

f. Inspect the valves. Replace if bent, galled, or burned. See Appendix A for clearance. If the stems are lightly scuffed, remove the marks by stoning and lapping. Clean the valve heads and stems with fine steel wool. Do not clean them with a hard scraper. Reface, removing as little material as possible to provide a clean face having a 45 degree angle. If the valve face is 5.6 mm (7/32 in) or wider, replace.

g. Lap the valve face to the seat, first using a coarse regrinding compound and following with a fine compound. Lapping should be continued until blueing shows the seats on both valve and insert to be concentric. The valves shall withstand 690 kPa (100 psi) air pressure without leakage.

h. Completely and thoroughly wash the head in kerosene to remove all ground metal and grinding compound.

i. Inspect the rocker arm thrust washers and replace if galled.

j. Inspect the rocker arm needle bearings and replace if necessary. Press them in, using the tool, Part No. 5710, always with the lettered side of the bearing out.

k. Inspect the rocker arm shafts and replace if any Brinelling is evident.

l. Inspect the rocker arm rollers and pins. Replace if worn or galled.

m. Check valve springs for fractures. See Appendix A for load limits.

n. Place the thrust washer, and the lower retaining washer over the valve guide in each rocker box.

o. Compress the valve springs, slip in the valve, and place the retaining keys on the valve stem, release the valve springs. Use the tool, part No. 5245.

p. Push the rocker shaft from the lower side of the rocker box, through the rocker arm, the thrust washer and the upper side of the rocker box. Use the drift, Part No. 5701.

q. Install rocker shaft covers, washers, and gaskets.

r. Check the thrust washer side clearance. If below 0.025 mm (0.001 in), lap the washer, if above 0.33 mm (0.013 in), replace it.

s. Check the mating surfaces between the head and cylinder housing and remove any nicks or burrs.

13.3.15 CYLINDER ASSEMBLY

a. Remove deposits with steel wool.
b. Inspect. See Appendix A for dimensions.
c. Check mating surfaces between sleeve flange and cylinder housing, removing all burrs and nicks.
d. Install copper gasket on sleeve outer ring sealing surface and 0.2 mm (0.007 in) thick nonmetallic gasket on sleeve flange.
e. Assemble sleeve and housing, using the two 3/8 in socket head cap screws.
f. Check all mating surfaces between cylinder housing and cylinder head, removing any burrs or nicks.
g. Install 0.2 mm (0.007 in) thick nonmetallic gasket between cylinder housing and cylinder head, and dry copper gasket between sleeve and cylinder head.
h. Fasten cylinder barrel assembly to cylinder head, using bolts with solid copper gaskets under the bolt heads, tighten the bolts evenly, using a torque wrench and 81.4 to 101.7 N·m (60 to 75 lb-ft) torque. Tighten bolts in an accepted sequence. Use a socket wrench with a torque measuring device.

13.3.16 PISTON ASSEMBLY

a. Remove deposits with a stiff brush, steel wool, or a scraper made by flattening the end of a copper tube. Do not use a wire brush or a buffing wheel.
b. Check piston for nicks and dents.

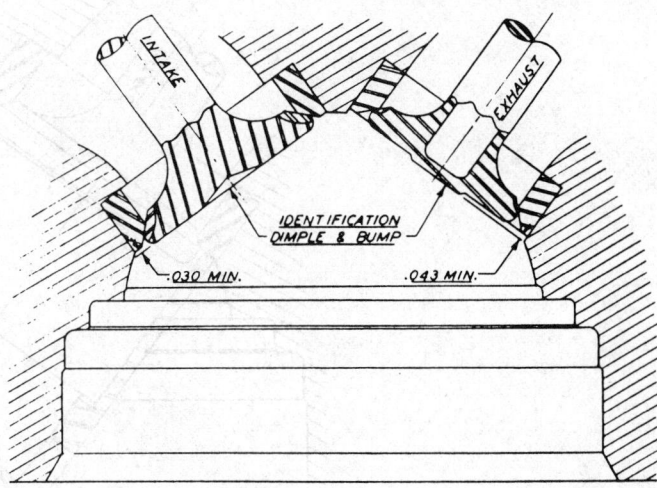

FIGURE 11—PROPER VALVE SEATING WITH REFACED VALVES AND REGROUND SEATS

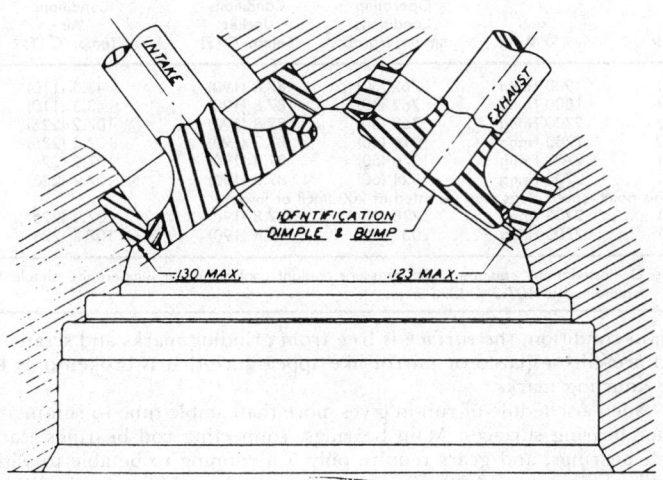

FIGURE 10—PROPER VALVE SEATING IN NEW CYLINDER HEAD

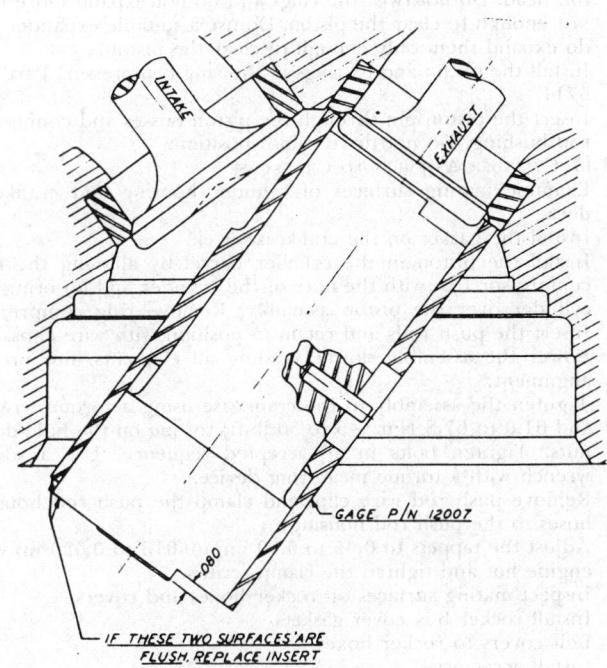

FIGURE 12—METHOD FOR CHECKING EXHAUST VALVE SEATING LIMITS

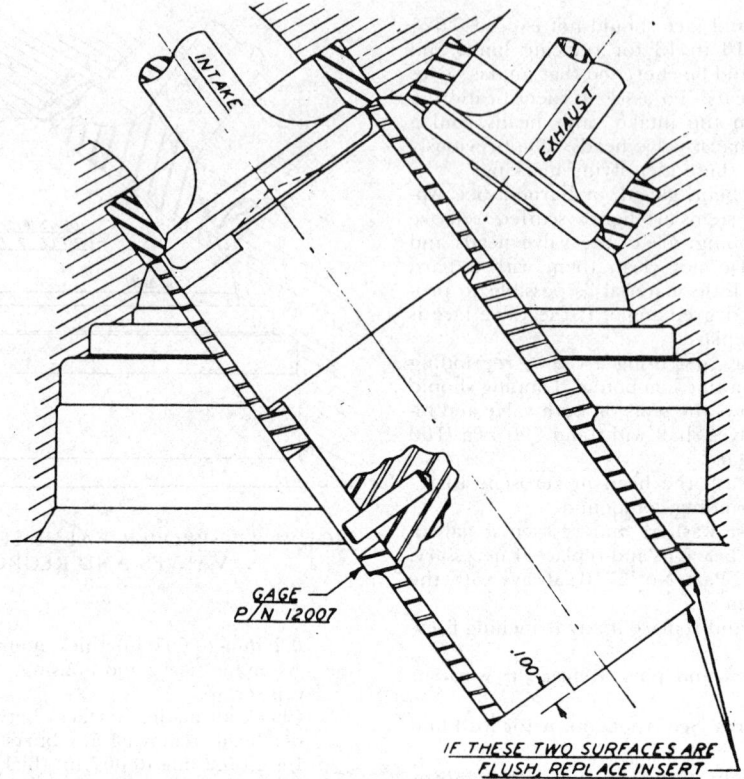

FIGURE 13—METHOD FOR CHECKING INTAKE
VALVE SEATING LIMITS

c. Check piston boss. Replace piston if scored or cracked.
d. Check for concentricity. Replace piston if skirt is more than 0.2 mm (0.007 in) out of round.
e. Inspect piston pin. Replace if worn or galled.
f. Remove ring deposits with steel wool.
g. See Appendix A for clearances.
h. If one ring is replaced, replace all rings.
i. Place the rings on the piston in the order removed. Install the compression rings with the inside bevel, if any, toward the piston head. Do not twist the rings and do not expand more than just enough to clear the piston. Do use a suitable expander and do expand them only enough to clear the piston.
j. Install the piston and its rings in the ring compressor, Part No. 5711.
k. Insert the piston pin through the piston bosses and connecting rod bushing. Do not drive it into position.

13.3.17 CYLINDER ASSEMBLY TO CRANKCASE
a. Examine mating surfaces of cylinder housing and crankcase deck.
b. Install the gasket on the crankcase deck.
c. Install the piston in the cylinder barrel by aligning the ring compressor I.D. with the bore of the cylinder and lowering the cylinder over the piston assembly. Remove ring compressor.
d. Insert the push rods and retain in position with wire clips.
e. Lower the assembly slowly, guiding all elements into proper alignment.
f. Tighten the assembly to the crankcase using a torque wrench and 61.0 to 67.8 N·m (45 to 50 lb-ft) torque on the hold-down nuts. Tighten bolts in an accepted sequence. Use a socket wrench with a torque measuring device.
g. Remove push rod wire clips and clamp the push rod housing hoses to the push rod housings.
h. Adjust the tappets to 0.45 to 0.50 mm (0.018 to 0.020 in) with engine hot and tighten the clamp screws.
i. Inspect mating surfaces on rocker boxes and covers.
j. Install rocker box cover gaskets.
k. Bolt covers to rocker boxes.
l. Install accessories.

14. Engine Run-In Schedule—The run-in schedule for a new or rebuilt engine is dictated by the rapidity with which the rings and cylinder sleeve bore run-in, provided bearing clearances follow recommendations and lubrication is adequate.

The engine schedule recommended for a new or rebuilt engine is shown in Table 2.

If the power absorption equipment is such that the speed schedules in Table 2 cannot be followed, the best compromise available is acceptable.

The schedule shown in Table 2 is conservative for most cases as run-in can be accomplished in less time. It is recommended that the engine be shut down before each change of operating conditions and the cylinder inspected with a light through the spark plug hole for any signs of scratching or scuffing. Any indication of distress is considered to be sufficient reason to drop back on the run-in schedule and operate longer at less severe conditions. When the barrel has reached a proper

TABLE 2—ENGINE RUN-IN SCHEDULE FOR NEW OR REBUILT ENGINES
PLUS USED CYLINDER AND NEW RINGS

Hr	RPM	Operating Conditions mm (in) Hg abs.	Operating Conditions Jacket Temp. °C (°F)	Operating Conditions Air Temp. °C (°F)
1	900 Firing	762 (30)	87.8 (190)	43.3 (110)
1	1800 Firing	762 (30)	87.8 (190)	43.3 (110)
2	2700 Firing	762 (30)	87.8 (190)	107.2 (225)
2	2700 Firing	1016 (40)	87.7 (190)	107.2 (225)
1	2700 Firing	1270 (50)	87.8 (190)	107.2 (225)
1	2700 Firing	1524 (60)	87.8 (190)	107.2 (225)
At this point, spark plugs may be rated at 200 IMEP or lower.				
1	2700 Firing	1778 (70)	87.8 (190)	107.2 (225)
1	2700 Firing	2032 (80)	87.8 (190)	107.2 (225)

NOTE: On installations equipped with pressure coolant control, jacket temperature should be maintained at 107.2 °C (225 °F).

run-in condition, the surface is free from grinding marks and scratches and presents a glazed or mirror-like appearance that is broken only by the knurling marks.

Such a schedule of run-in gives more than ample time to run-in any other bearing surfaces. Main bearings, connecting rod bearings, camshaft bearings, and gears require only 5 h running to be able to withstand severe duty and 1 h running is usually adequate for crankshaft oil seal rings, carriers and races. One hour's running at 2700 rpm with 762 mm (30 in) Hg abs. boost and 1 h running at 2700 rpm with 1524 mm (60 in) Hg abs. boost is considered adequate for the run-in of valve gear after a carbon and valve job.

It is not considered advisable to operate the engine at high power levels with cold engine oils, as the viscosity of the oil will be too high at low temperatures to ensure proper lubrication of the bearing surfaces. The following warmup schedule is recommended:

900 rpm firing with no boost until oil temperature has reached 50 °C (120 °F).
1800 rpm firing with no boost until oil temperature has reached 60 °C (140 °F).
2700 rpm firing with no boost until oil temperature has reached 70 °C (160 °F).

If the proper absorption equipment prevents following the warmup schedule, any suitable compromise is acceptable as long as high power and/or high rpm operation are avoided until the oil temperature has reached 70 °C (160 °F).

15. Operating Instructions

15.1 These operating instructions are for use with the 17.6 Spark Plug Rating Engine Standard Assembly as approved May 9, 1968. The operating conditions herein specified are subject to revision from time to time as approved by the majority of the members of the SAE Spark Plug Rating Engine Standardization Panel of the SAE Electrical Equipment Committee. It is recognized that individual laboratories will have slightly different operation requirements; however, the fundamental operations should follow these instructions as closely as possible so that the rating results will be comparable. Only those conditions that affect the rating of the spark plug or the safety of personnel and equipment are specified as required, but recommended procedures or conditions are indicated for information and guidance.

15.1.1 REQUIRED OPERATING CONDITIONS

Speed: 2700 + 65-30 rpm

Compression Ratio: 5.6 to 1

Ignition Timing: 30 degree btc (automotive types)
40 degree btc (aviation or others not rateable at 30 degrees)

Fuel: 98%—1 degree Benzene, 2%—Specification MIL-L-6082D Grade 1100 NONADDITIVE SAE 60 aviation oil, with 0.8 mL/L (3 cc/gal) T.E.L. added.

Fuel Injection Pressure: 5.17 Mpa (750 psi) minimum.

Fuel-Air Ratio: That which produces the maximum thermal plug temperature.

Air Inlet Temperature: 107.2 °C ± 2.8) (225 °F ± 5).

Air Inlet Humidity: 75 + grains of water per pound of dry air.

Bushing Outlet Coolant Temperature: 87.8 °C ± 1.0 (190 °F ± 2).

Engine Oil Temperature: 87.8 °C ± 5 (190 °F ± 10).

15.2 Step-by-Step Procedure

15.2.1 Select the proper spark plug insert (see Appendix B).

15.2.2 After being sure that the cylinder head insert seat, as well as the spark plug insert itself, is clean, install the insert carefully into the cylinder head, using new O-rings. The O-rings should be assembled onto the insert and may be held in place with a small amount of petroleum jelly or light grease.

15.2.3 The hold-down nuts are to be alternately tightened so that the insert will seat evenly in the cylinder head. The nuts are to be tightened finally to a uniform (0.92 kg-m (60 lb-in) torque.

15.2.4 Install a warmup spark plug similar to the type to be rated, using the specified installation torque (Appendix C). A new gasket is recommended for each installation, except conical, seating types.

15.2.5 Before starting the engine, be certain that it was the proper type and amount of lubricating oil in the crankcase, oil in the fuel injection pump base, cylinder jacket and spark plug insert cooling system filled, air pressure supply system filled to the proper water level, and the fuel supply tank filled with fuel. Be certain that the cylinder air pressure supply throttle valves are tightly closed and that the atmospheric air throttle valve is fully open. Turn on the electrical power to all units and set the switch on "high" for the base oil heaters. Turn on the water to the fuel cooler, the crankcase aspirator if used, and all temperature control valves. Turn on the control air supply.

15.2.6 Because there are many installations that use dynamometers other than the frequency changer synchronous type now furnished as part of the standard assembly in the United States, only typical directions will be given here for the starting of the dynamometer. However, most installations include certain safety interlocks that must be bypassed during the initial starting of the engine. This is normally accomplished by holding the "start" button depressed until the oil pressure increases to above the safety setting. As long as the fuel injection pump control is in the closed position, no fuel will be delivered to the injection valve and the engine will not fire. Under no conditions should a cold engine be motored or run at speeds above 900 rpm.

15.2.7 After any required motoring is completed, the engine can be fired by gradually increasing the displacement of the fuel injection control by means of the micrometer screw until firing takes place. The displacement should continue until the thermal plug temperature reaches a maximum. By this time, the dynamometer will have changed from motoring to absorbing. The dynamometer can now be unlocked for power and friction readings. No ratings are to be attempted until the oil temperature has been stabilized at 88 °C (190 °F). Much warmup time may be saved by leaving the crankcase oil heaters turned on—even overnight.

CAUTION: Always be certain that the dynamometer shell is locked and that the atmospheric inlet air throttle valve is fully open and the cylinder air pressure throttle valves are completely closed before the engine is started or stopped.

15.2.8 When the engine is ready for testing, it is stopped by disengaging the fuel injection pump control from the micrometer screw and then disconnecting the electrical power from the dynamometer. As soon as the engine comes to a complete stop, disconnect the ignition cable from the warmup spark plug and remove it. Install the spark plug to be rated and connect the ignition cable. Bring the engine up to speed with the dynamometer and connect the fuel injection pump control to the micrometer screw. The engine will start firing immediately. Unlock the dynamometer shell. With the speed maintained at 2700 rpm, slowly open the main cylinder air pressure throttle valve while slowly closing the atmospheric air throttle valve, constantly watching the intake boost gauge. During the transfer of air supplies, the intake boost should remain at about atmospheric pressure.

15.2.9 After the transfer to pressurized air is complete, the engine can be operated with increasing boost pressure in regular steps until preignition of the spark plug takes place. This is evidenced by a rapid rise in the thermal plug temperature.

CAUTION: As soon as preignition occurs, the fuel should be cut off by disconnecting the linkage between the fuel injector pump control and the micrometer screw. Failure to act promptly can cause damage to the engine and especially to the spark plug on test. As soon as the fuel is cut off, the thermal plug temperature will decrease.

15.3 The plug rating is that IMEP value obtained on the engine at a point where the supercharge pressure is 3.37 kPa (1 in) Hg below the preignition point. The following steps are recommended to attain this point:

15.3.1 The supercharge pressure is increased in 13.5 kPa (4 in Hg) increments until preignition occurs as indicated by a rapid rise in thermal plug temperature. At each setting, the mixture strength is adjusted such that a maximum thermal plug temperature is obtained.

15.3.2 When preignition occurs, the fuel supply is instantly cut off and the supercharge pressure is decreased 6.7 kPa (2 in Hg) at which point the fuel is turned on and again adjusted for maximum thermal plug temperature. This condition should be held for 3 min or until preignition again occurs.

15.3.3 If preignition occurs after 15.3.2, the supercharge pressure should be reduced by 3.37 kPa (1 in Hg), again adjusting for optimum thermal temperature until stable engine operation for 3 min is obtained or preignition occurs. If preignition occurs, refer to 15.3.5.

15.3.4 If after 15.3.2, stable engine operation is obtained, the supercharge pressure should be increased 3.37 kPa (1 in Hg), again adjusting for optimum thermal temperature until stable engine operation for 3 min is obtained or preignition occurs. If preignition occurs refer to 15.3.5.

15.3.5 Friction torque should be measured at supercharge pressure 3.37 kPa (1 in Hg) below the preignition point (or previous stabilized setting prior to preignition) and with 30 s after the engine ceases to fire.

15.3.6 Rating data may be verified using a plug that has a rating point at least 50 IMEP above the plugs that have been rated.

15.3.7 CALCULATION OF IMEP

Indicated HP = Friction HP + Brake HP (Eq.1)

$$IHP = \frac{2700}{5250} T_F + \frac{2700}{5252} T_B$$

IHP = 0.51 (T_F + T_B) = Plan/33,000
0.51(T_F + T_B) − (0.04)(0.01) P = IMEP
IMEP = 8.65 (T_F + T_B)

T_F—Friction Torque
T_B—Brake Torque
IMEP = Indicated Mean Effective Pressure

APPENDIX A
MANUFACTURING TOLERANCES AND REPLACEMENT LIMITS

All dimensions not otherwise indicated are in millimeters
(All dimensions in parentheses are in inches)

		As Manufactured	Condemning
TAPPET SETTING (INTAKE VALVE)		0.457(0.018) Hot	
TAPPET SETTING (EXHAUST VALVE)		0.457(0.018) Hot	
VALVE STEM TO GUIDE (INTAKE)(HAND REAM GUIDE AT ASSEMBLY)			
5835 Intake Valve	11.05-11.07 (0.435-0.436) Dia.	0.0127-0.0381 (0.0005-0.0015)	0.08 (0.003)
5134 Intake Valve Guide	11.07-11.087 (0.436-0.4365) Bore		
VALVE STEM TO GUIDE (EXHAUST)(HAND REAM GUIDE AT ASSEMBLY)			
5836 Exhaust Valve	11.010-11.036 (0.4335-0.4345) Dia.	0.051-0.076 (0.002-0.003)	0.1143 (0.0045)
5135 Exhaust Valve Guide	11.074-11.087 (0.436-0.4365) Bore		
VALVE GUIDES P.F. IN CYLINDER HEAD			
Cylinder Head	17.462-17.48 (0.6875-0.688)		
5134 Intake Valve Guide	17.488-17.495 (0.6885-0.6888)	0.0127-0.033 (0.0005-0.0013) P.F.	
5135 Exhaust Valve Guide			
VALVE SPRING LOAD			
5230 Inner Valve Spring	At 34.925 (1.375) Height	8.05-9.45 Kg (115-135 lb)	Under 7.7 Kg (110 lb)
	At 42.849 (1.687) Height	4.41-4.83 Kg (63-69 lb)	Under 4.06 Kg (58 lb)
5231 Outer Valve Spring	At 34.925 (1.375) Height	7.7-9.1 Kg (110-130 lb)	Under 7.35 Kg (105 lb)
	At 42.849 (1.687) Height	5.25-5.95 Kg (75-85 lb)	Under 4.90 Kg (70 lb)
VALVE ROCKER ARM SIDE CLEARANCE			
Cylinder Head	41.986-42.139 (1.653-1.659)		
Rocker Area	39.649-39.725 (1.561-1.564)	0.051-0.330 (0.002-0.013)	
5223 Thrust Washer	2.159-2.209 (0.085-0.087)		
VALVE ROCKER ARM SHAFT TO CYLINDER HEAD (5750 ENGINE)(HAND REAM AT ASSEMBLY)			
Cylinder Head	15.875-15.808 (0.625-0.6255) Bore	0.0127-0.0254 (0.0005-0.001)	
Rocker Arm Shaft	15.862-15.875 (0.6245-0.625) dia.		
VALVE PUSHROD HOUSING TO CYLINDER HEAD			
Cylinder Head	25.095-25.146 (0.988-0.990) Bore	0.000-0.076 (0.000-0.003) P.F.	
Intake, Upper Push Rod Housing	25.146-25.171 (0.990-0.991)		
Exhaust, Upper Push Rod Housing	25.146-25.171 (0.990-0.991)		
VALVE PUSHROD HOUSING TO CYLINDER HOUSING			
5544 Cylinder Housing	25.349-25.146 (0.998-0.990) Bore	0.000-0.076 (0.000-0.003) P.F.	
5164 Lower Push Rod Housing	25.146-25.171 (0.990-0.991) Dia.	0.000-0.076 (0.000-0.003) P.F.	
VALVE LIFTER ASSEMBLY TO GUIDE			
5501 Valve Lifter Guide	17.462-17.475 (0.6875-0.600) Bore	0.043-0.071 (0.0017-0.0028)	
5502 Tappet, Valve	17.404-17.419 (0.6852-0.6858) Dia.	0.043-0.071 (0.0017-0.0028)	

MANUFACTURING TOLERANCES AND REPLACEMENT LIMITS (CONTINUED)

All dimensions not otherwise indicated are in millimeters
(All dimensions in parentheses are in inches)

		As Manufactured	Condemning
5740 CYLINDER SLEEVE		66.662-66.700 (2.6245-2.626) Bore	0.050 (0.002) Out of Round 0.127 (0.005) Variation
PISTON TOP LAND			
5474 Piston (Top Land)		66.395-66.421 (2.614-2.615) Dia.	66.015 (2.599) Dia.
PISTON SKIRT AND OTHER LANDS			66.269 (2.609) Dia.
5474 Piston (Skirt and Other Lands)	66.523-66.548 (2.619-2.620) Dia.	66.523-66.548 (2.619-2.620) Dia.	0.178 (0.007) Out of Round
PISTON RING END CLEARANCE			
3296 Oil Control Ring	End Clearance at 66.675 (2.625)	0.178-0.181 (0.007-0.15)	0.686 (0.027) or
5863 Compression Ring	Gage Dia.		0.305 (0.012)
3387 Optional Compression Ring			Actual Increase
PISTON RING SIDE CLEARANCE NO. 1 (TOP)			
5474 Piston (Groove Width)	2.476-2.502 (0.0975-0.0985)	0.102-0.1397 (0.004-0.0055)	
5863 Ring	2.362-2.375 (0.093-0.0935)	0.102-0.1397 (0.004-0.0055)	
3387 Ring (Optional)(Tungsten)	2.362-2.375 (0.093-0.0935)	0.102-0.1397 (0.004-0.0055)	
PISTON RING SIDE CLEARANCE NOS. 2, 3, AND 4			
5474 Piston (Groove Width)	2.451-2.477 (0.0965-0.0975)	0.076-0.1143 (0.003-0.0045)	0.1397 (0.0055)
5863 Ring	2.362-2.375 (0.093-0.0935)	0.076-0.1143 (0.003-0.0045)	
PISTON RING SIDE CLEARANCE NO. 5 (BOTTOM)			
5474 Piston (Groove Width)	4.788-4.813 (0.1885-0.1895)	0.050-0.089 (0.002-0.0035)	
3296 Ring	4.724-4.737 (0.186-0.1865)	0.050-0.089 (0.002-0.0035)	0.127 (0.005)
PISTON PIN TO PISTON CLEARANCE			
5474 Piston	25.4025-25.4152 (1.0001-1.0006) Bore	0.0025-0.0203 (0.0001-0.0008)	
5120 Piston Pin	25.0000-25.3949 (1.0000-0.9998) Dia.	0.0025-0.0203 (0.0001-0.0008)	0.0381 (0.0015)
PISTON PIN TO CONNECTING ROD BUSHING CLEARANCE			
5120 Piston Pin	25.4000-25.3949 (1.0000-0.9998)	0.00635-0.0305 (0.00025-0.0012)	
5487 Connecting Rod Assembly	25.40635-25.4254 (1.00025-1.001) Bore	0.00635-0.0305 (0.00025-0.0012)	0.1143 (0.0045)
CRANKSHAFT TO CONNECTING ROD BEARING CLEARANCE			
5641 Crankshaft Connecting Rod Journal	57.1373-57.1500 (2.2495-2.250) Dia.		
5445 Connecting Rod Bearings		0.076 (0.003)	0.127 (0.005)
CRANKSHAFT TO MAIN BEARINGS CLEARANCE			
5641 Crankshaft Main Bearing Journal	57.1373-57.1500 (2.2495-2.250) Dia.	0.76-0.1143 (0.003-0.0045)	
5642 Front Main Bearing	57.2262-57.2389 (2.253-2.2535) Bore	0.76-0.1143 (0.003-0.0045)	
5677 Rear Main Bearing	57.2262-57.2389 (2.253-2.2535) Bore	0.76-0.1143 (0.003-0.0045)	0.1397 (0.0055)
5681 Rear Main (Front) Bearing	(at assembly)	0.76-0.1143 (0.003-0.0045)	

MANUFACTURING TOLERANCES AND REPLACEMENT LIMITS (CONTINUED)

All dimensions not otherwise indicated are in millimeters
(All dimensions in parentheses are in inches)

		As Manufactured	Condemning
CRANKSHAFT END CLEARANCE			
5641 Crankshaft (Between Surfaces)	101.041-101.092 (3.978-3.900)		
5681 Rear (Front) Main Bearing (Flange Thickness)			
5642 Front Main Bearing (Flange Thickness)	(Fitted at Assembly)	0.254-0.356 (0.010-0.014)	0.635 (0.025)
CRANKSHAFT BEARINGS P.F. IN CRANKCASE AND REAR ADAPTER			
5601 Crankcase	69.8627-69.8754 (2.7505-2.751) Bore		
5605 Rear Bearing Adapter	69.8627-69.8754 (2.7505-2.751) Bore		
5642 Front Main Bearing	Grind OD to Fit at Assembly		
5677 Rear Main Bearing	Grind OD to Fit at Assembly	0.0005-0.001 (0.0127-0.0254) P.F.	
5681 Rear (Front) Main Bearing	Grind OD to Fit at Assembly		
CAMSHAFT TO BEARINGS			
5641 Camshaft - Front Journal	45.2120-45.2245 (1.780-1.7805) Dia.	0.0254-0.0635 (0.001-0.0025)	
Rear Jornal	25.3746-25.3873 (0.999-0.9995) Dia.	0.0254-0.0635 (0.001-0.0025)	
5051 Front Bearing	45.2501-45.2755 (1.7815-1.7825) Bore	0.0254-0.0635 (0.001-0.0025)	0.102 (0.004)
5646 Rear Bearing	25.4127-25.4381 (1.0005-1.0015) Bore	0.0254-0.0635 (0.001-0.0025)	
CAMSHAFT END PLAY			
5640 Camshaft (Journal Length)	42.799-42.849 (1.685-1.687)	0.102-0.203 (0.004-0.008)	0.254 (0.010)
5051 Front Bearing (O.A.L.)	42.646-42.697 (1.679-1.681)	0.102-0.203 (0.004-0.008)	
CAMSHAFT FRONT BEARING TO CRANKCASE			
5601 Crankcase	57.150-57.175 (2.250-2.251) Bore	0.0127-0.0508 (0.0005-0.002)	
5051 Front Bearing	57.124-57.137 (2.249-2.2495)	0.0127-0.0508 (0.0005-0.002)	
CAMSHAFT REAR BEARING TO CRANKCASE (5750 ENGINE)			
5601 Crankcase	31.750-31.775 (1.250-1.251) Bore	0.0127-0.0508 (0.0005-0.002)	
5646 Rear Bearing	31.725-31.7373 (1.249-1.2495) Dia.	0.0127-0.0508 (0.0005-0.002)	
MAGNETO BREAKER POINT GAP		0.355-0.559 (0.014-0.022)	
MAGNETO CAM FOLLOWER PRESSURE (WITH CAM ON DWELL)		0.83-2.2 N (3-8 oz)	
MAGNETO-PRESSURE BETWEEN POINTS WITH FOLLOWER ON DWELL OF CAM		5.0-7.0 N (18-25 oz)	
COUNTERBALANCE SHAFTS TO BUSHINGS			
5608 Counterbalance Shafts (Bushing Journal)	31.699-31.711 (1.248-1.2485) Dia.	0.051-0.0889 (0.002-0.0035)	
5610 Rear Bushings	31.763-31.788 (1.2505-1.2515) Bore	0.051-0.0889 (0.002-0.0035)	0.1143 (0.0045)
5647 Front Bushings	31.763-31.788 (1.2505-1.2515) Bore		

MANUFACTURING TOLERANCES AND REPLACEMENT LIMITS (CONTINUED)

All dimensions not otherwise indicated are in millimeters
(All dimensions in parentheses are in inches)

		As Manufactured	Condemning
COUNTERBALANCE SHAFT BUSHINGS TO CRANKCASE AND BEARING ADAPTER			
5601 Crankcase	38.0873-38.1127 (1.4995-1.5005) Bore	0.000-0.0889 (0.000-0.0035)	
5609 Bearing Adapter	38.0873-38.1127 (1.4995-1.5005) Bore	0.000-0.0889 (0.000-0.0035)	
5610 Rear Bushings	38.075-38.0873 (1.499-1.4995) OD	0.000-0.0889 (0.000-0.0035)	
5647 Front Bearings	38.075-38.0873 (1.499-1.4995) OD	0.000-0.0889 (0.000-0.0035)	
COUNTERBALANCE SHAFT END PLAY			
5608 Counterbalance Shaft (Shoulder Length)	55.093-55.219 (2.169-2.174)	0.076-0.279 (0.003-0.011)	
5609 Bearing Adapter (Length Through Bore)	50.800-50.749 (2.000-1.998)	0.076-0.279 (0.003-0.011)	0.508 (0.020)
5647 Front Bushing (Flange Thickness)	4.191-4.216 (0.165-0.166)	0.076-0.279 (0.003-0.011)	
5656 OIL PUMP ASSEMBLY (5750 ENGINE)			
Driveshaft to Bushing in Body at Drive End			
5651 Shaft	12.725-12.7381 (0.501-0.5015) Dia.	0.038-0.076 (0.0015-0.003)	0.127 (0.005)
5002 Bushing	12.776-12.802 (0.503-0.504) Dia.	0.038-0.076 (0.0015-0.003)	
Driveshaft to Bushing in Cover			
5651 Shaft	12.6619-12.6746 (0.4985-0.499) Dia.	0.051-0.076 (0.002-0.003)	0.127 (0.005)
5652 Bushing	12.725-12.738 (0.501-0.5015) Bore	0.051-0.076 (0.002-0.003)	
Idler Shaft to Bushings			
5782 Shaft	12.6619-12.6746 (0.4985-0.499) Dia.	0.051-0.076 (0.002-0.003)	0.127 (0.005)
5652 Bushings	12.725-12.7381 (0.501-0.5015) Bore	0.051-0.076 (0.002-0.003)	
Bushings P.F. into Body and Cover			
Housing Bores	15.8623-15.8877 (0.6245-0.6255) Bore	0.0381-0.0889 (0.0015-0.0035) P.F.	
Bushing OD	15.925-15.951 (0.627-0.628) OD	0.031-0.0889 (0.0015-0.0035) P.F.	
Pump Gears Backlash		0.076-0.127 (0.003-0.005)	0.178 (0.007)

APPENDIX B
STANDARD SPARK PLUG INSERTS

B.1 See Table B1.

TABLE B1—STANDARD SPARK PLUG INSERTS

Spark Plug Size Thread	Spark Plug Size Reach mm (in)	Labeco Part Numbers (Std.) Meehanite	Labeco Part Numbers (Obs.) Stainless
8 mm	12.497 (0.492)	20001	
10 mm	6.350 (0.250)	16200	16151A
10 mm	12.497 (0.492)	16201	16163A
10 mm	17.780 (0.700)	16202	16166
12 mm	12.497 (0.492)	16204	16165A
12 mm	19.050 (0.750)	16205	16167
14 mm	9.525 (0.375)	16206	16152A
14 mm	11.100 (0.437)	16207	16153A
14 mm	12.700 (0.500)	16208	16154A
14 mm	11.684 (0.460) (conical seat)	16209	16171
14 mm	19.050 (0.750)	16210	16155A
14 mm	19.050 (0.750) (half thread)	16211	16156
14 mm	17.780 (0.700) (conical seat)	16219	
14 mm	17.272 (0.680)	16221	
14 mm	9.525 (0.375) (conical seat)	16223	
14 mm	17.780 (0.700) (conical seat) (1/2 Thread—Ford)	16232	
14 mm	23.241 (0.915) (conical seat) (1/2 Thread—Autolite)	16233	
14 mm	12.700 (0.500)	16154A	
18 mm	11.303 (0.445)	16212	16168
18 mm	12.700 (0.500)	16213	16157A
18 mm	11.684 (0.460) (conical seat)	16214	16158
18 mm	20.625 (0.812)	15215	16159A
18 mm	25.400 (1.000) (Special)	16226	
18 mm	29.718 (1.125)	16222	
18 mm	38.100 (1.500) (conical seat)	16231	
18 mm	29.718 (1.125) (3/16 thd relief)	16220	
0.875 in	15.900 (0.625)	16216	16160
0.875 in	20.625 (0.812)	16217	16161
0.875 in	20.625 (0.812) (Special)	16218	16162

APPENDIX C
SPARK PLUG INSTALLATION TORQUE

C.1 See Table C1.

TABLE C1—SPARK PLUG INSTALLATION TORQUE

Plug Size	Torque N·m (lb-ft)
10 mm	10-15 (7-11)
12 mm	15-25 (11-18)
14 mm	35-40 (26-30)
14 mm (conical seat)	9-20 (7-15)
18 mm	43-52 (32-38)
18 mm (conical seat)	20-27 (15-20)
0.875 in × 18	47-58 (35-43)

APPENDIX D
BILL OF MATERIAL FOR
5750 ENGINE ASSEMBLY, SAE SPARK PLUG
RATING W/INSERT TYPE HEAD

D.1 Control Items
REQUIRED
*SAE Control Item

Qty	Part No.	Description
1	5600	CRANKCASE ASSEMBLY, COUNTERBALANCED
1	5660	RETAINER, CRANKSHAFT OIL (REAR) ASSEMBLY (REAR) N. D.
1	5653	RETAINER, CRANKSHAFT OIL (REAR)
1	5661	SEAL OIL
1	5630	BEARING ADAPTER ASSEMBLY COUNTERBALANCE SHAFTS
1	5609	BEARING ADAPTER FOR COUNTERBALANCED SHAFT
2	5647	BUSHING, FRONT C-B SHAFT
2	4243	SCREW, SET, HEX SOCKET HEAD 0.375-16 UNC X 0.500 LONG CUP POINT
2	4260	PIN, DOWEL 0.1251/0.1253 DIA. X 0.375 LONG
1	5654	CAMSHAFT ASSEMBLY
1	5640	CAMSHAFT
1	5052	GEAR TIMING CAMSHAFT
1	5051	BEARING BUSHING CAMSHAFT FRONT
1	5053	NUT CAMSHAFT FRONT GEAR TO CAMSHAFT
1	4051	KEY, WOODRUFF 0.156 X 0.750 SAE #8, ANSI (506)
1	5054	WASHER, LOCK, GEAR TO CAMSHAFT
1	4076	KEY, WOODRUFF 0.250 X 0.875 SAE #A, ANSI (807)
1	5631	COVER, GEAR CASE ASSEMBLY
1	5602	COVER, GEAR CASE
1	5457	SEAL, OIL, FRONT CAMSHAFT
1	5111	FILLER CAP, WING TYPE
1	5112	GASKET FOR FILLER CAP
1	5661	SEAL OIL
4	5634	STUD, OIL PUMP TO CASE
1	4246	SCREW, SET, HEX SOCKET HEAD 0.375-16 UNC X 2.625 LONG CUP POINT
1	4247	SCREW, SET, HEX SOCKET HEAD 0.375-16 UNC X 1.000 LONG CUP POINT
1	5635	CRANKCASE SUB-ASSEMBLY
1	5601	CRANKCASE
1	5605	BEARING ADAPTER CRANKSHAFT REAR
1	5642	BEARING FRONT MAIN
1	5677	BEARING, CRANKSHAFT, REAR ADAPTER
1	5681	BEARING, CRANKSHAFT, FRONT ADAPTER
3	4329	PIN, TAPER, #3 X 2.500 LONG
15	35105	SCREW, CAP, HEX HEAD, 0.375-16 UNC X 1.250 LONG, GRADE 5 CADMIUM PLATED
15	4111	WASHER, PLAIN 0.375 0.391 I. D. X 0.625 O. D. X 0.062 THICK, CADMIUM PLATED AN960-616
2	4243	SCREW, SET, HEX SOCKET HEAD 0.375-16 UNC X 0.500 LONG CUP POINT
1	4060	FITTING, TUBE, INV FLARE, MALE 90 ELBOW 0.250 NPT X 0.375 TUBE BRASS WEATHERHEAD #402X6
1	5621	CRANKSHAFT ASSEMBLY
1	5641	CRANKSHAFT
2	5639	COUNTERWEIGHT— CRANKSHAFT
4	5019	CAPSCREW COUNTERWEIGHT TO CRANKSHAFT
1	4000	PIPE PLUG, SOCKET HEAD, STEEL 0.125 PT BLACK OXIDE FINISH
2	S11205	WIRE, SAFETY, 0.032 DIAMETER STAINLESS STEEL, MS20995C32
1	5673	OIL OUTLET FLANGE ASSEMBLY
1	5664	HEATER TERMINAL PLATE ASSEMBLY
1	5665	PLATE HEATER TERMINAL
1	5685	COVER HEATER TERMINAL PLATE (REAR)
4	4038	SCREW, MACHINE, FLAT HEAD #10-32 UNF X 0.750 LONG, PLATED
4	4030	WASHER, PLAIN #10 0.219 I. D. X 0.500 O. D. X 0.049 THICK, ZINC PLATED
4	4031	NUT, HEX, MACHINE SCREW #10-32 UNF, ZINC PLATED
1	5697	CONNECTOR TERMINAL
15	37729	TERMINAL, SOLDERLESS, RING, 16-14 AWG X #10 STUD
1	5696	COVER, HEATER TERMINAL PLATE FRONT
2	4179	SCREW, MACHINE, FLAT HEAD #6-32 UNC X 0.500 LONG 18-8 STAINLESS STEEL
2	5698	SEPARATOR, WIRE
1	200003-1	CONTROL ASSEMBLY, OIL SUMP HEATER
1	200005-1	ENCLOSURE, SWITCH, OIL SUMP HEATER
1	200006-1	COVER, SWITCH ENCLOSURE, OIL SUMP HEATER
1	200007-1	NAMEPLATE, SWITCH, OIL SUMP HEATER
1	37547	SWITCH, SELECTOR
1	37548	JUMPER, SELECTOR SWITCH
2	30741	SCREW, MACHINE, ROUND HEAD #10-32 UNF X 0.250 LONG, PLATED
1	5549	RECEPTACLE SUMP HEATER
4	35602	SCREW, CAP, HEX SOCKET HEAD #4-40 UNC X 0.313 LONG
1	5656	OIL PUMP ASSEMBLY
1	5649	BODY, OIL PUMP
1	5650	COVER, OIL PUMP
1	5651	SHAFT, OIL PUMP DRIVE
1	5782	SHAFT IDLER, OIL PUMP
1	5761	GEAR, OIL PUMP DRIVER
1	5762	GEAR, OIL PUMP DRIVEN
1	5643	COUPLING FLANGE
3	5652	BUSHING OIL PUMP COVER
1	5082	BUSHING OIL PUMP MAIN SHAFT
6	35105	SCREW, CAP, HEX HEAD 0.375-16 UNC X 1.250 LONG, GRADE 5 CADMIUM PLATED
6	4111	WASHER, PLAIN 0.375 0.391 I. D. X 0.625 O. D. X 0.062 THICK, CADMIUM

Qty	Part No.	Description
		PLATED AN960-616
2	4026	KEY, WOODRUFF 0.094 X 0.500 SAE #2 ANSI (304)
2	4054	PIN, DOWEL 0.2501/0.2503 DIA. X 0.625 LONG
1	5519	OIL PRESSURE REGULATOR ASSEMBLY A. C. SPEC
1	5923	HOUSING, RELIEF VALVE PRESSURE REGULATOR
1	5924	VALVE, RELIEF VALVE PRESSURE REGULATOR
1	5925	BOLT—RELIEF VALVE PRESSURE REGULATOR
1	5726	SCREW, OIL PRESSURE ADJUSTING
1	5938	SPRING—RELIEF VALVE, PRESSURE REGULATOR-SAE 17.6 3-7-62
2	4956	GASKET COPPER 0.750 I. D. 0.062 THICK SAME AS 5103 ("MCCORD" #511-A)
1	4339	NUT, HEX, LOCK, ELASTIC STOP
1	4340	GASKET 5/8 I. D. 7/8 O. D. COPPER CLAD ASBESTOS
1	5926	PLUG, RELIEF VALVE PRESSURE REGULATOR
8	35651	SCREW, CAP, HEX HEAD 0.313-18 UNC X 1.250 LONG GRADE 5 PLATED
6	35105	SCREW, CAP, HEX HEAD 0.375-16 UNC X 1.250 LONG GRADE 5 CADMIUM PLATED
3	4022	KEY, WOODRUFF 0.250 X 1.125 SAE #18, ANSI (809)
4	4030	WASHER, PLAIN #10 0.219 I. D. X 0.500 O. D. X 0.049 THICK, ZINC PLATED
4	4037	NUT, HEX 0.375-24 UNF ZINC PLATED
2	35066	SCREW, CAP, HEX HEAD 0.375-16 UNC X 0.750 LONG GRADE 5 CADMIUM PLATE
39	35024	SCREW, CAP, HEX HEAD 0.375-16 UNC X 0.875 LONG GRADE 5 CADMIUM PLATED
2	4047	WASHER, PLAIN 0.313 0.328 I. D. X 0.562 O. D. X 0.062 THICK, CADMIUM PLATED AN960-516
3	4051	KEY, WOODRUFF 0.156 X 0.750 SAE #8, ANSI (506)
8	35610	SCREW, CAP, HEX HEAD 0.250-20 UNC X 0.875 LONG GRADE 5 PLATED
74	4111	WASHER, PLAIN 0.375 0.391 I. D. X 0.625 O. D. X 0.062 THICK CADMIUM PLATED AN960-616
4	4129	SCREW, MACHINE, FILLISTER HEAD #10-32 UNF X 0.500 LONG CADMIUM PLATED
6	35106	SCREW, CAP, HEX HEAD 0.375-16 UNC X 1.750 LONG GRADE 5 CADMIUM PLATED
11	35153	SCREW, CAP, HEX HEAD 0.375-16 UNC X 3.250 LONG GRADE 5 CADMIUM PLATE
2	4256	SCREW, CAP, HEX SOCKET HEAD 0.313-18 UNC X 1.000 LONG
1	4267	SCREW, SET, HEX SOCKET HEAD #10-32 UNF X 0.375 LONG CUP POINT
4	4331	SCREW, MACHINE, FILLISTER HEAD #10-32 UNF X 1.250 LONG PLATED (AN501-10-20)
1	4556	VALVE, RELIEF, PESCO PROD. #3V, 195, SET FOR 2 in (25.8 mm) Hg
2	4861	SIGHT GAUGE, OIL
1	5013	GASKET FOR 5163 TO 5002 CYL. BARREL TO CRANKCASE
1	5023	WASHER, LOCK, FLYWHEEL NUT
1	5027	GEAR CRANKSHAFT TIMING
2	5035	WASHER CLAMP, CRANKSHAFT TIMING DISC
2	5042	NUT, LOCK (1 in to 20) CRANKSHAFT FRONT
1	5071	POINTER FLYWHEEL
1	5076	NUT, FLYWHEEL, HEX, 1.500—18
1	5603	PLATE, CRANKCASE COVER
2	5604	GASKET, CRANKCASE COVER PLATE
2	5608	SHAFT, COUNTERBALANCE
1	5611	SPROCKET CRANKSHAFT
2	5612	SPROCKET COUNTERBALANCE SHAFT
1	5613	SPROCKET, IDLER
1	5614	BUSHING, IDLER SPROCKET
1	5615	BOLT, IDLER SPROCKET BUSHING
1	5616	CHAIN, TIMING
1	5629	GASKET, GEAR CASE
4	5632	WASHER FOR REMOVING HEAT UNIT
4	5638	HEATER CARTRIDGE, OIL SUMP
1	5644	COUPLING CONNECTOR
1	5645	NUT, HEX DRIVING
1	5648-A	SLEEVE TIMING DISC SPACER
1	5670	KEY STRAIGHT SPECIAL FLYWHEEL TO CRANKSHAFT
2	5680	COVER, INSPECTION
2	5686	FLANGE BREATHER
2	5687	GASKET BREATHER FLANGE
1	5691	GASKET, REAR BEARING ADAPTER TO CASE
1	5692	GASKET, OIL PUMP TO COVER
1	5693	GASKET, OIL FILTER FLANGE
1	5694	GASKET, INSPECTION COVER TO CASE
1	5695	COVER PLATE DRILLED AND TAPPED FOR RELIEF VALVE
2	5837	BAFFLE, CRANKCASE, COVER
4	5838	SPACER, CRANKCASE COVER BAFFLE
6	35025	SCREW, CAP, HEX HEAD 0.375-16 UNC X 1.000 LONG GRADE 5 CADMIUM PLATED
3	4000	PIPE PLUG, SOCKET HEAD, STEEL 0.125 NPT BLACK OXIDE FINISH
1	4001	PIPE PLUG, SOCKET HEAD, STEEL 0.250 NPT BLACK OXIDE FINISH
1	5015	GASKET FOR 5014 TO 5008 RETAINER TO ADAPTER
2	36871	PIPE NIPPLE 0.250 NPTF X 0.375 NPTF WEATHERHEAD C3069X6X4
2	5610	BUSHING, REAR COUNTERBALANCE SHAFT
1	5646	BUSHING REAR CAMSHAFT
6	5003	STUD (CYLINDER TO CRANKCASE) CRANKCASE TO BARREL
4	4243	SCREW, SET, HEX SOCKET HEAD 0.375-16 UNC X 0.500 LONG CUP POINT
2	4002	PIPE PLUG, SOCKET HEAD, STEEL 0.500 NPT

		BLACK OXIDE FINISH
3	4318	SCREW, CAP, HEX SOCKET HEAD #10-32 UNF X 0.375 LONG
2	5500	VALVE LIFTER ASSEMBLY
1	5502	VALVE LIFTER, BODY
1	5503	VALVE LIFTER ROLLER
1	5504	VALVE LIFTER ROLLER PIN
1	5506	VALVE LIFTER PUSH ROD SOCKET
2	35100	SCREW, CAP, HEX HEAD 0.313-18 UNC X 1.000 LONG GRADE 5 CADMIUM PLATED
2	5501	VALVE LIFTER GUIDE
3	4242	PIN, DOWEL 0.3751/0.3753 DIA. X 1.000 LONG
*1	16018	CYLINDER HEAD ASSEMBLY, INSERT TYPE HEAD— SAE SPARK PLUG RATING ENGINE/SUB-
1	16001	CYLINDER HEAD, INSERTED TYPE
1	5136	VALVE SEAT INSERT INTAKE
1	5137	VALVE SEAT INSERT EXHAUST
1	5134	GUIDE, VALVE INTAKE
1	5135	GUIDE, VALVE EXHAUST
8	5138	STUD PORT FLANGE
1	16006	HOUSING, PUSH ROD, UPPER EXHAUST
1	16007	HOUSING, PUSH ROD, UPPER INTAKE
6	16013	STUD, 0.313 X 1.250, PLUG INSERT
10	16014	STUD, 0.250 X 1.250, ROCKER BOX
1	16020	ROCKER ARM, EXHAUST ASSEMBLY
1	5211	ROLLER, ROCKER ARMS
1	5212	HUB, ROCKER ARM ROLLER
1	5213	PIN, ROCKER ARM ROLLER AND HUB
1	5214	SCREW, ADJUSTING, ROCKER ARM
1	5215	LOCK FOR ROCKER ARM ADJUSTING SCREW
2	5216	BEARING, NEEDLE
1	16002	ROCKER ARM, EXHAUST
1	16021	ROCKER ARM, INTAKE ASSEMBLY
1	5211	ROLLER, ROCKER ARMS
1	5212	HUB, ROCKER ARM ROLLER
1	5213	PIN, ROCKER ARM ROLLER AND HUB
1	5214	SCREW, ADJUSTING, ROCKER ARM
1	5215	LOCK FOR ROCKER ARM ADJUSTING SCREW
2	5216	BEARING, NEEDLE
1	16003	ROCKER ARM, INTAKE, INSERTED HEAD
2	16011	PUSH ROD, ASSEMBLY— INSERTED HEAD
2	16004	COVER, ROCKER BOX— INSERTED HEAD
2	16005	GASKET, ROCKER BOX— INSERTED HEAD
1	5143	GASKET FOR 5142 TO 5133
2	5223	WASHER, THRUST
1	5524	ADAPTER, COOLANT
2	5227	RETAINER, VALVE SPRING LOWER
2	5230	SPRING VALVE INNER
2	5231	SPRING VALVE OUTER
4	3864	PLUG, DRAIN 0.625-18 THD.
6	4015	NUT, HEX 0.313-24 UNF ZINC PLATED
2	4111	WASHER, PLAIN 0.375 0.391 I. D. X 0.625 O. D. X 0.062 THICK CADMIUM PLATED AN960-616
2	35106	SCREW, CAP, HEX HEAD 0.375-16 UNC X 1.750 LONG GRADE 5 CADMIUM PLATED
8	35133	SCREW, CAP, HEX HEAD 0.313-18 UNC X 0.625 LONG GRADE 5 CADMIUM PLATE
10	4319	NUT, HEX 0.250-28 UNF ZINC PLATED
10	4341	WASHER, PLAIN 0.250 0.265 I. D. X 0.500 OC. D. X 0.063 THICK CADMIUM PLATE AN960-416
2	5574	ROCKER ARM, SHAFT
4	5575	PLATE ROCKER ARM SHAFT
4	5576	GASKET, ROCKER ARM SHAFT RETAINING PLATE
2	5831	SPACKER, VALVE SPRING
2	5832	KEY, VALVE SPRING RETAINING
2	5833	CAP, VALVE SPRING RETAINING
2	5834	RETAINER, VALVE SPRING (UPPER)
1	5835	VALVE, INTAKE
1	5836-2	VALVE, EXHAUST
1	36604	O-RING, 2-219, 1,296 I. D. X 0.139 W. RED SILICONE RUBBER
1	31328	O-RING, 2-228, 2.228 I. D. X 0.139 W.
1	16034-C	THERMOCOUPLE, INSERTED TYPE CYL. HD.
1	16048	THERMOCOUPLE GASKET
1	5587	CYLINDER HOUSING ASSEMBLY
2	5164	HOUSING LOWER FOR PUSH RODS IN CYLINDER BARREL
1	5544	HOUSING CYLINDER
2	35687	SCREW, CAP, HEX HEAD 0.375-16 UNC X 2.000 LONG GRADE 5 PLATED
2	35024	SCREW, CAP, HEX HEAD 0.375-16 UNC X 0.875 LONG GRADE 5 CADMIUM PLATED
4	4111	WASHER, PLAIN 0.375 0.391 I. D. X 0.625 O. D. X 0.062 THICK CADMIUM PLATED AN960-616
2	4293	SCREW, CAP, HEX SOCKET HEAD 0.375-16 UNC X 0.750 LONG
1	4359	GASKET, 3-7/16 I. D. 3-11/16 O. D. HcKIM #160 (SLEEVE TO CYL HOUSING)
1	5148	GASKET CYLINDER HEAD TO CYLINDER BARREL ("HEAD TO SLEEVE")
2	5207	HOSE FOR PUSH ROD HOUSING
4	3351	CLAMP, HOSE #16, 0.813 TO 1.500 CLAMP DIA
10	5546	CAPSCREW CYLINDER BARREL TO HEAD
2	5566	GASKET UPPER AND LOWER FOR 5567
1	5569	ADAPTER COOLANT INLET
10	5577	GASKET, COPPER 1/2 in
2	5578	GASKET, COPPER 3/8 in
1	5579	PLATE FOR 5544
2	5581	GASKET, CRANKSHAFT OIL RETAINER
*1	5740	SLEEVE, CYLINDER REVISED 10-3-60
1	5741	FLANGE, CYLINDER SLEEVE
*1	5939	PISTON ASSEMBLY
1	5474	PISTON & SODIUM CHAMBER ASSEMBLY

Qty	Part No.	Description
1	5120	PISTON PIN (SOLID TYPE)
1	3296	RING, PISTON, OIL CONTROL (0.187 WIDE)
4	5863	RING, PISTON, CHROME (0.094 WIDE)
6	5177	NUT, FLANGED 1/2-20; CLASS 5; HI-TEMP, 50 FT/LBS TORQUE MIN
2	5187	GASKET FOR PORT FLANGES
1	5233	EXHAUST PIPE ASSEMBLY WITH COOLANT JACKET
2	5303	FLANGE FOR EXHAUST PIPE ASSEMBLY WITH COOLANT JACKET
2	5304	SPACER FOR EXHAUST PIPE ASSEMBLY WITH COOLANT JACKET
1	5305	JACKET FOR EXHAUST PIPE ASSEMBLY WITH COOLANT JACKET
1	5306	PIPE FOR EXHAUST PIPE ASSEMBLY WITH COOLANT JACKET
2	5234	BOSS 3/8 in NPT FOR WELDING
1	17050	NAMEPLATE, LABECO—FOR GENERAL USE
1	5487	CONNECTING ROD ASSEMBLY
1	5472	CONNECTING ROD
2	5448	BOLT CONNECTING ROD
2	5117	NUT FOR CONNECTING ROD BOLT
2	4028	PIN, COTTER 0.094 DIA X 0.750 LONG
1	5445	SILVER GRID LINERS, HALF
1	5118	BUSHING, CONNECTING ROD, PISTON PIN END
*1	5496	INTAKE ASSEMBLY
*1	3128	INJECTION NOZZLE (BOSCH #ADN 12 SD12)
3	4002	PIPE PLUG, SOCKET HEAD, STEEL 0.500 NPT BLACK OXIDE FINISH
2	35100	SCREW, CAP, HEX HEAD 0.313-18 UNC X 1.000 LONG GRADE 5 CADMIUM PLATED
8	4015	NUT, HEX 0.313-24 UNF ZINC PLATED
2	36426	WASHER, LOCK HELICAL SPRING 0.313 MEDIUM PLATED
1	5179	AIR RECEIVER ASSEMBLY
1	5495	AIR RECEIVER SUB-ASSEMBLY
1	5180	AIR RECEIVER
6	5181	STUD FOR AIR RECEIVER COVER
10	5182	STUD FOR AIR RECEIVER FLANGES
1	4003	PIPE PLUG, SOCKET HEAD, STEEL 0.750 NPT BLACK OXIDE FINISH
1	4000	PIPE PLUG, SOCKET HEAD, STEEL 0.125 NPT BLACK OXIDE FINISH
1	5186	GASKET, AIR RECEIVER TO AIR RECEIVER COVER
1	5361	SCREEN ASSEMBLY
10	4015	NUT, HEX 0.313-24 UNF ZINC PLATED
1	5185	GASKET, AIR RECEIVER TO AIR RECEIVER CONNECTION FLANGE
1	5184	CONNECTION FLANGE AIR FOR AIR RECEIVER
1	5190	STANDPIPE FOR AIR RECEIVER
2	3386	THERMOCOUPLE, IRON CONSTANTAN
2	3191	THERMOCOUPLE CONNECTOR, FEMALE, CONSTANTAN "J"
1	5188	GASKET FOR INTAKE PIPE TO AIR RECEIVER ASSEMBLY
1	5189	PIPE INTAKE
1	5192	GASKET FOR 5193 TO 5347
1	5194	FLANGE FOR FUEL INJECTOR NOZZLE HOLDER ASSEMBLY
2	5195	STUD FOR NOZZLE HOLDER
*1	5347	NOZZLE HOLDER ASSEMBLY (BOSCH #AKB50S6777A OPENING PRESSURE Æ 1200 TO 1250 P. S. I.)
1	5493	BLANK TO COVER NOZZLE HOLDER FLANGE
1	5494	GASKET FOR BLANK COVER, NOZZLE HOLDER FLANGE
1	4000	PIPE PLUG, SOCKET HEAD, STEEL 0.125 NPT BLACK OXIDE FINISH
4	4099	SCREW, SELF-TAPPING, DRIVE #4 X 0.188 LONG TYPE "U" PLATED
8	5028	NUT, HEX, HIGH (BRASS) 0.375-16
1	5590	FLYWHEEL ASSEMBLY
1	5591	FLYWHEEL
1	5592	HUB, FLYWHEEL
6	35125	SCREW, HEX HEAD 0.500-13 UNC X 1.500 LONG GRADE 5
1	4322	TAPER PIN, #9 X 1.500 LONG
6	4362	WASHER, PLAIN 0.500 0.515 I. D. X 0.875 O. D. X 0.062 THICK CADMIUM PLATED AD960-816
1	5235	EXHAUST HOSE NOZZLE WELDMENT
1	5259	FUEL COOLER ASSEMBLY
1	5416	FUEL TUBE ASSEMBLY
1	5947	PIEZOMETER (EXHAUST BACK PRESSURE PICKUP)
1	5300	BRACKET MOUNTING MAGNETO & FUEL PUMP
4	4067	SCREW, CAP, HEX SOCKET HEAD 0.375-16 UNC X 0.875 LONG
8	35667	SCREW, CAP, HEX HEAD 0.313-24 UNF X 0.750 LONG GRADE 5 PLATED
1	5416	TUBE ASSEMBLY FUEL SHORT
2	5781	SPACER, FUEL PUMP AND MAGNETO MOUNTING BRACKET
4	35724	SCREW, CAP, HEX HEAD 0.500-13 UNC X 4.500 LONG GRADE 5 PLATED
1	4051	KEY, WOODRUFF 0.156 X 0.750 SAE #8 ANSI (506)
2	35150	SCREW, CAP, HEX HEAD 0.375-16 UNC X 0.500 LONG GRADE 5 CADMIUM PLATE
1	201016-1	COVER PLATE ASSEMBLY, MOUNTING BRACKET
6	3904	SCREW, MACHINE, ROUND HEAD #10-32 UNF X 0.500 LONG CADMIUM PLATED
6	36372	WASHER, LOCK, HELICAL SPRING #10 MEDIUM PLATED
1	202041	PUMP ASSEMBLY, FUEL INJECTION
1	5345	PUMP FUEL METERING, AMERICAN BOSCH #APE 1B-70P-300/3
1	4060	FITTING, TUBE, INV FLARE, MALE 90 ELBOW 0.250 NPT X 0.375 TUBE BRASS WEATHERHEAD #402X6
1	4023	NUT, TUBE, INV FLARE 0.375 TUBE STEEL WEATHERHEAD #105X6
1	5301	PLATE MOUNTING METERING PUMP

Qty	Part No.	Description
4	4065	SCREW, MACHINE, FLAT HEAD 0.375-16 UNC X 0.750 LONG, PLATED
1	4113	WASHER, LOCK, INTERNAL TOOTH 0.563 PLATED
1	5370	NUT, HEX SPECIAL FOR COUPLING BOSCH #NMU-2024/1X
1	202038	BRACKET ASSEMBLY, FUEL CONTROL
1	202039	TRIGGER, FUEL CONTROL
1	202040	BRACKET, FUEL CONTROL
1	4085	SCREW, CAP, HEX SOCKET HEAD #10-32 UNF X 0.500 LONG
1	5978-1	SUPPORT SHAFT FOR FUEL MICROMETER - (FUEL INJECTOR PUMP)
1	5978-2	FITTING FOR FUEL MICROMETER SUPPORT SHAFT
1	3005	BUSHING 0.250 I.P.T. X 0.750 I.P.T. BOSCH #WRV/2A1X
1	5978-3	SHAFT FOR FUEL MICROMETER SUPPORT SHAFT
1	202038*005	PIN, CLEVIS 0.250 DIA. X 0.750 LONG
1	4080	PIN, COTTER 0.094 DIA X 1.250 LONG
1	202038*007	SPRING
1	5409	COVER BLIND END FOR "BOSCH" PUMP
4	5413	SCREW, MACHINE, OVAL HEAD, AMERICAN BOSCH #NSR 734/27X
1	5790*020	MAGNETO BENDIX 10-518501-25
1	5790*021	COIL BENDIX 10-382080-1
1	5790*022	CONNECTOR KIT BENDIX
1	5404	COUPLING ASSEMBLY CAMSHAFT TO BOSCH PUMP
1	5394	COUPLING SPACER
2	5369	COUPLING, FLEXIBLE DISC ASSEMBLY
1	5372	FLANGE COUPLING TO CAMSHAFT
1	5371	FLANGE COUPLING TO "BOSCH" METERING PUMP
1	5418	FLANGE, ADJUSTABLE
2	35634	SCREW, CAP, HEX HEAD 0.250-28 UNF X 1.250 LONG GRADE 5 PLATED
2	35639	SCREW, CAP, HEX HEAD 0.250-28 UNF X 2.500 LONG GRADE 5 PLATED
6	4537	NUT, HEX, LOCK ELASTIC STOP 0.250-28 CADMIUM PLATED STEEL
2	4341	WASHER, PLAIN 0.250 0.265 I.D. X 0.500 OC.D. X 0.063 THICK CADMIUM PLATED AN960-416
2	35635	SCREW, CAP, HEX HEAD 0.250-28 UNF X 1.500 LONG GRADE 5 PLATED
2	4090	KEY, WOODRUFF 0.156 X 0.625 SAE #6, ANSI (505)
1	207001	INJECTOR PUMP-ACTUATOR ASSEMBLY
1	207003	BRACKET WELDMENT
1	207004	ROD, ACTUATOR
1	207005	ROD, ACTUATOR-BUSHING HOLDER
1	207006	BUSHING
1	207001*005	SPRING, TENSION
2	3552	NUT, HEX 0.250-20 UNC BRASS
1	201019	COVER INJECTOR PUMP ACTUATOR
1	201020	SPACER, SOLENOID
1	201021	ARM
1	201027	ARM, ACTUATOR
1	17050	NAMEPLATE, LABECO
1	207008	PIN
1	207002*01301	MOTOR, STEPPING, 12 VDC
1	35323	SOLENOID
1	207001*015	COVER
1	31310	TERMINAL BLOCK
1	207001*017	ROLL PIN, 1/4 DIA X 0.75 LG.
1	207001*018	ROLL PIN, 1/16 DIA. X 0.44 LG.
2	35325	SCREW, SHOULDER 0.250 DIA. X 0.375 LONG PIC 4330
12	32134	SCREW, MACHINE, ROUND HEAD #4-40 UNC X 0.250 LONG PLATED
4	31424	SCREW, MACHINE, ROUND HEAD #6-32 UNC X 0.750 LONG PLATED
3	3765	SCREW, MACHINE, ROUND HEAD #6-32 UNC X 0.250 LONG PLATED
4	4099	SCREW, SELF-TAPPING, DRIVE #4 X 0.188 LONG TYPE "U" PLATED
2	30097	SCREW, SET, HEX SOCKET HEAD #8-32 UNC X 0.375 LONG, CUT POINT
6	31306	SCREW, MACHINE, ROUND HEAD #8-32 UNC X 0.375 LONG PLATED
2	207001*026	SCREW, HEX HEAD, 1/4-20 UNC X 1.50 LG., BRASS
4	4804	SCREW, CAP, HEX SOCKET HEAD #8-32 UNC X 0.625 LONG
4	3547	WASHER, PLAIN #8 0.188 I.D. X 0.438 O.D. X 0.049 THICK, ZINC PLATED
7	3176	WASHER, LOCK, HELICAL SPRING #6 MEDIUM PLATED
6	36371	WASHER, LOCK, HELICAL SPRING #8 MEDIUM PLATED
2	4522	PIN, COTTER 0.032 DIA X 0.750 LONG
2	4826	NUT, HEX, LOCK, ELASTIC STOP #8-32 UNC PLATED
1	20013-1	WIRE ASSEMBLY, SPARK PLUG, 17.6 ENGINE
1	5403-2	COUPLING ASSEMBLY, CRANKSHAFT TO BENDIX #10-518501-25 MAGNETO
1	5394	COUPLING SPACER FOR "MORFLEX"
2	5369	COUPLING, FLEXIBLE DISC ASSEMBLY "MORFLEX" #302 (CENTER MEMBER ONLY)
1	5380	FLANGE COUPLING TO CRANKSHAFT ("MORSE" *302 X 1/2 in BORE—BLANK)
2	35634	SCREW, CAP, HEX HEAD 0.250-28 UNF X 1.250 LONG GRADE 5 PLATED
2	35639	SCREW, CAP, HEX HEAD 0.250-28 UNF X 2.500 LONG GRADE 5 PLATED
6	5403-2*008	NUT, HEX, THIN 0.250-38 UNF
2	4341	WASHER, PLAIN 0.250 0.265 I.D. X 0.500 O.D. X 0.063 THICK CADMIUM PLATED AN960-416
2	35635	SCREW, CAP, HEX HEAD 0.250-28 UNF X 1.500 LONG GRADE 5 PLATED

1	207015	FLANGE, COUPLING			X 1.000 LONG GRADE 5 CADMIUM PLATED
1	5801	OIL COOLER ASSEMBLY	10	4814	SCREW, MACHINE, ROUND HEAD 0.250-20 UNC X 0.500 LONG PLATED
1	5360	MAIN JACKET WELDMENT			
5	5313	JACKET, FLANGE			
1	5077	BASE PLATE			
1	5236	INTERMEDIATE TUBE WELDMENT	6	4248	NUT, HEX 0.375-16 UNC ZINC PLATED
1	5733-1	EXTENSION FLANGE WELDMENT	14	36375	WASHER, LOCK, HELICAL SPRING 0.375 MEDIUM PLATED
1	5362	CORE WELDMENT	10	36374	WASHER, LOCK, HELICAL SPRING 0.250 MEDIUM PLATED
1	5270	TAP WELDMENT			
6	35700	SCREW, HEX HEAD, CAP 0.375-24 UNF X 0.875 LONG GRADE 5	2	205010-1*038	VALVE, BALL SHUT OFF, 1/4 NPT
			3	35283	VALVE, INK BLEEDER BOSCH #F-I-7621
			1	31993	TWIST LOCK RECEPTACLE
12	35702	SCREW, HEX HEAD, CAP 0.375-24 UNF X 1.750 LONG GRADE 5	1	31994	TWIST LOCK PLUG
			1	3418	VALVE, GLOBE 0.500 NPT BRASS
			1	3395	VALVE, GLOBE, ANGLE 0.375 NPT BRONZE
18	4111	WASHER, PLAIN 0.375 0.391 I. D. X 0.625 O. D.	1	200012-1	COVER, THERMOCOUPLE JUNCTION BOX
12	4037	NUT, HEX 0.375 X 24 UNF STEEL, ZINC PLATED	1	32827	COVER FOR FS PYLET W/GASKET
3	4108	PIPE PLUG, SQUARE HEAD 0.375 NPT GALVANIZED STEEL	1	205010-1*084	STRAINER, PIPE, 3/4 in, MASONELILAN INT'L #16
			1	3500	VALVE, NEEDLE 0.375 NPT BRASS
2	5806	PIPE PLUG (SPECIAL)	1	205010-1*096	PUMP & MOTOR, CENTRIFUGAL—60 Hz, 1750 RPM, 1/3 HP, 115/230V, 1 PHASE, ALL BRONZE PUMP, NON DRIP PROOF, SERIES 1522, 3/4 AAB ITT BFLL & GOSSETT
1	201001	EXPANSION TANK			
1	201002	WELDMENT			
4	3393	ANGLE NEEDLE VALVE 0.250 N.P.T.			
1	3386	THERMOCOUPLE, IRON CONSTANTAN, 1-3/8 IMMERSION LENGTH, 0.250 NPT BRASS BUSHING			
			1	205010-1*100	VALVE GATE, 1/2, 125 LB. BRONZE RISING STEM, SOLID WEDGE DISC CRANE #428
1	34541	SIGHT GAUGE, OIL LINE			
1	3661	VALVE, SOLENOID			
1	34307	HEAT EXCHANGER			
2	200011-1	BOX, JUNCTION	1	207023	BOX, JUNCTION—THERMOCOUPLE WIRE
1	32827	COVER FOR FS PYLET W/GASKET			
1	12394	PRESSURE CAP, 7 PSI	1	207020	BRACKET, SUPPORT—WATER AND DRAIN LINE
2	205010-1*021	VALVE, ANGLE, 1/2 in NPT			
4	205010-1*031	U-BOLTS AND NUTS, 1/4-20 UNC X 0.75 INSIDE WIDTH	1	207021	BRACKET, MOUNTING—BALL VALVE
			1	207022	BRACKET, MOUNTING—BALL VALVE
			1	4600	VALVE, SWING CHECK K-105 STAND, 3/4 in PIPE ENDS 125# WORKING PRESSURE USED WITH 4557 (#37)
2	35101	SCREW, CAP, HEX HEAD 0.375-16 UNC X 1.250 LONG GRADE 5 CADMIUM PLATED			
12	35025	SCREW, CAP, HEX HEAD 0.375-16 UNC	1	4557	EJECTOR HYDRAULIC, 3/4 in

ELECTRIC FUSES (CARTRIDGE TYPE)—SAE J554 AUG87 — SAE Standard

Report of Electrical Equipment Division approved January 1914 and revised by the Electrical and Electronic Systems Technical Committee August 1987.

1. *Scope*—The fuses shown are for use in motor vehicles, boats, and trailers to protect electrical wiring and equipment. This standard is for the construction shown and is not intended to restrict the design and use of other configurations and materials capable of meeting the vehicle requirements.

2. *Definition*—A fuse is a device designed to open the electric circuit when subjected to overcurrents that could damage the circuit or equipment. This action is to be nonreversible, and the fuse is intended to be replaced after the circuit malfunction has been corrected.

φ 3. *Materials*—The fuses shown shall have clear glass tubes. End caps shall be of brass, copper, or other copper alloy and shall be plated with nickel or other suitable material having satisfactory electrical and corrosion protective properties.

φ 4. *Construction*—Fuse caps shall be tightly attached to the glass tube and the ends shall be square and free of solder externally. Fuse elements shall be clearly visible through the glass tube. Fuses shall be capable of being passed through a tubular gage having a length as long as the fuse and having a uniform inside diameter of 0.258-0.259 in (6.55-6.60 mm). Preferred and other fuse dimensions are shown in Fig. 1.

5. *Application*

φ 5.1 *General*—This standard applies to fuses of all lengths. However, the fuse derating chart shown in Fig. 2 applies specifically to the preferred length of 1.250 in (31.8 mm). The fuse manufacturer should be contacted for recommendations on other available lengths. (See Fig. 1.)

5.2 *Ampere Rating*—This standard covers ampere ratings up to and including 30 A. Preferred ampere ratings are shown in Table 1. These ratings are determined at 75°F (24°C) ambient temperature. Approximate capacity change with respect to temperature is shown in Fig. 3 for all length fuses. The use of fuses in ambient temperatures beyond the limits shown is not recommended without thorough testing experimentally in the vehicle. It is further recommended that fuses not be loaded to 100% of the adjusted capacity, according to ambient temperature, due to electrical system variances. See Fig. 2 for additional deratings when fuses are used on cable gage sizes other than the test gage wire.

5.3 *Voltage Rating*—Fuses shall be capable of interrupting any voltage up to and including 32 VDC.

5.4 *Maximum Voltage Drop*—The maximum voltage drop (in φ millivolts) at rated current across the fuse only, shall be as shown in Table 2, when measured across the fuse from ferrule to ferrule.

6. *Performance*—Tests shall be conducted within an ambient temperature range of 75 ± 9° F (24.0 ± 5°C) except for the overcurrent test which is to be conducted at 75 ± 2°F (23.9 ± 1.2°C).

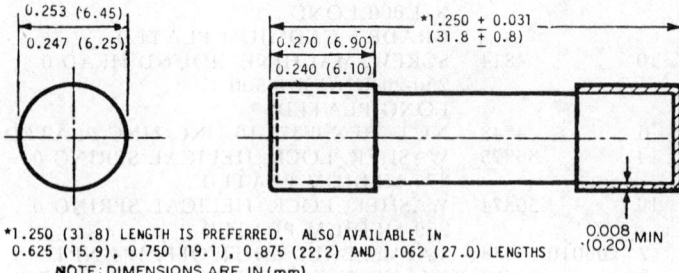

ΦFIG. 1—FUSE DIMENSIONS

TABLE 1—FUSE COLOR CODES

Ampere Rating	Color	Ampere Rating	Color
1	Dark green	9	Orange
2	Gray	10	Red
2½	Purple	14	Black
3	Violet	15	Light blue
4	Pink	20	Yellow
5	Tan	25	White
6	Gold	30	Light green
7½	Brown		

ΦFIG. 2—FUSE DERATING FOR VARIOUS WIRE SIZES (1.250 IN (31.8 MM) LENGTH)

6.1 Ampere Rating Tests—Fuses shall carry 110% of rated current continuously for 4 h, shall open at 135% of rated current in less than 1 h, and shall open at 200% of rated current in less than 10 s.

6.2 Cycling Test—Fuses shall perform satisfactorily for a 50 000 cycle load test. Each cycle shall consist of applying 70% of the rated current carrying capacity for 10 s followed by a 10 s interval of no applied current.

6.3 Vibration Test—Fuses shall perform satisfactorily after undergoing the following tests: Suitably mounted samples shall be subjected to a simple harmonic motion having an amplitude of 0.03 in (0.8 mm) travel (0.06 in (1.5 mm) max total excursion). The frequency shall be varied uniformly between the limits of 10 and 55 Hz. The entire range of 10-55 Hz and returning to 10 Hz shall be traversed in approximately 1 min. This motion shall be applied for a period of 2 h in each of the three mutually perpendicular directions (total of 6 h).

φ **6.4 Procedure**—The fuses, with the exceptions noted in the vibration test, shall be mounted horizontally. When testing two or more fuses in series, the fuses shall be mounted no less than 6 in (152 mm) apart and with no less than 24 in (609 mm) of interconnecting cable. All electrical tests shall be made using SAE No. 8[1] gage copper wire, and with fuse clip attachment terminals that have a max voltage drop of 4 mV per ampere when measured between points located on the wire 3 in (76 mm) from the attachment terminals. This determination shall be made by using a solid copper dummy 0.250 in (6.35 mm) in diameter and 1.250 ± 0.005 in (31.75 ± 0.13 mm) long, with suitably plated ferrules, installed in the fuse clips.

6.5 Marking—Fuses shall be permanently and legibly marked on the end caps with the ampere rating and the manufacturer's name or trademark. In addition, the ampere rating may be marked on the glass using numerals that are 0.150-0.200 in (4.0-5.0 mm) in height, or the fuses may be color coded with a permanent stripe around the interior or exterior of the glass tube. If a color stripe is used, the fuse element must still be clearly visible through the glass tube. This color coding shall be as shown in Table 1.

TABLE 2—MAXIMUM VOLTAGE DROP OF FUSE AT RATED CURRENT

Ampere Rating	Maximum Voltage Drop (Millivolts)
1	325.0
2	300.0
2½	275.0
3	250.0
4	235.0
5	215.0
6	200.0
7½	185.0
9	170.0
10	165.0
14	155.0
15	150.0
20	135.0
25	125.0
30	120.0

FIG. 3—EFFECT OF AMBIENT TEMPERATURE ON AMPERE RATING OF SAE SPECIFICATION FUSES

$t_{°C} = (t_{°F} - 32)/1.8$

[1] Conductor cross section area to be not less than 14.810 cir mil (7.23 mm²).

MINIATURE BLADE TYPE ELECTRICAL FUSES—SAE J2077 NOV90

SAE Standard

Report of the Circuit Protection & Switching Devices Standards Technical Committee approved November 1990.

1. Scope—The fuses shown in Figure 1 are for use in motor vehicles, boats, and trailers to protect electrical wiring and equipment. This SAE Standard is for the construction shown and is not intended to restrict the design and use of other configurations and materials capable of meeting the vehicle requirements.

2. References

2.1 Applicable Documents

2.1.1 SAE PUBLICATIONS—Available from SAE, 400 Commonwealth Drive, Warrendale, PA 15096-0001.

SAE J726—Air Cleaner Test Code
SAE J1034—Engine Coolant Concentrate—Ethylene-Glycol Type

2.1.2 ASTM PUBLICATIONS—Available from ASTM, 1916 Race Street, Philadelphia, PA 19103.

ASTM B 117—Method of Salt Spray (Fog) Testing

2.2 Definition

FUSE—A device designed to interrupt the electrical circuit when subjected to overcurrents. This action is to be nonreversible, and the fuse is intended to be replaced after the circuit malfunction has been corrected.

NOTE: ALL DIMENSIONS IN MILLIMETERS

FIGURE 1—OVERALL FUSE DIMENSIONS

3. Part 1—Design Parameters

3.1 Materials—The fuses shall have nonconductive bodies capable of withstanding vehicle environmental conditions as set forth in this document. Terminals shall have a suitable finish which will assure corrosion protection and satisfactory mechanical and electrical properties.

3.2 Construction—The fuse shall consist of a terminal and element combination securely attached to a nonconductive fuse body. Fuse terminals shall be exposed through the top portion of the fuse body to allow for inspection of the electrical continuity of the fuse while it is engaged in a fuseblock or fuseholder. Fuse elements shall be visible through the body. Typical overall dimensions are shown in Figure 1.

3.3 Marking (Initially and After Environmental Exposure)—Fuses shall be marked on the fuse body with the amperage, voltage rating, and manufacturer's name or trademark. In addition, the fuses shall be color coded as shown in Table 1. Markings shall be legible at the conclusion of all tests set forth in this document.

4. Part 2—Performance Requirements

4.1 Ampere Rating—This document covers ampere ratings up to and including 30 A. Preferred ampere ratings are shown in Table 1. These ratings were determined at 24°C ambient temperature using a test procedure as detailed in 5.1 of this document. The specific ampere capacity of the fuses is a function of the particular electrical system being utilized. To aid in determining the actual capacity change, several factors should be considered by the application engineer.

4.1.1 WIRE—Figure 2A represents the approximate ampere capacity change due to cable sizes other than 5 mm^2.

TABLE 1—PREFERRED AMPERE RATINGS AND FUSE COLOR CODE

Ampere Rating	Color
2	Grey
3	Violet
4	Pink
5	Tan
7-1/2	Brown
10	Red
15	Blue
20	Yellow
25	Natural (White)
30	Green

4.1.2 TEMPERATURE—Figure 2 represents the approximate capacity change with respect to ambient temperature. The use of fuses in ambient temperature beyond the limits shown is not recommended.

4.1.3 LOADING—It is recommended that the fuses not be loaded above 80% of their rerated ampere capacity based on ambient temperatures and the use of cable sizes other than 5 mm^2. In addition, it is further recommended that actual performance be verified through testing experimentally in the vehicle.

NOTE: The specific ampere capacity of a fuse is a function of the particular electrical system being utilized. To aid in determining the actual capacity change several factors should be considered by the application engineer. These factors include ambient temperature, wire size, insulating material, and connecting clips.

4.2 Voltage Rating—Fuses shall be capable of interrupting at any voltage up to and including 32 V.

4.3 Maximum Voltage Drop—The maximum voltage drop (in millivolts) at rated current across the fuse only shall be as shown in Table 2.

4.4 Performance—Tests shall be conducted within a temperature range of 24°C ± 5, which is considered room ambient within this specification.

4.4.1 AMPERE RATING TESTS—Fuses shall carry 110% of rated current continuously for a minimum of 100 h; shall open in not less than 0.75 s or more than 1800 s at 135% of rated current; shall open in not less than 0.15 s or more than 5 s at 200% of rated current; shall not open in less than 0.080 s or more than 0.50 s at 350% of rated current, and shall open in not less than 0.03 s or more than 0.10 s at 600% of rated current.

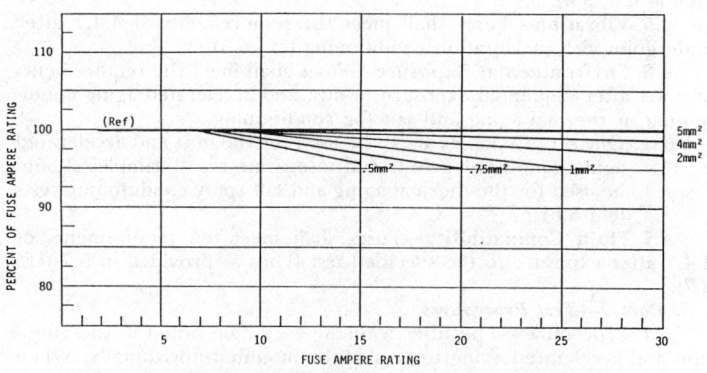

NOTE: THIS CURVE IS BASED ON 100% OF RATED CURRENT USING 5mm^2 COPPER CABLE AT 24°C IN A STANDARD TEST MODULE

FIGURE 2A—WIRE RERATING CURVE

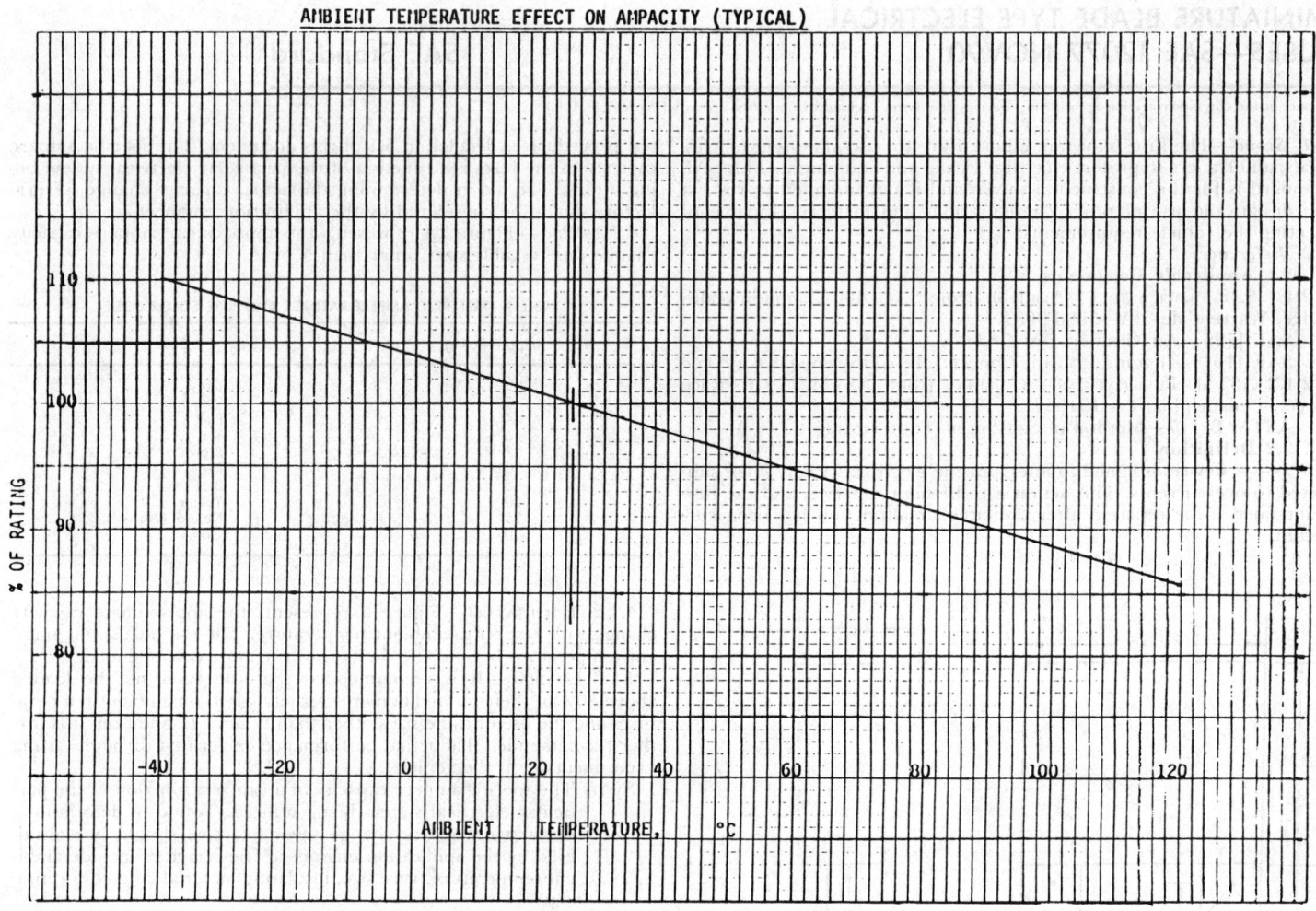

FIGURE 2—AMBIENT TEMPERATURE EFFECT ON AMPACITY (TYPICAL)

4.4.2 INTERRUPTING CAPACITY TEST—The fuses shall be capable of interrupting 1000 +5%/−0% A at 32 +5%/−0% V D.C. with any time constant up to 2.5 ms. After testing, there shall be no rupturing of the fuse body and no damage to the standard test module or the fuse such as to prevent removal of the fuse.

4.4.3 CURRENT CYCLING TEST—Fuses shall meet the requirements of 4.4.1 after current cycling for a minimum of 250 000 cycles (cf.; 5.3).

4.4.4 TRANSIENT CURRENT CYCLING—Fuses shall meet the requirements of 4.4.1 after a minimum of 50 000 cycles of transient current cycling (cf.; 5.4).

4.5 Vibration—Fuses shall meet the requirements of 4.4.1 after undergoing 6 h of vibration conditioning (cf.; 5.5).

4.6 Environmental Exposure—Fuses shall meet the requirements of 4.4.1 after sequential exposure to dust and accelerated aging conditioning or thermal aging and salt fog conditioning.

NOTE: One set of samples are to be used for the dust and accelerated aging conditioning tests and a separate set of samples should be used for the thermal aging and salt spray conditioning tests (cf.; 5.6).

4.7 Fluid Compatibility—Fuses shall meet the requirements of 4.4.1 after exposure to the specified test fluids as provided in 5.7 (cf.; 5.7).

5. *Part 3—Test Procedures*

5.1 Procedure—The fuses with the exception noted in the vibration and accelerated aging tests, shall be mounted horizontally. When testing two or more fuses in series, the fuses shall be mounted no less than 150 mm apart and with no less than 600 mm of interconnecting cable, except as noted for the transient current cycling test. All electrical tests shall be made with 5.0 mm² copper cable and a standard test module (shown in Figure 3). The interface voltage drop ($V_{CD}−V_{AB}$) of the fixture shall not exceed 30 mV. The total voltage drop (V_{EF}) should not exceed 60 mV. The voltage check shall be made using a solid copper dummy with silver plated terminals and with the dimensions as shown in Figure 4.

5.2 Voltage Drop—The voltage drop (in millivolts) after 15 min at rated current across the fuse shall be measured at the indicated points on the fuse terminals as shown in Figure 1.

5.3 Current Cycling Tests

5.3.1 Resistors should be employed as load(s) to adjust the current to 68 to 72% of the fuse rating as shown in Figure 5.

5.3.2 The test system voltage can be any convenient voltage up to and including 32 V.

5.4 Transient Current Cycling Test

5.4.1 Simulated or lamp loads should be used to adjust the initial peak transient current to the percent of the fuse rating as shown in Fig-

TABLE 2—MAXIMUM VOLTAGE DROP OF FUSE AT RATED CURRENT

Fuse Rating (Amps)	Max Voltage Drop (mV)
2	225
3	175
4	175
5	175
7-1/2	150
10	125
15	125
20	125
25	100
30	100

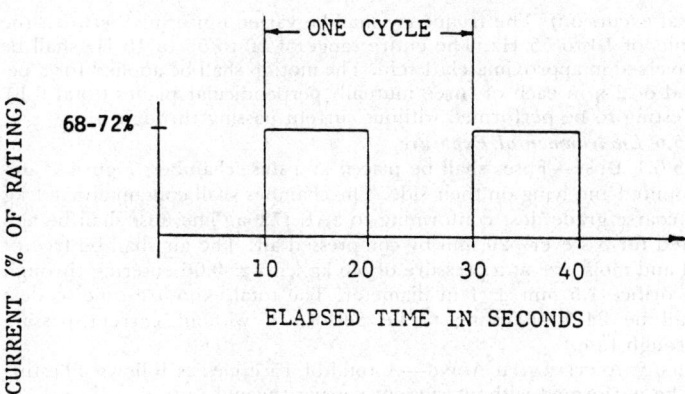

FIGURE 5—CURRENT CYCLING WAVEFORM

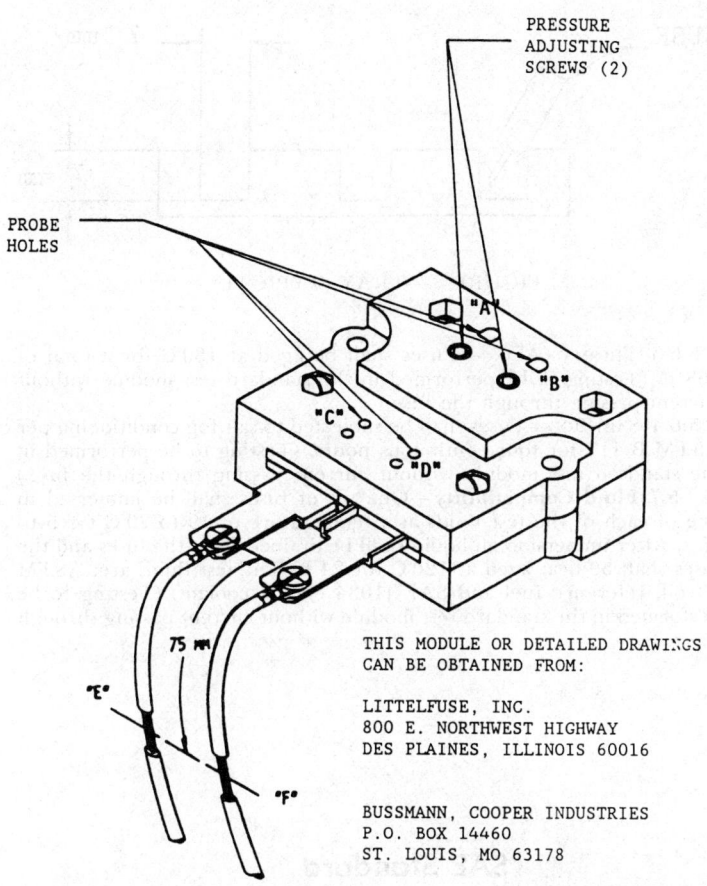

THIS MODULE OR DETAILED DRAWINGS CAN BE OBTAINED FROM:

LITTELFUSE, INC.
800 E. NORTHWEST HIGHWAY
DES PLAINES, ILLINOIS 60016

BUSSMANN, COOPER INDUSTRIES
P.O. BOX 14460
ST. LOUIS, MO 63178

FIGURE 3—STANDARD TEST MODULE

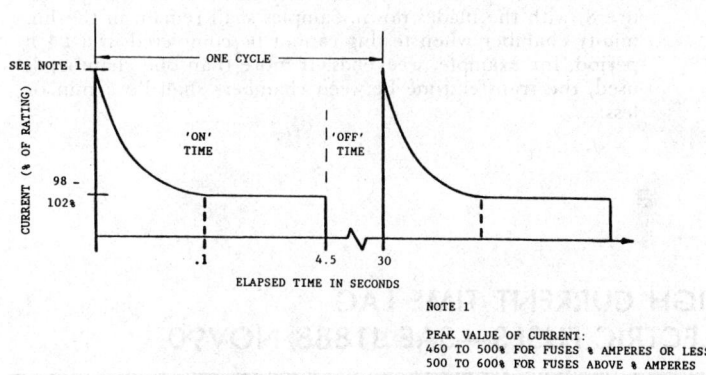

NOTE 1
PEAK VALUE OF CURRENT:
460 TO 500% FOR FUSES ％ AMPERES OR LESS
500 TO 600% FOR FUSES ABOVE ％ AMPERES

FIGURE 6—TRANSIENT CURRENT CYCLING WAVEFORM

ure 6 and the initial steady state current 98 to 102% of the fuse rating.
  5.4.2 The test system voltage can be any convenient voltage up to and including 32 V.
  5.4.3 A minimum of 300 mm of interconnecting cable should be used between each standard test module.
  5.5 Vibration Test—Suitably mounted samples shall be subject to a simple harmonic motion having an amplitude of 0.75 mm (1.50 mm

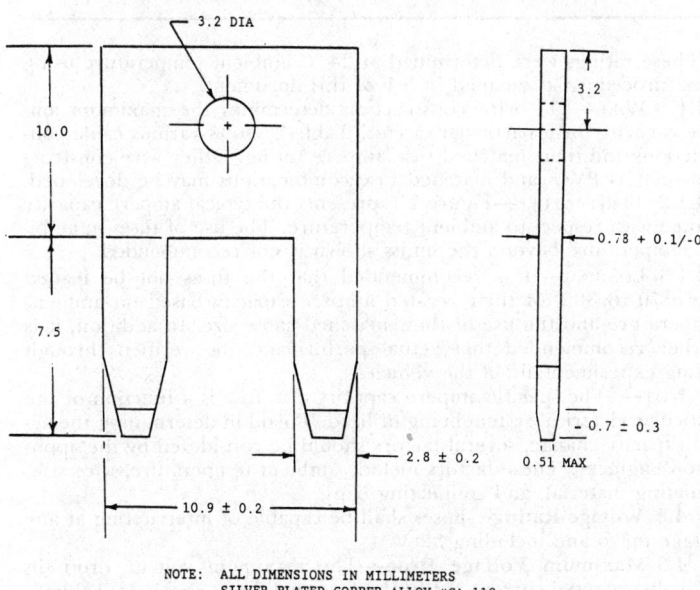

NOTE: ALL DIMENSIONS IN MILLIMETERS
SILVER PLATED COPPER ALLOY #CA 110

FIGURE 4—TEST SLUG

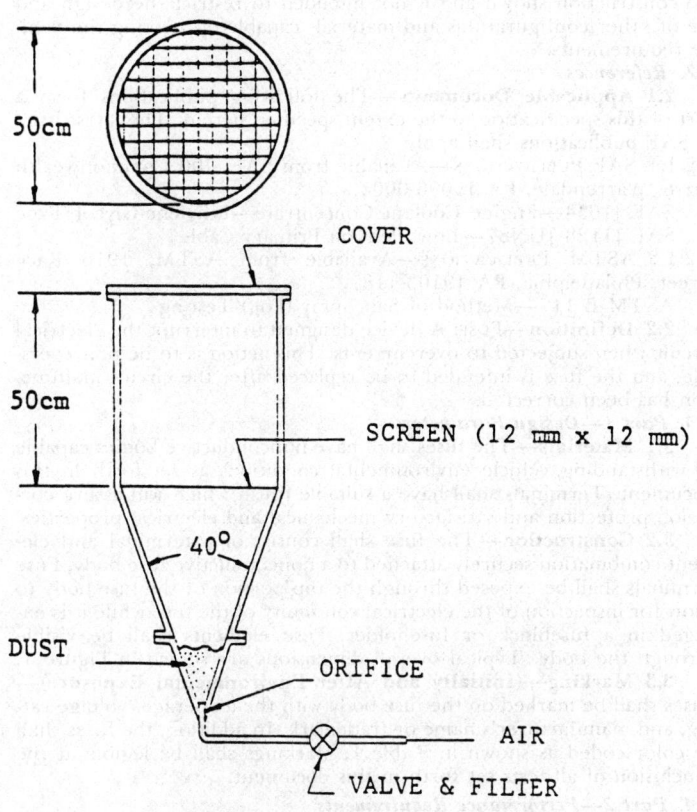

FIGURE 7—DUST CHAMBER (TYPICAL)

total excursion). The frequency shall be varied uniformly between the limits of 10 to 55 Hz. The entire range of 10 to 55 to 10 Hz shall be traversed in approximately 1 min. The motion shall be applied for a period of 2 h in each of three mutually perpendicular planes (total 6 h). (Testing to be performed without current passing through fuse.)

5.6 Environmental Exposure

5.6.1 DUST—Fuses shall be placed in a dust chamber, Figure 7, unmounted and lying on their side. The chamber shall contain about 1 kg of coarse grade dust conforming to SAE J726a. The dust shall be agitated for 3 s every 20 min by compressed air. The air shall be free of oil and moisture, at a pressure of 5.6 kg/cm^2 ± 0.06 entering through an orifice 1.5 mm ± 1 in diameter. The total exposure time to dust shall be 24 h. (Testing to be performed without current passing through fuse.)

5.6.2 ACCELERATED AGING—A total of 15 cycles as follows: (Testing to be performed without current passing through fuse.)

    16 h at 95 to 99% relative humidity and 37°C ± 1
    2 h at −40°C ± 1
    2 h at 70°C ± 1
    4 h at room ambient

NOTE: Fuses shall be placed in the humidity chamber on a tray, Figure 8, with the blades down. Samples shall remain in the humidity chamber when testing cannot be completed in a 24 h period, for example, weekends. If more than one chamber is used, the transfer time between chambers shall be 5 min or less.

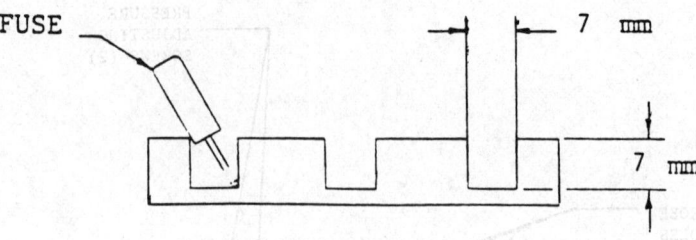

FIGURE 8—TRAY (TYPICAL)

5.6.3 THERMAL AGING—Fuses shall be aged at 180°C for a total of 168 h. (Testing to be performed in the standard test module without current passing through the fuse.)

5.6.4 SALT FOG—Fuses shall be subjected to salt fog conditioning per ASTM B 117 for four continuous hours. (Testing to be performed in the standard test module without current passing through the fuse.)

5.7 Fluid Compatibility—One set of fuses shall be immersed in one of each of the test fluids at a temperature of 19 to 29°C for 5 to 10 s. After immersion all liquid shall be drained from the fuses and the fuses shall be heat aged at 120°C for 24 h. The test fluids are: ASTM #3 oil, reference fuel and SAE J1034 engine coolant. (Testing to be performed in the standard test module without current passing through the fuse.)

HIGH CURRENT TIME LAG ELECTRIC FUSES—SAE J1888 NOV90

SAE Standard

Report of the SAE Circuit Protection & Switching Devices Standards Technical Committee approved November 1990.

1. Scope

The fuses shown in Figure 1 are for use in motor vehicles, boats, and trailers to protect electrical wiring. This SAE Standard is for the construction shown and is not intended to restrict the design and use of other configurations and materials capable of meeting the vehicle requirements.

2. References

2.1 Applicable Documents—The following publications form a part of this specification to the extent specified herein. The latest issue of SAE publications shall apply.

2.1.1 SAE PUBLICATIONS—Available from SAE, 400 Commonwealth Drive, Warrendale, PA 15096-0001.

    SAE J1034—Engine Coolant Concentrate—Ethylene-Glycol Type
    SAE J1128 JUN87—Low Tension Primary Cable

2.1.2 ASTM PUBLICATIONS—Available from ASTM, 1916 Race Street, Philadelphia, PA 19103-1187.

    ASTM B 117—Method of Salt Spray (Fog) Testing

2.2 Definition—FUSE: A device designed to interrupt the electrical circuit when subjected to overcurrents. This action is to be nonreversible, and the fuse is intended to be replaced after the circuit malfunction has been corrected.

3. Part 1—Design Parameters

3.1 Materials—The fuses shall have nonconductive bodies capable of withstanding vehicle environmental conditions as set forth in this document. Terminals shall have a suitable finish which will assure corrosion protection and satisfactory mechanical and electrical properties.

3.2 Construction—The fuse shall consist of a terminal and element combination securely attached to a nonconductive fuse body. Fuse terminals shall be exposed through the top portion of the fuse body to allow for inspection of the electrical continuity of the fuse while it is engaged in a fuseblock or fuseholder. Fuse elements shall be visible through the body. Typical overall dimensions are shown in Figure 1.

3.3 Marking—(Initially and After Environmental Exposure)—Fuses shall be marked on the fuse body with the amperage, voltage rating, and manufacturer's name or trademark. In addition, the fuses shall be color coded as shown in Table 1. Markings shall be legible at the conclusion of all tests set forth in this document.

4. Part 2—Performance Requirements

4.1 Ampere Rating—This document covers ampere ratings up to and including 120 amps. Preferred ampere ratings are shown in Table

TABLE 1—PREFERRED AMPERE RATINGS AND COLOR CODES

Ampere Rating	Color
20	Yellow
30	Green
40	Amber
50	Red
60	Blue
70	Tan
80	Natural
100	Violet
120	Pink

1. These ratings were determined at 24 °C ambient temperature using a test procedure as detailed in 5.1 of this document.

4.1.1 WIRE—The wire construction determines the maximum ampere capacity for a particular circuit. Table 2 shows various cable constructions and their matched fuse ampere rating. Other wire constructions such as PVCs and matched fuse combinations may be developed.

4.1.2 TEMPERATURE—Figure 2 represents the typical ampere capacity change with respect to ambient temperature. The use of fuses in ambient temperature beyond the limits shown is not recommended.

4.1.3 LOADING—It is recommended that the fuses not be loaded above 70 to 80% of their rerated ampere capacity based on ambient temperature and the use of their matched cable size. In addition, it is further recommended that actual performance be verified through testing experimentally in the vehicle.

NOTE—The specific ampere capacity of a fuse is a function of the particular electrical system being utilized. To aid in determining the actual capacity change, several factors should be considered by the application engineer. These factors include ambient temperature, wire size, insulating material, and connecting clips.

4.2 Voltage Rating—Fuses shall be capable of interrupting at any voltage up to and including 32 V.

4.3 Maximum Voltage Drop—The maximum voltage drop (in millivolts) at rated current across the fuse shall be as shown in Table 3.

4.4 Performance—Tests shall be conducted within a temperature range of 24 °C ± 5 °C, which is considered room ambient within this specification.

4.4.1 AMPERE RATING TESTS—Fuses shall carry 100% of rated current continuously for a minimum of 100 h. Opening times for 135%, 200%, 350%, and 600% of rated current are listed in Table 4.

4.4.2 INTERRUPTING CAPACITY TEST—The fuses shall be capable of interrupting 1000 + 5%/-0 A at 32 + 5%/-0 V D.C. with a circuit time constant up to 2.5 ms. After testing, the fuse body shall be intact and no damage to the standard test module or the fuse such as to prevent removal of the fuse shall be allowed.

4.4.3 CURRENT CYCLING TEST—Fuses shall meet the requirements of 4.4.1 after current cycling for a minimum of 250 000 cycles (cf.; 5.3).

4.4.4 TRANSIENT CURRENT CYCLING—Fuses shall meet the requirements of 4.4.1 after a minimum of 50 000 cycles of transient current cycling (cf.; 5.4).

4.5 Vibration Test—Fuses shall meet the requirements of 4.4.1 after undergoing 6 h of vibration conditioning (cf.; 5.5).

4.6 Environmental Exposure—Fuses shall meet the requirements of 4.4.1 after sequential exposure to dust and accelerated aging conditioning or thermal aging and salt fog conditioning (cf.; 5.6).

NOTE—One set of samples are to be used for the dust and accelerated aging conditioning tests and a separate set of samples should be used for the thermal aging and salt spray conditioning tests.

4.7 Fluid Compatibility—Fuses shall meet the requirements of 4.4.1 after exposure to specified test fluids (cf.; 5.7).

5. Part 3—Test Procedures

5.1 Procedure—The fuses with the exceptions noted in the vibration and accelerated aging tests, shall be mounted horizontally. When testing two or more fuses in series, the fuses shall be mounted no less than 150 mm apart and with no less than 600 mm of interconnecting cable, except as noted for the transient current cycling test. All electrical tests shall be made with matched cable gauge, as shown in Table 2 and a standard test module (shown in Figure 3). The interface voltage drop (V_{CD}—V_{AB}) of the fixture shall not exceed 50 mV. The total voltage drop (V_{FF}) should not exceed 100 mV. The voltage check shall be made using a solid copper dummy with silver plated terminals and with the dimensions as shown in Figure 4. The test current and matched cable gauge are shown in Table 2.

5.2 Voltage Drop—The voltage drop (in millivolts) at rated current across the fuse shall be measured at the indicated points on the

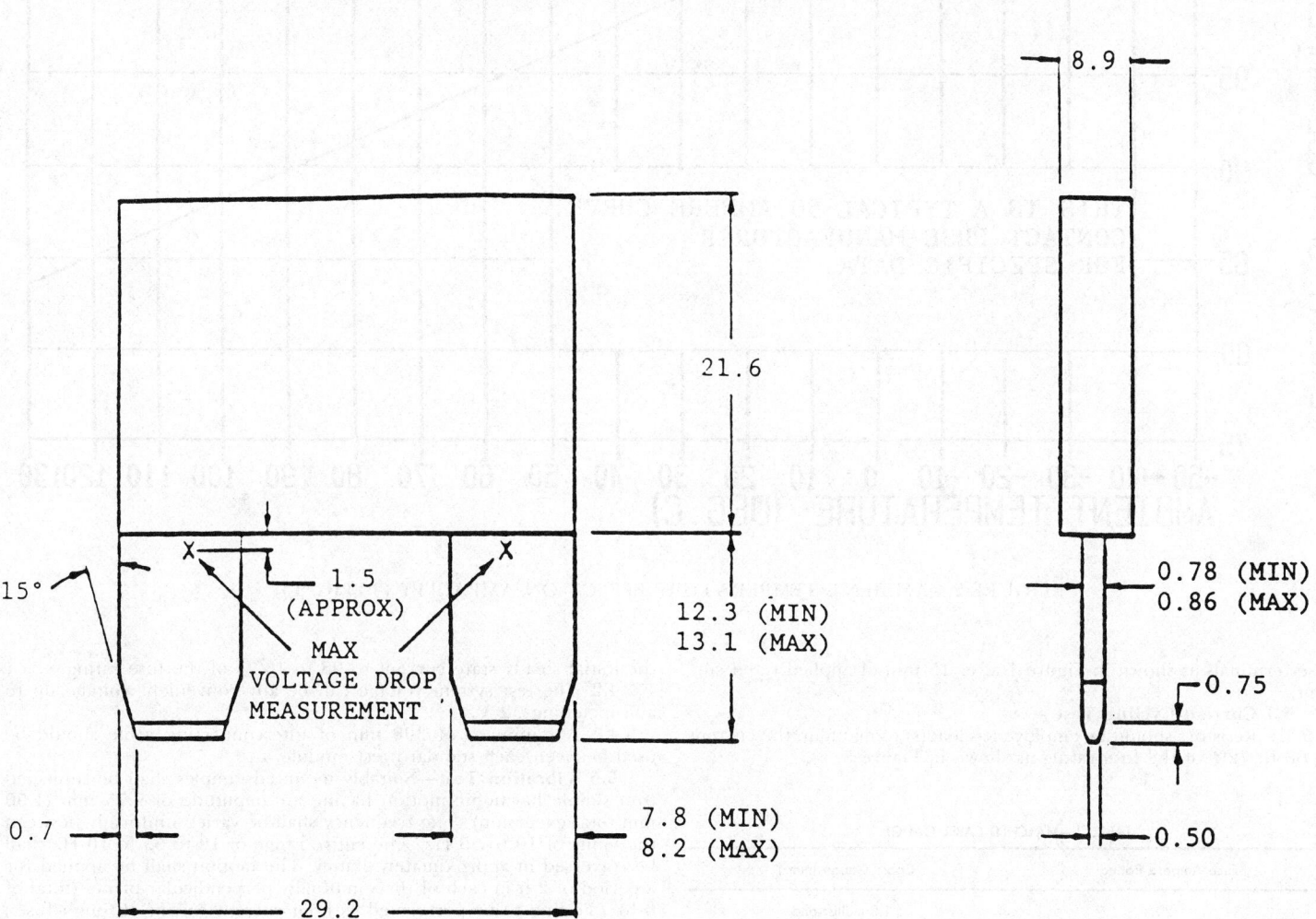

FIGURE 1—OVERALL FUSE DIMENSIONS

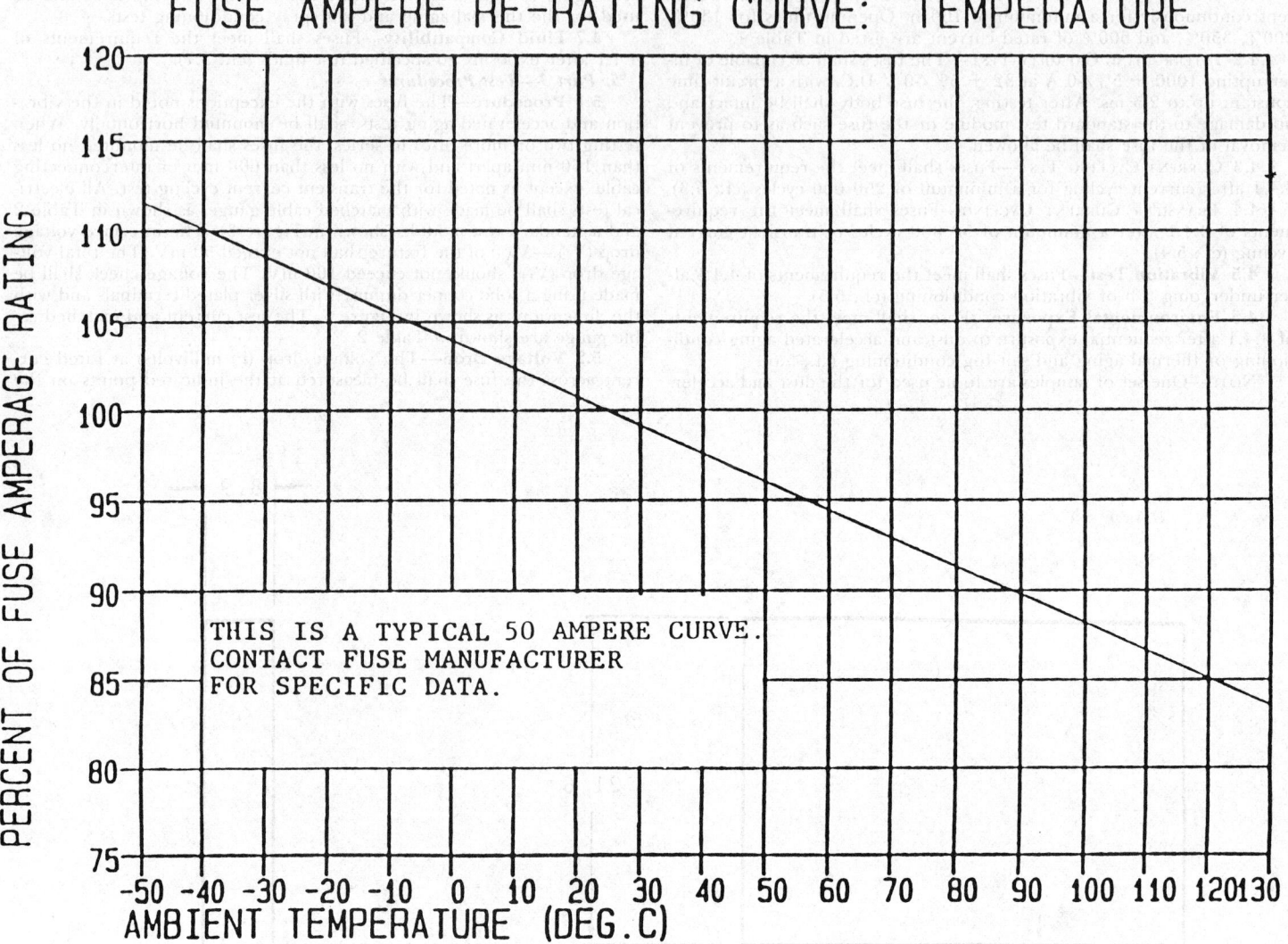

FIGURE 2—AMBIENT TEMPERATURE EFFECT ON AMPACITY (TYPICAL)

fuse terminals as shown in Figure 1 after 15 min of applied rated current.

5.3 Current Cycling Test

5.3.1 Resistors should be employed as load(s) to maintain the current to 68 to 72% of the fuse rating as shown in Figure 5.

TABLE 2—MATCHED CABLE GAUGE

Fuse Ampere Rating	Cable Gauge (mm²)
20	1.0 multistrand
30	2.0 multistrand
40	3.0 multistrand
50	5.0 multistrand
60	5.0 multistrand
70	8.0 multistrand
80	8.0 multistrand
100	13.0 multistrand
120	16.0 multistrand

NOTE—Wire type is cross-link polyethylene (GLX) as described in SAE specification J1128 (June, 1987).

5.3.2 The test system voltage can be any convenient voltage up to and including 32 V.

5.4 Transient Current Cycling Test

5.4.1 Simulated loads should be used to adjust the initial peak transient current to the percent of the fuse rating as shown in Figure 6 and the initial steady state current to 98 to 102% of the fuse rating.

5.4.2 The test system voltage can be any convenient voltage up to and including 32 V.

5.4.3 A minimum of 300 mm of interconnecting cable should be used between each standard test module.

5.5 Vibration Test—Suitably mounted samples shall be subjected to a simple harmonic motion having an amplitude of 0.75 mm (1.50 mm total excursion). The frequency shall be varied uniformly between the limits of 10 to 55 Hz. The entire range of 10 to 55 to 10 Hz shall be traversed in approximately 1 min. The motion shall be applied for a period of 2 h in each of three mutually perpendicular planes (total of 6 h). (Testing to be performed without current passing through fuse.)

5.6 Environmental Exposure

5.6.1 DUST—Fuses shall be placed in a dust chamber, Figure 7, unmounted and lying on their side. The chamber shall contain about 1 kg of coarse grade dust conforming to SAE J726a. The dust shall be agitated for 3 s every 20 min by compressed air. The air shall be free of oil and moisture, at a pressure of 5.6 kg/cm² ± 0.10 kg/cm² entering through an orifice 1.5 mm ± 1 mm in diameter. The total exposure time to dust shall be 24 h. (Testing to be performed without current passing through fuse.)

5.6.2 ACCELERATED AGING—A total of 15 cycles as follows: (Testing to be performed without current passing through fuse.)

16 h at 95 to 99% relative humidity and 37 °C ± 1 °C
2 h at −40 °C ± 1 °C
2 h at 70 °C ± 1 °C
4 h at room ambient

TEST MODULE OR DETAILED DRAWING
CAN BE OBTAINED FROM:

BUSSMANN
COOPER INDUSTRIES
P.O. BOX 14460
ST. LOUIS, MO 63178

OR

LITTELFUSE, INC.
800 E. NORTHWEST HIGHWAY
DES PLAINES, IL 60016

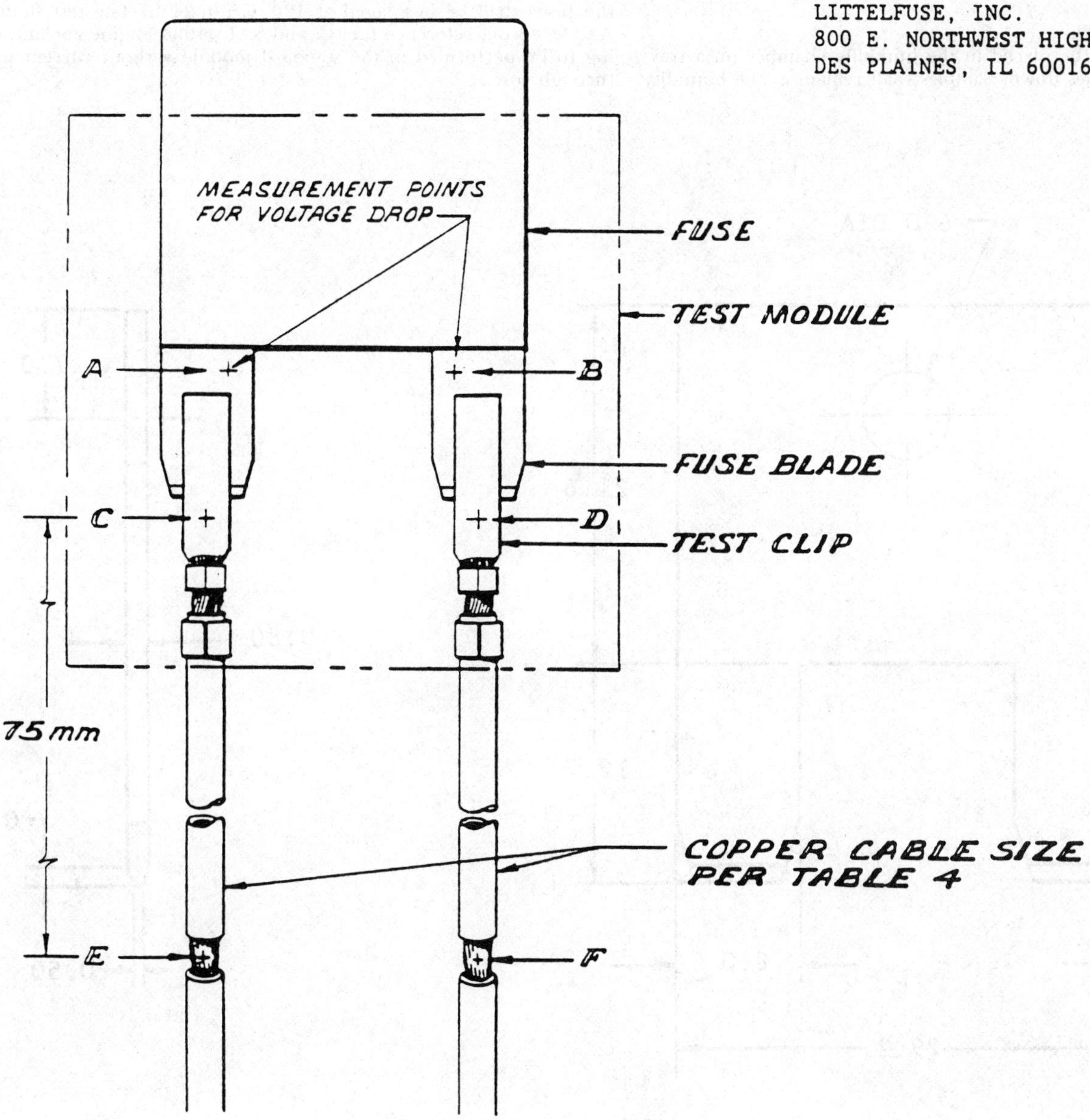

FIGURE 3—STANDARD TEST MODULE

TABLE 3—MAXIMUM VOLTAGE DROP OF FUSE AT RATED CURRENT

Fuse Rating (Amps)	Max. Voltage Drop (mV)
20	120
30	120
40	120
50	120
60	130
70	130
80	130
100	130
120	130

NOTE—Fuses shall be placed in the humidity chamber on a tray, Figure 8, with the blades down. Samples shall remain in the humidity chamber when testing cannot be completed in a 24 h period, for example weekends. If more than one chamber is used the transfer time between chambers shall be 5 min or less.

5.6.3 THERMAL AGING—Fuses shall be aged at 180 °C for a total of 168 h. (Testing to be performed in the standard module without current passing through fuse.)

5.6.4 SALT FOG—Fuses shall be subjected to salt fog conditioning per ASTM B 117 for 48 continuous hours. (Testing to be performed in the standard model without current passing through fuse.)

5.7 Fluid Compatibility—One set of fuses shall be immersed in one of each of the test fluids at a temperature of 24 °C ± 5 °C for 5 to 10 s. After immersion all liquid shall be drained from the fuses and the fuses shall be heat aged at 120 °C for 24 h. The test fluids are: ASTM #3 oil, reference fuel C, and SAE J1034 engine coolant. (Testing to be performed in the standard module without current passing through fuse.)

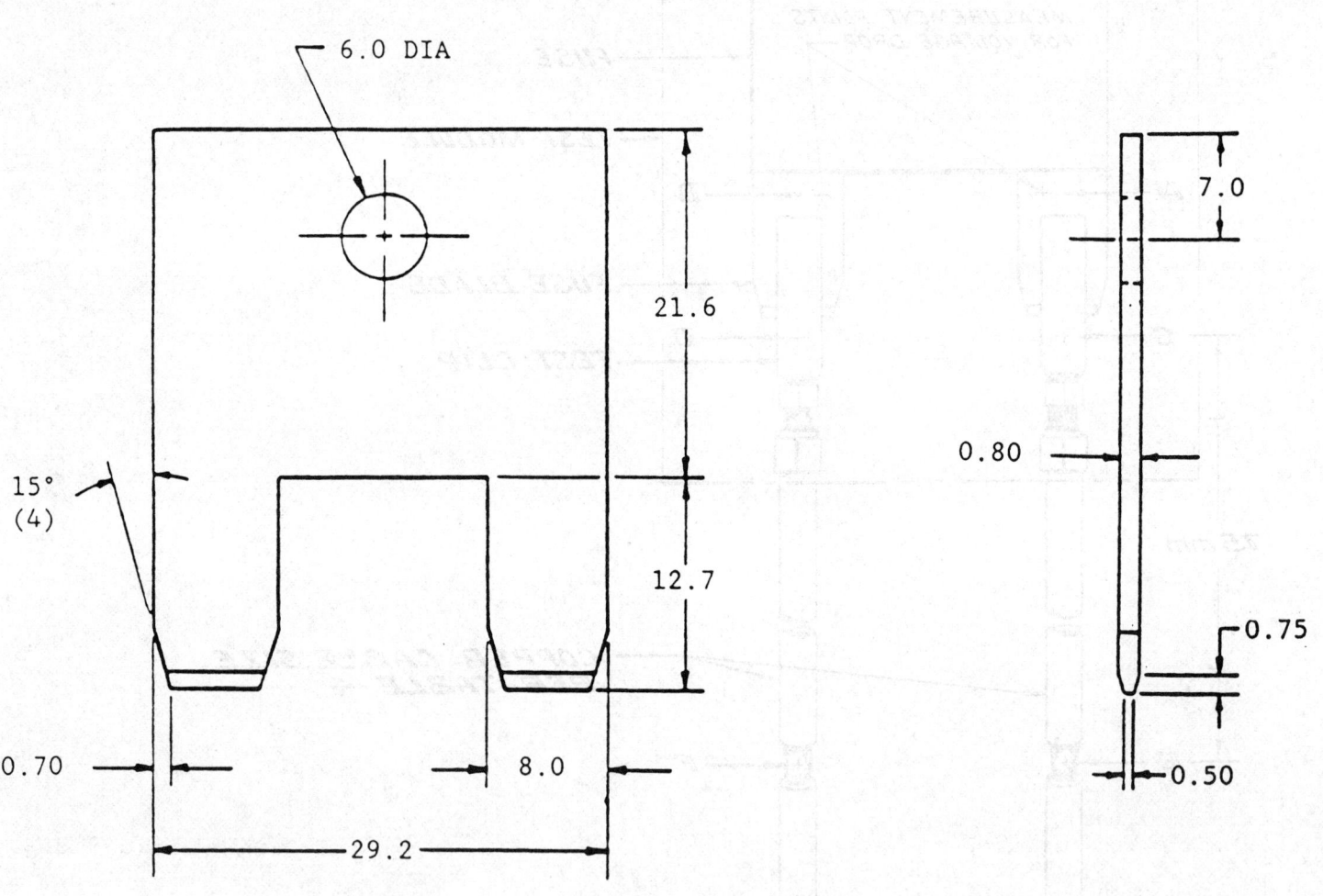

NOTE: ALL DIMENSIONS ARE IN MILLIMETERS

MATERIAL: COPPER ALLOY #CA110

PLATING: SILVER

FIGURE 4—TEST SLUG

TABLE 4—OPENING TIMES—MIN/MAX (SECONDS)

Fuse Rating (Amps)	Percent of Rated Current 135%	Percent of Rated Current 200%	Percent of Rated Current 350%	Percent of Rated Current 600%
20A Min	60	4	0.7	0.15
Max	1800	20	2.0	1.0
30A Min	60	6	1.0	0.2
Max	1800	30	4.0	1.0
40A Min	60	8	1.4	0.2
Max	1800	40	5.0	1.0
50A Min	60	10	1.7	0.2
Max	1800	50	6.0	1.0
60A Min	60	15	2.0	0.2
Max	1800	60	7.0	1.0
70A Min	60	4	0.2	0.04
Max	3600	60	2.0	0.15
80A Min	60	4	0.2	0.04
Max	3600	60	2.0	0.15
100A Min	60	4	0.2	0.04
Max	3600	60	2.0	0.15
120A Min	60	4	0.2	0.04
Max	3600	60	2.0	0.15

NOTE—The opening times shown are applicable to all ambients between −50 C and +130 C provided the fuse is rerated to the temperature in question in accordance with the manufacturer's rerating curve.

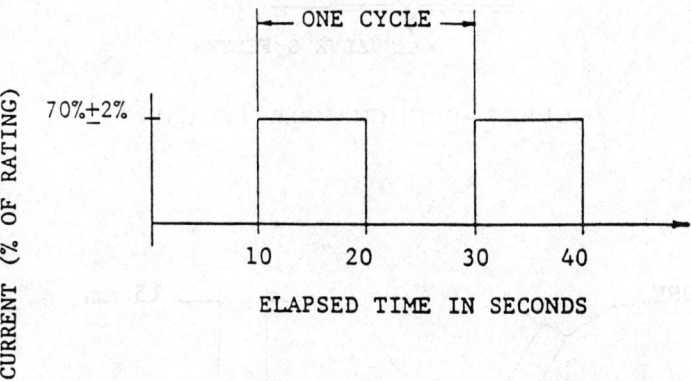

FIGURE 5—CURRENT CYCLING WAVEFORM

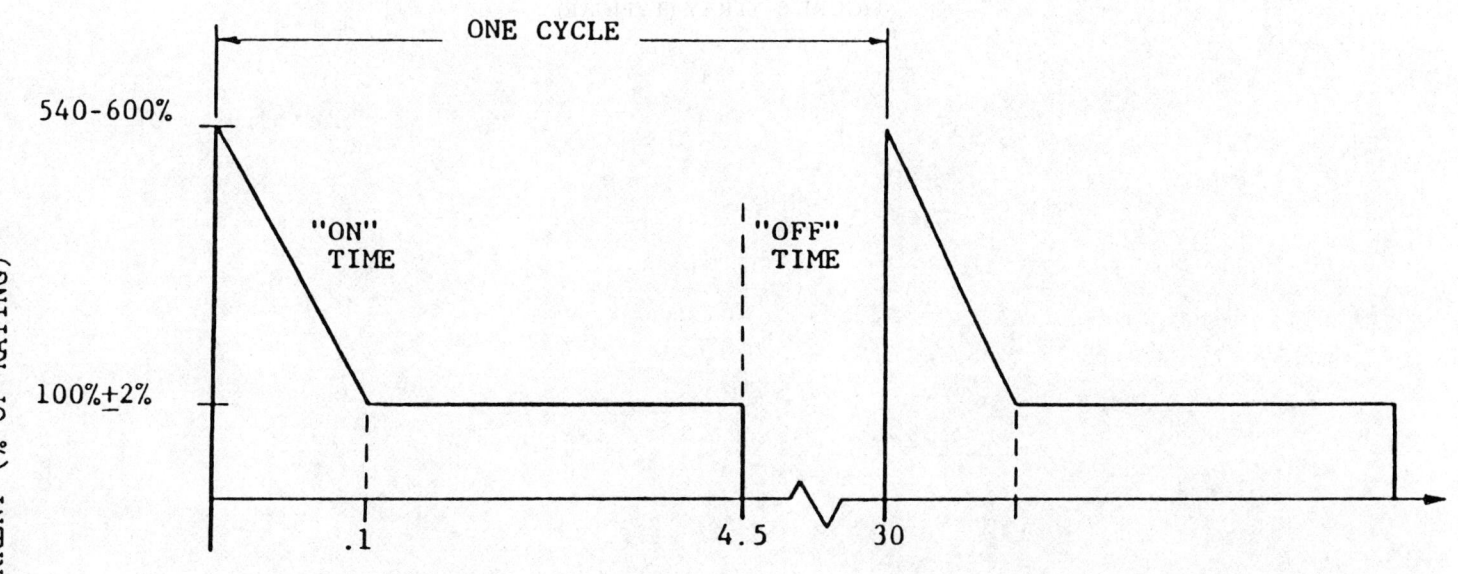

FIGURE 6—TRANSIENT CURRENT CYCLING WAVEFORM

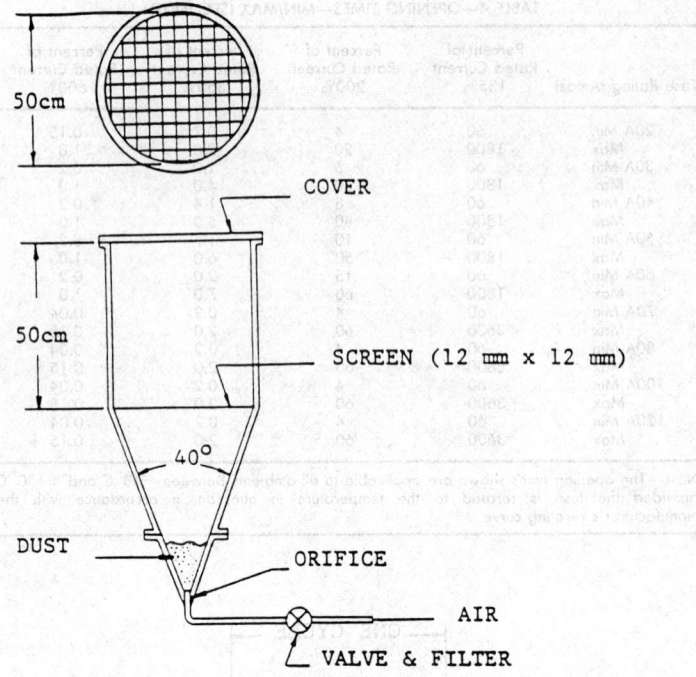

FIGURE 7—DUST CHAMBER (TYPICAL)

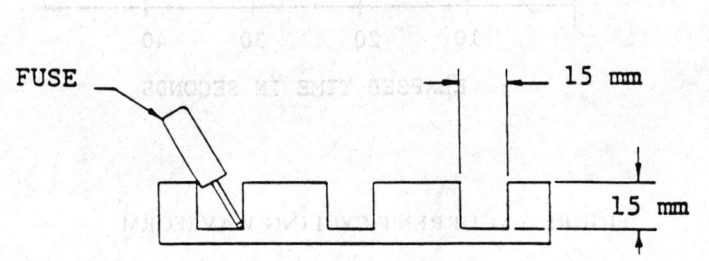

FIGURE 8—TRAY (TYPICAL)

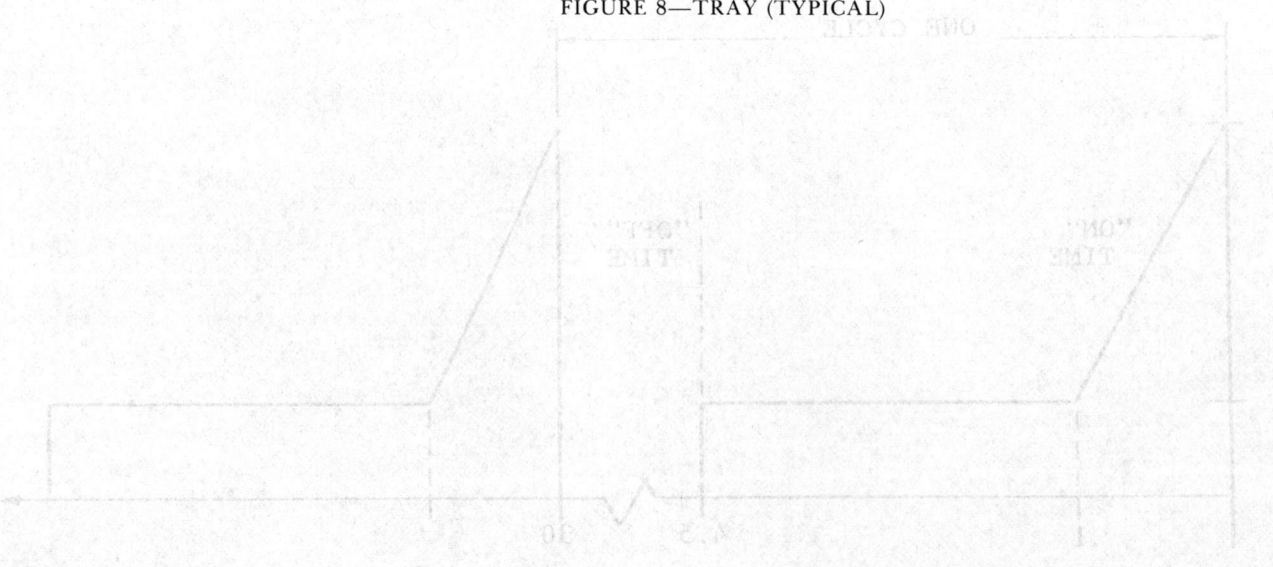

(R) CIRCUIT BREAKERS
—SAE J553 JUN92
SAE Standard

Report of the Electrical Equipment Committee approved November 1951 and completely revised by the Electrical and Electronic Systems Technical Committee June 1988. Rationale statement available. Completely revised by the SAE Truck and Bus Circuit Protection Subcommittee of the SAE Truck and Bus Electrical and Electronics Committee June 1992. Rationale statement available.

1. Scope—This SAE Standard defines the test conditions, procedures, and performance requirements for circuit breakers in ratings up to and including 50 A. The document includes externally or internally mounted automatic reset, modified reset, and manually reset types of circuit breakers for 12 V and 24 V DC operation. Some circuit breakers may have dual voltage ratings (AC and DC), however, this document evaluates DC performance only.

2. References

2.1 Applicable Documents—The following publications form a part of this specification to the extent specified herein. The latest issue of SAE publications shall apply.

2.1.1 SAE PUBLICATIONS—Available from SAE, 400 Commonwealth Drive, Warrendale, PA 15096-0001.

SAE J537—Storage Batteries
SAE J561—Electrical Terminals—Eyelet and Spade Type
SAE J858a—Electrical Terminals—Blade Type
SAE J1171—External Ignition Protection of Marine Electrical Devices
SAE J1211—Recommended Environmental Practices for Electronic Equipment Design
SAE J1428—Marine Circuit Breakers
SAE J1455—Joint SAE/TMC Recommended Environmental Practices for Electronic Equipment Design (Heavy- Duty Trucks)

2.2 Related Publications —The following publications are provided for information purposes only and are not a required part of this document.

2.2.1 SAE PUBLICATIONS—Available from SAE, 400 Commonwealth Drive, Warrendale, PA 15096-0001.

SAE J258—Circuit Breaker—Internal Mounted—Automatic Reset
SAE J554—Electric Fuses (Cartridge Type)
SAE J1284—Blade Type Electric Fuses
SAE J1888—High Current Time Lag Electric Fuses
SAE TSB 002—Preparation of SAE Technical Reports

2.2.2 OTHER PUBLICATIONS

CSA C22.2 No. 14-M1987—Industrial Control Equipment
CSA C22.2 No. 235-M89—Supplementary Protectors
MIL-STD-202F—Test Methods for Electronic and Electrical Component Parts
U.L. 1077—Standard for Supplementary Protectors for use in Electrical Equipment

2.3 Definitions

2.3.1 CIRCUIT BREAKERS are overcurrent protective devices, responsive to electric current and to temperature.

2.3.2 EXTERNALLY MOUNTED BREAKERS are defined as self-contained devices which are mounted individually or in combination via brackets, bus bars, plugged into terminal blocks, or panel mounted.

2.3.3 INTERNALLY MOUNTED BREAKERS are defined as subcomponent devices which are integral with a related unit such as a breaker/switch combination, within a motor housing for motor protection, etc. There are no implied restrictions on package design, provided the thermal circuit breaker exhibits performance characteristics and range of current rating as is covered by this document. While applications are generally found in motor vehicle electrical systems, other usages in DC circuit protection for accessories related or unrelated to the motor vehicle industry, may find this document of value for performance evaluation.

2.3.4 There are three general classes of breaker, defined as follows:

2.3.4.1 Type I—*Automatic Reset*—Automatic reset circuit breakers are cycling or continuously self resetting units which are opened by overcurrent.

2.3.4.2 Type II—*Modified Reset*—Modified reset circuit breakers are units which are opened by overcurrents and remain open as long as the power is on or until the load is removed. A number of cycles may occur prior to achieving the steady-state open condition.

2.3.4.3 Type III—*Manual Reset*—Manual reset circuit breakers are non-cycling units that are opened by overcurrents, but which remain open until manually reset.

3. Test Requirements

3.1 Test Equipment and Instrumentation

3.1.1 POWER SUPPLIES

3.1.1.1 A current and voltage regulated DC power supply shall be used for all tests except 3.2.7 Interrupt Test. The supply shall be capable of delivering 14 VDC and 28 VDC during open circuit portion of tests and have sufficient current output capacity to meet highest load requirements. Voltage and current settings shall be accurate to within ±1% of set point or better. Power transient response shall be such that a 30% step increase in power demanded by the load shall cause a transient in the regulation output which shall typically recover to within 3% of the final value within 100 ms or better. The power supply shall be operated with controlling circuitry to achieve all necessary test conditions. DC output shall have sufficient impedance via power resistors for buffering of load switching to prevent transitory output spikes.

3.1.1.2 Storage batteries specified in Table 1 shall be used as the power supply for 3.2.7 Interrupt Test. Open circuit voltage as specified shall be maintained by a battery charger or power supply with voltage regulated per Table 1 and current output restricted to 30 A or less.

TABLE 1—STORAGE BATTERIES

Voltage Rating	Minimum Battery Reserve Capacity[1]	Open Circuit Voltage
12 V	110 minutes	14.0 VDC ± 1% VDC
24 V[2]		28.0 VDC ± 1% VDC

[1] Reference SAE J537 (Types SAE 24-385/2, 24-410, 60-360, 74-410, or equivalent).
[2] Two 12 V batteries connected in series.

3.1.2 VOLTMETER—0 to 30 VDC maximum range, accuracy ±1/2%.

NOTE—A digital meter having at least a 3-1/2 digit readout with an accuracy of ±1% plus 1 digit is recommended for millivolt readings.

3.1.3 AMMETER—Capable of displaying full load current with an accuracy of ±1%. A calibrated shunt shall be used in series with the test circuit to minimize circuit resistance.

NOTE—Digital meter having at least a 3-1/2 digit readout with an accuracy of ±1% plus 1 digit is recommended for amperage readings when used in conjunction with a millivolt output calibrated shunt.

3.1.4 HIGH-VOLTAGE BREAKDOWN TESTER—Capable of providing 500 VAC RMS - 60 Hz, accuracy ±5%.

3.1.5 THERMOCOUPLE AND METER—0 to 150 °C minimum range, accuracy ±2%, maximum thermocouple wire size -0.22 mm^2 (#24 gage).

NOTE—A digital thermometer is recommended, with an accuracy of ±1 degree.

3.1.5.1 Two ambient observations are necessary during test cycles: ambient of test room and ambient of test chamber containing breakers under test.

3.1.5.2 Delta heat rise of terminations shall be calculated during 3.2.2.1 testing (at the tester's discretion).

3.1.6 OVEN—Variable controlled temperature oven able to vary temperature at a rate of 1 °C per minute and control temperature ±1 °C of set point accurate to ±2 °C.

3.1.7 TEST LOAD—Variable resistor(s) capable of varying circuit current to specified current requirements in conjunction with a power supply. Test circuit by-passes may be employed to verify current settings.

NOTE—Use of current regulated power supply would make it possible to use fixed resistors and achieve adjustment via the supply, however, voltage must be allowed to rise to either 14 VDC or 28 VDC during open circuit portion of tests when voltage rise is specifically required.

3.1.7.1 For transient current cycling tests, 12 VDC automotive lamps (such as sealed beams) shall be used in sufficient total wattage and quantity to meet test load requirements of 3.2.6.6.

3.1.8 TEST LEADS—Circuit breakers shall be tested using copper wire sizes listed in Table 2. The wire length shall be 1.22 m (48.0 in) for all voltages tested and insulation shall be rated 105 °C or better.

TABLE 2—TEST LEAD SIZES

Rated Current	SAE Metric Cable Size	SAE Wire Size
5 to 10A	1 mm²	#16
Greater than 10 to 15A	2 mm²	#14
Greater than 15 to 30A	3 mm²	#12
Greater than 30 to 40A	5 mm²	#10
Greater than 40 to 50A	8 mm²	#8

3.1.8.1 Termination of Test Leads—All test leads shall use standard commercially available terminals; ring terminals for threaded studs or screw type terminals, quick connect terminals for blade type terminations. To avoid secondary heat generation and/or adverse millivolt drop, it is recommended that test lead terminals be crimped and soldered; also, connections to breakers must be repeatable and uniform. Terminals shall be attached to breakers with screw threads to a specified torque value that is generally recommended for the particular thread size. Terminals applied to quick connect blades shall have an established minimum insertion and withdrawal force for test purposes to reduce the chance of marginal connections from fatigued test lead terminal materials. Secureness values shall be obtained from the terminal manufacturer. For custom terminations, consult the circuit breaker manufacturer.

NOTE—See SAE J858a and/or SAE J561 for a limited terminal listing.

3.1.9 TEST ENCLOSURE—Provide for a draft and convection air current free test chamber with a volume of approximately 5.66×10^4 cm³ (2.0 ft³). Chamber must allow for test lead access, internal chamber temperature monitoring, and indirect venting if needed to assure requirements of 3.2.1 are met.

3.2 Test Procedures

3.2.1 AMBIENT CONDITIONS—Environmental conditions have been selected for this document to help assure satisfactory operation under general customer use conditions. Circuit breakers shall be tested in still air at the temperatures indicated and allowed a 30 min soak without electrical load before testing (and repeated 30 min soaks for individual breakers that are involved in more than one test condition). Equipment listed in 3.1.9 fulfills the still air requirement. Where not otherwise specified, tests are to be run at 25 °C ± 2 °C (77 °F ± 3 °F). If room ambient is unstable or unregulated and an environmental chamber is employed, breakers under test must be isolated from chamber forced air currents. Paragraph 3.1.9 test enclosure shall be used within the chamber compartment and temperatures monitored per 3.1.5.1.

NOTE—Breakers stored in environments below 15 °C (59 °F) or above 35 °C (95 °F) shall be allowed a minimum of 1 h soak at the specified test temperature prior to initiation of any testing.

3.2.1.1 Test leads and terminations subject to thermal rise from test operations shall be allowed to restabilize to ambient conditions before starting a new test. Alternating between duplicate sets of leads is suggested.

3.2.2 CURRENT RATING TEST PROCEDURE—The circuit breaker shall be electrically connected with a pair of test leads described in 3.1.8 in series with the power supply as described in 3.1.1.1, a shunt with ammeter as described in 3.1.3, and an appropriate test load as described in 3.1.7 to provide the required current pass through the circuit breaker. Refer to Figure 1A for all tests except 3.2.6.6 and 3.2.7. Refer to Figure 1B for 3.2.6.6 only. Refer to Figure 1C for 3.2.7 only.

3.2.2.1 Maximum Voltage Drop Test Procedure—With the circuit breaker connected as described in 3.2.2, the voltage drop across the circuit breaker shall be measured while the breaker is passing full rated current and has achieved equilibrium (typically after 15 to 20 min of continuous operation at 100% of rated current, exhibited by no appreciable increase in voltage drop). If after 1/2 h of continuous operation at 100% of rated current, equilibrium has not been attained, continue testing until equilibrium has been attained and voltage drop is within acceptable limits, or, unit exceeds voltage drop limits and/or trips out.

NOTE—For applications sensitive to heat-rise at the circuit breaker terminations, thermocouple leads may be affixed to the terminations. General practice is to place the thermocouple lead on that portion of the terminal which is likely to come in contact with the wire lead insulation. Benchmark maximum values for delta heat rise (observed thermocouple temperature minus ambient temperature) are 65 °C delta for factory wired terminations and 50 °C delta for field wired terminations. Results of this testing will assist the user with circuit design and proximity considerations for breaker installation.

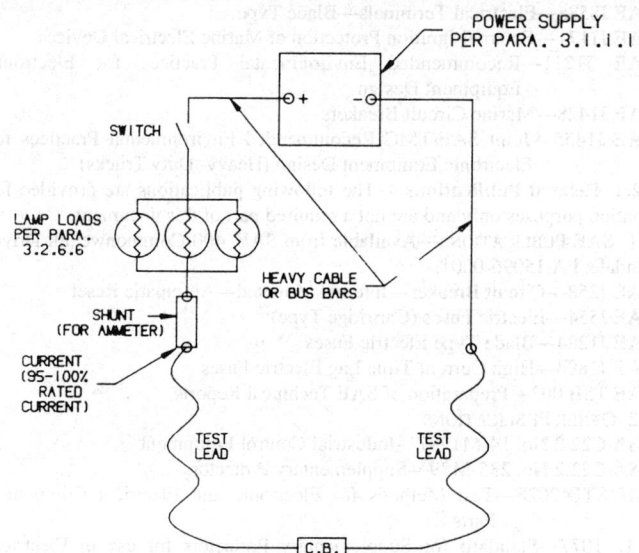

FIGURE 1B—TRANSIENT CYCLING TEST CIRCUIT

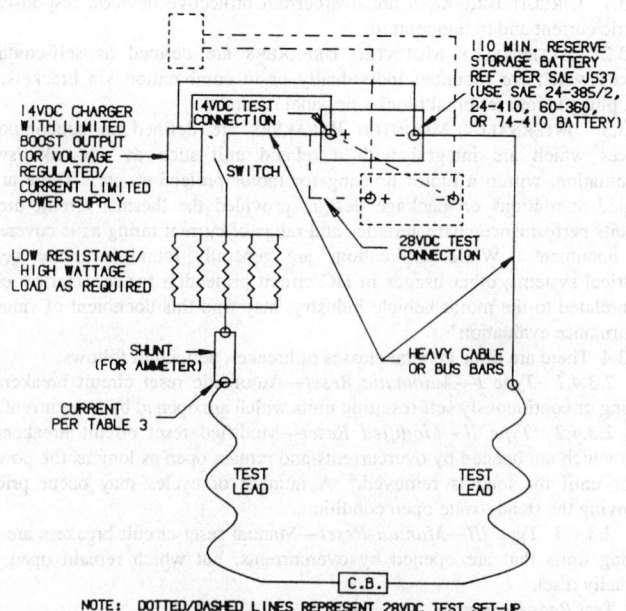

NOTE: DOTTED/DASHED LINES REPRESENT 28VDC TEST SET-UP. 14VDC CONNECTION AS SHOWN WOULD BE REMOVED.

FIGURE 1C—INTERRUPT CYCLING TEST CIRCUIT

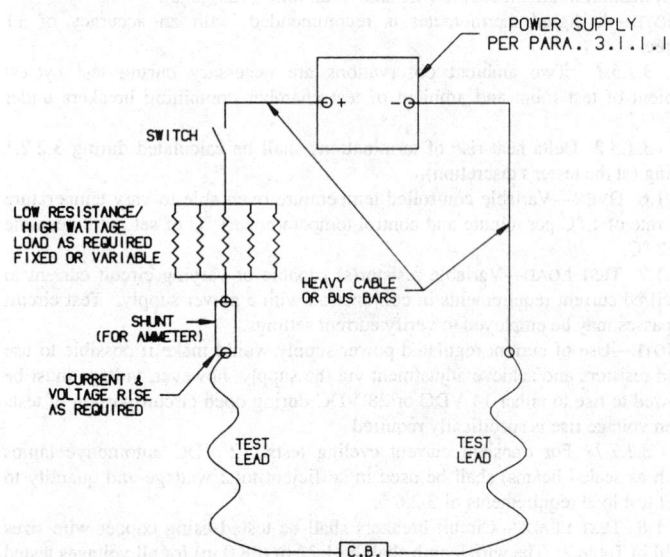

FIGURE 1A—CURRENT RATING TEST CIRCUIT

3.2.2.2 Overload Trip Rating Test—After a 30 min soak, reconnect the breaker as described in 3.2.2. Operate at 135% of rated current and record the elapsed time in seconds for the breaker to trip. If breaker has not opened after 1/2 h, discontinue the test. Repeat 30 min soak, operate at 200% of rated current and record elapsed time for the breaker to trip. If breaker has not opened after 60 s, discontinue the test.

3.2.3 EFFECTIVE CURRENT LIMITATION TEST PROCEDURE

3.2.3.1 Type I Only—With the circuit breaker connected as described in 3.2.2 and test current set at 200% of rated current, allow the breaker to cycle. At the end of 10 min (600 s), record the total elapsed time (in seconds) during which the breaker passed current. Multiply this figure by the 200% value of rated current used and divide the product by 600 s. The resulting quotient will represent the effective current for that particular breaker.

3.2.3.2 Type II Only—With the circuit breaker connected as described in 3.2.2 and test current set at 200% of rated current, allow the breaker to cycle. Begin timing the test from when the breaker initially trips and continue application of test current for 1 min (60 s). Count the number of cycles (one trip and one reset equals one cycle) that the breaker passed 200% of rated current. At the end of the 1 min (60 s) time period, observe if the breaker has stopped cycling and if it is passing a reduced current value. If so, record the reduced current value, expressed either in milliamps or tenths of an amp. If not, continue the cycling at 200% until cycling stops and a reduced current value is displayed for recording, or, terminate the test if 5 min elapses and the breaker is still cycling. For this particular test, the open circuit voltage must rise to 14 VDC ± 1% VDC for 12 VDC breakers 28 VDC ± 1% VDC for 24 VDC breakers and remain stable during reset portions of the cycling, and at the end of the 1 min (60 s) test duration when the reduced current value is measured (or until 5 min test termination limit if necessary).

3.2.3.3 Type III Only—Disable the trip indicator/reset mechanism in such a fashion as to allow the breaker to perform as a Type I style. Perform the same test instructions as in 3.2.3.1.

NOTE—If the Type III breaker is constructed in such a way that depression of the reset button does not allow the thermal element to cycle as if a Type I design, but rather trips and resets by definite mechanical action only, then this test is not required.

3.2.4 VOLTAGE BREAKDOWN TEST PROCEDURE (EXTERNALLY MOUNTED TYPE I AND TYPE II BREAKERS ONLY)

3.2.4.1 With the circuit breaker connected as described in 3.2.2, adjust the current to 400% of the circuit breaker rating and allow the breaker to cycle. At the end of 10 min, check the continuity at 440 VAC between each terminal of the circuit breaker individually and the cover of the breaker with the breaker in both an open and closed circuit condition.

NOTE—This test is not required on devices utilizing nonmetallic, nonconductive covers.

3.2.5 NO CURRENT TRIP AND RESET TEMPERATURE TEST PROCEDURE

3.2.5.1 Place the circuit breaker(s) in a variable temperature controlled environmental chamber heated to 10 °C below the minimum opening temperature (use 72 °C starting point for 10 A and below rated breakers and 102 °C starting point for above 10 A rated breakers) and soak at the starting temperature for 30 min. Utilize the test enclosure as described in 3.1.9 to shield the breaker(s) under test from environmental chamber convection currents (forced air models). After soak, raise temperature at a rate not exceeding 1 °C per minute. When the temperature has exceeded the minimum temperature the breaker(s) must endure without opening, continue elevating the temperature at the same rate of increase and record the temperature at which the breaker(s) opens. A test termination point at 200 °C is suggested. If a breaker under test fails to open by 200 °C, reevaluate performance per 3.2.2.2. Once the breaker has opened (or all breakers have opened if testing in multiples) decrease the temperature at a rate not exceeding 1 °C per minute and record the temperature at which the breaker(s) closes.

NOTE—If electrically operated indicators are employed to signal opened and closed states, voltage and current shall be at trace levels--6 V/100 mA maximum to prevent heating of breaker thermal elements if they are part of the indicator circuit loop, and to prevent operation of Type II heating circuits during ambient induced open cycles.

3.2.6 ENDURANCE TEST PROCEDURE

3.2.6.1 Test current for endurance tests shall be 600% of rated current, except where specified otherwise. Utilize Figure 1A test circuit except for 3.2.6.6.

3.2.6.2 With the circuit breaker connected as described in 3.2.2, Type I externally mounted circuit breakers shall be cycled for 30 min. The circuit breaker shall then be capable of passing 80% of rated current for 1/2 h, without tripping. Record voltage drop as in 3.2.2.1 at 80% current.

3.2.6.3 Using the circuit breaker from 3.2.6.2 (assuming 3.2.6.2 requirements were met), reconnect as described in 3.2.2 and cycle Type I breaker until failure as defined in 4.6.2.

NOTE—At tester's discretion, or as published by the manufacturer, an arbitrary minimum cycle time may be established as a milestone threshold for termination of the test (e.g.; 8 h, 12 h, etc.), to aid in test throughput.

3.2.6.4 With the circuit breaker connected as described in 3.2.2, Type II circuit breakers shall first be subjected to 30 on-off cycles. The "on" time of each cycle shall be 60 s, during which time the circuit breaker must open at least once, with repeated cycling possible and open circuit voltage rising to 14 VDC for 12 V breakers/28 VDC for 24 V breakers. The "off" time of each cycle shall be long enough to allow the circuit breaker to close by de-energizing the test circuit prior to initiating a subsequent "on" cycle. The "on" time of the thirtieth cycle shall be 24 h with voltage reduced to 11.3 V for 12 V breakers and 22.6 V for 24 V breakers once breakers are in a steady-state open circuit as induced by the heating circuit. During this time, the circuit breaker contacts must remain open. The circuit breaker shall then be allowed to reclose and again be subjected to the 30 cycle test, excluding the 24 h "on" time of the last cycle. The circuit breaker shall then be subjected to 80% of rated current for 1/2 h. Record voltage drop as in 3.2.2.1 at 80% current.

NOTE—For purposes of evaluating heater circuit endurance exclusive of cycling, a supplemental test is suggested in which the breakers are cycled until remaining open from heater circuit operation at 14 VDC for 12 V units and 28 VDC for 24 V units and are kept energized for 100 h elapsed time, observing for uninterrupted operation.

3.2.6.5 With the circuit breaker connected as described in 3.2.2, Type III circuit breakers shall be cycled for 100 on-off cycles utilizing the trip indicating/reset mechanism of the breaker. Means shall be provided to detect when the circuit breaker is capable of resetting in order to initiate the next cycle. The circuit breaker shall then be subjected to 80% of rated current for 1/2 h. Record voltage drop as in 3.2.2.1 at 80% current.

3.2.6.6 Transient Current Cycling Endurance Test—At the option of the tester, a transient current cycling endurance test shall be performed as is herein described. With the circuit breaker connected as described in 3.2.2 and utilizing Figure 1B test circuit, apply a transient current cycling waveform as shown in Figure 2 through the breaker for 25 000 cycles. The lamps employed for the test load as described in 3.1.7.2 shall be sufficient to create a current level of 95% to 100% of the circuit breaker rating. Lamp aging which diminishes load over time is acceptable, however, lamps that fail must be replaced immediately. This test procedure is applicable to all breaker types.

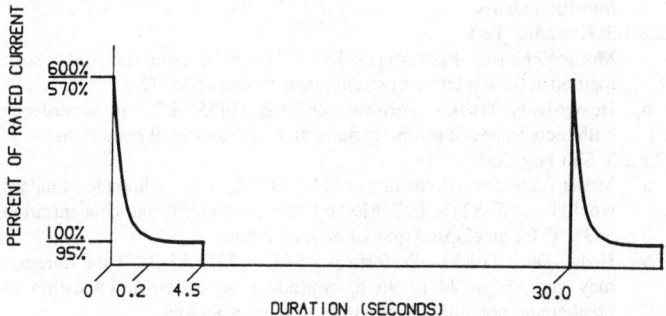

FIGURE 2—TRANSIENT CURRENT CYCLING WAVEFORM

3.2.7 INTERRUPT TEST PROCEDURE

3.2.7.1 Using the power source described in 3.1.1.2, test current shall be in accordance with Table 3. Refer to Figure 3 for interrupt cycle definition and Figure 1C test circuit.

TABLE 3—INTERRUPT TEST CURRENT REQUIREMENTS

Rated Current	12 VDC Amps	24 VDC Amps
5 to 10A	150	100
Greater than 10 to 15A	225	150
Greater than 15 to 20A	300	200
Greater than 20 to 30A	450	300
Greater than 30 to 40A	600	400
Greater than 40 to 50A	750	500

3.2.7.2 With the circuit breaker connected as described in 3.2.2 (utilizing 3.1.1.2 power source), Type I circuit breakers shall be subjected to 1-1/2 cycles of interrupt current.

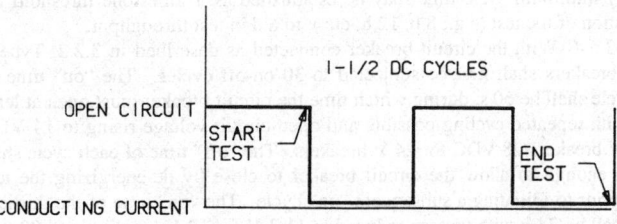

FIGURE 3—INTERRUPT CYCLE DEFINITION

3.2.7.3 With the circuit breaker connected as described in 3.2.2 (utilizing 3.1.1.2 power source), Type II circuit breakers shall be subjected to 1-1/2 cycles of interrupt current. If the breaker does not reset after 1/2 cycle due to its normal Type II function, terminate the test at that point.

3.2.7.4 With the circuit breaker connected as described in 3.2.2 (utilizing 3.1.1.2 power source), Type III circuit breakers shall be subjected to 1-1/2 cycles of interrupt current. Procedure shall be to apply fault current for first 1/2 cycle. The next 1/2 cycle shall consist of allowing the unit to come into the "ready to reset" mode. As soon as reset capability is enabled, the reset mechanism shall be activated to restore the circuit, at which time the last 1/2 cycle of interrupt current will be present.

3.2.8 ENVIRONMENTAL TESTS

3.2.8.1 Since end use applications may differ, the following tests are recommended, but not mandatory to determine general suitability of components. All tests shall follow the guidelines as set forth in SAE J1211 or SAE J1455 unless otherwise specified.

3.2.8.1.1 Temperature Test
  a. Motor Vehicles—Perform per SAE J1211, 4.1.3. Minimum temperature shall be -40 °C, maximum temperature shall be 105 °C. Cycle per SAE J1211 Figure 2B for a total elapsed time of 96 h.
  b. Heavy-Duty Trucks—Perform per SAE J1455, 4.1.3. Test as described for temperature cycling, thermal shock, and thermal stress at the specified test temperatures and in accordance with the temperature transition charts.

3.2.8.1.2 Humidity Test
  a. Motor Vehicles—Perform per SAE J1211, 4.2.3 using the 10 day soak method at 95% relative humidity, temperature at 38 °C.
  b. Heavy-Duty Trucks—Perform per SAE J1455, 4.2.3 in accordance with recommended test procedures and environmental conditions.

3.2.8.1.3 Salt Fog Test
  a. Motor Vehicles—Perform per SAE J1211, 4.3. Alternate standard would be MIL-STD-202F, Method 101D, with a 5% salt concentration at 35 °C for an elapsed time of 48 h minimum.
  b. Heavy-Duty Trucks—Perform per SAE J1455, 4.3.3. Time duration may vary, from 24 to 96 h, depending on anticipated location of breaker and potential for exposure to saline solutions.

3.2.8.1.4 Immersion and Splash Test—For general guidelines refer to SAE J1211, 4.4 (Motor Vehicles) or SAE J1455, 4.4 (Heavy-Duty Trucks).
  NOTE—Immersion testing shall apply only to devices which are stated as being "waterproof", "sealed", "watertight", etc. Test procedures per SAE J1171, Section 5 may be followed for basic test requirements.
  NOTE—Splash testing shall apply only to devices which are stated as being "splashproof", "water resistant", "weatherproof", etc. Test procedure in SAE J1428, 5.1.3 may be used. Devices passing immersion testing do not require splash testing.
  NOTE—Chemicals used for testing shall be restricted to water for immersion and splash. Evaluation of external identification marking shall be conducted by splash testing utilizing commonly encountered chemicals which shall include: engine oil, power steering fluid, windshield washer solvent, gasoline, diesel fuel, anti-freeze, steam, and salt water.

3.2.8.1.5 Mechanical Vibration Test
  a. Motor Vehicles—Perform per SAE J1211, 4.7.3. Test shall be for 1 h in each of three mutually perpendicular primary axes using the suggested current practice per SAE J1211, Figure 4.
  b. Heavy-Duty Trucks—Perform per SAE J1455, 4.9.4. Test shall be for 1 h in each of three mutually perpendicular primary axes using the suggested current practice per SAE J1455.

3.2.8.1.6 Drop Test
  a. Motor Vehicles—Test breakers shall be dropped onto a steel plate 6.35 mm (1/4 in) thick in one of six different directions along three mutually perpendicular primary axes from a height of 1.0 m ± 0.01 m.
  b. Heavy-Duty Trucks—Perform per SAE J1455, 4.10.3.1.

3.2.8.1.7 Environmental Extremes Test
  a. Motor Vehicles—For reference, see SAE J1211, 5.1 and 5.2. Actual test shall be to soak test breakers at 150 °C ± 2 °C for 240 h without any electrical load.
  b. Heavy-Duty Trucks—Generally handled in 3.2.8.1.1, but the motor vehicle test as described may be performed if deemed appropriate.

4. *Performance Requirements*

4.1 Current Rating—With the circuit breaker connected as described in 3.2.2, all circuit breakers shall pass 100% ± 1.5% of rated current continuously for a minimum of 1/2 h, shall open at 135% ± 1.5% of rated current within 1/2 h, and shall open at 200% ± 1.5% of rated current within 1 min. In addition, internally mounted circuit breakers shall pass 80% ± 1.5% of rated current at 52 °C ± 2 °C for 1/2 h without opening.

4.2 Maximum Voltage Drop—Using the procedure described in 3.2.3, the maximum voltage drop across the circuit breaker shall be within the limits shown in Figure 4.

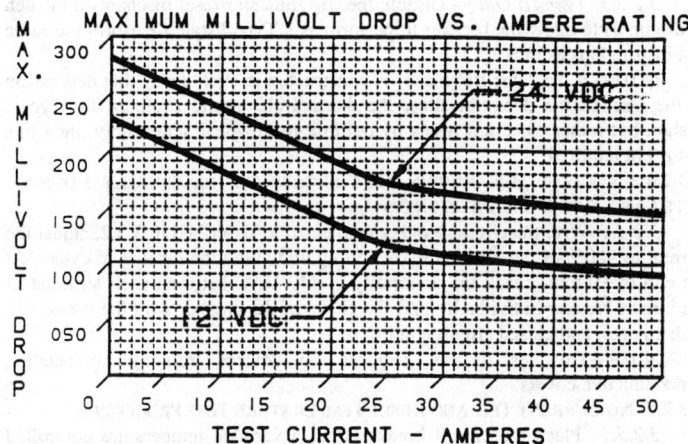

FIGURE 4—VOLTAGE DROP CURVES

4.3 Effective Current Limitation

4.3.1 TYPE I—Using the test procedure described in 3.2.3.1, the maximum value of effective current passed through the automatic reset circuit breaker shall not be greater than 135% of its rated current for an externally mounted breaker or greater than 150% of its rated current for an internally mounted breaker.

4.3.2 TYPE II—Using the test procedure described in 3.2.3.2, the current passing through the modified reset circuit breaker shall not exceed 1 A (1000 mA) after reaching a maintained open condition. The maintained open condition shall be reached within 60 s after the breaker initially opens. Breakers that exceed 60 s, but maintain open condition prior to 5 min test termination, shall be reported as variant. Suitability should be evaluated with regard to intended application.

4.3.3 TYPE III—Using the test procedure described in 3.2.3.3, the maximum value to effective current passed through the manual reset breaker (with disabled reset mechanism) shall not be greater than 135% of its rated current for an externally mounted breaker or greater than 150% of its rated current for an internally mounted breaker.

4.4 Voltage Breakdown—Using the test procedure described in 3.2.4, there shall be no continuity between either terminal of the circuit breaker and the cover.

4.5 No Current Trip and Reset Temperature—Using the procedure described in 3.2.5, all circuit breakers shall open and reclose in accordance with the following requirements:

NOTE—Recognizing device design variations as well as the possibility of ambient compensating mechanisms, it is recommended that manufacturers' temperature derating curves be consulted for application considerations. Consequently, the test procedure of 3.2.5 and performance requirements of 4.5 may, at the tester's discretion, be omitted from the test program.

4.5.1 Circuit breakers rated 10 A or less shall not open at less than 82 °C and shall reclose before the temperature is below 52 °C.

4.5.2 Circuit breakers rated above 10 A shall not open at less than 112 °C and shall reclose before the temperature is below 82 °C.

4.6 Endurance Test

4.6.1 Type I externally mounted circuit breakers shall be tested as described in 3.2.6.2 and then shall continuously pass 80% ± 1.5% of rated current for 1/2 h and the millivolt drop at 80% of rated current shall be within the limits specified in Figure 4 at the 80% rating value.

4.6.2 If failure occurs with the circuit breaker connected as described in 3.2.6.3, the ultimate failure of all circuit breakers shall result in an open circuit in the circuit breaker, and there shall be no damage to the associated wiring. Failure falls into three general categories: catastrophic failure—part of the electrical contacts and/or thermostatic material burns up and the circuit path is broken (contained within breaker housing/no external manifestations); operational fatigue—thermostatic material loses original form, no longer cycles or chatters (trip/reset excursions less than 1 s in duration), or loses contact pressure resulting in circuit discontinuity; contact failure—electrical contact material erodes or carbons to a level of nonconductance or high resistivity, causing inability to pass current (or trace levels only, below 1 A).

NOTE—In some instances, a high circuit resistance, and/or low current power source, may not provide enough fault current to assure that ultimate failure will always result in an open circuit breaker, or prevent wiring insulation breakdown near breaker terminations.

4.6.3 Type II externally mounted circuit breakers shall be tested as described in 3.2.6.4 and then shall continuously pass 80% ± 1.5% of rated current for 1/2 h and the millivolt drop at 80% of rated current shall be within the limits specified in Figure 4 at the 80% rating value. If performing suggested heater circuit endurance test, desired performance shall be 100 h of continuous heater circuit function, exhibiting no incidence of primary breaker circuit reclosure. Heater circuit performance may vary due to design, therefore, performance limits are advisory.

4.6.4 Type III externally mounted circuit breakers shall be tested as described in 3.2.6.5. The breaker shall trip and reset without failure. There shall be no measurable current passing through the breaker while in the tripped position for the 100 cycles. It shall then continuously pass 80% ± 1.5% of rated current for 1/2 h and the millivolt drop at 80% of rated current shall be within the limits specified in Figure 4 at the 80% rating value.

4.6.5 All breakers when tested as described in 3.2.6.6 shall maintain continuity for the 25 000 cycles.

4.7 Interrupt Test

4.7.1 When tested as described in 3.2.7, the preferred performance is for the circuit breaker to demonstrate continuity and functionality by passing 80% ± 1.5% of rated current for 1/2 h. Breakers which clear the circuit but cease to function shall be examined according to the guidelines of 4.6.2.

4.8 Environmental Tests

4.8.1 TEMPERATURE TEST (THERMAL SHOCK)

4.8.1.1 After completion of test as described in 3.2.8.1.1, the circuit breaker shall exhibit no signs of physical damage and be capable of passing 80% ± 1.5% of rated current for 1/2 h.

4.8.2 HUMIDITY TEST

4.8.2.1 After completion of test as described in 3.2.8.1.2, the circuit breaker shall perform in accordance with 4.1 and 4.2 at 100% ± 1.5% of rated current.

4.8.3 SALT FOG TEST

4.8.3.1 After completion of test as described in 3.2.8.1.3, the circuit breaker shall perform in accordance with 4.1 and 4.2 at 100% ± 1.5% of rated current. Physical corrosion shall not prevent proper fit and function of the breaker.

4.8.4 IMMERSION AND SPLASH TEST

4.8.4.1 Immersion Test—After completion of test as described in 3.2.8.1.4, pass/fail criteria of SAE J1171, Section 5 shall apply.

4.8.4.2 Splash Test—After completion of test as described in 3.2.8.1.4, pass/fail criteria of SAE J1428, 5.1.3 shall apply.

4.8.5 MECHANICAL VIBRATION TEST

4.8.5.1 While testing as described in 3.2.8.1.5, the circuit breaker shall continuously pass 80% ± 1.5% of rated current during the last 1/2 h with no loss in continuity. Loss of continuity is defined as a resistance across the circuit breaker terminals in excess of 100 Ω, or a voltage rise across the terminals exceeding 50% of test circuit unloaded voltage for longer than 5 ms.

4.8.6 DROP TEST

4.8.6.1 After completion of test as described in 3.2.8.1.6, the circuit breaker shall not exhibit any physical damage. It shall be capable of passing 80% ± 1.5% of rated current for 1/2 h minimum and comply with 4.2.

4.8.7 ENVIRONMENTAL EXTREMES TEST

4.8.7.1 After completion of testing described in 3.2.8.1.7, there shall be no significant degradation of product materials, such as softening of plastics, creep, or other deformations that could alter product performance or reliability. If test unit is suspect, perform tests per 3.2.2, 3.2.3, 4.1, and 4.2.

4.9 General Requirements

4.9.1 MARKING—Externally mounted circuit breakers shall be permanently and legibly marked with the current rating and voltage as well as any other identifying part numbers. Circuit breaker exterior package designs, which may appear identical in Type I or Type II versions, shall be marked in a consistent fashion to provide distinction between Type I or Type II. Date coding is strongly recommended. Marking shall be generally resistant to common contaminants and chemicals. Evaluate suitability during 3.2.8.1.4 testing.

NOTE—Specifying of marking information, use of color codes, or custom information shall be the responsibility of the O.E.M.

4.9.2 APPLICATION—The specific current capacity of the circuit breaker is a function of the particular electrical system being utilized. It is recommended that actual performance be verified through testing experimentally in the proposed application. To aid in determining the actual capacity change caused by variations in circuit parameters, several factors should be considered by the application engineer.

4.9.2.1 Voltage Rating—The voltage rating marked on the externally mounted circuit breaker is the maximum value recommended (system, not charging voltage). Use at higher voltages may significantly shorten the ultimate life under overload conditions and/or destroy Type II components.

4.9.2.2 Current Rating—The current rating marked on externally mounted circuit breakers is the maximum value/ultimate rating but is subject to redefinition based on the application analysis. It is generally not desirable to specify circuit protection where the breaker will pass 100% of rated current during normal continuous circuit load. Application engineers generally specify circuit protection such that normal continuous circuit loads are approximately 75 to 80% of the circuit breaker current rating. Paragraphs 4.9.2.3 and 4.9.2.4 explain why.

4.9.2.3 Ambient Temperature—The circuit breakers covered by this document are thermal devices. Changes in the ambient temperature will have an effect on the current carrying capacity and on the effective limitation of current during overload cycling. Therefore, the application engineer needs to consider environmental conditions to which the breaker will be subject during operation and make use of derating curve information if available.

4.9.2.4 Wire and Terminations—The connecting wires and their terminations will affect the heat dissipation characteristics of the circuit breaker. Deviations from the circuit breaker application specifications may affect the current carrying capacity or the effective limitation of current during overload cycling. Heat sources associated with poor interfacing terminations that connect with the circuit breaker may be a cause of abnormal circuit resistance, excessive millivolt drop, damage to associated wiring, and ultimately, significant derating.

5. Qualification Test Sequence

5.1 Test Programs—There shall be two separate test sequences; a basic test cycle which covers all core requirements, and an expanded test cycle, which in addition to core requirements, includes all possible tests (most of which are considered optional and are designated by a dotted line border). Figure 5 outlines the basic test cycle and Figure 6 the expanded test cycle.

5.2 Sample Sizes—Basic test cycle requires 15 samples. All 15 receive the first three tests. Afterwards the group is divided by 3 into 5 piece subgroups for the remaining tests. The expanded test cycle requires 45 samples. Fifteen are treated the same as in Figure 5. The other 30 are divided into six groups for the environmental tests.

23.140

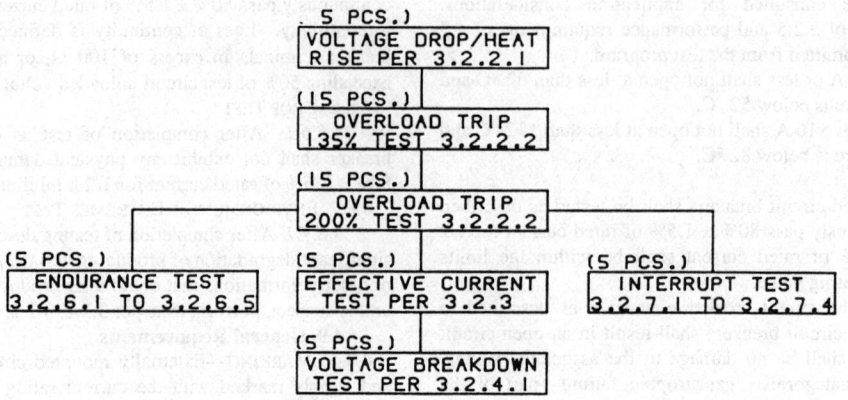

FIGURE 5—BASIC TEST CYCLE

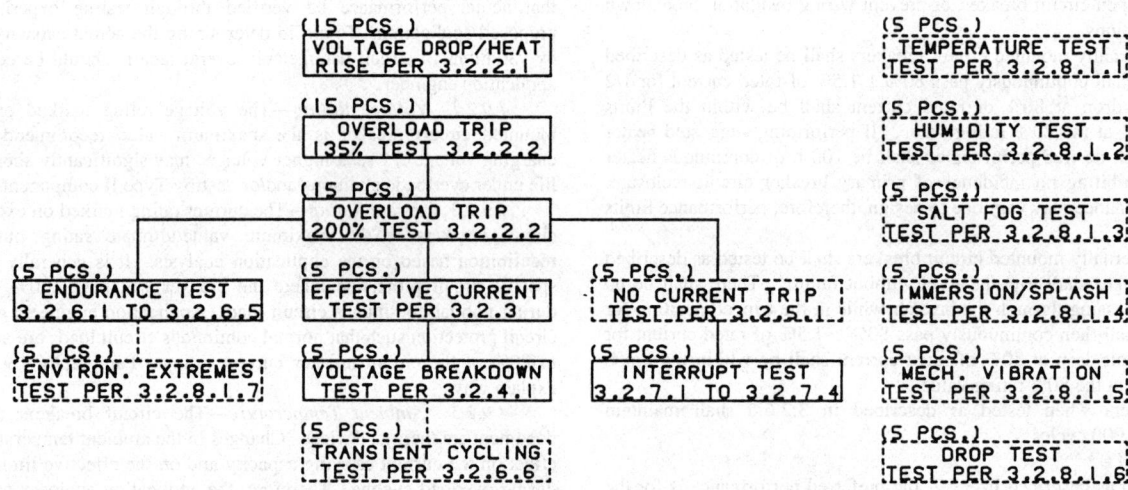

FIGURE 6—EXPANDED TEST CYCLE

CIRCUIT BREAKER—INTERNAL MOUNTED—AUTOMATIC RESET—SAE J258

SAE Recommended Practice

Report of Electrical Equipment Committee approved June 1971.

1. Definition—An internally mounted circuit breaker is one mounted within an automotive switch or other automotive device for protection against overload of the wiring. In a given application, this same circuit breaker may be designed to protect against overload of the electrically operable devices.

2. Test Procedure

2.1 The circuits containing the internally mounted circuit breakers shall have 3 ft leads connected to their terminals. The wire, size, and mating terminals specified must be the same as the intended application or, if there are a variety of applications, then the minimum wire size with which the circuit breaker will meet the requirements of this recommended practice.

2.2 The *circuits* containing the internally mounted circuit breakers shall be allowed a 0.20 V drop at rated current, when tested individually, in addition to the voltage drop allowed in the switch or other device specification.

3. Requirements—In a switch or other device with multiple circuit breakers, the following requirements are for testing each circuit breaker individually, with the design load applied to the other circuit breakers:

3.1 Continuously carry 100% of its rated current (as specified by the manufacturer for 1 h at 75 ± 3 F (24 ± 1.6 C).

3.2 After a 1 h soak at 125 ± 3 F (51.6 ± 1.6 C) and no load, the circuit breaker shall carry 80% of its rated current for 1 h at that temperature.

3.3 Effective current value shall not exceed 140% of its rated current when tested according to procedure as outlined in SAE J553 for Type I.

3.4 The circuit breaker must fail safe (open circuit) when tested as outlined in the life test of SAE J553 for 30 min or until destruction.

IGNITION SWITCH—SAE J259

SAE Recommended Practice

Report of Electrical Equipment Committee approved June 1971.

1. Definition—An ignition switch is that part of an electrical system by which the operator of a vehicle causes the ignition system to function. In addition to ignition circuitry, the switch may have circuits which control electrical accessories, engine starting, warning indicator lamp checking, etc.

2. Temperature Test

2.1 To insure basic function, the switch shall be manually cycled for 10 cycles at design electrical load at 75 ± 10 F (24 ± 5.5 C), $165 +0, -5$ F ($74 +0, -2.8$ C), and $-25 +5, -0$ F ($-32 +2.8, -0$ C) after a 1 h exposure at each of these temperatures. The switch shall be electrically and mechanically operable during each of these cycles.

2.2 This same switch shall be used for the endurance test described below.

3. Endurance Test Setup

3.1 The switch shall be set up to operate its design electrical load.

3.2 The test shall be set up to operate the switch for the prescribed number of complete cycles.

One complete cycle shall consist of sequencing through each position (with dwell in each position) and return without dwell in intermediate positions to the initial position.

The test equipment shall be so arranged as to provide the following mechanical time requirements:

Travel Time—0.1–0.5 s (time from one position to the next).
Dwell Time—1.0–2.0 s (time in each position).

3.3 During the test the switch shall be operated at:
6.4 V d-c for a 6 V system
12.8 V d-c for a 12 V system
25.6 V d-c for a 24 V system

These voltages shall be measured at the input termination on the switch.

The power supply shall not generate any adverse transients not present in motor vehicles and shall comply with the following specifications:

(a) *Output Current*—Capable of supplying the continuous current of the design electrical load.

(b) *Regulation*

Dynamic—The output voltage at the supply shall not deviate more than 1.0 V from zero to maximum load (including inrush current) and should recover 63% of its maximum excursion within 100 ms.

Static—The output voltage at the supply shall not deviate more than 2% with changes in static load from zero to maximum (not including inrush current), and means shall be provided to compensate for static input line voltage variations.

(c) *Ripple Voltage*—Maximum 300 mV peak to peak.

4. Endurance Requirements

4.1 The switch shall be capable of satisfactory operating for 25,000 complete cycles at a temperature of 75 ± 10 F (24 ± 5.5 C).

4.2 When the switch has a position which operates accessories only, the switch shall be cycled to that position during 25% of the endurance cycles.

4.3 The voltage drop from the input terminal(s) to the corresponding output terminal(s) shall be measured before and after the endurance test and shall not exceed 0.30 V (the average of three consecutive readings) at 10.0 A except in circuits such as warning indicator lamp checking and/or grounding type circuits which do not exceed 1 A load. In these circuits, the voltage drop shall not exceed 0.50 V (the average of three consecutive readings) at 1.0 A. If wiring is an integral part of the switch, the voltage drop measurement shall be made including 3 in. of wire on each side of the switch; otherwise, measurement shall be made at switch terminals.

DOOR COURTESY SWITCH—SAE J2108 DEC91

SAE Recommended Practice

Report of the SAE Circuit Protection and Switching Devices Standards Committee approved June 1991. Rationale statement available. Reaffirmed by the SAE Switch Task Force of the SAE Auxiliary Devices Standards Committee December 1991.

1. Scope—This SAE Recommended Practice defines the test conditions, procedures, and performance requirements for 6, 12, and 24 V Door Courtesy Switches which are intended for use in motor vehicles.

2. References

2.1 Applicable Documents—There are no referenced publications specified herein.

2.2 **Definitions**—The courtesy lamp switch is a door or door latch actuated device which controls the electrical operation of the courtesy lamp, ignition key alarm, and other related components.

2.3 **Types**

2.3.1 GROUNDED AND NON-GROUNDED—Grounded switches provide an electrical path to vehicle ground through their mounting attachment. Non-grounded switches have their electrical conductors insulated from vehicle ground.

2.3.2 SINGLE TERMINAL—Characterized by one wiring connection to the vehicle.

2.3.3 MULTI-TERMINAL—Any other terminal/connector configuration, other than single terminal.

2.3.4 SPECIAL—Switch types, which by their design, construction, and function, require separate definition.

2.4 Cycle—One cycle shall consist of allowing the actuation portion of the switch to move or be moved throughout its travel and to return to its initial position.

3. Test Requirements

3.1 Test Equipment and Instrumentation

3.1.1 POWER SUPPLY—The power supply shall comply with the following specifications:

3.1.1.1 *Output Current*—Capable of supplying the continuous and inrush currents of the design load (reference 3.2.1.1).

3.1.1.2 *Regulation*—Dynamic: The output voltage at the supply shall not deviate more than 1.0 V from zero to maximum load (including inrush current) and should recover 63% of its maximum excursion within 100 ms.

Static: The output voltage at the supply shall not deviate more than 2% with changes in static load from zero to maximum (not including inrush current), and means shall be provided to compensate for static input line variations.

3.1.1.3 *Ripple Voltage*—Maximum 300 mV peak-to-peak.

3.1.2 VOLTMETER—0 to 30 maximum full scale deflection, accuracy ± $1/2$%. (Note: a digital meter having at least a $3^{1}/_{2}$ digit readout with an accuracy of ± 1% plus 1 digit is recommended for millivolt readings.)

3.1.3 AMMETER—Capable of carrying full system load current, accuracy ± 3%.

3.2 Test Procedures—Environmental conditions have been selected for this document to help assure satisfactory operation under general customer use conditions. It is essential to duplicate the specific environmental conditions under which the device is expected to function.

3.2.1 ELECTRICAL LOADS

3.2.1.1 The design load applied to the switch is the electrical load defined by the number and type of bulbs (or other electrical load devices) to be operated by each circuit of the switch. For example, the design load for the courtesy lamp circuit may be four 1156 bulbs.

3.2.1.2 The switch shall be operated at 6.4 V DC ± 0.2 V DC for a 6 V system, 12.8 V DC ± 0.2 V DC for a 12 V system, or 25.6 V DC ± 0.2 V DC for a 24 V system. These voltages shall be the open circuit voltage measured at the input terminations of the switch.

3.2.2 TEMPERATURE TEST PROCEDURE

3.2.2.1 The switch shall be exposed for 1 h without electrical load to each of the following temperatures: 25 °C ± 5 °C, 74 °C +0, -3, -32 °C +3, -0. The switch shall be cycled at each temperature for ten cycles at design load.

3.2.2.2 The same switch shall be used for the endurance test described in 3.2.3.

3.2.3 ENDURANCE TEST PROCEDURE

3.2.3.1 The switch shall be electrically connected to operate its design load (both primary and secondary circuit function design electrical loads) at a temperature of 25 °C ± 5 °C.

3.2.3.2 The switch shall be operated for a minimum of 50 000 cycles. The speed and the incident angle of actuation shall be representative of the point of application ("A" pillar, "B" pillar) in the vehicle. The test equipment shall be designed to provide this timing: Travel Time: 0.1 to 1.0 s (time to travel from one extreme position to the other). Dwell Time: 2.0 to 5.0 s (time spent stationary at an extreme position). Make and Break Rate Range:

"A" Pillar application: 30 to 300 mm/s.

"B" Pillar application: 0.3 to 3.0 m/s.

3.2.3.3 At the conclusion of the endurance testing, the switch shall be operated for 1 h in each of its positions with the design load connected.

3.2.4 VOLTAGE DROP TEST PROCEDURE

3.2.4.1 Voltage drop from the input terminal(s) to the corresponding output terminal(s) shall be measured at design load before and after the completion of the endurance test. Three consecutive readings shall be taken and the average recorded. If wiring is an integral part of the switch, the voltage drop measurement shall be made by including 75 mm ± 6 mm of wire on each side of the switch. Otherwise, the measurement shall be made at the switch terminals.

4. Performance Requirements

4.1 During and after each of the cycles described in 3.2.2 and 3.2.3, the switch shall operate without hesitation mechanically, e.g., not more than 1.0 s, and shall be within its electrical design specifications.

4.2 The voltage drop shall not exceed 0.3 V when measured as in 3.2.4, either before or after the tests described in 3.2.3.

AUTOMOBILE, TRUCK, TRUCK-TRACTOR, TRAILER, AND MOTOR COACH WIRING—SAE J1292 OCT81

SAE Recommended Practice

Report of the Electrical Equipment Committee, approved June 1980, editorial change October 1981.

[This SAE Recommended Practice combines, revises, and replaces two previous recommended practices: SAE J555a and SAE J556.]

1. Scope—This SAE Recommended Practice covers the application of primary wiring distribution system harnesses to automotive, truck, and similar type vehicles. This is written principally for new vehicles but is also applicable to rewiring and service. It covers the areas of performance, operating integrity, efficiency, economy, uniformity, facility of manufacturing, and service. This practice applies to wiring systems of less than 50 V.

2. General Section

2.1 Definition—The systems of installation known as two wire or single wire are to be designated respectively as *insulated–return* and *ground–return* systems. Installations in which the frame and/or body of the vehicle are used as part of the return circuit are considered as *ground–return* systems.

2.2 Insulated Cable—All insulated cable shall conform to SAE Standards J1127 and J1128.

2.2.1 CONDUCTORS

2.2.1.1 All conductors are to be constructed in accordance with SAE J1127 and J1128 except when good engineering practice dictates special strand constructions.

2.2.1.2 Conductor materials and stranding other than copper can be used if all applicable requirements for physical, electrical, and environmental conditions are met as dictated by the end application.

2.2.2 CONDUCTOR INSULATION—Physical and dimensional values of conductor insulation are to be in conformance with the requirements of SAE J1127 or J1128 except when good engineering practice dictates special conductor insulations.

2.3 Insulated Cable Application

2.3.1 Select cable insulation in accordance with the vehicle's working environment. Consideration is given to physical and environmental factors such as flexing, heat, cold, bend, oil and fuel contact, dielectric, abrasion, short circuit, and pinch resistance among others.

NOTE—Most vehicle working environments permit the use of a thermoplastic insulated, SAE type GPT, general purpose cable. A cable of this type is generally used in static (non-flexing) applications when nominal abrasion, heat, cold, oil, dielectric, short circuit, and pinch resistance properties are desired.

2.3.2 Where vehicle working environments for cable require additional physical and environmental characteristics, upgraded insulations such as SAE types HDT, GPB, HDB, STS, HTS, and SXL shall be used as the severity of the applications dictate.

2.3.3 Specific continuous duty temperature limitations for each SAE cable type shall be observed. The total of the ambient temperature plus cable temperature rise, due to current flow, should not exceed the continuous duty guideline temperatures as shown in Table 1, unless extensive testing and/or evaluation has indicated that higher temperatures can be tolerated.

In addition, the maximum continuous duty temperature rating for any wire insulation shall be determined by an accelerated aging test conducted in accordance with ASTM D 573, with the samples of insulation being removed from the finished wire and aged 168 h. The test temperature shall be 30°C above the intended rated temperature. Tensile strength after aging shall be not less than 80% of the original tensile strength. The elongation after aging shall be at least 50% of the original elongation.

NOTE—Heavier conductors may be required to protect the carrying of current in wire bundles when all conductors are carrying maximum current. Temperature rise tests of the conductor bundle shall be run to determine the proper conductor size and insulation.

Resistance wire low tension cable may be used to limit the voltage applied to electrical devices. Since the nature of the wire is to limit the voltage applied to electrical devices, the distance of the device from the power source and the current demand of the device will determine the materials used. Because every application is different, no materials, conducting or insulating, can be specifically described as standard; thus the conductor and insulating materials must be carefully chosen for each application by the design engineer. It is desirable to identify resistance wire by printing the words *resistance wire* on the conductor.

Extreme care shall be used by the design engineer in choosing resistance wire as a conducting material to satisfy the current demand of the device and not create a temperature rise in the conductor that would deteriorate the insulating material even though the device is left on continuously.

Circuits using resistance wire shall be carefully placed in the vehicle so that their temperature rise will not create a hazard to, or malfunction of, any part of the vehicle. A general design guide would be that the conductor be required to dissipate no more than 5 W per insulated conductor foot.

2.3.4 FUSIBLE LINKS—A special section of low tension cable designed to open circuit when subjected to an extreme current overload shall conform to SAE J156.

2.3.5 It is desirable to color code each conductor in an electrical circuit to facilitate manufacture and service of a wire assembly. It is further desirable for all motor vehicle manufacturers to assign and use similar color code identifications for commonly used electrical circuits to promote ease of circuit analysis in service among the various manufacturers.

2.3.5.1 When feasible each circuit shall conform to a recommended color code by category of equipment as shown in Table 2. Otherwise, the color code may be a solid color (basic) and/or a basic color with secondary color stripes, dots, or hashes.

TABLE 1

SAE Cable (Ref. SAE J1128)	Temperature[a]
Type GPT, HDT, GPD, HDB	194°F (90°C)
Type STS, HTS	221°F (105°C)
Type SXL	275°F (135°C)

[a] Recommended maximum continuous duty temperature (ambient plus rise).

TABLE 2—CIRCUIT COLOR CODE—BASIC CIRCUITS (AUTOMOTIVE ONLY)

Function	Color
Left rear stop and turn	Yellow
Right rear stop and turn	Dark green
Auxiliary	Blue
Tail, side marker, license	Brown
Ground	White

NOTE: The above code is identical to the color code adopted for automotive type trailers—SAE J895.

2.3.5.2 Secondary color markings to be applied as to be visible throughout the entire length of the wire, or at each end of a lead.

2.3.5.3 Color combinations for special circuits not shown on Table 2 are to be selected by the user. As special circuit functions become standard with manufacturers, they shall be added to the recommended Color Code by category and shown in Table 2.

NOTE—It is desirable for the wire of any one circuit to be of uniform color code throughout the circuit regardless of the number of connections. A circuit is assumed to be continous until it can be interrupted by a relay or switch contacts, or when it reaches a load (such as bulbs, motors, etc.). Fusible links may differ in color from the circuits they are protecting as it could be advantageous to identify fusible link wire gauge size by insulation color.

2.3.5.4 Each circuit in the same wire assembly shall be distinguished from one another in some manner such as color code, or some substantial difference in insulation diameter (that is, two or more gauge sizes).

2.4 Conductor Termination

2.4.1 All stranded conductor stripped ends are to be fitted with terminals (exception—splices). Solid, precisely shaped conductors whose ends are the termination shall not have this fitting.

2.4.2 All terminal attachments to conductors shall conform with the physical and electrical performance requirements of SAE J163.

2.4.3 As a general practice, all terminations have integral and functional insulation grips, except where other secondary applications preclude their use. Special applications without insulation grips may be employed where other means of relieving strain are provided.

2.4.4 A terminal shall be attached to a conductor by a simple mechanical *crimp-type* process that will conform to the intent of paragraph 2.4.2. For maximum reliability and surety of connection, the *crimp* may also be soldered, swaged, brazed, or welded in a workmanlike manner. Care shall be taken to minimize wicking of solder in a stranded wire to avoid impairment of the strain relief or cable flexing.

2.4.5 CIRCUIT GROUNDING—Ground terminal lugs shall be solder dipped, cadmium, tin, or zinc plated. Ground terminals shall be accessible for service. A serrated paint cutting terminal may be utilized to make proper contact on painted surfaces. Ground terminal devices shall be cadmium, tin, or zinc plated. In special cases, plating may not be required for lugs and/or attaching devices.

Ground return connections shall be made to the vehicle structure, frame, or engine. In cases where the engine or body is mounted on rubber or other insulation, proper ground shall be provided.

2.4.6 Terminations used shall comply with the requirements of SAE J561, ring and spade types; SAE J858, blade type; and SAE J928, pin and receptacle type. Secondary applications will dictate the use of special terminations for special use or application.

NOTE—Terminations may be plated with a conductive and corrosion resistant material such as tin or silver to upgrade the current carrying capacity and to improve their resistance to corrosion.

2.5 Conductor Splicing

2.5.1 Conductors shall be mechanically crimped, soldered, swaged, brazed, or welded with other conductors to form a wire splice. All wire splices shall conform to the electrical specifications for splices per SAE J163.

2.5.2 Splices shall be mechanically secure to withstand all fabrication installation and vehicle environment abuse. The splice must be insulated.

2.6 Terminal and Connector Function

2.6.1 Single terminations shall be used only where there is no possibility of misconnections in assembly or service except when special applications may require otherwise.

2.6.2 Multiple terminal connect-disconnect connector bodies shall be used at all points where two or more conductors are terminated and where there is a possibility of misconnection in fabrication, assembly, or service; secondary applications may require a deviation from this practice.

2.6.3 All connections shall be designed to maintain surety of connections while subjected to vibration, shock, and the extreme temperatures that are normal environmental conditions for motor vehicles. Surety may be accomplished by employing the use of integral-molded lock devices, terminal to terminal interferences (detents), secondary locking clips, or attaching devices.

2.6.4 All multiple connect-disconnect connector bodies shall be polarized to prevent incorrect assembly unless circuitry permits use of a nonpolarized connector.

2.6.5 Connections shall be located in clean, dry areas when possible. Connections shall be designed to maintain circuit integrity regardless of environmental conditions (such as high humidity, road splash, rain, drainge, earth particles, fuels, lubricants, high and low temperatures, and solvent).

2.7 Conductor Grouping

2.7.1 Conductors are to be grouped together into multiple conductor assemblies whenever possible.

2.7.2 The number of wiring assemblies and electrical connections per vehicle shall be kept to a minimum with overlay or option wiring used only when justified by the economics of fabrication, vehicle installation, and service.

2.8 Wire Assembly Construction

2.8.1 Conductors are to be grouped, where practical, in cable or harness form.

NOTE—Suitable material such as braided cotton, braided paper and cotton, braided vinyl/nylon, flexible plastic conduit, friction or thermoplastic tape, extruded rubber and thermoplastic jackets, or woven loom may be used to form the assembly.

2.8.2 Wiring harness covering shall be adequate to protect the harness in the vehicle routing environment and shall furnish protection during all phases of vehicle assembly and operation.

NOTE—A general guideline to be used in the selection of coverings is specified in Table 3.

2.9 Wire Assembly Installation and Protection

2.9.1 Wiring and related devices shall be installed in a workmanlike manner, mechanically and electrically secure. Devices, lamps, and so forth requiring periodic service shall be serviceable and accessible by providing wire length sufficient to reasonably accomplish this.

2.9.2 In general, wire routing shall be such that maximum protection is provided by the vehicle sheet metal and structural components. Smooth protective channels especially designed for wiring and built into the vehicle body structure should be used when practicable. Avoid areas of excessive heat, vibration, and abrasion.

NOTE—Extra protection (such as braid, loom, conduit, etc.) should be provided when these areas cannot be avoided (Ref. Table 3).

2.9.3 All parts of the electrical system shall be adequately protected against corrosion.

2.9.4 If significant vibration levels exist, the edges of all metal members through which cables and harnesses pass shall be deburred, flanged, rolled, or bushed with suitable grommets. Suitable tubing or conduit over cables may be substituted for grommets if properly secured. Clips for retaining cables and harnesses shall be securely attached to body or frame member and cable or harness. Clips also assist in locating and routing at assembly.

2.9.5 Wiring shall be located to afford protection from road splash, stones, abrasion, grease, oil, and fuel. Wiring exposed to such conditions shall be further protected by either, or a combination of, the use of heavy wall thermoplastic insulated cable, (see SAE Standard J1128, Low Tension Primary Cable) additional tape application, plastic sleeving or conduit, nonmetallic loom, or metallic or other suitable shielding or covering.

2.9.6 Where cables must flex between moving parts, the last supporting clip shall be securely mounted and secure the cable in a permanent manner.

2.9.7 Wiring fasteners shall be non-conductive unless the wiring or fastener involved is provided with extra heavy outer covering such as nonmetallic conduit, tape, or dip.

NOTE—Overlay or option wiring should be routed in the same fasteners with standard wiring where practical, or should be fastened to the standard wiring with plastic straps or other mechanical means.

2.9.8 Electrical apparatus with integral wiring shall be supplied with grommets or other suitable mechanical fasteners for strain relief.

2.10 Wiring Overload Protective Devices

2.10.1 The current to all low-tension circuits, except starting motor and ignition circuits, shall pass through short circuit protective devices connected to the battery feed side of switches. Headlight systems shall be independently protected. Circuit protection shall be accomplished by utilizing fuses, circuit breakers, or fusible links which conform to SAE Standards.

2.10.2 The protective device shall be selected to prevent wire damage when subjected to extreme current overload.

2.10.2.1 *Fuses*—Fuse sizes shall be selected using guidelines presented in SAE J554, Electric Fuses.

2.10.2.2 *Circuit Breakers*—Fail-safe automatic reset circuit breakers shall be employed when it is necessary to quickly re-establish circuit continuity when that portion of the wiring has been subjected to an overload condition. Non-cycling type circuit breakers will not reset until the overload is removed, (unless they are the non-cycling manual-reset type). Circuit breakers shall conform to SAE J553 and SAE J258.

2.10.2.3 Fusible links shall be employed when heavy feed circuits exceed the continuous working limits of the fuses or circuit breakers. The link of wire, acting like a fuse, shall conform to the guidelines presented in SAE J156, Fusible Links.

3. Truck, Truck-Tractor Section

3.1 The following SAE Recommended Practice relates to wiring for exterior lamps, exclusive of head lamps, of commercial vehicles 80 in (203 cm) or more in width. Except as noted, the wiring system shall conform to the guidelines of Section 2.

3.1.1 LAMP—A lamp is a complete lighting unit. All lamps shall meet the requirements of SAE Standard J575, Lighting Equipment for Motor Vehicles. Lamps with pigtails not in excess of 12 in (30 cm) long shall have a minimum of 16-gauge wire; pigtails in excess of 12 in (30 cm) long shall have wire gauge conforming to the wiring requirements of the vehicle.

3.1.2 WIRE SIZE—To minimize voltage drops, the feed wire size for all circuits shall be a minimum of 12-gauge; branches or taps not in excess of 50 ft (15.2 m) in length shall be 14-gauge. The ground wire for insulated-return systems shall be equal to the respective feed wire. The main ground wire shall be a minimum of 10-gauge.

NOTE—In many cases 4 or 6 gauge may be required.

3.1.3 DESIGN VOLTAGE OF LAMPS—Reference SAE Standard J573, Lamp Bulb and Sealed Units for design voltage values applicable to various bulbs.

3.1.4 Truck-tractors shall conform to Section 2 and the following:

3.1.4.1 *Circuit Identification*—It is desirable to follow the SAE Recommended Practice J1067, Seven Conductor Jacketed Cable for Truck and Trailer Connections, for coding of truck-tractor jumper cable throughout the circuit. Where impractical, the coding is to be followed to a junction block or harness terminating point where visual inspection will identify the circuit coding change. The coding may also be numbers and/or letters printed on the wire insulation. Whatever coding system is chosen, the system shall facilitate in harness manufacturing and in service.

3.1.4.2 *Circuit Termination*—Wiring for trailer circuits shall terminate in:

(a) A connector socket conforming to SAE Recommended Practice J560, Seven-Conductor Electrical Connector for Truck-Trailer Jumper Cable, or

(b) A jumper cable with cable plug conforming to SAE J560.

4. Trailer Section

4.1 Trailers shall conform to Section 2 and the following:

4.1.1 WIRING—All wiring shall be installed in:

(a) Suitable conduit and boxes,
(b) Structure of the trailer, and
(c) Housings and/or raceways which provide equal protection.

Wiring shall be protected from stones, excess dirt, ice, moisture, chafing, and so forth, that will result in harmful effects.

All wiring for legally required lights shall be serviceable in a manner permitting removal and reinstallation from outside the trailer.

4.1.2 GROUNDING—The trailer shall be grounded to the tractor through the jumper cable.

NOTE—Contact of the trailer king pin or apron plate with the lower coupler or *grounding through the lower coupler* is not to be considered as providing a tractor-to-trailer ground.

TABLE 3

Type	Wire Harness Covering	General Application
1	Vinyl Plastic Tape—0.007 in (0.18 mm)	Primarily used for grouping cables into wire harnesses. Wiring not subject to damage from scuffing or scrubbing on rough metal edges.
2	Friction Tape, Cotton and Kraft Paper Braid	Generally optional; improved scuff and scrub resistance.
3	Vinyl/Nylon Braid	Improved abrasion resistance.
4	Non-Metallic Loom (Woven Asphalt, Impregnated Loom, Extruded Vinyl Plastic, or Elastomeric Tubing)	Improved scuff and abrasion resistance.
5	Rigid and Flexible Conduit	For maximum abrasion resistance and/or positive positioning for clearance to moving or heated vehicle components.

4.1.3 MARKING—The voltage of the lighting system shall be permanently or semi-permanently marked in a legible manner on a mounting surface, in proximity to the electrical connector receptacle. Preferably, the marking shall be in amber reflective letters.

4.1.4 TRAILER CONNECTOR SOCKET—The trailer connector socket for receiving the jumper cable plug shall conform to SAE J560.

4.1.5 CIRCUIT PROTECTION—Circuit protection independent of truck-tractor system shall be provided. Trailer circuit protective devices shall conform to SAE Standard J554, Electric Fuses or SAE Recommended Practice J553, Circuit Breakers and shall be located near the trailer wiring connector socket and be readily accessible for service.

5. Motor Coach Section

5.1 Motor coaches shall conform to Section 2 and the following:

5.1.1 WIRING—Where practical, wiring is to be located within the structure of the coach where it will not be subjected to damage by road splash, stones, grease, oil, fuel, or abrasion.

Wiring so located that it will be subjected to more than normal wear or hard usage shall be equipped with a means of disconnecting from the main harness and be easily removable for replacement or repair.

Wiring connections to lights mounted on the coach body shall be accessible from outside, with the light removed or through an access door in an interior trim panel.

6. Storage Battery Cables

6.1 Definition—Battery cables provide the link between the battery(s) and the balance of the starting/charging circuit. Items that dictate the design are:

6.1.1 ROUTING—Routing shall be established with the following guidelines:

6.1.1.1 Areas of excessive heat, abrasion, and vibration are to be avoided. Extra protection (such as loom, conduit, tubing, heat shield, etc.) shall be provided when these areas cannot be avoided.

6.1.1.2 Grommets or ferrules and nipples shall be provided when routed through holes in the frame or sheet metal.

6.1.1.3 Support at intervals of approximately 24 in (61 cm). Insulated or nonconductive supports shall be used.

6.1.1.4 Provide strain relief for the battery and starter motor terminals as close to terminals as practical.

6.1.1.5 Tailor such that the cables are not too loose nor too tight, considering engine rocking due to torque changes.

6.1.2 VOLTAGE DROP—Voltage drop for starting motor circuits as recommended in SAE J541, determines the maximum drop allowed for the total cranking circuits from the battery to the starter motor and the return to the battery.

6.1.3 CABLE SIZE—Cable size is determined by knowing the system parameters and subtracting their fixed resistances (such as connections, starter solenoid, ground path other than the battery cable, etc.) from the total specified in paragraph 6.1.2. This remaining resistance is the maximum allowed for the battery cables.

6.1.4 CABLE CONSTRUCTION—Cable construction is determined from the environment in which the battery cables must survive.

6.1.4.1 *Core Stranding*—Core stranding of conventional cable can be either bunched, concentric stranded, or rope lay. Bunched or concentric will suffice in most applications, except those requiring higher flex life. For larger cable sizes, rope stranding is needed for routing purposes as well. Battery strap is available for extreme flex requirements and restricted space or routing problems. Reference SAE J1127.

6.1.4.2 *Insulation*—Insulation provides electrical as well as environmental protection for the core. Polyvinyl chloride (PVC) can be used in most applications; cross-linked polyethylene, hypalon, neoprene, etc., may be needed for added protection against short circuit, high temperature, abrasion, etc. Reference SAE J1127 and Table 1.

6.1.4.3 *Terminals*—Terminals provide the connection to the battery, starter solenoid, junction blocks, switches, and grounding locations. A multitude of different types and styles are available for the variety of cable sizes. Also available are sleeves and covers which provide additional circuit and corrosion protection. Reference SAE J561 and SAE J163.

AUTOMOTIVE PRINTED CIRCUITS—SAE J771 APR86

SAE Standard

Report of Electrical Equipment Committee approved June 1961 and reaffirmed by the Electrical and Electronic Systems Technical Committee, April 1986.

1. Scope—This report relates to recommendations and specifications governing the classification, composition, test procedures, and properties of printed circuits commonly used to replace cable in automotive low voltage systems. It is not applicable to miniature circuits for solid state devices, high impedance or high voltage functions.

2. Base Materials—The base insulating materials fall into three categories: Phenolic Laminates, Plastic Molding Materials, and Flexible Plastic Films of the type which will meet the following flame retardancy requirement.

2.1 Flame Retardancy—Flame resistance of the material used for base or overlay shall be equal to or better than phenolic laminates, acrylonitrile butadiene, styrene, or polyester films from polyethylene terephthalate, respectively, as measured by ASTM D635, Tests for Flammability of Rigid Plastics Over 0.127 cm (0.050 in) in Thickness, or ASTM D 568, Test for Flammability of Plastics 0.127 cm (0.050 in) and Under in Thickness.

2.1.1 PHENOLIC LAMINATES—When plastic laminate base material is used, the quality characteristics and tolerances must conform to the following:

(a) Paper laminate thoroughly impregnated with thermosetting phenolic resin binder and properly cured.
(b) The material shall be opaque unless otherwise specified.
(c) Thickness shall be 0.062 in (1.58 mm) unless otherwise specified. (Refer to item E under Design Considerations.)
(d) Flexural strength shall not be less than 13,000 psi (89.6 MPa) with the grain or less than 11,000 psi (75.8 MPa) across the grain (ASTM D 229, Testing Rigid Sheet and Plate Materials Used for Electrical Insulation, and ASTM D 790, Method of Test for Flexural Properties of Plastics).

2.1.2 PLASTIC MOLDING MATERIALS—When plastic molding material is used, the quality characteristics and tolerances must conform to the following:

(a) The material must be opaque in cross section of 0.062 ± 0.005 in (1.58 ± 0.13 mm) thickness.
(b) Minimum physical properties:
 (1) Impact strength: 1.3 ft-lb/in (69.4 N · m/m) notch ¼ in (6.35 mm) bar sample notches at 73°F (22.8°C) (ASTM D 256, Methods of Test for Impact Resistance of Plastics and Electrical Insulating Materials).
 (2) Flexural strength: 11,220 psi (77.36 MPa) (ASTM D 790).
 (3) Tensile strength: 7750 psi (53.4 MPa) (ASTM D 638, Method of Test for Tensile Properties of Plastics).
 (4) Heat distortion temperature 225°F (107.2°C) at 264 psi (1.8 MPa) (ASTM D 648, Method of Test for Deflection Temperature of Plastics Under Load).

The materials in paragraphs 2.1.1 and 2.1.2 shall not be subjected to temperatures in excess of 225°F (107.2°C).

2.1.3 FLEXIBLE PLASTIC FILM—When flexible plastic film is used, the quality characteristics and tolerances must conform to the following:

(a) Tensile strength: 17,000 psi (117.2 MPa) (minimum), ASTM D 882, Methods of Test for Tensile Properties of Thin Plastic Sheeting (Method A-100% elongation per minute).

(b) Tensile modulus: 450,000 psi (3.1 GPa) (minimum), ASTM D 882 (Method A-100% elongation per minute).

(c) Tear initiation: 2000 lb/in (226 N · m) (minimum) Graves, ASTM D 1004, Method of Test for Tear Resistance of Plastic Film and Sheeting.

(d) Thermal coefficient of linear expansion: 15×10^{-6} (maximum) 70–120°F (21–49°C).

(e) Moisture absorption, water immersion: less than 0.5%, 24 h.

(f) Dielectric strength: 500 V/mil (19.7 kV/mm) minimum per ASTM D 149, Methods of Test for Dielectric Breakdown Voltage and Dielectric Strength of Electrical Insulating Materials at Commercial Power Frequencies.

(g) Melting point: more than 350°F (177°C).

(h) Thickness to be 0.003 in (0.08 mm) minimum.

3. Copper

3.1 Printed circuit grade electrodeposited copper foil with:

(a) 99.8% minimum purity per ASTM E 53-48.

(b) Maximum oxygen content 0.1% per ASTM E 53-48.

(c) Minimum elongation of 10%, both transverse and longitudinal. Measurement shall be made using a 2 in (50.80 mm) gage length with an extension rate of 2 in/min (0.85 mm/s).

(d) Tensile strength: Longitudinal—25,000 psi (172 MPa) minimum, transverse—25,000 psi (172 MPa) minimum.

(e) Bulge height: For test method, see Appendix. Bulge height shall be 0.180 in (4.57 mm) minimum.

(f) Resistivity: The resistivity shall not exceed 0.15940 Ω/m^2 at 68°F (20°C).

(g) The foil shall be free of any lead inclusions or other foreign materials.

(h) The foil shall have an antitarnish treatment which is compatible with the subsequent bonding or lamination process and which will offer no additional significant increase in the electrical resistance between the foil surface and any component.

3.2 Thicknesses employed may be 0.00135 in (0.034 mm), which is 1 oz/ft² (0.305 kg/m²); 0.0027 in (0.069 mm), which is 2 oz/ft² (0.610 kg/m²); or 0.004 in (0.102 mm), which is 3 oz/ft² (0.915 kg/m²). The choice of thickness is made on the basis of the thinnest copper foil that will meet the current-carrying requirements for any specific applications. (See Fig. 1.) Discoloration of the copper faces shall not be considered objectionable.

4. Coating—On phenolic laminate or molded plastics, a scratch resistant protective insulating coating film shall be bonded to all foil side surfaces which will not require contact with other components. The coating film or overlay used on circuits with flexible plastic base must be firmly bonded such that the fold endurance (see Physical Test Requirements) will be maintained.

5. Design Considerations

5.1 Practical Tolerances on Critical Dimensions

(a) Hole diameters:

Up to 0.500 ± 0.004 in (12.7 ± 0.10 mm)

0.500 to 1.000 ± 0.005 in ($12.7–25.4 \pm 0.13$ mm)

Over 1.000 ± 0.006 in (25.4 ± 0.15 mm)

For slots and notches, consider both length and width as hole diameters.

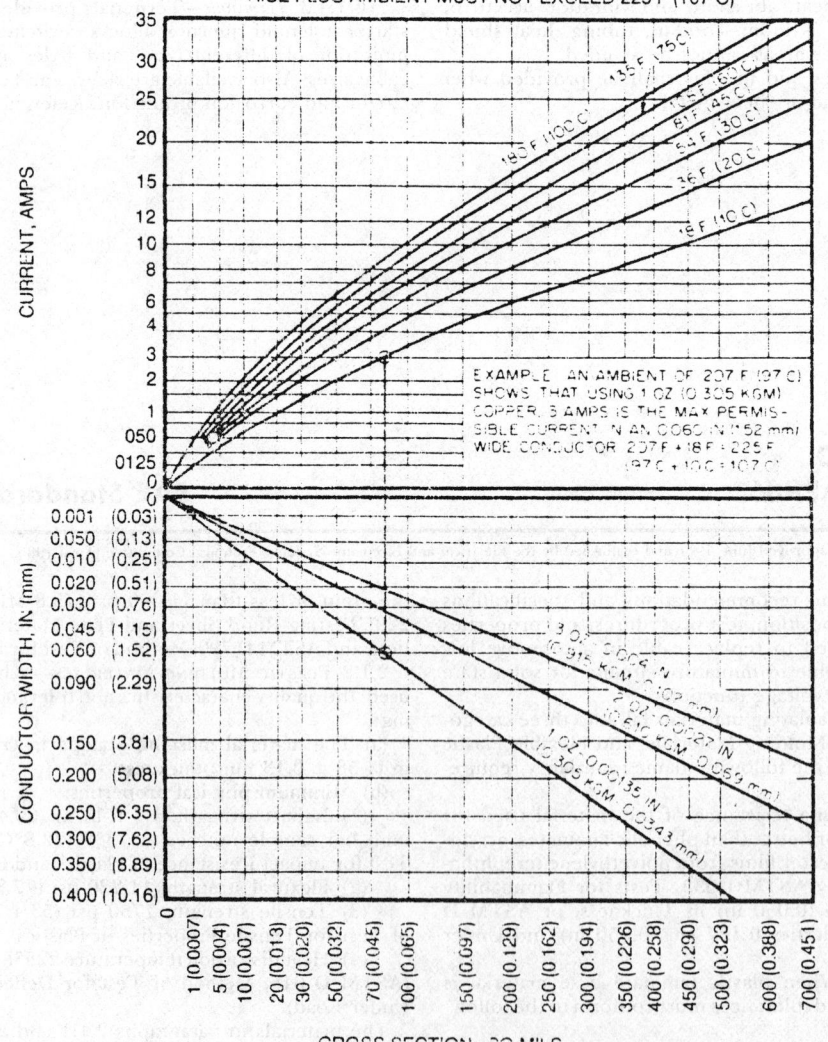

FOR USE IN DETERMINING CURRENT CARRYING CAPACITY AND SIZES OF COPPER CONDUCTORS FOR VARIOUS TEMPERATURE RISES ABOVE AMBIENT. CURVES ARE DERATED A NOMINAL 10% TO ALLOW FOR VARIATIONS IN PRODUCTION TECHNIQUES

FIG. 1—CONDUCTOR THICKNESS AND WIDTH

(b) Hole-to-hole centerline tolerance: ±0.005 in (±0.13 mm). Add ±0.001 in (±0.03 mm) for every inch (25.4 mm) over 1 in (25.4 mm).

(c) Line width and spacing tolerance: ±0.010 in (±0.25 mm). Line width tolerances do not include nicks, pin holes, and scratches. These imperfections are acceptable provided the line is not reduced by more than 20% in a local area.

(d) Warpage on phenolic laminate: Measured according to ASTM D 709, Specification for Laminated Thermosetting Materials; 0.025 in/in (0.025 mm/mm). Closer warp tolerances may limit the selection of raw materials or make necessary unusual manufacturing operations or shipping procedures.

(e) Panel thickness (0.062 in or 1.58 mm): Tolerance with copper foil:

oz/ft²	kg/m²	No. of Sides	Tolerance in	Tolerance mm
1	0.305	1	±0.0055	±0.14
1	0.305	2	±0.0060	±0.15
2	0.610	1	±0.0060	±0.15
2	0.610	2	±0.0065	±0.16

(f) Bond strength: Normal peel strength testing method to be used: a suitable universal testing machine (see Fig. 2) pulling the foil from the laminate within 5 deg (0.083 rad) of the perpendicular to the face of the sample. Jaws must grip foil across entire foil width. Convert results to pounds per inch width (newton/metre). Bond strength to base material 4 lb/in (0.45 N · m) of width minimum (6 lb/in (0.68 N · m) of width average) on 10 samples, one line each.

(g) Registration of circuit pattern to part: ±0.015 in (0.38 mm).

5.2 Good Practice (line widths, spacing, etc.[1])

[1] The 0.060 in (1.52 mm) minimum conductor widths and spacing refer to printed circuits commonly used to replace cable in automotive wiring. Miniature circuits for solid-state devices, etc., will normally have special requirements and specifications determined by the function.

(a) Minimum width of conductor: 0.060 in (1.52 mm). (See Figs. 1 and 3 for special application and typical pattern.)

(b) Minimum spacing between conductors: 0.060 in (1.52 mm).

(c) Minimum distance between copper conductors and edge of board: 0.060 in (1.52 mm).

(d) Minimum distance from a hole to the edge of the board should equal the diameter of the hole and should never be less than the thickness of the base material.

(e) Where conductor pad diameter is to be used for circuit mounting and grounding conductor, diameter should exceed screw head diameter by a minimum of 0.090 in (2.29 mm). Conductor fillet radii should be as large as possible, and in no case less than 0.03 in (0.76 mm) R.

(f) Maximum distance from edge of lamp socket hole to copper foil pad on phenolic laminate or plastic molded bases shall be 0.030 in (0.76 mm). (See Fig. 4.) Maximum distance from edge of lamp socket hole to copper foil pad on flexible plastic films shall be 0.065 in (1.65 mm). (See Fig. 5.)

(g) A radius shall be applied to all corners, notches, and slots, and shall not be less than 0.03 in (0.76 mm) R.

(h) Additional layout considerations. (See Figs. 3–5.)

6. Test Requirements
6.1 Electrical

(a) Dielectric: 200,000 Ω minimum resistance must exist between any two conductors after conditioning for 24 h at 170°F (77°C), and 100% relative humidity, when using a low voltage type resistance meter. Moisture present on sample surface to be wiped off or blown off with compressed air prior to the resistance check.

(b) Millivolt drop: Tests to be conducted at ambient room temperature of 75 ± 5°F (24 ± 2.8°C) in draft free area.

(1) Maximum permissible millivolt drop from any component contacting the foil on any printed circuit board or flexible plastic film shall be 5.0 mV at 2.0 A before conditioning.

(2) Maximum millivolt drop from any component to the foil after one humidity cycle of 100 h at 100% relative humidity at 100°F (37.8°C), must not exceed 10.0 mV at 2.0 A.

(c) Continuity of all electrical connections and circuits must be main-

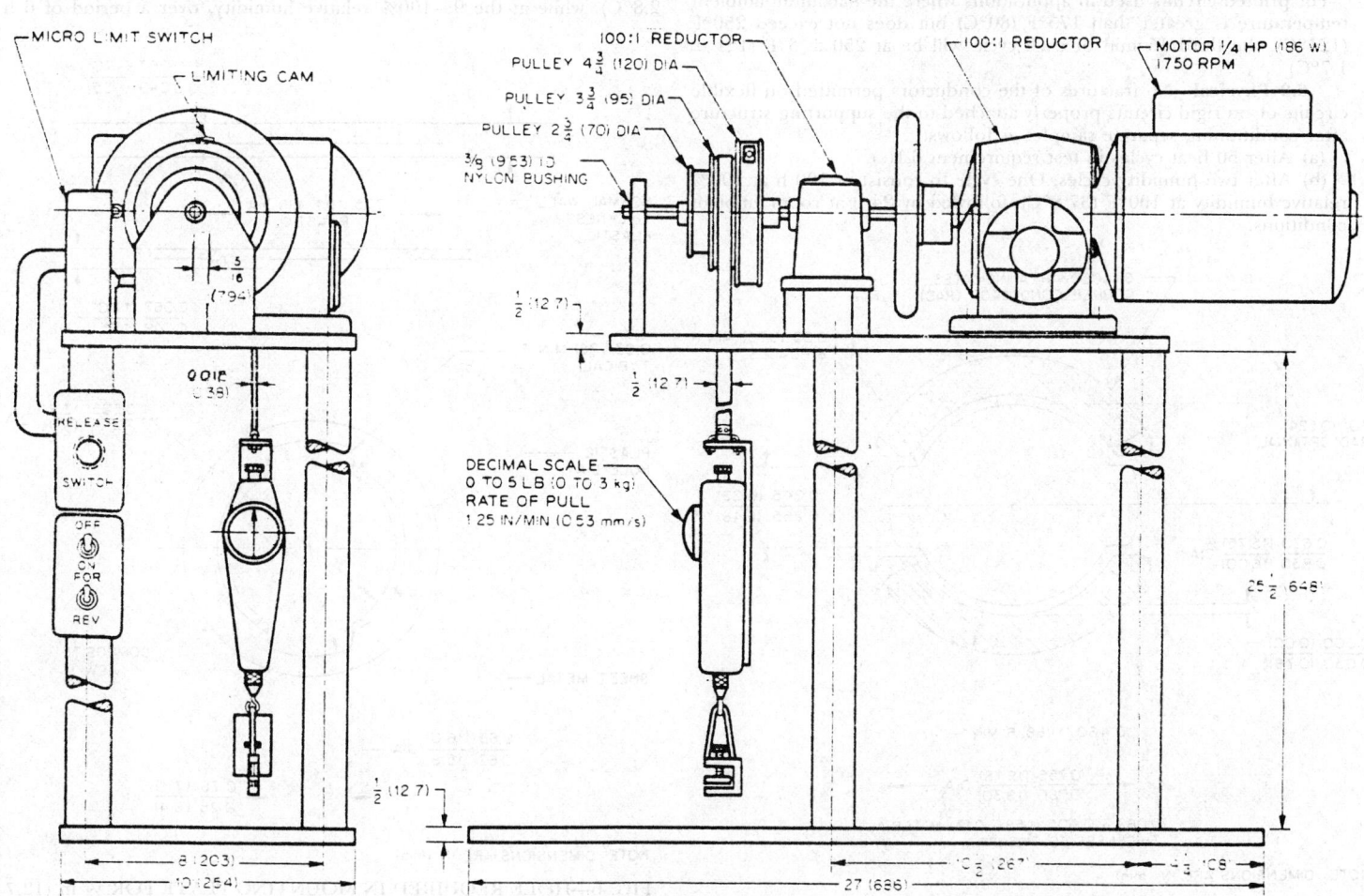

FIG. 2—BOND STRENGTH TESTER

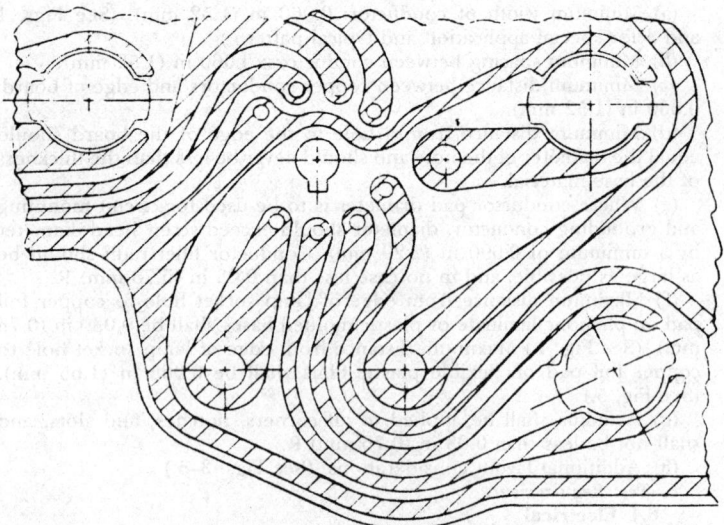

FIG. 3—CONDUCTOR PATTERNS: RECOMMENDED TYPICAL SECTION, WIDE CONDUCTOR LINES AND FILLETS IN AREA OF PADS AND LANDS

tained after 25 thermal shock cycles. For printed circuits used in applications where the maximum ambient temperature does not exceed 175°F (80°C), each cycle is to consist of:

15 min at $-30 \pm 3°F$ ($-34.5 \pm 1.7°C$)
15 min at $75 \pm 5°F$ ($24 \pm 2.8°C$)
15 min at $175 \pm 3°F$ ($80 \pm 1.7°C$)
15 min at $75 \pm 5°F$ ($24 \pm 2.8°C$)

For printed circuits used in applications where the maximum ambient temperature is greater than 175°F (80°C) but does not exceed 250°F (121°C), the third 15 min of each cycle will be at $250 \pm 3°F$ ($121 \pm 1.7°C$).

6.2 Physical—No fractures of the conductors permitted on flexible circuits or on rigid circuits properly attached to the supporting structure after conditioning separate samples as follows:

(a) After 50 heat cycles in test requirement 6.1(c).
(b) After two humidity cycles. One cycle to consist of 100 h at 100% relative humidity at 100°F (37.8°C), followed by 24 h at room ambient conditions.

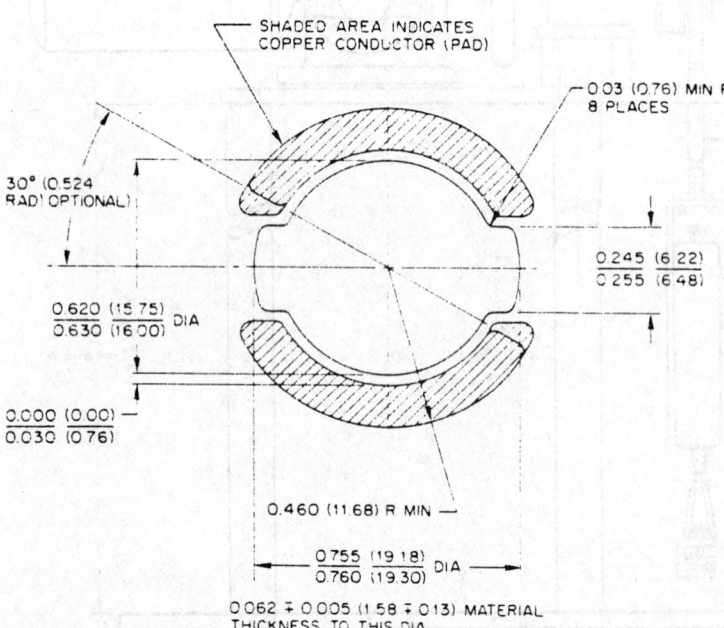

FIG. 4—⅝ in (15.9 mm) PRINTED CIRCUIT LAMP SOCKET HOLE STANDARD FOR RIGID PLASTIC OR PHENOLIC BASE

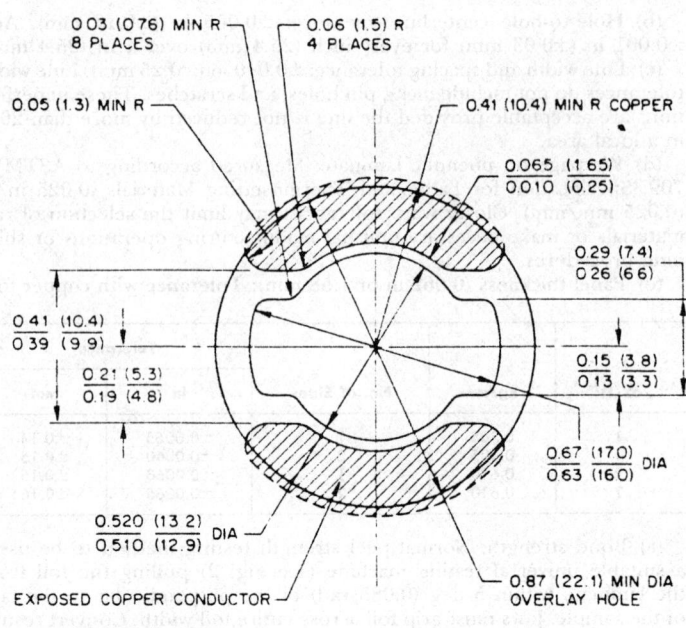

FIG. 5—½ in (12.7 mm) PRINTED CIRCUIT LAMP SOCKET HOLE STANDARD FOR FLEXIBLE PLASTIC FILM

(c) After being subjected to five continuous cycles as follows: $100 \pm 5°F$ ($37.8 \pm 2.8°C$) at 95–100% relative humidity for 2 h. Temperature then to be raised gradually to $160 \pm 5°F$ ($71 \pm 2.8°C$) at the same relative humidity over a 1 h period and then maintained for an additional 3 h. The temperature is then to be lowered gradually to $100 \pm 5°F$ ($37.8 \pm 2.8°C$), while at the 95–100% relative humidity, over a period of 6 h.

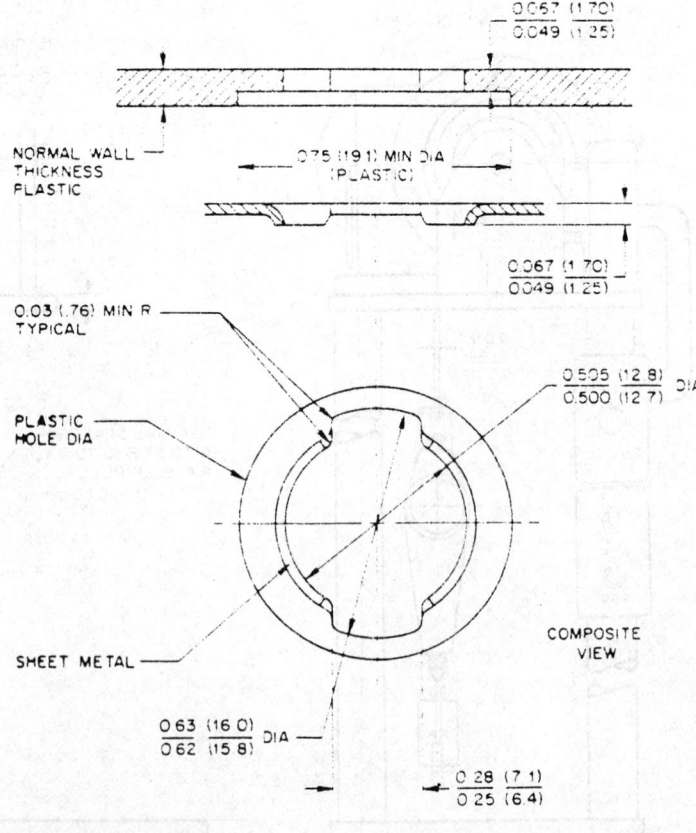

FIG. 6—HOLE REQUIRED IN MOUNTING PLATE FOR ½ in (12.7 mm) PRINTED CIRCUIT LAMP SOCKET WHEN BASE IN FLEXIBLE PLASTIC FILM

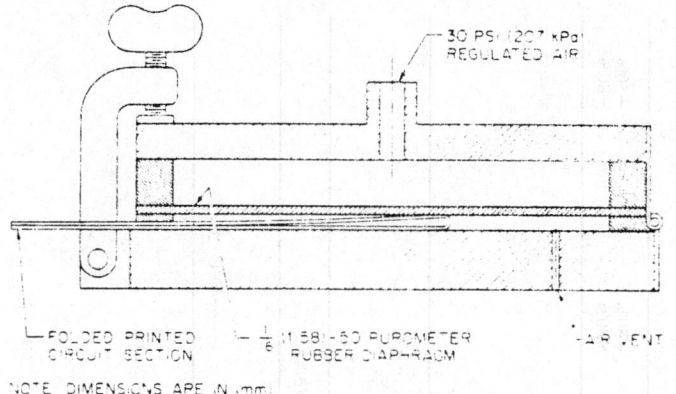

FIG. 7—FOLD TESTER FIXTURE

This would constitute one cycle. After completion of five cycles, parts examination should be initiated following 1 h minimum rest at room ambient conditions. If the printed circuit being tested is used in an application where the maximum ambient temperature is greater than 175°F (80°C) but does not exceed 250°F (121°C) at the completion of the preceding five cycles it shall be subjected immediately to an additional five continuous cycles as follows: −30 ± 3°F (−34.5 ± 1.7°C) for 1 h. Temperature then to be raised gradually to 250 ± 3°F (121 ± 1.7°C) over a 3 h period and then maintained for 1 h. The temperature is then to be lowered gradually to −30 ± 3°F (−34.5 ± 1.7°C) over a 3 h period and then maintained for 1 h. This would constitute one cycle. After completion of five cycles, the printed circuit would be placed in room ambient conditions for a minimum of 1 h before examination for fractures or delamination.

(d) After being subjected to one cycle of vibration in each of three mutually perpendicular directions, each cycle shall be as follows: The printed circuit shall be mounted or fastened as it normally would be in the application and contain all components and attachment devices. This assembly is then exposed to a temperature of −30 ±3°F (−34.5 ± 1.7°C) for 1 h. The assembly is then immediately subjected to a simple harmonic motion having an amplitude of 0.015 in (0.38 mm) with 0.03 in (0.76 mm) peak-to-peak maximum excursion. The frequency shall be varied uniformly between the limits of 10 and 55 Hz. The entire range from 10–55 Hz and returning to 10 Hz shall be traversed in approximately 1 min. This motion shall be applied for a period of 30 min. During this portion of the test the sample shall not be subjected to any temperature other than room ambient. Examination of the sample shall be after the third cycle.

If the sample being tested is used in an application where the maximum ambient temperature is greater than 175°F (80°C) but does not exceed 250°F (121°C), at the completion of the preceding three cycles, it shall be subjected immediately to an additional three cycles. These cycles will be similar to the first three except the conditioning temperature shall be 250 ± 3°F (121 ± 1.7°C) instead of −30°F (−34.5°C). Examination of the sample shall be after all six cycles are completed.

(e) Method for determining ductility of copper material intended for use in printed circuits: See Appendix.

APPENDIX
METHOD FOR DETERMINING DUCTILITY OF COPPER MATERIAL INTENDED FOR USE IN FLEXIBLE PLASTIC FILM PRINTED CIRCUITS

1. *Apparatus*—The apparatus used shall conform to that described in ASTM D 2210-64 Modified, Standard Method of Test for Bursting Strength of Leather, with Bulge Diameter of 1.250 in (32 mm), so that the output shaft turns at 10 rpm.

2. *Test Specimens*—From the samples selected to determine the ductility property, cut at least three specimens 3½ in (90 mm) square from sections located approximately at equal distances across the width of the material.

3. *Test Procedure:*

3.1 Set the bulge height indicator to zero using a flat rigid sheet between the plates.

3.2 Clamp the specimen securely between the plates of the circular hydraulic bulge tester with the uncoated (bright) side of the material facing downward.

3.3 Admit the hydraulic fluid to the bulging chamber between the plates.

3.4 Observe and record the maximum bulge height at burst.

HIGH TENSION IGNITION CABLE— SAE J2031 JAN90 — SAE Standard

Report of the Ignition Systems Standards Committee approved January 1990.

1. *Scope*—The specifications contained in this document pertain to high tension ignition cable used in road vehicle engine ignition systems.

2. *Cable Dimensions*—The average overall diameter of finished cable shall be either 7 mm or 8 mm. Allowable tolerance for either size shall be +0.3 mm. The average overall diameter shall be determined by taking the average of five sets of measurements along a 1 m length of finished cable. Each set of measurements shall consist of the determination of the maximum and minimum diameter at the point of measurement.

3. *Test Requirements*—When tested according to the methods outlined, the ignition cables covered by this document shall be capable of complying with the applicable requirements specified herein. Table 1 defines the applicability of the test and provides specific performance criteria.

NOTE: Wherever the term "room temperature" is used, it shall be defined as 23°C ± 5.

3.1 *Spark Test*—An AC spark test shall be performed on 100% of production of each ignition wire to which it is applicable (see Table 1). Apparatus shall consist of a voltage source, electrode, voltmeter, fault-current device or system, and the necessary electrical connections. The recommended apparatus and test method shall be that described in UL 1581, Section 900. Test potential shall be 25 kV for 7 mm cable and 30 kV for 8 mm cable.

NOTE: Alternative methods for this test may be considered provided that insulation faults are detected with the same degree of certainty.

3.2 *High Potential Test*—Refer to A1 for test apparatus. Immerse an approximate 1200 mm specimen for 4 h in a salt solution (3% m/m of NaCl in water) at room temperature with the ends twisted together and emerging approximately 400 mm above the surface of the solution. Apply a test voltage of 20 kV (rms) for 30 min between the conductor and the solution.

NOTE: The applied potential is to be increased from near zero at an essentially uniform rate not to exceed 500 V/s.

The cable shall not break down. Then increase the voltage at a rate not exceeding 500 V/s to the following levels:

7 mm cable to 30 kV (rms)
8 mm cable to 35 kV (rms)

Breakdown shall not occur.

3.3 *Capacitance*—Soak an approximate 1200 mm specimen in a salt solution (3% m/m of NaCl in water) at 70°C ± 2 for 24 h with each end of the cable emerging approximately 100 mm above the surface of the solution. Measure the capacitance between the conductor and the solution. Immerse the same specimen in tap water at 23°C ± 2 for 1 h with each end of the cable emerging approximately 100 mm above the surface of the water. Again, measure the capacitance between the conductor and the water. The frequency applicable to both measurements shall be 1000 Hz. Measured capacitance values shall not exceed those agreed on between the cable manufacturer and the user.

3.4 *Corona Resistance*—Affix an approximate 1200 mm specimen to a mandrel as specified in A2. Apply the test potential specified in Table 1 for a period of 8 h.

NOTE: The applied potential is to be increased from near zero at an essentially uniform rate not to exceed 500 V/s.

There shall be no breakdown of the cable nor shall the surface of the specimen exhibit any cracks, fractures, or other defects.

3.5 *Deformation Test*—Determine the average thickness of the covering(s) over the conductor of an approximate 100 mm specimen of

TABLE 1 — Test Applicability and Performance Criteria

Test		Class A				Class B				Class C				Class D				Class E				Class F			
	Type	1	2	3	4	1	2	3	4	1	2	3	4	1	2	3	4	1	2	3	4	1	2	3	4
3.1	Spark Test	applicable		not applicable		applicable		not applicable		applicable		not applicable		applicable		not applicable		applicable		not applicable		applicable		not applicable	
3.2	High-Potential Test	to be applied																							
3.3	Capacitance	Capacitance values shall be agreed on between cable manufacturer and user.																							
3.4	Corona Resistance																								
	7 mm cable	15 kV (rms), 50 or 60 Hz																							
	8 mm cable	18 kV (rms), 50 or 60 Hz																							
3.5	Deformation Test																								
	Test temperature	70°C ± 2				105°C ± 2				120°C ± 2				180°C ± 2				not applicable							
3.6	Thermal Overload Test																								
	Test Temperature	105°C ± 2				120°C ± 2				155°C ± 2				180°C ± 2				220°C ± 3				250°C ± 3			
	Max. change in res.	+50% / −80%		-	-	+50% / −80%		-	-	+50% / −80%		-	-	+50% / −80%		-	-	+50% / −80%		-	-	+50% / −80%		-	-
3.7	Test for Shrinkage																								
	Test temperature	155°C ± 2																							
	Max. shrinkage	2%																							
3.8	Res. to Flame Propagation																								
	Exposure Time	30 s				30 s								15 s											
	Extinction time																	70 s							
3.9	Low Temperature Test																								
	Test temperature	−20°C ± 3								−30°C ± 3								−40°C ± 3							

Test		Class	A				B				C				D				E				F			
		Type	1	2	3	4	1	2	3	4	1	2	3	4	1	2	3	4	1	2	3	4	1	2	3	4
3.10	Mech. Strength Test Max. change in res.		not applicable		+30%	—	not applicable		+30%	—	not applicable		+30%	—	not applicable		+30%	—	not applicable		+30%	—	not applicable		+30%	—
3.11	Stripping of Insulation										to be applied				to be applied				to be applied							
3.12	Resistance to Oil																									
3.13	Resistance to Fuel																									
3.14	Accelerated Life Test																									
	3.14.1 Res. to Salt Water										to be applied				to be applied				to be applied							
	3.14.2 Res. to Oil																									
	3.14.3 Res. to Fuel																									
	3.14.4 High Temp. Res. Test Temperature		90°C ± 2				105°C ± 2				120°C ± 2				155°C ± 2				180°C ± 2				220°C ± 3			
	3.14.5 Low Temp. Res. Test Temperature		−10°C ± 3				−15°C ± 3				−20°C ± 3				−30°C ± 3				−30°C ± 3				−40°C ± 3			

Type 1: Cables with copper conductor
Type 2: Cables with steel conductor
Type 3: Resistive cables
Type 4: Reactive cables

finished cable. Mount the specimen into the test apparatus shown in A3 and load with the following mass:

7 mm cable to 450 g (include mass of test frame)
8 mm cable to 510 g (include mass of test frame)

Place the test unit into a full draft, circulating air oven maintained at the temperature specified in Table 1 for a period of 4 h. Remove the test unit from the oven and cool within 10 s by immersing in cold water. Remove the specimen and measure the depth of the indentation in the area of the application of the load using a measuring microscope. The value of this measurement shall not exceed 50% of the original finding of the average thickness of the covering(s).

3.6 Thermal Overload Test—Suspend an approximate 500 mm specimen vertically for 48 h in a full draft, circulating-air oven maintained at the temperature specified in Table 1. Remove the specimen from the oven and allow to cool to room temperature. Subject the specimen to windings on a mandrel as defined in A4. On completion of the windings, there shall be no evidence of cracks, fractures, or other defects. For resistive cables, measure the resistance of the cable at room temperature before and after the test. The allowable change in resistance shall not exceed the values specified in Table 1.

3.7 Test for Shrinkage—This test shall apply where shrinkage of the conductor covering(s) is important with respect to the attachment of the connector. Measure the exact length (200 mm minimum) of a suitable specimen at room temperature. Place the specimen horizontally into a full draft, circulating-air oven maintained at the temperature specified in Table 1 for a period of 15 min. Remove the specimen, cool to room temperature, and measure the length again. Maximum length shrinkage shall not exceed the values shown in Table 1. In addition, there shall be no evidence of cracks, fractures, or other defects in the surface of the specimen.

3.8 Resistance to Flame Propagation—For this test, a Bunsen or Tirrill gas burner having a barrel of approximately 9.5 mm diameter shall be employed. While the barrel is vertical and the burner is well away from the specimen, the height of the flame is to be adjusted to approximately 100 to 125 mm. The blue inner cone is to be approximately 38 mm high and the temperature at its tip is to be a minimum of 816°C as measured using a chromel-alumel thermocouple. Suspend an approximate 500 mm specimen in a draft-free chamber and expose it to the tip of the inner cone of the test flame, as shown in A5, for the period of time specified in Table 1. Any combustion of the specimen must extinguish itself within the time specified in Table 1 following the removal of the test flame.

3.9 Low Temperature Test—Affix one end of an approximate 400 mm specimen to a 25 mm rotatable mandrel and attach a 4.5 kg mass to the free end. Subject the specimen, in a vertical position (i.e., with the mass freely hanging), to the temperature specified in Table 1 for a period of 4 h. Without removing the sample from the freezing chamber, wind it a minimum of three turns onto the rotatable mandrel at a speed of one turn/second. There shall be no evidence of cracks, fractures, or other defects.

NOTE: If the test device is precooled, a freezing time of 2 h is sufficient.

3.10 Mechanical Strength Test—An approximate 1200 mm specimen shall be suspended as defined by A6 and subjected to a force of 250 N for a period of 5 min. For resistive cables, measure the resistance prior to the test on the full length of the specimen and again after the test on a 250 mm straight portion of the specimen that was under stress. The change in the resistance per unit length shall not exceed the value shown in Table 1. For reactive cables, verify with approximately 12 V DC that there is no discontinuity in a 250 mm straight portion of the specimen that was under stress.

3.11 Stripping of Insulation—Where cables are required to be stripped, it shall be possible to remove at least 20 mm of insulation from the conductor cleanly and without difficulty. Specific stripping force values, when required, shall be agreed on between the manufacturer and the user.

3.12 Resistance to Oil—Measure the diameter of an approximate 400 mm specimen and then immerse for 48 h in ASTM No. 1 oil at a temperature of 90°C ± 2 with the cable ends emerging approximately 50 mm above the surface of the oil. (The oil shall be stirred during the test.) Remove the specimen, wipe off excess oil, and cool to room temperature. The maximum change in diameter shall not exceed +10%. Subject the specimen to windings as specified in A4. On completion of the windings, there shall be no evidence of cracks, fractures, or other defects.

3.13 Resistance to Fuel—Measure the diameter of an approximate 400 mm specimen and then immerse in ASTM fuel C at room temperature for 30 min with cable ends emerging approximately 100 mm above the surface of the fuel. Remove the sample and allow to dry for approximately 30 min. The maximum change in diameter shall not exceed +15%. Subject the specimen to windings as specified in A4. On completion of the windings, there shall be no evidence of cracks, fractures, or other defects.

3.14 Accelerated Life Test—Subject an approximate 1200 mm specimen to windings as specified in A4, and then as specified in A2. The specimen, while secured to the mandrel, shall then be subjected to the tests outlined in the sequence listed. Test voltages shall be 15 kV (rms) for 7 mm cable and 20 kV (rms) for 8 mm cable. Test voltage shall be applied for 30 min while the mandrel is contained within a close-fitting, nonmagnetic metallic sleeve, which may have flared ends, but conditioning prior to the application of the test potential may be conducted with or without the sleeve.

NOTE: The applied potential is to be increased from near zero at an essentially uniform rate not to exceed 500 V/s.

When immersed in liquids, the cable ends shall emerge approximately 100 mm above the surface of the liquid. There shall be no breakdown of the test specimen at any point in the sequential testing procedure.

3.14.1 RESISTANCE TO SALT WATER—Place test specimen in a full draft, circulating-air oven maintained at 90°C ± 2 for 4 h and then immediately immerse in a salt solution (3% m/m of NaCl in water) and maintain at 50°C ± 2 for 16 h. Remove the specimen from the solution, drain for 30 min at room temperature, and then apply the voltage test as specified.

3.14.2 RESISTANCE TO OIL—Place test specimen in a full draft, circulating-air oven maintained at 90°C ± 2 for 4 h and then immediately immerse in ASTM No. 1 oil and maintain at 90°C ± 2 for 16 h. Remove the specimen from the oil, drain for 30 min at room temperature, and then apply the voltage test as specified.

3.14.3 RESISTANCE TO FUEL—Immerse the test specimen in ASTM fuel C at room temperature for 30 min. Remove the specimen from the fuel, drain for 4 h without the sleeve, and then apply the voltage test (in the metallic sleeve) as specified.

3.14.4 HIGH TEMPERATURE RESISTANCE—Place the test specimen in a full draft, circulating-air oven maintained at the temperature shown in Table 1 for a period of 48 h. Remove the specimen from the oven, cool to room temperature, and then apply the voltage test as specified.

3.14.5 LOW TEMPERATURE RESISTANCE—Unwind the test specimen from the mandrel, leaving one end secured, and attach a 4.5 kg mass to the other end. With the mass fully supported by the test specimen, the entire arrangement shall be subjected to the temperature specified in Table 1 for a period of 4 h. Without removing the specimen from the freezing chamber, wind it onto the mandrel for five complete turns at the rate of one turn per 5 s. Remove the sample from the freezing chamber, allow it to return to room temperature, and then apply the voltage test as specified.

NOTE: If the test device is precooled, a freezing time of 2 h is sufficient.

APPENDIX A

A1. Test Apparatus for High Potential Test (see 3.2)

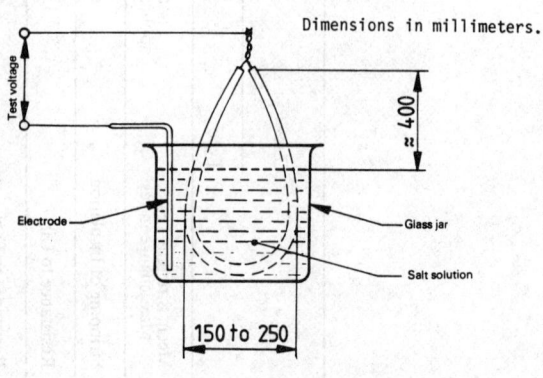

FIG. A1

A2. Test Apparatus for Corona Resistance Test (see 3.4)—Winding of the cable—attach on the cable specimen a mass of 2.5 kg. Fix the free end of the specimen to a mandrel so that the mass can hang freely. Rotate the mandrel against the force exerted by the mass so that the cable specimen is wound up in five complete turns at a pitch of approximately 19 mm. During winding, the specimen shall not be forced against the natural torsion. Then fix the ends of the cable, remove the mass, and push a closely fitting sleeve over the specimen.

The sleeve and the mandrel shall be of nonmagnetic metal. The sleeve may have flared ends.

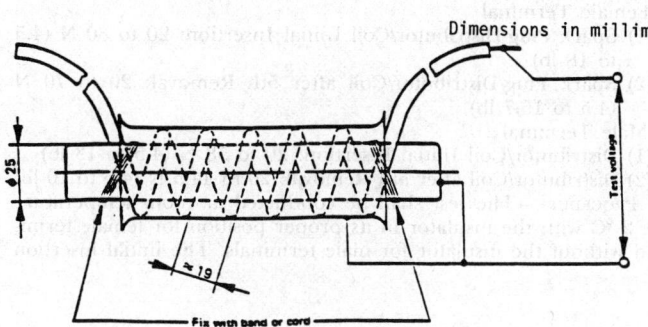

FIG. A2

A3. Test Apparatus for Deformation Test (see 3.5)

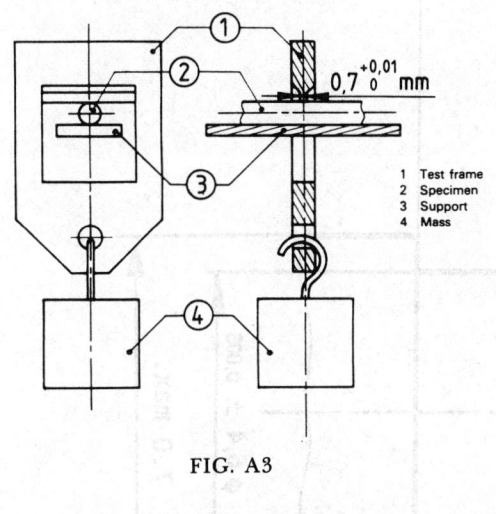

FIG. A3

A4. Winding on Mandrel (see 3.6, 3.12, 3.13 and 3.14)—Fix one end of the cable specimen to a rotatable mandrel of diameter 12.5 mm, and attach a mass of 4.5 kg to the insulation at the other end.

Wind the sample clockwise and then counterclockwise on the rotating mandrel in closely pitched turns to a minimum of four turns for each direction.

Speed of rotation—1 turn/s.

A5. Test Apparatus for Resistance to Flame Propagation Test (see 3.8)

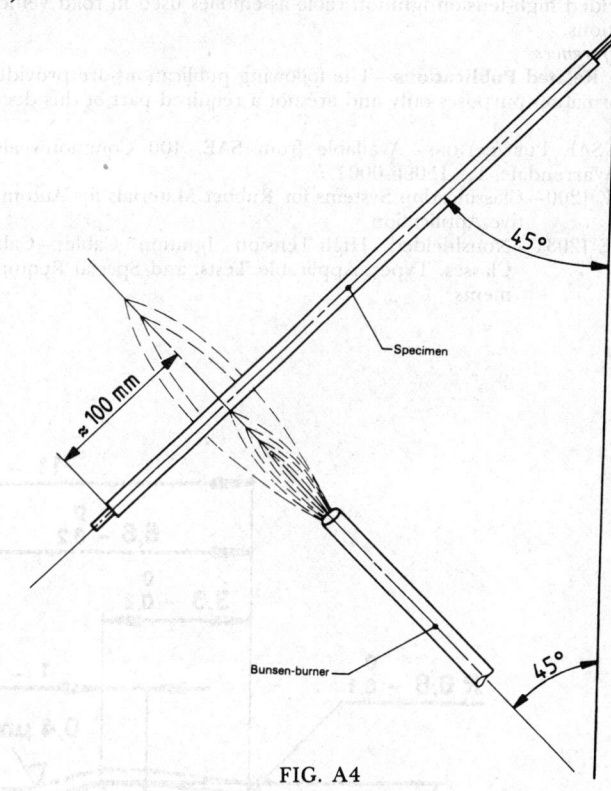

FIG. A4

A6. Test Apparatus for Mechanical Strength Test (see 3.10)

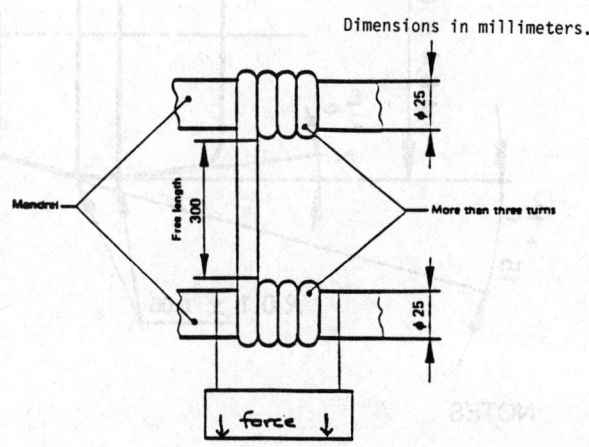

FIG. A5

IGNITION CABLE ASSEMBLIES —SAE J2032 NOV91

SAE Standard

Report of the SAE High Voltage Ignition Cable Task Force of the SAE Ignition Standards Committee approved November 1991.

1. Scope—This SAE Standard specifies the general requirements and test methods for nonshielded high-tension ignition cable assemblies.

1.1 Field of Application—This document applies to all types of nonshielded high-tension ignition cable assemblies used in road vehicle applications.

2. References

2.1 Related Publications—The following publications are provided for information purposes only and are not a required part of this document.

2.1.1 SAE PUBLICATIONS—Available from SAE, 400 Commonwealth Drive, Warrendale, PA 15096-0001.

  SAE J200—Classification Systems for Rubber Materials for Automotive Application
  SAE J2031—Nonshielded High-Tension Ignition Cables—Cable Classes, Types, Applicable Tests, and Special Requirements

3. Performance Requirements and Test Methods

3.1 Conductor Integrity—All finished assemblies will be tested for conductor continuity prior to testing.

3.2 Terminal Insertion and Removal Forces

3.2.1 REQUIREMENTS

  a. Female Terminal
   (1) Spark Plug/Distributor/Coil Initial Insertion: 20 to 80 N (4.5 to 18 lb)
   (2) Spark Plug/Distributor/Coil after 5th Removal: 20 to 70 N (4.5 to 15.7 lb)
  b. Male Terminal
   (1) Distributor/Coil Initial Insertion: 20 to 58 N (4.5 to 13 lb)
   (2) Distributor/Coil after 5th Removal: 20 to 44.5 N (4.5 to 10 lb)

3.2.2 PROCEDURE—The test shall be conducted at room temperature 23 °C ± 2 °C with the insulator in its proper position for female terminals and without the insulator for male terminals. The initial insertion

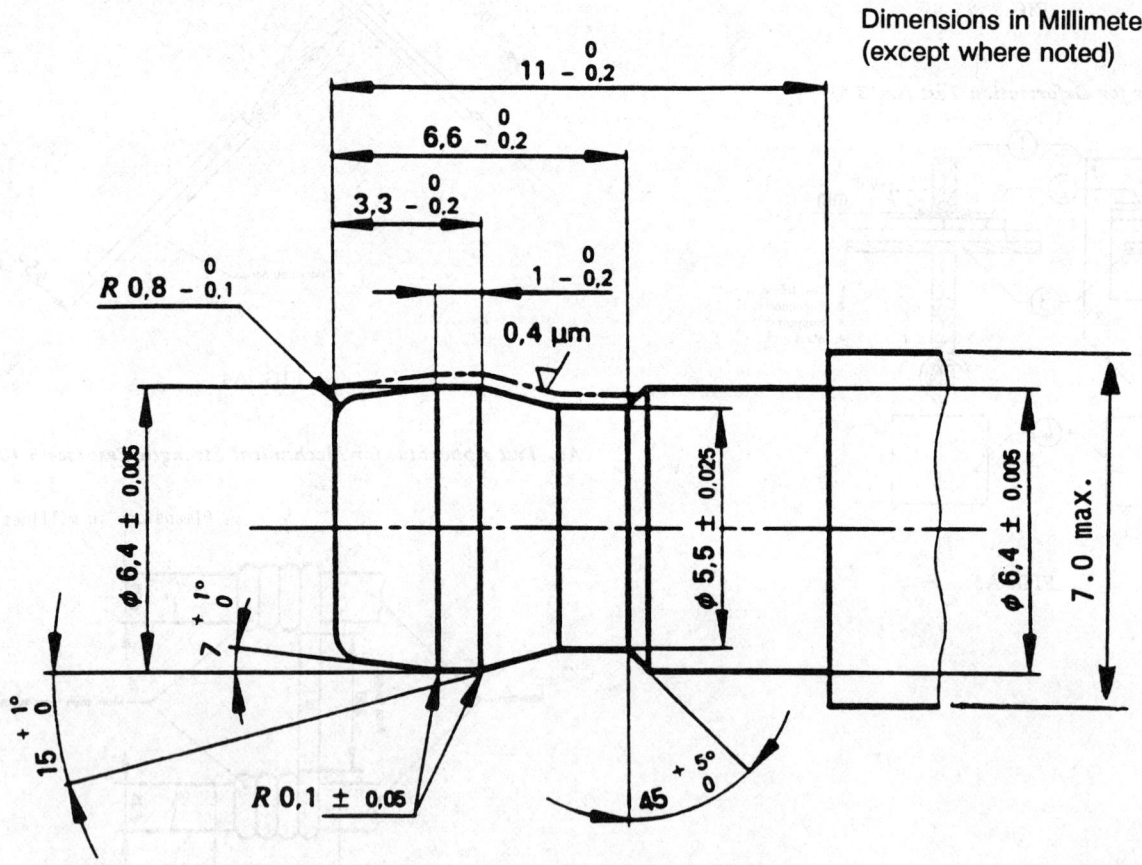

NOTES

1. The gage shall be of hardened steel.
2. The tolerances given for the gage dimensions include also the wear tolerances.

The dimensions 6.4 mm ± 0.005 mm and the angles of $7^{+1°}_{0}$ and $15^{+1°}_{0}$ are the most critical dimensions.

FIGURE 1—GAGE FOR MEASUREMENT OF INSERTION AND REMOVAL FORCES OF HIGH-TENSION CONNECTORS FOR SPARK-PLUGS WITH POST TERMINALS AND FOR IGNITION COIL AND DISTRIBUTORS WITH PLUG-TYPE HIGH-TENSION CONNECTIONS

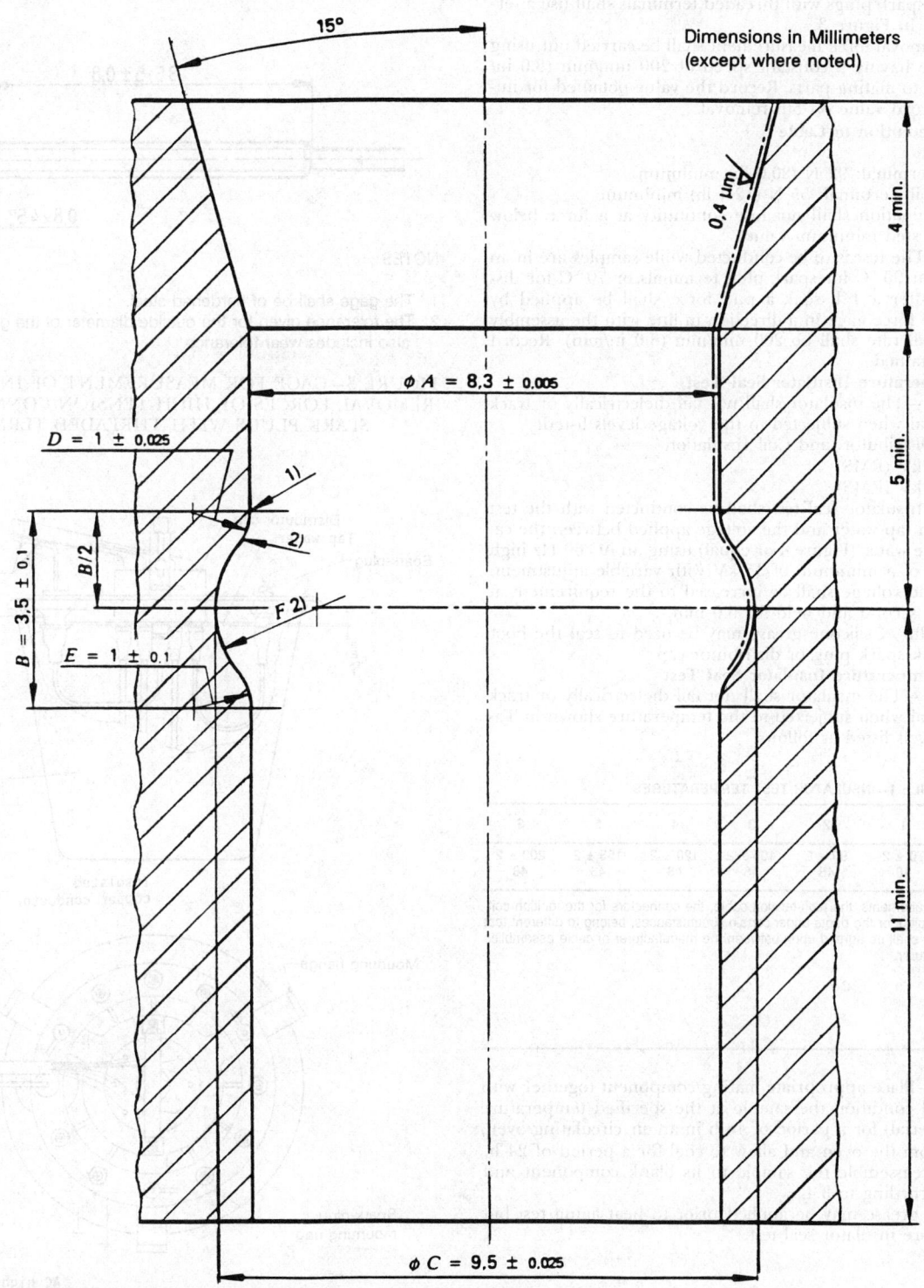

1. Tangential slope from diameter A to radius D.
2. Tangential slope from radius D to radius F. The value of F is implicitly determined by the values of dimensions A, B, C, D, and E.

NOTES

1. The gage shall be of hardened steel.
2. The tolerances given for the gage dimensions include also the wear tolerances. The dimensions A and D are the most critical dimensions.

FIGURE 2 – GAGE FOR MEASUREMENT OF INSERTION AND REMOVAL FORCES OF HIGH-TENSION CONNECTORS FOR SOCKET TYPE HIGH-TENSION CONNECTION FOR IGNITION COILS AND DISTRIBUTORS

and removal shall be done on a terminal gage for female terminals as shown in Figure 1 or terminal gage for male terminals as shown in Figure 2. Terminals for spark plugs with threaded terminals shall use a terminal gage as shown in Figure 3.

The insertion-removal force measurement shall be carried out using a suitable test fixture having a constant speed of 200 mm/min (8.0 in/min) aligned parallel to mating parts. Record the value obtained for initial insertion and record value on 5th removal.

3.3 Terminal Retention to Cable

3.3.1 REQUIREMENT

a. Spark Plug Terminal: 92 N (20.0 lb) minimum
b. Distributor/Coil Terminal: 55 N (12.4 lb) minimum

NOTE—The termination shall not lose continuity at a force below the suggested minimum value.

3.3.2 PROCEDURE—The test is to be conducted while samples are in an air circulating oven at 90 °C for spark plug terminals or 70 °C for distributor terminals. After a 1 h soak a pull force shall be applied by means of an accurate force gage in a direction in line with the assembly being tested. The pull rate shall be 200 mm/min (8.0 in/min). Record the highest value obtained.

3.4 Room Temperature Insulator Seal Test

3.4.1 REQUIREMENT—The insulator shall not fail dielectrically or track through the cable seal when subjected to the voltage levels listed:

a. Spark Plug, Distributor, and Coil Insulation
 (1) 7 mm: 15 kV (RMS)
 (2) 8 mm; 23 kV (RMS)

3.4.2 PROCEDURE—Insulator seal test shall be conducted with the test sample submerged in tap water and the voltage applied between the cable conductor and the water (Figure 4 or equal) using an AC 60 Hz high voltage unit capable of a minimum of 35 kV with variable adjustment. Beginning at 0 V, the voltage shall be increased to the requirement at a rate of 0.5 kV per second and held for 5.0 min.

NOTE—A thin film of silicone grease may be used to seal the boot to a blank spark plug or distributor cap.

3.5 Elevated Temperature Insulator Seal Test

3.5.1 REQUIREMENT—The insulator shall not fail dielectrically or track through the cable seal when subjected to the temperature shown in Table 1 and voltage levels listed as follows:

TABLE 1—INSULATOR TEST TEMPERATURES

Test Class	1	2	3	4	5	6
Test Temperature[1] °C	70 ± 2	90 ± 2	105 ± 2	120 ± 2	155 ± 2	200 ± 2
Test Time h	48	48	48	48	48	48

[1] According to differing requirements, the high-tension cable, the connectors for the ignition coil, the distributor, the spark plug, or the boots under certain circumstances, belong to different test classes. The test classes shall be agreed upon between the manufacturer of cable assemblies and the engine manufacturer.

Voltage
7 mm: 15 kV (RMS)
8 mm: 20 kV (RMS)

3.5.2 PROCEDURE—Place appropriate mating component together with the test sample and condition the sample at the specified temperature (depending on material) for a period of 48 h in an air circulating oven. Remove samples from the oven and allow to cool for a period of 24 h. Disassemble, then reassemble the sample to its blank component and run the seal test according to 3.4.

NOTE—Silicone grease may be applied prior to heat aging test but not before insulator seal test.

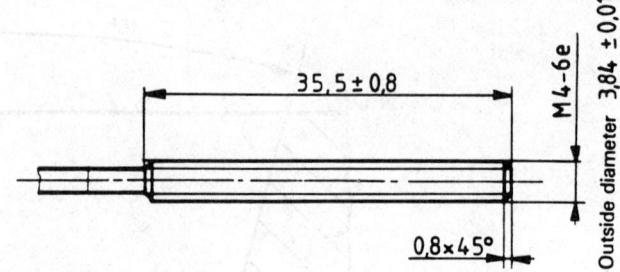

Dimensions in Millimeters

NOTES

1. The gage shall be of hardened steel.
2. The tolerance given for the outside diameter of the gage threaded part also includes wear tolerance.

FIGURE 3—GAGE FOR MEASUREMENT OF INSERTION AND REMOVAL FORCES OF HIGH-TENSION CONNECTORS FOR SPARK PLUGS WITH THREADED TERMINALS

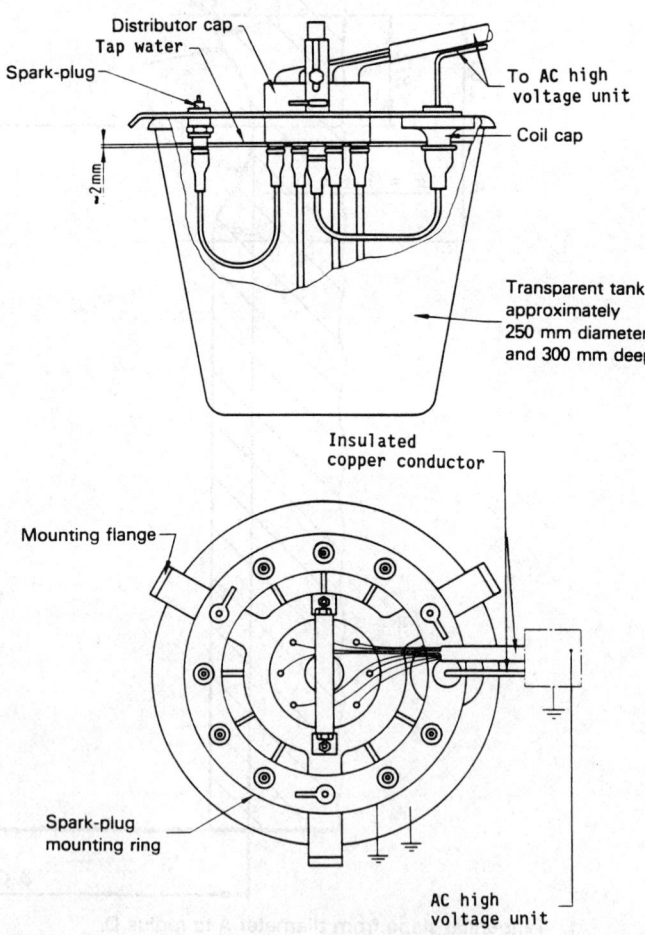

FIGURE 4—TEST APPARATUS FOR SEALING TEST

LOW TENSION WIRING AND CABLE TERMINALS AND SPLICE CLIPS—SAE J163

SAE Recommended Practice

Report of Electrical Equipment Committee approved January 1974.

Scope—This SAE Recommended Practice covers the application requirements for terminals and splice clips attached to stranded low tension wiring and cable as shown in J878 and J558. In addition, it covers maximum voltage drop limits for friction type connections.

Use of Terminals—Friction (quick disconnect) type brass connections should be used only where the maximum temperature (environmental ambient plus rise due to current), measured at the center of the terminal surface, does not exceed the capabilities of the physical properties of the material. Maximum temperatures for terminal materials other than brass should also be determined prior to using so as to be compatible with the physical properties of these materials.

Electrical connections and splices of standard types must be protected, as application dictates, from moisture, salt, soil accumulation, acid, or corrosive vapor which will deteriorate the connection beyond the limits of this recommended practice.

Performance Requirements (Electrical)—Terminals or splice clips shall be attached to wire or cable in such a manner that, following the humidity test, the voltage drop across the attachment shall not exceed the values in Table 1. Friction connections (terminal to terminal) shall be such that following four repeated insertions and the humidity test, the voltage drop across the connection shall not exceed the values in Table 2. For a terminal to be acceptable, all specimens tested must meet the requirements.

Test Procedure—Tests shall be conducted at 73 ± 5°F (23 ± 3°C). Test samples shall consist of terminals or splice clips attached to 12 in. (305 mm) of wire. It is suggested that at least 10 specimens of each wire size be subjected to each test.

Voltage Drop Test—Measurements shall be made after the temperature of the specimen has stabilized (2 h under test load).

Measurements across a wire to terminal attachment shall be made between the center of the conductor grip and a point on the cable core 3 in. (76.2 mm) behind the conductor grip. Probe point on the cable core shall be stripped and solder dapped. For preinsulated terminals, the measurements shall be made between a point in front of the conductor grip within 1/16 in. (1.6 mm) of the end of the insulation and a point on the cable core 3 in. (76.2 mm) behind the conductor grip. The voltage drop across the attachment is defined as the difference between this reading and the voltage drop through the 3 in. (76.2 mm) of wire.

Measurements across a splice clip connection shall be made between points on cable cores 3 in. (76.2 mm) behind the center of the conductor grip crimp. Probe points on the cable core shall be stripped and solder dapped. The voltage drop across the splice is defined as the difference between this reading and the voltage drop through the 6 in. (152 mm) of wire. Measurements shall be made across each combination of conductor pairs in the splice. The current value shall be selected according to the smaller gage cable in the cable pair being measured.

Measurements across a friction connection shall be made between the centers of the conductor grips of two joined line connector type terminals and from the center of the conductor grip of a line terminal to a similar point on a joined fixed terminal. For preinsulated terminals, the voltage drop across the wire to terminal attachment plus the 3 in. (76.2 mm) of cable shall be determined first per the above procedure. Then the terminals shall be connected and measurements across the connection shall be made from the stripped solder dapped points on the cable cores of two joined line connector type terminals and from the same point on the cable core to a point on a joined fixed terminal equivalent to that used for uninsulated terminals. The voltage drop across the friction connection is defined as the difference between this reading and the previous measurement(s) for the same specimen(s) (wire drop plus attachment drop).

Humidity Test—The humidity test shall consist of 100 h at 95–100% relative humidity at 100 ± 5°F (38 ± 3°C). (Demineralized water shall be used.) Specimens shall be prepared as follows:

1. Mounted at least 1 in. (25.4 mm) apart on test boards.
2. Placed in the humidity cabinet with the axis of each specimen in a horizontal plane and such that it has all surfaces of the terminal completely exposed and not in contact with other objects.
3. Removed from the cabinet after the prescribed exposure period and allowed to dry 24 h at room condition before final MVD test.

TABLE 1—WIRE TO TERMINAL OR WIRE TO WIRE (SPLICE CLIP METHOD) VOLTAGE DROP (AFTER HUMIDITY TEST)

Wire/Cable (SAE Gage)	Test Current, A	Drop, mV	
		Uninsulated Terminal	Preinsulated Terminal
20	5	3	3.5
18	10	5	5.5
16	15	8	9
14	20	10	11
12	30	15	17
10	40	20	22
8	50	25	—
6	60	15	—
4	70	18	—
2	80	20	—
0	90	23	—
00	100	25	—

TABLE 2—FRICTION VOLTAGE DROP (AFTER HUMIDITY TEST AND FOUR INSERTIONS)

Wire/Cable (SAE Gage)	Test Current, A	Drop, mV
20	5	7.5
18	10	15
16	15	22.5
14	20	30
12	30	45
10	40	60
8	50	75

φBATTERY CABLE—SAE J1127 JUN88

SAE Standard

Report of the Electrical Equipment Committee approved November 1975, and completely revised by the Electrical and Electronics Systems Technical Committee June 1988. J1127 NOV87 was not included in the SAE Handbook due to technical changes.

1. Scope—This standard covers battery cables intended for use at 50 V or less in surface vehicle electrical systems.

2. Specification Types

2.1 Type SGT—Starter or ground, thermoplastic insulated.

2.2 Type SGR—Starter or ground, thermoset elastomer insulated.

2.3 Type SGX—Starter or ground, crosslinked polyethylene insulated.

2.4 Type SGE—Starter or ground, thermoplastic elastomer insulated.

3. General Specifications

3.1 Conductors—The finished, uninsulated, conductor shall meet the elongation requirements specified in ASTM B-174. When tin, lead or lead alloy coated wires are used, they shall withstand the continuity of coating tests specified in paragraph 4.1. As required, a separator shall be used between uncoated conductors and insulation to comply with paragraph 3.2. The cross sectional area of stranded conductors shall not be less than the values specified in Table 1.

3.2 Insulation—Insulation shall be homogeneous and shall be placed concentrically within commercial tolerances about the conductor. Insulation shall adhere closely to, but strip readily from, the conductors leaving them reasonably clean and in suitable condition for terminating. The nominal wall thickness and maximum overall diameter of the finished cable shall be in accordance with Table 2 for the various cable types. Variations in insulation wall thickness are permissible due to eccentricity. However, the minimum wall thickness at any cross section of a test specimen shall not be less than 70% of the nominal wall thickness. The minimum wall thickness shall be measured with an optical device accurate to at least 0.001 in (0.01 mm) or with a pin dial micrometer that exerts a force of 25 ± 2 g (0.25 ± 0.02 N) on the specimen through a flat rectangular pressure foot. The dimensions at the pressure foot shall be 0.043 in × 0.312 in (1×8 mm) and the pin shall be 0.437 in (11 mm) long with a diameter of 0.020 in (0.5 mm). In case of dispute, the referee method shall be the optical device.

3.3 Finished Cable—The finished cable shall meet the requirements for all tests specified in Table 3 for each cable type.

TABLE 1—CONDUCTORS

SAE Wire Size		Minimum Conductor Area	
No.	mm²	cir cmils	mm²
6	13	24538	12.1
4	19	37360	18.3
2	32	62450	31.1
1	40	77790	38.1
0	50	98980	48.3
2/0	62	125100	59.8
3/0	81	158600	77.6
4/0	103	205500	98.5

NOTES:
1. SAE wire size number indicates that the cross sectional area of the conductors approximate the area of American Wire Gauge for equivalent sizes.
2. Metric wire size is the approximate nominal area of the stranded conductor. Metric dimensions are not direct conversion from circular mils.
3. See Appendix for various individual conductor constructions and nominal strand diameters.

TABLE 2—DIMENSIONS

Cable Type		SGT SGE SGR SGX			
SAE Wire Size		Nom Wall		Max Dia	
No.	mm²	in	mm	in	mm
6	13	0.060	1.52	0.340	8.59
4	19	0.065	1.65	0.420	10.21
2	32	0.065	1.65	0.505	11.93
1	40	0.065	1.65	0.557	12.87
0	50	0.065	1.65	0.600	14.05
2/0	62	0.065	1.65	0.655	16.00
3/0	81	0.078	1.98	0.750	18.29
4/0	103	0.078	1.98	0.810	19.79

NOTES:
1. Metric dimensions are not direct conversion from inches.
2. The 6 gage (13 mm²) and 4 gage (19 mm²) wall thickness can be the same as for GPT.

TABLE 3—TEST REQUIREMENTS

Tests	Cable Types			
	SGE	SGT	SGX	SGR
Conductor Area		X		X
Strand Coating		X		X
Physical Properties		X		X
Dimensions		X		X
Dielectric		X		X
Cold Bend		X		X
Flame		X		
Oil Absorption		X		X

4. Tests

4.1 Strand Coating—Continuity of coating test shall be conducted on individual strands prior to stranding and shall be conducted per ASTM B-33 or B-189.

4.2 Insulation Physical Properties—An accelerated aging test shall be conducted in accordance with ASTM D-573 except using specimens of insulation removed from finished cable. The original properties shall conform to the values shown in Table 4. Samples of insulation shall be aged 168 h in a forced air circulating air oven. The test temperature shall be as shown in Table 4. After aging, the tensile strength shall not be less than 80% of the original test value and the elongation shall not be less than 50% of the original test value.

Notes:
1. SGE samples may be preconditioned at test temperature for 24 h prior to taking original measurements. The samples will then be aged for 168 h.
2. The above accelerated aging test is appropriate for insulating materials currently specified in this standard; different test conditions may be necessary for other materials.
3. The temperature rating of the insulating material may be determined by conducting a long term test in accordance with IEC Publication 216-1 or equivalent using 0.030 in (0.76 mm) test specimens.

4.3 Dielectric Test—One in (25 mm) of insulation shall be removed from each end of a 24 in (600 mm) sample of finished cable and the two ends twisted together. The loop, thus formed, shall be immersed in water containing 5% salt by weight at room temperature so that not more than 6 in (150 mm) of each end of the sample protrudes above the solution. After being immersed for 5 h and while still immersed, the sample shall withstand the application of 1000 V at 60 Hz between the conductor and the solution for 1 min without puncture of the insulation.

4.4 Cold Bend Test—One in (25 mm) of insulation shall be removed from each end of a 24 in (600 mm) sample of finished cable. The sample shall be placed in a cold chamber at the specified temperature for a period of 3 h. While the sample is still at this low temperature, it shall be wrapped around a mandrel for a minimum of 180° at a uniform rate of one turn in 10 s. The temperature and mandrel size

TABLE 4—INSULATION PROPERTIES

Cable Type	Original Properties			Heat Aging Test[a] Temperature °C ± 2°C
	Min Tensile Strength		Min Elongation	
	psi	MPa	%	
SGT	1600	11	125	110
SGR	1000	7	150	110
SGX	1500	10	150	155
SGE	1600	11	200	150

[a] For higher temperature material, higher test temperatures may be used.

shall be as specified in Table 5. Either a revolving or stationary mandrel may be used. When a revolving mandrel is used, fasten one end of the sample to the mandrel and the specified weight to the other end. No weight is required when using a stationary mandrel. The sample is to be returned to room temperature and then subjected to the dielectric test specified in paragraph 4.3.

4.5 Flame Test—A bunsen burner having a ½ in (13 mm) inlet, a nominal bore of ⅜ in (10 mm), a length of approximately 4 in (100 mm) above the primary inlets, equipped with a wing top flame spreader having a 1/16 in × 2 in (1 × 50 mm) opening fitted to the top of the burner shall be used. A 24 in (600 mm) sample of finished cable shall be suspended, taut in a horizontal position within a partial enclosure which allows a flow of sufficient air for complete combustion but is free from drafts. The top of a 2 in (50 mm) gas flame with an inner cone one-third its height shall then be positioned vertically and applied to the center of the suspended cable with the flame spreader parallel to the axis of the cable. The time of application of the flame shall be 30 s. After removal of the bunsen burner flame, the sample shall not continue to burn for more then 30 s.

TABLE 5—COLD BEND TEST

Cable Type	SGT SGR SGX SGE 3h/−40° (−40°C)			
Test Conditions SAE Wire Size		Mandrel		Weight
No.	mm²	in	mm	lb	kg
6	13	10	254	6	2.72
4	19	10	254	6	2.72
2	32	10	254	6	2.72
1	40	18	457	6	2.72
0	50	18	457	10	4.53
2/0	62	18	457	10	4.53
3/0	81	18	457	10	4.53
4/0	103	18	457	10	4.53

NOTE: Metric dimensions and weights are direct conversion from inches and pounds.

APPENDIX

Recommended Conductor Constructions (AWG Strands)

SAE Wire Size No.	No. Strands/ AWG Size (in)	No. Strands/ AWG Size (in)
6	37/21 (0.0285)	7 × 19/27 (0.0142)
4	61/22 (0.0253)	7 × 19/25 (0.0179)
2	127/23 (0.0226)	7 × 19/23 (0.0226)
1	127/22 (0.0253)	7 × 37/25 (0.0179)
0	127/21 (0.0285)	7 × 37/24 (0.0201)
2/0	127/20 (0.0320)	7 × 37/23 (0.0226)
3/0		7 × 37/22 (0.0253)
4/0		19 × 22/23 (0.0226)

Recommended Conductor Constructions (Metric Strands)

SAE Wire Size mm²	No. Strands/ mm Size	No. Strands/ mm Size
13	37/0.66	7 × 19/0.36
19	61/0.63	7 × 19/0.45
32	127/0.57	7 × 19/0.57
40	127/0.63	7 × 19/0.63
50	127/0.71	7 × 19/0.71
62	127/0.79	7 × 19/0.79
81		7 × 37/0.63
103		19 × 18/0.63

NOTE: Stranding other than those shown above are acceptable providing they meet the minimum conductor area specified in Table 1.

4.6 Oil Absorption Test—One in (25 mm) of insulation shall be removed from each end of a 24 in (600 mm) sample of finished cable. The sample shall be immersed to within 1½ in (40 mm) from the end of the insulation in ASTM D-471 oil Number 3 at a temperature of 118-122°F (48-50°C) for a period of 20 h. The outside diameter of the cable shall not change more than 15% when measured 4 h after removal from the oil. The sample shall then be bent around a 10 in (250 mm) mandrel and then subjected to the dielectric test specified in paragraph 4.3.

5. Source for Referenced Specifications
  5.1 American Society of Testing and Materials (ASTM)
  1916 Race Street
  Philadelphia, PA 19103
  5.2 International Electrotechnical Commission (IEC)
  American National Standards Institute
  1430 Broadway
  New York, NY 10018

BATTERY BOOSTER CABLES—SAE J1494 JUN89

SAE Recommended Practice

Report of the Electrical Distribution Systems Standards Committee approved June 1989.

1. Scope—The purpose of this SAE Recommended Practice is to establish minimum performance and user information requirements for battery booster cable sets. Such sets may be used to provide a temporary connection of a surface vehicle battery to another similar battery to provide emergency power when required. This recommended practice DOES NOT ENDORSE NOR RECOMMEND the potentially hazardous procedure of jump starting a vehicle.

2. Specification Types—The battery booster cable sets covered by this specification shall have a minimum rating as shown below when tested in accordance with paragraph 3.3.
  2.1 Light duty - 125 A minimum
  2.2 Medium duty - 225 A minimum
  2.3 Heavy-duty - 350 A minimum
  2.4 Extra heavy-duty - 500 A minimum
  2.5 Super heavy-duty - 750 A minimum

3. General Specifications

3.1 Conductors—The conductors shall be bunched, concentric or rope stranded as specified in Table 1 and shall be annealed copper wire in accordance with ASTM B 3.

When tin-alloy-coated wires are used, they shall withstand the continuity of coating test as specified in ASTM B 33 or B 189. If a synthetic rubber insulation is used, a separator shall be placed between the uncoated conductor and the insulation. When coated conductors or chemically nonreactive insulation materials are used, no separator is required. The cross sectional area of stranded conductors shall not be less than the values specified in Table 1.

3.2 Insulation—Insulation shall be homogeneous in character and shall be placed concentrically within commercial tolerances around the conductor. Insulation shall adhere closely to, but strip readily from, the conductors, leaving them reasonably clean and in suitable condition for terminating.

TABLE 1—RECOMMENDED CONDUCTOR CONSTRUCTIONS AND INSULATION THICKNESS

SAE[a] Wire Size	Metric[b] Wire Size mm²	Minimum Conductor Area For Finished Cable		No. Strands	Gage		Minimum Wall Thickness	
		Cir Mil	mm²		in	mm	in	mm
10	5	9343	4.65	19/23	0.0226	0.574	0.042	1.07
8	8	14810	7.23	19/21	0.0285	0.724	0.042	1.07
6	13	25910	12.1	37/21	0.0285	0.724	0.042	1.07
4	19	37360	18.3	61/22	0.0253	0.643	0.046	1.17
2	32	62450	31.1	127/23	0.0226	0.574	0.046	1.17
1	40	77790	38.1	127/22	0.0253	0.643	0.046	1.17
0	50	98980	48.3	127/21	0.0285	0.724	0.046	1.17
2/0	62	125100	59.8	127/20	0.0320	0.813	0.046	1.17
3/0	81	158600	77.6	259/22	0.0253	0.643	0.055	1.40
4/0	103	205500	98.5	427/23	0.0226	0.574	0.055	1.40

[a]The SAE wire size number indicates that the cross sectional area of the conductors approximates the area of the American Wire Gage for equivalent sizes.
[b]The metric wire size is the approximate nominal area of the stranded conductor. The metric dimensions are not the direct conversion from the circular mils.
NOTE: Stranding other than those shown for SAE and metric wire sizes are acceptable provided they meet the minimum conductor area specifications in Table 1.

The minimum wall thickness of the finished cable shall be in accordance with Table 1. Variations in insulation wall thickness are permissible due to eccentricity. However, the minimum wall thickness at any cross section of a test specimen shall not be less than 70% of the nominal wall thickness (the nominal wall thickness is one half the difference between the nominal overall diameter of the finished cable and the nominal conductor diameter). The minimum wall thickness shall be measured with an optical device accurate to at least 0.001 in (0.01 mm) or with a pin dial micrometer that exerts a force of 25 g ± 2 (0.25 N ± 0.02) on the specimen through a flat rectangular pressure foot. The dimensions of the pressure foot shall be 0.043 in x 0.312 (1 mm x 8) and the pin shall be 0.437 in (11 mm) long with a diameter of 0.020 in (0.5 mm). In the case of a dispute, the referee method shall be the optical device.

The temperature rating of the cable insulation shall be a minimum of 140°F (60°C) as determined by an accelerated aging test conducted in accordance with ASTM D 573, except samples of insulation are to be removed from the finished cable and aged 168 h. The test temperature shall be 54°F (30°C) above the intended rated temperature.

The tensile strength after aging shall not be less than 80% of the original tensile strength. The elongation after aging shall be at least 50% of the original elongation. Samples may be preconditioned at the test temperature for a period not to exceed 24 h before conducting the accelerated aging test.

3.3 The Procedure for Determining Battery Booster Cable Set Rating—The following procedure should be performed at an ambient temperature of 73°F ± 9 (23°C ± 5), in order to determine the battery booster cable set rating as shown in section 2.

3.3.1 RECOMMENDED EQUIPMENT

  a. Source of constant DC current with sufficient capacity to allow a 2.5 V minimum drop on each cable with a current measuring capability of ±1% accuracy.
  b. Stainless steel electrodes per Fig. 1 attached to current source electrodes.
  c. Voltmeter (accurate to 0.01 V).
  d. Timing device.
  e. Thermocouple (iron/constantan type).
  f. Chart recorder with 250°F ± 5 (121°C ± 3) maximum reading. Accurate to 0.25% full scale.

3.3.2 TEST PROCEDURE

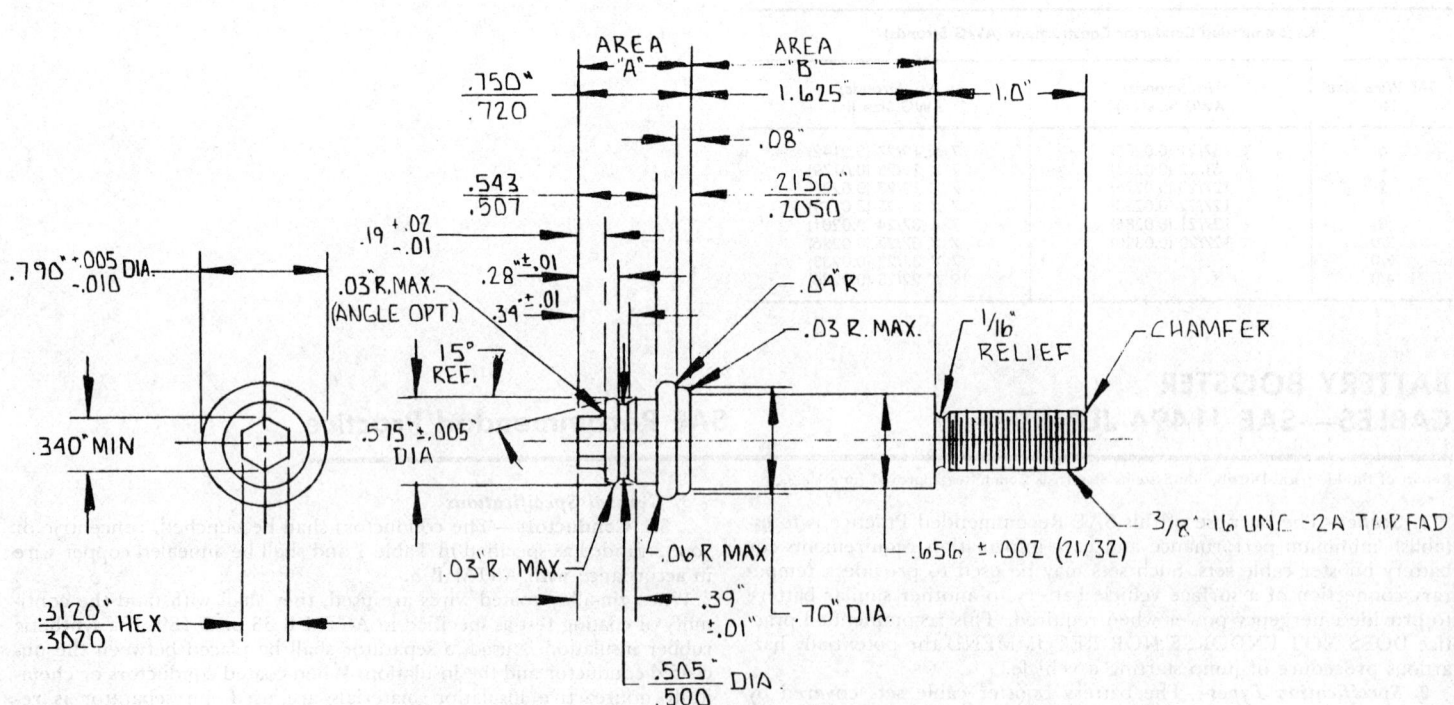

FIG. 1—STAINLESS STEEL ELECTRODE

a. Attach one of the cables of the battery booster cable set to the current source by clamping to the stainless steel electrode area A (see Fig. 1) in the manner the clamps would normally be used.

b. Attach the thermocouple to the battery booster cable clamp handle by taping tightly to the outer surface at a mid-point between the pivot and the rear of the current carrying clamp. If nonpermanently attached grips are used, slide the grip off the handle and apply thermocouple directly to the handle. (Thermocouple is attached in an area normally gripped by the user). Attach the other end of the thermocouple lead to the chart recorder.

c. Select a test current and apply to the cable.

d. After 10 s, measure and record the voltage drop from one stainless steel electrode to the other.

e. After 15 s, turn off the current.

f. Monitor the temperature of the handle for an additional 2 min and record the maximum temperature. The maximum temperature permitted is 150°F (66°C).

Note: Some clamps will be at a maximum temperature immediately after the current is turned off, others will continue to rise for various periods.

g. Repeat steps a through f with the other cable of the battery booster cable set.

h. Repeat steps a through g with successive greater test currents until the test current produces a combined total voltage drop of 5.0 V for both cables in the battery booster cable set. The cables should be allowed to cool for approximately 15 min before retesting.

i. Repeat steps a through h with the battery booster cable clamp attached to area B (see Fig. 1) of the stainless steel electrode.

j. Using the test data from area A and B, determine the largest of the test currents that does not cause a combined total voltage drop of more than 5.0 V or a temperature of more than 150°F (66°C) on the handle. The rating of the cable set shall be the lowest amperage rating of the results of tests performed using Clamping Area A and Clamping Area B.

4. Cable Requirements

4.1 Cold Bend Test—Remove 1 in (25 mm) of insulation from each end of a 24 in (610 mm) sample of finished cable. Attach the sample to either a revolving or a stationary mandrel. If a revolving mandrel is used, the sample is to be secured to the mandrel and a weight attached to the free end as specified in Table 2. If a stationary mandrel is used, the sample is to be conditioned at the low temperature without weights, and manually bent around the specified mandrel. Lower the temperature to -40°F (-40°C) and maintain for 3 h. While the sample is still at -40°F (-40°C), it shall be wrapped around the mandrel for 180 deg at a uniform rate of one turn in 10 s.

The cable shall then be looped and immersed in water containing 5% salt by weight for a period of 5 h at room temperature with 5 in (127 mm) to 6 in (152 mm) of each end of the sample protruding above the solution. The sample shall withstand the application of 1000 V at 60 Hz between the conductor and solution for 1 min without puncturing the insulation.

4.2 Flame Test—A bunsen burner having a 1/2 in (13 mm) inlet, a nominal bore of 3/8 in (10 mm), a length of approximately 4 in (102 mm) above the primary inlets equipped with a wing top flame spreader, is positioned parallel to the cable, and has a 1/16 x 2 in (1.6 x 51 mm) opening fitted to the top of the burner shall be used. A 24 in (610 mm) sample of finished cable shall be suspended taut in a horizontal position within a partial enclosure that allows a flow of sufficient air for complete combustion but is free from drafts. The top of a 2 in (51 mm) gas flame with an inner cone one-third its height shall then be applied to the center of the suspended cable. The time of application of the flame shall be 30 s for SAE wire gages 10 (5 mm^2) through 4/0 (103 mm^2). After removal of the bunsen burner flame, the sample shall not continue to burn for more than 30 s.

4.3 Deformation Test—The insulation thickness T_1 is to be determined from measurements made on a 24 in (610 mm) length of insulated cable. The ends of parallel cables must be separated. The difference method consists of obtaining five readings to determine the average diameter over the insulation and subtracting from it the diameter of the conductor plus any separator with the difference divided by two. Measurements are made with a machinist's caliper micrometer having flat surfaces on both the anvil and the end of the spindle and calibrated to read directly to at least 0.001 in (0.01 mm).

The insulation thickness T_2 at an elevated temperature shall be determined from measurements made by a dead-weight dial micrometer with a presser foot 0.375 in ± 0.010 (9.5 mm ± 0.2) in diameter and with graduations of 0.001 in (0.01 mm). The micrometer shall be actuated by a weight of a magnitude that causes the foot of the micrometer to press on a specimen positioned between the foot and the anvil with 500 gf (4.9 N) load.

With the weight in place on its spindle, the micrometer shall be placed beside the test specimen in a full-draft circulating air oven, which has been preheated to a temperature of 250°F ± 2.0 (121°C ± 1.0). The specimen and the micrometer shall remain side by side in the oven for 1 h of preliminary heating at full draft. At the end of the hour, the specimen shall be placed on the anvil of the micrometer. The loaded presser foot shall be gently brought to bear on the specimen and shall continue to bear on it while the micrometer and the specimen remain in the oven for an additional hour at full draft at a temperature of 250°F ± 2.0 (121°C ± 1.0). The entire surface of the presser foot shall be in contact with the specimen.

At the end of the second hour, the thickness T_2 of the specimen shall read directly from the dial on the loaded micrometer and shall be recorded to the nearest 0.001 in (0.01 mm). From this reading, subtract the core diameter and divide by two. The specimens shall not decrease more than 50% in thickness.

5. Connector Devices
—All connector devices shall provide a sound mechanical and good electrical connection to the point of attachment such as a battery terminal, stud, or metallic ground. The connector devices shall be free from burrs and sharp corners. The temper of the connectors shall be sufficiently soft to permit the connectors being assembled to the cable from showing any fracture or cracks that would impair the strength of the assembly. All connector devices shall be insulated to protect the user against cuts, burns, and scratches.

5.1 Connector Attachment—The connectors may be attached to the cables by crimping, swaging or a combination of both. Each end of a cable connected to a connector device must be able to withstand a tensile force of 100 lb (445 N) applied in an axial direction without affecting the cable/connector device interface or the integrity of the current carrying connection.

5.2 Connector Identification—The color black and any contrasting color, except white, connectors must be used at the battery contact point. If "+", "POS", and/or "POSITIVE" is marked on a connector, it must appear on the contrasting color clamp. If "-", "NEG", and/or "NEGATIVE" is marked on a connector, it must appear on the black clamp.

6. Labeling Information

6.1 Tags—The following information must be affixed to the battery booster cable set. It should be printed on a material that is capable of withstanding abuse during normal usage.

(a) Instructions for Jump Starting an Engine:

Warning - Batteries Contain Acid And Produce Explosive Gases

NOTE: Consult the owner's manual for complete instructions. SHIELD THE EYES AND FACE FROM THE BATTERIES AT ALL TIMES. Be sure the vent caps are tight and level. Place a damp cloth, if available, over any vent caps on both batteries. Be sure the vehicles do not touch and that both electrical systems are off and at the same voltage. These instructions are for negative ground systems only.

(1) Connect the positive (+) cable to the positive (+) terminal of the discharged battery that is wired to the starter or solenoid.

(2) Connect the other end of the positive cable to the positive terminal of the booster battery.

(3) Connect the black negative (-) cable to the other terminal (negative) of the booster battery.

TABLE 2—COLD BEND TEST
-40°F (-40°C)

SAE Wire Size	mm^2	Mandrel in	Mandrel mm	Weight lb	Weight kg
10	5	6	152	6	2.72
8	8	6	152	6	2.72
6	13	6	152	6	2.72
4	19	6	152	6	2.72
2	32	8	203	6	2.72
1	40	8	203	6	2.72
0	50	10	254	10	4.54
2/0	62	10	254	10	4.54
3/0	81	10	254	10	4.54
4/0	103	10	254	10	4.54

(4) MAKE THE FINAL CONNECTION ON THE ENGINE BLOCK OF THE STALLED VEHICLE (NOT TO THE NEGATIVE POST) AWAY FROM THE BATTERY. STAND BACK.

(5) Start the vehicle and remove the cables in the reverse order of connection (the engine block (black) connection is the first to disconnect).

6.2 Packaging—The tag information, see paragraph 6.1, must appear on the package in addition to the following warnings and safety procedures:

(a) Battery Booster Cable Instructions for Jump Starting an Engine: Warning - Batteries Contain Acid And Produce Explosive Gases.

These instructions are designed to minimize the explosion hazard. Keep sparks, flames, and cigarettes away from the batteries at all times. Wear safety glasses, and protect the eyes at all times. Do not lean over the batteries during this operation.

Both the battery to be jumped and the booster source must be of the same voltage (6 or 12 V, etc.). Power sources other than batteries should not exceed 16 V DC for use with 12 V systems, and 8 V DC for use with a 6 V DC system.

Position the vehicle with the booster battery, or other power source, adjacent to the vehicle with the discharged battery so that booster cables can be connected easily between both vehicles. Make certain that the vehicles do not touch each other.

(1) Turn off all electrical loads on all vehicles and set the parking brake. Place the automatic transmissions in "PARK" (manual transmission in "NEUTRAL").

(2) Determine whether the discharged battery has the negative (−) or the positive (+) terminal connected to the ground. The ground lead is connected to the engine block, the vehicle frame, or some other good metallic ground. The battery terminal connected to the starter is the one that is not grounded. All vehicles manufactured in the U.S.A. after 1955 have the negative battery terminal grounded. All European and Asian passenger vehicles manufactured after 1971 have the negative battery terminal grounded.

(3) On a negative ground system, connect the positive (+) cable to the positive (+) terminal of the discharged battery wired to the starter or solenoid. Do not allow the positive cable clamps to touch any metal other than the battery positive (+) terminals.

(4) Be sure that the vent caps are tight and level on both batteries. Place a damp cloth over any vent caps on each battery making certain it is clear of fan blades, belts, and other moving parts.

6.3 Supplemental Packaging Information—Additional supplemental information and instructions may also be included.

φLOW TENSION PRIMARY CABLE—SAE J1128 JUN88   SAE Standard

Report of the Electrical Equipment Committee approved November 1975 and completely revised by the Electrical and Electronic Systems Technical Committee June 1988. J1128 NOV87 was not included in the SAE Handbook due to technical changes.

1. Scope—This standard covers low tension primary cable intended for use at 50 V or less in surface vehicle electrical systems.

2. Specification Types

2.1 Type GPT—General purpose, thermoplastic insulated.

2.2 Type HDT—Heavy duty, thermoplastic insulated.

2.3 Type STS—Standard duty, thermoset elastomer, insulated.

2.4 Type HTS—Heavy duty, thermoset elastomer, insulated.

2.5 Type GXL—General purpose, crosslinked, polyethylene, insulated.

2.6 Type SXL—Special purpose, crosslinked polyethylene, insulated.

2.7 Type GTE—General purpose, thermoplastic elastomer insulated.

2.8 Type HTE—Heavy duty, thermoplastic elastomer insulated.

3. General Specifications

3.1 Conductors—The finished, uninsulated conductor, shall meet the elongation requirements specified in ASTM B-174. When tin, lead or lead alloy coated wires are used, they shall withstand the continuity of coating tests specified in paragraph 4.1. As required, a separator shall be used between uncoated conductors and insulation to comply with paragraph 3.2. The cross sectional area of stranded conductors shall not be less than the values specified in Table 1.

3.2 Insulation—Insulation shall be homogeneous and shall be placed concentrically within commercial tolerances about the conductor. Insulation shall adhere closely to, but strip readily from, the conductors leaving them reasonably clean and in suitable condition for terminating.

The nominal wall thickness and maximum overall diameter of the finished cable shall be in accordance with Table 2 for the various cable types. Variations in insulation wall thickness are permissible due to eccentricity. However, the minimum wall thickness at any cross section of a test specimen shall not be less than 70% of the nominal wall thickness. The minimum wall thickness shall be measured with an optical device accurate to at least 0.001 in (0.01 mm) or with a pin dial micrometer that exerts a force of 25 ± 2 g (0.25 ± 0.02 N) on the specimen through a flat rectangular pressure foot. The dimensions of the pressure foot shall be 0.043 in × 0.312 in (1 × 8 mm) and the pin shall be 0.437 in (11 mm) long with a diameter of 0.020 in (0.5 mm). In case of dispute the referee method shall be the optical device.

3.3 Finished Cable—The finished cable shall meet the requirements for all tests specified in Table 3 for each cable type.

4. Tests

4.1 Strand Coating—Continuity of coating test shall be conducted on individual strands prior to stranding and shall be conducted per ASTM B-33 or B-189.

4.2 Insulation Physical Properties—An accelerated aging test shall be conducted in accordance with ASTM D-573 except using specimens of insulation removed from finished cable. The original properties shall conform to the values shown in Table 4. Samples of insulation shall be aged 168 h in a forced air circulating oven. The test temperature shall be as shown in Table 4. After aging, the tensile strength shall not be less than 80% of the original test value and the elongation shall not be less than 50% of the original test value.

Note:
1. GTE and HTE samples may be preconditioned at test temperature for 24 h prior to taking original measurements. The samples will then be aged for 168 h.
2. The above accelerated aging test is appropriate for insulating materials currently specified in this standard; different test conditions may be necessary for other materials.
3. The temperature rating of the insulating material may be determined by conducting a long term test in accordance with IEC Publication 216-1 or equivalent using 0.030 in (0.76 mm) test specimens.

4.3 Dielectric Test—One in (25 mm) of insulation shall be removed from each end of a 24 in (600 mm) sample of finished cable and the two ends twisted together. The loop thus formed shall be immersed in water containing 5% salt by weight at room temperature so that not more than 6 in (150 mm) of each end of the sample protrudes above the solution. After being immersed for 5 h and while still immersed, the sample shall withstand the application of 1000 V at 60 Hz between the conductor and the solution for 1 min without puncture of the insulation.

4.4 Cold Bend Test—One in (25 mm) of insulation shall be removed from each end of a 24 in (600 mm) sample of finished cable. The sample shall be placed in a cold chamber at the specified temperature for a period of 3 h. While the sample is still at this low temperature, it shall be wrapped around a mandrel for a minimum of 180° at a uniform rate of one turn in 10 s. The temperature and mandrel size shall be as specified in Table 5. Either a revolving or stationary mandrel may be used. When a revolving mandrel is used, fasten one end of the sample to the mandrel and the specified weight to the other end. No weight is required when using a stationary mandrel. The sample is to be returned to room temperature and then subjected to the dielectric test specified in paragraph 4.3.

4.5 Flame Test—A Bunsen burner having a ½ in (13 mm) inlet, a nominal core of ⅜ in (10 mm), a length of approximately 4 in (100 mm) above the primary inlets, equipped with a wing top flame spreader having a 1/16 × 2 in (1 × 50 mm) opening fitted to the top of the burner shall be used. A 24 in (600 mm) sample of finished cable shall be suspended taut in a horizontal position within a partial enclosure which allows a flow of sufficient air for complete combustion but is free from drafts. The top of a 2 in (50 mm) gas flame with an inner cone one-third its height shall then be positioned vertically and applied to the center of the suspended cable with the flame spreader parallel to the axis of the cable. The time of application of the flame shall be 15 s for SAE size numbers 10 through 20 (5 through 0.5 mm^2) and 30 s for SAE size number 8 (8 mm^2) and larger. After removal of the bunsen burner flame, the sample shall not continue to burn for more than 30 s.

4.6 Oil Absorption Test—One in (25 mm) of insulation shall be removed from each end of a 24 in (600 mm) sample of finished cable. The sample shall be immersed to within 1½ in (40 mm) from the end of the insulation in ASTM D-471 oil Number 3 at a temperature of 118-122°F (48-50°C) for a period of 20 h. The outside diameter of the cable shall not change more than 15% when measured 4 h after removal from the oil. The sample shall then be bent around a mandrel as specified in Table 5. The sample shall then be subjected to the dielectric test specified in paragraph 4.3.

4.7 Overload Test—In an ambient temperature of 73 ± 9°F (23 ± 5°C), a 60 in (1500 mm) sample cable suspended in air or lying on a transite table top shall be subjected to an overload current sufficient to raise the conductor temperature to 400 ± 3°F (204 ± 2°C) and to hold it there for a period of 30 min (thermocouple to be inserted into sample conductor stranding 18 in (450 mm) from one end). After the overload test, cut 18 in (450 mm) from each end of the cable and discard. The remaining 24 in (600 mm) portion which was in the center of the original 60 in (1500 mm) shall then be subjected to the dielectric test, paragraph 4.3.

4.8 Short Circuit Test—(SAE Wire Size number 18 or 0.8 mm^2 only)—Using six 36 in (900 mm) lengths and one 48 in (1200 mm) length of SAE wire size 18 cable strip 1 in (25 mm) of insulation from each end of the 48 in (1200 mm) length. Twist the six 36 in (900 mm) lengths around the 48 in (1200 mm) length with approximately a 4 in (100 mm) lay. Position so that 6 in (150 mm) of the 48 in (1200 mm) cable extends beyond each end. Tape into position using woven glass tape with ⅓ lap. Apply a constant 55 amp current to the center conductor of the bundle for 3 min. Turn off current and allow bundle to cool to room temperature. Disconnect power source and test for short circuits between all conductors. Use 1000 V (rms) test voltage. There shall be no shorting of conductors and when the glass tape is removed from the bundle the individual wires shall be readily separated without tearing the insulation on the individual cables. This test is conducted to check the thermosetting properties of the insulation.

TABLE 1—CONDUCTORS

SAE Wire Size		Minimum Conductor Area for Finished Cable	
No.	mm^2	cmils	mm^2
20	0.5	1072	0.508
18	0.8	1537	0.760
16	1	2336	1.12
14	2	3702	1.85
12	3	5833	2.91
10	5	9343	4.65
8	8	14810	7.23
6	13	24538	12.1
4	19	37360	18.3

[a]SAE wire size number indicates that the cross sectional area of the conductors approximate the area of American Wire Gauge for equivalent sizes.

[b]Metric wire size is the approximate nominal area of the stranded conductor. Metric dimensions are not direct conversion from circular mils.

[c]See Appendix B for various individual conductor constructions and nominal strand diameters.

TABLE 3—TEST REQUIREMENTS

Tests	Cable Types							
	GPT	HDT	GTE	HTE	STS	HTS	SXL	GXL
Conductor Area	X	X	X	X	X	X	X	X
Strand Coating	X	X	X	X	X	X	X	X
Physical Properties	X	X	X	X	X	X	X	X
Dimensions	X	X	X	X	X	X	X	X
Dielectric	X	X	X	X	X	X	X	X
Cold Bend	X	X	X	X	X	X	X	X
Flame	X	X	X	X	X	X	X	X
Oil Absorption	X	X	X	X				
Overload					X	X	X	X
Short Circuit					X	X	X	X
Pinch	X	X	X	X	X	X	X	X
Abrasion Resistance	X	X	X	X	X	X	X	X

TABLE 2—DIMENSIONS

SAE Wire Size		GPT GXL GTE				STS		SXL		HDT		HTE		HTS	
		Nom Wall		Max Dia		Nom Wall		Max Dia		Nom Wall		Max Dia		Nom Wall		Max Dia	
No.	mm²	in	mm	in	mm	in	mm	in	mm	in	mm	in	mm	in	mm	in	mm
20	0.5	0.023	0.58	0.095	2.34	0.029	0.74	0.110	2.71	0.036	0.91	0.120	2.95	0.036	0.91	0.125	3.08
18	0.8	0.023	0.58	0.100	2.50	0.030	0.76	0.120	3.00	0.037	0.94	0.130	3.24	0.037	0.94	0.135	3.37
16	1	0.023	0.58	0.115	2.84	0.032	0.81	0.135	3.33	0.040	1.02	0.145	3.58	0.040	1.02	0.150	3.70
14	2	0.023	0.58	0.125	3.18	0.035	0.89	0.155	3.94	0.041	1.04	0.165	4.19	0.041	1.04	0.165	4.19
12	3	0.026	0.66	0.150	3.81	0.037	0.94	0.180	4.57	0.046	1.17	0.190	4.83	0.046	1.17	0.200	5.08
10	5	0.031	0.79	0.185	4.67	0.041	1.04	0.210	5.30	0.046	1.17	0.215	5.42	0.048	1.22	0.225	5.67
8	8	0.037	0.94	0.235	5.80	0.043	1.09	0.245	6.10	0.055	1.40	0.280	6.98	0.055	1.40	0.270	6.73
6	13	0.043	1.09	0.305	7.72	0.055	1.40	0.335	8.47	0.060	1.52	0.340	8.60	0.062	1.57	0.350	8.85
4	19	0.044	1.12	0.375	9.32					0.068	1.73	0.420	10.44				

Metric dimensions are not direct conversion from inches.

4.9 Pinch Test—One in (25 mm) of insulation shall be removed from one end of a 36 in (900 mm) sample of finished cable. The sample shall then be placed taut without stretching across a 1/8 in (3 mm) diameter steel bar and be subjected to the force of a weighted steel anvil. Increasing weight shall be applied to the steel anvil at a rate of 5 lb (2.3 kg) per min with a lever advantage of 10. At the moment the insulation is pinched through, the 1/8 in (3 mm) diameter rod will contact the conductor and the test shall stop. The weight in the receptacle shall then be recorded. After each reading, the sample shall be moved approximately 2 in (50 mm) and rotated clockwise 90 deg. Four readings shall be obtained for each sample. Obtain an average by calculating the arithmetic mean of all those readings. The average shall define the pinch resistance of the cable under test. The minimum values for each cable type and size shall be as shown in Table 6.

Note—The pinch test apparatus shall be equivalent to that shown in Fig. 1.

4.10 Abrasion Resistance—One in (25 mm) of insulation shown shall be removed from one end of a 36 in (900 mm) sample of finished cable. The sample shall then be placed taut, without stretching between the cable clamps as shown in military specification MIL-T-5438. Use the weight support bracket and weight specified in Table 8. The sample shall then be subjected to the abrasion test. After each reading, the sample shall be moved approximately 2 in (50 mm) and rotated clockwise 90 deg. Eight readings shall be obtained for each sample. Obtain an average by calculating the arithmetic mean of all readings. Discard all readings above the arithmetic mean and average the remaining readings. The average shall define the abrasion resistance of the cable under test. Values for individual cables are shown in Table 7.

5. Reference Information

5.1 Color Code

5.1.1 RECOMMENDED COLORS—The color of the cables should match as closely as possible the colors shown in Appendix A. These color limits are not applicable to type STS or HTS cable with respect to the lightness or darkness only.

5.1.2 STRIPES—When additional color coding is required, various colored stripes may be applied longitudinally, spirally, or by other manner agreed upon by the supplier and user. The color standards do not apply to stripes.

5.2 Source for Referenced Specifications

5.2.1 American Society of Testing and Materials (ASTM)
1916 Race Street
Philadelphia, PA 19103

5.2.2 International Electrotechnical Commission (IEC)
American National Standards Institute
1430 Broadway
New York, NY 10018

5.2.3 SAE
400 Commonwealth Dr.
Warrendale, PA 15096-0001

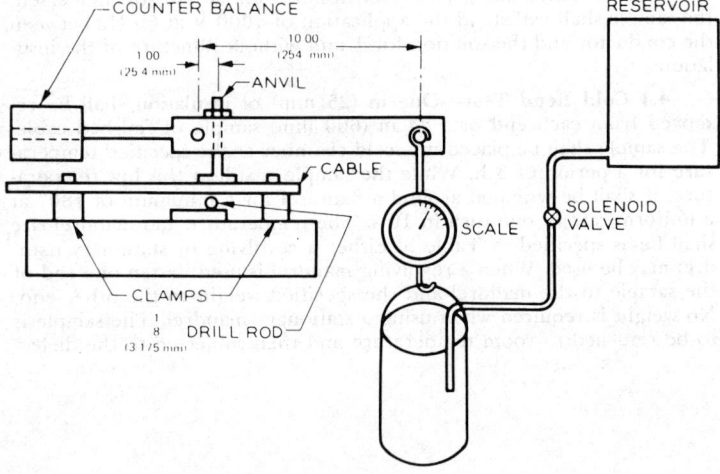

FIG. 1—PINCH TEST APPARATUS

TABLE 4—INSULATION PROPERTIES AND TEST TEMPERATURES

Cable Type	Original Properties			Heat Aging Test[a] Temperature °C ± 2°C
	Min Tensile Strength		Min Elongation	
	psi	MPa	Percent	
GPT HDT	1600	11	125	110
STS HTS	1600	11	250	110
GXL SXL	1500	10	150	155
GTE HTE	1600	11	200	150

[a]For higher temperature materials, higher test temperatures may be used.

TABLE 5—COLD BEND AND OIL ABSORPTION TEST

Cable Type Test Condition		GPT HDT 3h/−40°F (−40°C)				STS HTS 3h/−40°F (−40°C)				GXL SXL GTE HTE 3h/−60°F (−51°C)			
SAE Wire Size		Mandrel		Weight		Mandrel		Weight		Mandrel		Weight	
No.	mm²	in	mm	lb	kg	in	mm	lb	kg	in	mm	lb	kg
20	0.5	3.0	76	1.0	0.45	3.0	76	1.0	0.45	1.0	25	1.5	0.68
18	0.8	3.0	76	1.0	0.45	3.0	76	1.0	0.45	1.0	25	1.5	0.68
16	1	3.0	76	1.0	0.45	3.0	76	1.0	0.45	1.0	25	1.5	0.68
14	2	6.0	152	1.0	0.45	6.0	152	1.0	0.45	3.0	76	3.0	1.36
12	3	6.0	152	3.0	1.36	6.0	152	3.0	1.36	3.0	76	5.0	2.27
10	5	6.0	152	3.0	1.36	6.0	152	3.0	1.36	3.0	76	5.0	2.27
8	8	6.0	152	3.0	1.36	6.0	152	3.0	1.36	6.0	150	5.0	2.27
6	13	10.0	254	6.0	2.72	10.0	254	6.0	2.72	6.0	152	5.0	2.27
4	19	10.0	254	6.0	2.72					6.0	152	7.0	3.17

TABLE 6—PINCH TEST

SAE Wire Size		Minimum Pinch Resistance															
		GPT		HDT		STS		HTS		GXL		SXL		GTE		HTE	
No.	mm²	lb	kg	lb	kg	lb	kg	lb	kg	lb	kg	lb	kg	lb	kg	lb	kg
20	0.5	5	2.3	9	4.1	8	3.6	10	4.5	11	5.0	18	8.2	8	3.6	13	5.9
18	0.8	6	2.7	10	4.5	8	3.6	10	4.5	14	6.4	20	9.1	9	4.1	15	6.8
16	1	6	2.7	13	5.9	8	3.6	10	4.5	15	6.8	22	10.0	9	4.1	16	7.3
14	2	8	3.6	15	6.8	8	3.6	10	4.5	16	7.3	25	11.4	10	4.5	19	8.6
12	3	8	3.6	18	8.2	8	3.6	10	4.5	19	8.6	27	12.3	12	5.3	22	10.0
10	5	10	4.5	24	10.9	8	3.6	10	4.5	22	10.0	33	15.0	14	6.4	23	10.5
8	8	11	5.0	32	14.5	8	3.6	10	4.5	28	12.7	36	16.4	17	7.7	28	12.7
6	13	15	6.8	43	19.5	8	3.6	10	4.5			40	18.2				
4	19	27	12.3	54	24.5												

TABLE 7—ABRASION (REQUIREMENTS)

SAE Wire Size		Minimum Resistance Length of Tape															
		GPT		HDT		STS		HTS		GXL		SXL		GTE		HTE	
No.	mm²	in	mm	in	mm	in	mm	in	mm	in	mm	in	mm	in	mm	in	mm
20	0.5	18	460	16	410	18	460	30	760	18	460	27	700	18	460	16	410
18	0.8	21	530	22	560	21	530	35	890	21	530	33	850	21	530	22	560
16	1	22	560	29	740	22	560	40	1020	22	560	43	1100	22	560	29	740
14	2	7	190	27	700	7	190	18	460	7	190	20	500	7	190	27	700
12	3	11	290	43	1100	11	290	22	560	11	290	24	600	11	290	43	1100
10	5	20	510	51	1300	20	510	30	760	20	510	35	900	20	510	51	1300
8	8	36	920	87	2200	36	920	35	890	36	920	47	1200	36	920	87	2200
6	13	25	635	50	1270	25	635	60	1520	25	635	39	990	25	635	50	1270
4	19	30	760	60	1520	25	635	60	1520	30	760			30	760	60	1520

TABLE 8—ABRASION TEST (CONDITIONS)

SAE Wire Size		Test Conditions												
		GPT GTE GXL			HDT HTE			STS SXL			HTS			
No.	mm²	Br	lb	kg	Br	lb	kg	Br	lb	kg	Br	lb	kg	
20	0.5	A	1	0.45	B	3	1.36	A	1	0.45	B	1	0.45	
18	0.8	A	1	0.45	B	3	1.36	A	1	0.45	B	1	0.45	
16	1	A	1	0.45	B	3	1.36	B	1	0.45	B	1	0.45	
14	2	B	3	1.36	B	4.25	1.93	B	3	1.36	B	3	1.36	
12	3	B	3	1.36	B	4.25	1.93	B	3	1.36	B	3	1.36	
10	5	B	3	1.36	B	4.25	1.93	B	3	1.36	B	3	1.36	
8	8	B	3	1.36	C	4.25	1.93	B	3	1.36	C	3	1.36	
6	13	C	4.25	1.93	C	4.25	1.93	C	4.25	1.93	C	4.25	1.93	
4	19	C	4.25	1.93	C	4.25	1.93							

If the sandpaper is not available, this test may be omitted.

APPENDIX A—RECOMMENDED COLORS

Color	Light	Central	Dark
Red	2.5R 4.2/11.2	3.3R 3.8/11.0	4.4R 3.4/10.4
Orange	8.75R 6.0/11.5	8.75R 5.75/12.5	8.75R 5.5/13.5
Brown	10R 3.5/1.0	0.8YR 3.0/1.0	4.6YR 2.5/1.0
Tan	5YR 6.25/4.0	5YR 5.9/4.3	5YR 5.5/4.6
Yellow	8.4Y 8.5/8.3	8.2Y 8.5/9.8	8Y 8.5/11.2
Lt. Green	0.5G 6.25/6.3	0.5G 5.6/7.0	0.5G 5.1/7.5
Dk. Green	2.2BG 4.75/9.4	1.3BG 4.25/9.4	0.5BG 3.75/9.4
Lt. Blue	9B 5.4/5.0	9B 5.0/5.0	9B 4.7/5.0
Dk. Blue	4.6PB 3.8/10.2	5.2PB 3.3/9.8	5.6PB 2.75/9.4
Purple	4.4P 3.9/6.7	3.9P 3.4/6.7	3.4P 2.8/6.7
Pink	7RP 6.1/11.5	7.2RP 5.6/12.1	7.7RP 5.2/12.5
Gray	N6.3/(10GY,0.2)	N5.7/(10GY,0.2)	N5.2/(10GY,0.2)
White	—	5Y 9/1	5Y 8.5/1
Black	N4	N 2.25	—

a. Comparison must be made by a person with normal or corrected vision, under cool white fluorescent lighting. The surface being inspected and the tolerance set must be in the same plane. Cable samples must be placed flat, overlapping the color standard.

b. Color Tolerance Reference Sets are available from SAE, 400 Commonwealth Drive, Warrendale, PA 15096-0001.

APPENDIX B

Recommended Conductor Constructions (AWG Strands)

SAE Wire Size No.	No. Strands/ AWG Size (in)	No. Strands/ AWG Size (in)
20	7/28 (0.0126)	
18	16/30 (0.0100)	
16	19/29 (0.0113)	
14	19/27 (0.0142)	
12	19/25 (0.0179)	
10	19/23 (0.0226)	
8	19/21 (0.0285)	
6	37/21 (0.0285)	7 × 19/27 (0.0142)
4	61/22 (0.0253)	7 × 19/25 (0.0179)

Recommended Conductor Constructions (Metric Strands)

Metric Wire Size mm²	No. Strands/ mm Size	No. Strands/ mm Size
0.5	7/0.31	
0.8	16/0.26	
1	19/0.28	7 × 15/0.13
2	19/0.36	7 × 15/0.16
3	19/0.45	7 × 15/0.20
5	19/0.57	7 × 9/0.26
8	19/0.71	7 × 19/0.28
13	37/0.66	7 × 19/0.36
19	61/0.63	7 × 19/0.45

Stranding other than those shown above are acceptable providing they meet the minimum conductor area specified in Table 1.

LOW TENSION THIN WALL PRIMARY CABLE—SAE J1560 JAN92

SAE Standard

Report of the SAE Cable Task Force of the SAE Electrical Distribution Systems Standards Committee approved January 1992.

1. Scope
This SAE Standard covers thin wall, low tension primary cable intended for use at 50 V or less in surface vehicle electrical systems.

2. References

2.1 Applicable Documents
The following publications form a part of this specification to the extent specified herein.

2.1.1 ASTM PUBLICATIONS—Available from ASTM, 1916 Race Street, Philadelphia, PA 19103.

- ASTM B 1—Specification for Hard-Drawn Copper Wire
- ASTM B 33—Specification for Tinned Soft or Annealed Copper Wire for Electrical Purposes
- ASTM B 174—Specification for Bunch-Stranded Copper Conductors for Electrical Conductors
- ASTM B 189—Specification for Lead-Coated and Lead-Alloy-Coated Soft Copper Wire for Electrical Purposes
- ASTM D 471—Test Method for Rubber Property—Effect of Liquids
- ASTM D 573—Test Method for Rubber—Deterioration in an Air Oven

2.1.2 IEC PUBLICATIONS—Available from ANSI, 11 West 42nd Street, New York, NY 10036-8002.

- IEC 216.1—Part 1: General Guidelines for Aging Procedure and Evaluation of Test Results

3. Specification Types
3.1 Type TWP—Thinwall, thermoplastic insulated
3.2 Type TXL—Thinwall, crosslinked polyethylene insulated
3.3 Type TWE—Thinwall, thermoplastic elastomer, insulated

4. General Specifications

4.1 Conductors—The finished, uninsulated conductor, shall meet the elongation requirements specified in ASTM B 174. When tin, lead, or lead alloy coated wires are used, they shall withstand the continuity of coating tests specified in 5.1. A separator shall be used between uncoated conductors and insulation to comply with 4.2, as required. The cross-sectional area of stranded conductors shall not be less than the values specified in Table 1.

NOTE—Hard copper wire per ASTM B 1 may be used for special applications when agreed between the supplier and purchaser.

4.2 Insulation—Insulation shall be homogeneous and shall be placed concentrically within commercial tolerances about the conductor. Insulation shall adhere closely to, but strip readily from, the conductors leaving them reasonably clean and in suitable condition for terminating.

The nominal wall thickness, minimum wall thickness, and maximum overall diameter of the finished cable shall be in accordance with Table 2 for the various cable types. The minimum wall thickness shall be measured at five cross sections spaced approximately 50 mm (2 in) apart with an optical device accurate to at least 0.01 mm (0.001 in) or with a pin dial micrometer that exerts a force of 0.25 N ± 0.02 N (25 g ± 2 g) on the specimen through a flat rectangular pressure foot. The dimensions of the pressure foot shall be 1 × 8 mm (0.043 × 0.312 in) and the pin shall be 11 mm (0.437 in) long with a diameter of 0.5 mm (0.020 in). In case of dispute the referee method shall be the optical device.

4.3 Finished Cable—The finished cable shall meet the requirements for all tests specified in Table 3 for each cable type.

5. Tests

5.1 Strand Coating—Continuity of coating test shall be conducted on strands prior to stranding and shall be conducted per ASTM B 33 or ASTM B 189.

TABLE 1—CONDUCTORS

SAE Wire Size No.	SAE Wire Size mm²	Minimum Conductor Area cmils	Minimum Conductor Area mm²
24	0.22	405	0.205
22	0.35	681	0.324
20	0.50	1072	0.508
18	0.80	1537	0.760
16	1	2336	1.12
14	2	3702	1.85
12	3	5833	2.91
10	5	9343	4.65
8	8	14810	7.23

NOTES:
1. SAE wire size number indicates that the cross-sectional area of the conductors approximate the area of American Wire Gauge for equivalent sizes.
2. Metric wire size is the approximate nominal area of the stranded conductor. Metric dimensions and resistances are not direct conversions from circular mils.
3. See Appendix B for various individual conductor constructions and nominal strand diameters.

5.2 Insulation Physical Properties—An accelerated aging test shall be conducted in accordance with ASTM D 573 except using specimens of insulation removed from finished cable. The sample shall be stretched at a rate of 500 mm/min (20 in/min). The original properties shall conform to the values shown in Table 4. Samples of insulation shall be aged 168 h in a circulating air oven. The test temperature shall be as shown in Table 4. After aging, the tensile strength shall not be less than 80% of the original test value and the elongation shall not be less than 50% of the original test value.

5.2.1 NOTES
a. TWE samples may be preconditioned at test temperature for 24 h prior to taking original measurements. The samples will then be aged for 168 h. The TWE samples shall be stretched at a rate of 50 mm/min (2 in/min).
b. The previous accelerated aging test is appropriate for insulating materials currently specified in this document; different test conditions may be necessary for other materials.
c. The temperature rating of the insulating material may be determined by conducting a long-term test in accordance with IEC Publication 216-1 or equivalent using 0.40 mm (0.016 in) test specimens.

5.3 Dielectric Test—Twenty-five mm (1 in) of insulation shall be removed from each end of a 600 mm (24 in) sample of finished cable and the two ends twisted together. The loop thus formed shall be immersed in water containing 5% salt by weight at room temperature so that not more than 150 mm (6 in) of each end of the sample protrudes above the solution. After being immersed for 5 h and while still immersed, the sample shall withstand the application of 1000 V at 60 Hz between the conductor and the solution for 1 min without puncture of the insulation.

5.4 Cold Bend Test—Twenty-five mm (1 in) of insulation shall be removed from each end of a 600 mm (24 in) sample of finished cable. The sample shall be placed in a cold chamber at the specified temperature for a period of 3 h. While the sample is still at this low temperature, it shall be wrapped around a mandrel for a minimum of 180 degrees at a uniform rate of one turn in 10 s. The temperature and mandrel size

TABLE 2—DIMENSIONS (TWP/TXL/TWE)

SAE Wire No.	SAE Wire mm²	Minimum Wall Individual mm	Minimum Wall Individual in	Minimum Wall Average mm	Minimum Wall Average in	Nominal Wall mm	Nominal Wall in	Maximum O.D. mm	Maximum O.D. in
24	0.22	0.28	0.011	0.32	0.013	0.40	0.016	1.48	0.062
22	0.35	0.28	0.011	0.32	0.013	0.40	0.016	1.63	0.069
20	0.50	0.28	0.011	0.32	0.013	0.40	0.016	1.85	0.076
18	0.80	0.28	0.011	0.32	0.013	0.40	0.016	2.12	0.084
16	1	0.28	0.011	0.32	0.013	0.40	0.016	2.25	0.095
14	2	0.28	0.011	0.32	0.013	0.40	0.016	2.66	0.109
12	3	0.32	0.013	0.35	0.014	0.45	0.018	3.20	0.132
10	5	0.35	0.014	0.40	0.016	0.50	0.020	3.88	0.161
8	8	0.39	0.015	0.44	0.017	0.55	0.022	4.70	0.196

NOTE—Metric dimensions are not direct conversion from inches.

shall be as specified in Table 5. Either a revolving or stationary mandrel may be used. When a revolving mandrel is used, fasten one end of the sample to the mandrel and the specified weight to the other end. No weight is required when using a stationary mandrel. The sample is to be returned to room temperature and then subjected to the dielectric test specified in 5.3.

5.5 Flame Test—A Bunsen burner having a 13 mm (1/2 in) inlet, a nominal core of 10 mm (3/8 in), a length of approximately 100 mm (4 in) above the primary inlets, equipped with a wing top flame spreader having a 1 × 50 mm (1/16 × 2 in) opening fitted to the top of the burner shall be used. A 600 mm (24 in) sample of finished cable shall be suspended taut in a horizontal position within a partial enclosure which allows a flow of sufficient air for complete combustion but is free from drafts. The top of a 50 mm (2 in) gas flame with an inner cone one-third its height shall then be positioned vertically and applied to the center of the suspended cable with the flame spreader parallel to the axis of the cable. The time of application of the flame shall be 15 s. After removal of the Bunsen burner flame, the sample shall not continue to burn for more than 30 s.

5.6 Oil Absorption Test—Twenty-five mm (1 in) of insulation shall be removed from each end of a 600 mm (24 in) sample of finished cable. The original outside diameter shall be measured at five cross sections spaced approximately 50 mm (2 in) apart with the device described in 4.2. A minimum of two readings shall be taken at each cross section. The sample should be rotated approximately 90 degrees between readings. The sample shall be immersed to within 40 mm (1-1/2 in) from the end

TABLE 3—TEST REQUIREMENTS

Tests	Cable Types TWP	Cable Types TXL	Cable Types TWE
Conductor Area	X	X	X
Dimensions	X	X	X
Strand Coating	X	X	X
Physical Properties	X	X	X
Dielectric	X	X	X
Cold Bend	X	X	X
Flame	X	X	X
Oil Absorption	X	X	X
Pinch	X	X	X
Abrasion			
Crosslinking		X	

TABLE 4—INSULATION PROPERTIES

Cable Type	Minimum Tensile Strength MPa	Minimum Tensile Strength psi	Minimum Elongation %	Heat Aging Test[1] Temperature °C ± 2 °C
TWP	11	1600	125	110
TXL	10	1500	150	155
TWE	11	1600	200	150

[1] For higher temperature materials, higher test temperatures may be used.

TABLE 5—COLD BEND AND OIL ABSORPTION TEST

Cable Type Test Condition SAE Wire Size No.	Cable Type Test Condition SAE Wire Size mm²	TWP 3h/-40 °C (-40 °F) Mandrel mm	TWP 3h/-40 °C (-40 °F) Mandrel in	TWP 3h/-40 °C (-40 °F) Weight kg	TWP 3h/-40 °C (-40 °F) Weight lb	TXL 3h/-51 °C (-60 °F) Mandrel mm	TXL 3h/-51 °C (-60 °F) Mandrel in	TWE 3h/-51 °C (-60 °F) Weight kg	TWE 3h/-51 °C (-60 °F) Weight lb
24	0.22	25	1.0	0.45	1.0	25	1.0	0.68	1.5
22	0.35	25	1.0	0.45	1.0	25	1.0	0.68	1.5
20	0.50	76	3.0	0.45	1.0	25	1.0	0.68	1.5
18	0.80	76	3.0	0.45	1.0	25	1.0	0.68	1.5
16	1	76	3.0	0.45	1.0	25	1.0	0.68	1.5
14	2	152	6.0	0.45	1.0	76	3.0	1.36	3.0
12	3	152	6.0	1.36	3.0	76	3.0	2.27	5.0
10	5	152	6.0	1.36	3.0	76	3.0	2.27	5.0
8	8	152	6.0	1.36	3.0	152	6.0	2.27	5.0

TABLE 6—PINCH TEST

SAE Wire Size No.	SAE Wire Size mm²	Minimum Pinch Resistance TWP kg	Minimum Pinch Resistance TWP lb	Minimum Pinch Resistance TWE kg	Minimum Pinch Resistance TWE lb	Minimum Pinch Resistance TXL kg	Minimum Pinch Resistance TXL lb
24	0.22	1.2	2.6	1.5	3.3	2.3	5
22	0.35	1.2	2.6	1.5	3.3	2.7	6
20	0.50	1.5	3.3	2.0	4.4	2.7	6
18	0.80	1.5	3.3	2.0	4.4	3.2	7
16	1	2.0	4.4	2.5	5.5	3.6	8
14	2	2.0	4.4	2.5	5.5	4.1	9
12	3	2.5	5.5	3.0	6.6	4.5	10
10	5	2.5	5.5	3.0	6.6	5.5	12
8	8	3.0	6.6	3.5	7.7	5.9	13

of the insulation in ASTM D 471 #2 oil at a temperature of 48 to 50 °C (118 to 122 °F) for a period of 20 h. After removal from the oil, the sample shall be conditioned for 4 h at room temperature. After conditioning, the outside diameter of the cable shall be measured using the previous procedure. The grand average of the readings taken after conditioning shall not change by more than 15% from the grand average of the original readings. The conditioned sample shall be wrapped around a mandrel as specified in Table 5 for a minimum of 180 degrees at a uniform rate of one turn in 10 s. The sample shall then be subjected to the dielectric test specified in 5.3.

5.7 Pinch Test—Twenty-five mm (1 in) of insulation shall be removed from one end of a 900 mm (36 in) sample of finished cable. The sample shall then be placed taut without stretching across a 3 mm (1/8 in) diameter steel bar and be subjected to the force of a weighted steel anvil. Increasing weight shall be applied to the steel anvil at a rate of 2.3 kg (5 lb) per min with a lever advantage of 10. At the moment the insulation is pinched through, the 3 mm (1/8 in) diameter rod will contact the conductor and the test shall stop. The weight of the receptacle shall then be recorded. After each reading the sample shall be moved approximately 50 mm (2 in) and rotated clockwise 90 degrees. Four readings shall be obtained for each sample. Obtain an average by calculating the arithmetic mean of all those readings. The average shall define the pinch resistance of the cable under test. The minimum values for each cable type and size shall be as shown in Table 6.

NOTE—The pinch test apparatus shall be equivalent to that shown in Figure 1.

5.8 Abrasion Resistance—To be determined.

5.9 Crosslinking—Twenty-five mm (1 in) of insulation shall be removed from each end of a 600 mm (24 in) sample of finished cable. The sample shall be bent a minimum of 135 degrees around a 6 mm (1/4 in) mandrel. The cable and mandrel shall be placed against a hot plate which has been preheated to 250 °C ± 25 °C. A force of 5 to 7 N (1 to 1-1/2 lb) shall be applied for 5 to 6 s without rubbing or scraping the cable on the plate. After exposure, the cable core shall not be visible through the insulation. The sample is to be returned to room temperature and then subjected to the dielectric test specified in 5.3.

6. Reference Information
6.1 Color Code

6.1.1 RECOMMENDED COLORS—The color of the cables should match as closely as possible the colors specified in Appendix A.

6.1.2 STRIPES—When additional color coding is required, various colored stripes may be applied longitudinally, spirally, or by other manner agreed upon by the supplier and user. The color standards do not apply to stripes.

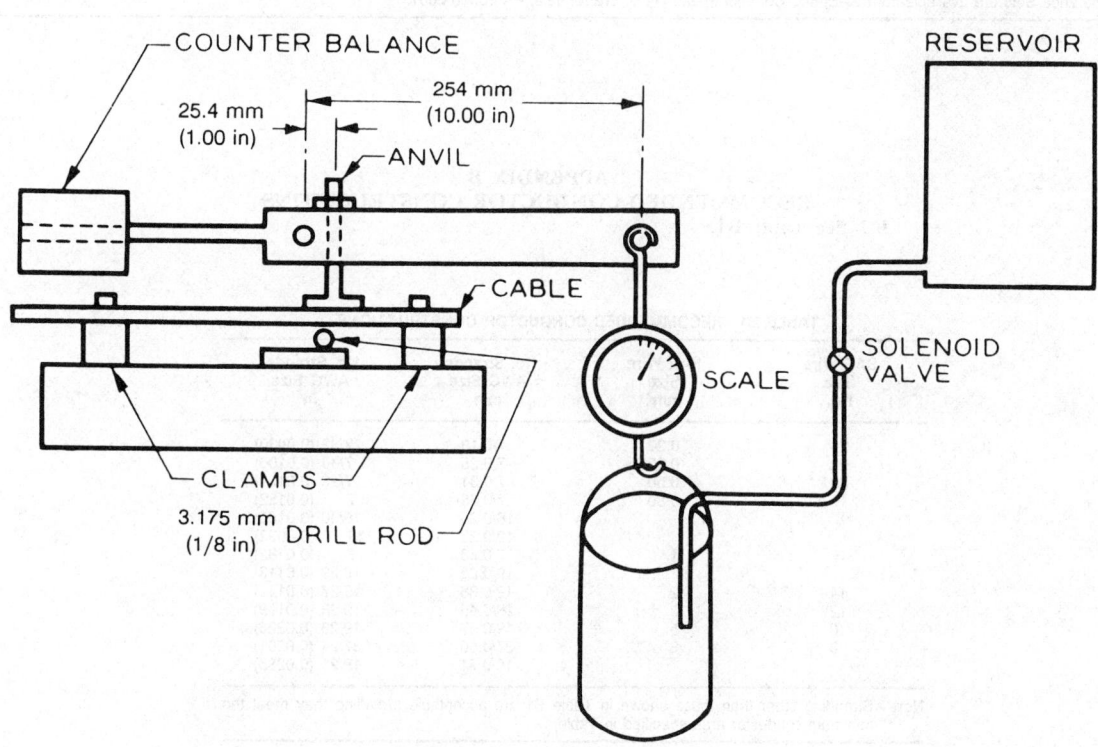

FIGURE 1—PINCH TEST APPARATUS

APPENDIX A
RECOMMENDED COLORS

A.1 See Table A1.

TABLE A1 — RECOMMENDED COLORS

Color	Light Munsell[1]	Light X, Y, Z[2]	Dark Munsell[1]	Dark X, Y, Z[2]
Red	2.5R 4.2/11.2	22, 13, 10	4.4R 3.4/10.4	15, 9, 5
Orange	8.75R 6.0/11.5	42, 30, 12	8.75R 5.5/13.5	38, 25, 7
Brown	10R 3.5/1.0	10, 9, 10	4.6YR 2.5/1.0	5, 5, 5
Tan	5YR 6.25/4.0	36, 33, 25	5YR 5.5/4.6	27, 25, 16
Yellow	8.4Y 8.5/8.3	62, 68, 23	8Y 8.5/11.2	61, 68, 13
Lt. Green	0.5G 6.25/6.3	25, 33, 22	0.5G 5.1/7.5	14, 21, 11
Dk. Green	2.2BG 4.75/9.4	10, 18, 21	0.5BG 3.75/9.4	5, 11, 12
Lt. Blue	9B 5.4/5.0	21, 24, 42	9B 4.7/5.0	15, 17, 32
Dk. Blue	4.6PB 3.8/10.2	11, 11, 38	5.6PB 2.75/9.4	6, 6, 24
Purple	4.4P 3.9/6.7	15, 12, 25	3.4P 2.8/6.7	8, 6, 16
Pink	7RP 6.1/11.5	45, 31, 36	7.7RP 5.2/12.5	35, 22, 24
Gray	N6.3/(10GY,0.2)	33, 34, 40	N5.2/(10GY,0.2)	21, 22, 26
White			5Y 8.5/1	66, 68, 70
Black	N4	12, 12, 14		

[1] Comparison must be made by a person with normal or corrected vision, under cool white fluorescent lighting. The surface being inspected and the tolerance set must be in the same plane. Cable samples must be placed flat, overlapping the color standard.
[2] FMCII, measured under CIE illuminant C, 2 degree observer.
NOTE — Color Tolerance Reference Sets are available from SAE, 400 Commonwealth Drive, Warrendale, PA 15096-0001.

APPENDIX B
RECOMMENDED CONDUCTOR CONSTRUCTIONS

B.1 See Table B1.

TABLE B1 — RECOMMENDED CONDUCTOR CONSTRUCTIONS

SAE Wire Size No.	SAE Wire Size mm²	No. Strands/ AWG Size mm	No. Strands/ AWG Size in
24	0.22	7/0.19	7/32 (0.0080)
22	0.35	7/0.25	7/30 (0.0100)
20	0.50	7/0.31	7/28 (0.0126)
18	0.80	7/0.38	7 (0.0152)
		16/0.26	16/30 (0.0100)
		19/0.23	19 (0.0092)
16	1	7/0.46	7 (0.0182)
		19/0.28	19/29 (0.0113)
14	2	19/0.36	19/27 (0.0142)
12	3	19/0.45	19/25 (0.0179)
10	5	19/0.57	19/23 (0.0226)
8	8	37/0.50	37/24 (0.0201)
		19/0.71	19/21 (0.0285)

NOTE — Stranding other than those shown in Table B1 are acceptable providing they meet the minimum conductor area specified in Table 1.

FUSIBLE LINKS—SAE J156 APR86

SAE Standard

Report of Electrical Equipment Committee, approved February 1970, and reaffirmed by the Electrical and Electronic Systems Technical Committee, April 1986.

1. Scope—This SAE Recommended Practice covers the details, use, and design evaluation testing of fusible links for motor vehicle electrical wiring protection. The specifications as listed are known good practice and are not intended to restrict new materials or construction.

2. Definition—A fusible link is a special section of low tension cable designed to open the circuit when subjected to an extreme current overload. Its purpose is to minimize wiring system damage when such an overload occurs accidentally in those circuits protected by the fusible link.

3. General Specifications

3.1 Conductors—Conductors shall conform to the specifications shown in Table 1 of SAE J1128.

3.2 Insulation—The insulating material shall meet the requirements shown in SAE J1128 Type HTS. A special insulation with a tensile strength of 1000 psi (6900 kPa) minimum and STS wall may also be used.

3.3 Wire Size—The fusible link must be of a smaller wire size than any connecting cable in the circuits being protected. Wire sizes are to be determined experimentally with the vehicle wiring system based on the type of harness wire insulation, circuit loads, and physical locations. This may be done either in the vehicle or with an equivalent laboratory set up.

3.4 Length—The length of each fusible link for effective protection is to be determined in the same manner as for the wire size.

3.5 Location—Fusible links shall be located such that any fumes generated during their destruction will not cause undue discomfort to any passenger, and no damage will occur to adjacent components, combustible material, or other circuits.

3.6 Terminations—The conductor and insulation at each end of a fusible link shall be securely fastened to its termination. If spliced to a connecting cable, the splice joint must either be welded or mechanically secured and soldered. The splice must then be properly insulated.

3.7 Identification—Each fusible link shall be permanently marked with the wire size and identification that it is a fusible link. After a link has fused and opened the circuit, sufficient identification shall still be present to establish this information for replacement.

4. Testing—Design evaluation testing is to be conducted to verify the ability of a specified fusible link to conduct the maximum design load of the electrical circuit and to ascertain that the link will open *under extreme current overload* without causing damage to the protected wiring, associated harness, or adjacent components.

4.1 Charging Circuit Protection—Fusible links located in circuits which conduct battery charging currents are to be tested either in the vehicle or in a duplicating laboratory set up. The specified generator and battery should be operating at maximum charge current and the battery shall have been completely discharged before the test began. Electrical accessory loads are to be such as would cause maximum current through the fusible link that could occur in a vehicle. In a laboratory set up, the generator may be duplicated by an equivalent current producing source.

At the start of the test, the generator temperature is to be $75 \pm 5°F$ ($24 \pm 3°C$) and the battery electrolyte temperature $110 \pm 5°F$ ($43 \pm 3°C$). The test shall be conducted for at least 5 min after the maximum fusible link core temperature attainable in the vehicle can be reached. After the test is completed, the fusible link insulation shall show no deterioration due to heat.

4.2 Short Circuit Protection—Fusible links are to be tested for design evaluation by grounding the conductor of the protected circuit at a point which is the most electrically remote from the fusible link. The point selected must not have any intervening circuit protecting devices, such as fuses or circuit breakers between it and the fusible link. Under extreme current overload conditions, the fusible link must open the circuit within a period of time such that no damage to the protected wiring, associated harness, or adjacent components occurs. After the link has opened, there shall be no exposed conductor in a location to cause subsequent short circuiting of the battery or generator.

4.3 Observations and Conclusions—At the conclusion of each test, visual inspection of the wiring is to be made. Other than opened fusible link sections, there shall be no evidence of cable insulation deformation or damage regardless of the type of insulation used. Any fusible link tested for maximum design current load shall show no insulation deterioration after the test.

FIVE CONDUCTOR ELECTRICAL CONNECTORS FOR AUTOMOTIVE TYPE TRAILERS—SAE J895 APR86

SAE Recommended Practice

Report of Electrical Equipment Committee approved June 1964 and reaffirmed by the Electrical and Electronic Systems Technical Committee, April 1986. This report is currently under revision by the Electrical Distribution Systems Subcommittee.

1. Scope—This SAE Recommended Practice covers the wiring and connector standards for nonpassenger carrying trailers, SAE Classes 1–3[1], with circuit loads not to exceed 7.5 amp per circuit. It provides the lighting circuits of these trailers with a universal connecting device, standard circuit coding and protection for the wiring from hazards and shorts.

2. Receptacle—The receptacle shall be of the design as shown in Fig. 1 and shall be attached to the towing vehicle as follows:
(a) White—Ground to frame
(b) Brown—Spliced to tail and license light circuit
(c) Yellow—Spliced to left turn and stop circuit
(d) Green—Spliced to right turn and stop circuit
(e) Blue—Auxiliary

The receptacle leads shall be attached to the vehicle wiring harness in a workmanlike manner, mechanically and electrically secure. Further, a well insulated strain relief shall be provided between the receptacle and the towing vehicle wiring harness connections so that there will be no strain on the vehicle harness in the event of an abnormal pull on the receptacle. The receptacle shall be placed in a location where it will not be exposed to road hazards either when connected or loose. The receptacle leads must be properly routed and protected against damage from cutting and pinching where they leave the vehicle body. No receptacle leads shall be smaller than 16 gage (single) or smaller than 18 gage (in multiconductor cables) heavy duty SAE insulated automotive primary wire.

3. Plug—The plug shall be as shown in Fig. 2, and wiring, shall be attached to the trailer so that the wires have the maximum protection against road splash, stones, abrasion, grease, oil, and fuel. The wiring shall be secured to the trailer frame at intervals not greater than 18 in so that the wiring does not shift or sag. The circuits used shall be color coded as follows:
(a) White—Ground
(b) Brown—The tail and license light
(c) Yellow—Left turn and stop light
(d) Green—Right turn and stop light
(e) Blue—Auxiliary

No plug leads shall be smaller than 16 gage (single) or smaller than 18 gage (in multiconductor cables) heavy duty SAE insulated automotive primary wire. Extra insulation should be provided between the strain relief at the trailer hitch and the wiring assembly so that an abnormal pull on the plug will not damage the wiring.

APPENDIX

A.1 Material Requirements—The receptacle and plug shall be made of an insulating material such that they can be processed to provide the spacing and splash protection indicated in Figs. 1 and 2. The material used shall have a hardness of Shore "A" 50 minimum and shall be compatible with the insulation used on the wire leads and/or jacket over the

[1] See SAE J684.

23.172

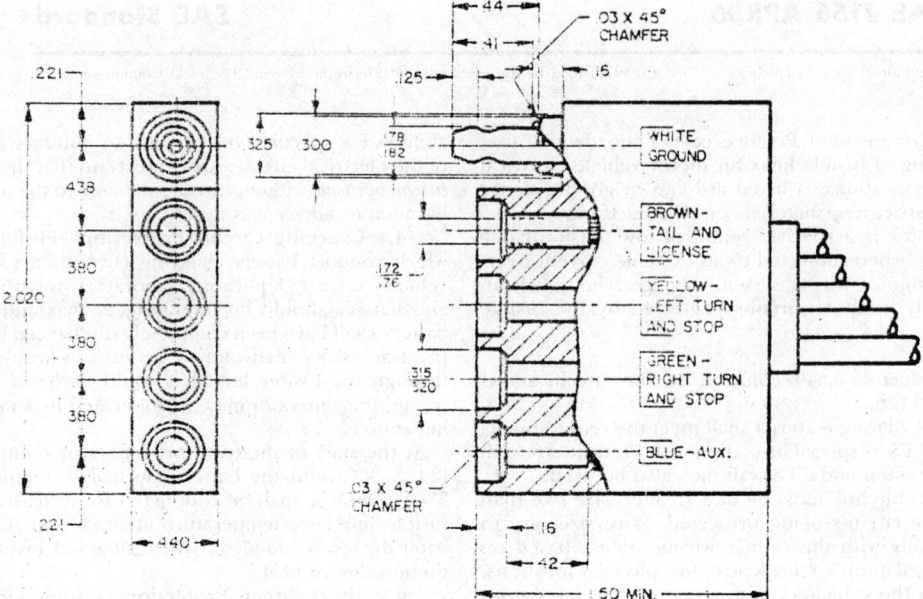

FIG. 1—RECEPTACLE

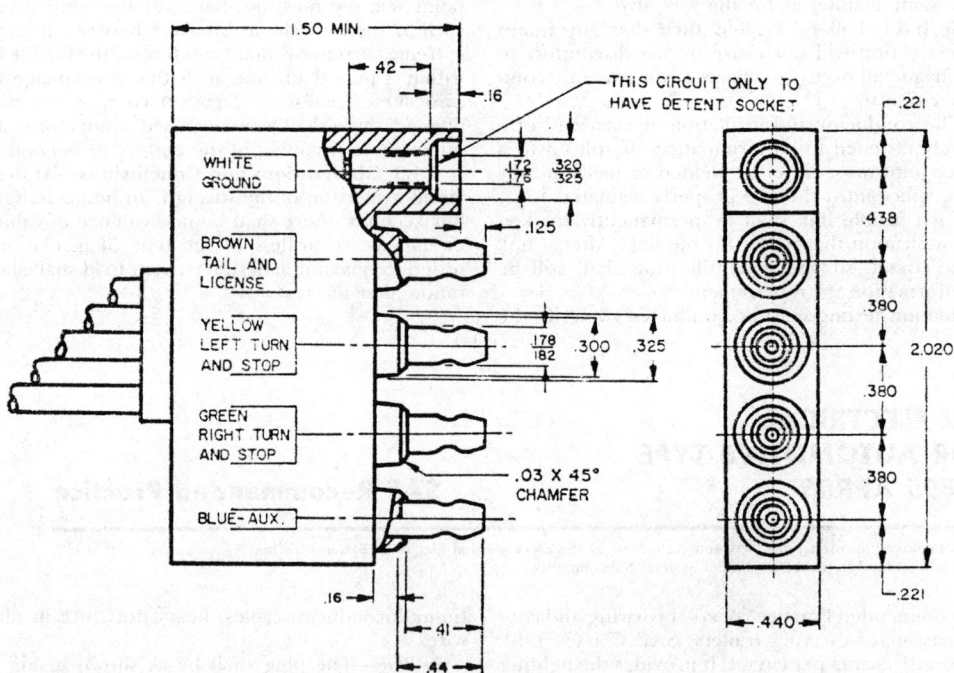

FIG. 2—PLUG

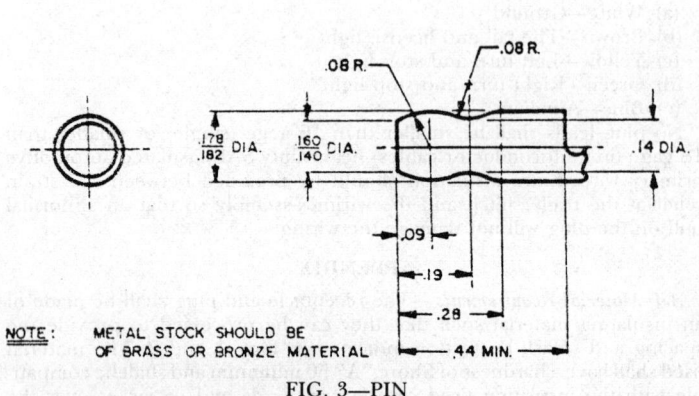

NOTE: METAL STOCK SHOULD BE OF BRASS OR BRONZE MATERIAL

FIG. 3—PIN

NOTE: THIS SOCKET TO BE USED ONLY ON THE GROUND CIRCUIT OF THE MOLDED PLUG. SOCKETS SIMILAR TO THIS, EXCEPT WITH THE DETENT OMITTED, ARE TO BE USED IN THE MOLDED RECEPTACLE.

METAL STOCK SHOULD BE OF 3/4 HARD TEMPER BRASS OR BRONZE MATERIAL.

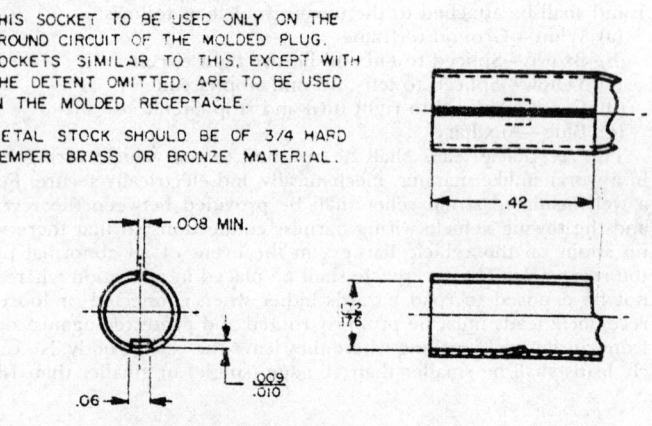

FIG. 4—SOCKET

leads. Where thermoplastic materials are used the hardness shall not exceed Shore "A" 70. The jacket on the leads shall be as deemed necessary to adequately protect the wiring. The metal pins and sockets shall be of the size and type shown in Figs. 3 and 4 made of either brass or bronze and suitably coated to protect against corrosion. The coating shall be smooth so as not to bind when the parts are engaged.

A.2 Assembly Requirements—The plug and receptacle assembly shall disengage with a minimum force of 3 lb per circuit and a maximum of 7 lb per circuit, except the ground circuit which can be 12 lb max. The mechanical force requirement of disengagement does not preclude the requirement for good electrical connections between the male and female connectors of the circuits.

FOUR- AND EIGHT-CONDUCTOR RECTANGULAR ELECTRICAL CONNECTORS FOR AUTOMOTIVE TYPE TRAILERS—SAE J1239

SAE Recommended Practice

Report of Electrical Equipment Committee approved June 1978.

1. Scope—This SAE Recommended Practice covers the wiring and rectangularly shaped connector standards for all types of trailers whose gross weight does not exceed 10 000 lb (4540 kg). These trailers are grouped in SAE classes 1 through 4 as delineated in SAE J684e (August, 1974), with running light circuit loads not to exceed 7.5 amps per circuit. This recommended practice provides circuits for lighting, electric brakes, trailer battery charging, and an auxiliary circuit color coding and protection for the wiring from hazards or short circuits. Color coding is compatible with SAE J560b (September, 1974) and ISO (1724-1975(E)).

2. Receptacle—The receptacle shall be of the configuration and design dimensions shown in Fig. 1 for four circuits and as shown in Fig. 2 for eight circuits.

2.1 The four-circuit receptacle (Fig. 1) shall be color coded and attached to the towing vehicle as follows:

2.1.1 WHITE—Ground to frame (SAE wire size 16 or metric size 1.2 minimum).

2.1.2 BROWN—Spliced to tail and license lamp circuit.

2.1.3 YELLOW—Spliced to left turn and stop circuit.

2.1.4 GREEN—Spliced to right turn and stop lamp circuit.

2.2 The eight-circuit receptacle (Fig. 2) shall be color coded and shall be attached to the towing vehicle as follows:

Left Bank of Receptacles:

2.2.1 RED—Independent stop.

2.2.2 BLUE—Brake circuit spliced to controller of brake.

2.2.3 OPTIONAL—Auxiliary (see Fig. 2, Note 1).

2.2.4 ORANGE—Battery charge circuit—connect to battery positive terminal through separate fuse or circuit breaker.

Right Bank of Receptacles:

2.2.5 WHITE—Direct to battery negative (SAE wire size 12 or metric size 3.0 minimum).

2.2.6 BROWN—Spliced to tail and license lamp circuit.

2.2.7 YELLOW—Spliced to left turn and stop lamp circuit.

2.2.8 GREEN—Spliced to right turn and stop lamp circuit.

2.3 The receptacle leads shall be attached to the vehicle wiring harness, brake controller, or battery in a workmanlike manner, mechanically and electrically secure.

2.4 The receptacle leads must be properly routed and protected against damage from cutting and pinching where they leave the vehicle body. No receptacle leads designated for lighting shall be smaller than SAE wire size 16 or metric size 1.2 if a single conductor, or smaller than SAE wire size 18 or metric size 0.8 if a multi-conductor cable.

2.5 No receptacle leads for brake circuits shall be smaller than SAE wire size 14 or metric size 2.0 and no circuits shall be smaller than SAE wire size 12 or metric size 3.0 for trailer battery charge circuit or battery return circuit.

The gauge of conductors for the auxiliary circuit shall be sized to provide at least the minimum ampacity for the load it will service with a voltage drop not exceeding 3%. The receptacle shall be placed in a location where it will not be exposed to road hazards when disconnected from trailer.

3. Plug—The plug shall be of the configuration and design dimensions shown in Fig. 1 for four circuits and as shown in Fig. 2 for eight circuits.

3.1 The four circuit plug (Fig. 1) shall be color coded and attached to the trailer harness as follows:

3.1.1 WHITE—Ground to frame (SAE wire size 16 or metric wire size 1.0 minimum).

3.1.2 BROWN—Spliced to tail and license lamp circuit.

3.1.3 YELLOW—Spliced to left turn and stop circuit.

3.1.4 GREEN—Spliced to right turn and stop lamp circuit.

3.2 The eight-circuit plug (Fig. 2) shall be color coded and attached to the trailer harness as follows:

Right Bank of Receptacles:

3.2.1 RED—Independent stop.

3.2.2 BLUE—Brake circuit spliced to controller of brake circuit.

3.2.3 OPTIONAL—Auxiliary (see Fig. 2, Note 1).

3.2.4 ORANGE—Battery charge circuit—connect to trailer battery positive terminal through separate fuse or circuit breaker.

Left Bank of Receptacles:

3.2.5 WHITE—Ground to frame and trailer battery negative terminal.

3.2.6 BROWN—Spliced to tail and license lamp circuit.

3.2.7 YELLOW—Spliced to left turn and stop lamp circuit.

3.2.8 GREEN—Spliced to right turn and stop lamp circuit.

4. Wiring

4.1 Exposed trailer wiring shall be run in conduits or secured at intervals not greater than 18 in (457 mm) to stop lateral movement and prevent rubbing or chafing.

4.2 So far as practicable wiring should be located to afford protection from road splash, stones, or abrasion. Wiring exposed to such conditions shall be further protected by the use of—or combination of—additional tape covering, plastic sleeving, non-metallic, or other suitable shielding or covering.

5. Appendix

5.1 Material Requirements

5.1.1 If the receptacles and plugs are fabricated of either compression molded or extruded plastic, the plastic material shall be stabilized for protection against exposure to ultra-violet light.

5.1.1.2 The hardness of a molded or extruded plastic receptacle or plug shall fall within the limits of Shore *A* 50 as minimum and Shore *A* 70 as a maximum.

5.1.2 The metal pins and sockets shall be of the size and type shown in Figs. 3 and 4.

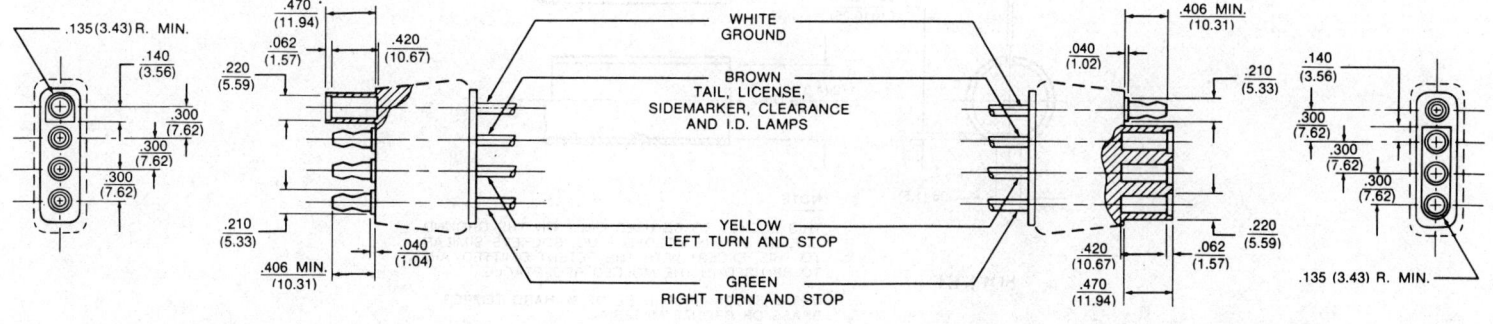

PLUG RECEPTACLE

FIG. 1

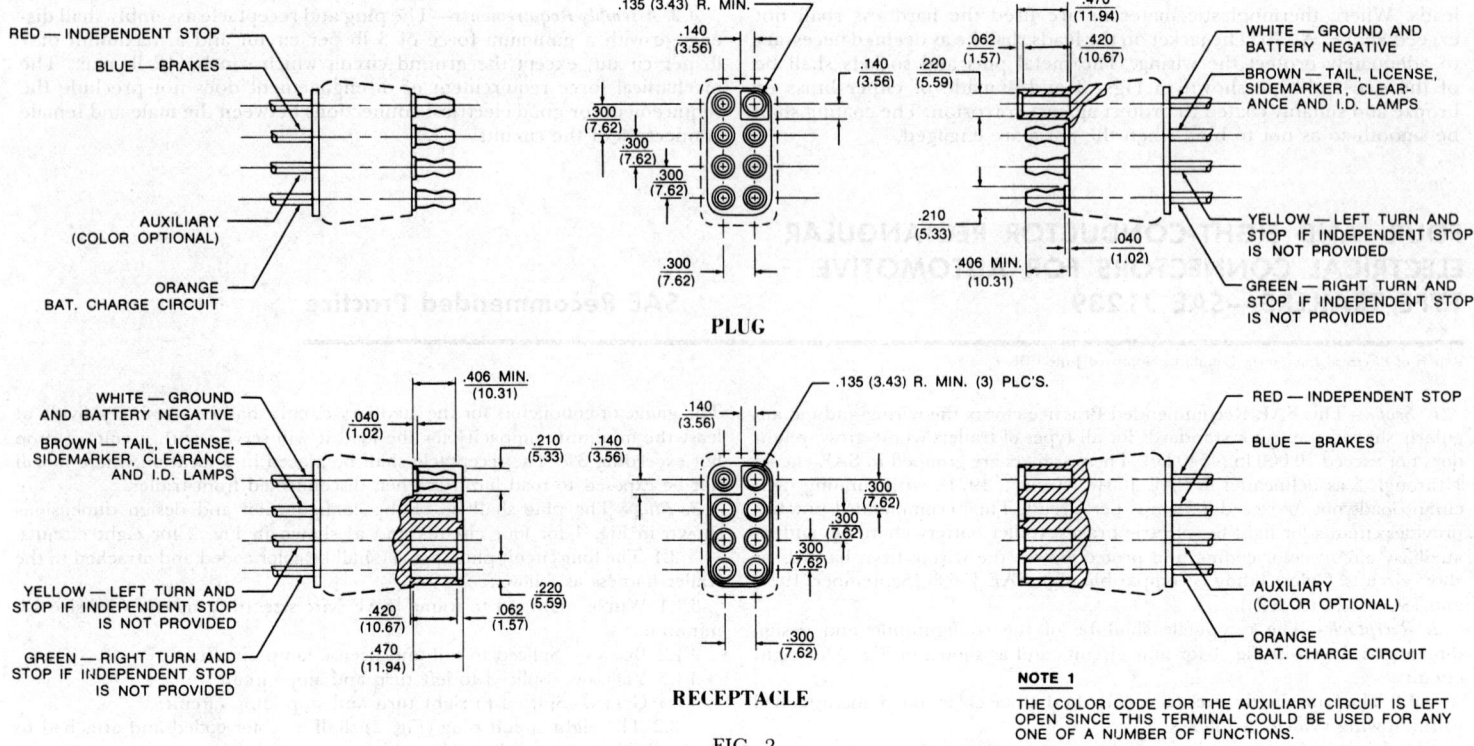

PLUG

RECEPTACLE

FIG. 2

NOTE
METAL STOCK SHOULD BE OF ¾ HARD TEMPER BRASS OR BRONZE MATERIAL.

NOTE 1
THE COLOR CODE FOR THE AUXILIARY CIRCUIT IS LEFT OPEN SINCE THIS TERMINAL COULD BE USED FOR ANY ONE OF A NUMBER OF FUNCTIONS.

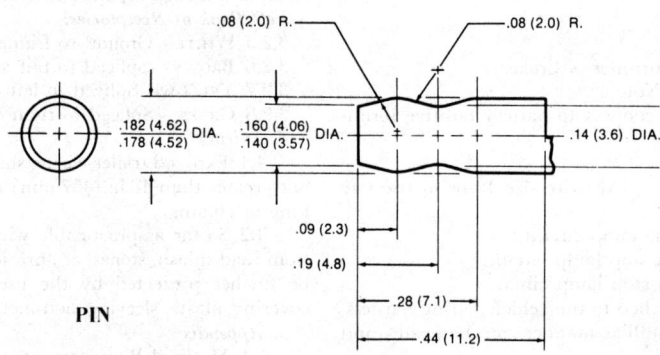

PIN

FIG. 3

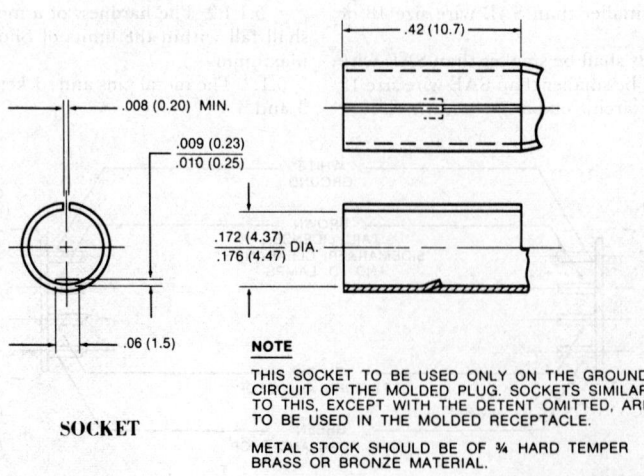

SOCKET

NOTE
THIS SOCKET TO BE USED ONLY ON THE GROUND CIRCUIT OF THE MOLDED PLUG. SOCKETS SIMILAR TO THIS, EXCEPT WITH THE DETENT OMITTED, ARE TO BE USED IN THE MOLDED RECEPTACLE.

METAL STOCK SHOULD BE OF ¾ HARD TEMPER BRASS OR BRONZE MATERIAL.

FIG. 4

5.1.2.1 Pins and receptacles shall be fabricated from brass or bronze and suitably coated to protect against corrosion. Finished surfaces of plugs and interior walls of sockets shall be smooth so as not to bind when the parts are engaged.

5.1.2.2 Pins and receptacles shall conform to TYPE 1—PIN TERMINALS, nominal diameter 0.180(5) specified as SAE J928a (April, 1970). Detailed pin and receptacle dimensions are illustrated in Figs. 3 and 4.

5.2 All wire and insulation shall conform to the requirements of SAE J1128 (November, 1975), Low Tension Primary Cable.

6. **Assembly Requirements**—The plug and receptacle of a 4-way connector assembly shall disengage with a minimum force of 5 lb (22.2 N) per assembly and a maximum force of 20 lb (89.0 N) per assembly. The plug and receptacle of an 8-way connector assembly shall disengage with a minimum of 8 lb (35.6 N) per assembly and a maximum force of 30 lb (133.6 N) per assembly.

STRANDED CONDUCTORS FOR 12 VOLT CIRCUITS
(PRIMARY CABLE DATA ABSTRACTED FROM J1128)
3% VOLTAGE DROP

SAE Wire Size	20	18	16	14	12	10
Stranding	7×28	16×30	19×29	19×27	19×25	19×23
Metric Wire Size	0.5	0.8	1.0	2.0	3.0	5.0
Min Cond Area Cir Mil	1072	1537	2336	3702	5833	9343
Min Cond Area mm²	0.508	0.760	1.12	1.85	2.91	4.65

MAXIMUM LENGTH OF CONDUCTOR IN FEET FROM POWER SOURCE TO LOAD

SAE Wire Size	20		18		16		14		12		10	
Circuit Current in AMPS	ft	m	ft	m	ft	m	ft	m	ft	m	ft	m
1	36.4	11.09	52.3	15.94	78.0	23.77						
2	18.2	5.55	26.1	7.96	39.0	11.89	63.0	19.20	99.0	30.17		
3	12.2	3.72	17.4	5.30	26.0	7.92	42.0	12.80	66.0	20.12		
4	9.1	2.77	13.1	3.99	19.5	5.94	31.5	9.60	49.5	15.09	78.8	24.02
5	7.3	2.22	10.4	3.17	15.6	4.75	25.2	7.68	39.6	12.07	63.0	19.20
6	6.1	2.65	8.7	2.65	13.0	3.96	21.0	6.40	33.0	10.06	52.5	16.00
7	5.2	1.58	7.4	2.26	11.1	3.38	18.0	5.49	28.2	8.60	45.0	13.72
8			6.5	1.98	9.8	2.99	15.8	4.82	24.8	7.56	39.4	12.00
9			5.8	1.77	8.6	2.62	14.0	4.27	22.0	6.71	35.0	10.67
10			5.2	1.58	7.8	2.38	12.6	3.84	19.8	6.04	31.5	9.60
15					5.2	1.58	8.4	2.56	13.2	4.02	21.0	6.40
20							6.3	1.92	9.9	3.02	15.8	4.82
20									6.6	2.01	10.5	3.20

FIG. 5

R) SEVEN CONDUCTOR ELECTRICAL CONNECTOR FOR TRUCK-TRAILER JUMPER CABLE—SAE J560 JUN93

SAE Standard

Report of Electrical Equipment Committee approved January 1951 and revised September 1974. Completely revised by the SAE Truck and Bus Electrical and Electronics Committee June 1993. Rationale statement available.

Foreword—The seven conductor electrical connector is used exclusively in the United States and Canada as the electrical interface between highway tractors and trailers. The exclusive use of this connector makes it possible to pull any trailer with any tractor without the use of adapters.

This connector is comparable to only one unit currently being considered as an ISO Standard. In addition to the seven conductor unit, ISO is considering twelve, thirteen, and fifteen conductor units. All of these may be included in any ISO Standard which will require a number of adapters to achieve universal compatibility of tractors and trailers.

1. *Scope*—This SAE Standard provides the minimum design requirements for the jumper cable plug and receptacle for the truck-trailer jumper cable system. It includes the test procedures, design, and performance requirements.

2. **References**

2.1 **Applicable Documents**—The following publications form a part of this specification to the extent specified herein. The latest issue of SAE publications shall apply.

2.1.1 SAE PUBLICATION—Available from SAE, 400 Commonwealth Drive, Warrendale, PA 15096-0001.

SAE J1067—Seven Conductor Jacketed Cable for Truck-Trailer Connections

2.1.2 ASTM PUBLICATION—Available from ASTM, 1916 Race Street, Philadelphia, PA 19103-1187.

ASTM B 117-73—Standard Method of Salt Spray (Fog) Testing

2.2 **Definitions**

2.2.1 RECEPTACLE—The receptacle consists of the connector socket, its housing, and a cover which latches the cable plug in place. The socket contains the male contacts. See Figures 1 and 2.

2.2.2 CABLE PLUG—The cable plug is part of the jumper cable assembly. The cable plug contains the female contacts. See Figure 3.

2.2.3 COUPLING CYCLE—Coupling and uncoupling the plug and receptacle is one coupling cycle.

3. *Identification Code Designation*—Devices conforming to this document shall be identified with the manufacturer's identification, model or part number, and shall be identified with SAE J560 and the revision (month and year) of the document to which the device conforms. For example:

XYZ Corp.
9999
SAE J560
Jun, 93

4. **Tests**

4.1 **Test Equipment and Instrumentation**

4.1.1 POWER SUPPLY—The power supply shall be capable of supplying the continuous current required to perform all tests.

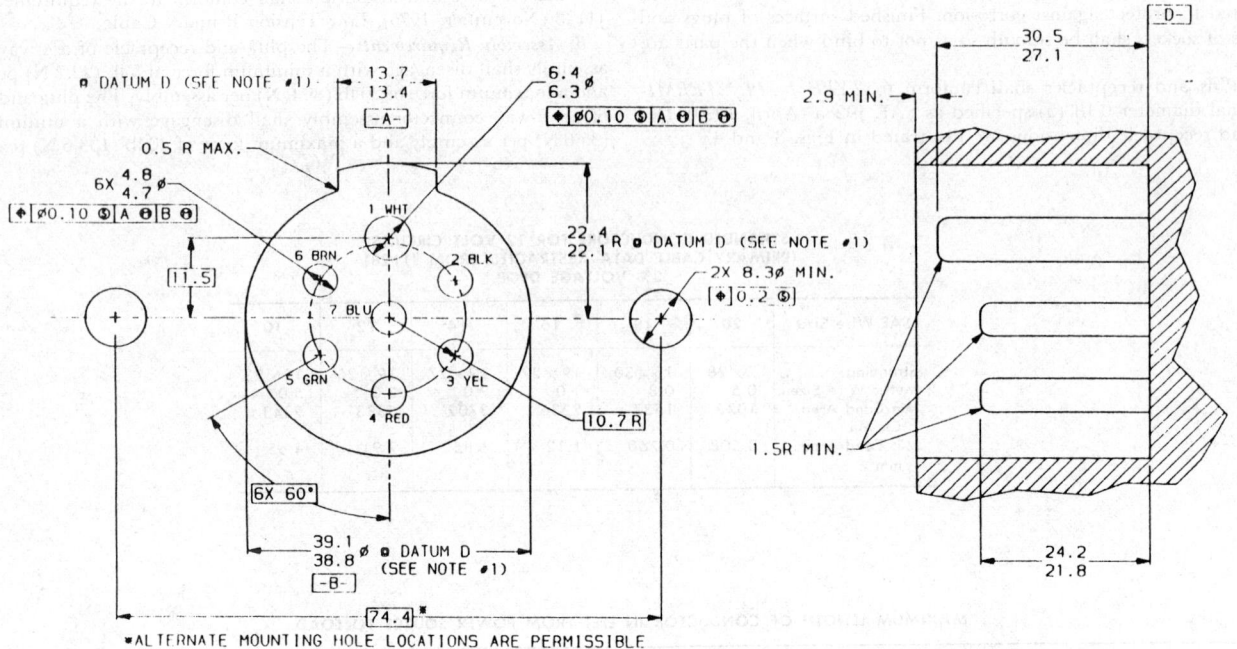

NOTES: 1. ALLOWABLE DRAFT ½ DEGREE PER SIDE.
2. REFER TO PARAGRAPH 7.7 FOR CIRCUIT IDENTIFICATION REQUIREMENTS.

FIGURE 1—RECEPTACLE SOCKET

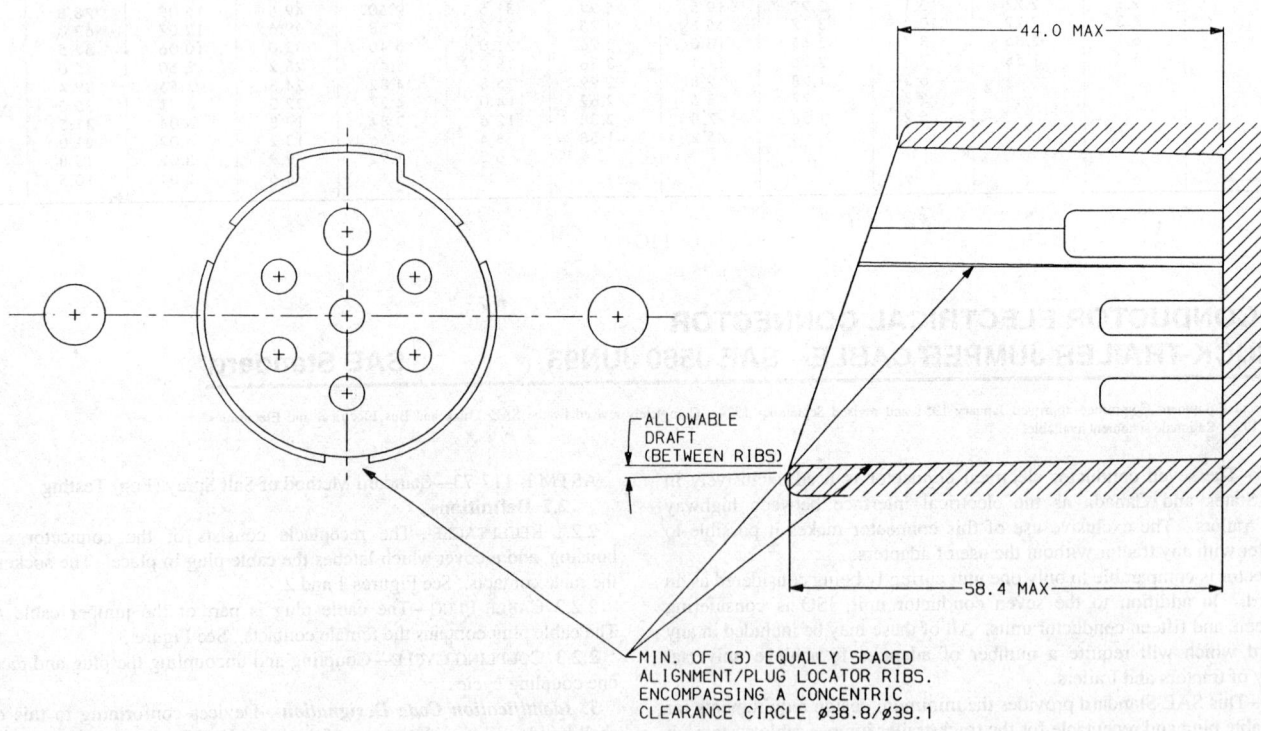

ALL DATUMS, NOTES, AND DIMENSIONS ON FIGURE 1 APPLY TO THIS FIGURE

FIGURE 2—ALTERNATE CONSTRUCTION RECEPTACLE SOCKET

4.1.2 VOLTMETER—A d-c voltmeter with an input resistance greater than 1000 Ω/V and with a resolution of 0.1 V shall be used. To achieve this resolution, the full-scale deflection shall be appropriate to the voltage rating of the system being tested.

A digital meter having at least a 3-1/2-digit readout with an accuracy of ±1% plus one digit is recommended for millivolt readings.

4.1.3 AMMETER—A d-c ammeter shall be used for current measurements. The meter range resolution shall be 0.1 A.

4.1.4 MILLIAMMETER—A d-c ammeter shall be used for current measurements. The meter range resolution shall be 1.0 mA.

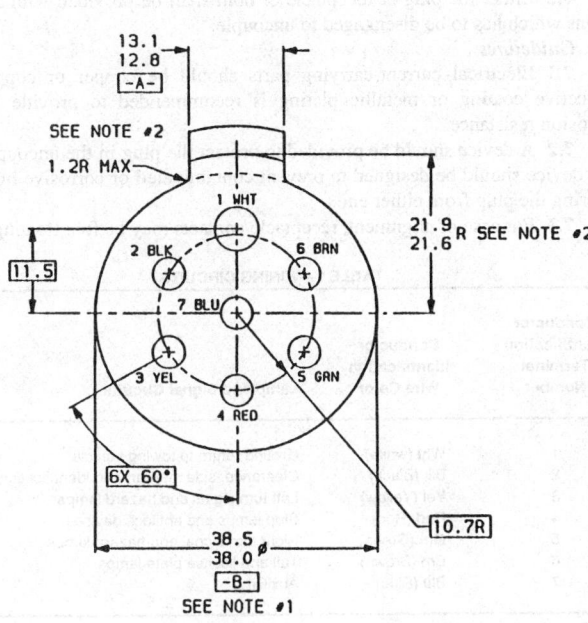
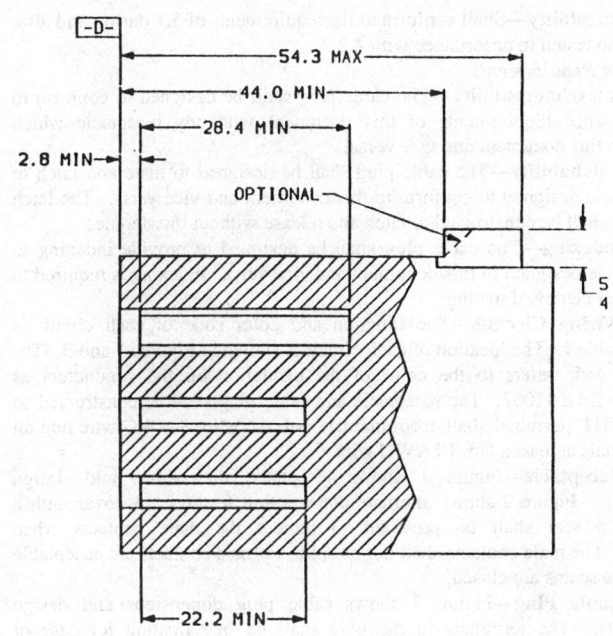

NOTES:
1. DIMENSION MUST BE MAINTAINED FOR 58.4 FROM DATUM "D"
2. DIMENSION MUST BE MAINTAINED FOR 44.0 FROM DATUM "D"

FIGURE 3—CABLE PLUG

4.2 Test Procedures

4.2.1 VOLTAGE DROP

4.2.1.1 Connectors Without Circuit Breakers—Connect a length of SAE J1067 type cable to the plug and receptacle terminals, then couple the mating parts. Connect to a power supply and apply a minimum of 35 A to each circuit which has a 4.75 mm (0.188 in) diameter terminal and a minimum of 70 A to each circuit which has a 6.35 mm (0.25 in) diameter terminal. After temperature stabilization, measure the voltage drop across each circuit of the assembly at a convenient point on the wire at least 25 mm (1 in) from the terminal. The test is to be conducted in a draft-free room maintained at an ambient temperature of 25 °C ± 5 °C.

4.2.1.2 Connectors With Circuit Breakers—Connect a length of SAE J1067 type cable to the plug and receptacle terminals, then couple the mating parts. Connect to a power supply and apply a minimum of 35 A to each circuit which has a 4.75 mm (0.188 in) diameter terminal and a minimum of 70 A to each circuit which has a 6.35 mm (0.25 in) diameter terminal. After temperature stabilization, measure the voltage drop across each circuit of the assembly at a convenient point on the wire at least 25 mm (1 in) from the terminal. Devices with circuit breakers may be certified using the noncircuit breaker version provided the construction is otherwise identical. If this is not possible, devices with circuit breakers may be tested by installing low-resistance shunts across the circuit breakers of the devices tested. The test is to be conducted in a draft-free room maintained at an ambient temperature of 25 °C ± 5 °C.

4.2.2 SHORT AND GROUNDED CIRCUIT—The test for shorts between circuits and ground is to be made with 70 V DC ± 5 V DC. Connect a milliammeter between the circuits to determine a circuit-to-circuit short and between each circuit and ground to determine a circuit-to-ground condition.

4.2.3 COUPLING FORCE—Measure the force to connect and disconnect the plug and receptacle.

4.2.4 STRAIGHT PULL—An assembled cable plug and jumper cable shall be securely mounted in a suitable fixture and a pull of 667 N (150 lb) exerted on the cable along the axis of the cable plug.

4.2.5 SALT SPRAY

4.2.5.1 With the plug inserted into the receptacle and with the assembly mounted in normal truck-trailer position, subject the normally exposed portion of the assembly to a 48 h salt spray test per ASTM B 117.

4.2.5.2 With the receptacle mounted in a normal position with the cover closed and with the open end of the plug pointed down, subject the uncoupled units to a 48 h salt spray test per ASTM B 117.

4.2.6 EXTREME TEMPERATURE—Use the same connector assembly for each extreme temperature condition test.

4.2.6.1 Connect a length of SAE J1067 type cable to the plug and receptacle.

4.2.6.2 Subject each assembly to a minimum ambient temperature of +82.2 °C. After the assembly has stabilized at +82.2 °C, perform a coupling cycle.

4.2.6.3 Subject each assembly to a maximum ambient temperature of -40 °C. After the assembly has stabilized at -40 °C, perform a coupling cycle.

4.2.7 DURABILITY—Connect a length of SAE J1067 type cable to the plug and receptacle and conduct the following test:

a. Conduct the voltage drop test per 4.2.1.1.
b. Perform 2500 coupling cycles.
c. Conduct salt spray test per 4.2.4.1.
d. Conduct voltage drop, short circuit, and grounded circuit tests per 4.2.1.
e. Conduct salt spray test per 4.2.4.2.
f. Repeat d.
g. Perform 2500 additional coupling cycles.
h. Repeat d.

5. Performance Requirements

5.1 Electrical

5.1.1 VOLTAGE DROP—After temperature stabilization, the voltage drop for each circuit shall not exceed 3 mV/A when tested in accordance with 4.2.1.

5.1.2 SHORT CIRCUIT AND GROUNDED CIRCUIT—The current flowing between any two circuits or between each circuit and ground shall not exceed 50 mA when tested in accordance with 4.2.2.1.

5.2 Coupling Force—The unlatched coupling force shall not exceed 223 N (50 lb) and the latched uncoupling force shall not be less than 110 N (25 lb) in accordance with 4.2.3.

5.3 Straight Pull—An assembled cable plug and trailer jumper cable shall not be damaged when tested in accordance with 4.2.4.

5.4 Extreme Temperature—Insulating materials shall not fracture and shall not deform when tested in accordance with 4.2.6.

5.5 Durability—Shall conform to the requirements of 5.1 during and after the test when tested in accordance with 4.2.7.

6. Design Requirements

6.1 Interchangeability—The cable plug shall be designed to conform to the performance requirements of this document with any receptacle which conforms to this document and vice versa.

6.2 Latchability—The cable plug shall be designed to mate and latch to any receptacle designed to conform to this document and vice versa. The latch mechanism shall be constructed to latch and release without interference.

6.3 Indexing—The cable plug shall be designed to provide indexing to any receptacle designed to this document and vice versa. Indexing is required to insure proper electrical mating.

6.4 Wiring Circuits—The function and color code of each circuit is shown in Table 1. The location of each circuit is shown in Figures 1 and 3. The wire color code refers to the color of the insulation on the conductors as specified in SAE J1067. The receptacle and cable plug shall be constructed so that the "WHT" terminal shall accommodate at least a No. 8 AWG wire and all other terminals at least a No. 10 AWG wire.

6.5 Receptacle—Figure 1 shows receptacle dimensions and design requirements. Figure 2 shows alternate construction features. A cover with a weather-tight seal shall be provided to protect the male contacts when uncoupled. The male contacts shall not be split. Formed contacts are acceptable provided the seams are closed.

6.6 Cable Plug—Figure 3 shows cable plug dimensions and design requirements. The terminals in the plug shall be free floating for ease of alignment with the receptacle during coupling. Cable plug assemblies shall incorporate a strain relief to relieve the tension on the electrical connection between the plug contacts and the jumper cable conductors.

6.7 Circuit identification by color or numeric is mandatory on the wire connection side of the cable plug and receptacle. It is recommended that circuit identification be on both the front and back sides of each.

6.8 Either the plug or receptacle or both shall be provided with a latching means which has to be disengaged to uncouple.

7. Guidelines

7.1 Electrical current-carrying parts should be copper or copper alloy. Protective coating or metallic plating is recommended to provide improved corrosion resistance.

7.2 A device should be provided to protect the plug in the uncoupled state. The device should be designed to prevent contaminated or corrosive liquid from entering the plug from either end.

7.3 For ease of alignment, receptacle contacts may be free floating.

TABLE 1—WIRING CIRCUITS

Conductor Identification Terminal Number	Conductor Identification Wire Color	Lamp and Signal Circuits
1	Wht (white)	Ground return to towing vehicle
2	Blk (Black)	Clearance, side marker, and identification lamps
3	Yel (Yellow)	Left turn signal and hazard lamps
4	Red (Red)	Stop lamps and antilock devices
5	Grn (Green)	Right turn signal and hazard lamps
6	Brn (Brown)	Tail and license plate lamps
7	Blu (Blue)	Auxiliary

SEVEN CONDUCTOR JACKETED CABLE FOR TRUCK TRAILER CONNECTIONS—SAE J1067

SAE Standard

Report of Electrical Equipment Committee approved October 1973.

1. Scope—This standard covers the minimum construction requirements and the configuration of the conductors in the cable to connect electrically a tractor to a trailer and/or trailer to trailer. This cable is used with the connector described in SAE J560.

2. Construction

2.1 Conductor

2.1.1 The conductors shall be made with tinned annealed copper wire according to ASTM B-33. Steel strands may be added to increase flexibility life.

2.1.2 The conductor stranding and lay shall be as shown in Table 1.

2.2 Insulation

2.2.1 The insulation on the No. 8 wire shall be a layer of .001 in. (.0254 mm) white polyethylene terephthalate film helically wrapped around the conductor with $1/3$–$1/2$ lap.

2.2.2 The insulation on the No. 10 wire and No. 12 wires shall be as specified in Table 2. The colors of the insulating compounds shall be as shown in Fig. 1. The nominal wall thickness of the insulation shall be 0.032 in. (.813 mm).

2.3 Cabling—The conductors shall be cabled together with a maximum lay of 6 in. (152.4 mm). The configuration of the conductors shall be as shown in Fig. 1. A suitable filler may be applied in twisting so as to fill the interstices between the conductors to produce a circular cross section. Use of a suitable separator over the conductors is required.

TABLE 1—CONDUCTORS

SAE Wire[a] Size	No. of Wires	Nominal Size of Strand AWG	Nominal Size of Strand in.	Lay in.	Conductor Area Cir Mils	Max Dia of Stranded Conductor, in.
12	65	30	.010 (.254 mm)	1.5 (38.1 mm)	6487	.100 (2.54 mm)
10	105	30	.010 (.254 mm)	1.5 (38.1 mm)	10479	.125 (3.18 mm)
8	168 or	30	.010 (.254 mm)	2.0 (50.8 mm)	16414	.175 (4.45 mm)
	427	34	.0063 (.160 mm)	2.0 (50.8 mm)		

[a] SAE wire size numbers indicate that the circular mil area of the stranded conductor approximates the circular mil area of American Wire Gage for equivalent gage size.

TABLE 2—INSULATION COMPOUND[a]

Recovery, in.	5 in. (127.0 mm) str, 1/2 (12.7 mm)
Elongation, %	250
Tensile strength, psi (MPa)	600 (4.12)
Oxygen-bomb aged (96 h at 70°C, 300 psi) elongation	70% of orig. min
Oxygen-bomb aged (96 h at 70°C, 300 psi) tensile	70% of orig. min

[a] Underwriters' Laboratories requirements for Class 3 Rubber—UL 62.

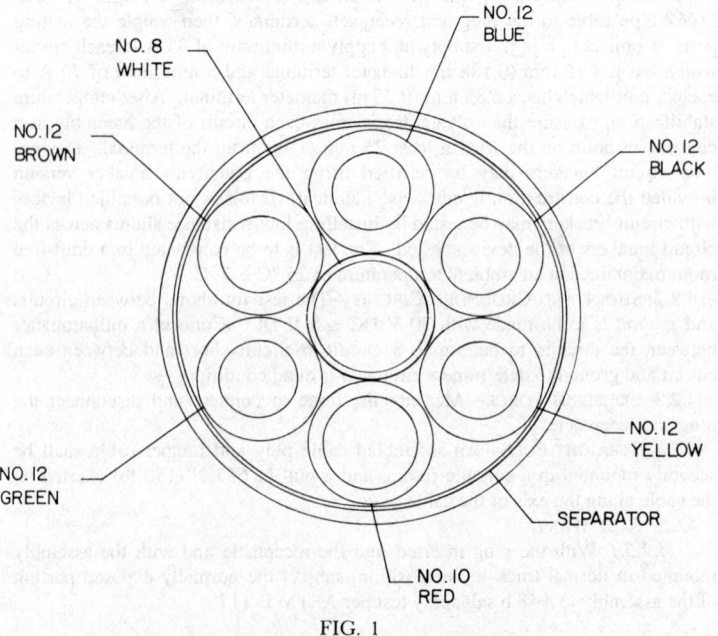

FIG. 1

2.4 Jacket—A 0.062 in. (1.57 mm) minimum thickness jacket shall be applied over the conductors and separator. The jacket compound shall meet the specifications in Table 3. The color of the jacket shall be red.

2.5 Cable Diameter and Finish—The finished outside diameter of the cable shall be 0.690 ± 0.020 in. (17.53 ± 0.508 mm). The finish of the cable shall be smooth and free from defects. Adjacent layers must not stick together when wound around a spool at any temperature below 120°F (49°C).

3. Tests

3.1 The cable shall be free from open circuits or twisted conductor splices.

3.2 Cold Test—A suitable length of cable shall be subjected to −20°F (−29°C) for a period of 4 h. While still at this temperature, it shall be bent 360 deg around a 3 in. (76.2 mm) diameter mandrel. The jacket shall not crack.

TABLE 3—JACKET COMPOUND[a]

Recovery, in.	6 in. (152.4 mm) str, 3/8 (9.53 mm)
Elongation, %	300
Tensile strength, psi (MPa)	1500 (10.35)
Oxygen-bomb aged (96 h at 70°C, 300 psi (2.06 MPa)) elongation	70% of orig. min[b]
Oxygen-bomb aged (96 h at 70°C, 300 psi (2.06 MPa)) tensile	70% of orig. min[b]
Air-oven aged (168 h at 70°C) tensile and elongation	70% of orig. min
Oil immersed (18 h at 121°C) tensile and elongation	60% of orig. min

[a] Underwriters' Laboratories requirements for Class 15 Neoprene—UL 62.
[b] 65% of result with unaged specimens if sum of tensile and elongation percentages is at least 140.

ELECTRICAL TERMINALS BLADE TYPE—SAE J858a

SAE Recommended Practice

Report of Electrical Equipment Committee approved June 1963 and last revised August 1969.

Blade terminals listed in this SAE Recommended Practice may be used for terminating wire ends, or for terminating circuits on devices other than wire.

When blade terminals are used for terminating wire, the temper of the terminals shall be sufficiently soft to permit the terminals being assembled to the wire and not show any fracture or cracks which would impair the strength of the assembly.

Terminals may be applied to wire by crimping, welding, swaging, soldering, or any combination at conductor grip.

Insulation grips must be used on all terminals, or some external means of relieving strain shall be provided.

When assembled to wire, the terminals shall fit, and securely grip, the conductor and when applicable, the insulation.

When blade terminals are used to terminate circuits on devices, they shall be of a temper that will permit the terminating section to be formed and attached to the device without fracturing or cracking. The temper should be high enough to resist displacement of the terminal and consequent misalignment to the mating connector.

TYPE 1A BLADE TERMINAL WITH DEPRESSION FOR USE WITH MATING SINGLE CONNECTORS

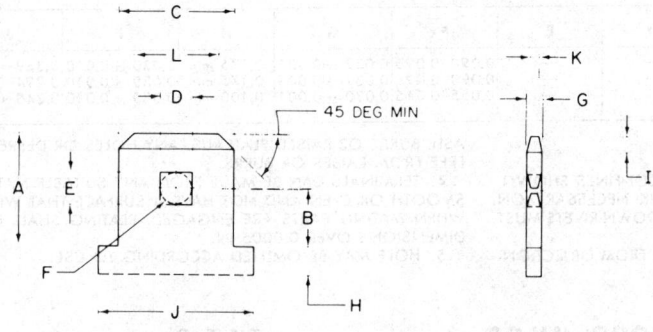

NOTES 1, 2, 3, 4 APPLY

SAE No.	Width	A	B	C	D	E	F	G	H	I	J	K	L
	5/16	0.308-0.318	0.160 ±0.003	0.304-0.320	0.080-0.100	0.070-0.080	0.070-0.080	0.032 ±0.001	0.075 min	0.030 ±0.010	0.369-0.383	0.013-0.019	0.240-0.260
	1/4	0.307-0.317	0.160 ±0.003	0.244-0.252	0.080-0.100	0.070-0.080	0.070-0.080	0.032 ±0.001	0.075 min	0.030 ±0.010	0.294-0.322	0.013-0.019	0.178-0.198
	3/16	0.245-0.260	0.140 ±0.003	0.183-0.192	0.055-0.075	0.045-0.055	0.045-0.055	0.020 ±0.001	0.075 min	0.030 ±0.010	0.240-0.270	0.006-0.012	0.117-0.138

TYPE 1B BLADE TERMINAL WITH HOLE FOR USE WITH MATING SINGLE CONNECTORS

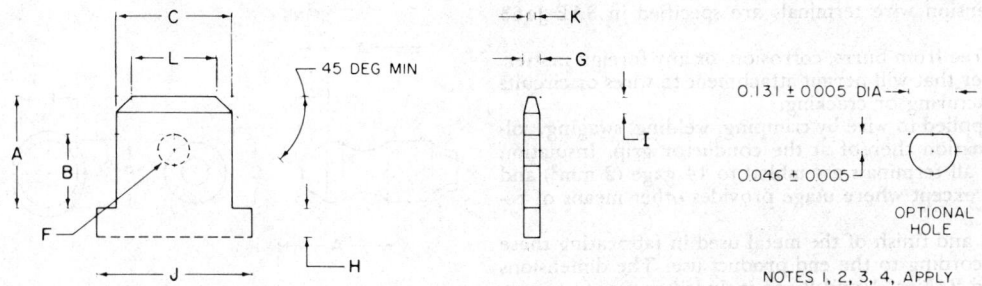

NOTES 1, 2, 3, 4, APPLY

SAE No.	Width	A	B	C	D	E	F	G	H	I	J	K	L
	5/16	0.308-0.318	0.194-0.200	0.304-0.320	—	—	0.090-0.096	0.032 ±0.001	0.075 min	0.030 ±0.010	0.369-0.383	0.013-0.019	0.240-0.260
	1/4	0.307-0.317	0.194-0.200	0.244-0.252	—	—	0.090-0.096	0.032 ±0.001	0.075 min	0.030 ±0.010	0.294-0.322	0.013-0.019	0.178-0.198
	3/16	0.245-0.260	0.150-0.160	0.183-0.192	—	—	0.055-0.065	0.020 ±0.001	0.075 min	0.030 ±0.010	0.240-0.270	0.006-0.012	0.117-0.138

23.180

TYPE 1C BLADE TERMINAL WITHOUT HOLE FOR USE WITH MATING MULTIPLE CONNECTOR PLUG

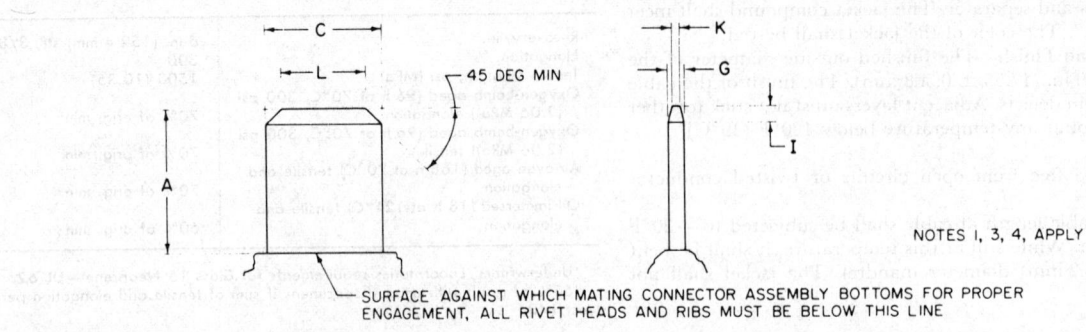

NOTES 1, 3, 4, APPLY

SAE No.	Width	A	B	C	D	E	F	G	H	I	J	K	L
	5/16	0.354–0.364	—	0.304–0.320	—	—	—	0.032 ±0.001	—	0.030 ±0.010	—	0.013–0.019	0.240–0.260
	1/4	0.353–0.363	—	0.244–0.252	—	—	—	0.032 ±0.001	—	0.030 ±0.010	—	0.013–0.019	0.178–0.198
	3/16	0.295–0.310	—	0.183–0.192	—	—	—	0.020 ±0.001	—	0.030 ±0.010	—	0.006–0.012	0.117–0.138

TYPE 2 BLADE TERMINAL

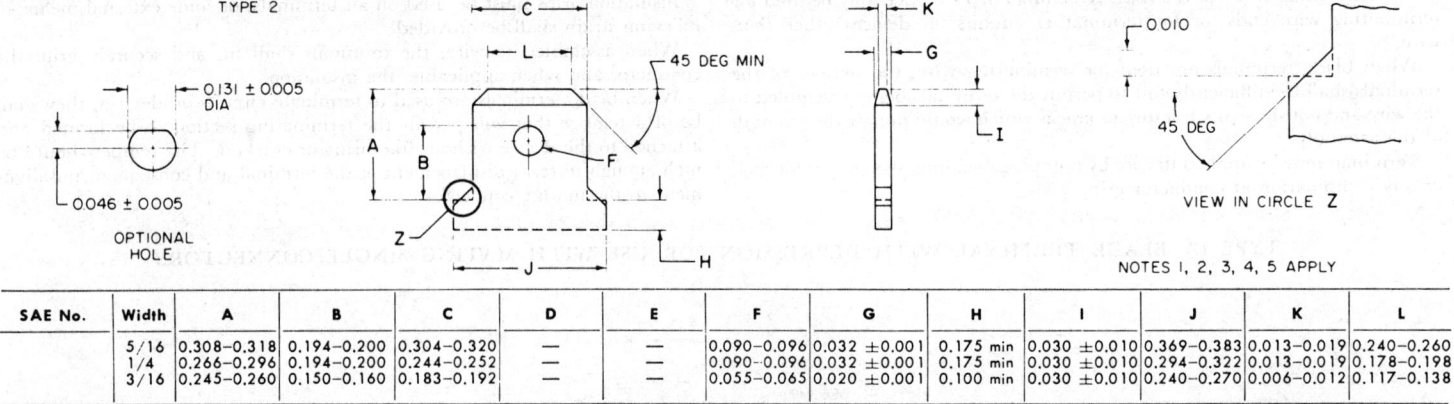

NOTES 1, 2, 3, 4, 5 APPLY

SAE No.	Width	A	B	C	D	E	F	G	H	I	J	K	L
	5/16	0.308–0.318	0.194–0.200	0.304–0.320	—	—	0.090–0.096	0.032 ±0.001	0.175 min	0.030 ±0.010	0.369–0.383	0.013–0.019	0.240–0.260
	1/4	0.266–0.296	0.194–0.200	0.244–0.252	—	—	0.090–0.096	0.032 ±0.001	0.175 min	0.030 ±0.010	0.294–0.322	0.013–0.019	0.178–0.198
	3/16	0.245–0.260	0.150–0.160	0.183–0.192	—	—	0.055–0.065	0.020 ±0.001	0.100 min	0.030 ±0.010	0.240–0.270	0.006–0.012	0.117–0.138

NOTES:
1. 45 DEG BEVEL NEED NOT BE A STRAIGHT LINE IF WITHIN THE CONFINES SHOWN.
2. H MINIMUM DIMENSION INDICATES THE AMOUNT OF SHANK NECESSARY ON TERMINAL TO CLEAR MATING PARTS. ALL PROTRUDING RIBS OR HOLD DOWN RIVETS MUST BE BELOW THIS LINE.
3. ALL PORTIONS OF TERMINAL SHOWN SHALL BE FLAT AND FREE FROM OBJECTIONABLE BURRS OR RAISED PLATEAUS. ANY HOLES OR DEPRESSION FOR DETENTS MUST BE FREE FROM RAISES OR BURRS.
4. TERMINALS CAN BE MADE FROM ANY SUITABLE MATERIAL. ANY PLATING MUST BE SMOOTH OR EVEN AND NOT HAVE A SURFACE THAT WILL INDUCE ADDITIONAL DRAG WHEN MATING PARTS ARE ENGAGED. PLATING SHALL NOT INCREASE THE TERMINAL DIMENSIONS OVER 0.0005 IN.
5. HOLE MAY BE OMITTED ACCORDING TO USE.

ELECTRICAL TERMINALS— PIN AND RECEPTACLE TYPE—SAE J928 JUL89

SAE Standard

Report of the Electrical Equipment Committee, approved July 1965, completely revised June 1980. Revised by the Electrical Distribution Systems Standards Committee July 1989.

1. Scope—This SAE Standard covers general requirements and terminal interface dimensions of various sizes of pin and receptacle type terminals.

(R) **2. General Requirements**—The pin and receptacle type terminals listed in this SAE Standard may be used for terminating wire ends, or for terminating circuits on devices other than wire. Performance requirements for low tension wire terminals are specified in SAE J163 JAN74.

Terminals shall be free from burrs, corrosion, or any foreign matter, and shall be of a temper that will permit attachment to wires or circuits on devices without fracturing or cracking.

Terminals may be applied to wire by crimping, welding, swaging, soldering, or any combination thereof at the conductor grip. Insulation grips shall be used on all terminals assembled to 14 gage (2 mm²) and smaller insulated wire except where usage provides other means of relieving strain.

The type, thickness, and finish of the metal used in fabricating these terminals may vary according to the end product use. The dimensions shown in Tables 1 and 2/Figs. 1 and 2 are included to assure proper fits between manufacturing sources.

Terminal sizes other than those listed are permissible, providing they meet the general requirements of this standard and the performance requirements of SAE J163.

Pin terminals fabricated from rod or bar stock must provide suitable stepped internal diameters to fit the wire conductors and insulation consistent with the method by which they are attached.

Insertion and removal forces are also variables that can be adjusted to fit the end use. It is recommended, however, that single connections with indentures should not exceed 15 lb (67 N) and multiple connections without indentures should not exceed 7 lb (31 N) per connection.

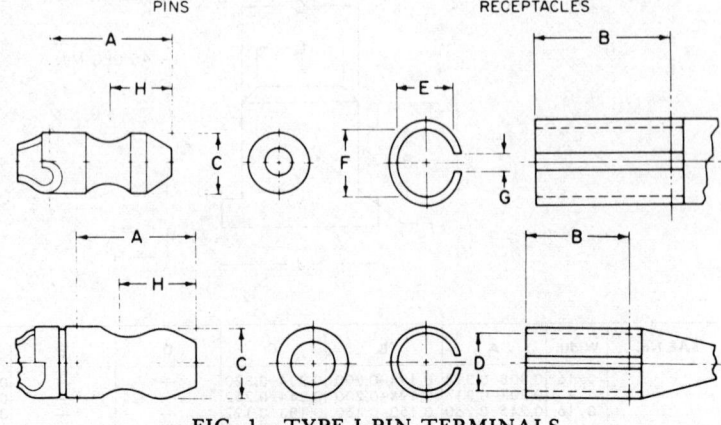

FIG. 1—TYPE I PIN TERMINALS

(R) TABLE 1—TYPE I PIN TERMINALS

Nominal Dia		SAE Wire Size Range	A Min[a]		B Min[a]		C		D Nominal		E Nominal		F Nominal		G Nominal		H	
in	mm		in	mm	in	mm	in	mm	in	mm	in	mm	in	mm	in	mm	in	mm
0.156	3.96	22-12 (0.35-3.0 mm^2)	0.34	8.7	0.34	8.7	0.159-0.155	4.04-3.94	0.150	3.81	0.147	3.73	0.181	4.60	0.034	0.86	0.219	5.56
0.180	4.57	23-10 (0.35-5.0 mm^2)	0.40	10.2	0.40	10.2	0.182-0.178	4.62-4.52	0.174	4.42	—	—	—	—	—	—	0.190	4.83

[a]Minimum insertion length.

NOTE 1: Detent Female—When a female detent is required, the detent of the receptacle must match the H dimension of the pin.

(R) NOTE 2: Metric dimensions are for reference purposes only.

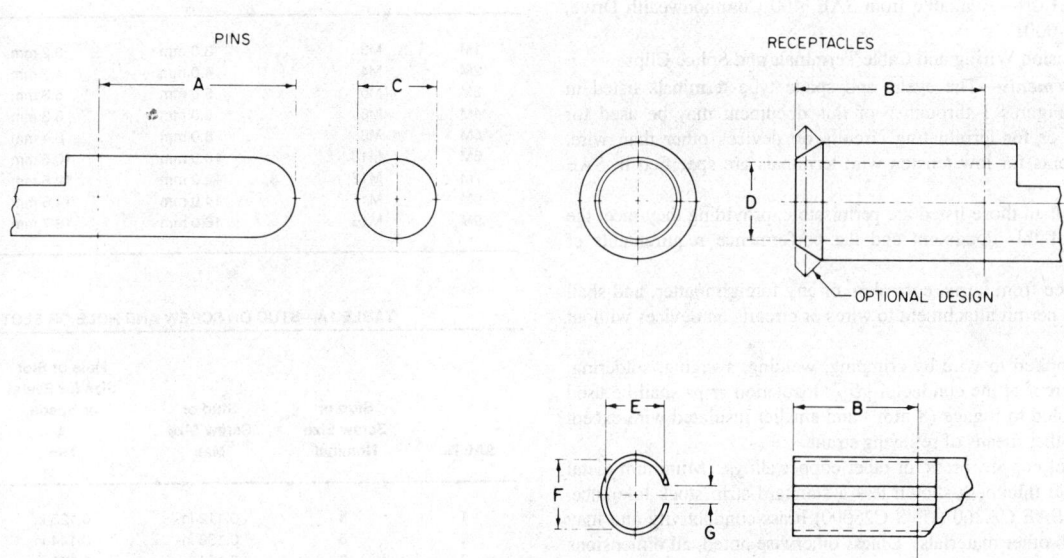

FIG. 2—TYPE II PIN TERMINALS

(R) TABLE 2—TYPE II PIN TERMINALS

Nominal Dia		SAE Wire Size Range	A Min[a]		B Min[a]		C		D Nominal		E Nominal		F Nominal		G Nominal	
in	mm		in	mm	in	mm	in	mm	in	mm	in	mm	in	mm	in	mm
0.086	2.18	22-14 (0.35-2.0 mm^2)	0.17	4.4	0.17	4.4	0.086-0.083	2.18-2.11	0.080	2.03	—	—	—	—	—	—
0.093	2.36	22-14 (0.35-2.0 mm^2)	0.21	5.4	0.21	5.4	0.093-0.091	2.36-2.31	0.086	2.18	0.086	2.18	0.134	3.40	0.022	0.56
0.156	3.96	22-12 (0.35-3.0 mm^2)	0.20	5.1	0.20	5.1	0.159-0.154	4.04-3.91	0.152	3.86	—	—	—	—	—	—

[a]Minimum insertion length.

(R) NOTE: Metric dimensions are for reference purposes only.

(R) ELECTRICAL TERMINALS—EYELET AND SPADE TYPE—SAE J561 JUN93

SAE Standard

Report of the Electrical Equipment Division approved August 1918. Completely revised by the Electrical Equipment Committee June 1980. Completely revised by the Electrical Distribution Systems Standards Committee June 1993. Rationale statement available.

1. **Scope**—This SAE Standard covers general requirements and dimensions of various sizes of eyelet and spade type terminals.

2. **References**

   2.1 **Applicable Document**—The following publications form a part of this specification to the extent specified herein. The latest issue of SAE publications shall apply.

   2.1.1 SAE PUBLICATION—Available from SAE, 400 Commonwealth Drive, Warrendale, PA 15096-0001.

   SAE J163—Low Tension Wiring and Cable Terminals and Splice Clips

3. **General Requirements**—The eyelet and spade type terminals listed in Tables 1 and 1a and Figures 1 through 5 of this document may be used for terminating wire ends or for terminating circuits on devices other than wire. Performance requirements for low tension wire terminals are specified in SAE J163.

   Terminal sizes other than those listed are permissible, providing they meet the general requirements of this document and the performance requirements of SAE J163.

   Terminals shall be free from burrs, corrosion, or any foreign matter, and shall be of a temper that will permit attachment to wires or circuits on devices without fracturing or cracking.

   Terminals may be applied to wire by crimping, welding, swaging, soldering, or any combination thereof at the conductor grip. Insulation grips shall be used on all terminals assembled to 8 gage (8 mm^2) and smaller insulated wire except where usage provides other means of relieving strain.

   Materials should be of copper, brass, or other copper alloys. Minimum metal thickness is the nominal thickness shown less a standard strip stock tolerance. Thickness is based on SAE CA260 (UNS C26000) brass conductivity and may be adjusted for use with other materials. Unless otherwise noted, all dimensions shall be held to a tolerance of ±0.25 mm (±0.010 in).

TABLE 1—METRIC STUD OR SCREW AND HOLE OR SLOT SIZES

SAE No.	Metric Stud or Screw Size Nominal	Metric Stud or Screw Size Max	Hole or Slot Size for Eyelet or Spade, A Min	Hole or Slot Size for Eyelet or Spade, A Max
1M	M3	3.0 mm	3.2 mm	3.4 mm
2M	M4	4.0 mm	4.2 mm	4.4 mm
3M	M5	5.0 mm	5.3 mm	5.5 mm
4M	M6	6.0 mm	6.3 mm	6.5 mm
5M	M8	8.0 mm	8.4 mm	8.6 mm
6M	M10	10.0 mm	10.5 mm	10.7 mm
7M	M12	12.0 mm	12.5 mm	12.9 mm
8M	M14	14.0 mm	14.6 mm	15.0 mm
9M	M16	16.0 mm	16.7 mm	17.1 mm

TABLE 1A—STUD OR SCREW AND HOLE OR SLOT SIZES

SAE No.	Stud or Screw Size Nominal	Stud or Screw Size Max	Hole or Slot Size for Eyelet or Spade, A Min	Hole or Slot Size for Eyelet or Spade, A Max
1	4	0.112 in	0.123 in	0.129 in
2	6	0.138 in	0.144 in	0.150 in
3	8	0.164 in	0.170 in	0.176 in
4	10	0.190 in	0.201 in	0.207 in
5	1/4	0.250 in	0.279 in	0.285 in
6	5/16	0.313 in	0.342 in	0.348 in
7	3/8	0.375 in	0.404 in	0.410 in
8	7/16	0.438 in	0.466 in	0.476 in
9	1/2	0.500 in	0.528 in	0.538 in

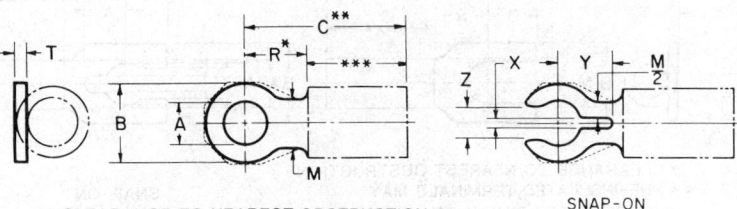

* CLEARANCE TO NEAREST OBSTRUCTION
** PRE-INSULATED TERMINALS MAY EXCEED THIS DIMENSION
*** DETAIL DESIGN IS MANUFACTURER'S OPTION

SAE No.	Screw Size	A Min mm	A Min in	A Max mm	A Max in	B Min mm	B Min in	C Max mm	C Max in	M Min mm	M Min in	R Min mm	R Min in	T Nom mm	T Nom in	X mm	X in	Y mm	Y in	Z mm ±0.13	Z in ±0.005
\multicolumn{22}{c}{To Use on SAE No. 18 and No. 20 (0.8 mm² and 0.5 mm²) Wire}
A001	4	3.13	0.123	3.27	0.129	6.1	0.24	16.0	0.63	3.9	0.15	4.9	0.19	0.64	0.025	---	---	---	---	---	---
A002	6	3.66	0.144	3.81	0.150	6.1	0.24	16.7	0.66	3.9	0.15	6.4	0.25	0.64	0.025	---	---	---	---	---	---
A003	8	4.32	0.170	4.47	0.176	8.7	0.34	17.7	0.70	3.9	0.15	7.7	0.30	0.64	0.025	1.5	0.06	4.6	0.18	3.56	0.140
A004	10	5.11	0.201	5.25	0.207	8.7	0.34	17.7	0.70	3.9	0.15	7.7	0.30	0.64	0.025	1.5	0.06	7.4	0.29	3.81	0.150
A005	1/4	7.09	0.279	7.23	0.285	11.0	0.43	21.5	0.85	3.9	0.15	9.4	0.37	0.64	0.025	---	---	---	---	---	---
A006	5/16	8.69	0.342	8.83	0.348	11.0	0.43	21.5	0.85	3.9	0.15	11.0	0.43	0.64	0.025	---	---	---	---	---	---
\multicolumn{22}{c}{To Use on SAE No. 14 and No. 16 (2.0 mm² and 1.0 mm²) Wire}
B101	4	3.13	0.123	3.27	0.129	6.1	0.24	19.3	0.76	4.4	0.17	4.9	0.19	0.72	0.028	---	---	---	---	---	---
B102	6	3.66	0.144	3.81	0.150	6.1	0.24	19.3	0.76	4.4	0.17	6.4	0.25	0.72	0.028	---	---	---	---	---	---
B103	8	4.32	0.170	4.47	0.176	8.7	0.34	20.8	0.82	4.4	0.17	7.7	0.30	0.72	0.028	1.5	0.06	4.6	0.18	3.56	0.140
B104	10	5.11	0.201	5.25	0.207	8.7	0.34	20.8	0.82	4.4	0.17	7.7	0.30	0.72	0.028	1.5	0.06	7.4	0.29	3.81	0.150
B105	1/4	7.09	0.279	7.23	0.285	11.0	0.43	22.3	0.88	4.4	0.17	9.4	0.37	0.72	0.028	---	---	---	---	---	---
B106	5/16	8.69	0.342	8.83	0.348	14.0	0.55	26.4	1.04	4.6	0.18	11.0	0.43	0.72	0.028	---	---	---	---	---	---
B107	3/8	10.27	0.404	10.41	0.410	14.0	0.55	26.4	1.04	4.6	0.18	12.7	0.50	0.72	0.028	---	---	---	---	---	---
\multicolumn{22}{c}{To Use on SAE No. 10 and No. 12 (5.0 mm² and 3.0 mm²) Wire}
B203	8	4.32	0.170	4.47	0.176	8.7	0.34	24.3	0.96	6.1	0.24	7.7	0.30	1.02	0.040	1.5	0.06	4.6	0.18	3.56	0.140
B204	10	5.11	0.201	5.25	0.207	8.7	0.34	24.3	0.96	6.1	0.24	7.7	0.30	1.02	0.040	1.5	0.06	7.4	0.29	3.81	0.150
B205	1/4	7.09	0.279	7.23	0.285	12.7	0.50	26.4	1.04	6.1	0.24	9.4	0.37	1.02	0.040	---	---	---	---	---	---
B206	5/16	8.69	0.342	8.83	0.348	17.3	0.68	29.4	1.16	6.1	0.24	11.0	0.43	1.02	0.040	---	---	---	---	---	---
B207	3/8	10.27	0.404	10.41	0.410	17.3	0.68	29.4	1.16	6.1	0.24	12.7	0.50	1.02	0.040	---	---	---	---	---	---
B208	7/16	11.84	0.466	12.09	0.476	17.3	0.68	29.4	1.16	7.7	0.30	15.8	0.62	1.02	0.040	---	---	---	---	---	---
B209	1/2	13.42	0.528	13.66	0.538	17.3	0.68	29.4	1.16	7.7	0.30	15.8	0.62	1.02	0.040	---	---	---	---	---	---
\multicolumn{22}{c}{To Use on SAE No. 8 (8.0 mm²) Wire}
B304	10	5.11	0.201	5.25	0.207	8.7	0.34	28.7	1.13	6.1	0.24	7.7	0.30	1.15	0.045	---	---	---	---	---	---
B305	1/4	7.09	0.279	7.23	0.285	12.7	0.50	28.7	1.13	6.1	0.24	9.4	0.37	1.15	0.045	---	---	---	---	---	---
B306	5/16	8.69	0.342	8.83	0.348	17.3	0.68	32.0	1.26	6.1	0.24	11.0	0.43	1.15	0.045	---	---	---	---	---	---
B307	3/8	10.27	0.404	10.41	0.410	17.3	0.68	32.0	1.26	6.1	0.24	12.7	0.50	1.15	0.045	---	---	---	---	---	---
B308	7/16	11.84	0.466	12.09	0.476	17.3	0.68	35.0	1.38	7.7	0.30	15.8	0.62	1.15	0.045	---	---	---	---	---	---
B309	1/2	13.42	0.528	13.66	0.538	17.3	0.68	35.0	1.38	7.7	0.30	15.8	0.62	1.15	0.045	---	---	---	---	---	---

FIGURE 1—STRAIGHT-TYPE EYELET AND SNAP-ON EYELET TERMINALS

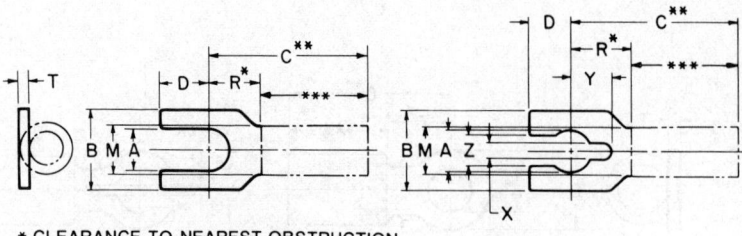

* CLEARANCE TO NEAREST OBSTRUCTION
** PRE-INSULATED TERMINALS MAY EXCEED THIS DIMENSION
*** DETAIL DESIGN IS MANUFACTURER'S OPTION

SAE No.	Screw Size	A Min mm	A Min in	A Max mm	A Max in	B Min mm	B Min in	C Max mm	C Max in	D mm	D in	M Min mm	M Min in	R Min mm	R Min in	T Nom mm	T Nom in	X mm	X in	Y mm	Y in	Z mm ±0.13	Z in ±0.005
								To Use on SAE No. 18 and No. 20 (0.8 mm² and 0.5 mm²) Wire															
H003	8	4.32	0.170	4.47	0.176	9.4	0.37	20.8	0.82	6.4	0.25	3.9	0.15	7.7	0.30	0.64	0.025	1.5-2.5	0.06-0.10	3.8-5.3	0.15-0.21	3.76	0.148
H004	10	5.11	0.201	5.25	0.207	9.4	0.37	20.8	0.82	6.4	0.25	3.9	0.15	7.7	0.30	0.64	0.025	1.5-2.5	0.06-0.10	3.8-5.3	0.15-0.21	4.44	0.175
								To Use on SAE No. 14 and No. 16 (2.0 mm² and 1.0 mm²) Wire															
H103	8	4.32	0.170	4.47	0.176	9.4	0.37	20.8	0.82	6.4	0.25	4.4	0.17	7.7	0.30	0.72	0.028	1.5-2.5	0.06-0.10	3.8-5.3	0.15-0.21	3.76	0.148
H104	10	5.11	0.201	5.25	0.207	9.4	0.37	20.8	0.82	6.4	0.25	4.4	0.17	7.7	0.30	0.72	0.028	1.5-2.5	0.06-0.10	3.8-5.3	0.15-0.21	4.44	0.175
								To Use on SAE No. 10 and No. 12 (5.0 mm² and 3.0 mm²) Wire															
H203	8	4.32	0.170	4.47	0.176	9.4	0.37	20.8	0.82	6.4	0.25	6.1	0.24	7.7	0.30	1.02	0.040	1.5-2.5	0.06-0.10	3.8-5.3	0.15-0.21	3.76	0.148
H204	10	5.11	0.201	5.25	0.207	9.4	0.37	20.8	0.82	6.4	0.25	6.1	0.24	7.7	0.30	1.02	0.040	1.5-2.5	0.06-0.10	3.8-5.3	0.15-0.21	4.44	0.175
								To Use on SAE No. 8 (8.0 mm²) Wire															
H304	10	5.11	0.201	5.25	0.207	9.4	0.37	28.7	1.13	6.4	0.25	6.1	0.24	7.7	0.30	1.15	0.045	1.5-2.5	0.06-0.10	3.8-5.3	0.15-0.21	4.44	0.175

FIGURE 2—STRAIGHT-TYPE SPADE TERMINALS

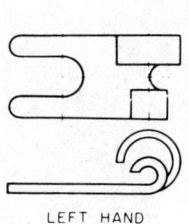 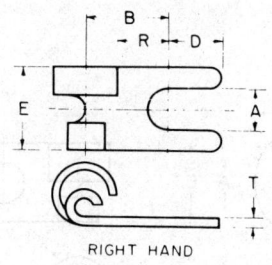 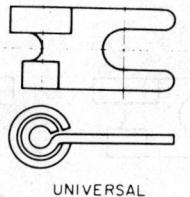 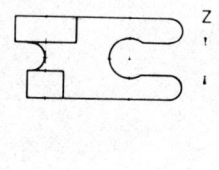

LEFT HAND   RIGHT HAND   UNIVERSAL   SNAP-ON

SAE No.	Screw Size	A Min mm	A Min in	A Max mm	A Max in	B mm	B in	D mm	D in	E mm	E in	R Min mm	R Min in	T Nom mm	T Nom in	Z mm ±0.13	Z in ±0.005
						To Use on No. 14 (2.0 mm^2) Wire and Smaller											
M103	8	4.32	0.170	4.47	0.176	8.7	0.34	6.4	0.25	9.7	0.38	7.7	0.30	0.72	0.028	3.76	0.148
M104	10	5.11	0.201	5.25	0.207	8.7	0.34	6.4	0.25	9.7	0.38	7.7	0.30	0.72	0.028	4.44	0.175
						To Use on SAE No. 10 and No. 12 (5.0 mm^2 and 3.0 mm^2) Wire											
M203	8	4.32	0.170	4.47	0.176	9.4	0.37	6.4	0.25	9.7	0.38	7.7	0.30	1.02	0.040	3.76	0.148
M204	10	5.11	0.201	5.25	0.207	9.4	0.37	6.4	0.25	9.7	0.38	7.7	0.30	1.02	0.040	4.44	0.175
						To Use on SAE No. 8 (8.0 mm^2) Wire											
M304	10	5.11	0.201	5.25	0.207	10.2	0.40	6.4	0.25	9.7	0.38	7.7	0.30	1.15	0.045	4.44	0.175

FIGURE 3—SIDE-TYPE SPADE TERMINALS

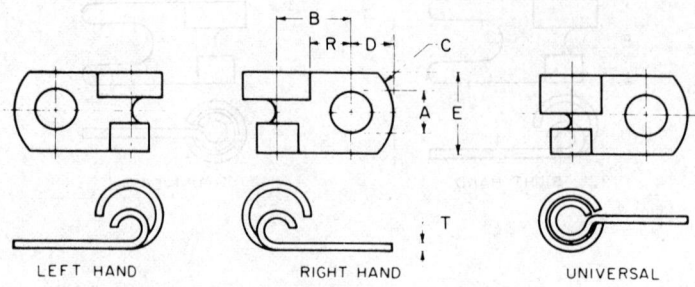

SAE No.	Screw Size	A Min mm	A Min in	A Max mm	A Max in	B Min mm	B Min in	C Max mm	C Max in	D mm	D in	E mm	E in	R Min mm	R Min in	T Nom mm	T Nom in
						To Use on SAE No. 14 (2.0 mm²) Wire and Smaller											
K101	4	3.13	0.123	3.27	0.129	6.9	0.27	9.6	0.38	3.9	0.15	7.9	0.31	4.9	0.19	0.72	0.028
K102	6	3.66	0.144	3.81	0.150	6.9	0.27	9.6	0.38	3.9	0.15	7.9	0.31	6.4	0.25	0.72	0.028
K103	8	4.32	0.170	4.47	0.176	8.7	0.34	9.6	0.38	4.8	0.19	9.7	0.38	7.7	0.30	0.72	0.028
K104	10	5.11	0.201	5.25	0.207	8.7	0.34	9.6	0.38	4.8	0.19	9.7	0.38	7.7	0.30	0.72	0.028
K105	1/4	7.09	0.279	7.23	0.285	9.4	0.37	16.0	0.63	6.4	0.25	12.7	0.50	9.4	0.37	0.72	0.028
K106	5/16	8.69	0.342	8.83	0.348	11.0	0.43	16.0	0.63	7.9	0.31	15.7	0.62	11.0	0.43	0.72	0.028
K107	3/8	10.27	0.404	10.41	0.410	12.0	0.47	19.3	0.76	9.5	0.37	19.0	0.75	12.7	0.50	0.72	0.028
						To Use on SAE No. 10 and No. 12 (5.0 mm² and 3.0 mm²) Wire											
K203	8	4.32	0.170	4.47	0.176	8.7	0.34	9.6	0.38	4.8	0.19	9.7	0.38	7.7	0.30	1.02	0.040
K204	10	5.11	0.201	5.25	0.207	8.7	0.34	9.6	0.38	4.8	0.19	9.7	0.38	7.7	0.30	1.02	0.040
K205	1/4	7.09	0.279	7.23	0.285	11.0	0.43	16.0	0.63	6.4	0.25	12.7	0.50	9.4	0.37	1.02	0.040
K206	5/16	8.69	0.342	8.83	0.348	12.5	0.49	16.0	0.63	7.9	0.31	15.7	0.62	11.0	0.43	1.02	0.040
K207	3/8	10.27	0.404	10.41	0.410	12.5	0.49	19.3	0.76	9.5	0.37	19.0	0.75	12.7	0.50	1.02	0.040
						To Use on SAE No. 6 and No. 8 (13.0 mm² and 8.0 mm²) Wire											
K304	10	5.11	0.201	5.25	0.207	10.2	0.40	16.0	0.63	4.8	0.19	9.7	0.38	7.7	0.30	1.15	0.045
K305	1/4	7.09	0.279	7.23	0.285	12.0	0.47	19.3	0.76	9.5	0.37	19.0	0.75	9.4	0.37	1.15	0.045
K306	5/16	8.69	0.342	8.83	0.348	14.0	0.55	19.3	0.76	9.5	0.37	19.0	0.75	11.0	0.43	1.15	0.045
K307	3/8	10.27	0.404	10.41	0.410	15.3	0.60	19.3	0.76	9.5	0.37	19.0	0.75	12.7	0.50	1.15	0.045
K308	7/16	11.84	0.466	12.09	0.476	17.3	0.68	19.3	0.76	9.5	0.37	19.0	0.75	15.8	0.62	1.15	0.045

FIGURE 4—SIDE-TYPE EYELET TERMINALS

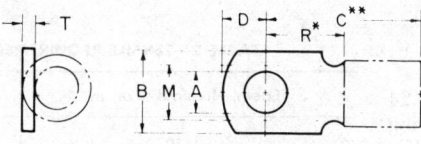
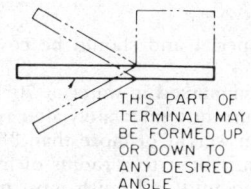

* CLEARANCE TO NEAREST OBSTRUCTION
** TERMINAL METAL LENGTH

THIS PART OF TERMINAL MAY BE FORMED UP OR DOWN TO ANY DESIRED ANGLE

SAE No.	Screw Size	A Min mm	A Min in	A Max mm	A Max in	B Min mm	B Min in	C Max mm	C Max in	D mm	D in	M Min mm	M Min in	R Min mm	R Min in	T Nom mm	T Nom in
To Use on SAE No. 4 and No. 6 (19.0 mm² and 13.0 mm²) Wire																	
N405	1/4	7.09	0.279	7.23	0.285	14.0	0.55	32.0	1.26	8.6-9.7	0.34-0.38	7.7	0.30	15.8	0.62	1.82	0.072
N406	5/16	8.69	0.342	8.83	0.348	17.3	0.68	32.0	1.26	8.6-9.7	0.34-0.38	7.7	0.30	15.8	0.62	1.82	0.072
N407	3/8	10.27	0.404	10.41	0.410	17.3	0.68	32.0	1.26	8.6-9.7	0.34-0.38	7.7	0.30	15.8	0.62	1.82	0.072
N408	7/16	11.84	0.466	12.09	0.476	17.3	0.68	32.0	1.26	8.6-9.7	0.34-0.38	7.7	0.30	15.8	0.62	1.82	0.072
N409	1/2	13.42	0.528	13.66	0.538	17.3	0.68	32.0	1.26	8.6-9.7	0.34-0.38	7.7	0.30	15.8	0.62	1.82	0.072
To Use on SAE No. 1 and No. 2 (40.0 mm² and 32.0 mm²) Wire																	
N505	1/4	7.09	0.279	7.23	0.285	14.0	0.55	42.9	1.69	8.6-9.7	0.34-0.38	12.5	0.49	15.8	0.62	2.05	0.080
N506	5/16	8.69	0.342	8.83	0.348	17.3	0.68	42.9	1.69	8.6-9.7	0.34-0.38	12.5	0.49	15.8	0.62	2.05	0.080
N507	3/8	10.27	0.404	10.41	0.410	17.3	0.68	42.9	1.69	8.6-9.7	0.34-0.38	12.5	0.49	15.8	0.62	2.05	0.080
N508	7/16	11.84	0.466	12.09	0.476	17.3	0.68	42.9	1.69	8.6-9.7	0.34-0.38	12.5	0.49	15.8	0.62	2.05	0.080
N509	1/2	13.42	0.528	13.66	0.538	17.3	0.68	42.9	1.69	8.6-9.7	0.34-0.38	12.5	0.49	15.8	0.62	2.05	0.080
To Use on SAE No. 00 and No. 0 (62.0 mm² and 50.0 mm²) Wire																	
N605	1/4	7.09	0.279	7.23	0.285	14.0	0.55	44.7	1.76	8.6-9.7	0.34-0.38	14.0	0.55	15.8	0.62	2.05	0.080
N606	5/16	8.69	0.342	8.83	0.348	17.3	0.68	44.7	1.76	8.6-9.7	0.34-0.38	14.0	0.55	15.8	0.62	2.05	0.080
N607	3/8	10.27	0.404	10.41	0.410	17.3	0.68	44.7	1.76	8.6-9.7	0.34-0.38	14.0	0.55	15.8	0.62	2.05	0.080
N608	7/16	11.84	0.466	12.09	0.476	17.3	0.68	44.7	1.76	8.6-9.7	0.34-0.38	14.0	0.55	15.8	0.62	2.05	0.080
N609	1/2	13.42	0.528	13.66	0.538	17.3	0.68	44.7	1.76	8.6-9.7	0.34-0.38	14.0	0.55	15.8	0.62	2.05	0.080
To Use on SAE No. 0000 and No. 000 (103.0 mm² and 81.0 mm²) Wire																	
N705	1/4	7.09	0.279	7.23	0.285	18.8	0.74	44.7	1.76	9.4-13.0	0.37-0.51	17.3	0.68	18.8	0.74	2.30	0.090
N706	5/16	8.69	0.342	8.83	0.348	18.8	0.74	44.7	1.76	9.4-13.0	0.37-0.51	17.3	0.68	18.8	0.74	2.30	0.090
N707	3/8	10.27	0.404	10.41	0.410	18.8	0.74	44.7	1.76	9.4-13.0	0.37-0.51	17.3	0.68	18.8	0.74	2.30	0.090
N708	7/16	11.84	0.466	12.09	0.476	18.8	0.74	44.7	1.76	9.4-13.0	0.37-0.51	17.3	0.68	18.8	0.74	2.30	0.090
N709	1/2	13.42	0.528	13.66	0.538	18.8	0.74	44.7	1.76	9.4-13.0	0.37-0.51	17.3	0.68	18.8	0.74	2.30	0.090

FIGURE 5—END-TYPE STARTING-CABLE TERMINALS

NONMETALLIC LOOM—SAE J562 APR86 — SAE Recommended Practice

Report of Electrical Equipment Division, approved March 1922, and reaffirmed by the Electrical and Electronic Systems Technical Committee, April 1986.

General Data—Nonmetallic flexible loom is recommended for use as an insulated covering giving mechanical protection over insulated wire, metal tubing, or other parts requiring a water-, oil-, and acid-proof covering resistant to fire or abrasion. It is also recommended for use as a covering for copper or other metal tubing to prevent crystallization and to eliminate rattles.

Construction—The loom shall be of single-wall construction, the material used to be strictly nonmetallic and of sufficient mechanical strength so that when formed or woven into a tubing it shall pass the tests for the size specified. Finished loom shall be free from obstruction and shall permit easy introduction of the maximum size wire or other part for which it is normally suited. Loom in any length shall slip freely over a polished mandrel 12 in long and equal in diameter to the minimum inside diameter specified. The dimensions of the standard sizes are listed in the accompanying table.

Saturation

Fire Resistant Loom—The loom shall be of such construction that when the asphaltic compound or equivalent water, acid, and fire resisting compound is applied, it will thoroughly impregnate the outside and lightly impregnate the inside of the loom.

Oil Proof Loom—The loom shall be thoroughly impregnated with a gum saturator or its equivalent. The saturator, when dry, shall be free from tackiness and gummy deposits. This impregnation is introduced to prevent absorption of moisture, oil or gasoline, to bind the material together to give the required wall strength and to prevent fraying.

Finish—For a fire resistant loom the outer surface shall be thoroughly covered by an asphaltic or equivalent water, acid, and fire resisting compound. Over the asphaltic or equivalent water, acid, and fire resisting compound, the loom may be coated with a thin coating of good paraffin wax or equivalent. For an oil-proof loom the outer surface shall be thoroughly covered with at least two coats of black pyroxylin lacquer or its equivalent, producing a good luster and a good bond to the fabric. The lacquer must be thoroughly dried before wrapping or boxing the loom for shipment. The use of heavy finishes or saturators to give artificial appearance is not permitted. The pyroxylin must be sufficiently plasticized so that it will not crack on a piece of loom kept three months at room temperature and then bent back sharply upon itself. Loom with finish of a higher luster than can be obtained with two coats of lacquer as

TABLE 1—DIMENSIONS OF NONMETALLIC LOOM, IN

Nominal Size	Inside Dia Min	Inside Dia Max	Outside Dia Min	Outside Dia Max	Nominal Size	Inside Dia Min	Inside Dia Max	Outside Dia Min	Outside Dia Max
5/32	0.156	0.176	0.245	0.265	5/8	0.625	0.645	0.785	0.805
3/16	0.187	0.207	0.287	0.307	11/16	0.687	0.707	0.847	0.867
1/4	0.250	0.270	0.350	0.370	3/4	0.750	0.770	0.934	0.954
5/16	0.312	0.332	0.412	0.432	13/16	0.812	0.832	0.996	1.016
3/8	0.375	0.395	0.505	0.525	7/8	0.875	0.895	1.079	1.099
7/16	0.437	0.457	0.567	0.587	15/16	0.937	0.957	1.141	1.161
1/2	0.500	0.520	0.630	0.650	1	1.000	1.020	1.204	1.224
9/16	0.562	0.582	0.722	0.742					

specified above shall be considered special and should be covered by other specifications when required.

Tests—A 6 in piece of loom totally immersed in water at 70°F for 24 h and then blown out with a mild air current immediately after removing the water shall not have an increase in weight of more than 35%. The wall must not collapse when the loom is bent to a radius of five times the inside diameter at 70°F. The compound and finish must not crack open in this test.

The material in the wall of the loom shall not crack or break when a 3 in length is flattened between two steel plates. When the inside diameter of loom is 3/8 in or smaller, the distance between plates is to be 11/64 in. When the inside diameter of loom is over 3/8 in, the distance between plates is to be 9/32 in. The finish shall not show excessive cracking when loom is subjected to this test.

The polished mandrels used for checking the inside diameters shall show no sticking or discoloration up to 150°F.

Loom shall be capable of standing a tension test for 5 min without breaking or opening at any point as required in Table 2.

Additional Test for Oil Proof or Fire Resistant Loom—When a piece of loom is totally immersed in an equal mixture of cylinder oil, kerosene, and gasoline at 70°F for 5 min and then subjected to a temperature not exceeding 250°F for 1 h, the saturating compound must not drip from the loom nor the finish show any appreciable defects.

Flame-resisting qualities shall be incorporated in the saturation or finish or both for the loom to pass the following test: the loom shall not convey fire nor support combustion for more than 1 min after five 15-sec applications of a standard test flame with intervals of 15 sec between applications. A standard test flame is the blue flame, about 5 in high, produced by a 1/2 in bunsen burner fed with ordinary illuminating gas at normal pressure. The loom shall be held vertically with either the lower or the upper end thoroughly sealed to prevent the passage of air, and the flame must be applied horizontally.

TABLE 2—TENSILE REQUIREMENTS OF NONMETALLIC LOOM[a]

Loom, Nominal Size, in	Min Tensile Requirement, lb
3/16	75
1/4	85
5/16 and 3/8	100
7/16 or larger	150

[a] The test piece shall be 6 in long between supports.

ELECTRIC WINDSHIELD WIPER SWITCH—SAE J112a — SAE Recommended Practice

Report of Electrical Equipment Committee approved July 1969 and last revised May 1971. Editorial change October 1977.

1. Definition—An electric windshield wiper switch is that part of an electric windshield wiper system by which the operator of a vehicle causes the windshield wipers to function.

2. Reference Standards

2.1 If the switch employs an internal circuit breaker(s), the circuit breaker shall comply with the requirements of SAE J258 (June, 1971).

2.2 If the switch employs the integral exterior circuit breaker, the circuit breaker shall comply with the requirements of SAE J553c (May, 1972).

3. Temperature Test

3.1 To insure basic function, the switch shall be manually cycled for 10 cycles at design electrical load at 75 ± 10 F (24 ± 5.5 C), 165, +0, −5 F (74, +0, −2.8 C), and −25, +5, −0 F (−32, +2.8, −0 C) after a 1 h exposure at each of these temperatures. The switch shall be electrically and mechanically operable during each of these cycles.

3.2 This same switch shall be used for the endurance test described below.

4. Endurance Test Setup

4.1 The switch shall be set up to operate its design electrical load.

4.2 The test shall be set up to operate the switch for the prescribed number of completed cycles.

One complete cycle shall consist of sequencing through each position (with dwell in each position) and return without dwell in intermediate positions to the initial position.

The test equipment shall be so arranged as to provide the following mechanical time requirements:

Travel Time—0.1–0.5 s (time from one position to the next).

NOTE: If the switch employs a rheostat, the travel time through the rheostat segment in each direction shall be 1.0–3.0 s.

Dwell Time—0.50–1.0 s (time in each position).

NOTE: After switching to OFF, if a motor is used, sufficient dwell time shall be provided to allow the motor to park. The dwell time in OFF can then be greater (if required) than the dwell time range indicated above.

4.3 During the test the switch shall be operated at 6.4 V d-c for a 6 V system, 12.8 V d-c for a 12 V system, or 25.6 V d-c for a 24 V system.

These voltages shall be measured at the input termination on the switch.

The power supply shall not generate any adverse transients not present in motor vehicles and shall comply with the following specifications:

(a) *Output Current*—Capable of supplying the continuous current of the design electrical load.

(b) *Regulation*

Dynamic—The output voltage at the supply shall not deviate more than 1.0 V from zero to maximum load (including inrush current) and should recover 63% of its maximum excursion within 100 ms.

Static—The output voltage at the supply shall not deviate more than 2% with changes in static load from zero to maximum (not including inrush current), and means shall be provided to compensate for static input line voltage variations.

(c) *Ripple Voltage*—Maximum 300 mV peak to peak.

5. Endurance Requirements

5.1 The switch shall be capable of satisfactorily operating for 10,000 complete cycles at a temperature of 75 ± 10 F (24 ± 5.5 C) followed by 1 h ON in low position at 75 ± 10 F (24 ± 5.5 C).

5.2 The average voltage drop from the input terminal(s) to the corresponding average output terminal(s) shall be measured before and after the endurance test and after the soak tests and shall not exceed 0.30 (excluding rheostat), the average of three consecutive readings, at design load. If wiring is an integral part of the switch, the voltage drop measurement shall be made including 3 in. (76 mm) of wire on each side of the switch; otherwise, measurement shall be made at switch terminals.

6. Combination Windshield Wiper and Washer Switch—The same combination switch shall be used for the test of each function. If the washer and wiper functions are mechanically coordinated, the functions shall be tested simultaneously. The wiper switch shall meet the requirements of this recommended practice. The washer switch shall meet the requirements of SAE J234 (May, 1971).

ELECTRIC WINDSHIELD WASHER SWITCH—SAE J234 — SAE Recommended Practice

Report of Electrical Equipment Committee approved May 1971. Editorial change October 1977.

1. Definition—An electric windshield washer switch is that part of an electric windshield washer system by which the operator of a vehicle causes the windshield washers to function.

2. Temperature Test

2.1 To insure basic function, the switch shall be manually cycled for 10 cycles at design electrical load at 75 ± 10 F (24 ± 5.5 C), 165 +0, −5 F (74, +0, −2.8 C), and −25, +5, −0 F (−32, +2.8, −0 C) after a 1 h exposure at each of these temperatures. The switch shall be electrically and mechanically operable during each of these cycles.

2.2 This same switch shall be used for the endurance test described below.

3. Endurance Test Setup

3.1 The switch shall be set up to operate its design electrical load.

3.2 The test shall be set up to operate the switch for the prescribed number of completed cycles.

One complete cycle shall consist of sequencing through each position (with dwell in each position) and return without dwell in intermediate positions to the initial position.

The test equipment shall be so arranged as to provide the following mechanical time requirements:

Travel Time—0.1–0.5 s (time from one position to the next).

Dwell Time—1.0–2.0 s (time in each position).

3.3 During the test the switch shall be operated at 6.4 V d-c for a 6 V system, 12.8 V d-c for a 12 V system, or 25.6 V d-c for a 24 V system.

These voltages shall be measured at the input termination on the switch.

The power supply shall not generate any adverse transients not present in motor vehicles and shall comply with the following specifications:

(a) Output Current—Capable of supplying the continuous current of the design electrical load.

(b) Regulation

Dynamic—The output voltage at the supply shall not deviate more than 1.0 V from zero to maximum load (including inrush current) and should recover 63% of its maximum excursion within 100 ms.

Static—The output voltage at the supply shall not deviate more than 2% with changes in static load from zero to maximum (not including inrush current), and means shall be provided to compensate for static input line voltage variations.

(c) Ripple Voltage—Maximum 300 mV peak to peak.

4. Endurance Requirements

4.1 The switch shall be capable of satisfactorily operating for 10,000 complete cycles at a temperature of 75 ± 10 F (24 ± 5.5 C).

4.2 The average voltage drop from the input terminal(s) to the corresponding output terminal(s) shall be measured before and after the endurance test and shall not exceed 0.30 V (the average of three consecutive readings) at design load. If wiring is an integral part of the switch, the voltage drop measurement shall be made including 3 in. of wire on each side of the switch; otherwise, measurement shall be made at switch terminals.

5. Combination Windshield Wiper and Washer Switch—The same combination switch shall be used for the test of each function. If the washer and wiper functions are mechanically coordinated, the functions shall be tested simultaneously. The washer switch shall meet the requirements of this recommended practice. The wiper switch shall meet the requirements of SAE J112a (November, 1971).

ELECTRIC BLOWER MOTOR SWITCH—SAE J235

SAE Recommended Practice

Report of Electrical Equipment Committee approved May 1971.

1. Definition—An electric blower motor switch is that part of a blower system by which the operator of a vehicle causes a blower motor to function.

2. Temperature Test

2.1 To insure basic function, the switch shall be manually cycled for 10 cycles at design electrical load at 75 ± 10 F (24 ± 5.5 C), 165, +0, −5 F (74, +0, −2.8 C), and −25, +5, −0 F (−32, +2.8, −0 C) after a 1 h exposure at each of these temperatures. The switch shall be electrically and mechanically operable during each of these cycles.

2.2 This same switch shall be used for the endurance test described below.

3. Endurance Test Setup

3.1 The switch shall be set up to operate its design electrical load.

3.2 The test shall be set up to operate the switches for the prescribed number of complete cycles.

One complete cycle shall consist of sequencing through each position (with dwell in each position) and return without dwell in intermediate positions to the initial position.

The test equipment shall be so arranged as to provide the following mechanical time requirements:

Travel Time—0.1–0.5 s (time from one position to the next).

Dwell Time—0.5–1.0 s (time in each position).

3.3 During the test the switch shall be operated at 6.4 V d-c for a 6 V system, 12.8 V d-c for a 12 V system, or 25.6 V d-c for a 24 V system.

These voltages shall be measured at the input termination on the switch.

The power supply shall not generate any adverse transients not present in motor vehicles and shall comply with the following specifications:

(a) Output Current—Capable of supplying the continuous current of the design electrical load.

(b) Regulation

Dynamic—The output voltage at the supply shall not deviate more than 1.0 V from zero to maximum load (including inrush current) and should recover 63% of its maximum excursion within 100 ms.

Static—The output voltage at the supply shall not deviate more than 2% with changes in static load from zero to maximum (not including inrush current), and means shall be provided to compensate for static input line voltage variations.

(c) Ripple Voltage—Maximum 300 mV peak to peak.

4. Endurance Requirements

4.1 The switch shall be capable of satisfactorily operating for 10,000 complete cycles at a temperature of 75 ± 10 F (24 ± 5.5 C) followed by 1 h ON in low position at 75 ± 10 F (24 ± 5.5 C).

4.2 The average voltage drop from the input terminal(s) to the corresponding average output terminal(s) shall be measured before and after the endurance test and shall not exceed 0.30 V (the average of three readings) at design load. If wiring is an integral part of the switch, the voltage drop measurement shall be made including 3 in. of wire on each side of the switch; otherwise, measurement shall be made at switch terminals.

SIX- AND TWELVE-VOLT CIGAR LIGHTER RECEPTACLES—SAE J563 MAR90

SAE Standard

Report of Electrical Equipment Committee approved June 1949 and last revised May 1978. Reaffirmed by the Circuit Protection and Switching Devices Standards Committee March 1990.

Foreword—This reaffirmed document has been changed to reflect the new SAE Technical Board format.

1. Scope—This SAE Standard covers the basic cigar lighter receptacle, which may optionally incorporate overload protective devices.

Fig. 1 and Table 1 show the cigar lighter receptacle and its pertinent dimensions.

Lighter plugs shall be permanently marked either "6 volt" or "12 volt".

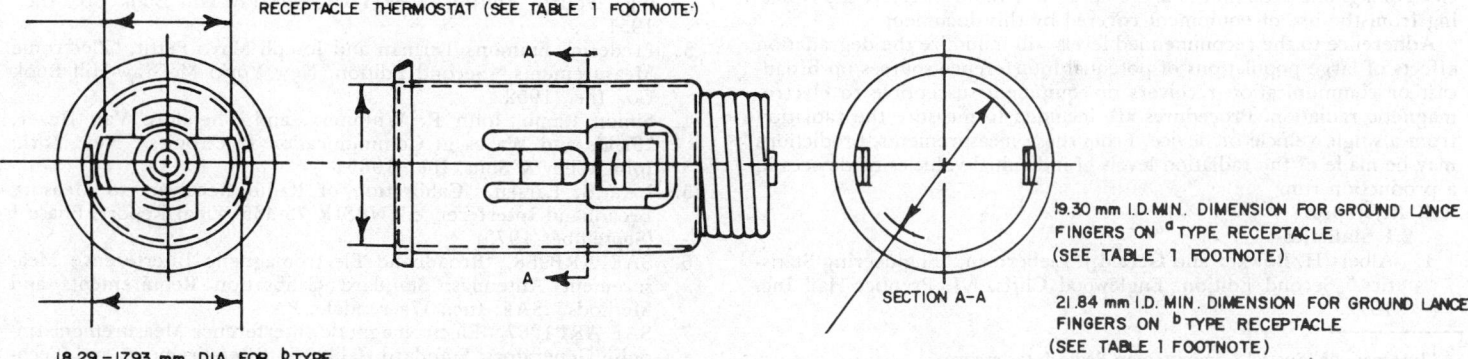

FIG. 1—CIGAR LIGHTER RECEPTACLE (TYPICAL)

Manufacturers of accessory plugs to engergize devices such as trouble lamps and razors are cautioned to conform to the provisions given in this document.

Live contact of any accessory plug shall be made with the center stud; the contacting member must not be less than 3.56 mm in diameter and have a minimum spherical radius of 2.54 mm at the contacting end.

To prevent damage, the body of an accessory plug back of the contact end should be large enough to serve as a centering guide when it is inserted into the receptacle. The contact end of any accessory plug should have sufficient taper so as not to interfere with the receptacle bimetal fingers. Any ground contact fingers should be so designed as not to interfere with, distort, or catch on the grounding lances in the receptacle when the plug is rotated.

Wiring capacity and overload protection, as provided by the automotive industry to the 12 V lighter receptacle, restricts accessory plug devices to a maximum current draw of 8.0 A.

Accessory plugs shall be permanently marked "6 volt" or "12 volt" in letters 4.7 mm high minimum.

TABLE 1—DIAMETERS OF RECEPTACLE AND LIGHTER PLUG

Volts	Inside Dia, B, mm	Plug Body Dia, mm
6	21.34-21.46	21.08-21.23
12	20.93-21.01[a]	20.73-20.88
12	21.41-21.51[b]	21.13-21.23

[a] Receptacles providing bimetal finger contact to the outer periphery of the heating element cup.
[b] Receptacles providing bimetal finger contact to the inner periphery of the heating element cup.

(R) PERFORMANCE LEVELS AND METHODS OF MEASUREMENT OF ELECTROMAGNETIC RADIATION FROM VEHICLES AND DEVICES (30 to 1000 MHz)—SAE J551 MAR90

SAE Standard

Report of the Electrical Equipment Committee, approved December 1947, tenth revision, Electromagnetic Radiation Subcommittee, October 1985. Rationale statement available. Completely revised by the Electromagnetic Radiation Standards Committee March 1990.

Foreword—International activity to control electromagnetic radiation from ignition systems of vehicles began in Paris, France in 1934. The fifty plus years following have led to the development of C.I.S.P.R.[1] Publication[2] 12, Third Edition.

C.I.S.P.R. limits are incorporated in the laws of European countries and are used in most of the world, except North America where the J551 method is used as a basis for control of broad band electromagnetic radiation. The latter is conceptually identical, but more rigorous. It is the policy of the Electromagnetic Radiation (EMR) Technical Committee to work toward a common worldwide standard. In the past ten years the U.S. has undertaken a concerted effort to arrive at a common document by changing SAE J551 and proposing changes to Publication 12. At this time, the two documents are about 95% harmonized.

This document adopts the vast majority of the wording of the third edition of C.I.S.P.R. Publication 12. The performance limits are identical to those of Publication 12. The changes that have been made broaden the applicability of the document, enhance its applicability to automated test methods, and increase the stringency of the test by requiring scanning over the entire frequency range.

Commencing with Section 3 changed or added paragraphs are indicated by **bold** print; otherwise the text and numbering of paragraphs is identical to Publication 12.

1. Scope—This SAE Standard covers the measurement of broadband electromagnetic radiation over the frequency range of 30 to 1000 MHz from a vehicle or other device powered by an internal combustion engine or electric motor. Operation of all engines (main and auxiliary) of a vehicle or device is included. All equipment normally operating when the engine is running is also included except operator-controlled equipment, which is excluded.

The recommended level applies only to complete vehicles or devices in their final manufactured form. Vehicle mounted rectifiers used for battery charging in electric vehicles are included in this document when operated in their charging mode.

1.1 Purpose—This document provides test procedures and recommended levels to assist engineers in the measurement of broadband electromagnetic radiation and the control of radio interference resulting from the use of equipment covered by this document.

Adherence to the recommended levels will minimize the degradation effects of large populations of potential interference sources on broadcast or communication receivers or equipment susceptible to electromagnetic radiation. Procedures are included to measure the radiation from a single vehicle or device. From these measurements, predictions may be made of the radiation levels of individual vehicles or devices in a production run.

2. References

2.1 Statistics

1. Albert H. Bowker and Gerald J. Lieberman, "Engineering Statistics." Second Edition, Englewood Cliffs, NJ: Prentice-Hall Inc, 1972.
2. D. B. Owen, "Factors for One-sided Tolerance Limits and for Variables Sampling Plans." Sandia Corporation Monograph, SCR-607 (March 1963).
3. D. B. Owen, "A Survey of Properties and Applications of the Non-Central t-Distribution." Technometrics, Vol. 10 (1968), pp. 445-478.
4. Edwin L. Crow, Francis A. Davis, Margaret W. Maxfield, "Statistics Manual." U.S. Naval Ordinance Test Station, China Lake, CA, (1955).
5. George J. Resnikoff and Gerald J. Liebermann, "Tables of the Non-Central t-Distribution." Stanford, CA: Stanford University Press, 1957.
6. Mary Gibbons Natrella, "Experimental Statistics." National Bureau of Standards Handbook 91, Issued August 1, 1963, Reprinted October 1966 with corrections.
7. Roy H. Wampler, "One-Sided Tolerance Limits for the Normal Distribution, $P=.80$, $\gamma=.80$." Journal of Research of the National Bureau of Standards Section B, Vol. 80B, No. 3, (July-September, 1976).
8. N. L. Johnson and F. C. Leone, "Statistics and Experimental Design I." New York: John Wiley & Sons, 1964, pp. 298-348.
9. H. J. Larson, "Introduction to Probability Theory and Statistical Inference." New York: John Wiley & Sons, 1969.
10. "The Statistical Consideration in the Determination of Limits of Radio Inference." International Electrotechnical Commission (International Special Committee on Radio Interference C.I.S.P.R.) Report No. 48 (Document CISPR (Secretariat) 952E, September 1973 as approved at West Long Branch meeting 1973).

2.2 Antennas, Electronics, and Fields

1. Henry Jasik and Richard C. Johnson, "Antenna Engineering Handbook." Second Edition, New York: McGraw-Hill Book Co., Inc. 1983.
2. Frederick Emmons Terman, "Electronic and Radio Engineering." Fourth Edition, New York: McGraw-Hill Book Co., Inc., 1951.
3. Frederick Emmons Terman and Joseph Mayo Pettit, "Electronic Measurements," Second Edition, New York: McGraw-Hill Book Co., Inc., 1952.
4. Simon Ramo, John R. Whinnery, and Theodore Van Duzer, "Fields and Waves in Communication Electronics." New York: John Wiley & Sons, Inc., 1965.
5. Ezra B. Larsen, "Calibration of Radio Receivers to Measure Broadband Interference." NBSIR 73-335, Final Report, Phase I (September 1973).
6. SAE ARP958, "Broadband Electromagnetic Interference Measurement Antennas; Standard Calibration Requirements and Methods." SAE, Inc., Warrendale, PA.
7. SAE ARP1267, "Electromagnetic Interference Measurement Impulse Generators; Standard Calibration Requirements and Techniques." SAE, Inc., Warrendale, PA.

---

[1] International-Special Committee on Radio Interference.
[2] Publication refers to C.I.S.P.R. documents.

8. ANSI C63.2, "Electromagnetic Noise and Field Strength Instrumentation, 10 kHz to 1 GHz." ANSI, New York.

2.3 Other—Available from American National Standards Institute, 1430 Broadway, New York, NY 10018.

1. International Electrotechnical Commission (IEC) Publication, International Electrotechnical Vocabulary (IEV), Chapter 161 Bureau Central of IEC, Geneva, Switzerland.
2. C.I.S.P.R. Publication 12 [1990-01], "Limits and Methods of Measurements of Radio Interference Characteristics of Vehicles, Motorboats and Spark-Ignited Engine-Driven Devices." Bureau Central of IEC, Geneva, Switzerland.
3. C.I.S.P.R. Publication 16, "Specification for Radio Interference Measuring Apparatus and Measurement Methods." Bureau Central of IEC, Geneva, Switzerland.

3. Definitions—For the purpose of this document, the definitions contained in IEC Publications 50 (161) and 50 (902) are applicable. **For additional definitions, refer to ANSI/IEEE Standard 100.**

The following definitions are specific for this document:

3.1 Impulsive Vehicular Noise—The unwanted emission of electromagnetic energy, predominantly impulsive in content, arising from sources within a vehicle or device.

3.2 Impulsive Ignition Noise—The unwanted emission of electromagnetic energy, predominantly impulsive in content, arising from the ignition system within a vehicle or device.

3.3 Ignition Noise Suppressor—That portion of a high voltage ignition circuit intended to limit the emission of impulsive ignition noise.

3.4 Noise Suppression Ignition Cable—High voltage ignition cable, which has a high impedance at radio frequencies.

3.5 Noise Suppression Ignition Cable Harness—A set of suppression ignition cables designed specifically for a given type of engine.

3.6 Distributed Ignition Noise Suppressors—An ignition cable having its suppressive element (resistive or reactive) distributed throughout its length.

3.7 Lumped Ignition Noise Suppressor—An ignition noise suppressor containing only discrete elements.

3.8 Spark Plug Ignition Noise Suppressor—A lumped suppression component designed for direct connection to a spark plug.

3.9 Sleeve Type Ignition Noise Suppressor—A lumped suppression component designed for insertion in series in a high voltage ignition cable.

3.10 Distributor Ignition Noise Suppressor—A lumped suppression component designed for direct connection to the high voltage terminals of a distributor cap.

3.11 Noise Suppression Spark Plug—A spark plug with a built-in noise suppressive element.

3.12 Noise Suppression Distributor Rotor—A rotor of an ignition distributor with a built-in suppressive element.

3.13 Resistive Distributor Brush—A resistive pickup brush in an ignition distributor cap.

3.14 Device—A machine equipped with an internal combustion engine but not self-propelled. Devices include, but are not limited to, chain saws, irrigation pumps, and air compressors.

3.15 Impulse—An impulse is a noise transient having a frequency spectrum that is instantaneously uniform over a specified frequency band.

3.16 Impulse Bandwidth—The maximum value of the output response envelope divided by the spectrum amplitude of an applied impulse.

3.17 Impulse Electric Field Strength—The root-mean-square value of the sinusoidally varying radiated electric field producing the same peak response in a bandpass system, antenna and bandpass filter, produced by the unknown impulse electric field.

3.18 Shall—Used to express a command. Conformance with the specific recommendation is mandatory and deviation is not permitted. The use of shall is not qualified by the fact that compliance with the document is considered voluntary.

3.19 Vehicle—A self-propelled machine (excluding aircraft and rail vehicles and boats over 10 m in length). Vehicles may be propelled by an internal combustion engine, electrical means, or both. Vehicles include, but are not limited to, automobiles, trucks, agricultural tractors, snowmobiles, mopeds, and small motorboats.

3.20 Characteristic Level—The maximum of the measured value shall be taken as the characteristic level at each measuring frequency.

NOTE: A characteristic level is used for the purpose of comparison with the recommended level. The characteristic level for each band shall be the maximum measurement obtained for that band for both polarizations and for all measurement positions of the vehicle or device as specified in Section 5 and shall be compared to the level at the arithmetic mean frequency of the band. Known ambient carriers shall be ignored in determining characteristic levels.

3.21 Tracking Generator—A narrowband radio frequency source synchronized to the instantaneous receive frequency of a scanning receiver or spectrum analyzer.

4. Limits of Interference—The limits for radiation are in the table and shown graphically in Figure 1. Only one of the bandwidths listed need to be chosen for testing. For more accurate determination, the table given in Figure 1 shall be used:

NOTES:
1. Limits for vehicles equipped with electric propulsion motors are under consideration.
2. For peak type measurements the limits given above may be related to bandwidths other than 1 KHz by adding a correction factor of 20 log bandwidth (kHz)/1 kHz]. For example, to relate the limit to 120 kHz bandwidth, the correction is 20 log (120 kHz/1 kHz) = 42 dB.
3. The correlation factor between quasi-peak and peak measurements is +20 dB at 120 kHz bandwidth.

5. Methods of Measurement

5.1 Measuring Apparatus Requirements

5.1.1 RECEIVER—Scanning receivers, which meet the requirements of ANSI C63.2 or C.I.S.P.R. Publication 16, are satisfactory for measurement of ignition noise. Either the peak or the quasi-peak detector may be used. Manual or automatic frequency scanning may be used.

5.1.1.1 *Detector Function*—The measuring instrument shall be a receiver capable of detecting the peak of the envelope of the response of a bandpass filter to an impulse type signal over the specified frequency range.

5.1.1.2 *Indicating Device*—The measuring instrument shall have an indicating device (meter, numerical display, graphical display, etc.) to determine the peak response as defined in Appendix G. This peak response is a function of the effective bandwidth of the measuring system (usually the impulse bandwidth of the tuned IF circuits of the measuring instrument). Hence, the effects of bandwidth shall be included in the calibration of the indicating device (see Appendix G). The indicating device readings shall be dB above 1 µV/kHz. The impulse bandwidth of the measuring instrument shall not exceed 10% of the frequency at which measurements are made.

5.1.2 ANTENNA TYPES

5.1.2.1 *Reference Antenna*—The reference antenna shall be a balanced half wave resonant dipole (see C.I.S.P.R. Publication 16).

5.1.2.2 *Broadband Antennas*—Although linearly polarized antennas are recommended, any receiving antenna is permitted, provided that it can be normalized to the reference antenna.

A broadband antenna is required when making measurements with an automated receiving system using a scanning receiver. Such a broadband antenna is usable for measuring radiation levels (over the frequency spectrum covered by this document), provided that its output can be normalized to the output of the reference antenna in the actual test environment at the actual test site.

NOTE: When broadband antennas are used, they shall meet the requirements for complex antennas given in C.I.S.P.R. Publication 16. Examples of factors to be considered include:
a. The effective aperture area of the antenna, including its polar response (horizontal and vertical planes).
b. The effect of a phase center, which moves with frequency.
c. The effect of ground reflection characteristics (including multiple ray reflections, which may arise at specific frequencies at about 500 MHz vertical polarization and 900 MHz horizontal polarization).

5.1.3 SCANNING PLOTTERS—The sine wave signal frequency response at 1.3 cm (0.5 in) peak-to-peak shall not be down by more than 3 dB at 10 Hz from the 1 Hz response.

5.1.4 ACCURACY—The measuring system, excluding source, shall be able to measure impulse electric field strength over the frequency range 30 to 1000 MHz with a maximum uncertainty of ±5 dB. The frequency uncertainty shall be less than ±3%.

NOTE: To insure that the measurements defined in this document are within the stated tolerances, consideration should be given to all pertinent measuring equipment characteristics such as: frequency and amplitude stability, image rejection, cross-modulation, overload levels, selectivity, time constants, and signal/noise ratio as well as those affecting antenna and lead-in cable.

5.1.5 REPEATABILITY—The repeatability of the measurement sys-

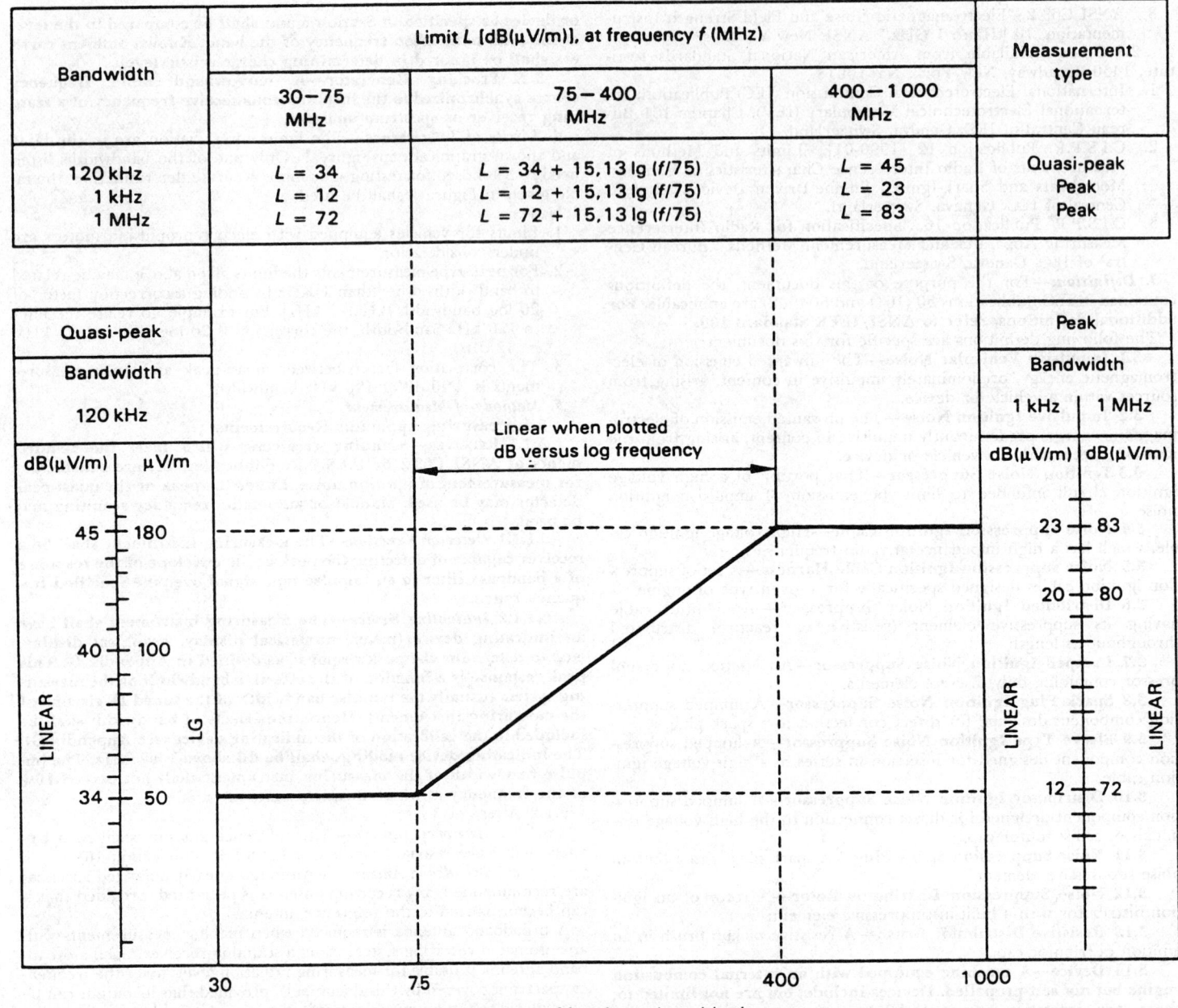

FIG. 1—LIMITS OF INTERFERENCE

tem shall be established and periodically checked to detect variability; the input/output characteristics of the measuring instrument shall be checked at shorter intervals of time.

NOTE: In view of the variations in ground conductivity and other factors that influence repeatability, it is reasonable to expect a standard deviation not to exceed 3 dB in the measurements made of an impulse electric field (see A.15) within the range of 30 to 1000 MHz.

5.1.6 CALIBRATION—See Appendix A.

5.2 Measuring Location Requirements

5.2.1 SITE REQUIREMENTS

5.2.1.1 The test site shall be a clear area free from electromagnetic reflecting surfaces within a circle of minimum radius 30 m measured from a point midway between the vehicle or device and the antenna.

5.2.1.2 The measuring set, test hut, or vehicle in which the measurement set is located may be within the test site, but only in the permitted region indicated by the crosshatched area of Figure 2.

NOTES:

1. The site requirements defined in 5.2.1 are the application of C.I.S.P.R. Publication 16 to large automotive objects.

2. Anechoic chambers may be used provided that the results obtained can be correlated with those obtained using the outdoor sites described. Such chambers have the advantages of all weather testing, controlled environment, and improved repeatability because of stable chamber electrical characteristics.

5.2.2 ANTENNA POSITION REQUIREMENTS—At each measuring frequency, measurements shall be taken for horizontal and vertical polarization (see Figures 3 and 4).

5.2.2.1 *Height*—The center of the antenna shall be 3.00 m ± 0.05 m above the ground or water surface.

5.2.2.2 *Distance*—The horizontal distance of the antenna to the nearest metal part of the vehicle or device shall be 10.0 m ± 0.2 m.

5.2.3 AMBIENT REQUIREMENTS—To ensure that there is no extraneous noise or signal of a magnitude sufficient to affect materially the measurement, measurements shall be taken before and after the main test, but without the engine under test running. In both of these measurements, the extraneous noise or signal shall be at least 10 dB below the limits of interference given in Section 4.

5.2.4 AUXILIARY ANTENNA—For simultaneous left and right measurement in automated test systems or for reference antenna purposes, the auxiliary antenna shall be located on the other side of the vehicle or device symmetrically opposite the antenna.

To minimize possible interaction when active, the antennas shall

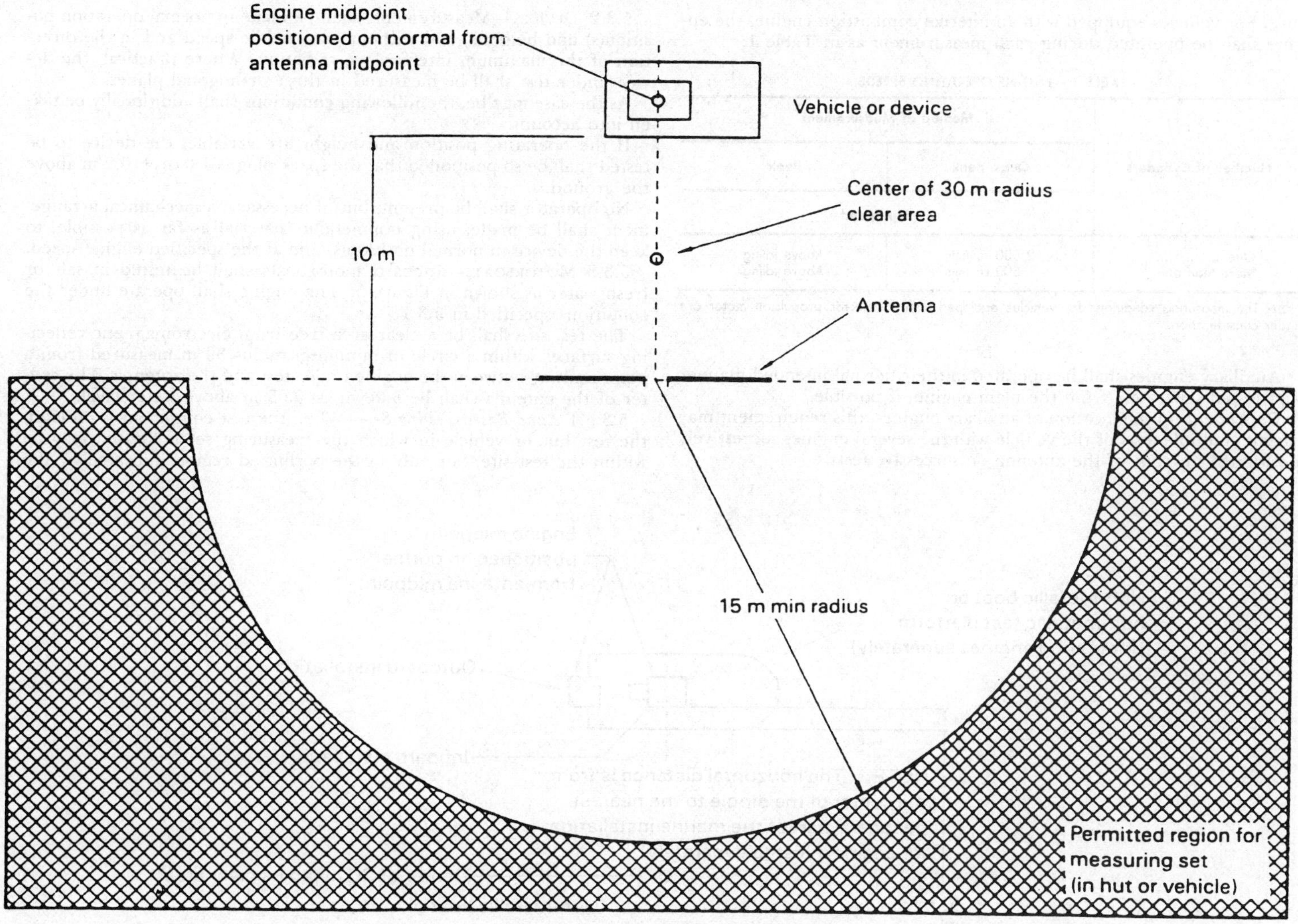

FIG. 2—MEASURING SITE *All dimensions ± 0.2 m*

be operated in opposite polarization modes to each other. For example, when making a simultaneous recording, one antenna would be in the horizontal mode while the other would be in the vertical mode and vice versa.

5.3 Test Object Conditions—Measurements shall not be made while rain is falling on a vehicle or boat, nor within 10 min after the rain has stopped. For outboard engines and devices, all surfaces other than those normally in contact with water shall be dry.

NOTE: Dew or light moisture may seriously affect readings obtained on devices having plastic enclosures.

5.3.1 VEHICLES—Measurements shall be made on the left and right sides of the vehicle (see Figures 3 and 4).

Only the ancillary electrical equipment necessary to run the engine shall be operating. The engine shall be at normal operating tempera-

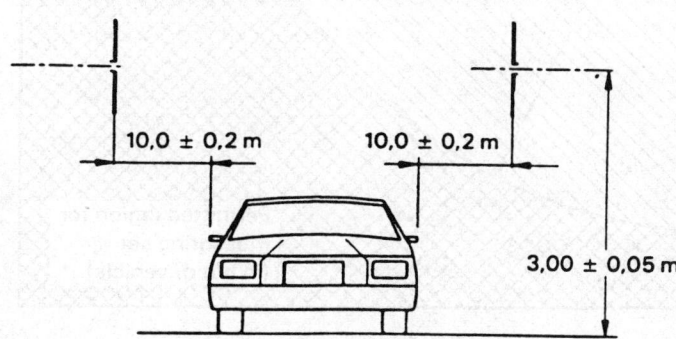

FIG. 3—ANTENNA POSITION TO MEASURE RADIATION: VERTICAL POLARIZATION

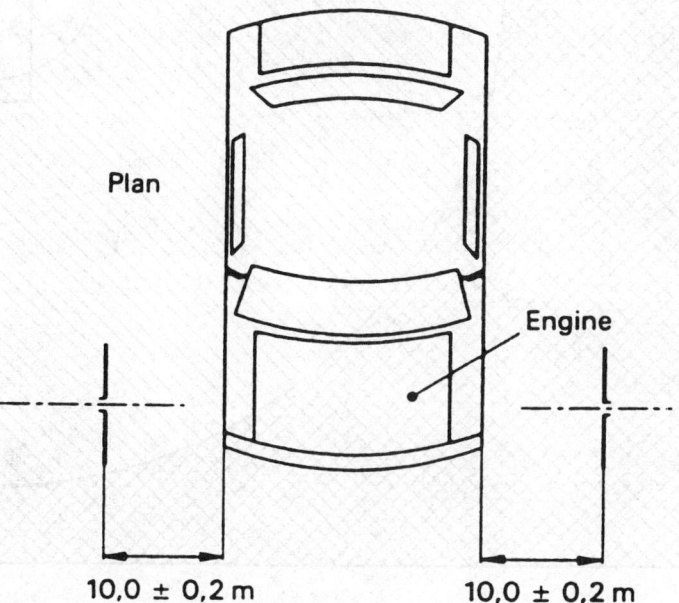

FIG. 4—ANTENNA POSITION TO MEASURE RADIATION: HORIZONTAL POLARIZATION

ture. For vehicles equipped with an internal combustion engine, the engine shall be operated during each measurement as in Table 1:

TABLE 1—ENGINE OPERATING SPEEDS

Number of Cylinders	Method of Measurement	
	Quasi-peak	Peak
	Engine Speed	
One	2 500 rev/min	Above idling
More than one	1 500 rev/min	Above idling

NOTE: The measuring conditions for vehicles equipped with an electric propulsion motor are under consideration.

Auxiliary engines shall be operated in their normal intended manner and tested separately from the main engine, if possible.

Dependent on the location of auxiliary engines, this requirement may dictate multiple tests of the vehicle with the several engines successively positioned in front of the antenna on successive tests.

5.3.2 DEVICES—Measurements shall be made in normal operation position(s) and height(s) and without load at idle speed and in the direction of the maximum interference radiation. Where practical, the device under test shall be measured in three orthogonal planes.

As the case may be, the following conditions shall additionally be taken into account:

If the operating position and height are variable, the device to be tested shall be so positioned that the spark plug is 1.0 m + 0.2 m above the ground.

No operator shall be present, but, if necessary, a mechanical arrangement shall be made, using nonmetallic material as far as possible, to keep the device in normal position(s) and at the specified engine speed.

5.3.3 MOTORBOATS—Inboard motorboats shall be tested in salt or fresh water as shown in Figure 5. The engine shall operate under the conditions specified in 5.3.1.

The test site shall be a clear area free from electromagnetic reflecting surfaces within a circle of minimum radius 30 m measured from a point midway between the engine under test and the antenna. The center of the antenna shall be 3.00 m ± 0.05 m above water level.

5.3.3.1 *Land-Based Testing Set*—When the test equipment is on land, the test hut or vehicle in which the measuring set is located may be within the test site, but only in the permitted region indicated by the

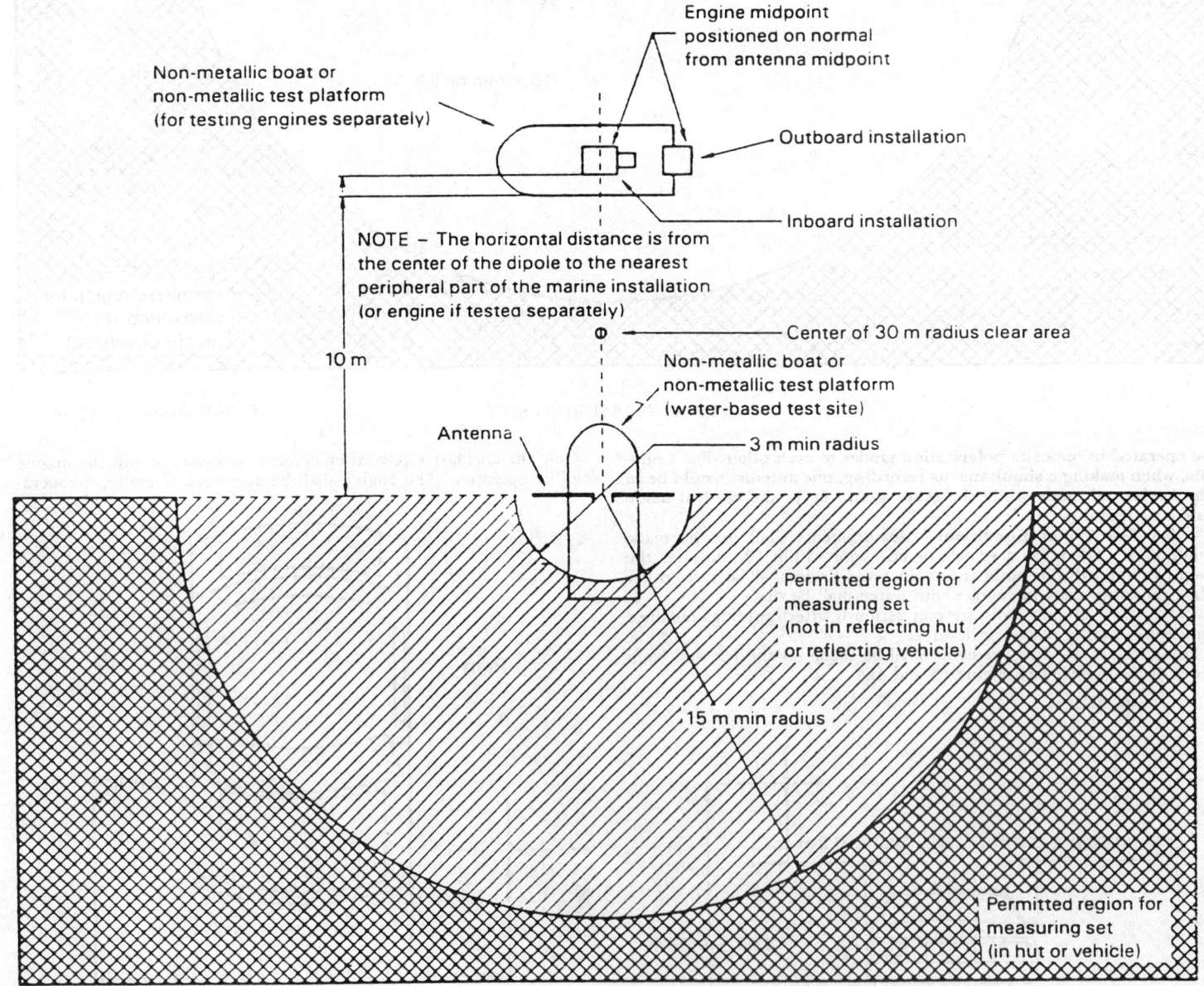

All dimensions ± 0,2 m

FIG. 5—MOTORBOAT MEASURING SITE

crosshatched area of Figure 5. If the measuring set is not in a hut or vehicle, it may be located within the test site in either the shaded or crosshatched area of Figure 5.

5.3.3.2 *Water-Based Testing Set*—The measuring set shall be installed in a nonmetallic boat or nonmetallic test fixture, which may be within the test site, but only within the permitted region indicated by the shaded area of Figure 5.

When tested separately, inboard, stern drive, and outboard engines shall be attached to a nonmetallic board or nonmetallic test fixture and tested in a similar way to that prescribed for inboard motorboats.

5.3.4 ELECTRIC VEHICLES—Electric vehicles shall be tested in accordance with Appendix H.

5.3.5 MOISTURE—During testing, all surfaces, other than those normally in contact with water, shall be dry. Immersible devices shall be tested at their normal operating depth.

NOTE: Dew or light moisture may seriously affect readings obtained on devices incorporating plastic enclosures. Data used to determine compliance with the performance levels shall comply with the requirement of 5.3.5. Relative data may be taken on wet vehicles or devices for development or engineering purposes, but such data shall not be used to determine compliance.

5.4 Test Frequencies

5.4.1 FREQUENCY RANGE—Measurements shall be made over the frequency range of 30 to 1000 MHz. For analysis, this range shall be divided into a minimum of 14 bands with approximately three bands in each octave (2:1 ratio) of frequency. For bands that include the frequency range of 75 to 400 MHz (that is where the recommended limit is not constant), the ratio of the highest frequency to the lowest frequency in each band shall be no greater than 1.33. Each band shall be scanned to determine the maximum radiation level for that band.

Example:

30 to 34 MHz	100 to 130 MHz	400 to 525 MHz
34 to 45	130 to 170	525 to 700
45 to 60	170 to 225	700 to 850
60 to 80	225 to 300	850 to 1000
80 to 100	300 to 400	

5.4.2 OBSERVATION TIME

5.4.2.1 *Peak Detected Measurements*—An analog swept instrument shall scan at a rate not to exceed: ten times the instrument IF bandwidth per second. An incrementally stepped instrument shall scan at an equivalent rate, taking into account the settling time and the control time of the instrument and its control system.

5.4.2.2 *Quasi-Peak Detected Measurements*—The time at each frequency shall not be less than the settling time plus the control time of the instrument and its control system, or 1 s, whichever is greater.

5.5 Expression of Results—The results of measurement shall be expressed in microvolts per meter for 120 kHz bandwidth. For statistical evaluation, the logarithmic unit dB (μV/m) shall be used. If the actual bandwidth (expressed in kilohertz) of the measuring apparatus is just outside 120 kHz for certain frequencies, the results measured shall be related to the 120 kHz bandwidth by applying the factor 120/bandwidth.

5.5.1 CHARACTERISTIC LEVEL—The maximum of the measured values shall be taken as the characteristic level at each measuring frequency.

6. *Methods of Checking for Compliance with C.I.S.P.R. Requirements*

6.1 Type Test

6.1.1 Compliance with the requirements given in Section 4 shall be checked as follows:

6.1.1.1 Measurements may be made on a prototype vehicle or a prototype device of a later production series. The results shall be at least 2 dB below the limits specified in Section 4.

6.1.1.2 Five or more additional samples may be tested and the results combined with the first test and evaluated statistically as defined in Appendix B; the result shall be below the specified limits of Section 4.

6.1.2 Some differences in the construction of vehicles or devices are unlikely to have a significant effect on the ignition noise radiation. For road vehicles, examples of such differences are given in Appendix C.

NOTE: For vehicles with spark-ignited internal combustion engines already in service and not yet equipped with ignition noise suppressors, suppression methods as shown in Appendix D are suggested. These methods can be expected to give effective compliance with C.I.S.P.R. requirements in the majority of cases. For devices, similar suppression methods are suggested.

6.2 Surveillance of Series Production

6.2.1 The results of the measurements on one vehicle or device may be 2 dB above the specified limits of Section 4.

6.2.2 Five or more additional samples may be tested and the results combined with the first test and evaluated statistically as defined in Appendix B; the overall result shall be below the specified limits of Section 4.

APPENDIX A

ANTENNA AND TRANSMISSION LINE CALIBRATION (C.I.S.P.R. REPORT 56)

Foreword—This appendix contains, for guidance, an example of an antenna and transmission line calibration procedure that complies with the intent of 5.1.2. Proper antenna and transmission line calibration is essential to account for transmission line loss and mismatch errors, and to characterize a broadband antenna, if used.

This report is intended to be tutorial in nature, as an aid for those who may not be familiar with antenna and transmission line calibration. Other methods, such as those using tracking generators, network analyzers, or narrowband signal sources may be equally satisfactory and nothing in this appendix should be interpreted as precluding their use.

A1. Impulse Electric Field Strength—Impulse electric field strength shall be expressed in units of decibels above 1 μV/m. The relationship expressing impulse electric field strength to the measurement system is:

$$F = R + AF + T \qquad (Eq.A1)$$

where:

F = Impulse electric field strength dBμV/m
R = Instrument reading dBμV
AF = Antenna factor, defined in A3 or A4
T = Transmission line factor, defined in A7

A2. Reference Antenna—The reference antenna for these measurements is the balanced half wavelength resonant dipole tuned to the measurement frequency. The reference point is the center of the two dipole elements.

A3. Antenna Factor—The factor relating the field strength to the loaded antenna terminal voltage[3] at the reference point of the antenna is called the antenna factor, designated AF, expressed in dB. The antenna factor shall include the effects of baluns, impedance matching devices, any mismatch losses, and operation off the resonant frequency of the antenna.

NOTE: This factor is a function of frequency and is usually provided by manufacturers of resonant dipoles. Knowledge of the antenna factor for free space operation for resonant dipoles is sufficiently accurate for purposes of this report. Greater accuracy can be obtained by knowing the antenna factor for the particular resonant dipole being used in the test environment. A method for determining antenna factor is described in SAE ARP958 (see Section 2).

A4. Alternate Antennas—The antenna factor for the alternate antenna is the antenna factor for the reference antenna (resonant dipole) minus the gain (dB) of the alternate antenna relative to the reference antenna.

A5. Antenna Support Structure—Electrical interaction between the antenna elements and the antenna support/guy system shall be avoided.

A6. Auxiliary Antenna—For simultaneous left and right measurement in automatic test systems or for calibration antenna purposes, the auxiliary antenna shall be located symmetrically opposite the antenna.

NOTE: To minimize possible interaction, the antennas **shall** be operated in opposite polarization modes to each other. Example: For simultaneous recording, one antenna would be in the horizontal mode while the other would be in the vertical mode and vice versa.

A7. Transmission Line—The transmission line factor (loss) as a function of frequency shall be known. The factor is designated T and is:

$$T = 20 \log\left(\frac{\text{input voltage}}{\text{output voltage}}\right) dB \qquad (Eq.A2)$$

NOTE: It is recommended that the transmission line be double braided or solid shielded coaxial cable to achieve proper shielding. It is preferable that transmission line loss and mismatch errors be accounted for by including the cable in the measuring instrument calibration. When this is done, T is dropped from the equation for F in A1.

Theoretical considerations of antenna and transmission line geometry demand that the transmission line not interact electrically with the antenna elements. One acceptable transmission line geometry for dipole antenna is to route the transmission line horizontally rearward for

---

[3] As this is a voltage ratio, calculations to convert to decibels should be made using the factor of 20.

a distance of 6 m at a height of 3 m before descending to ground level or below. Other geometries are acceptable if they can be shown not to affect the measurements, or if the effects can be included in equipment calibration.

A8. Reference Impulse Generator—The impulse generator output level (dB above 1 μV/unit bandwidth) shall be known to within ±1.0 dB.

NOTE: For convenience of testing, the reference instrument should be broadband impulse generator capable of producing a uniform spectrum to within ±3.0 dB in the frequency range 30 to 1000 MHz.

A9. Calibration Antenna—The prime function of the calibration instrument is to provide a repeatable RF field for the comparison of an alternate antenna to the dipole antenna. For ease in measurement and to assure freedom from variation caused by antenna adjustment, it is recommended that broadband antennas be used. Typical antennas are the biconical for up to 200 MHz, and the conical logarithmic spiral for 200 to 1000 MHz.

A10. Alternate Antenna Factor Determination—If an alternate antenna (see A4) is used, the antenna factor shall be determined by a substitution technique in the intended test environment. The reference shall be the dipole (A2). The radiated field to be measured for the substitution technique is generated by the calibration antenna and the impulse generator as specified in A8 and A9.

A10.1 Use of a Tracking Generator for Alternate Antenna Factor Determination—A tracking generator may be used in place of the impulse generator in A10 provided the output of the tracking generator is known to within ±1.0 dB and the calibration electric field causes the measured field strength to be at least 6 dB above the least measurable field strength of the measuring instrument (see A12).

A11. Test Geometry—The alternate antenna shall be located at its intended test position. When substitution occurs, the dipole shall be placed so that its reference point is at the same place that the reference point for the alternate antenna normally occupies. The calibration antenna shall be 10 m in horizontal distance from the alternate antenna reference point in Figure A1 (taking the place of the nearest vehicle periphery) and shall be 1 m high.

A12. Reference Impulse Electric Field Amplitude— For accurate measurements to be made, the calibration impulse electric field shall be at least 3 dB above the least measurable field of the measuring system. A value of at least 10 dB is preferred.

NOTE: Experience indicates that an impulse generator that meets A8 and has a nominal 100 dBμV/kHz level can produce a field of approximately 10 dBμV/m/kHz at the receiving antenna when an impedance matching attenuator of 10 dB is used at the output of the generator. This field strength varies depending on calibration antenna losses and radiation characteristics and on propagation anomalies. This approximate value is provided so that the antenna factor determination can be performed. It is then possible to estimate the required sensitivities and the tolerable losses in the measuring system.

A13. Test Procedure—The procedure to be used is to measure the reference field with the reference antenna positioned as in A11 to obtain a meter reading (usually voltage). Then the alternate antenna is substituted and a second reading is taken.

The antenna factor for the alternate antenna is calculated as discussed in A4. This procedure should be conducted for both horizontal and vertical polarizations to determine whether different antenna factors are required for each of the two cases.

NOTE: The antenna factor of the reference antenna may be assumed to be the same for both cases.

A14. Frequencies—The number of frequencies at which antenna factor values are required depends on the alternate antenna being evaluated. A sufficiently large number of frequencies shall be considered to describe the function adequately.

A15. Complete System Verification—The complete measurement system comprised of antenna, transmission cable, measuring instrument, and readout devices shall be verified by measuring an impulse electric field established with a wideband impulse generator and antenna(s) described in A8 and A9. This verification shall be made on a periodic basis so that any change in system performance can be detected (see Figure A1).

APPENDIX B

STATISTICAL ANALYSIS OF THE RESULTS OF MEASUREMENTS

B1. The following condition shall be fulfilled in order to ensure, with an 80% degree of confidence, that 80% of mass-produced vehicles/devices conform to a specified limit L:

$$\bar{x} + kS_n \leq L \quad \text{(Eq.B1)}$$

where:
$\bar{x}$ = arithmetical mean of the results on n vehicles/devices
k = statistical factor dependent on n, as given by Table B1:

TABLE B1

n = 6	7	8	9	10	11	12
k = 1,42	1,35	1,30	1,27	1,24	1,21	1,20

S_n = standard deviation of results on n assembly line units
$S_n^2 = \Sigma (x - \bar{x})^2/(n-1)$
x = individual result
L = specified limit
$S_n, x, \bar{x}$ and L are expressed in dBμV/m

If a first sample of n vehicles/devices does not meet the specifications, a second sample of n vehicles/devices shall be tested and all results assessed as coming from a sample of 2n vehicles/devices.

APPENDIX C

CONSTRUCTION FEATURES OF MOTOR VEHICLES AFFECTING THE RADIATION OF IGNITION NOISE
(C.I.S.P.R. REPORT 65)

Foreword—For guidance in testing and approval, it should be noted that some differences in vehicle construction are unlikely to have a significant effect on the ignition noise radiation. For this reason, measurements on one variant may be considered as being typical and such a variant may be used as the basis for the assessment of the design characteristics of road vehicles insofar as they affect the ignition noise radiation.

C1. The following construction differences[4] have little effect on ignition noise radiation:
a. Two-door or four-door vehicles or station wagons of similar overall length.
b. Differences in radiator grille construction, provided that grilles are of metal, offer approximately the same proportion of clear opening and have approximately the same mounting.
c. Shape of fenders or contour of hood/bonnet.
d. Different size wheels or tires.
e. Ordinary nonresistive spark plugs of different makes, provided they have equivalent electrical characteristics (capacitance, inductance, resistance).
f. Coils and distributors of different makes, provided they have equivalent electrical characteristics (capacitance, inductance, resistance).
g. Decorative ornamentations, heaters or air-conditioners, occupying the same location.
h. Ordinary resistive spark plugs of different heat ranges, provided they have equivalent electrical characteristics (capacitance, inductance, resistance).

C2. The following construction differences[5] can be expected to have a significant effect on ignition noise radiation:
a. Significant differences in compression ratio.
b. Use of plastic or metallic fenders, roofs, or body panels.

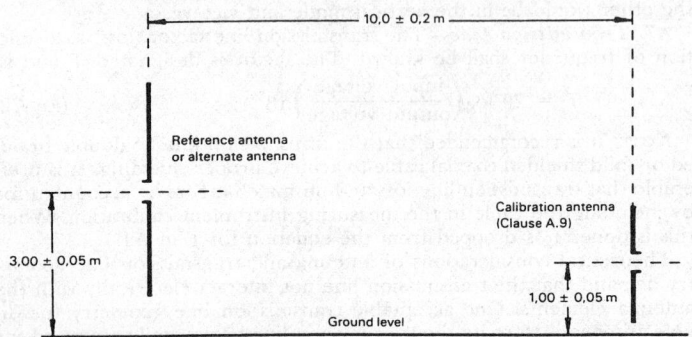

FIG. A1—ALTERNATE ANTENNA FACTOR DETERMINATION

---
[4] This is not all inclusive; it is a set of examples only.
[5] See Footnote 4.

c. Size, shape, and location of metallic air cleaners and use of plastic rather than metallic air cleaners or vice versa.
d. Location of distributor and coil on the engine or in the engine compartment.
e. Size and shape of the engine compartment and location of the high voltage harness.
f. Significant differences in the clear opening of engine compartment around the wheels.
g. Right or left hand steering as it may affect the position of the other components or parts.
h. Vehicles having auxiliary engine(s) for purposes other than propulsion.

APPENDIX D

GUIDANCE FOR NOISE SUPPRESSION EQUIPMENT
(C.I.S.P.R. REPORT 66)

D1. This appendix contains, for guidance, examples of suppression arrangements, which have been found satisfactory for many vehicles or devices.

It is not possible, however, to specify precise methods of suppression, which will be satisfactory for all types of motor vehicles (or devices) because features in the design of a vehicle or engine have a great affect on the magnitude of the noise generated or radiated. For example, the level of noise is dependent on the arrangement of the ignition components and the lengths of the connecting cables/wires. Such cables/wires should not run close to metallic body panels in which interference currents may be induced. The cables/wires should, as far as possible, follow paths close to the engine block.

In Table D1, vehicles and engines are divided into two groups for the purpose of suggesting suppression methods. Some assistance in suppression is often given by the metal body of a vehicle or a device enclosure, and more suppression may be needed where no metal body shielding exists. In all cases, final testing must be done on a complete vehicle or device (with all suppression components and/or body panels in place). Mere use of the components mentioned will not assure compliance with the limit specified in Section 4.

TABLE D1—EXAMPLES OF SUPPRESSION EQUIPMENT

	Engines with distributor	Engines without distributors
Vehicles/devices with metallic engine enclosures or special metallic ignition enclosures	A with 2 or 3 or 4 or B with 1 or 2 or 3 or 4 or C with 1 or 2 or 3 or 4 or D (all spark plug leads) with 1 or 2 or D (all leads) or E (all spark plug leads) with 1 or 2 or E (all leads)	A or B or C or D or E
Vehicles/devices without metallic engine enclosures, motor-cycles, mopeds	B with 3 or 4 or C with 3 or 4 or B with D (all leads) or B with E (all leads) or C with D (all leads) or C with E (all leads)	B or C

The letters and figures above are those shown in Figure D1. The metallic screen of screened spark plug ignition noise suppressors B must make firm electrical contact with the body of the spark plug.

APPENDIX E

MEASUREMENT OF THE INSERTION LOSS OF IGNITION NOISE SUPPRESSORS[6]
(C.I.S.P.R. REPORT 37/2)

E1. Introduction—Three methods of measurement of the insertion loss of ignition noise suppressors are used:

E1.1 "C.I.S.P.R. box method" (50/75 Ω laboraory method) described in E3.

E1.2 Model installation laboratory method ("earth current method") described in E4.

E1.3 Field comparison method. In this method, the insertion loss of the suppressor (or set of suppressors) is determined from the measurement of interference field intensity caused by the vehicle or device on the open test site. It is evaluated according to the formula:

$$A = E_1 - E_2 \qquad \text{(Eq.E1)}$$

[6] Not required by SAE, but permitted at the discretion of the user.

where: E_1 = intensity of the field caused by the ignition system without suppressors, expressed in dBμV/m
E_2 = intensity of the field caused by the same ignition system but with suppressors (or set of suppressors) expressed in dBμV/m

NOTE: Field intensity is to be measured in accordance with Section 5.

E2. Comparison of Test Methods

E2.1 C.I.S.P.R. Box Method—With the help of the "C.I.S.P.R. box method", it is possible to compare only the characteristics of single suppressors of the same kind under standard laboratory conditions. At present, this method is used in the frequency range from 30 to 300 MHz. Results obtained have no significant correlation with the efficiency of suppressors observed in practice (see E5). This method does not allow measurement of a set of suppressors consisting, for example, of four resistors and five cables with distributed attenuation. Nevertheless, it provides a means of quick control, for instance, of suppressors during manufacture after previous verification of their effectiveness in actual conditions.

E2.2 Model Installation Method—With the help of the model installation method, it is possible to compare the characteristics both of single suppressors and sets of suppressors taking into account the influence of environmental factors, for example, high voltage, more easily than in the box method. As with the C.I.S.P.R. box method, it may be used in a laboratory but results obtained have better correlation with the efficiency of suppressors observed in practice (see E5). At present, this method is used in the frequency range from 30 to 300 MHz.

E2.3 Field Comparison Method—The field comparison method may be considered the reference method since the results obtained give the insertion loss of suppressors observed in practice. It automatically takes into account all the factors influencing the insertion loss and it has no limitations in frequency range. Its main disadvantage is the need to perform measurements on an open test site (or in a large building of special construction) and the need to test the complete vehicle or device.

E2.4 Summary—Assessing the cost of instrumentation, additional equipment involved and time consumed in each method of measurement, it can be stated that the field comparison method is the most expensive, the box method and model installation method being much cheaper. The field comparison method is, however, to be considered the reference method. The remaining methods may be used only for guidance in design or for quality control of individual parts during production processing.

E3. C.I.S.P.R. Box Method (50/75 Ω Laboratory Method of Measurements of Insertion Loss of Ignition Noise Suppressors)

E3.1 General Conditions and Limitations of Measurement—The insertion loss of an ignition noise suppressor is measured with the test circuit shown in Figure E1. This method is intended to be used only as a comparative method for suppression devices of the same type and is not intended to give direct correlation with radiation measurements. The word "type" is understood to mean all suppression devices belonging to the same case of Figure D1.

E3.2 Test Procedure—In Figure E1 the coaxial switches (2) are adjusted so that the signal from the signal generator (1) is passed through the test box (4) and the specimen under test (5) giving an indication on the output indicator of the measuring instrument (7). Fixed "T" attenuators (3) have a loss of 10 dB.

The coaxial switches (2) are then turned so that the signal passes through the calibrated variable attenuator (6), which is adjusted to give the same indication on the output indicator of the measuring instrument (7). The insertion loss of the ignition noise suppressor is then given by the attenuation read on the calibrated variable attenuator (6) minus the attenuation of the fixed attenuators (3).

E3.3 Test Box Construction—Details of the usual test box are shown in Figures E2 to E4. For the majority of applications, this box is applicable; however, hole positions and box size may require modification for some applications. The arrangement of the suppressors in the test box is shown in Figures E5 to E11. All noncoaxial connecting leads within the C.I.S.P.R. box to the suppressors under measurement shall be kept as short as possible, or of specified length where shown. In all arrangements the spark plug is modified to accept a coaxial input and is constructed from a standard spark plug assembly having a direct connection between the spark plug terminal and the central electrode.

E3.4 Results—For ignition noise suppressors having a high impedance, the insertion loss a_1 in a circuit having a characteristic impedance z_1 can be converted to the insertion loss a_2 in a circuit having a characteristic impedance z_2; the following formula applies:

$$a_2 = a_1 + 20 \log (z_1/z_2). \qquad \text{(Eq.E2)}$$

23.198

A		Spark plug ignition noise suppressor
B		Screened spark plug ignition noise suppressor
C		Suppression spark plug
D		Resistive cable
E		Reactive cable
1		Distributor cap with inbuilt central resistor (resistive brush or distributor ignition noise suppressor or built-in resistor)
2		Suppression distributor rotor
3		Distributor cap with inbuilt resistors in the distributor cap outlets or in the cables near the distributor cap. The central resistor may be a resistive brush
4		Distributor cap with suppression distributor rotor and resistors in all spark plug outlets

FIG. D1—SUPPRESSION EQUIPMENT

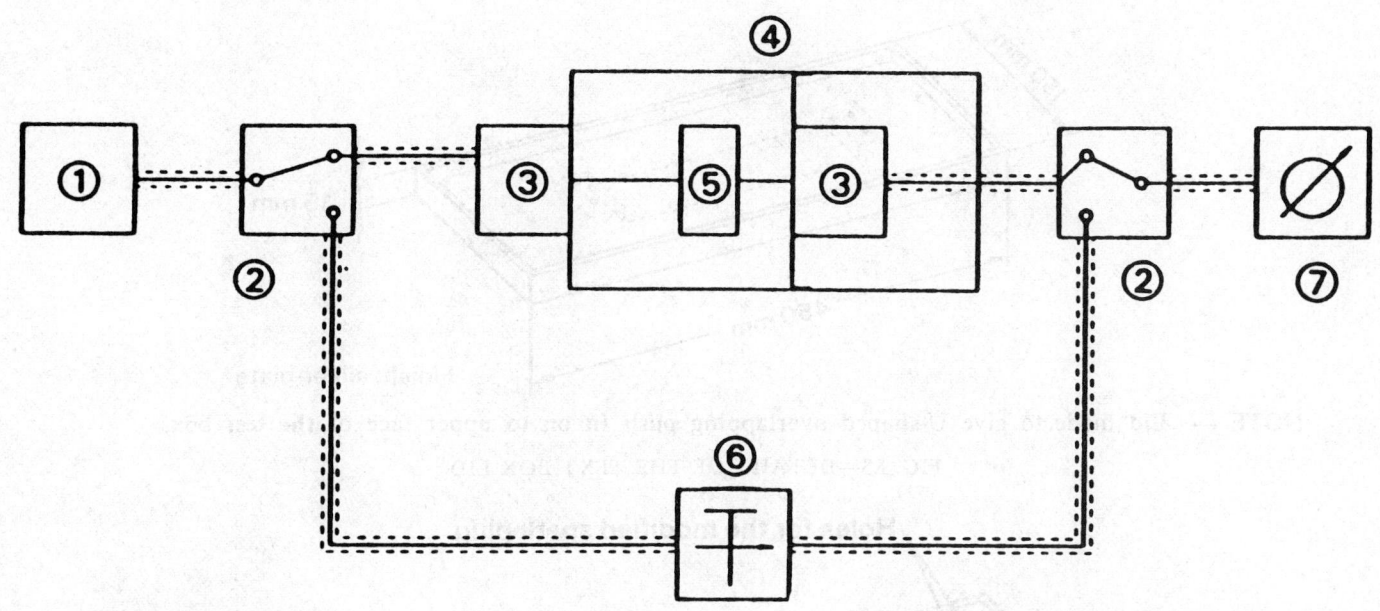

① signal generator
② coaxial switch
③ fixed "T" attenuator (10 dB)
④ test box
⑤ specimen under test
⑥ calibrated variable attenuator
⑦ measuring instrument

NOTE — Items ①, ②, ③, ⑥ and ⑦ must have the same characteristic impedance.

FIG. E1—TEST CIRCUIT

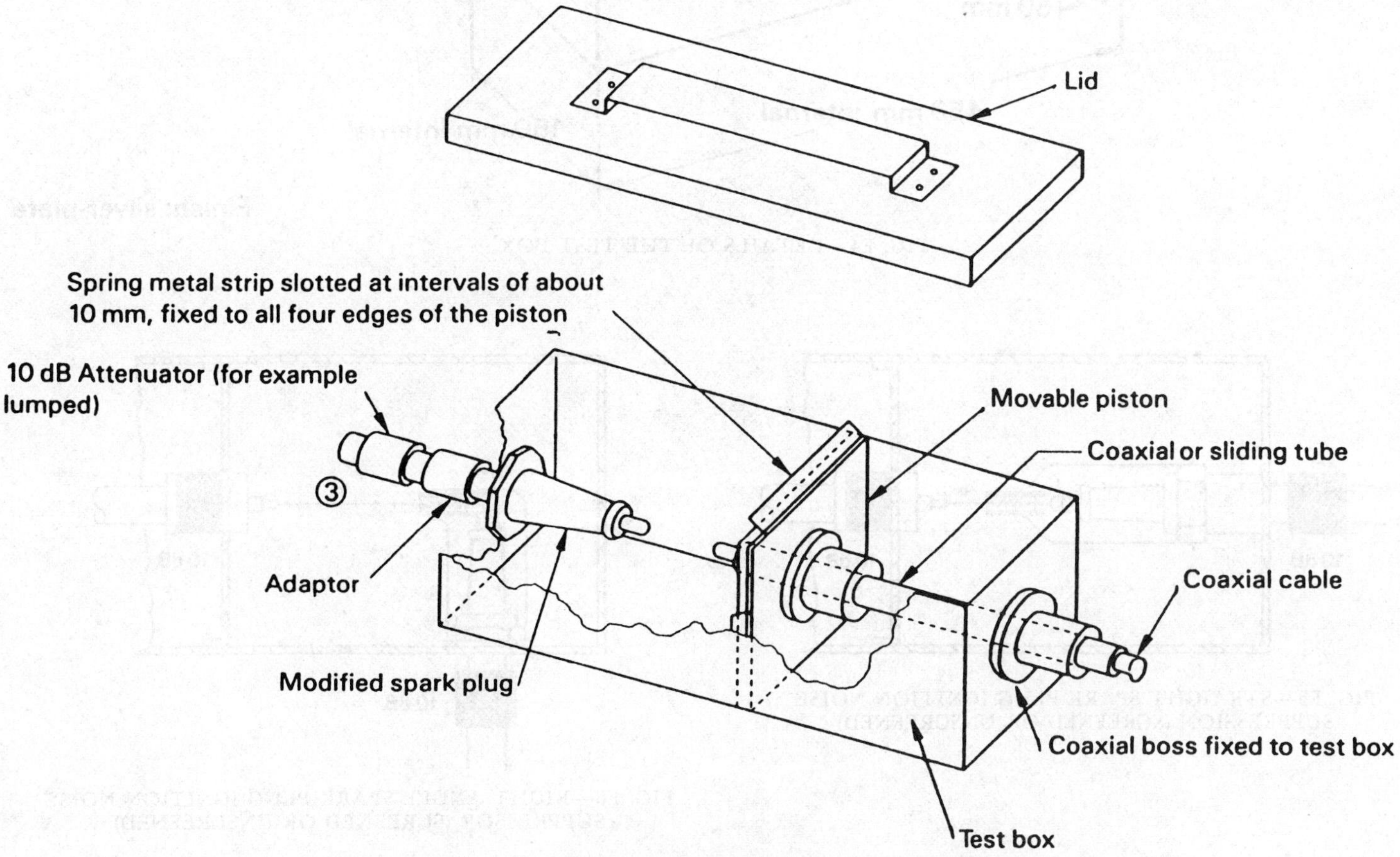

FIG. E2—GENERAL ARRANGEMENT OF THE TEST BOX

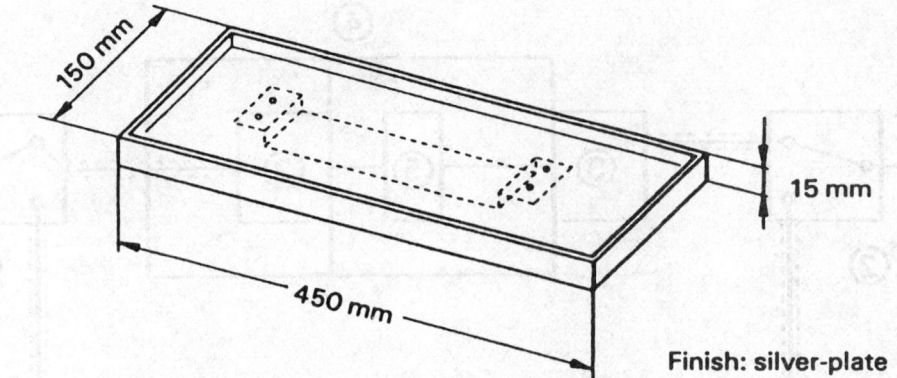

NOTE — Lid made to give U-shaped overlapping push fit on to upper face of the test box.

FIG. E3—DETAILS OF THE TEXT BOX LID

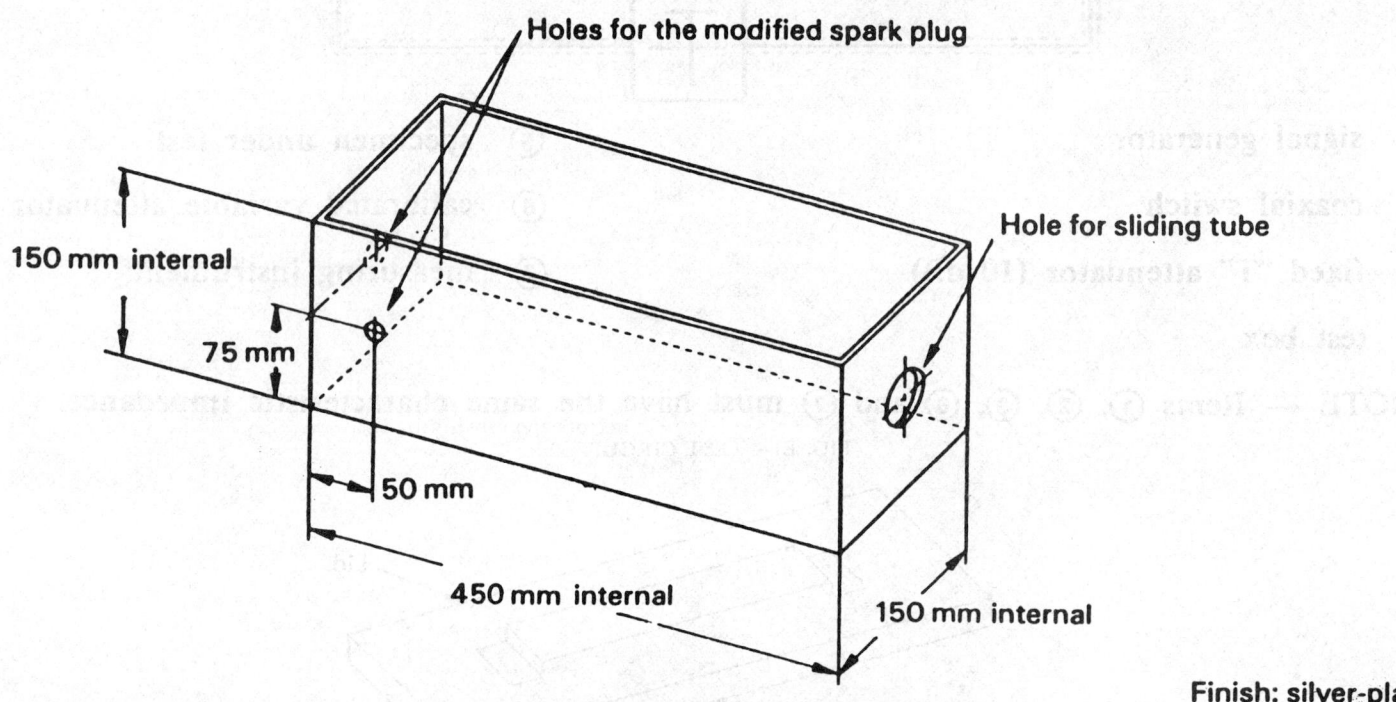

FIG. E4—DETAILS OF THE TEST BOX

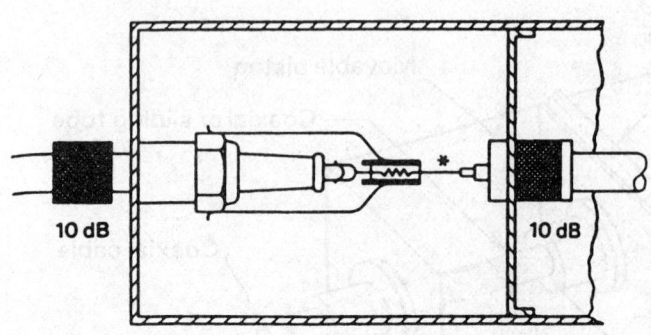

FIG. E5—STRAIGHT SPARK PLUG IGNITION NOISE SUPPRESSION (SCREENED OR UNSCREENED)

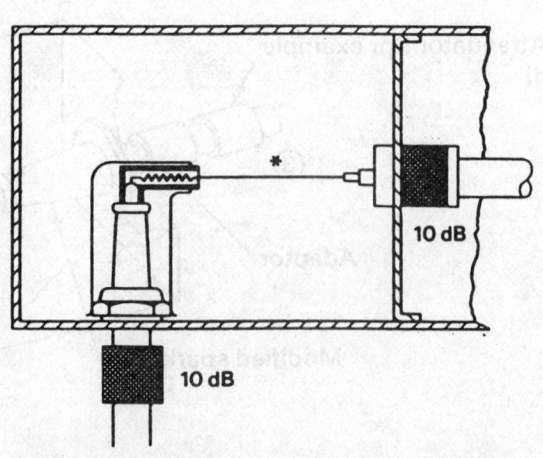

FIG. E6—RIGHT ANGLE SPARK PLUG IGNITION NOISE SUPPRESSOR (SCREENED OR UNSCREENED)

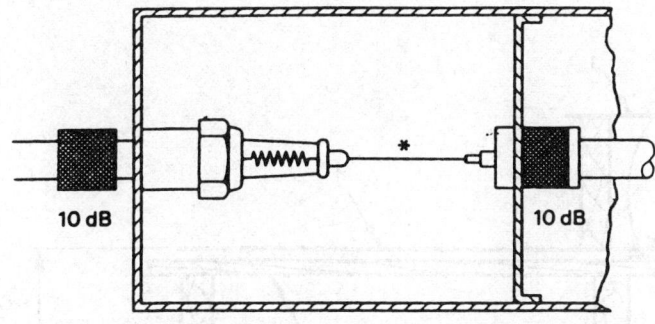

FIG. E7—NOISE SUPPRESSOR SPARK PLUG

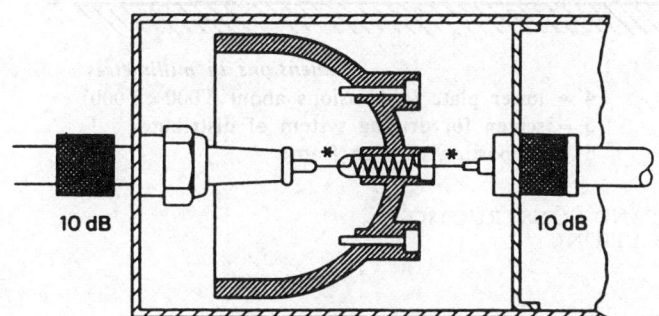

FIG. E8—RESISTIVE DISTRIBUTOR BRUSH

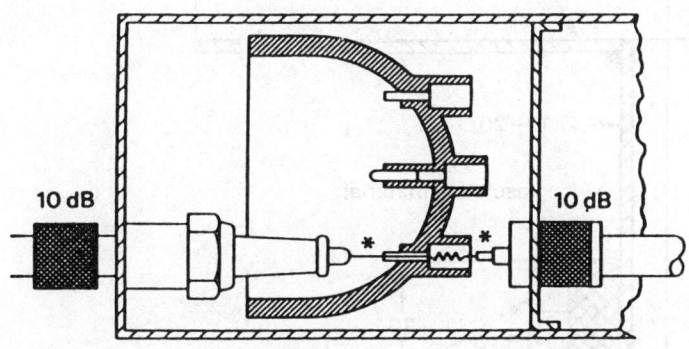

FIG. E9—NOISE SUPPRESSOR IN DISTRIBUTOR CAP

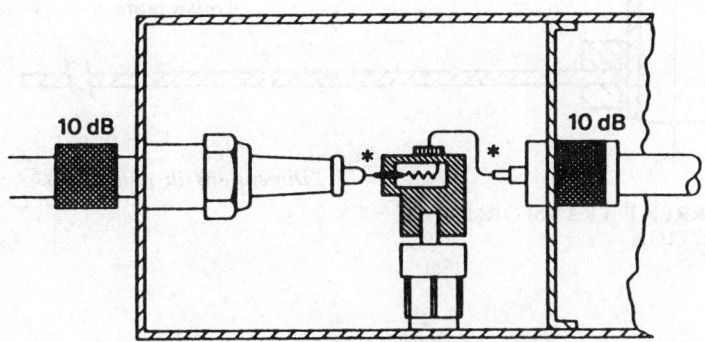

FIG. E10—NOISE SUPPRESSION DISTRIBUTOR ROTOR

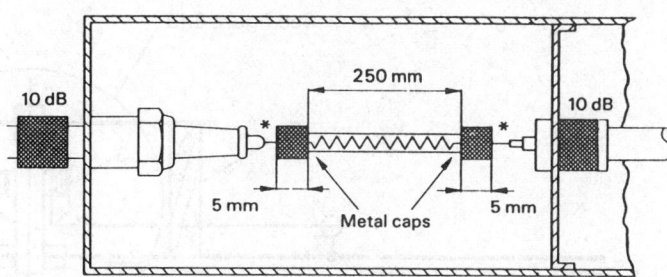

FIG. E11—NOISE SUPPRESSION IGNITION CABLE (RESISTIVE OR REACTIVE)

* All connecting leads to suppressors under measurement to be kept as short as possible or of specified length where shown.

E4. Model Installation Laboratory Method of Measurement of Insertion Loss of Ignition Noise Suppressors (Earth Current Method)

E4.1 General Conditions of Measurement—Measurements of a suppressor (or suppressors) are performed during its operation in a model installation of an ignition system of the type in which the suppressor shall be applied in practice.

The voltage proportional to the total earth current, induced by the interference field radiated by the installation, is measured.

E4.2 Test Setup—An example of the test stand construction for the frequency range 30 to 300 MHz is shown in Figures E12 and E13. The mounting method of a model installation is shown in Figure E14.

NOTE: Dimensions, materials, and construction shown in the figures are not critical, but they shall be chosen so that in the frequency range of interest, the test stand does not have self-resonances. The method of checking the test stand for the absence of resonances is given in E4.4.

E4.3 Test Procedure—Tests are performed in two stages:
 a. In the first stage, the voltage in the model installation without suppressors is measured. (Resistors are replaced by shorting connectors and cables with distributed impedance by ordinary non-lossy cables.
 b. In the second stage, the voltage is measured in the same model installation but with suppressors.

The insertion loss of the suppressors is determined according to the formula:

$$A = V_1 - V_2 \qquad (Eq.E3)$$

where: V_1 = interference voltage measured in the model installation without suppressors expressed in dBµV.
  V_2 = interference voltage measured in the model installation with suppressors investigated, expressed in dBµV.

NOTE: During the measurements in both stages, conditions must be the same in respect of:
1. Length and geometry of the ignition cables.
2. Distributor speed.
3. Pressure in the pressure box with the spark plugs.

E4.4 Method of Checking the Test Stand for Absence of Self-Resonances—Resonances can be detected from the shape of the frequency characteristic of the stand, i.e., from the shape of the curve of output voltage versus frequency when radiated power is maintained constant.

Checking is performed in the circuit given in Figure E15. The model installation is replaced by a quarter-wave rod antenna fed from the signal generator. Radiated power is given by:

$$P = \left(\frac{E}{R_g + R_a} \right)^2 R_a \qquad (Eq.E4)$$

where:

R_g = output resistance of the generator (equal to the characteristic impedance of the cable feeding antenna)
R_a = input resistance of the antenna
E = e.m.f. of the generator

The antenna length is adjusted at each frequency so that its input impedance is resistive. Then, knowing the input resistance of the antenna, the e.m.f. of the generator is set to keep radiated power constant at each frequency.

Absence of variation in the curve of voltage versus frequency indicates the absence of resonances.

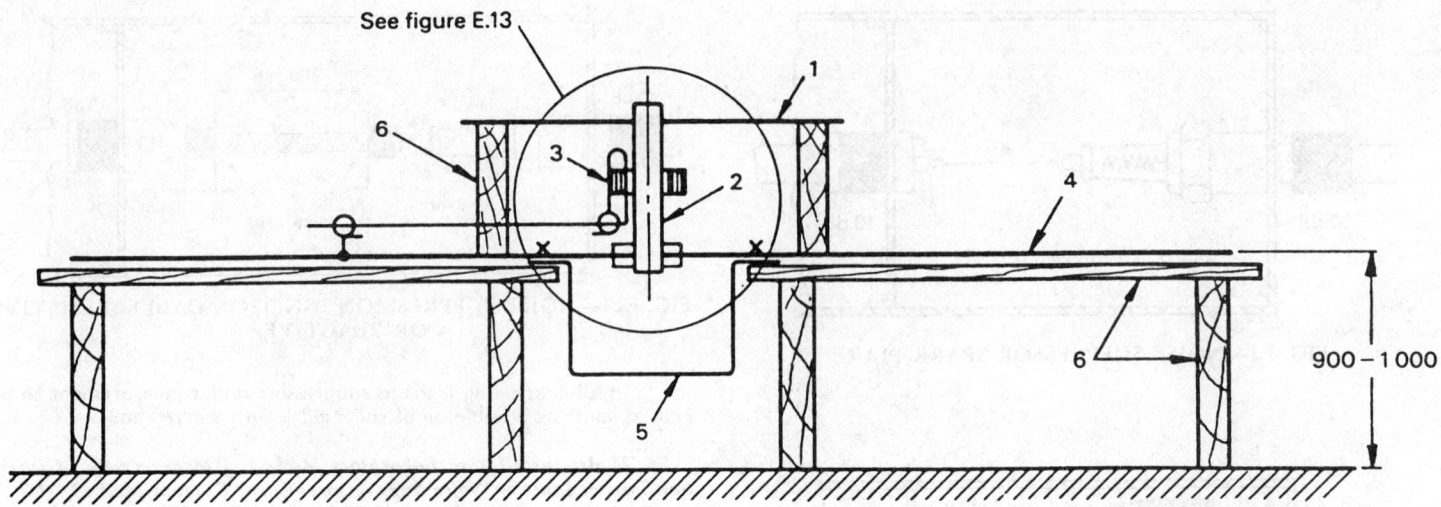

1 = upper plate
2 = metal tube connecting the plates (constituting the primary winding of the current transformer)
3 = current transformer
4 = lower plate (dimensions about 3 000 × 3 000)
5 = screen for driving system of distributor
6 = supporting wooden frame

Dimensions in millimetres

FIG. E12—EXAMPLE OF TEST STAND CONSTRUCTION
(SIMPLIFIED ELEVATION)

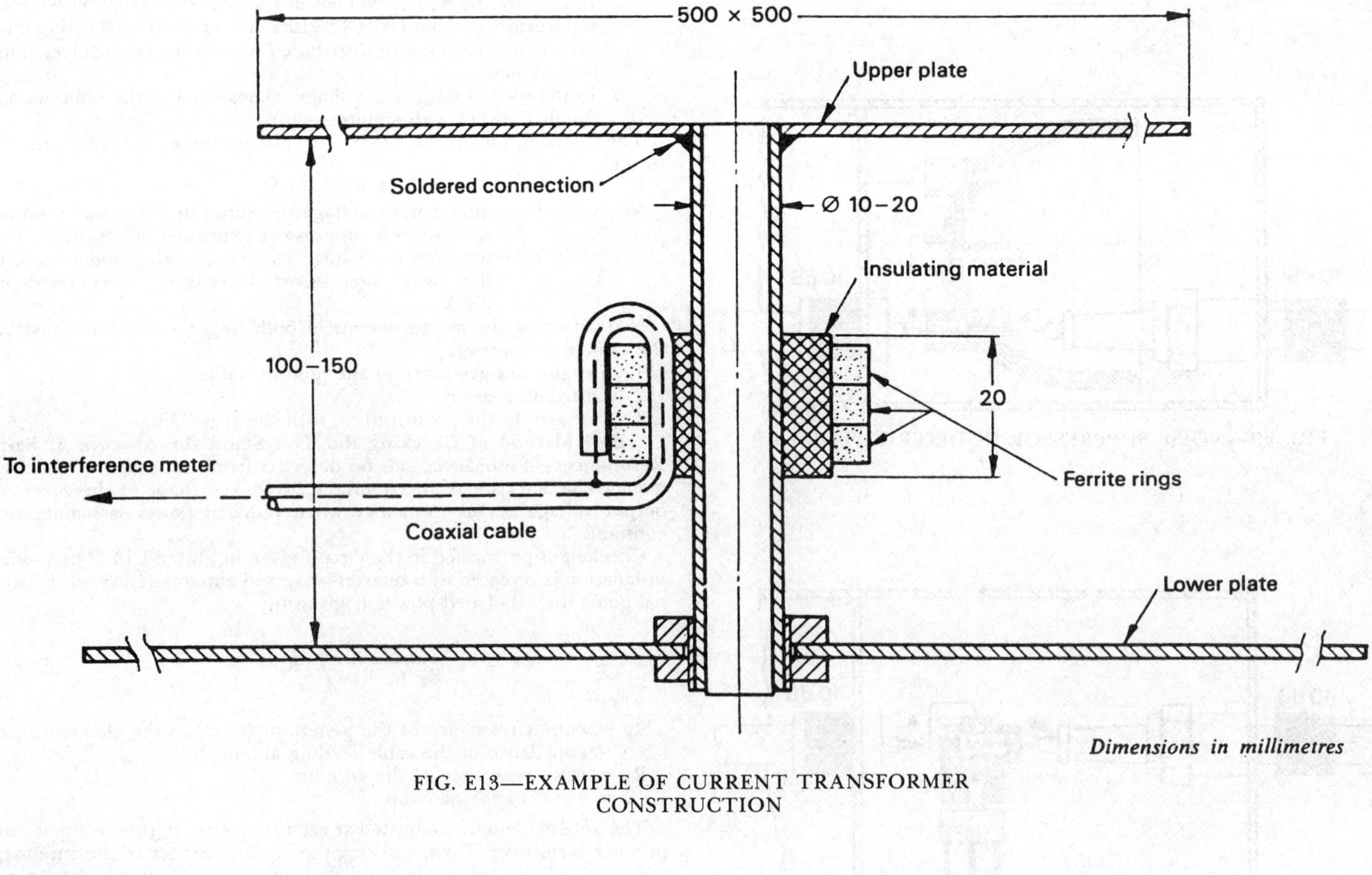

Dimensions in millimetres

FIG. E13—EXAMPLE OF CURRENT TRANSFORMER
CONSTRUCTION

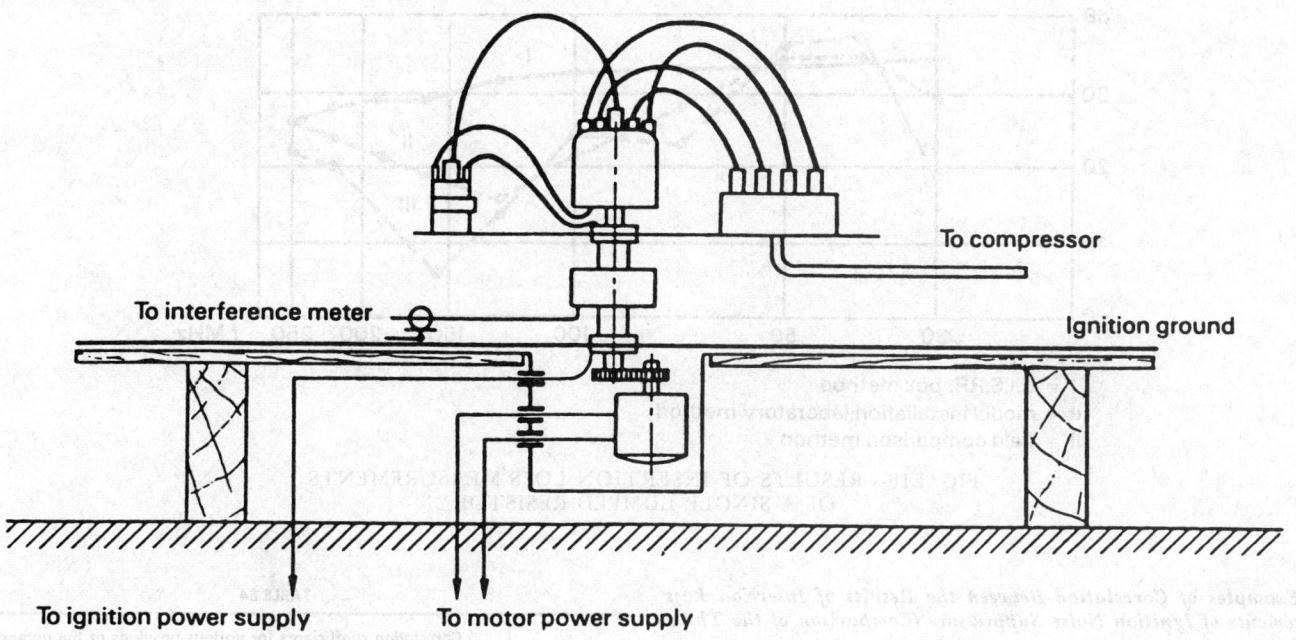

FIG. E14—EXAMPLE OF MODEL INSTALLATION MOUNTING

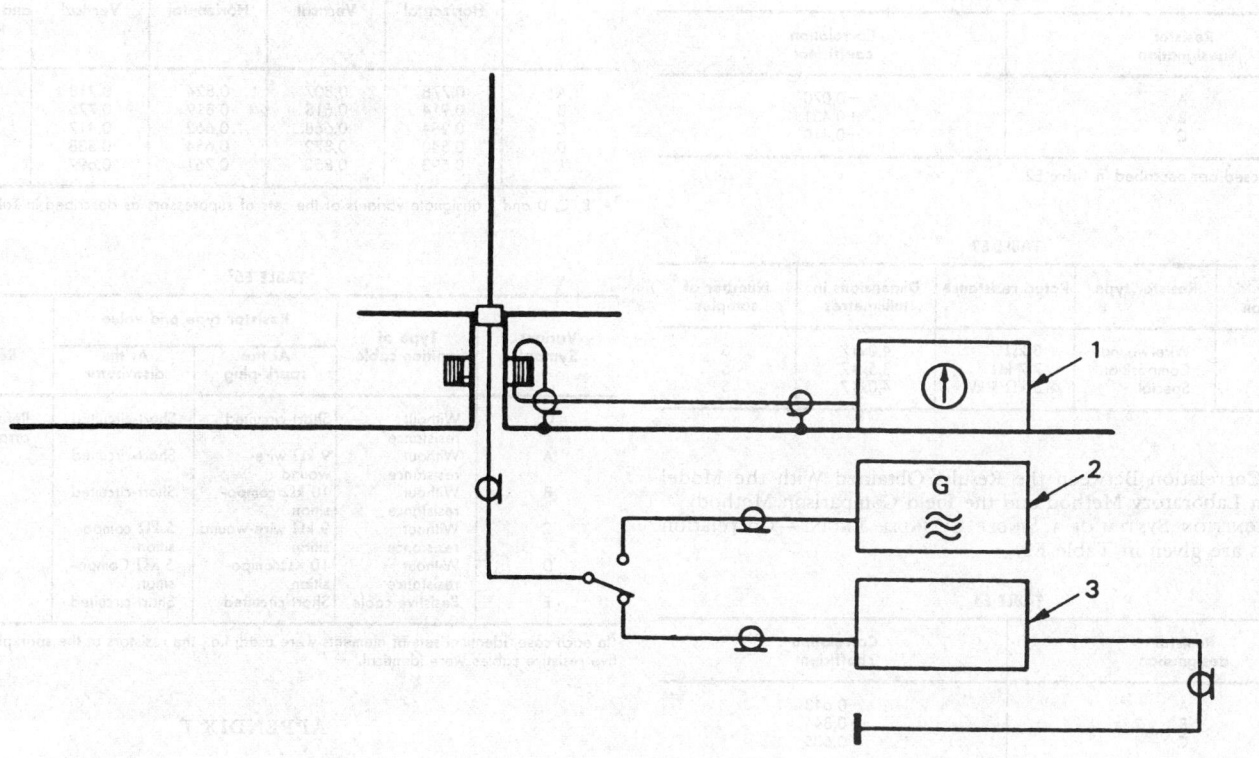

1 = interference meter
2 = signal generator
3 = impedance meter

FIG. E15—METHOD OF CHECKING THE TEST STAND FOR ABSENCE OF SELF-RESONANCES

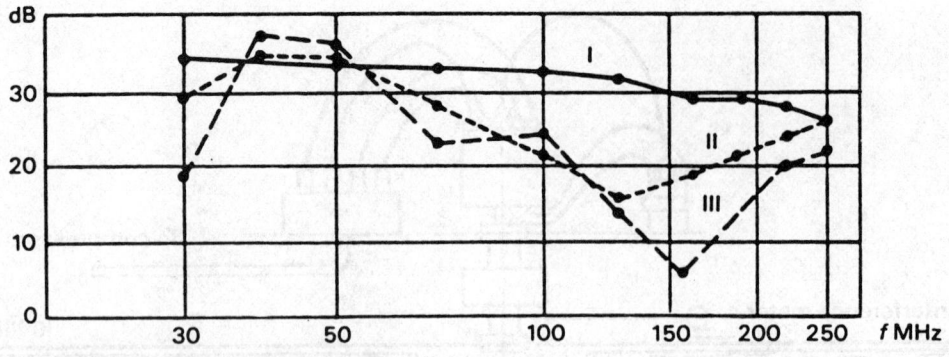

FIG. E16—RESULTS OF INSERTION LOSS MEASUREMENTS
OF A SINGLE LUMPED RESISTOR

E5. Examples of Correlation Between the Results of Insertion Loss Measurements of Ignition Noise Suppressors (Comparison of the Three Methods)

E5.1 Correlation Between the Results Obtained With the Field Comparison Method and the C.I.S.P.R. Box Method:

TABLE E1

Resistor designation	Correlation coefficient
A	−0,070
B	+0,431
C	+0,410

The resistors used are described in Table E2.

TABLE E2

Resistor designation	Resistor type	Rated resistance	Dimensions in millimetres	Number of samples
A	Wire-wound	8 kΩ	4,0×17	5
B	Composition	7,7 kΩ	3,5×17	5
C	Special	5 kΩ/5 kV	4,0×17	5

E5.2 Correlation Between the Results Obtained With the Model Installation Laboratory Method and the Field Comparison Method:

E5.2.1 IGNITION SYSTEM OF A SINGLE CYLINDER ENGINE—Correlation coefficients are given in Table E3.

TABLE E3

Resistor designation	Correlation coefficient
A	+0,643
B	+0,844
C	+0,605

The resistors used are described in Table E2.

For field comparison measurements, a motorcycle was used with the same ignition system and suppressors as in the model installation.

E5.2.2 IGNITION SYSTEM OF A FOUR CYLINDER ENGINE—Correlation coefficients are given in Table E4.

For the measurements, a car with an engine of 1300 cm^3 was used.

E5.3 Examples of the Results of Measurements— In Figures E16 and E17 examples of results of insertion loss measurements with the three methods described are shown.

For a multicylinder system (Figure E17), only results obtained with the model installation method and the field comparison method are compared, because the C.I.S.P.R. box method does not allow measurements of sets of suppressors.

TABLE E4

Variant Symbol	Correlation coefficients for various positions of the antenna in relation to the car and for various polarizations				Both positions and polarizations together
	Right side		Left side		
	Horizontal	Vertical	Horizontal	Vertical	
A	0,778	0,807	0,824	0,718	0,732
B	0,914	0,618	0,819	0,775	0,812
C	0,284	0,668	0,662	0,417	0,515
D	0,540	0,872	0,654	0,538	0,617
E	0,593	0,852	0,761	0,697	0,706

[1]A, B, C, D and E designate variants of the sets of suppressors as described in Table E5.

TABLE E5[7]

Variant Symbol	Type of ignition cable	Resistor type and value		Remarks
		At the spark-plug	At the distributor	
O	Without resistance	Short-circuited	Short-circuited	Reference arrangement
A	Without resistance	9 kΩ wire-wound	Short-circuited	
B	Without resistance	10 kΩ composition	Short-circuited	
C	Without resistance	9 kΩ wire-wound sition	5 kΩ composition	
D	Without resistance	10 KΩ composition	5 KΩ Composition	
E	Resistive cable	Short-circuited	Short-circuited	

[7]In each case, identical sets of elements were used, i.e., the resistors at the sparkplugs and the five resistive cables were identical.

APPENDIX F

REPORT ON ROADSIDE TESTING OF RADIATION FROM VEHICLES
(C.I.S.P.R. REPORT 62)

Foreword—Some national committees have initiated programs or have expressed interest in a method of measurement applicable to moving vehicles in a roadway. This document summarizes information presently available and the limitations of the roadside testing method.

F1. Summary—Roadside testing of moving traffic cannot be precise. It is possible only to determine in an approximate manner whether suppression is present on a given vehicle. Limited information can be gained, however, from static testing of vehicles chosen from a traffic stream. The roadside testing of moving vehicles may be used when performing statistical investigations of interference caused by the vehicles in use.

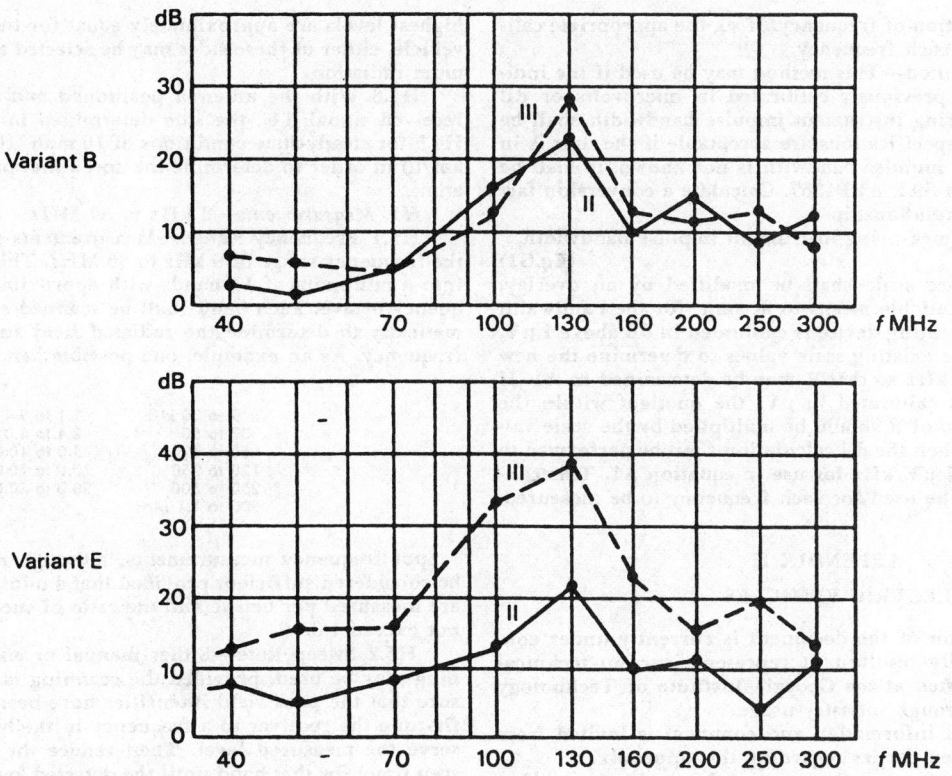

II = model installation laboratory method
III = field comparison method

FIG. E17—RESULTS OF INSERTION LOSS MEASUREMENTS OF SETS OF SUPPRESSORS

F2. Testing Moving Vehicles—In this case, measurement errors of significant magnitude arise from the total inability to reproduce laboratory practices or standard C.I.S.P.R. conditions:
  F2.1 Variables of the Measured Vehicle
  a. Varying and unequal distance to the vehicles being tested.
  b. Differing engine speeds at the same road speed caused by overall drive ratios, transmission slip, engine size, etc.
  c. Constantly varying speed of vehicles along the roadway, regardless of posted speed restrictions.
  d. Constantly varying engine/vehicle electrical system status (road load, air/fuel mixture, ignition voltage, electrical loads, etc.).
  F2.2 Variables of the Measurement Site
  a. Lack of correlation between C.I.S.P.R. site and the selected measurement site.
  b. Undefined condition of the near field emission pattern at low frequencies with reduced (5 m) antenna spacing.
  c. Undesired site electromagnetic ambient from narrow or broadband sources (superimposed short time radio frequency carriers, spurious interference, etc.).
  d. A method different from the C.I.S.P.R. method (measurement at only one or two frequencies, with only one antenna polarization and with the observation of the radiation of only one side of a vehicle.
  e. Inherent nonreproducibility of the data (among other causes: inadequate vehicle measurement time in front of the antenna, change in traffic flow, antenna half power beam width, measurement of multiple vehicles simultaneously, etc.).
  f. Undefined reflectivity of the site(s), dependent on adjacent signs, fences, utility poles, metal buildings, etc.
  F2.3 Expected Accuracy of Roadside Tests—For the reasons given, accuracy cannot be expected when measuring moving vehicles at a field site. Experience shows that with normal RFI suppression components, more than 95% of the vehicles will be below the applicable C.I.S.P.R. limit.

Vehicles with defective or missing RFI suppression components will be well above (10 to 20 dB) the limit. The difference between RFI suppressed/ unsuppressed vehicles will normally be at least 20 to 30 dB. This method will, therefore, be able to differentiate between such vehicles but cannot measure variations in each category.

F3. Testing Static Vehicles
  F3.1 C.I.S.P.R. Method
  F3.1.1 OPTIMUM CONDITIONS—Tests on a single vehicle shall be performed according to the C.I.S.P.R. test method, or shall be correlated back to the standard conditions of the C.I.S.P.R. test method.
  F3.1.2 PRACTICAL CONDITIONS—Most of the measurement errors mentioned in F.2.1 and F.2.2 will be encountered to some degree in any field test site. Correlation should be attempted (but will be difficult), and results should be used with caution, with doubts resolved in favor of the vehicle. A trained observer should be used.
  F3.2 Non-C.I.S.P.R. Methods—Specialized methods have been employed with success in particular situations. As mentioned in F.3.1.1, it is imperative that these be carefully correlated back to the C.I.S.P.R. standard and used with care in the specific situation for which the method was designed.

F4. Testing of Moving or Static Vehicles from a Moving Vehicle—The testing of moving or static vehicles from a moving vehicle should not be attempted.

APPENDIX G

INDICATING DEVICE CALIBRATION

G1. The calibration of the measuring instrument shall be accomplished in one of two ways

G1.1 Preferred Method—This method may be used for any indicating device whether previously calibrated or not. Use the reference impulse generator and appropriate RF attenuator or impulse generator output level controls to calibrate the indicating device directly in dB above 1 µV/kHz by injecting the impulse signal into the measuring instrument's RF input port and noting the indicating device readings at various known impulse signal levels. A scale overlay, look-up table, or other suitable technique can then be used to facilitate the determination of readings in dB above 1 µV/kHz (R in A1). The calibration shall be accomplished for at least one frequency in each band described in 5.4.1 to determine whether the indicating device calibration is a function of frequency. If so, the appropriate calibration shall be used at each frequency.

G1.2 *Acceptable Method*—This method may be used if the indicating device has been previously calibrated in microvolts or dB above 1 μV. The measuring instrument impulse bandwidth shall be known. Manufacturer's specifications are acceptable if the unit is in proper condition. If the impulse bandwith is not known, it shall be measured as discussed in SAE ARP1267. Calculate a conversion factor using the following relationship:

$$B = 20 \log (1 \text{ kHz/measuring instrument impulse bandwidth, kHz}) \quad \text{(Eq.G1)}$$

The indicating device scale shall be modified by an overlay, look-up table, or other suitable means to account for the bandwidth conversion B. If the indicating device is calibrated in dB above 1 μV, B should be added to the existing scale values to determine the new scale in dB above 1 μV/kHz so that R may be determined in A1. If the indicating device is calibrated in μV, the quotient within the brackets of the definition of B should be multiplied by the scale values to obtain μV/kHz. Then the dB calculation shall be performed to provide R in dB above 1 μV/kHz for use in equation A1. This calibration procedure shall be used for each frequency to be measured.

APPENDIX H
ELECTRIC VEHICLES

Foreword—This portion of the document is currently under consideration but not finally resolved. It represents current technical thought. It has been tested at the Georgia Institute of Technology (Project A-3089) and through industry usage.

Additional technical information and comment is invited from users and other interested parties regarding this appendix.

NOTE: Users of this proposed test procedure with its performance level are requested to test from 9 kHz to 30 MHz and report results to the SAE EMR Standards Committee for use in determining the final level and frequency range.

H1. Scope—The test procedures and performance levels in Appendix H cover the measurement of radiated magnetic and electric field strengths over the frequency range 9 kHz to 30 MHz. For measurements of electric vehicles at frequencies between 30 to 1000 MHz, follow the procedure in H.8 to arrange, operate, and test the vehicle using the measurement methods in Section 5 of this document.

H2. Equipment—The electromagnetic radiation receiver used for measuring absolute field intensity levels shall have peak detection capabilities over the frequency range 9 kHz to 30 MHz. The nominal input impedance shall be 50 ohms with a voltage standing wave ratio (VSWR) of less than 2.0:1 over the applicable frequency range. Quasi-peak detectors may be acceptable if the correlation factor between peak and quasi-peak detectors can be determined. The 20 dB correlation factor determined for spark-ignited engines is based on a pulse repetition rate of 50 to 100 Hz and may not apply to electric vehicles.

H3. Antennas—The receiving antennas to be used are the loop (electrostatically shielded) and the rod, above a ground plane as recommended by the antenna manufacturer, each equipped with switchable impedance matching networks.

H4. Preliminary Scan Procedure

H4.1 Elevate the drive wheels using jack stands as supports. If operation of the vehicles in the unloaded state would cause damage to the propulsion system, an absorption-type mechanical dynamometer shall be used.

H4.2 Use a rod antenna of 1.0 m nominal length mounted above a ground plane, as recommended by the antenna manufacturer. Use the antenna factors provided by the manufacturer to reduce the data to field strength.

H4.3 Establish steady-state conditions of 25 mph (40 km/h) in high gear.

H4.4 With the base of the rod antenna 1 m ± 0.05 m above the ground level and 1 m ± 0.05 m away from the nearest part of the front end of the vehicle, scan the radiated emission levels from 9 kHz to 30 MHz. Use a spectrum analyzer or a receiver.

H4.5 Record the data by taking photographs of the display, plotting the data on an X-Y plotter, or tabulating the data in sufficient detail in order to characterize the received signal levels over the 9 kHz to 30 MHz frequency range for the vertical electric field.

H4.6 Repeat for the other three sides of the vehicle.

H4.7 Determine the direction of maximum radiation based on the results of H4.1 to H4.6. This determination should be based on the highest level obtained from the four sides of the vehicle. If the highest levels are approximately equal for two different sides of the vehicle, either of these sides may be selected as the direction of maximum radiation.

H4.8 With the antenna positioned and oriented for maximum received signal, i.e., the side determined in H4.7, repeat H4.4 and H4.5 for steady-state conditions of 10 mph (16 km/h) and 40 mph (64 km/h) in order to determine the speed that produces maximum radiation.

H5. Measurements—9 kHz to 30 MHz

H5.1 *Frequency Range*—Measurements shall be performed over the frequency range of 9 kHz to 30 MHz. This range shall be divided into a minimum of 11 bands with approximately one band per frequency octave. Each band shall be scanned either manually or automatically to determine the radiated field strength as a function of frequency. As an example, one possible band selection would be:

9 to 30 kHz	1.1 to 2.4 MHz
30 to 60	2.4 to 5.0
60 to 120	5.0 to 10.0
120 to 250	10.0 to 20.0
250 to 500	20.0 to 30.0
500 to 1.1 MHz	

Spot frequency measurements, although not recommended, shall be considered sufficient provided that a minimum of two frequencies are measured per octave and the ratio of successive frequencies does not exceed 1.6.

H5.2 *Sweep Rate*—Either manual or automatic frequency scanning may be used, provided the scanning is sufficiently slow to ensure that the peak field intensities have been measured. As a check, fix-tune the receiver to a frequency in the band in question and observe the measured level. Then reduce the scan rate (increase the scan time) for that band until the detected level approximates (within 1 dB) the fix-tuned level at that particular frequency.

H5.3 *Operating Conditions*—All of the following radiated emission measurements shall be made with the drive wheels elevated and the vehicle supported by jack stands.

NOTE: If operation of the vehicle in the unloaded state would cause damage to the propulsion system, measurements may be made using a mechanical absorption dynamometer to load the vehicles at the zero grade road load for the particular speed determined in H4.8.

The vehicle shall be operated at the speed determined in H4.8 during all of the testing.

H5.4 *Vehicle Measurements*—The vehicle emissions shall be measured for each band as stated in H5.1, except that the vehicle shall be operated as specified in H5.3. Data shall be recorded in terms of received voltage (in either dBμV or dBμV/kHz versus frequency.

H6. Antenna

H6.1 *Antenna Position*—Position the receiving antenna (1) on the side of the vehicle found to emit maximum radiation as determined in H4.7, and (2) with the electrical center of the antenna (considered to be the base of the rod or the center of the loop) 10.0 m ± 0.2 m from the closest metallic part of the vehicle at a height of 3.0 m ± 0.05 m above ground level.

H6.2 *Antenna Polarization*—The strengths of the vertical component of the impulse electric field and both horizontal and vertical components of the impulse magnetic field shall be measured. The polarization for a magnetic loop antenna is referenced to an imaginary axis perpendicular to the plane of the loop. In the case of horizontal polarization, for example, the imaginary axis would be horizontally oriented in the plane transverse to the direction of propagation.

H7. Performance Levels

H7.1 Reduce the data by adding the antenna factors (in dB) provided by the antenna manufacturer to the received voltage. If the received voltage is in dBμV, convert it to dBμV/kHz as described in Appendix G.

H7.2 The recommended performance levels (9 kHz to 30 MHz) are given in Figures H1 (magnetic field strength) and H2 (electric field strength). To meet the intent of this document, the characteristic level for each band shall fall at or below the level at the arithmetic mean frequency of that band.

H8. Measurements—30 to 1000 MHz

H8.1 The vehicle shall be operated with the drive wheels elevated (unloaded) unless such operation is likely to cause damage to the vehicle. If operation in the unloaded state would cause damage, the vehicle shall be operated on a nonelectrical absorption dynamometer at a load corresponding to the zero grade load at a given speed.

H8.2 The preliminary test shall be made to determine the vehicle speed that produces maximum radiation.

H8.3 Final testing shall be conducted on all four sides of the vehicle at the speed that produces worst case conditions.

H8.4 Radiated emission testing of on board battery chargers is not required.

H8.5 Use the performance levels found in Section 4.

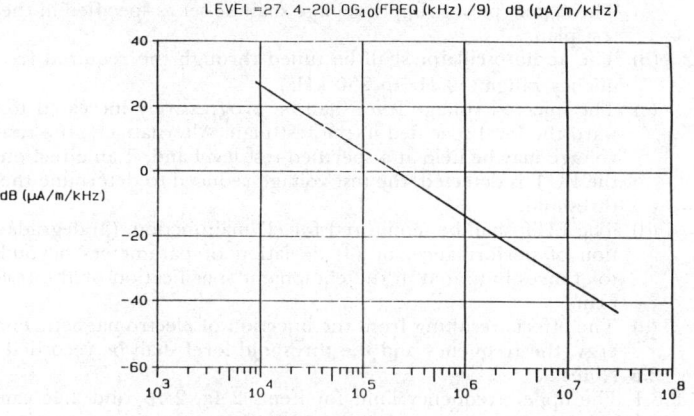

FIG. H1—RECOMMENDED PERFORMANCE LEVELS FOR PEAK IMPULSE MAGNETIC FIELD STRENGTH

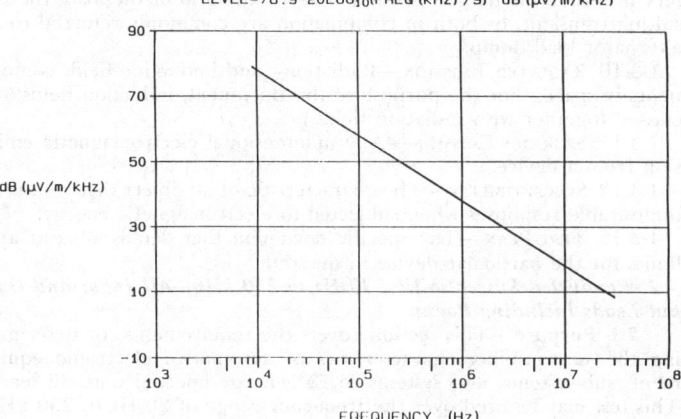

FIG. H2—RECOMMENDED PERFORMANCE LEVELS FOR PEAK IMPULSE ELECTRIC FIELD STRENGTH

φELECTROMAGNETIC SUSCEPTIBILITY MEASUREMENT PROCEDURES FOR VEHICLE COMPONENTS (EXCEPT AIRCRAFT)—SAE J1113 AUG87

SAE Recommended Practice

Report of the Subcommittee on EMI Standards and Test Methods, approved April 1975 and completely revised by the Electrical and Electronic Systems Technical Committee August 1987.

1. Introduction

1.1 Scope—This SAE Recommended Practice establishes uniform laboratory measurement techniques for the determination of the susceptibility to undesired electromagnetic sources of electrical, electronic, and electromechanical ground-vehicle components. It is intended as a guide toward standard practice, but may be subject to frequent change to keep pace with experience and technical advances, and this should be kept in mind when considering its use.

1.2 Measurement Philosophy—The need for measurement of the susceptibility of vehicle electronic components to electromagnetic sources has become more essential as more electronic components are introduced into motor vehicles. Electronic and electrical equipment may be susceptible to performance anomalies when subjected to electromagnetic sources, either of a transient or steady-state nature.

Electromagnetic interference (EMI) may be transient, intermittent, or continuous in nature arising from sources such as transmitters or other equipment located either on board or adjacent to the vehicle, or from component parts of the vehicle ignition or electrical power systems.

This recommended practice sets forth uniform procedures for establishing the susceptibility levels of individual vehicle components. It does not set limits on levels of EM energy in which vehicle components must perform; however, suggestions for developing functional performance status classifications for immunity are given in Appendix B.

A direct method of specifying the EM energy environment limits is to measure the actual fields, voltages, current, and impedances around the component or system of interest under all hazardous conditions. This will, of course, require a large enough sample of installations to determine possible variations. Some example data showing fields exists in NBS Technical Note 1014, "Electromagnetic Interference (EMI) Radiative Measurements for Automotive Applications."

It is recommended that a statistically valid number of components be tested using procedures adopted as standard by the testing organization. For destructive testing, such as transients on the power leads of Section 4, consult a handbook on statistical methods for details of the Karber method or the Bruceton (stair-step) method of sensitivity measurements. These methods eliminate the effects of cumulative degradation which often occurs during destructive testing.

It is suggested that only those portions of this recommended practice which are critical to the particular use of the component under test be applied, rather than subject the component to the provisions of the entire document. Thus, if the particular component under test is known to be susceptible mainly to transients, but otherwise well protected against conducted and radiated EMI, then only Section 4 need be applied. Or, if susceptibility to radiated energy is known to be a primary cause of malfunctions, then only Sections 6 through 9 need be applied.

Caution must be exercised in many portions of this procedure where high voltages or intense fields may be present.

ANSI and OSHA standards should be consulted concerning applicable limits on field exposure. For near field power density calculations, refer to paragraph 1.3.7.

1.3 Definitions and Terminology—The following definitions apply to the terms indicated as they are used in this recommended practice:

1.3.1 AMBIENT LEVEL—Those levels of radiated and conducted signal and noise existing at a specified test location and time when the test sample is in operation. Atmospherics, interference from other sources, and circuit noise or other interference generated within the measuring set compose the ambient level.

1.3.2 CONDUCTED EMISSION—Desired or undesired electromagnetic energy which is propagated along a conductor.

1.3.3 ELECTROMAGNETIC COMPATIBILITY (EMC)—Is the condition that enables equipment, subsystems, and systems (electronic, chemical, biological, etc.) to function without degradation from electromagnetic sources and without degrading the electromagnetic environment; that is, it is the condition which allows the coexistence of different electromagnetic sources without significant change in performance of any one in the presence of any or all of the others.

1.3.4 EMISSION—Electromagnetic energy propagated from a source by radiation or conduction.

1.3.5 EQUIPMENT UNDER TEST (EUT)—The device or system whose susceptibility is being checked.

1.3.6 FIELD DECAY (VOLTAGE)—The exponentially decaying negative voltage transient such as developed by an automotive alternator when the field excitation is suddenly removed, as when the ignition switch is turned off.

1.3.7 FIELD STRENGTH—The term field strength shall be applied to either the electric or the magnetic component of the field, and may be expressed as V/m or A/m. When measurements are made in the far field and in free space, the power density in W/m² may be obtained from field strengths approximately as $(V/m)^2/377$ or $(A/m)^2 \times 377$. When measurements are made in the near field and in free space, both the complex electric and magnetic vector components of the field must be fully defined. Power density may then be obtained by use of the Poynting vector.

1.3.8 GROUND PLANE—A metal sheet or plate used as a common

unipotential reference point for circuit returns and electrical or signal potential.

1.3.9 LOAD DUMP (VOLTAGE)—The exponentially decaying positive voltage transient developed by an automotive alternator when disconnected suddenly from its load, while operating without a storage battery or with a discharged storage battery. Removal of the load, the resulting transient, or both in combination are commonly referred to as alternator load dump.

1.3.10 RADIATED EMISSION—Radiation- and induction-field components in space. (For the purpose of this document, induction fields are classed together with radiation fields.)

1.3.11 SPURIOUS EMISSION—Any unintentional electromagnetic emission from a device.

1.3.12 SUSCEPTIBILITY—The characteristic of an object that results in undesirable responses when subjected to electromagnetic energy.

1.3.13 TEST PLAN—The specific document that details all tests and limits for the particular device in question.

2. Conducted Susceptibility, 30 Hz to 250 kHz—All Input and Output Leads Including Power

2.1 Purpose—This section covers the requirements for determining the susceptibility characteristics of automotive electronic equipment, sub-systems, and systems to EM energy injected onto all leads. This test may be used over the frequency range of 30 Hz to 250 kHz.

2.2 Measurement Philosophy—For the frequency range of this test, the impedances seen by the signal, load and power supply leads are generally known and can be treated as lumped constants. In this test, a wide range audio voltage source is coupled through a transformer to each specified pin of the EUT. The signal source impedance must be low in comparison to the impedance of the circuit being tested. Experience has shown that a signal source impedance of 0.5 ohms max is adequate for the test. The EUT should be connected so it will operate in its normal manner. Actual loads and sources should be used where appropriate or may be simulated. A capacitor shunt element should be used on a specific lead, if a large percentage of the test signal does not appear across the signal, load or power supply lead relative to ground. (See Fig. 1.)

2.3 Grounding and Shielding—For the stated frequency range there are no special grounding and shielding requirements. However, the requirements of paragraph 3.3 may be utilized here, if expedient.

2.4 Apparatus—To utilize the full frequency range of this test, the apparatus shall be as follows:

(a) Audio Oscillator—30 Hz to 250 kHz.
(b) Audio Power Amplifier—50 W or greater with output impedance equal to, or less than, 2.0 Ω (capable of delivering 50 W into a 0.5 Ω resistive load connected across an isolation-transformer secondary). 30 Hz to 250 kHz.
(c) Isolation Transformer—4:1 impedance ratio; secondary as connected shall be capable of handling the current flow without saturating the core. 30 Hz to 250 kHz.
(d) Measuring Instrument—Oscilloscope, Voltmeter, or EMI Meter.
(e) Power Supply—The power supply used for this test shall have the equivalent of 100 µf (min) capacitor across the output terminals.
(f) Capacitor—A 100 µf capacitor may be used to shunt the source end of the isolation transformer to ground, if difficulty is encountered in obtaining sufficient test voltage.

2.5 Test Setup and Procedures—The test setup is shown in Fig. 1.

(a) The system power supply voltage shall be set as specified in the test plan.
(b) The audio oscillator shall be tuned through the required frequency range (30 Hz to 250 kHz).
(c) The injected voltage level shall be progressively increased toward the level specified in the test plan. Alternatively, the test voltage may be held at a specified test level and, if an effect on the EUT is detected, the test voltage reduced to determine the threshold.
(d) The EUT shall be monitored for (1) malfunction, (2) degradation of performance, or (3) deviation of parameters beyond tolerances indicated in the equipment specification or the test plan.
(e) The effects resulting from the injection of electromagnetic energy, the frequency and the threshold level shall be recorded.

2.6 Notes

2.6.1 The upper frequency limit for items 2.4a, 2.4b, and 2.4c can be reduced in accordance with the user's frequency range requirements.

2.6.2 It is recognized that other types of equipment can produce equivalent signals, for example, a Power Oscillator can replace the oscillator and amplifier; a Power Operational Amplifier can replace the amplifier and power supply, etc.

2.6.3 The following procedure can be used to verify the signal source impedance at the isolation transformer secondary terminals.

(a) Set a voltage level at the primary terminals and measure the open circuit secondary voltage (V_{oc}).
(b) Connect a known load R_L across the secondary and measure the closed-circuit secondary voltage (V_{cc}).
(c) The impedance shall be calculated as follows:

$$Z = \frac{R_L (V_{oc} - V_{cc})}{V_{cc}} \, (\Omega)$$

(d) Repeat the above procedure at one frequency per decade from 30 Hz to 250 kHz (including 30 Hz and 250 kHz).
(e) The measured impedance shall be less than, or equal to 0.5 Ω.

3. Conducted Susceptibility, 50 kHz TO 100 MHz—All Input and Output Leads, Including Power

3.1 Purpose—This section covers the requirements for determining the susceptibility characteristics from 50 kHz to 100 MHz of automotive electronic equipment, subsystems, and systems to EMI injected onto all input leads, including signal and power.

3.2 Measurement Philosophy—Power-source RF impedance seen by a given type of electronic equipment depends upon this varying impedance and would render susceptibility measurements meaningless unless the impedance is also measured or controlled. In order to compare measurements made at various locations, powerline RF impedance seen by the equipment shall be controlled by line-impedance stabilization networks. The EUT should be connected so it will operate in its normal manner. Actual loads and sources should be used where appropriate or may be simulated.

3.3 Grounding and Shielding—To achieve uniform measurement conditions at radio frequencies requires that certain grounding practices be followed. Ground requirements are that EUT, LISNs, and terminating loads:

(a) be placed on a metallic ground plane having the following minimum dimensions:
　1. Thickness—1.5 mm (0.060 in) aluminum, copper, or brass sheet.
　2. Length—1 m or underneath entire equipment plus 0.5 m whichever is larger.
　3. Width—width of equipment plus 0.25 m on each side.
(b) be bonded to the ground plane as in its intended installation.
(c) not otherwise be grounded, unless required in installation instructions. The line-impedance stabilization networks shall be bonded to the ground plane as close as possible to the EUT ground. No shielding is to be used other than that called out in installation instructions.

3.4 Power Input Lead Test

3.4.1 APPARATUS

(a) Signal Source—A 50 Ω output-impedance source with an output of 100 V or greater into a matched load.
(b) One of the following to measure RF voltage.

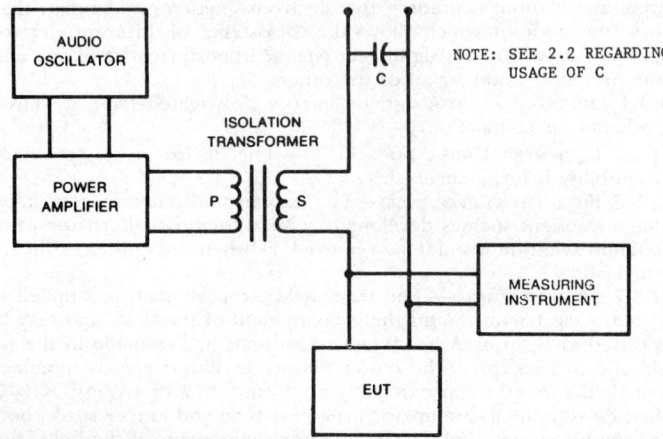

FIG. 1—TEST SETUP FOR MEASURING CONDUCTED SUSCEPTIBILITY, 30 HZ TO 250 KHZ, ALL LEADS

1. Calibrated Oscilloscope
2. High Impedance RF Voltmeter
3. EMI Meter
4. Spectrum Analyzer

(c) Line-Impedance Stabilization Networks (LISN's)—as specified in Figs. 2 and 3 with 50 Ω resistive RF terminations. (See Appendix A for suggestion of construction details.) When using a LISN, caution should be exercised to avoid load-current limiting due to series inductance in the LISN. This limiting may occur when loads switch between high-and low-impedance states. Use of a LISN may then result in increased susceptibility.

(d) Test-Source Injection Networks illustrated in Fig. 4.

(e) Power Supply—DC.

3.4.2 TEST SETUP AND PROCEDURE—The test setup is shown in Fig. 5. The procedure is as follows:

(a) Each control and signal lead shall be loaded with a termination impedance. At these frequencies, however, the impedance as seen by the control and signal leads may no longer be determined by the system designer, due to uncontrollable stray impedances. It may be possible to simulate these impedances with a simple capacitor and inductor added to the actual leads if the frequency in MHz does not greatly exceed $300/20\,\pi\ell$ where the ℓ is the characteristic lead length in meters. Above that frequency, the test designer should design the test plan and setup to given uniform results.

(b) The EUT shall be connected as shown in Fig. 5, observing the grounding and shielding requirements of paragraph 3.3.

(c) Signal sources and measuring instrumentation shall be connected to an LISN through test-source injection networks. Care shall be exercised to insure sufficiently short leads on the injection networks and LISN's to minimize loss of signal due to series inductance and shunt capacitance. A current probe on the injection lead right next to the EUT can be used to monitor the signal. For signal-source and measuring-instrument impedance equal to 50 Ω, use the signal-injection network of Fig. 4A. For a signal-source impedance of 50 Ω and a high-impedance measuring instrument, use the signal-injection network of Fig. 4B. Note the corresponding attenuation factors.

(d) Increase the level of the test signal while continuously scanning through the required frequency range (50 kHz to 100 MHz). Tests shall be conducted at not less than three frequencies per octave representing the maximum susceptibilities within that octave. Monitor the equipment under test for: (1) malfunction, (2) degradation of performance, or (3) deviation of parameters beyond tolerances indicated in the equipment specification or approved test plan (see paragraph 3.6). Record the highest level before degradation was observed.

(e) See paragraph 2.5c.

3.5 All Leads Except Power

3.5.1 APPARATUS

(a) Signal Sources—as for powerline measurements.
(b) Measuring Instruments—as for powerline measurements.
(c) Test Source Injection Networks—see Fig. 4.

3.5.2 TEST SETUP AND PROCEDURE

(a) Test setup is shown in Fig. 6. Note that the LISN remains in the powerline circuit and its RF injection terminal is loaded with 50 Ω.

(b) Each control and signal terminal not under test is loaded with its terminating impedance and test signals are injected into the test terminal as indicated in Fig. 6. At these frequencies, however, the impedances seen by control and signal leads may no longer be determined by the system designer. It may be possible to simulate the stray impedance as described in paragraph 3.4.2a. Care shall be exercised to insure sufficiently short leads on the injection networks and LISN's to minimize loss of signal due to series inductance and shunt capacity. A current probe on the injection lead right next to the EUT can be used to monitor the signal.

(c) Increase the level while continuously scanning through the required frequency range (50 kHz to 100 MHz). Tests shall be conducted at not less than three frequencies per octave representing the maximum susceptibilities within that octave. Monitor the EUT for: (1) malfunction, (2) degradation of performance, or (3) deviation of parameters beyond tolerances indicated in the equipment specification or approved test plan. (See paragraph 3.6.) The values at which these occur shall be recorded.

3.6 Notes

(a) Each LISN shall be tested over the range for which it is designed. The impedance should be within 20% of the curves in Figs. 2 and 3. If any discrepancies occur, then the network should be modified, for example, by adding ferrites to inductor leads to increase impedance at higher frequencies.

(b) Unless otherwise required in the equipment specifications or approved test plan, the test signals shall be modulated according to the following rules:
1. Test samples with audio channels/receivers.

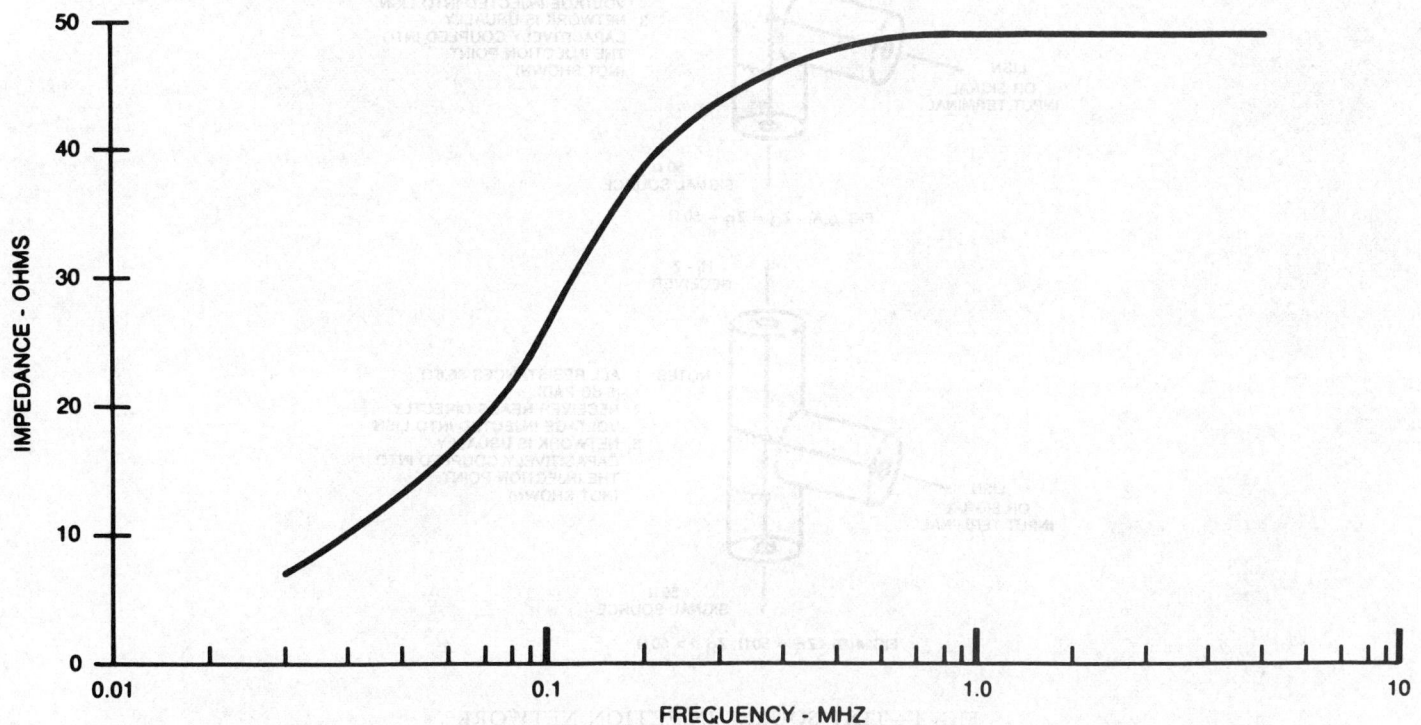

FIG. 2—LINE IMPEDANCE STABILIZATION NETWORK REFERENCE IMPEDANCE, 50 KHZ TO 5 MHZ, (50 µH)

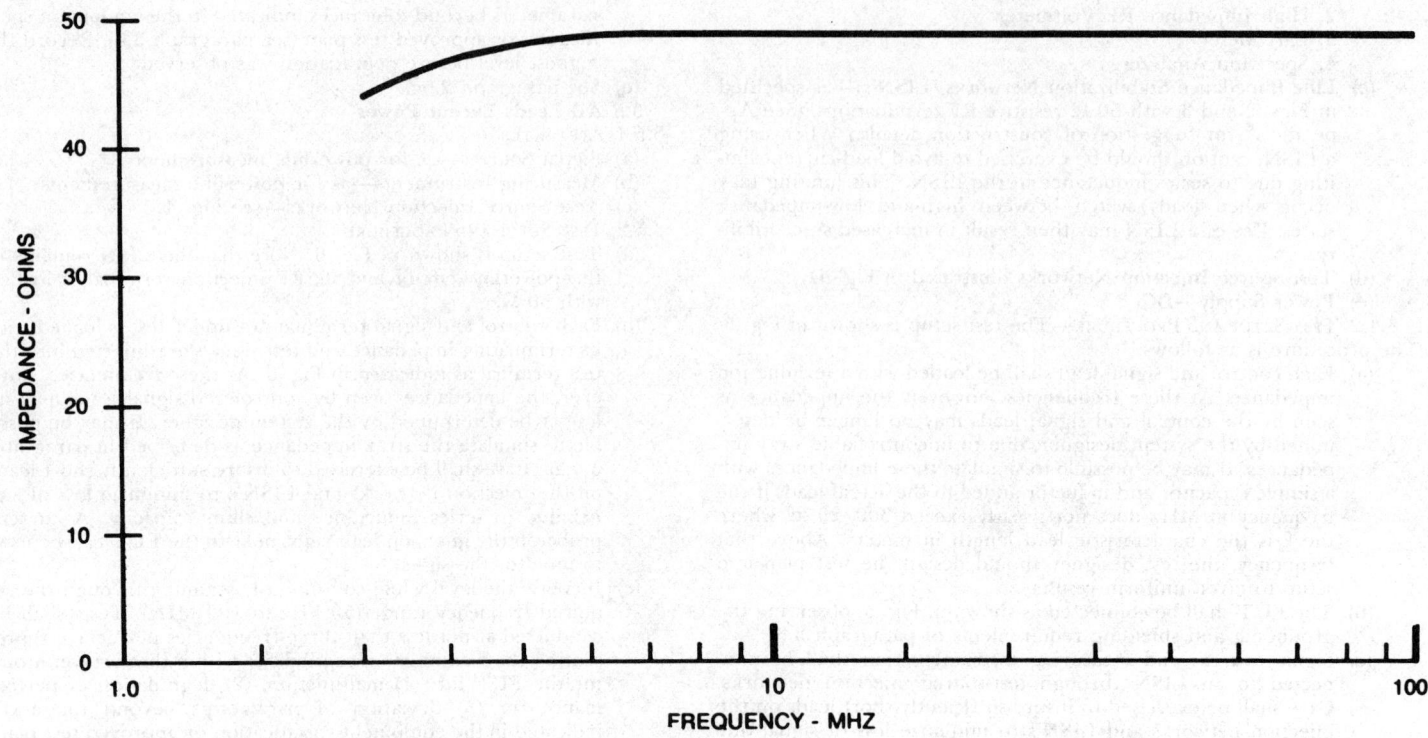

FIG. 3—LINE IMPEDANCE STABILIZATION NETWORK REFERENCE IMPEDANCE, 5 MHZ TO 100 MHZ, (5 µH)

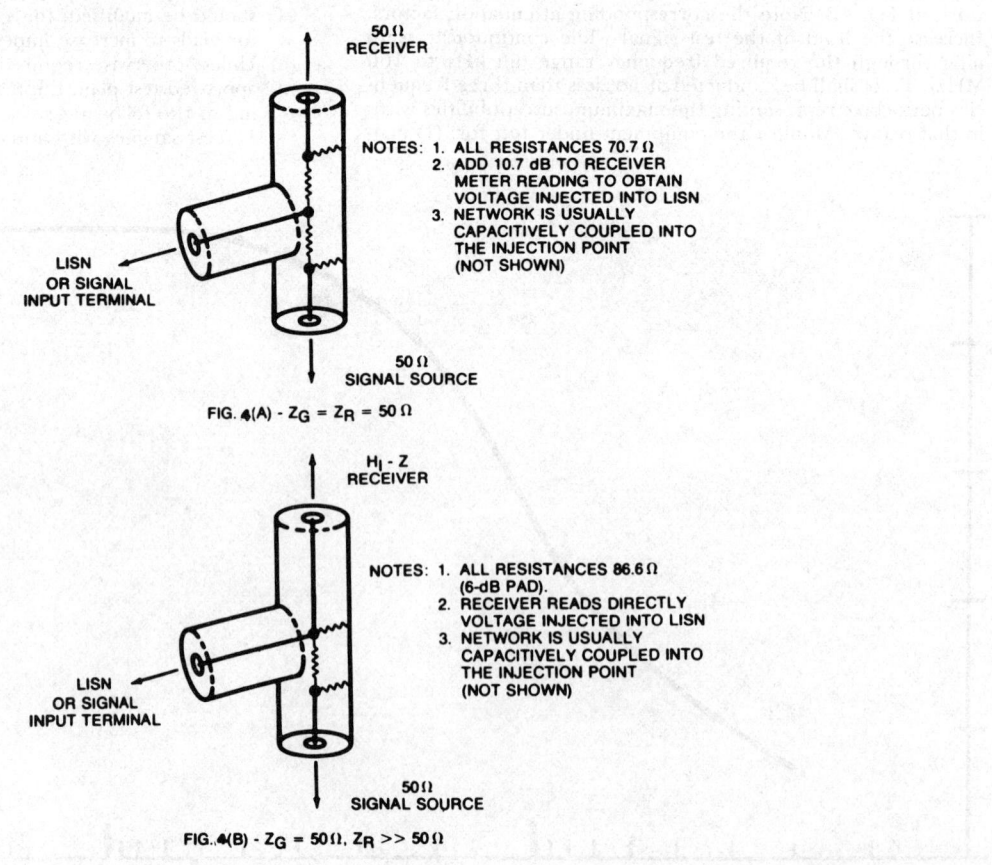

FIG. 4—TEST SOURCE INJECTION NETWORK

AM Receivers: Modulate 30% with 1000 Hz tone.
FM Receivers: Modulate with 1000 Hz signal using 30% modulation.
SSB Receivers: Use no modulation.
Other Equipments: Same as for AM receivers.
2. Test samples with video channels other than receivers. Modulate 90 - 100% with pulse of duration 2/BW and repetition rate equal to BW/1000, where BW is the video bandwidth (Hz).
3. Digital Equipment: Use pulse modulation as appropriate with pulse duration and repetition rate set equal to that used in the equipment under test or associated with other known external pulse sources that may be operating in close proximity.
4. Non-Tuned Equipment: Amplitude modulate 30% with 1000 Hz tone or as otherwise specified in the test plan.

4. Conducted Susceptibility, Transients, Power Leads 12 V Passenger Car Systems

4.1 Purpose—This section describes methods and apparatus to determine the capability of various electrical devices to withstand transients which normally occur in motor vehicles. Test apparatus specifications outlined in this section were developed for 12 V passenger cars. Similar specifications are being developed for 12 V trucks and 24 V vehicles.

Functional performance status classifications for immunity to transients are given in Appendix B. EUT performance requirements for each of the test pulses must be individually determined.

4.2 Measurement Philosophy—Installed equipment is powered from sources which contain, in addition to the desired electrical voltage, transients with peak values many times this value, caused by the release of stored energy during the operation of relay and other loads connected to the source and during start and turn off of vehicles. These tests are designed to determine the capability of equipment to withstand such transients. The tests may be made in the laboratory (bench tests) as well as on the vehicle. These tests are outlined in ISO/TR 7637, "Road Vehicles - Electrical Interference by Conduction and Coupling: Part 1, Vehicles with Nominal 12 V Supply Voltage -Electrical Transient Conduction Along Supply Lines Only." Bench test methods should give results which allow comparison between different laboratories and are intended to provide a basis for development of devices and systems. These tests may also be used later during production of these devices and systems.

These tests may not cover all types of transients which can occur in a vehicle. Therefore, the test pulses described in paragraph 4.3 are characteristic of typical pulses. To ensure proper operation of a vehicle in the electromagnetic environment, on-board testing must be performed in addition to bench testing:

4.3 Apparatus
(a) Pulse generators capable of generating the transient waveforms shown in Figs. 7A - 7H shall be adjustable up to the amplitudes indicated. The specification of rise time (T_r), duration (T), and internal resistance (R_i) represent fixed requirements unless otherwise specified.
(b) Oscilloscope (preferably storage): for monitoring the pulse generator band width: at least 100 MHz.
Writing Speed: at least 100 cm/μs.
Input Sensitivity: at least 5 mV/division.
(c) Voltage Probe: for use in conjunction with the oscilloscope.
Attenuation - 100/1.
Maximum Input Voltage - at least 1 kV.
Input Impedances as a function of the frequency, f:

f (MHz)	Z (kΩ)
1	40
10	4
100	0.40

maximum length of the probe cable - 3 m
maximum length of the ground cable - 0.13 m
Note: Any other cable lengths may influence the result of the measurement and should be stated in the test report.
(d) Switch S: Use standard production switch for the respective EUT. Current rating must be sufficient to handle the required loads.
(e) Equipment, as may be required, to perform a functional test of the component following transient application.

4.4 Test Setup and Procedure
(a) Connect a known functional EUT as shown in Fig. 8.
(b) Ensure that the ambient temperature and supply voltage V_s are maintained as required by the appropriate test specification.
(c) Set up the pulse generator to provide the polarity, amplitude, and pulse duration as specified in the appropriate test specification.
(d) Generate the test pulse.
(e) Perform the appropriate functional test to determine whether failure has occurred and record results.

4.5 Notes
(a) In determining the susceptibility level, care must be exercised to eliminate the effects of cumulative deterioration such as dielectric "punch through" in semi-conductor devices.
(b) When testing to a specified level, unnoticed failures may occur which be detected only by running life-cycle tests and comparing the results of tested components against those of untested components.
(c) Rise time requirements as specified in Figs. 7A - 7H represent a design objective. For short duration, high voltage pulse, some rise times may not be practical.
(d) The ambient temperature during the test shall be 23 ± 5°C.
(e) Power leads should be separated from all other leads for these tests.

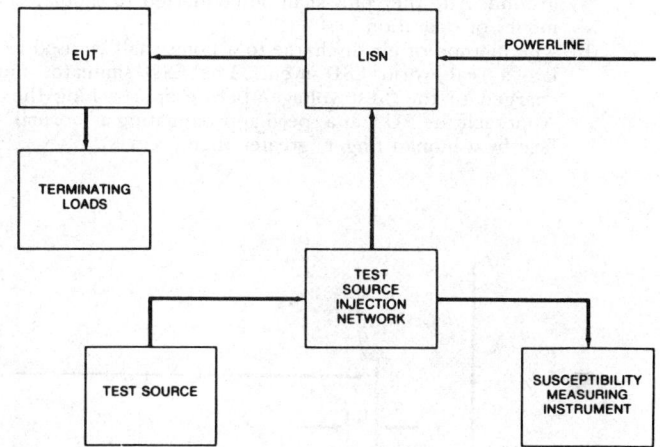

FIG. 5—EQUIPMENT BLOCK DIAGRAM FOR MEASURING CONDUCTED SUSCEPTIBILITY, 50 KHZ TO 100 MHZ, POWERLINE ONLY

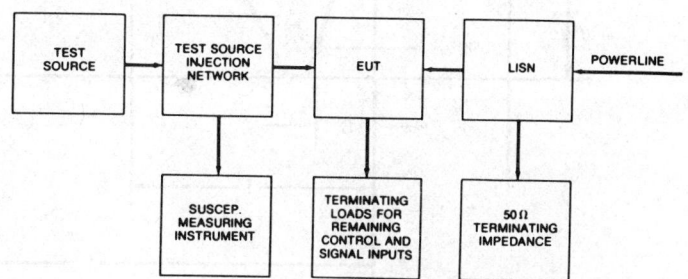

FIG. 6—EQUIPMENT BLOCK DIAGRAM FOR MEASURING CONDUCTED SUSCEPTIBILITY, 50 KHZ TO 100 MHZ, CONTROL AND SIGNAL INPUTS

5. Electrostatic Discharge

5.1 Purpose—This section covers the requirements for determining the susceptibility of automotive electronic components/subsystems to Electrostatic Discharge (ESD).

5.2 Measurement Philosophy—Occupants in vehicle can generate significant electrostatic potentials. Devices in the vehicle may fail when an electrostatic discharge occurs.

Electrostatic discharge simulator parameters for use in automotive applications are unique to the automotive environment and differ from values used in other applications.

5.3 Apparatus
(a) ESD Simulator characteristics:
- Voltage Range: variable from 1 kV to 15 kV (negative and positive polarity)
- Capacitor: 330 pF ($\pm$ 10%)
- Inductor: $<$ 1 μH
- Resistance: 2000 Ω ($\pm$ 10%)
- Tip: standard (IEC) finger model electrode. (See IEC 801-2.)

ESD simulators are commercially available. Appendix C defines a test method for evaluating the characteristics of the simulator to ensure the generation of accurate and reproducible discharge waveforms.

(b) Conductive Bench Top, at least 1 m^2.

5.4 Test Procedure
(a) The device shall be placed on a conductive-bench top as shown in Figs. 9A and 9B. All voltage supply pins should be connected to an appropriate power source. All ground pins should be connected together. The ground pins, bench top, ESD simulator, and power source shall be grounded together and to earth ground. All other pins shall be connected to simulate normal modes of operation.

(b) The method of air discharge to a point shall be used to simulate a real world ESD event. The ESD simulator must be charged to the "test voltage" before approaching the EUT. Approach the EUT at a speed approximating a "normal" interface by a human finger (greater than 5 cm/s).

(c) Twenty discharges shall be applied to specified test points at each voltage level; ten with positive polarity and ten with negative polarity. A minimum of five (5) s shall be maintained between pulse applications.

If significant variations are observed in the measurement data, 20 discharges may not be sufficient to obtain statistically significant results. If this is found to be the case, a larger number of discharges will be required.

(d) During and after each level of testing, the device shall be measured for performance which may indicate failure to meet design requirements.

6. Radiated Susceptibility, Power Line Magnetic Field 60 Hz to 30 kHz

6.1 Purpose—This section covers the recommended testing technique for determining the susceptibility of automotive electronic modules, subsystems, and systems to magnetic fields generated by power transmission lines and generating stations.

6.2 Measurement Philosophy—Electronic systems may be affected when immersed in a magnetic field. These fields are found near high power transmission lines and power generating stations that generate magnetic fields. The fields consist of the fundamental (60 Hz signal) and its odd harmonics with amplitudes approaching those shown in Fig. 10. Consequently, devices in vehicles that are driven near these sources may be subjected to these fields.

6.3 Test Specification
(a) Lower frequency: 60 Hz.
(b) Upper frequency: 30 kHz.
(c) Magnetic flux density of Helmholtz coil - see Fig. 10.

6.4 Test Apparatus—The following section describes a typical test setup that could be used to generate a uniform magnetic field for the testing as illustrated in Fig. 11.

(a) Helmholtz Coil—The radius of the coil will be determined by the size of the EUT. In order to obtain a uniform magnetic field ($\pm$ 10%), the relationship between the EUT and the coil is illustrated in Fig. 12. The coil should be capable of produc-

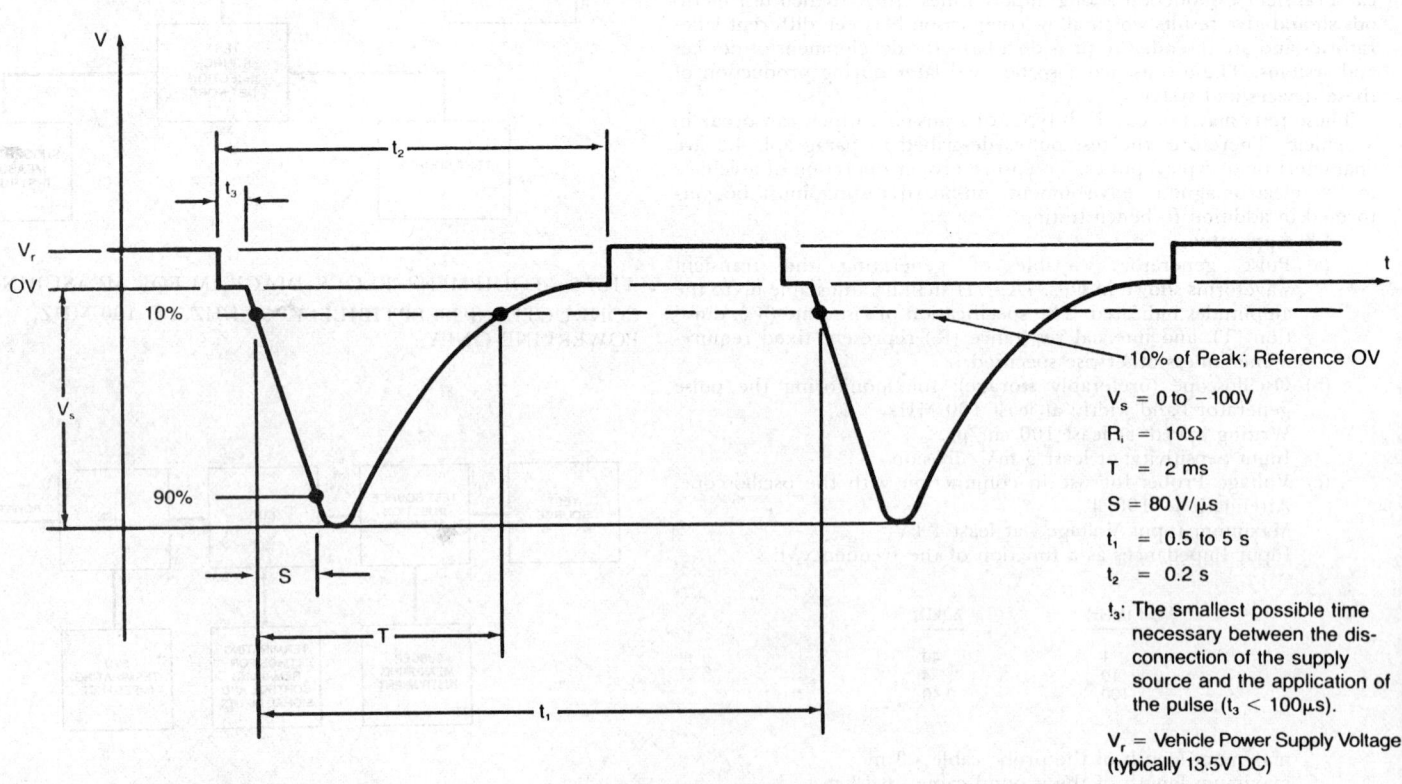

FIG. 7A—TEST PULSE 1 (DISCONNECTION FROM INDUCTIVE LOADS WITH DEVICE UNDER TEST REMAINING CONNECTED DIRECTLY IN PARALLEL WITH THIS INDUCTIVE LOAD)

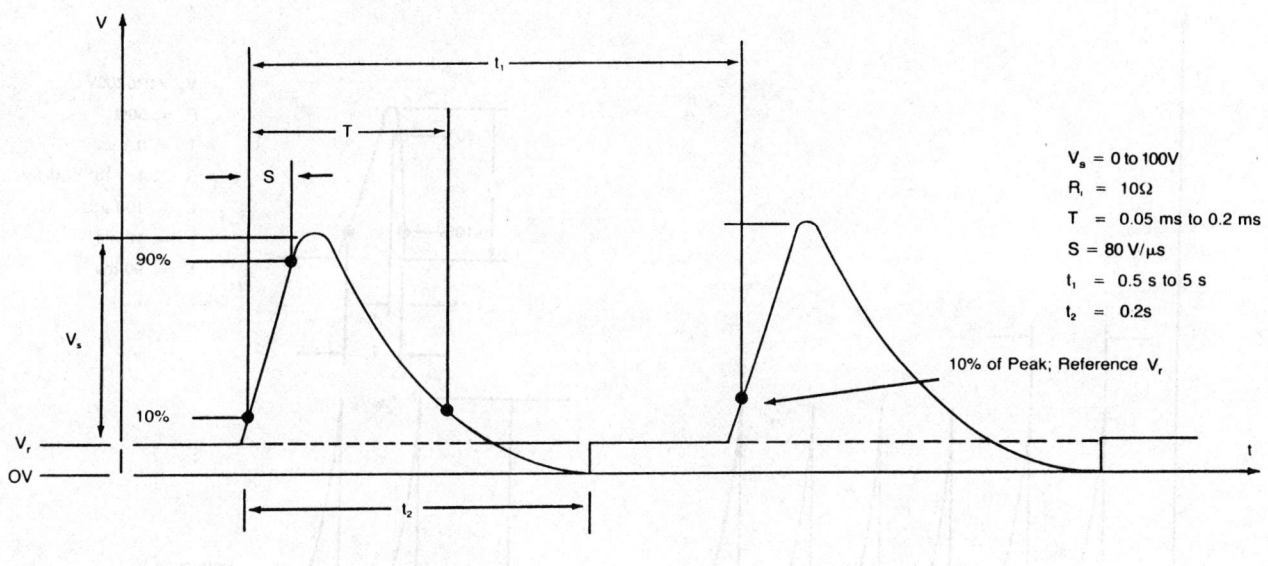

FIG. 7B—TEST PULSE 2 (SUDDEN INTERRUPTION OF A SERIES CURRENT)

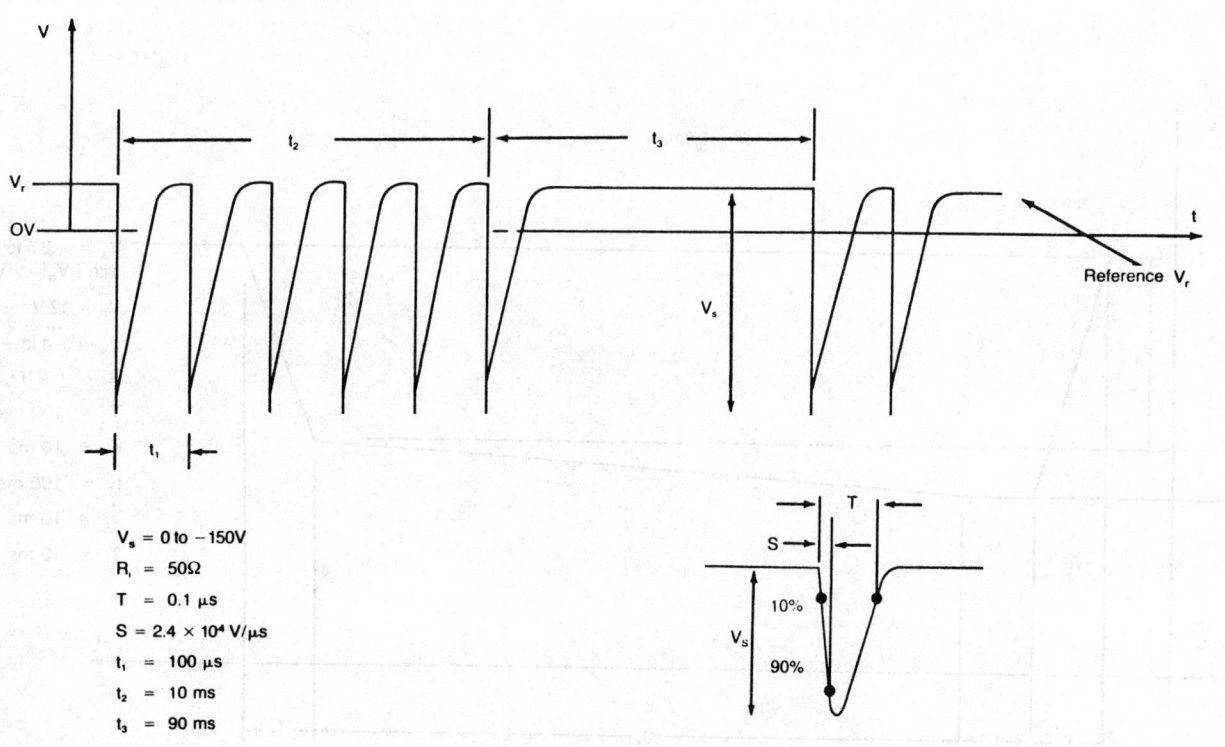

FIG. 7C—TEST PULSE 3A (SWITCHING SPIKES)

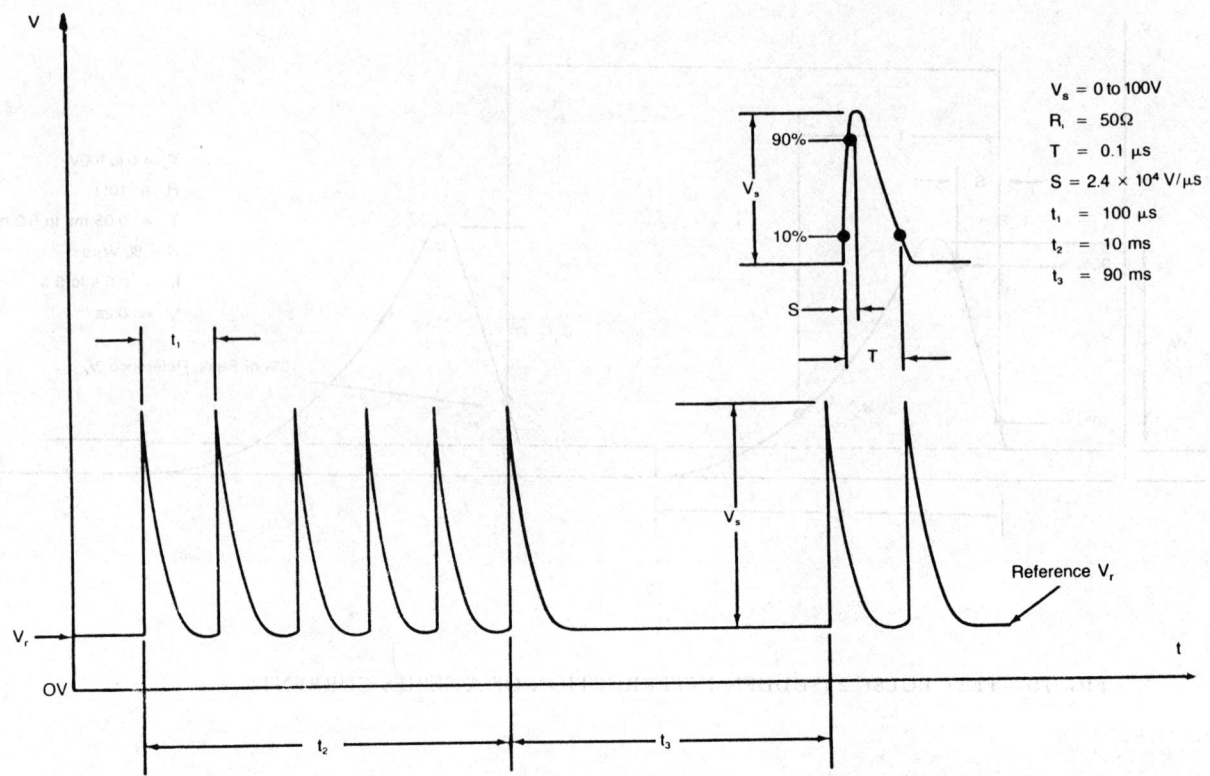

FIG. 7D—TEST PULSE 3B (SWITCHING SPIKES)

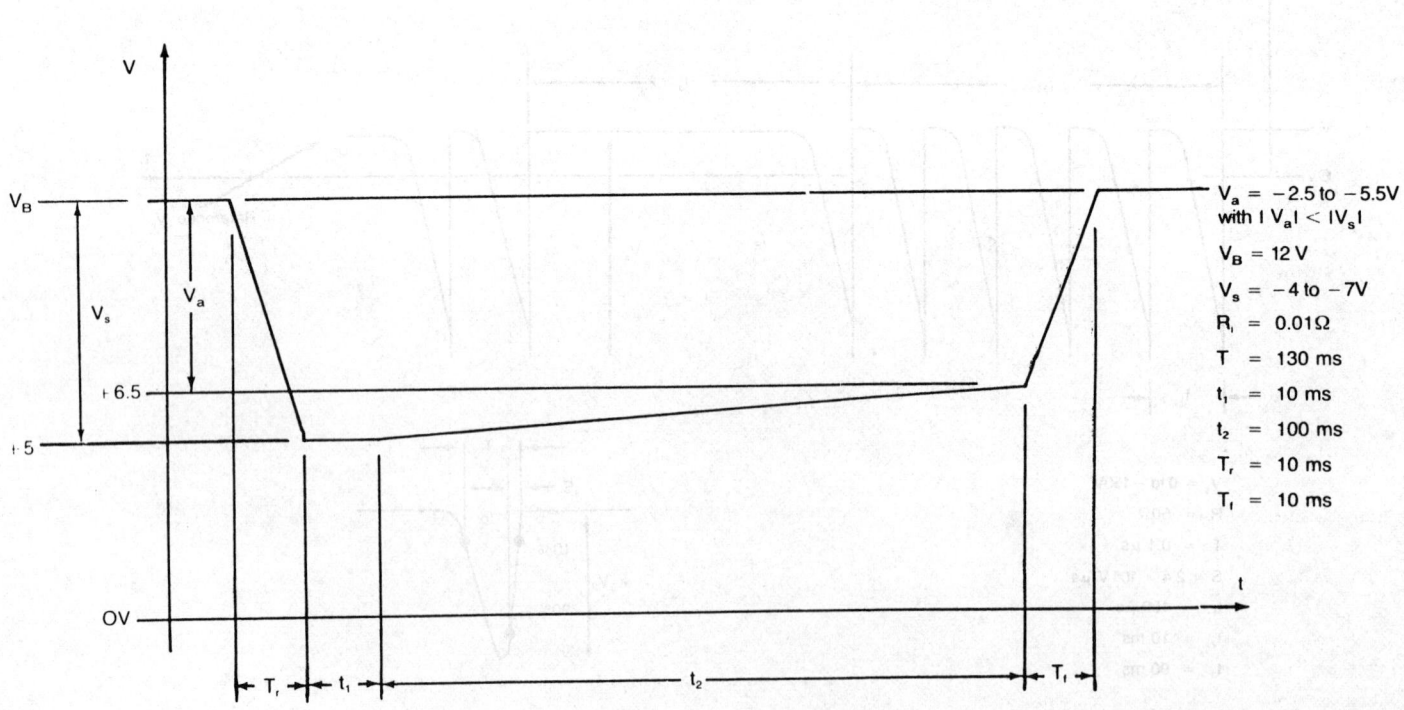

FIG. 7E—TEST PULSE 4 (SINGLE PULSE, FOR EXAMPLE, STARTER MOTOR ENGAGEMENT DISTURBANCE)

ing frequency dependent magnetic flux density of 160 dBpT at 60 Hz, decreasing at a rate of 12 dB/octave. For a pair of Helmholtz coils spaced one radius, the magnetic flux density at the center of the system is given by:

$$B = \mu_o H = (8.991 \times 10^{-7} N I)/R \text{ (Teslas)},$$

where N is the number of turns on each coil, R is the coil radius in meters, and I is the coil current in amperes. The coils are connected in series so that the magnetic fields add.

Or, the unperturbed magnetic field at the center of the system is given by:

$$H = \frac{0.7155 \, N \, I}{R} \text{ (A/m)},$$

The current carrying capability and turn ratio should be selected such that the test specification can be met. The coil should not have a self resonant frequency at or lower than the upper test frequency. Helmholtz coils can be purchased commercially.

(b) Function generator capable of producing 60 Hz to 30 kHz.
(c) Audio Power Amplifier—60 Hz to 30 kHz (approximately 200 W). Should be capable of delivering power to the coil to generate the specified magnetic field intensity at various frequencies as shown in Fig. 10.
(d) Current Monitor—60 Hz to 30 kHz.
(e) Magnetic Field Intensity Monitor—60 Hz to 30 kHz. Should be capable of measuring the specified magnetic field intensity.

6.5 Test Setup and Procedure
(a) Connect the test setup according to Fig. 11 without the EUT.
(b) Calibrate the system by generating the magnetic field, measuring the field using the intensity monitor and recording the current versus field values.

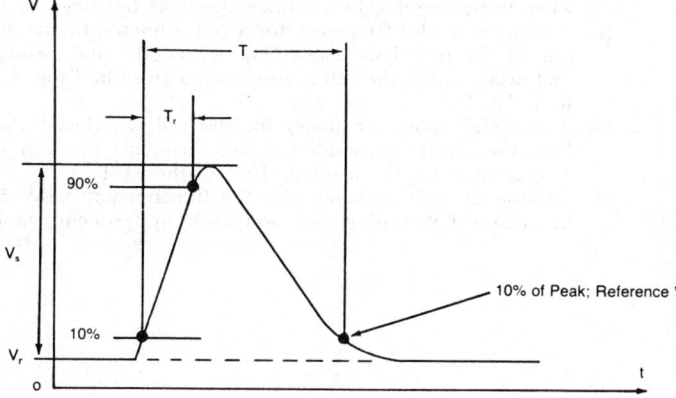

V_s = 25 to 120V
T = 40 to 400 ms
T_r = 5 to 10 ms
R_i = 0.5 to 4Ω

NOTES: General considerations of the dynamic behavior of alternators during load dump:
1) The internal resistance of an alternator in the case of load dump is mainly a function of speed and excitation current. The internal resistance of the load dump simulator shall be obtained from the following relationship:

$$R_i = \frac{10 \, V_{nom} \, N_{actual}}{0.8 \, I_{rated} \, 12000 \text{ min}^{-1}}$$

where V_{nom} = specified voltage of the generator, I_{rated} = specified current at a speed of 6000 min^{-1} (see ISO 8854), and N_{actual} = actual alternator speed (min^{-1}).
R_i shall represent a value corresponding to specified values of excitation current and speed.
2) The parameters are interrelated such that high values of V_s, R_i and T correspond to each other and low values vice versa.
3) Depending on the characteristics of the EUT (possibly including Zener diodes and varistors) the observed voltage waveform across the EUT may be quite different from the open circuit waveform.

FIG. 7F—TEST PULSE 5 (LOAD DUMP, SINGLE PULSE)

(c) Place the operating EUT in the central region of the Helmholtz coil. (Note: The Helmholtz coil criteria should be met.) Generate the desired magnetic field levels in accordance with the calibration established in paragraph 6.5b.
(d) Monitor the EUT and record the respective magnetic field intensity for: (1) malfunction, (2) degradation of performance, or (3) deviation of parameters beyond tolerances in the EUT specifications and approved test plan.

6.6 Notes—Caution must be exercised when operating high power amplifiers to avoid hazards to personnel and instrumentation. Instrumentation in the near vicinity of the coils must be shielded to prevent interference from radiated fields. Care should be exercised not to operate the coils near large metal objects or inside a shielded enclosure.

7. Radiated Susceptibility, 14 kHz - 200 MHz, Electric and Magnetic Fields Using a TEM Cell

7.1 Purpose—This section covers requirements for the determination, typically, of the electric-field susceptibility of equipment, subsystems, and systems (whose largest dimension is less than 15 cm) in the frequency range 14 kHz - 200 MHz using a TEM cell. However, since components of both the electric and magnetic fields are established in the TEM cell, the technique can be used for both electric and magnetic field susceptibility testing.

7.2 Measurement Philosophy—A TEM transmission cell is a rectangular adaptation of a coaxial line which sets up a region of uniform electric and magnetic fields in a traveling wave of essentially free-space impedance. The EUT is exposed to this electromagnetic source, but, typically, only the electric-field component is monitored. However, since components of both the electric and magnetic field are present, the exposure field can be calibrated in terms of either electric or magnetic field for either electric or magnetic field susceptibility testing.

This technique also prevents disturbance to equipment not under test since the RF field source and EUT are completely self-contained within the electromagnetic enclosure.

The TEM cell can be used to accurately generate absolute test fields when the EUT does not occupy an excessive portion of the test volume (see paragraph 7.5c). It is especially useful for diagnostic testing to determine, for example, frequencies of EUT susceptibility, some indication of how interference is coupled into the EUT, and the relative improvement in EUT immunity resulting from efforts to reduce EUT susceptibility. It cannot be used to determine EUT susceptibility to absolute test field levels if the EUT includes long wire harnesses that must be exposed to the test field. Only relative tests can be performed for this situation. The TEM cell lends itself to broadband automated testing and to continuous, swept frequency testing over the complete frequency range covered by the single TEM mode operation of the cell. TEM cell limitations due to EUT size and cell multimoding are discussed in SAE J1448 JAN84 (see Ref. 2). An alternate measurement approach is the use of open field tests as outlined in the SAE J1338 JUN81 (see Ref. 3). This approach however, is limited to discrete frequencies.

7.3 Apparatus
(a) Signal Source—Any commercially available signal source, power amplifier, and general-purpose amplifier capable of supplying at least 100 W of modulated and unmodulated power to develop the susceptibility levels specified in the test plan shall be used. Frequency accuracy shall be within ± 2%. Harmonics and spurious outputs shall not be more than −30 dB referred to the fundamental power.
(b) RF Voltmeter—A commercially available RF voltmeter capable of measuring 100 V over the frequency range 14 kHz - 200 MHz.
(c) Termination—One 200 W, 50 Ω load.
(d) Frequency Counter—A frequency counter capable of measuring frequencies up to 200 MHz.
(e) TEM Transmission Cell—A transverse electromagnetic transmission cell is shown in Fig. 13.
(f) Low-Pass Filter—Cutoff at 200 MHz, with the signal down 60 dB at frequencies greater than 300 MHz.
(g) Monitors—Required test equipment to monitor the operation of the EUT.
(h) Dual Directional Couplers—−30 dB or greater coupling ratio, 10 - 200 MHz.
(i) RF Power Meters with Sensors—Capable of measuring RF power levels up to 100 mW at frequencies of 10 - 200 MHz.
(j) Dual Channel XY Recorder

7.4 Test Setup and Procedure—A detailed, step-by-step measurement procedure, suggested as a systematic approach for evaluating the EM radiated susceptibility of EUT is contained in SAE J1448 JAN84.

Briefly, this includes:
(a) Place the EUT inside the cell as shown, for example, in Fig. 13.
(b) Access the EUT as required for operation and performance monitoring using appropriate shielded and fiber optic lines routed to filtered feed-through connectors mounted on the bottom outer shield as shown in Fig. 13. Care must be taken in routing the leads to obtain the most meaningful, repeatable results. Record lead positions for future reference.
(c) Connect up the measurement system as shown in Fig. 14A or 14B. Fig. 14A is used for frequencies below 10 MHz and Fig. 14B is used for frequencies above 10 MHz.
(d) Generate the test field as required. The field strength, E_o, at the center of the cell, midway between the septum and lower or upper wall is determined by:

$$E_o = \frac{V_{rf}}{b} \text{ (V/m)}$$

where V_{rf} is the input voltage to the cell in volts and b is the cell floor-to-septum separation in meters. At frequencies above 10 MHz, V_{rf} is determined from the expression:

$$V_{rf} = \sqrt{P_n/G_c} \text{ (V)}$$

where P_n is the net power flowing through the cell as measured by the power meters on the sidearm of the calibrated bi-directional coupler, and G_c is the real part of the cell's characteristics admittance (approximately = 0.2 mhos).
If the EUT is placed near the floor of the cell, the test field will be lower, relative to the field midway between the septum and floor. The correction is from 5-15% depending upon the cell form factor (width to height ratio, a/b) (that is, if a/b is 1.0, this correction factor is 0.85; if a/b is 1.5, this correction factor is 0.92; and if a/b is 1.67, the correction factor is 0.95).
(e) Operate the EUT as required while monitoring its response to the RF test fields. Scan the entire frequency range from 14 kHz to 200 MHz with particular emphasis made at the EUT's critical frequencies (local oscillator frequencies, intermediate frequencies, etc.) as specified in the test plan. Orient the EUT in each orthogonal plane within the cell to determine maximum susceptibility. Record the threshold of susceptibility as performance is observed or the maximum level specified in the test plan is achieved.

7.5 Notes
(a) Unless otherwise required in the equipment specification or approved test plan, the test signals shall be modulated according to the following rules:
1. EUTs with audio channels/receivers.
   AM Receivers: Modulate 30% with 1000 Hz tone.
   FM Receivers: When monitoring signal-to-noise ratio, modulate with 1000 Hz signal using 30% modulation. When monitoring receiver quieting, use no modulation.
   Other Equipment: Same as for AM receivers.
2. EUTs with video channels other than receivers. Modulate 90-100% with pulse duration of 2/BW and repetition rate equal to BW/1000, where BW is the video bandwidth.
3. Digital Equipment—Use pulse modulation with pulse duration(s) and repetition(s) equal to those used in the EUT.
4. Non-Tuned Equipment—Amplitude-modulate 30% with 1000 Hz tone, or as otherwise required in the test plan.

(b) Detailed considerations concerning EUT size, frequency limitations, and construction specifications can be found in SAE J1448 JAN84. In general, the device should be less than 1/3 the length (L), width (2a), and separation distance between cell septum and floor (b). Test samples of any size could be tested using a TEM cell modeled from Table 1 as long as the EUT size verses TEM cell size criteria defined above are satisfied. (These dimensions constraints prevent excessive impedance loading and test-field perturbation when inserting the EUT into the cell.) Thus, a small EUT could be tested at higher frequencies in small cells, and a large EUT could be tested at lower frequencies in a larger cell. The procedure for testing is the same for all sizes of cells except higher power signal sources and appropriate high-power terminations (50 Ω) are required when using larger cells to obtain the same test field levels.

(c) The upper useful frequency for a cell is limited by the distortion of the test signal caused by multimodes and resonances that occur within the cell at frequencies given in Table 1. (See Refs. 10, 11.)

(d) The useful upper frequency for the cell is reduced 10-20% from the cutoff-multimode resonant frequency given in Table 1 to account for the loading effect of the EUT.

(e) Because the cell operates with the fundamental TEM mode, broadband CW testing with amplitude or frequency modula-

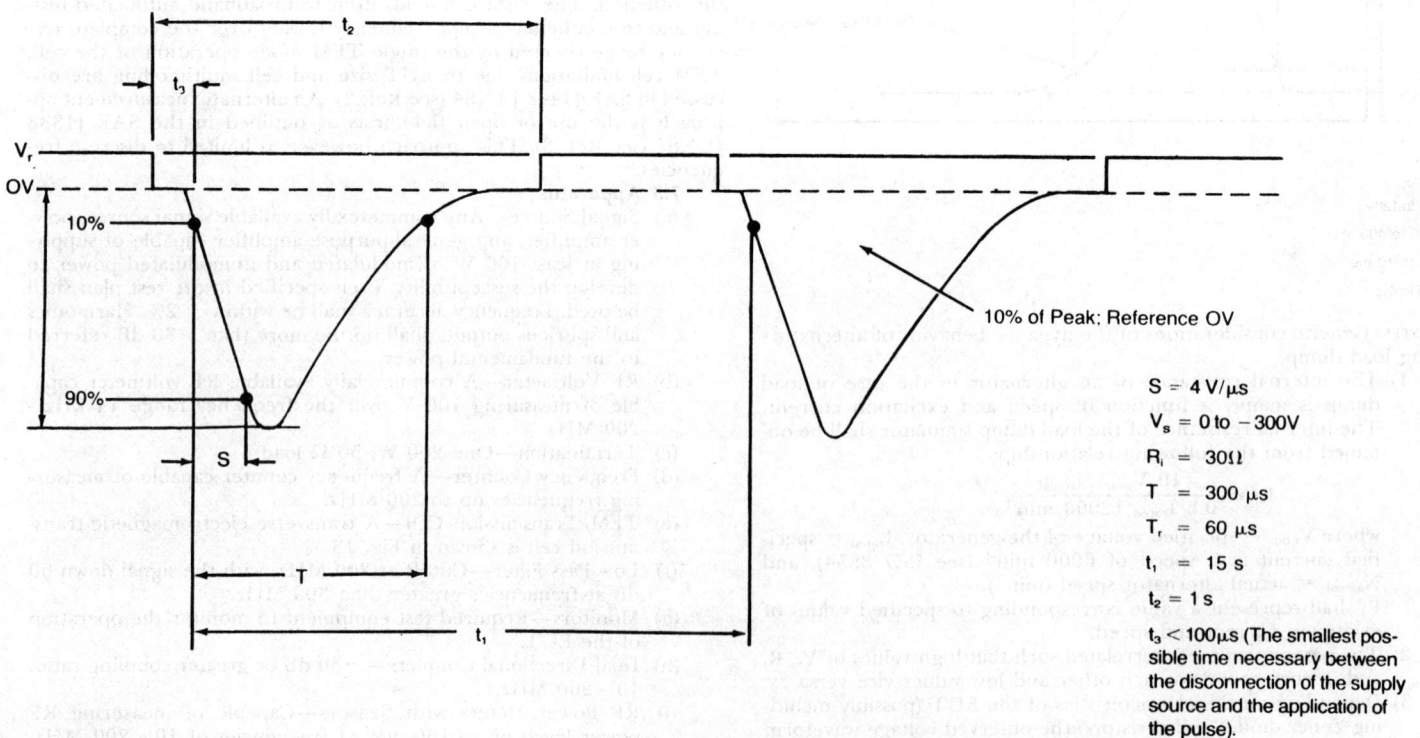

$S = 4 \text{ V/}\mu\text{s}$
$V_s = 0 \text{ to } -300\text{V}$
$R_i = 30 \Omega$
$T = 300 \mu\text{s}$
$T_r = 60 \mu\text{s}$
$t_1 = 15 \text{ s}$
$t_2 = 1 \text{ s}$
$t_3 < 100 \mu\text{s}$ (The smallest possible time necessary between the disconnection of the supply source and the application of the pulse).

FIG. 7G—TEST PULSE 6 (IGNITION COIL CURRENT INTERRUPTION)

tion is possible. In addition, the cell can be used to establish impulsive wave forms for testing by using an appropriate waveform generator connected to the cell's input port, assuming the frequency content of the wave form does not exceed the multimode cutoff frequency of the cell.

Not all the tests outlined in this measurement procedure may be required, and only those required by the test plan should be performed. For example, if the objective of the measurement program is to reduce the vulnerability (susceptibility) of the EUT, one EUT orientation with one input/output lead configuration could be tested in one particular operational mode to a preselected susceptibility test-field waveform and amplitude. Then, if corrective measures were made to the EUT and placement of the EUT and its leads inside the cell were carefully duplicated, repeat measurements could be made. These measurements could then be compared to determine the degree of improvement.

8. Radiated Susceptibility 14 kHz - 1 GHz, Electric Field and Magnetic Field Using a Strip Line

8.1 Purpose—This section covers requirements for the determination, typically, of the electric field susceptibility of equipment, subsystems, and systems (whose maximum height is less than 10 cm and maximum length is less than 2 m) in the frequency range 14 kHz to 1 GHz. However, since components of both the electric and magnetic fields are established in the line, the technique can be used for either electric or magnetic field susceptibility testing.

8.2 Measurement Philosophy—A strip line is a shieldless, unbalanced version of a TEM transmission line which sets up a region of uniform electric and magnetic fields in a traveling wave of essentially free-space impedance. The primary usage of the strip line is to couple RF onto the wire harness feeding the EUT. If desired, the EUT may also be tested using the strip line similar to using a TEM cell, Section 7, but with a proportionally smaller EUT. However, if the EUT is placed under the strip line, care must be exercised not to use the test setup above the multimode frequency of the strip line.

This technique is intended primarily for use in diagnostic testing to determine, for example, frequencies of EUT susceptibility, some indication of how interference is coupled into the EUT, and the relative improvement in EUT immunity resulting from efforts to reduce EUT susceptibility. It cannot be used to determine EUT susceptibility to absolute test field levels if the EUT includes long wire harnesses that must be exposed to the test field. This is because the wire harness is oriented longitudinally along the length of the strip line. Both the E and H fields are cross polarized with the wire harness for this configuration; hence, the coupling occurs mainly at the ends of the harness where it drops vertically to the EUT (aligned with the TEM E-field). It is also possible to induce RF onto the harness from differential mode coupling since the different wires are at different heights above the strip line's ground plate.

8.3 Apparatus
(a) Signal Source—Any commercially available signal source, power amplifier, and general-purpose amplifier capable of supplying at least 100 W of modulated and unmodulated power to develop the susceptibility levels specified in the test plan shall be used. Frequency accuracy shall be within ± 2%. Harmonics and spurious outputs shall not be more than −30 dB referred to the fundamental power.
(b) In-Line Wattmeter—A commercially available wattmeter capable of measuring 200 W over the frequency range 2 MHz to 1 GHz.
(c) RF Voltmeter—A commercially available RF voltmeter capable of measuring 100 V over the frequency range 14 kHz to 10 MHz.
(d) Frequency Counter—A frequency counter capable of measuring frequencies up to 1 GHz.
(e) Strip Line—A strip line is shown in Figs. 15A and 15B. The L dimension should be at least 2 m. The ratio of W to H determines the characteristic impedance according to the following equation:

$$Z_o = \frac{120\,\pi}{W/H + 2.42 - 0.44\,H/W + (1 - H/W)^6} \text{ for } W/H > 1$$

Typical strip line is generally constructed to be either 50 or 96 Ω with W/H equal to 5 and 1.75 respectively. The resistive load can be constructed of carbon resistors, conductive strips, etc. such that it matches the characterstic impedance of the strip line minimizing the standing waves.

8.4 Test Setup and Procedure
(a) Test setup should be as shown in Fig. 16.
(b) The EUT wire harness should be placed in non-conductive fixture and placed in the center of the line supported 5 cm off the ground plane.
(c) The EUT must not be grounded to the strip line, but may be placed on a non-conductive pad located on the strip line ground plane.
(d) Generate the test field as required. For frequencies below λ ≥ 10L, the field strength, E_v, can be measured by using an RF

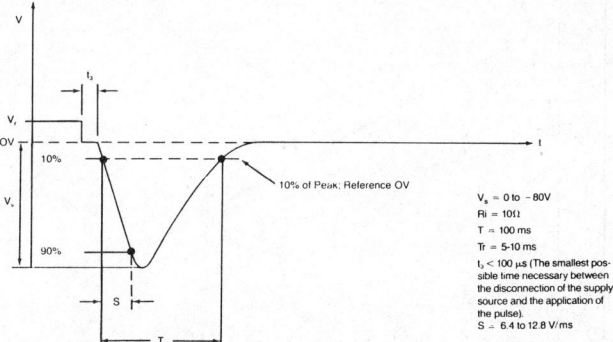

FIG. 7H—TEST PULSE 7 (SINGLE PULSE, FOR EXAMPLE, ALTERNATOR FIELD TRANSIENT AT ENGINE TURN-OFF)

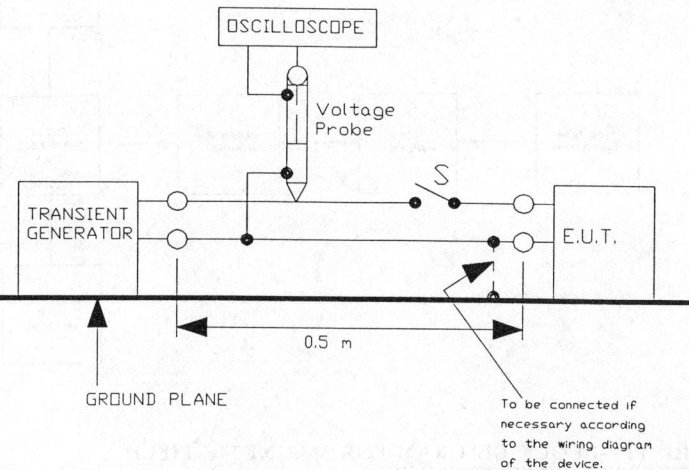

FIG. 8—TEST SETUP FOR CONDUCTED TRANSIENT SUSCEPTIBILITY

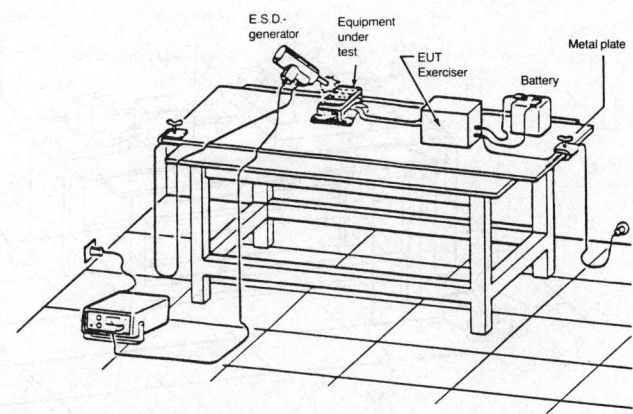

FIG. 9A—LABORATORY TESTING OF DEVICES INTENDED FOR INSTALLATION ON GROUND-CONNECTED CHASSIS PARTS OF VEHICLE

voltmeter where:

$$E_v = \frac{V_{rf}}{H} \text{ volts/meter}$$

For higher frequencies, a commercially available small field probe should be used to establish a calibration curve.

E_v can be monitored after establishing the calibration curve by:

$$E_v = \frac{\sqrt{PZ}}{H} \text{ volts/meter}$$

where P is the measurement forward power into the strip line in watts, Z is the strip line impedance in ohms, and H is the strip line spacing in meters.

(e) The EUT shall be operated by its normal inputs, where possible, and by simulators external to the strip line, filtered as required.

8.5 Notes

(a) Unless otherwise required in the equipment specification or approved plan, the test signals shall be modulated according to the following rules:
1. Test samples with audio channels/receivers.
   AM Receivers: Modulate 30% with 1000 Hz tone.
   FM Receivers: When monitoring signal-to-noise ratio, modulate with 1000 Hz signal using 30% modulation. When monitoring receiver quieting, use no modulation.
   Other Equipment: Same as for AM receivers.
2. EUT's with video channels other than receivers. Modulate 90 -100% with a pulse of duration 2/BW and repetition rate equal to BW/1000 where BW is the video bandwidth.
3. Digital Equipment—Use pulse modulation with pulse duration(s) and repetition rate(s) equal to those used in the equipment.
4. Non-Tuned Equipment—Amplitude modulate 30% with 1000 Hz tone or as otherwise required in the test plan.

(b) At frequencies over $\lambda = 2L$, and 2W, moding may occur which will reduce the accuracy of the fields generated by the strip line.

(c) At frequencies over $\lambda = 4H$, the strip line will radiate energy.

(d) If the EUT occupies a significant portion of the volume between the plates, the test field will be perturbed, resulting in a stronger field than that indicated by the measured forward power.

(e) Since the RF field is not self-contained, the test must be performed in a shielded room with the generating and monitoring equipment being outside the room. Energy radiated from the line into the enclosure will result in room resonance which can cause large errors. This effect can be reduced significantly by installing RF absorbing panels. These panels reduce the level of the reflections from the walls of the shielded room.

9. Radiated Susceptibility, 200 MHz - 18 GHz, Planar Field (Far-Field)

9.1 Purpose—This section covers the requirements for the determination of electric-field susceptibility of equipment, subsystems, and systems in the frequency range 200 MHz - 18 GHz.

9.2 Measurement Philosophy—At frequencies about 200 MHz, TEM cells become too small to test many types of equipment. However, RF absorbing material becomes effective and measurements can be made within an RF shielded enclosure provided RF absorbing material is used to make it anechoic. The use of such a facility is required if swept frequency testing is to be performed to contain the test field and prevent interference to equipment not under test. Alternately, open field tests as outlined in SAE J1338 (see Ref. 3) could be used but are limited to discrete frequencies.

9.3 Apparatus

(a) Anechoic Chamber—The size, shape, and construction of an anechoic chamber can vary considerably depending upon the tests to be performed, the size of the EUT and the frequency range to be covered. Basically, an anechoic chamber consists of a shielded room with RF absorbing material mounted on its internal surfaces. The minimum size of the room is determined by the size of the test region needed, the size of the transmitting antennas, the clearances needed between the absorber and antennas, and the separation distance required between the transmitting antenna and the EUT. To create the test region (quiet zone), the absorber, antenna systems, and chamber shape are selected to reduce the amount of extraneous energy in the test region below a minimum value which will give the desired measurement accuracy. Since the performance of the absorber is a function of its construction (thickness, shape, material, etc.) and the angle of incident energy, the determination of the optimum location and construction of the absorber in an anechoic chamber to meet specified reflectivity requirements can be complicated. Guidelines for designing and using an anechoic chamber for EMC measurements are available. (See Refs. 13-15.) Table 2 gives a general recommended minimum thickness of the absorber for covering the enclosure walls which will provide at least 20 dB attenuation of the reflected power.

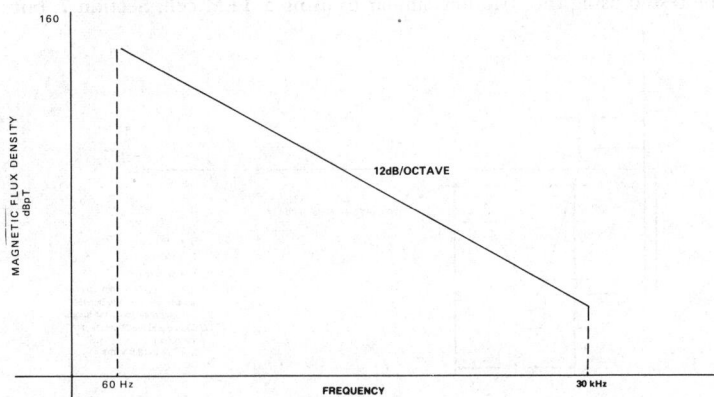

FIG. 10—MAGNETIC FLUX DENSITY VERSUS FREQUENCY

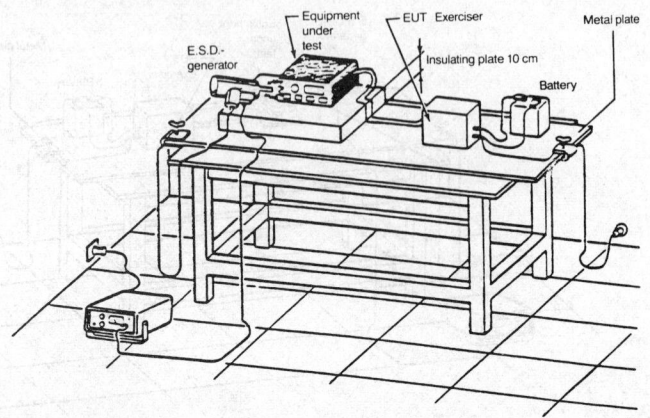

FIG. 9B—LABORATORY TESTING OF DEVICES INTENDED FOR INSTALLATION ON INSULATING PLASTIC PARTS ON VEHICLE

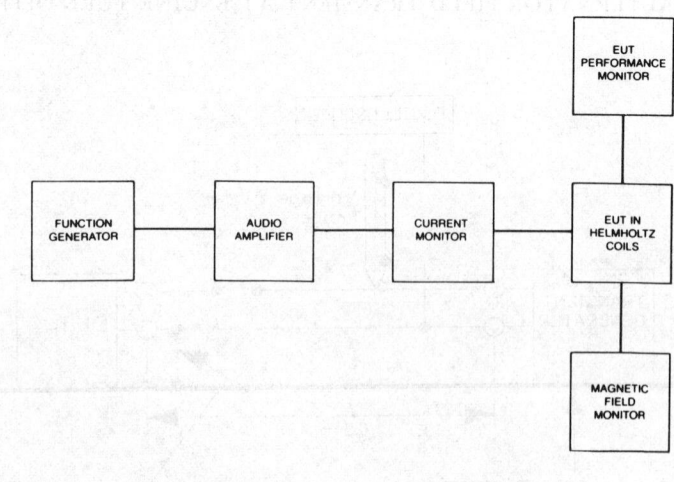

FIG. 11—BLOCK DIAGRAM FOR MAGNETIC FIELD SUSCEPTIBILITY TESTING

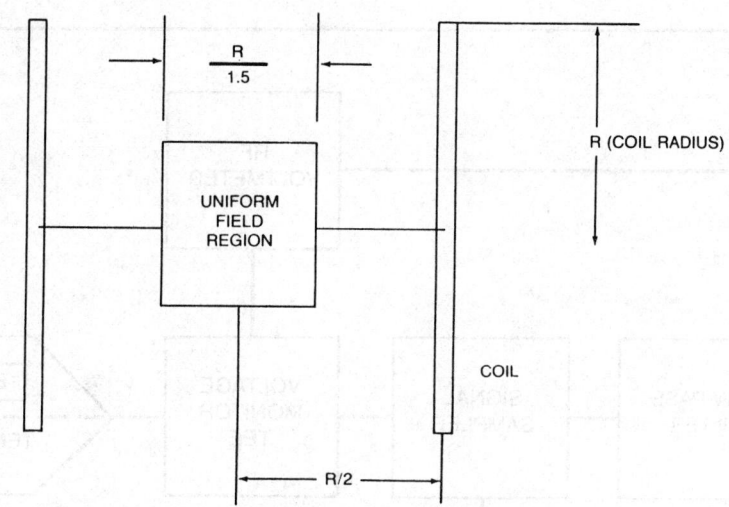

FIG. 12—HELMHOLTZ COIL CONFIGURATION

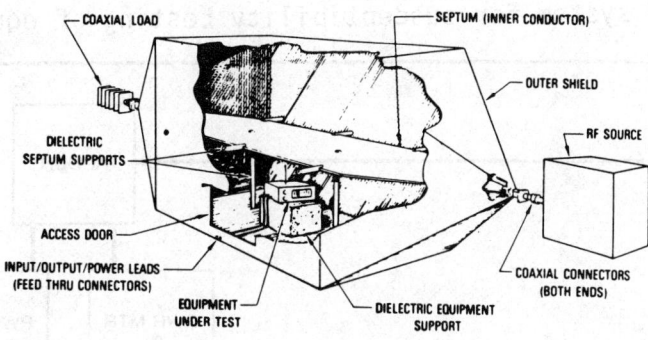

FIG. 13—CUT-AWAY VIEW OF TEM CELL USED FOR RADIATED SUSCEPTIBILITY TESTING. FIGURE SHOWS PLACEMENT OF EUT AND ASSOCIATED INPUT, OUTPUT AND MONITORING LEADS INSIDE THE CELL.

(b) Signal Source—Any commercially available signal source, power amplifier, and general purpose amplifier capable of supplying the necessary modulated and unmodulated power to develop the susceptibility test signals levels specified in the test plan may be used. Frequency accuracy shall be within ± 2%; Harmonics and spurious outputs shall not be more than −30 dB referred to the fundamental power.

(c) Radiating Antennas—Any standard antenna such as a log periodic (200 MHz - 1 GHz), and rectangular or ridged horns (1 - 8 GHz and 8 - 18 GHz bands) shall be used. Linear antennas yield more information about polarization if used.

(d) Sensing and Calibration Equipment—Any standard antenna with EMI meter or spectrum analyzer or calibrated field measuring probes can be used. Optimum results are obtained by using electrically small, isolated E-field probes, which remove many of the uncertainties associated with physically large receiving antennas. For a more detailed discussion and recommendations to avoid problems inherent in probing the fields around an EUT, see SAE Paper No. 830606 (see Ref. 16).

(e) Output Monitor—Appropriate instrumentation to monitor the performance of the EUT shall be used.

9.4 Test Setup and Procedure

(a) The block diagram is shown in Fig. 17.

(b) Distance from the EUT to the transmitting antenna is a function of frequency, EUT, and antenna size. If the EUT or the largest transmitting antenna dimension is less than a wavelength (λ) the far-field begins at $\lambda/2\pi$. A uniform test field then is obtained at separation distances greater than λ. For electrically large EUT and/or for antennas with electrically large apertures, the separation distance should exceed d^2/λ where d is the largest dimension of the antenna or EUT.

(c) Fields should be generated, as required, with the specified antenna. Care should be taken so that the test equipment is not affected by the test signals. Test equipment, except for the antenna, should be outside the shielded enclosure.

(d) The specified field strength and polarization shall be established prior to the actual testing by substituting a field-measuring antenna in place of the EUT and by adjusting and recording the transmitter power required to obtain a specified field intensity from the transmitting antenna. (This calibration may be used for all subsequent testing provided that exactly the same EUT location is used.)

(e) The EUT shall be oriented in each orthogonal plane to determine maximum susceptibility.

(f) The entire frequency range from 0.2 - 18 GHz shall be scanned. Tests shall be conducted at not less than three frequencies per octave, representing the maximum susceptibilities within that octave. In addition, tests shall also be made at the EUT critical frequencies (local oscillator frequency, intermediate frequency and others) as specified in the test plan. Determine the threshold of susceptibility by increasing the test signal until degradation of performance is observed or the

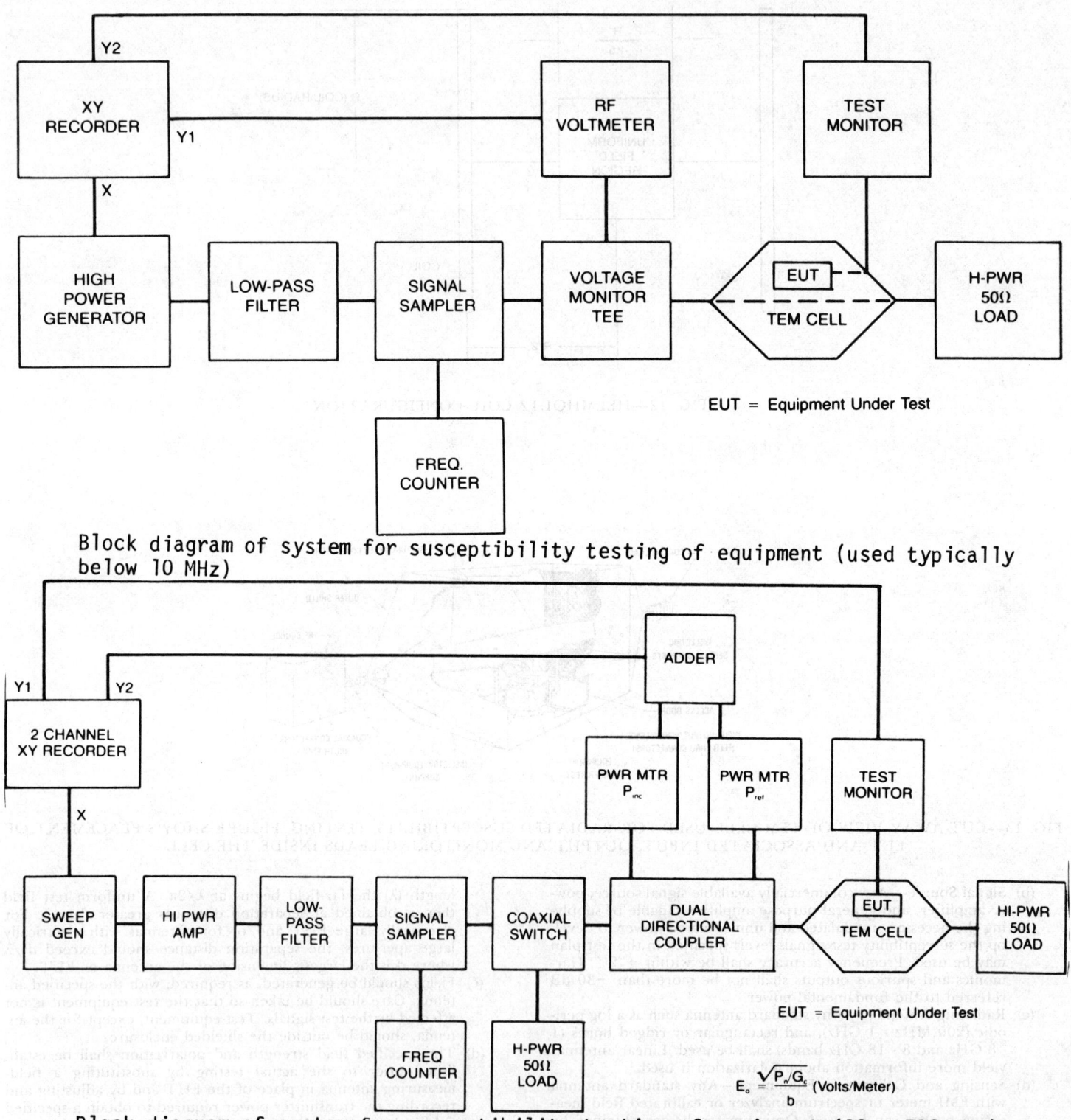

Block diagram of system for susceptibility testing of equipment (used typically below 10 MHz)

Block diagram of system for susceptibility testing of equipment (10 - 500 MHz)

FIG. 14—BLOCK DIAGRAMS OF TEM CELL SYSTEMS FOR SUSCEPTIBILITY TESTING OF EQUIPMENT

required test level is achieved.

9.5 Notes

(a) Unless otherwise required in the equipment specification or approved plan, the test signals shall be modulated according to the following rules:

1. EUT's with video channels other than receivers, modulate 90 -100% with a pulse of duration 2/BW and repetition rate equal to BW/1000, where BW is the video bandwidth.
2. Digital Equipment—Use pulse modulation with pulse duration(s) and repetition rate(s) equal to those used in the equipment.
3. Non-Tuned Equipment—Amplitude modulate 30% with the 1000 Hz tone or as otherwise required in the test plan.

(b) The EUT configuration shall simulate actual operating conditions as nearly as possible in terms of surrounding metal structure, lead length, and terminating impedances.

(c) RF power handling capability of absorber shall be adequate to insure safe operation.

(d) When required, a copper or brass ground plane (solid plate) shall be used that has a minimum thickness of 0.25 mm for copper or 0.63 mm for brass and is 2.25 m² or larger in area with the smaller side no less than 76 cm. The ground plane shall be bonded to the shielded room such that the DC bonding resistance shall not exceed 2.5 milliohms. In addition, the bonds shall be placed at distances no greater than 90 cm apart. For large equipment mounted on a metal test stand, the test stand shall be considered a part of the ground plane for testing purposes and shall be bonded accordingly. The faces of the test sample shall be located 10 ± 2 cm from the edge of the ground plane. All leads and cables shall be within 10 ± 2 cm from the edge of the ground plane and shall be approximately 5 cm above the ground plane.

10. References

1. [1]SAE J1595, Electrostatic Discharge Test For Vehicles. To be published.
2. [1]SAE J1448 JAN84, Electromagnetic Susceptibility Measurements of Vehicle Components Using TEM Cells, 14 kHz - 200 MHz.
3. [1]SAE J1338 JUN81, Open Field Whole-Vehicle Radiated Susceptibility, 10 kHz -18 GHz, Electric Field.
4. [1]SAE J1507 JAN87, Anechoic Test Facility Radiated Susceptibility, 20 MHz - 18 GHz, Electromagnetic Field.
5. Adams, J. W., Taggart, H. E., Kanda, M., and Shafer, J., "Electromagnetic Interference (EMI) Radiative Measurements for Automotive Applications," NBS Tech. Note 1014, June 1979.
6. Natrella, M. G., "Experimental Statistics," NBS Handbook 91.
7. Johnson, N. L. and Kotz, S., "Continuous Univariant Distributions," Vol. 2.
8. Resnikoft, "Non-Central t Distribution."
9. Owen, D. B., "Handbook of Statistical Tables."

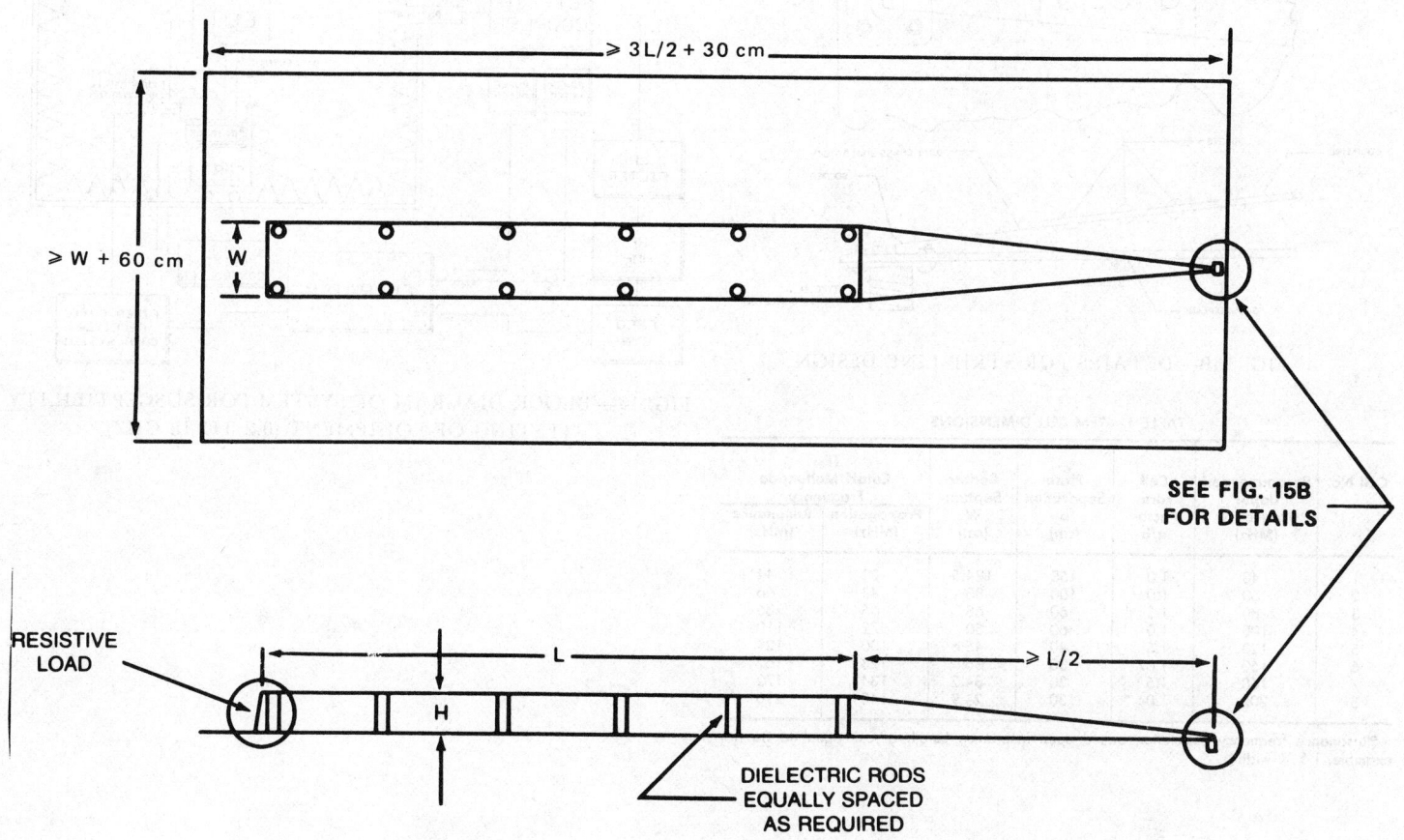

FIG. 15A—DESIGN FOR A STRIP LINE

10. ISO/TR 7637/0, 1-1984 (E), "Road Vehicles—Electrical Interference by Conduction and Coupling: Part 0, General, and Part 1, Vehicles with Nominal 12 V Supply Voltage—Electrical Transient Conduction Along Supply Lines Only," July 1, 1984.
11. Tippett, J. C., Chang, D. C., and Crawford, M. L., "An Analytical and Experimental Determination of the Cutoff Frequencies of Higher-Order TE Modes in a TEM Cell," NBSIR 76-841, June 1976.
12. Tippett, J. C., *Modal Characteristics of Rectangular Coaxial Transmission Line*, Thesis submitted June 1978 for degree of Doctor Philosophy to University of Colorado, Electrical Engineering Dept., Boulder, CO.
13. Nichols, F. J., and Hemming, L. H., "Recommendations and Design Guides for the Selection and Use of RF Shielded Anechoic Chamber in the 30-1000 MHz Frequency Range," IEEE Inter. Symp. on EMC., Boulder, CO, August 18-20, 1981, pp. 457-464.
14. [1]Kinderman, J. C. et al., "Implementation of EMC Testing of Automotive Vehicles," SAE Paper 810333, February 1981.
15. [1]Vrooman, "An Indoor 60 Hz to 40 GHz Facility for Total Vehicle EMC Testing," SAE Paper 821011, June 1983.
16. [1]Bronaugh, E. L., and McGinnis, W. H., "Whole-Vehicle Electromagnetic Susceptibility Test in Open-Area Test Sites: Applying SAE J1338," SAE Paper 830606.

TABLE 2—RECOMMENDED MINIMUM THICKNESS OF ABSORBER FOR COVERING ENCLOSURE WALLS

Frequency MHz	Absorber Thickness in (cm)
200	36 (91)
300	24 (61)
500-1000	18 (46)

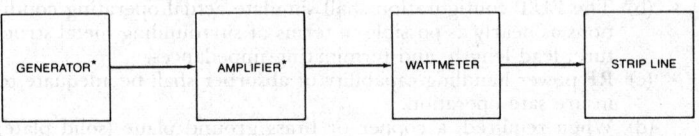

FIG. 16—BLOCK DIAGRAM FOR STRIP LINE

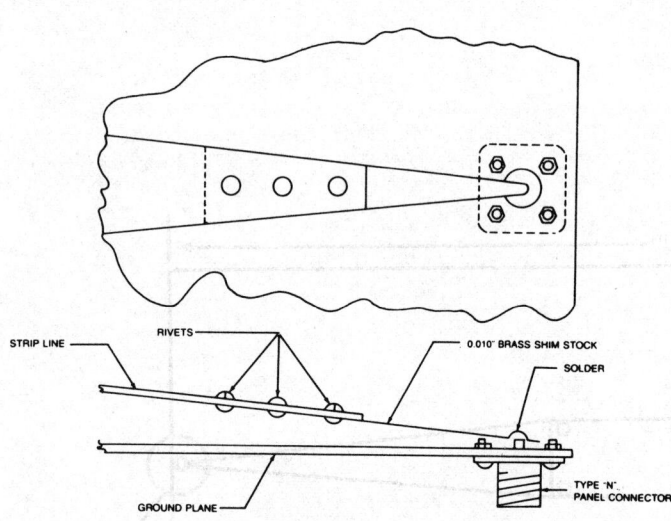

FIG. 15B—DETAILS FOR STRIP LINE DESIGN

FIG. 17—BLOCK DIAGRAM OF SYSTEM FOR SUSCEPTIBILITY TESTING OF EQUIPMENT (0.2 TO 18 GHZ)

TABLE 1—TEM CELL DIMENSIONS

Cell No.	Recommended Upper Frequency (MHz)	Cell Form Factor a/b	Plate Separation b (cm)	Center Septum W (cm)	TE_{01} Cutoff/Multimode Frequency Propagation (MHz)	TE_{01} Cutoff/Multimode Frequency Resonance (MHz)
1	40	1.0	150	124.5	29	44
2	60	1.0	100	83	43	66
3	80	1.5	60	68	66	86
4	100	1.0	60	50	72	110
5	120	1.5	40	45.6	100	129
6	150	1.67	30	36	128	162
7	160	1.5	30	34.2	134	172
8	200	1.0	30	24.9	143	220

aResonance frequency calculation based upon resonance length of cell equal to 3a (for example, 1.5 × width)

[1] Available from SAE, 400 Commonwealth Drive, Warrendale, PA 15096-0001.

APPENDIX A—LINE IMPEDANCE STABILIZATION NETWORKS

The following schematic diagrams are suggested for use in constructing the line impedance stabilization networks required for performing the conducted susceptibility measurements outlined in Section 3. Care must be taken in their use to insure that the impedances specified in Figs. 3 and 4 are obtained.

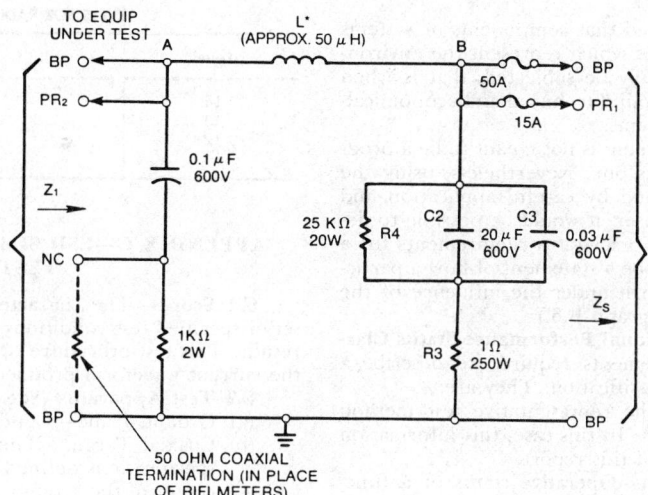

Z_1 - Impedance presented to the equipment when connected for measurements.
Z_s - Impedance of the power source used.
BP - Heavy duty binding posts (mfr. standard electric time co.)
PR_1 - Power receptacle, 115 V, 15 A (3-wire polarized twist lock, male base).
PR_2 - Power receptacle, 115 V, 15 A (3-wire non-polarized, "U" shaped grounding slot).

NC - Type "N" connector (UG-58/U) panel mounting.
L° - Coil - 26 turns of no. AWG-6 stranded wire with 600 V insulation wound on 5.5 in (14 cm) diameter coil form.
CASE - 17½"L × 17½"W × 8¾"H (44.4 cm L × 44.4 cm W × 22.2 cm H) brass (divided in two sections by a brass plate 17½" × 8⅝" × 1/16" (44.4 cm × 21.9 cm × 1.6 mm) thick).
NOTE—Dual line stabilization network consists of two of the above networks.

FIG. A1—LINE INPEDANCE STABILIZATION NETWORK (LISN), 50 KHZ - 5 MHZ

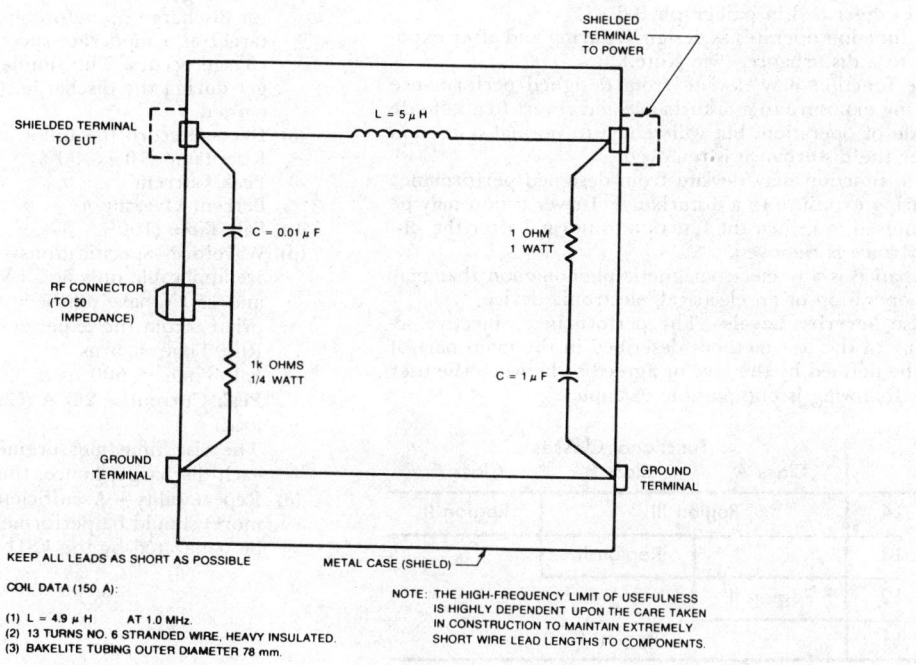

FIG. A2—LINE IMPEDANCE STABILIZATION NETWORK (LISN), 5 - 100 MHZ

APPENDIX B—FUNCTION PERFORMANCE STATUS CLASSIFICATION

B.1 Scope and Field of Application—The purpose of this appendix is to provide an example of a method to classify the functional status of automotive electronic devices upon application of the test conditions described in this report. The process outlined is applicable only to the bench testing of automotive electronic devices using the test methods described.

B.2 General—It must be emphasized that components or systems shall only be tested with those conditions which represent the environments to which the device would actually be subjected, that is when used in a vehicle. This will help assure sound technical and economically optimized designs for susceptible systems.

It should also be noted that this document is not meant to be a product specification and cannot function as one. Nevertheless, using the concepts described in this document and by careful application and agreement between manufacturer and user, it would be possible to develop a document to describe the functional status requirements for a specific device. This could then, in fact, be a statement of how a particular device could be expected to perform under the influence of the specified interference signals. (See paragraph B.6.)

B.3 Essential Elements of a Functional Performance Status Classification System—There are four elements required to describe a general function performance status classification. They are:
 (a) Test Method—Reference to the representative test method applied to the device under test. In this case, this information is contained in the main part of this report.
 (b) Functional Status Classification—Operative status of a function of a device during and after exposure to an electromagnetic environment.
 (c) Performance Region—For each Functional Class, three regions of Performance Objective levels are defined.
 (d) Test Pulse Severity Levels—Specification of severity levels for the three regions defined.

B.4 Functional Status Classification
Class A—Any function that provides a convenience (for example, entertainment radio, trip odometer).
Class B—Any function that enhances, but is not essential to the operation and/or control of the vehicle (for example, electronic climate control, fuel gauge).
Class C—Any function that is essential to the operation and/or control of the vehicle (for example, braking, steering).

B.5 Performace Region—For each functional class, the following performance regions are defined and used in conjunction with the performance objectives described in paragraph B.6.
Region I—The function operates as designed during and after exposure to a disturbance. (See Note.)
Region II—The function may deviate from designed performance during exposure to a disturbance and revert to a fail-safe mode of operation, but will return to normal operation after the disturbance is removed.
Region III—The function may deviate from designed performance during exposure to a disturbance. Driver action may be required to return the function to normal after the disturbance is removed.
Note—A disturbance is any electromagnetic phenomenon that may affect the proper operation of an electrical/electronic device.

B.6 Test Pulse Severity Levels—The performance objective severity levels for any of the test methods described in the main part of this report are to be defined by the user or agreed to between the user and supplier. The following is one possible example:

		Functional Classes		
		Class A	Class B	Class C
Performance	L4	Region III	Region III	Region II
Objective	L3		Region II	
Levels	L2	Region II		
	L1	Region I	Region I	Region I

LEVELS FOR TRANSIENT TESTING

Level	Pulse 1	Pulse 2	Pulse 3	---	---
L4	V1	V2			
L3	0.8 V1	0.75 V2			
L2	0.5 V1				
L1					

LEVELS FOR RADIATED SUSCEPTIBILITY TESTING

Level	Electric Field Strength
L4	X V/m
L3	0.6X V/m
L2	0.4X V/m
L1	

APPENDIX C—ESD SIMULATOR CHARACTERIZATION TEST PROCEDURE

C.1 Scope—The characterization of an ESD simulator requires a set of specified test conditions to ensure accurate and reproducible test results. This test procedure provides a standard method for measuring the current waveform produced by an ESD simulator.

C.2 Test Apparatus (See Fig. C1)
 (a) Ground Plane—Conductive top bench, at least 1 m².
 (b) Coaxial Target—The coaxial target is built according to the specifications defined in Fig. C2. The target is bonded to the center of the ground plane. An alternative to this design is the coaxial target defined in IEC 801-2. This target may be used if specified in the test plan.
 (c) Oscilloscope—Minimum Bandwidth - 400 MHz
  Storage Capability - Desirable
  Input Impedance - 50 ohms
  10X Attenuator (20 dB, 50 ohm)
 (d) Ground Cable—A ground cable is to be connected between the base of the coaxial target and the simulator (should minimize inductance).
 (e) Electrostatic Voltmeter—Input Impedance—greater than 10^{12} ohms.

C.3 Test Procedure (See Fig. C1)
 (a) Test Voltages—The ESD simulator is to be tested at 5 kV, 10 kV, and 15 kV. The voltage is measured at the discharge tip of the simulator using the electrostatic voltmeter.
 (b) Discharge Method—The test voltage is applied to the simulator discharge tip before approaching the target. Approach the target at a moderate speed (greater than 5 cm/s) until a discharge occurs. The simulator must not touch the coaxial target during the discharge. Only single discharges are to be recorded.
 (c) Data—Record the following data after each discharge:
Rise Time (10% - 90%)
Peak Current
Percent Overshoot
Fall Time (100% - 37%)
 (d) Waveform Specifications—The waveform specifications given are applicable only at 5 kV. Specifications applicable to 10 kV and 15 kV have not been defined as yet, but will be, as appropriate, from the experience gained in performing these tests.
Rise Time < 5 ns
Fall Time = 600 ns ± 20%
Peak Current ≥ 2.0 A (Current through 2 ohm coaxial target load)
The rise time measurements may not be reproducible with each discharge, hence, this is an average specification.
 (e) Repeatability—A sufficient number of measurements (20 or more) should be performed to ensure that repeatable data can be generated by the ESD simulator.

23.225

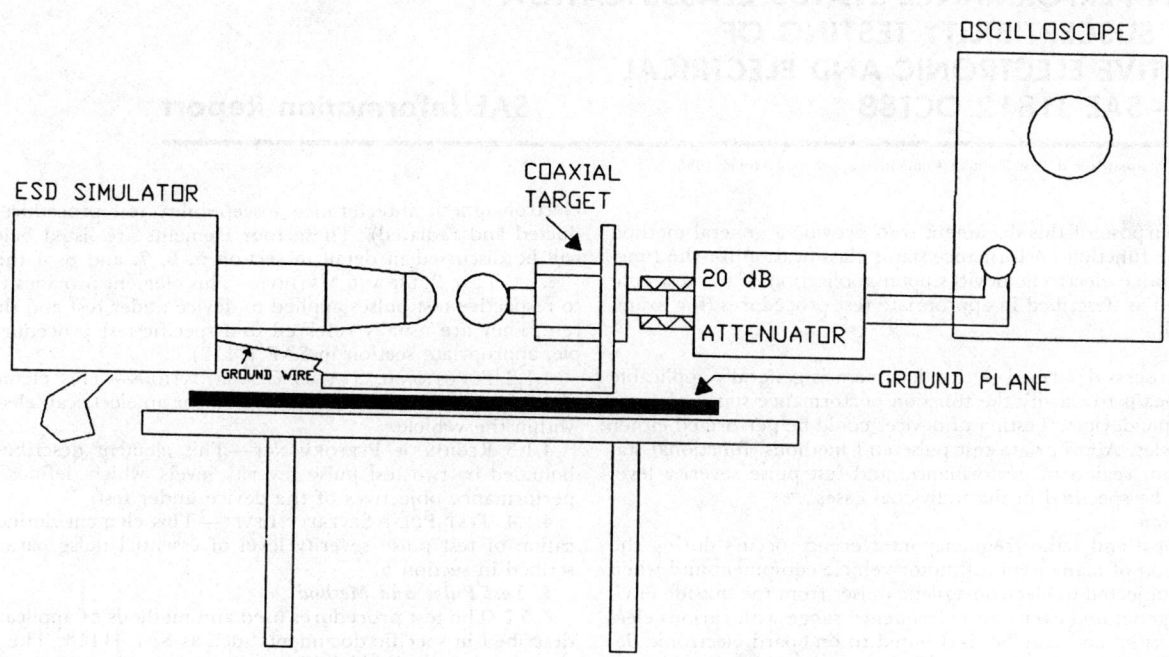

FIG. C1—ESD SIMULATOR CHARACTERIZATION TEST SETUP

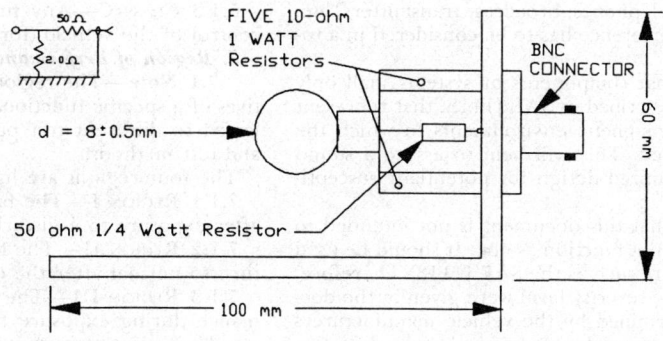

FIG. C2—COAXIAL TARGET

FUNCTION PERFORMANCE STATUS CLASSIFICATION FOR EMC SUSCEPTIBILITY TESTING OF AUTOMOTIVE ELECTRONIC AND ELECTRICAL DEVICES—SAE J1812 OCT88

SAE Information Report

Report of the EMI Standards and Test Methods Committee approved October 1988.

1. Purpose

1.1 The purpose of this document is to provide a general method for defining the function performance status classification for the functions of automotive electronic devices upon application of the test conditions specified as described in appropriate test procedures (for example SAE J1113).

2. Scope

2.1 The process described in this document is generally applicable to provide a means to classify the function performance status of automotive electronic devices. Testing of devices could be performed either on or off vehicles. Appropriate test pulse and methods, functional status classification, region of performance, and test pulse severity level would have to be specified in the individual cases.

3. Introduction

3.1 Electrical and radio frequency interference occurs during the normal operation of many items of motor vehicle equipment and when the vehicle is subjected to electromagnetic noises from the outside environment. It is generated over a wide frequency range with various electrical characteristics and may be distributed to on-board electronic devices and systems by conduction or radiation, or both.

3.2 During recent years, an increasing number of electronic devices has been introduced into vehicle designs in order to perform, control, monitor and display various functions including the engine management system. It has been necessary, therefore, to consider the electrical and electromagnetic environment in which these devices are required to operate. Interference can be generated in the vehicle electrical system itself by the normal operation of various power devices such as power window, power lock, air conditioning, etc. This interference can cause a temporary malfunction or even permanent damage to the electronic equipment. Significant number of performance deviations, resulting from this interference, have been reported.

3.3 Narrow band and broad band signals generated from sources inside or outside the vehicle could also be coupled into the electrical/electronic system, affecting the normal performance of electronic devices. These sources of electromagnetic interference are, for example, vehicle's ignition system, mobile telephones, broadcast transmitters, etc. Protection from this potential interference has to be considered in a total system validation.

3.4 It must be emphasized that components or systems shall only be tested with the conditions, as described in SAE J1113, that represent the simulated automotive electromagnetic environments to which the devices would actually be subjected. This will help to assure a sound technically and economically optimized design for potentially susceptible components and systems.

3.5 It should also be noted that this document is not intended to be a product specification and cannot function as one. It should be used in conjunction with a test procedure such as the SAE J1113. Therefore, no specific values for the test pulse severity level were given in the document since they should be determined by the vehicle manufacturers and the suppliers. Nevertheless, using the concepts described in this document and by careful application and agreement between manufacturer and supplier, this document could be used to describe the functional status requirements for a specific device. This could then, in fact, be a statement of how a particular device could be expected to perform under the influence of the specified interference signals.

3.6 Examples for the application of how the concept of function performance status classification could be applied to the conducted and radiated susceptibility testing are included in this document (see Appendix A and Appendix B).

4. Essential Elements of Function Performance Status Classification

4.1 There are four elements required to describe a function performance status classification. They can be generically applied to all electromagnetic interference susceptibility test procedures (both conducted and radiated). These four elements are listed below and they will be discussed in detail in section 5, 6, 7, and 8 of this document.

4.1.1 TEST PULSE AND METHOD—This element provides the reference to respective test pulses applied to device under test and the method of test. They are usually referred to a specific test procedure (for example, appropriate section in SAE J1113).

4.1.2 FUNCTIONAL STATUS CLASSIFICATIONS—This element describes the operational status of the function for an electrical/electronic device within the vehicle.

4.1.3 REGION OF PERFORMANCE—This element describes the region, bounded by two test pulse severity levels, which defines the expected performance objectives of the device under test.

4.1.4 TEST PULSE SEVERITY LEVEL—This element defines the specification of test pulse severity level of essential pulse parameters as described in section 5.

5. Test Pulse and Method

5.1 The test procedures used and methods of application are to be described in specific documents such as SAE J1113. The function performance status classification resulting from these tests would be applicable only to those particular test procedures.

6. Functional Status Classification

6.1 Note—All classifications are for the total device/system functional status. A given device or system may have several different functions and each individual function may have its own class of functional status. The classification of the function for any given device should be determined between the manufacturer and supplier. It is important to point out that, in many cases, only one or two classification(s) will apply to a particular product or function. For example, if the device has only one function, only part of Appendix A and Appendix B will apply (class A, B, or C).

6.1.1 CLASS A—Any function that provides a convenience (for example, entertainment, comfort).

6.1.2 CLASS B—Any function that enhances, but is not essential to the operation or control of the vehicle (for example, speed display).

6.1.3 CLASS C—Any function that is essential to the operation or control of the vehicle (for example, braking, engine management).

7. Region of Performance

7.1 Note—The region of performance defines performance objectives of a specific functional status classification when the device is subjected to different test pulse severity levels under various test pulses and test methods.

The four regions are listed below:

7.1.1 REGION I—The function shall operate as designed during and after exposure to a disturbance.

7.1.2 REGION II—The function may deviate from design but will return to normal after the disturbance is removed.

7.1.3 REGION III—The function may deviate from designed performance during exposure to a disturbance but simple operator action may be required to return the function to normal, once the disturbance is removed.

7.1.4 REGION IV—The device/function must not sustain any damage after the disturbance is removed.

8. Test Pulse Severity Level

8.1 The test pulse severity level is the stress level (voltage, volts per meter etc.) applied to the device under test for any given test method (section 5). The device should perform according to its functional status classification (section 6) and region of performance (section 7) during the test.

The test pulse severity level should be determined by the manufacturer and supplier (examples for how the test pulse severity level could be applied are included in Appendix A and B).

23.227

APPENDIX A
FUNCTIONAL STATUS CLASSIFICATION, REGION OF PERFORMANCE AND TEST PULSE SEVERITY LEVEL FOR CONDUCTED SUSCEPTIBILITY

A.1 Application of Functional Status Classification, Region of Performance and Test Pulse Severity Level:

FUNCTIONAL STATUS CLASSIFICATIONS

TEST PULSE SEVERITY LEVEL	CLASS A	CLASS B	CLASS C
LEVEL VI (L 6)	REGION IV	REGION IV	
LEVEL V (L 5)	REGION III	REGION III	
LEVEL IV (L 4)	REGION III		REGION II
LEVEL III (L 3)		REGION II	REGION II
LEVEL II (L 2)	REGION II		
LEVEL I (L 1)	REGION I	REGION I	REGION I

A.2 Test Pulse Severity Level Selection Table: Severity levels for each pulse or pulses corresponding to respective test pulse and method for Level I, II, III, IV, V, and VI are determined and entered in the table according to the test plan for the device under test (DUT). An example of a typical table (per SAE J1113) is listed below:

	PULSE 1	PULSE 2	PULSE 3a	PULSE 3b	PULSE 4	PULSE 5	PULSE 6	PULSE 7
L 6								
L 5								
L 4								
L 3								
L 2								
L 1								

APPENDIX B
FUNCTIONAL STATUS CLASSIFICATION, REGION OF PERFORMANCE AND TEST PULSE SEVERITY LEVEL FOR RADIATED SUSCEPTIBILITY

B.1 Functional Status Classification, Region of Performance and Test Pulse Severity Level

FUNCTIONAL STATUS CLASSIFICATIONS

TEST PULSE SEVERITY LEVEL	CLASS A	CLASS B	CLASS C
LEVEL VI (L 6)	REGION IV	REGION IV	
LEVEL V (L 5)	REGION III	REGION III	
LEVEL IV (L 4)	REGION III		REGION II
LEVEL III (L 3)		REGION II	REGION II
LEVEL II (L 2)	REGION II		
LEVEL I (L 1)	REGION I	REGION I	REGION I

B.2 Test Pulse Severity Level Selection Table: The pulse severity levels for the electric field (E) and magnetic field (H) strength corresponding to respective test pulse and method are determined and entered in the table according to the test plan for the device under test (DUT). An example of a typical table (per SAE J1113) is listed below:

PULSE SEVERITY LEVEL	E FIELD STRENGTH (VOLTS PER METER)	H FIELD STRENGTH (AMPERES PER METER)
L 6		
L 5		
L 4		
L 3		
L 2		
L 1		

RECOMMENDED ENVIRONMENTAL PRACTICES FOR ELECTRONIC EQUIPMENT DESIGN—SAE J1211

SAE Recommended Practice

Report of the Electronic Systems Committee approved June 1978. Editorial change November 1978.

1. Purpose—This guideline is intended to aid the designer of automotive electronic systems and components by providing material that may be used to develop environmental design goals.

2. Scope—The climatic, dynamic, and electrical environments from natural and vehicle-induced sources that influence the performance and reliability of automotive electronic equipment are included. Test methods that can be used to simulate these environmental conditions are also included in this document.

The information is applicable to vehicles that meet all the following conditions and are operated on roadways:

2.1 Front engine rear wheel drive vehicles.
2.2 Vehicles with reciprocating gasoline engines.
2.3 Coupe, sedan, and hard top vehicles.

Part of the information contained herein is not affected by the above conditions and has more universal application. Careful analysis is necessary in these cases to determine applicability.

3. Application

3.1 Environmental Data and Test Method Validity—The information included in the following sections is based upon test results achieved by major North American automobile manufacturers and automobile original equipment suppliers. Operating extremes were measured at test installations normally used by manufacturers to simulate environmental extremes for vehicles and original equipment components. They are offered as a design starting point. Generally, they cannot be used directly as a set of operating specifications because some environmental conditions may change significantly with relatively minor physical location changes. This is particularly true of vibration, engine compartment temperature, and electromagnetic compatability. Actual measurements should be made as early as practical to verify these preliminary design baselines.

The proposed test methods are either currently used for laboratory simulation or are considered to be a realistic approach to environmental design validation. They are not intended to replace actual operational tests under adverse conditions. The recommended methods, however, describe standard cycles for each type of test. The designer must specify the number of cycles over which the equipment should be tested. The number of cycles will vary depending upon equipment, location, and function. While the standard test cycle is representative of an actual short term environmental cycle, no attempt has been made to equate this cycle to an acceleration factor for reliability or durability. These considerations are beyond the scope of this guideline.

3.2 Organization of Test Methods and Environment Extremes Information—The data presented in this document is contained in Sections 4 and 5. Section 4, Environmental Factors and Test Methods, describes the 11 major characteristics of the expected environment that have an impact on the performance and reliability of automotive electronic systems. These descriptions are titled:

3.2.1 Temperature.
3.2.2 Humidity.
3.2.3 Salt Spray Atmosphere.
3.2.4 Immersion and Splash (Water, Chemicals, and Oils).
3.2.5 Dust, Sand, and Gravel Bombardment.
3.2.6 Altitude.
3.2.7 Mechanical Vibration.
3.2.8 Mechanical Shock.
3.2.9 Factors Affecting the Automotive Electrical Environment.
3.2.10 Steady State Electrical Characteristics.
3.2.11 Transient, Noise, and Electrostatic Characteristics.

They are organized to cover three facets of each factor:
(a) Definition of the factor.
(b) Description of its effect on control, performance, and long term reliability.
(c) A review of proposed test methods for simulating environmental stress.

Section 5, Environmental Extremes by Location, summarizes the anticipated limit conditions at five general control sites:
(a) Underhood
  1. Engine
  2. Bulkhead—dash panel
(b) Chassis
(c) Exterior
(d) Interior
  1. Instrument Panel
  2. Floor
  3. Rear Deck
(e) Trunk

3.3 Combined Environments—The automotive environment consists of many natural and induced factors. Combinations of these factors are present simultaneously. In some cases, the effect of a combination of these factors is much more serious than the effect of exposing samples to each environmental factor in series. For example, the suggested test method for humidity includes both high and low temperature exposure. This combined environmental test is very important to compments whose proper operation is dependent on seal integrity. Temperature and vibration is a second combined environmental test that can be significant to some components. During design analysis a careful study should be made to determine the possibility of design susceptibility to a combination of environmental factors that could occur at the planned mounting location. If the possibility of susceptibility exists, a combined environmental test should be considered.

3.4 Test Sequence—The optimum test sequence is a compromise between two considerations:

3.4.1 The order in which the environmental exposures will occur in operational use.

3.4.2 A sequence that will create a total stress on the sample that is representative of operation stress.

The first consideration is impossible to implement in the automotive case, since exposures occur in a random order. The second consideration prompts the test designer to place the more severe environments last. Many sequences that have been successful follow this general philosophy, except that temperature cycle is placed first in order to condition the sample mechanically.

4. Environmental Factors and Test Methods

4.1 Temperature

4.1.1 DEFINITION—Thermal factors are probably the most pervasive environmental hazard to automotive electronic equipment. Sources for temperature extremes and variations include:

4.1.1.1 The vehicle's climatic environment, including the diurnal and seasonal cycles. Additionally, variations in climate by geographical location must be considered. In the most adverse case, the vehicle that spends the winter in Canada may be driven in the summer in the Arizona desert. Temperature variations due to this source range from -40-85°C (-40-185°F).

4.1.1.2 Heat sources and sinks generated by the vehicle's operation. The major sources are the engine and drive train components, including the brake system. Very wide variations are to be found during operation. For instance, temperatures on the surface of the engine can range from the cooling system's 88-650°C (190-1200°F) on the surface of the exhaust system. This category also includes conduction, convection, and radiation of heat due to various modes of vehicle operation.

4.1.1.3 Self-heating of the equipment due to its own internal dissipation. A design review of the worst case combination of peak ambient temperature (due to 4.1.1.1 and 4.1.1.2 above), minimized heat flow away from the equipment and peak applied steady state voltage should be conducted.

4.1.1.4 Vehicle operational mode and actual mounting location. Measurements should be made at the actual mounting site during the following vehiclular conditions while subjected to the maximum heat generated by adjacent equipment and at the maximum ambient environment:

4.1.1.4.1 Engine start.
4.1.1.4.2 Engine idle.
4.1.1.4.3 Engine high speed.
4.1.1.4.4 Engine turn off—prior history important.
4.1.1.4.5 Various engine/road load conditions.

4.1.1.5 Ambient conditions before installation due to storage and transportation extremes. Shipment in unheated aircraft cargo compartments may lower the minimum storage (non-operating) temperature to −50°C (−58°F).

The thermal environmental conditions that are a result of these conditions can be divided into three categories:

4.1.1.5.1 Extremes—The ultimate upper and lower temperatures the equipment is expected to experience.

4.1.1.5.2 Cycling—The cumulative effects of temperatures cycling within the limits of the extremes.

4.1.1.5.3 Shock—Rapid change of temperature. Fig. 1 illustrates one form

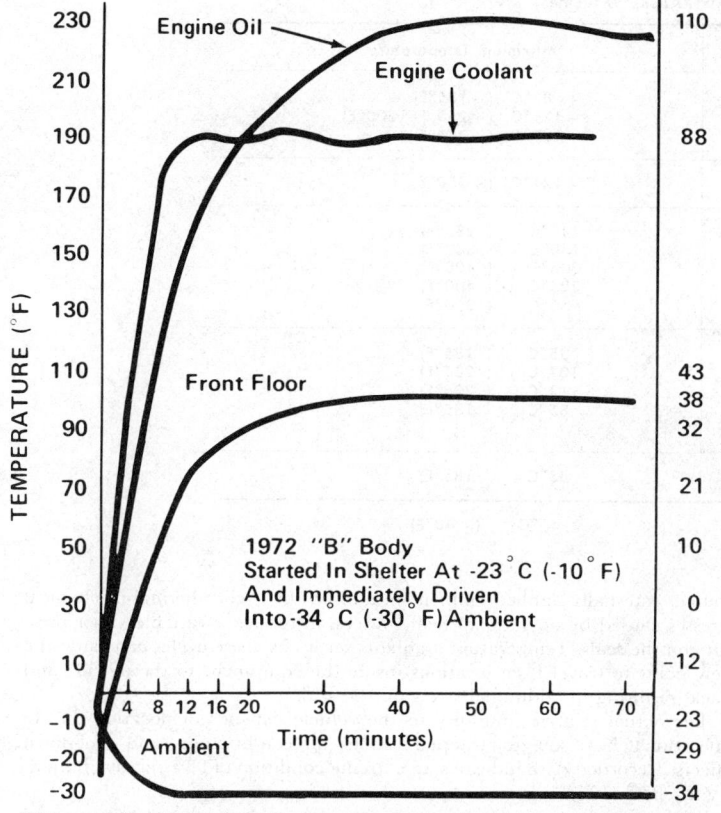

FIG. 1—VEHICLE COLD WEATHER WARM-UP CHARACTERISTICS

4.1.2.4 Seal failures, including the *breathing* action of some assemblies, due to temperature-induced dimensional variation which permit intrusion of liquid or vapor borne contaminants.

4.1.2.5 Failure of circuit components due to direct mechanical stress caused by differential thermal expansion.

4.1.2.6 The acceleration of chemical attack on interconnects, due to temperature rise, can result in progressive degradation of circuit components, printed circuit board conductors, and solder joints.

In addition to this, high temperature extremes can cause a malfunction by:

4.1.2.7 Exceeding the dissociation temperature of surrounding polymer or other packaging components.

4.1.2.8 Carbonization of packaging materials with eventual progressive failure of the associated passive or active components. This is possible in cases of extreme overtemperature. In addition, non-catastropic failure is possible due to electrical leakage in the resultant carbon paths.

4.1.2.9 Changes in active device characteristics with increased heat including changes in gain, impedance, collector-base leakage, peak blocking voltage, collector-base junction second breakdown voltage, etc. with temperature.

4.1.2.10 Changes in passive device characteristics such as permanent or temporary drift in resistor value and capacitor dielectric constants with increased temperature.

4.1.2.11 Changes in interconnect and relay coil performance due to the conductivity temperature coefficient of copper.

4.1.2.12 Changes in the properties of magnetic materials with increasing temperature, including Curie point effects and loss of *permanent* magnetism.

4.1.2.13 Dimensional changes in packages and components leading to separation of subassemblies.

4.1.2.14 Changes in the strength of soldered joints due to changes in mechanical characteristics of the solder.

Further, low temperature extremes can cause failure due to:

4.1.2.15 The severe mechanical stress caused by ice formation in moisture bearing voids or cracks.

4.1.2.16 The very rapid and extreme internal thermal stress caused by applying maximum power to semi-conductor or other components after extended cold soak under aberrant operating conditions such as 24-V battery jumper starts.

4.1.3 RECOMMENDED TEST METHODS

4.1.3.1 *Temperature Cycle Test*—A recommended thermal cycle profile is shown in Fig. 2 and recommended extreme temperatures in Table 1. The test method of Fig. 2A, a 24-h cycle, offers longer stabilization time and permits a convenient room ambient test period. Fig. 2B, and 8-h cycle, provides more temperature cycles for a given test duration. It is applicable only to modules whose temperatures will reach stabilization in a shorter cycle time. Stabilization should be verified by actual measurements. Thermocouples, etc.

Separate or single test chambers may be used to generate the temperature environment described by the thermal cycles. By means of circulation, the air temperature should be held to within ±2.8°C (±5°F) at each of the extreme temperatures. The test specimens should be placed in such a position, with respect to the air stream, that there is substantially no obstruction to the flow of air across the specimen. If two test specimens are used, care must be exercised to assure that the test samples are not subjected to temperature

of vehicle operation which induces thermal shock. Thermal shock is also induced when equipment at elevated temperature is exposed to sudden rain or road splash.

The automotive electronic equipment designer is urged to develop a systematic, analytic method for dealing with steady state and transient thermal analysis. The application of many devices containing semi-conductors will be temperature limited. For this reason, the potential extreme operating conditions for each application must be scrutinized to avoid later field failure.

4.1.2 EFFECT ON PERFORMANCE—The damaging effects of thermal shock and thermal cycling include:

4.1.2.1 Cracking of printed circuit board or ceramic substrates.

4.1.2.2 Thermal stress or fatigue failures of solder joints.

4.1.2.3 Delamination of printed circuit board and other interconnect system substrates.

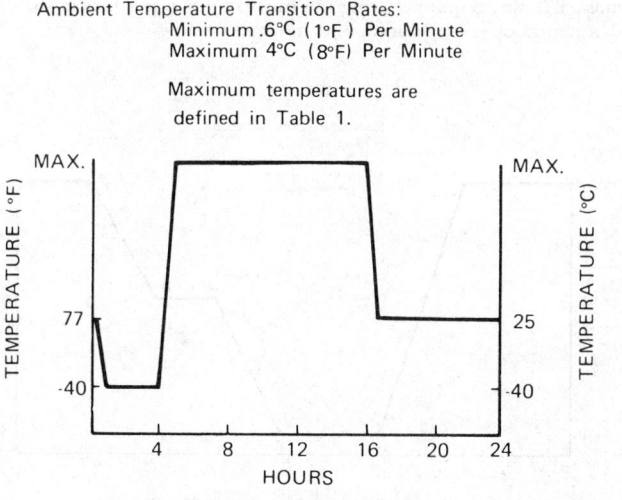

FIG. 2A—24-H CYCLE

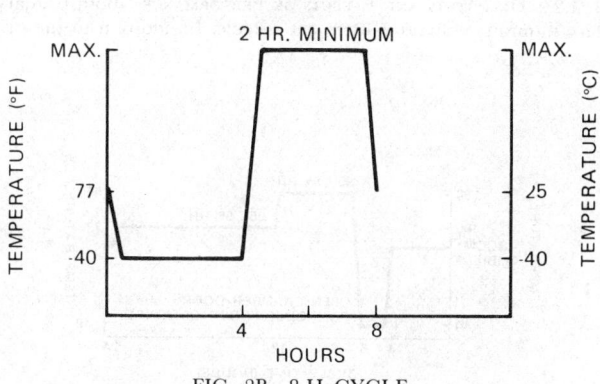

FIG. 2B—8-H CYCLE

FIG. 2—RECOMMENDED THERMAL CYCLES

TABLE 1—RECOMMENDED TEMPERATURE EXTREMES

Location		Maximum Temperature
Chassis	—Isolated Areas	+ 85°C (+185°F)
	—Exposed to Heat Sources	+121°C (+250°F–1200°F)
	—Exposed to Oils	+177°C (+350°F)
Exterior		+121°C (+250°F)
Underhood	—Dash Panel	140°C (285°F)
	—Engine (Typical)	150°C (300°F)
	—Choke Housing	205°C (400°F)
	—Starter Cable Near Manifold	205°C (400°F)
	—Exhaust Manifold	650°C (1200°F)
Interior	—Floor	85°C (185°F)
	—Rear Deck	107°C (225°F)
	—Instrument Panel (Top)	113°C (236°F)
	—Instrument Panel (Other)	85°C (185°F)
Door Interior	—No data available	
Trunk		85°C (185°F)
	Minimum Temperature	−40°C (−40°F)

transition rates greater than that defined in Fig. 2. Direct heat conduction from the temperature chamber heating element to the specimen should be minimized.

Electrical performance should be measured under the expected operational minimum and maximum extremes of excitation, input and output voltage and load at both the cold and hot temperature extremes. These measurements will provide insight into electrical variations with temperature.

Thermal shock normally expected in the automotive environment is simulated by the maximum rates of change shown on the recommended thermal cycle profile shown in Fig. 2. The proper thermal shock cycle should be determined by analysis of component power dissipation, expected rate of temperature change at its location in the system and the overall ambient operating temperature. In general, thermal shock is most severe when equipment is operated intermittently in low temperature environments. The effects of thermal shock include cracking and delamination of substrates, seal failures, wire bond breaks, and operating characteristic changes.

Thermal stress is caused by repeating cycling through the thermal profile of Fig. 2. The number of cycles required is a function of the equipment application. Functional electrical testing during temperature transitions or immediately after temperature transitions, is a means of detecting poor electrical connections. The effects of thermal stress are similar to thermal shock but are caused by fatigue.

NOTE: Although uniform oven temperatures are desirable, in some vehicle environments the only means of heat removal may be by special heat sinks or by free convection to surrounding air. It may be necessary to use conductive heat sinks with independent temperature controls in the former case and baffles or slow speed air stirring devices in the latter to simulate such conditions in the laboratory. (See Section 3.)

4.1.4 RELATED SPECIFICATIONS—A generally accepted method for small part testing is defined in MIL-STD-202E, Method 102A, Temperature Cycling. The short dwell periods at extreme temperature are satisfactory where temperature stabilization has been verified by actual measurements, thermocouples, etc.

4.2 Humidity

4.2.1, 4.2.2 DEFINITION AND EFFECTS ON PERFORMANCE—Both primary and secondary humidity sources exist in the vehicle. In addition to the primary source, externally applied ambient humidity, the cyclic thermal-mechanical stresses caused by operational heat sources, introduce a variable vapor pressure on the seals. Temperature gradients set up by these cycles can cause the dew point to travel from locations inside the equipment to the outside and back, resulting in additional stress on the seal.

The actual relative humidity in the vehicle depends on location due to operational heat sources, trapped vapors, air conditioning, and cool-down effects. Recorded data indicates an extreme condition of 98% relative humidity at 38°C (100°F).

Primary failure modes include corrosion of metal parts due to galvanic and electrolytic action, as well as corrosion due to interaction with water and due to adverse pH changes. Other failure modes include changes in electrical properties, surface bridging between circuits, and decomposition of organic matter.

4.2.3 RECOMMENDED TEST METHODS—The most common way to determine the effect of humidity on electronic equipment is to overtest and examine any failures for relevance to the more moderate actual operating conditions. Three general test methods are recommended. The most common is an active temperature humidity cycling under accelerated conditions. The second is a 10-day soak at 95% relative humidity and 38°C (100°F) temperature. A third method is an 8-24 h exposure at 103.4 kPa gauge pressure (15 psig) in a pressure vessel. This is a quick and effective method of uncovering defects in plastic encapsulated semi-conductors.

There are many acceptable accelerated humidity test cycles, including MIL-STD-202E, Method 103B; however, the test cycles in Fig. 3 are recommended as the most useful.

An optional frost condition may be incorporated during one of these humidity cycles (Fig. 3A). Electrical performance should be continuously monitored during these frost cycles to note erratic operation. Heat-producing and moving parts may require altering the frost condition portions of the cycle to allow a period of non-operation and induced frosting.

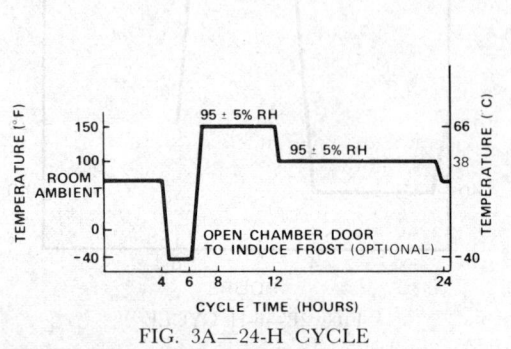

FIG. 3A—24-H CYCLE

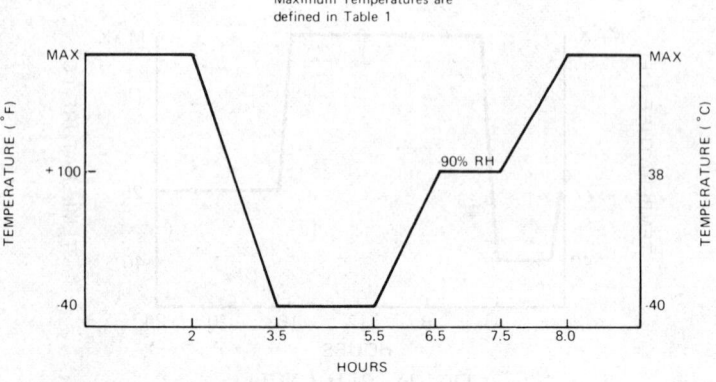

FIG. 3B—8-H CYCLE

FIG. 3—RECOMMENDED HUMIDITY CYCLES

The 10-day soak is normally conducted with equipment non-operating. Equipment that operates with standby voltage excitation and a low current drain when the ignition is off is a significant exception. Examples of this type include seat belt interlocks and electronic clocks. Samples of such equipment should be tested with normal standby conditions. Accelerated humidity effects should be expected under the conditions of high temperature, high humidity, and excitation voltage.

4.2.4 RELATED SPECIFICATIONS—A number of related humidity specifications are recommended for review and reference. The first: MIL-STD-810B, Method 507, Procedure 1, Humidity; is a system-oriented test method. The second; a modified version of MIL-STD-202E, Method 103B, Humidity (Steady State); is intended to evaluate materials. The third; MIL-STD-202E, Method 106D, Moisture Resistance; is a procedure for testing small parts.

4.3 Salt Atmosphere

4.3.1, 4.3.2 DEFINITION AND EFFECT ON PERFORMANCE—Electronic equipment mounted on the chassis, exterior, and underhood are often exposed to a salt spray environment. In coastal regions, the salt is derived from sea breezes and in colder climates, from road salt. Although salt spray is generally not found in the interior and trunk of the vehicle, it is advisable to evaluate the potential effects of saline solutions on the floor area as the result of transfer from the outside environment by vehicle operators, passengers, and transported equipment.

Failure modes due to salt spray are generally the same as those associated with water and water vapor. However, corrosion effects and alteration of conductivity are accelerated by the presence of saline solutions and adverse changes in pH.

4.3.3 RECOMMENDED TEST METHODS—The recommended test method for measuring susceptibility of electronic equipment to salt spray is the American Society for Testing and Materials (ASTM) Standard Method of Salt Spray (Fog) Testing-Number B 117-73.

The test consists of exposing the electronic equipment to a solution of 5 parts salt to 95 parts water atomized at a temperature of 35°C (95°F). The equipment being tested should be exposed to the salt spray for a period of from 24-96 h. The actual exposure time must be determined by analysis of the specific mounting location. When the tests have been concluded, the test specimens should be gently rinsed in clean running water, about 38°C (100°F) to remove salt deposits from their surface and then immediately dried. Drying should be done with a stream of clean, compressed dry air at about 241.3-275.8 kPa gauge pressure (35-40 psig). The equipment should then be tested under nominal conditions of voltage and load throughout the test.

NOTE: The Pascal (Pa) is the designated SI (metric) unit for pressure and stress. It is equivalent to 1 N/m^2.

Where leakage resistance values are critical, appropriate measurements in both the wet and dry states may be necessary.

4.3.4 RELATED SPECIFICATIONS—ASTM B 117-73, Salt Spray (Fog) Testing, is the recommended test method.

4.4 Immersion and Splash (Water, Chemicals, and Oils)

4.4.1 DEFINITION—Electronic equipment mounted on or in the vehicle is exposed to varying amounts of water, chemicals, and oil. A list of potential environmental chemicals and oils includes:

- Engine Oils and Additives
- Transmission Oil
- Rear Axle Oil
- Power Steering Fluid
- Brake Fluid
- Axle Grease
- Washer Solvent
- Gasoline
- Anti-Freeze Water Mixture
- Degreasers
- Soap and Detergents
- Steam
- Battery Acid
- Water and Snow
- Salt Water
- Waxes
- Freon
- Spray Paint
- Ether
- Vinyl Plasticizers
- Undercoating Material

The modified chemical characteristics of these materials when degraded or contaminated should also be considered.

4.4.2 EFFECT OF PERFORMANCE—Loss of the integrity of the container can result in corrosion or contamination of vulnerable internal components. The chemical compatibility can be determined by laboratory chemical analysis. Devices that may be immersed in fluids for a long period, such as sensors, should be subjected to laboratory life tests in these fluids.

4.4.3 RECOMMENDED TEST METHODS—The equipment designer should first determine whether the parts must withstand complete immersion or splash, and which fluids are likely to be present in the application. Immersion and splash tests are generally performed following other environmental tests because this sequence will tend to aggravate any incipient defects in seals, seams, and bushings which might otherwise escape notice.

Splash testing should be done with the equipment mounted in its normal operating position with all drain holes, if used, open. The sample is subjected to precipitation of 0.25 cm (0.1 in)/min delivered at an angle 45 deg below and above the sample with a nozzle having a solid cone spray.

During immersion testing, most commonly utilizing water as the fluid, the equipment ordinarily is not operated due to setup logistics and techniques of testing. Electrical tests should, therefore, be performed immediately before and after this test. In this test, the electronic equipment in its normal exterior package is immersed in tap water at about 18°C (65°F). The test sample should be completely covered by the wter. The sample is first positioned in its normal mounting orientation. It remains in this position for 5 min and then is rotated 180 deg. It should remain in that position for 5 min and then be rotated 90 deg about the other axis where it remains for 5 min. Immediately after removal, the sample should be exposed to some temperature below freezing until the entire mass is below freeezing. The sample is then returned to room temperature, air dried, functionally tested, and inspected for damage.

More severe tests such as combined temperature, pressure, and continuous fluid contact must be considered for equipment subjected to extreme environments as in the case of exposure to coolant water, brake fluid, and transmission oil. Caution must be used in specifying combined tests as they may be unrealistically severe for many applications.

4.5 Dust, Sand, and Gravel Bombardment

4.5.1 DEFINITION—Dust is a significant environment for chassis, underhood, and exterior-mounted devices; and can be a long-term problem in interior locations, such as under the dash and seats. Sand, primarily windblown, is an important environmental consideration for chassis, exterior, and underhood. Bombardment by gravel is significant for chassis and exterior-mounted equipment.

4.5.2 EFFECT ON PERFORMANCE—Exposure to fine dust can cause problems with moving parts, form conductive bridges, and act as an absorbent material for the collection of water vapor. Some electromechanical components may be able to tolerate fine dust, but larger particles may affect or totally inhibit their mechanical action. While the exposure in desert areas is severe, exposure to a reasonable amount of road dust is common to all areas.

4.5.3 RECOMMENDED TEST METHODS—Dust, sand, and gravel bombardment tests should be at room temperature and the sample need not be operating, although functional tests should be performed prior to and after testing.

Dust conforming to that defined in SAE J726b (November, 1976), coarse grade should be used. If this dust packs or seals openings in the test sample or if the sample contains exposed mechanical elements, the following alternate dust mixture may be used:

J726b Coarse or Equivalent	70%
120 Grit Aluminum Oxide	30%

Components should be placed in a dust chamber with sufficient dry air movement to maintain a concentration of 0.88 g/m^3 (0.025 g/ft^3) for a period of 24 h.

An alternate method is to place the sample about 15 cm (6 in) from one wall in a 3-ft cubical box. The box should contain (10 lb) 4.54 Kgm of fine powdered cement in accordance with ASTM C150-56, specification for Portland Cement. At intervals of 15 min, the dust must be agitated by compressed air or fan blower. Blasts of air for a 2-s period in a downward direction assure that the dust is completely and uniformly diffused throughout the entire cube. The dust is then allowed to settle. The cycle is repeated for 5 h.

The recommended test for susceptibility of equipment to damage from gravel bombardment is SAE J400 (July, 1968), Recommended Practice Test for Chip Resistance of Surface Coatings. This document is intended to detect susceptibility of surface coatings to chipping, but the basic test equipment and procedures are useful for evaluation of the electronic equipment. The test consists of exposing the test sample to bombardment by gravel 0.96-1.6 cm ($3/8$-$5/8$ in) in diameter for a period of approximately 2 min. The sample is positioned about 35 cm ($13 \frac{3}{4}$ in) from the muzzle of the gravel source. 470 cm^3, (approximately 1 pt) of gravel (250-300 stones) is delivered under a pressure of 483 kPa gauge pressure (70 psig) over an approximate 10-s period. The process is repeated 12 times for a total exposure of 2 min. Judgment must be used in determining which sides should be exposed to the bombardment. Certainly all forward-facing surfaces not shielded by other parts are included. In many cases, the bottom and sides should also be exposed.

4.5.4 RELATED SPECIFICATIONS—Three specifications are referenced. The first: MIL-STD-202E, Method 110A, Sand and Dust, is a piece part test and is included for information and comparison. The second is SAE J726b (November, 1976), Air Cleaner Test Code, which defines the recommended dust. It also describes some test apparatus. The third specification is SAE J400 (July, 1968), Test for Chip Resistance for Surface Coatings, which is recommended in part for a gravel bombardment guide. Continued integrity at the conclusion of the exposure is the passing criteria.

4.6 Altitude

4.6.1 DEFINITION—With the exception of air shipment of unenergized controls, operation in the vehicle should follow the anticipated operating limits. Completed controls are expected to be stressed over these limits of absolute pressure:

Condition	Altitude	Atmospheric Pressure
Operating	3.7 km (12 000 ft)	62.1 kPa absolute pressure (9 psia)
Non-operating	12.2 km (40 000 ft)	18.6 kPa absolute pressure (2.7 psia)

4.6.2 EFFECT ON PERFORMANCE—With increased altitude the following effects are generally observed:

4.6.2.1 Reduction in convection heat transfer efficiency.

4.6.2.2 Change in mechanical stress on packages which have internal cavities. The reference cavity of an absolute pressure sensor is an example of this.

4.6.2.3 A very noticeable reduction in the high voltage breakdown characteristics of systems with electrically stressed insulator, conductor or air surfaces; this may result in setup of surface tracking with eventual component failure.

4.6.3 RECOMMENDED TEST METHODS—The recommended test method is to operate equipment during the thermal cycles described in the Temperature Test Section, but with the added parameter of 62.1 kPa absolute pressure (9 psia) pressure. The equipment should operate under maximum load. Failure effects will be similar to those experienced with thermal cycle and shock. Non-operating tests should be done at a minimum temperature of $-51°C$ ($-60°F$) if possible.

4.7 Mechanical Vibration

4.7.1 DEFINITION—Vibration, which is prevalent whenever the vehicle engine or suspension system is in motion, is a key factor in the automotive environment. The intensity varies from low severity at smooth engine idle to extreme severity when traversing rough roads at high speed. Vibration also varies with location. Detailed data is included in Figs. 11–18.

4.7.2 EFFECT ON PERFORMANCE—A number of modes of degradation or failure are possible under applied vibration. A partial list includes:

4.7.2.1 Loss of wiring harness electrical connection due to improper connector design and/or assembly.

4.7.2.2 Excitation of tuned mass harmonic vibration within the equipment which eventually leads to failure due to metal fatigue at stress concentration points.

4.7.2.3 Failure of mounting structure due to the added acceleration forces acting on the mass of the equipment.

4.7.2.4 Mechanical flexure at seal and other interface areas which promotes the intrusion of other environmental factors, such as moisture, in a manner similar to the phenomena described under temperature cycling effects.

4.7.2.5 Temporary abberation of equipment performance due to acceleration forces on control component masses. Two examples illustrate this:

4.7.2.5.1 Sensor measurement error due to motion of the sense element. An example of this is a pressure sensor which gives incorrect information under some applied frequencies due to the mass of a diaphragm-spring mechanism.

4.7.2.5.2 False operation of electromechanical components—e.g., a relay whose contacts close or open, due to vibratory movement of its armature's mass.

The designer should be particularly alert to failures which are intermittent or which cause faulty operation during applied vibration. Many malfunctions of this type revert to normal operation after the vibration excitation is removed. It is, therefore, recommended that electronic performance tests be conducted during vibration tests for those functions which must perform under this condition. In most cases this is only practical under laboratory simulation of the road test situation.

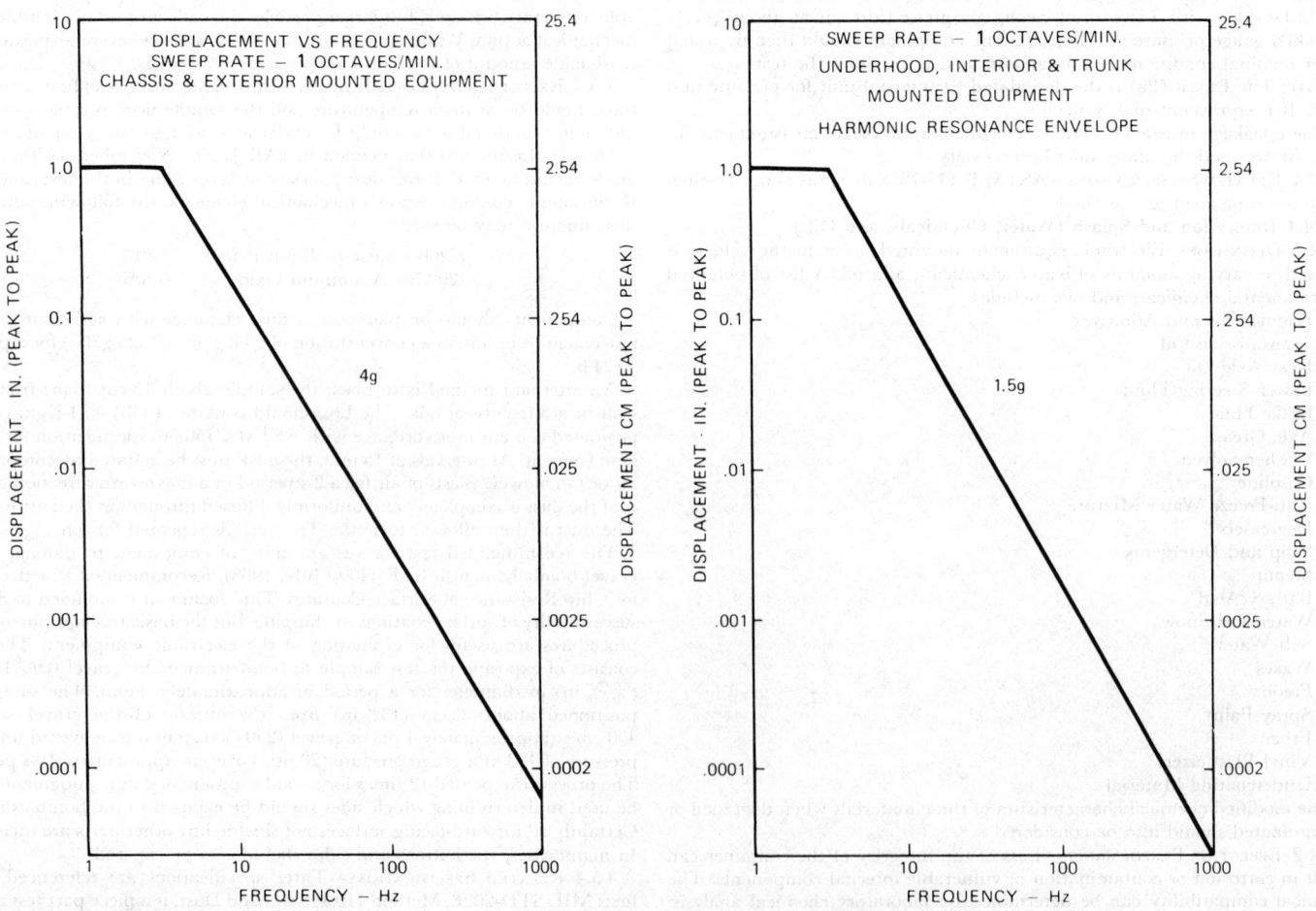

FIG. 4—RECOMMENDED RESONANT SEARCH VIBRATION PROFILE

4.7.3 RECOMMENDED TEST METHODS—A typical test for this environmental factor has been operation of a test vehicle over a group of severe road test track conditions. These include surfaces described as the Belgium Block Road, the Hop, the Tramp, the Square Block Test Course, and other complex surfaces. These courses are excellent test beds for complete transportation packages installed in the vehicle. Unfortunately, they are relatively inconvenient for electronic control evaluation during the design phase. In many cases electronic equipment exhibit intermittent or degraded performance during vibration and returns to normal operation when the excitation is removed.

Failure of electronic equipment in a vibratory environment may be the result of fixed frequency or random vibrations. Current practice within the industry is to conduct a resonant search up to 1000 Hz and then dwell at the major resonances if they are applicable to the operating environment spectrum.

Fig. 4 shows the recommended amplitude and sweep rate for this search. This profile is primarily gravity unit-oriented. A second recommended procedure is to sweep from 10–55 Hz and return in 1 min at an amplitude determined by measurements taken at the proposed mounting location. The test is conducted in each of three mutually perpendicular planes.

Experience has shown that in some cases random vibration may be a valuable approach in uncovering electronic equipment failure modes. While random testing is more difficult and costly, consideration should be given to this approach where required.

In the time sweep and resonant dwell, vibration must be conducted in each of three mutually perpendicular axes. Test duration must be determined by the equipment designer.

4.7.4 RELATED SPECIFICATIONS—Three specifications are referenced. The first, MIL-STD-202E, Method 201A. Vibration, and the second, MIL-STD-202E, Method 204C, Vibration, High Frequency, are concerned with sine vibration and offer procedural details and information on resonant dwell periods. The third, MIL-STD-202E, Method 214, Random Vibration, offers similar information on the random vibration approach.

4.8 Mechanical Shock

4.8.1, 4.8.2 DEFINITION AND EFFECT ON PERFORMANCE—The automotive shock environment is logically divided into four classes:
  Shipping and handling shock.
  Installation shock.
  Operational shock.
  Crash shock.

Shipping and Handling Shocks—These are similar to those encountered in non-automotive applications.

Installation Shock—It is common production-line practice to lift and carry equipment by its harness. Therefore, it is recommended that the harness design assure for secure fastening and suitable strain relief.

Operational Shocks—The shocks encountered during the life of the vehicle that are caused by curbs, pot holes, etc. can be very severe. These vary widely in amplitude, duration and number, and the test condition can only be generally simulated.

Crash Shock—This is included as an operating environment for safety systems. The operational requirements of these systems are limited to longitudinal shock at the present time.

4.8.3 RECOMMENDED TEST METHODS

Bench Handling Shock—The component shall be placed on a solid wooden bench top at least 3.4 cm (1 5/8 in) thick. The test shall be performed as follows: using one edge as a pivot, lift the opposite edge of the component until one of the following conditions shall first occur:

(a) The component forms a 45 deg angle with bench top.

(b) The lifted edge is just below the balance point. The component shall be allowed to drop to the bench top. Repeat using other practical edges of the same face as pivot points. The procedure is then repeated with the component resting on other faces until it has been dropped on each face that the component might normally be placed when bench handling or servicing.

Transit Drop Test—The drop shall be from a height of 122 cm (48 in) onto a solid 5 cm (2 in) thick plywood base backed by concrete or a rigid steel frame with the test sample properly installed in its shipping container. The drop shall be performed on each face, edge, and corner.

Installation Shock Test—A recommended test is to support the device and the far end of the harness at the same elevation, then release the device. Care should be taken to prevent the equipment from striking another object during this test. The drop should be repeated and the harness terminals or strain relief area inspected for damage.

Operation Shock—With the possible exception of collision, the most severe shock anticipated after production line installation is encountered when driving over complex road surfaces. The complex profile that was used to derive this test profile consists of a rise in the roadway followed by a depression or dip. Upon leaving the dip at 48 km/h (30 mph), the vehicle will often become airborne. The severe shock is experienced when the vehicle returns to the roadway. Fig. 5 shows the shock measured on a steering column just below

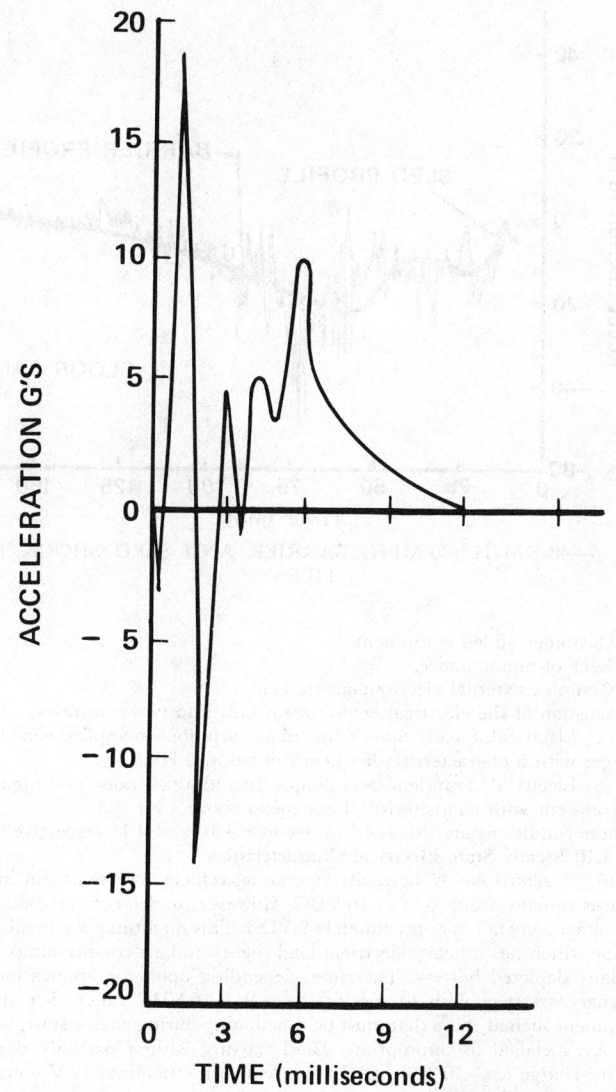

FIG. 5—OPERATIONAL SHOCK PROFILE

the steering wheel. The accelerometer was mounted with its sensitive axis perpendicular to the axis of the column and in the vertical plane.

While this location is not typical of component mounting locations, it probably represents the most severe operational shock environment. This information is provided for guidance only; there are no generally accepted test procedures at the present time.

Crash Shock Test—Only limited and preliminary data on the effects of crash shock on the electronic equipment environment are available. However, a representative deceleration profile for a 48 km/h (30 mph) barrier crash is shown in Fig. 6. The following factors vary with each installation and should be considered in pretest analysis:

(a) Equipment mass.

(b) Mounting system.

(c) Structure of the associated vehicle (crush distance, rate of collapse, etc.)

(d) Particular engine package.

(e) Direction of crash.

4.8.4 RELATED SPECIFICATIONS—Two specifications are recommended for consideration. The first, MIL-STD-202E, Method 203B, Random Drop, is designed to uncover failures that may result from the repeated random shocks that occur in shipping and handling. It is an endurance test. The second, MIL-STD-202E, Method 213B, Shock (Specified Pulse), is intended to measure the effect of known or generally accepted shock pulse shapes. It is intended that operational shock be reduced into a standard pulse shape to achieve a repeatable test method.

4.9 General Automotive Electrical Environment—Factors unique to the automobile that make the vehicular environment more severe than that encountered in most electrical equipment applications are:
  Interaction with other vehicular electronic/electrical systems.
  Voltage variations.

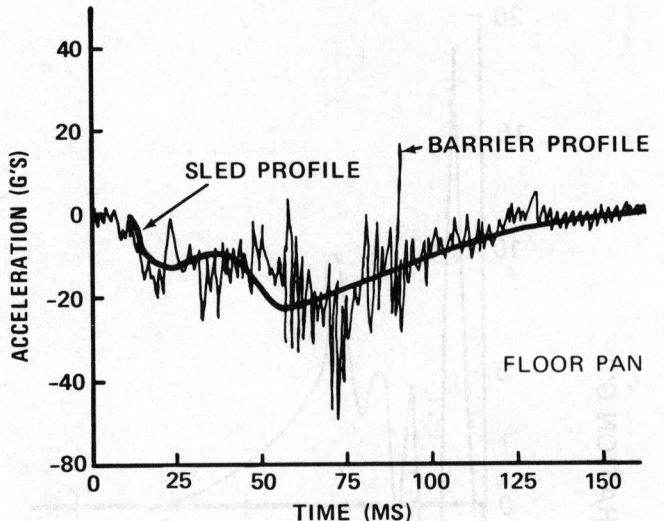

FIG. 6—48 KM/H (30 MPH) BARRIER AND SLED SHOCK PROFILES

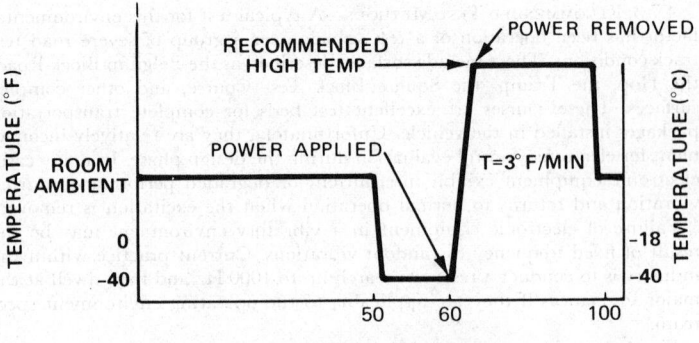

FIG. 7—COMBINED THERMAL AND ELECTRICAL STRESS PROFILE

Customer added equipment.
Lack of maintenance.
Complex external electromagnetic fields.
Discussion of the electrical environment falls into two categories:
  (a) Electrical, Steady-State—Including variations in applied vehicle DC voltages with a characteristic frequency of below 1 Hz.
  (b) Electrical, Transient, and Noise—Including all noise and high voltage transient with characteristic frequencies above 1 Hz.
These conditions are discussed in Sections 4.10 and 4.11 respectively.

4.10 Steady State Electrical Characteristics

4.10.1 DEFINITION—A normally operating vehicle will maintain supply voltages ranging from +11–+16 VDC. However, under certain conditions, the voltage may fall to approximately 9 VDC. This might happen in an idling vehicle which has a heavy electrical load (lights and air conditioning) and a partially depleted battery. Therefore, depending upon the application, the designer/user may wish to specify the +9–+16 VDC range. For specific equipment such as those that must be functioning during engine start, voltage may be specified as appropriate. Cold starting with a partially depleted battery charge at −40°C (−40°F) can reduce the nominal 12 V voltage to between 4.5 and 6.0 VDC.

Another condition affecting the DC voltage supply is developed when the vehicle voltage regulator fails, causing the alternator to drive the system 18 V. Extended 18 V operation will eventually cause boil-off of the battery electrolyte resulting in voltages as high as 75–130 V. Other charging system failures could result in lower than normal battery voltages. The general steady state voltage regulation characteristics are shown in Table 2.

Emergency starts by garages and emergency road services sometimes utilize 24 V sources, and there have been reports of 36 V being used for this purpose. High voltages such as these are applied for up to 5 min and sometimes even with reverse polarity. The use of voltages which are above the vehicle system voltage can damage components in a vehicular electrical system, and the higher the voltage, the greater the likelihood of damage. A designer cannot cope with ever-increasing excitation potentials, and the above values usually are not a part of his design criteria. The possibility of the use of voltages above system voltage is included here for information only.

4.10.2 EFFECT ON PERFORMANCE—Equipment that must operate during the starting conditon is generally designed to perform with slight degradation over a wide range of voltage. The designer is alerted to the possibility of failure from a combination of voltage and temperature variation. Over-voltage and high temperature, both from the external environment and internal dissipation, may cause excessive heat and result in failure. Under-voltage will probably result in degraded or non-performance. Conditions must be carefully examined to determine the true temperature and excitation voltage of the equiment.

4.10.3 RECOMMENDED TEST METHODS—Critical automotive equipment is performance-tested for operation within predetermined limits. Samples are also subjected to combinations of temperatures and supply voltage variation which are designed to represent the worst case stresses on control components. A typical cycle for this form of test is shown in Fig. 7.

The voltage applied and removed at the two points shown in Fig. 7 is generally 16 V, the maximum normal voltage. If the test is performed for the high voltage *booster battery* start condition of 24 V, a narrower temperature range is used. This is a destructive test which is often used as an indication of basic design environmental capability. The number of cycles expected before failure, the actual limit values for temperature and voltage, and the period of each cycle are dependent on the design goals for the equipment being considered.

Samples of finished units are generally tested for extended operation at the peak voltage/temperature combination expected at the equipment's location. In the absence of actual temperature measurements, the values in Table 1 are recommended. These tests often run for extended periods and are particularly stringent for equipment in the underhood environment.

4.11 Transient Noise and Electrostatic Characteristics

4.11.1, 4.11.2 DEFINITION AND EFFECT ON PERFORMANCE—Four principal types of transients are encountered on automobile wire harnesses. These are load dump, inductive switching transients, alternator field decay, and mutual coupling. Generally, they occur singly, but there are cases where the latter two could occur simultaneously. EMC characteristics vary considerably with type of vehicle and wiring harness. The equipment user and/or designer should determine the actual values of peak voltages, peak current, source impedance, repetition rate, frequency of occurrence at the interface between his equipment and the electrical distribution system, then design and test the electronic equipment to withstand values consistent with the expected use. Table 3 summarizes typical transient characteristics.

TABLE 2—AUTOMOTIVE VOLTAGE REGULATION CHARACTERISTICS

Condition	Voltage
Normal operating vehicle	16 V max
	14.2 V nominal
	9 V min[a]
Cold Cranking at −40°C (−40°F)	4.5–6.0 V
Jumper Starts	+24 V
Reverse Polarity	−12 V
Charging System Failure	<9–18 V
Battery Electrolyte Boil-Off	75–130 V

[a] See Section 4.10.1 for a definition of normal voltage.

TABLE 3—AUTOMOTIVE TRANSIENT VOLTAGE CHARACTERISTICS

Type	Max Amplitude (V)	Characteristic	Remarks
Load Dump	120	$106\epsilon^{-t/0.188} + 14$	Damage potential
Inductive Load Switching	−286	$-300\epsilon^{-t/0.001} + 14$ followed by +80 Volt excursion	Logic Errors
Alternator Field Decay	−90	$-90\epsilon^{-t/0.038}$	Occurs at Shutdown Only
Mutual Coupling	214	$+200\epsilon^{-t/0.001} + 14$	Logic Errors

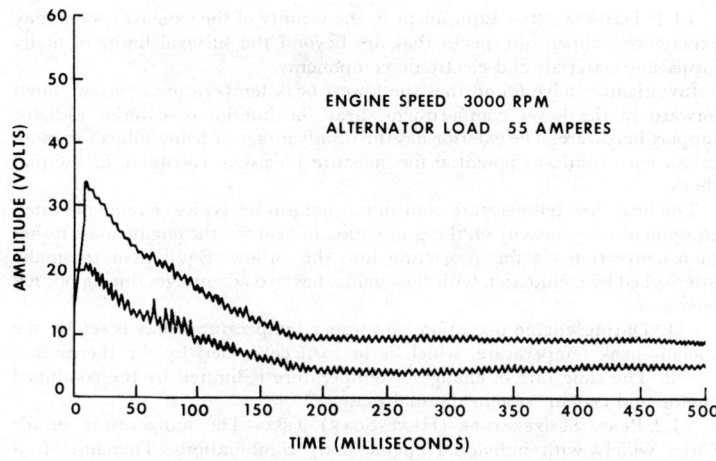

FIG. 8—LOAD DUMP TRANSIENT

TABLE 4—SUMMARY OF AUTOMOTIVE ELECTRICAL CONTINUOUS NOISE CHARACTERISTICS

Type	Max Amplitude	Duration	Repetition Rate	Remarks
Normal Accessory Noise	1.5 V Peak	Frequency	50 Hz–10 kHz	Total Pulse Height is 3 V-PP
Normal Ignition Pulses	3 V Peak	10–15 μs	Dependent on engine speed	Total Pulse Height is 6 V-PP
Abnormal Ignition Pulses	75 V Peak	~90 μs	Dependent on engine speed	
Transceiver Feedback	15–20 mV	Carrier	Frequency	Sinusoid

Load Dump Transient—Load dump occurs when the alternator load is abruptly reduced. This sudden reduction in current causes the alternator to generate a positive voltage spike. The worst case load dump is caused by disconnecting a discharged battery when the alternator is operated at rated load. Using the discharged battery load to create the load dump creates the worst situation for two reasons:

(a) The battery normally acts like a capacitor and absorbs transient energy when it is in the circuit.

(b) The partially discharged battery forms the single greatest load on the alternator and, therefore, disconnecting it creates the greatest possible step load change.

This transient may be the most severe encountered in the automobile and can result in component damage. In the practical case, it is most often initiated by defective battery terminal connections. Transient voltages of as high as 125 V or more have been reported with rise times of approximately 100 μs. Reports of decay time vary from 100 μs–4.5 s. The long duration decay occurs during vehicle turn off with a disconnected or dry vehicle battery. However, even the shortest time (100 ms) is relatively long, requiring that significant energy must be dissipated. Fig. 8 shows oscillograms of more typical load dump transients.

The load dump transient contains considerable electrical energy which must be safety dissipated to prevent damage to electronic equipment. This transient occurs randomly in time appearing as individual or repetitive pulses at random unknown rates due to vibration.

Inductive Load Switching Transient—Inductive transients are caused by solenoid, motor field, air conditioning clutch, and ignition system switching. These occur during vehicle operation whenever an inductive accessory is turned off. The severity is dependent on the magnitude of switched inductive load and line impedance. Unfortunately, measurements to date have not been taken with standardized procedures and were most probably observed with different loads.

These transients generally take the form of a large negative peak, followed by the smaller damped positive excursion. The highest reported by the data acquisition task force is −300/+80 V with an effective duration of 320 ms. Transients of this nature may cause component damage or introduce logic or functional computational errors.

Alternator Field Decay Transient—This is a special case of the inductive load switching transient. It is a negative pulse caused by alternator field decay and may occur when the field is disconnected from the battery as the ignition switch is turned to the *off* position. The amplitude is dependent on the voltage regulator cycle and load at the time of shutdown, varying from −40 to −100V and a duration of 200 ms.

Coupling—Coupling is not, strictly speaking, a generator of transients, but a mechanism which is capable of introducing transients into circuits not directly connected to the transient source. There are three general coupling modes in the automobile: magnetic, capacitive, and conducted. Briefly, the automobile coupling problems are caused by long harnesses, nonshielded conductors, and common ground return impedances. Long harnesses are one of the principal coupling media that distribute transients throughout the automobile (Ref. 2). When a number of wires are bundled into a harness and a step change in current or voltage occurs, inductive or capacitive coupling, between the conductor experiencing the change and the other wires, can result.

Other Effects—It is possible that inductive switching of certain solenoids and the alternator decay transient condition occur simultaneously. This hypothesis would account for the higher voltage transients that have been reported, but not explained. Measurement of 600 V transients on engine shutdown have been reported. Also to be considered are noise suppression capacitors that are sometimes placed on the fuse block, and some accessories that are applied to quiet interference on the entertainment radio. In some cases, these capacitors may form tuned circuits with automotive inductive loads, causing high voltage transient conditions.

Certain devices, with high levels of stored energy, such as coasting permanent magnetic motors, may maintain line voltage for a finite interval of time after the ignition is shut off. Some equipment may perform in an unsatisfactory mode of operation under such conditions.

NOTE: Direct conduction through common circuits constitutes the most frequent path by which transients are introduced into electronic equipment.

Electrical Noise—Noise will normally have a repetition rate which is dependent on the characteristics of the interferring device or engine speed. There are four general types as summarized in Table 4. A typical oscillogram of automotive electrical noise is shown in Fig. 9.

Normal Accessory Noise—Generally, the normal compliment of accessories contributes less than 1.5 V peak over a frequency range of 50 Hz–10 kHz.

Normal Ignition Pulses—Normal ignition pulses can cause 3 V peak pulses of 10–15 μs duration at a repetition rate dependent on engine speed.

Abnormal Ignition Pulses—Normally, the battery acts as a low impedance path to ground for the voltage pulse caused by the primary and secondary windings of the ignition coil. If the battery is disconnected, the repetitive voltage pulses will increase to a significant amplitude. Under this condition, there have been reports of voltages as high as 75 V peak and 90 ms duration with the repetition rate dependent on engine speed. The energy level is substantial and component damage is possible.

Since this condition can occur simultaneous with load dump, consideration should be given to testing both conditions together.

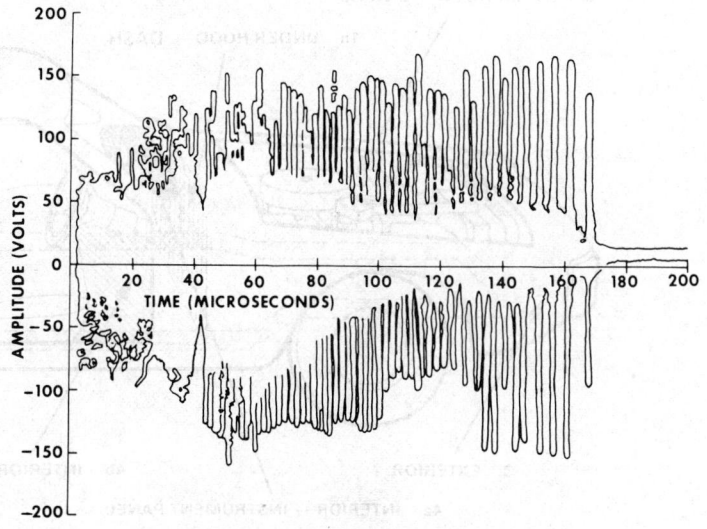

FIG. 9—POWER LINE ELECTRICAL NOISE

Transceiver Feedback—Some automotive transceivers feedback energy to the power line at carrier frequency when the transmitter is keyed. These potentials are small, 15–20 mV peak, and are mentioned here only because they are at a predictable frequency.

Electrostatic Discharge—The electrostatic charge stored by the human body and then discharged into a device may cause operating anomalies. Recent investigations indicate that discharging a 300 pF capacitor that has been charged to a potential of 15 kV through a 5 kΩ resistor is adequate to simulate this effect.

External Sources of Radiated Energy—The vehicle is exposed to radiated energy from a multitude of sources which have the potential to disrupt normal system operation.

A more detailed discussion of these transient and noise effects is available in Ref. 3 and 4.

NOTE: The mechanisms governing the introduction of transients into an electronic assembly or its interrelated components are very complex. The equipment designer/packager must, therefore, be familiar with the configuration of the total vehicle electrical system, e.g., wire routing, shielding, grounding, filtering and decoupling practices and equipment locations.

5. *Environmental Extremes by Location*—This section quantifies guidelines for the extreme operating conditions for five major in-vehicle equipment mounting sites:

(a) Underhood
  1. Engine Compartment
  2. Bulkhead—dash panel
(b) Chassis
(c) Exterior
(d) Interior
  1. Instrument Panel
  2. Floor
  3. Rear Deck
(e) Trunk

The physical locations of these sites are given in Fig. 10. Each site (denoted by shaded section) is individually discussed together with the following detail:

(a) A table listing extremes of temperature; humidity; salt spray; sand, dust, and gravel; oil and chemical; mechanical shock and vibration and electrical steady and transient; operating conditions.

(b) Comments germane to other operating conditions of interest.

(c) Charts and other information pertaining to the vibration environment.

This section contains data from environmental measurements made by North American vehicle manufacturers or automotive original equipment suppliers. Decisions concerning each environmental factor and the test methods used to determine equipment performance and durability, should only be arrived at after examining the information in Section 4 of this report. In addition, the designer should be satisfied, by referring to pertinent test data, that the particular application falls within the described operating extremes. See Section 3.

5.1 Underhood—Engine—Caution should be exercised in applying electronics equipment in the underhood region because of the wide range of environments. Data is summarized in Table 5.

5.1.1 TEMPERATURE—Equipment in the vicinity of the exhaust system may experience temperature peaks that are beyond the survival limits of many insulation materials and electronic components.

Investigators have found that the lowest peak temperature areas are often forward in the lower compartment, near the interior or exterior radiator support hardware. The exterior has the disadvantage of being subject to more splash with resultant potential for moisture intrusion, corrosion, or thermal shock.

The heat flow temperature control mechanism for typical engine-mounted equipment relies heavily on the conduction of heat via the engine mass rather than convection via fins projecting into the airflow. Equipment thermally interlocked by conduction with the engine, has two advantages during normal operation:

1. During engine operation, the upper temperature limit is set by the coolant peak temperature, which is in turn controlled by the thermostat.

2. The time rate of change of temperature is limited by the combined engine and coolant system thermal mass.

5.1.2 PEAK TEMPERATURE (HEAT SOAK) TEST—The temperature profile varies widely with individual engine/body combinations. Therefore, it is impossible to specify all possible operating conditions. Generally, worst case temperature operating conditions should be obtained by instrumenting a proposed location for the following operating conditions:

5.1.2.1 The largest engine installation expected in that body style.
5.1.2.2 Peak ambient temperature.
5.1.2.3 Air condition *ON*.

The vehicle is driven at highway speed for about 20 min and then parked. Underhood temperatures are monitored for the *heat soak* conditions as the thermal energy stored in the engine system is released in the absence of underhood airflow. Design modifications which contribute thermal energy to the underhood area, such as secondary air thermal reactors or catalytic reactors, should be in place and operating for this test.

Test procedures of this type have revealed that the region to the rear of the engine compartment, and the locations near radiated and conducted heat from the exhaust/reactor manifold tend to be much higher in temperature.

Present control practice has limited the location of electronic equipment to temperature situations similar to those shown for the intake manifold, although operation in the vicinity of the alternator heat source will probably add about 10°C (18°F) to the peak 121°C (250°F) shown for the intake manifold. Some experimenters expect the temperature near the radiator support structure to be no higher than 100°C (212°F).

Consideration should also be given to heat flow into the engine compartment from the front wheel suspension/brake and tire combination. Some consideration has been given to electronic equipment thermally interlocked with the engine cooling system, although the high pressure-temperature combination experienced during coolant boil-off may cause unacceptable catastrophic failure.

Rate of temperature change with time is also a consideration in this area, since cold starts will result in very rapid changes, as shown in Fig. 1.

5.1.3 VIBRATION—Vibration profiles recorded on the intake manifold are shown in Fig. 11, together with the equivalent power spectral density profiles.

5.2 Underhood—Dash Panel—Data is summarized in Table 6.

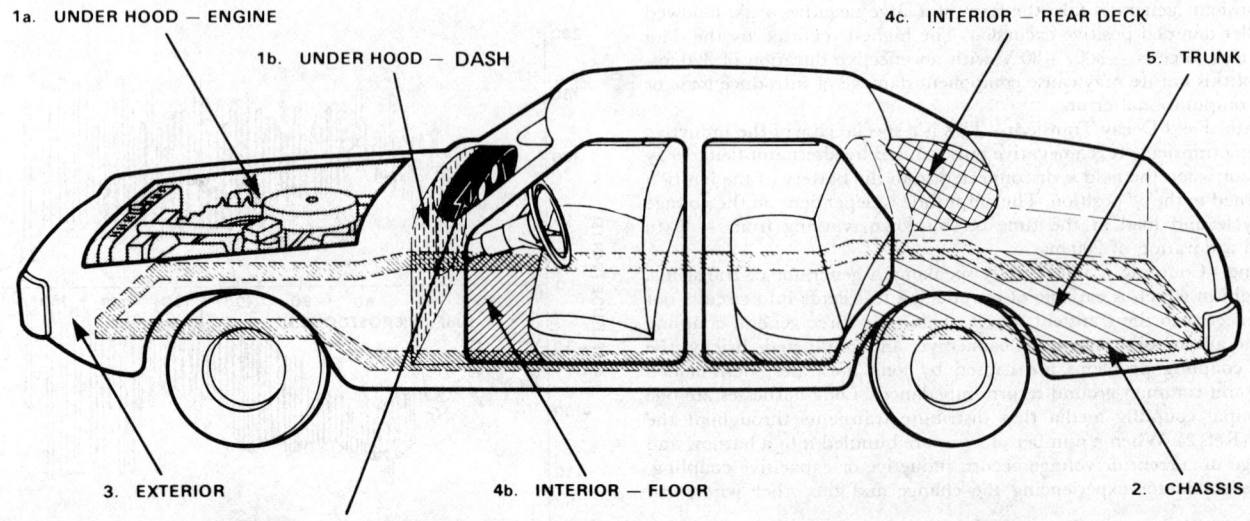

FIG. 10—VEHICLE ENVIRONMENTAL ZONES

UNDERHOOD-ENGINE ENVIRONMENTAL EXTREME DATA

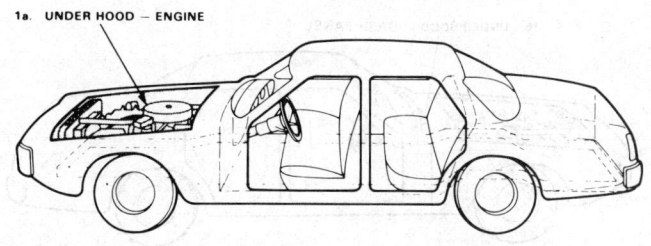

1a. UNDER HOOD – ENGINE

	TEMPERATURE			HUMIDITY (%RH)			SALT SPRAY	IMMERSION	SAND, DUST & GRAVEL	OIL & CHEMICAL	MECHANICAL SHOCK & VIBRATION	ELECTRICAL	
	LOW	HIGH	SLEW RATE	HIGH	LOW	FROST						STEADY-STATE	TRANSIENT
Choke Housing	-40°C (-40°F)	204°C (400°F)	-7°C/Min. (20°F/Min.)	95% at 38°C (100°F)	0	yes	Sect. 4.3	Splash present	Sect. 4.5	Sect. 4.4	Figure 11	Table 2 & Table 4	Table 3
Exhaust Manifold	-40°C (-40°F)	649°C (1200°F)											
Intake Manifold	-40°C (-40°F)	121°C (250°F)											

TABLE 5

5.2.1 TEMPERATURE—Temperature conditions are similar to the Underhood-Engine intake manifold, except that the primary method of heat flow is convection rather than conduction, and the resultant temperature slew rate is less. Equipment in this area generally relies heavily on convection due to the relatively low thermal conduction characteristics and unpredictable thermal interface between the equipment and the dash panel sheet metal. The rate of change in temperature is therefore set by the thermal mass of the equipment itself, and heat flow due to air movement in its vicinity rather than conduction via the mounting surface. Thermal shock due to the impact of cold mud, slush, etc., is not likely in the upper dash panel location. However, consideration should be given to melted snow and ice leakage from the hood/windshield area.

The majority of investigators have experienced peak temperatures of 121°C (250°F), although one data source expects this to be 140°C (285°F). Of course, locations on the dash panel near or just above the exhaust manifold(s) which is at 649°C (1200°F), will experience higher temperatures. The effects of underhood exhaust processing components (catalytic reactors, etc.) will also raise the peak temperatures.

5.2.2 HUMIDITY—This condition is similar to the associated engine condition, with the peak value shown in Table 6. The possibility of snow and ice intrusion, with hot ethylene glycol and water mixtures, due to cooling system failure, should also be considered.

5.2.3 SALT SPRAY—This condition is often a factor, particularly on the lower outboard portions where the dash panel joins the forward floor pan. Driving through salt slush can cause the entrance of salt spray through the radiator. The spray is then delivered to the engine compartment at high velocity by the fan. Spray due to this source is impacted on the dash panel, except for areas shielded by the engine or other underhood components.

5.2.4 IMMERSION—Not generally required.

5.2.5 SAND, DUST, AND GRAVEL—Gravel is not generally a problem, except at the lower dash panel near the transition into the forward floor pan.

5.2.6 OILS AND CHEMICALS—Commonly encountered components (with and without contaminants) are:
  Engine oils and additives (hot and cold)
  Brake fluid
  Gasoline
  Ethylene glycol and water (hot and cold)
  Water and snow
  Waxes
  Transmission oil
  Windshield washer solvent
  Degreasers and cleaning compounds
  Detergents
  Battery acid
  Steam
  Freon

5.2.7 MECHANICAL SHOCK AND VIBRATION—Vibration profiles are shown in Fig. 12.

5.3 Chassis—Data is summarized in Table 7.

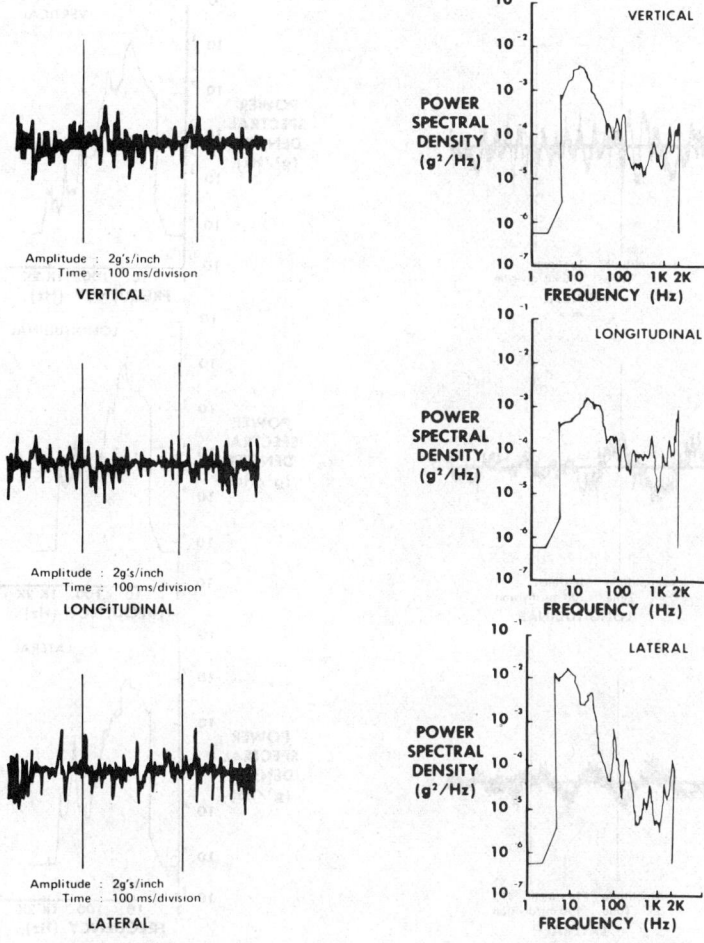

FIG. 11—ENGINE INTAKE MANIFOLD VIBRATION MEASUREMENTS

UNDERHOOD-DASH PANEL ENVIRONMENTAL DATA

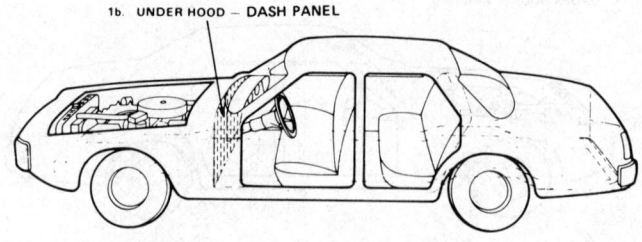

	TEMPERATURE			HUMIDITY (%RH)			SALT SPRAY	IMMERSION	SAND, DUST & GRAVEL	OIL & CHEMICAL	MECHANICAL SHOCK & VIBRATION	ELECTRICAL	
	LOW	HIGH	SLEW RATE	HIGH	LOW	FROST						STEADY-STATE	TRANSIENT
Normal	−40°C (−40°F)	121°C (250°F)	open	95% at 38°C (100°F)			Sect. 4.3	no	Sect. 4.5	Sect. 4.4	Figure 12	Tables 2 & 4	Table 3
Extreme	−40°C (−40°F)	141°C (285°F)	open	80% at 66°C (150°F)				no					

TABLE 6

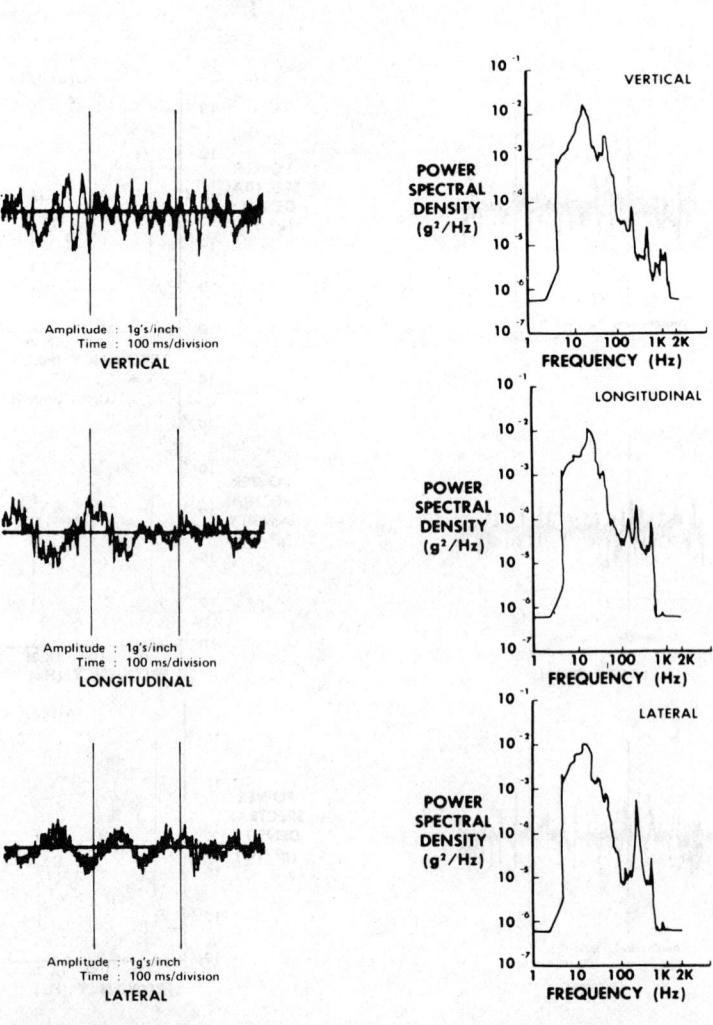

FIG. 12—PLENUM VIBRATION MEASUREMENTS

5.3.1 TEMPERATURE—The heat sources encountered in the chassis area include (in rank of decreasing surface temperature):

Source	Peak Temperature
a. Exhaust/catalytic reactor system	649°C (1200°F)
b. Brake system/tires and transmission/differential drivetrain components	177°C (350°F)
c. Engine	121°C (250°F)
d. Vehicle ambient peak temperature	85°C (185°F)

The practical limitations of equipment components (with the possible exception of sensors) will restrict the designer to locations with the peak temperatures given in c and d above. Again, the designer is urged to check his particular installation for the actual peak temperatures experienced under operating conditions.

5.3.2 HUMIDITY—As shown in Table 7.

5.3.3 SALT SPRAY—With the exception of a few shielded locations, all chassis components are subject to heavy salt spray.

5.3.4 IMMERSION—Typical chassis components are subject to immersion.

5.3.5 DUST, SAND, AND GRAVEL—All chassis components in line with the wheel track that are not shielded are subject to continuous bombardment during vehicle operation on gravel roads. In nontrack aligned portions of the chassis, some bombardment will be experienced by equipment mounted on forward-facing chassis surfaces. All chassis components are subject to heavy dust and sand environments.

5.3.6 OILS AND CHEMICALS—The chassis is subject to all of the oils and chemicals listed in Section 4.4.

5.3.7 MECHANICAL, SHOCK, AND VIBRATION—Vibration data collected on the frame bumper attachment, frame transmission mount, frame crossmember and wheel backplate with equivalent power spectral density profiles are shown in Figs. 13-15.

5.3.8 ELECTRICAL—Steady State (Refer to Section 4.10.1 for further information—Three operating conditions are recognized:

   a. Normal starting and running   9-16 V[1]
   b. Cold starting   4.5-6 V
   c. Booster battery starting   24 V

5.3.9 ELECTRICAL—TRANSIENT—This condition varies, depending upon the electrical distance of the equipment from the battery and the nearness of transient sources (e.g., inductive motors, solenoids, the alternator). Typical data is shown in Table 7. (Refer to Section 4.11 for further information.)

5.4 **Exterior**—The exterior consists of all outward and external vehicle surfaces above the chassis. This includes the forward grille area and potential mounting areas just above the bumpers. Data is summarized in Table 8.

[1] See Section 4.10.1 for a definition of normal voltage.

CHASSIS ENVIRONMENTAL EXTREME DATA

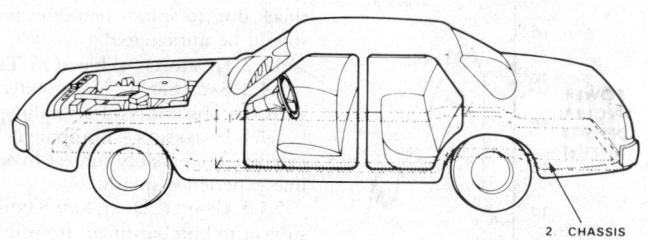

	TEMPERATURE			HUMIDITY (%RH)			SALT SPRAY	IMMERSION	SAND, DUST & GRAVEL	OIL & CHEMICAL	MECHANICAL SHOCK & VIBRATION	ELECTRICAL	
	LOW	HIGH	SLEW RATE	HIGH	LOW	FROST						STEADY-STATE	TRANSIENT
Isolated	-40°C (-40°F)	85°C (185°F)	NA	98% at 38°C (100°F)	0	yes	Sect. 4.3	Sect. 4.4	Sect. 4.5	Sect. 4.4	Figures 13, 14 & 15	Table 2 & 4	Table 3
Near Heat Source	-40°C (-40°F)	121°C (250°F)	NA	66°C (150°F)	0	yes							
At Drive Train High Temp Location	-40°C (-40°F)	177°C (350°F)	NA	80%	0	yes							

TABLE 7

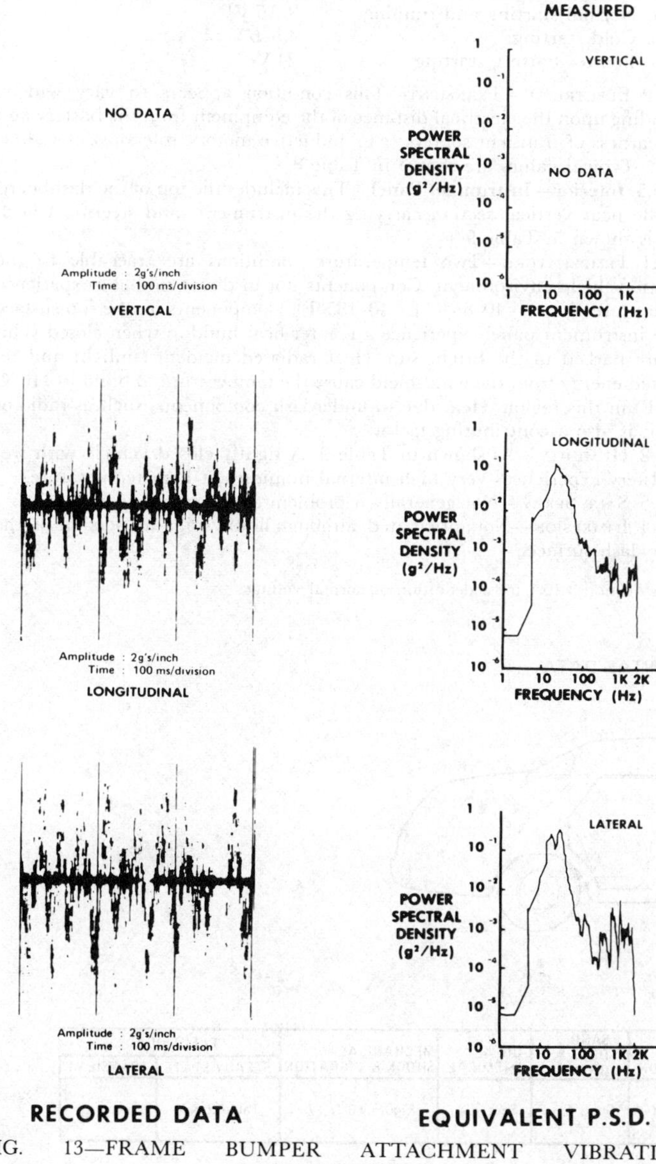

FIG. 13—FRAME BUMPER ATTACHMENT VIBRATION MEASUREMENTS

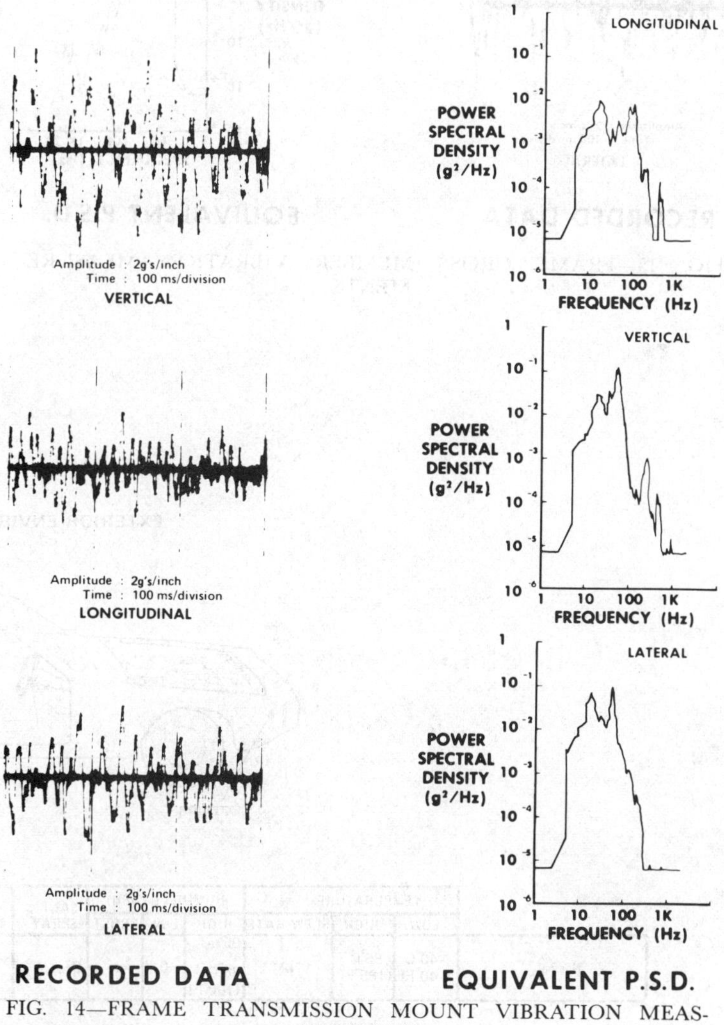

FIG. 14—FRAME TRANSMISSION MOUNT VIBRATION MEASUREMENTS

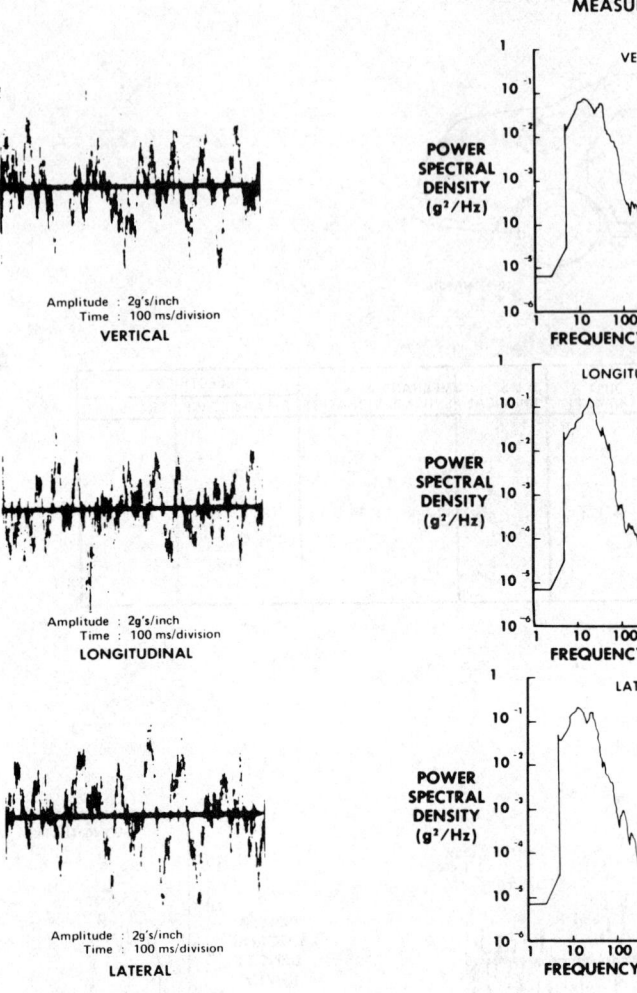

FIG. 15—FRAME CROSS MEMBER VIBRATION MEASUREMENTS

5.4.1 TEMPERATURE—Since all surfaces are away from internal vehicle heat sources, the temperature is primarily controlled by the climatic ambient conditions. These are discussed in Section 4.1 and shown in Table 8. Thermal shock due to splash or immersion, particularly on the front of the vehicle, should be anticipated.

5.4.2 HUMIDITY—Shown in Table 8.

5.4.3 SALT SPRAY—Most exterior surfaces are subject to heavy salt spray, with the possibility of crystalline salt buildup in some grill areas.

5.4.4 IMMERSION—Equipment mounted approximately below the vehicle axle line are possibly subject to occasional immersion. Components above this line experience splash.

5.4.5 GRAVEL, DUST, AND SAND—Components on the front of the vehicle are subject to bombardment from the vehicle ahead. Sand and dust impinges on all surfaces.

5.4.6 OILS AND CHEMICALS—Environmental chemicals include:
  Road tar
  Anti-freeze/water mixture
  Soaps and detergents
  Steam
  Salt spray
  Washer solvent
  Degreasers
  Waxes
  Water and snow

5.4.7 MECHANICAL SHOCK AND VIBRATION—Data collected at the wheel back plate is shown in Fig. 16, and center pillar data is shown in Fig. 17.

5.4.8 ELECTRICAL—STEADY STATE—Three operating conditions are recognized:

  a. Normal starting and running   9–16 V[1]
  b. Cold starting   4.5–6 V
  c. Booster battery starting   24 V

5.4.9 ELECTRICAL—TRANSIENT—This condition appears to vary widely, depending upon the electrical distance of the equipment from the battery and the nearness of transient sources (e.g., inductive motors, solenoids, the alternator). Typical values are shown in Table 8.

5.5 Interior—Instrument Panel—This includes the top of the dashboard and the near vertical section carrying the instruments and steering wheel. Data is shown in Table 9.

5.5.1 TEMPERATURE—Two temperature conditions are traceable to the climatic vehicle environment. Components not in direct sunlight experience temperatures from -40–$85°C$ (-40–$185°F$). Components on the top surface of the instrument panel experience a greater heat buildup when closed vehicles are parked in the bright sun. Heat radiated incident sunlight and re-radiated energy from the windshield cause the temperature to build to $113°C$ ($235°F$) in this region. Heat due to underdash components, such as radio or heater, is also a contributing factor.

5.5.2 HUMIDITY—As shown in Table 9. A tightly closed vehicle with wet upholstery experiences very high internal humidity at high temperature.

5.5.3 SALT SPRAY—Not generally a problem at the instrument panel.

5.5.4 IMMERSION—Not anticipated, although liquid spills are possible on the upper dash surface.

[1] See Section 4.10.1 for a definition of normal voltage.

EXTERIOR ENVIRONMENTAL DATA

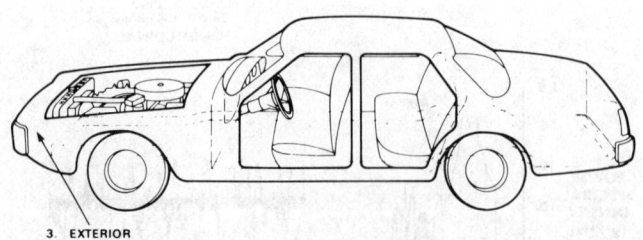

3. EXTERIOR

	TEMPERATURE			HUMIDITY (%RH)			SALT SPRAY	IMMERSION	SAND, DUST & GRAVEL	OIL & CHEMICAL	MECHANICAL SHOCK & VIBRATION	ELECTRICAL	
	LOW	HIGH	SLEW RATE	HIGH	LOW	FROST						STEADY-STATE	TRANSIENT
Normal	-40°C (-40°F)	85°C (185°F)	NA	95% at 38°C (100°F)	0	yes	Sect. 4.3	Sect. 4.4	Sect. 4.5	Sect. 4.4	Figure 16 & 17	Table 2 & 4	Table 3

TABLE 8

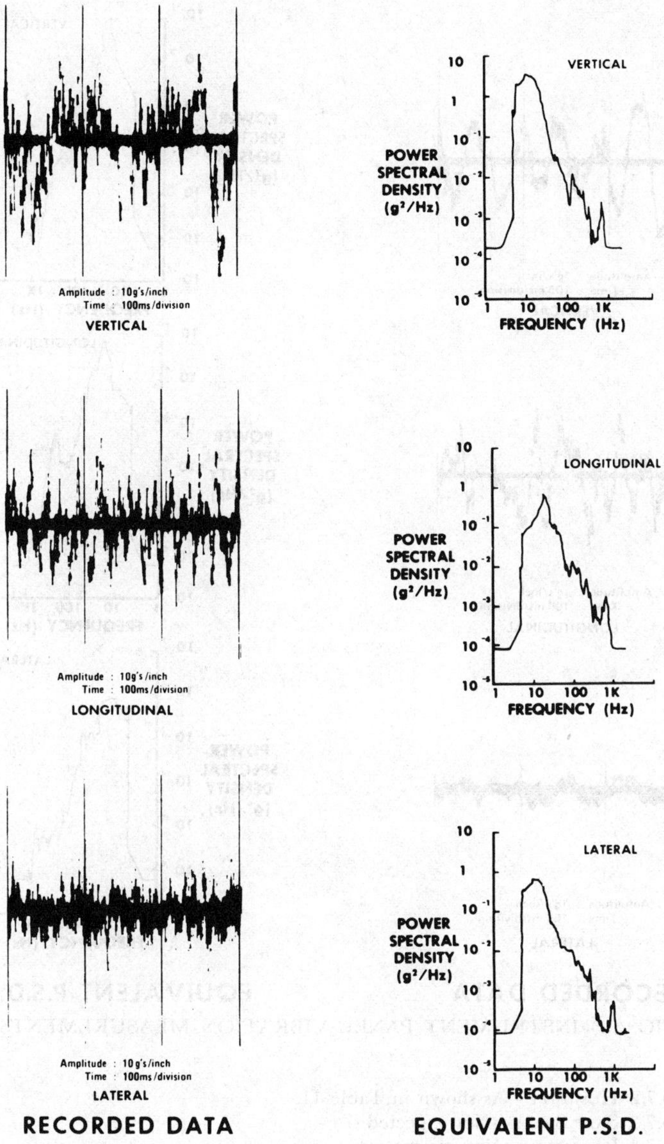

FIG. 16—WHEEL BACK PLATE VIBRATION MEASUREMENTS

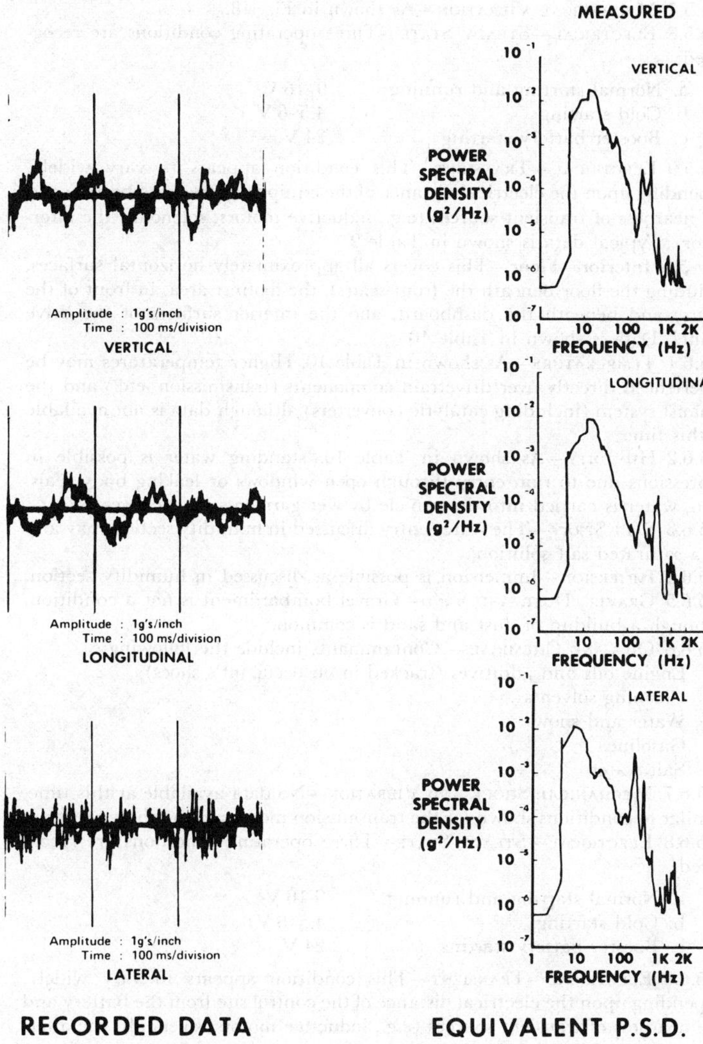

FIG. 17—CENTER PILLAR VIBRATION MEASUREMENTS

INTERIOR - INSTRUMENTAL PANEL ENVIRONMENTAL DATA

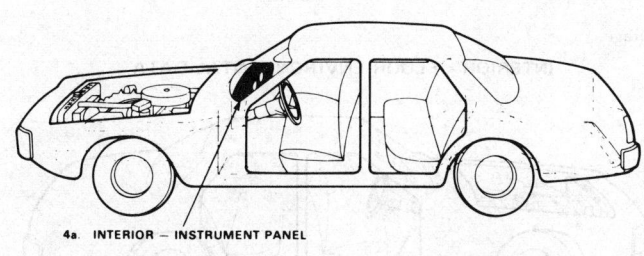

4a. INTERIOR – INSTRUMENT PANEL

	TEMPERATURE			HUMIDITY (%RH)			SALT SPRAY	IMMERSION	SAND, DUST & GRAVEL	OIL & CHEMICAL	MECHANICAL SHOCK & VIBRATION	ELECTRICAL	
	LOW	HIGH	SLEW RATE	HIGH	LOW	FROST						STEADY-STATE	TRANSIENT
Nominal	-40°C (-40°F)	85°C (185°F)	NA	98% at 38°C (100°F)	0	yes	no	Partial	Dust only	Sect. 4.4	Figure 18	Table 2 & 4	Table 3
Top Surface	-40°C (-40°F)	113°C (235°F)	NA	80% at 66°C (150°F)									

TABLE 9

5.5.5 GRAVEL, SAND, AND DUST—Gravel not anticipated. Coatings of sand and dust are expected on all horizontal surfaces.

5.5.6 OILS AND CHEMICALS—Mainly cleaning agents: waxes, soaps, and detergents.

5.5.7 MECHANICAL VIBRATION—As shown in Fig. 18.

5.5.8 ELECTRICAL—STEADY STATE—Three operating conditions are recognized:

   a. Normal starting and running   9-16 V[1]
   b. Cold starting   4.5-6 V
   c. Booster battery starting   24 V

5.5.9 ELECTRICAL—TRANSIENT—This condition appears to vary widely, depending upon the electrical distance of the equipment from the battery and the nearness of transient sources (e.g., inductive motors, solenoids, the alternator). Typical data is shown in Table 9.

5.6 Interior—Floor—This covers all approximately horizontal surfaces, including the floor beneath the front seat(s), the footrest areas in front of the seat(s) and beneath the dashboard, and the interior surfaces of the drive tunnel. Data is shown in Table 10.

5.6.1 TEMPERATURE—As shown in Table 10. Higher temperatures may be experienced directly over drivetrain components (transmission, etc.) and the exhaust system (including catalytic converters) although data is not available at this time.

5.6.2 HUMIDITY—As shown in Table 10, standing water is possible in depressions due to rain entry through open windows or leaking body seals. Also, water is carried into the vehicle by wet garments or packages.

5.6.3 SALT SPRAY—The water entry discussed in humidity section may also be a saturated salt solution.

5.6.4 IMMERSION—Immersion is possible as discussed in humidity section.

5.6.5 GRAVEL, DUST, AND SAND—Gravel bombardment is not a condition, although a buildup of dust and sand is common.

5.6.6 OILS AND CHEMICALS—Contaminants include the following:
   Engine oils and additives (tracked in on occupant's shoes)
   Cleaning solvents
   Water and snow
   Gasoline
   Salt water

5.6.7 MECHANICAL SHOCK AND VIBRATION—No data available at this time. Similar to conditions shown for the transmission mounts in the chassis section.

5.6.8 ELECTRICAL—STEADY STATE—Three operating conditions are recognized:

   a. Normal starting and running   9-16 V[1]
   b. Cold starting   4.5-6 V
   c. Booster battery starting   24 V

5.6.9 ELECTRICAL—TRANSIENT—This condition appears to vary widely, depending upon the electrical distance of the control site from the battery and the nearness of transient sources (e.g., inductive motors, solenoids, the alternator).

5.7 Interior—Rear Deck—This area includes horizontal surface extending from the top of the rear seat to the body work just below the bottom edge of the backlight. Data is shown in Table 11.

5.7.1 TEMPERATURE—The major heat source is climatic incident radiant energy from direct sunlight and sunlight reflected from the backlight. The peak temperature of 104°C (220°F) is slightly less than that given for the upper dashboard surface because of the absence of heat sources beneath the panel.

[1]See Section 4.10.1 for a definition of normal voltage.

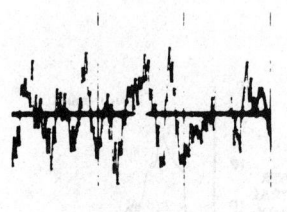
Amplitude : 1g's/inch
Time : 100 ms/division
VERTICAL

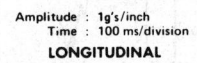
Amplitude : 1g's/inch
Time : 100 ms/division
LONGITUDINAL

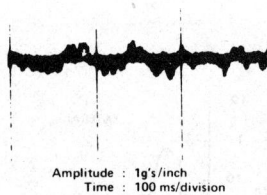
Amplitude : 1g's/inch
Time : 100 ms/division
LATERAL

RECORDED DATA

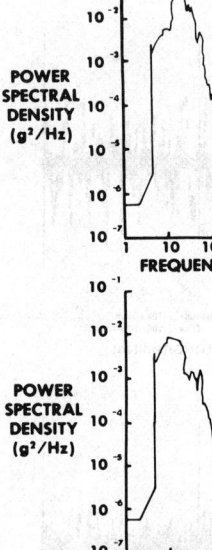

EQUIVALENT P.S.D.

FIG. 18—INSTRUMENT PANEL VIBRATION MEASUREMENTS

5.7.2 HUMIDITY—As shown in Table 11.

5.7.3 SALT SPRAY—Not expected.

5.7.4 IMMERSION—Not present.

5.7.5 SAND AND DUST—Light coatings of sand and dust are present.

5.7.6 OILS AND CHEMICALS—Only cleaning agents expected.

5.7.7 MECHANICAL SHOCK AND VIBRATION—No data available at this time. The condition is similar to conditions shown for the dashboard, with the exception that vibration at the vehicle's rear is a function of high variable trunk, rear seat, and bumper loading conditions.

5.7.8 ELECTRICAL—STEADY STATE—Three operating conditions are recognized:

INTERIOR - FLOOR ENVIRONMENTAL DATA

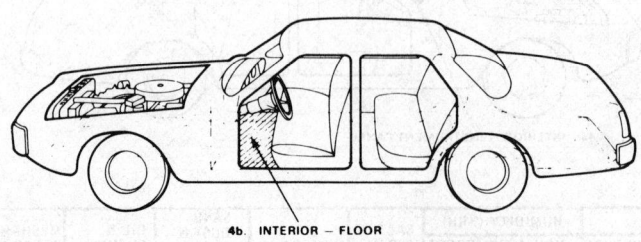

4b. INTERIOR - FLOOR

TEMPERATURE			HUMIDITY (%RH)			SALT SPRAY	IMMERSION	SAND, DUST & GRAVEL	OIL & CHEMICAL	MECHANICAL SHOCK & VIBRATION	ELECTRICAL	
LOW	HIGH	SLEW RATE	HIGH	LOW	FROST						STEADY-STATE	TRANSIENT
-40°C (-40°F)	85°C (185°F)	NA	98%RH 38°C (100°F)	0	-	NO	NO	Sect. 4.5	Sect. 4.4	Not measured	Table 2 & 4	Table 3

TABLE 10

INTERIOR - REAR DECK ENVIRONMENTAL DATA

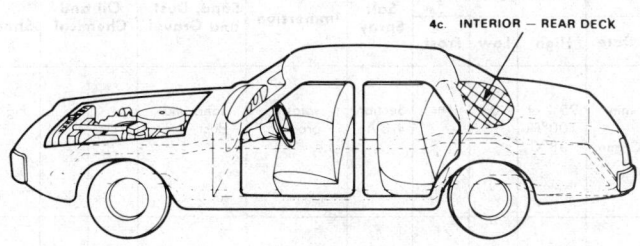

4c. INTERIOR – REAR DECK

TEMPERATURE			HUMIDITY (%RH)			SALT SPRAY	IMMERSION	SAND, DUST & GRAVEL	OIL & CHEMICAL	MECHANICAL SHOCK & VIBRATION	ELECTRICAL	
LOW	HIGH	SLEW RATE	HIGH	LOW	FROST						STEADY-STATE	TRANSIENT
-40°C (-40°F)	104°C (220°F)	NA	98% at 38°C (100°F) -------- 80% at 66°C (150°F)	0	no	no	no	Sect. 4.5	Sect. 4.4	Not Measured	Table 2 & 4	Table 3

TABLE 11

a. Normal starting and running 9–16 V[1]
b. Cold starting 4.5–6 V
c. Booster battery starting 24 V

5.7.9 ELECTRICAL—TRANSIENT—This condition appears to vary widely, depending upon the electrical distance of the equipment from the battery and the nearness of transient sources (e.g., inductive motors, solenoids, the alternator). Care should be taken to measure transients due to electrical rear window lifts, if present.

5.8 Trunk—Environmental data is defined in Table 12.

5.8.1 TEMPERATURE—The anticipated temperature limits are fairly similar to those expected for the interior. However, the thinner insulation may increase temperatures in some areas of the trunk floor due to radiated and conducted exhaust system heat. The inside of the trunk lid may also experience higher temperatures than those given due to direct sunlight heating.

5.8.2 HUMIDITY—The presence of stored liquids or wet clothing, etc., makes this the highest humidity enclosed volume in the car. Condensation is possible on all surfaces.

5.8.3 SALT SPRAY—A standing saturated salt solution is possible on the compartment floor.

5.8.4 IMMERSION—Standing liquid on the floor is anticipated unless the area is equipped with drains.

5.8.5 SAND AND DUST—A heavy buildup of sand and dust is anticipated.

5.8.6 OILS AND CHEMICALS—Spillage of all of the chemicals listed in Section 4.4 is possible.

5.8.7 MECHANICAL SHOCK AND VIBRATION—No data available at this time. Conditions are similar to those shown for the dashboard, with the exception that vibration at the vehicle's rear is a function of highly variable trunk, rear seat, and rear bumper loading conditions.

5.8.8 ELECTRICAL—STEADY STATE—This condition appears to vary widely, depending upon the electrical distance of the equipment from the battery and the nearness of transient sources (e.g., inductive motors, solenoids, the alternator). Typical values are shown in Table 12.

5.9 Environmental Extremes Summary—Table 13 summarizes the information provided in Sections 5.1–5.8.

ACKNOWLEDGMENTS

The subcommittee acknowledges the contribution of Mr. J. R. Morgan. Mr. Morgan provided oscillograms of the load dump transient and electrical noise, and critiqued the electrical section of this document.

REFERENCES

1. O. T. McCarter, "Environmental Guidelines for the Designer of Automotive Electronic Components." Paper 740017 presented at SAE Automotive Engineering Congress, Detroit, March 1974.
2. J. R. Morgan, "Transients in the Automotive Electrical System." Motorola CER-114, 1973.
3. G. B. Andrews, "Control of the Automotive Electrical Environment." Paper 730045 presented at SAE Automotive Engineering Congress, Detroit, January 1973.
4. SAE J1113a, Electromagnetic Susceptibility Procedures for Vehicle Components (Except Aircraft) (June, 1978).

[1] See Section 4.10.1 for a definition of normal voltage.

TRUNK & ENVIRONMENTAL DATA

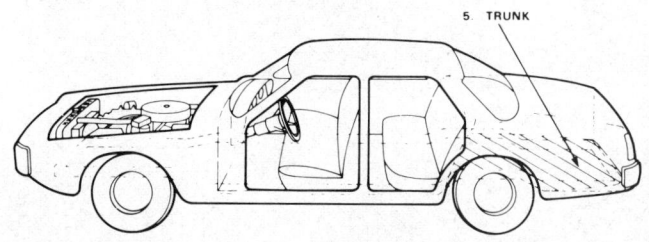

5. TRUNK

TEMPERATURE			HUMIDITY (%RH)			SALT SPRAY	IMMERSION	SAND, DUST & GRAVEL	OIL & CHEMICAL	MECHANICAL SHOCK & VIBRATION	ELECTRICAL	
LOW	HIGH	SLEW RATE	HIGH	LOW	FROST						STEADY-STATE	TRANSIENT
-40°C (-40°F)	85°C (185°F)	NA	98% at 38°C (100°F) -------- 80% at 66°F	0	yes	no	no	Sect. 4.5	Sect. 4.4	Not Measured	Tables 2 & 4	Table 3

TABLE 12

TABLE 13—ENVIRONMENTAL EXTREME SUMMARY

Location	Temperature			Humidity (%RH)			Salt Spray	Immersion	Sand, Dust, and Gravel	Oil and Chemical	Mechanical Shock and Vibration	Electrical	
	Low	High	Slew Rate	High	Low	Frost						Steady-State	Transient
1. Underhood—Engine Choke Housing	−40°F −40°C	400°F 204°C	20°F/min −7°C/min	95% at 100°F 38°C	0	yes	Section 4.3	splash present	sand and dust	↑	Fig. 11	↑	↑
Exhaust Manifold	−40°F −40°C	1200°F 649°C											
Intake Manifold	−40°F −40°C	250°F 121°C											
Underhood—Firewall Normal	−40°F −40°C	250°F 121°C	open	95% at 100°F 38°C	0		Section 4.3	no	sand and dust		Fig. 12		
Extreme	−40°F −40°C	285°F 141°C	open	80% at 150°F 66°C	0			no					
2. Chassis Isolated	−40°F −40°C	185°F 84°C	NA	98% at 100°F 38°C	0	yes	yes	yes	yes		See Section 4.10 and Tables 2 and 4		
Near Heat Source	−40°F −40°C	250°F 121°C	NA	80% at 150°F 66°C	0	yes		yes	yes		Figs. 13, 14, and 15		See Section 4 and Table 3
At Drivetrain High Temp Locations	−40°F −40°C	350°F 177°C	NA						yes				
3. Exterior Normal	−40°F −40°C	235°F 113°C	NA	95% at 100°F 38°C	0	yes	yes	yes	yes	See Section 4.4	Figs. 16 and 17		
4. Interior Instrument Panel	−40°F −40°C	185°F 84°C	NA	98% at 100°F 38°C	0	yes	no	no	dust only		Fig. 18		
Top		225°F 113°C											
Floor	−40°F −40°C	185°F 84°C	NA	98% at 100°F 38°C 80% at 150°F 66°C	0	no	no		dust and sand		Not measured		
Rear Deck	−40°F −40°C	220°F 104°C	NA	98% at 100°F 38°C	0	no	no	no	dust and sand		Not measured		
5. Trunk	−40°F −40°C	185°F 84°C	NA	98% at 100°F 38°C	0	yes	no	no	sand and dust	↓	Not measured	↓	↓

(R) DIAGNOSTIC CONNECTOR
—SAE J1962 JUN93

SAE Recommended Practice

Report of the SAE Vehicle E/E Systems Diagnostic Standards Committee approved June 1992 and completely revised June 1993.

Foreword—The purpose of this SAE Recommended Practice is to define a minimum set of diagnostic connector requirements that will promote the use of a common diagnostic connector throughout the motor vehicle industry.

It is intended that the connector specified herein be used on complex luxury vehicles as well as simple utility vehicles.

This SAE document is under the control and maintenance of the Vehicle E/E System Diagnostics Committee. This committee will periodically review and update this document as needs dictate.

NOTE—At the time of publication, SAE J1962 was in the revision process. Please consult with the SAE Troy, MI office for update.

Rationale relative to intent is provided, where applicable, to minimize ambiguity.

1. Scope—The SAE J1962 diagnostic connector consists of two mating connectors, the vehicle connector (see Figure 1) and the test equipment connector (see Figure 2).

This document:
a. Defines the functional requirements for the vehicle connector. These functional requirements are separated into three principal areas: connector location/access, connector design, and connector terminal assignments.
b. Defines the functional requirements for the test equipment connector. These functional requirements are separated into two principal areas: connector design and connector terminal assignments.

2. References

2.1 Applicable Documents—The following publications form a part of this specification to the extent specified herein. The latest issue of SAE publications shall apply.

2.1.1 SAE PUBLICATIONS—Available from SAE, 400 Commonwealth Drive, Warrendale, PA 15096-0001.

SAE J1850—Class B Data Communication Network Interface
SAE J1978—OBD II Scan Tool
SAE J2201—OBD II Scan Tool Universal Interface

2.1.2 ISO PUBLICATIONS—Available from ANSI, 11 West 42nd Street, New York, NY 10036-8002.

ISO 9141-2—Road vehicles—Diagnostic systems—Part 2: CARB requirements for interchange of digital information
ISO 8092—Road vehicles—Flat, quick connection terminations—Part 1: Tabs for single pole connections

2.1.3 CARB PUBLICATION—Available from California Air Resources Board. California Code of Regulations, Title 13, 1968.1: Malfunction and Diagnostic Systems Requirements, - 1994 and subsequent model year passenger cars, light-duty trucks, and medium-duty vehicles with feedback fuel control systems

2.2 Definitions, Abbreviations, Acronyms
N   Newtons
SAE J1850 Single Wire   SAE J1850 Single wire, 10.4 Kbps, VPW, CRC
SAE J1850 Two Wire   SAE J1850 Two wire, 41.6 Kbps, PWM, CRC

3. Vehicle Connector Location/Access

3.1 Consistency of Location—The vehicle connector shall be located in the passenger compartment in the area bounded by the driver's end of the instrument panel to 300 mm beyond the vehicle centerline, attached to the instrument panel, and accessible from the driver's seat. The preferred location is between the steering column and the vehicle centerline. The vehicle connector shall be mounted to facilitate mating and unmating.

3.2 Ease of Access—Access to the vehicle connector shall not require a tool for the removal of an instrument panel cover, connector cover, or any barriers. The vehicle connector should be fastened and located so as to permit a one-handed/blind insertion of the mating test equipment connector. Refer to Figure 1 for mated connector space requirements.

3.3 Visibility—The vehicle connector should be out of the occupant's (front and rear seat) normal line of sight but easily visible to a "crouched" technician.

3.4 Vehicle Operation—Attachment of any equipment to the vehicle connector should not preclude normal physical and electrical operation of the vehicle.

4. Vehicle and Test Equipment Connector Design

4.1 Number of Terminals—The vehicle and test equipment connectors shall each be capable of accommodating 16 terminals.

4.2 Terminal Requirements—The terminals shall be rated for currents not to exceed 10 A DC continuous.

4.2.1 TERMINAL TYPES—The vehicle connector shall consist of female terminals that will mate with the test equipment connector male blade terminals.

4.2.2 TERMINAL SPACING—Terminal spacing is shown in Figure 2.

4.3 Connector Mating—The test equipment connector contact mating shall be designed so that the signal ground and the chassis ground terminals of the test equipment connector will make electrical contact prior to any other test equipment connector terminals making electrical contact. On the disconnect cycle, these same two terminals will not lose electrical contact until all of the other terminals have been disconnected.

4.4 Connector Shape/Features—The mating portions of both connectors shall be "D" shaped. The connectors shall have easily discernible keying features to allow for easy connection in a one-handed/blind operation.

The vehicle connector and the test equipment connector shall have latching features that assure the test equipment connector will remain mated when properly connected. The latching feature will be designed to provide a positive feel when the test equipment connector is fully seated. The latching feature should not require the activation of any levers on either connector to mate or unmate. Pulling on the test equipment connector to separate the two mated halves shall not result in any damage to either connector.

4.5 Material Selection

4.5.1 CONNECTOR MATERIAL—Selection and specification of connector material shall be made by the appropriate vehicle and test equipment manufacturers. However, the minimum recommended temperature range for the selected material is -40 to 85 °C.

4.5.2 TERMINAL MATERIAL—Selection and specification of terminal base material shall be made by the appropriate vehicle and test equipment manufacturers.

4.6 Vehicle Connector Cycle Life—The vehicle connector shall meet the requirements of this document after 200 mating cycles.

4.7 Test Equipment Connector Cycle Life—The test equipment manufacturer shall specify the minimum number of mating cycles the test equipment connector is capable of while meeting the requirements of this document.

4.8 Strain Relief—The test equipment connector shall have strain relief features for the wires/cable connected to it.

4.9 Terminal and Connector Parameters and Performance Requirements

4.9.1 FUNCTIONAL PARAMETERS FOR TERMINALS
a. Blade Size for Test Equipment Connector: Shall conform to the dimensions shown in Figure 3
b. Maximum Current DC: 10 A DC continuous
c. Temperature Range: -40 to +85 °C
d. Voltage Range DC: -0.05 to 30.0 V DC
e. Suggested Maximum Cable Size: 0.8 mm^2 (18 AWG)

4.9.2 Performance Requirements for Terminal Pairs—The terminal system (i.e., mated terminal pairs) must meet the performance standards in (a) through (c) following each of the environmental exposures listed in 4.9.4. Performance measurements are to be taken at room temperature.
a. Resistance Interface (measured at 1 A): 3 mΩ maximum
b. Recommended Resistance Cable to Cable per Terminal Pair (measured at 1 A): 10 mΩ maximum
c. Recommended Low Energy Resistance: 100 mΩ maximum at 100 µA at 20 mV (open circuit voltage) at initial mating

4.9.3 CONNECTOR SYSTEM PERFORMANCE REQUIREMENTS—The connector system must meet the performance standards outlined in (a) through (f) following each of the environmental exposures listed in 4.9.4. Performance measurements must be taken at room temperature.
a. Isolation Resistance: Between adjacent terminals must exceed 20 MΩ at 16 V DC.
b. Retention Force: Terminal to connector retention force to exceed 80 N.

c. Connector Disengagement Force with 16 terminal pairs: Not to exceed 88 N.
d. Connector Mating Force with 16 terminal pairs:
   (1) without spring clip 110 N maximum
   (2) with spring clip 142 N maximum (see Figure 4)
e. Indexing Feature: Must prevent mismating of connectors when a force of 300 N is applied.
f. Mounting Feature: The vehicle connector mounting feature shall withstand a 300 N force applied to the connector mating area in the direction of the mating and unmating process.

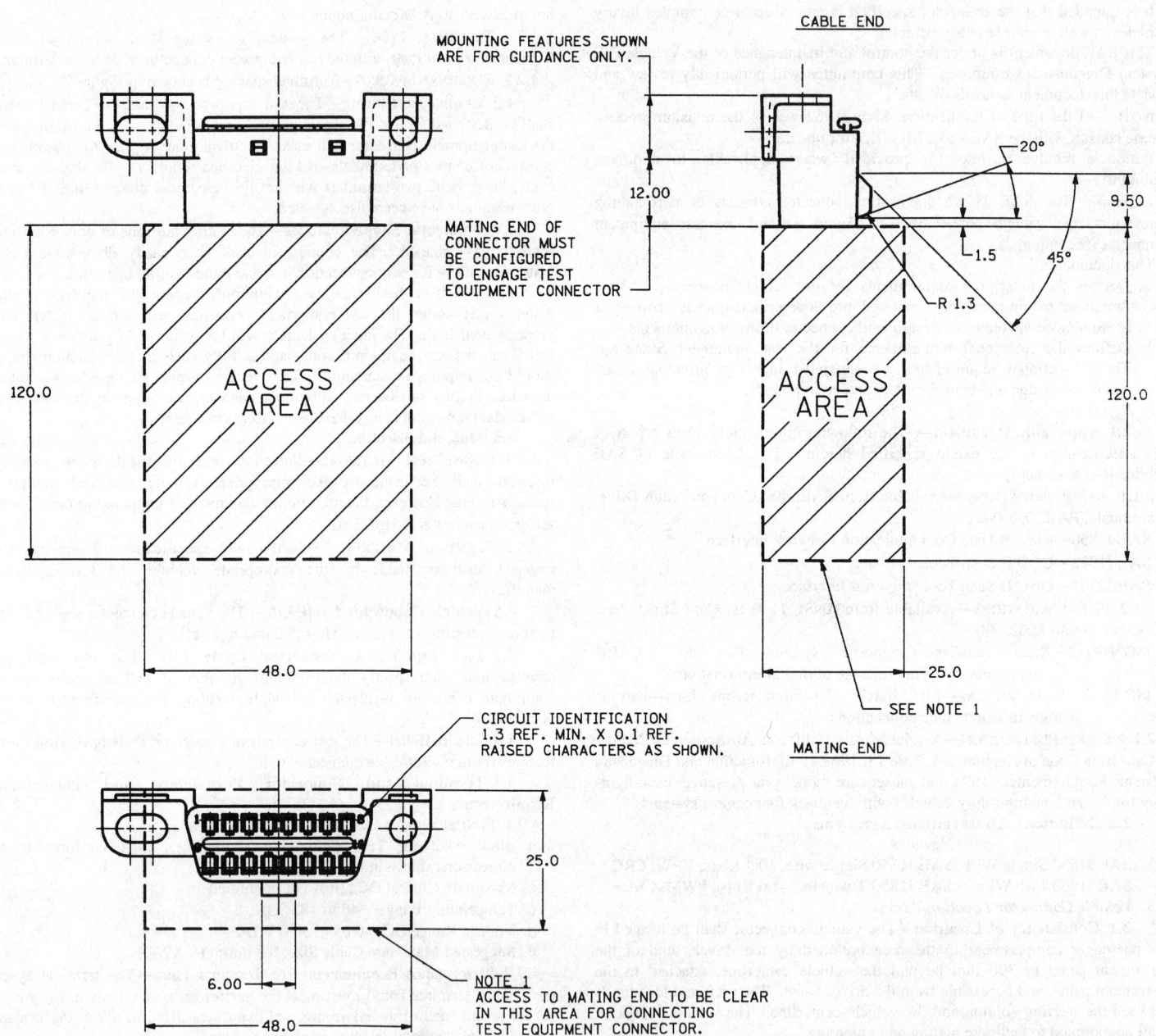

FIGURE 1—VEHICLE CONNECTOR

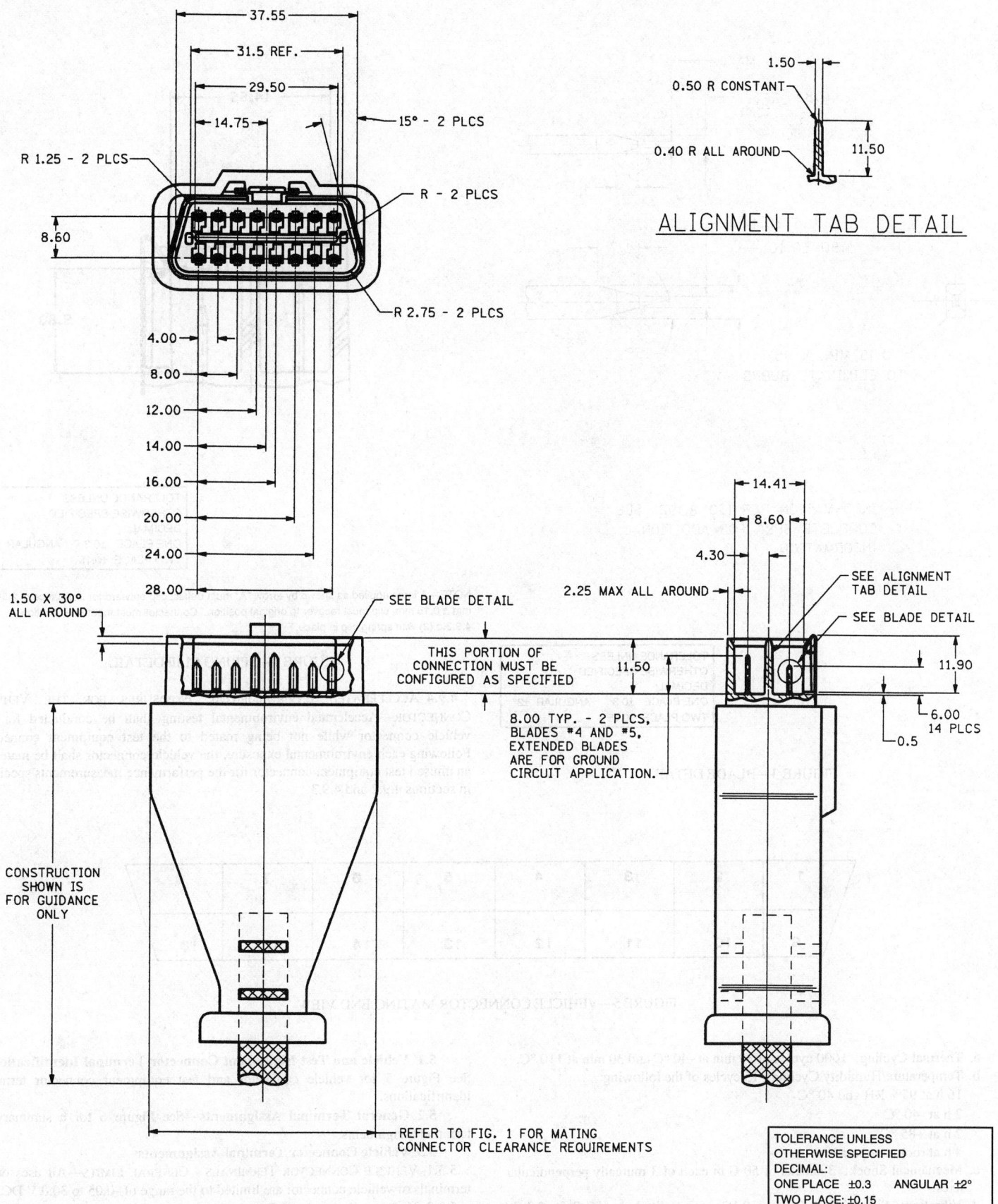

FIGURE 2—TEST EQUIPMENT CONNECTOR

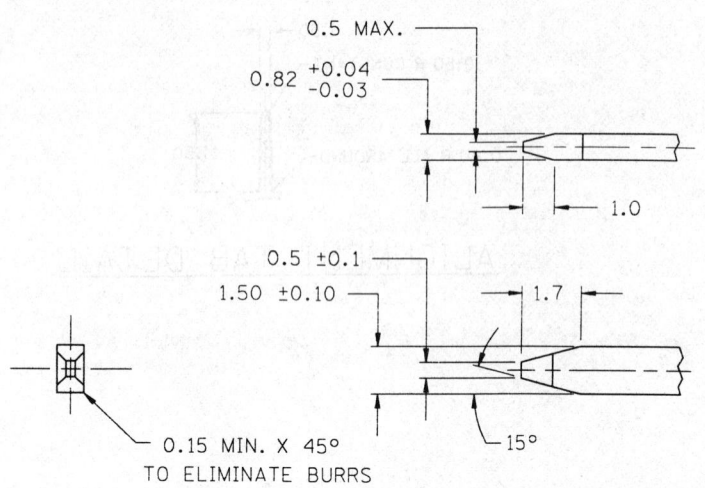

INFORMATION PER ISO 8092. SEE COMPLETE SPEC FOR ADDITIONAL INFORMATION.

TOLERANCE UNLESS OTHERWISE SPECIFIED
DECIMAL:
ONE PLACE ±0.3   ANGULAR ±2°
TWO PLACE: ±0.15

FIGURE 3—BLADE DETAIL

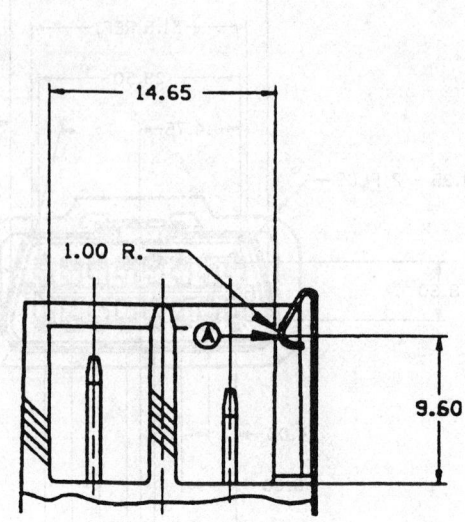

TOLERANCE UNLESS OTHERWISE SPECIFIED
DECIMAL:
ONE PLACE ±0.3   ANGULAR ±2°
TWO PLACE: ±0.15

NOTE—A force applied as shown by arrow "A" must deflect clip outward for a distance of 2.50 mm ± 0.15 mm, clip must recover to original position. Connector must meet specification 4.9.3.d.(2) with spring clip in place.

FIGURE 4—SPRING CLIP DETAIL

4.9.4 ACCELERATED ENVIRONMENTAL EXPOSURES FOR THE VEHICLE CONNECTOR—Accelerated environmental testing shall be conducted for the vehicle connector while not being mated to the test equipment connector. Following each environmental exposure, the vehicle connector shall be mated to an unused test equipment connector for the performance measurements specified in sections 4.9.2 and 4.9.3.

1	2	3	4	5	6	7	8
9	10	11	12	13	14	15	16

FIGURE 5—VEHICLE CONNECTOR MATING END VIEW

a. Thermal Cycling: 1000 cycles of 30 min at -40 °C and 30 min at 110 °C.
b. Temperature/Humidity Cycling: 15 cycles of the following:
   16 h at 95% RH and 40 °C
   2 h at -40 °C
   2 h at +85 °C
   4 h at room temperature
c. Mechanical Shock: 3 shocks of 50 G in each of 3 mutually perpendicular axes.
d. Vibration: Sinusoidal 1.5 mm ± 0.15 mm amplitude by 15 G for 2 h in each of 3 mutually perpendicular axes.

5. Vehicle and Test Equipment Connector Terminal Assignments

5.1 Vehicle and Test Equipment Connector Terminal Identification—See Figure 5 for vehicle connector and test equipment connector terminal identifications.

5.2 General Terminal Assignments—See Figure 6 for a summary of terminal assignments.

5.3 Vehicle Connector Terminal Assignments

5.3.1 VEHICLE CONNECTOR TERMINALS - GENERAL LIMITS—All uses of all terminals of vehicle connector are limited to the range of -0.05 to 30.0 V DC.

5.3.2 VEHICLE CONNECTOR TERMINALS 1, 3, 6, 8, 9, 11, 12, 13, AND 14—Assignment of vehicle connector terminals 1, 3, 6, 8, 9, 11, 12, 13, and 14 is left to the discretion of the vehicle manufacturer.

Terminal	General Assignment
1	Discretionary*
2	Bus + Line of SAE J1850**
3	Discretionary*
4	Chassis Ground
5	Signal Ground
6	Discretionary*
7	K Line of ISO 9141-2**
8	Discretionary*
9	Discretionary*
10	Bus - Line of SAE J1850**
11	Discretionary*
12	Discretionary*
13	Discretionary*
14	Discretionary*
15	L Line of ISO 9141-2**
16	Unswitched Vehicle Battery Positive

* Note, assignment of terminals 1, 3, 6, 8, 9, 11, 12, 13, and 14 in the vehicle connector is left to the discretion of the vehicle manufacturer.
** Note, for terminals 2, 7, 10, and 15 the related OBD II communication assignments are shown. These terminals may also be used for alternate assignments in the vehicle connector. See 5.3 and 5.5 for further information.

FIGURE 6—GENERAL TERMINAL ASSIGNMENTS

5.3.3 VEHICLE CONNECTOR TERMINAL 2—If SAE J1850 Single Wire is used in a vehicle to supply OBD II required communication services, then terminal 2 of the vehicle connector must be the SAE J1850 Single Wire Signal connection.

If SAE J1850 Two Wire is used in a vehicle to supply OBD II required communication services, then terminal 2 of the vehicle connector must be the Bus + Signal of the SAE J1850 Two Wire connection.

If neither SAE J1850 Single Wire nor SAE J1850 Two Wire is used in a vehicle to supply OBD II required communication services, then assignment of this terminal is left to the discretion of the vehicle manufacturer, provided this assignment does not interfere with the operation of, nor cause damage to, tools conforming to SAE J1978.

5.3.4 VEHICLE CONNECTOR TERMINAL 4—Vehicle connector terminal 4 is designated Chassis Ground and is further defined by SAE J2201. This terminal must be implemented in the vehicle connector.

5.3.5 VEHICLE CONNECTOR TERMINAL 5—Vehicle connector terminal 5 is designated Signal Ground and is further defined by SAE J2201. This terminal must be implemented in the vehicle connector.

5.3.6 VEHICLE CONNECTOR TERMINAL 7—If a two wire or a one wire ISO 9141-2 interface is used in a vehicle to supply OBD II required communication services, then terminal 7 of the vehicle connector must be the K Line of the ISO 9141-2 interface.

If neither a two wire nor a one wire ISO 9141-2 interface is used in a vehicle to supply OBD II required communication services, then assignment of this terminal is left to the discretion of the vehicle manufacturer, provided this assignment does not interfere with the operation of, nor cause damage to, tools conforming to SAE J1978.

5.3.7 VEHICLE CONNECTOR TERMINAL 10—If an SAE J1850 Two Wire interface is used in a vehicle to supply OBD II required communication services, then terminal 10 of the vehicle connector must be the Bus - Signal of the SAE J1850 Two Wire interface.

If an SAE J1850 Two Wire interface is not used in a vehicle to supply OBD II required communication services, then assignment of this terminal is left to the discretion of the vehicle manufacturer, provided this assignment does not interfere with the operation of, nor cause damage to, tools conforming to SAE J1978.

5.3.8 VEHICLE CONNECTOR TERMINAL 15—If a two wire ISO 9141-2 interface is used in a vehicle to supply OBD II required communication services, then terminal 15 of the vehicle connector must be the L Line of the ISO 9141-2 interface.

If a two wire ISO 9141-2 interface is not used in a vehicle to supply OBD II required communication services, then assignment of this terminal is left to the discretion of the vehicle manufacturer, provided this assignment does not interfere with the operation of, nor cause damage to, tools conforming to SAE J1978.

5.3.9 VEHICLE CONNECTOR TERMINAL 16—Vehicle connector terminal 16 is designated Unswitched Vehicle Battery Positive and must be implemented in the vehicle connector. This terminal must be connected directly (i.e., unswitched) to the DC Positive of the vehicle's battery. This connection does not preclude the use of a fuse or other circuit protection elements. This circuit may be grouped with other similar circuits. This terminal must be able to supply a minimum of 4.0 A at 14.4 V DC.

5.4 Vehicle Connector Terminal Protection—It is recommended that the vehicle manufacturer provide circuit protection in the event that the terminals of the vehicle connector are shorted together. This protection is limited to the ranges of voltages present at the vehicle connector before the test equipment connector is mated to it.

5.5 Test Equipment Connector Terminal Assignments

5.5.1 TEST EQUIPMENT CONNECTOR TERMINALS - GENERAL LIMITS—All uses of all terminals of test equipment connector are limited to the range of -0.05 to 30.0 V DC.

5.5.2 TEST EQUIPMENT CONNECTOR TERMINALS 1, 3, 6, 8, 9, 11, 12, 13, AND 14—The use of test equipment connector terminals 1, 3, 6, 8, 9, 11, 12, 13, and 14 is left to the discretion of the test equipment manufacturer.

These test equipment connector terminals shall normally be in a high impedance state, that is at greater than 500 kΩ impedance relative to Signal Ground and at greater than 500 kΩ impedance relative to Chassis Ground.

Before the condition of these test equipment connector terminals is changed from this high impedance state, the test equipment user and/or the test equipment must verify the proper usage of these vehicle connector terminals.

5.5.3 TEST EQUIPMENT CONNECTOR TERMINALS 2, 7, 10, AND 15—Assignment and use of test equipment connector terminals 2, 7, 10, and 15 must be compatible with the assignment and use of their mating terminal in the vehicle connector (see Vehicle Connector Terminal Assignments in 5.3) and SAE J2201.

5.5.4 TEST EQUIPMENT CONNECTOR TERMINAL 4—Test equipment connector terminal 4 is designated Chassis Ground and is defined by SAE J2201. Implementation of this terminal in the test equipment connector is optional.

5.5.5 TEST EQUIPMENT CONNECTOR TERMINAL 5—Test equipment connector terminal 5 is designated Signal Ground and is defined by SAE J2201. This terminal must be implemented in the test equipment connector for support of SAE J2201.

5.5.6 TEST EQUIPMENT CONNECTOR TERMINAL 16—Test equipment connector terminal 16 is designated as Unswitched Vehicle Battery Positive and is available to supply operating power and a reference voltage to test equipment.

5.6 Test Equipment Connector Terminal Protection—It is recommended that all circuits connected to the terminals of the test equipment connector be protected to the extent that no damage will come to these circuits if ANY terminal of the test equipment connector:

    a. Is connected to vehicle connector terminal 16 - Unswitched Vehicle Battery Positive for up to 10 A at 14.4 V DC,

    b. Is connected to vehicle connector terminal 4 - Vehicle Chassis Ground, or

    c. Is connected to vehicle connector terminal 5 - Vehicle Signal Ground.

6. Minimum Current Available—In order to ensure that adequate current is available to operate diagnostic scan tools, the minimum current available through terminal 16 of the vehicle connector shall be no less than 4.0 A at 14.4 V DC.

The vehicle manufacturer shall not be responsible for supplying more than 4.0 A at 14.4 V DC.

7. Liability for Devices That Draw More Than 4.0 A—Manufacturers of devices that include a connection to the vehicle connector, and which draw in excess of 4.0 A at 14.4 V DC from terminal 16 of the vehicle connector, may be responsible for any damage to the vehicle.

UNIVERSAL INTERFACE FOR OBD II SCAN—SAE J2201 JUN93

SAE Recommended Practice

Report of the Vehicle E/E System Diagnostics Standards Committee and the Vehicle Network for Multiplexing and Data Communications Standards Committee approved June 1993.

TABLE OF CONTENTS

1.	Scope
1.1	SAE Document Interrelationships
2.	References
2.1	Applicable Documents
2.1.1	SAE Publications
2.1.2	California Air Resource Board Documents
2.1.3	EPA Regulations
2.1.4	ISO Documents
2.2	Terms and Definitions
2.3	Acronyms and Abbreviations
3.	Requirements
3.1	Interface and Message Protocol Support
3.2	In-Frame Response
3.3	Signal Ground
3.4	Maximum Voltage Differentials
3.5	Chassis Ground
3.6	Minimum Connector Cable Length
3.7	Other Requirements
4.	Interface Functionality Evaluations
APPENDIX A	EXAMPLES
A.1	General Example Information
A.1.1	Transient Protection
A.1.2	Host Support Not Included
A.1.3	Supporting Documents
A.1.4	Common ISO 9141-2 Support
A.1.5	Electromagnetic Compatibility
A.1.6	No Responsibility Assumed By Contributors
A.1.7	Additional Capabilities of Examples
A.2	DLCS, HBCC, and ISO 9141-2 Interface Example
A.2.1	General Overview
A.2.2	DLCS Operation
A.2.2.1	Initialization
A.2.2.2	Message Transmission
A.2.2.3	Message Reception
A.2.3	DLCS Pin Names and Descriptions
A.2.4	HBCC Operation
A.2.4.1	Initialization
A.2.4.2	Sending a Message
A.2.4.3	Receiving a Message
A.2.5	HBCC Pin Descriptions
A.2.6	Additional Capabilities
A.3	PCI, HBCC, and ISO 9141-2 Interface Example
A.3.1	General Overview
A.3.2	Control Microcomputer
A.3.2.1	Control Microcomputer Software Structure
A.3.2.2	Control Microcomputer Interface to the Host Microcomputer
A.3.2.3	Control Microcomputer Interface to the SED
A.3.3	Symbol Encoder Decoder (SED)
A.3.3.1	Symbol Encoder
A.3.3.2	Symbol Decoder
A.3.3.3	Invalid Symbol
A.3.3.4	SED Inputs and Outputs
A.3.3.4.1	Transmit Clear (TRCLR~) Input
A.3.3.4.2	Reset (RST) Inputs
A.3.3.4.3	Oscillator (OSC) Input
A.3.3.4.4	Transmit Strobe (TRSTRB) Input
A.3.4	Integrated Driver/Receiver (IDR)
A.3.4.1	Bus Output Waveshaping
A.3.4.2	Transmitting Signals on the Bus
A.3.4.3	Processing Signals Received from the Bus
A.3.5	Additional Capabilities
A.4	SGS-Thomson Protocol Engine Interface Example
A.4.1	General Overview
A.4.2	Control Microcomputer
A.4.2.1	Control Microcomputer Software Structure
A.4.2.2	Control Microcomputer Interface to the Host
A.4.2.3	Automatic OBD II Interface Scan
A.4.3	GAL6001 Programmable Logic Array
A.4.4	Discrete Transceiver
A.4.5	Additional Capabilities
A.5	Silicon Systems F690/F691 Scan Tool Chipset Interface Example
A.5.1	General Overview
A.5.2	F690 Codec
A.5.3	F691 Transceiver
A.5.4	Pin Description
A.5.4.1	F690 Codec Pin Description
A.5.4.2	F691 Transceiver Pinout
A.5.5	Additional Capabilities
A.6	Motorola JCI and ISO 9141-2 Interface Example
A.6.1	General Overview
A.6.2	JCI Operation
A.6.2.1	Host Interface
A.6.2.2	Communication Mode Selection
A.6.2.3	In-Frame Response and I.D. Byte
A.6.2.4	Message Transmission
A.6.2.5	Message Reception
A.6.3	JCI Pin Names and Descriptions
A.6.4	Additional Capabilities
APPENDIX B	SUPPORTING DOCUMENTS

1. Scope—SAE J1978 defines the requirements of the OBD II scan tool. SAE J2201 defines the minimum requirements of the vehicle communications interface for the SAE J1978 OBD II scan tool. This interface connects the SAE J1962 test equipment connector to the hardware/software of the SAE J1978 OBD II scan tool that will use this interface to communicate with vehicles for the purpose of accessing required OBD II functions.

Included in this SAE Recommended Practice are several definitions relating to the interface, and interface functionality evaluation.

Appendix A - Examples include several example interface circuit implementations, which are believed to meet the requirements of this document and of SAE J1978. These examples are NOT requirements of this document. They are provided to assist circuit designers in developing interface circuits.

Appendix B - Supporting Documents includes a list of supporting documents for the examples shown in Appendix A.

1.1 SAE Document Interrelationships—Figure 1 shows the interrelationships between SAE J1978, SAE J1962, and this document.

NOTE—See SAE J1962 Diagnostic Connector for pin assignments.

Where any conflict may exist between the requirements contained in SAE J1978 and this document, SAE J1978 is the overriding document.

2. References—The terms, definitions, abbreviations, and acronyms contained in SAE J1930 are included by reference.

2.1 Applicable Documents—The following publications form a part of this specification to the extent specified herein. The latest issue of SAE publications shall apply.

2.1.1 SAE PUBLICATIONS—Available from SAE, 400 Commonwealth Drive, Warrendale, PA 15096-0001.

SAE J1850—Class B Data Communication Network Interface
SAE J1930—Electrical/Electronic Terms, Definitions, Abbreviations, and Acronyms
SAE J1962—Diagnostic Connector
SAE J1978—OBD II Scan Tool
SAE J1979—Diagnostic Test Modes
SAE J2205—Expanded Diagnostic Protocol

2.1.2 CALIFORNIA AIR RESOURCE BOARD DOCUMENTS

California Code of Regulation, Title 13, 1968.1 - Malfunction and Diagnostic System Requirements -- 1994 and Subsequent Model-Year Passenger Cars, Light-Duty Trucks, and Medium-Duty Vehicles With Feedback Fuel Control Systems

2.1.3 EPA REGULATIONS
Federal Register Tuesday September 24,1991, Part II Environmental Protection Agency 40 CFR Part 86 - Air Pollution Control; New Motor Vehicles and Engines: On-Board Diagnostic Systems on 1994 and Later Model Year Light-Duty Vehicles and Light-Duty Trucks; Proposed Rule

2.1.4 ISO DOCUMENTS—Available from ANSI, 11 West 42nd Street, New York, NY 10036-8002.

ISO 9141-2—Road vehicles—Diagnostic systems—CARB requirements for interchange of digital information ISO/TC 22/SC 3/WG 1 - N 425 E/REV April 1991

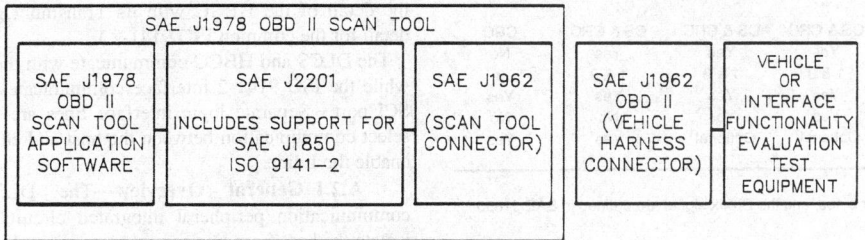

FIGURE 1—SAE DOCUMENT INTERRELATIONSHIPS

2.2 Terms and Definitions

2.2.1 APPLICATION SOFTWARE—As used in this document, this term refers to the microprocessor programming that controls the external equipment/scan tool hardware so as to perform required SAE J1978 OBD II Scan Tool functions.

2.2.2 DLCS—(Data Link Controller Serial) refers to the integrated circuit family developed by General Motors that supports SAE J1850 10.4 Kbps VPW CRC communication.

2.2.3 HBCC—(Hosted Bus Controller Circuit) refers to the integrated circuit developed by Ford Motor Co. that supports SAE J1850 41.6 Kbps PWM CRC communication.

2.2.4 OBD II—(On Board Diagnostics II) common term that refers to the requirements of the California legislation - California Code of Regulation, Title 13, 1968.1.

2.2.5 PCI—(Programmable Communications Interface) refers to a set of integrated circuits which includes a control microcomputer, the Symbol Encoder Decoder (SED) and the Integrated Driver Receiver (IDR). The PCI was developed by Chrysler Corporation and supports SAE J1850 10.4 VPW communication with both CRC and Checksum error techniques.

2.2.6 JCI—(SAE J1850 Communications Interface) refers to an integrated circuit manufactured by Motorola which supports both SAE J1850 10.4 VPW with CRC and SAE J1850 41.6 PWM with CRC communication.

2.3 Acronyms and Abbreviations

CRC - Cyclic Redundancy Check
CS - Checksum
EDP - Enhanced Diagnostic Protocol (SAE J2205)
IBS - Inter-Byte Separation
IFR - In-Frame Response
NRZ - Non Return to Zero
PWM - Pulse Width Modulation
SCI - Serial Communications Interface
SPI - Serial Peripheral Interface
UART - Universal Asynchronous Receiver/Transmitter
VDD - 5.0 VDC from Regulated Power Supply
VPW - Variable Pulse Width Modulation
"~" - Indicates Active Low Signals

3. Requirements—This section defines the required message structure support, signal ground, chassis ground, minimum scan tool connector cable length, and other requirements for the interface to be used by an SAE J1978 OBD II scan tool.

3.1 Interface and Message Protocol Support—The interface defined in this document must support the interface requirements and message protocol requirements of SAE J1978.

3.2 In-Frame Response—When a single byte in-frame response (IFR) is required during the reception of a 41.6 Kbps PWM SAE J1979 message, the interface will support the transmission of the node address as a single byte IFR.

3.3 Signal Ground—The Signal Ground pin of the scan tool side of the SAE J1962 diagnostic connector must be used as the signal ground reference for all interface transceivers required by this document.

3.4 Maximum Voltage Differentials—The interface described in this document includes any required interface transceivers and the cabling connecting the interface transceivers to the scan tool side of the SAE J1962 connector. Any interface connected to a vehicle through the SAE J1962 diagnostic connector is considered a part of the vehicle network and must operate within the limits of that network, which are described in SAE J1850 and ISO 9141-2.

The maximum voltage differential, e.g., due to load current, noise, etc., between any two nodes of a vehicle data communication network, where the interface described in this document is considered as a node on the vehicle network when connected to the vehicle, must be less than the limits specified in SAE J1850 and ISO 9141-2.

The following are maximum voltage differential values for the interface described in this document (as measured between the signal ground connection of all interface transceivers and the Signal Ground pin on the vehicle side of the SAE J1962 diagnostic connector) are:

0.25 V peak noise
0.1 VDC offset

3.5 Chassis Ground—The Chassis Ground pin of the scan tool side of the SAE J1962 diagnostic connector is available for any use, with the exception that the minimum DC impedance between the Chassis Ground and the Signal Ground is 1 MΩ.

3.6 Minimum Connector Cable Length—The minimum length between the ground reference connection of each interface transceiver and the Signal Ground connection of the vehicle side of the SAE J1962 diagnostic connector is 2 m.

3.7 Other Requirements—The interface must support the requirements of SAE J1850, SAE J1962, SAE J1978, SAE J1979, SAE J2205, and ISO 9141-2.

4. Interface Functionality Evaluations—The functionality of any proposed interface implementations must be evaluated by either of the following:

a. The use of multiplex bus interface equipment from I+ME, or equivalent equipment, that with appropriate software simulates vehicles that use the required implementations of SAE J1850 and ISO 9141-2, or

b. The use of representative vehicles of motor vehicle manufacturers that use the required implementations of SAE J1850 and ISO 9141-2.

APPENDIX A
EXAMPLES

A.1 General Example Information—This section shows example interface implementations that are believed to meet the requirements of this document.

These examples are NOT requirements of this document. These examples are intended as an assist to interface circuit designers.

Table A1 illustrates a summary of some of the capabilities available with these examples that are over and above the requirements of this document and some that are required.

TABLE A1—SUMMARY OF CAPABILITIES FOR THE FOLLOWING EXAMPLES

	DLCS & HBCC Section A.2	PCI & HBCC Section A.3	ST9 Section A.4	F690 & F691 Section A.5	JCI Section A.6
Error Checking	CRC	CS & CRC	CS & CRC	CS & CRC	CRC
Comprehends IBS[2]	No	Yes	Yes	Yes	No
Header Bytes	3	1 & 3	1 & 3	1 & 3	3
Generate Break	Yes	Yes	Yes	Yes	Yes
Control Function[3]	No	Yes	Yes	No	No
Host Required	Yes	Optional[1]	Optional[1]	Yes	Yes

Error checking (CS or CRC), IBS, header bytes, and the Break signal are defined in SAE J1850.

[1] Host Required refers to the need for a host microcomputer to perform the functionality of SAE J1978. Some interface implementations include a microcomputer as a part of the interface. In some of these implementations, this microcomputer may have enough excess capability so as to also be able to perform "host microcomputer" functions.

[2] IBS is part of only 10.4 Kbps VPW networks. It is not a part of 41.6 Kbps PWM networks, nor a part of ISO 9141-2 networks.

[3] Control Function refers to parts of an interface that are implemented in software and executed by some form of microcomputer.

A.1.1 Transient Protection—While some of the examples may show transient protection, this document does not specify or require transient protection.

A.1.2 Host Support Not Included—The examples shown here support a nominal type of host interface and are not meant to be directly applicable to any particular host microprocessor, host interface, and/or host software. The host application software (e.g., interface drivers, message building routines, message processing routines, etc.) and/or additional hardware (e.g., signal buffers, signal inverters, clock synchronization circuits, etc.) required to complete the interface to the example circuits shown in this document is the responsibility of the designer/implementer and is not shown here.

A.1.3 Supporting Documents—Appendix B - Supporting Documents identifies documentation that may be used to aid in the understanding of the examples shown in this document.

A.1.4 Common ISO 9141-2 Support—Many of the examples include a common implementation for ISO 9141-2. This common implementation is described here (see Figure 2-D ISO 9141-2).

The ISO 9141-2 document describes the requirements of the ISO 9141-2 interface. This interface is an asynchronous serial communication link using NRZ bit encoding. A 5 bps bit rate is used during module communication initialization and a 10.4 Kbps bit rate is used for all further communications.

This interface includes a K line which is bidirectional and used for all communication phases, and a L line which is output only from the scan tool and is only used during the 5 bps module communication initialization phase. When used to transmit data to a vehicle, both the L and K lines are driven to either ground or battery level, through a 510 Ω resistor, depending on the data value. When used to receive data from a vehicle, the K line is referenced to 1/2 the battery voltage, with a small amount of hysteresis for noise immunity, to determine the logic value of the data being transmitted by a vehicle module.

In all of the examples, the L and K lines are connected to the SCI port of either the host or a control microprocessor. A select signal from the host/control microprocessor is used to enable the L line when necessary.

A.1.5 Electromagnetic Compatibility—The Electromagnetic Compatibility (EMC) characteristics of these examples have not been investigated nor estimated. The ability of the examples to meet EMC or noise radiation requirements is unknown.

A.1.6 No Responsibility Assumed By Contributors—No responsibility is assumed by Chrysler Corporation, Ford Motor Co., General Motors Corporation, Delco Electronics Corporation, SAE, SGS-Thomson, Silicon Systems, Motorola, or any other company for use of any of the examples or products shown in this document nor for any infringements of patents and trademarks or other rights of other parties resulting from the use of these examples or products. No license is granted under any patents, patent rights, or trademarks of the previously named parties.

A.1.7 Additional Capabilities of Examples—Many of the examples support capabilities beyond the requirements of this document. Such capabilities are discussed in a separate section of each example.

A.2 DLCS, HBCC, and ISO 9141-2 Interface Example—This example (see Figure A1a) combines the use of the Data Link Controller Serial (DLCS) integrated circuit developed by Delco Electronics, the Hosted Bus Controller Circuit (HBCC) integrated circuit developed by Ford, a discrete 41.6 Kbps transmit driver, and the common ISO 9141-2 interface to provide the required interface support. (Figure A1b shows the detail of the DLCS, Figure A1c shows the detail of the HBCC with its Transmit Driver, and Figure A1d shows the detail for the common ISO 9141-2.)

The DLCS and HBCC communicate with the host through the host's SPI port, while the ISO 9141-2 interface communicates with the host through the host's SCI port. Separate logic interface lines are used by the host to individually select communication between the host and either the DLCS and HBCC, and to enable the L line.

A.2.1 General Overview—The DLCS and HBCC perform as communication peripheral integrated circuits. Following initialization, they isolate the host from bus communications tasks and are only serviced by the host when the host is loading a message for transmission or unloading a received message. The host receives an interrupt when a received message is available and when message transmission is completed. The host only handles communications as full messages. Several control and status registers are available in both the DLCS and HBCC to control their operation and indicate the status of their communications tasks.

This example uses chip select lines from the host to select which peripheral the host wishes to communicate with. Control data from the host enables and disables the DLCS and HBCC interfaces.

Because these devices are available from semiconductor suppliers who are also supplying them to major automotive manufacturers for use in production vehicles, compatibility with those vehicles should be readily verifiable.

The ISO 9141-2 interface is as described in A.1.4. The host uses its SCI interface to communicate with a vehicle through the ISO 9141-2 interface. A logic line is used by the host to enable the L line when necessary.

A.2.2 DLCS Operation—The DLCS contains a logic section and an analog section in a single integrated circuit. The logic section includes the SPI interface to the host, a transmit buffer, a receive buffer, status and control registers, bit timing, and symbol encoding and decoding logic. The analog section includes the transceiver circuitry. Data transfer between the host and the DLCS is done with two byte SPI transfers. The host sends to the DLCS a command byte and a data byte, while the DLCS sends to the host a status byte and a data byte. The content of the data bytes is dependent on the command.

a. Commands sent to the DLCS include:
   (1) General commands (e.g., send break, go to sleep, terminate transmit retry, load configuration data)
   (2) Transmit commands (e.g., load transmit data, transmit message)
   (3) Receive commands (e.g., unload receive data, flush byte, flush message)
b. Status data send from the DLCS indicate the following:
   (1) Receive status (e.g., buffer contains bytes or messages)
   (2) Transmit status (e.g., buffer contains bytes or messages)
   (3) Data link status (e.g., data link shorted)

Operation of the DLCS consists primarily of three functions: initialization, transmission of a message, and reception of a message. Each of these will be discussed in more detail as follows:

A.2.2.1 INITIALIZATION—Initialization is accomplished by sending the DLCS a command byte to load the accompanying data from the host as configuration data. The DLCS can be configured to enable or disable it to interrupt the host, thereby allowing either polling or interrupt based signalling schemes to be used, and to configure the oscillator divisor. Once the DLCS is initialized, normal transmission and reception can begin.

A.2.2.2 MESSAGE TRANSMISSION—Loading a message from the host to the DLCS for transmission on the bus is done by several 2 byte SPI transfers. Each transfer contains a command byte to load the accompanying data as transmit data. Loading the first and last byte of the message must be done with specific command bytes. Once the command to load the accompanying data as the last transmit data byte is received by the DLCS, the DLCS will begin an attempt to transmit the message on the bus. With each 2 byte SPI transfer with the DLCS, a status byte and data byte will be transferred to the host. The status byte is used to determine the status of the transmit buffer, i.e., empty, contains some bytes, full, etc. The status byte also contains receive status. When loading a transmit message into the DLCS, the data bytes returned by the DLCS to the host are ignored.

23.253

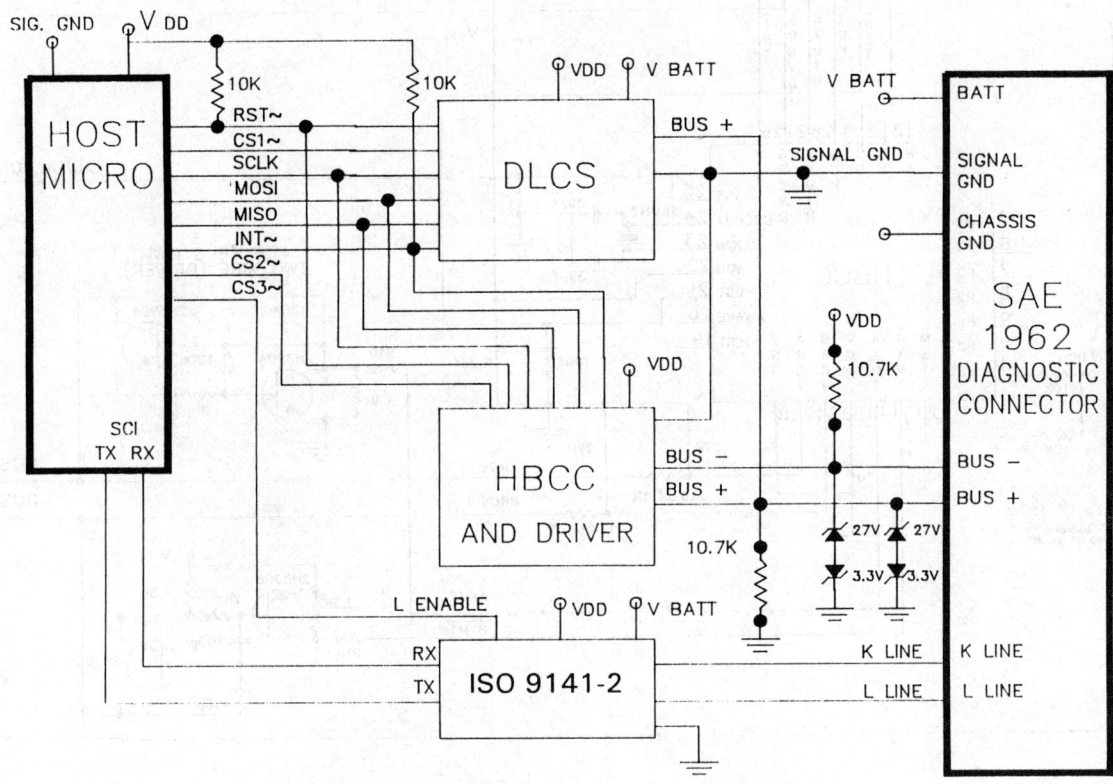

FIGURE A1a—DLCS, HBCC, AND ISO 9141-2 EXAMPLE

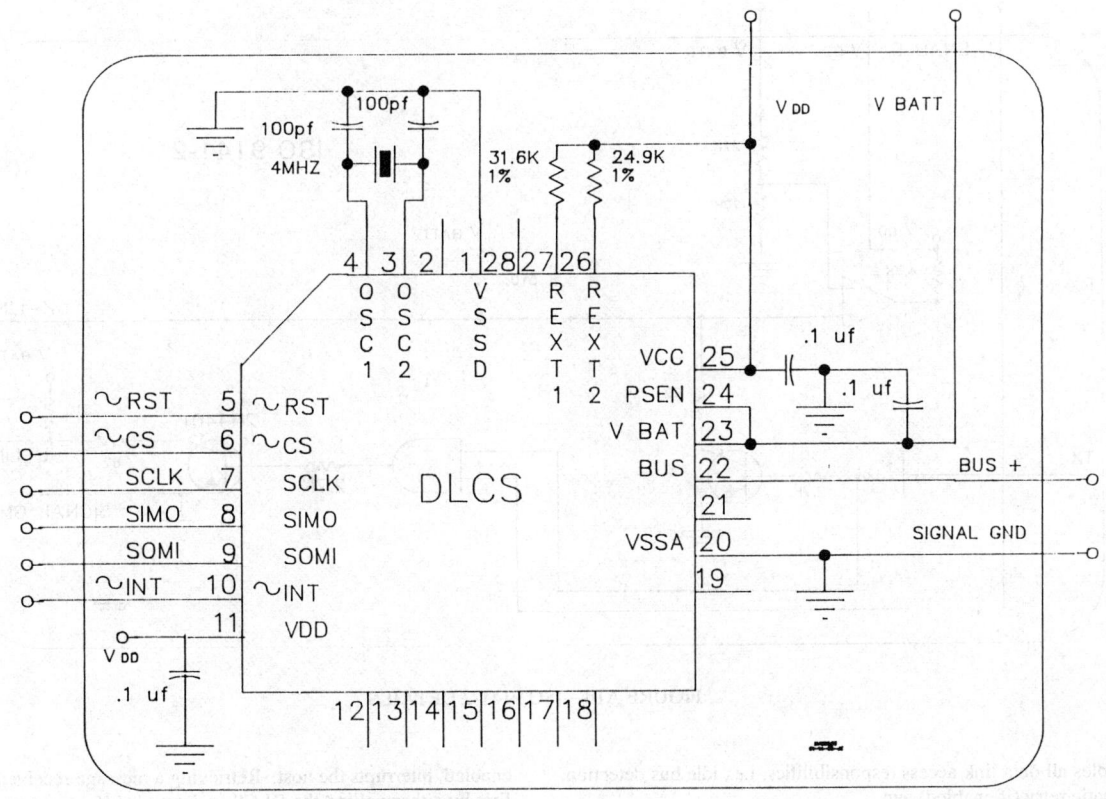

FIGURE A1b—DLCS DETAIL

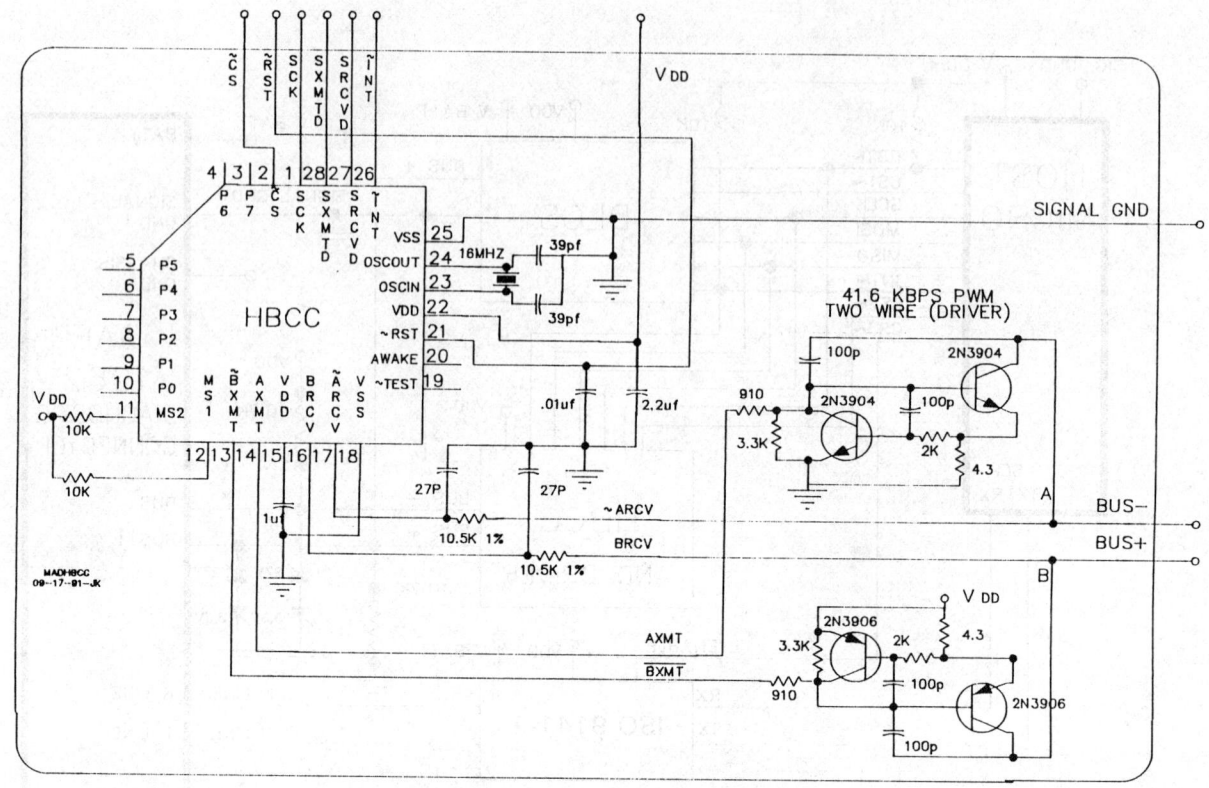

FIGURE A1c—HBCCC AND TRANSMIT DRIVER DETAIL

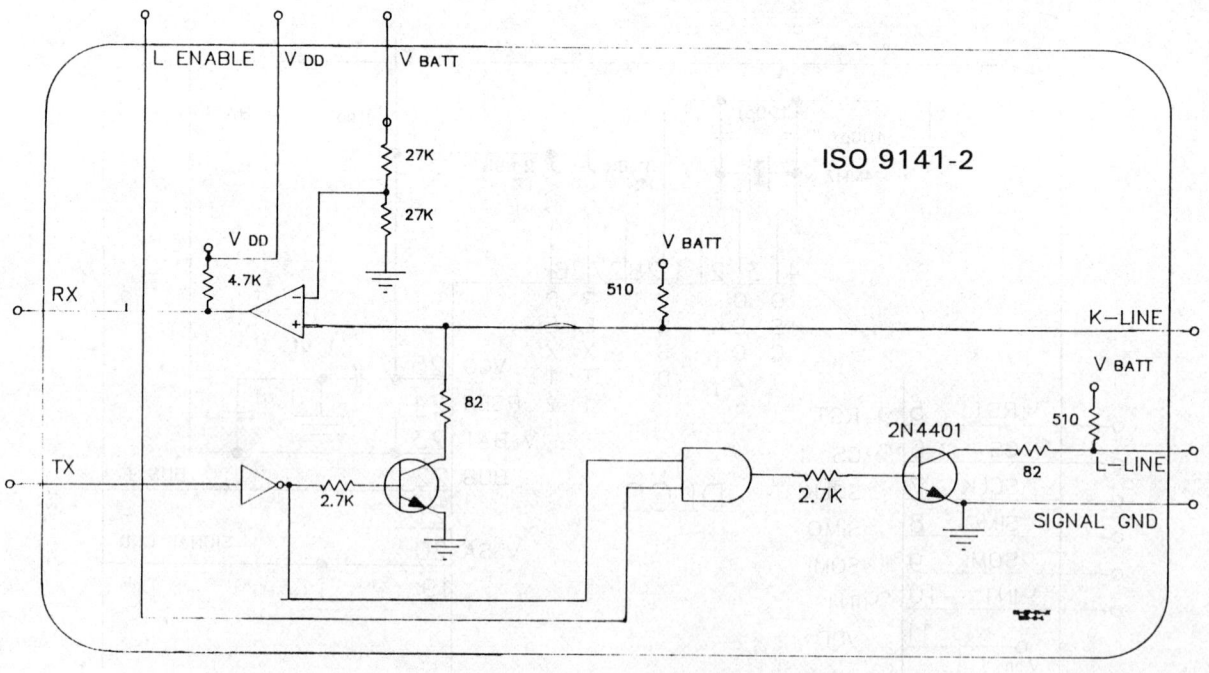

FIGURE A1d—ISO 9141-2 DETAILS

The DLCS handles all data link access responsibilities, i.e., idle bus detection, arbitration, automatic retry (if enabled), etc.

A.2.2.3 MESSAGE RECEPTION—The DLCS on its own receives and buffers a complete message from the bus, checks the received CRC, sets a status flag indicating a message has been received and if any errors occurred and, if enabled, interrupts the host. Retrieving a message received by the DLCS is done first by either polling the DLCS to determine if a message has been received, or by servicing an enabled interrupt that occurs upon reception of a message. (The remainder of this discussion assumes the interrupt method is used.)

Upon receiving an interrupt, performing a 2 byte SPI transfer with the DLCS will result in the transfer of status and data to the host. The status byte is used to determine if the receive buffer contains any bytes of a message or a complete message(s), and the status of the transmit buffer. Once all the bytes of a message have been received by the DLCS an indicator will be set in the status byte and, if enabled, an interrupt will be generated for the host. The status byte will also indicate whether the message was received correctly or that an error occurred, i.e., CRC error, incomplete byte, and bit timing error. If the message unloaded from the DLCS was received while the DLCS was attempting to transmit a message, the status byte will also indicate transmission status values such as overrun or underrun and whether bus arbitration was won or lost.

A.2.3 DLCS Pin Names and Descriptions—The names and descriptions of the pins of the DLCS are shown in Table A2.

TABLE A2—DLCS PIN NAMES AND DESCRIPTIONS

Pin Number	Pin Name	Description
1	Vssd	Digital Ground
2	LOTI	Logic Out Transmitter In (Test)
3	OSC2	Oscillator 2
4	OSC1	Oscillator 1
5	RST~	Reset~ (Active Low)
6	CS~	Chip Select~ (Active Low)
7	SCLK	SPI Serial Clock
8	SIMO	SPI Slave In Master Out
9	SOMI	SPI Slave Out Master In
10	INT~	Interrupt~ (Active Low)
11	Vdd	Digital Voltage Supply (+5 V)
12	N/A	Not Used
13	N/A	Not Used
14	N/A	Not Used
15	N/A	Not Used
16	N/A	Not Used
17	N/A	Not Used
18	N/A	Not Used
19	N/A	Not Used
20	Vssa	Analog Ground
21	LOAD	Bus Load
22	BUS	Bus Output
23	Vbatt	Battery Voltage
24	PSEN	Power Supply Enable
25	Vcc	Analog Voltage Supply (+5 V)
26	REXT2	External Resistor 2
27	REXT1	External Resistor 1
28	LITO	Login In Transmitter Out (Test)

A.2.4 HBCC Operation—The HBCC contains a logic section and an analog section in a single integrated circuit. The logic section of the HBCC supports two types of serial and two parallel host interfaces. This allows for the HBCC to interface with standard Intel or Motorola microprocessors or others which have compatible ports. For this example, the Motorola SPI interface has been used. The logic section also includes a transmit buffer, a receive buffer, status and control registers, bit timing, symbol encoding, and decoding logic. The analog section is limited to the receiver circuitry. The transmitter driver circuit and receiver input noise filter are implemented by external discrete devices.

Data transfer between a host and the HBCC is done with 2 byte SPI transfers. The host transfers an indirect address control byte (ACB) and a data byte to the HBCC. Generally the indirect address points to one of the internal registers of the HBCC and the data sent from the host is stored in the addressed register. During a 2 byte SPI transfer, the host receives a status byte and a data byte from the HBCC. Special address conventions also allow for block type data transfers between the host and the HBCC message buffers.

  a. Commands sent to the HBCC include:
    (1) General commands (e.g., initialize node address)
    (2) Transmit commands (e.g., load transmit buffer)
    (3) Receive commands (e.g., read receive buffer)
  b. Status data received from the HBCC include the following:
    (1) Receive status (e.g., received byte count)
    (2) Transmit status (e.g., transmission completed OK)
    (3) Network wire status (e.g., network wire shorted)

The operation of the HBCC consists primarily of three functions: initialization, transmission of a message, and reception of a message. Each of these will be discussed in more detail as follows:

A.2.4.1 INITIALIZATION—Initialization is accomplished by sending the HBCC commands to load the accompanying data as configuration information. The HBCC can be configured to enable or disable interrupts to the host, thereby allowing a polling scheme to be used, and to select the bus bit rate. Receive message filter look up tables can also be loaded to screen incoming messages, thus reducing host burden. Once the HBCC is initialized, normal transmission and reception can begin.

A.2.4.2 SENDING A MESSAGE—Loading a message into the HBCC for transmission on the bus is done by several SPI transfers. Each transfer contains an indirect address byte which defines where to store the accompanying data byte into the HBCC transmit buffer. Once the entire message has been loaded into the transmit buffer, the command to transmit the message is sent to the HBCC. The HBCC will then on its own begin to synchronize with the activity on the bus and transmit the message on the bus. The HBCC handles all data link access responsibilities, i.e., idle bus detection, arbitration, and automatic retry. When a message transmit attempt is completed, an interrupt is generated and the host can check the resultant status, including the status of an in-frame response.

A.2.4.3 RECEIVING A MESSAGE—Receiving bus messages using an HBCC is generally based on servicing the interrupt that occurs, if enabled, at the completion of message reception. The HBCC can be configured with interrupts either enabled or disabled, but for this example, interrupts are assumed to be enabled. Upon receiving an interrupt, the host will perform a two byte SPI transfer and receive from the HBCC a status byte and a data byte. The status byte can be used to determine how full the receive buffer is, and whether certain receive errors have occurred. A received message can subsequently be transferred from the HBCC to the host.

A.2.5 HBCC Pin Descriptions—The names and descriptions of the pins of the HBCC are shown in Table A3.

TABLE A3—HBCC PIN NAMES AND DESCRIPTIONS

Pin Number	Pin Name	Description
1	SCLK	Serial Clock
2	CS~	Chip Select~ (Active Low)
3	USER7	User Input 7
4	USER6	User Input 6
5	USER5	User Input 5
6	USER4	User Input 4
7	USER3	User Input 3
8	USER2	User Input 2
9	USER1	User Input 1
10	USER0	User Input 0
11	MS2	Mode Select 2
12	MS1	Mode Select 1
13	BXMT~	Network B Drive~ (Active Low)
14	AXMT	Network A Drive
15	Vdd	Analog Voltage Supply (+5 V)
16	BRCV	Network B Receive
17	ARCV~	Network A Receive (Active Low)
18	GND	Analog Ground
19	TESTENA~	Test Enable A (Active Low)
20	AWAKE	HBCC Awake
21	RST~	Reset~ (Active Low)
22	Vdd	Digital Voltage Supply (+5 V)
23	OSCIN	Oscillator 1
24	OSCOUT	Oscillator 2
25	GND	Digital Ground
26	INT~	Interrupt~ (Active Low)
27	SRCVD/SDAT	Serial Data Receive/Serial Data I/O
28	SXMTD/DIR	Serial Data Transmit/Data Direction

A.2.6 Additional Capabilities—Both the DLCS and HBCC include support for both transmit and receive message buffering and automatic message retransmission. The HBCC also includes support for receive message filtering, wake up, and 41.6 Kbps PWM bus physical layer fault tolerance. The DLCS also includes support to send and receive Break and for wake up.

A.3 PCI, HBCC, and ISO 9141-2 Interface Example—This example (see Figure A2a) combines the use of the Programmable Communication Interface (PCI) developed by Chrysler, the Hosted Bus Communication Controller (HBCC) developed by Ford, a discrete 41.6 Kbps transmit driver, and the common ISO 9141-2 interface to provide the required interface support. (Figure A2b shows the detail of the PCI.)

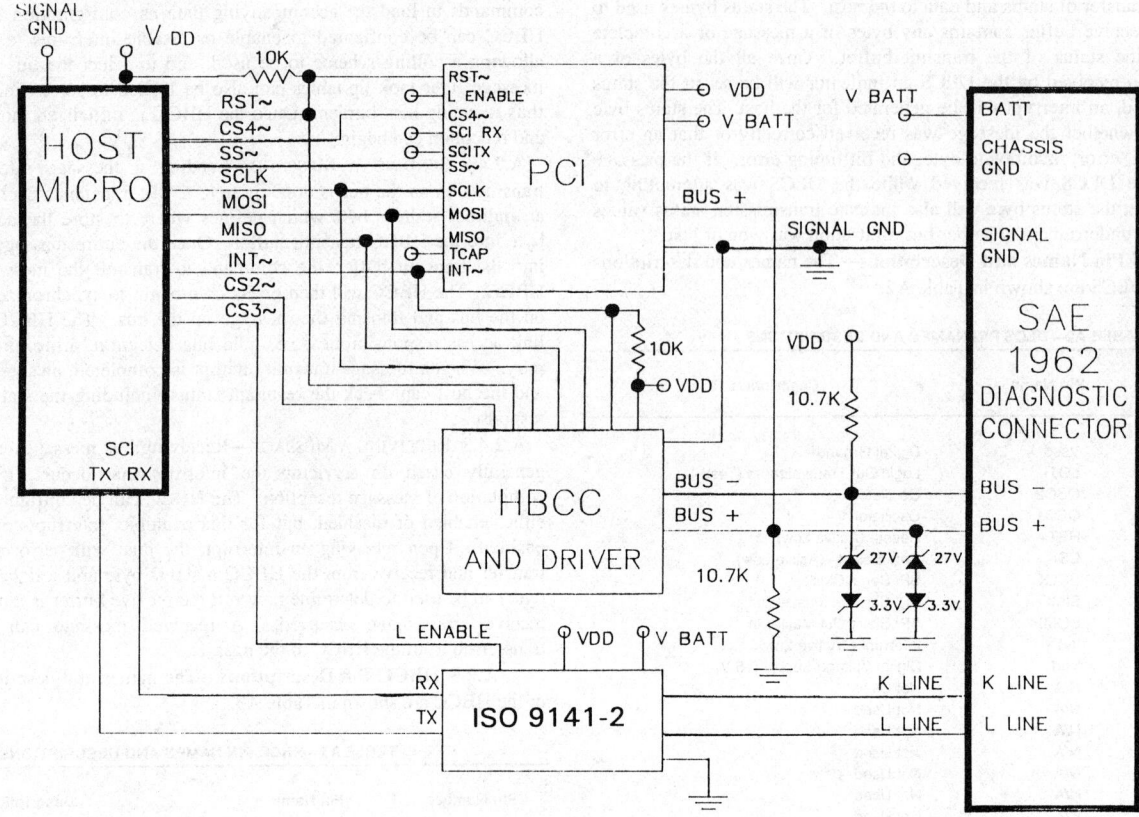

FIGURE A2a—PCI, HBCC, AND ISO 9141-2

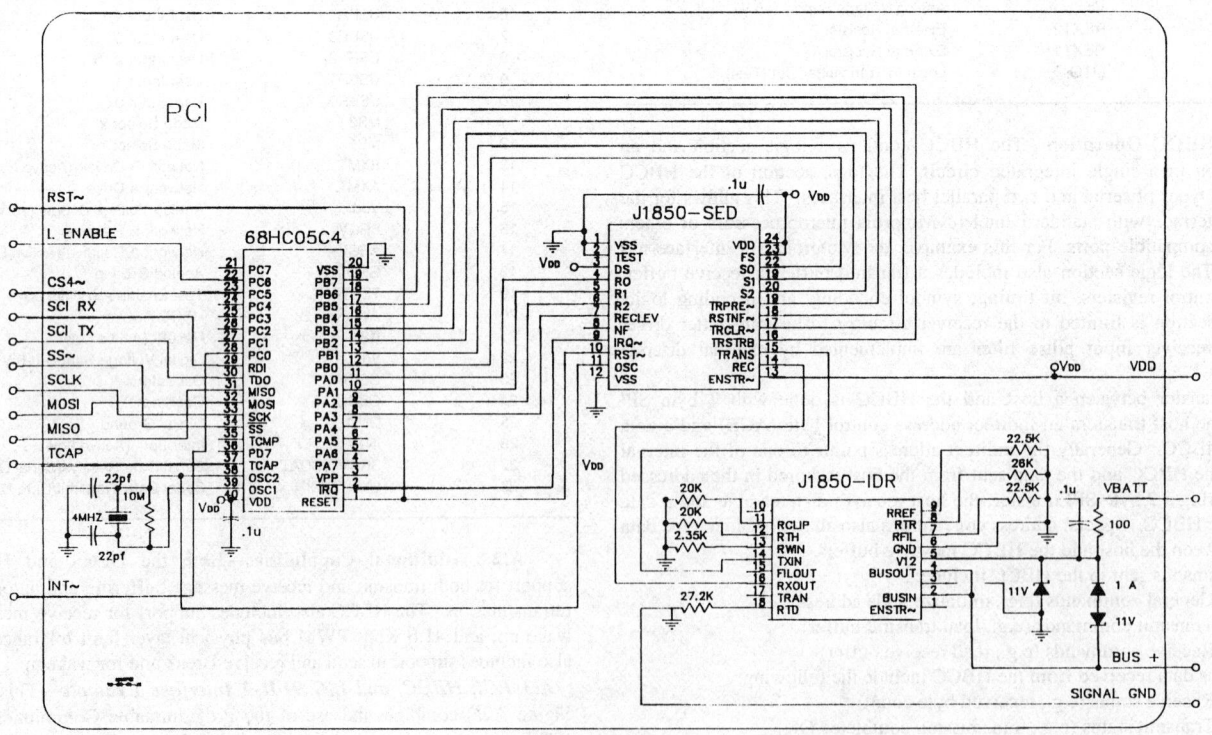

FIGURE A2b—PCI

A description of the HBCC interface is found in A.2.4.
A description of the ISO 9141-2 interface is found in A.1.4.
A description of the PCI is found in this section.

The host communicates with the PCI and the HBCC through the host's SPI interface. The host operates the SPI interface as the master and uses logic control lines to select either the PCI or the HBCC to communicate with. Control information sent from the host to the PCI and the HBCC cause the PCI and HBCC to be enabled or not. The ISO 9141-2 interface uses the SCI interface of the host and a logic line to enable the L line.

A.3.1 General Overview—The PCI operates at a 10.4 Kbps data rate using the symbols specified in SAE J1850.

The PCI is controlled by the coordinated interdependence of both hardware and software. The system consists of the interconnection of three devices:
 a. A Control Microcomputer with custom software
 b. A Symbol Encoder/Decoder (SED)
 c. An Integrated Driver Receiver (IDR)

The control microprocessor controls the operation of the SED and interfaces with the host.

The SED is a digital gate array device which performs symbol encoding and decoding, timing and bus synchronization.

The IDR is an analog ASIC which performs as a 10.4 Kbps VPW bus transceiver for the PCI.

A.3.2 Control Microcomputer—The control microcomputer has overall control of the PCI's operation and interfaces with the host and the SED. The control microcomputer, under software control, interprets host commands and bus symbols.

A.3.2.1 CONTROL MICROCOMPUTER SOFTWARE STRUCTURE—The control microcomputer's most critical task is to service the IRQ interrupts generated by the SED. Less critical tasks are handled on an as-needed basis by the interruptable main program.

The IRQ interrupt processing task services the following:
 a. Transmission of Frame Symbols
 b. Reception of Frame Symbols
 c. Bit-By-Bit Arbitration
 d. Error Detection
 e. Symbol encoding and decoding
 f. In-frame response

The main program services the following:
 a. Interface to the host
  (1) SPI (or Parallel) interface service
  (2) Chip selection of control microcomputer by the host
 b. Bus calculations
  (1) Checksum or CRC
  (2) Transmit and receive message buffering
  (3) Error interpretation
 c. Optional tasks when operating as host also
  (1) Keyboard inputs
  (2) Display drivers

A.3.2.2 CONTROL MICROCOMPUTER INTERFACE TO THE HOST MICROCOMPUTER—The PCI supports both a SPI or a parallel interface with the host. The host functions as interface master for both types of interfaces.

A.3.2.3 CONTROL MICROCOMPUTER INTERFACE TO THE SED—The interface between the control microcomputer and the SED shall be as defined by A.3.3 and Table A4.

TABLE A4—CONTROL MICROCOMPUTER AND SED INTERCONNECTION

Input Port	SED Output	Output Port	SED Input
PA0	R0	PB0	S0
PA1	R1	PB1	S1
PA2	R2	PB2	S2
PA3	RECLEV	PB3	TRPRE~
PA4	NF	PB4	RSTNF
IRQ~	IRQ~	PB5	TRCLR~
		PB6	TRSTRB

A.3.3 Symbol Encoder Decoder (SED)—The SED receives commands from the control microcomputer and converts these commands to SAE J1850 VPW symbols on a symbol-by-symbol bit-by-bit basis. Reception of messages is similarly accomplished, with the SED converting each received symbol into data that is fed to the control microcomputer for deciphering.

A.3.3.1 SYMBOL ENCODER—The control microcomputer initiates the transmission of a frame by first strobing Transmit Preset TRPRE~ to a logic "0" and strobing Transmit Strobe TRSTRB to a logic "1". The S0, S1, and S2 Inputs are then set to the logic levels defined in Table A5 in order to activate the transmit symbol generator circuit. During each symbol, the Transmit Strobe TRSTRB is strobed. The transmit symbol generator produces the required symbol on the Transmit TRANS Output of the SED. When the symbol has completed transmission, the SED Receive Circuit generates an IRQ~ interrupt to the control microcomputer to obtain the next symbol. This cycle is repeated until the entire frame is transmitted.

TABLE A5—TRANSMIT SYMBOL DEFINITION

Inputs S2	Inputs S1	Inputs S0	Output Description J1850 Xmit Symbol
0	0	1	Tv1 Short
0	1	0	Tv2 Long
0	1	1	Tv3 SOF or EOD
1	0	0	Tv6 IFS
1	0	1	Tv5 IBS
1	1	0	>Tv5 BREAK

A.3.3.2 SYMBOL DECODER—Similarly to the operation of the symbol encoder, the completion of the reception of a symbol from the IDR by the SED causes the SED to interrupt the control microcomputer. The control microcomputer reads the SED's R0, R1, and R2 outputs to determine which symbol was received, as defined by Table A6.

The meaning of each received VPW symbol is dependent on whether the symbol level is dominant or passive. This level is read by the control microcomputer on the Receive Level RECLEV Output of the SED.

TABLE A6—RECEIVE SYMBOL DEFINITION

Outputs R2	Outputs R1	Outputs R0	Input Description J1850 Rec. Symbol
0	0	0	Invalid Symbol
0	0	1	Tv1 Short
0	1	0	Tv2 Long
0	1	1	Tv3 SOF or EOD
1	0	0	Tv4 EOF
1	0	1	Tv6 IFS
1	1	0	Tv5 IBS

A.3.3.3 INVALID SYMBOL—The reception of an Invalid Symbol (Refer to Table A6) sets a Noise Flag NF Output. This Output remains at a logic "1" level until cleared by Reset Noise Flag RSTNF Input.

A.3.3.4 SED INPUTS AND OUTPUTS—All inputs and outputs from the SED are at standard CMOS voltage levels.

The following discusses some particular SED inputs and outputs.

A.3.3.4.1 Transmit Clear (TRCLR~) Input—The SED's TRCLR~ Input can be used by the control microcomputer to terminate the transmission of a IBS or Break symbol. IBS may be terminated after the minimum Tv5 period has been generated and a Break symbol may be terminated after a period greater than the maximum Tv5 has been generated.

A.3.3.4.2 Reset (RST) Inputs—The RST Input to the SED is used to initialize the SED.

A.3.3.4.3 Oscillator (OSC) Input—The OSC Input to the SED is the Time Base Clock Signal. When the DS Input is connected to VCC, a 4.0 MHz clock is required. When DS Input is grounded, an 8.0 MHz clock is required.

A.3.3.4.4 Transmit Strobe (TRSTRB) Input—The TRSTRB Input to the SED operates as a watchdog circuit to insure that the control microcomputer is functioning correctly. The TRSTRB Circuit requires that each transmit symbol must be strobed into the SED or the SED will transmit an idle Output. The TRSTRB Input is active when ENTRSTRB Input is grounded and disabled when the ENTRSTRB Input is connected to VCC.

A.3.4 Integrated Driver/Receiver (IDR)—The IDR is the interface between the vehicle bus wiring and the SED. The IDR integrates and simultaneously performs both the bus driver and receiver functions. The bus output waveform is specifically shaped to conform to SAE J1850 requirements and to minimize EMI.

A.3.4.1 BUS OUTPUT WAVESHAPING—The IDR bus output is wave shaped to have symmetrical rise and fall voltage waveforms within the range of allowed bus loading tolerances and battery operating voltages and during message arbitration. The waveshaping consists of a set rise/fall and corner rounding times. The maximum rise/fall time is 16 μs with the corner rounding of 4 μs. The typical rise/fall time is set to 14 μs with corner rounding of 3 μs.

A.3.4.2 TRANSMITTING SIGNALS ON THE BUS—The IDR receives 1's and 0's from the SED on the CMOS compatible Transmit pin (Tx) and translates these to wave shaped high and low signals at the bus out pin (Bout). The propagation delay between the transition on the Tx pin and the corresponding transition on the Bout pin measured at 3.875 V is a maximum of 12 μs, with a typical value of 10 μs.

A.3.4.3 PROCESSING SIGNALS RECEIVED FROM THE BUS—The IDR receives 1's and 0's on the bus input pin (Bin) and translates these to 1's and 0's at the CMOS compatible Receive pin (Rx). The propagation delay between the transition on the Rx pin and the corresponding transition on the Bin pin measured at 3.875 V is a maximum of 2 μs, with a typical value of 1 μs.

A.3.5 Additional Capabilities—The additional capabilities of the HBCC are shown in A.2.6.

The PCI is able to buffer both transmit and receive messages, send and receive Break, support both CRC and Checksum for message error checking, and comprehend IBS.

The MPU used in this example for the control microcomputer is a Motorola 68HC05C4. This particular version of the 68HC05 family is used for the designer's convenience. A much simpler 68HC05P7 microcomputer could also be used to perform the required PCI control tasks and in some cases the host tasks of a scan tool. The 68HC05C4's utilization during message transmission averages about 50% of the available CPU time. It is likely that with so much CPU time still available, the 68HC05C4 or the more powerful and faster 68HC11 could be used to perform both the control and host tasks.

A second example in this section (see Figure A3) shows the use of the control microprocessor associated with the PCI as the host also. As such the control microcomputer is connected to the HBCC and ISO 9141-2 interfaces. The external equipment manufacturer is left to design the interface to any required keyboard and display, which are not shown. Similarly to the case shown in Figure A2a, the control microcomputer externally selects the HBCC or the ISO 9141-2 interface when needed.

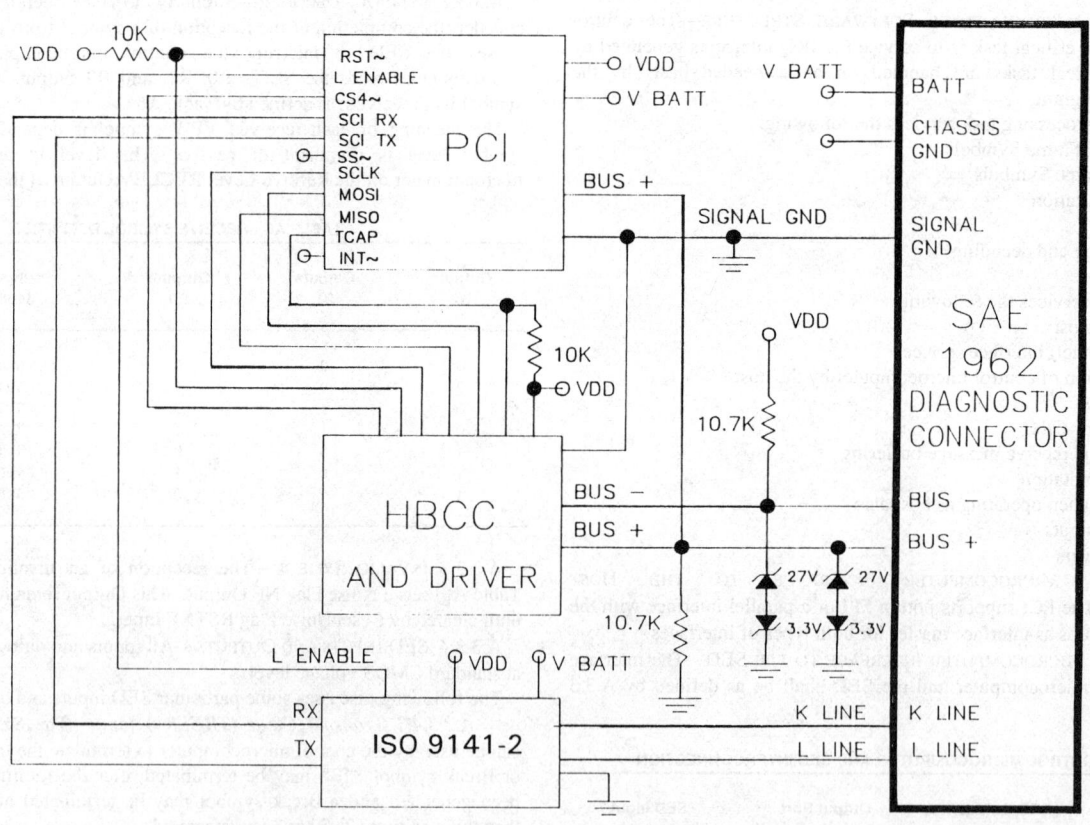

FIGURE A3—PCI, HBCC, AND ISO 9141-2 WITH PCI AS HOST ALSO

A.4 SGS-Thomson Protocol Engine Interface Example—This example (see Figure A4a) combines the use of an SGS-Thomson ST9 microprocessor as a control microcomputer, control microprocessor software, a GAL6001 programmable logic array device and discrete bus transceiver circuitry to provide the required interface support. Together these devices are identified as the SGS-Thomson Protocol Engine. (Figure A4b shows the detail of the ST9 and the GAL6001, Figure A4c shows the detail of the 10.4 Kbps Transceiver, Figure A4d shows the detail of the 41.6 Kbps Transceiver, and Figure A4e shows the detail of the ISO 9141 Transceiver.)

In this example the host microcomputer is connected to only one interface, the Protocol Engine. The Protocol Engine directly supports the required SAE J1850 and ISO 9141-2 interfaces and includes the ability to, on its own, determine the particular interface being used by a vehicle to support OBD II communication.

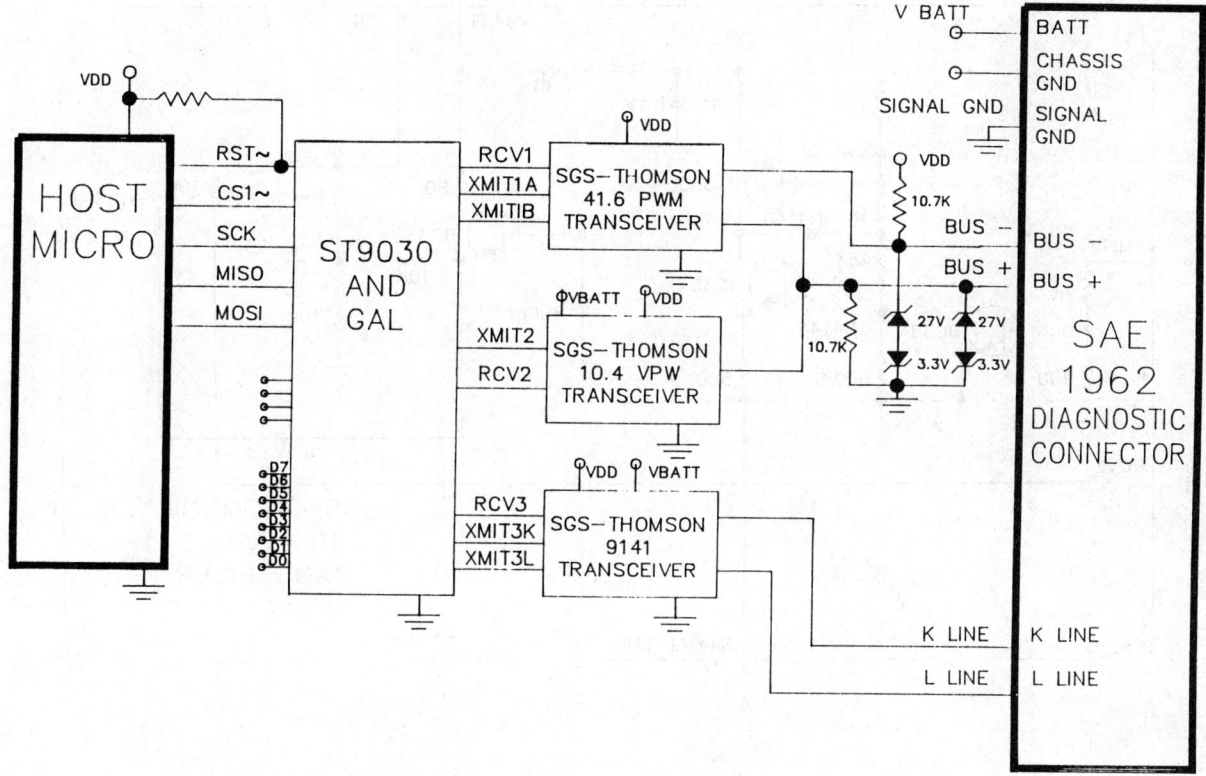

FIGURE A4a—SGS-THOMSON PROTOCOL ENGINE

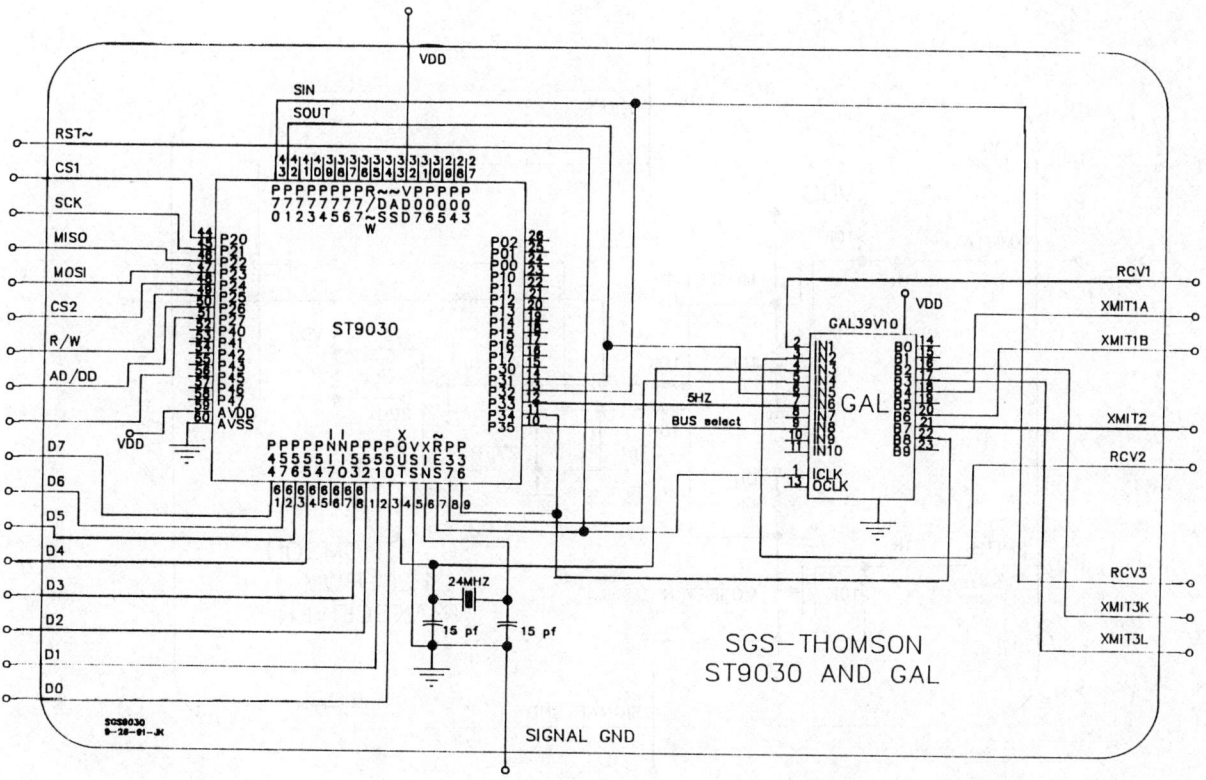

FIGURE A4b—SGS-THOMSON ST9030 AND GAL

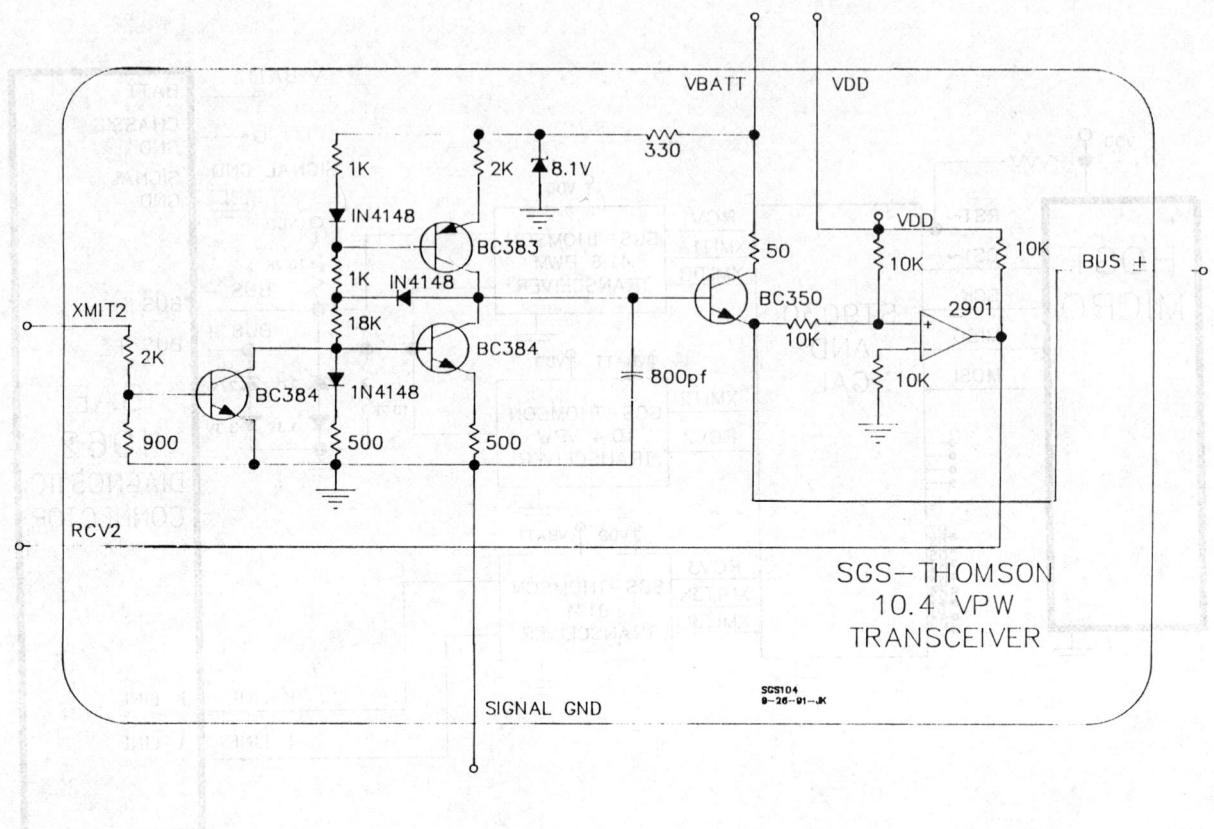

FIGURE A4c—SGS-THOMSON 10.4 VPW TRANSCEIVER

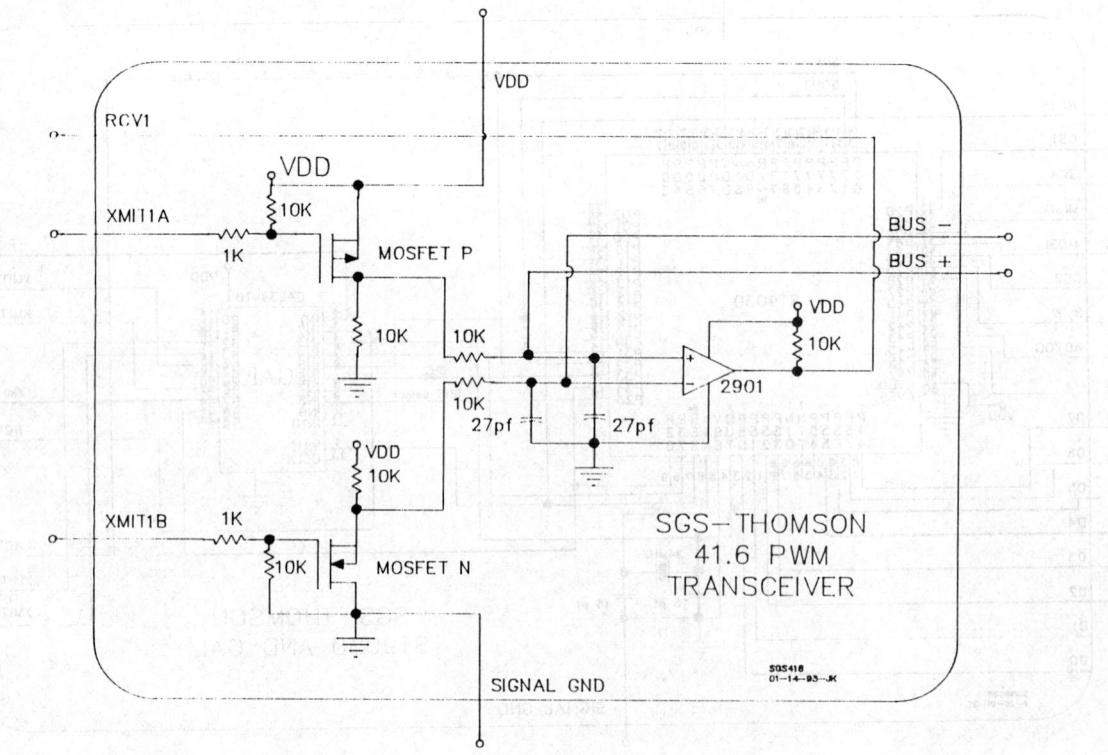

FIGURE A4d—SGS-THOMSON 41.6 PWM TRANSCEIVER

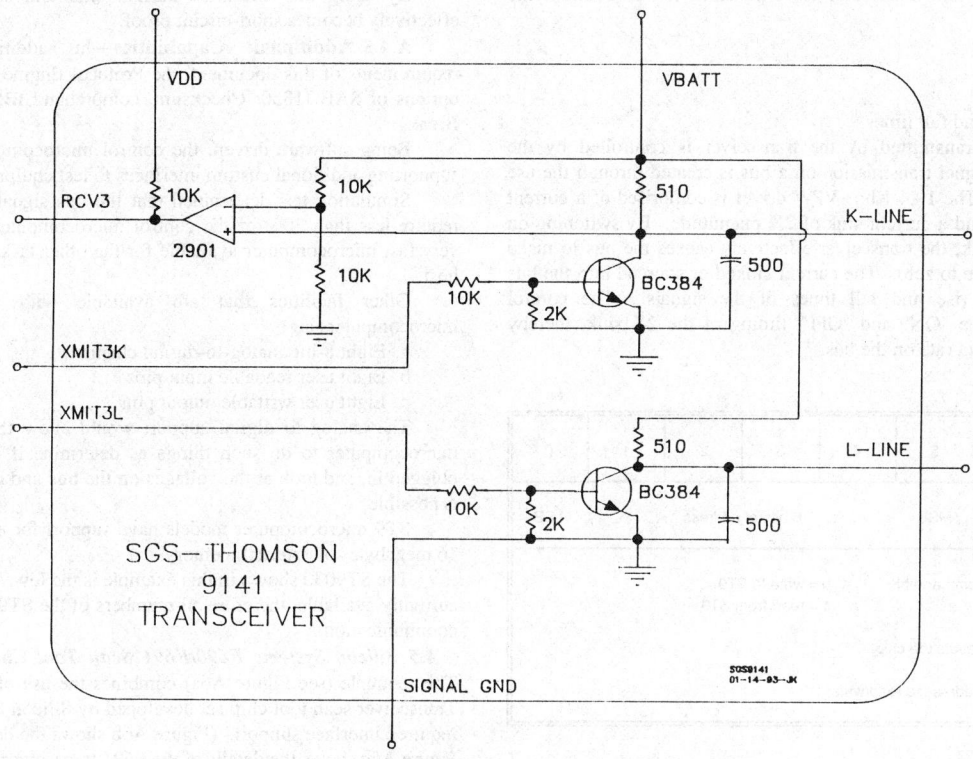

FIGURE A4e—SGS-THOMSON 9141 TRANSCEIVER

A.4.1 General Overview—As mentioned in Section A.4, the Protocol Engine is comprised of four parts: the ST9 control microcomputer, control microcomputer software, the GAL6001 programmable logic array device, and several discrete transceivers.

A.4.2 Control Microcomputer—A SGS-Thomson ST9 family microcomputer is used as an interface control microcomputer. The ST9 microcomputer is a general purpose 8/16 bit microcomputer. Its timer facilities are used extensively to provide the control of transmitted waveforms and the detection of received signals.

Both SPI and parallel interfaces to host microcomputers are supported.

The host can request the Protocol Engine to either scan the possible OBD II interfaces and determine the type used in a given vehicle or use a selected interface type.

A.4.2.1 CONTROL MICROCOMPUTER SOFTWARE STRUCTURE—The highest priority tasks in the control microcomputer support the symbol-by-symbol processing of each transmitted and received message frame. The control microcomputer, and in particular the ST9's timers, directly controls the rise and fall of transmitted signals and directly follows the rise and fall of received signals. When transmitting a message, it performs both of these processes simultaneously. All other control microcomputer tasks are processed on a time available basis.

The following further identifies the control microcomputer's tasks:
a. Highest Priority Tasks:
  (1) Bus synchronization
  (2) Transmission and reception of frame symbols
  (3) Bit-by-bit arbitration
  (4) Comprehension of IBS
b. Background tasks:
  (1) CRC or checksum calculation
  (2) Symbol encoding and decoding
  (3) Message buffering
  (4) In-frame response processing
  (5) Error checking
  (6) Host interface

A.4.2.2 CONTROL MICROCOMPUTER INTERFACE TO THE HOST—Communication between the control microcomputer and the host is based on the host reading from and writing to 64 bytes of memory in the control microcomputer. Sixteen bytes are reserved for status and control of SAE J1850 related operations, 13 bytes are reserved for status and control of ISO 9141-2 related operations, 3 bytes are reserved for general status and control, and 32 bytes are reserved for receive and transmit buffers and related pointers and counters.

Data transfers between the host and the control microcomputer are performed as 2 byte transfers, a control byte and a data byte, initiated by the host. The control byte from the host indicates whether the host wishes to read or write a byte in the control microcomputer and what the address of the byte is (see Figure A5). The second SPI byte transfer moves the byte read from the control microcomputer to the host or the byte to be written from the host to the control microcomputer.

The control microprocessor supports both an SPI and a parallel type of interface with the host. In order to properly service bus communications, the control microcomputer controls the actual data transfers between itself and the host.

A.4.2.3 AUTOMATIC OBD II INTERFACE SCAN—Upon host request, the Protocol Engine will, by itself, automatically scan all of the possible OBD II interfaces and determine the one supported by a given vehicle. The interface scan sequence used is as follows:
  a. SAE J1850 41.6 Kbps PWM
  b. SAE J1850 10.4 Kbps VPW
  c. ISO 9141-2

A.4.3 GAL6001 Programmable Logic Array—The GAL6001 Programmable Logic Array performs primarily as a digital filter of the signal received from the vehicle bus. It is a standard high-performance field programmable EEPROM logic device and is available in a variety of packages. The device chosen has a 20-year data retention specification. It is organized as a 78 X 64 X 36 FPLA.

A.4.4 Discrete Transceiver—The discrete transceiver circuity includes all the components necessary to create the required transmit signal waveforms on a bus and to receive bus signals. This includes support for 41.6 PWM, 10.4

VPW and ISO 9141-2. The transceiver is responsible for or controls the following:
  a. Voltage on the bus
  b. Current levels
  c. Media impedance
  d. Signal rise times and fall times

The overall signal transmitted by the transceiver is controlled by the control microcomputer. Signal transmission on a bus is created through the use of current sources/sinks. The 10.4 Kbps VPW driver is comprised of a current source of 1X magnitude and a current sink of 2X magnitude. By switching on and off the 2X current sink, the transceiver effectively causes the bus to make transitions from zero to one to zero. The current sinked or sourced into the bus capacitance controls the rise and fall times of the signals. The control microcomputer controls the "ON" and "OFF" timing of the 2X sink, thereby controlling the data and data rate on the bus.

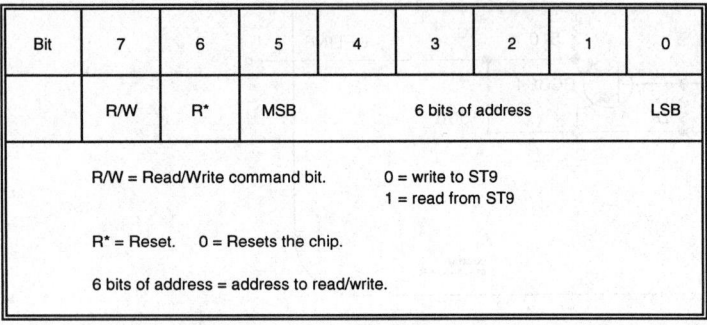

FIGURE A5—PROTOCOL ENGINE COMMAND BYTE

By using the controlled current sink and source approach this driver effectively becomes short-circuit proof.

A.4.5 Additional Capabilities—In addition to supporting the requirements of this document, the Protocol Engine also supports the following options of SAE J1850: Checksum, comprehend IBS, Single Byte Header, and Break.

Being software driven, the control microcomputer can be considered for supporting additional custom interfaces to test equipment host microcomputers.

Simulation has determined that the bus signal control and monitor tasks require less than 30% of the control microcomputer leaving at least 70% of a very fast microcomputer available for the other tasks such as performing as the host.

Other facilities that are available with various versions of ST9 microcomputers are:
  a. Eight 8-bit analog-to-digital channels
  b. Eight user readable input pins
  c. Eight user writable output pins

The analog to digital support would allow the host to have the control microcomputer to do such things as determine if the diagnostic connector is plugged in, and look at the voltages on the bus and determine if communication is possible.

ST9 microcomputer models have support for up to 32K of ROM or up to 16 megabytes of external memory.

The ST9030 shown in this example is the lowest member in the ST9 family currently available, therefore all members of the ST9 family can support OBD II communication.

A.5 Silicon Systems F690/F691 Scan Tool Chipset Interface Example—This example (see Figure A6a) combines the use of the F690 Codec and F691 Transceiver scan tool chip set developed by Silicon Systems (SSi) to provide the required interface support. (Figure A6b shows the detail of the F690 Codec, and Figure A6c shows the details of the F691 transceiver.)

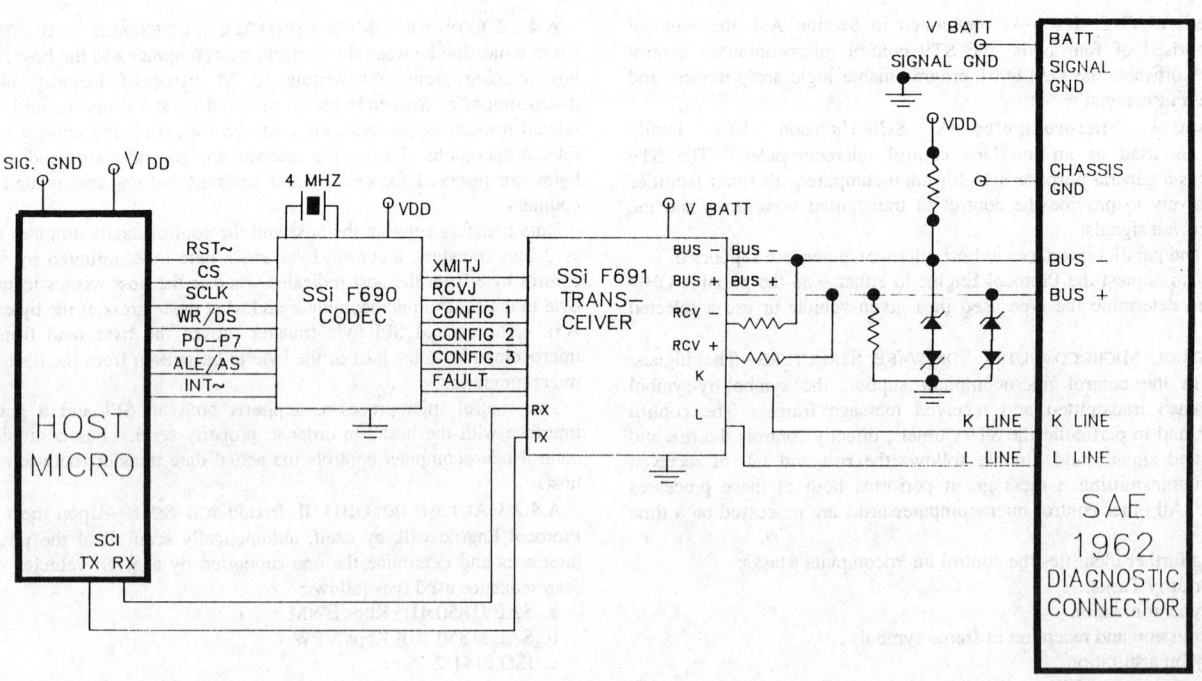

FIGURE A6a—SILICON SYSTEMS F690/F691 CHIPSET

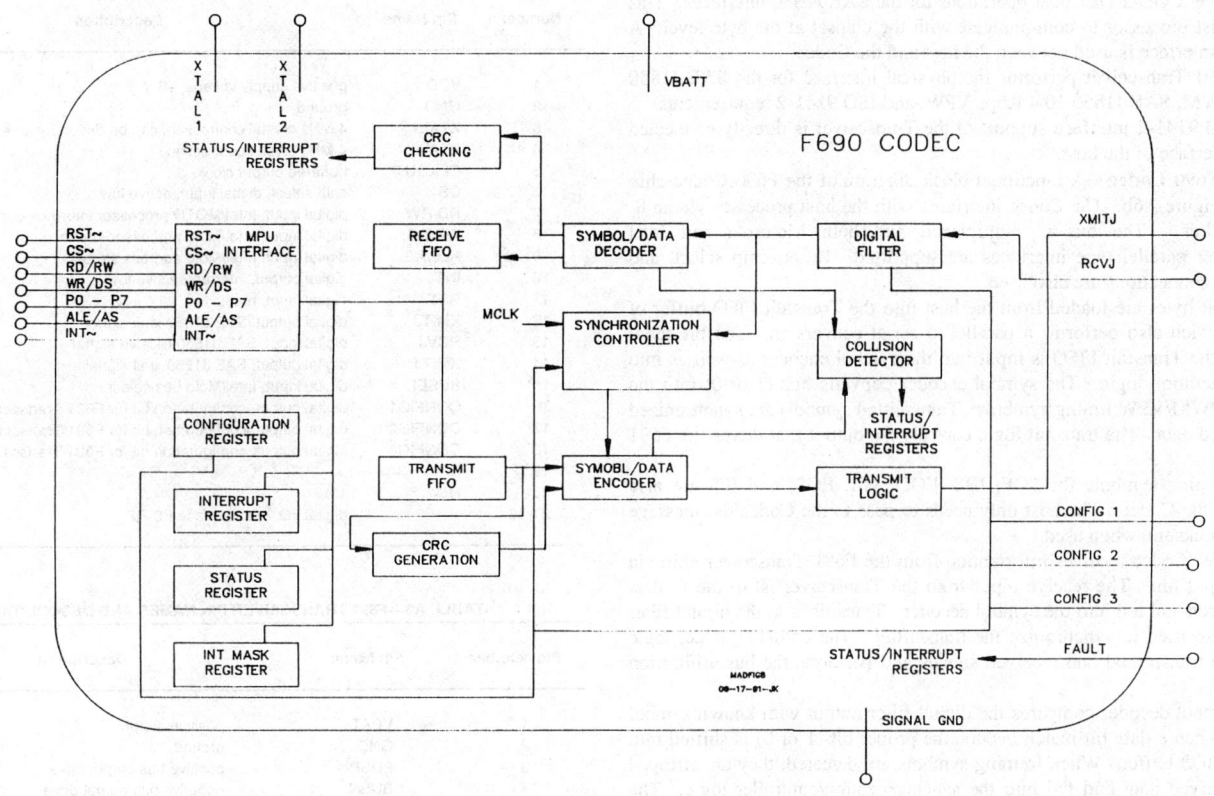

FIGURE A6b—F690 CODEC

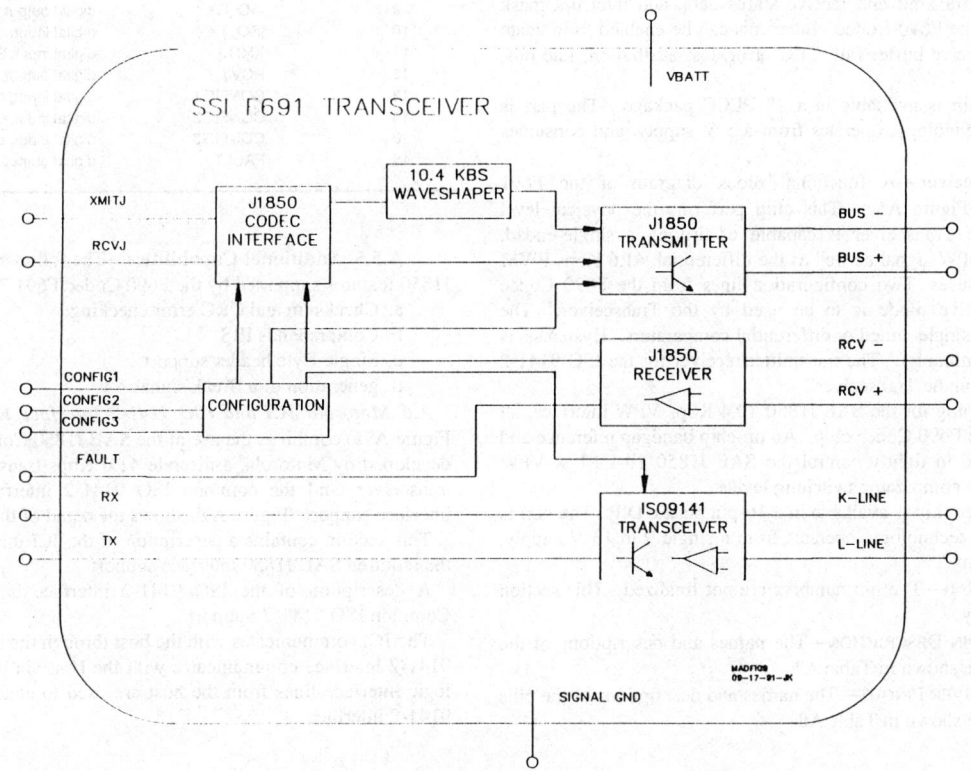

FIGURE A6c—F691 TRANSCEIVER

A.5.1 General Overview—The SSi Scan Tool chipset consists of the F690 Codec and the F691 Transceiver. The F690 Codec performs the bit/symbol level CODer/DECoder operations for the SAE J1850 interfaces. This allows the host processor to communicate with the chipset at the byte level. A parallel type interface is used between the host and the Codec.

The F691 Transceiver performs the physical interface for the SAE J1850 41.6 Kbps PWM, SAE J1850 10.4 Kbps VPW, and ISO 9141-2 requirements.

The ISO 9141-2 interface support of the Transceiver is directly connected to the SCI interface of the host.

A.5.2 F690 Codec—A functional block diagram of the F690 Codec chip is shown in Figure A6b. The Codec interfaces with the host processor via an 8-bit parallel bus. The bus is multiplexed and both Motorola and Intel microcomputer parallel type interfaces are supported. Reset, chip select, and interrupt line connections are also used.

Transmit bytes are loaded from the host into the Transmit FIFO buffer of the Codec, which also performs a parallel to serial conversion. The bit serial output from the Transmit FIFO is input into the symbol encoder as well as into the CRC generation logic. The symbol encoder converts bits (1 or 0) into the appropriate PWM/VPW timing symbols. Transmitted symbols are synchronized to the received data. The transmit logic controls the output that drives the F691 Transceiver.

The framing symbols for SOF, IBS, EOD, NB, EOF, and IFS are also generated by the Codec. The host only needs to pass to the Codec the message bytes, and checksum, when used.

The F690 Codec receives information from the F691 Transceiver chip via the RCVJ input pin. The receive input from the Transceiver is, in the Codec, digitally filtered and fed into the symbol decoder. Transitions at the digital filter output are also used to synchronize the transmitter. The collision detect logic compares the transmitted and received signals and performs the bus arbitration function.

The symbol decoder compares the digital filter output with known symbol templates. When a data bit match occurs, the proper bit (1 or 0) is shifted into the receive FIFO buffer. When framing symbols are detected, they are stripped from the received data and fed into the synchronization/controller logic. The data values from the symbol decoder output are also fed into the CRC checking logic. The receive FIFO buffer also performs a serial to parallel conversion such that the host processor can parallel read a received data byte.

Configuration data, transmit and receive status data, and interrupt mask registers are included on the F690 Codec. Interrupts can be enabled to indicate transmit buffer empty, receive buffer full, CRC error, lost attribution, idle bus, fault detection, etc.

The F690 Codec chip is available in a 28 PLCC package. The part is fabricated in a CMOS technology, operates from a 5 V supply, and consumes less than 100 mW max.

A.5.3 F691 Transceiver—A functional block diagram of the F691 Transceiver is shown in Figure A6c. This chip performs the physical level interface functions. The Transceiver is capable of driving a single-ended, waveshaped, 10.4 Kbps VPW signal as well as the differential, 41.6 Kbps PWM signal onto SAE J1850 buses. Two configuration lines from the F690 Codec determine what xmit/receive mode is to be used by the Transceiver. The receiver consists of either single-ended or differential comparators. Hysteresis is added to improve noise immunity. The transmitter/receiver for the ISO 9141-2 interface is also included on the Transceiver.

Other than waveshaping for the SAE J1850 10.4 Kbps VPW interface, all timing is controlled by the F690 Codec chip. An on-chip bandgap reference and prepackage trims are used to tightly control the SAE J1850 10.4 Kbps VPW waveshape parameters and comparator switching levels.

The F691 Transceiver chip is available in a 16-pin plastic DIP. The part is fabricated in a BIPOLAR technology, operates from a single 9 to 13 V supply, and dissipates 500 mW max.

A.5.4 Pin Description—The pin numbers are not finalized. This section describes pin function only.

A.5.4.1 F690 CODEC PIN DESCRIPTION—The names and descriptions of the pins of the F690 Codec are shown in Table A7.

A.5.4.2 F691 TRANSCEIVER PINOUT—The names and descriptions of the pins of the F691 Tranceiver are shown in Table A8.

TABLE A7—F690 CODEC PIN NAMES AND DESCRIPTIONS

Pin Number	Pin Name	Description
1	VDD	positive supply voltage, +5 V
2	GND	ground
3	XTAL1	4 MHz crystal connection; can be driven by an external clock
4	XTAL2	4 MHz crystal connection
5	CLKOUT	buffered output clock
6	CS	chip select, digital input, active low
7	RD/RW	digital input, Intel/MOTO processor interface control
8	WR/DS	digital input, Intel/Moto processor interface control
9	ALE/AS	digital input, Intel/Moto address strobe
10	INT	digital output, interrupt, active low
11	RST	digital input, reset
12	XMITJ	digital output, SAE J1850 xmit signal
13	RCVJ	digital input, SAE J1850 receive signal
14	XMITJ	digital output, SAE J1850 xmit signal
15	BUSEL	digital input, Intel/Moto bus select
16	CONFIG1	digital output, configuration bit for F691 Transceiver
17	CONFIG2	digital output, configuration bit for F691 Transceiver
18	CONFIG3	digital output, configuration bit for F691 Transceiver
19	N/A	
20	N/A	
21-28		digital I/O, bus interface P0-P7

TABLE A8—F691 TRANSCEIVER PIN NAMES AND DESCRIPTIONS

Pin Number	Pin Name	Description
1	VBAT	positive supply
2	GND	ground
3	BUSP	positive bus output drive
4	BUSN	negative bus output drive
5	RCVP	positive bus receiver input
6	RCVN	negative bus receiver input
7	K	ISO 9141-2 K line, input/output
8	L	ISO 9141-2 L line, output
9	ISO_RX	digital output, ISO 9141-2 receive signal
10	ISO_TX	digital input, ISO 9141-2 transmit signal
11	XMITJ	digital input, SAE J1850 xmit signal
12	RCVJ	digital output, SAE J1850 receive signal
13	CONFIG1	digital input, configuration bit
14	CONFIG2	digital input, configuration bit
15	CONFIG3	digital input, configuration bit
16	FAULT	digital output, fault detection

A.5.5 Additional Capabilities—The following is a list of additional SAE J1850 features supported by the F690 Codec/F691 Transceiver:

a. Checksum and CRC error checking
b. Comprehends IBS
c. Single Byte header support
d. generation of a Break signal

A.6 Motorola JCI and ISO 9141-2 Interface Example—This example (see Figure A7a) combines the use of the SAE J1850 Communications Interface (JCI) developed by Motorola, a discrete 41.6 Kbps transceiver, a discrete 10.4 Kbps transceiver, and the common ISO 9141-2 interface to provide the required interface support. Figure A7b shows the detail of the JCI.

This section contains a description of the JCI interface, which provides all of the required SAE J1850 interface support.

A description of the ISO 9141-2 interface is found in the section titled Common ISO 9141-2 Support.

The JCI communicates with the host through the host's SPI port, while the ISO 9141-2 interface communicates with the host via the host's SCI port. Separate logic interface lines from the host are used to enable either the JCI or the ISO 9141-2 interface.

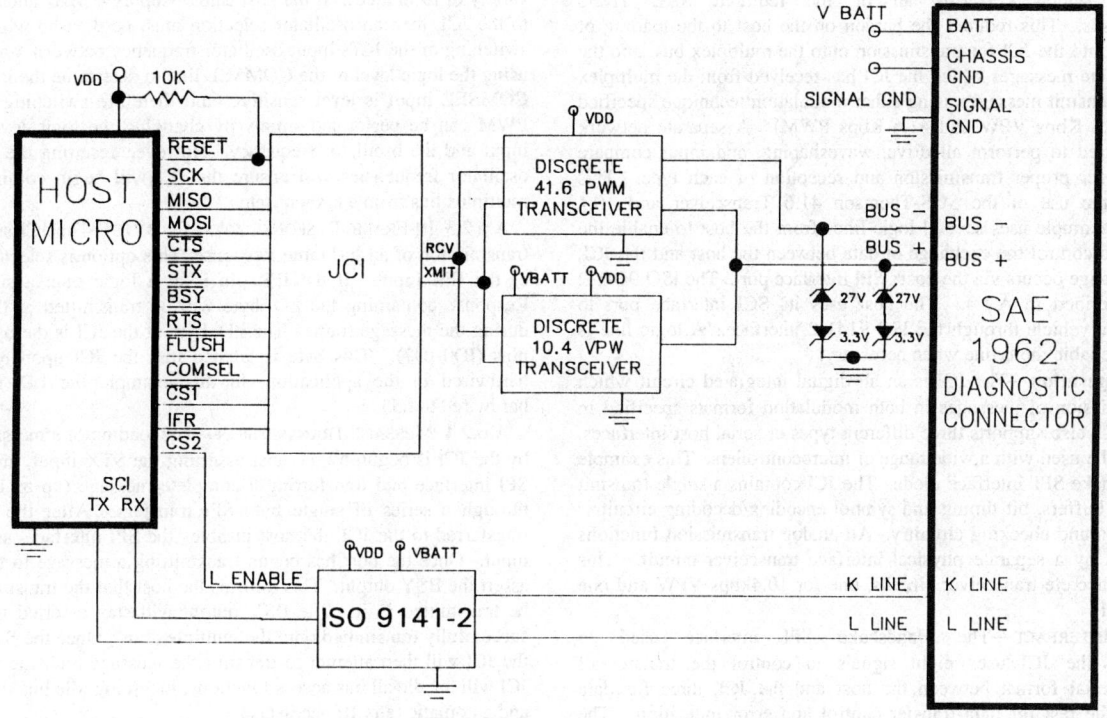

FIGURE A7a—JCI AND ISO 9141-2 EXAMPLE

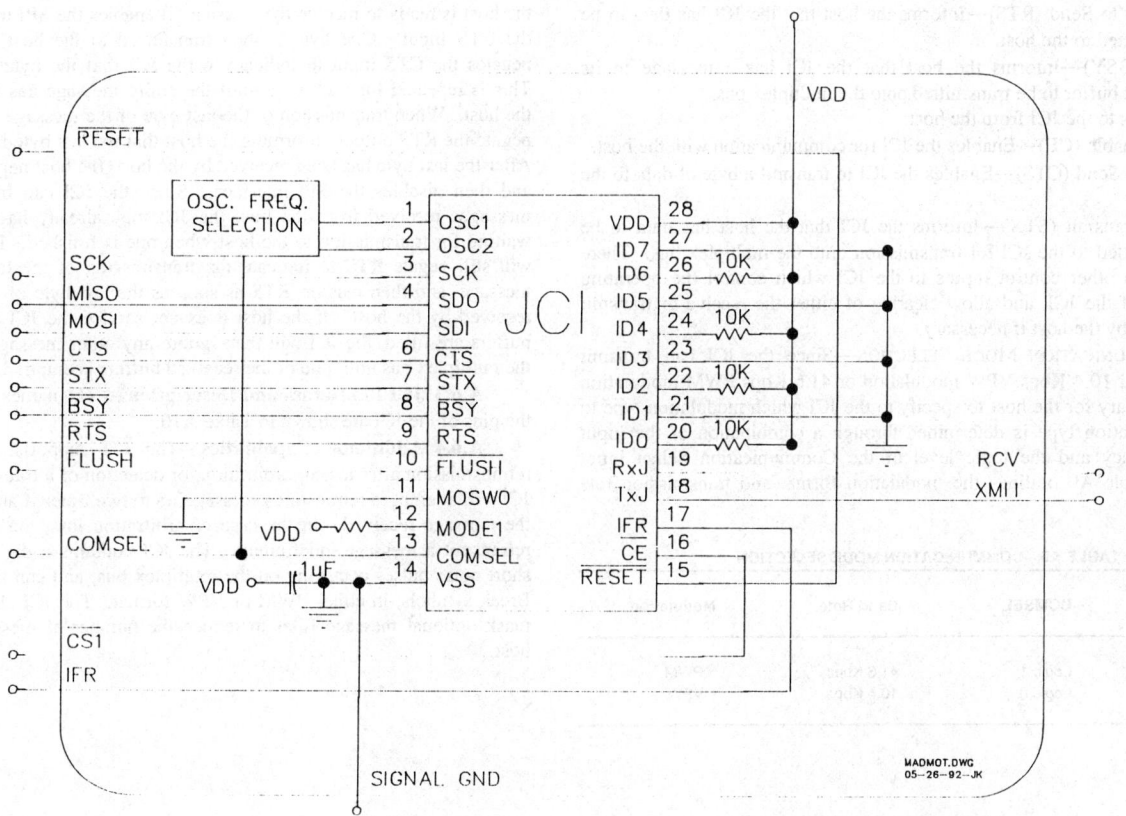

FIGURE A7b—MOTOROLA J1850 COMMUNICATIONS INTERFACE

A.6.1 General Overview—The JCI is an all-digital communications peripheral device which performs all of the required SAE J1850 communications tasks. This reduces the burden on the host to the loading of complete messages into the JCI for transmission onto the multiplex bus, and the unloading of complete messages which the JCI has received from the multiplex bus. The JCI can transmit messages using either modulation technique specified by SAE J1850 (10.4 Kbps VPW and 41.6 Kbps PWM). A separate network transceiver is required to perform all drive, waveshaping, and input compare functions required for proper transmission and reception of each type. This example includes the use of the SGS-Thomson 41.6 Transceiver and 10.4 Transceiver. This example uses several logic lines from the host to enable the JCI interface, and to control the exchange of data between the host and the JCI, while the data exchange occurs via the host's SPI interface port. The ISO 9141-2 interface is as described in A.1.4. The host uses its SCI interface port to communicate with a vehicle through the ISO 9141-2 interface. A logic line is used by the host to enable the L line when necessary.

A.6.2 JCI Operation—The JCI is an all-digital integrated circuit which can support transmission of messages in both modulation formats specified in SAE J1850. The JCI also supports three different types of serial host interfaces, allowing the JCI to be used with a wide range of microcontrollers. This example utilizes the `Handshake SPI' interface mode. The JCI contains a single transmit buffer, two receive buffers, bit timing and symbol encoding/decoding circuitry, and CRC generation and checking circuitry. All analog transmission functions must be performed by a separate physical interface transceiver circuit. This example uses two discrete transceiver circuits, one for 10.4kbps VPW and one for 41.6 Kbps PWM.

A.6.2.1 HOST INTERFACE—The Handshake SPI interface used to communicate with the JCI uses eight signals to control the transfer of information in a serial format between the host and the JCI, three for data transmission, and the rest for data transfer control and error indication. The actual data transfer is in the standard Motorola SPI format. The five control lines include two outputs from the JCI and three inputs to the JCI. The function of these lines are as follows:

a. Output signals from the JCI to the host:
 (1) Request to Send (RTS)—Informs the host that the JCI has data to be transmitted to the host.
 (2) Busy (BSY)—Informs the host that the JCI has a message in its transmit buffer to be transmitted onto the multiplex bus.

b. Input signals to the JCI from the host:
 (1) Chip Enable (CE)—Enables the JCI for communication with the host.
 (2) Clear to Send (CTS)—Enables the JCI to transmit a byte of data to the host.
 (3) Start Transmit (STX)—Informs the JCI that the host has data to be transmitted to the JCI for transmission onto the multiplex bus. There are also other control inputs to the JCI which control the operating mode of the JCI, and allow clearing of either the receive or transmit buffers by the host if necessary.

A.6.2.2 COMMUNICATION MODE SELECTION—Since the JCI can transmit messages in either 10.4 Kbps VPW modulation or 41.6 Kbps PWM modulation types, it is necessary for the host to specify to the JCI which modulation type to use. The modulation type is determined through a combination of the input oscillator frequency and the logic level of the Communication Select input (COMSEL). Table A9 outlines the modulation format and transmission rate selection.

TABLE A9—COMMUNICATION MODE SELECTION

Fosc	COMSEL	Baud Rate	Modulation
8 MHz	Logic 1	41.6 Kbps	PWM
4 MHz	Logic 0	10.4 Kbps	VPW

The different oscillator frequencies can be supplied to the JCI through a variety of techniques. If the host cannot supply 4 MHz and 8 MHz clock inputs to the JCI, then an oscillator selection must be devised which would allow the switching of the JCI's input oscillator frequency between 4 and 8 MHz, possibly using the logic level of the COMSEL line to determine the input frequency. The COMSEL input is level sensitive, and therefore switching between VPW and PWM can be performed simply by changing the logic level of the COMSEL input and the oscillator frequency. However, resetting the JCI when changing oscillator frequencies will ensure the JCI will begin communicating with the multiplex bus from a known state.

A.6.2.3 IN-FRAME RESPONSE AND I.D. BYTE—The JCI supports the optional transmission of an In-Frame Response. This option is selected by the logic level of the IFR input. If the IFR pin is at a logic one, a single byte In-Frame Response containing the I.D. byte will be transmitted at the appropriate time during the message frame. The I.D. byte of the JCI is the byte input on the I.D. pins (ID0-ID7). This byte is latched into the JCI upon reset, and should be hardwired in the application. In this example the I.D. byte of the JCI is hardwired to h55.

A.6.2.4 MESSAGE TRANSMISSION—The loading of a message for transmission by the JCI is begun by the host asserting the STX input, and then enabling the SPI interface and transferring a complete message (up to 11 bytes) to the JCI through a series of single byte SPI transfers. After the last byte has been transferred to the JCI, the host disables the SPI interface, and negates the STX input. Once the host has begun transmitting a message to the JCI, the JCI will assert the BSY output. This informs the host that the transmit buffer has data to be transmitted in it. The BSY output will stay asserted until the message is successfully transmitted onto the multiplex bus. Once the STX input is negated, the JCI will then attempt to transmit the message onto the multiplex bus. The JCI will handle all bus access functions, including idle bus detection, arbitration, and automatic retry (if necessary).

A.6.2.5 MESSAGE RECEPTION—The JCI will receive and buffer complete messages transmitted onto the multiplex bus. Once an entire message has been received with no errors, and the CRC byte verified as correct, the JCI will inform the host that a message is ready for delivery by asserting the RTS output. When the host is ready to receive the message, it enables the SPI interface, and asserts the CTS input. One byte is then transferred to the host, and the host then negates the CTS input to indicate to the JCI that the byte has been received. This is repeated for each byte until the entire message has been transmitted to the host. When transmission of the last byte of the message begins, the JCI will negate the RTS output, informing the host that the last byte is being transmitted. After the last byte has been received by the host, the host negates the CTS input, and then disables the SPI interface. Since the JCI can buffer two complete messages received from the bus, the JCI may already have another message waiting for transmission to the host when one is finished. In this case, the JCI will still negate RTS to indicate the transmission of the last byte of the first message, and then reassert RTS as soon as the last byte of the first message is received by the host. If the host does not service the JCI before both receive buffers are filled, the JCI will then ignore any other messages transmitted onto the multiplex bus until one of the received buffers is emptied by the host.

A.6.3 JCI Pin Names and Descriptions—The names and descriptions of the pins of the JCI are shown in Table A10.

A.6.4 Additional Capabilities—The JCI supports automatic message retransmission after loss of arbitration, or detection of a transmission error. The JCI will attempt to retransmit a message up to two times if an error is detected in the message received. In the case of arbitration loss, the JCI will attempt to retransmit a message indefinitely. The JCI contains a digital filter to remove short noise pulses occurring on the multiplex bus, and can transmit and receive Break symbols, in either PWM or VPW format. The JCI also contains a metal mask optional message filter to reduce the number of messages passed to the host.

TABLE A10—JCI PIN NAMES AND DESCRIPTIONS

Pin Number	Pin Name	Description
1	OSC1	Oscillator 1
2	OSC2	Oscillator 2
3	SCK	SPI Serial Clock
4	SDO	SPI Serial Data Out
5	SDI	SPI Serial Data In
6	CTS	Clear To Send (Active Low)
7	STX	Start Transmit (Active Low)
8	BSY	Busy (Active Low)
9	RTS	Request To Send (Active Low)
10	FLUSH	Flush Message Buffer (Active Low)
11	MODE0	Mode Select 0
12	MODE1	Mode Select 1
13	COMSEL	Communications Select
14	VSS	Ground
15	RESET	Reset (Active Low)
16	CE	Chip Enable (Active Low)
17	IFR	In Frame Response Enable
18	TxJ	Bus Transmit Output
19	RxJ	Bus Receive Input
20	ID0	I.D. Byte - bit 0
21	ID1	I.D. Byte - bit 1
22	ID2	I.D. Byte - bit 2
23	ID3	I.D. Byte - bit 3
24	ID4	I.D. Byte - bit 4
25	ID5	I.D. Byte - bit 5
26	ID6	I.D. Byte - bit 6
27	ID7	I.D. Byte - bit 7
28	VDD	Power Supply

APPENDIX B
SUPPORTING DOCUMENTS

The following is a list of documents that may be useful to the reader for a further understanding of the examples shown in Appendix A:

Ford Hosted Bus Controller Chip User's Guide
Ford Hosted Bus Controller Circuit
Delco Introduction to GM Class 2 Serial Data Bus XDE - 3100
Delco DLCS/P IC User's Guide XDE - 3101
Delco DLCS/P IC Data Sheets XDE - 3102
Silicon Systems, F691 Transceiver Data Sheet - 67F686
SGS-Thomson, ST9 Family Technical Manual - DBST9TMST/1
SGS-Thomson, ST9 Family 8/16 Bit MCU Programming - DBST9PMST/1
SGS-Thomson, ST9XXX Databook - DBST9ST/1
SGS-Thomson, ST9 Brochure - BRST9/1190
Motorola, MC68HC05C4/C8, Technical Data 8-Bit Microcomputer - AD1991R2
Motorola, M68HC05 Applications Guide - M68HC05AG/AD

(R) CLASS B DATA COMMUNICATION NETWORK INTERFACE—SAE J1850 AUG91

SAE Recommended Practice

Report of the Vehicle Network for Multiplexing and Data Communications Standards Committee approved November 1988. Completely revised by the SAE Vehicle Network for Multiplexing and Data Communications Standards Committee July 1990. Completely revised by the SAE Vehicle Network for Multiplexing and Data Communications Standards Committee August 1991.

Foreword—This SAE Recommended Practice constitutes the requirements for a vehicle data communications network. These requirements are related to the lowest two layers of the ISO Open System Interconnect (OSI) model (see ISO 7498). These layers are the Data Link Layer and the Physical Layer. This network has been described using the ISO conventions in ISO/TC 22/SC 3/WG1 N429 E, dated October, 1990. Both documents are intended to describe the same network requirements but using different descriptive styles. If any technical differences are identified, the very latest revision of these documents should be used.

This is an SAE Recommended Practice which has been submitted as an American National Standard. As such, its format is somewhat different from the formal ISO description in that descriptions have been expanded, but are in no way less precise. A more textual format has been adopted herein to allow explanations to be included.

The vehicle application for this class of data communication (Class B) network is defined (see SAE J1213/1 to allow the sharing of vehicle parametric information. Also per the definition, this Class B network shall be capable of performing Class A functions.

TABLE OF CONTENTS

Foreword
1. Scope
2. References and Related Documents
2.1 Applicable Documents
2.1.1 SAE Publications
2.1.2 ISO Documents
2.1.3 CISPR Documents
2.2 Related Publications
2.3 Definitions and Abbreviations
2.3.1 Definitions
2.3.1.1 Arbitration
2.3.1.2 Class A Data Communications
2.3.1.3 Class B Data Communications
2.3.1.4 Class C Data Communications
2.3.1.5 Dual Wire
2.3.1.6 Fault Tolerance
2.3.1.7 Frame
2.3.1.8 Functional Addressing
2.3.1.9 Message
2.3.1.10 Physical Addressing
2.3.1.11 Pulse Width Modulation (PWM)
2.3.1.12 Sleep-Mode
2.3.1.13 Variable Pulse Width (VPW) Modulation
2.3.2 Abbreviations/Acronyms
3. Description of the Architecture
3.1 General
3.2 Network Topology
3.2.1 Data Bus Topology
3.2.2 Data Bus Control
3.3 References to the OSI Model
3.3.1 Application Layer
3.3.2 Data Link Layer
3.3.3 Physical Layer
3.4 Network Implementation
4. Application Layer Details
4.1 Normal Vehicle Operation (Down the Road) Messages
4.2 Diagnostic Messages
4.2.1 Diagnostic Parametric Data
4.2.2 Diagnostic Malfunction Codes
4.3 Frame Filtering
5. Data Link Layer Details
5.1 Addressing Strategy
5.1.1 Physical Addressing
5.1.2 Functional Addressing
5.2 Network Access and Data Synchronization
5.2.1 Full Message Buffering
5.2.2 Byte Buffering
5.3 Network Elements and Structure
5.3.1 Frame Elements
5.3.2 Bit Ordering
5.3.3 Maximum Frame Length
5.3.4 Function of SOF, EOD, EOF, IFS, IBS, NB, and BRK
5.3.4.1 Start of Frame (SOF)
5.3.4.2 End of Data (EOD)
5.3.4.3 End of Frame (EOF)
5.3.4.4 Inter-Frame Separation (IFS)
5.3.4.5 Inter-Byte Separation (IBS)
5.3.4.6 Normalization Bit (NB)
5.3.4.7 Break (BRK)
5.3.5 Idle Bus (idle)
5.3.6 Data Byte(s)
5.3.7 In-Frame Response
5.3.7.1 Normalization Bit
5.4 Error Detection
5.4.1 Cyclic Redundancy Check (CRC)
5.4.2 Checksum
5.4.3 Frame/Message Length
5.4.4 Out-of-Range
5.4.5 Concept of Valid/Invalid Bit/Symbol Detection
5.4.5.1 Invalid Bit Detection
5.4.5.2 Invalid Frame Structure Detection
5.5 Error Response
5.5.1 Transmit
5.5.2 Receive
6. Physical Layer Details
6.1 Media
6.1.1 Single Wire
6.1.2 Dual Wires
6.1.3 Routing
6.2 Unit Load Specifications
6.3 Maximum Number of Nodes
6.4 Maximum Network Length
6.4.1 On-Vehicle / Off-Vehicle
6.5 Media Characteristics
6.6 Data Bit/Symbol Definition/Detection
6.6.1 Pulse Width Modulation (PWM)
6.6.1.1 The One "1" and Zero "0" Bits
6.6.1.2 Start of Frame (SOF)
6.6.1.3 End of Data (EOD)
6.6.1.4 End of Frame (EOF)
6.6.1.5 Inter-Frame Separation (IFS)
6.6.1.6 Break (BRK)
6.6.1.7 Idle Bus (Idle)
6.6.1.8 PWM Symbol Timing Requirements
6.6.2 Variable Pulse Width Modulation
6.6.2.1 The One "1" and Zero "0" Bits
6.6.2.2 Start Of Frame (SOF)
6.6.2.3 End Of Data (EOD)
6.6.2.4 End of Frame (EOF)
6.6.2.5 Inter-Byte Separation (IBS) Symbol
6.6.2.6 In-Frame Response Byte(s)/Normalization Bit
6.6.2.7 Inter-Frame Separation (IFS)
6.6.2.8 Break (BRK)
6.6.2.9 Idle Bus (Idle)
6.6.2.10 VPW Symbol Timing Requirements
6.7 Contention/Arbitration/Priority
6.7.1 Contention Definition
6.7.2 Contention Detection
6.7.3 Bit-by-Bit Arbitration
6.7.4 Arbitration Area
6.7.5 Frame Priority
6.8 Node Wake-Up Via Physical Layer
6.8.1 Network Media
6.8.2 Individual Nodes
6.8.3 Sleep State
6.8.3.1 Sleep State Exit (Wake-Up)
6.9 Physical Layer Fault Considerations
6.9.1 Failure Modes
6.10 EMC Requirements

6.10.1 Electromagnetic Compatibility (EMC)
7. Parameters
7.1 Application Layer
7.1.1 Wake-Up Requirements
7.1.2 Priority
7.2 Data Link Layer
7.2.1 Pulse Width Modulation (PWM) at 41.6 Kbps
7.2.2 Variable Pulse Width (VPW) at 10.4 Kbps
7.3 Physical Layer
7.3.1 General Network Requirements
7.3.2 Pulse Width Modulation (PWM)
7.3.2.1 PWM Timing Requirements
7.3.2.2 PWM DC Parameters
7.3.3 Variable Pulse Width (VPW) Modulation (See Appendix A for derivation)
7.3.3.1 VPW Timing Requirements
7.3.3.2 VPW DC Parameters
APPENDIX A.
APPENDIX B.
APPENDIX C.

1. Scope—This SAE Recommended Practice establishes the requirements for a Class B Data Communication Network Interface applicable to all On and Off-Road Land Based Vehicles. It defines a minimum set of data communication requirements such that the resulting network is cost effective for simple applications and flexible enough to use in complex applications. Taken in total, the requirements contained in this document specify a data communications network that satisfies the needs of automotive manufacturers.

This specification describes two specific implementations of the network, based on media/Physical Layer differences. One Physical Layer is optimized for a data rate of 10.4 Kbps while the other Physical Layer is optimized for a data rate of 41.6 Kbps. Additionally, this document outlines one Physical Layer alternative to the two fully defined implementations that may allow the network to operate at a 125 Kbps data rate.

Although devices may be constructed that can be configured to operate in either of the two primary implementations defined herein, it is expected that most manufacturers will focus specifically on either the 10.4 Kbps implementation or the 41.6 Kbps implementation depending on their specific application and corporate philosophy toward network usage. However, low volume users of network interface devices are expected to find it more effective to use a generic interface capable of handling either of the primary implementations specified in this document.

While two implementations are fully characterized here (i.e., 10.4 Kbps and 41.6 Kbps), one other implementation is being pursued and will be proposed in a future version of this document. This implementation may be generally characterized as 125 Kbps, with possibly different bit encoding, drive type, media, and redundancy characteristics than the current two implementations.

This SAE document is under the control and maintenance of the Vehicle Networks for Multiplexing and Data Communications (Multiplex) Committee. This committee will periodically review and update this document as needs dictate.

2. References and Related Documents
2.1 Applicable Documents—The following publications form a part of this specification to the extent specified herein. The latest issue of SAE publications shall apply.
2.1.1 SAE Publications—Available from SAE, 400 Commonwealth Drive, Warrendale, PA 15096-0001.

SAE J1113—Electromagnetic Susceptibility Measurements Procedures for Vehicle Components
SAE J1211—Recommended Environmental Procedure for Electronic Equipment Design
SAE J1213/1—Glossary of Vehicle Networks for Multiplexing and Data Communications
SAE J1879—General Qualification and Production Acceptance Criteria for Integrated Circuits in Automotive Applications
SAE J1962—Diagnostic Connector
SAE J1979—E/E Diagnostic Test Modes
SAE J2012—Diagnostic Codes/Messages
SAE J2178—Class B Data Communication Network Messages
SAE J2190—Enhanced E/E Diagnostic Test Modes

2.1.2 ISO Documents—Available from ANSI, 11 West 42nd Street, New York, NY 10036-8002.

ISO/TC22/SC3/WG1 N429E, OCT. 90—Road vehicles—Serial data communication for automotive applications, Low speed (125 Kbps and below)
ISO 7498—Data processing systems—Open systems interconnection—Standard reference model

2.1.3 CISPR Documents—Available from ANSI, 11 West 42nd Street, New York, NY 10036-8002.

CISPR/D/WG2 (19 Sept 1989)—Radiated Emissions Antenna and Probe Test

2.2 Related Publications—The following publications are provided for information purposes only and are not a required part of this document.
2.2.1 SAE Publications—Available from SAE, 400 Commonwealth Drive, Warrendale, PA 15096-0001.

SAE J1547—Electromagnetic Susceptibility Measurement Procedures for Common Mode Injection
SAE J1587—Joint SAE/TMC Electronic Data Interchange Between Microcomputer Systems In Heavy Duty Vehicle Applications
SAE J1930—Electrical/Electronic Systems Diagnostic Terms, Definitions, Abbreviations, & Acronyms
SAE J1978—OBD II Scan Tool
SAE J2008—Recommended Organization of Vehicle Service Information
SAE J2201—Universal Interface for OBD II Scan Tool
SAE J2205—Expanded Diagnostic Protocol for OBD II Scan Tools

2.3 Definitions and Abbreviations
2.3.1 Definitions
2.3.1.1 *Arbitration*—The process of resolving which frame, or In-Frame Response data, continues to be transmitted when two or more nodes begin transmitting frames, or In-Frame Response data, simultaneously.

2.3.1.2 *Class A Data Communications*—A system whereby vehicle wiring is reduced by the transmission and reception of multiple signals over the same signal bus between nodes that would have been accomplished by individual wires in a conventionally wired vehicle. The nodes used to accomplish multiplexed body wiring typically did not exist in the same or similar form in a conventionally wired vehicle.

2.3.1.3 *Class B Data Communications*—A system whereby data (e.g., parametric data values) is transferred between nodes to eliminate redundant sensors and other system elements. The nodes in this form of a multiplex system typically already existed as stand-alone modules in a conventionally wired vehicle. A Class B network shall also be capable of performing Class A functions.

2.3.1.4 *Class C Data Communications*—A system whereby high data rate signals typically associated with real time control systems, such as engine controls and anti-lock brakes, are sent over the signal bus to facilitate distributed control and to further reduce vehicle wiring. A Class C network shall also be capable of performing Class A and Class B functions.

2.3.1.5 *Dual Wire*—Two wires that are routed adjacently throughout the network and can be either a twisted or a parallel pair of wires.

2.3.1.6 *Fault Tolerance*—The ability of a system to survive a certain number of failures with allowance for possible down-graded performance while maintaining message transmission capability at the specified data rate.

2.3.1.7 *Frame*—One complete transmission of information, which may or may not include an "in-frame response." For this network, each frame contains one and only one message. A frame is delineated by the Start of Frame (SOF) and End of Frame (EOF) symbols.

2.3.1.8 *Functional Addressing*—Labeling of messages based on their operation code or data content.

2.3.1.9 *Message*—All of the data bytes contained in a frame. The message is what is left after the frame symbols have been removed from the frame. As such, the message is the sequence of bytes contained in the frame.

2.3.1.10 *Physical Addressing*—Labeling of messages for the physical location of their source and/or destination(s).

2.3.1.11 *Pulse Width Modulation (PWM)*—A data bit format, where the width of a pulse of constant voltage or current determines the value (typically one or zero) of the data transmitted.

2.3.1.12 *Sleep-Mode*—Node behavior in a low power consumption standby state waiting to be switched on by a frame or other activity. This is distinct from an off mode where there is no power consumption, disconnected from the power supply.

2.3.1.13 *Variable Pulse Width (VPW) Modulation*—A method of using both the state of the bus and the width of the pulse to encode bit information. This encoding technique is used to reduce the number of bus transitions for a given bit rate. One embodiment would define a "ONE"

(1) as a short dominant pulse or a long passive pulse while a "ZERO" (0) would be defined as a long dominant pulse or a short passive pulse. Since a frame is comprised of random 1's and 0's, general byte or frame times cannot be predicted in advance.

2.3.2 Abbreviations/Acronyms
BRK – Break
CRC – Cyclic Redundancy Check
CS – Checksum
E/E – Electrical and Electronic
EMC – Electromagnetic Compatibility
EMI – Electromagnetic Interference
EOD – End of Data
EOF – End of Frame
ERR – Error Detection Byte (CRC or CS)
IBS – Inter-Byte Separation
IFR – In-Frame Response (Byte/Bytes)
IFS – Inter-Frame Separation
ISO – International Standards Organization
Kbps – Kilo bits per second
NA – Not Applicable
NB – Normalization Bit
OSI – Open System Interconnect
SOF – Start of Frame

3. Description of the Architecture

3.1 General – It is the intent of this network to interconnect different electronic modules on the vehicle using an "Open Architecture" approach. An open architecture network is one in which the addition or deletion of one or more modules (data nodes) has minimal hardware and/or software impact on the remaining modules.

In order to support an open architecture approach, the Class B network utilizes the concept of Carrier Sense Multiple Access (CSMA) with nondestructive contention resolution. Additionally this network supports the prioritization of frames such that, in the case of contention, the higher priority frames will always win arbitration and be completed.

The architecture of this network is particularly well suited for nonperiodic, event driven, broadcast frames.

3.2 Network Topology

3.2.1 DATA BUS TOPOLOGY – Data bus topology is the map of physical connections of the data bus nodes to the data bus. It includes all nodes and data buses involved in the data bus integration of the vehicle.

A single-level bus topology, the simplest bus topology, is currently being used in several automotive applications. In a single-level bus topology, all nodes are interconnected via the same data bus. The redundancy requirements of a particular application may require a single-level topology to be implemented using multiple interconnecting cables operating in various modes (active or passive). However, the requirement to use multiple buses for redundancy purposes does not change the single-level bus topology definition if the following criteria are maintained:

a. All nodes/devices transmit and receive from a single path
b. All nodes/devices receive all frames at the same time
c. Communication on each data bus is identical

3.2.2 DATA BUS CONTROL – Although various methods of data bus control can be used, this Class B network is intended for "masterless" bus control.

The principal advantage of the masterless bus control concept is its ability to provide the basis for an open architecture data communications system. The masterless bus control concept is ideally suited for data that is characterized as: non-periodic and event driven. Since a master does not exist, each node has an equal opportunity to initiate a data transmission once an idle bus has been detected. However, not all nodes and/or data are of equal importance, prioritization of frames is allowed and the highest priority frame will always be completed. This also implies that frame/data contention will not result in lost data. Two disadvantages of the masterless bus concept are that data latency cannot be guaranteed, except for the single highest system priority frame, and bus utilization extremes are difficult to evaluate.

3.3 References to the OSI Model – Although this document focuses on the data link layer and the physical layer, references are included for the application layer since this needs to be included for emission related, diagnostic communication legislation requirements. The Class B network maps into the OSI model as described in the following paragraphs. This "mapping" is illustrated in Figure 1.

3.3.1 APPLICATION LAYER – At the top of the OSI reference model is the Application Layer. This layer establishes the relationship between the various application input and output devices, including what is expected of human operators. This layer documents the high level description of the function including control algorithms if appropriate. An example of an Application Layer functional description might be; "Pressing the head lamp button shall cause the low beam head lamp, marker, and tail lamp filaments to be energized." Legislated diagnostics is another area in which application layer requirements need to be specified.

3.3.2 DATA LINK LAYER – The primary function of the Data Link Layer is to convert bits and/or symbols to validated error free frames/data. Typical services provided are serialization (parallel to serial conversion) and clock recovery or bit synchronization. An important additional service provided by the Data Link Layer is error checking. When errors are detected, they may be corrected or higher layers may be notified.

3.3.3 PHYSICAL LAYER – The Physical Layer and its associated wiring form the interconnecting path for information transfer between Data Link Layers. Typical Physical Layer protocol elements include voltage/current levels, media impedance, and bit/symbol definition and timing.

3.4 Network Implementation – The network implementations based on this document have been reduced to commonize hardware, software, messages, and tools. The consolidation of messages has been documented in SAE J2178. The first byte or the first three bytes of these messages are called the "Header" byte(s). These header bytes fully define the associated requirements of this network interface, which previously had been optional. Figure 2 shows the general format for single byte header forms. Figure 3 shows the three byte header form. Figure 4 shows the specific bit assignments for priority, In-Frame Response, and Functional / Physical Address mapping in the three byte header format.

The combinations shown in Table 1 are the available sets which comply with this specification.

The KYZZ bits shown in Figure 4 have been further defined in Table 2. An SAE J1850 compatible network must use these definitions but may or may not include use or support of the full set of these defined headers. That is, a system designer may restrict his system to only support some of these forms, but those that are used must conform with these definitions.

4. Application Layer Details

The application of this communication network is the transfer of information from one node of the network to one or more other nodes. This transfer of information supports both operational and diagnostic needs. SAE has developed documents describing each of these types of applications, consistent with this document.

4.1 Normal Vehicle Operation (Down the Road) Messages – The messages sent during nondiagnostic operations are called normal vehicle operation messages. These normal vehicle operation messages are used for communication from a transmitter to one or more receivers across this network. The normal operation messages have been developed by the SAE for this communication network and are defined in SAE J2178. SAE defined messages and the "Reserved" messages of SAE J2178 shall remain specific to those definitions. In SAE J2178, there is also a set of "Reserved – Manufacturer" messages which, if used, will have meanings specific to a vehicle manufacturer but are likely to be different between manufacturers.

4.2 Diagnostic Messages – It is expected that this network will be used for diagnostics of the devices utilizing the network. These diagnostic procedures may include legislated diagnostics, industry standard diagnostics, or manufacturer specific diagnostic procedures.

Legislated diagnostics, and some level of voluntary industry standard diagnostics, that reference this document, should only specify procedures and frames that conform to this document. SAE J1979 and SAE J2190 define the set of recognized test modes that are available and have been reserved for diagnostic purposes.

Manufacturer specific test procedures utilizing this network may specify procedures that do not conform to the requirements of this document.

4.2.1 DIAGNOSTIC PARAMETRIC DATA – SAE J1979 and SAE J2190 define test modes and frame formats for use by off-vehicle test equipment to obtain diagnostic data from the vehicle. SAE J1979 and SAE J2190 messages conform to the requirements and limitations of this document.

4.2.2 DIAGNOSTIC MALFUNCTION CODES – SAE J2012 defines trouble codes to be assigned to various vehicle system malfunctions, and also assigns ranges of codes to be used for manufacturer specific codes. SAE J1979 and SAE J2190 include messages to be used to retrieve these codes from the on-vehicle systems. When trouble codes are to be assigned to system malfunctions, the code structure of SAE J2012 should be used.

4.3 Frame Filtering – The network interface device may be capable of filtering frames on the network to select those appropriate to a given node. Because this Class B protocol may use more than one type of frame addressing (e.g., functional and physical; see 5.1), the criteria for these filtering operations may include multiple byte comparisons occurring over the first several frame bytes. Regardless of the exact technique

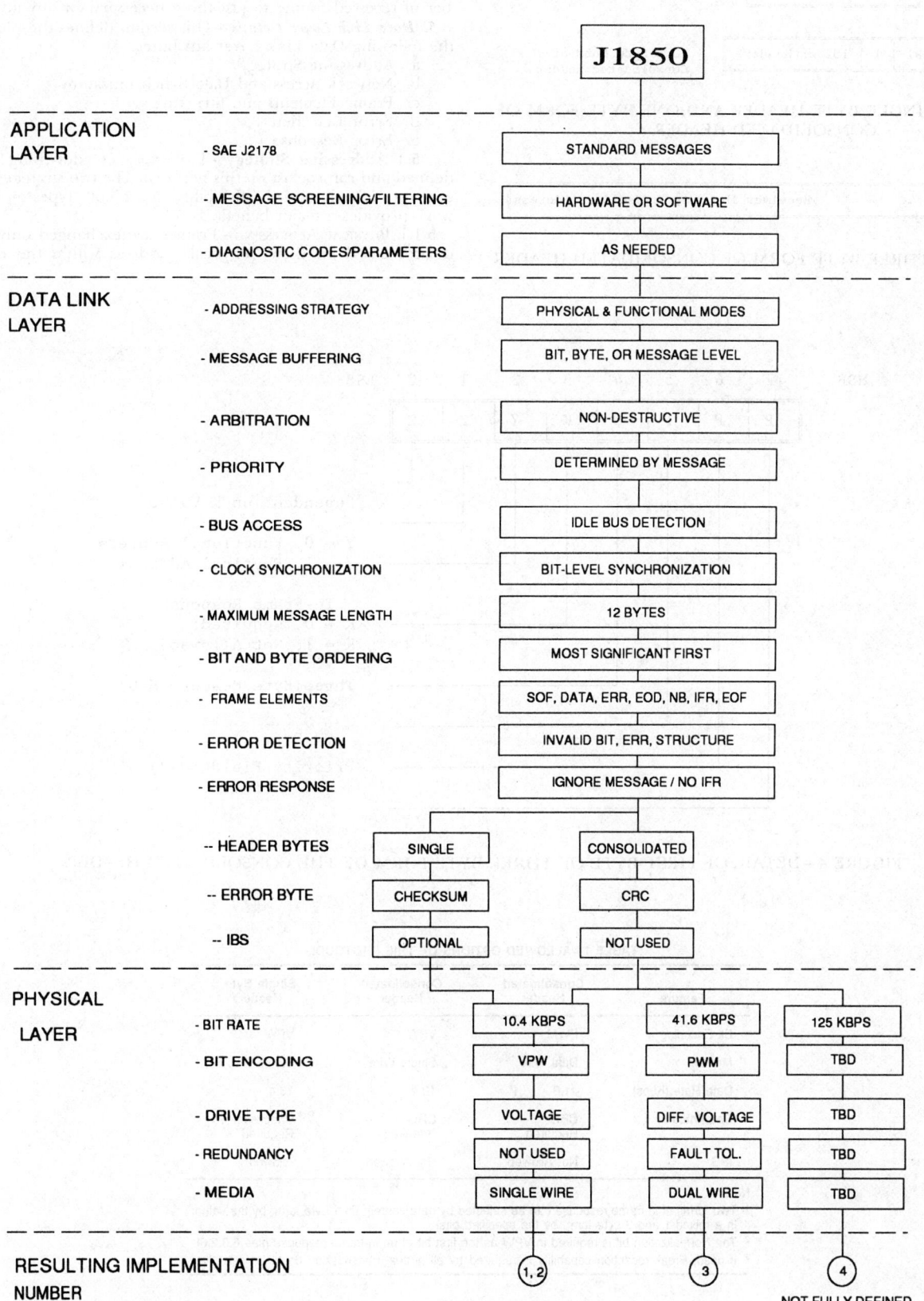

FIGURE 1 — MAP OF SAE J1850 TO THE ISO OSI MODEL

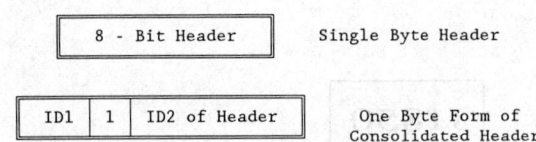

FIGURE 2 – SINGLE BYTE HEADER AND ONE BYTE FORM OF CONSOLIDATED HEADER

| Fig. 4 | Receiver ID | Transmitter Address |

FIGURE 3 – THREE BYTE FORM OF CONSOLIDATED HEADER

used for frame filtering, the objective is to reduce the software and processing burden associated with network operations by limiting the number of received frames to just those necessary for any given node.

5. Data Link Layer Details – This section defines the requirements on the following Data Link Layer attributes:
  a. Addressing Strategy
  b. Network Access and Data Synchronization
  c. Frame Elements and Structure
  d. Error Detection
  e. Error Response

5.1 Addressing Strategy – Two types of addressing strategies are defined and can coexist on this network. The two strategies serve different types of tasks and the flexibility to use both types on the same network provides a major benefit.

5.1.1 PHYSICAL ADDRESSING – Frames are exchanged only between two devices based on their "Physical" Address within the network. Each

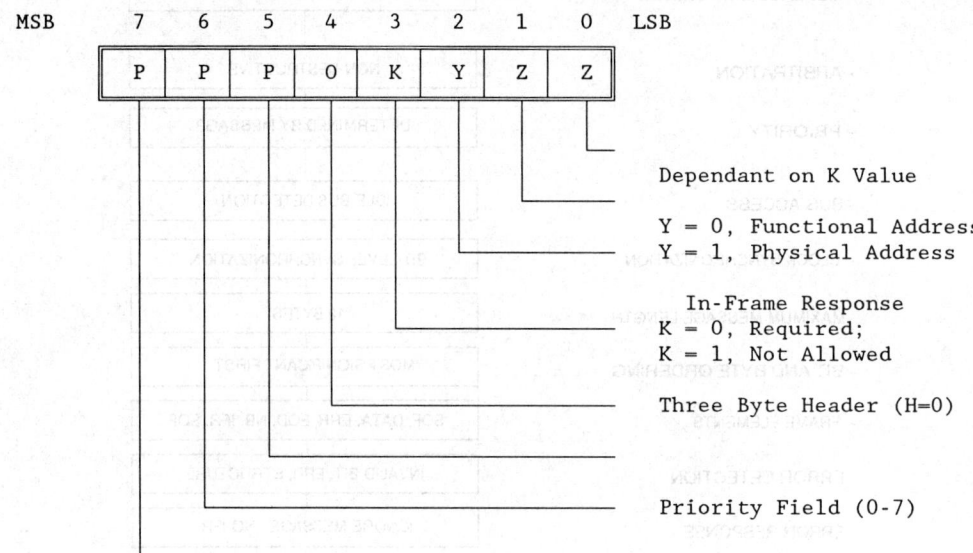

FIGURE 4 – DETAIL OF FIRST BYTE OF THREE BYTE FORM OF THE CONSOLIDATED HEADER

TABLE 1 – ALLOWED OPTIONS OF THIS PROTOCOL

Feature	Consolidated Header	Consolidated Header	Single Byte Header
Bit Encoding	PWM	VPW	VPW
Medium	Dual Wire	Single Wire	Single Wire
Data Rate (Kbps)	41.6	10.4	10.4
Data Integrity	CRC Required	CRC Required	Checksum Required
IBS	Not Allowed	Not Allowed	Allowed

NOTES:
1. Two forms of in-frame response can be selected by data content (in 1 byte form, by the value; in a mixed 1 and 3 byte form, by the specified bits).
2. The Normalization bit is required in VPW as the first bit of an in-frame response (see 6.6.2.6).
3. If used, Break reception capability is required by all nodes, transmission could be selective.

TABLE 2 – LOWER 4 BITS OF FIRST BYTE OF THE 3 BYTE FORM OF CONSOLIDATED HEADER

Lower 4 Bits	In-Frame Response	Addressing	Description
			Response Type
0 0 0 0	Required	Functional	1 Byte Multi Node
0 0 0 1	Required	Functional	1 Byte 1 Node
0 0 1 0	Required	Functional	1 Byte Multi Node (Function No-Op)
0 0 1 1	Required	Functional	Multi Byte 1 Node
0 1 0 0	Required	Physical	1 Byte 1 Node
0 1 0 1	Required	Physical	Multi Byte 1 Node
0 1 1 0	Required	Physical	Reserved – SAE
0 1 1 1	Required	Physical	Reserved – SAE
			Message Type
1 0 0 0	Not Allowed	Functional	Functional Command/Status
1 0 0 1	Not Allowed	Functional	Functional Request
1 0 1 0	Not Allowed	Functional	Function Query
1 0 1 1	Not Allowed	Functional	Reserved – SAE
1 1 0 0	Not Allowed	Physical	Node-to-Node
1 1 0 1	Not Allowed	Physical	BLOCK Transfer[1]
1 1 1 0	Not Allowed	Physical	Acknowledgement
1 1 1 1	Not Allowed	Physical	Reserved – MFG.

[1] BLOCK Transfer extends the normal maximum message length requirement.

node must be assigned a unique physical address within the network. This type of addressing strategy is used when the communications involve specific nodes and not the others that may be on the network. Diagnostic access would be one case where identification of a specific module is important.

5.1.2 FUNCTIONAL ADDRESSING – Frames can be transmitted between many devices based on the function of that frame on the network. Each node is assigned the set of functions that it cares about, either as transmitter or receiver, and can be located anywhere in the network. This type of addressing strategy is used when the physical location of the function is not important but could move around from one module to another. In the case of functional addressing, the function of the message is important and not the physical addresses of the nodes.

5.2 Network Access and Data Synchronization – The network interface shall implement a multiple access arbitration based protocol using nondestructive bit-by-bit arbitration to transparently resolve simultaneous access to the bus. Network access is allowed after detection of an idle bus. The definition of an idle bus is contained in 6.6.1.7.

Since a discrete clock wire is not used with this network, node synchronization can be derived from bit/symbol transitions on the bus.

5.2.1 FULL MESSAGE BUFFERING – One or more messages exist in their entirety in the interface device. This approach reduces software burden at the expense of hardware costs. Message filtering (or screening) is possible in such a device which reduces software burden even further.

5.2.2 BYTE BUFFERING – Each byte of a received message (or transmit message) is stored individually in the interface device. The controlling device is responsible for the timely servicing of the interface device to keep up with frame traffic.

5.3 Network Elements and Structure – The general format is:
(SOF),(data 0)..(IBS)..(data n),(ERR),(EOD),(NB),(IFR),(EOF),(IFS),(idle):

The preceding acronyms are defined as follows:
(SOF) Start of Frame
(data) Data bytes, each 8 bits long
(IBS) Inter-Byte Separation
(EOD) End of Data
(ERR) Error Byte, either CRC or Checksum
(NB) Normalization Bit
(IFR) In-Frame Response Byte(s)
(EOF) End of Frame
(IFS) Inter-Frame Separation
(idle) Idle Bus

NOTE – Break (BRK) can occur (be sent) on a network at any time. Requirements on the frame and/or its elements are as follows:

5.3.1 FRAME ELEMENTS – The frame elements other than the symbols SOF, IBS, EOD, NB, EOF, and BRK will be byte oriented and must end on byte boundaries. Each byte will be 8 bits in length.

5.3.2 BIT ORDERING – The first bit of each byte transmitted on the network shall be the most significant bit (i.e., MSB first).

5.3.3 MAXIMUM FRAME LENGTH – The maximum number of continuous bit times that a single node is able to control the bus shall not exceed the value specified in 7.2.

5.3.4 FUNCTION OF SOF, EOD, EOF, IFS, IBS, NB, AND BRK – In addition to actual data bytes (i.e., data, ERR, IFR) frame delimiter symbols are defined to allow the data bus to function properly in a multitude of different applications. An overview of these symbols is provided here. Detailed timing requirements on each symbol can be found in 7.3.

5.3.4.1 *Start of Frame (SOF)* – The SOF mark is used to uniquely identify the start of a frame. The SOF mark shall not be used in the calculation of the CRC or checksum error detection code.

5.3.4.2 *End of Data (EOD)* – End of Data (EOD) is used to signal the end of transmission by the originator of a frame. The in-frame response (IFR) section of the frame, if used, begins after the EOD bit but before the EOF. If the "In-Frame Response" feature (see 5.3.7) is not used, then the bus would remain in the passive state for an additional bit time, thereby signifying an "End of Frame."

If a frame includes an In-Frame Response, the originator of the frame will expect the recipient(s) of the frame to drive the network with one or more in-frame response bytes immediately following EOD.

5.3.4.3 *End of Frame (EOF)* – The completion of the EOF defines the end of a frame. After the last transmission byte (including in-frame response bytes where applicable), the bus will be left in a passive state. When EOF has expired, all receivers will consider the transmission complete.

5.3.4.4 *Inter-Frame Separation (IFS)* — Inter-Frame Separation is used to allow proper synchronization of various nodes during back-to-back frame transmissions. A transmitter must not initiate transmission on the bus before the completion of the IFS period. However, receivers must synchronize to any other SOF occurring during the IFS period in order to accommodate individual clock tolerances.

A transmitter that desires bus access must wait for either of two conditions before transmitting a SOF:
 a. EOF and IFS have expired
 b. EOF has expired and another rising edge has been detected

5.3.4.5 *Inter-Byte Separation (IBS)* — Only applicable to 10.4 Kbps implementation — IBS is allowed to accommodate unbuffered interface hardware and/or alleviate critical timing constraints. If used, it must adhere to the requirements as specified in 6.6.2.5.

5.3.4.6 *Normalization Bit (NB)* — Only applicable to 10.4 Kbps implementation — For Variable Pulse Width Modulation, the first bit of In-Frame Response data is also passive and therefore it is necessary to generate a Normalization Bit to follow the EOD symbol. This Normalization Bit shall define the start of the in-frame response. The Normalization Bit is defined in 6.6.2.6.

5.3.4.7 *Break (BRK)* — BRK is allowed to accommodate those situations in which bus communication is to be terminated and all nodes reset to a "ready-to-receive" state. If BRK is used, it must adhere to the requirements as specified in 6.6.

5.3.5 IDLE BUS (IDLE) — Idle bus is defined as any period of passive bus state occurring after IFS. During an idle bus, any node may transmit immediately. Contention may still occur when two or more nodes transmit nearly simultaneously; therefore, resynchronization to rising edges must continue to occur.

5.3.6 DATA BYTE(S). (DATA 0 ... DATA n) — A number of data bytes, each 8 bits in length, can be transmitted at the discretion of the system designer. However, the total message length (from SOF to EOF) shall not exceed the limit defined in 7.2. Typical uses include:
 a. Message header
 b. 8 bit data values
 c. 7 bit data values, 1 bit parity

5.3.7 IN-FRAME RESPONSE — For In-Frame Response, the response byte(s) are transmitted by the responders and begin after EOD.

If the first bit of the in-frame response byte does not occur at this point and the bus remains passive for a period of time defined as EOF, then the originator and all receivers must consider the frame complete.

In-Frame Response Byte(s) (IFR): In-frame response bytes may take one of the following forms (refer to Figure 5):
 a. None

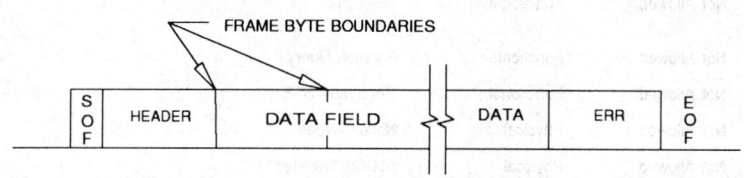

TYPE 0 - NONE

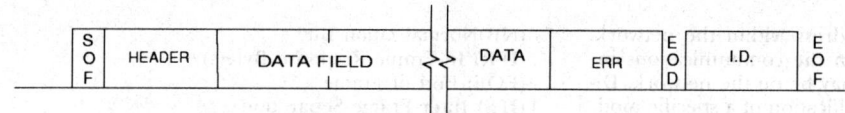

TYPE 1 - SINGLE BYTE FROM A SINGLE RESPONDER

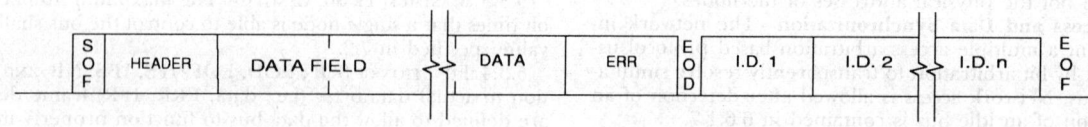

TYPE 2 - SINGLE BYTE FROM MULTIPLE RESPONDERS

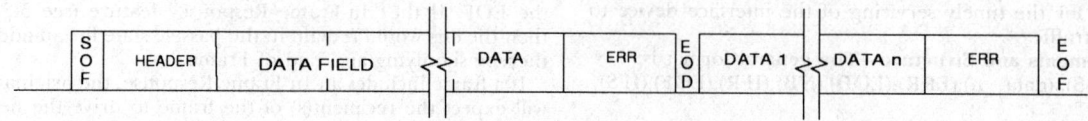

TYPE 3 - MULTIPLE BYTES FROM A SINGLE RESPONDER

FIGURE 5 — FORMS OF IN-FRAME RESPONSE

b. A single byte transmitted from a single recipient, typically a unique identifier (ID) or address.
c. Multiple bytes, each byte being transmitted from a single recipient. The effect is to concatenate the individual response bytes into a response "stream." The response byte from each recipient must be unique, typically a physical address (ID n). Arbitration takes place during the response process so that each recipient, if arbitration is lost during its response byte, will retransmit the single byte until the recipient observes its unique byte in the response stream. Once a given recipient observes its own unique response byte, it discontinues the transmission process to allow any remaining responders to transmit their byte.
d. One or more data bytes, all from a single recipient. A CRC or checksum byte may be appended to the data byte(s). The CRC or checksum byte is calculated as described in 5.4.1 or 5.4.2, respectively, except only the data in the response is used for the CRC or checksum calculation.

If in-frame response bytes are used, the overall frame/message length limit remains in effect. The sum total of data bytes, CRC or checksum bytes, and in-frame response bytes shall not exceed the frame length as specified in 7.2.

5.3.7.1 *Normalization Bit*—If the In-Frame Response is employed in the 10.4 Kbps implementation then a "Normalization Bit" is required. The Normalization Bit is described in 6.6.2.6.

5.4 Error Detection—Two forms of data integrity error detection have been defined, CRC and Checksum. The required form of error detection is associated with the header byte system used. The method of calculating and checking the error (ERR) byte are defined as follows:

5.4.1 CYCLIC REDUNDANCY CHECK (CRC)—The CRC field, if used, shall be calculated using the following method:
a. The CRC calculation and the CRC checker shift registers (or memory locations) will be located in the sender and receiver nodes, respectively, and shall be initially set to the "all ones" state during SOF. (The setting to "ones" prevents an "all zeros" CRC byte with an all zero data stream.)
b. All frame bits that occur after SOF and before the CRC field are used to form the Data Segment Polynomial which is designated as $D(X)$. For any given frame, this number can be interpreted as an "n-bit" binary constant, where n is equal to the frame length, counted in bits.
c. The CRC division polynomial is $X^8 + X^4 + X^3 + X^2 + 1$. This polynomial is designated as $P(X)$.
d. The remainder Polynominal $R(X)$ is determined from the following Modulo 2 division equation:

$$\frac{X^8 * D(X) + X^n + X^{n+1} + ... + X^{n+7}}{P(X)} = Q(X) + \frac{R(X)}{Q(X)} \quad \text{(Eq. 1)}$$

where:
$Q(X)$ is the quotient resulting from the division process.
e. The CRC byte is made equal to $\overline{R(X)}$, where $\overline{R(X)}$ is the ones complement of $R(X)$.
f. The Frame Polynomial $M(X)$ that is transmitted is:
$$M(X) = X^8 * D(X) + \overline{R(X)} \quad \text{(Eq. 2)}$$
g. The receiver checking process shifts the entire received frame, including the transmitted CRC byte, through the CRC checker circuit. An error free frame will always result in the unique constant polynomial of $X^7 + X^6 + X^2$ (C4 hex) in the checker shift register regardless of the frame content.
h. Examples of frames with the appropriate CRC bytes are listed in Table 3:
i. A status flag may be used to indicate the occurrence of a received CRC error.
j. When In-Frame Response data is protected by a CRC field, the previous rules are used to define the CRC, except that the sender and receiver nodes are interchanged. The CRC calculation only includes the in-frame response bytes. (Note that the SOF, EOD, EOF, and NB are not used in the CRC calculation and serve as data delimiters.)

5.4.2 CHECKSUM—The checksum byte, if used, is inserted at the end of the data block. It is defined as the simple 8 bit summation series of all the bytes in the frame, excluding the symbols and the checksum $<CS>$ itself.

If the frame is $SOF,<1><2><3>...<N>,<CS>,EOD$ where $<i>$ ($1 \le i \le N$) is the numeric value of the ith byte of frame.

Then $<CS> = <CS>_N$, where $<CS>_i$ ($i=2$ to N) is defined as $<CS>_i = \{<CS>_{i-1} + <i>\}_{\text{modulo }256}$ and $<CS>_1 = <1>$.

The checksum byte is then appended to the end of the data and is used to verify the received data validity. Examples of frames with the appropriate checksum bytes are listed in Table 3.

All frame bits that occur after SOF and before the CS field (except the IBS bits) are used to form the Checksum value.

When In-Frame Response data is protected by a CS field, the previous rules are used to define the CS, except that the sender and receiver nodes are interchanged. The CS calculation only includes the in-frame response bytes. (Note that the SOF, EOD, EOF, IBS, and NB are not used in the CS calculation and serve as data delimiters.)

5.4.3 FRAME/MESSAGE LENGTH—Frame/message length exceeding its defined limit may also constitute a detected error.

5.4.4 OUT-OF-RANGE—Data is corrupted in a vehicle network when transient interference is large enough to drive the receiver out of its dynamic range of operation. This out-of-range condition, where the receiver can no longer accurately decode the data, may be detected by an out-of-range detector. The following defines the operation of an out-of-range detector:
a. Data is recovered by holding the receiver output in the state it was prior to the out-of-range condition for the duration of the interference.
b. If the interfering transient is short enough and does not cross data state boundaries, accurate data recovery will result. Otherwise, a frame error has been detected.

5.4.5 CONCEPT OF VALID/INVALID BIT/SYMBOL DETECTION

5.4.5.1 *Invalid Bit Detection*—In some cases, data integrity may be increased by detecting the condition in the data stream where the received data bit does not match the specifications for either a "one" or a "zero" bit.

5.4.5.2 *Invalid Frame Structure Detection*—Regardless of data encoding, data integrity may be increased by detecting the condition when an EOD or EOF occurs on a non-byte boundary within the data stream, or the frame exceeds the maximum frame length.

5.5 Error Response

5.5.1 TRANSMIT—When an originator of a frame detects an abnormal situation on the network (e.g., received signal does not match desired transmit signal), the originator must discontinue the transmit operation immediately. After the specified period of EOF and IFS, the originator is allowed to retransmit the frame.

5.5.2 RECEIVE—If a frame is received which contains any error the frame is to be ignored. If "In-Frame Response" is being used, the receiver must not respond to a received frame containing any error. This lack of response serves as the signal to the originator to retransmit the frame.

6. Physical Layer Details—This section defines the requirements on the following physical layer attributes:
a. Media
b. Unit Load Specifications
c. Maximum Number of Nodes
d. Maximum Network Length
e. Media Characteristics
f. Data Bit/Symbol Definition/Detection
g. Network Wake-Up Via Physical Layer
h. Physical Layer Fault Considerations
i. EMC Requirements

Specific parametric values associated with the physical layer attributes are contained in Section 7.

6.1 Media—Physical Network Media—Although this specification focuses on the data carrying media, it is assumed that each node shall be supplied with appropriate power and ground.

6.1.1 SINGLE WIRE—The network medium for the single wire voltage drive shall be a single random lay wire.

6.1.2 DUAL WIRES—The network medium for the dual wire voltage drive shall be either a parallel wire pair separated by a constant distance, or a twisted pair of wires.

TABLE 3—EXAMPLES OF FRAMES WITH APPROPRIATE CRC 2ND CHECKSUM BYTES

Data Bytes (in hex)	CRC Byte (in hex)	Checksum Byte (in hex)
00,00,00,00	59	00
F2,01,83	37	76
0F,AA,00,55	79	0E
00,FF,55,11	B8	65
33,22,55,AA,BB,CC,DD,EE,FF	CB	A5
92,6B,55	8C	52
FF,FF,FF,FF	74	FC

NOTE: Figure 6 illustrates a typical CRC generator and Figure 7 illustrates a typical CRC checker. With appropriate gating, the two circuits may be combined to use only a single shift register for both CRC generation and CRC checking.

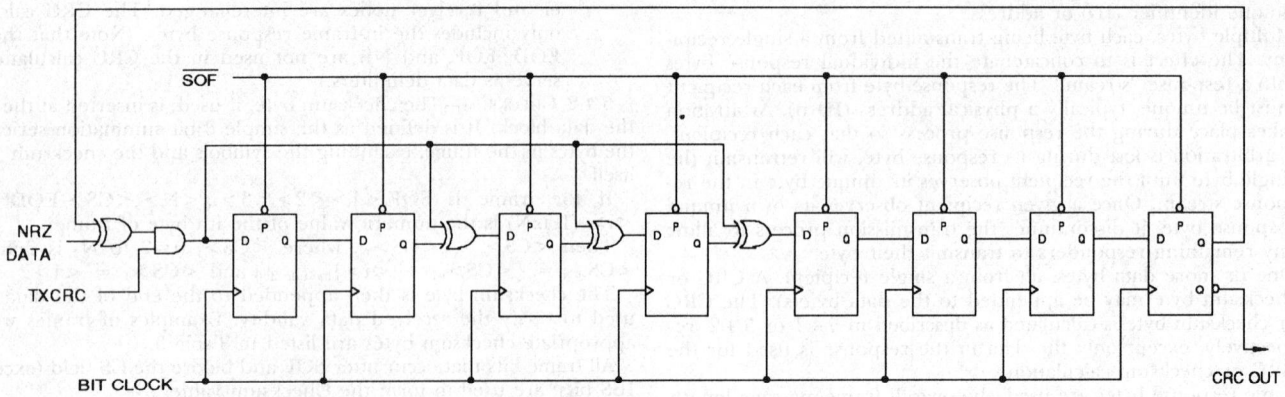

FIGURE 6 — CRC GENERATOR

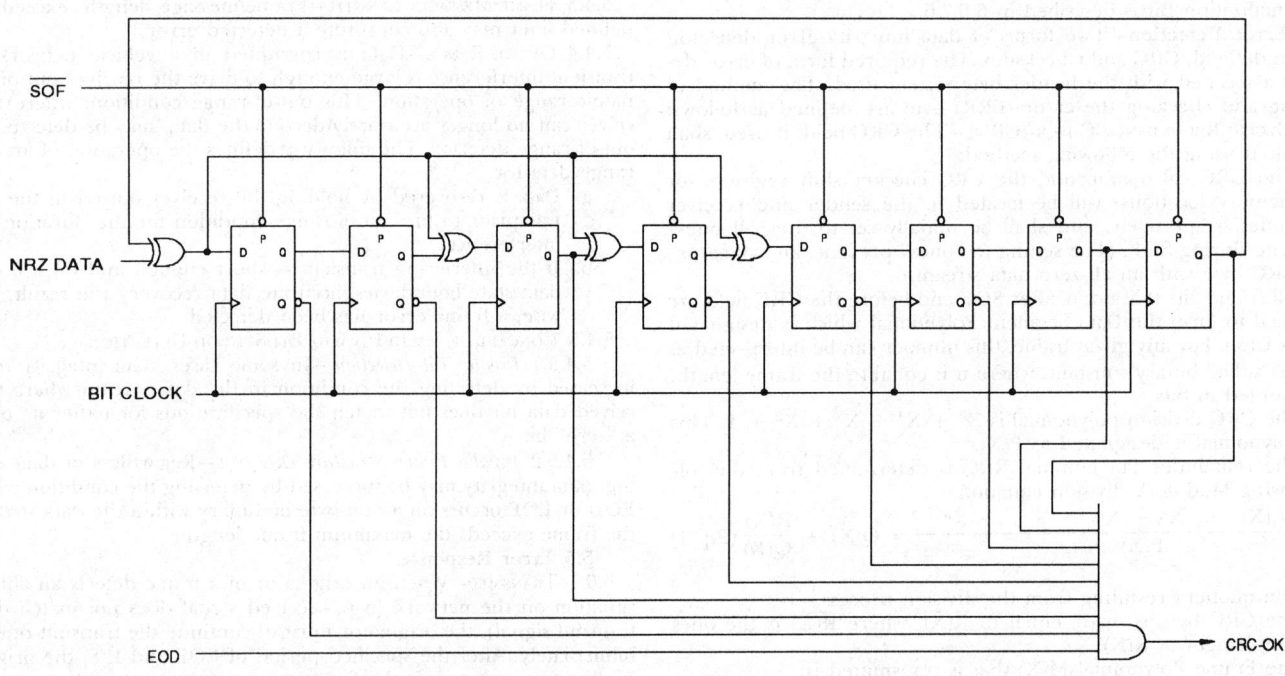

FIGURE 7 — CRC CHECKER

6.1.3 ROUTING — No Restrictions

6.2 Unit Load Specifications — The electrical loading effect of each device connected to this network will be measured in terms of unit loads. A unit load is a nominal value which, if all nodes correspond to one unit load, will allow the maximum specified number of nodes to be connected to the network. There is no requirement that a given node must be equal to a standard unit load, but the combination of all load values must not exceed the limits for any given system.

6.3 Maximum Number of Nodes — The maximum number of nodes, assuming each node is the equivalent of a standard unit load, is specified in Section 7.

6.4 Maximum Network Length — The maximum medium length between any two nodes shall not exceed the value specified in Section 7.

6.4.1 ON-VEHICLE/OFF-VEHICLE — The maximum network length, maximum capacitance value and minimum load/termination resistance values for any off-vehicle equipment have been specified in 7.3. Because all applications must allow for such off-vehicle equipment, the allowed maximum on-vehicle loads shall be limited to account for this level of off-vehicle loading.

6.5 Media Characteristics — The characteristics of the media are as specified in Section 7.

6.6 Data Bit/Symbol Definition/Detection — The data bus can be in one of two valid states, dominant or passive. For clarity in the following sections, a rising edge is a transition from the passive to dominant state, and a falling edge is a transition from the dominant to the passive state.

There are two methods of bit encoding specified in this document, Pulse Width Modulation (PWM) and Variable Pulse Width (VPW) modulation. The timing diagrams to follow all represent the requirements for the logical waveform.

It is the transmitter's responsibility to transmit bits/symbols which are valid (that is, meets these specifications). In some contention situations, the transmitter will have to resynchronize to ensure that the falling edge is within specification. The requirements associated with the reception of these bits and symbols (that is, bus receiver) are not stated explicitly in this specification, but are to be derived from transmitter specifications by the module or circuit designer. It is expected that the receiver will employ a simple clock-driven digital filter and digital integrator or majority vote sampling circuit for decoding data and maintaining "clock" synchronization. All timing requirements are specified in Section 7.

The following bits/symbols are defined for both PWM and VPW:
a. One "1" bit
b. Zero "0" bit
c. Start of Frame (SOF)
d. End of Data (EOD)
e. End of Frame (EOF)
f. Inter-Frame Separation (IFS)

Other symbols, Inter-Byte Separation (IBS), and Normalization Bit (NB), are only applicable for VPW implementations and are therefore only defined for VPW.

6.6.1 PULSE WIDTH MODULATION (PWM)—Figure 8 illustrates a detection approach for PWM encoded information and Figure 9 defines an approach for invalid bit detection.

Table 4 shows the interpretation of the bits based on these sample points.

6.6.1.1 The One "1" and Zero "0" Bits (See Figures 10 and 11)
  a. "1" Bit—A "1" bit is characterized by:
    (1) A rising edge that follows the previous rising edge by at least Tp3. Two rising edges shall never be closer than Tp3.
    (2) A falling edge that occurs Tp1 after the rising edge.
  b. "0" Bit—A "0" bit is characterized by:
    (1) A rising edge that follows the previous rising edge by at least Tp3. Two rising edges shall never be closer than Tp3.
    (2) A falling edge that occurs Tp2 after the rising edge.

6.6.1.2 *Start of Frame (SOF)*—The Start of Frame (SOF) mark has the distinct purpose of uniquely determining the start of a frame (see Figure 12). The SOF is characterized by:
  a. A reference rising edge that follows the previous rising edge by at least Tp5.
  b. A falling edge that occurs Tp7 after the reference rising edge.
  c. The rising edge of the first data bit will occur at Tp8 after the reference rising edge.

NOTE—Last bit of a frame may be the last data bit, last CRC bit, or last In-Frame Response bit.

6.6.1.3 *End of Data (EOD)*—End of Data is used to signal the end of transmission by the originator of a frame. The In-Frame Response (IFR) section of the frame, if used, begins immediately after the EOD bit (see Figure 13). If the In-Frame Response feature is not used, then the bus would remain in the passive state for an additional bit time, thereby signifying an End of Frame (EOF).

For In-Frame Response, the response byte(s) are driven by the responders and begin with the rising edge of the first bit of the response, Tp4 after the rising edge of the last bit sent from the originator of the frame.

If the first bit of the response byte does not occur at Tp4, and the bus remains passive for one additional bit time (Tp3), then the originator

TABLE 4—INTERPRETATION OF BITS BASED ON SAMPLE POINTS

SAMPLE 1	SAMPLE 2	SAMPLE 3	Result
Low	Low	Low	Possible EOD, EOF
High	Low	Low	Valid "One" Bit
Low	High	Low	Invalid Bit
High	High	Low	Valid "Zero" Bit
Low	Low	High	Invalid Bit
High	Low	High	Invalid Bit
Low	High	High	Invalid Bit
High	High	High	Possible SOF

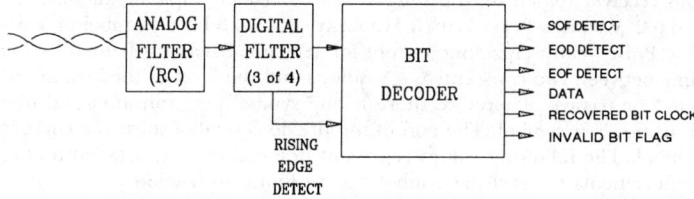

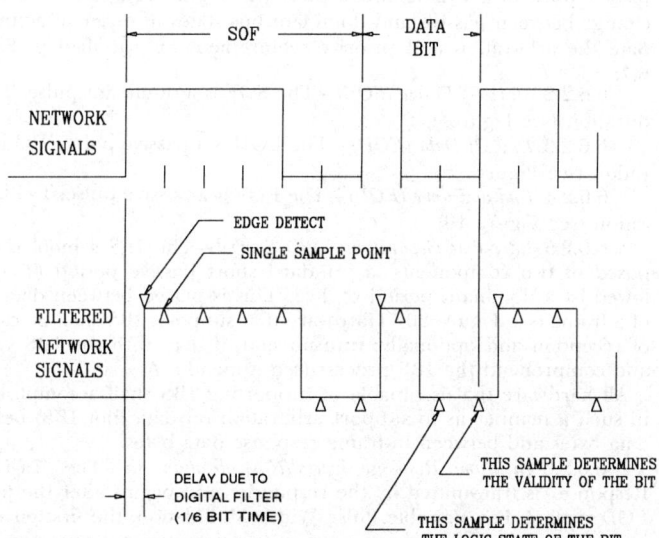

FIGURE 8—PWM BIT DECODER BLOCK DIAGRAM

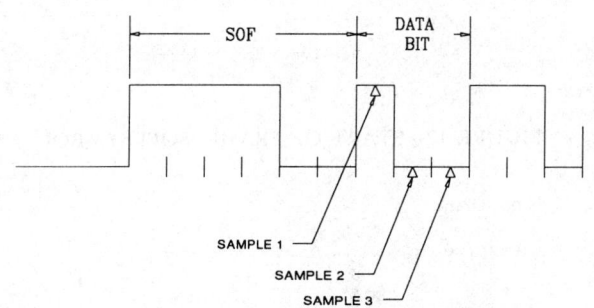

FIGURE 9—INVALID BIT DETECTION

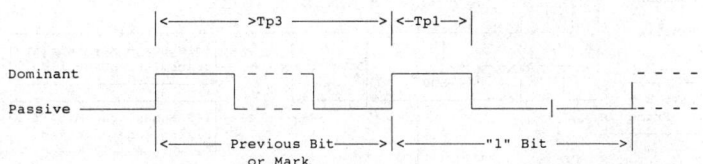

FIGURE 10—"1" BIT DEFINITION

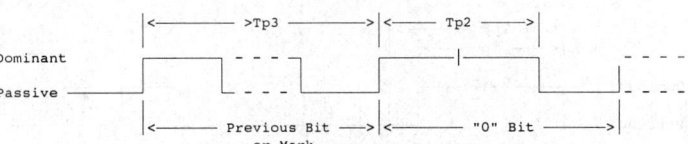

FIGURE 11—"0" BIT DEFINITION

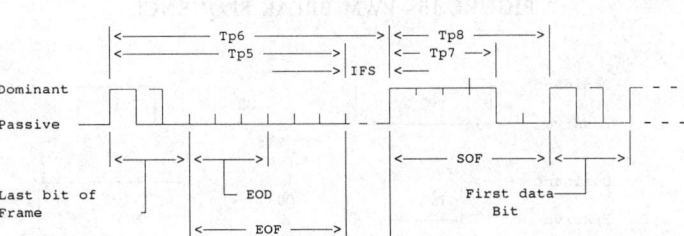

FIGURE 12—FRAME SYMBOLS

and all receivers must consider the frame complete (i.e., EOD has been transformed into an EOF).

6.6.1.4 *End of Frame (EOF)* — The completion of the EOF defines the end of a frame (by definition, an EOD forms the first part of the EOF — see Figure 14). After the last transmission byte (including in-frame response bytes where applicable), the bus will be left in a passive state. When EOF has expired (Tp5), all receivers will consider the transmission complete.

6.6.1.5 *Inter-Frame Separation (IFS)* — Inter-Frame Separation allows proper synchronization of various nodes during back-to-back frame operation.

A transmitter that desires bus access must wait for either of two conditions before transmitting a SOF:

a. EOF and IFS have expired (Last Data Bit + EOF + IFS = Tp6).

b. EOF after Last Data Bit (Tp5) has passed and another rising edge has been detected.

6.6.1.6 *Break (BRK)* — BRK is allowed to accommodate those situations in which bus communication is to be terminated and all nodes reset to a "ready-to-receive" state (see Figure 15). The PWM Break symbol is an extended SOF Symbol and will be detected as an "invalid" symbol to some devices, which will then ignore the current frame, if any. Following the break symbol, an EOF + IFS period (Tp6) is needed to resynchronize the receivers. For the "Breaking" device to obtain control, the highest priority frame must then be sent, otherwise, other frames will be sent under the normal rules of arbitration.

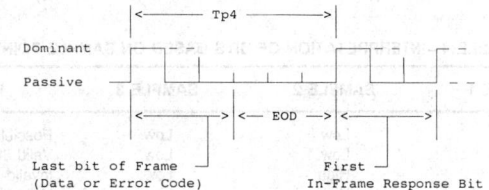

FIGURE 13 — END OF DATA SYMBOL

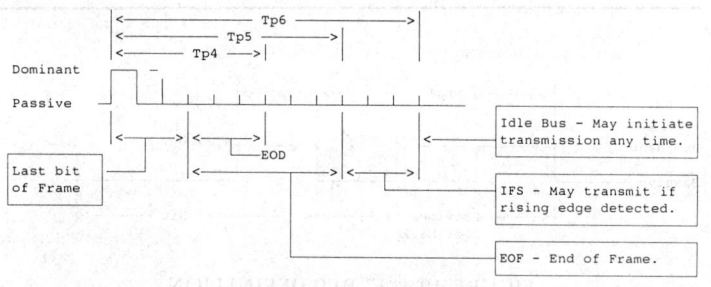

FIGURE 14 — EOF AND IDLE BUS DEFINITION

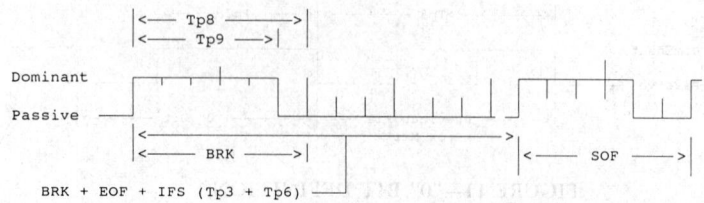

FIGURE 15 — PWM BREAK SEQUENCE

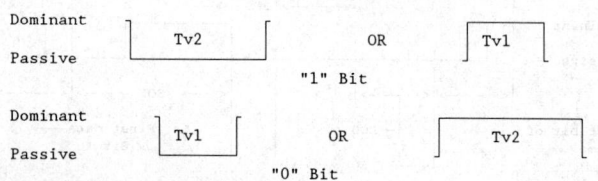

FIGURE 16 — ONE AND ZERO BIT DEFINITIONS

6.6.1.7 *Idle Bus (Idle)* — Idle bus is defined as any period of passive bus state occurring after IFS (see Figure 14). A node may begin transmission at any time during idle bus.

An idle bus will exist after the IFS (Tp6 after last rising edge). During an idle bus, any node may transmit immediately. Contention may still occur when two or more nodes transmit nearly simultaneously; therefore, resynchronization to rising edges must continue to occur.

6.6.1.8 *PWM Symbol Timing Requirements* — The symbol timing reference for PWM encoding is based on transitions from the passive state to the dominant state. The SOF and each data bit in PWM has a "leading edge" from which all subsequent timing is derived. The transition from dominant to passive (which occurs within the SOF or data bits) is not used as a timing reference. The leading edge is used as the only reference because the transition from passive to dominant appears on the bus wires as a fast clean edge while the transition from dominant to passive is slow and ambiguous due to variations in network capacitance.

The timing values specified in Section 7 are based on a "bit rate times 24" clock, which provides a convenient source for the suggested 3 out of 4 digital filter and the 3 phases of the PWM bit encoder/decoder. The "times 24" clock can be derived from 1.00, 2.00, 4.00, etc., MHz oscillators with simple binary dividers. Paragraph 7.3.2.1 defines the timing values for PWM at 41.6 Kbps. Values are provided for the transmitter and receiver (based on the suggested bit decoder implementation).

6.6.2 VARIABLE PULSE WIDTH MODULATION — Each bit or symbol in Variable Pulse Width encoding (except for IBS and Break) is defined by the time between two consecutive transitions and the level of the bus, dominant, or passive. Therefore there is one symbol per transition and one transition per symbol. The end of the previous symbol starts the current symbol. The following values represent nominal timing, detailed timing requirements for each bit/symbol can be found in Section 7.

6.6.2.1 *The One "1" and Zero "0" Bits* — A "1" bit is either a Tv2 passive pulse or a Tv1 dominant pulse. Conversely, a "0" bit is either a Tv1 passive pulse or a Tv2 dominant pulse (see Figure 16). The pulse widths change between passive and dominant bus states in order to accommodate the arbitration and priority requirements as specified in Section 6.7.

6.6.2.2 *Start Of Frame (SOF)* — The SOF is a dominant pulse, Tv3 in duration (see Figure 17).

6.6.2.3 *End Of Data (EOD)* — The EOD is a passive pulse, Tv3 in duration (see Figure 18).

6.6.2.4 *End of Frame (EOF)* — The EOF is a passive pulse, Tv4 in duration (see Figure 19).

6.6.2.5 *Inter-Byte Separation (IBS) Symbol* — The IBS symbol is composed of two components: a standard short passive period (Tv1) followed by a dominant period of Tv5. This is placed between data bytes of a frame (see Figure 20). Hardware that supports IBS shall be capable of reception and optionally, transmission of data with the IBS symbol and comprehend the IBS process (see Appendix A).

All hardware that is capable of supporting IBS shall accomplish IBS in such a manner as to support arbitration on data (not IBS) between data bytes and between in-frame response data bytes.

6.6.2.6 *In-Frame Response Byte(s)/Normalization Bit* — The "In-Frame Response" is transmitted by the responder and begins after the passive EOD symbol. For Variable Pulse Width Modulation, the first bit of the in-frame response data is also passive. It is necessary to generate a Normalization Bit to follow the EOD symbol. The responding device generates the normalization bit prior to sending the IFR data. This Normal-

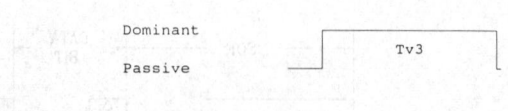

FIGURE 17 — START OF FRAME (SOF) SYMBOL

FIGURE 18 — END OF DATA (EOD) SYMBOL

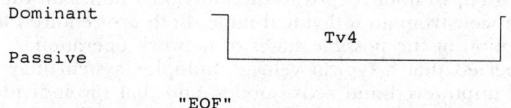

FIGURE 19—END OF FRAME (EOF) SYMBOL

ization Bit defines the start of the in-frame response and can take three forms. The first type is a dominant short period (Tv1). The second type shall be a dominant long period (Tv2) and may be used to define an in-frame response with a CRC or checksum. A third type of Normalization Bit, if used, is a dominant pulse of Tv5 to define an IBS period. Figure 21 illustrates the in-frame response using the normalization bit.

If Inter-Byte Separation is employed, then the in-frame response process shall comprehend the IBS process (see Appendix A).

If in-frame response bytes are used, the overall frame/message length limit remains in effect. The sum total of data bytes, CRC or checksum bytes, and in-frame response bytes shall not exceed the value specified in Section 7.

6.6.2.7 *Inter-Frame Separation (IFS)*—Inter-Frame Separation is used to allow proper synchronization of various nodes during back-to-back frame operation.

A transmitter that desires bus access must wait for either of two conditions before transmitting a SOF:
  a. IFS has expired (Tv6).
  b. At least EOF Min. after Last Data Bit falling edge (Tv4) has passed and another rising edge has been detected.

6.6.2.8 *Break (BRK)*—BRK is allowed to accommodate those situations in which bus communication is to be terminated and all nodes reset to a "ready-to-receive" state (see Figure 22).

6.6.2.9 *Idle Bus (Idle)*—Idle bus is defined as any period of passive bus state occurring after IFS. A node may begin transmission at any time during idle bus.

An idle bus will exist after IFS (Tv6). Contention may still occur when two or more nodes transmit nearly simultaneously; therefore, resynchronization to rising edges must continue to occur.

6.6.2.10 *VPW Symbol Timing Requirements*—The most important factor in symbol timing uncertainty is the edge position uncertainty due to the time taken to make a transition between $V_{ol,max}$ and $V_{oh,min}$ (in either direction). The maximum allowable transition time, $T_{t,max}$ (the area between $V_{ol,max}$ and $V_{oh,min}$), bounds the time span between the earliest and latest node to recognize a transition and is a key design parameter. As $T_{t,max}$ is reduced it becomes more and more difficult to design a driver which also meets the necessary EMI requirements. In this manner, the corners of the waveform, which are a major contributor to the radiated contents of the signal do not fall within the limits imposed by $T_{t,max}$ as found in Section 7.

Other factors that affect the transmitted pulse width are oscillator tolerance and variations in delay through the receiver, noise filter, and driver. For a fixed oscillator and a well designed digital filter, most of the variation is in the time it takes the driver to leave the current state and start the transition. The various factors (plus some guard band and allowance for alternative implementations) can be lumped into a single set of limits, $T_{x,min}$ and $T_{x,max}$, for each symbol. Each is measured from the trip point of the previous transition as seen by the transmitting node (assuming a step input) to the beginning or end of the transition at $V_{oh,min}$ or $V_{ol,max}$.

The pulse width as seen by another node, which may perceive the leading edge either sooner or later than the transmitting node, can range from $T_{x,min} - T_{t,max}$ to $T_{x,max} + T_{t,max}$ in absolute terms. This range represents a receiver's required acceptance range for a given symbol. The actual acceptance range must be wider to allow for receiver oscillator tolerance and any implementation dependent uncertainties or constraints. The necessary guard band limits how close together symbols can be defined. To keep the symbols short, there is no space allocated for an unambiguous (with normal tolerances) forbidden zone between symbols.

The time windows are not affected by multiple nodes trying to transmit at the same time during arbitration. This is because a single node effectively dominates each transition. It is either the first node to leave the passive state or the last node to leave the active state. Although the fastest or slowest node dominates a particular transition, the contention scheme assures that the highest priority frame always wins regardless of which node started transmitting first.

The purpose of the noise filter is to eliminate impulse noise which may exceed the steady-state noise margin and to minimize the need for hysteresis at the receiver input. The delay through the filter is compensated for in the timing logic. A specific implementation of the filter is not required for compatibility so long as the delay is properly compensated and the uncertainty bounds are not exceeded. In the presence of arbitrary noise during the transition period, the filter should respond as though a single step input had occurred at some point within that period.

The symbol limits defined in Section 7 are consistent with $T_{t,max}$ and the specified oscillator tolerance. T_{nom} is the nominal symbol time with no oscillator error and the receiver detecting the transition at V_t (see analysis in Appendix B).

6.7 Contention/Arbitration/Priority—A contention situation arises when more than one node attempts to access the bus at essentially the same time.

6.7.1 CONTENTION DEFINITION—While transmitting its flow of symbols and bits onto the bus, each transmitter monitors the flow of symbols and bits it receives. The bit-by-bit arbitration resolves the conflicting bus signalling requests. In this way a node detects a conflict to its access to the bus from one or more other nodes, a condition called contention. When a node receives a different symbol or bit from the bus than the one it is attempting to transmit, it immediately terminates the transmission of any further symbols or bits for the remainder of that frame. Since all nodes receive all frames, the received data is not lost by the contention resolution.

6.7.2 CONTENTION DETECTION—Contention detection is the recognition of conflicting symbols or bits. The process of bit-by-bit arbitration allows conflicting frame transmissions to be detected. A node that detects a difference between the symbol or bit it receives and the symbol or bit it is currently transmitting, has detected contention to the transmission of its frame. Only the one frame that wins all conflicts of different symbols and bits with all the other nodes that began transmitting during that frame will not detect any contention.

6.7.3 BIT-BY-BIT ARBITRATION—The bit-by-bit arbitration scheme described as follows settles the conflicts that occur when multiple nodes attempt to transmit frames simultaneously. This scheme is applied to each symbol/bit of the frame, starting with the SOF symbol and continuing until the end of the frame.

Bit-by-bit arbitration is based on the use, at the Physical layer, of two values of the bus, called the dominant state and the passive state. All symbols and bits are encoded on the bus by the Physical layer as combinations of dominant and passive state signals. During simultaneous transmission of dominant state and passive state signals on the bus, the resultant state on the bus is always dominant. See Figure 23 showing this operation on the physical layer, based on the two forms of modulation allowed.

As soon as the transmitting node detects a signal state on the bus that

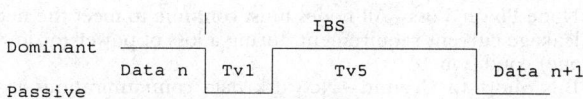

FIGURE 20—INTER-BYTE SEPARATION SYMBOL

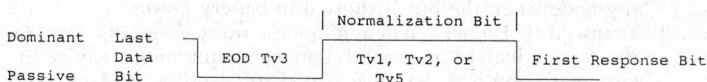

FIGURE 21—NORMALIZATION BIT

FIGURE 22—BREAK SIGNAL

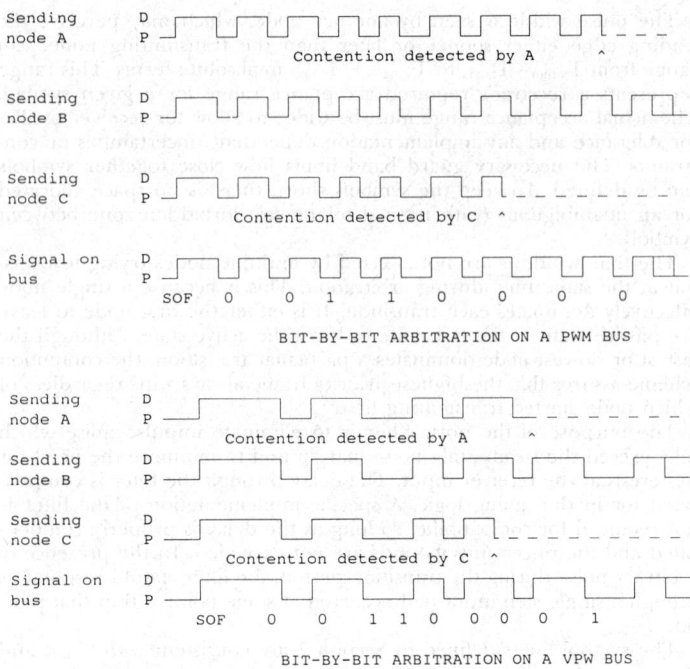

FIGURE 23 — ARBITRATION

is different from the state being transmitted by the node, the node ceases transmitting immediately. Priority is thus granted to nodes sending a dominant state signal over nodes sending a passive state signal.

The arbitration process begins with the SOF and continues throughout the bits of the frames being sent. As each symbol or bit of all the simultaneous frames is transmitted, all of the nodes still transmitting after the last symbol or bit will detect whether any contention is taking place. Nodes that detect contention have, by definition, lost arbitration and will immediately discontinue sending any further symbols or bits. Any remaining nodes will continue to send their symbols/bits until the next contention is detected. The node which obtains sole access to the medium is that which sends its symbols and bits without detecting any contention.

6.7.4 ARBITRATION AREA — The arbitration resolution described previously concerns all the symbols and bits between the SOF and the EOF including, in the case of an in-frame response, all the symbols and bits of the in-frame response. In other words, arbitration acts throughout the frame.

6.7.5 FRAME PRIORITY — The bit-by-bit arbitration mechanism allows the implementation of a frame priority system. When unique frames or, more particularly, when arbitration is resolved down to a single transmitter prior to any data values within the frame, a structure of relative frame priority is inherent.

Should two or more nodes attempt to access the bus within the same frame synchronization window (see IFS in 6.6.1.5 and 6.6.2.7), nondestructive arbitration will occur based on the bit value within each frame. The bus allows two states — dominant and passive. If a dominant state bit and a passive state bit are transmitted on the bus simultaneously, then the dominant state bit will override the passive state bit. Hence, frame arbitration occurs in a bit-wise manner. The result of the arbitration is that the node(s) sending lower priority frame(s) will recognize that they have lost arbitration and will discontinue transmitting before the next bit time. The node transmitting the highest priority frame will continue transmitting uninterrupted.

Based on the bit definitions contained in 6.6, the lowest value bytes immediately following the SOF will have the highest priority. Therefore, regardless of the number of bits allocated for "frame prioritization," the numerical value of zero will have the absolutely highest priority (that is, 000 has more priority than 001 or 111).

6.8 Node Wake-Up Via Physical Layer — The transition to a functional/operating network from an unpowered or standby state is a common occurrence in vehicle multiplex systems. In the context of this document the Session Layer controls the transitions between the operating and the standby states.

Two perspectives are used to define the session layer: (a) the view of the media itself, without regard to the individual nodes on the network, and (b) the view from an individual node. Both are required for a complete definition of the possible states of network operation.

It is expected that a typical vehicle multiplex system may contain a mixture of unpowered and active nodes, and that the individual nodes may themselves have both active and standby states. Therefore this document defines the following possible states for the network media and individual nodes:

6.8.1 NETWORK MEDIA
  a. Unbiased (all conductors at ground voltage, impedance of conductors not controlled)
  b. Biased (all conductors at "passive" voltage levels with appropriate impedances capable of supporting communication)

6.8.2 INDIVIDUAL NODES
  a. Unpowered (module not capable of network communication, nor wake-up from network signal transitions)
  b. Sleeping (an optional low power standby mode capable of detecting network signal transitions for the purpose of wake-up)
  c. Awake/operational (fully capable of receiving and transmitting frames on the network

In the case of an unbiased network no communication is possible. The media must first be brought to the biased state before communication can take place. The transition of the network from the unbiased to the biased state may serve as a node wake-up signal for certain applications, but this is not a requirement (i.e., it is possible for individual nodes to be in the sleep state on a fully biased network or in the awake state on an unbiased network).

In the case of the biased network the individual nodes on the network may exist in any of the three states described previously. Note that nodes which use the optional sleep state are required to wake up from the appropriate network signal within a defined wake-up period. In other words, all nodes which are capable of wake-up from network signals must do so; otherwise these nodes are considered "unpowered." This requirement assures a finite and limited delay to establish communication with nodes which may be in the sleep state.

The time required for the media to make the transition from the unbiased to the biased state is not defined in this document, since this parameter is largely application dependent, and, in fact, may not even occur (e.g., media always biased). For the same reason the time required for a node to make the transition between the unpowered state and the sleep or awake state is also not defined.

The maximum allowable time for a node to make the transition from the sleep state to the awake state is defined in 7.1.1. There is no specified requirement for the transition from the awake state to the sleep state, since this is an implementation specific issue.

6.8.3 SLEEP STATE — With regard to power modes, this document defines two types of nodes. The first type of node is not capable of entering a sleep state and when in the off state, shall not cause a network leakage current in excess of the amount specified in Section 7.

The second type of node is capable of entering a very low current standby mode (sleep state) in which the interface device monitors the communications bus (and hose microprocessor) for a "wake-up" signal. Any interface device in the sleep state can be awakened by any other interface device via the network or by its host via its interface to the hose. The following paragraph defines sleep state exit:

6.8.3.1 *Sleep State Exit (Wake-Up)* — Each device (node) which is capable of separate sleep and wake states shall wake up from the sleep state and become fully functional within the time specified in Section 7 for wake-up. A transition of the bus from the passive state to the dominant state shall be considered a wake-up signal. Proper bus bias must be supplied by all nodes designated to supply bias when the network is active.

6.9 Physical Layer Fault Considerations

6.9.1 FAILURE MODES — The network must meet the requirements as defined per the following failure modes:
  a. Node Power Loss — All nodes must continue to meet the network leakage current requirement during a loss of power (or low voltage) condition.
  b. Bus Short to Ground — Network data communications may be interrupted but there shall be no damage to any node when the bus is shorted to ground.
  c. Bus Short to Battery — Network data communications may be interrupted and it is desirable that there shall be no damage to any node when the bus is shorted to battery power.
  d. Transceiver Failure — When a node's transceiver I/O fails by shorting to battery or ground, data communications may be interrupted, and it is desirable that there shall be no damage to any other node.
  e. Loss of Termination/Bias — Biasing and termination resistors shall be redundant such that no single point failure will cause

the network to become inoperative.
  f. Loss of Connection to Network — When a node becomes disconnected from the network, the remaining nodes shall remain capable of communications.
  g. Loss of Node Connection to Ground — When a node loses its ground connection, the remaining nodes shall remain capable of communications.
  h. Optional Fault Tolerant Operation — Nodes on a dual wire bus may be capable of full communication (sending and receiving frames) at the specified bit rate in the presence of any one of the following faults occurring anywhere on the network:
    (1) Bus "+" wire Open Circuit
    (2) Bus "-" wire Open Circuit
    (3) Bus "+" wire Shorted to Ground
    (4) Bus "-" wire Shorted to Ground
    (5) Bus "+" wire Shorted to Battery + (V_{BATT})
    (6) Bus "-" wire Shorted to Battery + (V_{BATT})
    (7) Bus "+" wire Shorted to any voltage between Ground and Battery
    (8) Bus "-" wire Shorted to any voltage between Ground and Battery

Noise immunity and emissions may be somewhat degraded during the faulted period but shall return to normal after the fault is removed. All nodes shall be protected from damage during the presence of these faults such that recovery to normal operation occurs automatically upon removal of the fault (i.e., faults shall not propagate through the network).

Continued communication is not required for the "Double Fault" condition of the Bus "+" wire shorted to the Bus "-" wire.

6.10 EMC Requirements — The vehicle manufacturer shall specify a minimum EMC level of operation for a module that utilizes this network interface device. A philosophical method for classification of functional performance can be found in SAE J1113 Part 1 Appendix B (formerly SAE J1113 Appendix B). A component manufacturer that designs the interface device to conform to the most severe classification of Class C Region I, for example, could be assured of adequate performance for all conditions.

The modules that communicate with each other via this interface device shall not generate levels of noise emissions (EMI) large enough to interfere with each other. For reference the levels defined by CISPR/D/WG2 (Secretariat) 19 Sept 1989 may be used. In general the specifications on wave shaping and transition rise and fall times control the EMI levels.

The vehicle manufacturers may find SAE J1211 helpful in specifying their multiplex systems. SAE J1879 also may be utilized for specifying components.

6.10.1 ELECTROMAGNETIC COMPATIBILITY (EMC) — This document recommends the vehicle manufacturers use the following documents and the outlined EMC test plan defined in Appendix C. EMC and voltage "level" requirements specified by the Test Plan are given for reference only and the vehicle manufacturer specifications shall be used for compliance testing.
  a. Radiated Emissions Antenna & Probe Test [CISPR/D/WG2 (Secretariat) 19 Sept 1989]
  b. Transfer Function, Current probe monitoring (SAE J1113 Part 2)
  c. Transfer Function, Antenna monitoring (SAE J1113 Part 3)
  d. RF Susceptibility (SAE J1113 Part 13)
  e. Transient Susceptibility (SAE J1113 Part 10)

7. Parameters

7.1 Application Layer — The following Application Layer requirements shall exist whenever the network is used.

7.1.1 WAKE-UP REQUIREMENTS — The maximum amount of time from the wake-up stimulus until the node is capable of communicating is 1.0 ms.

7.1.2 PRIORITY — With regard to priority, a "0" dominates over a "1".

7.2 Data Link Layer

7.2.1 PULSE WIDTH MODULATION (PWM) AT 41.6 KBPS
  a. Maximum Frame Length (Time) (From SOF to EOF inclusive) — 101 bit times
  b. Maximum Frame Length (Excluding SOF, EOD, EOF) — 12 Bytes
  c. Inter-Frame Separation (IFS) minimum — 1 bit time

7.2.2 VARIABLE PULSE WIDTH (VPW) AT 10.4 KBPS
  a. Maximum Frame Length (Excluding SOF, EOD, EOF, and NB; but including IBS periods — 12 Bytes
  b. Inter-Frame Separation (IFS) minimum — TV6 (see 7.3.3.1)
  c. Inter-Byte Separation (IBS) Range — Tv1 + Tv5 (see 7.3.3.1)

7.3 Physical Layer

7.3.1 GENERAL NETWORK REQUIREMENTS
  a. Maximum Network Length — 40 m
    (1) On-Vehicle — 35 m
    (2) Off-Vehicle — 5 m
  b. Maximum Number of Standard Unit Loads — 32 nodes (including Off-Vehicle Equipment)
    (1) Off-Vehicle Load Resistance — 10 600 Ω min
    (2) Off-Vehicle Capacitance (Each Bus Wire to Signal Ground or to Chassis Ground, as measured at the SAE J1962 connector) — 500 pF max

7.3.2 PULSE WIDTH MODULATION (PWM)

7.3.2.1 *PWM Timing Requirements (Note all times in microseconds)* — The following requirements, Table 5, are for a nominal bit time of 24 μs or 41.6 Kbps:

TABLE 5 — PWM TIMING REQUIREMENTS

Symbol	Tx,nom	Tx,min	Tx,max	Rx,min	Rx,max
Tp1	7.00	6.86	7.14	≥4.00	≤12.00
Tp2	15.00	14.70	15.30	>12.00	≤20.00
Tp3	24.00	23.52	24.48	>20.00	≤28.00
Tp4	48.00	47.04	48.96	>44.00	≤52.00
Tp5	72.00	70.56	NA	≥68.00	≤76.00
Tp6	96.00	94.08	NA	NA	NA
Tp7	31.00	30.38	31.62	>28.00	≤36.00
Tp8	48.00	47.04	48.96	>44.00	≤52.00
Tp9	39.00	38.22	39.78	>36.00	≤44.00

7.3.2.2 *PWM DC Parameters*
Passive state (diff input V) — 0 — hysteresis
Dominant state (diff input V) — 0 + hysteresis
Hysteresis (diff input V) — 200 mV max
Dominant state current — 50 mA max
Driver voltage drop — 0.6 V max
Ground offset between any two nodes — 1.0 V (DC) max
Nominal Bus Driver Supply (V_{sup}) — 5 V (DC)
Receiver input range (each input maintains proper state) — -1 to 6 V
Receiver input range (common mode) — Gnd + 1.5 V to V_{sup} -1.5 V
Total bus termination resistance — 85 Ω min, 378 Ω max
Total bus capacitance — 500 pF min, 10 000 pF max
Unpowered State (standard unit load):
  Capacitance (each signal wire to ground) — 250 pf max
  Leakage Current (each signal wire, biased network — passive state) — 100 μA max
  Leakage Current (each signal wire, biased network — dominant state) — 100 μA max
Awake or Sleep State (standard unit load):
  Capacitance (each signal wire to ground) — 250 pF max
  Leakage Current (each signal wire, biased network — passive state) — 20 μA max
  Leakage Current (each signal wire, biased network — dominant state) — 100 μA max

7.3.3 VARIABLE PULSE WIDTH (VPW) MODULATION (SEE APPENDIX B FOR DERIVATION)

7.3.3.1 *VPW Timing Requirements (Note all times in microseconds)* — Table 6 shows the VPW timing values.

TABLE 6 — VPW TIMING REQUIREMENTS

Symbol	Tx,nom	Tx,min	Tx,max	Rx,min	Rx,max
Tv1	64	53	75	≥34	≤96
Tv2	128	116	141	>96	≤163
Tv3	200	186	214	>163	≤239
Tv4	280	265	NA	>239	NA
Tv5	>280 <768	265	794	>239	≤828
Tv6	300				

7.3.3.2 *VPW DC Parameters*
Minimum output high voltage ($V_{oh,min}$) — 6.25 V
Maximum output low voltage ($V_{ol,max}$) — 1.5 V
Minimum input high voltage ($V_{ih,min}$) — 4.25 V

Maximum input low voltage ($V_{il,max}$) — 3.5 V
Nominal receiver trip point (V_t) — 3.875 V
Ground offset between any two nodes — 2.0 V (DC) max
Maximum transition time ($T_{t,max}$) — 16 μs
Load bus resistance (R_{load}) — 240 Ω min, 1500 Ω max
Total bus capacitance (C_{tot}) — 3000 pF min, 15 000 pF max
Unpowered State:
  Nominal resistance (termination node) — 1500 Ω
  Capacitance (termination node) — 3300 pF max
  Nominal resistance (standard unit load) — 10 600 Ω
  Capacitance (standard unit load) — 470 pF max
  Leakage Current (V_{out} = 0 V) — 10 μA max
Awake or Sleep State:
  Nominal resistance (termination node) — 1500 Ω
  Capacitance (termination node) — 3300 pF max
  Nominal resistance (standard unit load) — 10 600 Ω
  Capacitance (standard unit load) — 470 pF max
  Leakage Current (V_{out} — 0 V) — 10 "A max

APPENDIX A
Inter-Byte Separation (IBS) Analysis

A.1 Inter-Byte Separation (IBS), when used, shall also adhere to these guidelines described in this section.

A.1.1 The IBS symbol (see Figure 20) shall not be present when it is not required and therefore shall only be present and consume bus bandwidth when required to support that message.

A.1.2 Hardware capable of supporting IBS shall support transmission and reception of data with the IBS Symbol between data bytes. Hardware capable of supporting IBS and in-frame response shall support transmission and reception of data with the IBS Symbol between data bytes and also between the EOD and (Normalization Bit = Tv5) the first in-frame response byte (see Figure 20) between response bytes in the in-frame response portion of the message. If longer IBS periods are required multiple IBS symbols may be employed.

A.1.3 Interface hardware that does not transmit IBS (because it contains additional buffering or is fast enough to not require IBS) must be capable of supporting the proper reception and the transmission of frames that arbitrate with nodes that use IBS. Attention must be given to insure proper arbitration of data through the IBS period. The interface shall not erroneously decode the passive short period (Tv1) as a "0" data bit and prematurely shift the data to the next bit. Nodes arbitrating with a node that transmits an IBS must resynchronize their transmission of the first bit of their current byte at the end of the IBS period to continue arbitration.

This interface type shall also support arbitration of frames as specified in 6.7.3 but at no time is required or needs to transmit the IBS Symbol.

APPENDIX B
VPW Waveform Analysis

B.1 Figure B1 shows a drawing of a bus waveform with various voltage levels and trip points. The voltage levels and trip points are defined as follows:

B.1.1 $V_{oh,min}$ — Minimum guaranteed output high voltage. This is also the highest trip point, with the source having no ground offset noise, and the receiver having 2 V of noise.

B.1.2 $V_{oh,max}$ — Maximum guaranteed output low voltage. This is also the highest trip point, with the receiver having no ground offset noise, and the source having 2 V of noise.

B.1.3 V_t — Ideal receiver trip point.

B.1.4 $V_{ih,min}$ — Minimum guaranteed input high voltage, and is also the highest trip point with no offset noise.

B.1.5 $V_{il,max}$ — Maximum guaranteed input low voltage, and is also the lowest trip point with no offset noise.

B.2 Factors that affect the transmitted pulse width are oscillator tolerance and variations in delay through the source driver and the medium. These factors, along with some guard banding, are lumped into a single set of limits called $T_{x,min}$ and $T_{x,max}$ for each bus symbol.

The equations used to generate the transmitted pulse values shown in Table B1 are:

$$T_{x,min} = (T_{nom} - T_{t,max}/2) * 0.98 - T_{x,mar} \quad \text{(Eq. B1)}$$
$$T_{x,max} = (T_{nom} + T_{t,max}/2) * 1.02 + T_{x,mar}$$

TABLE B1 — BUS SYMBOL TIMING REQUIREMENTS

Symbol	Tnom	Tx,min	Tx,max	Tr,min	Tr,max
Tv1	64	53	75	33	96
Tv2	128	116	141	95	163
Tv3	200	186	214	164	238
Tv4	280	265	NA	241	NA

The ($T_{nom} \pm T_{t,max}/2$)* 2% factor covers the 2% oscillator tolerance. The $T_{x,max}$ factor is assumed to be 2.0 which covers the rest of the factors (lumped variation in delay through the source driver and medium).

For VPW, mode synchronization is an integral part of symbol timing. All transmitting nodes reference their transmit timing from their receiver's perception of the previous edge, without regard to whether that edge was due to their own or another transmitter.

Figure B2 shows a plot of Tx, $T_{x,min}$, and $T_{x,max}$. Each is measured from the ideal trip point on the leading edge of the waveform. The fact that what is seen at the source's own receiver contributes to the transmitted pulse width, implies that the source device could be using echoed back information to shape what is transmitted.

The pulse width seen by another node (receiver) can range from $T_{x,min} - T_{t,max}$ to $T_{x,max} + T_{t,max}$. This range comes from the variations in the transmitted pulse width combined with possible noise offset affects on the transmitted and receiving node. The $T_{x,min} - T_{t,max}$ situation occurs when a $T_{x,min}$ pulse width is transmitted by a source node with no ground offset noise, and received by a receiving node with 2 V of ground offset noise. This sets up the possibility of the receiver tripping at $V_{oh,min}$ on the leading and trailing edge of the waveform. The $T_{x,max} + T_{t,max}$ situation occurs with a $T_{x,max}$ pulse being transmitted, the source node having 2 V of noise, and the receiver having no offset. This sets up the possibility of the receiving node tripping at $V_{ol,max}$. The receiving node may perceive the leading edge sooner or later than the transmitting node (source node). This range represents a receiver's required acceptance range for a given bus symbol. Figure B2 shows a drawing of this acceptance range.

The actual acceptance range must be wider to allow for the receiving node's oscillator tolerance and any implementation dependent uncertainties or constraints. The values for this range shown in Table B1 were arrived at using the formulas:

$$T_{r,min} = (T_{x,min} - T_{t,max})*0.98 - T_{r,mar} \quad \text{(Eq. B2)}$$
$$= ([(T_{nom} - T_{t,max}/2)* 0.98 - T_{x,mar}] - T_{t,max})*0.98 - T_{r,mar}$$

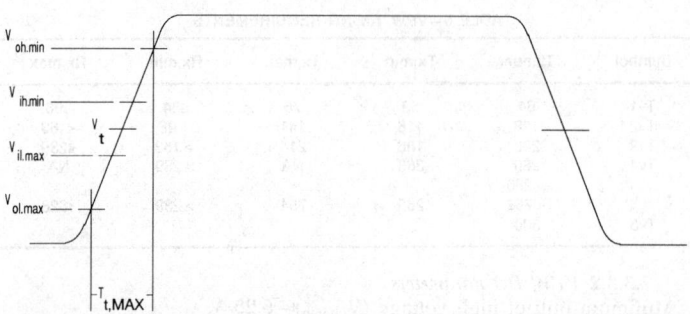

FIGURE B1 — VOLTAGE LEVELS AND TRIP POINTS

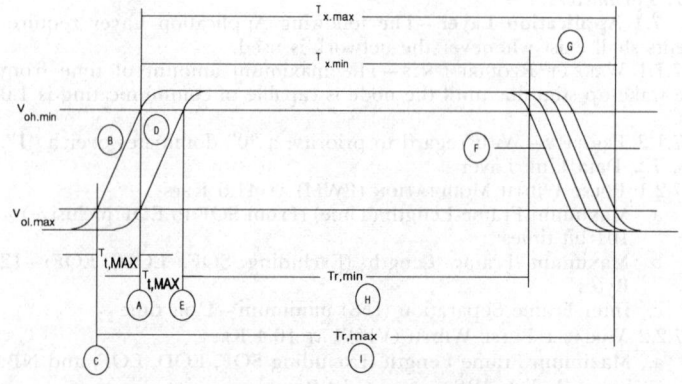

FIGURE B2 — WAVEFORM TIMING PARAMETERS

$$T_{r,max} = (T_{x,max} + T_{t,max})*1.02 + T_{r,mar} \quad (Eq.\ B3)$$
$$= ([(T_{nom} + T_{t,max}/2)*1.02 + T_{x,mar}] + T_{t,max})*1.02 + T_{r,mar}$$

The 2% guard bands cover the oscillator tolerance of both transmitter and receiver and $T_{r,max} = 3.0$ occurs the other uncertainties, including receiver delay and digital filter delay. VPW bus symbols allow no forbidden zones between symbols.

A—The trigger point on waveform "B" or "D" where the transmitting node sees the transition ending the previous symbol, and starts timing the current symbol. Take note that a node could trigger at any level of a transition between $V_{ol,max}$ and $V_{oh,min}$. The maximum and minimum trigger points of any given transition are illustrated by showing line A cutting through two different waveform transitions. This method of illustration allows separating the uncertainties of nodes sensing the beginning and ending of a symbol.

B—The earliest waveform that could trigger the transmitter at point A.

C—The earliest point on waveform "B" that could trigger any other node.

D—The latest waveform that could trigger the transmitter at point A.

E—The latest point on waveform "D" that could trigger any other node.

F—The waveform dispersion due to transmitter tolerances in oscillator, driver, etc.

G—The total transmitter spread when transition time limits are combined with the other tolerances in "F".

H—The shortest legitimate symbol time which can legitimately be seen under worst case conditions.

I—The longest legitimate symbol time which can legitimately be seen under worst case conditions.

NOTES—$T_{x,max}$ — Starts at $V_{oh,min}$ on wave "B"; $T_{x,min}$ — Starts at $V_{ol,max}$ on wave "D".

APPENDIX C
I/O EMC Test Plan

C.1 Radio Disturbance From Vehicle Components—(CISPR/D/WG2 (Secretariat) 19 Sept 1989) See Figure C1 for Diagram.

C.1.1 Radiated Emissions Antenna Test
C.1.1.1 FREQUENCY RANGE—10 KHz to 1.0 GHz
C.1.1.2 BANDWIDTH SETTING
 a. Frequency 10 KHz to 149 KHz Bandwidth 1 KHz
 b. Frequency 150 KHz to 1 GHz Bandwidth 3 KHz
C.1.1.3 TEST DATA
 C.1.1.3.1 *Ambient Reference*
  a. 10 KHz modulation off
  b. Both transmitter and receiver power off
  c. Sweep signal off
 C.1.1.3.2 *Idle State*
  a. 10 KHz modulation off
  b. Both transmitter and receiver power on
  c. Sweep signal off
 C.1.1.3.3 *Active Bus*
  a. 10 KHz modulation on
  b. Both transmitter and receiver power on
  c. Sweep signal off

C.1.2 Radiated Emissions Probe Test
C.1.2.1 FREQUENCY RANGE—10 KHz to 200 MHz
C.1.2.2 BANDWIDTH SETTING
 a. Frequency 10 KHz to 149 KHz Bandwidth 1 KHz
 b. Frequency 150 KHz to 1 GHz Bandwidth 3 KHz
C.1.2.3 PROBE TYPE—As required to make measurements
 a. Recommended probe type Tektronix A6302 0 to 29 MHz
 b. Recommended probe type Ailtech 94111-1 30 to 200 MHz
C.1.2.4 PROBE DISTANCE—5 cm from part
C.1.2.5 TEST DATA
 C.1.2.5.1 *Ambient Reference*
  a. 10 KHz modulation off
  b. Both transmitter and receiver power off
  c. Sweep signal off
 C.1.2.5.2 *Idle State*
  a. 10 KHz modulation off
  b. Both transmitter and receiver power on
  c. Sweep signal off
 C.1.2.5.3 *Active Bus*
  a. 10 KHz modulation on
  b. Both transmitter and receiver power on
  c. Sweep signal off

C.2 Transfer Function—Electromagnetic Susceptibility Measurement Procedures for Vehicle Components

C.2.1 Current Probe Monitoring—Adapted from SAE J1113 Part 2 (formerly SAE J1113 AUG87 Section 2).
 a. Frequency 30 Hz to 250 KHz

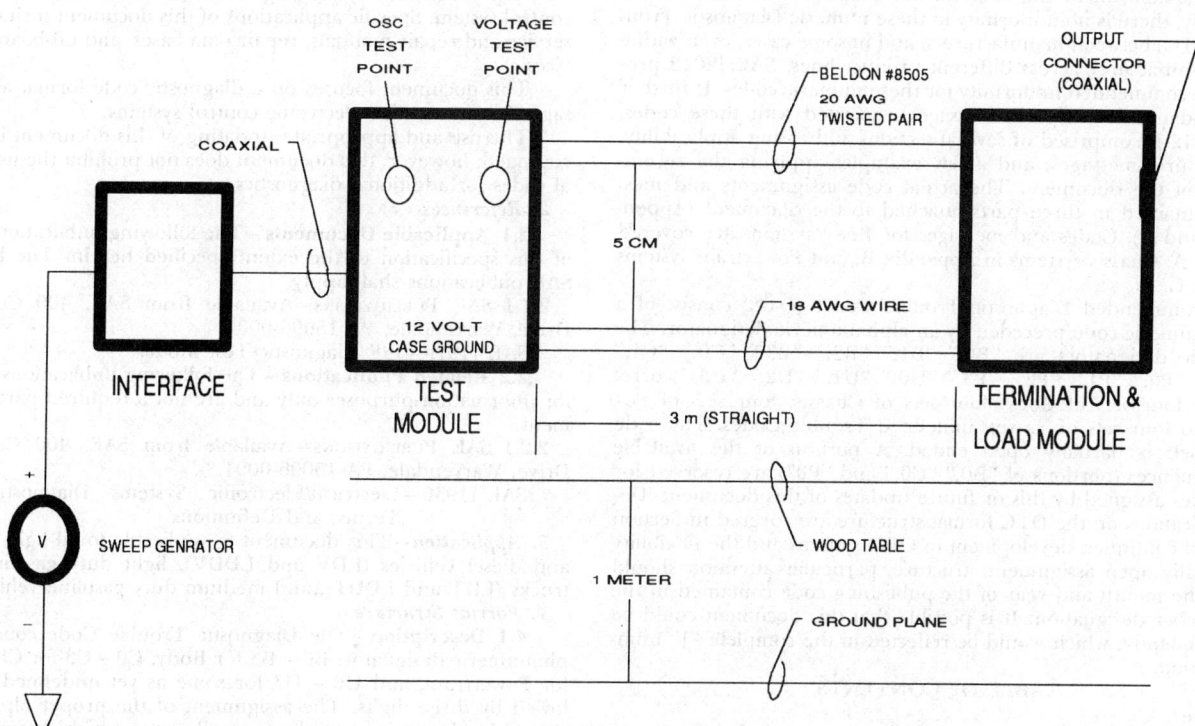

FIGURE C1—EMC TEST SETUP

b. Amplitude 1 V peak to peak at test point
C.2.2 Antenna Monitoring—Adapted from SAE J1113 Part 3 (formerly SAE J1113 AUG87 Section 3).
  a. Frequency 250 KHz to 200 MHz
  b. Attenuator input power ≈ 25 mW to generate 1 V peak to peak at test point
C.2.3 Test Data
C.2.3.1 TRANSFER FUNCTION FROM VCC TO BUS
  a. 10 KHz modulation on
  b. Both transmitter and receiver power on
  c. Sweep signal on
C.2.3.2 TRANSFER FUNCTION FROM GROUND TO BUS
  a. 10 KHz modulation on
  b. Both transmitter and receiver power on
  c. Sweep signal on

C.3 RF Susceptibility—Electromagnetic Susceptibility Measurement Procedures for Common Mode Injection for Compliance. Test per SAE J1113 Part 13 (formerly SAE J1547).
  C.3.1 Frequency Range—1 MHz to 200 MHz
  C.3.2 Max Level—100 mA RMS
  C.3.3 Procedure—Injection and monitoring to be applied to Bus pair only.
  C.3.4 Test Data—Record the nature of any interaction occurring below the test level along with the current and frequency.
C.4 Transient Susceptibility—Test per SAE J1113 Part 10 (adapted from ISO 7637/3).
  C.4.1 Test Pulse Amplitude
    a. Test Pulse 1 — −30 V (Reference)
    b. Test Pulse 2 — +30 V (Reference)
    c. Test Pulse 3a — −60 V (Reference)
    d. Test Pulse 3b — +40 V (Reference)

DIAGNOSTIC TROUBLE CODE DEFINITIONS—SAE J2012 MAR92

SAE Recommended Practice

Report of the SAE Vehicle E/E Systems Diagnostic Standards Committee approved March 1992.

Foreword—Most automobile manufacturers are equipping at least a portion of their product line with some On-Board Diagnostic (OBD) capability. Although currently confined to California, emission regulations sponsored by the California Air Resources Board are requiring the implementation of OBD systems across all manufacturers' product lines of cars and light and medium duty trucks offered for sale in California, beginning with Model Year 1988. These systems are required to provide an indication to the service technician as to the general location of the diagnosed malfunction so that the proper service procedures are used to resolve the problem. In most cases, this information is provided through a numeric code.

Additional on-board diagnostic requirements have been defined by the OBD-II legislation in California and the Federal Clean Air Act. These requirements include the need for standardization of various parts of the system, including some of the Diagnostic Trouble Codes.

Currently, there is no uniformity in these numeric Diagnostic Trouble Codes (DTC) between manufacturers, and in some cases, even within the same manufacturer across different product lines. SAE J2012 provides some recommended uniformity for these numeric codes. It further provides guidance for uniform messages associated with these codes.

SAE J2012 is comprised of several sections addressing applicability, format structure, messages, and a few examples applying the recommendations of the document. The actual code assignments and messages are contained in three parts attached to the document (Appendices A, B, and C). Codes and messages for Body systems are covered in Appendix A, Chassis systems in Appendix B, and Powertrain systems in Appendix C.

The recommended Diagnostic Trouble Codes (DTC) consist of a three-digit numeric code preceded by an alphanumeric designator. The alphanumeric designators are "B0," "B1," "B2," "B3," "C0," "C1," "C2," "C3," "P0," "P1," "P2," "P3," "U0," "U1," "U2," "U3," corresponding to four sets of Body, four sets of Chassis, four sets of Powertrain, and four sets of, as yet, undefined Trouble Codes. The code structure itself is partially open ended. A portion of the available numeric sequences (portions of "B0," "C0," and "P0") are reserved for uniform codes assigned by this or future updates of this document. Detailed specifications on the DTC format structure are covered in Section 4. Because of continued development in OBD systems and the flexibility of the partially open assignment structure, particular attention should be paid to the month and year of the publishing code contained in the full "J" number designation. It is possible that this document could be updated frequently, which would be reflected in the complete "J" number designation.

TABLE OF CONTENTS

1. Scope
2. References
2.1 Applicable Documents
2.2 Related Publications
3. Application
4. Format Structure
4.1 Description
4.2 Core DTCs
4.3 Non-Uniform DTC
4.4 Body System Groupings
4.5 Chassis System Groupings
4.6 Powertrain System Groupings
5. Messages
6. Examples
Appendix A—Body System Diagnostic Codes
Appendix B—Chassis System Diagnostic Codes
Appendix C—Powertrain System Diagnostic Codes

1. Scope—This SAE Recommended Practice is applicable to all light duty and medium duty passenger vehicles and trucks with feedback fuel control system. Specific applications of this document include diagnostic, service and repair manuals, repair data bases, and off-board readout devices.

This document focuses on a diagnostic code format and code messages for automotive electronic control systems.

The use and appropriate updating of this document is strongly encouraged; however, this document does not prohibit the use of additional codes for additional diagnostics.

2. References
  2.1 Applicable Documents—The following publications form a part of this specification to the extent specified herein. The latest issue of SAE publications shall apply.
  2.1.1 SAE PUBLICATIONS—Available from SAE, 400 Commonwealth Drive, Warrendale, PA 15096-0001.
  SAE J1979—E/E Diagnostic Test Modes
  2.2 Related Publications—The following publications are provided for information purposes only and are not a required part of this document.
  2.2.1 SAE PUBLICATIONS—Available from SAE, 400 Commonwealth Drive, Warrendale, PA 15096-0001.
  SAE J1930—Electrical/Electronic Systems Diagnostic Acronyms, Terms, and Definitions

3. Application—This document is applicable to all light duty gasoline and diesel vehicles (LDV and LDDV), light duty gasoline and diesel trucks (LDT and LDDT), and medium duty gasoline vehicles (MDGV).

4. Format Structure
  4.1 Description—The Diagnostic Trouble Code consists of an alphanumeric designator, B0 – B3 for Body, C0 – C3 for Chassis, P0 – P3 for Powertrain, and U0 – U3 for some as yet undefined catch-all, followed by three digits. The assignment of the proper alpha designator should be determined by the controller into which the particular function being diagnosed is being integrated, or in the case of multiple controllers, the area most appropriate for that function. In most cases, the

alpha designator will be implied since diagnostic information will be requested from a particular controller. In the cases where the source of the diagnostic information is not clear, the uppermost nibble of the two-byte code message as defined in J1979 (E/E Diagnostic Test Modes) will define the source system as follows:

- 0000 – P0 – Powertrain codes SAE controlled
- 0001 – P1 – Powertrain codes Manufacturer controlled
- 0010 – P2 – Powertrain codes Reserved
- 0011 – P3 – Powertrain codes Reserved
- 0100 – C0 – Chassis codes SAE controlled
- 0101 – C1 – Chassis codes Manufacturer controlled
- 0110 – C2 – Chassis codes Manufacturer controlled
- 0111 – C3 – Chassis codes Reserved
- 1000 – B0 – Body codes SAE controlled
- 1001 – B1 – Body codes Manufacturer controlled
- 1010 – B2 – Body codes Manufacturer controlled
- 1011 – B3 – Body codes Reserved
- 1100 – U0 – Undefined codes Reserved
- 1101 – U1 – Undefined codes Reserved
- 1110 – U2 – Undefined codes Reserved
- 1111 – U3 – Undefined codes Reserved

Within each code class, the first of the three digits identifies a particular grouping of codes. These particular groupings each contain a series of 100 sequence numbers for particular code definitions.

Codes have been defined to indicate a suspected trouble or problem area and are intended to be used as a directive to the proper service procedure. To minimize service confusion, fault codes should not be used to indicate the absence of problems or the status of parts of the system (e.g., Powertrain System O.K., or MIL illuminated), but should be confined to indicated areas in need of service attention.

4.2 Core DTCs—Core Diagnostic Trouble Codes are those codes where industry uniformity has been achieved. These codes were felt to be common enough across most manufacturers' applications that a common number and fault message could be assigned. All undefined numbers in each grouping of systems have been reserved for future growth. Although service procedures may differ widely among manufacturers, the fault being indicated is common enough to be assigned a particular fault code.

4.3 Non-Uniform DTC—Areas within each alpha designator have been made available for non-uniform DTCs. These are fault codes that will not generally be used by a majority of the manufacturers due to basic system differences, implementation differences, or diagnostic strategy differences. Each manufacturer will define their own code and error message definition for use in these assigned areas. It is understood that each manufacturer must remain consistent across their product line when assigning codes in the Manufacturer controlled areas. For Powertrain codes, the same groupings should be used as in the SAE controlled area, i.e., 100's and 200's for Fuel and Air Metering, 300's for Ignition System or Misfire, etc.

Code groupings for non-Powertrain codes will be defined at a later date.

4.4 Body System Groupings
4.4.1 — B0XXX SAE CONTROLLED
4.4.2 — B1XXX MANUFACTURER CONTROLLED
4.4.3 — B2XXX MANUFACTURER CONTROLLED
4.4.4 — B3XXX RESERVED

4.5 Chassis System Groupings
4.5.1 — C0XXX SAE CONTROLLED
4.5.2 — C1XXX MANUFACTURER CONTROLLED
4.5.3 — C2XXX MANUFACTURER CONTROLLED
4.5.4 — C3XXX RESERVED

4.6 Powertrain System Groupings
4.6.1 — P0XXX SAE CONTROLLED
4.6.1.1 — P01XX Fuel and Air Metering
4.6.1.2 — P02XX Fuel and Air Metering
4.6.1.3 — P03XX Ignition System or Misfire
4.6.1.4 — P04XX Auxiliary Emission Controls
4.6.1.5 — P05XX Vehicle Speed Control and Idle Control System
4.6.1.6 — P06XX Computer and Output Circuits
4.6.1.7 — P07XX Transmission
4.6.1.8 — P08XX Transmission
4.6.1.9 — P09XX Reserved for SAE
4.6.1.10 — P00XX Reserved for SAE
4.6.2 — P1XXX MANUFACTURER CONTROLLED
4.6.2.1 — P11XX Fuel and Air Metering
4.6.2.2 — P12XX Fuel and Air Metering
4.6.2.3 — P13XX Ignition System or Misfire
4.6.2.4 — P14XX Auxiliary Emission Controls
4.6.2.5 — P15XX Vehicle Speed Control and Idle Control System
4.6.2.6 — P16XX Computer and Output Circuits
4.6.2.7 — P17XX Transmission
4.6.2.8 — P18XX Transmission
4.6.2.9 — P19XX Category to be Determined by SAE
4.6.2.10 — P10XX Category to be Determined by SAE
4.6.3 — P2XXX SAE RESERVED
4.6.4 — P3XXX SAE RESERVED

5. Messages—Each defined fault code has been assigned a message to indicate the circuit or area that was determined to be at fault. The messages are organized such that different messages related to a particular sensor or system are grouped together. In cases where there are various fault messages for different types of faults, the group also has a **"generic"** message as the first Code/Message of the group. A manufacturer has a choice when implementing diagnostics, based on the specific strategy and complexity of the diagnostic, whether to use one **"generic"** code for any fault of that circuit or system or to use the more specific codes for better defining the type of fault that was detected. The manufacturer must determine what codes and messages best fit the diagnostics actually implemented. The intent is to have only one code stored for each fault detected.

Where messages are broken down into more specific fault descriptions for a sensor or system, the manufacturer should choose the code most applicable to their diagnosable fault. The messages are intended to be somewhat general to allow manufacturers to use them as often as possible yet still not conflict with their specific repair procedures. The terms "LOW" and "HIGH" when used in a message, especially those related to input signals, refer to the voltage, frequency, etc., at the pin of the controller. The specific level of "LOW" and "HIGH" must be defined by each manufacturer to best meet their needs.

6. Examples—For manufacturers choosing to implement basic diagnostics that provide general fault information but depend on service procedures (Service Charts) and off-board techniques to isolate the problem, general sensor and system codes would be used. If a fault is detected in the Throttle Position Sensor circuit, rather than burden the on-board system with determining the specific type of fault, a Code P0120 would be stored indicating some type of problem with that circuit, and the service procedure would then allow the technician to determine the type of fault and the specific location of the fault. On these types of systems, a shorted sensor input, an open sensor input, and even an out-of-range sensor input, would all set the same fault code.

Manufacturers choosing to allow the on-board diagnostics to better isolate the fault would not use the general fault code/message, but would use the more specific codes/messages associated with the particular sensor or circuit. For example, in diagnosing the Throttle Position Sensor, if the input signal was outside of the normal operating range of the sensor, Code P0122 would be stored for a low signal with the associated message being Throttle Position Sensor—Low Input, or Code P0123 would be stored for a high signal with the associated message being Throttle Position Sensor—High Input. For a signal that is within operating range, but is determined to be otherwise in error for the given test conditions, Code P0121 could be stored (Throttle Position Sensor—Range/Performance), or some other code defined in the Manufacturer Specific Code Area could be used with a message that better indicates the type or location of the problem.

APPENDIX A
Body System Diagnostic Codes

For future definition.

APPENDIX B
Chassis System Diagnostic Codes

For future definition.

APPENDIX C
Powertrain System Diagnostic Trouble Codes

C.1 Discussion—Following are the Recommended Industry Common Fault Codes for the Powertrain Control System. These include systems that might be integrated into an electronic control module that would be used for controlling Engine functions, such as Fuel, Spark, Idle Speed, and Vehicle Speed (Cruise Control) as well as those for Transmission control. The fact that a code is recommended as a Common Industry Code does not imply that it is a Required Code (Legislated), an Emission Related Code, or that it indicates a fault that will cause the Malfunction Indicator Light to be illuminated.

P01XX   Fuel and Air Metering

Code	Description
P0100	Mass or Volume Air Flow Circuit Malfunction
P0101	Mass or Volume Air Flow Circuit Range/Performance Problem
P0102	Mass or Volume Air Flow Circuit Low Input
P0103	Mass or Volume Air Flow Circuit High Input
P0105	Manifold Absolute Pressure/Barometric Pressure Circuit Malfunction
P0106	Manifold Absolute Pressure/Barometric Pressure Circuit Range/Performance Problem
P0107	Manifold Absolute Pressure/Barometric Pressure Circuit Low Input
P0108	Manifold Absolute Pressure/Barometric Pressure Circuit High Input
P0110	Intake Air Temperature Circuit Malfunction
P0111	Intake Air Temperature Circuit Range/Performance Problem
P0112	Intake Air Temperature Circuit Low Input
P0113	Intake Air Temperature Circuit High Input
P0115	Engine Coolant Temperature Circuit Malfunction
P0116	Engine Coolant Temperature Circuit Range/Performance Problem
P0117	Engine Coolant Temperature Circuit Low Input
P0118	Engine Coolant Temperature Circuit High Input
P0120	Throttle Position Circuit Malfunction
P0121	Throttle Position Circuit Range/Performance Problem
P0122	Throttle Position Circuit Low Input
P0123	Throttle Position Circuit High Input
P0125	Excessive Time to Enter Closed Loop Fuel Control
P0130	O_2 Sensor Circuit Malfunction (Bank 1* Sensor 1)
P0131	O_2 Sensor Circuit Low Voltage (Bank 1* Sensor 1)
P0132	O_2 Sensor Circuit High Voltage (Bank 1* Sensor 1)
P0133	O_2 Sensor Circuit Slow Response (Bank 1* Sensor 1)
P0134	O_2 Sensor Circuit No Activity Detected (Bank 1* Sensor 1)
P0135	O_2 Sensor Heater Circuit Malfunction (Bank 1* Sensor 1)
P0136	O_2 Sensor Circuit Malfunction (Bank 1* Sensor 2)
P0137	O_2 Sensor Circuit Low Voltage (Bank 1* Sensor 2)
P0138	O_2 Sensor Circuit High Voltage (Bank 1* Sensor 2)
P0139	O_2 Sensor Circuit Slow Response (Bank 1* Sensor 2)
P0140	O_2 Sensor Circuit No Activity Detected (Bank 1* Sensor 2)
P0141	O_2 Sensor Heater Circuit Malfunction (Bank 1* Sensor 2)
P0142	O_2 Sensor Circuit Malfunction (Bank 1* Sensor 3)
P0143	O_2 Sensor Circuit Low Voltage (Bank 1* Sensor 3)
P0144	O_2 Sensor Circuit High Voltage (Bank 1* Sensor 3)
P0145	O_2 Sensor Circuit Slow Response (Bank 1* Sensor 3)
P0146	O_2 Sensor Circuit No Activity Detected (Bank 1* Sensor 3)
P0147	O_2 Sensor Heater Circuit Malfunction (Bank 1* Sensor 3)
P0150	O_2 Sensor Circuit Malfunction (Bank 2 Sensor 1)
P0151	O_2 Sensor Circuit Low Voltage (Bank 2 Sensor 1)
P0152	O_2 Sensor Circuit High Voltage (Bank 2 Sensor 1)
P0153	O_2 Sensor Circuit Slow Response (Bank 2 Sensor 1)
P0154	O_2 Sensor Circuit No Activity Detected (Bank 2 Sensor 1)
P0155	O_2 Sensor Heater Circuit Malfunction (Bank 2 Sensor 1)
P0156	O_2 Sensor Circuit Malfunction (Bank 2 Sensor 2)
P0157	O_2 Sensor Circuit Low Voltage (Bank 2 Sensor 2)
P0158	O_2 Sensor Circuit High Voltage (Bank 2 Sensor 2)
P0159	O_2 Sensor Circuit Slow Response (Bank 2 Sensor 2)
P0160	O_2 Sensor Circuit No Activity Detected (Bank 2 Sensor 2)
P0161	O_2 Sensor Heater Circuit Malfunction (Bank 2 Sensor 2)
P0162	O_2 Sensor Circuit Malfunction (Bank 2 Sensor 3)
P0163	O_2 Sensor Circuit Low Voltage (Bank 2 Sensor 3)
P0164	O_2 Sensor Circuit High Voltage (Bank 2 Sensor 3)
P0165	O_2 Sensor Circuit Slow Response (Bank 2 Sensor 3)
P0166	O_2 Sensor Circuit No Activity Detected (Bank 2 Sensor 3)
P0167	O_2 Sensor Heater Circuit Malfunction (Bank 2 Sensor 3)
P0170	Fuel Trim Malfunction (Bank 1*)
P0171	System too Lean (Bank 1*)
P0172	System too Rich (Bank 1*)
P0173	Fuel Trim Malfunction (Bank 2)
P0174	System too Lean (Bank 2)
P0175	System too Rich (Bank 2)
P0176	Fuel Composition Sensor Circuit Malfunction
P0177	Fuel Composition Sensor Circuit Range/Performance
P0178	Fuel Composition Sensor Circuit Low Input
P0179	Fuel Composition Sensor Circuit High Input
P0180	Fuel Temperature Sensor Circuit Malfunction
P0181	Fuel Temperature Sensor Circuit Range/Performance
P0182	Fuel Temperature Sensor Circuit Low Input
P0183	Fuel Temperature Sensor Circuit High Input

P02XX   **Fuel and Air Metering**

Code	Description
P0201	Injector Circuit Malfunction—Cylinder 1
P0202	Injector Circuit Malfunction—Cylinder 2
P0203	Injector Circuit Malfunction—Cylinder 3
P0204	Injector Circuit Malfunction—Cylinder 4
P0205	Injector Circuit Malfunction—Cylinder 5
P0206	Injector Circuit Malfunction—Cylinder 6
P0207	Injector Circuit Malfunction—Cylinder 7
P0208	Injector Circuit Malfunction—Cylinder 8
P0209	Injector Circuit Malfunction—Cylinder 9
P0210	Injector Circuit Malfunction—Cylinder 10
P0211	Injector Circuit Malfunction—Cylinder 11
P0212	Injector Circuit Malfunction—Cylinder 12
P0213	Cold Start Injector 1 Malfunction
P0214	Cold Start Injector 2 Malfunction

P03XX   **Ignition System or Misfire**

Code	Description
P0300	Random Misfire Detected
P0301	Cylinder 1 Misfire Detected
P0302	Cylinder 2 Misfire Detected
P0303	Cylinder 3 Misfire Detected
P0304	Cylinder 4 Misfire Detected
P0305	Cylinder 5 Misfire Detected
P0306	Cylinder 6 Misfire Detected
P0307	Cylinder 7 Misfire Detected
P0308	Cylinder 8 Misfire Detected
P0309	Cylinder 9 Misfire Detected
P0310	Cylinder 10 Misfire Detected
P0311	Cylinder 11 Misfire Detected
P0312	Cylinder 12 Misfire Detected
P0320	Ignition/Distributor Engine Speed Input Circuit Malfunction
P0321	Ignition/Distributor Engine Speed Input Circuit Range/

---

* For systems with single O_2 sensors, use codes for Bank 1 sensor
Bank 1 contains cylinder #1

Sensor 1 is closest to engine

	Performance		Range/Performance
P0322	Ignition/Distributor Engine Speed Input Circuit No Signal	P0452	Evaporative Emission Control System Pressure Sensor Low Input
		P0453	Evaporative Emission Control System Pressure Sensor High Input
P0325	Knock Sensor 1 Circuit Malfunction		
P0326	Knock Sensor 1 Circuit Range/Performance	**P05XX**	**Vehicle Speed Control and Idle Control System**
P0327	Knock Sensor 1 Circuit Low Input		
P0328	Knock Sensor 1 Circuit High Input	P0500	Vehicle Speed Sensor Malfunction
		P0501	Vehicle Speed Sensor Range/Performance
P0330	Knock Sensor 2 Circuit Malfunction	P0502	Vehicle Speed Sensor Low Input
P0331	Knock Sensor 2 Circuit Range/Performance		
P0332	Knock Sensor 2 Circuit Low Input	P0505	Idle Control System Malfunction
P0333	Knock Sensor 2 Circuit High Input	P0506	Idle Control System RPM Lower Than Expected
		P0507	Idle Control System RPM Higher Than Expected
P0335	Crankshaft Position Sensor Circuit Malfunction		
P0336	Crankshaft Position Sensor Circuit Range/Performance	P0510	Closed Throttle Position Switch Malfunction
P0337	Crankshaft Position Sensor Circuit Low Input		
P0338	Crankshaft Position Sensor Circuit High Input	**P06XX**	**Computer and Output Circuits**
P0340	Camshaft Position Sensor Circuit Malfunction	P0600	Serial Communication Link Malfunction
P0341	Camshaft Position Sensor Circuit Range/Performance		
P0342	Camshaft Position Sensor Circuit Low Input	P0605	Internal Control Module (Module Identification Defined by J1979)
P0343	Camshaft Position Sensor Circuit High Input		
P04XX	**Auxiliary Emission Controls**	**P07XX**	**Transmission**
P0400	Exhaust Gas Recirculation Flow Malfunction	P0703	Brake Switch Input Malfunction
P0401	Exhaust Gas Recirculation Flow Insufficient Detected	P0705	Transmission Range Sensor Circuit Malfunction (PRNDL Input)
P0402	Exhaust Gas Recirculation Flow Excessive Detected	P0706	Transmission Range Sensor Circuit Range/Performance
P0405	Air Conditioner Refrigerant Charge Loss	P0707	Transmission Range Sensor Circuit Low Input
P0410	Secondary Air Injection System Malfunction	P0708	Transmission Range Sensor Circuit High Input
P0411	Secondary Air Injection System Insufficient Flow Detected	P0710	Transmission Fluid Temperature Sensor Circuit Malfunction
P0412	Secondary Air Injection System Switching Valve/Circuit Malfunction	P0711	Transmission Fluid Temperature Sensor Circuit Range/Performance
P0413	Secondary Air Injection System Switching Valve/Circuit Open	P0712	Transmission Fluid Temperature Sensor Circuit Low Input
P0414	Secondary Air Injection System Switching Valve/Circuit Shorted	P0713	Transmission Fluid Temperature Sensor Circuit High Input
P0420	Catalyst System Efficiency Below Threshold (Bank 1*)	P0715	Input/Turbine Speed Sensor Circuit Malfunction
P0421	Warm Up Catalyst Efficiency Below Threshold (Bank 1*)	P0716	Input/Turbine Speed Sensor Circuit Range/Performance
P0422	Main Catalyst Efficiency Below Threshold (Bank 1*)	P0717	Input/Turbine Speed Sensor Circuit No Signal
P0423	Heated Catalyst Efficiency Below Threshold (Bank 1*)		
P0424	Heated Catalyst Temperature Below Threshold (Bank 1*)	P0720	Output Speed Sensor Circuit Malfunction
		P0721	Output Speed Sensor Circuit Range/Performance
		P0722	Output Speed Sensor Circuit No Signal
P0430	Catalyst System Efficiency Below Threshold (Bank 2)		
P0431	Warm Up Catalyst Efficiency Below Threshold (Bank 2)	P0725	Engine Speed Input Circuit Malfunction
P0432	Main Catalyst Efficiency Below Threshold (Bank 2)	P0726	Engine Speed Input Circuit Range/Performance
P0433	Heated Catalyst Efficiency Below Threshold (Bank 2)	P0727	Engine Speed Input Circuit No Signal
P0434	Heated Catalyst Temperature Below Threshold (Bank 2)		
		P0730	Incorrect Gear Ratio
P0440	Evaporative Emission Control System Malfunction	P0731	Gear 1 Incorrect Ratio
P0441	Evaporative Emission Control System Insufficient Purge Flow	P0732	Gear 2 Incorrect Ratio
		P0733	Gear 3 Incorrect Ratio
P0442	Evaporative Emission Control System Leak Detected	P0734	Gear 4 Incorrect Ratio
P0443	Evaporative Emission Control System Purge Control Valve Circuit Malfunction	P0735	Gear 5 Incorrect Ratio
		P0736	Reverse Incorrect Ratio
P0444	Evaporative Emission Control System Purge Control Valve Circuit Open	P0740	Torque Converter Clutch System Malfunction
P0445	Evaporative Emission Control System Purge Control Valve Circuit Shorted	P0741	Torque Converter Clutch System Performance or Stuck Off
		P0742	Torque Converter Clutch System Stuck On
P0446	Evaporative Emission Control System Vent Control Malfunction	P0743	Torque Converter Clutch System Electrical
P0447	Evaporative Emission Control System Vent Control Open	P0745	Pressure Control Solenoid Malfunction
P0448	Evaporative Emission Control System Vent Control Shorted	P0746	Pressure Control Solenoid Performance or Stuck Off
		P0747	Pressure Control Solenoid Stuck On
P0450	Evaporative Emission Control System Pressure Sensor Malfunction	P0748	Pressure Control Solenoid Electrical
P0451	Evaporative Emission Control System Pressure Sensor	P0750	Shift Solenoid A Malfunction
		P0751	Shift Solenoid A Performance or Stuck Off
		P0752	Shift Solenoid A Stuck On
		P0753	Shift Solenoid A Electrical

* Bank 1 contains Cylinder #1

P0755	Shift Solenoid B Malfunction		P0765	Shift Solenoid D Malfunction
P0756	Shift Solenoid B Performance or Stuck Off		P0766	Shift Solenoid D Performance or Stuck Off
P0757	Shift Solenoid B Stuck On		P0767	Shift Solenoid D Stuck On
P0758	Shift Solenoid B Electrical		P0768	Shift Solenoid D Electrical
P0760	Shift Solenoid C Malfunction		P0770	Shift Solenoid E Malfunction
P0761	Shift Solenoid C Performance or Stuck Off		P0771	Shift Solenoid E Performance or Stuck Off
P0762	Shift Solenoid C Stuck On		P0772	Shift Solenoid E Stuck On
P0763	Shift Solenoid C Electrical		P0773	Shift Solenoid E Electrical

(R) ELECTRICAL/ELECTRONIC SYSTEMS DIAGNOSTIC TERMS, DEFINITIONS, ABBREVIATIONS, AND ACRONYMS—SAE J1930 JUN93

SAE Recommended Practice

Report of the Vehicle Electric and Electronic Systems Diagnostic Committee approved June 1988. Completely revised by the SAE Vehicle E/E Systems Diagnostic Standards Committee September 1991 and June 1993.

Foreword—As the number of sophisticated electrical and electronic (E/E) systems on motor vehicles has increased, the number of terms, abbreviations, and acronyms which describe various components of these systems has increased enormously. To bring some order to the proliferation of such terms, abbreviations, and acronyms, the Vehicle E/E Systems Diagnostic Committee has prepared this document.

The nomenclature used to convey automotive service information is being standardized in order to more accurately convey information to technicians faced with the diagnosis and repair of increasingly complex vehicles.

To be properly descriptive, each type of automotive nomenclature requires a consistent methodology. This document is concerned with a methodology for naming objects and systems and with the set of words from which names are built.

The methodology allows objects and systems to be completely described without ambiguity. It also is able to generate names which distinguish among similar objects or systems without confusion but with brevity. Using terms which are well-defined within the context of the automotive service industry, the methodology allows already existing imprecise names to be suitably changed and future names to be assigned in a predictable way which will reliably convey meaning to the technician.

The structure of this SAE document is open ended by design. As the need arises, additional entries can be added. Because of this flexibility, particular attention should be paid to the month and year publishing code contained in the full "J" number designation.

Table of Contents
Foreword
1. Scope
2. References
3. How to Use This Document
4. Methodology
4.1 Naming Objects
4.1.1 Base Words
4.1.2 Modifiers
4.1.3 Technological Terms
4.2 Naming Systems
4.3 Shortened Names
4.3.1 Acronyms
4.3.2 Abbreviations
4.4 Indexing of Names
5. Cross Reference and Look Up
6. Recommended Terms
7. Glossary of Terms
8. Revision Procedures
9. Notes
9.1 Marginal Indicia
APPENDIX A Request for Revision to SAE J1930 Electrical/Electronic Systems Diagnostic Terms, Definitions, Abbreviations, and Acronyms

1. Scope—This SAE Recommended Practice is applicable to all light-duty gasoline and diesel passenger vehicles and trucks, and to heavy-duty gasoline vehicles. Specific applications of this document include diagnostic, service and repair manuals, bulletins and updates, training manuals, repair data bases, under-hood emission labels, and emission certification applications.

This document focuses on diagnostic terms applicable to electrical/electronic systems, and therefore also contains related mechanical terms, definitions, abbreviations, and acronyms.

Even though the use and appropriate updating of this document is strongly encouraged, nothing in this document should be construed as prohibiting the introduction of a term, abbreviation or acronym not covered by this document.

Certain terms have already been in common use and are readily understood by manufacturers and technicians, but do not follow the methodology of this document. To preserve this understanding, these terms were included and have been identified with the footnote (2), "historically acceptable common usage," so they will not erroneously serve as a precedent in the construction of new names. These terms fall into three categories:
  a. Acronyms that do not logically fit the term.
  b. Acronyms existing at the component level, i.e., their terms contain the base word or noun that describes the generic item that is being further defined.
  c. Acronyms for terms that appear to contain the base word, but are frequently used as a modifier to another base word. (This use may possibly be thought of as following the methodology since the acronym is normally used as a modifier.)

2. References—There are no referenced publications specified herein.

3. How to Use This Document—To find the recommended term corresponding to an existing term, abbreviation or acronym, see Figure 1, Cross Reference and Look Up. See Figure 2, Recommended Terms, and Table 1, Glossary of Terms, for definitions of the recommended terms. Use Section 3.0, Methodology to construct a new name. Appropriate acceptable usages of Recommended Terms and Acronyms are contained in Figure 1.

4. Methodology—This naming methodology of describing objects and systems uses modifiers attached to base words. Appropriate modifiers are added to a base word until an object or system is uniquely specified within its context.

4.1 Naming Objects—When building names, select the most descriptive base word from the Glossary of Terms (Table 1). Add modifiers as necessary or as desirable within the context, in the order of most significance to least significance. The most significant word will be the base word, which denotes the basic function of the object. The most significant modifier will be adjacent to the base word, the second most significant will be next to that modifier, and so on until the least significant modifier is added. For the sake of future clarity, an additional modifier can be added to a name at any time, even if there is no present conflict with another object name. Figure 3 illustrates how modifiers can be added to build the name, "Instrumentation Engine Coolant Temperature Sensor."

When naming an object, it is tempting to choose the first modifiers according to the initial purpose for which the object was designed, but this will not always result in the name which is most helpful in the long run to a service technician. The information a technician needs is most often supplied by a term which describes a functional attribute, not purpose.

To ensure accuracy, always check the Glossary definitions of base words and modifiers before including them in a name. The Glossary is intended for diagnostic purposes, but provides only electrical/electronic terms for base words. Base words which describe nonelectrical objects (e.g., bolt, screw, bumper) should be used as in the past. Often, names for these objects are created by attaching the appropriate electrical/electronic object name to the mechanical base word. When using a common multiple word modifier, see Figures 1 and 2 to be sure that the modifier is acceptable or if it should be replaced with a more precise term.

4.1.1 BASE WORDS—The base word is the most generic term in a name. Simply stated, it answers the question, "What is this object?" In answering this question, the base word does not include information about the location or function of an object within a particular system. Specific information like this is provided by modifiers that are added to the base word. The following are examples of base words: diode, engine, module, motor, pump, relay, sensor, solenoid, switch, valve. The base word is always a noun and the last term in a name.

4.1.2 MODIFIERS—Modifiers provide functional/applicational meaning, system differentiation, and locational/directional information. Modifiers usually express nonelectrical ideas to describe base words which, in turn, convey electrical/electronic meaning. The range of modifiers is not limited and is used as necessary to uniquely describe an object in light of present knowledge, past experience, and potential future conflicts.

Although modifiers are used as adjectives, they are not necessarily terms which would normally be classified as adjectives. While neither "Air" or "Flow" are adjectives, the meaning of "Air Flow Valve" is clear to technicians; it is the name of a valve which regulates the flow of air. Both modifiers are nouns functioning as adjectives because of their position.

System modifiers can be added to object names to describe an object's purpose. When using a system name as a modifier in an object name, the word "System" is not included. For example, the device that directs the exhaust gases in the Exhaust Gas Recirculation (EGR) System is named "Exhaust Gas Recirculation (EGR) Valve."

4.1.3 TECHNOLOGICAL TERMS—Technologically specific terms tend to lengthen names without adding a corresponding level of useful service information about the function of an object. Add an appropriate technological modifier to a name only when it describes the primary difference between two objects. For example, the "thick film" technology used to construct the internal circuit of an Air Flow Sensor should not be identified in the object's name. However, if necessary for clarity, it would be appropriate to differentiate the relation to a specific external provision by adding "Hot Wire" to "Air Flow Sensor."

A technological term should be the first modifier conversationally (farthest from the base word, the position of least significance), unless a directional modifier is also present.

4.2 Naming Systems—When constructing a name for a system, consider it to be a combination of a "concept" and the word "System." Develop the concept name according to the rules for object naming and add the word "System." Keep in mind that a concept's most basic attribute is its purpose and that this attribute is described by the term closest to the word "System." For example, "recirculation" is the basic attribute of the Exhaust Gas Recirculation (EGR) concept. The group of components that embody the concept are together named the "EGR System."

4.3 Shortened Names—Techniques of shortening, including acronyms and abbreviations, are often necessary when space is limited and when names become awkwardly long. It is preferable to create a name first and its shortened form later, rather than the other way around.

Abbreviations and acronyms may be constructed not only of the letters of the alphabet, but of numbers, space characters, punctuation marks (such as "/" and "—"), subscripts, and any other ASCII characters. Treat the individual acronyms, modifier abbreviations, and base word abbreviations as words, separating them by space characters.

4.3.1 ACRONYMS—Specific definitions of acronyms vary, but for the purpose of this document, an acronym is a memorable combination of the first letters of the words of a name. While abbreviations are useful in text where space is limited, acronyms are particularly convenient for shortening verbal communication in addition to written materials. For this reason, acronyms are often pronounceable, which also makes them easy to remember. They are especially useful if a name is long and bulky both on paper and in conversation.

Use acronyms as modifiers or base words within names, such as "EGR System" and "Primary ECM." Do not use them as entire names, like "EGRS." Acronyms and other modifiers may be combined in any meaningful order to modify a base word. The following are examples of acceptable uses of acronyms:

EGR System EGRT Sensor Low Speed FC Switch High Speed FC Switch

Because there are a limited number of useful letter combinations for acronyms, new acronyms should be created for only the most commonly used terms. Also, avoid creating new acronyms by adding letters to those that already exist. For example, when using the acronym "FC" (Fan Control), do not add "H" or "L" to indicate "High Speed" or "Low Speed." Instead, use additional modifiers.

Usually, the first letters of each word of a name are used to build an acronym, but if a particular word is of little significance, it may be omitted ("United States of America" becomes "USA"). Also, more than the first letter of each word may be used ("Radio Detecting And Ranging" becomes "RADAR"). An acronym like "USA" which contains three letters or fewer may have its letters spoken separately, but a longer acronym such as "RADAR" must be pronounceable or its purpose will be defeated.

All of the letters of an acronym should be capitalized. Acronyms should not contain periods. Until an acronym is widely well-known, it should be accompanied by the spelled-out form when necessary for accurate reader comprehension in any given context.

In the very rare cases of strong historical meaning across all manufacturers, the rules for naming and acronym usage may be broken. For example, "AIR" is the approved acronym for "Secondary Air Injection," instead of "SAI." In fact, because there is no approved name "Primary Air Injection," the term "Secondary Air Injection" would be considered inappropriate. Despite this, historical precedent renders "AIR" and "Secondary Air Injection" the most easily understood terms. "AIR" originally meant "Air Injection Reactor." However, vehicles no longer necessarily use a separate air injector reactor, but instead might have additional air injected to the catalytic converter. Because of the similarity to the previous system, technicians have expressed a strong desire to retain "AIR" rather than "SAI."

Before using a new acronym, be sure to check Figures 1 and 2 for any conflicts with acronyms already in use.

4.3.2 ABBREVIATIONS—Use abbreviations to shorten base words and directional modifiers in written materials. Unlike an acronym, an abbreviation should have only its first letter capitalized and should end with a period. Wire colors are an exception to the rules of capitalization and punctuation. As in the past, they should continue to be completely capitalized in text and not followed by a period (for example, "a BLK wire"). Currently identified abbreviations for base words and modifiers are found in Figure 1.

4.4 Indexing of Names—Service information index designers consider the importance of each term in a name, and select the most appropriate word(s) to index. They most frequently index base words; following each by its modifier(s) to enhance users' retrieval. This document allows the designer flexibility to choose the indexed word(s); while it describes, in detail, the methodology for the conversational word order in text and illustrations. For example, the designer can conform to the methodology of this document and provide the user with the effective retrieval of the conversational name "Left Front Wheel Speed Sensor" by indexing it as "Sensor, Left Front Wheel Speed."

5. *Cross Reference and Look Up* —See Figure 1. The left column lists existing terms, acronyms and abbreviations. The center column provides the corresponding acceptable usages constructed of recommended terms combined with other modifiers and/or base words. The acceptable acronized usage is shown in the right column.

For information about using acronyms and abbreviations, see 4.3.1 (Acronyms) and 4.3.2 (Abbreviations). For additional information about Recommended Terms, see Figure 2 and Table 1.

6. *Recommended Terms* —Figure 2 is an alphabetical listing of modifiers to be used in combination with base words.

7. *Glossary of Terms* —Table 1 is an alphabetical listing of base words and single word modifiers, together with their definitions.

8. *Revision Procedures* —It will be appropriate to revise the published J1930 on an ongoing basis. Requested revisions and updates will be controlled by the SAE Vehicle E/E Systems Diagnostic Standards Committee using the normal Recommended Practice Ballot process. This will ensure proper distribution of the changes.

As required by SAE standards, the J1930 document will be formally updated and balloted at least once every five years. When warranted by the number of requested modifications, J1930 will be updated as often as every three months.

Use Appendix A for submission of new information.

Existing Usage	Acceptable Usage	Acceptable Acronized Usage
3GR (Third Gear)	Third Gear	3GR
4GR (Fourth Gear)	Fourth Gear	4GR
A/C (Air Conditioning)	Air Conditioning	A/C
A/C Cycling Switch	Air Conditioning Cycling Switch	A/C Cycling Switch
A/T (Automatic Transaxle)	Automatic Transaxle	A/T
A/T (Automatic Transmission)	Automatic Transmission	A/T
AC (Air Conditioning)	Air Conditioning	A/C
ACC (Air Conditioning Clutch)	Air Conditioning Clutch	A/C Clutch
Accelerator	Accelerator Pedal	AP
ACCS (Air Conditioning Cyclic Switch)	Air Conditioning Cycling Switch	A/C Cycling Switch
ACH (Air Cleaner Housing)	Air Cleaner Housing[1]	ACL Housing[1]
ACL (Air Cleaner)	Air Cleaner[1]	ACL[1]
ACL (Air Cleaner) Element	Air Cleaner Element[1]	ACL Element[1]
ACL (Air Cleaner) Housing	Air Cleaner Housing[1]	ACL Housing[1]
ACL (Air Cleaner) Housing Cover	Air Cleaner Housing Cover[1]	ACL Housing Cover[1]
ACS (Air Conditioning System)	Air Conditioning System	A/C System
ACT (Air Charge Temperature)	Intake Air Temperature[1]	IAT[1]
Adaptive Fuel Strategy	Fuel Trim[1]	FT[1]
AFC (Air Flow Control)	Mass Air Flow[1]	MAF[1]
AFC (Air Flow Control)	Volume Air Flow[1]	VAF[1]
AFS (Air Flow Sensor)	Mass Air Flow Sensor[1]	MAF Sensor[1]
AFS (Air Flow Sensor)	Volume Air Flow Sensor[1]	VAF Sensor[1]
After Cooler	Charge Air Cooler[1]	CAC[1]
AI (Air Injection)	Secondary Air Injection[1]	AIR[1]
AIP (Air Injection Pump)	Secondary Air Injection Pump[1]	AIR Pump[1]
AIR (Air Injection Reactor)	Pulsed Secondary Air Injection[1]	PAIR[1]
AIR (Air Injection Reactor)	Secondary Air Injection[1]	AIR[1]
AIRB (Secondary Air Injection Bypass)	Secondary Air Injection Bypass[1]	AIR Bypass[1]
AIRD (Secondary Air Injection Diverter)	Secondary Air Injection Diverter[1]	AIR Diverter[1]
Air Cleaner	Air Cleaner[1]	ACL[1]
Air Cleaner Element	Air Cleaner Element[1]	ACL Element[1]
Air Cleaner Housing	Air Cleaner Housing[1]	ACL Housing[1]
Air Cleaner Housing Cover	Air Cleaner Housing Cover[1]	ACL Housing Cover[1]
Air Conditioning	Air Conditioning	A/C
Air Conditioning Sensor	Air Conditioning Sensor	A/C Sensor
Air Control Valve	Secondary Air Injection Control Valve[1]	AIR Control Valve[1]
Air Flow Meter	Mass Air Flow Sensor[1]	MAF Sensor[1]
Air Flow Meter	Volume Air Flow Sensor[1]	VAF Sensor[1]
Air Intake System	Intake Air System[1]	IA System[1]
Air Flow Sensor	Mass Air Flow Sensor[1]	MAF Sensor[1]
Air Management 1	Secondary Air Injection Bypass[1]	AIR Bypass[1]
Air Management 2	Secondary Air Injection Diverter[1]	AIR Diverter[1]
Air Temperature Sensor	Intake Air Temperature Sensor[1]	IAT Sensor[1]
Air Valve	Idle Air Control Valve[1]	IAC Valve[1]
AIV (Air Injection Valve)	Pulsed Secondary Air Injection[1]	PAIR[1]
ALCL (Assembly Line Communication Link)	Data Link Connector[1]	DLC[1]
Alcohol Concentration Sensor	Flexible Fuel Sensor[1]	FF Sensor[1]
ALDL (Assembly Line Diagnostic Link)	Data Link Connector[1]	DLC[1]
ALT (Alternator)	Generator	GEN
Alternator	Generator	GEN
AM1 (Air Management 1)	Secondary Air Injection Bypass[1]	AIR Bypass[1]
AM2 (Air Management 2)	Secondary Air Injection Diverter[1]	AIR Diverter[1]
APS (Absolute Pressure Sensor)	Barometric Pressure Sensor[1]	BARO Sensor[1]
ATS (Air Temperature Sensor)	Intake Air Temperature Sensor[1]	IAT Sensor[1]
Automatic Transaxle	Automatic Transaxle[1]	A/T[1]
Automatic Transmission	Automatic Transmission[1]	A/T[1]
B+ (Battery Positive Voltage)	Battery Positive Voltage	B+
Backpressure Transducer	Exhaust Gas Recirculation Backpressure Transducer[1]	EGR Backpressure Transducer[1]
BARO (Barometric Pressure)	Barometric Pressure[1]	BARO[1]
Barometric Pressure Sensor	Barometric Pressure Sensor[1]	BARO Sensor[1]
Battery Positive Voltage	Battery Positive Voltage	B+
Block Learn Matrix	Long Term Fuel Trim[1]	Long Term FT[1]
BLM (Block Learn Memory)	Long Term Fuel Trim[1]	Long Term FT[1]
BLM (Block Learn Multiplier)	Long Term Fuel Trim[1]	Long Term FT[1]
BLM (Block Learn Matrix)	Long Term Fuel Trim[1]	Long Term FT[1]

Recommended Terms and Recommended Acronyms See Figure 2
[1] Emission-Related Term

FIGURE 1—CROSS REFERENCE AND LOOK UP

Existing Usage	Acceptable Usage	Acceptable Acronized Usage
Block Learn Memory	Long Term Fuel Trim[1]	Long Term FT[1]
Block Learn Multiplier	Long Term Fuel Trim[1]	Long Term FT[1]
BP (Barometric Pressure) Sensor	Barometric Pressure Sensor[1]	BARO Sensor[1]
C3I (Computer Controlled Coil Ignition)	Electronic Ignition[1]	EI[1]
CAC (Charge Air Cooler)	Charge Air Cooler[1]	CAC[1]
Camshaft Position	Camshaft Position[1]	CMP[1]
Camshaft Position Sensor	Camshaft Position Sensor[1]	CMP Sensor[1]
Camshaft Sensor	Camshaft Position Sensor[1]	CMP Sensor[1]
Canister	Canister[1]	Canister[1]
Canister	Evaporative Emission Canister[1]	EVAP Canister[1]
Canister Purge Valve	Evaporative Emission Canister Purge Valve[1]	EVAP Canister Purge Valve[1]
Canister Purge Vacuum Switching Valve	Evaporative Emission Canister Purge Valve[1]	EVAP Canister Purge Valve[1]
Canister Purge VSV (Vacuum Switching Valve)	Evaporative Emission Canister Purge Valve[1]	EVAP Canister Purge Valve[1]
CANP (Canister Purge)[1]	Evaporative Emission Canister Purge[1]	EVAP Canister Purge[1]
CARB (Carburetor)	Carburetor[1]	CARB[1]
Carburetor	Carburetor[1]	CARB[1]
CCC (Converter Clutch Control)	Torque Converter Clutch[1]	TCC[1]
CCO (Converter Clutch Override)	Torque Converter Clutch[1]	TCC[1]
CDI (Capacitive Discharge Ignition)	Distributor Ignition[1]	DI[1]
CDROM (Compact Disc Read Only Memory)	Compact Disc Read Only Memory[1]	CDROM[1]
CES (Clutch Engage Switch)	Clutch Pedal Position Switch	CPP Switch
Central Multiport Fuel Injection	Central Multiport Fuel Injection[1]	Central MFI[1]
CFI (Continuous Fuel Injection)	Continuous Fuel Injection[1]	CFI[1]
CFI (Central Fuel Injection)	Throttle Body Fuel Injection[1]	TBI[1]
Charcoal Canister	Evaporative Emission Canister	EVAP Canister[1]
Charge Air Cooler	Charge Air Cooler[1]	CAC[1]
Check Engine	Service Reminder Indicator[1]	SRI[1]
Check Engine	Malfunction Indicator Lamp[1]	MIL[1]
CID (Cylinder Identification) Sensor	Camshaft Position Sensor	CMP Sensor[1]
CIS (Continuous Injection System)	Continuous Fuel Injection[1]	CFI[1]
CIS-E (Continuous Injection System-Electronic)	Continuous Fuel Injection[1]	CFI[1]
CKP (Crankshaft Position)	Crankshaft Position[1]	CKP[1]
CKP (Crankshaft Position) Sensor	Crankshaft Position Sensor[1]	CKP Sensor[1]
CL (Closed Loop)	Closed Loop[1]	CL[1]
Closed Bowl Distributor	Distributor Ignition[1]	DI[1]
Closed Throttle Position	Closed Throttle Position[1]	CTP[1]
Closed Throttle Switch	Closed Throttle Position Switch[1]	CTP Switch[1]
CLS (Closed Loop System)	Closed Loop[1]	CL[1]
Clutch Engage Switch	Clutch Pedal Position Switch[1]	CPP Switch[1]
Clutch Pedal Position Switch	Clutch Pedal Position Switch[1]	CPP Switch[1]
Clutch Start Switch	Clutch Pedal Position Switch[1]	CPP Switch[1]
Clutch Switch	Clutch Pedal Position Switch[1]	CPP Switch[1]
CMFI (Central Multiport Fuel Injection)	Central Multiport Fuel Injection[1]	Central MFI[1]
CMP (Camshaft Position)	Camshaft Position[1]	CMP[1]
CMP (Camshaft Position) Sensor	Camshaft Position Sensor[1]	CMP Sensor[1]
COC (Continuous Oxidation Catalyst)	Oxidation Catalytic Converter[1]	OC[1]
Condenser	Distributor Ignition Capacitor[1]	DI Capacitor[1]
Continuous Fuel Injection	Continuous Fuel Injection[1]	CFI[1]
Continuous Injection System	Continuous Fuel Injection System[1]	CFI System[1]
Continuous Injection System-E	Electronic Continuous Fuel Injection System[1]	Electronic CFI System[1]
Continuous Trap Oxidizer	Continuous Trap Oxidizer[1]	CTOX[1]
Coolant Temperature Sensor	Engine Coolant Temperature Sensor[1]	ECT Sensor[1]
CP (Crankshaft Position)	Crankshaft Position[1]	CKP[1]
CPP (Clutch Pedal Position)	Clutch Pedal Position[1]	CPP[1]
CPP (Clutch Pedal Position) Switch	Clutch Pedal Position Switch	CPP Switch[1]
CPS (Camshaft Position Sensor)	Camshaft Position Sensor[1]	CMP Sensor[1]
CPS (Crankshaft Position Sensor)	Crankshaft Position Sensor[1]	CKP Sensor[1]
Crank Angle Sensor	Crankshaft Position Sensor[1]	CKP Sensor[1]
Crankshaft Position	Crankshaft Position[1]	CKP[1]
Crankshaft Position Sensor	Crankshaft Position Sensor[1]	CKP Sensor[1]
Crankshaft Speed	Engine Speed[1]	RPM[1]
Crankshaft Speed Sensor	Engine Speed Sensor[1]	RPM Sensor[1]
CTO (Continuous Trap Oxidizer)	Continuous Trap Oxidizer[1]	CTOX[1]
CTOX (Continuous Trap Oxidizer)	Continuous Trap Oxidizer[1]	CTOX[1]
CTP (Closed Throttle Position)	Closed Throttle Position[1]	CTP[1]
CTS (Coolant Temperature Sensor)	Engine Coolant Temperature Sensor[1]	ECT Sensor[1]

Recommended Terms and Recommended Acronyms See Figure 2
[1] Emission-Related Term

FIGURE 1—CROSS REFERENCE AND LOOK UP (CONTINUED)

Existing Usage	Acceptable Usage	Acceptable Acronized Usage
CTS (Coolant Temperature Switch)	Engine Coolant Temperature Switch[1]	ECT Switch[1]
Cylinder ID (Identification) Sensor	Camshaft Position Sensor[1]	CMP Sensor[1]
D-Jetronic	Multiport Fuel Injection[1]	MFI[1]
Data Link Connector	Data Link Connector[1]	DLC[1]
Detonation Sensor	Knock Sensor[1]	KS[1]
DFI (Direct Fuel Injection)	Direct Fuel Injection[1]	DFI[1]
DFI (Digital Fuel Injection)	Multiport Fuel Injection[1]	MFI[1]
DI (Direct Injection)	Direct Fuel Injection[1]	DFI[1]
DI (Distributor Ignition)	Distributor Ignition[1]	DI[1]
DI (Distributor Ignition) Capacitor	Distributor Ignition Capacitor[1]	DI Capacitor[1]
Diagnostic Test Mode	Diagnostic Test Mode[1]	DTM[1]
Diagnostic Trouble Code	Diagnostic Trouble Code[1]	DTC[1]
DID (Direct Injection - Diesel)	Direct Fuel Injection[1]	DFI[1]
Differential Pressure Feedback EGR (Exhaust Gas Recirculation) System	Differential Pressure Feedback Exhaust Gas Recirculation System[1]	Differential Pressure Feedback EGR System[1]
Digital EGR (Exhaust Gas Recirculation)	Exhaust Gas Recirculation[1]	EGR[1]
Direct Fuel Injection	Direct Fuel Injection[1]	DFI[1]
Direct Ignition System	Electronic Ignition System[1]	EI System[1]
DIS (Distributorless Ignition System)	Electronic Ignition System[1]	EI System[1]
DIS (Distributorless Ignition System) Module	Ignition Control Module[1]	ICM[1]
Distance Sensor	Vehicle Speed Sensor[1]	VSS[1]
Distributor Ignition	Distributor Ignition[1]	DI[1]
Distributorless Ignition	Electronic Ignition[1]	EI[1]
DLC (Data Link Connector)	Data Link Connector[1]	DLC[1]
DLI (Distributorless Ignition)	Electronic Ignition[1]	EI[1]
DS (Detonation Sensor)	Knock Sensor[1]	KS[1]
DTC (Diagnostic Trouble Code)	Diagnostic Trouble Code[1]	DTC[1]
DTM (Diagnostic Test Mode)	Diagnostic Test Mode[1]	DTM[1]
Dual Bed	Three Way + Oxidation Catalytic Converter[1]	TWC+OC[1]
Duty Solenoid for Purge Valve	Evaporative Emission Canister Purge Valve[1]	EVAP Canister Purge Valve[1]
E2PROM (Electrically Erasable Programmable Read Only Memory)	Electrically Erasable Programmable Read Only Memory[1]	EEPROM[1]
Early Fuel Evaporation	Early Fuel Evaporation[1]	EFE[1]
EATX (Electronic Automatic Transmission/Transaxle)	Automatic Transmission Automatic Transaxle	A/T A/T
EC (Engine Control)	Engine Control[1]	EC[1]
ECA (Electronic Control Assembly)	Powertrain Control Module[1]	PCM[1]
ECL (Engine Coolant Level)	Engine Coolant Level	ECL
ECM (Engine Control Module)	Engine Control Module[1]	ECM[1]
ECT (Engine Coolant Temperature)	Engine Coolant Temperature[1]	ECT[1]
ECT (Engine Coolant Temperature) Sender	Engine Coolant Temperature Sensor[1]	ECT Sensor[1]
ECT (Engine Coolant Temperature) Sensor	Engine Coolant Temperature Sensor[1]	ECT Sensor[1]
ECT (Engine Coolant Temperature) Switch	Engine Coolant Temperature Switch[1]	ECT Switch[1]
ECU4 (Electronic Control Unit 4)	Powertrain Control Module[1]	PCM
EDF (Electro-Drive Fan) Control	Fan Control	FC
EDIS (Electronic Distributor Ignition System)	Distributor Ignition System[1]	DI System[1]
EDIS (Electronic Distributorless Ignition System)	Electronic Ignition System[1]	EI System[1]
EDIS (Electronic Distributor Ignition System) Module	Distributor Ignition Control Module[1]	Distributor ICM
EEC (Electronic Engine Control)	Engine Control[1]	EC[1]
EEC (Electronic Engine Control) Processor	Powertrain Control Module[1]	PCM[1]
EECS (Evaporative Emission Control System)	Evaporative Emission System[1]	EVAP System[1]
EEPROM (Electrically Erasable Programmable Read Only Memory)	Electrically Erasable Programmable Read Only Memory[1]	EEPROM[1]
EFE (Early Fuel Evaporation)	Early Fuel Evaporation[1]	EFE[1]
EFI (Electronic Fuel Injection)	Multiport Fuel Injection[1]	MFI[1]
EFI (Electronic Fuel Injection)	Throttle Body Fuel Injection[1]	TBI[1]
EGO (Exhaust Gas Oxygen) Sensor	Oxygen Sensor[1]	O2S[1]
EGOS (Exhaust Gas Oxygen Sensor)	Oxygen Sensor[1]	O2S[1]
EGR (Exhaust Gas Recirculation)	Exhaust Gas Recirculation[1]	EGR
EGR (Exhaust Gas Recirculation) Diagnostic Valve	Exhaust Gas Recirculation Diagnostic Valve[1]	EGR Diagnostic Valve[1]
EGR (Exhaust Gas Recirculation) System	Exhaust Gas Recirculation System[1]	EGR System[1]
EGR (Exhaust Gas Recirculation) Thermal Vacuum Valve	Exhaust Gas Recirculation Thermal Vacuum Valve[1]	EGR TVV[1]
EGR (Exhaust Gas Recirculation) Valve	Exhaust Gas Recirculation Valve[1]	EGR Valve[1]
EGR TVV (Exhaust Gas Recirculation Thermal Vacuum Valve)	Exhaust Gas Recirculation Thermal Vacuum Valve[1]	EGR TVV[1]
EGRT (Exhaust Gas Recirculation Temperature)	Exhaust Gas Recirculation Temperature[1]	EGRT
EGRT (Exhaust Gas Recirculation Temperature) Sensor	Exhaust Gas Recirculation Temperature Sensor[1]	EGRT Sensor[1]

Recommended Terms and Recommended Acronyms See Figure 2
[1] Emission-Related Term

FIGURE 1—CROSS REFERENCE AND LOOK UP (CONTINUED)

Existing Usage	Acceptable Usage	Acceptable Acronized Usage
EGRV (Exhaust Gas Recirculation Valve)	Exhaust Gas Recirculation Valve[1]	EGR Valve[1]
EGRVC (Exhaust Gas Recirculation Valve Control)	Exhaust Gas Recirculation Valve Control[1]	EGR Valve Control[1]
EGS (Exhaust Gas Sensor)	Oxygen Sensor[1]	O2S[1]
EI (Electronic Ignition) (With Distributor)	Distributor Ignition[1]	DI[1]
EI (Electronic Ignition) (Without Distributor)	Electronic Ignition[1]	EI[1]
Electrically Erasable Programmable Read Only Memory	Electrically Erasable Programmable Read Only Memory[1]	EEPROM[1]
Electronic Engine Control	Electronic Engine Control[1]	Electronic EC[1]
Electronic Ignition	Electronic Ignition[1]	EI[1]
Electronic Spark Advance	Ignition Control[1]	IC[1]
Electronic Spark Timing	Ignition Control[1]	IC[1]
EM (Engine Modification)	Engine Modification[1]	EM[1]
EMR (Engine Maintenance Reminder)	Service Reminder Indicator[1]	SRI[1]
Engine Control	Engine Control[1]	EC[1]
Engine Coolant Fan Control	Fan Control	FC
Engine Coolant Level	Engine Coolant Level	ECL
Engine Coolant Level Indicator	Engine Coolant Level Indicator	ECL Indicator
Engine Coolant Temperature	Engine Coolant Temperature[1]	ECT[1]
Engine Coolant Temperature Sender	Engine Coolant Temperature Sensor[1]	ECT Sensor[1]
Engine Coolant Temperature Sensor	Engine Coolant Temperature Sensor[1]	ECT Sensor[1]
Engine Coolant Temperature Switch	Engine Coolant Temperature Switch[1]	ECT Switch[1]
Engine Modification	Engine Modification[1]	EM[1]
Engine Speed	Engine Speed[1]	RPM[1]
EOS (Exhaust Oxygen Sensor)	Oxygen Sensor[1]	O2S[1]
EPROM (Erasable Programmable Read Only Memory)	Erasable Programmable Read Only Memory[1]	EPROM[1]
Erasable Programmable Read Only Memory	Erasable Programmable Read Only Memory[1]	EPROM[1]
ESA (Electronic Spark Advance)	Ignition Control[1]	IC[1]
ESAC (Electronic Spark Advance Control)	Distributor Ignition[1]	DI[1]
EST (Electronic Spark Timing)	Ignition Control[1]	IC[1]
EVAP CANP	Evaporative Emission Canister Purge[1]	EVAP Canister Purge[1]
EVAP (Evaporative Emission)	Evaporative Emission[1]	EVAP[1]
EVAP (Evaporative Emission) Canister	Evaporative Emission Canister[1]	EVAP Canister[1]
EVAP (Evaporative Emission) Purge Valve	Evaporative Emission Canister Purge Valve[1]	EVAP Canister Purge Valve[1]
Evaporative Emission	Evaporative Emission[1]	EVAP[1]
Evaporative Emission Canister	Evaporative Emission Canister[1]	EVAP Canister[1]
EVP (Exhaust Gas Recirculation Valve Position) Sensor	Exhaust Gas Recirculation Valve Position Sensor[1]	EGR Valve Position Sensor[1]
EVR (Exhaust Gas Recirculation Vacuum Regulator) Solenoid	Exhaust Gas Recirculation Vacuum Regulator Solenoid[1]	EGR Vacuum Regulator Solenoid[1]
EVRV (Exhaust Gas Recirculation Vacuum Regulator Valve)	Exhaust Gas Recirculation Vacuum Regulator Valve[1]	EGR Vacuum Regulator Valve[1]
Exhaust Gas Recirculation	Exhaust Gas Recirculation[1]	EGR[1]
Exhaust Gas Recirculation Temperature	Exhaust Gas Recirculation Temperature[1]	EGRT[1]
Exhaust Gas Recirculation Temperature Sensor	Exhaust Gas Recirculation Temperature Sensor[1]	EGRT Sensor[1]
Exhaust Gas Recirculation Valve	Exhaust Gas Recirculation Valve[1]	EGR Valve[1]
Fan Control	Fan Control	FC
Fan Control Module	Fan Control Module	FC Module
Fan Control Relay	Fan Control Relay	FC Relay
Fan Motor Control Relay	Fan Control Relay	FC Relay
Fast Idle Thermo Valve	Idle Air Control Thermal Valve[1]	IAC Thermal Valve[1]
FBC (Feed Back Carburetor)	Carburetor[1]	CARB[1]
FBC (Feed Back Control)	Mixture Control[1]	MC[1]
FC (Fan Control)	Fan Control	FC
FC (Fan Control) Relay	Fan Control Relay	FC Relay
FEEPROM (Flash Electrically Erasable Programmable Read Only Memory)	Flash Electrically Erasable Programmable Read Only Memory[1]	FEEPROM[1]
FEPROM (Flash Erasable Programmable Read Only Memory)	Flash Erasable Programmable Read Only Memory[1]	FEPROM[1]
FF (Flexible Fuel)	Flexible Fuel[1]	FF[1]
FI (Fuel Injection)	Central Multiport Fuel Injection[1]	Central MFI[1]
FI (Fuel Injection)	Continuous Fuel Injection[1]	CFI[1]
FI (Fuel Injection)	Direct Fuel Injection[1]	DFI[1]
FI (Fuel Injection)	Indirect Fuel Injection[1]	IFI[1]
FI (Fuel Injection)	Multiport Fuel Injection[1]	MFI[1]
FI (Fuel Injection)	Sequential Multiport Fuel Injection[1]	SFI[1]
FI (Fuel Injection)	Throttle Body Fuel Injection[1]	TBI[1]
Flash EEPROM (Electrically Erasable Programmable Read Only Memory)	Flash Electrically Erasable Programmable Read Only Memory[1]	FEEPROM[1]

Recommended Terms and Recommended Acronyms See Figure 2
[1] Emission-Related Term

FIGURE 1—CROSS REFERENCE AND LOOK UP (CONTINUED)

Existing Usage	Acceptable Usage	Acceptable Acronized Usage
Flash EPROM (Erasable Programmable Read Only Memory)	Flash Erasable Programmable Read Only Memory[1]	FEPROM[1]
Flexible Fuel	Flexible Fuel[1]	FF[1]
Flexible Fuel Sensor	Flexible Fuel Sensor[1]	FF Sensor[1]
Fourth Gear	Fourth Gear	4GR
FP (Fuel Pump)	Fuel Pump	FP
FP (Fuel Pump) Module	Fuel Pump Module	FP Module
FT (Fuel Trim)	Fuel Trim[1]	FT[1]
Fuel Charging Station	Throttle Body[1]	TB[1]
Fuel Concentration Sensor	Flexible Fuel Sensor[1]	FF Sensor[1]
Fuel Injection	Central Multiport Fuel Injection[1]	Central MFI[1]
Fuel Injection	Continuous Fuel Injection[1]	CFI[1]
Fuel Injection	Direct Fuel Injection[1]	DFI[1]
Fuel Injection	Indirect Fuel Injection[1]	IFI[1]
Fuel Injection	Multiport Fuel Injection[1]	MFI[1]
Fuel Injection	Sequential Multiport Fuel Injection[1]	SFI[1]
Fuel Injection	Throttle Body Fuel Injection[1]	TBI[1]
Fuel Level Sensor	Fuel Level Sensor	Fuel Level Sensor
Fuel Module	Fuel Pump Module	FP Module
Fuel Pressure	Fuel Pressure[1]	Fuel Pressure[1]
Fuel Pressure Regulator	Fuel Pressure Regulator[1]	Fuel Pressure Regulator[1]
Fuel Pump	Fuel Pump	FP
Fuel Pump Relay	Fuel Pump Relay	FP Relay
Fuel Quality Sensor	Flexible Fuel Sensor[1]	FF Sensor[1]
Fuel Regulator	Fuel Pressure Regulator[1]	Fuel Pressure Regulator[1]
Fuel Sender	Fuel Pump Module	FP Module
Fuel Sensor	Fuel Level Sensor	Fuel Level Sensor
Fuel Tank Unit	Fuel Pump Module	FP Module
Fuel Trim	Fuel Trim[1]	FT[1]
Full Throttle	Wide Open Throttle[1]	WOT[1]
GCM (Governor Control Module)	Governor Control Module	GCM
GEM (Governor Electronic Module)	Governor Control Module	GCM
GEN (Generator)	Generator	GEN
Generator	Generator	GEN
Governor	Governor	Governor
Governor Control Module	Governor Control Module	GCM
Governor Electronic Module	Governor Control Module	GCM
GND (Ground)	Ground	GND
GRD (Ground)	Ground	GND
Ground	Ground	GND
Heated Oxygen Sensor	Heated Oxygen Sensor[1]	HO2S[1]
HEDF (High Electro-Drive Fan) Control	Fan Control	FC
HEGO (Heated Exhaust Gas Oxygen) Sensor	Heated Oxygen Sensor[1]	HO2S[1]
HEI (High Energy Ignition)	Distributor Ignition[1]	DI[1]
High Speed FC (Fan Control) Switch	High Speed Fan Control Switch	High Speed FC Switch
HO2S (Heated Oxygen Sensor)	Heated Oxygen Sensor[1]	HO2S[1]
HOS (Heated Oxygen Sensor)	Heated Oxygen Sensor[1]	HO2S[1]
Hot Wire Anemometer	Mass Air Flow Sensor[1]	MAF Sensor[1]
IA (Intake Air)	Intake Air	IA
IA (Intake Air) Duct	Intake Air Duct	IA Duct
IAC (Idle Air Control)	Idle Air Control[1]	IAC[1]
IAC (Idle Air Control) Thermal Valve	Idle Air Control Thermal Valve[1]	IAC Thermal Valve[1]
IAC (Idle Air Control) Valve	Idle Air Control Valve[1]	IAC Valve[1]
IACV (Idle Air Control Valve)	Idle Air Control Valve[1]	IAC Valve[1]
IAT (Intake Air Temperature)	Intake Air Temperature[1]	IAT[1]
IAT (Intake Air Temperature) Sensor	Intake Air Temperature Sensor[1]	IAT Sensor[1]
IATS (Intake Air Temperature Sensor)	Intake Air Temperature Sensor[1]	IAT Sensor[1]
IC (Ignition Control)	Ignition Control[1]	IC[1]
ICM (Ignition Control Module)	Ignition Control Module[1]	ICM[1]
IDFI (Indirect Fuel Injection)	Indirect Fuel Injection[1]	IFI[1]
IDI (Integrated Direct Ignition)	Electronic Ignition[1]	EI[1]
IDI (Indirect Diesel Injection)	Indirect Fuel Injection[1]	IFI[1]
Idle Air Bypass Control	Idle Air Control[1]	IAC[1]
Idle Air Control	Idle Air Control[1]	IAC[1]
Idle Air Control Valve	Idle Air Control Valve[1]	IAC Valve[1]
Idle Speed Control	Idle Air Control[1]	IAC[1]
Idle Speed Control	Idle Speed Control[1]	ISC[1]

Recommended Terms and Recommended Acronyms See Figure 2
[1] Emission-Related Term

FIGURE 1—CROSS REFERENCE AND LOOK UP (CONTINUED)

Existing Usage	Acceptable Usage	Acceptable Acronized Usage
Idle Speed Control Actuator	Idle Speed Control Actuator[1]	ISC Actuator[1]
IFI (Indirect Fuel Injection)	Indirect Fuel Injection[1]	IFI[1]
IFS (Inertia Fuel Shutoff)	Inertia Fuel Shutoff	IFS
Ignition Control	Ignition Control[1]	IC[1]
Ignition Control Module	Ignition Control Module[1]	ICM[1]
In Tank Module	Fuel Pump Module	FP Module
Indirect Fuel Injection	Indirect Fuel Injection[1]	IFI[1]
Inertia Fuel Shutoff	Inertia Fuel Shutoff	IFS
Inertia Fuel - Shutoff Switch	Inertia Fuel Shutoff Switch	IFS Switch
Inertia Switch	Inertia Fuel Shutoff Switch	IFS Switch
INT (Integrator)	Short Term Fuel Trim[1]	Short Term FT[1]
Intake Air	Intake Air	IA
Intake Air Duct	Intake Air Duct	IA Duct
Intake Air Temperature	Intake Air Temperature[1]	IAT[1]
Intake Air Temperature Sensor	Intake Air Temperature Sensor[1]	IAT Sensor[1]
Intake Manifold Absolute Pressure Sensor	Manifold Absolute Pressure Sensor[1]	MAP Sensor[1]
Integrated Relay Module	Relay Module	RM
Integrator	Short Term Fuel Trim[1]	Short Term FT[1]
Inter Cooler	Charge Air Cooler[1]	CAC[1]
ISC (Idle Speed Control)	Idle Air Control[1]	IAC[1]
ISC (Idle Speed Control)	Idle Speed Control[1]	ISC[1]
ISC (Idle Speed Control) Actuator	Idle Speed Control Actuator[1]	ISC Actuator[1]
ISC BPA (Idle Speed Control By Pass Air)	Idle Air Control[1]	IAC[1]
ISC (Idle Speed Control) Solenoid Vacuum Valve	Idle Speed Control Solenoid Vacuum Valve[1]	ISC Solenoid Vacuum Valve[1]
K-Jetronic	Continuous Fuel Injection[1]	CFI[1]
KAM (Keep Alive Memory)	NonVolatile Random Access Memory[1]	NVRAM[1]
KAM (Keep Alive Memory)	Keep Alive Random Access Memory[1]	Keep Alive RAM[1]
KE-Jetronic	Continuous Fuel Injection[1]	CFI[1]
KE-Motronic	Continuous Fuel Injection[1]	CFI[1]
Knock Sensor	Knock Sensor[1]	KS[1]
KS (Knock Sensor)	Knock Sensor[1]	KS[1]
L-Jetronic	Multiport Fuel Injection[1]	MFI[1]
Lambda	Oxygen Sensor[1]	O2S[1]
LH-Jetronic	Multiport Fuel Injection[1]	MFI[1]
Light Off Catalyst	Warm Up Three Way Catalytic Converter[1]	WU-TWC[1]
Light Off Catalyst	Warm Up Oxidation Catalytic Converter[1]	WU-OC[1]
Lock Up Relay	Torque Converter Clutch Relay[1]	TCC Relay[1]
Long Term FT (Fuel Trim)	Long Term Fuel Trim[1]	Long Term FT[1]
Low Speed FC (Fan Control) Switch	Low Speed Fan Control Switch	Low Speed FC Switch
LUS (Lock Up Solenoid) Valve	Torque Converter Clutch Solenoid Valve[1]	TCC Solenoid Valve[1]
M/C (Mixture Control)	Mixture Control[1]	MC[1]
MAF (Mass Air Flow)	Mass Air Flow[1]	MAF[1]
MAF (Mass Air Flow) Sensor	Mass Air Flow Sensor[1]	MAF Sensor[1]
Malfunction Indicator Lamp	Malfunction Indicator Lamp[1]	MIL[1]
Manifold Absolute Pressure	Manifold Absolute Pressure[1]	MAP[1]
Manifold Absolute Pressure Sensor	Manifold Absolute Pressure Sensor	MAP Sensor
Manifold Differential Pressure	Manifold Differential Pressure[1]	MDP[1]
Manifold Surface Temperature	Manifold Surface Temperature[1]	MST[1]
Manifold Vacuum Zone	Manifold Vacuum Zone[1]	MVZ[1]
Manual Lever Position Sensor	Transmission Range Sensor[1]	TR Sensor[1]
MAP (Manifold Absolute Pressure)	Manifold Absolute Pressure[1]	MAP[1]
MAP (Manifold Absolute Pressure) Sensor	Manifold Absolute Pressure Sensor[1]	MAP Sensor[1]
MAPS (Manifold Absolute Pressure Sensor)	Manifold Absolute Pressure Sensor[1]	MAP Sensor[1]
Mass Air Flow	Mass Air Flow[1]	MAF[1]
Mass Air Flow Sensor	Mass Air Flow Sensor[1]	MAF Sensor[1]
MAT (Manifold Air Temperature)	Intake Air Temperature[1]	IAT[1]
MATS (Manifold Air Temperature Sensor)	Intake Air Temperature Sensor[1]	IAT Sensor[1]
MC (Mixture Control)	Mixture Control[1]	MC[1]
MCS (Mixture Control Solenoid)	Mixture Control Solenoid[1]	MC Solenoid[1]
MCU (Microprocessor Control Unit)	Powertrain Control Module[1]	PCM[1]
MDP (Manifold Differential Pressure)	Manifold Differential Pressure[1]	MDP[1]
MFI (Multiport Fuel Injection)	Multiport Fuel Injection[1]	MFI[1]
MIL (Malfunction Indicator Lamp)	Malfunction Indicator Lamp[1]	MIL[1]
Mixture Control	Mixture Control[1]	MC[1]
Modes	Diagnostic Test Mode[1]	DTM[1]
Monotronic	Throttle Body Fuel Injection[1]	TBI[1]

Recommended Terms and Recommended Acronyms See Figure 2
[1] Emission-Related Term

FIGURE 1—CROSS REFERENCE AND LOOK UP (CONTINUED)

Existing Usage	Acceptable Usage	Acceptable Acronized Usage
Motronic	Multiport Fuel Injection[1]	MFI[1]
MPI (Multipoint Injection)	Multiport Fuel Injection[1]	MFI[1]
MPI (Multiport Injection)	Multiport Fuel Injection[1]	MFI[1]
MRPS (Manual Range Position Switch)	Transmission Range Switch	TR Switch
MST (Manifold Surface Temperature)	Manifold Surface Temperature[1]	MST[1]
Multiport Fuel Injection	Multiport Fuel Injection[1]	MFI[1]
MVZ (Manifold Vacuum Zone)	Manifold Vacuum Zone[1]	MVZ[1]
NDS (Neutral Drive Switch)	Park/Neutral Position Switch[1]	PNP Switch[1]
Neutral Safety Switch	Park/Neutral Position Switch[1]	PNP Switch[1]
NGS (Neutral Gear Switch)	Park/Neutral Position Switch[1]	PNP Switch[1]
Nonvolatile Random Access Memory	Nonvolatile Random Access Memory[1]	NVRAM[1]
NPS (Neutral Position Switch)	Park/Neutral Position Switch[1]	PNP Switch[1]
NVM (Nonvolatile Memory)	Nonvolatile Random Access Memory[1]	NVRAM[1]
NVRAM (Nonvolatile Random Access Memory)	Nonvolatile Random Access Memory[1]	NVRAM[1]
O2 (Oxygen) Sensor	Oxygen Sensor[1]	O2S[1]
O2S (Oxygen Sensor)	Oxygen Sensor[1]	O2S[1]
OBD (On Board Diagnostic)	On Board Diagnostic[1]	OBD[1]
OC (Oxidation Catalyst)	Oxidation Catalytic Converter[1]	OC[1]
Oil Pressure Sender	Oil Pressure Sensor	Oil Pressure Sensor
Oil Pressure Sensor	Oil Pressure Sensor	Oil Pressure Sensor
Oil Pressure Switch	Oil Pressure Switch	Oil Pressure Switch
OL (Open Loop)	Open Loop[1]	OL[1]
On Board Diagnostic	On Board Diagnostic[1]	OBD[1]
Open Loop	Open Loop[1]	OL[1]
OS (Oxygen Sensor)	Oxygen Sensor[1]	O2S[1]
Oxidation Catalytic Converter	Oxidation Catalytic Converter[1]	OC[1]
OXS (Oxygen Sensor) Indicator	Service Reminder Indicator[1]	SRI[1]
Oxygen Sensor	Oxygen Sensor[1]	O2S[1]
P/N (Park/Neutral)	Park/Neutral Position[1]	PNP
P/S (Power Steering) Pressure Switch	Power Steering Pressure Switch	PSP Switch
P- (Pressure) Sensor	Manifold Absolute Pressure Sensor[1]	MAP Sensor[1]
PAIR (Pulsed Secondary Air Injection)	Pulsed Secondary Air Injection[1]	PAIR[1]
Park/Neutral Position	Park/Neutral Position[1]	PNP
PCM (Powertrain Control Module)	Powertrain Control Module[1]	PCM[1]
PCV (Positive Crankcase Ventilation)	Positive Crankcase Ventilation[1]	PCV[1]
PCV (Positive Crankcase Ventilation) Valve	Positive Crankcase Ventilation Valve[1]	PCV Valve[1]
Percent Alcohol Sensor	Flexible Fuel Sensor[1]	FF Sensor[1]
Periodic Trap Oxidizer	Periodic Trap Oxidizer[1]	PTOX[1]
PFE (Pressure Feedback Exhaust Gas Recirculation) Sensor	Feedback Pressure Exhaust Gas Recirculation Sensor[1]	Feedback Pressure EGR Sensor[1]
PFI (Port Fuel Injection)	Multiport Fuel Injection[1]	MFI[1]
PG (Pulse Generator)	Vehicle Speed Sensor[1]	VSS[1]
PGM-FI (Programmed Fuel Injection)	Multiport Fuel Injection[1]	MFI[1]
PIP (Position Indicator Pulse)	Crankshaft Position[1]	CKP[1]
PNP (Park/Neutral Position)	Park/Neutral Position[1]	PNP[1]
Positive Crankcase Ventilation	Positive Crankcase Ventilation[1]	PCV[1]
Positive Crankcase Ventilation Valve	Positive Crankcase Ventilation Valve[1]	PCV Valve[1]
Power Steering Pressure	Power Steering Pressure	PSP
Power Steering Pressure Switch	Power Steering Pressure Switch	PSP Switch
Powertrain Control Module	Powertrain Control Module	PCM
Pressure Feedback EGR (Exhaust Gas Recirculation)	Feedback Pressure Exhaust Gas Recirculation[1]	Feedback Pressure EGR[1]
Pressure Sensor	Manifold Absolute Pressure Sensor[1]	MAP Sensor[1]
Pressure Transducer EGR (Exhaust Gas Recirculation) System	Pressure Transducer Exhaust Gas Recirculation System[1]	Pressure Transducer EGR System[1]
PRNDL (Park, Reverse, Neutral, Drive, Low)	Transmission Range	TR
Programmable Read Only Memory	Programmable Read Only Memory[1]	PROM[1]
PROM (Programmable Read Only Memory)	Programmable Read Only Memory[1]	PROM[1]
PSP (Power Steering Pressure)	Power Steering Pressure	PSP
PSP (Power Steering Pressure) Switch	Power Steering Pressure Switch	PSP Switch
PSPS (Power Steering Pressure Switch)	Power Steering Pressure Switch	PSP Switch
PTOX (Periodic Trap Oxidizer)	Periodic Trap Oxidizer[1]	PTOX[1]
Pulsair	Pulsed Secondary Air Injection[1]	PAIR[1]
Pulsed Secondary Air Injection	Pulsed Secondary Air Injection[1]	PAIR[1]
Radiator Fan Control	Fan Control	FC
Radiator Fan Relay	Fan Control Relay	FC Relay
RAM (Random Access Memory)	Random Access Memory[1]	RAM[1]

Recommended Terms and Recommended Acronyms See Figure 2
[1] Emission-Related Term

FIGURE 1—CROSS REFERENCE AND LOOK UP (CONTINUED)

Existing Usage	Acceptable Usage	Acceptable Acronized Usage
Random Access Memory	Random Access Memory[1]	RAM[1]
Read Only Memory	Read Only Memory[1]	ROM[1]
Recirculated Exhaust Gas Temperature Sensor	Exhaust Gas Recirculation Temperature Sensor[1]	EGRT Sensor[1]
Reed Valve	Pulsed Secondary Air Injection Valve[1]	PAIR Valve[1]
REGTS (Recirculated Exhaust Gas Temperature Sensor)	Exhaust Gas Recirculation Temperature Sensor[1]	EGRT Sensor[1]
Relay Module	Relay Module	RM
Remote Mount TFI (Thick Film Ignition)	Distributor Ignition[1]	DI[1]
Revolutions per Minute	Engine Speed[1]	RPM[1]
RM (Relay Module)	Relay Module	RM
ROM (Read Only Memory)	Read Only Memory[1]	ROM[1]
RPM (Revolutions per Minute)	Engine Speed[1]	RPM[1]
SABV (Secondary Air Bypass Valve)	Secondary Air Injection Bypass Valve[1]	AIR Bypass Valve[1]
SACV (Secondary Air Check Valve)	Secondary Air Injection Control Valve[1]	AIR Control Valve[1]
SASV (Secondary Air Switching Valve)	Secondary Air Injection Switching Valve[1]	AIR Switching Valve[1]
SBEC (Single Board Engine Control)	Powertrain Control Module[1]	PCM[1]
SBS (Supercharger Bypass Solenoid)	Supercharger Bypass Solenoid[1]	SCB Solenoid[1]
SC (Supercharger)	Supercharger[1]	SC[1]
Scan Tool	Scan Tool[1]	ST[1]
SCB (Supercharger Bypass)	Supercharger Bypass[1]	SCB[1]
Secondary Air Bypass Valve	Secondary Air Injection Bypass Valve[1]	AIR Bypass Valve[1]
Secondary Air Check Valve	Secondary Air Injection Check Valve[1]	AIR Check Valve[1]
Secondary Air Injection	Secondary Air Injection[1]	AIR[1]
Secondary Air Injection Bypass	Secondary Air Injection Bypass[1]	AIR Bypass[1]
Secondary Air Injection Diverter	Secondary Air Injection Diverter[1]	AIR Diverter[1]
Secondary Air Switching Valve	Secondary Air Injection Switching Valve[1]	AIR Switching Valve[1]
SEFI (Sequential Electronic Fuel Injection)	Sequential Multiport Fuel Injection[1]	SFI[1]
Self Test	On Board Diagnostic[1]	OBD[1]
Self Test Codes	Diagnostic Trouble Code[1]	DTC[1]
Self Test Connector	Data Link Connector[1]	DLC[1]
Sequential Multiport Fuel Injection	Sequential Multiport Fuel Injection[1]	SFI[1]
Service Engine Soon	Service Reminder Indicator[1]	SRI[1]
Service Engine Soon	Malfunction Indicator Lamp[1]	MIL[1]
Service Reminder Indicator	Service Reminder Indicator[1]	SRI[1]
SFI (Sequential Fuel Injection)	Sequential Multiport Fuel Injection[1]	SFI[1]
Short Term FT (Fuel Trim)	Short Term Fuel Trim[1]	Short Term FT[1]
SLP (Selection Lever Position)	Transmission Range	TR
SMEC (Single Module Engine Control)	Powertrain Control Module[1]	PCM[1]
Smoke Puff Limiter	Smoke Puff Limiter[1]	SPL[1]
SPI (Single Point Injection)	Throttle Body Fuel Injection[1]	TBI[1]
SPL (Smoke Puff Limiter)	Smoke Puff Limiter[1]	SPL[1]
SRI (Service Reminder Indicator)	Service Reminder Indicator[1]	SRI[1]
SRT (System Readiness Test)	System Readiness Test[1]	SRT[1]
ST (Scan Tool)	Scan Tool[1]	ST[1]
Supercharger	Supercharger[1]	SC[1]
Supercharger Bypass	Supercharger Bypass[1]	SCB[1]
Sync Pickup	Camshaft Position[1]	CMP[1]
System Readiness Test	System Readiness Test[1]	SRT[1]
TAB (Thermactor Air Bypass)	Secondary Air Injection Bypass[1]	AIR Bypass[1]
TAD (Thermactor Air Diverter)	Secondary Air Injection Diverter[1]	AIR Diverter[1]
TB (Throttle Body)	Throttle Body[1]	TB[1]
TBI (Throttle Body Fuel Injection)	Throttle Body Fuel Injection[1]	TBI[1]
TBT (Throttle Body Temperature)	Intake Air Temperature[1]	IAT[1]
TC (Turbocharger)	Turbocharger[1]	TC[1]
TCC (Torque Converter Clutch)	Torque Converter Clutch[1]	TCC[1]
TCC (Torque Converter Clutch) Relay	Torque Converter Clutch Relay[1]	TCC Relay[1]
TCM (Transmission Control Module)	Transmission Control Module	TCM
TFI (Thick Film Ignition)	Distributor Ignition[1]	DI[1]
TFI (Thick Film Ignition) Module	Ignition Control Module[1]	ICM[1]
Thermac	Secondary Air Injection[1]	AIR[1]
Thermac Air Cleaner	Air Cleaner[1]	ACL[1]
Thermactor	Secondary Air Injection[1]	AIR[1]
Thermactor Air Bypass	Secondary Air Injection Bypass[1]	AIR Bypass[1]
Thermactor Air Diverter	Secondary Air Injection Diverter[1]	AIR Diverter[1]
Thermactor II	Pulsed Secondary Air Injection[1]	PAIR[1]
Thermal Vacuum Switch	Thermal Vacuum Valve[1]	TVV[1]
Thermal Vacuum Valve	Thermal Vacuum Valve[1]	TVV[1]

Recommended Terms and Recommended Acronyms See Figure 2
[1] Emission-Related Term

FIGURE 1—CROSS REFERENCE AND LOOK UP (CONTINUED)

Existing Usage	Acceptable Usage	Acceptable Acronized Usage
Third Gear	Third Gear	3GR
Three Way + Oxidation Catalytic Converter	Three Way + Oxidation Catalytic Converter[1]	TWC+OC[1]
Three Way Catalytic Converter	Three Way Catalytic Converter[1]	TWC[1]
Throttle Body	Throttle Body[1]	TB[1]
Throttle Body Fuel Injection	Throttle Body Fuel Injection[1]	TBI[1]
Throttle Opener	Idle Speed Control[1]	ISC[1]
Throttle Opener Vacuum Switching Valve	Idle Speed Control Solenoid Vacuum Valve[1]	ISC Solenoid Vacuum Valve[1]
Throttle Opener VSV (Vacuum Switching Valve)	Idle Speed Control Solenoid Vacuum Valve[1]	ISC Solenoid Vacuum Valve[1]
Throttle Position	Throttle Position[1]	TP
Throttle Position Sensor	Throttle Position Sensor[1]	TP Sensor[1]
Throttle Position Switch	Throttle Position Switch[1]	TP Switch[1]
Throttle Potentiometer	Throttle Position Sensor[1]	TP Sensor[1]
TOC (Trap Oxidizer - Continuous)	Continuous Trap Oxidizer[1]	CTOX[1]
TOP (Trap Oxidizer - Periodic)	Periodic Trap Oxidizer[1]	PTOX[1]
Torque Converter Clutch	Torque Converter Clutch[1]	TCC[1]
Torque Converter Clutch Relay	Torque Converter Clutch Relay[1]	TCC Relay[1]
TP (Throttle Position)	Throttle Position[1]	TP[1]
TP (Throttle Position) Sensor	Throttle Position Sensor[1]	TP Sensor[1]
TP (Throttle Position) Switch	Throttle Position Switch[1]	TP Switch[1]
TPI (Tuned Port Injection)	Multiport Fuel Injection[1]	MFI[1]
TPS (Throttle Position Sensor)	Throttle Position Sensor[1]	TP Sensor[1]
TPS (Throttle Position Switch)	Throttle Position Switch[1]	TP Switch[1]
TR (Transmission Range)	Transmission Range	TR
Transmission Control Module	Transmission Control Module	TCM
Transmission Position Switch	Transmission Range Switch	TR Switch
Transmission Range Selection	Transmission Range	TR
TRS (Transmission Range Selection)	Transmission Range	TR
TRSS (Transmission Range Selection Switch)	Transmission Range Switch	TR Switch
Tuned Port Injection	Multiport Fuel Injection[1]	MFI[1]
Turbo (Turbocharger)	Turbocharger[1]	TC[1]
Turbocharger	Turbocharger[1]	TC[1]
TVS (Thermal Vacuum Switch)	Thermal Vacuum Valve[1]	TVV[1]
TVV (Thermal Vacuum Valve)	Thermal Vacuum Valve[1]	TVV[1]
TWC (Three Way Catalytic Converter)	Three Way Catalytic Converter[1]	TWC[1]
TWC + OC (Three Way + Oxidation Catalytic Converter)	Three Way + Oxidation Catalytic Converter[1]	TWC+OC[1]
VAC (Vacuum) Sensor	Manifold Differential Pressure Sensor[1]	MDP Sensor
Vacuum Switches	Manifold Vacuum Zone Switch	MVZ Switch[1]
VAF (Volume Air Flow)	Volume Air Flow[1]	VAF[1]
Vane Air Flow	Volume Air Flow[1]	VAF[1]
Variable Fuel Sensor	Flexible Fuel Sensor	FF Sensor[1]
VAT (Vane Air Temperature)	Intake Air Temperature[1]	IAT[1]
VCC (Viscous Converter Clutch)	Torque Converter Clutch[1]	TCC[1]
Vehicle Speed Sensor	Vehicle Speed Sensor[1]	VSS[1]
VIP (Vehicle In Process) Connector	Data Link Connector[1]	DLC[1]
Viscous Converter Clutch	Torque Converter Clutch[1]	TCC[1]
Voltage Regulator	Voltage Regulator	VR
Volume Air Flow	Volume Air Flow[1]	VAF[1]
VR (Voltage Regulator)	Voltage Regulator	VR
VSS (Vehicle Speed Sensor)	Vehicle Speed Sensor[1]	VSS[1]
VSV (Vacuum Solenoid Valve) (Canister)	Evaporative Emission Canister Purge Valve[1]	EVAP Canister Purge Valve[1]
VSV (Vacuum Solenoid Valve) (EVAP)	Evaporative Emission Canister Purge Valve[1]	EVAP Canister Purge Valve[1]
VSV (Vacuum Solenoid Valve) (Throttle)	Idle Speed Control Solenoid Vacuum Valve[1]	ISC Solenoid Vacuum Valve[1]
Warm Up Oxidation Catalytic Converter	Warm Up Oxidation Catalytic Converter[1]	WU-OC[1]
Warm Up Three Way Catalytic Converter	Warm Up Three Way Catalytic Converter[1]	WU-OC[1]
Wide Open Throttle	Wide Open Throttle[1]	WOT[1]
WOT (Wide Open Throttle)	Wide Open Throttle[1]	WOT[1]
WOTS (Wide Open Throttle Switch)	Wide Open Throttle Switch[1]	WOT Switch[1]
WU-OC (Warm Up Oxidation Catalytic Converter)	Warm Up Oxidation Catalytic Converter[1]	WU-OC[1]
WU-TWC (Warm Up Three Way Catalytic Converter)	Warm Up Three Way Catalytic Converter[1]	WU-TWC[1]

Recommended Terms and Recommended Acronyms See Figure 2
[1] Emission-Related Term

FIGURE 1—CROSS REFERENCE AND LOOK UP (CONTINUED)

Recommended Terms	Acronym/Abbrev.	Definitions
Accelerator Pedal	AP[2]	See Glossary Entry "ACCELERATOR PEDAL."
Air Cleaner[1]	ACL	See Glossary Entry "CLEANER."
Air Conditioning	A/C	See Glossary Entry "AIR CONDITIONING."
Automatic Transaxle	A/T	See Glossary Entry "TRANSAXLE."
Automatic Transmission	A/T	See Glossary Entry "TRANSMISSION."
Barometric Pressure[1]	BARO[2]	See Glossary Entry "PRESSURE."
Battery Positive Voltage	B+[2]	See Glossary Entry "BATTERY."
Camshaft Position[1]	CMP	See Glossary Entry "CAMSHAFT."
Canister[1]	---	See Glossary Entry "CANISTER."
Carburetor[1]	CARB[2]	See Glossary Entry "CARBURETOR."
Charge Air Cooler[1]	CAC[2]	A device which lowers the temperature of the pressurized intake air.
Closed Loop[1]	CL	See Glossary Entry "CLOSED LOOP."
Closed Throttle Position[1]	CTP	See Glossary Entry "THROTTLE."
Clutch Pedal Position[1]	CPP[1]	See Glossary Entry "CLUTCH."
Continuous Fuel Injection[1]	CFI	A fuel injection system with the injector flow controlled by fuel pressure.
Continuous Trap Oxidizer[1]	CTOX	A system for lowering diesel engine particulate emissions by collecting exhaust particulates and continuously burning them through oxidation.
Crankshaft Position[1]	CKP	See Glossary Entry "CRANKSHAFT."
Data Link Connector[1]	DLC[2]	Connector providing access and/or control of the vehicle information, operating conditions, and diagnostic information.
Diagnostic Test Mode[1]	DTM	A level of diagnostic capability in an On Board Diagnostic (OBD) system. This may include different functional states to observe signals, a base level to read diagnostic trouble codes, a monitor level which includes information on signal levels, bidirectional control with on/off board aids, and the ability to interface with remote diagnosis.
Diagnostic Trouble Code[1]	DTC	An alphanumeric identifier for a fault condition identified by the On Board Diagnostic System.
Direct Fuel Injection[1]	DFI	Fuel injection system that supplies fuel directly into the combustion chamber.
Distributor Ignition[1]	DI	A system in which the ignition coil secondary circuit is switched by a distributor in proper sequence to various spark plugs.
Early Fuel Evaporation[1]	EFE	Enhancing air/fuel vaporization during engine warm up.
EGR Temperature[1]	EGRT	Sensing EGR function based on temperature change. Primarily in systems with mechanical flow control devices.
Electrically Erasable Programmable Read Only Memory[1]	EEPROM	An electronic device named electrically erasable programmable read only memory.
Electronic Ignition[1]	EI	A system in which the ignition coil secondary circuit is dedicated to specific spark plugs without the use of a distributor.
Engine Control[1]	EC	See Glossary Entries "ENGINE" and "CONTROL."
Engine Control Module[1]	ECM[2]	See Glossary Entries "ENGINE," "CONTROL," "MODULE."
Engine Coolant Level	ECL	See Glossary Entries "ENGINE," "COOLANT," "LEVEL."
Engine Coolant Temperature[1]	ECT	See Glossary Entries "ENGINE," "COOLANT."
Engine Modification[1]	EM	A method of lowering engine emissions through changes in basic engine construction or in fuel and spark calibration.
Engine Speed[1]	RPM[2]	See Glossary Entries "ENGINE," "SPEED."
Erasable Programmable Read Only Memory[1]	EPROM	An electronic device named erasable programmable read only memory.
Evaporative Emission[1]	EVAP[2]	A system used to prevent fuel vapor from escaping into the atmosphere. Typically includes a charcoal canister to store fuel vapors.
Exhaust Gas Recirculation[1]	EGR	Reducing NOx emissions levels by adding exhaust gas to the incoming air/fuel mixture.
Fan Control	FC	See Glossary Entries "FAN," "CONTROL."
Flash Electrically Erasable Programmable Read Only Memory[1]	FEEPROM	An electronic device named flash electrically erasable programmable read only memory.
Flash Erasable Programmable Read Only Memory[1]	FEPROM	An electronic device named flash erasable programmable read only memory.
Flexible Fuel[1]	FF	A system capable of using a variety of fuels for vehicle operation.
Fourth Gear	4GR[2]	Identifies the gear in which the transmission is operating in at a particular moment (e.g., the Transmission Range (TR) switch may indicate that "drive" was selected, but the transmission is operating in 4th gear as indicated by 4GR switch).
Fuel Level Sensor	---	See Glossary Entries "FUEL," "SENSOR."
Fuel Pressure	---	See Glossary Entries "FUEL," "PRESSURE."
Fuel Pump	FP[2]	See Glossary Entries "FUEL," "PUMP."
Fuel Trim[1]	FT	A fuel correction term.
Generator	GEN[2]	See Glossary Entry "GENERATOR."
Governor	---	See Glossary Entry "GOVERNOR."
Governor Control Module	GCM[2]	See Glossary Entries "GOVERNOR," "CONTROL" and "MODULE."
Ground	GND	See Glossary Entry "GROUND."
Heated Oxygen Sensor[1]	HO2S[2]	An oxygen sensor (02S) that is electrically heated.
Idle Air Control[1]	IAC	Electrical or mechanical control of throttle bypass air.
Idle Speed Control[1]	ISC	Electronic control of minimum throttle position.
Ignition Control[1]	IC	See Glossary Entries "IGNITION" and "CONTROL."

--- Use Recommended Term Only
[1] Emission-Related Term
[2] Historically acceptable common usage

FIGURE 2—RECOMMENDED TERMS

Recommended Terms	Acronym/Abbrev.	Definitions
Ignition Control Module[1]	ICM[2]	See Glossary Entries "IGNITION," "CONTROL," "MODULE."
Indirect Fuel Injection[1]	IFI	An injection system that supplies fuel into a combustion pre-chamber (Diesel).
Inertia Fuel Shutoff	IFS	An inertia system that shuts off the fuel delivery system when activated by predetermined force limits.
Intake Air[1]	IA	See Glossary Entry "INTAKE AIR."
Intake Air Temperature[1]	IAT	See Glossary Entry "INTAKE AIR."
Knock Sensor[1]	KS[2]	See Glossary Entries "KNOCK," "SENSOR."
Malfunction Indicator Lamp[1]	MIL[2]	A required on-board indicator to alert the driver of an emission-related malfunction.
Manifold Absolute Pressure[1]	MAP	See Glossary Entries "MANIFOLD," "PRESSURE."
Manifold Differential Pressure[1]	MDP	See Glossary Entries "MANIFOLD," "PRESSURE."
Manifold Surface Temperature[1]	MST	See Glossary Entry "MANIFOLD."
Manifold Vacuum Zone[1]	MVZ	See Glossary Entries "MANIFOLD," "VACUUM."
Mass Air Flow[1]	MAF	A system which provides information on the mass flow rate of the intake air to the engine.
Mixture Control[1]	MC	A device which regulates bleed air, fuel, or both, on carbureted vehicles.
Multiport Fuel Injection[1]	MFI	A fuel delivery system in which each cylinder is individually fueled.
Nonvolatile Random Access Memory[1]	NVRAM	An electronic device named nonvolatile random access memory.
Oil Pressure	---	Pressure in the lubrication system.
On Board Diagnostic[1]	OBD	A system that monitors some or all computer input and control signals. Signal(s) outside of the predetermined limits imply a fault in the system or in a related system.
Open Loop[1]	OL	See Glossary Entry "OPEN LOOP."
Oxidation Catalytic Converter[1]	OC	A catalytic converter system that reduces levels of HC and CO.
Oxygen Sensor[1]	O2S[2]	A sensor which detects oxygen (O2) content in the exhaust gases.
Park/Neutral Position[1]	PNP	See Glossary Entry "PARK/NEUTRAL."
Periodic Trap Oxidizer[1]	PTOX	A system for lowering diesel engine particulate emissions by collecting exhaust particulates and periodically burning them through oxidation.
Positive Crankcase Ventilation[1]	PCV	Positive ventilation of crankcase emissions.
Power Steering Pressure	PSP	See Glossary Entry "POWER STEERING."
Powertrain Control Module[1]	PCM	See Glossary Entries "POWERTRAIN," "CONTROL," "MODULE."
Programmable Read Only Memory[1]	PROM	An electronic device named programmable (by the manufacturer) read only memory.
Pulsed Secondary Air Injection[1]	PAIR[2]	A pulse driven system for providing secondary air without an air pump by using the engine exhaust system pressure fluctuations or pulses.
Random Access Memory[1]	RAM	An electronic device named random access memory.
Read Only Memory[1]	ROM	An electronic device named read only memory.
Relay Module	RM[2]	See Glossary Entries "RELAY," "MODULE."
Scan Tool[1]	ST[2]	See Glossary Entry "SCAN TOOL."
Secondary Air Injection[1]	AIR[2]	A pump driven system for providing secondary air.
Sequential Multiport Fuel Injection[1]	SFI	A multiport fuel delivery system in which each injector is individually energized and timed relative to its cylinder intake event. Normally fuel is delivered to each cylinder once per two crankshaft revolutions in four cycle engines and once per crankshaft revolution in two cycle engines.
Service Reminder Indicator[1]	SRI[2]	An indicator used to identify a service requirement.
Smoke Puff Limiter[1]	SPL	A system to reduce diesel exhaust smoke during vehicle acceleration or gear changes.
Supercharger[1]	SC[2]	See Glossary Entry "SUPERCHARGER."
Supercharger Bypass[1]	SCB	See Glossary Entry "SUPERCHARGER."
System Readiness Test[1]	SRT	System readiness test as applicable to OBDII scan tool communications.
Thermal Vacuum Valve[1]	TVV[2]	A valve that controls vacuum levels or routing based on temperature.
Third Gear	3GR[2]	Identifies the gear in which the transmission is operating in at a particular moment (e.g., the Transmission Range (TR) switch may indicate that "drive" was selected, but the transmission is operating in 3rd gear as indicated by 3GR switch).
Three Way + Oxidation Catalytic Converter[1]	TWC+OC	A catalytic converter system that has both Three Way Catalyst (TWC) and Oxidation Catalyst (OC). Usually secondary air is introduced between the two catalysts.
Three Way Catalytic Converter[1]	TWC	A catalytic converter system that reduces levels of HC, CO and NOx.
Throttle Body[1]	TB[2]	See Glossary Entries "THROTTLE," "BODY."
Throttle Body Fuel Injection[1]	TBI	An electronically controlled fuel injection system in which one or more fuel injectors are located in a throttle body.
Throttle Position[1]	TP	See Glossary Entry "THROTTLE."
Torque Converter Clutch[1]	TCC[2]	See Glossary Entry "CONVERTER."
Transmission Control Module	TCM[2]	See Glossary Entries "TRANSMISSION," "CONTROL," "MODULE."
Transmission Range	TR	See Glossary Entries "TRANSMISSION," "RANGE."
Turbocharger[1]	TC[2]	See Glossary Entry "TURBOCHARGER."
Vehicle Speed Sensor[1]	VSS[2]	A sensor which provides vehicle speed information.
Voltage Regulator	VR[2]	See Glossary Entry "REGULATOR."
Volume Air Flow[1]	VAF	A system which provides information on the volume flow rate of the intake air to the engine.
Warm Up Oxidation Catalytic Converter[1]	WU-OC	A catalytic converter system designed to lower HC and CO emissions during engine warm up. Usually located in or near the exhaust manifold.
Warm Up Three Way Catalytic Converter[1]	WU-TWC	A catalytic converter system designed to lower HC, CO, and NOx emissions during engine warm up. Usually located in or near the exhaust manifold.
Wide Open Throttle[1]	WOT	See Glossary Entry "THROTTLE."

--- Use Recommended Term Only
[1] Emission-Related Term   [2] Historically acceptable common usage

FIGURE 2—RECOMMENDED TERMS (CONTINUED)

MODIFIERS					BASE WORD	
What is its purpose?	Where is it?	Which Temp?	What does it sense?	What is it?		
				Sensor		Most generic
			Temperature	Sensor		
		Coolant	Temperature	Sensor		
	Engine	Coolant	Temperature	Sensor		Most specific
Instrumentation	Engine	Coolant	Temperature	Sensor		
Least <------		------SIGNIFICANCE------			------> Most	

FIGURE 3—MODIFIER USAGE EXAMPLE

TABLE 1—GLOSSARY OF TERMS

Baseword/Single Word Modifier	Definition
Accelerator Pedal	A foot-operated device, which, directly or indirectly, controls the flow of fuel and/or air to the engine, thereby, controlling engine speed.
Accumulator	A vessel in which fluid is stored, usually at greater than atmospheric pressure.
Actuator	A mechanism for moving or controlling something indirectly instead of by hand. Compare: Solenoid, relay, and valve.
Air Conditioning	A vehicular accessory system that modifies the passenger compartment air by cooling and drying the air.
Alternator	See Generator.
Battery	An electrical storage device designed to produce a DC voltage by means of an electrochemical reaction.
Blower	A device designed to supply a current of air at a moderate pressure. A blower usually consists of an impeller assembly, a motor and a suitable case. The blower case is usually designed as part of a ventilation system. Compare: Fan.
Brake	A device for retarding motion, usually by means of friction.
Body	(1) The assembly of sheet-metal components, windows, doors, seats, etc., that provide enclosures for passengers and/or cargo in a motor vehicle. It may or may not include the hood and fenders. (2) The primary, central or key part of a feature.
Camshaft	A shaft on which phased cams are mounted. The camshaft is used to regulate the opening and closing of the intake and exhaust valves.
Canister	An evaporative emission canister contains activated charcoal which absorbs fuel vapors and holds tnem until the vapors can be purged at an appropriate time.
Capacitor	An electrical device for accumulating and holding a charge of electricity.
Carburetor	A mechanism which automatically mixes fuel with air in the proper proportions to provide a desired power output from a spark ignition internal combustion engine.
Catalyst	A substance that can increase or decrease the rate of a chemical reaction between substances without being consumed in the process.
Chassis	The suspension, steering, and braking elements of a vehicle.
Circuit	A complete electrical path or channel, usually includes the source of electric energy. Circuit may also describe the electrical path between two or more components. May also be used with fluids, air or liquid.
Cleaner	A device used in the intake system of parts that require clean air. An air cleaner usually has a filter in it to trap particulates and only pass clean air through.
Closed Loop (Engine)	An operating condition or mode which enables modification of programmed instructions based on a feedback system.
Clutch	A mechanical device which uses mechanical, magnetic or friction type connections to facilitate engaging or disengaging of two shafts or rotating members.
Code	A system of symbols (as letters, numbers, or words) used to represent meaning of information.
Coil (Ignition)	A device consisting of windings of conductors around an iron core, designed to increase the voltage, and for use in a spark ignition system.
Control	A means or a device to direct and regulate a process or guide the operation of a machine, apparatus or system.
Converter (Catalytic)	An in-line, exhaust system device used to reduce the level of engine exhaust emissions.
Converter (Torque)	A device which by its design multiplies the torque in a fluid coupling between an engine and transmission/transaxle.
Coolant	A fluid used for heat transfer. Coolants usually contain additives such as rust inhibitors and antifreeze.
Cooler	A heat exchanger that reduces the temperature of the named medium.
Crankshaft	The part of an engine which converts the reciprocating motion of the pistons to rotary motion.
Data	General term for information, usually represented by numbers, letters, symbols.
Device	A piece of equipment or a mechanism designed for a specific purpose or function. DO NOT use "Device" in naming.
Diagnostics	The process of identifying the cause or nature of a condition, situation or problem. To determine corrective action in repair of automotive systems.
Differential	(1) A device with an arrangement of gears designed to permit the division of power to two shafts. (2) See Pressure.
Distributor	A mechanical device designed to switch a high voltage secondary circuit from an ignition coil to spark plugs in the proper firing sequence.
Drive	A device which provides a fixed increase or decrease ratio of relative rotation between its input and output shafts.
Electrical	A type of device or system using resistors, motors, generators, incandescent lamps, switches, capacitors, batteries, inductors or wires. Compare: Electronic.
Electronic	(1) A type of device or system using solid-state devices or thermionic elements such as diodes, transistors, integrated circuits, vacuum fluorescent displays and liquid crystal displays. (2) The storage, retrieval and display of information through media such as magnetic tape, laser disc, electronic read only memory (ROM) and random access memory (RAM). Compare: Electrical.
Engine	A machine designed to convert thermal energy into mechanical energy to produce force or motion.
Fan	A device designed to supply a current of air. A fan may also have a frame, motor, wiring harness and the like. Compare: Blower.
Fuel	Any combustible substance burned to provide heat or power. Typical fuels include gasoline and diesel fuel. Other types of fuel include ethanol, methanol, natural gas, propane or in combination.
Generator	A rotating machine designed to convert mechanical energy into electrical energy.

TABLE 1—GLOSSARY OF TERMS (CONTINUED)

Baseword/Single Word Modifier	Definition
Governor	A device designed to automatically limit engine speed.
Ground	An electrical conductor used as a common return for an electric circuit(s) and with a relative zero potential.
Idle	Rotational speed of an engine with vehicle at rest and accelerator pedal not depressed.
Ignition	System used to provide high voltage spark for internal combustion engines.
Indicator	A device which visually presents vehicle condition information transmitted or relayed from some other source.
Injector	A device for delivering metered pressurized fuel to the intake system or the cylinders.
Intake Air	Air drawn through a cleaner and distributed to each cylinder for use in combustion.
Knock (Engine)	The sharp, metallic sound produced when two pressure fronts collide in the combustion chamber of an engine.
Level	The magnitude of a quantity considered in relation to an arbitrary reference value.
Link (Electrical/Electronic)	General term used to indicate the existence of communication facilities between two points.
Manifold	A device designed to collect or distribute fluid, air or the like. Compare: Rail.
Memory	A device in which data can be stored and used when needed.
Mode	One of several alternative conditions or methods of operating a device or control module.
Module (Electrical/Electronic)	A self-contained group of electrical/electronic components, which is designed as a single replaceable unit.
Motor	A machine that converts kinetic energy, such as electricity, into mechanical energy. Compare: Actuator.
Open Loop	An operating condition or mode based on programmed instructions and not modified by a feedback system.
Park/Neutral	The selected non-drive modes of the transmission.
Power Steering	A system which provides additional force to the steering mechanism, reducing the driver's steering effort.
Powertrain	The elements of a vehicle by which motive power is generated and transmitted to the driven axles.
Pressure	Unless otherwise noted, is gage pressure.
Pressure (Absolute)	The pressure referenced to a perfect vacuum.
Pressure (Atmospheric)	The pressure of the surrounding air at any given temperature and altitude. Sometimes called Barometric Pressure.
Pressure (Barometric)	Pertaining to atmospheric pressure or the results obtained by using a barometer.
Pressure (Differential)	The pressure difference between two regions, such as between the intake manifold and the atmospheric pressures.
Pressure (Gage)	The amount by which the total absolute pressure exceeds the ambient atmospheric pressure.
Pump	A device used to raise, transfer, or compress fluids by suction, pressure or both.
Radiator	A radiator is a liquid-to-air heat transfer device having a tank(s) and core(s) specifically designed to reduce the temperature of the coolant in an internal combustion engine cooling system.
Rail	A manifold for fuel injection fuel. Compare: Manifold.
Refrigerant	A substance used as a heat transfer agent in an air conditioning system.
Relay	A generally electromechanical device in which connections in one circuit are opened or closed by changes in another circuit. Compare: Actuator, Solenoid, and Switch.
Regulator (Voltage)	A device that automatically controls the functional output of another device by adjusting the voltage to meet a specified value.
Scan Tool	A device that interfaces with and communicates information on a data link.
Secondary Air	Air provided to the exhaust system.
Sensor	The generic name for a device that senses either the absolute value or a change in a physical quantity such as temperature, pressure or flow rate, and converts that change into an electrical quantity signal. Compare: Transducer.
Signal (Electrical/Electronic)	A fluctuating electric quantity, such as voltage or current, whose variations represent information.
Solenoid	A device consisting of an electrical coil which when energized, produces a magnetic field in a plunger, which is pulled to a central position. A solenoid may be used as an actuator in a valve or switch. Compare: Actuator, Relay, and Switch.
Solid State	Crystalline circuit structures used to perform electronic functions. Examples of such structures include transistors, diodes, integrated circuits and other semiconductors.
Speed	The magnitude of velocity (regardless of direction).
Supercharger	A mechanically driven device that pressurizes the intake air, thereby increasing the density of charge air and the consequent power output from a given engine displacement.
Switch	A device for making, breaking, or changing the connections in an electrical circuit. Compare: Relay, Solenoid, and Valve.
System	A group of interacting mechanical or electrical components serving a common purpose.
Test	A procedure whereby the performance of a product is measured under various conditions.
Throttle	A valve for regulating the supply of a fluid, usually air or a fuel/air mix, to an engine.
Transaxle	A device consisting of a transmission and axle drive gears assembled in the same case. Compare: Transmission.
Transducer	A device that receives energy from one system and retransmits (transfers) it, often in a different form, to another system. For example, the cruise control transducer converts a vehicle speed signal to a modulated vacuum output to control a servo. Compare: Sensor.
Transmission	A device which selectively increases or decreases the ratio of relative rotation between its input and output shafts. Compare: Transaxle.
Troubleshooting	See Diagnostics.
Turbocharger	A centrifugal device driven by exhaust gases that pressurize the intake air, thereby increasing the density of charge air and the consequent power output from a given engine displacement.
Ultraviolet	The portion of the electromagnetic spectrum between violet visible light and X-Rays.
Vacuum	A circuit in which pressure has been reduced below the ambient atmospheric pressure.
Valve	A device by which the flow of liquid, gas, vacuum, or loose material in bulk may be started, stopped or regulated by a movable part that opens, shuts or partially obstructs one or more ports or passageways. A "Valve" is also the moveable part of such a device. Compare: Actuator and Switch.
Vapor	A substance in its gaseous state as distinguished from the liquid or solid state.
Volatile	(1) Vaporizable at normal temperatures. (2) Not permanent.
Wheel	A circular frame of hard material that may be solid, partially solid, or spoked and that is capable of turning on an axle.

23.303

APPENDIX A
REQUEST FOR REVISION TO SAE J1930 ELECTRICAL/ELECTRONIC SYSTEMS DIAGNOSTIC TERMS, DEFINITIONS, ABBREVIATIONS, AND ACRONYMS

To insure that your request is accepted for ballot and incorporation into J1930, please supply the following information consistent with the Methodology of Section 4 through 4.3.2:

Please send completed form to:
SAE J1930 Task Force, 3001 West Big Beaver Rd, Ste. 320, Troy MI 48084-3174 U.S.A. Fax number: (313) 649-0425

PURPOSE or RATIONALE FOR REQUEST: _____
_____
_____

SECTION 5
EXISTING USAGE(S) _____ RECOMMENDED TERMS _____
_____   _____
_____   _____

SECTION 6
RECOMMENDED TERM   circle one:
    EXISTING: _____   Add   Delete   Change
    SUGGESTED: _____   Add   Delete   Change
ACRONYM / ABBREVIATIONS
    EXISTING: _____
    SUGGESTED: _____   Add   Delete   Change
DEFINITION
    EXISTING: _____
    _____
    SUGGESTED: _____   Add   Delete   Change
    _____

EMISSION RELATED? YES _____   NO _____

SECTION 7
GLOSSARY of TERMS _____
_____
_____

REQUESTOR: _____ Phone: _____ Fax: _____

Signature: _____ Address: _____
Date: _____   _____
                                     _____

COMMITTEE USE ONLY

RECOMMEND FOR BALLOT? YES _____   NO _____   BALLOT TARGET DATE _____
COMMENTS _____
_____

J1930 CHAIRPERSON _____   DATE _____

OFF-BOARD DIAGNOSTIC MESSAGE FORMATS—SAE J2037 NOV90

SAE Information Report

Report of the SAE Vehicle E/E Systems Diagnostic Standards Committee approved November 1990.

Foreword—This SAE Information Report is an example of how J1850 might be used to implement diagnostics in a vehicle. Related SAE Information Reports are J2054, "E/E Diagnostic Data Communication," J2062 "A Class B Serial Bus Diagnostic Protocol," and J2086 "An Applied Layered Protocol for a Generic Scan Tool."

1. Scope—The utilities defined for J2037 are designed to facilitate manufacturing and service diagnosis requirements.

Definition of the capability includes definition of standard messages and the dialogue necessary to provide the capability. The standard messages will be distinguished by the contents of the first data byte which specifies the diagnostic operation. Note that some vehicle applications will not require the implementation of all the defined diagnostic capabilities, and consequently, these applications will not support all message modes.

2. References

2.1 Applicable Documents—The following publications form a part of this specification to the extent specified herein. The latest issue of SAE publications shall apply.

2.1.1 SAE PUBLICATIONS—Available from SAE, 400 Commonwealth Drive, Warrendale, PA 15096-0001.

SAE J1850—Class B Data Communication Network Interface
SAE J2054—E/E Diagnostic Data Communication
SAE J2062—A Class B Serial Bus Diagnostic Protocol
SAE J2086—An Applied Layer Protocol for a Generic Scan Tool

3. J1850 has been advanced by the SAE as a Recommended Practice for vehicle multiplex applications. Government emissions regulatory agencies have indicated that adoption of the J1850 format may be an acceptable step forward toward standardization of vehicle diagnostics.

Further definition beyond J1850 is necessary in order to accomplish the desired standardization. The specifications contained in J1850 address only the lower layers of the seven layer ISO Open System model. Agreement in the content of the message fields defined by J1850 needs to be established. This document proposes a set of message definitions which are both compatible with J1850 and suitable for accomplishing the general requirements of off-board diagnostics.

The messages use physical addressing. The off-board unit sends messages to a specific physically addressed module; and the module may send return messages directly to the off-board. No broadcast or functionally addressed messages are included.

The general format of the J1850 message is shown as follows:

SOM	Start of Message
DATA	Data Byte(s) Field
ERR	CRC Error Detection byte
EOD	End of Data
RSP	Response Bytes
EOM	End of Message

The proposed Diagnostic Message adheres to this general format and provides special definition for the contents of the Data Bytes Field. The format for the proposed Diagnostic Message expands the J1850 definition as shown as follows:

SOM	Start of Message
DATA BYTE 1	Four Bits: Priority Four Bits: Code 4 hex
DATA BYTE 2	Target Module Physical Address
DATA BYTE 3	Source Module Physical Address
DATA BYTE 4	Diagnostic Operation Code
DATA BYTES	Additional Data Bytes as needed
ERR	CRC Error Detection byte
EOD	End of Data
RSP	Physical Address of Receiving Module
EOM	End of Message

The first three data bytes are used to specify priority, target, and source in all Diagnostic messages. Additionally, the lower four bits of the first data byte must be 0100 to indicate that the message is a Diagnostic Type. The fourth byte indicates the operation code. Seven bits (128 codes) are reserved for specification of the diagnostic operation. The MSB is always zero for this standard set of diagnostic Messages. Setting the MSB to logic one can define another set 128 operation which could be proprietary to each manufacturer.

3.1 Off-Board Diagnostic requirements can be categorized as follows:

a. Messages which request and report module I.D. and status.
b. Messages which invoke and report results for on-board diagnostic routines.
c. Messages which request and report parametric information.
d. Messages which transfer blocks of data.

The Diagnostics Message contains from 1 to 7 data bytes. The message always used the first data byte as an operation code as shown in Figure 1.

This allows for definition of 256 distinct operations. The First 128 Codes (00 to 7F hex) are dedicated to diagnostic operations. The other 128 codes (80 to FF) hex are reserved for future assignment. A list of current diagnostic operation code assignments is contained in Table 1.

3.2 CODE 00 hex Request Module Identification—The OFF-BOARD DIAGNOSTIC UNIT issues this command to request transmission of the Module Identification Report message.

This message contains 6 bytes as shown in Figure 2.

3.3 CODE 01 hex Request Diagnostic Status—The OFF-BOARD DIAGNOSTIC UNIT issues this command to request transmission of the Module Diagnostic Status Report message.

This message contains 6 bytes as shown in Figure 3.

3.4 CODE 09 HEX REQUEST MODULE TO OFF LINE—The OFF-BOARD DIAGNOSTIC UNIT issues this command to request a vehicle module to enter Off Line mode.

This message contains 6 bytes as shown in Figure 4.

3.5 CODE 10 hex Report Module Identification—This message is used by a vehicle module to report the Module Identification code(s).

This message is variable in length (7 to 12 bytes) depending upon the number of data bytes required to identify the module uniquely. The format is shown in Figure 5.

3.6 CODE 11 hex Report Diagnostic Status—This message is used by a vehicle module to report the Module Diagnostic Status.

This message is variable in length (7 to 12 bytes) depending upon the number of data bytes required to report the module status. The format is shown in Figure 6.

3.7 CODE 19 hex Report Module is Off Line—This message is used by a vehicle module to report that the Module has entered Off Line mode.

b7	b6	b5	b4	b3	b2	b1	b0
0 = DIAG	ENCODED (00 TO FF hex) AS 1 OF 128 DIAGNOSTIC OPERATIONS						
0	?	?	?	?	?	?	?

FIGURE 1—DIAGNOSTICS MESSAGE OPERATION CODE

TABLE 1—DIAGNOSTIC OPERATION CODE ASSIGNMENTS

OPERATION CODE	FUNCTION	ADDITIONAL DESCRIPTION
0000 0000 = 00 hex	Request Module Identification	Version, Revision etc.
0000 0001 = 01 hex	Request Module Status	Go/No 00 hex = OK
0000 0010 = 02 hex	NOT ASSIGNED	
0000 0011 = 03 hex	NOT ASSIGNED	
0000 0100 = 04 hex	NOT ASSIGNED	
0000 0101 = 05 hex	NOT ASSIGNED	
0000 0110 = 06 hex	NOT ASSIGNED	
0000 0111 = 07 hex	NOT ASSIGNED	
0000 1000 = 08 hex	NOT ASSIGNED	
0000 1001 = 09 hex	Request Module to Off-Line	
0000 1010 = 0A hex	NOT ASSIGNED	
0000 1011 = 0B hex	NOT ASSIGNED	
0000 1100 = 0C hex	NOT ASSIGNED	
0000 1101 = 0D hex	NOT ASSIGNED	
0000 1110 = 0E hex	NOT ASSIGNED	
0000 1111 = 0F hex	NOT ASSIGNED	
0001 0000 = 10 hex	Report Module Identification	Version, Revision etc.
0001 0001 = 11 hex	Report Module Status	Go/No 00 hex = OK
0001 0010 = 12 hex	NOT ASSIGNED	
0001 0011 = 13 hex	NOT ASSIGNED	
0001 0100 = 14 hex	NOT ASSIGNED	
0001 0101 = 15 hex	NOT ASSIGNED	
0001 0110 = 16 hex	NOT ASSIGNED	
0001 0111 = 17 hex	NOT ASSIGNED	
0001 1000 = 18 hex	NOT ASSIGNED	
0001 1001 = 19 hex	Report Module is Off Line	
0001 1010 = 1A hex	NOT ASSIGNED	
0001 1011 = 1B hex	NOT ASSIGNED	
0001 1100 = 1C hex	NOT ASSIGNED	
0001 1101 = 1D hex	NOT ASSIGNED	
0001 1110 = 1E hex	NOT ASSIGNED	
0001 1111 = 1F hex	NOT ASSIGNED	
0010 0000 = 20 hex	Request Parameter by PID	Single Parameter
0010 0001 = 21 hex	Request Parameter by DMR	Single Byte Parameter
0010 0010 = 22 hex	Request Parameter by DMR	Double Byte Parameter
0010 0011 = 23 hex	NOT ASSIGNED	
0010 0100 = 24 hex	Request Parameter Packet	Multiple Parameters
0010 0101 = 25 hex	NOT ASSIGNED	
0010 0110 = 26 hex	NOT ASSIGNED	
0010 0111 = 27 hex	NOT ASSIGNED	
0010 1000 = 28 hex	Set up Parameter Packet	Assigns Packet Content
0010 1001 = 29 hex	NOT ASSIGNED	
0010 1010 = 2A hex	NOT ASSIGNED	
0010 1011 = 2B hex	NOT ASSIGNED	
0010 1100 = 2C hex	NOT ASSIGNED	
0010 1101 = 2D hex	NOT ASSIGNED	
0010 1110 = 2E hex	NOT ASSIGNED	
0010 1111 = 2F hex	NOT ASSIGNED	
0011 0000 = 30 hex	Report PID Parameter	Single Parameter
0011 0001 = 31 hex	Report DMR Parameter	Single Byte Parameter
0011 0010 = 32 hex	Report DMR Parameter	Double Byte Parameter
0011 0011 = 33 hex	NOT ASSIGNED	
0011 0100 = 34 hex	NOT ASSIGNED	
0011 0101 = 35 hex	NOT ASSIGNED	
0011 0110 = 36 hex	NOT ASSIGNED	
0011 0111 = 37 hex	NOT ASSIGNED	
0011 1000 = 38 hex	NOT ASSIGNED	
0011 1001 = 39 hex	NOT ASSIGNED	
0011 1010 = 3A hex	NOT ASSIGNED	
0011 1011 = 3B hex	NOT ASSIGNED	
0011 1100 = 3C hex	NOT ASSIGNED	
0011 1101 = 3D hex	NOT ASSIGNED	
0011 1110 = 3E hex	NOT ASSIGNED	
0011 1111 = 3F hex	NOT ASSIGNED	
0100 0000 = 40 hex	Enter Diagnostic Routine	Invokes on board diag
0100 0001 = 41 hex	Exit Diagnostic Routine	cancel on-board diag
0100 0010 = 42 hex	NOT ASSIGNED	
0100 0011 = 43 hex	NOT ASSIGNED	
0100 0100 = 44 hex	NOT ASSIGNED	
0100 0101 = 45 hex	NOT ASSIGNED	
0100 0110 = 46 hex	NOT ASSIGNED	

TABLE 1—DIAGNOSTIC OPERATION CODE ASSIGNMENTS (CONTINUED)

OPERATION CODE	FUNCTION	ADDITIONAL DESCRIPTION
0100 0111 = 47 hex	NOT ASSIGNED	
0100 1000 = 48 hex	NOT ASSIGNED	
0100 1001 = 49 hex	NOT ASSIGNED	
0100 1010 = 4A hex	NOT ASSIGNED	
0100 1011 = 4B hex	NOT ASSIGNED	
0100 1100 = 4C hex	NOT ASSIGNED	
0100 1101 = 4D hex	NOT ASSIGNED	
0100 1110 = 4E hex	NOT ASSIGNED	
0100 1111 = 4f hex	NOT ASSIGNED	
0101 0000 = 50 hex	Confirm Diag Routine Entry	
0101 0001 = 51 hex	Confirm Diag Routine Exit	
0101 0010 = 52 hex	NOT ASSIGNED	
0101 0011 = 53 hex	NOT ASSIGNED	
0101 0100 = 54 hex	NOT ASSIGNED	
0101 0101 = 55 hex	NOT ASSIGNED	
0101 0110 = 56 hex	NOT ASSIGNED	
0101 0111 = 57 hex	NOT ASSIGNED	
0101 1000 = 58 hex	Report Diag Results	
0101 1001 = 59 hex	NOT ASSIGNED	
0101 1010 = 5A hex	NOT ASSIGNED	
0101 1011 = 5B hex	NOT ASSIGNED	
0101 1100 = 5C hex	NOT ASSIGNED	
0101 1101 = 5D hex	NOT ASSIGNED	
0101 1110 = 5E hex	NOT ASSIGNED	
0101 1111 = 5F hex	NOT ASSIGNED	
0110 0000 = 60 hex	substitution of parameter	
0110 0001 = 61 hex	NOT ASSIGNED	
0110 0010 = 62 hex	NOT ASSIGNED	
0110 0011 = 63 hex	NOT ASSIGNED	
0110 0100 = 64 hex	NOT ASSIGNED	
0110 0101 = 65 hex	NOT ASSIGNED	
0110 0110 = 66 hex	NOT ASSIGNED	
0110 0111 = 67 hex	NOT ASSIGNED	
0110 1000 = 68 hex	NOT ASSIGNED	
0110 1001 = 69 hex	NOT ASSIGNED	
0110 1010 = 6A hex	NOT ASSIGNED	
0110 1011 = 6B hex	NOT ASSIGNED	
0110 1100 = 6C hex	NOT ASSIGNED	
0110 1101 = 6D hex	NOT ASSIGNED	
0110 1110 = 6E hex	NOT ASSIGNED	
0110 1111 = 6F hex	NOT ASSIGNED	
0111 0000 = 70 hex	Report Packet #0	
0111 0001 = 71 hex	Report Packet #1	
0111 0010 = 72 hex	Report Packet #2	
0111 0011 = 73 hex	Report Packet #3	
0111 0100 = 74 hex	Report Packet #4	
0111 0101 = 75 hex	Report Packet #5	
0111 0110 = 76 hex	Report Packet #6	
0111 0111 = 77 hex	Report Packet #7	
0111 1000 = 78 hex	Report Packet #8	
0111 1001 = 79 hex	Report Packet #9	
0111 1010 = 7A hex	Report Packet #10	
0111 1011 = 7B hex	Report Packet #11	
0111 1100 = 7C hex	Report Packet #12	
0111 1101 = 7D hex	Report Packet #13	
0111 1110 = 7E hex	Report Packet #14	
0111 1111 = 7F hex	Report Packet #15	

23.306

Field	Value	Size
START DELIMITER		1 BIT
PRIORITY/TYPE FIELD	DIAG PRIORITY 3h NODE TO NODE 4h	8 BITS
TARGET FIELD	??h	8 BITS
SOURCE FIELD	OFF-BOARD DIAGNOSTIC MODULE – 00h	8 BITS
DATA FIELD	REQUEST MODULE I.D. CODE – 00h	8 BITS
CRC FIELD		8 BITS
ACKNOWLEDGE FIELD	??h	8 BITS

FIGURE 2—CODE 00 HEX REQUEST MODULE IDENTIFICATION

Field	Value	Size
START DELIMITER		1 BIT
PRIORITY/TYPE FIELD	DIAG PRIORITY 3h NODE TO NODE 4h	8 BITS
TARGET FIELD	??h	8 BITS
SOURCE FIELD	OFF-BOARD DIAGNOSTIC MODULE – 00h	8 BITS
DATA FIELD	REQUEST STATUS CODE – 01h	8 BITS
CRC FIELD		8 BITS
ACKNOWLEDGE FIELD	??h	8 BITS

FIGURE 3—CODE 01 HEX REQUEST DIAGNOSTIC STATUS

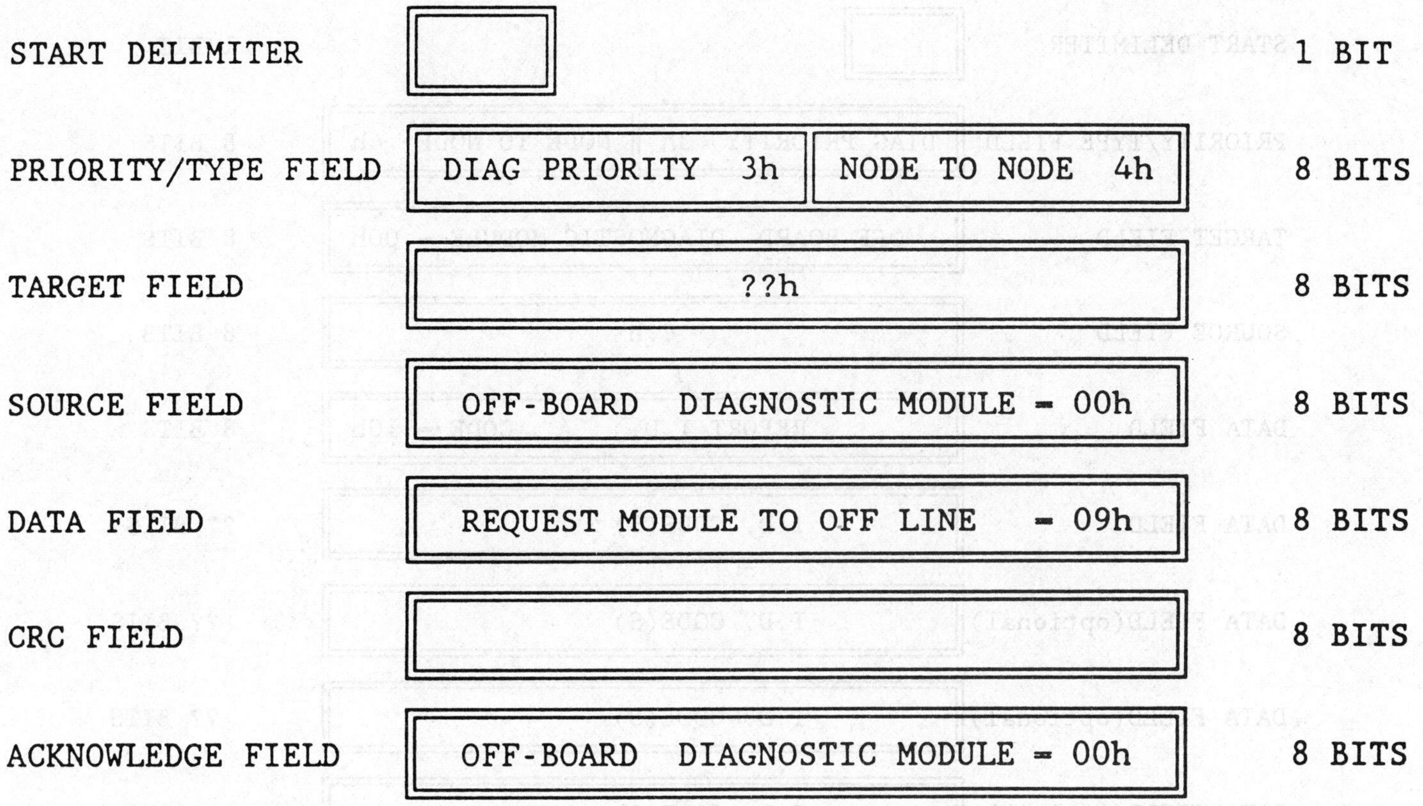

START DELIMITER			1 BIT
PRIORITY/TYPE FIELD	DIAG PRIORITY 3h	NODE TO NODE 4h	8 BITS
TARGET FIELD	??h		8 BITS
SOURCE FIELD	OFF-BOARD DIAGNOSTIC MODULE = 00h		8 BITS
DATA FIELD	REQUEST MODULE TO OFF LINE = 09h		8 BITS
CRC FIELD			8 BITS
ACKNOWLEDGE FIELD	OFF-BOARD DIAGNOSTIC MODULE = 00h		8 BITS

FIGURE 4—CODE 09 HEX REQUEST MODULE TO OFF LINE

This message contains 6 bytes as shown in Figure 7.

3.8 CODE 20 hex Request Parameter by PID—The OFF-BOARD DIAGNOSTIC UNIT issues this message to request a return transmission containing a parameter value. The parameter of interest is referenced by a preassigned index number which provides unique identification of the parameter within the module.

This message contains 7 bytes as shown in Figure 8.

3.9 CODE 21 hex Request Single Byte Parameter by DMR—The OFF-BOARD DIAGNOSTIC UNIT issues this message to request a return transmission containing a single byte parameter value. The parameter of interest is referenced by its memory location within the module.

This message contains 8 bytes as shown in Figure 9.

3.10 CODE 22 hex Request Double Byte Parameter by DMR—The OFF-BOARD DIAGNOSTIC UNIT issues this message to request a return transmission containing a double byte parameter value. The parameter of interest is referenced by its memory location within the module.

This message contains 8 bytes as shown in Figure 10.

3.11 CODE 24 hex Request Multiple Parameter Packet Report—The OFF-BOARD DIAGNOSTIC UNIT issues this message to request a return transmission containing multiple parameters within a single message. A Multiple Parameter Packet contains either 6 single byte or 3 double byte (word) parameters. Either type may be specified with this request. This request message also specifies whether the packet is to contain parameters which are addressed by index or direct memory reference. This message may be used to start, stop, or specify a fixed rate of transmission for the return message.

This message contains 7 bytes as shown in Figure 11.

3.12 CODE 28 hex Setup Parameter Report Packet—The OFF-BOARD DIAGNOSTIC UNIT issues this message to configure the contents of a Report Packet which contains multiple parameters within a single message. This message must be sent once for each parameter to be positioned within the Report Packet. That is, a 6 parameter Report Packet will require 6 Setup messages; a 3 parameter Report Packet requires only 3 iterations of the Setup message to complete the configuration.

This message contains 8 bytes when configuring Report Packets with PID referenced parameters; and requires 9 bytes when configuring DMR type Report Packets, as shown in Figure 12.

3.13 CODE 30 hex Report PID Parameter—This message is used by a vehicle module to report the value of a specific parameter value. The parameter of interest is referenced by a preassigned index number which provides unique identification. The parameter may be a single byte or a double byte (word).

This message contains 8 or 9 bytes as shown in Figure 13.

3.14 CODE 31 hex Report Single Byte DMR Parameter—This message is used by a vehicle module to report the value of a specific parameter. The parameter of interest is referenced by its memory location within the module. The parameter value is described with a single byte.

This message contains 9 bytes as shown in Figure 14.

3.15 CODE 32 hex Report Double Byte Parameter—This message is used by a vehicle module to report the value of a specifc parameter value. The parameter of interest is referenced by its memory location within the module. The parameter value is described with two bytes (word).

This message contains 10 bytes as shown in Figure 15.

3.16 CODE 40 hex Enter Diagnostic Routine—The OFF-BOARD DIAGNOSTIC UNIT issues this message to request that a specific on-board test diagnostic routine be entered and that the results, if any, be reported in a return message. The second data byte specifies which on-board diagnostic routine is being invoked.

This message contains 7 bytes as shown in Figure 16.

3.17 CODE 41 hex Exit Diagnostic Routine—The OFF-BOARD DIAGNOSTIC UNIT issues this message to request that a specific on-board test diagnostic routine be terminated and that the results, if any, be reported in a return message. The second data byte specifies which

23.308

Field	Content	Size
START DELIMITER		1 BIT
PRIORITY/TYPE FIELD	DIAG PRIORITY 3h | NODE TO NODE 4h	8 BITS
TARGET FIELD	OFF-BOARD DIAGNOSTIC MODULE - 00h	8 BITS
SOURCE FIELD	??h	8 BITS
DATA FIELD	REPORT I.D. CODE - 10h	8 BITS
DATA FIELD	I.D. CODE(S)	?? BITS
DATA FIELD (optional)	I.D. CODE(S)	?? BITS
DATA FIELD (optional)	I.D. CODE(S)	?? BITS
DATA FIELD (optional)	I.D. CODE(S)	?? BITS
DATA FIELD (optional)	I.D. CODE(S)	?? BITS
DATA FIELD (optional)	I.D. CODE(S)	?? BITS
CRC FIELD		8 BITS
ACKNOWLEDGE FIELD	OFF-BOARD DIAGNOSTIC MODULE - 00h	8 BITS

FIGURE 5—CODE 10 HEX REPORT MODULE IDENTIFICATION

Field	Value	Size
START DELIMITER		1 BIT
PRIORITY/TYPE FIELD	DIAG PRIORITY 3h	NODE TO NODE 4h	8 BITS
TARGET FIELD	OFF-BOARD DIAGNOSTIC MODULE - 00h	8 BITS
SOURCE FIELD	??h	8 BITS
DATA FIELD	REPORT STATUS CODE - 11h	?? BITS
DATA FIELD (optional)	STATUS CODE(S)	8 BITS
DATA FIELD (optional)	STATUS CODE(S)	8 BITS
DATA FIELD (optional)	STATUS CODE(S)	8 BITS
DATA FIELD (optional)	STATUS CODE(S)	8 BITS
DATA FIELD (optional)	STATUS CODE(S)	8 BITS
CRC FIELD		8 BITS
ACKNOWLEDGE FIELD	OFF-BOARD DIAGNOSTIC MODULE - 00h	8 BITS

FIGURE 6—CODE 11 HEX REPORT DIAGNOSTIC STATUS

23.310

START DELIMITER			1 BIT
PRIORITY/TYPE FIELD	DIAG PRIORITY 3h	NODE TO NODE 4h	8 BITS
TARGET FIELD	OFF-BOARD DIAGNOSTIC MODULE – 00h		8 BITS
SOURCE FIELD	??h		8 BITS
DATA FIELD	REPORT MODULE IS OFF LINE – 19h		8 BITS
CRC FIELD			8 BITS
ACKNOWLEDGE FIELD	OFF-BOARD DIAGNOSTIC MODULE – 00h		8 BITS

FIGURE 7—CODE 19 HEX REPORT MODULE IS OFF LINE

on-board diagnostic routine is being referenced.

This message contains 7 bytes as shown in Figure 17.

3.18 CODE 50 hex Confirm Diagnostic Routine Entry—This message is used to confirm that a specific on-board diagnostic routine has been entered.

This message contains 7 bytes as shown in Figure 18.

3.19 CODE 51 hex Confirm Diagnostic Routine Exit—This message is used to confirm that a specific on-board diagnostic routine has been exited. This message also contains a byte which specifies the number of bytes which will be sent to report results using the Report Diagnostic Routine Results message (code 58).

This message contains 8 bytes as shown in Figure 19.

3.20 CODE 58 hex Report Diagnostic Routine Results—This message is used to report the results of executing a specific on-board diagnostic routine. A maximum of 6 bytes of results information may be transmitted with each message.

This message contains from 7 to 12 bytes as shown in Figure 20.

3.21 CODE 60 hex Substitute Value

TBD

3.22 Code 7? hex Report Parameter Packet #?—This message is used to report the value of multiple parameters within a single message. A packet containing either 6 single byte or 3 double byte (word) parameters values may be transmitted with message. Codes 70 thru 7F correspond to Report Parameter Packet #'s 0 thru 15.

This message contains 12 bytes as shown in Figure 21.

START DELIMITER			1 BIT
PRIORITY/TYPE FIELD	DIAG PRIORITY 3h	NODE TO NODE 4h	8 BITS
TARGET FIELD	??h		8 BITS
SOURCE FIELD	OFF-BOARD DIAGNOSTIC MODULE – 00h		8 BITS
DATA FIELD	REQUEST PARAMETER BY PID – 20h		8 BITS
DATA FIELD	PARAMETER INDEX VALUE – ??		8 BITS
CRC FIELD			8 BITS
ACKNOWLEDGE FIELD	??h		8 BITS

FIGURE 8—CODE 20 HEX REQUEST PARAMETERS BY PID

23.312

Field	Value	Size
START DELIMITER		1 BIT
PRIORITY/TYPE FIELD	DIAG PRIORITY 3h | NODE TO NODE 4h	8 BITS
TARGET FIELD	??h	8 BITS
SOURCE FIELD	OFF-BOARD DIAGNOSTIC MODULE - 00h	8 BITS
DATA FIELD	REQUEST PARAMETER BY ADDRESS - 21h	8 BITS
DATA FIELD	PARAMETER ADDRESS LOW BYTE - ??	8 BITS
DATA FIELD	PARAMETER ADDRESS HIGH BYTE - ??	8 BITS
CRC FIELD		8 BITS
ACKNOWLEDGE FIELD	??h	8 BITS

FIGURE 9—CODE 21 HEX REQUEST SINGLE BYTE PARAMETER BY DMR

START DELIMITER			1 BIT
PRIORITY/TYPE FIELD	DIAG PRIORITY 3h	NODE TO NODE 4h	8 BITS
TARGET FIELD	??h		8 BITS
SOURCE FIELD	OFF-BOARD DIAGNOSTIC MODULE - 00h		8 BITS
DATA FIELD	REQUEST PARAMETER BY ADDRESS - 22h		8 BITS
DATA FIELD	PARAMETER ADDRESS LOW BYTE - ??		8 BITS
DATA FIELD	PARAMETER ADDRESS HIGH BYTE - ??		8 BITS
CRC FIELD			8 BITS
ACKNOWLEDGE FIELD	??h		8 BITS

FIGURE 10—CODE 22 HEX REQUEST DOUBLE BYTE PARAMETER BY DMR

23.314

Field		Size
START DELIMITER		1 BIT
PRIORITY/TYPE FIELD	DIAG PRIORITY 3h | NODE TO NODE 4h	8 BITS
TARGET FIELD	??h	8 BITS
SOURCE FIELD	OFF-BOARD DIAGNOSTIC MODULE – 00h	8 BITS
DATA FIELD	REQUEST PARAMETER PACKET – 24h	8 BITS
DATA FIELD	PID/DMR, BYTE/WORD, RATE, PACKET#	8 BITS
CRC FIELD		8 BITS
ACKNOWLEDGE FIELD	??h	8 BITS

FIGURE 11—CODE 24 HEX REQUEST MULTIPLE PARAMETER PACKET REPORT

Field	Content	Size
START DELIMITER		1 BIT
PRIORITY/TYPE FIELD	DIAG PRIORITY 3h | NODE TO NODE 4h	8 BITS
TARGET FIELD	??h	8 BITS
SOURCE FIELD	OFF-BOARD DIAGNOSTIC MODULE - 00h	8 BITS
DATA FIELD	SETUP PARAMETER REPORT PACKET - 28h	8 BITS
DATA FIELD	PACKET POSITION / PACKET #	8 BITS
DATA FIELD	PARAMETER ADDR LOW BYTE (PID) - ??	8 BITS
DATA FIELD (optional)	PARAMETER ADDRESS HIGH BYTE - ??	8 BITS
CRC FIELD		8 BITS
ACKNOWLEDGE FIELD	??h	8 BITS

FIGURE 12—CODE 28 HEX SETUP PARAMETER REPORT PACKET

23.316

Field	Value	Size
START DELIMITER		1 BIT
PRIORITY/TYPE FIELD	DIAG PRIORITY 3h | NODE TO NODE 4h	8 BITS
TARGET FIELD	OFF-BOARD DIAGNOSTIC MODULE – 00h	8 BITS
SOURCE FIELD	??h	8 BITS
DATA FIELD	REPORT PARAMETER BY PID – 30h	8 BITS
DATA FIELD	PARAMETER INDEX VALUE – ??	8 BITS
DATA FIELD	PARAMETER VALUE (LOW BYTE)	8 BITS
DATA FIELD (optional)	PARAMETER VALUE (HIGH BYTE)	8 BITS
CRC FIELD		8 BITS
ACKNOWLEDGE FIELD	OFF-BOARD DIAGNOSTIC MODULE – 00h	8 BITS

FIGURE 13—CODE 30 HEX REPORT PID PARAMETER

START DELIMITER			1 BIT
PRIORITY/TYPE FIELD	DIAG PRIORITY 3h	NODE TO NODE 4h	8 BITS
TARGET FIELD	OFF-BOARD DIAGNOSTIC MODULE – 00h		8 BITS
SOURCE FIELD	??h		8 BITS
DATA FIELD	REPORT PARAMETER BY DMR – 31h		8 BITS
DATA FIELD	PARAMETER ADDRESS (LOW BYTE)		8 BITS
DATA FIELD	PARAMETER ADDRESS (HIGH BYTE)		8 BITS
DATA FIELD	PARAMETER VALUE – ??		8 BITS
CRC FIELD			8 BITS
ACKNOWLEDGE FIELD	OFF-BOARD DIAGNOSTIC MODULE – 00h		8 BITS

FIGURE 14—CODE 31 HEX REPORT SINGLE BYTE DMR PARAMETER

23.318

START DELIMITER			1 BIT
PRIORITY/TYPE FIELD	DIAG PRIORITY 3h	NODE TO NODE 4h	8 BITS
TARGET FIELD	OFF-BOARD DIAGNOSTIC MODULE – 00h		8 BITS
SOURCE FIELD	??h		8 BITS
DATA FIELD	REPORT PARAMETER BY DMR – 32h		8 BITS
DATA FIELD	PARAMETER ADDRESS (LOW BYTE)		8 BITS
DATA FIELD	PARAMETER ADDRESS (HIGH BYTE)		8 BITS
DATA FIELD	PARAMETER VALUE (LOW BYTE)		8 BITS
DATA FIELD	PARAMETER VALUE (HIGH BYTE)		8 BITS
CRC FIELD			8 BITS
ACKNOWLEDGE FIELD	OFF-BOARD DIAGNOSTIC MODULE – 00h		8 BITS

FIGURE 15—CODE 32 HEX REPORT DOUBLE BYTE PARAMETER

START DELIMITER			1 BIT
PRIORITY/TYPE FIELD	DIAG PRIORITY 3h	NODE TO NODE 4h	8 BITS
TARGET FIELD	??h		8 BITS
SOURCE FIELD	OFF-BOARD DIAGNOSTIC MODULE - 00h		8 BITS
DATA FIELD	ENTER ON-BOARD DIAGNOSTIC - 40h		8 BITS
DATA FIELD	DIAGNOSTIC ROUTINE SPECIFIER - ??h		8 BITS
CRC FIELD			8 BITS
ACKNOWLEDGE FIELD	??h		8 BITS

FIGURE 16—CODE 40 HEX ENTER DIAGNOSTIC ROUTINE

23.320

START DELIMITER			1 BIT
PRIORITY/TYPE FIELD	DIAG PRIORITY 3h	NODE TO NODE 4h	8 BITS
TARGET FIELD	??h		8 BITS
SOURCE FIELD	OFF-BOARD DIAGNOSTIC MODULE — 00h		8 BITS
DATA FIELD	EXIT ON-BOARD DIAGNOSTIC — 41h		8 BITS
DATA FIELD	DIAGNOSTIC ROUTINE SPECIFIER — ??h		8 BITS
CRC FIELD			8 BITS
ACKNOWLEDGE FIELD	??h		8 BITS

FIGURE 17—CODE 41 HEX EXIT DIAGNOSTIC ROUTINE

START DELIMITER			1 BIT
PRIORITY/TYPE FIELD	DIAG PRIORITY 3h	NODE TO NODE 4h	8 BITS
TARGET FIELD	OFF-BOARD DIAGNOSTIC MODULE - 00h		8 BITS
SOURCE FIELD	??h		8 BITS
DATA FIELD	CONFIRM DIAGNOSTIC ENTRY - 50h		8 BITS
DATA FIELD	DIAGNOSTIC ROUTINE SPECIFIER - ??h		8 BITS
CRC FIELD			8 BITS
ACKNOWLEDGE FIELD	OFF-BOARD DIAGNOSTIC MODULE - 00h		8 BITS

FIGURE 18—CODE 50 HEX CONFIRM DIAGNOSTIC ROUTINE ENTRY

23.322

START DELIMITER			1 BIT
PRIORITY/TYPE FIELD	DIAG PRIORITY 3h	NODE TO NODE 4h	8 BITS
TARGET FIELD	OFF-BOARD DIAGNOSTIC MODULE - 00h		8 BITS
SOURCE FIELD	??h		8 BITS
DATA FIELD	CONFIRM DIAGNOSTIC EXIT - 50h		8 BITS
DATA FIELD	DIAGNOSTIC ROUTINE SPECIFIER - ??h		8 BITS
DATA FIELD	ROUTINE RESULTS BYTE COUNT - ??h		8 BITS
CRC FIELD			8 BITS
ACKNOWLEDGE FIELD	OFF-BOARD DIAGNOSTIC MODULE - 00h		8 BITS

FIGURE 19—CODE 51 HEX CONFIRM DIAGNOSTIC ROUTING EXIT

Field	Value	Size
START DELIMITER		1 BIT
PRIORITY/TYPE FIELD	DIAG PRIORITY 3h | NODE TO NODE 4h	8 BITS
TARGET FIELD	OFF-BOARD DIAGNOSTIC MODULE — 00h	8 BITS
SOURCE FIELD	??h	8 BITS
DATA FIELD	REPORT DIAG ROUTINE RESULTS — 58h	8 BITS
DATA FIELD	DIAGNOSTIC ROUTINE RESULTS — ??h	8 BITS
DATA FIELD (optional)	DIAGNOSTIC ROUTINE RESULTS — ??h	8 BITS
DATA FIELD (optional)	DIAGNOSTIC ROUTINE RESULTS — ??h	8 BITS
DATA FIELD (optional)	DIAGNOSTIC ROUTINE RESULTS — ??h	8 BITS
DATA FIELD (optional)	DIAGNOSTIC ROUTINE RESULTS — ??h	8 BITS
DATA FIELD (optional)	DIAGNOSTIC ROUTINE RESULTS — ??h	8 BITS
CRC FIELD		8 BITS
ACKNOWLEDGE FIELD	OFF-BOARD DIAGNOSTIC MODULE — 00h	8 BITS

FIGURE 20—CODE 58 HEX REPORT DIAGNOSTIC ROUTINE RESULTS

23.324

Field	Content	Size
START DELIMITER		1 BIT
PRIORITY/TYPE FIELD	DIAG PRIORITY 3h | NODE TO NODE 4h	8 BITS
TARGET FIELD	OFF-BOARD DIAGNOSTIC MODULE − 00h	8 BITS
SOURCE FIELD	??h	8 BITS
DATA FIELD	REPORT PARAMETER PACKET − 7?h	8 BITS
DATA FIELD	PARAMETER 1 or PARAMETER 1 (LO BYTE)	8 BITS
DATA FIELD	PARAMETER 2 or PARAMETER 1 (HI BYTE)	8 BITS
DATA FIELD	PARAMETER 3 or PARAMETER 2 (LO BYTE)	8 BITS
DATA FIELD	PARAMETER 4 or PARAMETER 2 (HI BYTE)	8 BITS
DATA FIELD	PARAMETER 5 or PARAMETER 3 (LO BYTE)	8 BITS
DATA FIELD	PARAMETER 6 or PARAMETER 3 (HI BYTE)	8 BITS
CRC FIELD		8 BITS
ACKNOWLEDGE FIELD	OFF-BOARD DIAGNOSTIC MODULE − 00h	8 BITS

FIGURE 21—CODE 7? HEX REPORT PARAMETER PACKET #?

OBD II SCAN TOOL—SAE J1978 MAR92

SAE Recommended Practice

Report of the SAE E/E Systems Diagnostic Standards Committee approved March 1992.

Table of Contents

Foreword
1. Scope
2. References
3. Application
4. Diagnostic Terms, Definitions, and Acronyms
5. Required Functions and Support
6. Vehicle Interface
6.1 Communication Data Link and Physical Layers
6.1.1 SAE J1850 Recommended Practice Class B Data Communication Network Interface
6.1.2 ISO 9141 - 2 Road Vehicles—Diagnostic Systems—Part 2: Carb Requirements For Interchange of Digital Information
6.2 Connector
6.3 Messages
6.4 Expanded Diagnostic Protocol
6.5 Automatic Hands off Determination of the Communication Interface Used in a Given Vehicle
6.5.1 General
6.5.2 Initialization Details
6.6 On-Board Diagnostic Evaluations
6.6.1 Completed On-Board System Readiness Tests
6.6.2 Supported On-Board System Readiness Tests
6.6.3 Malfunction Indicator Light
6.7 Use of SAE J2201 Universal Interface for OBD II Scan Tools
7. System Interaction Capability
7.1 Obtain and Display OBD II Emissions Related Diagnostic Trouble Codes
7.2 Obtain and Display OBD II Emissions Related Current Data, Freeze Frame Data, and Test Parameters and Results
7.3 Responses from Multiple Modules
7.4 Code Clearing
7.5 Oxygen Sensor Monitoring Tests
8. General Characteristics
8.1 Display
8.2 User Input
9. Power Requirements if Powered by the Vehicle Through the SAE J1962 Diagnostic Connector
10. Electromagnetic Compatibility (EMC)
11. Conformance Testing
11.1 General
11.2 Determine OBD II Communication Type
11.3 On-Board System Readiness Tests
11.4 Select Functions
11.5 Select and Display Items
11.6 Verify Requests to Clear Codes
11.7 General Diagnostic Communication Tests
11.8 Expanded Diagnostic Protocol
11.9 Capacitance and Impedance at the SAE J1962 Connector
11.10 Operating Voltage and Current Draw
11.11 Protocol Check
11.12 Alphanumeric Display
11.13 User Manual and Help Facility

Foreword—Title 13, California Code of Regulations, Section 1968.1 "Malfunction and Diagnostic System Requirements 1994 and Subsequent Model Year Passenger Cars, Light Duty Trucks, and Medium Duty Vehicles With Feedback Control Systems," more commonly known as OBD II (On-Board Diagnostics version II), defines diagnostic functions required to be supported by vehicles and functions to be supported by test equipment that interface with the vehicle diagnostic functions.

While a range of test equipment (e.g. handheld scan tools, PC based diagnostic computers, etc.) may be used to perform the required interface support functions, the term OBD II Scan Tool is used in this document to refer to any test equipment that meets the requirements of this document.

1. Scope—This SAE Recommended Practice defines the requirements of OBD II Scan Tools, i.e. test equipment that will interface with vehicle modules in support of the OBD II diagnostic requirements. It covers the required capabilities of and conformance criteria for OBD II Scan Tools.

2. References

2.1 Applicable Documents—The following publications form a part of this specification to the extent specified herein. The latest issues of SAE publications shall apply.

2.1.1 SAE PUBLICATIONS—Available from SAE, 400 Commonwealth Drive, Warrendale, PA 15096-0001.

SAE J1113—Electromagnetic Susceptibility Measurements Procedures for Vehicle Components (Except Aircraft)
SAE J1850—Class B Data Communication Network Interface
SAE J1930—Electrical/Electronic Systems Diagnostic Acronyms, Terms and Definitions
SAE J1962— Diagnostic Connector
SAE J1979—E/E Diagnostic Test Modes
SAE J2012—Recommended Format and Messages for Diagnostic Trouble Codes
SAE J2201 DRAFT—Universal Interface for OBD II Scan Tools
SAE J2205 DRAFT—Expanded Diagnostic Protocol

2.1.2 ISO PUBLICATIONS—Available from ANSI, 11 West 42nd Street, New York, NY 10036-8002.

ISO 9141-2 Road vehicles—Diagnostic systems—Part 2: Carb requirements for interchange of digital information

2.1.3 OTHER PUBLICATIONS

California Code of Regulations, Section 1968.1, Title 13—Malfunction and Diagnostic System Requirements 1994 and Subsequent Model Year Passenger Cars, Light Duty Trucks, and Medium Duty Vehicles with Feedback Control Systems

3. Application—The requirements contained in this document apply to all test equipment (e.g. handheld scan tool, PC based test equipment, inspection and maintenance equipment, etc.) that will be used to access the OBD II functions supported in vehicles.

Only tools passing the conformance test(s) specified in this document may claim or advertise that they meet or exceed the requirements of this document.

Nothing in this document precludes the inclusion of additional capabilities or functions in test equipment.

4. Diagnostic Terms, Definitions, and Acronyms—SAE J1930 is hereby referenced as the basis for all such terms in this document, with the following additions:

PID—Parameter Id (see SAE J1979 for PID definitions)

5. Required Functions and Support—The following are the basic functions that the OBD II Scan Tool is required to support or provide:

a. Automatic hands-off determination of the communication interface used
b. Obtaining and displaying the status and results of vehicle on-board diagnostic evaluations
c. Obtaining and displaying OBD II emissions related diagnostic trouble codes (DTC's)
d. Obtaining and displaying OBD II emissions related current data
e. Obtaining and displaying OBD II emissions related freeze frame data
f. Clearing the storage of OBD II emissions related diagnostic trouble codes, OBD II emissions related freeze frame data storage and OBD II emissions related diagnostic tests status
g. Ability to perform Expanded Diagnostic Protocol functions as described in SAE J2205
h. Obtaining and displaying OBD II emissions related test parameters and results as described in SAE J1979, mode 5
i. Provide a user manual and/or help facility

6. Vehicle Interface—The following specifies the minimum vehicle interfaces to be supported by an OBD II Scan Tool.

6.1 Communication Data Link and Physical Layers—The OBD II Scan Tool must be able to communicate with vehicle control modules using the communication interfaces as described as follows. The OBD II Scan Tool must perform the communications required to support the State of California requirements referenced in the introduction to this document. SAE J2205 describes a set of functions and communication criteria. The OBD II Scan Tool need not support any SAE J2205 functions or communications beyond the interfaces described in this section.

6.1.1 SAE J1850 Class B Data Communication Network Interface—SAE J1850 describes the data link and physical layers of a class B vehicle serial multiplex bus network. The two implementations that must be supported by the OBD II Scan Tool are the 41.6 Kbps PWM, and the 10.4 Kbps VPM with CRC. The OBD II Scan Tool must support both of the SAE J1850 protocols in a manner that is transparent to the user.

6.1.2 ISO 9141-2 Road Vehicles—Diagnostic Systems—Part 2: CARB Requirements for Interchange of Digital Information—ISO 9141 CARB describes the physical and data link layers of a vehicle serial diagnostic bus.

6.2 Connector—The OBD II Scan Tool must use a male SAE J1962 Diagnostic Connector to mate with the female SAE J1962 Diagnostic Connector required in vehicles. The OBD II Scan Tool must support the Standard Pin Assignments defined in SAE J1962. The electrical interface in the OBD II Scan Tool for the manufacturer discretionary pin assignments, shall be effectively open circuit as a default condition or state.

6.3 Messages—SAE J1979 describes the request messages to be sent by the OBD II Scan Tool to the vehicle and the response messages to be sent by the vehicle to the OBD II Scan Tool in order to perform the services required by OBD II.

6.4 Expanded Diagnostic Protocol—The OBD II Scan Tool must allow the user to enter and send vehicle specific messages defined and supplied in motor vehicle manufacturer documents and display the related response messages.

6.5 Automatic Hands-off Determination of the Communication Interface used in a Given Vehicle

6.5.1 General—While there are three types of communication interfaces that could be used to access the OBD II functions in a given vehicle (i.e. SAE J1850 41.6 Kbps PWM, SAE J1850 10.4 Kbps VPW with CRC, ISO 9141 CARB), only one is allowed to be used in any one vehicle to access all supported OBD II functions.

When connected to a vehicle and/or when the OBD II support is selected when such a selection is necessary, the OBD II Scan Tool will automatically attempt to determine which of the possible communication interfaces is being used in the vehicle to support OBD II related functions. The tool will continue to try to determine which interface is being used until it is successful in doing so. No user input will be required, nor allowed, to determine the appropriate interface.

Indications or messages will be displayed during this process informing the user that initialization is taking place and, if all interface types have been tested and none is responding properly to the request for OBD II services, the OBD II Scan Tool must indicate to the user:
a. To verify that the ignition is on.
b. To check the Emissions Label to verify that the vehicle is OBD II equipped.

If the ignition is on and the label indicates the vehicle is OBD II equipped, then there is a Data Link fault.

6.5.2 Initialization Details—Only the following steps may be used by an OBD II Scan Tool to attempt to determine the type of communications interface used in a given vehicle to support OBD II functions.
a. Test for SAE J1850 41.6 Kbps PWM
  (1) Step 1—Enable the SAE J1850 41.6 Kbps PWM interface
  (2) Step 2—Send a mode 1 PID 0 request message
  (3) Step 3—If a mode 1 PID 0 response message is received then SAE J1850 41.6 Kbps PWM is the type of interface used in a vehicle for OBD II support.
b. Test for SAE J1850 10.4 Kbps VPW with CRC
  (1) Step 1—Enable the SAE J1850 10.4 Kbps VPW with CRC interface
  (2) Step 2—Send a mode 1 PID 0 with CRC request message
  (3) Step 3—If a mode 1 PID 0 with CRC response message is received then SAE J1850 10.4 Kbps VPW with CRC is the type of interface used in a vehicle for OBD II support.
c. Test for ISO 9141 CARB
  (1) Step 1—Enable the ISO 9141 CARB interface
  (2) Step 2—If the initialization sequence defined in ISO 9141 CARB is completed successfully, then ISO 9141 CARB is the type of interface used in a vehicle for OBD II support.

The previous tests may be performed in any order and where possible may be performed in parallel.

The mode 1 PID 0 request and response messages are defined in SAE J1979.
SAE J1850 defines the requirements of SAE J1850 interfaces.
ISO 9141 CARB defines the requirements of an ISO 9141 CARB interface.

6.6 On-Board Diagnostic Evaluations

6.6.1 Completed On-board System Readiness Tests—Immediately after initial communications are established, the OBD II Scan Tool shall obtain the status of the on-board system readiness tests. If any tests have not been completed (i.e. any bits of the SAE J1979 mode 1 PID 1 data byte 6 are non-zero), the OBD II Scan Tool shall indicate to the user: "Not all supported on-board system readiness tests have been completed." or equivalent. The OBD II Scan Tool shall also allow the user to identify which readiness tests (if any) have not been completed.

6.6.2 Supported On-Board System Readiness Tests—The OBD II Scan Tool must be capable of indicating to the user which of the tests defined by SAE J1979 mode 1 PID 1 data bytes 4 and 5 are supported and which are completed.

6.6.3 Malfunction Indicator Light—The OBD II Scan Tool must be capable of indicating if the MIL has been commanded ON and if so, by which module or modules.

6.7 Use of SAE J2201 Universal Interface for OBD II Scan Tools—The OBD II Scan Tool shall use the interface described in SAE J2201, or an equivalent, as the interface to vehicles.

7. System Interaction Capability

7.1 Obtain and Display OBD II Emissions Related Diagnostic Trouble Codes—The OBD II Scan Tool must be capable of obtaining, converting, and displaying all of the OBD II emissions related diagnostic trouble codes from a vehicle that can be transmitted by a response to a SAE J1979 mode 3 request. Either the diagnostic trouble code, its descriptive text, or both must be displayed. Diagnostic Trouble Codes and their descriptive text are defined in SAE J2012. When "diagnostic trouble code data" are selected for display, the OBD II Scan Tool will continuously request of the vehicle its diagnostic trouble code data and display the data received in the corresponding response messages.

7.2 Obtain and Display OBD II Emissions Related Current Data, Freeze Frame Data, and Test Parameters and Results—The OBD II Scan Tool must be capable of obtaining, converting, and displaying (a) OBD II emissions related current data, (b) freeze frame data, and (c) test parameters and results data as described in SAE J1979. SAE J1979 details what data is available, the messages to be used to request the data, the messages to be used to return the data, the conversion values for the data and the format to be used to display the data. When current data items are selected for display, the OBD II Scan Tool will continuously request of the vehicle the data to be displayed and display the data received in the corresponding response messages. When freeze frame or test parameters and results items are selected for display, the OBD II Scan Tool does not need to continuously request and display those items.

7.3 Responses from Multiple Modules—The OBD II Scan Tool must be capable of interfacing with a vehicle in which multiple modules may be used to support OBD II requirements.

The OBD II Scan Tool must alert the user when multiple modules respond to the same request.

The OBD II Scan Tool must advise/alert the user when multiple modules respond with different values for the same data item.

The OBD II Scan Tool must provide the user with the ability to select for display as separate display items the responses received from multiple modules for the same data item.

7.4 Code Clearing—The OBD II Scan Tool must be capable of sending a request to clear OBD II emissions related diagnostic trouble codes, OBD II emissions related freeze frame data and OBD II emissions related diagnostic tests status information. The OBD II Scan Tool must require the user to validate a user's request to send this request before sending it (i.e. are you sure?).

7.5 Oxygen Sensor Monitoring Tests—The OBD II Scan Tool must be capable of requesting and displaying the results of the vehicle on-board oxygen sensor monitoring test results if the vehicle supports these optional tests.

8. General Characteristics

8.1 Display—The OBD II Scan Tool must be capable of displaying simultaneously at least two items of OBD II emissions related current data items, OBD II emissions related freeze-frame data items, or OBD II emissions related diagnostic trouble codes.

A list of the OBD II emissions related current data and freeze frame data items, their parameter id's, data resolution and data conversion information, units and display formats is provided in SAE J1979. The display units shall be the Standard International (SI) and English units as specified in SAE J1979. A user shall be able to select between English and SI values. The unit conversions specified in SAE J1979 shall be used.

The display of each OBD II emissions related current data or freeze frame data shall include the following:
a. Data value
b. Data Parameter id or name
c. The module id of the module that supplied the data

The display of each OBD II emissions related diagnostic trouble code shall include the module id of the module that supplied the code.

As a minimum the data values of two data items must be displayed simultaneously. A display of the parameter id's of the data items and the id's of the modules that supplied the data items must be easily accessible if not displayed with the data values.

The units of measure associated with the data items displayed must either be displayed with the data values, easily accessible on another display, or otherwise readily available to the user (e.g. on the tester body, as a part of the tester on a printed card etc.). Having this information available in a user's manual separate from the body of the tool does not satisfy this requirement.

The display must be capable of showing alphanumeric characters.

8.2 User Input—The OBD II Scan Tool must include some form of user input that would allow the user to:
  a. Select between the basic functions required by OBD II (i.e. display current data, display freeze frame data,   display trouble codes, clear emissions related data and display test parameters and results),
  b. Select for simultaneous display at least two items of any one of the following:
    (1) OBD II emissions related current data
    (2) OBD II emissions related diagnostic trouble codes
    (3) OBD II emissions related freeze frame data
    (4) OBD II emissions related test parameters and results
    Responses from multiple modules to requests for a current data item or a freeze frame data item are treated as separate data items for selection and display purposes.
  c. To verify a request to clear and/or reset OBD II emissions related diagnostic information as defined by   mode 4 of SAE J1979.
  d. Enter and send Expanded Diagnostic Protocol messages.

9. Power Requirements if Powered by the Vehicle Through the SAE J1962 Diagnostic Connector
  a. Voltage
    (1) Must operate normally within a range of 8.0 to 18.0 V D.C.
    (2) Must survive a steady state voltage of up to 24.0 V D.C. for at least 10.0 min
    (3) Must survive a steady state reverse voltage of up to 24.0 V D.C. for at least 10 min
  b. Current
    (1) Must not draw more than 4.0 A at 14.4 V D.C.

10. Electromagnetic Compatibility (EMC)—The tool must not interfere with the normal operation of vehicle modules.

The normal operation of the tool must be immune to conducted and radiated emissions present in a service environment and when connected to a vehicle.

The tool must be immune to reasonable levels of Electrostatic Discharge (ESD).

EMC and ESD measurements and limits will be according to SAE J1113.

11. Conformance Testing

11.1 General—Conformance testing defines the tests required to be passed in order for tools to be typed approved as "SAE J1978 MAR92 OBD II COMPATIBLE." Tools that do not pass these tests are not to be so labeled. Validation of conformance test is the responsibility of the Scan Tool manufacturer and the Scan Tool Manufacturer may elect to self-certify.

The tests in this section must be performed successfully five consecutive times in order to be considered passed.

Three examples of at least production intent level tools must pass all these tests in order for a given version of tool hardware and software to be considered passed.

Any changes to the hardware or software used in a tool for the functions described in this document will require a retest of these tests or an explanation from the tool manufacturer as to why the change should not require a retest. Where an explanation is submitted in lieu of a retest due to a change, the organization originally performing these tests will determine whether the explanation is acceptable or whether a retest is required. Reasonable normal engineering criteria will be used when determining whether to accept an explanation.

The Scan Tool Manufacturer shall make available to the buying public:
  a. The methods used to make these tests
  b. The results of the tests
  c. Clear indication of the versions of hardware and software that conform (i.e. labeled as conforming to or are compatible with the requirements of SAE J1978 OBD II Scan Tool or other labeling to that effect).

Both proper and improper response messages will be employed during these tests. Improper responses are those that have incorrect first, second, or third bytes of the header, an incorrect mode, an incorrect PID id, an incorrect length of the response message, or with an incorrect CRC. The tool must ignore all improper response messages and perform as if no response was received.

Situations involving multiple modules responding to a single request, single modules responding with multiple responses to a single request and multiple modules responding with multiple responses to a single request will be tested.

The interval between the end of the request message and the beginning of the response message(s) will be varied from 0 ms up to the delay required to show a no response indication on the OBD II Scan Tool. The delay that causes the no response indication will be compared to the value defined in SAE J1979.

The format, content, and order of messages transmitted on the SAE J1850 and ISO 9141 CARB buses will be observed and reviewed for correctness.

The ability to obtain and report the results of the on-board system readiness tests shall be verified. The ability to report which tests the vehicle supports and which have been completed shall be verified.

The requirements described in 11.3 through 11.7 (inclusive) shall be verified on each protocol specified in 6.1.

When performing these tests, observation of the indications and displays provided to the user and the signals on the SAE J1850 (bus +) and (bus -) lines and the ISO 9141 CARB K and L lines will be the criteria for proper performance.

These tests will be executed in an environment of 25 °C ± 3 °C and between 30% and 80% relative humidity.

The hardware and software used in the OBD II Scan Tool version being tested must be identified.

11.2 Determine OBD II Communication Type—Items to be tested:

Automatic hands-off determination of interface type.

That it is automatic when the SAE J1962 connector is plugged into its mating connector in the vehicle and/or OBD II support is selected, where such a selection is necessary.

That a test of all OBD II communication interfaces is performed at least once per scan.

That the scan of all interfaces continues until successful or until terminated by the user.

That some indication is provided to the user that the scan of interfaces is being performed.

That a failure to successfully find an OBD II interface during a scan of all the possible interfaces is indicated to the user at the completion of each and every scan.

That when an OBD II interface is successfully found, the tool automatically prompts the user for function selection.

That the tool provides and uses the facilities and/or messages defined in SAE J2201 (or equivalent), SAE J1979, ISO 9141 CARB and SAE J1850, and SAE J2012.

That the tool does not exceed the polling rates specified in SAE J1979.

That the tool provides the proper bias for the K and L lines as specified in ISO 9141 CARB.

That the tool performs the initialization tests according to 6.5.2 and indicates the information according to 6.5.1.

The interface determination tests will be performed: (a) with no modules available, (b) with an ISO 9141 CARB module available, (c) with a SAE J1850 41.6 Kbps PWM module available, and (d) with a SAE J1850 10.4 Kbps VPW with CRC module available.

11.3 On-Board System Readiness Tests—Item to be tested:

That the tool automatically requests and reports the results of the supported on-board system readiness tests.

11.4 Select Functions—Items to be tested:

That the tool supports the functions described in Section 5 of this document.

That the user is able to move back and forth between these functions.

The criteria for successfully passing this test is to be able to easily move back and forth between all functions and observe the results. The support provided should allow the user to easily move between functions.

11.5 Select and Display Items—Items to be tested:

That the user is able to select and display simultaneously at least two items from any one of:
  a. Available Diagnostic Trouble Codes
  b. Current data items
  c. Available freeze frame data items
  d. Test parameters and results

That the module id's and the parameter id's or PID names associated with all the items mentioned previously can also be displayed either (a) simultaneously with the displayed items or (b) in some alternate method (printed material, etc.)

That the units of measure information associated with all the possible current data items and freeze frame data items is easily available either as a part of the data display, displayed separately, or otherwise available on, or with the tool body itself.

That the tool is able to handle multiply responses from the same module due to one request.

That the tool is able to handle responses from multiple modules due to one request.

That the tool is able to handle multiple responses from multiple modules due to one request.

That the tool alerts the user that responses from multiple modules due to one request were received. Responses from multiple modules to a request are to be made available to the user as separate items for display.

That the tool advises/alerts the user that different responses from multiple modules due to one request were received.

The criteria for successfully passing this test is to select back and forth between all the items and observe the results. The support provided should allow the user to easily move between display items.

11.6 Verify Requests to Clear Codes—Items to be tested:

That the selection of the Clear Codes function incorporates a request to the user to verify the request.

That both yes and no responses to the request to the user to verify the selection of the Clear Codes function are processed appropriately.

This test should involve situations where there are some DTC's to clear and other situations where there are no DTC's to clear. When making this test, the presence or absence of DTC's must be verified both before and after the Clear Codes function is selected.

11.7 General Diagnostic Communication Tests—When performing tests involving diagnostic messages, tests are to be made of the tool's ability to handle an immediate response, a slow response and a response delayed longer than the maximum allowed in SAE J1979.

The tool should be able to process all responses that are received within the maximum time allowed by SAE J1979 and indicate a no response condition to the user when the response is delayed longer than the maximum allowed by SAE J1979.

The tool must support the transmission of its node address as an in-frame-response during the transmission of any response messages from modules on a SAE J1850 bus and must be able to handle both the presence and the absence of an in-frame-response during the tool's transmission of request messages.

11.8 Expanded Diagnostic Protocol—Items to be tested:

That the user is reasonably able to enter Expanded Diagnostic Protocol (EDP) input and that the OBD II Scan Tool correctly executes the entered EDP input.

A full range of EDP facilities should be exercised via the EDP data entered (see SAE J2205 DRAFT Expanded Diagnostic Protocol).

11.9 Capacitance and Impedance at the SAE J1962 Connector—Items to be tested:

That the capacitance and impedance of the OBD II Scan Tool, connecting cables and the male SAE J1962 connector, as seen at the connector, are within the limits defined in SAE J1850 and ISO 9141 CARB.

Measurement of these parameters will be made by the testing agency at their discretion following generally acceptable engineering practices.

11.10 Operating Voltage and Current Draw—Items to be tested:

That the OBD II Scan Tool will correctly operate throughout the voltage range specified in Section 9 of this document and will not require more than the maximum current specified in Section 9 of this document.

That the tool will survive the use of supply voltages of up to the maximum survival voltage and survival reverse voltage specified in Section 9 of this document.

During other conformance tests, the voltage supplied to the OBD II Scan Tool is to be varied throughout the specified range and the results observed. Also the current draw is to be compared with the limit specified.

11.11 Protocol Check—Items to be tested:

That all the request messages defined by SAE J1979 are properly and appropriately used by the OBD II Scan Tool.

That the OBD II Scan Tool is able to correctly process all forms of response messages defined by SAE J1979.

That the OBD II Scan Tool is able to perform the previously mentioned, the SAE J1850 and ISO 9141 CARB buses described in Section 7.

11.12 Alphanumeric Display—Items to be tested:

That the OBD II Scan Tool is able to display alphanumeric characters.

The results of the previous tests will be observed to determine the ability of the OBD II Scan Tool to Display alphanumeric characters.

11.13 User Manual and Help Facility—Items to be tested:

That a useful user manual and/or HELP facility is provided with the OBD II Scan Tool.

That the user manual and/or HELP facility includes:
a. SAE J1979 definitions
b. How to select the functions
c. How to select items for simultaneous display
d. How to determine the PID id, item name, and module id of data returned for display
e. How to verify the selection of the Clear Codes function
f. How to obtain and display OBD II emissions related test parameters and results as described in SAE J1979, mode 5
g. How multiple responses from one request is indicated
h. How different responses to the same request is indicated
i. What current and freeze frame data items are available through OBD II
j. How to enter requests for the Expanded Diagnostic Protocol and interpret the results

The OBD II Scan Tool will be tested for a HELP facility and/or the contents of the materials to be furnished to the user as a complete OBD II Scan Tool will be observed to determine the availability of these items.

E/E DIAGNOSTIC DATA COMMUNICATIONS—SAE J2054 NOV90

SAE Information Report

Report of the SAE Vehicle E/E Systems Diagnostic Standards Committee approved November 1990. Rationale statement available.

TABLE OF CONTENTS

1. Scope
1.1 Purpose
2. References
2.1 Applicable Documents
2.1.1 SAE PUBLICATIONS
3. Diagnostic Message Summary
4. Diagnostic Test Mode General Message Format
5. Device IDS
6. Diagnostic Test Mode General Conditions
7. Diagnostic Test Mode Functional Description
7.1 Mode 0—Return to Normal Mode
7.1.1 FUNCTIONAL DESCRIPTION
7.1.2 MODE 0 REQUEST
7.1.3 MODE 0 RESPONSE
7.2 Mode 1—Transmit Diagnostic Data
7.2.1 FUNCTIONAL DESCRIPTION
7.2.2 MODE 1 REQUEST
7.2.3 MODE 1 RESPONSE
7.3 Mode 2—Memory Dump
7.3.1 FUNCTIONAL DESCRIPTION
7.3.2 MODE 2 REQUEST
7.3.3 MODE 2 RESPONSE
7.4 Mode 3—Examine Memory
7.4.1 FUNCTIONAL DESCRIPTION
7.4.2 MODE 3 REQUEST
7.4.3 MODE 3 RESPONSE
7.5 Mode 4—Device Control Functions
7.5.1 FUNCTIONAL DESCRIPTION
7.5.2 MODE 4 REQUEST
7.5.3 MODE 4 RESPONSE
7.6 Mode 5—RAM Download Request
7.6.1 FUNCTIONAL DESCRIPTION
7.6.2 MODE 5 REQUEST
7.6.3 MODE 5 RESPONSE
7.7 Mode 6—RAM Download and Execute
7.7.1 FUNCTIONAL DESCRIPTION
7.7.2 MODE 6 REQUEST
7.7.3 MODE 6 RESPONSE (OPTIONAL)
7.8 Mode 7—Transmit Normal Message
7.8.1 FUNCTIONAL DESCRIPTION
7.8.2 MODE 7 REQUEST
7.8.3 MODE 7 RESPONSE
7.9 Mode 8—Disable Normal Communications
7.9.1 FUNCTIONAL DESCRIPTION
7.9.2 MODE 8 REQUEST
7.9.3 MODE 8 RESPONSE
7.10 Mode 9—Enable Normal Communications
7.10.1 FUNCTIONAL DESCRIPTION
7.10.2 MODE 9 REQUEST
7.10.3 MODE 9 RESPONSE
7.11 Mode 10—Clear Malfunction Codes
7.11.1 FUNCTIONAL DESCRIPTION
7.11.2 MODE 10 REQUEST
7.11.3 MODE 10 RESPONSE
7.12 Mode 11—Suspend Normal Message
7.12.1 FUNCTIONAL DESCRIPTION
7.12.2 MODE 11 REQUEST
7.12.3 MODE 11 RESPONSE
7.13 Modes 12 & 13—Define Diagnostic Message
7.13.1 FUNCTIONAL DESCRIPTION
7.13.2 MODE 12 REQUEST
7.13.3 MODE 12 RESPONSE
7.13.4 MODE 13 REQUEST
7.13.5 MODE 13 RESPONSE

1. Scope—This SAE Information Report describes the diagnostic data communications required for implementation of a set of diagnostic test modes for all electronic systems on the vehicle's serial data link. These test modes can be used by off-board test equipment for both service and assembly plant testing.

The goal of this document is to provide standard methods to perform common functions for all electronic systems. This standard set of procedures will aid development, production, and field service of those systems. Use of the standard data communications in this specification will potentially result in the following benefits:

Common methods and procedures for developers to use, without the need to invent methods to perform these functions

Common programming techniques for system programmers, with increased sharing of software procedures

Common hardware, software and test procedures for assembly plant testing

Common hardware, software and service procedures for service diagnostics across different vehicle manufacturers and systems

This specification includes:

Diagnostic Message Formats

Device ID's

Functional descriptions of all diagnostic test modes

Message and response formats for all diagnostic test modes

1.1 Purpose—The diagnostic test modes (DTM) provide off-board test equipment with communication access to the on-board vehicle electronic systems. The off-board equipment shall be able to interrogate the electronic systems on the vehicle, and exercise control over these systems for the purposes of verifying system operation and diagnosing malfunction conditions.

Diagnostic test modes are predefined and standardized for all systems on the vehicle. Each device on the vehicle will implement only those DTMs which are appropriate for that device. Not all modes will be implemented on each device. If use of these modes is not appropriate for a given application, or during some operating modes, then that device is responsible to verify safe and proper operation and not respond to the request.

2. References

2.1 Applicable Documents—The following publications form a part of this specification to the extent specified herein. The latest issue of SAE publications shall apply.

2.1.1 SAE PUBLICATIONS—Available from SAE, 400 Commonwealth Dr., Warrendale, PA 15096-0001.

SAE J1850—Class B Data Communication Network Interface

SAE J2037—Off-Board Diagnostic Message Formats

SAE J2062—A Class B Serial Bus Diagnostic Protocol

SAE J2086—An Application Layer Protocol for a Generic Scan Tool

3. Diagnostic Message Summary—This specification defines 14 Diagnostic Test Modes.

Mode 0 —Return to Normal Mode
Mode 1 —Transmit Diagnostic Data
Mode 2 —Memory Dump
Mode 3 —Examine RAM Memory
Mode 4 —Device Control Functions
Mode 5 —RAM Download Request
Mode 6 —RAM Download and Execute
Mode 7 —Command Normal Message
Mode 8 —Disable Normal Communications
Mode 9 —Enable Normal Communications
Mode 10—Clear Malfunction Codes
Mode 11—Suspend Normal Message
Mode 12—Define Diagnostic Message by Data Position
Mode 13—Define Diagnostic Message by Memory Address

4. Diagnostic Test Mode General Message Format—DTM messages all conform to the following general format:

BYTE	DESCRIPTION
1	Priority—first 4 bits Message Length—second 4 bits
2	Device ID for Target
3	Device ID for Source
4	Message type / format—first 4 bits Diagnostic Mode Number—second 4 bits

The first byte of the message is a combination of priority (4 bits) and message length (4 bits). Message length is used as one check for data integrity. This also allows shorter messages with the same priority as longer messages to be transmitted first, which can improve the efficiency of the data link.

Each device will be assigned an ID unique to that device. The second byte of all DTM messages is the Device ID for the intended receiver of the message, and the third byte is the Device ID for the source of the message.

The fourth byte contains a combination of Message type / format (4 bits) and diagnostic mode number (4 bits). Various types of messages will be defined for normal mode communications and for diagnostic modes. This value indicates the structure for the remaining bytes of the message. The mode number indicates the action the device shall take in response to the message. Those actions are defined in the functional description section for each mode.

SAE J1850 defines additional message elements that may be included in diagnostic messages. Use of these message elements is beyond the scope of this specification, but need to be considered when defining total diagnostic messages. Those elements are:

Start of Message (SOM)
Error Detection Byte (ERR)
End of Data (EOD)
Response Byte(s) (RSP)
End of Message (EOM)
Inter-Message Separation (IMS)
Inter-Byte Separation (IBS)

5. Device IDs—A unique device ID should be assigned for each type of electronic system that will be used on different vehicles (See Table 1). Wherever possible, those codes should remain constant for that type of device on all vehicles. This will enable that same device to be utilized on other vehicles without software modifications to alter the device code. If those codes are unique, it will also eliminate any message conflict with other devices trying to use the same device code.

Device ID $FF applies to all devices on the serial data link. This ID can be used to send the same request to all devices on the vehicle for purposes such as retrieving malfunction codes, disabling normal communications, or returning all devices to normal operation. All devices on the data link should then respond to this request with a response that includes their unique device ID. This enables the test device to determine which devices are on the data link and able to communicate.

TABLE 1—ASSIGNED DEVICE ID'S
(ALL IDS SHOWN IN HEX NOTATION)

Device ID	Device/Function
FF	All Devices
F0	DTM Test Device
F1	Body/Central Control Module
F2	Primary Display Device
F3	Secondary Display Device
F4	Engine Control
F5	Transmission Control
F6	Compass
E6	Engine Oil Life Monitor
F7	Ride Control
F8	Steering Control
F9	Brake/Traction Control
FA	Supplemental Inflatable Restraint
EA	HVAC Control
FB	Audio Warning (Voice Module/Chime)
EB	Remote Accessory Control
FC	Throttle Control
EC	Calculator
FD	Cellular Phone
ED	Memory Seat/Mirror

6. Diagnostic Test Mode General Conditions—If any on-board device is in a requested diagnostic mode, and if there is no other diagnostic message on the data link for a period of more than 5 s, that device shall terminate any active diagnostic mode and resume normal system operation. This solves the condition of a test device requesting a nonnormal mode of operation for a device and then disconnecting from the vehicle without returning operation of on-board devices to normal operation. If there is any diagnostic activity on the serial data link, then the on-board devices should assume that the test device may be controlling operation of some or all systems, and should remain in the present state of operation until instructed to change.

Diagnostic mode operation requires that test equipment be aware of previously requested modes, and accommodate for any desired changes. The automatic return to normal also requires that a test device transmit a diagnostic message at least once every 5 s if it needs to prevent on-board devices from returning to normal operation. That message does not need to be directed at any particular device on the link.

The potential exists in response messages for the contents of a multibyte variable to change after the first byte is sent, but before the last byte is sent, resulting in bad data. To maintain data integrity under this condition, hardware or software provisions must be included in each application to ensure that all bytes of multibyte variables are read simultaneously.

7. Diagnostic Test Mode Functional Description

7.1 Mode 0—Return to Normal Mode

7.1.1 FUNCTIONAL DESCRIPTION—The on-board device will return to the normal mode of operation when this command is received. All normal algorithms and normal communications will be resumed.

7.1.2 MODE 0 REQUEST

BYTE DESCRIPTION

1-4 Message Header (includes Mode = 0)

7.1.3 MODE 0 RESPONSE

BYTE DESCRIPTION

1-4 Message Header (includes Mode = 0)

7.2 Mode 1—Transmit Diagnostic Data

7.2.1 FUNCTIONAL DESCRIPTION—The on-board device will respond to this message by transmitting from 1 to 8 requested diagnostic data messages specific to that device. The message number bytes in the request allow multiple Mode 1 messages to be requested. Up to 32 messages can be defined for each device. These can either be predefined or defined during the diagnostic procedure using Mode 12 or 13. Messages should be defined with group data that is commonly used together or changes together.

Uses for these different messages include data such as wheel speeds for an ABS system. Data can be returned quickly with a minimal length request and response. Another use for these different messages is to return values that do not change, such as VIN and option content. These values need to be known one time only during testing. Other data bytes may contain present values of analog and discrete device I/O, device software flags and status words, and failure codes.

7.2.2 MODE 1 REQUEST

BYTE DESCRIPTION

1-4	Message Header (includes Mode = 1)
5	Transmit Message #1
6	Transmit Message #2—Optional
7	Transmit Message #3—Optional
8	Transmit Message #4—Optional
9	Transmit Message #5—Optional
10	Transmit Message #6—Optional
11	Transmit Message #7—Optional
12	Transmit Message #8—Optional

Message number is in the range from $00 to $1F.

7.2.3 MODE 1 RESPONSE

BYTE DESCRIPTION

1-4	Message Header (includes Mode = 1)
5	Message Number
6	Data Byte #1
7	Data Byte #2—Optional
8	Data Byte #3—Optional
9	Data Byte #4—Optional
10	Data Byte #5—Optional
11	Data Byte #6—Optional
12	Data Byte #7—Optional
13	Data Byte #8—Optional

Message number is in the range from $00 to $1F.

7.3 Mode 2—Memory Dump

7.3.1 FUNCTIONAL DESCRIPTION—The on-board device will respond to this message by transmitting the contents of eight memory locations beginning at the address specified in the request.

When using this mode to request data from EEPROM, data integrity will require that the software include provisions to either ensure that the controller is not writing to EEPROM, or that allows extra time for a response to this request. This is necessary because EEPROM cannot be read while a write to EEPROM is in progress.

7.3.2 MODE 2 REQUEST

BYTE DESCRIPTION

1-4	Message Header (includes Mode = 2)
5	Address 1 (High Order Byte)
6	Address 1 (Low Order Byte)

7.3.3 MODE 2 RESPONSE

BYTE DESCRIPTION

1-4	Message Header (includes Mode = 2)
5	Address #1 Contents
.	
.	
12	Address #8 Contents

7.4 Mode 3—Examine Memory

7.4.1 FUNCTIONAL DESCRIPTION—The on-board device will respond to this message by transmitting the contents of the memory locations requested by the test device.

The test equipment will include up to four independent two-byte memory addresses in the Request. The addressed device will respond with the one byte contents for each specified address.

When using this mode to request data from EEPROM, data integrity will require that the software include provisions to either ensure that the controller is not writing to EEPROM, or that allows extra time for a response to this request. This is necessary because EEPROM cannot be read while a write to EEPROM is in progress.

7.4.2 MODE 3 REQUEST

BYTE DESCRIPTION

1-4	Message Header (includes Mode = 3)
5	Address #1 (High Order Byte)
6	Address #1 (Low Order Byte)
7	Address #2 (High Order Byte)—Optional
8	Address #2 (Low Order Byte)—Optional
9	Address #3 (High Order Byte)—Optional
10	Address #3 (Low Order Byte)—Optional
11	Address #4 (High Order Byte)—Optional
12	Address #4 (Low Order Byte)—Optional

7.4.3 MODE 3 RESPONSE

BYTE DESCRIPTION

1-4	Message Header (includes Mode = 3)
5	Address #1 Contents
6	Address #2 Contents—Optional
7	Address #3 Contents—Optional
8	Address #4 Contents—Optional

7.5 Mode 4—Device Control Functions

7.5.1 FUNCTIONAL DESCRIPTION—The device control function mode provides the ability to define a totally device dependent set of functions. This is a general purpose mode used to communicate those messages which are unique to each device.

The test device will transmit a message to the selected device which may be used to temporarily alter software flags and status words, supply input values, or to exercise displays and outputs. The operation of the device control function mode may be restricted in order to ensure safe operation of the vehicle and to prevent equipment damage.

After a Mode 4 message is issued to change the system condition, other mode messages may be used to observe the effect of the changes.

If any device has received a Mode 4 command and has modified normal operation, and if there is no DTM message directed to any device, other than an FO device request, for a period of more than 5 s, that device shall terminate the Mode 4 and resume normal operation.

7.5.2 MODE 4 REQUEST

BYTE DESCRIPTION

1-4	Message Header (includes Mode = 4)
5	Data/Control Byte #1
.	
.	
N+4	Data/Control Byte #N

where:
N = Number of Data/Control bytes must be in the range from 1 to 8.

7.5.3 MODE 4 RESPONSE

BYTE DESCRIPTION

1-4	Message Header (includes Mode = 4)
5	Data/Status Byte #1—Optional
.	
.	
N+4	Data/Status Byte #N

where:
N = Number of Data/Status bytes must be in the range from 0 to 8.

The minimum response is three header bytes plus the message type / mode number.

7.6 Mode 5—Ram Download Request

7.6.1 FUNCTIONAL DESCRIPTION—The purpose of this message is to allow a test device to request an on-board device to prepare to receive executable code via the serial data link. The test device can be either an on-board device or an off-board test device.

The download request message includes the starting address to store the code to be downloaded and the number of bytes of code to be downloaded. This allows the on-board device to reject the request if that memory location is not acceptable.

This request gives the on-board device the necessary time to safely exit normal operations and prepare to receive data. Also, the on-board device can refuse the request if certain conditions are not satisfied due to other operations. The conditions required to allow or disallow a download will be dependent upon the on-board system, and are application dependent. Usually systems necessary for safe vehicle operation will not allow a download because of the possibility that the downloaded code could cause improper system operation.

Once a positive response has been sent by the on-board device, it will wait for a Message 6 from the tester to download code. A second Mode 5 command may be sent to the controller in the case of a garbled acceptance to the tester. The controller should always accept additional Mode 5 requests if currently waiting for a Mode 6 command. Other messages on the data link are allowed for other devices, but the on-board device will return to normal operation if a message is received for that device that is not a Mode 5 or Mode 6. The controller will also return to normal operation if there is no diagnostic test message on the data link for a period of 5 s.

Some systems will execute a monitor program after accepting the Mode 5 request. During the monitor program, normal software routines will not be executed, and the system may not be updated to allow normal operation. Some provision needs to be made to ensure that the system will operate normally after exiting from this monitor. This precaution may require a system reset to return to a known condition.

7.6.2 MODE 5 REQUEST

BYTE DESCRIPTION

1-4	Message Header (includes Mode = 5)
5	Starting Address (High Byte)
6	Starting Address (Low Byte)
7	Number of bytes to be downloaded

7.6.3 Mode 5 Response

BYTE DESCRIPTION

1-4 Message Header (includes Mode = 5)
5 Response Data—Accept = $AA
   Reject = any other
   response

7.7 Mode 6—RAM Download and Execute

7.7.1 FUNCTIONAL DESCRIPTION—The RAM Download and Execute message will always follow a RAM Download Request Message 5. The device receiving the message must verify that it can accept a download. This requires that the on-board device has responded to a Message 5 request and is now ready to accept code. If this condition has not been met, the on-board device should ignore the download message.

The on-board device will store the code at the address specified in the Mode 5 request message from the test device. At the end of this message, the onboard device should begin executing code at that starting address.

Normally data will be sent back to the test device during program execution. Multiple messages may be returned to the tester, depending on device specific criteria. If this is the case, then one of the data bytes should be used to distinguish between the possible messages. If messages are returned to the tester, then those messages should conform to the general format used in this document.

At the completion of the program, some provision needs to be made to return the operation of the device to normal. One possible action is for the on-board device to send a final Message 6 response indicating the end of the program, and then execute a Reset. Some systems will return back to a Mode 5 monitor program and wait for another Mode 6. In this case, care must be taken to ensure normal operation of the system after the return from the Mode 5 routine. In some applications, it may be necessary for the technician to power down the system and then turn power back on for the device to operate normally. The main caution to be observed is that while in the Mode 6 routine, the downloaded program has total control of registers and memory, and the controller must be returned to a known condition, which is most easily accomplished with a Reset.

Because of the wide variety of possible uses for this mode, and the wide variety of device hardware, a standard means cannot be specified to always return to normal operation. The caution is noted here only to ensure that this requirement is not overlooked.

This mode requires a block transfer mode that allows transmission of messages which are approximately 128 to 256 bytes long. These are considered special case modes which would not be used for standard diagnostics, but are required for assembly line testing and complex service procedures performed by high function diagnostic equipment.

7.7.2 MODE 6 REQUEST

BYTE DESCRIPTION

1-4 Message Header (includes Mode = 6)
5 First byte of code
.
.
.
N+4 Nth byte of code

where:
  N = Number of bytes of code to be downloaded.
    Must be in the range from 1 to 128 (256?).
This mode requires a block transfer mode.

7.7.3 MODE 6 RESPONSE (OPTIONAL)

BYTE DESCRIPTION

1-4 Message Header (includes Mode = 6)
5 Data Byte #1—Optional
.
.
.
N+4 Data Byte #N

where:
  N = Number of Data Bytes to be returned.
    Must be in the range from 0 to 8.
A Mode 6 response is optional, depending on the application.

7.8 Mode 7—Transmit Normal Message

7.8.1 FUNCTIONAL DESCRIPTION—The purpose of this mode is to allow the test equipment to request from 1 to 4 specific normal mode messages to be transmitted by the on-board device. The on-board device will respond to this message by transmitting the requested messages. The requested message will be the only response from the on-board device. This mode is usually only applicable if a previous Mode 8 command has suspended all normal communications.

7.8.2 MODE 7 REQUEST

BYTE DESCRIPTION

1-4   Message Header (includes Mode = 7)
5-6   Transmit Message #1
7-8   Transmit Message #2 (Optional)
9-10 Transmit Message #3 (Optional)
11-12 Transmit Message #4 (Optional)

1 to 4 normal mode messages can be requested.

7.8.3 MODE 7 RESPONSE—Normal mode messages specified in the Mode 7 Request are transmitted.

7.9 Mode 8—Disable Normal Communications

7.9.1 FUNCTIONAL DESCRIPTION—The purpose of this mode is to inhibit the on-board device from transmitting data on the link while still performing other functions normally. The device will continue to operate in whatever conditions were set by a previous Mode 4 command.

One use of this mode is to allow the test equipment to emulate the on-board device for diagnostic purposes. In this scenario, the test device would send a Mode 8 Request to the device to be emulated. The test device would then respond with all normal communication messages transmitted by that device, most likely with data intended to cause a known response by a system that uses the information in the response. The test device can then observe the actions of those systems.

If any device has received a Mode 8 command and has suspended normal communications, and if there is no diagnostic test message for a period of more than 5 s, that device shall terminate the Mode 8 and resume normal operations.

7.9.2 MODE 8 REQUEST

BYTE DESCRIPTION

1-4 Message Header (includes Mode = 8)

7.9.3 MODE 8 RESPONSE

BYTE DESCRIPTION

1-4 Message Header (includes Mode = 8)

7.10 Mode 9—Enable Normal Communications

7.10.1 FUNCTIONAL DESCRIPTION—The purpose of this mode is to cause an on-board device to resume normal communications after previously disabling these messages by a Mode 8 command. The device will continue to operate in whatever conditions were set by a previous Mode 4 command.

7.10.2 MODE 9 REQUEST

BYTE DESCRIPTION

1-4 Message Header (includes Mode = 9)

7.10.3 MODE 9 RESPONSE

BYTE DESCRIPTION

1-4 Message Header (includes Mode = 9)

7.11 Mode 10—Clear Malfunction Codes

7.11.1 FUNCTIONAL DESCRIPTION—The purpose of this mode is to provide a means for the external test device to command an on-board device to clear all malfunction codes.

7.11.2 MODE 10 REQUEST

BYTE DESCRIPTION

1-4 Message Header (includes Mode = 10)

7.11.3 MODE 10 RESPONSE

BYTE DESCRIPTION

1-4 Message Header (includes Mode = 10)

7.12 Mode 11—Suspend Normal Message

7.12.1 FUNCTIONAL DESCRIPTION—The purpose of this mode is to provide a means for the external test device to command the on-board communications scheduler to suspend some normal mode communications message.

The increased data communications required for diagnostics may require very high utilization of the serial data link. In order to allow time for the desired messages, this test mode is included to suspend scheduling of selected normal messages which use significant amounts of link time and are unnecessary during diagnostics. The test device sends a list of the messages to suspend.

When a mode 11 message is received, any normal messages previously suspended, but not included in the list of messages to suspend, should be scheduled for transmission. Each Mode 11 message can, therefore, be considered as a complete list of the messages to suspend. This will allow both disabling some messages and enabling others with a single message to the on-board device. All normal messages can be transmitted by sending a Mode 11 with no messages to suspend.

7.12.2 MODE 11 REQUEST

BYTE DESCRIPTION

1-4   Message Header (includes Mode = 11)
5-6   Suspend Message #1
7-8   Suspend Message #2 (Optional)
9-10   Suspend Message #3 (Optional)
11-12 Suspend Message #4 (Optional)

1 to 4 normal mode messages can be suspended.

7.12.3 MODE 11 RESPONSE

BYTE DESCRIPTION

1-4 Message Header (includes Mode = 11)

Normal mode messages specified in the Mode 11 request are suspended.

7.13 Modes 12 & 13—Define Diagnostic Message

7.13.1 FUNCTIONAL DESCRIPTION—These modes allow the test equipment to define a diagnostic message during the diagnostic procedure. This message can then be requested by the test device using a Mode 1 request. Messages typically should be defined which group data that is commonly used together or changes together.

Uses for these different messages include data such as wheel speeds for an ABS system. Data can be returned quickly with a minimal length request. Another use for these different messages is to return values that do not change, such as VIN and option content. These values need to be known one time only during testing. Other data bytes may contain present values of analog and discrete device I/O, device software flags and status words, and failure codes.

Data to be included in the message being defined by Mode 12 must already exist in one of the previously defined 32 messages. The individual values are specified by message number and position within that message.

Data to be included in the message being defined by Mode 13 are specified by the memory addresses for the data values.

7.13.2 MODE 12 REQUEST

BYTE DESCRIPTION

1-4 Message Header (includes Mode = 12)
5   Message Number defined
6   Data Value #1
     Message Number—5 bits
     Data Value Position—3 bits
.
.
.
N+5 Data Value #N
     Message Number—5 bits
     Data Value Position—3 bits

Message number is in the range from $00 to $1F.

Data values specified by this request must already exist in previously defined messages.

7.13.3 MODE 12 RESPONSE

BYTE DESCRIPTION

1-4 Message Header (includes Mode = 12)
5   Message Number defined

7.13.4 MODE 13 REQUEST

BYTE DESCRIPTION

1-4 Message Header (includes Mode = 13)
5   Data value range to fill (first bit)
     Message number defined (last 7 bits)
6   Data Value Address #1 (High Order Byte)
7   Data Value Address #1 (Low Order Byte)
8   Data Value Address #2 (High Order Byte)
     —Optional
9   Data Value Address #2 (Low Order Byte)
     —Optional
10   Data Value Address #3 (High Order Byte)
     —Optional
11   Data Value Address #3 (Low Order Byte)
     —Optional
12   Data Value Address #4 (High Order Byte)
     —Optional
13   Data Value Address #4 (Low Order Byte)
     —Optional

Message number is in the range from $00 to $1F.

This mode only allows 4 data values to be specified for a message. Two of these requests are required to fill a message with 8 bytes. If the first bit of the message number defined byte (Byte 5) is a 0, then the four values specified are to be Data Values 1 through 4. If the first bit of the message number defined byte (Byte 5) is a 1, then the four values specified are to be Data Values 5 through 8.

7.13.5 MODE 13 RESPONSE

BYTE DESCRIPTION

1-4 Message Header (includes Mode = 13)
5   Message Number defined

CLASS B DATA COMMUNICATION NETWORK MESSAGES: DETAILED HEADER FORMATS AND PHYSICAL ADDRESS ASSIGNMENTS—SAE J2178/1 JUN92

SAE Recommended Practice

Report of the SAE Vehicle Network for Multiplexing and Data Communications Standards Committee approved June 1992.

TABLE OF CONTENTS

1. Scope
2. Reference and Related Documents, Terms, Definitions, Abbreviations, and Acronyms
   2.1 Applicable Documents
   2.1.1 SAE Publications
   2.2 Terms and Definitions
   2.3 Abbreviations and Acronyms
3. General Information
   3.1 Part 1 Overview
   3.2 In-Frame Response Field Formats
   3.2.1 In-Frame Response Type 0
   3.2.2 In-Frame Response Type 1
   3.2.3 In-Frame Response Type 2
   3.2.4 In-Frame Response Type 3
4. Single Byte Header Messages and Format
5. Consolidated Header Messages and Format
   5.1 One Byte Form of the Consolidated Header Format
   5.2 Three Byte Form of the Consolidated Header Format
   5.2.1 Priority/Type—Byte 1
   5.2.1.1 Priority (PPP Bits)
   5.2.1.2 In-Frame Response (K Bit)
   5.2.1.3 Addressing Type (Y Bit)
   5.2.1.4 Type Modifier (ZZ Bits)
   5.2.1.5 In-Frame Response (IFR) Types
   5.2.2 Target Address—Byte 2
   5.2.3 Source Address—Byte 3
6. Data Fields
   6.1 Functional Data Field Formats
   6.1.1 Functional Data Field Format 0
   6.1.2 Functional Data Field Format 1
   6.1.3 Functional Data Field Format 2
   6.1.3.1 The Parameter/Quantity Bit "Q"
   6.1.3.2 The Control Bit "C"
   6.1.3.3 The Secondary Identification Address
   6.1.4 Functional Data Field Format 3
   6.1.5 Functional Data Field Format 4
   6.2 Physical Data Field Formats
   6.2.1 Physical Data Field Format 0
   6.2.2 Physical Data Field Format 1
   6.2.2.1 The Industry/Manufacturer Bit "I"
   6.2.2.2 The Request/Response Bit "R"
   6.2.2.3 The Test Mode Identification Reference
   6.2.3 Physical Data Field Format 2
   6.2.4 Physical Data Field Format 3
   6.2.4.1 The Transfer Type Byte
   6.2.4.2 The Block Length Bytes
   6.2.4.3 The Starting Address Bytes
   6.2.4.4 The Data Bytes
   6.2.4.5 The Block Checksum Bytes
   6.3 Extended Addressing
7. Physical Address Assignments
APPENDIX A
   A.1 Architectures of Networks
   A.2 Header Selection
APPENDIX B

1. Scope—This SAE Recommended Practice defines the information contained in the header and data fields of both diagnostic and non-diagnostic messages for automotive serial communications based on SAE J1850 Class B networks. This document describes and specifies the header fields, data fields, field sizes, scaling, representations, and data positions used within messages.

The general structure of a SAE J1850 message frame without in-frame response is shown in Figure 1. The structure of a SAE J1850 message with in-frame response is shown in Figure 2. Figures 1 and 2 also show the scope of frame fields defined by this document for non-diagnostic messages. Refer to SAE J1979 for specifications of emissions related diagnostic message header and data fields. Refer to SAE J2190 for the definition of other diagnostic message header and data fields. The description of the network interface hardware, basic protocol definition, the electrical specifications, and the CRC and Checksum fields (shown in the figures as ERR) are given in SAE J1850.

SAE J1850 defines two and only two formats of message headers. They are the Single Byte header format and the Consolidated header format. The consolidated header format has two forms, a single byte form and a 3 byte form. This document covers all of these formats and forms to identify the contents of messages which could be sent on a SAE J1850 network.

This document consists of four parts, each published separately.

Part 1 of SAE J2178 (this part, Titled: Detailed Header Formats and Physical Address Assignments) describes the two allowed forms of message header formats, single byte and consolidated. It also contains the physical node address range assignments for the typical sub-systems of an automobile. The details of this part are more fully described in paragraph 3.1 of this document.

Part 2 of SAE J2178 (Titled: Data Parameter Definitions) defines the standard parametric data which may be exchanged on SAE J1850 (Class B) networks. The parameter scaling, ranges, and transfer functions are specified.

Part 3 of SAE J2178 (Titled: Target Address for Single Byte Forms of Headers) defines the message assignments for the single byte header format and the 1 byte form of the consolidated header format.

Part 4 of SAE J2178 (Titled: Target Address (Second Byte) for Three Byte Headers) defines the message assignments for the second byte in the 3 byte form of the consolidated header format.

2. Reference and Related Documents, Terms, Definitions, Abbreviations, and Acronyms

2.1 Applicable Documents—The following publications form a part of this specification to the extent specified herein. The latest issue of SAE publications shall apply.

2.1.1 SAE PUBLICATIONS—Available from SAE, 400 Commonwealth Drive, Warrendale, PA 15096-0001.

SAE J1213/1—Glossary of Vehicle Networks for Multiplex and Data Communication
SAE J1850—Class B Data Communication Network Interface
SAE J1930—Electrical/Electronic Systems Diagnostic Terms, Definitions, Abbreviations, and Acronyms
SAE J1979—E/E Diagnostic Test Modes
SAE J2190—Enhanced E/E Diagnostic Test Modes

2.2 Terms and Definitions

DATA [DATA FIELD]—Data and data field are used interchangeably in this document and they both refer to a field within a frame that may include bytes with parameters pertaining to the message and/or secondary ID and/or extended addresses and/or test modes which further defines a particular message content being exchanged over the network.

EXTENDED ADDRESS—The extended address is a means to allow a message to be addressed to a specific geographical location or zone of the vehicle, independent of any node's physical address.

FRAME—A frame is one complete transmission of information which

<--- SAE J2178 --->			
SOF	Header Field	Data Field	ERR	EOF

FIGURE 1—SCOPE OF SAE J2178 FOR A SAE J1850 FRAME WITHOUT IN-FRAME RESPONSE (IFR)

<--- SAE J2178 --->				<-J2178->	
SOF	Header Field	Data Field	ERR	EOD	IFR	EOF

FIGURE 2—SCOPE OF SAE J2178 FOR A SAE J1850 FRAME WITH IN-FRAME RESPONSE (IFR)

may or may not include an In-Frame Response. The frame is enclosed by the start of frame and end of frame symbols. For Class B networks, each frame contains one and only one message (see "message" definition).

FUNCTIONAL ADDRESSING—Functional addressing allows a message to be addressed or sent to one or more nodes on the network interested in that function. Functional addressing is intended for messages that may be of interest to more than a single node. For example, an exterior lamp "off" message could be sent to all nodes controlling the vehicle exterior lamps by using a functional address. The functional address consists of a primary ID and may include a secondary ID.

HEADER [HEADER FIELD]—The header (or header field, used interchangeably) is a 1 or 3 byte field within a frame which contains information about the message priority, message source and target addressing, message type, and in-frame response type.

IN-FRAME RESPONSE (IFR) TYPE—The IFR type identifies the form of the in-frame response which is expected within that message.

LOAD—The load command indicates the operation of directly replacing the current/existing value of a parameter with the parameter value(s) contained in the message.

MESSAGE—A message consists of all of the bytes of a frame excluding the delimiter symbols (SOF, EOD, EOF, NB, IBS).

MODIFY—The modify command indicates the operation of using the message data parameter value to change (e.g., increment, decrement, or toggle) the current/existing value.

PARAMETER—A parameter is the variable quantity included in some messages. The parameter value, scaling, offset, units, transfer function, etc., are unique to each particular message. (The assigned parameters are contained herein.)

PHYSICAL ADDRESSING—Physical addressing allows a message to be addressed to a specific node or to all nodes or to a nonexistent, null node. The information in this message is only of relevance to a particular node, so the other nodes on the bus should ignore the message, except for the case of the "all nodes" address.

PRIMARY ID—The primary ID identifies the target for this functional message. This is the primary discriminator used to group functions into main categories.

PRIORITY—The priority describes the rank order and precedence of a message. Based upon the SAE J1850, Class B arbitration process, the message with the highest priority will win arbitration.

REPORT—A report indicates the transmission of parametric data values, based on: a change of state; a change of value; on a periodic rate basis; or as a response to a specific request.

REQUEST—A request is a command to, or a query for data, or action from another node on the network.

RESPONSE DATA—The response data is the information from a node on the network in response to a request from another node on the network. This may be an in-frame response or a report type of message.

SECONDARY ID—The secondary ID (along with the primary ID) identifies the functional target node for a message. The purpose of the secondary ID field within the frame is to further define the function or action being identified by the primary ID.

2.3 Abbreviations and Acronyms

CRC—Cyclic Redundancy Check
CS—Checksum
EOD—End of Data
EOF—End of Frame
ERR—Error Detection (CRC or CS)
IBS—Inter-Byte Separation
ID—Identifier
IFR—In-Frame Response
LSB—Least Significant Bit / Byte
MSB—Most Significant Bit / Byte
NB—Normalization Bit
PID—Parameter Identification number
SLOT—Scaling, Limit, Offset, and Transfer function
SOF—Start of Frame

3. General Information

3.1 Part 1 Overview—The messages defined by this four part document are specified for networks using single byte headers or consolidated 1 and 3 byte headers as specified in SAE J1850. Sections 4 and 5 of Part 1 provide the system architecture for the different possible headers used in Class B network communication (see Appendix A for supporting discussion). Section 6 of Part 1 defines the data fields used by the different header byte formats. Section 7 of Part 1 defines the physical address assignments.

Figure 3 shows an overview of this part (Part 1) which encompasses

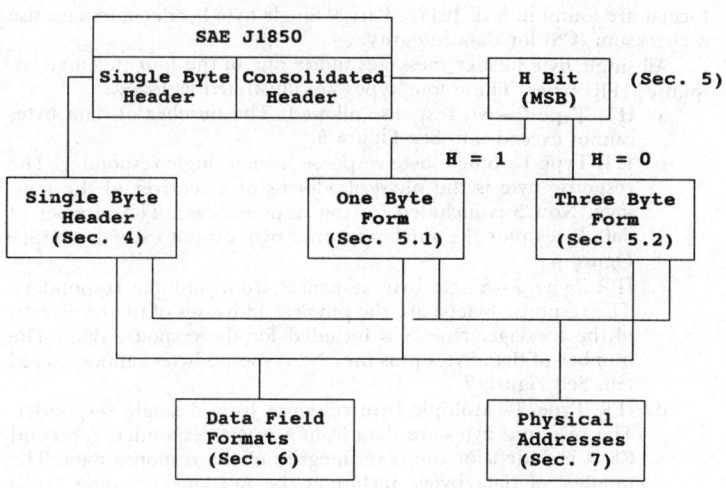

FIGURE 3—PART 1 OVERVIEW

the different possible messages and their component parts.

SAE J1850 defines two and only two formats of message headers. They are the Single Byte header format and the Consolidated header format. The consolidated header format has two forms, identified by the value of the H Bit. The two forms are: a 1 byte form and a 3 byte form. The information in the header field for both formats contains target, source, priority and message type information, while the data field contains additional addressing and/or parametric information and/or diagnostic test modes. This information is explicitly defined in some headers and implicitly defined in others. Messages defined by this document (Parts 1, 2, 3, and 4) are classified generally into two types:

a. Requests, that is, commands (load or modify) or queries for data, and

b. Responses, that is, reports or acknowledgements.

When a node generates a request, the target nodes which are responsible for the requested data or function must respond by reporting the requested information or by performing the requested function. For responses (that is, reports or acknowledgements), data information that a node responds with may have been previously requested by another node, or reported by the node when the desired information has changed, or reported by the node on a periodic basis.

Part 1 of this document describes the overall structure of messages. In total, parts 1, 2, 3, and 4:

a. Fully defines SAE (automotive industry) standard messages.
b. Reserves messages for future SAE standardization.
c. Reserves messages for manufacturers to allocate, which are typically unique or proprietary to that manufacturer.

In order to comply with this document, implementations need to use the defined messages on SAE J1850 networks in the exact way that they are defined here. However, there are a large number of message codes which are reserved for each manufacturer to independently allocate.

3.2 In-Frame Response Field Formats

3.2.1 IN-FRAME RESPONSE TYPE 0—The in-frame response type 0 is used to indicate the form without any in-frame response.

3.2.2 IN-FRAME RESPONSE TYPE 1—The in-frame response type 1 is a single byte response from a single responder. The response byte shall be the physical address of a receiver of the message. No CRC or CS checking byte is included for this response data.

3.2.3 IN-FRAME RESPONSE TYPE 2—The in-frame response type 2 is a single byte response from multiple responders. The response byte(s) shall be the physical address of the receiver(s) of the message. No CRC or CS checking is included for the response data.

3.2.4 IN-FRAME RESPONSE TYPE 3—The in-frame response type 3 is a multiple byte response from a single responder. The response bytes shall be data (generally not its address) from a single responder. The second CRC or CS checking byte is included for the data integrity of the response data. The actual in-frame response data shall be 1 byte in length, as a minimum.

4. Single Byte Header Messages and Format—For single byte header messages, the entire byte is used to define the message identifier (ID) as shown in Figure 4. Standard message identifiers that utilize this header format are found in SAE J2178; Part 3. Single byte header messages use a checksum (CS) for data integrity.

All single byte header messages utilize one of the four in-frame response (IFR) types. These four types are illustrated as follows:

a. IFR Type 0—No response allowed. The number of data bytes cannot exceed ten. See Figure 5.
b. IFR Type 1—Single byte response from a single responder. The response byte is the physical address of a receiver of the message. No CS is included for the response data. The number of data bytes plus the single response byte cannot exceed ten. See Figure 6.
c. IFR Type 2—Single byte responses from multiple responders. The response byte(s) are the physical addresses of the receiver(s) of the message. No CS is included for the response data. The number of data bytes plus the "N" response bytes cannot exceed ten. See Figure 7.
d. IFR Type 3—Multiple byte response from a single responder. The response bytes are data from a single responder. A second CS is included for the data integrity of the response data. The number of data bytes, including the in-frame response (IFR) data, cannot exceed nine. The actual in-frame response data must be one byte in length, as a minimum. See Figure 8.

5. Consolidated Header Messages and Format—The consolidated header format includes both a 1 byte form and a 3 byte form. All consolidated header format messages use a CRC code for data integrity.

In order to accommodate a 1 byte header form and a 3 byte header form on the same network, bit 4 (H bit) in the first byte of the message has been defined to indicate the header form. This bit is defined in Table 1.

TABLE 1—"H" BIT ASSIGNMENT

H Bit	Header Byte Form
0	3 Byte Form
1	1 Byte Form

5.1 One Byte Form of the Consolidated Header Format—The 1 byte form utilizes 7 bits for the message identifier (ID) and bit 4 (the H bit) = 1 to indicate the 1 byte form. Figure 9 illustrates this message header form. Standard message identifiers that utilize this header form are defined in SAE J2178; Part 3.

All 1 byte header form messages of the consolidated header format utilize one of the four in-frame response (IFR) types. These four types are illustrated as follows:

a. Type 0—No response allowed. The number of data bytes cannot exceed ten. See Figure 10.
b. IFR Type 1—Single byte response from a single responder. The response byte is the physical address of a receiver of the message. No CRC is included for the response data. The number of data bytes plus the single response byte cannot exceed ten. See Figure 11.
c. IFR Type 2—Single byte responses from multiple responders. The response byte(s) are the physical address of the receiver(s) of the message. No CRC is included for the response data. The number of data bytes plus the N response bytes cannot exceed ten. See Figure 12.
d. IFR Type 3—Multiple byte response from a single responder.

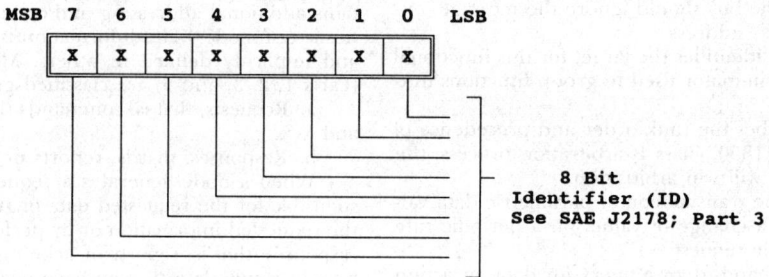

FIGURE 4—SINGLE BYTE HEADER FORMAT

| SOF | Header | Data | CS | EOF |

FIGURE 5—SINGLE BYTE HEADER, IFR TYPE 0

| SOF | Header | Data | CS | EOD | ID | EOF |

FIGURE 6—SINGLE BYTE HEADER, IFR TYPE 1

| SOF | Header | Data | CS | EOD | ID 1 | - | ID N | EOF |

FIGURE 7—SINGLE BYTE HEADER, IFR TYPE 2

| SOF | Header | Data | CS | EOD | IFR DATA | CS | EOF |

FIGURE 8—SINGLE BYTE HEADER, IFR TYPE 3

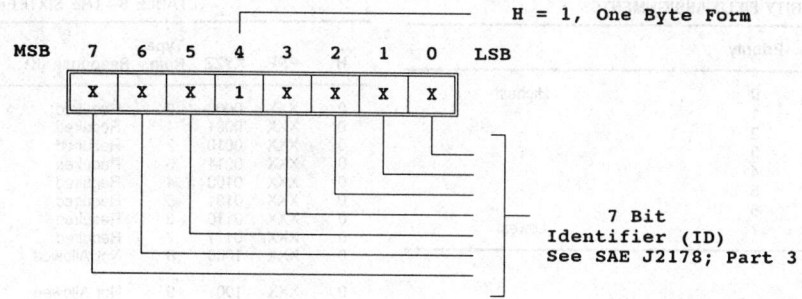

FIGURE 9 – CONSOLIDATED HEADER, SINGLE BYTE FORM

| SOF | Header | Data | CRC | EOF |

FIGURE 10 – ONE BYTE CONSOLIDATED HEADER, IFR TYPE 0

| SOF | Header | Data | CRC | EOD | ID | EOF |

FIGURE 11 – ONE BYTE CONSOLIDATED HEADER, IFR TYPE 1

| SOF | Header | Data | CRC | EOD | ID 1 | - | ID N | EOF |

FIGURE 12 – ONE BYTE CONSOLIDATED HEADER, IFR TYPE 2

The response bytes are data from a single responder. The second CRC is included for the data integrity of the response data. The number of data bytes, including the in-frame response (IFR) data, cannot exceed nine. The actual in-frame response data must be 1 byte in length as a minimum. See Figure 13.

5.2 Three Byte Form of the Consolidated Header Format — This header form utilizes the first 3 bytes of the message. In this form, the "H" bit is a zero (0), signifying that the 3 byte header form is used for this message. The remaining 7 bits of the first byte contain information about priority and message type (KYZZ). The second byte contains the target address information. The target can be either functionally addressed or physically addressed. The third byte contains the physical address of the source of the message. Arbitration is always resolved by the end of the third byte, since the source address must be unique. Figure 14 illustrates the 3 byte header form.

5.2.1 PRIORITY/TYPE – BYTE 1 – The priority/type byte contains information about priority, in-frame response, addressing type, and type modifier. Each of the field definitions are described in the following paragraphs.

5.2.1.1 *Priority (PPP Bits)* – The priority field is 3 bits in length which are the most significant bits of the first byte. The priority bit assignments are shown in Table 2.

5.2.1.2 *In-Frame Response (K Bit)* – The necessity for in-frame response is encoded into a single bit in the priority/type byte. This bit definition is shown in Table 3.

5.2.1.3 *Addressing Type (Y Bit)* – Message addressing is encoded with a single bit in the priority/type byte. This bit definition is shown in Table 4.

5.2.1.4 *Type Modifier (ZZ Bits)* – The type modifier is encoded as a 2 bit field and is used in conjunction with the in-frame response bit (K) and the address type bit (Y) to define sixteen message types, see Table 5.

5.2.1.5 *In-Frame Response (IFR) Types* – All 3 byte form messages of the consolidated header format listed in Table 5 utilize one of four in-frame response (IFR) types. These four types are illustrated as follows:
a. IFR Type 0 – No response allowed. The number of data bytes cannot exceed eight. See Figure 15.
b. IFR Type 1 – Single byte response from a single responder. The response byte is the physical address of a receiver of the message. No CRC is included for the response data. The number of data bytes plus the single response byte cannot exceed eight. See Figure 16.
c. IFR Type 2 – Single byte responses from multiple responders. The response byte(s) are the physical address of the receiver(s) of the message. No CRC is included for the response data. The number of data bytes plus the N response bytes cannot exceed eight. See Figure 17.
d. IFR Type 3 – Multiple byte response from a single responder. The response bytes are data from a single responder. The second CRC is included for the data integrity of the response data. The number of data bytes, including the in-frame response (IFR) data, cannot exceed seven. The actual in-frame response data must be 1 byte in length, as a minimum. See Figure 18.

5.2.2 TARGET ADDRESS – BYTE 2 – The second byte of the 3 byte form of the consolidated header format contains either a functional or physical

| SOF | Header | Data | CRC | EOD | IFR DATA | CRC | EOF |

FIGURE 13 – ONE BYTE CONSOLIDATED HEADER, IFR TYPE 3

TABLE 2 — PRIORITY FIELD ASSIGNMENTS

PPP Bits	Priority	
000	0	Highest
001	1	.
010	2	.
011	3	.
100	4	.
101	5	.
110	6	.
111	7	Lowest

TABLE 3 — "K" BIT ASSIGNMENT

K Bit	In-Frame Response
0	Required
1	Not Allowed

TABLE 4 — "Y" BIT ASSIGNMENT

Y Bit	Addressing Type
0	Functional
1	Physical

TABLE 5 — THE SIXTEEN MESSAGE TYPES

H	PPP	KYZZ	Type Num.	Response (K)	Address Type (Y)	IFR Type	Message Type (Name)
0	XXX	0000	0	Required	Functional	2	Function
0	XXX	0001	1	Required	Functional	1	Broadcast
0	XXX	0010	2	Required	Functional	2	Function Query
0	XXX	0011	3	Required	Functional	3	Function Read
0	XXX	0100	4	Required	Physical	1	Node-to-Node
0	XXX	0101	5	Required	Physical	Reserved	Reserved - MFG.
0	XXX	0110	6	Required	Physical	Reserved	Reserved - SAE
0	XXX	0111	7	Required	Physical	Reserved	Reserved - MFG.
0	XXX	1000	8	Not Allowed	Functional	0	Function (Command/Status)
0	XXX	1001	9	Not Allowed	Functional	0	Function (Request/Query)
0	XXX	1010	10	Not Allowed	Functional	0	Function Extended (Command/Status)
0	XXX	1011	11	Not Allowed	Functional	0	Function Extended (Request/Query)
0	XXX	1100	12	Not Allowed	Physical	0	Node-to-Node
0	XXX	1101	13	Not Allowed	Physical	0	Block Transfer
0	XXX	1110	14	Not Allowed	Physical	0	Reserved - MFG.
0	XXX	1111	15	Not Allowed	Physical	0	Reserved - MFG.

NOTES
1. Functional Addresses for the 3 byte header form of header are defined in SAE J2178; Part 4. Physical Address Ranges are defined in Section 7.
2. Message types 8 and 9 use the functional data format #2 only; Message types 10 and 11 use the functional data format #3 only.
3. Reserved-SAE indicates reserved for future SAE Committee action. Reserved-MFG indicates that manufacturers are allowed to allocate these definitions without requiring any commonality between motor vehicle manufacturers.

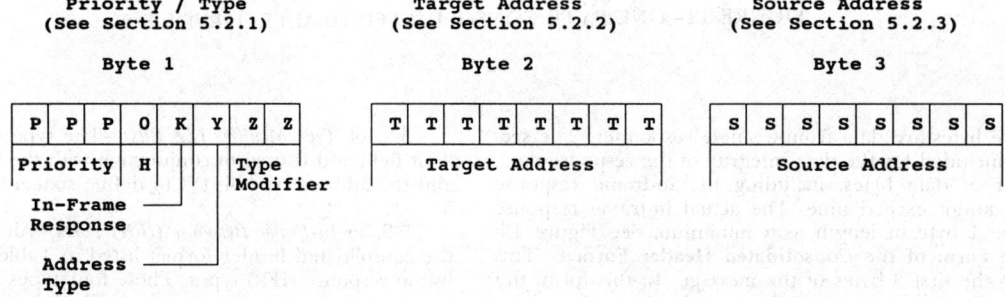

FIGURE 14 — CONSOLIDATED HEADER, THREE BYTE FORM

SOF	Header	Data	CRC	EOF

FIGURE 15 — THREE BYTE CONSOLIDATED HEADER, IFR TYPE 0

SOF	Header	Data	CRC	EOD	ID	EOF

FIGURE 16 — THREE BYTE CONSOLIDATED HEADER, IFR TYPE 1

SOF	Header	Data	CRC	EOD	ID 1	-	ID N	EOF

FIGURE 17 — THREE BYTE CONSOLIDATED HEADER, IFR TYPE 2

target address. The physical address assignments are found in section 7 of this part, while functional message address assignments are in Part 4 (of SAE J2178).

5.2.3 SOURCE ADDRESS—BYTE 3—The third byte of the 3 byte format of the consolidated header format is the physical address of the source of the message. Physical address assignments are found in Section 7. These physical address assignments shall be unique for each node of a network.

6. Data Fields—In both message header formats, single byte and consolidated, the data field can usually be encoded in the same way. This section briefly describes the different ways that information can be formatted in the data field. The data field immediately follows the header field. The number of bytes in this field will vary, based upon the content of the header field. The maximum data field length is limited by the requirements of SAE J1850. Because of differences in functionally and physically addressed or with in-frame response data, these cases are defined separately.

6.1 Functional Data Field Formats

6.1.1 FUNCTIONAL DATA FIELD FORMAT 0—One of the functional data field formats is, in fact, no additional bytes of data (an empty data field). The message consists of the header, error checking byte, and in-frame response bytes. This is the format used for functional message types 2 and 3. Because the data field does not actually exist but to allow referencing in other parts of this document, it has been identified as the functional data field format 0.

6.1.2 FUNCTIONAL DATA FIELD FORMAT 1—In the simplest case including data (format 1), the data field contains only parametric data. The first byte of the data field in this case contains the most significant byte of data. The data field must contain 1 byte as a minimum. Figure 19 illustrates this message format. This data field format may be used for message types 0 and 1 only.

6.1.3 FUNCTIONAL DATA FIELD FORMAT 2—In a data field format 2 message, the data field contains a byte used as an identifier which further defines the target function being addressed. In this format type, the data field would appear as shown in Figure 20. This data format may be used for message types 0, 1, 8 and 9.

The secondary ID consists of a parameter / quantity bit Q (see 6.1.3.1) in which binary data (such as on/off or yes/no) can be encoded. There is a listing of the possible uses for this bit in other parts of this document. The control bit C (see 6.1.3.2) is used to distinguish between an immediate load of data or a modify of the current data for command messages. For message types 9 and 11, the control bit distinguishes between request and query. The remaining 6 bits specify an ID address (see 6.1.3.3). The order of these bits within the secondary ID byte is shown in Figure 21.

In many cases, a single data bit (the Q bit) is not adequate to define the parameter being sent. In this case, the identifier field is followed by a data field as illustrated in Figure 22. The combination of primary and secondary ID define if additional data is used by that message.

6.1.3.1 *The Parameter / Quantity Bit "Q"*—The parameter / quantity bit represents the data or information value for single, binary values. If the information being transmitted has one of two opposite values, it is reflected as shown in Table 6. The specific meaning or relevance of this bit is defined for each message in other parts of this document.

6.1.3.2 *The Control Bit "C"*—The control bit represents an action to be taken with the associated data values. Table 7 shows the two possible states for this bit. If the "C" bit indicates a load, the associated data value replaces the current/existing value. If the "C" bit indicates a modify or toggle, the current/existing data value is manipulated by the transmitted data value, based on the definition of each specific message (the 6 secondary identification address bits). For example, if the message indicates a toggle, the current/existing value is changed to the opposite state, independent of what that current/existing data value had been.

| SOF | Header | Data | CRC | EOD | IFR DATA | CRC | EOF |

FIGURE 18—THREE BYTE CONSOLIDATED HEADER, IFR TYPE 3

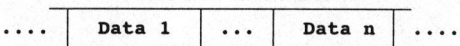

FIGURE 19—FUNCTIONAL DATA FIELD FORMAT 1

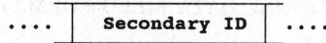

FIGURE 20—FUNCTIONAL DATA FIELD FORMAT 2

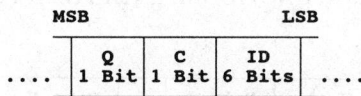

FIGURE 21—SECONDARY ID BYTE FORMAT

FIGURE 22—FUNCTIONAL DATA FIELD FORMAT 2, WITH DATA BYTES

TABLE 6—"Q" BIT ASSIGNMENT

Q Bit	Parameter/Quantity
0	No/False/Off/Down/..
1	Yes/True/On/Up/..

TABLE 7—"C" BIT ASSIGNMENT

C Bit	Command	Request/Query
0	Load	Request
1	Modify/Toggle	Query

For message types 9 and 11, the control bit distinguishes between request and query. The relevance of this bit is defined by the primary and secondary ID values and are specified in other parts of this document.

6.1.3.3 *The Secondary Identification Address*—The secondary identification address field is used to further identify the particular function or operation being addressed by this message. It is used to distinguish a function when the primary ID is not sufficient, or to define a specific operation to be performed by the function addressed by the primary ID. These secondary identification addresses are assigned in parts 3 and 4 of this document.

6.1.4 FUNCTIONAL DATA FIELD FORMAT 3—In a data field format 3 message, depending on the particular primary ID, the data field may also contain an extended address byte which defines the geographical location within the vehicle of the target function being addressed following the secondary ID byte. As an example, in a functional message whose primary ID target is window motion, an extended address byte is needed to identify which window, e.g., driver, passenger, all; along with the secondary ID byte needed to identify an operation such as Up or Down. This data field format may be used for message types 0 and 1 and must be used for message types 10 and 11. In this data field type, the data field appears as shown in Figure 23.

The extended address byte is used to determine where, geographically on a vehicle, a particular function is located. The exact definition of the extended address values can be found in 6.3. The secondary ID in Figure 23 is used as defined in 6.1.3. As an example, to roll the rear driver's side window down, the header would contain information that identifies window motion as the functional target. The first data byte would contain a 6-bit ID identifying the function Down with the Q-bit used to indicate enable the down or disable the down command, followed by the extended address, which would indicate that the particular window being addressed is on the rear driver's side. The C-bit is unused (default value = 0) in this particular message.

As in the other data field formats, a parameter data field may also be needed and in this case it is then appended to the end of the other identifiers. This format is shown in Figure 24.

6.1.5 FUNCTIONAL DATA FIELD FORMAT 4—In this data field format message, the data field contains a byte which defines the diagnostic test mode of the target function being addressed. In this type, the data field would appear as shown in Figure 25. This data format may be used for message types 0, 1, 8, and 9.

The test mode byte is used to determine which diagnostic function is involved. It may then be followed by a "Parameter Identification" (PID) number or other parameter data fields. In this case, the additional data bytes follow the test mode byte. This format is shown in Figure 26. This is the format for functionally addressed diagnostic messages such as those found in SAE J1979.

6.2 Physical Data Field Formats—The previous section (6.1) defined the data field formats for functionally addressed messages. This section (6.2) defines the data field formats for physically addressed messages.

6.2.1 PHYSICAL DATA FIELD FORMAT 0—One of the physical data field formats is, in fact, no additional bytes of data (an empty data field). The message consists of the header, error checking byte, and may be with or without in-frame response bytes. This is the format used for the acknowledgement message type. This message type simply confirms to another node that this node has correctly received the message it is acknowledging. Because this data field does not actually exist but to allow

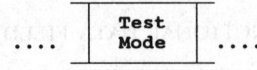

FIGURE 23—FUNCTIONAL DATA FIELD FORMAT 3, WITH EXTENDED ADDRESS

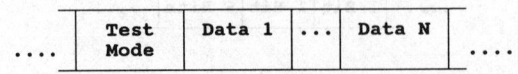

FIGURE 24—FUNCTIONAL DATA FIELD FORMAT 3, WITH DATA BYTES

.... | Test Mode | ....

FIGURE 25—FUNCTIONAL DATA FIELD FORMAT 4

.... | Test Mode | Data 1 | ... | Data N | ....

FIGURE 26—FUNCTIONAL DATA FIELD FORMAT 4, WITH DATA BYTES

referencing in other parts of this document, it has been identified as the physical data field format 0.

6.2.2 PHYSICAL DATA FIELD FORMAT 1—(Node to Node)

Physical data field format 1 is generally associated with the node-to-node message types. This message type is the one utilized for enhanced E/E diagnostic test modes (see SAE J2190).

This format assumes the 3 byte form of the consolidated header format. The single byte format and the 1 byte form of the consolidated header format are covered in physical data field format 2. Many of the specific diagnostic messages are defined in SAE J2190. This description of the format is consistent with that document but expands the definition to allow the format to be used in other messages as well. In particular, the manufacturer specific applications of this node-to-node message are expected to follow this format. The basic format is similar to the functional data field format 4. The format is shown in Figure 27.

The format will often include data bytes as shown in Figure 28. The content of the test mode byte indicates if additional data bytes are used.

The physical test mode byte details are shown in Figure 29. The description of the individual parts of this byte are described in the following paragraphs.

6.2.2.1 *The Industry/Manufacturer Bit "I"*—The industry/manufacturer bit represents the organization responsible for the definition of this test mode. The information that has been defined by SAE (I = 0) can be interpreted in the same way for all vehicles that use those definitions. The bit assignment is shown in Table 8. The specific meaning for the SAE (industry) messages are defined in SAE J2190. SAE has also reserved for future definition some of the I = 1 test mode bytes. Manufacturer defined test modes can be independently defined by each vehicle manufacturer and there is no attempt made to make these definitions common between manufacturers.

NOTE—See SAE J2190 for specific assignments.

6.2.2.2 *The Request/Response Bit "R"*—The request/response bit represents the direction of information flow contained in this message. Table 9 shows the two possible states for this bit. If the "R" bit indicates a request, the associated data bytes, if any, indicate the information being requested. If the "R" bit indicates a response, the data, if any, represent the response information previously requested.

6.2.2.3 *The Test Mode Identification Reference*—The test mode identification reference field is used to further identify the particular operation being addressed by this message. It is used to define a specific test mode to be performed by, or reported to, the node addressed by the target byte. These test mode identification addresses are assigned in SAE J2190 for the industry standard case.

6.2.3 PHYSICAL DATA FIELD FORMAT 2—(Diagnostic Test Modes)

TABLE 8—"I" BIT ASSIGNMENT

I Bit	Industry / Manufacturer
0	Industry (SAE Defined)
1	Industry (SAE Reserved/ Manufacturer Defined)

TABLE 9—"R" BIT ASSIGNMENT

R Bit	Request/Response
0	Request
1	Response

.... | Physical Test Mode | ....

FIGURE 27—PHYSICAL DATA FIELD FORMAT 1

.... | Phy. Test Mode | Data 1 | .... | Data N | ....

FIGURE 28—PHYSICAL DATA FIELD FORMAT 1, WITH DATA BYTES

```
            MSB              LSB
         ┌─────┬─────┬──────┐
    .... │  I  │  R  │  ID  │ ....
         │1 Bit│1 Bit│6 Bits│
         └─────┴─────┴──────┘
```

FIGURE 29—PHYSICAL TEST MODE BYTE FORMAT

.... | Physical Target Address | Physical Test Mode | ....

FIGURE 30—PHYSICAL DATA FIELD FORMAT 2

.... | Physical Target Address | Physical Test Mode | Data 1 | ... | Data N | ....

FIGURE 31—PHYSICAL DATA FIELD FORMAT 2, WITH DATA BYTES

In order to maximize commonality between the different header byte formats, the physical data field format 2 is essentially the same as for physical data field format 1 with the insertion of the physical target address byte ahead of the physical test mode byte. This allows the single byte header format and the 1 byte form of the consolidated header format to operate consistently with the 1 byte header form. The format is shown in Figures 30 and 31 for the two cases, with or without additional data bytes. The number of data bytes, if used, are not fixed length, unless so specified in other documents.

6.2.4 PHYSICAL DATA FIELD FORMAT 3 — (Block Transfer)

This physical data field format is unique and dedicated to the transfer of large blocks of data from one physical node to another node of the network. It would have very limited application if it were restricted to the maximum message length requirements of SAE J1850. Therefore, this data format is allowed to exceed that limitation. The physical data field format 3 is shown in Figure 32.

Since the message length limitation of SAE J1850 has been extended, the data field format includes additional error checking bytes. Each group of bytes in this format is described in the following paragraphs.

NOTE — This format is normally limited to use for service diagnostics or module/vehicle manufacture. It should not normally be used during vehicle operation because of the potentially long interruption of other vehicle message traffic.

6.2.4.1 *The Transfer Type Byte* — The first byte of this physical data field format defines the type of block transfer that is being transmitted. The transfer type byte is reserved for future definition. It has been included as a place holder to allow future definition by SAE or manufacturers.

6.2.4.2 *The Block Length Bytes* — The next 2 bytes of this physical data field format define the length of the data block (N) included in the block. The value of N represents the number of data bytes, not the total number of bytes in the data field. The maximum value of N is 0FFF Hex for a data block maximum length of 4095 bytes. The upper nibble (upper 4 bits) of the MSB of the block length is thus reserved for future definition and should be set to zero as its default value.

6.2.4.3 *The Starting Address Bytes* — The next 3 bytes identify the starting address of the data block being transmitted. In other words, it is the memory location where data byte 1 is located. The other data bytes of the block (data 2 to data N) will then represent, sequentially, in ascending address order, the additional data bytes of the block.

6.2.4.4 *The Data Bytes* — The data itself contains from 1 to N data bytes being transferred between nodes.

6.2.4.5 *The Block Checksum Bytes* — Because the block length is most likely longer than most other messages on the network, additional error checking is required. The last 2 bytes are a calculated checksum value based on the contents of the data block only. The calculation begins with the first data byte (Transfer Type) and continues to the last data byte (data N). The checksum is a 16 bit sum of the values included. The general checksum calculation is comparable to the SAE J1850 checksum extended to 16 bits. The transmitted checksum should match the received, calculated checksum to verify correct reception.

NOTE — The header is not included in this checksum calculation.

NOTE — The data field described previously is followed by the error checking byte of the SAE J1850 format (ERR byte, CS or CRC) to verify the integrity of the whole message, the same as all other Class B messages.

6.3 **Extended Addressing** — The extended addressing defines an extended (geographical) location in the vehicle. The extended address field (RR XXX YYY) is divided into three sub-fields:

a. The two most significant bits (RR) are reserved for future use. (00 is the defined value for geographic location.)

b. The three next most significant bits (XXX) indicate rows from front (001) to rear (111).

TABLE 10 — NODE PHYSICAL ADDRESS ASSIGNMENTS

Node	Address (Hex)
Powertrain Controllers:	
Integration / Manufacturer Expansion	00 - 0F
Engine Controller	10 - 17
Transmission Controller	18 - 1F
Chassis Controllers:	
Integration / Manufacturer Expansion	20 - 27
Brake Controller	28 - 2F
Steering Controller	30 - 37
Suspension Controller	38 - 3F
Body Controllers:	
Integration / Manufacturer Expansion	40 - 57
Restraints	58 - 5F
Driver Information / Displays	60 - 6F
Lighting	70 - 7F
Entertainment / Audio	80 - 8F
Personal Communications	90 - 97
HVAC	98 - 9F
Convenience (Doors, Seats, Windows, etc.)	A0 - BF
Security	C0 - C7
Future Expansion:	C8 - CF
Manufacturer Specific:	D0 - EF
Off-Board Testers / Diagnostic Tools:	F0 - FD
All Nodes:	FE
Null Node:	FF

c. The three least significant bits (YYY) indicate columns from left (001) to right (111).

The codes XXX = 000 (for rows) and YYY = 000 (for columns) indicate that ALL rows and/or ALL columns are indicated. In other words, to address all of the items in a specific row, regardless of which column, use XXX = 000. For all items in the same column, independent of which row, use YYY = 000. This can be useful if needed to address all headlamps, for example, to turn them all on, but for lamp outage, the particular location of the headlamp may be desirable.

A map illustrating 49 zones of the vehicle is shown in Figure 33. It generally represents a top view of the vehicle and supports both Left/Right and Driver/Passenger side references.

Appendix B shows some examples of this extended addressing as it identifies some common items in passenger vehicles.

7. Physical Address Assignments — The physical address assignment ranges for nodes on a network are shown in Table 10. It is important to note that node address assignments on any network must be unique. In other words, no two nodes can have the same physical address assignment.

NOTE — One address has been reserved to address all nodes on a network. The all nodes address, however, may or may not be recognized in a system. Additionally, one address has been reserved to address a "null node" which is never assigned to a real node. This "null node" allows some forms of testing network performance.

NOTE — The address ranges assigned to "Integration" are intended for modules which have more than one specific function. If this integration includes functions which involve crossing between Powertrain, Chassis, and Body, the highest priority (lowest address) range shall be selected, for example, if powertrain and chassis are integrated in one module, the 00 - 0F address range shall be used.

MSB							
...	Test Mode -SAEJ2190	Transfer Type	MSB Block Length	LSB Block Length	MSB-Start Address	Middle Address	LSB-Start Address

	DATA 1	...	Data N	MSB Checksum	LSB Checksum	...

(LSB)

FIGURE 32 — PHYSICAL DATA FIELD FORMAT 3 — BLOCK TRANSFER

LEFT SIDE FRONT SURFACE (RR 001 001)	DRIVER'S SIDE FRONT SURFACE (RR 001 010)	LEFT CENTER FRONT SURFACE (RR 001 011)	CENTER POINT FRONT SURFACE (RR 001 100)	RIGHT CENTER FRONT SURFACE (RR 001 101)	PASSENGER SIDE FRONT SURFACE (RR 001 110)	RIGHT SIDE FRONT SURFACE (RR 001 111)
LEFT SIDE UNDER HOOD (RR 010 001)	DRIVER'S SIDE UNDER HOOD (RR 010 010)	LEFT CENTER UNDER HOOD (RR 010 011)	CENTER POINT UNDER HOOD (RR 010 100)	RIGHT CENTER UNDER HOOD (RR 010 101)	PASSENGER SIDE UNDER HOOD (RR 010 110)	RIGHT SIDE UNDER HOOD (RR 010 111)
LEFT SIDE AFT BULKHEAD (RR 011 001)	DRIVER'S SIDE AFT BULKHEAD (RR 011 010)	LEFT CENTER AFT BULKHEAD (RR 011 011)	CENTER POINT AFT BULKHEAD (RR 011 100)	RIGHT CENTER AFT BULKHEAD (RR 011 101)	PASSENGER SIDE AFT BULKHEAD (RR 011 110)	RIGHT SIDE AFT BULKHEAD (RR 011 111)
LEFT SIDE AFT "A" PILLAR (RR 100 001)	DRIVER'S SIDE AFT "A" PILLAR (RR 100 010)	LEFT CENTER AFT "A" PILLAR (RR 100 011)	CENTER POINT AFT "A" PILLAR (RR 100 100)	RIGHT CENTER AFT "A" PILLAR (RR 100 101)	PASSENGER SIDE AFT "A" PILLAR (RR 100 110)	RIGHT SIDE AFT "A" PILLAR (RR 100 111)
LEFT SIDE MID VEHICLE (RR 101 001)	DRIVER'S SIDE MID VEHICLE (RR 101 010)	LEFT CENTER MID VEHICLE (RR 101 011)	CENTER POINT MID VEHICLE (RR 101 100)	RIGHT CENTER MID VEHICLE (RR 101 101)	PASSENGER SIDE MID VEHICLE (RR 101 110)	RIGHT SIDE MID VEHICLE (RR 101 111)
LEFT SIDE TRUNK (RR 110 001)	DRIVER'S SIDE TRUNK (RR 110 010)	LEFT CENTER TRUNK (RR 110 011)	CENTER POINT TRUNK (RR 110 100)	RIGHT CENTER TRUNK (RR 110 101)	PASSENGER SIDE TRUNK (RR 110 110)	RIGHT SIDE TRUNK (RR 110 111)
LEFT SIDE REAR SURFACE (RR 111 001)	DRIVER'S SIDE REAR SURFACE (RR 111 010)	LEFT CENTER REAR SURFACE (RR 111 011)	CENTER POINT REAR SURFACE (RR 111 100)	RIGHT CENTER REAR SURFACE (RR 111 101)	PASSENGER SIDE REAR SURFACE (RR 111 110)	RIGHT SIDE REAR SURFACE (RR 111 111)

```
XXX = ROW -- FROM FRONT TO REAR        000 = ALL ROWS
YYY = COLUMN -- FROM SIDE TO SIDE      000 = ALL COLUMNS
RR  = 00 (GEOGRAPHIC LOCATION)
```

FIGURE 33 – EXTENDED ADDRESS MAP

APPENDIX A
NETWORK ARCHITECTURES AND HEADER SELECTION

A.1 Architectures of Networks — A wide variety of network topologies can be envisioned by network designers. The message structure described in this document is very flexible and useful in exchanging information between network nodes. The following discussion describes two network architectures which are likely configurations that can use this message definition set:

a. A single network architecture
b. A multiple network architecture

Which of the different message definitions to use for these two network architectures is left as the system designer's choice, since the selection would be application specific. It should be noted that the hardware that supports these two message structures is generally not interchangeable. It is recommended that care be taken in choosing which message definition to use because the selection is generally irreversible because of these hardware limitations.

Consideration must be given by the network designer as to whether a single network architecture or a multiple network architecture, within the same vehicle, is preferable for that application. For example, a multiple network architecture could be based on one network optimized around data communication (Class B) protocol requirements, and another network optimized around sensor type (Class A) multiplexing requirements. The Class B network may be characterized such that time is a significant characteristic of the protocol where the functional "broadcast" type of message strategy can most effectively be used. The functional messages strategy can, if required, define the source or destination of a particular message by making it part of the message. However, it is unnecessary to designate or specify the source or destination of functional broadcast type messages. Reception becomes the exclusive responsibility of the receiving node.

A Class A network could handle the vehicle's event-driven multiplexing requirements. See SAE J2057; Part 5, for more information on Class A multiplexing considerations. A single network architecture is based on the concept of a network which handles both Class A and B requirements. It should be clear that both network architectures must be cost effective for the application and the specific nodes on each network.

In Class B communications, the network consists of the interconnection of intelligent nodes such as: an engine controller, a body computer, a vehicle instrument cluster, and other modules. Such a network normally does not significantly reduce the base vehicle wiring but provides an inter-module data communications capability for distributed processing. The data shared between modules may be repetitive in nature and sometimes requires handshaking between modules or acknowledgement of data reception. As a result of handling the repetitive data and response type data, a network can be optimized around functional addressing. Functional addressing sends data on the network which can be received by one or more nodes without regard to the physical location of the module but only by their "interest" in those specific functions. In general, the transmitting node doesn't care which, if any, nodes receive the data it is sending. When physical addressing is required in a data communications (Class B) network, it is usually for vehicle diagnostic purposes and can be easily handled without reducing network bandwidth. See Section 6 for a discussion of these two types of message addressing.

In Class A communications, the network generally consists of the interconnection of limited intelligence nodes, often simply sensors or actuators. These Class A networks can significantly reduce the base vehicle wiring as well as potentially remove redundant sensors from the vehicle. The data shared between nodes in this case are generally event driven in nature. In most vehicles, the number of event driven signals predominate but are only needed infrequently. The message to "turn headlamps on," for example can be easily seen as event driven. Because these messages are infrequent (only sent once when the signal changes), they generally require acknowledgement, either within the same message or a separate handshake/response message. Both of these approaches are supported by SAE J1850 and in this document.

The single network architecture carries both the Class A and Class B messages on one network. The characteristics of both time critical and event driven messages must be accommodated on a message by message basis. In general, this level of complexity will need the flexibility of the consolidated header structure. For some applications, however, the single byte header may be adequate and is in no way limited by this specification.

The multiple network architecture tends to separate the Class A messages from the Class B messages and optimize each network and node interface for the specific characteristics of each network Class. The time critical messages could be exchanged on one network while the event driven messages sent on another. For example, the data communication (Class B) repetitive messages can be handled on one network and the sensor and control (Class A) multiplexing requirements on another network. This architecture requires both networks to work together to achieve the total vehicle network requirements. If information is needed between the multiple networks, care must be exercised to meet the needs of each of the networks. This concept of multiple networks is not limited to two, but can be extended to several separate networks, if desired.

A.2 Header Selection—The selection of headers is dependent on a number of factors. Hardware cost is related to the complexity of interface hardware required, as opposed to level of software required to implement the system. A CRC, for example, is most efficiently calculated in hardware. A Checksum (CS) is easily calculated in software.

The 3 byte form of the consolidated header offers systems designers advantages over the single byte only header format. More bits are available for use in assigning message identifiers, priorities, message types, etc. In addition, the 3 byte header is not restricted to 256 addresses. The most significant bit of the consolidated header format is used to indicate the use of 1 byte or 3 byte header messages which provides flexibility for future applications. This consolidated format generally requires increased hardware implementations because of the variety of message forms allowed. However, because of the increased hardware definition, software complexity and coding error problems are reduced. The consolidated header format offers the flexibility to readily handle both Class A and Class B requirements, matching header length to each specific message, as needed.

The single byte only header format is a simple and efficient method of identifying messages in a Class B data communications optimized network. This type of header can be simply called the Message Identifier or ID. The ID becomes the name of the message that is broadcast to all the other nodes on the network. The message ID concept supports a very easily understood message protocol leading to the utilization of a simple hardware interface circuit. The single byte only header format in a Class B data communications environment is required to support a repetitive message strategy and manage network bandwidth efficiently. The results can yield an average message length of less than 4 bytes. Message overhead can be kept small and message latency minimized to improve the network's capability for handling time critical data. Another resulting feature of this method of message identification is that arbitration events are reduced (resolved within the first byte) and the need to carefully assign message priority is reduced.

As in the case of network architectures above, the choice of single byte header format or consolidated header format is left to the system designer. The choice must be made for each separate network for each application but this offers the ability to select which characteristic is important for the specific case.

APPENDIX B
EXAMPLES OF EXTENDED ADDRESSING

B.1 Seat locations in the passenger compartment are shown in Figure B1.

B.2 Typical lighting locations and external access examples are shown in Figure B2.

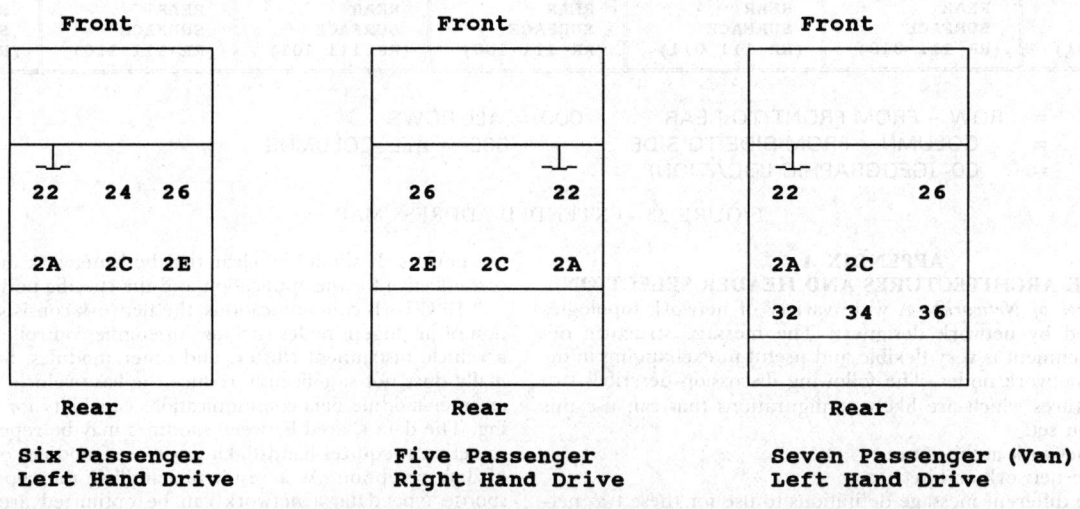

FIGURE B1—TYPICAL SEAT LOCATIONS

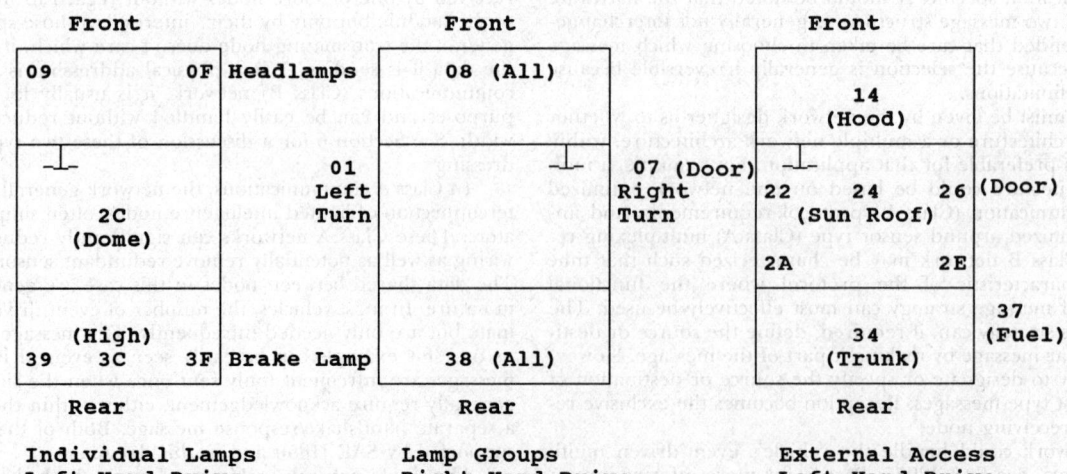

FIGURE B2—TYPICAL LIGHTING AND EXTERNAL ACCESS LOCATIONS

CLASS B DATA COMMUNICATION NETWORK MESSAGES
PART 2: DATA PARAMETER DEFINITIONS—
SAE J2178/2 JUN93

SAE Recommended Practice

Report of the Vehicle Network for Multiplexing and Data Communications Standards Committee approved June 1993.

TABLE OF CONTENTS

1. Scope
1.1 Standardized Parameter Definitions
2. References
2.1 Applicable Documents
2.1.1 SAE Publications
2.1.2 ANSI Publication
2.2 Related Publications
2.3 Terms and Definitions
2.4 Abbreviations and Acronyms
3. General Information
3.1 Part 2 Overview
3.2 How To Use This Document
4. Parameter Reference Number (PRN) Structure
5. Parameter Formats
5.1 Bit Mapped Parameters
5.1.1 Bit Mapped Data Without Mask Byte(s)
5.1.2 Bit Mapped Data With Mask Byte(s)
5.1.3 Bit Values
5.2 Byte (8 Bit) Parameters
5.3 Word (16 Bit) Parameters
5.4 Multi-Byte (>16 Bit) Parameters
5.5 Multiple Parameter Packets
6. Specific Parameter (PRN) Assignments
6.1 Specific Parameters
7. Scaling, Limit, Offset, and Transfer Function (SLOT) Definitions
7.1 Multiple Parameter Packeted (PKT) SLOTs
7.1.1 Multiple Parameter Packeted Assignments
7.2 Bit Mapped Without Mask (BMP) SLOTs
7.2.1 Bit Mapped Without Mask Parameter Assignments
7.3 Bit Mapped With Mask Bytes (BMM) SLOTs
7.3.1 Bit Mapped With Mask Parameter Assignments
7.4 Unsigned Numeric (UNM) SLOTs
7.4.1 Unsigned Numeric Variable Assginments
7.5 2's Complement Signed Numeric (SNM) SLOTs
7.5.1 2's Complement Signed Numeric Variable Assignments
7.6 State Encoded (SED) SLOTs
7.6.1 State Encoded Variable Assignments
7.7 ASCII Encoded (ASC) SLOTs
7.7.1 ASCII Encoded Variable Assignments
7.7.2 ASCII Character Set
7.8 Binary Coded Decimal (BCD) SLOTs
7.8.1 Binary Coded Decimal (BCD) Variable Assignments
7.9 Signed Floating Point [Scientific Notation] (SFP) SLOT
7.9.1 Signed Floating Point Variable Assignment
8. Multiple Frame, Single Parameter Format

1. Scope—This SAE Recommended Practice defines the information contained in the header and data fields of non-diagnostic messages for automotive serial communications based on SAE J1850 Class B networks. This document describes and specifies the header fields, data fields, field sizes, scaling, representations, and data positions used within messages.

The general structure of a SAE J1850 message frame without in-frame response is shown in Figure 1. The structure of a SAE J1850 message with in-frame response is shown in Figure 2. Figures 1 and 2 also show the scope of frame fields defined by this document for non-diagnostic messages. Refer to SAE J1979 for specifications of emissions-related diagnostic message header and data fields. Refer to SAE J2190 for the definition of other diagnostic data fields. The description of the network interface hardware, basic protocol definition, the electrical specifications, and the CRC byte are given in SAE J1850.

SAE J1850 defines two and only two formats of message headers. They are the Single Byte header format and the Consolidated header format. The consolidated header format has two forms, a single byte form and a three byte form. This document covers all of these formats and forms to identify the contents of messages which could be sent on a SAE J1850 network.

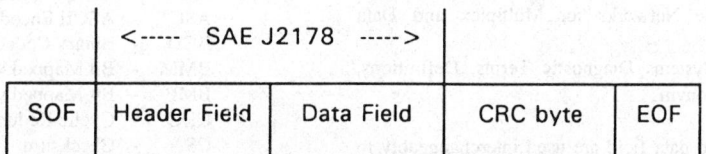

FIGURE 1—SCOPE OF SAE J2178 FOR A SAE J1850
FRAME WITHOUT IN-FRAME RESPONSE (IFR)

FIGURE 2—SCOPE OF SAE J2178 FOR A SAE J1850
FRAME WITH IN-FRAME RESPONSE (IFR)

This document consists of four parts, each published separately.

Part 1 of SAE J2178 (Titled: Detailed Header Formats and Physical Address Assignments) describes the two allowed forms of message header formats, single byte and consolidated. It also contains the physical node address range assignments for the typical subsystems of an automobile.

Part 2 of SAE J2178 (this part, Titled: Data Parameter Definitions) defines the standard parametric data which may be exchanged on SAE J1850 (Class B) networks. The parameter scaling, ranges, and transfer functions are specified. Messages which refer to these parametric definitions shall always adhere to these parametric definitions. It is intended that at least one of the definitions for each parameter in this part match the SAE J1979 definition. The details of this part are fully described in 3.1.

Part 3 of SAE J2178 (Titled: Frame IDs for Single Byte Forms of Headers) defines the message assignments for the single byte header format and the one byte form of the consolidated header format.

Part 4 of SAE J2178 (Titled: Message Definition for Three Byte Headers) defines the message assignments for the three byte form of the consolidated header format.

1.1 Standardized Parameter Definitions—The parameters used to describe data variables are one of the most important functions of this document. To achieve commonality of messages in Class B networks, the data parameters must become standardized. This applies to data parameter definitions for use during normal vehicle operations as well as during diagnostic operations. By using common parameter definitions for non-diagnostic and diagnostic functions on the network, the modules which form the network can maintain one image or description of a data parameter.

Where parameters have been defined in the Diagnostic Test Modes documents (SAE J1979 and J2190), such as Parameter Identifiers for diagnostic purposes, the definitions in Part 2 of this document match the diagnostic definition.

SAE J2178, Part 2 defines the parameters to be used for non-diagnostic and diagnostic data format definitions. For new parameter definitions which are needed in the future, the new definitions, if they are expected to become widely used, must be integrated into this document for commonality across these types of applications. Of course, manufacturers are free to assign their own definitions to data parameters which are unique or proprietary to their products. They are, however, restricted to using the "Manufacturer Reserved" message header assignments in Parts 3 and 4 of this document when using these unique or proprietary data parameter definitions.

2. References

2.1 Applicable Documents—The following publications form a part of this specification to the extent specified herein. The latest issue of SAE publications shall apply.

2.1.1 SAE PUBLICATIONS—Available from SAE, 400 Commonwealth Drive, Warrendale, PA 15096-0001.

SAE J1850—Class B Data Communication Network Interface
SAE J1979—E/E Diagnostic Test Modes
SAE J2190—Enhanced E/E Diagnostic Test Modes

2.1.2 ANSI PUBLICATION—Available from ANSI, 11 West 42nd Street, New York, NY 10036-8002.

ANSI/IEEE Std 754-1985, August 12, 1985—IEEE Standard for Binary Floating-Point Arithmetic

2.2 Related Publications—The following publications are provided for information purposes only and are not a required part of this document.

2.2.1 SAE PUBLICATIONS—Available from SAE, 400 Commonwealth Drive, Warrendale, PA 15096-0001.

SAE J1213/1—Glossary of Vehicle Networks for Multiplex and Data Communication
SAE J1930—Electrical/Electronic Systems Diagnostic Terms, Definitions, Abbreviations, and Acronyms

2.3 Terms and Definitions

2.3.1 DATA [DATA FIELD]—Data and data field are used interchangeably in this document and they both refer to a field within a frame that may include bytes with parameters pertaining to the message and/or secondary ID and/or extended addresses and/or test modes which further defines a particular message content being exchanged over the network.

2.3.2 EXTENDED ADDRESS—The extended address is a means to allow a message to be addressed to a specific geographical location or zone of the vehicle, independent of any node's physical address.

2.3.3 FRAME—A frame is one complete transmission of information which may or may not include an In-Frame Response. The frame is enclosed by the start of frame and end of frame symbols. For Class B networks, each frame contains one and only one message (see "message" definition).

2.3.4 FRAME ID—The Frame ID is the header byte for the Single Byte Header format and the one byte form of the consolidated header format. The definition of the frame ID is found in SAE J2178, Part 3. This header byte defines the target and source and content of the frame.

2.3.5 FUNCTIONAL ADDRESSING—Functional addressing allows a message to be addressed or sent to one or more nodes on the network interested in that function. Functional addressing is intended for messages that may be of interest to more than a single node. For example, an exterior lamp "off" message could be sent to all nodes controlling the vehicle exterior lamps by using a functional address. The functional address consists of a primary ID and may include a secondary ID and may also include an extended address.

2.3.6 HEADER [HEADER FIELD]—The header (or header field, used interchangeably) is a one or three byte field within a frame which contains information about the message priority, message source and target addressing, message type, and in-frame response type.

2.3.7 IN-FRAME RESPONSE (IFR) TYPE—The IFR type identifies the form of the in-frame response which is expected within that message.

2.3.8 LOAD—The load command indicates the operation of directly replacing the current/existing value of a parameter with the parameter value(s) contained in the message.

2.3.9 MESSAGE—A message consists of all of the bytes of a frame excluding the delimiter symbols (SOF, EOD, EOF, NB).

2.3.10 MODIFY—The modify command indicates the operation of using the message data parameter value to change (e.g., increment, decrement, or toggle) the current/existing value.

2.3.11 PARAMETER—A parameter is the variable quantity included in some messages. The parameter value, scaling, offset, units, transfer function, etc., are unique to each particular message. (The assigned parameters are contained herein.)

2.3.12 PHYSICAL ADDRESSING—Physical addressing allows a message to be addressed to a specific node or to all nodes or to a non-existent, null node. The information in this message is only of relevance to a particular node, so the other nodes on the bus should ignore the message, except for the case of the "all nodes" address.

2.3.13 PRIMARY ID—The primary ID identifies the target for this functional message. This is the primary discriminator used to group functions into main categories.

2.3.14 PRIORITY—The priority describes the rank order and precedence of a message. Based upon the SAE J1850, Class B arbitration process, the message with the highest priority will win arbitration.

2.3.15 REPORT—A report indicates the transmission of parametric data values, based on: a change of state; a change of value; on a periodic rate basis; or as a response to a specific request.

2.3.16 REQUEST—A request is a command to, or a query for data, or action from another node on the network.

2.3.17 RESPONSE DATA—The response data is the information from a node on the network in response to a request from another node on the network. This may be an in-frame response or a report type of message.

2.3.18 SECONDARY ID—The secondary ID (along with the primary ID or Frame ID) identifies the functional target node for a message. The purpose of the secondary ID field within the frame is to further define the function or action being identified by the primary ID.

2.4 Abbreviations and Acronyms

A/C	-	Air Conditioning
ASC	-	ASCII Encoded SLOT
BCD	-	Binary Coded Decimal (BCD) SLOT
BMM	-	Bit Mapped with Mask SLOT
BMP	-	Bit Mapped without Mask SLOT
CRC	-	Cyclic Redundancy Check
CS	-	Checksum
DTC	-	Diagnostic Trouble Code
EOD	-	End of Data
EOF	-	End of Frame
ERR	-	Error Detection
HVAC	-	Heating, Ventilation, Air Conditioning
ID	-	Identifier
IFR	-	In-Frame Response
LSB	-	Least Significant Bit/Byte
MSB	-	Most Significant Bit/Byte
NB	-	Normalization Bit
PID	-	Parameter IDentification (number, NOT the primary ID, See Section 6)
PKT	-	Multiple Parameter Packet SLOT
PRN	-	Parameter Reference Number
SED	-	State Encoded SLOT
SFP	-	Signed Floating Point (Scientific Notation) SLOT
SLOT	-	Scaling, Limit, Offset, and Transfer Function (See Section 7)
SNM	-	2's Complement Signed Numeric SLOT
SOF	-	Start of Frame
UNM	-	Unsigned Numeric SLOT
VIN	-	Vehicle Identification Number

3. General Information

3.1 Part 2 Overview—Section 4 provides a description of the parameter reference number (PRN) number groupings used for assigning PRN numbers to individual parameters. Section 5 defines the formats used to define all standard parameters to be used in SAE J2178 messages. Section 6 defines the specific parameter assignments in terms of names, units, and scale factor reference. Section 7 defines the actual parameter specifications (SLOT), in terms of the length, bit resolution, range, scale factor details, etc. Section 8 describes the case of very long parameters which cannot be transmitted in a single message. Appendix A provides a numerical cross reference to assist in finding the correct name of a parameter if the parameter identification number is known. Appendix B is an alphabetical cross reference to assist in finding the correct parameter identification number if the parameter name is known.

The messages contain header fields and data fields, described in Part 1 of SAE J2178. The header field contains target, source, priority, and message type information, while the data field contains optional additional addressing and parametric information. This document defines the parametric information.

For some applications, it is desirable to include multiple parameters in a single message. The multiple parameter format is called a packet in this document. For example, some diagnostic messages consist of combinations of these parameters to improve information density or to insure simultaneous readings of different variables. A very limited set of these combinations are defined here as industry standards but individual manufacturers are free to use this form in manufacturer specific messages, as needed.

3.2 How To Use This Document—This part (Part 2 of SAE J2178) provides the definition of parameters which are commonly found, or could be expected in vehicle Electrical/Electronic Systems today. These parameters have been defined to allow messages on a Class B communication system to have consistent meaning between manufacturers and over time. The parameter definition consists of two parts, the "PRN" and the "SLOT." The "PRN" (Parameter Reference Number) is a number used to identify a specific parameter by name, unit of measure, and its associated "SLOT." The "SLOT" defines the mathematical characteristics of parameters in terms of its representation (Binary, Unsigned Numeric, ASCII, BCD, etc.), its scaling (1 Bit =), its limits and offsets, and its transfer function.

To find a parameter by name or PRN number, Appendices A and B provide cross references to the page of this document where the PRN can be found. The PRN numbers have also been grouped by subsystem to enable the reader to look for parameters if the name is not known.

If the parameter has not as yet been included in the list of PRNs, users can define new parameters in terms of the SLOTs which have been defined.

4. Parameter Reference Number (PRN) Structure—Parameter Reference Numbers (PRNs) are used to simplify documentation. They do not, in themselves, have particular significance. PRNs do allow simplification of reference, particularly for diagnostic purposes. To this end, a structure for PRN number assignments has been developed. The structure is described in the following paragraphs. Some of the PRN numbers are assigned or reserved for SAE definitions but the majority are reserved for individual motor vehicle manufacturers to coordinate and assign.

All PRN addresses are two bytes long, with the first byte identifying a grouping or classification reference (refer to Tables 1 and 2). The second byte is then a sequence number pointing to the specific parameter used (refer to Tables 4 through 19). There has not been any attempt made to group or commonize the meaning in the second byte. The specific assignments are found in Section 6. Note that all PRN addresses are listed as hexadecimal numbers throughout this document.

SAE J1979 refers to PID numbers which are a single byte reference number. The first 256 PRNs defined here (first byte = 00), are identical with the SAE J1979 PID definitions.

Figure 3 shows the basic structure of PRNs and Tables 1 and 2 show bit assignments and address ranges based on these assignments.

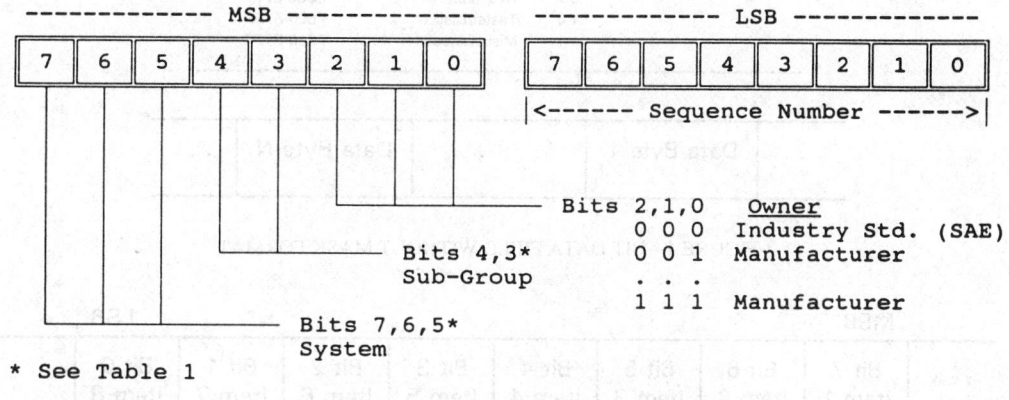

FIGURE 3—PRN STRUCTURE

5. Parameter Formats—Parameter values are represented in bit, byte, word, and multi-byte forms. The length of the parameter is uniquely associated with the message header and any included secondary ID field. These parameter definitions are referenced by one or more messages. The parameters allow a wide variety of variables, data definitions, and representations to provide the ability to use this definition for all messages, even when the industry standard messages are not useful for particular applications.

5.1 Bit Mapped Parameters—The bit mapped parameters, that is, those which have only two logical values (for example: True/False), are handled in one of two ways. The "Q" bit described in Section 6 of Part 1 of SAE J2178, is used if there is only one bit of information. In many cases, however, the bit values are associated together and form byte groupings, generally around common functional characteristics. Bit mapped data are transmitted either with or without corresponding mask bytes as described as follows.

5.1.1 BIT MAPPED DATA WITHOUT MASK BYTE(S)—In some cases, such as configuration identification, there is information which can be grouped as binary bits which represent whether, for example, a function or test is supported in a system. This form does not allow the bits to be supplied from different nodes in the network. If the bits potentially come from more than one node, the form with mask bytes described in 5.1.2, is used. The general form of the bit data bytes without mask is shown in Figure 4.

For these bit data cases, the bit names (that is: items) are somewhat different to emphasize the difference in format. These names are shown in Figure 5. If there is more than one byte of data of this form, the item numbers are incremented by one, sequentially, beginning at one from the MSB of the first byte.

For definition purposes, it is also possible to describe a single or multiple bit group which is smaller than a byte but which are combined into a byte or multiple bytes when the complete message is defined. This is a convenience used in this document to define some parameters. This definition notation applies equally to byte(s) with or without masks.

5.1.2 BIT MAPPED DATA WITH MASK BYTE(S)—For bit mapped data value groups that may come from several nodes or may not be valid for an application, a special format has been defined. Figures 6 and 7 show the data byte formats used. The data format allows two options in the number of bits in a group; 8 or 16. Since the defined bits for such groupings may not always come from a single network node, the format includes mask bytes indicating if an individual bit is valid for this message or if it should be ignored. The mask bytes map directly to the data bits with which they are associated. Figures 8 and 9 show how these bits are mapped within each byte. Table 3 shows the bit value for the mask bits.

TABLE 1—PRN GROUPINGS

7 6 5	System	4 3	Subgroup	Address (Hex)
0 0 0	Powertrain	0 0	J1979 Compatible	0000-00FF
		0 0	Reserved	0100-07FF
		0 1	Reserved	0800-0FFF
		1 0	Engine	1000-17FF
		1 1	Transmission	1800-1FFF
0 0 1	Chassis	0 0	Reserved	2000-27FF
		0 1	Brakes/Tires/Wheels	2800-2FFF
		1 0	Steering	3000-37FF
		1 1	Suspension	3800-3FFF
0 1 0	Body 1	0 0	Reserved	4000-47FF
		0 1	Reserved	4800-4FFF
		1 0	Reserved	5000-57FF
		1 1	Restraints	5800-5FFF
0 1 1	Body 2	0 0	Driver Info.	6000-67FF
		0 1	Reserved	6800-6FFF
		1 0	Lighting	7000-77FF
		1 1	Reserved	7800-7FFF
1 0 0	Body 3	0 0	Audio	8000-87FF
		0 1	Reserved	8800-8FFF
		1 0	Pers. Comm.	9000-97FF
		1 1	HVAC	9800-9FFF
1 0 1	Body 4	0 0	Convenience	A000-A7FF
		0 1	Reserved	A800-AFFF
		1 0	Reserved	B000-B7FF
		1 1	Reserved	B800-BFFF
1 1 0	Other 1	0 0	Security	C000-C7FF
		0 1	Reserved	C800-CFFF
		1 0	Reserved	D000-D7FF
		1 1	Reserved	D800-DFFF
1 1 1	Other 2	0 0	Config. Codes	E000-E7FF
		0 1	Reserved	E800-EFFF
		1 0	Tester/Diag.	F000-F7FF
		1 1	Miscellaneous	F800-FFFF

...	Data Byte 1	...	Data Byte N	...

FIGURE 4—BIT DATA FIELD WITHOUT MASK FORMAT

	MSB							LSB	
...	Bit 7 Item 1	Bit 6 Item 2	Bit 5 Item 3	Bit 4 Item 4	Bit 3 Item 5	Bit 2 Item 6	Bit 1 Item 7	Bit 0 Item 8	...

FIGURE 5—BITS WITHOUT MASK BYTE

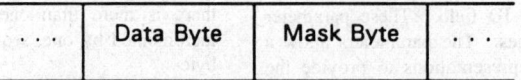

FIGURE 6—FORMAT FOR 8-BIT DATA WITH MASK

...	Data Byte 1	Mask Byte 1	Data Byte 2	Mask Byte 2	...

FIGURE 7—FORMAT FOR 16-BIT DATA WITH MASK

	MSB							LSB	
...	Bit 7 Item 1	Bit 6 Item 2	Bit 5 Item 3	Bit 4 Item 4	Bit 3 Item 5	Bit 2 Item 6	Bit 1 Item 7	Bit 0 Item 8	...

FIGURE 8—BIT MAPPED DATA BYTE

TABLE 2—SAE/MFG PRN RANGES

System	Subgroup	Address Map	Size
Powertrain	J1979 Compatible - SAE	0000 - 00FF	0.25K
	Reserved - MFG	0100 - 07FF	1.75K
	Reserved - SAE	0800 - 08FF	0.25K
	Reserved - MFG	0900 - 0FFF	1.75K
	Engine - SAE	1000 - 10FF	0.25K
	Engine - MFG	1100 - 17FF	1.75K
	Transmission - SAE	1800 - 18FF	0.25K
	Transmission - MFG	1900 - 1FFF	1.75K
Chassis	Reserved - SAE	2000 - 20FF	0.25K
	Reserved - MFG	2100 - 27FF	1.75K
	Brakes/Tires/Wheels - SAE	2800 - 28FF	0.25K
	Brakes/Tires/Wheels - MFG	2900 - 2FFF	1.75K
	Steering - SAE	3000 - 30FF	0.25K
	Steering - MFG	3100 - 37FF	1.75K
	Suspension - SAE	3800 - 38FF	0.25K
	Suspension - MFG	3900 - 3FFF	1.75K
Body 1	Reserved - SAE	4000 - 40FF	0.25K
	Reserved - MFG	4100 - 47FF	1.75K
	Reserved - SAE	4800 - 48FF	0.25K
	Reserved - MFG	4900 - 4FFF	1.75K
	Reserved - SAE	5000 - 50FF	0.25K
	Reserved - MFG	5100 - 57FF	1.75K
	Restraints - SAE	5800 - 5800	0.25K
	Restraints - MFG	5900 - 5FFF	1.75K
Body 2	Driver Information - SAE	6000 - 60FF	0.25K
	Driver Information - MFG	6100 - 67FF	1.75K
	Reserved - SAE	6800 - 68FF	0.25K
	Reserved - MFG	6900 - 6FFF	1.75K
	Lighting - SAE	7000 - 70FF	0.25K
	Lighting - MFG	7100 - 77FF	1.75K
	Reserved - SAE	7800 - 78FF	0.25K
	Reserved - MFG	7900 - 7FFF	1.75K
Body 3	Audio - SAE	8000 - 80FF	0.25K
	Audio - MFG	8100 - 87FF	1.75K
	Reserved - SAE	8800 - 88FF	0.25K
	Reserved - MFG	8900 - 8FFF	1.75K
	Personal Communications - SAE	9000 - 90FF	0.25K
	Personal Communications - MFG	9100 - 97FF	1.75K
	HVAC - SAE	9800 - 98FF	0.25K
	HVAC - MFG	9900 - 9FFF	1.75K
Body 4	Convenience - SAE	A000 - A0FF	0.25K
	Convenience - MFG	A100 - A7FF	1.75K
	Reserved - SAE	A800 - A8FF	0.25K
	Reserved - MFG	A900 - AFFF	1.75K
	Reserved - SAE	B000 - B0FF	0.25K
	Reserved - MFG	B100 - B7FF	1.75K
	Reserved - SAE	B800 - B8FF	0.25K
	Reserved - MFG	B900 - BFFF	1.75K
Other 1	Security - SAE	C000 - C0FF	0.25K
	Security - MFG	C100 - C7FF	1.75K
	Reserved - SAE	C800 - C8FF	0.25K
	Reserved - MFG	C900 - CFFF	1.75K
	Reserved - SAE	D000 - D0FF	0.25K
	Reserved - MFG	D100 - D7FF	1.75K
	Reserved - SAE	D800 - D8FF	0.25K
	Reserved - MFG	D900 - DFFF	1.75K
Other 2	Configuration Codes - SAE	E000 - E0FF	0.25K
	Configuration Codes - MFG	E100 - E7FF	1.75K
	Reserved - SAE	E800 - E8FF	0.25K
	Reserved - MFG	E900 - EFFF	1.75K
	Tester/Diagnostics - SAE	F000 - F0FF	0.25K
	Tester/Diagnostics - MFG	F100 - F7FF	1.75K
	Miscellaneous - SAE	F800 - F8FF	0.25K
	Miscellaneous - MFG	F900 - FFFF	1.75K

MSB									LSB
...	Mask Bit 7 Item 1	Mask Bit 6 Item 2	Mask Bit 5 Item 3	Mask Bit 4 Item 4	Mask Bit 3 Item 5	Mask Bit 2 Item 6	Mask Bit 1 Item 7	Mask Bit 0 Item 8	...

FIGURE 9—BIT MAPPED MASK BYTE

TABLE 3—MASK BIT ASSIGNMENT

Mask Bit	Bit Mapped Masking
0	Not A Valid Bit
1	Valid Bit

5.1.3 BIT VALUES—The general form of binary data is that 1 = true and 0 = false. One bit can carry a wide variety of interpretations, depending on the subject that it describes.

5.2 Byte (8 Bit) Parameters—Data parameters which can be expressed in 8 bits or less are expressed in a byte format. Byte parameters are the most common format. Figure 10 shows this format.

5.3 Word (16 Bit) Parameters—Data parameters which can be expressed in 9 to 16 bits are expressed in word format with the most significant byte transmitted first (high byte/low byte). Figure 11 shows this format.

5.4 Multi-Byte (>16 Bit) Parameters—Data parameters which can be expressed in more than 16 bits are expressed in multi-byte format with the most significant byte transmitted first (highest byte /.../ lowest byte). Figure 12 shows this format.

5.5 Multiple Parameter Packets—It is often useful to group parameters together into a packet to increase the information density of messages. For example, sending a fluid level and maximum capacity value in the same message, or a packet of single bit(s) or multiple bit definitions that may be smaller than a full byte, can be combined in this way. Parameter packets will be defined in this document in the same way as any other parameter, having a packet PRN number and SLOT reference assignment. If bit mapped parameters are included in a packet, the bit mapped byte(s) and the associated mask byte(s) are grouped together as a set representing a parameter. Thus, the data bytes and mask bytes for each parameter are together within the packet. Note that packets may be made of other packets, but each will be uniquely defined by combining each subpart into a message. Figure 13 shows the general form of a parameter packet.

A few examples, such as some of the SAE J1979 compatible PRN definitions in Section 6 will make this format clear.

FIGURE 10—BYTE PARAMETERS

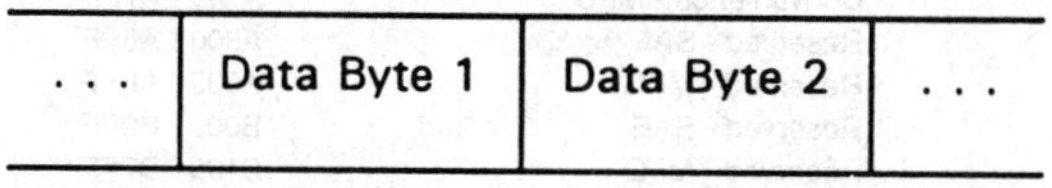

FIGURE 11—WORD PARAMETERS

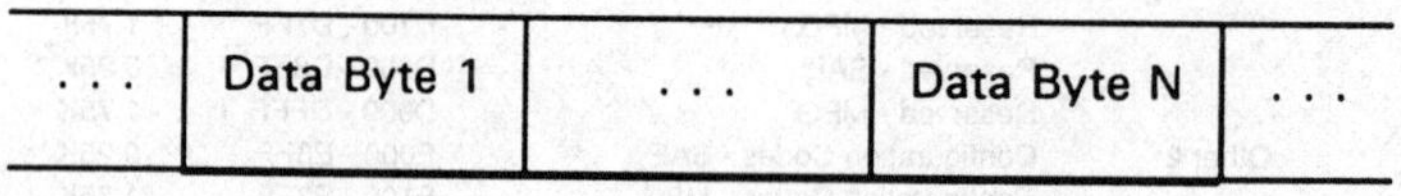

FIGURE 12—MULTI-BYTE PARAMETERS

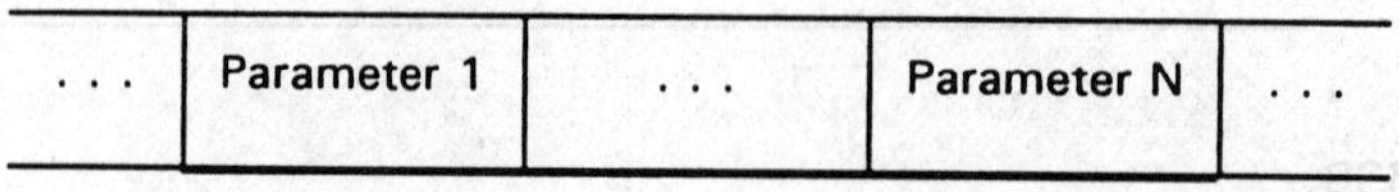

FIGURE 13—MULTIPLE PARAMETER PACKETS

6. Specific Parameter (PRN) Assignments—This section defines industry standard parameters used in messages found in parts 3 and 4. These parameters have been listed in random order based on when they were defined. Note that any PRN address in the SAE range that is not currently defined is reserved for SAE use. Cross references are provided in Appendices A and B of this part of SAE J2178 to assist the reader in finding the specific definition for each parameter.

6.1 Specific Parameters—The following definitions include the Parameter Reference Number (PRN), parameter name, units of measure, and associated SLOT number. The SLOT number is a reference to the Scaling, Limit, Offset, and Transfer function (SLOT) definition found in Section 7 as follows. The SLOT reference numbers have a format of: F-N-#, where F is a three-letter mnemonic indicating the format code type (see Section 7), N is the number of bits in the parameter, and # is the sequence number for that type of

SLOT. There is no relationship between sequence numbers for different format codes or number of bytes. Note that the sequence number is randomly assigned. Decimal sequence numbers of 1000 and above are available for manufacturers to assign. All SAE-assigned sequence numbers will be in the range from 1 to 999 exclusively and are not to be used by manufacturers. The sequence number ZERO (0) has the special purpose of a fill sequence which is used in packet definitions. The ZERO sequence number is all zeros, for the number of bits specified. The Parameter Reference Numbers (PRNs) are the index reference used by the message definitions found in Parts 3 and 4 of this document. The specific parameters are grouped as follows:
- Table 4 - SAE J1979 Compatible PRN Assignments
- Table 5 - Engine PRN Assignments
- Table 6 - Transmission PRN Assignments
- Table 7 - Brakes/Tires/Wheels PRN Assignments
- Table 8 - Steering PRN Assignments
- Table 9 - Suspension PRN Assignments
- Table 10 - Restraints PRN Assignments
- Table 11 - Driver Information PRN Assignments
- Table 12 - Lighting PRN Assignments
- Table 13 - Audio PRN Assignments
- Table 14 - Personal Communication PRN Assignments
- Table 15 - HVAC PRN Assignments
- Table 16 - Convenience PRN Assignments
- Table 17 - Security PRN Assignments
- Table 18 - Configuration Codes PRN Assignments
- Table 19 - Tester/Diagnostics PRN Assignments
- Table 20 - Miscellaneous PRN Assignments

TABLE 4—SAE J1979 COMPATIBLE PRN ASSIGNMENTS
(PRNs 0000 - 00FF)

PRN	Parameter Name	Resolution (1 Bit =)	Units Of Measure	Slot # (F-N-#)
0000	PIDs Supported (01h - 20h)	-	Bit Mapped	BMP-32-1
0001	Number of Emission-Related Trouble Codes & MIL Status	-	Packeted	PKT-32-1
0002	Trouble Code That Caused Freeze Frame Storage	-	Packeted	PKT-16-1
0003	Fuel System Status	-	Packeted	PKT-16-2
0004	Calculated Load Value	100/255	% Full Load	UNM-08-61
0005	Engine Coolant Temperature	1	Degrees Centigrade	UNM-08-102
0006	Short Term Fuel Trim - Bank 1	100/128	% Enrichment	UNM-08-92
0007	Long Term Fuel Trim - Bank 1	100/128	% Enrichment	UNM-08-92
0008	Short Term Fuel Trim - Bank 2	100/128	% Enrichment	UNM-08-92
0009	Long Term Fuel Trim - Bank 2	100/128	% Enrichment	UNM-08-92
000A	Fuel Pressure (Gage)	3	kPaG	UNM-08-131
000B	Intake Manifold Absolute Pressure	1	kPaA	UNM-08-101
000C	Engine RPM - High Resolution	1/4	RPM	UNM-16-31
000D	Vehicle Speed - Low Resolution - Metric	1	km/h	UNM-08-101
000E	Ignition Timing Advance (#1)	1/2	Degrees before TDC	UNM-08-72
000F	Intake Air Temperature	1	Degrees Centigrade	UNM-08-102
0010	Air Flow Rate from MAF	1/100	grams/sec	UNM-16-11
0011	Absolute Throttle #1 Position	100/255	% Full Throttle	UNM-08-61
0012	Commanded Secondary Air	-	Bit Mapped	BMP-08-5
0013	Oxygen Sensor Location	-	Bit Mapped	BMP-08-6
0014	Oxygen Sensor - BANK 1 - Sensor 1	-	Packeted	PKT-16-3
0015	Oxygen Sensor - BANK 1 - Sensor 2	-	Packeted	PKT-16-3
0016	Oxygen Sensor - BANK 1 - Sensor 3	-	Packeted	PKT-16-3
0017	Oxygen Sensor - BANK 1 - Sensor 4	-	Packeted	PKT-16-3
0018	Oxygen Sensor - BANK 2 - Sensor 1	-	Packeted	PKT-16-3
0019	Oxygen Sensor - BANK 2 - Sensor 2	-	Packeted	PKT-16-3
001A	Oxygen Sensor - BANK 2 - Sensor 3	-	Packeted	PKT-16-3
001B	Oxygen Sensor - BANK 2 - Sensor 4	-	Packeted	PKT-16-3
001C	Reserved SAE	-	-	-
001D	Reserved SAE	-	-	-
001E	Reserved SAE	-	-	-
001F	Reserved SAE	-	-	-
0020	PIDs Supported (21h - 40h)	-	Bit Mapped	BMP-32-2
0021-003F	Reserved SAE	-	-	-
0040	PIDs Supported (41h - 60h)	-	Bit Mapped	BMP-32-3
0041-00FF	Reserved SAE	-	-	-

TABLE 5—ENGINE PRN ASSIGNMENTS
(PRNs 1000 - 10FF)

PRN	PARAMETER NAME	RESOLUTION (1 Bit =)	UNITS OF MEASURE	SLOT # (F-N-#)
1000	MIL Status	-	Bit Mapped	BMP-01-1
1001	Number of Emissions Related DTCs	1	Quantity	UNM-07-1
1002	Continuous Evaluation Supported	-	Bit Mapped	BMP-08-1
1003	Trip Evaluation Supported	-	Bit Mapped	BMP-08-2
1004	Trip Evaluation Complete	-	Bit Mapped	BMP-08-3
1005	Sub-System Category of DTC	-	State Encoded	SED-02-1
1006	Most Significant Digit of DTC	-	State Encoded	SED-02-2
1007	Lower 3 Digits of DTC	-	BCD	BCD-12-1
1008	Fuel System Status - Bank 1	-	Bit Mapped	BMP-08-4
1009	Fuel System Status - Bank 2	-	Bit Mapped	BMP-08-4
100A	Oxygen Sensor Voltage	1/200	volts	UNM-08-11
100B	Short Term Fuel Trim	100/128	% Enrichment	UNM-08-92
100D	Most Significant Digit of VIN Number	-	ASCII	ASC-08-01
100E	A/C Clutch Load	25	Watts	UNM-08-165
1015	Injector On Time	2048	microseconds	UNM-08-231
1019	Crankshaft Torque - Absolute	4	N.m Torque	UNM-08-141
1020	Crankshaft Torque - Percent	100/255	% Maximum Torque	UNM-08-61
1021	Engine Boost	100/255	% Full Boost	UNM-08-61
1022	Engine RPM - Low Resolution	32	RPM	UNM-08-171
1023	Engine Idle RPM	16	RPM	UNM-08-161
1024	Engine Revolutions	2	Quantity	UNM-08-121
1025	Barometric Pressure	1	kPaA	UNM-08-101
1026	Engine Coolant Level - Percent	1/2	% Full	UNM-08-71
1027	Engine Coolant Level - Volume	1/10	liters	UNM-08-41
1028	Engine Coolant Capacity	1/10	liters	UNM-08-41
1029	Engine Coolant Pressure	4	kPaG	UNM-08-141
102A	Engine Coolant Fan #1 Speed	100/255	% Full On	UNM-08-61
102B	Engine Oil Temperature	1	Degrees Centigrade	UNM-08-102
102C	Engine Oil Level - Percent	1/2	% Full	UNM-08-71
102D	Engine Oil Level - Volume	1/10	liters	UNM-08-41
102E	Engine Oil Capacity	1/10	liters	UNM-08-41
102F	Engine Oil Pressure	4	kPaG	UNM-08-141
1030	Engine Oil Remaining Life	100/255	% Remaining Life	UNM-08-61
1031	Hydraulic Fan Speed	100/255	% Full On	UNM-08-61
1032	Methanol Content	100/255	% Methanol	UNM-08-61
1033	Maximum Crankshaft Torque	4	N.m Torque	UNM-08-141
1034	Accelerator Pedal Position	100/255	% Pressed Down	UNM-08-61
1035	Absolute Throttle #2 Position	100/255	% Full Throttle	UNM-08-61
1036	Absolute Throttle #3 Position	100/255	% Full Throttle	UNM-08-61
1037	Bank #1 - Converter #1 Temperature	8	Degrees Centigrade	UNM-08-151
1038	Bank #1 - Converter #2 Temperature	8	Degrees Centigrade	UNM-08-151
1039	Bank #2 - Converter #1 Temperature	8	Degrees Centigrade	UNM-08-151
103A	Bank #2 - Converter #2 Temperature	8	Degrees Centigrade	UNM-08-151
103B	Engine Coolant #2 Fan Speed	100/255	% Full On	UNM-08-61
103C	Engine Coolant Temperature - Low Range	1	Degrees Centigrade	UNM-08-104
103D	Engine Coolant Remaining Life	100/255	% Remaining Life	UNM-08-61
103F	Engine Oil Viscosity	1/10	Centistokes (cSt)	UNM-08-41
1040	Number of Engine Cylinders	1	Quantity	UNM-08-101
1041	Number of Valves per Cylinder	1	Quantity	UNM-08-101
1043	Engine Displacement	1/10	liters	UNM-08-41
1044	Fuel Temperature	1	Degrees Centigrade	UNM-08-102
1047	Ignition Switch Position	-	State Encoded	SED-08-5
1048	Engine Redline - Low Resolution	32	RPM	UNM-08-171
1049	Engine Redline - High Resolution	1/4	RPM	UNM-16-31

TABLE 6—TRANSMISSION PRN ASSIGNMENTS
(PRNs 1800 - 18FF)

PRN	Parameter Name	Resolution (1 Bit =)	Units Of Measure	Slot # (F-N-#)
1801	Transmission Fluid Level - Percent	1/2	% Full	UNM-08-71
1802	Transmission Fluid Level - Volume	1/10	liters	UNM-08-41
1803	Transmission Fluid Capacity	1/10	liters	UNM-08-41
1804	Transmission Oil Life	100/255	% Remaining Life	UNM-08-61
1805	Transmission Gear & Lockup Status	-	Packeted	PKT-08-1
1806	Transmission Range Actual (PRNDL)	-	State Encoded	SED-08-4
1807	Transmission Lockup Status	-	State Encoded	SED-02-3
1808	Transmission Actual Gear	-	State Encoded	SED-06-1
1809	Transmission Range Selected (PRNDL)	-	State Encoded	SED-08-4
180A	Transmission Transfer Case (4WD)	-	State Encoded	SED-08-6
180B	Transmission Fluid Temperature	1	Degrees Centigrade	UNM-08-102
180C	Transmission Fluid Pressure	8	kPaG	UNM-08-151
180D	Transmission Commanded Gear	-	State Encoded	SED-08-4
180E	Transmission Actual Gear	-	State Encoded	SED-08-4

TABLE 7—BRAKES/TIRES/WHEELS PRN ASSIGNMENTS
(PRNs 2800 - 28FF)

PRN	Parameter Name	Resolution (1 Bit =)	Units Of Measure	Slot # (F-N-#)
2801	Wheel Speed - Low Resolution	1	km/h	UNM-08-101
2802	Wheel Speed - High Resolution	1/128	km/h	UNM-16-5
2809	Wheel Slip	1/255	Dimensionless	UNM-08-6
2819	Hydraulic Brake Fluid Supply Pump Pressure	32	kPaG	UNM-08-171
281A	Hydraulic Brake Fluid Temperature	1	Degrees Centigrade	UNM-08-102
281B	Hydraulic Brake Fluid Recirculation Pump Pressure	1	kPaG	UNM-08-101
2821	Wheel Rate		SAE Reserved	SAE Reserved
2829	Wheel Angular Velocity		SAE Reserved	SAE Reserved
2831	Wheel Angular Acceleration		SAE Reserved	SAE Reserved
2839	Wheel Load	100/255	% Full Load	UNM-08-61
2841	Brake Fluid Level - Percent	1/2	% Full	UNM-08-71
2842	Brake Fluid Level - Volume	1/100	liters	UNM-08-15
2843	Brake Fluid Remaining Life	100/255	% Remaining Life	UNM-08-61
2844	Brake Fluid Capacity	1/100	liters	UNM-08-15
2849	Tire Temperature	1	Degrees Centigrade	UNM-08-102
2851	Tire Pressure	4	kPaG	UNM-08-141
2859	Tire Type		SAE Reserved	SAE Reserved
2861	Tire Tread Wear Level	100/255	% tread remaining	UNM-08-61

TABLE 8—STEERING PRN ASSIGNMENTS
(PRNs 3000 - 30FF)

PRN	Parameter Name	Resolution (1 Bit =)	Units Of Measure	Slot # (F-N-#)
3001	Steering Wheel Angle	6	Degrees CW from center	SNM-08-61
3005	Power Steering Fluid Temperature	1	Degrees Centigrade	UNM-08-102
3006	Power Steering Fluid Pressure	100	kPaG	UNM-08-185
3007	Power Steering Fluid Level - Percent	1/2	% Full	UNM-08-71
3008	Power Steering Fluid Level - Volume	1/100	liters	UNM-08-15
3009	Power Steering Fluid Remaining Life	100/255	% Remaining Life	UNM-08-61
300B	Power Steering Fluid Capacity	1/100	liters	UNM-08-15
300C	Steering Wheel Rate	1	RPM	UNM-08-101
300D	Steering Wheel Torque	1	N.m Torque	UNM-08-101
300E	Wheel Steer Angle	1/2	Degrees CW from Center	SNM-08-11

TABLE 9—SUSPENSION PRN ASSIGNMENTS
(PRNs 3800 - 38FF)

PRN	Parameter Name	Resolution (1 Bit =)	Units Of Measure	Slot # (F-N-#)
3801	Lateral Acceleration		SAE Reserved	SAE Reserved
3802	Longitudinal Acceleration		SAE Reserved	SAE Reserved
3803	Yaw Acceleration		SAE Reserved	SAE Reserved
3804	Suspension Ride Setting	100/255	% Stiff Setting	UNM-08-61
3805	Suspension Fluid Temperature	1	Degrees Centigrade	UNM-08-102
3806	Suspension Fluid Pressure	100	kPaG	UNM-08-185
3807	Suspension Fluid Level - Percent	1/2	% Full	UNM-08-71
3808	Suspension Fluid Level - Volume	1/32	liters	UNM-08-26
3809	Suspension Fluid Remaining Life	100/255	% Remaining Life	UNM-08-61
380A	Suspension Fluid Capacity	1/32	liters	UNM-08-26
380B	Vehicle Lateral Velocity		SAE Reserved	SAE Reserved
380C	Vehicle Longitudinal Velocity		SAE Reserved	SAE Reserved
380D	Vehicle Yaw Velocity		SAE Reserved	SAE Reserved

TABLE 10—RESTRAINTS PRN ASSIGNMENTS
(PRNs 5800 - 58FF)

PRN	Parameter Name	Resolution (1 Bit =)	Units Of Measure	Slot # (F-N-#)
5801	Shoulder Belt Position	1	Unscaled A/D Counts	UNM-08-101

TABLE 11—DRIVER INFORMATION PRN ASSIGNMENTS
(PRNs 6000 - 60FF)

PRN	Parameter Name	Resolution (1 Bit =)	Units Of Measure	Slot # (F-N-#)
6001	Vehicle Speed - High Resolution - Metric	1/128	km/h	UNM-16-5
6002	Vehicle Speed - High Resolution - English	1/128	mph	UNM-16-5
6003	Compass Direction	3/2	Degrees CW from North	SNM-08-51
6004	Odometer - Vehicle - Metric	1/64	kilometers	UNM-32-31
6005	Fuel Level - Percent	100/255	% Full	UNM-08-61
6006	Fuel Level - Volume	1/100	liters	UNM-16-11
6007	Fuel Capacity	1/100	liters	UNM-16-11
600A	Battery Voltage - Low Resolution	1/16	volts	UNM-08-32
600B	Battery Temperature	1	Degrees Centigrade	UNM-08-102
600C	Electrical Energy Load	1	amps	UNM-08-101
600D	Date (Dw$_8$:DD:MM:YY)	-	Packeted	PKT-32-3
600E	Year (YY)	-	BCD	BCD-08-1
600F	Year (Yr)	1	year	UNM-08-103
6010	Month (Mn)	-	State Encoded	SED-04-2
6011	Month (MM)	-	BCD	BCD-08-1
6012	Day of Week (Dw$_4$)	-	State Encoded	SED-04-1
6013	Day of Week (Dw$_8$)	-	State Encoded	SED-08-2
6014	Day of Month (Dm)	-	State Encoded	SED-08-3
6015	Day of Month (DD)	-	BCD	BCD-08-1
6016	Time of Day (HH:MM:SS)	-	Packeted	PKT-24-1
6017	Hours (HH)	-	BCD	BCD-08-1
6018	Minutes (MM)	-	BCD	BCD-08-1
6019	Seconds (SS)	-	BCD	BCD-08-1
601A	Battery Voltage - High Resolution	1/128	volts	UNM-16-5
601B	Distance Traveled - English	1/8000	miles	UNM-08-1
601C	Fuel Used - Metric	1/64	liters	UNM-16-8
601D	Distance To Empty - English	1/10	miles	UNM-16-21
601E	Vehicle Speed - Low Resolution - English	1	MPH	UNM-08-101
601F	Hours (Hr) - 0 to 23 numeric	1	hour	UNM-08-101
6020	Average Fuel Economy - Low Resolution - Metric	1	liters/100 kilometers	UNM-08-101
6021	Average Fuel Economy - Low Resolution - English	1	MPG	UNM-08-101
6022	Elapsed Time - Seconds	1	Seconds	UNM-08-101
6023	Date (Dw$_4$\Mn:Dm)	-	Packeted	PKT-16-6

Continued

TABLE 11—DRIVER INFORMATION PRN ASSIGNMENTS
(PRNs 6000 - 60FF) (CONTINUED)

PRN	Parameter Name	Resolution (1 Bit =)	Units Of Measure	Slot # (F-N-#)
6024	Elapsed Time - Minutes	1	Minutes	UNM-08-101
6025	Accumulated Ignition On Time	-	Packeted	PKT-24-2
6026	Fuel Used - English	1/64	gallons	UNM-16-8
6027	Distance To Empty - Metric	1/10	kilometers	UNM-16-21
6028	Average Fuel Economy - High Resolution - Metric	1/10	liters/100 kilometers	UNM-16-21
6029	Average Fuel Economy - High Resolution - English	1/10	MPG	UNM-16-21
602A	Elapsed Time - Hours	1	Hours	UNM-08-101
602B	Display Brightness	100/255	% Full On	UNM-08-61
602C	Ignition Off Duration	1	Minutes	UNM-08-101
602D	Outside Air Temperature - High Resolution	1/256	Degrees Centigrade	UNM-16-3
602E	Outside Air Temperature Display	1/2	Degrees Centigrade	UNM-08-73
602F	Minutes (Mn) 0 - 59 numeric	1	minute	UNM-08-101
6030	Time (Hr:Mn)	-	Packeted	PKT-16-5
6031	Odometer - Vehicle - High Resolution - English	1/8000	miles	UNM-32-11
6032	Odometer - Trip - High Resolution - English	128/8000	miles	UNM-24-21
6033	Odometer - Vehicle - Low Resolution - English	1/10	miles	UNM-24-11
6034	Odometer - Trip - Low Resolution - English	1/10	miles	UNM-16-21
6035	Charging Voltage - Low Resolution	1/16	volts	UNM-08-32
6036	Charging Voltage - High Resolution	1/128	volts	UNM-16-5
6037	Charging Current	1	amps	UNM-08-101
6038	Battery Current	1	amps	SNM-08-21
6039	Odometer - Trip - Metric	1/64	kilometers	UNM-24-41
603A	Instantaneous Fuel Economy - Low Resolution - Metric	1	liters/100 kilometers	UNM-08-101
603B	Fuel Used - Percent	100/255	% Used	UNM-08-61
603C	Fuel Used - Volume	1/100	liters	UNM-16-11
603D	Audible Signal Volume	100/255	% Full Volume	UNM-08-61
603E	Audible Signal Type		SAE Reserved	SAE Reserved
603F	Instantaneous Fuel Economy - High Resolution - Metric	1/10	liters/100 kilometers	UNM-16-21
6040	Instantaneous Fuel Economy - Low Resolution - English	1	MPG	UNM-16-21
6041	Instantaneous Fuel Economy - High Resolution - English	1/10	MPG	UNM-08-101
6042	Seconds (Sc) 0 - 59 numeric	1	seconds	UNM-08-101
6047	Alarm Time (HH:MM:SS)	-	Packeted	PKT-24-1
6049	Elapsed Years	1	years	UNM-08-101
604A	Elapsed Months	1	months	UNM-08-101
604B	Elapsed Days	1	days	UNM-08-101

TABLE 12—LIGHTING PRN ASSIGNMENTS
(PRNs 7000 - 70FF)

PRN	Parameter Name	Resolution (1 Bit =)	Units Of Measure	Slot # (F-N-#)

TABLE 13—AUDIO PRN ASSIGNMENTS
(PRNs 8000 - 80FF)

PRN	Parameter Name	Resolution (1 Bit =)	Units Of Measure	Slot # (F-N-#)

TABLE 14—PERSONAL COMMUNICATION PRN ASSIGNMENTS
(PRNs 9000 - 90FF)

PRN	Parameter Name	Resolution (1 Bit =)	Units Of Measure	Slot # (F-N-#)

TABLE 15—HVAC PRN ASSIGNMENTS
(PRNs 9800 - 98FF)

PRN	Parameter Name	Resolution (1 Bit =)	Units Of Measure	Slot # (F-N-#)
9801	HVAC Fan Speed	100/255	% Full On	UNM-08-61
9803	HVAC Door Position	100/255	% Open	UNM-08-61
9804	Electric Defrost Temperature	1	Degrees Centigrade	UNM-08-102
9808	HVAC High-Side Fluid Temperature	1	Degrees Centigrade	UNM-08-102
9809	HVAC Low-Side Fluid Temperature	1	Degrees Centigrade	UNM-08-102
980A	HVAC Low-Side Pressure	5/2	kPaG	UNM-08-125
980B	HVAC Fluid Charge - % Full Charge	100/255	% Full	UNM-08-61
980C	HVAC Fluid Charge - Absolute Weight	10	grams	UNM-08-155
980D	HVAC Fluid Charge Remaining Life	100/255	% Remaining Life	UNM-08-61
980E	HVAC Fluid Charge Capacity	10	grams	UNM-08-155
9810	HVAC Intake Temperature	1/2	Degrees Centigrade	UNM-08-73
9813	HVAC High-Side Pressure	14	kPaG	UNM-08-159
9815	Interior Humidity Level	100/255	% Relative Humidity	UNM-08-61
9816	Interior Air Filter Remaining Life	100/255	% Remaining Life	UNM-08-61
9817	Heat Load Sensor	1/2	mW/cm²	UNM-08-71
9820	Interior Set Temperature	1/2	Degrees Centigrade	UNM-08-73
9830	HVAC Zone Temperature	1/2	Degrees Centigrade	UNM-08-73

TABLE 16—CONVENIENCE PRN ASSIGNMENTS
(PRNs A000 - A0FF)

PRN	Parameter Name	Resolution (1 Bit =)	Units Of Measure	Slot # (F-N-#)
A001	Seat Temperature	1/2	Degrees Centigrade	UNM-08-73
A003	Wiper Mode	-	State Encoded	SED-08-1
A004	Wiper Delay	1/4	seconds	UNM-08-51
A006	Washer Fluid Temperature	1	Degrees Centigrade	UNM-08-102
A007	Washer Fluid Pressure	4	kPaG	UNM-08-141
A008	Washer Fluid Level - Percent	100/255	% Full	UNM-08-61
A009	Washer Fluid Level - Volume	1/10	liters	UNM-08-41
A00A	Washer Fluid Capacity	1/10	liters	UNM-08-41
A00C	Mirror Dimming Level	100/255	% Full Dim	UNM-08-61
A00D	Mirror Horizontal Position	1	Unscaled A/D Counts	UNM-08-101
A00E	Mirror Vertical Position	1	Unscaled A/D Counts	UNM-08-101
A00F	Window Position	1	Unscaled A/D Counts	UNM-08-101
A010	Door Lock Cylinder State	-	State Encoded	SED-08-7
A011	Steering Column Horizontal Position	1	Unscaled A/D Counts	UNM-08-101
A012	Steering Column Vertical Position	1	Unscaled A/D Counts	UNM-08-101
A014	Autolamp Off Delay Time	1	Seconds	UNM-08-101
A015	Vehicle Speed Setting - Low Resolution - Metric	1	km/h	UNM-08-101
A016	Vehicle Speed Setting - High Resolution - Metric	1/128	km/h	UNM-16-5
A017	Vehicle Speed Setting - Low Resolution - English	1	MPH	UNM-08-101
A018	Vehicle Speed Setting - High Resolution - English	1/128	MPH	UNM-16-5

TABLE 17—SECURITY PRN ASSIGNMENTS
(PRNs C000 - C0FF)

PRN	Parameter Name	Resolution (1 Bit =)	Units Of Measure	Slot # (F-N-#)
C001	Remote Transmitter ID	1	ID Number	UNM-08-101

TABLE 18—CONFIGURATION CODES PRN ASSIGNMENTS
(PRNs E000 - E0FF)

PRN	Parameter Name	Resolution (1 Bit =)	Units Of Measure	Slot # (F-N-#)
E021	Vehicle Id Number (VIN) #1	-	VIN character 1	PKT-32-2
E022	Vehicle Id Number (VIN) #2	-	VIN characters 2 - 5	ASC-32-1
E023	Vehicle Id Number (VIN) #3	-	VIN characters 6 - 9	ASC-32-1
E024	Vehicle Id Number (VIN) #4	-	VIN characters 10 - 13	ASC-32-1
E025	Vehicle Id Number (VIN) #5	-	VIN characters 14 - 17	ASC-32-1
E026	Vehicle Id Number (VIN) #6 (reserved for future use)	-	Reserved - SAE	Reserved - SAE
E027	Vehicle Id Number (VIN) #7 (reserved for future use)	-	Reserved - SAE	Reserved - SAE

TABLE 19—TESTER/DIAGNOSTICS PRN ASSIGNMENTS
(PRNs F000 - F0FF)

PRN	Parameter Name	Resolution (1 Bit =)	Units Of Measure	Slot # (F-N-#)

TABLE 20—MISCELLANEOUS PRN ASSIGNMENTS
(PRNs F800 - F8FF)

PRN	Parameter Name	Resolution (1 Bit =)	Units Of Measure	Slot # (F-N-#)
F801	One Byte Zero Fill	0	Zero	UNM-08-0
F802	Two Byte Zero Fill	0	Zero	UNM-16-0
F803	Three Byte Zero Fill	0	Zero	UNM-24-0
F804	Four Byte Zero Fill	0	Zero	UNM-32-0
F805	Five Byte Zero Fill	0	Zero	UNM-40-0
F806	Six Byte Zero Fill	0	Zero	UNM-48-0
F807	Seven Byte Zero Fill	0	Zero	UNM-56-0

7. Scaling, Limit, Offset, and Transfer Function (SLOT) Definitions—This section defines the parameter scaling, limit(s), offset value, and transfer function for bit, byte, or larger variables. These SLOT definitions have been grouped together to avoid duplication in this document and to offer a common list of definitions for use in assigning new parameter definitions. The wide range of these definitions are expected to cover a large number of applications, and should be used for most new definitions as well. Each of these definitions has been assigned a SLOT number for reference purposes but are formatted to include a three-letter mnemonic representing the format, the parameter length in bits, followed by a random sequence number.

The transfer function is shown in two forms to allow use in implementing messages in modules and in interpreting messages found on a network. The two forms are identical but are solved for each of the two variables. The transfer function defines the relationship between computer units (N) in decimal, and engineering units (E) of the data.

The format of the SLOT, identified by the three-letter mnemonic, indicates the category of bit representation. These formats include:
- Multiple Parameter Packeted   PKT
- Bit Mapped with Mask   BMM
- Bit Mapped without Mask   BMP
- Unsigned Numeric   UNM
- 2's Complement Signed Numeric   SNM
- State Encoded   SED
- ASCII Encoded   ASC
- Binary Coded Decimal (BCD)   BCD
- Signed Floating Point (Scientific Notation)   SFP

These formats for each are described in the following paragraphs.

7.1 Multiple Parameter Packeted (PKT) SLOTs—The multiple parameter packet is used to define PRNs which refer to more than one parameter as a group that are related. By grouping multiple parameters, the network bandwidth can be improved, if the groupings are done such that the individual parameters need to be associated such that the each need to be transmitted together. These packets are defined as a sequence of other PRN numbers, in the order that they appear in the message (MSB first). In some cases, a PRN may be assigned to fill bits or bytes with zeros. Each zero fill PRN is an unsigned numeric SLOT with zero as the sequence number.

7.1.1 MULTIPLE PARAMETER PACKETED ASSIGNMENTS

PKT-08-1   Transmission Gear & Lockup Status (PRN 1805)
              MSB   PRN 1807   Transmission Lockup Status .... 2 bits
              LSB   PRN 1808   Transmission Gear Engaged .... 6 bits

PKT-16-1   Trouble Code That Caused Freeze Frame Storage (PRN 0002)
              MSB   PRN 1005   Sub-System Category .... 2 bits
                       PRN 1006   MSB of Trouble Code .... 2 bits
              LSB   PRN 1007   Lower Bytes of Trouble Code - BCD .... 12 bits

PKT-16-2   Fuel System Status (PRN 0003)
              MSB   PRN 1008   Fuel System Status - Bank 1 .... 8 bits
              LSB   PRN 1009   Fuel System Status - Bank 2 .... 8 bits

PKT-16-3   Oxygen BANK 1, 2 - Sensor 1, 2, 3, or 4 (PRN 0014 - 001B)
              MSB   PRN 100A   Oxygen Sensor Voltage .... 8 bits
              LSB   PRN 100B   Short Term Fuel Trim .... 8 bits

PKT-16-5   Time (Hr:Mn) (PRN 6030)
              MSB   PRN 601F   Hours (0 - 23) .... 8 bits
              LSB   PRN 602F   Minutes (0 - 59) .... 8 bits

PKT-16-6   Date (Dw$_4$/Mn:Dm) (PRN 6023)
              MSB   PRN 6012   Day of Week (1 - 7) .... 4 bits
                       PRN 6010   Month (1 - 12) .... 4 bits
              LSB   PRN 6014   Day of Month (1 - 31) .... 8 bits

PKT-24-1   Time of Day (HH:MM:SS) (PRN 6016)
      MSB   PRN 6017   Hours (HH) - BCD .... 8 bits
               PRN 6018   Minutes (MM) - BCD .... 8 bits
      LSB   PRN 6019   Seconds (SS) - BCD .... 8 bits

PKT-24-2   Accumulated Ignition On Time (PRN 6025)
      MSB   PRN 602A   Elapsed Time-Hours (0-99) .... 8 bits
               PRN 6024   Elapsed Time-Minutes (0-59) .... 8 bits
      LSB   PRN 6022   Elapsed Time-Seconds (0-59) .... 8 bits

PKT-32-1   Number of Emission-Related Trouble Codes & MIL Status (PRN 0001)
      MSB   PRN 1000   MIL Status .... 1 bit
               PRN 1001   Number of Emission-Related Trouble Codes .... 7 bits
               PRN 1002   Continuous Evaluation Supported .... 8 bits
               PRN 1003   Trip Evaluation Supported .... 8 bits
      LSB   PRN 1004   Trip Evaluation Complete .... 8 bits

PKT-32-2   Vehicle Id Number (VIN) #1 (PRN 0021)
      MSB   PRN F803   Three Byte - Zero Fill .... 24 bits
      LSB   PRN 100D   MSB of VIN Number .... 8 bits

PKT-32-3   Date (Dw$_8$:DD:MM:YY) (PRN 600D)
      MSB   PRN 6013   Day of Week (1 - 7) .... 8 bits
               PRN 6015   Day of Month (DD) - BCD .... 8 bits
               PRN 6011   Month (MM) - BCD .... 8 bits
      LSB   PRN 600E   Year (YY) - BCD .... 8 bits

7.2 Bit Mapped Without Mask (BMP) SLOTs—Bit mapped (BMP) SLOTs are used to encode data that typically contains several binary parameters, such as status bits or flags, grouped into a single byte or several bytes. Bit mapped SLOTs can also be used for discrete output control such as warning lamps where each bit would indicate the state of a particular lamp. The data in these bit mapped SLOTs is not followed by a MASK byte. There can be up to 4 bytes of data without a mask.

7.2.1 BIT MAPPED WITHOUT MASK PARAMETER ASSIGNMENTS

BMP-01-1 MIL Status (PRN 1000)

		0	1
Length:	1 bit		
Item 1:	Malfunction Indicator Lamp (MIL)	Not Commanded On	Commanded On

BMP-08-1 Continuous Evaluation Supported (PRN 1002)

		0	1
Length:	8 bits		
Item 1:	Not Used		
Item 2:	Not Used		
Item 3:	Not Used		
Item 4:	Not Used		
Item 5:	Not Used		
Item 6:	Comprehensive Component Monitoring	Not Supported	Supported
Item 7:	Fuel System Monitoring	Not Supported	Supported
Item 8:	Misfire Monitoring	Not Supported	Supported

BMP-08-2 Trip Evaluation Supported (PRN 1002)

		0	1
Length:	8 bits		
Item 1:	EGR System	Not Supported	Supported
Item 2:	Oxygen Sensor Heater	Not Supported	Supported
Item 3:	Oxygen Sensor	Not Supported	Supported
Item 4:	A/C System Refrigerant	Not Supported	Supported
Item 5:	Secondary Air System	Not Supported	Supported
Item 6:	Evaporative Purge System	Not Supported	Supported
Item 7:	Heated Catalyst	Not Supported	Supported
Item 8:	Catalyst	NotSupported	Supported

BMP-08-3 Trip Evaluation Complete (PRN 1004)

		0	1
Length:	8 bits		
Item 1:	EGR System	Test Complete	Test Not Complete
Item 2:	Oxygen Sensor Heater	Test Complete	Test Not Complete
Item 3:	Oxygen Sensor	Test Complete	Test Not Complete
Item 4:	A/C System Refrigerant	Test Complete	Test Not Complete
Item 5:	Secondary Air System	Test Complete	Test Not Complete
Item 6:	Evaporative Purge System	Test Complete	Test Not Complete
Item 7:	Heated Catalyst	Test Complete	Test Not Complete
Item 8:	Catalyst	Test Complete	Test Not Complete

BMP-08-4 Fuel System Status (PRN 1008 & 1009)

		0	1
Length:	8 bits		
Item 1:	Reserved		
Item 2:	Reserved		
Item 3:	Reserved		
Item 4:	Closed Loop, Faulty O_2 Sensor	False	True
Item 5:	Open Loop, Detected Fault	False	True
Item 6:	Open Loop, Driving Conditions	False	True
Item 7:	Closed Loop, Using O_2 Sensor	False	True
Item 8:	Open Loop, Not Ready for Closed	False	True

BMP-08-5 Commanded Secondary Air Status (PRN 0012)

		0	1
Length:	8 bits		
Item 1:	Reserved		
Item 2:	Reserved		
Item 3:	Reserved		
Item 4:	Reserved		
Item 5:	Reserved		
Item 6:	Atmosphere/Off	Not Supported	Supported
Item 7:	Downstream - First Catalyst	Not Supported	Supported
Item 8:	Upstream - First Catalyst	Not Supported	Supported

BMP-08-6 O_2 Sensor Location (PRN 0013)

		0	1
Length:	8 bits		
Item 1:	Bank 2 - Sensor 4 (B4 - S2)	Not Present	Present
Item 2:	Bank 2 - Sensor 3 (B4 - S1)	Not Present	Present
Item 3:	Bank 2 - Sensor 2 (B3 - S2)	Not Present	Present
Item 4:	Bank 2 - Sensor 1 (B3 - S1)	Not Present	Present
Item 5:	Bank 1 - Sensor 4 (B2 - S2)	Not Present	Present
Item 6:	Bank 1 - Sensor 3 (B2 - S1)	Not Present	Present
Item 7:	Bank 1 - Sensor 2	Not Present	Present
Item 8:	Bank 1 - Sensor 1	Not Present	Present

BMP-32-1 PIDs Supported 01h - 20h (PRN 0000)

		0	1
Length:	32 bits		
Byte 1:			
Item 1 - 8:	PID 01h - 08h Supported	Not Supported	Supported
Byte 2:			
Item 9 - 16:	PID 09h - 10h Supported	Not Supported	Supported
Byte 3:			
Item 17 - 24:	PID 11h - 18h Supported	Not Supported	Supported
Byte 4:			
Item 25 - 32:	PID 19h - 20h Supported	Not Supported	Supported

BMP-32-2   PIDs Supported 21h - 40h (PRN 0020)

		0	1
Length:	32 bits		
Byte 1: Item 1 - 8:	Byte 1: PID 21h - 28h Supported	Not Supported	Supported
Byte 2: Item 9 - 16:	Byte 2: PID 29h - 30h Supported	Not Supported	Supported
Byte 3: Item 17 - 24:	Byte 3: PID 31h - 38h Supported	Not Supported	Supported
Byte 4: Item 25 - 32:	Byte 4: PID 39h - 40h Supported	Not Supported	Supported

BMP-32-3   PIDs Supported 41h - 60h (PRN 0040)

		0	1
Length:	32 bits		
Byte 1: Item 1 - 8:	PID 41h - 48h Supported	Not Supported	Supported
Byte 2: Item 9 - 16:	PID 49h - 50h Supported	Not Supported	Supported
Byte 3: Item 17 - 24:	PID 51h - 58h Supported	Not Supported	Supported
Byte 4: Item 25 - 32:	PID 59h - 60h Supported	Not Supported	Supported

7.3 Bit Mapped With Mask Bytes (BMM) SLOTs—Bit mapped with mask (BMM) SLOTs are used to encode data that typically contains several binary parameters, such as status bits or flags, grouped into a single byte or several bytes. Bit mapped SLOTs can also be used for discrete output control such as warning lamps where each bit would indicate the state of a particular lamp. The data in these bit mapped SLOTs are always followed by a MASK byte which is used to indicate which bits of the data byte are valid. There can be up to 4 bytes of data including the mask bytes. Valid combinations include up to 2 data bytes each with mask.

7.3.1 BIT MAPPED WITH MASK PARAMETER ASSIGNMENTS—None Defined

7.4 Unsigned Numeric (UNM) SLOTs—Unsigned numeric (UNM) SLOTs are used to encode data that is typically associated with information such as temperature, speed, or percent. The SLOT can be 8, 16, 24, ... 56 bits in length (1 to 7 bytes) and may or may not have an offset. Unsigned numeric SLOTs can also be used for sequential data such as month (1 - 12) or day of month (1 - 31). Each SLOT definition contains a field for: resolution per bit; minimum and maximum value; and transfer function. The transfer function defines the relationship between computer units (N) in decimal, and engineering units (E) of the data.

7.4.1 UNSIGNED NUMERIC VARIABLE ASSIGNMENTS—The unsigned numeric variables have been grouped as follows:
- Table 21 - ZEROs SLOT Assignments
- Table 22 - UNM-xx, Short (< 8 Bit) SLOTs
- Table 23 - UNM-08, 8 Bit SLOTs
- Table 24 - UNM-16, 16 Bit SLOTs
- Table 25 - UNM-24, 24 Bit SLOTs
- Table 26 - UNM-32, 32 Bit SLOTs Assignments

TABLE 21—ZEROs SLOT ASSIGNMENTS

Slot #	Description	Length
UNM-01-0	Zero (0)	1 Bit
UNM-02-0	Zero (0)	2 Bits
UNM-03-0	Zero (0)	3 Bits
UNM-04-0	Zero (0)	4 Bits
UNM-05-0	Zero (0)	5 Bits
UNM-06-0	Zero (0)	6 Bits
UNM-07-0	Zero (0)	7 Bits
UNM-08-0	Zero (0)	8 Bits
UNM-16-0	Zero (0)	16 Bits
UNM-24-0	Zero (0)	24 Bits
UNM-32-0	Zero (0)	32 Bits
UNM-40-0	Zero (0)	40 Bits
UNM-48-0	Zero (0)	48 Bits
UNM-56-0	Zero (0)	56 Bits

TABLE 22 - UNM-xx, SHORT (<8 BIT) SLOTs

Slot #	Scaling (Resolution; 1 Bit =)	Minimum Limit	Maximum Limit	Invalid Range	Transfer Function N =	Transfer Function E =	Comment
01-1	1	0	1	-	E	N	
02-1	1	0	3	-	E	N	
03-1	1	0	7	-	E	N	
04-1	1	0	15	-	E	N	
05-1	1	0	31	-	E	N	
06-1	1	0	63	-	E	N	
07-1	1	0	127	-	E	N	

TABLE 23 - UNM-08, 8 BIT SLOTs

Slot #	Scaling (Resolution; 1 Bit =)	Minimum Limit	Maximum Limit	Invalid Range	Transfer Function N =	Transfer Function E =	Comment
1	1/8000	0	0.031875	-	E * 8000	N / 8000	=0.000125
2	1/4000	0	0.06375	-	E * 4000	N / 4000	=0.00025
3	1/2000	0	0.1275	-	E * 2000	N / 2000	=0.0005
4	1/1000	0	0.255	-	E * 1000	N / 1000	=0.001
5	1/511	0	0.499	-	E * 511	N / 511	=0.001957
6	1/255	0	1	-	E * 255	N / 255	=0.003922
7	1/100	0	+1.27	-	(E+1.28) * 100	(N / 100) -1.28	=0.01 w/ offset
11	1/200	0	1.275	-	E * 200	N / 200	=0.005
15	1/100	0	2.55	-	E * 100	N / 100	=0.01
21	1/64	0	3.98	-	E * 64	N / 64	=0.0156
26	1/32	0	7.969	-	E * 32	N / 32	=0.03125
31	1/25	0	10.2	-	E * 25	N / 25	=0.04
32	1/16	0	15.94	-	E * 16	N / 16	=0.0625
41	1/10	0	25.5	-	E * 10	N / 10	=0.1
51	1/4	0	63.75	-	E * 4	N / 4	=0.25
55	1/3	0	85	-	E * 3	N / 3	=0.333333
61	100/255	0	100	-	E * 2.55	N / 2.55	=0.39215
71	1/2	0	127.5	-	E * 2	N / 2	=0.5
72	1/2	-64	+63.5	-	(E+64) * 2	(N / 2) -64	=0.5 w/ offset
73	1/2	-40	+87.5	-	(E+40) * 2	(N / 2) -40	=0.5 w/ offset
76	2/3	0	170	-	E * 1.5	N / 1.5	=0.666667
77	2/3	0	100	100.3 to 170	E * 1.5	N / 1.5	=0.666667 w/ limits
81	3/4	0	191.25	-	E / 0.75	N * 0.75	=0.75
82	3/4	-90 (00h)	+90 (F0h)	90.75 (F1h - FFh)	(E+90) /0.75	(N*0.75) -90	=0.75 w/ offset
91	100/128	0	199.22	-	E * 1.28	N / 1.28	=0.78125
92	100/128	-100	+99.22	-	(E+100) * 1.28	(N/1.28) -100	=0.78125 w/ Offset
101	1	0	255	-	E	N	=1
102	1	-40	+215	-	E + 40	N - 40	=1 w/ Offset
104	1	-128	+127	-	E + 128	N - 128	=1 w/ Offset
111	3/2	0 (00h)	360 (F0h)	361.5 (F1h - FFh)	E / 1.5	N * 1.5	-
121	2	0	510	-	E / 2	N * 2	-
125	5/2	0	637.5	-	E * 2 / 5	N * 5 / 2	-
131	3	0	765	-	E / 3	N * 3	-
141	4	0	1020	-	E / 4	N * 4	-
151	8	0	2040	-	E / 8	N * 8	-
155	10	0	2550	-	E / 10	N * 10	-
159	14	0	3570	-	E / 14	N * 14	-
161	16	0	4080	-	E / 16	N * 16	-
165	25	0	6375	-	E / 25	N * 25	-
171	32	0	8160	-	E / 32	N * 32	-
181	64	0	16320	-	E / 64	N * 64	-
185	100	0	25500	-	E / 100	N * 100	-
191	128	0	32640	-	E / 128	N * 128	-
201	256	0	65280	-	E / 256	N * 256	-
211	512	0	130560	-	E / 512	N * 512	-
221	1024	0	261120	-	E / 1024	N * 1024	-
231	2048	0	522240	-	E / 2048	N * 2048	-
241	4096	0	1044480	-	E / 4096	N * 4096	-

TABLE 24—UNM-16, 16 BIT SLOTs

Slot #	Scaling (Resolution; 1 Bit =)	Minimum Limit	Maximum Limit	Invalid Range	Transfer Function N =	Transfer Function E =	Comment
2	1/256	0	255.99	-	E * 256	N / 256	=0.003906
3	1/256	-70	185.99	-	(E + 70) * 256	(N / 256) - 70	=0.003906 w/ offset
5	1/128	0	511.99	-	E * 128	N / 128	=0.007813
8	1/64	0	1023.984	-	E * 64	N / 64	=0.015625
11	1/100	0	655.35	-	E * 100	N / 100	=0.01
21	1/10	0	6553.5	-	E * 10	N / 10	=0.1
31	1/4	0	16383.75	-	E * 4	N / 4	=0.25
41	1	0	65535	-	E	N	-

TABLE 25—UNM-24, 24 BIT SLOTs

Slot #	Scaling (Resolution; 1 Bit =)	Minimum Limit	Maximum Limit	Invalid Range	Transfer Function N =	Transfer Function E =	Comment
11	1/10	0	1677721.5	-	E * 10	N / 10	=0.1
21	128/8000	0	268435.44	-	(E/128) * 8000	(N * 128) / 8000	=0.016
31	128/4000	0	536870.88	-	(E/128) * 4000	(N * 128) / 4000	=0.032
41	1/64	0	262143.98	-	E * 64	N / 64	=0.01563

TABLE 26—UNM-32, 32 BIT SLOTs

Slot #	Scaling (Resolution; 1 Bit =)	Minimum Limit	Maximum Limit	Invalid Range	Transfer Function N =	Transfer Function E =	Comment
11	1/8000	0	536870.9	-	E * 8000	N / 8000	=0.000125
21	1/4000	0	1073741.8	-	E * 4000	N / 4000	=0.00025
31	1/64	0	67108864	-	E * 64	N / 64	=0.01563

7.5 2's Complement Signed Numeric (SNM) SLOTs—Signed numeric (SNM) SLOTs are represented in 2's complement notation. If the most significant bit of the number is set (one, 1), then the number is negative and the absolute value of the number is found by taking the 2's complement of the number. The 2's complement is found by inverting each bit of the number and then adding a binary one (1) to the result. For example, the number FFh, which has its most significant bit set, corresponds to -1.

7.5.1 2'S COMPLEMENT SIGNED NUMERIC VARIABLE ASSIGNMENTS—The 2's complement signed numeric variables are grouped as follows:
- Table 27 - SNM-08, 8 Bit SLOTs

TABLE 27—SNM-08, 8 BIT SLOTs

Slot #	Scaling (Resolution; 1 Bit =)	Minimum Limit	Maximum Limit	Invalid Range	Transfer Function N =	Transfer Function E =	Comment
11	1/2	-64	+63.5	-	E * 2	N / 2	=0.5
21	1	-128	+127	-	E	N	-
41	4	-512	+508	-	E / 4	N * 4	-
51	3/2	-192	+190.5	-	E * 2/3	N * 3/2	-
61	6	-768	+762	-	E / 6	N * 6	-

7.6 State Encoded (SED) SLOTs—State encoded (SED) SLOTs are used for data that can take one of several states such as Day of Week or Wiper Mode. Each SLOT definition contains a field for describing each state within the SLOT. There can be between 1 and 8 bits in any given SLOT with 2^n possible states where n is the number of bits in the SLOT.

7.6.1 STATE ENCODED VARIABLE ASSIGNMENTS

SED-02-1   Sub-System Category Reference Letter (PRN 1005)
Length: 2 bits (0 - 3)
0   "P" = Powertrain
1   "C" = Chassis
2   "B" = Body
3   "U" = Undefined

SED-02-2   Most Significant Digit of Trouble Code (PRN 1006)
Length: 2 bits (0 - 3)
0   "0"
1   "1"
2   "2"
3   "3"

SED-02-3   Transmission Lock-up Status (PRN 1807)
Length: 2 bits (0 - 3)
0   Unlock
1   Partial Lock
2   Full Lock
3   Invalid

SED-04-1   Day of the Week (PRN 6012)
Length: 4 bits (0 - F)
0   Unknown
1   Sunday
2   Monday
3   Tuesday
4   Wednesday
5   Thursday
6   Friday
7   Saturday
8 - F   Invalid

SED-04-2   Month (PRN 6010)
Length: 4 bits (0 - F)
0   Unknown
1   January
2   February
3   March
4   April
5   May
6   June
7   July
8   August
9   September
A   October
B   November
C   December

	D - F	Invalid

SED-06-1   Transmission Gear Engaged (PRN 1808)
- Length: 6 bits (00 - 3F)
- 00   Neutral
- 01   Reverse
- 02   Forward 1
- 04   Forward 2
- 08   Forward 3
- 10   Forward 4
- 20   Forward 5
- Others   Invalid

SED-08-1   Wiper Mode (PRN A003)
- Length: 8 bits (00 - FF)
- 00   Invalid
- 01   Off
- 02   Intermittent
- 03   Low Speed
- 04   Medium Speed
- 05   High Speed
- 06   Pulse
- 07 - FF   Invalid

SED-08-2   Day of the Week (PRN 6013)
- Length: 8 bits (00 - FF)
- 00   Unknown
- 01   Sunday
- 02   Monday
- 03   Tuesday
- 04   Wednesday
- 05   Thursday
- 06   Friday
- 07   Saturday
- 08 - FF   Invalid

SED-08-3   Day of Month (PRN 6014)
- Length: 8 bits (00 - FF)
- 00   Invalid
- 01   1st
- 02   2nd
- 03   3rd
- 04   4th
- 05   5th
- 06   6th
- 07   7th
- 08   8th
- 09   9th
- 0A   10th
- 0B   11th
- 0C   12th
- 0D   13th
- 0E   14th
- 0F   15th
- 10   16th
- 11   17th
- 12   18th
- 13   19th
- 14   20th
- 15   21st
- 16   22nd
- 17   23rd
- 18   24th
- 19   25th
- 1A   26th
- 1B   27th
- 1C   28th
- 1D   29th
- 1E   30th
- 1F   31st
- 20 - FF   Invalid

SED-08-4   Transmission Range (PRN 1806, 1809, 180D, & 180E)
- Length: 8 bits (00 - FF)
- 00   Invalid
- 01   Reverse
- 02   Forward 1
- 04   Forward 2
- 08   Forward 3
- 10   Forward 4
- 20   Forward 5
- 40   Forward 6/Park
- 80   Neutral
- Others   Invalid

SED-08-5   Ignition Switch Position (PRN 1047)
- Length: 8 bits (00 - FF)
- 00   Invalid
- 01   ACC
- 02   OFF/LOCK
- 04   OFF/UNLOCK
- 08   RUN
- 10   START
- Others   Invalid

SED-08-6   Transmission Transfer Case (4WD) (PRN 180A)
- Length: 8 bits (00 - FF)
- 00   Invalid
- 01   Neutral
- 02   Two Wheel Drive
- 03   Four Wheel Drive - Low
- 04   Four Wheel Drive - High
- 05 - FF   Invalid

SED-08-07   Door Lock Cylinder State (PRN C002)
- Length: 8 bits (00 - FF)
- 00   Key Out
- 01   Key In Lock Position
- 02   Key In Unlock Position
- 03 - FF   Invalid

7.7 ASCII Encoded (ASC) SLOTs—ASCII (ASC) SLOTs are used to encode ASCII data. The least significant 7 bits represent the standard ASCII codes from 0 to 127. The most significant bit is reserved at this time but may be assigned a special function in the future. Several ASCII characters can be included in a single message by placing up to 7 ASCII characters sequentially in the data field. This allows a character string of up to 7 characters to be sent in a single message.

The ASCII character set is included in Table 27.

7.7.1 ASCII ENCODED VARIABLE ASSIGNMENTS

ASC-08-1   One ASCII Character
- Length: 8 bits
- Byte 1: One ASCII Character

ASC-16-1   Two ASCII Characters
- Length: 16 bits
- Byte 1: Left Most Character
- Byte 2: Right Most Character

ASC-24-1   Three ASCII Characters
- Length: 24 bits
- Byte 1: Left Most Character
- Byte 2: Middle Character
- Byte 3: Right Most Character

ASC-32-1   Four ASCII Characters
- Length: 32 bits
- Byte 1: Left Most Character
- Byte 2: Middle Left Character
- Byte 3: Middle Right Character
- Byte 4: Right Most Character

ASC-40-1   Five ASCII Characters
  Length: 40 bits
  Byte 1:   Left Most Character
  Byte 2:   Middle Left Character
  Byte 3:   Middle Character
  Byte 4:   Middle Right Character
  Byte 5:   Right Most Character

ASC-48-1   Six ASCII Characters
  Length: 48 bits
  Byte 1:   Left Most Character
  Byte 2:   Middle Left Character
  Byte 3:   Middle Middle Left Character
  Byte 4:   Middle Middle Right Character
  Byte 5:   Middle Right Character
  Byte 6:   Right Most Character

ASC-56-1   Seven ASCII Characters
  Length: 56 bits
  Byte 1:   Left Most Character
  Byte 2:   Middle Left Character
  Byte 3:   Middle Middle Left Character
  Byte 4:   Middle Character
  Byte 5:   Middle Middle Right Character
  Byte 6:   Middle Right Character
  Byte 7:   Right Most Character

7.7.2 ASCII Character Set—The conversion chart, Table 28, can be used to convert from a two-digit (one byte) hexadecimal number to an ASCII character or from an ASCII character to a two-digit hexadecimal number.

TABLE 28—ASCII CONVERSION CHART

Bits 0 to 3 (LSB)	Bits 4 to 6 / First Hex Digit (MSB)							
	0	1	2	3	4	5	6	7
0	NUL	DLE	SP	0	@	P	`	p
1	SOH	DC1	!	1	A	Q	a	q
2	STX	DC2	"	2	B	R	b	r
3	ETX	DC3	#	3	C	S	c	s
4	EOT	DC4	$	4	D	T	d	t
5	ENQ	NAK	%	5	E	U	e	u
6	ACK	SYN	&	6	F	V	f	v
7	BEL	ETB	'	7	G	W	g	w
8	BS	CAN	(	8	H	X	h	x
9	HT	EM	)	9	I	Y	i	y
A	LF	SUB	*	:	J	Z	j	z
B	VT	ESC	+	;	K	[	k	{
C	FF	FS	,	<	L	\	l		
D	CR	GS	-	=	M	]	m	}
E	SO	RS	.	>	N	^	n	~
F	SI	US	/	?	O	_	o	DEL

7.8 Binary Coded Decimal (BCD) SLOTs—BCD SLOTs are similar to ASCII SLOTs and are used to encode BCD data. The upper nibble is used to represent the most significant digit in a two-digit number and the lower nibble the least significant digit. For example, the data value 38h would represent 38 decimal. Invalid data ranges include any value that has either nibble greater than 9 (i.e., A, B, ..., F in either nibble). Therefore, a valid value can represent a number from 0 to 9. Several BCD numbers can be included in a single message by placing up to 14 BCD digits sequentially in the data field. This allows any number from 0 to 99,999,999,999,999 to be sent in a single message.

7.8.1 Binary Coded Decimal (BCD) Variable Assignments

BCD-04-1   1 digit
  Length: 4 bits
  Scaling (Resolution): 1 bit = 1 (BCD)
  Limit:
    Minimum: 0
    Maximum: 9

BCD-08-1   2 digits
  Length: 8 bits
  Scaling (Resolution): 1 bit = 1 (BCD)
  Limit:
    Minimum: 00
    Maximum: 99

BCD-12-1   3 digits
  Length: 12 bits
  Scaling (Resolution): 1 bit = 1 (BCD)
  Limit:
    Minimum: 000
    Maximum: 999

BCD-16-1   4 digits
  Length: 16 bits
  Scaling (Resolution): 1 bit = 1 (BCD)
  Limit:
    Minimum: 0000
    Maximum: 9999

BCD-20-1   5 digits
  Length: 20 bits
  Scaling (Resolution): 1 bit = 1 (BCD)
  Limit:
    Minimum: 00000
    Maximum: 99999

BCD-24-1   6 digits
  Length: 24 bits
  Scaling (Resolution): 1 bit = 1 (BCD)
  Limit:
    Minimum: 000000
    Maximum: 999999

BCD-28-1   7 digits
  Length: 28 bits
  Scaling (Resolution): 1 bit = 1 (BCD)
  Limit:
    Minimum: 0000000
    Maximum: 9999999

BCD-32-1   8 digits
  Length: 32 bits
  Scaling (Resolution): 1 bit = 1 (BCD)
  Limit:
    Minimum: 00000000
    Maximum: 99999999

BCD-36-1   9 digits
  Length: 36 bits
  Scaling (Resolution): 1 bit = 1 (BCD)
  Limit:
    Minimum: 000000000
    Maximum: 999999999

BCD-40-1   10 digits
  Length: 40 bits
  Scaling (Resolution): 1 bit = 1 (BCD)
  Limit:
    Minimum: 0000000000
    Maximum: 9999999999

BCD-44-1   11 digits
  Length: 44 bits
  Scaling (Resolution): 1 bit = 1 (BCD)
  Limit:
    Minimum: 00000000000
    Maximum: 99999999999

BCD-48-1   12 digits
  Length: 48 bits
  Scaling (Resolution): 1 bit = 1 (BCD)
  Limit:
    Minimum: 000000000000
    Maximum: 999999999999

BCD-52-1   13 digits
Length: 52 bits
Scaling (Resolution): 1 bit = 1 (BCD)
Limit:
               Minimum: 0000000000000
               Maximum: 9999999999999

BCD-56-1   14 digits
Length: 56 bits
Scaling (Resolution): 1 bit = 1 (BCD)
Limit:
               Minimum: 00000000000000
               Maximum: 99999999999999

7.9 Signed Floating Point [Scientific Notation] (SFP) SLOT—The Signed Floating Point (SFP) SLOT is used to encode data that needs to be represented in floating point arithmetic, and always includes a leading sign character. The format exactly follows the ANSI/IEEE Standard (Std 754-1985) Single format. Please note that the data byte boundaries of the transmitted frame do not align with the boundaries of this format. The floating point parameter is sent as a 32 bit (4 byte) variable. The bit order is shown in Figure 14.

7.9.1 SIGNED FLOATING POINT VARIABLE ASSIGNMENT

SFP-32-1   Floating Point

	1 bit	8 bits	23 bits	
...	Sign Bit	Exponent	Fractional Part	...
	MSB		LSB	

FIGURE 14—SIGNED FLOATING POINT (SCIENTIFIC NOTATION) (SFP) SLOT

8. Multiple Frame, Single Parameter Format—This section defines the method of encoding long parameters that require more than one frame to complete. The first data byte for this format is a sequence number as shown in Figure 15. The sequence byte consists of an upper nibble that identifies this frame by number in the sequence. The lower nibble is the total number of frames that make up this grouping. The sequence byte is followed by up to the maximum number of data bytes allowed by SAE J1850, for each frame. The entire parameter is thus built up from the total number of frames shown in the sequence tracking byte, in the order carried in that byte. The data field format will thus have an order indicated by Figure 16.

A simple example of this format is shown in Figures 17, 18, and 19, sending an ASCII string of characters.

	MSB	LSB	
...	Frame # (4 bits)	# of Frames (4 bits)	...

FIGURE 15—SEQUENCE TRACKING BYTE

	Sequence Byte	Data Byte 1	Data Byte 2	Data Byte 3	
...					...

FIGURE 16—MULTIPLE FRAME, SINGLE PARAMETER FIELD

	Sequence 1 of 3	54h T	68h h	65h e	20h SP	51h Q	75h u	69h i	
...									...

FIGURE 17—FIRST SEQUENCE FRAME

	Sequence 2 of 3	63h c	6Bh k	20h SP	42h B	72h r	6Fh o	77h w	
...									...

FIGURE 18—SECOND SEQUENCE FRAME

	Sequence 3 of 3	6Eh n	20h SP	54h T	72h r	6Fh o	75h u	74h t	
...									...

FIGURE 19—LAST SEQUENCE FRAME

CLASS C APPLICATION REQUIREMENT CONSIDERATIONS—SAE J2056/1 JUN93

SAE Recommended Practice

Report of the SAE Vehicle Network for Multiplexing and Data Communications Standards Committee approved June 1993.

Foreword—Three classes of vehicle communications have been defined by the SAE Vehicle Networking for Multiplexing & Data Communications Standards Committee. One of those classes, Class C applications, represents those communications which are intended for real-time control systems such as engine controls and anti-lock brakes in order to facilitate distributed control and further reduce vehicle wiring. The requirements for these applications are different from those required for either Class A or Class B applications. This paper describes those requirements specific to a Class C application. An example system is provided for consistency of discussion.

Table of Contents

1. Scope
1.1 Background
2. References
2.1 Applicable Document
2.1.1 SAE Publication
2.1.2 Other Documents
3. Example Description
4. Application Requirements
4.1 Regularity of Information Transfer
4.2 Performance
4.3 Dependability
4.4 Impact on Vehicle System
4.5 Physical Implementation
4.5.1 Openness
4.5.2 Level of Hardware Implementation
4.5.3 Transmission Media and EMC Considerations
5. Summary
Appendix A Reasons For Standardization

1. Scope—This SAE Recommended Practice will focus on the requirements of Class C applications. The requirements for these applications are different from those required for either Class A or Class B applications. An overall example system is provided for consistency of discussion.

Figure 1 is a block diagram of a typical network node. The hardware and software for both the communication interface and the application itself are shown. For the purposes of this discussion the communication hardware and software will be considered as the communication interface. This paper will discuss the requirements for the communication interface (both hardware and software) without necessarily determining whether it will be accomplished by hardware or software. This choice is left as a subsequent trade-off. Thus requirements are presented from the perspective of the application.

The examples provided are for discussion purposes only and are in no way intended to be an endorsement or recommendation of how a specific application should be designed.

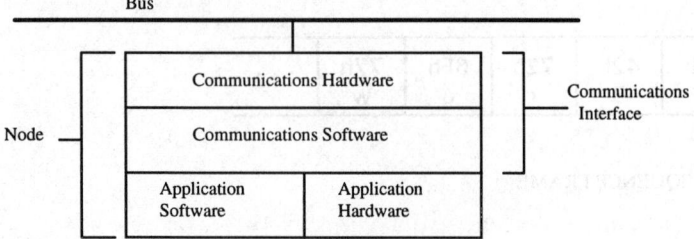

FIGURE 1—TYPICAL NODE BLOCK DIAGRAM

1.1 Background—Three classes of vehicle communications have been identified by the SAE Vehicle Networking for Multiplexing & Data Communications Standards Committee. These classes are defined as follows:

a. Class A—A potential multiplex system usage whereby vehicle wiring is reduced by the transmission and reception of multiple signals over the same signal bus between nodes that would have been accomplished by individual wires in a conventionally wired vehicle. The nodes used to accomplish multiplexed body wiring typically did not exist in the same or similar form in a conventionally wired vehicle.

b. Class B—A potential multiplex system usage whereby data is transferred between nodes to eliminate redundant sensors and other system elements. The nodes in this form of a multiplex system typically already existed as stand-alone modules in a conventionally wired vehicle.

c. Class C—A potential multiplex system usage whereby high data rate signals typically associated with real-time control systems, such as engine controls and anti-lock brakes, are sent over the signal bus to facilitate distributed control and further reduce vehicle wiring.

These three classes describe the various applications of communication that are anticipated to exist within a vehicle. Each class is intended to be able to support the lower level classes of applications also. That is, Class A systems are designed for basic low level switch multiplexing. Class B introduces the aspect of parametric data sharing while still providing for Class A applications. Class C introduces the aspect of real-time closed-loop feedback machine control but still allows Class B and Class A tasks to be performed. Issues such as cost, reliability, and performance will determine which link or combination of links are most appropriate for a given application.

It is believed there are significant benefits available to the automotive, component, and semiconductor manufacturers in developing a standard Class C communication network. The work performed toward standardization of the Class B network has provided insight into the magnitude of and potential methods for this effort, and has shown that this is a significant undertaking—one which must be initiated early. A discussion of the benefits of standardization are presented in Appendix A.

2. References

2.1 Applicable Document—The following publication forms a part of this specification to the extent specified herein. The latest issue of SAE publications shall apply.

2.1.1 SAE PUBLICATION—Available from SAE, 400 Commonwealth Drive, Warrendale, PA 15096-0001.

SAE J1850—Class B Data Communication Network Interface

2.1.2 OTHER DOCUMENTS

Ford Motor Company, "ETX-I Final Report," Volume I

Patil, P. B., et. al., "Electric Transaxle System Design for an Advanced Electric Vehicle Powertrain," EVC Expo 83, Paper No. 8324, Dearborn, Michigan, October, 1983

Bates, B., et. al., "A Vehicle Control System for an Electric Vehicle," 19th IECEC, Paper No. 849441, San Francisco, California, August, 1984

Landman, R. G., et. al., "Control System Architecture for an Advanced Electric Vehicle Powertrain," SAE Paper No. 871552, Future Transportation Technology Meeting, Seattle, Washington, August, 1987

3. Example Description—To clarify the communication requirements of a distributed control system, an electric vehicle drive- and brake-by-wire system is described. One version of the system was implemented in an advanced electric vehicle powertrain called ETX-I (ETX-Electric Trans-Axle). The system consisted of seven modules: the vehicle controller (V/C), the inverter/motor (I/M) controller, the instrument panel display, the transmission, the traction battery, brakes, and driver inputs.

The vehicle controller is the command center of the system. It electronically interprets all driver demands by monitoring the accelerator and brake pedals and the shift lever and provides the desired wheel torque response by appropriately controlling the inverter, motor, transmission, and brake operation. It also provides fault management and diagnostics. In the ETX implementation, two dedicated serial data links were used: one between the vehicle controller and I/M controller and the other between the vehicle controller and the display. All other signals, such as between the vehicle controller and the transmission or the I/M controller and the inverter, used hard wires. However, a network-based system configuration with intelligent nodes for each major subsystem as shown in Figure 2 is feasible and is used to illustrate the communication requirements of a distributed control system.

The operation of the system can best be described with reference to Figure 3 which shows the state diagram for the ETX-I system. The circles represent states and the interconnecting arrows represent possible transition paths. The logical expressions associated with each transition arrow describe the conditions necessary for the transition to occur. Tables 1 and 2 provide a list of the states and the events responsible for state transitions. Detailed description of the propulsion system, its operation and the design of the control system can be found in references 1 to 4. What follows is a brief description of certain aspects of system operation to illustrate the relationship between system operation and communication requirements.

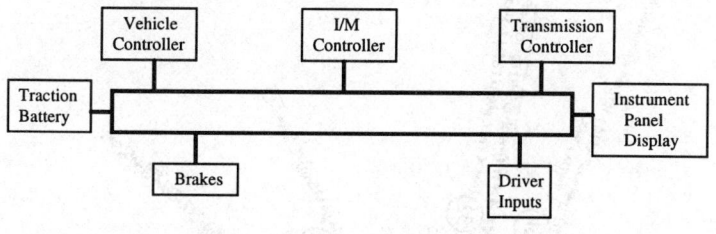

FIGURE 2—EXAMPLE BLOCK DIAGRAM

TABLE 1—ETX SOFTWARE STATE DIAGRAM DEFINITIONS

State	Definition
0	Initialize Vehicle Controller
1	Display "SHIFT NOT IN PARK OR NEUTRAL" Message
2	Close Inverter/Motor Controller Power Relay
3	Send Initialization Record to I/MC
4	Send PTR (Prepare to Run) message to I/MC
5	PARK or NEUTRAL - Output 0 torque
6	Close main contactor negative side. Start timeout for MLA.
7	Close main contactor positive side. Start timeout for MHA.
8	Send MCA
9	Send NOT(MCR) and start timeout for NOT(MCC)
10	Open Main Contactor and start timeout for NOT(MHNA)
11	Send NOT(MCA).
12	Send NOT(FRV) to I/MC
13	Energize 1st Gear Friction Clutch
14	DRIVING - Process Torque Schedule and output to I/MC continuously.
15	Start time delay 2 for regen to drive shift
16	Stop time delay 2
17	Start time delay 3 for clutch to clutch shift
18	De-energize 1st gear friction clutch
19	Energize 2nd gear clutch
20	De-energize 2nd gear clutch
21	Start timeout 1 for measured torque to equal 0
22	SHUTDOWN - Open inverter/motor controller power relay, main contactors, and clutch solenoids
23	Display diagnostic status
24	Open vehicle controller power relay
25	Send FRV to I/MC
26	Send MCR and start timeout for MCC
27	Send NOT(MCR)
28	Set motor overspeed warning flag
29	Display "CHARGER STILL PLUGGED IN" message
30	Clear motor overspeed warning flag
31	Decrement maximum power available to motor
32	Increment maximum power available to motor
33	Increase negative torque to limit motor speed
34	Decrease negative torque

State 0 in Figure 3 is the state to which the control system is initialized when the ignition key is turned on. Several messages are initiated as a result of this driver action: the vehicle controller needs to know the status of the shift lever, the friction clutches in the transmission, and the relay that locks the power on to the vehicle controller itself. If the transmission is not in "park" or "neutral," for example, the system goes to (fault) State 1 and displays an appropriate message. On the other hand, if the transmission is in "park" or "neutral," the friction clutches are disengaged and the power relay is locked on. When the key switch goes to the "start" position, the system transitions to State 2 and energizes the relay that provides power to the I/M Controller. Thus the driver action of turning the ignition key to "on" and "start" positions results in a "burst" of messages. These messages are not repetitive. However, they should have low, worst case latencies to prevent a perceptible delay between key turning and system initialization.

When the system is in State 14, the V/C is continuously sending a torque command to the I/M Controller at a 5 ms update rate and needs to receive the pedal position and the calculated torque value at the same rate. This is an example of repetitive data which must be updated at a rate fast enough to provide a smooth response. If the time required for acknowledgement of such data is a significant fraction of the update period, it may be desirable not to require an acknowledgement and use the old data until the next error-free data packet arrives. In State 14, the vehicle controller is also monitoring several other signals which are event driven and non-repetitive and whose receipt in certain combinations can lead to state transitions. These messages may have to be acknowledged to prevent faulty state transitions. Some of these messages need very short latencies to provide proper operation and fault management.

TABLE 2—ETX STATE TRANSITION EVENTS

C2A	Second Gear Friction Gear Solenoid Acknowledge
CFA	First Gear Friction Gear Solenoid Acknowledge
DEC	Rate of change of accelerator pedal position (APP) function: (dAPP/dt < -Y and (APP < X)
DRI	Gear Shift in Drive Position
FLT	Fault Detected
FRA	Forward/Reverse Acknowledge from I/MC
FRV	Forward/Reverse Request to I/MC
INR	Initialization Record Received
ICA	Inverter Power Relay Acknowledged
KSR	Key Switched On (in Run or Start Position)
KSW	Key Switched to Start Position
LOW	Gear Shift in Low Position
MCC	I/MC ready for contactor closure
MCA	Contactor acknowledge sent to I/MC
MS1	Motor Speed for Scheduled Downshift
MS2	Motor Speed for Scheduled Upshift
NEU	Gear Shift in Neutral Position
OSP	Motor Over Speed Detected
PAR	Gear Shift in Park Position
PTR	Prepare to Run Signal sent to I/MC
REV	Gear Shift in Reverse Position
SDN	Shutdown signal from Inverter Motor Controller
TD1	Time Delay 1 Timed Out (Time for Measured Torque to Reach 0)
TR1	Time Delay 1 Running
TD2	Time Delay 2 Timed Out (Time to Shift from Regen to Drive)
TR2	Time Delay 2 Running
TD3	Time Delay 3 Timed Out (Delay for clutch to Clutch Shift)
TR3	Time Delay 3 Running
TG0	Torque Request - 0 + delta
TH0	Torque Request < 0 - delta
TL0	Torque Request < 0 - delta
TQ0	0 - delta < Measured Torque < 0 + delta
VCA	Vehicle Controller Power Relay Acknowledge
VSM	Maximum Allowable Vehicle Speed for Downshift
VSR	Allowable Vehicle Speed for Shift Into Reverse
INT	Charger Interlock (charger is plugged in)
MLA	Main Contactor Neg. Contact Closed
MHA	Main Contactor Pos. Contact Closed
MLNA	Main Contactor Neg. Contact Open
MHNA	Main Contactor Pos. Contact Open
MS0	Motor Speed Equals Zero
SIP	Shift in Progress
SIN	Shift Inhibit
WRN	Motor Near Overspeed
TBV	Traction Battery Voltage
TBTM	Traction Battery Temperature
ITO	Inverter Overtemp
TOTEMP	Transaxle Overtemp

One of the ETX-I Control system state transitions is a gear shift from 1st to 2nd gear, State 19. Since the gear shift initiation and execution depend on the occurrence of certain conditions in a specific sequence within a given time interval, it is necessary for the communication system to meet requirements for data consistency, sequence, and latency. Also during this period the clutch pressure and motor speed signals need to be updated at a much faster rate (5 to 10 ms) than during normal driving in a particular gear. System design, must, therefore, address the issue of whether to allow for this high update rate and low latencies all of the time (high bandwidth) or to provide the means to dynamically change priorities and update rates.

In the ETX system the serial communication between the V/C and the I/M Controller was handled as described in the following. The controller was implemented in software. Information was passed as message packets of variable length from 3 to 7 bytes long. Each message packet began with a sync byte and ended with a checksum byte. These two features allowed the processors to determine with a high degree of confidence that they were interpreting the right string of bytes as a packet and described the information contained in the rest of the packet. Two types of data were passed. Non-periodic data, such as status information, were updated as required and had to be acknowledged by the receiving processor. If the acknowledgement was not received within a given period, the transmitting processor retransmitted the message packet. Other data were periodically updated. This type was not acknowledged. Since a packet with an incorrect checksum is ignored, the processor will continue using the previous value of the data until it is updated. This saves time on the data that are updated often (e.g., the torque command is updated every 5 ms) yet assures that valid information is always used.

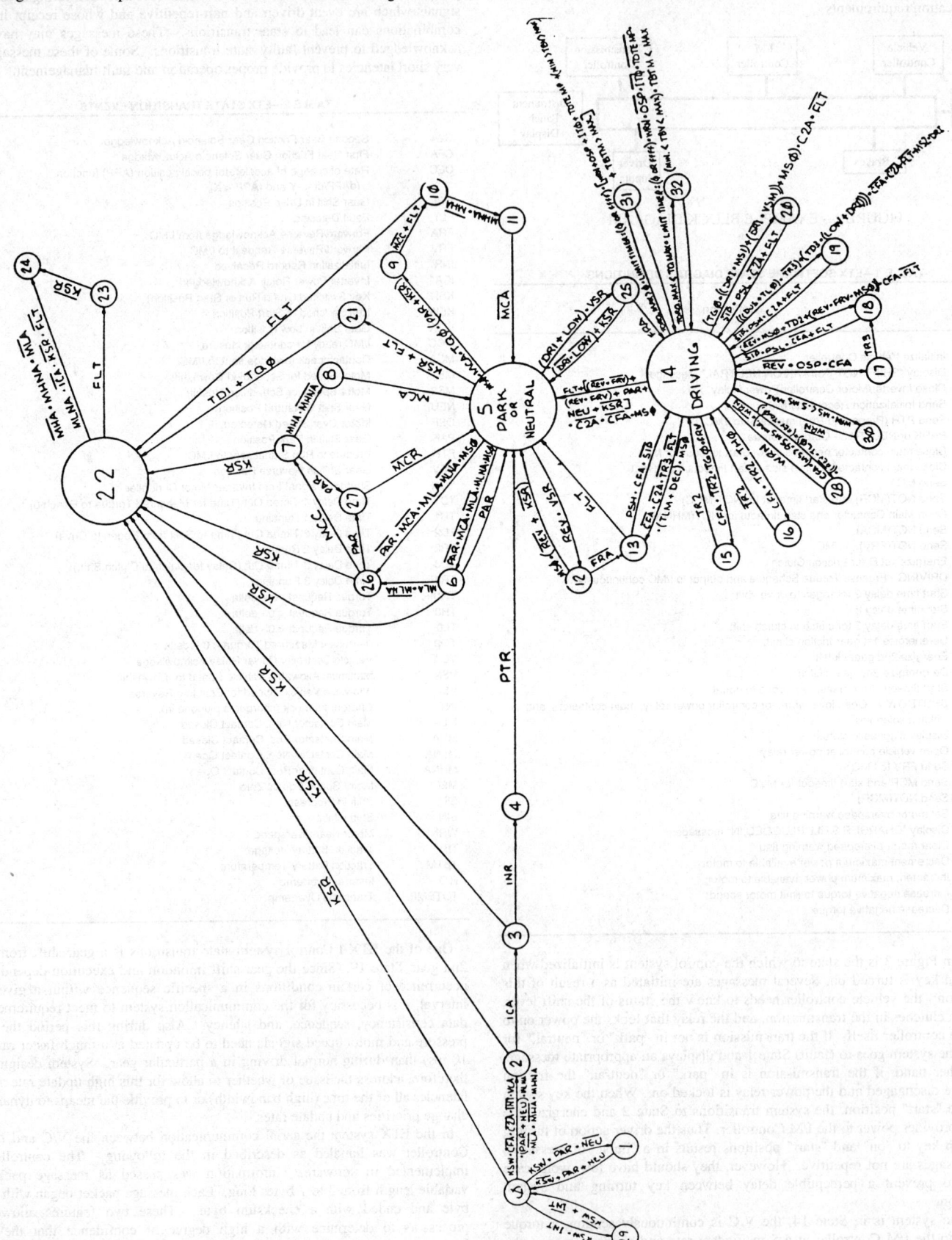

FIGURE 3—EXAMPLE STATE DIAGRAM

Table 3 provides a list of signals for the ETX-I control system. The first part of the table lists the signals which used individual wires in the ETX implementation. However, these signals could be sent using the communication bus in a fully multiplexed system. In this case, the analog signals would have to be converted to digital ones as indicated by the dual entries under the "type" column in Table 3. The second part of the table shows signals that were sent using the serial link between the vehicle and the inverter/motor controllers described previously. The signals which do not have an update rate associated with them in Table 3 are "event driven" (i.e., driver action or a change in state) and their update rate can only be described statistically. Some of the signals in Table 3 can cause a state change and therefore cause other signals to be generated as was described for the key-on signal earlier.

TABLE 3—ETX VEHICLE CONTROL SYSTEM SIGNALS

	Type	Update Interval (ms)		Symbol		From	To
Traction Battery Voltage	A/D8	100		TBV		TBat[1]	V/C
Traction Battery Current	A/D8	100		TBI		TBat	V/C
Traction Battery Temp, Avg	A/D8	1000		TBTA		TBat	V/C
Traction Battery Temp, Max	A/D8	1000		TBTM	TBat		V/C
Auxiliary Battery Voltage	A/D8	100		ABV		ABat	V/C
Auxiliary Battery Current	A/D8	100		ABI		ABat	V/C
Accelerator Position	A/D8	5		APP		Driver	V/C
Brake Pressure, Master Cylinder	A/D8	5		BPM		Brkes	V/C
Brake Pressure, Line	A/D8	5		BPL		Brkes	V/C
Transaxle Lubrication Pressure	A/D8	100		PLT		Trans	V/C
Transaxle Clutch Line Pressure	A/D8	5		PCT		Trans	V/C
Vehicle Speed	Pulse train	100		WHS		Brake	V/C
Traction Battery Ground Fault	D1	1000		TGF		TBat	V/C
Hi & Lo Contactor Open/Close	D4	-		Table 2	TBat		V/C
Key Switch Run	D1	-		KSR		Driver	V/C
Key Switch Start	D1	-		KSW		Driver	V/C
Accelerator Switch	D2	-		ASW		Driver	V/C
Brake Switch	D1	-		BSW		Brake	V/C
Emergency Brake	D1	-		PBK		Driver	V/C
Shift Lever (PRNDL)	D3	-		Table 2	Driver		V/C
Motor/Trans Over Temperature	D2	1000		TOTEMP	Trans		V/C
Speed Control	D3	-		SPC		Driver	V/C
12 V Power Acknowledge							
Vehicle Controller	D1	-		VCA		ABat	V/C
Inverter	D1	-		ICA		ABat	V/C
I/M Controller	D1	-		IMCA		ABat	V/C
Brake Mode (Parallel/Split)	D1	-		BMD		Driver	V/C
SOC Reset	D1	-		SOCR	Driver		V/C
Interlock	D1	-		INT		TBat	V/C
High Contactor Control	Pulse train	10		MHC		V/C	TBat
Low Contactor Control	Pulse train	10		MLC		V/C	TBat
Reverse & 2nd Gear Clutches	D2	-		PC1/2	V/C		Trans
Clutch Pressure Control	Pulse train	5		PC1/2	V/C		TBat
DC/DC Converter	D1	1000		DDC		V/C	TBat
DC/DC Converter Current Control	Pulse train	-		DIC		V/C	TBat
12 V Power Relays	D1	-		APC/TPC	V/C		TBat
Traction Batt. Ground Fault Test	D2	1000		TGF		V/C	Brake
Brake Solenoid	D1	-		BSL		V/C	Brake
Back Up Alarm	D1	-		BUA		V/C	Brake
Warning Lights	D7	-		-		V/C	InstPnl
Key Switch	D1	-		KSW		V/C	I/M C
Main Contactor Close	D1	-		MCC		I/M C	V/C
Torque Command	D8	5 ms		TQC		V/C	I/M C
Torque Measured	D8	5 ms		TQM		I/M C	V/C
FWD/REV	D1	-		FRA		V/C	I/M C
FWD/REV Acknowledge	D1	-		FRA		I/M C	V/C
Idle	D1	-		IDL		V/C	I/M C Shift
Inhibit	D1	-		SIN		I/M C	V/C
Shift in Progress	D1	-		SIP		V/C	I/M C
Processed Motor Speed	D8	5 ms		PMS		I/M C	V/C
Inverter Temperature Status	D2	-		ITS		I/M C	V/C
Shutdown	D1	-		SDN		I/M C	V/C
Status/Malfunction (TBD)	D8	-		SML		I/M C	V/C
Main Contactor Acknowledge	D1	-		MCA		V/C	I/M C

[1] Abbreviations:
    TBat - Traction Battery
    ABat - Auxiliary Battery
    Trans - Transmission
    A/D8 - Analog with 8 bits resolution
    D4 - Digital 4 bit value

4. Application Requirements—The requirements of Class C applications associated with the communication characteristics will be described in the following sections. This section is grouped into five major areas: (a) regularity of information transfer, (b) performance, (c) dependability, (d) system issues, and (e) implementation. Within each of these sections, the specific requirements affecting the communication network will be discussed.

4.1 Regularity of Information Transfer—As defined, Class C applications are control-oriented. By the fundamentals of control system design, there is a high level of repetitiveness inherent in the system. This occurs in particular with digital control systems where sample times and processing loops are a fundamental part of the design. Periodic events also arise from periodic physical phenomena such as crankshaft position in an engine.

For the control applications currently being considered, parameters such as coolant temperature, throttle position, engine load level, and vehicle speed are typically updated at regular intervals. This information must then be communicated to other subsystems within the vehicle. Update intervals may range from milliseconds to seconds depending on the rate of change of the data and its intended use. In the example in Section 2.0, the discussion of State 14, where V/C is continuously sending a torque command to the I/M C at 5 msec update rate and needs to receive pedal position and the calculated torque value at the same rate, is an example of repetitive data (reference Table 3).

Despite the high degree of regularity of information, it will also be necessary to accommodate the transfer of irregular information and information bursts. This includes the transfer of information associated with irregular events such as driver-initiated mode changes, failures, or command information being transferred between command modules which triggers multiple messages. An example of such information was described in Section 3. In that case, in State 0 (initialization) a burst of messages is set once at start-up. These are considered to be irregular data transmissions in this case.

It is anticipated that a large percentage of the data transferred on a Class C network will be repetitive in nature. This data will represent sampled information and values calculated in control loops. This is in contrast to a Class B application, where bursty information is more prevalent.

4.2 Performance—The performance of a communication system is typically discussed in terms of communications speed—how long it takes to get messages through the data link. The most common measures of communication speed are throughput and latency. Throughput rates are selected by the system designer within limits determined by cost and EMI considerations. Latency is the time delay between queuing a message for transmission and receipt at the destination. Latency is somewhat affected by bit rate choices but is more significantly affected by the protocol's logical structure—particularly the media access method.

The latency of data exchange may be considered from a number of different perspectives. From an application perspective, latency relates to the time delay associated with the transfer of information from one application program to another. In this respect it is necessary for the latency to be minimal (in some cases less than 1 ms) and predictable (as defined by the systems exchanging data). Predictability has to do with how the message latencies change over time—whether they follow a statistical distribution or are deterministic.

As a control system is designed, the delay resulting from information transfer is accounted for in such a way that the control algorithm is able to compensate for that fact. From this perspective the necessity of predictable latency (latency defined within a given tolerance) can be understood. Obviously, the need to minimize latency is also a function of the algorithms and acceptable limits for each application must be defined. In some cases such as chassis control systems, latencies less than 2 ms may be necessary.

Priority can be used as a means of reducing the latency associated with a particular message. In that sense, priority can be used to assure that critical messages remain within necessary latency bounds. In some cases, priority may be used to override normal operation of the system to allow certain messages to be sent immediately. Many forms of priority may be used, including dynamic alteration of the priority as a function of time in the message transmission queue. Class B systems may also utilize priority for a similar reason (in some cases it is considered an additional identifier field). In any case, the use of priorities can only improve latencies for the highest priority message at the expense of other messages. If the media access method is statistical by nature, priority methods cannot insure absolute guarantees on latencies.

For example, a message containing the vehicle speed and engine speed (rpm) may be sent at a normal priority level, where as the command from the brake system to the throttle system to reduce throttle for a given traction condition may be sent at the highest priority. An application of dynamic prioritization may be used for messages involving wheel speed. In this instance, the wheel speed message would be sent at normal priority, until the system entered a critical situation. At that time the priority for the wheel speed message might be raised to a higher level.

4.3 Dependability—Since a Class C data link is likely to be used in safety-critical systems, it must be designed so as to dependably perform its function. This need for dependability includes traditional component-oriented reliability requirements, but goes beyond this in ways that affect system reliability and safety. For example, a fault that could cause unsafe operation should be detected and the system put into a safe mode, perhaps with degraded performance, whenever possible. This implies that there should be means for defeating and recovering from communication failures such as corrupted messages, media faults, and failed nodes.

The communications protocol must be able to provide guaranteed delivery of critical messages. Some types of messages might not need this, but for those that do, various acknowledgement schemes can help. Immediate acknowledgement can be used to indicate correct receipt of a valid message. This might be automatically returned by the interface hardware itself or it might be generated by low-level communications software. In either case, the acknowledgement is based on the validity of the message framing, bit encoding, CRC codes, etc., that define syntactically correct messages.

Beyond this, the sending task may need to know that the message was successfully acted upon by the application program at the destination. In the example system, this occurs when the vehicle controller issues a torque request to the Inverter/Motor Controller and can observe the measured torque output to verify correct operation. This can be accomplished by returning status information, but may require special provisions in the protocol to ensure timely acknowledgement. If a fast acknowledgement is needed, it may not be acceptable to wait for normal media access methods. The initiating node might need to retain control of the medium and allow the destination node to transmit without arbitrating for access to the network.

The data link must also deliver the message intact. Various techniques such as message framing, checksums, and CRC fields can be used to validate messages. If a message can be sent to more than one recipient, then it is also necessary that the recognition of an invalid message at one node force all other receiving nodes to reject that message. Otherwise, data and parameters can have inconsistent values throughout the system. However, one node must not be able to coninuously invalidate messages used by other nodes. A receiving task may also need the ability to detect missed messages (e.g., sensor values).

This need for communications integrity applies to the time order of messages as well. If a sequence of messages is sent in a particular order, it should be received in that order as well. This requirement affects queuing disciplines used to buffer messages, particularly in bridge or gateway nodes. However, the same problems can arise in networks where nodes function as active relays with store-and-forward capacity. This can be accomplished by time stamping and sequence numbering the messages.

Obtaining adequate system dependability may require the use of fault tolerant computing techniques. This places special requirements on the communications interface. One such technique may be the use of special bit encoding techniques to allow the regeneration of the data in messages containing errors. The data link itself must also be fault tolerant with graceful degradation. The failure of individual nodes must not cause total loss of the communications capability. If the physical medium is damaged or severed, the remaining portions should be able to communicate within themselves.

If true fault tolerance is to be implemented, the system will need the capability to recognize faulty behavior, identify the faulty node, and reconfigure the system to provide the most acceptable performance possible. This implies that the nodes can monitor the network itself and identify other nodes that are not responding or transmitting correctly. There must then be some method for ignoring or isolating the failed nodes in order to reduce their impact on the system.

Functional reconfiguration means that the critical functions which were being performed at a node that failed, are shifted to remaining good nodes. To do this, the task-to-task communications must use functional addressing rather than physical. Additionally, it may be necessary to dynamically reconfigure the network in some cases.

4.4 Impact on Vehicle System—From the perspective of the vehicle or subsystem engineer, a data link is not a function in itself. Rather, it is a means to providing the features and functions desired for a control system. The decision to incorporate a data link is driven by the earlier decision to partition the system functionality among a number of nodes. The resulting need that data be shared between nodes having interrelated functions is the precursor that leads to adopting the use of a data link. However, the data link must also enable the system to meet the requirements that originally lead to the partitioning choice. These requirements fall into three general categories: cost, flexibility, and fault tolerance.

Clearly, any technical choices must be cost effective. The most tangible cost impact of a data link comes from the medium, the connectors (including assembly costs), and the hardware/software interfaces required to support the protocol. One immediate impact of the protocol design comes through requirements for timing accuracy and synchronization. If the timing requirements are strict enough to force use of tight-tolerance components such as crystal clocks, this will add to the cost of every node on the data link. These are easily measured, but represent only part of the costs.

The choice of a data link also impacts the overall costs of a vehicle system in less direct ways. The communications link itself places an overhead burden on the processor(s) at a node. This reduces the node's available computation capacity and may require a faster (more expensive) processor in order to meet the time-critical aspects of control. Similarly, the interface circuitry places demands on the node's electrical power.

The data link also impacts cost by affecting the design cycle. The hardware and software interfaces must be easy to use. It must support modular partitioning on the software and easy reconfiguration so that the system designer is not artificially constrained by the communications. In general, this can be achieved by making the communications link as transparent as possible to the application programs. Ideally, two tasks should be able to share data as easily over a data link as if they were running on the same processor. In particular, this implies the use of functional addresses rather than physical addresses.

The data link must also support test instrumentation used during prototype development. On the physical side, the instrumentation tools will need to be connected to the data link medium. This places extra drive loads on the electronics and may require automatic reconfiguration of the network to accommodate the instrumentation equipment as an additional node. There must be communications bandwidth to carry debugging and diagnostic messages in addition to the normal traffic. The development process may also require testing partial systems in a bench-top prototype form. This will be easier if the protocol does not require a master node in order to operate.

Similar capabilities are needed for service diagnostics. In addition, the diagnostics may need to monitor all data link traffic and verify correct operation. To do this effectively, the diagnostic equipment may need to determine the source of every message. One way to facilitate this is to have a source ID field in every message.

4.5 Physical Implementation—There are a number of issues associated with the implementation of a Class C network which may also impose requirements on any network solutions. These issues focus primarily on the physical interface between the application and the network. It is generally expected that between 2 and 30 nodes will reside on a Class C network. This number is dependent on the extent of control system distribution and the number of vehicle options incorporated. It is also expected that bus rates in excess of 100 kbits/s will be required and that they will likely reach 1 Megabit/s or greater. This, obviously, is dependent on issues discussed earlier associated with the performance requirements.

4.5.1 OPENNESS—An open system is defined as a system consisting of nodes interconnected by a common communications medium (signal bus) according to established standards, which will support temporary connections to manufacturing networks, diagnostics, and other local area networks. This may be extended to allow devices to be added or deleted to the network without significant physical alterations. Class C applications will require an open system allowing nodes to be added and deleted as freely as possible. As a minimum this will require a standard protocol for the physical and data link layers. This will allow nodes to communicate effectively on the network. It may also be desirable to establish an open interconnect (i.e., standardized interconnect) at higher levels (e.g., priority definitions, specific diagnostic messages); however, this is not essential.

4.5.2 LEVEL OF HARDWARE IMPLEMENTATION—The nature of the applications and performance requirements imply constraints on the hardware implementation. First, it is essential that the hardware implementation minimize the burden to the processor of the communication system. Thus, a controller attached to the communication bus should not be burdened with a complicated, processor-intensive hardware and/or software interface for communication with other devices connected to the bus.

Performance and application requirements discussed earlier lead to the conclusion that information transfer should occur on a message level rather than the bit level. Data rates make it awkward if not impossible to interpret bits individually and form the messages. Thus, an adequate hardware interface is necessary to collect the bits and buffer them as messages to be retrieved by the application. In addition, sufficient logic should be provided to allow consistent and common message handling where possible. The hardware and software interface to the serial bus should also be as transparent, to the host processor, as possible for ease of use. This will tend to make the level of hardware interface more substantial in comparison to other classes of applications.

4.5.3 TRANSMISSION MEDIA AND EMC CONSIDERATIONS—It is anticipated that the electromagnetic compatibility issues (emitted and susceptibility to noise interference) will be a critical design factor which will impact the media selection. As a minimum, the EMC requirements must be consistent with SAE practices and methods. EMC will be largely dependent on the bit rate of the data transfer. The impact of the EMC requirements, in turn will directly impact the transmission media selection (which will also be influenced by cost). The media must also meet the reliability requirements outlined previously. Alternatives such as twisted, shielded pair cable and fiber optics are most likely candidates.

5. Summary—The application requirements of Class C applications have been presented with regard to five major areas: (a) regularity of information transfer, (b) performance, (c) dependability, (d) impact on vehicle systems, and (e) physical implementation. The real-time control system orientation of Class C applications has a significant influence on the requirements in each of these areas.

The Class C application requirements discussed are also different from typical Class B application requirements. Information transfer associated with Class C applications tends to be more regular (periodic updates of information). Performance factors related to latency and priority must be clearly defined to establish predictable communication for controllability purposes. Class C applications are often safety-critical requiring the information transfer to be dependable. This includes issues associated with acknowledgements, message integrity, and data consistency. The communication system also has a larger influence in Class C applications with regard to the impact on the overall vehicle system and the physical implementation. These must be accomplished in ways that do not detract from the overall goals of the system.

These requirements will form the basis for evaluating protocols to determine an appropriate data communication recommendation for Class C applications. As such, each protocol considered will be evaluated for its ability to achieve each of the requirements described. Additional requirements definition will be necessary to identify the priority of the requirements such that protocols with different strengths and weaknesses can be more readily compared. Additional technical investigations may be necessary to achieve a single recommendation. These will be performed as required.

APPENDIX A
REASONS FOR STANDARDIZATION

This appendix addresses the reasons for pursuing a standard communication network for Class C applications. The benefits of standardization listed as follows must be considered in light of two factors which influence the impact the standardization will have. These are: (a) the timing of standard development relative to the status of individual corporation internal efforts and (b) the ability to share information jointly between peer competitive corporations and suppliers.

It is important to begin the standardization process early so a standard may be developed before it is required. When individual automotive manufacturers have made significant financial and resource investments on internal efforts based on their need for a product, it is difficult to merge their developments (as was learned in the early SAE J1850 activity). Beginning early on standardizing a communications network for Class C applications provides the potential to avoid this problem. A sense of awareness and urgency must be created to encourage this to occur.

The success of the standardization effort will also depend on cooperation, sharing of information and philosophies, joint efforts, and openness to others' ideas. This will require understanding information shared within the SAE Vehicle Data Networking and Multiplexing Committee, other related SAE Committees (e.g., Truck and Bus, and Construction and Agriculture applications), and world organizations. Working together on technical investigations and resolving philosophical issues will lead to the possibility of creating a better standard than might otherwise have been achieved.

Presuming these factors are achieved, the benefits to such standardization are manyfold with few or no measurable disadvantages. If, after a dedicated effort, a Class C standard is not pursued to completion, the worst that will have happened is that the representatives within SAE will have consciously and knowledgeably decided not to pursue a standard—and have reasons to support their direction. In contrast, the benefits are discussed as follows.

a. Resource savings are possible in areas where industry commonization is beneficial. Areas of vehicle design which do not offer vehicle manufacturers special opportunities to create unique, innovative products can be common among the manufacturers. One example is the serial data links. By establishing a common serial data network, resources which may have been required to design the serial network can then be used for development of other aspects of the vehicle.

b. Semiconductor companies will be able to proliferate protocol cells onto multiple architectures providing more application alternatives. With an industry standard protocol, rather than one for each vehicle manufacturer, semiconductor manufacturers may focus their efforts on proliferating one protocol onto several architectures rather than several protocols onto one architecture. This allows the system and automotive manufacturer to have a much broader selection of architectures and microcontroller configurations to use, resulting in a system more finely tuned in performance and cost.

c. Aftermarket additions will be less likely to compromise the system. Although aftermarket additions will not be a major concern with Class C applications based on the level of complexity, performance, and type of systems that are addressed in this network, in the case that aftermarket additions are pursued, it will be less likely to compromise the system. Adding a system to a network with a restricted protocol would require reverse engineering and increase opportunity for errors resulting in network failure, whereas, adding a system to a network with a standardized protocol would allow a compatible interface to be used and additionally provide better integrity through a greater number of manufacturers proving out the design.

d. Knowledge and resources will be shared. There is significant benefit in combining efforts and knowledge for a given area (as demonstrated with SAE J1850). The information shared in terms of encoding techniques, physical bus interfaces, error detection techniques, synchronization, etc., may allow definition of a better protocol than what an individual effort perhaps might attain.

e. There will potentially be cost and volume advantages. There is a potential volume benefit to cost. This would need to be measured since automotive volumes in the U.S. may be significantly high even within individual manufacturers.

f. Philosophical and technical information is available through ISO. There is an opportunity to take advantage of the significant philosophical and technical research already completed as a result of ISO investigations of data networks for Class C applications. In addition, methodologies for standardization have been developed for this effort.

SURVEY OF KNOWN PROTOCOLS—SAE J2056/2 APR93   SAE Information Report

Report of the SAE Vehicle Network for Multiplexing and Data Communication Standards Committee approved April 1993.

Table of Contents

1. Scope
1.1 Background
2. References and Definitions
2.1 Selected References
2.2 Definitions
3. Protocol Technical Summary
3.1 Aircraft Internal Time Division Command/Response Multiplex Data Bus (MIL-STD-1533)
3.2 Automotive Bit-Serial Universal-Interface System (ABUS)
3.3 Auto Local Area Network (AutoLAN)
3.4 Class B Data Communication Network Interface (SAE J1850)
3.5 Controller Area Network (CAN)
3.6 Digital Data Bus (D2B)
3.7 Ethernet (IEEE 802.3)
3.8 Joint Integrated Avionics Working Group (JIAWG)
3.9 Mini-Manufacturing Automation Protocol (Mini-MAP)
3.10 Synchronous Data Link Control/High-Level Data Link Control (SDLC/HDLC)
3.11 Token Ring (IEEE 802.5)
3.12 Token Slot Network
3.13 VAN (Vehicle Area Network)

1. Scope—This SAE Information Report is a summary comparison of existing protocols found in manufacturing, automotive, aviation, military, and computer applications which provide background or may be applicable for Class C application (see Figure 1). The intent of this report is to present a summary of each protocol, not an evaluation. This is not intended to be a comprehensive review of all applicable protocols. The form for evaluation of a protocol exists in this paper and new protocols can be submitted on this form to the committee for consideration in future revisions of this report.

This report contains a table which provides a side-by-side comparison of each protocol considered. The subsequent section provides a more detailed examination of the protocol attributes. Many of the protocols do not specify a method for one or more of the criteria. In these circumstances `user defined' or `not specified' will appear under the heading.

Certain protocol specifics or details are omitted for the sake of brevity. Every attempt is made to provide the interested reader with the necessary references for further research.

1.1 Background—Three classes of vehicle communications have been identified by the SAE Vehicle Networking Subcommittee. These classes are defined as follows:

a. Class A: A potential multiplex system usage whereby vehicle wiring is reduced by the transmission and reception of multiple signals over the same signal bus between nodes that would have been accomplished by individual wires in a conventionally wired vehicle (i.e., low-speed body wiring and control functions, for example, control of exterior lamps).

b. Class B: A potential multiplex system usage whereby data is transferred between nodes to eliminate redundant sensors and other system elements (i.e., data communications, for example, sharing of vehicle parametric data).

c. Class C: A potential multiplex system usage whereby high data rate signals typically associated with real-time control systems, such as engine controls and anti-skid brakes, are sent over the signal bus to facilitate distributed control and to further reduce vehicle wiring (i.e., high-speed real time control, for example, distributed engine control).

These three classes describe the various applications of communication that are anticipated to exist within a vehicle. Each class is intended to be able to support the lower level Class applications. That is, Class A systems are designed for basic low level switch multiplexing. Class B introduces the aspect of parametric data sharing while still providing for Class A applications. Class C introduces the aspect of real-time control but still allows Class B and Class A tasks to be performed. Issues such as cost, reliability, and bus bandwidth will determine which link or combination of links are most appropriate for a given application.

For definition of other terms used in this report, please refer to SAE J1213.

2. References and Definitions
2.1 Selected References

a. Aircraft Internal Time Division Command / Response Multiplex Data Bus (MIL-STD-1553): "MIL-STD-1553B Designer's Guide" published by ILC Data Device Corporation.

b. Automotive Bit-serial Universal-interface System (ABUS):
   (1) ABUS presentation by Volkswagen to SAE Class C Task Force 8/89.
   (2) Press Information and data sheet by Telefunken Electronic 8/88.

c. Auto Local Area Network (AutoLAN):
   (1) John D. H. Harris, Nigel M. Bailey "General Instrument AUTOLAN - A High Speed Multiplexing System for Automotive Body Wiring Systems" SAE 890541
   (2) "AUTOLAN CONTROLLER IC 616759 DRAFT 05-04-89"

	Aircraft MIL STD 1553	ABUS	AUTOLAN	CAN	DDB	ETHERNET	HDLC/SDLC	J1850	JIAWG	MINI-MAP	TOKEN RING	TOKEN SLOT DEVICE	VAN
AFFILIATION	Military Standard	VW	General Instrument	Proposed ISO Bosch	Philips	IEEE 802.3	ISO(HSLC) IBM(SDLC)	SAE J1850	SAE	IEEE	IEEE 802.5	GM	Proposed ISO
APPLICATION	Aircraft Network (fighter...)	Auto In-Vehicle	Auto In-Vehicle	Auto In-Vehicle	Audio/Video	Lab and Business	Industrial	Auto In-Vehicle	Avionics	Factory Communications	Computer	Auto In-Vehicle	Auto In-Vehicle
TRANSMISSION MEDIA	Shielded TW. Pair Fiber Optic	Single Wire	Twisted Pair	Twisted Pair Fiber Optic	Twisted Pair	"Thick" Coaxial Cable		Single Wire Twisted Pair	Fiber Optics	Twisted Pair, Coax Fiber Optics	Dual Twisted Pair	Twisted Pair Fiber Optics	Twisted Pair
BIT ENCODING	Manchester II Biphase	NRZ	Alternate Pulse Inversion	NRZ With Bit Stuffing	PWM	Manchester	NRZ With Inversion And Zero Bit Insrtn	VPW PWM	Manchester	Phase Shift Keying (PCFSK, AMPSK)	Manchester	NRZ With Bit Stuffing	Manchester
MEDIA ACCESS	Master/Slave or Token	Contention	Master/Slave	Contention	Contention	Contention	Master/Slave	Contention	Token Passing	Token Passing	Token	Token Slot	Contention
ERROR DETECTION	Parity	Bit Only	CRC	CRC	Parity	CRC	CRC	Optional CRC	CRC	Optional CRC	CRC	CRC	CRC
DATA FIELD LENGTH	2 Bytes Minimum	2 Bytes	2 Bytes	0-8 Bytes	2-128 Bytes	46-1500 Bytes	Any Number of Bytes	0-8 Bytes	32-65536 Bytes	2-256 Bytes	1-4 K Bytes	0-256 Bytes	0-8 Bytes
IN-MESSAGE ACK				Yes				Yes				Yes	
MAXIMUM BIT RATE	1 Mb/S	500 Kb/S	4 Mb/S	1 Mb/S	100 Kb/S	10 Mb/S	375 Kb/S 2.4 Mb/S	10 Kb/S 40 Kb/S	50 Mb/S	5,10,20 Mb/S	4 Mb/S 16 Mb/S	2 Mb/S	User Definable
MAXIMUM BUS LENGTH	Not Specified	Not Specified Typical 30 m	>40 m	Not Specified Typical >40 m	150 m	500 m		40 m	>40 m	>400 m	1500 m	Not Specified Typical 30 m	20 m
MAXIMUM NUMBER OF NODES	Not Specfied	Not Specified Typical 32	127	Not Specified Typical >16	50	1024		Not Specified	128	100	256	32 (Transmit Capability)	16
HARDWARE AVAILABLE	Yes	Yes	Yes	Yes	Yes	Yes	Yes	No	No	Yes	Yes	No	No

FIGURE 1—PROTOCOL DESCRIPTION

- d. Class B Data Communication Network Interface (SAE J1850): Recommended Practice J1850, Revised 9/19/88.
- e. Controller Area Network (CAN):
  - (1) Intel 1989 Automotive Handbook
  - (2) SAE Information Report J1583
- f. Digital Data Bus (D2B): Philips Single-Chip 8-Bit Microcontroller User's Guide
- g. Ethernet (IEEE 802.3):
  - (1) IEEE 802.3 Summary by M.R. Stepper of the SAE Truck and Bus Control and Communications Subcommittee
  - (2) Intel 1989 Microcommunications Handbook
  - (3) Intel Local Area Networking (LAN) Tutorial
- h. Joint Integrated Avionics Working Group (JIAWG):
  - (1) James H. Nelson, Larry T. Shafer, Daryle B. Hamlin, James J. Herrmann "A Candidate for Linear Token-Passing, High-Speed Data Bus Systems" SAE 872494
  - (2) Draft "Standard Joint Integrated Avionics Working Group Linear Token Passing Multiplex Data Bus Protocol" DOCUMENT J88-M5
  - (3) "Linear Token Passing Multiplex Data Bus Standard" Unisys Document 7340679
- i. Mini-Manufacturing Automation Protocol (Mini-MAP):
  - (1) "MiniMAP/MAP Controller/Carrierband Modem Interface" Motorola Inc. Rev. 1
  - (2) "MicroMAP1-7 Manufacturing Automation Protocol Software" Motorola Inc. Rev.1
  - (3) "IEEE 802.4 Token Bus"
- j. Synchronous Data Link Control/High-Level Data Link Control (SDLC/HDLC):
  - (1) Microcommunications Handbook, Intel, 1988
  - (2) Microprocessor, Microcontroller and Peripheral Data Volume II, Motorola, 1988
- k. Token Ring (IEEE 802.5):
  - (1) ANSI/IEEE std 802.5-1985, Token Ring Access Method
  - (2) ANSI/IEEE std 802.5-1985, Logical Link Control
- l. Token Slot Protocol: SAE Information Report J2106 "Token Slot Network for Automotive Control"
- m. Vehicle Area Network: VAN Specification, Version 1.2, ISO/TC22/SC3/WG1

2.2 Definitions—The descriptions of the protocol subsections follow. These descriptions appear in the order in which they are discussed for each protocol.

2.2.1 INTENT OF PROTOCOL SUBSECTIONS—The following describes the intent of the various subsections of the protocol characteristics.

2.2.2 APPLICATION/AFFILIATION—The application section briefly identifies the applications for which the protocol was designed to serve (e.g., military, aircraft, industrial, land vehicles, trucks). The affiliation section identifies the organization(s) that originally developed or specified the protocol or which now endorse the protocol.

2.2.3 TRANSMISSION MEDIA—The transmission media section describes the physical medium generally associated or required by the given protocol (e.g., single wire, dual (parallel) wire, twisted pair, twisted pair with shield, dual twisted pair, fiber optics).

2.2.4 PHYSICAL INTERFACE—This section describes the basic circuitry used to connect the nodes to the network. In some cases the schematic of a typical interface may be shown. In others, a reference to a generally known interface technique may be made. This section may also include additional data about aspects of the interface not readily shown. An example would be that receiver nodes synchronize to the signal from a transmitting node, or that receiver nodes adjust their receiver clock to the received data signal.

2.2.5 BIT ENCODING—The bit encoding section describes the way in which the logical bits, 1's and 0's, are translated into signals on the transmission medium by the physical interface (e.g., NRZ, PWM, MANCHESTER).

2.2.6 NETWORK ACCESS—The network access section describes the method used to award the communication network to one of the nodes for the transmission of a message (e.g., master slave, token passing, CSMA/CD).

2.2.7 MESSAGE FORMAT—The message format section describes the fields that make up the basic message(s) used in the protocol. This includes the order, name, and size of the fields.

2.2.8 HANDSHAKING—The handshaking section describes the interaction of nodes within a network in order to effect a transfer of data. This may include such things as negative and positive acknowledgement, and in-message acknowledgement.

2.2.9 ERROR DETECTION MANAGEMENT—The error detection management section describes the types of errors the protocol detects and recovery techniques it uses (e.g., wrong message length, CRC).

2.2.10 FAULT TOLERANCE—The fault tolerance section describes the ability of the protocol to continue operation, possibly at a degraded level, when various parts of the physical layer or medium of the network on which the protocol is operating fails (e.g., node connections are broken, bus wires are opened, bus wires are shorted to ground or to vehicle battery voltage).

2.2.11 DATA RATE—The data rate section identifies the maximum data rate supported by the protocol.

2.2.12 FRAMING OVERHEAD—The framing overhead section briefly shows the amount of non-data overhead (i.e., framing overhead) associated with the given protocol. If possible the calculation of overhead is shown. Because some protocols offer significantly different message formats and/or message sizes several overhead calculations may be necessary to give an accurate picture of the range of the protocol's overhead requirements.

2.2.13 LATENCY—The latency section describes the factors that affect the delay between the availability of a message to be transmitted and the beginning of the reception of that message by the intended receiver.

2.2.14 POWER REDUCTION MODE—The power reduction section has general information about any modes of operation that require less power than normal operation. As a minimum, this section identifies the lower power level(s). It also includes a brief description of the criteria used in transitioning to the lower power mode(s) and to return to normal power mode. Some of this information may be device version specific and will be so identified.

2.2.15 SELECTED REFERENCES—The selected references section identifies the source of the information used to create the summary of a given protocol (e.g., manuals, standards, presentations).

2.2.16 CONTRIBUTORS—The contributors section recognizes the person(s) who supplied information on the particular protocol and/or who reviewed the information.

3. Protocol Technical Summary

3.1 Aircraft Internal Time Division Command/Response Multiplex Data Bus (MIL-STD-1553)

3.1.1 APPLICATION/AFFILIATION—Military standard (presently in Revision B). Used in aircraft (fighter jet, other) and tank networks. One of the basic tools used by the Department of Defense for integration of weapon systems. In the future, it is intended to be used to integrate flight controls, to integrate propulsion controls, and to store management subsystems.

3.1.2 TRANSMISSION MEDIA—Standard defines specific characteristics for a twisted pair shielded cable. Other media may be used. A related protocol, MIL-STD-1773, supports fiber optics. A summary of Data Bus and Coupling Requirements may be found in Table 1.

TABLE 1—SUMMARY OF DATA BUS AND COUPLING REQUIREMENTS

Parameter	MIL-STD-1553B
Transmission line	
Cable type	Twisted-shielded pair
Capacitance (wire to wire)	30 pf/ft, maximum
Twist	Four per foot (0.33/in), minimum
Characteristic impedance (Z_0)	70 to 85 Ω at 1.0 MHz, maximum
Attenuation	1.5 dB/100 ft at 1.0 MHz, maximum
Length of main bus	Not specified
Termination	Two ends terminated in resistors equal to $Z(0) \pm 2\%$
Shielding	75% coverage minimum
Cable coupling	
Stub definition	Short stub < 1 ft
	Long stub > 1 to 20 ft (may be exceeded)
Coupler requirement	Direct coupled - short stub; transformer coupled - long stub (ref. Figure 1-1.7)
Coupler transformer	
Turns ratio	1:1.41
Input Impedance	3000 Ω, minimum (75 kHz to 1.0 MHz)
Droop	20% maximum (250 kHz)
Overshoot and ringing	± 1.0 V peak (250 kHz square wave with 100 ns maximum rise and fall time)
Common mode rejection	45.0 dB at 1.0 MHz
Fault protection	Resistor in series with each connector equal to $(0.75 Z_0) \pm 2.0\% \Omega$
Stub voltage	1.0 V to 14.0 V p-p, 1-1, minimum signal voltage (transformer coupled); 1.4 V to 20.0 V, p-p, 1-1, minimum signal voltage (direct coupled)

3.1.3 PHYSICAL INTERFACE—Direct coupling and transformer coupling based on length of connected stub.

3.1.4 BIT ENCODING—Valid Manchester II biphase encoding. Zero crossing deviation of ± 23 ns and rise and fall time of greater than or equal to 100 ns specified. 1-1/2 bits positive volts, 1-1/2 bits negative volts for Command and Status Words, the opposite for Data Word. Valid Manchester II for bits other than the word sync.

3.1.5 NETWORK ACCESS—In the "stationary master bus control system," MIL-STD-1553 operates as Master/Slave with respect to the bus controller and the remote terminal. The bus controller is in control of all communication and it is the sole device allowed to transmit command words. All messages are initiated by the bus controller using command word(s). The network implements "Internal Time Division Command/Response" to control access between the bus controller and the remote terminals.

In the "non-stationary master bus control system," multiple bus controllers may (one at a time) control the single data bus system. The protocol provides a method for issuing a bus controller offer, allowing a potential bus controller to accept or reject control via a bit in the returning status word. Two methods have been used: time based and round robin.

3.1.6 MESSAGE FORMAT—There are 10 message types (see Figure 2). The 10 message types may be broken into 2 groups -"information transfer formats" and "broadcast information transfer formats."

Each message consists of 3 primary sections or "words." They are the "Command Word," the "Data Word," and the "Status Word" (see Figure 3).

3.1.7 HANDSHAKING—Message acknowledgement is provided by the "Status Word." For a single receiver, the structure is shown in Figure 3. The broadcast message format requires the receiving remote terminals to suppress their status responses. Therefore, to analyze the status word after a broadcast message has occurred requires a mode code message to a unique terminal to be sent (and the message must be transmitted before any other transmission to that unique terminal).

All accesses to the bus are controlled through the bus controller. For transmissions to a specific remote terminal, the remote terminal address is used.

For broadcast messages, the bus controller transmits commands using a remote terminal address of "31 hex."

Acknowledgement is provided by the command/response philosophy, which requires that all error-free messages received by a remote terminal be followed by the transmission of a remote terminal status word.

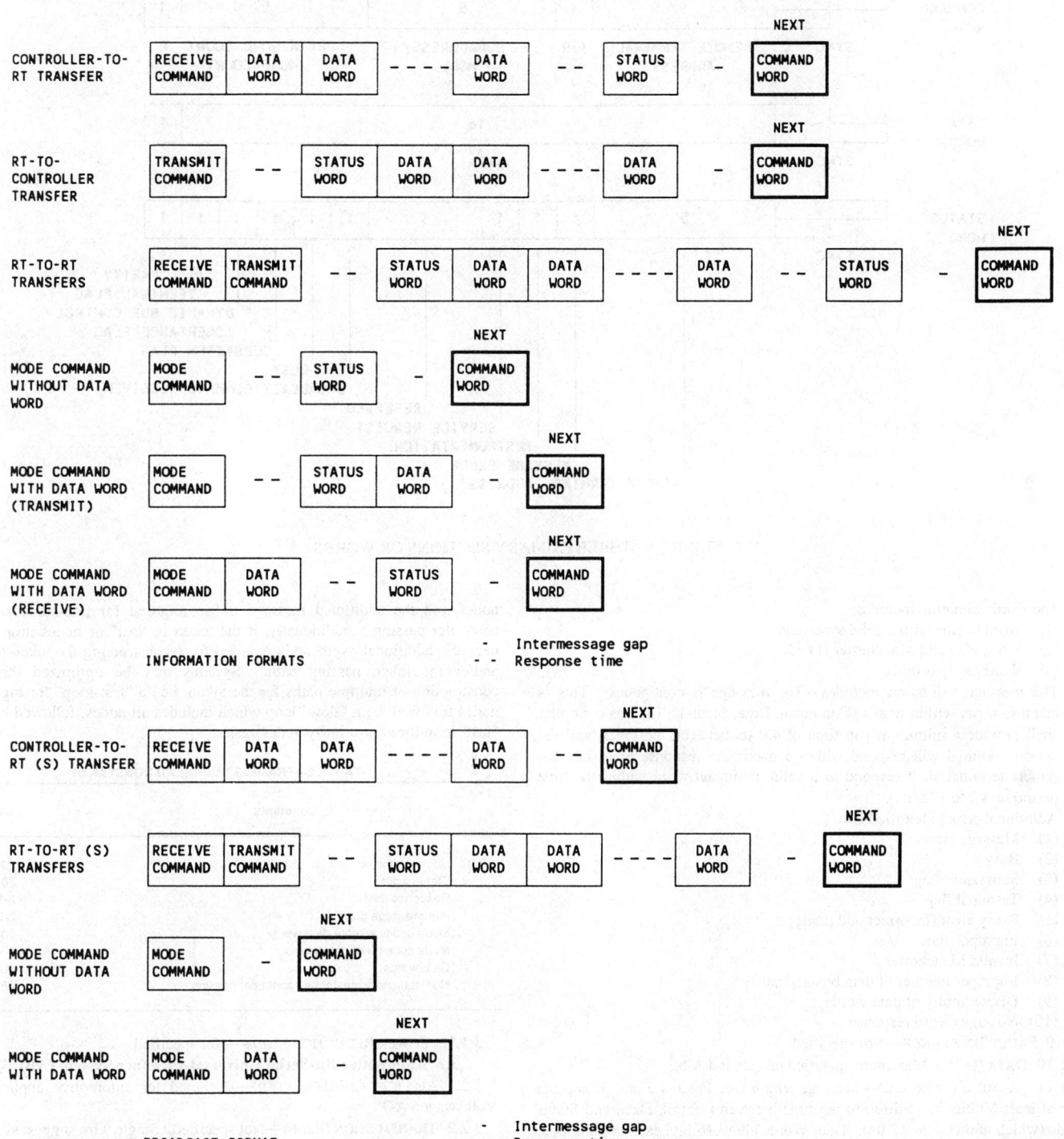

FIGURE 2—TRANSFER FORMATS

3.1.8 ERROR DETECTION MANAGEMENT—The message error bit is the only required status bit and it is used to identify messages which do not pass the word or message validation tests. This bit is set if a message fails to pass the tests and the status word is suppressed (i.e., not transmitted). All messages that are not error-free will not have a responding status word—allowing the bus controller to time-out on the no status response, alerting the bus controller of a failure condition. Parity is used at the end of each word. Odd parity is established for all words based on the 16 bits of data plus parity.

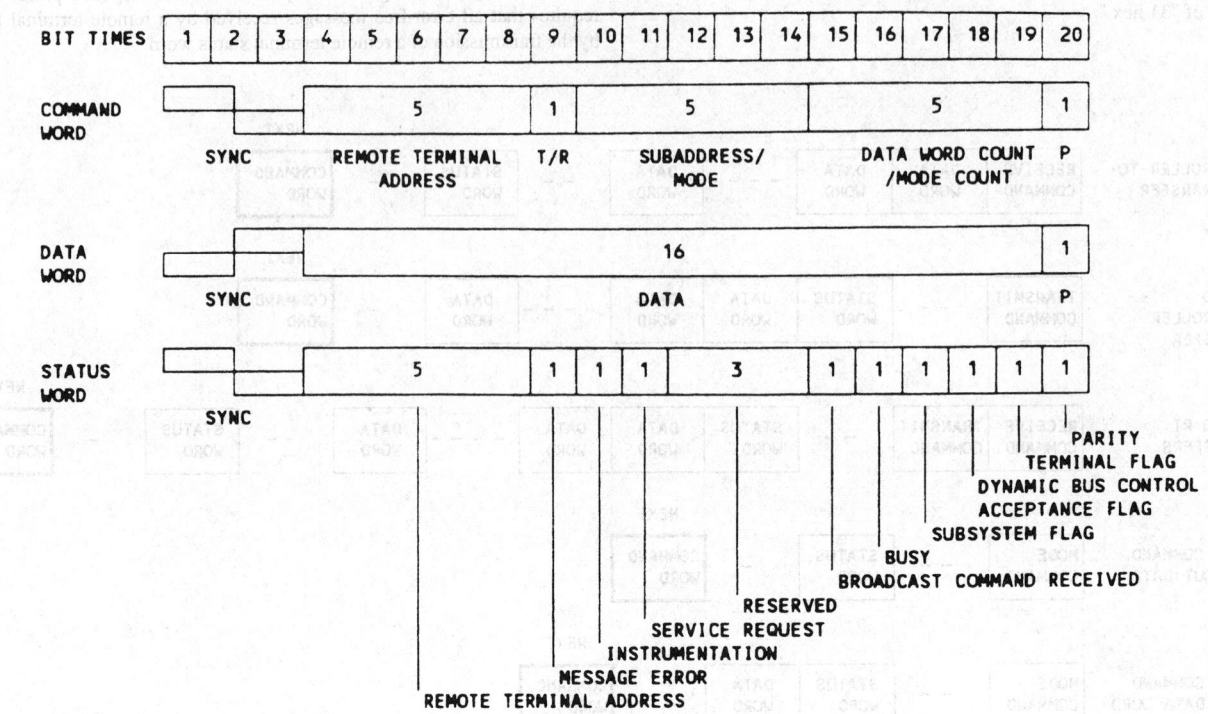

FIGURE 3—THREE PRIMARY SECTIONS OR WORDS

a. The word validation includes:
  (1) word begins with a valid sync field
  (2) bits are a valid Manchester II code
  (3) word parity is odd
b. The message validation includes—The message is contiguous. This is taken to mean within words (Command, Data, Status). The bus controller shall provide a minimum gap time of 4.0 μs between messages, and the remote terminal will respond within a maximum response time (i.e., the remote terminal shall respond to a valid command word within the time period of 4.0 to 12.0 μs).
c. Additional errors identified are:
  (1) Message error
  (2) Busy
  (3) Subsystem flag
  (4) Terminal flag
  (5) Parity error (incorrect odd parity)
  (6) Improper sync
  (7) Invalid Manchester
  (8) Improper number of data bits and parity
  (9) Discontinuity of data words
  (10) No status word response

3.1.9 FAULT TOLERANCE—Not specified.
3.1.10 DATA RATE—Maximum specified bit rate is 1 Mb/s.
3.1.11 FRAMING OVERHEAD—Message length (see Figures 2 and 3) appears to be at least 60 bits in addition to the time between Control, Data, and Status Words (which allows 4 to 12 μs). Data Words allow 16 bits each and multiple Data Words may be used within a message. Each Data Word is 20 bits. Using the formula: OVERHEAD/MESSAGE = FRAMING / (FRAMING + DATA) approximate framing overhead per message calculated is 60/(60+40)=60%.
3.1.12 LATENCY—The total length for the frame is network latency which is dependent on several system factors. Framing overhead constants are included in Table 2.

Stationary versus non-stationary bus controller systems may see different latency results.

The bus access method used may be either Master/Slave or a form of Token. In either case there are latency factors to be considered. In a master/slave system all communication goes through the master, effectively doubling overhead and/or latency. In a token system, latency is dependent in part on the number of nodes and the additional message traffic required for non-stationary master controller passing. Additionally, if the token is "lost" or nodes drop from the network, additional overhead is required for reconstructing the token ownership and/or the token passing path. Systems may be optimized through the construction of multiple paths for the token, i.e., a "fast loop" for high-priority nodes followed, by a "slow" loop which includes all nodes, followed by another "fast" loop for high-priority nodes again.

TABLE 2—FRAMING OVERHEAD CONSTANTS

Constants	Bits
Command word	20
Status word	20
Response time	2-10
Intermessage gap	2-30
Mode codes without data words	20
Mode codes with data words	40
Data words	20
Non-stationary master bus controller passing	48 min

3.1.13 POWER REDUCTION MODE—Not specified.

3.2 Automotive Bit-Serial Universal-Interface System (ABUS)

3.2.1 APPLICATION/AFFILIATION—Designed for automotive applications by Volkswagen AG.
3.2.2 TRANSMISSION MEDIA—Not specified. Single wire suggested.
3.2.3 PHYSICAL INTERFACE—Not specified.
3.2.4 BIT ENCODING—NRZ with eight samples / bit, which means each sample is 1/8 the length of the bit. A valid bit has to have either four or six samples the same (see Figure 4). The requirement of either four or six required common samples is user selectable.
3.2.5 NETWORK ACCESS—Contention using nondestructive bitwise arbitration. Arbitration occurs through the end of the identifier field.
3.2.6 MESSAGE FORMAT—Message length is constant and consists of 1 start bit, 1 bit (NC/DAT) indicating whether it is a data message or a command message, an 11-bit identifier, 16 bits of data, and 2 stop bits (STP0, STP1). Further detail is provided in Figure 5. Two different types of messages are possible, the data message and the command message. Command messages may

be used to insure that a following command will not be accepted on the bus until the receiver has read the command. The ABUS IC can be programmed in a way that every following command will not be accepted on the bus until the Master Controller Unit has read the command received last. In this case, no commands are lost.

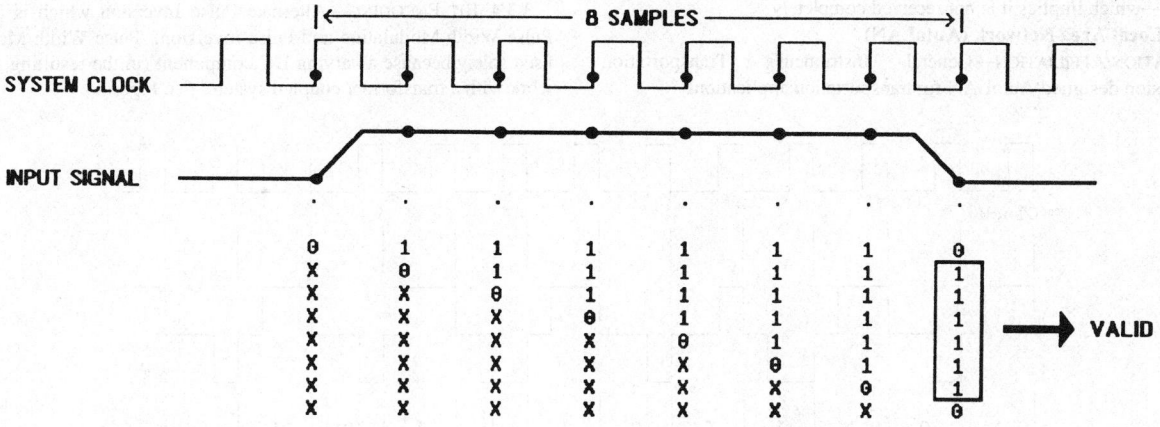

FIGURE 4—NRZ WITH EIGHT SAMPLES / BIT

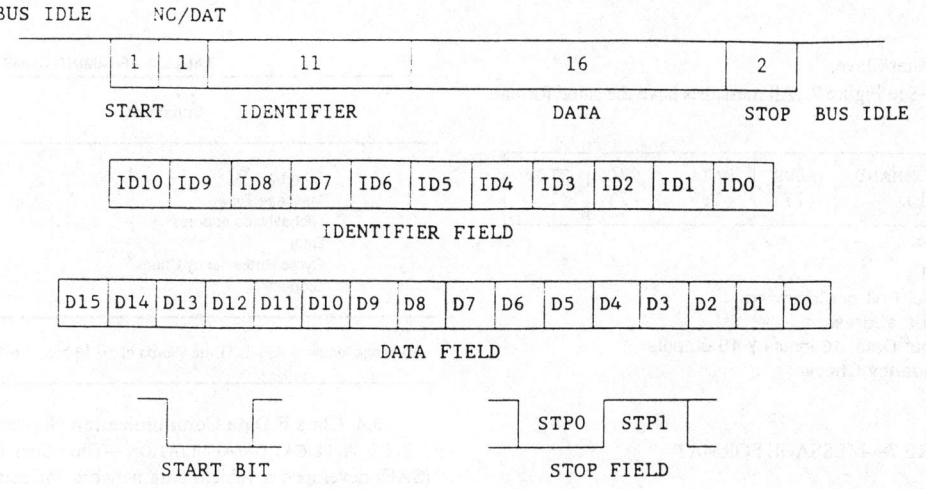

FIGURE 5—MESSAGE FORMAT

3.2.7 HANDSHAKING—Negative acknowledgement by receiver after error is detected. Any receiving device that notices a protocol error notifies all other bus members by pulling down the STP1 bit. If the STP1 bit is low, the message is regarded as invalid by all members. There is no required or dedicated positive acknowledgement.

3.2.8 ERROR DETECTION MANAGEMENT—The transmitter and receiver monitor the bus for errors. The transmitter reads its own message back from the bus and compares it with the message that it intended to send. The receiver detects code errors by sampling the bus and by comparing the samples. Four different types of errors are monitored:

  a. Start-bit Error—it has not been possible to generate a start bit at the beginning of the transmission process.
  b. Transmit Error—during transmission the value read back was not equal to the value sent out Y times consecutively between the arbitration and STP1 bit.
  c. Receive Error—during reception an error has been detected X times consecutively prior to the STP1 bit.
  d. Short Circuit—no logical "1" has been read for a period of 256 clock pulses after a high to low transition.

NOTE—X and Y are programmable values (to 8, 16, and 32). In the present implementation these error occurrences are stored in a status register and cause an interrupt. Before they are recognized as an error, the sender tries to send a message for 8 to 32 times.

3.2.9 FAULT TOLERANCE—Not specified.

3.2.10 DATA RATE—Maximum specified bit rate = 500 Kb/s.

3.2.11 FRAMING OVERHEAD—The message length is 31 bits, 16 of these are data. To transmit 4 bytes of data, 2 messages must be sent. Using Equation 1:

$$\text{OVERHEAD/MESSAGE} = \text{FRAMING} / (\text{FRAMING} + \text{DATA}) \quad (\text{Eq.1})$$

Approximate framing overhead/message calculated is: $(15+15)/((15+16)+(15+16))=48\%$.

3.2.12 LATENCY—The ABUS protocol uses nondestructive bitwise arbitration in contention to determine bus access. In the case of two or more nodes beginning transmission simultaneously, the message with the highest priority will win the arbitration and continue transmission. As a result, the maximum latency for the highest priority message is the number of bits in the maximum length message multiplied by the time per bit. A 31-bit message will require 64 ms at 500 Kb/s. Lower priority messages may encounter additional delay in the event that they lose arbitration. Their latency may be determined based on a statistical analysis of the system (bus load, priority, other).

3.2.13 POWER REDUCTION MODE—The power consumption of the ABUS IC can be reduced by using the sleep mode. In this mode, the oscillator is turned off and the power consumption is reduced to 10 to 100 µA. The sleep mode is initiated by the host microprocessor. The IC will wake up either by a reset or by bus activity, but will go through a reset in either case. After the reset it will take about 50 ms for the oscillator to work properly, e.g., the first message works as an `alarm clock'—which implies it is not received completely.

3.3 Auto Local Area Network (AutoLAN)

3.3.1 APPLICATION/AFFILIATION—General Instrument's Transportation Electronics Division designed AutoLAN for transportation applications.

3.3.2 TRANSMISSION MEDIA—The transmission media is a dual twisted pair. One pair carries data and idle bits from the Master to all Slaves. The second pair carries data from the Slaves to the Master.

3.3.3 PHYSICAL INTERFACE—The Physical Interface utilizes a custom 68 pin PLCC package operating at 5 V (±0.5 V).

3.3.4 BIT ENCODING—Alternate Pulse Inversion which is a combination of Pulse Width Modulation and Pulse Inversion. Pulse Width Modulation was not used solely because a varying DC component (in the resulting signal) would not work with a transformer coupled system. See Figure 6.

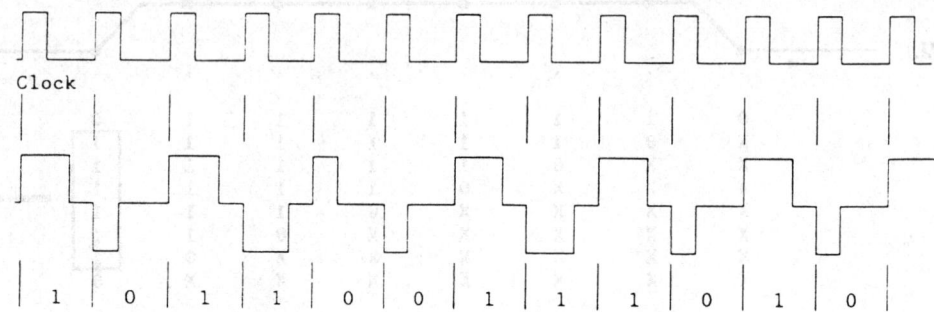

FIGURE 6—ALTERNATE PULSE INVERSION BIPOLAR DATA

3.3.5 MEDIA ACCESS—Master/Slave

3.3.6 MESSAGE FORMAT—See Figure 7. All messages have the same format.

| START (1) | POLL/COMMAND (1) | ADDR (7) | DATA (16) | CRC (7) | STOP (1) |

START : Unique signal
P/C : Message Type, Poll or Command
ADDR : Global or Node address
DATA : Input or Output Data, 16 inputs / 16 outputs
CRC : Cyclic Redundancy Check
STOP : Unique signal

FIGURE 7—MESSAGE FORMAT

3.3.7 HANDSHAKING—A Poll Request from the Master is transmitted to each Slave. The addressed Slave sends a Poll Reply which is the Slave's input status. From the Poll Reply, if the input status of the Slave has changed, the Master Polls the Slave and Broadcasts the information to all the Slaves in a Command Message (using a Global Address). All the Slaves read the Broadcast Message and determine what action, if any, should be taken. A Command Message does not require any acknowledgement from the Slaves.

3.3.8 ERROR DETECTION MANAGEMENT—If a Poll Request fails Framing, CRC, or Address Match, the message will be ignored. If a Poll Request message was received and the Poll Reply message failed the Framing, CRC, or Address Match test, the Master will generate a NACK interrupt to its processor.

3.3.9 FAULT TOLERANCE—If a transmission line is cut, all slaves beyond the cut are not accessible by the Master.

3.3.10 FRAMING OVERHEAD/LATENCY—Bit rates of up to 4 Mbps.
Dependent on the time it takes the Master to finally poll the particular slave; then, the time for the slave to reply with its status which, if changed since last poll, will be broadcasted to other slaves. See Table 3.

3.3.11 POWER REDUCTION MODE—When Slaves stop receiving clock pulses from the Master, Slaves enter Rest Mode with all output circuits maintained. If the Master completely powers down (no DC current to Slaves), all Slaves enter Sleep Mode with less than 100 µs current draw per node.

3.3.12 CONTRIBUTORS
  a. Abe Jacobs, General Instruments, El Paso
  b. Whit Leverett, General Instruments, El Paso
  c. Mike Thomas, General Instruments, Farmington Hills, Michigan

TABLE 3—FRAMING CONSTANTS

Constants	Bits
Start bit	1
Message Type	1
Global/Node address	7
Data	16
Cyclic Redundancy Check	7
Stop bit	1

Message length is 33 bits. Data Words allow 16 bits. So 16 data bits / 33 bits = 48%.

3.4 Class B Data Communication Network Interface (SAE J1850)

3.4.1 APPLICATION/AFFILIATION—The Society of Automotive Engineers (SAE) developed a vehicle data network for communications. A new revision (August 1991) of SAE J1850 has recently been approved by SAE for publication.

3.4.2 TRANSMISSION MEDIA—Three physical layer approaches have been defined: single wire voltage drive, dual wire voltage drive, and balanced current drive. For each of these approaches a different transmission media is specified. The medium for single wire is a single random lay wire. The medium for dual wire and balanced current drive is either a twisted or parallel wire pair.

3.4.3 PHYSICAL INTERFACE—A composite circuit diagram for the receiver/transmitter represents all three bus interfaces. This diagram is shown in Figure 8.

3.4.4 BIT ENCODING—Pulse Width Modulation. A "1" bit and "0" bit are shown in Figure 9. All bits are encoded in this manner except for a few unique symbols differentiated by the pulse timing. Some of the symbols include Start of Frame (SOF), End of Data (EOD), and End of Frame (EOF.) See Figure 9 for bit representations.

3.4.5 NETWORK ACCESS—Nondestructive prioritized bitwise arbitration.

3.4.6 MESSAGE FORMAT—The maximum length for a message frame is 101 bit times. The error detection byte is included in the data field at the discretion of the designer. Another possible use of the data field could be a message/address identifier. The response byte is explained more fully in 3.4.7. Interframe spacing is nominally 2 bit times. The message frame is shown in Figure 10.

3.4.7 HANDSHAKING—Acknowledgement is provided in the message frame using the response bytes. The response byte appears after the EOD. If an acknowledgement is not expected, a response byte will not be sent, and the bus will remain in the passive state signifying an EOF. If the "In-Message Acknowledge/Response" feature is active, then the response byte is an 8-bit acknowledge identifier, or one or more response bytes followed by an ERR byte. One or more nodes may attempt to respond to the requesting node and arbitration will occur during the response time period.

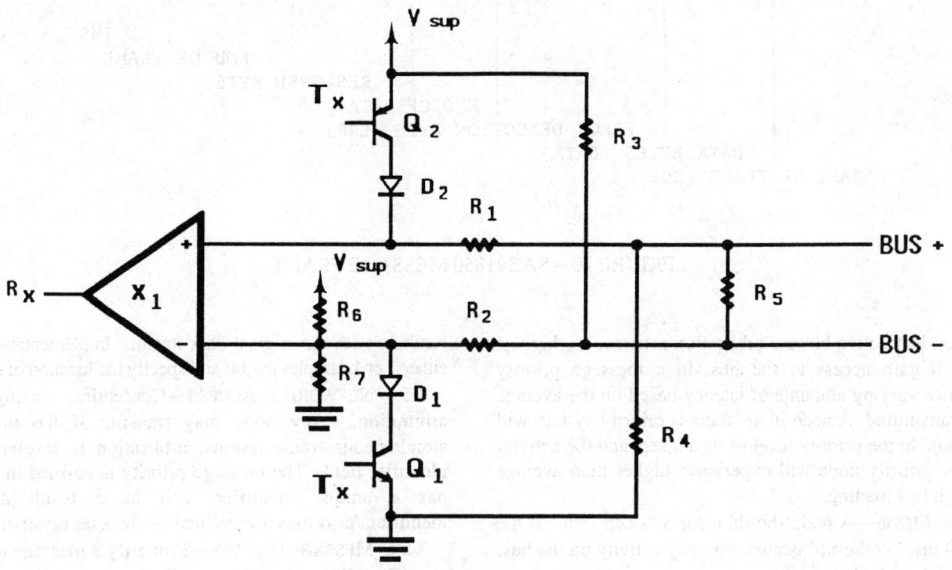

FIGURE 8—DIAGRAM OF SAE J1850 PHYSICAL INTERFACE

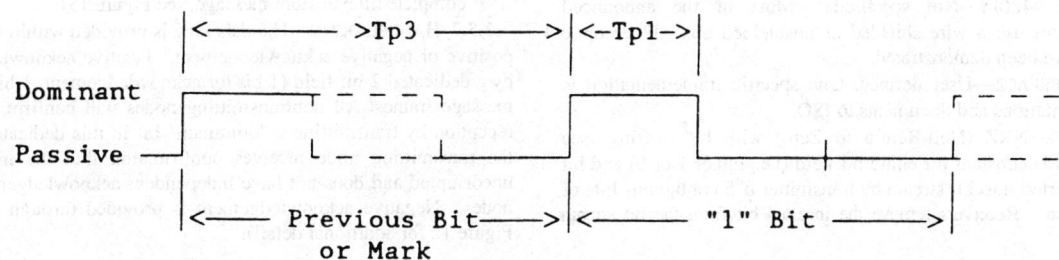

"1" Bit Definition

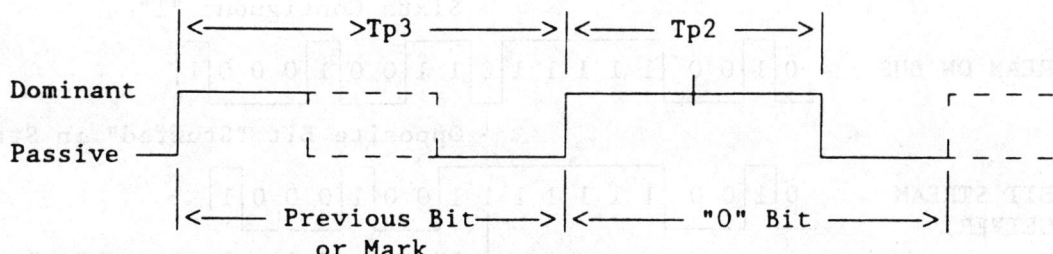

"0" Bit Definition

FIGURE 9—SAE J1850 BIT REPRESENTATION

3.4.8 ERROR DETECTION MANAGEMENT—Includes detection of bus out-of-range, invalid bit value, and invalid message structure. An invalid bit or an invalid message will cause the receive process to be terminated until the next SOF. The in-message Error Detection field (ERR) uses an 8-bit CRC based on the polynomial: $x^8 + x^4 + x^3 + x^2 + 1$.

3.4.9 FAULT TOLERANCE—Continued network operation when a node loses power, bus short to ground, bus short to battery, transceiver failure, loss of connection to network is optional.

3.4.10 DATA RATE—The bit rates specified are 10 Kb/s and 40 Kb/s.

3.4.11 FRAMING OVERHEAD—The total length for the frame is 101 bit times. For the overhead calculation add 2 bits for IMS. The total length for the calculation becomes 103. The total allowed data is 80 bits (assume one byte for address/message identifier). The percentage of overhead is therefore 23/(88 + 23)=27.2%.

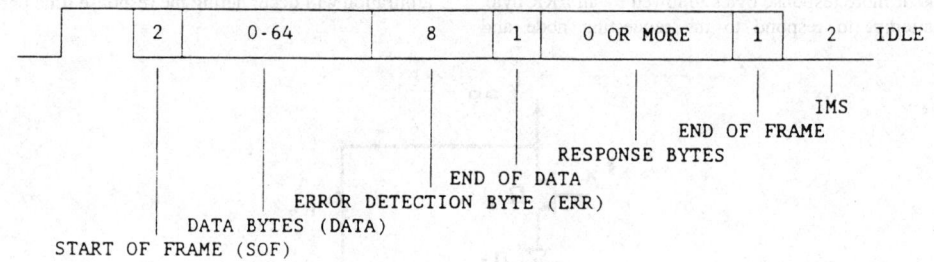

FIGURE 10—SAE J1850 MESSAGE FRAME

3.4.12 LATENCY—In a nondestructive bitwise arbitration scheme, the highest priority message/address will gain access to the bus. In a message priority scheme, a node will experience varying amounts of latency based on the average priority of messages to be transmitted. A node in an address priority system will experience a delay proportional to the priority level of its address and the activity on the bus. Example, a low priority node will experience higher than average latency during periods of high bus loading.

3.4.13 POWER REDUCTION MODE—A node should enter a "sleep state" if the bus is idle for more than 500 ms. "Wake-up" occurs with any activity on the bus.

3.5 Controller Area Network (CAN)

3.5.1 APPLICATION/AFFILIATION—Both Standard (S) and Extended (E) message format is intended for automotive in-vehicle applications. A draft standard within ISO. Published within SAE via Information Report SAE J1583.

3.5.2 TRANSMISSION MEDIA—Not specified. Most of the announced production-intent systems use a wire shielded or unshielded bus. Fiber optic systems using CAN have been demonstrated.

3.5.3 PHYSICAL INTERFACE—User defined. One specific implementation is defined in Bosch presentations and documents to ISO.

3.5.4 BIT ENCODING—NRZ (Non-Return to Zero) with bit stuffing (see Figure 11). Logic level is constant for entire bit field (i.e., either 1 or 0) and bit of opposite state is inserted into bit stream by transmitter if 5 contiguous bits of the same state are seen. Receivers remove the inserted bit from the bit stream resulting with the original data stream. Implementations are programmable to allow either 3 or 1 samples per bit and specify the location of samples within a bit.

3.5.5 NETWORK ACCESS—Contention using nondestructive bitwise arbitration. Any node may transmit if the bus is idle. In the case of simultaneous transmissions, arbitration is resolved through the value in the identifier field. The message priority is defined in the identifier. Each message has a unique identifier, and as a result a unique priority. These identifiers/priorities are defined by the user (system designer).

3.5.6 MESSAGE FORMAT—Primarily 3 message types:
a. Data Frame (see Figure 12)
b. Remote Frame (see Figure 13)
c. Error Frame (see Figure 14)
d. Overload Frame (used in events where individual node has not had complete time to store message, see Figure 15)

3.5.7 HANDSHAKING—Handshaking is provided within the message via either positive or negative acknowledgement. Positive acknowledgement is provided by a dedicated 2-bit field (1 bit for acknowledgement, 1 bit for delimiter) in the message frame. All nontransmitting nodes will confirm uncorrupted message reception by transmitting a "dominant" bit in this dedicated field. As a result, the transmitting node receives confirmation that the message was received uncorrupted and does not have independent acknowledgements from individual nodes. Negative acknowledgement is provided through the Error Frame (see Figure 12 for additional detail).

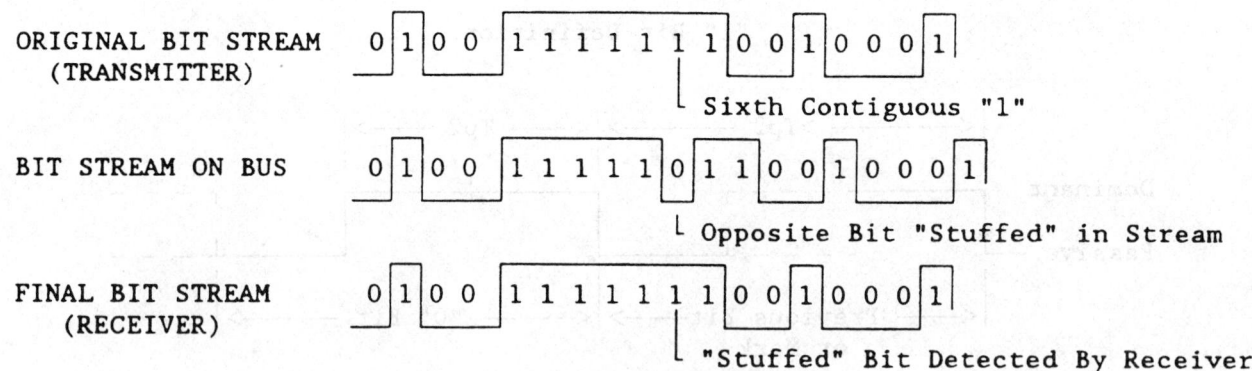

FIGURE 11—NRZ BIT "STUFFING"

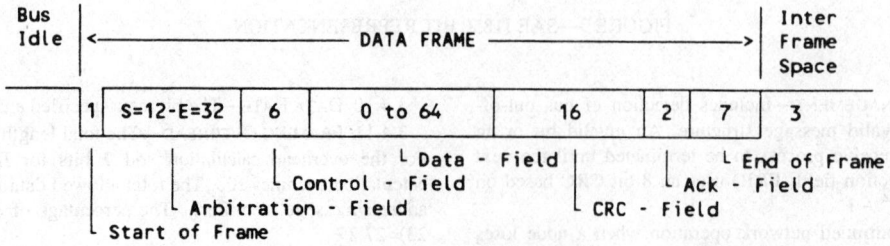

FIGURE 12—DATA FRAME

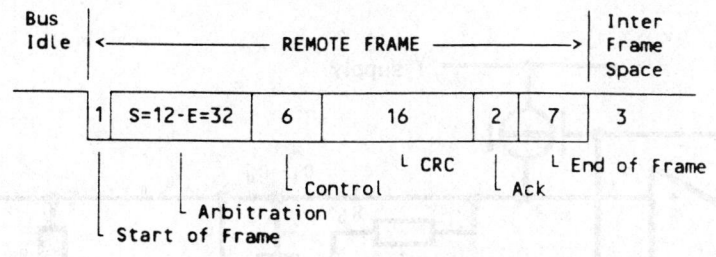

FIGURE 13—REMOTE FRAME

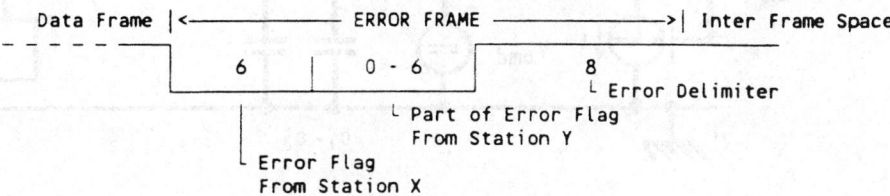

FIGURE 14—ERROR FRAME

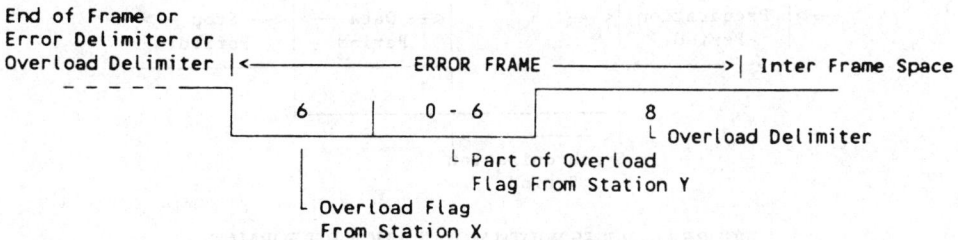

FIGURE 15—OVERLOAD FRAME

3.5.8 ERROR DETECTION MANAGEMENT—All nodes monitor all messages. If an error is detected within a message then the node(s) detecting that error destroys that message by transmitting an error frame. The result of this error frame is that all nodes (including the transmitting node) know that an error has been detected within the present message. The transmitter will retransmit the message at its next opportunity (through normal bus access arbitration). Error checking is provided on CRC, message length (message length is specified in the control field), message format, and bit level and timing.

3.5.9 FAULT TOLERANCE—Protocol is intended to be orthogonal, i.e., all nodes address faults in the same manner. Fault confinement is provided by each node constantly monitoring its performance with regard to successful and unsuccessful message transactions. Each node will act on its own bus status based on its individual history. As a result, graceful degradation allows a node transmitter to disconnect itself from the bus. If the bus media is severed or shorted, the ability to continue communications is dependent upon the condition and the physical interface used.

3.5.10 DATA RATE—Bit rate of up to 1 Mbits/s.

3.5.11 FRAMING OVERHEAD—Maximum message length (maximum time between messages) is 111 bit times for Standard Format and 131 bit times for Extended Format, i.e., 111 and 131 μs at 1 Mb/s. For the highest priority message, if a message has just begun and the message in question is queued up the latency will be 111/131 μs, and 222/262 μs maximum until its transmission is complete.

Maximum time between messages with four bytes data is 79 bits for Standard Format and 99 bits for Extended Format (please note: this includes Interframe Space). For a message transmitting 4 bytes of data, using the formula: OVERHEAD/MESSAGE = FRAMING / (FRAMING + DATA) the approximate framing overhead/message calculated for Standard Format is: 47/(47+32)=59%. For 8 data bytes the approximate framing overhead/message calculated is 47/(47+64)=42%. For Extended Format is 67/(67+32)=68% and 67/(67+64)=51%, respectively, (does not include Bit Stuffing).

3.5.12 LATENCY—The CAN protocol uses nondestructive bitwise arbitration in contention to determine bus access. In the case of two or more nodes beginning transmission simultaneously, the message with the highest priority will win the arbitration and continue transmission. As a result, the maximum latency for the highest priority message is the number of bits in the maximum length message multiplied by the time per bit (in other words 111 bits times or 111 μs at 1 Mbit/s for Standard Format and 131 bits times or 131 μs at 1 Mbit/s for Extended Format). Lower priority messages may encounter additional delay in the event that they lose arbitration. Their latency may be determined based on a statistical analysis of the system (bus load, priority, other).

3.5.13 POWER REDUCTION MODE—Not specified.

3.6 Digital Data Bus (D2B)

3.6.1 APPLICATION/AFFILIATION—Digital Data Bus is a product of Philips for use in Audio/Video communications, computer peripherals and automotive.

3.6.2 TRANSMISSION MEDIA—Twisted pair.

3.6.3 PHYSICAL INTERFACE—Differential floating pair. See Figure 16.

3.6.4 BIT ENCODING—Pulse Width Modulation (PWM). The general bit format is comprised of four sections:
 a. The preparation period
 b. The sync period
 c. The data period
 d. The stop period

The duration of the periods and the bit is dependent on the speed of the bus and the type of the bit. The speed of the bus is determined during contention. Low speed is dominant. There are three speeds possible. The general bit format is shown in Figure 17.

3.6.5 NETWORK ACCESS—Access is achieved by contention using nondestructive prioritized bitwise arbitration. Competing nodes arbitrate first on the mode in which the node will operate (3-bit field), where low mode is dominant, then all nodes in a common mode arbitrate based on the unique address bits of the competing masters. Low address is dominant. The mode designates the speed at which the bus will operate during the message transfer. A unit may use the bus for one time slot. The amount of data transferred in the time slot depends on the speed mode determined during arbitration.

3.6.6 MESSAGE FORMAT—The message frame consists of 6 fields. A parity bit follows the master, slave, control, and data fields. An acknowledge bit follows the slave field, control field, and the data field. An end-of-data bit follows each data byte. The total length of the frame is 47 bits. See Figure 18 for the frame.

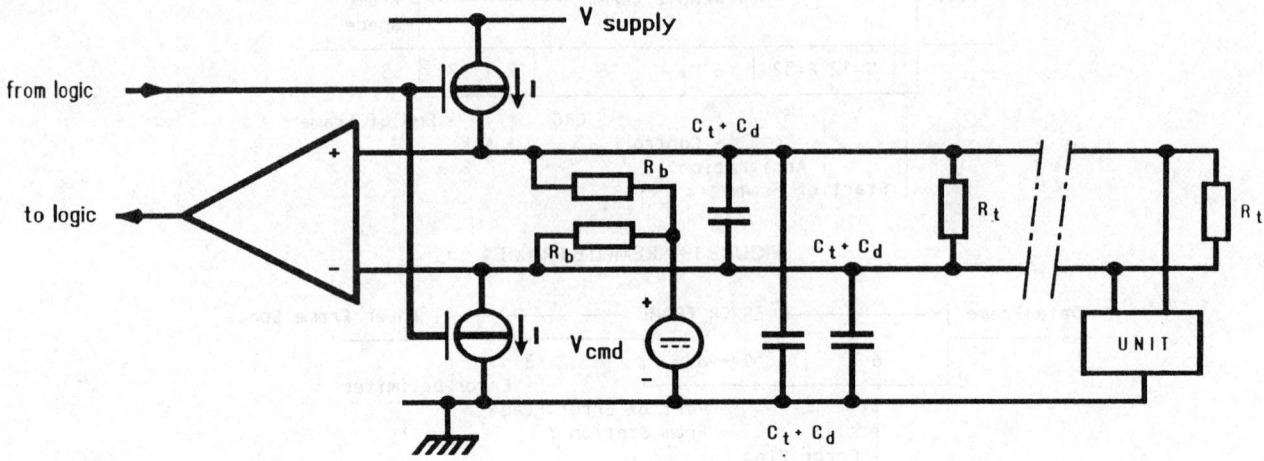

FIGURE 16—D2B PHYSICAL INTERFACE

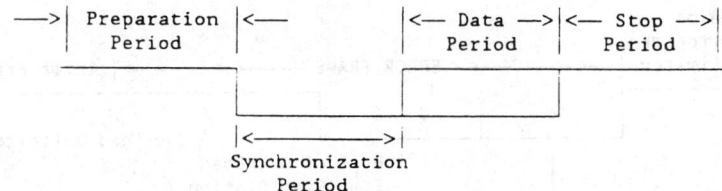

FIGURE 17—PULSE WIDTH MODULATION BIT FORMAT

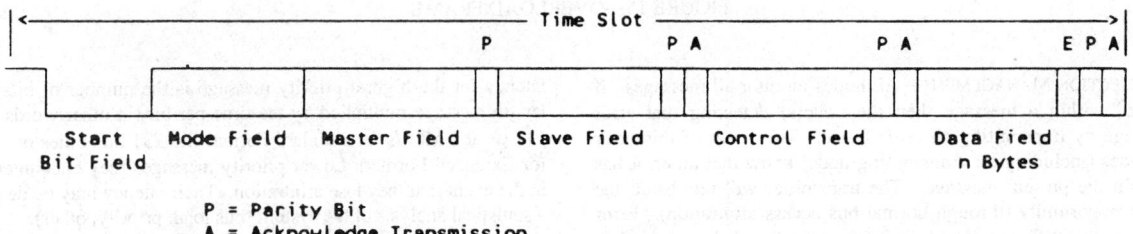

P = Parity Bit
A = Acknowledge Transmission
E = End-of-Data Bit

FIGURE 18—SIX FIELD MESSAGE FORMAT

3.6.7 HANDSHAKING—Handshaking is accomplished with positive acknowledgement in the transfer message. No reply from the slave is interpreted as a negative acknowledgement. The master can retry the message provided time remains in the slot. During every transfer there are three different acknowledge bits:
 a. After the slave address
 b. After the control bits
 c. After each data byte
A master has the ability to lock a slave node to its address having the effect of disabling the node from communicating with any other master on the network. This is done when a data transfer exceeds the time slot and the master must arbitrate again for the bus to complete the data transfer.

3.6.8 ERROR DETECTION MANAGEMENT—Error checking is performed through odd parity on the slave address, control field, and after each data byte. The acknowledge bit in the transfer message will not be transmitted by the addressed slave if there is a parity error, the speed mode is too high, timing error, slave locked to another master, or the receive buffer is full.

3.6.9 FAULT TOLERANCE—Fault tolerance for nodes is not specified.

3.6.10 DATA RATE—The maximum bit rate is 1 Mb/s. Three different transmission speeds are allowed.

3.6.11 FRAMING OVERHEAD—The total frame size is 34 bits including Inter Frame Separation. In speed mode 1, 32 data bytes can be transferred from master to slave. Therefore the percentage of overhead is 34/34+256=11.7%.

3.6.12 LATENCY—D2B allows for three different speed modes for transmission and arbitrates on the address of the competing nodes once in a speed mode. Therefore a low priority node may experience high latency times vs. the average. Latency is also affected by the ability of a master to lock a slave node; a locked node will not respond to any messages, and in certain situations this could degrade the overall performance of the system.

3.6.13 POWER REDUCTION MODES—Power reduction modes are available for the bus controllers.

3.7 Ethernet (IEEE 802.3)

3.7.1 APPLICATION/AFFILIATION—Ethernet is specified by IEEE 802.3. While not restricted to specific applications, Ethernet was developed as a laboratory network. Additionally it is used in CAE/CAD workstation clusters, office or business type network or low-cost PC environments. Ethernet falls into the 10BASE5 category which shows the characteristics of 10 Mb/s, baseband, and segments up to 500 m. Other protocols or implementations falling within the IEEE 802.3 specification include Cheapernet, Thinnet, Thinwire ENET, and Starlan. This summary will focus on Ethernet.

3.7.2 TRANSMISSION MEDIA—"Thick" coaxial cable defined as a 0.4-in. heavily shielded, 50-Ω coaxial cable which extends in segments of up to 500 m with passive terminators at either end.

3.7.3 PHYSICAL INTERFACE—Nodes are connected to the segment with a cable tap, a transceiver, and a shielded, twisted pair transceiver cable which can extend up to 50 m. The transceiver is isolated from the main cable by use of AC/DC converters and pulse transformers or optoisolators. Three signals—transmit, receive, and collision presence—are carried between the transceiver and the node.

3.7.4 BIT ENCODING—Baseband signalling and Manchester phase encoding at a 10 Mb/s rate.

3.7.5 NETWORK ACCESS—CSMA/CD (Carrier Sense Multiple Access / Collision Detect) is the medium access method. Each transmitting node that detects a collision (by sensing an abnormal voltage level on the cable) sends a short "jamming" signal to ensure collision detection by the entire network. The period that each station must wait after detecting a collision before attempting to transmit again is governed by a truncated binary exponential backoff algorithm, which specifies a random waiting time. The random time is chosen from a window whose size is a binary multiple of the round-trip propagation delay time of the network, called the slot time. The window size increases by a power of two after each collision until the tenth collision, after which it remains the same. Each transmission attempt is allowed a maximum 16 collisions before the higher layers are signalled that the message is undeliverable.

3.7.6 MESSAGE FORMAT—An Ethernet frame is made up of seven fields (see Figure 19). The first field is a 56-bit preamble used for hardware synchronization. This field is followed by an 8-bit start-of-frame delimiter field, two 6-byte address fields for destination and source addresses, and a 2-byte length field indicating the actual length of the data. This is followed by the data or information field, which must be a minimum 46 bytes and a maximum 1500 bytes. The last field is a 32-bit frame check sequence, which contains the cyclic redundancy check (CRC) for the frame.

7 BYTES	1 BYTE	6 BYTES	6 BYTES	2 BYTES	46-1500 BYTES	4 BYTES	
Preamble	Start Frame Delimiter	Dest Addr	Source Addr	Length Field	Data Field	Frame Check Sequence	End Frame Delimiter

FIGURE 19—SEVEN FIELD MESSAGE FORMAT

3.7.7 HANDSHAKING—Not specified.

3.7.8 ERROR DETECTION MANAGEMENT—Information reported on transmitted message events includes:
a. Transmission successful
b. Transmission unsuccessful due to lost Carrier Sense
c. Transmission unsuccessful due to lost Clear-to-Send
d. Transmission unsuccessful due to DMA underrun because the system bus did not keep up with the transmission
e. Transmission unsuccessful due to number of collisions exceeding the maximum allowed.

Information reported on incoming message events includes:
a. CRC error due to incorrect CRC in a well-aligned frame
b. Alignment error due to incorrect CRC in a misaligned frame
c. Frame too short—The frame is shorter than the configured value for minimum frame length
d. Overrun since the frame was not completely placed in memory because the system bus did not keep up with incoming data
e. Out of buffers, i.e., no memory resources to store the frame, so part of the frame was discarded

3.7.9 FAULT TOLERANCE—Specific implementations may provide a set of network-wide diagnostics that can serve as the basis for a network management entity. The primary reference used for the following information is the Intel 82586. Networked activity information provided includes number of collision and deferred transmissions. Statistics information includes number of CRC errors, alignment errors, no-resources (correct frames lost due to lack of memory resources) and overrun errors (number of frame sequences lost due to DMA overrun).

Additional diagnostics are provided by the external and internal loopback modes where the node can be configured to verify the transceivers collision detection circuitry, the transmit circuitry, the receive circuitry, internal memory, and the exponential backoff random number generator internal to the device.

3.7.10 DATA RATE—Bit rate of 10 Mb/s.

3.7.11 FRAMING OVERHEAD—Overhead is approximately 24 bytes (though some of the fields are specified in bit lengths). The data field is 46 bytes minimum to 1500 bytes maximum. Using the formula: OVERHEAD/MESSAGE = FRAMING / (FRAMING + DATA) the approximate worst case overhead calculated is 192 bits/(192 bits+368 bits)=34%. The approximate best case calculated is 192/(192+12000)=2%. Although a message with only 4 bytes of data is not allowed, its overhead would be 192/(192+32)=86%.

3.7.12 LATENCY—Ethernet is a nondeterministic contention protocol. Since collisions are resolved through a back-off scheme, latency may be significant for very low priority messages relative to a nondestructive bitwise arbitration protocol. Latency may be determined based on a statistical analysis of the system (bus load, priority, other).

3.7.13 POWER REDUCTION MODE—Not specified.

3.8 Joint Integrated Avionics Working Group (JIAWG)

3.8.1 APPLICATION/AFFILIATION—The High Speed Data Bus protocol was used in a demonstration project and will be succeeded by a protocol developed by the Joint Integrated Avionics working group (JIAWG). This protocol may be the next generation MIL-STD-1553. This survey encompasses the protocol information which the JIAWG has defined. The current applications for this protocol are military avionics, particularly the ATF, ATA, and LHX programs.

3.8.2 TRANSMISSION MEDIA—Fiber optics.

3.8.3 PHYSICAL INTERFACE—Requirements for the Physical Layer are defined in 'The Linear Token Passing Multiplex Bus Protocol (J88-M5).'

3.8.4 BIT ENCODING—The encoding format is Manchester II biphase level. A logic one is transmitted as a unipolar coded signal 1/0. A logic zero is transmitted as a unipolar coded signal 0/1.

3.8.5 NETWORK ACCESS—Token passing.

3.8.6 MESSAGE FORMAT—There is no interframe gap between frames generated from the same station for a given possession of the token. When multiple frames are transmitted for a given possession of the token, only the first frame shall be preceded by a preamble. Refer to Figures 20, 21, and 22.

A node will transmit a claim token frame immediately after the 'Bus Activity Timer' times out.

3.8.7 HANDSHAKING—A message may be received by all nodes or any subset of nodes except the transmitting node. Nodes accept messages based on physical addressing, logical addressing, or broadcast addressing.

3.8.8 FAULT TOLERANCE—The bus is dual redundant (each node has two transmitters and two receivers). When a station transmits a message, it goes onto both redundant buses at the same time. The receiving node(s) only takes the message off of one bus, until there is an error, and then it switches to the other bus. If both buses are bad in the same place in the data, the message is discarded.

When a node failure occurs, the reconfiguration will occur when the failed station's predecessor in the logical ring attempts to pass the token to it. After two unsuccessful attempts, the node will pass the token to the failed node's successor.

If a node fails while possessing the token, the network will reinitialize itself due to no activity on network.

3.8.9 DATA RATE—Bit rate of up to 50 Mbps.

3.8.10 FRAMING OVERHEAD—Maximum message length is 65608 bit times. Maximum time between message with four bytes data is 104 bits.

Framing / Framing + Data

$$2\text{-word Data Message} = \frac{72 \text{ bits overhead}}{104 \text{ overall bits}} = 70\% \qquad (\text{Eq.2})$$

$$256\text{-word Data Message} = \frac{72 \text{ bits overhead}}{4168 \text{ overall bits}} = 2\% \qquad (\text{Eq.3})$$

4096 - word Data Message = $\frac{72 \text{ bits overhead}}{65608 \text{ overall bits}}$ = 0.1%   (Eq.4)

3.8.11 POWER REDUCTION MODE—Not specified.

3.8.12 CONTRIBUTORS—James H. Nelson, Northrop Aircraft Division, Hawthorne, California

SD (4)	0 (1)	TDA (7)	TFCS (8)	ED (4)

SD   : Start Delimiter - Identifies the start of message.

TDA   : Token Destination Address - Specifies Physical Address of where Token is being sent.

TFCS : Token Frame Check Sequence - Checks for bit errors within each token frame.

ED   : End Delimiter - Identifies the end of message.

FIGURE 20—TOKEN FRAME

SD (4)	FC (8)	0 (1)	SA (7)	FW (1024)	ED (4)

SD : Start Delimiter - Identifies the start of message.

FC : Frame Control - Identifies Frame Type and Priority.

SA : Source Address - Identifies the physical address of node sending the message.

FW : Fill Words

ED : End Delimiter - Identifies the end of message.

FIGURE 21—CLAIM TOKEN FRAME

SD (4)	FC (8)	0 (1)	SA (7)	MDA (15)	WC (16)	INFO (16-65536)	MFCS (16)	ED (4)

SD   : Start Delimiter - Identifies the start of message.

FC   : Frame Control - Identifies Frame Type and Priority.

SA   : Source Address - Physical address of node sending the message.

MDA   : Message Destination Address - Identifies a Physical or Logical Address of where message is being sent.

WC   : Word Count - Specifies the number of 16-bit fields in the INFO field.

INFO : Information - Contains 1 to 4096 words of message data.

MFCS : Message Frame Check Sequence - Checks for bit errors within each message frame.

ED   : End Delimiter - Identifies the end of message.

FIGURE 22—MESSAGE FRAME

3.9 Mini-Manufacturing Automation Protocol (Mini-MAP)

3.9.1 APPLICATION/AFFILIATION—Mini-MAP was designed for Factory Floor applications, specifically automation. The three ISO layer Mini-MAP carrierband has an IEEE 802.4 Standard Interface.

3.9.2 TRANSMISSION MEDIA—The transmission media is unshielded twisted pair (broadband), shielded twisted pair, coaxial cable, or fiber optics.

3.9.3 PHYSICAL INTERFACE—The Physical Interface for a mini-MAP network is product specific. The Physical Layer restrictions are described in IEEE 802.4 (1989 Chapters 12,13).

3.9.4 BIT ENCODING—5 Mbps carrierband:
a. 5 MHz represents zero's
b. 10 MHz represents one's

3.9.5 NETWORK ACCESS—Token passing. Refer to Figure 23.

| SD (8) | DA (16) | SA (16) | MAC (8 - 65776) | FCS (32) | ED (8) |

SD (Start Delimiter) : Indicates Start of Message
DA (Destination Address) : Specifies address of where token is being passed
SA (Source Address) : Specifies transmitter address
MAC (Media Access Control) : Contains LLC (Logical Link Control) which holds the Function Code.
Contains message data (if appropriate)
Contains control code (i.e., acknowledgement message, a read command, a write command)
FCS (Frame Control Sequence) : Code for passing Token
ED (End Delimiter) : Indicates End of Message

FIGURE 23—MESSAGE FORMAT

3.9.6 HANDSHAKING—Acknowledgements are optional. An acknowledgement will be placed in the Media Access Control Field of a new frame.

3.9.7 ERROR DETECTION MANAGEMENT—If a node fails to receive or pass the token, that node will be patched out of the network. If a node fails while possessing the token, the network will reset itself to recover the token.

3.9.8 FAULT TOLERANCE—A network is protected by the ratio of taps. A node can be removed from the network without affecting the operation of the other nodes.

3.9.9 DATA RATE—Bit rate of up to 20 Mbps.

3.9.10 FRAMING OVERHEAD—Maximum message length is 65856 bit times.

3.9.11 LATENCY—The latency a node encounters when needing to get on the bus depends on the time it takes for the Token to get passed to the node.

3.9.11 POWER REDUCTION MODE—Not specified.

3.9.12 CONTRIBUTORS
a. Robert Yee, Ford Motor Company, Dearborn, Michigan
b. Robert Crowder, Ship Star Associates, Newark, Delaware
c. Keith McNab, Industrial Technology Institute, Ann Arbor, Michigan

3.10 Synchronous Data Link Control/High-Level Data Link Control (SDLC/HDLC)

3.10.1 APPLICATION/AFFILIATION—SDLC was originally designed for computer-to-computer and computer-to-terminal applications by IBM. HDLC is a standard communication link protocol established by International Standards Organization (ISO).

3.10.2 TRANSMISSION MEDIA—Not specified. (SDLC/HDLC is a data link layer protocol only.)

3.10.3 PHYSICAL MEDIA—Not specified. (SDLC/HDLC is a data link layer protocol only.)

3.10.4 BIT ENCODING—NRZ and NRZI (NRZ Inverted). NRZI is used to ensure that within a frame, data transitions will occur at least every five bit times.

Zero insertion and deletion is performed automatically. A binary "0" is inserted by the transmitter after any succession of five "1's" within a frame. The receiver deletes the binary "0" that follows any five continuous "1's" within a frame.

3.10.5 NETWORK ACCESS—Not specified. (SDLC/HDLC is a data link layer protocol only.)

3.10.6 MESSAGE FORMAT—Transmit and receive data are contained in a format called a frame. All frames start with an opening flag and end with a closing flag. Between the opening flag and closing flag, a frame contains an address field, a control field, an optional information field and a frame check sequence field.

The fields that make up a frame and their sizes are shown in Figure 24.

The flag is a unique binary pattern (01111110). It provides the frame boundary and a reference for the position of each field of the frame. Optionally two successive frames can share one flag as the closing flag of one frame and the opening flag for the next frame. The receiver searches for a flag on a bit by bit basis and recognizes a flag at any time. The receiver establishes the frame synchronization with every flag. The flags mark the frame boundary and reference for each field.

The 8 bits following the opening flag are the address field. The address field can be extended if selected as an option. The 8 bits following the address field are the control field. The control field can be extended if selected as an option. The information field follows the control field and precedes the Frame Check Sequence field (FCS). The information field contains the "data" to be transferred, if any, but is not always contained in every frame. The information field will continue until it is terminated by the 16 bit FCS and the closing flag. The information field may be subdivided into 5, 6, 7, or 8 bit words. The 16 bits preceding the closing flag form the Frame Check Sequence field. The FCS is a cyclic redundancy check character.

3.10.7 Handshaking—The SDLC/HDLC protocol allows the user to select between full duplex, half duplex, synchronous, asynchronous, point to point, and loop communication modes.

In loop mode, a master station sends messages which all slave stations relay from one to another, until the message is received back by the master. Any slave station finding its address in the message receives the frame.

3.10.8 ERROR DETECTION MANAGEMENT—The following errors are identified: FCS (i.e., CRC) errors, more than six `1' bits consecutively, and minimum frame size errors are detected when receiving.

3.10.9 FAULT TOLERANCE—Not specified.

3.10.10 DATA RATE—Data rates are not specified for SDLC/HDLC as it is only a data link layer protocol.

3.10.11 FRAMING OVERHEAD—There are a minimum of six (6) bytes of overhead in every message. The length of the information field has no specified maximum. The actual sizes of the address field, control field, and the information field for a given SDLC/HDLC network implementation determines the amount of framing.

3.10.12 LATENCY—The network topology (e.g., point to point, loop, master/slave, full duplex, half duplex), network size and message sizes determines message latency.

3.10.13 POWER REDUCTION MODE—Not specified.

3.11 Token Ring (IEEE 802.5)

3.11.1 APPLICATION/AFFILIATION—Token Ring is specified in IEEE standard 802.5. The protocol was created, "for the purpose of compatible interconnection of data processing equipment."(1)

3.11.2 TRANSMISSION MEDIA—Shielded twisted pair is the media of choice but the specification allows for the use of coaxial cable and optical fiber.

3.11.3 PHYSICAL INTERFACE—The node is connected to the trunk cable that serves as the ring through two balanced twisted pair cables to a trunk coupling unit (TCU). Insertion into the ring is achieved by a "phantom circuit technique." The node impresses a DC voltage on the cables that connect it to the TCU. This DC voltage is transparent to the passage of signals on the ring and allows the TCU to insert the node onto the ring. This connection technique provides for node detection of open circuits, shorts, and node self-test. Refer to Figure 25 for more details.

FLAG	address	control	information	FCS	FLAG
8	X * 8	Y * 8	Z * WS	16	8

where:

    FLAG = <01111110>

    FCS = a 16 bit CRC

    WS   = word size (5, 6, 7 or 8 bits)

    X   = number of bytes in the address field

    Y   = number of bytes in the control field

    Z   = number of words (see WS above)

FIGURE 24—MESSAGE FRAME FORMAT

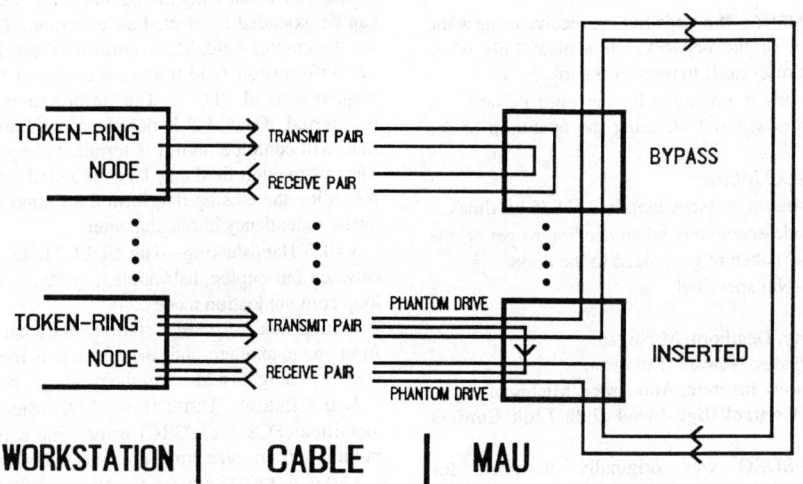

FIGURE 25—TOKEN RING PHYSICAL INTERFACE

3.11.4 BIT ENCODING—Coding is accomplished with differential Manchester-type.

It includes 4 symbols:
  a. Binary zero (0)
  b. Binary one (1)
  c. Non-data-J (J)
  d. Non-data-K (K)

3.11.5 NETWORK ACCESS—Any node may gain access by capturing the "token" that circulates the ring. The token consists of a 24-bit frame with unique signals that characterize the frame as a token. The token will have a priority assigned. A node may use the token only if its message is of equal or greater priority. Priorities are common for all nodes on the ring. Upon receiving the token, a node will transform it to a start-of-frame sequence by changing the token bit. A token holding timer controls the amount of time a node can use the token. When a node is not active, it will receive and retransmit each bit that is on the bus.

3.11.6 MESSAGE FORMAT—The token consists of three 8-bit fields. See Figure 26. The frame can range from 13 bytes (no information or data) to 21 bytes with the information field limited only by the maximum frame time. The maximum frame time is set for each ring. See Figure 27 for the frame. The configuration of the starting delimiter (SD), access control (AC), and frame control (FC) are shown in more detail in Figure 28.

3.11.7 HANDSHAKING—A transmitting node will append information to the token (the token becomes part of the frame) and send. A node(s), upon recognition of its address or a broadcast address, will copy the information into its receiving buffer and change the bits in the Frame Status (FS) field. These bits are Address-Recognized and Frame-Copied. The transmitting node receives the frame back after the message has passed through all the nodes on the ring and strips the information off and releases the token. The node will be able to determine from the FS if: the destination node is nonexistent/nonactive, the destination node exists but the frame was not copied, or the frame was copied.

SD	AC	ED

SD - Starting Delimiter (1 octet)
AC - Access Control (1 octet)
ED - Ending Delimiter (1 octet)

FIGURE 26—TOKEN THREE FIELD FORMAT

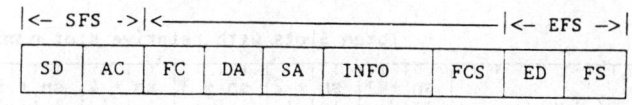

```
SFS - Start-of-Frame Sequence          INFO - Information (0 or more octets)
 SD - Starting Delimiter (octet)       FSC  - Frame-Check Sequence (4 octets)
 AC - Access Control (1 octet)         EFS  - End-of-Frame Sequence
 FC - Frame Control (1 octet)          ED   - Ending Delimiter (1 octet)
 DA - Destination Address (2 or 6 octets)  FS - Frame Status (1 octet)
 SA - Source Address (2 or 6 octets)
```

FIGURE 27—FRAME DEFINITION

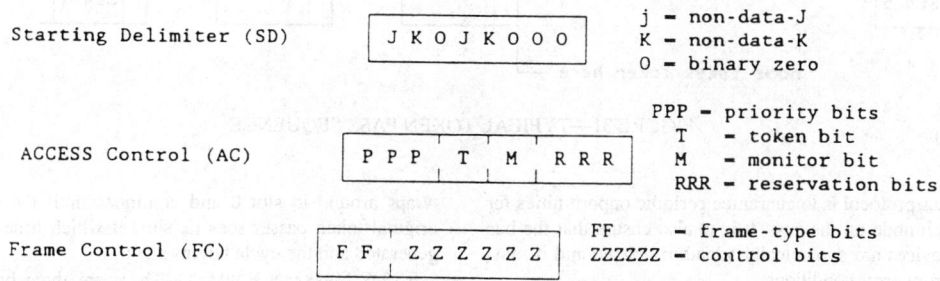

FIGURE 28—FRAME CONFIGURATION

3.11.8 ERROR DETECTION MANAGEMENT—Error and detection mechanisms exist to detect both hard and soft errors. Soft errors are logged by the individual nodes to determine the ring's service priority. Hard failures are acted upon immediately by the node upstream from the failure. A MAC (Medium Access Control) beacon is sent to all nodes suspending operation of the token ring until the disruption is removed. Most rings have a dedicated node to monitor the performance of the ring.

3.11.9 FAULT TOLERANCE—Each node can detect open and shorts on the ring and can monitor its own performance via closed-loop (deinsertion from the TCU) methods. Detection of hard faults is performed by individual nodes and system degradation/recovery is performed by a specific node.

3.11.10 DATA RATE—The data rate can be 4 Mb/s or 16 Mb/s.

3.11.11 FRAMING OVERHEAD—The largest frame size, no information or data, is 21 bytes. Assume the ring will allow the transmission of 4 Kbytes of data for a best case. The percentage of overhead is therefore 168/(168+32)=0.004%. Now assume the ring allows only 1 Kb of data to be transmitted. The percentage of overhead is therefore 168/(168+1024)=14%.

3.11.12 LATENCY—In a token pass protocol, latency is affected by the number of nodes on the ring, the maximum length of message allowed and bus loading. In IEEE 802.5, the token has an associated priority, this allows quick access to the bus for higher priority messages. Also, there is a token holding timer that monitors the amount of time a node has the token.

3.11.13 POWER REDUCTION MODE—Not specified.

3.12 Token Slot Network

3.12.1 APPLICATION/AFFILIATION—General Motors developed protocol for high-performance vehicle control and general information sharing.

3.12.2 TRANSMISSION MEDIA—Not specified. Electrical twisted pair or fiber optic media is recommended. Fiber optic Token Slot Networks operating at 1 Mb/s have been demonstrated.

3.12.3 PHYSICAL INTERFACE—Not specified. Multiple access to a logical common bus is required.

3.12.4 BIT ENCODING—NRZ (Non Return to Zero) with opposite logic level bit insertion (stuffing) after five contiguous bits of the same state. Receiving nodes detect and remove inserted bits.

3.12.5 NETWORK ACCESS—The token passing bus network is open, peer oriented, and multimaster. It is non-contention and uses a time slot token passing technique. See Figures 29, 30, and 31.

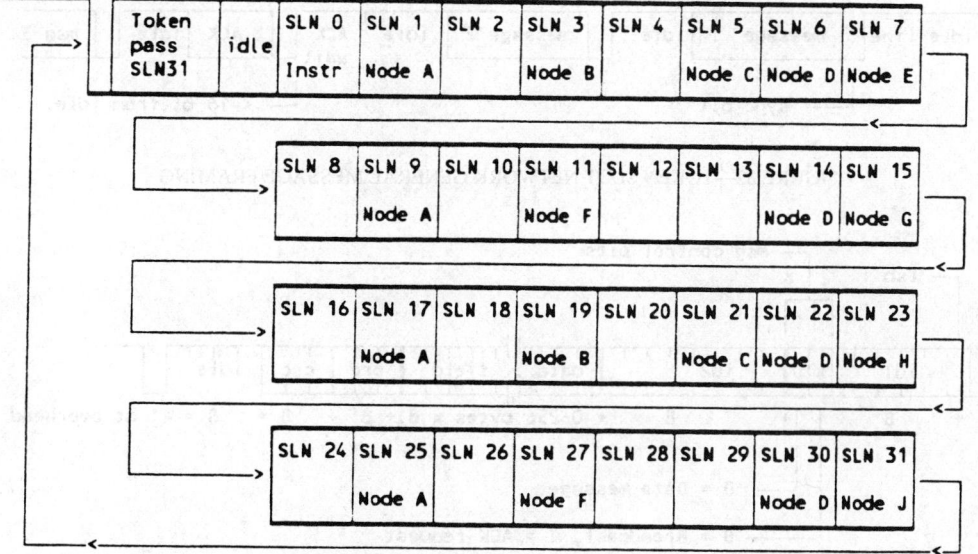

FIGURE 29—A TYPICAL TOKEN SLOT NODE ASSIGNMENT AND SLOT SEQUENCE CYCLE PATTERN

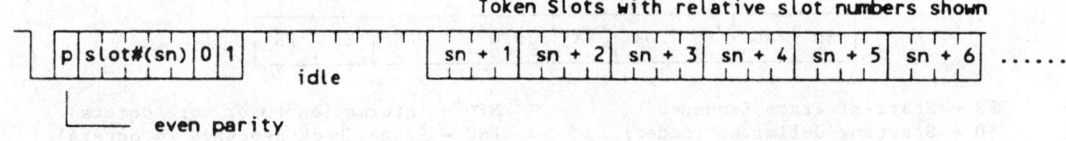

FIGURE 30—TOKEN PASS MESSAGE FORMAT

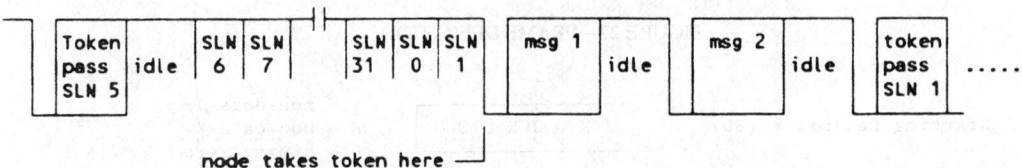

FIGURE 31—TYPICAL TOKEN PASS SEQUENCE

The intent of this bus access protocol is to guarantee periodic opportunities for message transmission by each node on the bus. It is to also ensure that the bus remains operational when devices are dynamically added or deleted and it must provide for quick recovery from error conditions.

After a node has completed sending its message traffic, a sequenced scan of short, equal time intervals (slots) offer bus transmit privileges to the node slot owners as follows (see Figure 30): A token pass message (or a bus jam) instructs all nodes to begin the token slot timing mode. Each node is assigned one or more specific time slots and will activate its transmitter to send a message during its slot only if it is operational and has message traffic to send. Otherwise the token slot interval is allowed to pass. When the transmitter is activated, all other nodes recognize that the token has been taken and they enter the receive mode.

The new token owner next proceeds to send its message traffic (see Figure 31). Token hold times are individually assigned to each node and are strictly limited to assure a system maximum message latency limit. Individual message transmit priorities are determined by each node's application and are not restricted by the communications data link.

A node concludes a transmit session by sending the token pass message which contains the current slot number (see Figure 31). In the ensuing token slot sequence, the node which owns the next sequential slot number may take the token (or let it pass). When the maximum slot number is reached, the sequence wraps around to slot 0 and continues until the slot is picked up or until the original token passer sees its slot, at which time a new token pass message is generated and the cycle begins again.

3.12.6 MESSAGE FORMAT—There are three basic message types which are distinguished by the 2-bit control field which is found in the first byte:

a. Token pass (see Figures 30 and 31)—This is a single byte message which contains the message control field (2 bits), the current slot number (5 bits), and a single parity (even) bit. It is followed by an idle line (8 bits) delimiter.

b. Data (see Figures 32 and 33)—This includes the message control field (2 bits), a message I.D. field (14 bits), up to 256 bytes of data, a 16-bit Cyclic Redundancy Check (CRC) field, and a message delimiter bus idle line (8 bits). The message control field is used to request an Acknowledge message response.

c. Acknowledge (see Figure 34)—This is a fixed, single byte ($D5) message plus an 8-bit idle line delimiter (16 total bits). It is initiated by the previous data message control field.

3.12.7 HANDSHAKING—Any data message can be dynamically programmed to command an immediate returned acknowledge message from one receiving node. If after a short wait interval, the requested acknowledge has not been received, the sender may retry and/or proceed to send other messages. Note that the responding node does not possess the token.

FIGURE 32—TOKEN SLOT NETWORK GENERAL MESSAGE FRAMING

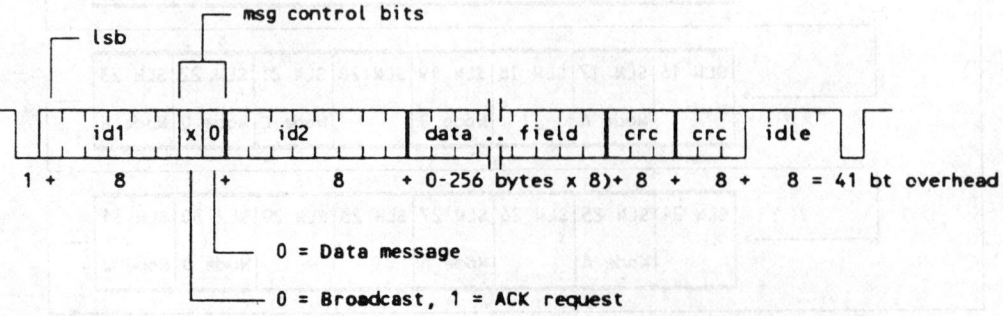

FIGURE 33—TOKEN SLOT NETWORK DATA MESSAGE FORMAT

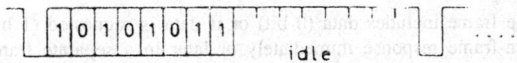

FIGURE 34—TOKEN SLOT NETWORK ACK MESSAGE FORMAT

3.12.8 ERROR DETECTION MANAGEMENT—Both the receiving and the transmitting nodes independently monitor transmissions on the bus. The transmitting node checks messages for a 1:1 received to transmitted bit correspondence. All nodes check for correct timing, CRC, intermessage gaps, and (if requested) an acknowledge message. Receiving nodes do not acknowledge erroneous messages.

The 16-bit CRC conforms to the CCITT standard and detects all single bit errors, all parity errors, and all burst errors less than 17 bits long. For burst errors longer than 16 bits, the CRC misses 0.0015% of errors.

A transmitter detected Bus Time Out (BTO) error occurs when the bus is idle for more than a complete token slot sequence period. When a BTO is detected, all nodes start a new token pass slot sequence beginning at slot 0.

Bus errors or collisions cause the detecting node to generate a bus jam signal (a dominant line for 8 bit times) before the end-of-message idle line. This declares the current operation or message invalid and instructs all nodes to start a new token pass slot sequence beginning at slot 0.

3.12.9 FAULT TOLERANCE—Each transmitting node monitors its own bus performance and fault history. Appropriate degraded mode operations are controlled by the node. The loss of any node or even a separated bus will not affect the continued bus operation by the remaining nodes.

3.12.10 DATA RATE—Data rate limit is specified at 2 MHz. However, the data rate is only limited by bus media bandwidth and future data rate growth is possible.

3.12.11 FRAMING OVERHEAD—The message framing overhead is summarized in Figures 35 and 36. See Figure 37 for total message latency calculation methods.

See Figures 30 and 31.

Slot width (assume xmtr ON at mid slot = 2 bt)	2
Token Pass Message	8
Delimiter gap	8
Token pass overhead per node:	18 bit

FIGURE 35—TOKEN PASSING SLOT OVERHEAD IN BIT TIMES (bt)

See Figures 32, 33, and 34.

Synchronization bit	1
ID	16
Data (system determined limit e.g. 0 - 256 bytes)	var
CRC	16
Intermessage delimiter gap	8
Acknowledge response (if requested) + gap	16
Per message overhead - with ACK	57 bit
or Per message overhead - without ACK	41 bit

FIGURE 36—TOKEN SLOT DATA MESSAGE OVERHEAD

3.12.12 LATENCY—The protocol is non-contention and deterministic. As such, message latencies in the Token Slot Network are both predictable and bounded—a requirement for feedback control systems.

Factors which affect message latency times are discussed below. See Figure 37 for methods of latency calculation and prediction.

The token loop time determines the interval between opportunities to transmit a message. It is defined as the total elapsed time between token possessions by a particular node. It includes all message traffic, token pass slot times, and all token hold times.

Token Slot Time Length—During the token pass sequence, each time slot must provide sufficient time for worst case signal propagation delays in order to allow nodes to detect that the token has been taken.

Token Hold Time is the maximum number of bit times that each node is allowed to hold the token. All message IDs, data fields, CRCs, intermessage gaps, message synchronization bits, NRZ5 bit insertions, acknowledge messages or acknowledge time outs, and token pass messages must not exceed this limit. Each node monitors and controls this time to stay within its assigned limit.

Token Slot Message Overhead (in bit times = bt):

Synchronization bit	1
ID	16
Data (system determined limit - bytes)	(0-256 bytes)
CRC	16
Intermessage delimiter gap	8
Acknowledge response (if requested) + gap	16
Per message overhead - with ACK:	57 bt
or Per message overhead - without ACK:	41 bt

Token Passing Slot Overhead:

Slot width (assume xmtr on at mid slot = 1 bt)	1
Token Pass Message	8
Delimiter gap	8
Token pass overhead per node:	17 bt

Summary of Loop Time Calculations:

For P nodes sending an N message loop with M total message data bytes:

Total message overhead (no ACK)	41N
Total token overhead	17P
Total message data time (m x 8 bits)	8M
Unused slots = slot width x (max #slots - #used slots) = 1(32-P)	
Total Loop Time	41N + 17P + 8M + 4(32-P)

Example Token Slot Timing Calculation

For 8 nodes sending a 16 message loop with 32 total message data bytes (2 msgs x 2 bytes per node x 8 nodes):

Total message overhead (no ACK)	41x16	=	656
Total token overhead	17x8	=	+ 136
			792
Total message data time (32 x 8 bits)	=		+ 256
Unused slot time	(1x24)	=	+ 24
Total Loop Time (bt)			1072

Data-to-total time overhead efficiency: 256/1072 = 24.0%

NOTE—See Figures 30 through 34

FIGURE 37—DETERMINATION OF TOKEN SLOT MESSAGE LATENCIES

3.12.13 POWER REDUCTION MODE—Not specified. Could be implemented.

3.13 VAN (Vehicle Area Network)

3.13.1 APPLICATION/AFFILIATION—VAN (Vehicle Area Network) is a multiplex bus protocol proposal being considered by the ISO Technical Committee 22/SC3/WG1.

3.13.2 TRANSMISSION MEDIA—Twisted pair.

3.13.3 PHYSICAL INTERFACE—Transmission is on differential pair using current sources. An optional analog filter is provided to increase noise immunity. See Figure 38.

3.13.4 BIT ENCODING—Two different bit representations are allowed: L-MANCHESTER and E-MANCHESTER. The bit representation is selected by the user. When L-MANCHESTER is selected, all the bits in the frame are Manchester encoded. E-MANCHESTER will be NRZ encoding except for the last bit of each nibble. With E-MANCHESTER: IDLE, IMS, and SOF are NRZ encoded, and ACK is MANCHESTER encoded. See Figure 39 for an example of E-MANCHESTER encoding.

3.13.5 NETWORK ACCESS—Prioritized nondestructive bitwise arbitration. Arbitration is resolved using the unique 12-bit identifier field at the start of the message. The format of this field is user defined. VAN allows in-frame access based on a media access rank "R." A node can access the frame provided the previous R-1 bits of the arbitration field have been emitted and the previous time slot was recessive.

3.13.6 MESSAGE FORMAT—The message frame consists of 8 fields. The start of frame (SOF) provides a common time reference that allows receiving nodes to correct their local clock. The start message bit initializes the frame. The two-byte frame identification field includes the unique 12-bit message identifier followed by three control bits and the remote transmission request bit (RTR). The RTR specifies if the frame includes data (0 bit) or if data is requested (1 bit). The RTR allows in-frame response immediately or later in a separate frame. The message frame is shown in Figure 40.

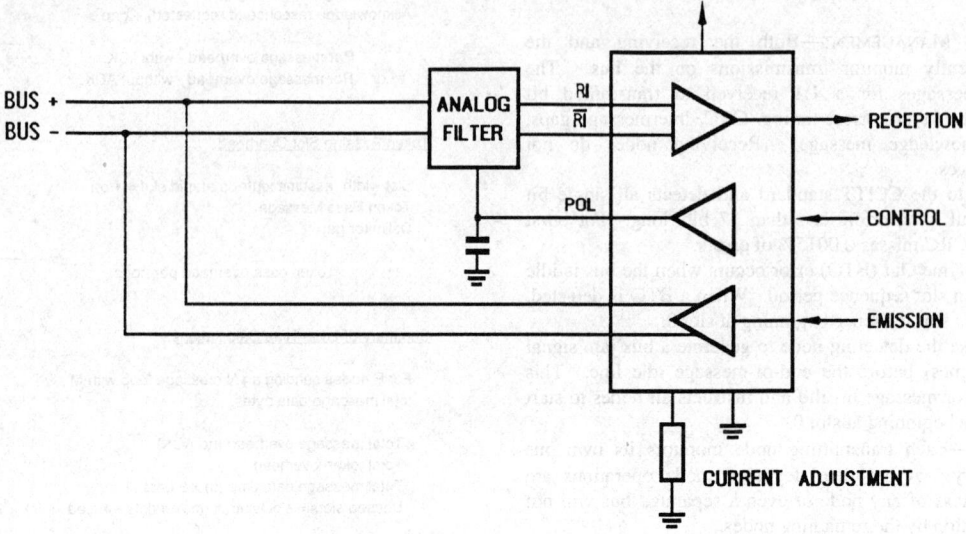

FIGURE 38—FAULT TOLERANT BUS INTERFACE

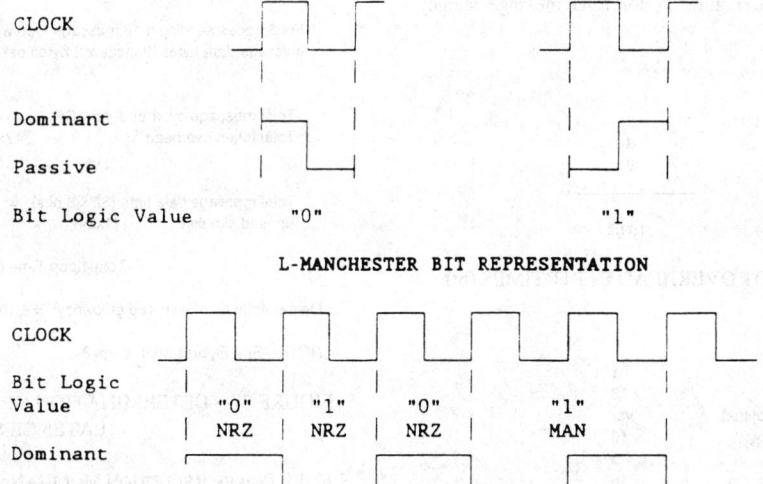

FIGURE 39—ILLUSTRATION OF VAN BIT ENCODING

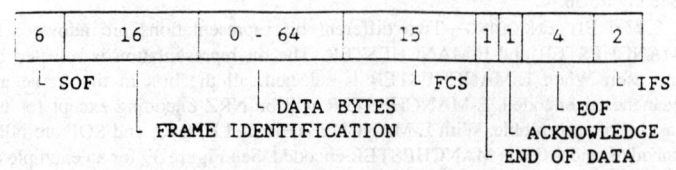

FIGURE 40—SINGLE TYPE FRAME FORMAT

3.13.7 HANDSHAKING—Message acknowledgement is achieved in one of two ways: no acknowledge or positive acknowledge, meaning at least one station has received and accepted the message. If the receiving station cannot make an in-frame acknowledge, a separate acknowledge message must be sent later.

3.13.8 ERROR DETECTION MANAGEMENT—Methods of error detection include: level monitoring, CRC, code violation detection and message frame check. Level monitoring is performed by the transmitting node. The transmitted bit levels are compared with the bit levels detected on the bus. Frame checking is performed through a 15-bit CRC code:

$$(x^8 + x^4 + x^3 + x^2 + 1)/(x^7 + 1) \qquad (Eq.5)$$

3.13.9 FAULT TOLERANCE—Single wire operation is possible with the differential drive scheme due to the AC coupling if the other wire is shorted to ground, shorted to Vbatt, or open circuited.

3.13.10 DATA RATE—The data rate is not specified.

3.13.11 FRAMING OVERHEAD—The total frame size (including data) is 109 bits. The total frame size includes start, stop and idle bits. The maximum amount of data allowed in a frame is 64 bits. The percentage of overhead is therefore 45/(45+64)=41.3%.

3.13.12 POWER REDUCTION MODE—Not specified.

SELECTION OF TRANSMISSION MEDIA
—SAE J2056/3 JUN91

SAE Information Report

Report of the SAE Vehicle Network for Multiplexing and Data Communications Standards Committee approved June 1991.

TABLE OF CONTENTS

1. Scope
   1.1 Background
   1.2 Interrelationship of Classes A, B and C
   1.3 Electromagnetic Susceptibility (EMS) Considerations
   1.4 Electromagnetic Interference (EMI) Considerations
2. References
   2.1 Applicable Documents
       2.1.1 SAE PUBLICATIONS
       2.1.2 OTHER PUBLICATIONS
3. Twisted Pair
   3.1 Inherent Advantages/Disadvantages of Twisted Pair Networks
       3.1.1 FAMILIARITY OF TWISTED PAIR NETWORKS
       3.1.2 RADIATED LINE LOSSES
       3.1.3 RECEIVER SUSCEPTIBILITY
       3.1.4 DRIVE PROBLEMS AND LINE LOSSES
   3.2 Network Architecture Options
       3.2.1 DATA ENCODING OF COMMUNICATION PROTOCOLS
       3.2.2 MFM ENCODING APPLIED TO VEHICLE MULTIPLEXING
       3.2.3 SIEFRIED ENCODING
       3.2.4 ARCNET ENCODING
       3.2.5 I/O HARDWARE CONFIGURATION
   3.3 Key Concerns of Twisted Pair Networks
       3.3.1 COMPUTER SIMULATION OF EMI LEVELS
       3.3.2 FOUR MEDIA DRIVING TECHNIQUES CONSIDERED
           a. Voltage Drive I/O
           b. Balanced Current I/O
           c. Transformer Coupling
           d. Optical Coupling
       3.3.3 MEDIUM DRIVING AND ENCODING TECHNIQUES CONCLUSIONS
4. Shielded/Coaxial Cable
   4.1 Inherent Features of Shielded/Coaxial Cable Networks
   4.2 Network Architecture Options
   4.3 Key Concerns of Shielded/Coaxial Cable Networks
5. Fiber Optic
   5.1 Inherent Features of Fiber Optic Systems
       5.1.1 PRINCIPAL ADVANTAGES
           a. EMI/RFI Immunity
           b. Data Bus Termination Impedance Elimination
           c. Immunity From Cross-Talk
           d. Ground Loop Elimination
       5.1.2 GENERAL ADVANTAGES
           a. Data Rate or Bandwidth
           b. Weight and Space Savings
           c. Long-Term Reliability
           d. Immunity from Short Circuits
           e. Data Integrity
           f. Lack of Familiarity
       5.1.3 DISADVANTAGES
           a. Sharp Bending Radii for Fibers
           b. Unidirectional Transfer of Data
           c. Lack of Standards
           d. Lack of Automotive-Grade Connectors
           e. Availability of Reliable and Cost Effective Automotive Components
   5.2 Network Architecture Options
       a. Cost
       b. Physical Complexity
       c. Power Moding
       d. Fault Tolerance
       e. Expandability
       f. Serviceability
       g. Latency
       5.2.1 ACTIVE STAR
       5.2.1.1 *Advantages*
           a. Cost
       5.2.1.2 *Disadvantages*
           a. Cost
           b. Complexity
           c. Power Moding
           d. Fault Tolerance
           e. Expandability
           f. Serviceability
           g. Latency
       5.2.2 PASSIVE STAR
       5.2.2.1 *Advantages*
           a. Cost
           b. Power Moding
           c. Fault Tolerance
           d. Latency
           e. Serviceability
       5.2.2.2 *Disadvantages*
           a. Cost
           b. Complexity
           c. Expandability
       5.2.3 SINGLE RING
       5.2.3.1 *Advantages*
           a. Cost
           b. Complexity
           c. Expandability
       5.2.3.2 *Disadvantages*
           a. Power Moding
           b. Fault Tolerance
           c. Serviceability
           d. Latency
       5.2.4 DOUBLE RING
       5.2.4.1 *Advantages*
           a. Fault Tolerance
           b. Expandability
           c. Serviceability
       5.2.4.2 *Disadvantages*
           a. Cost
           b. Complexity
           c. Power Moding
           d. Latency
       5.2.5 LINEAR TAPPED BUS
       5.2.5.1 *Advantages*
           a. Power Moding
           b. Latency
       5.2.5.2 *Disadvantages*
           a. Cost
           b. Complexity
           c. Power Moding
           d. Fault Tolerance
           e. Expandability
           f. Serviceability
           g. Latency
       5.2.6 NETWORK ARCHITECTURE CONCLUSIONS
   5.3 I/O Hardware Configuration
       5.3.1 TIME DIVISION MULTIPLEX (TDM)
       5.3.2 FREQUENCY (WAVELENGTH) DIVISION MULTIPLEX
       5.3.3 SPACE DIVISION MULTIPLEX (MULTISTRAND FIBER CABLE)
   5.4 Communication Protocols
       5.4.1 BIT WISE CONTENTION RESOLUTION BASED PROTOCOLS
       5.4.2 NON-CONTENTION BASED PROTOCOLS
   5.5 Key Concerns of Fiber Optic Systems
       5.5.1 NEW CULTURE/EDUCATION FOR AUTOMOTIVE ENVIRONMENT
           a. Manufacturing/Assembly
           b. Maintenance/Repair
           c. Perceived Cost
       5.5.2 LENGTH OF LINK
       5.5.3 DATA RATES
       5.5.4 FAILURE MODES
           a. LED
           b. LED Drive Circuit
           c. Fiber
           d. Connection System
           e. Receiver
           f. Passive Networking Components
           g. Active Networking Components
           h. Secondary Failure Effects
       5.5.4.1 *Consequences of Failure*
           a. Passive Networks

b. Active Networks
6. Summary & Conclusions
Appendix A

Foreword—It has been commonly accepted by most automotive RF engineers that a Class C Network at a transmission rate above 100 kilobits per second (kbps) will require either a fiber optic or a shielded cable for the transmission medium. Some communications engineers have proposed that transformer coupling to a twisted pair may be an acceptable alternative to a fiber optic or a shielded cable.

It has also been generally recognized that the EMI levels available in a vehicle to corrupt data transmission are very high and cannot be filtered out of the data. The employment of a fiber optic or a shielded cable for the transmission medium would also solve this EMI problem.

1. Scope—This SAE Information Report studies the present transmission media axioms and takes a fresh look at the Class C transmission medium requirements and also the possibilities and limitations of using a twisted pair as the transmission medium.

The choice of transmission medium is a large determining factor in choosing a Class C scheme.

1.1 Background—The Vehicle Network for Multiplexing and Data Communications (Multiplex) Committee has defined three classes of vehicle data communication Networks:

  a. Class A—Low-Speed Body Wiring and Control Functions, i.e., Control of Exterior Lamps
  b. Class B—Data Communications, i.e., Sharing of Vehicle Parametric Data
  c. Class C—High-Speed Real-Time Control, i.e., High-Speed Link for Distributed Processing

1.2 Interrelationship of Classes A, B, and C—The Class B Network is intended to be a functional superset of the Class A Network. That is, the Class B Bus must be capable of communications that would perform all of the functions of a Class A Bus. This feature protects the use of the same bus for all Class A and Class B functions or an alternate configuration of both buses with a "gateway" device. In a similar manner, the Class C Bus is intended as a functional superset of the Class B Bus.

1.3 Electromagnetic Susceptibility (EMS) Considerations—Inherent with the high data rates of a Class C Bus is a higher probability of electromagnetic interference (EMI) corrupting data. There has been a lot of research on Class B Networks that use twisted pair operating at data rates below 50 kbps and methods have been found to overcome the communication problems (SAE J1850). But, it is commonly agreed that the corruption of serial data by EMI will be an issue if a twisted pair or any other kind of conventional wiring and connector design is used at the higher data rates. Also, if data communication requirements dictate transmission rates above 50 kbps, another technique may be required because 50 kbps is the practical upper limit of these Class B Networks (SAE J1850) that use twisted pairs and conventional bus drivers.

1.4 Electromagnetic Interference (EMI) Considerations—A key concern is the generation of EMI when the Class C Vehicle Multiplexing Network is utilizing twisted pair for the transmission medium operating at data rates above 50 kbps. It is because of this EMI concern that most automotive RF engineers commonly accept that either a fiber optic or a shielded cable will be required for the transmission medium at data rates above 100 kbps.

It is expected that the growth of data communications on vehicles, the issue of shielding cost requirements, and electromagnetic compatibility of copper-based systems, will drive future development. These factors and other, as yet undefined, needs for Class C communication will eventually drive the implementation of automotive fiber optic systems for higher data transfer rates.

2. References

2.1 Applicable Documents—The following publications form a part of this specification to the extent specified herein. The latest issue of SAE publications shall apply.

2.1.1 SAE Publications—Available from SAE, 400 Commonwealth Drive, Warrendale, PA 15096-0001.

SAE J1850—Class B Data Communications Network Interface
SAE J2056/1—Class C Multiplexing Applications/Definition
SAE J2056/2—Class C Multiplexing Survey of Known Protocols

2.1.2 Other Publications

2.1.2.1 Henry W. Ott, Bell Laboratories, Noise Reduction Techniques in Electronic Systems, A Wiley-Interscience Publications. Second Edition, 1988.

2.1.2.2 CISPR/D/WG2 (Secretariat)19, September 1989, International Electrotechnical Commission, International Special Committee on Radio Interference (CISPR), Subcommittee D: Interference Relating to Motor Vehicles and Internal Combustion Engines, Working Group 2, Test Limits and Methods of Measurement of Radio Disturbance from Vehicle Components and Modules: Conducted Emissions, 150 kHz to 108 MHz and Radiated Emissions, 150 kHz to 1000 MHz

2.1.2.3 A. L. Harmer, SPIE Vol. 468 Fibre Optics '84, pp. 174-185 (1984).

2.1.2.4 W. A. Rogers, D. R. Kimberlin, and R. A. Meade, Soc. Automotive Eng. 88, pp. 50-56 (1980).

2.1.2.5 M. W. Lowndes and E. V. Phillips, 4th Int. Conf. Automotive Electronics IEE Vol. 229, pp. 154-159, 1983.

2.1.2.6 P. G. Duesbury and R. S. Chana, 4th Int. Conf. Automotive Electronics IEE Vol. 229, pp. 160-164, 1983.

2.1.2.7 K. Sekiguchi, Int. Fibre Optics and Commun. 3, pp. 56-60, 1982.

2.1.2.8 T. Sasayama, Hirayama, S. Oho, T. Shibata, A. Hasegawa, and Y. Minai, 4th Int. Conf. Automotive Electronics IEE Vol. 229, Nov. 1983.

2.1.2.9 K. Sasai, Sitev Conf., pp. i-ii, May 1983.

2.1.2.10 R. E. Steele and H. J. Schmitt, SPIE Vol. 840, Fiber Optic Systems for Mobile Platforms '87.

2.1.2.11 G. D. Miller, SPIE Vol. 989, Fiber Optic Systems for Mobile Platforms II '88, pp. 124-132 (1988).

2.1.2.12 D. A. Messuri, G. D. Miller and R. E. Steele, A Fiber Optic Connection System Designed for Automotive Applications, Soc. Automotive Eng. #890202, Feb. '89.

2.1.2.13 T. W. Whitehead, Du Pont Electronics Private Communications (1989).

2.1.2.14 T. Sasayama and A. Hideki, SPIE Vol. 989, Fiber Optic Systems for Mobile Platforms II, '88.

2.1.2.15 M. Kitazawa, Mitsubishi Rayon, Private Communications (1989).

3. Twisted Pair—A Twisted Pair is defined to be a transmission line consisting of two similar conductors that are insulated from each other and are twisted around each other to form a communication channel. The purpose for twisting the conductors around each other is to reduce the electric and magnetic field interaction with other conductors. In recent years there has been a lot of research on Class B Networks that use twisted pair operating at data rates below 50 kbps. At Class C data rates (>100 kbps) many new problems need development and attention.

3.1 Inherent Advantages/Disadvantages of Twisted Pair Networks—As the result of widespread Class B network development a lot of research has been completed on the use of copper-based twisted pair for a transmission media. Class C development is an extension of that activity.

3.1.1 Familiarity of Twisted Pair Networks—The desire to use twisted pair for the transmission medium of a Class C Network by the automotive industry is universal. This desire is twisted pair's biggest advantage. At lower data rates the automotive wiring requirements for twisted pair and connector techniques are well known and developed. The failure modes such as shorts to ground and battery have been extensively studied. The use of proper techniques for termination have been developed. An effective I/O can be easily achieved by integrating the transmission hardware, used for driving the twisted pair, into an interface device that also contains the receiver and some external discrete filter components for EMI rejection. Bidirectional data transfer is easily obtainable using the same twisted pair for both reception and transmission. Statistical studies have provided data so that the reliability of a twisted pair network is known. The connector industry is currently developing insulation displacement type connectors so that in the future automated machines can be programmed to place bus connector drops as required, further reducing the cost of the wiring harness. Of course at Class C data rates many of these and other factors such as the maintenance of twist uniformity and the harness interconnection requirements are likely to change. A large investment in research and development must be completed in order to demonstrate feasibility. The magnitude of the task could easily be underestimated even though this development is an extension of familiar work.

In most communications systems the length of line is a large factor in determining the upper limit of data rates. However, line length in automotive networking is relatively small and does not play a major role, but the number of connectors and losses due to impedance mismatching at the connector is a concern. Perhaps developments in ribbon cabling techniques and insulation displacement connectors could improve this impedance matching situation.

3.1.2 Radiated Line Losses—The biggest problem to overcome is the fact that for data rates above 100 kbps the radiated line losses are very high (2.1.2.1). These radiated line losses cause transmitter line driver problems and generate large amounts of EMI. The work at Class B data rates demonstrated that the transition rise time was responsible for most of the EMI. The present automotive quality of a twisted pair network medium does not exhibit good transmission line characteristics. Also the capacitance load to the output driver from the twisted pair was measured to be approximately 2000 pfd. At Class C data rates this capacitance loading, impedance mismatching at the connector, maintenance of twist uniformity, and drive symmetry match requirements between bus outputs make it very difficult, if not impossible, to design an output driver. The challenge will be to achieve a low enough output impedance to drive a twisted pair without incurring excessive losses or spectral distortion of the transitions especially for data rates above 1 Megabit per second (Mbps).

3.1.3 RECEIVER SUSCEPTIBILITY—The receiver is very susceptible to coupled (capacitive/inductive) and longitudinal noise interference (see 3.3.2 for details on longitudinal noise). At Class C data rates it is much more difficult to devise a filter that could eliminate the coupled line noise. The severity of this problem can be understood by realizing that the vehicle wiring harness appears to be resonant around 25 to 30 MHz which is approximately a quarter wavelength in length. Switching noise and spikes are broadband and excite the wiring harness to resonate at high levels. At Class B data rates this broadband noise is coupled into the circuit but is effectively eliminated by the filter. For Class C multiplexing the data rates required may be at 1 to 10 Mbps. This wire harness resonance is too close to the filter cutoff frequency for traditional filtering techniques to be very effective.

3.1.4 DRIVE PROBLEMS AND LINE LOSSES—The transmitter drive problem and line losses cause many experts to conclude that twisted pair and shielded twisted pair are not usable for data rates above 100 kbps. The filtering techniques for receiver susceptibility would also leave the network highly susceptible to data corruption and thus require very sophisticated error detection or reconstruction techniques.

3.2 Network Architecture Options—The suitable topology configurations of twisted pair is a very strong advantage. It can accommodate any configuration from a Star, Tee, Bus, Ring, Daisy Chain, or various Hybrids. Many data encoding techniques have been employed with twisted pair as the transmission medium with a variety of I/O hardware configurations.

3.2.1 DATA ENCODING OF COMMUNICATION PROTOCOLS—The data encoding technique has a significant effect on the radiated EMI. To achieve the highest possible data rate it is important to choose a data encoding method that has the fewest transitions per bit with the maximum of time between transitions and is bit synchronized so that invalid bit testing can be effective. Invalid bit testing has proven to play a large role in providing data integrity in a high EMI environment. PWM, for example, has two transit ions per bit with 1/3 bit times between transitions. NRZ has a maximum of one transition per bit but without the added overhead of synchronizing transitions is not suitable. Some of the disk drive encoding techniques such as Modified Frequency Modulation (MFM) or Run Length Limited (RLL) are synchronous with fewer than one transition per bit (see Table 1 for a comparison chart of a selection of encoding techniques). The variable column in the table describes an attribute whereby the transmission time for data byte is a variable quantity depending on the data value.

TABLE 1—COMPARISON OF DATA ENCODING TECHNIQUES

	Arbitrates	Compression	Synchronizing	Trans/Bit	Variable
PWM	Yes	Base	Yes	Two	No
VPWM	Yes	2 to 1	Yes	One	Yes
Manchester	Yes	1.5 to 1	Yes	1.5 Avg.	No
NRZ	Yes	3 to 1	No	One	No
MFM	Yes	3 to 1	Yes	0.75 Avg.	No
RLL	No	>6 to 1	Yes	<0.5 Avg.	Yes
Siefried	Yes	>8 to 1	Yes	<0.1	Yes
Arcnet	No	>16 to 1	Yes	<0.1	No

3.2.2 MFM ENCODING APPLIED TO VEHICLE MULTIPLEXING—A modulation technique developed during the latter 1960s called Modified Frequency Modulation (MFM) used in disk drives could be adapted to vehicle multiplexing. The advantage of using the MFM encoding technique is that it would be synchronous with an average of 0.75 transitions per bit. The encoding technique permits a transition rise time that can be maximized and wave shaped to significantly reduce EMI. Disk drives have a similar requirement where the modulation technique allows pulses to be recorded on a disk at maximum density. The diagram shown in Figure 1 demonstrates one way of applying MFM encoding technique to a communication data link.

The rule for encoding simply causes a transition at the data time slot when the data at that time slot is a logic "1." A transition is also generated at the clock time slot when the data before and after that clock time slot was a logic "0" (or two "0's" in a row).

3.2.3 SIEFRIED ENCODING—The Siefried Patented Encoding Process may be especially useful for Class C Multiplexing. This data compression encoding technique is based on digitally varying the pulse width proportionally to the data byte value and then passing the modulated square wave signal through a narrow band filter to change the signal to a near sine wave. The high frequency content of the signal is limited and can be transmitted across ordinary twisted pair network medium. The Siefried technique generates a redundant negative portion of the waveform that is a mirror image of the positive. The data decode circuit should extract two identical values from both the positive and negative portion of the signal. If not, this inconsistency can be used for detection of data corruption due to EMI. The data is recovered by employing a zero crossing detector and converting the sine wave back into square waves. The encoding technique defines a minimum duration square wave that corresponds to a zero data value. All other data values increase the square wave duration proportionally to the data value. This variation should be small in proportion to the zero value square wave duration and easily converted to a near sine wave by the filter. Assume a square wave bias value of 256 μs to illustrate the technique. In this example assume the data varies the square wave by 128 one-half microsecond steps corresponding to an 8 bit byte of data values thus varying the square wave pulse width from 256 to 384 μs. In this example one 8 bit byte of data was compressed into a single square wave, which duration varied from 256 to 384 μs.

The Siefried Encoding Technique could be used in a medium access scheme that requires bit-by-bit arbitration but, as discussed, arbitration requires nonsymmetrical output drives. Token passing or time division multiplexing, on the other hand, would be a better choice for the medium access method. These access methods would take advantage of the Siefried Compression Encoding Technique and allow balanced current or transformer coupling to achieve the highest possible data rates. It would seem realistic that the equivalent of 500 kbps to 1 Mbps of the Siefried encoded data could be communicated across a Class C Network that uses only twisted pair for the network medium. However, there are a number of factors such as cost, resolution requirements and signal distortion that need to be proven. These factors will require a large investment in research and development to demonstrate feasibility.

3.2.4 ARCNET ENCODING—The Arcnet Encoding Technique is similar to Siefried except instead of digitally varying the pulse width proportionally to the data byte value the process digitally varies the sine wave amplitude proportionally to the data byte value. The technique is reported to generate a 16 to 1 data compression level by encoding the positive sine wave separately from the negative. An 8 to 1 data compression for both halves of the sine wave is achieved.

3.2.5 I/O HARDWARE CONFIGURATION—The driving difficulties are increased if we require an arbitration based protocol because an arbitration based I/O such as that used with J1850 requires nonsymmetrical driving capabilities. The more difficult driving requirements of higher data rates are achievable if the transmitter could drive symmetrically both negatively and positively because better impedance matching is possible. Communications protocols such as Time Division Multiplexing, Master/Slave, or Token Pass may be a better choice than a bit-by-bit arbitration based protocol. A case can also be made for symmetrical driving by employing transformer coupling. However, the cost of transformer coupling may be an issue.

3.3 Key Concerns of Twisted Pair Networks—The EMI levels generated by a Class C Vehicle Multiplex Network and the susceptibility to externally generated EMI are the main concerns with utilizing twisted pair for transmission medium. A large factor in EMI levels is determined by both the data encoding methods and medium driving techniques.

3.3.1 COMPUTER SIMULATION OF EMI LEVELS—The EMI levels generated by a Class C Vehicle Multiplex Network has not been sufficiently studied or documented. A recent computer simulation was completed using Fourier Spectrum Modeling Technique. Appendix A details the method used for this EMI level modeling and the results of these studies are documented herein. The computer study predicts the EMI levels radiated by a single wire in a vehicle wiring harness and the effect of various data modulation methods. The four modulation techniques studied were PWM, VPWM, Manchester, and MFM. NRZ modulation would generate the same EMI levels as MFM and therefore was not simulated.

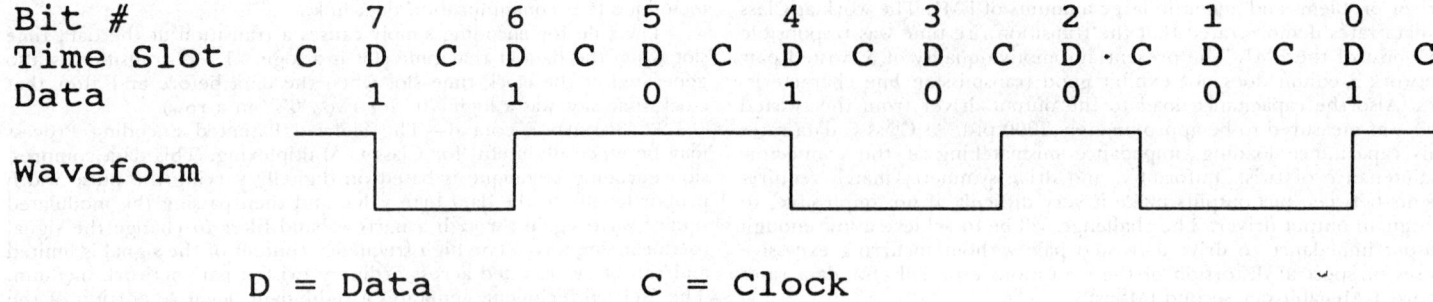

FIGURE 1—MFM ENCODED BYTE OF DATA

The first step in the computer simulation was to determine the worst case EMI levels for the four modulation methods studied. A constant 0.5 µs transition rise time was simulated for all the modulation techniques and all permutations of data were considered in order to establish the worst case examples of EMI levels for each data encoding method. Figure 2 illustrates the resulting predicted worst case EMI levels for these four modulation techniques. The curve plotted for each modulation is the maximum level generated by the fundamental frequency and each harmonic. All simulated modulation examples were for 100 kbps data rates. Notice all the curves are parallel to each other and only a few dbV apart with Manchester Encoding being worse and MFM predicting the least EMI.

It is a well-known fact that the level of EMI generated by the harmonics is a function of the transition rise times. The second step in the computer simulation was to model these four examples using a variable transition rise time in order to reduce these harmonic EMI levels. The simulation was generated using a 25% factor of the minimum feature size (minimum pulse width) to determine the rise time. The resulting graph is illustrated by Figure 3 and clearly shows the effect of rise time on EMI levels predicted. As for the first simulation all modulation examples used the same data rates of 100 kbps. Take note that the EMI levels predicted for MFM are significantly less than the other modulation methods as would be expected because it has proportionally the longest rise time.

The third step in simulation is illustrated by Figure 4. The resulting graph is for MFM where the data rate was varied from 41.6 kbps to 125 kbps. The simulation was generated using the same 25% factor of the minimum feature size for the rise time. Take note that the break point at 500 kHz for the 125 kbps data rate is approximately −32 dbV.

The EMI level at the 500 kHz point has special significance. The computer study predicts the EMI levels radiated by a single wire in a vehicle wiring harness. A simulation was generated for VPWM at 10.4 kbps data rate and a 16 µs rise time (see Figure 5). This simulation is of special significance because one implementation of SAE J1850 specifies this configuration and extensive vehicle verification tests show that it does not generate excessive EMI levels. Notice that the break point at 500 kHz is approximately −60 dbV.

CISPR/D/WG2 (Secretariat)19 Sept 1989 Radiated Emissions Antenna & Probe Test Document (2.1.2.2) has been generally interpreted by most automotive RF engineers to specify a break point at 500 kHz of −60 dbV. Another way of demonstrating the validity of this is to compare the predicted levels illustrated by Figure 5 to the levels referenced by this CISPR document.

3.3.2 FOUR MEDIA DRIVING TECHNIQUES CONSIDERED—Since the computer model simulates a single wire, a twisted pair transmission medium would provide generated EMI cancellation. The amount of EMI cancellation obtained is dependent on the performance of the balancing effectiveness of the medium driving technique employed. Drivers integrated into a single device can be carefully designed to have balanced source impedances. The receivers can be designed to have high common mode rejection ratios. It should be possible to achieve up to 60 dB of EMI cancellation particularly at AM band frequencies.

The susceptibility to interference of the network is a different issue. The source of coupled line noise is as was previously discussed in 3.1.3. To gain a better understanding as to the source of the longitudinal noise, consider the simplified circuit diagram shown in Figure 6. The transmitter is referenced at a different ground point than the receiver. The noise source at each ground point is directly (not capacitively or inductively) coupled to that ground point and normally significantly different. Voltage V_G represents the difference in ground noise potential or longitudinal transient noise between these two ground points. The magnitude of V_G is usually very large (i.e., < 200 V). The frequency content of longitudinal transient noise can go as low as the audio range thus making traditional low pass filters ineffective. I_A and I_B represent the two currents that flow as a result of V_G into the differential receiver. If these two currents were of equal magnitude and phase they would then be cancelled across the load resistor R_L. This situation usually never exists as illustrated. I_A takes the path passing through R_S, and I_B takes the more direct route to R_L. The receiver amplifies the difference current across R_L thus making it susceptible to longitudinal noise.

a. Voltage Drive I/O: A computer simulation was completed of a PWM encoded 41.6 kbps signal, assuming a 2 µs transition rise time. One implementation of SAE J1850 specifies this drive configuration and empirical tests have validated the EMI cancellation levels predicted by the study. The results of the simulation predict an approximately −37 dbV at the 500 kHz break point. This simulation can be interpreted to mean that in order to guarantee EMI-free operation greater than a modest 23 dbV of EMI cancellation must be achieved. If the same driving technique was used with MFM encoding, higher data rates would be achievable (see Figure 4), but operation at 100 kbps or 125 kbps would be within about 5 dbV of the PWM encoding 41.6 kbps signal. If required this 5 dbV improvement of EMI cancellation should be achievable.

The other point to consider with voltage drive I/O is the susceptibility to longitudinal noise. Figure 7 shows a simplified circuit diagram of the voltage drive I/O. Again the longitudinal noise currents do not cancel even though the source impedances R_S are balanced. I_A takes the path passing through Z_C, and I_B takes the more direct route to R_L. Z_C is of the same order of magnitude as R_S and, therefore, the longitudinal noise current match is poor.

b. Balanced Current I/O: The balanced current I/O can also be integrated into a single device and can achieve EMI cancellation similar to voltage drive.

The main advantage of balanced current I/O is that it can be designed to achieve a significant improvement in balancing the two longitudinal noise currents particularly at audio frequencies. Figure 8 shows a simplified circuit diagram of the balanced current I/O. Even though I_A takes the path passing through Z_C, and I_B takes the more direct route to R_L both currents must pass through the constant current sources I_S. These current sources present a much higher impedance than Z_C and significantly improves the current matching of the longitudinal noise currents. It is claimed that this factor is responsible for as much as 40 dB improvement in susceptibility to vehicle longitudinal noise.

The main concern in using balanced current I/O is that the sources and sink currents may be too low to allow the use of conventional automotive connectors.

c. Transformer Coupling: Transformer coupling is also a very effective method for matching load impedance and generated EMI cancellation. Transformer coupling also plays an important role by isolating the ground connection at both ends of the transmission line and effectively blocking the longitudinal noise current. It is claimed that this factor can be responsible for a greater than 40 dB improvement in susceptibility to data cor-

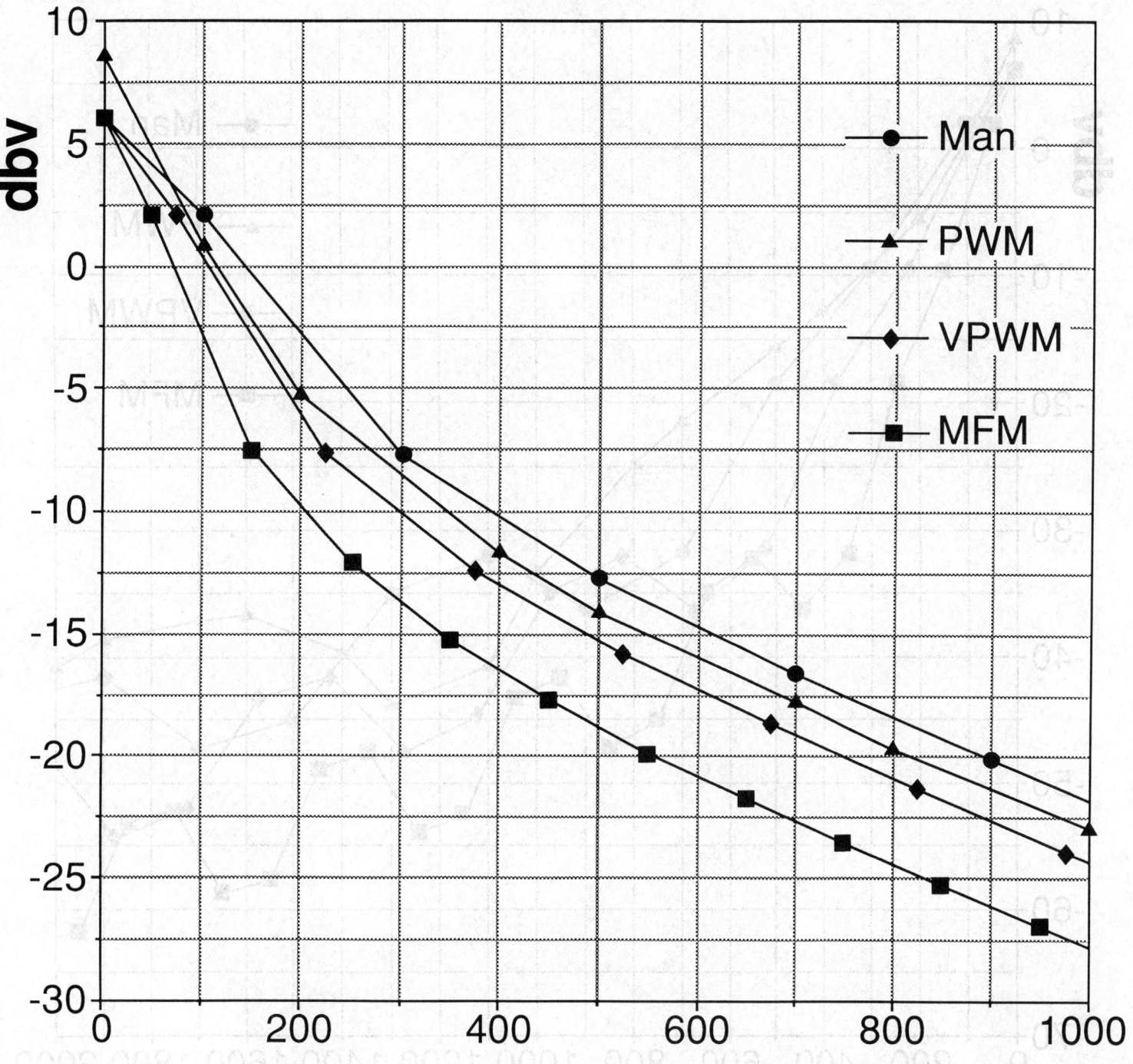

FIGURE 2—WORST CASE EMI LEVELS FOR PWM, VPWM, MANCHESTER, AND MFM

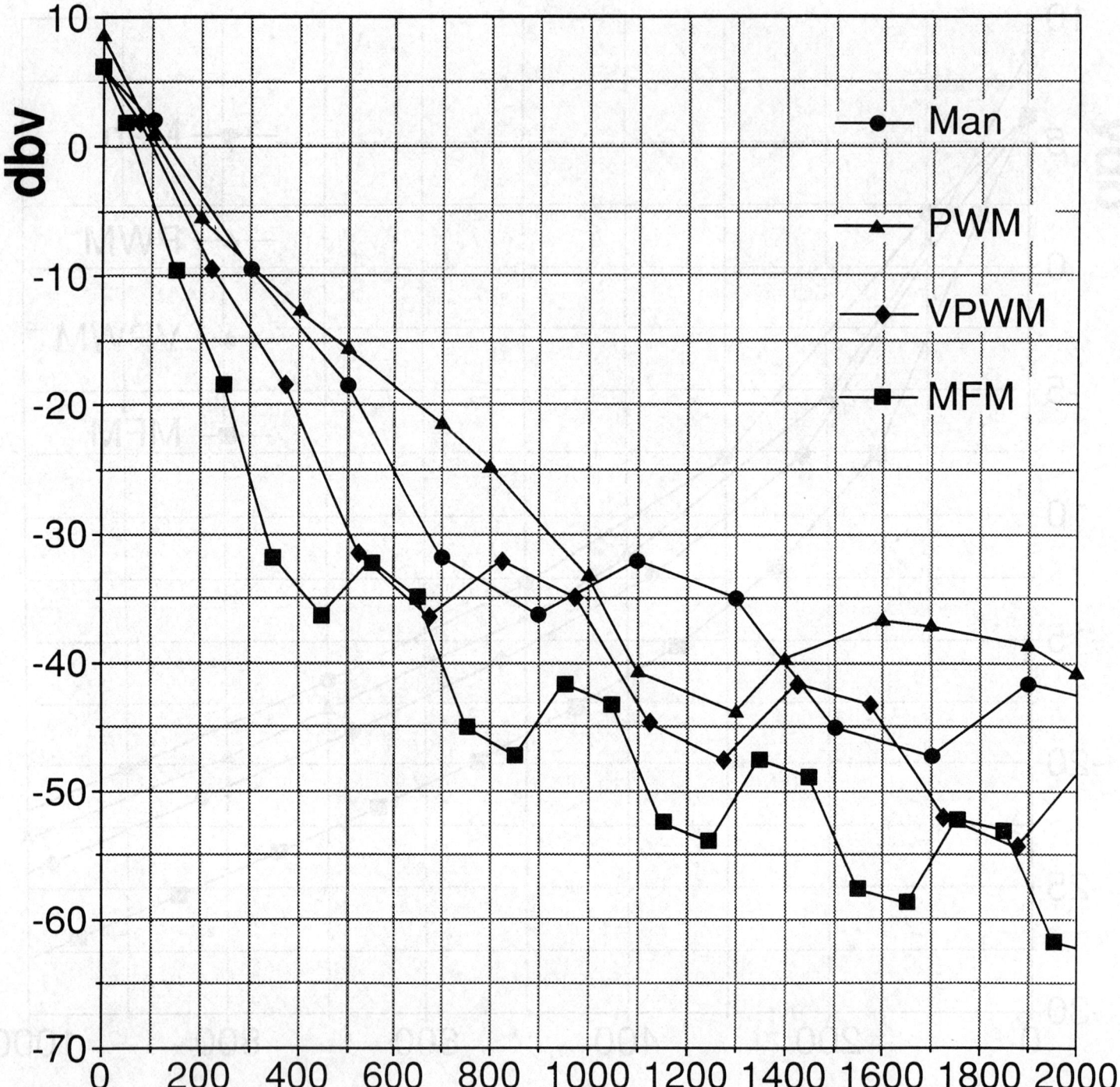

FIGURE 3—EMI LEVELS USING A 25% OF MINIMUM FEATURE TO ESTABLISH RISE TIME

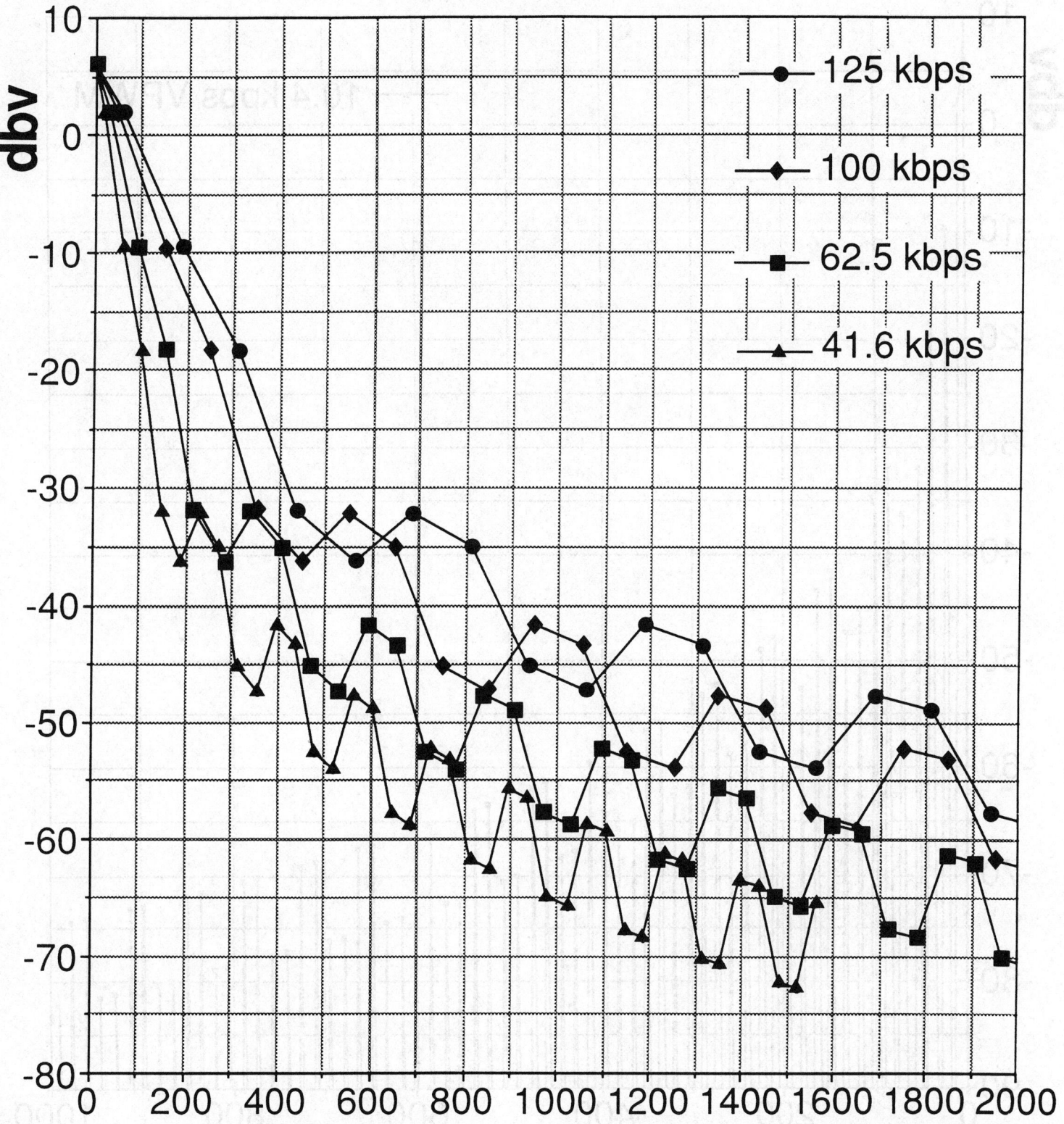

FIGURE 4—MFM FOR DATA RATES FROM 41.6 KBPS TO 125 KBPS

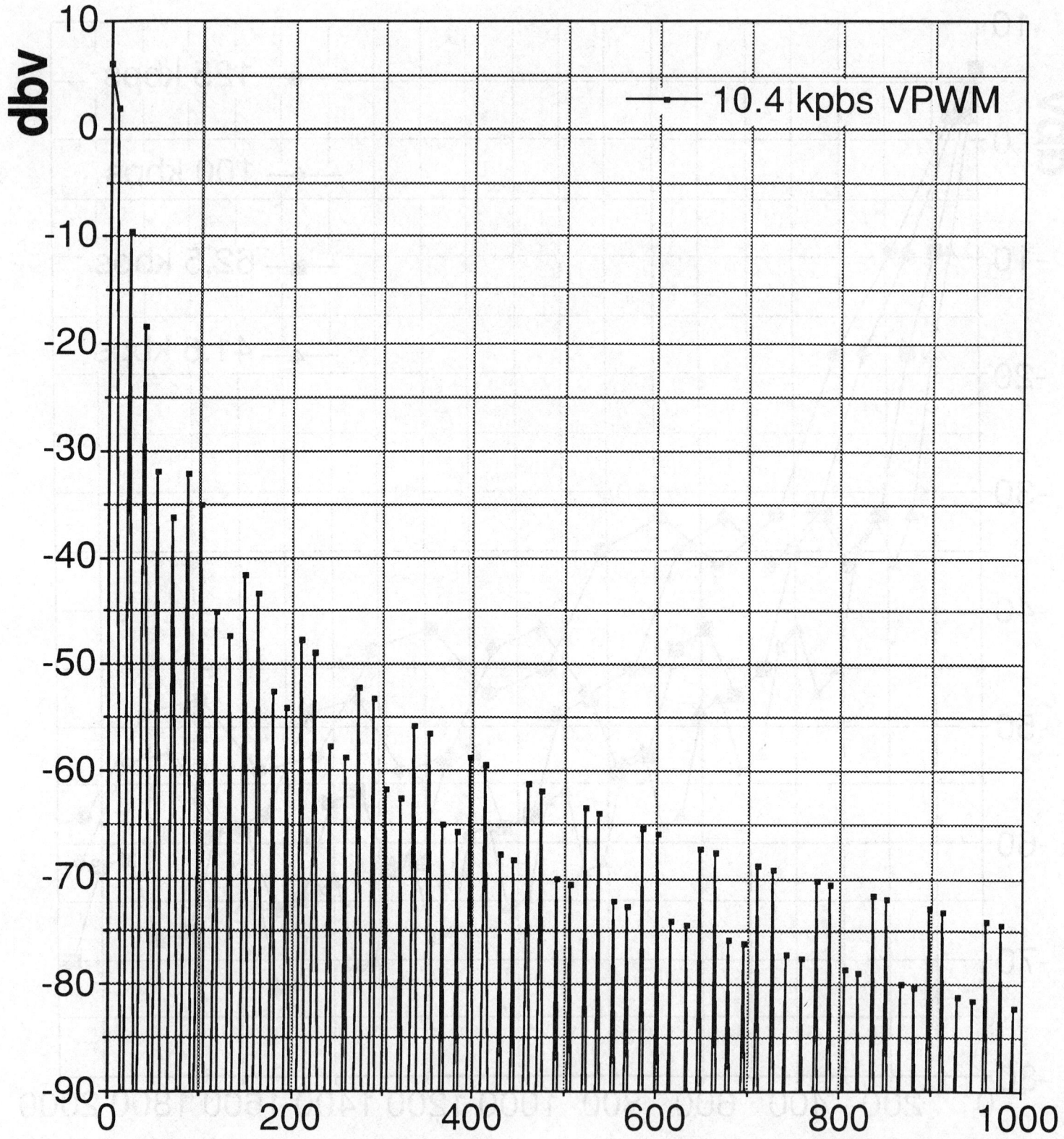

FIGURE 5—10.4 KBPS VPWM SIMULATION

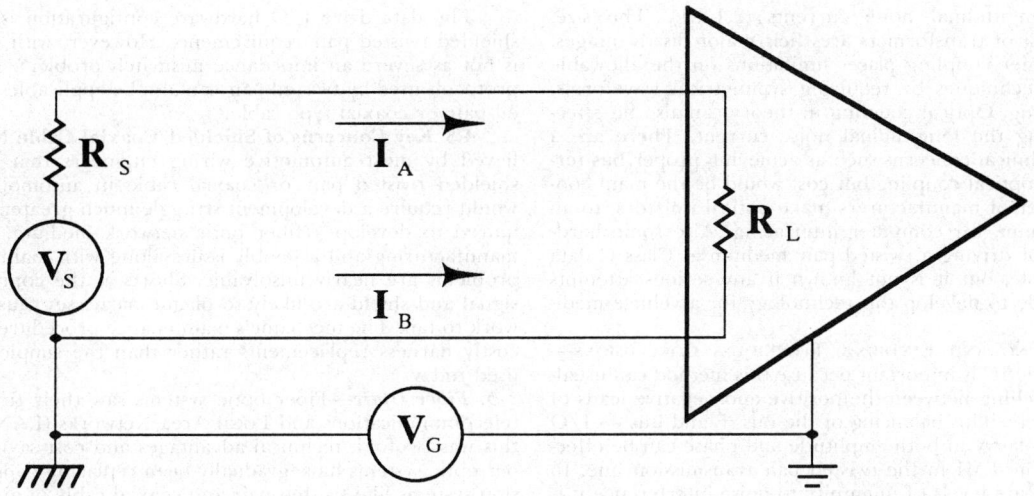

FIGURE 6—SIMPLIFIED CIRCUIT DIAGRAM OF DRIVER AND RECEIVER

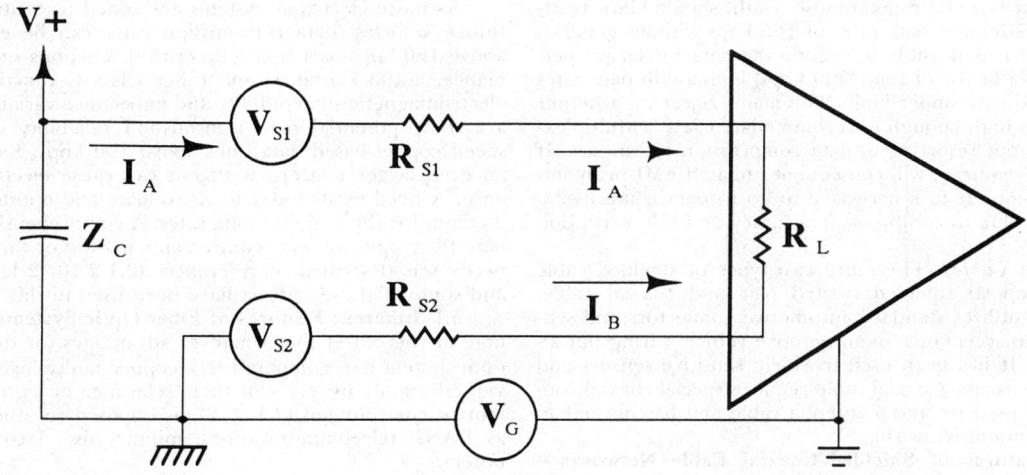

FIGURE 7—SIMPLIFIED CIRCUIT DIAGRAM OF VOLTAGE DRIVE I/O

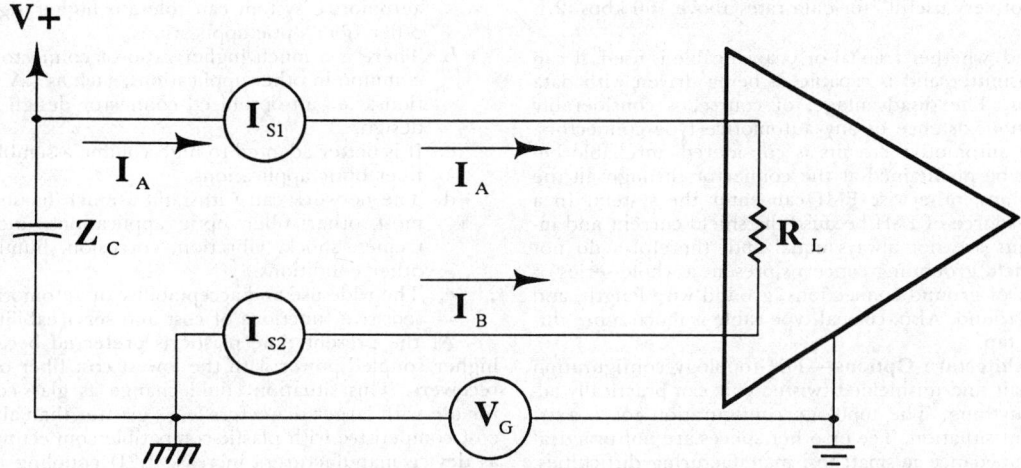

FIGURE 8—SIMPLIFIED CIRCUIT DIAGRAM OF BALANCED CURRENT I/O

ruption by longitudinal noise currents (2.1.2.1). The size, weight and cost of transformers are their major disadvantages. Also, transformer coupling places limitations on the allowable bit encoding techniques by requiring symmetrical waveforms.

d. Optical Coupling: Optical coupling in theory can also be effective in blocking the longitudinal noise current. There are a number of technical concerns such as achieving proper bus termination with optical coupling but cost would be the main concern. A number of manufacturers make optical isolators, to an electrical medium, for computer interfacing. Also some hardware capable of driving a twisted pair medium at Class C data rates is available, but it is not known if any serious attempts have been made to develop this technology for a vehicle medium driver.

3.3.3 MEDIUM DRIVING AND ENCODING TECHNIQUES CONCLUSIONS—Integration into a single IC is important because this method can usually result in better matching between the positive and negative leads of the drivers and receivers. This balancing of the Bus + and Bus — I/O signal at the proper polarity in both amplitude and phase can be effectively used to cancel the EMI in the twisted pair transmission line. In order to achieve adequate levels of immunity to noise interference it is important to choose a medium driving and receiving technique that either cancels or blocks the longitudinal noise current.

Some encoding methods arbitrate more naturally than others if arbitration is required. This encoding factor affects cost and reliability of the network. Encoding selection is not the subject of this report and will depend on the final protocol selection (2.1.2.2). The choice of a proper data encoding technique with wave shaping the rise time to 25% of the minimum feature size (minimum pulse width) should allow twisted pair for the bus medium at data rates of 125 kbps without generating too much EMI. If it is possible to decode the data for larger percentages of rise times a factor of four (500 kbps) increase in data rates establishes the approximate upper limit. The main concern is whether a 125 kbps data rate is high enough to accommodate Class C Multiplexing requirements without resorting to data compression techniques. If higher data rates are required, will consequent radiated EMI problems dictate data compression? If it is necessary to go to data compression will the data integrity due to compression accuracy or EMI corruption be acceptable?

4. Shielded/Coaxial Cable—There are two types of shielded cable considered in this report: shielded twisted pair and coaxial cable. Shielded twisted pair utilizes standard automotive connectors and sensors and has had some acceptance in automotive vehicle wiring but always in special cases. It has been used to shield sensitive sensors and low-level radio audio circuits. Coaxial cable requires special coaxial connectors such as those used for radio antenna cable and has never had wide acceptance in automotive wiring.

4.1 Inherent Features of Shielded/Coaxial Cable Networks—Shielded twisted pair for the transmission medium of a Class C Network does yield only about 15 dB of EMI and EMS improvements over twisted pair. This improvement is not significant because the shield must be broken at the ends where connection is made to the standard automotive connector or sensor and, unlike coaxial cable networks, allows the introduction of EMI noise. Shielded twisted pair does not offer any improvement in ability to drive and it is considered, along with the twisted pair, to be "not very useful" for data rates above 100 kbps (2.1.2.1).

On the other hand, whether triaxial or coaxial cable is used, it can be matched to a transmitter and is capable of being driven with data rates above 100 kbps. The disadvantage, of course, is considerably higher cost and the nonexistence of any automotive type connectors. Coaxial type cable in automotive circuits is considered unreliable because the shield must be maintained at the connector through all the environmental conditions, otherwise EMI can enter the system. In a practical sense, it is a source of EMI because the shield current and inner conductor currents are not always equal and, therefore, do not completely cancel. Shield grounding concerns present a whole series of issues such as number of ground connections, ground wire length, and shield distance from ground. Also, coaxial type cable is much more difficult to splice or tee tap.

4.2 Network Architecture Options—The topology configuration for shielded twisted pair and unshielded twisted pair can practically accommodate almost anything. The topology configuration for coaxial type cable is a different situation. Tee or other splices are not practical because they cause impedance mismatches, manufacturing difficulties, and increased cost. Coaxial cable is practically limited to accommodating only point-to-point or daisy chain type systems. The other alternative is to employ costly matching tee networks or active type repeaters.

The data drive I/O hardware configuration is similar to the unshielded twisted pair requirements. However, with coaxial cable there is not as severe an impedance mismatch problem. All communication protocols used by twisted pair can also be applicable with shielded twisted pair or coaxial type cable.

4.3 Key Concerns of Shielded/Coaxial Cable Networks—It is believed by most automotive wiring engineers that widespread use of shielded twisted pair or coaxial cable in automotive wire harnesses would require a development struggle much greater than the effort required to develop a fiber optic network medium. They believe that manufacturing and assembly issues along with maintenance and repair problems are nearly unsolvable. Shorts at the connector between the signal and shield are likely to plague harnesses causing the whole network to fail. The mechanic's maintenance procedure is likely to require costly harness replacements rather than the simpler repair technique used today.

5. Fiber Optic—Fiber optic systems saw their first widescale use in telecommunications and Local Area Networks (LANs). The reason for this was twofold; technical advantages and cost savings. Since then, fiber optic systems have gradually been replacing copper-based transmission systems like twisted pair and coaxial cable in aircraft, military use, factory automation, and others.

Applications of fiber optics to passenger cars started in the mid-sixties and were mostly for illumination and sensors (2.1.2.3). Over the past decade, several fiber optic multiplexing systems for experimental and test cars were described in the literature (2.1.2.4, 2.1.2.5, 2.1.2.6), and some isolated subsystems were implemented in low volume, high end cars (2.1.2.7, 2.1.2.8, 2.1.2.9).

As more electronic systems are added to control new functions in future vehicles, data transmission rates can be expected to increase above 150 kbps particularly in critical functions such as engine performance, antilock brakes, and other Class C functions. The effects of electromagnetic susceptibility and impedance variation on bit error rate are some potential data transmission reliability concerns with high-speed copper-based data links above 150 kbps. Experience has shown an even larger concern is that it can cause electromagnetic interference. A need exists today to re-evaluate and optimize the transmission medium for these higher data rates. A comprehensive approach to evaluate fiber optic system requirements in view of the distinct automotive needs was described in references (2.1.2.10, 2.1.2.11, and 2.1.2.12), and some of those criteria have been used in this report.

5.1 Inherent Features of Fiber Optic Systems—It is important to note at the outset that whatever advantages (or disadvantages) a fiber optic system has, compared to a copper multiplexed system, have been viewed mainly in terms of their relevance or significance to the automotive environment (2.1.2.12), as opposed to other applications, such as LANs, telecommunications, military use, factory automation, and others.

A fiber optic system could be designed for a glass fiber, a plastic fiber, or a combination of the two, depending on the application. The special needs of the automotive environment impose several requirements on any fiber optic system that is designed to operate in it. The main characteristics of an automotive fiber optic system are:

a. Automotive networks normally require short (an average of 1.5 m) distances between nodes. This short distance means that an automotive system can tolerate higher signal attenuation than other fiber optic applications.
b. There is a much higher ratio of connector to fiber use than is common in other applications, such as LANs or telecommunications; i.e., an optimized connector design will drive the system design.
c. It is better adapted to high volume assembly and test than other fiber optic applications.
d. The network can withstand a much harsher environment than most other fiber optic applications under temperature extremes, shock, vibration, corrosion, humidity, road salts, and other conditions.
e. The wide use and acceptability of automotive networks are very sensitive functions of cost and serviceability.

At the present time plastic is preferred because one can achieve higher coupled power with the lowest cost fiber optic transmitters and receivers. This situation could change as glass-compatible connectors for use with larger diameters (e.g., greater than about 250 µm) become cost competitive with plastic-compatible connectors going forward, and as device manufacturers increase LED coupling efficiency. This technology is rapidly evolving, forcing designers to constantly review the glass vs plastic decision. Unless otherwise specified, the advantages and disadvantages listed as follows will apply to both glass and plastic fiber systems.

5.1.1 PRINCIPAL ADVANTAGES—The main practical consequence in the four following advantages over copper-based transmission media, particularly at the higher data rates, is the minimization or virtual elimination of bus shielding and impedance matching requirements. However, one still has to shield the emitter/receiver components like any other electrical box.

   a. EMI/RFI Immunity: Optical signals are immune to EM and RF interference. Thus, a fiber optic bus can be placed next to high voltage or radio receiver lines without distorting their outputs or being affected by them. This will become more important as dielectric materials replace metals for car bodies.
   b. Data Bus Termination Impedance Elimination: Bus termination and biasing networks are not required.
   c. Immunity From Cross-Talk: Optical signals, by their nature, do not radiate energy that interferes with electronic signals. Therefore, cross-talk between a fiber optic data line and other electronic signal lines is nonexistent.
   d. Ground Loop Elimination: In a copper-based system different ground locations present different impedance values to the same high frequency signal and thus give rise to a potential difference or noise at the receiver circuit. Since optical fibers do not conduct electrical current this "ground loop" is broken and the noise associated with it is eliminated.

5.1.2 GENERAL ADVANTAGES

   a. Data Rate or Bandwidth: Although a fiber optic system has bandwidths of several hundred MHz this advantage is irrelevant for Class C automotive applications at present because it is well beyond present defined Class C data rate requirements. However, as more and more video displays are designed into future cars the need for larger bandwidths will become more important and fiber optics will be the obvious choice.
   b. Weight and Space Savings: There is a sizeable weight reduction when several twisted pairs or coaxial cables are replaced by a single fiber optic cable, as is the case in telecommunications, but the weight difference is insignificant if a single 22 gauge twisted pair is replaced by a fiber optic cable. The same argument could be applied to cable sizes as well, but if shielding requirements are taken into consideration there will be overall space savings and increased design flexibility in a fiber optic system. Furthermore, fiber is easier to proliferate across different vehicle makes because of its immunity to routing-induced EMC problems. Copper-based systems sometimes have to be rerouted when placed into a different vehicle due to unanticipated interference problems.
   c. Long-Term Reliability: Present long-term experience in automotive fiber optic systems indicates that these systems will not be prone to connector erosion or conductor hardening over long periods of time, as is the case for copper-based systems. It is expected that recently developed low-cost fiber optic data transmission components will be reliable in automotive applications, based on their successful performance in laboratory tests (2.1.2.8, 2.1.2.9).
   d. Immunity from Short Circuits: Since fibers do not carry electrical current, shorts to ground are eliminated. The network signal cannot be shorted to power, ground, or other signal wires. This means that depending on network architecture the network cannot be disrupted by a short. Also, there is no need for transient suppression components for shorts to power or Electro-Static Discharge (ESD).
   e. Data Integrity: It is anticipated that, because of the EMI immunity of optical signals, data integrity would be higher than in a copper-based system unless elaborate shielding is used in the latter. This needs to be explored further (see Section 4).
   f. Lack of Familiarity: One possible advantage to the use of fiber optics over the electrical alternatives is the respect it would be given due to the lack of familiarity. The high-speed electrical alternatives may be treated the same as their low frequency power and signal carrying cousins. The high frequency shield ground is not the same as any other ground. The shorting to ground of a high frequency ground could cause problems such as ground loops. A 6 in lead could mean the shield is not grounded. The practice of splicing or doubling at high frequencies is not acceptable. A wire at these frequencies is not just a wire anymore.

5.1.3 DISADVANTAGES

   a. Sharp Bending Radii for Fibers: Sharp bends have an inherent disadvantage in fibers compared to copper conductors. If the bend radius decreases below a critical value, which is related to the fiber radius, light transmission in the fiber is reduced. This effect is not a severe limitation but it implies that a routing-induced attenuation has to be anticipated as a system design parameter.
   b. Unidirectional Transfer of Data: This condition implies that one fiber is used to send data out from a transmitter and another fiber is used to receive incoming data. Thus a duplex cable is required instead of a coax or a twisted pair. With the advent of monolithic detector/transmitter devices (2.1.2.13), the need for a duplex cable could be eliminated and one fiber could be used.
   c. Lack of Standards: The intermittent use of fiber optic systems in automobiles, so far, has not called for standardization and vice versa. As fiber use gains more acceptance, this limitation will slowly disappear.
   d. Lack of Automotive-Grade Connectors: Most of the connectors on the market are designed for a glass fiber with applications for other than the automotive market. There are also connectors that are low cost, are not labor intensive, and require little training or skill to assemble, but most of these are not truly of automotive-grade. There are industry activities working to address this concern (2.1.2.12).
   e. Availability of Reliable and Cost Effective Automotive Components: This is probably the most important reason why fiber optics has not gained a wider automotive application. As mentioned earlier, a plastic system is believed to be more suitable for the automotive environment and at the heart of this system is the fiber. Although fibers and other components have been available for temperatures between -40 °C and 85 °C (passenger compartment) and fibers that operate under the hood ($+125$ °C) are also currently available, fibers for the engine compartment ($+150$ °C) have not been available on a commercial basis let alone at a competitive price. Some recent developments (2.1.2.14) indicate that this is likely to change soon. Quite recently, prototype fibers with operating temperatures of 135 °C (2.1.2.15) have been advertised. It is believed that a wider use of automotive fiber optic systems will spur more competition and drive the component price down. This could be accelerated if fiber optics is adopted as a standard for a transmission medium for Class C systems.

5.2 Network Architecture Options—There are a number of possible architectures that could be implemented in an automotive network. This report considers only the most likely candidate topologies the single and double ring, the active and passive star, and the linear tapped bus. All comparisons have been discussed on an equal basis. These five network architectures have been evaluated using cost, physical complexity, power moding, fault tolerance, expandability, serviceability, and latency as the seven criteria for comparison. The issues used to evaluate the architectures are discussed in the following:

   a. Cost: Cost comparisons should include all necessary hardware, beginning with the input to the encoding and transmitter drive circuitry through the receiver and decoding circuitry. Total system cost, not just initial piece cost, will determine which network architecture will be selected for production. Initial cost will be the combination of the fiber ends, receiver transmitter pairs, and fiber length. Total cost will include assembly, warranty, and service costs. A relative comparison has been made to determine system advantage or disadvantage. With respect to cost, the best system will have the fewest components; one receiver transmitter pair per node and one fiber per node.

   Cost factors that can be readily quantified are fiber ends, receiver transmitter pairs, fiber length, and any other significant system components.

   (1) Fiber ends: Due to the large number of fiber ends to finish in an automotive network, the number of fiber ends will be proportional to a significant part of the system cost and be inversely proportional to reliability. In other words the more mated ends, the more opportunities for a fault to occur.
   (2) Receiver/transmitter pairs: Active devices are costly components. Minimizing receiver transmitter pairs will minimize cost and also result in improved system reliability.
   (3) Fiber length: Although the amount of fiber per vehicle will be small it still represents a cost factor that is easily quantified.

   b. Physical Complexity: Physical complexity is an issue that will impact the vehicle assembly plant operation. A physically complex system will be difficult to assemble reliably. This could add to assembly costs, dealer warranty cost, or owner maintenance cost. The best system will not have concentrations of complexity or have extra components such as fibers, nodes, or receiver transmitter pairs.

c. Power Moding: Power Moding is defined to be the selective control of node power consumption and its level of activity. The electronic systems on the vehicle are not powered up at all times. Some systems may be purposely powered down during a portion of the time. Some systems will be maintained on for a time after the vehicle is turned off. Times such as ignition off, crank, or accessory operation require only a portion of the vehicle system to be operational to conserve battery charge. The network should allow for this reduced level of network operation. The best system would allow network operation with any combination of nodes powered.

d. Fault Tolerance: Continued network operation under a fault condition is desirable. The tolerance of a system to faults could impact the type of functions that could be included on the network. The more tolerant the network is to faults, the more critical functions that can be implemented on it. A fault tolerant system will continue to function with any single point fault.

e. Expandability: Expandability for vehicle variability, future vehicle improvements, and dealer installed options must be allowed for in the initial design. Easy network expansion and contraction is a desirable feature. The best system will be expandable without cost to the lesser system.

f. Serviceability: Second only to initial functionality, serviceability is a very important issue. The vehicle must be serviceable within the existing service network including local service stations. The best system will be self-diagnosing.

g. Latency: The time between data updates will impact the type of function performed on the network. Control applications will require rapid data exchange updates. The best system will have minimum transmission latency.

5.2.1 ACTIVE STAR—The active star network consists of a group of network nodes connected to the active star as shown in Figure 9. Each node consists of a receiver/transmitter (Rx/Tx) pair and a pair of optical fibers. The network also contains a central node that receives and retransmits all network communications. The central node contains an Rx/Tx pair for each network node and in its simplest form, an electrical bus. The active star is an electrical bus packaged in the central node and connected to the network nodes by point to point fiber optic links.

5.2.1.1 *Advantages*

a. Cost: The active star allows the use of simple point to point fiber optic links which could minimize the cost/complexity of the electro-optical components and the optical connection system.

5.2.1.2 *Disadvantages*

a. Cost: The active star requires an additional active node with its associated electrical content including a second Rx/Tx pair per node, power supply, and external electrical content (power and ground feeds).

b. Complexity: The active star is more complex because of the added node and its associated electrical complexity with two fibers per node connecting to it.

c. Power Moding: The star must be powered up at all times so that data can be exchanged on the network.

d. Fault Tolerance: The central node offers many internal (electrical bus) and external (power and ground wiring) fault opportunities that could cause the complete network to be inoperative. A fiber fault will affect only a single node.

e. Expandability: Expansion will be limited by the original hardware design and may require several design levels to minimize cost impact.

f. Serviceability: Some fault identification is possible. A fault will be isolated to a specific spoke on the star or the central node.

g. Latency: Central node retransmission will increase network latency. The added latency could be on the order of 140 ns.

5.2.2 PASSIVE STAR—The passive star network consists of a series of network nodes connected to the passive star as shown in Figure 10. Each node consists of an Rx/Tx pair and a pair of optical fibers. The network also contains a central element that acts like an optical splice and distributes the optical signal to all network nodes. The passive transmissive star contains a means to receive optical energy from any transmitting node and divide the optical energy evenly among the output fibers.

5.2.2.1 *Advantages*

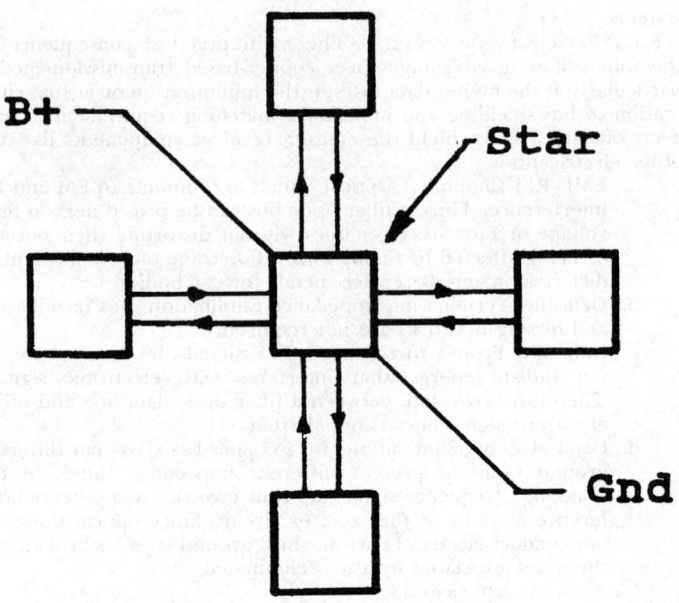

FIGURE 9—ACTIVE STAR

a. Cost: The passive star has a limited number of receiver/transmitter pairs and no electrical connections to the star.
b. Power Moding: All nodes can be powered in any order or combination.
c. Fault Tolerance: A fault will be limited to an individual node or its associated fibers. The star could be a single point fault, but because it is totally passive its failure modes are limited to the physical design and with sufficient hardening this design should not be a problem.
d. Latency: All nodes will monitor the data when it is transmitted with only the delay caused by fiber length.
e. Serviceability: Some fault identification is possible since a fault will be isolated to a specific spoke or node on the network. The remainder of the network is still operational.

5.2.2.2 *Disadvantages*

a. Cost: The passive star element will add cost to the network.
b. Complexity: The passive star requires two fibers per node to

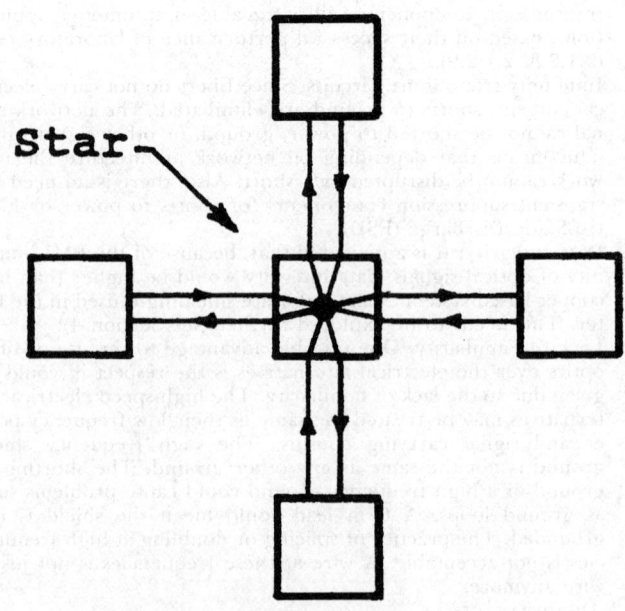

FIGURE 10—PASSIVE STAR

route to the star.
  c. Expandability: It is limited by star port count and optical flux budget. Currently available devices limit the number of ports to seven.

5.2.3 SINGLE RING—The single ring network, illustrated in Figure 11, consists of an Rx/Tx pair and a single fiber for each node connected in series. Data is transferred around the ring through a continuous series of retransmissions by each node until it completes a circuit around the network. Each data message is received and retransmitted by each node. The amount of logic required depends on the particular technique used. The logic could be as simple as a series of shift registers through the capability to receive, decode, and retransmit a complete message.

5.2.3.1 *Advantages*
  a. Cost: The single ring network contains the fewest fiber ends and requires no special hardware.
  b. Complexity: The complexity is minimized since there are no added nodes and the complexity is evenly distributed.
  c. Expandability: The single ring network is easily expanded by adding nodes into the ring.

5.2.3.2 *Disadvantages*
  a. Power Moding: All modules must be powered up for the network to operate.
  b. Fault Tolerance: A fault anywhere in the network will cause the network to be inoperative.
  c. Serviceability: Since any one fault causes the network to be inoperative, faults are not easily isolated.
  d. Latency: Each module must retransmit the data adding to the latency between the first and last node.

5.2.4 DOUBLE RING—The double ring or dual counter rotating ring, illustrated in Figure 12, consists of two single rings connected such that data flows in opposite directions. Data is passed around in two directions on two sets of fibers. Each node consists of two Rx/Tx pairs and two fibers, and some additional logic to allow for fault isolation and bypass. If a fault occurs in one or both rings, the nodes on either side of the fault wrap the data around the ring in the other direction on the second loop, as shown in Figure 13, completing the communication. Faulty nodes or fiber interconnects can be bypassed in this manner.

5.2.4.1 *Advantages*
  a. Fault Tolerance: The double ring network is very tolerant of single faults (nodes or fiber interconnect). The remainder of the network continues to function.
  b. Expandability: The network is easily expanded by inserting a node into the ring.
  c. Serviceability: The network is capable of determining the location of a fault.

5.2.4.2 *Disadvantages*
  a. Cost: The double ring network employs four Rx/Tx pairs, two fibers per node, and additional logic required for fault isolation and bypass will have significant cost impact.
  b. Complexity: Complexity is evenly distributed, but each node requires four optical connections and two receiver/transmitter pairs.
  c. Power Moding: One module could be powered down and the rest of the network will continue to function. For multiple modules to be powered down they would have to be positioned in line on the ring, which would impact the physical complexity.
  d. Latency: Each module must retransmit the data. This action adds to the latency between the first and last node.

5.2.5 LINEAR TAPPED BUS—The linear tapped bus can be implemented in many forms depending on the capability of the tap and the available optical flux budget. If the tap is assumed to be unidirectional, asymmetrical, and variable, the simplest configuration would consist of a single fiber that would route past each module twice as shown in Figure 14. The single fiber would be divided into a transmit leg and a receive leg. Each node has a launch tap on the transmit leg and a receive tap on the receive leg. The optical flux budget and the tap capabilities will determine the maximum number of nodes that can be handled passively. If a larger number of nodes is required, an active head end as shown in Figure 15 would be needed. Currently, with reasonably low-cost electro-optical devices and available tap technology, the maximum number of nodes that can be accomplished passively is ten.

5.2.5.1 *Advantages*
  a. Power Moding: Passive implementation will allow all nodes to power mode independently.
  b. Latency: All nodes will see the data when it is transmitted with only the delay caused by fiber length unless a head end is required.

5.2.5.2 *Disadvantages*
  a. Cost: The linear tapped bus requires a large number of taps, two per node, and possible need for a head end node will increase the cost.
  b. Complexity: The packaging of taps and head end node will impact complexity.
  c. Power Moding: The head end node must be powered up continuously if required.
  d. Fault Tolerance: A fiber or tap fault will always cut off some portion of the network. The farther down the bus fiber the

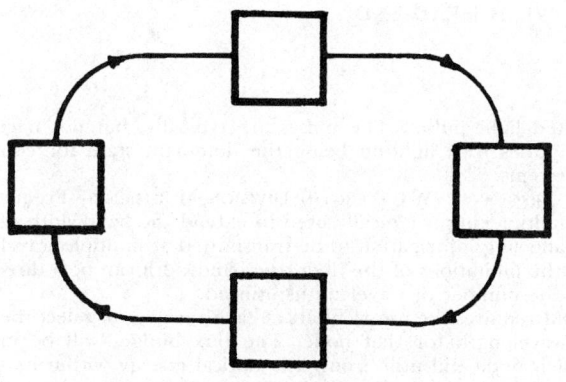

FIGURE 11—SINGLE RING

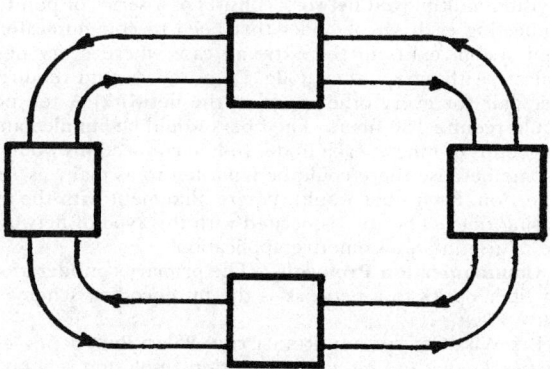

FIGURE 12—DOUBLE RING

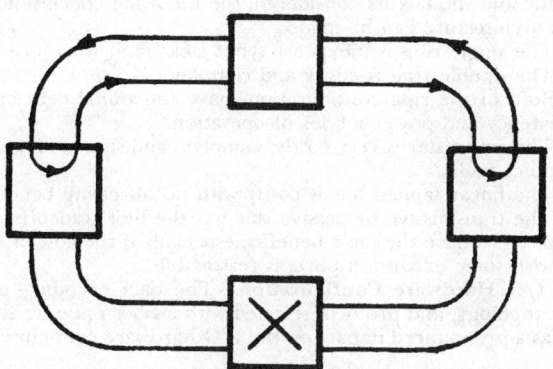

FIGURE 13—DOUBLE RING WITH FAULT

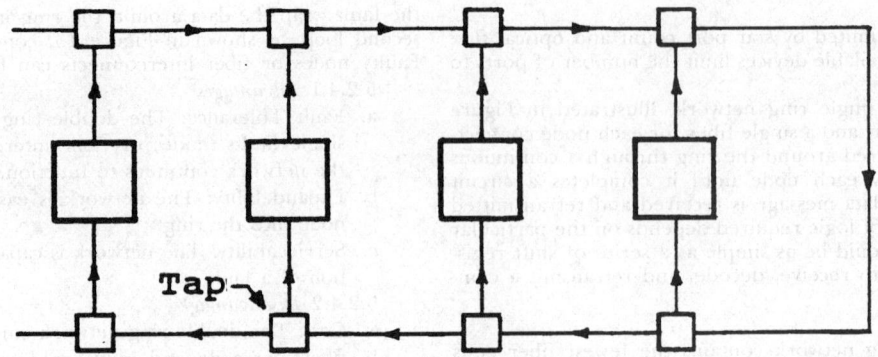

FIGURE 14—LINEAR TAPPED BUS

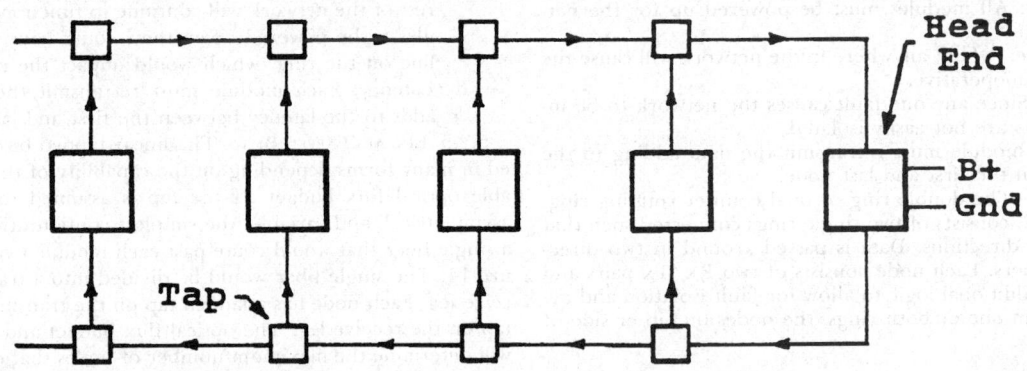

FIGURE 15—LINEAR TAPPED BUS WITH HEAD END

fault occurs, the more nodes that will be impacted until all nodes would be inoperative. A fault in the head end module or its power and ground feeds will incapacitate the whole network.
  e. Expandability: The expandability of the network will be limited by the tap characteristics and the original optical flux budget. The network would have to be designed and built with the capability for the addition of taps later. A tap could not be randomly applied anywhere within the network. It can only be applied where sufficient optical energy existed and where the downstream taps were designed to operate with the reduction in optical energy. Currently available taps and electro-optical devices limit the passive linear tapped bus to ten nodes.
  f. Serviceability: A number of conditions can cause the network to be inoperative making failure diagnosis difficult.
  g. Latency: The head end retransmission will add some latency.

5.2.6 NETWORK ARCHITECTURE CONCLUSIONS—Based on the assumptions made and the factors considered, the following conclusions about network architecture can be made:
  a. The single ring is inexpensive but lacks fault tolerance.
  b. The double ring is costly and complex.
  c. Both of the ring configurations have functional deficiencies in latency and power modes of operation.
  d. The active star is very costly, complex, and susceptible to single point faults.
  e. The linear tapped bus is costly with no offsetting benefits.
  f. The transmissive or passive star has the best tradeoffs and appears to have the most benefits, especially if the cost of the star with some expansion ports is reasonable.

5.3 I/O Hardware Configuration—The data encoding method, network topology, and protocol selected with a fiber optic transmission media has a pronounced impact on the I/O hardware configuration design.

5.3.1 TIME DIVISION MULTIPLEX (TDM)—Time division multiplex is the most commonly discussed method of data transmission for automotive applications. In TDM, the data is encoded in a series of amplitude modulated light pulses. The pulses are typically, but not necessarily, on/off pulses with light-on being the dominant state for contention based systems.

5.3.2 FREQUENCY (WAVELENGTH) DIVISION MULTIPLEX—Frequency division multiplexing is typically used to extend the bandwidth of the fiber by allowing information to be transmitted at multiple wavelengths. Within the limitations of the fiber, the bandwidth can be a direct function of the number of wavelengths utilized.

FDM requires the use of filters at each receiver to select the appropriate wavelength for that node. The flux budget will be impacted since each node will utilize only the optical energy within its selected wavelength and discard all optical energy in the unwanted wavelengths. The energy in the unwanted wavelengths is lost.

Since the bandwidth of the fiber with a single wavelength is sufficient, FDM would probably not be cost effective.

5.3.3 SPACE DIVISION MULTIPLEX (MULTISTRAND FIBER CABLE)—A space division multiplexed network consists of a series of point to point links connecting each set of nodes that need to communicate. A fully connected system exists in the extreme case where every node must communicate with every other node. Each node would require a separate fiber pair for every other node in the network. A ten node network would require 180 fibers. The fibers would be bundled and terminated in groups of nine at each node. In-line connections would be very cumbersome because there could be from ten to as many as 180 fibers per connection. Each fiber would require alignment with the appropriate opposing fiber. The cost associated with this type of network would preclude its use in an automotive application.

5.4 Communication Protocols—The primary considerations when applying fiber optics to a network is the bit encoding scheme and the protocol.

5.4.1 BIT WISE CONTENTION RESOLUTION BASED PROTOCOLS—Multiple access networks that use bit wise contention resolution rely on the network transport time being much shorter than the bit time. This means that in effect all nodes in the network monitor the same information at the same time. Bit wise contention allows the nodes to contend for control of the bus in a nondestructive manner and the highest priority message is not corrupted. If the transport time gets too long, two contending nodes are not aware of the other's presence on the bus until some time beyond a bit time and the data will be corrupted. In long distance LANs that use contention the data is allowed to be corrupted and lost. Bus control is then resolved through techniques such as random-back-off time.

Passive networks, such as the passive transmissive star or tapped linear bus, result in the shortest data transport delays. In passive buses, only the fiber delay of approximately 1.5 ns/ft exists. Active networks on the other hand use data retransmission to distribute the information throughout the network. In the active star, data is received, amplified, shaped, and then retransmitted. This process can add a significant delay in the network. Typical data reception and retransmission delays are on the order of 140 ns for devices in the 1 to 5 MBaud range. Shorter delays are possible but require more expensive devices.

Bit wise arbitration resolution protocols also limit the possible Rx design options. DC coupled receivers can be readily applied but have lower sensitivity than do receivers that utilize A/C coupling or AGC. A/C coupled receivers are difficult to apply to nonsymmetrical bit encoding schemes (NRZ, etc.), where due to contention multiple transmitters can be on simultaneously and where long bus idle times can exist (Multiple Access Protocols). Receivers with AGC require a preamble which precludes their use on bit wise contention buses.

5.4.2 NON-CONTENTION BASED PROTOCOLS—Non-contention based protocols such as master slave or token passing do not have the bus transport delay restriction. Bus ownership is predetermined and only one node will be transmitting on the bus at any one time. For these protocols, both active and passive optical networks are acceptable. Also, with the addition of a preamble, higher sensitivity receivers using AGC and A/C coupling can be applied to passive networks thus expanding the maximum node count. The addition of a preamble will have a negative impact on bus throughput efficiency.

5.5 Key Concerns of Fiber Optic Systems—Over the past ten years or so many predictions have been made that multiplex wiring (including fiber optics) would be in wide use by the early 1990s. Based on automotive market realities today it is not hard to see that these expectations will have to be pushed further out into the future. Several factors so far have slowed the implementation of fiber optic systems in the automotive market. Some of these factors were mentioned in 5.1.3 under "Disadvantages," and other factors are discussed below.

5.5.1 NEW CULTURE/EDUCATION FOR AUTOMOTIVE ENVIRONMENT—This concern covers three main areas: manufacturing/assembly, maintenance/repair, and perceived cost.
  a. Manufacturing/Assembly: Unlike fiber optic systems in telecommunications and LANs where the fibers and connectors are assembled on-site in a labor intensive fashion, the viable implementation of fiber optics in automotive systems requires automated production and assembly of components and systems on- and off-line. While these methods are well developed by the automakers and their OEMs for conventional wire harnesses and electric components and could be adapted for fiber, they have not been developed by the traditional fiber optic system suppliers that serve the LAN and other markets. This concern will gradually diminish as automotive-specific F/O connectors and cable assemblies are designed for automated high volume production and on-line assembly into the vehicle.
  b. Maintenance/Repair: This is a very legitimate concern at present. Since fiber optic communication is a fairly recent development, most technicians, let alone service shop mechanics, are not familiar with nor are trained in fiber optic tools and repair techniques. However, this is also true, to a lesser degree, for multiplexed systems and associated digital circuitry in general. Training programs that are understandable for service shop personnel and shop floor must be developed. Also diagnostic and repair kits have to be developed and put in place for these personnel, along with instruction manuals for car owners if full fiber optic benefits are to be realized.
  c. Perceived Cost: A fair cost comparison between fiber and copper is very difficult to make. First of all, it is very important to include all system costs and their impact on performance of other subsystems and to include long-term repair and warranty benefits. Second, most cost comparisons that have been made to date are done on the basis of a glass fiber system. Third, a fair comparison has to be based on equal performance and should, in the least, include the cost of EMI shielding for the copper-based system.

5.5.2 LENGTH OF LINK—In a glass fiber, attenuation is very small, and long links of fibers (a few km) are quite common. For an automotive plastic fiber system, however, the links have to be short (a few meters maximum), since signal attenuation is about 0.2 to 0.5 dB/m for plastic fibers. Particularly in a passive star topology the length of free fiber links should not exceed 2 to 3 m. Although this is not an ideal situation, it is not a limitation, since the distances between most nodes in the automobile are generally below 3 m anyway.

5.5.3 DATA RATES—The data rate capability of both copper and optical transmission media is limited by the amount of money invested in the components. Currently available plastic fiber is bandwidth limited to $\approx$ 60 MHz-km, at 10 m length, the fiber optic transmission medium is capable of 600 MHz. Visible red LEDs are capable of data rates in excess of 50 Mbps. Currently available receiver technology compatible with 1000 fiber is limited to 10 Mbps due more to industry needs than technology shortcomings. Higher data rate devices are possible in the future.

5.5.4 FAILURE MODES
  a. LED: The typical LED failure mechanism is reduced light output. The optical flux budget should contain margin for normal LED degradation.
  b. LED Drive Circuit: The electronic drive circuitry could fail in such a manner that the LED is forced on continuously. This condition can be relieved by designing in a reasonable amount of redundancy.
  c. Fiber: An optical fiber cannot be shorted out, it can only be pinched or opened. The light transmission will be reduced in proportion to the reduction in cross-sectional area.
  d. Connection System: Optical connections, like electrical connections, can be degraded by the introduction of foreign materials analogous to dirt or ice. The light transmission will be reduced accordingly.
  e. Receiver: The receiver will have similar failure modes as the other electronic components, including loss of sensitivity, shorts, opens, etc.
  f. Passive Networking Components: Passive networking components will have mostly mechanical failure modes which will physically interrupt or reduce the light level.
  g. Active Networking Components: Active networking components will contain active and passive electronic devices, electro-optical components (LEDs and Receivers), and have external power and ground circuits. These components will be subject to opens, shorts, and degradation.
  h. Secondary Failure Effects: Similar to the copper-based counterpart, fiber optic media have secondary failure effects. An inadvertent event such as a nick in the cable jacket will allow moisture/chemical intrusion that in the case of copper causes corrosion, in optical fiber could cause an increase in attenuation.

5.5.4.1 *Consequences of Failure*—The consequence of a component failure depends on the particular network architecture.
  a. Passive Networks: Low levels of light transmitted due to LED, transmitter fiber, or networking component degradation will eventually cause the signal to not be received by the node with the weakest receiver. In a contention based system this could cause collisions and data to be corrupted and lost. Low levels of received data due to receiver fiber or receiver degradation will cause the signal from the weakest transmitter LED to not be received. In a contention based system this could cause collisions and data to be corrupted and lost.
  b. Active Networks: In active networks, which are made up of point to point optical links, there may be more optical margin. This would allow larger component degradation before a system fault occurs.

6. Summary and Conclusions—A Class C Multiplex Network is intended to be a functional superset of a Class B and Class A Multiplex Network. Being a superset means that the Class C Network must be capable of communications that would perform all of the functions of a Class B and Class A Bus. However, it has been generally agreed that the main additional requirement a Class C Multiplex Network must be capable of producing is high-speed real-time control (SAE J2056/1). The most important aspect to this real-time control requirement that directly impacts transmission medium is very low message latencies. This relationship is true because message latency is often inversely proportional to data rate.

This document has provided an overview of twisted pair, shielded twisted pair, coaxial, and fiber optic transmission media (see Table 2 for a comparison matrix). The focus of discussion has been on the advantages and disadvantages of each of these media and their ability to achieve anticipated data rates. Other factors such as data encoding or medium driving techniques also affect the effective data rates achievable for a given medium. These have also been discussed.

The selection of a transmission medium appropriate for a Class C Multiplex Network will mainly depend on the factors presented in this report and the designer's resolution of the issues identified. Resolution of these issues, and the subject for future committee activity, must be consistent with the specific application requirements. These Class C application requirements will influence many aspects of the network in addition to the transmission medium and these in turn may then impact the transmission medium choice.

The particular bus access method selected will influence the transmission medium selected. This effect, caused by the network protocol, is because different access methods will produce different maximum and minimum amounts of bus bandwidth utilization. The imposition of different bandwidth loading caused by the choice of network protocols in turn imposes different transmission medium requirements, i.e., a network utilizing an arbitration-based protocol where multiple nodes contend for access to the network will have the effect of increasing the bus bandwidth loading while a token-passing protocol may require in-

TABLE 2—SUMMARY MATRIX OF AUTOMOTIVE TRANSMISSION MEDIA

	Twisted Pair	Shielded/Coaxial Twisted Pair	Fiber Optic
Main Advantage	Familiarity	Low EMI	EMI Immunity
Key Concern	High EMI Above 100 kbps	Development Struggle	Availability of Components
Interconnect Options	Accommodates all Media Topologies	Limited, no Tees or Splices	Limited, no Tees or Splices
Useful Data Rates	< 100 kbps	< 10 Mbps	< 10 Mbps

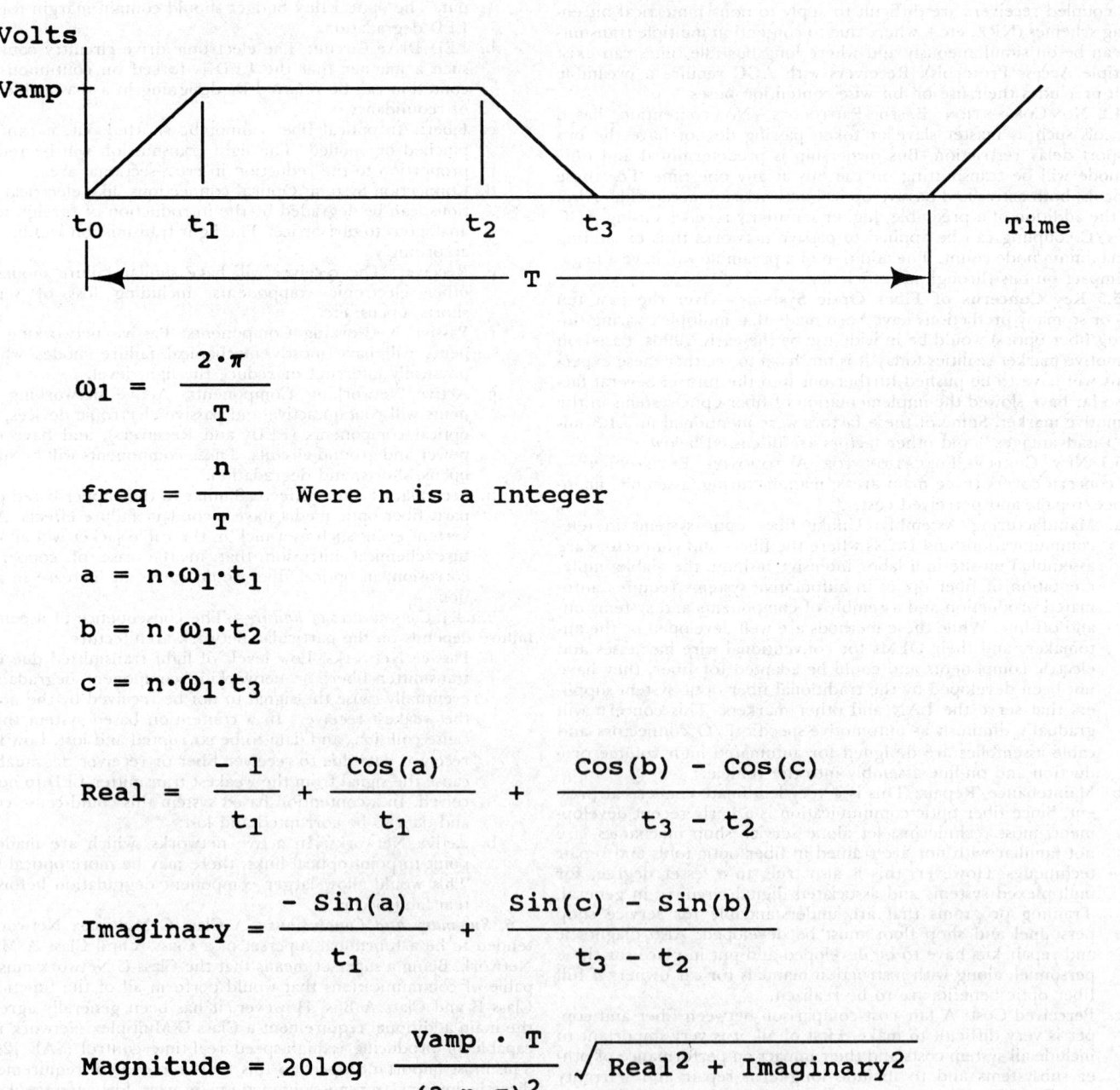

FIGURE A1—TRAPEZOIDAL WAVE SHAPING

creased data rates in order to produce the same effective data rate. Notice that in this example the key to utilizing an arbitration-based protocol is adequate bus bandwidth that is produced by higher data rates, if possible, in order to obtain acceptable Class C real-time control operation because latency cannot be guaranteed for all messages. The token-passing protocol can guarantee the latency for all messages but in a predominately bursty data environment may be less bus bandwidth efficient due to protocol overhead because time is lost passing the token from one network node to the next in order to find the next node that has data to transmit.

Special encoding techniques may also be used, such as Siefried encoding, to increase the bus bandwidth. This encoding technique may allow transmission media such as twisted pair to be utilized to achieve higher effective data rates (thus reducing the potential need for coax or optics). It is possible that Siefried encoding could increase the effective data rate by 8 to 10 times (see 3.2.3); however, this technique may also introduce a potential decrease in data transfer accuracy. A vehicle real-time control strategy such as fuzzy logic may accommodate this type of accuracy reduction and should be investigated as future committee activity. Fiber optics for high-speed data transmission (>100 kbps) in automotive applications is in about the same development state as shielded twisted pair or coaxial cable. These systems have been used for nonautomotive applications, but automotive applications have been limited due to cost. There is a need to develop automotive grade components, design guidelines, trained technicians, servicing procedures, and service equipment/tools for any of the three that are proliferated.

Fiber optic systems offer the potential benefit of stabilizing the present vehicle to vehicle EMC variability. Optical signals are immune to EM and RF interference and a fiber optic network is free from susceptibility to ground longitudinal noise. However, copper-based systems are susceptible to longitudinal noise.

As specific requirements for Class C applications mature so that they can be defined, and the specific Class C network begins to take form, it will be possible to utilize the factors and address the issues identified herein. At that time, it will be necessary to perform a thorough analysis of the attributes of the network which influence the medium to identify an appropriate choice. The basis for selection should involve many factors including application requirements, performance needs and system cost. This selection process is the challenge for further work.

APPENDIX A
TRAPEZOIDAL WAVE SHAPING

A.1 The purpose of this appendix is to outline the method used for EMI level modeling of a trapezoidal wave. The model should predict the EMI levels radiated by a single wire in a vehicle wiring harness. The technique used was to calculate and plot a Fourier series of a sample trapezoidal wave. Consider the trapezoidal wave shown in Figure A1.

CLASS A APPLICATION/DEFINITION—
SAE J2057/1 JUN91

SAE Information Report

Report of the SAE. Vehicle Network for Multiplexing and Data Communication Standards Committee approved June 1991.

Foreword—The Vehicle Network for Multiplexing and Data Communication Subcommittee has defined three classes of communication networks. Perhaps the least understood with respect to function and implementation is the Class A network. A clear understanding of Class A functions is necessary before any standards for protocol can be established.

1. Scope—This SAE Information Report will explain the differences between Class A, B, and C networks and clarify through examples the differences in applications. Special attention will be given to a listing of functions that could be attached to a Class A communications network.

2. References

2.1 Applicable Documents—The following publications form a part of this specification to the extent specified herein. The latest issue of SAE publications shall apply.

2.1.1 SAE PUBLICATIONS—Available from SAE, 400 Commonwealth Drive, Warrendale, PA 15096-0001.

SAE J1213—Glossary of Automotive Electronic Terms

2.2 Classification Definitions—The SAE Recommended Practice, J1213, defines three classes of communication networks, Class A, Class B, Class C.

2.2.1 CLASS A—The Class A network is defined as "A potential multiplex system usage whereby vehicle wiring is reduced by the transmission and reception of multiple signals over the same signal bus between nodes that would have been accomplished by individual wires in a conventionally wired vehicle. The nodes to accomplish multiplexed body wiring typically did not exist in the same or similar form in a conventionally wired vehicle."

2.2.2 CLASS B—The Class B network is defined as "A potential multiplex system usage whereby data is transferred between nodes to eliminate redundant sensors and other system elements. The nodes of this form of a multiplex system typically already existed as stand-alone modules in a conventionally wired vehicle."

2.2.3 CLASS C—The Class C network is defined as "A potential multiplex system usage whereby high data rate signals typically associated with real-time control systems, such as engine controls and anti-lock brakes, are sent over the signal bus to facilitate distributed control and to further reduce vehicle wiring."

3. Interrelationship of the Three Classes—A hierarchical relationship exists between the classes of networks. By definition, Class C is a superset of Class B. Also, Class B is a superset of Class A. It should be noted that this is a functional relationship only. Therefore, it is important to distinguish between the function and the application of the multiplex network.

3.1 System Speed vs Functional Speed—Most discussions on multiplexing focus on two issues, system speed and system complexity. Confusion arises from associating functional speed with system speed and complexity. As described in 2.2.3, Class C is defined as high-speed and real-time control. Intuitively, high function speed requires high system speeds and perhaps complexity. The Class B definition also makes no reference to the speed of the network or the function but places an emphasis on the type of function, "data communications." Class A defines the functions as being individually wired and not normally connected to intelligent nodes within the vehicle. Here again, no mention is made about the system speed or complexity required to achieve the function. Networks operating at high or medium speeds, therefore, must not be excluded from consideration as a Class A network.

4. Typical Applications of the Classes—Table 1 lists some characteristics of the three classes of multiplex networks. The real functional purpose is shown for each. In addition, the type of information and the timeliness of its distribution is noted.

4.1 Class C Typical Applications—Systems that require real-time, high-speed control normally require a significant amount of information to function properly. This information must be available within a narrow time window that cannot vary. A delay of information longer than the specified time window may cause reduced operation or, in extreme cases, could result in the vehicle becoming inoperable. Figure 1

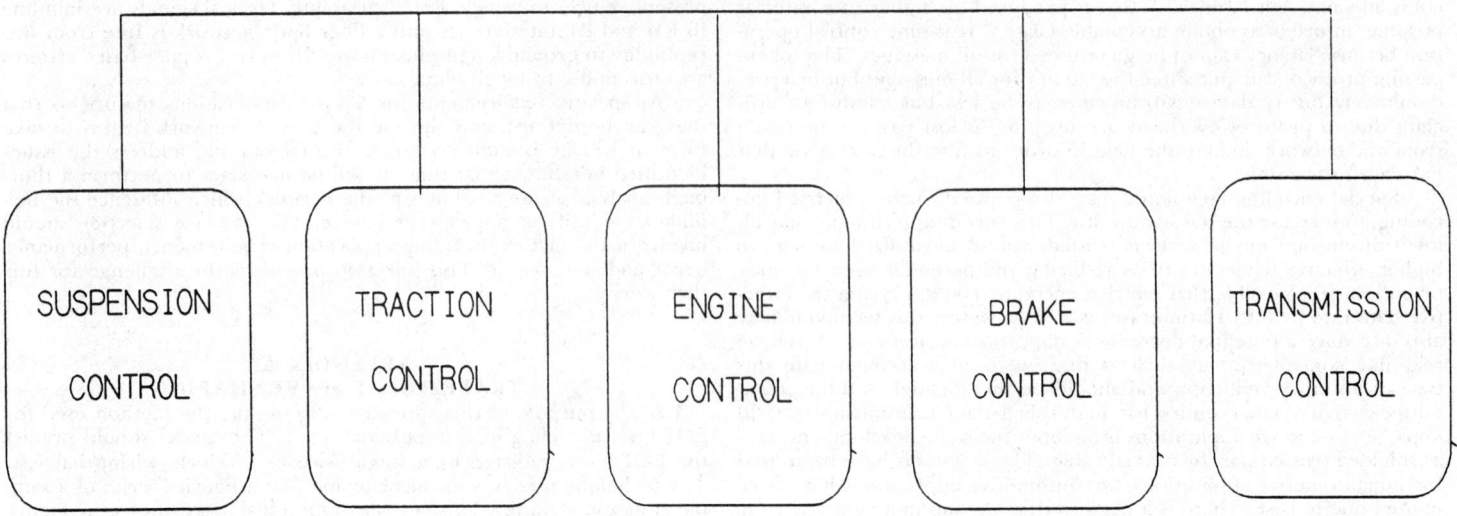

FIGURE 1—TYPICAL CLASS C APPLICATION

TABLE 1—CHARACTERISTICS OF MULTIPLEX NETWORKS

	CLASS A	CLASS B	CLASS C
Purpose	Sensor/Actuator Control	Information Sharing	Real-Time Control
Information	Real Time	Occasional	Real Time
Time Window	Wide Window	Varying Window	Narrow Window
System	Multiple Systems	Multiple Systems	System Specific
Information Lost or Corrupted	Nuisance	Nuisance or Failure	Failure

illustrates a Class C application.

4.1.1 EXAMPLES OF CLASS C APPLICATIONS—Anti-lock brakes, steer by wire, traction control.

4.2 Class B Typical Applications—Many systems within the vehicle require information that is common to other systems. While redundant sensors and actuators as well as parallel circuitry would support acceptable operation, multiplex data sharing of this information could result in simpler, more reliable systems. The shared information on a Class B network is not critical to the operation of all of the systems to which it is connected. The delay of a specific bit of information will not cause a critical failure in any of the systems. Therefore, the response window in the Class B network is not nearly as narrow as in the Class C. In fact, the response time may be variable, depending on the application. Another characteristic of Class B network is its interconnection of dissimilar systems. Figure 2 illustrates a Class B application.

4.2.1 EXAMPLES OF CLASS B APPLICATIONS—A typical Class B network could connect engine control modules, body computers, and system diagnostic modules.

4.3 Class A Typical Applications—Figure 3 illustrates the zone locations referenced in Table 2. Tables 2.1–2.4 list typical applications that could be considered for Class A networks. The chart is by no means complete, and will vary from application to application. It serves, however, to illustrate the numerous devices that can be serviced through a Class A network. It contains information on the device type, its anticipated latency requirements, and the severity of damage should the device fail.

4.3.1 EXAMPLES OF CLASS A APPLICATIONS—Class A applications include the control of lights, power convenience features, and information diagnostics. Figure 4 illustrates an application of a Class A network.

5. Parameters of Class A Devices—Class A devices can be divided into two main categories, inputs (sensors) and outputs (controls).

5.1 Input Device Definition—An input device is an information-gathering device for other devices or systems. This information can be binary (on or off, open or closed, up or down) or analog (several positions). Examples of input devices are switches and sensors.

5.2 Output Device Definition—An output device does work that causes a desired result. An output device responds to information received from an input device.

5.2.1 OUTPUT DEVICE WITH FEEDBACK (STATUS)—In addition to the requirements of 5.2, these devices also need to communicate information to other nodes. This information is typically associated with the status of the output device, e.g., is it on or off.

6. Class A System Criteria—Many items affect the performance of a Class A network. Depending on the application each item will have a different amount of importance. It is useful, however, to be able to judge each potential Class A system with respect to each of the following criteria. (Refer to Table 3.)

6.1 Latency—For Class A applications, latency is the total time required for a system to respond to a request. This includes the time to sense an input, process the information, and energize the controlled device. This does not include the additional time for the output device being controlled to reach its final state.

6.2 Reliability—This criterion is for the entire system including redundancy of any kind as specified by the system designer.

6.3 Bus and Node Failures—This notes what happens to the operation of the devices on the network in the event of bus or node failures.

6.4 EMC Susceptibility and Radiation—This criterion is for the systems generation of and susceptibility to EMI RFI noise.

6.5 Diagnostics—This is the ability of a system to determine if failures are present in the system and relay this information for appropriate action.

6.6 Cost—This is the total system cost for the Class A network. It should be relative to the anticipated cost, per function, of a non-multiplexed vehicle.

6.7 Open System—This criterion indicates the proprietary position of the network being evaluated.

6.8 Sensitivity to Environments—This is the measure of the entire system's ability to withstand the various environments within a motor vehicle. It must consider all of the hardware associated with the network.

6.9 Communications to Other Systems—This is the ability of the network to communicate with other networks that are likely to be in the motor vehicle.

6.10 Electrical Media—This defines the requirements for transmission media for the Class A network at the EMI RFI levels specified in 6.4.

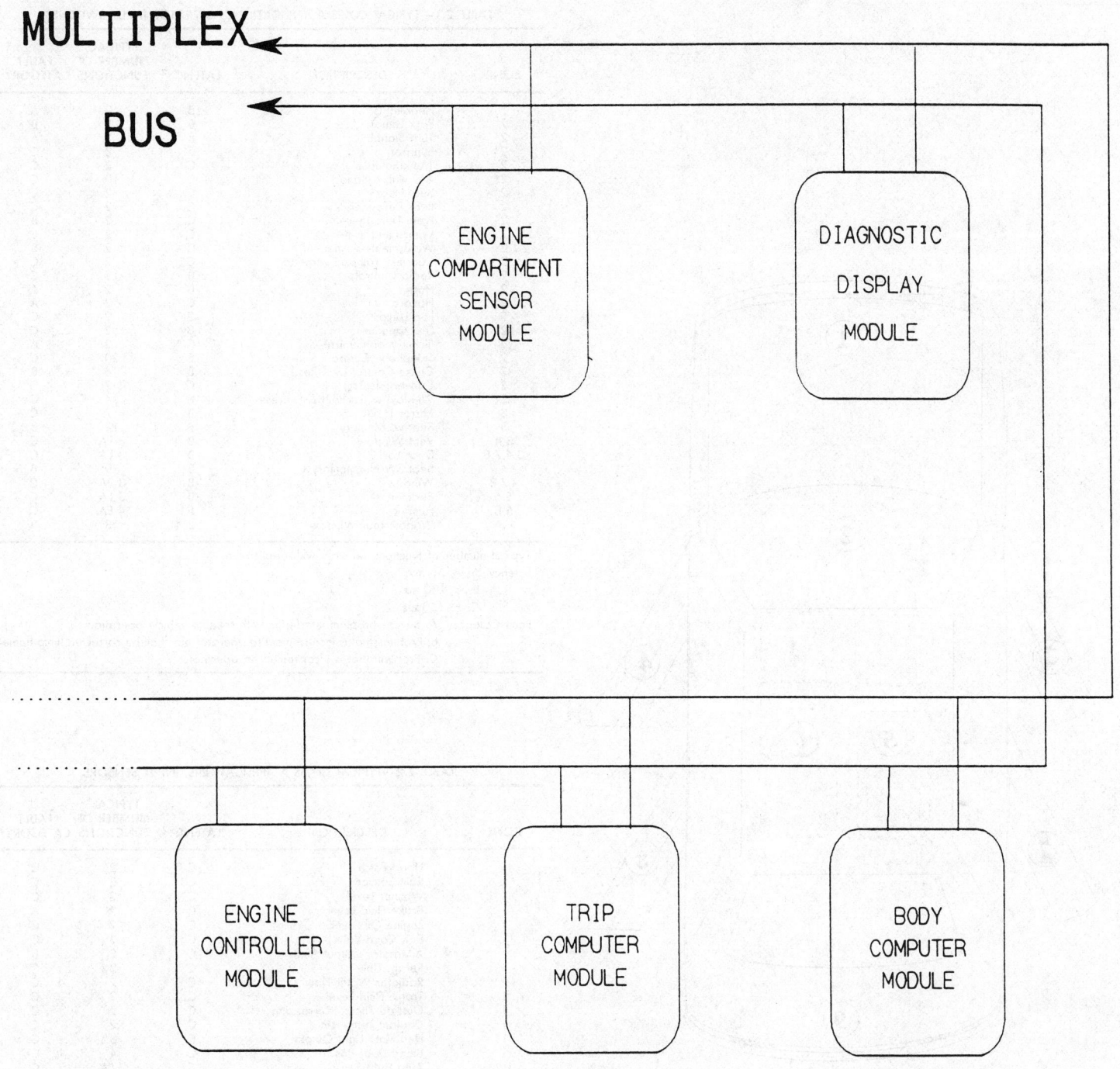

FIGURE 2—CLASS B MULTIPLEX APPLICATION

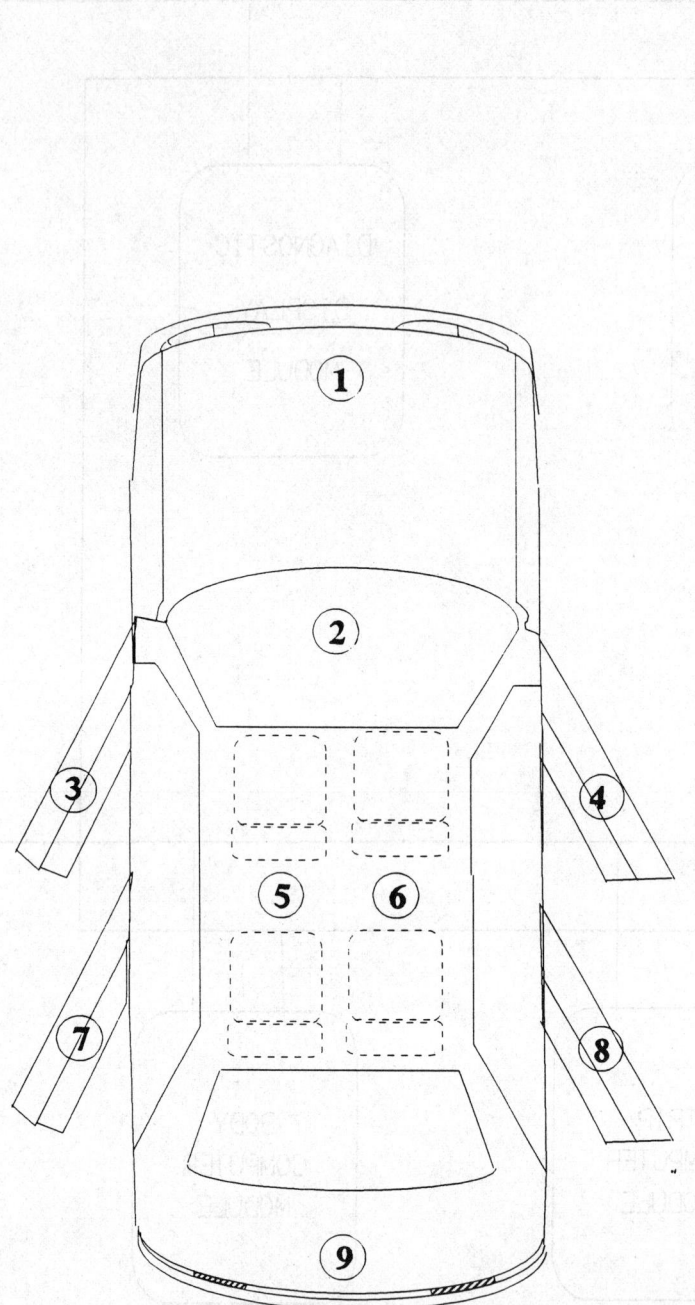

FIGURE 3—APPLICATIONS ZONES

TABLE 2.1—TYPICAL CLASS A APPLICATIONS, OPERATOR INPUT SWITCHES

ZONE	DESCRIPTION	LATENCY[2]	TYPICAL[1] NUMBER OF FUNCTIONS	FAULT CATEGORY[3]
2	Headlights	B	2	A
2	Park Lamps	B	2	B
2	Turn Signal	B	3	A
2	Sunroof	C	3	C
2	Trunk Release	C	2	C
2	A/C with 4 Speed	C	10	C
2	Seat Heat	C	2	C
2	Rear Defogger	C	2	C
2	Front Defogger	C	2	B
2	Windshield Wiper	C	2	B
2	Windshield Washer	C	2	B
2	Courtesy Lamps/Overhead	C	2	C
2	Radio Controls	B	12	C
2	Horn	B	2	A
2	Hazard	C	2	A
2	Fog Lamps	C	2	C
2	Fuel Opener	B	2	C
2	Illumination Control	C	2	A
2	Telephone Control	B	20	C
2	Cruise Control (Set.-Res.)	B	4	B
2	Convertible Top	C	2	C
3	Window w/ Lockout & Express	B	14	C
3	Mirror Lt./Rt.	B	7	C
3	Seat w/ Memory	B	12	B
3,4	Vent Window	B	3 EA	C
3,4,7,8	Door Lock	B	3 EA	C
4	Seat SW No Memory	B	9	B
4,7,8	Window	B	3 EA	C
5,6	Recliner	B	3 EA	C
5,6	Lumbar	B	4 EA	C
9	Wagon Rear Window	B	3	B

[1] Typical number of functions will vary with application.
[2] Latency, A: < 50 ms
    B: < 100 ms
    C: < 150 ms
[3] Fault Category, A: Severe problem interfering with reliable vehicle operation.
    B: Problem that is inconvenient to operator but shall be corrected (limp home).
    C: Problem that is inconvenient to operator.

TABLE 2.2—TYPICAL CLASS A APPLICATIONS, INPUT SENSORS

ZONE	DESCRIPTION	LATENCY[2]	TYPICAL[1] NUMBER OF FUNCTIONS	FAULT CATEGORY[3]
1	Hood Latch	C	2	C
1	Rain Sensor	C	2	C
1	Washer Level	C	2	C
1	Brake Fluid Level	C	2	A
1	Engine Oil Level	C	2	C
1	Bat. Cond-Volts	C	2	C
1	Alternator Output-Amps	C	2	B
1	Refrig. Flow	C	2	C
1	Radiator Water Flow	C	2	C
1	Trans. Fluid Level	C	2	C
1	Outside Temp. < Freezing	C	2	C
1	Coolant Temp. Limit	C	2	C
1	Headlight Light Output	C	8	C
1	Blend Door Pos.	C	4	C
1,2	Auto Light Sensor	C	2 EA	C
2	Park Brake Set	C	2	C
2	Brake Pedal Depressed	B	2	A
2	Clutch Depr.	C	2	C
3,4,7,8	Door Lock	C	2 PER	C
3,4,7,8	Door Latch	C	2 PER	A
3,4,7,8	Door Handle	B	2 PER	C
3,4	Window Limit-Express	C	2	C
5,6	Seat Temp.	C	2 PER	C
9	Fuel Level	C	2	C
9	Axle Oil Level	C	2	C
9	Park Lamp Output	C	2	C
9	Stop Lamp Output	C	2	C
9	Auto Level	C	3	C
9	Fuel Door	C	2	C
9	License Lamp Output	C	2	C
9	Trunk Latched	C	2	B

[1] Typical number of functions will vary with application.
[2] Latency, A: < 50 ms
    B: < 100 ms
    C: < 150 ms
[3] Fault Category, A: Severe problem interfering with reliable vehicle operation.
    B: Problem that is inconvenient to operator but shall be corrected (limp home).
    C: Problem that is inconvenient to operator.

TABLE 2.3—TYPICAL CLASS A APPLICATIONS, OUTPUT CONTROL

ZONE	DESCRIPTION	LATENCY[2]	TYPICAL[1] NUMBER OF FUNCTIONS	FAULT CATEGORY[3]	STATUS FEEDBACK DESIRED
1	Headlamp	B	2	A	Yes
1	Radiator Fan	C	2	B	Yes
1	Refrig. Flow	C	2	C	Yes
1	Cruise Cont. On-Off	B	3	A	Yes
1	Wipers Hi-Lo	C	2	A	No
1	Horn	A	2	B	No
1	Wind. Wash	C	2	B	No
1,9	Corner Lamps	B	2	B	Yes
1,9	Park Lamp	B	2	B	Yes
1,9	Turn Signal	B	2	B	Yes
2	Defog. Fr.	C	2	B	No
2	Sunroof Motor	C	2	C	No
2	Heater Blower	C	4	B	No
2	Heat Mode Doors	C	4	B	No
2,3,4,7,8	Courtesy Lamp	C	2	C	No
3	Defog Mirror	C	2	C	No
3	Mirror Motors	C	4	C	No
3,4	Vent Window	C	2	C	No
3,4,7,8	Window Motor	B	2	B	No
3,4,7,8	Door Lock	B	2 PER	C	Yes
5,6	Seat Recline Motor	B	8 PER	B	No
5,6	Heated Seat	C	2 PER	C	No
5,6	Lumbar Valves 2 Bag	C	2 PER	C	No
5,6	Lumbar Compr.	C	2 PER	C	Yes
6	Meter Illumination	B	2	B	Yes
9	Conv. Top Motor	C	2	B	No
9	Backup Lamp	C	2	B	Yes
9	Level Control Compr.	B	2	B	No
9	Brake Light	A	2	B	Yes
9	Defog. Rear	C	2	C	No
9	Trunk Rel.	B	2	C	No

[1] Typical number of functions will vary with application.
[2] Latency, A: < 50 ms
     B: < 100 ms
     C: < 150 ms
[3] Fault Category, A: Severe problem interfering with reliable vehicle operation.
     B: Problem that is inconvenient to operator but shall be corrected (limp home).
     C: Problem that is inconvenient to operator.

TABLE 2.4—CLASS A TYPICAL APPLICATIONS, ANALOG TO "B" BUS CONTROLLER

ZONE	DESCRIPTION	LATENCY[2]	TYPICAL[1] NUMBER OF FUNCTIONS	FAULT CATEGORY[3]
2	In Car Temp.	C	2	B
2	Wiper Delay	C	2	C
2	Panel Lamp Delay	C	2	C
2	Panel Lamp Dimmer	B	2	B
5	Seat Position	B	3	B

[1] Typical number of functions will vary with application.
[2] Latency, A: < 50 ms
     B: < 100 ms
     C: < 150 ms
[3] Fault Category, A: Severe problem interfering with reliable vehicle operation.
     B: Problem that is inconvenient to operator but shall be corrected (limp home).
     C: Problem that is inconvenient to operator.

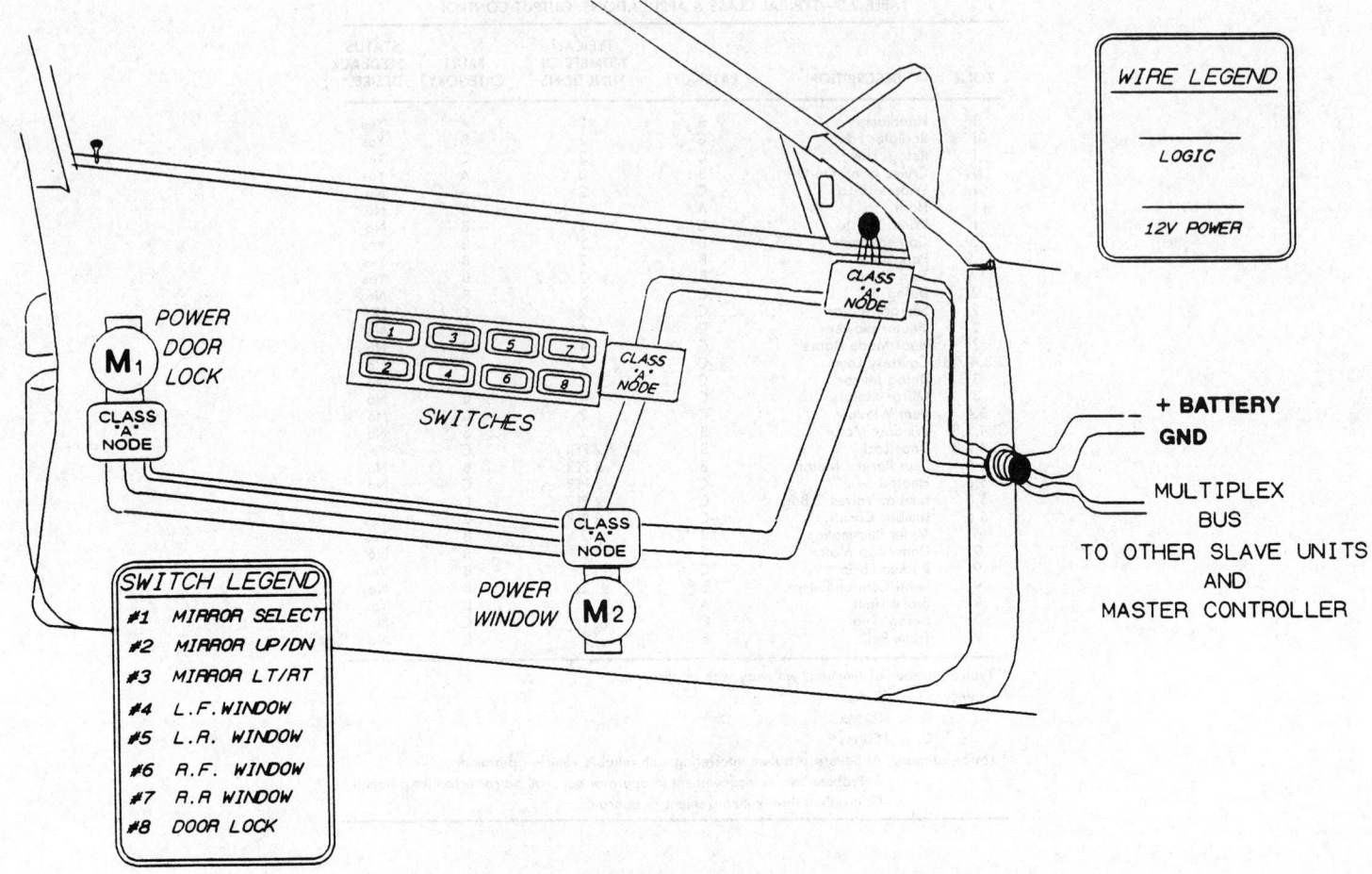

FIGURE 4—CLASS A MULTIPLEX APPLICATION

TABLE 3—CLASS A MULTIPLEX NETWORK CRITERIA

CRITERION	BEST	EXPECTED	LEAST ACCEPTABLE
Latency	< 50 ms	T.B.D.	> 150 ms
Reliability	T.B.D.	T.B.D.	= Non-Multiplexed Systems
Bus Failure	No Effect on Driveability	Limited Effect on Some Functions	Requires Redundancy
Node Failure	No Effect on Function	Function Fails Safe	Function Fails, Other Fcn's not Affected
EMC Susceptibility	30 dBm (Field Strength) Margin over OEM Specs	T.B.D.	0 dBm (Field Strength) Margin over OEM EMC Specs
EMC Radiation	Ambient	T.B.D.	0 dBm (Field Strength) Margin over OEM EMC Specs
Node Diagnostics	Automatic Continuous	Automatic Periodic	Only on Controls Only on Demand
Bus Diagnostics	Automatic Continuous	Self-Regulation	None
Cost	Decrease from Non-Multiplexed System	Same as Non-Multiplexed System	10% Increase Additional Functions
Open System	Public Domain	Free License	License at Minimal Fee
Sensitivity to Environments	Insensitive to Environment	85% of Applications	Separate Networks for Each Car Environment
Communications to Other Networks	Direct Subset of J1850	Via J1850 Micro.	Separate Gateway
Electrical Media	1 Wire & Gnd.	1 Tw. Pr.	Shielded Tw. Pr.
Software Requirements	Central CPU; < 10K	T.B.D.	Distributed > 5K per Node
Node Capabilities	Multiple "I" & "O" per Node	One "I" & "O" per Node	Only One "I" or One "O" per Node
System Capability	> 250 Functions per Network	T.B.D.	64 Functions per Network
Sleep State Current Drain	0 Amps	T.B.D.	= OEM Specs

T.B.D. = To be determined

6.11 Software Requirements—This is defined as the size of memory required to operate the number of functions indicated in 6.12. Specifications must be given for distributed software as well as a centrally controlled system.

6.12 Node Capabilities—This is the maximum number of nodes and functions that can be controlled by the network.

6.13 Sleep State Current Drain—This is the amount of current drain of each node of the system when the system is in its mode of least activity. This mode is commonly referred to as the "sleep state."

7. Preferred Class A System Criteria—Table 3 lists the preferred characteristics for each of the criteria listed in Section 6. Because it is unlikely that any one Class A network can satisfy all of the criteria, this table will serve only as a benchmark from which all potential Class A multiplex systems can be judged.

CLASS A MULTIPLEXING SENSORS—SAE J2057/3 JUN93

SAE Information Report

Report of the SAE Vehicle Network for Multiplexing and Data Communications Standards Committee approved June 1993.

Foreword—This SAE Information Report is the third in a series of Class A Multiplexing Information Reports. This sensors document is not a sensor definition report but intended to be a sensor multiplexing information report. The purpose of this document is to provide information about the types of sensors that can typically be used to meet Class A Bus system requirements. These sensors fall into two general categories; analog sensors and digital sensors, including the operator controlled switches. This document is not all inclusive but is meant to be used as a tool for the system engineer designing and developing a multiplexing network application.

Table of Contents

1. Scope
1.1 Three Classes of Multiplex Networks
2. References
2.1 Applicable Documents
2.2 Related Publications
2.3 Definitions
2.3.1 Analog Sensor
2.3.2 Digital Sensor
2.3.3 Engineering Units
2.3.4 Binary Resolution
2.3.5 Engineering Resolution
3. Typical Applications
3.1 Analog Sensors
3.2 Digital Sensors
4. Requirements
4.1 Network Requirements
4.2 Electrical Requirements
4.3 Latency
4.4 EMC Susceptibility and Radiation
4.5 Reliability
4.6 Sensor Failure
4.7 Diagnostics
5. Sensor Types and Parameters
6. Conclusions
7. Key Words
Appendix A

1. Scope—The Class A Task Force of the Vehicle Network for Multiplexing and Data Communications Subcommittee is providing information on sensors that could be applicable for a Class A Bus application. Sensors are generally defined as any device that inputs information onto the bus. Sensors can be an input controlled by the operator or an input that provides the feedback or status of a monitored vehicle function. Although there is a list of sensors provided, this list is not all-inclusive. This SAE Information Report is intended to help the network system engineer and is meant to stimulate the design thought process.

1.1 Three Classes of Multiplex Networks—The Vehicle Network for Multiplexing and Data Communications Committee has previously identified three classes of vehicle data communication networks.

1.1.1 CLASS A MULTIPLEXING—Class A Multiplexing contains many of the operator-controlled functions and the monitored vehicle function status inputs. Some examples of sensor inputs would be the operator control of powered convenience features (power window switches) or the status of a fluid level (windshield washer fluid).

1.1.2 CLASS B MULTIPLEXING—Class B Multiplexing provides the data communications between different modules, internal and external to the vehicle, for the purpose of sharing common data about the vehicle. An example of this is the diagnostic information shared between an internal (on-vehicle) module and an external (hand-held) module for service repair.

1.1.3 CLASS C MULTIPLEXING—Class C Multiplexing contains systems that require real time, high-speed control, and normally require a significant amount of information to function properly. An example is the wheel speed sensor for the Anti-Lock Brakes System.

2. References

2.1 Applicable Document—The following publication forms a part of this specification to the extent specified herein. The latest issue of SAE publications shall apply.

2.1.1 SAE PUBLICATION—Available from SAE, 400 Commonwealth Drive, Warrendale, PA 15096-0001.

SAE J2057/1—Class A Multiplexing Application/Definition

2.2 Related Publication—The following publication is provided for information purposes only and is not a required part of this document.

2.2.1 SAE PUBLICATION—Available from SAE, 400 Commonwealth Drive, Warrendale, PA 15096-0001.

SAE J1930—Electrical/Electronic Systems Diagnostic Terms, Definitions, Abbreviations, and Acronyms

2.3 Definitions—Class A sensors fall into the areas of operator convenience, vehicle status, and vehicle message information for a monitored function. They are characterized by moderate to slow times of being read and are non-time critical.

2.3.1 ANALOG SENSOR—A sensor that converts some measured continuously varying input characteristic as a continuously varying output value or magnitude. The sensor has a maximum and minimum measurable input range that corresponds to a maximum and minimum output represented value.

2.3.2 DIGITAL SENSOR—A sensor that converts some measured input characteristic as discrete output states. The sensor has a maximum and minimum measurable input range that corresponds to a fixed number of discrete output states.

2.3.3 ENGINEERING UNITS—Referred to as the units of measure detected by the sensor and processed by the measuring system. For example, Volume, Voltage, Displacement, Volume/Time, etc.

2.3.4 BINARY RESOLUTION—The number of digits, in base 2, required to represent the full-scale numerical value measurable by a sensor. A bit is a single unit of information which has only two states, On/Off, 1/0, HI/LO, or True/False. Binary bits may be combined into serial bits of data.

2.3.5 ENGINEERING RESOLUTION—The smallest subdivision to which a sensor's output must be resolved.

3. Typical Applications

3.1 Analog Sensors—Analog sensors are used where continuously varying measured data is required for display or mathematical calculations. Analog sensors continuously measure quantities such as voltages, resistances, pressures, etc., by representing the measured quantity with another type of continuously variable quantity, voltage or current. For example, in temperature measurement, input temperature is represented by an electric voltage or current output. The output signal is solely dependent upon the input signal and the sensor's transfer function to obtain a value or magnitude to express the measured information. To extract the information, it is necessary to compare the value or magnitude of the signal to a standard. For analog data to be transmitted on the Class A Bus, it is usually first converted to a digital format and then transmitted.

3.2 Digital Sensors—Digital sensors are used where measured data is required for status information. Digital sensors measure a variable and represent it by coded pulses or states based on discrete numerical techniques. The discrete states can be a representation of numerical values; for example, the number of motor turns can represent a seat position distance from a maximum or minimum travel point.

The discrete states could represent other information based on the various combinations of the states. The information can represent an ON/OFF state, for example, is the door locked or not; or the information can represent a status, for example, if the fuel level is at FULL, 7/8, 3/4, ..., EMPTY. The discrete states can be transmitted on the Class A Bus directly and no conversion is needed.

Digital sensors can also be switch inputs that can be closed by the operator. For example the power mirror directional switches.

4. Requirements—This is only a general list of requirements and is not meant to be specific for any one application; that would be defined by the user. The requirements in this report are for informational purposes only, the actual requirements for each specific sensor would be determined by the application and by the manufacturer.

4.1 Network Requirements—The sensor will be capable of interfacing to the Class A Bus through integral interface circuitry or through a stand-alone interface module. Reference SAE J2057/1 for specific requirements.

4.2 Electrical Requirements—The sensor must operate at all standard automotive voltages and survive the abnormal conditions, such as reverse voltage and load dump, as required by each user.

4.3 Latency—Refer to Table 2.2, Typical Class A Applications, included in SAE J2057/1.

4.4 EMC Susceptibility and Radiation—The sensor's generation and susceptibility to EMI RFI noise must meet the requirements of the user and SAE J2057/1, paragraph 7.4.

4.5 Reliability—The reliability of the sensor and its Class A Bus interface should not degrade the performance of the function or the network as compared to non-multiplexed vehicles. The actual sensor reliability requirement will be determined by the application and by the manufacturer.

4.6 Sensor Failure—The failure of the sensor must not affect operation of the Class A Bus and should provide a known default value when appropriate. Reference SAE J2057/1, paragraph 7.3.

4.7 Diagnostics—The sensor should have the ability to be interrogated by a system to determine if failures are present in the sensor and transmit this information for appropriate action.

5. Sensor Types and Parameters—Figure A1 in Appendix A contains two lists of sensor types: analog sensors and digital sensors. Refer to SAE J2057/1 for additional switches. The operator-controlled (actuated digital) switches are a subset of the digital sensors. The lists are not all-inclusive for all applications. Each sensor has some information associated with it. This information is not stated to be recommended practice but only as useful information. The specifics for each sensor will be determined by the application and by the manufacturer.

6. Conclusions—The use of Class A sensors on a vehicle network should offer the manufacturer and the customer several benefits in several key areas: customer confidence, vehicle design, assembly operations, and service.

a. To the customer, the confidence that each system or function is working properly and if it does not, the vehicle can provide warning information.
b. To the design engineer, a minimized number of wiring harness variations to mechanize and validate.
c. To the assembly line worker, the installation the wiring harnesses in the vehicle should be made simpler and easier due to minimized wiring harness size.
d. To the service personnel, any problems for which the vehicle is brought back, can be diagnosed and repaired efficiently.

7. Key Words—Multiplexing, Class A, Sensor

APPENDIX A

DESCRIPTION	DISPLAYED UNITS	SENSITIVITY	MAX/MIN READINGS	ENVIRONMENT
ANALOG SENSORS				
Washer Level, Continuous	Volume	16 parts	MFG	E,U,F
Engine Oil Level, Continuous	Volume	0.1 Liter	MFG	U,F
Fuel Level	Volume	0.4 Liters	MFG	E,I,F
Battery Condition Volts	Volts	0.1 Volts	MFG	I,U
Alternator Output Amps	Amperes	0.5 Amps	MFG	U
Transmission Fluid Temperature	Deg. C	0.5 Degree	MFG	U,F
Coolant Temperature, Continuous	Deg. C	1.0 Degree	MFG	U,F
Air Blend Door Position	Position	1 Degree	MFG	I
DIGITAL SENSORS				
Hood Latch	N/A	ON/OFF	O/C	E
Trunk Latched	N/A	ON/OFF	O/C	E,I
Door Locked, per door	N/A	ON/OFF	O/C	I
Washer Level, Low Limit	N/A	ON/OFF	L/H	E,U,F,
Brake Fluid, Low Limit	N/A	ON/OFF	L/H	U,F
Engine Oil Level, Low Limit	N/A	ON/OFF	L/H	U,F
Transmission Oil Level, Low Limit	N/A	ON/OFF	L/H	U,F
Coolant Temperature Limit	N/A	ON/OFF	L/H	U,F
OPERATOR SWITCHES				
Power Door Lock Switch	N/A	On/Off	O/C	I
Power Window, Up Switch	N/A	On/Off	O/C	I
Power Mirror, In Switch	N/A	On/Off	O/C	I
Power Seat, Backward Switch	N/A	On/Off	O/C	I
Power Seat, Recline Switch	N/A	On/Off	O/C	I
HVAC Fan, Low Speed Switch	N/A	On/Off	O/C	I
HVAC Fan, High Speed Switch	N/A	On/Off	O/C	I
Rear Window Defogger Switch	N/A	On/Off	O/C	I

NOTE: Metric Units Assumed

LEGEND:
O/C - Open/Closed
L/H - Low/High
MFG - Manufacturer Dependent
E - Exterior
I - Interior
U - Under Hood
F - Fluid Immersion

FIGURE A1—SENSOR TYPES AND TYPICAL PARAMETERS

CLASS A MULTIPLEXING ARCHITECTURE STRATEGIES—SAE J2057/4 JUN93

SAE Information Report

Report of the SAE Vehicle Network for Multiplex Data Standards Committee approved June 1993.

Foreword—There are generally three classes of multiplex application requirements within the vehicle. To cover these applications two prevalent multiplex architecture strategies have developed. The most popular is the Single Network Architecture. This architectural strategy sizes the network hardware to meet the requirements of the highest level application while maintaining the capability, where possible, of handling the lowest level application. The second strategy, Multiple Network Architecture, is to develop as many types of specialized network hardware components as required to efficiently handle each application and then gateway them together to have only one diagnostic service port. These two differing strategies are studied in detail and presented in this SAE Information Report.

Table of Contents

1. Scope
1.1 Three Classes Multiplex Networks
2. References
2.1 Applicable Documents
2.2 Related Publications
2.3 Terms and Definitions
3. Multilex Wiring System Architecture Strategies
3.1 Multiple Network Architecture Background
3.2 Single Network Architecture Background
4. Role of Class A Multiplexing
4.1 Other Driving Forces
4.2 Example Class A Systems
5. Proposed Vehicle Architecture
5.1 Engine Compartment Node
5.2 Door Nodes
5.3 General Node Concerns
5.4 Multiple Network Architecture
6. Requirements for Class A Sensors & Actuators
7. Summary and Conclusions
7.1 Advantages and Disadvantages
Appendix A

1. Scope—The subject matter contained within this SAE Information Report is set forth by the Class A Task Force of the Vehicle Network for Multiplexing and Data Communications (Multiplex) Committee as information the network system designer should consider. The Task Force realizes that the information contained in this report may be somewhat controversial and a consensus throughout the industry does not exist at this time. The Task Force also intends that the analysis set forth in this document is for sharing information and encouraging debate on the benefits of utilizing a multiple network architecture.

1.1 Three Classes Multiplex Networks—The Vehicle Network for Multiplexing and Data Communications (Multiplex) Committee has defined three classes of vehicle data communication networks.

1.1.1 CLASS A—Low-Speed Body Wiring and Control Functions, e.g., Control of Exterior Lamps

1.1.2 CLASS B—Data Communications, i.e., Sharing of Vehicle Parametric Data

1.1.3 CLASS C—High-Speed Real Time Control, e.g., High-Speed Link for Distributed Processing

1.1.4 INTERRELATIONSHIP OF CLASSES A, B, AND C—The Class B Network is intended to be a functional superset of the Class A Network. That is, the Class B Bus must be capable of communications that would perform all of the functions of a Class A Bus. This feature protects the use of the same bus for all Class A and Class B functions or an alternate configuration of both buses with a "gateway" device. In a similar manner, the Class C Bus is intended as a functional superset of the Class B Bus.

2. References

2.1 Applicable Documents—The following publications form a part of this specification to the extent specified herein. The latest issue of SAE publications shall apply.

2.1.1 SAE PUBLICATIONS—Available from SAE, 400 Commonwealth Drive, Warrendale, PA 15096-0001.

SAE J1850—Class B Data Communication Network Interface
SAE J2057/1—Class A Application/Definition
SAE J2058—Chrysler Sensor and Control (CSC) Bus Multiplexing for Class `A' Applications
SAE J2178/1/2/3/4—Class B Data Communication Network Messages

2.2 Related Publications—The following publications are provided for information purposes only and are not a required part of this document.

Thomas R. Wrobleski, "A Multiplexed Automotive Sensor System," Sensors Magazine dated February 1989, Volume 6, No. 2

Thomas R. Wrobleski, "A CSC Bus Multiplexing Technique for Sensors and Actuators Which Allows Common Vehicle Electronic Control Modules," Paper #89123, 20th International Symposium on Automotive Technology and Automation, Florence, Italy, May 1989

2.3 Terms and Definitions

2.3.1 EVENT-BASED—The attribute of transmission of data on a manually triggered event or on change of parametric value.

2.3.2 EVENT-DRIVEN—The attribute of event-based network protocol.

2.3.3 RESPONSE-TYPE MESSAGES—Messages that require Acknowledgement.

2.3.4 T-TAP—A splice in a wiring harness forming a "T" connection. Sometimes this configuration is associated with automated insulation displacement type connection at a connector.

2.3.5 TIME-BASED—The attributes of repetitive parametric data in a Class B Multiplex Network.

3. Multiplex Wiring System Architecture Strategies—It is a well-known fact that the cost of electronics is decreasing. More functions can now be integrated into fewer modules. The availability of Class B multiplexing now avails the automotive system designer with many new architecture partitioning options. The availability of customer-specific ICs to accomplish a function at a substantially lower cost is becoming a reality. On the other side of the equation is rising wiring and labor costs. Vehicle manufacturers have, in some instances, gone to off-shore or other countries to offset these labor-intensive assembly costs. However, the growth in size and complexity of wiring harnesses causes an ever-increasing investment in assembly facilities that overshadows these cost-containment efforts. These basic trends are projected to apply in the future and become our base assumptions.

3.1 Multiple Network Architecture Background—Initially, the Vehicle Network for Multiplexing and Data Communications Committee recognized the three different requirements for vehicle networking. A chart of these three vehicle multiplex networking typical characteristics is shown in Figure 1. This chart was presented late in 1986 to the SAE Truck and Bus Committee as the state of consensus by the Multiplexing Committee.

The chart shown in Figure 1 does not mean that three networks are needed to cover the multiplexing requirements, but that there are three different characteristic requirements within the vehicle that must be considered. Some of the entries in the chart such as "Status" and "Data Consistency" were not totally understood as noted by the missing entry under Class A. For example the "Status" entry was eventually recognized as the need for acknowledgement in a Class A Network. Some entries had slightly different meanings between committee members and, therefore, further development was dropped. The purpose of this document is not to explain all the characteristic requirements of vehicle networking, but to focus on the Class A and Class B interrelationship and, therefore, there will be no further discussion on Figure 1.

The decision to pursue a multiple network architecture strategy requires a careful study of the alternatives. This investigation of multiple network architectures begins with the assumption that a lower total vehicle system cost would result if one network were optimized around a data communications requirement and other networks were optimized around the sensor and control requirements.

Consider first the ramifications of optimizing a network around the data communications requirements, i.e., Class B multiplexing. Class B multiplexing interconnects intelligent modules such as the engine controller, body computer, vehicle instrument cluster, and other electronic modules. It normally does not affect the base vehicle wiring such as lighting, but it does affect wiring in that it may reduce the connections between sensors and modules. Class B Multiplexing in this case provides an intermodule data communications link for distributed processing. The parametric data shared between modules is almost exclusively repetitive in nature and rarely do these modules require handshaking or acknowledgment of data with other modules. Therefore, a network can be optimized around functional addressing. This is the result of handling the dominance of repetitive data and only a small amount of response type data, e.g., diagnostic data. It is consistent with this strategy to define a multiplex application so optimized by handling the dominance of time-based data communications requirements (see SAE J1850).

REQUIREMENTS OF VEHICLE NETWORKS			
	Class A	Class B	Class C
Repetitive	Allowed	Yes	Yes
Bursty	Yes	Yes	Yes
Handshaking	Yes	Yes	Yes
Status	------	Yes	Yes
Data Consistency	------	Allowed	Beneficial
Number of Nodes	> 100	> 10	> 10
Reliability	Better	Better	Better
Open	Truly	Qualified	Qualified
Priority	Allowed	Yes	Yes
Latency	< 50 ms	< 50 ms	~< 5 ms
Hardware Level	Complete	Flexible	Independent
EMC	*	*	*
TYPICAL ATTRIBUTES OF VEHICLE NETWORKS			
	Class A	Class B	Class C
Bus Rate	1 K bits/sec	10 K bits/sec	1 M bits/sec
Trans. Media	Single Wire	Dif Twist Pr	Coax/Fib Opt

* Note: Use SAE test methods to insure compatibility with automotive environment.

FIGURE 1—CHART OF TYPICAL VEHICLE MULTIPLEXING CHARACTERISTICS

When physical addressing is required in a data communication optimized network, usually for vehicle diagnostics, it can be handled without reducing efficiency. The amount of safety type data that a data communications optimized network has to handle is negligible and can be very effectively handled by other means such as discrete hard wiring. Hard wiring of sensitive functions is considered an advantage because it is consistent with the present conservative method of handling these functions.

When the encumbrance of handling safety type data and most Class A (sensor and control multiplexing) functions are eliminated from the network requirements, a significantly simpler data communications network is the result. This multiple network architectural philosophy also results in a simpler and more effective method of handling sensor and control multiplexing.

The logistical size and complexity of the vehicle manufacturers' systems organization is another factor to be considered. The multiple network architecture is better suited to development and production by multiple sources: a situation that may be important to some vehicle manufacturers. The multiple network architecture requires only a moderate systems organization to insure compatibility because fewer messages would be supported by the data communication network. The sensor and control subsystem requirements can be handled by the product development organization with less direction from the system engineering group. This direction is possible because most of the subsystem would interface with their relevant sensors via their own dedicated Class A Network.

The multiple network architecture strategy should not be a hindrance to multiplex standardization because the Class B Network would be used to support diagnostics.

This multiplexing strategy does have many other advantages. For example, the software required for system control is simplified because it is not required to support timers, counters, or other response types of communications or control. The interfacing hardware is simplified and a less-complex microcomputer is normally required. The data communication rates are consistent with SAE J1850 single wire interfacing, which also supports a lower-cost solution. Data communication multiplexing requirements are consistent, and tend to be associated with the cost-proven technique of integration of body feature modules.

As system designers choose between 4-bit, 8-bit, and 16-bit microcomputers and apply them to their requirements, one would similarly think that Class A is most likely to be bit oriented, Class B is likely to be byte oriented and Class C is likely to be message oriented. The multiple network architectural philosophy also conforms with the reasoning where a number of optimized and simpler solutions can be developed to handle the many and differing requirements of the vehicle multiplex spectrum.

3.2 Single Network Architecture Background—The single network architecture strategy alternative that meets the requirements of both Class A and B classifications leads to a more complex and costly solution. In order to handle data communications or time-based type messages (which are repetitive in nature) and control-type messages (which are event-driven by nature), the control-type message dominates in hardware and software complexity. Control-type messages, such as turn headlights on, are easily understood as event-based. A more complex situation is where parametric data such as vehicle speed, which is defined to be time-based type messages must transmit only on change in parametric value, e.g., change in vehicle speed from 45 to 46 mph to represent an event for transmission as an event-based message.

In this event-based protocol a loss of message is much more critical than a loss of a message in a time-based protocol where the data is naturally repetitive for the message being transmitted is generally current status. The addition of an acknowledgement to the protocol is a possible solution and the following complexity ramifications should be considered:

  a. RAM is required to save message(s) and control data for messages.
  b. Messages may need to be saved while they await acknowledgement.
  c. Timer data associated with a given message may need to be maintained while messages wait for acknowledgement.
  d. A retry counter data associated with a given message may need to be maintained while messages await acknowledgement.

In a time-based protocol, acknowledgement of data may add unnecessary complexity. This condition can be easily understood by considering the situation where a given copy of a message is not received, for whatever reason, and another more current copy will be coming shortly. There is no reason to keep a copy of the current message so that it can be retransmitted if the previous transmission is not acknowledged.

Generally, there is nothing that a module can do if one of its messages is not acknowledged aside from the exception of possibly creating a fault condition indication. There are exceptions where modules may be carrying on a particular dialogue, but these are the exception and not the general rule. The correct acknowledgement response may enhance the probability of detecting message corruption, but basic network communications capability is not improved by acknowledgement techniques. It remains the transmitting module's responsibility to transmit its messages and should register a fault if it is unable to do so. The modules that receive messages should monitor whether they are correctly receiving them and, if not, should register a fault. Acknowledgement techniques do not generally improve this capability.

Unlike the time-based optimized protocol where only the functional broadcast-type messages strategy can be effectively used, an event-driven protocol may need other message types to be most effective. For example, an event-driven protocol needs response-type messages, and the use of in-message response as the acknowledgement mechanism does not change the basic scenario. Distinct from functional broadcast-type messages that do not care where they came from or where they are going, the event-driven protocol does sometimes care.

4. Role of Class A Multiplexing—Class A Multiplexing is most appropriate for low-speed body wiring and control functions. The example most often used to illustrate the benefits of Class A Multiplexing is the base exterior lighting circuit. However, this example is the hardest function to cost justify. The base exterior lighting system is extremely simple and very low cost. A multiplex network applied to this lighting system could result in increased wiring complexity and cost. Data integrity in the lighting system can be a stringent requirement for Class A Multiplexing, e.g., a single bit error that results in Headlights "Off" when they should be "On." Adequate data integrity in a Class A Multiplex network is a constraint and bit error checking may be required.

In the future, the results could change if new features such as low current switching or lamp outage warning became a requirement or new lamp technology such as smart bulbs became a reality. In general the addition of new features, as just illustrated, will play a major role as to when and how multiplexing will become a cost-effective solution.

4.1 Other Driving Forces—The design of vehicles to minimize manufacturing complexity is a major force that will lead to architecture partitioning development. The properly developed multiplex architecture can be very effective in reducing the number of parts in the assembly plants and built-in diagnostics can substantially reduce build test time.

4.2 Example Class A Systems—To illustrate how a Class A Multiplex Network could be used to simplify the vehicle wiring situation, first consider the Vehicle Theft Alarm system shown in Figure 2. Although this example does not represent the epitome in theft alarm features, it does illustrate the nonmultiplexed condition. The horn actuator and the sensor switches are all wired directly to the theft alarm module. The module is then armed by activating the Dash Arm Switch. The module can be disarmed by either the driver door key switch, passenger door key switch, or the trunk key switch. The horn is sounded when either the hood, door, or trunk is tampered with when the module is armed.

The Vehicle Theft Alarm system shown in Figure 3 illustrates a near optimal configuration of a Class A Network. The sensors and actuators are integrated with the multiplexing electronics so that they can communicate over a single wire to the theft alarm module. The integration of electronics into the sensors and actuator improve sensor diagnostics because the sensor status and condition can be reported back to the controlling module. The integrity of the sensor status/condition can be linked to the mechanical operation of the sensor. This level of switch integrity cannot be achieved with normal switch biasing methods. In a theft alarm system there is an added benefit: the sensor condition can be used to set off the alarm and foul the tampering of a would-be thief.

The I/O requirements support T-tap connections which can be highly automated in the production of wiring harnesses, reduce bundle size, and eliminate dual crimps. The configuration also supports the concept of adding sensors or actuators as the option requires without changing the Theft Alarm Module configuration to support the optional features. This expandability feature allows the cost of the option to drive the system cost.

To show how this configuration is flexible and easily expandable consider the example condition where some versions of Theft Alarms are built as originally described, but an upscaled version is offered as an option where the unit is armed by the driver locking the doors. To support this option the Dash Arm Switch would be eliminated and the Driver Door Lock Switch would be configured with the integrated switch multiplex at a different address. The same Theft Alarm Module's software could then reconfigure itself without hardware modifications.

Statistically speaking, there are approximately seven sensors to every actuator in a real vehicle body system (See SAE J2057 Part 1). This Theft Alarm System is typical with ten sensors (switches) to one actuator (horn).

The sensors and multiplexing electronics can be integrated into the switch component. This configuration eliminates separate wiring and mounting of the multiplex module. Some component manufacturers have even been working on two wire (signal and ground) sensors where the power to run the sensor has been supplied by the multiplex signal (for an example component see SAE J2058). These sensors have been designed to include the multiplex circuit integrated with the Hall Effect Device in the same TO92 size package. The multiplexer portion is very small and requires approximately 300 logic gates.

The actuator driver and multiplexer can similarly be integrated into the horn or motor. This configuration also reduces wiring and mounting complexity. Actuators normally require more power than sensors and, therefore, usually require three wires: signal, power, and ground. Refer to SAE J2058 for an example component. However, some manufacturers are developing a method to eliminate one of these wires by placing the signal on the power wire.

Cost is perhaps the biggest factor and criteria of judgment as to whether a Class A Multiplexing Network will be successful. Ideally the cost of the sensor or actuator component should be less than or equal to the part it replaces. This cost requirement should not be too big a challenge for the aggressive sensor/actuator supplier. (For an explanation see Appendix A.)

Even with all the advantages of the Class A Network, the overall system cost must be competitive. The cost of the sensors is controlled by their mature volume. With many different sensors being required, the logistics involved with developing each part can significantly affect the final price. An effective means of programming the distinct address on a common part must be available.

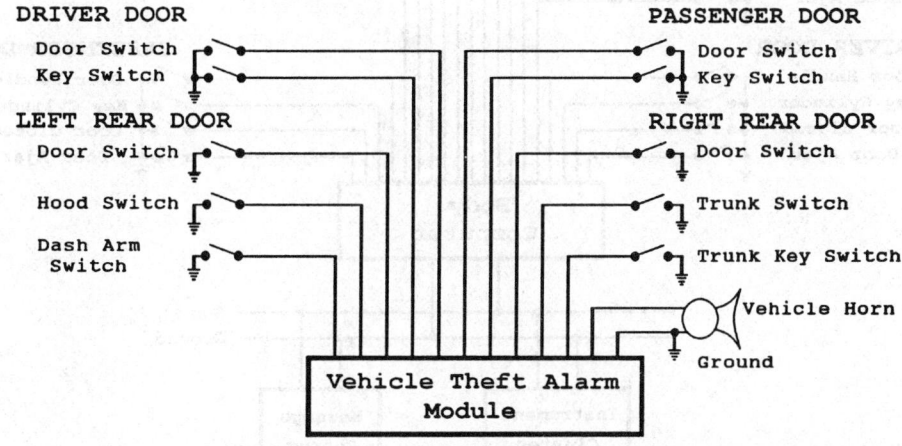

FIGURE 2—EXAMPLE VEHICLE THEFT ALARM SYSTEM

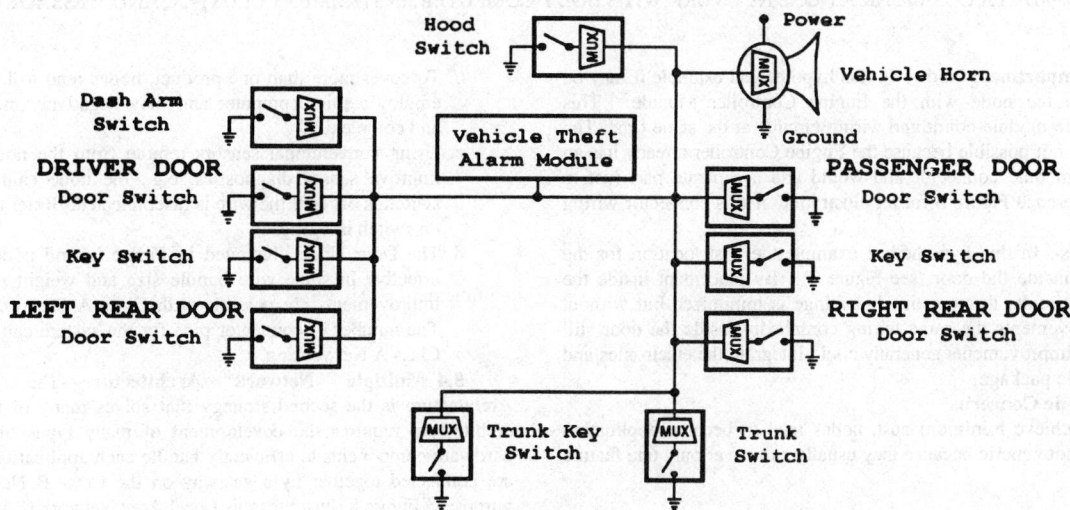

FIGURE 3—MULTIPLEXED VEHICLE THEFT ALARM SYSTEM

5. Proposed Vehicle Architecture—The vehicle system designer now has many architecture partitioning options. A strategy of when to integrate many features into a module or when to employ a dedicated node is a prime example. Care must be taken or the partitioning strategy may not achieve optimal results. The issue is much more complex when vehicle multiplexing is involved in this partitioning strategy. The most popular networking strategy is the Class B Single Network Architecture. However, Class B Multiplexing does not always result in an optimal solution.

A hypothetical vehicle will be described to illustrate this point. Figure 4 illustrates the part of a Data Communications Network that contains a Body Computer, an Instrument Cluster, and a Message Center. In this example all the sensors that feed the network enter through the Body Computer.

As illustrated in Figure 4, all sensors are wired directly to the Body Computer. This example shows that a base vehicle with only a small amount of electronic content, where all the sensors are directly wired to the body computer, the wire bundle size, and number of connector pins is attainable. As additional features are made standard, either by consumer demands or government regulations, it becomes more and more difficult to implement the required system. This added complexity is due to the tremendous number of interconnecting wires from sensors to the modules. The build complexity and trouble-shooting problems make this option a limited solution for this partitioning strategy.

The Class B Single Network Architecture strategy would solve this complexity problem by adding a sensor node and reduce the number of interconnecting wires. By this strategy conventional sensors are connected directly to the node which serves as a gateway to the other modules over the Class B data link. Figure 5 illustrates the dramatic reduction in the number of circuits required. This method is effective in reducing the number of sensor wires connected through "crunch points" such as the bulkhead or door hinge. However, this reduction in wiring is obtained at the expense of three added sensor nodes.

Class B Multiplexing is a very useful technique for reducing many of the problems encountered by the automotive system engineers. However, this report will demonstrate that in many situations the multiplex strategy, shown by Figure 5, leads to a less than optimum system architecture. It is highly desirable to have a multiplexing architecture which would:

a. Permit the use of smaller module connectors
b. Reduce the number of wires crowding through the congested areas
c. Accomplish without introducing more modules to mount, wire, and service

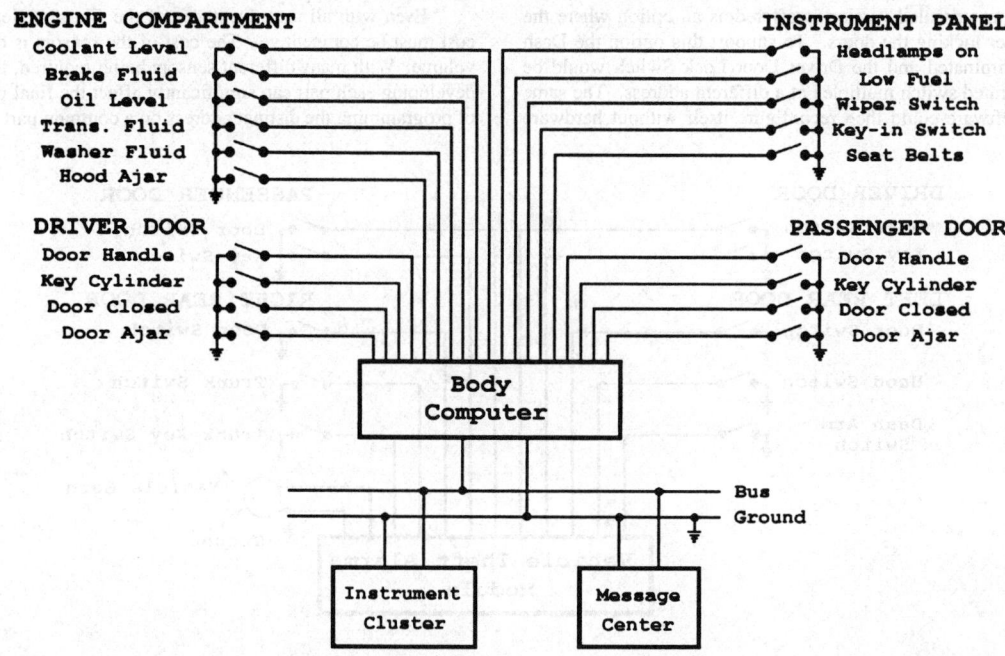

FIGURE 4—DATA COMMUNICATIONS NETWORK WITH BODY COMPUTER, INSTRUMENT CLUSTER, AND MESSAGE CENTER

5.1 Engine Compartment Node—In this hypothetical example it may be desirable to integrate the node with the Engine Controller Module. This operation would reduce module count and wiring circuits at the same time. The integration solution is not possible because the Engine Controller already has an uncontrollably large module connector and would add a separate part just to cover an option. Reference Figure 6 for an illustration of this connector wiring complexity.

5.2 Door Nodes—In this hypothetical example the best location for the door node would be inside the door (see Figure 7). By placement inside the door, the number of circuits through the door hinge is minimized but without making further improvements the same wiring complexity inside the door still exists. These further improvements generally could integrate the electronics and mechanics into a single package.

5.3 General Node Concerns

a. In order to achieve minimum cost, nodes tend to become application specific and not generic because they usually can cover only one feature product.

b. To cover more than one product, nodes tend to become intelligent and employ a microcomputer and may negatively impact the system cost and complexity.
c. Using conventional sensors remote from the node does not normally improve sensor diagnostics, e.g., the node cannot tell if the sensor switch is off or if the wire is disconnected. Refer to 4.2 for a discussion on switch integrity.
d. The Door Node illustrated in Figure 7, and nodes in general, can be effective in some wire bundle size and weight reductions but further improvements are possible with Class A Sensor/Actuator Networking. The number of connector pins for the system can also be reduced with Class A Networking.

5.4 Multiple Network Architecture—The Multiple Network Architecture is the second strategy that solves many of these concerns. This architecture requires the development of many types of specialized network hardware components to efficiently handle each application. These components are connected together by a gateway on the Class B Network for diagnostics purposes. Figure 8 illustrates this Local Area Network (LAN) solution.

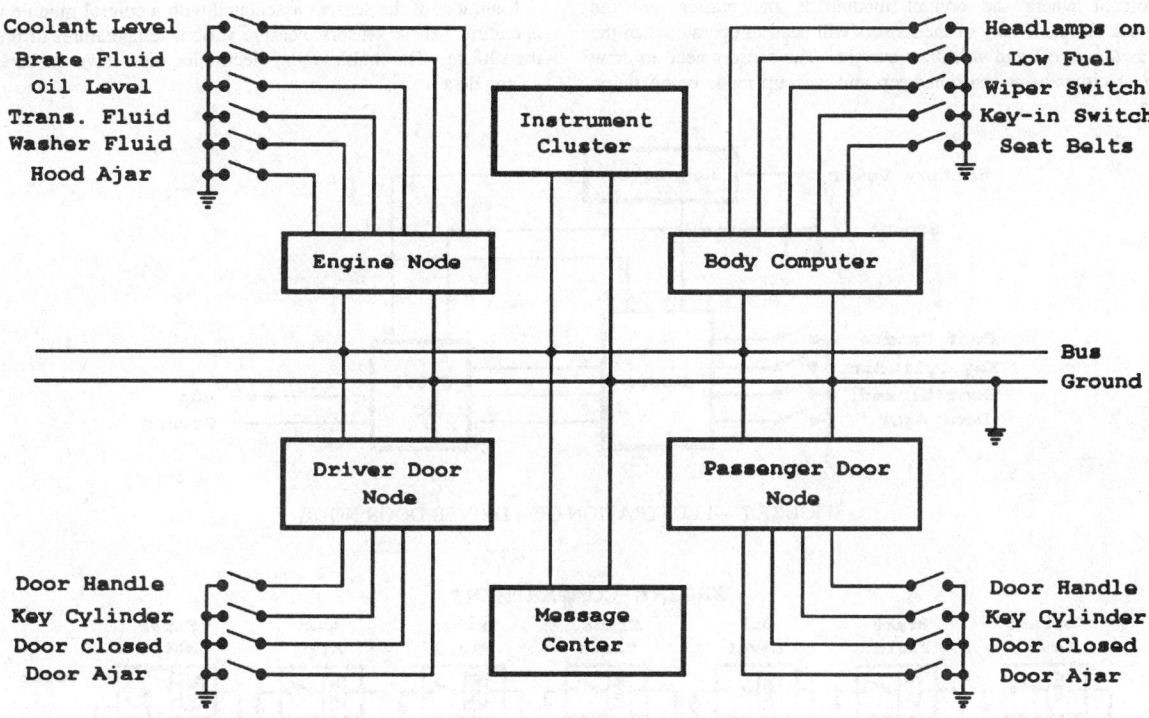

FIGURE 5—DATA LINK WITH BODY COMPUTER, INSTRUMENT CLUSTER, MESSAGE CENTER AND THREE SENSOR NODES

FIGURE 6—ENGINE CONTROLLER CONNECTOR AND WIRE BUNDLE

Multiple network architecture strategy requires the integration of electronics into the sensors, actuators, and motors so that they can communicate over a single wire into the module that utilizes them. Since the sensor and actuator components contain the added multiplex electronics the separate installation and wiring of the multiplex module is eliminated. Unlike the single network architecture strategy the integration of electronics into the sensors, actuators, and motors normally does improve sensor diagnostics because the sensor status and condition can be reported back to the controlling module. Also, the method makes use of a Class A LAN without adding components to the vehicle system.

Figure 8 shows that the Class A LAN eliminates the need for the Engine Compartment Node and two Door Sensor Modules while still reducing wiring at the crunch points. The Multiplex Architecture shown in Figure 8 significantly simplifies the same system shown in Figure 5. This simplification is made possible by separating the Class B intermodule communications network from the Class A sensor-to-module communications. The cost of adding multiplexing directly to the sensors is significantly less than the cost of adding the three sensor modules. Appendix A discusses how these conclusions could be achieved.

The Class A LAN connects all the multiplexed components in parallel. The I/O requirements support T-tap connections, which can be highly automated in the production of wiring harnesses. This reduces bundle size and eliminates dual crimps. The configuration also supports the concept of adding sensors or actuators as the option requires without changing the Body Computer configuration to support the option. This add-on feature allows the option to dictate the cost, not the cost of the added node dominating.

The two different Class A Networks using multiplexed sensors are shown in Figure 8 and Figure 9 illustrating a Body Computer flexibility commonality which is not available in the other architectural approaches. The typical Base Vehicle using the Class A Network is shown in Figure 9. The Body Computer in the Base, Medium, and Premium Vehicle Systems all have the Wiper Switch, Seat Belt Switch, Headlamps on, Low Fuel Level, Key-in Ignition Switch, Class A Interface, and the Class B Multiplex Interface. This commonality across option rates allows additional inputs to be connected without modifying the hardware in the Body Computer.

Unlike the hypothetical situation just used for illustration purposes, a real vehicle will have several actuators as well as sensors connected to the Body Computer. As it was previously discussed in 4.2 there are approximately seven sensors to every actuator in a real vehicle body system. For illustrative purposes 14 sensors and one actuator (Wiper Motor) were shown in this example. The principles shown, however, apply similarly any number of actuators.

6. Requirements for Class A Sensors and Actuators—As shown in Figures 8 and 9 a sensor/actuator network is best demonstrated when associated with a discrete sensor or actuator that is distributed throughout the vehicle. The multiplexing electronics can then be contained within the sensor or actuator and, therefore, eliminate the need to create a gateway node between the sensor or actuator and the controlling module as shown by Figure 5.

The multiplexing electronics must be simple and low cost. To facilitate manufacturing and service, it should contain diagnostic capability. The data on the Class A Network normally tends to be event-driven and requires acknowledgement or handshake with the control module. Although the latency time requirements for Class A sensors and actuators are normally much longer than Class B Multiplexing latencies. Some Class A sensors and actuator latencies may exceed the Class B requirements and sometimes require a guaranteed maximum latency. A possible low-cost solution is to utilize a master-slave protocol where the control module is the master and the sensors/actuators are slaves. Many of the sensors will need to operate when the vehicle ignition switch is off and will draw current. The sensors need to draw very low current or either a low current sleep and wakeup mode of operation must be provided.

A number of the sensors associated with a control module will require analog operation. These sensors measure various temperatures or pressures throughout the vehicle. The multiplexing electronics must have the capability of handling analog data.

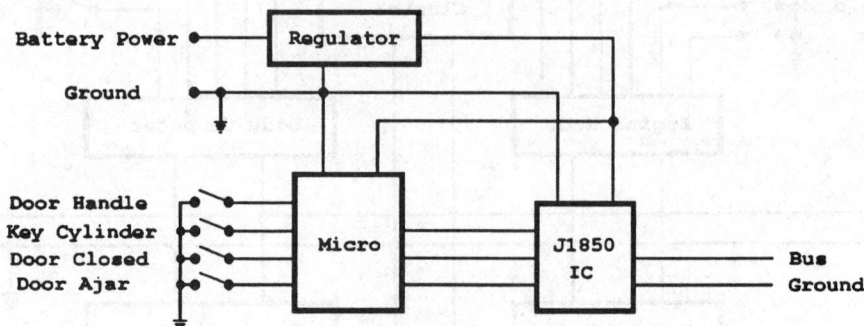

FIGURE 7—ILLUSTRATION OF A DRIVER DOOR NODE

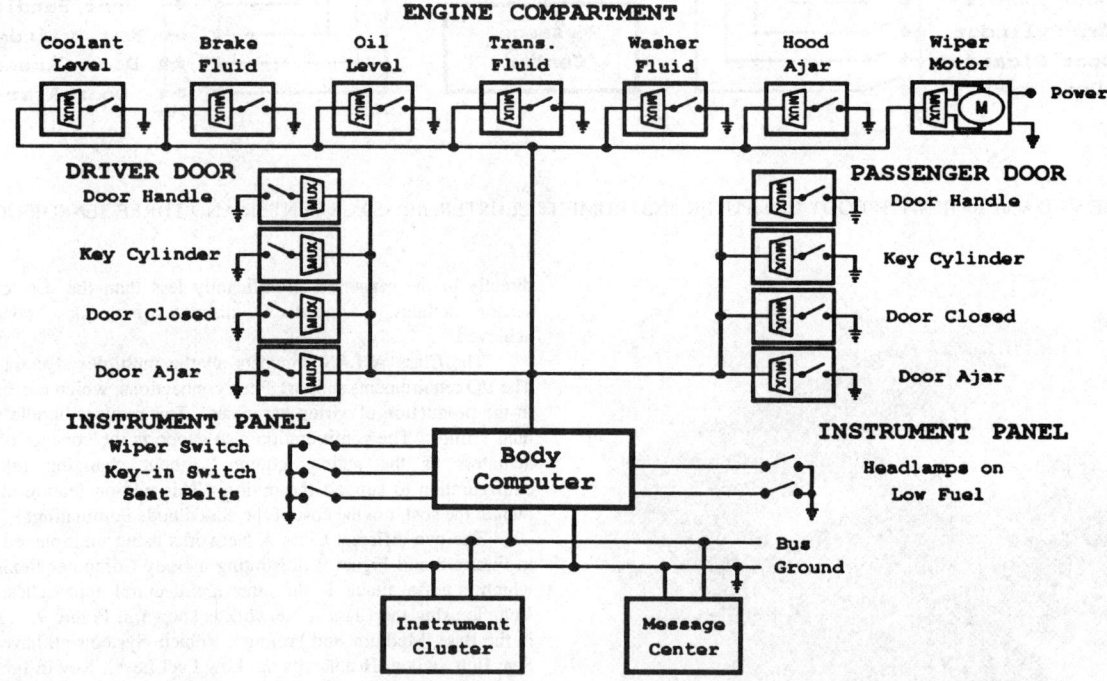

FIGURE 8—DATA LINK WITH BODY COMPUTER, INSTRUMENT CLUSTER, MESSAGE CENTER, AND CLASS A NETWORK FOR SENSORS

7. Summary and Conclusions—With the development of SAE J1850 Class B Network and the message strategies defined in SAE J2178, many of the potential vehicle multiplexing applications, for what was once considered a Class A application, are now resolved. This Class B solution should function very well and perform all of the functions of a Class A Network as defined by 1.1.4. However, this solution is considerably more costly and does not solve much of the body wiring problem.

This report on Architecture Strategies began with the assumption that a lower total vehicle system cost would result if one network were optimized around a data communication requirement and other networks were optimized around the sensor and control requirements. The Class A Sensor/Actuator Network addressed by this document does provide insight into a possible solution to these cost, complexity, and wiring problems.

7.1 Advantages and Disadvantages—There are many advantages to a Class A Sensor/Actuator Network Architecture. The sensors and actuators can be integrated with the multiplexer electronics, and will eliminate the need to mount separate control node(s) or module(s). This architecture promotes simplified wiring harnesses that are easy to manufacture and can be validated as a system in vehicle production. Vehicle service is greatly simplified because the sensor and actuator diagnostic capability can economically be built into the component.

The disadvantages of the Class A Sensor/Actuator Network is that it is inefficient for the sensor or actuator to be connected with a nonassociated node or module. If, for example, in the rare instance where an actuator is required to be controlled by more than one module, the module would have to coordinate this activity through a higher level network or through hard wires. This is the presently common direction for sensors on a Class B Multiplexing Network where sensor data is distributed across the network to be used by other modules. This consideration places an architectural constraint on the system designer. However, this is not the only constraint. As with any multiplex system, attention must also be given to the need to standardize on sensor/actuator addresses. The need to standardize component addresses complicates the supplier black/gray box programs with the vehicle manufacturer.

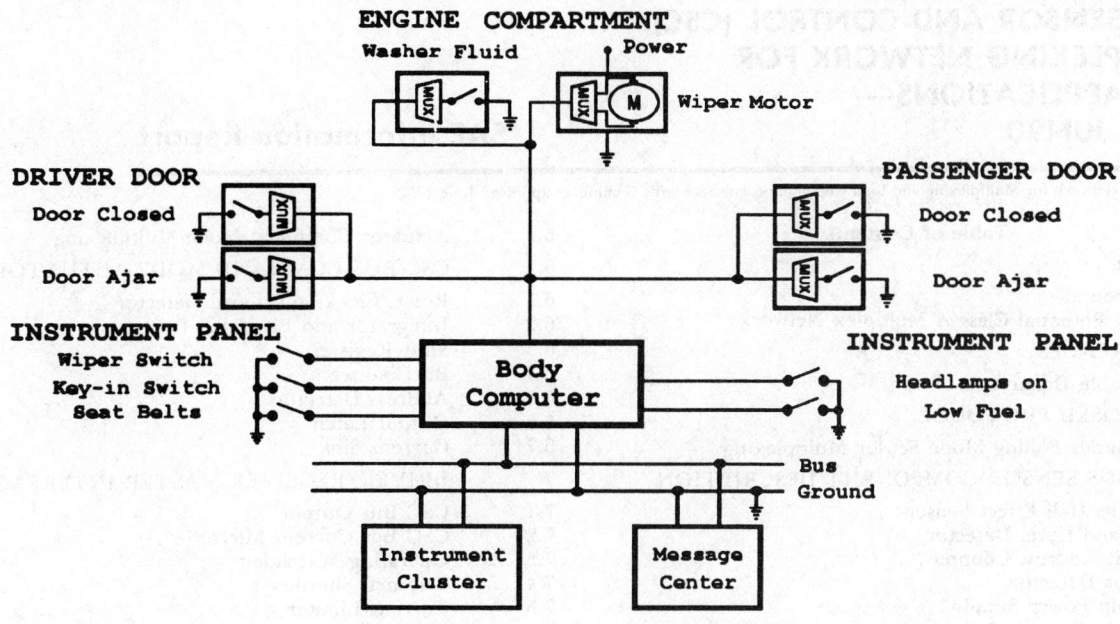

FIGURE 9—BASE VEHICLE WITH CLASS B DATA LINK, BODY COMPUTER, INSTRUMENT CLUSTER, MESSAGE CENTER, AND CLASS A NETWORK FOR SENSORS

APPENDIX A

A.1 Sensor/Actuator Technology—The theory is to find a method of utilizing an integrated sensor or actuator with the multiplexing circuitry. Sensors presently use other mechanical mechanisms such as floats, levers, or springs. Actuators use high-power drivers that require short-circuit protection or mechanical devices such as commutator or rare earth magnets. These elements can be eliminated to offset the cost of the integrated electronics. The cost savings in some devices is much more substantial than others. If on the average the savings of eliminating these devices plus the system savings in complexity and labor listed in Table A1 offset these electronics cost, then the challenge has been met.

TABLE A1—EFFECTS OF CLASS A SENSOR/ACTUATOR MULTIPLEXING

	Figure 4 Directly Wired	Figure 5 Class B Nodes	Figure 8 Class A Network
Number of Circuits		Fewer	Fewer
No. Connector Pins		Same	A lot Fewer
Auto T-Tap Support		No	Yes
Wire Bundle Size		Smaller	A lot Smaller
Wire Weight		Lighter	A lot Lighter
No. of Components		Larger	Same
Installation Time		Longer	Shorter
Build Test Time		Same	Shorter
Components Weight		A lot Heavier	Equal or Lighter
Sensors/Actuators Diagnostic		Not Usually	Usually
Post Build Upgrade		Same	Easy
Self Configuring		No	Yes
Req Std Addresses		No	Yes
Sensors/Actuators Cost		Same	See Section A.1

CHRYSLER SENSOR AND CONTROL (CSC) BUS MULTIPLEXING NETWORK FOR CLASS 'A' APPLICATIONS— SAE J2058 JUN90

SAE Information Report

Report of the Vehicle Network for Multiplexing and Data Communications Standards Committee approved June 1990.

Table of Contents

1. SCOPE
1.1 Background
1.2 CSC A Potential Class A Multiplex Network
2. REFERENCES
2.1 Applicable Documents
3. PROPOSED PROTOCOL
3.1 Continuous Polling Mode Sensor Multiplexing
4. CSC BUS SENSOR COMPONENT DESCRIPTION
4.1 CSC Bus Hall Effect Sensor
4.1.1 Clock and Level Detector
4.1.2 Five-Bit Address Counter
4.1.3 Address Detector
4.1.4 On-Chip Power Supply
4.1.5 Constant Current Sink
4.1.6 Sensing Element
5. ACTUATOR POLLING MULTIPLEXING
5.1 Actuator Command Mode Multiplexing
6. CSC BUS COMMAND MODE ACTUATOR
6.1 Reset, Clock, and Level Detector
6.2 Integrator and Bit Value Detector
6.3 Shift Register
6.4 Bit Counter
6.5 Address Detector
6.6 Output Latch
6.7 Current Sink
7. DRIVER/RECEIVER MASTER INTERFACE
7.1 CSC Bus Output
7.2 CSC Bus Current Mirror
7.3 Operating Watchdog
7.4 Thermal Shutdown
7.5 Current Limiter
7.6 Output Wave Shaper
8. ELECTROMAGNETIC COMPATIBILITY
9. CONCLUSION AND FUTURE DIRECTION

Foreword—A proprietary multiplexing technique called the Chrysler Sensor and Control (CSC) Bus has been designed which gives the systems engineer an effective means to meet the future demands of automotive customers. The CSC Bus allows the design of the base vehicle, while attaching the complete cost of optional features to the option. This paper will describe the protocol for this CSC Bus multiplexing technique and show that it leads to the simplest vehicle electronic system.

1. Scope—The CSC Bus components defined herein were developed to provide simple, yet reliable, communication between a host master module and its sensors and actuators. The scheme chosen provides the ability to communicate in both polling mode and direct addressing modes.

1.1 Background—Vehicle applications for multiplexing can be broken into three broad classifications:
a. Class A–Sensor and Control Multiplexing
b. Class B–Data Communications
c. Class C–High Speed Real Time Control

Sensor and control multiplexing is the more classical form. An example of this kind would be fluid level sensors, door/hatch switches, and control of the vehicle headlamps, tail lamps, stop lamps, horn, wipers, etc., through a multiplex wiring network. Normally, control multiplexing affects the base vehicle wiring system, which directly affects the base cost. On the other hand data communications multiplexing (Class B) interconnects intelligent modules such as the engine controller, body computer, vehicle instrument cluster, and other electronic modules. It normally does not affect the base vehicle wiring but provides an inter-module data communications link for distributed processing and, if properly partitioned, will reduce costs on all but the base vehicle. Finally, there is a class of medium to very fast real time control modules such as the engine controller, automatic transmission, and antiskid controller.

Class B multiplexing is a very useful technique for reducing the problems encountered by the automotive electronic system engineer. However, in many situations Class B multiplexing leads to a less than optimum system. This Class B network can be an effective enabler to vehicle integration but it does not solve the sensor and actuator connection problem. After a careful study of the current Class B Network, it was obvious that a different system needed to be developed. Statistically speaking, it was determined that there are approximately seven sensors to every actuator in a potential Class A multiplex network. A form of multiplexing which would permit smaller module connectors and reduce the number of wires crowding through the congested areas without introducing more modules is highly desirable. The CSC Bus is a style of multiplexing which meets these objectives.

Another problem presented to the system designer consists of implementing different levels of option content. Base vehicles receive limited electronic feature content, medium vehicles receive some optional electronics, and premium vehicles contain an extensive amount of electronic features. CSC multiplexing has the ability to solve this variable content uncertainty as well as the wiring congestion problem.

1.2 CSC A Potential Class A Multiplex Network—The proposed solution defined herein integrates a multiplexer with a Hall Effect Sensing Element and replace the reed switch used in current sensors. When standard switches are given multiplexing capability they are really sensors. They no longer 'switch' as they do today. Instead they 'sense' some parameter and report back the status including sensor diagnostics. The CSC Bus allows for increased feature content in existing modules by connecting additional sensors. More exotic sensors are also practical using multiple CSC sensors.

Perhaps the biggest factor and criteria of judgment as to whether a Class A multiplexing network will be successful or not is cost. The CSC sensors have been designed to include the multiplex circuit integrated with the Hall Effect Device in the same TO92 size package. The multiplexer portion is very small and requires approximately 300 logic gates.

2. References
2.1 Applicable Documents
SAE J1850, Class B Data Communications Network

R. Vig and T. Wroblewski, "Addressable Single Wire Magnetic Sensor IC for Multiplexed Control," 3rd International Conference "Automobile Electronics–Interface and Environment," SIA. Toulouse, France; April 1988

T. Wroblewski, "A Multiplexed Automotive Sensor System," Sensors, February 1989

U. Kiencke, S. Dais, and M. Litshel, "Automotive Serial Controller Area Network," SAE Technical Paper Series, International Congress and Exposition, February 1986

D. J. Arnett, "A High Performance Solution for In-House Networking Controller Area Network (CAN)," SAE Technical Paper Series, Earth-Moving Industry Conference, April 1987

T. Wroblewski, "A SC Bus Multiplexing Technique Which Allows for Common Electronic Modules," Paper 89123, 20th ISATA, Florence, Italy; May 1989

3. Proposed Protocol—The network method for this communication system is a master-slave polling and/or direct address method. The master uses a voltage waveform to communicate to the sensors and actuators. A sensor is addressed through successive, ordered polling of each address (time slot or period) in ascending order.

<SOM>, Wake-Up Period, Addr 1, Addr 2, ..., Addr 32

An actuator may be controlled by either the polling mode above or the direct-address, command-mode message below. The 'start of message', (SOM), is defined as the rising edge from 0 V to 6 V in the voltage waveform of Figure 1.

<SOM>, <Five-Bit address>, <1 to N data bits>, [<parity bit>]

  a. CSC Bus Sensor Polling Mode Voltage Waveform
  b. Voltage Representing Current Drawn by the CSC Sensors

3.1 Continuous Polling Mode Sensor Multiplexing—Sensor multiplexing was addressed first because sensors outnumber actuators in the vehicle. Figure 1 contains the typical voltage waveform used to communicate with multiplexed sensors. As shown, the sensors use current to respond back to the master which generates the voltage waveform. In this scheme, each sensor has an internal preprogrammed address. The voltage begins at a reset (0 V) level and climbs to 6 V. This initial 6 V level provides power to the sensors. During this time, the master reads the amount of current required to keep the sensors powered, called the sensor power current. At each change from 6 to 9 V, a counter contained in the sensor is incremented. The sensors are addressed consecutively so this mode of CSC Bus communication is called the 'Continuous Polling Mode'.

4. CSC Bus Sensor Component Description—Two CSC components are available as of this publication;
  a. 1: CSC Bus Two-Pin Hall-Effect Sensor
  b. 2: Driver/Receiver Master Interface

4.1 CSC Bus Hall Effect Sensor—(Reference Sprague P/N UGN/UGS3055U, Data Sheet #27680)

While the voltage is at approximately 9 V, the sensors compare the value in their counter to their preprogrammed address. If a sensor detects a match between these two values, the sensor will increase the current drain on the CSC Bus. This 'response current' informs the master that a sensor has recognized its address. This condition remains until the voltage falls to about 6 V.

When the sensor being addressed detects that the voltage is below its threshold of about 7.5 V, it will determine the status of the sensing element (magnetic field detecting Hall effect sensor, optical sensor, or mechanical switch to ground). If the sensing element is active (a magnetic field is detected, light is detected, or a mechanical switch to ground is closed), the sensor continues to draw the response current so that the master can sense the sensor's status. If the sensing element is not active, response current will cease and only the sensor power current will be drawn.

Each sensor is addressed in this manner until sensor address 32 has been addressed. After address 32, the voltage waveform returns to re-

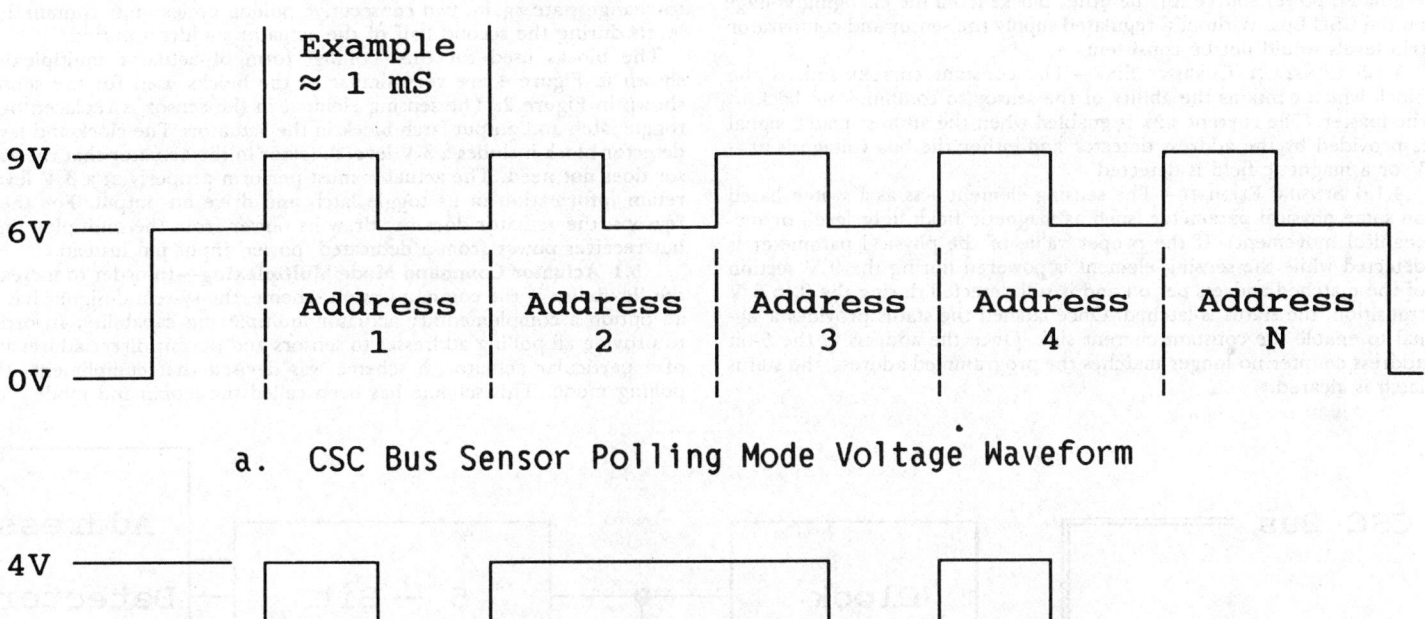

a. CSC Bus Sensor Polling Mode Voltage Waveform

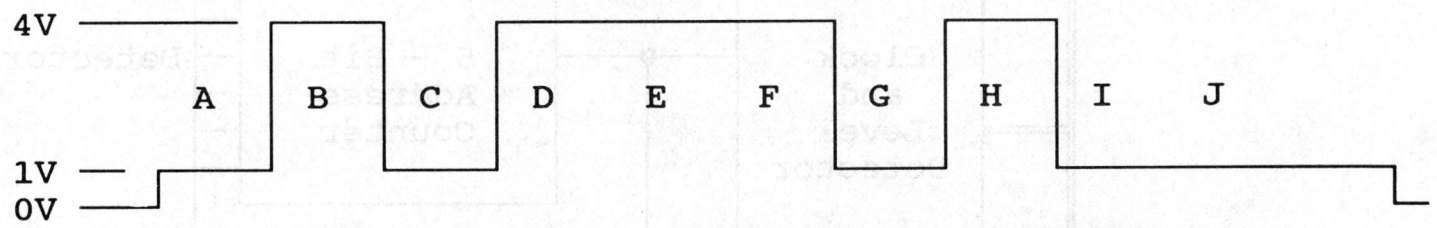

b. Voltage Representing Current Drawn by the CSC Sensors

A: Sensor Power Current   B: Sensor Presence Current
C: Switch Open   D: Sensor Presence Current
E: Switch Closed   F: Sensor Presence Current
G: Switch Open   H: Sensor Presence Current
I: Switch Open   J: No Sensor Present

FIG. 1—CSC BUS SENSOR POLLING MODE WAVEFORM

set. The following blocks shown in Figure 2 are required to implement the CSC Bus addressable hall effect sensor:
a. Clock and Level Detector
b. Five-Bit Address Counter
c. Address Detector
d. On-Chip Power Supply
e. Constant Current Sink
f. Sensing Element

4.1.1 CLOCK AND LEVEL DETECTOR—The clock and level detector block contains a comparator which detects the voltage level on the CSC Bus. When the voltage rises to approximately 8 V, the comparator output becomes high and a positive signal increments the 5-bit address counter. This signal is also used in the logic to enable the constant current sink. Due to the comparator's 1-V hysteresis, the comparator waits until the voltage falls to about 7 V before its level returns low. The 1 V hysteresis prevents incorrect changes in the output of the comparator when low level oscillations are coupled to the CSC Bus line.

4.1.2 FIVE-BIT ADDRESS COUNTER—The 5-bit address counter monitors the address period being queried by the master. The clock and level detector provides the signal which increments the counter. This counter resets each time the CSC Bus returns to the reset level (approximately 0 V).

4.1.3 ADDRESS DETECTOR—The address detector determines when the sensor is being addressed by the master. The particular address is determined by programming during wafer probe. When the value in the 5-bit address counter matches the programmed value, a signal is provided to enable the constant current sink.

4.1.4 ON-CHIP POWER SUPPLY—The on-chip power supply provides a regulated power source for the other blocks from the changing voltage on the CSC Bus. Without a regulated supply the sensor and comparator trip levels would not be consistent.

4.1.5 CONSTANT CURRENT SINK—The constant current sink is the block which contains the ability of the sensor to communicate back to the master. The current sink is enabled when the address match signal is provided by the address detector and either the bus voltage is at 9 V or a magnetic field is detected.

4.1.6 SENSING ELEMENT—The sensing element acts as a switch based on some physical parameter (such as magnetic field, light level, or mechanical movement). If the proper value of the physical parameter is detected while the sensing element is powered during the 9 V section of the matched address period and is still detected during the 9 to 6 V transition, the status is latched. Once latched the status provides a signal to enable the constant current sink. Once the address in the 5-bit address counter no longer matches the programmed address, the status latch is cleared.

5. Actuator Polling Multiplexing

The first effort to control actuators using the CSC Bus multiplexing technique used an extension of the sensor polling mode. A particular actuator is assigned an address, just as the sensor is assigned an address (see 4.1). Each actuator monitors the CSC Bus to count the 6 to 9 V transitions in the same way as the sensor does. When the value in the counter of the actuator matches the actuator's address during the 6 V portion of the address, the actuator draws current to tell the master that the actuator is recognizing its address.

To activate an actuator output the status must first be monitored by checking the current drawn during the second half of the address cycle. In contrast to the sensor polling scheme which uses 6 V and 9 V levels only, the actuator multiplexing scheme adds a third 3 V level (see Figure 3). During the second half of the address cycle, the level is driven to 3 V by the master when the output of the actuator is to be toggled. The actuator monitors the CSC Bus during its address. If the actuator detects the 3 V level, a latch is set. The actuator does not change its output after the first 3 V level is detected because of noise considerations.
a. CSC Bus Actuator Polling Mode Voltage Waveform
b. Voltage Representing Current Drawn by Sensors and Actuators

The actuator monitors the following polling cycle. If the second half of the actuator's address period in the very next polling cycle is 3 V, then the actuator will toggle its output. Every subsequent polling cycle must contain the 6 V level during the actuator's address period for the output to remain constant. The current drawn by the actuator during this second half indicates the status of the output. When the output is to change state again, two consecutive polling cycles must contain 3 V levels during the second half of the actuator's address period.

The blocks used for this 'polling' form of actuator multiplexing shown in Figure 4 are very similar to the blocks used for the sensor shown in Figure 2. The sensing element in the sensor is replaced by a toggle latch and output latch block in the actuator. The clock and level detector block includes a 3 V level detector in the actuator that the sensor does not need. The actuator must perform properly at a 3 V level, retain information in its toggle latch and drive an output. For these reasons, the actuator does not draw its power from the multiplex link, but receives power from a dedicated 'power' input pin instead.

5.1 Actuator Command Mode Multiplexing—In order to increase the flexibility of the communication scheme, the system designer has as an option a complementary actuator multiplexing capability. In order to provide all polling addresses to sensors and permit direct addressing of a particular actuator, a scheme was devised that compliments the polling mode. This scheme has been called the 'command mode', 'di-

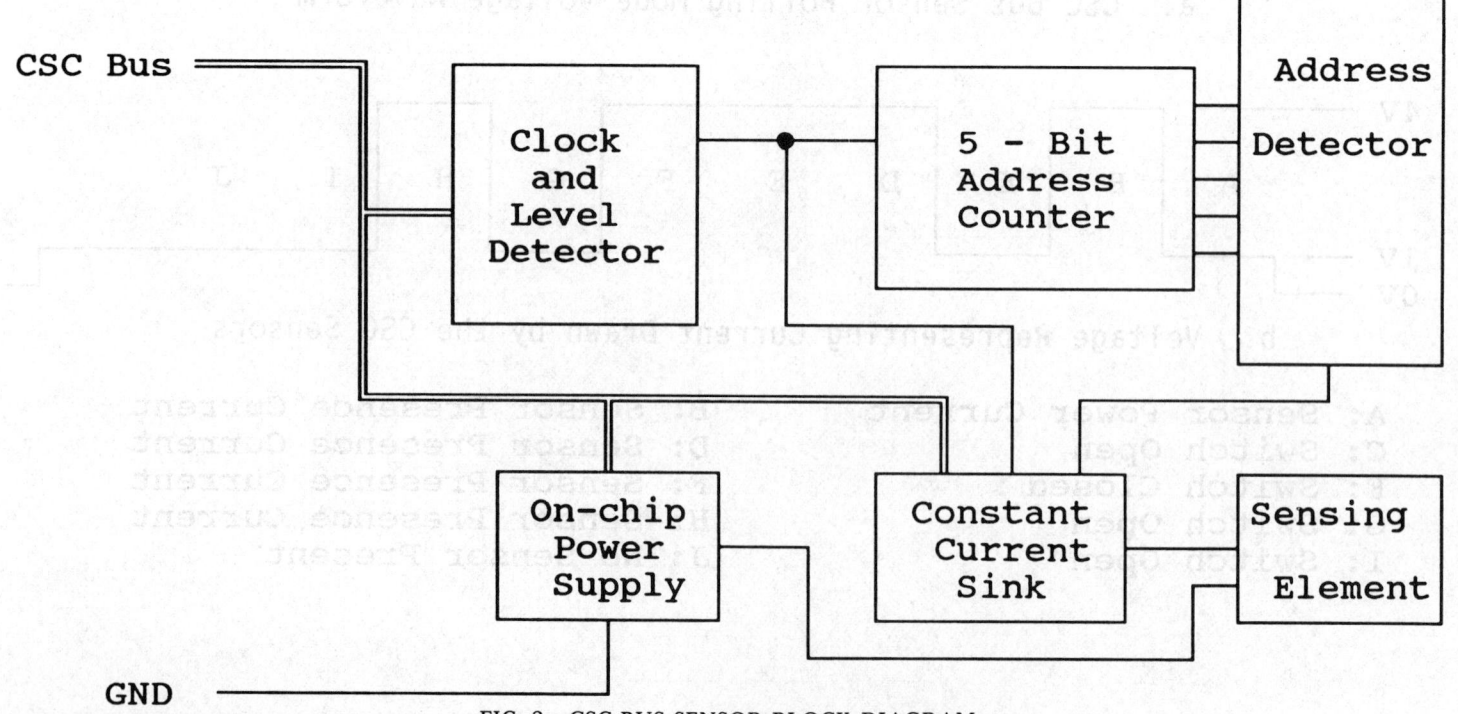

FIG. 2—CSC BUS SENSOR BLOCK DIAGRAM

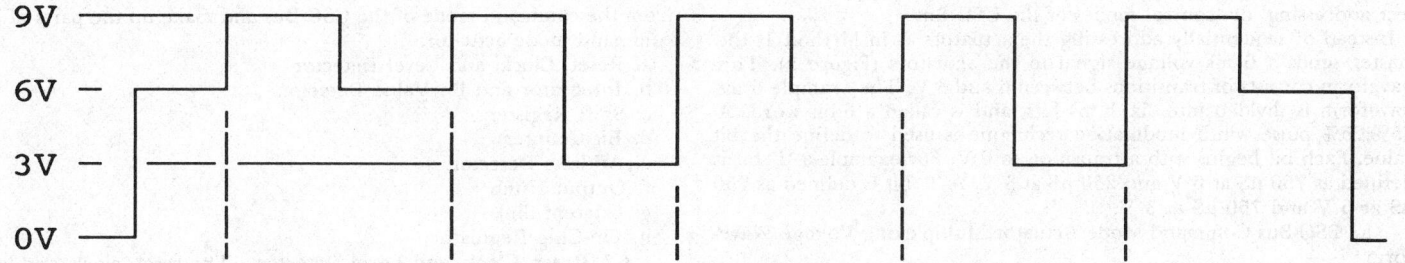

a. CSC Bus Actuator Polling Mode Voltage Waveform

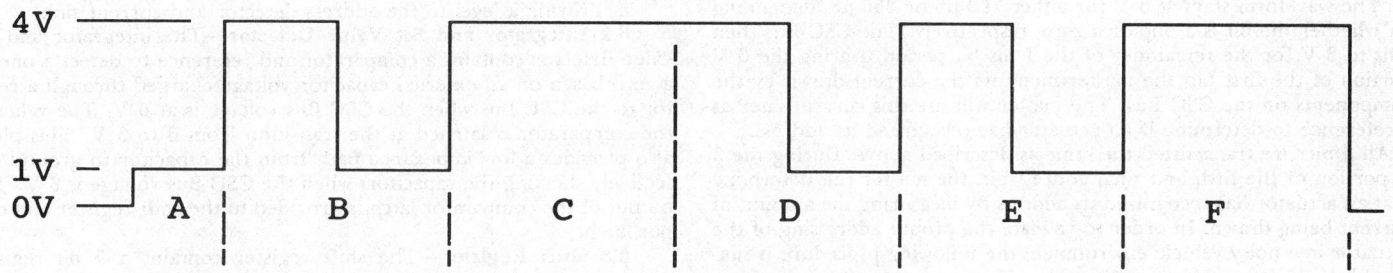

b. Voltage Representing Current Drawn by Sensors and Actuators

A: Sensor Power Current B: Sensor Present, not Active
C: Actuator Toggled, High D: Actuator Status Presently High
E: Actuator Toggled, Low F: Sensor Present, Active

FIG. 3—CONTINUOUS POLLING ACTUATOR MULTIPLEXING WAVEFORMS

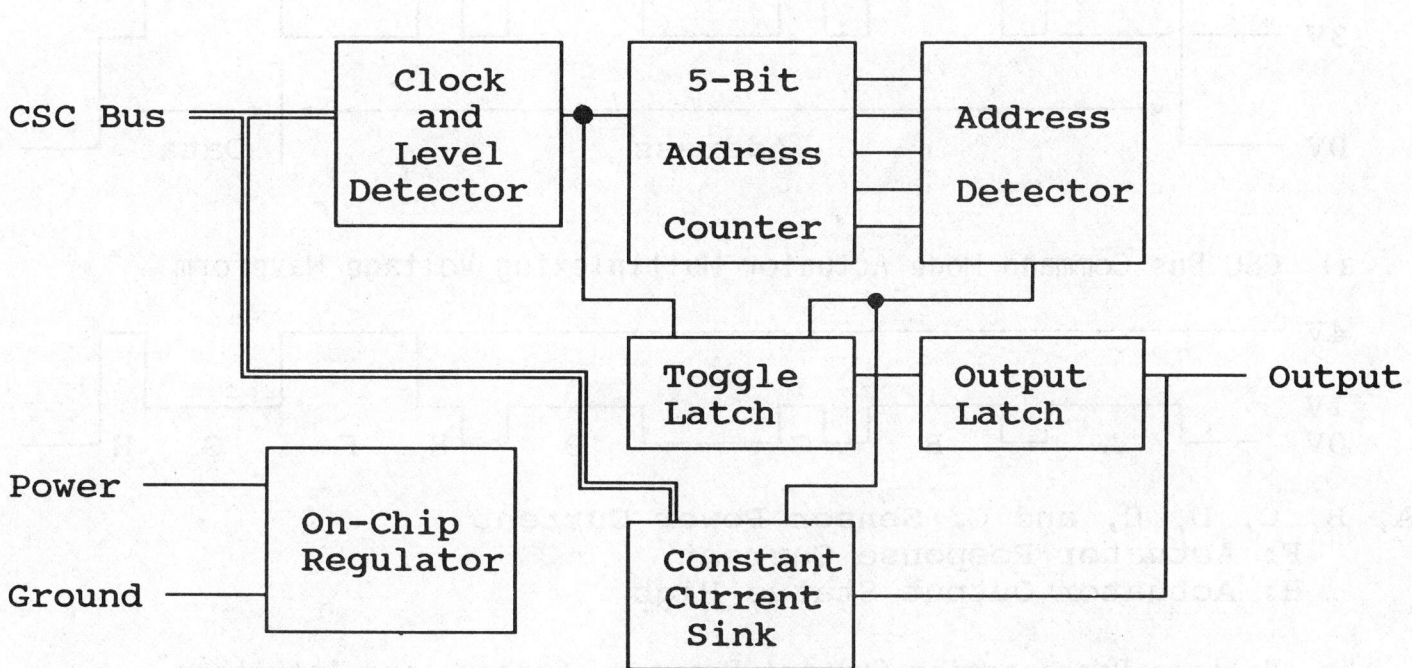

FIG. 4—CSC BUS ACTUATOR POLLING BLOCK DIAGRAM

rect addressing' or 'control mode' of the CSC Bus.

Instead of sequentially addressing the actuators as in Method 1, the master sends a 6 ms voltage signal to the actuators (Figure 5). This waveform consists of transitions between 6 and 3 V. The example 6 ms waveform is divided into six 1 ms bits, and is called a 6-bit word. A 75%:25% pulse width modulation technique is used to define the bit value. Each bit begins with a transition to 6 V. For example a '1' bit is defined as 750 µS at 6 V and 250 µS at 3 V. A '0' bit is defined as 250 µS at 6 V and 750 µS at 3 V.

    a. CSC Bus Command Mode Actuator Multiplexing Voltage Waveform

    b. Voltage Representing Current Drawn by Sensors and Actuators

The first 5 bits of the 6-bit word are used to address the particular actuator and the sixth bit is used to control the state of the actuator's output. The master provides the voltage waveform. The CSC Bus is initially at reset (≈ 0 V). From reset, the voltage waveform is driven to 6 V. The waveform stays at 6 V for either 750 µS or 250 µS, depending on whether the bit is a one or a zero, respectively. The CSC Bus then falls to 3 V for the remainder of the 1 ms bit period. During the 6 V portion of the first bit, the master monitors the current drawn by the components on the CSC Bus. The master will use this current later as a reference to determine if an actuator has recognized its address.

All 6 bits are transmitted the same as described above. During the 3 V portion of the fifth and final address bit, the master can determine that an actuator has recognized its address by measuring the amount of current being drawn. In order to validate the proper addressing of the actuator in a noisy vehicle environment the following procedure is suggested. If this current is not above the reference current measured, no actuator is listening and there is no need to send the data bit. If two or more actuators are listening (determined by double the expected response current during the 3 V portion of the fifth address bit), the master can reset (output 0 V) the CSC Bus instead of sending the data bit to an incorrect actuator. If one actuator is listening, the master would then send the data bit (sixth and final bit in the word). During the 3 V portion of the data bit, the actuator returns the status of its output to the master by sinking the proper amount of current. If the master detects that the actuator output is not in the correct state, the command can be resent.

6. CSC Bus Command Mode Actuator—The following blocks (Figure 6) are necessary to implement an actuator that can receive commands from the command mode of the CSC Bus and make up the parts of the command mode actuator:

    a. Reset, Clock, and Level Detector
    b. Integrator and Bit Value Detector
    c. Shift Register
    d. Bit Counter
    e. Address Detector
    f. Output Latch
    g. Current Sink
    h. On-Chip Regulator

6.1 Reset, Clock, and Level Detector—The reset, clock, and level detector block contains the circuitry to

    a. Provide a power-on reset of the output latch upon application of power
    b. Provide a clock signal to the counter, shift register, and integrator and bit value detector
    c. Provide a level to the address detector and current sink

6.2 Integrator and Bit Value Detector—The integrator and bit value detector contains a comparator and reference to detect a one or a zero based on an external capacitor voltage charged through a resistor to the CSC Bus when the CSC Bus voltage is at 6 V. The value of the comparator is latched at the transition from 6 to 3 V. This block also provides a low impedance path from the capacitor to ground (effectively shorting the capacitor) when the CSC Bus voltage is 3 V. The output of the comparator latch is provided to the shift register and output latch.

6.3 Shift Register—The shift register contains a 5 bit register which gets its data from the integrator and bit value detector. Data is shifted during the transition to 6 V.

6.4 Bit Counter—The bit counter contains a 3 bit counter which is clocked during the 6 to 3 V transition. When the value is five, all address bits have been received and the address detector can compare the bits in the shift register to an internal address. If the value is six and an address match has been detected, the output of the integrator and bit value detector is latched into the output latch.

6.5 Address Detector—The address detector detects a match between the address detected in the shift register and an internal address. If a match is detected, the current sink is enabled and the data bit from the integrator and bit value detector is clocked into the output latch.

6.6 Output Latch—The output latch is used to latch the value of

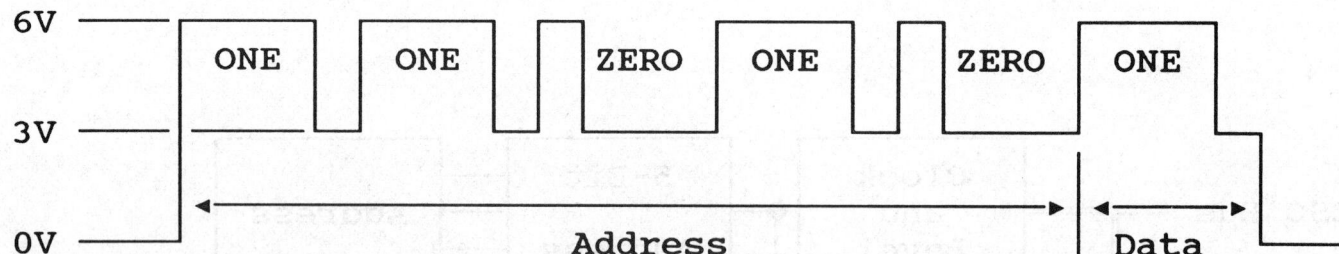

a. CSC Bus Command Mode Actuator Multiplexing Voltage Waveform

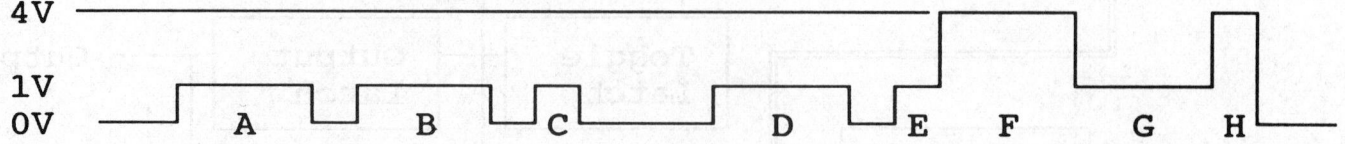

A, B, C, D, E, and G: Sensor Power Current
    F: Actuator Response Current
    H: Actuator Output Status High

b. Voltage Representing Current Drawn by Sensors and Actuators

FIG. 5—CSC BUS COMMAND MODE ACTUATOR MULTIPLEXING WAVEFORM

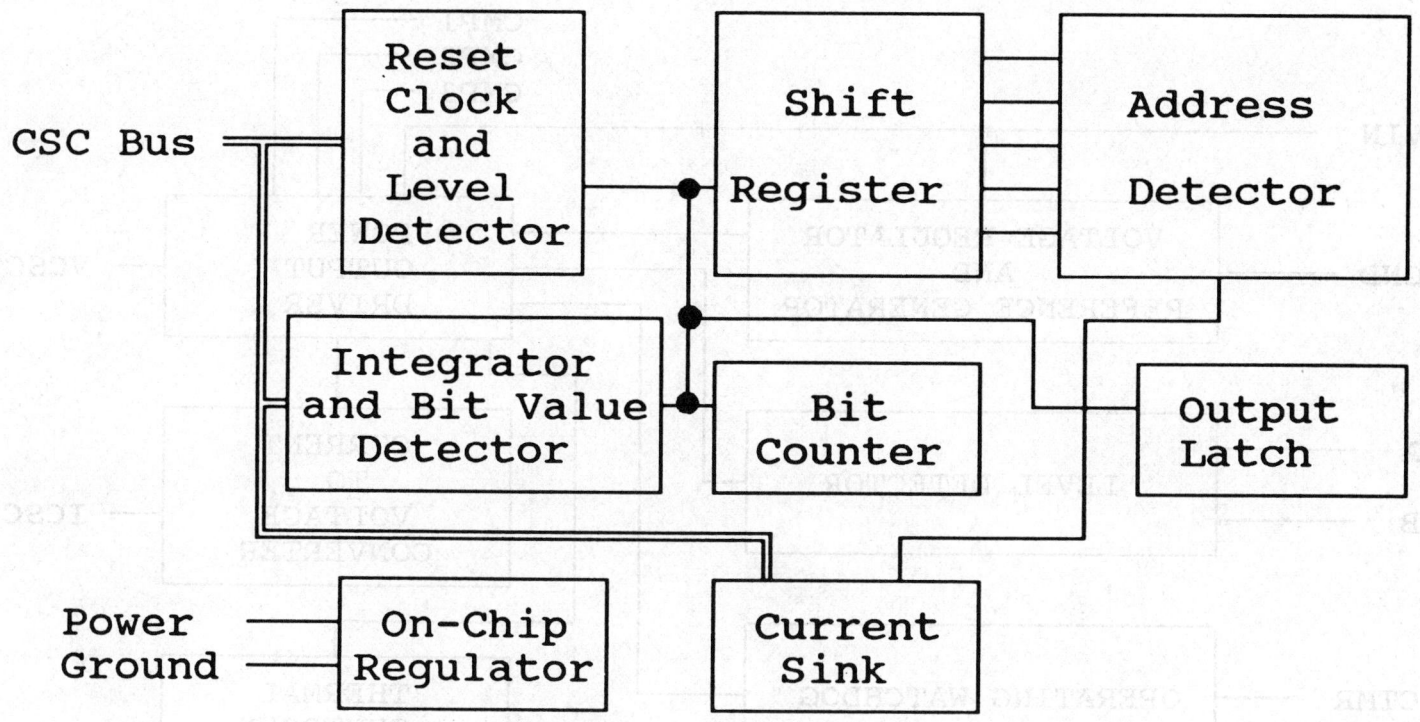

FIG. 6—CSC BUS COMMAND MODE ACTUATOR BLOCK DIAGRAM

the data bit from the integrator and bit value detector at the transition from 6 to 3 V when the bit counter is six.

6.7 Current Sink—The current sink is used to report the actuator's status to the master. If the address sent by the master matches an internal address, the current sink is enabled during the 3 V portion of the last address (fifth) bit. The current sink is also enabled during the 3 V portion of the data (sixth) bit if the output is high and the address sent matches the internal address.

7. Driver/Receiver Master Interface—(Referance Cherry Semiconductor P/N CS-4252)

The CSC Bus utilizes a Master-Slave protocol. This protocol is appropriate because the master is usually the present major feature module and its associated sensors and actuators are the slaves. The feature module microcomputer is the host to the driver/receiver interface IC. In many applications the module connector size is significantly reduced and the host microcomputer ROM size associated with the sensors and actuators can be reduced because software efficiency is improved. In some instances Class B multiplex slave modules can be replaced by simpler CSC Bus sensor/actuator nodes.

The CSC Bus driver/receiver master interface integrated circuit contains the circuitry required to provide a DC offset square wave output (VCSC). This output is controlled by two digital CMOS inputs, A and B. These two inputs are provided by a host microcomputer acting as the 'brains' of the master control module. The IC is able to sense the CSC Bus current and convert it to an analog voltage. This voltage is provided by the IC output (ICSC) for use by the microcomputer analogue input. The driver/receiver master interface IC has the blocks required to perform the following functions (See Figure 7):

7.1 CSC Bus Output—When A and B are both logic LOW, VCSC is OFF (RESET, approximately 0 V). When B is logic HI and A is logic LOW, VCSC is approximately 3 V. When B is logic LOW and A is logic HI, VCSC is approximately 6 V. When A and B are both logic HI, VCSC is approximately 9 V. This output is able to source 125 mA. If VIN falls below 9 V, the voltages track VIN. This output has sufficient filtering and waveshaping to keep the frequency emissions below that level which would interfere with RF communications (AM, FM, CB, etc.) through appropriate external passive networks connected to CMP1, CMP2, and CMP3. Figure 8 shows a typical continuous polling mode waveform with the A and B input pattern required to achieve this waveform. Figure 9 shows a typical command mode waveform with the A and B inputs required to realize this waveform.

7.2 CSC Bus Current Mirror—The ICSC output provides a voltage proportional to the current on the single wire CSC Bus when connected to a resistor to ground. The relationship is shown in Table 1.

TABLE 1—CURRENT TO VOLTAGE CONVERTER

CSC Bus Current	ICSC Output (with 2.7K to Ground)		
	MIN	TYP	MAX
10 mA	0.356 V	0.375 V	0.394 V
20 mA	0.712 V	0.750 V	0.788 V
30 mA	1.069 V	1.125 V	1.181 V
40 mA	1.425 V	1.500 V	1.575 V
50 mA	1.781 V	1.875 V	1.969 V
60 mA	2.137 V	2.250 V	2.363 V
70 mA	2.494 V	2.625 V	2.756 V
80 mA	2.850 V	3.000 V	3.150 V
90 mA	3.206 V	3.375 V	3.544 V
100 mA	3.562 V	3.750 V	3.938 V
110 mA	3.919 V	4.125 V	4.331 V
120 mA	4.275 V	4.500 V	4.725 V
> 133 mA	4.730 V	5.000 V	5.300 V

This output is driven to 5 V during thermal shutdown or external short circuits.

7.3 Operating Watchdog—To eliminate the possibility of destroying any of the sensors if the bus gets locked at 9 V while a sensor's current sink is on, an operating watchdog disables the bus voltage. The watchdog is a timer that resets each time AB=00 is received. The timer length is determined by the charging of an external capacitor.

7.4 Thermal Shutdown—The IC has a built-in thermal shutdown capability that disables the output if the chip temperature reaches an unsafe operating temperature.

7.5 Current Limiter—The output device current supply is limited to 150 mA by an on-board current-limiter to prevent the destruction of the IC and a CSC Bus multiplexed smart sensor connected backwards.

7.6 Output Wave Shaper—Passive networks connected to pins CMP1, CMP2, and CMP3 shapes the VCSC waveform to limit the amplitude of frequency components that would interfere with RF communication (AM, FM, CB, etc.). The networks limits the rise and fall times to between 10 µS and 50 µS. This network does not affect the drive capability or source impedance of the VCSC stage.

8. Electromagnetic Compatibility—As was previously discussed the output waveform of the driver/receiver master interface must be wave shaped to limit the rise and fall time. Empirical vehicle testing has confirmed acceptable EMI levels if the waveform transients exceed 20 µS at data rates of 1 Kbps. The susceptibility to EMI is of greater concern. First is should be remembered that in the vehicle CSC devices replaced sensors that usually employed reed switches as the sensing element, these sensors were originally very susceptible EMI. The normal solution for eliminating EMI susceptibility that caused false actuation was to software filtering the signal in the control module.

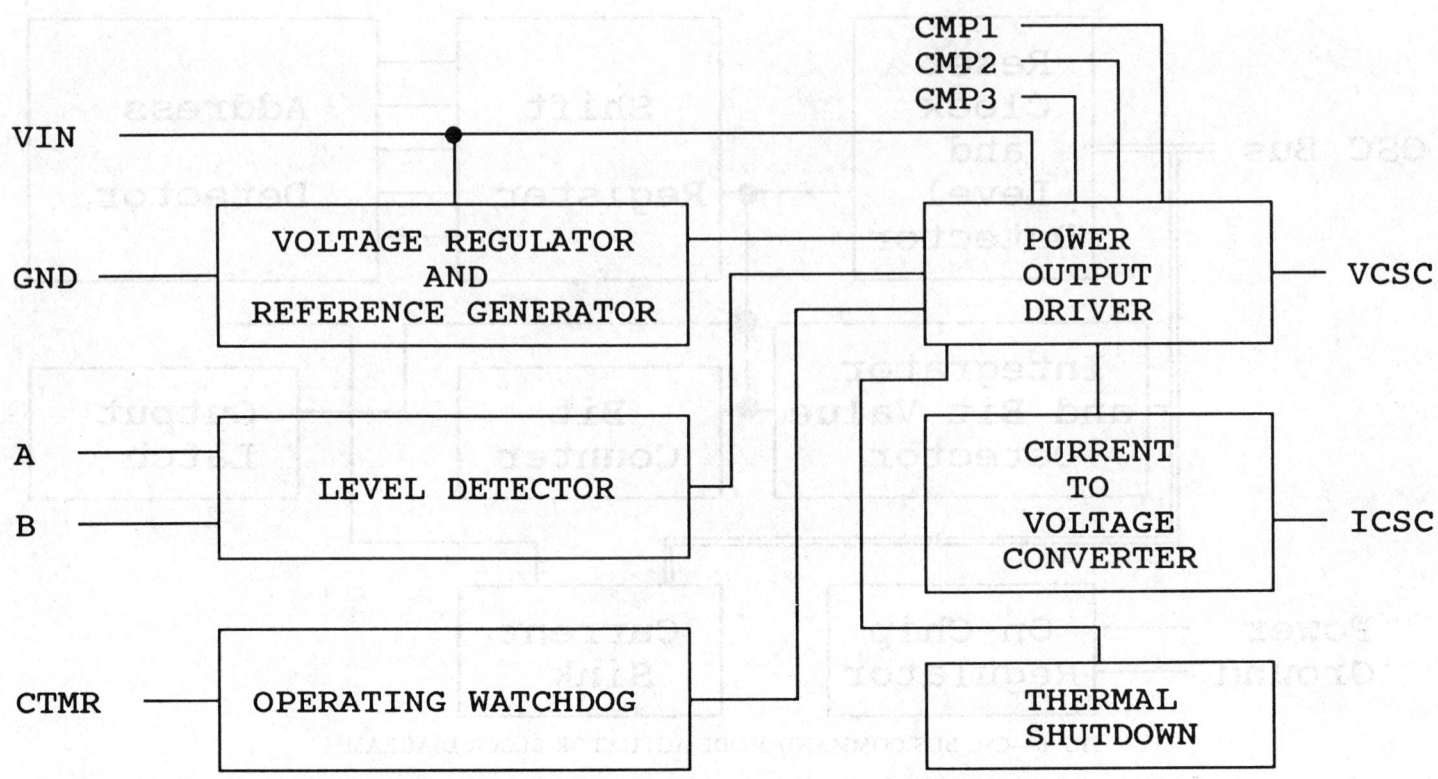

FIG. 7—DRIVER/RECEIVER MASTER INTERFACE INTEGRATED CIRCUIT BLOCK DIAGRAM

To gain a better understanding as to the source of the interference consider a network where the CSC sensor is referenced at a different ground point than the driver/receiver master interface. A noise voltage is developed between these two different ground points. This noise potential is known as longitudinal noise. The magnitude of this longitudinal noise voltage can be as large as 50 to 100 V and can simply wipe out total network communications for the duration of the transient. In such a system the longitudinal noise is in series with the sensor signal and overdrives the receiver input. The only protection the receiver has for this longitudinal noise susceptibility is the input filter which must also be designed to withstand the transient voltage. This filter can only function properly while driven within its dynamic range. When overdriven the output becomes indeterminate.

Another solution tested with outstanding success was to reference the CSC sensor at the driver/receiver master interface ground. This solution virtually eliminates this longitudinal noise current and at the same time significantly reduces the effects of ground offset voltage. In a production vehicle returning the CSC sensor references to the master module usually does not add a wire circuit because these sensors originally had a independent ground wire return to the body or chassis. Further improvement to EMI susceptibility can be achieved by placing a small bypass capacitor across the CSC Bus at the sensor.

9. Conclusion and Future Direction—After a careful study of the current electronic systems, it became obvious that a revolutionary system needed to be developed. A technique for reducing connector size, increasing module commonality, and adding sensor and relay diagnostics was developed and exists in the CSC Bus. Add-a-feature, add-a-module design strategies may be discarded and module commonality between base vehicle requirements and premium vehicle requirements will be retained. The premium vehicle requirements may be met while significantly reducing the number of wiring circuits and wiring bundle size.

The CSC Bus allows for increased feature content in existing modules by connecting up to 32 additional sensors and 32 additional actuators. This CSC Bus scheme will allow for the continued expansion and sophistication of electronic features demanded by tomorrow's customer while improving performance and reliability.

The complimentary modes of the CSC Bus yield flexibility that would not be realized if the polling mode was the only method of communication between the master and both sensors and actuators. By including the Command Mode, more addresses are available and outputs may be controlled with less latency.

A rigorous analysis of module requirements, wiring bundle sizes, total cost and system complexity will yield the best applications for the CSC Bus system. The system designer now has a new technology to apply when partitioning a new system.

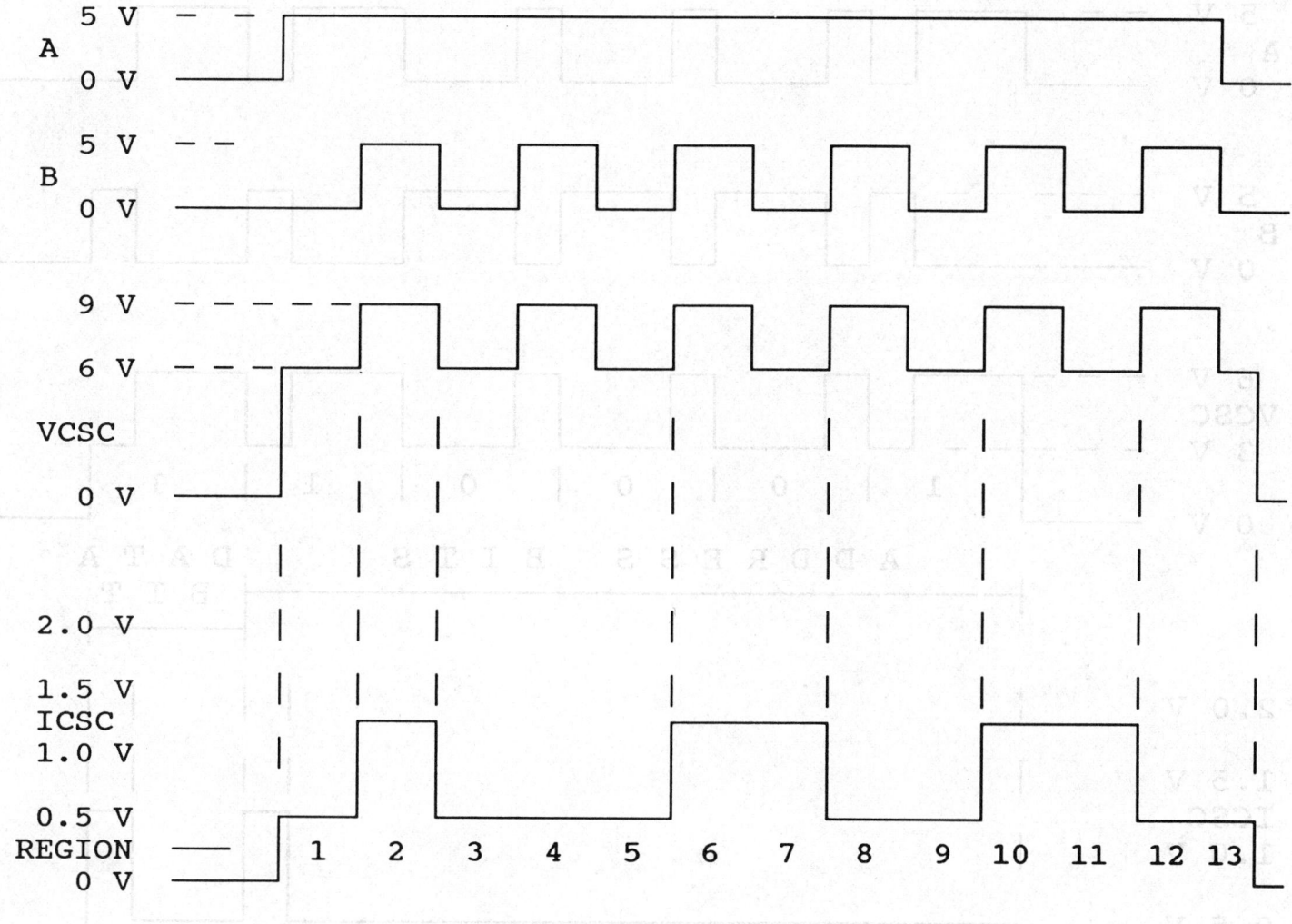

REGION	EXPLANATION
1	Initial voltage representing quiescent current
2	Sensor #1 PRESENT
3	Sensor #1 DOES NOT DETECT a magnet
4	Sensor #2 NOT PRESENT
5	Sensor #2 NO MEANING since sensor #2 not detected
6	Sensor #3 PRESENT
7	Sensor #3 DOES DETECT a magnet
8	Sensor #4 NOT PRESENT
9	Sensor #4 NO MEANING since sensor #4 not detected
10	Sensor #5 PRESENT
11	Sensor #5 DOES DETECT a magnet
12	Sensor #6 NOT PRESENT
13	Sensor #6 NO MEANING since sensor #6 not detected

FIG. 8—UNFILTERED BUS WAVEFORM FOR THE CONTINUOUS POLLING MODE

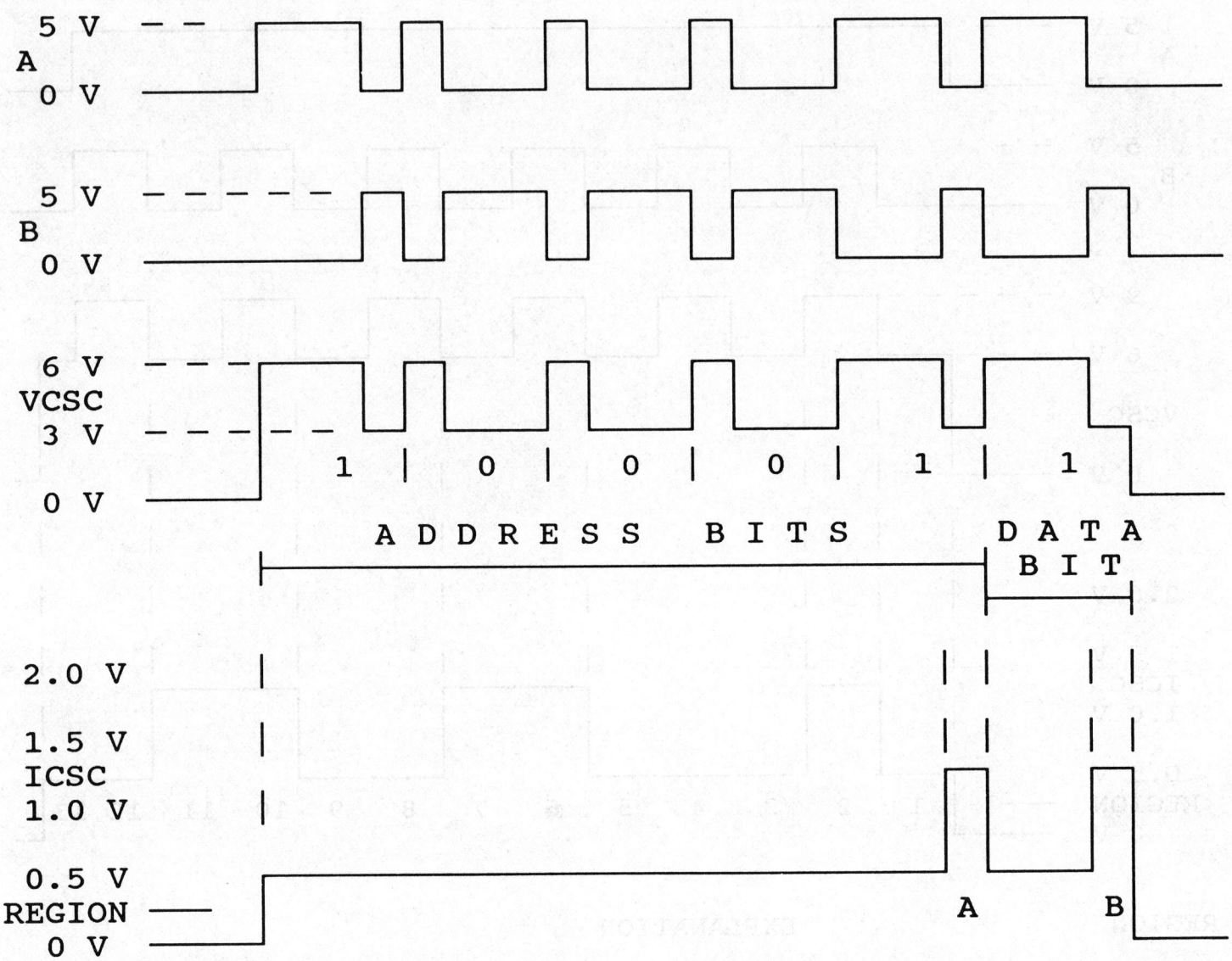

FIG. 9—UNFILTERED BUS WAVEFORM FOR THE COMMAND ADDRESSING MODE

TOKEN SLOT NETWORK FOR AUTOMOTIVE CONTROL—SAE J2106 APR91

SAE Information Report

Report of the SAE Class C Task Force of the SAE Vehicle Network for Multiplexing and Data Communications Standards Committee approved April 1991.

TABLE OF CONTENTS

Foreword—Data Links in Vehicle Control
  —Introducing the Token Slot Network Protocol (TSN)
  —Token Slot Protocol Background
  —Token Slot Network Application Characteristics
  —Token Slot Network Design Philosophy

1. Scope
1.1 Purpose
1.2 Background
1.3 Design Objectives
2. References
2.1 Applicable Documents
  2.1.1 SAE PUBLICATIONS
  2.1.2 OTHER PUBLICATIONS
3. Technical Requirements
3.1 Bus Access Procedure
  3.1.1 TOKEN SLOT OPERATION
  3.1.2 TOTAL NUMBER OF SLOTS AVAILABLE
  3.1.3 NUMBER OF SLOTS PER NODE
  3.1.4 INSTRUMENTATION SLOT NUMBER
  3.1.5 TIME SLOT SEQUENCE DELAY CALCULATION
  3.1.6 TIME SLOT WIDTH
  3.1.7 TOKEN POSSESSION GUIDELINES
    3.1.7.1 *Decision to Transmit*
    3.1.7.2 *Maintaining Synchronization*
    3.1.7.3 *Maximum Token Possession Time*
  3.1.8 TOKEN SLOT BUS EXCEPTION CONDITIONS
    3.1.8.1 *Initialization*
    3.1.8.2 *Bus Jam Signal*
    3.1.8.3 *Response to Bus Jam*
    3.1.8.4 *Collisions*
    3.1.8.5 *Bus Time Out*
    3.1.8.6 *BTO During Token Possession*
    3.1.8.7 *BTO at Initialization*
    3.1.8.8 *Transmit Error Counter*
3.2 Bit Encoding and Message Delimiting
  3.2.1 NRZ5 BIT ENCODING RULES
  3.2.2 ASYNCHRONOUS OPERATION
3.3 Message Framing
  3.3.1 DATA MESSAGE FRAMES
    3.3.1.1 *Data Message Identification Field*
    3.3.1.2 *Data Field*
    3.3.1.3 *Frame Check Sequence (CRC) Field*
  3.3.2 ACKNOWLEDGE FRAME
  3.3.3 TOKEN PASS MESSAGE FRAME
    3.3.3.1 *Token Pass Message Control Bits*
    3.3.3.2 *Token Pass Slot Number Bits*
    3.3.3.3 *Token Pass Parity Bit*
3.4 Message Delineation
  3.4.1 IDLE LINE (NORMAL MESSAGE DELIMITATION)
    3.4.1.1 *Idle Line (Inter-Message Gap)*
  3.4.2 SYNCHRONIZATION BIT
  3.4.3 ACKNOWLEDGE TIME OUT
  3.4.4 BUS TIME OUT
3.5 Network Structure
  3.5.1 BUS TOPOLOGY
    3.5.1.1 *Bus Length*
  3.5.2 PHYSICAL MEDIUM
    3.5.2.1 *Transmission Media*
    3.5.2.2 *Data Bit Rates*
    3.5.2.3 *Bit Rate Tolerance*
    3.5.2.4 *Logic Zero Dominance*
3.6 Node Device Implementation Considerations
  3.6.1 MESSAGE TRANSMISSION
    3.6.1.1 *Required Resources for Transmission*
    3.6.1.2 *Transfer of Transmitting Privileges*
    3.6.1.3 *Message Selection*
    3.6.1.4 *Message Update Rate*
    3.6.1.5 *Maximum Token Possession Time*
    3.6.1.6 *Transmission Monitoring*
    3.6.1.7 *Transmit Message Validation*
    3.6.1.8 *Error Handling and Recovery*
    3.6.1.9 *Transmitted and Received Data Differences*
    3.6.1.10 *Transmitter Underrun*
    3.6.1.11 *Acknowledge Frame Errors*
    3.6.1.12 *Transmitter Error Counter*
  3.6.2 RECEIVE MESSAGE VALIDATION
    3.6.2.1 *Resources for Message Reception*
    3.6.2.2 *Receive Message Selection*
    3.6.2.3 *Token Message Reception*
    3.6.2.4 *Broadcast Data Messages*
    3.6.2.5 *Data Messages with Acknowledge Request*
    3.6.2.6 *Acknowledge Frame*
    3.6.2.7 *Message Delimiting*
    3.6.2.8 *Bit Encoding Checks*
    3.6.2.9 *Error Check Field*
    3.6.2.10 *Minimum Message Length*
    3.6.2.11 *Receive Message Validation Summary*
APPENDIX A: DETERMINATION OF MESSAGE LATENCIES
APPENDIX B: EXAMPLE OF SLOT ASSIGNMENTS
APPENDIX C: SUMMARY OF MESSAGE TYPES
APPENDIX D: TOKEN SLOT OVERHEAD COMPARISON WITH J1850 & CAN

Foreword—Data Links in Vehicle Control

Introducing the Token Slot Network Protocol (TSN): This document describes a network protocol called the Token Slot Network that has been developed by General Motors for use in real-time vehicle control applications. This protocol offers a number of technical advantages such as controllable maximum latency, efficiency, and reliability. In addition, it is cost competitive with other protocols which lack these advantages.

Token Slot Protocol Background: The need exists for data links to be used in real-time control of vehicle systems. These systems require message delivery capabilities beyond those of the SAE Class B protocol that is now known as the J1850 protocol. After examining the communications requirements of the real-time control domain, no existing protocol was found that would completely meet those needs. Therefore, a new protocol was developed with the basic features of token passing schemes, but which included a time slot scheme for performing the token pass.

The following sections will first discuss the application needs that drove the protocol development, then describe the protocol itself, and finally present methods for time analyses of systems which use the protocol.

Token Slot Network Application Characteristics: The Token Slot Network is intended for use in distributed, real-time control applications in land based vehicles. The network nodes are typically intelligent (i.e., microprocessor based) components that cooperate in order to perform some control task. They may be functionally independent nodes, such as an engine control module and a transmission control module, that each have relatively autonomous functions that need to be coordinated for better total vehicle behavior. Alternatively, there may be a hierarchical control structure where some nodes implement a local low-level control algorithm, but are subservient to a higher level control running in another node. In either case, rapid and predictable communication is needed.

Real-time control communication systems must provide:
  —Rapid control loop feedback via data link
  —Regular data updates for periodic or event driven data
  —Minimal, predictably bounded message delays (latencies)
  —System specified and limited transmit privilege hold time

Data transfer efficiency requirements:
  —Low message overhead
  —No physical limits to higher data rates

Fault tolerance requirements:
  —Rapid fault detection and data error rejection
  —Quick initialization and recovery from bus errors
  —Masterless—unaffected by node drop off

Token Slot Network Design Philosophy: The need for bounded latency characteristics in applications with both bursty and periodic traffic require us to reject all protocols that use a contention mechanism for media access control (such as J1850 and CAN). While such protocols can be fashioned to provide statistical "guarantees" that message latency will not exceed required limits, there can always be cases where traffic loading will cause latencies to grow beyond any predefined limit. This is unacceptable for applications controlling real-time physical processes.

For this reason, it was decided to pursue protocols with more deterministic media access methods. A token passing scheme was chosen because it offers determinable and bounded latencies and it can be implemented efficiently with low data overhead.

In token passing protocols, the single node in possession of the token is a temporary bus master and can broadcast messages or initiate message transmission to other nodes as it chooses. When this node either finishes sending its messages or approaches its token hold (transmit privilege) time limit, it passes the token to another node which then becomes the master and the process repeats. The network can be configured so as to guarantee orderly access to the data link by all nodes.

One problem area in such schemes has to do with the token passing mechanism itself. Typically, the current token owner has a "next" node in a logical sequence order. When the token is passed, it is to be explicitly passed to the "next" node. If that node has no network traffic to send, it must still accept the token, and then pass it on to its successor. This procedure can lead to substantial inefficiencies especially if the "next" node has dropped off the bus. In this case, the current node must follow an exception procedure for determining to which node it is to pass the token. This usually involves polling to determine which nodes are still active, building an "active node" list, and broadcasting the list to all nodes. This total procedure is often complicated and is always time consuming.

Similarly, following system power-up initialization or if noise causes corruption of the token message, the token may become lost. This also requires a time-consuming reinitialization process for reorganization of the virtual node sequence ring formed by each node's idea of the logical "next" node in the sequence.

To solve this problem, it was decided to use a different token passing method based on rotating time slots. This allows transparent skipping of nodes which have no pending traffic or which are inactive. This protocol also provides for quick system initialization and for error recovery. The details of this method will be described in the following sections.

1. Scope—The Token Slot Data Link is intended to provide periodic, broadcast communications (communication that must occur on a regular, predetermined basis) within a vehicle system.

The Token Slot protocol achieves this by implementing a masterless, deterministic, non-contention Token Slot sequence which is designed to offer a transmit token to all devices (or nodes) without requiring that they respond. After acquiring the token, messages may be sent and verified using a variety of built-in techniques. The token passing slot sequence is then reinitiated by the current token holder.

1.1 Purpose—This SAE Information Report describes the Token Slot Data Link Communication Protocol. It is intended to cover the attributes of the network structure, network management, bus access procedures, message framing, bit encoding, and message delimiting required to communicate on the Token Slot Data Link. These aspects relate to those items that must be standardized excluding the physical implementation. The physical details of the network such as media, line driver/receivers, and semiconductor implementation are discussed generally but are not specified by this document.

1.2 Background—In recent years it has become apparent that future electronic applications within vehicles will need to communicate critical control information among distributed controllers. This control information to be transmitted is typically periodic in nature and must arrive on a regular, predetermined basis. In many cases it is changing at rapid rates and only sampled information is provided at appropriate intervals (rather than updating the information only when it changes). This must be done at speeds that are sufficiently fast to meet the requirements of the interactive control applications. The Token Slot Network Data Link is designed to provide this capability.

1.3 Design Objectives—The Token Slot Network was defined with the following goals in mind:
  a. It must meet the needs for on-vehicle, rapid, periodic communication between computer assemblies.
  b. Target applications for this link will be higher speed control applications (typically above 1 Mbit/s).
  c. Message maximum latency time delays must be minimal and bounded to assure stability in closed loop control applications. See Appendix A for methods of calculating and predicting message latencies.
  d. There must be an open data rate growth path to accommodate future media technologies and demands for increased data traffic.
  e. The architecture must be an open system to allow for the addition and deletion of nodes (devices) both in new designs and during dynamic operation. Note that open architecture does not mean that devices can be added to the data link without regard to their affect on message latency times.

It is intended that this link will be used primarily in vehicle applications requiring feedback and control data to be shared among various distributed devices. For example, the link could be used between distributed engine, transmission, anti-lock braking, and power steering controllers. Flexibility has been built into the protocol to also include event driven and block memory transfer data types. This allows vehicles to be equipped with a single data link to cover all communications to avoid gateways and reduce costs, when system and reliability conditions permit.

2. References

2.1 Applicable Documents—The following publications form a part of this specification to the extent specified herein. The latest issue of SAE publications shall apply.

2.1.1 SAE Publications—Available from SAE, 400 Commonwealth Drive, Warrendale, PA 15096-0001.

SAE J1850 Part 3—Road Vehicles—Serial Data Communication
SAE J2056 Part 1—Class C Application Requirements
SAE J2056 Part 2—Survey of Known Protocols
SAE J2056 Part 3—Selection of Transmission Media

2.1.2 Other Publications

82526 Serial Communications Controller Architectural Overview—INTEL Corporation

"Data Transmission Over the Telephone Network: Series V Recommendations" Section V41, The Orange Book VIII.1, International Telecommunications Union, Geneva (1977)

3. Technical Requirements

3.1 Bus Access Procedure—This section describes the token slot bus access method of providing transmit privileges to devices communicating on the Token Slot Data Link. It is the primary difference between Token Slot and contention protocols (e.g., J1850, CAN, etc.).

The intent of this bus access protocol is to guarantee periodic opportunities for message transmission by each node on the bus. It is to ensure that the bus remains operational when devices are dynamically added or deleted and it must provide for quick recovery from error conditions.

The token passing bus protocol is open, peer oriented, and multimaster. It is non-contention and uses a time slot token passing technique. The bus access procedure is performed on a distributed, masterless basis, with all devices behaving identically to maintain an orderly rotation of transmit privileges.

The token slot method described here has several features which help achieve the goals of a deterministic data link. The principal advantage of this method of bus access is that the worst-case delay of information exchange due to the data link can be determined and bounded. Requirements for line driver/receiver circuitry are less stringent than would be required with a contention based system since only one transmitter is normally on at a time and since the protocol is not data rate limited by signal propagation delay. The protocol has provisions for rapid initialization of the link and for handling cases in which the number of devices on the bus changes during operation (due to power moding sequences, failure or power loss of individual devices, or intermittent software operation of devices on the bus). Fast error recovery is a primary feature of this approach.

3.1.1 Token Slot Operation—The token slot method offers transmit privileges to all devices on the bus in an orderly manner. After a node has completed sending its message traffic, a sequenced scan of short, equal time intervals (slots) offer bus transmit privileges to the node slot owners as follows: A token pass message (see Figure 1) instructs all nodes to begin the token slot timing mode. Each slot interval following a token pass message has a sequential Slot Number (SLN). Control is offered to the slot owning node devices by time progressing through successively higher Slot Numbers. Each node is assigned one or more specific time slots (Slot Numbers) and will activate its transmitter to send a message during its slot only if it is operational and has message traffic to send. Otherwise the token slot interval is allowed to pass. When the transmitter is activated, all other nodes recognize that the token has been taken and they enter the receive mode. See Appendix B for a summary of token pass messages. See Appendices C and D for a comparison with contention based protocols.

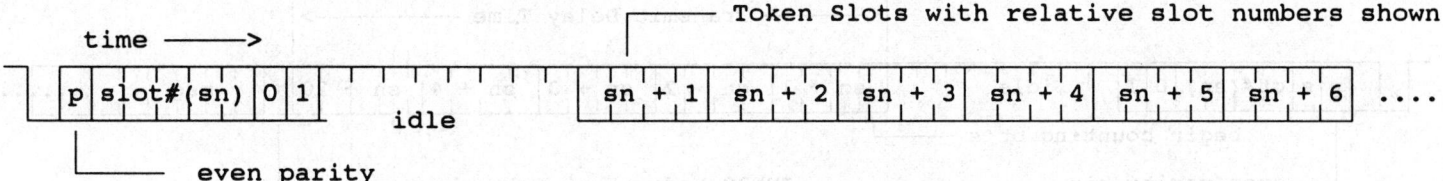

FIGURE 1—TOKEN PASS MESSAGE FORMAT

The token slot method differs from time-division multiplexing, which provides a "time slot" wide enough for an entire message, to each device on the bus. If the device is not present on the bus or has no message to transmit, the bus must remain inactive during that entire time slot.

The Token Slot protocol does not send a directly addressed token pass message to hand off control of the bus to a specific device which may or may not be active. This avoids the requirement that all devices must maintain a prioritized list of all currently active nodes to which the token must be passed.

Figure 2 depicts the bus waveform of a token message sent by the device which owns SLN 5, followed by the owner of time slot 1 (SLN1) taking the token, sending two messages, and then initiating its own token pass message.

In this example, the devices holding slot numbers 6, 7, ... through 31 and 0 were either not present or did not have a data message ready for transmission. Thus, the bus remained in the idle state during those time slots.

As shown, when the maximum slot number is reached, the sequence wraps around to slot 0 and continues until the slot is picked up or until the original token passer sees its slot, at which time a new token pass message is generated and the token slot cycle begins again.

Once a device takes the token, it proceeds to send its message traffic. Token hold times are individually assigned to each node by the system designer and are strictly limited to assure a system maximum message latency limit. Individual message transmit priorities are determined by each node's application and are not restricted by the communications data link.

3.1.2 TOTAL NUMBER OF SLOTS AVAILABLE—All applications of the Token Slot bus have thirty-two time slots available. Thus, there can be a maximum of thirty-two transmitting type devices present on the bus in any particular application.

3.1.3 NUMBER OF SLOTS PER NODE—A particular device may have zero (for a "receive only" device), one, two, or more Slot Numbers, depending on how often that device needs to transmit messages compared to the other devices on the bus. A device which owns multiple Slot Numbers must take care to include the Slot Number for the time slot it actually took during that token possession in any token messages it sends. The number of time slots that a particular device may occupy is to be defined by the system application designer who is responsible for message latency assurance.

3.1.4 INSTRUMENTATION SLOT NUMBER—SLN 0 is ideally reserved for use by instrumentation in all applications to provide an external access by standard test equipment. It may be prudent to require that the vehicle be placed in a special instrumentation interactive mode before SLN 0 be allowed to take the token.

3.1.5 TIME SLOT SEQUENCE DELAY CALCULATION—All devices use the Slot Number contained in a received token message to determine how long they must wait until they should take control of the bus if no other node intervenes. When a token message is received, each device, including the one that sent the token, calculates its delay using the following equation:

Transmit Delay Time (TDT) = $\{[(A-B)-1]\}$ modulo32 * T (Eq.1)

where:
A = Slot Number of calculating device
B = Slot Number in last received token message
T = Time Slot Width, expressed as an integral number of bit times (bt)

Modulo32 means that:
If $[(A-B)-1] < 0$
then TDT $\{= [(A-B)-1] + 32\} * T$
Else TDT $= [(A-B)-1] * T$
Therefore: $0 < TDT < 32*T$

Any device holding multiple Slot Numbers will need to calculate a transmit delay time for each of its Slot Numbers, then take the token on the earliest TDT.

If another device begins a transmission in an earlier slot before this transmit delay time has elapsed, the bus access calculation must be restarted by waiting for a new token message to be received. However, if the bus remains inactive for this calculated transmit delay time following a token message, it means that this device's token slot interval has arrived. If the node has messages to send, it may begin a transmission at this point and thus take control of the token and the bus.

The transmit delay time begins immediately following the 8 bit time (bt) idle line delimiting the token message. This means that the transmit delay time counter should begin counting at the beginning of the 9th bit time following a token message. Figure 3 illustrates this situation.

3.1.6 TIME SLOT WIDTH—The time slot width is determined by a timing analysis of the system bus propagation delay and is shown as an integral number of bit times (bt). The time slot must be long enough to allow a device to start a transmission and for all other devices to recognize that event. Its width must also take into account the accumulated time base drifts that occur as each device is counting and waiting for its time slot to occur. A typical slot width might be 1 to 5 bit times when operating at 1 Mb/s for propagation delay characteristics on the order of 5 ns/m for either a properly terminated electrical or a plastic fiber optic medium.

3.1.7 TOKEN POSSESSION GUIDELINES

3.1.7.1 *Decision to Transmit*—When the bus remains inactive for a device's transmit delay time following a token message, all devices except the one that sent the last token message have two choices. These devices have the option of taking possession of the token and beginning transmission of a message or of remaining silent. A device may send one or more messages once it takes token possession, and it then relinquishes control of the bus by sending a token pass as its last message.

3.1.7.2 *Maintaining Synchronization*—The device that sent the last token has the duty to maintain a minimum level of activity on the bus.

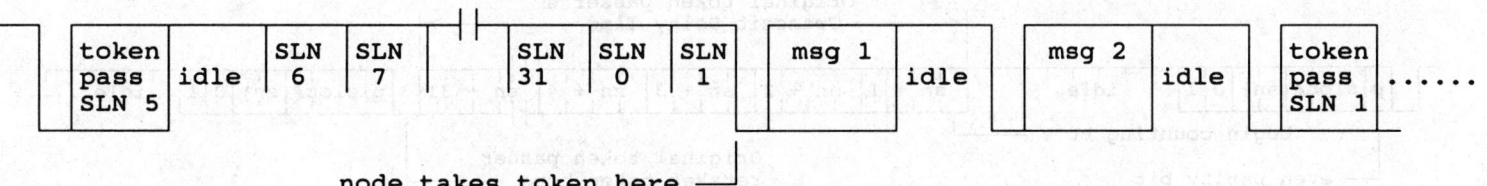

FIGURE 2—TYPICAL TOKEN PASS SEQUENCE

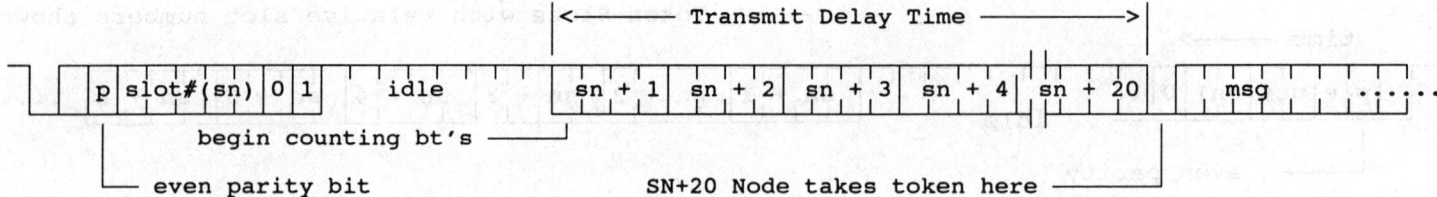

FIGURE 3—TOKEN PASS MESSAGE FORMAT

In the event that no other device initiates a transmission during the thirty-one slots following a token, the device that sent the last token must take the token and send either a data message if one is ready or only a token message if no data message is ready. The purpose of this is to maintain synchronization of the time slots during periods of low bus utilization. Accumulated time base differences among all the devices causes blurring of the time slot boundaries as the time since the last token message increases. This explains the need for periodic synchronization. Figure 4 illustrates the case where no device takes the token and the original token passer must re-take it.

3.1.7.3 *Maximum Token Possession Time*—Each device on the bus is assigned a maximum time that it may hold the token on any individual possession. This maximum token possession time helps to bound the token rotation time for the system and is used to avoid continuous transmission caused by a device failure. A device must not begin transmission of a particular data message unless completion of the message can be accomplished without exceeding its maximum token possession time. If that message is too long to be sent during the present token possession, the device must send a token message, then cease transmission. The time that the token has been held and time required to send the next message may be calculated by the message lengths + bit insertions + message acknowledges + idle lines between messages. There are various methods of calculating and/or estimating the maximum token possession time. See the section on Maximum Token Possession Time Rationale for further discussion. See Appendix A for Methods of calculating and predicting message latencies.

The system designer may assign different maximum token possession times to different devices in a particular application.

3.1.8 Token Slot Bus Exception Conditions—This section describes exception conditions that may occur on the bus and how these conditions should be handled.

3.1.8.1 *Initialization*—When a device requires initialization following a power up or internal reset condition, it monitors the bus for a token message or for a Bus Time Out to occur. The Bus Time Out (BTO) occurs when the bus is idle for thirty-two time slots (see 3.1.8.5). If a token message is received, the device calculates its transmit delay time and begins normal operation.

3.1.8.2 *Bus Jam Signal*—The Bus Jam is defined as the receipt of six or more consecutive logic zeros by a receiving device. A transmitter communicates error situations to all devices on the bus with the Bus Jam. It is intended to cause all devices on the bus to cancel any current message and begin recovery procedures.

The bus jam must be a dominant signal level and must last at least 6 bit times which is longer than any normal data signal condition. This will prevent a legal bit stuffing signal from causing a bus jam. The bus jam signal must be started before a post-message 8 bit time idle line is allowed to occur on the bus. This ensures that the jam signal causes all devices to discard the corrupted message. If a complete idle line were allowed to occur, then devices receiving the subsequent bus jam would not necessarily reject the corrupted message.

3.1.8.3 *Response to Bus Jam*—When a device recognizes a bus jam condition, it rejects the message in progress (if any) as invalid and calculates a transmit delay time as if a token message containing a Slot Number of thirty-one was just received. This results in slots for devices beginning with SLN 0, 1, 2, ... etc., to take the token following a bus jam. Once the jam condition ends and an 8 bit time (bt) idle line is detected, all devices begin waiting for their time slot to occur. If a device finds that the bus remains idle for its transmit delay time following any jam, it is to transmit either a data message or a token message if it has no data messages to send. This ensures that normal bus activity resumes properly under light bus load conditions. Figure 5 shows the bus activity following a bus jam.

In this example, no device on the bus holds SLNs 0, 1, or 2. The device holding SLN 3 transmits either a data message or a token message if it has no data messages ready for transmission.

3.1.8.4 *Collisions*—A collision is the result of two or more devices attempting to transmit at the same time. Collisions may occur during operation of the bus due to corrupted token messages, noise during the time slots following a token message, or due to other noise-induced conditions. Each transmitting device must monitor the data it receives during message transmission and compare it to the data it transmitted. This comparison may be done on a bit or byte level. If a discrepancy is detected between transmitted and received data, the transmit device is to jam the bus for 8 bit times (bt) and relinquish control of the bus (see 3.1.8.3).

3.1.8.5 *Bus Time Out*—The Bus Time Out (BTO) occurs when the bus is idle for thirty-two time slots. This condition may occur due to failure of a transmitting device to send a token message or loss of the token message due to noise. The reaction to a BTO is similar to the response to a message error. When a BTO occurs, all devices assume that a token message containing a Slot Number of thirty-one was just received and calculate the appropriate transmit delay time. If any device finds that the bus remains idle for this transmit delay time, it transmits either a data message or a token message if it has no data messages to transmit. Figure 6 depicts this situation.

In this example, no device on the bus holds SLNs 0, 1, or 2. The device holding SLN 3 begins either a data message if it has one to send, or a token message if it does not, in the appropriate time slot.

3.1.8.6 *BTO During Token Possession*—It may be possible for a Bus Time Out to occur while a particular device is holding the token. For example, a device may transmit one message properly and then experience a long delay in preparing the next message for transmission during that token possession. Once the device senses a BTO, it must immediately relinquish possession of the bus, and begin waiting for its time slot to occur just as if it had not had token possession prior to the BTO.

3.1.8.7 *BTO at Initialization*—When a device requires initialization

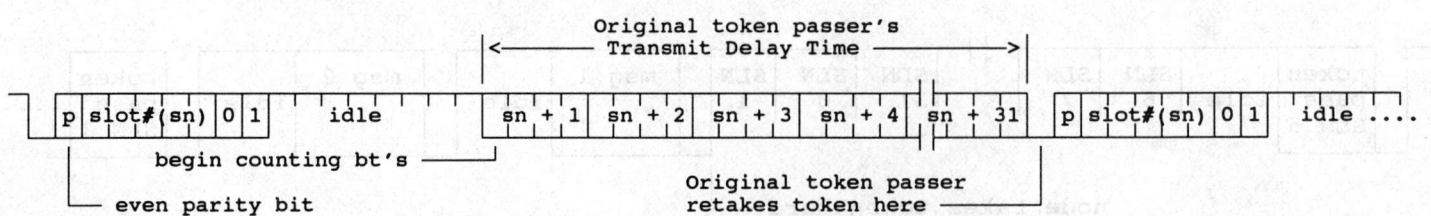

FIGURE 4—TOKEN PASS MESSAGE FORMAT

following a power up or internal reset condition, it monitors the bus for a token message or for a BTO to occur. If a BTO is detected, it calculates its transmission delay time as if a token message with a Slot Number of thirty-one had been received and continues as if the BTO had occurred during normal operation.

3.1.8.8 *Transmit Error Counter*—Each device maintains a transmit error counter to monitor the frequency of errors that occur while that device is transmitting. The transmit error counter should be initialized to zero and incremented by some system assigned algorithm consistent with maintaining bus message latency boundaries in the face of a number of bus failures.

3.2 Bit Encoding and Message Delimiting—This section describes the mechanism for transmitting information on the Token Slot Data Link. It includes the bit encoding and the rules governing synchronization and delineation of message frames.

3.2.1 NRZ5 BIT ENCODING RULES—All messages on the Token Slot data link are encoded as Non-Return-to-Zero with opposite level bit insertion after the fifth consecutive bit remains at a constant one or zero logic level. This technique will be called NRZ5 encoding. NRZ5 is applied regardless of transmission media used.

NRZ5 is similar to standard NRZ in that the logic level on the bus is a logic one for the entire bit period to transmit a binary one and is a logic zero for the entire bit period to transmit a binary zero. In addition, synchronization information is added to the bit stream at the transmitter by inserting a bit of the opposite state into the bit stream whenever five consecutive bits of the same state are to be sent. For example, when the transmitter detects that a string of five zeros is to be transmitted, it inserts a one into the bit stream, then continues the data transmission, again watching for a string of five consecutive zeros or ones including any inserted bits. Upon reception of this string of five zeros, the receiver checks that the next bit to arrive is a one (the one inserted by the transmitter), and deletes this one from the bit stream. If a sixth zero is received, the receiver treats these six zeros as a bus jam signal and takes the proper recovery steps. A corresponding situation occurs for five ones in a row except that six received logic ones in a row indicate the start of an idle line, not a bus jam. An example of this procedure is shown in Figure 7.

If a message should happen to end in five zeros or five ones, the transmitter must still perform the one or zero insertion. The receiver will check for the presence of the inserted bit and delete it as required.

3.2.2 ASYNCHRONOUS OPERATION—Serial data transfers are done asynchronously, meaning that a single data stream with imbedded pre-message synchronization is the only signal among the devices on the bus. This avoids sending a separate clock and data signal which would increase the cost of the transmission medium and bus transceivers. Receivers decode the incoming data by synchronizing on the logic one-to-zero transitions in the data stream. During normal data transmission these transitions will never be more than 10 bit times apart due to zero and one insertion in NRZ5 encoding. When used with closely matched clock sources in all devices, this synchronization information is sufficient for proper data decoding.

3.3 Message Framing—All information transmitted on the link must be consistently framed for correct interpretation. For the Token Slot Data Link four message types are identified, distinguished by two control bits in the first byte of each message. The four message types are broadcast data messages, data messages requiring an acknowledge, single byte acknowledge messages, and single byte token messages.

Data bytes are transmitted least significant bit first to conform to traditional serial data link standards. See Figure 8.

The two most significant bits in the first byte of every message identify the type of message being transmitted. These bits are labelled Control Bits 0 and 1 (CB0 and CB1) and are defined in Table 1:

TABLE 1—DEFINITIONS—CONTROL BITS 0 AND 1 (CB0 AND CB1)

CB0	CB1	Message type
0	0	Broadcast data message
1	0	Data message with acknowledge request
0	1	Token message
1	1	Acknowledge message

Note that CB1 distinguishes data message frames from the two single byte message frames, while CB0 indicates that this message is either requesting or supplying an acknowledge. See Figure 9.

3.3.1 DATA MESSAGE FRAMES—The data message is used for the actual transfer of information on the data link. There are two types of data messages, one which requires no acknowledgement, and another which requires that an acknowledge message be returned by a single receiving device.

The data message frame consists of a two byte message control (2 bits) and identification (14 bits) field, an optional data field, and a two byte CRC frame check sequence (FCS) field. These fields are described in the following sections. The shortest possible data message frame is four bytes long, consisting of only the ID and FCS fields. Message frames shorter than this minimum length are to be rejected by the receiving device.

3.3.1.1 *Data Message Identification Field*—The Identification Field identifies the message using a 14 bit message ID number which is contained in the first two bytes of a data message. The format of the ID field for a data message is in Figure 10.

In the data message, CB1 is set to zero to indicate a data message. CB0 is set to zero for a broadcast data message and set to one for a data message which requires an acknowledgement. ID13 through ID0 form the rest of the message ID.

3.3.1.2 *Data Field*—The optional data field contains any information beyond the message ID that needs to be transferred by the message. The information transmitted in the data field is sent in byte for-

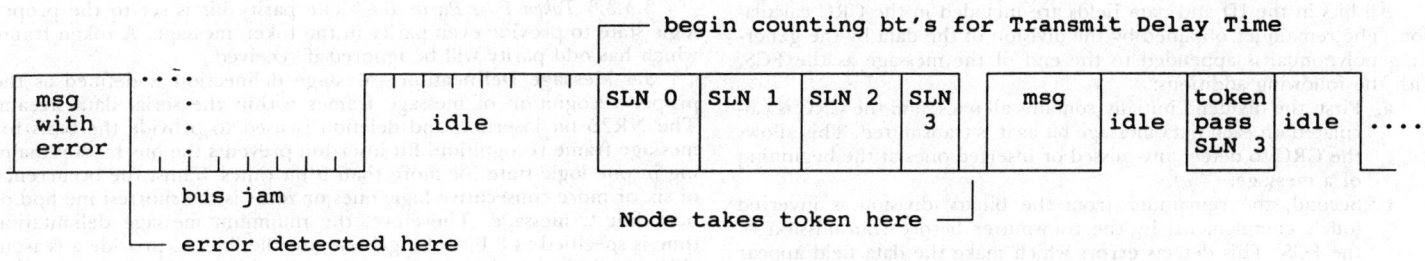

FIGURE 5—BUS ACTIVITY FOLLOWING BUS JAM

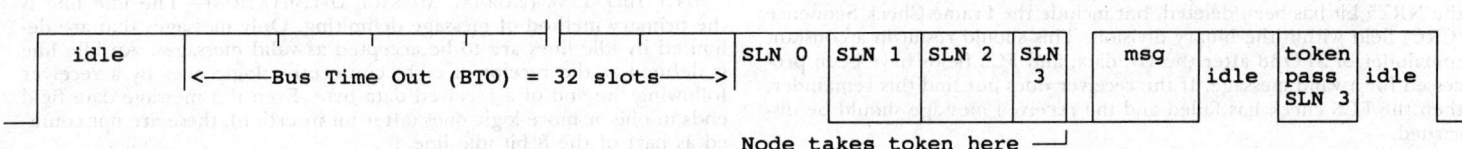

FIGURE 6—BUS ACTIVITY FOLLOWING A BUS TIME OUT

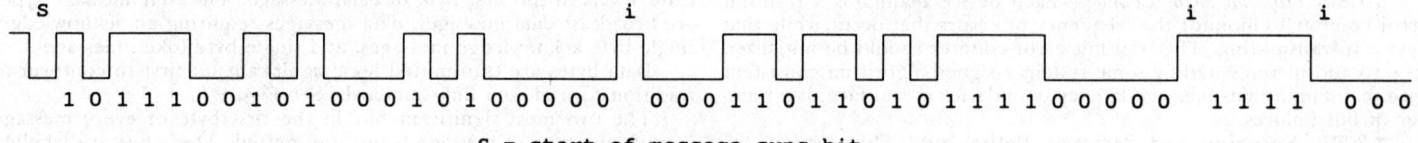

FIGURE 7—NRZ5 EXAMPLE

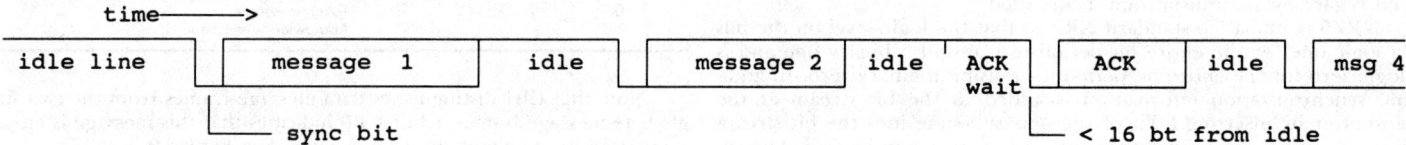

FIGURE 8—GENERAL MESSAGE FRAMING

mat, meaning that the data field is always an integral number of bytes. The data field may be from 0 to 256 bytes in length. The contents, scaling, order of appearance, and processing of any data in the data field is user defined and is a function of the message ID number.

3.3.1.3 *Frame Check Sequence (CRC) Field*—The 16 bit CRC conforms to the CCITT standard and detects all single bit errors, all parity errors, and all burst errors less than 17 bits long. For burst errors longer than 16 bits, the CRC misses 0.0015% of errors.

The Frame Check Sequence (FCS) field consists of two bytes of CRC code. The FCS field is used by receiving devices to detect errors in incoming data messages. Generation and decoding of this field is as follows:

The CRC is calculated by performing a binary division of the entire contents of a message (before NRZ5 bit insertion) by a generating polynomial G(X). The following CCITT-CRC (see Section 2, "Data Transmission Over the Telephone Network: Series V Recommendations") generating polynomial is used where $X = 2$ (for modulo 2 operations):

$$G(X) = X^{**}16 + X^{**}12 + X^{**}5 + 1 \qquad (Eq.2)$$

All bits in the ID and data fields are included in the CRC calculation. The remainder obtained by the division of the data by the generating polynomial is appended to the end of the message as the FCS, with the following additions:

a. First, the dividend initially contains all ones, and the CRC is calculated on each data message bit as it is transmitted. This allows the CRC to detect any missed or inserted ones at the beginning of a message.

b. Second, the remainder from the binary division is inverted (one's complement) by the transmitter before transmission of the FCS. This detects errors which make the data field appear to be rotated by a number of bits.

c. Third, the FCS contains the CRC remainder, transmitted most significant bit first. Note that NRZ5 bit insertion is performed on the FCS after the calculation and prior to transmission.

Receivers perform a similar computation on incoming data after the NRZ5 bit has been deleted, but include the Frame Check Sequence (CRC) field within the binary division. This should result in a constant remainder of $F0B8 after the ID, data, and FCS fields have been processed for a valid message. If the receiver does not find this remainder, then the FCS check has failed and the received message should be discarded.

3.3.2 A<small>CKNOWLEDGE</small> F<small>RAME</small>—The acknowledge frame has the fixed format as in Figure 11.

The control bits CB1 and CB0 are both set to ones to indicate an acknowledge message. Thus, the valid acknowledge frame is always $D5.

The system designer must assign only one node the responsibility for returning the acknowledge response to a given message.

3.3.3 T<small>OKEN</small> P<small>ASS</small> M<small>ESSAGE</small> F<small>RAME</small>—The token frame is used to pass control of the link from one device to the next. This frame indicates the Slot Number of the device that currently has transmitting privileges on the bus. This frame is transmitted by the device currently in control of the link after it has completed sending its current messages. The token message is always a single byte in length, and the format of this field is as in Figure 12.

3.3.3.1 *Token Pass Message Control Bits*—The CB0 and CB1 bits are set to zero and one, respectively, to indicate that this frame is a token message.

3.3.3.2 *Token Pass Slot Number Bits*—The five slot number bits contain the slot number that the device used to gain access to the bus during this token possession. All devices use the slot number to determine their transmit delay time following the idle line at the end of the token message.

3.3.3.3 *Token Pass Parity Bit*—The parity bit is set to the proper logic state to provide even parity in the token message. A token frame which has odd parity will be ignored if received.

3.4 Message Delineation—Message delineation is defined as the proper recognition of message frames within the serial data stream. The NRZ5 bit insertion and deletion is used to provide the basis for message frame recognition. Bit insertion prevents the bus from remaining in one logic state for more than 5 bit times. Thus, the occurrence of six or more consecutive logic ones or zeros is the shortest method of bounding a message. Therefore, the minimum message delimitation time is specified as 8 bits of logic one (or idle line) to provide a reasonable signal propagation delay margin. Longer periods of bus inactivity may occur in several situations: the end-of-message (EOM) idle line message delimiter, the start-of-message (SOM) synchronization bit, the acknowledge time out (ATO), the bus time out (BTO), and the bus jam.

3.4.1 I<small>DLE</small> L<small>INE</small> (N<small>ORMAL</small> M<small>ESSAGE</small> D<small>ELIMITATION</small>)—The idle line is the primary method of message delimiting. Only messages that are delimited by idle lines are to be accepted as valid messages. An idle line is defined as the receipt of eight consecutive logic ones by a receiver following the end of a received data byte. Even if a message data field ends in one or more logic ones (after bit insertion), these are not counted as part of the 8 bit idle line.

An idle line is to be recognized only after the bus has remained at a logic one state for eight entire bit times past the boundary of the last data byte in a message. This is important when passing the token. If a device were to declare an idle line when it actually samples an eighth one, it could start transmitting immediately, possibly before all devices had actually sampled that 8th bit. Delaying idle line recognition beyond six until eight full bit times have elapsed eliminates that possibility, since all devices would have had time to properly sample the 8th bit.

3.4.1.1 *Idle Line (Inter-Message Gap)*—Messages from the same device during a token possession time should be transmitted with a minimum of idle time between them. The shortest idle time is the 8 bits necessary to properly delimit the messages, and this should be achieved whenever practical. However, exception cases may exist which cause this inter-message gap to exceed the 8 bit time minimum. As a goal, consecutive messages should not have more than 16 bit times of idle between them to preserve the data throughput on the bus. This figure is subject to change if the constraints it places on transmitting devices is too severe. The maximum message gap must always be less than a Bus Time Out in any case. The occurrence of a BTO will force a device to relinquish token possession and respond to the BTO error.

3.4.2 SYNCHRONIZATION BIT—The synchronization bit is a single binary zero which is transmitted for one bit time immediately preceding any message on the bus. This bit is similar to a start bit used in standard UARTs. Before the start of any message, the bus will be at the logic one (idle) state for at least an idle line time (8 bt) and the logic zero sync bit both informs all receivers that a message is starting and provides the first synchronizing edge for NRZ5 bit decoding purposes.

Figures 13 and 14 depict the idle line, sync bit, and start of message.

3.4.3 ACKNOWLEDGE TIME OUT—Messages which have requested an acknowledge must allow an Acknowledge Time Out (ATO) period of 20 bit times beyond the request message's idle line delimiter before declaring a no-response and beginning to transmit the next message.

The Acknowledge Time Out (ATO) is the maximum number of bit times that a device is to wait for an expected acknowledge message to begin. The ATO is defined as the time from the end of the idle line delimiting the message which requires an acknowledge to the start of the acknowledge frame being returned by the destination device. The required ATO time must be less than a Bus Time Out (see 3.4.4) and should be as short as practical for minimal impact on bus throughput. The ATO time is therefore defined to be 20 bit times in duration. This allows adequate time for the device transmitting the acknowledge to decide if the incoming message requires an acknowledge and time to begin its transmission. To provide a timing margin for the ATO, a device should not begin the acknowledge if it is unable to do so within 16 bit times from the idle line at the end of the previous message. See Figure 15.

Another important, though less critical, time associated with the acknowledge function is the time for the device holding the token to resume transmission. This could be either a retransmission of that message if an ATO occurred or another data message or a token message if a valid acknowledge was received. The transmission start time should be as short as practical, but in general should not exceed 20 bit times following either an ATO or a valid acknowledge message. This requirement is important to keep the bus throughput burden of the acknowledge function to a definable minimum.

3.4.4 BUS TIME OUT—A Bus Time Out (BTO) is declared when the bus has remained at the logic one state for 32 time slots. The BTO does not include the idle line at the end of a message. During normal operation, the requirement that one device continue to send tokens in the absence of other bus traffic prevents idle periods of more than 31 time slots. See Figure 16.

In general, a device should start counting time slots whenever it finds that an idle line has occurred. Following any bus activity except for a valid token message, the device should begin waiting for the 32 time slots which define a BTO. When a valid token message is received, the device should instead calculate and wait for its transmit delay time and decide whether to begin transmission as described in the "Token Slot Operation" section.

3.5 Network Structure—This section describes the physical structure of the Token Slot Data Link. This includes the connection of devices to the bus, transmission media, and bit rate. Each application may have different objectives which affect the final configuration of the system. For instance, a requirement for easy addition and subtraction of devices on the bus will affect the method of coupling to the transmission medium. The guidelines given here represent the basic characteristics which all applications of the Token Slot bus are likely to meet.

3.5.1 BUS TOPOLOGY—The Token Slot Data Link is functionally configured as a bus, such that all devices transmit and receive over a single common path, and all devices receive information transmitted on the link simultaneously (within the bounds imposed by finite propagation

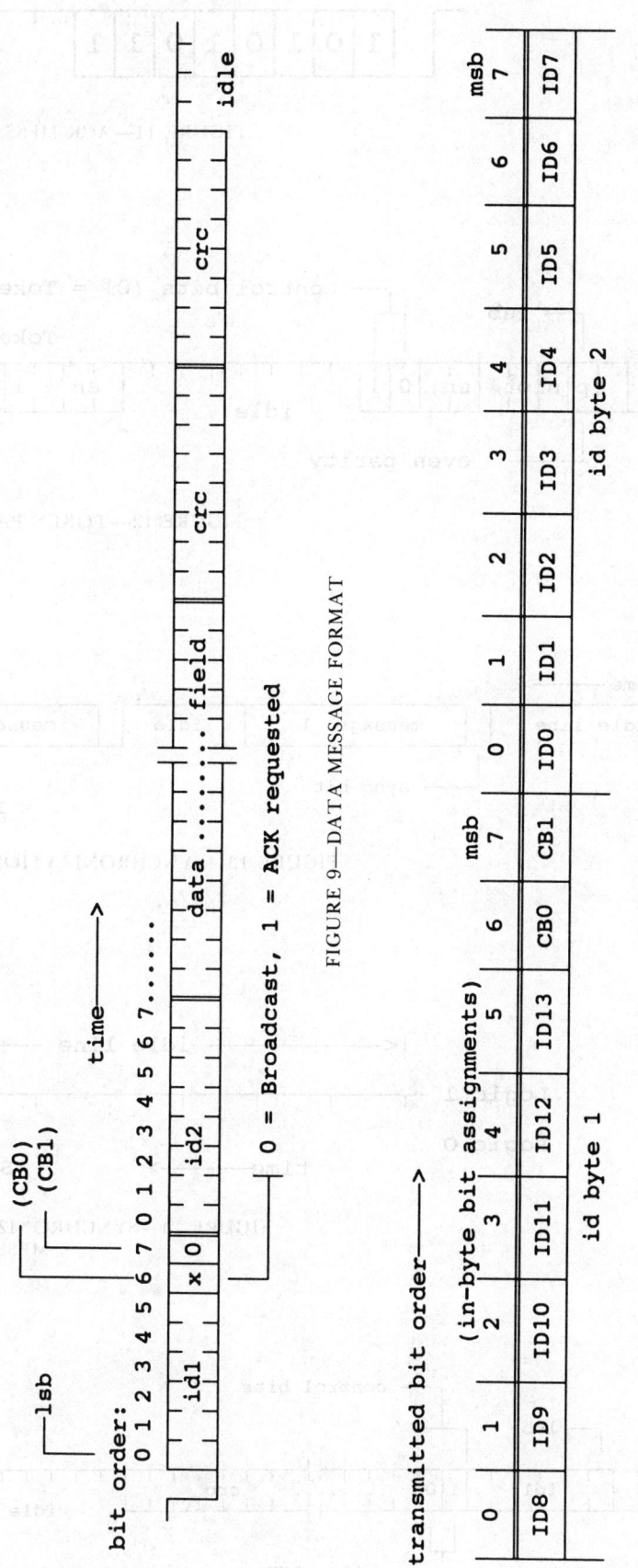

FIGURE 9—DATA MESSAGE FORMAT

FIGURE 10—DATA MESSAGE ID FIELD

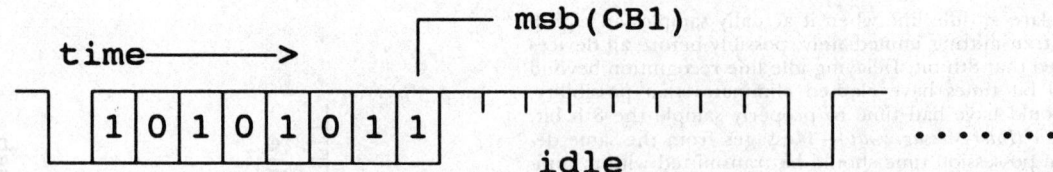

FIGURE 11—ACK MESSAGE FORMAT (FIXED)

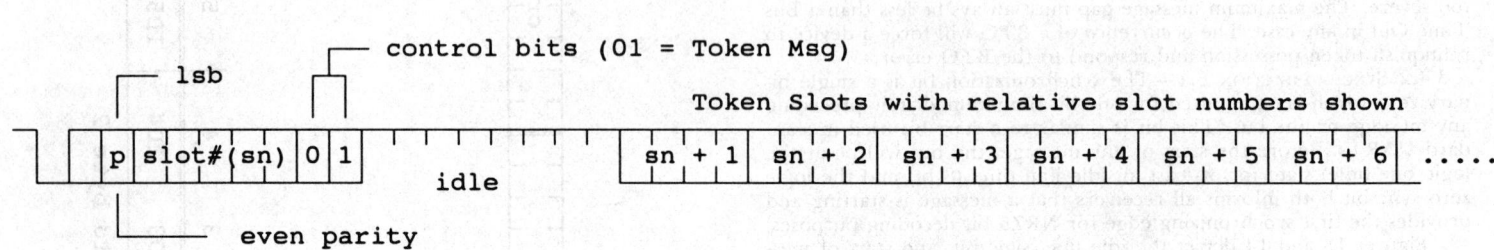

FIGURE 12—TOKEN PASS MESSAGE FORMAT

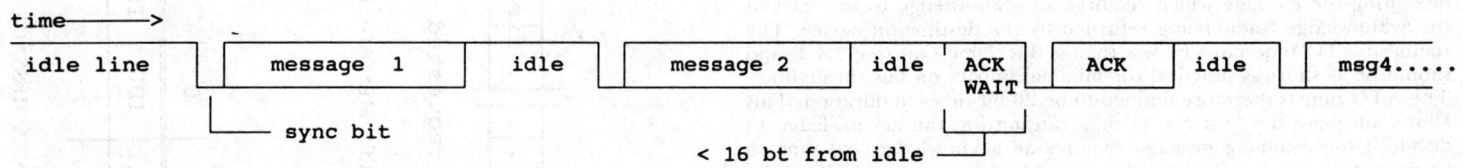

FIGURE 13—SYNCHRONIZATION AND GENERAL MESSAGE FRAMING

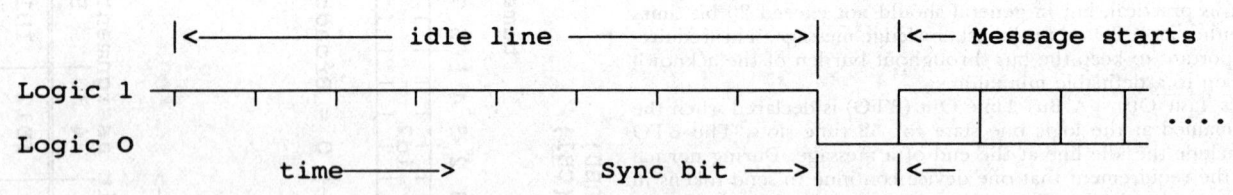

FIGURE 14—SYNCHRONIZATION BIT STARTING A MESSAGE

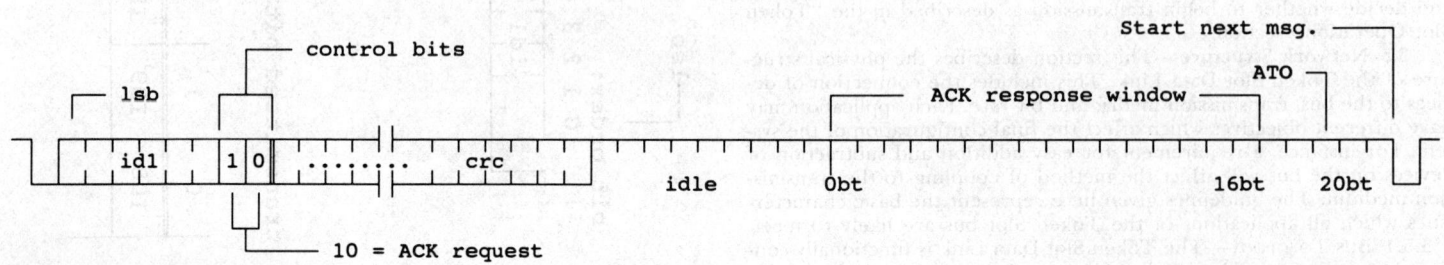

FIGURE 15—ACKNOWLEDGE MESSAGE TIMING

delays in the wiring, receive circuitry, etc.).

A discussion of key considerations associated with the physical bus configuration is given in the following paragraph. Consideration of these factors during the physical bus design will produce a proper determination of the time width of the token slot. This is necessary to assure that all nodes can detect that the token has been taken. The Token Slot Protocol is specially adept at accommodating the bus propagation delay in that the token slot periods can be adjusted to varying time widths to accommodate a given vehicle platform design without affecting either the ability of the protocol to operate at the highest practical media data rate or the ability to determine the maximum message latency time.

Figure 17 describes the propagation delay considerations for the Token Slot Network. It is most important that the slot width time is long enough to include two worst-case maximum propagation delays (2*Tpd max = a round trip), an adequate receive node sample window time, and a transmitter turn-on delay. The two nodes in question are presumed to be the furthest time distance apart (Tpd max). Note that many of these considerations also apply to the contention based protocols—and on each bit in the contention field.

The physical layer configuration is not specified here, since transmission medium, physical location of devices on the bus, and failure mode concerns dictate these requirements at the system level. However, the method of coupling devices to the bus must ensure that individual device loss of power, transmitter turn-off, or other device failures do not render the bus inoperative.

3.5.1.1 *Bus Length*—The maximum length of the bus will be application determined by the characteristics of the bus media components. Consideration of such factors as transmitter output energy, receiver sensitivity and dynamic range, bus losses, terminations, propagation delays, environmental effects on components, etc., determines the greatest acceptable length of wire or optical fiber between the bus transceivers of any two devices on the bus. The connection to any off-vehicle instrumentation must be included in this length budget.

3.5.2 PHYSICAL MEDIUM—The Token Slot Data Link offers implementation flexibility to accommodate various transmission media at a variety of bit rates. Each medium supports a maximum bit rate which can be achieved with acceptable electromagnetic interference (EMI) and compatibility (EMC) performance in on-vehicle applications. Thus the maximum bit rate is limited by the range allowed by the transmission media.

3.5.2.1 *Transmission Media*—The following media are candidates for use in systems using the Token Slot bus: single random lay wire, twisted pairs, twisted and shielded pairs, and fiber optics. The actual choice depends on the required bit rate and economic trade-off studies for each application.

3.5.2.2 *Data Bit Rates*—Data bit rate is only limited by the available technology of bus media, receivers, transmitters, etc. Future improvements in components and media will allow a virtually unlimited future data rate growth path beyond the 1 or 2 Mbits/s capability.

3.5.2.3 *Bit Rate Tolerance*—Each device must maintain the selected bit rate within a tolerance maximum consistent with reliable bit sampling and token acceptance detection. The Bus Time Out condition is the most vulnerable to this parameter with 32 token slot times to run without a clock synchronization signal. Factors which influence the bit rate tolerance include variations due to part tolerances, operating temperature range, and component aging.

3.5.2.4 *Logic Zero Dominance*—The transmission medium and the bus transceivers must be designed such that a logic zero being transmitted by one device will override logic ones being transmitted by one or more other devices. Use of a dominant logic state allows both better collision detection and the ability to force a jam signal onto the bus.

3.6 Node Device Implementation Considerations—This section describes the duties of devices which transmit and receive information on the Token Slot Data Link. The actual partitioning of these duties between the microcomputer software, protocol support hardware, and the line driver/receiver hardware is not specified here, although the indicated timing constraints will affect that partitioning.

3.6.1 MESSAGE TRANSMISSION—The required capabilities and duties of a device transmitting a message include acquisition of transmitting privileges, message selection, actual transmission of the message, message validation, and proper recovery from error conditions. These requirements are detailed in the following sections.

3.6.1.1 *Required Resources for Transmission*—A device must have

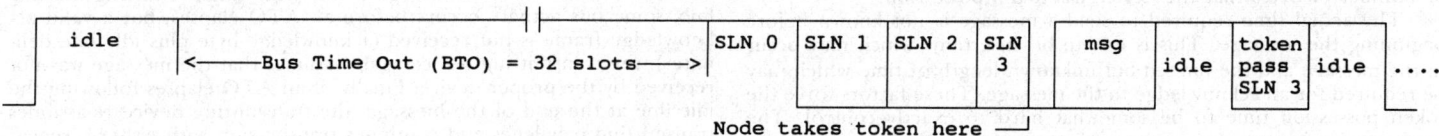

FIGURE 16—BUS TIME OUT

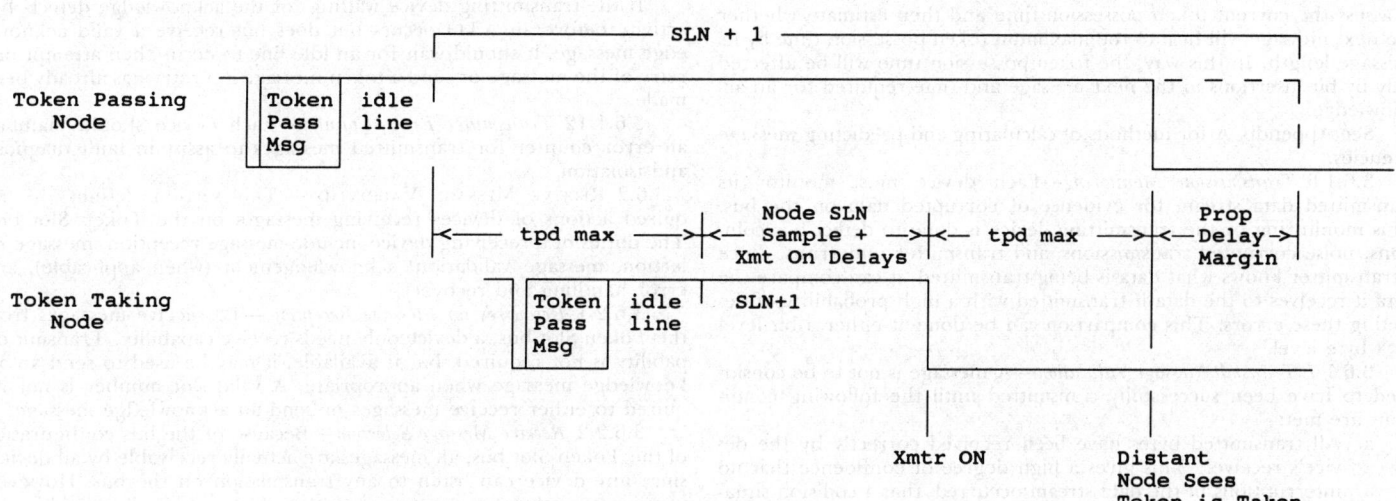

FIGURE 17—PROPAGATION DELAY CONSIDERATIONS

both transmit and receive capabilities and a valid slot number to transmit messages on the Token Slot bus. Receive capabilities are needed so that it can acquire the token and monitor its own transmissions. The slot number is needed so that the proper time slot can be used to take the token before beginning transmission. Devices without these resources are not to be allowed to acquire token possession.

3.6.1.2 *Transfer of Transmitting Privileges*—A device obtains transmitting privileges by determining the proper transmit delay time following activity on the bus and then beginning a transmission if the bus has remained at the logic one state for that time. The transmit delay time is calculated using the Slot Number in a received token message or using a Slot Number of 31 after a bus jam or bus time out has occurred. The device must begin its transmission (that is, drive the bus to the dominant state) within a short time after its transmit delay time elapses. This time includes the delays in the serial hardware and the bus transceiver and is compensated by adjusting the system token slot width. In normal operation, the last message sent by a device will be a token message containing that device's Slot Number.

3.6.1.3 *Message Selection*—When a device obtains transmitting privileges, it may transmit one or more messages on the bus. Each device is responsible for deciding which message(s) to transmit.

3.6.1.4 *Message Update Rate*—In general, it is undesirable for a device to send every possible message during every token possession. Such an approach would use excessive bus bandwidth and tend to limit the growth potential of the network. Each message should be transmitted only at the update rate required by other devices in the system or when an event causes a change in status. Background status summary messages may be scheduled and sent periodically to update receiving nodes which may have missed the initial event driven message. Decisions to send or withhold messages based on current bus "business" and maximum hold time limits may be implemented by a node if authorized by the system designer.

3.6.1.5 *Maximum Token Possession Time*—Each device on the bus should be assigned a maximum time that it may hold the token on any individual possession. This maximum token possession time helps to bound the token rotation time for the system. A device must look ahead and may not begin transmission of a data message which would cause the maximum token possession time to be exceeded. The token message must be sent as the final message even if this maximum time has been reached or exceeded. While this function is defined as a time period, the actual implementation may count the number of bit times or number of bytes that the device has token possession.

The actual time required to send a message is not known before beginning the message. This is due to bit insertions which may occur in the message and the limited but unknown length of time which may be required for an acknowledge to the message. These factors force the token possession time to be somewhat hard to exactly control. The minimum implementation of the maximum token possession time monitor is for the device to add the number of bytes already transmitted to the length of the next message, and compare this sum to a fixed maximum. If the sum is less than the maximum, the message is transmitted. This method does not account for bit insertions, inter-message gaps, and acknowledges which may have occurred during the current token possession. A better method is for the device to use a timer to accurately assess the current token possession time and then estimate whether the next message will fit into the maximum token possession time by its message length. In this way, the token possession time will be affected only by bit insertions in the next message and time required for an acknowledge.

See Appendix A for methods of calculating and predicting message latencies.

3.6.1.6 *Transmission Monitoring*—Each device must monitor its transmitted data stream for evidence of corrupted data on the bus. This monitoring by the transmitting device is done to detect bus collisions, noise corrupted transmissions, and transmitter underruns. Since a transmitter knows what data is being transmitted, it can compare the data it receives to the data it transmitted with a high probability of detecting these errors. This comparison can be done at either a bit level or a byte level.

3.6.1.7 *Transmit Message Validation*—A message is not to be considered to have been successfully transmitted until the following conditions are met:
   a. All transmitted bytes have been received correctly by the device's receiver. This gives a high degree of confidence that no interruptions in the data stream occurred, that a collision situation did not exist during transmission, and to a lesser degree that noise did not corrupt the transmission.
   b. An idle line condition was sensed by the receiver both preceding and following the message. This ensures that the message was properly delimited.
   c. If the message was a data message (CB1 bit = zero in ID field) and contained a request for acknowledgement (CB0 bit set in the ID field), a valid acknowledge frame must have been received following the idle line at the end of that message. This acknowledge frame must have been started before an Acknowledge Time Out (ATO) condition is reached.

3.6.1.8 *Error Handling and Recovery*—Proper response to error conditions is essential for consistent, reliable operation of the Token Slot bus. The following error recovery procedures must be observed by all devices:

3.6.1.9 *Transmitted and Received Data Differences*—When a discrepancy between transmitted data and received data is detected, the current message must be terminated and a jam signal sent on the bus. The jam signal is intended to communicate the error condition to all devices by forcing bit encoding errors to occur in their received data. This jam signal must be started before an idle line occurs which would delimit this message to the receiving devices. Otherwise, the jam would not cause other devices to reject the corrupted message. For NRZ5 encoding, the transmitted jam signal should be 8 bits in duration to ensure that all receivers sense at least six consecutive zeros.

The jam signal is also sent if a transmitter detects that the idle line following the current message is corrupted. Once the message has been delimited by eight consecutive logic ones, a continuation of logic ones may extend the idle line. The transmitting device is to continue to monitor the bus for an idle condition (logic ones received) during this inter-message gap and issue a bus jam if a logic zero is received during this time.

3.6.1.10 *Transmitter Underrun*—A transmit underrun typically occurs when the transmitter hardware runs out of data to transmit before a message is completed, causing a gap in the bit stream. If a transmitter should detect a transmit underrun condition, it must immediately issue the bus jam signal and cease transmission. As discussed in the previous section, this jam must begin before a valid idle line is completed following the message data.

3.6.1.11 *Acknowledge Frame Errors*—When a device sends a message requesting an acknowledge, an acknowledge frame must be started before an Acknowledge Time Out (ATO) is declared. Any activity on the bus before the idle condition is achieved will cause a transmission error in the transmitting device, since an early start of the acknowledge frame will cause the acknowledge to be ignored. If, following the idle line, some bus activity occurs before an ATO elapses, but a valid acknowledge frame is not received (acknowledge byte plus idle line delimiter), the transmitting device again assumes that the message was not received by the proper device. Finally, if an ATO elapses following the idle line at the end of the message, the transmitting device re-assumes transmitting privileges and continues transmission with either a repeat of that message, a new data message, or a token message. This subsequent transmission must begin before a BTO can be detected by any receiver on the bus. The only condition that will cause a transmitting device to validate a message containing an acknowledge request is that a valid acknowledge frame is started within one ATO of the idle line delimiter.

If the transmitting device waiting for the acknowledge detects bus activity before an ATO occurs but does not receive a valid acknowledge message, it should wait for an idle line to occur then attempt one retry of the message or send a token message if a retry has already been made.

3.6.1.12 *Transmitter Error Counter*—Each device should maintain an error counter for transmitted messages to assist in fault diagnosis and isolation.

3.6.2 RECEIVE MESSAGE VALIDATION—This section defines the required actions of devices receiving messages on the Token Slot bus. The duties of a receiving device include message reception, message selection, message validation, acknowledgement (when applicable), and error handling and recovery.

3.6.2.1 *Resources for Message Reception*—To receive messages from the Token Slot bus, a device only needs receive capability. Transmit capability is not required, but if available, it may be used to send an acknowledge message when appropriate. A valid slot number is not required to either receive messages or send an acknowledge message.

3.6.2.2 *Receive Message Selection*—Because of the bus configuration of the Token Slot bus, all messages are actually receivable by all devices since any device can listen to any transmission on the bus. However, certain restrictions on message reception exist and are described in this section.

3.6.2.3 *Token Message Reception*—All devices that require transmitting privileges on the bus must accept and interpret all token messages before assuming the transmit mode.

3.6.2.4 *Broadcast Data Messages*—Messages identified as broadcast data messages can be received and used by any device on the bus that knows the format and processing rules of that message. This implies that changes to this message must be communicated to all users of the bus.

3.6.2.5 *Data Messages with Acknowledge Request*—These messages are used when only one device on the bus has been designated to receive and acknowledge the message. The receiving device is required to transmit an acknowledge frame in response to that message if it meets the requirements in 3.6.2.11.

All devices except the one requesting an acknowledge to a message should disregard acknowledge messages which may occur on the bus. This restriction does not apply to diagnostic equipment.

3.6.2.6 *Acknowledge Frame*—If a received data message containing a request for acknowledge (CB0 bit in ID = 1) passes the preceding validation tests, an acknowledge frame must be transmitted. This acknowledge frame is to begin within 16 bit times from the end of the idle line delimiting the message to allow the original data message sender time to detect the response before it declares an ATO. This 16 bit time limit may need to be adjusted to accommodate physical bus propagation properties. If the message did not pass all the validation tests, or if the receiving device does not have transmit capability, no transmission shall take place in response to the acknowledge request. A device that has transmit capability must return the acknowledge message when required if it is functional and capable of acting on the message.

Any errors encountered during transmission of the acknowledge frame are to be handled as if the device were transmitting a data message or token message. It is appropriate for the receiver to issue a bus jam when these errors are detected.

3.6.2.7 *Message Delimiting*—All messages are to be delimited with an 8 bit time idle condition that will be seen in all receivers as an idle line. Receiving devices use idle lines to define the boundaries of individual messages on the bus.

3.6.2.8 *Bit Encoding Checks*—During reception of a message, each receiving device is to monitor the data stream for bit encoding errors. An NRZ5 bit encoding error is the occurrence of six consecutive bit times of the same logic state. Six or more consecutive logic zeros is interpreted as a bus jam, while six or more consecutive ones may be seen as either an encoding error or the beginning of an idle line.

3.6.2.9 *Error Check Field*—All data messages contain a 16 bit CCITT CRC at the end of the message. All receiving devices should use this CRC to check the validity of the received data.

3.6.2.10 *Minimum Message Length*—The minimum length for data messages is 4 bytes not including the idle line delimiter (2 ID bytes + 2 CRC bytes, if no data bytes). A receiver is to reject any data message that is not at least 4 bytes long. The length of token messages and acknowledge messages is always 1 byte. Note that Control Bit CB1 = 0 in the ID field distinguishes data messages from tokens and acknowledges.

3.6.2.11 *Receive Message Validation Summary*—A message will not be considered to have been successfully received until the following conditions are met:
a. The message was properly delimited by idle lines (this is a basic requirement for defining an individual message).
b. No bit encoding errors were detected between the idle lines.
c. The CRC or parity check revealed no data errors.
d. The received message contained an integral number of bytes (counted after the bit deletion performed during decoding), and if it was a data message, was at least the minimum 4 bytes in length.
e. No receiver overflow conditions occurred during reception of the message.

Failure of a message to meet these criteria will cause the receiving node to reject (ignore) the message and, if assigned Acknowledge responsibility, it will not return the ACK message.

APPENDIX A
DETERMINATION OF MESSAGE LATENCIES

A.1 Token Slot Message Data Overhead Summary (in bit times = bt):

Synchronization bit	1
ID	16
Data (system determined limit, e.g., 0 to 256 bytes)	var
CRC	16
Intermessage delimiter gap	8
Acknowledge response (if requested) + gap	16
Per message overhead—with ACK:	57 bt
or Per message overhead—without ACK:	41 bt

A.2 Token Passing Slot Overhead:

Slot width (assume xmtr on at mid slot = 1 bt)	1
Token Pass Message	8
Delimiter gap	8
Token pass overhead per node:	17 bt

A.3 Summary of Loop Time Calculations:

For P nodes using R slots sending an N message loop with M total message data bytes:

Total message overhead (no ACK)	=	41N
Total token overhead	=	17R
Total message data time (m × 8 bits)	=	8M
Unused slots = slot width × (max #slots—#used)	=	1(32-R)
Total Loop Time Formula:		41N + 17R + 8M + 1(32-R)

A.4 Example Token Slot Timing Calculation:

For a 32 node system using 8 slots to send a 16 message loop with 32 total message data bytes (2 msgs × 2 bytes) per slot × 8 slots:

Total message overhead (no ACK)	41×16 = 656
Total token overhead	17×8 = 136
	792
Total message data time (16 msgs × 2 bytes × 8 bits)	+256
Unused slot time (1 bt per slot × 24 unused slots)	+ 24
Total Loop Time	1072 bt

Data efficiency = Total Data/Total Loop Time = 256/1072 = 24.0%

APPENDIX B
EXAMPLE OF SLOT ASSIGNMENTS

B.1 See Figures B1, B2, and B3.

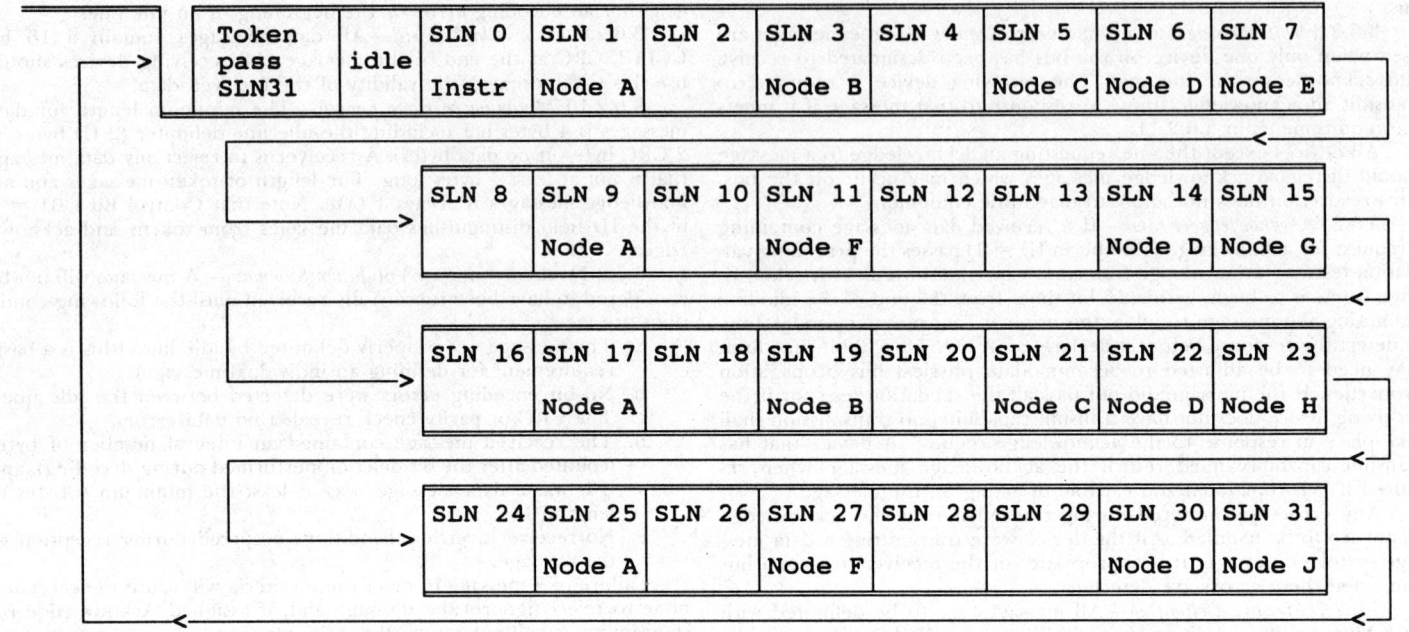

FIGURE B1—A TYPICAL TOKEN SLOT NODE ASSIGNMENT AND SLOT SEQUENCE CYCLE PATTERN

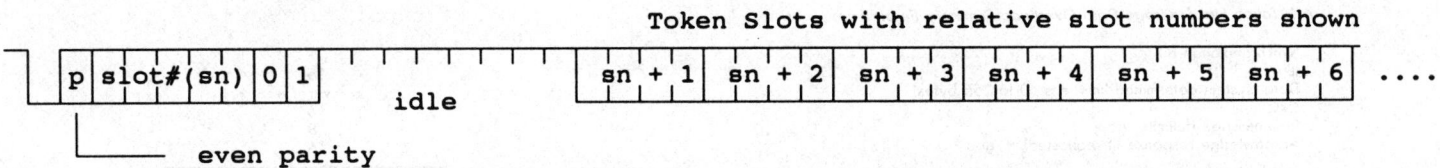

FIGURE B2—TOKEN PASS MESSAGE FORMAT

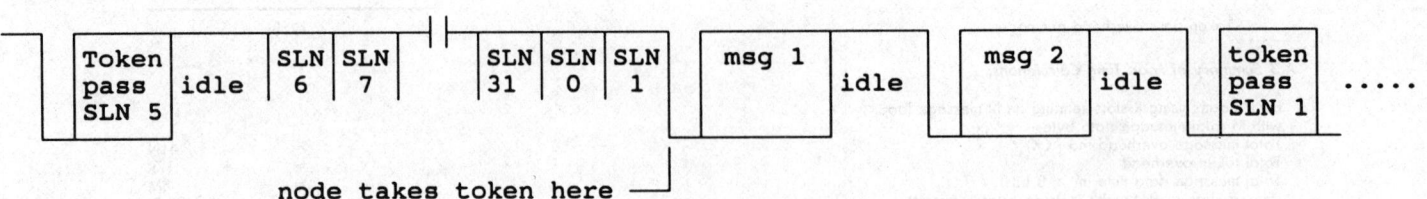

FIGURE B3—TYPICAL TOKEN PASS SEQUENCE

APPENDIX C
SUMMARY OF MESSAGE TYPES

C.1 See Figures C1, C2, C3, and C4.

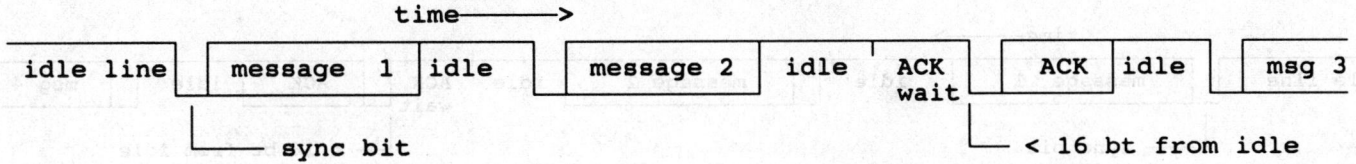

FIGURE C1—TOKEN SLOT NETWORK GENERAL MESSAGE FRAMING

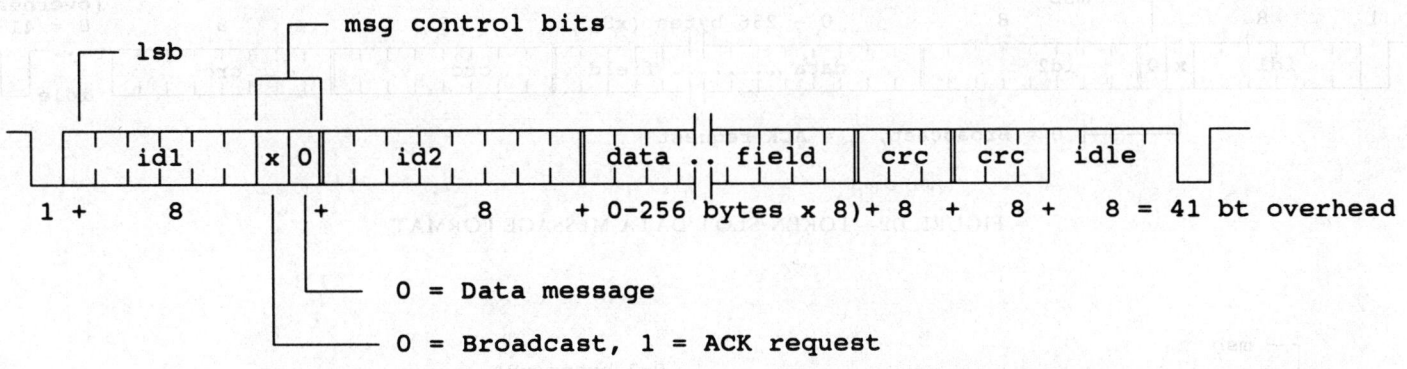

FIGURE C2—TOKEN SLOT NETWORK DATA MESSAGE FORMAT

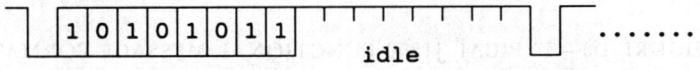

FIGURE C3—TOKEN SLOT NETWORK ACK MESSSAGE FORMAT

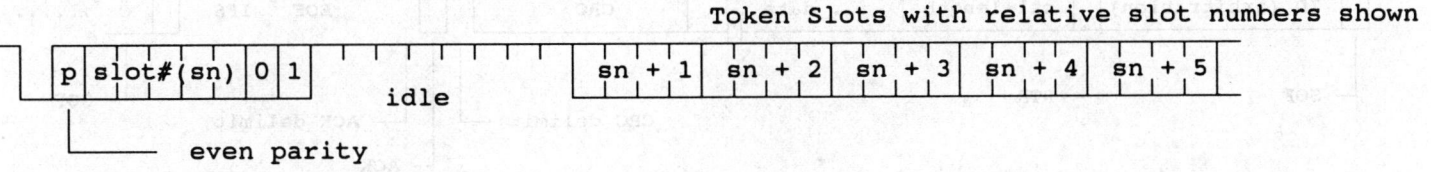

FIGURE C4—TOKEN SLOT NETWORK TOKEN PASS MESSAGE FORMAT

APPENDIX D
TOKEN SLOT OVERHEAD COMPARISON WITH J1850 & CAN

D.1 See Figures D1, D2, D3, and D4.

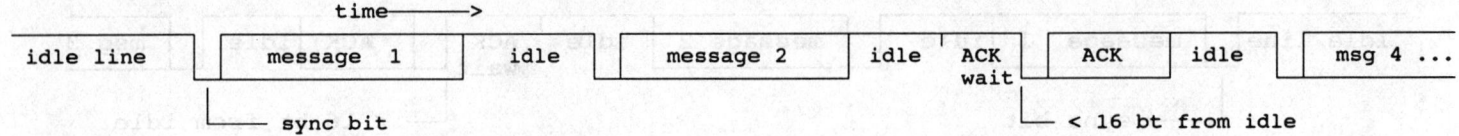

FIGURE D1—TOKEN SLOT GENERAL MESSAGE TIMING

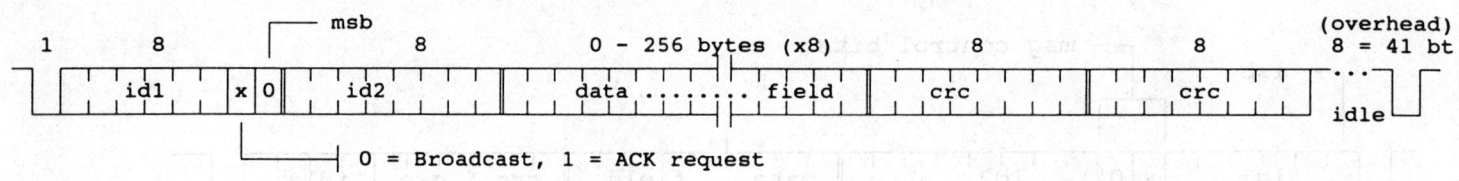

FIGURE D2—TOKEN SLOT DATA MESSAGE FORMAT

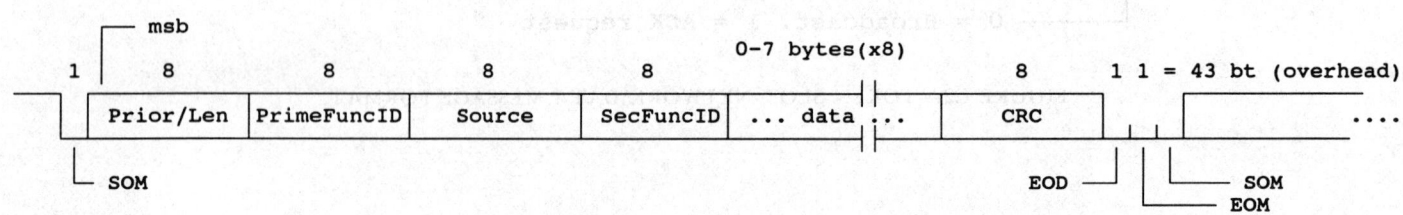

FIGURE D3—TYPICAL J1850 FUNCTIONAL MESSAGE FORMAT

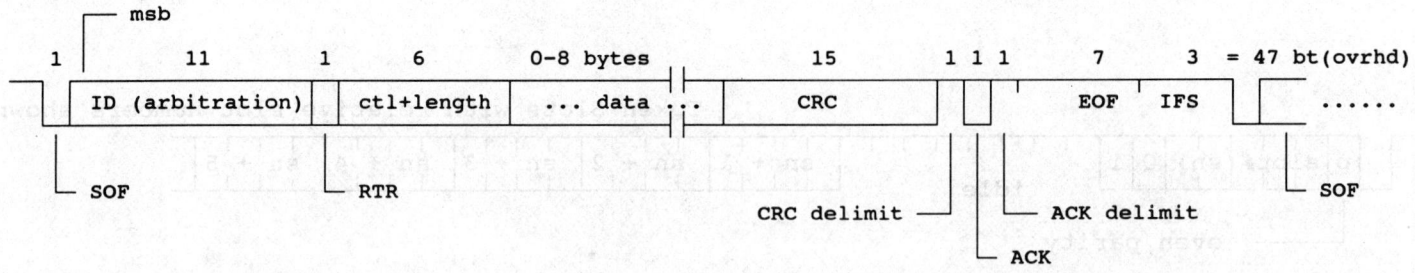

FIGURE D4—CAN MESSAGE FORMAT

APPENDIX D
OVERHEAD COMPARISON WITH J1850 AND CAN
(continued)
D.2 Overhead Comparison Calculation

D.2.1 Message Overhead

	TOKEN SLOT		J1850		CAN	
Synchronization bit		1	SOM	1	SOF	1
ID		16	Priority/length	8	ID(Arb)	11
Data (system determined limit - bytes)	(0-256 bytes)		Pimary Function	8	RTR	1
CRC		16	Secondary Func.	8	CTL Field	6
Intermessage delimiter gap		8	Source	8	DATA	(0-64 bits)
Acknowledge response (if requested) + gap		16	DATA	(0-56 bits)	CRC	16
			ACK	(0-57)	ACK	2
			CRC	8	EOF	7
			EOD/EOM	3	InterFrame	3
Per message overhead - with ACK:		57 bt		var		47 bt
or Per message overhead - without ACK:		41 bt		43 bt		na

D.2.2 Token Passing Slot Overhead

```
Slot width (assume xmtr on at mid slot = 1 bt)   1
Token Pass Message                               8
Delimiter gap                                    8
                                               ____
            Token pass overhead per node:   17  bt
```

D.2.3 Summary of Loop Time Calculation—For P nodes sending an N message loop with M total message data bytes:

	TOKEN SLOT	J1850	CAN
Total message overhead (no ACK)	41N	43N	47N
Total token overhead	17P	0	0
Total message data time (m x 8 bits)	8M	8M	8M
Unused slots = slot width x (max #slots - #used slots) =	1(32-P)	0	0
Total Loop Time	41N + 17P + 8M + 4(32-P)	43N + 8M	47N + 8M

D.2.4 Example Timing Calculation with Comparison to J1850—
For 8 nodes sending a 16 message loop with 32 total message data bytes
(2 msgs × 2 bytes per node × 8 nodes):

```
Total message overhead (no ACK)     41x16 =      656      43x16 = 688     47x16 = 752
Total token overhead                17x8  =    + 136              0               0
                                                 792            688             752
Total message data time (32 x 8 bits) =       + 256            256             256
Unused slot time         (1x24)       =       +  24              0               0

                    Total Loop Time (bt)       1072            944 *          1008 *
```

* Does not include time added when messages lose arbitration.

Data-to-total time overhead efficiency: $256/1072 = 24.0\%$
$256/944 = 27.1\%$
$256/1008 = 25.3\%$

D.3 J1850 Messages Contained within Token Slot Format—Figure D5 shows a possible method of retaining existing J1850 message formats within the Token Slot message structure. This would allow a smooth transition for users to retain their established message IDs and formats while taking advantage of the features of the Token Slot protocol. Note that the contention based J1850 message priorities do not apply to the Token Slot protocol. This allows stripping of the Priority field from the message before transmission.

Figure D6 shows the placement of typical J1850 functional message fields (lower line) within the Token Slot data message structure (upper line).

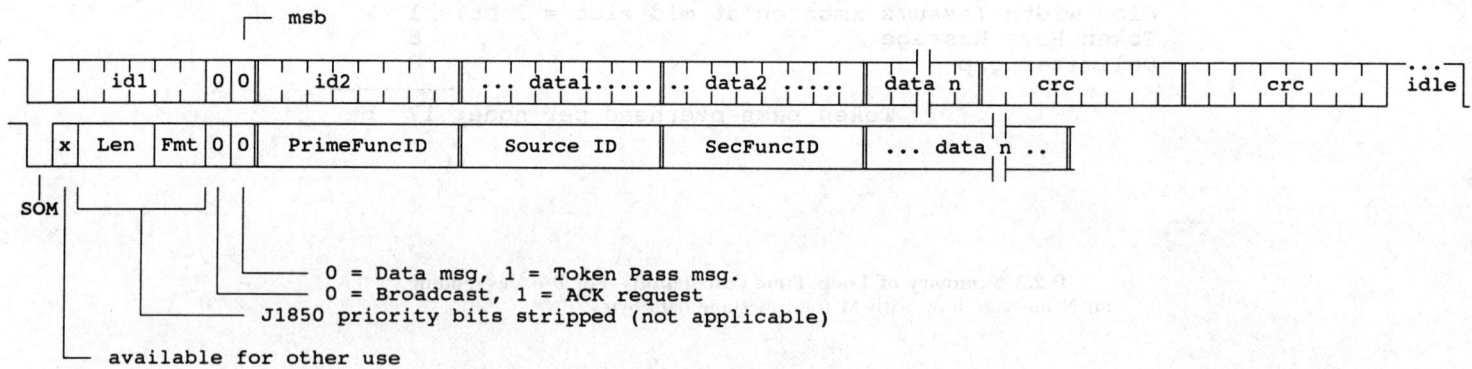

FIGURE D5—J1850 FUNCTIONAL MESSAGE FIELDS CONTAINED IN A TOKEN SLOT DATA MESSAGE FORMAT

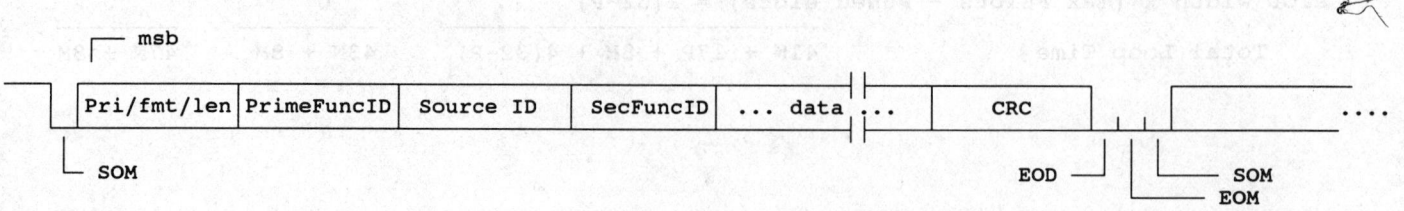

FIGURE D6—TYPICAL J1850 FUNCTIONAL MESSAGE FORMAT

OPEN FIELD WHOLE-VEHICLE RADIATED SUSCEPTIBILITY 10 kHz–18 GHz, ELECTRIC FIELD—SAE J1338 JUN81

SAE Information Report

Report of the Electronic Systems Committee, approved June 1981.

1. *Purpose*—This report covers the open field requirements for the determination of system susceptibility in the presence of an electric field in the frequency range 10 kHz–18 GHz.

2. *Measurement Philosophy*—The increasing use of electronic control systems for automotive application dictates a need to ensure electromagnetic compatibility (EMC) of these systems within their intended environment. Testing techniques are needed for complete systems (whole-vehicle) since component systems, which may not be susceptible to electromagnetic interference (EMI) when tested by themselves, may be susceptible when installed as part of a larger or whole-vehicle system. Open-field testing offers an important possible test environment since it can provide both a reference and perturbation-free environment in which actual operating conditions of the equipment under test (EUT) can be simulated. Tests can be performed at discrete frequencies from 10 kHz to 18 GHz provided proper experimental licensing is first obtained from the Federal Communications Commission (FCC). Since tests must be limited to only specific FCC-approved frequencies, alternative sites, such as anechoic chambers, buried low-Q chambers, TEM cells, etc., may be more appropriate in some instances.

The scope of this report is limited to susceptibility tests of a whole vehicle as a system. Susceptibility tests on subsystems or components should be carried out as indicated in SAE J1113a (15) using laboratory facilities such as TEM cells, anechoic enclosures, parallel plate lines, or other test facilities. In such laboratory facilities, the operating parameters or variables of the device under test can be better controlled, and swept frequency testing can be performed to pin-point specific problems in the device under test.

3. *Licensing*—An FCC experimental license for open field radiated susceptibility tests of whole vehicles may be obtained in accordance with Part 5 of the Rules and Regulations of the FCC. The procedure and notes for this are contained in Appendix A. An FCC license *must* be obtained *before* performing open field radiated susceptibility tests.

4. *Grounding and Shielding*—It is recommended that a ground plane over which to perform tests be established. Such a ground plane will improve the repeatability of the tests, make establishing the test fields easier, and make the test simulate the 'worst-case' conditions in the vehicle's normal environment. The ground plane may be made of wire mesh, pierced aluminum planking, pierced steel planking, or other similar material laid on the surface, buried 1–2 in (2.5–5 cm) below the surface or cast in pavement 1–2 in (2.5–5 cm) below the surface. The ground plane must encompass the setup, including the external field generating antenna when used. It should preferably extend as far as is practicable beyond the perimeter of the vehicle under test and any test antenna used. If the extent of the ground plane around the perimeter of the vehicle under test and the test antenna is reduced to about 1 m or less, the amount of ground plane is not large enough to provide an adequate return path for displacement currents. This modifies the physical shape and extent of the electric fields and changes the coupling between the antennas and the vehicle under test in a manner which is very difficult to predict.

The test equipment must be shielded and all power and control lines to it must be shielded and filtered with sufficient effectiveness to prevent instability and false indications of susceptibility. If the test equipment is not designed to be used in high-strength fields, external shielding and filtering must be provided. The test equipment shields must be bonded to the ground plane using care not to create any ground loops or other undesired coupling paths. The test equipment operators and any other personnel involved in the tests should be provided shielding from the test fields or be located at a distance from the test setup.

An example of a ground plane layout and test facility is shown in Fig. 1 in an elevation sketch of the National Bureau of Standards (NBS) facility. Such a facility is more elaborate than is necessary, but serves as a good example of one approach to the problem.

5. *Apparatus*—The following listed test apparatus and equipment are needed. Typical examples of equipment and references, for example, (1), (5), etc. are listed in Section 7. The apparatus discussed in this section is recommended assuming that the highest electric field strengths to which the test vehicles are to be exposed are on the order of 200 V/m. This level was selected based on the premise that the maximum EM field exposure of the occupants of a vehicle should be 10 mW/cm². The national safety standard for nonionizing radiation gave this as the limit until late 1979. Also, there have been studies by the Environmental Protection Agency (EPA) and others (12), (13) which would generally support this level for exposure of automobiles. If the maximum field strength to be used in the tests is much less than 200 V/m, then lower-powered generators and other apparatus may be used.

Frequency Range	Antenna Type
10 kHz–1.6 MHz	Modified Beverage antenna (1), (5), (6), (9), parallel plate (2), (7), or 3-meter whip (transformer coupled) (3), (8)
1.6–10 MHz	3-Meter whip (transformer coupled) (3), (8)
10–25 MHz	2-Meter whip (transformer coupled or base-loaded) (3), (4), (8)
26–50 MHz	Mobile base-loaded or λ/4 whip
25–200 MHz	High-power biconical dipole
20–200 MHz	Log periodic array (Note 1)
148–174 MHz	Mobile 5λ/8 or λ/4 whip
200 MHz–1 GHz	Ridged waveguide horn or log periodic array
200 MHz–1 GHz	Conical log spiral (Note 2)
450–470 MHz	Mobile colinear or λ/4 whip
800–915 MHz	Mobile colinear or λ/4 whip
1–8 GHz	Ridged waveguide horn (Note 3)
1–10 GHz	Conical log spiral (Note 2)
8–18 GHz	Rectangular waveguide horn (Note 3)

NOTE 1: The log periodic arrays for the frequency range of 20–200 MHz are not as effective as the high-power biconical dipoles when used near a ground plane and near the equipment under test.

NOTE 2: The conical log spiral antennas do not have as much gain as the ridged waveguide horns and log periodic arrays. Also, they emit a circularly polarized wave. These two factors tend to make them less desirable as susceptibility test field radiators, since more RF power is required to achieve a given linear (equivalent) field. Their VSWR also tends to be higher, making them more difficult to drive. However, since they are circularly polarized, they tend to excite all responses of the vehicle without being repositioned to provide both vertical and horizontal polarizations as is necessary with the linearly polarized antennas.

NOTE 3: Ridged waveguide horns could be used up to 18 GHz but most are coaxial cable fed—too lossy for frequencies above 8 GHz at the power levels necessary to achieve 200 V/m.

5.2 Generators and Transmitters—The generators and transmitters needed depend on the desired field strength and frequency for external field generation. It is assumed that tests for on-board generated fields, that is, from on-board mobile transmitters, will be performed with the maximum FCC authorized power to the antenna of 110 W for the business, public service, police, and mobile telephone radio services. The approximate generator power required to produce external fields of 200 V/m under the conditions discussed in Section 6 is given in the following table.

In the frequency range of 10 kHz–1.6 MHz, the power level is antenna height dependent. All other power level variations are versus frequency. These approximate levels are based on driving the external field generating (fixed) antennas to produce 200 V/m at the vehicle under test when set up as specified in paragraph 6.1. (These are estimates which will be refined as the technique develops.)

Frequency Range	Power Level
10 kHz–1.6 MHz	0.3–2 kW
1.6–30 MHz	0.6–2 kW
30–200 MHz	2–0.7 kW
200 MHz–1 GHz	0.5 kW
1–18 GHz	0.2 kW

5.3 Field Sensing and Calibration Equipment—The fields may be set using a standard EMI meter with calibrated dipole and rod antennas, but under some of the conditions of irradiation of the vehicle, a considerable uncertainty will exist with this method of field calibration. The magnitude of this uncertainty is generally unknown. Electrically small, isolated E-Field probes which can be used near the vehicle to sense the fields

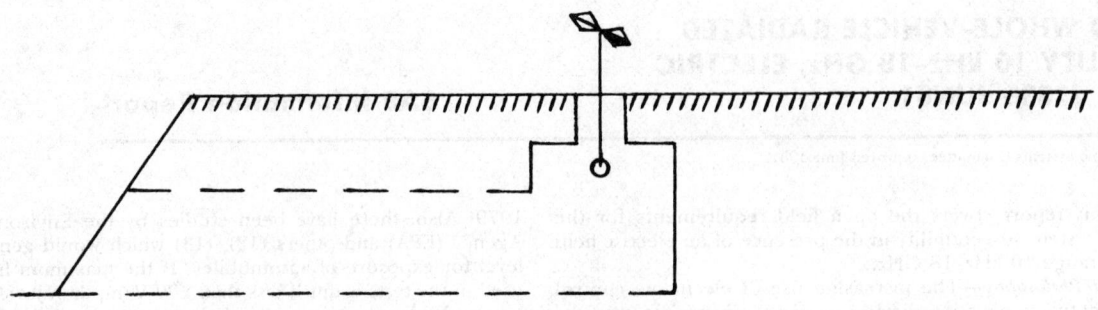

FIG. 1—NBS GROUND PLANE FACILITY

reduce the uncertainty to a known and acceptable amount provided great care is used in their placement. This is discussed more in paragraph 6.1.2. Other needed calibration equipment is listed below:
(1) Accurate frequency meter.
(2) Directional wattmeters 10 MHz-18 GHz.
(3) High-impedance RF voltmeter 10 kHz-30 MHz.
(4) Vehicle susceptibility monitoring equipment.

6. Test Setup and Procedures—This section is divided into two parts. The first part, paragraph 6.1, covers the generation of fields external to the vehicle, and the second part, paragraph 6.2, covers the generation of fields from on-board transmitters.

Climatic conditions may affect the tests and should be considered. Tests must not be performed during periods of precipitation or with the surface of the vehicle under test wet. Precipitation in the air and moisture on the vehicle will cause anomalies in the electromagnetic fields and their coupling to the vehicle. Moisture collected on tents and shelters for the test site may also cause anomalous test results. The test equipment should be protected from direct sunlight and extreme temperatures, hot and cold. Temperatures that are much higher or lower than are typical for indoor laboratory environments may degrade the performance of the test equipment.

6.1 External Field Source
6.1.1 Antenna Setup
6.1.1.1 *Modified Beverage or Parallel Plate Line*—This antenna is set up over the ground plane so that the single-wire flat top of the modified Beverage or the upper plate of the parallel plate line is 1–2 m above the top of the vehicle. (The ground plane forms the lower plate of the parallel plate line. See References 2 and 7 on the parallel plate line for its application.) The following refers solely to the modified Beverage antenna. (See References 1 and 5.) The 50-Ω transmission line from the test signal generator must be connected to the feed end of the antenna through one or more matching transformers. The impedance transformation ratio (N^2) should be nearly equal to the ratio of the antenna impedance to 50 Ω. Fig. 2 depicts the antenna configuration. The antenna should be sufficiently long to accommodate the maximum dimension of the vehicle that is to be parallel to the antenna during the test. The flat top part of the antenna should be long enough so that the field is uniform over this dimension of the vehicle without the vehicle present. The height of the antenna and the slope of the end sections affect the uniformity of the field near the end of the flat top, so some experimentation will probably be needed to determine the necessary length.

6.1.1.2 *Other Antennas*—These antennas are set up near the edge of the ground plane 1 m from the vehicle under test. The height above the ground plane depends upon the type of antenna and its size. A rod or whip antenna may use the ground plane as its counterpoise, or if the antenna has its own counterpoise, it should be on or near the ground plane and electrically bonded to it. Dipole and log periodic linear antennas should be placed at a height above the ground plane as near the level of potentially susceptible circuits as possible. However, such antennas must be at least a half-length of their longest element above the ground

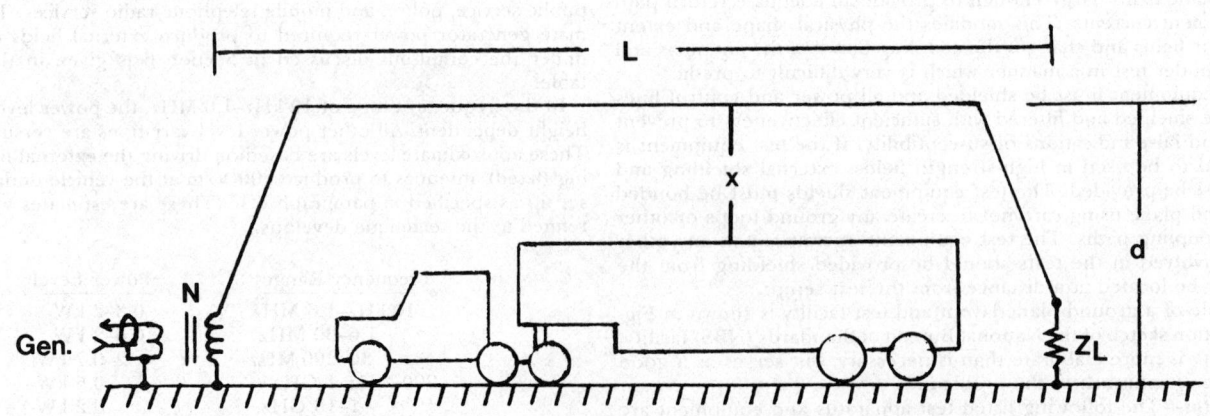

x = space above vehicle, ideally, x > d/2
d = height of antenna above ground plane
L = antenna length including end sections, L < 23 m
a = radius of the wire

Z_a = characteristic impedance of the antenna
Z_L = load resistance, calculated below:

$$P_{in} \simeq \frac{E^2 d^2}{Z_L}, \quad Z_L = Z_a \simeq 138 \log 2d/a)$$

FIG. 2—MODIFIED BEVERAGE ANTENNA

plane where horizontally polarized and at least 0.3 m more than their longest element's half-length when vertically polarized. (This choice in vertical polarization is somewhat arbitrary, but it is accepted by the developers of MIL-STD-462 (6), (10) as minimally adequate.) The spiral antennas and the horn antennas should be placed at the level of potentially susceptible circuits, with the limit that the spiral antennas not be closer than 1 m (6), (10) to the ground plane. (The effect of the ground plane on the antenna impedance is relatively small at this and greater distances.) The position of the antenna relative to the vehicle must be changed and the test repeated as many times as necessary to assure that all critical areas of the vehicle are exposed to fields of the required strength. The area of 'uniform' exposure for the whip, dipole, and log periodic antennas is a surface of the approximate length calculated from the following equation (1) from Reference 6:

$$L = 2\sqrt{Rd - (d/2)^2} \quad (1)$$

where: R is the distance between the antenna and the vehicle, and d is the thickness of the vehicle measured perpendicular to the surface directly exposed to the antenna (see Fig. 3).

The area of 'uniform' exposure for the horn antennas varies with the design of the horn. Refer to the manufacturer's literature to find the 3 dB beam width and apply equation (2) to find the approximate length of the area.

$$L = 2R \tan \frac{\theta}{2} \quad (2)$$

where: θ is the 3 dB beam width of the antenna and R and L are as for equation (1).

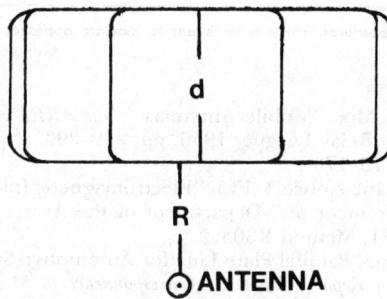

FIG. 3—PLAN VIEW OF TEST SETUP

The higher frequency antennas which tend to have narrower beams of radiation may need to be mounted at elevations and angles that will maximize their radiation through windows or other openings. This may be needed for adequate tests of cluster or dash-mounted electronics. In planning the test, the effects of the vehicle body must be considered.

The whip antennas are used vertically polarized. Details on their design and construction can be found in References 3, 4, and 8. The dipole, log periodic, and horn antennas are used both vertically and horizontally polarized.

6.1.2 Field Calibration Procedure

6.1.2.1 *Standard EMI Meter Method*—If a standard EMI meter is used[1], place its calibrated field measuring antenna for the appropriate frequency range in the position which the vehicle is to occupy during the tests. Using the RF voltmeter below 25 MHz and the directional wattmeter above 25 MHz to monitor the RF drive to the transmitting antenna, establish the required field strength at each test frequency and record the RF voltage or the incident power to the transmitting antenna. Do not use a loop antenna to measure the field strength produced. Position the vehicle to be tested in its proper place in the test setup and re-establish the RF fields by driving the antennas with the same voltage or power at each frequency determined during calibration.

This calibration procedure is based upon the faulty premise that the field radiated by the antenna (the primary field) is unperturbed by the presence of the vehicle under test so close to the transmit antenna. This perturbation may be very large and practically unquantifiable. However, the primary field is *always* proportional to the RF voltage across the terminals of the antenna as long as the antenna is much shorter than a wavelength ($l \leq 0.15\lambda$). Thus, for frequencies below 25 MHz, the primary field may be reset by monitoring the RF voltage at or near (within about 30 cm at 25 MHz) the antenna terminals. Above 25 MHz, one way to counteract the interaction of the antenna and vehicle under test is to monitor both forward and reflected power and adjust the transmitter for the same net power (forward power minus reflected power) to the antenna both before and after locating the vehicle under test 1 m from it. This is not a very accurate means for maintaining the primary field, but it can be used. If a base-loaded resonant (having an electrical length of $\lambda/4$) antenna is used above 10 MHz, the directional wattmeter method will have to be used along with it. The voltmeter method can still be used with a base-loaded antenna that is electrically shorter than $\lambda/4$.

If an indication of susceptibility is observed at any test frequency, reduce the generator level to determine the threshold of susceptibility and record the RF drive voltage or power. The field strength is proportional to the drive voltage directly or the square root of the drive power.

6.1.2.2 *Isolated Probe Method*—If isolated probes[2] are used to calibrate and monitor the susceptibility test fields, arrange the transmitting antenna and the vehicle as described in paragraph 6.1.1. Place one or more isolated probes near the surface or area of the vehicle to be irradiated. Follow the probe manufacturer's instructions as to exact placement of the probes. If the probes are not properly placed, the field produced by the reflections from the vehicle under test may cause indications considerably different from those that would be produced by the primary field. For example, if the probe were placed near a discontinuity, such as a sharp bend or corner of the vehicle, or near a resonant cable or component, the reflected or scattered field could be much larger than the primary field produced by the transmit antenna. Generally, the reflections from approximately flat surfaces parallel to the direction of propagation (perpendicular to the field) will be elliptically polarized in the direction of propagation. If an electrically small probe is placed parallel to the primary field above such surfaces, the scattered field will cause only a small increase or decrease in the indication of the primary field. Again, the probe manufacturer's directions should be obtained and followed. At each test frequency, increase the generator output level until the probe (or probes) indicates the desired field strength. If more than one probe is used, the indicated field strength from each probe will probably be different, depending on its placement. The lowest indicating probe should be used as the field strength level standard for worst-case tests. Be sure that the probe polarization is appropriate for the field transmitted, that is, vertical for vertically polarized fields and horizontal[3] for horizontally polarized fields. If susceptibility is found, determine and record the threshold of susceptibility.

6.1.3 Susceptibility Monitoring—Susceptibility of automotive systems and equipment must be monitored by methods appropriate to their particular characteristics. The susceptibility monitoring method must not include metallic or highly conductive wires attached to the vehicle. Such wires may perturb the test causing spurious indications of susceptibility or failure to indicate actual susceptibility. Fiber-optic links are considered best to carry susceptibility indications from the vehicle to a remote point. However, high-resistance wires may also be used at frequencies where they will cause no significant field perturbations. This condition is typically obtained for wires having a linear resistance of 670 Ω/cm or greater when used at frequencies higher than approximately 100 kHz.

The operating bias point of certain equipment on a vehicle, and therefore its susceptibility to EM fields may be influenced by the speed of the vehicle. The vehicle may have to be in motion during the test to establish the susceptibility, or lack thereof, of such equipment. At the least, a dynamometer may be needed to exercise some of the vehicular equipment. In planning the tests and the test site, these needs should be considered.

6.2 On-Board Field Source

6.2.1 Location of Antennas—The mobile transmitting antennas should be mounted in locations that are typically used by the mobile radio industry. Suggested locations are listed below versus frequency band.

Band	Antenna Location
26–50 MHz	Rear Deck, Bumper, Trunk Lip
148–174 MHz	Rear Deck, Trunk Lip, Cowl, Bumper
420–470 MHz	Roof, Cowl, Rear Deck, Trunk Lip
851–866 MHz	Roof, Cowl, Rear Deck, Trunk Lip

Ideally, tests should be performed with each combination of antenna and mounting location that is typically used. There are about 20 such combinations, the exact number depending on engineering judgment as to what is typical.

---

[1] Be sure that the calibrated field strength measuring antennas are rated for use in the high-strength fields that are intended to be produced.

[2] Battery operated optically isolated monopole E-Field probes are available from Instruments for Industry and several types of optically and high-resistance-transmission-line isolated dipole probes are available from Southwest Research Institute.

[3] There is disagreement among the experts as to the accuracy of the field strength indication of unbalanced probes, e.g., monopoles, when used horizontally polarized. Balanced dipole probes are preferred for horizontally polarized fields.

TABLE 1—TABLE OF TYPICAL TEST EQUIPMENT[a]

Frequency Range	Typical Generators/Amplifiers	Radiating Antennas	Field Calibrating/Monitoring Equip.	Remarks
10 kHz–1.6 MHz	Amplifier Research Model 1000L Instruments For Industry M410	Modified Beverage Ant. Parallel Plate Line 3-Meter Whip	Hewlett-Packard Frequency Counter Boonton Model 92S RF Voltmeter Instruments For Industry EFS-1, etc. Southwest Research Institute Spherical Dipole & Isolated Dipole FS Meter Electro Metrics EMC-25 w/accessories	The radiating antennas can be fabricated by the user or can be obtained on special order from research and development activities.
1.6–10 MHz		3-Meter Whip		
10–25 MHz		2-Meter Whip		
26–50 MHz		Decibel Products DB 712–1 Antenna Specialists ASP-3BL EMCO 3109 2 kW Biconical Decibel Products DB 702 General Electric EY-12A	Bird Model 43 Directional Wattmeter (with other equipment listed above)	Where a standard EMI meter is called for, any instrument meeting ANSI C63.2, "American National Standard Specification for Electromagnetic Noise and Field Strength Instrumentation 10 kHz to 1 GHz," or any instrument satisfactory for use in MIL-STD-461/462 tests may be used. Instruments conforming to American National Standards or to other standards coordinated by ANSI are preferred.
25–200 MHz				
148–174 MHz				
200–500 MHz	Microwave Power Devices MPD ENA2240–52	EMCO 3106 Ridged Guide Horn Antenna Research Associates LPD-2010/RAK Log Periodic Decibel Products DB-705 General Electric EY-12A	Hewlett-Packard Frequency Counter Bird Model 43 Directional Wattmeter Electro Metrics EMC-25 w/accessories Southwest Research Institute Resistively Isolated Dipole FS Meter	
450–470 MHz				
500 MHz–1 GHz	Logimetrics A 680-P	EMCO 3106 ARA LPD-2010/RAK		
1–8 GHz	Logimetrics Series A600L, S, C, X, and Xu.	EMCO 3105	AILTECH NM-62 w/accessories General Microwave Directional Wattmeters Hewlett-Packard Frequency Counter Southwest Research Institute Resistively Isolated Dipole FS Meter	
8–18 GHz		Systron-Donner DBG-520 and DBH-520 W.G. Horns with waveguide		

[a] The equipment listed is representative of equipment required to implement the open field susceptibility test procedures. There is no intent to indicate approval or disapproval of any manufacturer by the fact that his equipment is or is not included.

6.2.2 CALIBRATING ANTENNA DRIVE—The RF drive to the antenna should be adjusted and monitored using a directional wattmeter. The transmitter tuning and coupling should be adjusted to maximize forward power to the antenna, and if the antenna is adjustable, it should be normally adjusted to maximize forward power and minimize reflected power. These adjustments will interact somewhat. The adjustments should be made at reduced transmitter output both for safety to personnel and to prevent possible damage to the transmitter. After tuning and coupling adjustments are complete, the transmitter should be adjusted for maximum power output (as limited by FCC rules) for the susceptibility tests. The forward power indicated on the directional wattmeter should be recorded along with the operating frequency and any indications of susceptibility that are found. In addition to the above adjustments and tests, further tests may be desirable with the antenna misadjusted and/or with different cable routing arrangements. Practical experience with mobile transmitters has shown that the antenna feed cable routing and VSWR sometimes affect the susceptibility of on-board electronics. It is suggested that further testing be performed with as high a VSWR as the transmitter manufacturer allows, and with alternate cable routing.

6.2.3 SUSCEPTIBILITY MONITORING—If these tests are performed from a remote location, that is, if the operator is not in the vehicle, the same consideration for susceptibility monitoring discussed in paragraph 6.1.3 should be observed. If these tests are performed with the operator in the vehicle, most subsystems can be monitored without extra equipment. When extra equipment that is not a normal part of the vehicle is used for susceptibility monitoring, care must be exercised in its use to assure that any susceptibility indicated has not arisen in the extra equipment or as a result of its being there.

7. *Notes*—This section contains a list of references intended to assist the user of this document in implementing test equipment that is not normally available commercially, a list of typical test equipment (Table 1), and a discussion of FCC experimental licensing for susceptibility tests (Appendix). Reference 11 is listed as a source for further study of antenna theory. This is only one of numerous references on antenna theory. Any other text or handbook on antenna theory would probably serve as well.

References

1. H. H. Beverage, C. W. Rice, and E. W. Kellog, "The Wave Antenna," Trans. AIEE, 42, 1923, p. 215.
2. B. E. Roseberry and R. B. Schulz, "A Parallel-Strip Line for Testing RF Susceptibility," *IEEE Trans. on Electromagnetic Compatibility*. Vol. 7, No. 2, June 1965, pp. 142–150.
3. W. A. Stirrat, "Analysis of the High-Impedance Field in Susceptibility Tests," *Proceedings of the Ninth Tri-Service Conference on Electromagnetic Compatibility*. October 1963, pp. 747–774.
4. Donald H. Mix, "Mobile Antennas," *The ARRL Antenna Book*. The American Radio Relay League, 1956, pp. 279–293.
5. Ibid, pp. 175–176.
6. MIL-STD-462 Notice 3 (EL), "Electromagnetic Interference Characteristics, Measurement of." Department of the Army, Washington, DC, February 9, 1971, Method RS03.
7. J. J. Laggan, "Parallel Plate Line for Automotive Susceptibility Testing," *Information Report, Committee Correspondence to M. Crawford*. May 3, 1979.
8. E. L. Bronaugh and D. R. Kerns, "Rod Antennas for High-Strength Radiated Susceptibility Tests for VLF to HF," *IEEE 1979 Electromagnetic Compatibility Record* 79CH1383–9 EMC. October 9–11, 1979, pp. 413–416.
9. R. W. P. King, "Electromagnetic Waves and Antennas Above and Below the Surface of the Earth," *Radio Science*. Vol. 14, No. 2, March–April 1979, pp. 189–196.
10. MIL-STD-462 Notice 1, "Electromagnetic Interference Characteristics, Measurement of." Department of Defense, Washington, DC, August 1, 1968.
11. John D. Kraus, "Antennas." McGraw-Hill Book Company, Inc., NY, 1950.
12. D. L. Lambdin, "An Investigation of Energy Densities in the Vicinity of Vehicles with Mobile Communications Equipment and Near a Hand-Held Walkie Talkie," *ORP/EAD 79-2*. U.S. Environmental Protection Agency, Office of Radiation Program, Electromagnetic Radiation Analysis Branch, P.O. Box 15027, Las Vegas, NV 89114, March 1979.
13. J. W. Adams, H. E. Taggart, M. Kanda, and J. Shafer, "Electromagnetic Interference (EMI) Radiative Measurements for Automotive Applications," *NBS Technical Note 1014*. National Bureau of Standards, Department of Commerce, Boulder, CO 80303, June 1979.
14. ANSI C63.2, "Specification for Electromagnetic Noise and Field Strength Instrumentation 10 kHz to 1 GHz." American National Standards Institute, New York, NY, 1979.
15. SAE J1113a, "Electromagnetic Susceptibility Procedures for Vehicle Components (Except Aircraft)," SAE, 400 Commonwealth Drive, Warrendale, PA 15096, (1978).

APPENDIX

Obtaining an Experimental License

A1. Cognizant FCC Section—The cognizant section of the Federal Communications Commission (FCC) is:

Frequency Liaison Branch
Experimental Radio Services
Office of Science and Technology
Washington, DC 20544

Present contact (as of April 1981):
Mr. Frank Wright
(202) 653-8137

A2. References—*Volume II* of the Rules and Regulations of the Federal Communications Commission, August 1976, contains the following four parts that are applicable:

(a) Part 2—Frequency allocation and radio treaty matters; general rules and regulations.
(b) Part 5—Experimental radio services (other than broadcast).
(c) Part 15—Radio frequency devices.
(d) Part 18—Industrial, scientific, and medical equipment.

Forms to be used in application:
(a) FCC Form 442, August 1976 (6 pp.).
(b) FCC Form 440A (to be used when license involves a government contract).

A3. Procedure

A3.1 Well in advance of need, 120–180 days prior to first scheduled test, do the following:

(a) Obtain Volume II of the FCC Rules and Regulations from the Superintendent of Documents, U.S. Government Printing Office, Washington, DC 20402.
(b) Request FCC Form 442. Do this by writing or calling the cognizant FCC section or the nearest FCC field office. (An information copy of this form is found at the end of this report.)
(c) Study Part 5 carefully.
(d) Compile details for FCC Form 442. Use Part 5, Part 2, and the notes of this section along with Section A4 of this document as guides.
(e) File FCC Form 442. Expect a time lapse of 90–120 days after date of application until approval of license.

A3.2 Notes—A special temporary authorization (STA) may be obtained on the basis of an existing experimental license (paragraph 5.5.6, Part 5). This may be obtained by letter, and the response time is usually less than 30 days. The STA is valid for 30 days of operation, but allows adding new frequencies temporarily if needed for special tests.

The equipment licensed under the experimental license must be operated only under the supervision of the holder of an FCC commercial first or second class operator's license (either radiotelephone or radiotelegraph); thus, if a licensed operator is not already a member of the testing staff, a staff member should obtain such an operator's license (see Vol. I, Part 13, Rules and Regulations of FCC).

Test signals should be unmodulated, if possible. This will speed up license processing by the FCC. If modulation must be used, it must be requested in the application. The length of time on each frequency and the length of time for modulation to be used must be stated. Any type of modulation authorized by Part 2 may be used. It may speed up license processing if the type of modulation is the same as that authorized for the sub-band in the Table of Frequency Allocations in Part 2 (paragraph 2.106 of Part 2), but this should be considered on a case-by-case basis since it is not a basic requirement under Part 5. The main factor to consider is potential interference to other users of the sub-band.

When renewing the experimental license, a statement and facts must be included in the application for renewal indicating what has been accomplished. This experimental report must justify renewal of the license (see paragraph 5.204, Part 5).

A4. Selecting Frequencies

A4.1 Study the Table of Frequency Allocations in Part 2 (paragraph 2.106) to select frequencies that are most likely to be approved. In general, frequencies where no interference to other licensed radio services will likely occur will probably be approved. Always state the length of time each frequency will be used. If frequencies are to be used intermittently, that is, for periods of only a few minutes at a time only a few times per hour, they are more likely to be approved. Under intermittent use, interference tends to be minimized, and the FCC may approve intermittent use of frequencies for which continuous use could not be approved. Try to keep request in the business, industrial, and petroleum radio service frequencies.

A4.2 Frequencies to Avoid—In general, try to avoid the Domestic Public Radio Service frequencies, since this service is protected, and police and fire frequencies.

A4.2.1 Do not request the exact frequency of a commercial broadcast station (MF-AM, VHF-FM, VHF/UHF-TV) if there is a reasonable chance of interference occurring.

A4.2.2 Do not request the following frequencies: on or within the guard bands of any emergency frequencies; in any of the VLF, LF, MF, or HF Radio Navigation Aid sub-bands; or the VHF, UHF, or SHF Radio Navigation channels active at or near the test locations. See Part 2 for frequency allocations.

A4.2.3 Avoid government frequencies if possible. Requesting government frequencies (especially), and frequencies in the maritime service will slow the license processing. If government frequencies are needed, the local 'Area Frequency Coordinator' can be quite helpful. He can usually be contacted through the nearest military base communications officer. It is suggested that early establishment of good rapport with the Area Frequency Coordinator may be very beneficial. If he is satisfied that there will be no harmful interference to the government radio services for which he is responsible, he will very likely be of considerable help in obtaining license authorization for government frequencies.

A4.2.4 Avoid Standards frequencies such as WWV, Canadian Time, U.S. Naval Observatory, etc. The FCC cannot authorize their use in the experimental radio service. Also, avoid radio astronomy frequencies that are active in or near the test area. See Part 2 for frequency allocations.

A4.2.5 All requests should be for discrete frequencies. A request for a band of frequencies should include a justification of why discrete frequencies cannot be used.

A4.3 Suggested frequencies for susceptibility test use are shown in two lists. The first list consists of authorized frequencies for ISM from Part 18 and Field Disturbance Sensors from Part 15. The second list was developed on the basis of establishing three test frequencies per octave of spectrum from 10 kHz–18 GHz consistent with apparently usable frequencies from paragraph 2.106 of Part 2.

A4.3.1 ISM and FDS Frequencies

kHz	MHz
13560 ± 6.78	915 ± 13
27120 ± 160	2450 ± 50
40680 ± 20	5800 ± 75
	10525 ± 25

A4.3.2 Three Frequencies Per Octave from 10 kHz–18 GHz—These 63 test frequencies are not spaced evenly throughout the spectrum. This is done to avoid obviously unapprovable frequencies as indicated in Part 2. Some of these suggested frequencies may not receive FCC approval in all parts of the country.

kHz	kHz	MHz	MHz	MHz	GHz
10.000	111	1.0[a]	13.560	130	1.29[c]
14.000	130	1.3[a]	16.000	160	1.86
16.000	160	1.995	20.02	209[a]	2.1
20.5	200	2.6	27.120	260[b]	2.45
25	250	3.2	33.3	327[b]	3.29[c]
32	326	4.06	40.680	415[b]	4.19
40	400	5.1	52	523[a]	5.800
50	520	6.525	65[a]	661[a]	6.6
64	640[a]	8.1	81[a]	830	8.4[c]
80	810[a]	10.1	100[a]	915	10.495[c]
					13.22
					18

[a] These frequencies are in the broadcasting bands. Check to see if the listed frequencies are occupied and, if so, select a nearby locally unallocated frequency.

[b] These frequencies are a part of the government frequency band from 225–420 MHz and, therefore, might not be assignable or authorizable by the FCC. The local Area Frequency Coordinator may be able to assist in selecting frequencies.

[c] These are shared government/non-government frequencies; thus, some problem in assignment of these frequencies may occur.

A4.3.3 If mobile transmitters on board the test vehicle are to be used, the above list should be slightly modified to include at least one frequency in each of the standard mobile communication bands within the following frequency ranges:

MHz
26–50
148–174
420–470
851–866

23.452

Approved by GAO
B-180227(R0376)

FEDERAL COMMUNICATIONS COMMISSION
Washington, D. C. 20554

**APPLICATION FOR NEW OR MODIFIED RADIO STATION AUTHORIZATION UNDER PART 5 OF FCC RULES
EXPERIMENTAL RADIO SERVICES (OTHER THAN BROADCAST)**

These Instructions should be removed from the attached body of the form before the form is submitted to the Commission.

General Information and Instructions.

1. Submit application direct to the Federal Communications Commission, Washington, D.C. 20554.

2. No fee is required with this application.

3. Before this application is prepared applicant should refer to Part 5 of the Rules and Regulations of the Commission, copies of which may be obtained by purchasing Volume II of FCC Rules and Regulations from the Superintendent of Documents, Government Printing Office, Washington D. C. 20402.

4. If a corporation, state corporate name; if a partnership, state names of all partners and the name under which the partnership does business, if an unincorporated association, state the name of an executive officer, the office held by him, and the name of the association. If this application involves a station that is now authorized, the name herein shown must correspond exactly with that shown on current authorization.

5. Applicants should make every effort to file complete applications. Failure to do so may result in a rejection and return of the application or a delay in the processing of the application.

6. Each document required to be filed as an exhibit should be current as of the date of filing. If reference is made to information already on file with the Commission see item 7 below.

7. To refer to the material on file it is necessary that the applicant provide sufficient information to locate the previously filed material. If filed with a station application, furnish name of applicant or licensee, call sign, file number and date of filing. In other cases furnish date of filing and describe the matters in connection with which filed.

8. Each document or statement required to be filed as an exhibit should be numbered separately. Exhibit numbers should be shown in the blank space provided for this purpose in the individual items of the application form. Where the space left in the application for narrative answers is insufficient a separate statement, bearing an exhibit number in sequence with other exhibits numbered in the application, should be attached to the application and reference to the statement's exhibit number should be made in the answer space.

9. When the antenna structures proposed to be erected will extend more than 6 meters above the ground or above an existing building, each application shall be accompanied by FCC Form 714 indicating that notification has or has not been submitted to the FAA. Landing area, as defined in Part 17 of the Commission's Rules "Landing Area" means any locality, either of land or water, including airports and intermediate landing fields, which is used, or approved for use, for the landing and take-off of aircraft, whether or not facilities are provided for the shelter, servicing, or repair of aircraft, or for receiving or discharging passengers or cargo.

10. Applications proposing a developmental program of experimentation related to an established service which involves the assignment of frequencies not included in the rules governing such service or involves the use of allocated frequencies in a manner contrary to the rules governing that service must be accompanied by a petition requesting the amendment of the rules governing the service involved to provide for the proposed operation.

11. Applications proposing to develop a new service for which no frequencies have been allocated and for which rules have not been promulgated must be accompanied by a petition requesting the allocation of frequencies for the proposed service and setting forth the reasons in support of the petition.

12. An applicant may request that the Commission withhold from public inspection any information or material filed with this application. Requests shall be filed in accordance with Section 0.459 of the Federal Communications Commission Rules and Regulations.

(This form supersedes FCC Forms 440 & 441.)

FCC Form 442
July 1979

23.453

Approved by GAO
B-180227(R0376)

Federal Communications Commission
Washington, D. C. 20554

APPLICATION FOR NEW OR MODIFIED RADIO STATION AUTHORIZATION UNDER PART 5 OF FCC RULES EXPERIMENTAL RADIO SERVICES (OTHER THAN BROADCAST)

1. Applicant's Name and Post Office address
(Give street, city, state, and ZIP Code, see Instruction No. 4)

DO NOT WRITE IN THIS BLOCK
File No.

2.(a) Class of station applied for

2.(b) Nature of Service

3.(a) Application for (check only one box)
☐ New Station   ☐ Modification of existing authorization

3.(b) For Modification indicate below
File No.   Call Sign

4. Application for modification indicate change in (check all that apply)
☐ Frequency   ☐ Emission   ☐ Power   ☐ Location
☐ Other particulars (describe below or in attached Exhibit No. _____)

5. Particulars of Operation see instructions below)

Frequency (State Whether kHz or MHz) (A)	POWER			EMISSION (E)	MODULATING SIGNAL (F)	NECESSARY BANDWIDTH (kHz) (G)
	(B)	(C)	(D)			

(A) List each frequency or frequency band separately. (If more space is required, attach as Exhibit No. _____).
(B) Insert maximum R.F. output power at the transmitter terminals. Specify units.
(C) Insert maximum effective radiated power from the antenna (If pulsed emission specify peak power).
(D) Insert "MEAN" or "PEAK" (See definitions in Part 5).
(E) List each type of emission separately for each frequency. (See Section 2.201 FCC Rules.)
(F) Insert the following information for each type of emission:
　　for A1, insert maximum speed of keying in bauds
　　for A2, insert maximum audio modulating frequency and maximum speed of keying in bauds
　　for A3, insert maximum audio modulating frequency
　　for F1, insert mark and space frequencies, maximum speed of keying in bauds
　　for F2, insert maximum audio modulating frequency, frequency deviation of carrier, maximum speed of keying in bauds
　　for F3, insert maximum audio modulating frequency and frequency deviation of carrier
　　for P0, insert pulse duration and repetition rate
　　for other emissions describe in detail in space provided below.
(G) Describe how the necessary bandwidth was determined in space provided below.

(This form supersedes FCC Forms 440 & 441.)

FCC Form 442
July 1979

FCC Form 442　　Page 2

6(a). Proposed location of transmitter and transmitting antenna (Check only one box)

　　[ ] FIXED BASE　　　　　　[ ] MOBILE　　　　　　[ ] BASE & MOBILE

(b) If permanently located at a fixed location, give below

State	County	City or Town

(c) If mobile, describe the exact area of operation

Number and street (or other indication of location)

Geographical coordinates exact to the nearest second

Geographical coordinates of the approximate center of proposed area of operation (mobile applications)

North Latitude ° ' "	West Longitude ° ' "	North Latitude ° ' "	West Longitude ° ' "

Place an "X" in the appropriate column	YES	NO
7. Is a directional antenna *(other than radar)* used? If "YES", give the following information: (a) Width of beam in degrees at the half-power point. _____ (b) Orientation in horizontal plane _____ (c) Orientation in vertical plane _____		
8. Is this authorization to be used for fulfilling the requirements of a government contract with an agency of the United States Government? If "Yes", attach FCC Form 440-A, Supplemental Information for Applications in the Experimental Radio Service Involving Government Contract, in triplicate to this application.		
9. Is this authorization to be used for the exclusive purpose of developing radio equipment for export to be employed by stations under the jurisdiction of a foreign government? If "Yes", attach as EXHIBIT No. _____, the following information: (a) The contract number and the name of the foreign government concerned. (b) The daily hours of operation and the estimated date of the beginning and end of the specific time period for which the authorization is required.		
10. Is this authorization to be used for providing communications essential to a research project. (The radio communication is not the objective of the research project). If "Yes", attach as EXHIBIT No. _____, a narrative statement providing the following information (a) A description of the nature of the research project being conducted. (b) A showing that the communications facilities requested are necessary for the research project involved. (c) A showing that existing communications facilities are inadequate.		

11. If all the answers to Items 8, 9, and 10 are "No", attach as EXHIBIT NO. _____, a narrative statement describing in detail the following:

　(a) The complete program of research and experimentation proposed including description of equipment and theory of operation.
　(b) The specific objectives sought to be accomplished.
　(c) How the program of experimentation has a reasonable promise of contribution to the development, extension, expansion, or utilization of the radio art, or is along lines not already investigated.

12. (a) Give an estimate of the length of time that will be required to complete the program of experimentation proposed in this application
　　(b) If less than 2 years, give the length of time in months that the authorization requested in this application will be required.

FCC Form 442		Page 3

13. List below transmitting equipment to be installed (if experimental, so state):

MANUFACTURER	TYPE	NO. OF UNITS

14. State provisions that will be made for measurement and periodic checking of the station's technical characteristics (power, frequency, bandwidths) to insure compliance with terms of license.

Place an "X" in the appropriate column	YES	NO
15. Will the antenna extend more than 6 meters above the ground, or if mounted on an existing building will it extend more than 6 meters above the building, or will the proposed antenna be mounted on an existing structure other than a building? If "Yes", give the following (See Instruction 9): (a) Overall height above ground to tip of antenna is _____ meters. (b) Elevation of ground at antenna site above mean sea level is _____ meters. (c) Distance to nearest aircraft landing area is _____ kilometers. (d) List any natural formations or existing man-made structures (hills, trees, water tanks, towers, etc.) which, in the opinion of the applicant, would tend to shield the antenna from aircraft and thereby minimize the aeronautical hazard of the antenna. (e) Submit as EXHIBIT No. _____, a vertical profile sketch of total structure (including supporting building, if any, giving heights in meters above ground for all significant features. Clearly indicate existing portion, noting particulars of aviation obstruction lighting already available.		
16. Does this application propose a developmental program of experimentation which requires the submission of a petition? (See Instructions 10 and 11) If "Yes", attach the petition as EXHIBIT No. _____.		
17. Applicant is *(check only one box)* ☐ Individual ☐ Association ☐ Partnership ☐ Corporation ☐ Other *(describe below)*		
18. Is applicant a foreign government or a representative of a foreign government?		
19. Has applicant or any party to this application had any FCC station license or permit revoked or had any application for permit, license or renewal denied by this Commission? If "Yes", attach as EXHIBIT No. _____, a statement giving call sign of license or permit revoked and relate circumstances.		

FCC Form 442   Page 4

Place "X" in the appropriate column.	YES	NO
20. Will applicant be owner and operator of station? If "No", explain applicant's relation to station.		
21. Will applicant have absolute control of station, both as to physical operation and service conducted? If "No", explain circumstances and give name of person who will have control of station.		

22. Will the transmitter be operated with licensed operator on duty at a remote control point?
 	
    If "Yes", the following information must be provided. (If there is to be a licensed operator on duty at the transmitter, data required by this item may be omitted.)
 	
    (a) Location of remote control point. (If more than one remote control point is involved, attach as EXHIBIT No. _____, a statement giving location of each remote control point, plan of operation, etc.)		

State	County
City or Town	Street & No.

	YES	NO
(b) By what means will the transmitter be rendered inaccessible?		
(c) Can transmitter be placed in an inoperative condition from the remote control point?		
(d) Describe below the equipment to be used to enable the operator at the control point to determine when there is a deviation from the terms of the station license or when operation is not in accordance with the Commission's Rules governing the class of station involved.		

23. Give name, title, and telephone number (include area code) of person who can best handle inquiries pertaining to this application.

24. List below all exhibits in numerical sequence and the item number of form requiring the exhibit identified.

EXHIBITS AND ITEM NO. OF FORM

Exhibit Number	Item No. of Form	Exhibit Number	Item No. of Form	Exhibit Number	Item No. of Form

FCC Form 442   Page 5

25. CERTIFICATION

ATTENTION: Read this certification carefully before signing this application.

THE APPLICANT CERTIFIES THAT:

(a) Copies of the FCC Rules Parts 2 and 5 are on hand; and

(b) Adequate financial appropriations have been made to carry on the program of experimentation which will be conducted by qualified personnel; and

(c) All operations will be on an experimental basis in accordance with Part 5 and other applicable rules, and will be conducted in such a manner and at such a time as to preclude harmful interference to any authorized station; and

(d) Grant of the authorization requested herein will not be construed as a finding on the part of the Commission

    (1) that the frequencies and other technical parameters specified in the authorization are the best suited for the proposed program of experimentation, and

    (2) that the applicant will be authorized to operate on any basis other than experimental, and

    (3) that the Commission is obligated by the results of the experimental program to make provision in its rules including its table of frequency allocations for applicant's type of operation on a regularly licensed basis.

APPLICANT CERTIFIES FURTHER THAT:

(e) All the statements in the application and attached exhibits are true, complete and correct to the best of the applicant's knowledge; and

(f) The applicant is willing to finance and conduct the experimental program with full knowledge and understanding of the above limitations; and

(g) The applicant waives any claim to the use of any particular frequency or of the ether as against the regulatory power of the USA.

Signed and dated this _____ day of _____, 19____.

Name of Applicant _____
                                    correspond with name given on page 1)

By _____
           (print)   *(signature)*

Title _____

```
WILLFUL FALSE STATEMENTS MADE ON THIS
FORM ARE PUNISHABLE BY FINE AND IMPRISON-
MENT. U. S. CODE, TITLE 18. SECTION 1001.
```

Check Appropriate Classification
- [ ] Individual Applicant
- [ ] Member of Applicant Partnership
- [ ] Officer of Applicant Corporation or Association
- [ ] Authorized Employee

NOTIFICATION TO INDIVIDUALS UNDER PRIVACY ACT OF 1974

Information requested through this form are authorized by the Communications Act of 1934, as amended, and specifically by Section 308 therein. The information will be used by Federal Communications Commission staff to determine eligibility for issuing authorizations in the use of the frequency spectrum and to effect the provisions of regulatory responsibilities rendered the Commission by the Act. Information requested by this form will be available to the public unless otherwise requested pursuant to Section 0.459 of FCC Rules and Regulations.

THE FOREGOING NOTICE IS REQUIRED BY THE PRIVACY ACT OF 1974, P.L. 93-579, DECEMBER 31, 1974, 5 U.S.C. 552a(e)(3).

ELECTROMAGNETIC SUSCEPTIBILITY PROCEDURES FOR COMMON MODE INJECTION (1 — 400 MHz), MODULE TESTING—SAE J1547 OCT88

SAE Information Report

Report of the EMI Standards and Test Methods Committee approved October 1988.

1. Introduction

1.1 Scope—This document establishes a method of determining the relative level of susceptibility of electronic modules and the improvement or degradation which results from design changes to those modules. Further work needs to be done to correlate this module test to whole vehicle coupling in an electromagnetic field.

1.2 Measurement Philosophy—The Common Mode Injection Test method induces radio frequency (rf) energy into the electronic module under test. The test is designed to simulate conducted energy from transient sources in the vehicle as well as energy coupled to the vehicle wiring harnesses from on- and/or off-board radiating sources.

2. Apparatus—See Fig. 1 for Equipment Connection Diagram.

(a) Function Generator: 20 Hz – 50 kHz (Sine, Square, Pulse)
(b) RF Signal Generator: 1 – 400 MHz
(c) Broadband Power Amplifier(s): 1 – 400 MHz
(d) Directional Power Meter: 50Ω, Coaxial
(e) Injection Probe: High power broadband ferrite current transformer, Low Z to 50Ω
(f) Detector Probe: Same as above (low power)
(g) Wideband Voltmeter, 50Ω input
(h) Frequency Selective Voltmeter (Spectrum Analyzer): 50Ω input
(i) Ground Plane
(j) Power Divider
(k) Line Impedance Stabilization Network (LISN): use the five µH LISN defined in 3.4.1 (c) and Fig. 3 of SAE J1113 AUG87.

3. Test Signals
Each module type to be tested should be analyzed to determine its probable frequencies of susceptibility. Modules should be tested at selected frequencies over the range of 1 - 400 MHz. Appropriate modulation should be used to attempt to stimulate interaction. Example: An rf carrier amplitude modulated to 100% by a 20 Hz square wave has been used to produce interaction similar to that caused by conducted transients.

Care should be taken to ensure that the fundamental and desired sideband energy is being measured and not harmonic and spurious energy.

4. Test Set Up and Procedures

4.1 General Test Conditions—The module shall be connected using a 1 m long harness. Each power lead shall be connected through a 5 µH LISN. Signal and control leads shall be terminated as is typical in the system. Unless otherwise specified, representative loads from the system shall be used as loads for the component test. Unless otherwise specified, the equipment under test (EUT), harness, LISN, and control/load box shall be mounted on a ground plane as shown in Fig. 2. If the EUT is normally mounted to the metallic body of the vehicle, it should be mounted directly to the ground plane. If mounted to a non-conducting or isolated metallic surface, then the EUT should be insulated from and mounted 5 cm (or other specified distance) above the ground plane. In the latter case, the ground wire from the EUT should be routed in a similar fashion as in the vehicle installation, that is, connected close to the EUT or routed with the harness and grounded by the LISN. Wire used for the harness should be of the gauge and insulation type intended for production.

4.2 The equipment shall be connected as shown in Fig. 2.

4.3 At each selected frequency, the power should be increased until a malfunction occurs or the user's selected limit (may be equipment limited) is reached. Different frequencies and forms of modulation should be considered at each test frequency. For the purposes of finding the most efficient coupling, the injection probe should be moved from the nearest to the farthest position along the harness guide relative to the EUT.

4.4 Fig. 1 shows two alternatives for monitoring the detector output. The transfer impedance of the detector probe is typically a func-

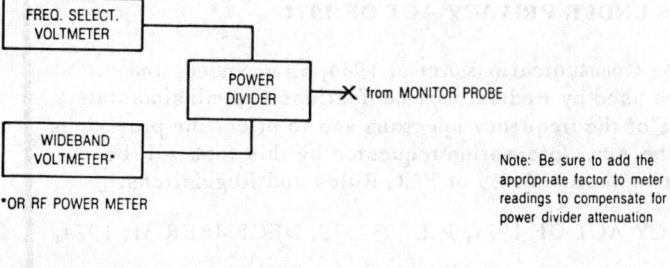

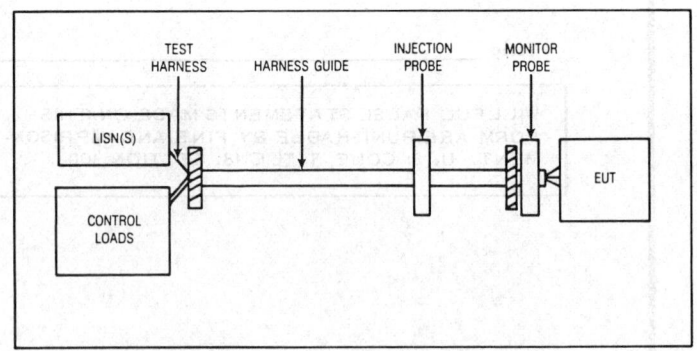

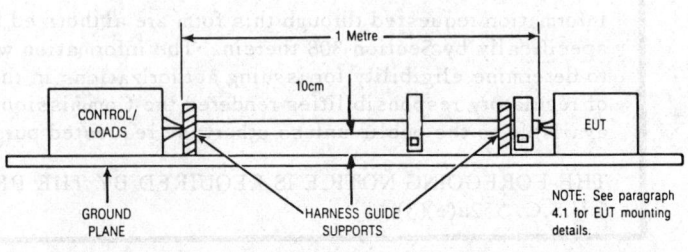

FIG. 1—EQUIPMENT CONNECTION DIAGRAM

FIG. 2—PHYSICAL LAYOUT OF TEST APPARATUS

tion of frequency and correlates the output voltage to the monitored current.

$$I_{mon} = \frac{V_{out}}{Z \text{ transfer } (f)}$$

Some manufacturers provide the probe calibration in terms of transfer admittance Y_t (or conductance G_t) which is the reciprocal of transfer impedance Z_t. Also, these data are usually shown on the calibration charts in decibels; either $dB(\Omega)$ for Z_t or $dB(S)$ for Y_t (or G_t). The above relationship between the monitored current and output voltage becomes:

$$I_{mon} [dB(\mu A)] = V_{out} [dB(\mu V)] - Z_t [dB(\Omega)]$$
or
$$I_{mon} [dB(\mu A)] = V_{out} [dB(\mu V)] + Y_t [dB(S)]$$

With a suitable conversion factor, a 50Ω rf power meter can be used in place of the wideband voltmeter for unmodulated or frequency modulated test signals. A suitable conversion factor for a 50Ω rf power meter calibrated in dBm (decibels referred to one milliwatt) is +107 dB. The equation then becomes:

$$I_{mon} [dB (\mu A)] = P_{meter} [db (m)] + 107 \text{ dB} - Z_t [dB(\Omega)]$$

Where pulse or amplitude modulation is used, a frequency selective voltmeter or spectrum analyzer should be used to measure the individual components of the current signal coupled to the harness.

5. Notes

5.1 The test should be performed in a screen room to minimize the effects of any extraneous radiation.

5.2 Due to the uncertainty of the load impendance, the rf amplifier should be designed to drive a wide range of mismatch without damage. The injection probe used should be of sufficient size to enable it to dissipate a large portion of the amplifier output power without damage. During extended test periods, the probe may become hot and care should be exercised to avoid overheating. The probe's characteristics will change if it is operated at high temperatures.

5.3 Each user should evaluate his product to determine appropriate test levels and frequencies.

This should take into account rf currents induced on the vehicle harnesses by external as well as internal sources. Transient, continuous and intermittent repetitive noise sources (that is, motors) should be considered.

5.4 Because of the possible high standing wave ratio on the coax between the amplifier and the injection probe, the length of the cable may affect the power delivered to the injection probe. Changing the cable length may overcome an apparent amplifier power deficiency.

5.5 To obtain proper output voltage, the current probe must be terminated in the manufacturer's specified impedance.

5.6 For repeatable measurements, the harness wires must be centered in both probes.

6. Open Questions—Readers of this information report are requested to submit comments regarding the following question raised during Subcommittee discussion of this information report:

6.1 Is the 1 m length of the harness sufficient to allow representative interwire coupling effects to occur?

6.2 A phenomenon has been observed during one user's experimentation that is not explained. With a given EUT being tested, the injection probe was moved along the length of the harness guide. This caused a change in the monitor probe output while holding the injection probe input constant and has been attributed to a change in coupling due to SWR on the harness. The unexplained phenomenon is that the threshold of susceptibility varies as a function of the injection probe location. For this part of the testing, the amplifier output was adjusted to maintain constant monitor probe output level. Why does this change in susceptibility threshold exist?

VEHICLE ELECTROMAGNETIC RADIATED SUSCEPTIBILITY TESTING USING A LARGE TEM CELL—SAE J1407 MAR88

SAE Information Report

Report of the Electronic Systems Committee, approved August 1982, and reaffirmed by the Electrical and Electronic Systems Technical Committee March 1988.

1. Scope—This information report gives the procedures for use and operation of a large transverse electromagnetic (TEM) mode cell for the determination of electromagnetic (EM) radiated susceptibility of equipment, subsystems and systems (whose dimensions are less than 3 m × 6 m × 18 m) in the frequency range 10 kHz–20 MHz. Several large TEM cells have been designed and constructed by various organizations for EMP and high power CW testing. Two cell designs and associated instrumentation are included for example purposes in this report. Other cell configurations have also been constructed. Users should consult the literature before undertaking a project of this magnitude for other cell and instrumentation designs.

2. Measurement Philosophy—A reasonably uniform (variations ≤ ±2 dB[1]) TEM field can be established over a large region using a TEM transmission cell at frequencies below approximately 20 MHz. The TEM field simulates a planar field in free space with the electric (E) and magnetic (H) fields orthogonal to each other and related by a wave impedance, nominally of 377 Ω or $\eta_0 \cong E/H$; typically, only the electric field component is monitored.

A typical TEM transmission cell is simply a rectangular 50-Ω transmission line constructed with transitions at each end to adapt from stripline (flat center plate) to coaxial transmission line with standard connectors[2]. Susceptibility test fields are established inside the cell by connecting an rf generator to the cell's input port and coupling rf energy through the cell to a 50-Ω termination connected to the cell's output port. Tests are performed on equipment under test (EUT) by placing the EUT inside the cell and exposing it to the test field while monitoring the functional parameters of the EUT. This technique prevents disturbance to equipment not under test since the rf test field and EUT are completely self-contained within the EM enclosure.

3. Apparatus—The test apparatus consists of the following:

(a) Signal Source—Any commercially available signal source, power amplifier, capable of supplying up to 5000 W of modulated and unmodulated power to develop the susceptibility levels specified in the test plan may be used.[3] The harmonic and spurious outputs from the rf source should be less than −20 dB referred to the fundamental. Specific tests may require additional harmonic suppression requiring the use of appropriate low pass filters as specified in (f). The frequency accuracy shall be within ±2%, if a frequency counter is not used.

(b) Rf Voltmeter—A commercially available rf voltmeter capable of measuring up to 500 V over the frequency range 10 kHz—20 MHz.

(c) Termination—One 5000-W, 50-Ω load with maximum VSWR ≤ 1.10.

(d) Frequency Counter—A frequency counter capable of measuring frequencies up to 20 MHz, or an equivalent frequency determination built into the rf signal source.

(e) TEM Transmission Cell—A TEM transmission cell similar to the examples shown in Figs. 1 (Ref. 1) or 2. The cell's outer wall can be fabricated using a lining of screen wire mesh or metal sheets welded or soldered together to form a continuous electrical conductor. The small ends of the cell may be fabricated from brass, electroformed copper, or other appropriate metal material and attached to complete the cell transmission line transitions to conventional coaxial connectors. The cell's center plate or septum may also be fabricated from screen wire mesh or metal sheets welded or soldered together into the appropriate shape. All supporting members within the cell which are not a part of the effective cell conductance are constructed from material with the lowest possible dielectric constant to minimize the effects of these members on the characteristic impedance of the cell. A fabrication procedure with detailed drawings for constructing a medium-sized symmetric cell (2.8 m × 2.8 m × 5.6 m) is available (Ref. 2) and may be useful in designing a full-sized cell.

(f) Low-Pass Filters—As required and as appropriate with cut-off frequencies, Fc, of from 15 kHz–30 MHz and with the signal down as required at frequencies greater than 2 Fc.

(g) Signal Samplers and Monitor Tees—As required by frequency and rf voltage monitoring equipment.

(h) EUT Monitors—Test equipment required to monitor the operation of the EUT.

4. Test Setup and Procedure—The TEM cell susceptibility measurement technique has been described in great detail (Ref. 3). The setup and measurement procedure is summarized below.

(a) The test setup is shown in Fig. 3A or Fig. 3B.

(b) The EUT may be placed on the bottom of the cell, centered with respect to the width and length, or it can be placed on an inclined ramp as shown in Fig. 2. If the EUT is not normally grounded, it should be insulated from the cell. The EUT would be placed on the inclined ramp if exposure (polarization matched proportional to sin θ to the vertical TEM field) of the long dimension of the EUT was required. The field strength, E_v, is determined by

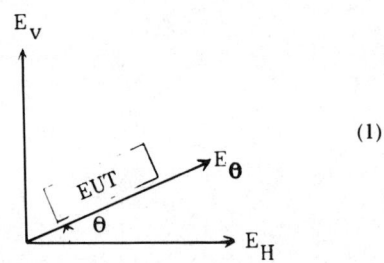

FIG. A

$$E_v = V_{rf}/b \text{ volts/meter, or} \quad E_\theta = E_v \sin\theta \quad (1)$$

where E_θ and θ are defined in the above diagram, V_{rf} is the input voltage to the cell in volts, and b is the cell bottom-to-septum separation in meters. If the test frequency is high enough (> 1 MHz) that the wavelength, λ, of the test frequency is an appreciable part of the cell length, the cell input voltage may be different than the voltage potential at the center of the cell. This would result in an error proportional to the VSWR or cell mismatch impedance. A more accurate determination of the test-field amplitude at these frequencies can be made by measuring the net input power to the cell, P_{net}, and its complex admittance. The vertical E-field is then given as

$$E_v = \frac{\sqrt{P_{net}/G_c}}{b} \quad (2)$$

where G_c is the real part of the cell's complex characteristic admittance (≃ 0.02 mhos) referenced to the center of the cell (i.e., halfway along to length of the cell transmission line). The value of E_v determined from either equation (1) or (2) is the field approximately midway between the septum and bottom of the cell. If the EUT is placed on the bottom of the cell, the test field is approximately 3% lower for the cell of Fig. 1 and 20% lower for the cell of Fig. 2 as compared to the field midway between the floor and septum as determined by expression (2). This decrease in the field amplitude is due to the E-field gradient that exists between the septum and floor.

(c) The EUT should be operated as required. Connection to monitoring instrumentation may be accomplished by using fiber optics, by high resistance (carbon impregnated teflon) leads or by leads from the

---

[1] Obtaining a uniform test field over a test volume as large as shown in the examples of Figs. 1 and 2 at high field levels is very difficult to obtain especially at frequencies below 20 MHz. Hence variations ≤ ±2 dB are considered very reasonable.

[2] Other cell configurations have been developed that eliminate the need for the second output taper.

[3] 5000 W of power applied to the inputs of the cells shown in Figs. 1 and 2 would result in test E-field strengths approximately 125 V/m and 83 V/m, respectively. Other rf power generation equipment may be substituted, as required to generate higher or lower field strengths.

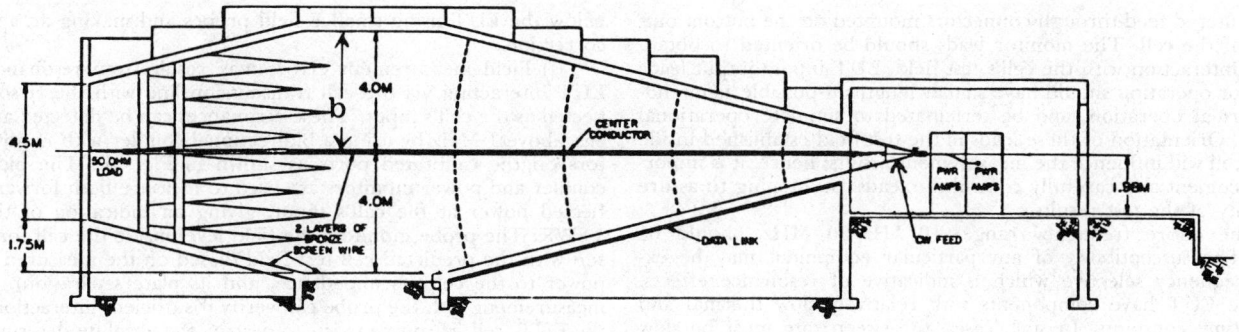

FIG. 1A—SIDE VIEW—EMES

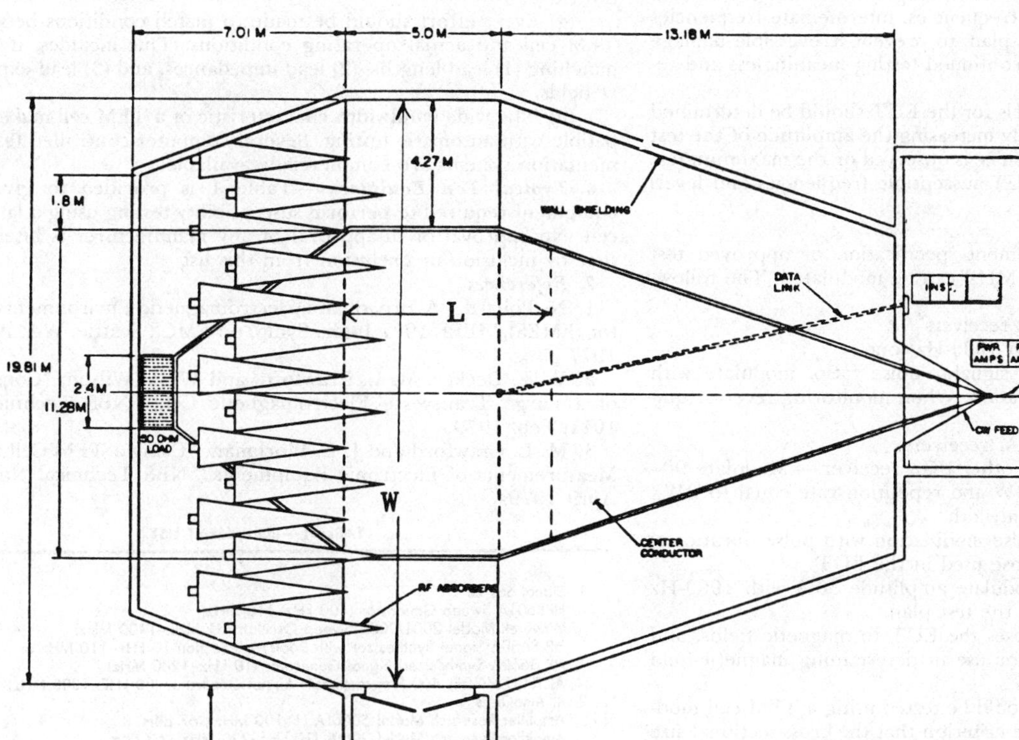

FIG. 1B—FLOOR PLAN—EMES FACILITY

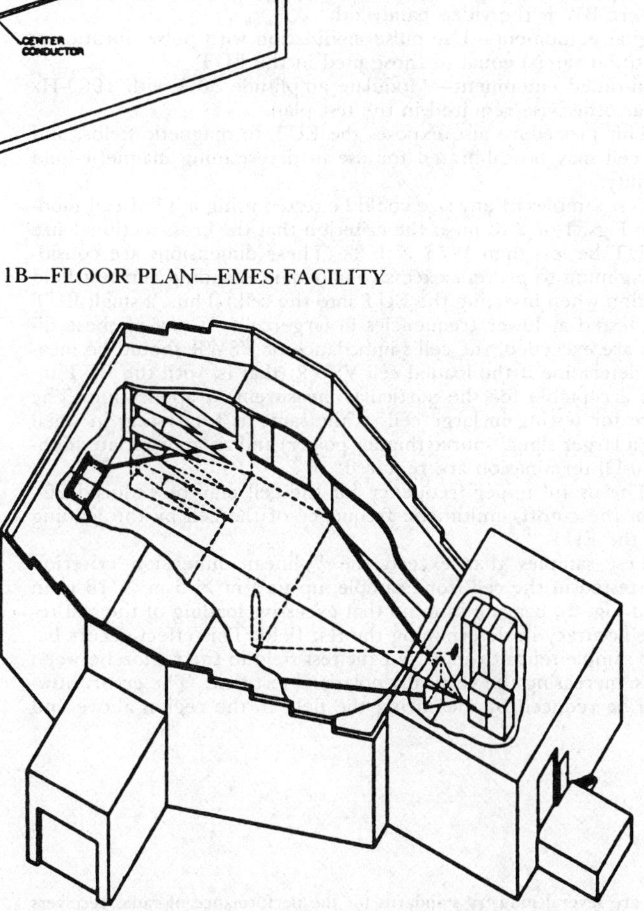

FIG. 1C—ELECTROMAGNETIC ENVIRONMENTS SIMULATOR (EMES)

FIG. 1—EXAMPLE OF END LOADED TEM CELL FOR RADIATED EM SUSCEPTIBILITY MEASUREMENTS

device to filtered feed-through connectors mounted on the bottom outer shield of the cell. The monitor leads should be oriented to obtain minimum interaction with the cell's test field. EUT input/output leads required for operation should have similar lengths if possible as intended for normal operation, and be terminated to simulate operational conditions. Orientation of these leads in the test field established inside the TEM cell will influence the measurement results; hence, it is important to document and carefully control the leads' positioning to assure repeatability of the test results.

(d) The entire frequency range, 10 kHz–20 MHz, should be scanned. The susceptibility of any particular equipment may be extremely frequency selective which is indicative of resonance effects. Also, some EUT have components with relatively slow thermal and electrical time constants. In such cases, the sweep rate must be slow enough to allow reaction/interaction of the EUT to the test field. In addition, care must be exercised when testing the EUT at critical frequencies, such as local oscillation frequencies, intermediate frequencies and others as specified in the test plan, to prevent irreversible damage to the EUT which would render continued testing meaningless and require costly repairs.

The susceptibility threshold levels for the EUT should be determined as a function of frequency by slowly increasing the amplitude of the test field until degradation in performance is observed or the maximum test level required is reached. The EUT susceptible frequencies and levels are then recorded.

5. Notes

(a) If required in the equipment specification or approved test plan, the test signals (10 kHz–20 MHz) can be modulated. The following guidelines are suggested:

(1) EUT's with audio channels/receivers[4]
AM receivers: Modulate 30% with 1000-Hz tone.
FM receivers: When monitoring signal-to-noise ratio, modulate with 1000-Hz signal using 10-kHz deviation. When monitoring receiver quieting, use no modulation.
Other Equipment: Same as for AM receivers.

(2) EUT's with video channels other than receivers—Modulate 90–100% with pulse duration of 2/BW and repetition rate equal to BW/1000, where BW is the video bandwidth.

(3) Digital equipment—Use pulse modulation with pulse duration(s) and repetition rate(s) equal to those used in the EUT.

(4) Non-tuned equipment—Modulate amplitude 30% with 1000-Hz tone, or as otherwise required in the test plan.

(b) This procedure also exposes the EUT to magnetic fields, and thus the cell may be calibrated for use in determining magnetic-field susceptibility.

(c) Test samples of any size could be tested using a TEM cell modeled from Figs. 1 or 2 to meet the criterion that the cross-sectional size of the EUT be less than $W/3 \times b/3$. (These dimensions are considered a maximum to prevent excessive impedance loading and test-field perturbation when inserting the EUT into the cell.) Thus, a small EUT could be tested at lower frequencies in large cells. If any of these dimensions are exceeded, the cell's impedance or VSWR should be measured to determine if the loaded cell VSWR (that is, with the EUT inserted) is acceptable for the particular measurement application. The procedure for testing in large cells is the same as for testing in small cells, but a larger signal source (higher power) and an appropriate high-power (50-Ω) termination are required.

(d) The useful upper frequency for the cell may be reduced 20–30% from the cutoff/multimode frequency of the cell by the loading effect of the EUT.

(e) Test samples that exceed the ⅓-linear dimension criterion could be tested in the cell (for example, up to 3 m × 6 m × 18 m in the cell of Fig. 2), bearing in mind that excessive loading of the cell reduces the accuracy in determining the test field. This effect occurs because the sample tends to short out the test field in the region between the plates, increasing the vertically polarized test field. The error, however, can be reduced by measuring the field in the region above and below the EUT using small E-field probes and making an appropriate correction.

(f) Field measurement errors may result from resonances in the EUT interacting via the cell transmission line with the rf source connected at the cell's input. These resonances can be detected at frequencies above 1 MHz by using a bidirectional coupler with rf power monitors on the calibrated ports as shown in Fig. 3B. The bidirectional coupler and power monitors are used to measure both forward and reflected power at the cell's input, giving an indication of the system VSWR. The probe monitors the field level inside the cell for comparison with the predicted cell test field (based on the measured input net power to the cell, its impedance, and its plate separation). The field measurement with the probe can verify if sufficient interaction between the EUT/cell/rf source exists to destroy the absolute determination of the test-level amplitude inside the cell that stimulates the resonances in the EUT.

(g) Every effort should be made to match conditions between the TEM cell and actual operating conditions. This includes, if possible, matching (1) lead lengths, (2) lead impedances, and (3) lead exposure to rf fields.

(h) The wide bandwidth characteristic of a TEM cell makes it compatible with automatic testing. Several computer-controlled test instrumentation systems are commercially available.

6. Typical Test Equipment—Table 1 is provided to give typical equipment required to perform susceptibility testing using a large TEM cell. No approval or disapproval of any manufacturer is intended, either by inclusion or exclusion from this list.

7. References

1. N. Pollard, "A Broadband Electromagnetic Environments Simulator (EMES)," IEEE 1977 Inter. Symp. on EMC., Seattle, WA, Aug. 2–4, 1977.

2. W. F. Decker, M. L. Crawford, and W. A. Wilson, "Construction of a Large Transverse Electromagnetic Cell," NBS Technical Note 1011, Feb. 1979.

3. M. L. Crawford and J. L. Workman, "Using a TEM Cell for EMC Measurements of Electronic Equipments," NBS Technical Note 1013, April 1979.

TABLE 1—EQUIPMENT LIST

1. Signal Source
   HP8601A Sweep Generator (100 kHz–110 MHz)
   Wavetek Model 2001/2002 Sweep Oscillator (1 MHz–1400 MHz)
   HP 8660A Signal Synthesizer with 86601A F Section 10 kHz–110 MHz or
   HP 8662A Synthesized Signal Generator (10 kHz–1280 MHz)
   AILTECH MODEL 460 Programmable Signal Generator (30 kHZ–1300 MHz)
2. Rf Amplifiers
   Amplifier Research Model 5000LA (1–100 MHz), 67 dBm
   Amplifier Research Model 2000L (10 kHz–220 MHz), 63 dBm
   Instruments for Industry Model M466 (100 kHz–100 MHz), 66 dBm
3. Frequency Counter
   HP 5305A (DC–1100 MHz)
   Fluke Model 1910/1911A/1912A (DC–125 MHz/250 MHz/520 MHz)
4. Power Meters
   HP 436A with HP8482B Sensors (100 kHz–4.28 GHz)
5. Low Pass Filters—cut-off frequencies specified as required
   BIRD Model 5323
   Lark Engineering
   MDI Model FCL A35
6. Directional Couplers
   BIRD Model 4310 or 4311 with appropriate elements
   BIRD Model 4712 with appropriate element (2 MHz–30 MHz)
   MDI Model 568A8 4 kW or Model 568A9, 12 kW
7. Coaxial Termination (high power)
   BIRD Model 8720 (dc–4 GHz), 4 kW
   BIRD Model 8736 (dc–1400 MHz), 10 kW
   MDI 638A (dc–1 GHz), 6 kW
8. XY Recorder
   HP 7046A Dual Channel
9. Test Monitor
   Instrument for Industries EFS-1—E-field monitor, 10 kHz–200 MHz
   Radiation Devices Fiber Optic Model FAT-6PB (Fiberoptic analog transmitter and receiver system with optical fiber for telemetry of data).
   –3 dB bandwidth = 15 Hz–30 MHz.
10. RF Voltmeter
    HP400GL (20 Hz–4 MHz)
    HP410C (20 Hz–700 MHz), (0.5 V–300 V), 10 MΩ with HP 455A Probe T connector
11. RF Signal Samples
    BIRD Model 4273, 5 kW (1.5 MHz–35 MHz)

---

[4] There are several industry standards for the performance of radio receivers that may be helpful for determining the modulation to be used on the susceptibility test signal. Two of them are IEEE 185 and EIA RS-204-A, others may be more appropriate for certain equipment.

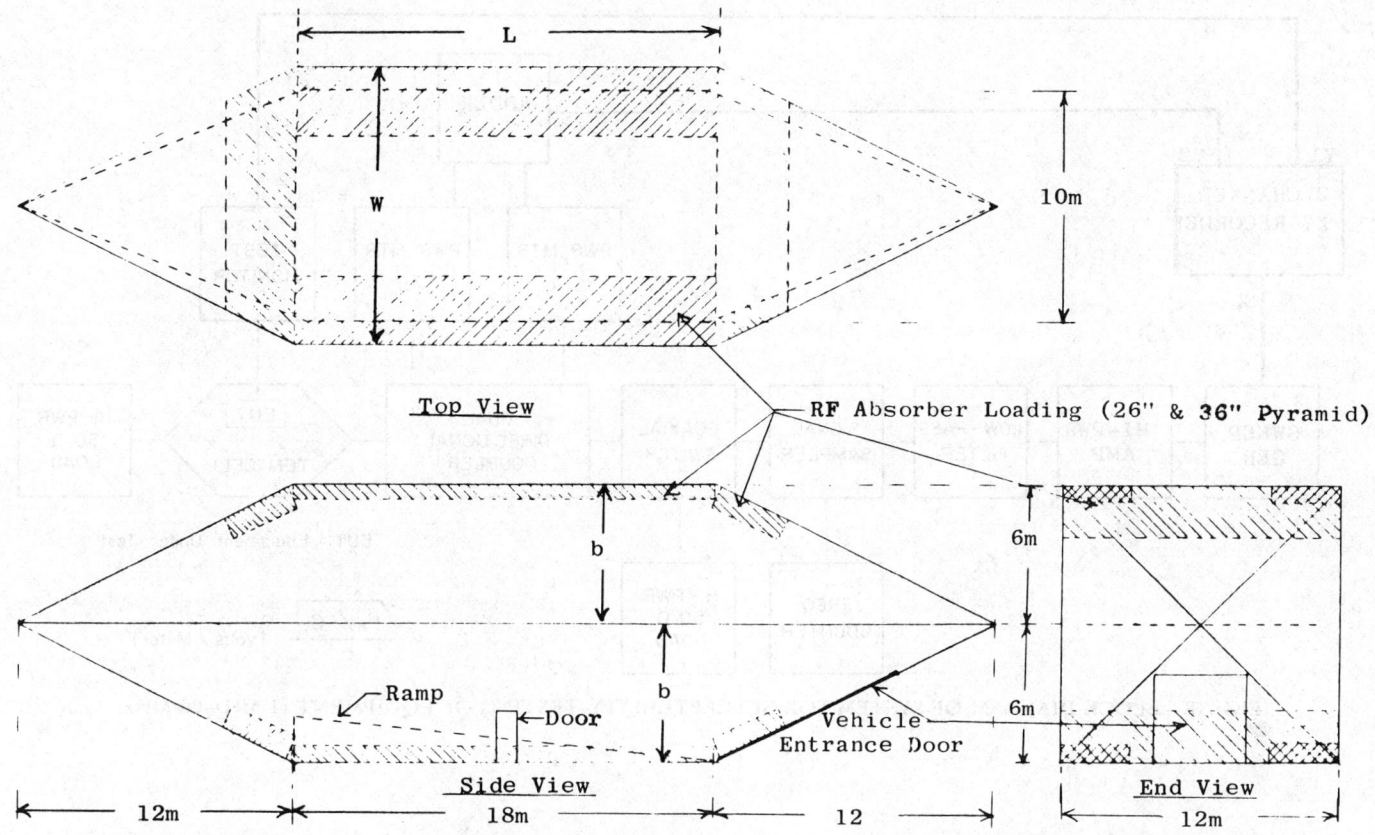

FIG. 2—EXAMPLE OF DISTRIBUTED LOADED TEM CELL FOR RADIATED EM SUSCEPTIBILITY MEASUREMENTS

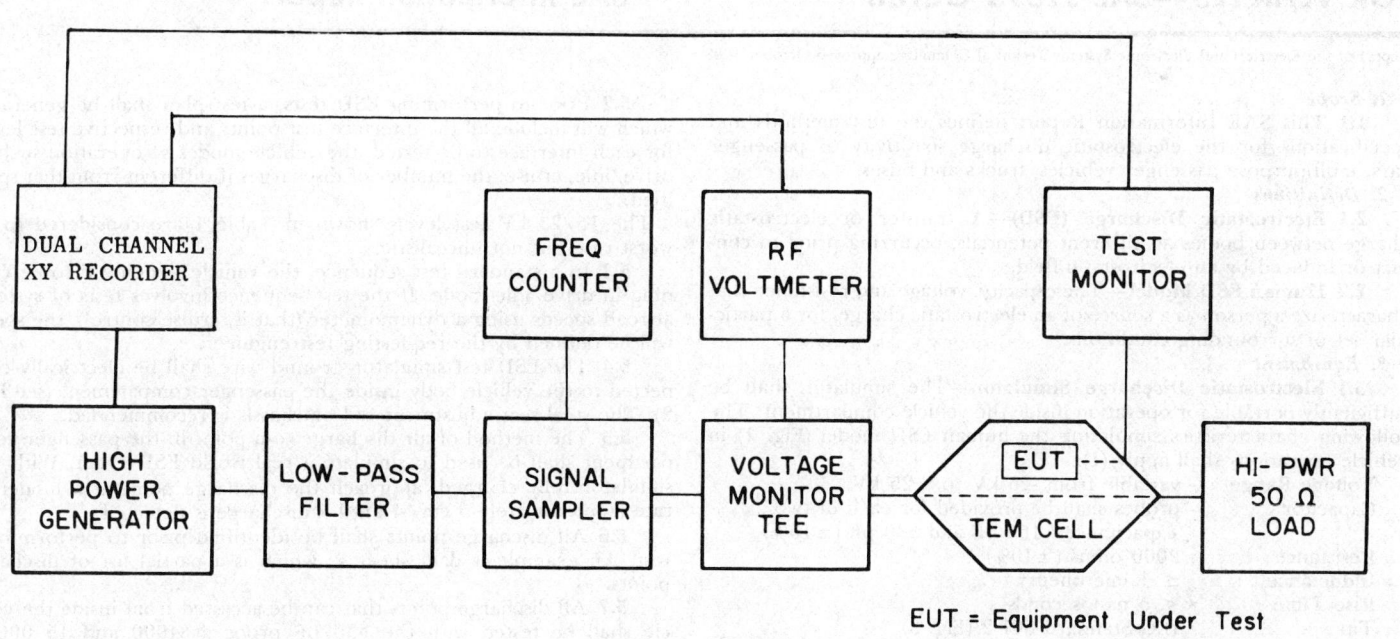

EUT = Equipment Under Test

$$E_v = \frac{V_{rf}}{b} \text{ (Volts / Meter)}$$

FIG. 3A—BLOCK DIAGRAM OF SYSTEM FOR SUSCEPTIBILITY TESTING OF EQUIPMENT (USED TYPICALLY BELOW 1 MHz)

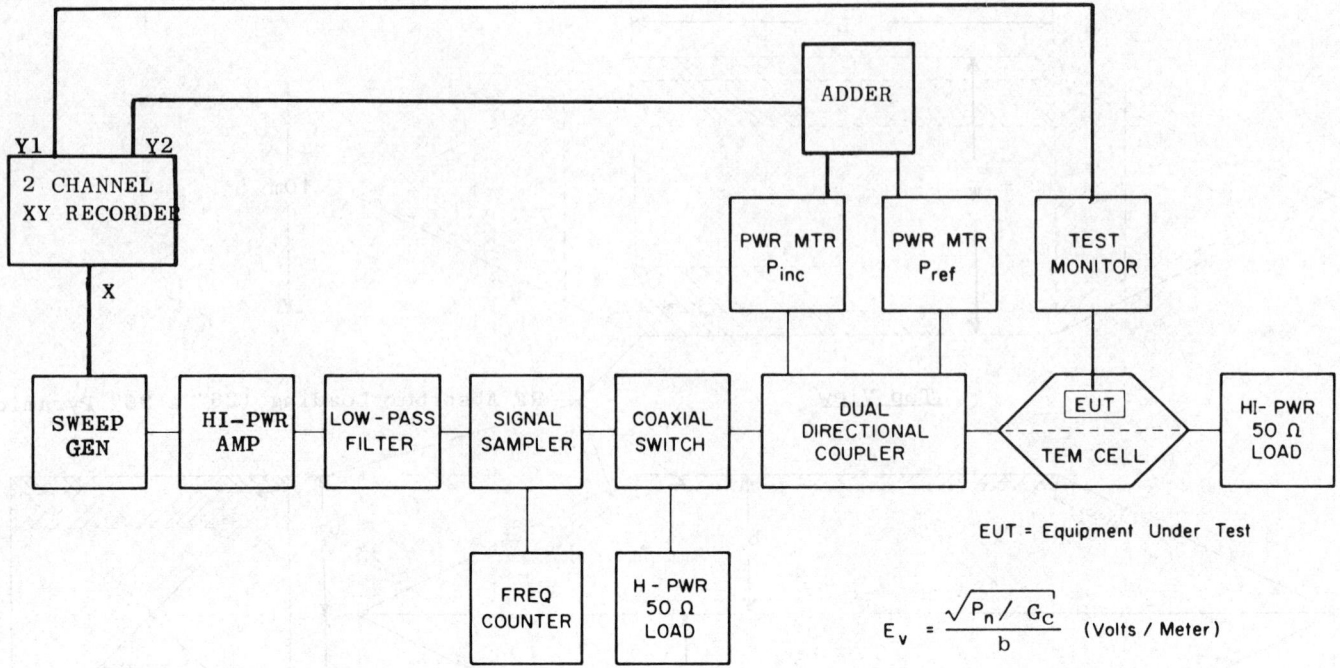

FIG. 3B—BLOCK DIAGRAM OF SYSTEM FOR SUSCEPTIBILITY TESTING OF EQUIPMENT (1 MHz–20 MHz)

$$E_v = \frac{\sqrt{P_n / G_c}}{b} \text{ (Volts / Meter)}$$

EUT = Equipment Under Test

ELECTROSTATIC DISCHARGE TEST FOR VEHICLES—SAE J1595 OCT88

SAE Information Report

Report of the Electrical and Electronic Systems Technical Committee approved October 1988.

1. Scope

1.1 This SAE Information Report defines the test methods and specifications for the electrostatic discharge sensitivity of passenger cars, multipurpose passenger vehicles, trucks and buses.

2. Definitions

2.1 Electrostatic Discharge (ESD)—A transfer of electrostatic charge between bodies at different potentials, occurring prior to contact or induced by an electrostatic field.

2.2 Human ESD Model—The capacity, voltage and resistance that characterize a person as a source of an electrostatic charge, for a particular set of surrounding conditions.

3. Equipment

3.1 Electrostatic Discharge Simulator—The simulator shall be sufficiently portable for operation inside the vehicle compartment. The following characteristics simulating the human ESD model (Fig. 1) in vehicle operations shall apply (1):

Voltage Range	- variable from -25 kV to +25 kV.
Capacitor	- probes shall be provided for each of two capacitances: 150 pF and 330 pF (±10%)
Resistance	- 2000 ohms (±10%)
Inductance	- ≤ 1 microhenry
Rise Time	- < 5 nanoseconds
Tip	- IEC Standard 801-2 (Fig. 3)

The simulator shall be designed so that the discharge capacitance is fully charged to the desired voltage before the energy can be switched to the device under test.

Simulator equipment is commercially available.

4. Test Vehicle

4.1 Record the pertinent test vehicle information or any specific test conditions on a data sheet such as the example data sheet 1.

5. Test Procedure

5.1 Before the application of any discharges to the vehicle, the ESD Simulator Discharge Verification procedure of section 6 shall be performed.

5.2 Prior to performing ESD tests, a test plan shall be generated which will include all the interface test points and respective test levels for each interface to be tested; the vehicle modes of operation such as drive/idle, cruise; the number of discharges if different from that specified.

The 15/25 kV test levels shown in Table 1 are considered to be worst case but not unrealistic.

5.3 In a standard test sequence, the vehicle's engine is to be running in drive/idle mode. If the test sequence involves tests of systems at road speeds using a dynamometer (that is, cruise control), the speed will be defined by the requesting test engineer.

5.4 The ESD test simulator ground wire shall be electrically connected to the vehicle body inside the passenger compartment (see Fig. 2). The steel seat adjustment rail or chassis is recommended.

5.5 The method of air discharge to a point in the passenger compartment shall be used to simulate a real world ESD event. With the simulator fully charged, approach the discharge point at a moderate rate (approximately 5 cm/s) until a discharge is detected.

5.6 All discharge points shall be identified prior to performing a test. An example is data sheet 2, which is a partial list of discharge points.

5.7 All discharge points that can be accessed from inside the vehicle shall be tested with the 330 pF probe at 4000 and 15 000 V through 2000 ohms. Only those discharge points that can be conveniently accessed when standing outside the vehicle (this, in effect, lowers a person's capacitance) and reaching inside, such as the headlight switch or the ignition switch, shall be tested with the 150 pF probe at 25 000 V through 2000 ohms. This distinction shall be at the discretion of the test engineer. Any other discharge point not listed on data sheet 2, but deemed pertinent to this test, shall be included. These points shall, at a minimum, include all electrical switches and controls that can be interfaced by an occupant within the passenger compartment. Also, any knobs, levers or handles which would be actuated in the normal operation of the vehicle should be included.

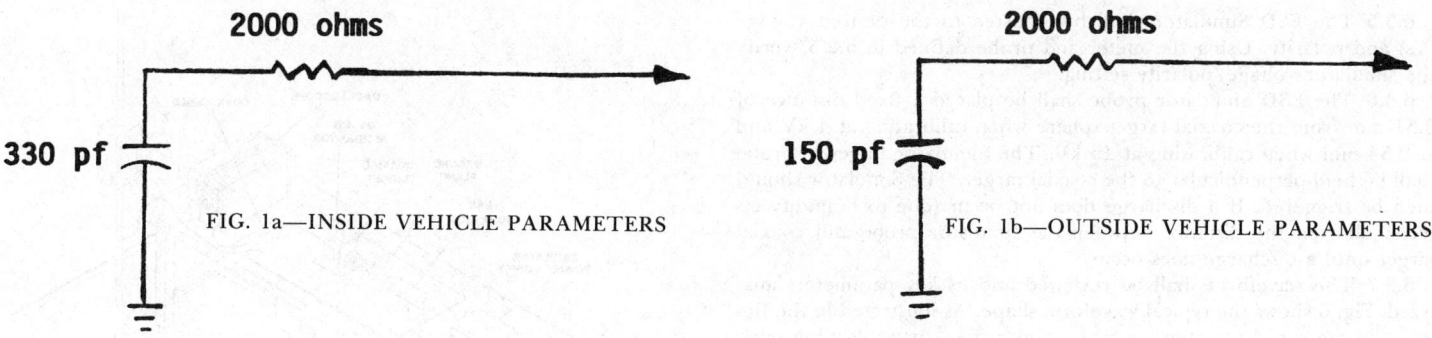

FIG. 1—HUMAN ESD MODELS FOR VEHICLE TESTING

TABLE 1—VEHICLE ESD TEST PARAMETERS

	METHOD	TEST LEVELS	NO. OF[1] PULSES	TEST PERIOD INTERVALS
Discharge points in vehicle accessed from inside or outside of vehicle	Fig. 1a	±4 kV ±15 kV	3 minimum 3 minimum	5 s minimum 5 s minimum
Discharge points in vehicle accessed only from outside of vehicle	Fig. 1b	±25 kV	3 minimum	5 s minimum

[1]Each discharge point shall be subjected to a minimum of three positive and three negative polarity discharges at each voltage level.

5.8 Each discharge point shall be subjected to a minimum of three positive polarity and three negative polarity discharges at each voltage level. The time duration between discharges shall be a minimum of five seconds. Table 1 gives a summary of the test parameters.

5.9 Visual, audible and instrumented deviations from normal vehicle operation will be entered into the data log. Various operating systems such as heater controls, air conditioner controls, radio controls, and digital displays will be exercised periodically during the test to demonstrate normal response.

6. ESD Simulator Discharge Verification

6.1 This section defines a test method for verifying the operation of the ESD Simulator.

6.2 Equipment for Verification

6.2.1 GROUND PLANE—The ground plane shall be a conductive metallic sheet (copper or aluminum) 0.25 mm thick, with an area of at least one square meter. The ground plane shall be connected to the facility earth ground by a braided strap which should be less than 1 m long and at least 1 cm wide.

6.2.2 COAXIAL TARGET—A 50 ohm coaxial target is defined by the International Electrotechnical Commission (IEC) Standard 801-2 (see Fig. 4). The target shall be bonded to the center of the ground plane defined in 6.2.1.

6.2.3 20 dB ATTENUATOR—A 50 ohm, 20 dB, wideband attenuator is to be attached to the output of the coaxial target.

6.2.4 MEASUREMENT INSTRUMENTATION—A measurement device with a minimum bandwidth of 250 MHz (1 GHz bandwidth preferred) and a 50 ohm input impedance shall be used. (An analog oscilloscope is preferred. If, however, a sampling oscilloscope is used, an instrument having a digitizing rate of 1 Gigasamples per second may be required to provide a reliable 250 MHz single-shot bandwidth.) The measurement instrument may have to be shielded against electromagnetic interference.

6.2.5 The ESD Simulator shall be verified using a meter (or oscilloscope). The probe used must have a high impedance (100 Megohm minimum).

6.3 Verification Procedure

6.3.1 The test setup shall be configured in accordance with Fig. 5.
6.3.2 The coaxial target shall be located on and bonded to the center of the ground plane.

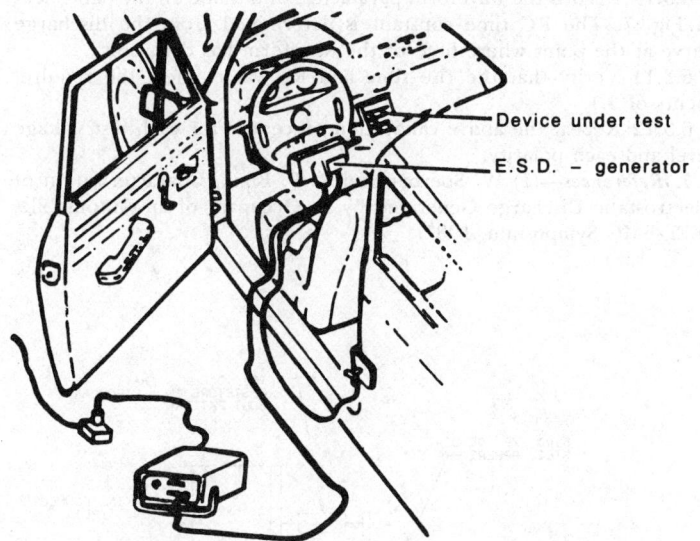

FIG. 2—VEHICLE ESD TEST SETUP

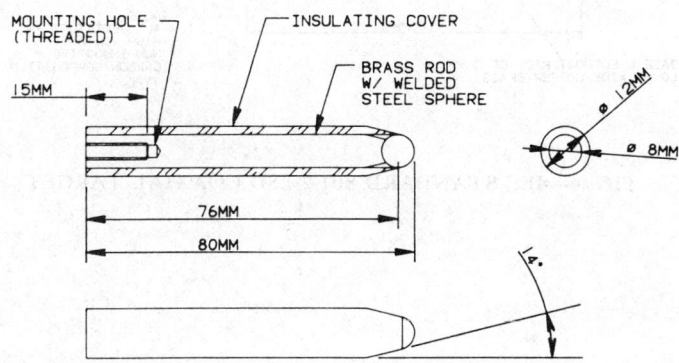

FIG. 3—ESD SIMULATOR FINGER TIP PROBE

6.3.3 The horizontal time base and vertical amplifier level of the measurement instrument shall be set as necessary to enable viewing of the complete ESD waveform. The horizontal sweep shall be set to single event trigger. (If a Gigasamples per second digitizing oscilloscope is used and the bandwidth of the vertical amplifier is greater than 250 MHz, use an anti-aliasing filter.)

6.3.4 The ESD Simulator high voltage ground shall be directly connected to the ground plane defined in 6.2.1. The ESD Simulator shall be set up and turned ON per its instruction manual.

6.3.5 The ESD Simulator shall be adjusted to the desired voltage (Vs) and polarity. Using the meter and probe defined in 6.2.5, verify the simulator voltage/polarity setting.

6.3.6 The ESD Simulator probe shall be placed a fixed distance of 0.51 mm from the coaxial target sphere when calibrating at 4 kV and at 2.54 mm when calibrating at 15 kV. The Simulator fingertip probe shall be held perpendicular to the coaxial target. The Simulator should then be triggered. If a discharge does not occur (due to humidity effects), shorten the distance between the simulator probe and coaxial target until a discharge does occur.

6.3.7 The waveform shall be reviewed and its key parameters analyzed. Fig. 6 shows the typical waveform shape. As illustrated in the figure, the measured waveform may contain a high speed, leading edge transient.

6.3.8 Only single event discharge waveforms are acceptable.

6.3.9 Acceptable ESD waveforms shall be repeatable 6 times minimum in 10 attempts.

6.3.10 Record the waveform parameters and shape on the data sheet in Fig. 7. The RC time constant is determined from the discharge curve at the point where 63% of the waveform has decayed.

6.3.11 Verify that the rise time and RC values meet the requirements of 3.1.

6.3.12 Repeat the above calibration procedure for each test voltage level and each polarity.

7. References—(1) W. Sperber and R. P. Blink, Characterization of Electrostatic Discharge Generated by an Occupant of an Automobile, IEEE-EMC Symposium, 1987.

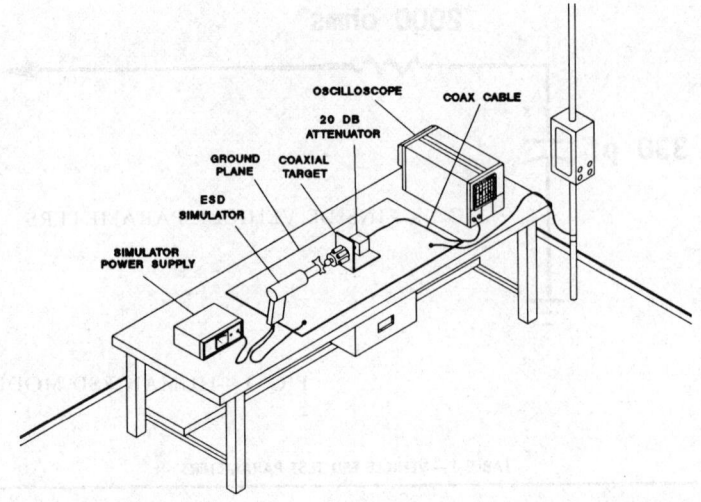

FIG. 5—ELECTROSTATIC DISCHARGE SIMULATOR VERIFICATION TEST CONFIGURATION

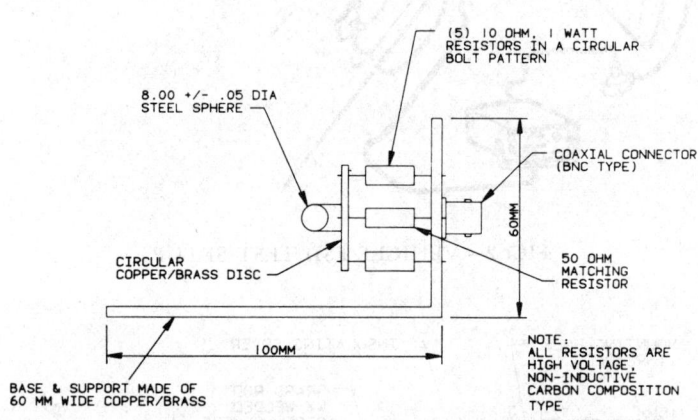

FIG. 4—IEC STANDARD 801-2 ESD COAXIAL TARGET

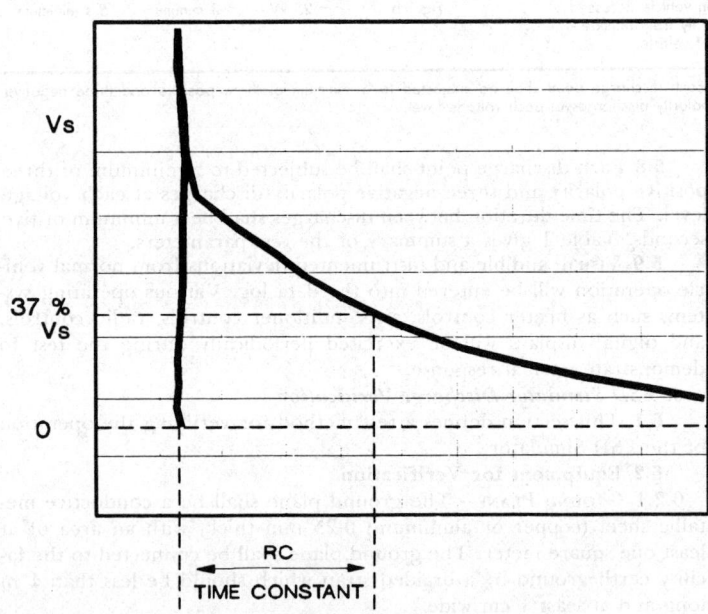

FIG. 6—DISCHARGE VERIFICATION WAVEFORM USING A COAXIAL TARGET

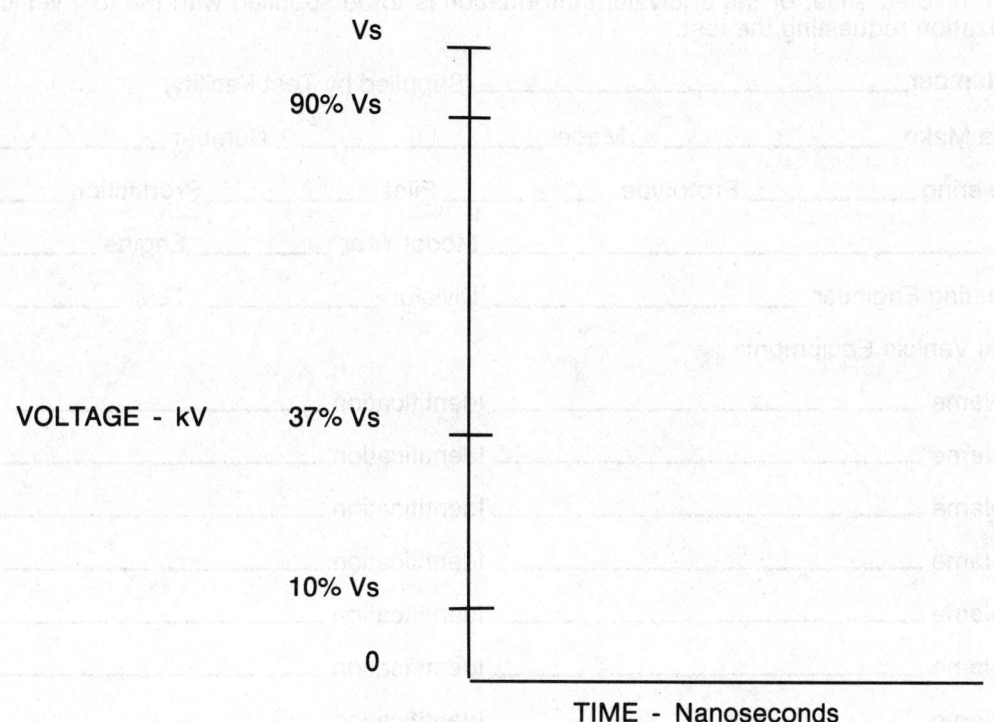

FIG. 7—ESD SIMULATOR WAVESHAPE RECORD

DATA SHEET 1 - Electrostatic Discharge Test for Vehicles

This completed sheet or the equivalent information is to be supplied with the test vehicle by the organization requesting the test.

Test Number_____ (Supplied by Test Facility)

Vehicle Make_____ Model _____ Number _____

Engineering _____ Prototype _____ Pilot _____ Production _____

VIN _____ Model Year _____ Engine _____

Requesting Engineer _____ Division _____ Tele. _____

Special Vehicle Equipment:

    Name _____ Identification _____

    Name _____ Identification _____

    Name _____ Identification _____

    Name _____ Identification _____

    Name _____ Identification _____

    Name _____ Identification _____

    Name _____ Identification _____

Special Test Conditions or Variations

_____

_____

_____

_____

23.469

DATA SHEET 2 - Electrostatic Discharge Test for Vehicles

The following information shall become part of the data log. Additional discharge points shall be added as directed by the requesting engineer. The 25kV test should include only those controls that can be conveniently accessed when standing outside the vehicle.

Electrostatic Discharge Test.   Date Tested _____
Performed at _____
Test Number _____ Vehicle Number _____
Temperature _____ Humidity _____ (at test start)

Functions or Accessories Monitored:
_____
_____
_____
_____

Test Equipment:

Name	Manufacturer	Model
_____	_____	_____
_____	_____	_____

Test Data:

Discharge Point	Voltage	Vehicle Performance Deviation
POWER DOOR LOCK SWITCH	4 kV	
	15 kV	
	25 kV	
POWER WINDOW SWITCH	4 kV	
	15 kV	
	25 kV	
POWER SEAT CONTROL	4 kV	
	15 kV	
	25 kV	
TILT WHEEL LEVER	4 kV	
	15 kV	
	25 kV	
MULTI-FUNCTION LEVER	4 kV	
	15 kV	
	25 kV	
GEAR SHIFT LEVER	4 kV	
	15 kV	
	25 kV	
STEERING WHEEL	4 kV	
	15 kV	
	25 kV	
IGNITION SWITCH	4 kV	
	15 kV	
	25 kV	
HEADLIGHT SWITCH	4 kV	
	15 kV	
	25 kV	
HOOD RELEASE LEVER	4 kV	
	15 kV	
	25 kV	
BRAKE RELEASE LEVER	4 kV	
	15 kV	
	25 kV	

DATA SHEET 2 (Continued)

Discharge Point	Voltage	Vehicle Performance Deviation
DOOR HANDLE	4 kV	
	15 kV	
	25 kV	
WINDOW CRANK	4 kV	
	15 kV	
	25 kV	
DRIVER SEAT BELT	4 kV	
	15 kV	
	25 kV	
RADIO/TAPE PLAYER CONTROLS	4 kV	
	15 kV	
	25 kV	
REAR DEFOGGER SWITCH	4 kV	
	15 kV	
	25 kV	
A/C CONTROLS	4 kV	
	15 kV	
	25 kV	
CIGARETTE LIGHTER	4 kV	
	15 kV	
	25 kV	
SIDE MIRROR CONTROLS	4 kV	
	15 kV	
	25 kV	
TRIP ODOMETER RESET	4 kV	
	15 kV	
	25 kV	
HAZARD FLASHER SWITCH	4 kV	
	15 kV	
	25 kV	
AUTOMATIC REAR VIEW MIRROR CONTROL	4 kV	
	15 kV	
	25 kV	

Comments and Unusual Events During Test:

_____

_____

_____

NOTE: The phrase 'No Discharge' shall be entered under deviations where this situation occurs. 'No Deviation' shall be entered where a discharge is evident and no deviation observed.

ELECTROMAGNETIC SUSCEPTIBILITY MEASUREMENTS OF VEHICLE COMPONENTS USING TEM CELLS (14 kHz–200 MHz)—SAE J1448 JAN84

SAE Information Report

Report of the Electrical and Electronic Systems Technical Committee, approved January 1984.

1. Purpose—This report covers requirements for the determination of electric-field susceptibility of equipment, subsystems, and systems (whose largest dimension is less than 15 cm) in the frequency range 14 kHz–200 MHz, using a TEM cell [1]. This report outlines a systematic approach, using a TEM cell, for evaluating the electromagnetic (EM) radiated susceptibility of electronic equipment. The purpose of the report is to provide guidelines, for those using TEM cells for performing EM susceptibility measurements, to improve the repeatability and, hence, the value of their test results. The report describes the test setup, details the step-by-step procedures to use in performing susceptibility measurements, and discusses pertinent information related to the range of application and limitations associated with the use of TEM cells.

2. Measurement Philosophy—A transverse electromagnetic (TEM) transmission cell is a rectangular adaptation of a coaxial line which sets up a region of uniform electric and magnetic fields in a traveling wave of essentially freespace impedance. The equipment under test (EUT) is exposed to this electromagnetic source, but only the electric-field component is monitored. This technique also prevents disturbance to equipment not under test since the rf field source and EUT are completely self-contained within the electromagnetic enclosure. This technique is intended primarily for use in diagnostic testing to determine, for example, frequencies of EUT susceptibility, some indication of how electromagnetic interference (EMI) is coupled into the EUT, and the relative improvement in immunity that may result from efforts to reduce an EUT's susceptibility (that is, A-B measurements). It is not intended for use in determining EUT susceptibility to absolute test field levels if, for example, the EUT includes long wiring harnesses that must be exposed polarization matched to the test field.

3. Apparatus—The test apparatus shall consist of the following:

(a) *Signal Source*—Any commercially available signal source, power amplifier, and general-purpose amplifier capable of supplying at least 100 W of modulated and unmodulated power to develop the susceptibility levels specified in the test plan shall be used, provided the following requirements are met: Frequency accuracy shall be within ±2% and harmonics and spurious outputs shall not be more than −30 dB referred to the fundamental power.

(b) *rf Voltmeter*—A commercially available rf voltmeter capable of measuring 100 V over the frequency range 14 kHz–200 MHz.

(c) *Termination*—One 100 W, 50 Ω load.

(d) *Frequency Counter*—A frequency counter capable of measuring frequencies up to 200 MHz.

(e) *TEM Transmission Cell*—A transverse electromagnetic transmission cell is shown in Figs. 1 and 2. Typical dimensions for cells are given in Table 1. The dimension for a cell suggested for use in the frequency range 14 kHz–200 MHz is underlined on the table.

(f) *Low-Pass Filter*—Cutoff at 200 MHz, with the signal down 60 dB at frequencies greater than 300 MHz.

(g) *Signal Samplers or Monitor Tees*—Frequency and rf-voltage monitoring equipment.

(h) *Monitors*—Required test equipment to monitor the operation of the test sample.

(i) *Dual Directional Coupler(s)*—−30 dB or greater coupling ratio, 10–200 MHz.

(j) *rf Power Meters with Sensors*—Capable of measuring rf power levels up to 100 mW at frequencies of 10–200 MHz.

(k) *Dual Channel xy Recorder.*

4. Test Setup—Figs. 3A and 3B show two block diagrams of systems using the TEM cell for susceptibility measurements. Fig. 3A is used for frequencies below 10 MHz and Fig. 3B is used for frequencies above 10 MHz. Both systems are configured for swept frequency testing; however, either system could be used for discrete frequency testing without the use of the xy recorder or a sweep-type generator.

The high-power generator generally consists of a variable frequency generator and high-power linear amplifier. Most of today's linear amplifiers have built-in protection circuits which protect them in the event the cell system becomes impedance mismatched. The low-pass filter is needed to keep unwanted higher frequency components (second, third, etc., harmonics) from being introduced into the cell. This is especially important if the harmonics or spurious frequencies are above the multimode or resonant frequency limit of the cell. A frequency counter is used to tell precisely at what frequency a failure occurs. This can be important if the signal generator frequency calibration is not sufficiently accurate since EUT susceptibility often is characterized by high Q-resonances which have very narrow frequency responses. The dual coupler and power meters are used to measure the incident and reflected power at the input of the cell. A test monitor is shown with the EUT and is used to determine failure of the EUT. A 50 Ω, high-power load is used to terminate the system.

The main difference between Figs. 3A and 3B is the use of an rf voltmeter with monitor tee in place of the directional coupler and power meters. At frequencies below 10 MHz, the cell is electrically short (that is, much less than one wavelength), and accurate voltmeter measurements, which can be referred to the EUT location at the center of the cell, can be made.

Automated, interactive susceptibility measurements can be made by using a minicomputer controlled system as shown in Fig. 4. This system places the testing operation under control of the computer which progressively increases the test-field level in the cell at selected frequencies while monitoring the test-field level around the EUT and the performance of the EUT. If degradation occurs as determined from preestablished, programmed criteria, the computer can respond by limiting the test-field level to prevent damage to the EUT. The computer will also print out the susceptibility information according to software instruction and format.

5. Measurement Procedures—The following measurement procedures are suggested as a systematic approach for evaluating the EM radiated susceptibility of equipment using a TEM cell.

Step 1. Place the EUT inside the cell. Normally, the EUT is placed in one of two locations in the lower half space (below the septum) centered in the cell. The first location (location A) for placing the EUT is on the floor of the cell (Fig. 5), but insulated from it (unless grounding the EUT case to the cell is desired). This position is used to evaluate field-to-EUT case coupling by minimizing the exposure of the EUT's input/output leads and monitor leads to the test field as explained in Step 2. If the EUT is placed near the floor of the cell, the test field will be lower, relative to the field at the center (midway between the septum and floor) of the cell test zone—from 5 to 15% depending upon the cell form factor (width to height ratio). (That is, if a/b is 1.0, this correction factor is 0.85; if a/b is 1.5, this correction factor is 0.92; and if a/b is 1.67, the correction factor is 0.95).

The second common location (location B, Fig. 6) is halfway between the septum and cell bottom. The EUT is supported on dielectric material with as low a dielectric constant as possible. For example, plastic foams with a dielectric constant of approximately 1.04 to 1.08 are almost invisible electrically and make good support material. This position increases the exposure of the EUT's leads, thus giving some indication, when comparing measurement results obtained with the EUT in the two positions, of how energy is coupled into the EUT (that is, some indication of field-to-EUT leads coupling). After placing the EUT in either location A or B, orient the EUT as desired relative to the cell's TEM field. Normally, the first orientation position is with the EUT in its flat, upright, normal operating position.

Step 2. Access the EUT as required for operation and performance monitoring. The EUT input/output leads should be as nearly the same as in anticipated use. Leads should be the same length, if possible, and be terminated into their equivalent operational impedances so as to simulate the EUT in its operation configuration. Care must be taken in routing the leads including monitor leads (if nontransparent to the rf field) inside the cell for the most meaningful, repeatable results (that is, desired lead exposure, minimum field perturbation/interaction, etc.). To minimize exposure of the leads to the test field, the EUT should be lowered as close to the cell floor as possible and its leads should be routed along the floor of the cell in appropriate holders,[1] to the bulkhead panel and ac receptacle, as shown in Fig. 6. Shield the leads by either taping the leads to the cell floor with 5 cm wide conductive tape or by using braided wire slipped over the lead. Note, keep the leads separate; do not bundle

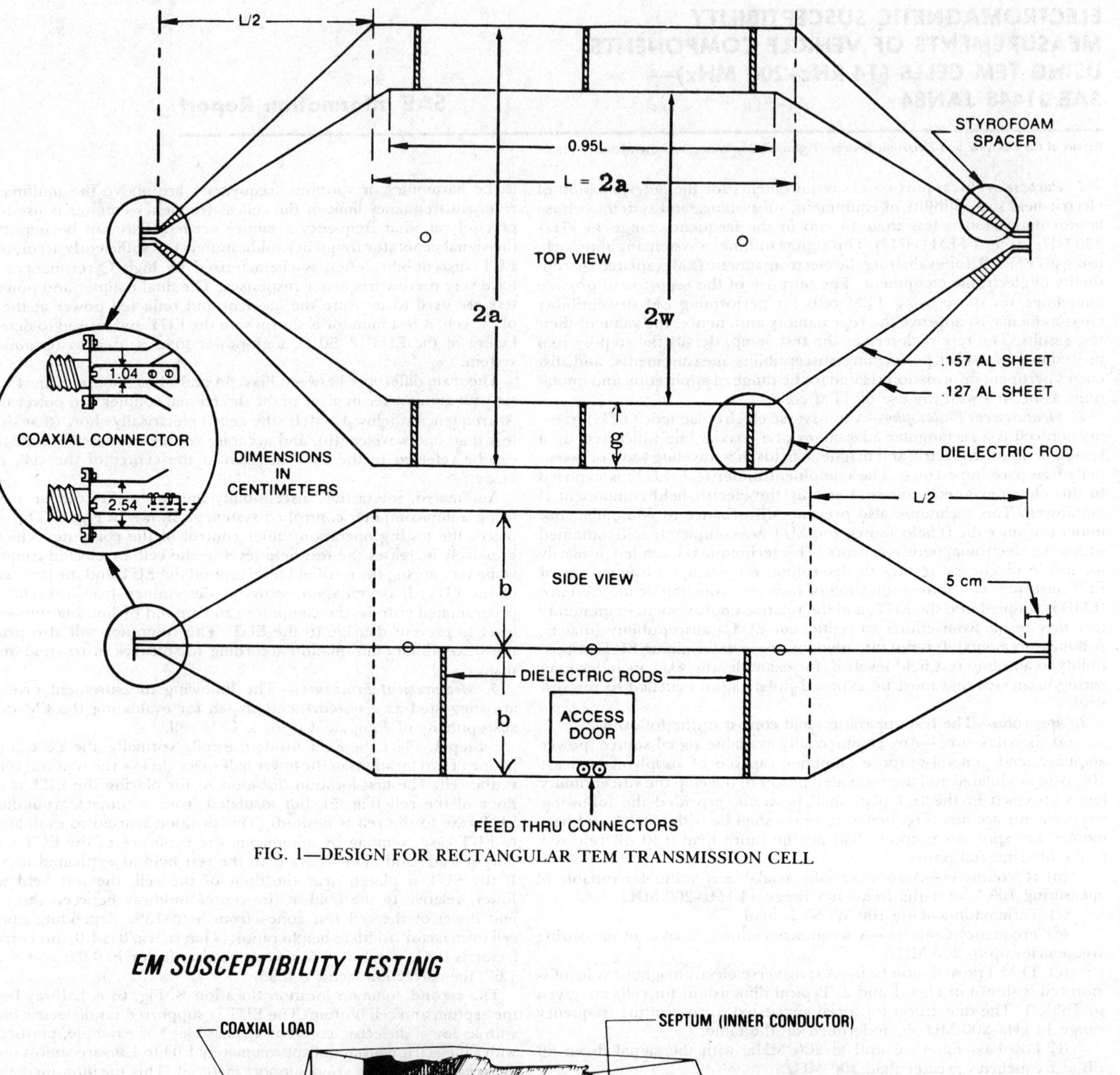

23.472

FIG. 1—DESIGN FOR RECTANGULAR TEM TRANSMISSION CELL

EM SUSCEPTIBILITY TESTING

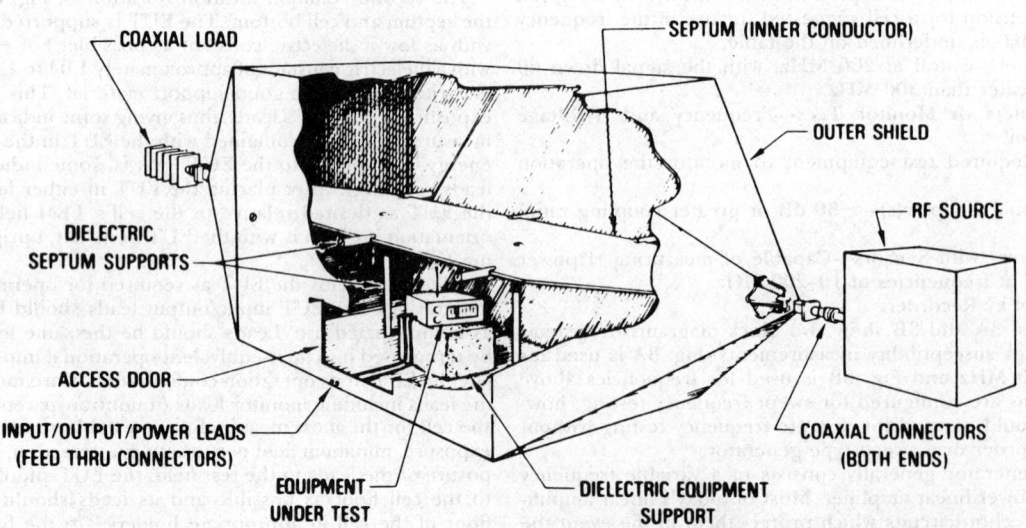

FIG. 2—CUT-AWAY VIEW OF TEM CELL BEING USED FOR EM SUSCEPTIBILITY TESTING. FIGURE SHOWS PLACEMENT OF EUT AND ASSOCIATED INPUT, OUTPUT, AND MONITORING LEADS INSIDE THE CELL

TABLE 1—TEM CELL DIMENSIONS

Cell No.	Recommended Upper Frequency (MHz)	Cell Form Factor a/b	Plate Separation b (cm)	Center Septum W (cm)	TE_{01} Cutoff/Multimode Frequency	
					Propagation MHz	Resonance MHz
1	46	1.0	150	124.5	29	54
2	70	1.0	100	83	43	81
3	85	1.5	60	68	66	101
4	120	1.0	60	50	72	135
5	130	1.5	40	45.6	100	150
6	160	1.67	30	36.0	128	187
7	170	1.5	30	34.2	134	200
8	230	1.0	30	24.9	143	268
9	260	1.5	20	22.8	204	300
10	320	1.67	15	18.0	256	374
11	350	1.50	14	16.0	272	410
12	460	1.0	15	12.5	287	538

$$f_{res_{mn\ell}} = \sqrt{(f_{cmn})^2 + \left(\frac{c}{2\ell}\right)^2}$$

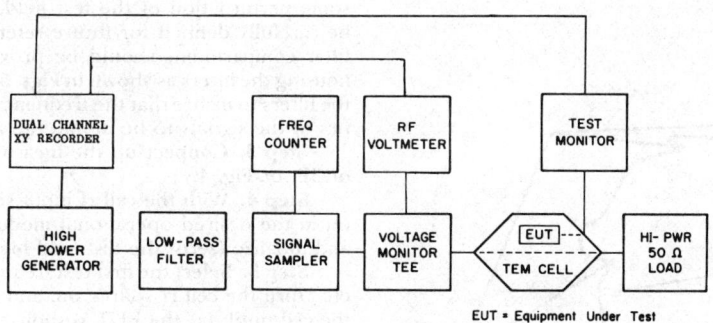

$$E_v = \frac{V_{rf}}{b} \text{ (Volts / Meter)}$$

FIG. 3A—BLOCK DIAGRAM OF SYSTEM FOR SUSCEPTIBILITY TESTING OF EQUIPMENT (USED TYPICALLY BELOW 10 MHz)

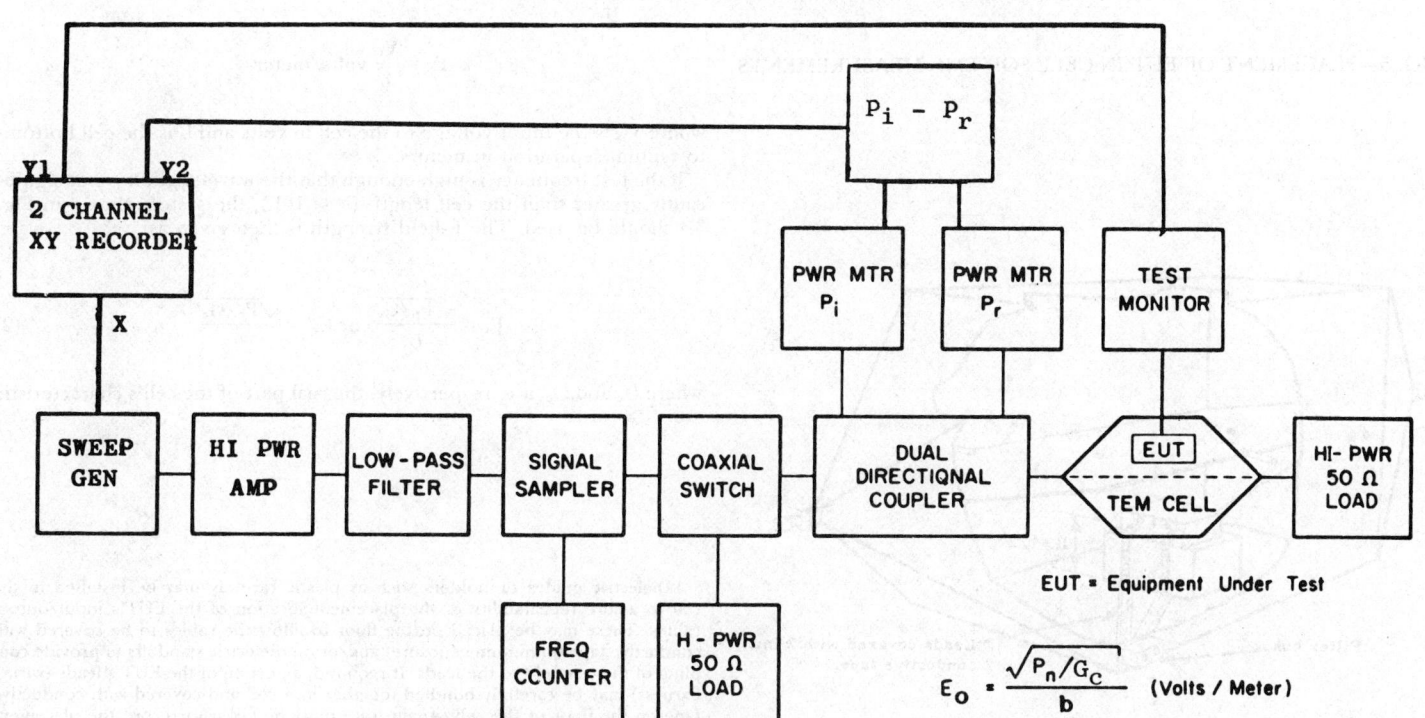

$$E_o = \frac{\sqrt{P_n/G_c}}{b} \text{ (Volts / Meter)}$$

FIG. 3B—BLOCK DIAGRAM OF SYSTEM FOR SUSCEPTIBILITY TESTING OF EQUIPMENT (10–200 MHz)

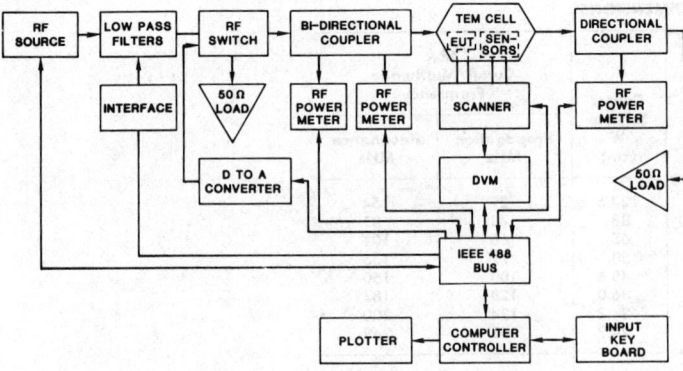

FIG. 4—BLOCK DIAGRAM OF AUTOMATED TEM CELL SUSCEPTIBILITY MEASUREMENT SYSTEM

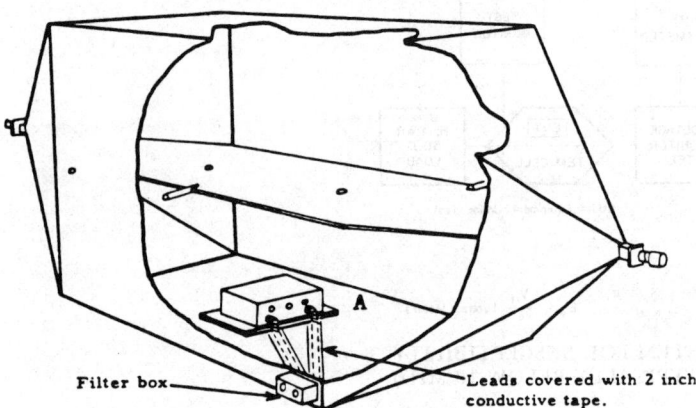

FIG. 5—PLACEMENT OF EUT IN CELL FOR EMC MEASUREMENTS

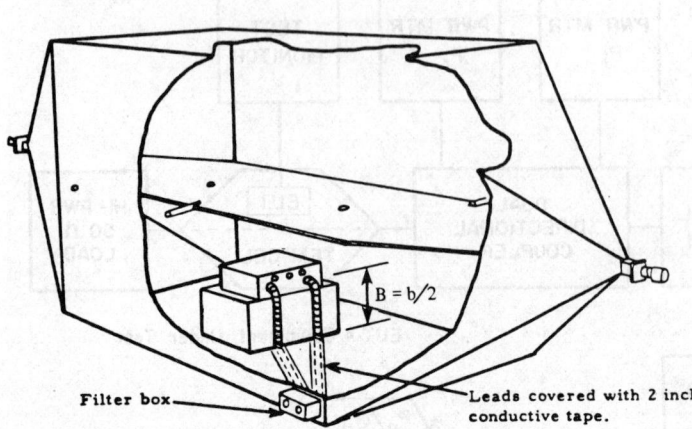

FIG. 6—PLACEMENT OF EUT IN CELL FOR MINIMUM EXPOSURE OF LEADS TO THE TEM FIELD

input/output, monitor, and ac power line leads together. Twist the leads if they cannot be kept separated. If braided wire is used, be sure the braid is in electrical contact with the cell floor. However, care must be taken to prevent the braided shield or tape from contacting the case of the EUT unless, once again, a common ground between the EUT and cell is required. Grounding the two together will obviously influence the results of the susceptibility measurements. Connect the EUT input/output and monitor leads to appropriate feed-through, filtered connectors for accessing and operating the EUT. Input/output leads should be filtered to prevent rf leakage into or out of the cell, which would reduce the shielding integrity of the measurement system. The monitor leads must also be accessed through the bulkhead as appropriate for the measurement requirement. These leads, which are used for sensing and telemetering the performance of the EUT while exposed to the test fields, may require special "invisible," high-resistance leads made of carbon-impregnated plastic [2] or fiber optics lines to prevent perturbation of or interaction with the test environment. dc signals or signals with frequency components below 1 KHz, may be monitored via the high-resistance lines. rf signals should be monitored via fiber optic lines.

If the monitored signal is at a frequency or frequencies sufficiently different from the susceptibility test frequency or frequencies, conductive (hard wire) leads may be used with appropriate filtering (high-pass, low-pass, band-pass, etc.) at the bulkhead. Such leads, however, will cause some perturbation of the test field, thus their placement location must be carefully defined for future reference. Note that a separate, shielded filter compartment should be provided on the outside of the cell for housing the filters as shown in Figs. 5 and 6. Care must be used in selecting the filters to insure that the frequency and amplitude response characteristics of the signals to be monitored are not significantly changed.

Step 3. Connect up the measurement system as shown in Figs. 3A or 3B, or Fig. 4.

Step 4. With the cell rf input source turned *off* and the EUT turned on in the desired operational mode, record the EUT monitor response and intialize (zero) the test-field measurement instrumentation.

Step 5. Select the first test frequency, modulation rate, test waveform, etc., turn the cell rf source on, and slowly bring up the field level inside the cell until: (a) the EUT response monitor(s) indicate susceptibility or (b) the maximum required test level is obtained.

NOTE: Do not increase the test level too fast. Sufficient time must be spent at each test level to allow the EUT to respond. Also, the rate of change of the field used should be specified in relation to the response time of the particular function of the EUT being monitored. Record the monitored response. If the block diagram of Fig. 3A is used, the E-field strength, E_v, in the cell is determined by the expression:

$$E_v = \frac{V_{rf}}{b} \text{ volts/meter} \qquad (1)$$

where V_{rf} is the input voltage to the cell in volts and b is the cell bottom-to-septum separation in meters.

If the test frequency is high enough that the wavelength λ is not significantly greater than the cell length ($\lambda \leq 10L$), the system shown in Fig. 3B should be used. The E-field strength is then given as:

$$E_v = \frac{\sqrt{P_n/G_c}}{b}, \text{ or } E_v' = \frac{\sqrt{P_n/G_c'}}{b} \qquad (2)$$

where G_c and G_c' are, respectively, the real part of the cell's characteristic

---

[1] Dielectric guides or holders such as plastic raceway may be installed in the cell to assure repeatability of the placement location of the EUT's input/output cables. These may be placed on the floor to allow the cables to be covered with conductive tape (minimum exposure) and/or on dielectric standoffs to provide coupling of the test field to the leads. If required, a portion of the EUT's leads (wiring harness) may be carefully bundled together in a coil and covered with conductive tape on the floor of the cell. Again, care must be taken to record the placement location and how this is done so that it can be repeated if necessary. It may be helpful to mark the bottom of the cell with a uniform array of scribe marks to assist in determining placement locations precisely.

admittance referenced to the center of the cell with or without the EUT[2] inserted in the cell.

P_n is determined from the power meter readings on the sidearms of a calibrated bi-directional coupler using the following equation:

$$P_n = CR_f \cdot P_i - CR_R \cdot P_r \qquad (3)$$

where CR_f and CR_R are the forward and reverse coupling ratios of the bi-directional coupler and P_i and P_r are the indicated incident and reflected coupler sidearm power meter readings.

Step 6. Select the next test frequency, modulation rate, or test waveform, etc., and repeat procedures of Step 5 until all frequencies, modulation levels and rates, and waveforms required by the test plan are complete.

NOTE: It may be desirable to select specific test levels and sweep the frequency range of interest at these levels while monitoring the EUT response. If this procedure is used, the following precautions must be exercised: (a) the sweep rate must be slow enough to allow the EUT to respond (remember, susceptibility often is due to resonance in the EUT circuits, leads, or apertures in its case); and (b) the selected test level must not be too high or damage may occur to the EUT.

Step 7. If exposure to interference fields of the EUT input/output and ac power line leads occurs in actual use, these leads should be raised from the floor of the cell and extended inside the cell, matching polarization as much as allowed. Care must be taken to clamp the leads with dielectric supports so they can be placed in the same location (exposure) if the EUT is taken out and then returned to the cell. Both the EUT placement locations and orientation, and the input/output ac power line leads placement locations and orientations must be carefully recorded in order to obtain repeatability when reevaluating the susceptibility characteristics of an EUT. This is true no matter what the test environment (for example, shielded enclosure, TEM cell, anechoic chamber, etc.) since EUT susceptibility obviously is a function of how the interference is coupled into the EUT (for example, EUT exposure aspect angle relative to interference polarization, etc.). With the EUT leads now exposed to the test field, Steps 4 through 6 should be repeated.

NOTE: Performing Step 7 provides some information on how the interference fields are coupled into the EUT.

Step 8. Next, the EUT should be reoriented from position 1 (flat, normal operating position) by laying it on its side or on its end. All three orthogonal orientations of the EUT may need to be tested in the cell. This is required to expose each surface of the EUT, matching polarization to the TEM field of the cell. After changing the EUT orientation, Steps 2, 4, 5, 6, and 7 should be repeated.

NOTE: If the field close to the EUT is monitored, using, for example, small calibrated electric and/or magnetic field probes, the measured results must be interpreted carefully. This is because such measurements are made in the near field of the scattered field from the EUT and its leads. This field can be stronger than the test field (TEM field) launched inside the cell, and erroneous results or conclusions may result. If possible, it is preferable to mount the field monitoring probe in the opposite half space (from where the EUT is located) in the mirror image location of the EUT.

The entire frequency range of interest should be scanned (but not exceeding the upper frequency limit of the cell), with particular emphasis made at the EUT's critical frequencies (local oscillator frequency, intermediate frequency, and others) as specified in the test plan. Experience in performing susceptibility tests indicate high-Q resonances can occur in some EUT, hence, discrete frequency testing may not be adequate to completely characterize the EUT's response.

---

[2] If the EUT is not small, the EUT will effectively short out part of the vertical separation, b, between the plates resulting in an increase in the exposure field. This new E_v' (with the EUT in the cell) can be found by determining the new effective separation distance, b', given as: $b' = b-h$, where h is the effective height of the EUT between the plates. E_v' then is equal to V/b'. Actually, unless the EUT's length and width occupied the full length and width dimensions of the cell, and the EUT case was metal (highly conductive), the effective height of the EUT is not as large as its physical height. The effective separation distance b' is then determined by measuring the distributed impedance of the TEM cell as a transmission line after inserting the EUT inside. This measurement is made using a time domain reflectometer (TDR). The value of the cell transmission line impedance in the section occupied by the EUT is then used as shown in Appendix A to compute b' and hence E_v'. An example computation of b' and E_v' based on TDR measurements is contained in Appendix A. If a TDR is not available or it is not convenient to make this impedance measurement, b' can be estimated by measuring the exposure field level at the center of the cell test zone, without the EUT in the cell, using a small dipole probe [2,3]. The probe's output voltage is then recorded and used as a reference to compare measurements of the exposure field made using the probe with the EUT in the cell.

Not all the tests outlined in this measurement procedure may be required, and only those required by the test plan should be performed. For example, if the objective of the measurement program is to reduce the vulnerability (susceptibility) of the EUT, one EUT orientation with one input/output lead configuration could be tested in one particular operational mode to a preselected susceptibility test-field waveform and amplitude. Then, if corrective measures were made to the EUT and placement of the EUT, and its leads inside the cell were carefully duplicated, repeat measurements could be made. These measurements could then be compared to determine degree of improvement.

6. Notes

(a) Because the cell operates with the fundamental TEM mode, broadband CW testing with amplitude or frequency modulation is possible. In addition, the cell can be used to establish impulsive wave forms for testing by using an appropriate wave-form generator connected to the cell's input port, assuming the frequency content of the wave form does not exceed the multimode cutoff frequency of the cell.

(b) This procedure also exposes the EUT to magnetic fields, and thus the cell may be calibrated for use in determining magnetic-field susceptibility.

(c) Test samples of any size could be tested using a TEM cell modeled from Table 1 to meet the criteria that the size of the EUT be less than $L/3 \times 2W/3 \times b/3$. (These dimensions are considered a maximum to prevent excessive impedance loading and test-field perturbation when inserting the EUT into the cell.) Thus, a small EUT could be tested at higher frequencies in small cells, and a large EUT could be tested at lower frequencies in large cells. The procedure for testing in large cells is the same as for testing in small cells, but a larger signal source (higher power) and an appropriate high-power termination (50 Ω) are required.

(d) The upper useful frequency for a cell is limited by the distortion of the test signal caused by multimodes and resonances that occur within the cell at frequencies given in Table 1 [4,5]. The frequencies, fc_{mn}, for the first few modes can be determined using Fig. 7. The frequencies of resonances, $fres_{mn}$, associated with these modes can be found from the following expression:

$$fres_{mn} = \sqrt{(fc_{mn})^2 + \left(\frac{c}{2\ell}\right)^2} \text{ (MHz)}$$

where c is the wave propagation velocity (3.0×10^8 m/s), ℓ is the resonant length on the cell, and m and n are integers corresponding to the particular

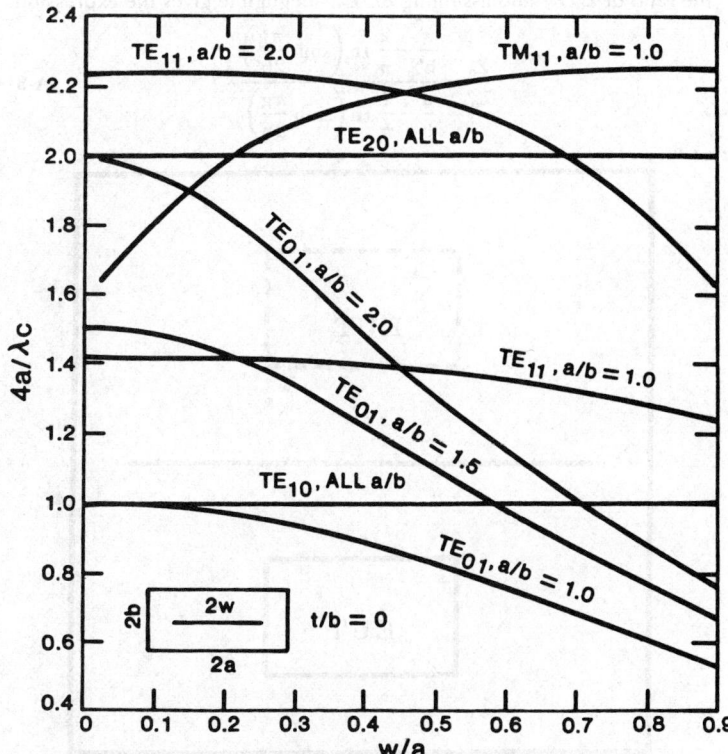

FIG. 7—NORMALIZED CUTOFF FREQUENCY VERSUS w/a FOR DIFFERENT MODE IN RECTANGULAR TRANSMISSION LINES WITH a/b = 1.0, 1.5, AND 2.0

waveguide mode. Note that the resonant length of the cell is only slightly longer than the main body length, L, for the first resonance. It is also important to note (i) that the influence of the first order TE modes do not become significant until approaching their resonant frequencies; and (ii) since the septum (center conductor) of the cell is centered symmetrically, the odd order TE modes are not excited in the cell, except, of course, by the presence of the EUT place in the cell. Thus, the recommended upper frequencies exceed the multimode cutoff frequency of the first higher-order mode (TE_{01}) but are less than this mode's resonant frequency.

(e) The useful upper frequency for the cell is reduced (10–20%) from the cutoff/multimode resonant frequency given in Table 1 to account for the loading effect of the EUT.

(f) Test samples that exceed the ⅓-linear dimension criterion could be tested in the cell, bearing in mind that excessive loading of the cell reduces the accuracy in determining the test field. This effect occurs because the sample tends to short out the test field in the region between the plates, increasing the vertically polarized test field. The error, however, could be partially corrected by measuring the field in the region above and below the EUT using miniature E-field probes and making an appropriate correction.

7. References

1. Crawford, M. L., and Workman, J. L., "Using a TEM Cell for EMC Measurements of Electronic Equipment," NBS Technical Note 1013, Revised July 1981, 65 pp.
2. Green, F. M., "Development of Electric and Magnetic Near-Field Probes," NBS Technical Note 658, January 1975.
3. Larsen, E. B., and Ries, F. X., "Design and Calibration of the NBS Isotropic Electric-Field Monitor (EFM-5), 0.2 to 1000 MHz," NBS Technical Note 1033, March 1981, 97 pp.
4. Tippet, J. C., Chang, D. C., and Crawford, M. L., "An Analytical and Experimental Determination of the Cutoff Frequencies of Higher-Order TE Modes in a TEM Cell," NBSIR 76–841, June 1976.
5. Tippet, J. C., Modal Characteristics of Rectangular Coaxial Transmission Line. Thesis submitted June 1978 for degree of Doctor of Philosophy to University of Colorado, Electrical Engineering Dept., Boulder, CO.

APPENDIX A—SAMPLE CALCULATIONS OF b' AND E_v' FOR SUSCEPTIBILITY TESTING
(Correcting Test Field Level for Loading Effect of EUT)

The following is a sample calculation for determining the corrected, effective separation distance b' and hence the electric field, E_v', inside a TEM cell after inserting an EUT.

From Eq. (1) of Ref. 1, NBS Technical Note 1013, the empty cell's impedance was given as:

$$Z_o \cong \frac{376.7}{4\left[\frac{a}{b} - \frac{2}{\pi} \ln\left(\sinh \frac{\pi g}{2b}\right)\right] - \frac{\Delta c}{\epsilon_o}}. \quad (A\text{-}1)$$

The equation parameters a, b, and g were defined in Fig. 1.

The cell's impedance after inserting an EUT with its image in the opposite half space of the cell as shown, for example, in Fig. A-1 is:

$$Z_o' \cong \frac{376.7}{4\left[\frac{a}{b'} - \frac{2}{\pi} \ln\left(\sinh \frac{\pi g}{2b'}\right)\right] - \frac{\Delta c}{\epsilon_o}}. \quad (A\text{-}2)$$

The parameter, b', is the effective separation distance between the cell septum and outer wall, and is the parameter to be determined. Taking the ratio of Z_o/Z_o' and assuming $\Delta c/\epsilon_o$ is negligible gives the expression:

$$\frac{Z_o}{Z_o'} = \frac{\frac{a}{b'} - \frac{2}{\pi} \ln\left(\sinh \frac{\pi g}{2b'}\right)}{\frac{a}{b} - \frac{2}{\pi} \ln\left(\sinh \frac{\pi g}{2b}\right)}. \quad (A\text{-}3)$$

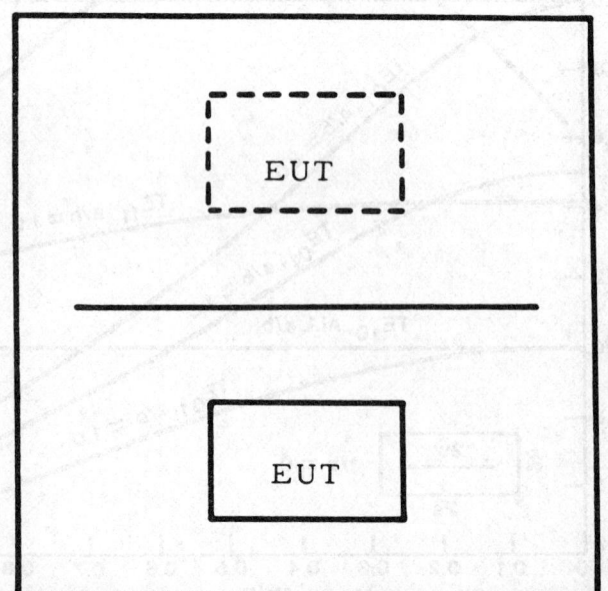

FIG. A-1—PLACEMENT OF EUT AND ITS IMAGE (SECOND IDENTICAL EUT) IN CELL FOR MEASUREMENT OF LOADED IMPEDANCE, Z_o'

To find b', measure the distributed impedance of both the empty cell and the cell with the EUT and its image using a time domain reflectometer (TDR). Assume for the purpose of this example that the EUT is to be susceptibility tested in the cell #4 of Table 1. Assume that the impedance of this cell, measured by the TDR in the section of the cell to be occupied by the EUT, is 50.5 Ω with the cell empty and 48.3 Ω with the EUT and its image inside the cell. We can compute the effective separation distance b' as follows:

For this cell, a = 0.6 m, b = 0.6 m, and g = 0.104 m

Substituting these values along with the measured value for Z_o and Z_o' into equation (A-3) we obtain:

$$\frac{50.5}{48.3} = \frac{\frac{0.6}{b'} - \frac{2}{\pi} \ln\left(\sinh \frac{0.104\pi}{2b'}\right)}{1 - \frac{2}{\pi} \ln\left(\sinh \frac{0.104\pi}{1.2}\right)}$$

or

$$1.902 = \frac{0.6}{b'} - 0.637 \left(\ln \sinh \frac{0.163}{b'}\right) \quad (A\text{-}4)$$

Equation (A-4) can now be solved numerically to obtain the value of b', which is approximately 0.50.

If it is inconvenient to place a second equivalent EUT inside the cell to serve as an image for the principle EUT when measuring Z_o', the following approximation can be used. Measure the distributed impedance Z_o'', with only the EUT (no image) inside the cell. Then:

$$Z_o = 2\Delta Z_o \cong Z_o' \quad (A\text{-}5)$$

where

$$\Delta Z_o = Z_o - Z_o''$$

or

$$Z_o' \cong 2Z_o'' - Z_o$$

The approximate value, Z_o', the symmetrically loaded impedance, can then be calculated and used to obtain the corrected equivalent separation distance b'. For example, Z_o'' measured, for the sample calculation of b' given above was 49.4 Ω, or computing the approximate value of Z_o' gives:

$$Z_o' \cong 2(49.4) - 50.5 = 48.3 \ \Omega$$

If the EUT is centered in the test zone of the cell, midway between the septum and floor, position A, see Fig. 5, we can obtain the value of E_v' from equation (2) as:

$$E_v' = \frac{\sqrt{P_n/G_c'}}{b'} \tilde{E}_{x,y} = \frac{\sqrt{48.3\,P_n}}{0.5} = 13.9\sqrt{P_n} \text{ V/m} \quad (A\text{-}6)$$

where $1/G_c' \cong Z_o' = 48.3$ Ω, and $\tilde{E}_{x,y}$ is the relative magnitude of the E field at the location of the EUT inside the cell. $\tilde{E}_{x,y}$ is equal to 1 at this location in the cell. For the empty cell, E_v is given as:

$$E_v = \frac{\sqrt{P_n R_c}}{b} \tilde{E}_{x,y} = \frac{\sqrt{50.5\,P_n}}{0.6} = 11.8\sqrt{P_n} \text{ V/m} \quad (A\text{-}7)$$

the ratio of:

$$\frac{E_v'}{E_v} = 1.18$$

This represents again for this particular EUT located at the center of the test zone, inside this particular cell, a 1.44 dB correction to the E-field, E_v, computed from the power measured at the cell's input port.

If the EUT was located in the center of the cell, but near the cell floor (position B, see Fig. 6), and if the impedance loading caused by the EUT inside the cell is the same as the above example, the corrected electric field strength, E_v' is:

$$E_v' = 13.9 \sqrt{P_n} (0.853) = 11.86 \sqrt{P_n} \text{ V/m} \quad (A-8)$$

where $\tilde{E}_{x,y}$ is equal to 0.853 at this location in the cell. This value for $\tilde{E}_{x,y}$ was obtained from Table 3 of NBS Technical Note 1013 [1], which gives the magnitude of the normalized electric field as a function of cross section location in a square TEM cell.

The value of the electric field strength, E_v at position B in the empty cell is:

$$E_{x,y} = 11.8 \sqrt{P_n} (0.853) = 10.07 \sqrt{P_n} \text{ V/m} \quad (A-9)$$

The ratio $E_{x,y}'/E_{x,y}$ then is $11.86\sqrt{P_n}/10.07\sqrt{P_n} = 1.18$. This correction factor, for the loading effect of the EUT, is the same for either test location of the EUT.

ANECHOIC TEST FACILITY RADIATED SUSCEPTIBILITY 20 MHz—18 GHz ELECTROMAGNETIC FIELD—SAE J1507 JAN87

SAE Information Report

Report of the Electrical and Electronic Systems Technical Committee approved January 1987.

1. *Purpose*—This information report gives typical requirements for an anechoic chamber in which the system susceptibility of an operating motor vehicle to electromagnetic fields can be determined in the frequency range of 20 MHz to 18 GHz. Because of the large cone sizes required for 20 MHz cut-off, several anechoic facilities have been designed with lower cut-off frequencies of 200 MHz or greater. Testing below cut-off is then accomplished using customized antennas at reduced accuracy. Users should carefully review their testing requirements before undertaking the construction of a test facility the magnitude of an anechoic chamber. Other test approaches include, but are not limited to, open field testing per SAE J1338 and mode stirred reverberation chambers.

2. *Measurement Philosophy*—The design objective of an anechoic chamber is to create an indoor electromagnetic compatibility testing facility which simulates open field testing without the weather and Federal Communications Commission radiation constraints. Typical chambers have absorbing material lining as many surfaces in the chamber as possible to minimize reflections and resonances. The design objective is to reduce the reflectivity in the test area to −10dB or less.

The test consists of generating radiated electromagnetic fields by using antenna sets with rf sources capable of producing the desired field strengths over the range of test frequencies. Typically the fields are monitored with small probes to ensure proper leveling. To reduce testing error, the vehicle under test is usually monitored by optical couplers.

3. *Apparatus*—Typical test apparatus consists of the following:

(a) **Anechoic Chamber**—The size, shape, and construction of an anechoic chamber can vary considerably. (See Figs. 1 or 2.) The chamber shape is a function of the types of tests to be performed, the size of vehicle to be tested and the frequency range to be covered. Basically, an anechoic chamber consists of a shielded room with absorbing material mounted on its internal surfaces. The minimum size of the room is determined by the size of the test region needed, the size of the transmitting antennas and the clearances needed between the absorber, antennas, and the largest vehicle that is to be tested. To create the test region (quiet zone), the absorber, antenna systems, and chamber shape are selected to reduce the amount of extraneous energy in the test region below a minimum value which will give the desired measurement accuracy. Since the performance of absorber is a function of its construction (thickness, shape, material, etc.) and the angle of incident energy, the determination of the optimum location and construction of the absorber in an anechoic chamber to meet the reflectivity requirements is a complex process.

The use of scale models can greatly reduce both the cost and time involved in establishing an acceptable design. In addition, the model may reveal room resonances which are not predicted by a theoretical study. The test area in the chamber may have a dynamometer and/or a vehicle turntable. Additional guidelines for designing an anechoic chamber are available in Ref. 9. Specific design requirements and construction details for an anechoic chamber are available in Refs. 1, 2 and 3. For typical anechoic chamber specification, see the Appendix.

(b) **Signal Sources**—Any commercially available signal source and power amplifier, as required, capable of supplying adequate power (modulated and unmodulated) to develop the susceptibility levels specified in the test plan may be used. Typically for the lower MHz bands, these power levels range from 10,000 watts for five meter test distances to 2,000 watts for one meter. For the GHz bands, these powers drop to 100 or 200 watts. The harmonic and spurious outputs from the rf source should be less than −20 dB referred to the fundamental. Specific tests and frequencies near the band edges of the generators may require additional harmonic suppression with appropriate filter prior to final amplification. The frequency accuracy of the signal generators shall be within ± 2%, if a frequency counter is not used.

(c) **Signal Samplers and Monitor Tees**—As required by frequency and rf voltage monitoring equipment.

(d) **Power Monitoring Equipment**—Commercially available power meters and directional couplers capable of measuring the maximum generated powers given in (a) over the frequency range of 20 MHz - 18 GHz. Several different meters will be required. To protect the generators in the high frequency band, the power meters and couplers must be able to measure VSWR. Electric field probes may be used over most of the frequency range as an option if due care is exercised. See Section 4(c).

(e) **Frequency Counter**—Frequency counters capable of measuring frequencies from 20 MHz - 18 GHz or an equivalent frequency determination built into the rf signal sources. If a frequency counter is not used, the frequency accuracy of the rf signal sources shall be within ± 2%.

(f) **Antenna Sets**—Any commercial available antenna set including high power baluns which is capable of radiating the specified field strengths at the specified distance with power available. A typical set consists of several horn or parabolic antennas, a log periodic, and a biconical dipole.[1] For the 20 MHz - 150 MHz band, specialized antennas may be needed. Ref. 15 outlines the construction of a high power rod antenna which could be used in the lower part of this band.

(g) **EUT Monitor**—Test equipment, hardened to the test power levels, required to monitor the operation of the equipment under test (EUT) coupled to the control center by commercially available fiber optic links or high resistance leads.

---

[1] High power balun is especially needed for this antenna.

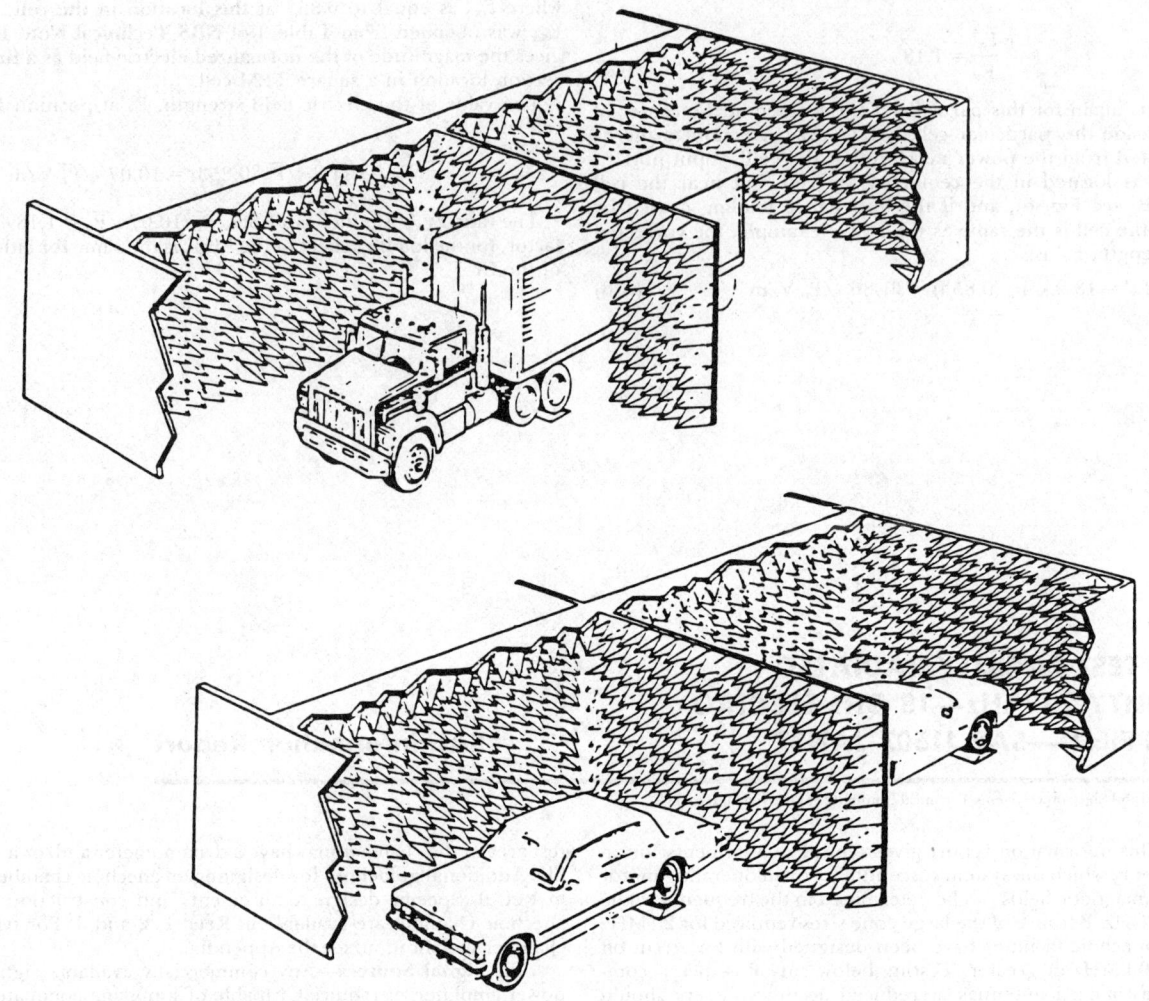

FIG. 1—EXAMPLE OF CONDUCTIVE FLOOR ANECHOIC CHAMBER

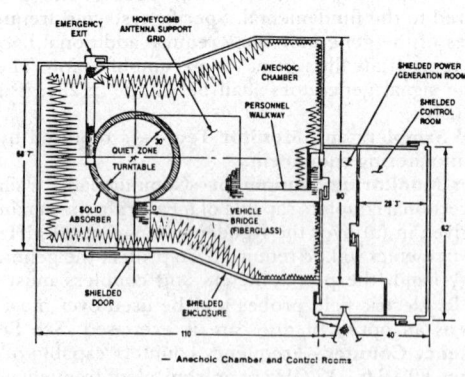

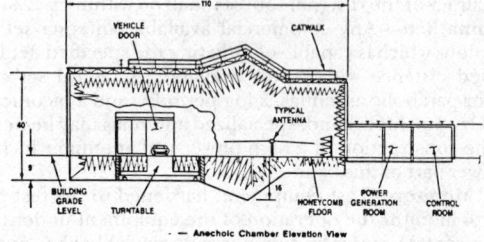

FIG. 2—EXAMPLE OF ANECHOIC CHAMBER WITH A QUIET ZONE

(h) **System Controller**—If desired, process controllers or small computers compatible with the IEEE 488 Bus can be used to control the testing and test equipment.

4. Test Setup Procedure—Anechoic chamber susceptibility measurement techniques have been described in detail in several papers (Refs. 4, 1 and 2.) The setup and measurement procedure is summarized below.

(a) Typical test setups are shown in Figs. 3A and 3B.

(b) The vehicle may be placed in the chamber centered with respect to the width and length, or placed in the chamber's "quiet zone" if the chamber is so designed. The test area may contain a dynamometer and/or a turntable. The surface under the vehicle may be conductive or absorptive depending on the test plan and facility capabilities.

If direct comparisons between the anechoic chamber and an open field site are desired, a conducting floor in the anechoic chamber may be needed since the open field site has a reflective ground plane.

(c) The determination of the field strength at the vehicle can be done several ways. Field probes placed near the vehicle can be used to measure the field directly. Measurement of the forward RF power into the antenna set and use of the antenna factors to calculate the field strength can be used. Measurement of the field at test area, without the vehicle present, with a field probe in conjunction with the forward rf power necessary to maintain the field level is another. These and others can be used exclusively or in combination. Each method has its uncertainties.

Field probes, if used, should not perturb the field, therefore, they should be electrically small and electrically isolated. The transmission lines from the probes should be either very high resistance lines or fiber optic links. Placement of the probe is critical to reducing measurement errors. If the probe is placed near a discontinuity, such as a sharp bend or corner of the vehicle, or near a resonant cable or component, the resulting measurement could be in error by a substantial amount. Unless otherwise specified by the probe's manufacturer, the probe should be placed parallel to the generated field above an approximately flat surface. For additional information on the use of field probes, see Ref. 8.

If the field strength is determined by the use of antenna factors, the vehicle should be in the antenna's far field[2] and the site attenuation must be known over the frequency band. Details of determining both site attenuation and errors associated with antenna factors are given in several sources (Refs. 10, 11, 12, 13, 14 and 16). In particular, Pate (Ref. 16) reported errors of greater than 1.5dB in the antenna factor if a dipole antenna is placed closer than 0.2 of a wave length from the absorber and if a biconical antenna is closer than 2 m.

If the field strength is determined by the use of either the antenna factor or a field probe in the absence of the vehicle, a potentially large error can be caused by the vehicle interacting with the transmitting antenna. To minimize this error, both the forward and reflected powers should be monitored and the transmitter adjusted for the same net power (forward less reflected) to the antenna both before and after the vehicle is placed in the test area. Discussions of additional errors in generating a "known field" are given in Refs. 5, 6 and 7. J1338 and Ref. 8 further discuss establishing the test field.

(d) The EUT should be operated as required in the test plan by actuators which have a minimum effect on the EUT; i.e., plastic blocks on the accelerator, pneumatic actuators with plastic tubes. Connections to equipment monitoring for electromagnetic interference reactions of the EUT may be accomplished by using fiber optics, or by high resistance leads. Other types of leads may be used but require extreme care to minimize interactions. The orientation, length, and location of the latter leads must be carefully documented to assure repeatability of the test results.

(e) Because of the multiple power amplifiers and antennas required, the vehicle is generally scanned in bands over the test frequency range of 20 MHz - 18 GHz. Since some EUT's may have relatively slow thermal and/or electrical time constants or long cycle times, the sweep rate or dwell time (depending on the frequency generation equipment) must be compatible with reaction/interaction times of the EUT to the test field. This reaction has two parts: 1) the time required for the EUT to respond to a given frequency and 2) the frequency band width over which the response can occur. The latter quantity can be used to define a "Q" of the response in the normal manner (i.e. $Q = F/B$ where B is the $-3dB$ band width and F is the center frequency). Typical values for this Q range from about 2-40.

If the response is the result of a resonant circuit which is not affected by other vehicle conditions, then the response time (Tr) can be approximated by $Tr \approx 0.3F$ and the maximum scan speed (SSm) by $SSm \leq 1.73 \ (F/Q)^2$.

If, however, the response can only occur during part of an electrical system cycle (i.e., while an engine controller is sampling an external sensor), the response time is hard to predict and the scan speed or dwell time must be much slower than SSm. If a field probe is used to level the field during testing, the system will require time to stabilize at each frequency.

[2] For antennas less than a wave length (λ), the far field begins at $\lambda/2\pi$. For antennas with electrically large apertures, it begins at d^2/λ where d is the largest dimension of the antenna.

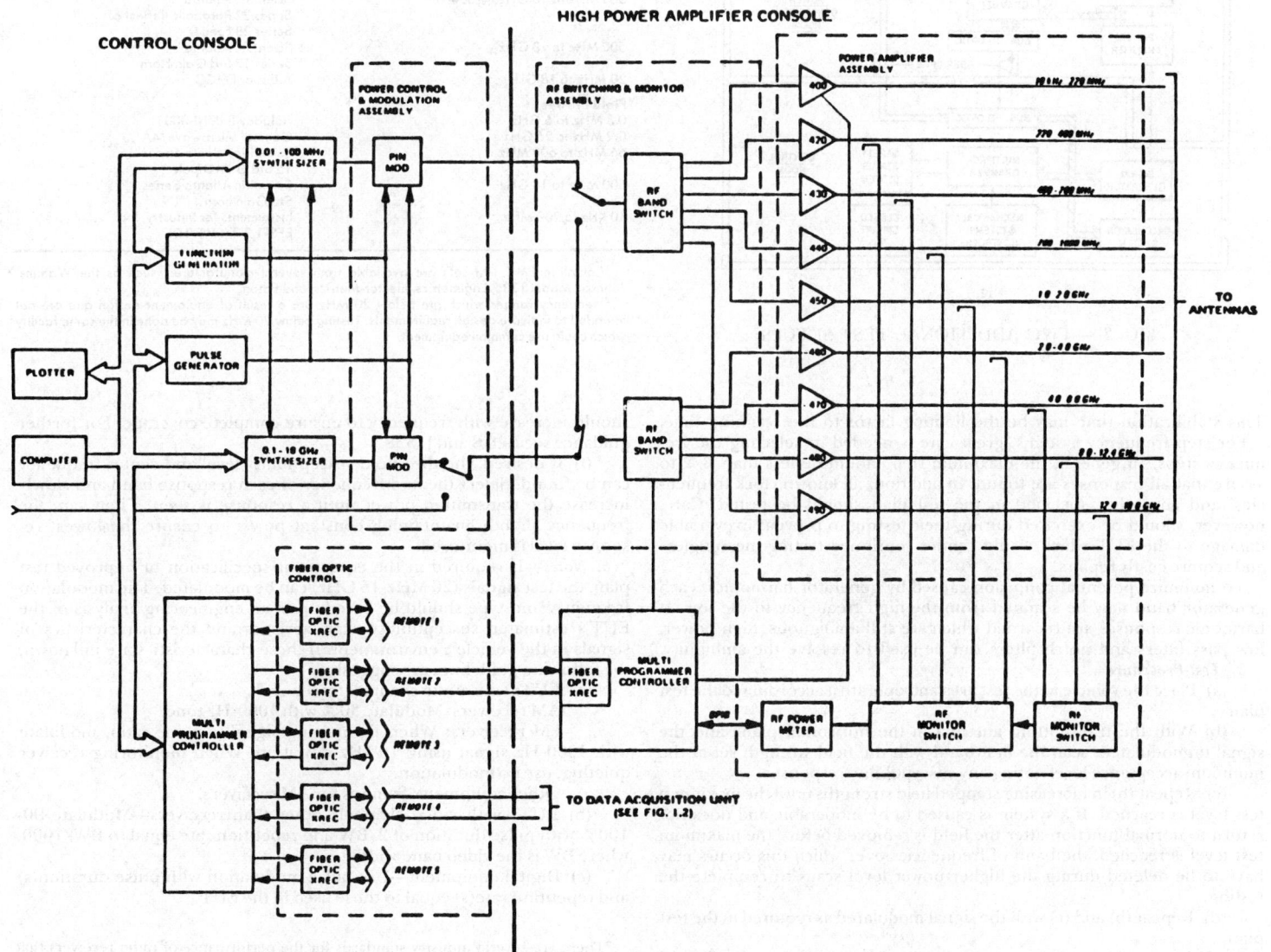

FIG. 3a—BLOCK DIAGRAM OF A TYPICAL EMC TEST SYSTEM

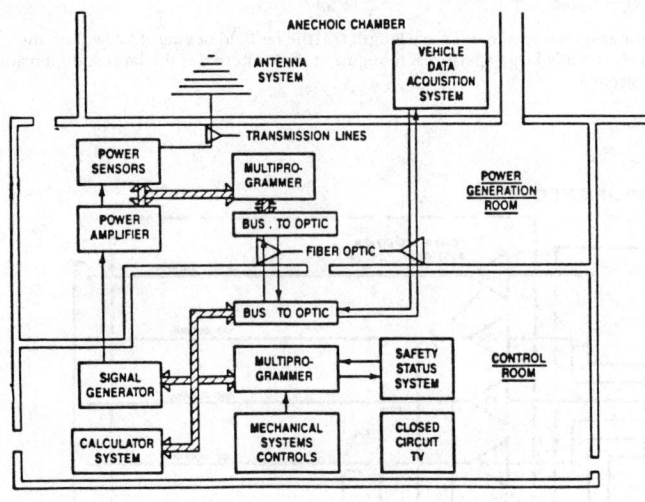

FIG. 3b—TWO ADDITIONAL TEST SETUPS

TABLE 1—EQUIPMENT LIST[1]

Signal Generation	Manufacture/Model
10 kHz to 1 GHz[2]	HP 8660C
100 MHz to 26 GHz	WJ 1204, HP 8340A
Power Amplifiers	
10 KHz to 100 MHz @ 10 KW	Amplifier Research 10,000L
10 kHz to 220 MHz @ 2 KW	AR 2000L
220 MHz to 400 MHz @ 1250 W	AR 1000
400 MHz to 700 MHz @ 500 W	MPD
700 MHz to 1000 MHz @ 500 W	MPD
1 GHz to 2 GHz @ 200 W	HP 489A plus Varian TWT Series
2 GHz to 4 GHz @ 200 W	HP 491C plus Varian TWT Series
4 GHz to 8 GHz @ 200 W	HP 493A plus Varian TWT Series
8 GHz to 12.4 GHz @ 200 W	HP 495A plus Varian TWT Series
12.4 GHz to 18 GHz @ 100 W	HP (Special Order) TWT Series
1 GHz to 2 GHz @ 200 W	Logimetrics 200 Watt TWT Series
2 GHz to 4 GHz @ 200 W	Logimetrics 200 Watt TWT Series
4 GHz to 8 GHz @ 200 W	Logimetrics 200 Watt TWT Series
8 GHz to 12.4 GHz @ 200 W	Logimetrics 200 Watt TWT Series
12.4 GHz to 18 GHz @ 100 W	Logimetrics 200 Watt TWT Series
Power Meters—Directional Couplers	
10 kHz to 100 MHz @ 10 KW	AR DC4000
10 kHz to 1000 MHz (below 2 KW)	Internal amplifier directional coupler and detector
0.5 MHz to 1000 MHz (power as required)	Bird (thru line Series)
1 GHz to 18 GHz @ 200 W	Narda Directional Couplers (High Power Series)
10 kHz to 18 GHz	HP 436
Antenna System	
20 MHz to 150 MHz @ 5 to 10 KW	Electrospace (special order)
150 MHz to 1 GHz @ 2 KW	Electrospace (special order)
500 MHz to 18 GHz @ 500 W	Scientific Atlanta Series 22 Parabolic Reflect & Series 28 Feed
500 MHz to 18 GHz	Scientific Atlanta Series 12 Std Gain Horn
20 MHz to 18 GHz	A.E.L. or EMCO
Field Probes	
0.5 MHz to 6 GHz	Holaday Ind. HI-3001
0.2 MHz to 26 GHz	General Microwave, 4A
66 MHz to 600 MHz	Scientific Atlanta Series 15 Std Gain Dipole
500 MHz to 18 GHz	Scientific Atlanta Series Std Gain horn
10 KHz to 200 MHz	Instruments for Industry, Inc. EFS-1,2,3/LMT/LDI

[1]Complete EMC test sets are available from several manufacturers such as the Watkins-Johnson Model 1235, Logimetrics, Elector-Metrics, and Eaton.

[2]Frequency ranges which are below 20 MHz are a result of equipment design and are not intended to indicate design requirements. Testing below 20 MHz may be done in the same facility which could use common equipment.

This stabilization time may be the limiting factor to the scanning time.

For step frequency systems, great care is needed in selecting the frequency steps. In general, the maximum step should be less than B/2 to assure that all responses are found. In addition, all known clock frequencies, and any other specified in the test plan should be tested. Care, however, should be exercised during their testing to prevent irreversible damage to the EUT which would render continued testing meaningless and require costly repairs.

To minimize potential confusion caused by generator harmonics, each generator band may be scanned from the high frequency to the low. If harmonic responses are recorded which are still ambiguous, high power, low pass filters and notch filters can be used to resolve the ambiguity.

5. Test Procedure

(a) Place the vehicle in the test area and operate it according to the test plan.

(b) With the transmitting antenna in the horizontal plane and the signal unmodulated, scan the first band with the field strength set at the minimum acceptance level noting any susceptibility responses.

(c) Repeat (b) in increasing stepped field strengths until the maximum test level is reached. If a system is caused to be inoperable and does not return to normal function after the field is removed before the maximum test level is reached, the band of frequencies over which this occurs may have to be deleted during the higher power level scans to complete the testing.

(d) Repeat (b) and (c) with the signal modulated as required in the test plan.

(e) Repeat (b), (c), and (d), for each of the frequency bands and vehicle orientations specified in the test plan. Because of the narrow beam width of the high frequency antennas, the number of vehicle positions should increase with frequency to ensure complete coverage. For further guidance see Ref. 8 and J1338.

(f) If desired, the threshold susceptibility level and center frequency can be found. Select the mid frequency of each response band and slowly increase the transmitted power until a response is seen. Then vary the frequency slightly maintaining constant power to ensure the lowest response level is monitored.

6. Notes—If required in the equipment specification or approved test plan, the test signals (20 MHz-18 GHz) can be modulated. The modulation frequency and type should be based upon an engineering analysis of the EUT's estimated susceptibility characteristics and the characteristics of signals in the vehicle's environment. If these characteristics are unknown, the following guidelines are suggested:

(a) EUTs with audio channels/receivers[3]

AM receivers: Modulate 30% with 1000-Hz tone.

FM receivers: When monitoring signal-to-noise ratio, modulate with 1000-Hz signal using 10-kHz deviation. When monitoring receiver quieting, use no modulation.

Other equipment: Same as for AM receivers.

(b) EUTs with video channels other than receivers—Modulate 90-100% with pulse duration of 2/BW and repetition rate equal to BW/1000, where BW is the video band width.

(c) Digital equipment—Use pulse modulation with pulse duration(s) and repetition rate(s) equal to those used in the EUT.

[3]There are several industry standards for the performance of radio receivers that may be helpful for determining the modulation to be used on the susceptibility test signal. Two of them are IEEE 185 and EIA RS-204-A; others may be more appropriate for certain equipment.

(d) Nontuned equipment—Modulate amplitude 30% with 1000-Hz tone, or as otherwise required in the test plan.

7. *Typical Test Equipment*—Table 1 is provided to give typical equipment required to perform susceptibility testing using an anechoic chamber. No approval or disapproval of any manufacturer is intended, either by inclusion or exclusion from this list, nor is the list intended to be all inclusive.

8. *References*

1. J. C. Kinderman, et al., "Implementation of EMC Testing of Automotive Vehicles," SAE Paper 810333, SAE, 400 Commonwealth Drive, Warrendale, PA 15096 (Feb. 1981).

2. Gary F. E. Vrooman, "An Indoor 60 Hz to 40 GHz Facility for Total Vehicle EMC Testing," SAE Paper 831011, SAE, 400 Commonwealth Drive, Warrendale, PA 15096 (June 1983).

3. Donald A. Weber, et al., "Development of a Vehicle Electromagnetic Test Environment," SAE Paper 831015, SAE, 400 Commonwealth Drive, Warrendale, PA 15096 (June 1983).

4. Wayne L. Minten, "Vehicle Testing for Electromagnetic Compatibility," SAE Paper 810332, SAE, 400 Commonwealth Drive, Warrendale, PA 15096 (Feb. 1981).

5. H. E. Taggart, "Radiated EMI Instrumentation Errors," EMC Technology, Vol. 1, No. 4, October 1982, pp. 26-35.

6. A. A. Smith, Jr., "Control of Errors on Open Area Test Sites," EMC Technology, Vol. 1, No. 4, October 1982, pp. 50-58.

7. W. S. Bennett, "Error Control in Radiated Emission Measurements," EMC Technology, Vol. 1, No. 4, October 1982, pp. 38-47.

8. E. L. Bronaugh and W. H. Mcginnis, "Whole-Vehicle Electromagnetic Susceptibility Tests in Open-Area Test Sites: Applying SAE J1338," SAE Paper 830606, SAE, 400 Commonwealth Drive, Warrendale, PA 15096 (Feb. 1983).

9. F. J. Nichols and L. H. Hemming, "Recommendations and Design Guides for the Selection and Use of RF Shielded Anechoic Chambers in the 30-1,000 MHz Frequency Range," IEEE International Symposium on Electromagnetic Compatibility, Boulder, CO, August 18-20, 1981, pp. 457-464.

10. A. A. Smith, R. F. German, J. B. Pate, "Calculation of Site Attenuation from Antenna Factors," IEEE Transactions on Electromagnetic Compatibility, Vol. EMC-24, No. 3 (Aug. 1982) pp. 301-316.

11. A. A. Smith, "Standard-Site Method of Determining Antenna Factors," IEEE Transactions on Electromagnetic Compatibility, Vol. EMC-24, No. 3 (Aug. 1982) pp. 316-322.

12. R. F. German, "Comparison of Semi-Anechoic Chamber and Open-Field Site Attenuation Measurements," IEEE International Symposium on Electromagnetic Compatibility, Santa Clara, CA (Sept. 8-10, 1982) pp. 260-265.

13. S. D. Bloom, "Correlation and Calibration of Anechoic Chambers," IEEE National Symposium on Electromagnetic Compatibility, San Antonio, TX (April 24-26, 1984) pp. 21-25.

14. L. Farber and D. H. Chapman, "Antenna Factors Determined in an Absorber-Lined Shielded Enclosure," IEEE National Symposium on Electromagnetic Compatibility, San Antonio, TX (April 24-26, 1984) pp. 1-7.

15. E. L. Bronaugh and D. R. Kerns, "Rod Antennas for High-Strength Radiated Susceptibility Tests from VLF and HF," IEEE International Symposium on Electromagnetic Compatibility, San Diego, CA (Oct. 9-11, 1979) pp. 413-416.

16. J. B. Pate, "Potential Measurement Errors Due to Mutual Coupling Between Dipole Antennas and Radio Frequency Absorbing Material in Close Proximity," IEEE National Symposium on Electromagnetic Compatibility, San Antonio, TX (April 24-26, 1984) pp. 13-19.

APPENDIX

The tabulation below summarizes the salient electromagnetic and mechanical features of a typical anechoic chamber less its instrumentation.

ELECTRICAL

· Frequency Range	20 MHz to 20 GHz[1]
· Quiet Zone Size	Cylinder, h = 12'6", d-30
· Quiet Zone Reflectivity	−10db @ 20 MHz
	−40db @ 640 MHz
	−40db @ 18 GHz
	−20db @ 40 GHZ
· Floor Options	Anechoic
	Dielectric (RFI)
	Ground Plane
· Shielding	11 GA Steel, MIG Weld
· Closed Circuit TV	2 Camera, Tilt, Pan, Zoom
· Data Link	Fiber Optic

MECHANICAL

· Turntable	± 180° Vehicle Rotation
	22,000 Max. Load
	30' Diameter
	12' × 30' Platform
· Test Volume Envelope	36'2" Length
	10' Width
	12'6" Height
· Overall Size	110' Long
	48' High
	90' Width (Max.)
· Dynamometer	Passenger Car—100 HP
	Heavy Truck—250 HP
· Vehicle Cooling & Exhaust	11,500 CFM Cooling
	2,025 CFM Exhaust

SAFETY

· High EM Fields	Shielding & Door Interlocks
· Gas Detection	Hydrocarbon & Co.
· Fire	
—Detection	Cross Zone UV & Smoke
—Suppression	Low Pressure CO_2
· Vehicle	Tie-Down & Shut-off
· CCTV	Monitor Test Area
· Warning	Key Lockout & Signs

[1]Upper frequency limit determined by absorber reflectivity characteristics

COLLISION DETECTION SERIAL DATA COMMUNICATIONS MULTIPLEX BUS—SAE J1567 AUG87

SAE Information Report

Report of the Electrical and Electronic Systems Technical Committee approved August 1987.

1. *Introduction*—The serial communications bus and bus interface special function integrated circuit defined herein was intended to provide a simple, yet reliable, data communications link between members of a distributed processing vehicle multiplex system. The communications protocol chosen minimizes the software support overhead requirement of the modules on the multiplex bus. Appendix A contains a detailed technical summary of the Differential Serial Communications Bus Protocol and Interface Integrated Circuit.

2. *Proposed Protocol*—The network access method that best meets the requirements is known as CSMA (Carrier Sense Multi Access). A deterministic priority access method of resolving contention, by non-destructive bit-by-bit arbitration, was chosen instead of the nondeterministic random backoff procedure associated with classic collision detection. The message format chosen is shown below and will support a number of higher level protocols. Take note that idle periods are allowed between each byte of data. This permits the use of firmware control and direct connection to the host microcomputer's asynchronous serial I/O port.

(SOM), (id 1), (id 2), (data 1), (data 2) ... (data n), (EOM)

    a— SOM, defines start of message
    b— id 1, 8 bit firmware addressing scheme
    c— id 2, optional identifier
    d— Data, may take any form, that is, data value, CRC, checksum, number of bytes in message, acknowledgement, etc.
    e— EOM, defines end of message

Standard byte synchronization NRZ was chosen as the bit encoding modulation technique. This data byte synchronizing method also provides message integrity by framing error detection. Three bit encoding methods NRZ, PWM, and Bi Phase were originally considered. The choice to use byte sync NRZ encoding as the modulation method was because it is believed to have the potential of providing the best compromise in bandwidth efficiency and data recovery in a noisy vehicle environment while maintaining the required high network data rates without generating excessive EMI.

The first thing considered was the data recovery performance in a noisy vehicle environment. The best performing data decoders should digitally approximate an integration over the bit period. In a noisy environment up to ½ of the samples can be corrupted for the NRZ encoding method before data is lost. PWM encoding method suffers a significantly smaller fraction dependent on the duty factor. This reduction in noisy environment performance cannot be compensated by just reducing the data rate.

The second thing considered was the EMI generated by the different data bit encoding method. Some empirical data was collected in a vehicle experiment. The data, although not completely conclusive, indicated that NRZ could operate at a significantly higher bit rate than PWM or Bi Phase encoding methods. It was also felt that the main advantage to using PWM or Bi Phase encoding is that the data bit rate is also encoded in the signal and, therefore, operation can be maintained using a larger clock tolerance. The NRZ requirement for a more precision clock was not a large advantage because all module applications already demanded a crystal or ceramic resonator.

The need for an open network so that minimal effect would result when nodes are added or removed, led to the selection of a broadcast type system. The identification (ID) byte of the message data content is unique and is broadcast across the network. The collision detection hardware within the interface IC determines which transmitting ID's wins without losing bus time when contention occurs. This action satisfies the need for establishing access priority and, therefore, gaining control of the latency for important messages. The message format and the data integrity method is determined by firmware. Maximum flexibility is the result and, thus, the software developer's creativity is not restricted for future applications.

3. Functional Description—Interfacing between a microcomputer and the bus is fairly simple. There are basically three methods; asynchronous serial, synchronous serial, and parallel to the address and data I/O port. The cost of the parallel interface was avoided by development of a low cost serial interface IC. To make it universally interfaceable to most microcomputers three modes of operation are supported in one device: (1) SCI, (2) SPI, and (3) Buffered SPI.

The circuits of the interface IC used when connected to the microcomputer SCI are basic to the operation of all the other modes of serial communications. Therefore, this mode will be explained first. The components of the device (Fig. 1) include the following:

a— a contention permitting differential transceiver
b— a collision detector
c— an arbitration detector
d— a digital filter
e— a bus idle detector
f— a timing and synchronizing circuit

3.1 Contention Permitting Differential Transceiver—The differential transceiver is a serial interface device which accepts digital signals and translates this information for transmission on a two-wire differential bus. The transmitter section, when transmitting, shall provide matched constant current sources to the bus "+" and bus "−" drivers (Fig. 1), sourcing and sinking current respectively. When a logic zero is supplied to the "transmit data" input, the differential amplifier shall cause the bus "+" driver to provide source current and the bus "−" driver to provide a matched current sink. A logic one at the "transmit data" input must cause the bus "+" and bus "−" drivers to simultaneously provide a high impedance state.

The wired "OR" action of the transmitting section allows more than one device to transmit at the same time thus permitting data collisions. The nonsymmetrical action of the bus drivers will allow a transmission of a logic zero, from one device, to overpower the transmission of a logic one from other devices. In this manner two or more devices can simultaneously transmit and contend for the bus, each using a unique message ID byte. The winner is determined by the value (or priority) of the ID byte, without losing bus time. A logic zero bit in one message ID byte has priority over a logic one bit in another message ID byte.

The bus shall depend on external resistor and other components for bias and termination. Clamping diodes may be added to provide a high level of transient protection.

In addition to the transmission of data, the differential transceiver receives data at its bus "+" and bus "−" terminals. The received data is translated back into the standard digital logic levels by a differential amplifier. The microcomputer always receives the actual transmitted data and in this manner can test for loss of arbitration.

3.2 Collision Detector—Data collision detection occurs because the transceiver output is reflected back into its input. The data collision detector samples the transmitted signal and the received signal. The timing of this sampling is determined by the timing circuit. When the collision detector determines that a logic one bit is being transmitted, but a logic zero bit is being received, the collision detector blocks the transmitted signal. (See waveform shown in Fig. 2.) In this way the data collision detector will permit only the interface with the highest priority to continue transmitting. The collision detector action of blocking transmission is also reset by detection of bus idle (10 consecutive idle bits).

3.3 Arbitration Detector—The arbitration detector works in conjunction with the timing and synchronizing circuit to arbitrate between the start of data to be transmitted with the start of a received message from the bus. The arbitration detector blocks a transmission that could corrupt a message that is already in progress. Also, it allows the device that starts transmission first, after a bus idle, to pass its data through the interface and out onto the bus. In all other nontransmitting devices the arbitration detector blocks transmission of data until the detection of a bus idle condition. When more than one device wants to transmit at about the same time, (greater than 1/4 bit time) the arbitration detector will allow transmission on a first-come, first-serve basis. If data transmission from more than one device on the bus is attempted in

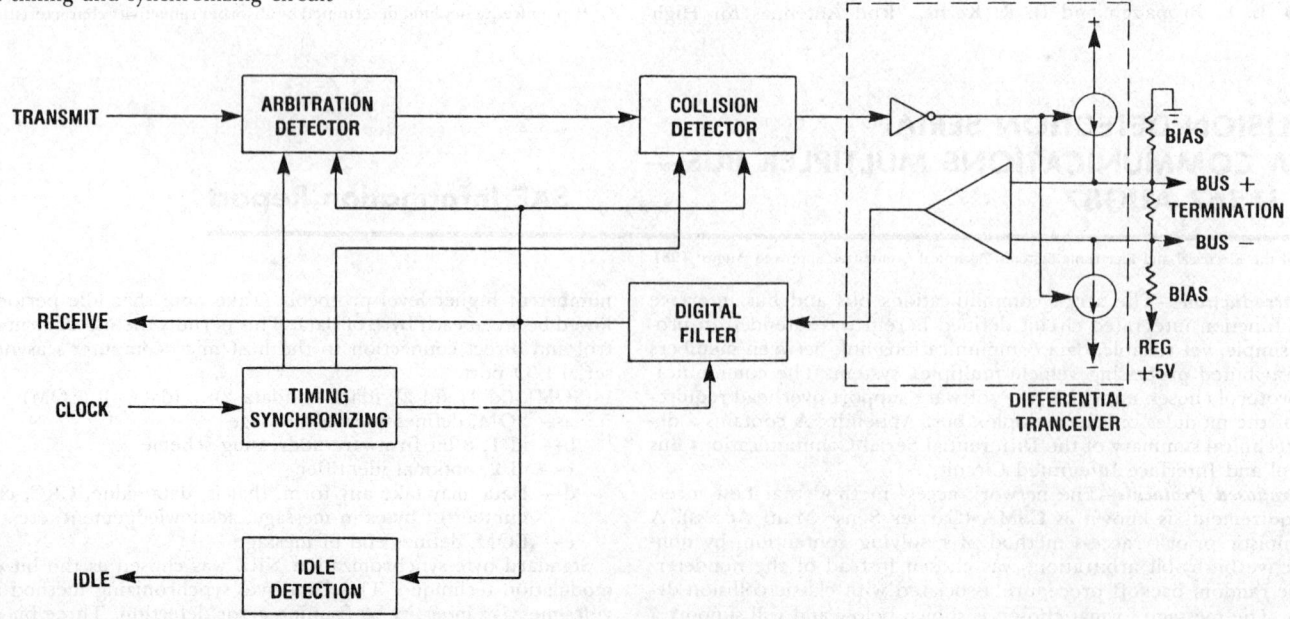

FIG. 1—SIMPLIFIED BLOCK DIAGRAM OF BUS INTERFACE

near synchronism, (less than or equal 1/4 bit time) the arbitration detector will allow the transmission. However, the data collision detector will permit only the one with the highest priority to continue transmission. The arbitration detector is also reset by the detection of a bus idle.

a valid start bit, it causes the Word Counter to synchronize itself to the timing of the received data word (8 bit byte). A valid start bit is determined by the arbitration detector to be negative signal that remains negative for 1/4 bit time after an idle period or a valid stop bit. The

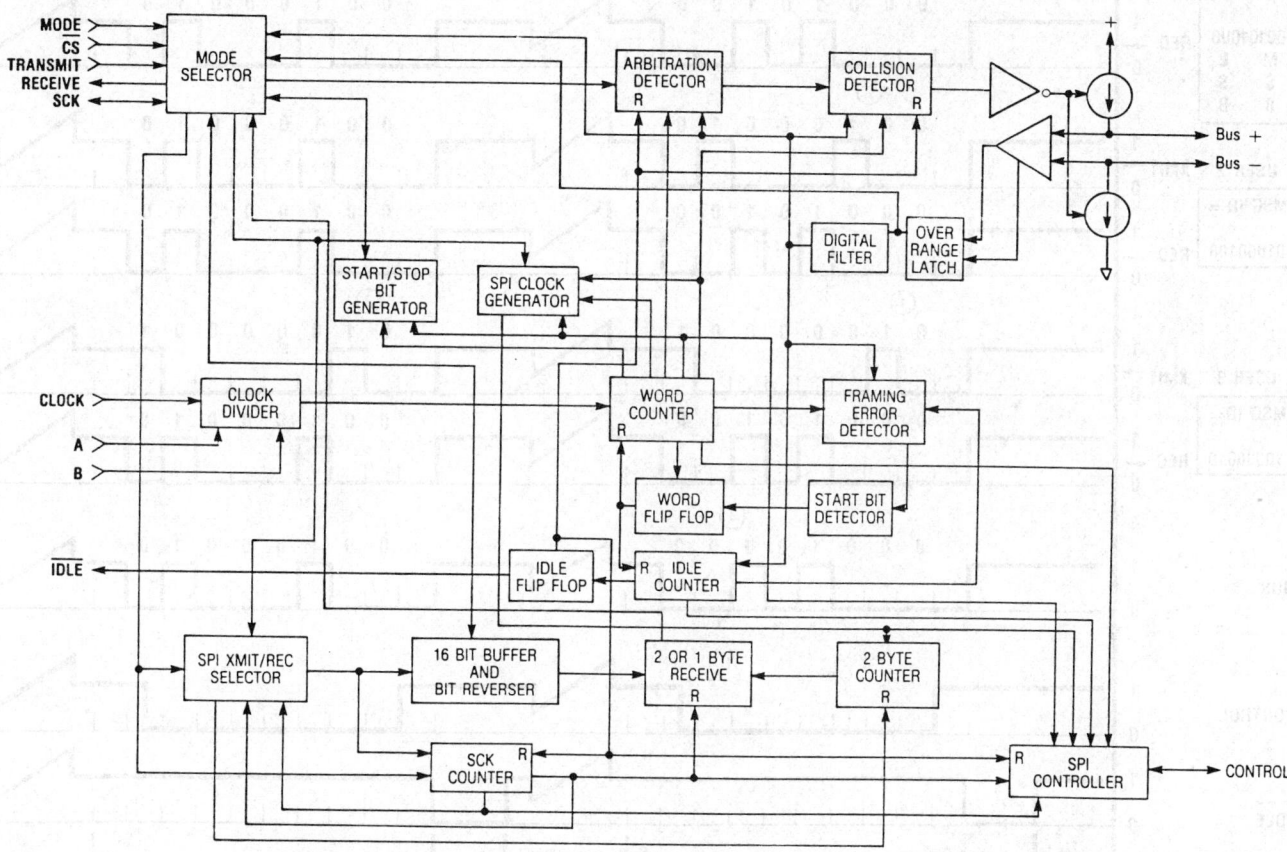

FIG. 2—ARBITRATION OF THREE SINGLE BYTE MESSAGES

3.4 Digital Filter—A low pass digital filter is placed between the transceiver and the received data output. This circuit functions to filter out any received EMI from the desired digital data signal before being processed by the other circuits of the interface.

3.5 Bus Idle Detector—The function of the bus idle detector circuit is to detect when the bus is idle (not active) or busy (active) and then feed this information back to the microcomputer. It accomplishes idle detection by sensing a received stop bit followed by ten (10) bits of continuous idle or logic ones. Normally an active or busy period follows an idle period. The sensing of an active or busy bus is accomplished by detecting a start bit. During unusual conditions such as node start up, any transition from a logic one to a logic zero that is maintained for a period longer 1/4 bit time sets the active or busy flag.

3.6 Timing and Synchronizing Circuit—The timing and synchronizer circuit uses an external clock and establishes the synchronizing and baud rate timing signal. This generator circuit first synchronizes on the negative edge of a start bit and then generates a timing signal at the center (1/2 bit time) of each bit. This timing signal is used by the arbitration and collision detector circuit for sampling received data. This timing signal is also used by the idle detection circuit to determine an idle bus.

4. Universal Interface IC—A block diagram of the integrated circuit developed that supports all three modes: (1) SCI, (2) SPI, and (3) Buffered SPI is shown in Fig. 3. Notice the additions to the diagram shown in Fig. 1. An Over Range Latch has been added between the differential receiver and the Digital Filter to hold the received data at the logic level it was before the out of dynamic range signal occurred. Also the timing and Synchronizing circuit is shown in greater detail.

4.1 SCI Mode of Operation—When the Start Bit Detector senses word counter is used to generate the 1/4 bit time (32 μs for 7812.5 baud) pulse for the arbitration detector and 1/2 bit time (64 μs for 7812.5 baud) pulse for the data collision detector. The word counter triggers the framing error detector at the stop bit time. If the stop bit is not detected, the Idle counter is extended by the framing error circuit until 10 consecutive idle bit periods are received.

When a microcomputer connected to the bus is ready for transmission, it should use the following procedure. First it reads the $\overline{\text{IDLE}}$ line and waits until it goes to a logic "0", which indicates the bus is idle. Then the microcomputer tries to transmit the 8 bit ID word associated with the data to be transmitted. If it started transmitting first or has the highest priority ID, the collision detector circuit will permit transmission. The microcomputer must confirm transmission by reading the received ID word and comparing it with the ID it wanted to transmit. If there is confirmation that the same ID was transmitted, then data can be transmitted. If not, then the microcomputer needs to compare the received ID with the list of required ID's in order to determine if it needs to process the received message. Please take note, the microcomputer need not cut off the data transmission because the collision detector will accomplish that.

It is important to realize that data collision may result due to outside interference. The microcomputer that is transmitting data can compare the transmitted data with the received data for this type of data collision. Appropriate action should then be taken by the microcomputer.

The function of the idle counter and idle flip flop circuit is to detect when the bus is in the idle (not active) condition. It does this by sensing a received stop bit and then delays a short idle period of ten (10) bit times. The idle detector output is then set to a logic "0". The idle detector output is set to a logic "1" by receiving a start bit.

4.2 SPI Mode of Operation—In the SPI mode the word counter generates the timing signals to drive the SPI clock generator and the start/stop bit generator. The microcomputer operates as an SPI slave. The interface IC operates as an SPI master. When the microcomputer

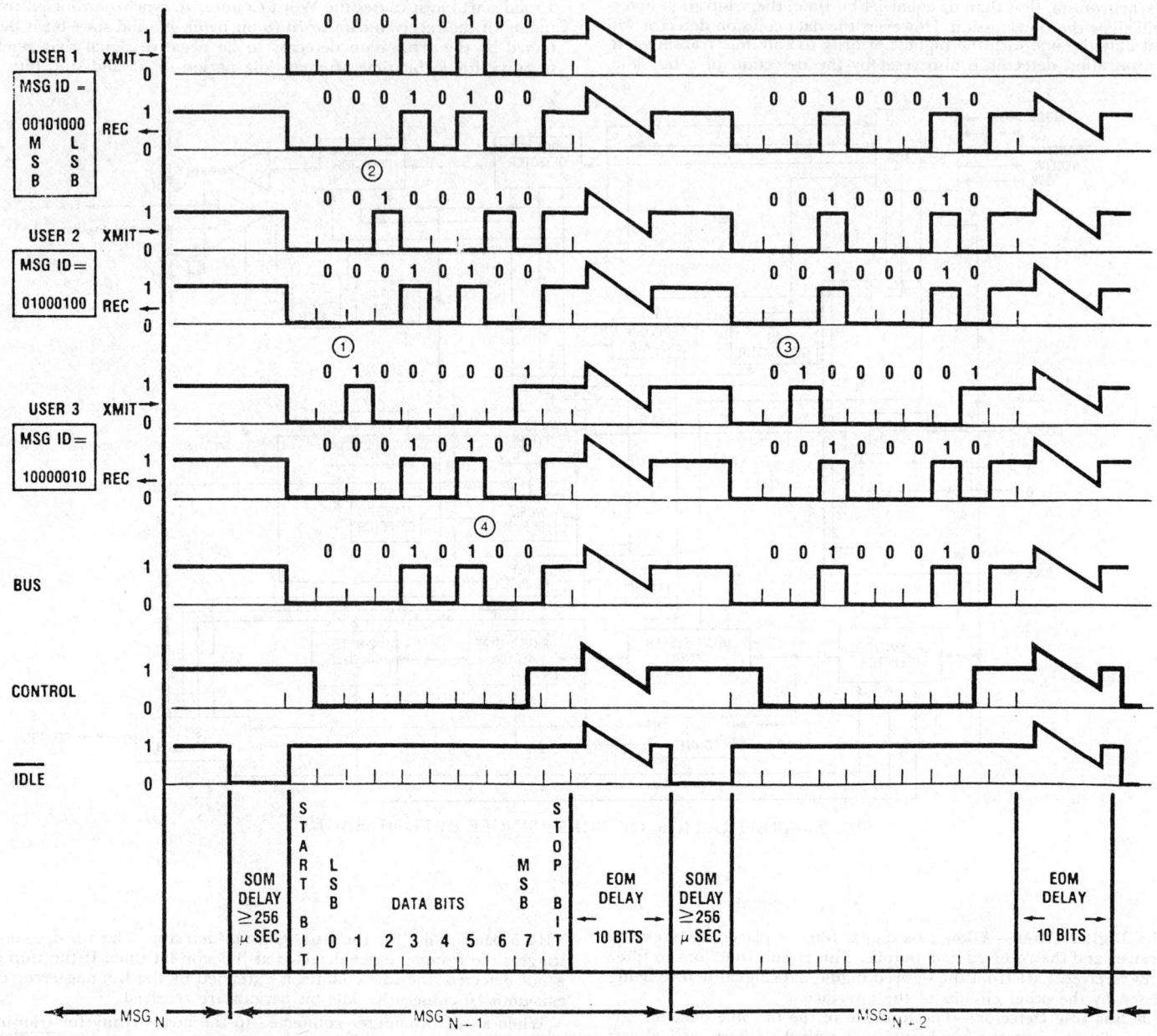

FIG. 3—BLOCK DIAGRAM OF UNIVERSAL BUS INTERFACE IC

needs to transmit a message, it loads the ID of that message into its SPI data register. The microcomputer should first monitor the $\overline{\text{IDLE}}$ pin to determine when the bus goes idle. The microcomputer then pulses the CONTROL pin to a logic "0". This sets the SPI controller circuit to schedule a transmission at the appropriate delay time after idle. The interface IC will: (1) then generate a start bit, (2) supply the microcomputer with 8 SCK shift pulses in synchronization with the start pulse and, (3) then generate a stop bit. If the transmitted ID collides with a higher priority ID transmitted by another module, the interface IC will stop transmitting immediately and transfer to the microcomputer the data it receives from the bus. If no collision is detected, the module has won arbitration and the microcomputer is free to complete data transmission. The microcomputer must monitor the CONTROL pin of the bus IC. The SPI controller circuit will set its output to a logic "1" when it completes the transmission of each ID or data word.

The SPI data buffer can then be loaded with the next word and the CONTROL pin must be momentarily pulled low to transmit that data. The SPI controller circuit works in conjunction with the start/stop bit generator and produces synchronized shift clock pulses for both receiving and transmitting of data. However, the start/stop bit generator does not output shift pulses for the start or stop bits.

The data on the bus must be transmitted least significant bit (LSB) first and most significant bit (MSB) last (Fig. 2). Generally, SPI transfers are done MSB first and LSB last. The data transmitted and received may have to be reversed within the microcomputer for the SPI mode of operation.

4.3 Buffered SPI Mode of Operation—In the buffered SPI mode, a 16 bit buffer and bit reverser circuit is provided for both receiving and transmitting data. This buffer is accessed by the microcomputer through the mode selector circuit. The order of the data bits is reversed as the data is transferred between the microcomputer and the buffer. The microcomputer SPI generally transfers data MSB first and LSB last while the data on the bus is transferred LSB first and MSB last. This insures that the order of the data on the bus while operating in the SCI, SPI, and buffered SPI modes is the same (LSB first). The 2 or 1 byte receive circuit works in conjunction with the buffer system and the 2 or 1 bytes counter to allow the reception of odd numbers of bytes in a message. The control flip flop in the SPI xmit/rec selector circuit is used to determine whether the buffer is connected to the microcomputer input or to the bus transmit circuitry. When powered up, the buffer is connected to the microcomputer.

The microcomputer is the SPI master and the IC is an SPI slave. In a system where there are multiple SPI slave units, the bus IC can be selected by using the $\overline{CS}$ pin. When the microcomputer wants to transmit, it selects the bus interface $\overline{IC}$ by outputting a "0" to the $\overline{CS}$ pin and then monitors the CONTROL pin. Before the microcomputer can transmit, it must wait for the completion of all bus activity. The SPI controller circuit monitors the idle flip flop and the SCK counter circuit to generate the control signal at the proper time. When the CONTROL pin goes to a logic "1" signifying that the buffer register is full of received data and can be read by the microcomputer, the microcomputer then supplies 16 shift pulses (two 8 bit SPI transfer cycles) and reads the received data at the same time it loads the SPI buffer with the ID and data it wants to transmit. The microcomputer then momentarily pulls the CONTROL pin to a logic "0" and the data will be transmitted at the proper time. If the microcomputer just needs to access the received data it just does the two 8 bit SPI transfers and does not pulse the CONTROL pin low. The interface IC contains the circuitry to hold the received data in the buffer register and ignore receiving new data until after the old data has been read. The maintenance of the received data in the buffer register until it is accessed by the microcomputer insures that the transmitted data can be tested even in the case of a slow data transfer. It is important to test all attempts at bus access to verify when correct transmission has been completed and when it will be necessary to retransmit the message.

5. An Item For Future Development—The normal method used to detect message data error is a firmware checksum scheme. Based on the following statements, a good alternative to this practice may be the detection of Out of Common Mode Range.

   a) All electrical interference that would corrupt data in a vehicle network is single ended; for example, it affects both Bus+ and Bus− equally.
   b) Maintaining a sufficiently high ratio between single ended interference to differential interference using a properly terminated twisted pair transmission media may be feasible.
   c) Therefore, if the above paragraphs are attainable then all electrical interference that would corrupt data would first generate an out-of-range condition. Also, out-of-range integrity checking of the data is much simpler than the firmware generated method and does not have the bandwidth reduction.

6. Conclusions—The quality of a serial communications bus protocol should be judged by the following factors:

6.1 Low Cost—Cost is contained by achieving a very simple hardware interface IC design and the handling of the bus protocol in microcomputer ROM firmware.

6.2 Universally Microcomputer Friendly—This goal is achieved by being capable of directly connecting to the microcomputer's SCI or SPI serial communications I/O ports.

6.3 Low Generation of EMI while Maintaining High Bus Message Throughout—Byte sync NRZ data modulation was chosen because it has the advantage of being the most efficient modulation method, that is, has fewer transitions for data byte. Empirical testing shows that this translates to low generation of EMI.

6.4 Low Data Error Rate—This goal was achieved by using differential current mode NRZ data modulation. NRZ data demodulation has been shown to achieve superior data recovery performance in a noisy environment and can be accomplished using simple data sampling techniques.

6.5 Reasonably Fault Tolerant—The Interface IC is protected to survive bus shorts to battery and ground. Microcomputer watchdog circuits can be implemented to protect against bit streaming failures. These measures are not sufficient to guarantee data communications. The known methods of overcoming or partially overcoming many of these failure modes are either too complex or generates too much EMI to be practical. The best protection is for the system designer to realize this when partitioning the vehicle system. The few failure conditions that could cause a safety problem can be handled using other partitioning options or with limp in modes. Another option, of course, is to separately hard wire these few safety related situations.

APPENDIX A—TECHNICAL SUMMARY
General Bus Interface IC Functionality

A1. Bus Message Format

<msg id 1> <msg id 2> ... <data byte 1> <data byte 2> ... <data byte n>

The msg id and data bytes are individual bytes that are asynchronously transmitted with 1 start bit, 8 data bits and 1 stop bit using a typical NRZ format with a '1' level signal for a bus idle and stop bit value and a '0' level signal for a start bit value.

Interbyte Delay—The maximum delay between bytes is <10 bit times. The minimum delay between bytes is 0 bit times.

Idle—A continuous period 10 bit times long of a '1' level signal after the detection of a proper stop bit is interpreted as an IDLE condition. IDLE is used to synchronize all stations to the condition that: 1) the last message on the bus is complete, 2) the next byte on the bus will be the first msg id byte of a message, 3) any stations that have messages to transmit can begin the transmission 2 bit times later.

Use of Msg Id byte(s)—The msg id byte(s) are used to determine the winner in message collision situations. The msg id can be 1 or more bytes.

Use of Data Bytes—The data bytes can be anything (that is, CRC byte, Checksum, etc.).

Minimum Message Size—The minimum message size is 1 byte (that is, 1 msg id byte).

Maximum Message Size—The maximum message size is unlimited (Defined by host microcomputer ROM firmware).

A2. Control Procedure—The control procedure determines when a station may attempt to transmit on the bus and who wins the use of the bus if multiple stations attempt to transmit on the bus at the same time.

A collision occurs on the bus if more than 1 station attempts to transmit a msg id byte within $\frac{1}{4}$ bit time of each other. Stations that attempt to transmit a msg id byte later than $\frac{1}{4}$ bit time after the first station began to transmit have already lost bus arbitration.

If multiple stations transmit the same msg id byte at the same time, then these stations will continue to arbitrate with each other for the use of the bus using the next msg id byte just like the first msg id byte (that is, the winner is the station that transmits the next msg id byte first or whose next msg id byte has the highest priority value).

The priority of msg id bytes is determined by the rule that a '0' bit will win out over a '1' bit. As soon as a given station loses during bit by bit comparison, it is blocked from transmitting any more bits onto the bus. Bit by bit comparison includes all bits of a byte (that is, start bit, data bits and stop bit) but normally is determined by the value of the data bits. Data bit transmission begins with the LSB and continues on to the MSB.

All stations receive the data that is actually transmitted on the bus. Thus, a station that is attempting to begin transmitting a message will receive the msg id of whatever message won the use of the bus.

Abort—Stations that receive a msg id different from the one that they were attempting to transmit have collided with the transmission of a higher priority message from another station and have lost. Stations that detect this should stop attempting the transmission of their message and should consider switching over to receiving the winning message if it is something that they want to receive.

Defer—A station that loses at message arbitration should immediately stop further transmission of message bytes but it does not have to. The bus interface IC will automatically cut off its transmitter from affecting the bus as soon as that station has lost to another station and will not allow the losing station to transmit onto the bus until IDLE has occurred again. EMI that causes message corruption has the same effect and the IC will automatically cut off transmission.

Retransmission—A station that loses at bus arbitration can attempt to retransmit its message as soon as IDLE occurs again.

Number of Stations—Any number of stations may be connected to that same network.

Transmission of Start Time—The transmission of the first msg id byte shall be such that at least a nominal 2 bit times delay exists between IDLE and the beginning of the transmission of the start bit.

Bus Driver Interface—The bus driver drives the bus through a 'wired OR' type of connection where any single station transmitting a '0' level signal forces the bus to a '0' level even if multiple stations attempt to simultaneously transmit a '1' level signal.

Hardware Collision Detection and Bus Arbitration—The hardware collision detection function is reset at IDLE. Bus arbitration begins when one or more stations begin to transmit a message after IDLE. Collision begins when more than one station attempts to transmit a message on the bus at the same time. Bus arbitration and any collisions end: 1) at $\frac{1}{4}$ bit time after the beginning of a start bit (that is, the first '0' level signal that lasts for at least $\frac{1}{4}$ bit time and occurs after a full stop bit) if the 'O' level signal lasts for longer than ¼ bit time and if only one station is attempting to transmit or 2) when the last remaining station(s) lose a bit by bit priority comparison with the resulting winning station due to the 'wired OR' nature of the hardware bus drivers, if multiple stations began transmitting a start bit within the ¼ bit time of each other.

A3. Data Rates
Frequency Range—Any value defined by input clock; recommend 7812.5 Baud
Accuracy Requirements—+ or − 2.0% for best results < + or − 0.5%

A4. Cable Requirements
Impedance—120 ohm conventional automotive
Loss—Negligible
Physical—Differential twisted pair > 1 twist/inch

A5. Connector Requirements and Terminators
The connectors are of standard automotive quality as defined by the vehicle manufacturer. The wire shall be constructed as twisted differential pair with termination resistors at both physical ends of the wire. T taps may be used to connect other modules to the network than those connected at the terminated ends, but the length of the pair of wires between the tap and the module must be short (< a meter) to minimize EMI.

A6. Transceiver
Connection Rules—A large number of transceivers may be placed on the bus. The limit is a function of the number and value of bias resistors.
Cable Bias Rules—The differential Bus+ shall be connected to a bias resistor to ground and Bus− shall be connected to an identical resistor to transceive power.
Bias Resistor—< 13K ohms
Cable Interface—(5 volt operation 2.5 volt bias)
Input Leakage Shut Down Current—10 microamps source or sink from Bus+ or Bus−
Signal Output Range—>2.75 mA Source and Sink for logic level '0', <1 Microamp for logic level '1', < 5% Current match
Receiver Differential Sensitivity—> 120 mV logic level '0', < 20 mV logic level '1', > 20 mV hysteresis
Signal Rise and Fall Time—< 1.5 microsec with ± 50 nanosec Transition Match (50% point)
Receiver Dynamic Range—1.2 − 3.8 V
Transceiver IC to Microcomputer Interface
Power—Vcc = 4.0 - 7.0 V power supply
Power Down Mode—I/O in the high impedance state
Failure Modes Protection—Protected against shorts to battery and ground. Protected against SCR latch up during power up or down.
Signal Definitions
Input Voltage—0.7 × Vcc and 0.3 × Vcc V
Output Voltage—Vcc 0.05 and 0.05 V

CONTROLLER AREA NETWORK (CAN), AN IN-VEHICLE SERIAL COMMUNICATION PROTOCOL—SAE J1583 MAR90

SAE Information Report

Report of the Vehicle Network for Multiplexing and Data Communications Standards Committee approved March 1990.

Foreword—Controller Area Network (CAN) is an advanced serial communication protocol which efficiently supports distributed real-time control with a very high level of security.

Its domain of applications range from high speed networks (greater than 125K bits/s) to low speed multiplex wiring (10K bits/s or less).

1. Scope
The scope of this specification is to define the transfer layer and the consequences of the Controller Area Network (CAN) protocol on the surrounding layers.

2. General Features
CAN has the following general features:

MULTIMASTER ARCHITECTURE—Assurance that any node can have access to the bus for transmission of messages.

PRIORITIZATION OF MESSAGES—Bus access is determined by the selectable predetermined priority of the 2032 different messages allowable.

GUARANTEE OF LATENCY TIME—Assurance that the maximum latency time for the highest priority message is less than 150 μs (at 1M bits/s).

BROADCAST MESSAGE TRANSFER—Any number of nodes can receive and simultaneously act upon the same message.

MESSAGE ARBITRATION—If two or more senders start transmitting messages at the same time, the bus access conflict is resolved by nondestructive prioritized bitwise arbitration via the message identifier.

DATA CONSISTENCY—Assures that a message is simultaneously accepted either by all nodes or by no node.

PROGRAMMABLE TRANSFER RATE—The bit-rate is programmable up to 1m bit/s; once set it is uniform and fixed in the system.

POWERFUL ERROR HANDLING—Includes automatic retransmission of corrupted messages and distinction between temporary errors and permanent failures of nodes.

CONFIGURATION FLEXIBILITY—Nodes may be added without requiring any change in the software or hardware of any node or application layer.

FAULT DETECTION—Distinction between temporary errors and permanent failures of node and autonomous switching off of defect nodes.

3. CAN Architecture
To achieve design transparency and implementation flexibility CAN has been developed structurally in layers. Different silicon implementations will integrate these layers to different depths in hardware. This information report will address the protocol, or bus communication rules, that all implementations must support or have the capability of supporting. One implementation example already in production is the Intel 82526. An overview of its functionality and use is provided in Appendix A. Another such implementation is BasicCAN. It is not within the scope of this document to cover BasicCAN in detail. Future updates may provide additional information.

The layers are:
a. Object Layer—The scope of the object layer includes:
  1) Deciding which messages received by the transfer layer are actually to be used (acceptance filtering)
  2) Determining which messages are to be transmitted (prioritized message handling)
  3) Providing an interface to the host CPU if there is a host in the system.
b. Transfer Layer—The scope of the transfer layer includes:
  1) Error checking
  2) Error signalling and confinement
  3) Message validation and acknowledgement
  4) Performing arbitration
  5) Controlling message framing
  6) Fault confinement
  7) General features of bit timing
c. Physical Layer—The scope of the physical layer is the actual transfer of the message bits between the different nodes in the network with respect to all electrical properties. This includes: Signal level, bit representation, and transmission medium. The protocol does not specify the signal level and transmission medium. This is done intentionally to allow the user or system designer to adapt CAN to the specific application and application environment. Within one network the physical layer, of course, has to be the same for all nodes. There may be, however, much freedom in selecting a physical layer.

4. Bit Representations, Coding/Decoding, and Stuffing

4.1 Bit Representations—There are two logical bit representations defined: "dominant" and "recessive". A recessive bit on the bus line appears only if all connected nodes at that time send a recessive bit. If one or more nodes send a dominant bit, the bus line reflects a dominant bit. In other words, a dominant bit always overwrites a recessive bit if emitted on the bus at the same time by a different node. If the electrical bus is implemented as an npn-open-collector bus, a dominant bit corresponds to a low level signal and a recessive bit corresponds to a high level signal.

4.2 Bit Coding/Decoding and Stuffing—Bit encoding is performed with an NRZ bit waveform representation using bit stuffing. Bit stuffing is used in the following frame segments:
  a. Start of frame
  b. Arbitration field
  c. Control field

d. Data field
e. CRC sequence

Whenever the transmit logic of CAN device detects five consecutive bits of identical levels to be transmitted, it inserts a complement bit in the transmitted bit stream. Whenever a CAN device receives five identical consecutive bit levels in the bit stream, its receive logic will automatically delete the next bit from the data stream (destuffing). See Figure 1 for further detail.

5. Arbitration—The CAN protocol defines that each single message used in the communication network has a unique identifier. Using this method, the identifier assigns a name to the data frame and automatically implies the priority of the message. As a result, the identifier during bus access represents not only the message name but, more important, the priority of each specific message. Since the most significant bit (MSB) of an identifier is transmitted first, the identifier with the smallest digital value has the highest priority for bus access.

When bus idle is detected, any node may start to transmit a message. In a case where two or more CAN devices start a message transfer concurrently, the bus access conflict is solved by prioritized bitwise arbitration performed during the transmission of the arbitration field.

The transmit logic of a given node compares the bit level transmitted to the level monitored on the serial bus. The transmit logic immediately will stop a current message transfer if a recessive bit was sent but a dominant bit was monitored. This method guarantees the data transfer of the message with the highest priority even if there is a collision during the arbitration field of one or more messages.

An identifier can never be used for more than one specific message (total available number of identifiers = 2032). This guarantees that two or more nodes never simultaneously start a transmission of a data frame with the same priority of data. One exception would be a frame transfer simultaneously initiated by a transmitter and receiver. If one CAN device generates a request for actual data of a certain type by transmitting a remote frame and the CAN device normally transmitting that data transmits simultaneously, the CAN device responsible for this type of data will win the arbitration and continue the transmission. Hence, arbitration can not be solved by the identifier itself. For this reason, the remote transmission request (RTR-BIT) is included in the arbitration field. The RTR-BIT of the transmitter is always set dominant and, therefore, has a higher priority than the requesting RTR-BIT recessive. In this case, the remote transmission request by the receiver gets an immediate response by the transmitter.

6. CAN Message Frame Types— The CAN device supports four different frame types:
a. Data frame
b. Remote frame
c. Error frame
d. Overload frame

6.1 Data Frame—A Data Frame is composed of seven different fields:
a. Start of frame
b. Arbitration field
c. Control field (Identifier)
d. Data field (Data segment)
e. CRC field
f. Acknowledgement field
g. End of frame

6.1.1 START OF FRAME—Signals the start of a data or remote frame. It consists of a single dominant bit used for a synchronization of receiving nodes. The CAN device will start a transmission only if a bus-idle state is detected. All nodes must synchronize to the leading edge of the node starting transmission first.

6.1.2 ARBITRATION FIELD—Consists of the message identifier and one

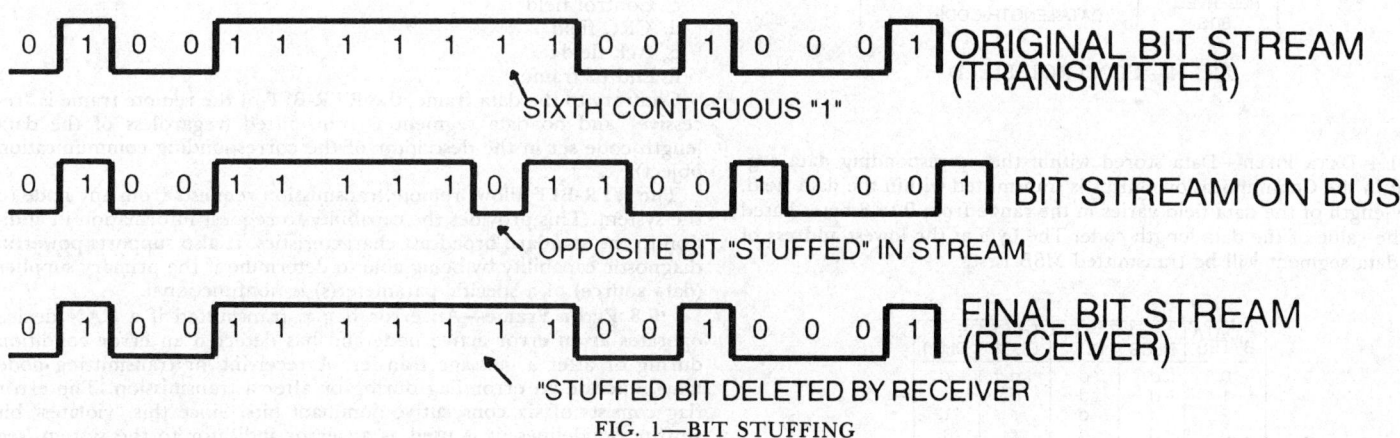

FIG. 1—BIT STUFFING

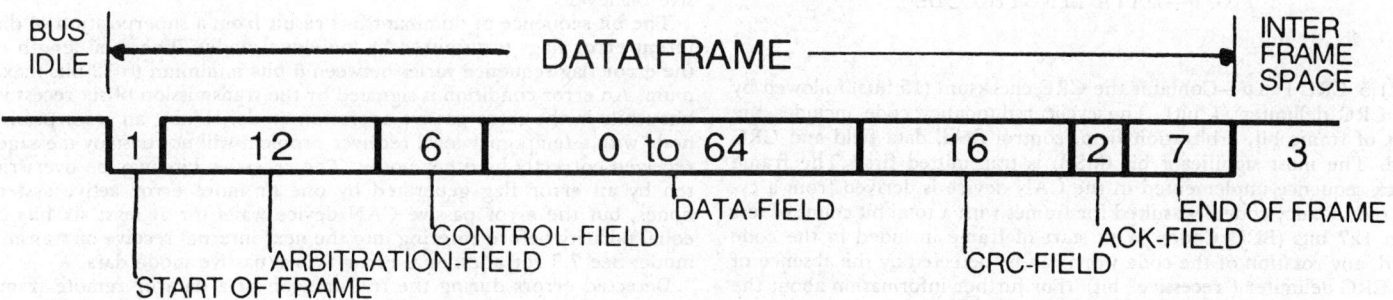

FIG. 2—DATA FRAME

additional control bit RTR-BIT. The simultaneous message transmission start of two or more nodes is solved by bitwise arbitration during the transmission of the arbitration field (for more details see Section 5).

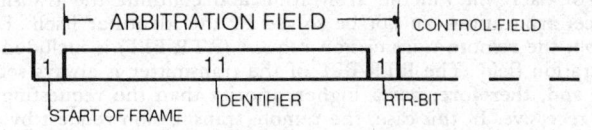

FIG. 3—ARBITRATION FIELD

6.1.2.1 *Identifier*—This 11-bit field provides information about each individual message as well priority for the message. The identifier defines the corresponding specific message (e.g., engine speed, temperature, pressure, etc.) of a communication object. It's digital value represents the message priority for bus access.

NOTE: An identifier may be a physical or functional address or a combination of the two to determine a receiver of a frame on a multi-drop line. The receiving CAN device decides, based on the acceptance filter process of a received identifier, whether data being received within a correct frame is to be accepted or not.

6.1.2.2 *RTR-BIT*—(Remote Transmission Request) A station, active as a receiver may initiate the transmission of the data by transmission of a remote frame to the network, addressing the data source via the identifier. Within a data frame, the RTR-BIT is transmitted as a dominant bit level (for more information see 6.2).

6.1.3 CONTROL FIELD—This field consists of 6 bits. It includes the data length code and two reserved bits. The reserved bits are automatically transmitted with a dominant bit level. The length of the data field (data segment) is coded in bytes.

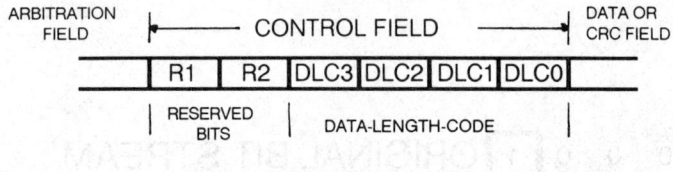

FIG. 4—CONTROL FIELD

6.1.4 DATA FIELD—Data stored within the corresponding data segment in the Communication Buffer is transmitted within the data field. The length of the data field varies in the range from 0 to 8 bytes based on the value of the data length code. The byte at the lowest address of the data segment will be transmitted MSB first.

# DATA BYTES	DATA-LENGTH-CODE			
	DLC3	DLC2	DLC1	DLC0
0	d	d	d	d
1	d	d	d	r
2	d	d	r	d
3	d	d	r	r
4	d	r	d	d
5	d	r	d	r
6	d	r	r	d
7	d	r	r	r
8	r	d	d	d

"d" = dominant, "r" = recessive

FIG. 5—DATA LENGTH CODE

6.1.5 CRC FIELD—Contains the CRC checksum (15 bits) followed by the CRC delimiter (1 bit). The cyclic redundancy code includes the start of frame bit, arbitration field, control field, data field and CRC field. The most significant bit (MSB) is transmitted first. The frame check sequence implemented in the CAN device is derived from a cyclic redundancy code best suited for frames with a total bit count of less than 127 bits (BCH Code). With start of frame included in the code word, any rotation of the code word can be detected by the absence of the CRC delimiter ("recessive" bit). (For further information about the CRC, see 7.6.)

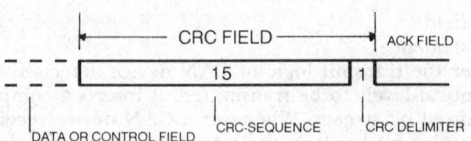

FIG. 6—CRC FIELD

6.1.6 ACKNOWLEDGEMENT FIELD—The ACK-field consists of two bits, the ACK-slot and ACK-delimiter which are sent with a "recessive" level by the transmitter. Any receiving CAN device acknowledges to the transmitting CAN device a match of the complete/correct CRC check sum within the ACK-slot by superscribing this recessive bit with a dominant bit. A transmitter monitoring the bus level recognizes that at least one receiver within the system has received a complete and correct message (no error was found). The ACK-SLOT is surrounded by two recessive bits—The CRC delimiter and the ACK delimiter.

6.1.7 END OF FRAME—Each data frame or remote frame is delimited by the end of frame bit sequence which consists of seven "recessive" bits (exceeding the bit stuff width by 2 bits). A receiver detects the end of a data frame independent of a previous transmission error because the

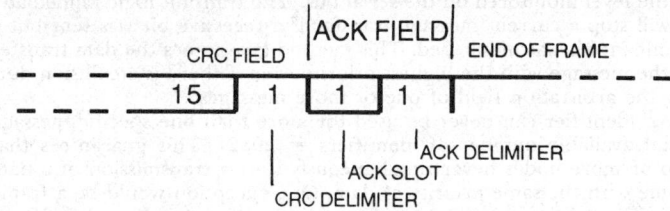

FIG. 7—ACKNOWLEDGEMENT FIELD

receiver expects all bits up to the end of CRC-field to be coded by the bit stuffing method and bit stuffing is not used in end of frame.

6.2 Remote Frame—A remote frame is composed of six different fields:
a. Start of frame
b. Arbitration field
c. Control field
d. CRC field
e. Ack field
f. End of frame

Contrary to the data frame, the RTR-BIT of the remote frame is "recessive" and no data segment is transmitted (regardless of the data length code set in the descriptor of the corresponding communication object).

The RTR-BIT allows remote transmission requests from any node to the system. This provides the capability to request information in addition to the standard broadcast characteristics. It also supports powerful diagnostic capability by being able to determine if the primary supplier (data source) of a specific parameter(s) is nonfunctional.

6.3 Error Frame—An error flag is transmitted if a CAN device operates as an error active node and has detected an error condition during or after a message transfer. A receiving or transmitting node may generate an error flag during or after a transmission. The error flag consists of six consecutive dominant bits. Since this "violates" bit stuffing guidelines, it is used as an error indicator to the system (see 4.2). If an error flag is generated by a transmitter, or a receiver, all other nodes interpret the error flag as a bit stuffing rule violation. As a consequence, the transmission of an error flag occurs.

The error-delimiter consists of seven recessive bits generated by the CAN device after the end of an error flag on the serial bus line. This is monitored by detection of a transition from the dominant to recessive bit level.

The bit sequence of dominant bits result from a superposition of different error flags transmitted by individual nodes. The total length of the error flag sequence varies between 6 bits minimum to 12 bits maximum. An error condition is signaled by the transmission of six recessive bits while in the error passive operation mode. Hence, an error passive node with a temporary local receiver problem will not destroy messages received correctly by other nodes. The recessive bits may be overwritten by an error flag generated by one or more error active system nodes, but the error passive CAN device waits for at least six bits of equal polarity before entering into the next internal receive or transmit mode. See 7.7 for additional error active/passive mode data.

Detected errors during the transmission of a data or remote frame can be signaled within the transmission time of the respective frame.

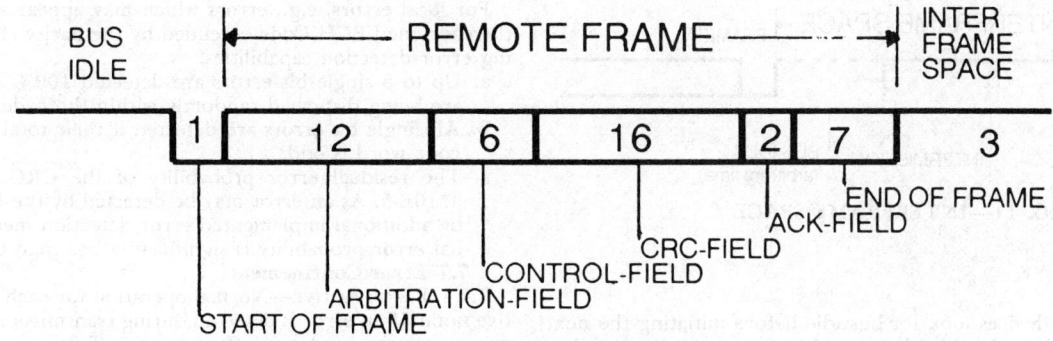

FIG. 8—REMOTE FRAME

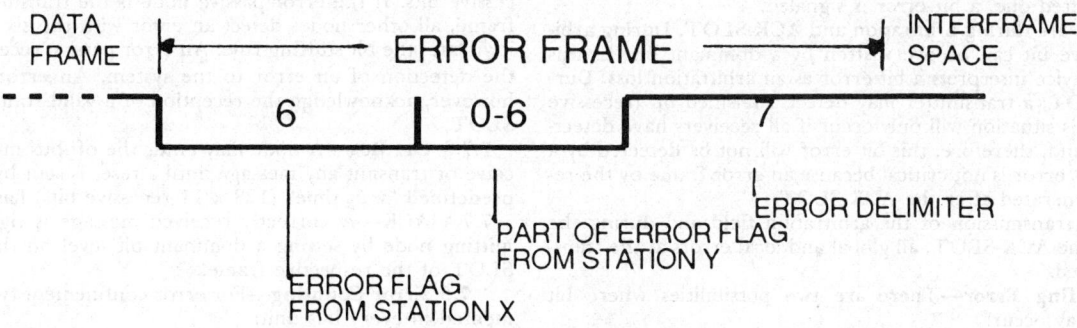

FIG. 9—ERROR FRAME

This procedure associates an error flag to the frame and initiates an automatic retransmission of the frame. If the CAN device detects any deviation of its error frame it will start retransmitting an error frame. If this occurs several times in a sequence the CAN device will become error passive.

6.4 Overload Frame—The overload frame consists of 2 bit fields: The Overload Flag and the Overload Delimiter.

There are two cases of overload conditions which result in the transmission of an overload flag:
  a. Internal conditions of the receiver circuitry of the CAN device which require a delay time before receiving the next frame (receiver not ready).
  b. Detection of a "dominant" bit during intermission.

At most, two overload frames may be generated to lengthen one data frame or remote frame.

The overload flag consists of six dominant bits that correspond to the error flag and destroy the fixed form of the intermission field. As a result, all other stations also detect an overload condition and start transmission of an overload-flag. If a "dominant" bit is detected during the 3rd bit of intermission by some but not all nodes, the other nodes will interpret the first of these six 'dominant' bits as start-of-frame. The sixth "dominant" bit will then violate the bit stuffing rule and cause an error condition. The overload delimiter consists of seven recessive bits generated by the CAN device.

NOTE: An overload frame can be transmitted earliest at the first bit time of the interframe space field. This is contrary to the error frame and allows the CAN device to differentiate between the two frame types.

6.5 Interframe Space—Data frames and remote frames are separated from preceding frames by an interframe space consisting of the intermission bit field and a possible bus idle time. An error frame is not preceded by an interframe space.

Intermission consists of three recessive bits. During intermission time the CAN device will not start transmission of a frame. Intermission requires a fixed time period for the CAN device to execute internal processes prior to the next receive or transmit task.

Data received within a data frame will be stored in the communication buffer and the control bits are updated if no error condition has occurred through last bit of the end of frame field.

The bus idle time may be of arbitrary length. After the interframe

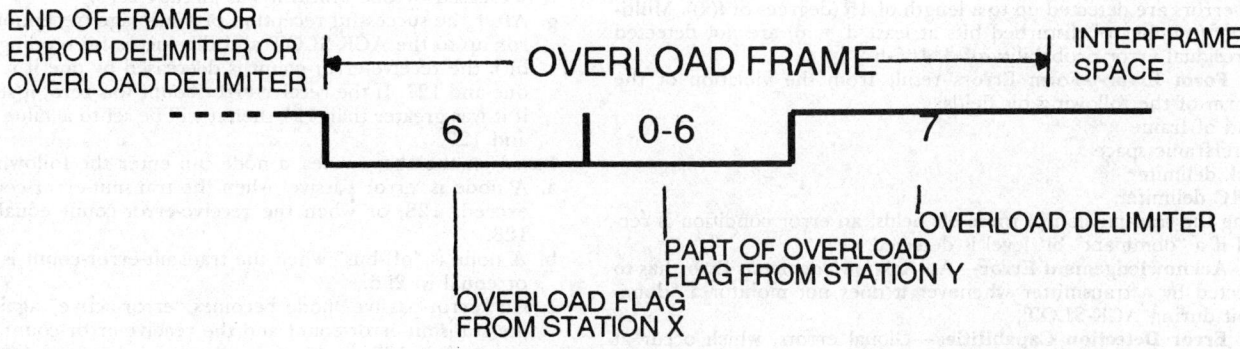

FIG. 10—OVERLOAD FRAME

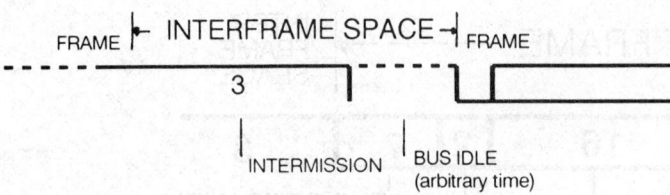

FIG. 11—INTERFRAME SPACE

space period, CAN devices look for bus idle before initiating the next transmissions. The detection of a dominant bit after intermission or bus idle is interpreted by the CAN device as start of frame.

7. Error Detection

7.1 Bit Error—During a transmit operation, the CAN device monitors the bus on a bit-by-bit basis. If the bit level monitored is different from the transmitted one, a bit error is signaled.

The exceptions are during arbitration and ACK-SLOT. During arbitration, a recessive bit can be overwritten by a dominant bit. In this case, the CAN device interprets a bit error as an arbitration loss. During the ACK-SLOT, a transmitter may detect a falsified bit (recessive to dominant). This situation will only occur if all receivers have detected a CRC error and, therefore, this bit error will not be detected by a CAN device. This error is not critical because an error frame by the receivers will be generated after the ACK-SLOT.

Except during transmission of the arbitration field and during the time window of the ACK-SLOT, all global and local errors at the transmitter are detected.

7.2 Bit Stuffing Error—There are two possibilities where bit stuffing errors may occur:
 a. A disturbance generates more consecutive bits of equal level than allowed by the rule of bit stuffing. These errors are detected by all nodes.
 b. A disturbance falsifies one or more of the 5 bits preceding the stuff bit. This error is not recognized by a receiver, but if an error appears at the transmitter as well, it will be detected as a bit error.

Otherwise, the error is detected by a receiver either by the bit stuffing mechanism (the stuff bit of transmitter is not dropped but taken as an information bit) or by the CRC check.

7.3 CRC Error—To ensure the validity of a transmitted message, all receivers perform a CRC check. In addition to the information digits, any code word includes control digits used for error detection.

7.3.1 DESCRIPTION OF THE CRC CODE—The code used for the CAN device is a (shortened) BCH Code, extended by a parity check and has the following attributes:
 a. 127 bits as maximum length of the code word
 b. 113 bits as maximum number of information digits
 c. Length of the CRC SEQUENCE 15 bits
 d. Hamming distance d = 6
  d = min A (x EXOR y) / x,y different code words
  A(x) = number of "recessive" bits in the code word x

As a result, d-1 random errors are detectable (some exceptions exist).

The CRC sequence is determined by the request that the code word, if interpreted as polynomial with coefficients 0 or 1 is devisable by the polynomial.

$$f(x) = (x^{E14} + x^{E9} + x^{E8} + x^{E6} + x^{E5} + x^{E4} + x^{E2} + x + 1)$$
(Eq. 1)
$$= 1100010110011001 = 0C599 \text{ HEX}$$

Burst errors are detected up to a length of 15 (degrees of f(x)). Multiple errors (number of disturbed bits at least d = 6) are not detected with a residual error probability of $3*10E-5$.

7.4 Form Error—Form Errors result from the violation of the fixed form of the following bit fields:
 a. End of frame
 b. Interframe space
 c. Ack delimiter
 d. CRC delimiter

During the transmission of these bit fields, an error condition is recognized if a "dominant" bit level is detected.

7.5 Acknowledgement Error—An Acknowledgement Error has to be detected by a transmitter whenever it does not monitor a "dominant" bit during ACK-SLOT.

7.6 Error Detection Capabilities—Global errors, which occur at all fully functional nodes, are 100% detectable.

For local errors, e.g., errors which may appear at some nodes only, the shortened BCH Code extended by the parity check has the following error detection capabilities:
 a. Up to 5 single bit errors are detected 100% even if those errors are being disturbed randomly within the code word.
 b. All single bit errors are detected if their total number within the code word is odd.
 c. The residual error probability of the CRC check is 2E-15 or $3*10E-5$. As an error may be detected by the CRC check, and/or by additional implemented error detection mechanisms, the residual error probability is significantly less than the above.

7.7 Error Confinement

7.7.1 ERROR ACTIVE—Normal operation for each node is an error active node. If an error is detected during transmission of a frame, a node enters into the send error flag state (see 6.3).

7.7.2 ERROR PASSIVE—An error passive node, like an error active node, may function as a receiver and/or transmitter, but actions based on transmit and/or receive error conditions are different. As an example, after detection of an error, an error passive node will send six recessive bits. If the error passive node is the transmitter of a disturbed frame, all other nodes detect an error with the six recessive bits since it violates the bit stuffing-rule. An error passive receiver does not signal the detection of an error to the system. An error passive node will, however, acknowledge the reception of a valid frame during the ACK-SLOT.

7.7.3 OFF BUS—A node may enter the off-bus mode and will not receive or transmit any message until a reset is sent by the host CPU and predefined "wait time" (128 x 11 recessive bits) has passed.

7.7.4 ACK—A correctly received message is signaled to the transmitting node by setting a dominant bit level on the bus in the ACK-SLOT of the respective frame.

7.8 Error Counting—For error confinement two counts are implemented in every bus unit:
 a. Transmit-Error-Count
 b. Receive-Error-Count

These counts are modified according to the following rules: (note that more than one rule may apply during a given message transfer)
 a. When a receiver detects an error the receive-error-count will be increased by one, except when the detected error was a bit error during the sending of an active-error flag.
 b. When a receiver detects a 'dominant' bit as the first bit after sending an error flag the receive-error-count will be increased by eight.
 c. When a transmitter sends an error flag the transmit-error-count is increased by eight.
  Exception 1: If the transmitter is 'error-passive' and detects an acknowledgement error because of not detecting a 'dominant' ACK and does not detect a 'dominant' bit while sending its passive-error flag.
  Exception 2: If the transmitter sends an error flag because a stuff error occurred during the arbitration whereby the stuff bit should have been 'recessive' and has been sent as 'recessive' but monitored as 'dominant'.
  In exception 1 and 2 the transmit-error-count is not changed.
 d. If an 'error-active' transmitter detects a bit-error while sending an error flag.
 e. If an 'error-active' receiver detects a bit error while sending an error flag, the receive-error-count is increased by eight.
 f. After the successful transmission of a message (getting ACK and no error until end of frame is finished) the transmit-error-count is decreased by one unless it was already zero.
 g. After the successful reception of a message (reception without error up to the ACK-SLOT and the successful sending of the ACK bit), the receive-error-count is decreased by one if it was between one and 127. If the receive-error-count was zero, it stays zero, and if it was greater than 127 then it will be set to a value between 119 and 127.

Based on the above rules, a node can enter the following states:
 a. A node is 'error passive' when the transmit-error-count equals or exceeds 128, or when the receive-error-count equals or exceeds 128.
 b. A node is "off-bus" when the transmit-error-count is greater than or equal to 256.
 c. An "error-passive" node becomes "error-active" again when both the transmit-error-count and the receive-error-count are less than or equal to 127.

d. A node which is "off-bus" is permitted to become "error-active" (no longer "off-bus") after reset with its error counters both set to 0 after 128 occurrences of 11 consecutive "recessive" bits have been monitored on the bus.

An error count value greater than about 96 indicates a disturbed bus. It may be of advantage to provide means to test for this condition.

If during system start-up only one node is on-line, and if this node transmits some message, it will get no acknowledgement, detect an error and repeat the message. It can become "error-passive" but not "off-bus" due to this reason. If a wake-up message is used, a node which sends this message can become "error-passive" for the same reason.

8. Bit Timing—A bit time is subdivided in a number of BTL cycles. This number results from an addition of the segments SJW 1, SJW 2, TSEG 1 and TSEG 2, plus the general segment INSYNC.

8.1 INSYNC—The incoming edge of a bit is expected during this state; this segment corresponds to one BTL cycle.

8.2 SJW 1 & 2—Both segments determine the maximum synchronization jump width and are programmable from 1 to 4 BTL cycles. The width of SJW 1 is increased to a maximum of two times the bit time during resynchronization. The width of SJW 2 is reduced or canceled to shorten the bit time during resynchronization.

8.3 TSEG 1—Determines the sampling point based on the number of BTL cycles programmed by TSEG 1 (4 bits). The sampling point is located at the end of TSEG 1 (if sampling only once per bit). TSEG 1 is used to compensate delay times on the bus and to have some reserved time to tolerate one or more nonsynchronization pulses caused by spikes on the bus line. TSEG 1 is programmable from 1 to 16 BTL cycles.

8.4 TSEG 2—Defines the time between the sampling point and the end of the bit time; programmable from 1 to 8 BTL cycles. This segment is necessary to tolerate one or more nonsynchronization spikes on the busline. Also necessary to guarantee sufficient time to generate a transmit signal dependent on the sampled bus level. The transmit point is determined internally in such a way that with zero delay the generated transmit signal will appear within the INSYNC state. For example, no transmit signal would be generated if an arbitration was lost. This guarantees that the transmit logic immediately stops and immediately enters into the receive mode.

8.5 Bit Time Calculation—The number of clock cycles at every bit time determines (together with the oscillator frequency and the baud-rate-prescaler) the period of each bit time, and as a consequence, the baud rate.

bit time = (INSYNC + SJW1 + TSEG1 + TSEG2 + SJW2) BTL cycles (Eq. 2)

BTL cycle = 2 * $T_{oscillator}$ * (baud-rate-prescaler + 1)

BAUD rate = 1 / bit time

9. Synchronization—Synchronization is performed by comparing the incoming edges with the intended bit timing, and adapts the bit timing by hard synchronization or resynchronization.

Hard synchronization occurs only at the beginning of a frame. The CAN device synchronizes to the first incoming recessive to dominant edge of a frame.

Resynchronization occurs during the message bit stream to compensate differences in the oscillator frequencies of individual CAN devices as well as changes introduced by switching from one to another transmitter; e.g., after arbitration of two or more nodes. SJW 1 and SJW 2 define the maximum number of clock cycles a bit time may be shortened or lengthened by resynchronization.

There are two modes of sampling.
a. Resynchronization may take place on both edges. In this case, resynchronization is always done on the edge of a bus level which is different from the one read at the last sample point.
b. Resynchronization may take place on the edge of a dominant level only. In this case, resynchronization is done if the bus level monitored at the last sample point was a recessive level.

These modes are necessary due to physical bus characteristics at the maximum cable length and high baud rates.

Synchronization is done once during a bit time and is released after the sample point dependent on the level of the actual bit. Thereafter, the CAN device synchronizes to the next relevant edge and synchronization is suppressed until the following sample point.

By resynchronization, the bit time can be lengthened or shortened based on the following conditions:
a. If an edge appears in the segment SJW 1 or TSEG 1, then lengthen.
b. If an edge appears in the segment SJW 2 or TSEG 2, then shorten.

10. Sleep-mode/wake-up—In order to wake up other nodes of the system, which are in sleep-mode, a special wake-up message with the dedicated lowest possible identifier (rrr rrrd rrr where r = 'recessive' and d = 'dominant') may be used.

APPENDIX A

A1. 82526 Functional Overview—The CAN major functional blocks are:
a. Interface management processor (IMP)
b. Quasi dual port RAM
c. Transceiver control logic (TCL)
d. Bit stream processor (BSP)
e. Error management logic (EML)
f. Bus timing logic (BTL)
g. Processor interface unit (PIU)
h. Clock generator (CG)

A block diagram is shown Figure A1.

A1.1 IMP (Interface Management Processor)—The IMP executes commands from the CPU and controls data transmissions on the serial bus. This processor computes addresses for the many communication buffer accesses required during normal operation. Global status and control register bits, as well as the control bits of the descriptor are used primarily by the interface management processor. The IMP consists of a data path section with various local registers including an ALU with an accumulator and a control section for receive and transmit processes.

A1.2 Quasi Dual Port RAM—The host CPU communicates with the CAN device through a quasi dual port RAM memory which includes global status registers and control registers. This memory serves as the communication buffer interface between the host CPU and the IMP. The host CPU initializes the global status and control registers and creates data structures called communication objects within the communication buffer for reception and transmission of defined messages. The memory size is 64 bytes.

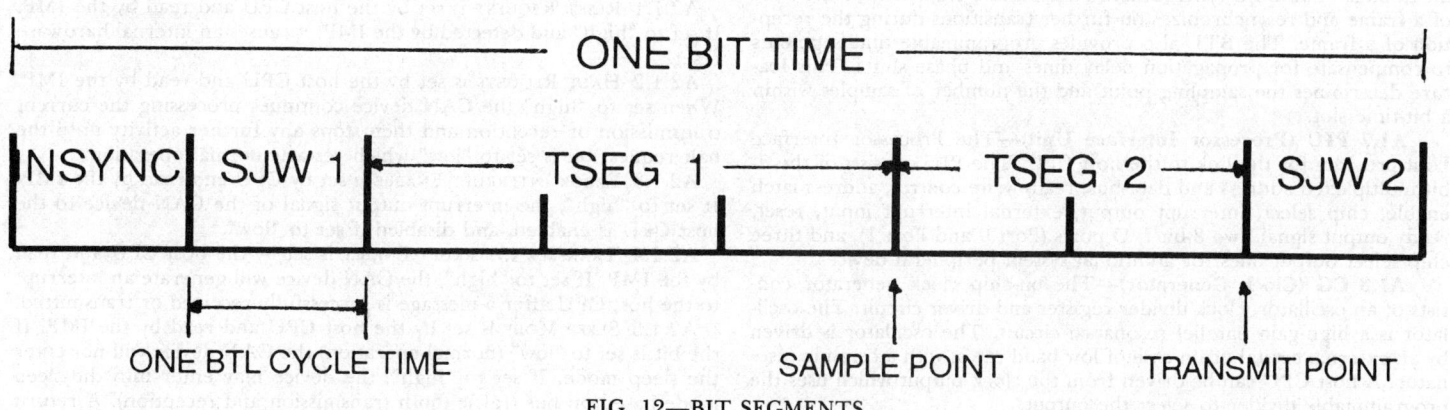

FIG. 12—BIT SEGMENTS

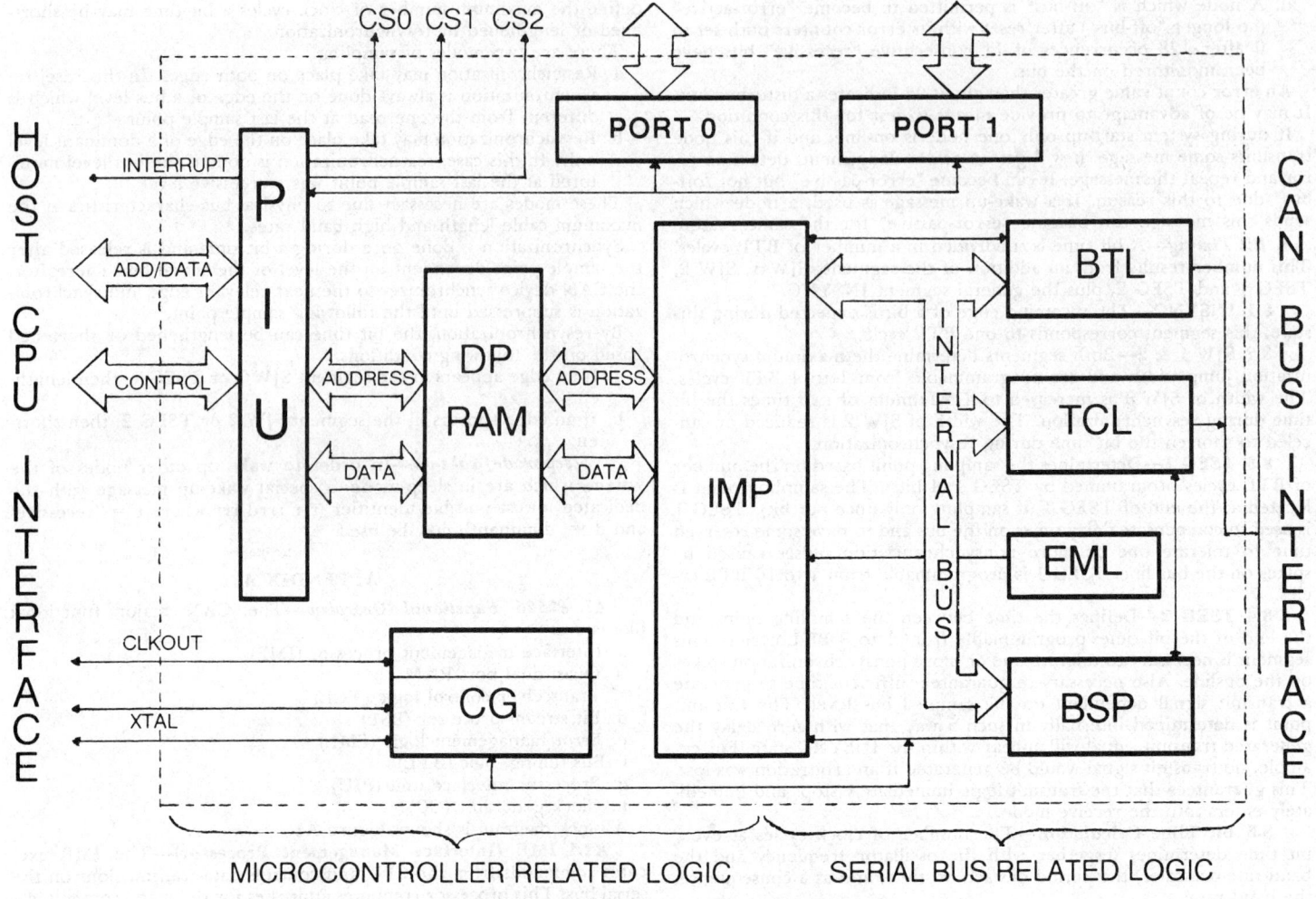

FIG. A1—FUNCTIONAL BLOCK DIAGRAM

A1.3 TCL (Transceiver Control Logic)—The transceiver control logic consists of bit stuffing logic, programmable output driver logic, CRC logic and shift registers.

A1.4 BSP (Bit Stream Processor)—The Bit Stream Processor controls the data stream between the IMP (parallel data) and the bus line (serial data). Message reception and transmission, bus arbitration, error signaling and control of the TCL are duties of the BSP.

A1.5 EML (Error Management Logic)—Errors are reported to the EML by the bit stream processor (BSP). The EML informs the BSP, TCL and IMP of error statistics.

A1.6 BTL (Bus Timing Logic)—This logical block monitors the busline through an input comparator and performs the busline related bit timings. The BTL synchronizes on a busline transition at the start of a frame and resynchronizes on further transitions during the reception of a frame. The BTL also provides programmable time segments to compensate for propagation delay times and phase shifts. This feature determines the sampling point and the number of samples within a bit-time slot.

A1.7 PIU (Processor Interface Unit)—The Processor Interface Unit provides for the link to the host CPU. The PIU consists of the 8-bit multiplexed address and data bus, read/write control, address latch enable, chip select, interrupt output, external interrupt input, reset, ready output signal, two 8-bit I/O ports (Port 0 and Port 1), and three chip select output lines for additional system peripheral devices.

A1.8 CG (Clock Generator)—The on-chip clock generator consists of an oscillator, clock divider register and driver circuit. The oscillator is a high-gain parallel resonance circuit. The oscillator is driven by an external crystal or, in case of low baud rates, with a ceramic resonator. A host CPU can be driven from the clock output which uses the programmable divider to select the output.

A2. Quasi Dual Port RAM Memory Layout and Control Registers—The on-chip quasi dual port RAM (communication buffer) is seen as part of the host CPU's memory map and may be defined as a "decoupling device" between the CPU and the "Bus-Interface" of the CAN device. The buffer memory area used for communication is configured by the user during an initialization download after power-up. It consists of dedicated control registers, communication objects, and an end mark signifying the end of RAM area used for communication. Each individual communication object is composed of the identifier, control bits, and data. The CPU programmer operates only on this communication buffer to perform message transfers. The buffer memory layout is shown in Figure A2.

A2.1 Control Register (Address 00H)

A2.1.1 RESET REQUEST is set by the host CPU and read by the IMP. If set to "high" and detected by the IMP, it causes an internal hardware reset.

A2.1.2 HALT REQUEST is set by the host CPU and read by the IMP. When set to "high" the CAN device continues processing the current transmission or reception and then stops any further activity until the halt request bit is set to "low" which restarts normal operation.

A2.1.3 ERROR INTERRUPT ENABLE is set by CPU and read by the IMP. If set to "high", the interrupt output signal of the CAN device to the host CPU is enabled, and disabled if set to "low".

A2.1.4 TRANSFER INTERRUPT ENABLE is set by the host CPU and read by the IMP. If set to "high", the CAN device will generate an interrupt to the host CPU after a message is successfully received or transmitted.

A2.1.5 SLEEP MODE is set by the host CPU and read by the IMP. If the bit is set to "low" (normal operation) the CAN device will not enter the sleep mode. If set to "high", the device may enter into the sleep mode based on bus traffic (both transmission and reception). A return

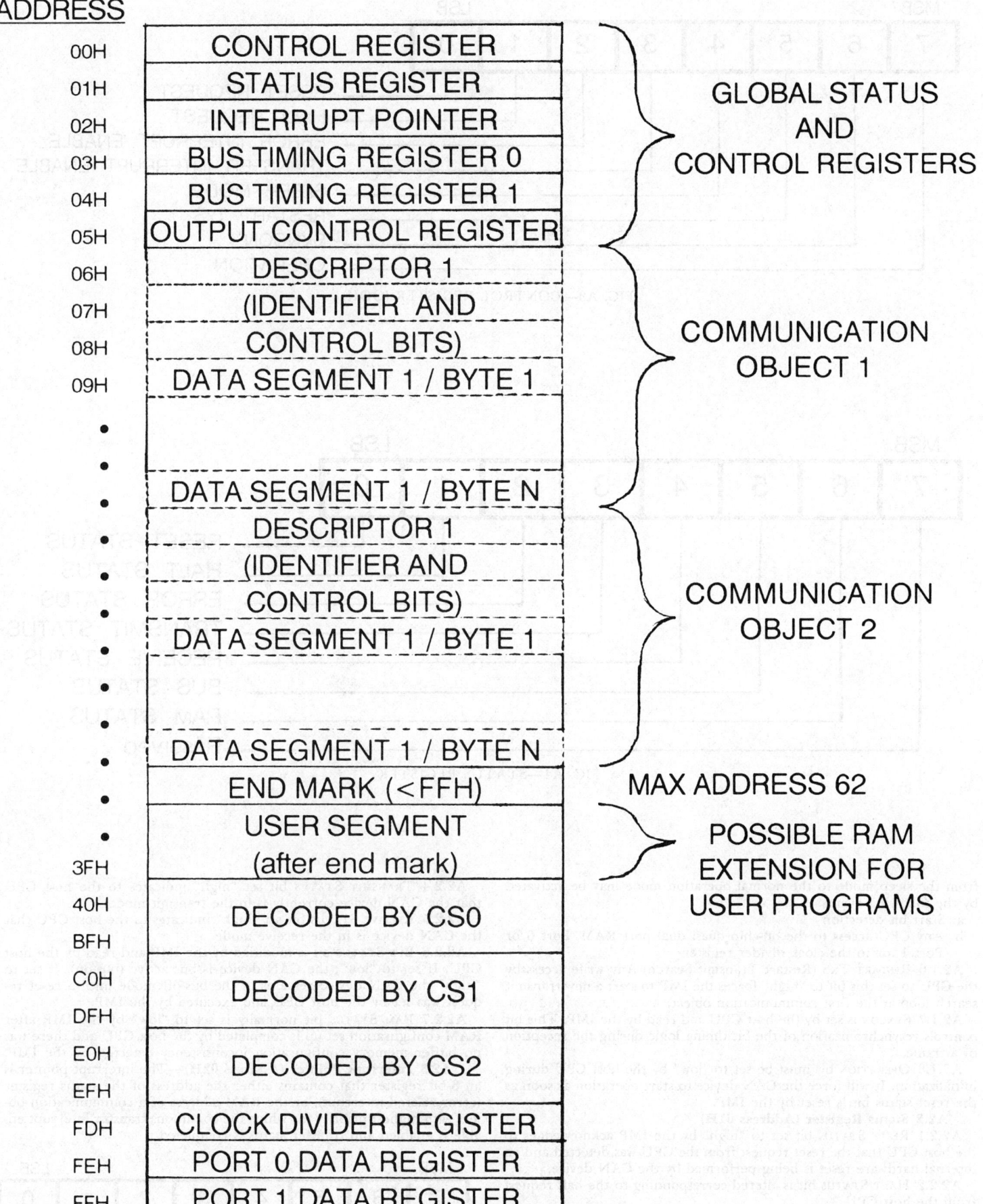

FIG. A2—BUFFER MEMORY LAYOUT

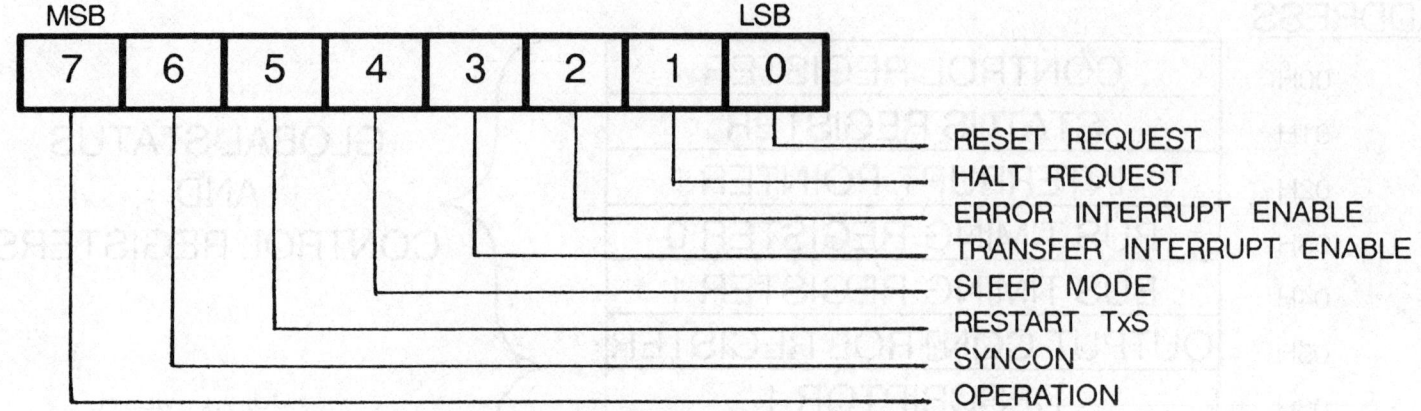

FIG. A3—CONTROL REGISTER (00H)

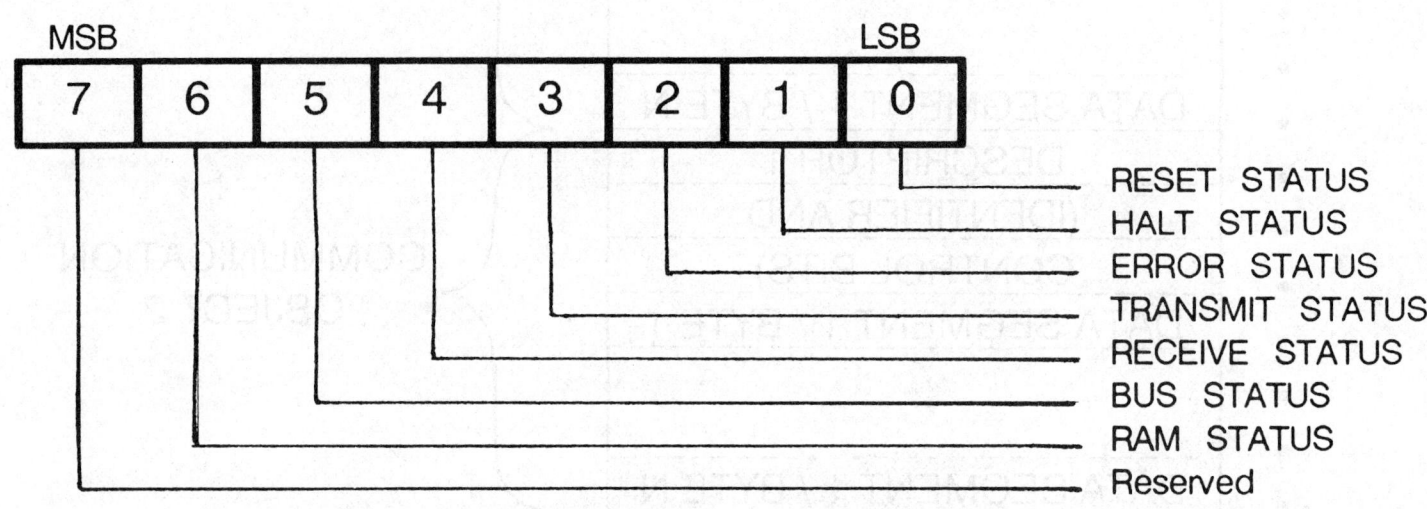

FIG. A4—STATUS REGISTER

from the sleep mode to the normal operation mode may be activated by the following events:
  a. Start bit detection
  b. Any CPU access to the on-chip quasi dual port RAM, Port 0 or Port 1, or to the clock divider register

A2.1.6 RESTART TxS (Restart Transmit Search) Any write access by the CPU to set this bit to "high" forces the IMP to start a new transmit search loop at the first communication object.

A2.1.7 SYNCON is set by the host CPU and read by the IMP. This bit controls resynchronization of the bit timing logic during the reception of a frame.

A2.1.8 OPERATION bit must be set to "low" by the host CPU during initialization. It will force the CAN device to start operation as soon as the reset status bit is reset by the IMP.

A2.2 Status Register (Address 01H)

A2.2.1 RESET STATUS bit set to "high" by the IMP acknowledges to the host CPU that the reset request from the CPU was detected and an internal hardware reset is being performed by the CAN device.

A2.2.2 HALT STATUS bit is altered corresponding to the halt request from the host CPU.

A2.2.3 ERROR STATUS bit set to "high" by the IMP indicates to the host CPU that the error management logic (EML) has detected a major problem either with transmission or reception of messages.

A2.2.4 TRANSMIT STATUS bit set "high" indicates to the host CPU that the CAN device currently is in the transmit mode.

A2.2.5 RECEIVE STATUS bit set "high" indicates to the host CPU that the CAN device is in the receive mode.

A2.2.6 BUS STATUS bit is modified by the IMP and read by the host CPU. If set to "low", the CAN device is bus active (bus-on). If set to "high" the CAN device remains in the bus-off mode until a reset request was set by the host CPU and executed by the IMP.

A2.2.7 RAM STATUS bit normally is set to "low" by the IMP after RAM configuration set up is completed by the host CPU and there was no buffer memory configuration inconsistency detected by the IMP.

A2.3 Interrupt Pointer (Address 02H)—The interrupt pointer is an 8-bit register that contains either the address of the status register (error related interrupt), or the RAM address of a communication object with highest priority for which the conditions transfer interrupt enable is enabled and transfer status is completed.

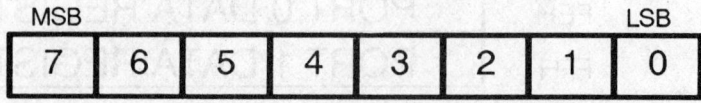

FIG. A5—INTERRUPT POINTER (02H)

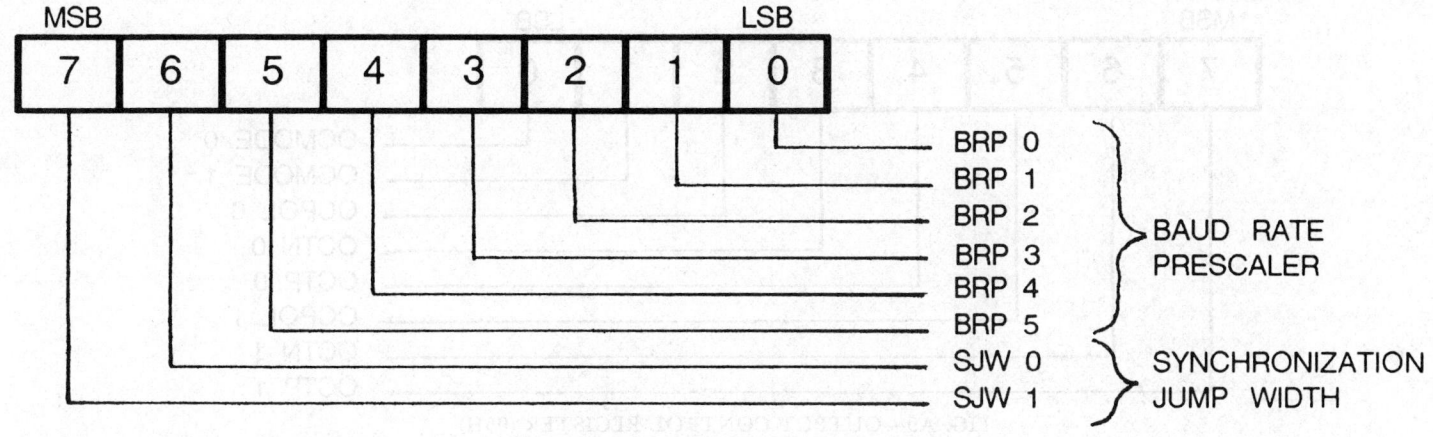

FIG. A6—BUS TIMING REGISTER 0 (03H)

A2.4 Bus Timing Register 0 (Address 03H)

A2.4.1 BAUD RATE PRESCALER—By programming the six bits of the baud rate prescaler, the BTL cycle time is determined. The BTL cycle time is derived from the system cycle time. The desired baud rate is determined by the BTL cycle time and the programmable bit timing segments.

BAUD RATE PRESCALER BRP0-BRP5	BIT CYCLE TIME
000000	1 * System Cycle Time
000001	2 * System Cycle Time
000010	3 * System Cycle Time
:	:
:	:
111111	64 * System Cycle Time

FIG. A7—BAUD RATE PRESCALER

A2.4.2 SYNCHRONIZATION JUMP WIDTH—These two bits are introduced to compensate for phase shifts between clock oscillators of different bus nodes. Any node may resynchronize to a relevant transition of the bus bitstream of a transmitter. The synchronization jump width defines the maximum number of BTL cycles. A bit may be shortened or lengthened by one resynchronization during transmission of a data frame or remote frame.

A2.5 Bus Timing Register 1 (Address 04H)—The number of samples per bit, the delay time and the phase shift buffer are programmed by bus timing register 1.

A2.5.1 BUS SAMPLE—This bit determines whether a bit is sampled directly by the BTL or if each bit first is filtered by the majority logic. If sample is set to "low", a bit is sampled once by the BTL, and if set to "high", three samples per bit are taken.

A2.5.2 TIME SEGMENT 1/TIME SEGMENT 2—Time segments within a bit time are introduced to determine the number of BTL cycles per bit time and the location of the sample point within a bit time.

A2.6 Output Control Registers (Address 05H)—The output control register allows set up of different output driver configurations under software control. The user may select active pull-up, pull-down, or push-pull output.

A2.6.1 NORMAL MODE—Output signal levels are assigned to the logical bit levels "dominant" and "recessive" depending on the output characteristics (pull-up, pull-down, push-pull).

A2.6.2 BIPOLAR MODE—If galvanically decoupled from the busline by a transformer, the bit stream has to be coded in such a way that there is no resulting DC-current transferred from the transformer. With "recessive" bits, all outputs are deactivated. "Dominant" bits are sent alternately on Tx0 and Tx1; e.g., the first "dominant" bit is sent on Tx0, the second on Tx1, the third on Tx0, and so on.

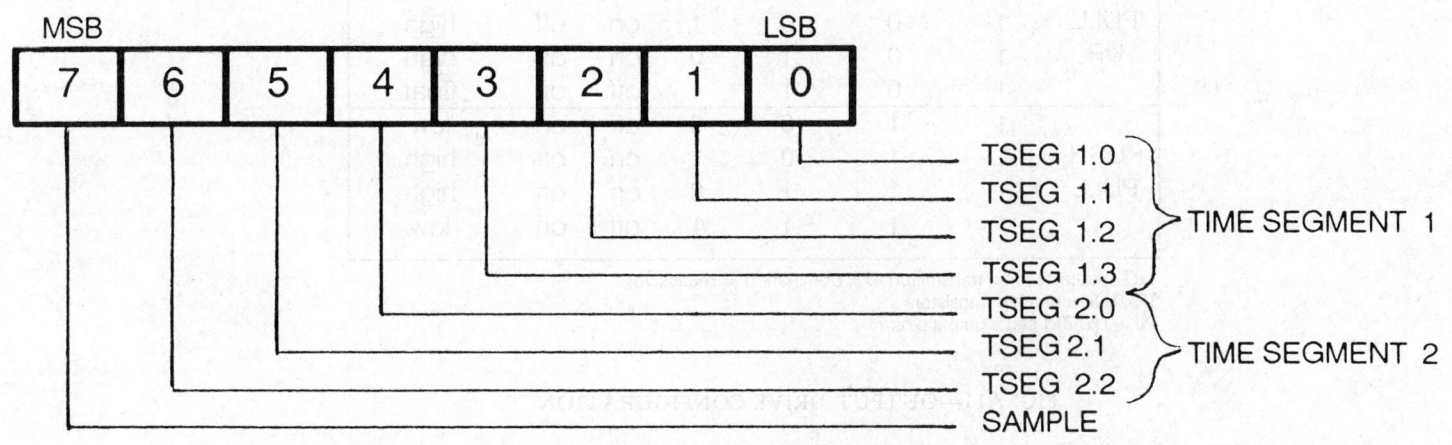

FIG. A8—BUS TIMING REGISTER 1 (04H)

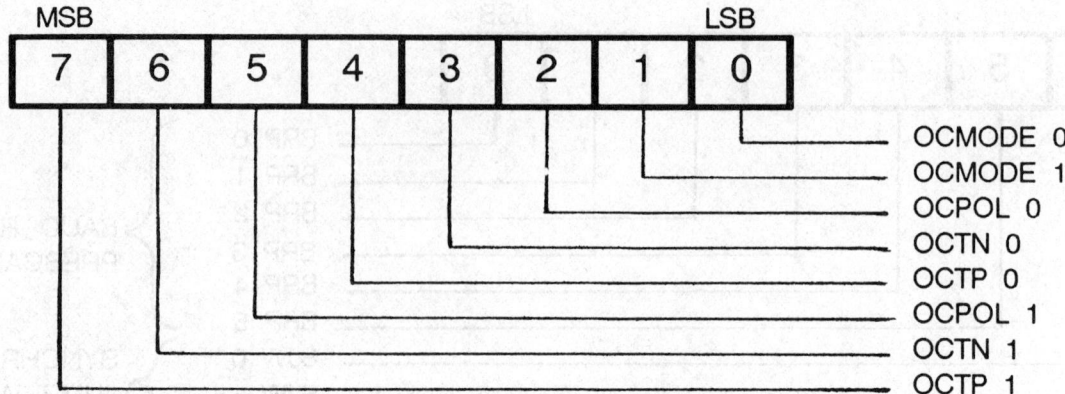

FIG. A9—OUTPUT CONTROL REGISTER (05H)

OCMODE 1	OCMODE 0	FUNCTION
1	0	Normal Mode, bit stream transmitted on both Tx0 and Tx1
1	1	Normal Mode, Tx0 = bit stream, Tx1 = bus clock (txclk)
0	0	Bipolar Mode
0	1	Not Used

FIG. A10—PROGRAMMABLE FEATURES

MODE	OCTPi	OCTNi	OCTLi	TxD	TPx	TNx	TXi OUPUT
FLOAT	0	0	0	0	off	off	float
	0	0	0	1	off	off	float
	0	0	1	0	off	off	float
	0	0	1	1	off	off	float
PULL DOWN	0	1	0	0	off	on	low
	0	1	0	1	off	off	float
	0	1	1	0	off	off	float
	0	1	1	1	off	on	low
PULL UP	1	0	0	0	off	off	float
	1	0	0	1	on	off	high
	1	0	1	0	on	off	high
	1	0	1	1	off	off	float
PUSH PULL	1	1	0	0	off	on	low
	1	1	0	1	on	off	high
	1	1	1	0	on	off	high
	1	1	1	1	off	on	low

TxD = data bit to be transmitted (0 = dominant, 1 = recessive)
TPx/TNx = on-chip transistors
Txi = Tx0, Tx1 serial output pins

FIG. A11—OUTPUT DRIVE CONFIGURATION

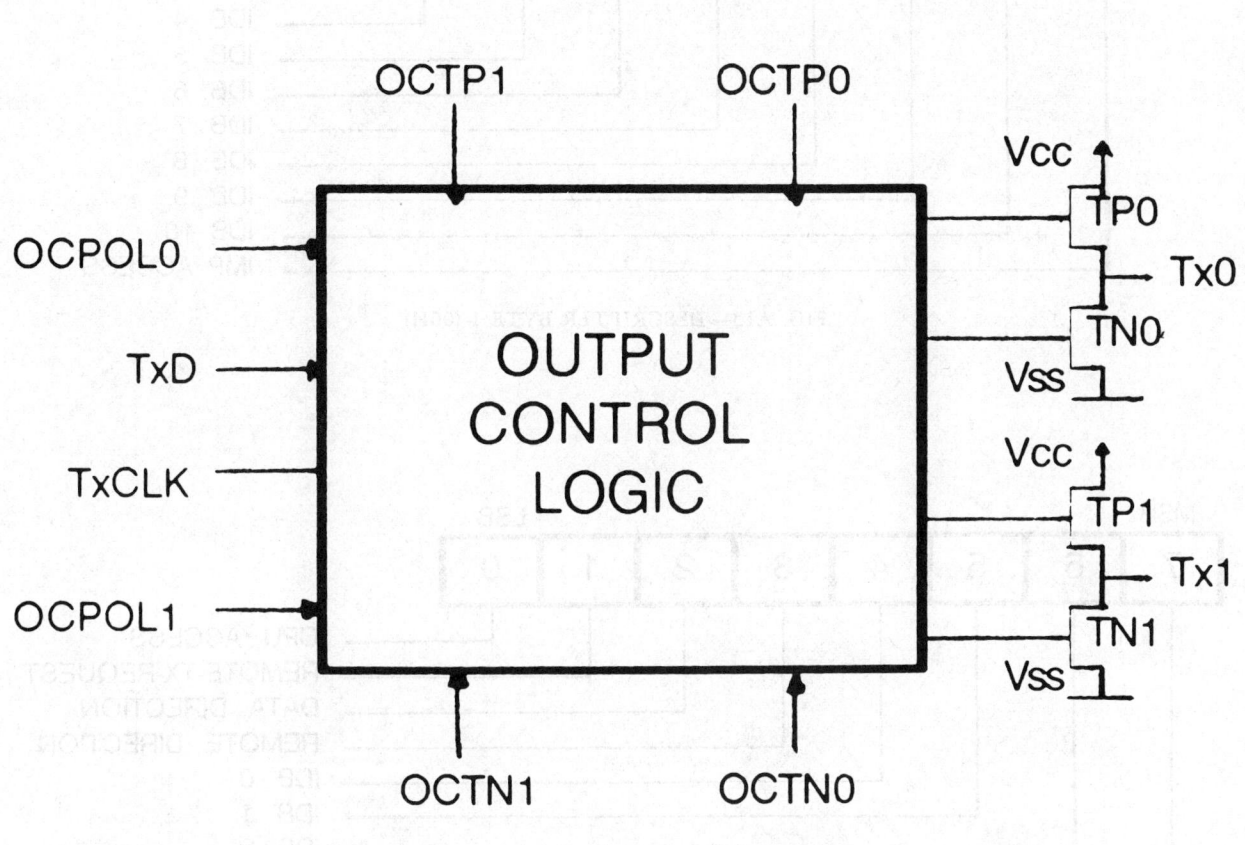

FIG. A12—OUTPUT CONTROL LOGIC

A2.7 Communication Object (Starting at Address 06H) Communication Objects are created by the user for every message, to perform a transmission or a reception of the respective message. This allows message transfer and an integrity check to be handled without user interference. Each communication object consists of a descriptor and data segment. The number of communication objects which may be configured within the buffer memory is dependent on the number of bytes in the data field (data segment) since communication objects are placed contiguously.

A2.7.1 DESCRIPTOR (BYTES 1, 2, AND 3)—The descriptor consists of three bytes, assigned to each communication object by the user. The three bytes include the message identifier and a control segment which contains semaphore bits to guarantee mutual exclusion of access to the communication buffer, if required.

 a. Identifier—An Identifier is a unique name for each communication object and defines the corresponding, specific message. It's digital number also determines the priority of each single message for bus access (see Section 5).

 b. Control Bits—IMP access is set by the host CPU and read by the IMP. If set to "low", IMP access to the data of this communication object is locked out. If reset to "high", IMP access is released. CPU access is set by the IMP and read by the host CPU. If set to

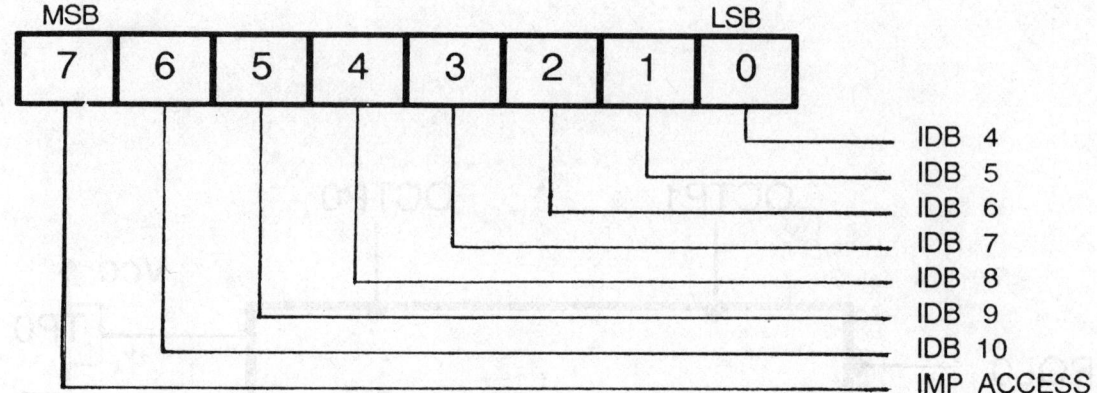

FIG. A13—DESCRIPTER BYTE 1 (06H)

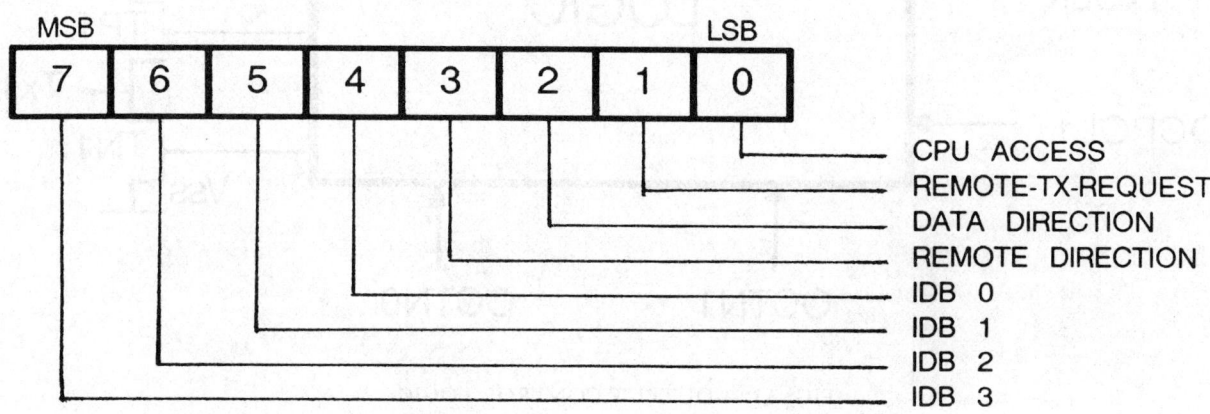

FIG. A14—DESCRIPTER BYTE 2 (07H)

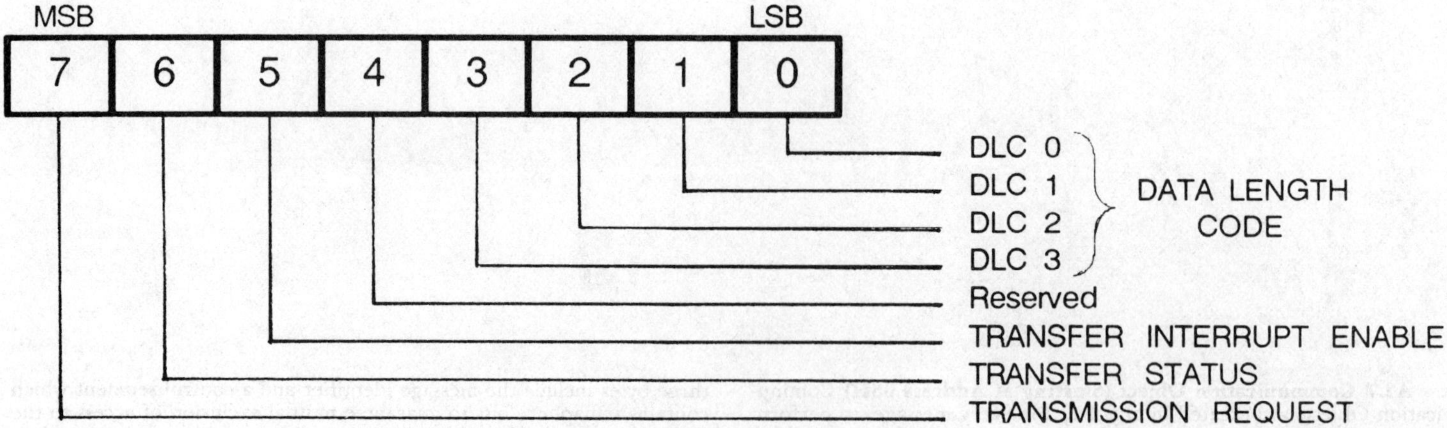

FIG. A15—DESCRIPTER BYTE 3 (08H)

"low", CPU access to the data of this communication object is locked out. If reset to "high" CPU access is released. In case of a conflict, the priority is assigned to the CPU access.

Data direction is set by the host CPU and read by the IMP. If set to "low", a communication object is assigned to receive a message of the respective Identifier. If set to "high", a communication object is assigned for transmission.

Remote direction set by the host CPU and read by the IMP. If set to "high" the data direction is set to "low", a CPU may initiate the transmission of a remote frame when transmission request is set "high" (see Table A4).

Transmission request is modified by the host CPU and IMP. If set to "high" by the CPU, a data frame or remote frame will be transmitted based on which bit (data direction or remote direction) is set to "transmit".

Transfer status is modified by the IMP and the host CPU. This bit is set "high" by the IMP after a successful transmission or reception. The transfer status bit is reset by the host CPU in order to signal to the IMP that the CPU transfer is complete.

Transfer interrupt enable is modified by the host CPU. If set "high" (enable), the interrupt signal will be activated if the transfer status of a corresponding communication object was found complete by the IMP.

Data length code is set by the host CPU during the initialization routine and determines the number of bytes (data byte count) of a respective message located in the data segment.

A2.7.2 DATA SEGMENT—The data segment holds the data of the corresponding communication objects. The number of data bytes per segment is defined by the data length code in each descriptor.

A2.7.3 ENDMARK—The endmark is set to FFH by the host CPU indicates to the IMP that there is no communication object stored beyond the endmark (maximum address = 62). The endmark indicates to the IMP the end of the linked list of communication objects.

A2.7.4 USER SEGMENT—The user segment may follow the endmark since that portion of RAM is not used by the CAN device. It is available to the user as general data memory and does not affect the function of the device.

A VEHICLE NETWORK PROTOCOL WITH A FAULT TOLERANT MULTIPLEX SIGNAL BUS—SAE J1813 AUG87

SAE Information Report

Report of the Electrical and Electronic Systems Technical Committee approved August 1987. Rationale statement available. The draft, SAE Recommended Practice J1850 currently under development, is a generalized approach (making extensive use of software) which can support VNP (described in SAE J1813) and similar protocols. Working Group I of the International Standards Organization (ISO) is currently giving strong consideration to maximizing the compatibility and similarity between any standard they propose for adoption and SAE J1850.

1. A Vehicle Network Protocol (VNP)

1.1 Introduction—This Vehicle Network Protocol is intended to service all vehicle communication applications which do not require a latency of less than 20 ms. This assumes a 10 Kbit/s bus speed operation. The fault tolerant multiplex signal bus complements the VNP by providing successful operation despite component failures.

1.2 Design Goals—The design goals of the Vehicle Network Protocol include:

a. Message integrity—assurance that a transmitted message has arrived at the intended receiver(s) as transmitted; an error checking code, such as cyclic redundancy check (CRC), and an acknowledgment of correct reception of data are required.

b. Bit integrity—assurance that a logic one is distinct from a logic zero; and that noise corruption does not readily cause a logic one to be decoded as a logic zero and vice versa.

c. Asynchronous bus access—The ability to transmit on an unscheduled or demand basis.

d. Priority contention—The ability to gain access to the bus based upon a predetermined priority.

e. Contention resolution—The assurance that the dominant contender maintains control of the bus and will continue message transmission while the losing contender assumes the role of receiver such that contention is resolved in a deterministic manner.

f. Assignable priority—The ability to prioritize a message independent of its content.

g. Message content arbitration—The ability to resolve bus access contention based upon message content.

h. Expandability—The ability to accommodate future communication needs by:
   1. Providing features which afford flexibility to the system designer.
   2. Incorporating a mechanism by which other data formats can be defined which will operate on the same bus.

i. Functional addressing—The ability to communicate on the basis of functions independent of physical node partitioning and the ability to communicate the function information to multiple receivers simultaneously.

j. Node to node addressing—The ability to communicate information between two specific physical nodes, a transmitter and a receiver.

k. Data transfer—The ability to transmit multiple bytes of data to one or more nodes.

l. Low latency time—Guaranteed access (with maximum priority) to the bus within 20 ms.

m. Optimized baud rate—High enough to provide minimum required message latency, but low enough to allow simple transmission line media and exhibit acceptable automotive radiation.

n. Microprocessor compatible—Assurance that message format is such that encoding and decoding of the message is easily accomplished by eight-bit microprocessors. Byte-orientation is required.

o. Fault diagnosis—The ability to isolate faults to physical nodes.

p. Self-clocking waveforms—The ability to synchronize message transmissions on a bit by bit basis.

1.3 Bit Waveform Representation—The proposed protocol utilizes a pulse width modulated bit waveform to represent the logic state as well as unique start and end-of-data bits. The waveforms are shown in Fig. 1.1.

1.4 General Message Format—In the interest of universal applicability of the VNP, the message format has been left flexible. All users must subscribe to basic rules to access the bus. All messages must begin with a start bit followed by a field which determines bus priority as well as defining the message type. All messages end with a stop bit field. No message may exceed 101 bits in length. Within these constraints, new message types may be invented for the unused message type codes. Fig. 1.2 illustrates the basic VNP format.

1.5 Message Types—All messages are composed of fields which have predefined significance. Although the field format for all currently defined message types is consistent, the message type is required to indicate to the target node(s) how the remainder of the message is to be interpreted. All messages must begin with a start bit, contain a priority/type field, and terminate within 101 bit cell times of initiation of the start bit. The general message format is shown in Fig. 1.2.

There are currently five defined message types. They are Functional, Functional Read, Broadcast, Physical, and Physical Read.

1.6 Functional Message Type—The Functional Message Type is used to communicate data or commands to functional target groups; for example, "coolant temperature = 179" or "windows up." The targets of this type message are indicated by a functional address field which follows immediately after the priority/type field. In the second

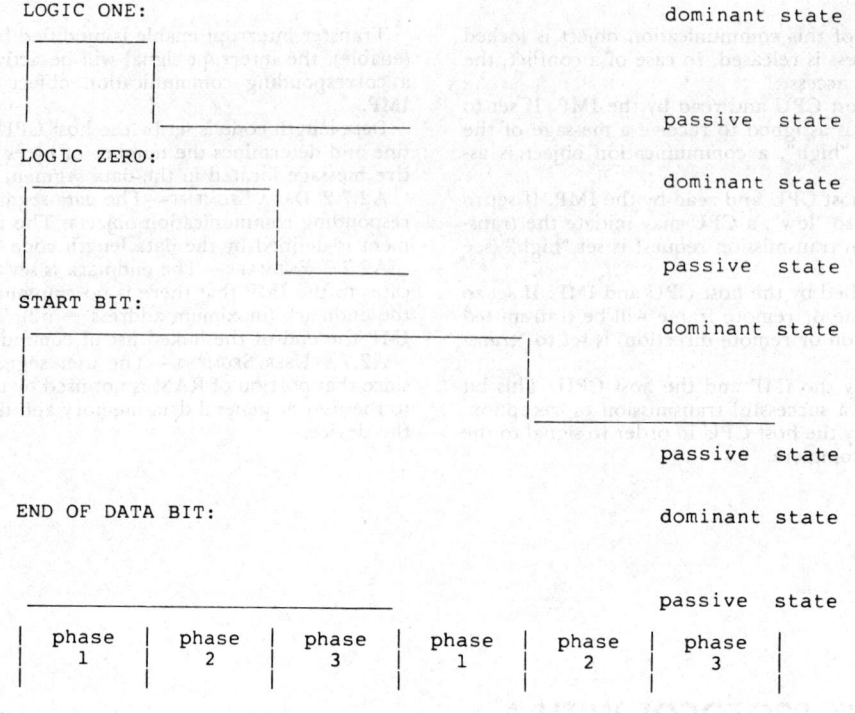

FIG. 1—BIT WAVEFORMS

```
| S | PRIORITY | ?????? | ?????? | ?????????? / / ????? |
|   | & TYPE   |        |        |                      |
```

FIG. 1.2—GENERAL MESSAGE FORMAT

example above, the functional address field would contain the eight-bit code for the "windows" function. The next field transmitted in this message type contains the initiating node address. The data field which may have a variable length follows. This field will contain data for information transfer, but in the case of functional commands, this field is used to provide specifics of the request: that is, up, down, on, off, left, right, etc. In either case, the length of the data field plus the length of the acknowledge field must not exceed eight bytes total. The data field is followed by the eight-bit Cyclic Redundancy Check (CRC) Field. An End-Of Data bit, after the CRC, signals the receiving unit that the CRC has been received and must be compared with the calculated value. The initiating node ceases transmission at this point. Acknowledging node(s) respond using an arbitration scheme. The format for the Functional Message Type is shown in Fig. 1.3.

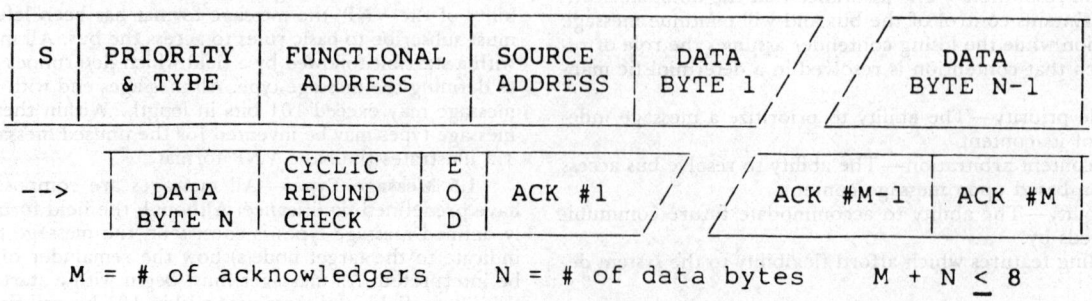

FIG. 1.3—FUNCTIONAL MESSAGE TYPE FORMAT

1.7 Functional Read Message Type—The Functional Read Message is an interrogative message type. It may be used to elicit information from nodes which normally initiate functional commands and/or data transfers. Through this message type, the request and the response are both contained in a single bus transaction. Fig. 1.4 shows the format for this message type. The message begins with a start bit and the priority/type field as do all messages. The functional address field contains the descriptor of the function which is being interrogated. The next field specifies the address of the initiating node. Note that there is no data field transmitted by the initiating node. Consequently, the next field is the Cyclic Redundancy Check (CRC) field which is followed by the EOD. At this point, all nodes which are normally initiators of this function are expected to contend for the bus using their relevant transmission of the single data byte response and then append a single byte CRC for data security. Potential respondents which lose arbitration do not attempt a second time to reply. Only the node with the "highest priority data" will successfully respond to this message type. Because of this, the system designer will be required to judiciously encode the status reply bytes of potential functional read respondents in order to insure desired information transfer. The message length for a Functional Read Message type is fixed at 53 bit cell times.

1.8 Broadcast Message Type—The Broadcast Message type has been specified to provide the possibility of initiating the action of a Functional Message type without incurring the message length penalty associated with the multiple acknowledge bytes. The multiple acknowledges have been replaced by a single acknowledge from the potential respondent with the highest priority node address. The Broadcast Message type consequently does not have the level of message security inherent with the Functional Message type. With the Broadcast Message type, it can be ascertained that the message has been properly received by at least one other node on the network. The format for this message type is represented in Fig. 1.5. It is identical to the format for the Functional Message type, except that the acknowledging receivers arbitrate for one byte only then give up. Message length varies with the number of bytes in the data field. The number of data bytes may vary from one to seven. The maximum message length is fixed at 101 bit cell times.

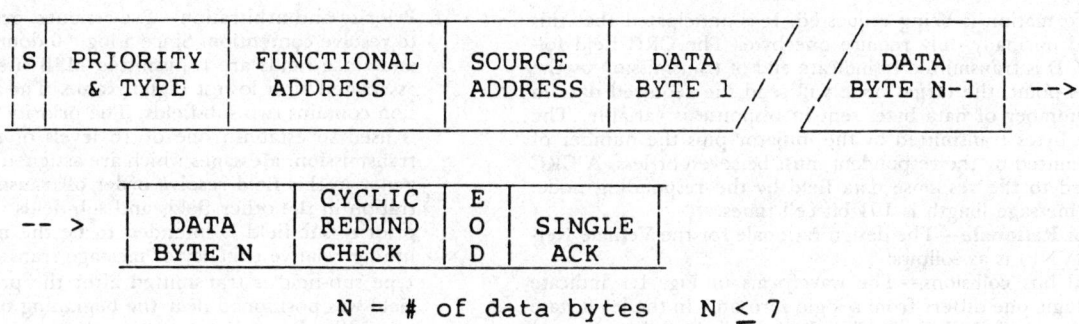

FIG. 1.4—FUNCTIONAL READ MESSAGE TYPE FORMAT

1.9 Physical Message Type—The Physical Message type is used to communicate data or commands to a specific physical location (node). It is anticipated that this message type will be useful in performing system diagnostics. The format for this message type is shown in Fig. 1.6. The message begins with a start bit and the priority/type field as do all messages. The physical address field contains the descriptor of the single node which is being addressed. The next field specifies the address of the initiating node. The variable length data field follows. The number of bytes may vary from one to seven. A Cyclic Redundancy Check (CRC) byte is transmitted next to insure message security. The CRC is followed by an End of Data bit (EOD) to signal that the initiator portion of the message is complete and that the CRC should be in the receiver's buffer. At this point, the target node is required to respond with transmission of its node address as confirmation of proper message reception. The maximum message length is 101 bit cell times.

FIG. 1.5—BROADCAST MESSAGE TYPE FORMAT

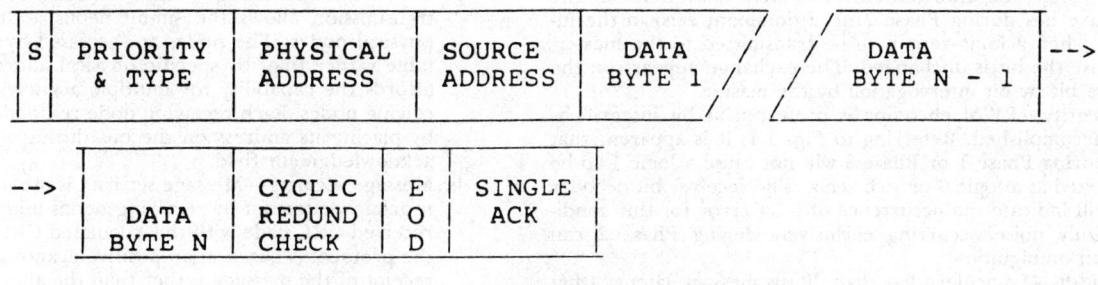

FIG. 1.6—PHYSICAL MESSAGE TYPE FORMAT

1.10 Physical Read Message Type—The Physical Read Message is an interrogative message type. It may be used to elicit the status of individual nodes. Through the use of this message type, the request and response may both be contained in a single transaction. Fig. 1.7 illustrates the Physical Read Message format. The message begins with a start bit and the priority/type field as do all messages. The address field contains the address of the node which is being interrogated. Following the physical address is the address of the initiating node. The next field is a data field. The purpose of this variable length field is to specify what information is being requested. It is anticipated that this data field would normally only require one byte. The CRC field follows, and the EOD is transmitted to indicate end of transmission by the initiator. At this point, the target node will send the required data in response. The number of data bytes sent in response is variable. The number of data bytes transmitted by the initiator plus the number of data bytes transmitted by the respondent must be seven or less. A CRC Byte is appended to the response data field by the responding node. The maximum message length is 101 bit cell times.

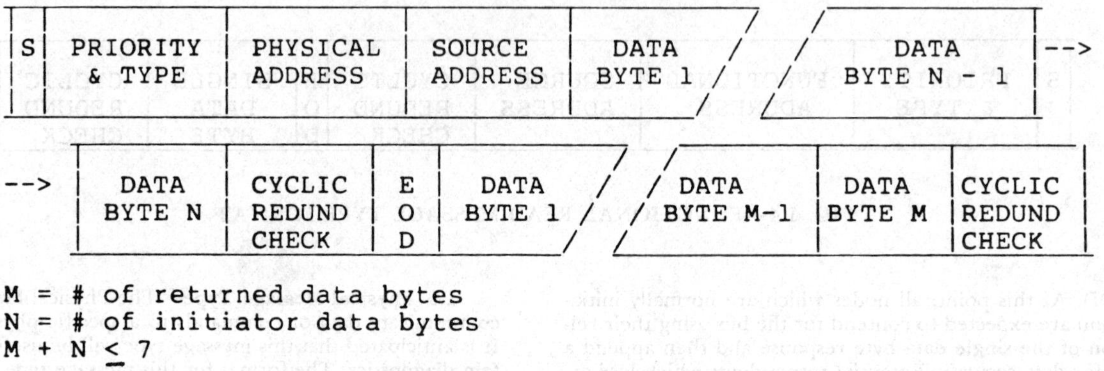

1.11 Design Rationale—The design rationale for the Vehicle Network Protocol (VNP) is as follows:

a. Physical bus collisions—The waveforms in Fig. 1.1 indicate that a logic one differs from a logic zero only in the level state of the bus during Phase 2. The physical bus is designed to exhibit a passive state and an active state such that the active state dominates in the case that distinct drivers impress both states simultaneously. In this manner, a logic one is dominated on the bus by a logic zero since the bus will be driven active during Phase 2. A start bit is also shown. The start bit is a unique waveform which provides message synchronization among the nodes.

b. Bit encoding—The logical waveforms are pulse width modulated (PWM) rather than traditional NRZ type. This allows nodes to receive data using the leading edge of the bit waveform to generate a sample strobe to clock the data bit during the critical second phase of the bit waveform.
Transmission by the slave node is also accommodated by the selected PWM waveform. In this case, the master node "clocks" logic one onto the bus. The slave node modifies the bit (active bus during Phase 2) to a dominant zero in the instances when a logic zero is to be transmitted to the master, otherwise the bit is unchanged. The exchange appears on the bus as a bit by bit interrogation by the master.

c. Bit integrity—PWM encoding is preferred as bit integrity is easily accomplished. Referring to Fig. 1.1, it is apparent that noise during Phase 1 or Phase 3 will not cause a logic 1 to be interpreted as a logic 0 or vice versa. The receiver bit decoder logic will indicate the occurrence of a bit error for this condition. Only noise occurring exclusively during Phase 2 can cause bit ambiguity.

d. Bandwidth—To achieve less than 20 ms message latency (that is, two longest message types back to back), a bit rate of 10 Kbit/s is required. This latency could be reduced by operating at higher bus speeds.

e. Contention resolution—The physical bus is designed such that multiple nodes can simultaneously impress conflicting logic states on the differential pair with a predictable predetermined result. Any single logic 0 will dominate the transmission of logic 1 by a number of nodes. With this feature, it is possible to resolve bus contention without message collision and destruction. When two or more nodes begin transmission simultaneously, contention for bus access occurs on a bit-by-bit basis. Each node is required to monitor its own transmission for integrity. Upon detection that its transmission of a bit was dominated by a transmission of another node, a transmitting node will revert to the status of a receiving node.

f. Priority and arbitration—Pre-assigned priority fields are used to resolve contention. Since a logic 0 dominates over a logic 1, and since fields are transmitted MSB first, highest priority is associated with lowest digital value. The first field of arbitration contains two sub-fields. The priority sub-field of four bits is used to establish one of 16 levels of urgency for message transmission. Messages which are assigned the same level of urgency in this field resolve order of transmission through arbitration in the other fields and sub-fields which follow; but the priority sub-field is intended to be the mechanism for establishing relative urgency of message transmission. The message type sub-field is transmitted after the priority sub-field. This field was positioned near the beginning of the message, not as a priority issue, but to allow those receivers not concerned with the message type being transmitted to ignore the remaining bytes of the message. Perhaps more important, it allows the remainder of the message to be custom formatted to the common format for all VNP messages. Messages with the same priority and message type must resolve contention in fields which are specific to the message type format. The Functional Message Type (FMT) uses a function field followed by the receiver address followed by the transmitter field, and the Physical Message Type (PMT) uses the receiver address followed by the transmitter address to resolve contention. It is important to note that while these fields resolve contention and determine order of transmission in some cases, it is not the intention of the protocol design to use these fields to establish priority.

g. Functional addressing—The Functional Message Type (FMT) transmission allows the simultaneous addressing of multiple physical nodes. The nodes are accessed by a function look-up table rather than by specific physical address. The FMT also affords the capability for multiple acknowledgments from receiving nodes. Each receiving node responds to the transmitter by placing its address on the bus during a priority arbitrated acknowledgment field.

h. Message security—Message security is provided by an automatic acknowledgment by receiving nodes upon correlation of the received CRC code with the calculated CRC code. Security for the protocol is based upon positive acknowledgment of proper receipt of the message rather than the alternative: negative acknowledgment for messages corrupted in transmission. While negative acknowledgment has desirable characteristics, it was rejected for the VNP protocol because it does not allow the creation of new message types which deviate from the currently prescribed methods of checking message integrity (CRC). Negative acknowledgment requires that all nodes check all

```
| S | PRIORITY | PHYSICAL | SOURCE  | DATA   //    DATA     |-->
|   | & TYPE   | ADDRESS  | ADDRESS | BYTE 1 //  BYTE N - 1  |

--> | DATA   | CYCLIC | E | DATA   // DATA    | DATA   | CYCLIC |
    | BYTE N | REDUND | O | BYTE 1 // BYTE M-1| BYTE M | REDUND |
    |        | CHECK  | D |        //         |        | CHECK  |
```

M = # of returned data bytes
N = # of initiator data bytes
M + N ≤ 7

FIG. 1.7—PHYSICAL READ MESSAGE TYPE FORMAT

messages for CRC integrity. Any new message not conforming to the currently defined message type integrity checks would be "negative acknowledged." The negative acknowledgment is too restrictive for the expandability desired in the VNP protocol.

The positive acknowledgment given by the receiving nodes is an indication that the message has been successfully transmitted; it is not a verification of command actuation. It is anticipated that CRC/acknowledgment handshake will be used at each node continuously to monitor communication system integrity. A counter could be implemented to flag failure at some predetermined number of occurrences. Verification of proper device actuation as a result of command messages must occur through multiple messages.

2. A Fault Tolerant Multiplex Signal Bus

2.1 Introduction—The fault tolerant multiplex signal bus described herein is intended to permit system operation despite individual node failures or a failure of a portion of the signal bus itself. The design specifically addressed shorts at the node or one of the two signal bus wires to ground or some other dc potential. The objective is to permit continued successful communication between remaining operational units over remaining operational link elements. The fault tolerant multiplex signal bus is intended for use in conjunction with the Vehicle Network Protocol (VNP).

2.2 Functional Description—Key elements of the fault tolerant multiplex signal bus include:
   a. Multiplex signal bus wires.
   b. Node isolation resistors.
   c. Signal bus termination networks.
   d. Pulse width modulated differential waveforms.
   e. Transmitter.
   f. Receiver.
   g. Input/output logic.

Fig. 2.1 is a schematic of the fault tolerant multiplex bus arrangement.

2.3 Multiplex Signal Bus Wires—A two-wire, multiplex signal bus is used to minimize susceptibility to and generation of electromagnetic interference when used in conjunction with a differential waveform. The use of this form of communications medium also complements other aspects of the fault tolerant design.

2.4 Isolation Resistors—The fault tolerant multiplex signal bus employs resistors which partially isolate individual nodes from the multiplex signal bus and thus limit the propagation and effect of individual failures. These resistors are of a type guaranteed not to fail in a short circuit mode or at a resistance significantly less than their nominal value. Similar resistors are used to terminate the signal bus wires and thus provide the multiplex signal bus voltage and current levels needed for proper operation.

2.5 Signal Bus Termination Networks—The signal bus termina-

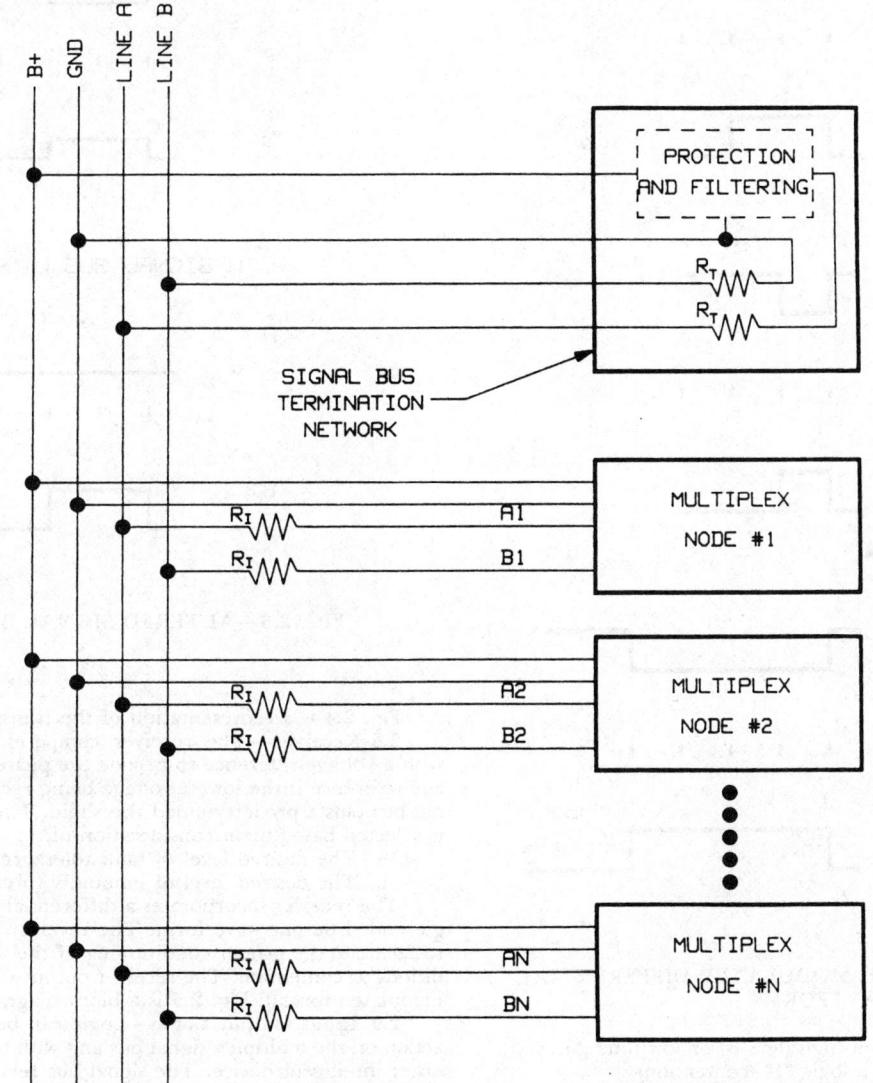

FIG. 2.1—FAULT TOLERANT MULTIPLEX BUS ARRANGEMENT

tion networks perform the following functions:
  a. Sourcing one wire of the multiplex signal bus (designated line A) from battery voltage through a bus termination resistor.
  b. Termination of the other multiplex signal bus wire (designated line B) to ground through a bus termination resistor.
  c. Filtering of the line A multiplex signal bus wire from possible noise present on the battery line.
  d. Protection of the line A multiplex signal bus and nodes tied to it from possible transients present on the battery line.
  e. Providing a circuit arrangement which does not draw significant battery current when the key is off and no transmitters are in operation.

2.6 Pulse Width Modulated Differential Waveforms—The waveforms shown in Fig. 2.2 contain the following characteristics:
  a. Three phases per bit cell.

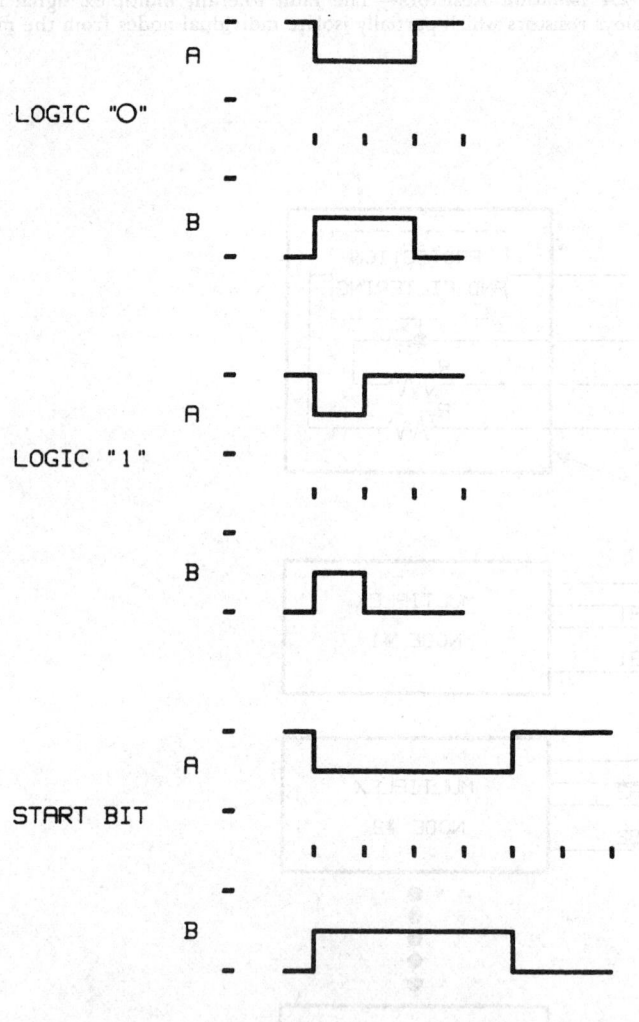

FIG. 2.2—PULSE WIDTH MODULATED DIFFERENTIAL WAVEFORMS

  b. Each line is only driven high (line B) or low (line A).
  c. A logic "0" overrides a logic "1" (contention).
  d. Self-clocking.
  e. Inclusion of reference information enhances fault tolerant detection of the received waveform.

Fig. 2.3 shows the effect of contention and failures on the signal bus waveform.

2.7 Transmitter—The transmitter drives the multiplex signal bus through the node isolation resistors. The transmitter must provide a low voltage drop when on and high internal impedance when off to provide a usable signal which permits:
  a. Decoding of the multiplex signal bus waveforms under contention and failure conditions.
  b. Sufficient link margins to provide adequate electrical noise immunity.

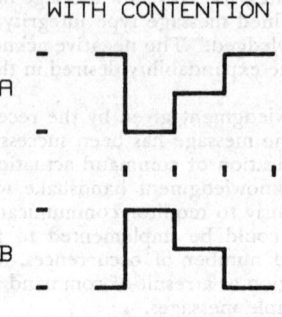

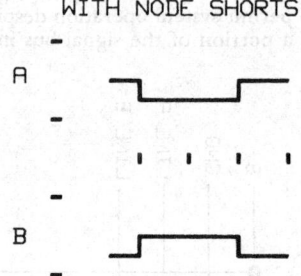

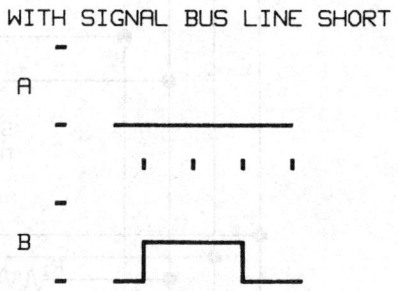

FIG. 2.3—ALTERED SIGNAL BUS WAVEFORMS

Fig. 2.4 is a representation of the transmitter.

2.8 Receiver—The receiver compares the received bus voltage with a voltage reference to decode the phases of the bit cell. The voltage reference is the lowest voltage being received on the multiplex signal bus plus a predetermined threshold. The predetermined threshold is selected based upon consideration of:
  a. The desired level of fault tolerance.
  b. The desired level of immunity to electrical noise.

The receiver incorporates a differential amplifier to produce a single ended output wave form. The receiver also employs level shifting to facilitate the proper conditioning of the waveform during fault conditions or contention. The receiver output waveform is in a pulse width modulated format. Fig. 2.5 is a block diagram of the receiver.

2.9 Input/Output Logic—Logic can be used to coordinate interaction on the multiplex signal bus and with the node microcomputer or other intelligent device. The signal bus related functions may include:
  a. Start bit generation and detection.
  b. Signal arbitration of bus data.
  c. Error detection.
  d. Message acknowledgment.

The Microcomputer related functions may include:
  a. Input of data from the multiplex signal bus.
  b. Output of data destined for the multiplex signal bus.
  c. Input of diagnostic information relative to bus messages.

Alternative implementations of this logic in hardware and software can be designed which are consistent with this proposal. These alternative implementations must, however, interact properly with the multiplex signal bus and properly in conjunction with signals placed on the bus by other nodes.

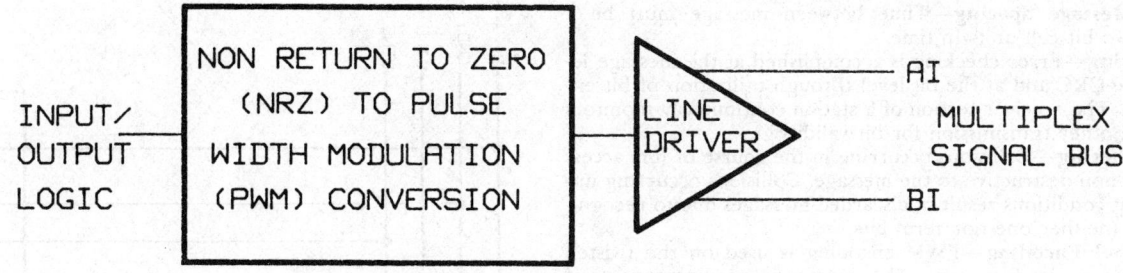

FIG. 2.4—TRANSMITTER

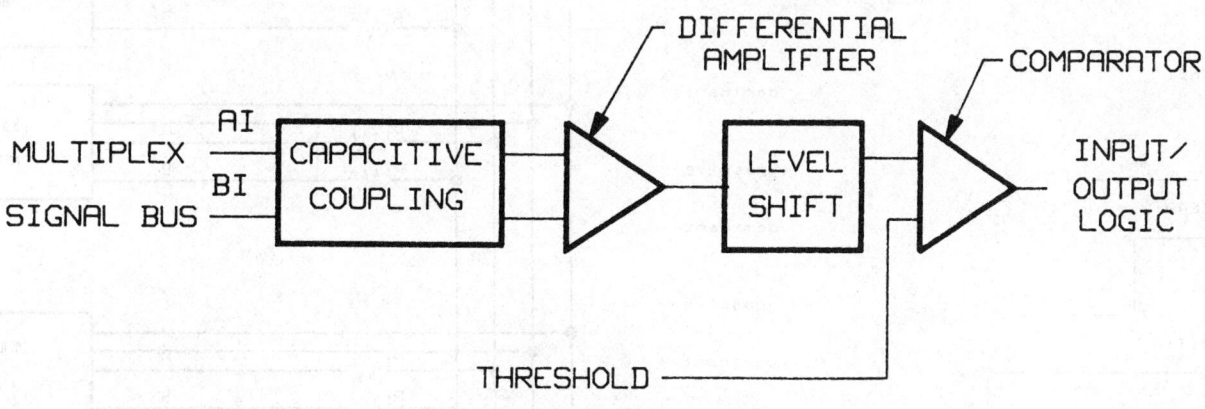

FIG. 2.5—RECEIVER

APPENDIX A—TECHNICAL SUMMARY FOR VNP PROTOCOL

A1. Packet Format

```
 _____
| S |PRIORITY|DESTINATION| SOURCE | DATA   /
|   |& TYPE  | ADDRESS   |ADDRESS |BYTE 1 /
|___|_____|_____|_____|_____/
    <----------Header------------>

   /        |CYCLIC  |E|              |
-->/ DATA   |REDUN   |O| ACKNOWLEDGE  |
  / BYTE N  |CHECK   |D| FIELD        |
 /_____|_____|_|_____|
```

Stations must be able to receive messages on the common twisted pair bus with indicated format. Each message should be viewed as a sequence of 8-bit bytes. The most significant bit of each byte is transmitted first to facilitate arbitration.

Maximum Message Length—101 bits (start bit + 3 bytes header + 8 bytes data and acknowledgment combined + 1 byte CRC + end-of-data-bit) with total message time not to exceed 10 ms.

Minimum Message Length—53 bits (start bit + 3 bytes header + 1 byte data + 1 byte CRC + 1 byte acknowledgment + end-of-data-bit).

Start Bit—A unique occurrence extending over two bit cell times to synchronize all stations to the beginning of a message.

Priority/Type Field—The first 4 bits transmitted as part of this field specify the assigned message priority. The second 4 bits identify which type of transmission is in process: this subfield defines how the remainder of the message is to be interpreted.

Destination Address—This 8-bit field specifies either the physical address of the station to which the message is directed or a functional group which may reside in one or more unspecified physical stations.

Source Address—This 8-bit field contains the unique address of the station which is transmitting the message.

Data Field—This field contains an integral number of bytes from 1 to 7.

A2. Control Procedure—The control procedure defines how and when a station may transmit messages over the common bus.

Defer—A station must not transmit onto the bus until the station has ascertained that there has been no bus activity for at least two bit cell units of time.

Transmit—A station may transmit if it is not deferring. It may continue to transmit until either the last transmit field of the message (CRC) or until the station detects a bit-level discrepancy between what it is transmitting and what is being detected by the receiver section of the station.

Abort—If the receiver section of the station detects a bit-level discrepancy between its transmission and its reception, the station aborts transmission beginning with the next bit. Other bus participants are not informed of this occurrence.

Re-transmit—Since transmissions are non-destructive in that a single dominant message always completes, re-try may occur immediately without danger of continuous unresolved collision. The re-try function is enabled by a higher level of protocol. Re-trys are discontinued after a preset number of attempts.

Back-off—As stated above, back-off is required while the dominant message continues, but re-try may occur immediately without concern for re-collision with the dominant node.

CRC Field—This 1-byte field contains a redundancy check defined by the generating polynomial: $x^8 + x^4 + x^3 + x^2 + x + 1$. The CRC covers the priority/type, destination address, source address, and data fields.

End-of-Data Bit—The end-of-data bit is defined as an idle bus occurrence for a duration of one bit cell time.

Minimum Message Spacing—Time between message must be a minimum of two bit cell units in time.

Error Checking—Error checking is accomplished at the message level through the CRC and at the bit level through utilization of bit encoding (PWM). The receiver section of a station continuously monitors its own and all other transmission for bit validity.

Collision Filtering—Collisions occurring in the course of line access arbitration are non-destructive to the message. Collisions occurring under system fault conditions result in discarded messages due to recognition of invalid (neither one nor zero) bits.

A3. Channel Encoding—PWM encoding is used on the twisted pair. There is no carrier frequency. The waveform encoding consists of three phases. During the first phase, the bus is unconditionally driven to the active state: during the last phase the bus must remain in the passive (non-dominant) state. The state of the bus during the second phase determines whether the bit is a logic one or a logic zero. The bit encoding is return-to-zero (RZ) type. The bus transitions from passive to active state at the beginning of each bit and returns to passive state at the end of phase one or phase two.

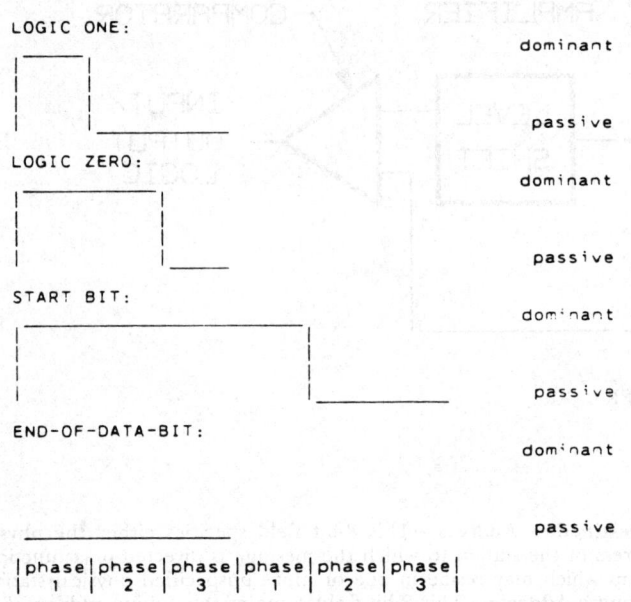

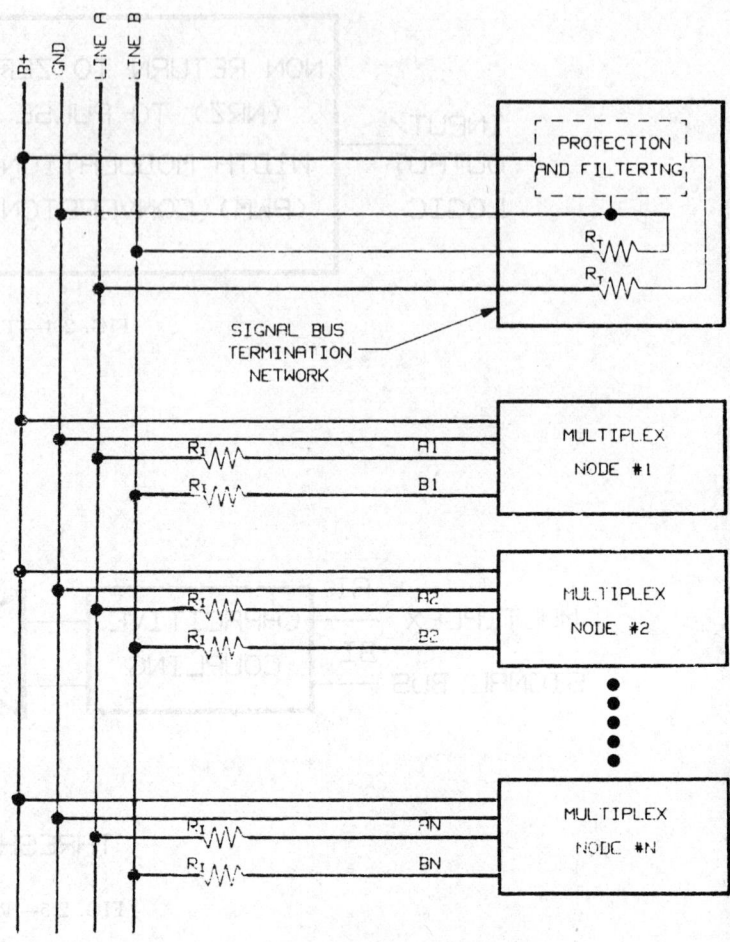

FIG. B1—FAULT TOLERANT MULTIPLEX BUS ARRANGEMENT

A4. Data Rate—10 Kbits/s is a typical data rate.

A5. Carrier—The presence of data transitions indicates carrier is active (that is, the bus is not idle).

APPENDIX B—TECHNICAL SUMMARY FOR VNP-COMPATIBLE FAULT TOLERANT BUS

B1. Schematic of Network Architecture—A distributed (not centralized) architecture with a ring configuration topology is preferred to complement the fault tolerant approach and to permit a more complete realization of its advantages. See Fig. B1, Fault Tolerant Multiplex Bus Arrangement.

B2. Cable Requirements
Impedance—approximately 120 ohm characteristic impedance.
Loss—consistent with 2.0 amperes or greater capacity.
Grounding—See Fig. B1 schematic.
Physical Dimensions and Materials—copper, twisted pair consistent with above requirements.

B3. Connector Requirements and Terminators—Connector on bus (female shell with male pins) with internal 200 ohm, 3 watt flameproof resistors is desired on signal lines. (Male pins with a female shell configuration is also acceptable.) Rating of power and ground line pins should be consistent with 10 or higher AWG wires. Insulation displacement connectors should be used if proven sufficiently reliable in the automotive environment and cost effective to ease the installation of nodes to the system in manufacturing, the dealership, and in aftermarket applications.

B4. Transceiver
Connection Rules—TBD
Cable Interface
Input Impedance—See Fig. B1 schematic.
Signal Swing Ranges and Accuracies—Potential swing is between 0 and 9-16 volts. Actual swing is between 0 and ½ of 9-16 volts on one line and between ½ of 9-16 volts and 9-16 volts on the other line. Voltage swings and bias points can change under fault conditions.
AC Switch Parameters—At approximately 0.6 volts of network swing on either line from idle (passive) condition, the dominant condition will be recognized. (A threshold in this range is needed to continue to operate successfully despite two worst case fault conditions.)
Signal Symmetry Requirements—One line switches between 0 volts and approximately ½ of battery voltage while the other line switches between battery voltage and approximately ½ of battery voltage (except under fault conditions).

B5. Isolation—The Fault Tolerant Multiplex Signal Bus is designed to provide effective isolation and protection for individual nodes. There is also protection for loss of one of the two signal bus wires—open or grounded (two opens can be tolerated with the preferred ring configuration). By using distributed control and the Fault Tolerant Multiplex Signal Bus, the system is designed to be relatively immune from individual failures bringing down the system and rendering it unavailable to aid in diagnosis.

B6. Transceiver Cable and Connectors
Insertion Loss Specification—TBD
List of Signal Names Versus Pin Number—TBD
Sketch of Wire Harness Detail—TBD

E/E DATA LINK SECURITY
—SAE J2186 SEP91

SAE Recommended Practice

Report of the SAE Vehicle E/E Systems Diagnostics Standards Committee approved September 1991.

TABLE OF CONTENTS
1. Scope
2. References
2.1 Applicable Documents
2.1.1 SAE Publications
2.1.2 California Air Resources Board Regulations
2.2 Related Publications
2.2.1 SAE Publications
2.2.2 ISO Publications
3. Definitions
4. Technical Requirements
4.1 Data Link Security Strategy
4.2 Data Link Access Function Examples
4.2.1 Unsecured Read Data
4.2.2 Unsecured Service Alteration
4.2.3 Unsecured Permanent Alteration
4.2.4 Secured Read Data
4.2.5 Secured Service Alteration
4.2.6 Secured Permanent Alteration
4.3 Characteristics of Security
4.4 Functional Requirements

1. Scope—This SAE Recommended Practice establishes a uniform practice for protecting vehicle modules from "unauthorized" intrusion through a vehicle diagnostic data communication link. The security system represents a recommendation for motor vehicle manufacturers and provides flexibility for them to tailor their system to their specific needs. The vehicle modules addressed are those that are capable of having solid-state memory contents altered external to the electronic module through a diagnostic data communication link. Improper memory content alteration could potentially damage the electronics or other vehicle modules; risk the vehicle compliance to government legislated requirements; or risk the vehicle manufacturer's security interests. This document is intended to meet the "tampering protection" provisions of California Air Resources Board OBD II regulations and does not imply that other security measures are not required nor possible.

2. References

2.1 Applicable Documents—The following publications form a part of this specification to the extent specified herein. The latest issue of SAE publications shall apply.

2.1.1 SAE PUBLICATIONS—Available from SAE, 400 Commonwealth Drive, Warrendale, PA 15096-0001.

J1978 MAR92 OBD II Scan Tool

2.1.2 CALIFORNIA AIR RESOURCES BOARD REGULATIONS—Available from Air Resources Board, Mobile Source Division, 9528 Telstar Avenue, El Monte, CA 91731.

Mail out #91-18 (OBD II)—Title 13, California Code of Regulations, Section 1968.1 Malfunction and Diagnostic System Requirements—1994 and Subsequent Model Year, Passenger Cars, Light Duty Trucks, and Medium-Duty vehicles with Feedback Fuel Control Systems.

2.2 Related Publications—The following publications are provided for information purposes only and are not a required part of this document.

2.2.1 SAE PUBLICATIONS—Available from SAE, 400 Commonwealth Drive, Warrendale, PA 15096-0001.

J1850 AUG91—Class B Data Communication Network Interface
J1930 SEP91—Terms, Definitions, Abbreviations and Acronyms
J1979 DEC91—E/E Diagnostic Test Modes

2.2.1.1 The following publications are under development:

J1962 Draft Diagnostic Connector
J2012 Draft Format and Messages for Diagnostic Trouble Codes
J2190 Draft Enhanced E/E Diagnostic Test Modes
J2201 Draft Universal Interface for OBD II Scan Tool
J2205 Draft Expanded Diagnostic Protocol for OBD II Scan Tool

2.2.2 ISO PUBLICATIONS—Available from ANSI, 11 West 42nd Street, New York, NY 10036.

ISO/CD 9141-1 Road vehicles—Diagnostic systems—CARB requirements for April 1991 interchange of digital information.

3. Definitions

STANDARD SERVICE TOOL—A service tool providing basic emission related power train diagnostic tool functions, as required by CARB OBD II regulations. This service tool complies with SAE J1978 and is not necessarily limited to only OBD II functions.

ENHANCED SERVICE TOOL—A service tool providing expanded emission related power train diagnostic functions in excess of the standard service tool and/or non-power train systems diagnostic tool functions.

SECURED SERVICE TOOL—A service tool containing a feature for accessing secured data link functions. Use of this feature is limited by the vehicle manufacturer.

UNSECURED FUNCTIONS—Standard diagnostic functions that are provided by vehicle manufacturers. These are controlled and protected by the on-vehicle controller. The unsecured capability includes reprogramming of selected items for which the reprogrammer is liable.

SECURED FUNCTIONS—Manufacturer restricted functions that require "Unlocking" the on-vehicle controller to gain access. Typical functions include programming of vehicle emission systems.

READ DATA FUNCTION—Operation which reads parameters and codes via the data link.

SERVICE ALTERATION FUNCTION—Operation which temporarily alters the vehicle control system for the purpose of diagnosing the system's operation.

PERMANENT ALTERATION FUNCTION—Operation which permanently alters the operating characteristics of a vehicle module or system.

SEED—The data value sent from the on-board controller to the secured service tool.

KEY—The data value sent from the secured service tool to the on-board controller.

4. Technical Requirements—Proper "Unlocking" of the controller shall be a prerequisite to access certain critical on-board controller functions; the only access to the on-board controller permitted while in a "Locked" mode is through the product-specific software. This permits the product-specific software to protect itself and the rest of the vehicle control system from unauthorized intrusion.

This document does not attempt to define capability or information that is under security; this is left to the vehicle manufacturer. The security system shall not prevent basic diagnostic communications between the external tool and the on-board controller.

4.1 Data Link Security Strategy—As shown in Table 1, the data link access of function is divided into two classifications: Unsecured and Secured. Each of these functions are subclassified as: Read Data, Service Alteration, and Permanent Alteration. These classifications are allocated to three types of service tools: Standard Tools, Enhanced Tools, and Secured Tools. The allocation of access classification by service tool type should provide for normal service capability of the vehicle and protect the vehicle from "unauthorized" intrusion of certain critical functions.

**TABLE 1—SECURITY AUTHORITY ACCESS
DATA LINK ACCESS FUNCTION (REFER TO SECTION 4.2)**

Service Tool Type	Unsecured Read Data (4.2.1)	Unsecured Service Alteration (4.2.2)	Unsecured Permanent Alteration (4.2.3)	Secured Read Data (4.2.4)	Secured Service Alteration (4.2.5)	Secured Permanent Alteration (4.2.6)
Standard (SAE J1978)	X	B	B	NA	NA	NA
Enhanced	X	E	E	NA	NA	NA
Secured	X	X	X	X	X	X

X — Function Allowed
NA — Not Allowed
B — Basic Diagnostic Capability Per Vehicle Manufacturer's Specification
E — Enhanced Diagnostic Capability

4.2 Data Link Access Function Examples

4.2.1 UNSECURED READ DATA
Read emission related data.
Read emission related trouble codes.

4.2.2 UNSECURED SERVICE ALTERATION
Cycle device on/off.
Substitute sensor value.

4.2.3 UNSECURED PERMANENT ALTERATION
Change vehicle option/configuration data (i.e., tire size).
Reset electronic module.

4.2.4 SECURED READ DATA

Read keyless entry parameters.
Read executable code.

4.2.5 SECURED SERVICE ALTERATION
Vehicle assembly plant verification tests involving parameters not normally used in service.

4.2.6 SECURED PERMANENT ALTERATION
Alteration of a vehicle emission calibration.
Alteration of executable code.

4.3 Characteristics of Security—A special communications mode shall be provided to "Unlock" the on-board controllers which have secured or restricted functions. The security system is intended to make the emission related controller more immune to:
  a. Unauthorized intrusion into the controller without full control of the product specific software.
  b. Unauthorized alteration of the on-board control system or control parameters.

Disclosure of the "Seed/Key" relationship shall be limited to those persons, authorized by the vehicle manufacturer, who are responsible for the production of the secured service tool. The security system shall not prevent basic diagnostic or vehicle communications between external devices and the on-board controller.

There shall be three parameters which control the security access of the on-board controller:
  a. The "Seed" and "Key" shall each be a minimum of 2 bytes in length. The relationship between the "Seed" and "Key" is the responsibility of the vehicle manufacturer. Multiple "Seed/Key" relationships may exist for access to different functions within a controller or systems within a vehicle.
  b. The Delay Time (DT) shall be a minimum of 10 seconds; the vehicle manufacturer may specify an increased delay time to suit its specific requirements.
  c. The Number of False Access Attempts (NFAA) shall be a maximum of two; the vehicle manufacturer may specify a reduced number of false attempts to suit its specific requirements. When the "Key" received by the controller is not correct, it shall be considered as a false access attempt; if access is rejected for any other reason, it shall not be considered a false access attempt.

4.4 Functional Requirements—Two request/response communication message pairs (Request #1/Response #1, Request #2/Response #2) shall be used to "Unlock" the on-board controller. The specific message content is not specified by this document and is the responsibility of the vehicle manufacturer.
  a. Step 1—The external device shall request the on-board controller to "Unlock" itself by sending Request #1. The controller shall respond by sending a "Seed" using Response #1. A seed value of zero shall indicate that the controller is currently unlocked.
  b. Step 2—The external device shall respond by returning a "Key" number back to the controller using Request #2. The controller shall compare this "Key" to one internally determined and issue Response #2.

If the two numbers agree, then the controller shall enable ("Unlock") the external device's access to secured communication modes.

If, upon "NFAA" attempts, the two keys do not compare (false attempt), then the controller shall insert the DT before allowing further attempts. The DT shall also be required at each controller power-on.

Three on-board controller responses shall be decoded by the external device:
  a. Accept—The controller has "Unlocked" its access.
  b. Invalid Key—The access attempt was rejected because the key was determined to be invalid by the controller; the access attempt was false.
  c. Process Error—The access attempt was rejected for reasons other than receiving the wrong key; this shall not be counted as a false access attempt.

Termination of security access, "Locking" the product, shall result after any of the following conditions:
  a. Each time the product is "powered up."
  b. Upon commanding the product to a normal operational mode.
  c. Conditions at the vehicle manufacturers discretion.

If an attempt is made to communicate with a "Locked" on-board controller and access a "Secured" function, the controller may return a special response indicating that the controller is "Locked" and cannot respond as requested.

PERFORMANCE LEVELS AND METHODS OF MEASUREMENT OF ELECTROMAGNETIC RADIATION FROM VEHICLES AND DEVICES NARROWBAND, 10 kHz to 1000 MHz— SAE J1816 OCT87

SAE Recommended Practice

Report of the Electrical and Electronic Systems Technical Committee approved October 1987.

1. Foreword—With the advent of computer based electronics being utilized in automobiles, the Electromagnetic Radiation Subcommittee has deemed it prudent that a new test method be written to provide a common test for the measurement of narrowband radiation from vehicles and devices.

2. Purpose—This SAE Standard covers methods of measuring incidental narrowband radiation from vehicles and devices. The standard also establishes performance levels intended to protect nearby communication and broadcast receivers. It is intended to serve as an alternate method of measuring electromagnetic radiation which is analogous to the FCC Part 15 methodology but adapted to measuring vehicles. The equivalent procedures for broadband emissions are set forth in SAE J551.

3. Scope—This standard covers narrowband emissions in the frequency range of 10 kHz to 1000 MHz. An example of such radiation is the unintended emission from on-board logic and computer modules. In particular, the standard is intended to provide protection to adjacent mobile communication and broadcast receivers (vertical E-Field test) and portable broadcast radios (magnetic field test). In the future horizontal E-Field measurement requirements will be considered.

4. Definitions—Terms used in this Standard are defined either in SAE J551 or in the IEEE Dictionary of Electrical Terms.

5. Measuring Equipment—The measuring equipment shall conform to the requirements of CISPR Pub 16 or American National Standard C63.2. Because of near-field interaction, the following antennas shall be used in order to obtain repeatable results between different test sites:

Frequency Range	Antenna
10 kHz - 30 MHz	1.1 metre monopole with appropriate ground plane, and with a loop antenna whose size will allow it to be completely enclosed by a square having a side of 60 cm in length
30 MHz - 200 MHz	Broadband biconical dipole
200 MHz - 1 GHz	Log-periodic dipole array with a maximum length of 1.2 metres

6. Method of Measurement—The test method to be used shall be based on either CISPR Publication 16 or American National Standard C63.4, except:
  a. The electric field antenna shall be vertically polarized.
  b. Measurements using the loop antenna shall be made in three orthogonal planes.

6.1 The test antenna shall be positioned at 1.0 ± 0.1 metre height above the ground as measured to the ground plane of the monopole or the midpoint of the elements of the dipole or log-periodic antennas. The measurement point for the loop antenna shall be its center point.

6.2 The test antenna shall be placed at a distance of 3.0 ± 0.1 metres from the test vehicle or device [measured from the center of the antenna elements (tip for a log-periodic) to the perimeter of the vehicle or device]. The measurement point for the loop antenna shall be at its center point.

6.3 Measurements shall be made using a peak or quasipeak detector.

Note—FCC Part 15 permits the use of either a peak or quasipeak detector with a single limit.

The following bandwidths shall be used (or the closest available bandwidth on the measuring equipment used with appropriate correction factor, if required):

Frequency Range	Quasipeak Bandwidth	Peak Bandwidth
10 kHz - 150 kHz	200 kHz	1 kHz
150 kHz - 30 MHz	9 kHz	10 kHz
30 MHz - 1 GHz	120 kHz	100 kHz

6.4 Note—While open site measurements are satisfactory, an anechoic shieldroom at higher frequencies and a shieldroom at lower frequencies will be desirable to reduce ambient signals allowing the low level radiated emissions to be easily measured.

At low frequencies, there will be standing waves even in an anechoic shieldroom. Correlation must be established between indoor and outdoor test sites. In any case, free space (outdoor) shall be the reference condition. For additional information on anechoic shieldrooms, refer to SAE J1507.

7. Test Procedure
7.1 Calibrate Equipment

7.2 The vehicle or device shall be positioned adjacent to the antenna and each long-term operator-controlled device (such as the blower motor) shall be turned on to its maximum noise producing condition. For some subsystems, this may require running the engine. Equipment, such as thermostatically controlled devices, shall be caused to remain in a running condition, unless doing so would cause an unsafe operating condition.

7.3 Measurements shall be made at the following locations:
  a. Center front
  b. Center rear
  c. Right and left side in line with major sources (such as instrument panel and power plant)

8. Performance Levels

Note—For magnetic field measurement, apply the rationale in ANSI C63.12-1984 Section 7.1, revised as follows:

It is tacitly understood that a radiation guideline expressed in uV/m implies a magnetic field guideline related by the free space characteristic impedance of 377 ohms. Nevertheless, the free space impedance does not hold in the near field, for example at frequencies where the measurement distance is less than $\lambda/2\pi$.

Applying the free space magnetic field impedance factor in the near field, however, provides equivalent protection for receivers (and other devices) whether magnetic or electric fields are the source of interference.

9. References

SAE J551, Performance Levels and Methods of Measurement of Electromagnetic Radiation from Vehicles and Devices (30-1000 MHz).

SAE J1507, Anechoic Test Facility Radiated Susceptibility 20 MHz to 18 GHz Electromagnetic Field.

Obtainable from the Society of Automotive Engineers, 400 Commonwealth Drive, Warrendale, PA 15096 USA.

IEEE Dictionary of Electrical Terms.

Obtainable from IEEE Service Center, 445 Hoes Lane, Piscataway, NJ 08854 USA.

CISPR Pub 16.

International Special Committee on Radio Interference, obtainable from American National Standards Institute, 1430 Broadway, New York, NY 10018 USA.

ANSI C63.12-1984.
ANSI C63.2.

Obtainable from American National Standards Institute, 1430 Broadway, New York, NY 10018 USA.

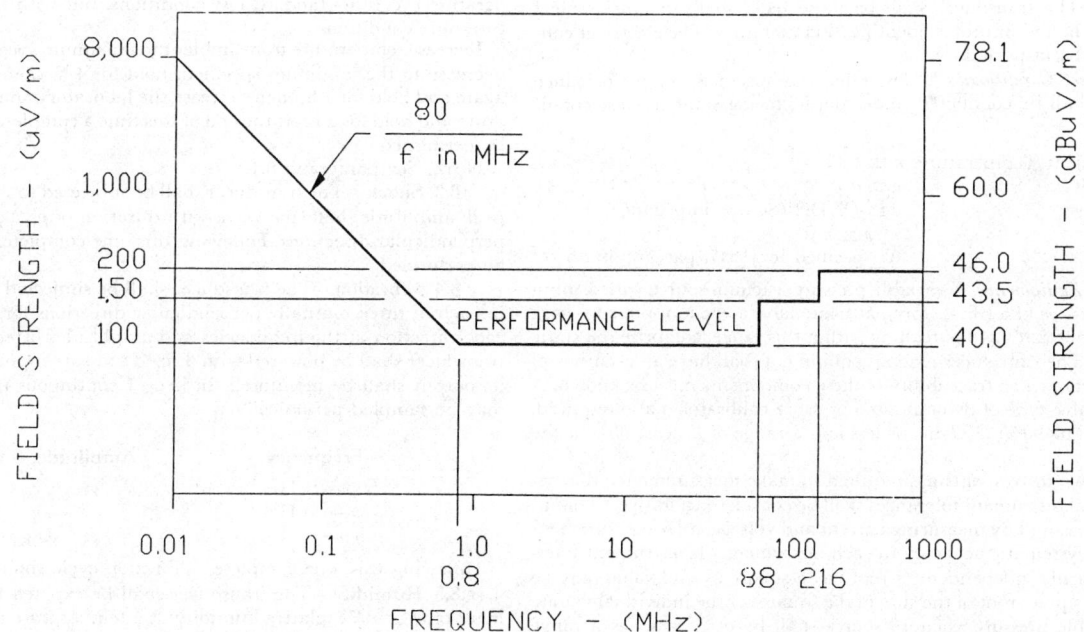

FIG. 1

GUIDE TO MANIFOLD ABSOLUTE PRESSURE TRANSDUCER REPRESENTATIVE TEST METHOD—SAE J1346 JUN81

SAE Information Report

Report of the Electronic Systems Committee, approved June 1981.

Introduction—This document is intended as a guide for technical personnel of both using and supplier firms whose duties include specifying, calibrating, testing, developing, or demonstrating the performance characteristics of Manifold Absolute Pressure (MAP) transducers. By basing users' specifications as well as supplier technical advertising and reference literature on this document, or by referencing portions thereof, as applicable, a clear understanding of the methods used for evaluating or proving performance will be provided. Adhering to the exemplary test method outline, terminology, and procedures shown will result in simple, complete test method documents; it will reduce design time, procurement lead time, labor, and materials costs. Of major importance will be the reduction of qualification test lead time and duration resulting from the use of commonly accepted test procedures.

The guide may also serve a useful purpose as a general example for test procedures on other types of transducers.

The MAP Transducer Representative Test Method has a companion dependent document, SAE J1347, Guide to Manifold Absolute Pressure Transducer Representative Specification, which is referenced to it.

Scope—This guide is intended to cover test procedures applicable to MAP transducers; it is also applicable to transducers such as Barometric (Ambient) Absolute Pressure transducers, Manifold Vacuum transducers, and similar pressure transducers used in automotive systems. Although oriented towards active devices (those using internal signal conditioning), it can be applied to passive devices with minor modifications.

Values—The guide is intended to be general in nature. Specific values for test data are not included in order to maintain generality. Exemplary values are contained in the Appendix in an attempt to clarify the text. *The SAE does not imply any recommendation regarding these values.*

TEST METHOD
MANIFOLD ABSOLUTE PRESSURE TRANSDUCER
A. Qualification Tests

1. Inspection—The transducer shall be inspected visually for mechanical defects, poor finish, and improper identification markings. The electrical connector shall also be inspected.

2. Standard Test Conditions—Unless otherwise specified, all performance test procedures shall be conducted under the following standard test conditions:

Laboratory Ambient Temperature:	$a \pm b\,°C$[1]
Relative Humidity:	$c \pm d\%$
Excitation Voltage:	$e \pm f$ V DC (Source impedance $< g \pm h\,\Omega$)
Output Load:	As specified in J1347, paragraph 2.6

3. Calibration Equipment—A variable pressure/vacuum source with a minimum range of pp–qq kPa (rr–ss torr), absolute, and a vacuum-pressure gage with an accuracy of $\pm tt\%$ of the reading within this range, comprise the static pressure system. The transducer readout equipment shall have an accuracy of $\pm uu\%$ of the reading. The traceability of these components must be known. A quick-opening-valve-type of dynamic step pressure calibrator is also required, with a rise time (10–90%) of kk ms or less and a range of at least 80% of the transducer range.

3.1 When two instruments are required to make measurements, the systematic individual instrument tolerances shall be considered additive. (That is, if resistance is measured by measuring current and voltage, a 1% measurement requires ±0.5% systematic accuracy in each instrument.) If instrument tolerances are statistically independent, a root-mean-square overall value may be computed as the square root of the sum of the squares of the individual values.

3.2 A variable pressure/vacuum source shall be used that has a range extending above and below the range of the transducer to be measured, and not less than 110% of that range.

4. Installation Checks—The transducer shall be connected to the pressure source (manifold) and secured with the recommended force or torque. The excitation source, signal conditioner, and read-out instrument shall also be connected to the transducer and turned on.

A warm-up time of 1/2 h shall be allowed for the test equipment, prior to energizing the transducer under test.

The pressure source, connecting tubing, and transducer system shall pass a leak test: the transducer system shall be pressurized to $i \pm j$ kPa (absolute)

[1]The letters referenced in place of numeric values are keyed to the Appendix.

after which the system shall be sealed off from the pressure source. During the next 30 min, the system pressure shall not change by more than $\pm k$ kPa. Alternative leak tests may be specified.

5. Initial Static Calibrations—The transducer shall be energized for vv min, after which the initial calibration shall be performed.

The transducer output shall be recorded at standard test conditions and eleven input pressures representing 20% increments of the range of the transducer in ascending and descending directions (or as otherwise specified) approaching each pressure slowly in a monotonic manner to avoid overshoot. In the case of overshoot, the pressure shall be backed up by approximately a 10% increment and the step repeated. Three complete calibration cycles shall be performed consecutively to establish sensitivity, linearity (least squares line of regression is recommended), hysteresis, output noise, and repeatability.

6. Excitation-Voltage Effects—A complete calibration cycle (as specified in Section 5) shall be performed at laboratory standard conditions but with excitation voltages of these percentages of nominal: 70%, 85%, 100%, or as otherwise specified.

7. Warm-Up Characteristics—The transducer shall be subjected to abbreviated calibration cycles. The transducer output shall be recorded at standard test conditions and four input pressures of 0%, 50%, 100%, and 0% of the range of the transducer. These abbreviated calibration cycles shall be performed at specified time intervals, starting ww min after the transducer is energized, and continuing for approximately xx min.

8. Environmental Effects

8.1 *Temperature*—A complete calibration cycle (as specified in Section 5) shall be performed at standard test conditions, but at the following temperatures and in this sequence: from laboratory ambient to specified minimum, up through laboratory ambient to maximum and returning to laboratory ambient. Readings shall not be taken until the output value at constant pressure input has stabilized to within $\pm l\%$ of the value of m min.

8.2 *Temperature Cycling*—The transducer shall be subjected to 200 temperature cycles at standard test conditions, but with the following temperature-time conditions:

Increase temperature from ambient to maximum specified, hold for 1 h, then decrease to the minimum specified, hold for 1 h, then increase to maximum again and hold for 1 h, then decrease the laboratory ambient standard temperature and hold for 2 h. At the end of this time a complete calibration cycle shall be performed.

NOTE: See paragraph 8.1.

8.3 *Shock*—The transducer shall be subjected to shock stimuli of $n \pm o$ g_n peak amplitude, half sine wave with duration of $p \pm q$ ms in three mutually perpendicular directions. Following this, one complete calibration cycle shall be performed.

8.4 *Vibration*—The transducer shall be subjected to sinusoidal vibration in each of three mutually perpendicular directions for a duration of r min in each direction at the frequencies and amplitudes listed (within $\pm 10\%$). The transducer shall be powered with a fixed pressure input of $aaa \pm bbb$ kPa, and its output shall be monitored. In lieu of continuous monitoring, the output may be sampled periodically.

Frequency	Amplitude, 0 to Peak
$s - t$ Hz	u g_n
$v - w$ Hz	x g_n
$y - z$ Hz	aa g_n

Following this, one complete calibration cycle shall be performed.

8.5 *Humidity*—The transducer shall be exposed to a humidity environment of $bb \pm cc\%$, relative humidity at a temperature of $dd\,°C$, for a period of ee h. The transducer shall be powered, with a fixed pressure input of $aaa \pm bbb$ kPa, and its output shall be monitored. In lieu of continuous monitoring, the output may be sampled periodically.

8.6 *Pressure Cycling*—The transducers shall be subjected to ff pressure cycle from ambient pressure to a specified minimum at standard test conditions and at a rate of gg cycles/second or slower.

Following this, one complete calibration cycle shall be performed.

9. Environmental Pressure—The transducer shall be subjected to an external (case) pressure of hh kPa (ii torr) corresponding to an elevation of jj m, and one complete calibration cycle shall be performed.

10. Dynamic Response—The transducer shall be subjected to a step pressure change from yy—zz kPa. The rise time of this step pressure (10–90%) shall be a maximum of *kk* ms. Three pressure steps shall be applied and the average transducer rise time determined.

11. Electromagnetic Susceptibility—The transducer and associated cabling shall be tested for electromagnetic susceptibility, conducted and radiated, in accordance with specified, pertinent parts of the procedures of SAE Recommended Practice J1113a.

12. Life Test—The transducer shall be exercised for *ccc* pressure cycles from ambient pressure to minimum range at standard test conditions and at a rate of *ddd* cycles/second or slower, while being subjected to the temperature-humidity cycle shown in Table 1. The sensor shall have power applied throughout this test. Following this, one complete calibration cycle shall be performed.

13. Other Tests—Additional tests may be specified if required to establish the effects of: overpressure, case isolation, output loading, excitation voltage ripple, over-voltage, voltage transients, radiation, dust, watersplash, oil splash, gasoline vapor exposure, thermal shock, and others. Tests which establish the sensor performance against the standard pressure calibration procedure while the sensor is being subjected to a variation in another parameter or in its vehicle operational environment may be specified. An example is a test of pressure calibration during vibration.

B. Individual Acceptance Tests

Individual acceptance tests are performed on transducers of previously qualified models and types for the purposes of incoming inspection on delivery. These tests will be similar to some of the qualification tests.

14. Inspection—As in Qualification Tests: Section 1.
15. Standard Test Conditions—As in Qualification Tests: Section 2.
16. Calibration Equipment—As in Qualification Tests: Section 3.
17. Installation Checks—As in Qualification Tests: Section 4.
18. Initial Static Calibrations—The transducer shall be energized for *mm* min and then a single complete calibration cycle shall be performed as in Section 5 of the Qualification Tests.
19. Pressure Cycling—The transducer shall be exercised for *nn* pressure cycles from ambient pressure to minimum range at standard test conditions and at a rate of *oo* cycles/second or slower. Following this, one complete calibration cycle shall be performed.

References

1. ANSI MC 6.1—1975 (ISA S37.1), Standard, "Electrical Transducer Nomenclature and Terminology", October 1975.
2. ANSI MC 6.2—1975 (ISA S37.3), Standard, "Specifications and Tests for Strain Gage Pressure Transducers", January 1976.
3. ANSI MC 6.5—1976 (ISA S37.6), Standard, "Specifications and Tests of Potentiometric Pressure Transducers", July 1976.
4. ANSI B88.1—1972, Standard, "A Guide for the Dynamic Calibration of Pressure Transducers", August 1972.
5. P. S. Lederer, "Methods for Performance-Testing of Electromechanical Pressure Transducers", NBS Technical Note 411 (February 1967).
6. SAE Recommended Practice J1113a, "Electromagnetic Susceptibility Procedures for Vehicle Components (Except Aircraft)", 1975.

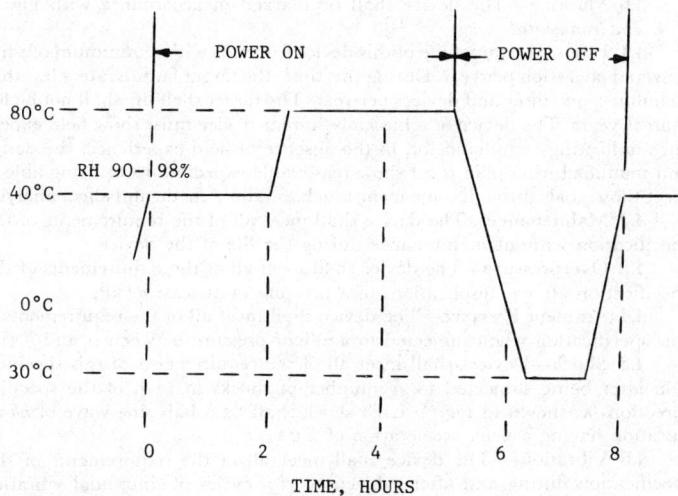

ESTABLISH 90-98% RH AT 40°C AND MAINTAIN MOISTURE IN CHAMBER FOR THE REMAINDER OF THE CYCLE. MONITOR FOR LOS CONTINUOUSLY.

LIFE TEST CYCLE

TABLE 1

Temperature-Humidity Cycle

Twenty-four (24) hour cycle—2 h allowed for transition

Temperature (°C) ± 3°C	Relative Humidity (%) ± 3%	Hours ± 0.2 h	Power
−29 (−20°F)	0	3	off
27 (80°F)	20	2	on
65 (150°F)	20	7	on
110 (230°F)	20	1	on
65 (150°F)	95	9	off

APPENDIX
GUIDE TO MANIFOLD ABSOLUTE PRESSURE TRANSDUCER REPRESENTATIVE TEST METHOD

NOTE: The SAE does not imply any recommendation concerning these values.

Code	Typical Value(s)
a	23
b	±3
c	50
d	15
e	12
f	0.5
g	0.1
h	0.01
i	20
j	±5
k	5
l	0.25
m	1
n	100
o	10
p	10
q	2
r	20
s	2
t	20
u	0.5 in
v	20
w	200
x	15g
y	200
z	2000
aa	10g
bb	95
cc	3
dd	80
ee	72
ff	10^6
gg	1
hh	76
ii	570
jj	4270
kk	5
mm	1
nn	100
oo	1
pp	10
qq	120
rr	75
ss	900
tt	0.1
uu	0.1
vv	30
ww	0.5
xx	30
yy	40
zz	100
aaa	70
bbb	0.1
ccc	10^6–10^7
ddd	4

GUIDE TO MANIFOLD ABSOLUTE PRESSURE TRANSDUCER REPRESENTATIVE SPECIFICATION— SAE J1347 JUN81

SAE Information Report

Report of the Electronic Systems Committee, approved June 1981.

Introduction—This document is intended as a guide for technical personnel of both using and supplier firms whose duties include specifying, calibrating, testing, developing, or demonstrating the performance characteristics of Manifold Absolute Pressure (MAP) transducers. By basing users' specifications as well as suppliers' technical advertising and reference literature on this document, or by referencing portions thereof, as applicable, a clear understanding of the users' needs and of the transducers' performance capabilities will be provided. Adhering to the specification outline, terminology, and procedures shown will result in simple, complete specifications; it will also reduce design time, procurement lead time, and labor, as well as material costs.

This guide is also intended for use as a general example for specifying other types of transducers.

The MAP Transducer Representative Specification is referenced to and dependent upon SAE J1346, Guide to Manifold Absolute Pressure Transducer Representative Test Method.

Scope—This guide is intended to cover specifications applicable to MAP transducers. It is also applicable to transducers such as Barometric (Ambient) Absolute Pressure transducers, Manifold Vacuum transducers, and similar pressure transducers used in automotive systems. Although this guide is oriented towards active devices (those using internal signal conditioning), it can be applied to passive devices with minor modifications.

Values—The guide is intended to be general in nature. Specific values for performance data are not included in order to maintain generality. Exemplary values are contained in the Appendix in an attempt to clarify the text. *The SAE does not imply any recommendations regarding these values.*

SPECIFICATION
MANIFOLD ABSOLUTE PRESSURE TRANSDUCER

1. Operating Pressure Range—The operating pressure range is from a–b kPa exclusive of abnormal pressure excursions. (All pressures in this document are given as absolute pressures.)

2. Electrical

2.1 Power Supply Voltage—The normal operating supply voltage is $c \pm d$ V DC.

2.2 Reference Voltage—The nominal reference voltage is $e \pm f$ V DC.

2.3 Overvoltage Protection
EITHER: The supplied voltage is free from transients and voltage reversals.
OR: Devices must survive the following without degradation of performance.
LIST: (The list shall contain a short description, the name and number of times it may be encountered with reference to a figure showing voltage versus time. A recommended test circuit should be included if available.)

2.4 Current Draw
2.4.1 POWER SUPPLY CURRENT—Maximum allowable current draw shall be h mA.

2.5 Transducer Input/Output Functional Relationship
NOTE: Y may be in units of volts, ohms, henries, farads, hertz, seconds, etc.
EITHER: The output parameter, Y_o, shall correspond to the following:
$$Y_o = AP + B$$
where: P is the pressure in kilopascals
A is i Y units/kPa
B is j Y units
OR: The output parameter, Y_o, shall correspond to the following:
$$Y_o = V_r(AP + B)$$
where: P is the pressure in kilopascals
V_r is either the power supply voltage, or the reference voltage
A is k Y/volt-kPa
B is l Y/volt

2.5.1 Sensors which exhibit a non-linear input-output relationship shall be specified in a similar manner as determined by the purchaser. Where possible, a power series expression is preferred.

2.6 Output Load—The transducer must meet the accuracy requirements while driving a load of m Ω returned to the n supply, or as otherwise specified.

2.7 Output Parameter Range—The maximum and minimum output parameter range over the full required pressure range (and supply voltage range) are from o X–p X units (see paragraph 2.5–Y units).

2.8 Output Error—The allowable output error will be expressed as an equivalent change in the measurand and will be within the limits shown in Figs. 1 and 2. The error limits shown in the figures include all sources of error. (The figures shall show error in kPa versus applied pressure.)

2.9 Warm-Up Time—The transducer shall be operating within the allowable error band no more than q s after the application of power.

2.10 Output Noise and Ripple—The output ripple shall not exceed that shown in Fig. 3. (The figure shall show allowable output peak-to-peak ripple versus frequency.) The output noise shall be as specified by the purchaser.

2.11 Generated and Radiated Noise—(As specified by the purchaser. Adherence to SAE Recommended Practice J1113a is urged.)

2.12 Susceptibility to Conducted and Radiated Noise—(As specified by the purchaser. Adherence to SAE Recommended Practice J1113a is urged.)

3. Mechanical

3.1 Measurement Cavity Leakage (Body Leak)—The measurement cavity shall not leak at a rate greater than r cm^3/min when subjected to a pressure of s kPa.

3.2 Pressure Response Time—The device shall have a response time to a step input of pressure of no more than t ms. The input pressure step shall be from u–v kPa with a 10–90% rise time not to exceed w ms. The response time of the device is defined as the time duration from the 0.1 time until the time it takes the output to settle to within $\pm x$ kPa of its final value.

3.3 Location—The device shall be located _____.

3.4 Materials—Selections of materials are the prerogative of the supplier. However, certain goals of the purchaser of the device may make the use of certain classes of material preferable. These goals, listed as follows, shall be discussed with supplier.

LIST OF GOALS:

3.5 Packaging and Termination—The device shall meet the critical dimensions noted in Fig. 4 and shall have electrical terminations as shown in that drawing.

3.6 Marking—The device shall be marked in accordance with Fig. 4.

4. Environmental

4.1 Life—The target life of this device is y years with a minimum of z h of powered operation per year. During this time, the target failure rate is less than aa failures/one thousand devices per year. The target shelf life shall not be less than bb years. The device or a basically similar device must show field experience indicating compliance or, in the absence of field experience, the design and manufacturing plan must show reasonable expectations of being able to meet those goals through some means such as failure mode and effect analysis.

4.2 Maintenance—The device shall meet all of the requirements of this specification without maintenance during the life of the device.

4.3 Overpressure—The device shall meet all of the requirements of this specification after cc applications of a pressure of at least dd kPa.

4.4 Ambient Pressure—The device shall meet all of the requirements of this specification when subjected to ambient pressure between ee and ff kPa.

4.5 Shock—Devices shall meet all of the requirements of this specification after being subjected to gg number of shocks in each of the specified directions as shown in Fig. 4. Each shock shall be a half sine wave of hh ms duration having a peak acceleration of ii g's.

4.6 Vibrations—The device shall meet all of the requirements of this specification during and after subjection to jj cycles of sinusoidal vibration applied on each axis as specified in Fig. 4. Each cycle shall consist of a frequency sweep from kk–ll Hz and back at a constant acceleration, velocity, or displacement over the appropriate portion of the frequency range. Each cycle of vibration shall last mm min.

4.7 Humidity—The device shall meet all of the requirements of this specification during and after being subjected to nn h of $oo\%$ relative humidity at a temperature of pp °C.

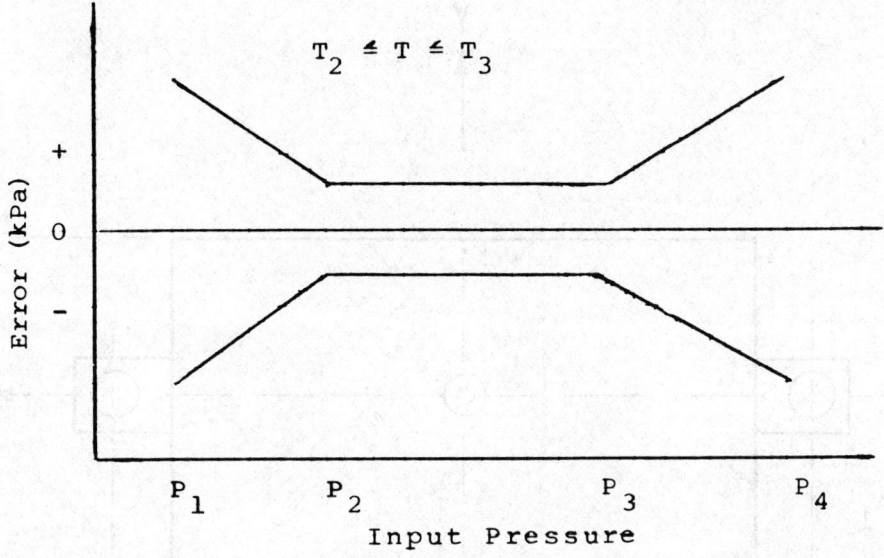

Error = Indicated Pressure-Actual Pressure
FIG. 1

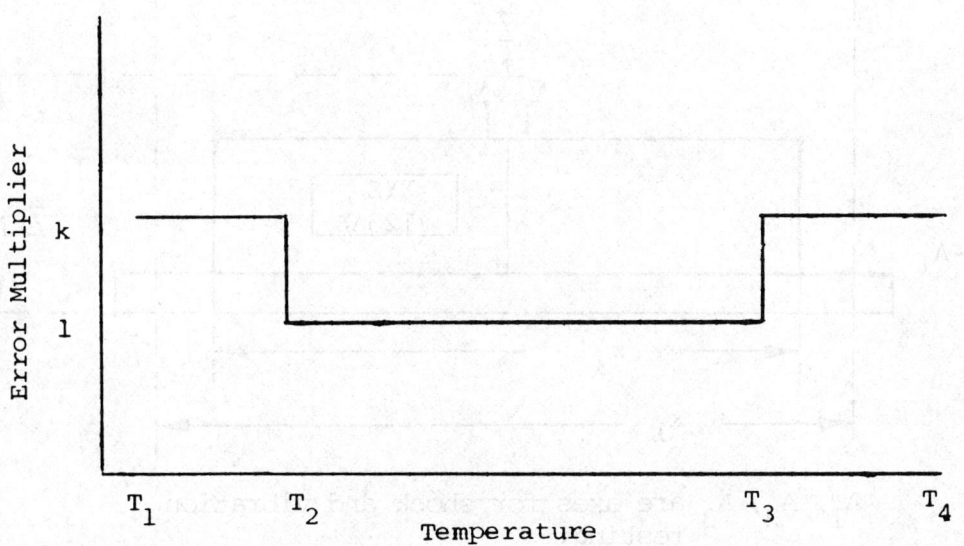

Error Multiplying Factor over Temperature Range
FIG. 2

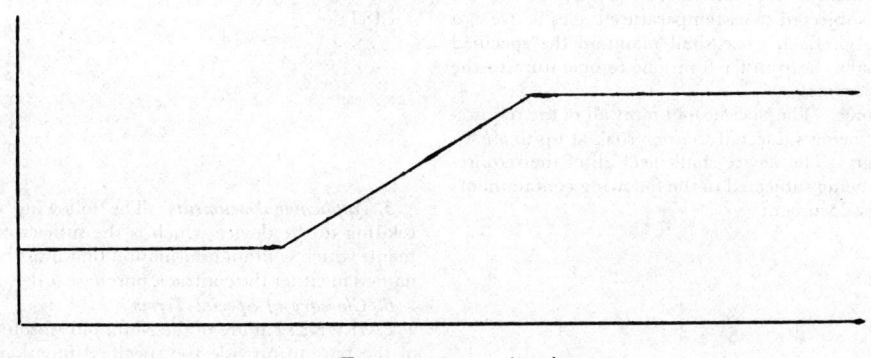

FIG. 3

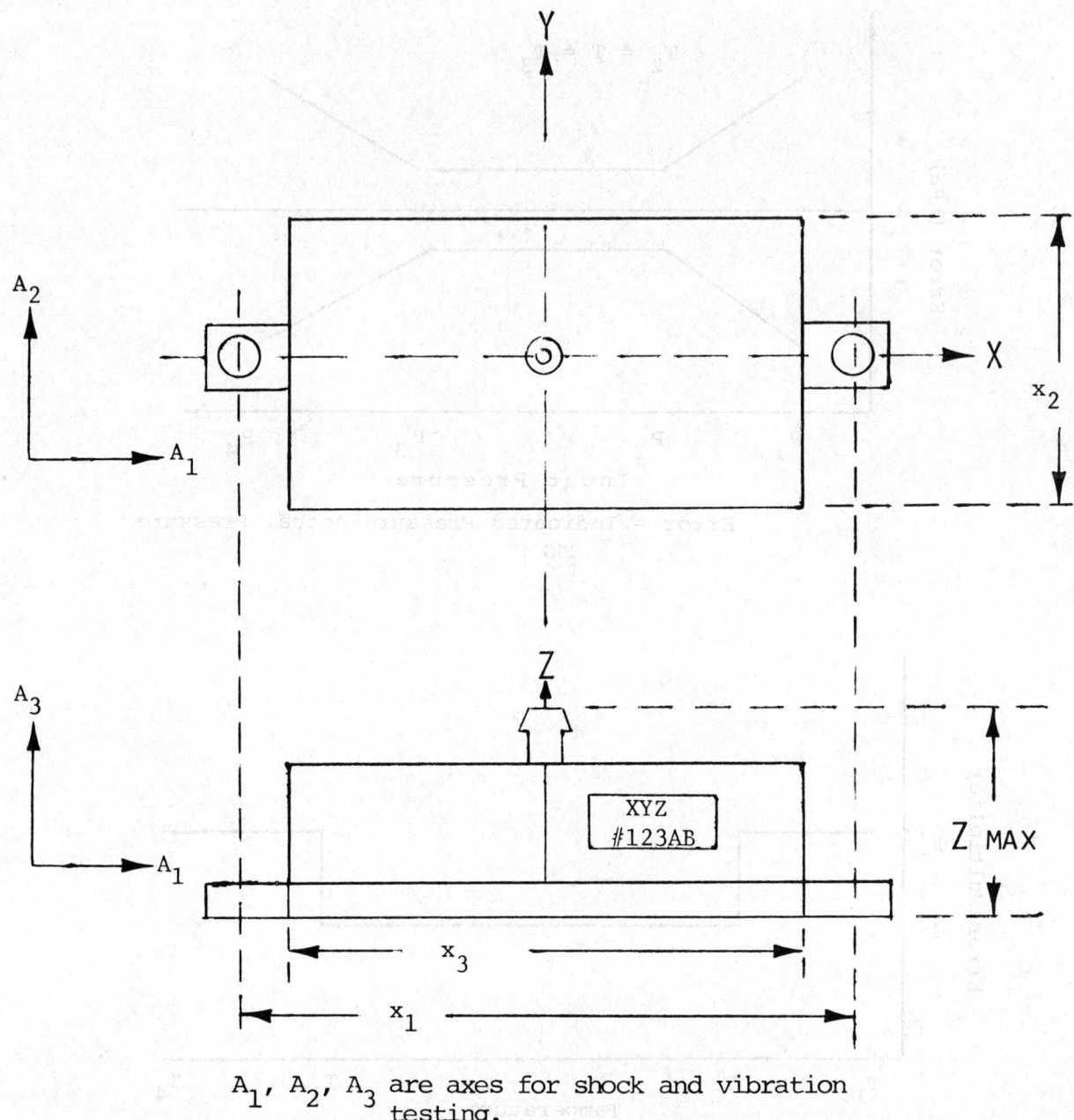

A_1, A_2, A_3 are axes for shock and vibration testing.

FIG. 4

4.8 Pressure Cycle—The device shall meet all of the requirements of this specification after being subjected to qq pressure cycles from $\leq rr$ kPa–$\geq ss$ kPa at a rate $\leq tt$ cycles/h.

4.9 Temperature Cycling—The device shall meet all of the requirements of this specification after being subjected to uu temperature cycles between a temperature $\leq vv°$C and $\geq ww°$C. Each cycle shall maintain the specified temperatures for xx min and make the transfer from one temperature to the next in a time between yy and zz min.

4.10 Temperature Endurance—The device shall meet all of the requirements of this specification after being subjected to aaa h soak at up to $bbb°$C.

4.11 External Contaminants—The device shall meet all of the requirements of this specification after being subjected to the following contaminants per the applicable test methods document:
LIST:

4.11.1 M<small>EASURAND</small> C<small>ONTAMINANTS</small>—The device shall meet all of the requirements of this specification after being subjected to the following contaminants which could be contained in the measurand gas per the applicable test method document:
LIST:

5. Applicable Documents—The following documents contain information relating to the device which is the subject of this specification. Those documents which contain information that must be complied with shall be directly named in either the contract, purchase order, or on the specific device drawing.

6. Glossary of Special Terms
FAILURE: Failure of the Manifold Absolute Pressure sensor is the inability of the part to provide the specified function within the specified tolerances when properly installed in the specified environment, including all interfaces, for a continuous period of time equal to or greater than 10 s.

APPENDIX
GUIDE TO MANIFOLD ABSOLUTE PRESSURE TRANSDUCER REPRESENTATIVE SPECIFICATION

NOTE: The SAE does not imply any recommendation concerning these values.

Code	Typical Value(s)
a	16
b	107
c	12
d	$+2, -7$
e	9
f	0.01
g	100
h	10
i	0.016 V/kPa
j	0.5 V
k	0.0018/kPa
l	0.05
m	1000
n	+12 or ground
o	0.5 V
p	6.5 V
q	10
r	1
s	20
t	20
u	40
v	100
w	5
x	0.4
y	5
z	500

APPENDIX (CONT'D)

Code	Typical Value(s)
aa	0.3
bb	10
cc	1000
dd	500
ee	120
ff	60
gg	10
hh	10
ii	100
jj	10
kk	20
ll	2000
mm	5
nn	72
oo	95
pp	50
qq	10^6 (10^{10})
rr	40 (60)
ss	100 (63)
tt	3600 (1.5×10^6)
uu	100
vv	120
ww	-50
xx	60
yy	1
zz	5
aaa	72
bbb	120
Fig. 1	± 0.4 kPa, 40–80 kPa
Fig. 2	$K = 2(3); T_1 = -50°C, T_2 = 25°C, T_3 = 80°C, T_4 = 120°C$
Fig. 4	$x_1 = 5$ cm, $x_2 = 3$ cm, $x_3 = 4$ cm, $Z = 3$ cm

(R) ACCELERATOR PEDAL POSITION SENSOR FOR USE WITH ELECTRONIC CONTROLS IN MEDIUM- AND HEAVY-DUTY VEHICLE APPLICATIONS—SAE J1843 APR93

SAE Recommended Practice

Report of the Truck and Bus Electrical and Electronics Committee approved June 1989. Completely revised by the Truck and Bus Diesel Engine Electronic Controls Subcommittee of the SAE Truck and Bus Electrical and Electronics Committee April 1993.

Foreword—Many electronic controls used in medium- and heavy-duty vehicles require an electrical indication of accelerator pedal position. A common accelerator pedal position sensor function and performance criterion is desired to minimize the number of different designs that would have to be stocked by those who service the many different types and brands of vehicles. A single universal electrical interface has not been defined. Two electrical interface types are defined in this SAE Recommended Practice. The intent of providing a choice of two signal types is to allow the industry time to prove by actual application the best selection.

While a common mechanical definition of the size, shape, etc., of the accelerator pedal and accelerator position sensor is desirable, it is realized that vehicles are not designed around the accelerator pedal. The present variations in vehicle configurations and design requirements cannot be satisfied by a single mechanical interface specification for the accelerator pedal. The intent of this specification is to limit sensor variations to one physical mounting interface and one of two electrical signal types.

The specification to outline portions of the physical interface between an accelerator pedal and the accelerator pedal position sensor should serve two purposes. First, it will minimize the number of base mechanical pedals required to mount the appropriate sensor(s) for different applications. Second, this specification would encourage the use of the sensors with the same mechanical interface for floor-mounted, suspended, and remote accelerator applications. For the remainder of this document, the term "pedal" can be construed to mean any physical means of converting operator motion into an acceleration or deceleration command.

1. Scope—The purpose of this SAE Recommended Practice is to provide a common electrical and mechanical interface specification that can be used to design electronic accelerator pedal position sensors and electronic control systems for use in medium- and heavy-duty vehicle applications.

2. References

2.1 Applicable Document—The following publication forms a part of this specification to the extent specified herein. The latest issue of SAE publications shall apply.

2.1.1 SAE PUBLICATION—Available from SAE, 400 Commonwealth Drive, Warrendale, PA 15096-0001.

SAE J1455—Joint SAE/TMC Recommended Environmental Practices for Electronic Equipment Design Heavy-Duty Trucks)

2.1.2 The APS assembly shall comply with all appropriate Federal Motor Vehicle Safety Standards.

2.2 Definitions

2.2.1 ACCELERATOR POSITION SENSOR (APS)—The sensor portion of the physical device used to convert the accelerator position into an electrical signal.

2.2.2 DIAGNOSTIC RANGES—The ranges of APS outputs between the maximum allowable output span during normal operation and the APS output values specified as an indication of an absolute fault condition. APS outputs in the diagnostic ranges may be used by the controller(s) as an out-of-range indication, but do not necessarily indicate an absolute fault.

2.2.3 DUTY CYCLE—The ratio of signal time high to signal period (Figure 1).

2.2.4 ELECTRICAL INTERFACE—The electrical signals to be passed from the APS to other electronic/electrical devices.

2.2.5 FAULT RANGES—The ranges of the APS output values beyond the diagnostic range(s) that indicate an absolute fault condition in the accelerator pedal assembly.

2.2.6 FULL SCALE—The difference between the theoretical maximum and minimum signal outputs (i.e., 100% of analog supply voltage or 100% duty cycle).

2.2.7 MECHANICAL INTERFACE—The physical boundaries of the APS.

2.2.8 OUTPUT HYSTERESIS—The maximum output signal difference for a given input pedal position due to previous history of pedal motion in either the increasing or decreasing direction.

2.2.9 OUTPUT LINEARITY—The maximum deviation of the actual output transfer function from a straight line defined by the best fit linear regression straight line through the actual values (Figure 2).

2.2.10 OUTPUT SMOOTHNESS—Any spurious variation in the output not present in the input is measured as the difference between the actual output transfer function and the end points of a 2.0% of total pedal travel long line parallel to the output linearity function that passes through the actual output value for any APS position. The difference between the actual output values and the parallel line end points located ±1.0% of total travel from the APS position should be less than the output smoothness specification (Figure 3).

2.2.11 PULSE WIDTH MODULATED (PWM)—A system of modulation where the duty cycle of discrete pulses are varied by controlling the leading, trailing, or both edges to represent an output signal where the duty cycle of the pulse is proportional to the value represented.

2.2.12 SENSING ELEMENT—The portion of, or discrete device contained within, the APS that converts physical motion into a usable electrical signal.

2.2.13 SUPPLY VOLTAGE—The voltage measured between the +V supply and -V supply leads with the APS device connected.

2.2.14 TREADLE—The lever operated by the foot.

2.2.15 IDLE VALIDATION SIGNAL (IVS)—A signal generated by the accelerator pedal assembly to indicate that the assembly is in the idle position.

2.2.16 TRANSMISSION SHIFT POINT TRANSITION SIGNAL—The electrical signal used by an automatic transmission to provide early shift points at low throttle and higher shift points at increased throttle positions.

2.2.17 KICK DOWN SIGNAL—The electrical signal used by an automatic transmission to raise the shift points to provide maximum performance at full throttle.

3. *Mechanical Interface*—The following specifications are for an accelerator pedal to accelerator position sensor interface. It is intended to allow the design of sensors that are interchangeable for different electronic applications. The driveshaft configuration and APS mounting pattern are the critical areas for commonality.

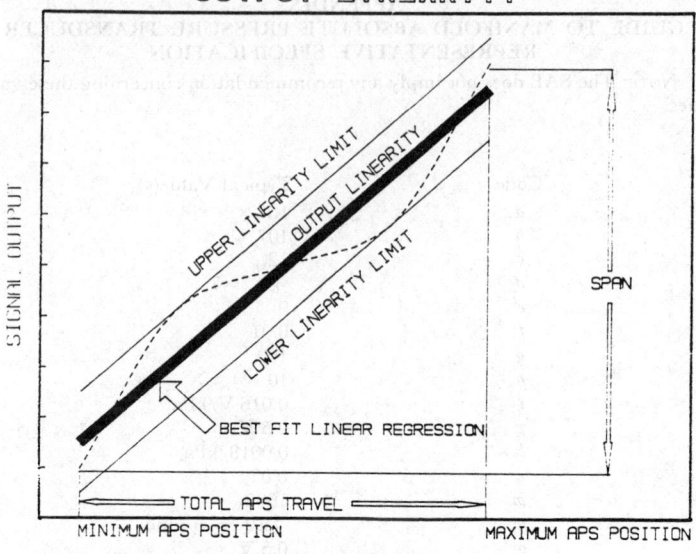

FIGURE 2—OUTPUT LINEARITY DEFINITION

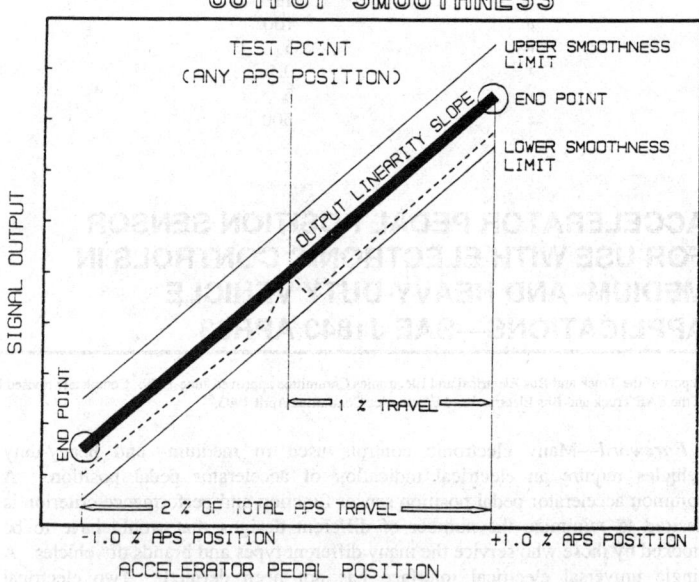

FIGURE 3—OUTPUT SMOOTHNESS DEFINITION

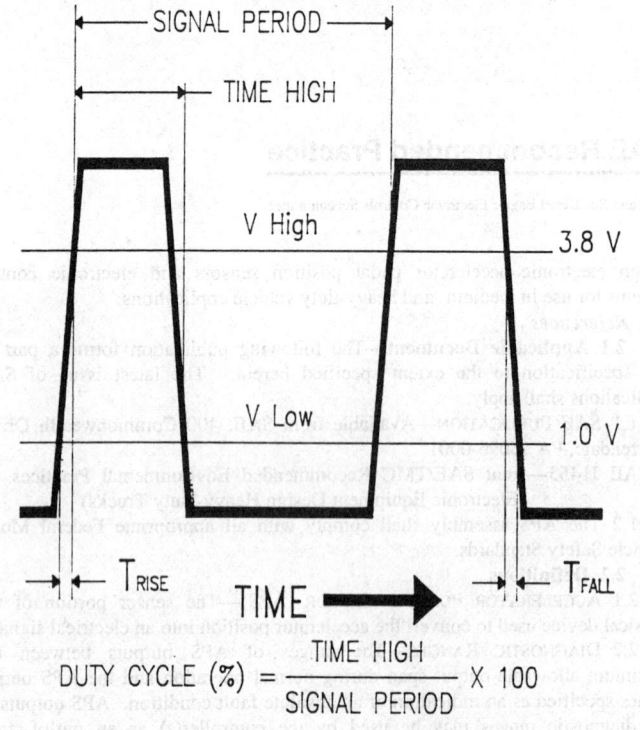

FIGURE 1—PULSE WIDTH MODULATED SIGNAL WAVEFORM

Figure 4 outlines the mounting pattern and driveshaft orientation.

Figure 5 outlines the APS mechanical interface in the area around the APS driveshaft. Figure 5 is a view from section A-A of Figure 4.

Overall drift of the minimum accelerator pedal position driveshaft to be ±3 degrees over the operating life of the accelerator pedal.

Overall drift of the maximum accelerator pedal position driveshaft to be ±3 degrees over the operating life of the accelerator pedal.

Due to variations in actual pedal designs and applications, sources of auxiliary signals, as defined in Section 7, may utilize, but are not required to utilize, this APS-to-pedal mechanical interface.

If the APS has an optional cutaway driveshaft receptacle as illustrated in Figure 6, then the APS must not contain an integral IVS. In this case, if the IVS is required, it must be located and operated independently of the APS.

The APS shall contain an internal source of energy capable of returning the internal portions of the APS to the end of travel nearest the idle state. It is not intended that the APS be capable of returning the entire accelerator pedal assembly to an idle condition. The accelerator pedal assembly is expected to accomplish return of the treadle through other sources of energy.

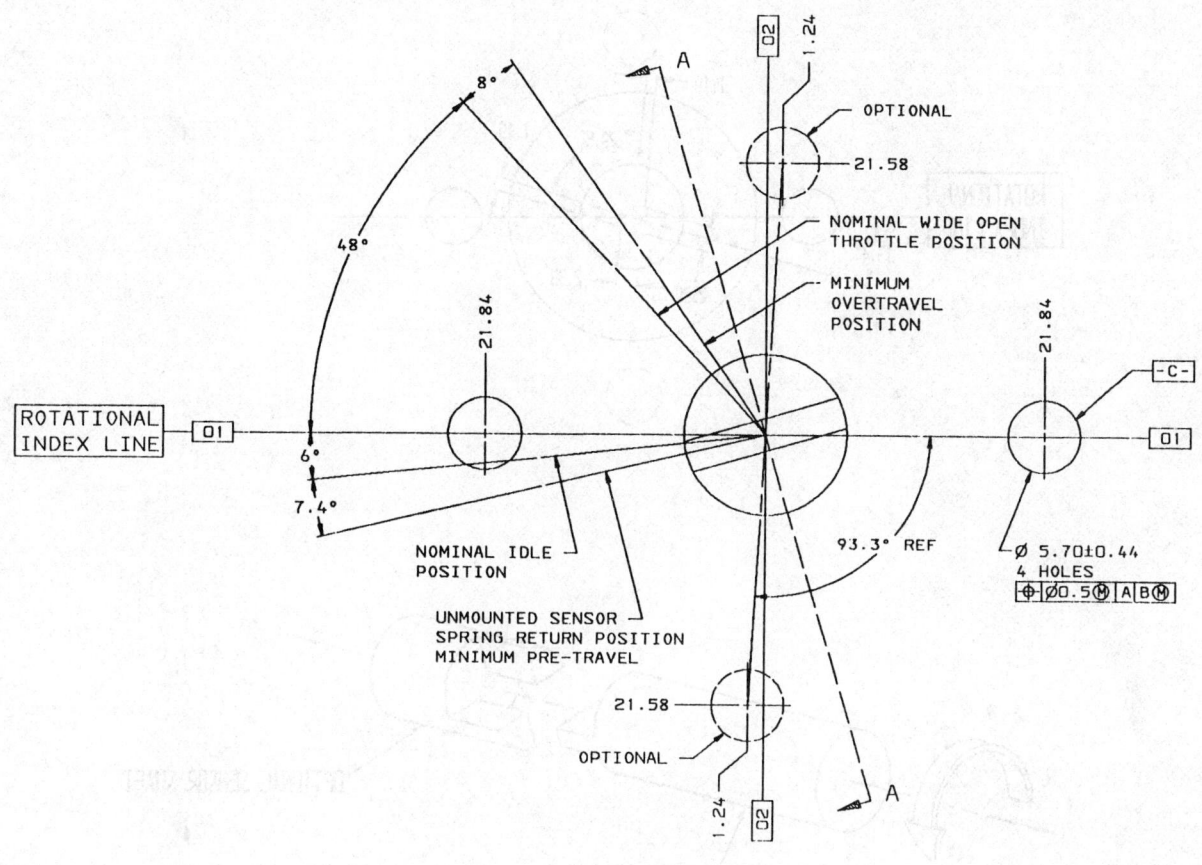

FIGURE 4—APS MOUNTING SPECIFICATIONS SENSOR MOUNTING HOLE PATTERN AND ANGULAR ROTATION REQUIREMENTS

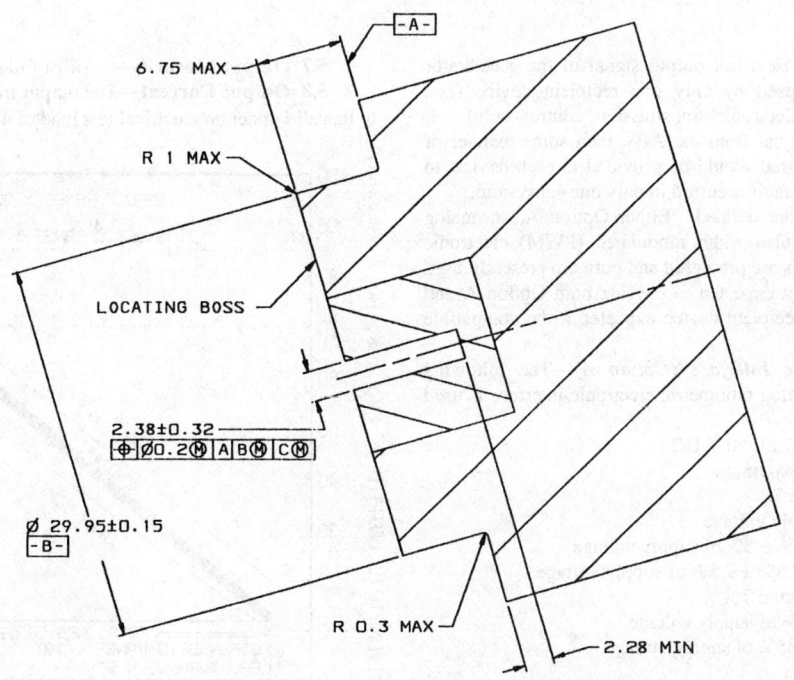

FIGURE 5—APS MOUNTING SPECIFICATIONS SECTION A-A

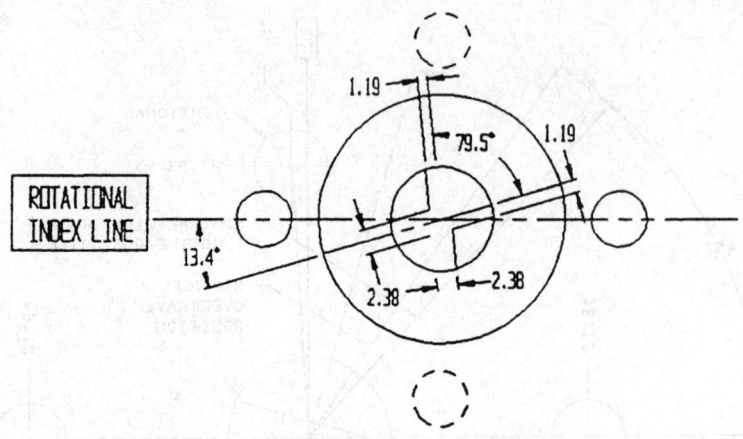
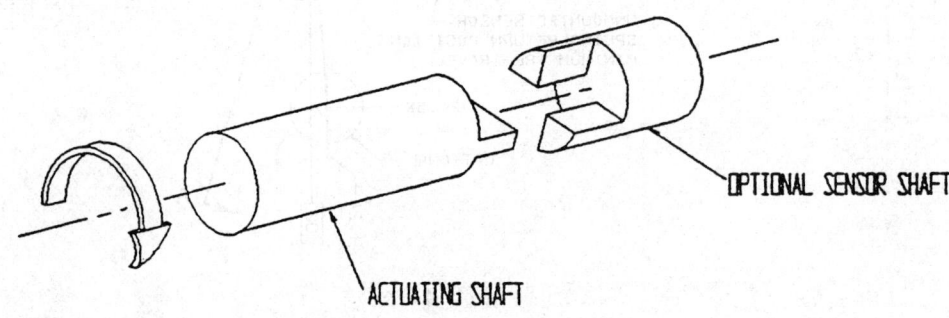

FIGURE 6—OPTIONAL SENSOR CUTOUT

4. Electrical Interface—Any one electrical output signal of the accelerator pedal assembly is intended to be used by only one recipient device (i.e., electronic engine control only, or electronic transmission control only). If multiple devices require a reliable signal from the APS, then some manner of isolation and buffering of the APS signal should be provided to each device, to prevent the loss of the APS signal if a fault occurred in only one subsystem.

Two optional electrical interfaces are defined. Either Option A, an analog ratiometric signal or Option B, a pulse width modulated (PWM) electronic interface can be used. The two options are presented and both are presently used in the industry today. An APS is not expected to provide both Option A and Option B output signals, nor is the recipient device expected to be compatible with both.

5. Analog Ratiometric Electronic Interface (Option A)—The following specifications shall apply when an analog ratiometric electronic interface is used in the APS.

    5.1 Supply Voltage—5.0 V DC ± 0.50 V DC
    5.2 Supply Current—20 mA maximum
    5.3 Output Range—See Figure 7.
    a. Span = 67.5% ± 7.5% of supply voltage
    b. Minimum APS Position = 15% ± 5% of supply voltage
    c. Maximum APS Position = 77.5% ± 7.5% of supply voltage
    5.4 Diagnostic Range—See Figure 7.
    a. Lower Range = Less than 10% of supply voltage
    b. Upper Range = Greater than 85% of supply voltage
    5.5 Fault Range—See Figure 7.
    a. Lower Range = Less than 5% of supply voltage
    b. Upper Range = Greater than 90% of supply voltage
    5.6 Output Smoothness—0.5% of full scale output for any 2% interval of total travel over the output range (Figure 3).

    5.7 Output Linearity—±5% of full scale output over the output range.
    5.8 Output Current—The output transfer function defined in Figure 7 is to be valid under an electrical test load of 47 kΩ ± 5% (see Figure 8).

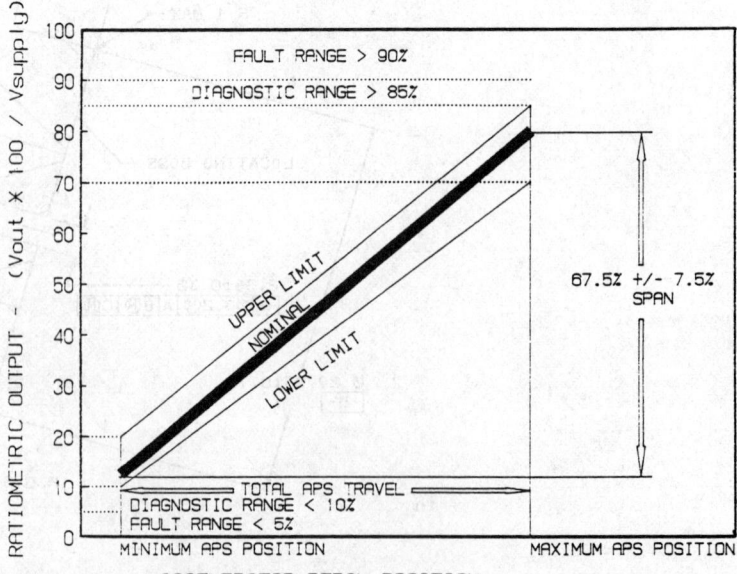

FIGURE 7—ANALOG RATIOMETRIC OUTPUT TRANSFER FUNCTION

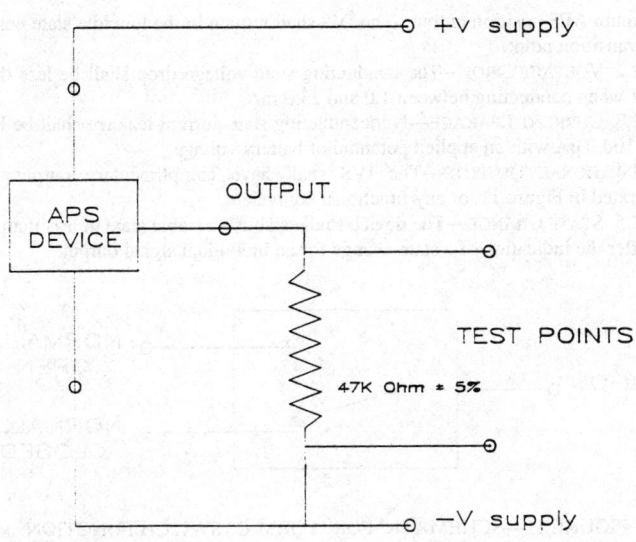

FIGURE 8—RATIOMETRIC APS OUTPUT TEST CIRCUIT

5.9 Output Hysteresis—The sensing device must not exhibit output hysteresis greater than 2% of full scale output when measured at mid-travel. Output hysteresis is measured at the direct mechanical input to the accelerator pedal position sensing element. Hysteresis of the linkages between the treadle and the sensing element of the APS is not included.

5.10 Open Circuit Response—An open circuit of any lead to the APS shall result in a signal as measured across the test points (47 kΩ ± 5% test lead as per Figure 8) within a specified fault range as shown in Figure 7 and within a maximum time of 1.0 s. The signal shall transit from a specified fault range signal to the correct reading at any APS position in less than 0.1 s (signal slew rate only) upon return to a normal operation.

5.11 Short Circuit Response—A short circuit between any two leads of the APS shall result in a signal as measured across the test points (47 kΩ ± 5% load as per Figure 8) within a specified fault range as shown in Figure 7 and within a maximum time of 1.0 s. The signal shall transit from a specified fault range signal to the correct reading at any APS position in less than 0.1 s (signal slew rate only) upon return to a normal operation.

6. Pulse Width Modulated (PWM) Electronic Interface (Option B)—The following specifications shall apply when a pulse width modulated electronic interface, Figure 1, is used in the APS.

6.1 Supply Voltage—Positive battery voltage, 12 V DC or 24 V DC nominal, regulated 8 V DC ± 0.4 V DC, or regulated 5 V DC ± 0.25 V DC.

6.2 Supply Current—100 V DC maximum.

6.3 Output Range—See Figure 9.
 a. Minimum APS Position = 6% duty cycle
 b. Maximum APS Position = 94% duty cycle
 c. Minimum Accelerator Assembly Position = 16% ± 6% duty cycle
 d. Maximum Accelerator Assembly Position = 82.5% ± 7.5% duty cycle

6.4 Fault Range—See Figure 9.
 a. Lower Range = Less than 5% duty cycle
 b. Upper Range = Greater than 95% duty cycle

6.5 Output Smoothness—0.5% of full scale output for any 2% interval of total travel over the output range (Figure 3).

6.6 Output Linearity—±5% of full scale output over the output range.

6.7 Output Frequency
 a. Minimum = 200 Hz
 b. Maximum = 1100 Hz

6.8 Output Current—The output transfer function as defined in Figure 9 is to be valid under an electrical test load impedance of 47 kΩ ± 5% and 0.001 µF capacitance. See Figure 10 for test schematic. The output voltage across the test points high shall be greater than 3.8 V while sourcing a minimum 8.0 mA current. The output voltage low shall be less than 1.0 V while sinking a maximum 10 mA current.

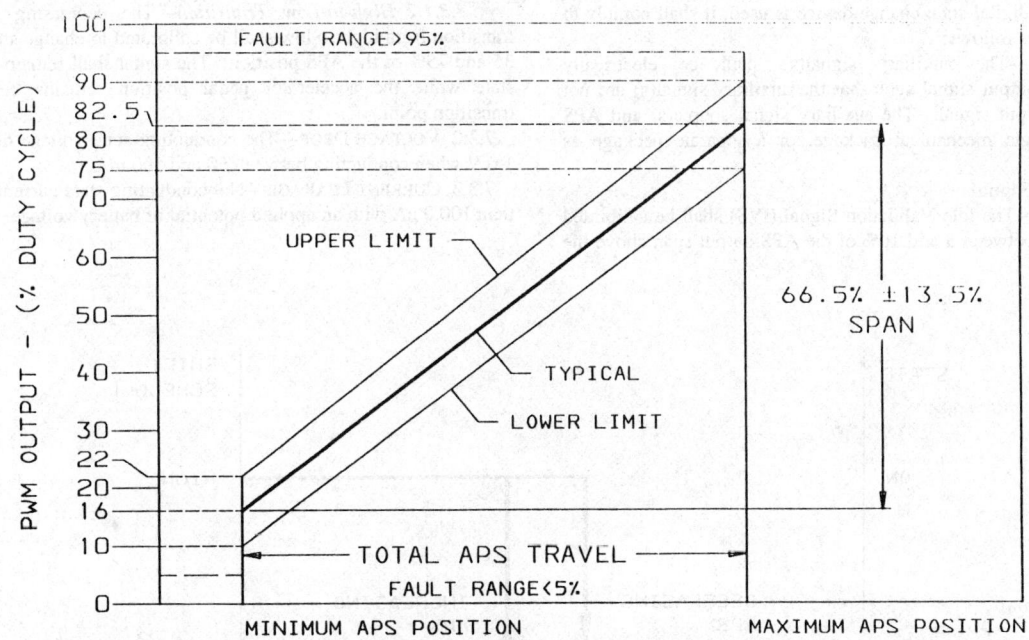

FIGURE 9—PULSE WIDTH MODULATED OUTPUT TRANSFER FUNCTION

6.9 Output Waveform—The pulse width modulated signal shall have the wave shape shown in Figure 1 while connected to the electrical test load impedance as shown in Figure 10.

6.10 Output Hysteresis—The sensing device must not exhibit output hysteresis greater than 2% of full scale output when measured at mid-travel. Output hysteresis is measured at the direct mechanical input to the accelerator pedal position sensing element. Hysteresis of the linkages between the treadle and the sensing element of the APS are not included.

6.11 Open Circuit Response—An open circuit of any lead to the APS shall result in a signal as measured across the test points (47 kΩ ± 5% and 0.001 µF test load per Figure 10) within a specified fault range as shown in Figure 9 and within a maximum time of 1.0 s. The signal shall transit from a specified fault range signal to the correct reading at any APS position in less than 0.1 s (signal slew rate only) upon return to a normal operation.

6.12 Short Circuit Response—A short circuit between any two leads to the APS shall result in a signal as measured across the test points (47 kΩ ± 5% and 0.001 µF test load per Figure 10) within a specified fault range as shown in Figure 9 and within a maximum time of 1.0 s. The signal shall transit from a specified fault range signal to the correct reading at any APS position in less than 0.1 s (signal slew rate only) upon return to a normal operation.

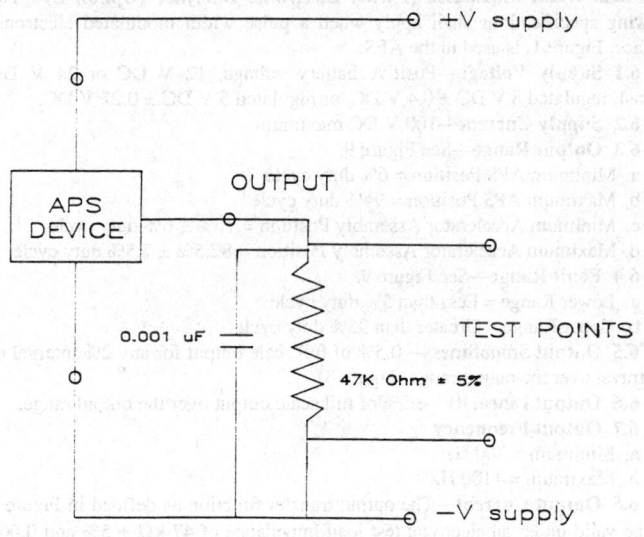

FIGURE 10—PWM APS OUTPUT TEST CIRCUIT

7. Auxiliary Signals—Some applications may require auxiliary "low idle," "shift point transition," and/or "kickdown" accelerator pedal position signal functions. If an auxiliary digital state change device is used, it shall comply to the appropriate section(s) as follows:

7.1 Signal Source—The auxiliary signal(s) shall be electrically independent of the APS output signal such that the auxiliary signal(s) are not derived from the APS output signal. The auxiliary signal source(s) and APS may be housed in a single mechanical package, or a separate package as required by the application.

7.2 Idle Validation Signal

7.2.1 LOW IDLE STATE—The Idle Validation Signal (IVS) shall be calibrated to change state at a point between 3 and 10% of the APS output span above the minimum APS position output. The IVS shall remain in the low idle state below this transition point.

7.2.2 VOLTAGE DROP—The conducting state voltage drop shall be less than 1.2 V when conducting between 1.0 and 25.0 mA.

7.2.3 CURRENT LEAKAGE—Nonconducting state current leakage shall be less than 100.0 µA with an applied potential of battery voltage.

7.2.4 SIGNAL OUTPUTS—The IVS shall have complimentary outputs as illustrated in Figure 11, or any functional equivalent.

7.2.5 STATE CHANGE—The device shall establish a stable state in less than 50 ms after the indication of a state change for an individual signal output.

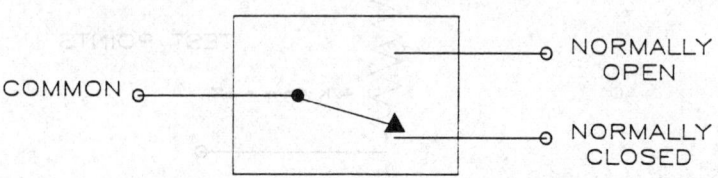

FIGURE 11—SCHEMATIC FOR "FORM-C" SWITCH FUNCTION

7.3 Transmission Shift Point Transition Signal

7.3.1 TRANSMISSION SHIFT POINT TRANSITION STATES—A single APS assembly position signal to provide a transition signal to change from low to high and from high to low transmission shift schedules. The low-to-high transition point is always at a greater APS assembly percentage of position than the high-to-low transition point by at least 10% of the APS position. See Figure 12.

7.3.1.1 *Low-to-High Transition*—The increasing APS position signal transition from low to high shall be calibrated to change state at a point between 50 and 90% of the APS position but not less than the high-to-low transition point described in 7.3.1.2. The signal shall remain in the high shift point state while the accelerator pedal position remains above the high-to-low transition point.

7.3.1.2 *High-to-Low Transition*—The decreasing APS position signal transition from high to low shall be calibrated to change state at a point between 35 and 75% of the APS position. The signal shall remain in the low shift point state while the accelerator pedal position remains below the low-to-high transition point.

7.3.2 VOLTAGE DROP—The conducting state voltage drop shall be less than 1.0 V when conducting between 50 and 500 mA.

7.3.3 CURRENT LEAKAGE—Nonconducting state current leakage shall be less than 100.0 µA with an applied potential of battery voltage.

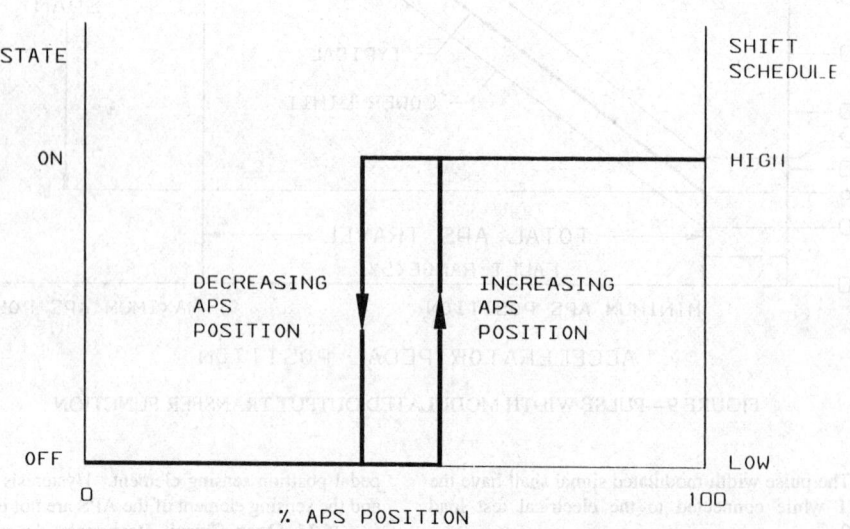

FIGURE 12—TRANSMISSION SHIFT POINT TRANSITION SIGNAL

7.4 Kickdown Signal

7.4.1 KICKDOWN STATE—The APS kickdown position signal shall be calibrated to change state at a point between 90 and 97% of the APS output span above the minimum APS position output and remain in the kickdown state above the point at which the state change occurs.

7.4.2 VOLTAGE DROP—The conducting state voltage drop shall be less than 1.2 V when conducting between 1.0 and 25.0 mA.

7.4.3 CURRENT LEAKAGE—Nonconducting state current leakage shall be less than 100.0 µA with an applied potential of battery voltage.

7.4.4 SIGNAL OUTPUTS—The kickdown signal shall have complimentary outputs as illustrated in Figure 11, or any functional equivalent.

7.4.5 STATE CHANGE—The device shall establish a stable state in less than 50 ms after the indication of a state change.

8. Environmental Requirements—The APS and auxiliary signal source(s) must meet these requirements over all applicable environmental specifications contained in SAE J1455.

9. Durability Requirements—Performance degradation over the life of the entire accelerator pedal assembly shall not result in operation of the APS or auxiliary signal source(s) outside of these requirements.

OEM/VENDOR INTERFACE SPECIFICATION FOR VEHICLE ELECTRONIC PROGRAMMING STATIONS —SAE J1924 DEC92

SAE Information Report

Report of the SAE Truck and Bus Vehicle Electronic Components Programming Subcommittee of the SAE Truck and Bus Electrical and Electronic Committee approved December 1992. Rationale statement available.

Foreword—In our previous meeting, we tried to define an OEM to Vendor communications protocol. There were too many differences between what each vendor needed to establish a common RS-232 based protocol. We decided to have each vendor supply its own communications program. The OEM would host the vendor communications programs on the OEM production line computers.

1. Scope—The purpose of the SAE Information Report is to address the method of loading vehicle electronic controllers with chassis and customer specific parameters. This specification shall establish an interface definition. The interface definition must be mutually agreeable to truck OEMs and vendors. The purpose of this specification is not to answer the large protocol issues raised by systems such as GM's MAP.

1.1 SAE Standard—In the future, SAE may use this specification as a basis for an OEM/Vendor interface specification standard.

2. References—There are no referenced publications specified herein.

3. Programming System

3.1 Plant System—Each OEM will create its own system of programming vehicle control modules. This specification only addresses the programming station setup and the OEM/vendor program interface. It does not suggest any specifics concerning the OEM's manufacturing system or production facilities.

3.2 Station Setup

3.2.1 PROPOSED STATION—The link between the OEM computer and the vendor interface tool is RS232c. The OEM host communications port shall support the 9 pin standard. The OEM host computer may have multiple ports or one common port. The common port is possible if the OEM provides an external switch. See the example in Figure 1.

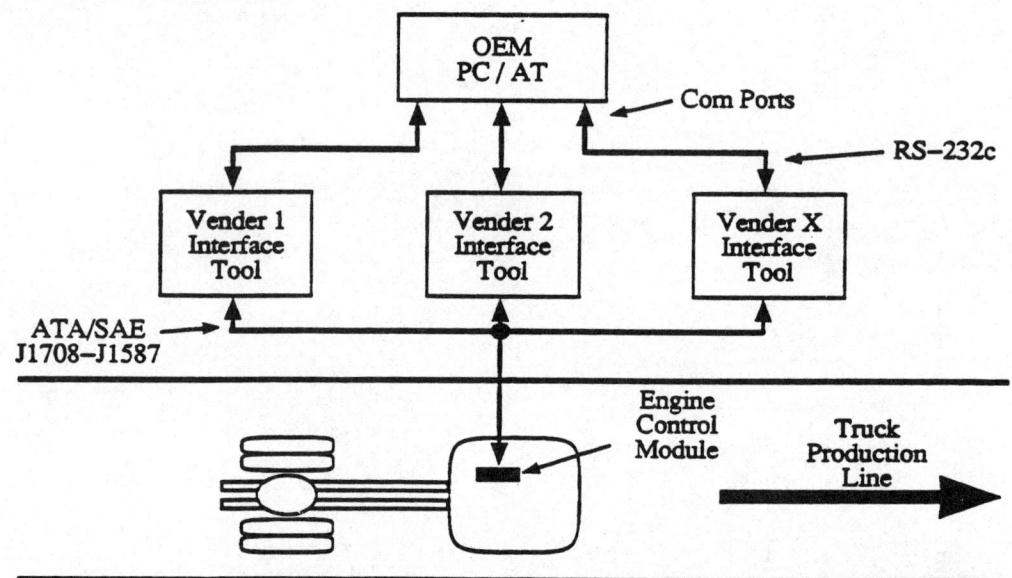

FIGURE 1—PRODUCTION LINE

3.2.2 PROPOSED SOFTWARE—The OEM production line PC shall execute an OEM production line program and the vendor communications programs. The OEM program shall gather customer and chassis specific parameter information from its manufacturing system and invoke the vendor program. The vendor program shall down-load the parameters via the host computer's communications port to the vehicle control module. The vendor program shall also return parameter verification and warranty data to the OEM program. The vendors shall supply the communication program to the OEMs as an executable program file (.exe). See example in Figure 2.

3.2.3 FUTURE DIRECTION—The OEMs want to phase out vendor tools used on the production line. In the future, an ATA communications card will replace the vendor interface tools. The OEMs intend to evolve to one ATA/SAE communications card which will plug into the host computer. This card will replace the vendor interface tools. The OEMs anticipate that the definition process will begin in the Spring of 1989. This should lead to a production line prototype in the early 1990s.

4. Interface Files

4.1 File Characteristics—Each interface file is an ASCII text file with multiple records. An end of line marker terminates each record and delimiters separate each field. A maximum character count limits the length of each field. If a larger field is needed, a field may contain a file name pointing to a separate file. An end of file marker terminates the file.

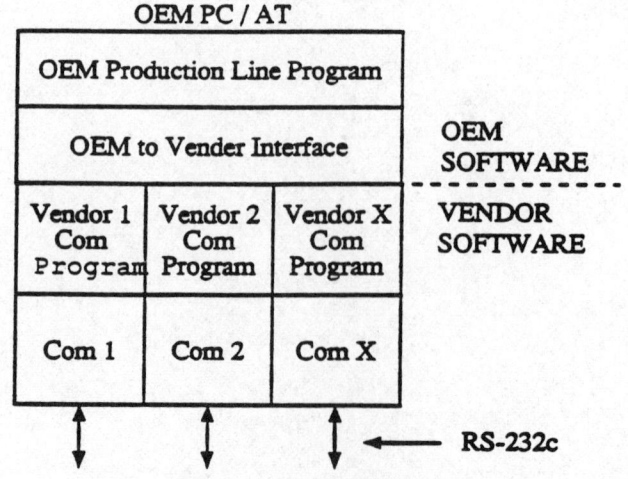

FIGURE 2—PRODUCTION LINE COMPUTER

4.1.1 TEXT FILE FORMAT—The record fields may contain any printable (20H to 7EH) ASCII characters except the field delimiter. The comma shall delimit record fields and a carriage return line feed shall terminate each record. A control-Z shall mark the end of file.

4.1.2 CASE SENSITIVITY—The files may contain either upper or lower case alpha characters. However, the OEM and vendor programs are not required to make "case sensitive" tests. The programs shall interpret the upper and lower case alpha characters (A through Z) as being the same. For example, the programs shall interpret "COMx" as equivalent to "COMX".

4.1.3 FIELD LENGTH—The field lengths shall be limited to 64 characters. Any field requiring more than 64 characters shall be passed as a file. Field may contain a file name which points to a separate file.

4.1.4 FILE NAMES—All vendor-supplied files shall conform to the following naming conventions. The file name shall include the vendor's company name, the version number, and the type of file. The first few letters of the file name shall be the vendor's company mnemonic and the following letters the version number. The file qualifier shall identify the type of file.

4.1.4.1 Vendor Mnemonic—The first 5 characters of the file name shall identify vendor company name.

4.1.4.2 Program Version—The next 3 characters following the vendor mnemonic shall identify the version of the vendor file.

4.1.4.3 File Type—The 3-character file qualifier shall identify the file type. The following list itemizes the possible file types. See the example in Figure 3.

  a. Executable file = .EXE
  b. Verification file = .VER
  c. Definition file = .DEF
  d. Parameter file = .PAR
  e. Remarks file = .REM

4.1.4.4 Examples
  a. CECOM120.DEF—Definition file version 1.20 for Cummins Electronics Company
  b. CATPL100.VER—Verification file version 1.00 for Caterpillar
  c. DDALS210.EXE—Executable file version 2.10 for Detroit Diesel Allison

4.2 Definition Files—The definition file provides descriptive information for both the OEM and vendor programs. It contains a set of records necessary to describe vehicle controller parameters and tattletale warranty information. The definition file shall also contain setup configuration data for the OEM and vendor programs. Each definition file record will hold different sets of data and, therefore, have a unique record structure. The OEM and vendor shall jointly create and maintain the definition file. They shall perform these operations off-line. See Figures 4 through 7 as examples.

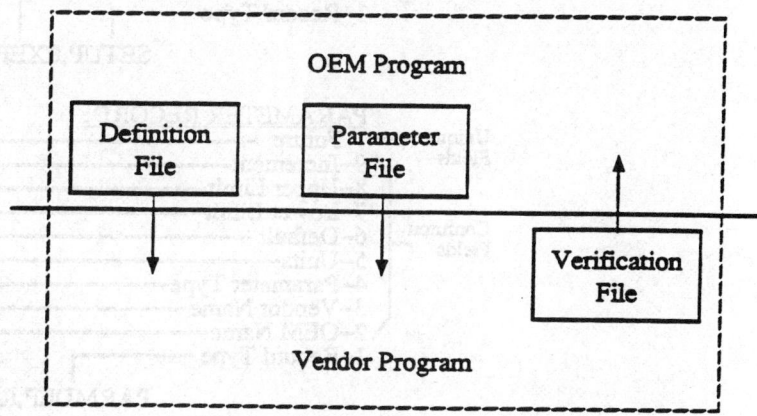

FIGURE 3—OEM/VENDOR INTERFACE FILES

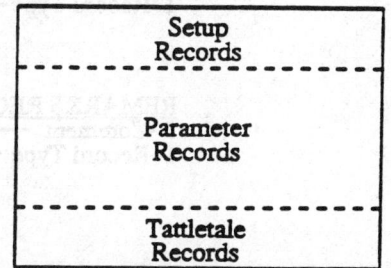

Records grouped by Record Type with remark records placed where needed

FIGURE 5—DEFINITION FILE STRUCTURE

	OEM Name	Vendor Name	Type	Units	Default Value	Lower Limit	Upper Limit	Increment
	SpeedCal	BC00	BOTH	RPM	3000	0	3000	.5
	SpeedLim	ADFF	BOTH	MPH	127	0	127	.5
2**	RPMSpeed	92CF	BOTH	RPM	3000	0	3000	10
	LowCruise	ADC0	BOTH	MPH	127	0	127	.5

FIGURE 4—SAMPLE DEFINITION TABLE

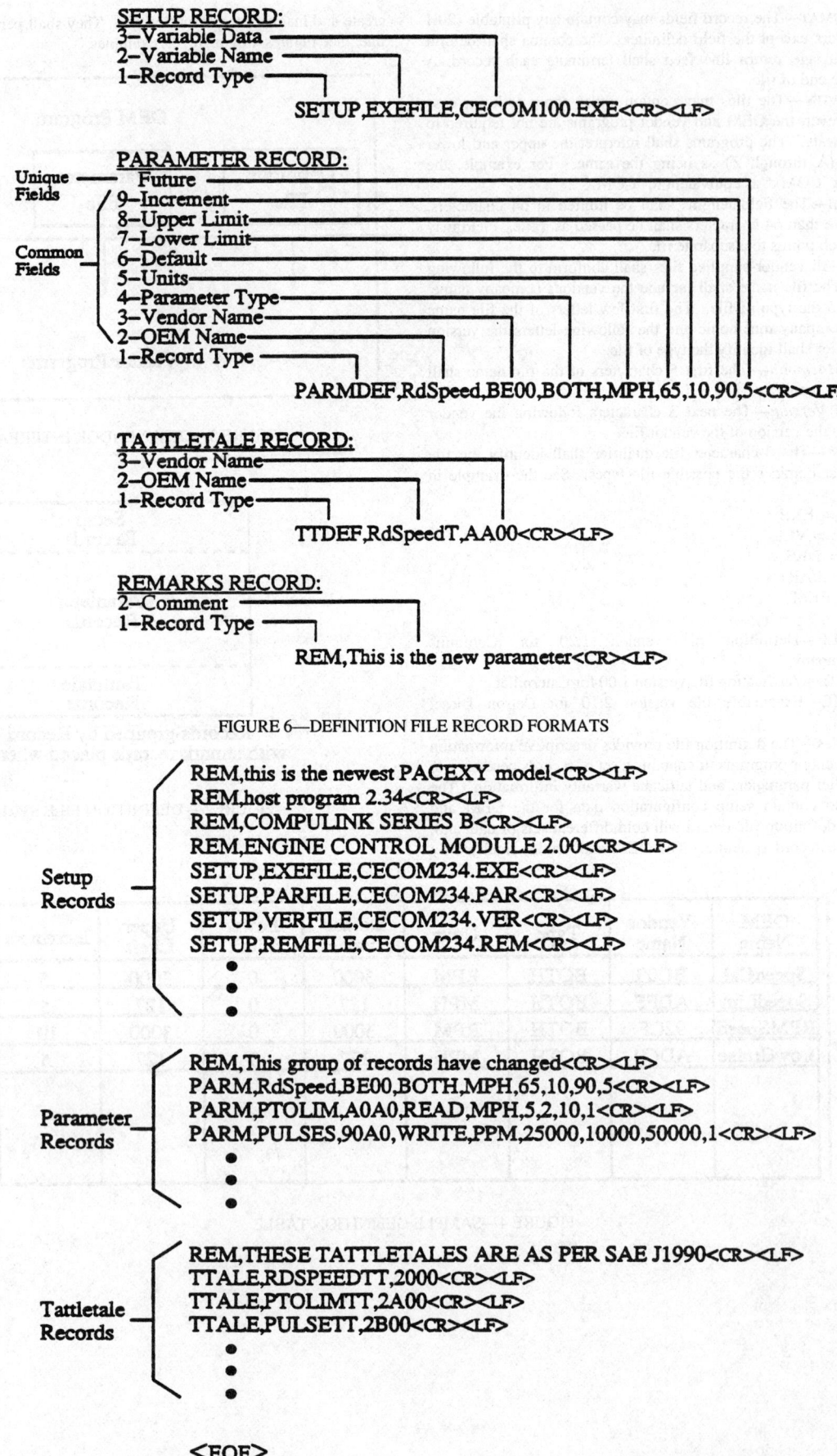

FIGURE 6—DEFINITION FILE RECORD FORMATS

FIGURE 7—SAMPLE DEFINITION FILE

4.2.1 DEFINITION FILE RECORD TYPES—Each of the records shall contain a record type field. The record type field shall be the first field of each record. It shall indicate what the remaining fields of the record will contain. For example, if the record type field contains the ASCII character "P" then the following fields of that record will contain parameter definition information. The first character of this field shall be one of the following:
  a. "S"—setup definition type record
  b. "P"—parameter definition type record
  c. "T"—tattletale definition type record
  d. "R"—remarks type record
Other characters may follow the first character but only the first is significant to the OEM and vendor programs.

4.2.2 DEFINITION FILE SETUP RECORDS—Setup records shall provide configuration data for the OEM and vendor programs. These programs shall read the setup records to find such things as interface file names. These records shall contain a field for a variable name and variable data. The variable name indicates the type of data contained in the record. The variable data field shall contain the record data. The following define types of setup records.
  a. Executable File Name—(vendor supplied communications program)
     Purpose: enable the vendor to define a version specific executable program file name for the OEM program.
     Variable Name: "EXEFILE"
     Variable Data: executable file name
     Example: EXEFILE,CECOM234.EXE
  b. Parameter File Name
     Purpose: enable the vendor to define a version specific parameter file name for the OEM program.
     Variable Name: "PARFILE"
     Variable Data: parameter file name
     Example: PARFILE,CATPL002.PAR
  c. Verification File Name
     Purpose: enable the vendor to define a version specific verification file name for the OEM program.
     Variable Name: "VERFILE"
     Variable Data: verification file name
     Example: VERFILE,DDALS001.VER
  d. Remarks File Name
     Purpose: enable the vendor to define a version specific comments file name for the OEM.
     Variable Name: "REMFILE"
     Variable Data: remarks or "readme" file name
     Example: REMFILE,BENDX100.REM

4.2.3 DEFINITION FILE PARAMETER RECORDS—The parameter records contain the information necessary to translate and describe vehicle controller settings. Each record contains data required to define one setting. Each record contains groups of common and unique fields. The common fields define the basic attributes of each parameter which are common to all vendors and OEMs. The unique fields shall define parameter attributes which are unique to the vendor.

4.2.3.1 Parameter Record—Common Fields—The common fields define parameter attributes which are common to all vendors and OEMs. The following is a list of common fields:
  a. OEM Name—The OEM name identifies the parameter which may be changed or verified.
  b. Vendor Name—The vendor name identifies a label or address used by the vendor program.
  c. Parameter Type—The parameter type indicates the accessibility of the parameter to the OEM program. The first character of the third field shall be one of the following:
     "R"—read only
     "W"—write only
     "B"—both read and write
     "S"—special types of parameters
     Other characters may follow the first character but only the first character is significant to the OEM and vendor programs.
  d. Units—The units field contains the parameter value units. As an example, the units field may have the following values:
     "PPM"—pulses per mile
     "MPH"—miles per hour
     "RPM"—revolutions per minute
     " "—(blank character)—no units
  e. Default—The default field contains the value of the parameter as supplied by the vendor unchanged.
  f. Lower Limit—The lower limit field contains the lowest value to which the parameter may be set.
  g. Upper Limit—The upper limit field contains the highest value to which the parameter may be set.
  h. Increment—The increment field contains the finest resolution to which the parameter may be set.

4.2.3.2 Parameter Record—Unique Fields—The unique fields define items specific to the vendor. These fields shall be defined by the vendor for the OEM as needed.

4.2.4 DEFINITION FIELD TATTLETALE RECORDS—The tattletale records provide a definition of data items which the OEM may use for warranty purposes. Each record shall contain OEM name and a vendor name. These records shall contain additional fields in the future when a new standard for tattletale information is established.
  a. OEM Tattletale Name—The OEM tattletale name identifies the parameter tattletale information which the OEM program may read for warranty purposes.
  b. Vendor Name—The vendor name identifies a label or address used by the vendor program.

4.2.5 DEFINITION FILE REMARK RECORDS—Remark records shall provide text information such as a version description of a new vehicle control module or vendor interface tool. This information shall identify the components of the programming station used to set the vehicle control parameters. As a minimum, the vendor shall include the following identification items:
  a. Host-resident vendor program identification
  b. Vendor interface tool identification
  c. Vehicle control module identification
The vendor may also add remark records to clarify other sections of the file as needed.

4.2.6 DEFINITION FILE CREATION—The vendors shall supply definition files to each OEM. The vendors shall fill in all fields except for the OEM Name. The OEMs shall fill in the OEM name for each parameter and tattletale record.

4.2.7 DEFINITION FILE MAINTENANCE—The vendor shall update or replace the definition file for each new version of the vendor's program. The distribution of this new information shall be specified by the OEM.

4.3 Parameter File—The parameter file shall contain a set of records which the OEM program passes to the vendor program. These records shall communicate to the vendor program: setup data, controller parameter settings, and requests for tattletale information. The program setup records shall contain special OEM to vendor communications data such as the communication port assignment. The parameter setting records shall provide the information necessary to change the vehicle controller settings. The tattletale records shall request a specific tattletale data item. Each of these records shall contain the data required to modify and verify one setting. See Figures 8 through 10 as examples.

4.3.1 PARAMETER FILE RECORD TYPES—Each of the records shall contain a record type field. The record type field shall be the first field of each record. It shall indicate what the remaining fields of the record will contain. For example, if the record type field contains the ASCII character "P" then the following fields of that record shall contain parameter data. The first character of this field shall be one of the following:
  a. "S"—setup variables type record
  b. "P"—parameter data type record
  c. "T"—tattletale request type record
Other characters may follow the first character but only the first is significant to the OEM and vendor programs.

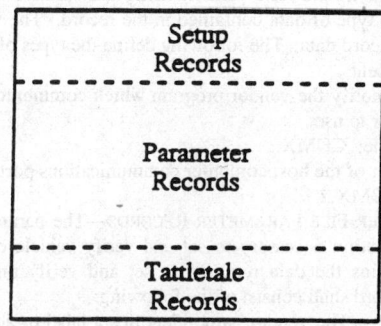

Records grouped by Record Type

FIGURE 8—PARAMETER FILE STRUCTURE

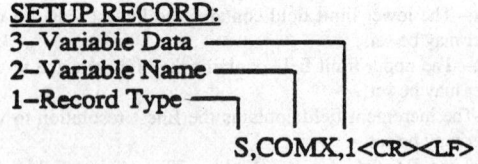

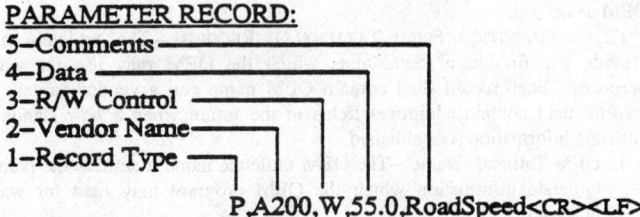

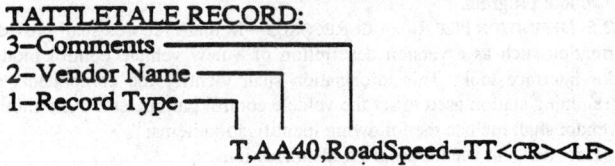

FIGURE 9—PARAMETER FILE RECORD FORMATS

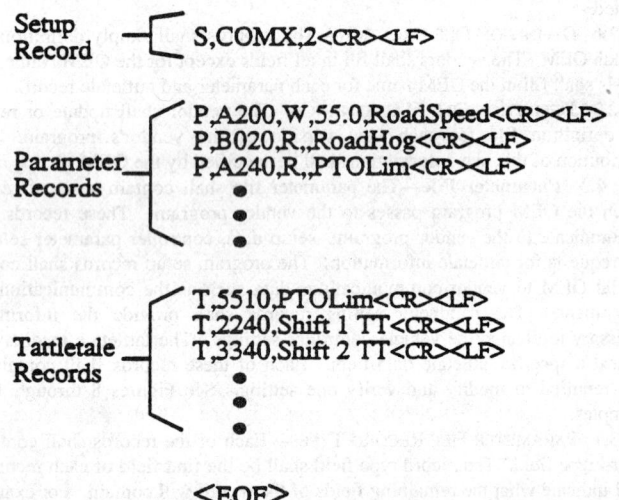

FIGURE 10—SAMPLE PARAMETER FILE

4.3.2 PARAMETER FILE SETUP RECORDS—Setup records shall provide information for the vendor program. The vendor program shall read the setup records to find such things as the communications port assignment. These records shall contain a field for a variable name and variable data. The variable name indicates the type of data contained in the record. The variable data field shall contain the record data. The following define the types of setup records:
  a. Port Assignment
     Purpose: to notify the vendor program which communications port of the host computer to use.
     Variable Name: COMX
     Data: number of the host computer communications port (1-8)
     Example: COMX,2

4.3.3 PARAMETER FILE PARAMETER RECORDS—The parameter data records contain the information necessary to set and verify vehicle controller settings. Each record contains the data required to set and verify one parameter. The parameter data record shall consist of the following:
  a. Vendor Name—The vendor name identifies a label or address used by the vendor program. This name shall identify for the vendor program the parameter which the OEM program needs to read, write, or verify.
  b. Read/Write Control—The Read/Write control field contains an OEM command. The command shall instruct the vendor program to read or write/verify a controller setting. The first character of the read/write control field shall contain one of the following:
     "R"—the vendor program shall read the parameter
     "W"—the vendor program shall write and verify the parameter
     Other characters may follow the first character but only the first is significant to the OEM and vendor programs.
  c. Parameter Data—The parameter data field contains the new parameter values. These values represent the customer and chassis specific settings which shall be written to the vehicle controller memory.
  d. Comments—The comments field may supply text information. This field will enhance the field readability. For example, it may contain the OEM parameter name.

4.3.4 PARAMETER FILE TATTLETALE RECORDS—The OEM program shall pass a request for tattletale information to the vendor program. The tattletale records in the parameter shall contain the following fields:
  a. Vendor Name—The vendor name identifies a label or address used by the vendor program. This name shall identify tattletale information which the vendor program shall read for the OEM program.
  b. Comments—The comments field may supply text information. This field is intended to enhance the file readability. For example, it may contain the OEM parameter name.

4.4 Verification File—The verification file shall contain a set of records which the vendor program passes to the OEM program. These records shall communicate to the OEM program: setup, parameter verification data, and tattletale information. The program setup records shall contain special vendor to OEM communications data such as error codes for component failures. The parameter verification and tattletale records shall provide feedback information for OEM reliability and warranty reports. Each of these records shall contain the data required to verify one setting. See Figures 11 through 13 as examples.

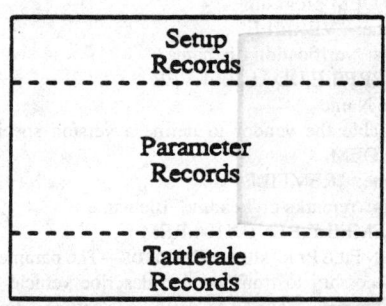

FIGURE 11—VERIFICATION FILE STRUCTURE

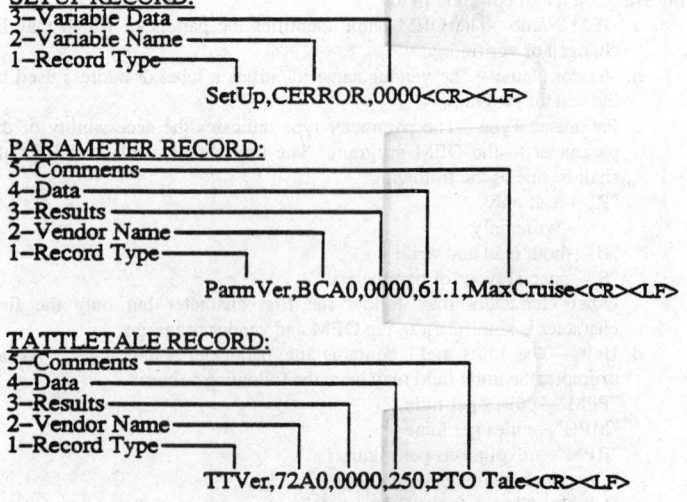

FIGURE 12—VERIFICATION FILE RECORD FORMATS

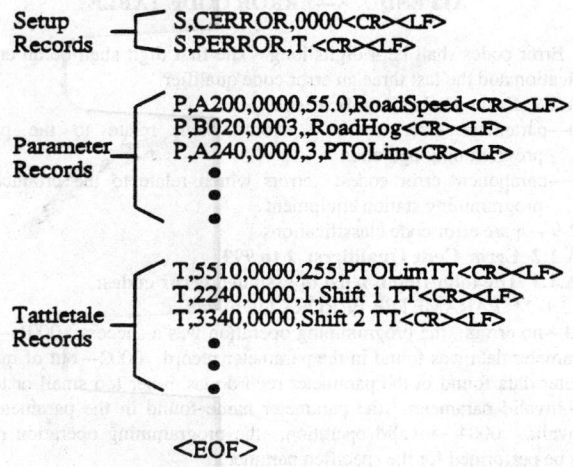

FIGURE 13—SAMPLE VERIFICATION FILE

4.4.1 VERIFICATION FILE RECORD TYPES—Each of the records shall contain a record type field. The record type field shall be the first field of each record. It shall indicate what the remaining fields of the record will contain. For example, if the record type field contains the ASCII character "P" then the following fields of that record shall contain parameter verification data. The first character of this field shall be one of the following:
 a. "S"—a setup program variable type record
 b. "P"—a parameter verification type record
 c. "T"—a tattletale type record

Other characters may follow the first character but only the first is significant to the OEM and vendor programs.

4.4.2 VERIFICATION FILE SETUP RECORDS—Setup records shall provide information for the OEM program. The OEM program shall read the setup records to find such things as program error codes. These records shall contain a field for a variable name and variable data. The variable name indicates the type of data contained in the record. The variable data field shall contain the record data. The following define the types of setup records:
 a. Component Error
  Purpose: to notify the OEM program that a component of the programming station has failed.
  Variable Name: "CERROR"
  Variable Data: component error codes as listed in the error code table found in APPENDIX A
  Example: CERROR,0000
 b. Parameter Error
  Purpose: to notify the OEM program that one of the parameter records of the verification file contains an error code.
  Variable Name: "PERROR"
  Data: shall contain one of the following characters:
   "T"—true, there is an error present
   "F"—false, there is not an error present
  Example: PERROR,T

4.4.3 VERIFICATION FILE PARAMETER RECORDS—The parameter verification data shall provide the information necessary to validate vehicle controller settings. Each record contains the data required to validate one parameter. The verification record shall contain the following fields:
 a. Vendor Name—The vendor name identifies a label or address used by the vendor program. This name shall identify to the OEM program that parameter which the vendor program read or wrote/verified.
 b. Programming Results—The programming results field shall indicate the success or failure of the programming operation. This field shall contain one of the parameter error codes listed in the error code table. See the error code table in Appendix A.
 c. Verification Data—The verification data field contains the new parameter values as stored in the vehicle control module memory. These values represent the actual customer and chassis specific settings which were written to the vehicle controller memory.
 d. Comments—The comments field may supply text information. This field is intended to enhance the file readability. For example, it may contain the OEM parameter name.

4.4.4 VERIFICATION FILE TATTLETALE RECORDS—The tattletale records shall provide warranty information for the OEM program. When the OEM requests tattletale information, the vendor program shall return tattletale information. The vendor program shall return the tattletale information in the same manner as other parameters in the verification file. The tattletale records in the verification file shall contain the following fields:
 a. Vendor Name—The vendor name identifies a label or address used by the vendor program. This name shall identify to the OEM program that tattletale information which the vendor program read.
 b. Read Results—The read results field shall indicate the success or failure of the tattletale read operation. This field shall contain one of the parameter error codes listed in the error code table. The error code table may be found in Appendix A.
 c. Tattletale Data—The tattletale data field contains the tattletale values as stored in the vehicle control module. These values represent parameter history data which the OEM may use to process warranty claims.
 d. Comments—The comments field may supply text information. This field is intended to enhance the file readability. For example, it may contain the OEM parameter name.

4.4.5 PARAMETER ANOMALIES—The vendor program shall return verification data which will represent the actual data that it stored in the controller's middle memory. Sometimes, the vendor program may return verification data which is different from the parameter data sent by the OEM program. If the vendor returns data different from the parameter data, this may be acceptable. It is acceptable if the difference is due to vendor algorithms or round-off errors. When this is the case, the vendor program shall set the results field to OK. An example: the OEM sends 62.0 and the vendor returns 61.9. If the results field indicates OK, then 61.9 was the actual controller setting and not 62.0.

5. Program Requirements

5.1 Program Scenario—The vendor shall supply a MS-DOS executable file (.exe). This file shall be executable from a DOS command line or as a "child" program. The OEM may choose to invoke the vendor program from a DOS batch file (.bat) or from another DOS executable (.exe) program.

5.1.1 OEM SETUP

5.1.1.1 The OEM shall gather and validate the customer specified and chassis specific parameter information.

5.1.1.2 The OEM shall perform the processing necessary to transform the customer and chassis parameter information. The OEM program shall format this information as specified by the parameter file definitions.

5.1.1.3 The OEM program shall place all the vendor interface files on the default disk drive.

5.1.1.4 The OEM program shall load the parameter file with the parameter information.

5.1.1.5 The OEM program shall load the parameter file with the communications port assignment.

5.1.1.6 The OEM program shall invoke the vendor program to transfer parameters to the vehicle control module.

5.1.2 VENDOR INITIALIZATION

5.1.2.1 The vendor program shall read the communications port assignment from the parameter file.

5.1.2.2 The vendor program shall configure its assigned communications port for operation.

5.1.3 VENDOR EXECUTION

5.1.3.1 The vendor program shall read the parameter information from the parameter file.

5.1.3.2 The vendor program shall execute the commands specified in the read/write control field of each parameter field record.

5.1.3.3 The vendor's program shall perform all host to vehicle communication functions via the assigned communications port.

5.1.3.4 The vendor program shall write the results of each command to a verification file record.

5.1.4 VENDOR TERMINATION

5.1.4.1 The vendor program shall disable its assigned communication port USART interrupts before returning program control to the OEM program.

5.1.4.2 The vendor program shall restore the CPU registers to the values they were when called by the OEM program.

5.2 Resource Allocations

5.2.1 MEMORY ALLOCATION—The OEM, vendor, and MS-DOS programs must be co-resident during the programming operation. The following allocations assure that there will be enough memory available.

a. OEM program—384k
b. Vendor program—182k
c. MS-DOS—128k
d. ---   640k

5.2.2 TIME ALLOCATION—The OEMs recommend that the vendor program will take no more than 30 s to execute. The vendor execution time includes the time from when it is invoked until it returns control to the OEM program.

5.2.3 HARDWARE ALLOCATION

5.2.3.1 The vendor shall only access the host computer communications port and disk drive hardware.

5.2.3.2 The vendor shall not access any other of the host computer resources such as the keyboard and video monitor.

5.2.3.3 The vendor shall not enable or disable any interrupts other than those interrupts directly related to the host assigned com port USART.

5.3 Host Operating System

5.3.1 It is recommended that the host operating system be compatible with MS-DOS version 3.1 or higher.

5.4 Program Portability

5.4.1 To promote portability, the vendor programs should use the BIOS service routines to access the RS-232 com ports. If the vendor cannot (due to execution speed) use the DOS comport BIOS, they should isolate the com port I/O routine. This routine should be programmed so that it can be easily replaced.

5.4.2 The vendor shall use DOS routines to access the host computer files.

APPENDIX A—ERROR CODE TABLE

A.1 Error codes shall be 4 digits long. The first digit shall be an error code classification and the last three an error code qualifier.

A.1.1 Error Code Classifications
0—parameter error codes: errors which relate to the parameter programming operation
1—component error codes: errors which relate to the production line programming station equipment
2-9—spare error code classifications

A.1.2 Error Code Qualifiers: 1 to 999

A.1.3 The following is a list of assigned error codes:

A.1.3.1 ***PARAMETER ERROR CODES***

0000—no errors: the programming operation was a success. 0001—no data: no parameter data was found in the parameter record. 0002—out of range: the parameter data found in the parameter record was either too small or too large. 0003—invalid parameter: the parameter name found in the parameter record was invalid. 0004—invalid operation: the programming operation requested cannot be performed for the specified parameter.

A.1.3.2 ***COMPONENT ERROR CODES***

1001—no parameter file. 1002—no definition file. 1003—DOS file error. 1004—no end of file marker. 1005—Tool failure: either the RS-232C link or the vendor interface tool failed. 1006—Module failure: either the ATA link or the vehicle control module failed. 1007—no communications port defined.

A TILT TABLE PROCEDURE FOR MEASURING THE STATIC ROLLOVER THRESHOLD FOR HEAVY TRUCKS—SAE J2180 APR93

SAE Recommended Practice

Report of the Truck and Bus Safety Dynamics Subcommittee of the SAE Truck and Bus Total Vehicle Systems Committee approved April 1993.

Foreword—This SAE Recommended Practice is intended as a guide toward a standard practice and is subject to change to keep pace with experience and technical advances.

The term "tilt table" refers to a device that rolls (rotates) the surface ("table") supporting a vehicle about a longitudinal axis. These devices may have one table that is larger than the vehicle's wheelbase (2.1.1) or a number of smaller tables, each large enough to support the wheels on an axle (2.1.2). In the case of multiple tables, the pivot axes should be aligned to fall on a common line. The important quality of the device is to maintain equal angles of tilt (within 0.1 degree if possible) under the wheels of all axles.

The test is conceptually very simple. The vehicle is driven onto the table (or tables) and then one side of the table is gradually elevated thereby placing the vehicle at a sequence of roll angles due to the tilting of the table.

When the table is at a tilt angle, the test simulates a nonvibratory steady turn. The "simulated" weight of the vehicle is the load perpendicular to the table surface, that is, the actual weight of the vehicle times the cosine of the angle of tilt. The "simulated" lateral acceleration force is the component of load horizontal to the table surface, that is, the actual weight of the vehicle times the sine of the tilt angle. The simulated lateral acceleration is the simulated lateral force divided by the simulated weight, that is, the tangent of the tilt angle. Thus, the static rollover threshold of the vehicle, in g's (1 g = the acceleration of gravity) of lateral acceleration, can be measured by determining the tangent of the tilt angle at which the vehicle just becomes unstable in roll.

1. Scope—The test procedure applies to roll coupled units such as straight trucks, tractor semitrailers, full trailers, B-trains, etc. The test is aimed at evaluating the level of lateral acceleration required to rollover a vehicle or a roll-coupled unit of a vehicle in a steady turning situation. Transient, vibratory, or dynamic rollover situations are not simulated by this test. Furthermore, the accuracy of the test decreases as the tilt angle increases, although this is a small effect at the levels of tilt angle used in testing heavy trucks. The test accuracy is accepted for vehicles that will rollover at lateral acceleration levels below 0.5 g corresponding to a tilt table angle of less than approximately 27 degrees. Even so, the results for heavy trucks with rollover thresholds greater than 0.5 g could be used for comparing their relative static roll stability.

1.1 Purpose—The purpose of this SAE Recommended Practice is to provide an interim test procedure for using tilt tables to measure a static rollover threshold for heavy trucks.

2. References

2.1 Applicable Documents—The following publications form a part of this specification to the extent specified herein.

2.1.1 L. Laird, "Measurement of Heavy Vehicle Suspension Roll-Stability Properties, and a Method to Evaluate Overall Stability Performance," SAE Paper No. 881869.

2.1.2 C. Winkler, "Experimental Determination of the Rollover Threshold of Four Tractor-Semitrailer Combination Vehicles," University of Michigan Transportation Research Institute report No. UMTRI-87-31.

2.1.3 G. Box, W. Hunter, and J. Hunter, "Statistics for Experimenters," Part I: Comparing Two Treatments, Wiley Interscience.

2.1.4 C. Winkler and M. Hagan, "A Test Facility for the Measurement of Heavy Vehicle Suspension Parameters," SAE Paper No. 800906.

3. Vehicle Identification, Test Setup, and Instruments—When preparing the test vehicle for testing, several vehicle factors (that play an important role in determining rollover threshold) need to be considered. These factors are:

a. Payload—its weight, center of gravity location, and how it is attached to the vehicle.

b. Tires—size, model, construction type, and pressure setting and wear state (Nomenclature and DOT identification number).

c. Suspension—model, size, type, and characteristics such as air spring height. (Height regulation valves should be deactivated [held at static values] during the actual tilt to avoid inflation/deflation of the air bag during the tilt. Cross coupling air lines from side to side may need to be deactivated. The investigation of active suspensions is beyond the scope of this procedure.)

In addition, the experimenter should note that the trailer(s) and the tractor comprise the "test vehicle." For articulated vehicles, all units that are roll coupled should be tested together. Each vehicle unit that is free to roll independently is tested separately. For example, since a fifth wheel provides roll coupling, a tractor and semitrailer combination comprise a single roll unit to be tested. Full trailers, of the type which are connected to their towing unit without roll coupling (such as through a pintle hitch), comprise a single roll unit to be tested.

(Individual testing of tractors and semitrailers is conceptually possible, but beyond the scope of this procedure. The roll stability properties of roll coupled units such as tractors and semitrailers generally have a highly nonlinear, synergistic relationship with one another. For example, it is very possible for two trailers, which show very similar results when coupled to one particular tractor, to show very dissimilar results when coupled to different tractors.)

All of the factors discussed previously can affect the rollover threshold. Care should be exercised in choosing realistic and repeatable test conditions. The test vehicle, including payload, tire, and suspension characteristics as listed previously, shall be identified and an adequate description shall be included with the results. This documentation is needed so that users of the results will understand the test conditions and not be misled since there are many possible choices of loading states, suspensions, and tires for heavy trucks.

(Also, the influences of stick-slip in the vehicle's compliant and coupling components should be taken into account. It can generally be expected that typical levels of hysteresis in suspensions and other elements of the vehicle will have minimal influence on measured roll stability [that is, the rollover threshold]. However, hysteresis may significantly influence other "events," such as initial wheel lift, which occurs prior to rollover at relatively low levels of tilt angle. Since hysteresis is difficult or practically impossible to control directly, the vehicle may be removed from the table in between tests and driven around to "randomize" or "equalize" the influences of hysteresis and stick-slip. Any procedures used to control stick-slip and hysteresis should be documented with the test results so that the test conditions can be understood. The number and sequence of left and right turns and the roughness of the road surface may influence the results at test angles below the rollover threshold. However, the influence on the rollover threshold is expected to be small and probably negligible.)

In addition to the previous test vehicle parameters, care should be taken to insure that the test vehicle is placed on the tilt table as straight as possible. The vehicle should be tested with all units in a straight line parallel to the tilt axis such that no axle centerline is off line by more than 25 mm (1 in). This provides a uniform test arrangement, even though various amounts of articulation may exist during turning maneuvers.

Rollover tests should be conducted at very low roll rates. The dynamic response of the test vehicle as it transitions the various "events" of the tilt table procedure is typically very slow. For example, when the vehicle begins to "fall" as the roll stability limit is reached, it accelerates very slowly. If table speed is too fast, the table can "chase" the vehicle, making precise identification of the moment of instability difficult. Similar problems can arise at each of the "events" occurring in the procedure. To avoid measurement errors which can result from the table "chasing" the vehicle, roll rates of 0.25 degree/s or less are desirable in the immediate vicinity of any "event" of interest, such as wheel lift off or the occurrence of suspension lash.

If the tilt table is located outdoors, note the wind conditions. Record wind condition (speed and direction). Do not test when wind speed exceeds 6 km/h (10 mph) and wind speeds less than 3 km/h (5 mph) are desirable. (A 6 km/h side wind on a stationary 14.4 m [48 ft] van semitrailer could produce a side force of approximately 1300 N [300 lb] which would be equivalent to an error of approximately 0.005 g in the rollover threshold for a semitrailer with a gross weight of 267 000 N [60 000 lb]. Since there is a velocity squared relationship between side force and wind velocity, a 3 km/h side wind would result in approximately 0.001 g of error in rollover threshold for this van semitrailer.)

Clearly, the vehicle needs to be restrained to keep it from rolling over or sliding down the table. Straps, chains, etc., can be used to catch the vehicle once rollover has started (see Figure 1). Restraints which can be adjusted under load are particularly desirable. A high-friction, ridged surface should also be used to provide enough friction force at the tire/table interface to keep the tires from sliding sideways. Even though the tire should not be resting against a vertical surface during the tests, it is good safety practice to have a constraining surface in place to hold the wheel in case it does slide sideways. An initial clearance of 75 to 100 mm (3 to 4 in) between the tires and the restraint is generally adequate to prevent contact in the normal conduct of a test.

Table 1 lists accuracy requirements for the instrumentation and the test set-up that are aimed at achieving 0.01 g accuracy in determining rollover threshold.

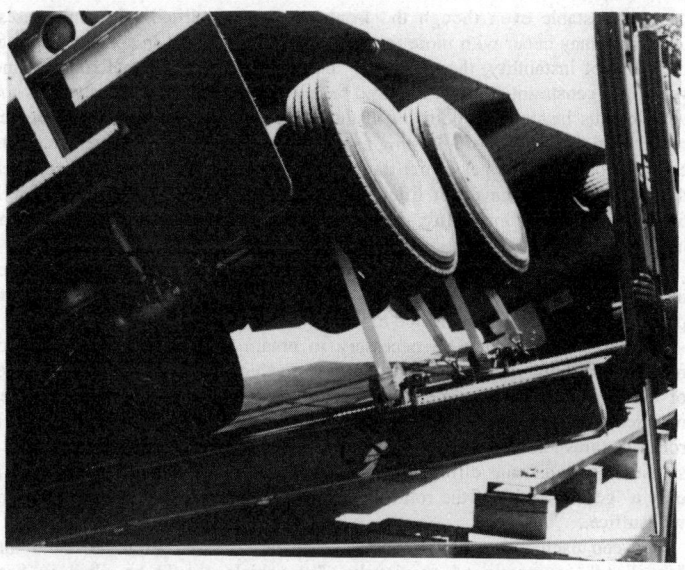

FIGURE 1—ILLUSTRATION OF STRAPS USED TO RESTRAIN THE VEHICLE FROM ROLLOVER

TABLE 1—TILT TABLE ACCURACY REQUIREMENTS

Tilt angle measurement	±0.1 degree
Vehicle-to-pivot axis alignment	Vehicle centerline parallel to pivot axis with ±25 mm at each axle
Tilt rate	≤0.25 degree/s in the vicinity of any event of interest
Pivot axis alignment	
Overall:	Horizontal ±0.25 degree
Multiple axle tables:	Co-linear ±2.5 mm
Tilt angle alignment at each axle[1]	±0.1 degree
Pretest suspension setting (Hysteretic effects)	Nominally centered; no special requirement for rollover threshold measurement
Wind disturbance acceleration	≤0.003 g magnitude (nominally 4.5 km/h wind speed)

[1] Reflects on required table stiffness and/or the alignment of individual axle tables.

4. Test Procedure—A basic rollover test consists of very gradually increasing the angle of the table, at a rate not to exceed 0.25 degree/s, and recording the angles at which:

a. The vehicle becomes unstable and starts to rollover
b. Wheels lift off
c. Suspension lash is encountered
d. The fifth wheel separates (if it does before rollover)

To begin testing, a preliminary test should be performed to (1) adjust the anti-rollover constraints, and (2) learn approximately when wheels lift, lash is encountered, and rollover occurs.

As the tilt table test proceeds, the various axles of the vehicle will lift off at different times. As the test proceeds further, axles which lift early may clear the table surface by distances exceeding 0.3 m prior to the vehicle becoming unstable. To establish proper adjustment of constraints, start the initial tilt with all constraints adjusted to allow only a few centimeters of wheel lift. As the tilt proceeds and individual constraints become taught, lower the table and adjust constraints to provide freedom to roll as required. Continue this procedure up to the tilt angle at which the vehicle becomes unstable. Allow only the minimum unconstrained roll motion needed to clearly identify the occurrence of instability.

(SAFETY NOTE—Most loaded heavy trucks will become unstable when all axles other than the front axle have lifted off. That is, usually the vehicle will become unstable even though the front axle has not lifted. In some cases, instability may occur with more than just the front axle still in contact. Right at the point of instability, the vehicle is "balanced" on the verge of rollover and very little constraint force is required to restrain the vehicle. However, as the vehicle rolls beyond the point of instability, required restraint force increases rapidly. In the process of adjusting axle constraints, care should be exercised to carefully observe whether or not the point of instability has been reached, even though some axles have not lifted off. If the vehicle is allowed to roll well beyond the point of instability, danger of failure of the restraint straps or chains increases quickly.)

It may be possible to identify the approximate tilt angle for each event of interest in the same tilt in which the constraints are adjusted. If not, a second preliminary tilt may be conducted to identify these angles.

Several test runs may be necessary to obtain a statistical distribution of rollover threshold values for a particular test vehicle (2.1.3). The total number of test runs will be dependent on the test objective as well as the accuracy required. As an example, multiple test vehicles may be tested to compare their rollover thresholds. Sufficient test turns will be needed to determine if a statistically significant difference exists between the test vehicles. However, if only a "general idea" of the rollover threshold value is required, three test runs will suffice.

Between each test run, the vehicle may be removed from the table and "equalized" as mentioned previously. The vehicle should be tilted in both directions. Because of asymmetries in mass, geometry, and system stiffness, the results for rolling to the left and to the right can be expected to differ.

5. Data Presentation and Analysis—The fundamental result from this test is the "rollover threshold." This performance measure is equal to the tangent of the table angle corresponding to the initiation of rollover. For example, if the vehicle had a rollover threshold of 0.4 g, the angle of the tilt table at the initiation of rollover would be the arctangent of 0.4 or 21.8 degrees. The rollover threshold shall be reported for all test runs.

Other factors such as table angles when wheel lift off, suspension lash, and fifth wheel separation occur are useful to help determine their relative influence on the test vehicle's rollover threshold. It is recommended that these events be tabulated as a function of the tangent of the table angle for all of the tests run.

Also, experimenters may want to measure and record angles at various cross members of the vehicle such as at the axles, at the front bumper of the tractor, or at points on the bed of the trailer. (See 2.1.1 for discussion of these types of measurements.) These angles can be useful for comparisons with experimental or theoretical results obtained in other tests or analyses. For comparisons with theoretical concepts, it is useful to plot the simulated lateral acceleration (i.e., the tangent of the table angle) versus the roll angle measured from the angle of the tilt table to the angle of the rolled lateral axis of the main sprung mass of the unit undergoing the test. (See Figure 2 for an example test result in which the rollover threshold is 0.34 g.)

Inclinometers are convenient instruments for measuring both table and vehicle component roll angles. However, especially for those inclinometers mounted on the vehicle, care should be taken in interpreting these data signals. Inclinometers measure tilt angle by sensing the angular orientation of the total acceleration vector to which they are subject. When the instrument is static, the acceleration vector is due only to gravity and the sensed angle is, indeed, inclination. But if the instrument is subject to any other acceleration in addition to gravity, an error is likely to result. As the vehicle transits the various "events" of the rollover process, it does indeed accelerate and decelerate in roll. The levels of linear acceleration experienced by inclinometers mounted on the vehicle can be significant, particularly if they are mounted far from the "roll center" of the various roll motions of the vehicle. It is generally good practice, therefore, to mount inclinometers low and toward the "downhill" side of the vehicle.

6. Discussion—In the tilt table method, the roll plane behavior of a steady-state turn is simulated by tilting a test vehicle to some angle ϕ on a table inclined in the roll direction. In this state, one component of gravity ($g \sin \phi$) acts laterally while the other component ($g \cos \phi$) acts perpendicular to the simulated road surface (the table surface). Assuming ($g \cos \phi$) simulates gravity, then simulated lateral acceleration (in "simulated g's") is ($g \sin \phi$)/($g \cos \phi$) or $\tan \phi$. Thus, if the table angle is slowly increased, the tangent of the tilt angle at which the vehicle rolls over can estimate the lateral acceleration (in g's) at which the static roll stability limit of the vehicle is reached.

The quality of the physical simulation involved in this method depends, in large part, on how closely $\cos \phi$ approximates unity. In the tilt table experiment, both the vertical and lateral loading of the vehicle are reduced by the factor, $\cos \phi$, relative to the "real" loads they represent. That is, on the tilt table ($g \cos \phi$) represents gravitational acceleration of one g and ($g \sin \phi$) represents a lateral acceleration of ($g \tan \phi$). Because of the reduced vertical loading, the vehicle may rise on its compliant tires and suspensions relative to its normal ride height, resulting in a higher center of gravity position and, possibly, an unrealistically low static rollover threshold. At the same time, simulated lateral acceleration is also reduced. This reduced loading may result in compliant lateral and roll motions of the vehicle which are unrepresentatively small, tending to make the vehicle appear more stable than it actually is. For loaded commercial vehicles, these error sources are generally small, since rollover will usually occur at a simulated lateral acceleration of less than 0.5 g; that is, at a tilt angle (ϕ) of less than 27 degrees or at a condition where $\cos (\phi) \geq 0.9$.

Other error sources include (a) artificial articulation angles for combination vehicles, and (b) artificial distribution of lateral tire loads for vehicles whose axles number more than 1+n where n is the number of yaw units. Item (a) is probably the larger error source. Item (b) may be large or small depending on the mix of suspensions involved.

Tilt tables may also be used to obtain various mechanical properties of suspensions, etc., or a separate suspension measurement facility may be used (2.1.4). The procedures for component measurements are not included here. They are facility and user dependent. (Perhaps standardized procedures for these auxiliary measurements will be forthcoming in the future.)

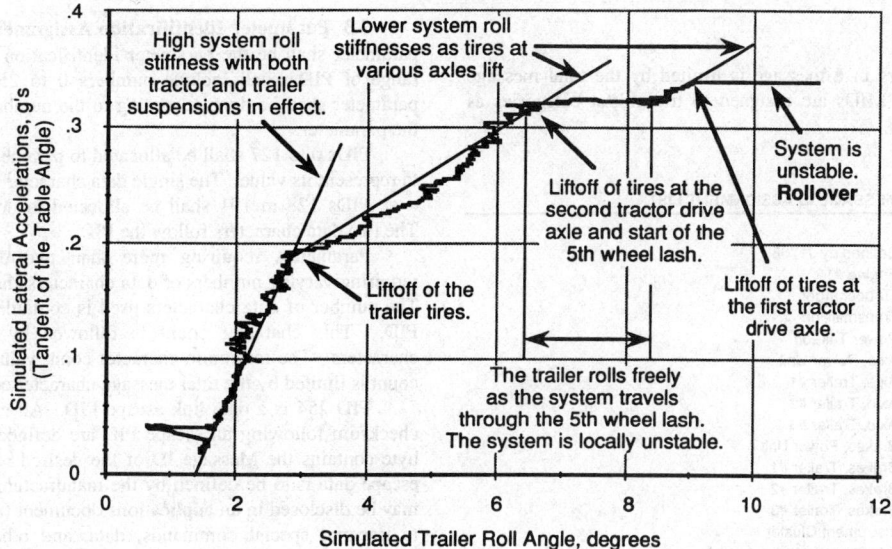

FIGURE 2—A PLOT OF TRAILER ROLL ANGLE VERSUS SIMULATED LATERAL ACCELERATION FROM A TILT TABLE TEST OF A LOADED, FIVE-AXLE TRACTOR SEMITRAILER COMBINATION

(R) JOINT SAE/TMC ELECTRONIC DATA INTERCHANGE BETWEEN MICROCOMPUTER SYSTEMS IN HEAVY-DUTY VEHICLE APPLICATIONS—SAE J1587 AUG92

SAE Recommended Practice

Report of the Truck and Bus Electrical Committee approved January 1988. Completely revised by the Truck and Bus Electrical and Electronics Committee November 1989. Completely revised by the Truck and Bus Electrical and Electronics Committee November 1990. Completely revised by the SAE Truck and Bus Data Format Diagnostics Subcommittee of the SAE Truck and Bus Electrical and Electronics Committee August 1992.

Foreword—This SAE/TMC Joint document has been developed by the Truck and Bus Data Format Diagnostics Subcommittee of the Truck and Bus Electrical and Electronics Committee and by the S.1 Electrical & Electronics Study Group of the Maintenance Council. The objectives of the subcommittee are to develop information reports, recommended practices, and standards concerned with the format of electronic signals and information transmitted among Truck and Bus electronic components.

1. Scope—This SAE Recommended Practice defines a document for the format of messages and data that is of general value to modules on the data communications link. Included are field descriptions, size, scale, internal data representation, and position within a message. This document also describes guidelines for the frequency of and circumstances in which messages are transmitted.

In order to promote compatibility among all aspects of electronic data used in heavy-duty applications, it is the intention of the Data Format Subcommittee (in conjunction with other industry groups) to develop recommended message formats for:

a. Vehicle and Component Information: This includes all information that pertains to the operation of the vehicle and its components (such as performance, maintenance, and diagnostic data).
b. Routing and Scheduling Information: Information related to the planned or actual route of the vehicle. It includes current vehicle location (for example, geographical coordinates) and estimated time of arrival.
c. Driver Information: Information related to driver activity. Includes driver identification, logs (for example, DOT), driver expenses, performance, status, and payroll data.
d. Freight Information: Provides data associated with cargo being shipped, picked up, or delivered. Includes freight status, overage, shortage and damage reporting, billing and invoice information as well as customer and consignee data.

This document represents the recommended formats for basic vehicle and component identification and performance data. This document is intended as a guide toward standard practice and is subject to change to keep pace with experience and technical advances.

1.1 Purpose—The purpose of this document is to define the format of the messages and data being communicated between microprocessors used in heavy-duty vehicle applications. It is meant to serve as a guide toward a standard practice to promote software compatibility among microcomputer based modules. This document is to be used with SAE J1708. SAE J1708 defines the requirements for the hardware and basic protocol that is needed to implement this document.

The primary use of the communications link and message format is expected to be the sharing of data among stand-alone modules. It is anticipated that this document (when used in conjunction with SAE J1708) will reduce the cost and complexity associated with developing and maintaining software for heavy-duty vehicle microprocessor applications.

2. References

2.1 Applicable Documents—The following publications form a part of this specification to the extent specified herein. The latest issue of SAE publications shall apply.

2.1.1 SAE PUBLICATIONS—Available from SAE, 400 Commonwealth Drive, Warrendale, PA 15096-0001.

SAE J1708—Serial Data Communications Between Microcomputer Systems in Heavy-Duty Vehicle Applications.

SAE J1455—Recommended Environmental Practices for Electrical Equipment Design (Heavy-Duty Trucks)

2.1.2 OTHER PUBLICATIONS

EIA RS-485—"Standard for Electrical Characteristics of Generators and Receivers for Use in Balanced Digital Multipoint Systems," Electronics Industries Association, Washington, DC, April 1983

ANSI/IEEE STANDARD 754-1985—"IEEE Standard for Binary Floating-Point Arithmetic"

3. Electronic Data Interchange—All data transmitted on the communication link, defined by SAE J1708, using message identification (MID) in the range 128 to 255, shall follow this document.

3.1 Message Format—The message shall consist of the following:
Message ID

One or More Parameters
Checksum

The number of parameters in a message is limited by the total message length defined in SAE J1708. MIDs are assigned to transmitter categories as identified in Table 1.

TABLE 1—MESSAGE ID ASSIGNMENT LIST

0-127	Defined by J1708
128	Engine #1
129	Turbocharger
130	Transmission
131	Power Takeoff
132	Axle, Power Unit
133	Axle, Trailer #1
134	Axle, Trailer #2
135	Axle, Trailer #3
136	Brakes, Power Unit
137	Brakes, Trailer #1
138	Brakes, Trailer #2
139	Brakes, Trailer #3
140	Instrument Cluster
141	Trip Recorder
142	Vehicle Management System
143	Fuel System
144	Cruise Control
145	Road Speed Indicator
146	Cab Climate Control
147	Cargo Refrigeration/Heating, Trailer #1
148	Cargo Refrigeration/Heating, Trailer #2
149	Cargo Refrigeration/Heating, Trailer #3
150	Suspension, Power Unit
151	Suspension, Trailer #1
152	Suspension, Trailer #2
153	Suspension, Trailer #3
154	Diagnostic Systems, Power Unit
155	Diagnostic Systems, Trailer #1
156	Diagnostic System, Trailer #2
157	Diagnostic System, Trailer #3
158	Electrical Charging System
159	Proximity Detector, Front
160	Proximity Detector, Rear
161	Aerodynamic Control Unit
162	Vehicle Navigation
163	Vehicle Security
164	Multiplex
165	Communication Unit—Ground
166	Tires, Power Unit
167	Tires, Trailer #1
168	Tires, Trailer #2
169	Tires, Trailer #3
170	Electrical
171	Driver Information Center
172	Off-board Diagnostics #1
173	Engine Retarder
174	Cranking/Starting System
175	Engine #2
176	Transmission, Additional
177	Particulate Trap System
178	Vehicle Sensors to Data Converter
179	Data Logging Computer
180	Off-board Diagnostics #2
181	Communication Unit—Satellite
182	Off-board Programming Station
183	Engine #3
184	Engine #4
185	Engine #5
186	Engine #6
187-255	Reserved—to be assigned

3.2 MID Assignment List Additions—No two transmitters in the system shall have the same MID. System manufacturers may request additions be made to the MID list. The Data Format Subcommittee will review the value of any additional MIDs for general interest and/or purpose and may or may not add it to the list.

3.3 Parameter Identification Assignments—The first character of every parameter shall be the parameter identification character (PID). The permitted range of PIDs shall include numbers 0 to 255. Assignment of a PID to a parameter shall be done according to the number of data characters required by the parameter.

PIDs 0 to 127 shall be allocated to parameters using a single data character to represent its value. The single data character follows the PID.

PIDs 128 to 191 shall be allocated to double data character parameters. The two data characters follow the PID.

Parameters requiring more than two data characters and parameters requiring varying numbers of data characters shall be allocated PIDs 192 to 253. The number of data characters used is contained in the first character after the PID. This character count is followed by the specified number of data characters. The minimum character count value is 0. The maximum character count is limited by the total message character count permitted by SAE J1708.

PID 254 is a data link escape PID. All characters excluding the message checksum following an escape PID are defined as escape data. The first data byte contains the Message ID of the desired receiving device. The remaining escape data is to be defined by the manufacturer of the transmitting device and may be disclosed in an applications document (reference SAE J1708). It is used to transmit special commands, data, and other proprietary information to a specified component.

Parameter ID 255 is an extension PID reserved for future use.

The PID assignment list is shown in Table 2.

The procedure for assigning new PIDs is contained in 3.9.

3.4 Parameter Data-Types—Parameter data shall use one or more of the following data-types as in Table 3.

Alphanumeric data will be transmitted with the most significant character first. All other data will be transmitted least significant character first.

Signed integer values will use two's complement notation.

Alphanumeric characters will conform to the ASCII character set for values 0 to 127. Values from 128 to 255 may be used but are not defined in this document.

Floating-Point values will conform to the IEEE Floating-Point Standard.

3.5 Parameter Transmission Update Period and Message Priority—The update period and message priority at which a parameter is transmitted on the data link is primarily the responsibility of the transmitting electronic device. Because overloading the data link and providing compatible update rates are major concerns, a recommended transmission update period and message priority for each parameter is included in Appendix A. Variations from the listed update periods shall be included in the application document (reference SAE J1708).

If multiple parameters are grouped into one message, the message assignment would be based on the highest message priority associated with the group parameters. All requested parameters were assigned the lowest message priority, priority 8, so that the messages would not disrupt the regularly broadcast data.

3.6 Parameter Definitions—See Appendix A for parameter definitions.

3.7 Subsystem Identification Assignments—Subsystem Identification Numbers (SIDs) are numbers assigned by the SAE staff or the Data Format Subcommittee. There are 255 SIDs definable for each controller or MID. SIDs are numbers that can be used to identify a section of a control system without a related PID. SIDs should only be assigned to field-repairable or replaceable subsystems for which failures can be detected and isolated by the controller (MID). SIDs 1 to 150 are assigned by SAE staff using the procedure in 3.9. SIDs 156 to 255 are assigned by the Data Format Subcommittee using the procedure in 3.9. MID related SIDs start with number 1 and sequentially increase. Common SIDs start at 254 and sequentially decrease.

SIDs 151 through 155 are defined as "System Diagnostic Codes" and are used to identify failures that cannot be tied to a specific field replaceable component. Specific subsystem fault isolation is the goal of any diagnostic system, but for various reasons this cannot always be accomplished. These SIDs allow the manufacturer some flexibility to communicate non-"specific component" diagnostic information. PID 194 SID/FMI format of SIDs 151-155 permit the use of standard diagnostic tools, electronic dashboards, satellite systems and other advanced devices that scan for PID 194. Because manufacturer defined codes are not desirable in terms of standardization, the use of these codes should only be used when diagnostic information cannot be communicated as a specific component and failure mode.

TABLE 2—PARAMETER IDENTIFICATION ASSIGNMENT LIST

PID	Parameter
	Single Data Character Length Parameters
0	Request Parameter
1[1]	Invalid Data Parameter (see Appendix A)
2[1]	Transmitter System Status (see Appendix A)
3[1]	Transmitter System Diagnostic (see Appendix A)
4	Reserved—to be assigned
5[1]	Underrange Warning Condition (see Appendix A)
6[1]	Overrange Warning Condition (see Appendix A)
7-52	Reserved—to be assigned
53	Transmission Synchronizer Clutch Value
54	Transmission Synchronizer Brake Value
55	Shift Finger Positional Status
56	Transmission Range Switch Status
57	Transmission Actuator Status #2
58	Shift Finger Actuator Status
59	Shift Finger Gear Position
60	Shift Finger Rail Position
61	Parking Brake Actuator Status
62	Retarder Inhibit Status
63	Transmission Auxiliary Actuator Status #1
64	Direction Switch Status
65	Service Brake Switch Status
66	Vehicle Enabling Component Status
67	Shift Request Switch Status
68	Torque Limiting Factor
69	Two Speed Axle Switch Status
70	Parking Brake Switch
71	Idle Shutdown Timer Status
72	Blower Bypass Value Position
73	Auxiliary Water Pump Pressure
74	Maximum Road Speed Limit
75	Steering Axle Temperature
76	Axle Lift Air Pressure
77	Forward Rear Drive Axle Temperature
78	Rear Rear-Drive Axle Temperature
79	Road Surface Temperature
80	Washer Fluid Level
81	Particulate Trap Inlet Pressure
82	Air Start Pressure
83	Road Speed Limit Status
84	Road Speed
85	Cruise Control Status
86	Cruise Control Set Speed
87	Cruise Control High-Set Limit Speed
88	Cruise Control Low-Set Limit Speed
89	Power Takeoff Status
90	PTO Oil Temperature
91	Percent Accelerator Pedal Position
92	Percent Engine Load
93	Output Torque
94	Fuel Delivery Pressure
95	Fuel Filter Differential Pressure
96	Fuel Level
97	Water in Fuel Indicator
98	Engine Oil Level
99	Engine Oil Filter Differential Pressure
100	Engine Oil Pressure
101	Crankcase Pressure
102	Boost Pressure
103	Turbo Speed
104	Turbo Oil Pressure
105	Intake Manifold Temperature
106	Air Inlet Pressure
107	Air Filter Differential Pressure
108	Barometric Pressure
109	Coolant Pressure
110	Engine Coolant Temperature
111	Coolant Level
112	Coolant Filter Differential Pressure
113	Governor Droop
114	Net Battery Current
115	Alternator Current
116	Brake Application Pressure
117	Brake Primary Pressure
118	Brake Secondary Pressure
119	Hydraulic Retarder Pressure
120	Hydraulic Retarder Oil Temperature
121	Engine Retarder Status
122	Engine Retarder Percent
123	Clutch Pressure
124	Transmission Oil Level
125	Transmission Oil Level High/Low
126	Transmission Filter Differential Pressure
127	Transmission Oil Pressure
	Double Data Character Length Parameters
128	Component-specific request
129-154	Reserved—to be assigned
155	Auxiliary Input and Output Status
156	Injector Timing Rail Pressure
157	Injector Metering Rail Pressure
158	Battery Potential (Voltage)—Switched
159	Gas Supply Pressure
160	Main Shaft Speed
161	Input Shaft Speed
162	Transmission Range Selected
163	Transmission Range Attained
164	Lubricant Rail Injection Control Pressure
165	Compass Bearing
166	Rated Engine Power
167	Alternator Potential (Voltage)
168	Battery Potential (Voltage)
169	Cargo Ambient Temperature
170	Cab Interior Temperature
171	Ambient Air Temperature
172	Air Inlet Temperature
173	Exhaust Gas Temperature
174	Fuel Temperature
175	Engine Oil Temperature
176	Turbo Oil Temperature
177	Transmission Oil Temperature
178	Front Axle Weight
179	Rear Axle Weight
180	Trailer Weight
181	Cargo Weight
182	Trip Fuel
183	Fuel Rate
184	Instantaneous Fuel Economy
185	Average Fuel Economy
186	Power Takeoff Speed
187	Power Takeoff Set Speed
188	Idle Engine Speed
189	Rated Engine Speed
190	Engine Speed
191	Transmission Output Shaft Speed
	Variable and Long Data Character Length Parameters
192	Multimessage Parameter
193[1]	Transmitter System Diagnostic Table
(see Appendix A)	
194	Transmitter System Diagnostic Code and Occurrence Count Table
195	Diagnostic Data Request/Clear Count
196	Diagnostic Data/Count Clear Response
197-234	Reserved—to be assigned
235	Total Idle Hours
236	Total Idle Fuel Used
237	Vehicle Identification Number
238	Velocity Vector
239	Vehicle Position
240	Change Reference Number
241	Tire Pressure
242	Tire Temperature
243	Component Identification
244	Trip Distance
245	Total Vehicle Distance
246	Total Vehicle Hours
247	Total Engine Hours
248	Total PTO Hours
249	Total Engine Revolutions
250	Total Fuel Used
251	Clock
252	Date
253	Elapsed Time
	Special Parameters
254	Data Link Escape
255	Extension

[1] NOTE: These PIDs are superseded by PIDs 194, 195, and 196.

TABLE 3—PARAMETER DATA TYPES

Data-Type	Characters
Binary Bit-Mapped (B/BM)	1
Unsigned Short Integer (Uns/SI)	1
Signed Short Integer (S/SI)	1
Unsigned Integer (Uns/I)	2
Signed Integer (S/I)	2
Unsigned Long Integer (Uns/LI)	4
Signed Long Integer (S/LI)	4
Alphanumeric (ALPHA)	1
Single-Precision Floating-Point (SP/FP)	4
Double-Precision Floating-Point (DP/FP)	8

Possible reasons for using a System Diagnostic Code include:
a. Cost of specific component fault isolation is not justified, or
b. New concepts in Total Vehicle Diagnostics are being developed, or
c. New diagnostic strategies that are not component specific are being developed.

Due to the fact that SIDs 151-155 are manufacturer defined and are not component specific, FMIs 0-13 have little meaning. Therefore, FMI 14, "Special Instructions," will usually be used. The goal is to refer the service personnel to the manufacturer's troubleshooting manual for more information on the particular diagnostic code.

The SID assignment list is shown in Table 4.

3.8 Failure Mode Identifier Assignments—The Failure Mode Identifier, FMI, describes the type of failure detected in the subsystem identified by the PID or SID. The FMI, and either the PID or SID combine to form a given diagnostic code (see PID 194 for added clarification). The remaining failure mode identifiers would be assigned by the Data Format Subcommittee if additional common failure modes become detectable.

The failure mode identifier assignment list is shown in Table 5.

3.9 SAE Procedure for MID, PID, and SID Assignment
a. Purpose—To outline the procedure for the assignment of MID, PID, and SID elements within the documents established in the SAE Data Format Subcommittee.
b. General—MIDs, PIDs, and SIDs will be requested using the request form (Figure 1). All requests for MIDs, PIDs, and common SIDs will be forwarded to the chairperson of the SAE Data Format Subcommittee for action at the next scheduled committee meeting. All requests for MID related SIDs will be processed by the SAE staff. A confirmation for MID, PID, and common SID requests will be sent to the requestor stating the date the request will be reviewed to ensure the requestor has the opportunity to be present at that meeting. MID related SID requests will be handled by SAE staff with copies of the request form sent to the chairperson of the SAE Data Format Subcommittee.
c. Verification of Request—The request form will be reviewed to ensure all required fields are provided by the requestor. If information is missing, the request form shall be returned to the requestor asking for the additional information. If the information is complete, either the MID/PID/Common SID process or the MID related SID process shall be followed depending on the type of request.

MID/PID/Common SID Process—SAE will complete the request form by filling in the date and time of the next SAE Data Format Subcommittee meeting. They will make two copies of the request form. File one copy in a SAE staff maintained file of requests. Send the original to the chairperson of the SAE Data Format Subcommittee for review and approval by the committee. Send the second copy of the request back to the requestor.

The chairperson of the SAE Data Format Subcommittee will present to the committee all MID, PID, and common SID requests since the last meeting. An approval or disapproval vote is required during the committee meeting. The chairperson of the SAE Data Format Subcommittee will document the approval or disapproval by completing the review section of the request form. These completed request forms for all MIDs, PIDs, and common SIDs will be sent to the SAE staff.

The SAE staff will verify that all requests were handled and notify the requestor by sending a copy of the completed form to the requestor. The original form should be filed in a completed request file. The copy of the request form that is in the request file should be removed.

d. MID Related SID Process—The SAE staff will keep records of SIDs allocated to each MID. This will be accomplished by maintaining a control log for each MID. If the requestor is asking for a new SID that is similar to an existing SID, the SAE staff will document the current SID on the request form and return it to the requestor. If the request is for a new MID related SID which is not currently assigned, the SAE staff will assign the next sequential number. This will be documented on the request form (Figure 1). The SAE staff will make two copies of the request form. The original will be returned to the requestor. The first copy will be sent to the SAE Data Format Subcommittee chairperson. The second copy will be filed in the assigned SID file by MID. The new SID number will be logged on the MID/SID control log for that MID. If the total number of SIDs assigned reaches 100 for an MID, the SAE staff is required to notify the chairperson of the SAE Data Format Subcommittee.

NOTE—Parameters considered to be of a data link command or control nature should be added to the parameter list at the lowest PID value available within the appropriate data size grouping. All other parameters should be added at the highest PID value available within the appropriate data grouping.

Requestor Name   _____
Requestor Address _____
                  _____
                  _____

Company Name   _____

Request Type   MID _____   PID _____

                  SID _____   Requested FOR MID # _____

Description of MID/PID/SID

_____
_____
_____
_____
_____
_____
_____
_____
_____

For PIDs Only (See SAE J1587 for Description)

    Parameter Data Length   _____
    Data Type   _____
    Resolution   _____
    Maximum Range   _____
    Transmission Update Period _____
    Message Priority   _____
    Format   _____

_____

For Use By SAE Only

Approved _____ Disapproved _____ Signature _____
New MID Number _____ New PID Number _____   New SID Number _____
Current MID _____ Current PID _____   Current SID _____

Date of next SAE Data Format Committee meeting _____
Location _____   Time _____

Incomplete Information _____   (Please complete items marked)

Please mail completed form to:

        SAE
        Data Format Subcommittee
        400 Commonwealth Drive
        Warrendale, PA 15096

FIGURE 1—SAE DATA FORMAT SUBCOMMITTEE MID, PID, SID REQUEST FORM

TABLE 4—SUBSYSTEM IDENTIFICATION (SID) ASSIGNMENT LIST

SIDs 1 to 150 are not common with other systems and are assigned by SAE.
SIDs 151 to 255 are common among other systems and are assigned by the Data Format Subcommittee.

SID	Description
151	System Diagnostic Code #1
152	System Diagnostic Code #2
153	System Diagnostic Code #3
154	System Diagnostic Code #4
155	System Diagnostic Code #5
156-231	Reserved for future assignment by SAE Data Format Subcommittee
232	5 Volts DC Supply
233	Controller #2
234	Parking Brake On Actuator
235	Parking Brake Off Actuator
236	Power Connect Device
237	Start Enable Device
238	Diagnostic Lamp—Red
239	Diagnostic Light—Amber
240	Program Memory
241[2]	Set aside for Systems Diagnostics
242	Cruise Control Resume Switch
243	Cruise Control Set Switch
244	Cruise Control Enable Switch
245	Clutch Pedal Switch #1
246	Brake Pedal Switch #1
247	Brake Pedal Switch #2
248	Proprietary Data Link
249	SAE J1922 Data Link
250	SAE J1708 (J1587) Data Link
251	Power Supply
252	Calibration Module
253	Calibration Memory
254	Controller #1
255	Reserved

Engine SIDs (MID = 128, 175, 183, 184, 185, 186)

SID	Description
0	Reserved
1	Injector Cylinder #1
2	Injector Cylinder #2
3	Injector Cylinder #3
4	Injector Cylinder #4
5	Injector Cylinder #5
6	Injector Cylinder #6
7	Injector Cylinder #7
8	Injector Cylinder #8
9	Injector Cylinder #9
10	Injector Cylinder #10
11	Injector Cylinder #11
12	Injector Cylinder #12
13	Injector Cylinder #13
14	Injector Cylinder #14
15	Injector Cylinder #15
16	Injector Cylinder #16
17	Fuel Shutoff Valve
18	Fuel Control Valve
19	Throttle Bypass Valve
20	Timing Actuator
21	Engine Position Sensor
22	Timing Sensor
23	Rack Actuator
24	Rack Position Sensor
25	External Engine Protection Input
26	Auxiliary Output Driver #1
27	Variable Geometry Turbocharger Actuator #1
28	Variable Geometry Turbocharger Actuator #2
29	External Fuel Command Input
30	External Speed Command Input
31	Tachometer Signal Output
32	Wastegate Output Device Driver
33	Fan Clutch Output Device Driver
34	Exhaust Back Pressure Sensor
35	Exhaust Back Pressure Regulator Solenoid
36	Glow Plug Lamp
37	Electronic Drive Unit Power Relay
38	Glow Plug Relay
39	Engine Starter Motor Relay
40	Auxiliary Output Device Driver #2
41	ECM 8 Volts DC Supply
42	Injection Control Pressure Regulator
43	Autoshift High Gear Actuator
44	Autoshift Low Gear Actuator
45	Autoshift Neutral Actuator
46	Autoshift Common Low Side (Return)
47-150	Reserved for future assignment by SAE

Transmission SIDs (MID = 130, 176)

SID	Description
0	Reserved
1	C1 Solenoid Valve
2	C2 Solenoid Valve
3	C3 Solenoid Valve
4	C4 Solenoid Valve
5	C5 Solenoid Valve
6	C6 Solenoid Valve
7	Lockup Solenoid Valve
8	Forward Solenoid Valve
9	Low Signal Solenoid Valve
10	Retarder Enable Solenoid Valve
11	Retarder Modulation Solenoid Valve
12	Retarder Response Solenoid Valve
13	Differential Lock Solenoid Valve
14	Engine/Transmission Match
15	Retarder Modulation Request Sensor
16	Neutral Start Output
17	Turbine Speed Sensor
18	Primary Shift Selector
19	Secondary Shift Selector
20	Special Function Inputs
21	C1 Clutch Pressure Indicator
22	C2 Clutch Pressure Indicator
23	C3 Clutch Pressure Indicator
24	C4 Clutch Pressure Indicator
25	C5 Clutch Pressure Indicator
26	C6 Clutch Pressure Indicator
27	Lockup Clutch Pressure Indicator
28	Forward Range Pressure Indicator
29	Neutral Range Pressure Indicator
30	Reverse Range Pressure Indicator
31	Retarder Response System Pressure Indicator
32	Differential Lock Clutch Pressure Indicator
33	Multiple Pressure Indicators
34	Reverse Switch
35	Range High Actuator Indicator
36	Range Low Actuator Indicator
37	Splitter Direct Actuator
38	Splitter Indirect Actuator
39	Shift Finger Rail Actuator 1
40	Shift Finger Gear Actuator 1
41	Upshift Request Switch
42	Downshift Request Switch
43	Torque Converter Interrupt Actuator
44	Torque Converter Lockup Actuator
45	Range High Indicator
46	Range Low Indicator
47	Shift Finger Neutral Indicator
48	Shift Finger Engagement Indicator
49	Shift Finger Center Rail Indicator
50	Shift Finger Rail Actuator 2
51	Shift Finger Gear Actuator 2
52	Hydraulic System
53	Defuel Actuator
54-150	Reserved for future assignment by SAE

Brake SIDs (MID = 136, 137, 138, 139)

SID	Description
0	Reserved
1	Wheel Sensor ABS Axle 1 Left
2	ABS Axle 1 Right
3	ABS Axle 2 Left
4	ABS Axle 2 Right
5	ABS Axle 3 Left
6	ABS Axle 3 Right
7	Pressure Modulation Valve ABS Axle 1 Left
8	ABS Axle 1 Right
9	ABS Axle 2 Left
10	ABS Axle 2 Right
11	ABS Axle 3 Left
12	ABS Axle 3 Right
13	Retarder Control Relay
14	Relay Diagonal 1
15	Relay Diagonal 2
16	Mode Switch ABS
17	Mode Switch ASR
18	DIF 1—ASR Valve
19	DIF 2—ASR Valve

TABLE 4—CONTINUED

20	Pneumatic Engine Control
21	Electronic Engine Control (Servomotor)
22	Speed Signal Input
23	Warning Light Bulb
24	ASR Light Bulb
25	Wheel Sensor, ABS Axle 1 Average
26	Wheel Sensor, ABS Axle 2 Average
27	Wheel Sensor, ABS Axle 3 Average
28	Pressure Modulator, Drive Axle Relay Valve
29	Pressure Transducer, Drive Axle Relay Valve
30	Master Control Relay
31-150	Reserved for Future Assignment by SAE

Instrument Panel SIDs (MID = 140)

0	Reserved
1	Left Fuel Level Sensor
2	Right Fuel Level Sensor
3	Fuel Feed Rate Sensor
4	Fuel Return Rate Sensor
5-150	Reserved for future assignment by SAE

Fuel System SIDs (MID =143)

0	Reserved
1	Injector Cylinder #1
2	Injector Cylinder #2
3	Injector Cylinder #3
4	Injector Cylinder #4
5	Injector Cylinder #5
6	Injector Cylinder #6
7	Injector Cylinder #7
8	Injector Cylinder #8
9	Injector Cylinder #9
10	Injector Cylinder #10
11	Injector Cylinder #11
12	Injector Cylinder #12
13	Injector Cylinder #13
14	Injector Cylinder #14
15	Injector Cylinder #15
16	Injector Cylinder #16
17	Fuel Shutoff Valve
18	Fuel Control Valve
19	Throttle Bypass Valve
20	Timing Actuator
21	Engine Position Sensor
22	Timing Sensor
23	Rack Actuator
24	Rack Position Sensor
25	External Engine Protection Input
26	Auxiliary Output Device Driver
27-150	Reserved for future assignment by SAE

Vehicle Navigation SIDs (MID = 162)

0	Reserved
1	Dead Reckoning Unit
2	Loran Receiver
3	Global Positioning System (GPS)
4	Integrated Navigation Unit
5-150	Reserved for future assignment by SAE

Particulate Trap System SIDs (MID = 177)

1	Heater Circuit #1
2	Heater Circuit #2
3	Heater Circuit #3
4	Heater Circuit #4
5	Heater Circuit #5
6	Heater Circuit #6
7	Heater Circuit #7
8	Heater Circuit #8
9	Heater Circuit #9
10	Heater Circuit #10
11	Heater Circuit #11
12	Heater Circuit #12
13	Heater Circuit #13
14	Heater Circuit #14
15	Heater Circuit #15
16	Heater Circuit #16
17	Heater Regeneration System
18-150	Reserved for future assignment by SAE

[2] Superseded by SIDs 151-155.

TABLE 5—FAILURE MODE IDENTIFIERS (FMIS)

0	Data valid but above normal operational range (that is, engine overheating)
1	Data valid but below normal operational range (that is, engine oil pressure too low)
2	Data erratic, intermittent, or incorrect
3	Voltage above normal or shorted high
4	Voltage below normal or shorted low
5	Current below normal or open circuit
6	Current above normal or grounded circuit
7	Mechanical system not responding properly
8	Abnormal frequency, pulse width, or period
9	Abnormal update rate
10	Abnormal rate of change
11	Failure mode not identifiable
12	Bad intelligent device or component
13	Out of Calibration
14	Special Instructions
15	Reserved for future assignment by the SAE Data Format Subcommittee

APPENDIX A
PARAMETER DEFINITIONS

A.0 Request Parameter—Used to request parameter data transmission from other components on the data link.

Parameter Data Length: 1 Character
Data Type: Unsigned Short Integer
Resolution: Binary
Maximum Range: 0 to 255
Transmission Update Period: As needed
Message Priority: 8
Format:

    PID Data
    0   a
    a - Parameter ID of the requested parameter

Any and all components measuring or calculating the specified parameter should transmit it if possible.

A.1 Invalid data parameter—Used to notify other components on the data link that invalid data has been detected in a parameter that is normally available and will not be transmitted.

The SAE Truck and Bus Data Format Subcommittee established PIDs 194 to 196 in May 1988; therefore, this Parameter ID should no longer be used by manufacturers in the design of new components. However, this parameter is being reserved for use by manufacturers who have developed systems prior to January 1989 and are, therefore, unable to accommodate the new diagnostic formats as defined in PIDs 194 to 196. It is recommended that manufacturers using this parameter fully define the contents and circumstances under which it is used in the application document.

A.2 Transmitter System Status—Used to notify other components on the data link of the present status of the transmitting electronic component.

Parameter Data Length: 1 Character
Data Type: Unsigned Short Integer
Resolution: Binary
Maximum Range: 0 to 255
Transmission Update Period: As needed
Message Priority: 8
Format:

    PID Data
    2   a
    a - Status code defined by the component manufacturer in an application document.

The SAE Truck and Bus Data Format Subcommittee established PIDs 194 to 196 in May 1988; therefore, this Parameter ID should no longer be used by manufacturers in the design of new components. However, this parameter is being reserved for use by manufacturers who have developed systems prior to January 1989 and are, therefore, unable to accommodate the new diagnostic formats as defined in PIDs 194 to 196. It is recommended that manufacturers using this parameter fully define the contents and circumstances under which it is used in the application document.

A.3 Transmitter System Diagnostic—Used to notify other components on the data link of the diagnostic condition of the transmitting electronic component.
    Parameter Data Length: 1 Character
    Data Type: Unsigned Short Integer
    Resolution: Binary
    Maximum Range: 0 to 255
    Transmission Update Period: As needed
    Message Priority: 8
    Format:
        PID Data
        3   a
        a - Status code defined by the component manufacturer in an application document.

The SAE Truck and Bus Data Format Subcommittee established PIDs 194 to 196 in May 1988, therefore, this Parameter ID should no longer be used by manufacturers in the design of new components. However, this parameter is being reserved for use by manufacturers who have developed systems prior to January 1989 and are therefore unable to accommodate the new diagnostic formats as defined in PIDs 194 to 196. It is recommended that manufacturers using this parameter fully define the contents and circumstances under which it is used in the application document.

A.4 Reserved—To be assigned

A.5 Under Range Warning Condition—Used to notify other components on the data link that the transmitter's internal monitoring process has declared the data transmitted by this PID is below or less than the acceptable operating level.
    Parameter Data Length: 1 character
    Data Type: Unsigned Short Integer
    Resolution: Binary
    Maximum range: 0 to 255
    Transmission Update Period: Transmitted as frequently as, and immediately prior to, the offending PID
    Message Priority: Parameter specific
    Format:
        PID Data
        5   a—Where a is the value of the offending PID

a. Example—The Monitoring device (perhaps the engine controller) determines oil pressure is below acceptable operating range. The portion of the transmitted message would read as shown in Figure A1:

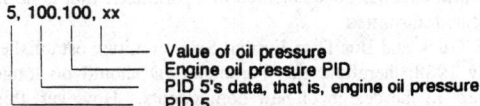

FIGURE A1—EXAMPLE UNDER RANGE WARNING CONDITIONS

The SAE Truck and Bus Data Format Subcommittee established PIDs 194 to 196 in May 1988, therefore, this Parameter ID should no longer be used by manufacturers in the design of new components. However, this parameter is being reserved for use by manufacturers who have developed systems prior to January 1989 and are, therefore, unable to accommodate the new diagnostic formats as defined in PIDs 194 to 196. It is recommended that manufacturers using this parameter fully define the contents and circumstances under which it is used in the application document.

A.6 PID Over Range Warning Condition—Used to notify other components on the data link that the transmitter's internal monitoring process has declared the data transmitted by this PID is above or greater than the acceptable operating level.
    Parameter Data Length: 1 Character
    Data Type: Unsigned Short Integer
    Resolution: Binary
    Maximum range: 0 to 255
    Transmission Update Period: Transmitted as frequently as, and immediately prior to, the offending PID.

a. Example: The monitoring device (perhaps the engine controller) determines coolant temperature is above the acceptable operating range. The portion of the transmitted message would read as shown in Figure A2:

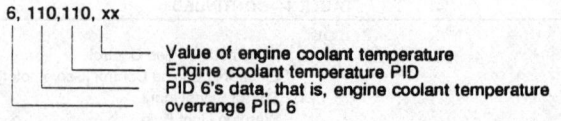

FIGURE A2—EXAMPLE PID OVER RANGE WARNING CONDITION

The SAE Truck and Bus Data Format Subcommittee established PIDs 194 to 196 in May 1988, therefore, this Parameter ID should no longer be used by manufacturers in the design of new components. However, this parameter is being reserved for use by manufacturers who have developed systems prior to January 1989 and are, therefore, unable to accommodate the new diagnostic formats as defined in PIDs 194 to 196. It is recommended that manufacturers using this parameter fully define the contents and circumstances under which it is used in the application document.

A.7 to A.52 Reserved—To be assigned

A.53 Transmission Synchronizer Clutch Value—The current modulation value for the air supply to the synchronizer clutch.
    Parameter Data Length: 1 Character
    Data Type: Unsigned Short Integer
    Bit Resolution: 0.4%
    Maximum Range: 0.0 to 102.0%
    Transmission Update Period: On request
    Message Priority: 8
    Format:
        PID Data
        53   a
        a - Transmission synchronizer clutch value

A.54 Transmission Synchronizer Brake Value—The current modulation value for the air supply to the synchronizer brake.
    Parameter Data Length: 1 Character
    Data Type: Unsigned Short Integer
    Bit Resolution: 0.4%
    Maximum Range: 0.0 to 102.0%
    Transmission Update Period: On request
    Message Priority: 8
    Format:
        PID Data
        54a
        a - Transmission synchronizer brake value

A.55 Shift Finger Positional Status—Identifies the current status of the switches that represent the position of the shift finger.
    Parameter Data Length: 1 Character
    Data Type: Binary Bit-Mapped
    Resolution: Binary
    Maximum Range: 0 to 255
    Transmission Update Period: On request
    Message Priority: 8
    Format:
        PID Data
        55   a
        a - Shift finger positional status
            Bits 1-2: Neutral sense
            Bits 3-4: Fore/aft sense
            Bits 5-6: Center rail sense
            Bits 7-8: Reserved—both bits set to 1
    NOTE: Each status will be described using the following nomenclature:
        00 Off
        01 On
        10 Error condition
        11 Not available

A.56 Transmission Range Switch Status—Identifies the current status of the switches that represent range position.
    Parameter Data Length: 1 Character
    Data Type: Binary Bit-Mapped
    Resolution: Binary
    Maximum Range: 0 to 255
    Transmission Update Period: On request
    Message Priority: 8
    Format:
        PID Data

56 a
 a - Transmission range switch status
  Bits 1-2: High range sense
  Bits 3-4: Low range sense
  Bits 5-8: Reserved—all bits set to 1
NOTE: Each status will be described using the following nomenclature:
 00 Off
 01 On
 10 Error condition
 11 Not available

A.57 Transmission Actuator Status #2—Identifies the current status of the actuators that control the torque converter interrupt clutch, lockup clutch, and the engine defuel mechanism.
Parameter Data Length: 1 Character
Data Type: Binary Bit-Mapped
Resolution: Binary
Maximum Range: 0 to 255
Transmission Update Period: On request
Message Priority: 8
Format:
 PID Data
 57 a
 a - Transmission actuator status #2
  Bits 1-2: Interrupt clutch actuator status
  Bits 3-4: Lockup clutch actuator status
  Bits 5-6: Defuel actuator status
  Bits 7-8: Reserved—all bits set to 1
NOTE: Each status will be described using the following nomenclature:
 00 Off
 01 On
 10 Error condition
 11 Not available

A.58 Shift Finger Actuator Status—Identifies the current status of the actuators that move the shift finger.
Parameter Data Length: 1 Character
Data Type: Binary Bit-Mapped
Resolution: Binary
Maximum Range: 0 to 255
Transmission Update Period: On request
Message Priority: 8
Format:
 PID Data
 58 a
 a - Shift finger actuator status
  Bits 1-2: Rail actuator #1 status
  Bits 3-4: Gear actuator #1 status
  Bits 5-6: Rail actuator #2 status
  Bits 7-8: Gear actuator #2 status
NOTE: Each status will be described using the following nomenclature:
 00 Off
 01 On
 10 Error condition
 11 Not available

A.59 Shift Finger Gear Position—The current position of the shift finger in the gear direction.
Parameter Data Length: 1 Character
Data Type: Unsigned Short Integer
Bit Resolution: 0.4%
Maximum Range: 0.0 to 102.0%
Transmission Update Period: On request
Message Priority: 8
Format:
 PID Data
 59 a
 a - Shift finger gear position

A.60 Shift Finger Rail Position—The current position of the shift finger in the rail direction.
Parameter Data Length: 1 Character
Data Type: Unsigned Short Integer
Bit Resolution: 0.4%
Maximum Range: 0.0 to 102.0%
Transmission Update Period: On request
Message Priority: 8
Format:
 PID Data
 60 a
 a - Shift finger rail position

A.61 Parking Brake Actuator Status—Identifies the current status of the actuators that control the parking brakes.
Parameter Data Length: 1 Character
Data Type: Binary Bit-Mapped
Resolution: Binary
Maximum Range: 0 to 255
Transmission Update Period: On request
Message Priority: 8
Format:
 PID Data
 61 a
 a - Parking brake actuator status
  Bits 1-2: Parking brake on actuator status
  Bits 3-4: Parking brake off actuator status
  Bits 5-8: Reserved—all bits set to 1
NOTE: Each status will be described using the following nomenclature:
 00 Off
 01 On
 10 Error condition
 11 Not available

A.62 Retarder Inhibit Status—Identifies the current state of the device that inhibits use of the engine retarder.
Parameter Data Length: 1 Character
Data Type: Binary Bit-Mapped
Resolution: Binary
Maximum Range: 0 to 255
Transmission Update Period: On request
Message Priority: 8
Format:
 PID Data
 62 a
 a - Retarder inhibit status
  Bits 1-2: Retarder inhibit status
  Bits 3-8: Reserved—all bits set to 1
NOTE: Each status will be described using the following nomenclature:
 00 Off
 01 On
 10 Error condition
 11 Not available

A.63 Transmission Actuator Status #1—Identifies the current status of the actuators used to control the functions of the auxiliary unit.
Parameter Data Length: 1 Character
Data Type: Binary Bit-Mapped
Resolution: Binary
Maximum Range: 0 to 255
Transmission Update Period: On request
Message Priority: 8
Format:
 PID Data
 63 a
 a - Transmission actuator status #1
  Bits 1-2: Range high actuator status
  Bits 3-4: Range low actuator status
  Bits 5-6: Splitter direct actuator status
  Bits 7-8: Splitter indirect actuator status
NOTE: Each status will be described using the following nomenclature:
 00 Off
 01 On
 10 Error condition
 11 Not available

A.64 Direction Switch Status—Identifies the current state of the switches that indicate the direction of the transmission.
Parameter Data Length: 1 Character
Data Type: Binary Bit-Mapped
Resolution: Binary
Maximum Range: 0 to 255
Transmission Update Period: On request

Message Priority: 8
Format:
   PID Data
   64   a
     a - Direction switch status
        Bits 1-2: Reverse switch status
        Bits 3-4: Neutral switch status
        Bits 5-6: Forward switch status
        Bits 7-8: Reserved—both bits set to 1

NOTE: Each status will be described using the following nomenclature:
   00 Off
   01 On
   10 Error condition
   11 Not available

A.65 Service Brake Switch Status—Identifies the current state of the switch that indicates the status of the service brakes.
Parameter Data Length: 1 Character
Data Type: Binary Bit-Mapped
Resolution: Binary
Maximum Range: 0 to 255
Transmission Update Period: On request
Message Priority: 8
Format:
   PID Data
   65   a
     a - Service brake switch status
        Bits 1-2: Service brake switch status
        Bits 3-8: Reserved—all bits set to 1

NOTE: Each status will be described using the following nomenclature:
   00 Off
   01 On
   10 Error condition
   11 Not available

A.66 Vehicle Enabling Component Status—Identifies the current state of the components that enable the vehicle to start and operate properly.
Parameter Data Length: 1 Character
Data Type: Binary Bit-Mapped
Resolution: Binary
Maximum Range: 0 to 255
Transmission Update Period: On request
Message Priority: 8
Format:
   PID Data
   66   a
     a - Vehicle enabling component status
        Bits 1-2: Ignition switch status
        Bits 3-4: Start enable device status
        Bits 5-6: Power connect device status
        Bits 7-8: Reserved—both bits set to 1

NOTE: Each status will be described using the following nomenclature:
   00 Off
   01 On
   10 Error condition
   11 Not available

A.67 Shift Request Switch Status—Identifies the current state of the switches used to request an upshift or downshift.
Parameter Data Length: 1 Character
Data Type: Binary Bit-Mapped
Resolution: Binary
Maximum Range: 0 to 255
Transmission Update Period: On request
Message Priority: 8
Format:
   PID Data
   67   a
     a - Vehicle enabling component status
        Bits 1-2: Upshift switch status
        Bits 3-4: Downshift switch status
        Bits 5-8: Reserved—all bits set to 1

NOTE: Each status will be described using the following nomenclature:
   00 Off
   01 On
   10 Error condition
   11 Not available

A.68 Torque Limiting Factor—Ratio of current output torque allowed (due to adverse operating conditions) to the maximum torque available at the current engine speed (under normal operating conditions).

$$\text{Torque Limiting Factor} = 100 \times \frac{\text{Allowed Max. Torque at current engine speed}}{\text{Max. Torque Available at current engine speed}} \quad \text{(Eq. A1)}$$

Parameter Data Length: 1 Character
Data Type: Unsigned Short Integer
Bit Resolution: 0.5 %
Maximum Range: 0.0 to 127.5%
Transmission Update Period: 1.0 s
Message Priority: 4
Format:
   PID Data
   68   a
     a - Torque Limiting Factor

A.69 Two Speed Axle Switch Status—Identifies the commanded range for a two speed axle.
Parameter Data Length: 1 Character
Data Type: Binary Bit-Mapped
Resolution: Binary
Maximum Range: 0 to 255
Transmission Update Period: 1.0 s
Message Priority: 6
Format:
   PID Data
   69   a
     a - Two speed axle switch status
        Bit 8:   0=high range is commanded
                    1=low range is commanded
        Bits 1 to 7: undefined

A.70 Parking Brake Switch Status—Identifies the state (active/inactive) of the parking brake switch.
Parameter Data Length: 1 Character
Data Type: Binary Bit-Mapped
Resolution: Binary
Maximum Range: 0 to 255
Transmission Update Period: 1.0 s
Message Priority: 5
Format:
   PID Data
   70   a
     a - Parking brake switch status
        Bit 8: 1=active / 0=inactive
        Bits 1 to 7: Undefined

A.71 Idle Shutdown Timer Status—State of the idle shutdown timer system (active, not active) for the various modes of operation.
Parameter Data Length: 1 Character
Data Type: Binary Bit-Mapped
Resolution: Binary
Maximum Range: 0 to 255
Transmission Update Period: 1.0 s
Message Priority: 5
Format:
   PID Data
   71   a
     a - Idle shutdown timer status
        Bit 8: Idle shutdown timer status   1=active / 0=inactive
        Bits 5 to 7: Undefined
        Bit 4: Idle shutdown timer function   1=enabled in calibration
                                                       0=disabled in cal.
        Bit 3: Idle shutdown timer override   1=active / 0=inactive
        Bit 2: Engine has shutdown by idle timer   1=yes / 0=no
        Bit 1: Driver alert mode   1=active / 0=inactive

A.72 Blower Bypass Valve Position—Relative position of the blower bypass valve.
Parameter Data Length: 1 Character
Data Type: Unsigned Short Integer
Bit Resolution: 0.4%
Maximum Range: 0.0 to 102.0%

Transmission Update Period: 0.5 s
Message Priority: 3
Format:
    PID Data
    72   a
    a - Blower bypass valve position

A.73 Auxiliary Water Pump Pressure—Gage pressure of auxiliary water pump driven as a PTO device.
Parameter Data Length: 1 Character
Data Type: Unsigned Short Integer
Bit Resolution: 13.8 kPa (2 lbf/in^2)
Maximum Range: 0.0 to 3516 kPa (0.0 to 510 lbf/in^2)
Transmission Update Period: 1.0 s
Message Priority: 4
Format:
    D Data
    73   a
    a - Auxiliary water pump pressure

A.74 Maximum Road Speed Limit—Maximum vehicle velocity allowed.
Parameter Data Length: 1 Character
Data Type: Unsigned Short Integer
Bit Resolution: 0.5 mph (0.805 km/h)
Maximum Range: 0.0 to 205.2 km/h (0.0 to 127.5 mph)
Transmission Update Period: On request
Message Priority: 8
Format:
    PID Data
    74   a
    a - Maximum road speed limit

A.75 Steering Axle Temperature—Temperature of lubricant in steering axle.
Parameter Data Length: 1 Character
Data Type: Unsigned Short Integer
Bit Resolution: 0.667 °C (1.2 °F)
Maximum Range: -17.8 to +152.2 °C (0.0 to 306.0 °F)
Transmission Update Period: 1.0 s
Message Priority: 5
Format:
    PID Data
    75   a
    a - Steering axle temperature

A.76 Axle Lift Air Pressure—Gage pressure of air in system that utilizes compressed air to provide force between axle and frame.
Parameter Data Length: 1 Character
Data Type: Unsigned short Integer
Bit Resolution: 4.14 kPa (0.6 lbf/in^2)
Maximum Range: 0.0 to 1055 kPa (0.0 to 153.0 lbf/in^2)
Transmission Update Period: 1.0 s
Message Priority: 5
Format:
    PID Data
    76   a
    a - Axle lift air pressure

A.77 Forward Rear Drive Axle Temperature—Temperature of axle lubricant in forward rear drive axle.
Parameter Data Length: 1 Character
Data Type: Unsigned Short Integer
Bit Resolution: 0.667 °C (1.2 °F)
Maximum Range: -17.8 to +152.2 °C (0.0 to 306.0°F)
Transmission Update Period: 1.0 s
Message Priority: 5
Format:
    PID Data
    77   a
    a - Forward rear drive axle temperature

A.78 Rear Rear Drive Axle Temperature—Temperature of axle lubricant in rear rear drive axle.
Parameter Data Length: 1 Character
Data Type: Unsigned Short Integer
Bit Resolution: 0.667 °C (1.2 °F)
Maximum Range: -17.8 to +152.2 °C (0.0 to 306.0 °F)
Transmission Update Period: 1.0 s
Message Priority: 5
Format:
    PID Data
    78   a
    a - Rear rear drive axle temperature

A.79 Road Surface Temperature—Indicated temperature of road surface over which vehicle is operating.
Parameter Data Length: 1 Character
Data Type: Signed Short Integer
Bit Resolution: 1.39 °C (2.5 °F)
Maximum Range: -196.0 to +159.0 °C (-320.0 to +317.5 °F)
Transmission Update Period: 10.0 s
Message Priority: 7
Format:
    PID Data
    79   a
    a - Road surface temperature

A.80 Washer Fluid Level—Ratio of volume of liquid to total container volume of fluid reservoir in windshield wash system.
Parameter Data Length: 1 Character
Data Type: Unsigned Short Integer
Bit Resolution: 0.5%
Maximum Range: 0.0 to 127.5%
Transmission Update Period: 10.0 s
Message Priority: 7
Format:
    PID Data
    80   a
    a - Washer fluid level

A.81 Particulate Trap Inlet Pressure—Exhaust back pressure as a result of particle accumulation on filter media placed in the exhaust stream.
Parameter Data Length: 1 Character
Data Type: Unsigned Short Integer
Bit Resolution: 0.169 kPa (0.05 in Hg)
Maximum Range: 0.0 to 43.1 kPa (0.0 to 12.75 in Hg)
Transmission Update Period: 10.0 s
Message Priority: 7
Format:
    PID Data
    81   a
    a - Particulate trap inlet pressure

A.82 Air Start Pressure—Gage pressure of air in an engine starting system that utilizes compressed air to provide the force required to rotate the crankshaft.
Parameter Data Length: 1 Character
Data Type: Unsigned Short Integer
Bit Resolution: 4.14 kPa (0.6 lbf/in^2)
Maximum Range: 0.0 to 1055 kPa (0.0 to 153 lbf/in^2)
Transmission Update Period: On request
Message Priority: 8
Format:
    PID Data
    82   a
    a - Air start pressure

A.83 Road Speed Limit Status—State (active or not active) of the system used to limit maximum vehicle velocity.
Parameter Data Length: 1 Character
Data Type: Binary Bit-Mapped
Resolution: Binary
Maximum Range: 0 to 255
Transmission Update Period: 1.0 s
Message Priority: 4
Format:
    PID Data
    83   a
    a - Road speed limit status
        Bit 8: 1=active / 0=not active
        Bits 1 to 7: undefined

A.84 Road Speed—Indicated vehicle velocity.
Parameter Data Length: 1 Character
Data Type: Unsigned Short Integer
Bit Resolution: 0.805 km/h (0.5 mph)
Maximum Range: 0.0 to 205.2 km/h (0.0 to 127.5 mph)
Transmission Update Period: 0.1 s

Message Priority: 1
Format:
   PID Data
   84   a
       a - Road speed

A.85 Cruise Control Status—State of the vehicle velocity control system (active, not active), and system switch (on, off), for various system operating modes.
Parameter Data Length: 1 Character
Data Type: Binary Bit-Mapped
Resolution: Binary
Maximum Range: 0 to 255
Transmission Update Period: 0.2 s
Message Priority: 3
Format:
   PID Data
   85   a
       a - Cruise control status
          Bit 1: cruise control switch 1=on / 0=off
          Bit 2: set switch 1=on / 0=off
          Bit 3: coast switch 1=on / 0=off
          Bit 4: resume switch 1=on / 0=off
          Bit 5: accel switch 1=on / 0=off
          Bit 6: brake switch 1=on / 0=off
          Bit 7: clutch switch 1=on / 0=off
          Bit 8: cruise mode 1=active / 0=not active

A.86 Cruise Control Set Speed—Value of set (chosen) velocity of velocity control system.
Parameter Data Length: 1 Character
Data Type: Unsigned Short Integer
Bit Resolution: 0.805 km/h (0.5 mph)
Maximum Range: 0.0 to 205.2 km/h (0.0 to 127.5 mph)
Transmission Update Period: 10.0 s
Message Priority: 6
Format:
   PID Data
   86   a
       a - Cruise control set speed

A.87 Cruise Control High Set Limit Speed—Maximum vehicle velocity allowed at any cruise control set speed.
Parameter Data Length: 1 Character
Data Type: Unsigned Short Integer
Bit Resolution: 0.805 km/h (0.5 mph)
Maximum Range: 0.0 to 205.2 km/h (0.0 to 127.5 mph)
Transmission Update Period: On request
Message Priority: 8
Format:
   PID Data
   87   a
       a - Cruise control high set limit speed

A.88 Cruise Control Low Set Limit Speed—Minimum vehicle velocity allowed by cruise control before a speed adjustment is called for.
Parameter Data Length: 1 Character
Data Type: Unsigned Short Integer
Bit Resolution: 0.805 km/h (0.5 mph)
Maximum Range: 0.0 to 205.2 km/h (0.0 to 127.5 mph)
Transmission Update Period: On request
Message Priority: 8
Format:
   PID Data
   88   a
       a - Cruise control low set limit speed

A.89 Power Takeoff Status—State of the system used to transmit engine power to auxiliary equipment. Status indication is for system (active, not active), and system switch (on, off), for various operating modes.
Parameter Data Length: 1 Character
Data Type: Binary Bit-Mapped
Resolution: Binary
Maximum Range: 0 to 255
Transmission Update Period: 1.0 s
Message Priority: 5
Format:
   PID Data
   89   a
       a - Power takeoff status
          Bit 1: PTO control switch 1=on / 0=off
          Bit 2: set switch 1=on / 0=off
          Bit 3: coast switch 1=on / 0=off
          Bit 4: resume switch 1=on / 0=off
          Bit 5: accel switch 1=on / 0=off
          Bit 6: brake switch 1=on / 0=off
          Bit 7: clutch switch 1=on / 0=off
          Bit 8: PTO mode 1=active / 0=not active

A.90 Power Takeoff Oil Temperature—Temperature of lubricant in device used to transmit engine power to auxiliary equipment.
Parameter Data Length: 1 Character
Data Type: Unsigned Short Integer
Bit Resolution: 0.667 °C (1.2 °F)
Maximum Range: -17.8 to +152.2 °C (0.0 to 306.0 °F)
Transmission Update Period: 1.0 s
Message Priority: 5
Format:
   PID Data
   90   a
       a - Power takeoff oil temperature

A.91 Percent Accelerator Pedal Position—Ratio of actual accelerator pedal position to maximum pedal position.
Parameter Data Length: 1 Character
Data Type: Unsigned Short Integer
Bit Resolution: 0.4%
Maximum Range: 0.0 to 102.0%
Transmission Update Period: 0.1 S
Message Priority: 3
Format:
   PID Data
   91   a
       a - Percent accelerator pedal position

A.92 Percent Engine Load—Ratio of current output torque to maximum torque available at the current engine speed.
Parameter Data Length: 1 Character
Data Type: Unsigned Short Integer
Bit Resolution: 0.5%
Maximum Range: 0.0 to 127.5%
Transmission Update Period: 0.1 s
Message Priority: 3
Format:
   PID Data
   92   a
       a - Percent engine load

A.93 Output Torque—Amount of torque available at the engine flywheel.
Parameter Data Length: 1 Character
Data Type: Signed short Integer
Bit Resolution: 27.1 Nm (20 lbf/ft)
Maximum Range: -3471 to +3444 Nm (-2560 to +2540 lbf/ft)
Transmission Update Period: 1.0 s
Message Priority: 5
Format:
   PID Data
   93   a
       a - Output torque

A.94 Fuel Delivery Pressure—Gage pressure of fuel in system as delivered from supply pump to the injection pump.
Parameter Data Length: 1 Character
Data Type: Unsigned Short Integer
Bit Resolution: 3.45 kPa (0.5 lbf/in^2)
Maximum Range: 0.0 to 879.0 kPa (0.0 to 127.5 lbf/in^2)
Transmission Update Period: 1.0 s
Message Priority: 4
Format:
   PID Data
   94   a
       a - Fuel delivery pressure

A.95 Fuel Filter Differential Pressure—Change in fuel delivery pressure, measured after the filter, due to accumulation of solid or semisolid matter on the filter element.
Parameter Data Length: 1 Character

Data Type: Unsigned Short Integer
Bit Resolution: 1.724 kPa (0.25 lbf/in^2)
Maximum Range: 0.0 to 439.5 kPa (0.0 to 63.75 lbf/in^2)
Transmission Update Period: 10.0 s
Message Priority: 7
Format:
PID Data
95 a
a - Fuel filter differential pressure

A.96 Fuel Level—Ratio of volume of fuel to the total volume of fuel storage container.
Parameter Data Length: 1 Character
Data Type: Unsigned Short Integer
Bit Resolution: 0.5%
Maximum Range: 0.0 to 127.5%
Transmission Update Period: 10.0 s
Message Priority: 6
Format:
PID Data
96 a
a - Fuel level

A.97 Water in Fuel Indicator—Indication (yes/no) of presence of unacceptable amount of water in fuel system.
Parameter Data Length: 1 Character
Data Type: Binary Bit-Mapped
Resolution: Binary
Maximum Range: 0 to 255
Transmission Update Period: 10.0 s
Message Priority: 7
Format:
PID Data
97 a
a - Water in fuel indicator
   Bit 8: 1=yes / 0=no
   Bits 1 to 7: undefined

A.98 Engine Oil Level—Ratio of current volume of engine sump oil to maximum required volume.
Parameter Data Length: 1 Character
Data Type: Unsigned Short Integer
Bit Resolution: 0.5%
Maximum Range: 0.0 to 127.5%
Transmission Update Period: 10.0 s
Message Priority: 6
Format:
PID Data
98 a
a - Engine oil level

A.99 Engine Oil Filter Differential Pressure—Change in engine oil pressure, measured after filter, due to accumulation of solid or semisolid material on or in the filter.
Parameter Data Length: 1 Character
Data Type: Unsigned short Integer
Bit Resolution: 0.431 kPa (0.0625 lbf/in^2)
Maximum Range: 0.0 to 109.9 kPa (0.0 to 15.9375 lbf/in^2)
Transmission Update Period: 10.0 s
Message Priority: 6
Format:
PID Data
99 a
a - Oil filter differential pressure

A.100 Engine Oil Pressure—Gage pressure of oil in engine lubrication system as provided by oil pump.
Parameter Data Length: 1 Character
Data Type: Unsigned Short Integer
Bit Resolution: 3.45 kPa (0.5 lbf/in^2)
Maximum Range: 0.0 to 879.0 kPa (0.0 to 127.5 lbf/in^2)
Transmission Update Period: 1.0 s
Message Priority: 2
Format:
PID Data
100 a
a - Engine oil pressure

A.101 Crankcase Pressure—Gage air pressure inside engine crankcase.
Parameter Data Length: 1 Character
Data Type: Signed Short Integer
Bit Resolution: 0.862 kPa (0.125 lbf/in^2)
Maximum Range: -110.0 to +109.5 kPa (-16.00 to +15.875 lbf/in^2)
Transmission Update Period: 1.0 s
Message Priority: 4
Format:
PID Data
101 a
a - Crankcase pressure

A.102 Boost Pressure—Gage pressure of air measured downstream on the compressor discharge side of the turbocharger.
Parameter Data Length: 1 Character
Data Type: Unsigned Short Integer
Bit Resolution: 0.862 kPa (0.125 lbf/in^2)
Maximum Range: 0.0 to 219.8 kPa (0.0 to 31.875 lbf/in^2)
Transmission Update Period: 1.0 s
Message Priority: 4
Format:
PID Data
102 a
a - Boost pressure

A.103 Turbo Speed—Rotational velocity of rotor in turbocharger.
Parameter Data Length: 1 Character
Data Type: Unsigned Short Integer
Bit Resolution: 500 rpm
Maximum Range: 0 to 127 500 rpm
Transmission Update Period: 1.0 s
Message Priority: 4
Format:
PID Data
103 a
a - Turbo speed

A.104 Turbo Oil Pressure—Gage pressure of oil in turbocharger lubrication system.
Parameter Data Length: 1 Character
Data Type: Unsigned Short Integer
Bit Resolution: 4.14 kPa (0.6 lbf/in^2)
Maximum Range: 0.0 to 1055 kPa (0.0 to 153 lbf/in^2)
Transmission Update Period: 1.0 s
Message Priority: 4
Format:
PID Data
104 a
a - Turbo oil pressure

A.105 Intake Manifold Temperature—Temperature of precombustion air found in intake manifold of engine air supply system.
Parameter Data Length: 1 Character
Data Type: Unsigned short Integer
Bit Resolution: 0.556 °C (1.0°F)
Maximum Range: -17.8 to +123.9 °C (0.0 to 255.0 °F)
Transmission Update Period: 1.0 s
Message Priority: 5
Format:
PID Data
105 a
a - Intake manifold temperature

A.106 Air Inlet Pressure—Absolute air pressure at inlet to intake manifold or air box.
Parameter Data Length: 1 Character
Data Type: Unsigned Short Integer
Bit Resolution: 1.724 kPa (0.25 lbf/in^2)
Maximum Range: 0.0 to 439.5 kPa (0.0 to 63.75 lbf/in^2)
Transmission Update Period: 1.0 s
Message Priority: 5
Format:
PID Data
106 a
a - Air inlet pressure

A.107 Air Filter Differential Pressure—Change in engine air system pressure, measured after the filter, due to accumulation of solid foreign matter on or in the filter.
    Parameter Data Length: 1 Character
    Data Type: Unsigned Short Integer
    Bit Resolution: 0.0498 kPa (0.2 in H_2O)
    Maximum Range: 0.0 to 12.7 kPa (0.0 to 51.0 in H_2O)
    Transmission Update Period: 10.0 s
    Message Priority: 7
    Format:
        PID Data
        107 a
        a - Air filter differential pressure

A.108 Barometric Pressure—Absolute air pressure of the atmosphere.
    Parameter Data Length: 1 Character
    Data Type: Unsigned Short Integer
    Bit Resolution: 0.431 kPa (0.0625 lbf/in^2)
    Maximum Range: 0.0 to 109.9 kPa (0.0 to 15.9375 lbf/in^2)
    Transmission Update Period: 1.0 s
    Message Priority: 5
    Format:
        PID Data
        108 a
        a - Barometric pressure

A.109 Coolant Pressure—The gage pressure of liquid found in engine cooling system.
    Parameter Data Length: 1 Character
    Data Type: Unsigned Short Integer
    Bit Resolution: 0.862 kPa (0.125 lbf/in^2)
    Maximum Range: 0.0 to 219.8 kPa (0.0 to 31.875 lbf/in^2)
    Transmission Update Period: 10.0 s
    Message Priority: 6
    Format:
        PID Data
        109 a
        a - Coolant pressure

A.110 Engine Coolant Temperature—The temperature of liquid found in engine cooling system.
    Parameter Data Length: 1 Character
    Data Type: Unsigned Short Integer
    Bit Resolution: 0.556 °C (1.0 °F)
    Maximum Range: -17.8 to +123.9 °C (0.0 to 255.0 °F)
    Transmission Update Period: 1.0 s
    Message Priority: 4
    Format:
        PID Data
        110 a
        a - Engine coolant temperature

A.111 Coolant Level—Ratio of volume of liquid found in engine cooling system to total cooling system volume.
    Parameter Data Length: 1 Character
    Data Type: Unsigned Short Integer
    Bit Resolution: 0.5%
    Maximum Range: 0.0 to 127.5%
    Transmission Update Period: 10.0 s
    Message Priority: 7
    Format:
        PID Data
        111 a
        a - Coolant level

A.112 Coolant Filter Differential Pressure—Change in coolant pressure, measured after the filter, due to accumulation of solid or semisolid matter on or in the filter.
    Parameter Data Length: 1 Character
    Data Type: Unsigned Short Integer
    Bit Resolution: 0.431 kPa (0.0625 lbf/in^2)
    Maximum Range: 0.0 to 109.9 kPa (0.0 to 15.9375 lbf/in^2)
    Transmission Update Period: 10.0 s
    Message Priority: 6
    Format:
        PID Data
        112 a
        a - Coolant filter differential pressure

A.113 Governor Droop—The difference between full load rated engine speed and maximum no-load governed engine speed.
    Parameter Data Length: 1 Character
    Data Type: Unsigned Short Integer
    Bit Resolution: 2.0 rpm
    Maximum Range: 0.0 to 510.0 rpm
    Transmission Update Period: On request
    Message Priority: 8
    Format:
        PID Data
        113 a
        a - Governor droop

A.114 Net Battery Current—Net flow of electrical current into/out of the battery or batteries.
    Parameter Data Length: 1 Character
    Data Type: Unsigned Short Integer
    Bit Resolution: 1.2 A
    Maximum Range: -153.6 to +152.0 A
    Transmission Update Period: 1.0 s
    Message Priority: 5
    Format:
        PID Data
        114 a
        a - Net battery current

A.115 Alternator Current—Measure of electrical flow from the alternator.
    Parameter Data Length: 1 Character
    Data Type: Unsigned Short Integer
    Bit Resolution: 1.2 A
    Maximum Range: 0.0 to 306 A
    Transmission Update Period: 1.0 s
    Message Priority: 5
    Format:
        PID Data
        115 a
        a - Alternator current

A.116 Brake Application Pressure—Gage pressure of compressed air or fluid in vehicle braking system measured at the brake chamber when brake shoe (or pad) is placed against brake drum (or disc).
    Parameter Data Length: 1 Character
    Data Type: Unsigned Short Integer
    Bit Resolution: 4.14 kPa (0.6 lbf/in^2)
    Maximum Range: 0.0 to 1055 kPa (0.0 to 153.0 lbf/in^2)
    Transmission Update Period: 0.2 s
    Message Priority: 1
    Format:
        PID Data
        116 a
        a - Brake application pressure

A.117 Brake Primary Pressure—Gage pressure of air in the primary, or supply side, of the air brake system.
    Parameter Data Length: 1 Character
    Data Type: Unsigned Short Integer
    Bit Resolution: 4.14 kPa (0.6 lbf/in^2)
    Maximum Range: 0.0 to 1055 kPa (0.0 to 153.0 lbf/in^2)
    Transmission Update Period: 1.0 s
    Message Priority: 1
    Format:
        PID Data
        117 a
        a - Brake primary pressure

A.118 Brake Secondary Pressure—Gage pressure of air in the secondary, or service side, of the air brake system.
    Parameter Data Length: 1 Character
    Data Type: Unsigned Short Integer
    Bit Resolution: 4.14 kPa (0.6 lbf/in^2)
    Maximum Range: 0.0 to 1055 kPa (0.0 to 153.0 lbf/in^2)
    Transmission Update Period: 1.0 s
    Message Priority: 1
    Format:
        PID Data
        118 a
        a - Brake secondary pressure

A.119 Hydraulic Retarder Pressure—Gage pressure of oil in hydraulic retarder system.
    Parameter Data Length: 1 Character
    Data Type: Unsigned Short Integer
    Bit Resolution: 4.14 kPa (0.6 lbf/in^2)
    Maximum Range: 0.0 to 1055 kPa (0.0 to 153.0 lbf/in^2)
    Transmission Update Period: 1.0 s
    Message Priority: 5
    Format:
        PID Data
        119 a
        a - Hydraulic retarder pressure

A.120 Hydraulic Retarder Oil Temperature
    Parameter Data Length: 1 Character
    Data Type: Unsigned Short Integer
    Bit Resolution: 1.11 °C (2 °F)
    Maximum Range: -17.8 to 265 °C (0.0 to 510°F)
    Transmission Update Period: 1.0 s
    Message Priority: 5
    Format:
        PID Data
        120 a
        a - Hydraulic retarder oil temperature

A.121 Engine Retarder Status—State of device used to convert engine power to vehicle retarding (stopping) force.
    Parameter Data Length: 1 Character
    Data Type: Binary Bit-Mapped
    Resolution: Binary
    Maximum Range: 0 to 255
    Transmission Update Period: 0.2 s
    Message Priority: 3
    Format:
        PID Data
        121 a
        a - Engine retarder status
            Bit 8: 1=on / 0=off
            Bit 7: undefined
            Bit 6: undefined
            Bit 5: 1=8 cylinder active / 0=8 cylinder not active
            Bit 4: 1=6 cylinder active / 0=6 cylinder not active
            Bit 3: 1=4 cylinder active / 0=4 cylinder not active
            Bit 2: 1=3 cylinder active / 0=3 cylinder not active
            Bit 1: 1=2 cylinder active / 0=2 cylinder not active

A.122 Engine Retarder Percent—Ratio of current engine retard force to maximum retard force available.
    Parameter Data Length: 1 Character
    Data Type: Unsigned Short Integer
    Bit Resolution: 0.5%
    Maximum Range: 0.0 to 127.5%
    Transmission Update Period: 1.0 s
    Message Priority: 5
    Format:
        PID Data
        122 a
        a - Engine retarder percent

A.123 Clutch Pressure—Gage pressure of oil within a wet clutch.
    Parameter Data Length: 1 Character
    Data Type: Unsigned Short Integer
    Bit Resolution: 13.8 kPa (2.0 lbf/in^2)
    Maximum Range: 0.0 to 3516 kPa (0.0 to 510.0 lbf/in^2)
    Transmission Update Period: 1.0 s
    Message Priority: 4
    Format:
        PID Data
        123 a
        a - Clutch pressure

A.124 Transmission Oil Level—Ratio of volume of transmission sump oil to recommended volume.
    Parameter Data Length: 1 Character
    Data Type: Unsigned Short Integer
    Bit Resolution: 0.5%
    Maximum Range: 0.0 to 127.5%
    Transmission Update Period: 10.0 s
    Message Priority: 7
    Format:
        PID Data
        124 a
        a - Transmission oil level

A.125 Transmission Oil Level High/Low—Amount of current volume of transmission sump oil compared to recommended volume.
    Parameter Data Length: 1 Character
    Data Type: Unsigned Short Integer
    Bit Resolution: 0.473 L (1.0 pt)
    Maximum Range: -60.6 to 60.1 L (-128 to +127 pt)
    Transmission Update Period: 10.0 s
    Message Priority: 6
    Format:
        PID Data
        125 a
        a - Transmission oil level High/Low

A.126 Transmission Filter Differential Pressure—Change in transmission fluid pressure, measured after the filter, due to accumulation of solid or semisolid material on or in the filter.
    Parameter Data Length: 1 Character
    Data Type: Unsigned Short Integer
    Bit Resolution: 1.724 kPa (0.25 lbf/in^2)
    Maximum Range: 0.0 to 439.5 kPa (0.0 to 63.75 lbf/in^2)
    Transmission Update Period: 10.0 s
    Message Priority: 7
    Format:
        PID Data
        126 a
        a - Transmission filter differential pressure

A.127 Transmission Oil Pressure—Gage pressure of lubrication fluid in transmission, measured after pump.
    Parameter Data Length: 1 Character
    Data Type: Unsigned Short Integer
    Bit Resolution: 13.8 kPa (2.0 lbf/in^2)
    Maximum Range: 0.0 to 3516 kPa (0.0 to 510.0 lbf/in^2)
    Transmission Update Period: 1.0 s
    Message Priority: 4
    Format:
        PID Data
        127 a
        a - Transmission oil pressure

A.128 Component Specific Parameter Request—Used to request parameter data transmissions from a specified component on the data link.
    Parameter Data Length: 2 Characters
    Data Type: Unsigned Short Integer (both characters)
    Resolution: Binary (both characters)
    Maximum Range: 0 to 255 (both characters)
    Transmission Update Period: As needed
    Message Priority: 8
    Format:
        PID Data
        128 a b
        a - Parameter number of the requested parameter
        b - MID of the component from which the parameter data is requested

Only the specified component should transmit the specified parameter. If the specified component is in the MID range 0 to 127, its response is not defined in this document.

A.129 to A.154 Reserved—To be assigned.

A.155 Auxiliary Input and Output Status—Identifies the current status of auxiliary input and output functions that are configured uniquely per application. Not to be used in place of existing PIDs.
    Parameter Data Length: 2 Characters
    Data Type: Binary Bit-Mapped
    Bit Resolution: Binary
    Maximum Range: 0 to 65535
    Transmission Update Period: On request
    Message Priority: 8
    Format:
        PID Data

155 a b
- a - Auxiliary input status
  - Bits 1-2: Auxiliary Input #1
  - Bits 3-4: Auxiliary Input #2
  - Bits 5-6: Auxiliary Input #3
  - Bits 7-8: Auxiliary Input #4
- b - Auxiliary output status
  - Bits 1-2: Auxiliary Output #1
  - Bits 3-4: Auxiliary Output #2
  - Bits 5-6: Auxiliary Output #3
  - Bits 7-8: Auxiliary Output #4

NOTE: Each status will be described using the following nomenclature:
- 00 Off
- 01 On
- 10 Error condition
- 11 Not available

A.156 Injector Timing Rail Pressure—The gage pressure of fuel in the timing rail as delivered from the supply pump to the injector timing inlet.
Parameter Data Length: 2 Characters
Data Type: Unsigned Integer
Bit Resolution: 0.689 kPa (0.1 lbf/in^2)
Maximum Range: 0.0 to 45 153.6 kPa (0.0 to 6553.5 lbf/in^2)
Transmission Update Period: 1.0 s
Message Priority: 4
Format:
  PID Data
  156 a a
  a a - Injector timing rail pressure

A.157 Injector Metering Rail Pressure—The gage pressure of fuel in the metering rail as delivered from the supply pump to the injector metering inlet.
Parameter Data Length: 2 Characters
Data Type: Unsigned Integer
Bit Resolution: 0.689 kPa (0.1 lbf/in^2)
Maximum Range: 0.0 to 45 153.6 kPa (0.0 to 6553.5 lbf/in^2)
Transmission Update Period: 1.0 s
Message Priority: 4
Format:
  PID Data
  157 a a
  a a - Injector metering rail pressure

A.158 Battery Potential (Voltage)— Switched—Electrical potential measured at the input of the electronic control unit supplied through a switching device.
Parameter Data Length: 2 Characters
Data Type: Unsigned Integer
Bit Resolution: 0.05 V
Maximum Range: 0.0 to 3276.75 V
Transmission Update Period: On request
Message Priority: 8
Format:
  PID Data
  158 a a
  a a - Battery potential (voltage)—switched

A.159 Gas Supply Pressure—Gas supply pressure (gage) to fuel metering device.
Parameter Data Length: 2 Characters
Data Type: Unsigned Integer
Bit Resolution: 0.345 kPa (0.05 lbf/in^2)
Maximum Range: 0.0 to 22 609.6 kPa (0.0 to 3276.75 lbf/in^2)
Transmission Update Period: 1.0 s
Message Priority: 4
Format:
  PID Data
  159 a a
  a a - Gas supply pressure

A.160 Main Shaft Speed—Rotational velocity of the first intermediate shaft of the transmission.
Parameter Data Length: 2 Characters
Data Type: Unsigned Integer
Bit Resolution: 0.25 rpm
Maximum Range: 0.0 to 16383.75 rpm
Transmission Update Period: On request
Message Priority: 2
Format:
  PID Data
  160 a a
  a a - Main shaft speed

A.161 Input Shaft Speed—Rotational velocity of the primary shaft transferring power into the transmission. When a torque converter is present, it is the output of the torque converter.
Parameter Data Length: 2 Characters
Data Type: Unsigned Integer
Bit Resolution: 0.25 rpm
Maximum Range: 0.0 to 16383.75 rpm
Transmission Update Period: On request
Message Priority: 2
Format:
  PID Data
  161 a a
  a a - Input shaft speed

A.162 Transmission Range Selected—Range selected by the operator. Characters may include P, R2, R1, R, N, D, D1, D2, L, L1, L2, 1, 2, 3, ... If only one character is required, the second character shall be used and the first character shall be a space (ASCII 32).
Parameter Data Length: 2 Characters
Data Type: Alphanumeric
Resolution: ASCII
Maximum Range: 0 to 255 (each character)
Transmission Update Period: 0.5 s
Message Priority: 4
Format:
  PID Data
  162 a a
  a a - Transmission range selected

A.163 Transmission Range Attained—Range currently being commanded by the transmission control system. Characters may include P, R2, R1, R, N, D, D1, D2, L, L1, L2, 1, 2, 3, ... If only one character is required, the second character shall be used and the first character shall be a space (ASCII 32).
Parameter Data Length: 2 Characters
Data Type: Alphanumeric
Resolution: ASCII
Maximum Range: 0 to 255 (each character)
Transmission Update Period: 0.5 s
Message Priority: 4
Format:
  PID Data
  163 a a
  a a - Transmission range attained

A.164 Injection Control Pressure—The gage pressure of the hydraulic accumulator that powers fuel injection.
Parameter Data Length: 2 Characters
Data Type: Unsigned Integer
Bit Resolution: 1/256 MPa
Maximum Range: 0 to 255.996 MPa
Transmission Update Period: 1.0 s
Message Priority: 5
Format:
  PID Data
  164 a a
  a a - Injection control pressure

A.165 Compass Bearing—Present compass bearing of vehicle
Parameter Data Length: 2 Characters
Data Type: Unsigned Integer
Bit Resolution: 0.01 degree
Maximum Range: 0.00 to 655.35 degree
Transmission Update Period: On request
Message Priority: 6
Format:
  PID Data
  165 a a
  a a - Present compass bearing

A.166 Rated Engine Power—Net brake power that the engine will deliver continuously, specified for a given application at a rated speed.
Parameter Data Length: 2 Characters
Data Type: Unsigned Integer
Bit Resolution: 0.745 kW (1.0 hp)

Maximum Range: 0.0 to 48 869.4 kW (0.0 to 65 535.0 hp)
Transmission Update Period: On request
Message Priority: 8
Format:
    PID Data
    166 a a
    a a - Rated engine power

A.167 Alternator Potential (Voltage
Parameter Data Length: 2 Characters
Data Type: Unsigned Integer
Bit Resolution: 0.05 V
Maximum Range: 0.0 to 3276.75 V
Transmission Update Period: 1.0 s
Message Priority: 5
Format:
    PID Data
    167 a a
    a a - Alternator potential

A.168 Battery Potential (Voltage)—Measured electrical potential of the battery.
Parameter Data Length: 2 Characters
Data Type: Unsigned Integer
Bit Resolution: 0.05 V
Maximum Range: 0.0 to 3276.75 V
Transmission Update Period: 1.0 s
Message Priority: 5
Format:
    PID Data
    168 a a
    a a - battery potential (voltage)

A.169 Cargo Ambient Temperature—Temperature of air inside vehicle container used to accommodate cargo.
Parameter Data Length: 2 Characters
Data Type: Signed Integer
Bit Resolution: 0.139 °C (0.25 °F)
Maximum Range: -4568.9 to +4533.2 °C (-8192.00 to +8191.75 °F)
Transmission Update Period: 10.0 s
Message Priority: 6
Format:
    PID Data
    169 a a
    a a - Cargo ambient temperature

A.170 Cab Interior Temperature—Temperature of air inside the part of the vehicle that encloses the driver and vehicle operating controls.
Parameter Data Length: 2 Characters
Data Type: Signed Integer
Bit Resolution: 0.139 °C (0.25 °F)
Maximum Range: -4568.9 to +4533.2 °C (-8192.00 to +8191.75 °F)
Transmission Update Period: 10.0 s
Message Priority: 7
Format:
    PID Data
    170 a a
    a a - Cab interior temperature

A.171 Ambient Air Temperature—Temperature of air surrounding vehicle.
Parameter Data Length: 2 Characters
Data Type: Signed Integer
Bit Resolution: 0.139 °C (0.25 °F)
Maximum Range: -4568.9° to +4533.2 °C (-8192.00 to +8191.75 °F)
Transmission Update Period: 10.0 s
Message Priority: 7
Format:
    PID Data
    171 a a
    a a - Ambient air temperature

A.172 Air Inlet Temperature—Temperature of air entering vehicle air induction system.
Parameter Data Length: 2 Characters
Data Type: Signed Integer
Bit Resolution: 0.139 °C (0.25 °F)
Maximum Range: -4568.9 to 4533.2 °C (-8192.00 to +8191.75 °F)
Transmission Update Period: 1.0 s
Message Priority: 5
Format:
    PID Data
    172 a a
    a a - Air inlet temperature

A.173 Exhaust Gas Temperature—Temperature of combustion by-products leaving the engine.
Parameter Data Length: 2 Characters
Data Type: Signed Integer
Bit Resolution: 0.139 °C (0.25 °F)
Maximum Range: -4568.9 to +4533.2 °C (-8192.00 to +8191.75 °F)
Transmission Update Period: 1.0 s
Message Priority: 5
Format:
    PID Data
    173 a a
    a a - Exhaust gas temperature

A.174 Fuel Temperature—Temperature of fuel entering injectors.
Parameter Data Length: 2 Characters
Data Type: Signed Integer
Bit Resolution: 0.139 °C (0.25 °F)
Maximum Range: -4568.9 to +4533.2 °C (-8192.00 to +8191.75 °F)
Transmission Update Period: 1.0 s
Message Priority: 4
Format:
    PID Data
    174 a a
    a a - Fuel temperature

A.175 Engine Oil Temperature—Temperature of engine lubricant.
Parameter Data Length: 2 Characters
Data Type: Signed Integer
Bit Resolution: 0.139 °C (0.25°F)
Maximum Range: -4568.9 to +4533.2 °C (-8192.00 to +8191.75 °F)
Transmission Update Period: 1.0 s
Message Priority: 4
Format:
    PID Data
    175 a a
    a a - Engine oil temperature

A.176 Turbo Oil Temperature—Temperature of turbocharger lubricant.
Parameter Data Length: 2 Characters
Data Type: Signed Integer
Bit Resolution: 0.139 °C (0.25 °F)
Maximum Range: -4568.9 to +4533.2 °C (-8192.00 to +8191.75 °F)
Transmission Update Period: 1.0 s
Message Priority: 4
Format:
    PID Data
    176 a a
    a a - Turbo oil temperature

A.177 Transmission Oil Temperature—Temperature of transmission lubricant.
Parameter Data Length: 2 Characters
Data Type: Signed Integer
Bit Resolution: 0.139 °C (0.25 °F)
Maximum Range: -4568.9 to +4533.2 °C (-8192.00 to +8191.75 °F)
Transmission Update Period: 1.0 s
Message Priority: 4
Format:
    PID Data
    177 a a
    a a - Transmission oil temperature

A.178 Front Axle Weight—Total force of gravity imposed by the front tires on the road surface.
Parameter Data Length: 2 Characters
Data Type: Unsigned Integer
Bit Resolution: 4.448 N (1.0 lbf)
Maximum Range: 0.0 to 291 514.2 N (0.0 to 65 535.0 lbf)
Transmission Update Period: On request
Message Priority: 8
Format:
    PID Data

178 a a
a a - Front axle weight

A.179 Rear Axle Weight—Force of gravity imposed on the road surface by all the tires on each individual rear axle.
Parameter Data Length: 2 Characters
Data Type: Unsigned Integer
Bit Resolution: 4.448 N (1.0 lbf)
Maximum Range: 0.0 to 291 514.2 N (0.0 to 65 535.0 lbf)
Transmission Update Period: On request
Message Priority: 8
Format:
PID Data
179 a a
a a - Rear axle weight

A.180 Trailer Weight—Total force of gravity of freight-carrying vehicle designed to be pulled by truck, including the weight of the contents.
Parameter Data Length: 2 Characters
Data Type: Unsigned Integer
Bit Resolution: 17.792 N (4.0 lbf)
Maximum Range: 0.0 to 1 166 056.9 N (0.0 to 262 140.0 lbf)
Transmission Update Period: On request
Message Priority: 8
Format:
PID Data
180 a a
a a - Trailer weight

A.181 Cargo Weight—The force of gravity of freight carried.
Parameter Data Length: 2 Characters
Data Type: Unsigned Integer
Bit Resolution: 17.792 N (4.0 lbf)
Maximum Range: 0.0 to 1 166 056.9 N (0.0 to 262 140.0 lbf)
Transmission Update Period: On request
Message Priority: 8
Format:
PID Data
181 a a
a a - Cargo weight

A.182 Trip Fuel—Fuel consumed during all or part of a journey.
Parameter Data Length: 2 Characters
Data Type: Unsigned Integer
Bit Resolution: 0.473 L (0.125 gal)
Maximum Range: 0.0 to 31 009.6 L (0.0 to 8191.875 gal)
Transmission Update Period: 10.0 s
Message Priority: 7
Format:
PID Data
182 a a
a a - Trip fuel

A.183 Fuel Rate—Amount of fuel consumed by engine per unit of time.
Parameter Data Length: 2 Characters
Data Type: Unsigned Integer
Bit Resolution: 16.428×10^{-6} L/s (4.34×10^{-6} gal/s)
Maximum Range: 0.0 to 1.076 65 L/s (0.0 to 0.284 421 90 gal/s)
Transmission Update Period: 0.2 s
Message Priority: 3
Format:
PID Data
183 a a
a a - Fuel rate

A.184 Instantaneous Fuel Economy—Current fuel economy at current vehicle velocity.
Parameter Data Length: 2 Characters
Data Type: Unsigned Integer
Bit Resolution: $1.660\ 72 \times 10^{-3}$ km/L (1/256 mpg)
Maximum Range: 0.0 to 108.835 km/L (0.0 to 255.996 mpg)
Transmission Update Period: 0.2 s
Message Priority: 3
Format:
PID Data
184 a a
a a - Instantaneous fuel economy

A.185 Average Fuel Economy—Average of instantaneous fuel economy for that segment of vehicle operation of interest.
Parameter Data Length: 2 Characters
Data Type: Unsigned Integer
Bit Resolution: $1.660\ 72 \times 10^{-3}$ km/L (1/256 mpg)
Maximum Range: 108.835 km/L (0.0 to 255.996 mpg)
Transmission Update Period: 10.0 s
Message Priority: 7
Format:
PID Data
185 a a
a a - Average fuel economy

A.186 Power Takeoff Speed—Rotational velocity of device used to transmit engine power to auxiliary equipment.
Parameter Data Length: 2 Characters
Data Type: Unsigned Integer
Bit Resolution: 0.25 rpm
Maximum Range: 0.0 to 16383.75 rpm
Transmission Update Period: 0.1 s
Message Priority: 2
Format:
PID Data
186 a a
a a - Power takeoff speed

A.187 Power Takeoff Set Speed—Rotational velocity selected by operator for device used to transmit engine power to auxiliary equipment.
Parameter Data Length: 2 Characters
Data Type: Unsigned Integer
Bit Resolution: 0.25 rpm
Maximum Range: 0.0 to 16383.75 rpm
Transmission Update Period: 10.0 s
Message Priority: 6
Format:
PID Data
187 a a
a a - Power takeoff set speed

A.188 Idle Engine Speed—Minimum nontransient rotational velocity of crankshaft while engine is supplying power to itself and its attendant support systems.
Parameter Data Length: 2 Characters
Data Type: Unsigned Integer
Bit Resolution: 0.25 rpm
Maximum Range: 0.0 to 16383.75 rpm
Transmission Update Period: On request
Message Priority: 8
Format:
PID Data
188 a a
a a - Idle engine speed

A.189 Rated Engine Speed—The maximum governed rotational velocity of the engine crankshaft under full load conditions.
Parameter Data Length: 2 Characters
Data Type: Unsigned Integer
Bit Resolution: 0.25 rpm
Maximum Range: 0.0 to 16383.75 rpm
Transmission Update Period: On request
Message Priority: 8
Format:
PID Data
189 a a
a a - Rated engine speed

A.190 Engine Speed—Rotational velocity of crankshaft.
Parameter Data Length: 2 Characters
Data Type: Unsigned Integer
Bit Resolution: 0.25 rpm
Maximum Range: 0.0 to 16383.75 rpm
Transmission Update Period: 0.1 s
Message Priority: 1
Format:
PID Data
190 a a
a a - Engine speed

A.191 Transmission Output Shaft Speed—Rotational velocity of shaft transferring force from transmission to driveshaft.

    Parameter Data Length: 2 Characters
    Data Type: Unsigned Integer
    Bit Resolution: 0.25 rpm
    Maximum Range: 0.0 to 16383.75 rpm
    Transmission Update Period: 0.1 s
    Message Priority: 2
    Format:
        PID Data
        191 a a
        a a - Transmission output shaft speed

A.192 Multisection Parameter—Used to transmit parameters that are longer than what is limited by SAE J1708. A specified parameter can be broken into sections with each section being transmitted in a different message.

    Parameter Data Length: Variable
    Data Type: Defined by specified sectioned parameter
    Resolution: Defined by specified sectioned parameter
    Maximum Range: Defined by specified sectioned parameter
    Transmission Update Period: Defined by specified sectioned parameter
    Message Priority: Parameter specific
    Format:
        PID Data
        192 n, a, b, c/d, c, c, c, c, c, c, c
        n- Byte count of data that follows this character. This excludes characters MID, PID 192, and n, but it includes a , b, c, or d type characters.
        a- PID specifying parameter that has been selected.
        b- The last section number (total number of sections minus ONE) and the current section number. The upper nibble contains the last section number (1 to 15). The lower nibble contains the current section number and is limited to the range 0 to 15. Section numbers are assigned in ascending order.
        c- Data portion of sectioned parameters. May be 1 to 14 characters in the first packet. May be 1 to 15 characters in the middle and ending packets.
        d- Byte count of the total data portion. This character is sent only in the first packet. The values are limited to 239 or less but must be greater than 17.

Application Notes -
1. Single sections of data are not allowed to be sent alone. Message packets must be sent in sequence from the transmitting device.
2. Receiver devices should have the capacity to receive concurrent PID 192 type messages from different transmitters.
3. Caution must be taken in interpreting data. The value of a parameter with multiple sections may have been updated during the time between which the packets are sent.
4. Other PID's and associated parameters can be incorporated in the message packet if character count limitations are not violated.

A.193 Transmitter System Diagnostic Table—Used to notify other components on the data link of the diagnostic condition of the transmitting electronic component. The parameter contains a list of diagnostic codes.

    Parameter Data Length: Variable
    Data Type: Defined by manufacturer application document
    Resolution: Defined by manufacturer application document
    Maximum Range: Defined by manufacturer application document
    Transmission Update Period: Defined in application document
    Message Priority: 8
    Format:
        PID Data
        193 n a a a a a a a
        n- Byte count of data that follows this character
        a- Diagnostic codes defined by the component manufacturer in an application document.

The SAE Truck and Bus Data Format Subcommittee established PIDs 194 to 196 in May 1988; therefore, this Parameter ID should no longer be used by manufacturers in the design of new components. However, this parameter is being reserved for use by manufacturers who have developed systems prior to January 1989 and are, therefore, unable to accommodate the new diagnostic formats as defined in PIDs 194 to 196. It is recommended that manufacturers using this parameter fully define the contents and circumstances under which it is used in the application document.

A.194 Transmitter System Diagnostic Code and Occurrence Count Table—Used to notify other components on the data link of the diagnostic condition of the transmitting electronic component. The parameter contains a list of diagnostic codes and occurrence counts.

    Parameter Data Length: Variable
    Data Type: Binary Bit-Mapped
    Resolution: Binary
    Maximum Range: 0 to 255
    Transmission Update Period: The diagnostic code is transmitted once whenever the fault becomes active and once whenever the fault becomes inactive but never more than once per second. All diagnostic codes are also available on request. All active diagnostic codes are retransmitted at a rate greater than or equal to the refresh rate of the associated PID but not greater than once per second. Active diagnostic codes for on-request PIDs and SIDs are transmitted at a rate of once every 15 s.
    Message Priority: 8
    Format:
        PID Data
        194 n a b c a b c a b c a b c a b c ...
        n- Byte count of data that follows this character. This excludes characters MID, PID 194, and n but includes a, b, and c type characters.
        a- SID or PID of a standard diagnostic code.
        b- Diagnostic code character.
            Bits 1-4: Failure mode identifier (FMI) of a standard diagnostic code.
            Bit 5: Low character identifier for a standard diagnostic code
                1=low character is subsystem identifier (SID)
                0=low character is parameter identifier (PID)
            Bit 6: Type of diagnostic code
                1=standard diagnostic code
                0=reserved for expansion diagnostic codes
            Bit 7: Current Status of fault
                1=fault is inactive
                0=fault is active
            Bit 8: Occurrence Count included
                1=count is included
                0=count not included
        c- Occurrence count for the diagnostic code defined by the preceding 2 characters. The count is optional and bit 8 of the first character of the diagnostic code is used to determine if it is included.

Using the MID, FMI, and PID or SID associated with a diagnostic code, the control system which has the fault, which subsystem of the control system is failing, and how the subsystem is failing can be determined. The text used in J1587 to describe the FMIs and SIDs should be used whenever a standard diagnostic code is being described. The use of common descriptions for the FMIs and SIDs is needed to allow the diagnostic codes to be interpreted consistently. The subsystem identification assignment list is shown in Table 3. The failure mode identifier assignment list is shown in Table 4.

1. If the diagnostic code PID is requested and there are no diagnostic codes, the response would be a PID 194 with the n set to o.

2. If the length of the message would exceed the maximum message length allowable, PID 192 would be used and the data would be sent in a multisection transmission.

3. Application designers should be aware that at a later date the zero state of bit 6 of character b may be utilized to expand the number of diagnostic codes. This means that the current definition of character a and bits 1 to 5 of character b may change.

4. In the event the data is invalid, for example, the case of a shorted sensor, the PID at fault will not be broadcast. (However, a PID 194 with the offending PID will be broadcast per the above.)

Example—Normal broadcast of engine speed (PID 190) and oil pressure (PID 100) prior to oil pressure sensor failure

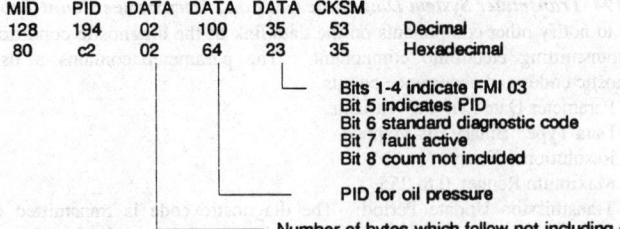

Diagnostic broadcast, Oil Pressure Sensor Shorted High

MID	PID	DATA	DATA	DATA	CKSM	
128	194	02	100	35	53	Decimal
80	c2	02	64	23	35	Hexadecimal

Next scheduled broadcast of engine speed (PID 190) oil pressure (PID 100) is not broadcast due to a failed sensor

MID	PID	DATA	DATA	CKSM	
128	190	32	28	134	Decimal
80	be	20	1c	86	Hexadecimal

A.195 Diagnostic Data Request/Clear Count—Used to request additional information about a given diagnostic code or clear its count.

Parameter Data Length: 3 Characters
Data Type: Binary Bit-Mapped
Resolution: Binary
Maximum Range: 0 to 255
Transmission Update Period: As needed
Message Priority: 8
Format:

PID Data
195 n a b c

n- Number of parameter data characters = 3
a- MID of device to which request is directed.
b- SID or PID of a standard diagnostic code.
c- Diagnostic code character
   Bits 1-4: Failure mode identifier (FMI) of a standard diagnostic code
   Bit 5: Low character identifier for a standard diagnostic code
    1=low character is subsystem identifier (SID)
    0=low character is parameter identifier (PID)
   Bit 6: Type of diagnostic code
    1=standard diagnostic code
    0=reserved for expansion diagnostic codes
   Bits 7-8: (00) -Request an ASCII descriptive message for the given SID or PID
    (01) -Request count be cleared for the given diagnostic code on the device with the given MID.
    (10) -Request counts be cleared for all diagnostic codes on the device with the given MID. The diagnostic code given in this transmission is ignored.
    (11) -Request additional diagnostic information for the given diagnostic code, the content of which is defined in a manufacturer's application document.

A.196 Diagnostic Data/Count Clear Response—Used to acknowledge the clearing of diagnostic codes or supply additional information about a diagnostic code as requested by PID 195.

Parameter Data Length: Variable
Data Type: Binary Bit-Mapped
Resolution: Binary
Maximum Range: 0 to 255
Transmission Update Period: As needed
Message Priority: 8
Format:

PID Data
196 n a b c c c c c c c c c c ....

n- Byte count of data that follows this character. This excludes characters MID, PID 196, and n, but includes a, b, c type characters.
a- SID or PID of a standard diagnostic code.
b- Diagnostic code character
   Bits 1-4: Failure mode identifier (FMI) of a standard diagnostic code.
   Bit 5: Low character identifier for a standard diagnostic code
    1=low character is subsystem identifier (SID)
    0=low character is parameter identifier (PID)
   Bit 6: Type of diagnostic code
    1=standard diagnostic code
    0=reserved for expansion diagnostic codes
   Bits 7-8: (00) -Message is an ASCII descriptive message for the given SID or PID.
    (01) -The count has been cleared for the given diagnostic code.
    (10) -All clearable diagnostic counts have been cleared for this device.
    (11) -Message is additional diagnostic information for the given diagnostic code, the content of which is defined in a manufacturer's application document.
c- If bits 7 and 8 of character b are (00), the data in field C are an ASCII string, which describes the given SID or PID. If bits 7 and 8 of character b are (11), the data in field C are defined by the manufacturer's application document with the exception that the first five characters of the data define the make of the component, which is responding. The five characters defining the make correspond to the codes defined in the American Trucking Association Vehicle Maintenance Reporting Standard (ATA/VMRS). It is suggested that spaces (ASCII 32) are used to fill the remaining characters if the ATA/VMRS make code is less than five characters in length. Data type c would be omitted if bits 7 and 8 of character b are either (01) or (10) or if no data of the type requested is available.

Application Note—If the length of the message would exceed the maximum message length allowable, PID 192 would be used and the data would be sent in a multisection transmission.

A.197 to A.234 Reserved—To be assigned.

A.235 Total Idle Hours—Accumulated time of operation of the engine while under idle conditions.

Parameter Data Length: 4 Characters
Data Type: Unsigned Long Integer
Bit Resolution: 0.05 h
Maximum Range: 0.0 to 214 748 364.8 h
Transmission Update Period: On request
Message Priority: 8
Format:

PID Data
235 n a a a a

n- Number of parameter data characters = 4
a a a a- Total idle hours

A.236 Total Idle Fuel Used—Accumulated amount of fuel used during vehicle operation while under idle conditions.

Parameter Data Length: 4 Characters
Data Type: Unsigned Long Integer
Bit Resolution: 0.473 L (0.125 gal)
Maximum Range: 0.0 to 2 032 277 476 L (0.0 to 536 870 911.9 gal)
Transmission Update Period: On request
Message Priority: 8
Format:

PID Data
236 n a a a a

n- Number of parameter data characters = 4
a a a a- Total idle fuel used

A.237 Vehicle Identification Number—Vehicle Identification Number (VIN) as assigned by the vehicle manufacturer.

Parameter Data Length: Variable
Date Type: Alphanumeric
Resolution: ASCII
Maximum Range: 0 to 255 (each character)
Transmission: On request
Message Priority: 8
Format:

PID Data
237 n a a a a

n- number of parameter data character
a- character specifying VIN

A.238 Velocity Vector—Any combination of the velocity, heading, and pitch, as calculated by the navigation device(s).
    Parameter Data Length: 5 Characters
    Data Type: Character 1 = Unsigned Short Integer
                Character 2 = Unsigned Integer
                Character 3 = Unsigned Integer
                Character 4 = Signed Integer
                Character 5 = Signed Integer
    Bit Resolution: Character 1 = 0.805 km/h (0.5 mph)
                    Character 2 = 0.01 degree/bit
                Character 3 = 0.01 degree/bit
                Character 4 = 0.01 degree/bit
                Character 5 = 0.01 degree/bit
    Maximum Range: Character 1 = -36.2 to +90.6 km/h (-45 to +112.5 mph)
                    (range is offset to acknowledge backward motion)
                    91 km/h (113 mph) indicates "Data Not Available"
                    Characters 2-3 = 0 to 655.34 degree
                    655.35 degree indicates "Data Not Available"
                    Characters 4-5 = -327.67 to +327.67 degree
                    -327.68 degree indicates "Data Not Available"
    Transmission Update Period: On request
    Message Priority: 6
    Format:

        PID   Data
        238   n a b b c c
        n-   Number of parameter data characters
        a-   Calculated vehicle speed
        b-   Present vehicle heading
        c-   Pitch, positive = ASCENT, negative = DESCENT

A.239 Position—The three-dimensional location of the vehicle.
    Parameter Data Length: 10 Characters
    Data Type: Characters 1-4 = Signed Long Integer
                    Characters 5-8 = Signed Long Integer
                Characters 9-10 = Signed Integer
    Resolution: Characters 1-4 = (10^{-6}) degree/bit
                  Characters 5-8 = (10^{-6}) degree/bit
                  Characters 9-10 = 0.15 m/bit (0.5 ft/bit)
    Maximum Range: Characters 1-4 = -2147.483 648 to +2147.483 647 degree
                    Characters 5-8 = -2147.483 648 to +2147.483 647 degree
                    Characters 9-10 = -2497 to 4993.7 m (16 384 to +16 383.5 ft)
    Transmission Update Period: On request
    Message Priority: 6
    Format:

        PID   Data
        239   n a a a a b b b b c c
        n-   Number of parameter data characters
                8 = latitude and longitude only (a a a a b b b b)
                2 = altitude only (c c)
                10 = latitude, longitude, and altitude
        a-   Latitude, positive = NORTH, negative = SOUTH
        b-   Longitude, positive = EAST, negative = WEST
        c-   Altitude referenced to sea level at standard atmospheric pressure and temperature

A.240 Change Reference Number—Used to indicate that a change has occurred in the calibration data.
    Parameter Data Length: Variable
    Data Type: Defined by manufacturer
    Resolution: Defined by manufacturer
    Maximum Range: Defined by manufacturer
    Transmission Update Period: On request
    Message Priority: 8
    Format:

        PID   Data
        240   n a a a a
        n-   Byte count of data that follows this character.
        a-   Change reference number

A.241 Tire Pressure—Pressure at which air is contained in cavity formed by tire and rim.
    Parameter Data Length: 3 Characters
    Data Type :   Character 1 = Unsigned Short Integer
                  Character 2 = Unsigned Short Integer
                  Character 3 = Unsigned Short Integer
    Resolution: Character 1 = Binary
                  Character 2 = Binary
                Character 3 = 4.14 kPa/bit (0.6 lbf/in^2/bit)
    Maximum Range: 0.0 to 1055 kPa (0.0 to 153.0 lbf/in^2)
    Transmission Update Period: 10.0s
    Message Priority: 6
    Format:

        PID   Data
        241   n a b c
        n-   Number of parameter data characters = 3
        a-   Trailer or power unit MID
        b-   Tire position = (axle number * 16) = wheel number
        c-   Tire pressure

Axle number is incremented from front to back with the front most axle being number 1. Wheel numbers on the axle are assigned as follows:
    Outer left tire = 1
    Inner left tire   = 2
    Inner right tire = 3
    Outer right tire = 4

The outer numbers are used when only one tire is on either side of an axle.

A.242 Tire Temperature—Temperature at the surface of the tire sidewall.
    Parameter Data Length: 3 Characters
    Data Type: Character 1 = Unsigned Short Integer
                  Character 2 = Unsigned Short Integer
                Character 3 = Unsigned Short Integer
    Resolution: Character 1 = Binary
                Character 2 = Binary
              Character 3 = 1.39 °C/bit (-2.5 °F/bit)
    Maximum Range: -17.8 to +366 °C (0.0 to 637.5 °F)
    Transmission Update Period: 10.0s
    Message Priority: 6
    Format:

        PID   Data
        242   n a b c
        n-   Number of parameter data characters = 3
        a-   Trailer of power unit MID
        b-   Tire position = (axle number * 16) + wheel number
        c-   Tire temperature

Axle number is incremented from front to back with the front most axle being number 1. Wheel numbers on the axle are assigned as follows:
    Outer left tire = 1
    Inner left tire   = 2
    Inner right tire = 3
    Outer right tire = 4

The outer numbers are used when only one tire is on either side of an axle.

A.243 Component Identification Parameter—Used to identify the Make, Model, and Serial Number of any component on the vehicle.
    Parameter Data Length: Variable
    Data Type: Alphanumeric
    Resolution: ASCII
    Maximum Range: 0 to 255 (each character)
    Transmission Update Period: On request
    Message Priority: 8
    Format:

        PID   Data
        243   n b c c c c c * d d d d d d d d d * e e e e e e e e e
        n-   Number of parameter data characters
        b-   MID of component being identified
        c-   Characters specifying component Make
        d-   Characters specifying component Model
        e-   Characters specifying component Serial Number

When used, the Make is five characters long and shall correspond to the codes defined in the American Trucking Association Vehicle Maintenance Reporting Standard (ATA/VMRS). It is suggested that spaces (ASCII 32) are used to fill the remaining characters if the ATA/VMRS make code is less than five characters in length. The Model and Serial Number fields are variable in length and separated by an ASCII "*". It is not necessary to include all three fields; however, the delimiter ("*") is always required.

A.244 Trip Distance—Distance traveled during all or part of a journey.
    Parameter Data Length: 4 Characters
    Data Type: Unsigned Long Integer

Bit Resolution: 0.16 km (0.1 mile)
Maximum Range: 0.0 to 691 207 984.6 km (0.0 to 429 496 729.5 mile)
Transmission Update Period: 10.0 s
Message Priority: 7
Format:
    PID Data
    244 n a a a a
    n- Number of parameter data characters = 4
    a a a a- Trip distance

A.245 Total Vehicle Distance—Accumulated distance travelled by vehicle during its operation.
Parameter Data Length: 4 Characters
Data Type: Unsigned Long Integer
Bit Resolution: 0.161 km (0.1 mile)
Maximum Range: 0.0 to 691 207 984.6 km (0.0 to 429 496 729.5 mile)
Transmission Update Period: 10.0 s
Message Priority: 7
Format:
    PID Data
    245 n a a a a
    n- Number of parameter data characters = 4
    a a a a- Total vehicle distance

A.246 Total Vehicle Hours—Accumulated time of operation of vehicle.
Parameter Data Length: 4 Characters
Data Type: Unsigned Long Integer
Bit Resolution: 0.05 h
Maximum Range: 0.0 to 214 748 364.8 h
Transmission Update Period: On request
Message Priority: 8
Format:
    PID Data
    246 n a a a a
    n- Number of parameter data characters = 4
    a a a a- Total vehicle hours

A.247 Total Engine Hours—Accumulated time of operation of engine.
Parameter Data Length: 4 Characters
Data Type: Unsigned Long Integer
Bit Resolution: 0.05 h
Maximum Range: 0.0 to 214 748 364.8 h
Transmission Update Period: On request
Message Priority: 8
Format:
    PID Data
    247 n a a a a
    n- Number of parameter data characters = 4
    a a a a- Total engine hours

A.248 Total PTO Hours—Accumulated time of operation of power takeoff device.
Parameter Data Length: 4 Characters
Data Type: Unsigned Long Integer
Bit Resolution: 0.05 h
Maximum Range: 0.0 to 214 748 364.8 h
Transmission Update Period: On request
Message Priority: 8
Format:
    PID Data
    248 n a a a a
    n- Number of parameter data characters = 4
    a a a a- Total PTO hours

A.249 Total Engine Revolutions—Accumulated number of revolutions of engine crankshaft during its operation.
Parameter Data Length: 4 Characters
Data Type: Unsigned Long Integer
Bit Resolution: 1000 r
Maximum Range: 0 to 4 294 967 295 000 r
Transmission Update Period: On request
Message Priority: 8
Format:
    PID Data
    249 n a a a a
    n- Number of parameter data characters = 4
    a a a a- Total engine revolutions

A.250 Total Fuel Used—Accumulated amount of fuel used during vehicle operation.
Parameter Data Length: 4 Characters
Data Type: Unsigned Long Integer
Bit Resolution: 0.473 L (0.125 gal)
Maximum Range: 0.0 to 2 032 277 476 L (0.0 to 536 870 911.9 gal)
Transmission Update Period: On request
Message Priority: 8
Format:
    PID Data
    250 n a a a a
    n- Number of parameter data characters = 4
    a a a a- Total fuel used

A.251 Clock—Accumulated number of revolutions of engine crankshaft during its operation.
Parameter Data Length: 3 Characters
Data Type: Each Character Unsigned Long Integer
Bit Resolution: Character 1 = 0.25 s/bit
                Character 2 = 1 min/bit
                Character 3 = 1 h/bit
Maximum Range: Character 1 = 0 to 63.75 s
                 Character 2 = 0 to 255 min
                 Character 3 = 0 to 255 h
Transmission Update Period: On request
Message Priority: 8
Format:
    PID Data
    251 n a b c
    n- Number of parameter data characters = 3
    a- Seconds
    b- Minutes
    c- Hours

A.252 Date
Parameter Data Length: 3 Characters
Data Type: Each Character Unsigned Short Integer
Resolution: Character 1 = 0.25 day/bit
              Character 2 = 1 month/bit
              Character 3 = 1 year/bit
Maximum Range: Character 1 = 0 to 63.75 day
                 Character 2 = 0 to 255 month
                 Character 3 = 0 to 255 year
Transmission Update Period: On request
Message Priority: 8
Format:
    PID Data
    252 n a b c
    n- Number of parameter data characters = 3
    a- Day
    b- Month
    c- Year - 1985

A.253 Elapsed Time
Parameter Data Length: Variable
Data Type: Each Character Unsigned Short Integer
Bit Resolution: Character 1 = 0.25 s/bit
                Character 2 = 1 min/bit
                Character 3 = 1 h/bit
                Character 4 = 1 day/bit
Maximum Range: Character 1 = 0 to 63.75 s
                 Character 2 = 0 to 255 min
                 Character 3 = 0 to 255 h
                 Character 4 = 0 to 255 day
Transmission Update Period: 10.0 s
Message Priority: 7
Format:
    PID Data
    253 n a b c d
    n- Number of parameter data characters = 3
    a- Seconds
    b- Minutes
    c- Hours
    d- Days

This parameter can be shortened by dropping days, days and hours, or days, hours, and minutes.

A.254 Data Link Escape—This PID allows transmission of information on the data bus in a nonstandard (per the protocol outlined in SAE J1587) but specific electronic module vendors proprietary fashion. The intent of this PID is to allow a means to use the data bus for vendor specific transmissions that do not benefit the general purpose nature of the communication data link.

Parameter Data Length: Variable
Data Type: Variable
Resolution : Variable
Maximum Range: Variable
Transmission Update Period: Variable up to 10 times per second
Message Priority: Parameter specific
Format:

PID   Data
254   a b

a- Receiving module's MID
b- Data

A.255 Extension—This PID allows for usage of PIDs with a length greater than a single byte. The subcommittee will need to review and approve any use of this PID.

Parameter Data Length: Variable
Data Type: Variable
Resolution : Variable
Maximum Range: Variable
Transmission Update Period: Variable up to 10 times per second
Message Priority: Parameter specific
Format:

PID   Data
255   Variable

(R) SERIAL DATA COMMUNICATIONS BETWEEN MICROCOMPUTER SYSTEMS IN HEAVY DUTY VEHICLE APPLICATIONS— SAE J1708 OCT90

SAE Recommended Practice

Report of the Truck and Bus Electrical Committee approved January 1986 and revised June 1987. Completely revised by the Truck and Bus Electrical Committee November 1989. Completely revised by the Truck and Bus Electrical and Electronics Committee October 1990.

Foreword—This SAE/TMC Joint Recommended Practice has been developed by the Truck and Bus Electronic Interface Subcommittee of the Truck and Bus Electrical Committee and by the S.1 Study Group of the Maintenance Council. The objectives of the subcommittee are to develop information reports, recommended practices, and standards concerned with the interface requirements and connecting devices required in the transmission of electronic signals and information among truck and bus components.

Objectives: Some of the goals of the subcommittee in developing this document were to:
a. Minimize hardware cost and overhead;
b. Provide flexibility for expansion and technology advancements with minimum hardware and software impact on in-place assemblies;
c. Utilize widely accepted electronics industry standard hardware and protocol to give designers flexibility in parts selection;
d. Provide a high degree of electromagnetic compatibility;
e. Provide original equipment manufacturers, suppliers, and aftermarket suppliers the flexibility to customize for product individuality and for proprietary considerations.

1. Scope—This document defines a recommended practice for implementing a bidirectional, serial communication link among modules containing microcomputers. This document defines those parameters of the serial link that relate primarily to hardware and basic software compatibility such as interface requirements, system protocol, and message format. The actual data to be transmitted by particular modules, which is an important aspect of communications compatibility, is not specified in this document. These and other details of communication link implementation and use should be specified in the separate application documents referenced in Section 2.

1.1 Purpose—The purpose of this document is to define a general-purpose serial data communication link that may be utilized in heavy duty vehicle applications. It is intended to serve as a guide toward standard practice to promote serial communication compatibility among microcomputer-based modules. The primary use of the general-purpose communications link is expected to be the sharing of data among stand-alone modules to cost effectively enhance their operation. Communication links used to implement functions that require a dedicated communication link between specific modules may deviate from this document.

2. References—It is recommended that a separate applications document be published by the manufacturer for each device using the serial link. These documents should define the data format, message I.D.'s, message priorities, error detection (and correction), maximum message length, percent bus utilization, and methods of physically adding/removing units to/from the line for the particular application.

2.1 Applicable Documents

SAE J1455—Joint SAE/TMC Recommended Environmental Practices for Electronic Equipment Design (Heavy Duty Trucks)

SAE J1587—Joint SAE/TMC Recommended Practice for Electronic Data Interchange Between Microcomputer Systems in Heavy Duty Vehicle Applications

SAE J1992—Powertrain Control Interface for Electronic Controls Used in Medium and Heavy Duty Diesel on Highway Vehicle Applications

Electronics Industries Association Standard RS-485 (EIA RS-485) "Standard for Electrical Characteristics of Generators and Receivers for Use in Balanced Digital Multipoint Systems," April 1983

2.2 Definitions

2.2.1 Access Time—Two bit times multiplied by the message priority (which ranges from 1 to 8) added to the idle line time.

2.2.2 Baud—The maximum number of analog signal transitions per second that can occur on a channel. In this coding system, this is the reciprocal of the bit time.

2.2.3 BIT TIME—Duration or period of one unit of information.

2.2.4 CHARACTER TIME—The duration of one character. The character must start with a low logic bit, then 8 bits of data (least significant bit first) followed by a high logic level stop bit.

2.2.5 CONTENTION—A state of the bus in which two or more transmitters are turned on simultaneously to conflicting logic states.

2.2.6 DIFFERENTIAL SIGNAL—A two-wire process in which both lines are switches as opposed to a single-ended signal wherein one line is grounded and the signal line is switched between logic states.

2.2.7 IDLE STATE—The state that produces a high logic level on the input of the bus receiver when all transmitters on the network are turned off.

2.2.8 IDLE LINE—The condition that exists when the bus has remained in a continuous high logic state for at least 10 bit times after the end of the last stop bit.

NOTE: The idle line serves as the delimiter between messages on the bus. A receiver that cannot distinguish between a stop bit and any other high logic state may become synchronized with the bus by noting the receipt of 12 consecutive high logic bits. In the absence of errors, the first low logic bit (0) following 12 consecutive high logic bits (1) is the start bit of a message identification character (MID) (that is, the first character of a message).

2.2.9 MESSAGE PRIORITY—A measure of message criticality assigned on a scale of 1 to 8 by the appropriate applications document. The most critical message has a priority of one.

2.2.10 NODE—A receiver or transceiver circuit connected to the bus.

2.2.11 START BIT—Initial element of a character defined as a low logic level of 1 bit time duration as viewed at the output of the bus receiver.

2.2.12 STOP BIT—Final element of a character defined as a high logic level of 1 bit time duration as viewed at the output of the bus receiver.

3. Electrical Parameters—The electrical parameters of this serial data link are a modification of the EIA RS-485 standard. In some areas this document conflicts with RS-485. This document shall serve as the guiding document in such cases. Appendix A details a serial data bus standard node which defines the interface circuit parameters. Operation of this standard node is detailed in this section.

3.1 Logic State—Positive true logic will be used when referring to the states of transmitted inputs and received outputs. Referring to Appendix A, the input of the transmitter (marked as point Tx) and the output of the receiver (marked as Rx) will be in logic 1 state when driven or passively pulled to +V, and will be at a logic 0 state when driven to ground.

3.2 Bus State—The bus is in a logic 1 (high) state whenever Point A is at least 0.2 V more positive than Point B. The bus is in a logic 0 (low) state whenever Point A is at least 0.2 V more negative than Point B (Points A and B, refer to Figure A1). The bus state is indeterminate when the differential voltage is less than 0.2 V.

3.2.1 LOGIC HIGH STATE—The bus will be in a logic 1 (high) state when all connected transmitters are idle or sending logic 1. An idle state is produced when all transmitters on the network are turned off. All nodes shall include means to pull the bus to a logic 1 (high) when all transmitters are off (see Appendix A).

3.2.2 LOGIC LOW STATE—The bus will be in a logic 0 (low) state when one or more transmitters are sending logic 0, which guarantees that logic 0 (low) dominates when the bus is in contention.

3.3 Network Capacity—The bus will support a minimum of 20 standard nodes where each node is comprised of the circuit defined in Appendix A. Deviations from this circuit must be carefully analyzed to determine impact on bus loading and noise margins over the common mode range.

3.4 Bus Termination—Bus termination resistors as referenced in RS-485 are not required and shall not be used.

3.5 Ground—All assemblies using the link must have common ground reference.

3.6 Wire—A minimum of 18 gauge twisted pair wire, with a minimum of one twist (360 degrees) per inch (2.54 cm), is required. The twists shall be distributed evenly over the length of the wire.

3.7 Length—This recommended practice is intended for, but not limited to, applications with a maximum length of 130 ft (40 m).

4. Network Parameters

4.1 Network Topology—The network interconnect shall use a common or global bus.

4.2 Network Access—The method of access to the network is random.

4.2.1 BUS ACCESS—A transmitter shall begin transmitting a message only after an idle state has continuously existed on the bus for at least a bus access time. The transmitter must verify that the idle state continues to exist immediately prior to initiating a transmission (that is, within one-half bit time).

4.2.1.1 *Bus Access Time*—Bus access time is a time duration equal to the minimum time of an idle line plus the product of 2 bit times and the message priority. This relationship can be expressed as follows:

$$T_a = T_i + [2 * T_b] * P \qquad (Eq.1)$$

where:

T_a = Bus access time
T_b = Bit time, or period of one unit of information
P = Message priority
T_i = Minimum time duration of an idle line

NOTE: The minimum time duration of an idle line is defined in 2.2.9. However, a transmitter that cannot distinguish between a stop bit and any other high logic state may not assume that T_i has elapsed until it has received 19 consecutive high logic bits.

4.2.1.2 *Message Priority Assignment*—All messages will be assigned a priority from 1 to 8 as indicated in Table 1:

TABLE 1

Priority	Message Assignment
1 and 2	Reserved for messages that require immediate access to the bus
3 and 4	Reserved for messages that require prompt access to the bus in order to prevent severe mechanical damage
5 and 6	Reserved for messages that directly affect the economical or efficient operation of the vehicle
7 and 8	All other messages not fitting into the previous priority categories should be assigned a priority 7 or 8

The applications document shall define the priority associated with each message. In the event that more than one priority could be assigned to a particular message, the application document shall define each priority and the circumstances in which the priority is assigned.

4.2.2 BUS CONTENTION—All transmitters shall monitor the message identification portion of their message to determine if another transmitter has attempted to gain access to the bus at the same time. If a transmitter detects a collision, the transmitter shall relinquish control of the bus after completing the transmission of the current character or sooner if possible. After relinquishing control, it is recommended that the transmitter become a receiver, using the received MID as the beginning of the incoming message. The transmitter may attempt to regain access to the bus after a bus access time has elapsed. An example bus reaccess procedure is shown in Appendix B.

5. Protocol

5.1 Bit Time—A bit time shall be 104.17 μs ± 0.5% (±500 ns). This is equivalent to a baud rate of 9600 bits per second.

5.2 Character Format—A character shall consist of 10 bit times. The first bit shall always be a low logic level and is called the start bit. The last (tenth) bit shall always be a high logic level and is called the stop bit.

This convention is consistent with standard UART operation. The remaining eight center bits are data bits that are transmitted least significant bit (LSB) first.

5.3 Message Format

5.3.1 MESSAGE CONTENT—A message appearing on the communication bus shall consist of the following:

  a. Message Identification Character (MID);
  b. Data Characters;
  c. Checksum.

As indicated in 4.2.1, a message shall always be preceded by an idle state of duration equal to or greater than the appropriate bus access time. The length of time between characters within a message shall not exceed 2 bit times.

5.3.2 MESSAGE IDENTIFICATION CHARACTER (MID)—The first character of every message shall be a MID. The permitted range of MIDs shall include the numbers 0 to 255. The MIDs 0 to 68 shall be assigned to transmitter categories as identified in Table 2. These assignments have been made to accommodate existing systems, or systems that may presently be under development, and to avoid conflicts, which otherwise might arise if indiscriminate use of MIDs were permitted.

MIDs 69 to 86 have been set aside for use by the SAE J1922.

MIDs 87 to 110 shall be allocated as reserved MIDs for transmitter categories beyond those that are identified in Table 2. These MIDs shall be individually assigned by the SAE Electronics Interface Subcommittee of the SAE Truck and Bus Electrical Committee on petition by a manufacturer at the time a new transmitter category is identified, or when additional MIDs are required within a previously identified category. The content and format of the messages using the assigned MIDs (0 to 110) is the responsibility of the transmitter. Content of format of the data within these messages is not defined in this document but should be identified in an appropriate applications document as described in Section 2.

MID 111 shall be used exclusively for factory test of electronic modules. Since it is possible that during factory test the normal control software is bypassed, giving the tester direct control of module I/o, several precautions should be observed:

a. Entry into factory test should be granted by the module control software only after ensuring that it is safe to do so.
b. This MID should not be transmitted by any on-board module.

MIDs 112 to 127 are not assigned to any category and are not reserved for future assignment. These MIDs are available to any manufacturer or user for any message identification purpose outside the scope of this document.

TABLE 2—MESSAGE IDENTIFICATION CHARACTER ALLOCATION

Mid Range	Transmitter Category
00-07	ENGINE
08-09	BRAKES, TRACTOR
10-11	BRAKES, TRAILER
12-13	TIRES, TRACTOR
14-15	TIRES, TRAILER
16-17	SUSPENSION, TRACTOR
18-19	SUSPENSION, TRAILER
20-27	TRANSMISSION
28-29	ELECTRICAL CHARGING SYSTEM
30-32	ELECTRICAL
33-35	CARGO REFRIGERATION/HEATING
36-40	INSTRUMENT CLUSTER
41-45	DRIVER INFORMATION CENTER
46-47	CAB CLIMATE CONTROL
48-55	DIAGNOSTIC SYSTEMS
56-61	TRIP RECORDER
62-63	TURBOCHARGER
64-68	OFF-BOARD DIAGNOSTICS
69-86	SET ASIDE FOR SAE J1922
87-110	RESERVED—TO BE ASSIGNED BY ELECTRONIC INTERFACE SUBCOMMITTEE (see 5.3.2, Section 3).
111	RESERVED—FACTORY ELECTRONIC MODULE TESTER (OFF VEHICLE)
112-127	UNASSIGNED—AVAILABLE FOR USE
128-255	TO BE ASSIGNED BY DATA FORMAT SUBCOMMITTEE (see 5.3.2.)

MIDs in the 0 to 68 and 87 to 127 ranges shall be defined in the manufacturer's applications document. It shall be the responsibility of the systems integrator or user to ensure that a particular MID is not used by more than one device on the same vehicle.

MIDs in the range of 128 to 255 shall be reserved for applications using formatted data as set forth in a document issued by the SAE Truck and Bus Electrical Committee Data Format Subcommittee. These MIDs shall only be used when the data format set forth within that document is strictly followed. See SAE J1587.

5.3.3 DATA CHARACTERS—Data characters shall be characters that convey the intelligence of the message and shall conform to the character format as defined in 5.2. The 8 bit data character may be given any value from 0 to 255. The data characters shall be defined in an appropriate applications document at the option of the supplier. The application document shall define parameters, parameter order, scaling and error detection/correction coding if applicable.

5.3.4 CHECKSUM—The last character of each message shall be the two's complement of the sum of the MID and the data characters. Simple message error detection may be implemented by adding the checksum to the sum of all previous message characters (including the MID). The 8 bit sum will be zero, neglecting the CARRY, for a correctly received message.

5.3.5 MESSAGE LENGTH—Total message length, including MID and checksum, shall not exceed 21 characters. Exceptions to this length limitation may be made when the engine is not running and the vehicle is not moving. Messages longer than 21 characters may also be broken up into several separate messages of 21 or fewer characters and may then be transmitted while the engine is running and/or the vehicle is moving by conforming to the 21 character message length limitation of SAE J1708.

APPENDIX A
Serial Data Bus Standard Node
(Unipolar Drive With Passive Termination in Each Module)

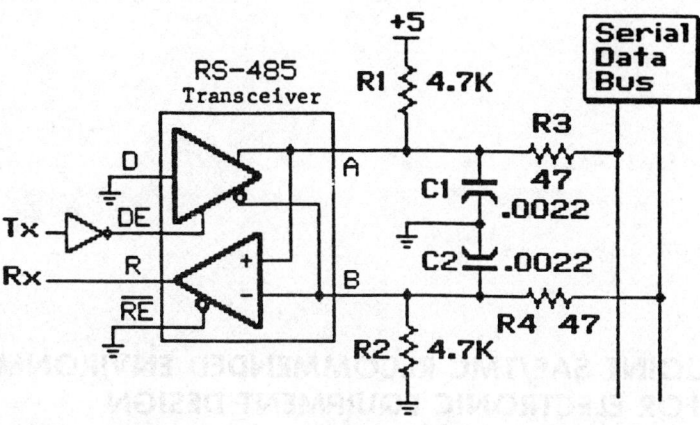

FIG. A1—SERIAL DATA BUS STANDARD NODE DIAGRAM

A1. This circuit utilizes standard RS-485 transceivers (less than or equal to one RS-485 unit load) connected to drive the differential data bus to the logic zero state only (unipolar drive). In the above circuit, a standard RS-485 receiver may be used in place of a transceiver in applications where data need not be placed on the bus (that is, receive only).

A2. The logic one state (also idle state) is controlled by pull-up resistor R1 and pull-down resistor R2.

A3. The transceiver output impedance, C1 and C2, form the transmit filter for transient and EMI suppression (approximately 6 MHz low pass).

A4. R3, C1, R4, and C2 form the receive filter for EMI suppression (approximately 1.6 MHz low pass). These parts also form a pseudo line termination at high frequencies.

A5. The active (high-to-low) transition delay is approximately 0.6 μs at the receiver with two nodes on the bus and 2.3 μs with 20 nodes on the bus.

A6. The passive (low-to-high) transition delay at the receiver remains at 10 μs with any number of loads on the bus (up to 20).

A7. The values shown were chosen for use with commercially available RS-485 drivers to provide maximum fan-out, EMI suppression, and bus termination. Remaining nodes may be in either the powered or unpowered state.

A8. This method of unipolar drive prevents unresolved contention (logic zero always wins).

A9. The resistors shown should be 5% parts to assure sufficient noise margin under worst case conditions. R3, R4, C1, and C2 should be balanced within 10% on each side of the data bus to minimize common mode electromagnetic radiation.

APPENDIX B
Example Bus Reaccess Procedure

B1. A method for reaccessing the bus can be described by the following example:

B1.1 Sequence of Events

a. First crash occurs for the current attempt to access the bus.
b. Each device wishing to access the bus then waits their predefined bus access time (as described in 4.2.2).
c. Second crash occurs for the same attempt to access the bus.
d. Any device that has experienced two consecutive crashes in its attempt to transmit the same message shall follow the bus access procedure defined in 4.2.1 but with the bus access time calculated as follows:

$$T_a = T_i + 2 + P_2 \times T_b \qquad (Eq. B1)$$

where:

T_i and T_b are defined as in 4.2.1.1.

P_2 = A three bit psuedo random number such as the three least significant bits of the stack pointer.

For example, if 18 is the location of stack pointer register, the contents of this register is the stack pointer. This value will be ANDED with 0007, which results in a number from 0 to 7. P2 would, therefore, be a value from 0 to 7.

e. If any more consecutive crashes occur, the procedure described in d is repeated.

B2. This example addresses the recognized possibility that two or more devices could continue to crash if their priorities were the same. The above method would greatly reduce the possibility of a third crash with the same device or devices.

JOINT SAE/TMC RECOMMENDED ENVIRONMENTAL PRACTICES FOR ELECTRONIC EQUIPMENT DESIGN (HEAVY-DUTY TRUCKS)—SAE J1455 JAN88 — SAE Recommended Practice

Report of the Truck and Bus Electrical Committee approved January 1988.

1. Purpose—This guideline is intended to aid the designer of automotive electronic systems and components by providing material that may be used to develop environmental design goals.

2. Scope—The climatic, dynamic, and electrical environments from natural and vehicle-induced sources that influence the performance and reliability of vehicle electronic equipment are included in this document. Test methods that can be used to simulate these environmental conditions are also included. This information is applicable to diesel powered trucks in Classes 6, 7, and 8.

3. Application

3.1 Environmental Data and Test Method Validity—The information included in the following sections is based upon test results achieved by major North American truck manufacturers and component equipment suppliers. Operating extremes were measured at test installations normally used by manufacturers to simulate environmental extremes for vehicles and original equipment components. They are offered as a design starting point. Generally, they cannot be used directly as a set of operating specifications because some environmental conditions may change significantly with relatively minor physical location changes. This is particularly true of vibration, engine compartment temperature, and electromagnetic compatibility. Actual measurements should be made as early as practicable to verify these preliminary design baselines.

The proposed test methods are currently being used for laboratory simulation or are considered to be a realistic approach to environmental design validation. They are not intended to replace actual operational tests under adverse conditions. The recommended methods describe standard cycles for each type of test. The designer must specify the number of cycles over which the equipment should be tested. The number of cycles will vary depending upon equipment, location, and function. While the standard test cycle is representative of an actual short term environmental cycle, no attempt is made to equate this cycle to an acceleration factor for reliability or durability. These considerations are beyond the scope of this document.

3.2 Organization of Test Methods and Environmental Extremes Information—The data presented in this document are contained in Sections 4 and 5. Section 4, Environmental Factors and Test Methods, describes the 13 characteristics of the expected environment that have an impact on the performance and reliability of truck and bus electronic systems. These descriptions are titled:

(a) Temperature
(b) Humidity
(c) Salt Spray Atmosphere
(d) Immersion and Splash (Water, Chemicals, and Oils)
(e) Steam Cleaning and Pressure Washing
(f) Fungus
(g) Dust, Sand, and Gravel Bombardment
(h) Altitude
(i) Mechanical Vibration
(j) Mechanical Shock
(k) General Heavy Duty Truck Electrical Environment
(l) Steady-State Electrical Characteristics
(m) Transient, Noise, and Electrostatic Characteristics

They are organized to cover three facets of each factor:
(a) Definition of the factor.
(b) Description of its effect on control, performance, and long term reliability.
(c) A review of proposed test methods for simulating environmental stress.

In Section 5, Environmental Extremes by Location, summaries are presented of the anticipated limit conditions at the following five general control sites:

(a) Underhood
  (1) Engine (Lower Portion)
  (2) Engine (Upper Portion)
  (3) Bulkhead
(b) Interior (Cab)
  (1) Floor
  (2) Instrument Panel
  (3) Head Liner
  (4) Inside Doors
(c) Interior (Aft of Cab)
  (1) Bunk Area
  (2) Storage Compartment
(d) Chassis
  (1) Forward
  (2) Rear
(e) Exterior of Cab
  (1) Under Floor
  (2) Rear
  (3) Top
  (4) Doors

3.3 Combined Environments—The vehicle environment consists of many natural and induced factors. Combinations of these factors are present simultaneously. In some cases, the effect of a combination of these factors is more serious than the effect of exposing samples to each environmental factor in series. For example, the suggested test method for humidity includes high- and low-temperature exposure. This combined environmental test is important to components when proper operation is dependent on seal integrity. Temperature and vibration is a second combined environmental test method that can be significant to components. During design analysis a careful study should be made to determine the possibility of design susceptibility to a combination of environmental factors that could occur at the planned mounting location. If the possibility of susceptibility exists, a combined environmental test should be considered.

3.4 Test Sequence—The optimum test sequence is a compromise between two considerations:

3.4.1 The order in which the environmental exposures will occur in operational use.

3.4.2 A sequence that will create a total stress on the sample that is representative of operation stress.

The first consideration is impossible to implement in vehicle testing since exposures occur in a random order. The second consideration prompts the test designer to place the most severe environments last. Many sequences that have been successful follow this general philosophy, except that the temperature cycle is placed first in order to condition the sample mechanically.

4. Environmental Factors and Test Methods
4.1 Temperature

4.1.1 DEFINITION—Thermal factors are probably the most pervasive environmental hazard to vehicle electronic equipment. Sources for temperature extremes and variations include:

4.1.1.1 The vehicle's climatic environment, including the diurnal and seasonal cycles. Variations in climate by geographical location must be considered. In the most adverse case, the vehicle that spends the winter in Canada may be driven in the summer in the Arizona desert. Temperature variations due to this source range from −54 to +85°C (−65 to +185°F).

4.1.1.2 Heat sources and sinks generated by the vehicle's operation. The major sources are the engine and drive train components, including the brake system. Wide variations are found during operation. For instance, temperatures on the surface of the engine can range from the cooling system's 88°C (190°F) to 816°C (1500°F) on the surface of the exhaust system. This category also includes conduction, convection, and radiation of heat because of the various modes of the vehicle's operation.

4.1.1.3 Self-heating of the equipment due to its internal dissipation. A design review of the worst case combination of peak ambient temperature (see paragraphs 4.1.1.1 and 4.1.1.2), minimized heat flow away from the equipment, and peak-applied steady-state voltage should be conducted.

4.1.1.4 Vehicle operational mode and actual mounting location. Measurements should be made at the actual mounting site during the following vehicular conditions while they are subjected to the maximum heat generated by adjacent equipment, and while they are at the maximum ambient environment:

(a) Engine Start.
(b) Engine Idle.
(c) Engine High Speed.
(d) Engine Turn Off. (Prior history important.)
(e) Various Engine/Road Conditions.

4.1.1.5 Ambient conditions before installation due to storage and transportation extremes. Shipment in unheated aircraft cargo compartments may lower the minimum storage (non-operating) temperature to −50°C (−58°F).

The thermal environmental conditions that are a result of these conditions can be divided into three categories:

(a) Extremes—The ultimate upper and lower temperatures the equipment is expected to experience.
(b) Cycling—The cumulative effects of temperature cycling within the limits of the extremes.
(c) Shock—Rapid change of temperature. Fig. 1 illustrates one form of vehicle operation that induces thermal shock and is derived from an actual road test of two vehicles. Thermal shock is also induced when equipment at elevated temperature is exposed to sudden rain or road splash, or when it is moved from a heated shelter into a low (−40°C/−40°F) ambient temperature environment.

The vehicle electronic equipment designer is urged to develop a systematic, analytic method for dealing with steady-state and transient thermal analysis. The application of many devices containing semiconductors is temperature limited. For this reason, the potential extreme operating conditions for each application must be scrutinized to avoid later field use failure.

4.1.2 EFFECT ON PERFORMANCE—The damaging effects of thermal shock and thermal cycling include:

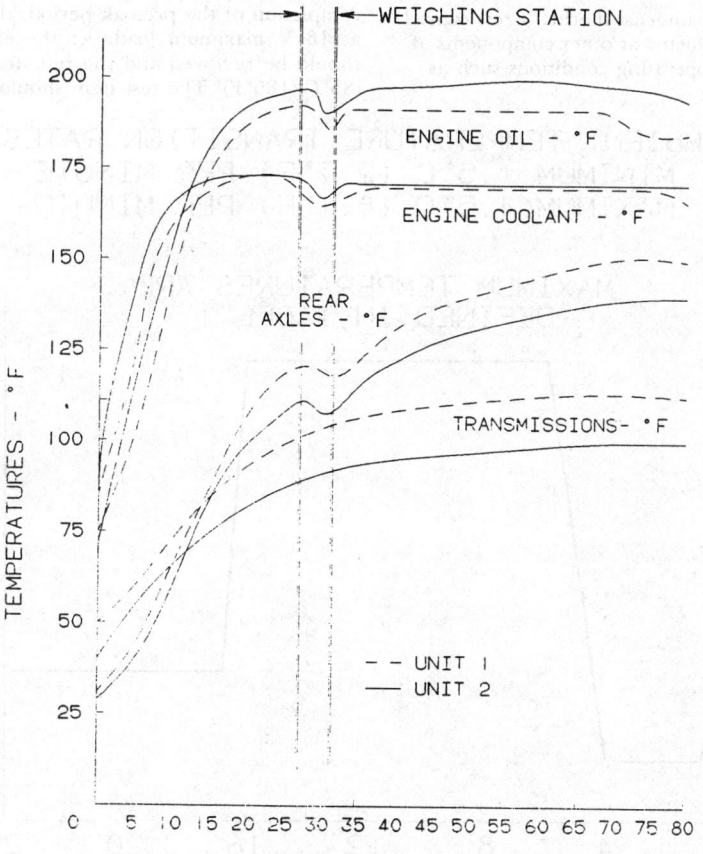

FIG. 1—TIME INTO RUN-MINUTES VEHICLE WARM-UP CHARACTERISTICS

4.1.2.1 Cracking of printed circuit board or ceramic substrates.
4.1.2.2 Thermal stress or fatigue failures of solder joints.
4.1.2.3 Delamination of printed circuit boards and other interconnect system substrates.
4.1.2.4 Seal failures, including the breathing action of some assemblies, due to temperature-induced dimensional variation that permits intrusion of liquid or vapor borne contaminants.
4.1.2.5 Failure of circuit components due to direct mechanical stress caused by differential thermal expansion.
4.1.2.6 The acceleration of chemical attack on interconnects, due to temperature rise, can result in progressive degradation of circuit components, printed circuit board conductors, and solder joints.

In addition to these phenomena, high temperature extremes can cause a malfunction by:

4.1.2.7 Exceeding the dissociation temperature of surrounding polymer or other packaging components.
4.1.2.8 Carbonizing of the packaging materials resulting in the eventual progressive failure of the associated passive or active components. This is possible in cases of excessively high temperature. In addition, noncatastrophic failure is possible because of electrical leakage in the resultant carbon paths.
4.1.2.9 Changing the active device characteristics with increased heat, including changes in gain, impedance, collector-base leakage, peak blocking voltage, collector-base junction second breakdown voltage, etc.
4.1.2.10 Changing the passive device characteristics, such as permanent or temporary drift in resistor value and capacitor dielectric constants, with increased temperature.
4.1.2.11 Changing the interconnect and relay coil performance due to the conductivity temperature coefficient of copper.
4.1.2.12 Changing the properties of magnetic materials with increasing temperature, including Curie point effects and loss of permanent magnetism.
4.1.2.13 Changing the dimensions of packages and components leading to the separation of subassemblies.
4.1.2.14 Changing the strength of soldered joints because of changes in the mechanical characteristics of the solder.
4.1.2.15 The severe mechanical stress caused by ice formation in moisture bearing voids or cracks.
4.1.2.16 The very rapid and extreme internal thermal stress caused by applying maximum power to semiconductor or other components after extended cold soak under aberrant operating conditions such as 36 V battery jumper starts.

4.1.3 RECOMMENDED TEST METHODS

4.1.3.1 Temperature Cycle Test—Recommended thermal cycle profiles are shown in Figs. 2A, B, and C, and recommended extreme temperatures in Tables 1(A) and (B). The test method of Fig. 2A, a 24 h cycle, offers longer stabilization time and permits a convenient room ambient test period. Fig. 2B, an 8 h cycle, provides more temperature cycles for a given test duration. It is applicable only to modules whose temperatures will reach stabilization in a shorter cycle time. Stabilization should be verified by actual measurements; thermocouples, etc. It is important that all parts of the test specimen be held at the specified maximum and minimum temperatures for at least 30 min during each cycle, after reaching stability at that temperature. This is to maintain thermal or pressure stresses generated in the test specimen for a reasonable period of time. Fig. 2C illustrates a test method for thermal shock.

Separate or single test chambers may be used to generate the temperature environment described by the thermal cycles. By means of circulation, the air temperature should be held to within ±3°C (±5°F) at each of the extreme temperatures. The test specimens should be placed in a position, with respect to the air stream, that there is substantially no obstruction to the flow of air across the specimen. If two test specimens are used, care must be exercised to assure that the test samples are not subject to temperature transition rates greater than that defined in Figs. 2A and 2B. Direct heat conduction from the temperature chamber heating element to the specimen should be minimized.

NOTE: Airflow is a function of actual equipment location. Simulation of actual airflow and thermal transfer operation conditions should be considered in test design.

Electrical performance should be measured under the expected operational minimum and maximum extremes of excitation, input and output voltage and load at both the cold and hot temperature extremes. These measurements provide insight into electrical variations with temperature.

4.1.3.2 Thermal Shock Test—Thermal shock that can be expected in the vehicle environment is simulated by the maximum rates of change shown on the recommended thermal cycle profile portrayed in Fig. 2C. The thermal shock test should begin within a 3½ h presoak at −40°C (−40°F) with power off. Approximately 30 min before the completion of the presoak period, the test item should be powered-up at 16 V maximum load. At the end of the presoak period, power should be removed and the test item transferred to the hot chamber (85°C/185°F). The test item should remain in the hot chamber for 2

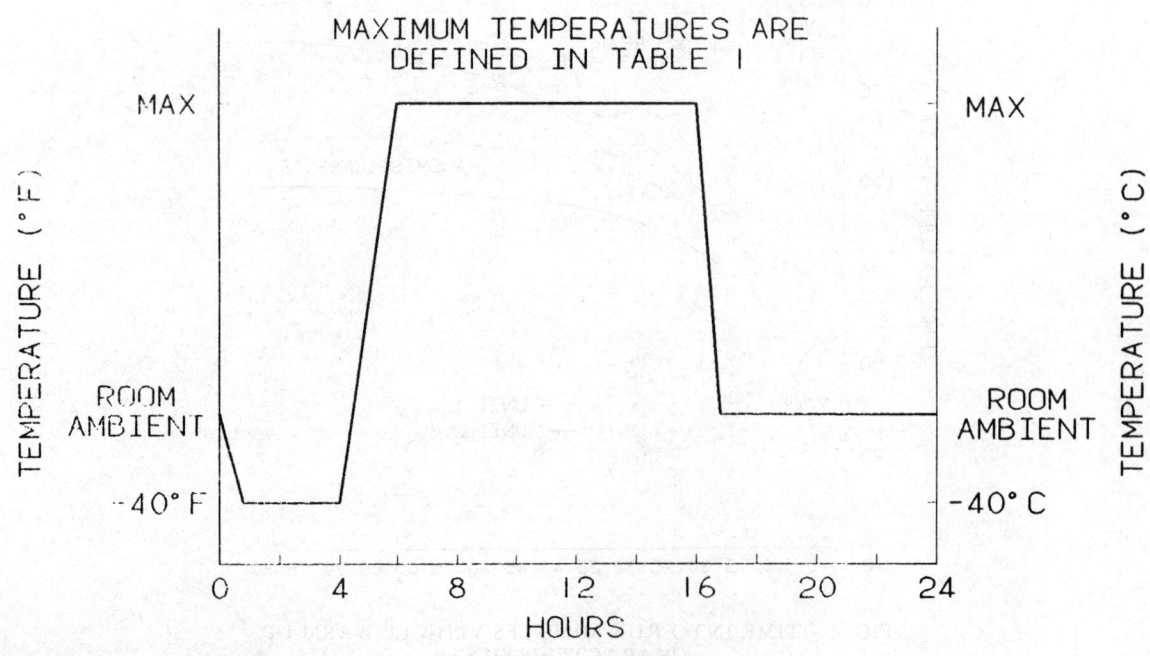

FIG. 2A— 24 H CYCLE

h, then returned to the cold chamber (−40°C/−40°F) for 2 h. This cycle should be repeated at least five times. Each transfer should be accomplished in 1 min or less. The manufacturer may wish to consider a "power-up" sequence to occur during each soak time to create internal stress at low temperature and maximum internal heating at high temperature.

4.1.3.3 Thermal Stress—Thermal stress is caused by repeat cycling through the thermal profiles of Figs. 2A, B and C. Many failures are due to fatigue. Slow cycles not repeated often will not demonstrate this. The number of cycles required is a function of the equipment application. Functional electrical testing during temperature transitions or immediately after temperature transitions is a means of detecting poor electrical connections. The effect of thermal stress is similar to thermal shock but is caused by fatigue.

NOTE: Although uniform oven temperatures are desirable, the only means of heat removal in some vehicle environments may be by special heat sinks or by free convection to surrounding air. It may be necessary to use conductive heat sinks with independent temperature controls in the former case and baffles or slow speed air stirring devices in the latter to simulate such conditions in the laboratory. (See Section 3.)

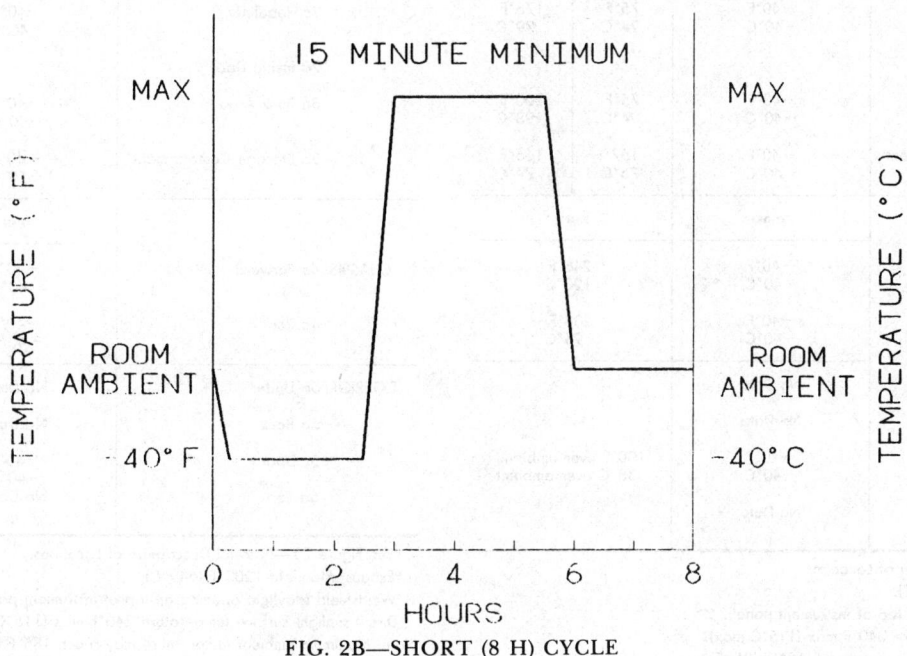

FIG. 2B—SHORT (8 H) CYCLE

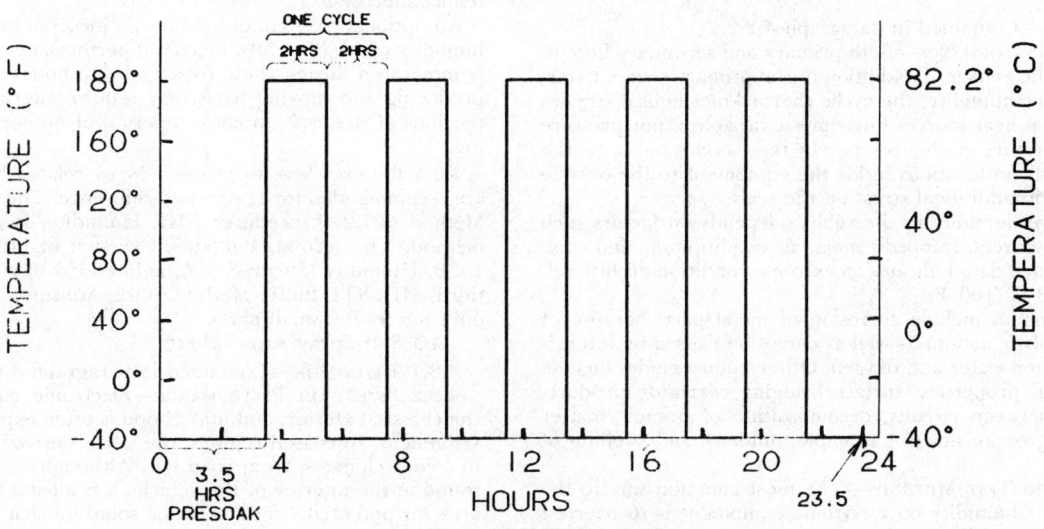

FIG. 2C—THERMAL SHOCK

TABLE 1A—ENVIRONMENTAL EXTREME SUMMARY
HEAVY DUTY CAB OVER ENGINE TRUCK/TRACTOR

Location[a]	Temperatures		
	min	max	
ENGINE: 1a Underhood - Lower	−40°F −40°C	100°F over ambient 38°C over ambient	
1b[b] Underhood - Upper	−40°F −40°C	400°F over ambient 204°C over ambient	
1c Underhood - Bulkhead	−40°F −40°C	100°F over ambient 38°C over ambient	
	min	operating	max
INTERIOR: 2a Floor	−40°F −40°C	80°F 27°C	150°F 66°C
2b[c] Instrument Panel	−40°F −40°C	75°F 24°C	185°F 85°C
2c Headliner	−40°F −40°C	75°F 24°C	175°F 79°C
2d Inside Door			
3a Bunk Area	−40°F −40°C	75°F 24°C	200°F 93°C
3b Storage Compartment	−40°F −40°C	167°F 75°C	165°F 74°C
	min	max	
CHASSIS: 4a Forward	−40°F −40°C	248°F 120°C	
4b Rear	−40°F −40°C	203°F 95°C	
EXTERIOR: 5a Under	No Data		
5b Back	No Data		
5c Door	−40°F −40°C	100°F over ambient 38°C over ambient	
5d Top	No Data		

[a]See Fig. 23—Pictorial Description of Locations.
[b]Exhaust Manifold 1200°F (649°C).
[c]Windshield (daylight opening on top of instrument panel).
Direct sunlight surface temperature 240°F max (115°C max).
Note: Maximum ambient temperature may reach 185°F (85°C).

TABLE 1B—ENVIRONMENTAL EXTREME SUMMARY
HEAVY DUTY CONVENTIONAL ENGINE TRUCK/TRACTOR

Location[a]	Temperatures		
	min	max	
ENGINE: 1a Underhood - Lower	−40°F −40°C	100°F over ambient 38°C over ambient	
1b[b] Underhood - Upper	−40°F −40°C	400°F over ambient 204°C over ambient	
1c Underhood - Bulkhead	−40°F −40°C	100°F over ambient 38°C over ambient	
	min	operating	max
INTERIOR: 2a Floor	−40°F −40°C	80°F 27°C	165°F 75°C
2b[c] Instrument Panel	−40°F −40°C	75°F 24°C	185°F 85°C
2c Headliner	−40°F −40°C	75°F 24°C	175°F 79°C
2d Inside Door			
3a Bunk Area	−40°F −40°C	75°F 24°C	200°F 93°C
3b Storage Compartment	−40°F −40°C	167°F 75°C	165°F 74°C
	min	max	
CHASSIS: 4a Forward	−40°F −40°C	250°F 121°C	
4b Rear	−40°F −40°C	200°F 93°C	
EXTERIOR: 5a Under	No Data		
5b Back	No Data		
5c Door	−40°F −40°C	100°F over ambient 38°C over ambient	
5d Top	No Data		

[a]See Figure 23—Pictorial Description of Locations.
[b]Exhaust Manifold 1200°F (649°C).
[c]Windshield (daylight opening on top of instrument panel).
Direct sunlight surface temperature 240°F max (115°C max).
Note: Maximum ambient temperature may reach 185°F (85°C).

4.1.4 RELATED SPECIFICATIONS—A generally accepted method for small part testing is defined in MIL-STD-202F, Method 107F, Thermal Shock, Method A or B, Alternately MIL-STD-810D, Method 503.2. The short dwell periods at high temperature are satisfactory where temperature stabilization is verified by actual measurements.

4.2 Humidity

4.2.1 DEFINITIONS—Contained in paragraph 4.2.2.

4.2.2 EFFECTS ON PERFORMANCE—Both primary and secondary humidity sources exist in the vehicle. In addition to the primary source externally applied ambient humidity, the cyclic thermal-mechanical stresses caused by operational heat sources introduce a variable vapor pressure on the seals. Temperature gradients set up by these cycles can cause the dew point to travel from locations inside the equipment to the outside and back, resulting in additional stress on the seals.

The actual relative humidity in the vehicle depends on factors such as operational heat sources, trapped vapors, air conditioning, and cool-down effects. Recorded data indicates an extreme condition of 98% relative humidity at 38°C (100°F).

Primary failure modes include corrosion of metal parts because of galvanic and electrolytic action, as well as corrosion caused by interaction with contaminated water and oxygen. Other failure modes include changes in electrical properties, surface bridging corrosion products and condensation between circuits, decomposition of organic matter because of attacking organisms (for example, mildew), and swelling of elastomers.

4.2.3 RECOMMENDED TEST METHODS—The most common way to determine the effect of humidity on electronic equipment is to overtest and examine any failure for relevance to the more moderate actual operating conditions. It is most important that the equipment or component be allowed to "breathe" during testing. The most common test is an 8 h active temperature humidity cycling under accelerated conditions (Fig. 3A). A second test is an 8-24 h exposure at 103.4 kPa gage pressure (15 lbf/in^2 gage) in a pressure vessel (Fig. 3B). This is a quick and effective method for uncovering defects in plastic encapsulated semiconductors.

An optional frost condition may be incorporated during one of these humidity cycles (Fig. 3B). Electrical performance should be continuously monitored during these frost cycles to note erratic operation. Heat-producing and moving parts may require altering the frost condition portions of the cycle to allow a period of non-operation induced frosting.

4.2.4 RELATED SPECIFICATIONS—Many related humidity specifications are recommended for review and reference. The first: MIL-STD-810D, Method 507.2, Procedures I-III, Humidity, is a system-oriented test method. The second, a modified version of MIL-STD-202F, Method 103B, Humidity (Steady-State), is intended to evaluate materials. The third, MIL-STD-202F, Method 106E, Moisture Resistance, is a procedure for testing small parts.

4.3 Salt Spray Atmosphere

4.3.1 DEFINITION—Contained in paragraph 4.3.2.

4.3.2 EFFECT ON PERFORMANCE—Electronic equipment mounted on the chassis, exterior, and underhood is often exposed to a salt spray environment. In coastal regions, the salt is derived from sea breezes, and in colder climates, from road salt. Although salt spray is generally not found in the interior of the vehicle, it is advisable to evaluate the floor area for potential effects of saline solutions that were transferred from the outside environment by vehicle operators, passengers, and transported equipment.

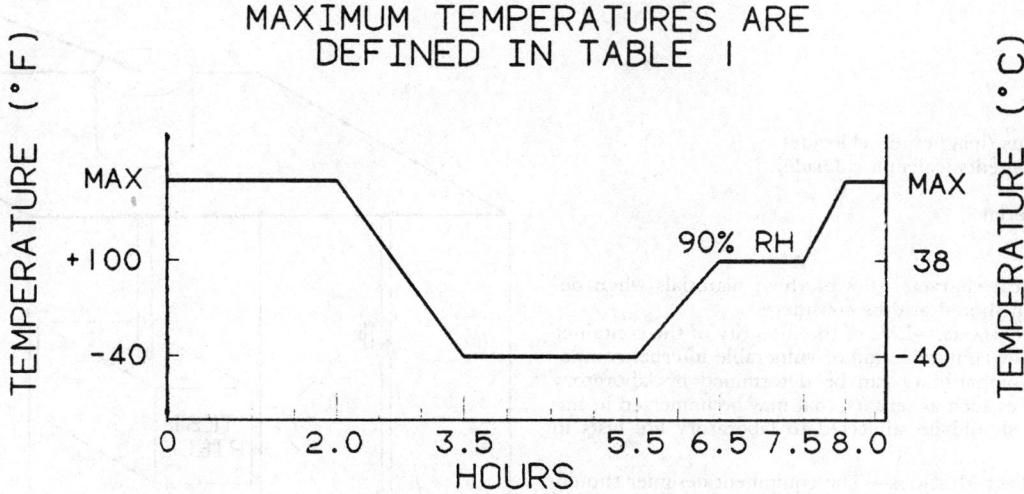

FIG. 3A—RECOMMENDED HUMIDITY 8 H CYCLE

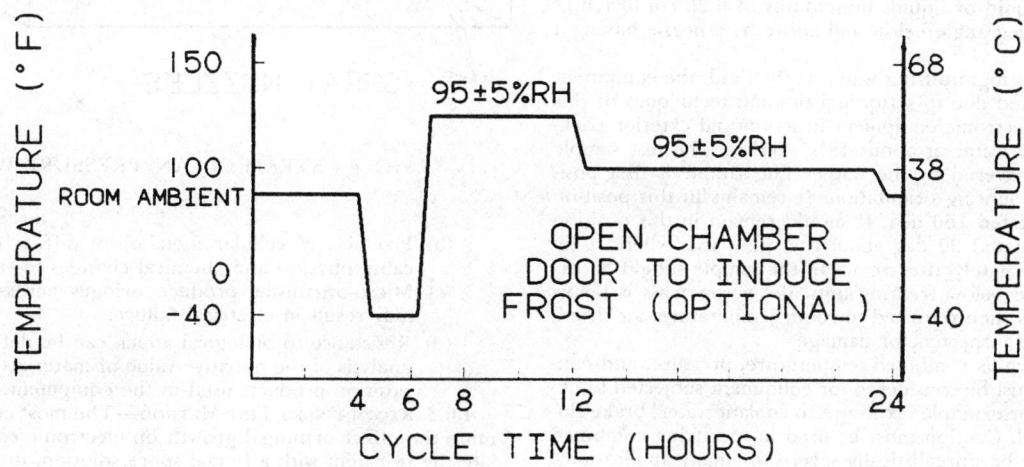

FIG. 3B—RECOMMENDED HUMIDITY 24 H CYCLE

Failure modes due to salt spray are generally the same as those associated with water and water vapor. However, corrosion effects and alteration of conductivity are accelerated by the presence of saline solutions and adverse changes in pH.

4.3.3 RECOMMENDED TEST METHODS—The recommended test method for measuring susceptibility of electronic equipment to salt spray is the American Society for Testing and Materials (ASTM) Standard Method of Salt Spray (Fog) Testing Number B 117-73. Similar test methods are found in MIL-STD-202, Method 101D, and MIL-STD-810D, Method 509.2.

The test consists of exposing the electronic equipment to a solution of 5 parts salt to 95 parts water, atomized at a temperature of 35°C (95°F). The equipment being tested should be exposed to the salt spray for a period of from 24-96 h. The actual exposure time must be determined by analysis of the specific mounting location. When the tests are concluded, the test specimens should be gently rinsed in clean running water, about 38°C (100°F), to remove salt deposits from the surface, and then immediately dried. Drying should be done with a stream of clean, compressed dry air at about 175.8-241.3 kPa gage pressure (35-40 lbf/in^2 gage). The equipment should then be tested under nominal conditions of voltage and load throughout the test.

NOTE: The Pascal (Pa) is the designated SI (metric) unit for pressure and stress. It is equivalent to 1 N/m^2.

Where leakage resistance values are critical, appropriate measurements under wet and dry conditions may be necessary.

4.3.4 RELATED SPECIFICATIONS—ASTM B 117-73, Salt Spray (Fog) Testing, is the recommended test method.

4.4 Immersion and Splash (Water, Chemicals, and Oils)

4.4.1 DEFINITION—Electronic equipment mounted on or in the vehicle are exposed to varying amounts of water, chemicals, and oil. A list of potential environmental chemicals and oils includes:

Engine Oils and Additives
Transmission Oil
Rear Axle Oil
Power Steering Fluid
Brake Fluid
Axle Grease
Washer Solvent
Gasoline
Diesel Fuel
Fuel Additives
Alcohol
Anti-Freeze Water Mixture
Degreasers
Soap and Detergents
Steam
Battery Acid
Water and Snow
Salt Water
Waxes

Kerosene
Freon
Spray Paint
Paint Strippers
Ether
Dust Control Agents (magnesium chloride)
Moisture Control Agents (calcium chloride)
Vinyl Plasticizers
Undercoating Material
Mercuric Acid
Ammonia

The modified chemical characteristics of these materials when degraded or contaminated should also be considered.

4.4.2 EFFECT ON PERFORMANCE—Loss of the integrity of the container can result in corrosion or contamination of vulnerable internal components. The chemical compatibility can be determined by laboratory chemical analysis. Devices such as sensors, that may be immersed in fluids for a long period, should be subjected to laboratory life tests in these fluids.

4.4.3 RECOMMENDED TEST METHODS—The equipment designer should first determine whether the parts must withstand complete immersion or splash, and which fluids are likely to be present in the application. Immersion and splash tests are generally performed following other environmental tests because this sequence tends to aggravate incipient defects in seals, seams, and bushings that might otherwise escape notice.

Splash testing should be done with the equipment mounted in a normal operating position with drain holes, if used, open. The sample is subjected to the test liquid or liquids in amounts of 0.25 cm (0.1 in)/min delivered at a 45 deg angle below and above by a nozzle having a solid cone spray.

During immersion testing, utilizing water as the fluid, the equipment ordinarily is not operated due to setup logistics and techniques of this test. In this test, the electronic equipment in its normal exterior package is immersed in tap water at about 18°C (65°F). The test sample should be completely covered by the water. The sample is first positioned in its normal mounting orientation. It remains in this position for 5 min and then rotated 180 deg. It should remain in this position for 5 min and then rotated 90 deg about the other axis where it remains for 5 min. Immediately after removal, the sample should be exposed to a temperature below freezing until the entire mass is below freezing. The sample is then returned to room temperature, air dried, functionally tested, and inspected for damage.

More severe tests such as combined temperature, pressure, and continuous fluid contact must be considered for equipment subjected to extreme environments; for example, exposure to coolant water, brake fluid, and transmission oil. Caution must be used in specifying combined tests because they may be unrealistically severe for many applications.

4.4.4 RELATED SPECIFICATIONS—None.

4.5 Steam Cleaning and Pressure Washing

4.5.1 DEFINITION—Contained in paragraph 4.5.2.

4.5.2 EFFECTS ON PERFORMANCE—The intense heat from cleaning sprays and the caustic nature of chemical agents in washing solutions create a severe environment for devices and associated wiring and connectors mounted in the engine, chassis and exterior areas. This exposure can cause a degradation of insulation and seals as well as cracking of vinyl connectors and component packaging. High pressure washdown may produce results similar to salt spray in many truck interiors.

4.5.3 RECOMMENDED TEST METHOD—The test item should be mounted in its normal operating position with drain holes, if used, open. The test apparatus should be designed to provide 100% coverage of the exposed surface of the test item using flat fan spray nozzles. This apparatus should provide an impact pressure of at least 31.0 kPa (4.5 lbf/in^2) at the test item. Water temperature should be 93°C (200°F). The test item should be exposed to the spray for 3 s of a 6 s period for a total of 375 cycles.

A sample test device is illustrated in Fig. 4.

For pressure washing with water/detergent, the above test should be run at 40°C (104°F) and 700 kPa gage (102 lbf/in^2 gage) with a flow rate of approximately 9460 cm^3/min (150 gal/min).

4.5.4 RELATED SPECIFICATIONS—None.

4.6 Fungus

4.6.1 DEFINITION—The fungus test is used to determine the resistance of equipment to fungi and to determine if it is adversely affected by fungi under conditions favorable for their development; for example, high humidity, warm atmosphere, and inorganic salts.

4.6.2 EFFECTS ON PERFORMANCE
  (a) Micro-organisms digest organic materials; thus, degrading the substrate, reducing the surface tension, and increasing moisture penetration.

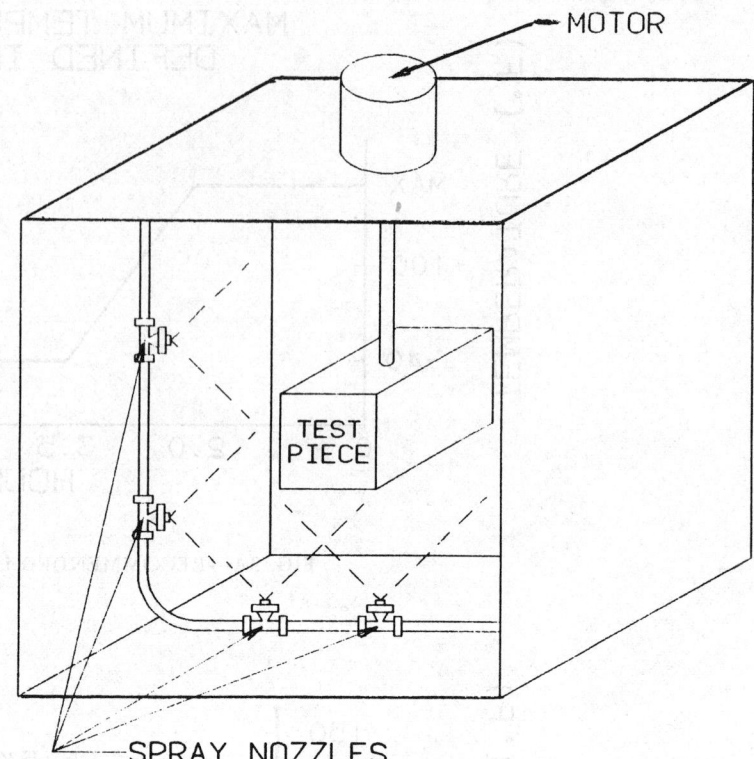

FIG. 4—STEAM CLEAN/PRESSURE WASH CHAMBER

  (b) Products of cellular metabolism diffuse out of the cells and cause physical and chemical changes to the materials.
  (c) Micro-organisms produce bridges across components which may result in electrical failure.
  (d) Resistance to biological attack can be determined by chemical analysis of the nutritive value of materials and material decomposition products used in the equipment.

4.6.3 RECOMMENDED TEST METHOD—The most common way to determine the effect of fungal growth on electronic equipment is to inoculate the test item with a fungal spore solution, incubate the inoculated item to permit fungal growth, and examine and test the item. Incubation normally takes place under cyclic temperature and humidity conditions that approximate environmental conditions and assure suitable fungal growth. A recommended test is MIL-STD-810D, Method 508.3.

NOTE: Conductive solutions used as a spore media and growth accelerator may affect operational tests.

4.6.4 RELATED SPECIFICATIONS—None.

4.7 Dust, Sand, and Gravel Bombardment

4.7.1 DEFINITION—Dust creates a harsh environment for chassis, underhood, and exterior-mounted devices, and can be a long-term problem in interior locations, such as under the dash and seats. Sand, primarily windblown, is an important environmental consideration for items mounted in the chassis, exterior, and underhood areas. Bombardment by gravel is significant for chassis, lower engine, and exterior-mounted equipment.

4.7.2 EFFECT ON PERFORMANCE—Exposure to fine dust causes problems with moving parts, forms conductive bridges, and acts as an absorbent material for the collection of water vapor. Some electromechanical components may be able to tolerate fine dust, but larger particles may affect, or totally inhibit, their mechanical action. While the exposure in desert areas is severe, exposure to a reasonable amount of road dust is common to all areas.

4.7.3 RECOMMENDED TEST METHODS—Dust, sand, and gravel bombardment tests should be at room temperature. The sample need not be operating, although functional tests should be performed prior to and after testing.

Dust conforming to that defined in SAE J726 MAY81 as coarse grade should be used. If this dust packs or seals openings in the test sample or if the sample contains exposed mechanical elements, the following alternate dust mixture may be used:

  (a) J726 Coarse or Equivalent 70%
  (b) 120 Grit Aluminum Oxide 30%

Components should be placed in a dust chamber with sufficient dry air movement to maintain a concentration of 0.88 g/m³ (0.025 g/ft³) for a period of 24 h.

An alternate method is to place the sample about 15 cm (6 in) from one wall in a 91.4 cm (3 ft) cubical box. The box should contain 4.54 kg (10 lb) of fine powdered cement in accordance with ASTM C 150-56, Specification for Portland Cement. At intervals of 15 min, the dust must be agitated by compressed air or fan blower. Blasts of air for a 2 s period in a downward direction assure that the dust is completely and uniformly diffused throughout the entire cube. The dust is then allowed to settle. The cycle is repeated for 5 h.

Condensation may be induced on electronic equipment following dust testing for a combined environment and operational test.

The recommended test for susceptibility of equipment to damage from gravel bombardment is SAE J400 JUN80, Recommended Practice Test for Chip Resistance of Surface Coatings. This document is intended to detect susceptibility of surface coatings to chipping, but the basic test equipment and procedures are useful for evaluation of the electronic equipment. The test consists of exposing the test sample to bombardment by gravel 0.96-1.6 cm (³⁄₈–⅝ in) in diameter for a period of approximately 2 min. The sample is positioned about 35 cm (13¾ in) from the muzzle of the gravel source. 470 cm³, (approximately 1 pt) of gravel (250-300 stones), is delivered under a pressure of 483 kPa gage (70 lbf/in² gage) over an approximate 10 s period. The process is repeated 12 times for a total exposure of 2 min. Judgment must be used in determining which sides should be exposed to the bombardment. Certainly all forward-facing surfaces not shielded by other parts are included. In many cases, the bottom and sides should also be exposed.

4.7.4 RELATED SPECIFICATIONS—Three specifications are referenced. The first, MIL-STD-202F, Method 110A, Sand and Dust, is a piece part test and is included for information and comparison. MIL-STD-810D, Method 510.2, is another reference. The second is SAE J726 MAY81, Air Cleaner Test Code, which defines the recommended dust. It also describes test apparatus. The third specification is SAE J400 JUN80, Test for Chip Resistance for Surface Coatings, which is recommended in part as a gravel bombardment guide. Continued integrity at the conclusion of the exposure is the passing criteria.

4.8 Altitude

4.8.1 DEFINITION—With the exception of air shipment of unenergized controls, operation in a vehicle should follow the anticipated operating limits. Completed controls are expected to be stressed over these limits of absolute pressure:

Condition	Altitude	Atmospheric Pressure
Operating	3.6 km (12 000 ft)	62.0 kPa absolute pressure (9 lbf/in² absolute)
Non-Operating	12.2 km (40 000 ft)	18.6 kPa absolute pressure (2.7 lbf/in² absolute)

4.8.2 EFFECT ON PERFORMANCE—With increased altitude, the following effects are generally observed:

4.8.2.1 Reduction in convection heat transfer efficiency.

4.8.2.2 Change in mechanical stress on packages that have internal cavities. The reference cavity of an absolute pressure sensor is an example of this.

4.8.2.3 A noticeable reduction in the high voltage breakdown characteristics of systems with electrically stressed insulator, conductor or air surfaces. This may result in surface cracking with eventual component failure.

4.8.3 RECOMMENDED TEST METHODS—The recommended test method is to operate equipment during the thermal cycles described in the Temperature Test Section, but with the added parameter of 62.0 kPa absolute pressure (9 lbf/in² absolute pressure). The equipment should operate under maximum load. Failure effects will be similar to those experienced with thermal cycle and shock. Non-operating tests should be done at a minimum temperature of −50°C (−58°F), if possible.

4.9 Mechanical Vibration

4.9.1 DEFINITION—Mechanical vibration is another key factor in electronic equipment design for the truck environment. For diesel powered trucks, mechanical vibration is likely the most important factor to consider in truck electronics design. Vibration levels may vary during vehicle operation from low severity to high severity when traversing rough roads at high speeds. The vibration characteristics may vary with the mounting location in addition to the vehicle mode of operation. Vibration levels and frequency content are significantly different for various mounting locations as shown in Figs. 5-15. Note that most of these examples are low frequency. Significant "G" levels above these frequencies are common.

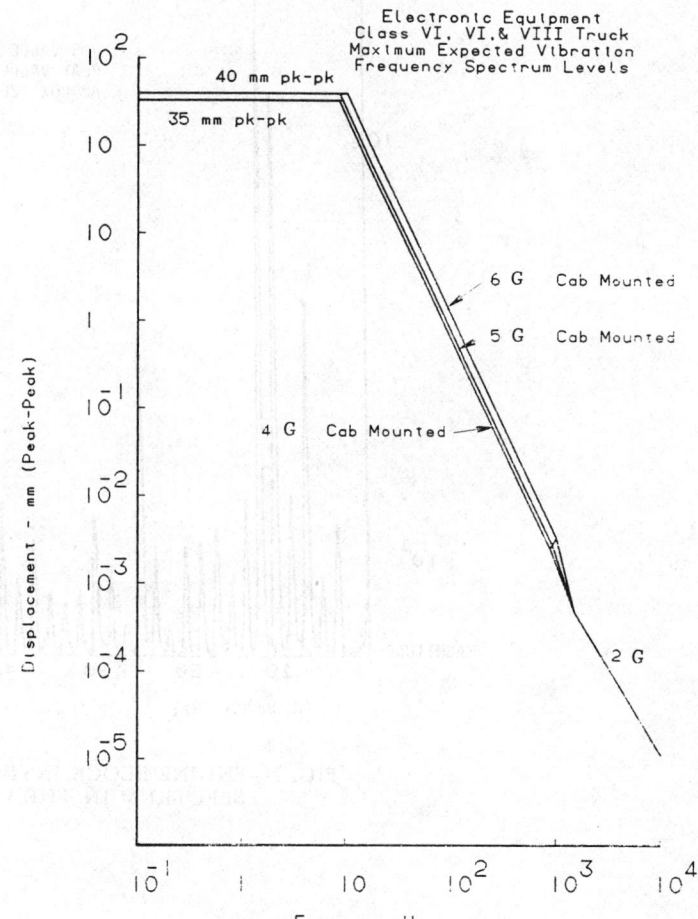

FIG. 5—MAXIMUM EXPECTED TRUCK VIBRATION LEVELS

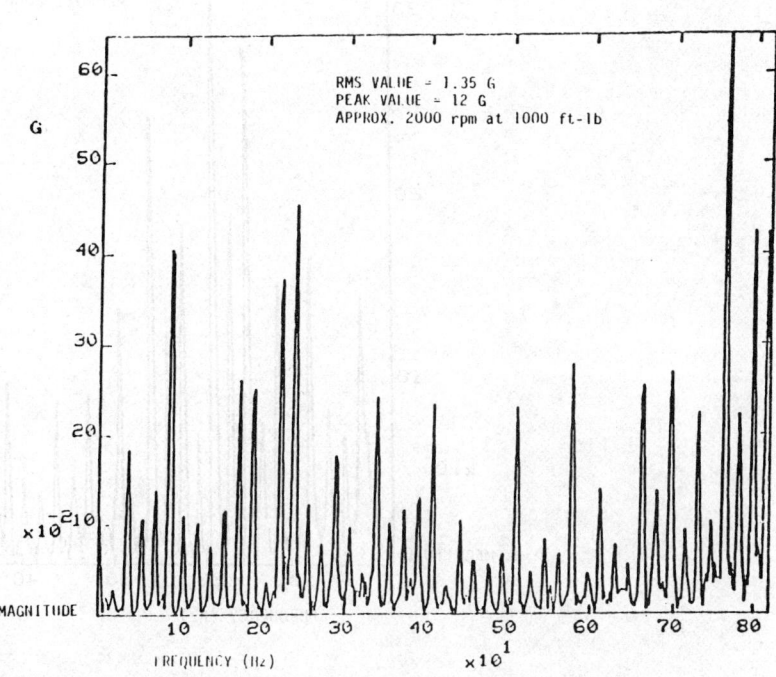

FIG. 6—ENGINE BLOCK INSTANTANEOUS ACCELERATION SPECTRUM IN THE HORIZONTAL DIRECTION

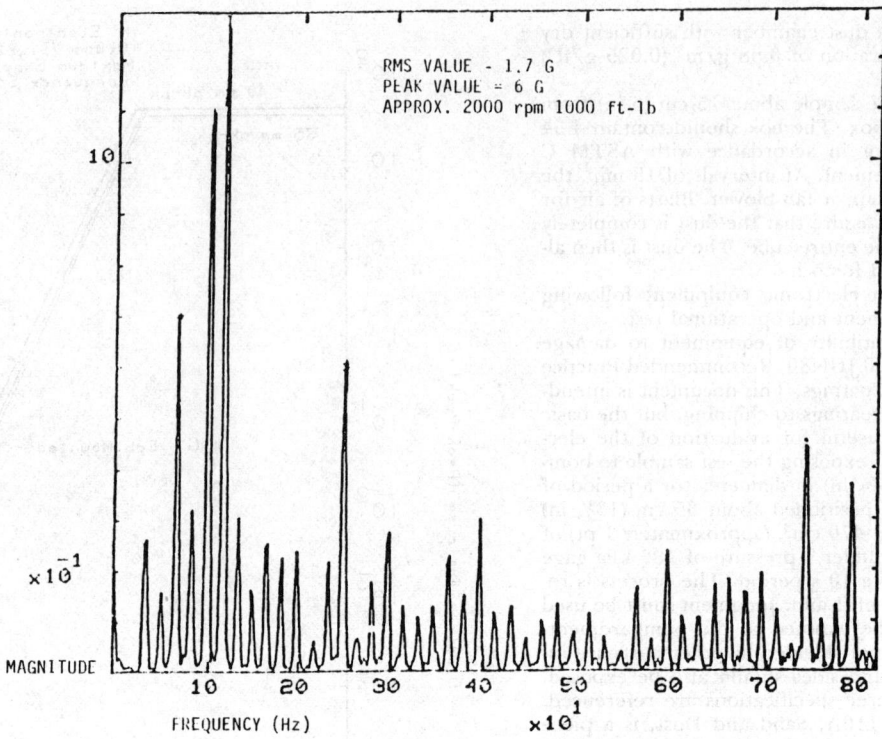

FIG. 7—ENGINE BLOCK INSTANTANEOUS ACCELERATION SPECTRUM IN THE VERTICAL DIRECTION

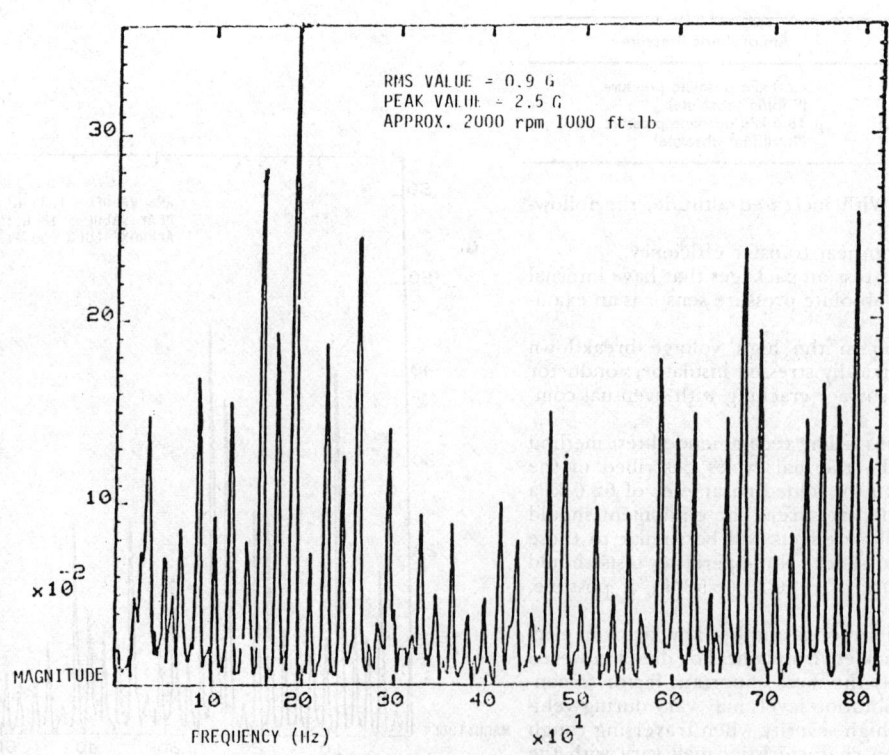

FIG. 8—ENGINE BLOCK INSTANTANEOUS ACCELERATION SPECTRUM IN THE AXIAL DIRECTION

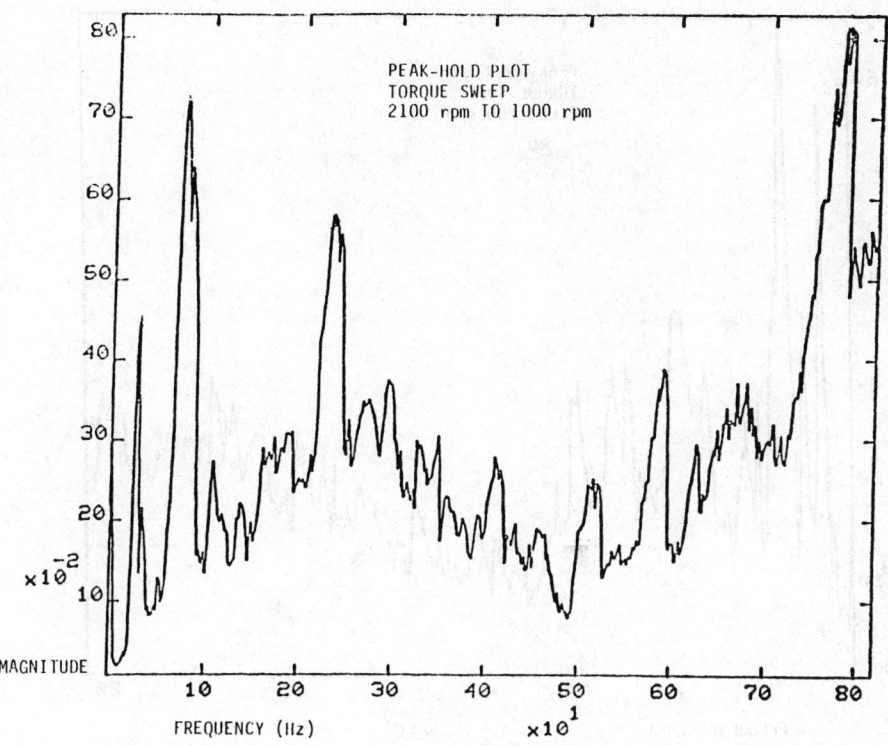

FIG. 9—ENGINE BLOCK PEAK ACCELERATION SPECTRUM
IN THE HORIZONTAL DIRECTION

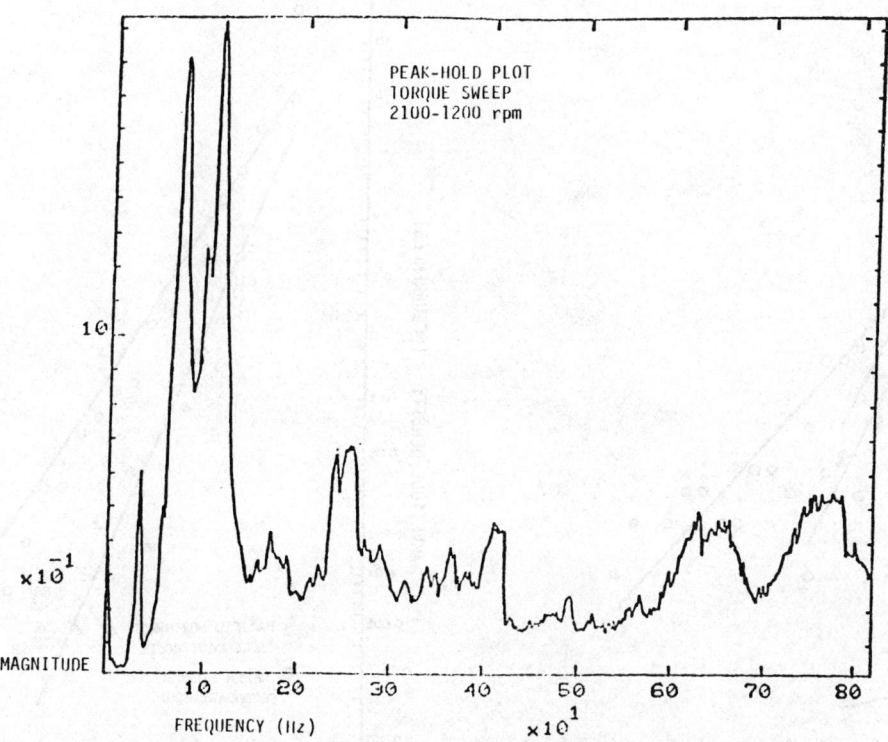

FIG. 10—ENGINE BLOCK PEAK ACCELERATION SPECTRUM
IN THE VERTICAL DIRECTION

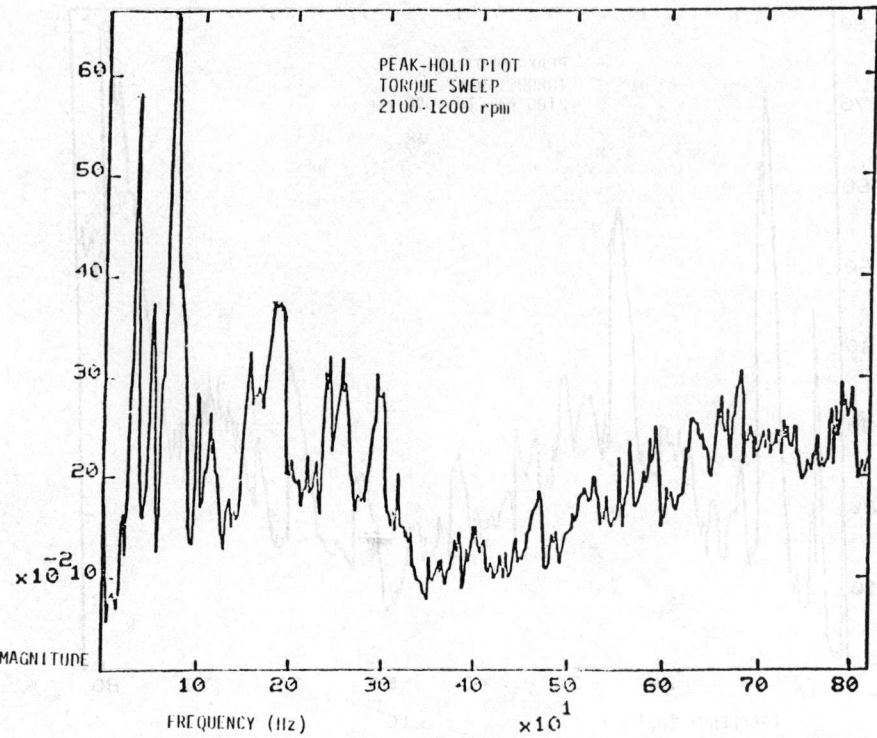

FIG. 11—ENGINE BLOCK PEAK ACCELERATION SPECTRUM IN THE AXIAL DIRECTION

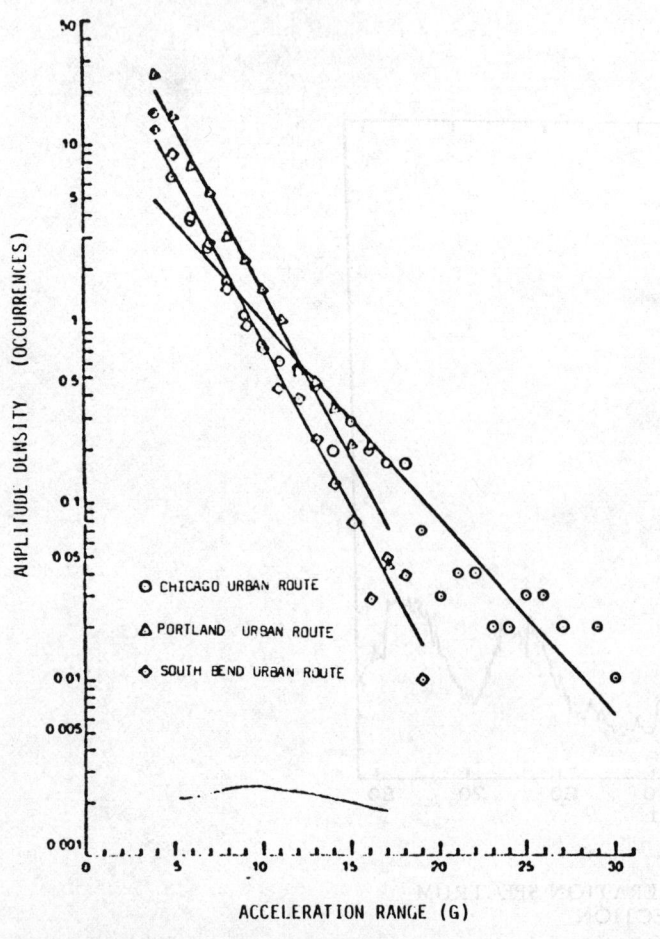

FIG. 12—DATA FROM FRONT AXLE ACCELEROMETER FOR THREE URBAN ROUTES

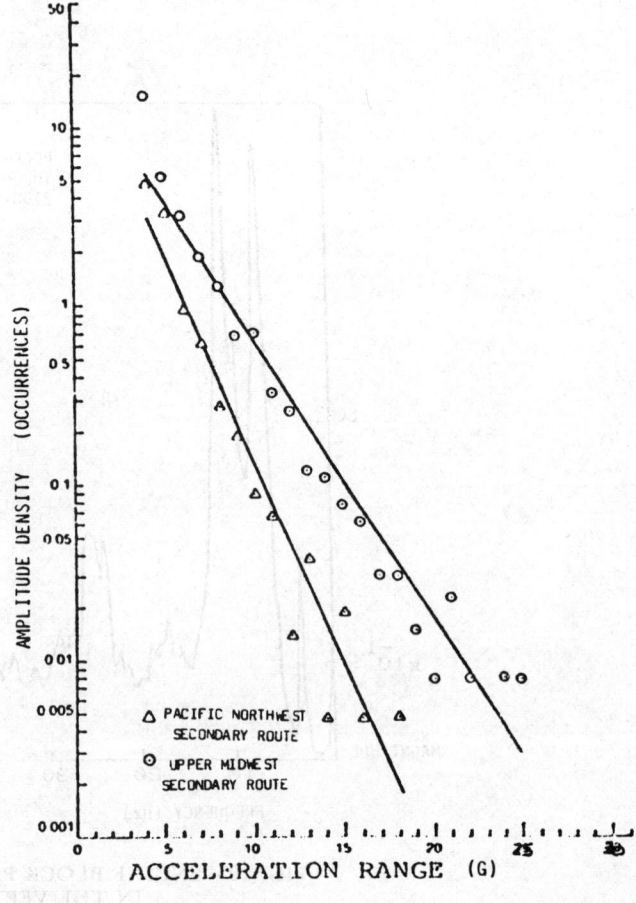

FIG. 13—DATA FROM FRONT AXLE ACCELEROMETER FOR TWO SECONDARY ROUTES

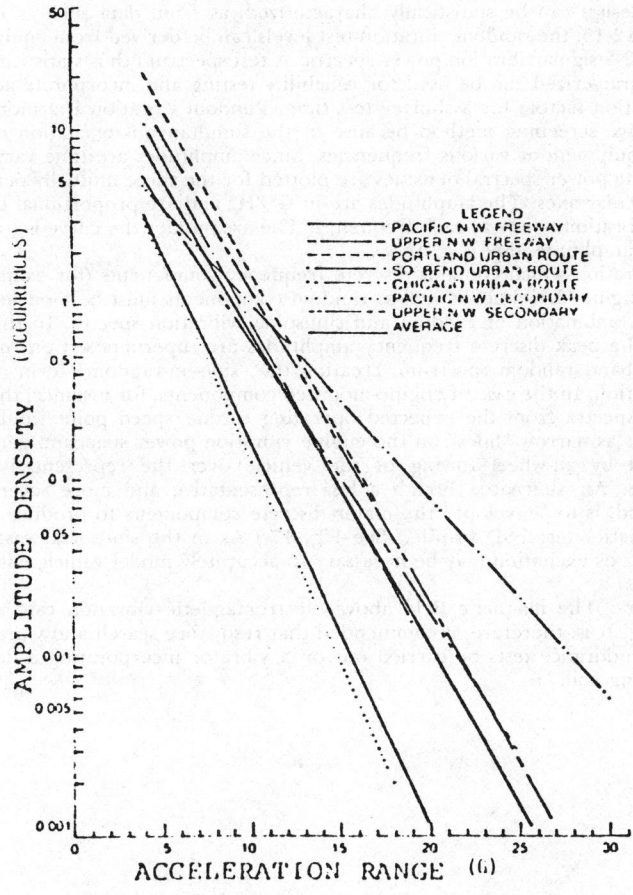

FIG. 14—RESPONSE HISTOGRAM DATA SUMMARY, FRONT AXLE ACCELEROMETER

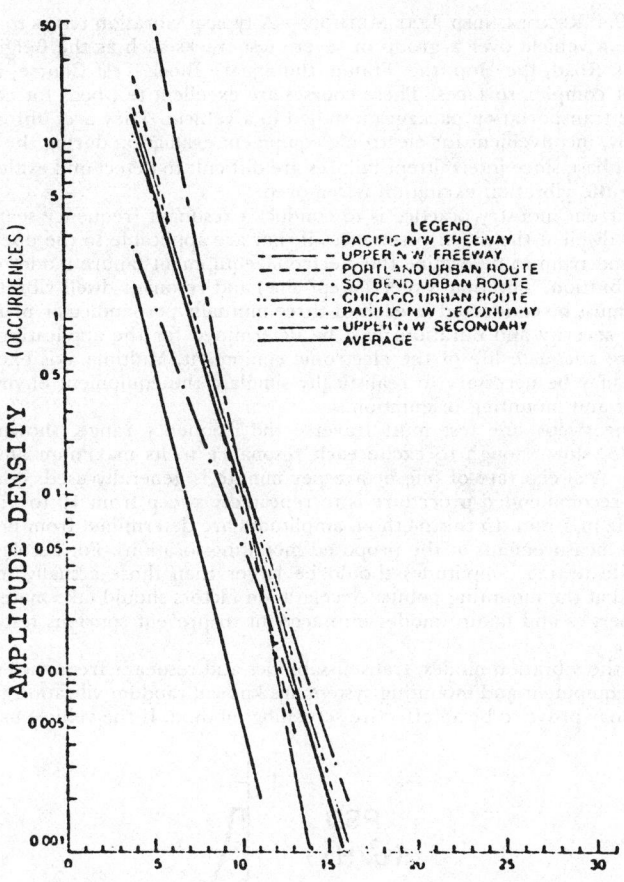

FIG. 15—RESPONSE HISTOGRAM DATA SUMMARY, INTERAXLE CROSSMEMBER ACCELEROMETER

4.9.2 Effect on Performance—A number of electronic equipment failure modes or performance degradations are possible during applied vibration. A partial list includes:

4.9.2.1 Loss of wiring harness electrical connection due to improper connector design or assembly, or both, or due to fretting corrosion.

4.9.2.2 Metal fatigue failure at stress concentration points due to resonant excitation of tuned mass structures in the electronic equipment.

4.9.2.3 Mount structure failures due to acceleration forces acting on the equipment mass.

4.9.2.4 Seal leakage due to mechanical flexing at the seal or other interface areas, which promotes intrusion of unwanted environmental factors such as moisture, in a similar phenomenon as described under temperature cycling effects.

4.9.2.5 Temporary abberation of equipment performance due to acceleration forces on control component masses. Examples are as follows:

4.9.2.5.1 Sensor measurement error due to motion of the sensor element, such as a pressure sensor, which gives incorrect information at applied frequencies because of the acceleration of the diaphragm and spring mechanism masses.

4.9.2.5.2 False operation of electro-mechanical components, such as a relay whose contacts open or close due to vibration-induced motion of the armature mass. The designer should be particularly alert to intermittent failures or faulty operation during applied vibration that may revert to normal operation after the vibration excitation is removed. Electronic performance tests conducted during vibration tests are recommended for functions that must perform during these conditions. In most cases, this is only practical under laboratory simulation of road tests.

4.9.3 Vibration Levels

4.9.3.1 Predicted Levels—The expected maximum vibration frequency spectrum profiles for various electronic equipment mounting locations are shown in Fig. 5. The expected cab mounted acceleration levels are ±4 G in any direction limited to a maximum peak to peak displacement of 35 mm (1.3 in). Truck chassis mounted electronic components may be subjected to acceleraton levels of ±5 G and engine mounted components to ±6 G with peak to peak displacements up to 40 mm (1.6 in). The expected acceleration levels should decrease linearly between 1000-1500 Hz to 2 G. The displacement roll off above 1500 Hz should be determined experimentally. An assumed value of 2 G above 1500 Hz should be used for unknown environments. Each mounting location must be considered independently since frequency spectrum levels can exceed those shown in Fig. 5. Engine mounted components may have acceleration levels twice the chassis levels at the mount resonant frequencies for some mounting system designs.

Note: Careful consideration should be given to the method used in mounting equipment so that it is not subject to major resonance input vibration. This may be achieved by either ensuring that the major resonance is outside the operating frequency range or by incorporating adequate damping techniques.

4.9.3.2 Example Levels—The acceleration spectrum of an engine is composed of numerous components, all of which are multiples of the cylic frequency of the engine. The components combine when the data is viewed as a function of time. Even though all of the spectral components are small, the time domain signal can be over 10 G. This can be seen in Figs. 6-8, which are the instantaneous values for an engine block in the horizontal, vertical, and axial directions, respectively. Since the spectrum changes with operating conditions, the peak-hold plots, which are the envelopes of the maximum spectral components, are included in Figs. 9-11 for the engine for a torque curve sweep. The RMS Value and Peak Value, shown in Figs. 6-8, are the voltages that would be seen if the instantaneous G level waveforms were fed into a broadband volt meter and oscilloscope.

Figs. 12-15 represent chassis data from a truck driven over many routes and shown as amplitude density versus G levels. Amplitude density is a useful unit that may be interpreted to be the number of occurrences of an amplitude of acceleration encountered on a per unit distance basis. This allows one to relate the environment to the distance driven instead of time as when only frequency units are used. RMS amplitudes per distance traveled is more appropriate in this method since it implies the vibration power level.

4.9.4 RECOMMENDED TEST METHODS—A typical vibration test is to operate a vehicle over a group of severe test tracks such as the Belgium Block Road, the Hop, the Tramp, the Square Block Test Course, and other complex surfaces. These courses are excellent test beds for complete transportation packages installed in a vehicle. They are, unfortunately, inconvenient for electronic equipment evaluation during the design phase since intermittent failures are difficult to detect and evaluate once the vibration excitation is removed.

Current industry practice is to conduct a resonant frequency search, then dwell at the major resonances if they are applicable to the operating spectrum to determine the electronic equipment failure modes due to vibration. The time sweep (swept sine) and resonant dwell vibration test must be conducted in each of three mutually perpendicular planes. Test severity and duration must be determined for the application to assure adequate life of the electronic equipment. Multiple axis excitation may be necessary to realistically simulate the equipment environment and mounting orientation.

The swept sine test must traverse the frequency range, shown in Fig. 5, slow enough to excite each resonance to its maximum amplitude. A sweep rate of one octave per minute is generally used. A second recommended procedure is to repeatedly sweep from 10 to 55 to 10 Hz in 1 min. In this method, amplitudes are determined from peak-hold measurements at the proposed mounting locations. For accelerated life testing, amplitudes should be larger than those actually measured at the mounting points. Acceleration factors should take material properties and failure modes into account to prevent spurious test results.

If the vibration modes, transmissabilities and resonant frequencies of the equipment and mounting system are known, random vibration testing may prove to be an effective screening method. If the vehicle usage and design can be statistically characterized, as from data such as in Figs. 12-15, the random vibration test levels can be derived from equivalent 2-3 sigma vibration power spectra. A test spectrum thus statistically characterized can be used for reliability testing and incorporate acceleration factors for a shorter test time. Random vibration is a more effective screening method because of the simultaneous excitation of the equipment at various frequencies. Since amplitudes are time varying, the power spectral densities are plotted for the three mutually perpendicular axes. The amplitudes are in G^2/Hz and are proportional to the vibration power at each frequency. The area under the curve is the RMS amplitude. (See Fig. 16.)

Vibration which contains discrete frequency components (for example, engine/driveline) as well as random components must be modeled by a combination of random and sinusoidal vibration spectra. In this case the peak discrete frequency amplitudes are superimposed on the broadband random spectrum, creating the "sine-on-random" form of excitation. In the case of engine-mounted components, for instance, the peak spectra from the expected operating torque/speed point would appear as narrow "lines" on the engine vibration power spectrum generated by all-wheel towing of the vehicle over the representative course. An alternate, though a less representative and more severe method, is to "envelope" the major discrete components to produce a "synthetic" test PSD profile. (See Fig. 17.) As in the sinusoidal case, multi-axis excitation may be necessary to accurately model vehicle conditions.

NOTE: The magnetic field above electromagnetic vibration can be strong. It is, therefore, recommended that resonance search and vibration endurance tests be carried out on a vibrator incorporating a degaussing coil.

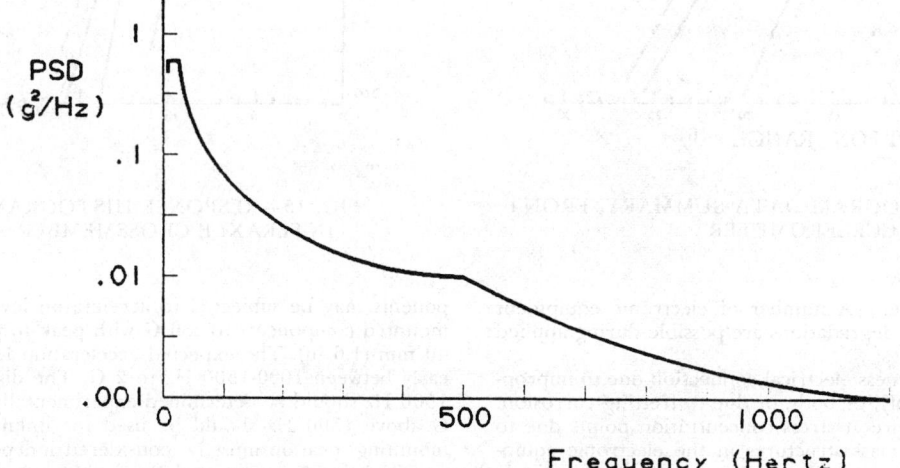

FIG. 16—FENDER PROFILE[1]

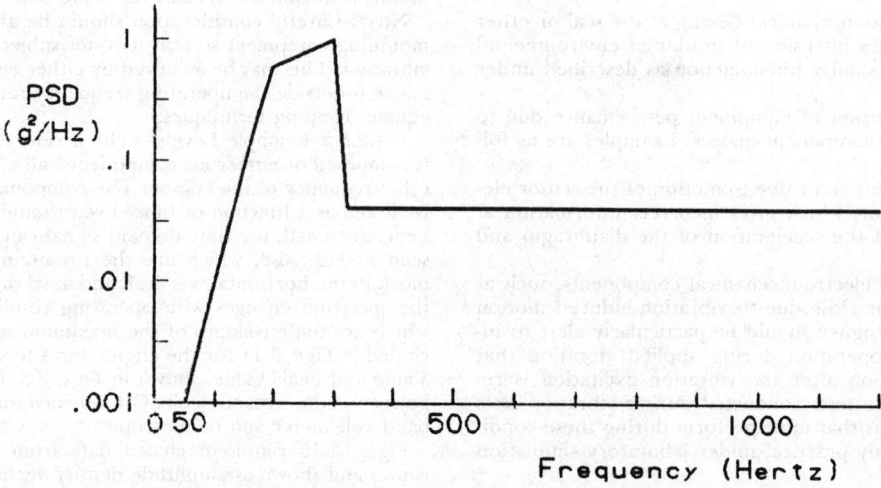

FIG. 17—SIX CYLINDER ENGINE PROFILE[1]

[1]Plots in Figs. 16 and 17 are excerpted from SAE Paper 840501, and do not necessarily represent truck environments.

4.9.5 RELATED SPECIFICATIONS—Three methods in MIL-STD-202F relate to vibration testing. Method 201A refers to tests between 10 and 55 Hz. Method 204D Vibration, High Frequency covers the ranges 10-500 Hz, 10-2000 Hz, and 10-3000 Hz, with several levels selected to suit expected service conditions. Both tests use swept sine vibration and offer procedural details and information on resident dwell periods. Method 214 Random Vibration, in MIL-STD-202F covers a wide range of test conditions that may be appropriate to body, chassis and axle mounted equipment. Guidance to test procedures and acceleration factors can be found in MIL-STD-810D, Method 514.3.

4.10 Mechanical Shock

4.10.1 DEFINITIONS—Contained in Section 4.11.2.

4.10.2 EFFECT ON PERFORMANCE—The automotive shock environment is logically divided into four classes:

4.10.2.1 Shipping and Handling Shocks—These are similar to those encountered in non-vehicle applications.

4.10.2.2 Installation Harness Shock—It is common production-line practice to lift and carry equipment by its harness. Therefore, it is recommended that the harness design incorporate secure fastening and suitable strain relief.

4.10.2.3 Operational Shock—The shocks encountered during the life of the vehicle that are caused by curbs, pot holes, etc., can be very severe. These vary widely in amplitude, duration, and number, and test conditions can only be generally simulated.

4.10.2.4 Crash Shock—This is included as an operating environment for safety systems. The operational requirements for these systems are limited to longitudinal shock at the present time.

4.10.3 Recommended Test Methods

4.10.3.1 Handling Shock Test—The equipment shall survive a single 0.9 m (3 ft) drop on a concrete floor, regardless of the point of impact, with only cosmetic damage. This test is performed with power off. Performance shall be checked before and after the test.

4.10.3.2 Transit Drop Test

4.10.3.2.1 Test Equipment—Shall comply with ASTM D 775 and D 880; TAPPI T-801 and T-802.

  (a) Drop tester, or hoist with suitable sling and tripping device, for packaged-products weighing less than 27 kg (60 lb). (The surface on which the packaged-product is to be dropped must provide a flat, firm, non-yielding base such as steel, concrete, etc.)

  (b) Incline impact tester, or alternative equipment, for packaged-products weighing from 28-45 kg (61-100 lb).

4.10.3.2.2 Test Procedure

Step 1—With the packaged-product in its normal shipping position, face one end of the container and identify the surfaces as follows:
  top, 1
  right side, 2
  bottom, 3
  left side, 4
  near end, 5
  far end, 6

Step 2—Identify edges by the numbers of those surfaces forming that edge, for example, the edge formed by the top and right side is identified as 1-2.

Step 3—Identify the corners by the numbers of three surfaces that meet to form that corner; for example, the corner formed by the right side, bottom and near end is identified as 2-3-5.

Step 4—The drop height shall be as follows:
  (1) Packaged-products up to 45 kg (100 lb).
    0.45-9.52 kg (1.00-20.99 lb) - 76 cm (30 in)
    9.53-18.59 kg (21.00-40.99 lb) - 61 cm (24 in)
    18.60-27.66 kg (41.00-61.99 lb) - 46 cm (18 in)
    27.67-53.31 kg (62.00-100 lb) - 31 cm (12 in)

  (2) As an alternative, when the packaged-product's configuration is such that dropping is impractical, ten incline impacts from a height necessary to achieve a minimum impact velocity of 1.75 m/s (5.75 f/s) may be performed in lieu of the 31 cm (12 in) drops. The impact sequence is delineated under Step 5.

Step 5—Drop or impact the packaged-product as specified under Step 4 in the following sequence:
  (1) the 2-3-5 corner
  (2) the shortest *edge* radiating from that corner
  (3) the next longest *edge* radiating from that corner
  (4) the longest *edge* radiating from that corner
  (5) flat on one of the smallest faces
  (6) flat on the opposite small face
  (7) flat on one of the medium faces
  (8) flat on the opposite medium face
  (9) flat on one of the largest faces
  (10) flat on the opposite large face

Step 6—Inspect both the package and the product. The packaged-product shall be considered to have satisfactorily passed the test if, upon examination, the product is free from damage and the container still provides reasonable protection to the contents.

4.10.3.3 Installation Harness Shock Test—A recommended test is to support the device and the far end of the installation harness at the same elevation, then release the device. Care should be taken to prevent the equipment from striking another object during this test. The drop should be repeated and the harness terminals or main relief area inspected for damage.

4.10.3.4 Operational Shock—With the possible exception of collision, the most severe vertical shock anticipated after production line installation may occur when driving over complex road surfaces. Trailer coupling or low speed loading dock collision provide the most severe horizontal shock in truck operation. The complex profile used to derive an operational shock test consists of a rise in the roadway followed by a depression or dip. Upon leaving the dip at 48 km (30 mph), the vehicle can become airborne. Severe shock may be experienced when the vehicle returns to the roadway. Another severe vertical shock is encountered in dump body trucks when loaded with rock and soil. Fig. 18 illustrates the shock measured on a steering column just below the steering wheel.

While this location is not typical of component mounting locations, it represents the most severe operational shock environment. This information is provided for guidance only; there are no generally accepted test procedures at the present time.

4.10.3.5 Crash Shock Test—Only limited and preliminary data on the effects of crash shock on the electronic equipment environment are available. However, a representative deceleration profile for a 48 km/h (30 mph) barrier crash is shown in Fig. 19. The following factors vary with each installation and should be considered in pretest analysis:
  (a) Equipment mass
  (b) Mounting system
  (c) Structure of the associated vehicle (crash distance, rate of collapse, etc.)
  (d) Particular engine package
  (e) Direction of crash

4.10.4 RELATED SPECIFICATIONS—Two specifications are recommended for consideration. The first, MIL-STD-202F, Method 203B, Random Drop, is designed to uncover failures that may result from the repeated random shocks that occur in shipping and handling. It is an endurance test. The second, MIL-STD-202F, Method 213B, Shock (Specified Pulse), is intended to measure the effect of known or generally accepted shock pulse shapes. It is intended that operational shock be reduced into a standard pulse shape to achieve a repeatable test method. Other valuable guidance can be found in MIL-STD-810D, Method 516.3.

4.11 General Heavy Duty Truck Electrical Environment—Factors unique to the truck/tractor that makes the vehicular environment more severe than those encountered in most electrical equipment applications are:
  (a) Interaction with other vehicular electronic/electrical systems on the truck
  (b) Voltage variations
  (c) Customer added equipment
  (d) Lack of maintenance
  (e) Complex external electromagnetic fields

Discussion of the electrical environment falls into two categories:
  (a) Electrical, Steady-State—Including variations in applied vehicle DC voltages with a characteristic frequency at or below 1 Hz.
  (b) Electrical Transient, and Noise—Including all noise and high voltage transients with characteristic frequencies above 1 Hz.

These conditions are discussed in Sections 4.11.1 and 4.11.2, respectively.

4.11.1 STEADY-STATE ELECTRICAL CHARACTERISTICS

4.11.1.1 Twelve Volt Systems

4.11.1.1.1 Definition—A normally operating vehicle will maintain supply voltages ranging from +11 to +16 V DC. However, under certain conditions, the voltage may fall to approximately 9 V DC. This might happen in an idling vehicle with a heavy electrical load (lights and air conditioning) and a fully discharged battery. Therefore, depending upon the application, the designer/user may specify the +9 to +16 VDC range. For specific equipment, such as those that must func-

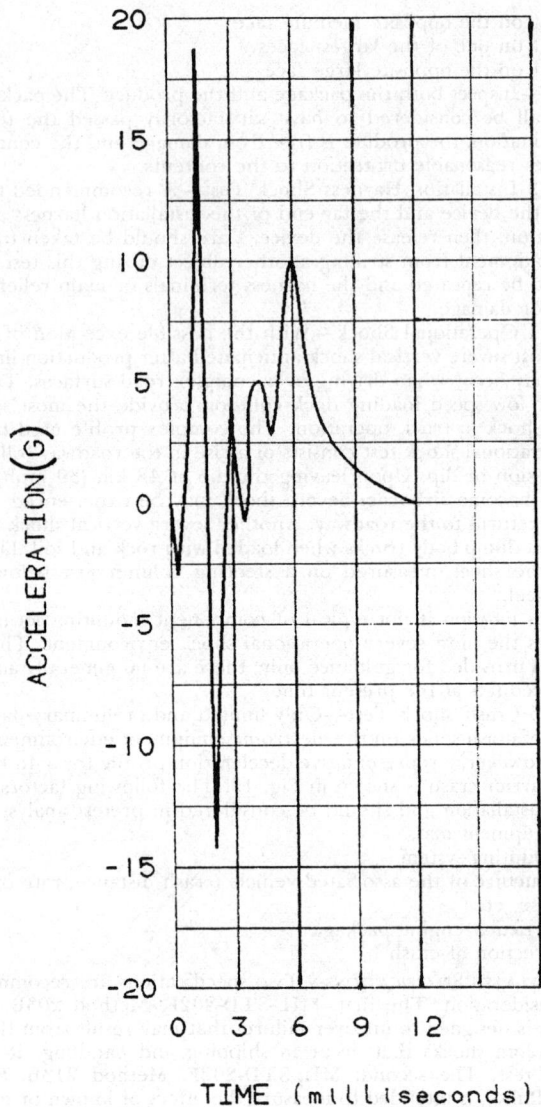

FIG. 18—OPERATIONAL SHOCK PROFILE

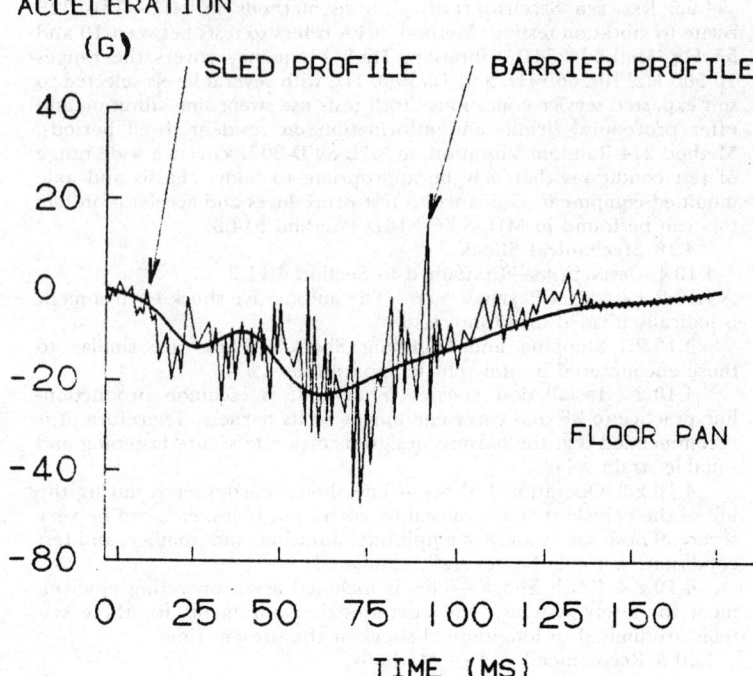

FIG. 19—48 KM/H (30 MPH) BARRIER AND SLED SHOCK PROFILES

tion during engine start, voltage may be specified as appropriate. Cold cranking of the engine with a partially depleted battery at −40°C (−40°F) can reduce the nominal 12 V to between 4.5 and 6 V DC.

Another condition affecting the DC voltage supply occurs when the voltage regulator fails, causing the alternator to drive the system at 18 V or higher. Extended 18 V operation will eventually cause boil-off of the battery electrolyte, resulting in voltages as high as 75-130 V. Other charging system failures can result in lower than normal battery voltages. General steady-state voltage regulation characteristics are shown in Table 2A.

Garages and emergency road services have been known to utilize 24 V sources for emergency starts, and there are reports of 36 V being used for this purpose. High voltages such as these are applied for up to 5 min and sometimes with reverse polarity. The use of voltages that exceed the vehicle system voltage can damage electrical components, and the higher the voltage, the greater the likelihood of damage.

NOTE: Since a design cannot preclude every contingency, this discussion of the application of voltage above normal system voltage is included for information only.

4.11.1.1.2 Effect on Performance—Equipment that must operate during the starting condition is generally designed to perform with slight degradation over a wide range of voltage. The designer is alerted to the possibility of failure from a combination of voltage and temperature variation. Over-voltage and high temperature, both from the external environment and internal dissipation, may cause excessive heat and result in failure. Under-voltage will probably result in degraded or non-performance. Conditions must be carefully examined to determine the true temperature and excitation voltage of the equipment.

4.11.1.1.3 Recommended Test Methods—Critical automotive equipment is performance-tested for operation within predetermined limits. Samples are also subjected to combinations of temperatures and supply voltage variations that are designed to represent the worst case stresses on control components. A typical cycle for this form of test is shown in Fig. 20.

The voltage applied and removed at the two points shown in Fig. 20 is generally 16 V, the maximum normal voltage. If the test is performed for the high voltage battery jump start condition of 24 V, a narrower temperature range is used. This is a destructive test that is often used as an indication of basic design environmental capability. The number of cycles expected before failure, the actual limit values for temperature and voltage, and the period of each cycle are dependent on the design goals for the equipment being considered.

TABLE 2A—TRUCK/TRACTOR (12 V SYSTEM) VOLTAGE REGULATION CHARACTERISTICS

Condition	Voltage
Normal operating vehicle	16 V max 14.2 nominal 9 V min[a]
Cold cranking at −40°C (−40°F)	3.5 - 6.0 V
Jumper starts	+24 V
Reverse polarity	−12 V
Voltage regulator failure	9 - 18 V
Battery electrolyte boil-off	75 - 130 V

[a]See Section 4.11.1 for a definition of normal voltage.

Samples of finished units are generally tested for extended operation at the peak voltage/temperature combination expected at the equipment's location. In the absence of an actual temperature combination, the values in Table 1 are recommended. These tests often run for extended periods and are particularly stringent for equipment in the underhood environment.

4.11.1.2 Twenty-Four Volt System

4.11.1.2.1 Definition—A normally operating vehicle will maintain supply voltages ranging from +22 to +32 V DC. However, under certain conditions, the voltage may fall to approximately 18 V DC. This

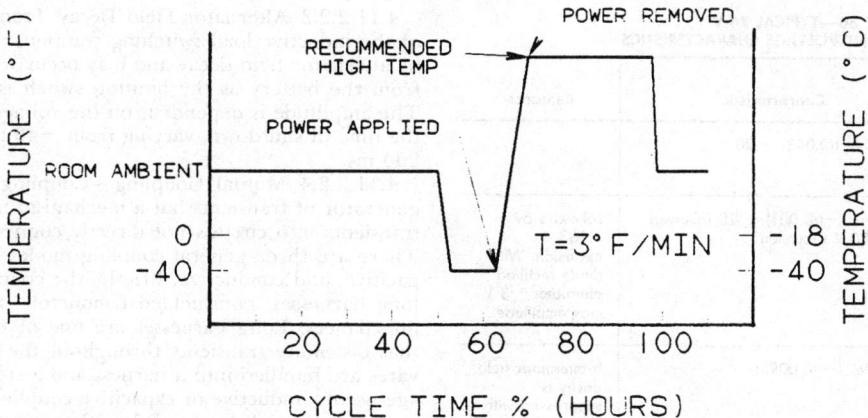

FIG. 20—COMBINED THERMAL AND ELECTRICAL STRESS PROFILE

can happen in an idling vehicle that has a heavy electrical load (lights and air conditioning) and a fully discharged battery. Therefore, depending upon the application, the designer/user may specify the +18 to +32 V DC range. For specific equipment that must function during engine start, voltage may be specified. Cold cranking of the engine with a partially depleted battery at −40°C (−40°F) can reduce the nominal 24 V to between 9 and 12 V DC.

Another condition affecting the DC voltage supply occurs when the voltage regulator fails, causing the alternator to drive the system at 36 V. Extended 36 V operation will eventually cause boil-off of the battery electrolyte, resulting in voltages as high as 75-250 V. Other charging system failures could result in lower than normal battery voltages. General steady-state voltage regulation characteristics are shown in Table 2B.

4.11.1.2.2 Effect on Performance—Equipment that must operate during the starting condition is generally designed to perform with slight degradation over a wide range of voltage. The designer is alerted to the possibility of failure from a combination of voltage and temperature variation. Over-voltage and high temperature, both from the external environment and internal dissipation, may cause excessive heat and result in failure. Under-voltage will probably result in degraded or non-performance. Conditions must be carefully examined to determine the true temperature and excitation voltage of the equipment.

TABLE 2B—TRUCK/TRACTOR (24 V SYSTEM) VOLTAGE REGULATION CHARACTERISTICS

Condition	Voltage
Normal operating vehicle	32 V max 28.4 nominal 18 V min[a]
Cold cranking at −40°C (−40°F)	9 - 12 V
Jumper starts	+48 V
Reverse polarity	−24 V
Voltage regulator failure	18 - 36 V
Battery electrolyte boil-off	75 - 130 V

[a] See Section 4.11.1 for a definition of normal voltage.

4.11.1.2.3 Recommended Test Methods—Critical automotive equipment is performance-tested for operation within predetermined limits. Samples are also subjected to combinations of temperatures and supply voltage variation that are designed to represent the worst case stresses on control components. A typical cycle for this form of test is shown in Fig. 20.

The voltage applied and removed at the two points shown in Fig. 20 is generally 32 V, the maximum normal voltage. If the test is performed for the high voltage battery jump start condition of 48 V, a narrower temperature range is used. This is a destructive test that is often used as an indication of basic design environmental capability. The number of cycles expected before failure, the actual limit values for temperature and voltage, and the period of each cycle are dependent on the design goals for the equipment being considered.

Samples of finished units are generally tested for extended operation at the peak voltage/temperature combination expected at the equipment's location. In the absence of actual temperature measurements, the values in Tables 1A and 1B are recommended. These tests often run for extended periods and are particularly stringent for equipment in the underhood environment.

4.11.2 Transient, Noise and Electrostatic Characteristics for 12 and 24V

4.11.2.1 Definition—Contained in paragraph 4.11.2.2.

4.11.2.2 Effect on Performance—Four principal types of transients are encountered on truck/tractor wire harnesses. These are load dump, inductive switching transients, alternator field decay, and mutual coupling. Generally, they occur singly, but there are cases where the latter two could occur simultaneously. Electromagnetic compatibility (EMC) characteristics vary considerably with type of vehicle and wiring harness. The equipment user and/or designer should determine the actual values of peak voltages, peak current, source impedance, repetition rate, frequency of occurrence at the interface between his equipment and the electrical distribution system, then design and test the electronic equipment to withstand values consistent with the expected use. Tables 3A and 3B summarize typical transient characteristics for 12 and 24 V systems, respectively.

4.11.2.2.1 Load Dump Transient—Load dump occurs when the alternator load is abruptly reduced. This sudden reduction in current causes the alternator to generate a positive voltage spike. The most severe case load dump is caused by disconnecting a discharged battery when the alternator is operated at rated load. Using the discharged battery load to create the load dump creates the worst situation for two reasons:

(a) The battery normally acts like a capacitor and absorbs transient energy when it is in the circuit.
(b) The partially discharged battery forms the single greatest load on the alternator and, therefore, disconnecting it creates the greatest possible step load change.

TABLE 3A—TYPICAL 12 V VEHICLE TRANSIENT VOLTAGE CHARACTERISTICS

Type	Max Amplitude (V)	Characteristic	Remarks
Load Dump	150	$171e^{(-t/0.400)} + 14$	Damage Potential
Inductive Load Switching	−600	$-614e^{(-t/0.001)} + 14$ followed by +80 V excursion	Logic Errors
Field Decay	−90	$-90e^{(-t/0.038)}$	Occurs at Shutdown Only for Separately Regulated Generators
Mutual	214	$+200e^{(-t/0.001)} + 14$	Logic Errors

TABLE 3B—TYPICAL 24 V VEHICLE TRANSIENT VOLTAGE CHARACTERISTICS

Type	Max Amplitude (V)	Characteristic	Remarks
Load Dump	250	$222e^{(-t/0.048)} + 28$	
Inductive Load Switching	−1000	$-106e^{(-t/0.001)} + 28$ followed by +80 V excursion	Followed by +133 V excursion. With diode rectified alternator −3 V max amplitude.
Alternator Field Decay	150	$-150e^{(-t/0.0095)}$	If alternator field decay is suppressed with a diode, this is not necessary.
Mutual	356	$-333e^{(-t/0.001)} + 28$	

This transient may be the most severe encountered in the vehicle and can result in component damage or fuse opening. It is most often initiated by defective battery terminal connections. On 12 V systems, transient voltages as high as 185 V or more have been reported with rise times of approximately 100 μs. Reports of decay time vary from 100 μs to 4.5 μs. The long duration decay occurs during vehicle turn off with a disconnected or dry vehicle battery. However, even the shortest time (100 μs) is relatively long, requiring that significant energy must be dissipated. Fig. 21 shows typical load dump transients for both 12 V and 24 V systems.

The load dump transient contains considerable electrical energy which must be safely dissipated to prevent damage to electronic equipment. This transient occurs randomly in time appearing as individual or repetitive pulses at random unknown rates due to vibration.

4.11.2.2.2 *Inductive Load Switching Transient*—Inductive transients are caused by solenoid, motor field, air conditioning clutch, and ignition system switching. These occur during vehicle operation whenever an inductive accessory is turned off. Severity is dependent on the magnitude of switched inductive load and line impedance. Unfortunately, measurements to date have not been taken with standardized procedures and were most probably observed with different loads.

These transients generally take the form of a large negative peak, followed by the smaller damped positive excursion. Transients of this nature may cause component damage or introduce logic or functional computational errors. Table 3A illustrates typical values for 12 V systems and Table 3B shows typical values for 24 V systems.

4.11.2.2.3 *Alternator Field Decay Transient*—This is a special case of the inductive load switching transient. It is a negative pulse caused by alternator field decay and may occur when the field is disconnected from the battery as the ignition switch is turned to the off position. The amplitude is dependent on the voltage regulator cycle and load at the time of shutdown, varying from −40 to −100 V and a duration of 200 ms.

4.11.2.2.4 *Mutual Coupling*—Coupling is not, strictly speaking, a generator of transients but a mechanism that is capable of introducing transients into circuits not directly connected to the transient source. There are three general coupling modes in the vehicle: magnetic, capacitive, and conductive. Briefly, the coupling problems are caused by long harnesses, nonshielded conductors, and common ground return impedances. Long harnesses are one of the principal coupling media that distribute transients throughout the vehicle. When a number of wires are bundled into a harness and a step change in current or voltage occurs, inductive or capacitive coupling between the conductor experiencing the change and the other wires can result. Multiple ground returns with different potentials cause "ground loops" and result in conductive coupling of transients.

4.11.2.2.5 *Other Effects*—It is possible that inductive switching of certain solenoids and the alternator decay transient condition can occur simultaneously. This hypotheses would account for the higher voltage transients that have been reported, but not explained. Measurement of 600 V transients on engine shutdown have been reported. Also to be considered are noise suppression capacitors that are sometimes placed on the fuse block, and some accessories that are applied to quiet radio interference. In some cases, capacitors may form tuned circuits with inductive loads, causing high voltage transient conditions.

Certain devices with high levels of stored energy, such as coasting permanent magnetic motors, may maintain line voltage for a finite interval after the ignition is shut off. Some equipment may perform in an unsatisfactory mode of operation under such conditions.

NOTE: Direct conduction through common circuits constitutes the most frequent path by which transients are introduced into electronic equipment.

4.11.2.3 *Electrical Noise*—Noise will normally have a repetition rate which is dependent on the characteristics of the interfering device or engine speed. There are two general types, as summarized in Table 4. A typical oscillogram of vehicle electrical noise is shown in Fig. 22.

4.11.2.3.1 *Normal Accessory Noise*—Generally, the normal compliment of accessories contributes less than 1.5 V peak over a frequency range of 50 Hz-10 KHz.

4.11.2.3.2 *Transceiver Feedback*—Some vehicle transceivers feedback energy to the power line at carrier frequency when the transmitter is keyed. These potentials are small, 15-20 mV peak, and are mentioned here only because they are at a predictable frequency.

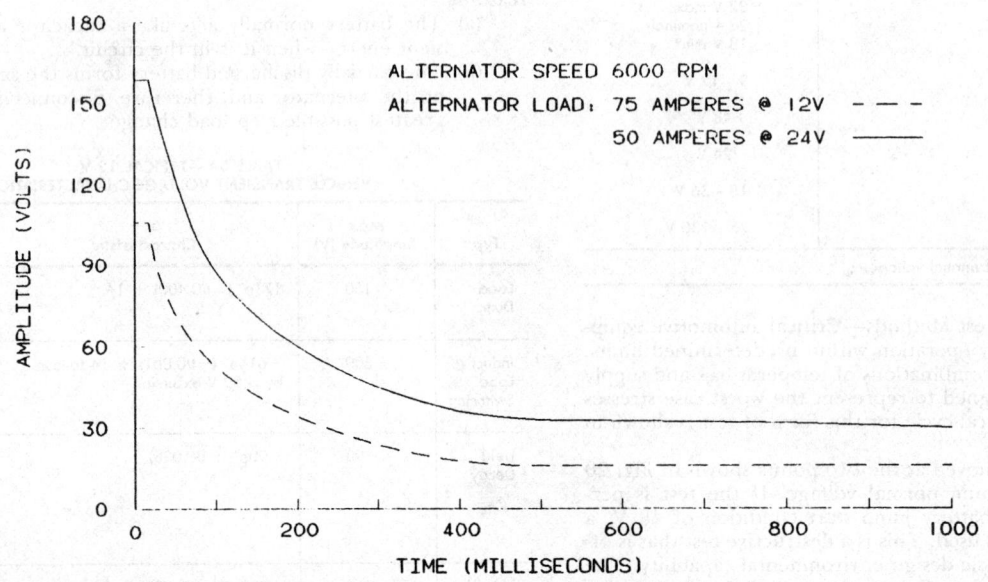

FIG. 21—LOAD DUMP TRANSIENT

4.11.2.3.3 Electrostatic Discharge—The electrostatic charge stored by the human body and then discharged into a device may cause operating anomalies. International Electrotechnical Commission (IEC) investigation indicates that discharging a 150 pF capacitor that has been charged to a potential of 15 kV through a 150 ohms resistor is adequate to stimulate this effect at the terminals or other areas subjected to handling. (See Ref. 6.4.) For in-vehicle wiring harness operation, apply 15 kV from 300 pF through 5 k ohm to exposed surfaces. (See SAE J1211.)

TABLE 4—SUMMARY OF HEAVY TRUCK ELECTRICAL CONTINUOUS NOISE CHARACTERS

Type	Max Amplitude	Duration	Repetition Rate	Remarks
Normal Accessory Noise	1.5 V Peak	Frequency	50 Hz to 10 KHz	Total Pulse is 3 V-PP
Transceiver	15 - 20 mV	Carrier	Frequency	Sinusoid

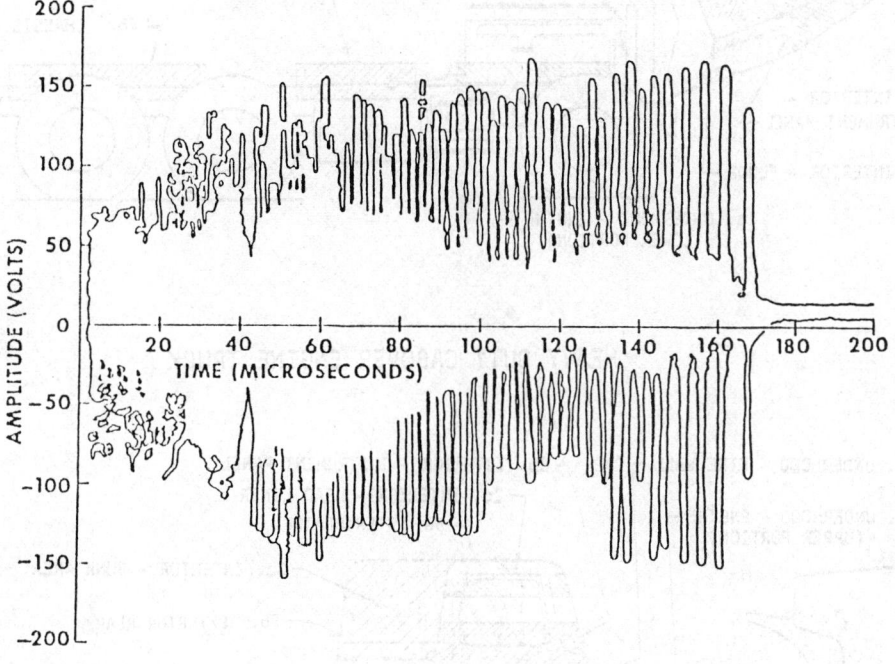

FIG. 22—12 V POWER LINE ELECTRICAL NOISE

4.11.2.3.4 External Sources of Radiated Energy—The vehicle is exposed to radiated energy from a multitude of sources that have the potential to disrupt normal system operation.

A more detailed discussion of these transient and noise effects is available in Ref. 6.3.

NOTE: The mechanisms governing the introduction of transients into an electronic assembly or its interrelated components are very complex. Thus, the equipment designer/packager must be familiar with the configuration of the total vehicle electrical system; for example, wire routing, shielding, grounding, filtering and decoupling practices, and equipment locations.

5. *Environmental Extremes by Location*—This section qualifies guidelines for the extreme operating conditions for five major vehicle equipment mounting sites:

(a) Underhood
  (1) Engine (Lower Portion)
  (2) Engine (Upper Portion)
  (3) Bulkhead
(b) Interior (Forward)
  (1) Floor
  (2) Instrument Panel
  (3) Head Liner
  (4) Inside Doors
(c) Interior (Rear)
  (1) Bunk Area
  (2) Storage Compartment
(d) Chassis
  (1) Forward
  (2) Rear
(e) Exterior of Cab
  (1) Under Floor
  (2) Rear
  (3) Top
  (4) Doors

The physical location of these sites is given in Fig. 23. Each site is individually discussed together with the following detail:

(a) A table listing extremes in operating conditions.
(b) Comments germane to other operating conditions of interest.
(c) Charts and other information not provided in Section 4.

This section contains data from environmental measurements made by North American vehicle manufacturers or original equipment suppliers. Decisions concerning each environmental factor and the test methods used to determine equipment performance and durability, should only be arrived at after examining the information in Section 4 of this document. In addition, the designer should be satisfied, by referring to pertinent test data, that the particular application falls within the described operating extremes. (See Section 3.)

5.1 Underhood-Engine and Bulkhead—Caution should be exercised in applying electronics equipment in the underhood region because of the wide range of environments. Data is summarized in Table 5.

5.1.1 TEMPERATURE

5.1.1.1 Temperature-Engine—Equipment in the vicinity of the exhaust system may experience temperature peaks that are beyond the survival limits of many insulation materials and electronic components.

Investigators have found that the lowest peak temperature areas are often forward in the lower compartment, near the interior or exterior radiator support hardware. The exterior has the disadvantage of being subject to more splash with resultant potential for moisture intrusion, corrosion, or thermal shock.

The temperature control mechanism for typical engine-mounted equipment relies heavily on the conduction of heat via the engine mass rather than convection via fins projecting into the airflow. Units which have a built-in source of heat energy may operate at temperatures above the highest coolant temperature. Equipment thermally interlocked by conduction with the engine has two advantages during normal operation:

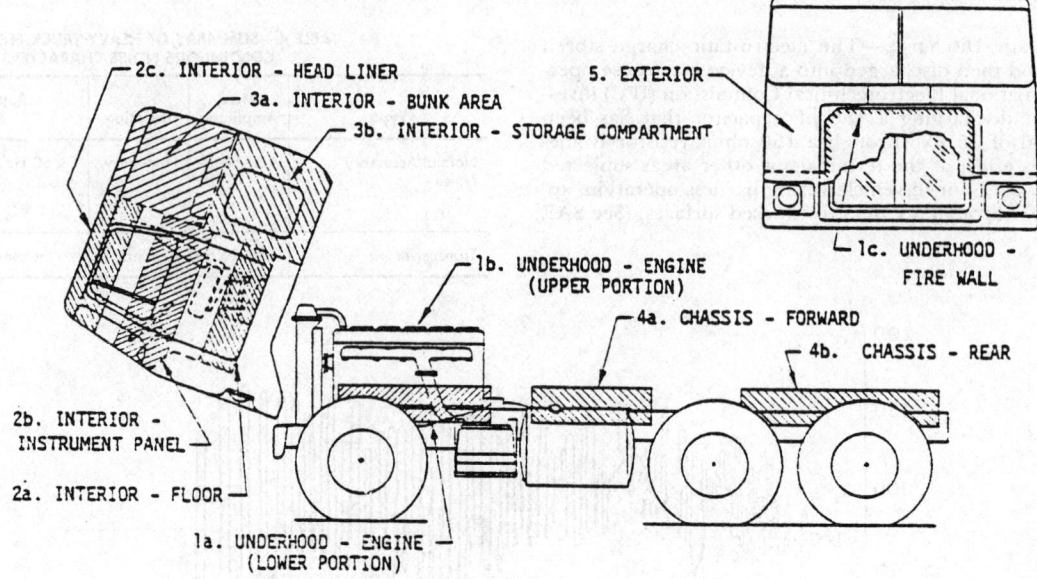

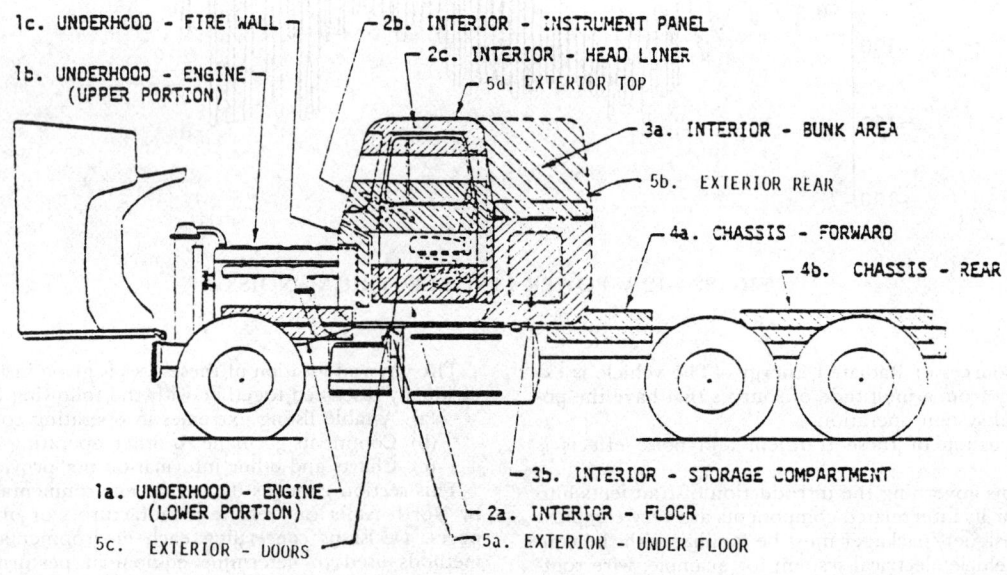

FIG. 23—VEHICLE ENVIRONMENTAL ZONES

(a) During engine operation, the upper temperature limit is set by the coolant peak temperature, which is in turn controlled by the thermostat.

(b) The time rate of change of temperature is limited by the combined engine and coolant system thermal mass.

Thermal shock as it is usually thought of for electronic components would not occur in engine-mounted electronics (assuming good thermal connection to the engine) in normal vehicle operation (normal engine warm-up and cool down) because of the large mass except for transient phenomena such as ice water splash in the winter or steam cleaning.

Consideration should also be given to the applications where fuel cooling is used. Fuel temperature can remain at ambient temperature plus 22°C (40°F) fairly constant throughout operation. Stabilization rates shown in Figs. 2A, 2B, and 2C are not precise and must be experimentally or analytically determined for a given application.

5.1.1.2 Temperature—Bulkhead—Temperature conditions are similar to the underhood-engine intake manifold, except that the primary method of heat flow is convection rather than conduction, and the resultant temperature slew rate is less. Equipment in this area generally relies heavily on convection due to the relatively low thermal conduction characteristics and unpredictable thermal interface between the equipment and the bulkhead sheet metal. The rate of change in temperature is therefore set by the thermal mass of the equipment itself, and convection due to air movement rather than conduction via the mounting surface. Thermal shock due to the impact of cold mud, slush, etc., is not likely in the upper bulkhead location. Consideration should be given to melted snow and ice leakage from the hood/windshield area.

The majority of investigators has experienced peak temperatures of 121°C (250°F), although one data source expects this to be 140°C (285°F). Locations on the bulkhead near or just above the exhaust manifold(s) which is at 649°C (1200°F), will experience higher temperatures. The effects of underhood exhaust processing components (catalytic reactors, etc.) will also raise the peak temperatures.

5.1.2 PEAK TEMPERATURE (HEAT SOAK) TEST—The temperature profile varies widely with individual engine/body combinations. Therefore, it is impossible to specify all conditions. Generally, worst case temperature operating conditions should be obtained by instrumenting a proposed location for the following operating conditions:

(a) The largest engine installation expected in that body style.
(b) Peak ambient temperature.
(c) Air condition ON.

The vehicle is driven at highway speed at rated GCVW for about 60 min and then parked. Underhood temperatures are monitored for the heat soak conditions as the thermal energy stored in the engine system is released in the absence of underhood airflow. Design modifications that contribute thermal energy to the underhood area, such as secondary air thermal reactors, engine charge air coolers (aftercoolers) or catalytic reactors, should be in place and operating for this test.

Test procedures of this type have revealed that the region to the rear of the engine compartment, and locations near radiated and conducted heat from the exhaust/reactor manifold tend to be high temperature areas. Present control practice has limited the location of electronic equipment to temperature situations similar to those shown for the intake manifold, although operation in the vicinity of the alternator heat source will probably add about 10°C (18°F) to the peak 121°C (250°F) shown for the intake manifold. Some experimenters expect the temperature near the radiator support structure to be no higher than 93°C (200°F).

Consideration should also be given to heat flow into the engine compartment from the front wheel suspension/brake and tire combination. Some consideration has been given to electronic equipment thermally interlocked with the engine cooling system, although the high pressure-temperature combination experienced during coolant boiling off may cause unacceptable catastrophic failure.

Rate of temperature change with time is also a consideration in this area, since cold starts will result in very rapid changes, as shown in Fig. 1.

High speed running with power take-off loads can cause maximum cooling system temperatures due to lack of ram air.

5.1.3 MECHANICAL SHOCK AND VIBRATION—As shown in Table 5.

5.1.4 HUMIDITY—This condition is similar to the associated engine condition, with the peak value shown in Table 5. The possibility of snow and ice intrusion, and hot ethylene glycol and water mixtures, due to cooling system failure, should also be considered.

5.1.5 SALT FOG AND SPRAY—This condition is often a factor, particularly on the lower outboard portions where the bulkhead joins the forward floor pan. Driving through salt slush can cause the entrance of salt spray through the radiator. The spray is then delivered to the engine compartment at high velocity by the fan. Spray due to this source is impacted on the dash panel, except for areas shielded by the engine and other underhood components.

5.1.6 IMMERSION AND SPLASH—Immersion is not generally required; however, this does provide a controlled way of checking for the effects of water and chemicals. Engine-mounted equipment will be exposed to many chemicals that are found in and around vehicles, including those listed in Section 4.4.

5.1.7 SAND, DUST, AND GRAVEL—Gravel is not generally a problem, except at the lower dash panel near the transition into the forward floor pan.

5.1.8 PRESSURE WASH/STEAM CLEAN—Many vehicles are exposed to high pressure washing and steam cleaning, or both. This could be directed at any point under the hood and probably represents the most likely chance for leakage into sealed equipment.

5.2 Interior-Forward—This area includes the floor, instrument panel and head liner.

5.2.1 INSTRUMENT PANEL—This includes the top of the dashboard and the near vertical section carrying the instruments and steering wheel. Data is shown in Table 6.

5.2.1.1 Temperature—Two temperature conditions are traceable to the climatic vehicle environment. Components not in direct sunlight experience temperatures from −40 to 85°C (−40 to 185°F). Components on the top surface of the instrument panel experience a greater heat buildup when closed vehicles are parked in the bright sun. Heat radiated, incident sunlight, and reradiated energy from the windshield can cause the temperature to build to 116°C (240°F) in this region. Heat due to underdash components, such as radio or heater, is also a contributing factor.

5.2.1.2 Humidity—As shown in Table 6. A tightly closed vehicle with wet upholstery experiences high internal humidity at high temperature.

5.2.1.3 Salt Spray—Not generally a problem at the instrument panel.

5.2.1.4 Immersion—Not anticipated, although liquid spills are possible on the upper dash surface.

5.2.1.5 Gravel, Sand, and Dust—Gravel not anticipated. Coatings of sand and dust are expected on all horizontal surfaces.

5.2.1.6 Oils and Chemicals—Mainly cleaning agents: waxes, soaps, and detergents.

5.2.1.7 Mechanical Shock and Vibration—As shown in Table 6.

5.2.1.8 Electrical-Steady State—Three operating conditions are recognized:

(a) Normal starting and running
(b) Cold starting
(c) Battery jump starting

NOTE: See Tables 2A and 2B.

5.2.1.9—Electrical-Transient—This condition appears to vary widely, depending upon the electrical distance of the equipment from the battery and the nearness of transient sources (for example, inductive motor, solenoids, the alternators). Typical data is shown in Table 3.

5.2.2 FLOOR—This covers all approximately horizontal surfaces, including the floor beneath the front seat(s), the footrest areas in front of the seat(s) and beneath the dashboard, and the interior surfaces of the drive tunnel. Data is shown in Table 7.

5.2.2.1 Temperature—As shown in Table 7, higher temperatures may be experienced directly over drivetrain components (transmission, etc.) and the exhaust system (including catalytic converters), although data is not available at this time.

5.2.2.2 Humidity—As shown in Table 7, standing water is possible in depressions due to rain entry through open windows or leaking seals. Also, water is carried into the vehicle by wet garments and packages.

5.2.2.3 Salt Spray—The water entry discussed in the humidity section may also be a saturated salt solution.

5.2.2.4 Immersion—Immersion is possible as discussed in the humidity section.

5.2.2.5 Gravel, Dust, and Sand—Gravel bombardment is not a condition, although a buildup of dust and sand is common.

5.2.2.6 Oils and Chemicals—Contaminants include the following:
Engine oils and additives (tracked in on occupants' shoes)
Cleaning solvents
Water and snow
Gasoline
Diesel fuel
Salt water
Water/ethylene glycol (from leaking heater)

5.2.2.7 Mechanical Shock and Vibration—No data available at this time.

5.2.2.8 Electrical-Steady State—Three operating conditions are recognized:

(a) Normal starting and running
(b) Cold starting
(c) Battery jump starting

NOTE: See Tables 2A and 2B.

5.2.2.9 Electrical-Transient—This condition appears to vary widely, depending on the electrical distance of the control site from the battery and the nearness of transient sources (for example, inductive motors, solenoids, the alternator).

5.2.3 HEADLINER—This covers all approximately horizontal surfaces over the cab floor. Data are shown in Table 8. The temperature does not include localized heat sources such as dome lights where temperatures up to 163°C (325°F) may be reached.

5.3 Interior-Rear—This area includes the bunk area and storage compartment.

5.3.1 BUNK AREA—See Table 9.

5.3.2 STORAGE COMPARTMENT—See Table 10.

5.4 Interior-Inside Doors—See Table 11.

5.5 Chassis—Data are summarized in Table 12.

5.5.1 TEMPERATURE—The heat sources encountered in the chassis area include, but are not limited to, the following:

Source	Peak Temperature
a. Exhaust	650°C (1200°F)
b. Brake System, Tires & Transmission, Differential Drive Train Components	180°C (350°F)
c. Engine	120°C (250°F)
d. Vehicle Ambient Peak Temperature	85°C (185°F)

The practical limitations of equipment components (with the possible exception of sensors) restricts the designer to locations with the peak temperatures given in c and d above. Again, the designer is urged to check his particular installation for the actual peak temperatures experienced under operating conditions.

5.5.2 HUMIDITY—As shown in Table 11.

5.5.3 SALT SPRAY—Assume all chassis components are subject to heavy salt spray unless otherwise shielded.

5.5.4 IMMERSION—Typical chassis components are subject to immersion.

5.5.5 DUST, SAND, AND GRAVEL—All chassis components in line with the wheel track that are not shielded are subject to continuous bombardment during vehicle operation on gravel roads. In non-track aligned portions of the chassis, some bombardment is experienced by equipment mounted on chassis surfaces. All chassis components are subject to heavy dust and sand environments.

5.5.6 OILS & CHEMICALS—The chassis is subject to all the oils and chemicals listed in Section 4.4, plus spilled cargo.

5.5.7 MECHANICAL SHOCK AND VIBRATION—Vibration data should be collected as required for chassis configuration and locations under study.

5.5.8 ELECTRICAL—STEADY STATE—Four operating conditions are recognized:

	12 V System	24 V System
a. Nominal Regulator Setting	14 V	28 V
b. Normal Range	12-15 V	18-32 V
c. Starting Low −18°C (0°F)	7 V	9-12 V
d. Booster Starting—Extreme High	24 V	48 V

5.5.9 ELECTRICAL TRANSIENT—Typical transient electrical signals experienced by chassis equipment will be due to load dumping and inductive load switching. (See Table 11.) In addition radio frequency interference can affect chassis-mounted equipment. Equipment should be subjected to radio signals which might be expected to be encountered. (See Ref. 6.3.)

5.6 Exterior—The exterior consists of all outward and external vehicle surfaces (top, rear, doors, under floor) above the chassis. This includes the forward grill area and potential mounting areas just above the bumpers. Data are summarized in Table 13.

5.6.1 TEMPERATURE—Since all surfaces are away from internal vehicle heat sources, the temperature is controlled primarily by the climatic ambient conditions and solar heating of exposed surfaces. These are discussed in Section 4.1 and shown in Table 13. Thermal shock due to splash or immersion, particularly on the front of the vehicle, should be anticipated. Under floor temperatures are also influenced by heat radiated or convected from exhaust system and engine/driveline components.

5.6.2 HUMIDITY—Shown in Table 13.

5.6.3 SALT SPRAY—Most exterior surfaces are subject to heavy salt spray, with the possibility of crystalline salt build-up.

5.6.4 IMMERSION—Equipment mounted approximately below the vehicle axle line is possibly subject to occasional immersion. Components above this line experience splash.

5.6.5 GRAVEL, DUST, AND SAND—Components on the front and underside of vehicles are subject to bombardment from vehicles ahead. Dust and sand impinge on all surfaces.

5.6.6 ELECTRICAL—STEADY STATE—Three operating conditions are recognized:
  (a) Normal starting and running
  (b) Cold starting
  (c) Battery jump starting

5.6.7 ELECTRICAL—TRANSIENT—This condition appears to vary widely, depending upon the electrical distance of the equipment from the battery and the nearness of the transient sources (for example, inductive motors, solenoids, the alternator). (See Table 13.)

5.6.8 EXTERIOR APPEARANCE/WASHABILITY—Exterior mounted equipment must resist degradation in appearance due to solar radiation, ozone, air pollutants and vehicle washing. Mounting and orientation of drain/vent holes should allow for mechanical wash rack service.

5.6.9 MECHANICAL SHOCK AND VIBRATION—Location of components near the center of large, edge supported areas of exterior panels can result in large displacements at the resonant frequency of the panel.

6. References

6.1 SAE J400 JAN85—Test for Chip Resistance of Surface Coatings.

6.2 SAE J726 MAY81—Air Cleaner Test Code.

6.3 SAE J1113 JUN84—Electromagnetic Susceptibility Procedures for Vehicle Components (Except Aircraft).

6.4 SAE J1211—Recommended Environmental Practices for Electronic Equipment Design.

6.5 MIL-STD-810D, 19 July 1983—Environmental Test Methods and Engineering Guidelines.

6.6 International Electrotechnical Commission (IEC), Publication 801 (1984)—Electrostatic Discharge (for Industrial Process Measurement and Control).

6.7 MIL-STD-202F, 01 April 1980 Test—Methods for Electronic and Electrical Component Parts.

6.8 ASTM C 150-84—Specification for Portland Cement.

6.9 ASTM B 117-73—Standard Method of Salt Spray (Fog) Testing.

6.10 ASTM D 775-80.

6.11 ASTM D 880-79.

6.12 TAPPI T801 OM-83.

6.13 TAPPI T802 OM-81.

6.14 SAE Paper 840501.

TABLE 5—ENGINE—ENVIRONMENTAL DATA

	Temperature		Humidity	Salt Spray	Immersion and Splash	Steam Clean Pressure Wash Direct Spray	Sand & Dust	Shock Vibration	Altitude	Electrical
	Min	Max	High Low Frost							
Underhood	(over ambient)									
Lower	−40°F (−40°C)	100°F (38°C)	Sec. 4.2	Sec. 4.3	Sec. 4.4	Sec. 4.5	Sec. 4.7	Figs. 5-16	Sec. 4.8	Sec. 4.11
Upper	−40°F (−40°C)	401°F (205°C)								
Bulkhead	−40°F (−40°C)	100°F (38°C)								

TABLE 6—INTERIOR—INSTRUMENTAL PANEL ENVIRONMENTAL DATA

	Temperature			Humidity (%RH)			Salt Spray	Immersion	Sand Dust & Gravel	Oil & Chemical	Mechanical Shock & Vibration	Electrical	
	Low	High	Slew Rate	High	Low	Frost						Steady-State	Transient
Nominal	−40°C (−40°F)	85°C (185°F)	NA	98% at 38°C (100°F)	0	yes	no	Partial	Dust only	Sec. 4.4	Figs. 11-16	Tables 2A, 2B	Table 3
Top Surface	−40°C (−40°F)	113°C (235°F)	NA	80% 66°C (150°F)									

TABLE 7—INTERIOR—FLOOR ENVIRONMENTAL DATA

Temperature			Humidity (%RH)			Salt Spray	Immersion	Sand Dust & Gravel	Oil & Chemical	Mechanical Shock & Vibration	Electrical	
Low	High	Slew Rate	High	Low	Frost						Steady-State	Transient
−40°C (−40°F)	85°C (185°F)	NA	98% at 38°C (100°F)	RH 0	—	no	no	Sec. 4.7	Sec. 4.4	Not measured	Tables 2A, 2B	Table 3

TABLE 8—INTERIOR—HEADLINER ENVIRONMENTAL DATA

Temperature			Humidity (%RH)			Salt Spray	Immersion	Sand Dust & Gravel	Oil & Chemical	Mechanical Shock & Vibration	Electrical	
Low	High	Slew Rate	High	Low	Frost						Steady-State	Transient
−40°C (−40°F)	104°C (220°F)	NA	98% at 38°C (100°F); 80% at 66°C (150°F)	RH 0	—	no	no	Sec. 4.7	Sec. 4.4	Not measured	Tables 2A, 2B	Table 3

TABLE 9—INTERIOR—BUNK AREA ENVIRONMENTAL DATA

Temperature			Humidity (%RH)			Salt Spray	Immersion	Sand Dust & Gravel	Oil & Chemical	Mechanical Shock & Vibration	Electrical	
Low	High	Slew Rate	High	Low	Frost						Steady-State	Transient
−40°C (−40°F)	85°C (185°F)	NA	98% at 38°C (100°F); 80% at 66°C	RH 0	yes	no	no	Sec. 4.7	Sec. 4.4	Not measured	Tables 2A, 2B	Table 3

TABLE 10—INTERIOR—STORAGE COMPARTMENT ENVIRONMENTAL DATA

Temperature			Humidity (%RH)			Salt Spray	Immersion	Sand Dust & Gravel	Oil & Chemical	Mechanical Shock & Vibration	Electrical	
Low	High	Slew Rate	High	Low	Frost						Steady-State	Transient
−40°C (−40°F)	104°C (220°F)	NA	98% at 38°C (100°F); 80% at 66°C (150°F)	RH 0	no	no	no	Sec. 4.7	Sec. 4.4	Not measured	Tables 2A, 2B	Table 3

TABLE 11—INTERIOR—INSIDE DOORS ENVIRONMENTAL EXTREMES DATA

Temperature		Thermal Shock	Humidity (%RH)			Salt Spray	Immersion	Sand Dust & Gravel	Oil & Chemical	Mechanical Shock & Vibration	Electrical	
Low	High		High	Low	Frost						Steady-State	Transient
−40°F (−40°C)	175°F (78°C)	no	98% at (100°F) (38°C)	0	yes	yes	yes	yes	Sec. 4.4	See Fig. 5	Tables 2A, 2B	Table 3

TABLE 12—CHASSIS—ENVIRONMENTAL DATA

	Temperature			Humidity (%RH)			Salt Spray	Immersion	Sand Dust & Gravel	Oil & Chemical	Electrical	
	Low	High	Slew Rate	High	Low	Frost						Transient
Isolated	−40°C (−40°F)	85°C (185°F)	NA	98% at 38°C (100°F)	0	yes	Sec. 4.3	Sec. 4.4	Sec. 4.7	Sec. 4.4		Table 12
Near Heat Source	−40°C (−40°F)	121°C (250°F)	NA	66°C (150°F)	0	yes						
At Drive Train High Temp Location	−40°C (−40°F)	177°C (350°F)	NA	80%	0	yes	—	—	—	—	—	—

TABLE 13—EXTERIOR—ENVIRONMENTAL DATA

	Temperature			Humidity (%RH)			Salt Spray	Immersion	Sand Dust & Gravel	Oil & Chemical	Mechanical Shock & Vibration	Electrical	
	Low	High	Slew Rate	High	Low	Frost						Steady-State	Transient
Normal	−40°C (−40°F)	85°C (185°F)	NA	95% at 38°C (100°F)	0	yes	Sec. 4.3	Sec. 4.4	Sec. 4.7	Sec. 4.4	Figs. 11-16	Tables 2A, 2B	Table 3

DESIGN/PROCESS CHECKLIST FOR VEHICLE ELECTRONIC SYSTEMS—SAE J1938 OCT88

SAE Information Report

Report of the Electrical and Electronic Systems Technical Committee approved October 1988.

1. Introduction—To obtain a high degree of quality and reliability, a wide variety of subjects need to be addressed when designing a vehicle electronic system. No single designer can be expected to have the experience necessary to consider all aspects of a design. Such experience is often spread throughout an organization and not concentrated on any one project.

2. Purpose—The main purpose of this checklist is to provide a systematic approach to insuring that all aspects of an electronic systems design are addressed. Such a list would be useful for design reviews, "fresh eyes" reviews, and for education/training.

3. Scope—The following subjects reflect the automotive environment and are based on good engineering practices and past "lessons learned" experiences. Since it is impossible to be all inclusive and cover every aspect of quality and reliability, this document should be used as a guide for preparation of a checklist that reflects the accumulated "lessons learned" at a particular company.

4. Format—To keep in a form that will be readily used, each subject will be addressed in an abbreviated format using short, direct, to-the-point phrases. It is not the intent of this document to give a lot of detail, but to point out those subjects that need to be investigated and acted upon.

5. Design Checklist

5.1 Component Selection/Application—One of the first major concerns for a reliable design is part selection and application. Efforts to use best-in-class suppliers cannot be overemphasized. Much of the input for this topic will come from the corporate electronic components department.

References: 11, 12, 18, 19, 20, 21, 22, data bases of Reliability Analysis Center of IIT Research Institute.

— Resistors: types, tolerances, packages, reliability concerns, failure modes (for example, opens most common), power/temperature derating.
— Capacitors: types, tolerances, packages, reliability concerns, failure modes (for example, shorts, value change most common), power/temperature derating.
— Transistors/Diodes: types, packages, reliability concerns, failure modes, voltage/current/temperature derating.
— I.C.'s: types, packages, reliability concerns, failure modes, voltage/current/temperature derating.
— I.C. Sockets: types, reliability concerns, failure modes.
— Connectors/Interconnects:
  . Types, reliability concerns, failure modes.
  .. Between PCB's (for example, individual wires, flat cable, flex cable).
  .. Pin/socket connector to wiring harness.
  .. Blade/socket connector to wiring harness.
  . Stress relief.
  . "Dry" circuits - low voltage, film buildup.
— Printed Circuit Boards (PCB's):
  . Reliability concerns, failure modes.
  .. Opens, shorts, warpage.
  .. Edge connector (if applicable) to wiring harness - reliability subject to many parameters, for example, plating uniformity, tolerances.
  . Material selection.
  . Copper thickness (1, 2, and 3 oz).
  . Tolerances.
  . Thermal considerations (that is, matching of thermal expansions).
  . Manufacturability criteria (see Section 6).
  . EMC criteria (see paragraph 5.7.7).
— Thick film substrates:
  . Reliability concerns, failure modes.
  . Material selection.
  . Tolerances.
  . Thermal considerations (that is, matching of thermal expansions).
  . Manufacturability criteria (see Section 6).
  . EMC criteria (see paragraph 5.7.7).
— Potting, conformal coating: where used, types, limitations.

- Identification of critical reliability components: for example, power transistors, power zeners, etc.
- Special requirements for these critical components: derating, screening, handling, failure mode response.
- Components to avoid: for example, variable resistors if fixed can be used, hand inserted parts if auto insertion viable.
- Part availability.
- Part specifications: for example, MIL-STD-883, MIL-STD-202.
- Testing sample size: statistical significance, attribute or variable data, cost/time/test facility limitations.
- Electrostatic Discharge: most sensitive components, precautions, handling.
- Vendor quality/reliability control program.
- Acceptable Quality Level (AQL), in parts/million (ppm), required: how verified.
- Process flow/control plans.
- Process change procedures.
- Closed loop failure analysis, corrective action plan.
- Degree of component Statistical Process Control (SPC) used: where used, adherence, effect on AQL.

5.2 Thermal Considerations (Components/Assemblies)—Temperature has a major effect on reliability. In fact, as the temperature of a system rises, thermal failures almost completely overweigh failures from other causes.

References: 3, 11, 12, 13, 14, 15, 16, 17, 22.
- Conduct thermal survey of environment (under hood, passenger compartment, etc.): start temperature (heat, cold soak), warm up time, operational temperature (range, rate of temperature change, frequency of change), number of cycles, cooling effects.
- Assembly (module) temperature environment vs. reliability: field experiences.
- Component (resistor, capacitor, transistor, diode, etc.) thermal analysis: worst case analysis (electrical loading, environment), heat sinking, derating (safety margins).
- Assembly (module) thermal analysis: worst case analysis (electrical loading, environment), heat sinking, derating (safety margins).
- Thermal analysis using thermal resistance values is best case: does not consider nonlinearity (hot spots), interface bonds <100% of area.
- Thermal testing evaluation: for example, thermocouple critical areas in module and test under worst case electrical loading and environment in temperature chamber. Vehicle evaluation shall also be done (temperature chamber, wind tunnel, etc.).
- Different expansion coefficient stresses: potting, conformal coating, Surface Mount Devices (SMD's), Leadless Chip Carriers (LCC's), PCB interfaces, etc.
- Rules for mounting components.
- Rules for mounting assemblies (modules).
- Thermal shock (splash, cold start): typical failure modes.
- Identification of critical components and special requirements.
- Thermal stress test for design verification: tailored to find defects in new design, should be failure oriented (overstressed). Temperature cycling profile: extremes, number of cycles, rate of change, when powered, parameters monitored.
- Thermal stress test for qualification: mission life oriented. Temperature cycling profile: extremes, number of cycles, rate of change, when powered, parameters monitored.
- Thermal stress test for production acceptance: should include Environmental Stress Screening (ESS) tailored to reduce infant mortality and precipitate process problems. Temperature cycling profile: extremes, number of cycles, rate of change, when powered, parameters monitored.
- Combined thermal stress with other tests (for example, thermal, vibration, humidity, voltage) more realistic.
- Testing sample size: statistical significance, attribute or variable data, cost/time/test facility limitations.

5.3 Vibration/Shock Considerations (Components/Assemblies) References: 3, 11, 13, 14, 16, 17, 22.
- Conduct vibration/shock survey of environment (under hood, passenger compartment, etc.): conditions (bumps/potholes, road vibration, handling, rail shock), type (sine, random, complex), frequency range, amplitude/Power Spectral Density (PSD), axis, duration.
- Stresses on components, bonds, mounting brackets, etc.: concerns, typical failure modes.
- Rules for mounting components: for example, part size/mass vs. mounting technique.
- Module mounting techniques: consider mounting bracket effects (for example, resonances).
- Resonances: conduct resonant search, failure modes, solutions.
- Vibration test for design verification: tailored to find defects in new design, should be failure oriented (overstressed). Type (sine, random, complex), frequency range, amplitude/PSD, axis, duration, monitored to detect intermittents.
- Vibration test for qualification: mission life oriented. Type (sine, random, complex), frequency range, amplitude/PSD, axis, duration, monitored to detect intermittents.
- Vibration test for production acceptance: should include Environmental Stress Screening (ESS) tailored to reduce infant mortality and precipitate process problems. Type (sine, random, complex), frequency range, amplitude/PSD, axis, duration, monitored to detect intermittents.
- Shock test for design verification, qualification and production acceptance: similar to vibration above.
- Conduct vibration/shock testing before climatic testing (if done separately).
- Vibration/shock combined with temperature cycling, humidity more realistic.
- Testing sample size: statistical significance, attribute or variable data, cost/time/test facility limitations.

5.4 Humidity/Splash/Dust Considerations (Components/Assemblies)—References: 3, 11, 13, 14, 22.
- Component/Assembly sealing: gasketing, potting, etc.
- Connector integrity: type of connector (open, sealed, greased, etc.).
- Failure modes: shunt resistance, series impedance.
- Test procedure for design verification: similar to paragraphs 5.2, 5.3.
- Test procedure for qualification: similar to paragraphs 5.2, 5.3.
- Test procedure for production acceptance: similar to paragraphs 5.2, 5.3.
- More realistic if combined with temperature cycling, vibration.
- Testing sample size: statistical significance, attribute of variable data, cost/time/test facility limitations.

5.5 Burn In
- Determine need, component vs. assembly or both: field correlation, experiences, cost analysis.
- Component burn-in: which ones, more stress than assembly, minimizes rework. If ppm failure rates low, burn-in may make worse (handling, ESD).
- Assembly burn-in: thermal mass test considerations.
- Test conditions: elevated temperature and voltage accelerates failure modes (different times for different failure modes). Static, dynamic operation.
- Determine optimum burn-in empirically: time vs. temperature/voltage failure rates.
- Combined powered thermal cycle and burn-in.
- Testing sample size: statistical significance, attribute or variable data, cost/time/test facility limitations.

5.6 Electromagnetic Compatibility (EMC)—References: 1, 2, 4, 5, 6, 7, 8, 9, 10, 22, 23, 24.

5.6.1 COMPONENT LEVEL
- Radiated Susceptibility:
  . Moderate Radio Frequencies (RF) fields (50 V/m)—represents nearby transmitters, low power on-board transmitters.
  . High RF fields (100 V/m)—represents high power (100 W) on-board transmitters.
  . Test procedures—SAE J1113, SAE J1448, and SAE J1547.
- Radiated Emissions:
  . On board entertainment/communications antennas and radio sensitivities determine specification limit.
  . Test procedures—MIL-STD-461.
- Conducted Susceptibility:
  . Supply Voltage:
  .. Normally 10 to 16 V.
  .. Reverse battery.
  .. Overvoltage—failed regulator (17 V), double voltage jump start.
  .. Cold start—5 to 6 V.
  .. Ignition switch rotation—voltage dropouts.
  . Vehicle electrical system noise (for example, load dump, switch arcing, inductive transients).
  . Test procedure SAE J1113.

— Conducted Emissions: Test procedure — VDE 0879, part 3.
— Electrostatic Discharge (ESD):
  . ±15 kV.
  . Test procedure—SAE J1113, SAE J1595.

5.6.2 VEHICLE LEVEL—Vehicle EMC test procedures.
  . Internally generated EMI—check interactions of subsystems under various conditions.
  . Radiated susceptibility—10 kHz to 1 GHz or higher, powerlines, nearby lightning: SAE J1338, SAE J1507, SAE J1407.
  . Radiated emissions—SAE J551, SAE J1816.
  . Esd—SAE J1595.
  . Charging system anomalies—disconnected battery (engine running), load dump, malfunctioning regulator, reverse battery.

5.7 Circuit Design Guidelines—References: 12, 15, 22.

5.7.1 GENERAL
— Minimize number of parts.
— Maximize use of proven circuits.
— Maximize use of standard parts and widest tolerances.
— Use slowest speed technology consistent with function.
— Where possible, include hysteresis on analog/digital circuits.
— High gain circuits with differential inputs (for example, op-amps, comparators) should use filter capacitor across inputs and/or capacitor from each input to ground.
— Breadboards may aggravate problems: long leads, poor ground(s).
— Discrete vs. custom circuits: cost, reliability, volume tradeoffs. For custom circuits, use pessimistic cost/timing.
— Redundancy: critical circuits.
— Designed for manufacturability: see process guidelines Section 6.
— Repairability, remanufacturing (if applicable), rework considerations.
— Diagnostics considered in design.
— Terminate unused inputs to I.C.'s.
— Relay precautions: for example, diode increases dropout time, contact arcing, transients, contact sticking.

5.7.2 COMPONENTS - SPECIFIC DEVICES
— Transistors/Diodes/MOSFET's:
  . Consider diode/zener response time to transients.
  . Transistor within safe operating area (SOA). Consider temperature, loading, signal input.
  . Use transistor base to emitter resistor.
  . Collector to housing stray capacitance for switching circuits—may cause radiated emissions.
  . Limit base/gate drive—fast drive into saturation creates noise, balance Electromagnetic Interference (EMI) with heat dissipation.
  . MOSFET's—ESD one of major failure modes.
— Linear (Op Amps, etc):
  . Consider single supply limitations—input range usually does not include power supply rails, output loading determines voltage range.
  . Overdriven inputs—may drive output to power supply rail (transmits power supply noise).
  . High gain amplifiers—stability, oscillations, stray capacitance and inductance varies with temperature/sample.
  . Use Bode gain/phase plot analysis to determine stability margin.
  . Differential amps—consider all sources of unbalance to ground, DC and AC (for example, capacitors, source impedance).
  . Differential amps limited in rejection of common mode signals at higher frequencies.
  . Use op amp internal compensation capacitor, if accessible, for filtering (acts as nonlinear filter).
  . Voltage follower latch up—input levels too high.
  . Avoid high impedances.
— Digital (microprocessors, etc.):
  . Fanout limitations—for example, loading affects propagation delays especially for CMOS.
  . Logic levels compatible over minimum and maximum temperature/specification limits.
  . Maximize logic levels margins (for example, V low max and V high min).
  . CMOS—latchup when input > power supply or < ground.
  . Microprocessor clock—operates, including start up, under all temperature and power supply transitions.

5.7.3 MODULE INPUTS
— Protection for shorts to ground/power.
— Switch requirements: contact material/pressure, type of connector, minimum voltage/current for oxidation burn through (dry circuits).
— Allow for contact resistance, shunt resistance.
— Maximize input thresholds for noise immunity.
— Maximize input filtering considering maximum signal information delay, minimum signal pulse width to be recognized, fastest signal rate of change (dV/dT).
— Shared sensors compatible: that is, one sensor to multiple modules.

5.7.4 MODULE OUTPUTS
— Protection for shorts to ground/power.
— Inductive driver transient protection.
— Output driver current source vs. current sink considerations: current source has same failure mode (wiring short or open).
— Limit high current actuator transition times (without overheating): generates noise, wiring harness ringing.
— H-Bridge driver: insure both drivers in each leg not on simultaneously during transitions.

5.7.5 POWER SUPPLY RELATED
— Circuits compatible with run - start - run cycle (starting) et al.
— Power up/down sequence.
— Overvoltage, undervoltage, reverse voltage, load dump.
— Power supply protection schemes.
— Power supply regulator response time: not too fast or may be noise sensitive.
— Power supply capacitor design: for example, aluminum electrolytic voltage/temperature, ripple calculations.
— Avoid voltage divider circuits. If used, use worst case power supply voltages.
— Minimum current draw may deregulate power supply.
— Two power supplies may cause latch up if not tracking.
— For mixed technologies (for example, CMOS, TTL), power up/down may produce errors due to different valid/invalid levels.

5.7.6 ELECTRICAL OVERSTRESS
— Ignition arcover design considerations for under hood applications.
— Transient protection: Resistor, Capacitor, R/C, clamps (for example, zener, diode).
— All circuits connected to main power, or through loads to main power, must withstand electrical overstress.
— Shutdown circuits must have fast response.
— ESD protection:
  . Often misanalyzed as electrical overstress.
  . Part (for example, IC) protection limited—too slow (ESD <5ns.).
  . Use Resistor, Capacitor, R/C, clamps—consider high peak voltage.

5.7.7 PCB LAYOUT RULES FOR EMC
— Ground plane interconnecting circuit grounds: ideally greater than 50% of PCB area.
— Common impedance: sensitive circuits not shared with high rate of current change (dI/dT) circuits. $E = L * dI/dT$ (typically, L=25 nh/in).
— Decoupling capacitors very near IC's, especially microprocessors and high dI/dT circuits. Use ground plane between capacitor and IC ground.
— Input/output filtering configurations: near entry of Input/Output, grounded via ground plane.

5.7.8 CIRCUIT TOLERANCE/ANALYSIS
— Sneak circuit analysis.
— Failure Modes and Effects, criticality (severity/probability) analysis.
— AQL for assembly (module) levels required, how determined.
— Reliability Prediction models used for assembly:
  . MIL-STD-217—not directly applicable, assumes exponential failure distribution (no infant mortality). Need automotive data base. Consider dormancy.
  . Field experience—reliability growth model. Similar equipment, complexity/function.
— Degree of circuit/tolerance analysis used:
  . Worst case combinations of part tolerances/inputs over temp. range.
  . Design of Experiments, Taguchi methods.
  . Monte Carlo method.
  . Component aging considered—value change with stress, time.

5.8 Software
— Use modularization.
— Optimize decoupling with other software modules.
— Designed for testability.
— Documentation.
— Sufficient time allowed for testing and debug: can be as much as design and coding.
— Software module testing: simulation on mainframe.
— Static (change inputs manually) bench testing of system (total program).
— Dynamic (many combinations of inputs) design verification on mainframe simulator.
— Fault tolerance: for example, inputs within realistic limits (for example, dV/dT, dFreq./dT, edges, change in A/D counts).
— Watchdog timer strategy and implementation.
— Low voltage reset.
— Fault tolerance strategy: revert to old data, ignore, try again.
— Software noise immunity strategies and limitations.
— Switch contact bounce strategy.
— Software development tools: portable engineering and calibration consoles.
— Vehicle testing program.

5.9 Diagnostics
— What functions to check: assign probability/severity index.
— Diagnostic troubleshooting procedures: philosophy, documentation.
— Diagnostics considered in warranty analysis.
— Built in monitor circuits.
— Warning indicators: for example, instrument panel light.
— Intermittents: how to precipitate, store in nonvolatile memory.
— Self Test Methodology: factory and field service.
— Test equipment requirements.
— Software memory allocation.

5.10 Miscellaneous Design Guidelines:
— Identify critical characteristics from customer perspective: Quality Functional Deployment (QFD).
— Vehicle wiring guidelines:
  . Critical modules that must operate during engine cranking (for example, electronic engine controls) may require power and ground feeds directly to battery (minimizes common impedance, voltage drops).
  . Low level signals—do not use sheet metal return.
  . Maintain low resistance between body panels/structures.
  . Avoid unterminated wires—act as antennas.
  . Sensitive wiring >5 cm from secondary ignition parts (for example, ignition wires, coil, plugs, distributor).
  . Test for wiring crosscoupling and correct (separation, twisting, etc).
  . Twisted wires—effective low cost option for noise reduction.
  . Wire shielding—verify need (usually needed for wire crosscoupling), insure coverage in area of noise, single point ground (drain wire short).
  . Maximize wire routing near sheet metal.
  . Analyze multiple grounds—ground loops.
  . Inductive loads—test for noise and, if excessive, suppress.
  . Insure reliability of ground connections to sheet metal.
  . Ignition arcover—maintain separation; "non conductors" may be conductive (for example, carbon loaded hoses, wet plastic).
— Allow for impedance buildup during system life.
— Vehicle durability, field testing program.
— Failure Modes and Effects analysis: module, subsystem, system level.
— Criticality analysis (severity/probability of occurrence): module, subsystem, system level.
— Limited operation strategy.
— Design reviews, audits at various stages.
— Design change procedures.
— Closed loop failure analysis and corrective action plan: concern description, define root cause, containment, corrective actions, verification of containment/corrective actions, prevent recurrence.
— Analyze warranty returns: "non defective returns" often >50% of returns, functional testing at temperature limits often will not identify problem. Use Environmental Stress Screening to precipitate problems.
— Factor accessibility: harder to replace item within system will have better warranty.
— Wiring harness reliability affects module warranty.

6. Process Checklist—The manufacture of electronic components/modules is a process that is continually being changed. These guidelines are basically intended for the module design engineer so that a greater appreciation of what is involved can be obtained and considered in the early design stages.

6.1 Through Hole (TH) Technology
6.1.1 GENERAL TH DESIGN GUIDELINES
— For single sided PCB, conductor width = 0.020 in minimum.
— For double sided PCB, power/ground conductor width = 0.016 in minimum, signal conductor width = 0.012 in minimum.
— Plated through hole (PTH) pad diameter = or >2 x hole diameter.
— PTH diameter = or >40% of material thickness.
— PTH double sided: hole diameter 0.010 to 0.028 in > lead diameter.
— PTH single sided: hole diameter 0.005 to 0.020 in > lead diameter.
— Do not use sharp corners for conductor traces.
— Web between holes: for punched holes = 0.060 in minimum or 1.5 times hole diameter whichever is greater. For drilled holes = 0.035 in minimum.
— Warpage: balance copper density within 30% both sides. For large copper areas (for example, ground plane), use voids at random intervals.
— Use thermal relief around component holes in large copper areas.
— Auto insertion: hole diameter >0.015 in over lead diameter minimum.
— Lead formed double kink parts for PTH: euroform type preferred, stress relief, trapped gas in PTH area.
— For clinched leads, use tear drop pads: more bonding area.
— Solder mask: types (screened, dry film).

6.1.2 PROCESS STEPS
— Component packaging: for example, bulk and vibration feeders (avoid both), reel/tape.
— Parts insertion, clinching: automatic, hand.
— Wave solder process: types, variables.
— Solder/flux chemistry.
— Cleaning: for example, flux residues, solvents, migration.
— Solder inspection criteria.
— Potting/conformal coating.
— Testing/inspection: during assembly, final assembly, automatic, high density concerns.
— Repair/rework philosophy for PTH technology.
— SPC: where used, adherence, effect on AQL.
— ESD practices unique to TH technology.

6.2 Thick Film Hybrid (TFH) Technology
6.2.1 GENERAL TFH DESIGN GUIDELINES
— Resistors: 10 to 200 ppm/C, ± 15% tolerance typical (untrimmed). Top hat resistors more susceptible to high voltage arcover.
— Conductor: paladium silver = 0.035 ohms/square, 0.08 with solder coating, 250 - 800 ppm/C.
— Conductor signal width/spacing normally 0.5 mm.
— Component attachment pad configuration.
— Minimize crossovers and number of resistive paste values.
— Conductor concerns.
  . Silver migration—silver + electrolyte + moisture = dendritic growth of silver (0.020 in gap in minutes with 0.5 V bias). Preventive measures = cleaning, potting.
  . Galvanic action—Alum + electrolyte + moisture = battery with paladium or silver (2.5 to 2.8 V). Preventive measures = cleaning, potting.
  . Intermetallics—Paladium + tin = poor adhesion of pal silver films (conductor adhesion). Control solder.

6.2.2 PROCESS STEPS
— Screen/print, fire conductors.
— Double screen/print, fire dielectric (crossovers).
— Screen/print, fire resistors.
— Trim: for example, laser, scribe.
— Solder screen.
— Conductive epoxy vs. solder paste.
— Solder/flux chemistry.
— Component packaging: for example, bulk and vibration feeders (avoid both), reel/tape.
— Pick/place components.
— Reflow solder: variables = component mass, specific heat, furnace set point, belt speed.

- Cleaning: for example, flux residues, solvents.
- Solder inspection criteria.
- Break substrate apart.
- Wire bonding (if applicable).
  . Materials—gold, aluminum (most common).
  . Thermocompression.
  . Ultrasonic—most common.
  . Thermosonic—0.001 to 0.005 gold wires.
- Potting/coating.
- Testing/inspection: during assembly, final assembly, automatic, high density concerns.
- Repair/rework philosophy for TFH technology.
- SPC: where used, adherence, effect on AQL.
- ESD practices unique to TFH technology.

6.3 Surface Mount Device (SMD) Technology—There are three types of SMD technology:

Type 1 - All SMD's

Type 2 - SMD's and/or TH on top, SMD's and/or TH on bottom of PCB

Type 3 - TH on top, SMD's on bottom of PCB.

At this time, Type 3 is most common in the automotive world, so it will be used as an example.

6.3.1 GENERAL SMD DESIGN GUIDELINES
- See paragraph 6.1 for items similar to TH technology.
- Vias, test pads not part of attachment pad.
- Lead coplanarity (flatness): <0.004
- Underside leaded component clearance: >0.010 for solvent action.
- Component orientation relative to soldering direction.
- Tombstoning prevention.

6.3.2 PROCESS STEPS
- Component packaging: for example, bulk and vibration feeders (avoid both), reel/tape.
- Parts insertion/clinch TH devices.
- Flip PCB.
- Apply adhesive: type, concerns.
- SMD component packaging: for example, bulk and vibration feeders (avoid both), reel/tape.
- Place SMD's.
- Cure adhesive.
- Flip PCB.
- Wave solder process.
- Solder/flux chemistry.
- Cleaning: for example, flux residues, solvents, migration.
- Solder inspection criteria.
- Potting/conformal coating.
- Testing/inspection: during assembly, final assembly, automatic, high density concerns.
- Repair/rework philosophy for SMD technology.
- SPC: where used, adherence, effect on AQL.
- ESD practices unique to SMD technology.

6.4 Miscellaneous Process Guidelines
- Design engineer interface with manufacturing throughout project.
- Conduct feasibility study.
- Compatibility of parts/tooling: assessed early in design phase.
- Incoming inspection: vendor quality #1 reliability concern.
- In-process and end of line test requirements.
- Repair/rework methods.
- Layout for EMC: see paragraph 5.7.7.
- Electrostatic discharge (ESD) control program, training.
- Vision technology: type, where used.
- Optimize utilization of PCB, TFH area: use standard sizes.
- Tolerance stackup: registration, hole location, artwork, camera repeatability, etc.
- Consider dimensional variances: samples, vendors.
- Placement machine accuracy/repeatability.
- Board positioning accuracy/repeatability.
- Component lead finish, materials, thickness.
- Position components to provide clearance for assembly, handling operations.
- Process flow plans.
- Process change procedures.
- Closed loop failure analysis and corrective action plans.

7. References

1. SAE J551, Performance Levels and Methods of Measurement of Electromagnetic Radiation from Vehicles and Devices (30 - 1000 MHz).

2. SAE J1113 AUG87, Electromagnetic Susceptibility Measurement Procedures for Vehicle Components (Except Aircraft).

3. SAE J1211 NOV78, Recommended Environmental Practices for Electronic Equipment Design.

4. SAE J1338 JUN81, Open Field Whole-Vehicle Radiated Susceptibility, 10 KHz - 18 GHz, Electric Field.

5. SAE J1407 AUG82, Vehicle Electromagnetic Radiated Susceptibility Using a Large TEM Cell.

6. SAE J1448 JAN84, Electromagnetic Susceptibility Measurements of Vehicle Components Using TEM Cells (14 KHz - 200 MHz).

7. SAE J1507 JAN87, Anechoic Test Facility Radiated Susceptibility, 20 MHz to 18 GHz, Electromagnetic Field.

8. SAE J1547 (Draft) MAY86, Electromagnetic Susceptibility Procedures for Common Mode Injection (1 - 400 MHz) Module Testing.

9. SAE J1595 (Draft), Electrostatic Discharge Test for Vehicles.

10. SAE J1816 OCT87, Performance Levels and Methods of Measurement of Electromagnetic Radiation from Vehicles and Devices, Narrowband, 10 KHz - 1000 MHz.

11. MIL-STD-202F, Test Methods for Electronic and Electrical Component Parts. 1 Apr., 1980.

12. MIL-STD-217D, Reliability Prediction of Electronic Equipment. 15 Jan., 1982.

13. MIL-STD-781C, Reliability Design Qualification and Production Acceptance Tests. Exponential Distribution. 21 Oct., 1977.

14. MIL-STD-810D, Environmental Test Methods and Engineering Guidelines. 19 July, 1983.

15. MIL-HDBK-251, Reliability Design Thermal Applications. 19 Jan., 1978.

16. DOD-HDBK-344, Environmental Stress Screening (ESS) of Electronic Equipment. 20 Oct., 1986.

17. NAVMAT P-9492, Navy Manufacturing Screening Program. May, 1979.

18. MIL-STD-S-19500G, Semiconductor Devices, General Specification For. 16 Feb., 1984.

19. MIL-STD-750C, Test Methods for Semiconductor Devices. 23 Feb., 1983.

20. MIL-STD-38510F, Microcircuits, General Specification For. 31 Oct., 1983.

21. MIL-STD-883C, Test Methods and Procedures for Microelectronics. 25 Aug., 1983.

22. Automotive Electronics Reliability Handbook. SAE publication AE-9. Feb., 1987.

23. VDE 0879 - Part 3, Radio Interference Suppression of Vehicles, of Vehicle Equipment and of Internal Combustion Engines. Interference Suppression for "On-board" Radio Reception; Measurements of Vehicle Equipment.

24. MIL-STD-461C - Aug 86, Electromagnetic Emission and Susceptibility Requirements for the Control of Electromagnetic Interference.

GENERAL QUALIFICATION AND PRODUCTION ACCEPTANCE CRITERIA FOR INTEGRATED CIRCUITS IN AUTOMOTIVE APPLICATIONS—SAE J1879 OCT88

SAE Recommended Practice

Report of the Electrical and Electronic Systems Technical Committee approved October 1988.

1. Introduction

1.1 Purpose—The overall purpose of this document is to provide a common document (procedure) to specify automotive integrated circuits that will improve supplier-user interface relationships with resulting benefits such as:
 a) Improved quality
 b) Reduced cost
 c) Reduced development time
 d) Improved availability

This SAE Recommended Practice is intended as a guide towards standard practice and is subject to change to keep pace with experience and technical advances. The Electronics Reliability Subcommittee solicits comments to further enhance the usefulness of this document. Comments received will be reviewed and the document revised when appropriate (annually, if required).

Comments should be addressed to: SAE Electronics Reliability Standards Committee, c/o SAE, Troy, Michigan

Users are encouraged to compare their existing specifications with this document to determine whether their requirements are covered. In those cases where the requirements are different, they can be addressed in the user's purchase document.

1.2 Scope—This document establishes the general environmental, mechanical and quality requirements pertaining to IC's. It also describes the qualification and production acceptance procedures performed by the supplier.

It is to be used as a guideline which device buyers can apply wholly or in part as appropriate for their needs and applications.

This document is essentially part number independent. The detail specification particular to a device part number will be given separately, containing all detailed electrical and mechanical parameters.

2. Quality Assurance

2.1 General—In general, the manufacturer must demonstrate with the help of a Quality Assurance Program that design, manufacture, inspection and testing of integrated circuits are adequate to assure compliance with state-of-the-art quality standards. Such a Quality Assurance Program should be structured along MIL-M-38510 (all references to class S removed).

Upon previous request, the buyer shall have access to production areas in order to verify proper implementation of the procedures agreed upon.

2.2 Statistical Process Control (SPC)—A SPC program shall be followed assuring control of device parameters specified in the specific device Engineering Specification (ES).

2.2.1 CHARACTERIZATION REPORT—Prior to qualification approval, the supplier shall provide a Characterization report for Product Engineering approval. This report shall include all tests indicated in the specific device ES. These tests typically contain device parameters at various supply voltages. For qualification, the tests shall be run at five or more specific temperatures including (Ta min - 15°C), (Ta max + 15°C), 25°C, Ta min and Ta max.

The data will be collected from 300 pieces (100 pieces from each of three qualification lots).

During production, the supplier shall do a limited characterization on 10 parts per lot. This limited characterization is also outlined in the specific device ES. The tests indicated in the ES shall be run at Ta min, 25°C and Ta max.

2.2.2 REPORT CONTENTS (QUALIFICATION)—For each parameter, if user has no other reporting format, the data shall be presented using the following format:
 a) For the combination of three lots, the supplier shall generate a plot of the mean parameter value for each supply voltage versus temperature. This should be presented on one graph.
 b) For each supply voltage, the supplier shall generate a separate graph showing the ±6 sigma, ±3 sigma and mean parameter values versus temperature.
 c) For each of the three lots, the supplier shall generate a bar chart which plots the individual lot versus the range of measured parameters. For this graph, only data collected in the temperature range of Ta min $\leq$ Ta $\leq$ Ta max shall be included.

2.2.3 REPORT CONTENTS (PRODUCTION)—The production report shall consist of:
 a) A control chart for each parameter in the table from the specific device ES at the three temperatures. Each chart will track the mean parameter value and standard deviation from lot to lot.
 b) Corrective actions or out-of-control readings.
 c) Process changes and effective dates.

3. Qualification And Production Acceptance (PA) Tests

—The table of tests defines the requirements and acceptance criteria for all qualification and production level tests.

Compliance with all the tests in this section must be demonstrated as follows:

3.1 Qualification—These tests are used to demonstrate the potential of the process to produce parts that meet engineering requirements. They must be completed satisfactorily with initial parts from production tooling and processes before approval and authorization for shipment of parts can be effected.

3.2 Production Acceptance 1 (PA1)—These tests are used to demonstrate the capability and ongoing performance of the process and must be completed using parts from production tooling and processes prior to first production shipment approval. These tests are to continue in effect until process capability is demonstrated.

NOTE: In the event that qualification and PA1 tests could be performed concurrently and would be redundant, that is, if samples from production tooling and processes are available, and if a test plan is developed to accomplish the purpose of both qualification and PA1 test programs, it is acceptable to perform one series of tests for both qualification and PA1.

3.3 Production Acceptance 2 (PA2)—The PA2 test program may be implemented only after process capability to meet PA1 requirements has been established and only if the suppliers in-line quality and reliability monitoring procedures are accepted by the buyer. Samples for these tests must be randomly selected from production parts on a continuing basis to represent the entire production population.

3.4 General Notes—Approval for use of the PA2 test sequence will be granted if appropriate, based on the results of qualification testing and the availability of relevant specific and generic data attesting to the stability of the manufacturing process.

Alternate procedures for the tests defined in Section 4 and sample plans contained in the table of tests may be proposed by the vendor. These can be approved by the buyer. The requirements for one or more of these tests may be deleted on the basis of commonality with previously approved parts or on the basis of acceptable part family data.

4. Test Procedures And Requirements

—The following paragraphs describe the tests identified in Section 3, the table of tests. These functional and durability tests may be conducted in any sequence except for the electrical tests in paragraphs 4.3.3, 4.3.4, and 4.3.5. Some of these tests are made up of a series of steps which must be carried out in the sequence indicated. Parts subjected to these tests may not be delivered as production parts unless specifically indicated to the contrary in the test description. Temperature tolerance is ±3°C.

NOTE: For tests which require multiple lots, the number of lots required may be reduced by the submission of generic data.

4.1 Environmental Tests

4.1.1 HERMETIC DEVICE WATER VAPOR CONTENT—Performed on hermetic devices only.

Step 1 - Internal Water Vapor Content: MIL-STD-883, method 1018. Criteria: 5000 ppm maximum water vapor content at 100°C.

4.1.2 TEMPERATURE CYCLING—Performed on molded devices only.

Step 1 - Electrical Parameters: Verification of functionality and measurements of DC parameters to be made per paragraphs 4.3.4 and 4.3.5. Datalogs to be retained for review.

Step 2 - Temperature Cycling: MIL-STD-883, method 1010, modified condition C (-55 to 150°C), 1000 cycles.

Step 3 - Electrical Parameters: Verification of functionality and measurements of DC parameters to be made per paragraphs 4.3.4 and 4.3.5. Datalogs to be retained for review.

TABLE OF TESTS 4.1[1] — ENVIRONMENTAL

TEST NAME	TEST NUMBER	CORRESPONDING STANDARD	QUALIFICATION MINIMUM SAMPLE SIZE	QUALIFICATION STATISTICAL ACCEPTANCE CRITERIA	PROD. ACCEPT. – 1 (PA1) MINIMUM SAMPLE SIZE	PROD. ACCEPT. – 1 (PA1) STATISTICAL ACCEPTANCE CRITERIA	PROD. ACCEPT. – 2 (PA2) MINIMUM SAMPLE SIZE	PROD. ACCEPT. – 2 (PA2) STATISTICAL ACCEPTANCE CRITERIA
HERM. DEVICE WATER VAPOR CONTENT	4.1.1	MIL-STD 883 1018	3	0 REJECTS	3 ONCE PER QUARTER	0 REJECTS	3 ONCE PER QUARTER	0 REJECTS
TEMP. CYCLING MOLDED DEVICES	4.1.2	MIL-STD 883 1010C	45 DEVICES FROM EA. OF 3 LOTS TOT-135	LTPD-5 0 FAIL PER LOT	45 PER QUARTER		45 PER QUARTER	
THERMAL SHOCK QUALIFICATION	4.1.3	MIL-STD 883 1011C	45 DEVICES FROM EA. OF 3 LOTS TOT-135	LTPD-5 0 FAIL PER LOT	45 ONCE PER QUARTER	LTPD-5 (0 REJ)	45 ONCE PER 6 MOS.	LTPD-5 (0 REJ)
UNBIASED AUTOCLAVE QUALIFICATION	4.1.4	JEDEC 22 A102A	77 DEVICES FROM EA. OF 3 LOTS TOT-231	LTPD-5 1 FAIL PER LOT	77 ONCE PER QUARTER	LTPD-5 (1 REJ)	77 ONCE PER 6 MOS.	LTPD-5 (1 REJ)
UNBIASED AUTOCLAVE PRODUCTION LEVEL	4.1.5	JEDEC 22 A102A			45 PER LOT	LTPD-5 (0 REJ)	45 PER LOT	LTPD-5 (0 REJ)
BIASED HUMIDITY	4.1.6	JEDEC 22 A101 COND. A	45 DEVICES FROM EA. OF 3 LOTS TOT-135	LTPD-5 0 FAIL PER LOT	45 ONCE PER QUARTER	LTPD-5 (0 REJ)	45 ONCE PER 6 MOS.	LTPD-5 (0 REJ)
SALT ATMOSPHERE	4.1.7	MIL-STD 883 1009A	38	LTPD-10 (1 REJ)				
OPERATING LIFE QUALIFICATION	4.1.8	JEDEC 22 A108	76 DEVICES FROM EA. OF 3 LOTS TOT - 228	LTPD-3 0 FAIL PER LOT				
BURN - IN	4.1.9	JEDEC 22 A108	100%		REFER TO TEXT		REFER TO TEXT	
HIGH TEMP REVERSE BIAS	4.1.10	JEDEC 22 A108	76	LTPD-3 (0 REJ)				
POWER AND TEMP. CYCLE	4.1.11	JEDEC 22 A105	76	LTPD-3 (0 REJ)				

[1]REQUIREMENTS:
[a]When subgroup sample sizes are greater than the minimum specified and permit acceptance with one or more rejects, the buyer must be notified prior to the start of testing.
[b]When a sample size is tied to a period of time longer than that which is normally controlled and shipped as a lot, e.g., X/quarter; one sample of the specified number of pieces must be selected from one lot manufactured in that specified time period.

Step 3 (alternate) - Procedures and sampling criteria per manufacturer's specification may be substituted for PA2 tests after approval by the buyer.

4.1.3 THERMAL SHOCK—Performed on molded devices only.

Step 1 - Electrical Parameters: Verification of functionality and measurement of DC parameters to be made per paragraphs 4.3.4 and 4.3.5. Datalogs to be retained for review.

Step 2 - Thermal Shock: MIL-STD-883, method 1011, modified condition C (-55 to 150°C), 1000 cycles.

Step 3 - Electrical Parameters: Verification of functionality and measurement of DC parameters to be made per paragraphs 4.3.4 and 4.3.5. Datalogs to be retained for review.

Step 3 (alternate) - Procedures and sampling criteria per manufacturer's specification may be substituted for PA2 tests after approval by the buyer.

4.1.4 UNBIASED AUTOCLAVE (QUALIFICATION) - Performed on molded devices only.

Reference: JEDEC 22B, A102A

Step 1 - Electrical Parameters: Verification of functionality and measurement of DC parameters to be made per paragraph 4.3.4. Leakage current measurements must be made. Datalogs and data to be retained for review.

Step 2 - Place parts in a stainless steel autoclave and maintain at 121°C, 15 ± 2 psig for 96 h. Use deionized water with resistivity no less than 1 x 10 E 6 ohm-cm at 25°C. Parts to be dried at room temperature for 24 h prior to electrical testing. Testing is to be completed within 48 h after drying.

Step 3 - Electrical Parameters: Verification of functionality and measurements of DC parameters to be made per paragraph 4.3.4. Leakage current measurements shall be made. Datalogs and data to be retained for review. Lack of functionality or parametric deviations beyond specification limits shall indicate lack of sufficient durability.

4.1.5 UNBIASED AUTOCLAVE (PRODUCTION ACCEPTANCE)—Performed on molded devices only.

Reference: JEDEC 22B, A102A

Step 1 - Electrical Parameters: Verification of functionality and measurements of DC parameters to be made per paragraph 4.3.4. Leakage current measurements shall be made.

Step 2 - Place parts in stainless steel autoclave and maintain at 121°C, 15 ± 2 psig for 48 h. Use deionized water with resistivity no less than 1 x 10 E 6 ohm-cm at 25°C. Parts to be dried at room temperature for 24 h prior to electrical testing.

Step 3 - Electrical Parameters: Verification of functionality and measurements of DC parameters to be made per paragraph 4.3.4. Leakage current measurements shall be made. Lack of functionality or parametric deviations beyond specification limits shall indicate inability to adequately sustain stress.

Step 4 - Evaluation: If the LTPD criteria of the table of tests is not met, Steps 1, 2 and 3 may be repeated one time. If the stated LTPD criteria is not met the second time, the lot is rejected.

TABLE OF TESTS 4.2[1] — MECHANICAL

TEST NAME	TEST NUMBER	CORRESPONDING STANDARD	QUALIFICATION		PROD. ACCEPT. – 1 (PA1)		PROD. ACCEPT. – 2 (PA2)	
			MINIMUM SAMPLE SIZE	STATISTICAL ACCEPTANCE CRITERIA	MINIMUM SAMPLE SIZE	STATISTICAL ACCEPTANCE CRITERIA	MINIMUM SAMPLE SIZE	STATISTICAL ACCEPTANCE CRITERIA
INTERNAL EXAM A	4.2.1	MIL-STD 883 2014	3	(0 REJ)	3 ONCE PER QUARTER	(0 REJ)	3 ONCE PER QUARTER	(0 REJ)
INTERNAL EXAM B	4.2.2	MIL-STD 883 2010-B	100%	REMOVE DEFECTIVES	100%	REMOVE DEFECTIVES		
PHYSICAL DIMENSION	4.2.3	MIL-STD 883 2016	3	(0 REJ)	3 PER LOT	(0 REJ)	3 PER LOT	(0 REJ)
EXTERNAL VISUAL	4.2.4	MIL-STD 883 2009	76	LTPD-3 (0 REJ)	4 PER LOT	(0 REJ)	4 PER LOT	(0 REJ)
RESISTANCE TO SOLVENTS	4.2.5	MIL-STD 883 2015	76	LTPD-3 (0 REJ)	4 PER LOT	(0 REJ)	4 PER LOT	(0 REJ)
VISUAL INSPECTION	4.2.6	MIL-STD 883 2009	100%	REMOVE DEFECTIVES	100%	REMOVE DEFECTIVES	100%	REMOVE DEFECTIVES
BOND STRENGTH	4.2.7	MIL-STD 883 2011 C OR D	76 BONDS FROM EA. OF 3 LOTS TOT-228	LTPD-3 (0 REJ)	76 BONDS PER LOT	LTPD-3 (0 REJ)	76 BONDS PER LOT	LTPD-3 (0 REJ)
TERMINAL STRENGTH & DURABILITY	4.2.8	MIL-STD 883 2004B2, 1014 JEDEC22, A102A	22	LTPD-10 (0 REJ)				
SOLDERABILITY	4.2.9	MIL-STD 883 2003	22	LTPD-10 (0 REJ)	3 LEADS ON 7 DEV. PER LOT	LTPD-10 (0 REJ)	3 LEADS ON 7 DEV. PER LOT	LTPD-10 (0 REJ)
HERM. DEVICE PACKAGE INTEGRITY - TEST 1	4.2.10	MIL-STD 883 1010C, 2001E 1014	45	LTPD-5 (0 REJ)	45 ONCE PER QUARTER	LTPD-5 (0 REJ)	45 ONCE PER 6 MOS.	LTPD-5 (0 REJ)
HERM. DEVICE PACKAGE INTEGRITY - TEST 2	4.2.11	MIL-STD 883 2001B, 2002B 1014	22	LTPD-10 (0 REJ)	22 ONCE PER QUARTER	LTPD-10 (0 REJ)	22 ONCE PER 6 MOS.	LTPD-10 (0 REJ)

[1]REQUIREMENTS:

[a]When subgroup sample sizes are greater than the minimum specified and permit acceptance with one or more rejects, the buyer must be notified prior to the start of testing.

[b]When a sample size is tied to a period of time longer than that which is normally controlled and shipped as a lot, e.g., X/quarter; one sample of the specified number of pieces must be selected from one lot manufactured in that specified time period.

4.1.6 BIASED HUMIDITY—Performed on molded devices only.
Reference: JEDEC 22B, A101 Condition A
Step 1 - Electrical Parameters: Verification of functionality and measurements of DC parameters to be made per paragraph 4.3.4. Leakage current measurements must be made. Datalogs and data to be retained for review.
Step 2 - Maintain parts in ambient of 85°C, 85% RH for 1000 h. Monitor part condition at 168, 504 and 1008 h. Use deionized water with resistivity no less than 1 x 10 E 6 ohm-cm at 25°C. The chamber must be sequenced during start up and shut down so as to avoid forming condensation on the test devices. Bias circuit to be specified in detailed part specification, to be configured for minimum power dissipation at 2/3 of minimum specified breakdown voltages. Parts to be dried at room temperature for 24 h prior to electrical testing. Testing to be completed within 48 h after drying.
Step 3 - Electrical Parameters: Verification of functionality and measurements of DC parameters to be made per paragraph 4.3.4. Leakage current measurements must be made. Datalogs and data to be retained for review. Lack of functionality or parametric deviations beyond specification limits shall indicate lack of sufficient durability.
4.1.7 SALT ATMOSPHERE—This test may be performed using electrical rejects.
Step 1 - Salt atmosphere (corrosion): MIL-STD-883, method 1009, condition A.
4.1.8 OPERATING LIFE (QUALIFICATION)—Reference: JEDEC 22B, A108

This test is intended for use on complex digital or digital and analog components. The high temperature reverse bias test may be used instead of this test for certain SSI digital and analog devices.
Step 1 - Electrical Parameters: Verification of functionality and measurements of DC parameters shall be made per paragraphs 4.3.4 and 4.3.5. Full datalogs are to be retained for analysis. Twenty parts from each wafer fabrication run shall be tested and data retained to establish initial values of AC parameters and other parameters defined in the specific device ES which are not included in paragraphs 4.3.4 and 4.3.5.
Step 2 - Subject all devices to the operating life test. The detailed test conditions will be defined in the specific device ES. In general, these conditions will determine load networks and operating frequencies or clock rates so as to maintain a maximum junction temperature of 150°C or an ambient temperature of 125°C.
All samples will be tested per paragraph 4.3.4 after 96, 168, 504 and 1008 h of operation. In addition, test all samples per paragraph 4.3.5 after 1008 h of operation.
The parts will be cooled to room temperature with power applied prior to removal from the burn-in oven. Electrical tests will be conducted within 96 h of removal of power except for the monitoring at 1008 h which shall be within 24 h.
Step 3 - Data Analysis: Parts which successfully pass this test will be those which pass paragraph 4.3.4 at each monitoring point and 4.3.5 at 1008 h.
In addition, leakage current measurements in paragraph 4.3.4 and any other parameters must remain within specified delta limits of zero

TABLE OF TESTS 4.3[1] — ELECTRICAL

TEST NAME	TEST NUMBER	CORRESPONDING STANDARD	QUALIFICATION		PROD. ACCEPT. – 1 (PA1)		PROD. ACCEPT. – 2 (PA2)	
			MINIMUM SAMPLE SIZE	STATISTICAL ACCEPTANCE CRITERIA	MINIMUM SAMPLE SIZE	STATISTICAL ACCEPTANCE CRITERIA	MINIMUM SAMPLE SIZE	STATISTICAL ACCEPTANCE CRITERIA
HIGH TEMP. CONTINUITY	4.3.1	JEDEC 22 C100	100%	REMOVE DEFECTIVES	100%	REMOVE DEFECTIVES	100%	REMOVE DEFECTIVES
ELECTROSTATIC DISCHARGE	4.3.2	MIL-STD 883 3015	10 DEVICES	0 REJECTS				
FINAL ELECTRICAL	4.3.3		100%	REMOVE DEFECTIVES	100%	REMOVE DEFECTIVES	100%	REMOVE DEFECTIVES
ELECTRICAL QUALITY	4.3.4		100%	0 REJECTS	200 PER LOT	0 REJECTS	200 PER LOT	0 REJECTS
ELECTRICAL QUALITY TEMP. LIMITS	4.3.5		100%	0 REJECTS	200 PER LOT	0 REJECTS	200 PER LOT	0 REJECTS

[1]REQUIREMENTS:
[a]When subgroup sample sizes are greater than the minimum specified and permit acceptance with one or more rejects, the buyer must be notified prior to the start of testing.
[b]When a sample size is tied to a period of time longer than that which is normally controlled and shipped as a lot, e.g., X/quarter; one sample of the specified number of pieces must be selected from one lot manufactured in that specified time period.

hour data specified in the specific device ES. Any parameter drifts must have demonstrably "leveled off" by the final reading.

4.1.9 BURN-IN (QUALIFICATION, PRODUCTION ACCEPTANCE)—For production level, parts subjected to this test are deliverable as production parts.

Reference: JEDEC 22B, A108

Step 1 - Electrical Parameters: Verification of functionality and measurement of DC parameters is to be made using automatic test equipment. Though it is preferable that the test equipment be identical to the electric test of paragraph 4.3.3, the supplier may use an alternate test if the availability of suitable test equipment or other circumstances so dictates. In this case, the test program conditions and limits must be approved by the buyer.

Step 2 - Subject the components to burn-in at 125°C using the operating life test circuitry, procedures and device specification. In most cases, this will be paragraph 4.1.8. For some components, such as analog or simple digital devices, this will consist of the high temperature reverse bias circuit defined for use in conjunction with paragraph 4.1.10.

During qualification, the failure rate versus burn-in time will be empirically determined by testing at intervals of 24, 48, 96, and 168 h.

This data shall be used to determine the optimized ratio between costs and effectiveness of burn-in. A reasonable decision can be made with data from approximately 2000 devices from at least 10 wafer lots. The decision can be, depending on the results, to shorten the burn-in life or to introduce sample burn-in. The criteria for this decision shall be negotiated between the buyer and supplier.

Step 2 (alternate) - The supplier may substitute other time, temperature and circuit conditions if equivalent stress is demonstrated and approval is granted by the buyer.

Step 3 - Electrical Parameters: The testing described in Step 1 will be repeated.

4.1.10 HIGH TEMPERATURE REVERSE BIAS:

Reference: JEDEC 22B, A108

Step 1 - Electrical Parameters: Verification of functionality and measurement of DC parameters is to be made using automatic test equipment. Though it is preferable that the test be identical to the test of paragraphs 4.3.4 and 4.3.5, the supplier may use an alternate test if the availability of suitable test equipment or other circumstances so dictate. In this case, the test program conditions and limits must be approved by the buyer.

Step 2 - Parts are to be operated in a reverse biased condition. Details of the bias circuit, temperature and time will be specified in the part specification. The parts will be operated for 250 h at a junction temperature of 175°C or 1000 h at a junction temperature of 150°C, or as specified in the part specification.

Step 3 - Electrical Parameters: Repeat Step 1.

4.1.11 POWER AND TEMPERATURE CYCLE:

Reference: JEDEC 22B, A105

Step 1 - Electrical Parameters: Verification of functionality and measurements of DC parameters to be made at 25°C per paragraph 4.3.4. Datalogs and data to be retained for review.

Step 2 - The components shall be operated in the same circuit utilized for the Operating Life test. The tests shall be conducted in an environmental chamber which maintains the selected ambient temperatures through air circulation. Power and appropriate drive signals shall be applied to the device in a cycle of 5 min on and 5 min off during the test. The environmental chamber shall provide a 1 h temperature cycle as follows:

- 1/4 h min at -40°C
- Transition within 1/4 h
- 1/4 h min at 125°C
- Transition within 1/4 h

The test shall be conducted for 1008 h.

Step 3 - Electrical Parameters: Verification of functionality and measurements of DC parameters to be made at 25°C per paragraph 4.3.4. Datalogs and data to be retained for review. This test shall monitor Step 2 at 504 and 1008 h.

4.2 Mechanical Tests

4.2.1 INTERNAL EXAM A—Step 1 - Internal Visual and Mechanical: MIL-STD-883, method 2014. Requirements for materials, design and construction to be those of PA1 devices. 8 x 10 in photograph (per method 2014) of each die is to be supplied to the buyer.

4.2.2 INTERNAL EXAM B—Parts inspected to these criteria are deliverable as production parts.

Step 1 - Internal Visual: MIL-STD-883, method 2010, test condition B. Other details to be specified in detailed part print.

Step 1 (alternate) - Internal visual examination to manufacturer's specification may be performed after approval of the procedure by the buyer.

4.2.3 PHYSICAL DIMENSION—This test may be performed using electrical rejects.

Step 1 - Physical Dimension: MIL-STD-883, method 2016. Other details to be specified in detailed part print.

Step 1 (alternate) - Insert and check for proper fit in specified gauge block.

4.2.4 EXTERNAL VISUAL—This test may be performed using electrical rejects.

Step 1 - External Visual: MIL-STD-883, method 2009.

4.2.5 RESISTANCE TO SOLVENTS—This test may be performed using electrical rejects.

Step 1 - Resistance to Solvents: MIL-STD-883, method 2015.

4.2.6 VISUAL INSPECTION—Parts subjected to this test are deliverable as production parts.

Step 1 - External Visual: MIL-STD-883, method 2009.

Step 1 (alternate) - External visual per manufacturer's specification after approval by the buyer.

4.2.7 BOND STRENGTH—Step 1 - Bond Strength: MIL-STD-883, method 2011, conditions C or D.

Step 1 (alternate) - Bond strength procedure and sampling criteria per manufacturer's specification may be utilized after approval by the buyer. SPC data may waive this requirement.

4.2.8 TERMINAL STRENGTH AND DURABILITY—Samples for this test may be drawn from those selected from paragraph 4.1.4.

Reference: JEDEC 22B, A102A

Step 1 - Lead Integrity: MIL-STD-883, method 2004, condition B2, two leads per device.

Step 2a - Hermetic Devices Only: Seal (fine and gross), MIL-STD-883, method 1014.

Step 2b - Molded Devices Only: Pressure cooker 121°C, 15 ± 2 psig, for 96 h. Note that for convenience these devices may share the autoclave with devices being tested to requirements of paragraph 4.1.4.

Step 3 - Molded Devices Only: Electrical parameters: Verification of functionality and measurement of DC parameters per paragraph 4.3.5. Lack of full functionality and/or parametric deviations due to mechanical damage or corrosion shall indicate lack of sufficient terminal strength and durability.

4.2.9 SOLDERABILITY—Except for qualification, this test may be performed using electrical rejects.

Step 1 - Solderability: MIL-STD-883, method 2003. Solder all leads. For qualification, examine all leads per the method. For PA, view all leads at low magnification, then select the 3 worst leads per device and examine at higher magnification to the criteria of the method. Record the identification of the three leads examined for each device.

Step 2 - For qualification only, verification of functionality to be made per paragraph 4.3.4.

Step 1-2 (alternate) - Solderability testing to MIL-STD-883, method 2003, may be done per the manufacturer's sampling criteria upon approval of the buyer.

4.2.10 HERMETIC DEVICE PACKAGE INTEGRITY (TEST 1): Performed on hermetic devices only.

Step 1 - Electrical Parameters: Verification of functionality and measurements of DC parameters to be made per paragraphs 4.3.4 and 4.3.5. Datalogs to be retained for review.

Step 2 - Seal (fine and gross): MIL-STD-883, method 1014.

Step 3 - Temperature Cycling: MIL-STD-883, method 1010, condition C, 1000 cycles.

Step 4 - Constant Acceleration: MIL-STD-883, method 2001, condition E, Y1 plane only (condition D for packages = or > 40 pins).

Step 5 - Seal (fine and gross): MIL-STD-883, method 1014.

Step 6 - Electrical Parameters: Verification of functionality and measurements of DC parameters to be made per paragraphs 4.3.4 and 4.3.5.

Step 1-6 (alternate) - Procedures and sampling criteria per manufacturer's specification may be substituted for PA2 tests after approval by the buyer.

4.2.11 HERMETIC DEVICE PACKAGE INTEGRITY (TEST 2): Performed on hermetic devices only.

Step 1 - Electrical Parameters: Verification of functionality and measurements of DC parameters to be made per paragraphs 4.3.4 and 4.3.5. Datalogs to be retained for review.

Step 2 - Seal (fine and gross): MIL-STD-883, method 1014.

Step 3 - Vibration, Variable Frequency: MIL-STD-883, method 2007, condition B.

Step 4 - Mechanical Shock: MIL-STD-883, method 2002, condition B, Y1 plane only.

Step 5 - Seal (fine and gross): MIL-STD-883, method 1014.

Step 6 - Electrical Parameters: Verification of functionality and measurements of DC parameters to be made per paragraphs 4.3.4 and 4.3.5.

4.3 Electrical Tests

4.3.1 HIGH TEMPERATURE CONTINUITY—This test is performed on molded devices only. Parts which pass this test are deliverable as production parts.

Reference: JEDEC 22B, C100

The parts shall be placed in a chamber which permits the parts to stabilize at 125°C prior to testing. Lead integrity will then be checked to verify the presence of current flow when a voltage is sequentially applied to each pin, with respect to another pin such as Vss. This test may be combined with 4.3.2 if the specific device specification requires that test is to be conducted at 85°C or above.

4.3.2 ELECTROSTATIC DISCHARGE (ESD)—Reference: MIL-STD-883, 3015.2

Step 1 - Electrical Parameters: Verification of functionality and measurement of DC and AC parameters are to be made per paragraphs 4.3.4 and 4.3.5. Datalogs are to be retained for review.

Step 2 - Reference the specific device ES for details such as between which pins the ESD should be applied, number of positive and/or negative discharges, failure criteria.

Step 3 - Apply the following exposure levels:
. Human Body Model—R = 1.5 K ohms, C = 100 pf, 2 kV
. Charged Device Model—R = 0 ohms, C = 200 pf, 200 V

Step 4 - Repeat Step 1.

Step 5 - Evaluation: Successful completion of this test will include functional verification (paragraphs 4.3.4 and 4.3.5). Also, unless otherwise specified in specific device ES, Icc and Idd will change by no more than 15% and input/output leakage currents shall not have increased by more than a factor of ten.

4.3.3 FINAL ELECTRICAL (TEMPERATURE PER SPECIFIC DEVICE ES)—Parts subjected to this test are deliverable as production parts.

This test shall consist of those electrical tests which are required to guarantee conformance of all production parts to the electrical requirements in the detailed part specification. Functionality, DC and AC parameters will be tested to the extent required for such a guarantee. The individual device specification may require this test to be performed at elevated temperature.

4.3.4 ELECTRICAL QUALITY (25°C)—Parts subjected to this test are deliverable as production parts.

This test shall consist of those electrical tests which are required to verify conformance of all production parts to the electrical requirements in the detailed part specification. Functionality, DC and AC parameters will be tested to the extent required for such a guarantee.

If the acceptance criteria of the table of tests is not met, a 100% screen is required.

4.3.5 ELECTRICAL QUALITY (TEMPERATURE LIMITS)—Parts subjected to this test are deliverable as production parts.

This test is the same as paragraph 4.3.4 except that it is run twice, first at the low and then at the high operating temperature limits specified in the individual device specification. When running the low temperature test, adequate "soak" time will be provided to allow the part to stabilize at the cold temperature, before power is applied. When running the high temperature test, the ambient temperature will be adjusted above the specification limit to account for the thermal time constant of the device, and adequate "soak" time will be allowed, with the intent that the test shall be run as close to normal operating maximum junction temperature as possible.

The parameter limits used in this test will reflect the specification limits established for high and low temperature operation.

If the acceptance criteria of the table of tests is not met, a 100% screen is required.

5. Revalidation Requirements - Subsequent Year Revalidation—For parts which have met all production level test requirements for at least six months prior to revalidation, the revalidation shall consist of the Internal Examination A and the Operating Life (qualification) test. All other parts will be subjected to the entire qualification series unless monitor data can be compiled to demonstrate conformance.

6. Lot Definition—This section establishes the size of acceptance lots.

Acceptance lot formations	Maximum number of devices	
	Single wafer lots	Multiple wafer lots
	30 000	25 000

Multiple wafer lots shall consist of no more than 20 wafer lots, fabricated over a period of not greater than 30 days between starts.

7. Failure Analysis, Corrective Actions—Stress failures shall invoke a failure analysis of the root cause. The failure analysis is to be performed on representative devices of each failure mode unless commonality to a previously analyzed failure can be demonstrated.

Corrective action plans shall be formulated for these failures by the supplier.

8. Applicable Documents
  8.1 Reference Documents—
  a) Purchase order
  b) Detail specification of applicable part
  c) Data sheet of supplier
  d) MIL-STD-883C, Test Methods and Procedures for Microelectronics
  e) MIL-STD-105D, Inspection Level 2
  f) MIL-M-38510F, including appendices
  g) JEDEC STD 22B, Test methods and procedures for solid state devices used in automotive transportation applications

8.2 Priority of Documents—In the event of a conflict between this document and other documents (paragraph 8.1), the precedence in which requirements shall govern, in decreasing order, is as follows:

a) The purchase order
b) The detail specification of the applicable part
c) This document
d) The MIL-STD, JEDEC documents designated in paragraph 8.1
e) The data sheet of the supplier

24 Lighting

φ TERMINOLOGY—MOTOR VEHICLE LIGHTING—SAE J387 OCT88

SAE Information Report

Report of the Lighting Committee, approved March 1969 and completely revised October 1988. Rationale statement available.

1. Scope—This report provides definitions of common terms used in SAE Technical Reports pertaining to motor vehicle lighting. It covers not only basic lighting terms but also terms which identify major segments of technical reports.

2. Lighting Basics and Equipment

2.1 Bulb—An indivisible assembly which contains a source of light and which is normally used in a lamp.

2.1.1 ACCURATE RATED BULB—A seasoned bulb operated at design mean spherical candle power and having its filament(s) positioned within ±0.25 mm, and with its pins oriented within ±5 deg of nominal design positions. Rated bulbs shall be seasoned at rated voltage for 1% of their design life or 10 h maximum.

2.2 Device—Any piece of equipment or mechanism designed to serve a specific purpose or perform a specific function.

2.3 Lamp—A divisible assembly which contains a bulb or other light source and generally an optical system such as a lens or a reflector, or both, and which provides a lighting function.

2.3.1 MULTIPLE COMPARTMENT LAMP—A lamp which provides its lighting function using two or more separately lighted areas which are joined by one or more common parts, such as a housing or lens.

2.3.2 MULTIPLE LAMP ARRANGEMENT— An array of two or more separate lamps on each side of the vehicle which operate together for a particular lighting function.

2.3.3 OPTICALLY COMBINED—A lamp shall be deemed to be "optically combined" if both of the following conditions exist:
  a. It has a single or two filament light source or two or more separate light sources that operate in different ways.
  b. Its optically functional lens area is wholly or partially common to two or more lamp functions.

2.4 Light—Visible radiant energy.

2.5 Light Source

2.5.1 LOCATION—The geometric center of the light emitting element.

2.5.2 LIGHT EMITTING SURFACE—"Light emitting surface" means all or part of the exterior surface of the transparent or translucent lens that encloses the lighting or light signalling device and allows conformance with photometric and colorimetric requirements.

2.5.3 EFFECTIVE PROJECTED LUMINOUS AREA—"Effective projected luminous area" is that area of the light emitting surface projected on a plane at right angles to the axis of a lamp, excluding reflex reflectors (but including congruent reflexes), which is not obstructed by opaque objects such as mounting screws, mounting rings, bezels or trim, or similar ornamented feature areas. Areas of optical or other configurations, for example, molded optical rings or markings, shall be considered part of the total "effective projected luminous area," even if they do not contribute significantly to the total light output. The axis of the lamp corresponds to the H-V axis used for photometric requirements.

2.5.4 CENTROID OF A LENS AREA— The geometric centroid of a plane area which is perpendicular to the axis of reference of the vehicle and upon which the projection of the light emitting lens area falls. As an example: The axis of reference for lamps mounted on the front and rear of a vehicle is the longitudinal axis of the vehicle.

2.6 Unit—An indivisible assembly which provides a mechanical, electrical, or lighting function, for example, sealed beam unit or flasher.

3. Technical Report Content

3.1 Guidelines—Advisory, informational or instructional statements to assist designers, installers, laboratory personnel or manufacturers in meeting the requirements in the evaluation and use of a device or component.

3.2 Requirements—Objectives to be attained in the evaluation and use of new and unused devices, manufactured using production tooling and assembled by production processes.

3.2.1 PERFORMANCE REQUIREMENTS— Characteristics of a device which are essential for its proper functioning, for example, color, luminous intensity, and ability to withstand vibration.

3.2.2 DESIGN REQUIREMENTS— Dimensional or physical characteristics to be attained.

PREFERRED CONVERSION VALUES FOR DIMENSIONS IN LIGHTING—INCH-POUND UNITS/SI—SAE J1322 JUN85

SAE Information Report

Report of the Lighting Committee, approved May 1981, editorial change May 1981, reaffirmed without change June 1985.

1. Scope—This information report is being issued to cover conversions[1] from inch-pound units to those in the Système International (SI) approach. Since all conversions cannot possibly be covered, the list contained in Section 4 is composed of the lighting oriented items which are found in SAE or Motor Vehicle Safety Standards (MVSS) in North America. This reference is to FMVSS in the United States and to CMVSS in Canada.

2. Purpose—The purpose of this report is to give guidance for uniformity particularly to those organizations, governmental or otherwise, who may in the future have occasion to express present inch-pound units in metric or SI terms.

---

[1] N.B. These preferred conversion values do not alter the obligations imposed for compliance purposes by NHTSA and Transport Canada.

ed.

3. Background—The unique character of standards for lighting is aptly illustrated by the requirement of MVSS 108, which provides that various devices "shall be designed to conform" to stipulated specifications. At one and the same time this requirement recognizes, among other things, that lighting items are more subjective than those in any other safety field and that the eye differentiates intensities, for example, on a gross rather than a narrow basis. Additionally, and realistically, some provisions have been derived empirically. A good example is the provision of SAE J586 that visibility of a stop lamp must include the device's having an "unobstructed projected illuminated area of outer lens surface, . . . at least 2 in² in extent, measured at 45 deg . . . .". Some time ago, J586 provided merely that stop lamps must be visible from 45 deg; thus, the addition of a 2 in² requirement was an attempt to define the conditions under which a lamp could be considered as being visible. The 2 in² was determined by viewing a great number of satisfactory lamps already in existence and then picking, in a rounded-off form, a characteristic common to all of them. It is likely that a standard of 1.8 in² or 2.2 in² may in fact have been equally satisfactory. This is the reason that a slightly more rounded-off figure of 12.5 cm² (rather than the literal conversion of 12.9) is used in SAE J586d.

ed. Similarly, 8 in² may more logically be expressed as 50 cm² (instead of the more exact conversion of 51.6) on that day when inch-pound units may be eliminated entirely and initial designs done from a metric standpoint. In the same terms, a guideline of 10 ft for the minimum distance between a lamp filament and a photometer screen may be stated as 3 m.

A metric conversion generally should not increase the severity of a perfectly satisfactory standard which does not demand a strict *go* or *no go* in deciding whether to round up or down.

Yet another factor which argues for the adoption of preferred conversion values is the inexact nature of some commonly accepted techniques. 80°F by using common procedures converts to 27°C; yet, 27°C by applying the same principles becomes 81°F.

There are, however, instances in which the importance of a number is such that a most literal conversion is indicated. An example is that of the 80 in vehicle width. This dividing line currently occurs in so many laws, regulations, standards, etc. that a conversion to 2032 mm is the only reasonable path to follow at this time.

The dimensions contained in SAE specifications such as J571, J573, J602, J760, J1049, and J1132 demand more exact conversions, by their very nature. Such dimensions, therefore, are not covered in this information report. Similarly, units quoted from other sources—such as ASTM—are not covered here.

ed. Although SAE J916 sets forth conversion procedures that may apply to the vast majority of instances, this information report is intended to be an extension to J916 which recognizes that in some cases the use of rationalized numbers may be preferable to exact mathematical conversions.

To summarize, the factors governing establishment of a conversion result should include consideration of the following:

(a) How important or exact is the number being converted?

(b) Does the rounding-up or rounding-down of a converted number outlaw items already in existence and performing satisfactorily?

(c) Is the conversion result *logical* if subsequently used as an original design parameter? In most instances 400 mm, for example, would be a better form than 401 mm or 402 mm.

(d) In what manner may international harmonization be affected by the choice of SI conversion values?

4. Preferred Conversion Values
 4.1 General Rules
  4.1.1 LINEAR DIMENSIONS
   (a) Less than 1 in expressed in whole millimeters, except for numbers of 0.1 in or less.
   (b) Numbers from 1 in to 7 ft expressed to nearest 5 mm.
   (c) Numbers greater than 7 ft expressed to 0.1 m.
  4.1.2 FORCE—Expressed to nearest 10 N.
  4.1.3 TORQUE—Expressed to nearest 0.1 N · m.

ed. **4.2 Spacings, Mounting Heights, and General Linear Dimensions**

Inch-Pound Units	SI
0.01 in	0.25 mm
1/16 in	1.5 mm
0.08 in	2 mm
0.1 in	2.5 mm
1/8 in	3 mm
3/16 in	5 mm
0.465 in	12 mm
0.5 in	13 mm
7/8 in	20 mm
1 in	25 mm
1.25 in	30 mm
2 in	50 mm
2.5 in	60 mm
3 in	75 mm
4 in	100 mm
4.5 in	110 mm
5 in	130 mm
6 in	150 mm
9 in	230 mm
10 in	255 mm
12 in	300 mm
14 in	355 mm
15 in	380 mm
16 in	410 mm
18 in	450 mm
22 in	560 mm
24 in or 2 ft	600 mm
27 in	685 mm
30 in	765 mm
32 in	800 mm
36 in or 3 ft	900 mm
40 in	1000 mm
48 in or 4 ft	1200 mm
54 in	1350 mm
60 in	1525 mm
72 in or 6 ft	1800 mm
80 in	2032 mm
83 in	2100 mm
10 ft	3 m
15 ft	4.6 m
25 ft	7.6 m
30 ft	9.1 m
60 ft	18.3 m
100 ft	30 m

4.3 Areas ed.

Inch-Pound Units	SI
0.028 in²	18 mm²
0.1 in²	60 mm²
2 in²	12.5 cm²
3.5 in²	22 cm²
5 in²	32 cm²
6 in²	37.5 cm²
8 in²	50 cm²
12 in²	75 cm²

4.4 Force ed.

Inch-Pound Units	SI
8 lbf	40 N
14 lbf	60 N
50 lbf	220 N

4.5 Torque ed.

Inch-Pound Units	SI
2 lb-in	0.3 N · m
5 lb-in	0.6 N · m
20 lb-in	2.3 N · m

4.6 Terms of Reference
 4.6.1 HEADLAMPS ed.

Inch-Pound Units	SI
5-3/4 in	146 mm
7 in	178 mm
4 × 6-1/2 in	100 × 165 mm
—	142 × 200 mm²
—	92 × 150 mm²

4.6.2 License plates

Inch-Pound Units	SI
4 × 7 in	100 × 175 mm
6 × 12 in	150 × 300 mm

APPENDIX

It is recommended that this document be updated on a timely basis, possibly even at shorter intervals than the 5-year period currently provided by SAE. Exact frequency should be made a function of progress toward metrication occurring in North America. Such updating should give attention to further *rationalizing* numbers, including the combining of numbers which are somewhat close together. An outstanding example of rational, longer-range future action could be to express as 2 m the *dividing line* for cars and trucks, now expressed as 80 in. Also included could be the expression of linear dimensions to the nearest 10 mm (rather than 5) and areas to the nearest 5 cm².

One regulatory plan to be used could be to proceed in stages:
(a) show the preferred conversion values parenthetically following present inch-pound units system values;
(b) after a period of time, reverse the order;
(c) finally, drop the inch-pound units values entirely.

While numbers are given in both systems, compliance could be demonstrated to either system; one system, however, would have to be used throughout.

[2] This was metric from its inception.

LIGHTING IDENTIFICATION CODE —SAE J759 JUN91

SAE Recommended Practice

Report of the Lighting Committee, approved November 1960, and revised December, 1987. Completely revised by the SAE Lighting Coordinating Committee June 1991. Rationale statement available.

1. Scope—This SAE Recommended Practice provides the lighting function identification codes for use on all passenger cars and trucks.

2. References

2.1 Applicable Documents—The following publications form a part of this specification to the extent specified herein.

2.1.1 GOVERNMENT PUBLICATIONS—Available from the Superintendent of Documents, U.S. Government Printing Office, Washington, D.C. 20402.

FMVSS 108 Lamps, Reflective Devices, and Associated Equipment

2.2 Definition—A lighting identification code is a series of standardized markings for lighting devices which a manufacturer or a supplier may use to mark his product to indicate the SAE Lighting Specification or Specifications to which the device is designed. The code is not intended to limit the manufacturer or supplier in applying other markings to the devices.

NOTE—SAE does not approve products; hence, the use of markings in accordance with this code should not be interpreted to mean that a device so marked has SAE approval.

3. Requirements

3.1 Code Location—The identification code should be permanently marked on the lens or body where it can be observed with the device mounted in its normal position on the vehicle, except that headlamps, turn signal switches, hazard warning switches, and flashers may have the markings located where they can be observed by removing other parts. This exception is granted because many of these devices are not visible as installed on the vehicle. However, when these same devices are externally mounted, the markings must be visible. The manufacturer's identification and model designation (or part number) is required on both the housing and lens of separable devices.

3.2 Size of Markings—Identification numerals and letters shall be at least 3 mm high, except that raised molded markings 2 mm high may be used on lenses containing less than 12.5 cm² of area. The smaller markings are permitted when they are raised, molded markings (not stamped, etched, or lettered in indelible ink).

3.3 Flashers—Flashers, because of their small size, may use 2 mm high markings and these may be permanently stamped, etched, or lettered in indelible ink. The markings required on flashers are:
a. The manufacturer's identification and model number (or part number).
b. The appropriate SAE Specification.
EXAMPLE: Flasher Co. 200
SAE J590e

3.4 Multicompartment and Multiple Lamp Arrangements—Multicompartment lamps or individual lamps designed to use more than one compartment or lamp to comply with the appropriate SAE requirements shall be marked to indicate the number of compartments or lamps that are to be operated together to meet this requirement. Lamps which do not carry a number within a circle or parentheses to indicate that more than one lamp is to be used are intended to be used alone unless exceptions are noted in the appropriate SAE Specification. See example in Section 4.

3.5 Content of Identification Code—The identification code should consist of a series of letters and numbers in the following sequence:

TABLE 1—LETTERS INDICATING DEVICE FUNCTIONS

A	Reflex reflectors
C	Motorcycle auxiliary front lamps
D	Motorcycle and motor driven cycle turn signal lamps
E	Side turn signal lamps—vehicles 9.1 m or more in length
E2	Side turn signal lamps—vehicles less than 9.1 m in length
F	Front fog lamps
F2	Fog tail lamps
G	Truck cargo lamp
H	Sealed beam headlamp
HH	Sealed beam headlamp housing
HR	Replaceable bulb headlamp
I	Turn signal lamps
I3	Turn signal lamps spaced from 75 mm to less than 100 mm from headlamp
I4	Turn signal lamps spaced from 60 mm to less than 75 mm from headlamp
I5	Turn signal lamps spaced less than 60 mm from headlamp
I6	Rear mounted turn signal lamps and front mounted turn signal lamps mounted 100 mm or more from the headlamp, for use on vehicles 2032 mm or more in overall width
I7	Front mounted turn signal lamps mounted less than 100 mm from the headlamp, for use on vehicles 2032 mm or more in overall width
I8	Truck turn signal lamps spaced less than 60 mm from headlamp
J590	Turn signal flasher
J945	Hazard warning signal flasher
J1054	Warning lamp alternating flasher
K	Front cornering lamps
K2	Rear cornering lamps
L	License plate lamps
L2	License plate lamps for use on vehicles of 2032 mm or more in overall width
M	Motorcycle and motor driven cycle headlamps—motorcycle type
N	Motorcycle and motor driven cycle headlamps—motor driven cycle type
O	Spot lamps
P	Parking lamps
P2	Clearance, sidemarker, and identification lamps
P3	Clearance, sidemarker, and identification lamps for use on vehicles 2032 mm or more in overall width
PC	Combination clearance and sidemarker lamps
PC2	Combination clearance and sidemarker lamps for use on vehicles 2032 or more in overall width
Q	Turn signal operating units—Class A
QB	Turn signal operating units—Class B
QC	Vehicular hazard warning signal operating unit
R	Backup lamps
S	Stop lamps
S2	Stop lamp for use on vehicles 2032 mm or more in overall width
T	Tail lamps (rear position lamps)
T2	Tail lamps (rear position lamps) for use on vehicles 2032 mm or more in overall width
U	Supplemental high mounted stop and turn signal lamps
U2	Supplemental center high mounted stop lamp for trucks
U3	Center high mounted stop lamp for passenger cars, light trucks, and MPVs
W	Warning lamps for emergency, maintenance and service vehicles
W2	Warning lamps for school buses
W3	360 degree emergency warning lamps
W4	Emergency warning device
W5-1	360 degree gaseous discharge lamp—Class 1
W5-2	360 degree gaseous discharge lamp—Class 2
W5-3	360 degree gaseous discharge lamp—Class 3
Z	Auxiliary low beam lamps

3.5.1 SAE.

3.5.2 A number within a circle or parentheses indicating the number of compartments in a multicompartment device or the number of separate lamps when more than one compartment or one lamp is needed to satisfy the requirements of the applicable SAE Specification. No number is required for a single compartment device.

3.5.3 One or more letters identifying the function or functions for which the device was designed. Multipurpose devices shall be marked to cover each function for which the device was originally designed. Such devices may be used to carry out one or more of these functions. Table 1 lists the identifying designations for SAE Specifications.

3.5.4 The last two numbers of a year which means that the code letters refer to SAE Specifications listed in the SAE Handbook current in the year indicated, or the applicable requirements of Federal Motor Vehicle Safety Standard 108 specified for the device function in the year indicated. To denote that a function meets the requirements of FMVSS 108, but not the current SAE Specification, a dash line shall be placed under the function letter. Example: AIRST 75.

Devices marked prior to the adoption of this SAE Specification need not be remarked.

3.6 Content of Manufacturer's Identification—The manufacturer's lettered identification and model designation (or part number). Pictorial trademarks are permitted but they must be in addition to the required lettered identification. The lettered identification must be reproducible on a typewriter or computer. The manufacturer's identification does not have to be located in sequence with the identification code markings but may be located on a different area of the lens or body.

4. Examples—Examples of code and manufacturer's identification markings:

SAE AIST 75

XYZ Corp. 400

Translated, this means SAE Specifications current in 1975 for stop lamps, tail lamps, turn signals, and reflectors. The manufacturer is the XYZ Corp. (the word corporation is shown for example only, since this is generally not used) and their model number is Model 400.

SAE A (2) I (2) S (3) T 75

Translated, this means a three-compartment or three-lamp arrangement with two of the compartments needed to meet optically the stop and turn signal requirements, and all three of the compartments needed to meet the tail lamp requirements. If, for example, each of the compartments fully complied with the tail lamp requirements, the 3 preceding the letter T would be omitted. This would signify that all of the compartments were not necessary to meet the tail lamp requirement and that each individual compartment met it as a separate single function. If the SAE Specifications referenced in the examples were not revised and would still be current in the year 1976, for example, the manufacturer could change the marking on this device to 76, if he so desired. This coding change regarding the year might also be made if the SAE Specifications have been revised in the interim and the device meets the new requirements.

PHOTOMETRY LABORATORY ACCURACY GUIDELINES—SAE J1330 JUN88

SAE Information Report

Report of the Lighting Committee, approved August 1982 and revised June 1988.

1. Scope—The purpose of this report is to list and explain major equipment, instrumentation, and procedure variables which can affect inter-laboratory differences and repeatability of photometric measurements of various lighting devices listed in SAE Technical Reports. The accuracy guidelines listed in the report are for the purpose of controlling variables that are not a direct function of the lighting device being measured. The control of these individual variables is necessary to control the overall accuracy of photometric measurements. These accuracy guidelines apply to the measurement of the luminous intensities and reflected intensities of devices at the specified geometrically distributed test points and areas. These guidelines do not apply to photometric equipment used to measure license plate lamps or lighting devices using gaseous discharge lamps.

2. Use of Guidelines—The accuracy limit guidelines suggested in this report are intended as a reference guide to photometric laboratories of various accuracy parameters to help maintain correlation of photometric measurements between laboratories. The guidelines are not intended as specifications to be applied to all photometric equipment, test fixtures, and measurements. Actual photometric performance of various functions and the designs of lighting devices and test fixtures may vary considerably. The use of the guideline information in this report as rigid specifications applied to all types of photometric measurements would be impractical and in some cases would result in equipment with unnecessary accuracy restrictions. These guidelines should be used to aid the photometric laboratory in the awareness of the major variables and to provide information on equipment, instrumentation, and procedure accuracies which may affect overall laboratory differences and repeatability.

3. Accuracy Guidelines

3.1 Test Fixtures

3.1.1 DEVICE POSITIONING—The lighting device to be photometered should be mounted on a rigid test fixture in a position corresponding to the designed nominal operating position of the device on the vehicle. For devices designed for a specific vehicle, the designed nominal position should be determined from the vehicle manufacturer's specifications. For devices designed for multiple vehicle use, the designed nominal position should be determined from the device manufacturer's specifications or instructions. Multiple use devices should be tested in each position in which they are designed for use, or the equivalent, by mathematically translating axis angles and test points.

One of the factors which can significantly affect the device mounting attitude is the torque used to fasten the device to the test stand. This is particularly important when the device floats on a compression type gasket. Mounting torques should be specified for all devices, and these torques should be sufficient to compress the specified gaskets so that "floating" of parts does not occur, unless certain parts are so designed as a means of absorbing shock and vibration.

3.1.2 TEST FIXTURE POSITIONING—Numerous factors affect the ability of the test fixture to position the test device in its designed nominal position. Some of these accuracy factors are the rigidity and flatness of the base, the rigidity of the test fixture structure, and the length of the machined alignment edge or the spacing between alignment pins. Each test fixture should be built from a manufacturer's test fixture design standard to minimize these errors. One suggested example of a test fixture design guide is shown in Appendix A. Other test fixture designs may be equally satisfactory (for example, specialized fixtures for sealed beam units) if they provide proper positioning accuracy.

3.1.3 POSITIONING TOLERANCE—Tolerance guidelines for positioning the test device are listed below.

3.1.3.1 *Lighting Devices Except Headlamp Units*—The tolerance for positioning the device in the test fixture should be ±0.1 deg in each axis.

3.1.3.2 *Headlamp Units*—The positioning of headlamp units is generally more critical than other lighting devices. The tolerance for positioning of headlamp units in the test fixture should be ±0.05 deg in each axis.

3.2 Measurement of Spatial Distributions of Luminous Intensity

—The photometric test point patterns specified in the SAE technical reports are based on measurement of luminous intensity as a function of angle when using a Type A goniometer (as shown in Appendix B and recommended in SAE J575) to position the device with the configuration of horizontal rotation over elevation. The other goniometer configuration is elevation over horizontal rotation (Type B). The use of a Type B goniometer configuration may require the use of a conversion table as given in Appendix B. Methods other than a two axis goniometer, as described above, may be used. For example, a fixed test device with fixed photometer sensors mounted at every specified test point at some suitable distance, or other configurations of fixed and/or moveable sensors may be used. It should be noted that specified luminous intensity maximums or values between test points cannot be measured using sensors at fixed test points.

3.2.1 POSITIONER ACCURACY DETERMINATION—The accuracy of positioners or goniometers used to measure spatial distributions of luminous intensities can be determined by using a checking procedure such as outlined in Appendix C.

3.2.2 POSITIONER TOLERANCES—The repeatability of photometric measurements stated as a percent difference between laboratories when measuring the same device cannot be solely determined as a function of the accuracy of the positioner system or goniometer. For example, a difference of 0.1 deg due to mounting/positioner accuracy will result in no difference between measurements (other factors being equal) if the light distribution is uniform and does not vary significantly with angle in the area of interest. On the other hand, if the area of interest is in a high gradient (rate of change of luminous intensity with angle), one system may, for example, measure 500 cd and another system with a difference of 0.1 deg may measure 550–600 cd. The percent difference between the two measurements could then be 10–20% even with the small orientation difference. This example demonstrates that it is not possible to state a specific accuracy for a photometric measurement system as a function of angle accuracy alone, as both the positioner angular accuracy and the luminous intensity gradient are involved.

Two axis goniometers are available with accuracies of ±0.05 deg and with a resolution of ±0.01 deg. However, in most photometric measurements, a positioner system accuracy deviation of ±0.1 deg with a resolution of ±0.03 deg is considered adequate. (See paragraph 3.1.3.2 above.)

3.3 Power Supplies

—Unless otherwise specified, a regulated DC power supply should be used for all photometric measurements. The following are suggested specifications for the DC power supply:

3.3.1 LINE REGULATION—±0.1%

3.3.2 RIPPLE AND NOISE—0.4% max

3.3.3 STABILITY—Any power supply that remains stable within ±0.1% during the photometric measurement period is satisfactory.

3.4 Voltage Measurements

—A 4½ digit digital voltmeter (DC) with a 10 meg ohm minimum impedance and with an accuracy of 0.05% of the reading is recommended for measurements up to at least 20 V. Voltage measurements should be taken as close to the device input terminations as practicable.

3.5 Current Measurements

—A 4½ digit digital voltmeter reading the output of a precision current shunt is recommended for measurements up to 100 W. The size of the shunt should be sufficient to prevent error due to excessive heat loading. The minimum accuracy of this system should be 0.09% of the value of the current being measured.

3.6 Accurate Rated Bulbs

—Unless otherwise specified, accurate rated bulbs should be used for all photometric measurements. When applicable ratings are available, these bulbs should be rated at the current to produce the designed luminous flux (lumens) or mean spherical intensity (candela) in the attitude in which they are intended to be used with respect to gravity (any attitude for vacuum bulbs). Yellow glass bulbs should also be rated for current to produce their designed luminous flux. Because there are several filament parameters in addition to those controlled in accurate rated bulbs such as coil length, pitch, diameter, and color temperature, some lighting devices may produce significantly different luminous intensity measurements with two different accurate rated bulbs, particularly lighting devices with exceptionally short focal lengths. The light source should be allowed sufficient warmup period for the luminous flux to stabilize. Yellow glass bulbs should also have a sufficient warmup period for any color change to stabilize.

3.7 Sensor/Photometer System

3.7.1 COLOR RESPONSE—The spectral response of the photometer sensor system should be such that color corrections in the yellow color being tested are less than 2% between the yellow-green limit and the yellow-red limit. Likewise, the red correction should not exceed 3% from the red-yellow limit to medium red. One method to determine this color response requires a 2856°K (C.I.E. Illuminant A) standard lamp and a set of at least four glass color filters calibrated for transmittance at 2856°K. Suggested filters are:

Color Region	Chromaticity Coordinates	
	x	y
SAE Yellow-Green Limit Area	0.56	0.44
SAE Yellow-Red Limit Area	0.61	0.39
SAE Red-Yellow Limit Area	0.67	0.33
Medium Red Area	0.69	0.31

In addition, blue filters can be used to check the blue response such as ones with 2856°K chromaticity coordinates of approximately $x = 0.16$, $y = 0.13$; and approximately $x = 0.12$, $y = 0.30$.

Other methods for determinations of color response are acceptable. For example, a calibrated filter may be used whose color matches the color of the light emitted by the device(s) being measured.

3.7.2 RANGE LINEARITY—The linearity of the sensor/photometer system should be verified at least over the range of luminous intensities used for the specific type of device being measured. A deviation from linear response over the range from the calibration level to the extreme luminous intensity value measured should not exceed 2.5%. Measurements made over a narrow range of intensities should have a smaller deviation from linear response.

3.7.3 PHOTOMETER SENSOR APERTURE SIZE—Since photometric measurements are allowed at various distances with different lighting device functions, the area of the photosensor aperture determines the intercepted solid angle of the light flux at the measured test point. Use of large diameter photodetectors should be avoided at shorter measurement distances. Unless otherwise specified, the actual effective area of the sensor used for making the photometric measurements should fit within a circle whose diameter is approximately 0.009 times the distance from the measured light source to the sensor. A 0.009 ratio is equivalent to a solid angle formed by a radius of 0.26 deg. At test distances greater than 5 or 10 m it is common practice for sensor areas to be considerably smaller than the 0.009 maximum. Unless otherwise required, a minimum size is not specified for a sensor; however, some differences in measurements due to large differences in sensor sizes may be experienced, particularly in areas of high luminous intensity gradients.

3.7.4 PHOTOMETER SYSTEM CALIBRATION—The preferred method of photometric calibration is with a calibrated photometric standard lamp. When a calibrating lamp is used for the procedure, it should be calibrated using standard lamps of a high order of accuracy and be traceable to the National Bureau of Standards or other qualified laboratories, through no more than four steps. A minimum of three photometric standard lamps or calibrating lamps should be maintained by the testing laboratory. These lamps should be intercompared periodically and records of the results maintained by the laboratory.

3.8 Instrumentation Calibration—Maintenance of instrumentation accuracy is necessary for the photometric laboratory to make measurements with consistent precision. The following are recommendations of calibration intervals on various instrumentation and standards:

3.8.1 GENERAL INSTRUMENTATION—All instrumentation including voltmeters, current shunts, standard lamps, and accurate rated bulbs should be recalibrated at least annually by comparison with standards of a higher order of accuracy whose calibration is traceable to the National Bureau of Standards or other national laboratories, or by comparison to other like calibrated instrumentation. Records of these checks should be maintained by the laboratory.

3.8.2 SENSOR/PHOTOMETER SYSTEM—The system should be checked for color response and linearity (see paragraphs 3.7.1 and 3.7.2) at least annually. Records of these checks should be maintained by the laboratory.

3.8.3 STANDARD LAMPS AND BULBS—In addition to the annual recalibration, operating time records should be kept on standard lamps, calibrating lamps, and accurate rated bulbs. Secondary and working calibrating lamps should be recalibrated or replaced at the interval recommended by the calibrating laboratory.

3.9 Environmental Variables—The following environmental variables should be controlled so that their effect on photometric measurement accuracy is minimal.

3.9.1 TEMPERATURE AND HUMIDITY—Temperature and humidity can have a significant influence on electrical and photometric measuring instrumentation. Unless special steps have been taken to negate the effects of these factors, such as temperature control (or compensation) and humidity protection on the sensors and amplifiers, the photometry test area should be maintained at a calibrated room temperature within a sufficient temperature and humidity range so as not to have a significant effect on the instrumentation. Typical acceptable tolerances are ±5°C and a maximum of 80% relative humidity. Large temperature gradients and turbulent air can cause fluctuations in luminous intensity measurements over long test distances of 20 m or more. Uniform temperatures should be maintained in long, enclosed photometric tubes or tunnels. Likewise, high concentrations of smoke or dust in the atmosphere may also influence results.

3.9.2 CONDITIONING PERIODS—Injection molded plastic optical components may tend to change dimensionally after molding. Final critical photometric measurements should be made after the parts have stabilized. Similar precautions should be taken with devices which have been exposed to extreme temperatures or humidities in shipping or storage. For maximum accuracy and repeatability, parts should be conditioned in the laboratory environment for a minimum of 24 h, immediately prior to testing.

4. References

SAE J575—Tests for Motor Vehicle Lighting Devices and Components.

5. Bibliography

SAE J387—Terminology—Motor Vehicle Lighting.
SAE J573—Lamp Bulbs and Sealed Beam Headlamp Units.
I.E.S. Lighting Handbook, Sixth Edition, 1981, Illuminating Engineering Society.
Journal of I.E.S., October 1971—Practical Guide to Photometry.
Illuminating Engineering, March and April 1955—I.E.S. General Guide to Photometry.

APPENDIX A
TEST FIXTURE DESIGN GUIDE

To assist the photometric laboratory in using test fixtures which control the main factors influencing the overall positioning accuracy (see paragraph 3.2.1), an example of a test fixture design is shown in Fig. A-1. Other designs, suitable for specific devices such as sealed beam units, or for multipurpose use, and sufficient to provide proper positioning accuracy, may be equally satisfactory.

APPENDIX B
GONIOMETER POSITION CONVERSIONS

To assist the laboratory in converting the test position settings from the recommended goniometer configuration in SAE J575 (Horizontal Rotation Over Elevation, Type A) to the opposite configuration (Elevation Over Horizontal Rotation, Type B), calculated equivalent positions are given in Table B-1. Other positions, not given in the table may be calculated as follows:

$$V_B \text{ deg} = \tan^{-1} (\tan V_A \text{ deg}/\cos H_A \text{ deg})$$
$$H_B \text{ deg} = \sin^{-1} ((\cos V_A \text{ deg})/(\sin H_A \text{ deg}))$$

It should be noted that test position settings on both the horizontal and the vertical axes are identical for both goniometer configurations. Differences in the coordinates are also insignificant within 5 deg in any direction from H-V. A sketch of both goniometer configurations is shown in Fig. B-1.

APPENDIX C
GONIOMETER ACCURACY CHECKING PROCEDURE

To aid the photometric laboratory in determining the positioning deviation of a goniometer (see paragraph 3.2), the following procedure is presented.

C1. Scope—The purpose of this procedure is to check the positioning accuracy of a goniometer used to position lighting devices for photometric measurements.

C2. Equipment—A telescopic or laser transit is required. If the goniometer to be checked is of the rotation over elevation configuration (recommended goniometric configuration, Type A, referenced in SAE J575), the transit should be of the elevation over rotation configuration for the direct measurement of compound angles. For a Type B goniometer with an elevation over rotation configuration, a rotation over elevation configured transit must be used for direct measurement of compound angles. If the proper configured transit is not available, the angular displacement for compound angles may be calculated. (See Appendix B.)

C3. Checking Procedure

C3.1. Set the goniometer to the 0 deg vertical and horizontal positions.

C3.1.1. Position a laser so that its beam is located on the goniometer/photometer sensor or goniometer/projector lamp H-V axis.

C3.1.2. Mount on the goniometer a precision angle plate (flat and square within 0.1 mm over a 150 mm span) with a mirrored surface perpendicular to the H-V axis of the goniometer.

C3.1.3. Adjust the goniometer controls until the laser beam is reflected back onto itself as precisely as possible.

C3.1.4. Adjust the position indicators of the goniometer to read 0, 0 if necessary.

C3.2. Install the transit on the goniometer positioner so that the rotational axes of the transit is on the rotational axes of the goniometer.

C3.3. Set the goniometer to the 0 deg vertical and horizontal positions.

C3.4. Align the optical center of the transit to the center of the photometer sensor, and set the angular indicators of the transit to 0 deg for both axes.

C3.5. Rotate the goniometer along the horizontal or vertical axis to the test point angle to be checked.

C3.6. Align the optical center of the transit to the center of the photometer sensor and record the angular displacement. The deviation of the goniometer angle from the actual angle is the recorded angular displacement minus the accuracy of the transit.

C3.7. Return the goniometer and the transit to the 0 deg horizontal and vertical position.

C3.8. Repeat steps C3.5 and C3.6 to check additional test point positioner accuracy along the horizontal or vertical axes.

C3.9. In a similar manner, the accuracy of compound angle test points may be checked by use of a correctly configured transit (see paragraph C2 above) and by moving the positioner and transit to the compound angle location.

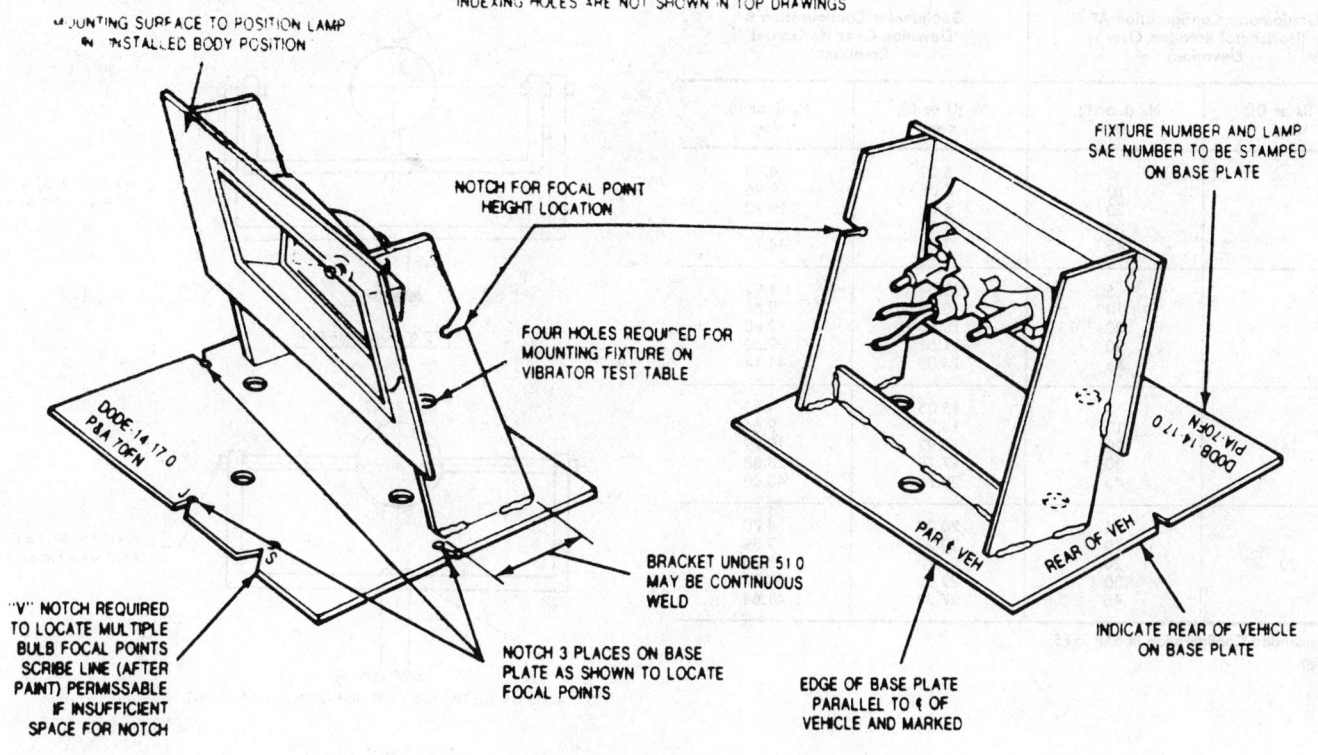

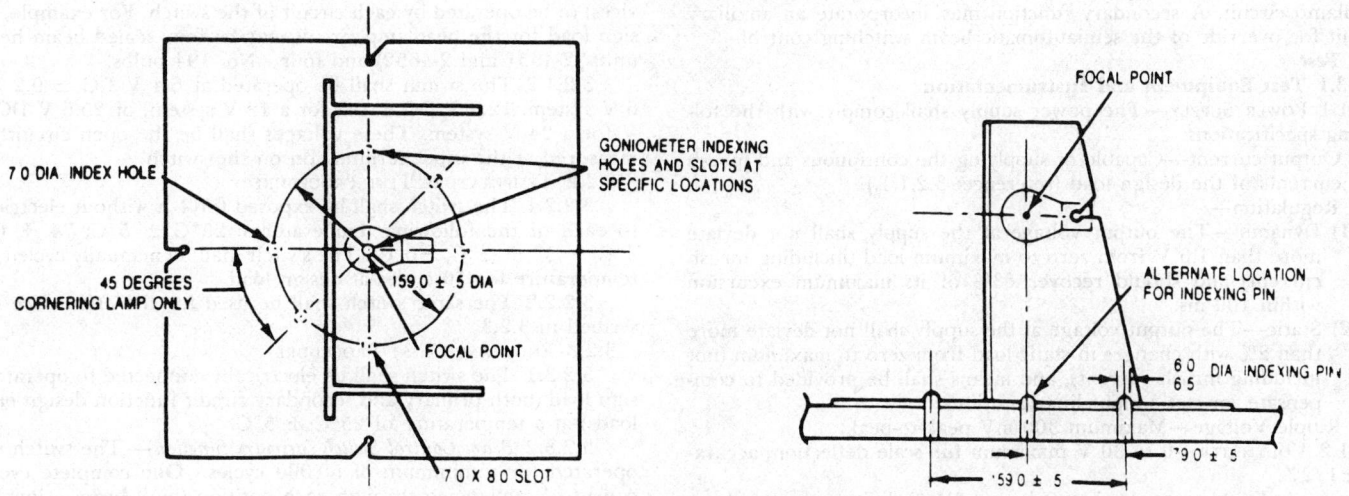

FIG. A-1—LAMP TEST FIXTURE DESIGN

TABLE B-1—GONIOMETER COORDINATE CONVERSIONS

Goniometer Configuration A[a] (Horizontal Rotation Over Elevation)		Goniometer Configuration B (Elevation Over Horizontal Rotation)	
V_A (U or D), deg	H_A (L or R), deg	V_B (U or D), deg	H_B (L or R), deg
5	5	5.02	4.98
	10	5.08	9.96
	20	5.32	19.92
	30	5.77	29.87
	45	7.05	44.78
10	5	10.04	4.92
	10	10.15	9.85
	20	10.63	19.68
	30	11.51	29.50
	45	14.00	44.14
15	5	15.05	4.83
	10	15.22	9.66
	20	15.92	19.29
	30	17.19	28.88
	45	20.75	43.08
20	5	20.07	4.70
	10	20.28	9.39
	20	21.17	18.75
	30	22.80	28.02
	45	27.24	41.64

[a] Recommended Configuration in SAE J575.

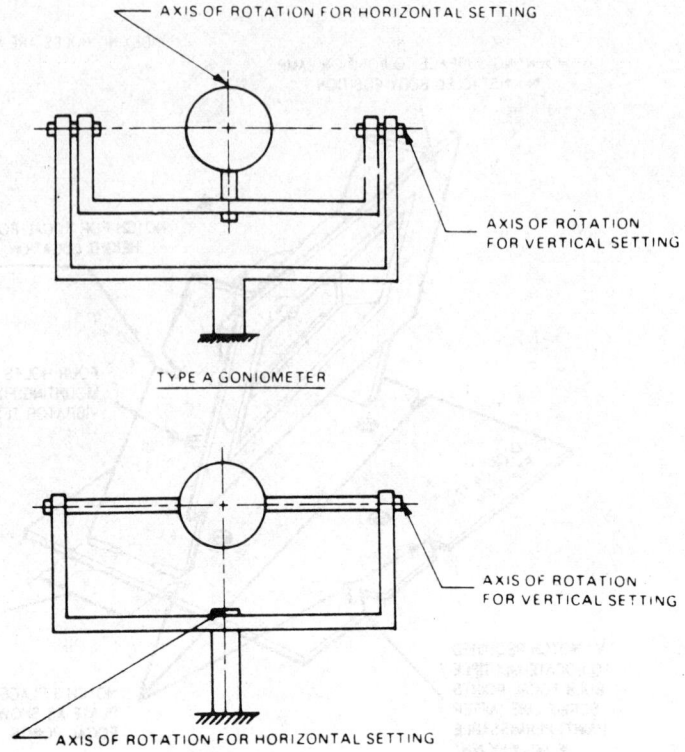

FIG. B-1—GONIOMETER CONFIGURATIONS

(R) HEADLAMP BEAM SWITCHING— SAE J564 MAR90

SAE Standard

Report of Lighting Division approved January 1934 and last revised by Lighting Committee June 1971. Editorial change October 1977. Completely revised by the Auxiliary Devices Standards Committee March 1990. Rationale statement available.

1. Scope—This SAE Standard defines the test conditions, procedures and performance specification for 6, 12, and 24 V manually actuated headlamp beam control switches.

2. Definition—The headlamp beam control switch is an operator activated device intended primarily to select the high or low beam headlamp circuit. A secondary function may incorporate an auxiliary circuit for override of the semiautomatic beam switching control.

3. Test

3.1 Test Equipment and Instrumentation

3.1.1 POWER SUPPLY—The power supply shall comply with the following specifications:
  a. Output current—Capable of supplying the continuous and inrush currents of the design load (reference: 3.2.1.1.).
  b. Regulation—
  (1) Dynamic—The output voltage at the supply shall not deviate more than 1.0 V from zero to maximum load (including inrush current) and should recover 63% of its maximum excursion within 100 ms.
  (2) Static—The output voltage at the supply shall not deviate more than 2% with changes in static load from zero to maximum (not including inrush current), and means shall be provided to compensate for static input line variations.
  c. Ripple Voltage—Maximum 300 mV peak-to-peak.

3.1.2 VOLTMETER—0 to 30 V maximum full-scale deflection, accuracy ±1/2%.

NOTE: A digital meter having at least a 3-1/2 digit readout with an accuracy of ±1% plus 1 digit is recommended for mV readings.

3.1.3 AMMETER—Capable of carrying full system load current, accuracy ±3%.

3.2 Test Procedures—Environmental conditions have been selected for this document to help assure satisfactory operations under general use conditions. It is essential to duplicate specific environmental conditions under which the device is expected to function.

3.2.1 ELECTRICAL LOADS

3.2.1.1 The design load applied to the switch is the electrical load specified by the number and type of bulbs (or other electrical load devices) to be operated by each circuit of the switch. For example, the design load for the headlamp circuit may be four sealed beam headlamp units (2-4651 and 2-4652) and four - No. 194 bulbs.

3.2.1.2 The switch shall be operated at 6.4 V DC ± 0.2 V for a 6 V system, 12.8 V DC ± 0.2 for a 12 V system, or 25.6 V DC ± 0.2 V for a 24 V system. These voltages shall be the open circuit voltage measured at the input termination on the switch.

3.2.2 TEMPERATURE TEST PROCEDURE

3.2.2.1 The switch shall be exposed for 1 h without electrical load to each of the following temperatures: 25°C ± 5°C; 74 + (0°C − 3°C); −32 + (3°C, −0°C). The switch shall be manually cycled at each temperature for 10 cycles at design load.

3.2.2.2 The same switch shall be used for the endurance test described in 3.2.3.

3.2.3 ENDURANCE TEST PROCEDURE

3.2.3.1 The switch shall be electrically connected to operate its design load (both primary and secondary circuit function design electrical loads) at a temperature of 25°C ± 5°C.

3.2.3.2 *Beam Control Switch (primary function)*—The switch shall be operated for a minimum of 50 000 cycles.[1] One complete cycle shall consist of sequencing through each position (high beam – low beam – high beam with dwell in each position) and return without dwell in any of the intermediate positions to the initial position.

---

[1] 50 000 cycles represents 14 cycles of headlamp switch operation every day for approximately 10 years, or 1 cycle for each 2 miles driven for 100 000 miles with 50% night driving.

The test equipment shall be arranged to provide the following switch operating time requirements:
a. Travel Time: 0.1 to 0.5 s
   (time from one position to the next)
b. Dwell Time: 0.5 to 2.0 s
   (time in each position)
c. Make and
   Break Rate: 130 to 150 mm per s

3.2.3.3 *Semiautomatic Beam Control Switch (secondary function):* The switch shall meet all the requirements of the beam control switch except as follows:

One complete cycle shall consist of sequencing through each position (with dwell in each beam position):
a. High beam
b. Override mechanism
c. Low beam
d. Override mechanism
e. High beam

The test equipment shall be arranged to provide the operating time requirements.

The semiautomatic beam control switch shall be operated for 25 000 cycles.[2]

[2] 25 000 cycles represents 14 manual override beam changes every day for approximately 10 years or cycle for each 4 miles driven for 100 000 miles with 50% night driving.

3.2.3.4 At the completion of the cycle testing, the switch shall be operated for 1 h in the headlamp position with the design load(s) connected.

3.2.4 VOLTAGE DROP TEST PROCEDURE:

3.2.4.1 The voltage drop from the input terminal(s) to the corresponding output terminal(s) shall be measured at design load before and after the completion of the endurance test and shall be the average of three consecutive readings. These voltage drop readings should exclude the voltage drop across the circuit breaker(s). If wiring is an integral part of the switch, the voltage drop measurement shall be made including 75 mm ± 6 mm of wire on each side of the switch; otherwise the measurement shall be made at the switch terminals.

4. Performance Requirements

4.1 During and after each of the cycles described in 3.2.2 and 3.2.3, the switch shall be electrically and mechanically operable.

4.2 The voltage drop shall not exceed 0.3 V when measured as in 3.2.4, before and after completion of the tests described in 3.2.3.

φSEMIAUTOMATIC HEADLAMP BEAM SWITCHING DEVICES—SAE J565 JUN89

SAE Standard

Report of the Lighting Committee, approved August 1954, completely revised by the Lighting Coordinating Committee June 1989.

1. Scope—This SAE Standard provides test procedures, performance requirements, and guidelines for semiautomatic headlamp beam switching devices.

2. Definitions

2.1 Semiautomatic Headlamp Beam Switching Device—A device that provides either automatic or manual control of beam switching at the option of the driver. When the control is automatic, the headlamps switch from the upper beam to the lower beam when a photosensor is illuminated by the headlamps of an approaching car and switch back to the upper beam when the light from the road ahead is at an appropriate intensity. When the control is manual, the driver may obtain either beam manually regardless of the condition of light ahead of the vehicle.

2.2 "Dim" Sensitivity Light Intensity—The emitted candela of the light source at the time the device switches from upper beam to lower beam as the intensity of the light source is gradually increased.

2.3 "Hold" Sensitivity Light Intensity—The emitted candela of the light source at the time the device switches from lower beam to upper beam as the intensity of the light source is gradually decreased.

3. Identification Code Designation—Not applicable.

4. Tests

4.1 Test Facilities and Environment

4.1.1 LABORATORY FACILITIES—See SAE J575.

4.1.2 LIGHT SOURCES—The light sources used for sensitivity testing shall be of the incandescent tungsten filament type operating as C.I.E. Illuminant A (2856 K).

4.2 Equipment and Instrumentation

4.2.1 TESTING DISTANCE—The testing distance for sensitivity shall not be less than 3 m with the lamp intensity and test distance adjusted to give the same results as that obtained when using 30 m.

4.2.2 AIM—The device shall be mounted and operated in the laboratory in the same environment as that encountered on the vehicle, that is behind tinted glass, grille work, etc. The H-V position/axis shall be taken as parallel to the longitudinal axis of the vehicle.

4.3 Test Samples

4.3.1 See SAE J575.

4.4 Test Procedures—The following sections describe individual tests, which need not be performed in any particular sequence unless otherwise specified. The completion of the tests may be expedited by performing the tests simultaneously on separately mounted samples. For all tests unless otherwise specified, the device shall be set up, adjusted, and operated in accordance with paragraphs 4.4.3.1 and 4.2.2 in an ambient temperature of 25°C ± 5 with an input of 13 V ± 0.1. All light intensity values should normally be measured at 30 m unless otherwise specified.

4.4.1 DUST TEST—The device shall be subjected to the dust test of SAE J575. The device shall not be operated during the dust test.

4.4.2 CORROSION TEST—The device and all components, which are located outside the driver compartment of the vehicle, shall be subjected to corrosion testing in accordance with SAE J575 with the device not operating. (Water should not be allowed to collect on any connector socket).

4.4.3 SENSITIVITY TEST

4.4.3.1 With the device in the design position and aimed with the H-V position per paragraph 4.2.2 facing a light source of 15 cd, the voltage output of the sensitivity control shall be gradually increased until the device switches from upper beam to lower beam. The voltage at this setting of the sensitivity control shall be noted and maintained throughout the remainder of the sensitivity test. The "dim" sensitivity and the "hold" sensitivity, shall be determined at each test position.

4.4.3.2 To provide more complete information on sensitivity, a "dim" curve shall be made radially in intervals not exceeding 10 deg around the H-V axis at a light level of 25 cd ± 0.5. The sensitivity control output voltage shall be maintained as found in paragraph 4.4.3.1 for establishing the "dim" curve. The device shall be aimed a sufficient amount from the H-V position that it will remain on upper beam. The aim of the device shall gradually be changed along a radial line towards the H-V position until the device switches from upper beam to lower beam to establish a point on the "dim" curve. As the change in aim of the device continues to the H-V axis, the device shall remain on lower beam. A switch from lower beam to upper beam automatically within the "dim" curve would constitute a void in sensitivity and shall be so identified.

4.4.4 VOLTAGE REGULATION—The "dim" sensitivity at H-V shall be determined with the voltage input to the device at 11 V ± 0.1 and at 15 V ± 0.1.

4.4.5 MANUAL OVERRIDE OF AUTOMATIC CONTROL—The test light shall be turned on to cause the device to be on lower beam. The manufacturer's instructions shall be followed to cause the device to override the test light and switch to upper beam. In a similar manner the test light shall be turned off to cause the device to be on upper beam. Again the manufacturer's instructions shall be followed to cause the device to switch to the lower beam.

4.4.6 WARMUP—The device shall be aimed on the H-V axis at a light source emitting 25 cd ± 0.5.

4.4.7 TEMPERATURE REGULATION—The device shall not be operated while being exposed for 1 h to an ambient temperature of 98°C ± 2 if the device is mounted in the passenger or engine compartment, or 65°C ± 2 if mounted elsewhere. After the high temperature exposure, the device shall be operated at the same conditions as initially adjusted in ambient temperatures of -30°C ± 2, 25°C ± 5, and 40°C ± 2 for 1 h each.

4.4.8 VIBRATION—The device shall not be operated while being subjected to sinusoidal vibration of 5 g ± 0.2 constant acceleration as follows:

a. The device shall be mounted in proper vehicle position and vibrated for 0.5 h in each of three directions; vertical, horizontal and parallel to the vehicle axis, and horizontal and normal to the vehicle axis.

b. The vibration frequency shall be varied from 30 to 200 Hz and back to 30 Hz over a period of approximately 1 min.

4.4.9 SUNLIGHT EXPOSURE—The device shall not be operated while being exposed for 1 h in bright noonday sunlight (54 000 lx minimum illumination with a clear sky) with the photounit aimed as it would be on a car and facing an unobstructed portion of the horizon in the direction of the sun. After resting for 1 h in normal room temperature, the device shall be operated at the same conditions as initially adjusted.

4.4.10 DURABILITY—The device shall be subjected to the following test:

a. The photounit shall be actuated by a light source which impresses 0.07 lx ±0.01 (equivalent to approximately 60 cd ± 10 at 30 m) on the lens surface and which is cycled on and off four times per minute with a duty cycle of 50% on time.

b. The device shall be operated at 13 ± 0.V input for 90 min on and 30 min off for 200 h total time.

After resting for 2 h at room temperature in a lighted area of 500-1500 lx, the device shall be operated at the same conditions as initially adjusted.

4.4.11 RETURN TO UPPER BEAM TIME—An illumination of 1000 lx ± 100 intensity shall be impressed on the lens of the device for 10 s ± 1. The light shall then be extinguished.

5. Requirements—A device, when tested in accordance with the test procedures specified in Section 4, shall meet the following requirements:

5.1 Dust—After the dust test, the photounit lens shall be wiped clean. The device shall be operated at the same conditions as intially adjusted. The H-V "dim" sensitivity shall be between 8 and 25 cd.

5.2 Corrosion—After the test, the device shall be operated at the same conditions as initially adjusted. The H-V "dim" sensitivity shall be between 8 and 25 cd.

5.3 Sensitivity

5.3.1 The "dim" and "hold" sensitivities shall be within the limits of Table 1.

5.3.2 There shall be no sensitivity voids within the dim curve where the headlamps return to upper beam automatically.

TABLE 1—OPERATING LIMITS
(candela at 30 m)

Test Position (deg)	Dim Intensity (cd)	Hold Intensity (cd)
H-V	15 - adjust	1.5 min to 3.75 max
H-2L	25 max	1.5 min
H-4L	40 max	1.5 min
H-6L	75 max	1.5 min
H-2R	25 max	1.5 min
H-5R	40 min to 150 max	1.5 min
ID-V	30 max	1.5 min
IU-V	30 max	1.5 min

5.4 Voltage Regulation—The H-V "dim" sensitivity shall be between 8 and 25 cd at 11 V ± 0.1 and 15 V ± 0.1.

5.5 Manual Override of Automatic Control—When the device is in the upper beam mode, the override shall be capable of switching to lower beam. When the device is in the lower beam mode, the override shall be capable of switching to upper beam.

5.6 Warmup—The warmup time after being energized to maintain lower beam shall not exceed 3 s.

5.7 Temperature Regulation—After a 1 h soak in each of the operating ambient temperatures, the H-V "dim" sensitivity shall be between 8 and 25 cd.

5.8 Vibration—At the conclusion of the test, the device shall be operated at the same conditions as initially adjusted. The H-V "dim" sensitivity shall be between 8 and 25 cd and the mechanical aim of the photounit shall not change more than 0.3 deg.

5.9 Sunlight Exposure—The H-V "dim" sensitivity shall be between 8 and 25 cd.

5.10 Durability—The H-V "dim" sensitivity shall be between 8 and 25 cd.

5.11 Return to Upper Beam Time—The time to return to upper beam shall not exceed 2 s.

6. Guidelines

6.1 Installation Requirements—The following requirements apply to the device as used on the vehicle and are not part of the laboratory test requirements and procedures.

6.1.1 LENS ACCESSIBILITY—The device lens shall be accessible for cleaning when the unit is mounted on a vehicle.

6.1.2 MOUNTING HEIGHT—The center of the device lens shall be mounted not less than 600 mm above the road surface.

6.1.3 MALFUNCTION OVERRIDE—In the event that there is a malfunction of the automatic control (light sensor) portion of the system, which would maintain the system on upper beam, means must be provided for manual switching to low beam. Malfunction of the automatic control portion of the system may be simulated by placing the photounit in darkness (light level below 0.001 lx) where the device should effectively be locked on high beam.

6.1.4 AUTOMATIC DIMMING INDICATOR—There shall be a convenient means of informing the driver when the device is controlling the headlights automatically. The manufacturer's instructions shall be followed to determine the means of indication.

6.1.5 UPPER BEAM INDICATOR—The device shall not affect the function of the upper beam indicator light.

6.1.6 SENSITIVITY CONTROL—A sensitivity control shall be provided for the driver.

6.2 Installation Recommendations—The following design parameters should be considered in mounting the light sensor portion of the device on the vehicle for satisfactory performance.

6.2.1 For optimum performance, the light sensor portion of the device should face forward and be located and mounted, whenever possible, to sense the equivalent perception of the driver's field of view ahead of the vehicle.

6.2.2 As a minimum mounting height, the light sensor portion of the device should be above the top edge of the highest mounted headlamp unit on the vehicle.

6.2.3 The lens of the light sensor portion of the device should be mounted for an unobstructed field of view ahead of the vehicle within 15 deg L, 10 deg R, 5 deg D, and 10 deg U, relative to the vehicle longitudinal axis.

6.3 Operating Instructions—A set of operating instructions shall be included to permit a driver to operate the device correctly. The following items shall be covered:

a. How to turn the automatic control on and off.
b. How to adjust sensitivity.
c. Any other instructions applicable to the particular device.

LAMP BULB RETENTION SYSTEM—SAE J567 NOV87   SAE Standard

Report of Electrical Equipment Division approved August 1915 and revised by the Lighting Committee November 1987. Rationale statement available.

1. Scope—This SAE Standard covers the performance and functional requirements of the lamp bulb retention system applicable for use in motor vehicles.

2. Definition—The lamp bulb retention system is a device which retains a lamp bulb in its intended application and provides electrical continuity.

3. Requirements

3.1 The lamp bulb retention system shall accept and provide for the retention and removal of the maximum and minimum bulb gages. See Table 1 for bulb gages.

3.2 The bulb retention system shall provide required electrical connections.

3.3 Bulb retention systems employing multiple contacts shall have them spaced so that they will not contact (electrically insulated from) each other or short to ground.

3.4 When the bulb retention system is assembled in its intended application, the insertion and rotational forces required to lock the maximum bulb gage in its final seating position shall not exceed the values shown below:

Maximum Insertion Force
B and C base sockets—60 N
A base sockets—40 N

Maximum Torque
C base sockets—0.6 N·m
B base sockets—0.6 N·m
A base sockets—0.3 N·m

NOTE: Bulb retention systems designed to be removed from their intended application for bulb service may be checked while removed.

3.5 When the bulb retention system is assembled in its intended application, the B and C base bulb retention systems shall provide a minimum bulb support as measured with the bulb support gage. (See Table 2 for bulb support gages and Fig. 3 for gage.)

NOTES:

1. Bulb retention systems designed to be removed from their intended application for bulb service may be checked while removed.

2. Bulb retention systems which provide alternative equivalent bulb supporting means may be used and need not be checked with the bulb support gage.

⏀TABLE 1—MAXIMUM AND MINIMUM BULB RETENTION SYSTEM GAGES (MM)

Dimension[a]	C-2 Base Max Gage	C-2 Base Min Gage	B-2 Base Max Gage	B-2 Base Min Gage	B-1 Base Max Gage	B-1 Base Min Gage
A	15.30 +0.01/−0.00	15.05 +0.00/−0.01	15.30 +0.01/−0.00	15.05 +0.00/−0.01	15.30 +0.01/−0.00	15.05 +0.00/−0.01
C	16.97 +0.01/−0.00	16.32 +0.00/−0.01	16.97 +0.01/−0.00	16.32 +0.00/−0.01	16.97 +0.01/−0.00	16.32 +0.00/−0.01
D	—	0.64 +0.00/−0.01	—	0.64 +0.00/−0.01	—	0.64 +0.00/−0.01
E	2.03 +0.01/−0.00	1.88 +0.00/−0.01	2.03 +0.01/−0.00	1.88 +0.00/−0.01	2.03 +0.01/−0.00	1.88 +0.00/−0.01
F	8.03 +0.01/−0.00	6.32 +0.00/−0.01	8.03 +0.01/−0.00	6.32 +0.00/−0.01	8.03 +0.01/−0.00	6.32 +0.00/−0.01
G	1.27 +0.00/−0.13	0.38 +0.00/−0.01	1.27 +0.00/−0.13	0.38 +0.00/−0.01	1.27 +0.00/−0.13	0.38 +0.00/−0.01
H	4.32 +0.03/−0.00	3.51 +0.00/−0.03	4.32 +0.03/−0.00	3.51 +0.00/−0.03	4.32 +0.03/−0.00	3.51 +0.00/−0.03
K	—	8.89 +0.00/−0.01	—	8.89 +0.00/−0.01	—	8.89 +0.00/−0.01
L	16.38 +0.01/−0.00	—	16.38 +0.01/−0.00	—	16.38 +0.01/−0.00	—
N	5.21 +0.03/−0.00	4.50 +0.00/−0.03	5.21 +0.03/−0.00	4.50 +0.00/−0.03	5.21 +0.03/−0.00	4.50 +0.00/−0.03
P	3.38 +0.03/−0.00	2.97 +0.00/−0.05	0.00 +0.03/−0.03	0.00 +0.00/−0.03	0.00 +0.03/−0.03	0.00 +0.00/−0.03
S	6.78 +0.01/−0.00	6.78 +0.00/−0.01	6.78 +0.01/−0.00	6.78 +0.00/−0.01	NA	NA
T	1.52 +0.13/−0.00	1.52 +0.13/−0.00	1.52 +0.13/−0.00	1.52 +0.13/−0.00	1.52 +0.13/−0.00	1.52 +0.13/−0.00
U	—	—	—	—	—	—

ΦTABLE 1—MAXIMUM AND MINIMUM BULB RETENTION SYSTEM GAGES (continued)

Dimension[a]	A-1 Base Max Gage		A-1 Base Min Gage		Wedge Base Max Gage		Wedge Base Min Gage	
A	9.30	+0.01 / -0.00	9.07	+0.00 / -0.01	4.45	+0.01 / -0.00	3.68	+0.00 / -0.01
C	10.97	+0.01 / -0.00	10.26	+0.00 / -0.01	6.10	+0.01 / -0.00	3.68	+0.00 / -0.01
D	NA		0.64	+0.00 / -0.01	—		—	
E	1.70	+0.01 / -0.00	1.55	+0.00 / -0.01	9.50	+0.01 / -0.00	9.50	+0.00 / -0.01
F	6.48	+0.01 / -0.00	4.57	+0.00 / -0.01	3.05	+0.01 / -0.00	NA	
G	1.27	+0.00 / -0.13	0.38	+0.00 / -0.01	4.06	+0.01 / -0.00	3.30	+0.00 / -0.01
H	3.33	+0.01 / -0.00	2.41	+0.00 / -0.03	—		—	
K	—		4.57	+0.00 / -0.01	—		—	
L	10.41	+0.01 / -0.00	—		—		—	
N	4.32	+0.03 / -0.00	3.73	+0.00 / -0.03	1.65	+0.00 / -0.01	—	
P	0.00	+0.03 / -0.03	0.00	+0.03 / -0.03	2.41	+0.01 / -0.00	1.91	+0.00 / -0.01
S	NA		NA		—		—	
T	1.52	+0.13 / -0.00	1.52	+0.13 / -0.00	—		—	
U	6.40	+0.03 / -0.00	—		—		—	

[a] See Figs. 1 and 2.

ΦTABLE 2—BULB SUPPORT GAGE (SEE FIG. 3)

Dimension	B-1 and B-2 Base Bulb Support Gage		C-2 Base Bulb Support Gage		Dimension	B-1 and B-2 Base Bulb Support Gage		C-2 Base Bulb Support Gage	
A	15.05	+0.01 / -0.00	15.05	+0.01 / -0.00	N	4.85	+0.13 / -0.00	4.85	+0.13 / -0.00
B	19.05	+0.25 / -0.25	19.05	+0.25 / -0.25	P	10.36	+0.01 / -0.00	10.36	+0.01 / -0.00
C	16.32	+0.01 / -0.00	16.32	+0.01 / -0.00	R	18.90	+0.00 / -0.25	26.44	+0.00 / -0.25
D	0.64	+0.01 / -0.00	0.64	+0.01 / -0.00	S	4.75D	S/F	4.75D	S/F
E	1.88	+0.01 / -0.00	1.88	+0.01 / -0.00	T	6.35D	S/F	6.35D	S/F
F	6.32	+0.01 / -0.00	6.58	+0.01 / -0.00	U	0.89	+0.00 / -0.13	0.89	+0.00 / -0.13
G	0.00	+0.03 / -0.03	2.97	+0.05 / -0.00	V	15.48	+0.01 / -0.00	15.48	+0.01 / -0.00
H	3.51	+0.01 / -0.00	3.51	+0.01 / -0.00	W	1.52	+0.13 / -0.00	1.52	+0.13 / -0.00
J	30.18	+0.13 / -0.00	44.45	+0.13 / -0.00	X	B-1 0.00 / B-2 3.45	+0.00 / -0.08	3.45	+0.00 / -0.08
K	6.86	+0.00 / -0.13	6.86	+0.00 / -0.13					
L	3.18	+0.01 / -0.00	3.18	+0.01 / -0.00	Y	B-1 0.00 / B-2 6.78	+0.13 / -0.00	6.78	+0.13 / -0.00
M	0.76R	+0.03 / -0.03	0.76R	+0.03 / -0.03					

NOTE: Dim. P − L = 7.19 mm Dim.

24.13

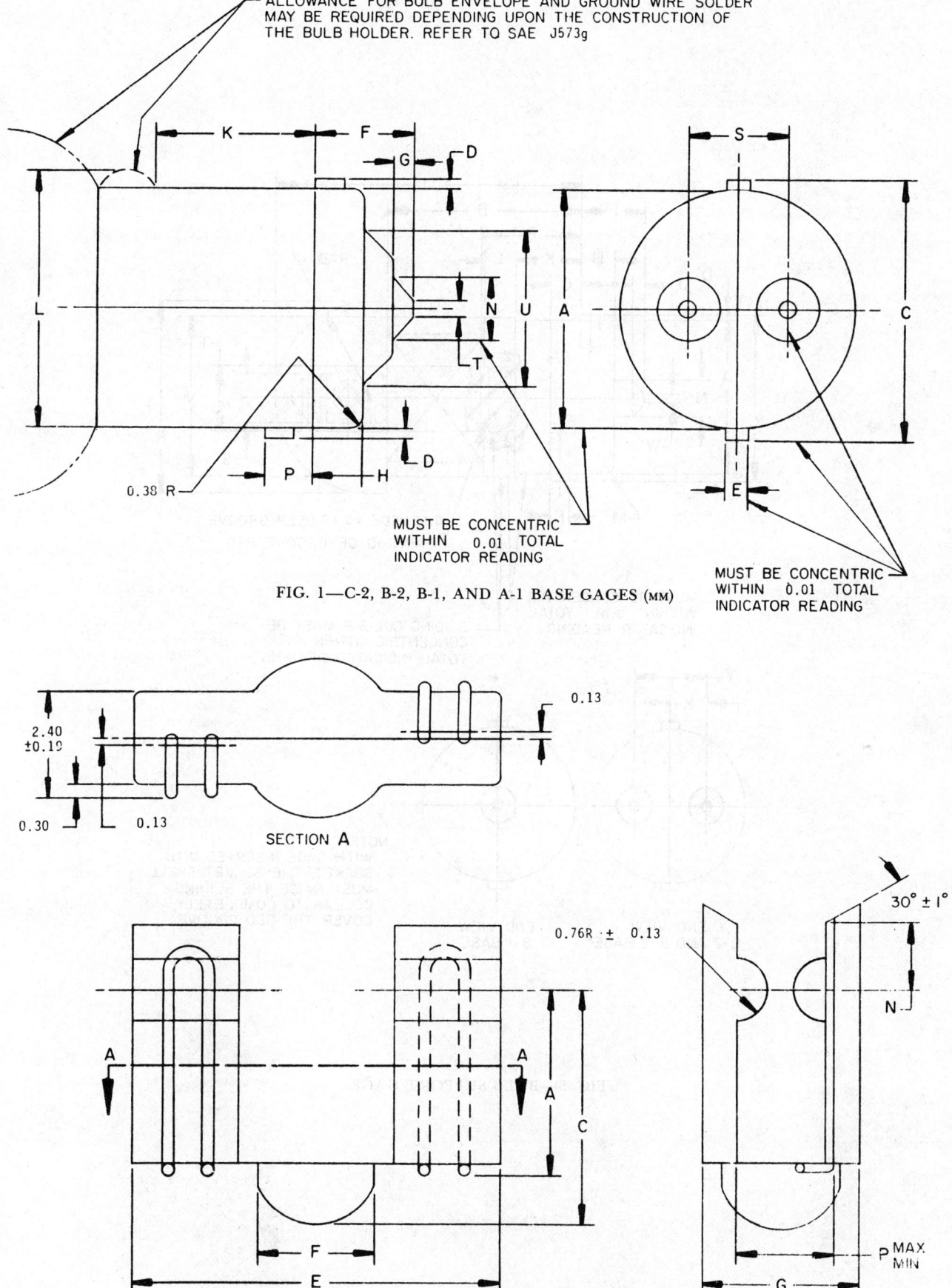

FIG. 1—C-2, B-2, B-1, AND A-1 BASE GAGES (mm)

FIG. 2—WEDGE BASE GAGES (mm)

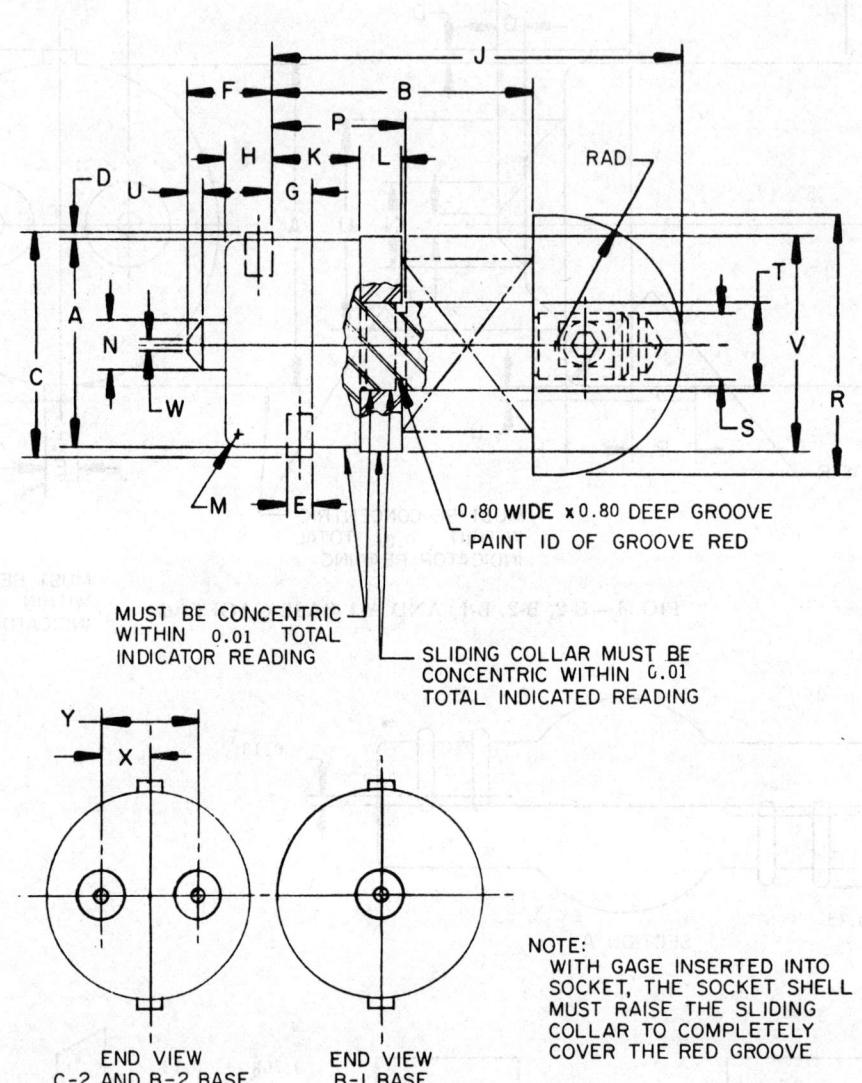

FIG. 3—BULB SUPPORT GAGE

CONNECTORS AND PLUGS—SAE J856

SAE Standard

Report of Electric Equipment Division approved June 1936 and last revised by Lighting Committee April 1963.

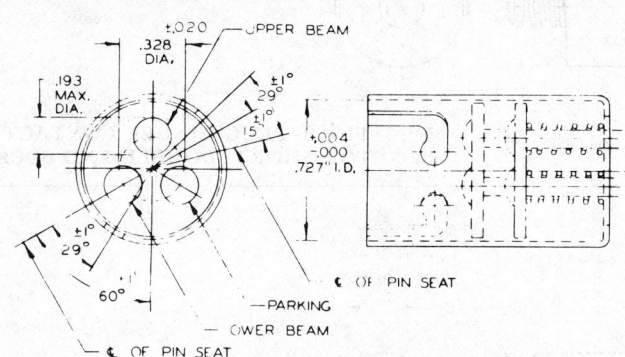

FIG. 1—SOCKET, PLUG, THREE WAY OFFSET PIN, LARGE
THREE WAY OFFSET PIN, LARGE FOR PLUG BASE TYPE SEE FIG. 11

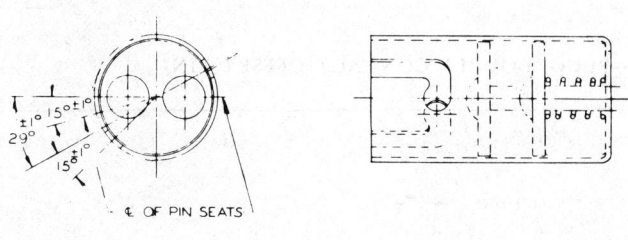

FIG. 2—SOCKET, PLUG, DOUBLE CONTACT OFFSET PIN
DOUBLE CONTACT OFFSET PIN FOR PLUG BASE TYPE SEE FIG. 8

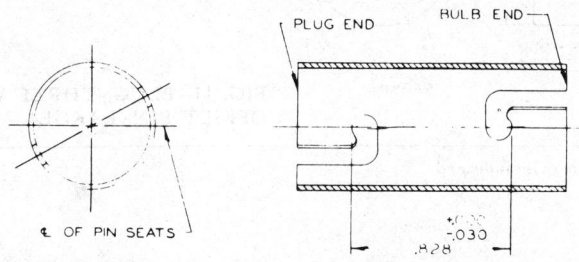

FOR PLUG BASE TYPE SEE FIG. 9, FOR BULB BASE TYPES B-1 AND B-2
FIG. 3—SOCKET, PLUG-BULB, DOUBLE END, SHORT[1]

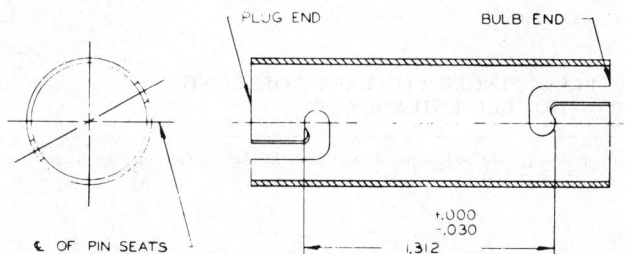

FOR PLUG BASE TYPE SEE FIG. 10, FOR BULB BASE TYPES B-1 AND B-2
FIG. 4—SOCKET, PLUG-BULB, DOUBLE END, LONG[1]

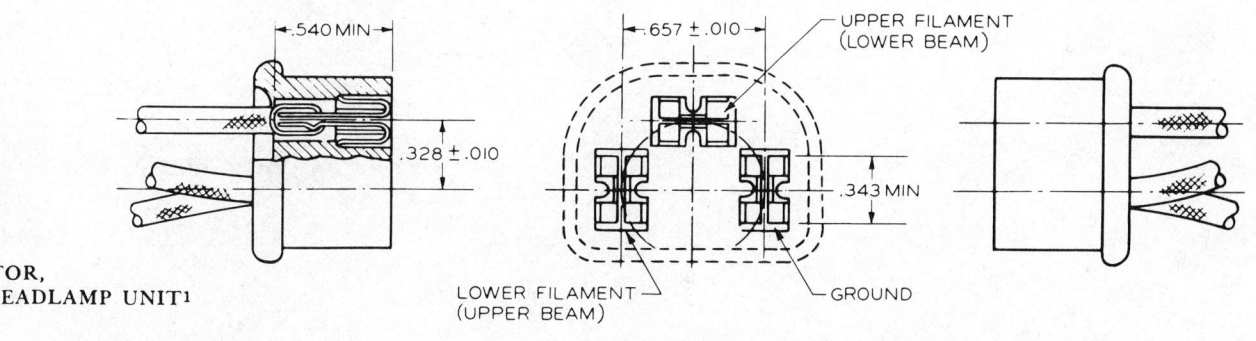

FIG. 5—CONNECTOR, SEALED BEAM HEADLAMP UNIT[1]

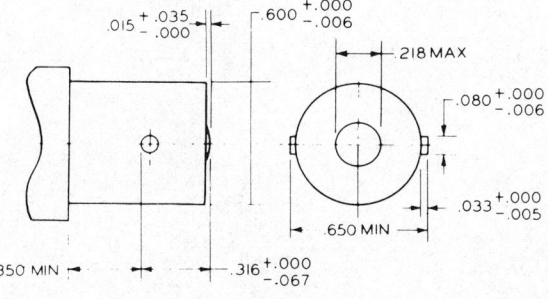

FIG. 6—PLUG, SINGLE CONTACT

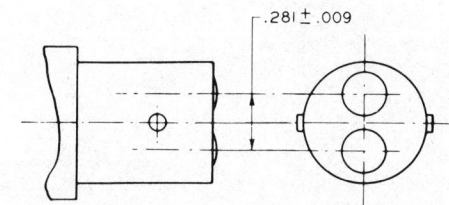

FIG. 7—PLUG, DOUBLE CONTACT

---

[1]All dimensions for the following are the same as those given for Figs. 1 and 2, except where otherwise indicated.

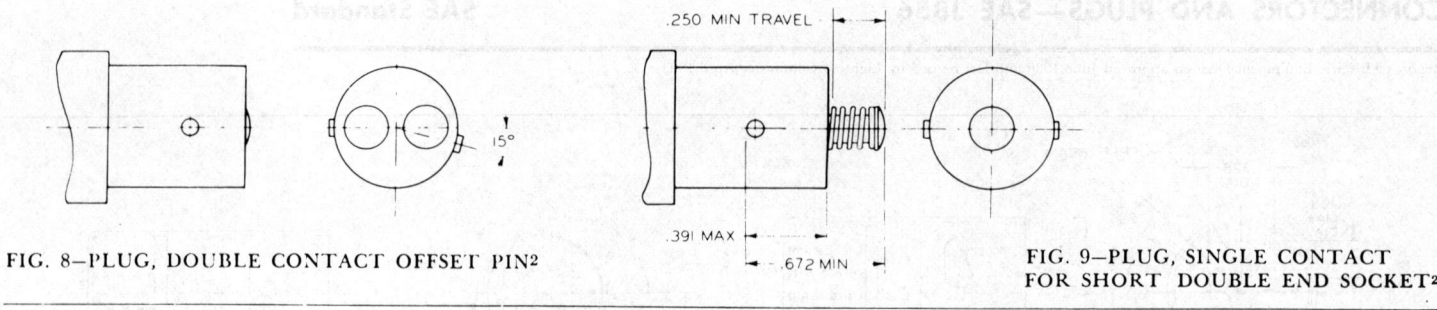

FIG. 8—PLUG, DOUBLE CONTACT OFFSET PIN[2]

FIG. 9—PLUG, SINGLE CONTACT FOR SHORT DOUBLE END SOCKET[2]

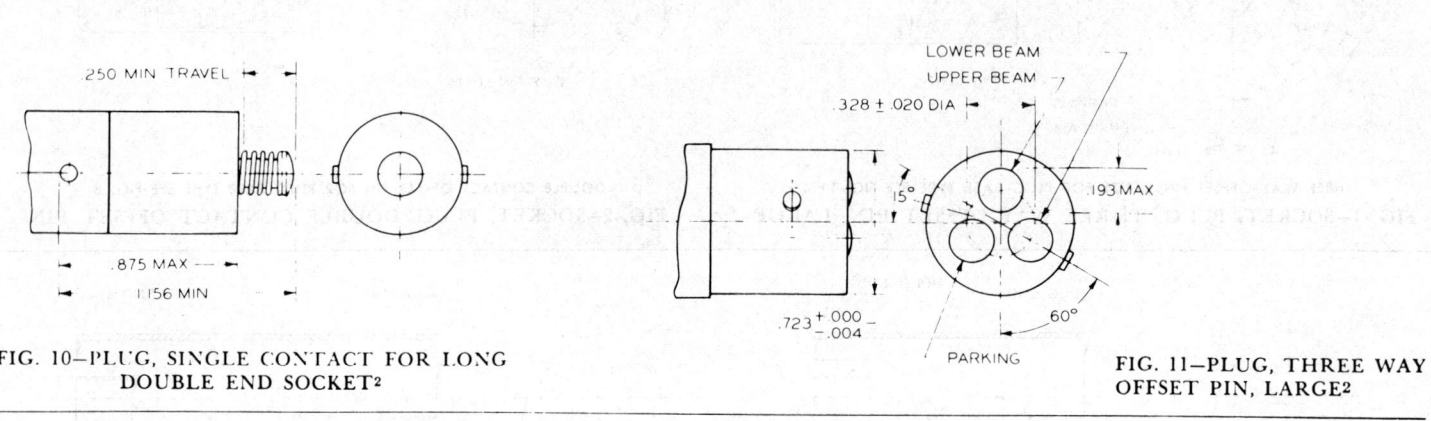

FIG. 10—PLUG, SINGLE CONTACT FOR LONG DOUBLE END SOCKET[2]

FIG. 11—PLUG, THREE WAY OFFSET PIN, LARGE[2]

[2] All dimensions for the following are the same as those given for Figs. 6 and 7, except where otherwise indicated.

DIMENSIONAL SPECIFICATIONS FOR GENERAL SERVICE SEALED LIGHTING UNITS—SAE J760a

SAE Recommended Practice

Report of Lighting Committee approved March 1961 and last revised December 1974.

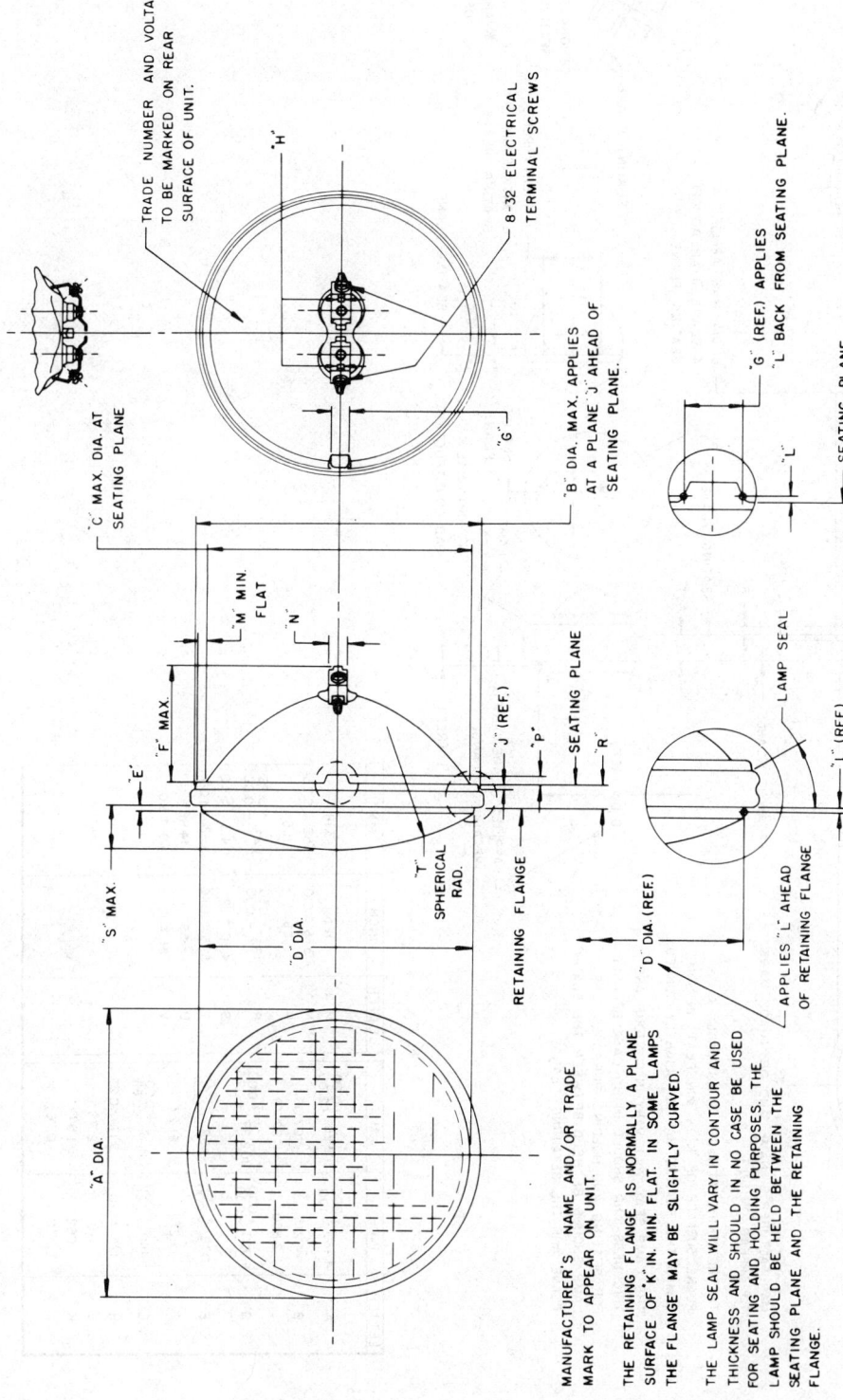

FIG. 1—DIMENSIONS OF 5¾ IN (146 mm) DIAMETER SEALED LIGHTING UNIT

LETTER	INCH	MM
A	5.70 +.00/-.10	144.8 +0.0/-2.5
B	5.475	139.06
C	5.100	129.54
D	5.265 +.000/-.030	133.7 +0.00/-0.76
E	.078 +.062/-.000	2.00 +1.50/-0.00
F	2.50	63.5
G	.440 +.000/-.025	11.17 +0.00/-0.63
H	.125 ±.010	3.17 ±0.25
J	.110	2.79
K	.062	1.57
L	.030	0.76
M	.125	3.18

LETTER	INCH	MM
N	.375 ±.010	9.52 ± 0.25
P	.135 +.030/-.000	3.43 +0.76/-0.00
R	.53 ± .04	13.5 ± 1.0
S	.92	23.36
T	5.00 ± .12	127.0 ± 3.0

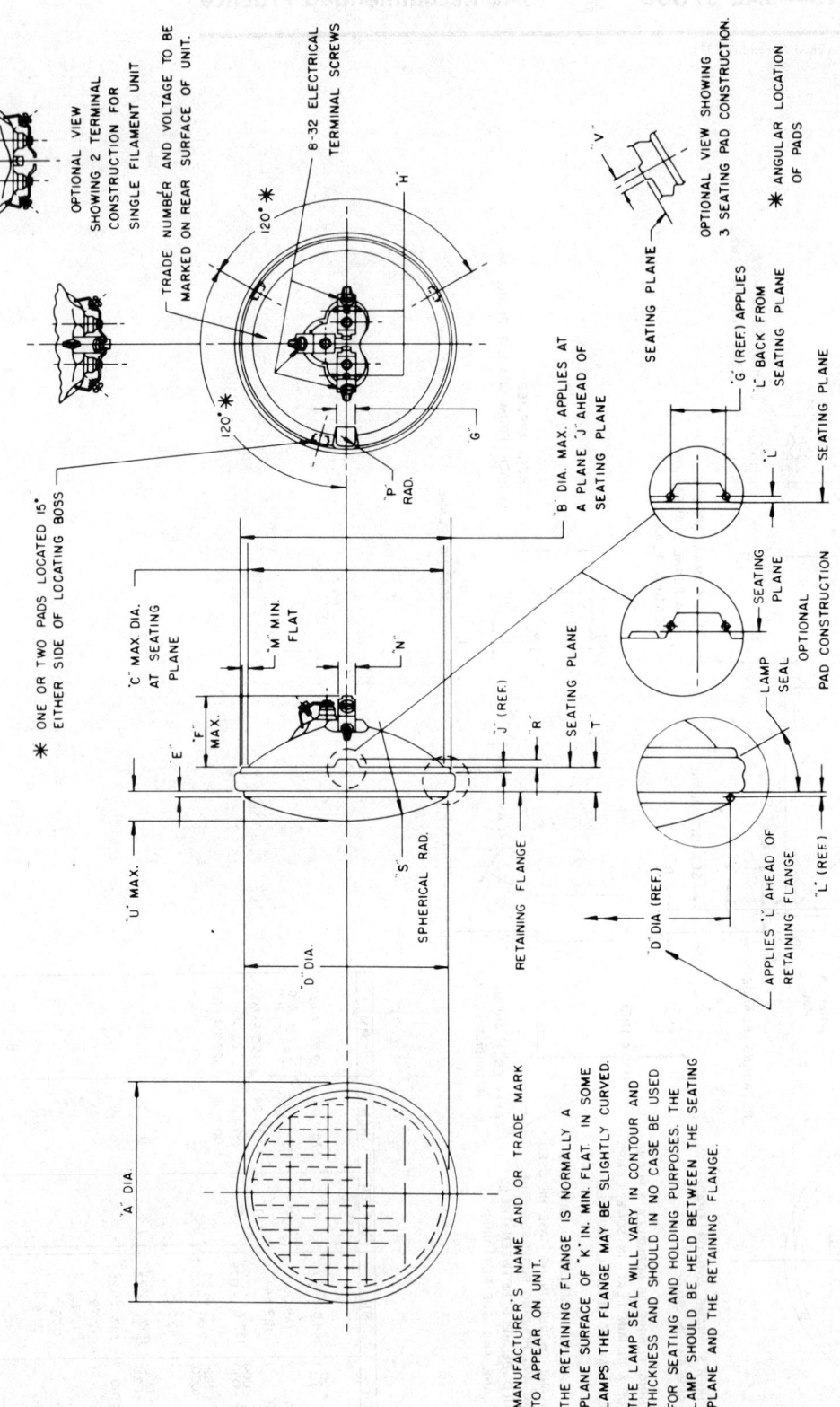

FIG. 2—DIMENSIONS OF 4½ IN (114 mm) DIAMETER SEALED LIGHTING UNIT

L.E.D. LIGHTING DEVICES—
SAE J1889 JUN88

SAE Recommended Practice

Report of the Lighting Committee approved June 1988. Rationale statement available.

1. Scope—This technical report applies to motor vehicle signalling and marking lighting devices which use light emitting diodes (L.E.D.) as light sources. This report provides test methods, requirements, and guidelines applicable to the special characteristics of L.E.D. lighting devices. These are in addition to those required for devices designed with incandescent light sources. This report is intended as a guide to standard practice and is subject to change to reflect additional experience and technical advances.

2. Definitions

2.1 Semiconductor—A material whose resistivity lies in the broad range between conductors and insulators.

2.2 L.E.D.—An indivisible, discrete light source unit containing a semiconductor junction in which visible light is non-thermally produced when a forward current flows as a result of applied voltage.

2.3 L.E.D. Lighting Device—A lighting device in which light is produced by an array of L.E.D. light sources.

2.4 Incandescence—The generation of light caused by heating a body to a high temperature. Generally this heating is obtained by passing an electric current through a wire filament. The resistance of the filament to the current causes the filament to heat up and emit radiant energy, some of which is in the visible range. Ordinary automotive bulbs have incandescent light sources.

2.5 L.E.D. Light Source Center—For a single L.E.D., the point that is located at the geometric center of the junction where the luminescence takes place.

2.6 Lighting Device Light Center—The geometric center of all the single L.E.D. light source centers within the L.E.D. array(s) used to illuminate the device function, or the geometric center of the illuminated area if the light output is produced indirectly.

3. Tests—The following section describes individual tests which need not be performed in any particular sequence. Testing may be expedited by performing two or more tests simultaneously on separate samples.

3.1 SAE J575 is a part of this report. Unless otherwise specified, the following tests are applicable with modifications as indicated.

3.1.1 Vibration Test—The evaluation of the sample at the completion of the test shall also include a functional lighting check. If a partial outage is observed, a photometry test (paragraph 3.1.5) shall be performed and the results recorded.

3.1.2 Moisture Test

3.1.3 Dust Test—If dust is found, the change in the maximum photometric luminous intensity of the sample shall be determined by using the photometric measurement procedures in paragraph 3.1.5.

3.1.4 Corrosion Test

3.1.5 Photometry Test—The photometric output (luminous intensity) of a L.E.D. lighting device typically decreases as the temperature of the L.E.D. light sources increases. In addition to the test procedures in SAE J575 the following shall apply:

3.1.5.1 Design Voltage—The device shall be operated at its design voltage during all photometric tests.

3.1.5.2 Photometric Maximums—For measurements to photometric maximum requirements, first allow the test device to stabilize at laboratory ambient temperature (23 ± 5°C) unenergized. After all the device components are at laboratory ambient temperature, energize the test device and record the maximum photometric value(s) within 60 s of the initial on-time.

3.1.5.3 Photometric Minimums—For measurements to photometric minimum requirements, the test device light output shall first be stabilized by energizing the device at laboratory ambient temperature (23 ± 5°C) until either internal heat buildup saturation has occurred or 30 min has elapsed, whichever occurs first.

3.1.6 Warpage Test on Devices With Plastic Components—Not required.

3.2 Color Test—SAE J578 is a part of this report.

3.3 Thermal Cycle Test

3.3.1 Scope—This test evaluates the ability of the sample device to resist optical, electrical, or physical malfunctions due to exposures to repeated changes from hot to cold temperature extremes. Devices installed in vehicle locations that could produce temperatures outside the test range specified may necessitate special test requirements.

3.3.2 Test Equipment—A thermal cycle chamber capable of providing the temperature extremes and rates of change of temperature in the temperature–time profile specified in Fig. 1.

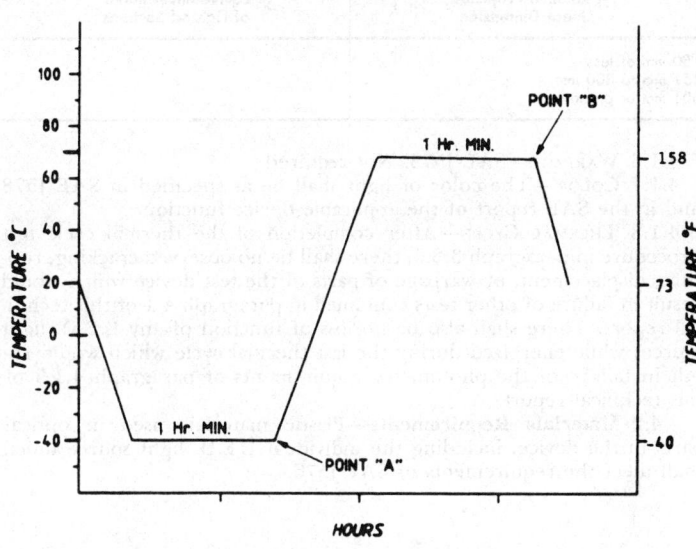

FIG. 1—THERMAL CYCLE PROFILE

3.3.3 Test Procedure—The sample device, mounted on a test fixture shall be subjected to thermal cycles as follows:

3.3.3.1 Thermal Cycle—The device shall be exposed to the thermal cycle profile shown in Fig. 1.

3.3.3.2 Device Operation—The device shall be energized at design voltage commencing at point "A" of Fig. 1 and de-energized at point "B" of each cycle. When energized, the lighting function(s) shall be cycled as specified in SAE J575, Table 1.

3.3.3.3 Test Duration—The test shall consist of 25 complete cycles of the thermal cycle profile shown in Fig. 1.

3.3.3.4 Sample Evaluation—During the final thermal cycle, the sample lighting function(s) shall be continuously checked for permanent or intermittent outages while energized from Point "A" (cold temperature) to Point "B" (hot temperature) on Fig. 1 and the results recorded. If partial outage is observed, a photometry test (paragraph 3.1.5) with the remaining functional L.E.D. segments lighted shall be performed and the results recorded. Upon completion of the thermal cycle exposure the sample device shall be visually examined for any cracking, rupture or warpage of parts and the results recorded. If any of the above changes are observed that could result in failure of the other tests contained in Section 3, these test(s) shall be performed on the same sample used for the thermal cycle test and the results recorded.

4. Requirements

4.1 Performance Requirements—A L.E.D. lighting device when tested in accordance with the test procedures specified in Section 3 shall meet the following requirements.

4.1.1 Vibration—SAE J575. The following requirements also apply:

4.1.1.1 After completion of test procedure paragraph 3.1.1, all L.E.D. light sources contained within the device shall function *or* the device shall comply with the photometric requirements in paragraph 4.1.5 of this report.

4.1.2 Moisture—SAE J575.

4.1.3 Dust—SAE J575.

4.1.4 Corrosion—SAE J575.

4.1.5 Photometry—SAE J575. The photometric performance requirements in the applicable SAE technical report for the lighting function being tested shall also apply. Specified photometric maximum and minimum test points shall be determined as specified in paragraphs 3.1.5.2 and 3.1.5.3 of this report. The following requirements shall also apply:

4.1.5.1 Lighted Sections—Applicable photometric requirements specified in other SAE technical reports which are based on the number of lighted sections shall instead be applied based on the dimensions of the L.E.D. lighting device function being tested. The maximum horizontal or vertical projected lighted linear dimension of the function shall be equivalent to the following number of lighted sections.

Maximun Projected Linear Dimension	Equivalent Number of Lighted Sections
150 mm or less	1
151 mm to 300 mm	2
301 mm or greater	3

4.1.6 Warpage—SAE J575. Not required.

4.1.7 Color—The color of light shall be as specified in SAE J578 and in the SAE report of the applicable device function.

4.1.8 Thermal Cycle—After completion of the thermal cycle test procedure in paragraph 3.3.3, there shall be no observed cracking, rupture, displacement, or warpage of parts of the test device which would result in failure of other tests contained in paragraph 4.1 of this technical report. There shall also be no loss of function of any L.E.D. light sources while energized during the last thermal cycle which would result in failure of the photometry requirements of paragraph 4.1.5 of this technical report.

4.2 Materials Requirements—Plastic materials used in optical parts in the device, including the individual L.E.D. light source units, shall meet the requirements of SAE J576.

4.3 Design Requirements

4.3.1 Reverse Voltage—Some L.E.D. light sources may be damaged by the application of a voltage of reverse polarity. Protection shall be provided in the device to prevent any damage when the voltage polarity to the lighting device is reversed.

5. Guidelines

5.1 Photometric Design Guidelines—The photometric design guidelines in the applicable SAE technical report for the lighting function design shall be required. Specified photometric maximum and minimum values shall be measured as specified in paragraphs 3.1.5.2 and 3.1.5.3 of this report. Requirements using the number of lighted sections shall apply as specified in paragraph 4.1.5.1 of this report.

5.2 Installation Guidelines—The following guidelines are provided due to the special characteristics of L.E.D. lighting devices:

5.2.1 The luminous intensity of L.E.D. lighting devices typically vary with applied voltage. The electrical system of a vehicle should, under normal operating conditions, provide design voltage to the device as closely as practicable bearing in mind the inherent variability of such systems.

5.2.2 The luminous intensity of a L.E.D. lighting device typically decreases as the temperature of the L.E.D. light sources increases. Installation of lamps on vehicles should be considered to minimize the effect of accumulating excessive temperatures in the device.

5.2.3 While L.E.D. light sources typically have a very long energized life, outage of a segment of a L.E.D. light source array may occur when one of the L.E.D. light sources within the array segment malfunctions. The user should be cautioned to replace or repair the device since the luminous intensity of the device is reduced by such an outage.

6. Appendix—As a matter of additional information, attention is called to SAE J387.

(R) MINIATURE LAMP BULBS—
SAE J573 DEC89

SAE Standard

Report of Lighting Division aproved march 1918 and last revised by Lighting committee December 1976. Rationale statement available. Completely revised by the Signalling and Marking Devices Standards Committee December 1989. Rationale statement available.

1. Scope—Many of the lighting devices on motor vehicles are required and essential to operation on public roadways. To assure field replacement, it is important that the bulb types employed be readily available, when needed, in normal service channels. Therefore, this document lists an assortment of current popular types, together with their design characteristics, which are recommended for use wherever practicable. It is recognized that because of constantly changing and improving technology, the list may be incomplete. Also, instances may arise in the design of some devices that require the employment of other types while achieving the desired performance.

Some of the design characteristics in this document are listed solely for the sake of standardization and have no bearing on how lamp bulbs perform in lighting devices on the highway.

2. Definition

2.1 Accurate Rated Miniature Bulb—A bulb operated at design mean spherical candela (Table 2) and having its filament(s) within ±0.25 mm of nominal design position. This applies to No. 1156, 1157, 1157NA, 2057, and 2057NA only. (See Figure 1 for the spacing between the major and minor filaments of these bulbs.)

2.2 Seasoned Bulb—A bulb that has been lighted at 1% of its average lab life, or 10 h maximum, whichever is shorter.

3. Tests

3.1 Samples for Test—Test samples shall be new, unused lamp bulbs fabricated from production processes.

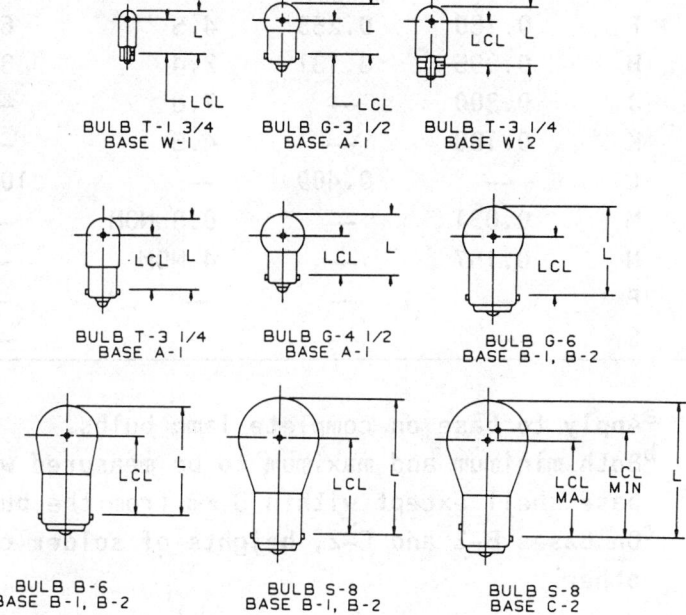

FIG. 1—BULB TYPES

TABLE 1—BULB DIMENSIONS (SEE FIGURE 1)

Bulb	Base	Max Bulb Dia (D)		Max Exposed Length (L)	
		in	mm	in	mm
G – 3-1/2	A–1	0.460	11.7	0.700	17.8
T – 1-3/4	W–1	0.230	5.8	0.598	15.2
T – 3-1/4	W–2	0.405	10.3	0.815	20.7
T – 3-1/4	A–1	0.433	11.0	0.941	23.9
G – 4-1/2	A–1	0.590	15.0	0.843	21.4
G – 6	B–1, B–2	0.748	19.0	1.189	30.2
B – 6	B–1, B–2	0.775	19.7	1.469	37.3
S – 8	B–1, B–2	1.043	26.5	1.772	45.0
S – 8	C–2	1.043	26.5	1.772	45.0

See Table 2 for LCL dimensions and tolerances.

TABLE 2—TYPICAL LAMP BULBS FOR MOTOR VEHICLES

Typical[a] Service	Trade No.	Design Mean Spherical Candela	Cd Tol. ±%	Volts	Design Amps	Amp Tol. ±%	Rated Average Lab Life. H.	Type[b]	Light Center Length (LCL) in	mm	LCL Tolerance ±in	±mm	Axial Alignment Tolerance ±in	±mm	Bulb[c] Type	Type[d]	Data Base Designation
C	74	0.7	30	14.0	0.1	15	500	C-2F	0.402	10.2	0.040	1.0	0.040	1.0	T 1 3/4	W1	Sub-Min-Wedge
C	53	1	20	14.4	0.12	10	1000	C-2V	0.500	12.7	0.090	2.3	0.090	2.3	G 3 1/2	A1	Min Bay
C.M	57	2	20	14.0	0.24	10	500	C-2V	0.560	14.2	0.090	2.3	0.090	2.3	G 4 1/2	A1	Min Bay
C.M	1895	2	20	14.0	0.27	10	1500	C-2F	0.560	14.2	0.090	2.3	0.090	2.3	G 4 1/2	A1	Min Bay
T.P.M.L	67	4	15	13.5	0.59	8	2000	C-2R	0.811	20.6	0.090	2.3	0.090	2.3	G6	B1	SC Bay
T.P.M.L	97	4	15	13.5	0.69	8	2000	C-2V	0.811	20.6	0.090	2.3	0.090	2.3	G6	B1	SC Bay
C	161	1	20	14.0	0.19	10	1500	C-2F	0.560	14.2	0.090	2.3	0.090	2.3	T 3 1/4	W2	Wedge
C.M	168	3	20	14.0	0.35	10	1500	C-2F	0.560	14.2	0.090	2.3	0.090	2.3	T 3 1/4	W2	Wedge
C.M.T.L	194	2	20	14.0	0.27	10	1500	C-2F	0.560	14.2	0.040	1.0	0.060	1.5	T 3 1/4	W2	Wedge
D.S.B	1156	32	10	12.8	2.10	5	600	C-6	1.252	31.8	0.040	1.0	0.040	1.0	S8	B1	SC Bay[f]
P.S.T.D	1157	32	10	12.8	2.10	5	600	C-6	1.252	31.8	0.040	1.0	0.040	1.0	S8	C2	DC Bay[f]
		3	12	14.0	0.59	8	2000	C-6									Index[f]
D.M.P	1157NA	24	30	12.8	2.10	5	600	C-6	1.252	31.8	0.040	1.0	0.040	1.0	S8	C2	DC Bay
		2.2	30	14.0	0.59	8	2000	C-6	e		e		e				Index[f]
P.S.T.D	2057	32	10	12.8	2.10	5	600	C-6	1.252	31.8	0.040	1.0	0.040	1.0	S8	C2	DC Bay
		2	12	14.0	0.48	8	5000	C-6	e		e		e				Index[f]
D.M.P	2057NA	24	30	12.8	2.10	5	600	C-6	1.252	31.8	0.040	1.0	0.040	1.0	S8	C2	DC Bay
		1.5	30	14.0	0.48	8	5000	C-6	e		e		e				Index[f]

[a]Letter designations are defined as follows: B–Backup; C–Indicator; D–Turn Signal; M–Marker, Clearance, Identification; P–Parking; S–Stop; T–Tail; L–License
[b]Filament types – see Figure 3
[c]Bulb types – see Figure 1
[d]Base types – see Figures 4, 5, and 6
[e]See Figure 2 for filament spacing and light center length
[f]Plane of pins with respect to filament is 90° ± 5

TABLE 3—BASE DIMENSIONS[a] (SEE FIGURE 4)

Dim	Bayonet (A-1)				Bayonet (B-1, B-2, C-2)			
	inches		mm		inches		mm	
	Min	Max	Min	Max	Min	Max	Min	Max
A[b]	0.357	0.366	9.0	9.3	0.593	0.602	15.0	15.3
B	0.383	0.400	9.7	10.1	0.616	0.636	15.6	16.1
C	--	0.431	--	10.97	--	0.668	--	16.9
D	0.025	--	0.6	--	0.025	--	0.6	--
E	0.060	0.067	1.5	1.7	0.071	0.087	1.8	2.2
F	0.180	0.255	4.5	6.48	0.248	0.316	6.3	8.0[c]
H	0.095	0.131	2.4	3.33	0.138	0.170	3.5	4.3
J	0.300	--	7.6	--	0.492	--	12.5	--
K	0.180	--	4.5	--	0.350	--	8.8	--
L	--	0.409	--	10.4	--	0.642	--	16.3
M	0.031	--	0.8 NOM	--	0.031	--	0.8 NOM	--
N	0.157	--	4 NOM	--	0.189	--	4.8 NOM	--
P	--	--	--	--	0.117	0.133	2.9	3.3
S	--	--	--	--	0.255	0.280	6.4	7.0

[a]Apply to base on complete lamp bulbs.
[b]Both minimum and maximum to be measured with a ring gauge. Applies to all parts of base shell except within 3 mm from the bulb and base junction.
[c]On bases B-2 and C-2, heights of solder contacts are to be within 0.5 mm of each other.

4. Requirements

The test samples shall comply with the following requirements:

4.1 Candela—Seasoned bulbs shall be measured at design volts in a properly calibrated photometer in accordance with accepted photometric procedures. See Table 2 for candela requirements. An acceptable seasoning schedule at rated volts is 1% of rated average lab life as shown in Table 1 or 10 h maximum, whichever is shorter. For lamp bulbs not listed in Table 1, use the manufacturer's published design life for rated average lab life.

4.2 Physical Dimensions:

4.2.1 Table 1 lists the bulb dimensions necessary to allow interchangeability.

4.2.2 Table 2 lists the electrical rating and physical locations of the filaments.

4.2.3 Table 3 lists the base dimensions considered important for metal based bulbs to insure that lamp bulbs will perform satisfactorily in a bulb retaining device (socket) made in accordance with SAE J567. Appendix A contains the following ANSI Standards:

Base Type/Description	ANSI Pub./Std. Sheet No.	IEC Designation
SAE A-1 Miniature Bayonet	C81.30,1-1	BA9s
SAE B-1 Candelabra Bayonet	C81.30,1-2	BA15s
SAE B-2 Candelabra Bayonet	C81.30,1-3	BA15d
SAE C-2 Candelabra Bayonet	C81.30,1-11	BAY15d

TABLE 4—WEDGE BASE DIMENSIONS (SEE FIGURE 5) TYPE W-2

Dimension	mm	
	Min	Max
A[a]	3.43	4.45
B	4.83	--
C	--	6.35
D	1.5 NOM	--
E	8.89	9.50
F	--	3.04
G	--	4.06
H	5.6 NOM	--
J	0.8 NOM	--
K[b]	0.8R NOM	--
P	1.90	2.41
N	1.65	--

[a]To be measured on longest side only with the wire in intimate contact with the bottom of the glass wedge.
[b]Optional construction, radius under wire not required, and dimension J becomes 1.2 nominal.

4.2.4 Table 4 lists the base dimensions considered important for wedge base (Type W-2) lamps to insure that the lamp bulbs will perform satisfactorily in a bulb retaining device (socket) made in accordance with SAE J567.

4.2.5 Table 5 lists the base dimensions considered important for subminiature wedge base (Type W-1) lamps to insure that lamp bulbs will perform satisfactorily in a bulb retaining device (socket) made in accordance with SAE J567.

TABLE 5—SUBMINIATURE WEDGE BASE DIMENSIONS (SEE FIGURE 6) TYPE W-1

Dimension	mm Min	mm Max
A[a]	2.03	3.04
B	3.05	5.08
C	--	5.08
E	4.83	5.08
G	--	3.10
H	3.3 NOM	--
P	1.78	2.20
N	1.65	--
M	1.5 NOM	--
Q	0.50	--

[a] To be measured on longest side only with the wire in intimate contact with the bottom of the glass wedge.

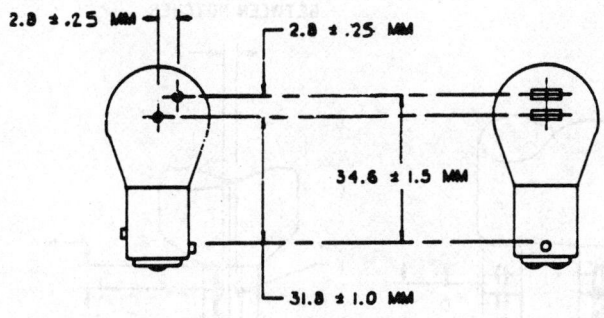

USE DIMENSIONS OF LOWER FILAMENT FOR 1156 BULB

FIG. 2—BULB FILAMENT DESIGN LOCATION

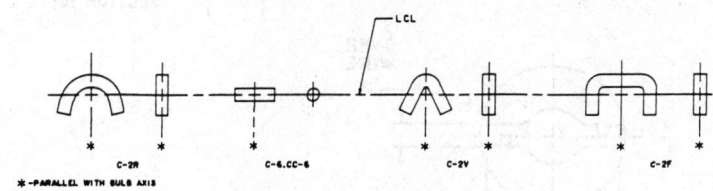

FIG. 3—FILAMENT TYPES

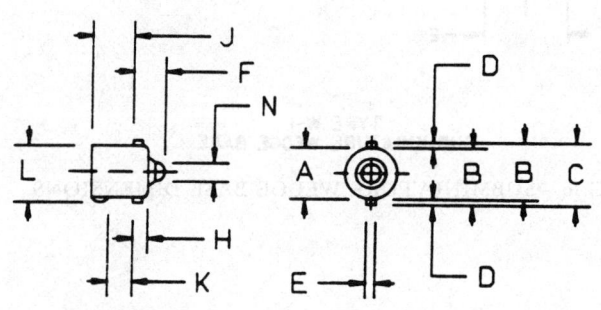

TYPE A-1
MINIATURE BAYONET
SINGLE CONTACT

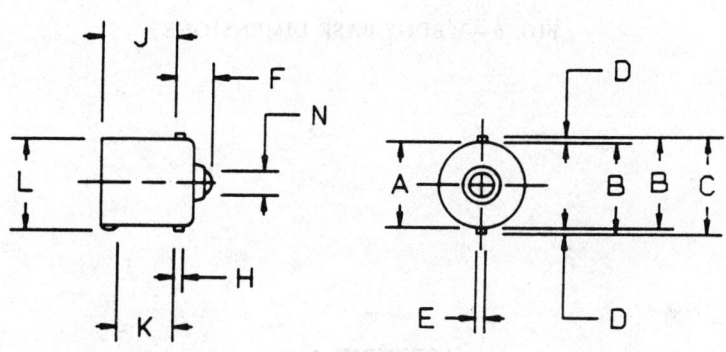

TYPE B-1
BAYONET
SINGLE CONTACT

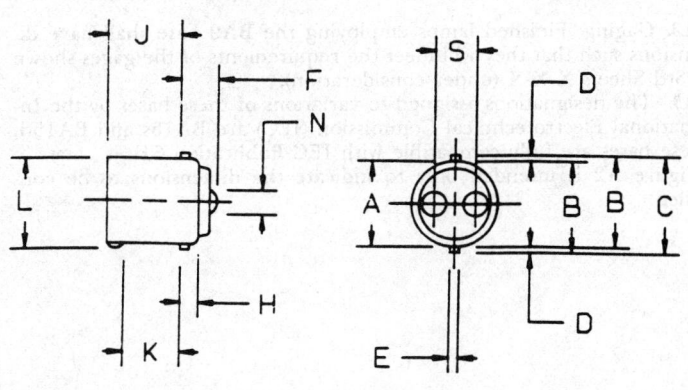

TYPE B-2
BAYONET
DOUBLE CONTACT

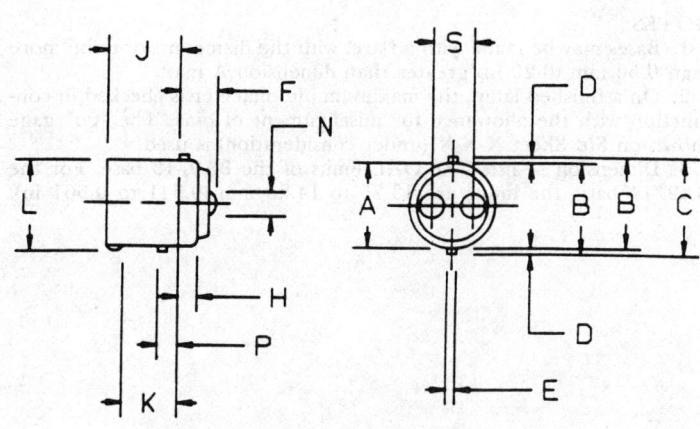

TYPE C-2
INDEXING BAYONET
DOUBLE CONTACT

FIG. 4—BASE TYPES

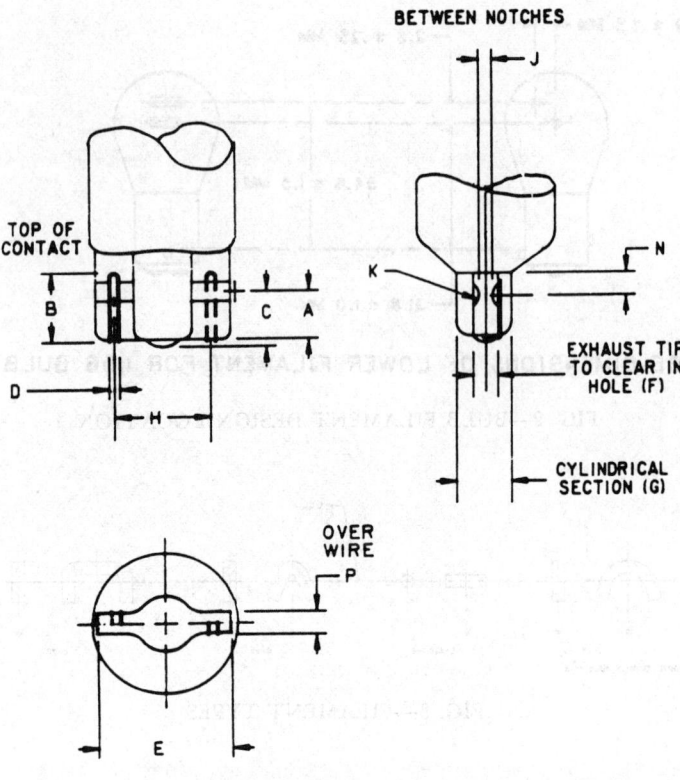

FIG. 5—WEDGE BASE DIMENSIONS

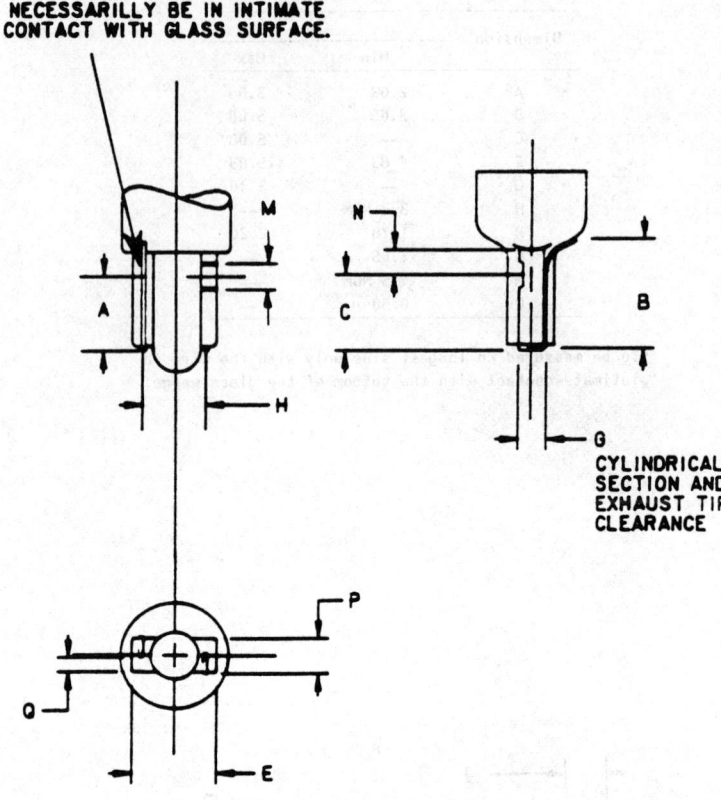

TYPE W-1
SUBMINIATURE WEDGE BASE

FIG. 6—SUBMINIATURE WEDGE BASE DIMENSIONS

APPENDIX A

A1. The designation assigned to this base by the International Electrotechnical Commission (IEC) is BA9. This base is fully compatible with IEC Publication 61.

Figure A1 is intended only to indicate the dimensions to be controlled.

NOTES:

1. Bases may be made with a flare, with the diameter not to be more than 0.50 mm (0.20 in) greater than dimension A max.

2. On a finished lamp, the maximum pin diameter is checked in conjunction with the allowance for misalignment of pins. The "go" gage shown on Std Sheet X-X-X (under consideration) is used.

3. Dimension M lists The OAL limits of the BA9/13 base. For the BA9/14 base, the limits are 13.75 to 14.25 mm (0.541 to 0.561 in).

4. Dimension N denotes the minimum length of shell that shall conform to the limits of dimension A.

5. Dimension U includes allowance for side-solder. It is measured from opposite the barrel or the flare, if present.

6. Rounded edge recommended to aid insertion.

A2. Gaging: Finished lamps employing the BA9 base shall have dimensions such that they will meet the requirements of the gages shown on Std Sheets X-X-X (under consideration).

A3. The designations assigned to variations of these bases by the International Electrotechnical Commission (IEC) are BA15s and BA15d. These bases are fully compatible with IEC Publication 61.

Figure A2 is intended only to indicate the dimensions to be controlled.

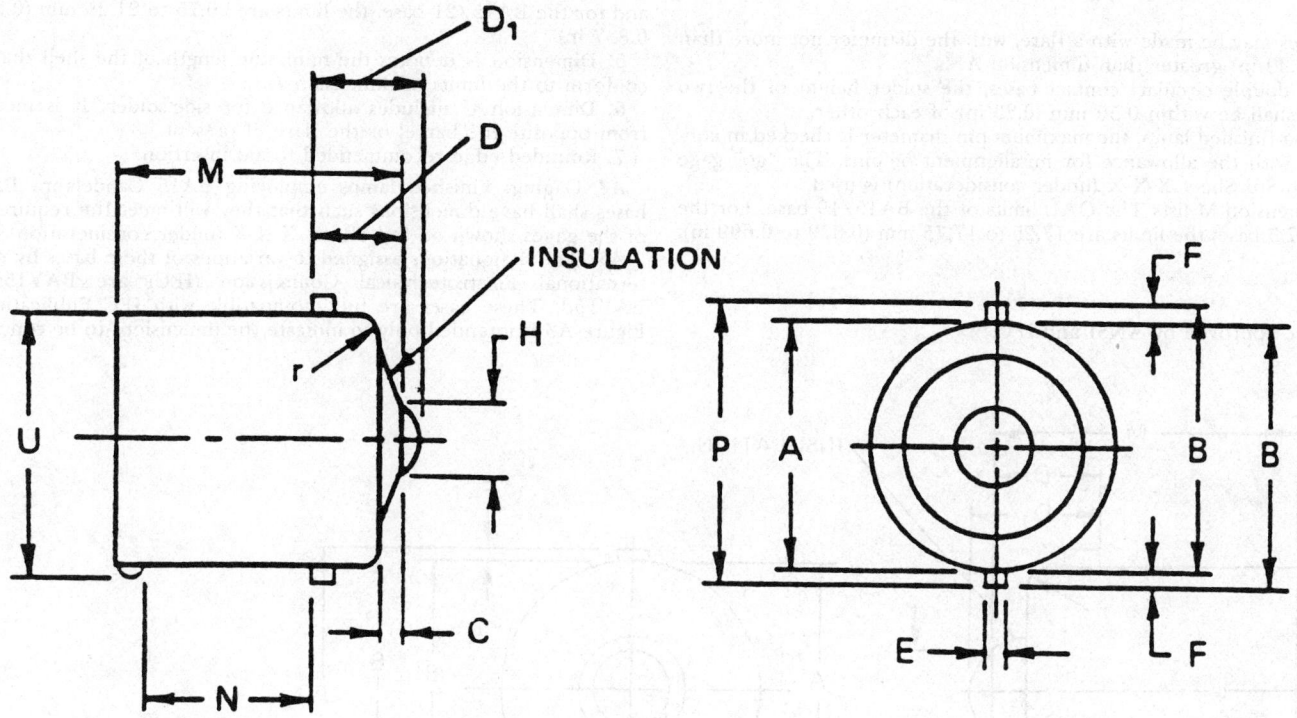

	Standard Dimension (millimeters)				Nearest Equivalent (inches)			
	Unmounted*		Finished Lamp		Unmounted*		Finished Lamp	
Reference	Min	Max	Min	Max	Min	Max	Min	Max
A (Note 1)	9.08	9.20	9.08	9.25	0.357	0.362	0.357	0.364
B	9.75	10.11	9.75	10.16	0.384	0.398	0.384	0.400
C	1.50	–	–	–	0.059	–	–	–
D	4.30	5.20	–	–	0.169	0.205	–	–
D_1	–	–	4.30	5.90	–	–	0.169	0.232
E (Note 2)	1.50	1.70	1.50	1.70	0.059	0.067	0.059	0.067
F	0.64	–	0.64	–	0.025	–	0.025	–
H	3.50	4.00	3.50	4.00	0.138	0.157	0.138	0.157
M (Note 3)	12.90	13.30	–	–	0.508	0.524	–	–
N (Note 4)	4.50	–	4.50	–	0.177	–	0.177	–
P	–	10.95	–	11.00	–	0.431	–	0.433
U (Note 5)	–	–	–	10.41	–	–	–	0.410
r (Note 6)	–	–	–	–	–	–	–	–

*These dimensions are solely for base design and are not to be gaged on the finished lamp.

FIG. A1

NOTES:

1. Bases may be made with a flare, with the diameter not more than 1 mm (0.39 in) greater than dimension A.

2. On double circular[1] contact bases, the solder height of the two contacts shall be within 0.50 mm (0.20 in) of each other.

3. On a finished lamp, the maximum pin diameter is checked in conjunction with the allowance for misalignment of pins. The "go" gage shown on Std Sheet X-X-X (under consideration) is used.

4. Dimension M lists The OAL limits of the BA15/19 base. For the BA15/17.5 base, the limits are 17.25 to 17.75 mm (0.679 to 0.699 in), and for the BA15/21 base, the limits are 20.75 to 21.25 mm (0.817 to 0.837 in).

5. Dimension N denotes the minimum length of the shell that shall conform to the limits of dimension A.

6. Dimension U includes allowance for side-solder. It is measured from opposite the barrel or the flare, if present.

7. Rounded edge recommended to aid insertion.

A4. Gaging: Finished lamps employing BA15 Candelabra Bayonet bases shall have dimensions such that they will meet the requirements of the gages shown on Std Sheets X-X-X (under consideration).

A5. The designations assigned to variations of these bases by the International Electrotechnical Commission (IEC) are BAY15s and BAY15d. These bases are fully compatible with IEC Publication 61. Figure A3 is intended only to indicate the dimensions to be controlled.

[1] Change approved by ANSI and IEC.

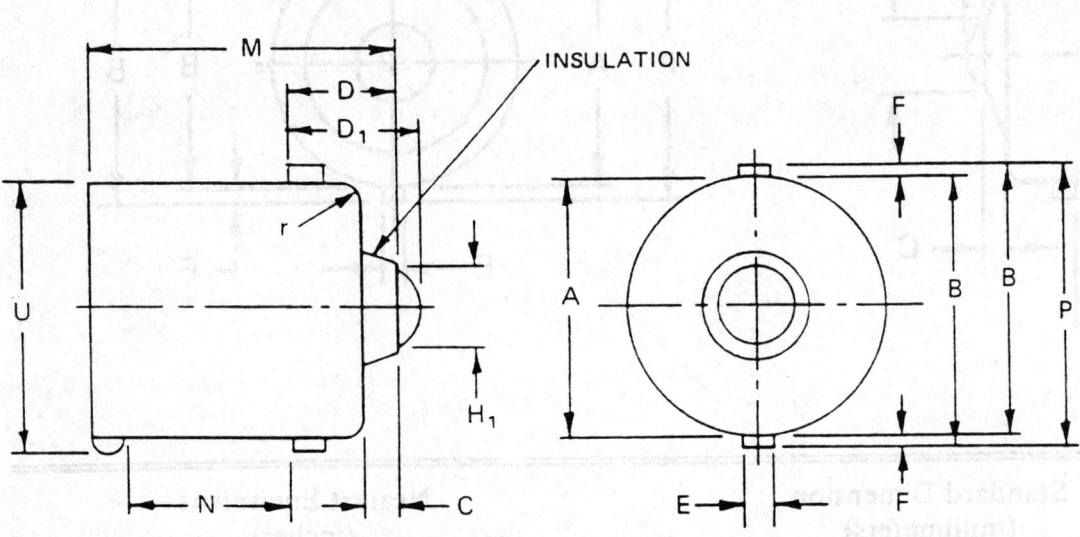

SINGLE CONTACT
BA15s

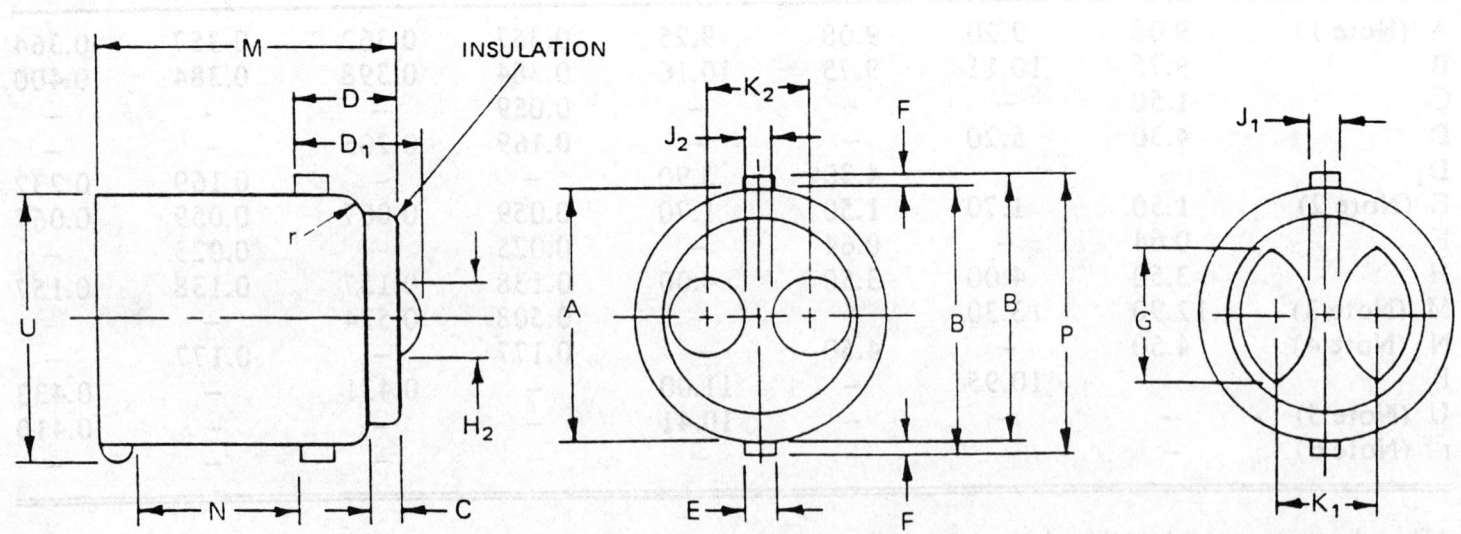

DOUBLE CONTACT
BA15d

OPTIONAL CONTACT
CONFIGURATION

FIG. A2

	Standard Dimension (millimeters)				Nearest Equivalent (inches)			
	Unmounted*		Finished Lamp		Unmounted*		Finished Lamp	
Reference	Min	Max	Min	Max	Min	Max	Min	Max
A (Note 1)	15.05	15.25	15.05	15.30	0.5925	0.6004	0.5925	0.6025
B	15.65	16.10	15.65	16.10	0.616	0.634	0.616	0.636
C	1.50	–	–	–	0.059	–	–	–
D	6.00	6.60	–	–	0.236	0.260	–	–
D_1 (Note 2)	–	–	6.32	7.50	–	–	0.249	0.295
E (Note 3)	1.80	2.20	1.80	2.20	0.071	0.087	0.071	0.087
F	0.64	–	0.64	–	0.025	–	0.025	–
G	9.00 Nom		–		0.354 Nom		–	
H_1	4.50	5.20	–	–	0.177	0.204	–	–
H_2	4.50	–	–	–	0.177	–	–	–
J_1	3.00	–	–	–	0.118	–	–	–
J_2	1.70	–	–	–	0.067	–	–	–
K_1	7.00	8.00	–	–	0.276	0.315	–	–
K_2	6.50	7.10	–	–	0.256	0.280	–	–
M (Note 4)	18.75	19.25	–	–	0.738	0.758	–	–
N (Note 5)	8.90	–	8.90	–	0.350	–	0.350	–
P	–	16.95	–	17.00	–	0.667	–	0.669
U (Note 6)	–	–	–	16.26	–	–	–	0.640
r (Note 7)	–	–	–	–	–	–	–	–

*These dimensions are for base design only, and are not to be gaged on the finished lamp.

FIG. A2 (Continued)

NOTES:
1. Bases may be made with a flare, with the diameter not more than 1 mm (0.39 in) greater than dimension A.
2. On double circular[1] contact bases, the solder height of the two contacts shall be within 0.50 mm (0.20 in) of each other.
3. On a finished lamp, the maximum pin diameter is checked in conjunction with the allowance for misalignment of pins. The "go" gage shown on Std Sheet X-X-X (under consideration) is used.
4. Dimension M lists the OAL limits of the BAY15/19 Base. For the BAY15/21 base, the limits are 20.75 to 21.25 mm (0.817 to 0.837 in).
5. Dimension N denotes the minimum length of the shell that shall conform to the limits of dimension A.
6. Dimension U includes allowance for side-solder. It is measured from opposite the barrel or the flare, if present.
7. Rounded edge recommended to aid insertion.

A6. Gaging: Finished lamps employing BAY15 Candelabra Bayonet bases shall have dimensions such that they will meet the requirements of the gages shown on Std Sheets X-X-X (under consideration).

[1] Change approved by ANSI and IEC.

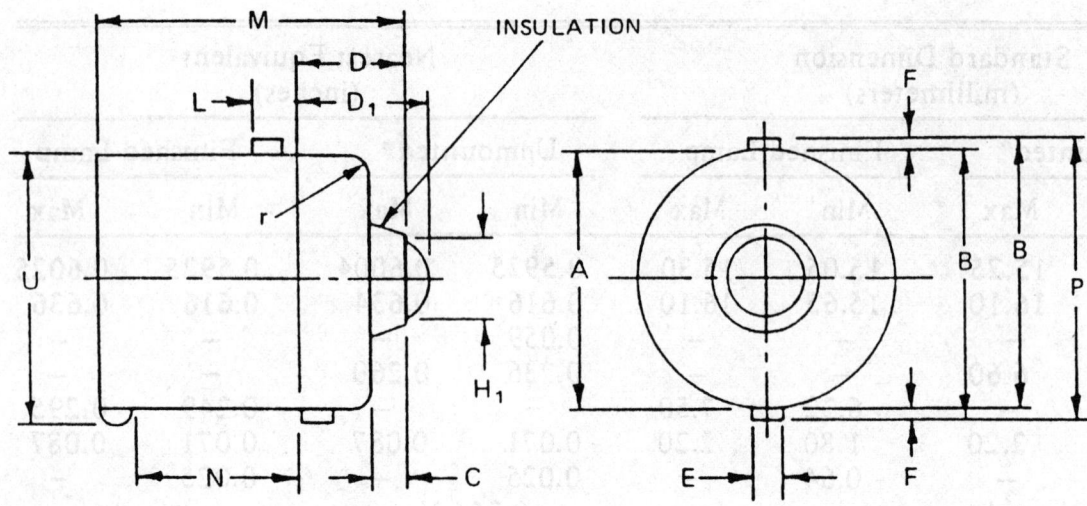

SINGLE CONTACT
BAY15s

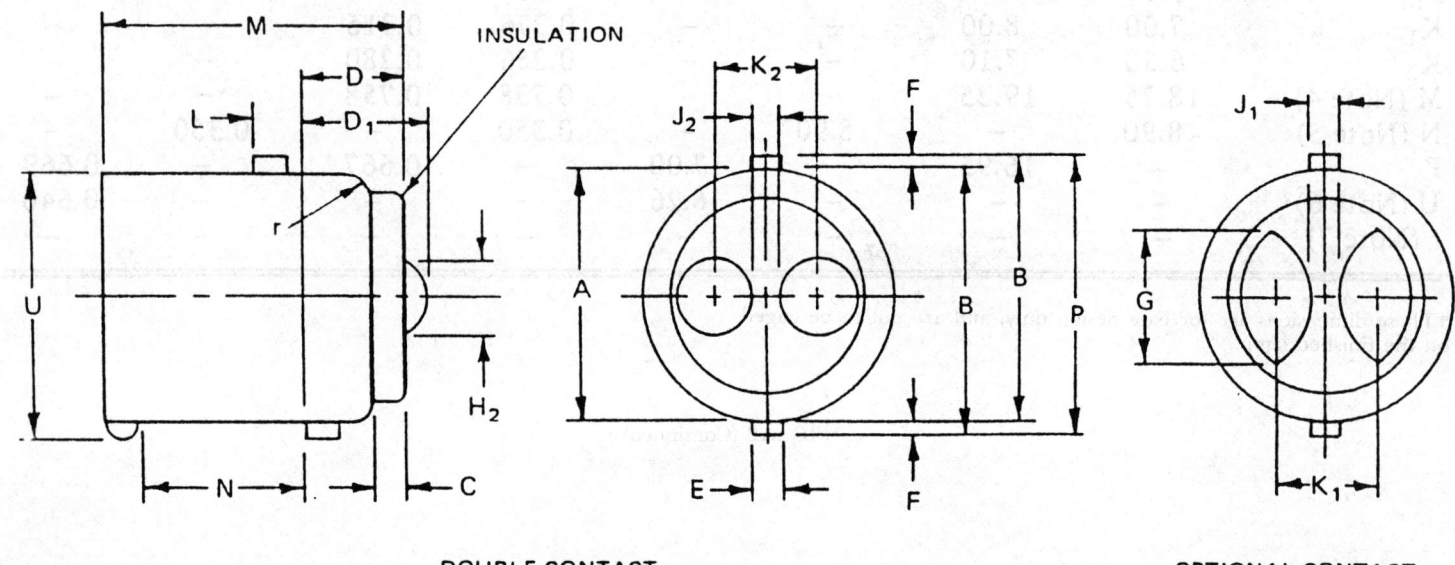

DOUBLE CONTACT
BAY15d

OPTIONAL CONTACT
CONFIGURATION

FIG. A3

| | Standard Dimension (millimeters) | | | | Nearest Equivalent (inches) | | | |
| | Unmounted* | | Finished Lamp | | Unmounted* | | Finished Lamp | |
Reference	Min	Max	Min	Max	Min	Max	Min	Max
A (Note 1)	15.05	15.25	15.05	15.30	0.5925	0.6004	0.5925	0.6025
B	15.65	16.10	15.65	16.15	0.616	0.634	0.616	0.636
C	1.50	–	–	–	0.059	–	–	–
D	6.00	6.60	–	–	0.236	0.260	–	–
D_1 (Note 2)	–	–	6.32	7.50	–	–	0.249	0.295
E (Note 3)	1.80	2.20	1.80	2.20	0.071	0.087	0.071	0.087
F	0.64	–	0.64	–	0.025	–	0.025	–
G	9.00 Nom		–	–	0.354 Nom		–	–
H_1	4.50	5.20	–	–	0.177	0.204	–	–
H_2	4.50	–	–	–	0.177	–	–	–
J_1	3.00	–	–	–	0.118	–	–	–
J_2	1.70	–	–	–	0.067	–	–	–
K_1	7.00	8.00	–	–	0.276	0.315	–	–
K_2	6.50	7.10	–	–	0.256	0.280	–	–
L	3.00	3.40	3.00	3.40	0.118	0.134	0.118	0.134
M (Note 4)	18.75	19.25	–	–	0.738	0.758	–	–
N (Note 5)	8.90	–	8.90	–	0.350	–	0.350	–
P	–	16.95	–	17.00	–	0.667	–	0.669
U (Note 6)	–	–	–	16.26	–	–	–	0.640
r (Note 7)								

*These dimensions are for base design only, and are not to be gaged on the finished lamp.

FIG. A3 (Continued)

(R) TEST METHODS AND EQUIPMENT FOR LIGHTING DEVICES AND COMPONENTS FOR USE ON VEHICLES LESS THAN 2032 mm IN OVERALL WIDTH—SAE J575 JUN92

SAE Recommended Practice

Report of the Lighting Division, approved May 1942, completely revised by the Lighting Committee July 1983, and reaffirmed by the Lighting Coordination Committee December 1988. Completely revised by the SAE Test Methods and Equipment Standards Committee June 1992. Rationale statement available.

1. Scope—This SAE Recommended Practice provides standardized laboratory tests, test methods, and requirements applicable to many of the lighting devices and components covered by SAE Recommended Practices and Standards and is intended for reference for devices used on vehicles less than 2032 mm in width, regardless of length, or 7620 mm in length regardless of width. Tests for vehicles larger than 2032 mm in overall width are covered in J2139.

2. References

2.1 Applicable Documents—The following publications form a part of this specification to the extent specified herein. The latest issue of SAE publications shall apply.

2.1.1 SAE PUBLICATIONS—Available from SAE, 400 Commonwealth Drive, Warrendale, PA 15096-0001.

SAE J387—Terminology—Motor Vehicle Lighting
SAE J1330—Photometry Laboratory Accuracy Guidelines
SAE J2139—Test Methods and Equipment for Lighting Devices and Components for Use on Vehicles More than 2032 mm in Overall Width

2.1.2 ASTM PUBLICATIONS—Available from ASTM, 1916 Race Street, Philadelphia, PA 19103.

ASTM B 117-73—Method of Salt Spray (Fog) Testing
ASTM C 150-84—Specification for Portland Cement
ASTM E 308-85—Standard Practice for Spectrophotometry and Description of Color in CIE 1931 System

2.2 Definitions

2.2.1 LIGHTING DEVICES—An assembly (divisible or indivisible) which contains a bulb or other light source and generally an optical system such as a lens or a reflector, or both, and which provides a lighting function. Lighting device samples submitted for test shall be representative of the device as regularly manufactured and marketed, unless otherwise identified. Each sample shall be securely mounted on a test fixture in its designed operating position and shall include all accessory equipment necessary to operate the device in its normal manner.

2.2.2 BULB—An indivisible assembly which contains a source of light and which is normally used in a lamp. Unless otherwise specified, bulbs used in the tests shall be supplied by the test facility and shall be representative of bulbs in regular production. Where special bulbs are specified, they shall be submitted with the sample devices and the same or similar bulbs shall be used in the tests. Lighting devices designed for use in 6 V, 12 V, or 24 V systems shall be tested with 12 V bulbs.

2.2.3 TEST FIXTURE—A device specifically designed to support the lighting device in its designed operating position during laboratory testing. This fixture, when used for the vibration test, shall not have a resonant frequency in the 10 to 55 Hz range with the sample installed.

3. Test Descriptions—The following sections describe individual tests which need not be performed in any particular sequence. The completion of the tests may be expedited by performing the tests simultaneously on separately mounted samples. However, it is recommended that the design of each device be evaluated to determine if vibration or warpage tests might affect the results of other tests, in which case the vibration and/or warpage test(s) should be performed first.

3.1 Vibration Test—This test evaluates the ability of the sample device to resist damage from vibration induced stresses. This test is not intended to test the vibration resistance of bulb filaments or headlamp light source filaments.

3.1.1 VIBRATION TEST EQUIPMENT—A vibration test machine capable of linear frequency variation at a constant unidirectional excursion shall be used. The vibrator table shall be of sufficient size to completely contain the test fixture base with no overhang. If this is not possible, a transition table shall be used to mechanically interface the large test fixture base to the smaller vibrator table. Precautions shall be taken to minimize the introduction of extraneous responses in the test setup. The vibration machine output wave form shall be sinusoidal with a maximum permissible harmonic distortion as shown in Figure 1, when measured as follows:

3.1.1.1 *Distortion Measurement*—The test machine output wave form shall be measured with an accelerometer having a flat frequency response (±5%) from 5 to 2200 Hz, attached to the unloaded vibrator table or to the transition table, if used. The acceleration component measured shall be in the direction of table travel.

3.1.1.2 *Harmonic Distortion Analysis*—The percent distortion shall be measured directly or shall be computed by taking the ratio (×100) of the rms (root mean squared) voltage of the distortion components to the rms voltage of the total signal (distortion plus fundamental) of the accelerometer.

3.1.2 VIBRATION TEST PROCEDURE—A sample device as mounted on a test fixture shall be securely bolted to the table of the vibration test machine and subjected to vibration according to the following test parameters:

3.1.2.1 *Frequency*—Varied from 10 to 55 Hz and return to 10 Hz at a linear sweep period of 2 min/complete sweep cycle.

3.1.2.2 *Excursion*—1.0 + 0.1/- 0.0 mm peak to peak over the specified frequency range.

3.1.2.3 *Direction of Vibration*—Vertical axis of the device as it is mounted on the vehicle.

3.1.2.4 Test Duration 60 +1/-0 minutes

3.2 Moisture Test—This test evaluates the ability of the sample device to resist moisture leakage from a water spray and determines the drainage capability of those devices with drain holes or other exposed openings in the device. This test is not intended to provide a complete test on the device seal (see Dust Exposure Test 3.3). A sample device as mounted on the test fixture shall be tested according to either Test 3.2.1 (Water Spray) or Test 3.2.3 (Water Submersion) as applicable. The purpose of the Water Submersion Test is to reduce the test time for sealed lighting devices. Devices which comply with the Water Submersion Test are considered to have complied with all requirements of the Moisture Test.

3.2.1 WATER SPRAY TEST EQUIPMENT—A water spray cabinet with the following characteristics shall be used:

3.2.1.1 *Cabinet*—The cabinet shall be equipped with a nozzle(s) which provides a solid cone water spray of sufficient angle to completely cover the sample device. The centerline of the nozzle(s) shall be directed downward at an angle of 45 degrees ± 5 degrees to the vertical axis of a rotating test platform.

3.2.1.2 *Rotating Test Platform*—Having a minimum diameter of 140 mm and rotating about a vertical axis in the center of the cabinet.

3.2.1.3 *Precipitation Rate*—The precipitation rate of the water spray at the device shall be 2.5(+1.6/-0) mm/min as measured with a vertical cylindrical collector centered on the vertical axis of the rotating test platform. The height of the collector shall be 100 mm and the inside diameter shall be a minimum 140 mm.

3.2.2 WATER SPRAY TEST PROCEDURE—The mounted sample device shall be subjected to a water spray as follows:

3.2.2.1 *Device Openings*—All drain holes and other openings shall remain open. Devices having a portion completely protected in service (i.e., trunk mounted lamps) shall have that part of the device covered to prevent moisture entry during the test. Drain wicks, when used, shall be tested in the device.

3.2.2.2 *Rotational Speed*—The device shall be rotated about its vertical axis at a rate of 4.0 ± 0.5 rpm.

3.2.2.3 *Test Duration*—The water spray test shall continue for 12 h.

3.2.2.4 *Drain Period*—The rotation and the water spray shall be turned off and the device allowed to drain for 1 h with the cabinet door closed.

3.2.2.5 *Sample Evaluation*—Upon completion of the drain period, the interior of the device shall be observed for moisture accumulation. If a standing pool of water has formed, or can be formed by tapping or tilting the device, the accumulated moisture shall be extracted and measured.

3.2.3 WATER SUBMERSION TEST PROCEDURE—The device shall be completely submerged under laboratory ambient temperature water at a depth of 150 to 175 mm as measured from the top of the device.

3.2.3.1 *Test Duration*—The device shall be submerged for 1 h.

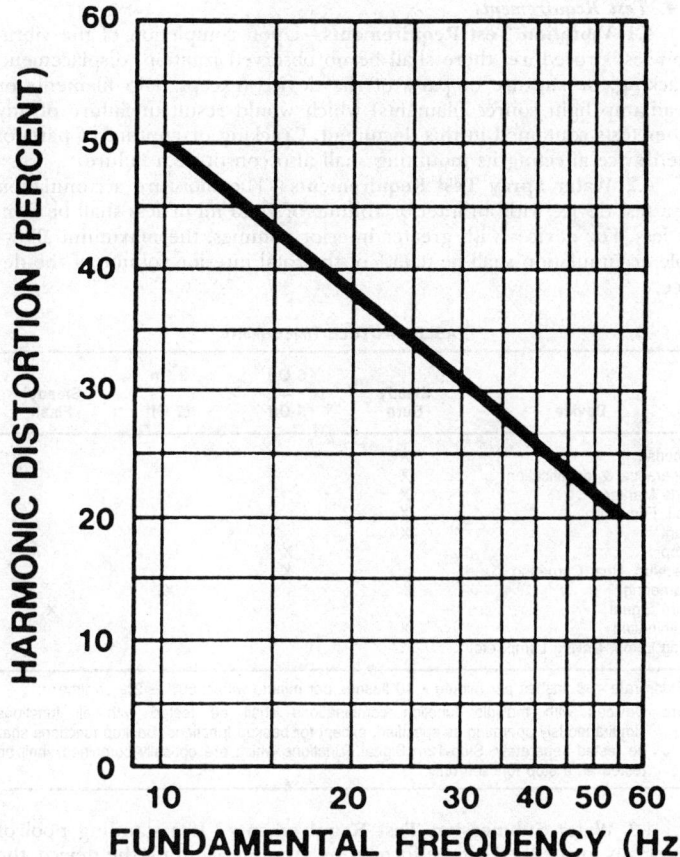

FIGURE 1—MAXIMUM PERMISSIBLE VIBRATION WAVE FORM HARMONIC DISTORTION

3.2.3.2 *Sample Evaluation*—Immediately after the device is removed from submersion, the interior of the test device shall be observed for water accumulation.

3.3 Dust Exposure Test—This test evaluates the ability of the sample device to resist dust penetration which could significantly affect the photometric output of the lamp device. This test is not intended to provide a complete test on the device seals. A sample device shall be tested to either the Dust Exposure Test or Water Submersion Test. Devices which comply with the water submersion requirements are considered to have complied with all requirements of the Dust Exposure Test. If the device does not comply with the water submersion requirements, it may still comply with all requirements of the Dust Exposure Test.

3.3.1 DUST EXPOSURE TEST EQUIPMENT—The following equipment shall be used to test for dust exposure:

3.3.1.1 *Dust Exposure Test Chamber*—The interior of the test chamber shall be cubical in shape with measurements of 0.9 to 1.5 m per side. The bottom may be "hopper shaped" to aid in collecting the dust. The internal chamber volume, not including a "hopper shaped" bottom, shall be 2 m^3 maximum and shall be charged with 3 to 5 kg of the test dust. The chamber shall have the capability of agitating the test dust by means of compressed air or blower fans in such a way that the dust is diffused throughout the chamber.

3.3.1.2 *Test Dust*—The test dust used shall be fine powdered cement in accordance with ASTM C 150-84.

3.3.2 DUST EXPOSURE TEST PROCEDURE—A sample device, mounted on a test fixture, with the initial maximum luminous intensity photometrically measured and recorded, shall be subjected to dust as follows:

3.3.2.1 *Device Openings*—All device openings shall be open. A device which has a portion completely protected in service (i.e., trunk mounted lamp) shall have that portion of the device covered to prevent dust entry during the dust exposure.

3.3.2.2 *Dust Exposure*—The mounted device shall be placed in the dust chamber no closer than 150 mm from a wall. Devices with a length exceeding 600 mm shall be horizontally centered in the test chamber. The test dust shall be agitated as completely as possible by compressed air or blower(s) at intervals of 15 min for a 2 to 15 s period. The dust shall be allowed to settle between the agitation periods.

3.3.2.3 *Test Duration*—5 h.

3.3.2.4 *Sample Evaluation*—Upon completion of the dust exposure test, the lamp exterior shall be cleaned and the maximum luminous intensity measured.

3.4 Corrosion Test—This test evaluates the ability of the sample device to resist salt corrosion which would impair the functional characteristics of the device.

3.4.1 CORROSION TEST EQUIPMENT—A salt spray (fog) cabinet, operating at the conditions specified by ASTM B 117-73, shall be used.

3.4.2 CORROSION TEST PROCEDURE—A sample device as mounted on the test fixture shall be subjected to salt spray (fog) as follows:

3.4.2.1 *Device Openings*—All device openings shall remain open. If a portion of the device is completely protected in service (such as a trunk mounted lamp) that portion shall be covered to prevent salt fog entry during the salt exposure.

3.4.2.2 *Salt Exposure*—The device shall be placed in the salt spray chamber for a period of 48 h.

3.4.2.3 *Sample Evaluation*—After removal from the chamber and after a 1 h drying period, the device shall be visually examined for corrosion which could affect other tests contained in this document.

3.5 Photometry Test—This test measures luminous intensities at test points throughout the light distribution pattern as specified by the applicable SAE report for the sample device.

3.5.1 PHOTOMETRIC TEST EQUIPMENT—Unless otherwise specified, the following equipment shall be used to make the photometric measurements:

3.5.1.1 *Positioner*—The positioner (goniometer) configuration shall be capable of positioning the sample device at the test point positions specified in 3.5.2.4 and in the applicable SAE report. (The recommended configuration is shown in SAE J1330.) Other systems may be used to achieve equivalent positioning, but it may be necessary at compound angles greater than 5 degrees from H-V to calculate the position which is equivalent to that of the recommended goniometer.

3.5.1.2 *Photometer*—The photometer system consists of a sensor, amplifier, and indicator instrument. The system shall be capable of providing the luminous intensity readings (candela) of the output of the device being tested. The sensor shall be located at the distance from the device specified in the applicable SAE report and shall have the following characteristics:

3.5.1.2.1 Maximum Size—Unless otherwise specified, the maximum effective area of the sensor shall fit within a circle whose diameter is equal to 0.009 times the actual test distance from the light source of the sample device to the sensor. The sensor effective area is the actual area of intercepted light striking the detector surface of the photometer. For systems with lens(es) that change the diameter of the intercepted light beam before it reaches the actual detector surface, the maximum size requirements shall apply to the total area of the light actually intercepted by the lens surface. The sensor shall be capable of intercepting all direct illumination from the largest illuminated dimension of the sample device at the test distance.

3.5.1.2.2 Photopic Response—The color response of the photometer sensor shall be corrected to that of ASTM E 308-85.

3.5.2 PHOTOMETRY TEST PROCEDURE—The sample device shall be mounted on a test fixture and luminous intensity measurements made as follows:

3.5.2.1 *Bulbs*—Unless otherwise specified, accurate rated bulbs (selected per SAE J387) shall be used and shall be operated at their rated luminous flux output. Where special bulbs are used, they shall be seasoned per SAE J387 and operated at their rated luminous flux output.

3.5.2.2 *Test Voltage*—If the rated luminous flux output is not available, or not applicable, operate the bulb at its specified design voltage. If the luminous flux output of the bulb is intentionally modified from specifications for the device through internal or external circuitry, operate the bulb at its modified voltage, or with the voltage modification circuitry attached and with the specified design voltage applied to the input of the modification circuitry.

3.5.2.3 *Test Distance*—The luminous intensity measurements shall be made at a distance equal to, or greater than, the minimum test distance between the center of the light source (or the face of a reflex reflector) and the photometer sensor as specified in the SAE Technical Report applicable to the function of the sample device. If no test distance is specified, the distance shall be at least 10 times the largest illuminated dimension of the sample device.

3.5.2.4 *Test Point Positions*—Test point positions are specified in the applicable SAE Technical Report. The following nomenclature shall also apply: The letters "V" and "H" designate the vertical and horizontal planes intersecting both the device light source (or center or a reflex reflector) and the goniometer axis. A device using a bulb with a major and minor light source shall be oriented with respect to its major light source. "H-V" designates the test point angle at the intersection of the H and V planes (H = O, V = O degrees). Unless otherwise specified, this intersection shall be parallel to the longitudinal axis of the vehicle in the case of the designed operating position of front or rear device functions and shall be horizontal and perpendicular to the longitudinal axis of the vehicle in the case of side function devices. The letters "U," "D," "L," and "R" (up, down, left, and right, respectively) designate the angular position in degrees from the H and V planes to the goniometer as viewed from a lamp, or to the source of illumination as viewed from a reflex reflector. This angular direction is defined as follows:

3.5.2.4.1 HORIZONTAL ANGLE (L AND R)—The angle between the vertical plane and the projection onto the horizontal plane of the ray from the center of the light source of the device to the center of the photometer sensor.

3.5.2.4.2 VERTICAL ANGLE (U AND D)—The true angle between the horizontal plane and the ray from the center of the light source of the device to the center of the photometer sensor.

3.5.2.4.3 The direction can be visualized where an observer stands behind the device and looks in the direction of the emanating light beam towards the photometer sensor when the device is properly aimed with respect to H-V. It should be noted that when rotating the device on a goniometer, it is necessary to move the aim of the device from the H-V point in the opposite direction of the test point being measured. For example, to read a 5U-V test point, the goniometer shall aim the device 5 degrees down. A similar reversal applies to the down (D), left (L), and right (R) test points.

3.5.2.5 *Photometric Measurements*—Photometric measurements shall be made with the light source(s) steady burning. The luminous intensity measurements, in candela, shall be recorded for each of the test points and zones specified for the function of the device being tested.

3.6 Warpage Test on Devices with Plastic Components—This test evaluates the ability of the plastic components of the sample device to resist damage due to ambient and light source heat.

3.6.1 WARPAGE TEST EQUIPMENT—A circulating air oven having a predominant air flow direction shall be used with the air flow inlet on one side of the interior test chamber and the exhaust air outlet on the opposite side of the chamber.

3.6.2 WARPAGE TEST PROCEDURE—A sample device as mounted on the test fixture shall be placed in the circulating air oven and tested to the following procedures:

3.6.2.1 *Oven Temperature*—The oven temperature shall be controlled between 46 to 49 °C.

3.6.2.2 *Sample Position*—The device shall be positioned at the center of the oven such that the predominant direction of air flow approximates that which the device will encounter in its installed position on the vehicle.

3.6.2.3 *Bulb Operation*—Unless otherwise specified, the light source(s) shall be operated at design voltage and cycled as specified in Table 1.

3.6.2.4 *Test Duration*—1 h.

3.6.2.5 *Sample Evaluation*—Upon completion of the test, the device shall be visually examined for warpage of the plastic components.

4. Test Requirements

4.1 Vibration Test Requirements—Upon completion of the vibration test procedure, there shall be no observed rotation, displacement, cracking, or rupture of parts of the device (except bulb filaments or headlamp light source filaments) which would result in failure of any other tests contained in this document. Cracking or rupture of parts of the device affecting its mounting shall also constitute a failure.

4.2 Water Spray Test Requirements—The moisture accumulation in a test device with an interior volume of 7000 ml or less shall be 2 ml or less. For devices with greater interior volumes, the maximum allowable accumulation shall be 0.03% of the total interior volume of the device.

TABLE 1 — CYCLE TIMES (MIN)

Device	Steady Burn	5 On — 5 Off	3 On — 12 Off	Steady[1] Flash
License	X			
Clearance & Identification	X			
Side Marker	X			
Tail, Fog Tail	X			
Park	X			
Stop		X		
Back-up, Rear Cornering		X		
Cornering			X	
Turn Signal				X
Illuminating (Fog Lamp, Driving Lamp, etc)	X			

[1] Flash rate—90 flashes per minute ± 10 flashes per minute with a 50% ± 2% on time.

NOTE—Devices with multiple function combinations shall be tested with all functions simultaneously operating as specified, except for backup functions. Backup functions shall be tested separately. Stop-Turn Signal Functions which are optically combined shall be tested as a stop function only.

4.3 Water Submersion Test Requirements—If a standing pool of water has formed, or can be formed by tapping or tilting the device, the accumulated moisture shall be extracted and measured. The moisture accumulation in the device, regardless of the volume of the device, shall not exceed 1 ml.

4.4 Dust Exposure Test Requirements—The maximum luminous measured intensity after the dust exposure test shall be at least 90% of the initial maximum luminous intensity measured before the test.

4.5 Corrosion Test Requirements—If corrosion is found that could affect other tests in this document, the test(s) shall be performed on the corrosion sample to ensure compliance to that test requirement.

4.6 Photometry Design and Performance Requirements—Upon completion of the test procedure, the luminous intensities at the test points or zones shall be within the limits specified in the applicable SAE Technical Report for the function being tested.

4.6.1 Unless otherwise specified in the applicable SAE Technical Report, the minimum luminous intensity requirements between the specified test points shall be no less than 60% of the lower specified design requirement minimum values for any two adjacent test points on a horizontal or vertical line.

4.7 Warpage Test Requirements—If warpage is observed that could result in failure of other tests contained in this document, the test(s) shall be performed on the warpage sample to insure compliance to that test requirement.

(R) PLASTIC MATERIALS FOR USE IN OPTICAL PARTS SUCH AS LENSES AND REFLEX REFLECTORS OF MOTOR VEHICLE LIGHTING DEVICES—SAE J576 JUL91

SAE Standard

Report of the Lighting and Nonmetallic Materials Committee approved January 1955, and revised by the Lighting Committee September 1986. Rationale statement available. Completely revised by the SAE Lighting Coordinating Committee and the Materials Standards Committee July 1991. Rationale statement available.

1. Scope—This SAE Recommended Practice provides test methods and requirements to evaluate the suitability of plastic materials intended for optical applications in motor vehicles. The tests are intended to determine physical and optical characteristics of the material only. Performance expectations of finished assemblies, including plastic components, are to be based on tests for lighting devices, as specified in SAE Standards and Recommended Practices for motor vehicle lighting equipment. Field experience has shown that plastic materials meeting the requirements of this document and molded in accordance with good molding practices will produce durable lighting devices.

2. References

2.1 Applicable Documents—The following publications form a part of this specification to the extent specified herein. The latest issue of SAE publications shall apply.

2.1.1 SAE PUBLICATIONS—Available from SAE, 400 Commonwealth Drive, Warrendale, PA 15096-0001.

SAE J578—Color Specification

2.1.2 ASTM PUBLICATIONS—Available from ASTM, 1916 Race Street, Philadelphia, PA 19103.

ASTM D 1003-61—Test for Haze and Luminous Transmittance of Transparent Plastics

ASTM D 4364—Standard Practice for Performing Accelerated Outdoor Weathering Using Concentrated Natural Sunlight Utilizing Night Cycle Water Spray

ASTM E 308-66—Recommended Practices for Spectrophotometry and Description Color in CIE 1931 System

2.2 Definitions

2.2.1 MATERIAL—The type and grade of plastics, composition, and manufacturer's designation (number) and color.

2.2.1.1 *Coated Materials*—A coated material is a material as defined in 2.2.1 which has a coating applied to the surface of the finished sample to impart some protective properties. Coating identification includes manufacturer's name, formulation designation (number), and recommendations for application.

2.2.2 MATERIAL EXPOSURE

2.2.2.1 *Exposed*—Material used in lenses or optical devices exposed to direct sunlight as installed on the vehicle.

2.2.2.2 *Protected*—Material used in inner lenses for optical devices where such lenses are protected from exposure to the sun by an outer lens made of materials meeting the requirements for exposed plastics.

2.2.3 WEATHERING EFFECTS

2.2.3.1 *Color Bleeding*—The migration of color out of a plastic part onto the surrounding surface.

2.2.3.2 *Crazing*—A network of apparent fine cracks on or beneath the surface of materials.

2.2.3.3 *Cracking*—A separation of adjacent sections of a plastic material with penetration into the specimen.

2.2.3.4 *Haze*—The cloudy or turbid appearance of an otherwise transparent specimen caused by light scattered from within the specimen or from its surface.

2.2.3.5 *Delamination*—A separation of the layers of a material including coatings.

3. Test Procedures

3.1 Materials to be Tested—Outdoor exposure tests shall be made on each material (as defined in 2.2.1 and 2.2.1.1) offered for use in optical parts employed in motor vehicle lighting devices. Concentrations of polymer components and additives such as plasticizers, lubricants, colorants, weathering stabilizers, and antioxidants in plastic materials and/or coatings may be changed without outdoor exposure testing if: the changes are within the limits of composition represented by higher and lower concentrations of these polymer components and additives have been tested in accordance with 3.3 and found to meet the requirements of Section 4.

3.2 Samples Required

3.2.1 GENERAL—Samples of plastic preferably should be injection molded into polished metal molds to produce test specimens with two flat and parallel faces. Alternative processing techniques may also be used to produce equivalent test specimens. Test specimen shape may vary, but each exposed surface shall contain a minimum uninterrupted area of 32 cm^2 (5.0 in^2).

3.2.2 THICKNESS—Samples shall be furnished covering the thickness range stated by the manufacturer. Recommended nominal thicknesses are: 1.6 mm (0.063 in); 3.2 mm (0.125 in); 6.4 mm (0.250 in). A 2.3 mm (0.090 in) sample is also suggested.

3.2.3 NUMBER OF SAMPLES REQUIRED—Outdoor Exposure Test—1 sample/each thickness/each site × 2 sites for each material = 2 samples/each thickness for each material. Control: 1 sample/each thickness for each material—1 sample each.

NOTE—The control sample must be kept properly protected from influences which may change its appearance and properties.

3.3 Outdoor Exposure Tests

3.3.1 EXPOSURE SITES—Florida (warm, moist climate) and Arizona (warm, dry climate).

3.3.2 SAMPLE MOUNTING—One sample of each thickness of each material at each test station shall be mounted so that the exposed upper surface of the samples is at an angle of 45 degrees to the horizontal facing south. The exposed surface of the sample shall contain a minimum uninterrupted area of 32 cm^2 (5.0 in^2). The sample shall be mounted in the open no closer than 30 cm (11.8 in) to its background.

3.3.3 EXPOSURE TIME AND CONDITIONS—The time of exposure shall be as noted in 3.3.3.1 for each type of material exposed. During the exposure time the samples shall be cleaned once every three months by washing with mild soap or detergent and water, and then rinsing with distilled water. Rubbing shall be avoided.

3.3.3.1 *Exposure Time Based on Material Usage*—Exposed—(defined in 2.2.2.1): 3 years. Protected—(defined in 2.2.2.2): 6 consecutive months starting in May.

3.3.3.2 *Accelerated Weathering*—After establishing and documenting correlation between accelerated and SAE outdoor exposure tests (3.3) for the plastic material and colorant under consideration, accelerated weathering may be used to evaluate minor changes in concentrations of polymer components and additives (3.1) previously found to be acceptable in the outdoor exposure tests. These tests may be used to establish acceptable high and low concentrations of the components and additives pending completion of 3 year weathering tests. These tests will serve as an indication that the plastic materials are capable of meeting the performance requirements of Section 4.

3.4 Optical Measurements

3.4.1 LUMINOUS TRANSMITTANCE AND COLOR MEASUREMENTS—Measurements shall be made in accordance with ASTM E 308-66 (1973).

3.4.2 HAZE MEASUREMENTS—Measurements shall be made in accordance with ASTM D 1003-61 (1977).

4. Material Performance Requirements—A material in the range of thickness as stated by the material manufacturer, and defined in 2.2.1 or 2.2.1.1, shall conform to the following requirements:

4.1 Before Exposure to Any Tests—The chromaticity coordinates shall conform with the requirements of SAE J578 in the range of thickness stated by the material manufacturer.

4.2 After Outdoor Exposure

4.2.1 LUMINOUS TRANSMITTANCE—The luminous transmittance of the exposed samples using CIE Illuminant A (2856K) shall not have changed by more than 25% of the luminous transmittance of the unexposed control sample when tested in accordance with ASTM E 308.

4.2.2 CHROMATICITY COORDINATES—The chromaticity coordinates shall conform with the requirements of SAE J578 in the range of thickness stated by the material manufacturer.

4.2.3 HAZE—The haze of plastic materials used for lamp lenses shall not be greater than 30% as measured by ASTM D 1003 (1977). The haze of plastic materials used for reflex reflectors and/or exposed cover lens materials used in front of reflex reflectors shall not be greater than 7% as measured by ASTM D 1003. Plastic materials used for forward road illumination devices, excluding cornering lamps, shall show no deterioration.

4.2.4 APPEARANCE—The exposed samples when compared with the unexposed controls shall not show physical changes affecting performance such as color bleeding, delamination, crazing, or cracking.

5. Detection of Coatings—In order to test for the presence of a coating, a trace quantity (100 ppm maximum in wet state) of an optical brightener should be added to a coating formulation. This should be checked by ultraviolet inspection against a known coated sample. Additionally, coating suppliers have the option of providing coatings without optical brighteners if they can provide an industry accepted method to detect the coating.

φCOLOR SPECIFICATION—SAE J578 MAY88   SAE Standard

Report of Lighting Committee approved January 1942 and completely revised May 1988. Rational statement available.

1. Scope—This standard defines and provides a means for the control of colors employed in motor vehicle external lighting equipment, including lamps and reflex reflectors. The standard applies to the overall effective color of light emitted by the device in any given direction and not to the color of the light from a small area of the lens. It does not apply to pilot, indicator, or tell-tale lights.

2. Test Methods

2.1 Method of Color Measurement—One of the methods listed in paragraphs 2.1.1, 2.1.2, or 2.1.3 shall be used to check the color of the light from the device or its optical components for compliance with the color specifications. The device shall be operated at the design test voltage. Components (bulbs, cap lenses, and the like) shall be tested in a fixture or in a manner simulating the intended application.

In measuring the color of reflex devices, precautions shall be made to eliminate the first surface reflections of the incident light.

Lighting devices that are covered with neutral density filters shall be tested for color with such filters in place.

2.1.1 Visual Method—In this method, the color of the emitted light from the device is visually compared to the light from a filter/source combination of known chromaticity coordinates. The filter/source combinations are generally chosen to describe the limits of chromaticity coordinates of the color being measured. The color of the filter/source combination is determined spectrophotometrically.

In making visual appraisals, the light from the device lights one portion of a comparator field and filter/source standard lights an adjacent area. The two fields should be in close proximity to each other.

To make valid visual comparisons, the two fields to be viewed must be of near equal luminance (photometric brightness). A means of mechanically adjusting the filter/source standard is generally used to accomplish this. See Appendix for measuring precautions.

2.1.2 Tristimulus Method—In this method, photoelectric detectors with spectral responses that approximate the 1931 CIE standard spectral tristimulus values are used to make the color measurements. These measured tristimulus values are used to calculate the chromaticity coordinates of the color of emitted light from the device. The instrument used for this type of measurement is a colorimeter. These instruments are generally used for production control of color and are satisfactory if calibrated against color filters of known chromaticity coordinates.

Visual tristimulus colorimeters can also be used for color evaluation. See Appendix for measuring precautions.

2.1.3 Spectrophotometric Method—The standard CIE method of color measurement is computing chromaticity coordinates from the spectral energy distribution of the device. This method should be used as a referee approach when the commonly used methods produce questionable results.

Refer to ASTM E308-66 for more details on spectrophotometric measurements (reprinted in the SAE Lighting Manual HS-34).

3. Definitions

3.1 Chromaticity Coordinates—The fundamental requirements for color are expressed as chromaticity coordinates according to the CIE (1931) standard colorimetric system (see Fig. 1). The following requirements shall apply when measured by the tristimulus or spectrophotometric methods.

3.1.1 Red—The color of light emitted from the device shall fall within the following boundaries:
$$y = 0.33 \text{ (yellow boundary)}$$
$$y = 0.98 - x \text{ (purple boundary)}$$

3.1.2 Yellow (Amber)—The color of light emitted from the device shall fall within the following boundaries:
$$y = 0.39 \text{ (red boundary)}$$
$$y = 0.79 - 0.67x \text{ (white boundary)}$$
$$y = x - 0.12 \text{ (green boundary)}$$

3.1.2.1 Selective Yellow (See Appendix, Section A2)—The color of light emitted from the device shall fall within the following boundaries:
$$y = 0.58 x + 0.14 \text{ (red boundary)}$$
$$y = 1.29 x - 0.10 \text{ (green boundary)}$$
$$y = 0.97 - x \text{ (white boundary)}$$

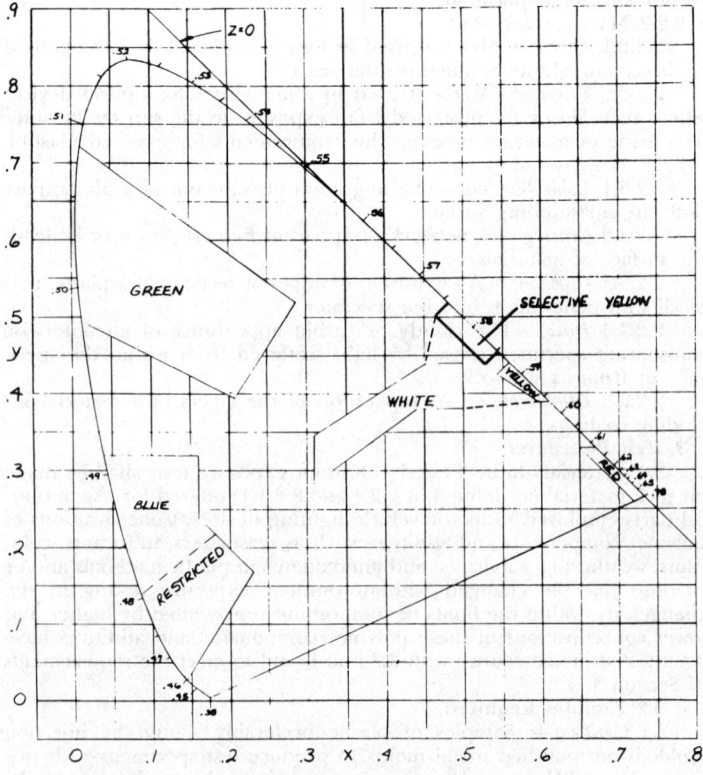

FIG. 1—CHROMATICITY DIAGRAM

3.1.3 WHITE (ACHROMATIC)—The color of light emitted from the device shall fall within the following boundaries:

$x = 0.31$ (blue boundary)
$x = 0.50$ (yellow boundary)
$y = 0.15 + 0.64x$ (green boundary)
$y = 0.05 + 0.75x$ (purple boundary)
$y = 0.44$ (green boundary)
$y = 0.38$ (red boundary)

3.1.3.1 White to Yellow—The color of light emitted from the device shall fall within one of the following areas:
(a) That defined in paragraph 3.1.2 Yellow.
(b) That defined in paragraph 3.1.2.1 Selective Yellow.
(c) That defined in paragraph 3.1.3 White.
(d) The area between Yellow, Selective Yellow, and White as shown by the dashed line in Fig. 1.

3.1.4 GREEN—The color of light emitted from the device shall fall within the following boundaries:

$y = 0.73 - 0.73x$ (yellow boundary)
$x = 0.63y - 0.04$ (white boundary)
$y = 0.50 - 0.50x$ (blue boundary)

3.1.5 BLUE—The color of light emitted from the device shall fall within the following boundaries:

3.1.5.1 Restricted Blue—This color should be elected when recognition of blue as such is necessary.

$y = 0.07 + 0.81x$ (green boundary)
$x = 0.40 - y$ (white boundary)
$x = 0.13 + 0.60y$ (violet boundary)

3.1.5.2 Signal Blue—This color may be elected when, due to other factors, it is not always necessary to identify blue as such.

$y = 0.32$ (green boundary)
$x = 0.16$ (white boundary)
$x = 0.40 - y$ (white boundary)
$x = 0.13 + 0.60y$ (violet boundary)

3.2 Visual Method—When checking by the visual method of paragraph 2.1.1, the following subjective guidelines shall be considered:

3.2.1 RED—Red shall not be acceptable if it is less saturated (paler), yellower, or bluer than the limit standards.

3.2.2 YELLOW (AMBER)—Yellow shall not be acceptable if it is less saturated (paler), greener, or redder than the limit standards.

3.2.3 WHITE—White shall not be acceptable if its color differs significantly from that of a blackbody source operating at a color temperature between CIE Illuminant A (2854K) and CIE Illuminant B (5000K).

3.2.4 GREEN—Green shall not be acceptable if it is less saturated (paler), yellower, or bluer than the limit standards.

3.2.5 BLUE—Blue shall not be acceptable if it is less saturated (paler), greener, or redder than the limit standards.

APPENDIX

A1. Precautions—The following are applicable to all methods of determining the color of light:

a) Some devices may emit a different color of light in one direction than another. Measurements should be made in as many directions as required to define the color characteristic of emitted light.

Some instruments (tristimulus and spectroradiometric) use an integrating sphere at the inlet port of the device to integrate all the light from the device. Care should be taken to assure that the integrating sphere is not combining different color light emitted in different directions from the device and thereby providing an erroneous reading.

b) The lamp and optical components should be allowed to reach operating temperature before any measurements are made. Lamps should be operated at design voltage.

If visually the device does not appear to be emitting light with a uniform color, additional precautions should be taken.

c) The distance between the test instrument and the device under test should be great enough so that further increases in distance do not affect the results. The visual field of the instrument should view the entire lighted area of the device.

A2. Color Application—Selective yellow is used on a limited basis primarily for fog lights and is not to be used in turn signal, parking, identification, clearance, sidemarker, and school bus warning lamps, or yellow reflex reflector applications as required by FMVSS 108.

A3. Neutral Density—Filtering materials are sometimes used over existing lighting devices to reduce the light intensity but not to change the fundamental color requirements as detailed in SAE J578.

A4. Orange Fluorescent Information Guideline—Definitions and Requirements for Orange Fluorescent color can be found in the appropriate SAE Recommended Practice or Standard. Refer to SAE J774, Emergency Warning Device, or SAE J943, Slow-Moving Vehicle Identification Emblem or to FMVSS No. 125, Warning Devices, 39 FR 28636, Aug. 9, 1974 as amended at 40 FR4, Jan. 2, 1975.

A5. Color Measurements of Gaseous Discharge Lighting Devices—Some laboratories cannot measure the color of light from the short pulses of lamps that use discharge tubes and, therefore, these lamps need a steady burning test source, operated at the color temperature of the gaseous discharge warning lamp. Use of CIE Illuminant C for strobe lights has been confirmed by independent testing laboratories.

A6. Cited ASTM Report—ASTM E 308-66, Standard Practice for Spectrophotometry and Description of Color in CIE 1931 System. Reprinted in SAE Ground Vehicle Lighting Manual, HS-34.

(R) PERFORMANCE REQUIREMENTS FOR MOTOR VEHICLE HEADLAMPS—SAE J1383 JUN90

SAE Recommended Practice

Report of the Lighting Committee, approved April 1985. Rationale statement available. Completely revised by the Road Illumination Devices Standards Committee June 1990. Rationale statement available.

TABLE OF CONTENTS

1. Scope
2. References
   2.1 Applicable Documents
   2.2 Definitions
3. Identification Code Designation
   3.1 SAE J759 Lighting Identification Code
   3.2 Headlamp Marking Requirements
   3.3 Replaceable Bulb Marking Requirements
4. Tests
   4.1 Bulbs
   4.2 SAE J575
   4.3 SAE J578 Color
   4.4 SAE J576 Plastic Materials
   4.5 Beam Pattern Location Test
   4.6 Wattage Test
   4.7 Luminous Flux Test
   4.8 Luminous Flux Maintenance Test
   4.9 Out-of-Focus Test
   4.10 Impact Test
   4.11 Aiming Adjustment Test
   4.12 Inward Force Test
   4.13 Torque Deflection Test
   4.14 Deflection Test—Replaceable Headlamp Bulbs
   4.15 Sealing Test—Replaceable Headlamp Bulbs
   4.16 Chemical Resistance Test
   4.17 Abrasion Test of Plastic Headlamp Lens Material
   4.18 Thermal Cycle Test
   4.19 Internal Heat Test
   4.20 Humidity Test
   4.21 Filament Rated Laboratory Life Test
5. Requirements
   5.1 Vibration Test
   5.2 Dust Test
   5.3 Corrosion Test
   5.4 Photometry
   5.5 SAE J578 Color
   5.6 SAE J576 Plastic Materials
   5.7 Beam Pattern Location Test
   5.8 Wattage Test
   5.9 Luminous Flux Test
   5.10 Luminous Flux Maintenance Test
   5.11 Out-of-Focus Test
   5.12 Impact Test
   5.13 Aiming Adjustment Test
   5.14 Inward Force Test
   5.15 Torque Deflection Test
   5.16 Deflection Test—Replaceable Headlamp Bulbs
   5.17 Sealing Test—Replaceable Headlamp Bulbs
   5.18 Chemical Resistance Test
   5.19 Abrasion Test of Plastic Headlamp Lens Material
   5.20 Thermal Cycle Test
   5.21 Internal Heat Test
   5.22 Humidity Test
   5.23 Retaining Ring Requirements
   5.24 Design Requirements
6. Guidelines
7. Bulb Filament Dimension and Location Test for the 9004 Replacement Bulb
   7.1 Low Beam Filament Location Test
   7.2 High Beam Filament Location Test
8. Low Beam Filament Location
   8.1 Axial
   8.2 Vertical
   8.3 Transverse
9. High Beam Filament Location
10. Bulb Filament Dimension and Location Test for the 9005 and 9006 Replacement Bulb
11. High Beam Filament and Location
12. Low Beam Filament Location
13. Viewing Direction for HB3 (9005) and HB4 (9006) Bulbs
14. Bulb Filament Dimension and Location Test for the 9007 Replacement Bulb
15. Low Beam Filament Location
16. High Beam Filament Location
17. Methods of Measuring Internal Elements of H4/HB2 Bulbs
18. Reference Plane, Reference Axis, and Planes for Measurements
19. Viewing Directions
20. Measuring Points (MP)

LIST OF FIGURES AND TABLES

Figure 1	Photometric Tables
Figure 2	Test Enclosure Dimensions
Figure 3	Inward Force Tester
Figure 4	Deflectometer
Figure 5	Deflectometer
Figure 6	Deflectometer
Figure 7	Deflectometer
Figure 8	Deflectometer
Figure 9	Bulb Deflection Test
Figure 10	Test for Airtight Seal
Figure 11	Abrasion Test Machine
Figure 12	Thermal Cycle Profile
Figure 13	Rectangular Headlamp Mounting and Retaining Ring
Figure 14	Rectangular Headlamp Mounting and Retaining Ring
Figure 15	Rectangular Headlamp Retaining Ring—146mm Diameter
Figure 16	Rectangular Headlamp Retaining Ring—178mm Diameter
Figure 17	Mounting and Retaining Rings (LF and UF)
Figure 18	Aiming Ring—55 × 135 UK/LK
Figure 19	Guideline Replaceable Bulb Filament End Coil Definition
Figures 20-32	Sealed Beam Headlamps
Figures 33-76	Replaceable Headlamp Bulbs
Table 1	Replaceable Headlamp Bulbs
Table 2	Replaceable Bulbs and Related Dimensional Figures
Table 3	Photometry
Table 4	Headlamp Type Identification
Table 5	Headlamp Photometric Performance Requirements
Table 6	Typical Headlamps
Table 7	Flange Thickness
Table 8	Test Classification
Table 9	Dimensions to be Measured

1. Scope—This SAE Recommended Practice is intended as a guide toward standard practice and is subject to change to keep pace with experience and technical advances. This document establishes performance requirements, material requirements, design requirements, and design guidelines for headlamps and replaceable bulbs for headlamps.

2. References

2.1 Applicable Documents

2.1.1 SAE PUBLICATIONS—Available from SAE, 400 Commonwealth Drive, Warrendale, PA 15096-0001.

SAE J575—Tests for Motor Vehicle Lighting Devices and Components

SAE J759—Lighting Identification Code

2.2 Definitions

2.2.1 HEADLAMP—A lighting device providing an upper and/or a lower beam used for providing illumination forward of the vehicle.

2.2.2 SEALED BEAM HEADLAMP ASSEMBLY—A major lighting assembly which includes one or more indivisible optical assemblies used to provide general illumination ahead of the vehicle.

2.2.3 REPLACEABLE BULB (BULB)—A light source with related envelope and mounting base which is removable from the headlamp for the purpose of replacement.

2.2.4 MECHANICALLY AIMABLE HEADLAMP—A headlamp having three pads on the lens, forming an aiming plane used for laboratory photometric testing and for adjusting and inspecting the aim of the headlamp when installed on the vehicle.

2.2.5 AIMING PLANE—A plane defined by the surface of the three aiming pads on the lens.

2.2.6 HEADLAMP MECHANICAL AXIS—The line formed by the intersection of a horizontal and a vertical plane through the light source parallel to the longitudinal axis of the vehicle. If the mechanical axis of the headlamp is not at the geometric center of the lens, then the location will be indicated by the manufacturer on the headlamp.

2.2.7 H-V AXIS—A line from the center of the principal filament (low beam filament of two filament bulbs) to the intersection of the horizontal (H) and vertical (V) lines on the screen (see Figure 1).

2.2.8 SEASONING—Process of energizing the filament of a bulb at design voltage for a period of time equal to 1% of design life or 10 h maximum, whichever is shorter.

TABLE 1—REPLACEABLE HEADLAMP BULBS

Design Designation	Number of Filaments	Luminous Wattage (Watts) at 12.8 V U.B./L.B.	Flux (Lumens) at 12.8 V U.B./L.B.	Design Life (Hours) Æ 14 V U.B./L.B.[2]	Filament Type
9004 (ANSI)	2	65/45	1200/700[1]	150/320	C6/C6
9005 (ANSI)	1	65	1700	150	C8
9006 (ANSI)	1	55	1000[1]	320	C8
9007 (ANSI)	2	65/55	1350/1000[1]	150/320	C8/C8
H1	1	60.5	1410	320	C8
H3	1	60.5	1310	320	C6
H4	2	67/60.5	1500/910[1]	150/320	C8/C8

[1] With opaque coating
[2] Guideline

TABLE 2—REPLACEABLE BULBS AND RELATED DIMENSIONAL FIGURES

Bulb Identification	Figures Relative To: Dimensional Specifications Standard Bulb	Figures Relative To: Specifications For Bulb Holders	Figures Relative To: Measurement Method—Bulb Filament Dimension and Tolerance	Figures Relative To: Dimensional Specifications for Accurate Rated Bulb
9004	Figures 33 to 36	37	Sections 7, 8, 9	38
9005	Figures 39 to 43	44	Sections 10, 11, 12, 13	Sections 10, 11, 12, 13, and Figure 46
9006	Figures 47 to 51	52	53	Sections 10, 11, 12, 13, and Figure 53
9007	Figures 54 to 57	58	Sections 14, 15, 16	59
H1	Figures 60 to 61	62	63	64
H3	Figures 65 to 66	67	68	69
H4	Figures 70 to 72	73	Sections 17, 18, 19, 20, 21 and Figure 74	71
HB2	Figures 70 to 72	73	Sections 17, 18, 19, 20, 21 and Figure 74	71

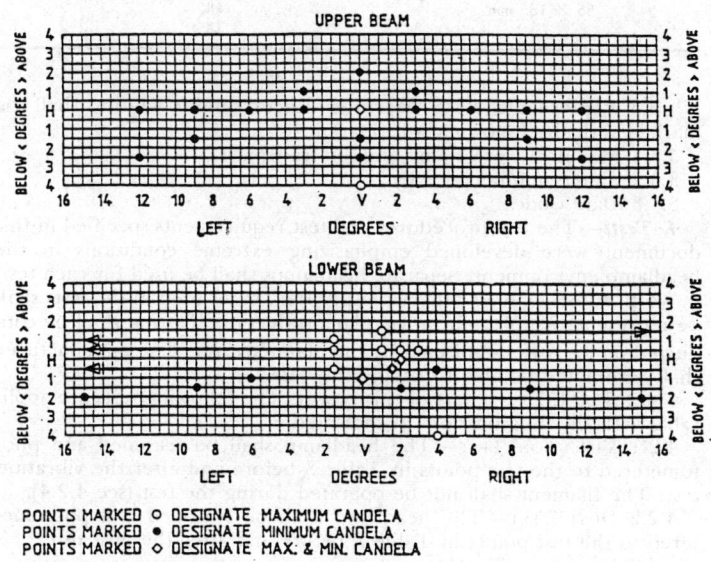

POINTS MARKED ○ DESIGNATE MAXIMUM CANDELA
POINTS MARKED ● DESIGNATE MINIMUM CANDELA
POINTS MARKED ◊ DESIGNATE MAX. & MIN. CANDELA

FIGURE 1—PHOTOMETRIC TABLES

2.2.9 DESIGN VOLTAGE—The voltage used for design purposes.

2.2.10 TEST VOLTAGE—The specified voltage and tolerance to be used when conducting a test.

2.2.11 RATED VOLTAGE—The nominal circuit or vehicle electrical system voltage classification. (Example: 12 V Headlamp)

2.2.12 HEADLAMP TEST FIXTURE—Device specifically designed to support a headlamp in the test position during laboratory testing. Mounting hardware and components shall be representative of those necessary to operate the headlamp in its normal manner.

2.2.13 MOUNTING RING—The adjustable ring upon which the sealed beam bulb is mounted and which forces the sealed beam bulb to seat against the aiming ring when assembled into a sealed beam headlamp assembly.

2.2.14 RETAINING RING—The clamping ring that holds the sealed beam bulb against the mounting ring.

2.2.15 AIMING RING—The clamping ring that retains the sealed beam bulb against the mounting ring, and that provides an interface between the bulb's aiming/seating pads and the headlamp aimer adapter (locating plate). It also describes and is coincident with the aiming plane.

2.2.16 AIMING SCREWS—Horizontal and vertical adjusting screws with self-locking features used to aim and retain the headlamp unit in the proper position.

2.2.17 INTEGRAL AIM—A vertical aiming system which is mounted to the headlamp and does not require a separate vertical mechanical aiming device.

2.2.18 ACCURATE RATED BULB—A seasoned bulb operated at design luminous flux shown in Table 1 and having its filaments located within the tolerances indicated in figures specified in Table 2. Separate bulbs may be used for high and low beams.

2.2.19 HIGH BEAM—A beam intended primarily for distant illumination and for use when not meeting or following other vehicles.

2.2.20 LOW BEAM—A beam intended to illuminate the road ahead of the vehicle when meeting or following another vehicle.

2.2.21 HIGH BEAM FILAMENT—Filament coil designed to provide high beam function.

2.2.22 LOW BEAM FILAMENT—Filament coil designed to provide low beam function.

2.2.23 FILAMENT ROTATION—Any nonparallelism of either coil with respect to the centerline of the design nominal filament location or any additional width of the end view of the filament in excess of the outside diameter of the first full turn.

2.2.24 RATED AVERAGE LAB LIFE—An average life in hours which is obtained by laboratory life testing of bulbs at the specified test voltage over a long period of production time. It is meant to partially describe a manufactured product recognizing that individual lifetimes vary greatly. It is not the same as service life which is generally shorter due to environmental conditions such as vibration, voltage fluctuations, and temperature.

2.2.25 DESIGN LIFE—An operational time objective in hours of a headlamp filament at the test voltage.

2.2.26 GAGING STANDARD—A gage produced for each bulb type with all critical tolerances affecting filament location one-tenth the stated tolerances in the design designation.

3. Identification Code Designation
3.1 SAE J759 Lighting Identification Code
3.2 Headlamp Marking Requirements—Headlamps shall be marked with the following markings:
3.2.1 Manufacturer's name and/or trademark shall appear on the lens.
3.2.2 Voltage and part number or trade number shall appear on the headlamp.
3.2.3 The face of letters, numbers, or other symbols molded on the surface of the lens shall not be raised more than 0.5 mm (0.020 in).
3.2.4 HEADLAMP TYPE IDENTIFICATION CODE
3.2.4.1 Headlamp lenses shall be marked with a two or three character code.

3.2.4.2 The marking shall be molded in the lens and shall be 6.35 mm (0.25 in) or greater in size.

3.2.4.3 The first character (a number) of the three character identification code indicates the number of beams in the headlamp. All headlamps marked with a "1" are aimed on the high beam and all headlamps marked with a "2" are aimed on the low beam.

3.2.4.4 The second character (a letter) stands for the size and number of headlamps used on the vehicle.

 A — 100 × 165 mm rectangular, four lamp system
 B — 142 × 200 mm rectangular, two lamp system
 C — 146 mm round, four lamp system
 D — 178 mm round, two lamp system
 E — 100 × 165 mm rectangular, two lamp system
 F — 92 × 150 mm rectangular, four lamp system
 G — 100 × 165 mm rectangular, four lamp system
 H — 100 × 165 mm rectangular, two lamp system
 J — 56 × 75 mm rectangular, eight lamp system
 K — 55 × 135 mm rectangular, four lamp system

3.2.4.5 The third character (a number) indicates the photometric specification which applies to the headlamp. Headlamps designed to Table 3 have "1" as the third character.

TABLE 3—PHOTOMETRY

	Low Beam Min	Low Beam Max
10U to 90U, 45R to 45L		125 cd
8L to 8R, H to 4U	64 cd	
4L to 4R, H to 2U	135	
1U—1-1/2L to L		700
1/2U—1-1/2L to L		1 000
1/2D—1-1/2L to L		3 000
1-1/2U—1R to R		1 400
1/2U—1R, 2R, 3R		2 700
1/2D—1-1/2R	8 000	20 000
1D—6L		750
1-1/2D—2R	15 000	
1-1/2D—9L & 9R		750
2D—15L & 15R		700
4D—4R		8 000
H—2R	4 000	10 000
1D—V	6 000	15 000

TABLE 3—PHOTOMETRY (CONTINUED)

	High Beam Min	High Beam Max
2U—V		1 500 cd
1U—3R & 3L		5 000
H—V	20 000	75 000 cd
H—3R & 3L	10 000	
H—6R & 6L	3 250	
H—9R & 9L	2 000	
H—12R & 12L	500	
1-1/2D—V		5 000
1-1/2D—9R & 9L		1 500
2-1/2D—V		2 000
2-1/2D—12R & 12L		750
4D—V		12 500
Maximum Beam Candela [1]—30 000 cd Min		

[1] The highest candela reading found in the beam pattern.

3.2.4.6 Headlamps designed to UF, UK, LF, and LK specifications shall meet the following criteria:

3.2.4.6.1 The first character indicates the upper (high) or low beam function.

3.2.4.6.2 The second character indicates the size and number of headlamps used on the vehicle.

3.2.4.7 *Headlamp Type Identification*—See Table 4.

TABLE 4—HEADLAMP TYPE IDENTIFICATION

Size	Type	Number of Headlamps
100 × 165 mm (4 × 6.5 in)	1A1	2
	2A1	2
	1G1	2
	2G1	2
142 × 200 mm	2B1	2
146 mm (5.75 in)	1C1	2
	2C1	2
178 mm (7.0 in)	2D1	2
100 × 165 mm (4 × 6.5 in)	2E1	2
	2H1	2
92 × 150 mm	UF	2
	LF	2
56 × 75 mm	UJ	4
	LJ	4
55 × 135 mm	UK	2
	LK	2

3.3 Replaceable Bulb Marking Requirements—Bulbs shall be marked with the following information.

3.3.1 Manufacturer's name and/or trademark
3.3.2 Trade number (ANSI)
3.3.3 Date Code

4. Tests—The test procedures and test requirements specified in this document were developed emphasizing extreme conditions in the headlamp environment. Separate headlamps shall be used for each test.

4.1 Bulbs—Unless otherwise specified, bulbs used in the tests shall be representative of bulbs in regular production. Testing shall be conducted on lot sizes established by the manufacturer. The manufacturer shall obtain and be able to supply the data.

4.2 SAE J575 is a part of this report. The following tests are applicable with the modifications as indicated.

4.2.1 VIBRATION TEST—The headlamp shall be seasoned and photometered to the test points in Table 3 before and after the vibration test. The filament shall not be operated during the test (see 4.2.4).

4.2.2 DUST TEST—The headlamp shall be seasoned and photometered to the test points in Table 3 before and after the dust test.

4.2.3 CORROSION TEST

4.2.3.1 The headlamp shall be seasoned and photometered to the test points in Table 3 before and after the corrosion test.

4.2.3.2 The test period shall be 240 h consisting of 10 cycles of 23 h exposure followed by 1 h drying.

4.2.4 PHOTOMETRY

4.2.4.1 Test samples shall be new, unused headlamps manufactured from production tooling and assembled by production processes.

4.2.4.2 The headlamp shall be seasoned and photometered at the appropriate test points as listed in Table 3. The headlamp shall be in operation a minimum of 5 min prior to photometry.

4.2.4.3 Photometric test shall be made with the photometer sensor at a distance of at least 18.3 m from the headlamp.

4.2.4.4 The headlamp shall be aimed mechanically with the aiming plane at the design angle(s) to the photometer axis and the mechanical axis of the headlamp on the photometer axis.

4.2.4.5 *Test Voltage*—The voltage for the photometric test shall be 12.8 V ± 20 mV, DC as measured at the terminals of the headlamp.

4.3 Color Test—SAE J578 is a part of this report.

4.4 Plastic Materials—SAE J576 is a part of this report except 4.2.1, Luminous Transmittance.

4.5 Beam Pattern Location Test

4.5.1 Headlamps designed to be aimed on high beam, shall be seasoned and photometered to find the location of maximum intensity (see 4.2.4).

4.5.2 Headlamps designed to be aimed on low beam, shall be seasoned and photometered (see 4.2.4) at the test points H-2R and 1D-V.

4.6 Wattage Test

4.6.1 The wattage of each filament shall be determined at 12.8 V ± 20 mV DC.

4.6.2 Filaments shall be seasoned prior to wattage measurement.

4.7 Luminous Flux Test

4.7.1 Each filament shall be seasoned and photometered at 12.8 V ± 20 mV DC to determine luminous flux.

4.7.2 The tests shall be conducted in accordance with IES Approved Method for Electrical and Photometric Measurements of General Service Incandescent Filament Lamps, IES Lighting Handbook, Reference Volume, Illuminating Engineering Society, New York, NY, Procedure LM-45.

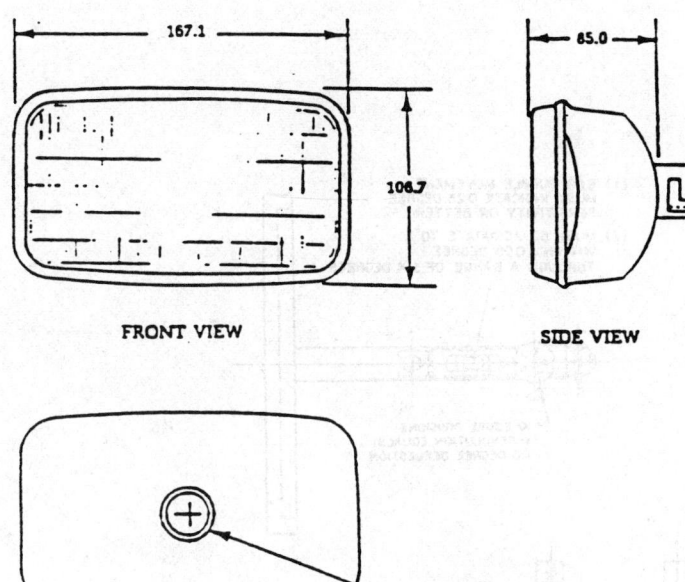

FIGURE 2—TEST ENCLOSURE DIMENSIONS

4.8 Luminous Flux Maintenance Test

4.8.1 The luminous flux for each filament shall be determined in accordance with 4.7.

4.8.2 The bulb shall then be energized in a horizontal or its normal burning position in the test enclosure shown in Figure 2.

4.8.3 The test voltage shall be 14.0 V ± 0.1 V DC. Two-filament bulbs shall be tested by cycle burning the high beam filament 12 min for each hour of testing.

4.8.4 The luminous flux of a single filament bulb shall be measured after burning for 70% of the design life.

4.8.5 The luminous flux of each filament of two-filament bulbs shall be measured in accordance with 4.7 after the low beam filament on-time equals 70% of the design life when tested according to 4.8.3.

4.9 Out-of-Focus Test

4.9.1 This test shall be conducted on headlamps with replaceable bulbs.

4.9.2 The headlamp shall be mounted in the goniometer with the mechanical axis coincident with the photometer axis.

4.9.3 The test voltage for the headlamp shall be 12.8 V ± 20 mV DC.

4.9.4 The headlamp shall be photometered at the appropriate test points as listed in Table 3.

4.9.5 Intensity measurements shall be made at six out-of-focus positions with the filament located at 2/3 of the tolerance value specified for the filament tolerance specifications referenced in Table 2.

4.10 Impact Test

4.10.1 The headlamp shall be rigidly mounted in a test fixture on the seating plane with the lens facing up.

4.10.2 The seating plane of the test fixture shall consist of 13 mm thick oak wood. The test fixture shall rest on an oak wood base.

4.10.3 One impact shall be delivered to the headlamp lens along the mechanical axis using a 23 mm diameter steel sphere (approximately 50 g) dropped freely, without side forces, from a distance of 40 cm above the lens.

4.11 Aiming Adjustment Test

4.11.1 When making the aiming adjustment test, an accurate measurement technique shall be used. This may consist of: (a) Attaching a device such as a spot projector to the headlamp, or (b) replacing the headlamp with a mirror along with a separate light source, or (c) other equally accurate means.

4.11.2 When conducting the test, the headlamp shall be mounted in the design position with the unit at nominal aim (0,0).

4.12 Inward Force Test—The mechanism, including the aiming adjusters, shall be subjected to an inward force of 222 N directed normal to the headlamp aiming plane and symmetrically about the center of the sealed beam unit face (see Figure 3).

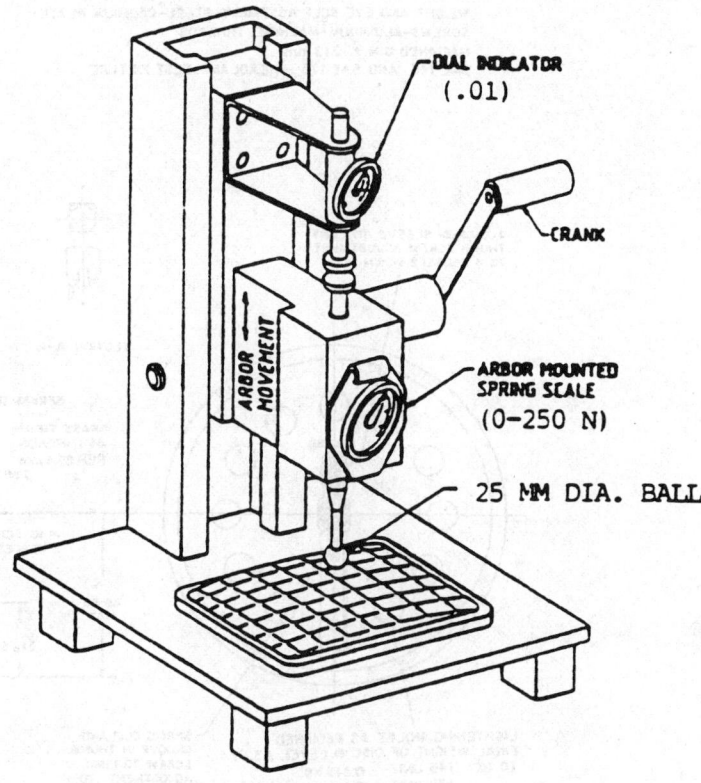

FIGURE 3—INWARD FORCE TESTER

4.13 Torque Deflection Test—Applies to headlamps which do not incorporate on-board headlamp aiming system.

4.13.1 The headlamp assembly to be tested shall be mounted in design vehicle position and set at nominal aim (0,0).

4.13.2 Sealed beam headlamps shall be replaced by the appropriate deflectometer (Figures 4 to 8).

4.13.3 Replaceable bulb headlamps shall be equipped with an appropriate fixture on the face of the lens with the applied load acting parallel to the aiming reference plane and in a downward direction. The force shall be applied through the aiming pads.

4.13.4 A torque of 2.25 Nm shall be applied to the headlamp assembly through the deflectometer and a reading on the thumbwheel shall be taken. The torque shall then be removed and a second reading on the thumbwheel shall be taken.

4.14 Deflection Test—Applies to replaceable headlamp bulbs.

4.14.1 The bulb shall be rigidly mounted in a fixture in a manner indicated in Figure 9.

24.40

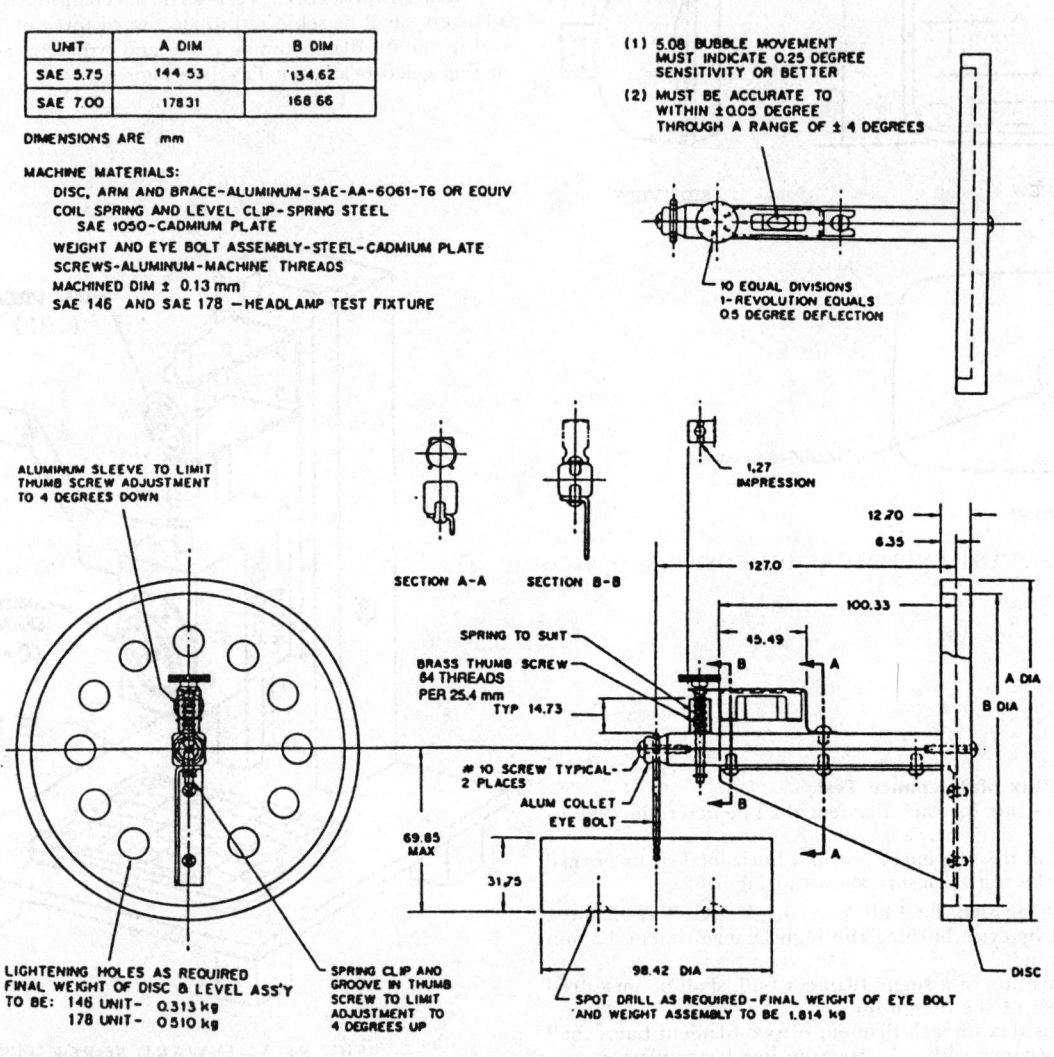

FIGURE 4—DEFLECTOMETER

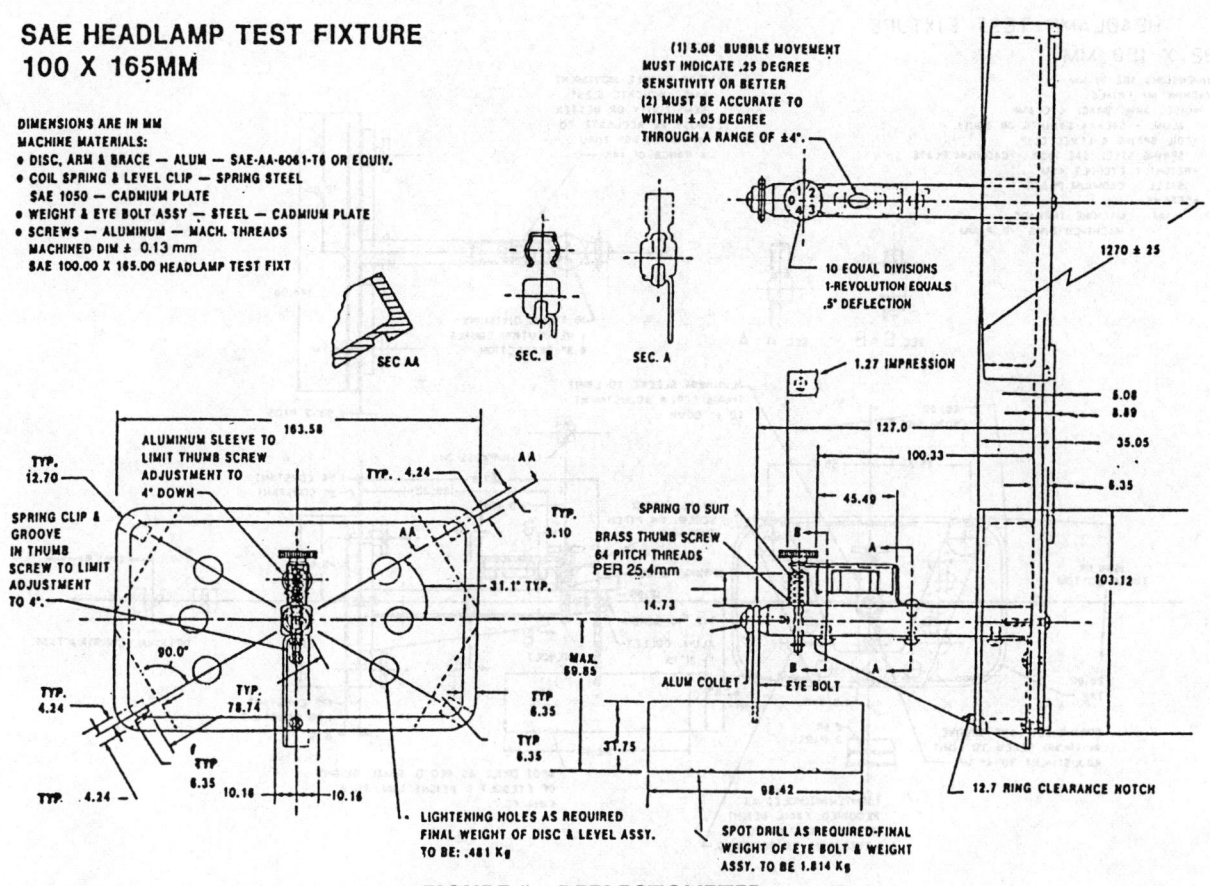

FIGURE 5—DEFLECTOMETER

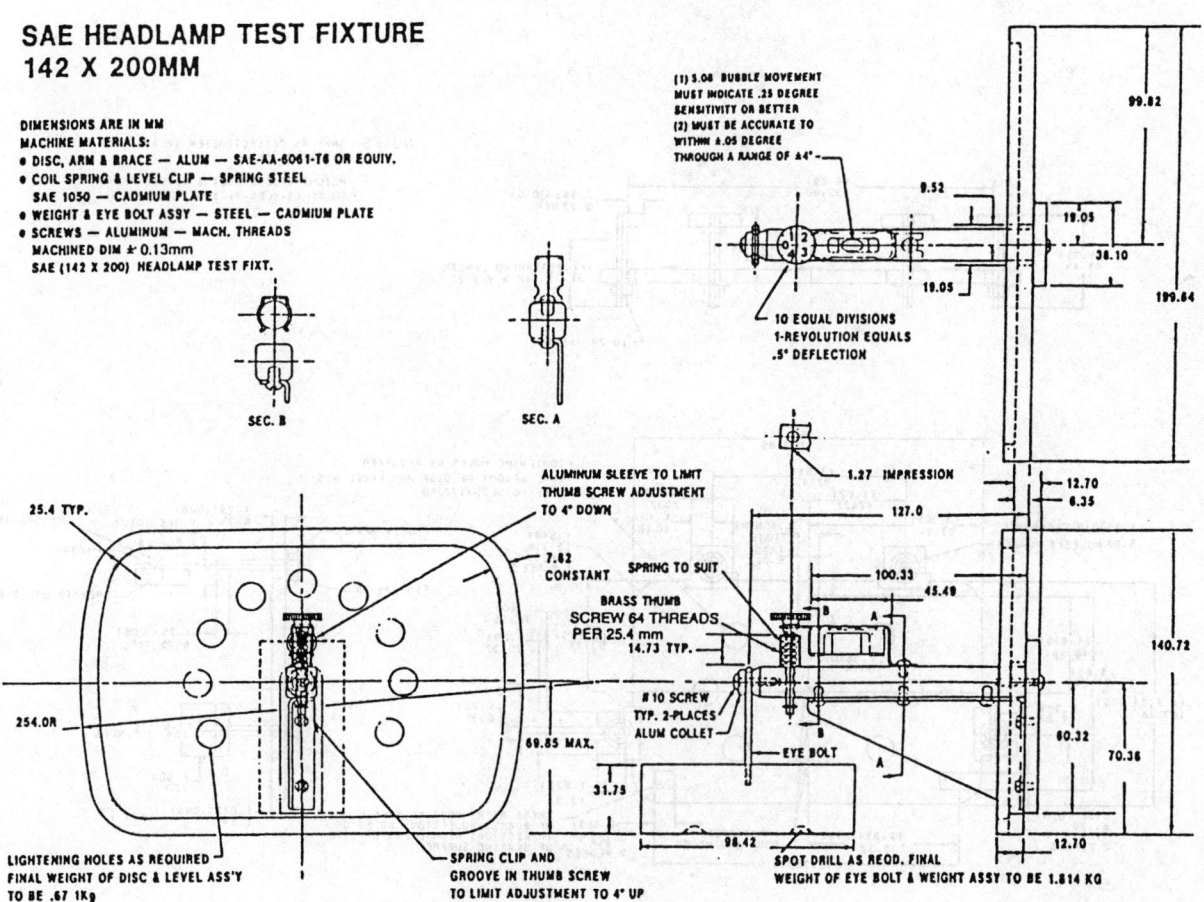

FIGURE 6—DEFLECTOMETER

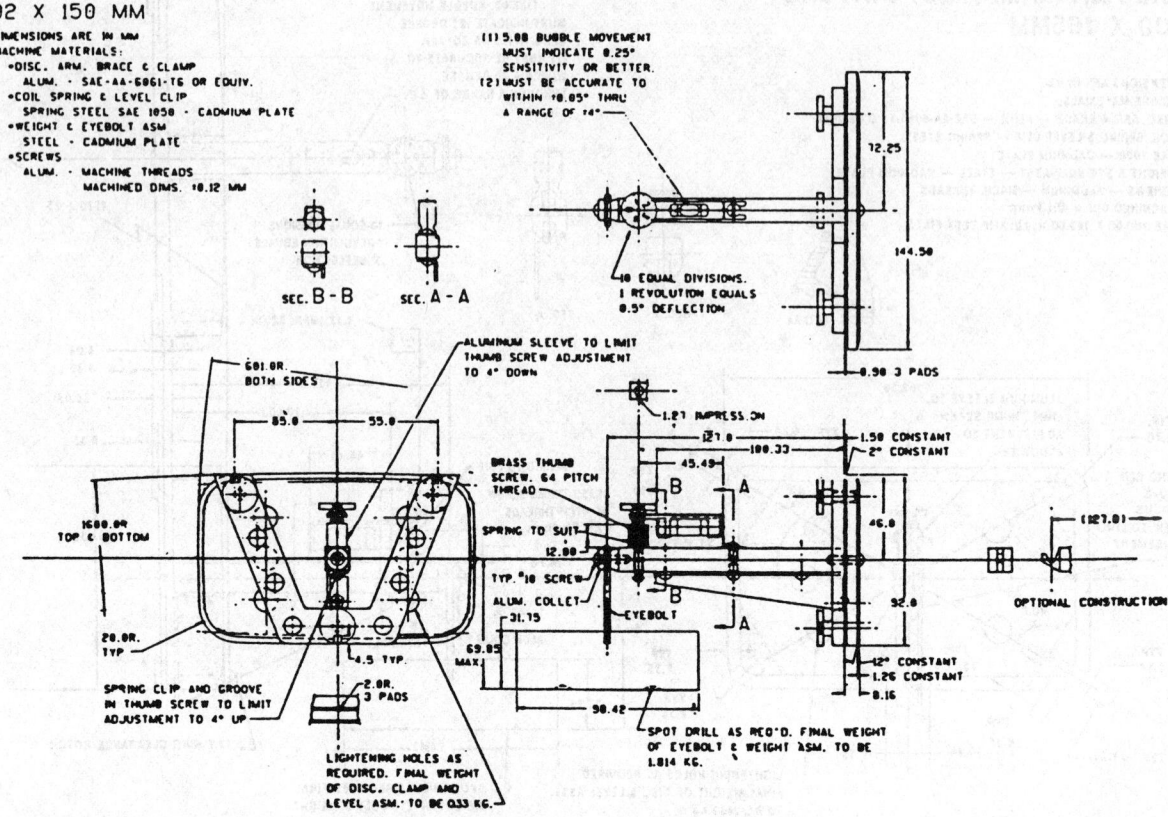

FIGURE 7—DEFLECTOMETER

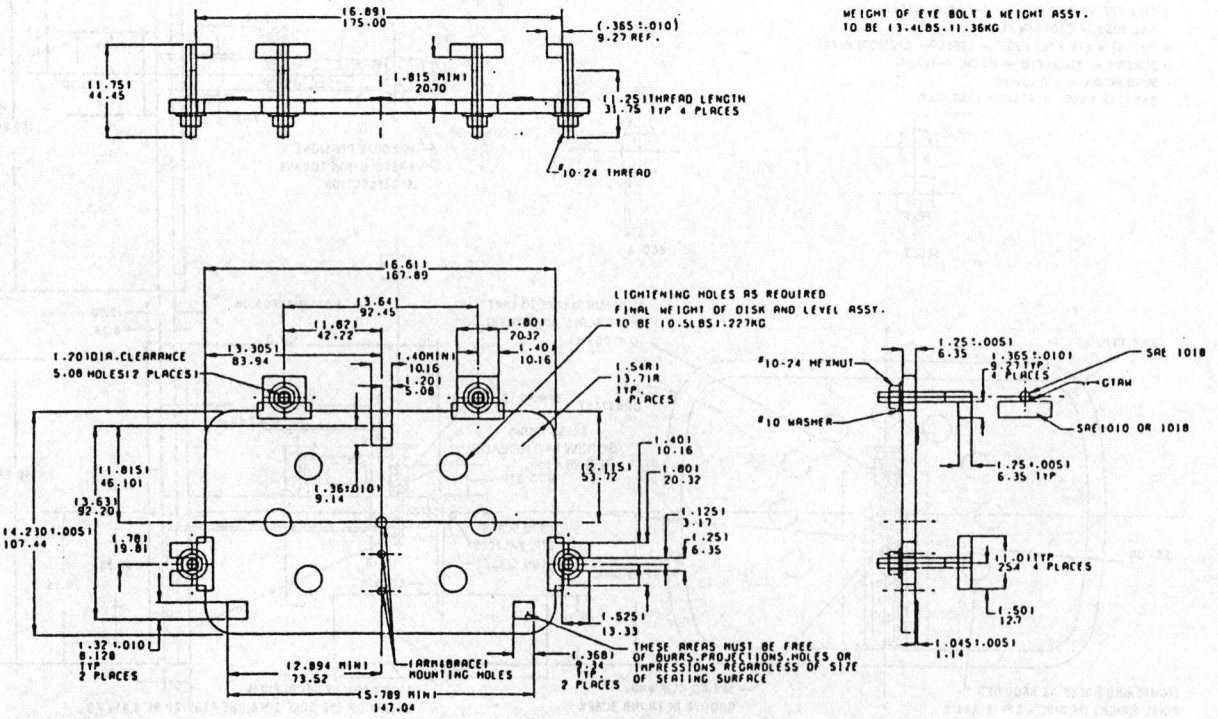

FIGURE 8—DEFLECTOMETER

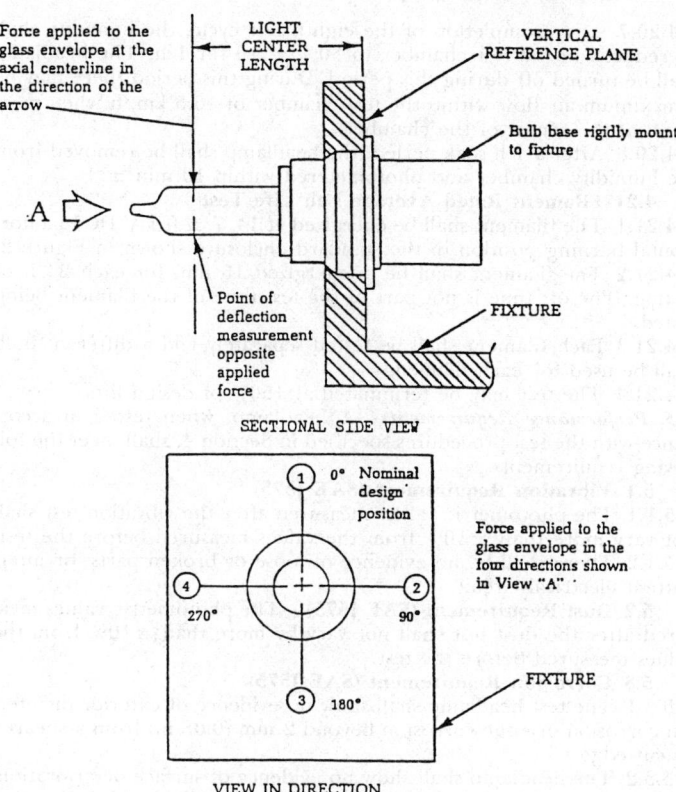

FIGURE 9—BULB DEFLECTION TEST

4.14.2 A force of 18.0 N ± 0.4 shall be applied for a maximum of 5.0 s at the locations shown in Figure 9 using a rod with a hard rubber tip with a minimum radius of 1.0 mm.

4.14.3 A separated bulb shall be used for each load application at 0, 90, 180, and 270 degrees.

4.15 Sealing Test—Applies to bulbs designed for an airtight fit to the headlamp.

4.15.1 The bulb shall be inserted into a fixture as shown in Figure 10 and retained by the same method intended for application, or equivalent.

4.15.2 The chamber shall be gradually pressurized to 70.0 kPa ± 1.0 gage while the fixture and terminal end of the bulb is completely submerged in water. The 70 kPa gage pressure shall be held for 60 s.

4.15.3 The bulb shall be observed for the presence of air bubbles during the 60 s time period.

4.16 Chemical Resistance Test

4.16.1 The test shall be conducted with the headlamps and the test fluids at an ambient temperature of 23 °C ± 4.

4.16.2 The test headlamps shall be seasoned and photometered to the test points in Table 3 before and after the chemical resistance test (see 4.2.4).

4.16.3 A separated headlamp shall be used for each of the test fluids.

4.16.4 The test fluids are:
   a. Windshield washer fluid (50% concentration by volume of methanol/detergent base, 0.16% ethanolamine)
   b. Antifreeze (50% concentration by volume of ethylene glycol in water)
   c. Simulated unleaded gasoline (test fluid ASTM D 471-79 Reference fuel "D")

4.16.5 An unfixtured headlamp in its design operating position and condition shall be used for the test.

4.16.6 A 6 in square cotton cloth shall be folded twice to form a 3 in square and placed onto the bottom of a beaker.

4.16.7 Meter 3 mL of the test fluid onto the folded cloth.

4.16.8 Remove the cloth from the beaker (5 s after completion of test fluid metering for Reference Fuel D and windshield washer fluid, and 60 s after completion of test fluid metering for antifreeze).

4.16.9 Within 5 s after removal of the cloth from the beaker, wipe the lens with that cloth surface which was uppermost in the beaker. The entire exterior optical surface of the lens of the fixtured headlamp shall be wiped in three horizontal cycles (one cycle consists of one back and forth motion). The first cycle shall apply the test fluid to the upper segment of the lens, the second cycle shall apply it to the center segment, and the third cycle shall apply it to the lower segment.

4.16.10 After applying the test fluid, the test headlamp shall be set aside for a period of 48 h where upon the headlamp shall be wiped clean with a soft, dry, cotton cloth.

4.17 Abrasion Test of Plastic Headlamp Lens Material

4.17.1 A 100 × 165 mm flat test specimen shall be measured for luminous transmittance before and after wiping clean after the abrasion test.

4.17.2 The test specimen shall be mounted in the abrasion test machine as indicated in Figure 11.

4.17.3 The size of the abrading pad shall be 25 × 100 mm constructed of 0000 steel wool and firmly attached to a pad support of equal size such that the "grain" of the pad is perpendicular to the direction of motion.

4.17.4 The abrading pad shall be loaded such that an average pad pressure of 14 kPa ± 1 exists normal to the surface of the test specimen.

4.17.5 The density of the abrading pad shall be such that when the abrading pad mounted to the pad support is resting unloaded on the test specimen, the pad support shall be no closer than 3.1 mm to the surface of the test specimen.

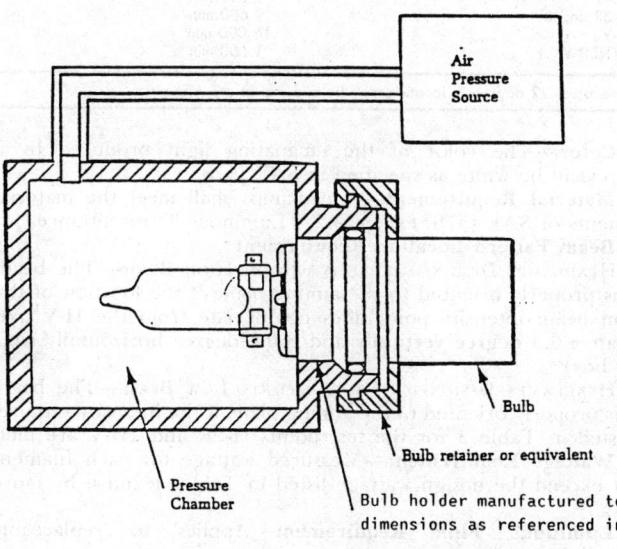

FIGURE 10—TEST FOR AIRTIGHT SEAL

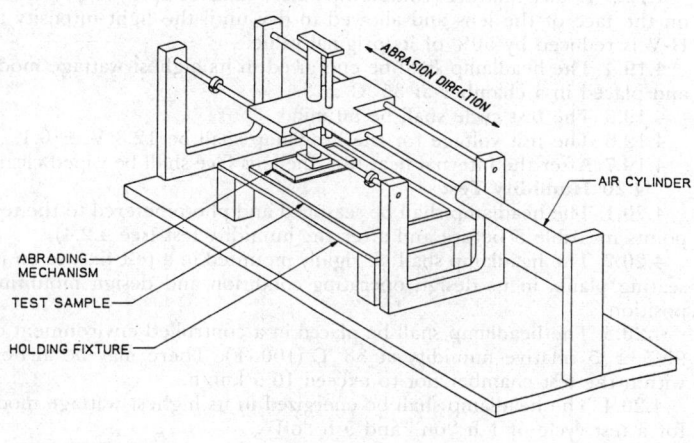

FIGURE 11—ABRASION TEST MACHINE

4.17.6 An abrasion cycle is one forward stroke 10 cm ± 2 and one rearward stroke of the same distance. The velocity of the abrading pad shall be 10 cm/s ± 2.

4.17.7 The test specimen shall be subjected to 20 abrasion cycles.

4.18 Thermal Cycle Test

4.18.1 The headlamp shall be seasoned and photometered to the test points in Table 3 before and after the thermal cycle test (see 4.2.4).

4.18.2 The headlamp shall be rigidly mounted in a test fixture on its seating plane in its design operating condition and design mounting position.

4.18.3 The headlamp shall be exposed to the thermal cycle profile shown in Figure 12.

4.18.4 Separate or single test chambers may be used to generate the temperature environment described by the thermal cycle.

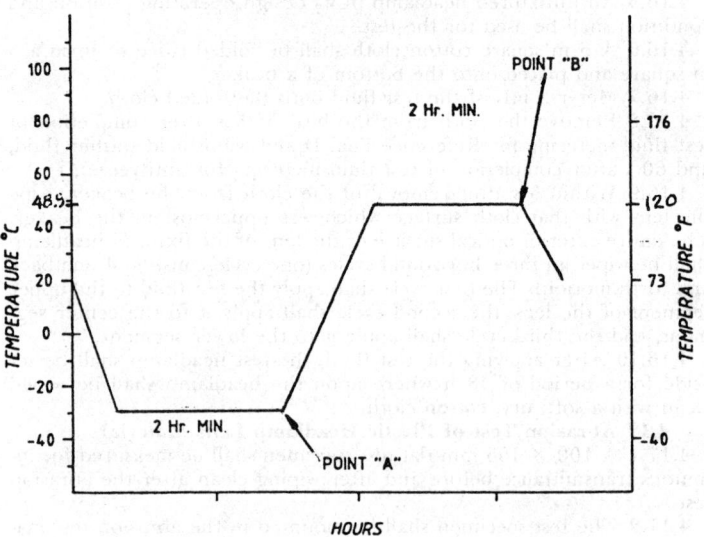

FIGURE 12—THERMAL CYCLE PROFILE

4.18.5 The headlamp shall be energized at 12.8 V ± 20 mV, its highest wattage mode commencing at point "A" of Figure 12 and de-energized at point "B" of each cycle.

4.18.6 The test period shall be 10 cycles of 8 h per cycle.

4.19 Internal Heat Test

4.19.1 The headlamp shall be seasoned and photometered to the test points in Table 3 before and after the internal heat test (see 4.2.4).

4.19.2 The headlamp shall be rigidly mounted in a test fixture on its seating plane in its design operating condition and design mounting position.

4.19.3 A dirt mixture, soluble in water, shall be sprayed uniformly on the face of the lens and allowed to dry until the light intensity at H-V is reduced by 50% of its original value.

4.19.4 The headlamp shall be energized in its highest wattage mode and placed in a chamber at 35 °C ± 3.

4.19.5 The test cycle shall be 30 min.

4.19.6 The test voltage for the headlamp shall be 12.8 V ± 0.1.

4.19.7 After the internal heat test, the lens face shall be wiped clean.

4.20 Humidity Test

4.20.1 The headlamp shall be seasoned and photometered to the test points in Table 3 before and after the humidity test (see 4.2.4).

4.20.2 The headlamp shall be rigidly mounted in a test fixture on its seating plane, in its design operating condition and design mounting position.

4.20.3 The headlamp shall be placed in a controlled environment of 95% ± 5 relative humidity at 38 °C (100 °F). There may be airflow within the test chamber not to exceed 16.5 km/h.

4.20.4 The headlamp shall be energized in its highest wattage mode for a test cycle of 1 h "on" and 5 h "off".

4.20.5 The test voltage for the headlamp shall be 12.8 V ± 0.1.

4.20.6 Test Duration—Eight Complete Cycles—The test is to end in the "off" cycle mode.

4.20.7 After completion of the eighth test cycle, the humidity shall be reduced in the test chamber to 30% ± 10 for 1 h. The headlamp shall be turned off during this period. During this period there may be a maximum air flow within the test chamber of 16.5 km/h when measured at the center of the chamber.

4.20.8 After a 1 h soak period, the headlamp shall be removed from the humidity chamber and photometered within 10 min ± 1.

4.21 Filament Rated Average Lab Life Test

4.21.1 The filament shall be energized at 14 V ± 0.1 V DC in a horizontal burning position in the standard enclosure shown in Figure 2.

4.21.2 The filament shall be unenergized 15 min for each 24 h of testing. The off time is not part of the test time of the filament being tested.

4.21.3 Each filament shall be tested separately and a different bulb shall be used for each filament.

4.21.4 The test may be terminated at 150% of design life.

5. Performance Requirements—A headlamp, when tested in accordance with the test procedures specified in Section 4, shall meet the following requirements.

5.1 Vibration Requirement (SAE J575)

5.1.1 The photometric values measured after the vibration test shall not vary more than ±10% from the values measured before the test.

5.1.2 There shall be no evidence of loose or broken parts, or intermittent electrical circuit.

5.2 Dust Requirement (SAE J575)—The photometric values measured after the dust test shall not vary by more than ±10% from the values measured before the test.

5.3 Corrosion Requirement (SAE J575)

5.3.1 The test headlamp shall show no evidence of exterior or internal corrosion or edge corrosion beyond 2 mm (0.08 in) from a sheared or cut edge.

5.3.2 The headlamp shall show no evidence of surface deterioration, fractures, color bleeding, or deterioration of bonding materials.

5.3.3 The photometric values measured after the corrosion test shall not vary more than ±10% from the values measured before the test.

5.4 Photometric Performance Requirement—Headlamps designed to meet the specifications of Table 3 shall meet the photometric requirements of Table 5.

TABLE 5—HEADLAMP PHOTOMETRIC PERFORMANCE REQUIREMENTS

Test Point[1]	Requirement, cd
Low Beam	
Type 2A1, 2B1, 2C1, 2D1, 2E1, or Equivalent	
10U-90U, 45'R-45'L	438 cd max permissible within 2 degree conical angle
1/2U—1-1/2L	1 100 max
1/2U—1R	3 240 max
1/2D—1-1/2R	6 400 min/24 000 max
1D—6L	600 min
High Beam	
Type 1A1 and 1C1, or Equivalent	
2U-V	800 min
H-3R and 3L	9 600 min
H-V	16 000 min
2-1/2D-V	1 600 min

[1] A tolerance of ±1/2 degree in location may be allowed at any test point.

5.5 Color—The color of the emanating light produced by a headlamp shall be white as specified in SAE J578.

5.6 Material Requirements—Headlamps shall meet the material requirements of SAE J576, except 4.2.1 Luminous Transmittance.

5.7 Beam Pattern Location Requirement

5.7.1 Headlamps Designed to be Aimed on High Beam—The beam pattern is properly oriented to the aiming plane if the location of the maximum beam intensity point does not deviate from the H-V axis more than ±0.5 degree vertically and ±0.8 degree horizontally (rectangular box).

5.7.2 Headlamps Designed to be Aimed on Low Beam—The beam pattern is properly oriented to the aiming plane if the intensity requirements listed in Table 3 for the test points H-2R and 1D-V are met.

5.8 Wattage Requirement—Measured wattage for each filament shall not exceed the design wattage listed in Tables 1 and 6 by more than 7.5%.

5.9 Luminous Flux Requirement—Applies to replaceable headlamp bulbs.

24.45

TABLE 6—TYPICAL HEADLAMPS

Headlamp Type and Identification Code[1]	Trade No.[2]	Design Watts at 12.8 V U.B.	Design Watts at 12.8 V L.B.	Design Life at 14 V[3] U.B.	Design Life at 14 V[3] L.B.	Max. Amps at 12.8 V U.B.	Max. Amps at 12.8 V L.B.	Size, mm	Dimensional Specs	Terminals No.
2C1	4000	37.5	60	200	320	3.14	5.02	146 Dia	Figure 24	3
2C1	4040[2]	37.5	60	200	320	3.14	5.02	146 Dia	Figure 24	3
2C1	H5006	35	35	200	320	2.94	2.94	146 Dia	Figure 24	3
1C1	4001	37.5		200		3.14		146 Dia	Figure 23	2
1C1	H4001	37.5		200		3.14		146 Dia	Figure 23	2
1C1	5001	50		200		4.20		146 Dia	Figure 23	2
1C1	H5001	50		200		4.20		146 Dia	Figure 23	2
2D1	6014	50	50	200	320	5.02	4.20	178 Dia	Figure 25	3
2D1	H6014	60	50	200	320	5.02	4.20	178 Dia	Figure 25	3
2D1	60152	60	50	200	320	5.02	4.20	178 Dia	Figure 25	3
2D1	60162	60	50	300	500	5.02	4.20	178 Dia	Figure 25	3
2D1	H6017	60	35	200	320	5.02	2.94	178 Dia	Figure 25	3
1A1	4651	50		200		4.20		100 X 165	Figure 20	2
1A1	H4651	50		200		4.20		100 X 165	Figure 20	2
2A1	4652	40	60	200	320	3.36	5.02	100 X 165	Figure 21	3
2A1	H4656	35	35	200	320	2.94	2.94	100 X 165	Figure 21	3
2A1	H4662	40	45	200	320	3.36	3.78	100 X 165	Figure 21	3
2A1	H4739	40	50	500	2000	3.36	4.20	100 X 165	Figure 21	3
2B1	6052	65	55	150	320	5.46	4.62	142 X 200	Figure 22	3
2B1	H6052	65	55	150	320	5.46	4.62	142 X 200	Figure 22	3
2B1	H6054	65	35	150	320	5.46	2.94	142 X 200	Figure 22	3
2E1	H4666	65	45	150	320	5.46	3.78	100 X 165	Figure 26	3
UF	H4701	65		150		5.46		92 X 150	Figure 27	2
LF	H4703	55		200		4.62		92 X 150	Figure 27	2
1G1		50	35	200	200	4.20	2.94	100 X 165	Figure 28	3
2G1		35	45	150	200	2.94	3.78	100 X 165	Figure 28	3
2H1		65	25	150	150	5.46	2.10	100 X 165	Figure 28	3
UJ		25		150		2.10		56 X 75	Figure 29	2
LJ		20			500	1.68		56 X 75	Figure 30	2
UK[4]	H4352	65		150		5.46		55 X 135	Figure 31	3
LK	H4351		55		500		4.62	55 X 135	Figure 32	2

[1] Headlamp identification codes are explained in 5.25.
[2] Heavy duty headlamps.
[3] All headlamps designs for 12.8 V usage are life tested at 14 V. In general, the life at vehicle voltage is longer.
[4] UK and LK photometric beam patterns are combined for high beam.

5.9.1 For bulbs with no opaque coating, the measured luminous flux shall be within ±12% of the design luminous flux listed in Table 1.

5.9.2 For bulbs with opaque coating, the measured luminous flux shall be within ±15% of the design luminous flux listed in Table 1.

5.10 Maintenance of Luminous Flux Requirement—When tested in accordance with 4.7—For samples from each lot tested, the average luminous flux value for single filament bulbs for each filament of two-filament bulbs after burning for 70% of design life shall be no less than 90% of the initial average luminous flux value.

5.11 Out-of-Focus Requirement—The headlamp shall meet the requirements of Table 5 for each of the out-of-focus test positions.

5.12 Impact Requirement—The headlamp shall show no evidence of broken, cracked, or chipped pieces of the headlamp, coating adhesion failure, or delamination of material, or visible loosening or breaking apart of headlamp parts.

5.13 Aiming Adjustment Requirement—When tested in accordance with 4.11, the headlamp shall meet the following requirements:

5.13.1 For headlamps with individual horizontal and vertical aim adjustments, tested in the laboratory, a minimum aiming adjustment of ±4.0 degrees shall be provided in the vertical plane and ±2.5 degrees in the horizontal plane.

5.13.2 On headlamp assemblies with independent vertical and horizontal aiming provision, the adjustments shall be such that when tested in the laboratory, neither the vertical nor horizontal aim shall deviate more than 100 mm from horizontal or vertical planes, respectively, at a distance of 7.6 m through an angle of ±4.0 degrees vertically and ±2.5 degrees horizontally.

5.13.3 On headlamps with integral aim tested in the laboratory, the headlamp shall be able to indicate variations in vertical aim within a range extending from 1.2 degrees above to at least 1.2 degrees below a longitudinal horizontal plane through the center of the headlamp system.

5.13.4 On headlamps with integral aim, photometric tests shall be performed with the vertical aiming system set to its specified design vertical aim, and with the headlamp assembly mounted to the test fixture in the same attitude as its design mounting position in the vehicle.

5.13.5 The self-locking devices used to hold aiming screws in position shall continue to operate satisfactorily for a minimum of 20 adjustments on each screw, over a length of screw thread of not less than 3 mm.

NOTE: 5.13.2 and 5.13.3 are not applicable to headlamps with ball and socket or equivalent adjusting means.

5.14 Inward Force Requirements—When subjected to the tests in 4.12, the headlamp shall meet the following requirements:

5.14.1 The headlamp shall not permanently recede by more than 2.5 mm.

5.14.2 The aim of the headlamp shall not permanently deviate by more than 3.2 mm at a distance of 7.6 m.

5.15 Torque Deflection Requirement—When subjected to the tests in 4.13, the difference between the two readings shall not exceed 0.30 degree.

5.16 Deflection Requirement—After the load application, the permanent deflection of the glass envelope of the bulb shall not exceed 0.13 mm.

5.17 Sealing Requirement—While the fixture and terminal end is submerged, no bubble(s) shall develop outside the test fixture.

5.18 Chemical Resistance Requirement

5.18.1 The exposed headlamp, when compared to an unexposed headlamp, shall not show surface deterioration, delamination, fractures, deterioration of bonding materials, color bleeding, or color pickup as a result of exposure to the test fluids.

5.18.2 The photometric values measured after the chemical resistance test shall not vary more than ±10% from the values measured before the test.

5.19 Abrasion of Plastic Headlamp Lens Material Requirements—The luminous transmittance of the abraded test specimen using CIE Illuminant A (2856D), shall show a maximum of 3% deterioration from the luminous transmittance of the unabraded control sample.

5.20 Thermal Cycle Requirement

5.20.1 The headlamp shall show no evidence of delamination, fractures, seal fractures, deterioration of bonding material, color bleeding, warp, or deforming.

5.20.2 The photometric values measured after the temperature cycle test shall not vary by more than ±10% from values measured before the test.

5.21 Internal Heat Requirement—The photometric values measured after the internal heat test shall not vary by more than ±10% from the values measured before the test.

5.22 Humidity Requirement

5.22.1 At the end of the 10 min test period (see 4.22), the headlamp shall be inspected immediately and show no evidence of condensed moisture or droplets inside the headlamp.

5.22.2 The headlamp shall show no evidence of delamination, bonding, material deterioration, or seal failure.

5.22.3 The photometric values measured after the humidity test shall not vary by more than ±10% from the values measured before the test.

5.23 Retaining Ring Requirements

5.23.1 Positive means shall be provided for holding the headlamp to the mounting ring.

5.23.2 The fastening means shall be capable of holding the headlamp securely in its proper position at the end of 20 replacements.

5.23.3 When a headlamp having a flange thickness (as shown in Table 7) is secured between the retaining ring and mounting ring, there shall be no evidence of looseness:

TABLE 7—FLANGE THICKNESS

Headlamp Type	Flange Thickness
146 mm	11.7 mm
178 mm	11.7 mm
100 × 165 mm	33.9 mm
142 × 200 mm	10.1 mm
92 × 150 mm	9.6 mm
55 × 135 mm	9.6 mm
56 × 75	3.6 mm

5.24 Design Requirements

5.24.1 Dimensions of sealed beam headlamp mounting-sealed beam headlamp mounting rings and retaining rings shall meet the dimensions marked "I" in the following figures to assure compatibility with the corresponding types of units.

Type 1A1—Figure 13
Type 2A1—Figure 13
Type 2B1—Figure 14
Type 1C1—Figure 15
Type 2C1—Figure 15
Type 2D1—Figure 16
Type 2E1—Figure 13
Type UF —Figure 17
Type LF —Figure 17
Type UK —Figure 18
Type LK —Figure 18

5.24.2 DIMENSIONS OF SEALED BEAM HEADLAMPS—Sealed beam headlamps shall meet the dimensions marked "I" in the following figures to assure interchangeability with other sealed beam headlamps of the same type.

Type 1A1—Figure 20
Type 2A1—Figure 21
Type 2B1—Figure 22
Type 1C1—Figure 23
Type 2C1—Figure 24
Type 2D1—Figure 25
Type 2E1—Figure 26
Type UF —Figure 27
Type LF —Figure 27
Type 1G1—Figure 28
Type 2G1—Figure 28
Type 2H1—Figure 28
Type UJ —Figure 29
Type UK —Figure 31
Type LK —Figure 32

5.24.3 DIMENSIONS FOR MECHANICAL AIMING OF HEADLAMPS—Headlamps shall meet the following requirements to assure compatibility with mechanical aimers.

5.24.3.1 Type 1C1, 2C1, and 2D1 headlamps shall have no raised letters or embossing on the outside surface of the lens between the diameters of 40 and 90 mm about the lens center.

5.24.3.2 Type 1A1, 2A1, 2B1, 2E1, UF, LF, 1G1, 2G1, and 2H1 headlamps shall have no raised letters or embossing on the outside surface of the lens within a diameter of 70 mm about the lens center.

5.24.3.3 Aiming pad design may vary, but shall meet limiting dimensions as shown on the figures specified in 5.24.1 and 5.24.2.

5.24.4 HEADLAMP MOUNTING ASSEMBLY—The headlamp mounting assembly shall meet the requirements of Figure 13, Dimensions of Sealed Beam Headlamp Mounting.

5.24.5 AIMER COMPATIBILITY—Headlamps which do not incorporate integral headlamp aim shall be designed and installed so that they may be inspected and aimed by mechanical aimers as specified in SAE J602 without the removal of any ornamental trim rings or other parts.

5.24.6 Bulbs and bulb holders shall meet the requirements referenced in Table 2 to ensure interchangeability.

5.24.7 Accurate rated bulbs shall meet the dimensional requirements shown in each applicable figure of Table 2.

5.24.8 Typical replaceable headlamp bulbs are listed in Table 1.

6. Guidelines

6.1 When in use, a headlamp shall not have any styling ornament or other feature, such as a glass cover or grille, in front of the lens.

6.2 Photometric Design Guidelines—Guidelines for the photometric design of headlamps are shown in Table 3.

6.3 Dimensional Guidelines—Guidelines for dimensions are shown in the following figures:

6.3.1 MOUNTING AND RETAINING RINGS—Figures 13 to 18
6.3.2 SEALED BEAM HEADLAMPS—See Table 6 and Figures 20 to 32
6.3.3 REPLACEABLE BULBS—See Table 2 and Figures 33 to 70
6.3.4 REFLECTOR BULB MOUNTING HOLE FOR REPLACEABLE BULBS—See Table 2.

6.4 Filament Rated Average Lab Life Guideline—Rated average lab life shall approximate design life. The design life for the filament(s) of each bulb type is shown in Table 1 or 5.

6.5 Replaceable Bulb Filament End Coil Definition—Shown in Figure 19.

6.6 Summary of Requirements and Guidelines—Table 8 summarizes the classification of the various sections of this report into requirements and guidelines.

6.7 Fixed Horizontal Aim Guideline—When horizontal aim adjusting screws are provided on fixed horizontal aim headlamps, they shall be of a tamperproof design or shall be difficult to access.

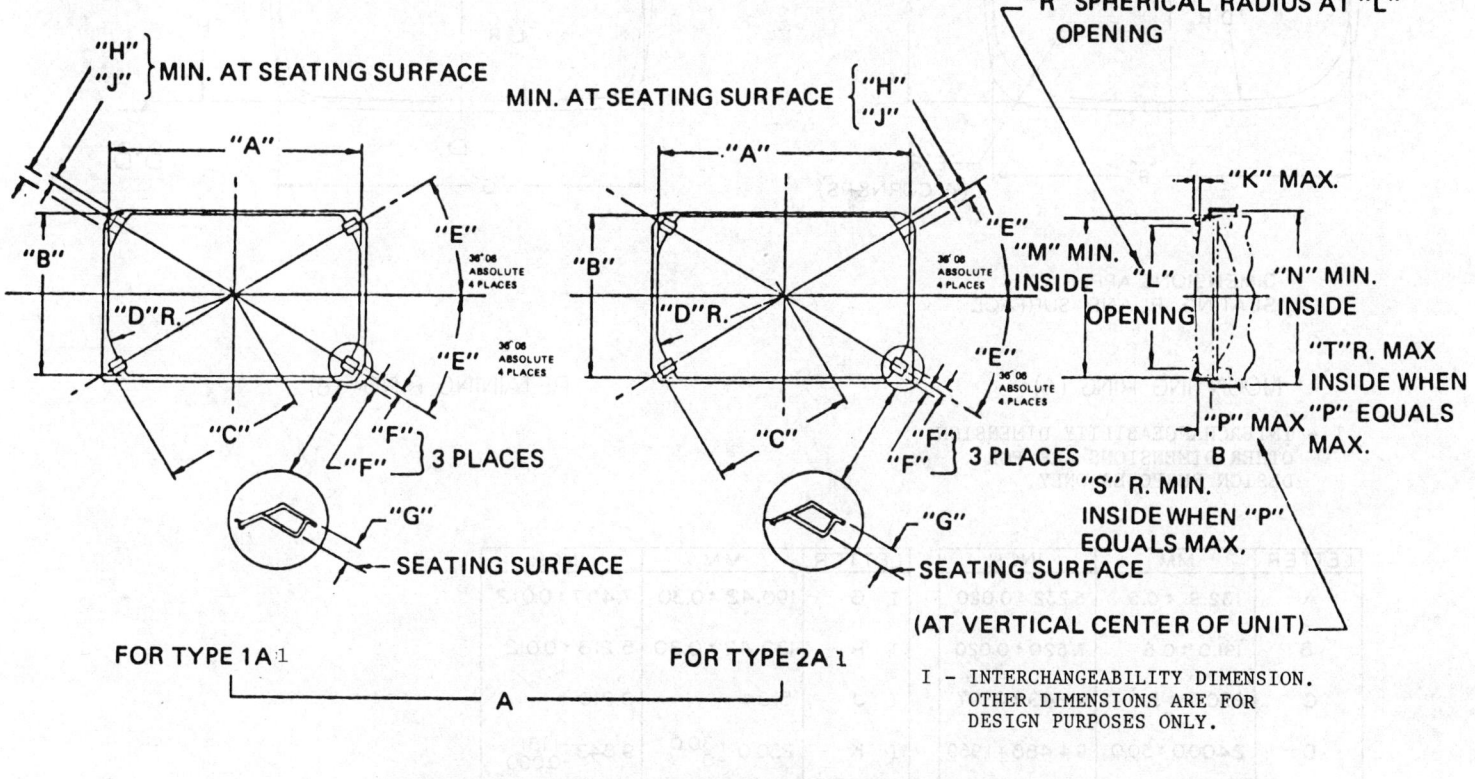

FIGURE 13—(A) FRONT VIEW OF SLOTS OR NOTCHES FOR 100 × 165 mm RECTANGULAR HEADLAMP MOUNTING RING OR LAMP BODY; (B) RECTANGULAR HEADLAMP RETAINING RING

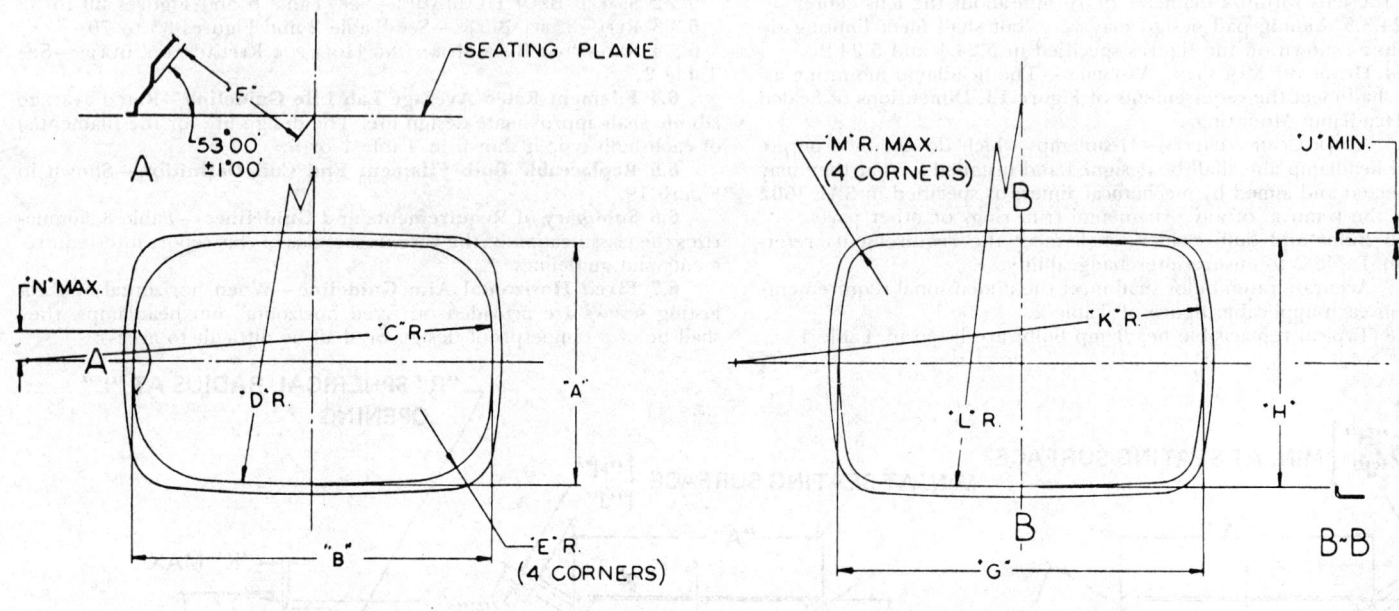

DIMENSIONS APPLY AT SEATING PLANE SURFACE

MOUNTING RING (A) RETAINING RING (B)

I - INTERCHANGEABILITY DIMENSION. OTHER DIMENSIONS ARE FOR DESIGN PURPOSES ONLY.

LETTER	MM	INCH
A	132.9 ± 0.5	5.232 ± 0.020
B	191.0 ± 0.5	7.520 ± 0.020
C	250.0 ± 5.0	9.843 ± 0.197
D	2400.0 ± 50.0	94.488 ± 1.969
E	41.0 ± 2.0	1.614 ± 0.079
F	79.90 ± 0.40	3.146 ± 0.016

	LETTER	MM	INCH
I	G	190.42 ± 0.30	7.497 ± 0.012
I	H	132.42 ± 0.30	5.213 ± 0.012
I	J	5.34	0.210
I	K	250.0 +30.0/-0	9.843 +1.181/-0.000
I	L	2402.0 +2250.0/-0	94.567 +88.583/-0.000
I	M	20.4	0.803
	N	19.0	0.748

FIGURE 14—(A) FRONT VIEW OF MOUNTING RING OR LAMP BODY FOR 142 × 200 mm RECTANGULAR HEADLAMP; (B) RETAINING RING

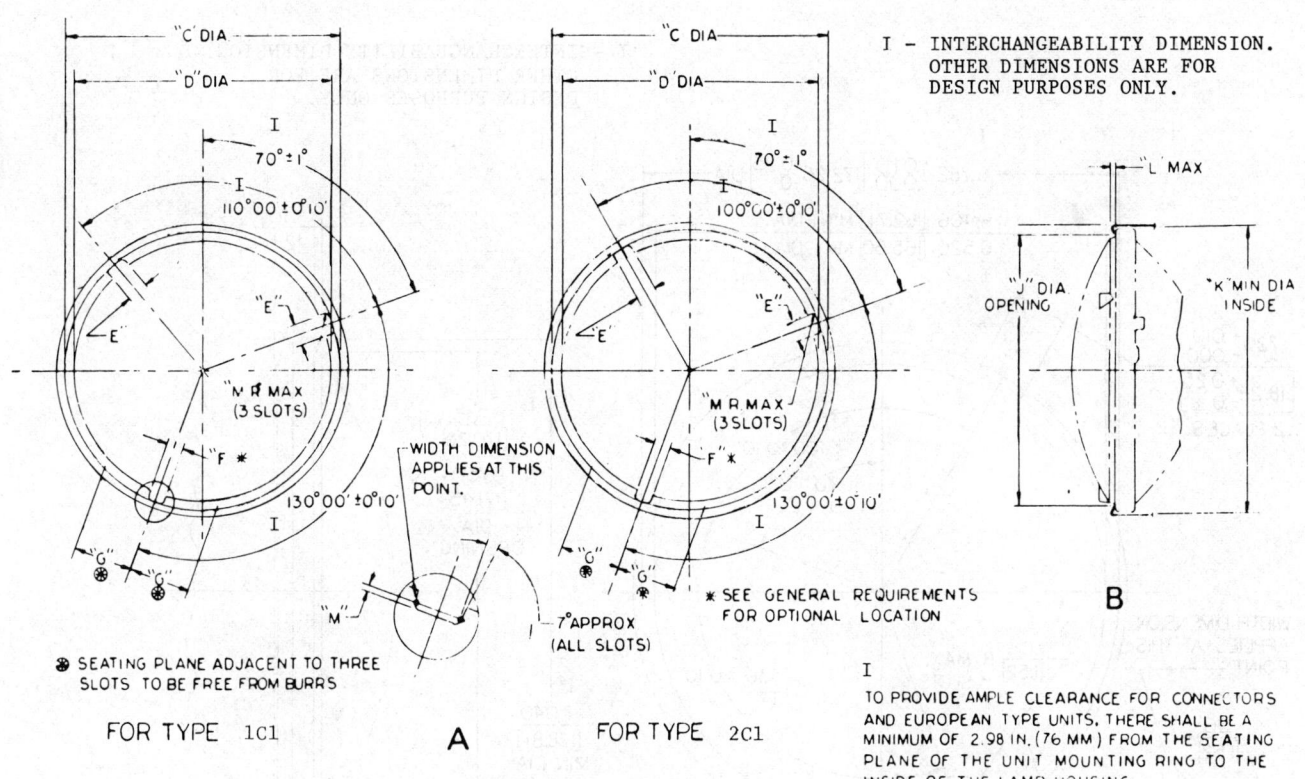

Letter		in	mm	Letter		in	mm
I	C	5.450 +0.010/-0.000	138.43 +0.25/-0		G	1.20	30.48
I	D	5.250 – 5.140	133.35 – 130.55	I	J	5.400 – 5.360	137.16 – 136.14
I	E	0.410 +0.010/-0.000	10.41 +0.25/-0	I	K	5.710	145.03
I	F	0.330 +0.005/-0.000	8.38 +0.12/-0		L	0.100	2.54
					M	0.06	1.52

FIGURE 15—(A) FRONT VIEW OF SLOTS OR NOTCHES FOR 146 mm DIAMETER HEADLAMP MOUNTING RING OR LAMP BODY; (B) 146 mm HEADLAMP RETAINING RING

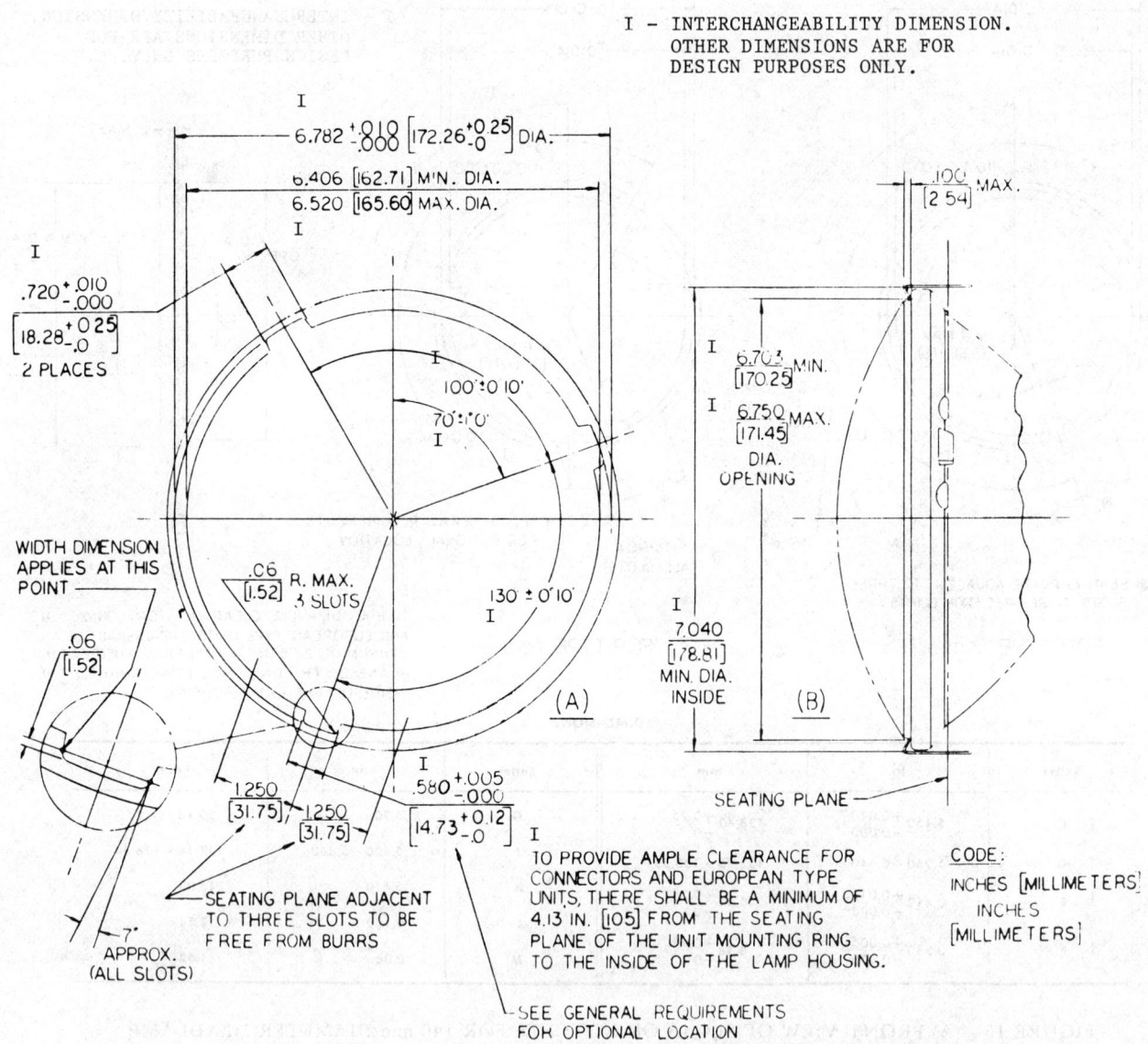

FIGURE 16—(A) FRONT VIEW OF SLOT OR NOTCHES FOR 178 mm DIAMETER HEADLAMP MOUNTING RING OR LAMP BODY; (B) 178 mm HEADLAMP RETAINING RING

24.51

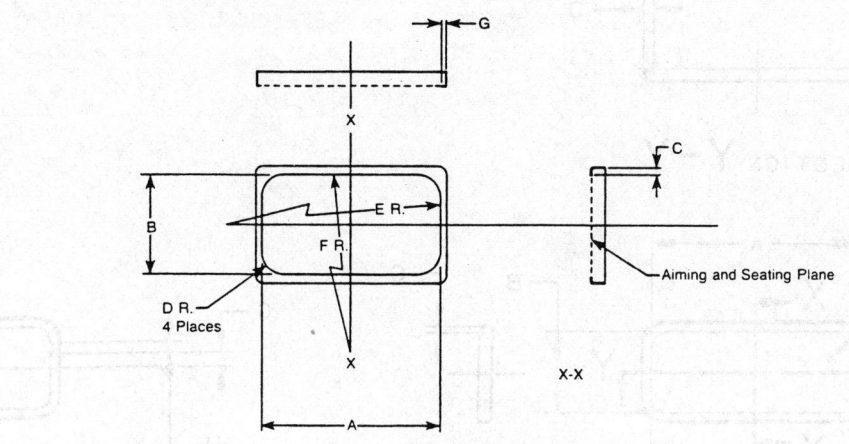

Aiming Ring

LETTER	INCH	MM
A	5.721 ± .006	145.30 ± 0.30
B	3.284 ± .006	83.40 ± 0.30
C	.213 MIN.	5.40 MIN.
D	.670 MAX.	17.00 MAX.
E	23.7 ± 2.0	602.2 ± 50.0
F	63.0 ± 3.93	1600.0 ± 100.5
G	.134 MIN.	3.40 MIN.

FIGURE 17—AIMING/SEATING RING FOR TYPE LF AND UF RECTANGULAR SEALED BEAM HEADLAMP UNITS

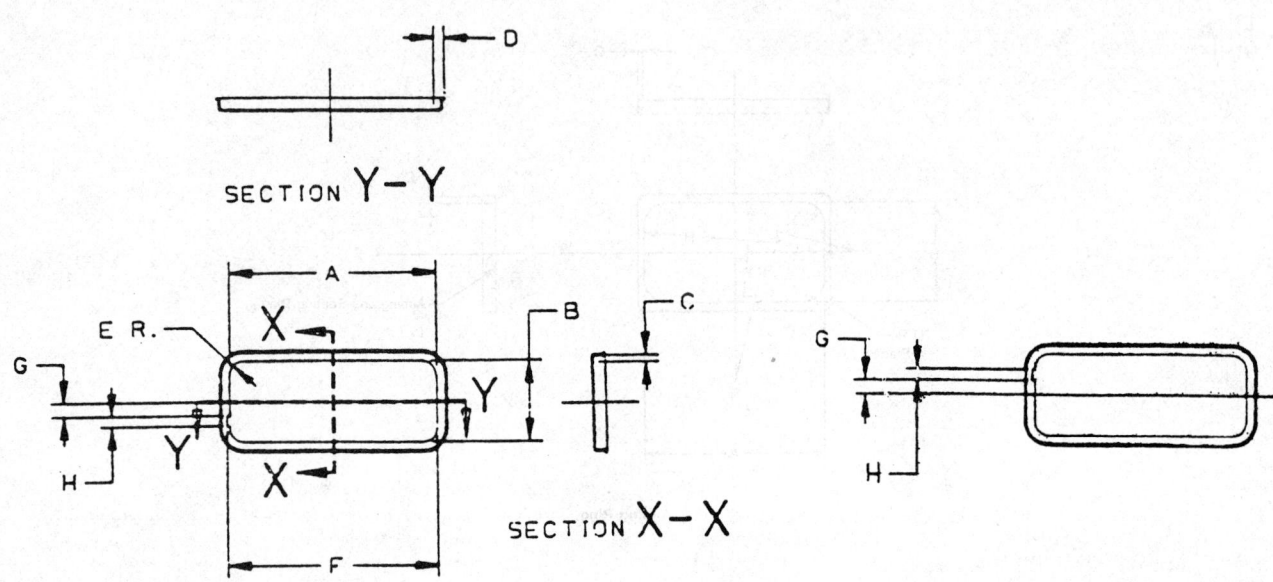

HIGH BEAM

LETTER	INCH	MM
A	5.38 ±.010	136.7 ±0.30
B	2.25 ±.010	57.0 ±0.30
C	.154 MIN	3.9 MIN
D	.197 MIN	5.0 MIN
E	.26 ±.010	6.5 ±0.30
F	5.47 ±.02	139.0 ±0.30
G	.33 ±.02	8.5 ±0.5
H	.39 ±.02	10.0 ±0.5

LOW BEAM

NOTE: SAME AS HIGH BEAM EXCEPT AS SHOWN

LETTER	INCH	MM
G	.33 ±.02	8.5 ±0.5
H	.39 ±.02	10.0 ±0.5

FIGURE 18—AIMING RING—55 × 135 UK/LK

The following guideline is intended to depict the current methods which are being used by the lighting industry for identifying the end of filaments. It is not possible to predict every filament leg configuration. When the filament legs are in some other configuration than those shown below, a guideline is: The end turns of the filament are defined as being the first luminous turn and the last luminous turn that are substantially at the correct helix angle. The ends of the filament would then be the beginning of the first turn and conclusion of the last turn that are substantially at the correct helix angle.

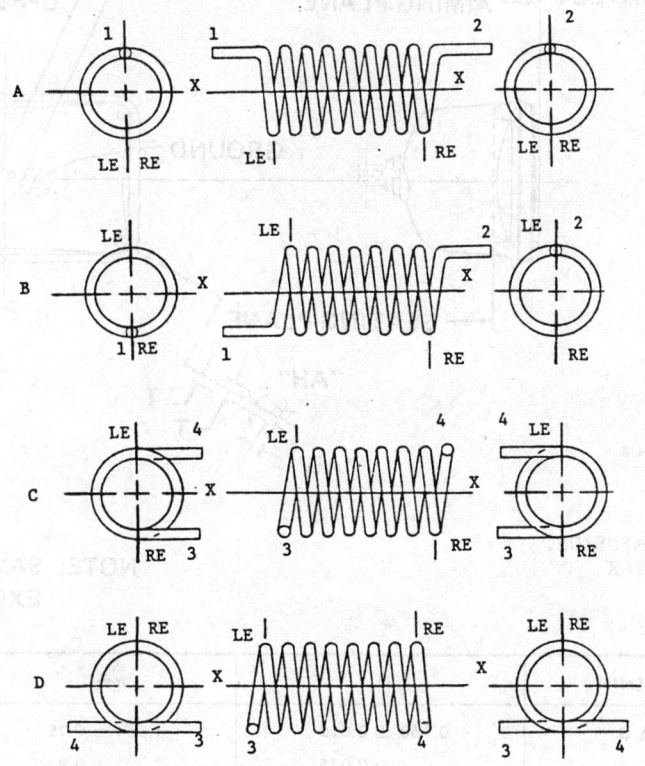

X-X axis of the filament
   LE - Left end of filament
   RE - Right end of filament

Filament configuration A and B:
   LE - is 180 degrees around circumference on the first turn from (1) leg of the filament, when looking parallel to X-X

   RE - is 180 degrees around circumference on the first turn from (2) leg of the filament, when looking parallel to X-X

Filament configuration C and D:

   LE - is 180 degrees around circumference on the first turn from (3) centerline of filament leg, when looking parallel to X-X

   RE - is 180 degrees around circumference on the first turn from (4) centerline of filament leg, when looking parallel to X-X

FIGURE 19—GUIDELINE REPLACEABLE BULB FILAMENT END COIL DEFINITION

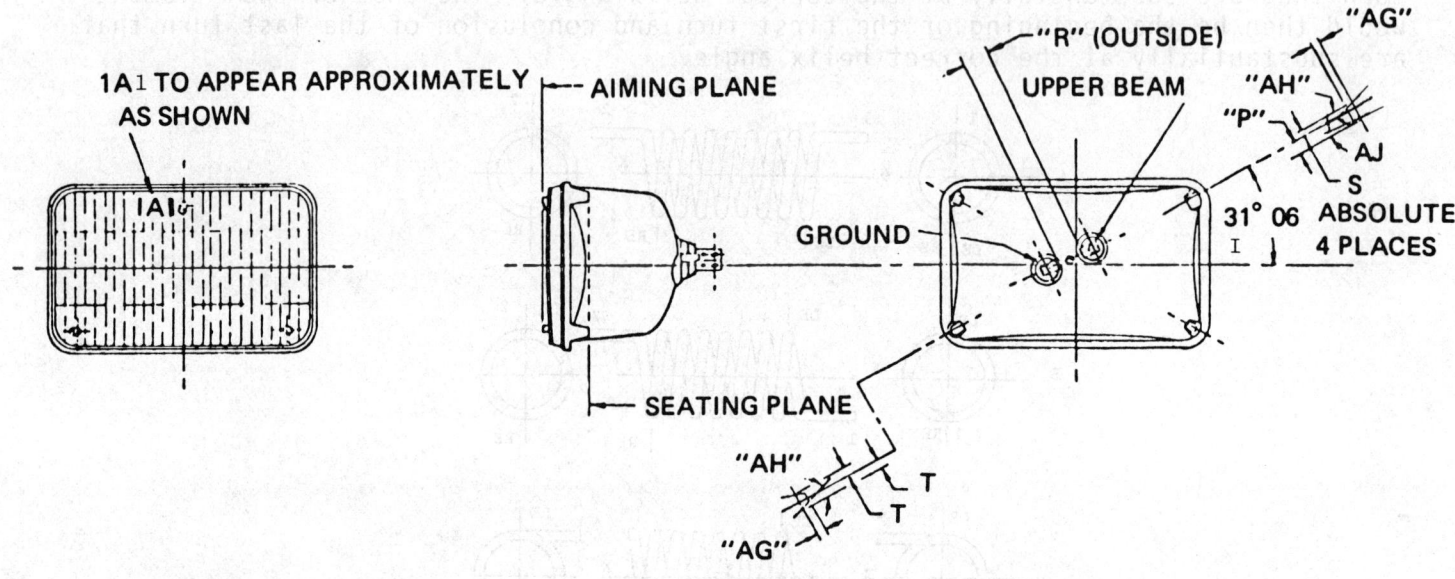

Letter		in	mm
	AG	0.160 ± 0.010	4.064 ± 0.25
I	R	0.669 +0.035 / −0.000	17.0 +0.9 / −0
	AH	15°00′ ± 3°00′	
I	T	0.167 ± 0.0100	4.24 ± 0.25
	AJ	4°20′ ± 1°00′	
I	P	0.313 +0.015 / −0.010	7.95 +0.38 / −0.25
I	S	0.122 +0.015 / −0.010	3.10 +0.38 / −0.25

FIGURE 20—TYPE 1A1 HEADLAMP 100 × 165 mm RECTANGULAR

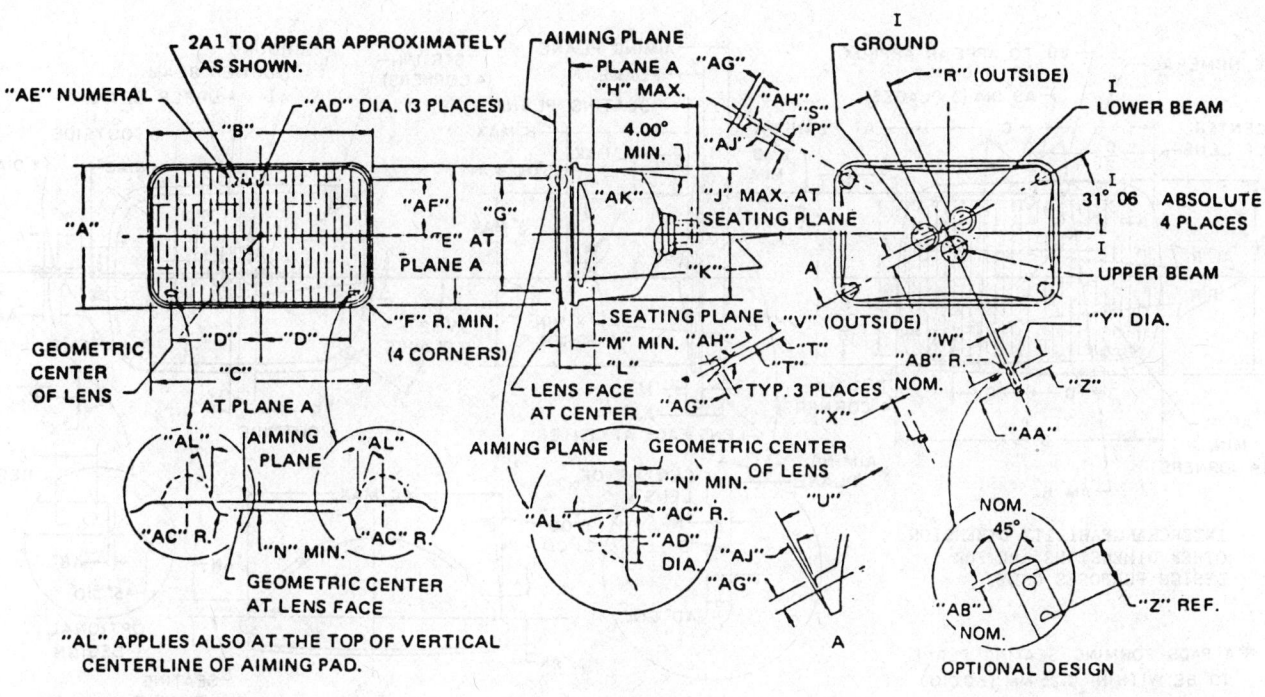

Letter	in	mm	Letter	in	mm	Letter	in	mm
I A	4.200 +0.030/−0.170	106.68 +0.76/−4.32	I P	0.313 +0.015/−0.010	7.95 +0.38/−0.25	AA	0.535 +0.000/−0.071	13.58 +0/−1.80
I B	6.580 +0.030/−0.170	167.13 +0.76/−4.32	I R	0.669 +0.035/−0.000	17.02 +0.88/−0.00	AB	0.060 ± 0.020	1.5 ± 0.5
I C	6.440 ± 0.030	163.58 ± 0.76	I S	0.122 ± 0.015	3.10 +0.38/−0.25	AC	0.060 ± 0.020	1.5 ± 0.5
D	2.700 ± 0.020	68.58 ± 0.51	I T	0.167 ± 0.010	4.24 ± 0.25	AD	0.200 ± 0.010	5.08 ± 0.25
I E	4.060 ± 0.030	103.12 ± 0.76	I U	3.640 ± 0.010	92.47 ± 0.25	AE	0.250 ± 0.030	6.35 ± 0.76
I F	0.540	13.71	I V	0.335 +0.020/−0.000	8.5 +0.5/−0	AF	1.660 ± 0.010	42.16 ± 0.25
G	3.320 ± 0.030	84.33 ± 0.76				AG	0.160 ± 0.010	4.06 ± 0.25
I H	3.350	85.09	I W	0.304 +0.016/−0.000	7.72 +0.40/−0.00	AH	15° max	
I J	4.01	101.85	I X	0.030 ± 0.002	0.76 ± 0.05	AJ	3.33° min	
K	50.000 +0.500/−2.00	1270.0 +13.0/−50.8	Y	0.120 +0.010/−0.000	3.05 +0.25/−0	AK	1.56° max	39.6 max
I L	1.375 ± 0.040	34.93 ± 1.02	Z	0.345 +0.059/−0.000	8.76 +1.50/−0	AL	16° max	
M	0.420	10.68						
N	0.020	0.51						

I – INTERCHANGEABILITY DIMENSION. OTHER DIMENSIONS ARE FOR DESIGN PURPOSES ONLY.

FIGURE 21—TYPE 2A1 HEADLAMP 100 × 165 mm RECTANGULAR

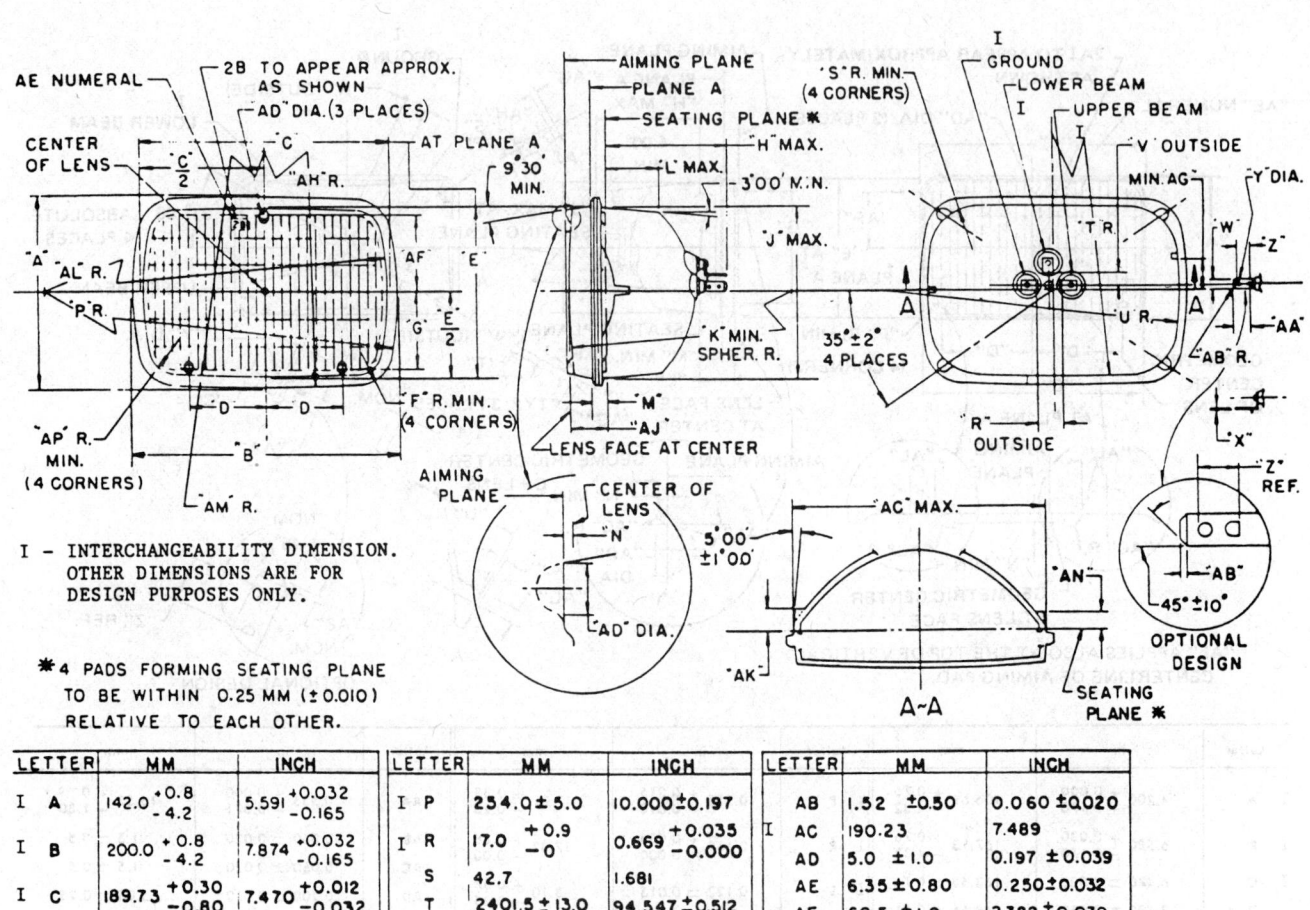

LETTER	MM	INCH	LETTER	MM	INCH	LETTER	MM	INCH
I A	142.0 +0.8 / -4.2	5.591 +0.032 / -0.165	I P	254.0 ±5.0	10.000±0.197	AB	1.52 ±0.50	0.060 ±0.020
I B	200.0 +0.8 / -4.2	7.874 +0.032 / -0.165	I R	17.0 +0.9 / -0	0.669 +0.035 / -0.000	I AC	190.23	7.489
I C	189.73 +0.30 / -0.80	7.470 +0.012 / -0.032	S	42.7	1.681	AD	5.0 ±1.0	0.197 ±0.039
D	64.0 ±1.0	2.520±0.039	T	2401.5 ±13.0	94.547 ±0.512	AE	6.35±0.80	0.250±0.032
I E	131.73 +0.30 / -0.80	5.186 +0.012 / -0.032	U	249.0 ±5.0	9.803 ±0.197	AF	60.5 ±1.0	2.382 ±0.039
I F	25.5	1.004	I V	8.5 +0.5 / -0	0.335 +0.020 / -0.000	I AG	19.7	0.776
G	59.6 ±1.0	2.346±0.039	I W	7.72 +0.40 / -0	0.304 +0.016 / -0.000	I AH	2406.5 ±13.0	94.744±0.512
I H	107.0	4.213	I X	0.76 ±0.05	0.030±0.002	AJ	26.7 +4.0 / -1.0	1.051 +0.157 / -0.039
I J	132.23	5.206	I Y	3.05 +0.25 / -0	0.120 +0.010 / -0.000	AK	16.0 +2.0 / -1.0	0.630 +0.079 / -0.039
K	1200.0	47.244	Z	8.76 +1.50 / -0	0.345 +0.059 / -0.000	AL	250.0 +0 / -25.0	9.843 +0.000 / -0.984
L	49.0	1.929	AA	13.58 +0 / -1.80	0.535 +0.000 / -0.071	AM	2402.0 +0 / -775.0	94.567 +0.000 / -30.512
I M	11.1 ±1.0	0.437 ±0.039				AN	12.0 +2.0 / -1.0	0.472 +0.079 / -0.039
N	0.5 +4.0 / -0.5	0.020 +0.157 / -0.020				AP	20.4	0.803

FIGURE 22—TYPE 2B1 HEADLAMP 142 × 200 mm RECTANGULAR

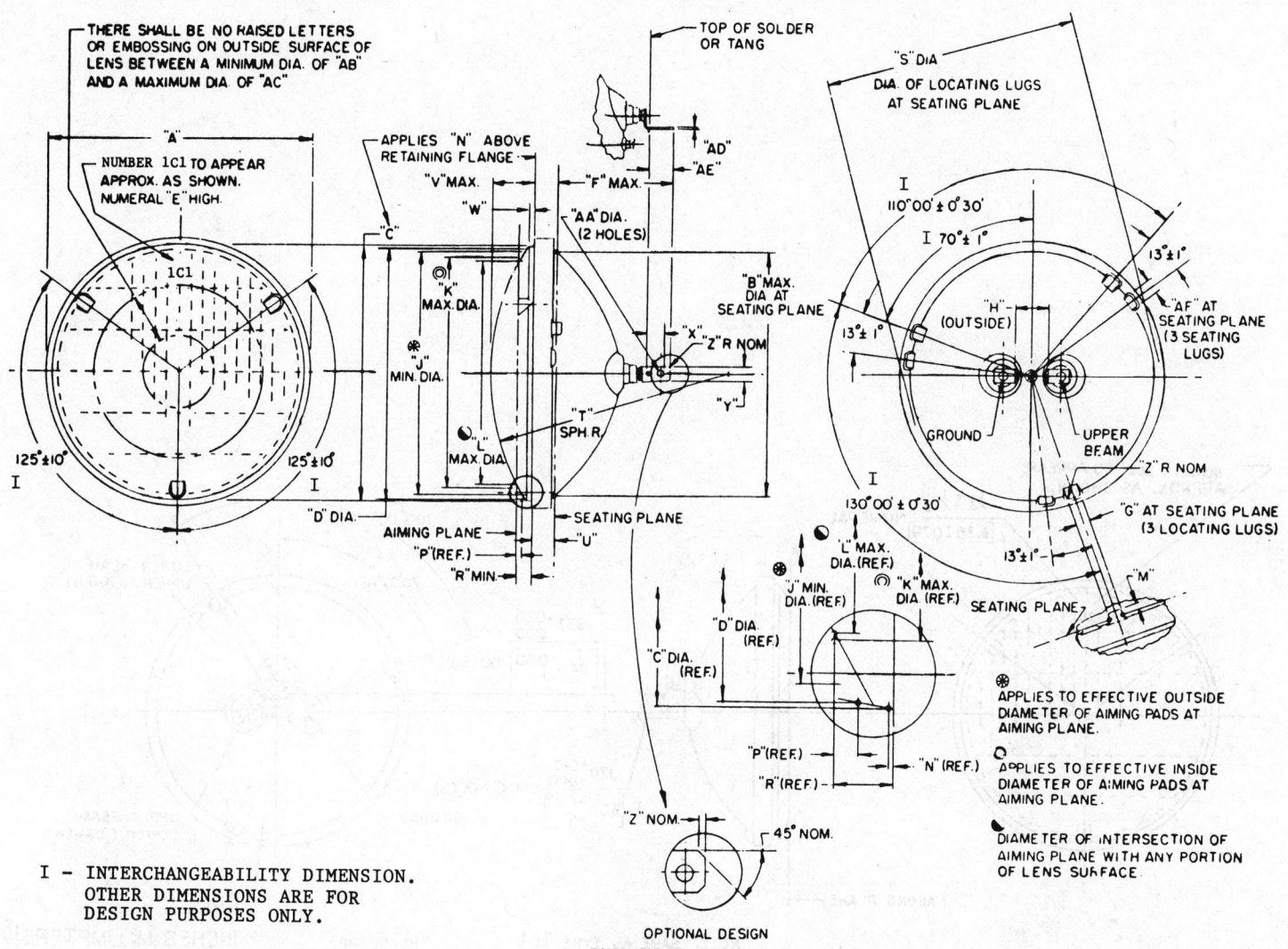

I - INTERCHANGEABILITY DIMENSION. OTHER DIMENSIONS ARE FOR DESIGN PURPOSES ONLY.

DIMENSIONS

Letter	in	mm	Letter	in	mm
I A	5.700 +0.000/−0.100	144.78 +0/−2.54	T	5.06 ±0.12	128.52 ±3.04
I B	5.120	130.04	I U	0.500 ±0.040	12.70 ±1.01
I C	5.355 +0.000/−0.030	136.01 +0/−0.76	V	0.92	23.36
D	5.280−5.340	134.11−135.63	W	0.078 +0.062/−0.000	1.98 +1.57/−0
E	1/4 ±1/32	6.35 ±0.79	I X	0.345 +0.060/−0.000	8.76 +1.52/−0
I F	2.60	66.04	I Y	0.304 +0.016/−0.000	7.72 +0.40/−0
I G	0.312 ±0.010	7.92 ±0.25	Z	0.06	1.52
I H	0.670 +0.035/−0.000	17.01 +0.88/−0	I AA	0.120 +0.010/−0.000	3.04 +0.25/−0
I J	5.060	128.52	AB	1.50	38.10
K	4.57	116.07	AC	3.60	91.44
L	4.53	115.06	I AD	0.030 ±0.002	0.76 ±0.05
I M	0.100 +0.050/−0.000	2.54 +1.27/−0	AE	0.535 +0.000/−0.070	13.58 +0/−1.77
N	0.030	0.76	AF	0.31 ±0.12	7.87 ±3.04
P	0.165	4.19			
I R	0.320	8.12			
I S	5.440 +0.000/−0.040	138.17 +0/−1.01			

FIGURE 23—TYPE 1C1 HEADLAMP 146 mm DIAMETER

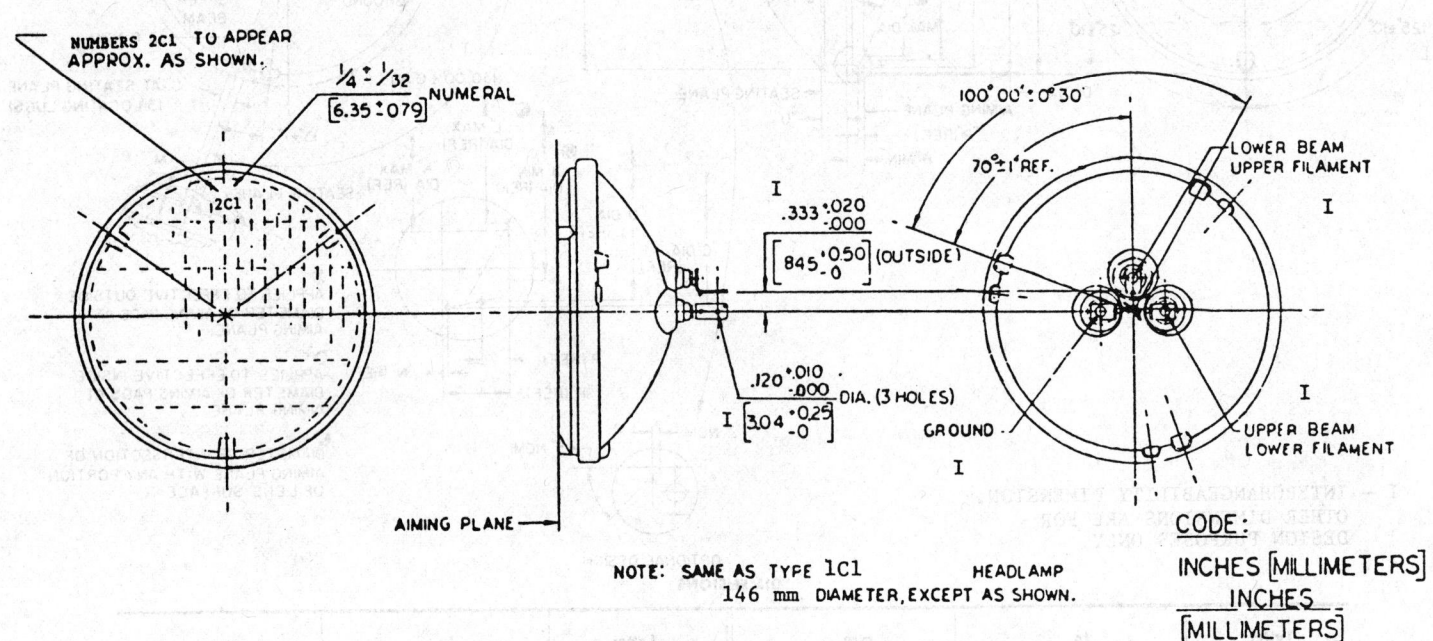

FIGURE 24—TYPE 2C1 HEADLAMP 146 mm DIAMETER

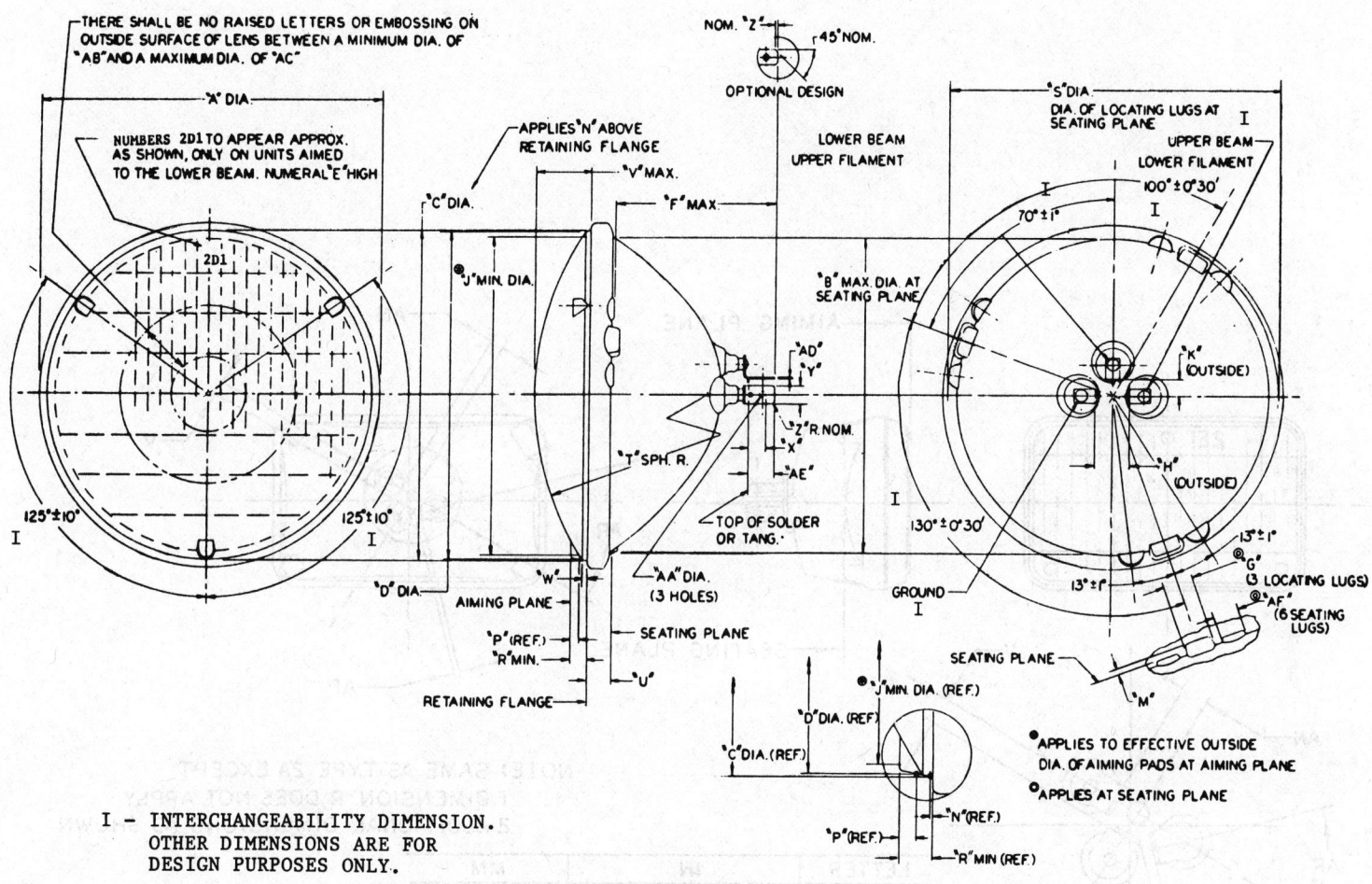

I - INTERCHANGEABILITY DIMENSION. OTHER DIMENSIONS ARE FOR DESIGN PURPOSES ONLY.

DIMENSIONS

	Letter	in	mm		Letter	in	mm
I	A	7.031 +0.000/−0.109	178.58 +0/−2.76	I	S	6.770 +0.000/−0.040	171.95 +0/−1.01
I	B	6.380	162.05		T	6.000 +0.250/−0.000	152.40 +6.35/−0
I	C	6.687 +0.000/−0.030	169.84 +0/−0.76	I	U	0.500 ±0.040	12.70 ±1.01
	D	6.595 − 6.675	167.52 − 169.54		V	1.150	29.21
	E	3/8 ±1/32	9.52 ±0.79		W	0.078 +0.062/−0.000	1.98 +1.57/−0
I	F	3.500	88.90	I	X	0.345 +0.060/−0.000	8.76 +1.52/−0
I	G	0.575 +0.000/−0.025	14.60 +0/−0.63	I	Y	0.304 +0.016/−0.000	7.72 +0.40/−0
I	H	0.670 +0.035/−0.000	17.01 +0.88/−0		Z	0.06	1.52
I	J	6.450	163.83	I	AA	0.120 +0.010/−0.000	3.04 +0.25/−0
I	K	0.333 +0.020/−0.000	8.45 +0.50/−0		AB	1.50	38.10
					AC	3.60	91.44
I	M	0.106 +0.100/−0.000	2.69 +2.54/−0	I	AD	0.030 ±0.002	0.76 ±0.05
	N	0.030	0.76		AE	0.535 +0.000/−0.070	13.58 +0/−1.77
	P	0.180	4.57		AF	0.50 ±0.25	12.70 ±6.35
I	R	0.350	8.89				

FIGURE 25—TYPE 2D1 HEADLAMP 178 mm DIAMETER

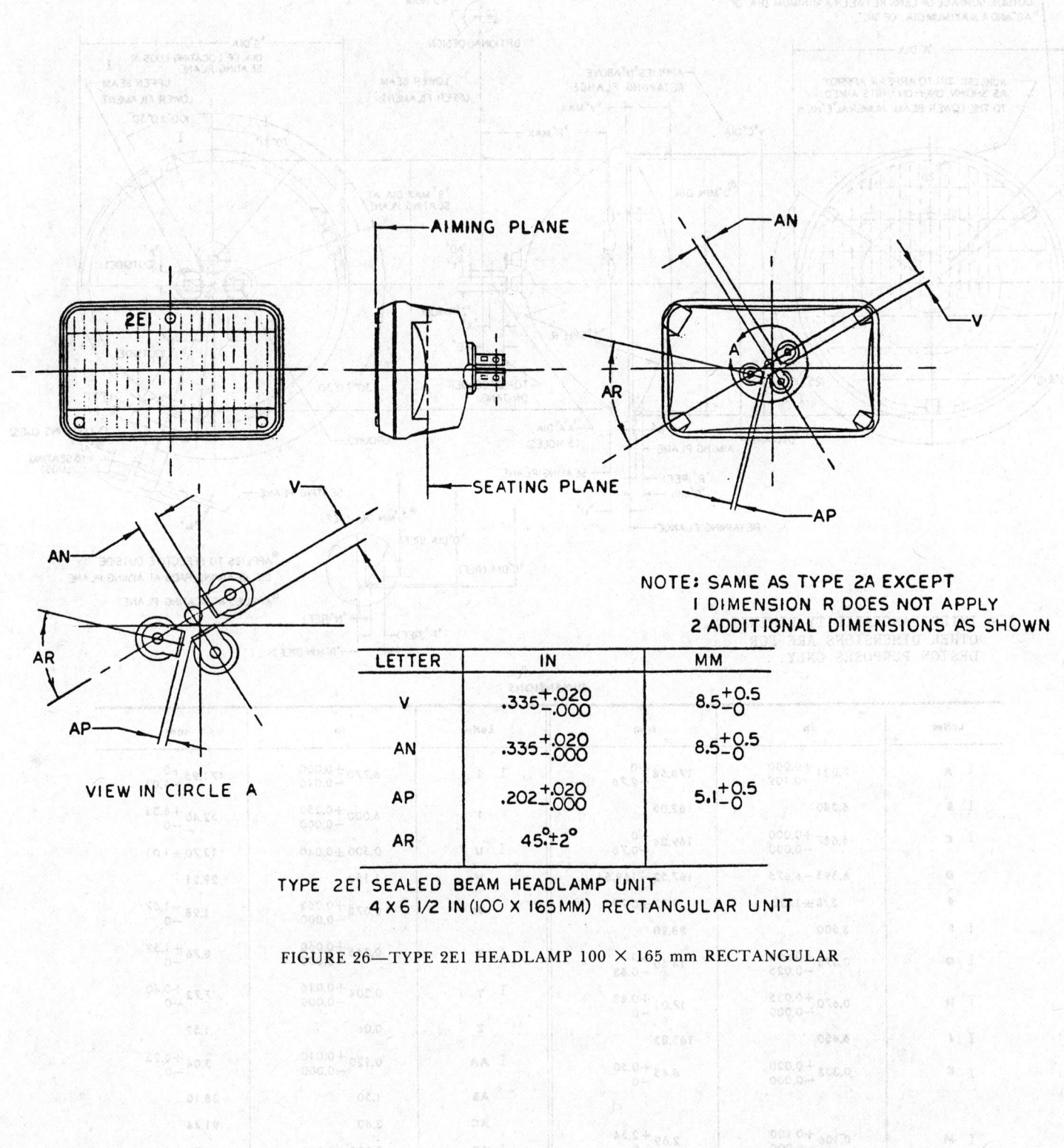

FIGURE 26—TYPE 2E1 HEADLAMP 100 × 165 mm RECTANGULAR

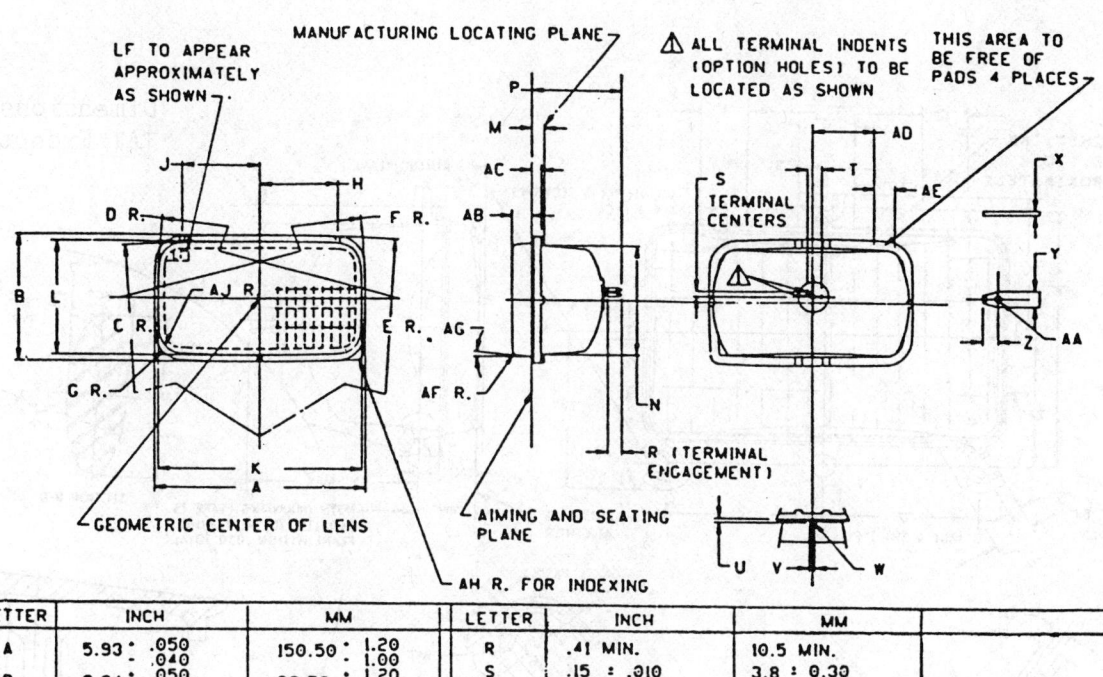

LETTER	INCH	MM	LETTER	INCH	MM	
A	5.93 +.050/-.040	150.50 +1.20/-1.00	R	.41 MIN.	10.5 MIN.	
B	3.64 +.050/-.040	92.50 +1.20/-1.00	S	.15 ± .010	3.8 ± 0.30	
C	63.0 ± 3.94	1600.0 ± 100.0	T	.41 ± .010	10.43 ± 0.30	
D	23.6 ± 1.97	600.0 ± 50.0	U	.024 MIN.	0.60 MIN.	TYPE LF
E	63.0 ± 3.94	1600.0 ± 100.0	V	.315 MAX.	8.0 MAX.	RECTANGULAR
F	23.8 ± 1.97	600.0 ± 50.0	W	RADIUS	RADIUS	SEALED BEAM
G	.787 ± .010	20.00 ± 0.30	X	.032 ± .002	0.82 ± 0.04	HEADLAMP UNIT
H	2.16 ± .010	55.0 ± 0.30	Y	.110 ± .004	2.80 ± 0.10	
J	2.16 ± .010	55.0 ± 0.30	Z	.104 ± .010	2.65 ± 0.30	
K	5.689 +.008/-.040	144.50 +0.20/-1.00	AA	.051 ± .010 DIA.	1.30 ± 0.30 DIA.	
L	3.252 +.008/-.040	82.60 +0.20/-1.00	AB	.56 ± .020	14.3 ± 0.50	
M	.46 MAX.	11.7 MAX.	AC	.295 MAX.	7.50 MAX.	
N	3.19 MAX.	81.0 MAX.	AD	1.77	45.0	
P	2.87 MAX.	73.0 MAX.	AE	.63	16.0	
			AF	.13 ± .02	3.2 ± 0.5	
			AG	5° ± 1°	5° ± 1°	
			AH	.24 ± .02	6.0 ± 0.5	
			AJ	.63 MIN.	16.0 MIN.	

TYPE UF
RECTANGULAR SEALED BEAM HEADLAMP UNIT

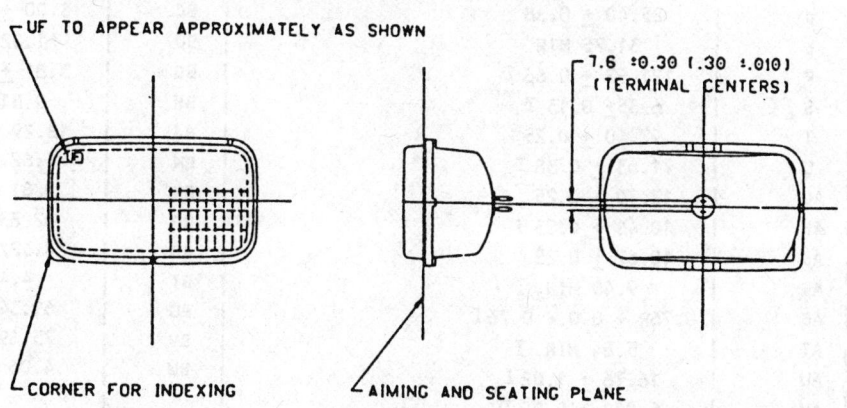

Note: Same as Type LF except as shown (.XX) Inch Dim.

FIGURE 27—TYPE "UF" AND "LF" HEADLAMPS 92 × 150 mm RECTANGULAR

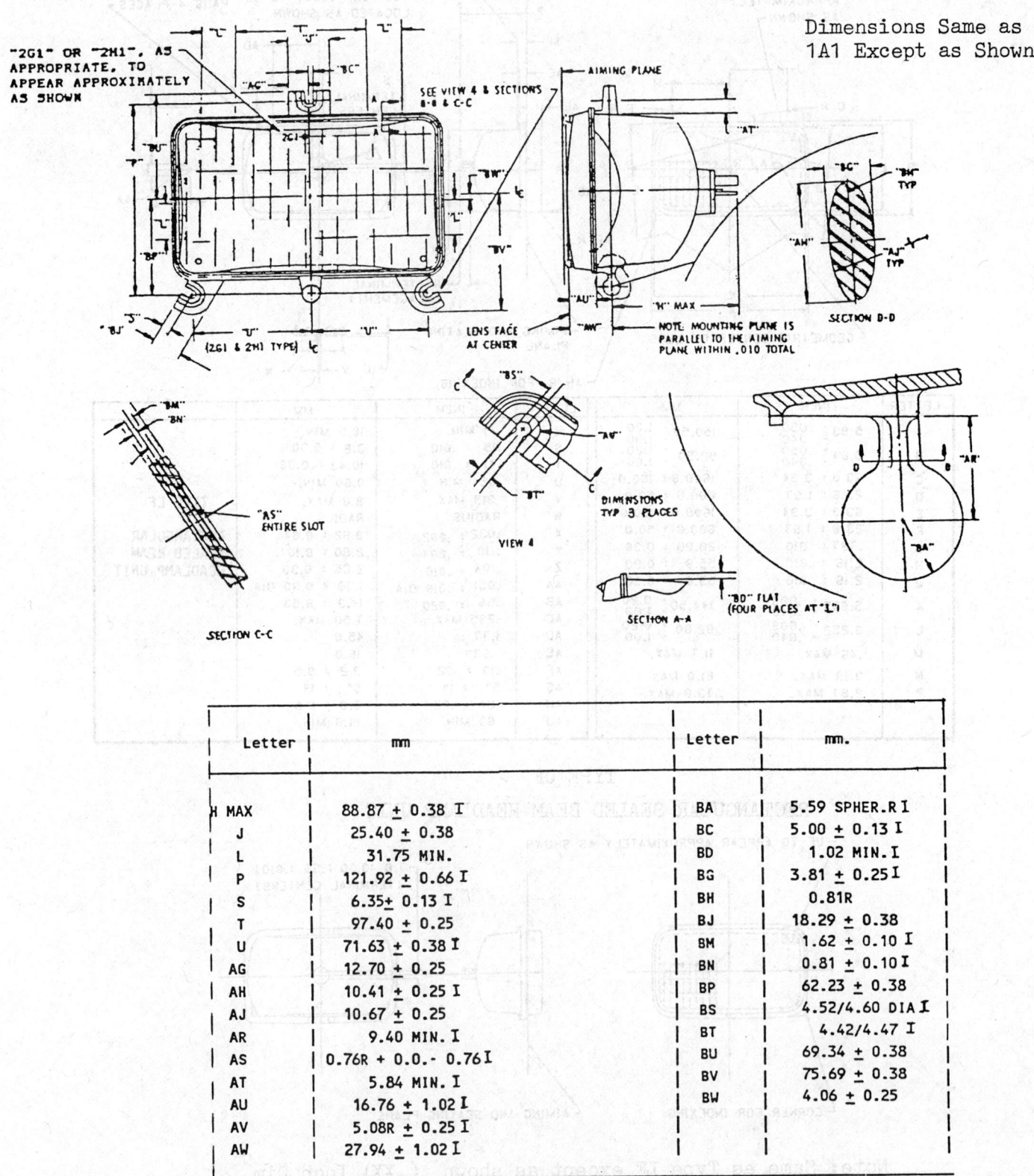

Letter	mm	Letter	mm.
H MAX	88.87 ± 0.38 I	BA	5.59 SPHER. R I
J	25.40 ± 0.38	BC	5.00 ± 0.13 I
L	31.75 MIN.	BD	1.02 MIN. I
P	121.92 ± 0.66 I	BG	3.81 ± 0.25 I
S	6.35 ± 0.13 I	BH	0.81R
T	97.40 ± 0.25	BJ	18.29 ± 0.38
U	71.63 ± 0.38 I	BM	1.62 ± 0.10 I
AG	12.70 ± 0.25	BN	0.81 ± 0.10 I
AH	10.41 ± 0.25 I	BP	62.23 ± 0.38
AJ	10.67 ± 0.25	BS	4.52/4.60 DIA I
AR	9.40 MIN. I	BT	4.42/4.47 I
AS	0.76R + 0.0. − 0.76 I	BU	69.34 ± 0.38
AT	5.84 MIN. I	BV	75.69 ± 0.38
AU	16.76 ± 1.02 I	BW	4.06 ± 0.25
AV	5.08R ± 0.25 I		
AW	27.94 ± 1.02 I		

FIGURE 28—TYPES 1G1, 2G1, AND 2H1 HEADLAMPS 100 × 165 mm RECTANGULAR

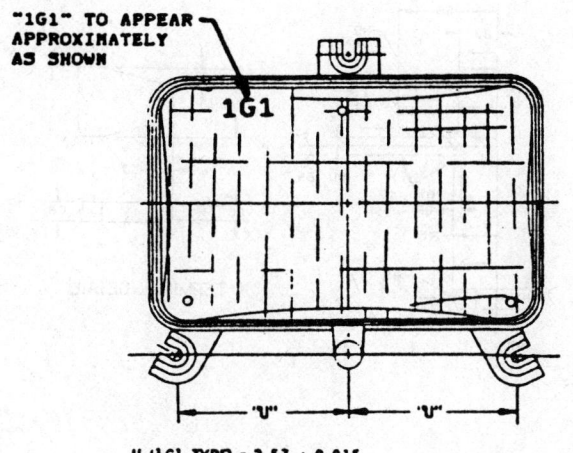
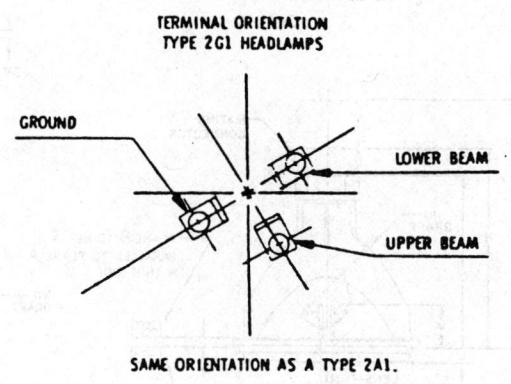
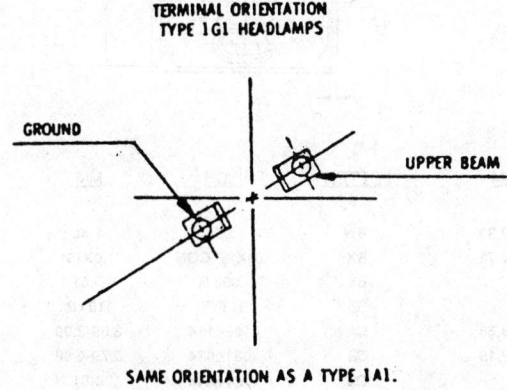
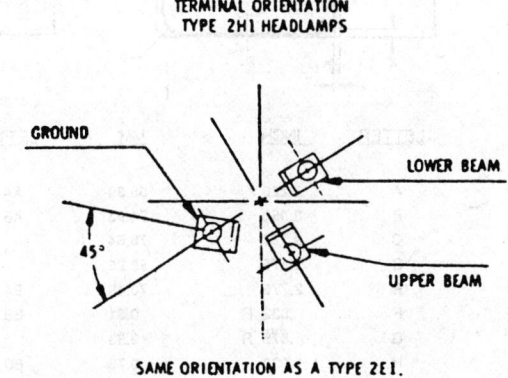

Noninterchangeability Configurations for Integral Mount Sealed Beam Headlamps, Type G and H

FIGURE 28 (CONTINUED)

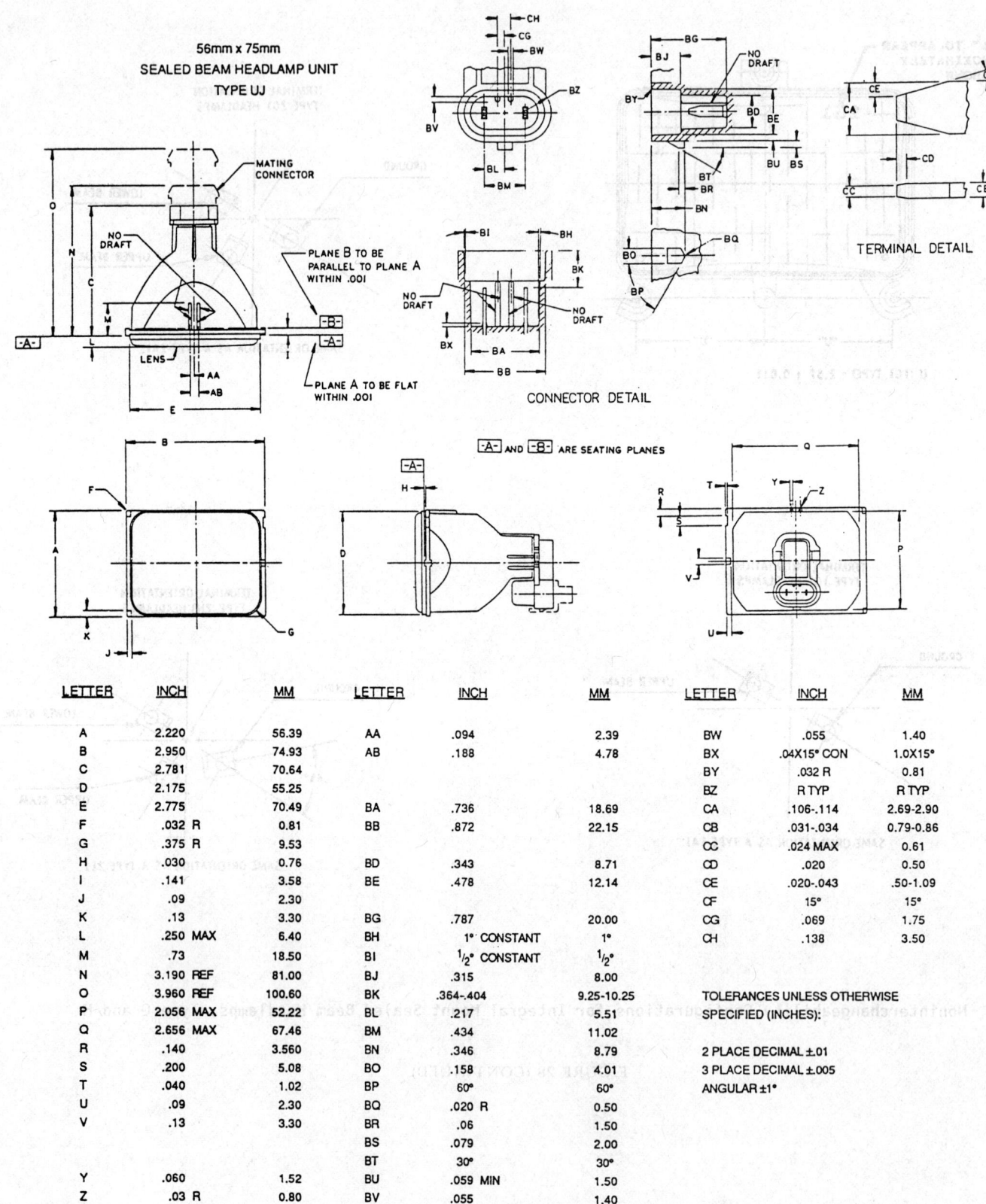

FIGURE 29—SEALED BEAM HEADLAMP UNIT
TYPE UJ

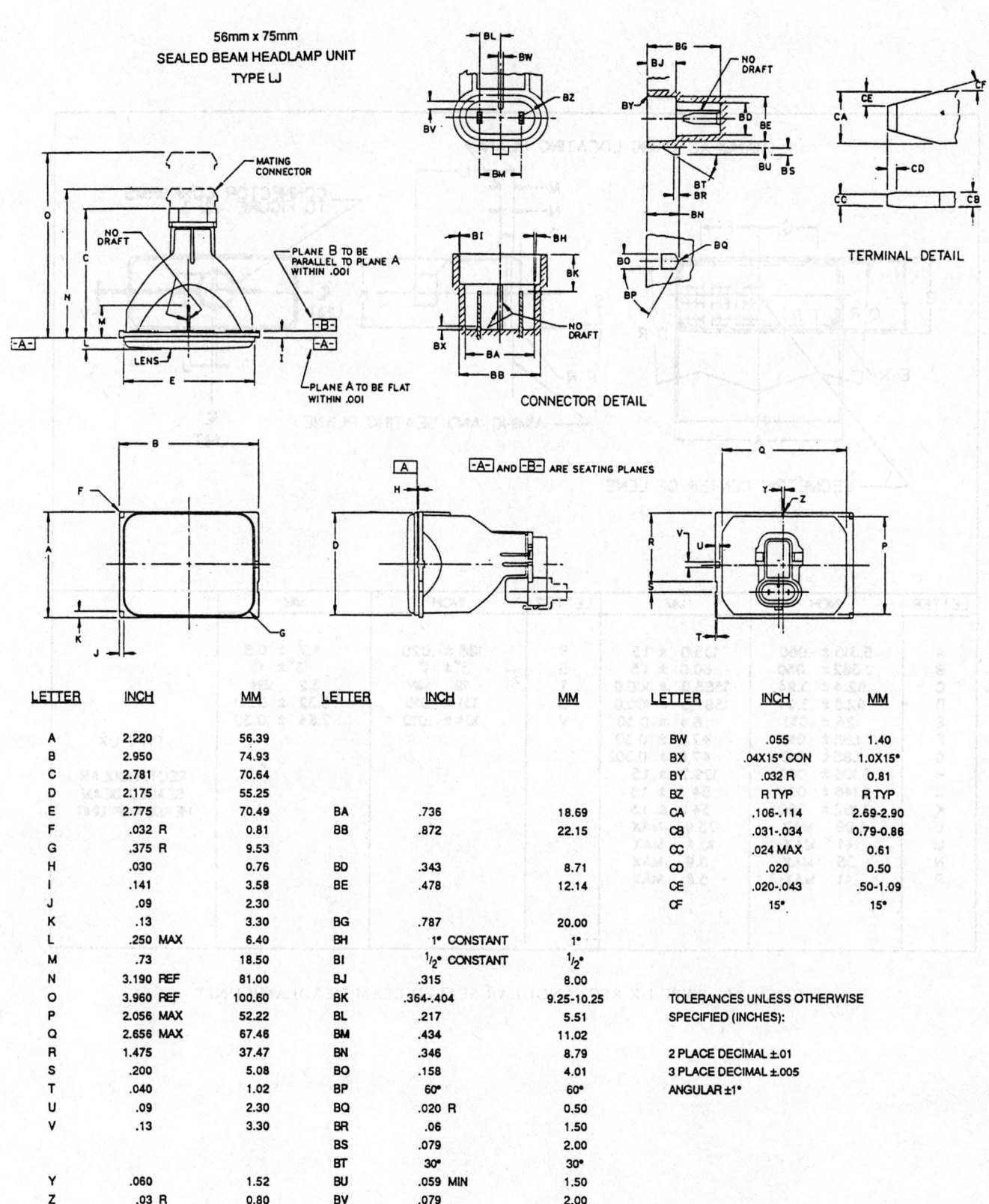

FIGURE 30—SEALED BEAM HEADLAMP UNIT
TYPE LJ

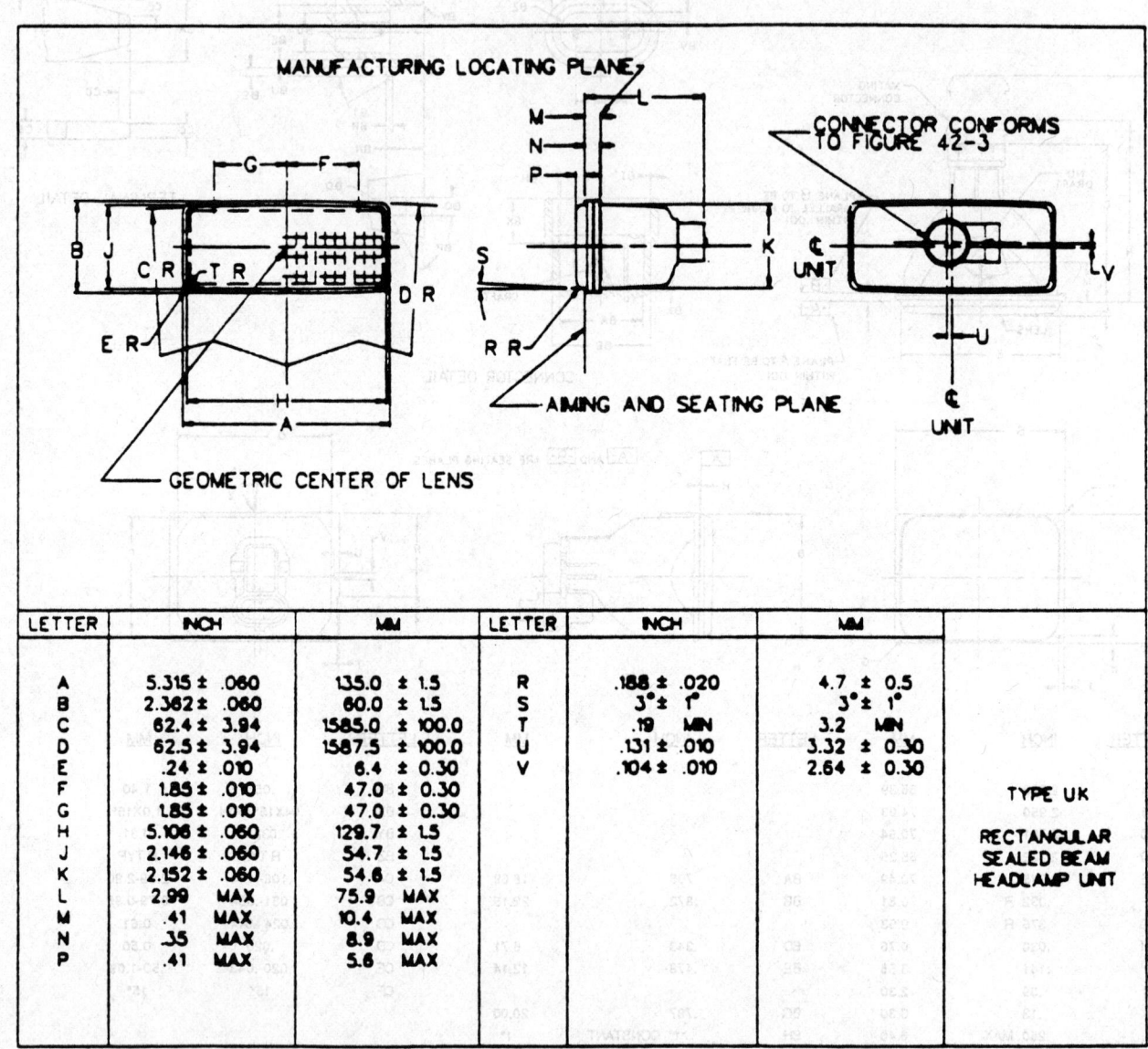

FIGURE 31—TYPE UK RECTANGULAR SEALED BEAM HEADLAMP UNIT

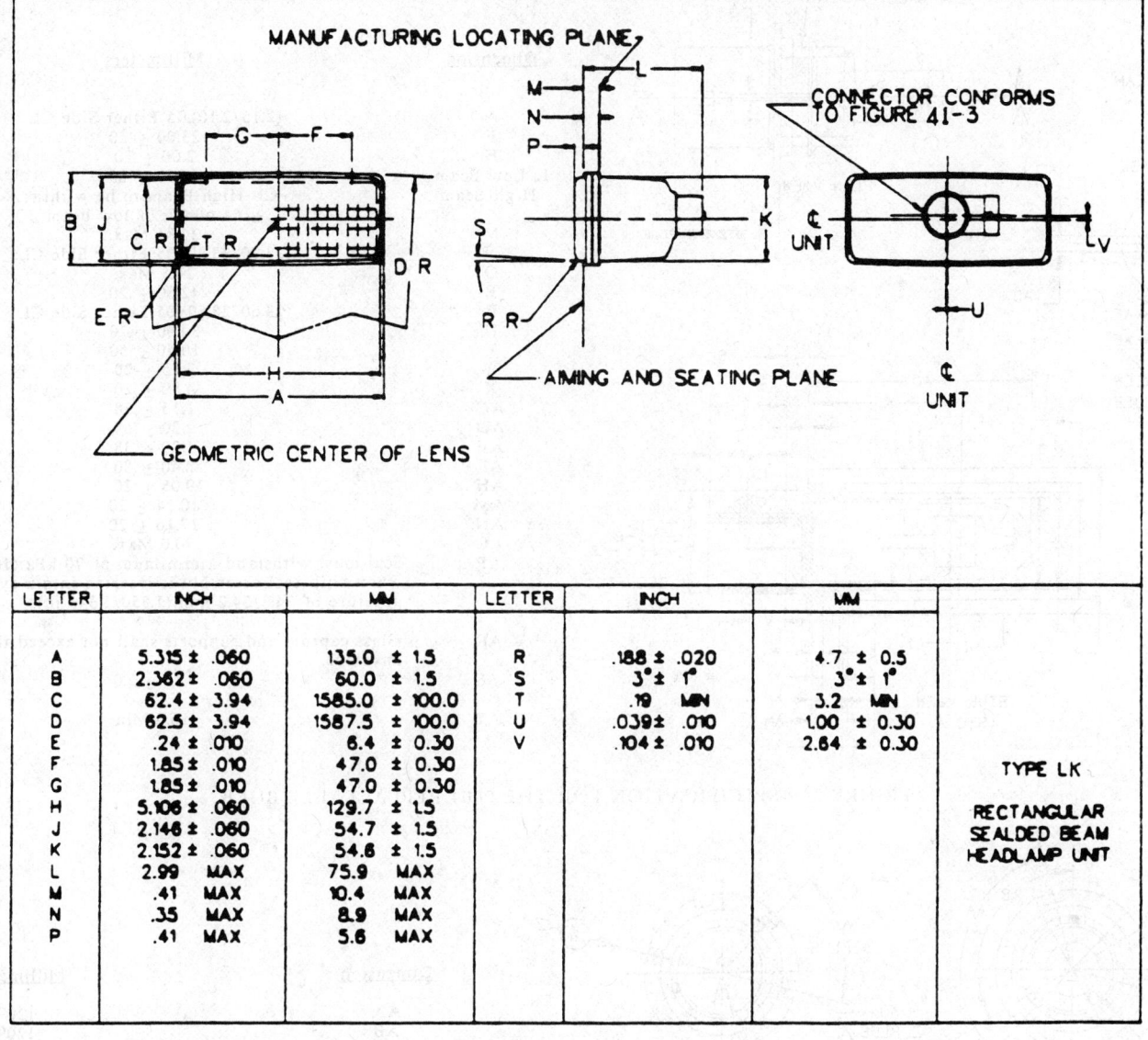

FIGURE 32—TYPE LK RECTANGULAR SEALED BEAM HEADLAMP UNIT

7. Bulb Filament Dimension and Location Test for the 9004 Replaceable Bulb—Filament locations relative to the bulb base (with O-ring removed) shall be determined for both production and accurate rated bulbs, as outlined below. For the actual conduct of these measurements, gaging standards shall be used for equipment calibration purposes.

7.1 Low Beam Filament Location Test—The location shall be determined by measuring from the midpoint of the smallest rectangle which encloses the filament image to the axial centerline of the base (see Figure 38):

a. Axially—in the right side view
b. Vertically—in the right side view
c. Transversely—in the plan view

7.2 High Beam Filament Location Test—The location shall be determined as indicated in 7.1.

8. Low Beam Filament Location—Production bulbs (refer to Figure 38).

8.1 Axial—The low beam filament axial or fore/aft location shall be measured in the right side view from the reference plane of the base to the center of the smallest rectangle which encloses the low beam filament image.

8.2 Vertical—The low beam filament vertical or side view from a horizontal plane through the base centerline to the center of the smallest rectangle which encloses the low beam filament image.

8.3 Transverse—The low beam filament transverse or left/right location shall be measured in the plan view from the vertical plane through the center of the base to the midpoint of the smallest rectangle which encloses the low beam filament image.

9. High Beam Filament Location—Production bulbs (refer to Figure 38).

9.1 Axial—The high beam filament axial location shall be measured in the right side view of the high beam filament from the centerline of the low beam filament to the centerline of the smallest rectangle which encloses the high beam filament image.

9.2 Vertical—The high beam filament vertical location shall be measured in the right side view of the high beam filament from the centerline of the low beam filament to the centerline of the smallest rectangle which encloses the high beam filament image.

9.3 Transverse—The high beam filament horizontal location shall be measured in the plan view of the high beam filament from the midpoint of the low beam filament to the midpoint of the smallest rectangle which encloses the high beam filament image.

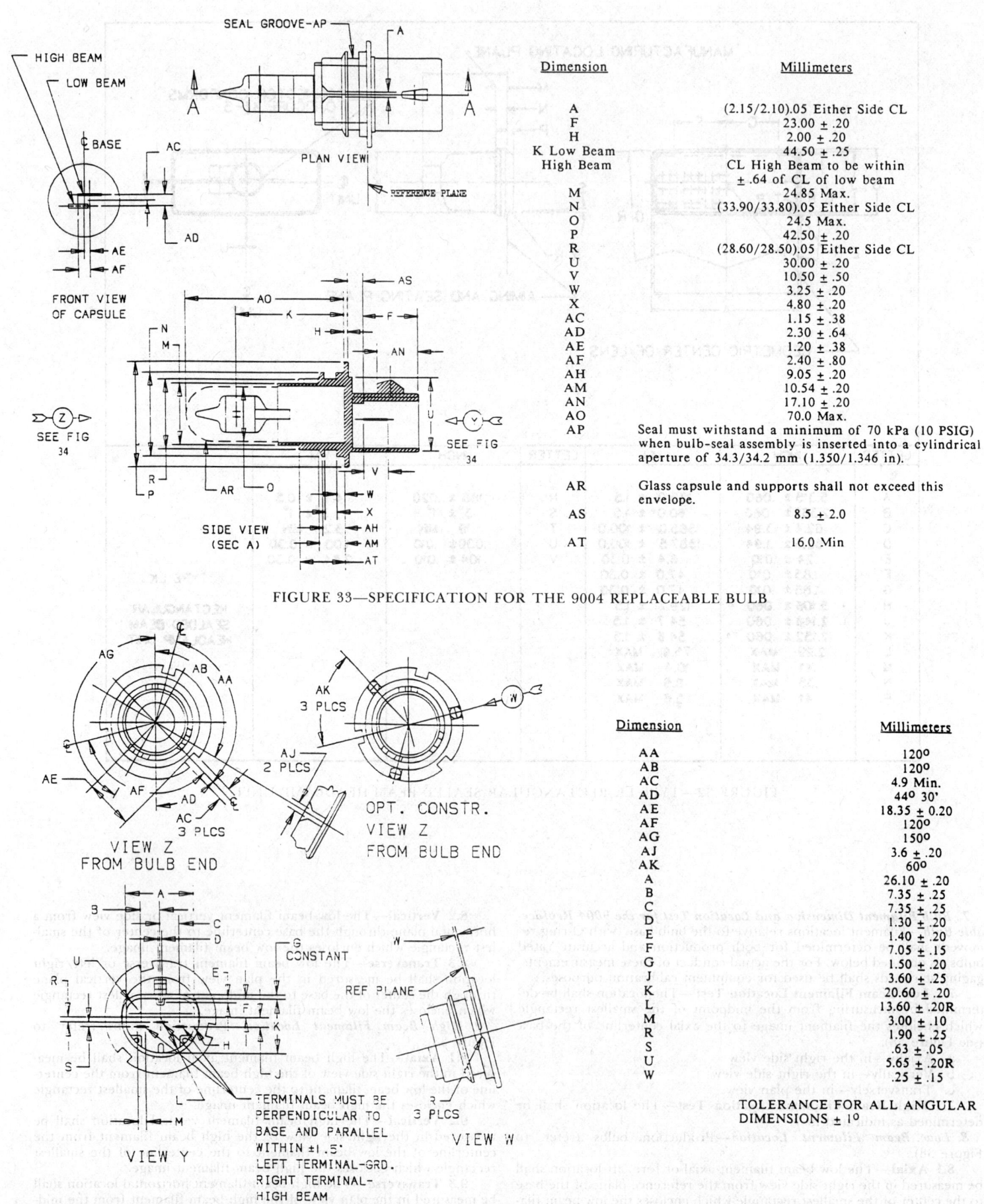

FIGURE 33—SPECIFICATION FOR THE 9004 REPLACEABLE BULB

FIGURE 34—SPECIFICATION FOR THE 9004 REPLACEABLE BULB

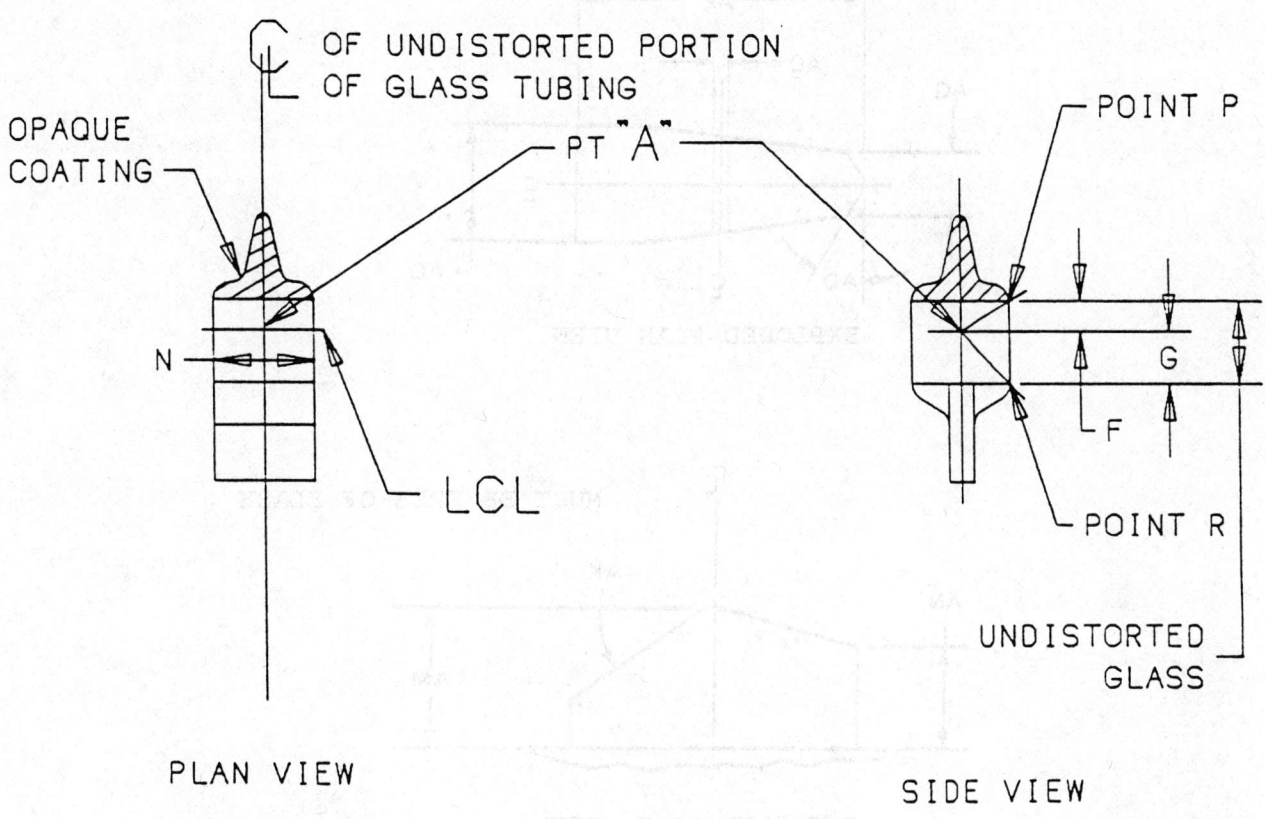

PLAN VIEW   SIDE VIEW

Dimensional Specifications
Figure 35

Dimension	Specification
F	$(N/2)\tan 38° \pm 1.0$ mm
G	$(N/2)\tan 43°$ MIN
N	Actual Capsule Dia. (To Be Established By Manufacturer)
P	Entire Radius and Distorted Glass Shall Be Covered to the Plane Passing Through Point "P", Perpendicular to the Glass Capsule Centerline.

FIGURE 35—SPECIFICATION FOR THE 9004 REPLACEABLE BULB

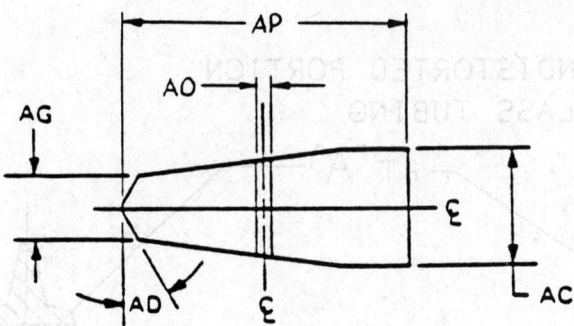

EXPLODED PLAN VIEW

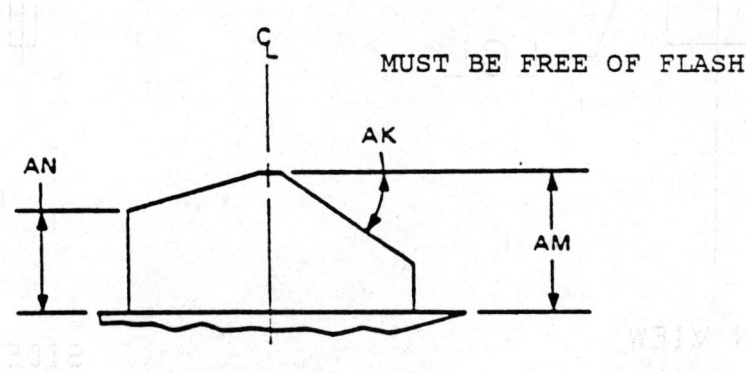

MUST BE FREE OF FLASH

EXPLODED SIDE VIEW

Dimension	Millimeters
AC	4.55 ± .20
AD	30° ± 3°
AG	2.50 ± .20
AK	35° ± 3°
AM	5.50 ± .20
AN	4.00 ± .20
AO	.5 ± .20
AP	11.4 ± .20

FIGURE 36—SPECIFICATION FOR THE 9004 REPLACEABLE BULB
LOCKING FEATURE

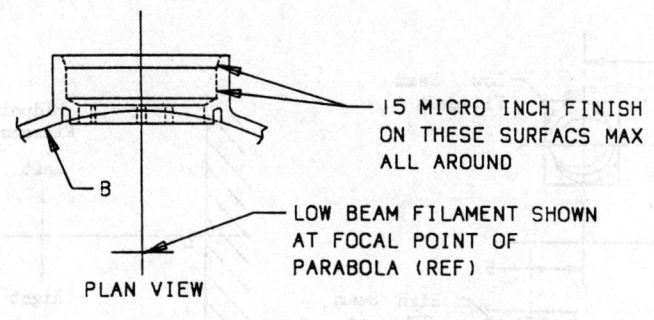
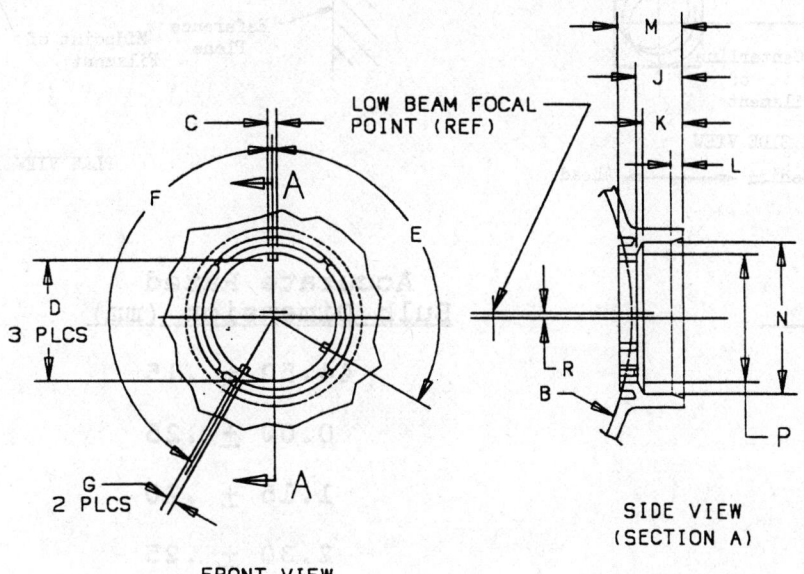

Dimension	Millimeters
B	Ref Line
	Lamp Parabola
C	2.00 ± .05
	.05 Either Side of CL
D	27.10 ± .20
E	120°
F	150°
G	2.00 ± .20
H	15.15 ± .20
J	11.10 ± .20
K	9.50 ± .20
L	2.75 ± .20
N	34.24 +.08/−.05
P	28.70 +.10/−.05
	Diameter P shall be concentric to diameter N within ± .05
R	1.15 ± .10

TOLERANCE FOR ALL ANGULAR
DIMENSIONS ± 1°

FIGURE 37—SPECIFICATION FOR THE 9004 REPLACEABLE BULB

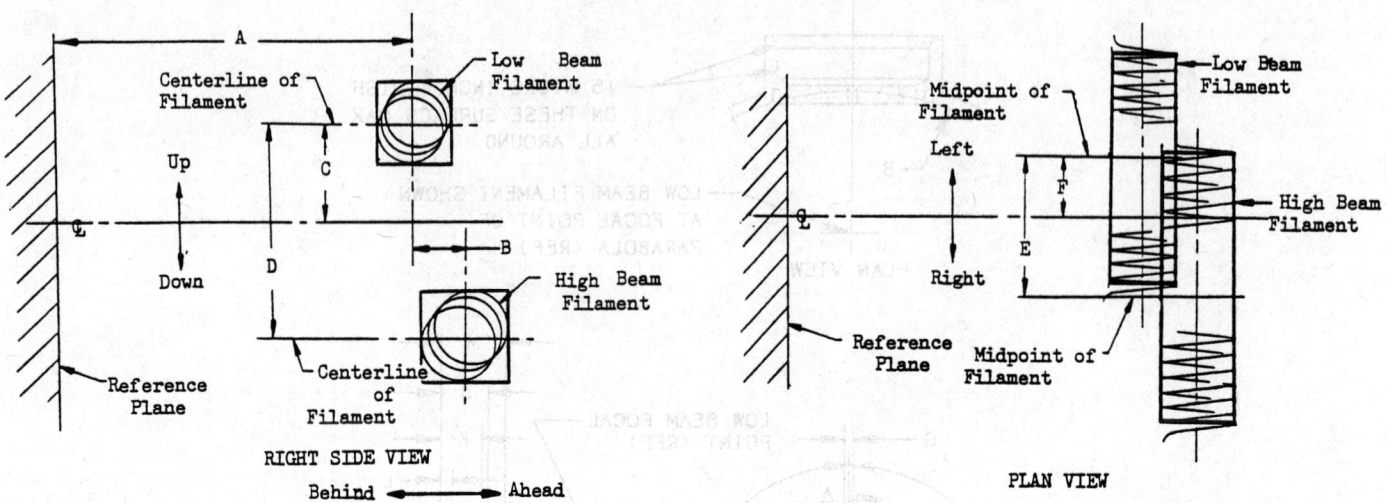

Letter	Accurate Rated Bulb Dimension (mm)
A	44.50 ± .15
B	0.00 ± .25
C	1.15 ± .20
D	2.30 ± .25
E	2.40 ± .40
F	1.20 ± .20
Low Beam Filament Length 1/3/	4.80 ± .40
High Beam Filament Length 2/3/	4.80 ± .40

1/ Low beam filament rotation shall not exceed 0.3 diameters of the coil.

2/ High beam filament rotation shall not exceed 0.4 diameters of the coil.

3/ Filament Length - The length of any filament shall be considered to be the length of the smallest rectangle which encloses the filament image in the plan view or right side view, as appropriate.

FIGURE 38—DIMENSIONAL SPECIFICATIONS FOR THE 9004 REPLACEABLE BULB
FILAMENT DIMENSION AND LOCATION—MEASUREMENT METHOD

24.73

10. Bulb Filament Dimension and Location Test for the 9005 and 9006 Replaceable Bulb—Filament locations relative to the bulb base (with O-ring removed) shall be determined for both production and accurate rated bulbs, as outlined below. For the actual conduct of these measurements, gaging standards shall be used for equipment calibration purposes.

10.1 High Beam Filament Location Test (see Figure 39)—The high beam filament location shall be determined by measuring:
   a. Axially—in the side view
   b. Vertically—in the side view
   c. Transversely—in the bottom view

10.2 Low Beam Filament Location Test (see Figure 47)—The low beam filament location shall be determined by measuring:
   a. Axially—in the side view
   b. Vertically—in the side view
   c. Transversely—in the bottom view

11. High Beam Filament Location
   11.1 Production Bulbs
   11.1.1 AXIAL—The end coil nearest to Plane A shall be within the volume "B" and the end coil farthest from Plane A shall be within the volume "C" as shown in Figure 43.
   11.1.2 VERTICAL—Same as 11.1.1.
   11.1.3 TRANSVERSE—Same as 11.1.1.
   11.2 Accurate Rated Bulbs
   11.2.1 AXIAL—The axial or fore/aft location shall be measured from Plane A to the beginning of the end coil nearest to Plane A and to the finish of the end coil farthest from Plane A. See Figure 46, Volume D and E.
   11.2.2 VERTICAL
      11.2.2.1 *End Coils*—The vertical or up/down location shall be measured from line A to the center of the smallest rectangle which encloses the end coil. See Figure 46, Volume D and E.
      11.2.2.2 *Center Section*—The vertical or up/down location shall be measured from line A to the center of the smallest rectangle which encloses the center coil. See Figure 46, Section F, Area G.
   11.2.3 TRANSVERSE
      11.2.3.1 *End Coils*—Same as 11.2.2.1.
      11.2.3.2 *Center Section*—Same as 11.2.2.2.

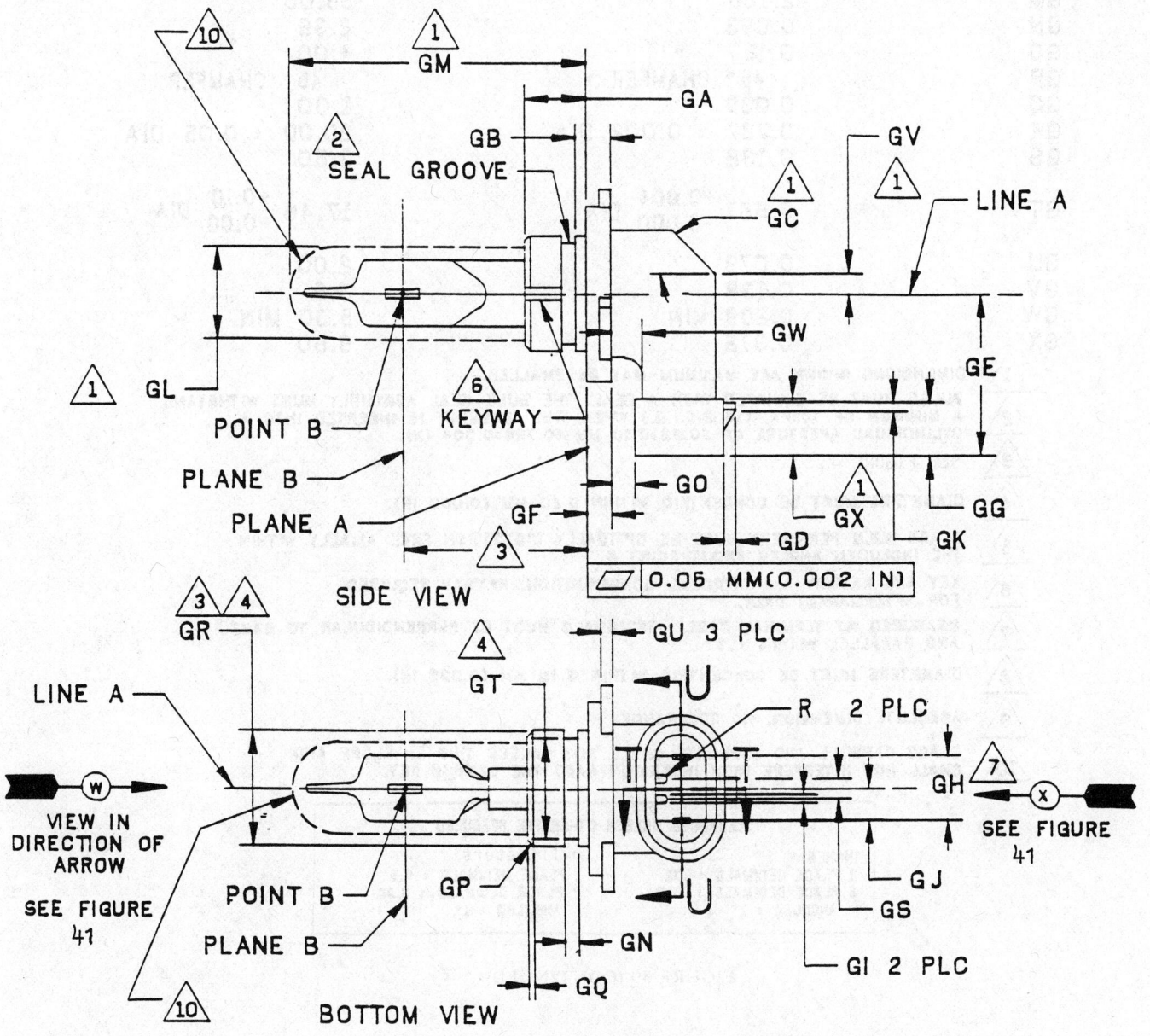

FIGURE 39—SPECIFICATION FOR THE 9005 REPLACEABLE BULB

DIMENSION	INCHES	MILLIMETERS
GA	0.591 MAX/0.217 MIN	15.00 MAX/5.50 MIN
GB	0.236	6.00
GC	45°	45°
GD	0.079	2.00
GE	1.09	27.8
GF	0.165	4.20
GG	0.346	8.80
GH	0.433	11.00
GI	0.055	1.40
GJ	0.217 ± 0.006	5.50 ± 0.15
GK	0.06	1.5
GL	0.775 DIA	19.68 DIA
GM	2.165	55.00
GN	0.093	2.36
GO	0.157	4.00
GP	45° CHAMFER	45° CHAMFER
GQ	0.039	1.00
GR	0.787 ± 0.002 DIA	20.00 ± 0.05 DIA
GS	0.138	3.50
GT	0.687 +0.004/-0.000 DIA	17.46 +0.10/-0.00 DIA
GU	0.079	2.00
GV	0.138	3.5
GW	0.209 MIN	5.30 MIN
GX	0.378	9.60

/1\ DIMENSIONS SHOWN ARE MAXIMUM-MAY BE SMALLER.

/2\ BULBS MUST BE EQUIPPED WITH A SEAL. THE BULB-SEAL ASSEMBLY MUST WITHSTAND A MINIMUM OF 70kPA. (10 P.S.I.G.) WHEN THE ASSEMBLY IS INSERTED INTO A CYLINDRICAL APERTURE OF 20.22±0.10 MM (0.796±0.004 IN).

/3\ SEE FIGURE 43

/4\ DIAMETERS MUST BE CONCENTRIC WITHIN 0.20 MM (0.008 IN).

/5\ GLASS BULB PERIPHERY MUST BE OPTICALLY DISTORTION FREE AXIALLY WITHIN THE INCLUDED ANGLES ABOUT POINT B.

/6\ KEY AND KEYWAY ARE OPTIONAL CONSTRUCTION. KEYWAY REQUIRED FOR AFTERMARKET ONLY.

/7\ MEASURED AT TERMINAL BASE. TERMINALS MUST BE PERPENDICULAR TO BASE AND PARALLEL WITHIN ±1.5°

/8\ DIAMETERS MUST BE CONCENTRIC WITHIN 0.20 MM (0.008 IN).

/9\ ABSOLUTE DIMENSION. NO TOLERANCE.

/10\ GLASS CAPSULE AND SUPPORTS SHALL NOT EXCEED THIS ENVELOPE AND SHALL NOT INTERFERE WITH INSERTION PAST THE LAMP'S KEY.

TOLERANCES UNLESS OTHERWISE SPECIFIED	
INCHES	millimeters
2 PLACE DECIMALS ± .02	1 PLACE DECIMALS ± 0.5
3 PLACE DECIMALS ± .010	2 PLACE DECIMALS ± 0.30
ANGULAR ± 1°	ANGULAR ± 1°

FIGURE 39 (CONTINUED)

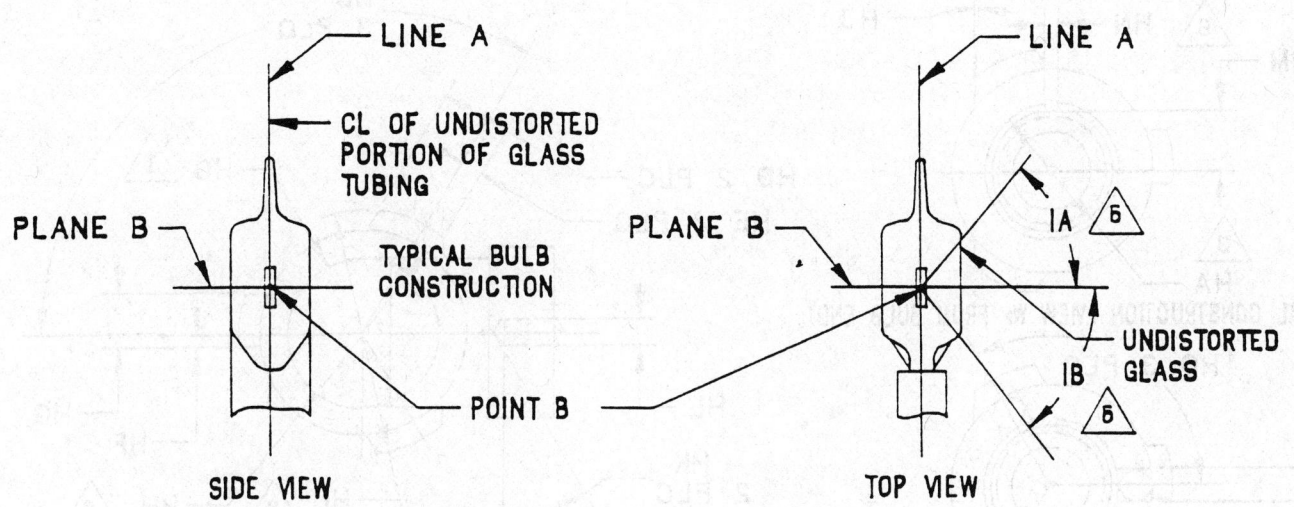

POINT B IS INTERSECTION OF PLANE B AND CENTERLINE OF UNDISTORTED GLASS TUBING

DIMENSION	INCHES	MILLIMETERS
1A	45° MIN	45° MIN
1B	52° MIN	52° MIN

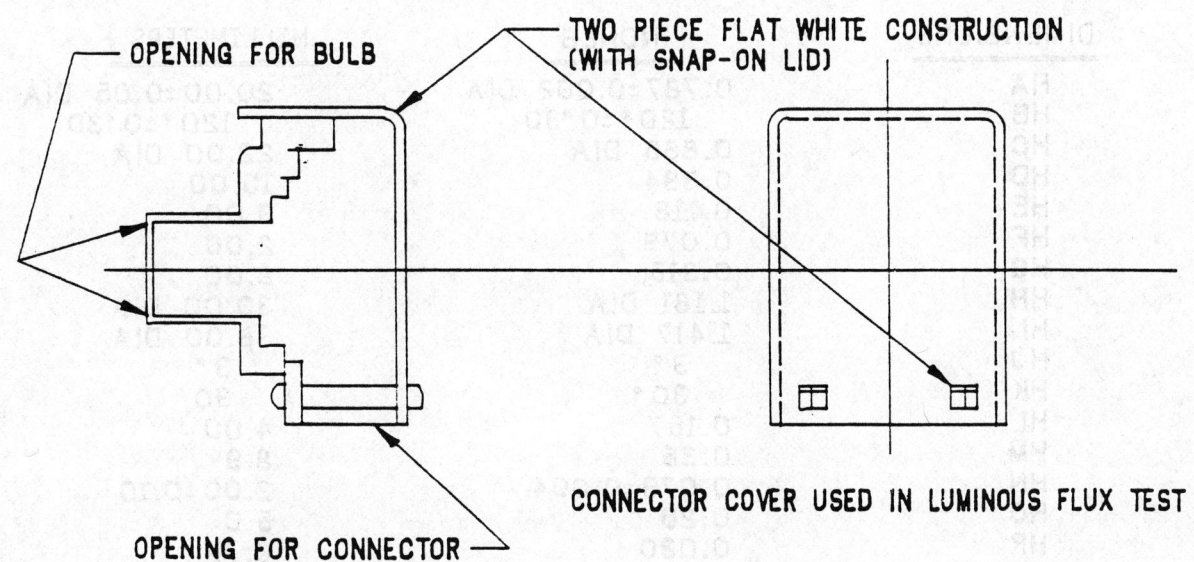

FIGURE 40—SPECIFICATION FOR THE 9005 REPLACEABLE BULB

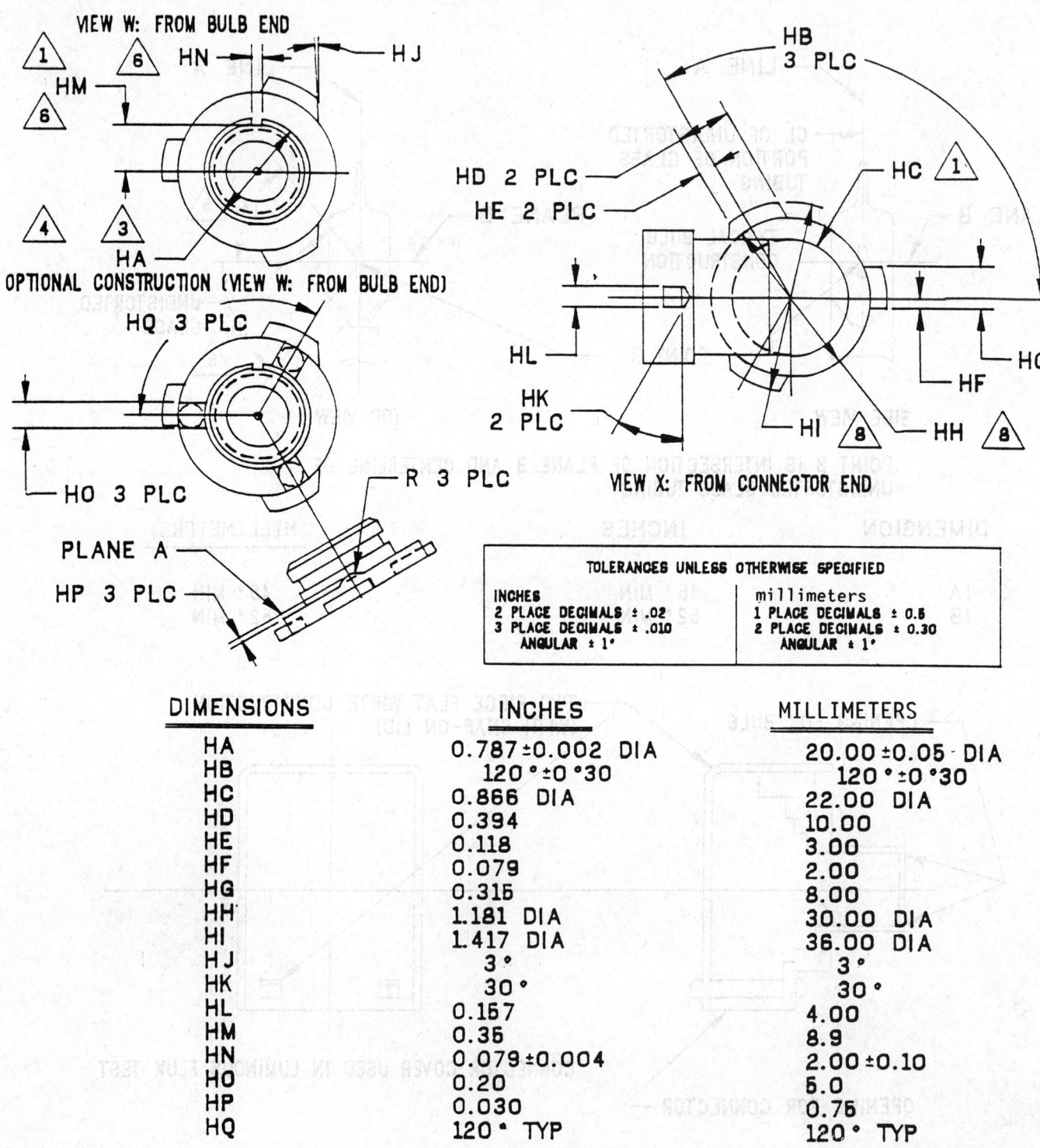

FIGURE 41—SPECIFICATION FOR THE 9005 REPLACEABLE BULB

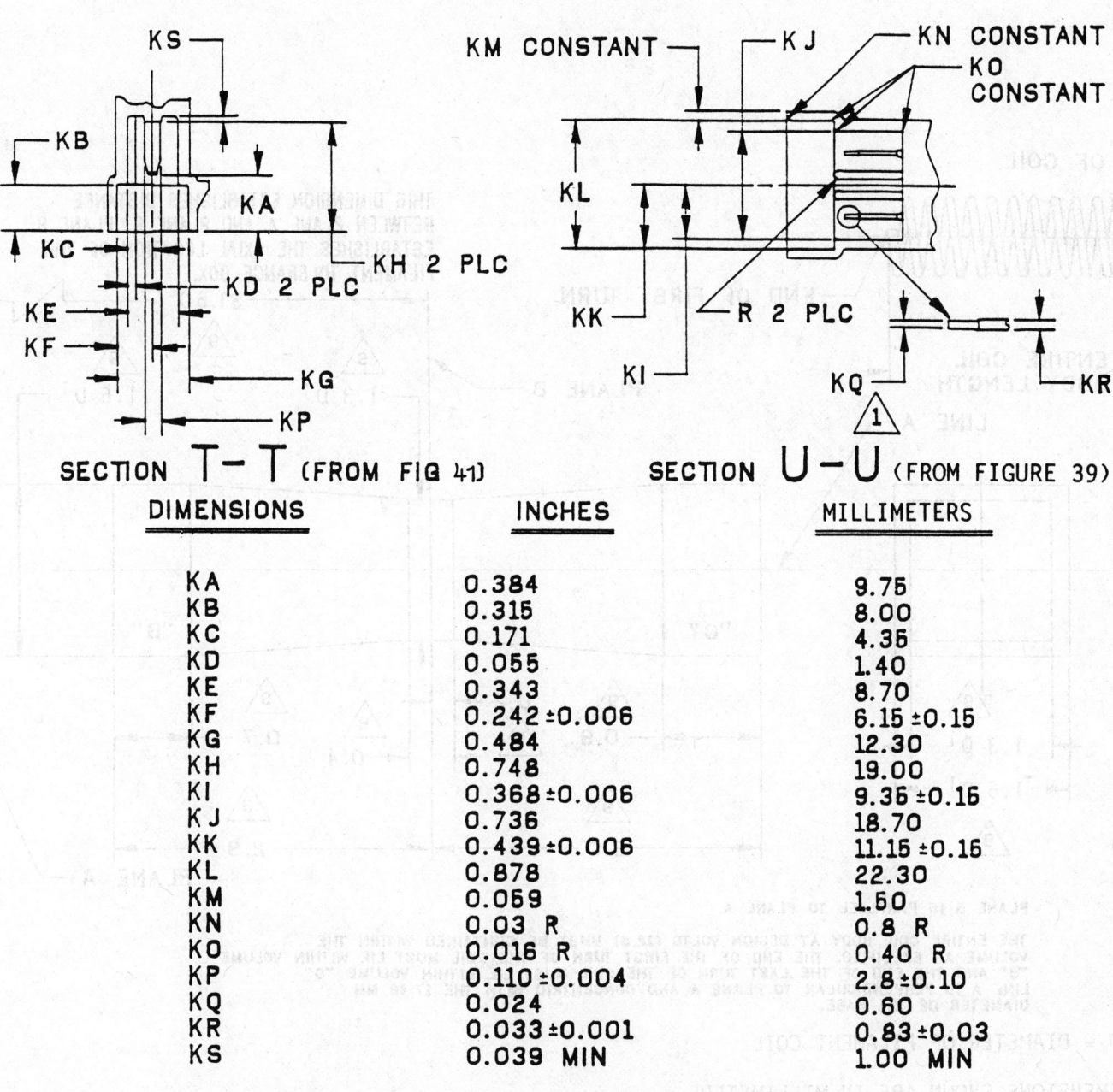

SECTION T-T (FROM FIG 41)   SECTION U-U (FROM FIGURE 39)

DIMENSIONS	INCHES	MILLIMETERS
KA	0.384	9.75
KB	0.315	8.00
KC	0.171	4.35
KD	0.055	1.40
KE	0.343	8.70
KF	0.242 ± 0.006	6.15 ± 0.15
KG	0.484	12.30
KH	0.748	19.00
KI	0.368 ± 0.006	9.35 ± 0.15
KJ	0.736	18.70
KK	0.439 ± 0.006	11.15 ± 0.15
KL	0.878	22.30
KM	0.059	1.50
KN	0.03 R	0.8 R
KO	0.016 R	0.40 R
KP	0.110 ± 0.004	2.8 ± 0.10
KQ	0.024	0.60
KR	0.033 ± 0.001	0.83 ± 0.03
KS	0.039 MIN	1.00 MIN

TOLERANCES UNLESS OTHERWISE SPECIFIED	
INCHES	millimeters
2 PLACE DECIMALS ± .02	1 PLACE DECIMALS ± 0.5
3 PLACE DECIMALS ± .010	2 PLACE DECIMALS ± 0.30
ANGULAR ± 1°	ANGULAR ± 1°

FIGURE 42—SPECIFICATION FOR THE 9005 REPLACEABLE BULB

24.78

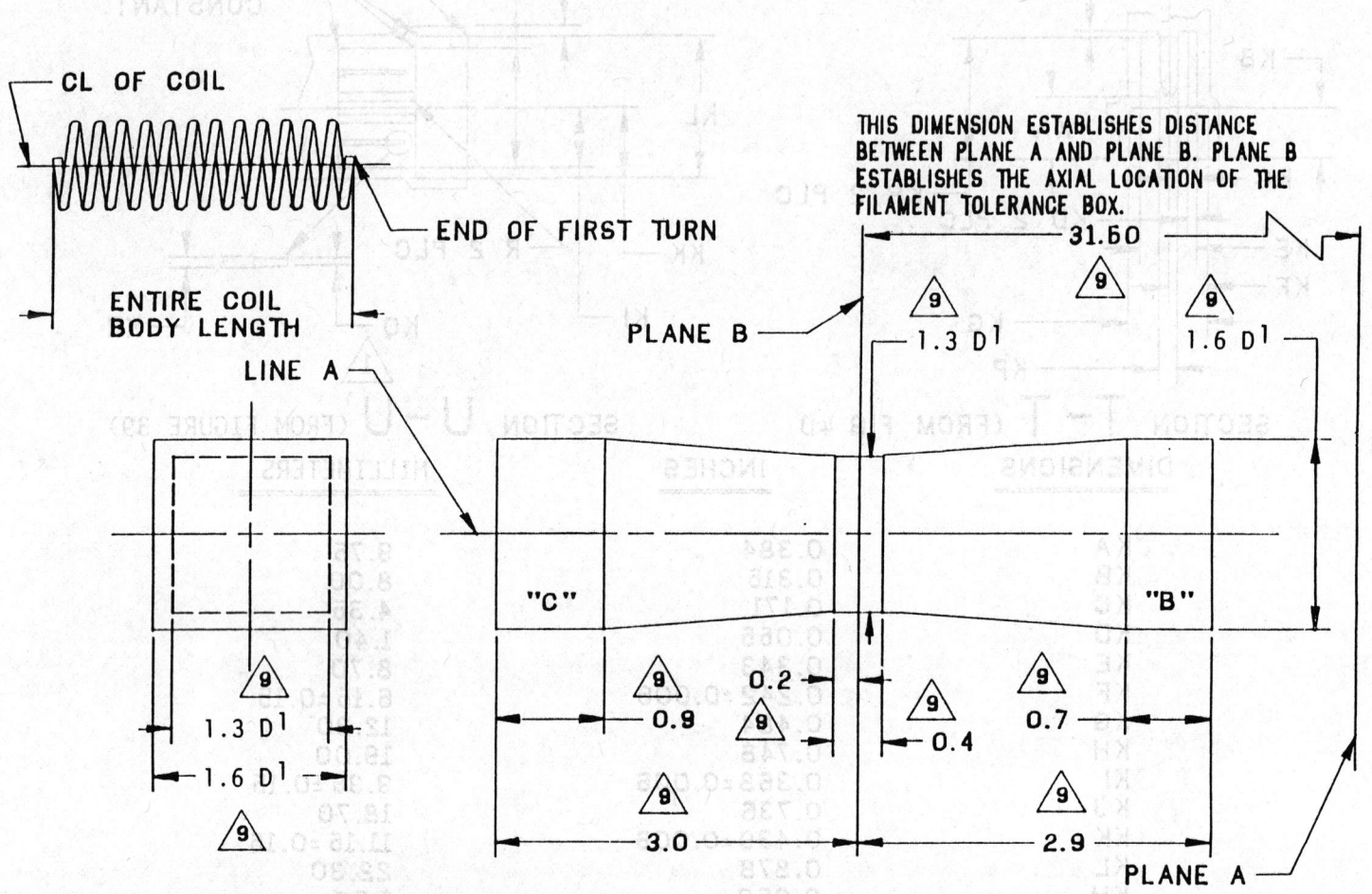

PLANE B IS PARALLEL TO PLANE A.

THE ENTIRE COIL BODY AT DESIGN VOLTS (12.8) MUST BE CONTAINED WITHIN THE VOLUME AS SPECIFIED. THE END OF THE FIRST TURN OF THE COIL MUST LIE WITHIN VOLUME "B" AND THE END OF THE LAST TURN OF THE COIL MUST LIE WITHIN VOLUME "C" LINE A IS PERPENDICULAR TO PLANE A AND CONCENTRIC WITH THE 17.46 MM DIAMETER OF THE BASE.

[1]D = DIAMETER OF FILAMENT COIL

DIMENSIONS SHOWN ARE IN MILLIMETERS

FIGURE 43—SPECIFICATION FOR THE 9005 REPLACEABLE BULB

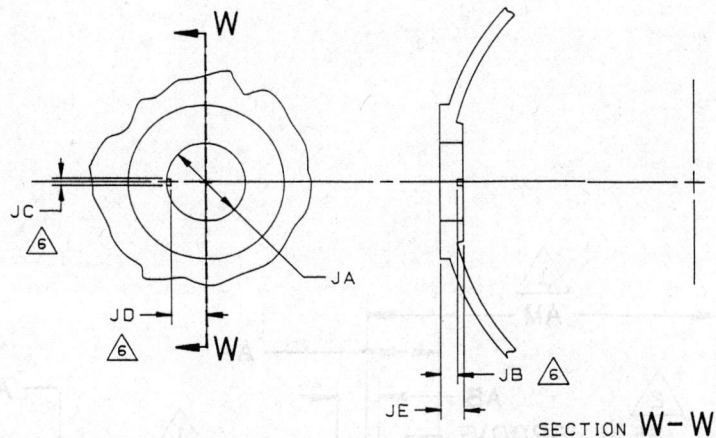

DIMENSIONS	INCHES	MILLIMETERS
JA	0.796 ±0.004 DIA	20.22 ±0.10 DIA
JB	0.172 +0.010/−0.000	4.36 +0.30/−0.00
JC	0.067 ±0.004	1.70 ±0.10
JD	0.352 +0.004/−0.000	8.95 +0.10/−0.00
JE	0.236 MIN	6.00 MIN

FIGURE 44—SPECIFICATION FOR THE 9005 REPLACEABLE BULB
BULB HOLDER

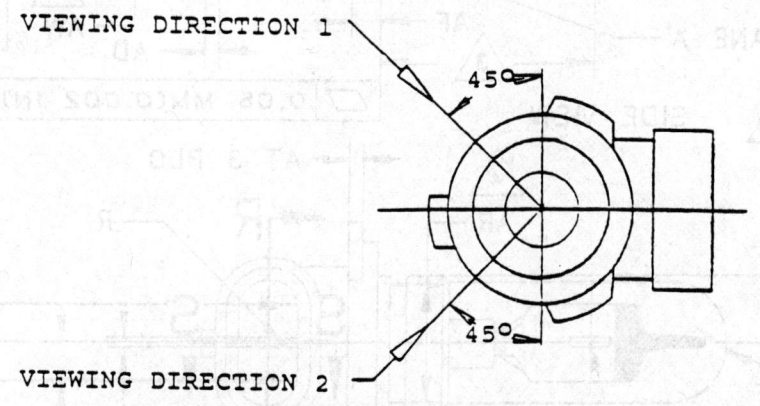

FIGURE 45—MODIFIED VIEW W FROM FIGURE 41,
SIMILAR FOR 49

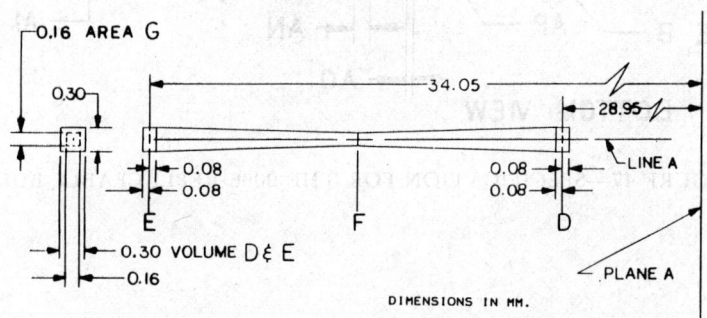

DIMENSIONS IN MM.

THE CENTROID OF THE FIRST TURN OF THE COIL MUST BE WITHIN
VOLUME D AND THE CENTROID OF THE LAST TURN OF THE COIL MUST
BE WITHIN VOLUME E. F IS AT THE MID-LENGTH OF THE COIL. THE
CENTROID AT F MUST BE WITHIN AREA G.

FIGURE 46—SPECIFICATION FOR THE 9005
REPLACEABLE BULB
ACCURATE RATED BULB

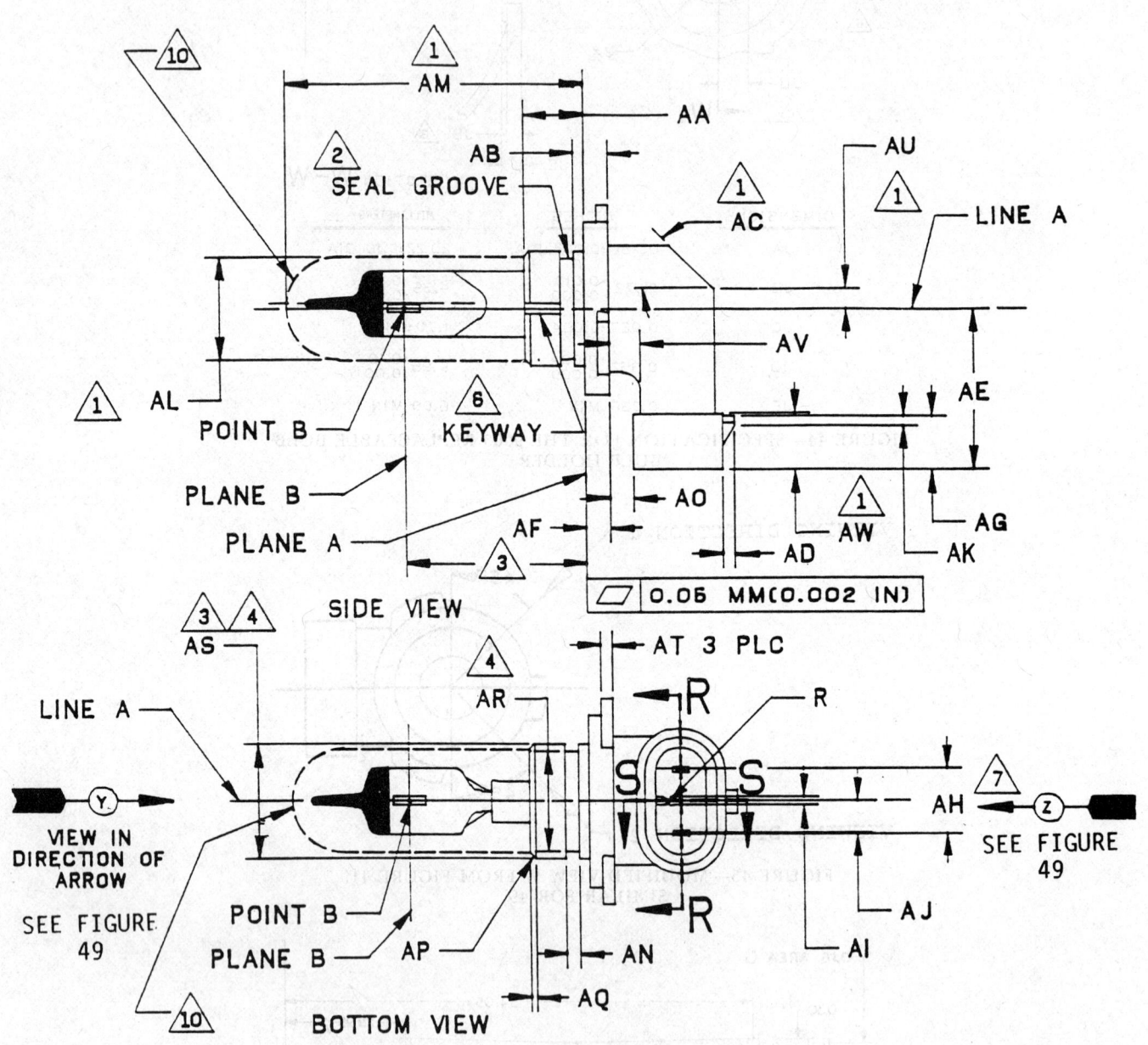

FIGURE 47—SPECIFICATION FOR THE 9006 REPLACEABLE BULB

DIMENSION	INCHES	MILLIMETERS
AA	0.591 MAX/0.217 MIN	15.00 MAX/5.50 MIN
AB	0.236	6.00
AC	45°	45°
AD	0.079	2.00
AE	1.09	27.8
AF	0.165	4.20
AG	0.346	8.80
AH	0.433	11.00
AI	0.055	1.40
AJ	0.217 ± 0.006	5.50 ± 0.15
AK	0.06	1.5
AL	0.780 DIA	19.81 DIA
AM	2.165	55.00
AN	0.093	2.36
AO	0.157	4.00
AP	45° CHAMFER	45° CHAMFER
AQ	0.039	1.00
AR	0.766 +0.004/−0.000 DIA	19.46 +0.10/−0.00 DIA
AS	0.866 ± 0.002 DIA	22.00 ± 0.05 DIA
AT	0.079	2.00
AU	0.138	3.5
AV	0.209 MIN	5.30 MIN
AW	0.378	9.60

/1\ DIMENSIONS SHOWN ARE MAXIMUM-MAY BE SMALLER.

/2\ BULBS MUST BE EQUIPPED WITH A SEAL. THE BULB-SEAL ASSEMBLY MUST WITHSTAND A MINIMUM OF 70kPA (10 P.S.I.G.) WHEN THE ASSEMBLY IS INSERTED INTO A CYLINDRICAL APERTURE OF 22.22±0.10 MM (0.875±0.004 IN).

/3\ SEE FIGURE 51

/4\ DIAMETERS MUST BE CONCENTRIC WITHIN 0.20 MM (0.008 IN).

/5\ GLASS BULB PERIPHERY MUST BE OPTICALLY DISTORTION FREE AXIALLY WITHIN THE INCLUDED ANGLES ABOUT POINT B.

/6\ KEY AND KEYWAY ARE OPTIONAL CONSTRUCTION. KEYWAY REQUIRED FOR AFTERMARKET ONLY.

/7\ MEASURED AT TERMINAL BASE. TERMINALS MUST BE PERPENDICULAR TO BASE AND PARALLEL WITHIN ±1.5°

/8\ DIAMETERS MUST BE CONCENTRIC WITHIN 0.20 MM (0.008 IN).

/9\ ABSOLUTE DIMENSION. NO TOLERANCE.

/10\ GLASS CAPSULE AND SUPPORTS SHALL NOT EXCEED THIS ENVELOPE.

TOLERANCES UNLESS OTHERWISE SPECIFIED	
INCHES 2 PLACE DECIMALS ± .02 3 PLACE DECIMALS ± .010 ANGULAR ± 1°	millimeters 1 PLACE DECIMALS ± 0.5 2 PLACE DECIMALS ± 0.30 ANGULAR ± 1°

FIGURE 47 (CONTINUED)

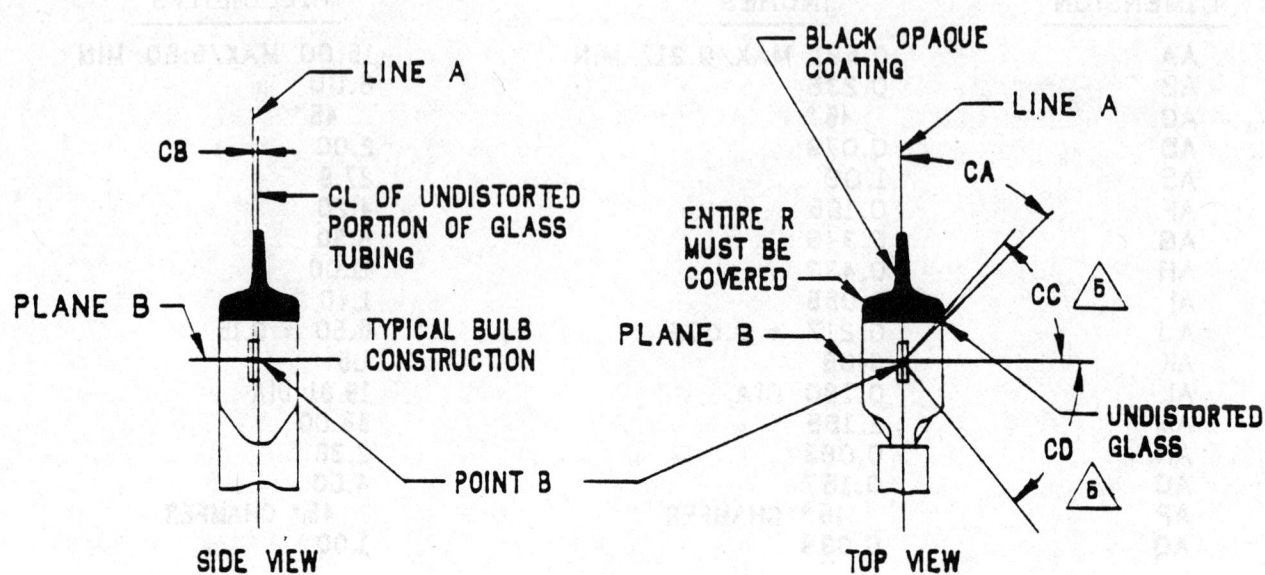

DIMENSION	INCHES	MILLIMETERS
CA	45°±5°	45°±5°
CB	0.030±0.020	0.75±0.50
CC	50° MIN	50° MIN
CD	52° MIN	52° MIN

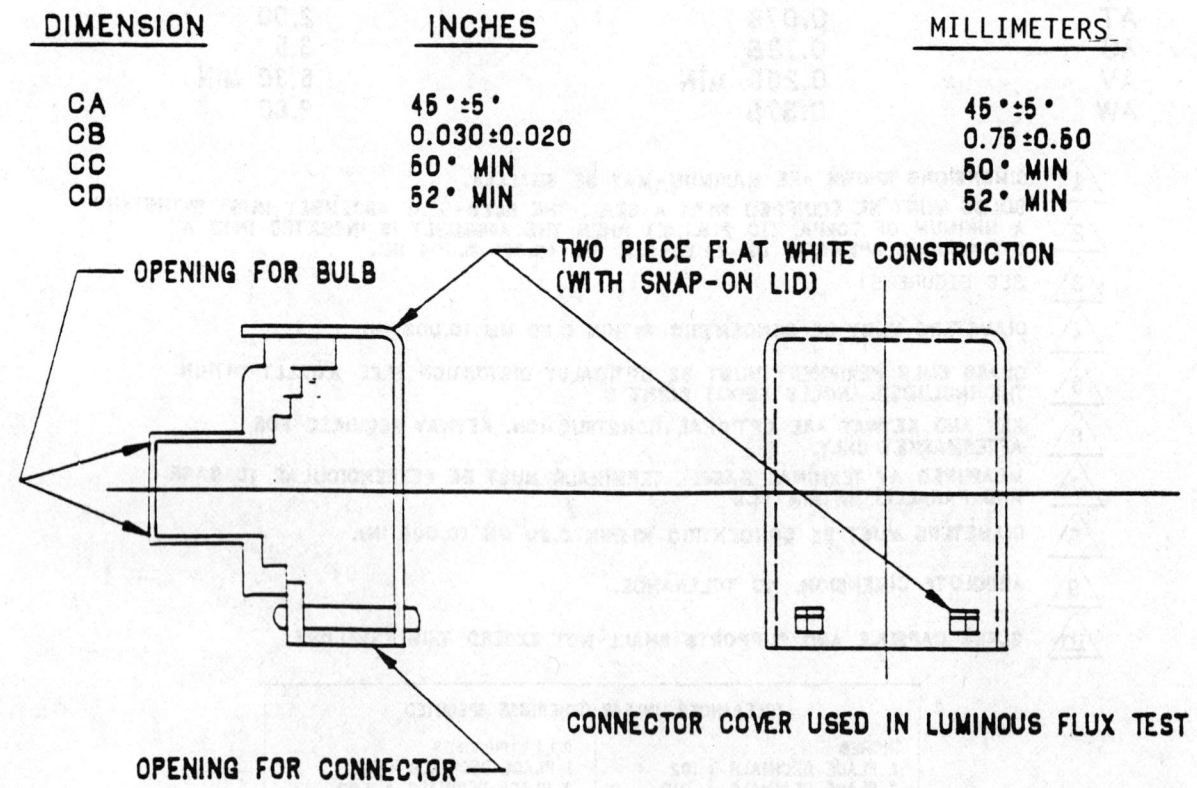

FIGURE 48—SPECIFICATION FOR THE 9006 REPLACEABLE BULB

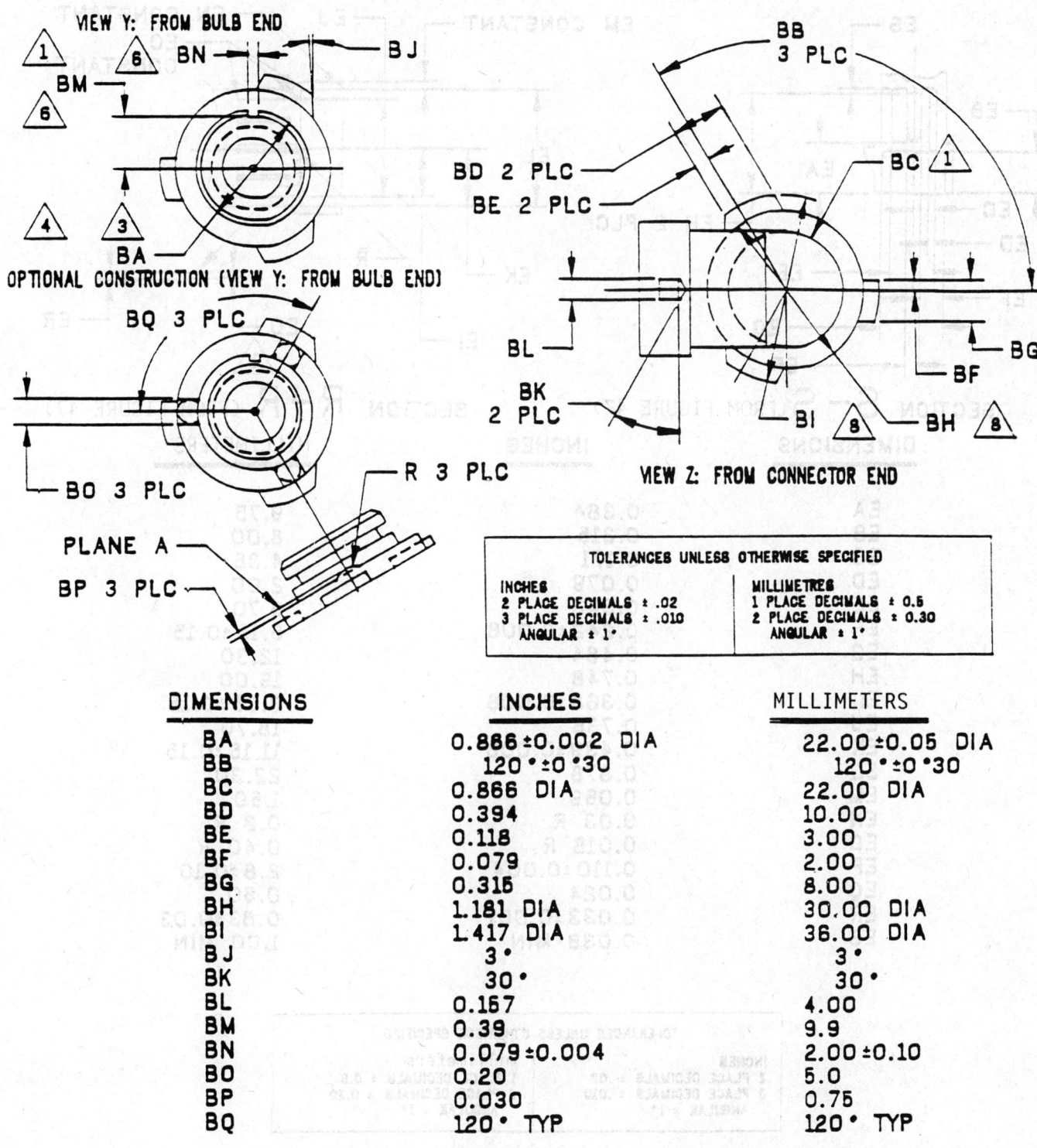

FIGURE 49—SPECIFICATION FOR THE 9006 REPLACEABLE BULB

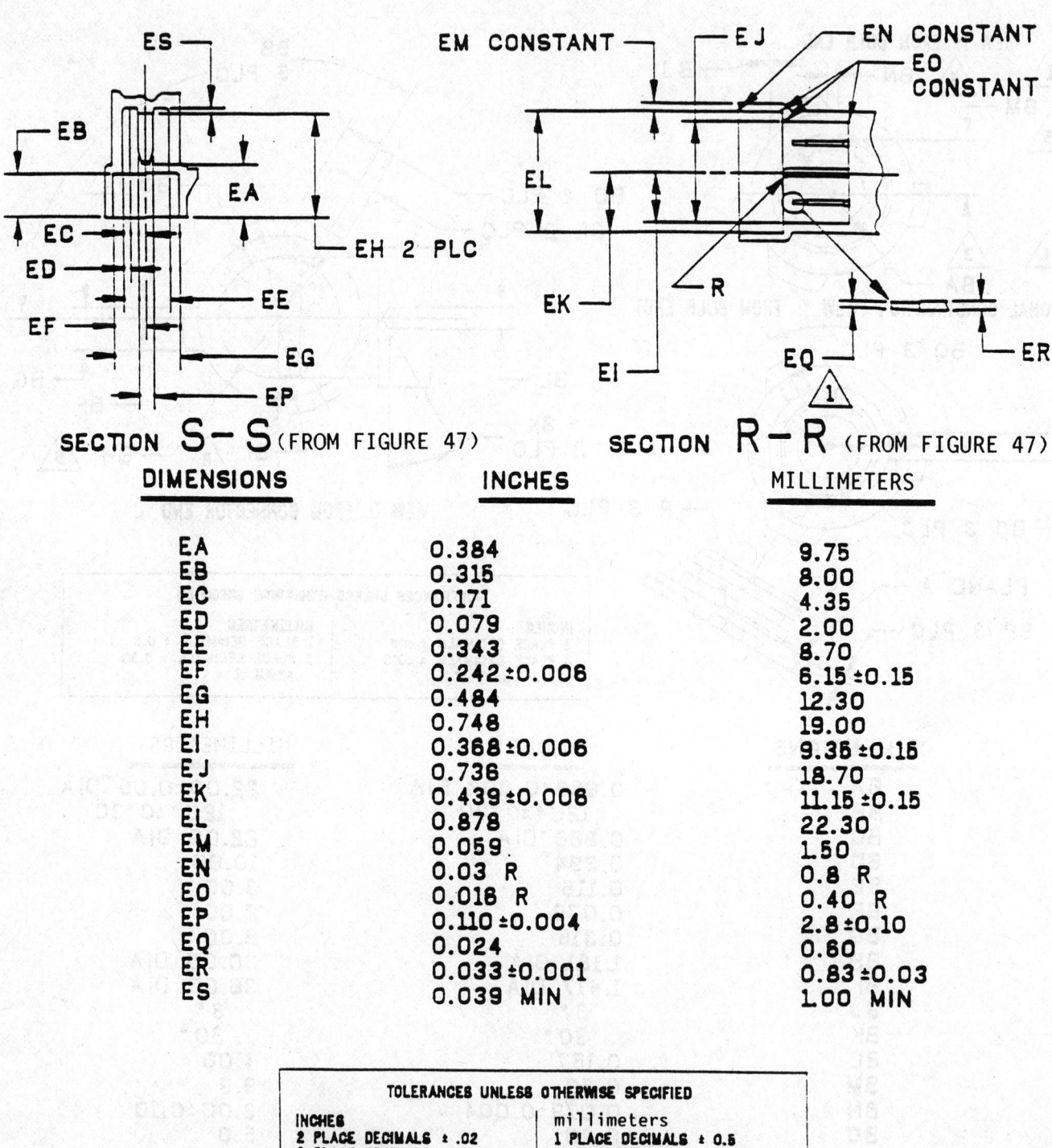

SECTION S-S (FROM FIGURE 47)　　　SECTION R-R (FROM FIGURE 47)

DIMENSIONS	INCHES	MILLIMETERS
EA	0.384	9.75
EB	0.315	8.00
EC	0.171	4.35
ED	0.079	2.00
EE	0.343	8.70
EF	0.242 ± 0.006	6.15 ± 0.15
EG	0.484	12.30
EH	0.748	19.00
EI	0.368 ± 0.006	9.35 ± 0.15
EJ	0.736	18.70
EK	0.439 ± 0.006	11.15 ± 0.15
EL	0.878	22.30
EM	0.059	1.50
EN	0.03 R	0.8 R
EO	0.016 R	0.40 R
EP	0.110 ± 0.004	2.8 ± 0.10
EQ	0.024	0.60
ER	0.033 ± 0.001	0.83 ± 0.03
ES	0.039 MIN	1.00 MIN

TOLERANCES UNLESS OTHERWISE SPECIFIED	
INCHES 2 PLACE DECIMALS ± .02 3 PLACE DECIMALS ± .010 ANGULAR ± 1°	millimeters 1 PLACE DECIMALS ± 0.5 2 PLACE DECIMALS ± 0.30 ANGULAR ± 1°

FIGURE 50—SPECIFICATION FOR THE 9006 REPLACEABLE BULB

24.85

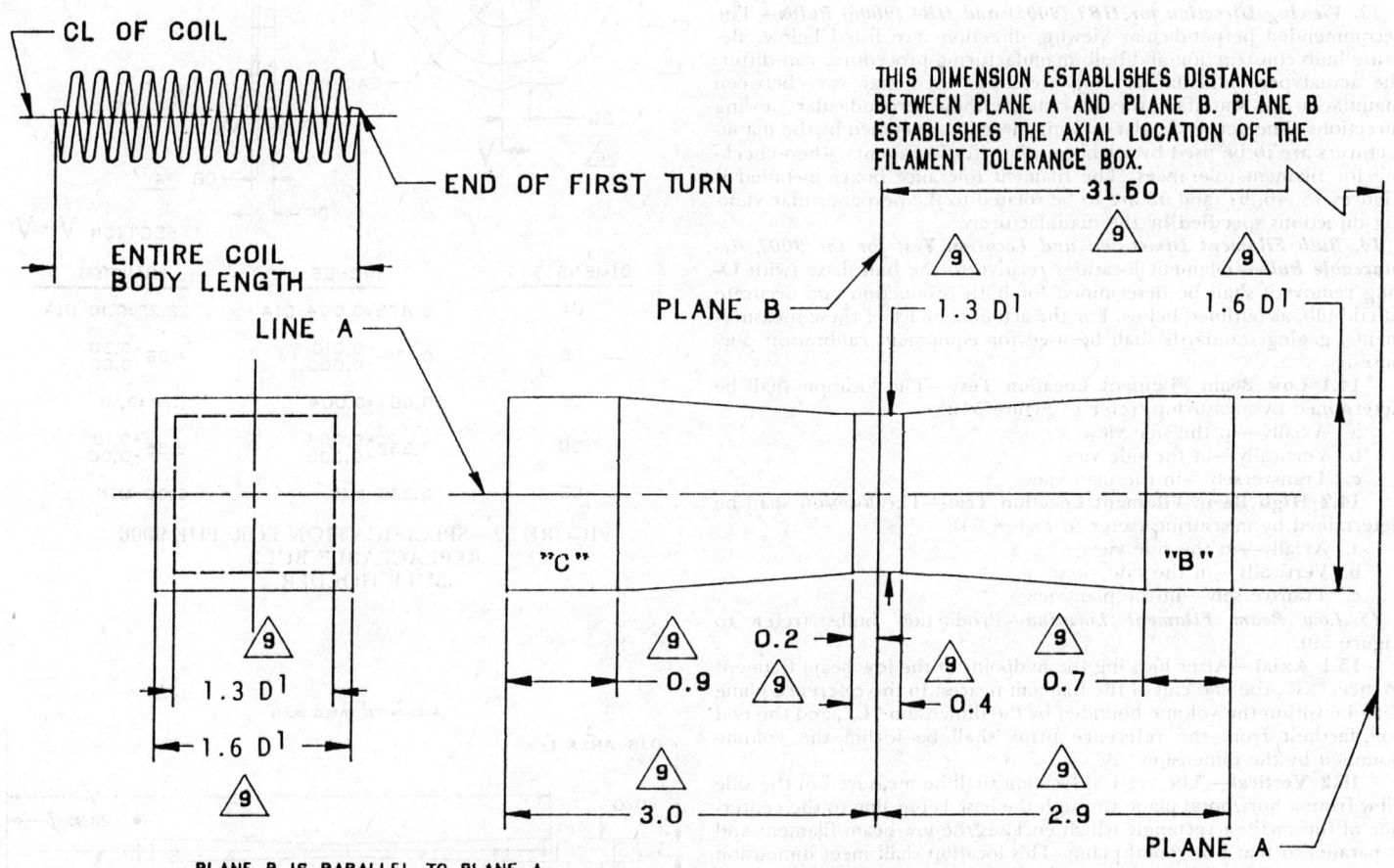

FIGURE 51—SPECIFICATION FOR THE 9006 REPLACEABLE BULB

12. Low Beam Filament Location
  12.1 Production Bulbs
  12.1.1 AXIAL—Same as 11.1.1 except Figure 51.
  12.1.2 VERTICAL—Same as 12.1.1.
  12.1.3 TRANSVERSE—Same as 12.1.1.
  12.2 Accurate Rated Bulbs
  12.2.1 AXIAL—Same as 11.2.1 except Figure 53.
  12.2.2 VERTICAL
    12.2.2.1 *End Coils*—Same as 11.2.2.1 except Figure 53.
    12.2.2.2 *Center Section*—Same as 11.2.2.2 except Figure 53.
  12.2.3 TRANSVERSE
    12.2.3.1 *End Coils*—Same as 12.2.2.1.
    12.2.3.2 *Center Section*—Same as 12.2.2.2.

13. Viewing Direction for HB3 (9005) and HB4 (9006) Bulbs—The recommended perpendicular viewing directions are listed below. Because bulb construction and bulb manufacturing procedures can differ, the actual perpendicular viewing directions used may vary between manufacturers. Manufacturers may choose their perpendicular viewing directions. The perpendicular viewing directions specified by the manufacturers are to be used by a laboratory or testing agency when checking for filament tolerances. The filament tolerance boxes included if Figures 43, 46, 51, and 53 are to be rotated to the perpendicular viewing directions specified by the manufacturer.

14. Bulb Filament Dimension and Location Test for the 9007 Replaceable Bulb—Filament locations relative to the bulb base (with O-ring removed) shall be determined for both production and accurate rated bulb, as outlined below. For the actual conduct of these measurements, gaging standards shall be used for equipment calibration purposes.

  14.1 Low Beam Filament Location Test—The location shall be determined by measuring (refer to Figure 54):
   a. Axially—in the side view
   b. Vertically—in the side view
   c. Transversely—in the plan view
  14.2 High Beam Filament Location Test—The location shall be determined by measuring (refer to Figure 54):
   a. Axially—in the side view
   b. Vertically—in the side view
   c. Transversely—in the plan view

15. Low Beam Filament Location—Production bulbs (refer to Figure 59).
  15.1 Axial—After locating the midpoint of the low beam filament to meet "G", the end coil of the filament nearest to the reference plane shall be within the volume bounded by the dimension "C", and the end coil farthest from the reference plane shall be within the volume bounded by the dimension "B".
  15.2 Vertical—The vertical location shall be measured in the side view from a horizontal plane through the base centerline to the centerline of the smallest rectangle which encloses the low beam filament and is parallel to that horizontal plane. This location shall meet dimension "A". The width of this rectangle shall not exceed 1.6X the diameter of the low beam coil.
  15.3 Transverse—The transverse location shall be measured in the plan view from a vertical plane through the center of the base to the centerline of the smallest rectangle which encloses the low beam filament and is parallel to that plane. This location shall meet the dimension "L". The width of this rectangle shall not exceed 1.6X the diameter of the low beam coil.

16. High Beam Filament Location—Production bulbs (refer to Figure 59).
  16.1 Axial—The filament location shall be measured from the midpoint of the low beam filament to the midpoint of the smallest rectangle which encloses the high beam filament image.
  16.2 Vertical—The location shall be measured from the centerline of the low beam filament to the centerline of the smallest rectangle which encloses the high beam filament image and is parallel to the horizontal plane referenced in 15.2. This location shall not exceed dimension "J" and the width of the rectangle shall not exceed 1.6X the diameter of the high beam filament coil.
  16.3 Transverse—The location shall be measured from the centerline of the low beam filament to the centerline of the smallest rectangle which encloses the high beam filament image and is parallel to that plane referenced in 15.3. This location shall not exceed dimension "H" and the width of the rectangle shall not exceed 1.6X the diameter of the high beam filament coil.

17. Methods of Measuring Internal Elements of H4/HB2 Bulbs
  17.1 These paragraphs specify the methods of measuring internal elements of H4 and HB2 bulbs.

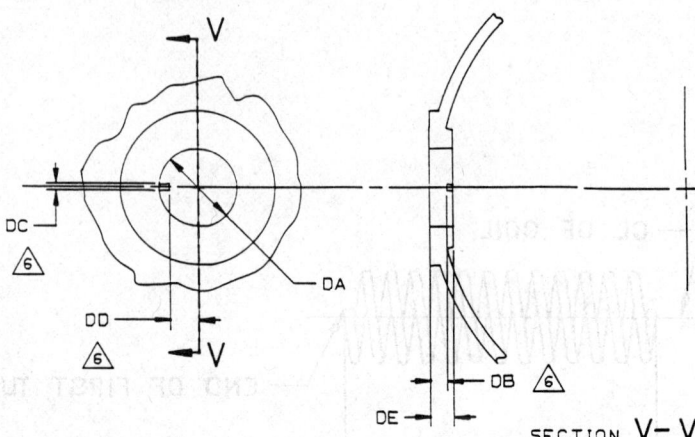

DIMENSIONS	INCHES	MILLIMETERS
DA	0.875 ± 0.004 DIA	22.22 ± 0.10 DIA
DB	0.172 +0.010/-0.000	4.36 +0.30/-0.00
DC	0.067 ± 0.004	1.70 ± 0.10
DD	0.392 +0.004/-0.000	9.95 +0.10/-0.00
DE	0.236 MIN	6.00 MIN

FIGURE 52—SPECIFICATION FOR THE 9006 REPLACEABLE BULB BULB HOLDER

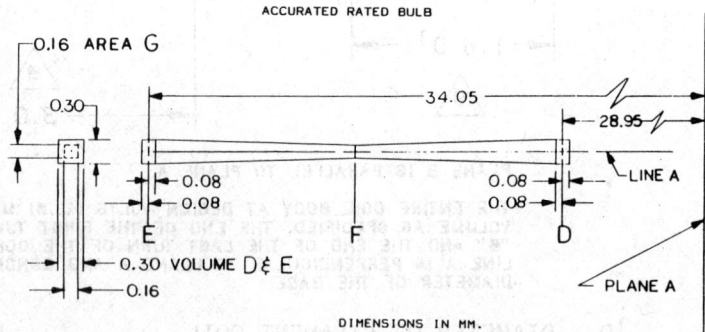

THE CENTROID OF THE FIRST TURN OF THE COIL MUST BE WITHIN VOLUME D AND THE CENTROID OF THE LAST TURN OF THE COIL MUST BE WITHIN VOLUME E. F IS AT THE MID-LENGTH OF THE COIL. THE CENTROID AT F MUST BE WITHIN AREA G.

FIGURE 53—SPECIFICATION FOR THE 9006 REPLACEABLE BULB ACCURATE RATED BULB

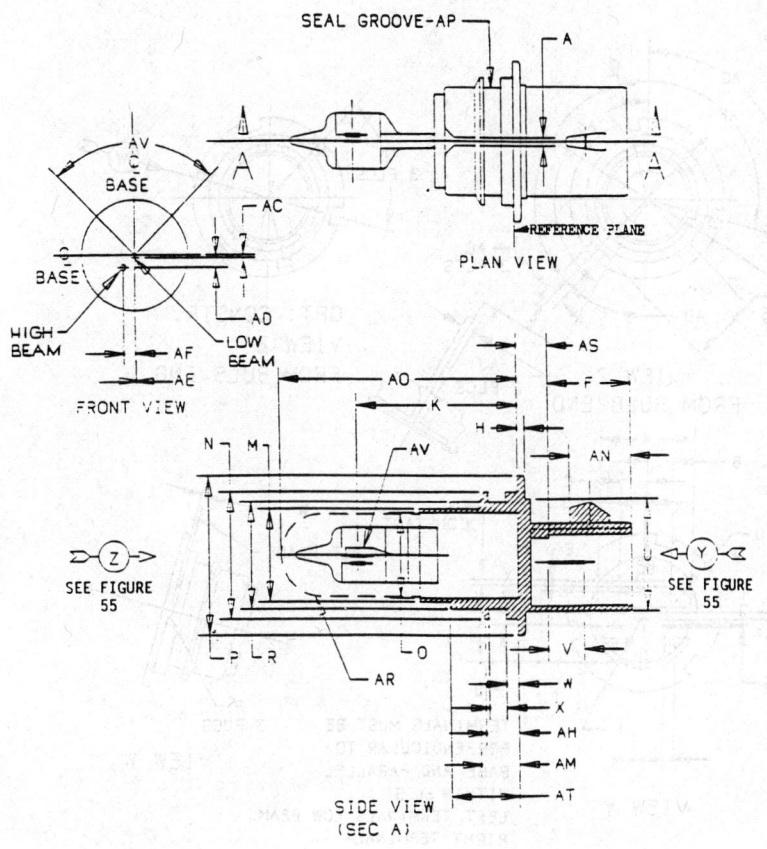

Dimension	Millimeters
A	(2.15/2.10) .05 Either Side CL
F	23.00 ± .20
H	2.00 ± .20
K Low Beam	44.50 ± .25
High Beam	CL High Beam to be within ± .64 of CL of low beam
M	24.85 Max.
N	(33.90/33.80) .05 Either Side CL
O	24.5 Max.
P	42.50 ± .20
R	(28.60/28.50) .05 Either Side CL
U	30.00 ± .20
V	10.50 ± .50
W	3.25 ± .20
X	4.80 ± .20
AC	0.38 ± .38
AD	1.60 ± .64
AE	.000 ± .38
AF	1.60 ± .81
AH	9.05 ± .20
AM	10.54 ± .20
AN	17.10 ± .20
AO	70.0 Max.
AP	Seal must withstand a minimum of 70 kPa (10 PSIG) when bulb-seal assembly is inserted into a cylindrical aperture of 34.3/34.2 mm (1.350/1.346 in).
AR	Glass capsule and supports shall not exceed this envelope.
AS	8.5 ± 2.0
AT	16.00 Min.
AV	Support wires extending forward of the filaments shall be within ± 45° of vertical.

FIGURE 54—SPECIFICATION FOR THE 9007 REPLACEABLE BULB

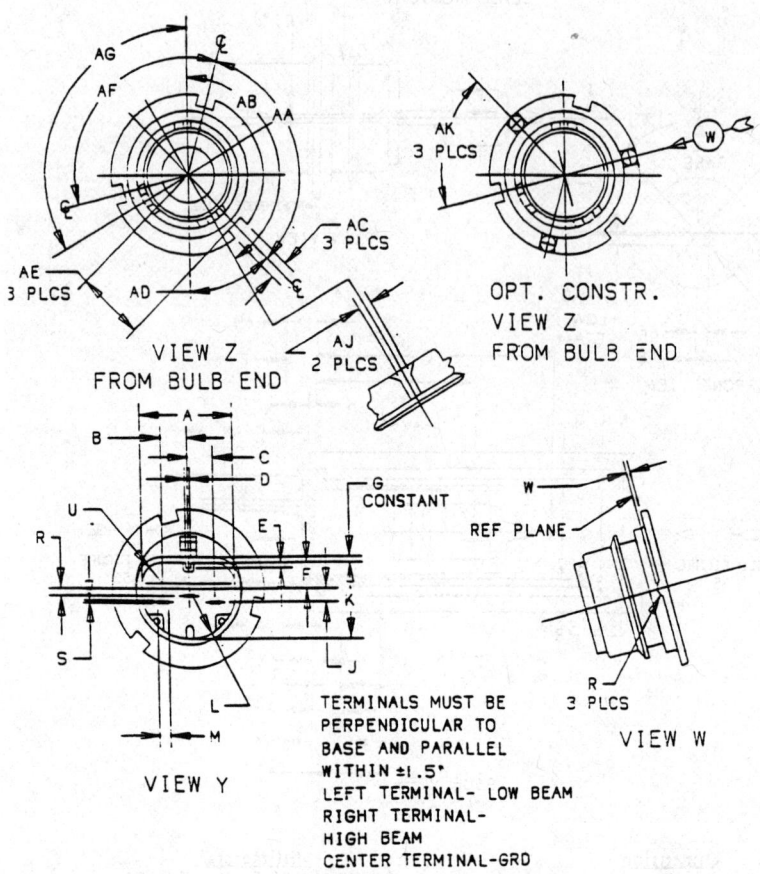

Dimension	Millimeters
AA	120°
AB	150°
AC	4.9 Min.
AD	44° 30'
AE	18.35 ± 0.20
AF	120°
AG	120°
AJ	3.6 ± .20
AK	60°
A	26.10 ± .20
B	7.35 ± .25
C	7.35 ± .25
D	1.30 ± .20
E	1.40 ± .20
F	7.05 ± .15
G	1.50 ± .20
J	3.60 ± .25
K	20.60 ± .20
L	13.60 ± .20R
M	3.00 ± .10
R	1.90 ± .25
S	.63 ± .05
U	5.65 ± .20R
W	.25 ± .15

TOLERANCE FOR ALL ANGULAR DIMENSIONS ± 1°

FIGURE 55—SPECIFICATION FOR THE 9007 REPLACEABLE BULB

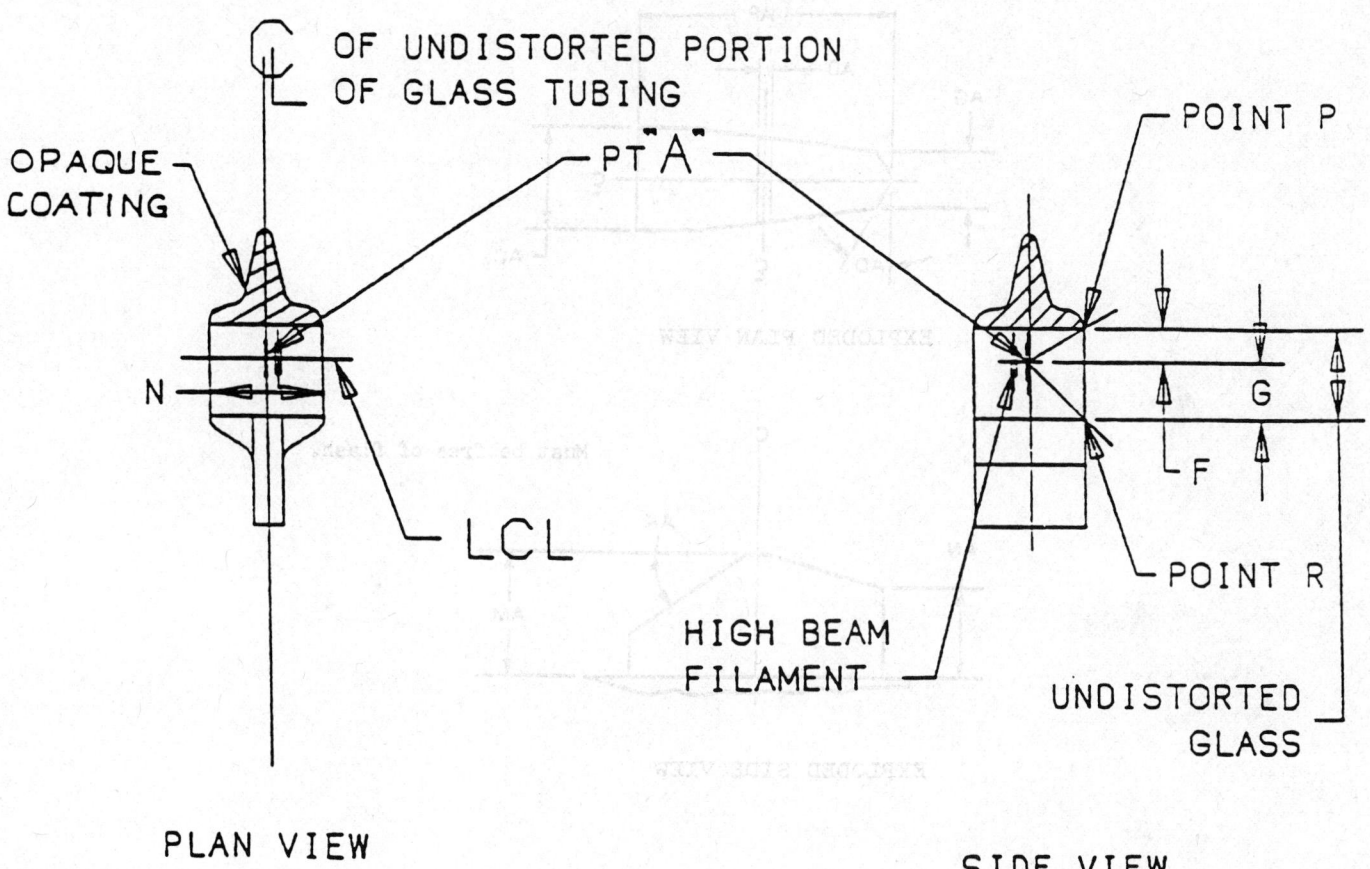

Dimensional Specifications
Figure 56

Dimension	
F	$(N/2)\tan 38° \pm 1.0$ mm
G	$(N/2)\tan 43°$ MIN
N	Actual Capsule Dia. (To Be Established By Manufacturer)
P	Entire Radius and Distorted Glass Shall Be Covered to the Plane Passing Through Point "P", Perpendicular to the Glass Capsule Centerline.

FIGURE 56—SPECIFICATION FOR THE 9007 REPLACEABLE BULB

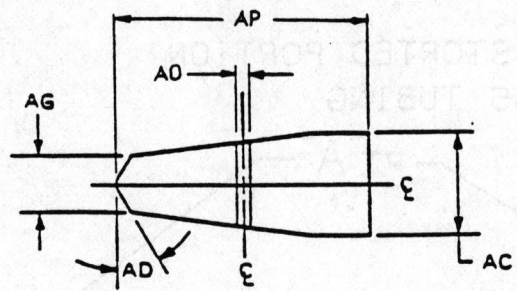

EXPLODED PLAN VIEW

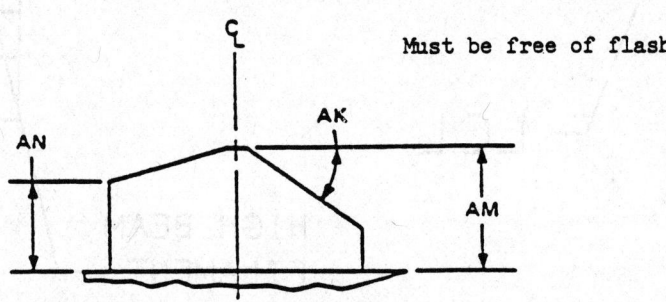

Must be free of flash.

EXPLODED SIDE VIEW

Dimension	Millimeters
AC	4.55 ± .20
AD	30° ± 3°
AG	2.50 ± .20
AK	35° ± 3°
AM	5.50 ± .20
AN	4.00 ± .20
AO	.5 ± .20
AP	11.4 ± .20

FIGURE 57—SPECIFICATION FOR THE 9007 REPLACEABLE BULB LOCKING FEATURE

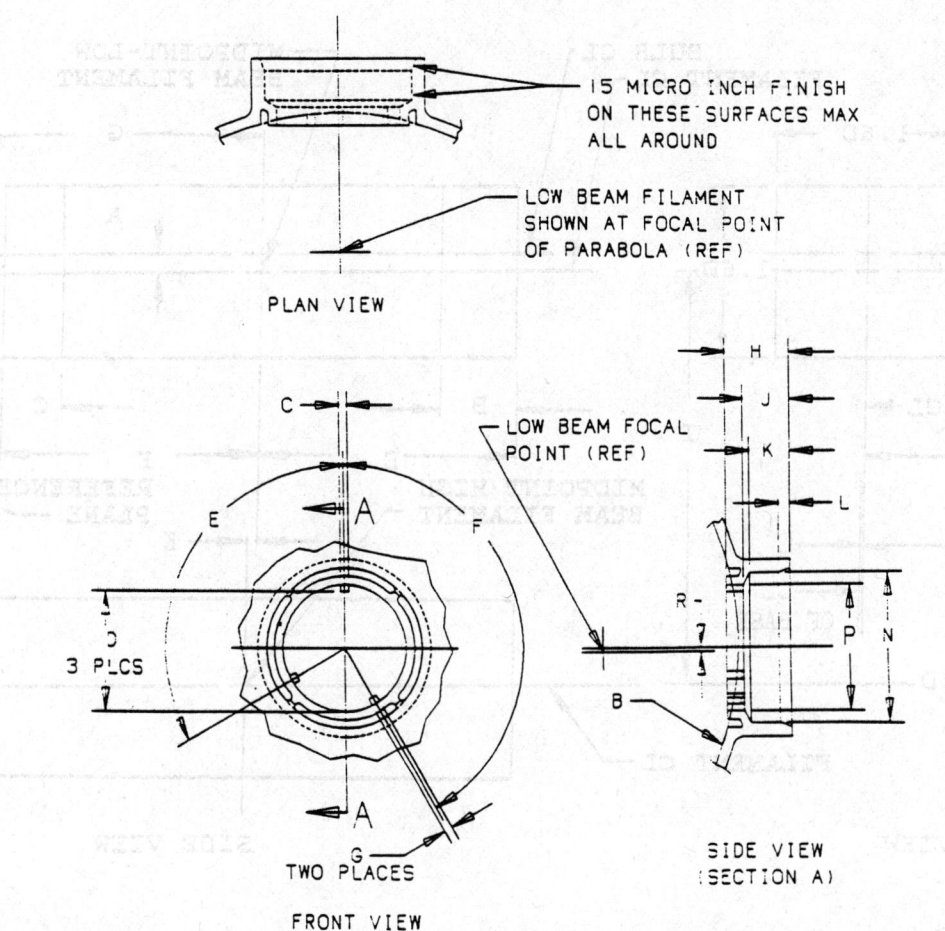

Dimension	Millimeters
B	Ref Line Lamp Parabola
C	2.00 ± .05
	.05 Either Side of CL
D	27.10 ± .20
E	120°
F	150°
G	2.00 ± .20
H	15.15 ± .20
J	11.10 ± .20
K	9.50 ± .20
L	2.75 ± .20
N	34.24 +.08/−.05
P	28.70 +.10/−.05
	Diameter P shall be concentric to diameter N within ± .05
R	0.38 ± 0.10
	TOLERANCE FOR ALL ANGULAR DIMENSIONS ± 1°

FIGURE 58—SPECIFICATION FOR THE 9007 REPLACEABLE BULB BULB HOLDER

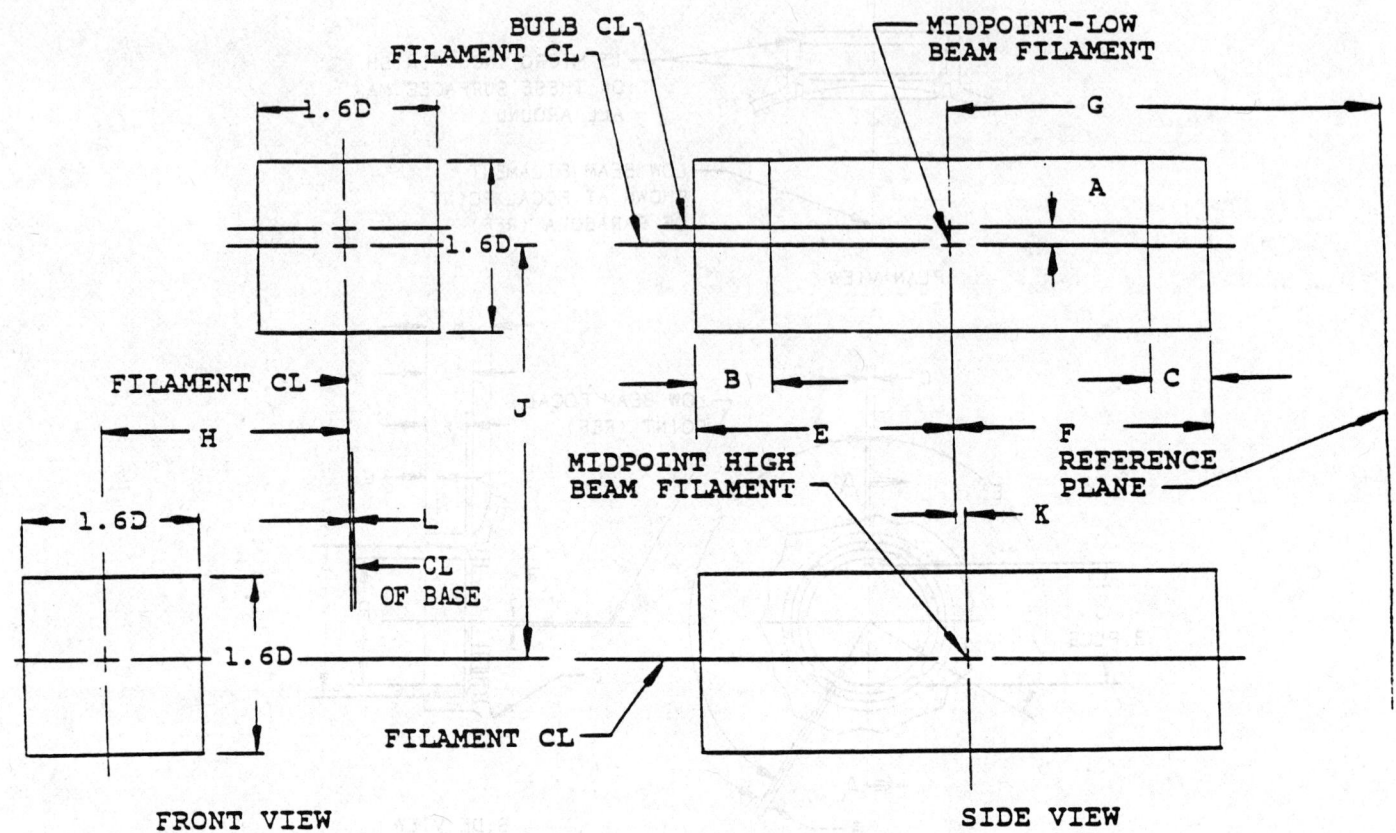

LETTER	STANDARD DIMENSION	ACCURATE RATED BULB
A	0.38 ± 0.38 mm	0.38 ± 0.20 mm
B	0.9 Basic	---
C	0.7 Basic	---
D	Actual Filament Diameter	---
E	3.0 Basic	---
F	2.9 Basic	---
G	44.50 ± 0.25	44.50 ± 0.15
H	1.60 ± 0.81	1.60 ± 0.25
J	1.60 ± 0.64	1.60 ± 0.25
K	000 ± 0.64	000 ± 0.40
L	000 ± 0.38	000 ± 0.25

FIGURE 59—DIMENSIONAL SPECIFICATIONS FOR THE 9006 REPLACEABLE BULB FILAMENT DIMENSION AND LOCATION—MEASUREMENT METHOD

The drawing is not mandatory, their sole purpose is to show which dimensions must be verified.

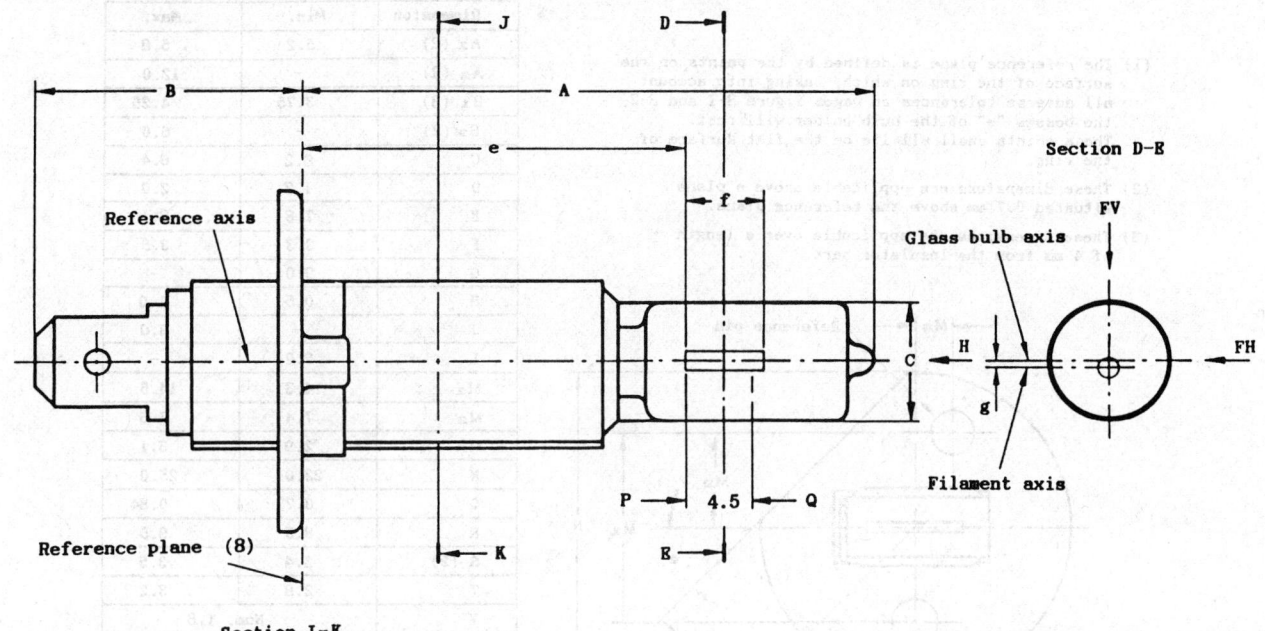

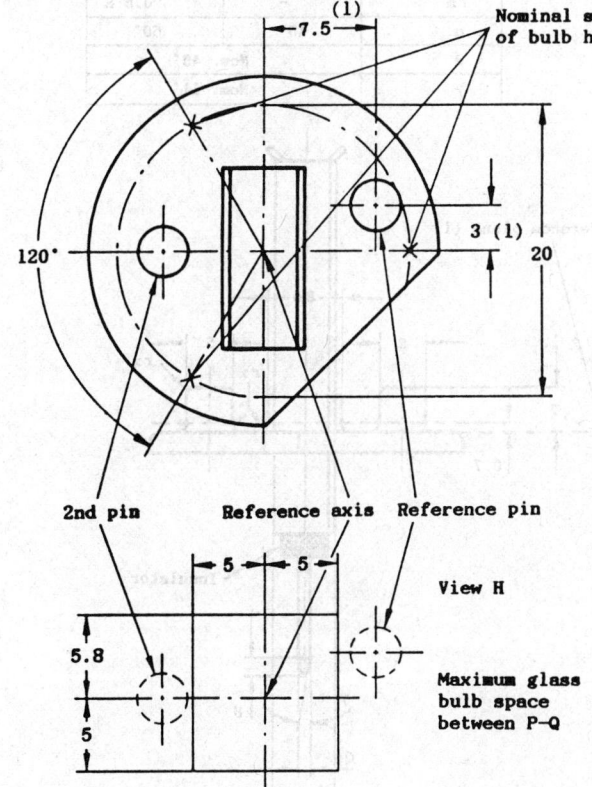

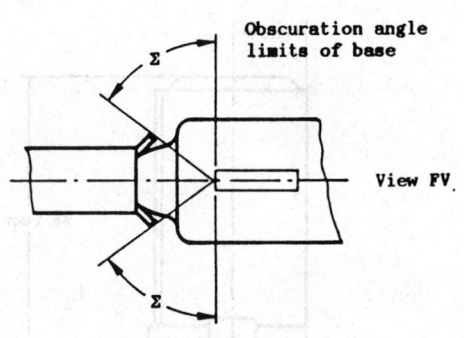

Dimensions in millimeters

Reference	Dimension	Tolerance
A	44 max.	—
B	18.5 max.	—
C	8.5 max.	—
e (6)	25	—
f (2)(3)(6)	5.0	± 0.5
g (4)(5)	0.5d	± 0.5d
Σ	45°	± 12°

FIGURE 60—SPECIFICATION FOR THE TYPE H1 REPLACEABLE BULB

The drawing is intended only to indicate
the dimensions essential for interchangeability.

(1) The reference plane is defined by the points on the surface of the ring on which, taking into account all adverse tolerances on pages Figure 8-1 and 8-2, the bosses "e" of the bulb holder will rest. These points shall all lie on the flat surface of the ring.
(2) These dimensions are applicable above a plane situated 0.7 mm above the reference plane.
(3) These dimensions are applicable over a length of 4 mm from the insulator part.

Dimensions in millimeters

Dimension	Min.	Max.
A_1 (2)	5.2	5.8
A_2 (3)	–	12.0
B_1 (3)	3.75	4.25
B_2 (2)	–	6.0
C	6.2	6.4
D	1.7	2.0
E	7.8	8.1
F	3.3	3.5
G	9.0	–
H	0.5	1.0
J	–	3.0
L	5.0	–
M_1	14.3	14.5
M_2	7.4	7.6
M_3	2.9	3.1
N	23.0	25.0
Q	0.77	0.84
R	8.5	9.5
S (2)	3.4	3.5
T	2.8	3.2
V	Nom. 1.6	
Y	–	18.5
r_1	–	0.6
r_2	–	0.5 S
α	40°	50°
β	Nom. 45°	
γ	Nom. 11°	

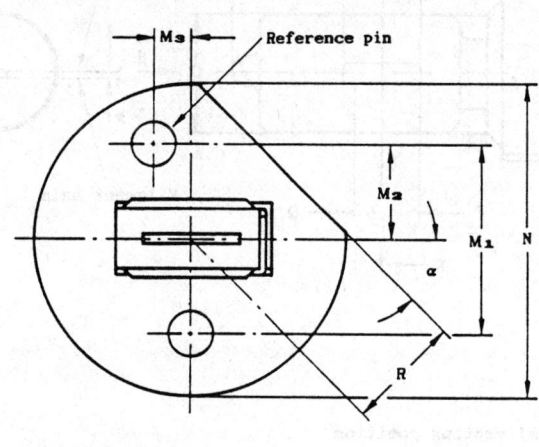

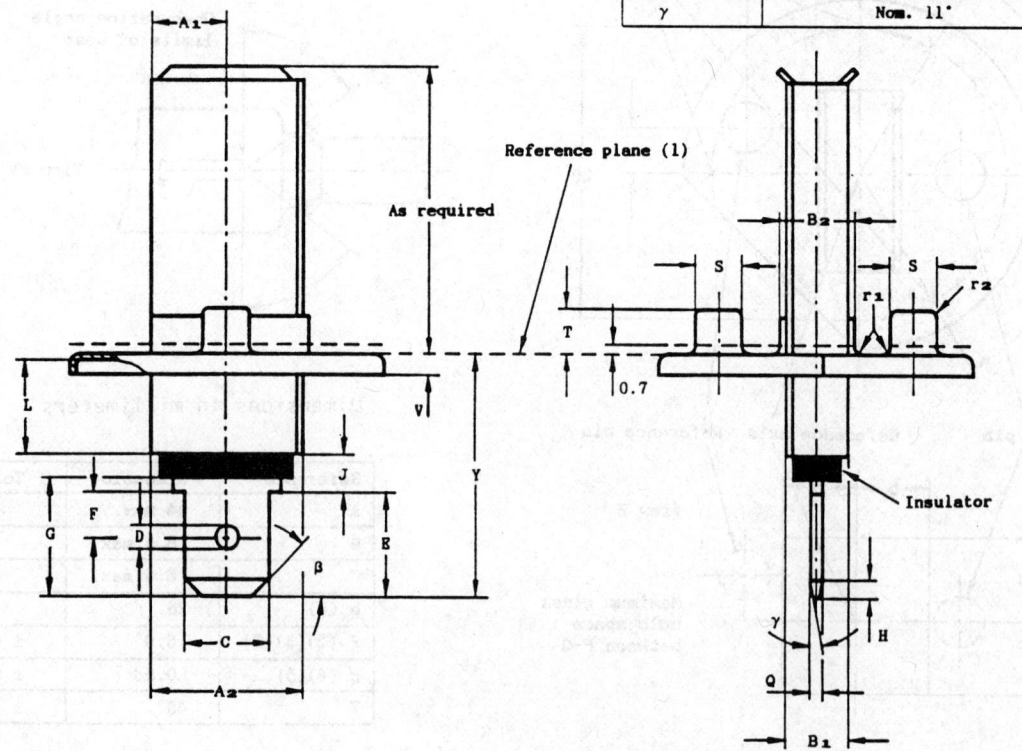

FIGURE 61—SPECIFICATION FOR THE TYPE H1 REPLACEABLE BULB
BASE P14.5s

The drawings are intended only to indicate the dimensions essential for interchangeability.

Section I-I

Dimensions in millimeters

Dimension	Min.	Max.
A_1	6.1	6.3
A_4	11.7	-
A_5	7.0	7.5
B_2	7.0	7.5
B_3	4.0	4.2
M_1	Nom. 14.5	
M_2	7.4	7.6
M_3	2.9	3.1
M_4	18.1	18.3
S	3.6	3.7
U_1	0.8	1.0
U_2	1.8	2.2
X	9.0	9.2
Z	19.5	20.5
δ	40°	45°
θ	59°	61°

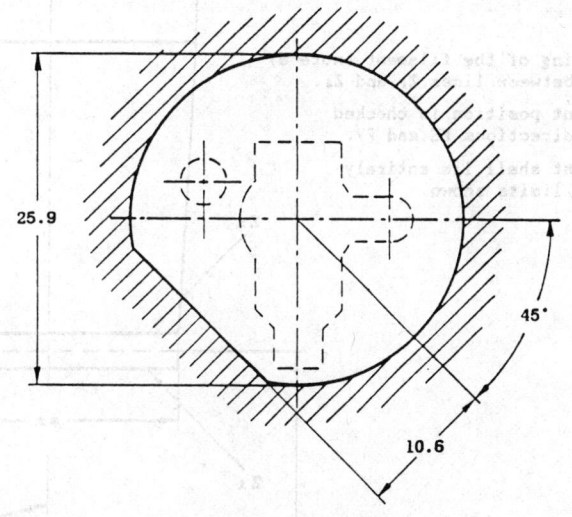

Minimum free space for the base ring

The correct orientation of the bulb is made by the apertures "t_1" and "t_2".
The three bosses "e" determine the reference plane.

The holder shall be so designed that the means of retention can be applied only when the bulb is in the correct position.

The means of retention shall make contact only with the prefocus ring of the base, and the total force exerted when the bulb is in position, shall be not less than 10 N and not greater than 60 N.

FIGURE 62—SPECIFICATION FOR THE TYPE H1 REPLACEABLE BULB
BULB HOLDER P14.5s

(1) These dimensions define the reference axis.

(2) The longer lead wire should be positioned above the filament (the bulb being viewed as shown in the figure).
The internal design of the bulb should then be such that stray light images and reflections are reduced to the minimum e.g. by fitting cooling jackets over the non-coiled parts of the filament.

(3) The cylindrical portion of the glass bulb over length "f" shall be such as not to deform the projected image of the filament to such an extend as appreciable to affect the optical results.

(4) Offset of filament in relation to glass bulb axis measured at 27.5 mm from the reference plane in direction FV.

(5) d = actual diameter of filament.

(6) The ends of the filament are defined as the points where, when the viewing direction as defined in foot-note 7, the projection of the outside of the end turns nearest to or furthest from the reference plane crosses the reference axis.

(7) The viewing direction is the perpendicular to the reference axis contained in the plane defined by the reference axis and the centre of the second pin of the base.

(8) The reference plane is the plane formed by the seating points of the three bosses of the bulb holder on the base ring.

Filament position requirements

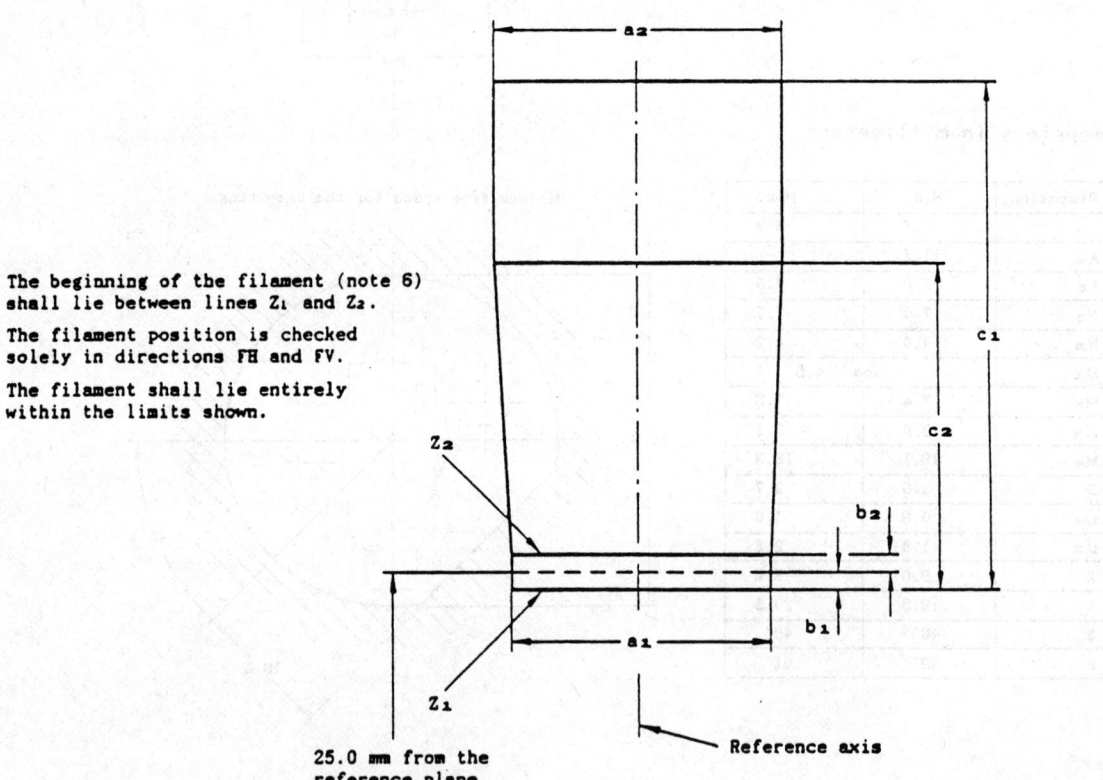

Dimensions in millimeters

Reference	Dimensions
a_1 (5)	1.4d
a_2 (5)	1.9d
b_1, b_2	0.25
c_1	7
c_2	4.5

FIGURE 63—SPECIFICATION FOR THE TYPE H1 REPLACEABLE BULB

Additional requirements for accurate rated bulbs

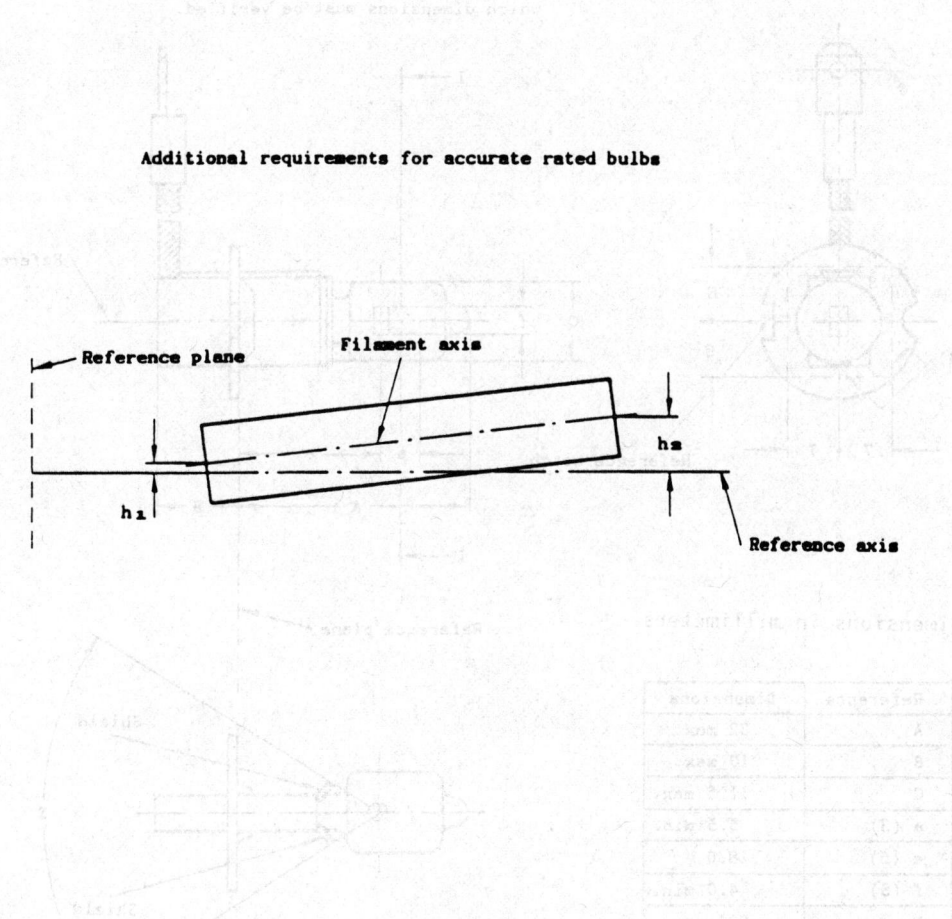

Dimensions in millimeters

Reference	Dimension	Tolerance
e (1)	25	± 0.15
f (1)	5.0	+ 0.5
g (1)	0.5d (2)	± 0.25d
Σ (1)	45°	± 3°
h₁	0	± 0.20
h₂	0	± 0.25

(1) See Figure 60
(2) d = actual filament diameter

FIGURE 64—SPECIFICATION FOR THE TYPE H1 REPLACEABLE BULB

24.98

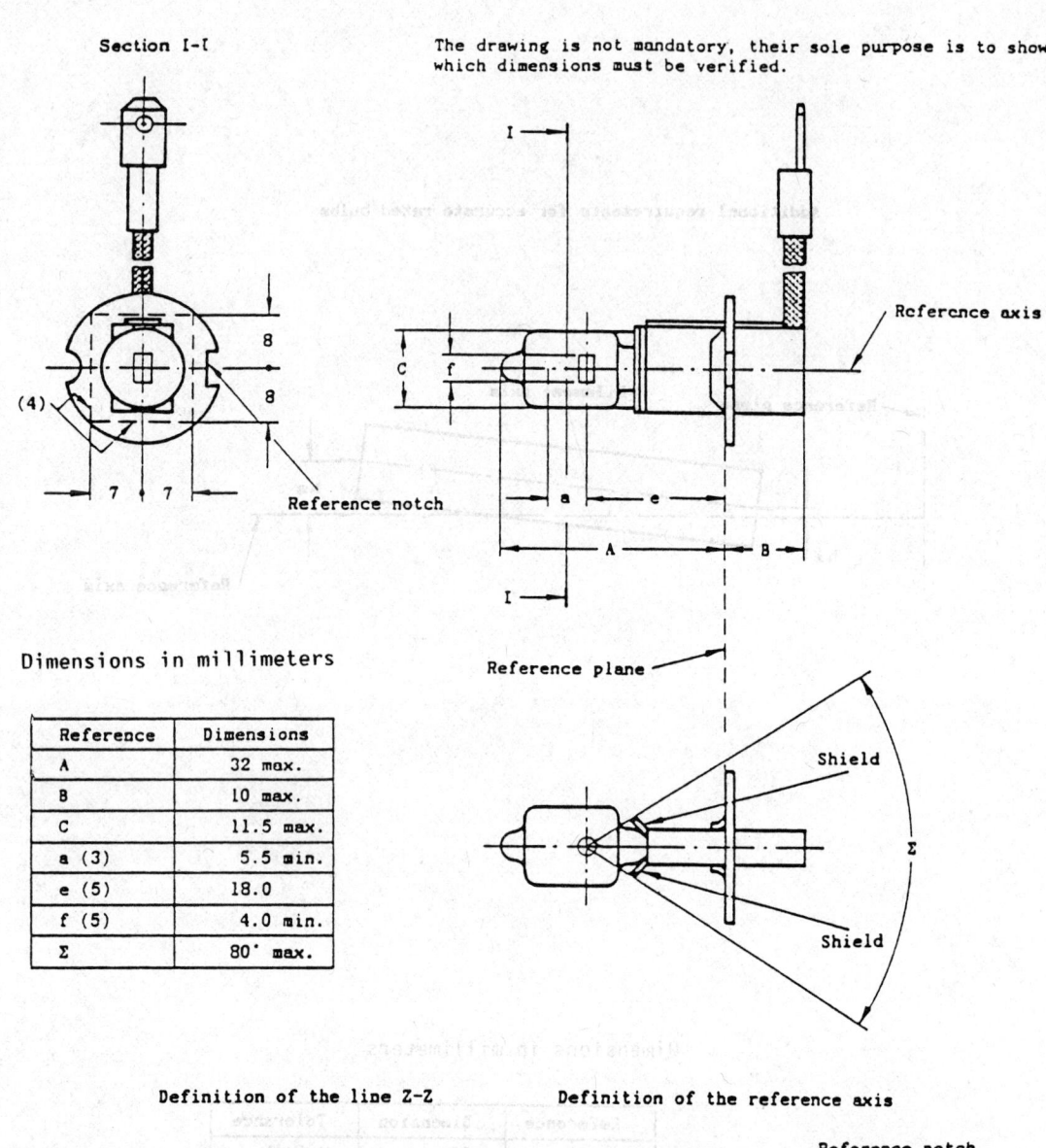

FIGURE 65—SPECIFICATION FOR THE TYPE H3 REPLACEABLE BULB

The drawings are intended only to indicate
the dimensions essential for interchangeability.

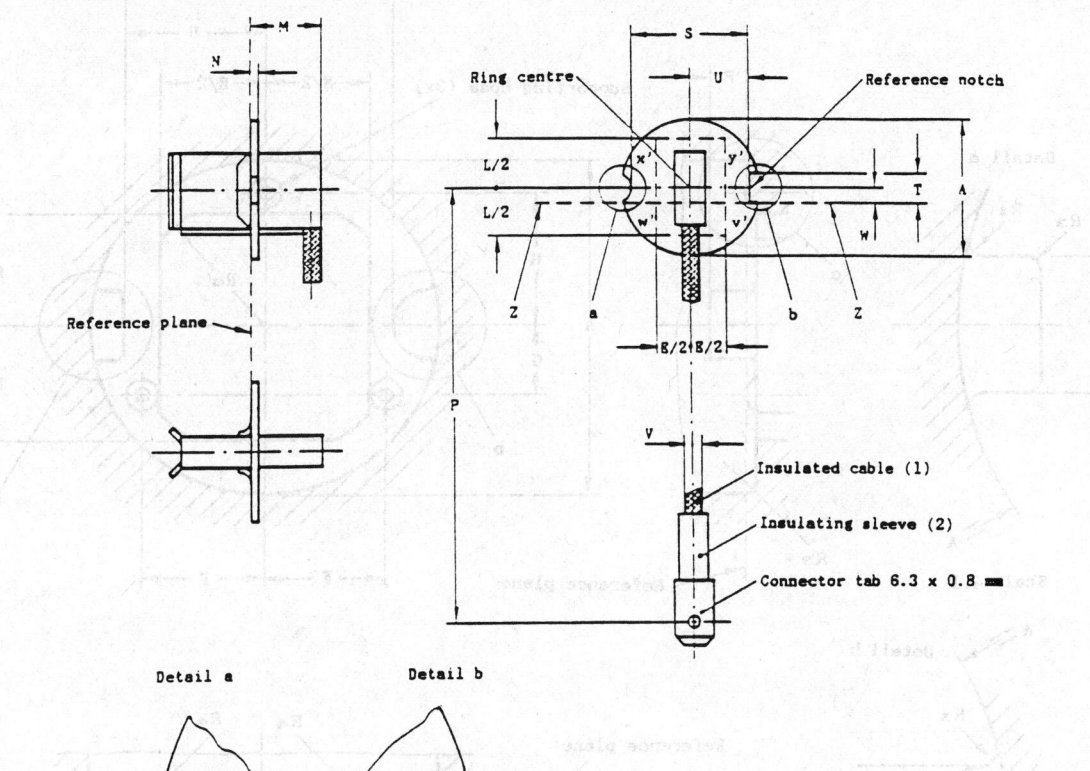

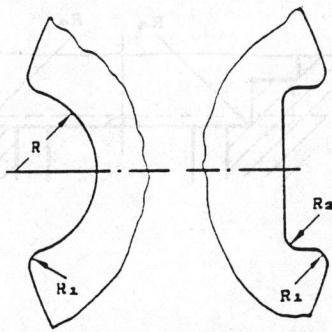

Detail a   Detail b

(1) It shall be possible to bend the cable within a cylinder of 22.2 mm diameter co-axial with the axis of the ring.
(2) The insulating sleeve shall be securely fastened, shall adequately overlap the wire insulation and shall cover all metal parts up to the shoulders of the tab.
(3) The space to be reserved for the parts of the base below the ring—with the exception of the cable outlet, is bounded by a rectangular box of x', y', v', w'.
(4) A reduction of the minimum value is under consideration.
(5) This dimension is not to be gauged.
(6) Outside the area defined by x', y', v' and w', the flatness of the ring, on the reference plane side, shall be within 0.25 mm (0.01 in).

Dimensions in millimeters

Dimension	Min.	Max.
A	22.15	22.25
E (3)(6)	11.0	
L (3)(6)	16.0	
M	–	10.0
N (4)	0.7	1.1
P	95	105
R	2.5	2.6
R_1	–	0.4
R_2	–	0.5
S	18.1	18.3
T	5.0	5.1
U	9.55	9.65
V (5)	1.75	2.75
W	2.0	3.0

FIGURE 66—SPECIFICATION FOR THE TYPE H3 REPLACEABLE BULB BASE PK22s

24.100

The drawings are intended only to indicate
the dimensions essential for interchangeability.

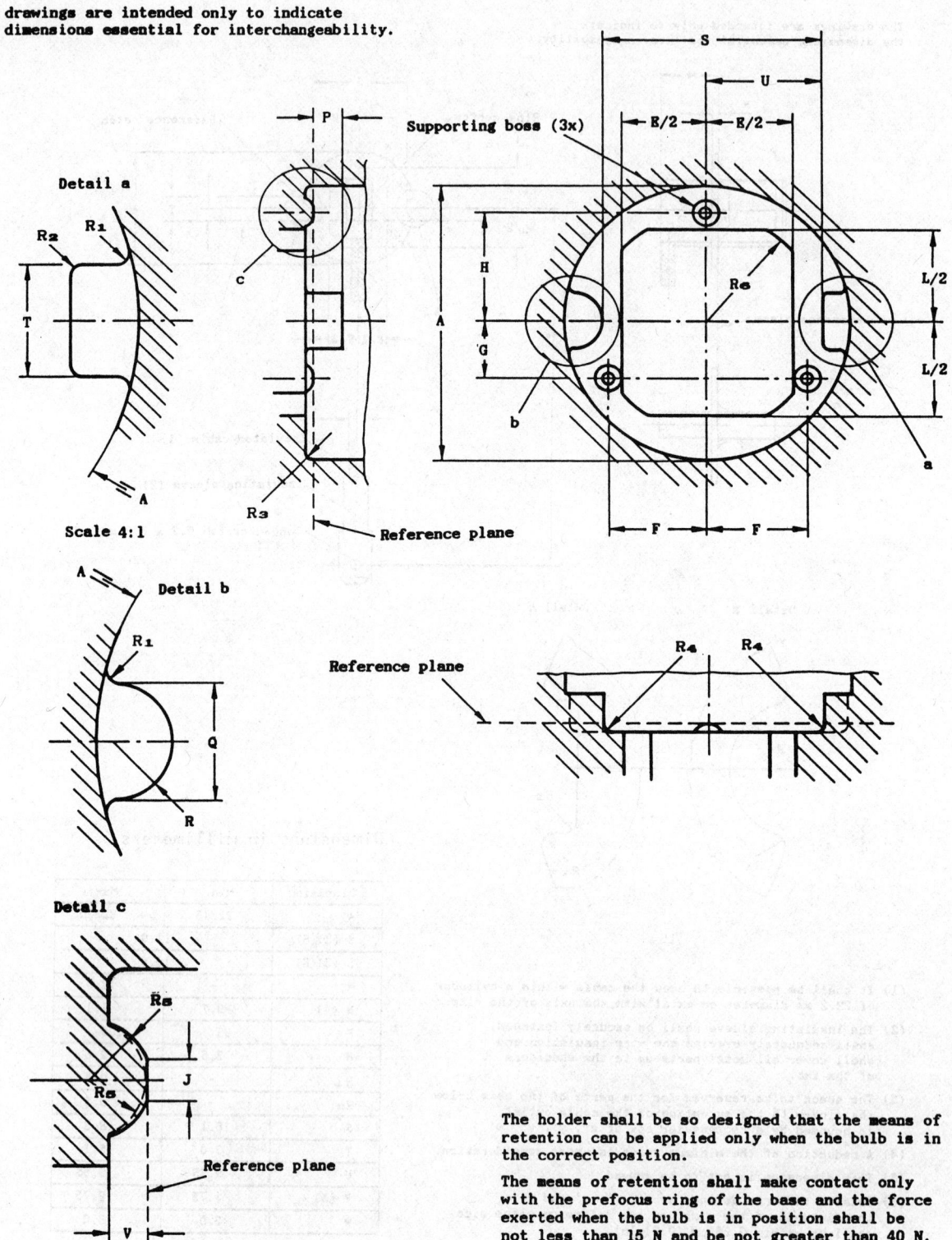

The holder shall be so designed that the means of
retention can be applied only when the bulb is in
the correct position.

The means of retention shall make contact only
with the prefocus ring of the base and the force
exerted when the bulb is in position shall be
not less than 15 N and be not greater than 40 N.

FIGURE 67—SPECIFICATION FOR THE TYPE H3 REPLACEABLE BULB
BULB HOLDER PK22s

(1) The maximum difference in height between the supporting bosses shall not exceed 0.1 mm when the value of dimension V of the smallest boss is 0.3 mm.

If this value exceeds 0.3 mm then the difference in height may be increased accordingly.
(2) If the value of dimension A exceeds A min., then the values of dimensions R_1 max. and R_4 max. may be increased accordingly.
(3) If the value of dimension V exceeds V min., then the values of dimensions R_3 max. and R_4 max. may be increased accordingly.
(4) Dimensions E, L and R_6 denote the minimum free space to be reserved for the bulb.
(5) Dimension J denotes the allowed flat area.

Dimensions in millimeters

Dimension	Min.	Max.
A (2)	24	-
E (4)	14	-
F	8.05	8.35
G	4.65	4.95
H	9.35	9.65
J (5)	-	1
L (4)	16	-
P	2.5	-
Q	4.75	4.95
R	Q/2	
R_1	-	0.5 (2)
R_2	0.3	0.5
R_3	-	0.3 (2)(3)
R_4	-	0.3 (3)
R_5	1.2	-
R_6 (4)	9.5	-
S	18.35	18.65
T	4.75	4.95
U	9.70	9.85
V (1)(3)	0.3	0.8

FIGURE 67 (CONTINUED)

(1) The distortion of the base end-portion of the glass bulb shall not be visible from any direction outside the obscuration angle of 80° max.
The shields shall produce no inconvenient reflections.
The angle between the reference axis and the plane of each shield, measured on the glass bulb side, should not exceed 90°.

(2) The permissible deviation of the ring centre from the reference axis is 0.5 mm in the direction perpendicular to the Z-Z line, and 0.05 mm in the direction parallel to the Z-Z line.

(3) Minimum length above the height of the actual light emitting centre over which the glass bulb shall be cylindrical.

(4) No part of the spring and no component of the bulb holder shall bear on the prefocus ring elsewhere than outside the rectangle shown in discontinuous outline.

(5) The positions of the first and the last turn of the filament are defined by the intersections of the outside of the first and the outside of the last light emitting turn, respectively, with the plane parallel to and 18 mm distance from the reference plane.

(6) The reference plane is the plane formed by the seating points of the three bosses of the bulb holder on the base ring.

Filament position requirements

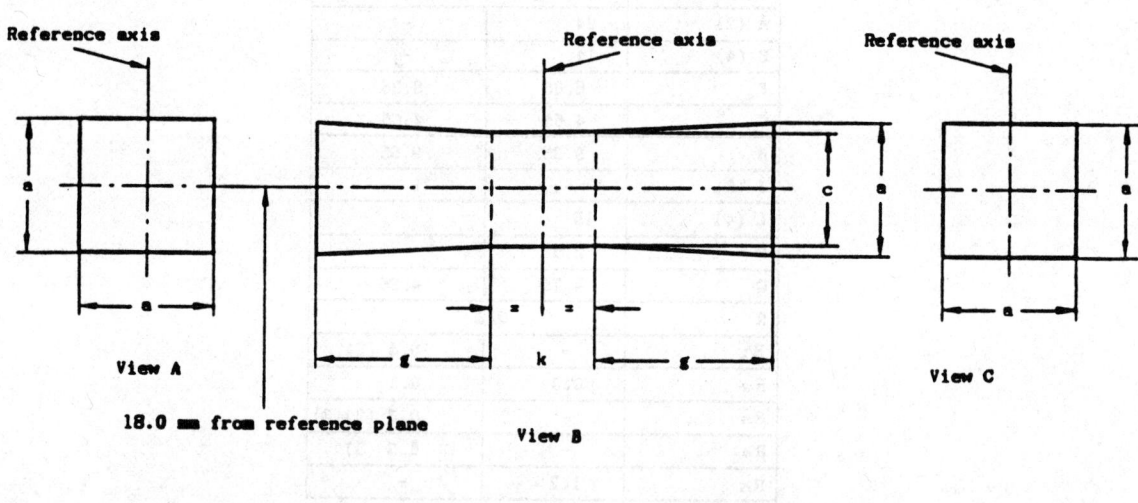

The first and last turn of the filament must lie entirely within the limits shown in respectively view A and C.
The transverse projection of the filament must lie within the limits shown in view B.
The centre of the filament shall lie within the limits of dimension k.

Dimensions in millimeters

(1) d = actual filament diameter

Reference	Dimension
a	1.8d
c	1.6d
g	2.8
k	1.0

FIGURE 68—SPECIFICATION FOR THE TYPE H3 REPLACEABLE BULB

17.2 General Test Conditions

17.2.1 The bulb shall be measured in a horizontal operating position.

17.2.2 Each filament shall be aged for approximately 1 h at test voltage. Immediately prior to a measurement the filament shall be operated for a minimum of 2 min at test voltage.

17.2.3 Measurements of filaments are carried out at test voltage.

18. Reference Plane, Reference Axis, and Planes for Measurements

18.1 Reference Plane—The reference plane is the plane formed by the seating points of the three lugs.

18.2 Reference Axis—The reference axis is perpendicular to the reference plane and passed through the center of the outer circle with diameter M of the base-ring.

18.3 Plane V-V—Plane V-V is the plane perpendicular to the reference plane and contains the reference axis and the center line of the reference lug.

18.4 Plane H-H—Plane H-H is the plane perpendicular to the reference plane and plane V-V and contains the reference axis.

18.5 Plane X-X—Plane X-X is the plane perpendicular to the reference plane, contains the reference axis, and has an angle of 15 degrees to plane H-H turned clockwise away from the reference lug.

18.6 Plane Y_1-Y_1—Plane Y_1-Y_1 is a plane parallel to the reference plane at a distance of 29.5 mm from it.

18.7 Plane Y_2-Y_2—Plane Y_2-Y_2 is a plane parallel to the reference plane at a distance of 33.0 mm from it.

18.8 Plane Y_3-Y_3—Plane Y_3-Y_3 is a plane parallel to the reference plane at a distance of 23.5 mm from it.

18.9 Plane Y_4-Y_4—Plane Y_4-Y_4 is a plane parallel to the reference plane at a distance of 26.0 mm from it.

18.10 Plane Y_5-Y_5—Plane Y_5-Y_5 is a plane parallel to the reference plane at a distance of 28.95 mm from it.

19. Viewing Directions (see Figure 74)

19.1 Viewing Direction 1—Perpendicular to plane V-V, seen from the side of the left-handed shield edge.

19.2 Viewing Direction 2—Perpendicular to plane H-H, seen from the side of the reference lug.

19.3 Viewing Direction 3—Parallel to plane X-X and reference plane, seen from the side of the right-handed shield edge.

20. Measuring Points (MP)—The following points as specified in Figures 75 and 76 shall be measured.

Measurements are to be made perpendicular to the viewing directions.

Additional requirements for accurate rated bulbs

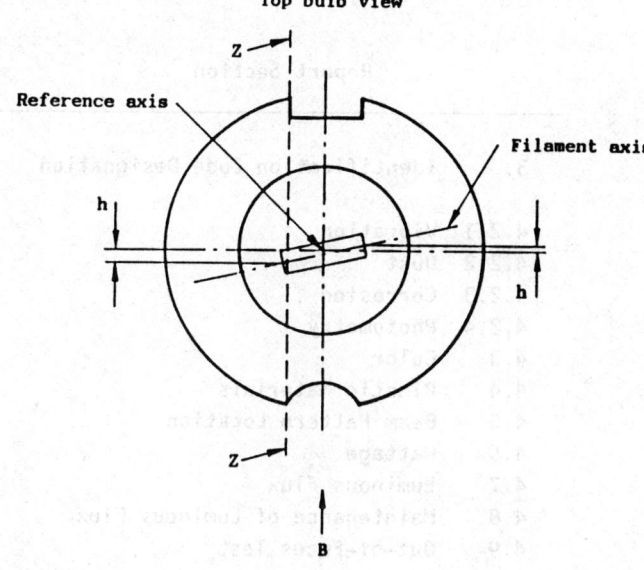

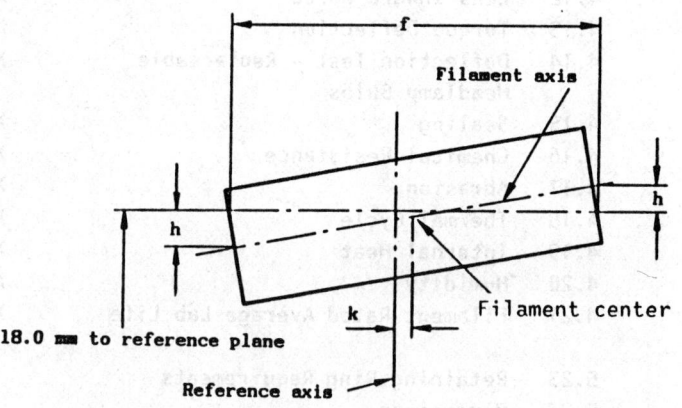

Dimensions in millimeters

Reference	Dimension	Tolerance
f	5.0	± 0.50
h	0	± 0.25
k	0	± 0.20

FIGURE 69—SPECIFICATION FOR THE TYPE H3 REPLACEABLE BULB

TABLE 8 – Test Classification

Report Section		Requirements Performance	Requirements Design	Requirements Material	Requirements Guidelines
3.	Identification Code Designation				X
4.2.1	Vibration	X			
4.2.2	Dust	X			
4.2.3	Corrosion	X			
4.2.4	Photometry	X			
4.3	Color	X			
4.4	Plastic Materials	X			
4.5	Beam Pattern Location	X			
4.6	Wattage	X			
4.7	Luminous Flux	X			
4.8	Maintenance of Luminous Flux	X			
4.9	Out-of-Focus Test	X			
4.10	Impact	X			
4.11	Aiming Adjustment	X			
4.12	Lens Inward Force	X			
4.13	Torque Deflection	X			
4.14	Deflection Test – Replaceable Headlamp Bulbs	X			
4.15	Sealing	X			
4.16	Chemical Resistance	X			
4.17	Abrasion	X		X	
4.18	Thermal Cycle	X			
4.19	Internal Heat	X			
4.20	Humidity	X			
4.21	Filament Rated Average Lab Life	X			
5.23	Retaining Ring Requirements		X		X
5.24	Dimensions		X		X
6.2	Photometric Design		X		X
6.4	Filament Life				X

The drawing is not mandatory, their sole purpose is to show which dimensions must be verified.

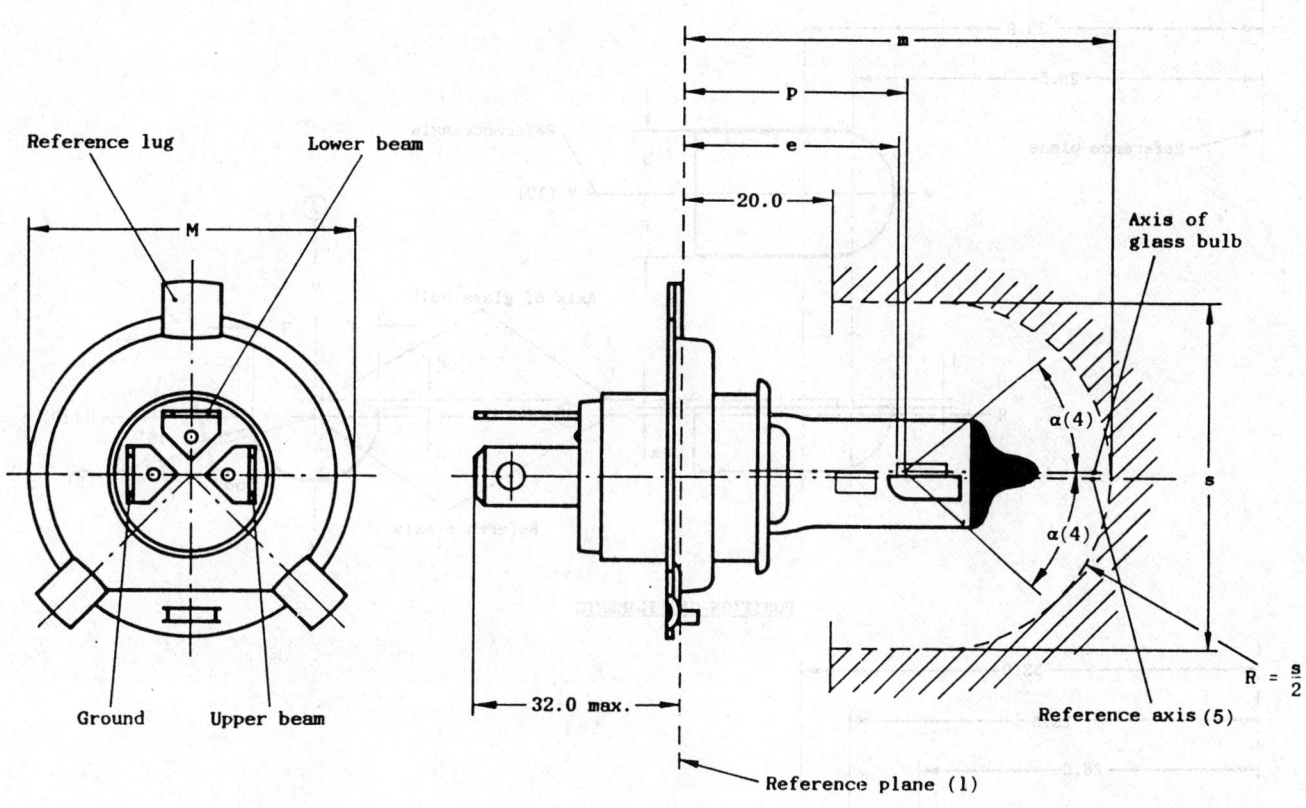

(1) The reference plane is the plane formed by the seating points of the three lugs of the base ring.
(2) "m" denotes the maximum length of the bulb.
(3) It must be possible to insert the bulb into a cylinder of diameter "s" concentric with the reference axis and limited at one end by a plane parallel to and 20 mm distance from the reference plane and at the other end by a hemisphere of radius s/2.
(4) The obscuration must extend at least as far as the cylindrical part of the glas bulb.
It must also overlap the internal shield when the latter is viewed in a direction perpendicular to the reference axis.
The effect sought by the obscuration may also be achieved by other means.*
(5) The reference axis is the line perpendicular to the reference plane and passing through the centre of the circle of diameter "M".
* Not applicable to HB2.

Dimensions in millimeters

Reference	Dimension	Tolerance
e	28.5	+ 0.45 - 0.25
p	28.95	-
m (2)	max. 60.0	-
s (3)	45.0	-
α (4)	max. 40°	-

FIGURE 70—SPECIFICATION FOR THE TYPE H4/HB2 REPLACEABLE BULB

The drawings are not mandatory with respect to the design of the shield.

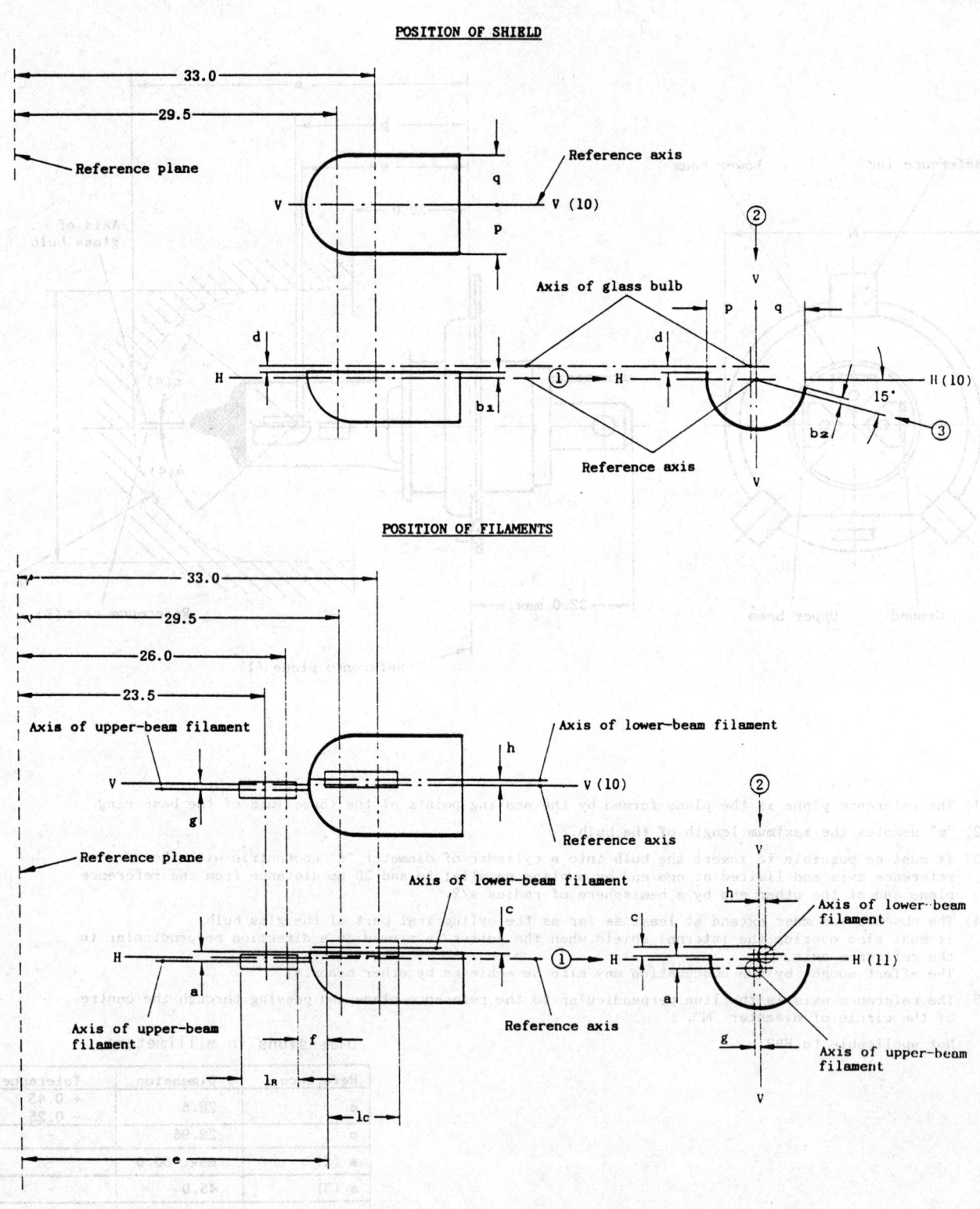

FIGURE 71—SPECIFICATION FOR THE TYPE H4/HB2 REPLACEABLE BULB

Dimensions indicated in Table are measured in three directions:

   Direction ① for dimensions a, b_1, c, d, e, f, l_R and l_C;
   Direction ② for dimensions g, h, p and q;
   Direction ③ for dimension b_2.

Dimensions p and q are measured in a plane parallel to and 33 mm away from the reference plane.

Dimensions b_1, b_2, c and h are measured in planes parallel to and 29.5 mm and 33 mm away from the reference plane.

Dimensions a and g are measured in planes parallel to and 26.0 mm and 23.5 mm away from the reference plane.

(6) The end turns of the filaments are defined as being the first luminous turn and the last luminous turn that are at substantially the correct helix angle.

(7) For the lower-beam filament the points to be measured are the intersections, seen in direction ①, of the lateral edge of the shield with the outside of the end turns defined under footnote (6).

(8) "e" denotes the distance from the reference plane to the beginning of the lower-beam filament as defined under footnote (7).

(9) For the upper-beam filament the points to be measured are the intersections, seen in direction ①, of a plane parallel to plane HH and situated at a distance of 0.8 mm below it, with the end turns defined under footnote (6).

(10) Plane VV is the plane perpendicular to the reference plane and passing through the reference axis and through the intersection of the circle of diameter "M" with the axis of the reference lug.

(11) Plane HH is the plane perpendicular to both the reference plane and plane VV and passing through the reference axis.

Dimensions in millimeters

Reference	Dimension	Tolerances		
		H4 ***		HB2
a/26 *	0.8	± 0.35	± 0.2	± 0.30
a/23.5 *	0.8	± 0.60	± 0.2	+ 0.40
b_1/29.5 *	0	± 0.30	± 0.2	± 0.25
b_1/33 *	b_1/29.5vm **	± 0.30	± 0.15	± 0.20
b_2/29.5 *	0	± 0.30	± 0.2	+ 0.25
b_2/33 *	b_2/29.5vm **	± 0.30	± 0.15	± 0.20
c/29.5 *	0.6	± 0.35	± 0.2	± 0.30
c/33 *	c/29.5vm **	± 0.35	± 0.15	± 0.30
d	min. 0.1	-	-	-
e (8)	28.5	+ 0.35 / - 0.25	+ 0.2 / - 0.0	+ 0.35 / - 0.15
f (6)(7)(9)	1.7	+ 0.50 / - 0.30	+ 0.3 / - 0.1	+ 0.30 / - 0.30
g/26 *	0	± 0.5	± 0.3	+ 0.4
g/23.5 *	0	± 0.7	± 0.3	± 0.5
h/29.5 *	0	± 0.5	± 0.3	± 0.5
h/33 *	h/29.5vm **	± 0.35	± 0.2	± 0.35
l_R (6)(9)	4.5	± 0.8	± 0.4	± 0.8
l_C (6)(7)	5.5	± 0.5	± 0.35	± 0.8
p/33 *	Depends on the shape of the shield	-	-	-
q/33 *	$\frac{p+q}{2}$	± 0.6	± 0.3	± 0.6
$b_1 - b_2$	-	-	-	± 0.25

* Dimension to be measured at the distance from the reference plane indicated in mm after the slash.

** ./29.5vm means the value measured at a distance of 29.5 mm from the reference plane.

*** Left column shows the tolerances for normal production bulbs.
Right column shows the tolerances for accurate rated bulbs.

FIGURE 71 (CONTINUED)

The drawings are intended only to indicate
the dimensions essential for interchangeability.

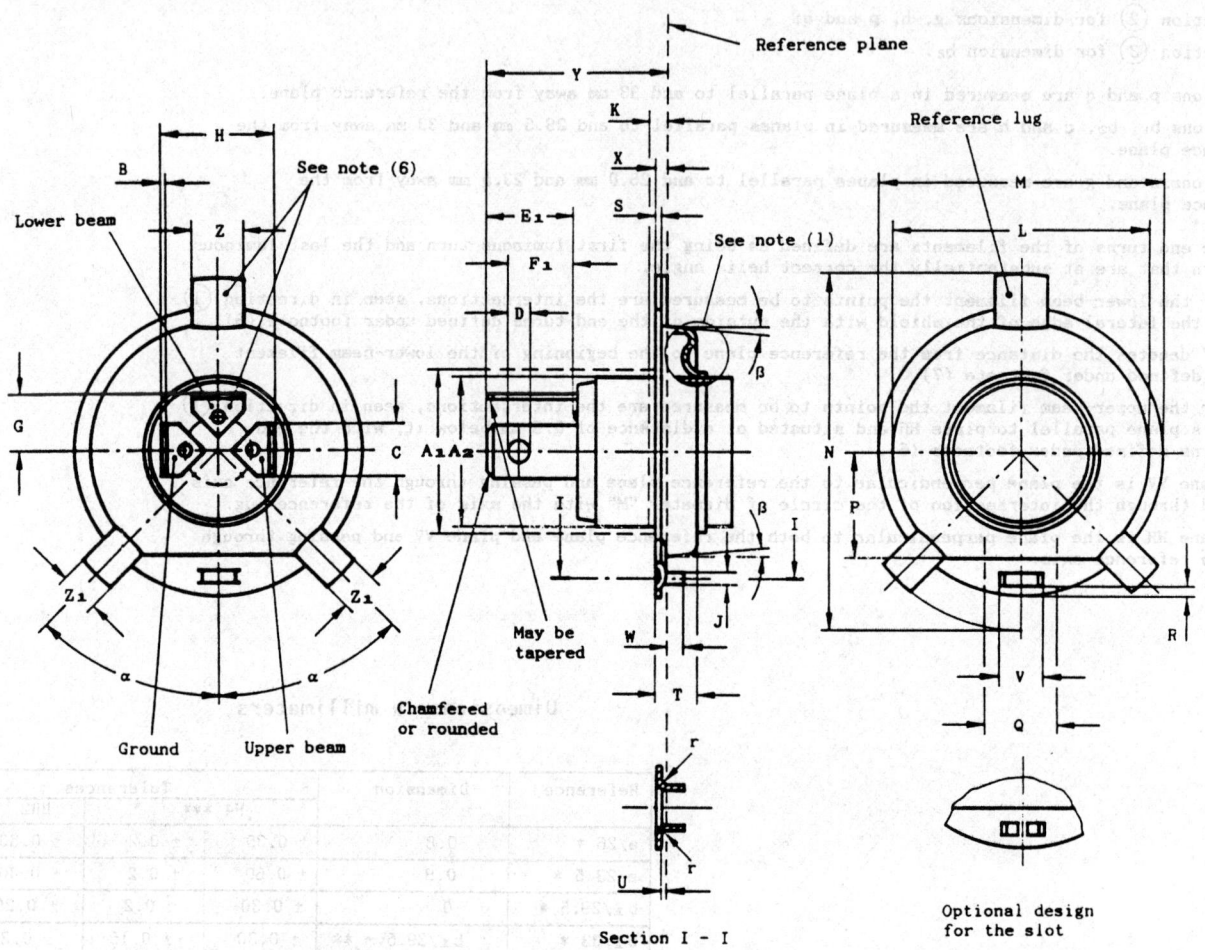

Dimensions in millimeters

Dimension	Min.	Max.
A_1 (8)	25.0	
A_2 (10)	Nom. 22*	
B	0.7	0.8
C	7.7	8.1
D	3.0	3.3
E_1	11.8	13.6
F_1	8.8	10.3
G	8.5	9.0
H	17.0	17.9
J	1.9	2.1
K (10)		2.0
L (2)(4)	37.8	38.0
M (3) H4	42.8	43.0
M (3) HB2	42.9	43.0
N	51.6	52.0

Dimension	Min.	Max.
P (2)(7)	15.3	15.5
Q (2)(7)	8.5	—
R	1.3	1.7
S	0.50	—
T	5.0	6.0
U	(9)	
V (2)(5)	6.3	6.5
W	1.8	2.2
X	1.1	1.3
Y	—	32.0
Z	7.9	8.0
Z_1	5.8	6.2
r	(9)	
α	44°	46°
β	—	5°

FIGURE 72—SPECIFICATION FOR THE TYPE H4/HB2 REPLACEABLE BULB
BASE P43t-38

(1) The form of this annular part of the ring is optional and may be flat or recessed. However, the form shall be such that it will not cause any abnormal glare from the lower beam filament when the bulb is in its normal operating position in the vehicle.

(2) This dimension is measured at the reference plane.

(3) Dimension M is the diameter on which the bulb is centred.

(4) The maximum allowable eccentricity of cylinder L with respect to the circle of diameter M is 0.05 mm.

(5) The maximum allowable displacement of the centre of the nose from the line running through the centre of the reference lug and the circle of diameter M is 0.05 mm.
The sides of the nose shall not bend outwards.

(6) The relative positions of the contact tabs and the reference lug shall not deviate from the position shown by more than ± 20°.

(7) Dimension Q denotes the minimum width over which both the minimum and maximum limits of dimension P shall be observed.
Outside dimension Q, the maximum limit for dimension P shall not be exceeded.

(8) The means of securing the ring in the headlamp shall not encroach on this cylindrical zone, which extends over the full length of the shell shown on this side of the ring.

(9) The radius r shall be equal to or smaller than dimension U.

(10) Beyond distance K, in the direction of the contact tabs, dimension A_2* shall be observed.

* This dimension is solely for base design and is not to be gauged on the finished lamp.

FIGURE 72 (CONTINUED)

20.1 Shield and Filaments
20.1.1 VIEWING DIRECTION 1
a. MP 1 and MP 2: The intersections of the high beam filament axis with planes Y_3-Y_3 and Y_4-Y_4.
b. MP 3 and MP 4: The intersections of the shield edge with plane Y_1-Y_1 and Y_2-Y_2.
c. MP 5 and MP 6: The intersections of the envelope of the low beam filament with planes Y_1-Y_1 and Y_2-Y_2, farthest from plane H-H.
d. MP 7: The intersection of the glass bulb axis with plane Y_1-Y_1.
e. MP 8 and MP 11: The intersections of the outer part of respectively the first and last luminous turn of the low beam filament with the shield edge.
f. MP 9 and MP 10: The intersections of the outer part of respectively the first and last luminous turn of the high beam filament with the center line (axis) of that filament.

20.1.2 VIEWING DIRECTION 2
a. MP 12 and MP 13: The intersections of the high beam filament axis with planes Y_3-Y_3 and Y_4-Y_4.
b. MP 14 and MP 15: The intersections of the low beam filament axis with planes Y_1-Y_1 and Y_2-Y_2.
c. MP 16 and MP 17: The intersections of the shield edges with plane Y_2-Y_2.

20.1.3 VIEWING DIRECTION 3
a. MP 18 and MP 19: The intersections of the shield edge planes Y_1-Y_1 and Y_2-Y_2.

20.2 Top Obscuration
20.2.1 VIEWING DIRECTION 2
a. MP 23: Intersection of the glass bulb axis with plane Y_5-Y_5.
b. MP 21 and MP 22: Intersections of the top shielding with a plane parallel to plane H-H and containing the glass bulb axis.

20.2.2 VIEWING DIRECTION 1
a. MP 20: Intersection of the top shielding with a plane parallel to plane V-V and containing the glass bulb axis.

21. Dimensions to be Measured—Table 9 states the dimensions to be measured. Values and tolerances are given in Figure 65.

TABLE 9—DIMENSIONS TO BE MEASURED

Distances	Measured Perpendicular to Plane	Viewing Direction	Reference 12 V
MP 2 to MP 3	H-H	1	a/26.0
MP 1 to MP 3	H-H	1	a/23.5
MP 3 to H-H	H-H	1	b1/29.5
MP 4 to H-H	H-H	1	b1/33.0
MP 18 to X-X	X-X	3	b2/29.5
MP 19 to X-X	X-X	3	b2/33.0
MP 3 to MP 5	H-H	1	c/29.5
MP 4 to MP 6	H-H	1	c/33.0
MP 7 to MP 3	H-H	1	d
MP 8 to ref. plane	ref. plane	1	e
MP 8 to MP 9	ref. plane	1	f
MP 13 to V-V	V-V	2	g/26.0
MP 12 to V-V	V-V	2	g/23.5
MP 14 to V-V	V-V	2	h/29.5
MP 15 to V-V	V-V	2	h/33.0
MP 9 to MP 10	ref. plane	1	1R
MP 8 to MP 11	ref. plane	1	1c
MP 16 to V-V	V-V	2	p/33.0
MP 17 to V-V	V-V	2	q/33.0
Angle			
MP 21 & 22 to MP 23	V-V	2	α
MP 20 to MP 23	H-H	1	α

24.110

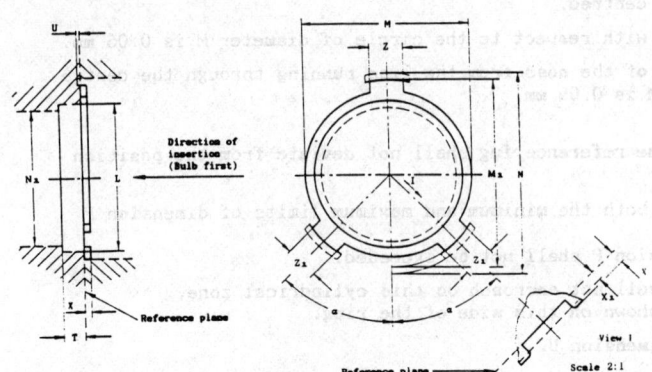
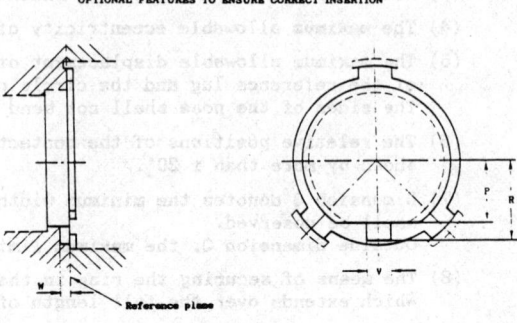

The holder shall be so designed that, without using undue force, the means of retention of the bulb can be applied only when it is in the correct position.

The means of retention shall make contact with the prefocus base ring only and the total force exerted, when the bulb is in position, shall be not less than 10 N and be not more than 60 N.

(1) This value shall be complied with between the rim of the holder and the reference plane (dimension X). However, it may be reduced to 38.5 mm within the dimensions Z and Z_1 which correspond with the support points for the lugs of the ring.

(2) Dimension X_1 denotes the minimum distance over which dimensions Z and Z_1 shall apply. Outside dimension X_1 the slots may be chamfered or rounded.

(3) Wrong adjustment of the bulb in the holder can be prevented in different ways e.g.:
  - by applying the additional optional features. (See figures).
  - By decreasing dimension Z_1 to 7.5 - 7.7 mm followed by a decrease of the tolerance for α to give values of 44°40' - 45°20'.
  - by using a sufficiently large value for X depending on the construction of the holder.

(4) If dimension L is smaller than 40.5 mm, dimension V, R and W shall apply.

(5) Dimension N delineates the minimum free space to be reserved for the three lugs of the ring.

(6) Dimension N_1 shall be not less than 35 mm diameter over a distance of 20 mm from the reference plane and shall be not less than 45 mm diameter at any distance greater than 20 mm from the reference plane.

Dimensions in millimeters

Dimension	Min.	Max.
L (4)	38.2	-
M	43.02 (1)	43.2
M_1	-	49.0
N (5)	52.5	
N_1	(6)	
P (3)	16.0	-
R (4)	20.5	-
T	5.5	-

Dimension	Min.	Max.
U	0.4	-
V (4)	6.8	-
W (4)	2.5	-
X (3)	1.8	-
X_1 (2)	1.4	-
Z (3)	8.05	8.15
Z_1 (3)	8.0	8.5
α	44°	46°

FIGURE 73—SPECIFICATION FOR THE TYPE H4/HB2 REPLACEABLE BULB BULB HOLDER P43t

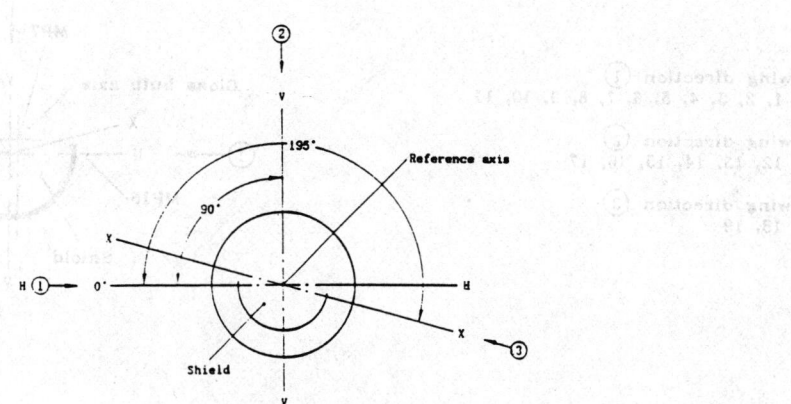

FIGURE 74—SPECIFICATION FOR THE TYPE H4/HB2
REPLACEABLE BULB
VIEWING DIRECTION SEEN FROM THE TOP OF THE BULB

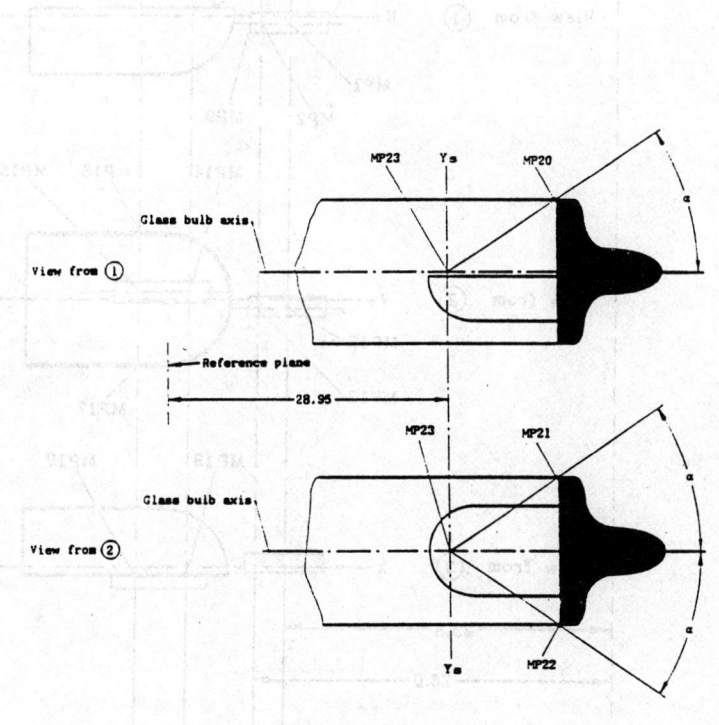

FIGURE 75—SPECIFICATION FOR THE TYPE
H4/HB2 REPLACEABLE BULB
TOP OBSCURATION

Viewing direction ①
MP 1, 2, 3, 4, 5, 6, 7, 8, 9, 10, 11

Viewing direction ②
MP 12, 13, 14, 15, 16, 17

Viewing direction ③
MP 18, 19

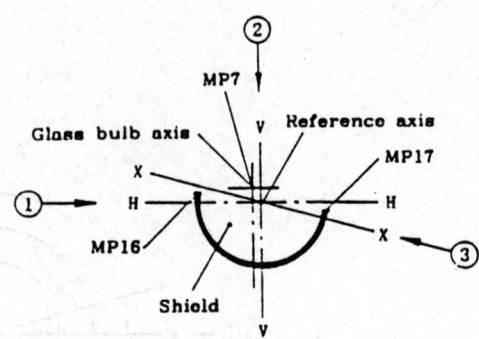

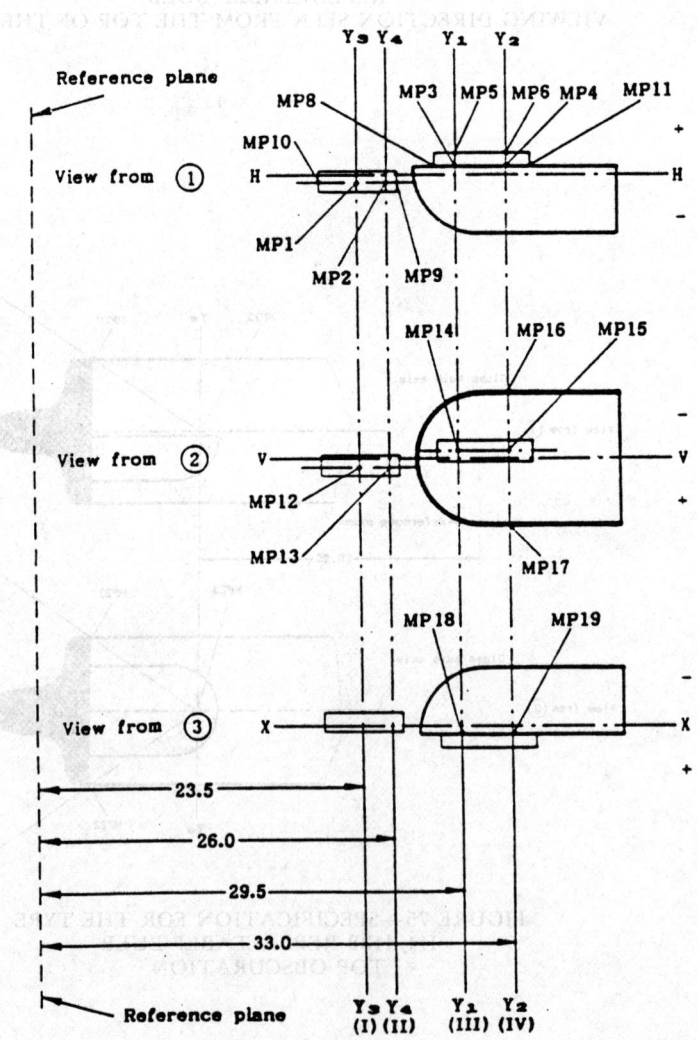

FIGURE 76—SPECIFICATION FOR THE TYPE H4/HB2 REPLACEABLE BULB
POSITIONS OF MEASURING POINTS OF H4 BULBS

DISCHARGE FORWARD LIGHTING SYSTEM —SAE J2009 FEB93

SAE Recommended Practice

Report of the SAE Road Illumination Devices Standards Committee approved February 1993. Rationale statement available.

1. Scope
This SAE Recommended Practice applies to motor vehicle Forward Illumination Systems which use light generated by discharge sources. It provides test methods, requirements, and guidelines applicable to the special characteristics of gaseous discharge lighting devices which supplement those required for forward illumination systems using incandescent light sources. This document is intended to be a guide to standard practice and is subject to change to reflect additional experience and technical advances.

2. References

2.1 Applicable Documents—The following publications form a part of this specification to the extent specified herein. The latest issue of SAE publications shall apply.

2.1.1 SAE PUBLICATIONS—Available from SAE, 400 Commonwealth Drive, Warrendale, PA 15096-0001.

SAE J575—Tests for Motor Vehicle Lighting Devices and Components
SAE J578—Color Specification
SAE J759—Lighting Identification Code
SAE J1113—Electromagnetic Susceptibility Measurement Procedures for Vehicle Components
SAE J1211—Recommended Environmental Practices for Electronic Equipment Design
SAE J1383—Performance Requirements for Vehicle Headlamps
SAE J1816—Performance Levels and Methods of Measurement of Electromagnetic Radiation From Vehicles and Devices (Narrow Band), 10 kHz - 1000 MHz

2.1.2 ANSI PUBLICATIONS—Available from American National Standards Institute, Inc., 11 West 42nd Street, New York, NY 10036.

ANSI Z311.1—Photobiological Safety for Lamps and Lighting Systems
ANSI C78.376—Spectroradiometrically Determined Assignments

2.1.3 FMVSS PUBLICATIONS—Available from the National Highway Traffic Safety Administration, 400 Seventh Street SW, Washington, DC 20024-0002.

FMVSS 108—Lamps, Reflective Devices, and Associated Equipment (Available as 49 CFR 571.108)
FMVSS 112—Headlamp Concealment Devices (Available as 49 CFR 571.112)

2.1.4 CIE PUBLICATION—Available from Commission Internationale de L'eclairage, 52 Bd Malesherbes, F-75008 Paris, France.

CIE Pub. 13.2—Method of Measuring and Specifying Color Rendering Properties of Light Sources (TC3.2) 1974

2.1.5 ACGIH PUBLICATIONS—Available from American Council of Governmental Industrial Hygienists, 6500 Glenway Avenue, Building D-7, Cincinnati, OH 45211.

Threshold Limit Values and Biological Exposure Indices for 1989-1990, American Conference of Governmental Industrial Hygienists

2.2 Definitions

2.2.1 DISCHARGE FORWARD LIGHTING (DFL) SYSTEM—An automotive lighting system, providing forward illumination, comprised of the headlamps, discharge source, ballast/starting system, and interconnecting wiring.

2.2.2 DISCHARGE SOURCE—An electric light source in which light is produced by a stabilized arc.

2.2.3 START-UP TIME—The period of time between the instant when the user operates a switch to power a lamp ON and the instant when the DFL system reaches a level within X% of "steady-state" output level.

2.2.4 RESTART—The ability of the "hot" DFL system to relight before its temperature has returned to initial ambient.

2.2.5 PHOTOMETRIC MAINTENANCE—Change in beam intensity of the test points of the beam pattern light output over time (life).

2.2.6 LIFE—Time in hours and starting cycles of a DFL system during which it meets specified operational characteristics under specified test conditions.

2.2.7 RATED LAB LIFE—Life specified by the manufacturer as the period during which the DFL system meets the performance specifications. (Rated lab life equals design life.)

2.2.8 COLOR RENDERING INDEX (LIGHT SOURCE—CRI)—Measure of the degree of color shift objects undergo when illuminated by the light source as compared with the color of those same objects when illuminated by a reference source of comparable color temperature.

2.2.9 ULTRAVIOLET RADIATION—Radiation in the spectral region between 200 and 400 nm. Definitions and terminology are adopted in accordance with proposed ANSI specification standard Z311.

a. UVA flux—Radiant energy flux between 320 and 400 nm
b. UVB flux—Radiant energy flux between 260 and 320 nm
c. UVC flux—Radiant energy flux between 200 and 260 nm

2.2.10 STEADY-STATE—A condition under which the light output of the device is considered to be stable or changing at such a slow rate as to be insignificant. A "Steady-state" condition would be generally measured in terms of a "maximum percent change per time period."

Steady-state light level (100%) is established by allowing the lamp to operate for 120 s after being switched "on." The average light level within the period from 120 to 140 s will be defined as the 100% level.

If the light output is not stable to within ±10% during the 120 to 140 s time interval, the test should be repeated on the system. If a system fails to stabilize after three attempts, a new system should be selected for a test sample.

2.2.11 AUTOMOTIVE BALLAST—A device for stabilizing the operating characteristics of a discharge lamp. The ballast contains all the necessary circuitry to ignite a lamp and cause it to operate within a specified power profile range. It controls the required light output characteristics of the automotive discharge lighting system. The ballast may consist of one or more separate components.

2.2.12 INTEGRAL BEAM—An "Integral Headlamp" produces a light pattern when normal vehicle voltage is applied. It cannot be disassembled by the user for the purpose of replacing any failed subassemblies within the lamp or housing package.

Discharge headlamps in which the ballast subunit is remote from the starter/lamp subunit, may be considered integral if the user cannot disconnect the two subunits. Such a lamp may be disassembled and serviced by the manufacturer for the purpose of recycling the assembly by replacing nonfunctioning parts. It may also be disassembled and serviced by a service factory or dealer service facility. In any case, there is the assumption that the servicing facility will certify that the performance of the serviced device will meet all standards applicable to the original equipment.

3. Lighting Identification Codes, Markings, and Notices

3.1 Headlamps shall be marked in accordance with SAE J759.

3.2 The DFL system shall contain a label indicating the presence of high voltage, e.g., the International electric shock hazard symbol ("lightning bolt").

4. Tests
All sample DFL systems shall be seasoned for 20 h prior to being subjected to the tests that follow and a new DFL system may be used for each test.

NOTE—The power supply used for all testing should have its output isolated from the input to prevent any potential danger to laboratory personnel when running test as required.

4.1 Lamp/System Starting Procedures—The headlamp shall be held in its normal operating position and mechanically aimed with a photocell or cells at the test points shown in Table 1. Tests shall be conducted at room temperature (23 °C ± 3 °C), at 12.8 VDC ± 0.1 VDC, and for a duration required to obtain a reading. The response time of the measurement instrument should be less than 100 ms.

TABLE 1—TEST POINTS FOR DFL HEADLAMP STARTING TESTS

Lower Beam Lamp	Upper Beam Lamp
1.5 D - 2 R	H - V

4.1.1 INITIAL START-UP—The DFL system (ballast/starter) shall be activated and the luminous intensity at the photometric test points of Table 1 sampled and recorded for each headlamp from initial actuation through the intervals specified in Figure 1 or Figure 2. The test lamp(s) is then turned off.

4.1.2 SWITCHING (COLD LAMPS)—The DFL system (ballast/starter) shall be activated and the luminous intensity at the photometric test points of Table 1 sampled and recorded for each headlamp from initial activation through the intervals specified in Figure 1 or Figure 2.

4.1.3 SWITCHING (HOT RESTART)—The lamp shall be energized for 5 min minimum. After this time period, a restart test shall be conducted once for every time interval as follows:

a. Cool down times for DFL hot restart test—1 s, 4 s, 10 s, 20 s, 30 s, 1 min

The system shall be switched off for the period of time shown as previously stated in order to simulate momentary switching to the alternate beam. The test lamp shall be energized and the luminous intensity at the applicable photometric test point shown in Table 1 sampled.

4.1.4 SWITCHING (CONTINUOUS LOW BEAM MODE)—For DFL systems designed to have the lower beam on continuously, the lower beam lamp shall be operated during the test. However, only the photometric characteristics of the upper beam switching shall be measured. The tests for "continuous low beam mode" are identical to those described in 4.1.3 except that only the upper beam lamp is tested.

4.2 Electrical Characteristics

4.2.1 SYSTEM OPERATING WATTAGE RANGE—DFL system wattage shall be measured at 12.8 VDC ± 0.1 VDC with all components in normal operating orientation using the equipment described in 4.2.3.

4.2.2 SYSTEM OPERATING VOLTAGE RANGE—The DFL system shall operate in the regulated mode from 9.0 to 18.0 VDC with all components in normal operating orientation using the equipment described in 4.2.3. Additional considerations for voltages between 9.0 and 4.5 V are presented in 6.10.

4.2.3 EQUIPMENT REQUIREMENTS—The input terminals of the DFL system shall be connected to a laboratory power supply which shall have a range of voltages from at least 4.0 to 18.5 VDC and which shall be capable of controlling voltage to ±0.1 VDC input voltage. In addition, the power supply shall be capable of satisfying the DFL system's current drain in all operational modes.

4.3 Photometric Maintenance—The DFL system shall be operated under nominal laboratory operating test conditions (23 °C ± 3 °C) at 12.8 VDC ± 0.1 VDC input voltage. The test cycle shall be the same as for the life test (6.2). The photometric maintenance test shall be performed after 70% of rated life of operation (e.g., 1400 h for a 2000 h design value of rated life).

Low or High Beam Non-Continuous Low		
Time(sec)	Min (%)	Max (%)
0.25	20	300
0.50	30	300
0.75	50	300
1.00	60	200
2.00	70	200
3.00	70	150
5.00	70	130
60.00	70	130

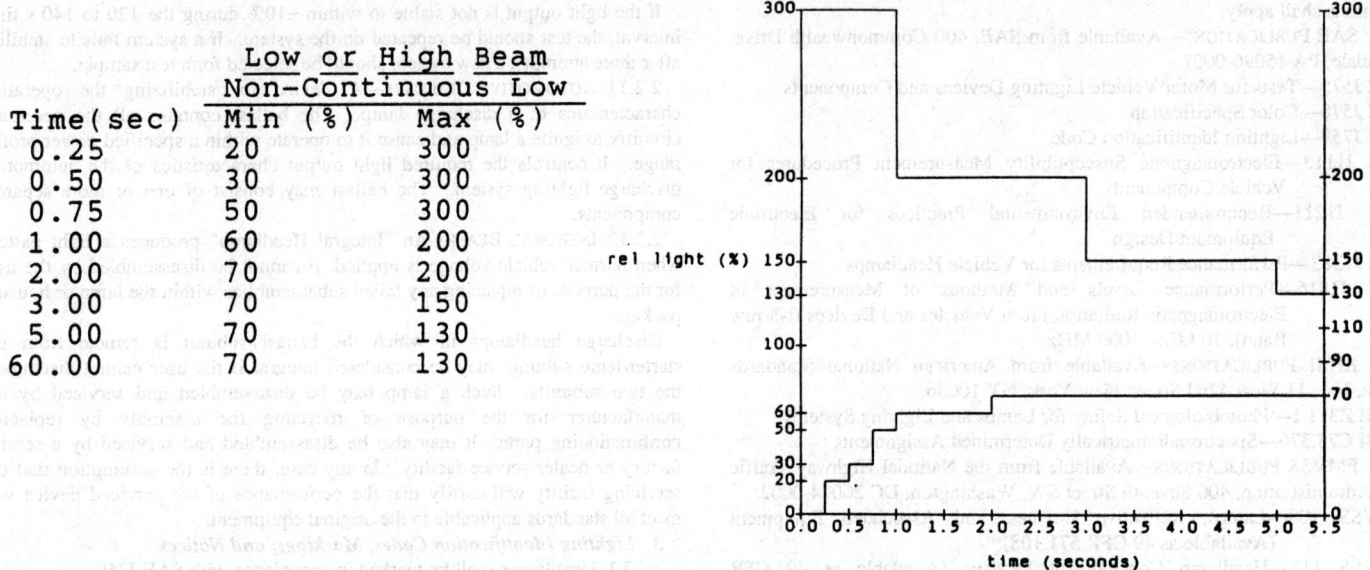

FIGURE 1—LAMP OUTPUT VS START-UP TIME LOW OR HIGH BEAM NONCONTINUOUS LOW

Low or High Beam Continuous Low		
Time(sec)	Min (%)	Max (%)
0.25	10	300
1.00	25	200
2.00	50	200
3.00	70	150
5.00	70	130
60.00	70	130

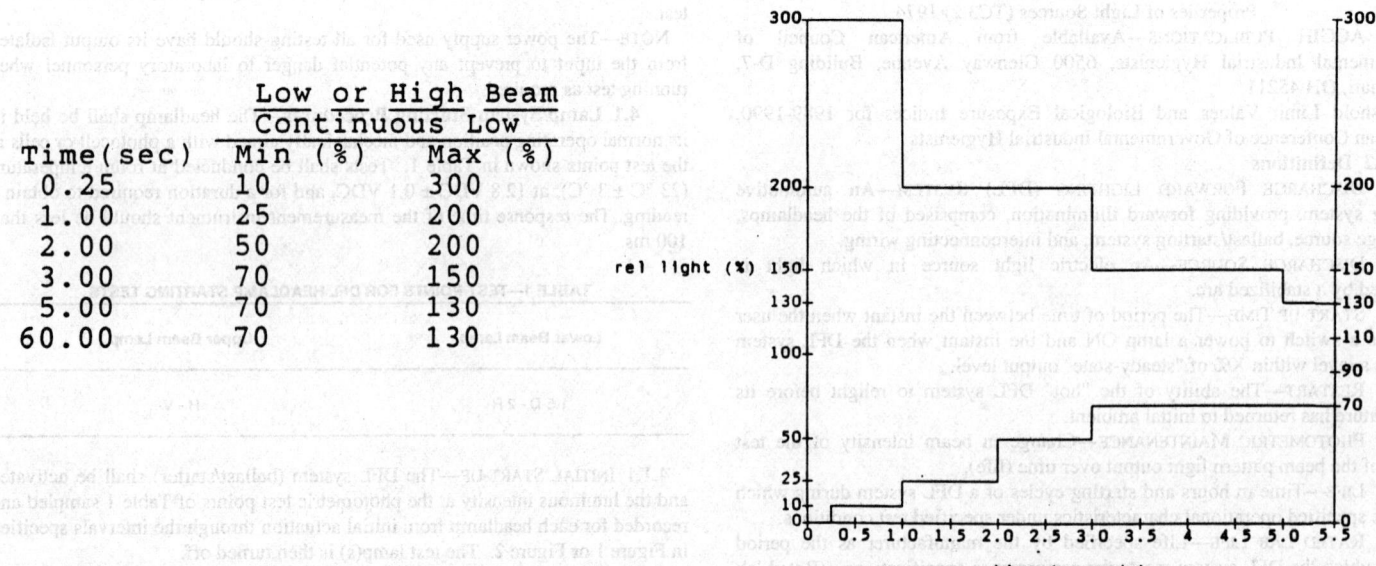

FIGURE 2—LAMP OUTPUT VS START-UP TIME LOW OR HIGH BEAM CONTINUOUS LOW

NOTE—(%) = Percent of steady-state light intensity. Each beam should be measured at the point prescribed in Table 1, with the other beam blocked or not operating.

4.4 Color and CRI

4.4.1 COLOR—The color coordinates shall be tested and fall within the chromaticity limits in SAE J578 for "White" light.

4.4.2 COLOR RENDERING—The color rendering properties of a DFL system shall be determined using the procedure outlined in CIE Publication Number 13.2 (TC 3.2) 1974, "Method of measuring and specifying color rendering properties of light sources." See 6.8 for performance criteria.

4.5 Environmental Tests—Testing shall be accomplished on a complete DFL system, i.e., ballast, interconnections, and headlamp unless otherwise specified in the specific test.

4.5.1 LEAKAGE CURRENT/BREAKDOWN TEST—The test shall be made on a system positioned in its design orientation by completely covering the exterior of the DFL system to be tested with aluminum foil. The foil is to be connected to a current-sensing device which terminates at the power source common (chassis ground). The sensing device shall be a noninductive resistor of 1000 _. The leakage current occurring during starting and operating (transient and steady-state) shall be measured using an oscilloscope with a bandwidth capability five times the bandwidth being measured for the observed frequencies and rise times. Current readings shall be recorded during the first 10 s of the initial start. The unit shall then continue to operate for 30 min, be turned off, and immediately restarted. The current readings shall again be recorded during the first 10 s after restart. After completion of this procedure, and without submitting the unit to any other tests, the environmental test shall be carried out on the unit. Within 30 min of the completion of the environmental test, the breakdown test shall be repeated. The final readings are then compared with the respective (initial and 30 min) readings made before the environmental test.

4.5.2 THERMAL CYCLE—A DFL system shall be mounted on a test fixture in its design orientation and shall be exposed to the test described in J1383 "Thermal Cycle Test." In addition, electronic components shall be subjected to the test in SAE J1211, Section 4.1 using conditions that are appropriate for the location of the DFL system components in the vehicle.

4.5.3 HUMIDITY—The DFL system shall be mounted on a test fixture in its design orientation and shall be subjected to the test described in SAE J1383 "Humidity Test." The DFL system shall be tested before and after the humidity test in accordance with the Breakdown Test in 4.5.1. Photometric testing shall begin at 10 min ±1 min following completion of the humidity test. In addition, electronic components shall be subjected to the test in SAE J1211 Section 4.2 using conditions that are appropriate for the location of the DFL system components in the vehicle.

4.5.4 INTERNAL HEAT TEST—The DFL system shall be subjected to the conditions specified in SAE J1383 "Internal Heat Test." The DFL system shall be tested before and after the internal heat test in accordance with the Breakdown Test in 4.5.1.

4.5.5 DUST TEST—Conducted per SAE J575 Section 4.3. "DFL System" replaces "Headlamp" in the specifications. The DFL system shall be tested before and after the dust test in accordance with the Breakdown Test in 4.5.1.

4.5.6 CORROSION TEST—Test the DFL system per J575 Section 4.4. "DFL System" replaces "Headlamp" in the test. The DFL system shall be tested before and after the corrosion test in accordance with the Breakdown Test in 4.5.1.

4.5.7 CHEMICAL RESISTANCE TEST—The DFL system shall be tested per SAE J1383 "Chemical Resistance Test," except "DFL System" replaces "Headlamp" in the test. In addition, electronic components shall be subjected to the test specified in SAE J1211 Section 4.4 using conditions that are appropriate for the location of the DFL system components in the vehicle. The DFL system shall be tested before and after the chemical test in accordance with the Breakdown Test in 4.5.1.

4.5.8 VIBRATION TEST—The DFL system shall be tested as specified in SAE J575 Section 4.1 except "DFL System" replaces "Headlamp" in the test. In addition, electronic components shall be subjected to SAE J1211 Section 4.7 and 4.8 using conditions that are appropriate for the location of the DFL system components in the vehicle. The DFL system shall be tested before and after the vibration test in accordance with the Breakdown Test in 4.5.1.

4.5.9 ALTITUDE TEST—Electronic components shall be subjected to the requirements of Section 4.6 of SAE J1211 using conditions that are appropriate for the location of the DFL system components in the vehicle. The DFL system shall also be tested before and after the altitude test in accordance with the Breakdown Test in 4.5.1.

4.6 Photometry—The DFL system shall first be seasoned (per Section 4) at the nominal ballast input voltage of 12.8 V ± 0.1 V. The seasoned DFL system shall be aimed and after attaining steady-state conditions as specified in 6.5, be photometered to SAE J1383 requirements. Photometric measurements shall be made at a minimum distance of 18.3 m (60 ft) from the unit.

4.7 Electromagnetic Susceptibility (EMS)—The DFL system shall be tested to SAE J1113 (guidelines and test methods Sections 2 through 9) to evaluate compatibility with potential sources of EMI.

4.8 Electromagnetic Radiation (EMR)—DFL systems shall be tested in accordance with the guidelines of SAE J1816.

4.9 Life—DFL system(s) shall be mounted in its design orientation and operated using the following cycle to determine DFL system life. The "life test cycle" is a 1 h cycle, starting with five 10 min lighted cycles (50 min total). The operating cycle is 9 min 45 s "On" and 15 s "Off." Following the fifth operating cycle, the lamp is allowed to cool in the "Off" state for 10 min (81.25% hot time). When the DFL system "hot time" reaches 70% of rated lab life, the DFL system shall be subjected to the tests of 4.1 (starting procedures), 4.4 (color), and 4.5.1 (breakdown). Upon completion of the previous tests, the DFL system is returned to the life test cycle listed in Table 2 until it fails to restart (meet Figure 1 or Figure 2) following any off period. Test result guidelines are covered in 6.2.

TABLE 2—LIFE TEST CYCLE (FOR BOTH HIGH AND/OR LOW BEAM)

Cycle	Cycle Time (Minutes)		Total (Min)
1 - Operating Cycle	(9:45 on, 0:15 off)	=	10
1 - Operating Cycle	(9:45 on, 0:15 off)	=	10
1 - Operating Cycle	(9:45 on, 0:15 off)	=	10
1 - Operating Cycle	(9:45 on, 0:15 off)	=	10
1 - Operating Cycle	(9:45 on, 0:15 off)	=	10
1 - Off Cycle		=	10
Total Life test cycle		=	**60 min**

4.10 UV Test—UV radiation refers to the radiation in the spectral region between 200 and 400 nm.
    a. UVA flux—is the energy flux between 320 and 400 nm
    b. UVB flux—is the energy flux between 260 and 320 nm
    c. UVC flux—is the energy flux between 200 and 260 nm

The measurement setup shall be as shown in Figure 1 of ANSI Z311 and the radiation sensor shall be located at a specified distance from the source (typically 50 cm). The source is defined as a lamp without any outermost lens(es). Energy levels shall be recorded at 10 nm intervals over the UV range.

UV weighting factors are defined in tables in ANSI Z311. (ANSI and NIOSH use the same tables, DIN values are also defined.)

5. Performance Requirements

5.1 Lamp/System Starting Procedures

5.1.1 INITIAL START-UP—Start-up intensities shall conform to Figure 1 or Figure 2. The test lamp photometric values shall meet the percent of the minimum values specified in J1383 for the test points in Table 1.

5.1.2 SWITCHING (COLD LAMPS)—Figure 1 or Figure 2 indicates the acceptable percent of steady-state light intensity versus time after a cold DFL lamp has been turned on, for DFL systems with noncontinuous and continuous low beam illumination, respectively. The lamp shall produce not less than the percent of the minimum light level specified in J1383 for the test points shown in Table 1. For DFL systems which are designed for "continuous" low beam operation, see 5.1.4.

5.1.3 SWITCHING (HOT RESTART)—After a thermally stabilized DFL headlamp has been allowed to cool for varying periods of time as shown in 4.1.3a, upon restart it shall produce not less than the percentage indicated in Figure 1 or Figure 2 of the luminous intensity values given in J1383 for the test points shown in Table 1. For DFL systems which are designed for "continuous" low beam operation, see 5.1.4.

5.1.4 For DFL systems which are designed for "continuous" low beam operation, the luminous intensity of the low beam at the upper beam test point is combined with that of the upper beam to determine conformance to this specification.

5.2 Electrical Characteristics

5.2.1 SYSTEM OPERATING WATTAGE RANGE—At an input voltage of 12.8 V, the power consumption of the DFL system shall stabilize at its rated value with a maximum deviation of ±7.0%.

5.2.2 SYSTEM OPERATING VOLTAGE RANGE—Ignition and hot reignition of the lamp shall occur for all voltage settings between 9.0 and 18.0 VDC, and the arc shall be maintained to a low voltage of 4.5 VDC. (Reduced lumen output is acceptable for the 4.5 to 8.9 VDC range.)

The DFL system manufacturer shall specify the DFL minimum system voltage and current required for DFL system start-up.

5.3 Photometric Maintenance—When tested in accordance with the described test procedure, the DFL system shall meet the appropriate photometric specifications of SAE J1383, except 85% initial luminous flux value replaces 90% at 70% of life.

5.4 Color—The color of light emitted from the DFL headlamp following seasoning and attaining steady-state shall fall within the white light chromaticity boundaries as defined in SAE J578. The color of light shall be within the chromaticity limits both initially and after the photometric maintenance test of 5.3. Also see 6.1.

5.5 Environmental Requirements

5.5.1 LEAKAGE CURRENT/BREAKDOWN TEST—The acceptance criteria for this test shall be based on a comparison of the initial value of leakage current measured before the environmental test and the value measured after the test. The leakage value after the environmental test shall not exceed 200% (twice) of the initial test value.

5.5.2 THERMAL CYCLE—After the test, the DFL system shall meet SAE J1383 "Thermal Cycle Requirements" without magnification. Lens warpage shall be less than 3 mm (0.118 in) when measured normal to the lens surface at the geometric center of the lens. No breakdown shall be detected when the DFL system is tested in accordance with the Breakdown Test. In addition, electronic components shall meet the requirements of Section 5 of SAE J1211 using conditions that are appropriate for the location of the DFL system components in the vehicle.

5.5.3 HUMIDITY TEST—After the test, the DFL system shall meet SAE J1383 "Humidity Requirements" without magnification, and meet the photometric requirements of SAE J1383. There shall be no evidence of breakdown during the Breakdown Test. In addition, electronic components shall meet the requirements of Section 5 of SAE J1211 using conditions that are appropriate for the location of the DFL system components in the vehicle.

5.5.4 INTERNAL HEAT—The DFL system shall meet J1383 photometry values after the internal heat test. There shall be no evidence of breakdown during the Breakdown Test. In addition, electronic components shall meet the requirements of Section 5 of SAE J1211 using conditions that are appropriate for the location of the DFL system components in the vehicle.

5.5.5 DUST TEST—The DFL system shall meet the requirements of 4.3 of SAE J575. There shall be no evidence of breakdown during the Breakdown Test. In addition, electronic components shall meet the requirements of Section 5 of SAE J1211 using conditions that are appropriate for the location of the DFL system components in the vehicle.

5.5.6 CORROSION TEST—The DFL system shall be evaluated in accordance with 4.4 of SAE J575, except "DFL System" replaces "Headlamp" in the specifications. There shall be no evidence of breakdown during the Breakdown Test. In addition, electronic components shall meet the requirements of Section 5 of SAE J1211 using conditions that are appropriate for the location of the DFL system components in the vehicle.

5.5.7 CHEMICAL RESISTANCE TEST—The DFL system shall meet SAE J1383 "Chemical Resistance Requirement," except that "DFL System" replaces "Headlamp" in the specifications. There shall be no evidence of breakdown during the Breakdown Test. In addition, electronic components shall meet the requirements of Section 5 of SAE J1211 using conditions that are appropriate for the location of the DFL system components in the vehicle.

5.5.8 VIBRATION TEST—The DFL system shall meet the requirements specified in 4.1 of SAE J575. "DFL System" replaces "Headlamp" in the specifications. There shall be no evidence of breakdown during the Breakdown Test. In addition, electronic components shall meet the requirements of Section 5 of SAE J1211 that are appropriate for vibration tests for the location of the DFL system components in the vehicle.

5.5.9 ALTITUDE TEST—The DFL system's electronics shall comply with the applicable requirements of J1211. There shall be no evidence of breakdown during the Breakdown Test. In addition, electronic components shall meet the requirements of Section 5 of SAE J1211 using conditions that are appropriate for the location of the DFL system components in the vehicle.

5.6 Photometry—Each High and Low Beam of the DFL system shall meet the photometry specified in SAE J1383 Table 3.

5.7 Electromagnetic Susceptibility (EMS)—The DFL system shall meet the test requirements as specifically determined for the user's application and the environment. See applicable Sections 2 through 9 of SAE J1113 for guidance. After exposure to the tests in SAE J1113, the DFL system shall meet the requirements specified in 5.1.1, 5.1.2, 5.1.3, and 5.2.1.

5.8 Electromagnetic Radiation (EMR)—The DFL system shall meet the test requirements as specifically determined for the user's application and environment. See applicable sections of SAE J1816 for guidance.

6. Guidelines

6.1 Colorimetric Characteristics—Until an ANSI Standard[1] for colorimetric characteristics is developed for mercury, sodium, xenon, and metal halide lamps, it is required that the color of the DFL beams be perceived as essentially white light by drivers (for the accurate perception of colors in order to interpret road signs and signals). For this purpose, a color rendering index value may be established.

The "WHITE" color of a DFL headlamp device is presumed to exhibit only minor localized variations from the integrated measurement. If significant color variations exist within the projected beam, or if color changes occur during a period of time when the device is energized, the manufacturer of the device must be assured that such color will not be confused with that of an emergency warning device.

This assurance may be realized by using a panel of observers or by comparison of colorimetric measurements to standards for signal colors.

6.2 Life—Following cycle operation per 4.9 to 70% of rated life (Example—1400 h for a 2000 h design life), the DFL system shall meet the requirements of 5.1 (starting procedures), 5.4 (color), and 5.5.1 (breakdown). Upon completion of the previous maintenance checks, the DFL system shall be returned to cycle operation.

6.3 Voltage Regulation—The DFL system electrical supply shall be designed such that an inoperative or removed lamp will not affect the operation and performance of the remaining lamp(s) in the DFL system.

6.4 Light and Near-Infrared Radiation Exposure Limits—Manufacturers and users of DFL systems should ensure that the DFL system does not exceed the maximum allowable limit value for three retinal hazards as specified in 4.3 of the ANSI Standard Z311.1 (titled Photobiological Safety). Those three hazards are retinal thermal injury from short-term viewing (Z311.1 paragraph 4.3.1), retinal photochemical injury from chronic exposure (Z311.1 paragraph 4.3.2), and long-term ocular exposure to infrared radiation (Z311.1 paragraph 4.3.3). The three hazards cover wavelengths between 400 to 1400 nm. The measurement of the exposure levels is strongly dependent on the value used for the angular subtense of the light source (the parameter, alpha, in Z311.1). The handbook "Safety With Lasers and Other Optical Sources" by Sliney and Wolbarsht (1980, Plenum Press) is recommended as a reference.

6.5 Steady-State—Steady-state light level (100%) shall be established by allowing the DFL system to operate for 120 s after being switched "on." The average light level within the period from 120 to 140 s will be defined as the 100% level. If the light output is not stable to within ±10% during the 120 to 140 s time interval, the process should be repeated. If the system selected does not stabilize within three attempts, another unit should be selected and subjected to the previous procedure. Steady-state is only used to define a system's baseline in order to evaluate test effects on the system.

6.6 High-Voltage Shock Safety—High-voltage shock from DFL systems is an important concern just as it is with other automotive components. Appropriate levels of safety must be designed in and other precautionary measures, such as use of caution labels, should be implemented to assure a sufficiently low level of risk. Since individual DFL products and vehicle applications will differ in regard to high voltage levels, power, and integrity of construction, each DFL system will require a specific evaluation with the system installed in the vehicle to assure a low problem potential. This is a vehicle design and testing issue and is beyond the scope of this document. Designs of replacement equipment will also need to be evaluated on specific vehicle models for which they are intended.

6.7 High-Voltage Vapor Ignition Safety—Protection from the possibility of vapor ignition from DFL high voltages is a concern. However, the concerns of vapor ignition are not too much different from those experienced from damaged high-voltage ignition systems, shorted and burning wiring, and exposed hot bulbs and filaments. Furthermore, the safety of a DFL system depends not only on its basic design, but also upon the design of the vehicle in which it is installed. Each vehicle application, therefore, needs to be individually evaluated as a complete DFL system in the vehicle to assure a low level of risk. This is a vehicle design and testing issue and is beyond the scope of this document.

6.8 CRI—The color rendering index shall meet the general criteria of Ra = 60. Each source manufacturer and user shall determine that the light produced shall readily allow the customer to distinguish between typical road sign colors.

---

[1] Current fluorescent lamp chromaticity standards in the United States are now based on spectroradiometrically determined assignments by the National Institute of Standards and Technology. Refer to ANSI C78.376-1969.

6.9 UV Test—UV weighting factors defined in ANSI Z311 (ANSI and NIOSH use the same tables, DIN tables are also defined) shall be used to determine time for minimum effect. Measurements shall be made in accordance with 4.10. (See Table 3 for examples.)

TABLE 3—EXAMPLE: 175 W MULTIVAPOR LAMP MEASURED AT 50 cm

Standard	Type	Time for Min. Effect
NIOSH (200-320)	ERYTH	124.4 h
DIN (240-325)	ERYTH	677.6
DIN (300-440)	PIGMT	26.0
DIN (220-305)	CONJT	29435.

The ANSI Z311 Standard shall be used (voluntary draft at present). Energy level to be determined by application.

It is recognized that ultraviolet radiation (UV) normally emitted by arc discharge sources may pose health hazards at certain levels and durations of exposure. The magnitudes of acceptable levels of exposure are outlined in documents such as those published by NIOSH or ACGIH (American Conference of Governmental Industrial Hygienists). These tables may be used as references when determining the exposure potential of DFL headlamps.

The concern with UV light may be addressed by the device manufacturer by using shields, coatings, and/or absorbing materials. It is anticipated that most lens materials will normally provide safe UV levels by absorption/reflection. However, where the possibility exists that lens protection may be lost while the arc source remains functional (stone damage or low-energy impact), it is the device manufacturer's responsibility to assure that protection from UV is provided to all who may be exposed to the light.

6.10 System Operating Voltage Range—Vehicles are designed to continue operating at voltage levels below 9.0 V. The vehicle manufacturers must define acceptable reductions in light output for voltages below 9.0 V and determine system dropout voltage.

φAUXILIARY DRIVING LAMPS—J581 JUN89

SAE Standard

Report of Lighting Committee approved March 1979 and completely revised by the Lighting Coordinating Committee June 1989. Rationale statement available.

1. Scope—This SAE Standard provides test procedures, performance requirements and guidelines for auxiliary driving lamps.

2. Definition

2.1 Auxiliary Driving Lamp—A lighting device mounted to provide illumination forward of the vehicle and intended to supplement the upper beam of a standard headlamp system. It is not intended for use alone or with the lower beam of a standard headlamp system.

3. Lighting Identification Code—The auxiliary driving lamps may be identified by the code "Y", in accordance with SAE J759.

4. Tests

4.1 SAE J575—The following test procedures in SAE J575 are a part of this report with the modifications indicated:

4.1.1 VIBRATION TEST
4.1.2 MOISTURE TEST
4.1.3 DUST TEST—(Dust test shall not be required for sealed units.)
4.1.4 CORROSION TEST
4.1.5 WARPAGE TEST—(Devices produced from plastic components.)
4.1.6 A photometric test.

4.1.6.1 The photometric tests for bulb replaceable units shall be made at a distance of at least 18.3 m (60 ft) from the photometer to the lamp.

4.1.6.2 Lamp Aim—A lamp or sealed beam unit, which is designed to be aimed mechanically, shall be centered on the photometric axis with the aiming planes normal to that axis. A lamp or sealed unit, not designed to be aimed mechanically, shall be photoelectrically aimed so that the test points in Fig. 1 designated by the squares have equal intensity and those designated by triangles have equal intensity.

4.2 Color Test—SAE J578 is a part of this report.

5. Requirements

5.1 Performance Requirements

5.1.1 SAE J575—A device when tested in accordance with the test procedures in Section 4, shall meet the following requirements in SAE J575, with the modifications indicated:

5.1.1.1 Vibration
5.1.1.2 Moisture
5.1.1.3 Dust
5.1.1.4 Corrosion
5.1.1.5 Warpage test on devices with plastic components.
5.1.1.6 Photometry—The lamp under test shall meet the photometric requirements contained in Table 1.

TABLE 1—PHOTOMETRIC REQUIREMENTS

Test Point Deg[a]	Candela, CD
2U – 3R and 3L	1600 min
1U – 3R and 3L	4000 min
H – V	20 000 min and 60 000 max
H – 3R and 3L	8000 min
1D – 6R and 6L	2960 min
2D – 6R and 6L	1600 min
4D – V	6000 max

[a] A tolerance of ±1/4 deg in location may be allowed at any test point

5.2 Color—The color of the emitted light shall be white as defined in SAE J578.

5.3 Plastic Materials—The plastic materials used in optical parts shall meet the requirements in SAE J576.

6. Guidelines

6.1 The photometric design guidelines for auxiliary driving lamps, when tested in accordance with paragraph 4.1.6 of this standard, are contained in Table 2.

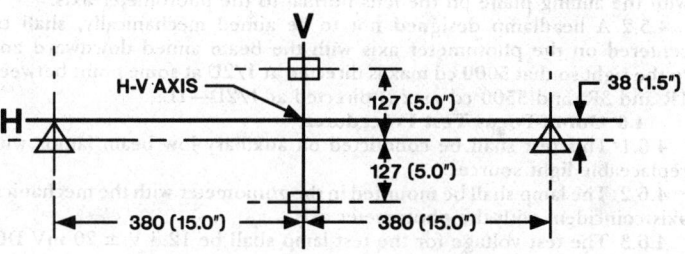

FIG. 1—TEST POINTS ON A SCREEN AT 7.6 m

TABLE 2—PHOTOMETRIC DESIGN GUIDELINES

Test Point Deg[a]	Candela, CD
2U – 3R and 3L	2000 min
1U – 3R and 3L	5000 min
H – V	25 000 min and 50 000 max
H – 3R and 3L	10 000 min
1D – 6R and 6L	3700 min
2D – 6R and 6L	2000 min
4D V	5000 max

[a] A tolerance of ±1/4 deg in location may be allowed at any test point.

6.2 These guidelines apply to the device as used on the vehicle and are not a part of the laboratory test procedures and requirements.

6.3 Lamp Aim—The lamp aim adjustments on the vehicle should be with mechanical aimers, if possible. Set the mechanical aim to 0-0, reference SAE J599.

6.4 Other Aiming Procedures—If the vehicle mounting or lamp design precludes mechanical aiming, the lamp shall be aimed photometrically (see paragraph 4.1.6.2), or visually aimed. The correct visual aim is with the high intensity zone of the beam symmetric about the H-V axis of the lamp on an aiming screen at 7.6 m (25 ft).

AUXILIARY LOW BEAM LAMPS—SAE J582 SEP84 — SAE Recommended Practice

Report of the Lighting Division, approved January 1941, completely revised by the Lighting Committee September 1984. Rationale statement available.

1. Scope—This SAE Technical Report provides general design and performance requirements, test procedures, and installation guidelines for auxiliary low beam lamps.

2. Definitions

2.1 An auxiliary low beam lamp supplements the lower beam of a standard headlamp system.

3. Lighting Identification Code—Auxiliary low beam lamps may be identified by the code "Z," in accordance with SAE J759, Lighting Identification Code.

4. Tests

4.1 Test Voltage—In conducting tests to the auxiliary low beam lamps, the test voltage used shall be 12.8 V ± 20 mV, DC as measured at the terminals of the lamp.

4.2 SAE J575—The following procedures in SAE J575, are a part of this report with modifications as indicated.

Lighting Devices
Bulbs
Test Fixture
Laboratory Facilities
Vibration Test
Moisture Test
Dust Test
Corrosion Test
Warpage Test on Devices with Plastic Components

4.3 Color Test—The test procedures in SAE J578 are a part of this report.

4.4 Plastic Materials—Plastic materials used in optical parts shall be tested in accordance with the procedures in SAE J576.

4.5 Photometric Test

4.5.1 Photometric tests shall be made with the photometer at a distance of at least 18.3 m (60 ft) from the headlamp. The headlamp shall be aimed mechanically by centering the headlamp on the photometer axis with the aiming plane on the lens normal to the photometer axis.

4.5.2 A headlamp designed not to be aimed mechanically, shall be centered on the photometer axis with the beam aimed downward and to the right so that 5000 cd max is directed at 1/2U at some point between 1R and 3R, and 3500 cd max is directed at 1/2D—1L.

4.6 Out-of-Focus Test Procedures

4.6.1 This test shall be conducted on auxiliary low beam lamps with replaceable light sources.

4.6.2 The lamp shall be mounted in the goniometer with the mechanical axis coincident with the photometer axis.

4.6.3 The test voltage for the test lamp shall be 12.8 V ± 20 mV DC.

4.6.4 The lamp shall be photometered at the appropriate test points as listed in Table 2.

4.6.5 Intensity measurements shall be made at six out-of-focus positions with the filament located at 2/3 of the tolerance values specified by the manufacturer above-below, ahead-behind, and right-left of the design position.

5. Requirements

5.1 Performance Requirements

5.1.1 LIGHTING DEVICES—The performance requirements apply only to new unused and undamaged lamps fabricated from production tools and assembled by production processes.

5.1.2 BULBS—Unless otherwise specified, bulbs used in the tests shall be supplied by the laboratory and be representative of standard bulbs in regular production. The rated standard bulbs shall be operated at their designed luminous intensity (MSCP); sealed units shall be seasoned and operated at their design voltage.

5.1.3 VIBRATION—Upon completion of the test, the same device shall be examined. There shall be no evidence of rotation, displacement, cracking or rupture of parts (except bulbs and sealed beam unit internal components) which would prevent the device from meeting the performance criteria of any of the tests contained in Section 4 of J575. Additionally, there shall be no evidence of cracking or rupture of parts of the device affecting its mounting.

5.1.4 MOISTURE—There shall be no moisture accumulation in excess of 2 mL.

5.1.5 DUST—The device shall be considered to have met the requirements if no dust is found on the interior surfaces of the device, or if the maximum beam intensity output is within 10% as compared with the condition after the device is cleaned inside and out. Sealed beam units shall be exempt from this test.

5.1.6 CORROSION—There shall be no evidence of internal or external corrosion or edge corrosion beyond 2.5 mm (0.100 in) from any sharp or cut edge. The lamps shall show no evidence of surface deterioration, fractures, color bleeding, or deterioration of bonding materials.

5.1.7 COLOR—The color light from auxiliary low beam lamps shall be white as specified in SAE J578.

5.1.8 PLASTIC MATERIALS—Any plastic materials used in optical parts shall conform to the requirements in SAE J576.

5.1.9 WARPAGE—This is a requirement only for devices with plastic components. There shall be no evidence of warpage, delamination, fractures, deterioration of bonding material, or deformation which would result in failure to meet the requirements of Section 5 of this technical report.

5.1.10 OUT-OF-FOCUS—The auxiliary low beam lamp shall meet the photometric requirements of Table 2 for each of the out-of-focus positions.

5.2 Design Guidelines

5.2.1 PHOTOMETRIC DESIGN—Table 1 establishes the desired auxiliary low beam lamp pattern. The beam from the device shall meet the beam candela distribution guidelines indicated in Table 1.

5.3 Service Performance Requirements

5.3.1 PHOTOMETRY—When tested in accordance with the procedures in paragraph 4.5, the device shall meet the requirements shown in Table 2.

6. Guidelines—The following recommendations and test procedures apply to the device as used on the vehicle and are not part of the laboratory test requirements and procedures.

6.1 Lamp Aim on Vehicle—Mechanical lamp aim adjustment and inspection may be performed in accordance with SAE J599 and SAE J602.

6.2 If vehicle mounting precludes mechanical aiming, the lamp may be visually aimed. The correct visual aim is with the top edge of the high intensity zone 25 mm (1 in) above horizontal at 7.6 m (25 ft) and the left edge of the high intensity zone 130 mm (5 in) left of vertical at 7.6 m (25 ft).

6.3 Means shall be provided to turn off the auxiliary low beam lamp independently of the lower beam lamps of the standard headlighting system.

6.4 Lamp Mounting—A single lamp shall be mounted at the front and to the left side (driver's side) of the center of the vehicle. If two lamps are used, they shall be mounted at the same mounting height level with respect to the standard headlamps.

TABLE 1—PHOTOMETRIC DESIGN GUIDELINES

Test Point Deg[a]	Candela-Max	Candela-Min
10U —90U[b]	175	—
1-1/2U —1L to L	800	—
1-1/2U —1R to R	2000	—
1/2U —1L to L	1000	—
1/2U —1R to 3R	5000	—
1/2D —1-1/2R to 3R	50 000	15 000
1/2D —1L to L	3500	—
1D —1R	—	10 000
1D —3R	—	15 000
4D —2R	8000	—

[a] A tolerance of ± 1/4 deg in location is allowed at any test point.
[b] From the normally exposed surface of the lens.

TABLE 2—PHOTOMETRIC SERVICE PERFORMANCE REQUIREMENTS

Test Point[a]	Requirement, cd
1-1/2U—1R	2000 max
1/2U —1L to L	1200 max
1/2U —1R	5000 max
1/2D —1-1/2R to 3R	10 000 min
1D —3R	15 000 min

[a] A tolerance of ± 1/4 deg in location may be allowed at any test point.

(R) FRONT FOG LAMPS
—SAE J583 JUN93 — SAE Standard

Report of the Lighting Division, approved May 1937. Completely revised by the Lighting Committee July 1977, editorial change May 1981. Rationale statement available. Completely revised by the SAE Road Illumination Devices Standards Committee June 1993. Rationale statement available.

1. Scope—This SAE Standard provides performance requirements, test procedures, and design and installation guidelines for front fog lamps.

2. References

2.1 Applicable Documents—The following publications form a part of this specification to the extent specified herein. The latest issue of SAE publications shall apply.

2.1.1 SAE PUBLICATIONS—Available from SAE, 400 Commonwealth Drive, Warrendale, PA 15096-0001.

SAE J575—Tests for Motor Vehicle Lighting Devices and Components
SAE J576—Plastic Materials for Use in Optical Parts Such as Lenses and Reflectors of Motor Vehicle Lighting Devices
SAE J578—Color Specification
SAE J599—Lighting Inspection Code
SAE J759—Lighting Identification Code

2.2 Definition

2.2.1 A FRONT FOG LAMP is a lighting device providing illumination forward of the vehicle under conditions of fog, rain, snow, or dust. Principally, the front fog lamp supplements the lower beam of a standard headlamp system.

3. Lighting Identification Code—Front fog lamps may be identified by the code "F" in accordance with SAE J759.

4. Tests

4.1 Test Voltage—In conducting tests on front fog lamps, the test voltage shall be 12.8 V ± 20 mV, DC as measured at the terminals of the lamp.

4.2 SAE J575—SAE J575 is a part of this report. The following test procedures are applicable with modifications indicated:

4.2.1 TESTS
 4.2.1.1 Vibration Test
 4.2.1.2 Moisture Test
 4.2.1.3 Dust Test
 4.2.1.4 Corrosion Test
 4.2.1.5 Photometry Test

4.2.1.5.1 Photometric tests shall be made with the photometer at a distance of at least 18.3 m (60 ft) from the fog lamp. A front fog lamp designed to be aimed mechanically shall be centered on the photometer axis with the aiming plane on the lens normal to the photometer axis.

4.2.1.5.2 A front fog lamp not intended to be aimed mechanically shall be centered on the photometer axis with the beam aimed downward so that 500 cd is directed at some point on the horizontal between 6L and 6R, whereby the horizontal beam distribution is being kept symmetrical with respect to the photometer axis. This can also be accomplished by balancing the beam.

4.2.1.5.3 A front fog lamp that has no provisions for mechanical lateral aim, shall be mounted in the design position of the goniometer, as it is prescribed to be mounted on the vehicle. The beam is aimed downward so that 500 cd is directed at some point on the horizontal between 6L and 6R.

4.2.1.6 Warpage Test (on devices with plastic components)

4.3 Color Test—SAE J578 is a part of this report.

4.4 Plastic Materials—Plastic materials used in optical parts shall be tested in accordance with the procedures in SAE J576.

4.5 Sealed Beam Unit Tests—Sealed beam units designed for use as front fog lamps, when tested without other parts of the lamp assembly, are not subject to moisture, dust, and corrosion tests.

5. Requirements

5.1 Performance Requirements

5.1.1 SAE J575 REQUIREMENTS—A device, when tested in accordance with the test procedures specified in Section 4, shall meet the following requirements in SAE J575:

5.1.1.1 Vibration
5.1.1.2 Moisture
5.1.1.3 Dust
5.1.1.4 Corrosion
5.1.1.5 Warpage (This is a requirement only for devices with plastic components.)

5.1.2 COLOR—The color of the light from a front fog lamp shall be white to yellow within the limits specified in SAE J578.

5.1.3 PHOTOMETRY—The beam of the front fog lamp shall be designed to conform to the light intensity distribution (candela) values as shown in Table 2. The lamp shall meet the photometric performance requirements contained in Table 1.

TABLE 1—PHOTOMETRIC REQUIREMENTS

Test Point[1] Degree	Candela (Cd) Max	Candela (Cd) Min
8U - 90U[2]	90	—
4U - 6L and 6R	150	—
2U - 6L and 6R	300	—
1U - 6L and 6R	420	—
H - 6L and 6R	600	—
1-1/2D - 3L and 3R	12 000	1600
1-1/2D - 9L and 9R	—	800
3D - 15L and 15R	—	800

[1] A tolerance of ±1/4 degree in location is allowed at any test point.
[2] From the normally exposed surface of the lens.

5.1.4 PLASTIC MATERIALS—The plastic materials used in optical parts shall meet the material performance requirements as listed in SAE J576.

6. Guidelines

6.1 Photometric Design Guidelines—The photometric design guidelines for front fog lamps when tested in accordance with 4.2.5 of this document are contained in Table 2.

TABLE 2—PHOTOMETRIC DESIGN GUIDELINES

Test Point[1] Degree	Candela (Cd) Max	Candela (Cd) Min
8U - 90U[2]	75	—
4U - 6L and 6R	125	—
2U - 6L and 6R	250	—
1U - 6L and 6R	350	—
H - 6L and 6R	500	—
1-1/2D - 3L and 3R	10 000	2000
1-1/2D - 9L and 9R	—	1000
3D - 15L and 15R	—	1000

[1] A tolerance of ±1/4 degree in location is allowed at any test point.
[2] From the normally exposed surface of the lens.

6.2 Installation Guidelines—These guidelines apply to the device as used on the vehicle and are not part of the design guidelines, performance requirements, or the test procedures.

6.2.1 LAMP AIM ON THE VEHICLE

6.2.1.1 Lamp aim adjustments and inspection should be made with mechanical aimers. The correct mechanical aim is 0-0 based on SAE J599.

6.2.1.2 If vehicle mounting or lamp design preclude mechanical aim, the lamp should be aimed visually. The correct visual aim is made with the top of the beam 100 mm (4 in) below the lamp center at 7.6 m (25 ft). The beam should be centered laterally about a vertical line directly ahead of the lamp.

6.2.1.3 Fog lamps designed for universal mounting applications to fit various vehicle makes and models should be provided with means for vertical and horizontal aim adjustment. Lamps and composite units, which are designed to be integrated into one specific vehicle, need only have means for vertical adjustment.

CARGO LAMPS FOR USE ON VEHICLES UNDER 12 000 LB GVWR—SAE J1424 JUN93

SAE Recommended Practice

Report of the Signalling and Marking Devices Standards Committee and The Sae Lighting Coordinating Committee approved August 1991 and revised June 1993. Rationale statements available.

1. Scope—This SAE Recommended Practice provides test procedures, performance requirements, design guidelines, and installation requirements for cargo lamps that are mounted on the exterior of vehicles weighing under 12 000 lb GVWR (Gross Vehicle Weight Rating).

2. References

2.1 Applicable Documents—The following publications form a part of this specification to the extent specified herein. The latest issue of SAE publications shall apply.

2.1.1 SAE PUBLICATIONS—Available from SAE, 400 Commonwealth Drive, Warrendale, PA 15096-0001.

SAE J567—Lamp Bulb Retention System
SAE J575—Tests for Motor Vehicle Lighting Devices and Components
SAE J576—Plastic Material for Use in Optical Parts Such as Lenses and Reflectors of Motor Vehicle Lighting Devices
SAE J578—Color Specification
SAE J759—Lighting Identification Code

2.2 Definition

2.2.1 A CARGO LAMP(S) is a supplemental lamp mounted on the exterior of a vehicle weighing under 12 000 lb GVWR for the purpose of providing illumination to load and unload cargo in an environment of otherwise insufficient light.

3. Lighting Identification Code—A cargo lamp for use on vehicles weighing less than 12 000 lb GVWR may be identified with the code "G" in accordance with SAE J759.

4. Tests

4.1 SAE J575 is a part of this report. The following tests are applicable with the modifications as indicated:

4.1.1 VIBRATION TEST
4.1.2 MOISTURE TEST
4.1.3 DUST TEST
4.1.4 CORROSION TEST
4.1.5 PHOTOMETRY TEST—In addition to the test procedures in SAE J575, the following apply:

4.1.5.1 Photometric measurements shall be made with the light source of the lamp at a distance of at least 3 m from the photometer.

4.1.6 WARPAGE TEST ON DEVICES WITH PLASTIC COMPONENTS—The device shall be operated with the bulb burning steadily during the test period.

4.2 Color Test—SAE J578 is a part of this report.

5. Requirements

5.1 Performance Requirements—A device when tested in accordance with the test procedures specified in Section 4 shall meet the following requirements:

5.1.1 VIBRATION—SAE J575
5.1.2 MOISTURE—SAE J575
5.1.3 DUST—SAE J575
5.1.4 CORROSION—SAE J575
5.1.5 PHOTOMETRY—SAE J575

5.1.5.1 A single lamp mounted on a test stand to simulate mounting attitude on the vehicle shall meet the photometric performance requirements contained in Table 1 and its footnotes.

5.1.6 WARPAGE TEST—SAE J575

5.2 Color—The color of the light from a cargo lamp shall be white as specified in SAE J578.

5.3 Material Requirements—Plastic materials used in optical parts shall meet the requirements of SAE J576.

6. Guidelines

6.1 Photometric Design Guidelines for cargo lamps, when tested in accordance with 4.1.5 of this report, are contained in Table 2.

6.2 Installation Guidelines

6.2.1 These guidelines apply to the device as used on the vehicle and are not a part of the design requirements, performance requirements, or test procedures.

6.2.1.1 The cargo lamp shall be wired in one of the following ways:
 a. Into the vehicle dome light circuit so that it cannot be energized unless the dome light is also energized, or
 b. In such a manner that it can be turned on only when the vehicle is stopped, or
 c. Independent with a separate on-off switch and a dash-mounted telltale lamp with an appropriate label to indicate the lamp operation.

6.2.1.2 The cargo lamp shall be mounted on the rear of the vehicle, or the vehicle cab, or in the cargo bed. One or more lamps may be used.

TABLE 1—PHOTOMETRIC PERFORMANCE REQUIREMENTS[2]

Zone	Test Points (degrees)	Minimum Luminous Intensity[1] Zone (cd)
1	10D - V	
	40D - V	67
	40D - 10L	
	40D - 10R	
2	10D - 10L	
	10D - 30L	
	40D - 30L	69
	70D - 30L	
	70D - 10L	
3	10D - 10R	
	10D - 30R	
	40D - 30R	69
	70D - 30R	
	70D - 10R	

[1] The measured value at each test point shall not be less than 60% of the minimum value for that test point in Table 2.
(R) [2] The maximum per lamp at 1.5 degrees down and above shall be 300 cd.

TABLE 2—PHOTOMETRIC DESIGN GUIDELINES

Test Points (degrees)	Luminous Intensity (cd)
1.5D & Above	300 max
10D - V	22 min
10D - 10L	22 min
10D - 10R	22 min
10D - 30L	22 min
10D - 30R	22 min
40D - V	15 min
40D - 10L	15 min
40D - 10R	15 min
40D - 30L	15 min
40D - 30R	15 min
70D - 10L	5 min
(R) 70D - 10R	5 min
70D - 30L	5 min
70D - 30R	5 min

APPENDIX A

A.1 As a matter of information, attention is called to SAE J567 for requirements and gages used in socket design.

(R) MOTORCYCLE HEADLAMPS—SAE J584 DEC83

SAE Standard

Report of the Lighting Committee, approved January 1949, completely revised by the Motorcycle Committee December 1983. Rationale statement available.

1. Scope—This SAE Standard provides design parameters and general requirements for motorcycle headlamps.

2. Definition—A motorcycle headlamp is a major lighting device used to provide general illumination ahead of the vehicle. For definition and classes of motorcycles, see SAE J213.

3. Laboratory Requirements

3.1 The following sections from SAE J575 are a part of this standard:

3.1.1 Section 2 —Samples for Test
3.1.2 Section 2.2—Bulbs
3.1.3 Section 3 —Laboratory Facilities
3.1.4 Section 4.1—Vibration Test
3.1.5 Section 4.2—Moisture Test
3.1.6 Section 4.3—Dust Test
3.1.7 Section 4.4—Corrosion Test
3.1.8 Section 4.6—Photometry
3.1.9 Section 4.8—Warpage Test on Devices with Plastic Components

3.2 Plastic Materials—Any plastic material used in optical parts shall comply with the requirements set forth in SAE J576.

3.3 Color Test—Color of the light from a motorcycle headlamp shall be white, as defined in SAE J578.

3.4 Aiming Adjustment Tests

3.4.1 A minimum aiming adjustment of 4 deg in each direction from the vertical and horizontal planes shall be provided.

3.4.2 Headlamps with independent vertical and horizontal aiming adjusting mechanisms:

3.4.2.1 The headlamp unit mounting shall be provided with independent vertical and horizontal aiming adjustments. The adjustment mechanisms shall be designed so that neither the vertical nor horizontal aim will deviate more than 100 mm (4 in) from the horizontal or vertical planes, respectively, at a distance of 7.6 m (25 ft) through an angle of ±4 deg.

3.4.2.2 When adjusting screws are employed, they shall be equipped with self-locking devices which operate satisfactorily for a minimum of 10 adjustments on each screw, over a length of screw thread of ±3 mm (1/8 in).

3.4.3 Headlamps with ball and socket or equivalent adjustment means need not conform with 3.4.2.

3.5 Inward Force Test—The mechanism, including the aiming adjusters, shall be designed to prevent the unit from receding permanently by more than 2.5 mm (0.1 in) into the lamp body or housing when an inward force of 222 N (50 lbf) is exerted at the geometric center of the outer surface of the lens.

3.6 Clarity of Hot Spot Definition—The geometric center of the high intensity zone of the upper beam of the multiple beam headlamps shall be deemed sufficiently defined for the purpose of service aiming if it can be set by three experienced observers on a vertical screen at 7.6 m (25 ft) within a maximum vertical deviation of ±0.3 deg and within a maximum horizontal deviation of ±0.4 deg. The aim for each observer shall be taken as the average of at least three observations.

3.7 Beam Aim During Photometric Test

3.7.1 The upper beam of a multiple beam headlamp shall be aimed photoelectrically so that the center of the zone of highest intensity falls 0.4 deg vertically below the lamp axis and is centered laterally. The center of the zone of highest intensity shall be established by the intersection of a horizontal plane passing through the point of maximum intensity, and the vertical plane established by balancing the photometric values at 3 deg left and 3 deg right.

3.7.2 The beam of a single beam Class C (moped) lamp shall be aimed photoelectrically so that the center of the zone of highest intensity falls 1.5 deg vertically below the lamp axis and is centered laterally. The center of the zone of highest intensity shall be established by the intersection of a horizontal plane passing through the point of maximum intensity, and the vertical plane established by balancing the photometric values at 3 deg left and 3 deg right.

3.8 Photometric Design Requirements

3.8.1 TEST PROCEDURES—Photometric tests shall be made with photometer at a distance of at least 18.3 m (60 ft) from the unit. The bulb or unit shall be operated at 6.4 V for a 6 V system and 12.8 V for a 12 V system during the test.

3.8.2 DESIGN INTENSITY REQUIREMENTS— The beam or beams from the unit shall be designed to conform to the intensity specifications in Tables 1, 2, or 3. A tolerance of ±0.25 deg in location may be allowed for any test point.

4. Optional Systems—One or two 178 mm (Type 2D1) or 142 × 200 mm (Type 2B1) sealed beam units meeting the requirements, including aim, of SAE J579 may be used on Class A, B, C, and D motorcycles. One 146 mm (Type 1C1) and one 146 mm (Type 2C1), or one 100 × 165 mm (Type 1A1) and one 100 × 165 mm (Type 2A1) sealed beam units meeting the requirements, including aim, of SAE J579 may be used on Class A, B, C, and D motorcycles.

TABLE 1—CLASS A AND D MOTORCYCLE

Upper Beam

Test Points (Deg)	Min. cd	Max. cd
2U-V	1000	
1U-3L and 3R	2000	
H-V	12 500	
1/2D-V	20 000	
1/2D-3L and 3R	10 000	
1/2D-6L and 6R	3300	
1/2D-9L and 9R	1500	
1/2D-12L and 12R	800	
1D-V	17 500	
2D-V	5000	
3D-V	2500	
3D-9L and 9R	1500	
3D-12L and 12R	300	
4D-V	1500	7500
Anywhere		75 000

Lower Beam

Test Points (Deg)	Min. cd	Max. cd
1-1/2U-1R to R		1400
1U-1-1/2L to L		700
1/2U-1-1/2L to L		1000
1/2U-1R to 3R		2700
1-1/2D-9L and 9R	700	
2D-V	7000	
2D-3L and 3R	4000	
2D-6L and 6R	1500	
2D-12L and 12R	700	
3D-6L and 6R	800	
4D-V	2000	
4D-4R		12 500

TABLE 2—CLASS B, C AND E MOTORCYCLE

Upper Beam

Test Points (Deg)	Class B Min. cd	Class B Max. cd	Class C and E Min. cd	Class C and E Max. cd
1U-3L and 3R	2000		1000	
H-V	10 000		5000	
1/2D-V	20 000		7500	
1/2D-3L and 3R	5000		3000	
1/2D-6L and 6R	2000		800	
1D-V	15 000		5000	
2D-V	5000		3000	
3D-V	2500		1000	
3D-6L and 6R	800		500	
4D-V		7500		7500
Anywhere		75 000		75 000

Lower Beam

Test Points (Deg)	Class B Min. cd	Class B Max. cd	Class C and E Min. cd	Class C and E Max. cd
1-1/2U-1R to R		1400		1400
1U-1-1/2L to L		700		700
1/2U-1-1/2L to L		1000		1000
1/2U-1R to 3R		2700		2700
2D-V	5000		4000	
2D-3L and 3R	3000		3000	
2D-6L and 6R	1500		1500	
3D-6L and 6R	800		800	
4D-V	2000		2000	
4D-4R		12 500		12 500

TABLE 3—CLASS C AND E MOTORCYCLE

Single Beam

Test Points (Deg)	Min. cd	Max. cd
1-1/2U-1R to 3R		1400
1U-1-1/2L to L		700
1/2U-1-1/2L to L		1000
1/2U-1R to 3R		2700
2D-V	4000	
2D-3L and 3R	3000	
2D-6L and 6R	1500	
4D-V	1000	
4D-4R		12 500

NOTE: Although automotive headlamp units may be optionally used, it should be noted that they conventionally supply a lesser amount of low beam light on the left side.

5. Installation Requirements—The following requirements apply to the devices as used on the vehicle and are not part of laboratory test requirements and procedures:

5.1 Beam Switching—The switching of motorcycle headlamps between the upper and lower beams should be by means of a switch designed and located so that it may be operated conveniently by a simple movement of the operator's hand or foot. The switch shall have no dead point between upper and lower beam switch position.

5.2 Means shall be provided for indicating to the driver that the upper beam is on. The upper beam indicator shall be plainly visible to operators of all heights under normal driving conditions when headlights are required. See SAE J107 APR80 for recommended high beam indicator.

5.3 Semiautomatic headlamp beam switching devices are permitted. See SAE J565.

φMOTORCYCLE AUXILIARY FRONT LAMPS—SAE J1306 JUN89

SAE Recommended Practice

Report of the Motorcycle Committee, approved June 1980 and completely revised June 1989.

1. Scope—This engineering design specification provides parameters and general requirements for auxiliary front lamps to be used on motorcycles. It may be supplemented by a service performance requirement.

2. Definition—An auxiliary lamp as covered by this specification is a unit, including sealed beam, intended to supplement either the upper or the lower beam from motorcycle headlamps.

3. Laboratory Requirements

3.1 The following sections of SAE J575 are a part of this recommended practice:

Section 2   —Samples for Tests
Section 2.2—Bulbs
Section 3   —Laboratory Facilities
Section 4.1—Vibration Test
Section 4.2—Moisture Test
Section 4.3—Dust Test
Section 4.4—Corrosion Test
Section 4.6—Photometry Test
Section 4.8—Warpage Test on Devices with Plastic Components

3.2 Sealed beam units need to comply only with Sections 2, 3, and 4.6 of SAE J575.

3.3 Color Test—The color of the light from a motorcycle auxiliary front lamp shall be white. (See SAE J578.)

3.4 Plastic Materials—Any plastic materials used in exterior optical parts shall comply with the requirements set forth in SAE J576.

3.5 Photometric Tests—These shall be made with the photometer at a distance of at least 60 ft (18.3 m) from the lamp.

3.5.1 AT-FOCUS TESTS—The light source shall be located in the designed position as specified by the manufacturer.

The beam from the lamp shall be aimed with the left edge of the high intensity zone at a vertical line straight ahead of the lamp center and with the top edge of the high intensity zone at the level of the lamp center at a distance of 25 ft (7.6 m) from the lens.

The beam from the lamp shall meet the photometric specifications listed in Table 1 when it is aimed as specified.

3.5.2 OUT-OF-FOCUS TESTS ON UNSEALED UNITS— Similar tests shall be made for each of four out-of-focus filament positions, except that the completed distribution may be omitted. Where conventional bulbs with two-pin bayonet bases are used, intensity tests shall be made with the light source 0.060 in (1.5 mm) above, below, ahead, and behind the designed position. If prefocused bulbs are used, the limiting positions at which tests are made shall be 0.020 in (0.5 mm) above, below, ahead, and behind the designed position.

The beam from the lamp may be reaimed as specified in paragraph 3.5.1 for each of the out-of-focus positions of light source.

3.5.3 The lamp shall be designed to comply with the photometric requirements shown in Table 1 for the design filament position and the required out-of-focus filament position. An aiming tolerance of ±0.25 deg shall be allowed at each test point.

4. Installation and Usage Requirements—The following items apply to the device as used on the motorcycle and are not a part of the laboratory test requirements and procedures.

For greatest visibility, with reasonable limitation of glare to approaching drivers, the beam from the lamp shall be aimed in accordance with paragraph 3.5.1. The unit should be turned off when traveling in congested areas in cities. It may be wired so that it can be turned on or off with either beam of the regular headlamps.

APPENDIX A

A.1 As a matter of information, attention is called to SAE J567 for requirements and gages to be used in socket design for unsealed units.

TABLE 1—TEST POSITIONS AND LUMINOUS INTENSITY REQUIREMENTS

Position, deg[a]	Max Intensity (cd)
1U-1L to left and above	400
1/2U-1L to left	500
1/2D-1L to left	1000
1-1/2D-1L to left	3000
2U-1R to right and above	1000
1U-1R to right	3000
H-1R to right	7000
1-1/2D-2R to 4R	10 000 min

[a]An aiming tolerance of ±1/4 deg should be allowed on individual points.

REPLACEABLE MOTORCYCLE HEADLAMP BULBS—SAE J1577 JUN91

SAE Recommended Practice

Report of the SAE Motorcycle Electrical Systems Subcommittee of the SAE Motorcycle Committee approved June 1991. Rationale statement available.

1. Scope—This SAE Recommended Practice provides performance parameters and dimensional specifications for available light sources (replaceable bulbs) which are appropriate for motorcycle headlamps.

2. References

2.1 Applicable Documents—The following publications form a part of this specification to the extent specified herein.

2.1.1 SAE Publications—Available from SAE, 400 Commonwealth Drive, Warrendale, PA 15096-0001.

SAE J584 DEC83—Motorcycle Headlamps
SAE J1383 APR85—Performance Requirements for Motor Vehicle Headlamps

2.1.2 IEC Publications (see Table 1)—Available from International Electrotechnical Commission, 3, rue de Verambe, P.O. Box 131, 1211 Geneva 20, Switzerland.

2.2 Definitions

2.2.1 Replaceable Motorcycle Headlamp Bulb (hereinafter referred to as "bulb" or "bulbs")—A radiant energy source with related envelope and mounting base which is removable from the motorcycle headlamp for the purpose of replacement.

2.2.2 Seasoning—Process of energizing the filament of a bulb at design voltage for a period of time equal to 1% of design life or 10 h, whichever is shorter.

3. Bulb Type and Specifications

3.1 Table 2 lists each bulb type with its specifications. Bulbs may be added to Table 2 in the future after review.

3.2 All the bulbs shall satisfy the following vibration test requirements.

3.2.1 Bulb shall be seasoned and mounted in a relevant headlamp and photometered to the applicable test points in accordance with article 3.8 of SAE J584 DEC83 before the vibration test.

3.2.2 Bulbs shall be mounted on the vibration test machine in their designed operating position.

3.2.3 Conditions of vibration test are shown in Table 3.

3.2.4 During the vibration test, the upper beam filament and the lower beam filament shall be energized alternately at 1 h intervals at design voltage.

3.2.5 Filament(s) shall not fail throughout the test period.

3.2.6 After the vibration test, the bulb shall be photometered to the applicable test points in accordance with article 3.8 of SAE J584 DEC83. The values shall not vary by more than ±10% from the values measured before the test.

TABLE 1—IEC PUBLICATIONS

Type	Bulb	Base	Socket (Holder)
H4	Figures 1 and 2	Figure 3	Figure 4
Reference	Data sheet: 809-IEC-2120-1 (1985) of Doc. IEC Publication 809, 1st edition Title: Road Vehicle Lamp Data Sheet Category: H4 Cap: P43t-38	Data sheet: 7004-39-3 (1980) of Doc. IEC Publication 61-1J Title: Prefocus Cap for Automobile Lamps P43t-38 Assembly of Ring and Cap on Finished Lamps	Data sheet: 7005-39-2 (1983) of Doc. IEC Publication 61-2, 3rd edition Title: Lampholder for Automobile Lamps P43t-38
HS1	Figures 5 and 6	Figure 7	Figure 8
Reference	Data sheet: 809-IEC-2130-1 (1988) of Doc. IEC Publication 809, Amendment 2 Title: Road Vehicle Lamp Data Sheet Category: HS1 Cap: PX43t	Data sheet: 7004-34-1 (1987) of Doc. IEC Publication 61-1L Title: Prefocus Cap PX43t Assembly of Ring and Cap on Finished Lamps	Data sheet: 7005-34-1 (1987) of Doc. IEC Publication 61-2H Title: Lampholder PX43t

TABLE 2—TYPICAL REPLACEABLE MOTORCYCLE HEADLAMP BULBS

Bulb Type	Number of Filaments	Rated Watts	Rated Volts	Design Voltage (D.V.)	Luminous Flux (Lumens Approx.) at D.V.	Max. Wattage at D.V.	Rated Average Laboratory Life at D.V.[3]	Filament Type	Dimensions
H4	2 (U.B./L.B.)	60/55	12/12	13.2	1650[1,2] / 1000[1,2]	75/68	450	Axial/Axial	See Figures 1 to 4
HS1	2 (U.B./L.B.)	35/35	12/12	13.2	825[1,2] / 525[1,2]	36.75/36.75	[4]	Axial/Axial	See Figures 5 to 8

U.B. = upper beam
L.B. = lower beam

[1] With black cap.
[2] Tolerances are ±15% for both U.B./L.B.
[3] The filaments are operated alternately according to the following cycle and starting with the lower beam filament: lower beam filament 15 h on 45 min off, then the upper beam filament 7.5 h on 45 min off; repeat the cycle.
The end of bulb life is determined by failure of either filament. The off periods are not considered as part of the bulb life.
[4] Not determined at this time.

TABLE 3—CONDITIONS OF VIBRATION TEST

Wave Form	Sinusoidal
Frequency	50-500-50 Hz at a linear sweep period of two minutes
Acceleration and Direction	98 m/s² at bulb retaining portion. Vertical
Test Cycle	180 cycles

Dimensions in mm
The drawings are not mandatory; their sole purpose is to show which dimensions must be verified.

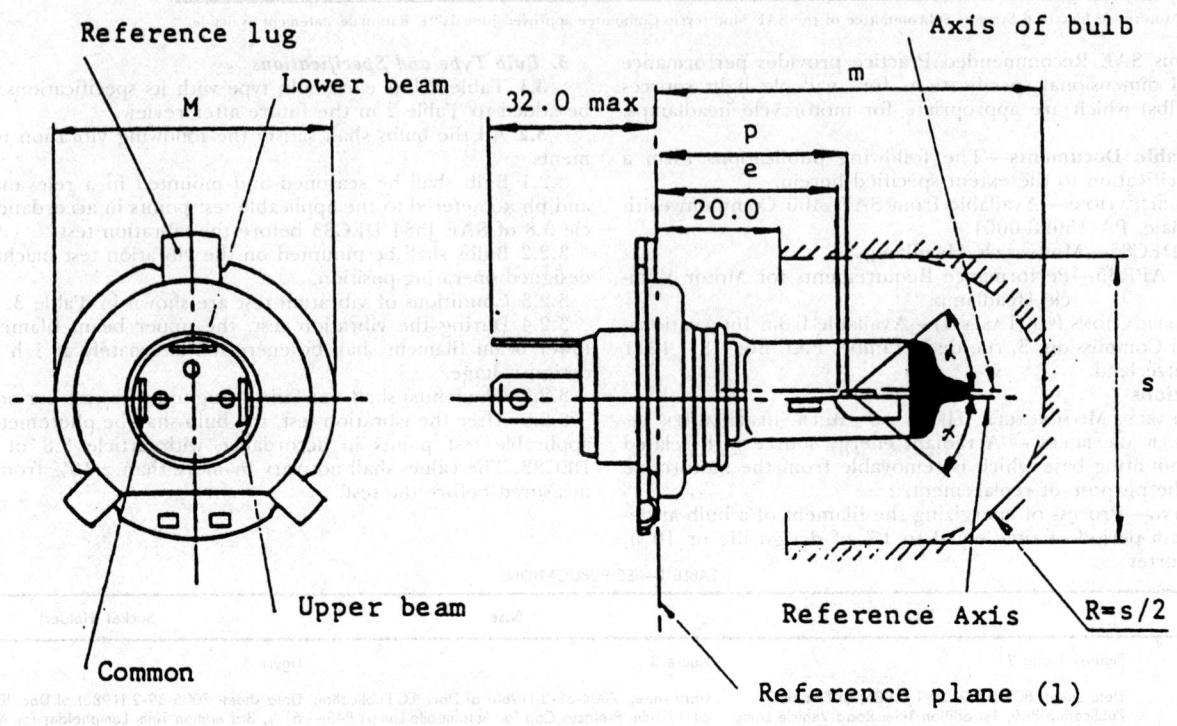

Reference	Dimension	Tolerance
e	28.5	+0.45/−0.25
p	28.95	—
m (2)	max. 60.0	—
s (3)	45.0	—
α (4)	max. 40°	—

(1) The reference plane is the plane formed by the seating points of the three lugs of the base.
(2) Dimension m denotes maximum length of the bulb.
(3) It must be possible to insert the bulb into a cylinder of diameter s concentric with the reference axis and limited at one end by a plane parallel to and 20 mm distant from the reference plane and at the other end by a hemisphere of radius s/2.
(4) The obscuration must extend at least as far as the cylindrical part of the capsule. It must also overlap the internal shield when the latter is viewed in a direction perpendicular to the reference axis. The effect sought by obscuration may also be achieved by other means.

FIGURE 1—TYPE H4 REPLACEABLE MOTORCYCLE HEADLAMP BULB—DIMENSIONAL SPECIFICATIONS

Dimensions in mm

The drawings are not mandatory with respect to the design of the shield.

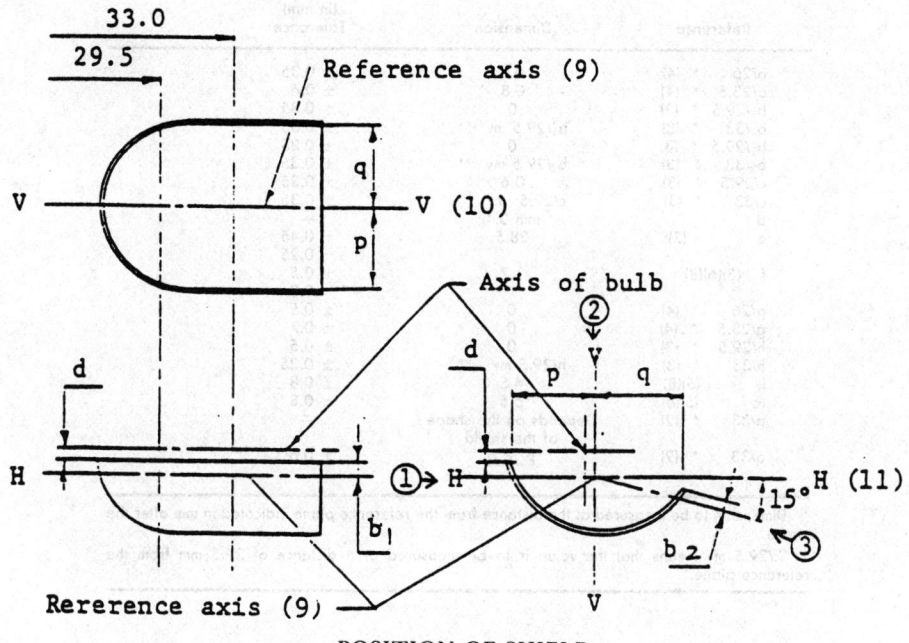

POSITION OF SHIELD

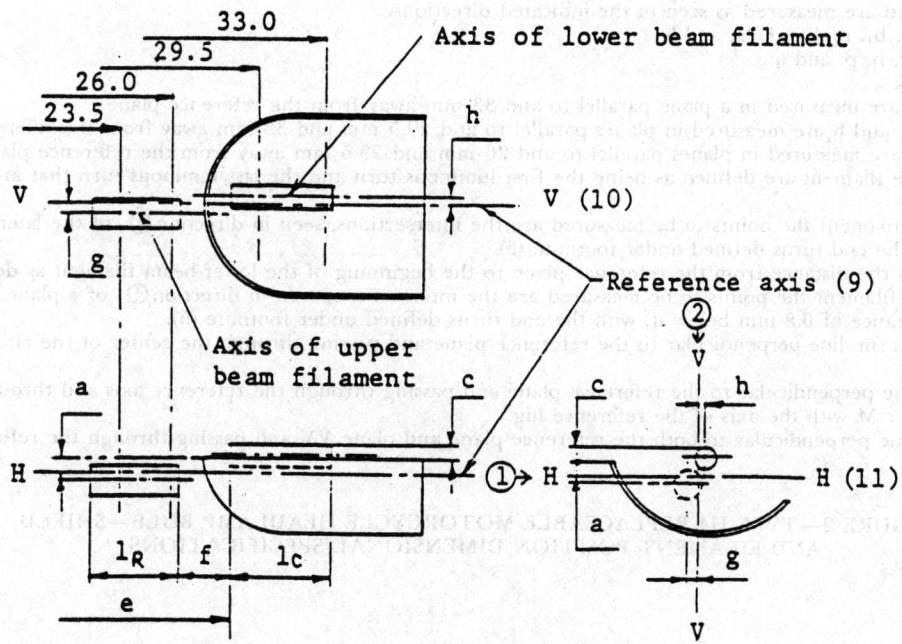

POSITION OF FILAMENTS

Reference		Dimension	(in mm) Tolerance
a/26	* (4)	0.8	± 0.35
a/23.5	* (4)	0.8	± 0.6
b_1/29.5	* (3)	0	± 0.35
b_1/33	* (3)	b_1/29.5 mv **	± 0.35
b_2/29.5	* (3)	0	± 0.35
b_2/33	* (3)	b_2/29.5 mv **	± 0.35
c/29.5	* (3)	0.6	± 0.35
c/33	* (3)	c/29.5 mv **	± 0.35
d		min 0.1	—
e	(7)	28.5	+ 0.45 / − 0.25
f	(5)(6)(8)	1.7	+ 0.5 / − 0.3
g/26	* (4)	0	± 0.5
g/23.5	* (4)	0	± 0.7
h/29.5	* (3)	0	± 0.5
h/33	* (3)	h/29.5 mv **	± 0.35
l_R	(5)(8)	4.5	± 0.8
l_c	(5)(6)	5.5	± 0.8
p/33	* (2)	Depends on the shape of the shield	—
q/33	* (2)	$\frac{p+q}{2}$	± 0.6

* Dimension to be measured at the distance from the reference plane indicated in mm after the stroke.

** "./29.5 mv means that the value is to be measured at a distance of 29.5 mm from the reference plane.

(1) The dimensions noted are measured as seen in the indicated directions.
  ① for dimensions a, b_1, c, d, e, f, l_R, and lc;
  ② for dimensions g, h, p, and q;
  ③ for dimension b_2.
(2) Dimensions p and q are measured in a plane parallel to and 33 mm away from the reference plane.
(3) Dimensions b_1, b_2, c, and h are measured in planes parallel to and 29.5 mm and 33 mm away from the reference plane.
(4) Dimensions a and g are measured in planes parallel to and 26 mm and 23.5 mm away from the reference plane.
(5) The end turns of the filament are defined as being the first luminous turn and the last luminous turn that are at substantially the correct helix angle.
(6) For the lower-beam filament the points to be measured are the intersections, seen in direction ①, of the lateral edge of the shield with the outside of the end turns defined under footnote (5).
(7) Dimension e denotes the distance from the reference plane to the beginning of the lower-beam filament as defined above.
(8) For the upper-beam filament the points to be measured are the intersections, seen in direction ①, of a plane, parallel to plane HH and situated at a distance of 0.8 mm below it, with the end turns defined under footnote (5).
(9) The reference axis is the line perpendicular to the reference plane and passing through the center of the circle of diameter M (see Figure 1).
(10) Plane VV is the plane perpendicular to the reference plane and passing through the reference axis and through the intersection of the circle of diameter M with the axis of the reference lug.
(11) Plane HH is the plane perpendicular to both the reference plane and plane VV and passing through the reference axis.

FIGURE 2—TYPE H4 REPLACEABLE MOTORCYCLE HEADLAMP BULB—SHIELD AND FILAMENT POSITION DIMENSIONAL SPECIFICATIONS

24.127

Dimensions in mm

The drawing is intended only to indicate the dimensions essential for interchangeability.

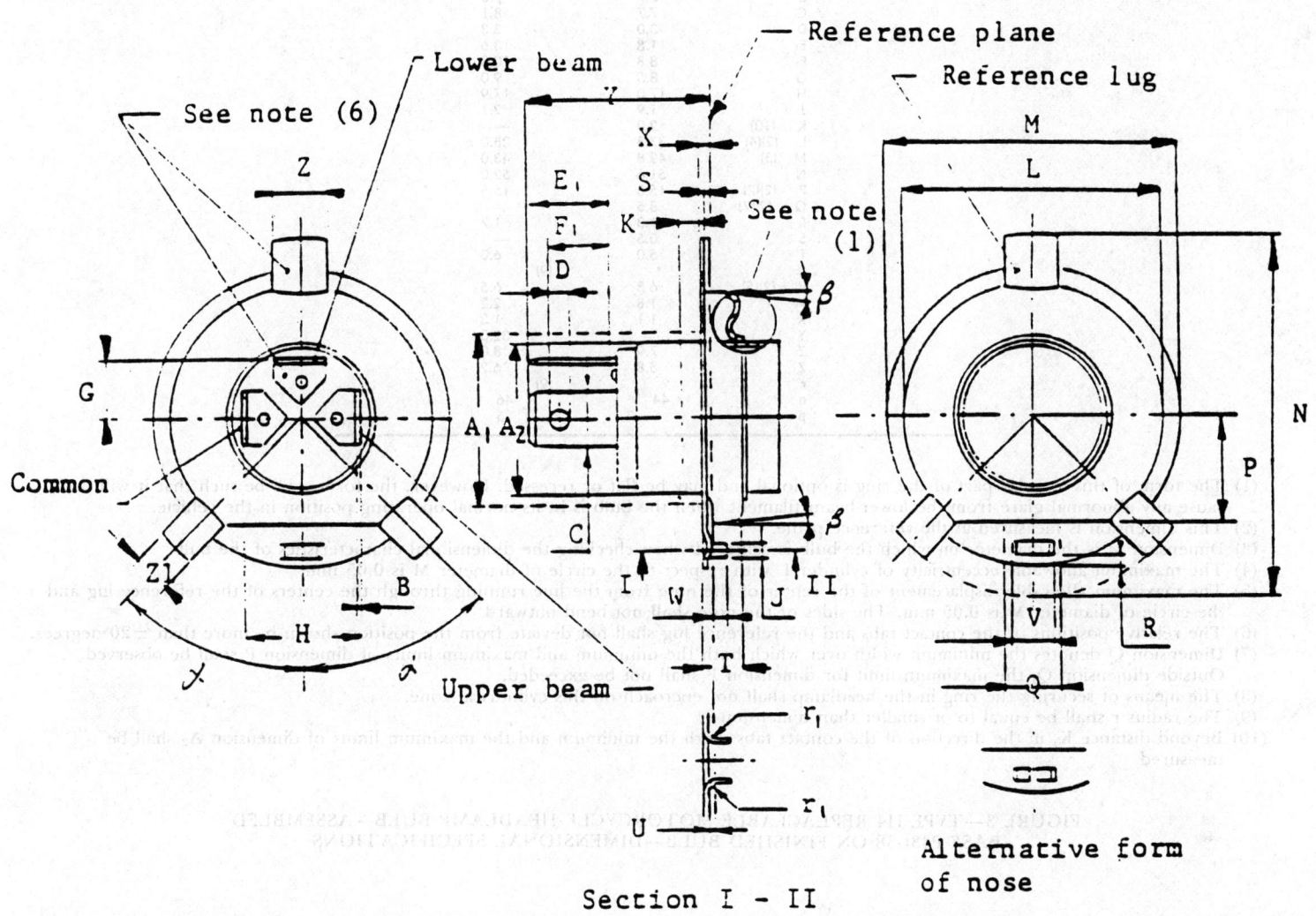

Dimension		Min.	Max.
A_1	(8)	25.0	—
A_2		21.94	22.0
B		0.7	0.8
C		7.7	8.1
D		3.0	3.3
E_1		11.8	13.6
F_1		8.8	10.3
G		8.5	9.0
H		17.0	17.9
J		1.9	2.1
K	(10)	2.0	—
L	(2)(4)	37.8	38.0
M	(3)	42.8	43.0
N		51.6	52.0
P	(2)(7)	15.3	15.5
Q	(2)(7)	8.5	—
R		1.3	1.7
S		0.5	—
T		5.0	6.0
U			(9)
V	(2)(5)	6.3	6.5
W		1.8	2.2
X		1.1	1.3
Y		—	32.0
Z		7.9	8.0
Z_1		5.8	6.2
r			(9)
α		44°	46°
β		—	5°

(1) The form of this annular part of the ring is optional and may be flat or recessed. However, the form shall be such that it will not cause any abnormal glare from the lower-beam filament when the bulb is in its normal operating position in the vehicle.
(2) This dimension is measured at the reference plane.
(3) Dimension M is the diameter on which the bulb is centered when checking the dimensional characteristics of the bulb.
(4) The maximum allowable eccentricity of cylinder L with respect to the circle of diameter M is 0.05 mm.
(5) The maximum allowable displacement of the center of the nose from the line running through the centers of the reference lug and the circle of diameter M is 0.05 mm. The sides of the nose shall not bend outward.
(6) The relative positions of the contact tabs and the reference lug shall not deviate from the position shown by more than ±20 degrees.
(7) Dimension Q denotes the minimum width over which both the minimum and maximum limits of dimension P shall be observed. Outside dimension Q, the maximum limit for dimension P shall not be exceeded.
(8) The means of securing the ring in the headlamp shall not encroach on this cylindrical zone.
(9) The radius r shall be equal to or smaller than dimension U.
(10) Beyond distance K, in the direction of the contact tabs, both the minimum and the maximum limits of dimension A_2 shall be measured.

FIGURE 3—TYPE H4 REPLACEABLE MOTORCYCLE HEADLAMP BULB—ASSEMBLED BASE P43t-38 ON FINISHED BULB—DIMENSIONAL SPECIFICATIONS

Dimensions in mm

The drawings are intended only to indicate the dimensions essential for interchangeability.

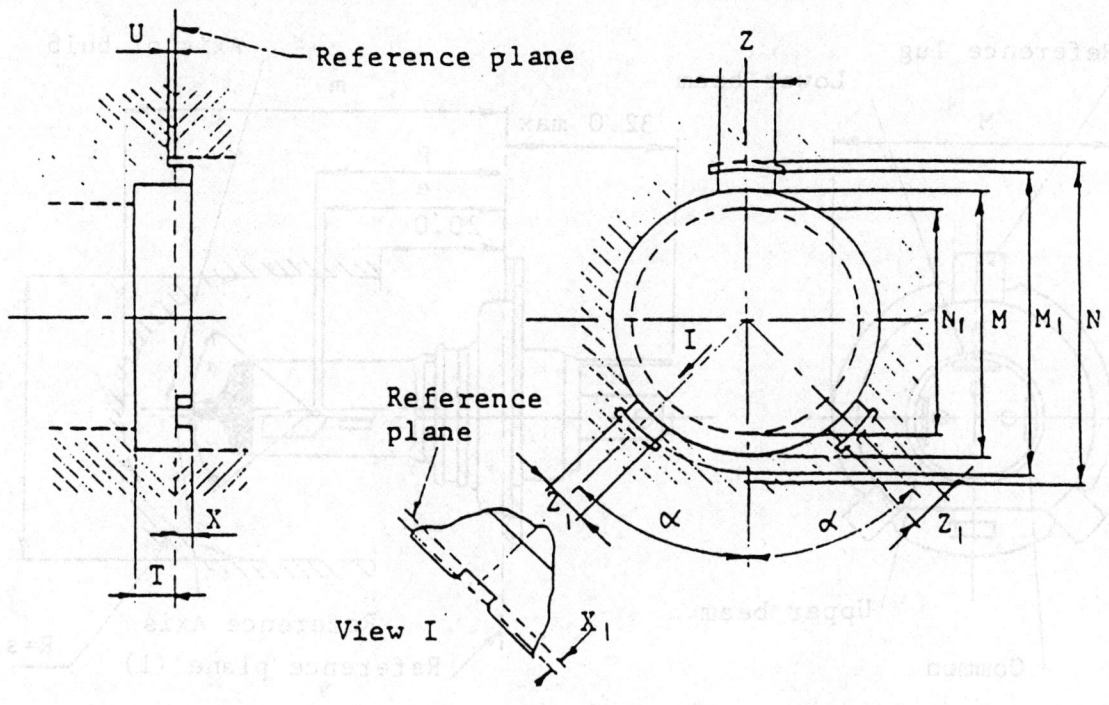

Dimension	Min.	Max.
M	43.02 (1)	43.2
M_1	—	49.0
N	52.2	—
N_1	35.0	—
T	5.5	—
U	0.4	—
X	1.8	—
X_1	1.4	—
Z	8.05	8.15
Z_1	8.0	8.5
α	44°	46°

The holder shall be so designed that the means of retention of the bulb can be applied only when the bulb is in the correct position.

The means of retention shall make contact only with the prefocus ring of the base and the total force exerted, when the bulb is in position, shall be not less than 10 N and be not greater than 60 N.
  (1) This value shall be complied with between the rim of the bulbholder and the reference plane (dimension X).
     However, it may be reduced to 38.5 mm within the dimensions Z and Z_1 which correspond with the support points for the lugs of the cap.
  (2) Dimension X_1 denotes the minimum distance over which dimensions Z and Z_1 shall apply.
     Outside dimension X_1 the slots may be chamfered or rounded.

FIGURE 4—TYPE H4 REPLACEABLE MOTORCYCLE HEADLAMP BULB—
BULBHOLDER P43t—DIMENSIONAL SPECIFICATIONS

Dimensions in mm

The drawings are not mandatory; their sole purpose is to show which dimensions must be verified.

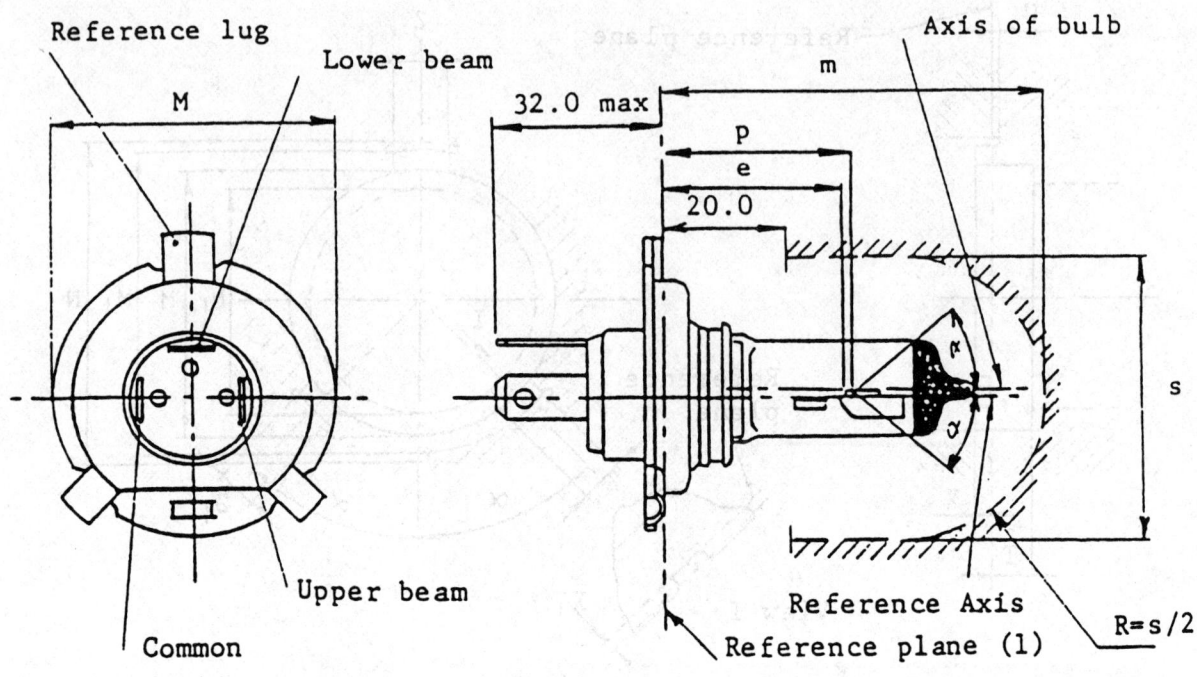

Reference	Dimension	Tolerance
e	28.5	+0.45/−0.25
p	28.95	—
m (2)	max. 60.0	—
s (3)	45.0	—
α (4)	max. 40°	—

(1) The reference plane is the plane formed by the seating points of the three lugs of the base.
(2) Dimension m denotes maximum length of the bulb.
(3) It must be possible to insert the bulb into a cylinder of diameter s concentric with the reference axis and limited at one end by a plane parallel to and 20 mm distant from the reference plane and at the other end by a hemisphere of radius s/2.
(4) The obscuration must extend at least as far as the cylindrical part of the capsule. It must also overlap the internal shield when the latter is viewed in a direction perpendicular to the reference axis. The effect sought by obscuration may also be achieved by other means.

FIGURE 5—TYPE HS1 REPLACEABLE MOTORCYCLE HEADLAMP
BULB—DIMENSIONAL SPECIFICATIONS

24.131

Dimensions in mm

The drawings are not mandatory with respect to the design of the shield.

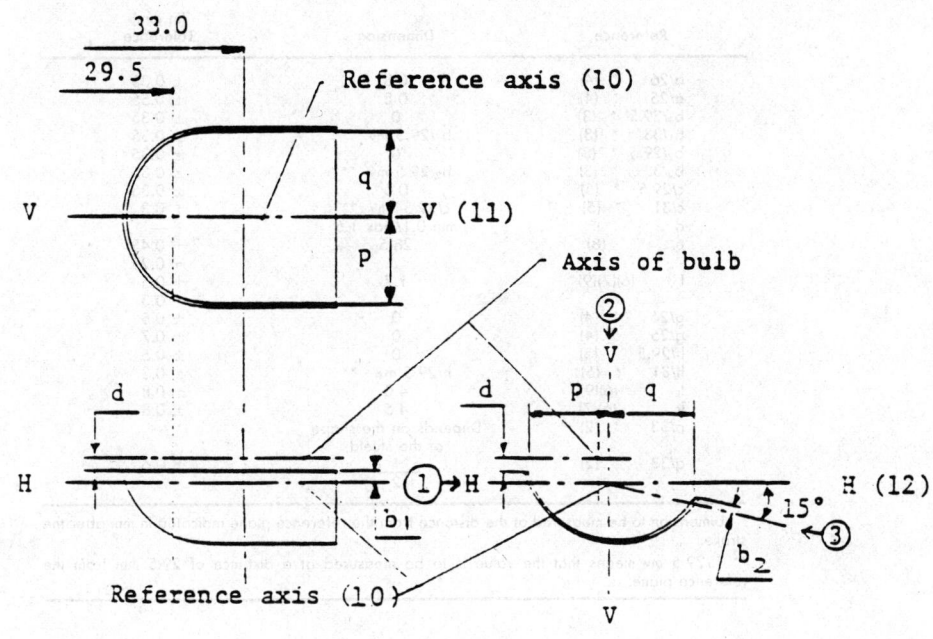

POSITION OF SHIELD

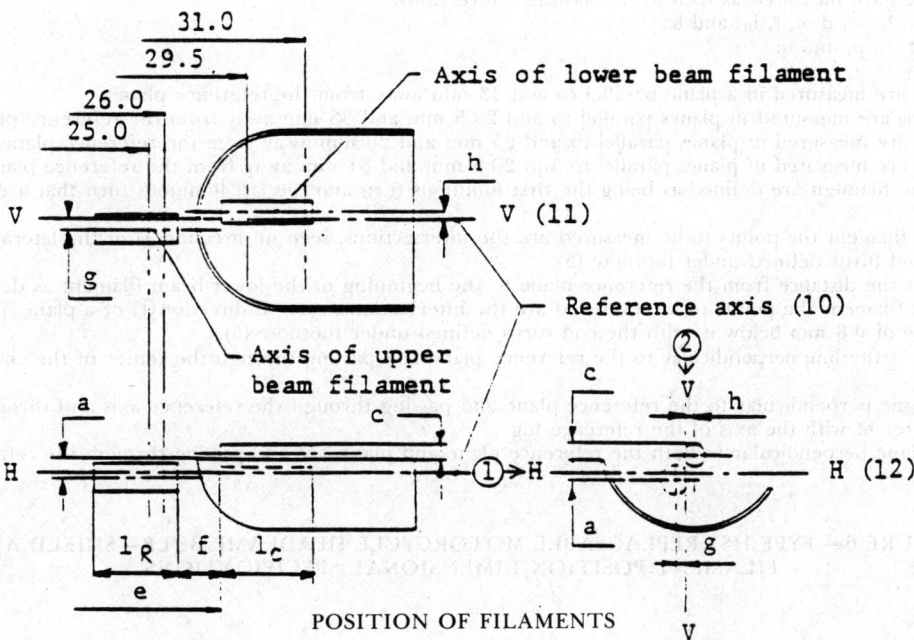

POSITION OF FILAMENTS

Reference			Dimension	(in mm) Tolerance
a/26	*	(4)	0.8	± 0.35
a/25	*	(4)	0.8	± 0.55
b_1/29.5	*	(3)	0	± 0.35
b_1/33	*	(3)	b_1/29.5 mv **	± 0.35
b_2/29.5	*	(3)	0	± 0.35
b_2/33	*	(3)	b_2/29.5 mv **	± 0.35
c/29.5	*	(5)	0.6	± 0.35
c/31	*	(5)	c/29.5 mv **	± 0.3
d			min 0.1/max 1.5	—
e		(8)	28.5	+ 0.45 / − 0.25
f		(6)(7)(9)	1.7	+ 0.5 / − 0.3
g/26	*	(4)	0	± 0.5
g/25	*	(4)	0	± 0.7
h/29.5	*	(5)	0	± 0.5
h/31	*	(5)	h/29.5 mv **	± 0.3
l_R		(6)(9)	4.0	± 0.8
lc		(6)(7)	4.5	± 0.8
p/33	*	(2)	Depends on the shape of the shield	—
q/33	*	(2)	$\frac{p+q}{2}$	± 0.6

* Dimension to be measured at the distance from the reference plane indicated in mm after the stroke.

** "./29.5 mv means that the value is to be measured at a distance of 29.5 mm from the reference plane.

(1) The dimensions noted are measured as seen in the indicated directions.
  ① for dimensions a, b_1, c, d, e, f, l_R, and lc;
  ② for dimensions g, h, p, and q;
  ③ for dimension b_2.
(2) Dimensions p and q are measured in a plane parallel to and 33 mm away from the reference plane.
(3) Dimensions b_1 and b_2 are measured in planes parallel to and 29.5 mm and 33 mm away from the reference plane.
(4) Dimensions a and g are measured in planes parallel to and 25 mm and 26 mm away from the reference plane.
(5) Dimensions c and h are measured in planes parallel to and 29.5 mm and 31 mm away from the reference plane.
(6) The end turns of the filament are defined as being the first luminous turn and the last luminous turn that are at substantially the correct helix angle.
(7) For the lower-beam filament the points to be measured are the intersections, seen in direction ①, of the lateral edge of the shield with the outside of the end turns defined under footnote (5).
(8) Dimension e denotes the distance from the reference plane to the beginning of the lower-beam filament as defined above.
(9) For the upper-beam filament the points to be measured are the intersections, seen in direction ① of a plane, parallel to plane HH and situated at a distance of 0.8 mm below it, with the end turns defined under footnote (5).
(10) The reference axis is the line perpendicular to the reference plane and passing through the center of the circle of diameter M (see Figure 5).
(11) Plane VV is the plane perpendicular to the reference plane and passing through the reference axis and through the intersection of the circle of diameter M with the axis of the reference lug.
(12) Plane HH is the plane perpendicular to both the reference plane and plane VV and passing through the reference axis.

FIGURE 6—TYPE HS1 REPLACEABLE MOTORCYCLE HEADLAMP BULB—SHIELD AND FILAMENT POSITION DIMENSIONAL SPECIFICATIONS

Dimensions in mm

The drawing is intended only to indicate the dimensions essential for interchangeability.

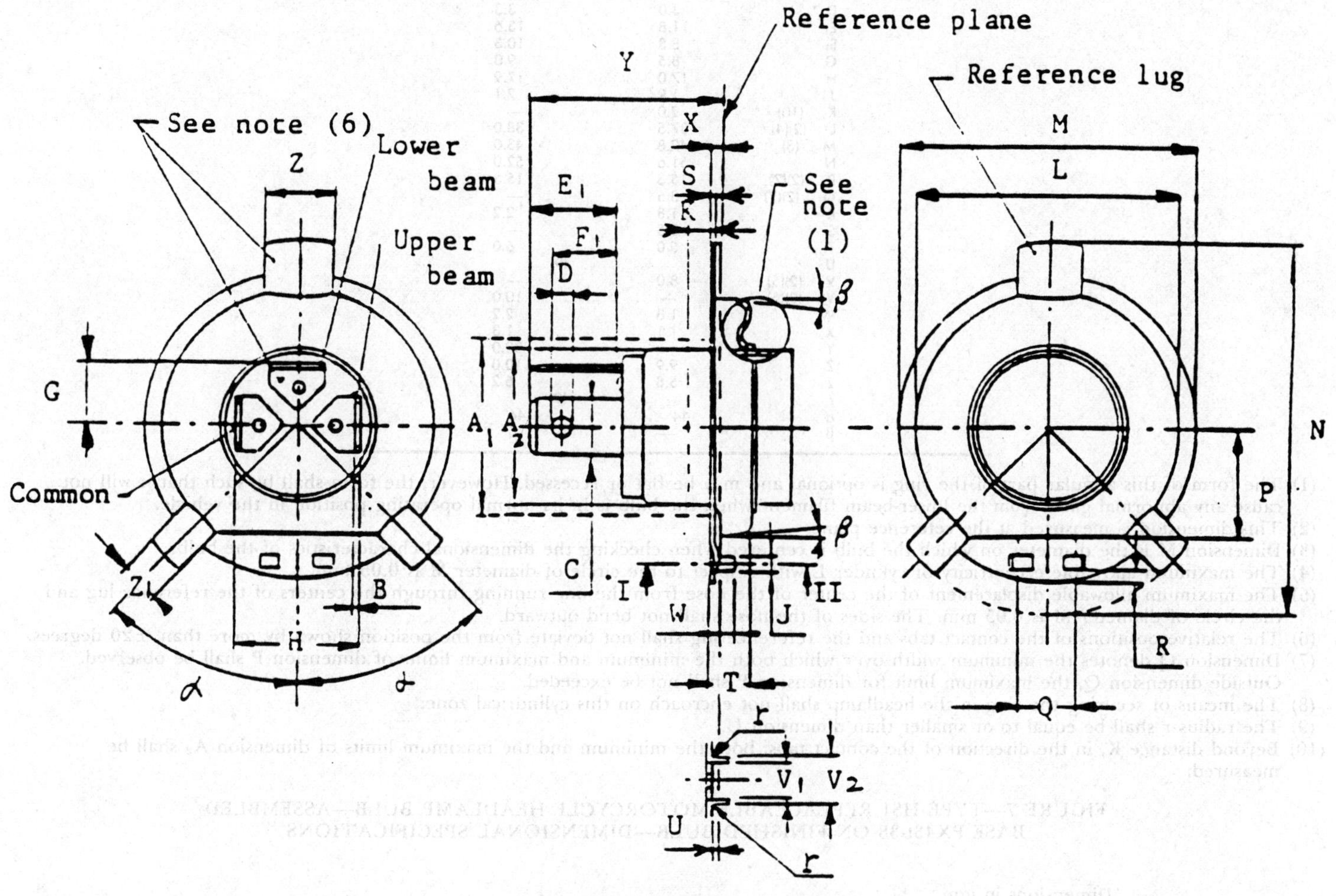

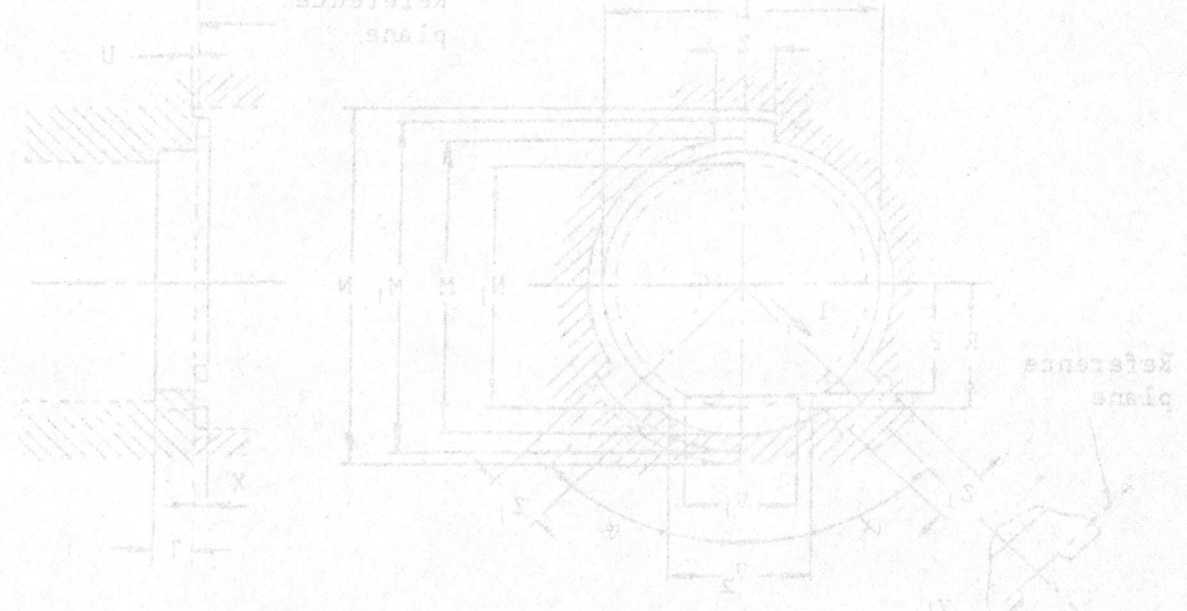

Section I - II

Dimension		Min.	Max.
A_1	(8)	25.0	—
A_2		21.94	22.0
B		0.7	0.8
C		7.7	8.1
D		3.0	3.3
E_1		11.8	13.6
F_1		8.8	10.3
G		8.5	9.0
H		17.0	17.9
J		1.9	2.1
K	(10)	2.0	—
L	(2)(4)	37.5	38.0
M	(3)	42.8	43.0
N		51.6	52.0
P	(2)(7)	15.3	15.5
Q	(2)(7)	8.5	—
R		1.8	2.2
S		0.5	—
T		5.0	6.0
U			(9)
V_1	(2)(5)	8.0	—
V_2	(2)(5)	—	10.0
W		1.8	2.2
X		1.1	1.3
Y		—	32.0
Z		9.9	10.0
Z_1		5.8	6.2
r			(9)
α		44°	46°
β		—	5°

(1) The form of this annular part of the ring is optional and may be flat or recessed. However, the form shall be such that it will not cause any abnormal glare from the lower-beam filament when the bulb is in its normal operating position in the vehicle.
(2) This dimension is measured at the reference plane.
(3) Dimension M is the diameter on which the bulb is centered when checking the dimensional characteristics of the bulb.
(4) The maximum allowable eccentricity of cylinder L with respect to the circle of diameter M is 0.05 mm.
(5) The maximum allowable displacement of the center of the nose from the line running through the centers of the reference lug and the circle of diameter M is 0.05 mm. The sides of the nose shall not bend outward.
(6) The relative positions of the contact tabs and the reference lug shall not deviate from the position shown by more than ±20 degrees.
(7) Dimension Q denotes the minimum width over which both the minimum and maximum limits of dimension P shall be observed. Outside dimension Q, the maximum limit for dimension P shall not be exceeded.
(8) The means of securing the ring in the headlamp shall not encroach on this cylindrical zone.
(9) The radius r shall be equal to or smaller than dimension U.
(10) Beyond distance K, in the direction of the contact tabs, both the minimum and the maximum limits of dimension A_2 shall be measured.

FIGURE 7—TYPE HS1 REPLACEABLE MOTORCYCLE HEADLAMP BULB—ASSEMBLED BASE PX43t-38 ON FINISHED BULB—DIMENSIONAL SPECIFICATIONS

Dimensions in mm

The drawings are intended only to indicate the dimensions essential for interchangeability.

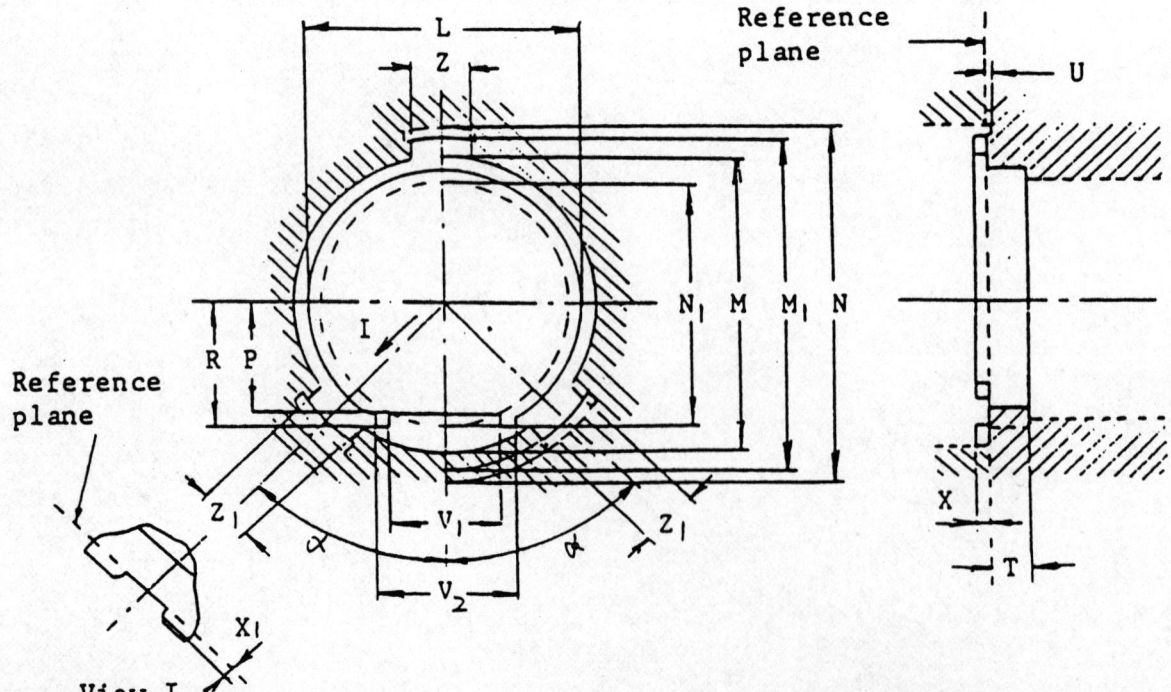

Dimension	Min.	Max.
M	43.02 (1)	43.2
M_1	—	49.0
N	52.2	—
N_1	35.0	—
L	38.2	38.5
P	15.7	16.7
R	20.0	—
T	5.5	—
U	0.4	—
V_1	7.5	7.8
V_2	10.2	—
X	1.8	—
X_1 (2)	1.4	—
Z	10.05	10.15
Z_1	8.0	8.5
α	44°	46°

The holder shall be so designed that the means of retention of the bulb can be applied only when the bulb is in the correct position. The means of retention shall make contact only with the prefocus ring of the base and the total force exerted, when the bulb is in position, shall be not less than 10 N and be not greater than 60 N.

(1) This value shall be complied with between the rim of the bulbholder and the reference plane (dimension X). However, it may be reduced to 38.5 mm within the dimensions Z and Z_1 which correspond with the support points for the lugs of the cap.

(2) Dimension X_1 denotes the minimum distance over which dimensions Z and Z_1 shall apply. Outside dimension X_1 the slots may be chamfered or rounded.

FIGURE 8—TYPE HS1 REPLACEABLE MOTORCYCLE HEADLAMP BULB—
BULBHOLDER PX43t—DIMENSIONAL SPECIFICATIONS

MOTORCYCLE AND MOTOR DRIVEN CYCLE ELECTRICAL SYSTEM MAINTENANCE OF DESIGN VOLTAGE
—SAE J392 FEB92   SAE Recommended Practice

Report of the Motorcycle Committee and Lighting Committee, approved December 1969, completely revised by the Motorcycle Committee May 1984. Rationale statement available. Reaffirmed by the SAE Motorcycle Electrical Systems Subcommittee of the SAE Motorcycle Committee February 1992.

Foreword—This reaffirmed document has been changed only to reflect the new SAE Technical Standards Board format.

1. Scope—This SAE Recommended Practice pertains to both battery-equipped and batteryless motorcycle electrical systems.

1.1 Purpose—This document provides minimum illumination voltage values for motorcycle and motor driven cycle electrical systems and accompanying test procedures.

NOTE: Wherever the word "motorcycle" appears in the report, it is understood to include "motor driven cycle."

2. References—There are no referenced publications specified herein.

3. Test Apparatus

3.1 Voltmeter—0 to 20 V maximum full-scale deflection, accuracy ±1/2% (two voltmeters required).

3.2 Ammeter—Capable of carrying full system load current. Accuracy ±3% FS.

3.3 Means for Measuring Engine rpm—Accuracy ±3%.

4. Test Procedure

4.1 Install fully charged original equipment or equal battery on the motorcycle (if motorcycle is battery equipped).

4.1.1 Battery temperature to be 26.7 °C ± 5.6 °C (80 °F ± 10 °F).

4.2 Connect one voltmeter between the headlamp low beam terminal and the ground; connect the other voltmeter between the taillamp terminal and the ground.

4.3 Connect the ammeter in series with the battery.

NOTE: Disregard 4.3 for batteryless machines.

4.4 Start engine and turn on headlamp(s).

4.4.1 Switch headlamp to the low beam position.

4.4.2 External fan cooling may be applied to the motorcycle engine.

4.5 Run the engine at an rpm equivalent to 48.3 km/h (30 mph) in top gear for 10 min.

4.5.1 Record the lowest and highest headlamp voltage and taillamp voltage observed during the 10 min period.

4.6 Run the engine at an rpm equivalent to 88.5 km/h (55 mph) in top gear for 10 min.

4.6.1 Record the lowest and highest headlamp voltage and taillamp voltage observed during the 10 min period.

4.7 Increase speed to manufacturer's suggested maximum rpm.

4.7.1 Record the highest and lowest headlamp and taillamp voltages observed during a 5 s period.

4.8 Run the engine at manufacturer's rated idle speed for 10 min.

4.8.1 Record the lowest and highest taillamp voltage observed during the 10 min period.

4.8.2 Record the lowest and highest headlamp voltage observed during the 10 min period.

4.9 Slowly increase the engine speed until generating equipment cancels the system load, indicated by "0" reading on the ammeter.

NOTE: Disregard 4.9 for batteryless motorcycles.

4.9.1 Record the engine rpm at ammeter zero point.

5. Test Limits

5.1 Voltages recorded in 4.5.1, 4.6.1, 4.7.1, and 4.8.1 shall be between 90 and 120% of the rated headlamp design voltage.

5.2 Voltages observed in 4.8.2 shall be between 60 and 120% of rated headlamp design voltage.

5.3 Engine rpm observed in 4.9.1 shall be less than the motorcycle equivalent speed at 48.3 km/h (30 mph) in top gear operation.

MOTORCYCLE TURN SIGNAL LAMPS—SAE J131 MAR83

SAE Standard

Report of the Lighting Committee and Motorcycle Committee, approved October 1969, completely revised by the Motorcycle Committee March 1983. Rationale statement available.

1. Scope—This Engineering Design Specification provides design parameters and general requirements for motorcycle turn signal lamps. It does not apply to mopeds.

2. Definition—Motorcycle turn signal lamps are the signalling elements of a turn signal system which indicate a change in direction by giving flashing lights on the side toward which the turn will be made.

3. General Requirements—The effective projected illuminated area measured on a plane at right angles to the axis of a lamp shall not be less than 22 cm² (3½ in²).

   3.1 The following sections from SAE J575 are a part of this standard:
   3.1.1 Section 2—Samples for Test
   3.1.2 Section 2.2—Bulbs
   3.1.3 Section 3—Laboratory Facilities
   3.1.4 Section 4.1—Vibration Test
   3.1.5 Section 4.2—Moisture Test
   3.1.6 Section 4.3—Dust Test
   3.1.7 Section 4.4—Corrosion Test
   3.1.8 Section 4.6—Photometry
     3.1.8.1 See Table 1 for Class A and D motorcycles and Table 2 for Class B and C motorcycles.
     3.1.8.2 All intensity measurements shall be made with the filament of the signal lamp at least 3 m (10 ft) from the photometer screen. The H-V axis shall be taken as parallel to the longitudinal axis of the vehicle.
   3.1.9 Section 4.8—Warpage Test on Devices with Plastic Components

4. Color Test—The color of the light emitted from turn signal lamps shall be red or yellow to the rear and yellow to the front of the vehicle. See SAE J578.

5. Plastic Materials—Any plastic materials used in optical parts shall comply with requirements set forth in SAE J576.

6. Installation Requirements—The following requirements apply to the device as used on the vehicle and are not a part of the laboratory requirements and test procedures.

   6.1 The filament center of each signal lamp on the front shall be symmetrically spaced a minimum of 200 mm (8 in) from the centerline of the vehicle and 100 mm (4 in) from the inside diameter of the retaining ring of the headlamp unit providing the lower beam. On the rear, the symmetrical spacing shall be a minimum of 110 mm (4½ in) from the centerline of the vehicle to the filament axis of the signal lamp.

   6.1.1 Visibility of the front turn signal to the front and the rear turn signal to the rear shall not be obstructed by any part of the vehicle throughout the test angles for the lamps, if such obstruction causes the lamp to fail to meet minimum photometric visibility requirements.

   6.1.2 The signals from each lamp shall be visible through a horizontal angle of 45 deg outboard. To be considered visible, the lamp must project a minimum unobstructed illuminated area of 12.5 cm² (2 in²), measured at all angles throughout the 45 deg requirement.

   6.2 **Turn Signal Pilot Indicator**—There shall be either a light or sound signal to give the operator a clear and unmistakable indication that the turn signal system is functioning correctly. The illuminated indicator shall consist of one or more lights flashing at the same frequency as the signal lamps, and shall be plainly visible to operators of all heights when seated in normal position in the operator's seat, while driving in bright sunlight.

If a sound signal is used, it shall be audible to the operator when the motorcycle is operating at 55 km/h (35 mph) or less. The signal shall cycle on and off at the same frequency as the signal lamps. Failure of one or more turn signal lamps to operate shall be indicated by a continuous tone or by failure of the signal to sound.

APPENDIX

A1. As a matter of information, attention is called to SAE J567 for requirements and gages to be used in socket design.

A2. For flashing rate and "on" time, see SAE J590.

A3. All motorcycle class designations are those given in SAE J213.

TABLE 1—MINIMUM INTENSITY REQUIREMENTS (cd)—CLASS A AND D MOTORCYCLES

Test Points (deg)		Front Signal	Rear Signal	
		Yellow	Red	Yellow
10 U and D	10 L and R	25	6	15
	V	60	15	40
5 U and D	20 L and R	25	6	15
	10 L and R	75	20	50
	5 L and R	125	30	80
	V	175	40	110
H	20 L and R	35	10	25
	10 L and R	100	25	65
	5 L and R	200	50	120
	V	200	50	130
Maximum Rear Lamp Only		—	300	750

TABLE 2—MINIMUM INTENSITY REQUIREMENTS (cd)—CLASS B AND C MOTORCYCLES

Test Points (deg)		Red	Yellow
10 U and D	10 L and R	5	15
	V	13	30
5 U and D	20 L and R	5	15
	10 L and R	15	40
	5 L and R	25	60
	V	35	90
H	20 L and R	8	20
	10 L and R	20	45
	5 L and R	40	120
	V	40	120
Maximum Rear Lamp Only		300	750

NOTES:
1. Specifications in Tables 1 and 2 are based on laboratories using accurate, rated bulbs during testing.
2. Lamps designed for use in both 6V and 12V systems shall be tested with 12V bulbs. Lamps designed to operate on the vehicle through a resistor or equivalent shall be photometered with the listed design voltage of the design source applied across the combination of resistance and filament.

(R) TAIL LAMPS (REAR POSITION LAMPS) FOR USE ON MOTOR VEHICLES LESS THAN 2032 mm IN OVERALL WIDTH—SAE J585 DEC91

SAE Standard

Report of the Lighting Division approved March 1918, revised by the Lighting Committee March 1986. Rationale statement available. Completely revised by the SAE Lighting Coordinating Committee and the SAE Signalling & Marking Devices Standards Committee December 1991.

1. Scope—This SAE Standard provides test procedures, requirements, and guidelines for tail lamps (rear position lamps).

2. References

2.1 Applicable Documents—The following publications form a part of this specification to the extent specified herein. The latest issue of SAE publications shall apply.

2.1.1 SAE PUBLICATIONS—Available from SAE, 400 Commonwealth Drive, Warrendale, PA 15096-0001.

SAE J567—Lamp Bulb Retention System
SAE J575—Tests for Motor Vehicle Lighting Devices and Components
SAE J576—Plastic Materials for Use in Optical Parts Such as Lenses and Reflectors of Motor Vehicle Lighting Devices
SAE J578—Color Specification
SAE J759—Lighting Identification Code
SAE J2040—Tail Lamps (Rear Position Lamps) for Use on Vehicles 2032 mm or More in Overall Width

2.2 Definitions

2.2.1 TAIL LAMPS—Lamps used to designate the rear of a vehicle by a steady burning low intensity light.

2.2.2 MULTIPLE COMPARTMENT LAMP—A device which gives its indication by two or more separately lighted areas which are joined by one or more common parts such as a housing or lens.

2.2.3 MULTIPLE LAMP ARRANGEMENT—An array of two or more separated lamps on each side of the vehicle which operate together or give a signal.

3. Lighting Identification Code—Tail lamps may be identified by the code "T" in accordance with SAE J759.

4. Tests

4.1 SAE J575 is a part of this report. The following tests are applicable with the modifications as indicated:

4.1.1 VIBRATION TEST
4.1.2 MOISTURE TEST
4.1.3 DUST TEST
4.1.4 CORROSION TEST
4.1.5 PHOTOMETRY TEST—In addition to the test procedures in SAE J575, the following apply:

4.1.5.1 Photometric measurements shall be made with the light source of the lamp at least 3 m from the photometer. The H-V axis shall be taken as parallel to the axis of reference of the lamp as mounted on the vehicle.

4.1.5.2 Photometric measurements shall be made with the bulb filament steadily burning. Photometric measurements of multiple compartment lamps or multiple lamp arrangements shall be made by either of the following methods:

4.1.5.2.1 All compartments or lamps shall be photometered together provided that a line from the light source of each compartment or lamp to the center of the photometer sensing device does not make an angle of more than 0.6 degree with the photometer H-V axis. When compartments or lamps are photometered together, the H-V axis shall intersect the midpoint between their light sources.

4.1.5.2.2 Each compartment or lamp shall be photometered separately by aligning the axis of each lamp or compartment with the photometer. The photometric measurement for the entire multiple compartment lamp or multiple lamp arrangement shall be determined by adding the photometric outputs from each individual lamp or component at corresponding test points.

4.1.6 WARPAGE TEST FOR DEVICES WITH PLASTIC COMPONENTS

4.2 Color Test—SAE J578 is a part of this report.

5. Requirements

5.1 Performance Requirements—A device when tested in accordance with the test procedures specified in Section 4 shall meet the following requirements:

5.1.1 VIBRATION—SAE J575.
5.1.2 MOISTURE—SAE J575.
5.1.3 DUST—SAE J575.

TABLE 1—PHOTOMETRIC REQUIREMENTS[1,2,3]

Zone	Test Points (Degrees)	Minimum Luminous Intensity (cd)	Minimum Luminous Intensity (cd)	Minimum Luminous Intensity (cd)
	Lighted Sections	1	2	3
1	10U- 5L 5U-20L 5D-20L 10D- 5L	1.4	2.4	3.5
2	5U-10L H-10L 5D-10L	2.4	4.2	6.0
3	5U- V H- 5L H- V H- 5R 5D- V	9.6	16.8	24.0
4	5U-10R H-10R 5D-10R	2.4	4.2	6.0
5	10U- 5R 5U-20R 5D-20R 10D- 5R	1.4	2.4	3.5
MAXIMUM LUMINOUS INTENSITY (cd) H AND ABOVE (See Note 2)	MAXIMUM LUMINOUS INTENSITY (cd) H AND ABOVE (See Note 2)	18	20	25

[1] The measured values at each test point shall not be less than 60% of the minimum values in Table 2.
[2] The listed maximum shall not be exceeded over any area larger than that generated by an 0.5 degree radius within the solid angle defined by the test points in Table 1.
[3] Ratio requirements of 5.1.5.3 apply.

5.1.4 Corrosion—SAE J575.
5.1.5 Photometry—SAE J575.
5.1.5.1 The lamp under test shall meet the photometric performance requirements contained in Table 1. The summation of the luminous intensity measurements at the specified test points in a zone shall be at least the value shown.
5.1.5.2 A multiple compartment lamp or multiple lamps may be used to meet the photometric requirements. If a multiple compartment lamp or multiple lamps are used and the distance between adjacent light sources does not exceed 560 mm for two compartments or lamp arrangements and does not exceed 410 mm for three compartments or lamp arrangements, then the combination of the compartments or lamps must be used to meet the photometric requirements for the corresponding number of lighted sections in Table 1. If the distance between adjacent light sources exceeds the previous dimensions, each compartment or lamp shall comply with the photometric performance requirements for one lighted section in Table 1.
5.1.5.3 When a tail lamp is combined with the turn signal or stop lamp, the signal lamp intensity shall not be less than three times the luminous intensity of the tail lamp at any test point, except that at H-V, H-5L, H-5R, and 5U-V, the turn signal or stop lamp intensity shall not be less than five times the luminous intensity of the tail lamp. If a multiple compartment or multiple lamp arrangement is used and the distance between optical axes for both the tail lamp and the turn signal or stop lamp is within the dimensions specified in 5.1.5.2, the ratio of the turn signal or stop lamp to the tail lamp shall be computed with all the compartments or lamps lighted. If a multiple compartment or multiple lamp arrangement is used and the distance between optical axes for one of the functions exceeds the dimensions specified in 5.1.5.2, the ratio shall be computed for only those compartments or lamps where the tail lamp and turn signal or stop lamp are optically combined. When the tail lamp is combined with the turn signal or stop lamp and the maximum luminous intensity of the tail lamp is located below horizontal and within an area generated by a 1.0 degree radius around a test point, the ratio for the test point may be computed using the lowest value of the tail lamp luminous intensity within the generated area.
5.1.6 Warpage—SAE J575.
5.1.7 Color—The color of the light from a tail lamp shall be red as specified in SAE J578.
5.2 **Material Requirements**—Plastic materials used in optical parts shall meet the requirements of SAE J576.
5.3 **Design Requirements**
5.3.1 If a turn signal or stop lamp is optically combined with the tail lamp and a two-filament replaceable bulb is used, the bulb shall have an indexing base and the socket shall be designed so that bulbs with non-indexing bases cannot be used. Removable sockets shall have an indexing feature so that they cannot be reinserted into lamp housings in random positions, unless the lamp will perform its intended function with random light source orientation.
5.4 **Installation Requirements**—The tail lamp shall meet the following requirements as installed on the vehicle:
5.4.1 Visibility of the tail lamp shall not be obstructed by any part of the vehicle throughout the photometric test angles for the lamp unless the lamp is designed to comply with all photometric and visibility requirements with these obstructions considered. Signals from lamps on both sides of the vehicle shall be visible through a horizontal angle from 45 degrees to the left to 45 degrees to the right. Where more than one lamp or optical area is lighted on each side of the vehicle, only one such area on each side need comply. To be considered visible, the lamp must provide an unobstructed view of the outer lens surface, excluding reflex, of at least 12.5 cm^2 measured at 45 degrees to the longitudinal axis of the vehicle.

6. **Guidelines**
6.1 **Photometric Design Guidelines** for tail lamps, when tested in accordance with 4.1.5 of this report, are contained in Table 2.
6.2 **Installation Guidelines**—The following guidelines apply to tail lamps as used on the vehicle and shall not be considered part of the requirements:
6.2.1 Tail lamps on the rear of the vehicle should be spaced as far apart laterally as practicable so that the signal will be clearly visible and its intent clearly understood.
6.2.2 The luminous intensity of incandescent filament bulbs will vary with applied voltage. The electrical wiring in the vehicle should be adequate to supply design voltage to the lamp filament.
6.2.3 Performance of lamps may deteriorate significantly as a result of dirt, grime, and/or snow accumulation on their optical surfaces. Installation of lamps on vehicles should be considered to minimize the effect of these factors.
6.2.4 Where it is expected that lamps must perform in extremely severe environments, such as in off-highway, mining, fuel haulage or where it is expected that they will be totally immersed in water, the user should specify lamps specifically designed for such use.

Appendix A

A.1 As a matter of additional information, attention is called to SAE J567, Lamp Bulb Retention System, for requirements and gages used in socket design.
A.2 For vehicles over 2032 mm wide see SAE J2040.

TABLE 2—PHOTOMETRIC DESIGN GUIDELINES[1,2]

Test Points (Degrees)	Test Points (Degrees)	Minimum Luminous Intensity (cd) Lighted Sections 1	Minimum Luminous Intensity (cd) Lighted Sections 2	Minimum Luminous Intensity (cd) Lighted Sections 3
10U, 10D	5L, 5R	0.4	0.7	1.0
	20L, 20R	0.3	0.5	0.7
5U, 5D	10L, 10R	0.8	1.4	2.0
	V	1.8	3.1	4.5
	10L, 10R	0.8	1.4	2.0
H	5L, 5R	2.0	3.5	5.0
	V	2.0	3.5	5.0
MAXIMUM LUMINOUS INTENSITY (cd) H AND ABOVE (See Note 1)	MAXIMUM LUMINOUS INTENSITY (cd) H AND ABOVE (See Note 1)	18	20	25

[1] The maximum design value of a lamp intended for the rear of the vehicle should not exceed the listed design maximum over any area larger than that generated by an 0.25 degree radius within the solid angle defined by the test points in this table.
[2] Ratio requirements of 5.1.5.3 apply.

TAIL LAMPS (REAR POSITION LAMPS) FOR USE ON VEHICLES 2032 mm OR MORE IN OVERALL WIDTH—SAE J2040 JUN91

SAE Standard

Report of the SAE Lighting Coordinating Committee and the Heavy-Duty Lighting Standards Committee approved June 1991. Rationale statement available.

1. Scope—This SAE Standard provides test procedures, requirements, and guidelines for tail lamps intended for use on vehicles 2032 mm or more in overall width. Tail lamps conforming to the requirements of this document may also be used on vehicles less than 2032 mm in overall width.

2. References

2.1 Applicable Documents—The following publications form a part of this specification to the extent specified herein. The latest issue of SAE publications shall apply.

2.1.1 SAE Publications—Available from SAE, 400 Commonwealth Drive, Warrendale, PA 15096-0001.

SAE J567—Lamp Bulb Retention System for Requirements and Gages Used in Retention System Design

SAE J576—Plastic Material for Use in Optical Parts Such as Lenses and Reflectors of Motor Vehicle Lighting Devices

SAE J578—Color Specification

SAE 585—Tail Lamps (Rear Position Lamps)

SAE J759—Lighting Identification Code

SAE J2139—Tests for Lighting Devices, Reflective Devices and Components Used on Vehicles 2032 mm or More in Overall Width

SAE Technical Paper 830566, "Motor Vehicle Conspicuity," R.L. Henderson, K. Ziedman, W.J. Burger, and K.E. Cavey, National Highway Traffic Safety Administration

2.1.2 Other Publications—Attention is called to the following documents for additional information on lamp design and installation requirements.

FMVSS 108
FHWA 393 Subpart B
TTMA #RP-9
TMC #RP-702

2.2 Definitions

2.2.1 A tail lamp is a lamp used to designate the rear of a vehicle by a steady burning low intensity light.

3. Lighting Identification Code—Tail lamps for use on vehicles 2032 mm or more in overall width may be identified by the code "T2" in accordance with SAE J759.

4. Tests

4.1 SAE J2139 is a part of this document. The following tests are applicable with modification as indicated.

4.1.1 Vibration
4.1.2 Moisture
4.1.3 Dust
4.1.4 Corrosion
4.1.5 Photometry

4.1.5.1 Photometric measurements shall be made with the light source of the device at least 3 m from the photometer.

4.1.5.2 The H-V axis of the device shall be taken to be parallel to the longitudinal axis of the vehicle, when the device is mounted in its design position.

4.1.5.3 Photometric measurements shall be made with the light source steadily burning. Photometric measurements of multiple compartment lamps or multiple lamp arrangements shall be made by either of the following methods.

4.1.5.3.1 All compartments or lamps shall be photometered together provided that a line from the light source of each compartment or lamp to the center of the photometer sensing device does not make an angle of more than 0.6 degrees with the photometer H-V axis. When compartments or lamps are photometered together, the H-V axis shall intersect the midpoint between their light sources.

4.1.5.3.2 Each compartment or lamp shall be photometered separately by aligning the axis of each lamp or compartment with the axis of the photometer. The photometric measurement for the entire multiple compartment lamp or multiple lamp arrangement shall be determined by adding the photometric outputs from each individual lamp or component at corresponding test points.

4.1.6 Warpage Test on Devices with Plastic Components

4.2 Color—SAE J578 is a part of this document.

4.3 Plastic Materials—SAE J576 is a part of this document.

5. Requirements

5.1 Performance Requirements—The device when tested in accordance with the test procedures of this document shall meet the requirements of SAE J2139 or as indicated.

5.1.1 Vibration
5.1.2 Moisture
5.1.3 Dust
5.1.4 Corrosion
5.1.5 Photometry—The device tested shall meet the photometric performance requirements of Table 1 and its footnotes.

The summation of the luminous intensity measurements at the specified test points in a zone shall be at least the value shown.

5.1.5.1 A multiple compartment lamp or multiple lamps may be used to meet the photometric requirements. If a multiple compartment lamp or multiple lamps are used and the distance between adjacent light sources does not exceed 560 mm for two compartments or lamp arrangements and does not exceed 410 mm for three compartments or lamp arrangements, then the combination of the compartments or lamps must be used to meet the photometric requirements of Table 1. If the distance between adjacent light sources exceeds the above dimensions, each compartment or lamp shall comply with the photometric requirements of Table 1.

5.1.5.2 When a tail lamp is combined with the stop lamp or turn signal lamp, the stop lamp or turn signal lamp intensity shall be not less than three times the luminous intensity of the tail lamp at any test point, except that at H-V, H-5L, H-5R, and 5U-V, the stop lamp or turn signal lamp intensity shall be not less than five times the luminous intensity of the tail lamp.

When a tail lamp is combined with the stop lamp or turn signal lamp, and the maximum luminous intensity of the tail lamp is located below the horizontal and is within an area generated by a 1.0 degree radius around the test point, the ratio for the test point may be computed using the lowest value of the tail lamp intensity within the generated area.

5.1.6 Warpage

5.2 Color—The color of the light from the tail lamp shall be red as specified in SAE J578.

5.3 Plastic Materials—The plastic materials used in the optical parts shall meet the requirements of SAE J576.

5.4 Design Requirements

5.4.1 If a tail lamp is combined with a stop lamp or a turn signal lamp and a replaceable multiple light source is used, the light source retention system shall be designed with an indexing means so that the light source is properly indexed. Removable light source retention systems shall have an indexing feature so that they cannot be reinserted into the lamp housing in a random position, unless the lamp will perform its intended function with random light source orientation.

5.4.2 The effective projected luminous lighted lens area of a single lamp shall be at least 75 cm^2.

5.4.3 If a multiple compartment lamp or multiple lamps are used to meet the photometric requirements, the effective projected luminous lens area of each compartment or lamp shall be at least 40 cm^2 provided the combined area is at least 75 cm^2.

5.4.4 A tail lamp shall not be combined with a clearance lamp.

5.5 Installation Requirements—The tail lamp shall meet the following requirements as installed on the vehicle.

5.5.1 The tail lamps shall be mounted on the permanent structure of the vehicle, facing rearward, at the same height and spaced as far apart laterally as practicable, so that the signal will be clearly visible.

5.5.2 Visibility of each tail lamp shall not be obstructed by any part of the vehicle throughout the photometric test pattern unless the lamp is designed to comply with all photometric and visibility requirements with the obstructions considered.

To be considered visible, the lamp must provide an unobstructed view of a portion of the lighted outer lens surface, excluding reflex reflector area, of at least 13 cm^2 measured at a horizontal angle of 45 degrees to the left and 45 degrees to the right of the longitudinal axis of the vehicle and a vertical angle from 20 degrees up to 10 degrees down.

See Table 1, Note 6. Where more than one lamp or optical area is lighted on each side of the vehicle, only one such area on each side need comply.

6. Guidelines

6.1 Design Guidelines

6.1.1 Photometric design guidelines are contained in Table 2 and its footnotes.

6.2 Installation Guidelines—The following guidelines apply to tail lamps as used on the vehicle and shall not be considered part of the requirements.

6.2.1 Performance of lamps may deteriorate significantly as a result of dirt, grime, snow, and ice accumulation on the optical surfaces. Installation of the device on the vehicle should be considered to minimize the effects of these factors.

6.2.2 Where it is expected that the device must perform in extremely severe environments, or where it is expected to be totally immersed in water, the user should specify devices specifically designed for such use.

7. Advance Requirements—This section of the document gives advance notice to manufacturers and users of the device of a pending change in the requirements for a tail lamp. The change in the requirements shall be effective on devices marketed and used on or after January 1, 1996.

See Tables 3 and 4 and the footnotes.

TABLE 1—TAIL LAMP PHOTOMETRIC PERFORMANCE REQUIREMENTS

Zone	Test Point Deg.	Zone Total Luminous Intensity, Candela, Red
1	10U— 5L 5U—20L 5D—20L 10D— 5L	1.4
2	5U—10L H—10L 5D—10L	2.4
3	5U— V H— 5L H— V H— 5R 5D— V	9.6
4	5U—10R H—10R 5D—10R	2.4
5	10U— 5R 5U—20R 5D—20R 10D— 5R	1.4
Maximum Luminous Intensity, Candela		18.0 H and above

[1] The maximum luminous intensity shall not be exceeded over any area larger than that generated by a 0.5 degree radius within the area defined by the test point pattern of Table 2.

[2] Unless otherwise specified, the lamp shall be considered to have failed the photometric requirements of this document if the luminous intensity at any test point is less than 60% of the values specified in Table 2.

[3] Unless otherwise specified, the lamp shall be considered to have failed the photometric requirements of this document if the minimum luminous intensity between test points is less than 60% of the lower design values of Table 2 for the closest adjacent test points on a horizontal and vertical line as defined by the test point pattern.

[4] The summation of the luminous intensity measurements at the specified test points in the zone shall be at least the values shown.

[5] When a tail lamp or a clearance lamp is combined with a stop lamp or a turn signal lamp, see 5.1.5.2 of this document for luminous intensity ratio requirements.

[6] Throughout the photometric pattern defined by the corner points of 20U-45L, 20U-45R, 10D-45R, and 10D-45L, the light intensity shall be not less than 0.10 candela in red.

TABLE 2—TAIL LAMP PHOTOMETRIC DESIGN GUIDELINES

Test Point Deg.	Luminous Intensity, Candela, Red
10U— 5L	0.4
5R	0.4
5U—20L	0.3
10L	0.8
V	1.8
10R	0.8
20R	0.3
H—10L	0.8
5L	2.0
V	2.0
5R	2.0
10R	0.8
5D—20L	0.3
10L	0.8
V	1.8
10R	0.8
20R	0.3
10D— 5L	0.4
5R	0.4
Maximum Luminous Intensity, Candela	18.0 H and above

[1] The maximum luminous intensity shall not be exceeded over any area larger than that generated by a 0.25 degree radius within the area defined by the test point pattern of Table 2.

[2] When a tail lamp is combined with a stop lamp or a turn signal lamp, see 5.1.5.2 of this document for luminous intensity ratio requirements.

[3] Throughout the photometric pattern defined by the corner points of 20U-45L, 20U-45R, 10D-45R, and 10D-45L, the light intensity shall be not less than 0.10 candela in red.

TABLE 3—TAIL LAMP PHOTOMETRIC PERFORMANCE REQUIREMENTS—ADVANCE REQUIREMENTS

Zone	Test Point Deg.	Zone Total Luminous Intensity, Candela, Red
1	20U—45L 20U—20L 10D—20L 10D—45L	1.0
2	10U— 5L 5U—20L 5D—20L 10D— 5L	1.4
3	5U—10L H—10L 5D—10L	2.4
4	5U— V H— 5L H— V H— 5R 5D— V	9.6
5	5U—10R H—10R 5D—10R	2.4
6	10U— 5R 5U—20R 5D—20R 10D— 5R	1.4
7	20U—20R 20U—45R 10D—45R 10D—20R	1.0
Maximum Luminous Intensity, Candela		18.0 H and above

[1] The maximum luminous intensity shall not be exceeded over any area larger than that generated by a 0.5 degree radius within the area defined by the test point pattern of Table 4.

[2] Unless otherwise specified, the lamp shall be considered to have failed the photometric requirements of this document if the luminous intensity at any test point is less than 60% of the values specified in Table 4.

[3] Unless otherwise specified, the lamp shall be considered to have failed the photometric requirements of this document if the minimum luminous intensity between test points is less than 60% of the lower design values of Table 4 for the closest adjacent test points on a horizontal and vertical line as defined by the test point pattern.

[4] The summation of the luminous intensity measurements at the specified test points in the zone shall be at least the values shown.

[5] When a tail lamp is combined with a stop lamp or a turn signal lamp, see 5.1.5.2 of this document for luminous intensity ratio requirements.

TABLE 4—TAIL LAMP PHOTOMETRIC DESIGN GUIDELINES—ADVANCE REQUIREMENTS

Test Point Deg.	Luminous Intensity, Candela, Red
20U—45L	0.25
20L	0.25
20R	0.25
45R	0.25
10U— 5L	0.4
5R	0.4
5U—20L	0.3
10L	0.8
V	1.8
10R	0.8
20R	0.3
H—10L	0.8
5L	2.0
V	2.0
5R	2.0
10R	0.8
5D—20L	0.3
10L	0.8
V	1.8
10R	0.8
20R	0.3
10D—45L	0.25
20L	0.25
5L	0.4
5R	0.4
20R	0.25
45R	0.25
Maximum Luminous Intensity, Candela—	18.0 H and above

[1] The maximum luminous intensity shall not be exceeded over any area larger than that generated by a 0.25 degree radius within the area defined by the test point pattern of Table 4.

[2] When a tail lamp is combined with a stop lamp or a turn signal lamp, see 5.1.5.2 of this document for luminous intensity ratio requirements.

FOG TAIL LAMP (Rear Fog Light) SYSTEMS—SAE J1319 JUN93

SAE Recommended Practice

Report of the Lighting Committee approved August 1987. Rationale statement available. Revised by the Signalling and Marking Devices Standards Committee June 1993. Rationale statement available.

1. **Scope**—This SAE Recommended Practice provides test procedures, requirements, and guidelines for fog tail lamp systems. See Appendices A and B.

2. **References**

   2.1 *Applicable Documents*—The following publications form a part of this specification to the extent specified herein. The latest issue of SAE publications shall apply.

   2.1.1 SAE PUBLICATIONS—Available from SAE, 400 Commonwealth Drive, Warrendale, PA 15096-0001.

   SAE J567—Lamp Bulb Retention System
   SAE J575—Test Methods and Equipment for Lighting Devices and Components for Use on Vehicles Less Than 2032 mm in Overall Width
   SAE J576—Plastic Materials for Use in Optical Parts Such as Lenses and Reflectors of Motor Vehicle Lighting Devices
   SAE J578—Color Specification
   SAE J585—Tail Lamps (Rear Position Lamps) for Use on Motor Vehicles Less Than 2032 mm in Overall Width
   SAE J759—Lighting Identification Code

   2.2 *Definitions*

   2.2.1 FOG TAIL LAMP—A lighting device providing a continuous red light of higher intensity than a tail lamp (SAE J585) for the purpose of marking the rear of a vehicle during fog or similar conditions of reduced visibility.

   2.2.2 FOG TAIL LAMP SYSTEM—One or two fog tail lamps with their respective wiring, connectors, switch, and a function indicator.

3. **Lighting Identification Code**—Fog tail lamps may be identified by the code F2 in accordance with SAE J759.

4. **Tests**

   4.1 SAE J575 is a part of this report. The following tests are applicable:
   4.1.1 VIBRATION TEST
   4.1.2 MOISTURE TEST
   4.1.3 DUST TEST
   4.1.4 CORROSION TEST
   4.1.5 PHOTOMETRY TEST
   4.1.5.1 Photometric measurements shall be made with light source of the lamp at least 3 m from the photometer. The H-V axis shall be taken as parallel to the axis of reference of the lamp as mounted on the vehicle.
   4.1.6 WARPAGE TEST FOR DEVICES WITH PLASTIC COMPONENTS
   4.2 *Color Test*—SAE J578 is a part of this report.

5. **Requirements**

5.1 Performance Requirements—A device, when tested in accordance with the test procedures specified in Section 4, shall meet the following requirements with the modifications indicated:
5.1.1 VIBRATION—SAE J575
5.1.2 MOISTURE—SAE J575
5.1.3 DUST—SAE J575
5.1.4 CORROSION—SAE J575
5.1.5 PHOTOMETRY—SAE J575

(R) 5.1.5.1 The lamp shall meet the photometric performance requirements contained in Table 1 and its footnotes. The summation of the luminous intensities at the test points specified for each zone in column 2 of Table 1 shall be at least the value shown for that zone in column 3.

TABLE 1—PHOTOMETRIC REQUIREMENTS

Zone	Test Points[1] (deg)	Minimum Luminous Intensity (candela)
1	10U-5L 5U-20L 5D-20L 10D-5L	50
2	5U-10L H-10L 5D-10L	100
3	5U-V H-5L H-V H-5R 5D-V	380
4	5U-10R H-10R 5D-10R	100
5	10U-5R 5U-20R 5D-20R 10D-5R	50
Maximum Luminous Intensity (candela[2])		300

[1] The measured values of each test point shall not be less than 60% of the minimum value in Table 2.
(R) [2] The listed maximum at any test point shall not be exceeded over any area larger than that generated by a 0.5 degree radius with the solid angle defined by the test points in Table 1.

5.1.6 WARPAGE—SAE J575
5.1.7 COLOR—The color of light from a fog tail lamp shall be red as specified in SAE J578.

5.2 Materials Requirements—Plastic materials used in the optical parts shall meet the requirements of SAE J576.

5.3 Design Requirements
5.3.1 A fog tail lamp shall not be optically combined with any lamp other than a tail lamp. If a fog tail lamp is optically combined with the tail lamp and a two-filament bulb is used, the bulb shall have an indexing base and the socket shall be designed so that bulbs with nonindexing bases cannot be used.

(R) **6. Guidelines**—The following guidelines are intended to provide optimal performance of the system and uniformity in use but shall not be considered part of the requirements.

6.1 Photometric design guidelines for a fog tail lamp, when tested in accordance with 4.1.5 of this document, are contained in Table 2.

6.2 Installation Guidelines—The user is cautioned that the mounting and use of fog tail lamps are specified by various regulatory agencies.

6.2.1 The illuminated edge of a fog tail lamp lens should be no closer than 100 mm from the illuminated edge of any stop lamp lens when projected on a vertical transverse plane.

6.2.2 The fog tail lamp system should consist of either: (a) one lamp mounted on or to the left of a vertical plane through the longitudinal centerline of the vehicle, or (b) two lamps symmetrically located about the vehicle centerline.

6.2.3 The fog tail lamp system should be wired so that it can be turned on only when the headlamps and/or front fog lamps are on, and should have a switch that allows the fog tail lamp to be turned off when headlamps are on.

6.2.4 Visibility of the fog tail lamp should not be obstructed by any part of the vehicle throughout the photometric test angles for the lamp unless the lamp is designed to comply with all photometric and visibility requirements with these obstructions considered. The signal from the lamp should be visible through a horizontal angle from 45 degrees to the left to 45 degrees to the right.

(R) 6.2.5 The fog tail lamp system should include a continuous yellow indicator that illuminates when the system is switched on that should be mounted in a location readily visible to the driver of the vehicle.

TABLE 2—PHOTOMETRIC DESIGN GUIDELINES

Test Points (deg)		Minimum Luminous Intensity (candela)
10U, 10D	10L, 10R	10
	5L, 5R	16
	V	25
5U, 5D	20L, 20R	10
	10L, 10R	30
	5L, 5R	50
	V	70
H	20L, 20R	15
	10L, 10R	40
	5L, 5R	80
	V	80
Maximum Luminous Intensity (candela)		300

APPENDIX A

As a matter of additional information, attention is called to SAE J567c for requirements and gages used in socket design.

(R) APPENDIX B

A fog tail lamp is a lighting device, required within the European Economic Community (EEC), which provides a steady burning marker on the rear of vehicles during conditions of reduced visibility. As a consequence of this regulation, many vehicle manufacturers already have vehicle designs which can accommodate fog tail lamps, but do not provide a completely operable system because of the absence of an SAE Standard; this Recommended Practice could be adopted by U.S. governmental agencies thereby permitting such a device through uniform, harmonized regulations.

The SAE Recommended Practice was developed to harmonize the existing EEC requirements and State of California Administrative Code, Title 13; it is believed that devices designed to comply with this SAE Recommended Practice will satisfy both the current California and the EEC requirements.

The EEC regulation was used as the basis for this document but differs from those European requirements in several areas:

a. Photometric test point location requirements and environmental test procedures were adopted from the SAE practice for stop lamps. Photometric design guidelines were adopted from the State of California requirements.
b. Minimum design candela requirements correspond to those for an SAE stop lamp and the California specifications. (Maximum permissible candela values, however, are the same as the EEC and California standards.)
c. The EEC mounting height requirements were not incorporated into this standard as this traditionally has been left to the discretion of government regulatory agencies. No other SAE Standard contains mounting height requirements.
d. Lateral visibility requirements were taken from the SAE stop lamp requirements which are more stringent than the EEC requirements (45 degrees versus 25 degrees).
e. The service performance requirements are identical to those for SAE stop lamps.
f. This SAE Recommended Practice specifies the installation of one fog tail lamp on the vehicle centerline or to the left of this position or two lamps displayed symmetrically about the centerline. The EEC requirements also specify that not more than two lamps may be installed.
g. Wiring of this lamp type is proposed so as to permit operation only with the headlamps to ensure compatibility with existing state governmental regulations for front fog lamps. Selectivity with beam switching was not included in order to correspond with the EEC requirements.

STOP LAMPS FOR USE ON MOTOR VEHICLES LESS THAN 2032 MM IN OVERALL WIDTH— SAE J586 DEC89

SAE Standard

Report of the Lighting Division, approved February 1927, completely revised by the Lighting Committee February 1984. Rationale statement available. Reaffirmed by the Signalling and Marking Devices Standards Committee December 1989.

1. Scope—This document provides test procedures, requirements, and guidelines for stop lamps intended for use on vehicles of less than 2032 mm in overall width.

2. Definitions

2.1 Stop Lamps—Lamps giving a steady light to the rear of a vehicle to indicate the intention of the operator of a vehicle to stop or diminish speed by braking.

3. Lighting Identification Code—Stop lamps for use on vehicles less than 2032 mm in overall width may be identified by the code "S" in accordance with SAE J759.

4. Tests

4.1 SAE J575 is a part of this document. The following tests are applicable with modifications as indicated.

4.1.1 Vibration Test
4.1.2 Moisture Test
4.1.3 Dust Test
4.1.4 Corrosion Test
4.1.5 Photometry Test

4.1.5.1 Photometric measurements shall be made with the light source of the signal lamp at least 3 m from the photometer. The H-V axis shall be taken as parallel to the longitudinal axis of the vehicle.

4.1.5.2 Photometric measurements shall be made with the bulb filament steadily burning. Photometric measurements of multiple compartment lamps or multiple lamp arrangements shall be made by either of the following methods by aligning the axis of each lamp or compartment with the photometer:

4.1.5.2.1 All compartments or lamps shall be photometered together provided that a line from the light source of each compartment or lamp to the center of the photometer sensing device does not make an angle of more than 0.6 deg with the photometer H-V axis. When compartments or lamps are photometered together, the H-V axis shall intersect the midpoint between their light sources.

4.1.5.2.2 Each compartment or lamp shall be photometered separately. The photometric measurement for the entire multiple compartment lamp or multiple lamp arrangement shall be determined by adding the photometric outputs from each individual lamp or component of corresponding test points.

4.1.6 WARPAGE TEST FOR DEVICES WITH PLASTIC COMPONENTS

4.2 Color Test—SAE J578 is a part of this document.

5. Requirements

5.1 Performance Requirements—A device when tested in accordance with the test procedures specified in Section 4, shall meet the following requirements:

5.1.1 Vibration—SAE J575
5.1.2 Moisture—SAE J575
5.1.3 Dust—SAE J575
5.1.4 Corrosion—SAE J575
5.1.5 Photometry—SAE J575

5.1.5.1 The lamp shall meet the photometric performance requirements contained in Table 1 and its footnotes. The summation of the luminous intensity measurements at the specified test points in a zone shall be at least the value shown.

5.1.5.2 A multiple compartment lamp or multiple lamps may be used to meet the photometric requirements of a stop lamp. If a multiple compartment or multiple lamps are used and the distance between adjacent light sources does not exceed 560 mm for two compartment or lamp arrangements and does not exceed 410 mm for three compartments or lamp arrangements, then the combination of the compartments or lamps must be used to meet the photometric requirements for the corresponding number of lighted sections (Table 1). If the distance between adjacent light sources exceeds the dimensions, each compartment or lamp shall comply with the photometric requirements for one lighted section (Table 1).

5.1.5.3 When a tail lamp is combined with the stop lamp, the stop lamp shall not be less than three times the luminous intensity of the tail lamp at any test point; except that at H-V, H-5L, H-5R, and 5U-V, the stop lamp shall not be less than five times the luminous intensity of the tail lamp. If a multiple compartment or multiple lamp arrangement is used and the distance between optical axis for both the tail lamp and stop lamp is within the dimensions specified in 5.1.5.2, the ratio of the stop lamp to the tail lamp shall be computed with all the compartments or lamps lighted. If a multiple compartment or multiple lamp arrangement is used and the distance between optical axes for one of the functions exceeds the dimensions specified in 5.1.5.2, the ratio shall be computed for only those compartments or lamps where the tail lamp and stop lamp are optically combined. When the tail lamp is combined with the stop lamp, and the maximum luminous intensity of the tail lamp is located below horizontal and within an area generated by a 0.5 deg radius around a test point, the ratio for the test point may be computed using the lowest value of the tail lamp luminous intensity within the generated area.

5.1.6 Warpage—SAE J575

5.1.7 Color—The color of light from the stop lamps shall be red as specified in J578.

5.2 Materials Requirements—Plastic materials used in the optical parts shall meet the requirements of SAE J576.

TABLE 1—PHOTOMETRIC REQUIREMENTS[c]

Zone	Test Points[a] (deg)	Minimum Luminous Intensity (candela) Lighted Sections[d]		
		1	2	3
1	10U-5L 5U-20L 5D-20L 10L-5L	50	60	70
2	5U-10L H-10L 5D-10L	100	115	135
3	5U-V H-5L H-V H-5R 5D-V	380	445	520
4	5U-10R H-10R 5D-10R	100	115	135
5	10U-5R 5U-20R 5D-20R 10D-5R	50	60	70
Maximum Luminous Intensity (candela)[b]		300	360	420

[a] The measured values at each test point shall not be less than 60% of the minimum value in Table 2.

[b] The listed maximum shall not be exceeded over any area larger than that generated by a 0.5 deg radius within the solid angle defined by the test points in Table 1.

[c] Ratio requirements of paragraph 5.1.5.3 apply.

[d] A multiple device signaling unit gives its indication by two or more separately lighted sections which may be separate lamps, or areas that are joined by common parts. The photometric values are to apply when all sections that provide the same signal are considered as a unit except when the dimensions between optical centers exceed those given in 5.1.5.2. For a separate lamp arrangement, where lamps are interchangeable, each lamp shall be of approximately the same performance.

TABLE 2—PHOTOMETRIC DESIGN GUIDELINES

Test Points (deg)		Minimum Luminous Intensity (candela)		
		Lighted Sections		
		1	2	3
10U,10D	5L,5R	16	19	22
	20L,20R	10	12	15
5U,5D	10L,10R	30	35	40
	V	70	82	95
	10L,10R	40	47	55
H	5L,5R	80	95	110
	V	80	95	110
Maximum Luminous Intensity[a] (candela)		300	360	420

[a] The maximum design value of a stop lamp should not exceed the listed design maximum over any area larger than that generated by 0.25 deg radius within the solid angle defined by the test points in Table 2.

5.3 Design Requirements

5.3.1 If a stop signal is optically combined with the tail lamp and a two-filament bulb used, the bulb shall have an indexing base and the socket shall be designed so that bulbs with nonindexing bases cannot be used. Removable sockets shall have an indexing feature so that they cannot be re-inserted into lamp housings in random positions, unless the lamp will perform its intended function with random light source orientation.

5.3.2 The functional lighted lens area of a single compartment lamp shall be at least 37.5 cm^2.

5.3.3 If a multiple compartment lamp or multiple lamps are used to meet the photometric requirements, the functional lighted lens area of each compartment or lamp shall be at least 22 cm^2 provided the combined area is at least 37.5 cm^2.

5.4 Installation Requirements—The stop lamp shall meet the following requirements as installed on the vehicle:

5.4.1 Visibility of the signal shall not be obstructed by any part of the vehicle throughout the photometric test angles for the lamps unless the lamp is designed to comply with all photometric and visibility requirements with these obstructions considered. Signals from lamps on both sides of the vehicle shall be visible through a horizontal angle of 45 deg to the left and to 45 deg to the right. Where more than one lamp or optical area is lighted on each side of the car, only one such area on each side need comply. To be considered visible, the lamp must provide an unobstructed view of the outer lens surface, excluding reflex reflectors, of at least 12.5 cm^2 measured at 45 deg to the longitudinal axis of the vehicle.

5.4.2 When a stop signal is optically combined with the turn signal, the circuit shall be such that the stop signal cannot be turned on if the signal is flashing.

6. Guidelines

6.1 Photometric design guidelines for stop lamps, when tested in accordance with 4.1.5 of this report, are contained in Table 2 and its footnotes.

6.2 Installation Guidelines—The following apply to stop lamps as used on the vehicle and shall not be considered part of the requirements:

6.2.1 Stop lamps on the rear of the vehicle should be spaced as far apart laterally as practicable, so that the signal will be clearly visible.

6.2.2 The luminous intensity of incandescent filament bulbs will vary with applied voltage. The electrical power system of the vehicle should, under normal running conditions, provide design voltage to the lamp as closely as practical bearing in mind the inherent variability of such systems.

6.2.3 Performance of lamps may deteriorate significantly as a result of dirt, grime, and/or snow accumulation on the optical surfaces. Installation of lamps on vehicles should be considered to minimize the effect of these factors.

6.2.4 Where it is expected that lamps must perform in severe environments, for example, be totally immersed in water periodically, the user should specify lamps designed for such use.

APPENDIX A

As a matter of additional information, attention is called to SAE J567 for requirements and gages used in socket design.

STOP LAMPS FOR USE ON MOTOR VEHICLES 2032 mm OR MORE IN OVERALL WIDTH— SAE J1398 JUN91

SAE Standard

Report of the Lighting Division, approved May 1985. Rationale statement available. Completely revised by the SAE Lighting Coordinating Committee and the Heavy-Duty Lighting Standards Committee June 1991. Rationale statement available.

1. Scope—This SAE Standard provides test procedures, requirements, and guidelines for stop lamps intended for use on vehicles 2032 mm or more in overall width. Stop lamps conforming to the requirements of this document may be used on vehicles less than 2032 mm in overall width.

2. References

2.1 Applicable Documents—The following publications form a part of this specification to the extent specified herein. The latest issue of SAE publications shall apply.

2.1.1 SAE PUBLICATIONS—Available from SAE, 400 Commonwealth Drive, Warrendale, PA 15096-0001.

SAE J567—Lamp Bulb Retention System for Requirements and Gages Used in Socket Design
SAE J576—Plastic Material for Use in Optical Parts Such as Lenses and Reflectors of Motor Vehicle Lighting Devices
SAE J578—Color Specification
SAE J586—Stop Lamps for Use on Motor Vehicles Less Than 2032 mm in Overall Width
SAE J759—Lighting Identification Code
SAE J2139—Tests for Lighting Devices, Reflective Devices and Components Used on Vehicles 2032 mm or More in Overall Width

SAE Technical Paper 830566, "Motor Vehicle Conspicuity," R.L. Henderson, K. Ziedman, W.J. Burger, and K.E. Cavey, National Highway Traffic Safety Administration

2.1.2 OTHER PUBLICATIONS—Attention is called to the following documents for additional information on lamp design and installation.

FMVSS 108
FHWA 393 Subpart B
TTMA #RP-9
TMC #RP-702

2.2 Definitions

2.2.1 A stop lamp is a lamp giving a steady light to the rear of a vehicle to indicate the intention of the operator of the vehicle to stop or diminish speed by braking.

3. Lighting Identification Code—Stop lamps for use on vehicles 2032 mm or more in overall width may be identified by the code "S2" in accordance with SAE J759.

4. Tests

4.1 SAE J2139 is a part of this document. The following tests are applicable with modification as indicated.

4.1.1 VIBRATION
4.1.2 MOISTURE
4.1.3 DUST
4.1.4 CORROSION
4.1.5 PHOTOMETRY

4.1.5.1 Photometric measurements shall be made with the light source of the device at least 3 m from the photometer.

4.1.5.2 The H-V axis of the device shall be taken to be parallel to the longitudinal axis of the vehicle, when the device is mounted in its design position.

4.1.5.3 Photometric measurements shall be made with the light source steadily burning. Photometric measurements of multiple compartment lamps or multiple lamp arrangements shall be made by either of the following methods by aligning the axis of each lamp or compartment with the photometer.

4.1.5.3.1 All compartments or lamps shall be photometered together provided that a line from the light source of each compartment or lamp to the center of the photometer sensing device does not make an angle of more than 0.6 degrees with the photometer H-V axis. When compartments or lamps are photometered together, the H-V axis shall intersect the midpoint between their light sources.

4.1.5.3.2 Each compartment or lamp shall be photometered separately. The photometric measurement for the entire multiple compartment lamp or multiple lamp arrangement shall be determined by adding the photometric outputs from each individual lamp or component at corresponding test points.

4.1.6 WARPAGE TEST ON DEVICES WITH PLASTIC COMPONENTS

4.2 Color—SAE J578 is a part of this document.

4.3 Plastic Materials—SAE J576 is a part of this document.

5. Requirements

5.1 Performance Requirements—The device when tested in accordance with the test procedures of this report shall meet the requirements of SAE J2139 or as indicated.

5.1.1 VIBRATION
5.1.2 MOISTURE
5.1.3 DUST
5.1.4 CORROSION
5.1.5 PHOTOMETRY—The device tested shall meet the photometric performance requirements of Table 1 and its footnotes.

The summation of the luminous intensity measurements at the specified test points in a zone shall be at least the value shown.

5.1.5.1 A multiple compartment lamp or multiple lamps may be used to meet the photometric requirements of a stop lamp. If multiple compartments or multiple lamps are used and the distance between adjacent light sources does not exceed 560 mm for two compartments or lamp arrangements and does not exceed 410 mm for three compartments or lamp arrangements, then the combination of the compartments or lamps must be used to meet the photometric requirements of Table 1. If the distance between adjacent light sources exceeds the above dimensions, each compartment or lamp shall comply with the photometric requirements of Table 1.

TABLE 1—STOP LAMP PHOTOMETRIC PERFORMANCE REQUIREMENTS

Zone	Test Point Deg.	Zone Total Luminous Intensity, Candela, Red
1	10U— 5L 5U—20L 5D—20L 10D— 5L	50
2	5U—10L H—10L 5D—10L	100
3	5U— V H— 5L H— V H— 5R 5D— V	380
4	5U—10R H—10R 5D—10R	100
5	10U— 5R 5U—20R 5D—20R 10D— 5R	50
Maximum Luminous Intensity, Candela		300.0

[1] The maximum luminous intensity shall not be exceeded over any area larger than that generated by a 0.5 degree radius within the area defined by the test point pattern of Table 2.

[2] Unless otherwise specified, the lamp shall be considered to have failed the photometric requirements of this document if the luminous intensity at any test point is less than 60% of the values specified in Table 2.

[3] Unless otherwise specified, the lamp shall be considered to have failed the photometric requirements of this document if the minimum luminous intensity between test points is less than 60% of the lower design values of Table 2 for the closest adjacent test points on a horizontal and vertical line as defined by the test point pattern.

[4] The summation of the luminous intensity measurements at the specified test points in the zone shall be at least the values shown.

[5] When a tail lamp or a clearance lamp is combined with a stop lamp, see 5.1.5.2 of this document for luminous intensity ratio requirements.

[6] Throughout the photometric pattern defined by the corner points of 20U-45L, 20U-45R, 10D-45R, and 10D-45L, the light intensity shall be not less than 0.4 candela in red.

5.1.5.2 When a tail lamp or a clearance lamp is combined with the stop lamp, the stop lamp intensity shall be not less than three times the luminous intensity of the tail lamp or clearance lamp at any test point, except that at H-V, H-5L, H-5R, and 5U-V, the stop lamp intensity shall be not less than five times the luminous intensity of the tail lamp or clearance lamp.

When a tail lamp is combined with the stop lamp, and the maximum luminous intensity of the tail lamp is located below the horizontal and is within an area generated by a 1.0 degree radius around the test point, the ratio for the test point may be computed using the lowest value of the tail lamp intensity within the generated area.

5.1.6 WARPAGE

5.2 Color—The color of the light from the stop lamp shall be red as specified in SAE J578.

5.3 Plastic Materials—The plastic materials used in the optical parts shall meet the requirements of SAE J576.

5.4 Design Requirements

5.4.1 If a stop lamp is combined with a tail lamp or a clearance lamp and a replaceable multiple light source is used, the light source retention system shall be designed with an indexing means so that the light source is properly indexed. Removable light source retention systems shall have an indexing feature so that they cannot be reinserted into the lamp housing in a random position, unless the lamp will perform its intended function with random light source orientation.

5.4.2 The effective projected luminous lighted lens area of a single lamp shall be at least 75 cm^2.

5.4.3 If a multiple compartment lamp or multiple lamps are used to meet the photometric requirements, the effective projected luminous lens area of each compartment or lamp shall be at least 40 cm^2 provided the combined area is at least 75 cm^2.

5.5 Installation Requirements—The stop lamp shall meet the following requirements as installed on the vehicle.

5.5.1 The stop lamps shall be mounted on the permanent structure of the vehicle, facing rearward, at the same height and spaced as far apart laterally as practicable, so that the signal will be clearly visible.

5.5.2 Visibility of each stop lamp shall not be obstructed by any part of the vehicle throughout the photometric test pattern unless the lamp is designed to comply with all photometric and visibility requirements with these obstructions considered.

To be considered visible, the lamp must provide an unobstructed view of a portion of the lighted outer lens surface, excluding reflex reflector area, of at least 13 cm^2 measured at a horizontal angle of 45 degrees to the left and 45 degrees to the right of the longitudinal axis of the vehicle and a vertical angle from 20 degrees up to 10 degrees down.

See Table 1, Note 6. Where more than one lamp or optical area is lighted on each side of the vehicle, only one such area on each side need comply.

6. Guidelines

6.1 Design Guidelines

6.1.1 Photometric design guidelines are contained in Table 2 and its footnotes.

6.2 Installation Guidelines—The following guidelines apply to stop lamps as used on the vehicle and shall not be considered part of the requirements.

6.2.1 Performance of lamps may deteriorate significantly as a result of dirt, grime, snow, and ice accumulation on the optical surfaces. Installation of the device on the vehicle should be considered to minimize the effects of these factors.

6.2.2 Where it is expected that the device must perform in extremely severe environments, or where it is expected to be totally immersed in water, the user should specify devices specifically designed for such use.

7. Advance Requirements—This section of the document gives advance notice to manufacturers and users of the device of a pending change in the requirements for a stop lamp. The change in the requirements shall be effective on devices marketed and used on or after January 1, 1996.

See Tables 3 and 4 and the footnotes.

TABLE 2—STOP LAMP PHOTOMETRIC DESIGN GUIDELINES

Test Point Deg.	Luminous Intensity, Candela, Red
10U— 5L	16.0
5R	16.0
5U—20L	10.0
10L	30.0
V	70.0
10R	30.0
20R	10.0
H—10L	40.0
5L	80.0
V	80.0
5R	80.0
10R	40.0
5D—20L	10.0
10L	30.0
V	70.0
10R	30.0
20R	10.0
10D— 5L	16.0
5R	16.0
Maximum Luminous Intensity, Candela	300.0

[1] The maximum luminous intensity shall not be exceeded over any area larger than that generated by a 0.25 degree radius within the area defined by the test point pattern of Table 2.

[2] When a tail lamp or a clearance lamp is combined with a stop lamp, see 5.1.5.2 of this document for luminous intensity ratio requirements.

[3] Throughout the photometric pattern defined by the corner points of 20U-45L, 20U-45R, 10D-45R, and 10D-45L, the light intensity shall be not less than 0.4 candela in red.

TABLE 3—STOP LAMP PHOTOMETRIC PERFORMANCE REQUIREMENTS—ADVANCE REQUIREMENTS

Zone	Test Point Deg.	Zone Total Luminous Intensity, Candela, Red
1	20U—45L 20U—20L 10D—20L 10D—45L	12
2	10U— 5L 5U—20L 5D—20L 10D— 5L	50
3	5U—10L H—10L 5D—10L	100
4	5U— V H— 5L H— V H— 5R 5D— V	380
5	5U—10R H—10R 5D—10L	100
6	10U— 5R 5U—20R 5D—20R 10D— 5R	50
7	20U—45R 20U—20R 10D—20R 10D—45R	12
Maximum Luminous Intensity, Candela		300 Red

[1] The maximum luminous intensity shall not be exceeded over any area larger than that generated by a 0.5 degree radius within the area defined by the test point pattern of Table 4.

[2] Unless otherwise specified, the lamp shall be considered to have failed the photometric requirements of this document if the luminous intensity at any test point is less than 60% of the values specified in Table 4.

[3] Unless otherwise specified, the lamp shall be considered to have failed the photometric requirements of this document if the minimum luminous intensity between test points is less than 60% of the lower design values of Table 4 for the closest adjacent test points on a horizontal and vertical line as defined by the test point pattern.

[4] The summation of the luminous intensity measurements at the specified test points in the zone shall be at least the values shown.

[5] When a tail lamp or a clearance lamp is combined with a stop lamp, see 5.1.5.2 of this document for luminous intensity ratio requirements.

TABLE 4—STOP LAMP PHOTOMETRIC DESIGN GUIDELINES—ADVANCE REQUIREMENTS

Test Point Deg.	Luminous Intensity, Candela, Red
20U—45L	1.0
20L	5.0
20R	5.0
45R	1.0
10U— 5L	16.0
5R	16.0
5U—20L	10.0
10L	30.0
V	70.0
10R	30.0
20R	10.0
H—10L	40.0
5L	80.0
V	80.0
5R	80.0
10R	40.0
5D—20L	10.0
10L	30.0
V	70.0
10R	30.0
20R	10.0
10D—45L	1.0
20L	5.0
5L	16.0
5R	16.0
20R	5.0
45R	1.0
Maximum Luminous Intensity, Candela	300 Red

[1] The maximum luminous intensity shall not be exceeded over any area larger than that generated by a 0.25 degree radius within the area defined by the test point pattern of Table 4.

[2] When a tail lamp or a clearance lamp is combined with a stop lamp, see 5.1.5.2 of this document for luminous intensity ratio requirements.

HIGH MOUNTED STOP LAMPS FOR USE ON VEHICLES 2032 mm OR MORE IN OVERALL WIDTH— SAE J1432 OCT88

SAE Information Report

Report of the Lighting Committee approved October 1988. Rationale statement available.

1. Scope—This SAE Information Report will provide a uniform arrangement with which to evaluate the concept of high-mounted lamps on large vehicles. The report provides test procedures, requirements, and guidelines for high-mounted stop lamps intended for use on certain vehicles 2032 mm (80 in) or more in overall width.

This information report applies to trucks, motor coaches, closed and open top van trailers and other vehicles with permanent structures greater than 2.8 m high. They are not intended for use on school busses, truck tractors, flat bed, pole, and boat trailers and all other trailers or trucks/truck bodies whose permanent structures are less than 2.8 m (approximately 112 in) high. These lamps are for the purpose of providing a signal over intervening vehicles to following drivers.

Additionally, four widely spaced lamps will make a more conspicuous stop lamp pattern, thus making it easier to identify a large vehicle as slowing or stopping when approaching it from the rear.

2. Definitions

2.1 High Mounted Stop Lamp—A lamp mounted high on the vehicle giving a steady light to the rear to indicate the intention of the operator to stop or diminish speed by braking. These lamps are supplemental and are in addition to the regular stop lamps.

3. Lighting Identification Code—High mounted stop lamps for use on vehicles 2032 mm or more in overall width may be identified by the code "U2" in accordance with SAE J759.

4. Tests

4.1 The device shall be tested according to the procedures specified in SAE J575. The following tests are applicable with the modifications as indicated:

4.1.1 VIBRATION TEST

4.1.2 MOISTURE TEST

4.1.3 DUST TEST

4.1.4 CORROSION TEST

4.1.5 PHOTOMETRY, with the following addition:

4.1.5.1 Photometric measurements shall be made with the light source of the signal lamp at least 3 m from the photometer. If the location and the intended orientation of the H-V axis of the lamp is not obvious from its physical configuration, the lamp manufacturer shall provide explicit instructions concerning the method of installing the lamp so that its H-V axis is horizontal and coincident with the O-O or H-V axis of the goniometer.

4.1.6 WARPAGE TEST ON DEVICES WITH PLASTICS COMPONENTS

TABLE 1 – PHOTOMETRIC REQUIREMENTS[a]

Zone	Test Results (deg)		Minimum Luminous Intensity Total for Zone (cd)
1	5U	V 10R – 10L 20R – 20L	150
2	H	V 5R – 5L 10R – 10L	320
3	5D	V 10R – 10L 20R – 20L	150
4	10D	5R – 5L 20R – 20L	52
Maximum Luminous Intensity (cd)[b]			300

[a] The measured value at each test point shall not be less than 60% of the minimum in Table 2.
[b] The maximum value shall not be exceeded over any area larger than that generated by a 0.5 deg radius within the solid cone angle defined by the test points in Table 2.

4.2 Color Test—The device shall be tested according to the procedures specified in SAE J578.

5. Requirements

5.1 Performance Requirements

5.1.1 SAE J575—A device, when tested in accordance with the test procedures specified in Section 4, shall meet the requirements indicated in the following sections of SAE J575:

5.1.1.1 Vibration
5.1.1.2 Moisture
5.1.1.3 Dust
5.1.1.4 Corrosion
5.1.1.5 Photometry

5.1.1.5.1 The lamp under test shall meet the photometric performance requirements contained in Table 1 - Photometric Requirements and its footnotes. The summation of the luminous intensity measurements at the specified test points in a zone shall be at least the value shown.

5.1.1.6 Warpage

5.1.2 COLOR—The color of the emitted light shall be red as specified in SAE J578.

5.2 Materials Requirements—Plastic materials used in optical parts shall meet the requirements of SAE J576.

5.3 Design Requirements

5.3.1 No other lamp or reflex reflector functions shall be combined with a high mounted stop lamp.

5.3.2 The effective projected luminous lens area of a single lamp shall be at least 50 cm^2.

5.4 Installation Requirements—The following requirements apply to the device as installed on the vehicle, and are not part of the laboratory test procedures and requirements.

5.4.1 Visibility of the lamp shall not be obstructed by any part of the vehicle throughout photometric test angles for the lamp unless the lamp is designed to comply with all photometric and visibility requirements with these obstructions considered.

5.4.2 Two high-mounted stop lamps are required. Both lamps must be mounted at the same height, as far apart laterally as practicable, at a minimum height of 2.7 m (approximately 108 in), measured from the road surface to the center of the lens with the vehicle unladen.

Only one lamp is required on vehicles, such as tankers, whose structure does not permit mounting two lamps at the same height as required above. Where only one lamp is required, it shall be mounted on vehicle centerline at a minimum height of 2.7 m (approximately 108 in).

5.4.3 The lamps shall be mounted so that their H-V axes are horizontal and parallel to the longitudinal axis of the vehicle with the vehicle unladen.

6. Guidelines:

6.1 Photometric Design Guidelines—for high-mounted stop lamps, when tested in accordance with paragraph 4.1.5 of this report, are contained in Table 2 - Photometric Design Guidelines and its footnotes.

6.2 Installation Guidelines—The following guidelines apply to high-mounted stop lamps as used on the vehicle and shall not be considered part of the requirements.

6.2.1 High-mounted stop lamps should be spaced as far apart laterally as practicable so that the signal will be clearly visible.

6.2.2 The luminous intensity of the light source will vary with applied voltage. The electrical wiring in the vehicle should be adequate to supply design voltage to the stop lamp.

6.2.3 Performance of lamps may deteriorate significantly as a result of dirt, grime or snow accumulation, or both, on their optical surfaces. Installation of lamps on vehicles should be considered to minimize the effect of these factors.

6.2.4 Where it is expected that lamps must perform in extremely severe environments, such as in off-highway, mining, fuel haulage, etc., the user should specify lamps designed for such use.

7. Appendix—As a matter of additional information, attention is called to SAE J567 for requirements and gages used in socket design.

TABLE 2 – PHOTOMETRIC DESIGN GUIDELINES

Test Points (deg)		Minimum Luminous Intensity (cd)
10D	5L and 5R 20L and 20R	16 10
5U and 5D	20L and 20R 10L and 10R V	10 30 70
H	10L and 10R 5L and 5R V	40 80 80
Maximum Luminous Intensity (cd)[a]		300

[a] The maximum design value should not be exceeded over any area larger than that generated by a 0.25 deg radius within the solid angle defined by the test points in Table 2.

CENTER HIGH MOUNTED STOP LAMP STANDARD FOR VEHICLES LESS THAN 2032 mm OVERALL WIDTH—SAE J1957 JUN93

SAE Standard

Report of the Signalling and Marking Devices Standards Committee approved June 1993. Rationale statement available.

1. Scope—This SAE Standard provides test procedures, requirements, and guidelines for center high mounted stop lamps (CHMSL) for use on vehicles less than 2032 mm in overall width.

2. References

2.1 Applicable Documents—The following publications form a part of this specification to the extent specified herein. The latest issue of SAE publications shall apply.

2.1.1 SAE PUBLICATIONS—Available from SAE, 400 Commonwealth Drive, Warrendale, PA 15096-0001.

SAE J575—Tests for Motor Vehicle Lighting Devices and Components
SAE J576—Plastic Materials for Use in Optical Parts Such as Lenses and Reflectors of Motor Vehicle Lighting Devices
SAE J578—Color Specification
SAE J759—Lighting Identification

2.2 Definitions

2.2.1 The center high mounted stop lamp (CHMSL) is an additional lamp of the stop lamp system, giving a brake actuated steady warning light to the rear of the vehicle. The CHMSL is intended to provide a signal to both the operator of the following vehicle as well as through intervening vehicles.

3. Lighting Identification Code

3.1 CHMSL for passenger vehicles may be identified with U3 code in accordance with SAE J759.

4. Tests

4.1 SAE J575—The following test procedures in SAE J575 are part of this document, with the modifications indicated:

4.1.1 VIBRATION TEST
4.1.2 MOISTURE TEST
4.1.3 DUST TEST
4.1.4 CORROSION TEST
4.1.5 PHOTOMETRY

4.1.5.1 Photometric tests shall be made with the photometer at least 3 m from the light source. The lamp axis shall be taken as the horizontal line through the light source and parallel to what would be the longitudinal axis of the vehicle if the lamp were mounted in its normal position on the vehicle.

4.1.6 WARPAGE TEST ON DEVICES WITH PLASTIC COMPONENTS—Stop lamp cycle time and temperature in Table 1 of SAE J575 shall be used for evaluating a CHMSL.

4.2 Color Test—SAE J578 is a part of this report.

5. Requirements

5.1 Performance Requirements—Center high mounted stop lamps, when tested in accordance with the test procedures specified in 4.1 shall meet the requirements indicated in the following sections of SAE J575.

5.1.1 VIBRATION TEST
5.1.2 MOISTURE TEST—Does not apply to CHMSLs mounted inside the vehicle.
5.1.3 DUST TEST—Does not apply to CHMSLs mounted inside the vehicle.
5.1.4 CORROSION TEST—Does not apply to CHMSLs mounted inside the vehicle.
5.1.5 PHOTOMETRY TEST—The lamp, when tested in accordance with 4.1 of this document, shall meet the photometric requirements contained in Table 1. For interior mounted CHMSLs, the photometry test shall include the vehicle manufacturer's specified glazing in the design position.

  a. The luminous intensity values at each test point shall not be less than 60% of the minimum value specified in Table 2.
  b. The listed maximum shall not be exceeded over any area larger than that generated by a 0.5 degree radius within the solid angle defined by the test points in Table 1.

5.1.6 WARPAGE TEST—There shall be no evidence of warpage which results in failure of any test contained in 4.1 of this document.

5.2 Color Test—The light emitted by the CHMSL shall be red.

5.3 Material Requirements—Plastic materials used in CHMSL optic parts shall conform to the requirements in SAE J576.

5.4 Dimensional Requirements—The effective projected luminous lens area measured on a plane at right angles to the lamp axis shall not be less than 29 cm^2 (4.5 in^2).

5.5 Installation Requirements

5.5.1 The CHMSL shall not be optically combined with any other signal lamp or reflective device other than with a cargo lamp.

5.5.2 The center of the CHMSL shall be mounted on the vertical centerline of the vehicle.

5.5.3 If the lamp is mounted below the rear window, no portion of the lens shall be lower than 152 mm (6 in) below the rear window on convertibles, or 76 mm (3 in) on other passenger cars.

5.5.4 CHMSL shall have a signal visible from 45 degrees to the left to 45 degrees to the right of the longitudinal axis of the vehicle.

5.5.5 The CHMSL shall be activated only upon application of the service brakes.

5.5.6 If the CHMSL is mounted inside the vehicle, means shall be provided to minimize reflections at the rear window glazing that might be visible to the driver when viewed directly, or indirectly in the rearview mirror.

6. Guidelines

6.1 Photometric design guidelines for center high mounted stop lamps, when tested in accordance with 4.1.5 of this document, are contained in Table 2.

6.2 Serviceability/Cleanability—The CHMSL shall be designed to be serviced and cleaned with either commonly available tools or no tools. The number of trim pieces that must be removed for this purpose should be minimized. This guideline applies to lamp servicing, bulb replacement, and access to the rear window glazing for cleaning purposes.

6.3 Replacement bulb identification should be permanently located on the lamp housing.

6.4 The vertical location is specified with the intent of positioning the lamp higher than the conventional stop lamps. The lamp shall be mounted high to insure its visibility through intervening vehicles. It may be located forward of the tail, stop, and rear turn signal lamps.

6.5 Heat test cycle time and temperature cycle may be increased to represent a more severe heat test based on CHMSL mounting environment and/or performance cycle.

TABLE 1—MINIMUM ZONAL PHOTOMETRIC REQUIREMENTS FOR CENTER HIGH MOUNTED STOP LAMPS

Group	Test Points (degrees)	Minimum Total Intensity (candela)
1	5U-V H-5L H-V H-5R 5D-V	125
2	5U-5R 5U-10R H-10R 5D-10R 5D-5R	98
3	5U-5L 5U-10L H-10L 5D-10L 5D-5L	98
4	10U-10L 10U-V 10U-10R	32

TABLE 2—DESIGN PHOTOMETRIC GUIDELINES FOR CENTER HIGH MOUNTED STOP LAMPS

Test Points		Minimum Intensity (candela)
10U	10L	8
	V	16
	10R	8
5U and 5D	10L	16
	5L	25
	V	25
	5R	25
	10R	16
H	10L	16
	5L	25
	V	25
	5R	25
	10R	16
Maximum[1]		130

[1] The lamp shall not exceed the listed maximum over an area larger than that generated by a 0.25 degree radius within a solid cone angle from 10 degrees L to 10 degrees R and from 10 degrees U to 5 degrees D.

φMECHANICAL STOP LAMP SWITCH—SAE J249 JUN88   SAE Standard

Report of Lighting Committee approved February 1972, and completely revised June 1988.

1. Purpose—This standard defines the test conditions, procedures, and performance specifications for 6-, 12-, and 24-V manually actuated mechanical stop lamp switches.

2. Definition—The mechanical stop lamp switch is an operator activated mechanical device intended primarily to control the functioning of the stop lamp and high mounted stop lamp circuits. Secondarily, the device may control the functioning of various accessories, such as disengaging cruise control, with operator actuation of brake pedal.

3. Test Requirements

3.1 Test Equipment and Instrumentation

3.1.1 POWER SUPPLY—The power supply shall comply with the following specifications:
  a. Output Current—capable of supplying the continuous and inrush currents of the design load (reference paragraph 3.2.1.1).
  b. Regulation:
    Dynamic—the output voltage at the supply shall not deviate more than 1.0 V from zero to maximum load (including inrush current) and should recover 63% of its maximum excursion within 100 ms.
    Static—the output voltage at the supply shall not deviate more than 2% with changes in static load from zero to maximum (not including in-rush current), and means shall be provided to compensate for static input line voltage variations.
  c. Ripple Voltage—maximum 300 mV peak-to-peak.

3.1.2 VOLTMETER—0 - 30 maximum full scale deflection, accuracy ± ½%.

NOTE—A digital meter having at least 3½ digit readout with an accuracy of ±1% plus one digit is recommended for millivolt readings.)

3.1.3 AMMETER—Capable of carrying full system load current, accuracy ±3%.

3.2 Test Procedures—Environmental conditions have been selected for this standard to help assure satisfactory operation under general customer use conditions. It is essential to duplicate specific environmental conditions under which the device is expected to function.

3.2.1 ELECTRICAL LOADS

3.2.1.1 The design load applied to the switch is the electrical load specified by the number and type of bulbs (or other electrical load devices) to be operated by each circuit of the switch. For example, the design load for the stop lamp circuit may be two #1157 (high current filament) and high mounted stop lamp circuit may be two #922 bulbs.

The switch shall be operated at 6.4 ± 0.2 V DC for a 6 V system, 12.8 ± 0.2 V DC for a 12 V system, or 25.6 ± 0.2 V DC for a 24 V system. These voltages shall be open circuit voltage measured at the input termination on the switch.

3.2.2 TEMPERATURE TEST PROCEDURES

3.2.2.1 The switch shall be exposed for 1 h without electrical load to each of the following temperatures: 25 ± 5°C; 74 + 0°C, −3°C; −32 + 3°C, −0°C. The switch shall be manually cycled at each temperature for ten cycles at design load.

3.2.2.2 The same switch shall be used for the endurance test described in Section 3.2.3.

3.2.3 ENDURANCE TEST PROCEDURE

3.2.3.1 The switch shall be electrically connected to operate its design load (both primary and secondary circuit function design electrical loads) at a temperature of 25 ± 5°C.

3.2.3.2 The switch shall be operated for a minimum of 300 000 cycles. One complete cycle shall consist of energizing and de-energizing the design load (with a dwell in each position).

The test equipment shall be arranged to provide the following switch operating time requirements:
Travel Time: 0.1 - 0.5 s (time from one position to next)
Dwell Time: 1.0 - 2.0 s (time in each position)
Make & Break Rate: 10 - 15 mm/s

NOTE—300 000 cycles represents 82 cycles of switch operation every day for approximately 10 years, or 3 cycles for each 1.0 mile driven 100 000 miles.)

3.2.4 VOLTAGE DROP TEST PROCEDURE

3.2.4.1 The voltage drop from the input terminal(s) to the corresponding output terminal(s) shall be measured at design load before and after the completion of the endurance test and shall be average of three consecutive readings. If wiring is an integral part of the switch, the voltage drop measurement shall be made including 75 ± 6 mm of wire on each side of the switch terminals.

4. Performance Requirements

4.1 During and after each of the cycles described in paragraph 3.2.2.1 and Section 3.2.3, the switch shall be electrically and mechanically operable.

4.2 The voltage drop shall not exceed 0.3 V when measured as in Section 3.2.4, at the beginning and after the test described in Section 3.2.3.

(R) LICENSE PLATE ILLUMINATION DEVICES (REAR REGISTRATION PLATE ILLUMINATION DEVICES)—SAE J587 MAR93

SAE Standard

Report of the Lighting Division, approved March 1918, completely revised, Lighting Committee, August 1985. Rationale statement available. Completely revised by the SAE Signalling and Marking Devices Standards Committee March 1993. Rationale statement available.

1. **Scope**—This SAE Standard provides test procedures, requirements, and guidelines for vehicular license plate illumination devices.

2. **References**

   2.1 **Applicable Documents**—The following publications form a part of this specification to the extent specified herein. The latest issue of SAE publications shall apply.

   2.1.1 SAE PUBLICATIONS—Available from SAE, 400 Commonwealth Drive, Warrendale, PA 15096-0001.

   SAE J567—Lamp Bulb Retention System
   SAE J575—Tests for Motor Vehicle Lighting Devices and Components
   SAE J576—Plastic Materials for Use in Optical Parts Such as Lenses and Reflectors of Motor Vehicle Lighting Devices
   SAE J578—Color Specification
   SAE J759—Lighting Identification Code

   2.1.2 ASTM PUBLICATION—Available from ASTM, 1916 Race Street, Philadelphia, PA 19103-1187.

   ASTM E 179—Selection of Geometric Conditions for Measurement of Reflection and Transmission Properties of Materials

   2.2 **Definitions**

   2.2.1 A LICENSE PLATE ILLUMINATION DEVICE is a device that illuminates the license plate on the rear of a vehicle.

3. **Lighting Identification Code**—License plate illumination devices may be identified by the code "L" in accordance with SAE J759.

4. **Tests**

   4.1 SAE J575 is a part of this document. The following tests are applicable with modifications as indicated.

   4.1.1 VIBRATION TEST
   4.1.2 MOISTURE TEST
   4.1.3 DUST TEST
   4.1.4 CORROSION TEST
   4.1.5 WARPAGE TEST ON DEVICES WITH PLASTIC COMPONENTS

   4.2 **Color Test**—SAE J578 is part of this document.

   4.3 **Photometry Test**

   4.3.1 TEST EQUIPMENT

   4.3.1.1 *Test Plate*—All luminance measurements shall be made on a rectangular test plate of clean, smooth, matte white blotting paper or an equivalent material with a diffuse white surface. The test plate shall have a total reflectance factor of 85% ± 5% when measured in accordance with ASTM E 179 (0/t illumination/viewing geometry). The size of the test plate is shown in Figure 1 or 2, as applicable. For devices used on vehicles other than motorcycles and motor-driven cycles, test stations shall be located on the face of the test plate as shown in Figure 1. For devices used on motorcycles and motor-driven cycles, the test stations shall be located on the face of the test plate as shown in Figure 2.

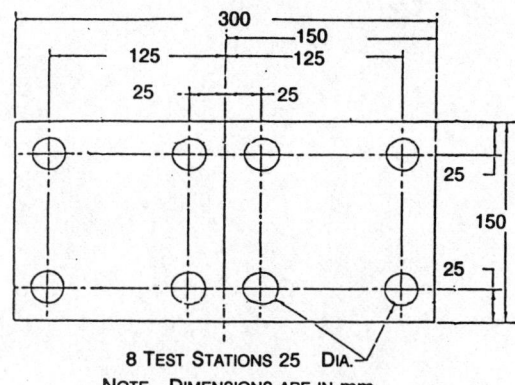

FIGURE 1—TEST PLATE FOR VEHICLES OTHER THAN MOTORCYCLES AND MOTOR-DRIVEN CYCLES

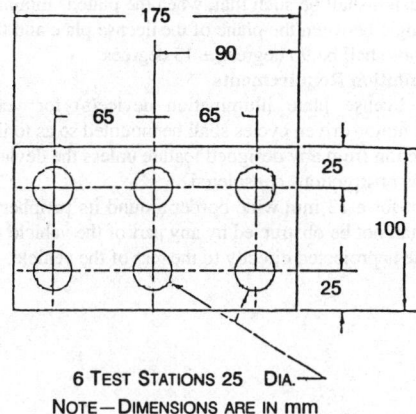

FIGURE 2—TEST PLATE FOR MOTORCYCLES AND MOTOR-DRIVEN CYCLES

4.3.1.2 *Luminance Meter*—A luminance meter shall be used to measure the luminance over the entire area of each circular test station on the test plate. The meter shall be calibrated to measure luminance in cd/m^2. Measurements shall not include any area beyond that of the test station.

4.3.2 TEST PROCEDURE

4.3.2.1 The test plate shall be mounted in the position ordinarily taken by the license plate. The face of the test plate shall be located 2 mm from the plane of the license plate holder toward the luminance meter.

4.3.2.2 Luminance measurements shall be made with the optical axis of the luminance meter perpendicular to the test plate surface within ±5 degrees. Measurements shall be recorded at each of the circular test station areas specified in Figure 1 or Figure 2, as applicable.

4.3.2.3 Calculate and record the ratio of the maximum to the minimum luminance over all specified test points. For test plates conforming to Figure 1, the average of the two highest and the two lowest luminance values recorded at the eight test stations shall be taken as maximum and minimum, respectively. For test plates conforming to Figure 2, the highest luminance value and the average of the two lowest luminance values recorded at the six test stations shall be taken as maximum and minimum, respectively.

5. **Requirements**

   5.1 **Performance Requirements**—A device, when tested in accordance with the test procedures specified in Section 4, shall meet the following requirements:

   5.1.1 Vibration—SAE J575
   5.1.2 Moisture—SAE J575
   5.1.3 Dust—SAE J575
   5.1.4 Corrosion—SAE J575
   5.1.5 Warpage—SAE J575

   5.2 **Color**—The color of the light from the license plate illumination device(s) shall be white as specified in SAE J578.

   5.3 **Photometry**—Upon completion of the photometry test procedure (paragraph 4.3), the following requirements shall apply:

   5.3.1 The luminance at each of the test station areas on the applicable test plate shall be at least 2.5 cd/m^2.

   5.3.2 For tests based on Figure 1, the ratio of maximum to minimum luminance shall not exceed 20/1. For tests based on Figure 2, the ratio of maximum to minimum luminance shall not exceed 15/1.

   5.3.3 If a tail or stop lamp is combined with a license plate illumination device, the combination shall also meet the requirements for these devices.

   5.4 **Materials Requirements**—Plastic materials used in the optical parts shall meet the requirements of SAE J576. Since some license plate illumination devices are mounted in shaded or protected locations, attention is called to the section of SAE J576 which covers exposure time and conditions.

5.5 Design Requirements

5.5.1 License plate illumination devices for vehicles other than motorcycle and motor-driven cycles shall be of such size and design as to provide illumination on all parts of a 150 × 300 mm test plate, except for a 13 mm wide border around the plate periphery. License plate illumination devices for motorcycle and motor-driven cycles shall be of such size and design as to provide illumination on all parts of a 100 × 175 mm test plate.

5.5.2 The design shall be such that, when the plate is mounted on a vehicle as intended, the angle between the plane of the license plate and the plane on which the vehicle stands shall be 90 degrees ± 15 degrees.

5.6 Installation Requirements

5.6.1 The license plate illumination device(s) for vehicles other than motorcycles or motor-driven cycles shall be mounted so as to illuminate the plate without obstruction from any designed feature unless the device(s) is designed to comply with the obstructions considered.

5.6.2 Except for a 13 mm wide border around its periphery, visibility of the license plate shall not be obstructed by any part of the vehicle when any point on the license plate is projected directly to the rear of the vehicle.

5.6.3 The license plate illumination device(s) shall be installed so that no white light is projected from the illumination device(s) directly to the rear of the vehicle.

5.6.4 The license plate illumination device(s) for vehicles other than motorcycles and motor-driven cycles shall be mounted so as to illuminate the plate from the top or sides. Illumination from the bottom of the plate is permitted provided other illumination is also provided from the top or sides of the plate.

6. Guidelines

6.1 **Installation Guidelines**—The following apply to license plate illumination devices as used on the vehicle and shall not be considered part of the requirements.

6.1.1 The license plate holding device shall be designed and constructed to provide a substantial plane surface on which to mount the plate.

APPENDIX A

A.1 As a matter of information, attention is called to SAE J567 for requirements and gages to be used in socket design.

TURN SIGNAL LAMPS FOR USE ON MOTOR VEHICLES LESS THAN 2032 MM IN OVERALL WIDTH—SAE J588 JUN91

SAE Standard

Report of the Lighting Division, approved February 1927, completely revised, Lighting Committee, November 1984. Rationale statement available. Revised by the SAE Lighting Coordinating Committee and the SAE Signaling and Marking Devices Standards Committee June 1991.

1. Scope—This SAE Standard provides test procedures, requirements, and guidelines for turn signal lamps intended for use on vehicles of less than 2032 mm in overall width.

(R) *2. References*

2.1 Applicable Documents—The following publications form a part of this specification to the extent specified herein. The latest issue of SAE publications shall apply.

2.1.1 SAE PUBLICATIONS—Available from SAE, 400 Commonwealth Drive, Warrendale, PA 15096-0001.

SAE J567—Lamp Bulb Retention System
SAE J575—Tests for Motor Vehicle Lighting Devices and Components
SAE J576—Plastic Materials for Use in Optical Parts such as Lenses and Reflectors of Motor Vehicle Lighting Devices
SAE J578—Color Specification
SAE J759—Lighting Code Identification
SAE J1050—Describing and Measuring the Driver's Field of View

2.2 Definitions

2.2.1 TURN SIGNAL LAMPS—The signalling elements of a turn signal system which indicate an intention to turn by giving a flashing light on the side toward which the turn will be made.

3. Lighting Identification Code—Turn signal lamps for use on vehicles less than 2032 mm in overall width may be identified by the codes I, I2, I3, I4, or I5 in accordance with SAE J759.

4. Tests

4.1 SAE J575 is a part of this document. The following tests are applicable with modifications as indicated.

4.1.1 VIBRATION TEST
4.1.2 MOISTURE TEST
4.1.3 DUST TEST
4.1.4 CORROSION TEST
4.1.5 PHOTOMETRY TEST

4.1.5.1 Photometric measurements shall be made with the light source of the signal lamp at least 3 m from the photometer. The H-V axis shall be taken as parallel to the longitudinal axis of the vehicle.

4.1.5.2 Photometric measurements shall be made with the bulb filament steadily burning. Photometric measurements of multiple compartment lamps or multiple lamp arrangements shall be made by either of the following methods by aligning the axis of each lamp or compartment with the photometer:

4.1.5.2.1 All compartments or lamps shall be photometered together provided that a line from the light source of each compartment or lamp to the center of the photometer sensing device does not make an angle of more than 0.6 degrees with the photometer H-V axis. When compartments or lamps are photometered together, the H-V axis shall intersect the midpoint between their light sources.

4.1.5.2.2 Each compartment or lamp shall be photometered separately. The photometric measurement for the entire multiple compartment lamp or multiple lamp arrangement shall be determined by adding the photometric outputs from each individual lamp or component at corresponding test points.

4.1.6 WARPAGE TEST FOR DEVICES WITH PLASTIC COMPONENTS

4.2 Color Test—SAE J578 is a part of this document.

5. Requirements

5.1 Performance Requirements—A device when tested in accordance with the test procedures specified in Section 4, shall meet the following requirements:

5.1.1 VIBRATION—SAE J575
5.1.2 MOISTURE—SAE J575
5.1.3 DUST—SAE J575
5.1.4 CORROSION—SAE J575
5.1.5 PHOTOMETRY—SAE J575

5.1.5.1 The lamp under test shall meet the photometric performance requirements contained in Table 1 and its footnotes. The summation of the luminous intensity measurements at the specified test points in a zone shall be at least the value shown.

5.1.5.2 A multiple compartment lamp or multiple lamps may be used to meet the photometric requirements of a turn signal lamp. If a multiple compartment or multiple lamps are used and the distance between adjacent light sources does not exceed 560 mm for two compartments or lamp arrangements and does not exceed 410 mm for three compartments or lamp arrangements, then the combination of the compartments or lamps must be used to meet the photometric requirements for the corresponding number of lighted sections (see Table 1). If the distance between adjacent light sources exceeds the above dimensions, each compartment or lamp shall comply with the photometric requirements for one lighted section (see Table 1).

5.1.5.3 When a tail lamp or parking lamp is combined with the turn signal lamp, the signal lamp shall not be less than three times the luminous intensity (a) of the tail lamp at any test point, or (b) of the parking lamp at any test point on or above horizontal except that at H-V, H-5L, H-5R, and 5U-V, the signal lamp shall not be less than five times the luminous intensity of the tail lamp or parking lamp. If a multiple compartment or multiple lamp arrangement is used and the distance between optical axis for both the tail lamp (parking lamp) and the turn signal is within the dimensions specified in 5.1.5.2, the ratio of the signal to the tail lamp (parking lamp) shall be computed with all the compartments or lamps lighted. If a multiple compartment or multiple lamp arrangement is used and the distance between optical axis for one of the functions exceeds the dimensions specified in 5.1.5.2, the ratio shall be computed for only those compartments or lamps where the tail lamp (parking lamp) and turn signal are optically combined. Where the tail lamp is combined with the turn signal lamp, and the maximum luminous intensity of the tail lamp is located below horizontal and within an area generated by a 0.5 degree radius around a test point, the ratio for the test point may be computed using the lowest value of the tail lamp luminous intensity within the generated area.

5.1.5.4 In the case where the front turn signal is mounted in close proximity to the low beam headlamp or any additional lamp used to supplement or used in lieu of the low beam, such as an auxiliary low beam or fog lamp, Table 2 shall be used to modify Table 1 as follows:

5.1.5.4.1 Spacing for a direct light source type design front turn signal lamp, that is, a lamp primarily employing a lens to meet photometric requirements (for example, a lamp that does not employ a reflector) shall be measured from the light source to the lighted edge of the low beam headlamp or any additional lamp used to supplement or used in lieu of the lower beam, such as an auxiliary low beam or fog lamp.

5.1.5.4.2 Spacing for a front turn signal lamp which primarily employs a reflector (for example, one of parabolic section) in conjunction with a lens to meet photometric requirements, shall be measured from the geometric centroid of the front turn signal functional lighted area to the lighted edge of the low beam headlamp or any additional lamp used to supplement or used in lieu of the lower beam, such as an auxiliary low beam or fog lamp.

5.1.6 WARPAGE—SAE J575

5.1.7 COLOR—The color of light from the turn signal lamps shall be red or yellow to the rear and yellow to the front of the vehicle as specified in J578.

5.2 Materials Requirements—Plastic materials used in the optical parts shall meet the requirements of SAE J576.

5.3 Design Requirements

5.3.1 If a turn signal is optically combined with the tail lamp and a two-filament bulb used, the bulb shall have an indexing base and the socket shall be designed so that bulbs with nonindexing bases cannot be used. Removable sockets shall have an indexing feature so that they cannot be reinserted into lamp housings in random positions, unless the lamp will perform its intended function with random light source orientation.

5.3.2 The functional lighted lens area of a single compartment lamp shall be at least 37.5 cm^2 for a rear lamp and at least 22 cm^2 for a front lamp.

5.3.3 If a multiple compartment lamp or multiple lamps are used to meet the photometric requirements of a rear turn signal lamp, the functional lighted lens area of each compartment or lamp shall be at least 22 cm^2 provided the combined area is at least 37.5 cm^2.

(R) TABLE 1—PHOTOMETRIC REQUIREMENTS[3]

Minimum Luminous Intensity (cd)[4]

Zone Lighted Sections	Test Points[1] (deg) Lighted Sections	Front Signals Yellow 1	Front Signals Yellow 2	Front Signals Yellow 3	Rear Signals Red 1	Rear Signals Red 2	Rear Signals Red 3	Rear Signals Yellow 1	Rear Signals Yellow 2	Rear Signals Yellow 3
1	10U—5L 5U—20L 5D—20L 10D—5L	130	155	180	50	60	70	80	100	120
2	5U—10L H—10L 5D—10L	250	295	340	100	115	135	165	185	220
3	5U—V H—5L H—V H—5R 5D—V	950	1130	1295	380	445	520	610	710	825
4	5U—10R H—10R 5D—10R	250	295	340	100	115	135	165	185	220
5	10U—5R 5U—20R 5D—20R 10D—5R	130	155	180	50	60	70	80	100	120
Maximum Luminous Intensity (cd)										
Rear Lamps Only[2]		—	—	—	300	360	420	750	900	1050

[1] The measured values at each test point shall not be less than 60% of the minimum value in Table 3.
[2] The listed maximum shall not be exceeded over any area larger than that generated by a 0.5 degree radius within the solid angle defined by the test points in Table 1.
[3] Ratio requirements of 5.1.5.3 apply.
[4] Multipliers of Table 2 are applicable per 5.1.5.4.

TABLE 2—LUMINOUS INTENSITY MULTIPLIERS FOR FRONT TURN SIGNAL SPACINGS

Spacing to Lighted Edge of Low Beam Headlamp[1]	Multiplier of Table 1 and 3 Values to Obtain Required Minimum Luminous Intensities
100 mm or greater	1.0
75 mm to less than 100 mm	1.5
60 mm to less than 75 mm	2.0
Less than 60 mm	2.5

[1] See 5.1.5 for methods to be used for measurements of spacings.

5.4 Installation Requirements—The turn signal lamp shall meet the following requirements as installed on the vehicle:

5.4.1 Visibility of the signal shall not be obstructed by any part of the vehicle throughout the photometric test angles for the lamps unless the lamp is designed to comply with all photometric and visibility requirements with these obstructions considered. Signals from lamps mounted on the left side of the vehicle shall be visible through a horizontal angle of 45 degrees to the left and signals from lamps mounted on the right side of the vehicle shall be visible through a horizontal angle of 45 degrees to the right. To be considered visible, the lamp must provide an unobstructed view of the outer lens surface, excluding reflex reflectors, of at least 12.5 cm^2 measured at 45 degrees to the longitudinal axis of the vehicle.

5.4.2 When a stop signal is optically combined with the turn signal, the circuit shall be such that the stop signal cannot be turned on if the turn signal is flashing.

5.4.3 TURN SIGNAL PILOT INDICATOR

5.4.3.1 If one right and one left turn signal are not readily visible to the driver, there shall be an illuminated indicator provided to give a clear and unmistakable indication that the turn signal system is activated. The illuminated indicator shall consist of one or more lights flashing at the same frequency as the signal lamps.

5.4.3.2 If the illuminated indicator is located inside the vehicle, it should emit a green colored light and have a mimimum area of 18 mm^2.

5.4.3.3 If the illuminated indicators are located on the outside of the vehicle, for example on the front fenders, they should emit a yellow colored light and have a minimum projected illuminated area of 60 mm^2.

5.4.3.4 The minimum required illuminated area of the indicators specified in 5.4.3.2 and 5.4.3.3 shall be visible according to the procedures described in SAE J1050. The steering wheel shall be turned to a straight-ahead driving position and in the design location for an adjustable wheel or column.

6. Guidelines

(R) **6.1 Photometric Design Guidelines**—Guidelines for turn signal lamps, when tested in accordance with 4.1.5 of this document, are contained in Table 3 and its footnotes. Depending on the spacing of the front turn signal relative to the forward illumination lamps as defined in 5.1.5 of this document, the multipliers specified in Table 2 are applicable to the values to Table 3.

6.2 Installation Guidelines—The following guidelines apply to front and/or rear signal lamps as used on the vehicle and shall not be considered part of the requirements.

6.2.1 Signal lamps on the front and rear of the vehicle should be spaced as far apart laterally as practicable, so that the direction of turn will be clearly understood.

6.2.2 The luminous intensity of incandescent filament bulbs will vary with applied voltage. The electrical power system of the vehicle should, under normal running conditions, provide design voltage to the lamp as closely as practical bearing in mind the inherent variability of such systems.

6.2.3 Performance of lamps may deteriorate significantly as a result of dirt, grime, and/or snow accumulation on the optical surfaces. Installation of lamps on vehicles should be considered to minimize the effect of these factors.

6.2.4 Where it is expected that lamps must perform in severe environments, e.g., be totally immersed in water periodically, the user should specify lamps designed for such use.

7. Additional Information
—As a matter of additional information, attention is called to SAE J567 for requirements and gages to be used in socket design.

24.155

(R) TABLE 3—PHOTOMETRIC DESIGN GUIDELINES

Test Points (deg) Lighted Sections		Front Signals Yellow 1	Front Signals Yellow 2	Front Signals Yellow 3	Rear Signals Red 1	Rear Signals Red 2	Rear Signals Red 3	Rear Signals Yellow 1	Rear Signals Yellow 2	Rear Signals Yellow 3
					Minimum Luminous Intensity (cd)					
10U, 10D	5L, 5R	40	48	55	16	19	22	26	30	35
	20L, 20R	25	30	35	10	12	15	15	20	25
5U, 5D	10L, 10R	75	88	100	30	35	40	50	55	65
	V	175	205	235	70	82	95	110	130	150
	10L, 10R	100	120	140	40	47	55	65	75	90
H	5L, 5R	200	240	275	80	95	110	130	150	175
	V	200	240	275	80	95	110	130	150	175
Maximum Luminous Intensity (cd)										
Rear Lamps Only[1]		—	—	—	300	360	420	750	900	1050

[1] The maximum design value of a lamp intended for the rear of the vehicle should not exceed the listed design maximum over any area larger than that generated by 0.25 degree radius within the solid angle defined by the test points in Table 3.

(R) SUPPLEMENTAL HIGH MOUNTED STOP AND REAR TURN SIGNAL LAMPS FOR USE ON VEHICLES LESS THAN 2032 MM IN OVERALL WIDTH—SAE J186 DEC89

SAE Recommended Practice

Report of the Lighting Committee, approved July 1970, completely revised November 1982. Rationale statement available. Completely revised by the Lighting Coordinating Committee December 1989. Rationale statement available.

1. Scope—This document provides design parameters, performance requirements, and general installation recommendations for supplemental high mounted stop and/or rear turn signal lamps, intended to supplement stop and/or rear turn signal lamps described in SAE J586 and SAE J588, for use on vehicles less than 2032 mm in overall width. Lamps for vehicles more than 2032 mm in width are covered in SAE J1432.

2. Definitions

2.1 Supplemental high mounted stop and rear turn signal lamps are additional lamps that are mounted high and possibly forward of the rear mounted tail, stop, and turn signal lamps. The supplemental stop and/or turn signals may be provided by separate lamps or both functions may be combined in, and provided by, a single lamp.

2.2 Supplemental high mounted stop lamps are additional lamps of a stop lamp system giving a brake-actuated, steady warning light to the rear of the vehicle. They are intended to provide a signal to both the operator of the next following vehicle as well as, through intervening vehicles, to the operators of the other following vehicles.

2.3 Supplemental high mounted rear turn signal lamps are additional lamps of a turn signal system which indicate a change in direction by giving a flashing warning signal on the side toward which the vehicle operator intends to turn. They are intended to provide a signal of the next following vehicle as well as, through intervening vehicles, to the operators of the other following vehicles.

3. Lighting Identification Code—May be U in accordance with SAE J759.

4. Test

4.1 SAE J575 is a part of this document. The following tests are applicable with the modifications as indicated:

  4.1.1 VIBRATION TEST
  4.1.2 MOISTURE TEST
  4.1.3 DUST TEST
  4.1.4 CORROSION TEST
  4.1.5 PHOTOMETRIC TEST

4.1.5.1 Photometric tests shall be made with the photometer a distance of at least 3 m from the light source. The lamp axis shall be taken as the horizontal line through the light source and parallel to what would be the longitudinal axis of the vehicle if the lamp were mounted in its normal position on the vehicle.

TABLE 1—PHOTOMETRIC PERFORMANCE ZONAL REQUIREMENTS

Test Points (degrees)	Total for Zone (cd)	
	Supplemental High Mounted Stop and Red Rear Turn Signal	Supplemental High Mounted Yellow Rear Turn Signal
5U-V H-5L H-V H-5R 5D-V	67	108
5R-5U 10R-5U H-10R 10R-5D 5R-5D	54	86
5L-5U 10L-5U H-10L 10L-5D 5L-5D	54	86
10L-10U 10U-V 10R-10U	18	29
MAXIMUM	75	145

1. An adjustment in lamp orientation from design position may be made in determining conformance to Table 1, provided such adjustment does not exceed 3 deg. All zones shall comply after final re-aim.
2. The measured values at each test point shall not be less than 60% of the minimum requirements in Table 2.
3. The maximum value shall not be exceeded over an area larger than that generated by a 1/4 deg radius within a solid cone from 10L to 10R and from 10U to 5D.

4.1.6 WARPAGE TEST ON DEVICES WITH PLASTIC COMPONENTS

4.1.6.1 Cycle times for stop and turn signal lamps listed in Table 1 of SAE J575 shall be employed for supplemental high mounted stop lamps and supplemental high mounted rear turn signals respectively.

4.2 Color Test—SAE J578 is a part of this document.

5. Requirements

5.1 Performance Requirements—Supplemental high mounted stop and rear turn signal lamps, when tested in accordance with the following tests, with modifications indicated, shall meet the requirements indicated in SAE J575.

5.1.1 VIBRATION TEST

5.1.2 MOISTURE TEST

5.1.3 DUST TEST

5.1.4 CORROSION TEST

5.1.5 PHOTOMETRIC TEST

5.1.5.1 The lamp under test, when tested in accordance with 4.1, shall meet the photometric requirements contained in Table 1. The summation of the luminous intensity measurements, at the specified test points in a zone, shall be at least the value shown.

5.1.6 WARPAGE TEST

5.2 Color Test—The light from the supplemental high mounted stop lamps shall be red and the light from the supplemental high mounted rear turn signal lamps shall be red or yellow in conformance with SAE J578.

5.3 Material Requirements—Plastic materials used in the optical parts shall conform to the requirements in SAE J576.

5.4 Dimensional Requirements—The effective projected luminous area measured on a plane at right angles to the lamp axis shall not be less than 29 cm^2.

6. Guidelines

6.1 Photometric design guidelines for supplemental high mounted stop and turn signal lamps, when tested in accordance with 4.1.5, are contained in Table 2.

TABLE 2—PHOTOMETRIC DESIGN GUIDELINES

Test Points (Degrees)		Red (cd)	Yellow (cd)
10U	10L	5	8
	V	10	16
	10R	5	8
5U and 5D	10L	10	16
	5L	15	24
	V	15	24
	5R	15	24
	10R	10	16
H	10L	10	16
	5L	15	24
	V	15	24
	5R	15	24
	10R	10	16
MAXIMUM		60	120

1. The listed maximum design value shall not be exceeded over an area larger than that generated by a 1/4 deg radius within a solid cone from 10L to 10R and from 10U to 5D.

6.2 Visibility of the signal shall not be obstructed by any part of the vehicle from 10U to 5D and from 10L to 10R, unless the lamp conforms with the cell requirements when obstruction is considered.

6.3 Supplemental turn signals shall flash simultaneously (not alternately) with the required turn signals.

6.4 No function other than red reflex reflectors shall be combined in the supplemental high mounted stop and/or turn signal lamps.

7. Notes—As a matter of additional information, attention is called to SAE J567 for requirements and gages to be used in the bulb retention system (socket) design.

(R) FRONT AND REAR TURN SIGNAL LAMPS FOR USE ON MOTOR VEHICLES 2032 mm OR MORE IN OVERALL WIDTH—SAE J1395 JUN91

SAE Standard

Report of the Lighting Committee, approved April 1985. Rationale statement available. Completely revised by the SAE Lighting Coordinating Committee and the Heavy-Duty Lighting Standards Committee June 1991. Rationale statement available.

1. Scope—This SAE Standard provides test procedures, requirements, and guidelines for turn signal lamps intended for use on vehicles 2032 mm or more in overall width. Front and rear turn signal lamps conforming to the requirements of this document may be used on vehicles less than 2032 mm in overall width.

2. References

2.1 Applicable Documents—The following publications form a part of this specification to the extent specified herein. The latest issue of SAE publications shall apply.

2.1.1 SAE PUBLICATIONS—Available from SAE, 400 Commonwealth Drive, Warrendale, PA 15096-0001.

SAE J567—Lamp Bulb Retention System for Requirements and Gages Used in Retention System Design

SAE J588—Turn Signal Lamps for Use on Motor Vehicles Less Than 2032 mm in Overall Width

SAE J576—Plastic Material for Use in Optical Parts Such as Lenses and Reflectors of Motor Vehicle Lighting Devices

SAE J578—Color Specification

SAE J759—Lighting Identification Code

SAE J1050—Describing and Measuring the Driver's Field of View

SAE J2139—Tests for Lighting Devices, Reflective Devices and Components Used on Vehicles 2032 mm or More in Overall Width

SAE Technical Paper 830566, "Motor Vehicle Conspicuity," R.L. Henderson, K. Ziedman, W.J. Burger, and K.E. Cavey, National Highway Traffic Safety Administration

2.1.2 OTHER PUBLICATIONS—Attention is called to the following documents for additional information on lamp design and installation.

FMVSS 108

FHWA 393 Subpart B

TTMA #RP-9

TMC #RP-702

2.2 Definitions

2.2.1 A turn signal lamp is the signaling element of a turn signal system which indicates a change in direction by giving a flashing light on the side toward which the turn or lane change will be made. See SAE J590 for flash rate and percent on time.

3. Lighting Identification Code—Turn signal lamps for use on vehicles 2032 mm or more in overall width may be identified by the code:

"I 6" for a rear mounted turn signal lamp and for a front mounted turn signal lamp mounted 100 mm or more from the headlamp,

"I 7" for a front mounted turn signal lamp mounted less than 100 mm from the headlamp, in accordance with SAE J759.

4. Tests

4.1 SAE J2139 is a part of this document. The following tests are applicable with modification as indicated.

4.1.1 VIBRATION

4.1.2 MOISTURE

4.1.3 DUST

4.1.4 CORROSION

4.1.5 PHOTOMETRY

4.1.5.1 Photometric measurements shall be made with the light source of the device at least 3 m from the photometer.

4.1.5.2 The H-V axis of the device shall be taken to be parallel to the longitudinal axis of the vehicle, when the device is mounted in its design position.

4.1.5.3 Photometric measurements shall be made with the light source steadily burning. Photometric measurements of multiple compartment lamps or multiple lamp arrangements shall be made by either of the following methods by aligning the axis of each lamp or compartment with the photometer.

4.1.5.3.1 All compartments or lamps shall be photometered together provided that a line from the light source of each compartment or lamp to the center of the photometer sensing device does not make an angle of more than 0.6 degrees with the photometer H-V axis. When compartments or lamps are photometered together, the H-V axis shall intersect the midpoint between their light sources.

4.1.5.3.2 Each compartment or lamp shall be photometered separately. The photometric measurement for the entire multiple compartment lamp or multiple lamp arrangement shall be determined by adding the photometric outputs from each individual lamp or component at corresponding test points.

4.1.6 WARPAGE TEST ON DEVICES WITH PLASTIC COMPONENTS

4.2 Color—SAE J578 is a part of this document.

4.3 Plastic Materials—SAE J576 is a part of this document.

5. *Requirements*

5.1 Performance Requirements—The device when tested in accordance with the test procedures of this document shall meet the requirements of SAE J2139 or as indicated.

5.1.1 VIBRATION

5.1.2 MOISTURE

5.1.3 DUST

5.1.4 CORROSION

5.1.5 PHOTOMETRY—The device tested shall meet the photometric performance requirements of Table 1 and its footnotes.

The summation of the luminous intensity measurements at the specified test points in a zone shall be at least the value shown.

5.1.5.1 A multiple compartment lamp or multiple lamps may be used to meet the photometric requirements of a turn signal lamp. If multiple compartments or multiple lamps are used and the distance between adjacent light sources does not exceed 560 mm for two compartments or lamp arrangements and does not exceed 410 mm for three compartments or lamp arrangements, then the combination of the compartments or lamps must be used to meet the photometric requirements of Table 1. If the distance between adjacent light sources exceeds the above dimensions, each compartment or lamp shall comply with the photometric requirements of Table 1.

5.1.5.2 When a tail lamp, clearance lamp, or a parking lamp is combined with the turn signal lamp, the turn signal lamp intensity shall be not less than three times the luminous intensity of the tail lamp, clearance lamp, or a parking lamp at any test point, except that at H-V, H-5L, H-5R, and 5U-V, the turn signal lamp intensity shall be not less than five times the luminous intensity of the tail lamp, clearance lamp, or parking lamp.

When a tail lamp or a clearance lamp is combined with the turn signal lamp and the maximum intensity of the tail lamp or clearance lamp is located below the horizontal and is within an area generated by a 1.0 degree radius around the test point, the ratio for the test point may be computed using the lowest value of the tail lamp or clearance lamp luminous intensity within the generated area.

5.1.5.3 Rear signals from a forward mounted double-faced turn signal lamp need only meet the performance requirements contained in Table 1 from directly to the rear to the left for left-hand lamp, and from directly to the rear to the right for a right-hand lamp. The intent is to permit the manufacturer to provide glare protection for the driver.

5.1.5.4 When a front turn signal lamp is mounted less than 100 mm from the low beam headlamp, the turn signal lamp luminous intensity shall be not less than 2.5 times the values specified in Table 1 for a front turn signal lamp.

5.1.5.5 Spacing for a direct light source type design front turn signal lamp, that is, a lamp primarily employing a lens to meet photometric requirements (for example, a lamp that does not employ a reflector), shall be measured from the center of the light source to the closest lighted edge of the low beam headlamp or any additional lamp used to supplement or used in lieu of the low beam, such as an auxiliary low beam or fog lamp.

5.1.5.6 Spacing for a front turn signal lamp which primarily employs a reflector (for example, a parabolic section) in conjunction with a lens to meet photometric requirements shall be measured from the geometric centroid of the front turn signal effective projected luminous lighted lens area to the closest lighted edge of the low beam headlamp or any additional lamp used to supplement or used in lieu of the low beam, such as an auxiliary low beam or fog lamp.

5.1.6 WARPAGE

5.2 Color—The color of the light from the front turn signal lamp shall be yellow and the color of the light from the rear turn signal lamp may be red or yellow as specified in SAE J578.

5.3 Plastic Materials—The plastic materials used in the optical parts shall meet the requirements of SAE J576.

TABLE 1—TURN SIGNAL LAMP PHOTOMETRIC PERFORMANCE REQUIREMENTS

Zone	Test Point Deg.	Zone Total Luminous Intensity, Candela, Yellow Front	Zone Total Luminous Intensity, Candela, Red Rear	Zone Total Luminous Intensity, Candela, Yellow Rear
1	10U— 5L 5U—20L 5D—20L 10D— 5L	130.0	50.0	84.0
2	5U—10L H—10L 5D—10L	250.0	100.0	165.0
3	5U— V H— 5L H— V H— 5R 5D— V	950.0	380.0	610.0
4	5U—10R H—10R 5D—10R	250.0	100.0	165.0
5	10U— 5R 5U—20R 5D—20R 10D— 5R	130.0	50.0	84.0
Maximum Luminous Intensity, Candela		—	300.0	750.0

[1] The maximum luminous intensity shall not be exceeded over any area larger than that generated by a 0.5 degree radius within the area defined by the test point pattern of Table 2.

[2] Unless otherwise specified, the lamp shall be considered to have failed the photometric requirements of this document if the luminous intensity at any test point is less than 60% of the values specified in Table 2.

[3] Unless otherwise specified, the lamp shall be considered to have failed the photometric requirements of this document if the minimum luminous intensity between test points is less than 60% of the lower design values of Table 2 for the closest adjacent test points on a horizontal and vertical line as defined by the test point pattern.

[4] The summation of the luminous intensity measurements at the specified test points in the zone shall be at least the values shown.

[5] When a tail lamp, clearance lamp, or a parking lamp is combined with a turn signal lamp, see 5.1.5.2 of this document for luminous intensity ratio requirements.

[6] Throughout the photometric pattern defined by the corner points of 20U-45L, 20U-45R, 10D-45R, and 10D-45L, the light intensity shall be not less than 0.4 candela in red or 1.0 candela in yellow for the rear turn signal lamp and 1.25 candela in yellow for the front turn signal lamp.

5.4 Design Requirements

5.4.1 If a turn signal lamp is combined with a tail lamp, a clearance lamp, or a parking lamp, and a replaceable multiple light source is used, the light source retention system shall be designed with an indexing means so that the light source is properly indexed. Removable light source retention systems shall have an indexing feature so that they cannot be reinserted into the lamp housing in a random position, unless the lamp will perform its intended function with random light source orientation.

5.4.2 The effective projected luminous lighted area of a single lamp shall be at least 75 cm^2.

5.4.3 If a multiple compartment lamp or multiple lamps are used to meet the photometric requirements, the effective projected luminous lighted area of each compartment or lamp shall be at least 40 cm^2 provided the combined area is at least 75 cm^2.

5.5 Installation Requirements—The turn signal lamp shall meet the following requirements as installed on the vehicle.

5.5.1 The turn signal lamps, facing rearward for the rear lamp and facing forward for the front lamp, shall be mounted on the permanent structure of the vehicle, at the same height, and spaced as far apart laterally as practicable, so that the signal will be clearly visible.

5.5.2 Visibility of each turn signal lamp shall not be obstructed by any part of the vehicle throughout the photometric test pattern unless the lamp is designed to comply with all photometric and visibility requirements with these obstructions considered.

To be considered visible, the lamp must provide an unobstructed view of a portion of the lighted outer lens surface, excluding reflex reflector area, having an area of at least 13 cm^2 measured at a horizontal angle of 45 degrees to the left and 45 degrees to the right of the longitudinal axis of the vehicle and a vertical angle from 20 degrees up to 10 degrees down.

See Table 1, Note 6. When more than one lamp or optical area is lighted on each side of the vehicle, only one such area on each side need comply.

5.5.3 TURN SIGNAL PILOT INDICATOR

5.5.3.1 If one right and one left turn signal lamp are not readily visible to the driver, there shall be an illuminated indicator provided to give a clear and unmistakable indication that the turn signal system is activated. The illuminated indicator shall consist of one or more lights flashing at the same frequency as the turn signal lamps.

5.5.3.2 If the illuminated indicator is located inside the vehicle, it shall emit a green colored light and have a minimum functional lighted area of 18 mm^2.

5.5.3.3 If the illuminated indicators are located on the outside of the vehicle, they shall emit a yellow colored light and have a minimum functional lighted area of 60 mm^2.

5.5.3.4 The minimum required illuminated lighted area of the indicators shall be visible according to the procedures described in SAE J1050.

The steering wheel shall be turned to a straight-ahead driving position and in the design location for an adjustable wheel or column.

6. Guidelines

6.1 Design Guidelines

6.1.1 Photometric design guidelines are contained in Table 2 and its footnotes.

6.2 Installation Guidelines—The following guidelines apply to turn signal lamps as used on the vehicle and shall not be considered part of the requirements.

6.2.1 Performance of lamps may deteriorate significantly as a result of dirt, grime, snow, and ice accumulation on the optical surfaces. Installation of the device on the vehicle should be considered to minimize the effects of these factors.

6.2.2 Where it is expected that the device must perform in extremely severe environments, or where it is expected to be totally immersed in water, the user should specify devices specifically designed for such use.

7. Advance Requirements—This section of the document gives advance notice to manufacturers and users of the device of a pending change in the requirements for a front and rear turn signal lamp. The change in requirements shall be effective on devices marketed and used on or after January 1, 1996.

See Tables 3 and 4 and the footnotes.

TABLE 2—TURN SIGNAL LAMP PHOTOMETRIC DESIGN GUIDELINES

Test Point Deg.	Luminous Intensity, Candela, Yellow Front	Luminous Intensity, Candela, Red Rear	Luminous Intensity, Candela, Yellow Rear
10U— 5L	40.0	16.0	27.0
5R	40.0	16.0	27.0
5U—20L	25.0	10.0	15.0
10L	75.0	30.0	50.0
V	175.0	70.0	110.0
10R	75.0	30.0	50.0
20R	25.0	10.0	15.0
H—10L	100.0	40.0	65.0
5L	200.0	80.0	130.0
V	200.0	80.0	130.0
5R	200.0	80.0	130.0
10R	100.0	40.0	65.0
5D—20L	25.0	10.0	15.0
10L	75.0	30.0	50.0
V	175.0	70.0	110.0
10R	75.0	30.0	50.0
20R	25.0	10.0	15.0
10D— 5L	40.0	16.0	27.0
5R	40.0	16.0	27.0
Maximum Luminous Intensity, Candela	—	300.0	750.0

[1] The maximum luminous intensity shall not be exceeded over any area larger than that generated by a 0.25 degree radius within the area defined by the test point pattern of Table 2.

[2] When a tail lamp, clearance lamp, or a parking lamp is combined with a turn signal lamp, see 5.1.5.2 of this document for luminous intensity ratio requirements.

[3] Throughout the photometric pattern defined by the corner points of 20U-45L, 20U-45R, 10D-45R, and 10D-45L, the light intensity shall be not less than 0.4 candela in red or 1.0 candela in yellow for the rear turn signal lamps or 1.25 candela in yellow for the front turn signal lamp.

TABLE 3—TURN SIGNAL LAMP PHOTOMETRIC PERFORMANCE REQUIREMENTS—ADVANCE REQUIREMENTS

Zone	Test Point Deg.	Zone Total Luminous Intensity, Candela, Yellow Front	Zone Total Luminous Intensity, Candela, Red Rear	Zone Total Luminous Intensity, Candela, Yellow Rear
1	20U—45L 20U—20L 10D—20L 10D—45L	30.0	12.0	20.0
2	10U— 5L 5U—20L 5D—20L 10D— 5L	130.0	50.0	84.0
3	5U—10L H—10L 5D—10L	250.0	100.0	165.0
4	5U— V H— 5L H— V H— 5R 5D— V	950.0	380.0	610.0
5	5U—10R H—10R 5D—10R	250.0	100.0	165.0
6	10U— 5R 5U—20R 5D—20R 10D— 5R	130.0	50.0	84.0
7	20U—20R 20U—45R 10D—45R 10D—20R	30.0	12.0	20.0
Maximum Luminous Intensity, Candela		—	300.0	750.0

[1] The maximum luminous intensity shall not be exceeded over any area larger than that generated by a 0.5 degree radius within the area defined by the test point pattern of Table 4.

[2] Unless otherwise specified, the lamp shall be considered to have failed the photometric requirements of this document if the luminous intensity at any test point is less than 60% of the values specified in Table 4.

[3] Unless otherwise specified, the lamp shall be considered to have failed the photometric requirements of this document if the minimum luminous intensity between test points is less than 60% of the lower design values of Table 4 for the closest adjacent test points on a horizontal and vertical line as defined by the test point pattern.

[4] The summation of the luminous intensity measurements at the specified test points in the zone shall be at least the values shown.

[5] When a tail lamp, clearance lamp, or parking lamp is combined with a turn signal lamp, see 5.1.5.2 of this document for luminous intensity ratio requirements.

TABLE 4—TURN SIGNAL LAMP PHOTOMETRIC DESIGN GUIDELINES—ADVANCE REQUIREMENTS

Test Point Deg.	Luminous Intensity, Candela, Yellow Front	Luminous Intensity, Candela, Red Rear	Luminous Intensity, Candela, Yellow Rear
20U—45L	2.5	1.0	2.0
—20L	12.5	5.0	8.0
—20R	12.5	5.0	8.0
—45R	2.5	1.0	2.0
10U— 5L	40.0	16.0	27.0
5R	40.0	16.0	27.0
5U—20L	25.0	10.0	15.0
10L	75.0	30.0	50.0
V	175.0	70.0	110.0
10R	75.0	30.0	50.0
20R	25.0	10.0	15.0
H—10L	100.0	40.0	65.0
5L	200.0	80.0	130.0
V	200.0	80.0	130.0
5R	200.0	80.0	130.0
10R	100.0	40.0	65.0
5D—20L	25.0	10.0	15.0
10L	75.0	30.0	50.0
V	175.0	70.0	110.0
10R	75.0	30.0	50.0
20R	25.0	10.0	15.0
10D—45L	2.5	1.0	2.0
20L	12.5	5.0	8.0
5L	40.0	16.0	27.0
5R	40.0	16.0	27.0
20R	12.5	5.0	8.0
45R	2.5	1.0	2.0
Maximum Luminous Intensity, Candela	—	300.0	750.0

[1] The maximum luminous intensity shall not be exceeded over any area larger than that generated by a 0.25 degree radius within the area defined by the test point pattern of Table 4.

[2] When a tail lamp, clearance lamp, or a parking lamp is combined with a turn signal lamp, see 5.1.5.2 of this document for luminous intensity ratio requirements.

TURN SIGNAL SWITCH—SAE J589b

SAE Standard

Report of Lighting Committee approved September 1950 and last revised June 1971. Editorial change October 1977.

1. Definition

1.1 A turn signal switch is that part of a turn signal system by which the operator of a vehicle causes the turn signal lamps to function.

1.2 A *Class A* turn signal switch may be used on any vehicle but is intended for use on multipurpose passenger vehicles, trucks, and buses that are 80 in. or more wide overall.

1.3 A *Class B* turn signal switch is intended for use in passenger cars, motorcycles, and multipurpose passenger vehicles, trucks, and buses of less than 80 in. overall width.

2. Reference Standards

2.1 The following sections from SAE J575f (April, 1975) are a part of this standard:

Section B—Samples for Test
Section C—Lamp Bulbs
Section D—Laboratory Facilities

2.2 Turn signal pilot indicators—See SAE J588e (September, 1970).

3. *Temperature Test*

3.1 To insure basic function, the switch shall be manually cycled for 10 cycles at design electrical load at: 75 ± 10 F (24 ± 5.5 C); 165 +0, −5 F (74 +0, −2.8 C); −25 +5, −0 F (−32 +2.8, −0 C). This to be done after a 1 h exposure at each of these temperatures. The switch shall be electrically and mechanically operable during each of these cycles.

3.2 This same switch shall be used for the endurance test described in paragraph 4.

4. *Endurance Test Setup*

4.1 The switch shall be operated with the maximum design bulb load stated by the switch manufacturer with the flasher not included in the circuit. Failed bulbs shall be replaced during the test.

4.2 When the switch is provided with a self-canceling mechanism, the test equipment shall be arranged so that the switch can be turned off by the self-canceling mechanism. Provision shall also be made for manual canceling.

4.3 The test shall be set up to operate the switch for the prescribed number of cycles.

One cycle shall consist of the following sequence of positions: off, left turn, off, right turn, off.

The test requirement shall function within the following mechanical timing requirements at a cycle rate of 12–20 cycles/minute:

Travel time—0.1–0.5 s max (time from one position to the next position)
Dwell time—0.4 s min (in each position)

4.4 During the test the switch shall be operated at 6.4 V d-c for a 6 V system, 12.8 V d-c for a 12 V system, or 25.6 V d-c for a 24 V system, measured at the input termination of the switch. The power supply shall not generate any adverse transients not present in motor vehicles and shall comply with the following specifications:

(a) Output current—Capable of supplying a continuous output current of the design load and inrush currents as required by the bulb load complement.

(b) Regulation—
Dynamic—The output voltage shall not deviate more than 1.0 V from zero to maximum load (including inrush current) and should recover 63% of its maximum excursion within 5 ms.
Static—The output voltage shall not deviate more than 2% with changes in static load from zero to maximum (not including inrush current), and means shall be provided to compensate for static input line voltage variations.

(c) Ripple voltage—Maximum 300 mV, peak to peak.

5. *Endurance Requirements*

5.1 Class A turn signal switches shall be capable of meeting the following endurance requirements:

(a) 165,000 cycles at 75 ± 10 F (24 ± 5.5 C).

(b) When the switch is provided with a self-canceling mechanism it shall be tested as follows: 155,000 cycles of self-canceling followed by 10,000 cycles of manual canceling.

(c) If the turn signal switch includes stop lamp circuitry, the stop lamp circuit shall be fed electrically for the first 100,000 cycles only.

5.2 Class B turn signal switches shall be capable of meeting the following endurance requirements:

(a) 100,000 cycles at 75 ± 10 F (24 ± 5.5 C).

(b) When the switch is provided with a self-canceling mechanism it shall be tested as follows: 95,000 cycles of self-canceling followed by 5000 complete cycles of manual canceling.

(c) If the turn signal switch includes stop lamp circuitry, the stop lamp circuit shall be fed electrically for the first 50,000 complete cycles only.

5.3 If the turn signal switch includes cornering light circuitry which is fed from the headlight switch, the cornering light circuit shall be fed electrically for the first 50,000 cycles only.

5.4 The voltage drop from the input terminal of each circuit to the lamp terminal of each circuit shall be measured at the beginning of the test and at intervals of 25,000 cycles.

This voltage drop shall not exceed:

0.25 V for 2 lamp load (or less) per side
0.30 V for 3 lamp load per side
0.35 V for 4 lamp load per side
0.40 V for 5 lamp load (or greater) per side

before, during, and after the endurance test.

If wiring is an integral part of the switch, the voltage drop measurement is to be made including 3 in. of wire on each side of switch; otherwise, measurement is to be made at switch terminals. Care shall be taken not to include the voltage drop of other devices in the circuit.

6. *Combination Turn Signal and Hazard Warning Signal Switches*

6.1 The same combination switch shall be used for the test of each function. The turn signal switch function shall meet the requirements of this standard. The hazard warning signal switch function shall meet the requirements of SAE J910b (June, 1971).

6.2 The operating motion of the hazard warning signal switch function shall differ from the actuating motion of the turn signal switch function.

φSIDE TURN SIGNAL LAMPS—SAE J914 NOV87 — SAE Standard

Report of Lighting Committee approved February 1965 and completely revised November 1987. Rationale statement available.

1. *Scope*—This SAE Standard provides installation requirements, test procedures, design guidelines, and performance requirements for side turn signal lamps.

2. *Definitions*—A side turn signal is a lighting device normally mounted on the side of a vehicle at or near the front, and used as part of the turn signal system to indicate a change in direction by means of a flashing warning signal on the side toward which the vehicle operator intends to turn or maneuver.

Note: Side turn signals, when used, are supplemental to, and should not be confused with turn signals described in SAE J588, which, in some cases, may be mounted on the side of the vehicle.

3. *Lighting Identification Code*—Turn signal lamps for use on vehicles 9.1 m or more in length and for vehicles less than 9.1 m in length may be identified respectively by the codes "E" and "E2" in accordance with SAE J759—Lighting Identification Code.

4. *Tests*

4.1 SAE J575—Tests for Motor Vehicle Lighting Devices and Components is a part of this report. The following tests are applicable with the modifications as indicated:

4.1.1 VIBRATION TEST
4.1.2 MOISTURE TEST
4.1.3 DUST TEST
4.1.4 CORROSION TEST
4.1.5 PHOTOMETRIC TEST—In addition to the test procedures in SAE J575, the following apply:

4.1.5.1 Photometric tests shall be made with the photometer at a distance of at least 3 m from the lamp. The H-V axis shall be taken as the horizontal line through the light source and normal to the longitudinal axis of the vehicle.

4.1.5.2 Photometric measurements shall be made with the bulb filament steadily burning.

4.1.6 WARPAGE TEST ON DEVICES WITH PLASTIC COMPONENTS—The device shall be operated with the bulb flashing steadily during the test period. The flash rate shall be in accordance with J590.

4.2 *Color Test*—SAE J578—Color Specification for Electric Signal Lighting Devices is part of this report.

5. *Requirements*

5.1 *Performance Requirements*—A device when tested in accordance with the test procedures specified in Section 4 shall meet the following requirements:

5.1.1 VIBRATION TEST—SAE J575
5.1.2 MOISTURE TEST—SAE J575
5.1.3 DUST TEST—SAE J575
5.1.4 CORROSION TEST—SAE J575
5.1.5 PHOTOMETRIC TEST—SAE J575—The photometric performance requirements are shown in Table 1. A single lamp mounted on a test stand to simulate mounting on the vehicle shall meet the photometric performance requirements.

5.1.5.1 When a sidemarker lamp is combined with a side turn signal lamp, the signal lamp shall not be less than three times the luminous intensity of the side marker lamp at any test point, except a right hand side turn signal lamp shall not be less than five times the luminous intensity of the sidemarker lamp at 5U-30R, 5U-70R, H-30R and H-70R, and the left hand side turn signal shall not be less than five times the luminous intensity of the sidemarker lamp at 5U-30L, 5U-70L, H-30L and H-70L.

5.1.5.2 When the sidemarker lamp is combined with the side turn signal lamp, and the maximum luminous intensity of the sidemarker lamp is within an area generated by a 0.5 deg radius around a test point, the ratio for the test point may be computed by using the lowest value of the sidemarker luminous intensity within the generated area.

5.1.6 WARPAGE—SAE J575

5.1.7 COLOR TEST—The color of the light from a side turn signal lamp shall be yellow, in accordance with SAE J578.

5.2 Material Requirements

5.2.1 Plastic material used in optical parts shall meet the requirements of SAE J576—Plastic Materials for Use in Optical Parts such as Lenses and Reflectors of Motor Vehicle Lighting Devices.

5.3 Installation Requirements

5.3.1 Visibility and photometric performance of the side turn signal lamp within the test angles shown in Tables 1 and 2 shall not be obstructed by any portion of the vehicle unless the lamp is designed to comply with all requirements when the obstruction is considered.

5.3.2 Side turn signal lamps shall flash simultaneously or alternately with the required front turn signal lamps.

5.3.3 MOUNTING HEIGHT

5.3.3.1 Side turn signal lamps shall be mounted on vehicles with a length of 9.1 m or more at a height of no more than 1830 mm and no less than 810 mm.

5.3.3.2 Side turn signal lamps shall be mounted on vehicles with a length of less than 9.1 m at a height of no more than 1220 mm and no less than 685 mm.

6. Guidelines

6.1 Photometric Design Guidelines for the side turn signal lamps, when tested in accordance with paragraph 4.1.5 of this report, are contained in Table 2—Photometric Design Guidelines and its footnotes.

6.2 **Installation Guidelines**—These guidelines apply to the device as used on the vehicle and are not a part of the design requirements, performance requirements, or test procedures.

6.2.1 Side turn signal lamps should be located as close to the front of the vehicle as practicable.

6.2.2 The luminous intensity of incandescent filament bulbs will vary with applied voltage. The electrical wiring in the vehicle should be adequate to supply design voltage to the lamp filament.

6.2.3 Performance of lamps may deteriorate significantly as a result of dirt, grime, and/or snow accumulation on their optical surfaces. Installation of lamps on vehicles should be considered to minimize the effect of these factors.

6.2.4 Where it is expected that lamps must perform in extremely severe environments, such as off-highway, mining, fuel haulage, or where it is expected that they will be totally immersed in water, the user should specify lamps specifically designed for such use.

TABLE 1—PHOTOMETRIC PERFORMANCE REQUIREMENTS[a,b]

Trucks, Buses and Multipurpose Vehicles 9.1 m or More in Length		Passenger Cars, Multipurpose Vehicles, Trucks and Buses Less Than 9.1 m in Length	
Position (deg)	Zone Total (Candela)[c]	Position (deg)	Zone Total (Candela)[d]
15U-V H-V 15D-V	100	15U-30L 5U-30L H-30L 5D-30L	40
10U-30L 5U-70L H-70L 5D-70L 10D-30L	130	15U-70L 5U-70L H-70L 5D-70L	40
5U-85L H-85L 5D-85L	130		

NOTES:
[a] Angles shown for lamps mounted on left hand side of vehicle. For lamps mounted on right hand side of vehicle, substitute right hand angles.
[b] The measured value for any test point shall not be less than 60% of the minimum value for that test point specified in Table 2.
[c] 300 Cd max at all points.
[d] 60 Cd max at all points.

TABLE 2—PHOTOMETRIC DESIGN GUIDELINES[a]

Trucks, Buses and Multipurpose Vehicles 9.1 m or More in Length		Passenger Cars, Multipurpose Vehicles, Trucks and Buses Less Than 9.1 m in Length	
Position (deg)	Candela[b] (min)	Position (deg)	Candela[c] (min)
15U-V	20	15U-30L	5
10U-30L	20	15U-70L	5
5U-85L	40		
5U-70L	30	5U-30L	15
		5U-70L	15
H-85L	50		
H-70L	30	H-30L	15
H-30L	30	H-70L	15
H-V	30		
		5D-30L	5
5D-85L	40	5D-70L	5
5D-70L	30		
10D-30L	20		
15D-V	20		

[a] Angles shown for lamps mounted on left hand side of vehicle. For lamps mounted on right hand side of vehicle, substitute right hand angles.
[b] 300 Cd max at all points.
[c] 60 Cd max at all points.

APPENDIX

As a matter of information, attention is called to SAE J567, for requirements and gauges to be used in socket design.

φFRONT CORNERING LAMPS FOR USE ON MOTOR VEHICLES—SAE J852 NOV87

SAE Recommended Practice

Report of Lighting Committee approved April 1963 and completely revised November 1987. Rationale statement available.

1. Scope—This SAE Recommended Practice provides test procedures, requirements, and guidelines for front cornering lamps that are mounted on the exterior of a vehicle.

2. Definition—Front cornering lamps are steady burning lamps used in conjunction with the turn signal system to supplement the headlamps by providing additional illumination in the direction of turn.

3. Lighting Identification Code—Front cornering lamps meeting the performance requirements of Section 5 of this recommended practice may be identified by the code K in accordance with SAE J759, Lighting Identification Code.

4. Tests

4.1 SAE J575, Tests for Motor Vehicle Lighting Devices and Components, is a part of this report. The following tests are applicable with the modifications as indicated:

4.1.1 VIBRATION TEST
4.1.2 MOISTURE TEST
4.1.3 DUST TEST
4.1.4 CORROSION TEST
4.1.5 PHOTOMETRY—In addition to the test procedures in SAE J575, the following apply:

4.1.5.1 Photometric measurements shall be made with the light source of the lamp at least 3 m from the photometer. The H-V axis shall be taken as the horizontal line through the light source and perpendicular to the longitudinal axis of the vehicle.

4.1.6 WARPAGE TEST ON DEVICE WITH PLASTIC COMPONENTS

4.2 Color Test—SAE J578, Color Specifications for Electric Signal Lighting Devices, is a part of this report.

5. Requirements

5.1 Performance Requirements—A device when tested in accordance with the test procedures specified in Section 4 shall meet the following requirements:

5.1.1 VIBRATION—SAE J575
5.1.2 MOISTURE—SAE J575
5.1.3 DUST—SAE J575
5.1.4 CORROSION—SAE J575
5.1.5 PHOTOMETRY—SAE J575—The lamp under test shall meet the photometric requirements contained in Table 1—Photometric Requirements. Test points shown are for a lamp mounted on the left side of the vehicle - left hand angles should be substituted for right hand angles for a lamp mounted on the right side of the vehicle.

TABLE 1—PHOTOMETRIC REQUIREMENTS

TEST POSITION	CANDLEPOWER
8°U to 90°U— 90°L to 90°R	150 max
4°U— 90°L to 90°R	240 max
2°U— 90°L to 90°R	360 max
1°U— 90°L to 90°R	480 max
H— 90°L to 90°R	600 max
2.5°D −30°R	240 min
2.5°D −45°R	400 min
2.5°D −60°R	240 min

5.1.6 WARPAGE—SAE J575
5.1.7 COLOR—SAE J578—The color of the light from a front cornering lamp shall be white or amber, as specified in SAE J578.

5.2 Material Requirements—Plastic materials used in optical parts shall meet the requirements of SAE J576, Plastic Material for Use in Optical Parts Such as Lenses and Reflectors of Motor Vehicle Lighting Devices.

6. Guidelines

6.1 Photometric Design Guidelines for front cornering lamps, when tested in accordance with paragraph 4.1.5 of this recommended practice, are contained in Table 2—Photometric Design Guidelines. Test points shown are for a lamp mounted on the left side of the vehicle - left hand angles should be substituted for right hand angles for a lamp mounted on the right side of the vehicle.

6.2 Operating Guidelines—The following guidelines apply to front cornering lamps as used on the vehicle and shall not be considered part of the requirements:

6.2.1 The front cornering lamps are primarily intended to be used during the times that headlamps are required.

6.2.2 Means should be provided to turn on the front cornering lamps with the turn signal lamps and they should turn off when the turn signal lamps are turned off. If the front cornering lamps are not turned off automatically, a visual or audible means should be provided to indicate to the driver when the lamps are on.

TABLE 2—PHOTOMETRIC DESIGN GUIDELINES

TEST POSITION	CANDLEPOWER
8°U to 90°U— 90°L 90°R	125 max
4°U— 90°L to 90°R	200 max
2°U— 90°L to 90°R	300 max
1°U— 90°L to 90°R	400 max
H— 90°L to 90°R	500 max
2.5°D −30°R	300 min
2.5°D −45°R	500 min
2.5°D −60°R	300 min

7. Appendix—As a matter of additional information, attention is called to SAE J567, Lamp Bulb Retention System, for requirements and gages used in socket design.

φREAR CORNERING LAMP—
SAE J1373 OCT87

SAE Recommended Practice

Report of the Lighting Committee, approved June 1982 and completely revised October 1987. Rationale statement available.

1. Scope—This SAE Recommended Practice provides test procedures, requirements, and guidelines for rear cornering lamps for use on vehicles less than 9.1 m in overall length.

2. Definitions

2.1 Rear Cornering Lamps—Supplemental lamps used to illuminate an area to the side and rearward of the vehicle to provide light when the vehicle is backing up.

3. Lighting Identification Code—Rear cornering lamps may be identified by the code "K2" in accordance with SAE J759, Lighting Identification Code.

4. Tests

4.1 SAE J575, Tests for Motor Vehicle Lighting Devices and Components, is a part of this recommended practice. The following tests are applicable with the modifications as indicated:

4.1.1 VIBRATION TEST
4.1.2 MOISTURE TEST
4.1.3 DUST TEST
4.1.4 CORROSION TEST
4.1.5 PHOTOMETRY TEST—In addition to the test procedure in SAE J575, the following apply:

4.1.5.1 Photometric measurements shall be made with the light source of the lamp at least 3 m from the photometer. The H-V axis shall be taken as parallel to the longitudinal axis of the vehicle.

4.1.6 WARPAGE TEST FOR DEVICES WITH PLASTIC COMPONENTS

4.2 Color Test—SAE J578, Color Specification for Electric Signal Lighting Devices, is a part of this recommended practice.

5. Requirements

5.1 Performance Requirements—A device, when tested in accordance with the test procedures specified in Section 4, shall meet the following requirements:

5.1.1 VIBRATION—SAE J575.
5.1.2 MOISTURE—SAE J575.
5.1.3 DUST—SAE J575.
5.1.4 CORROSION—SAE J575.
5.1.5 PHOTOMETRY—SAE J575.

5.1.5.1 The lamp under test shall meet the performance requirements contained in Table 1. (Test points shown are for a lamp mounted on the left side of the vehicle—right hand angles should be substituted for left angles for a lamp mounted on the right side of the vehicle.)

5.1.6 WARPAGE—SAE J575.

5.1.7 COLOR—The color of the light from a rear cornering lamp shall be white as specified in SAE J578.

The lamp may project incidental red, yellow, or white light through reflectors or lenses that are adjacent, close to, or a part of the lamp assembly. If a lamp has portions of its lens which project nonwhite light, that light shall be regarded as incidental if, quantitatively, it does not exceed 20% of the total device output at all specified test points; the lamp shall also meet the photometric requirements of this recommended practice with white light alone and the lamp shall meet the color requirements with the total output measured, including the incidental light.

5.2 Material Requirements—Plastic materials used in optical parts shall meet the requirements of SAE J576, Plastic Materials for Use in Optical Parts Such as Lenses and Reflectors of Motor Vehicle Lighting Devices.

6. Guidelines

6.1 Photometric Design Guidelines for rear cornering lamps, when tested in accordance with paragraph 4.1.5 of this recommended practice, are contained in Table 2. (Test points shown are for a lamp mounted on the left side of the vehicle—right hand angles should be substituted for left angles for a lamp mounted on the right side of the vehicle.)

6.2 Installation Guidelines—The following guidelines apply to rear cornering lamps as used on the vehicle and shall not be considered part of the requirements:

6.2.1 Rear cornering lamps should be mounted on each side, near or on the rear of the vehicle. These lamps may be combined with other lamps on the vehicle providing each function of the combined lamp meets its respective requirements.

6.2.2 The rear cornering lamp should be illuminated only when the ignition switch is in the run position and reverse gear is engaged.

6.2.3 The luminous intensity of the light source will vary with applied voltage. The electrical wiring in the vehicle should be adequate to supply design voltage to the lamp filament.

6.2.4 Performance of lamps may deteriorate significantly as a result of dirt, grime, and/or snow accumulation on their optical surfaces. Installation of lamps on vehicles should be considered to minimize the effect of these factors.

6.2.5 Where it is expected that lamps must perform in extremely severe environments, such as in off-highway, mining, fuel haulage or where it is expected that they will be totally immersed in water, the user should specify lamps specifically designed for such use.

7. Appendix—As a matter of additional information, attention is called to SAE J567, Lamp Bulb Retention System, for requirements and gages used in socket design.

TABLE 1—PHOTOMETRIC REQUIREMENTS

Test Points (deg) Lighted Sections	Minimum Luminous Intensity (cd)
2-½ D - 30L	30
2-½ D - 45L	60
2-½ D - 60L	30
Max Luminous Intensity (cd) H and Above	600

NOTE: Test points shown in Table 1 are for a lamp mounted on the left hand side of the vehicle.

TABLE 2—PHOTOMETRIC DESIGN GUIDELINES

Test Points (deg) Lighted Sections	Minimum Luminous Intensity (cd)
2-½ D - 30L	40
2-½ D - 45L	80
2-½ D - 60L	40
Max Luminous Intensity (cd) H and Above	500

NOTE: Test points shown in Table 2 are for a lamp mounted on the left hand side of the vehicle.

TURN SIGNAL FLASHERS—SAE J590 APR93

SAE Standard

Report of the Lighting Committee approved March 1960 and completely revised July 1986. Rationale statement available. Revised by the SAE Flasher Task Force of the SAE Auxiliary Devices Standards Committee April 1993.

(R) *Foreword*—The development of this SAE Standard was based on the premise that it described the requirements for all turn signal flashers regardless of the electrical load. Since this document was first introduced, the predominant load for passenger car use has been two lamp bulbs, one at the front and one at the rear of the vehicle and the vehicles have used fixed load flashers. The load for trucks can vary from two to as many as ten depending on the vehicle. Trucks with trailers generally use variable load flashers. Passenger cars with trailers should use a variable load flasher. Federal Motor Vehicle Safety Standard 108 requires vehicles to be equipped to indicate the loss of one or more turn signal lamps except when a variable load flasher is required.

(R) 1. *Scope*—This SAE Standard defines the test conditions, procedures, and minimum design requirements for nominal 6, 12, and 24 V turn signal flashers.

2. *References*

   2.1 **Applicable Documents**—The following publications form a part of this specification to the extent specified herein. The latest issue of SAE publications shall apply.

   2.1.1 SAE PUBLICATIONS—Available from SAE, 400 Commonwealth Drive, Warrendale, PA 15096-0001.

   SAE J588—Turn Signal Lamps for Use on Motor Vehicles Less Than 2032 mm in Overall Width
   SAE J759—Lighting Identification Code
   SAE J823—Flasher Test

(R) 2.2 **Definition**—The flasher is a device installed in a vehicle lighting system which has the primary function of causing the turn signal lamps to flash when the turn signal switch is actuated. Secondary functions may include the visible pilot indication for the turn signal system (required by SAE J588) an audible signal to indicate when the flasher is operating, and an indication of turn signal lamp outage.

(R) 3. *Flasher Identification Code*—Flashers conforming to this document may be identified by the code J590 in accordance with SAE J759.

4. *Tests*

   4.1 **Test Equipment**—The standard test equipment and circuitry for performing flasher tests shall conform with the specifications in SAE J823.

   4.2 **Test Procedures**—All of the following tests shall be performed at 12.8 V (or 6.4 V and/or 25.6 V) at the bulbs unless otherwise specified.

(R) 4.2.1 START TIME—The start time of a normally closed type flasher is the time to open the circuit after the voltage is applied, provided the closed circuit remains closed for a minimum of 0.10 s. If the closed circuit opens in less than 0.10 s, the flasher shall be considered a normally open type flasher for this test. The start time of a normally open type flasher is the time to complete one cycle (close the circuit then open the circuit) after voltage is applied. For a fixed-load flasher, the test shall be made with the specific ampere design load connected. For a variable-load flasher, the test shall be made with both the minimum and maximum ampere design load. The test shall be made in an ambient temperature of 24 °C ± 5 °C. The start time shall be measured and recorded for three starts, each of which is separated by a cooling interval of at least 5 min.

4.2.2 VOLTAGE DROP—The lowest voltage drop across the flasher shall be measured between the input and the load terminals at the flasher and during the "on" period. The voltage drop shall be measured and recorded during any three cycles after the flasher has been operating for five consecutive cycles. For fixed-load flashers, the voltage drop is measured with the specific ampere design load connected. For variable load flashers, the voltage drop shall be measured with the maximum ampere design load connected. The test shall be made in an ambient temperature of 24 °C ± 5 °C.

4.2.3 FLASH RATE AND PERCENT CURRENT ON TIME—The flash rate and percent current on time shall be measured and recorded after the flasher has completed five consecutive cycles and shall be an average of at least three consecutive cycles at each of the following bulb voltages and ambient temperature conditions.

a. 12.8 V (or 6.4 V or 25.6 V) and 24 °C ± 5 °C
b. 12.0 V (or 6.0 V or 24.0 V) and -17 °C ± 3 °C
c. 15.0 V (or 7.5 V or 30.0 V) and -17 °C ± 3 °C
d. 11.0 V (or 5.5 V or 22.0 V) and 50 °C ± 3 °C
e. 14.0 V (or 7.0 V or 28.0 V) and 50 °C ± 3 °C

The flashers shall be temperature stabilized before each test. For a fixed load flasher, the test shall be made with the specific ampere design load connected. For a variable load flasher, the test shall be made with both the minimum and maximum ampere design load connected.

4.2.4 EXTREME TEMPERATURE—The flasher shall be subjected to ambient temperatures of 63 °C ± 3 °C and -32 °C ± 3 °C until stabilized. The start time and flash rate shall be measured and recorded at each extreme temperature. The flash rate measurement must be completed within the first minute of energization. Otherwise the procedure shall be as specified in 4.2.1 and 4.2.3a.

4.2.5 DURABILITY—The durability test shall be conducted near the following conditions:

a. 24 °C ± 5 °C ambient temperature
b. 14.0 V (7.0 V or 28.0 V) applied to the input terminals of the test circuit
c. Specific ampere design load for fixed load flashers and maximum specified ampere design load for variable load flashers
d. 100 h of intermittent flashing (15 s on, 15 s off) followed by 50 h of continuous flashing

5. *Performance Requirements*

   5.1 **Start Time**—The average and maximum of the three start time measurements (see 4.2.1) for the flasher shall not exceed the values shown in Table 1.

TABLE 1—START TIME, s

Flasher Type	Average Time	Maximum Time
Normally closed	1.3	2.0
Normally open	1.5	2.0

   5.2 **Voltage Drop**—The average of the three voltage drop measurements (see 4.2.2) for the flasher shall not exceed 0.5 V. No single measurement shall exceed 0.8 V.

   5.3 **Flash Rate and Percent Current On Time** The average flash rate and percent current on time shall fall within 60 to 120 flashes per minute and 30 to 75% on under all conditions of 4.2.3.

   5.4 **Extreme Temperature**—At the extreme temperature conditions, start time shall not exceed 3 s and flash rate shall be 50 to 130 flashes per minute.

   5.5 **Durability**—The flasher shall conform to 5.1, 5.2, and 5.3 (under test condition 4.2.3a only) at the start and conclusion of the test.

(R) VEHICULAR HAZARD WARNING FLASHERS —SAE J945 JUN93

SAE Standard

Report of the Lighting Committee approved February 1966 and completely revised June 1987. Rationale statement available. Completely revised by the Auxiliary Device Standards Committee and the Flasher Task Force of the SAE Lighting Coordinating Committee June 1993. Rationale statement available.

1. *Scope*—This SAE Standard defines the test conditions, procedures, and minimum design requirements for nominal 6, 12, and 24 V hazard warning flashers.

2. *References*

   2.1 **Applicable Documents**—The following publications form a part of this specification to the extent specified herein. The latest issue of SAE publications shall apply.

   2.1.1 SAE PUBLICATIONS—Available from SAE, 400 Commonwealth Drive, Warrendale, PA 15096-0001.

SAE J759—Lighting Identification Code
SAE J823—Flasher Test
SAE J910—Hazard Warning Signal Switch

2.2 Definition

2.2.1 The HAZARD WARNING FLASHER is a device installed in a vehicle lighting system which has the primary function of causing the turn signal lamps to flash when the hazard warning switch is actuated. Secondary functions may include the visible pilot indication for the hazard system (required by SAE J910) and an audible signal to indicate when the flasher is operating.

3. Flasher Identification Code—Flashers conforming to this document may be identified by the code J945 in accordance with SAE J759.

4. Tests

4.1 Test Equipment—The standard test equipment and circuitry for performing flasher tests shall conform with the specifications in SAE J823.

4.2 Test Procedures—All of the following tests shall be performed at 12.8 V (or 6.4 V or 25.6 V) at the bulbs unless otherwise specified.

4.2.1 START TIME—The start time of a normally closed type flasher is the time to open the circuit after the voltage is applied, provided the closed circuit remains closed for a minimum of 0.10 s. If the closed circuit opens in less than 0.10 s, the flasher shall be considered a normally open type flasher for this test. The start time of a normally open type flasher is the time to complete one cycle (close the circuit and then open the circuit) after the voltage is applied. For a fixed load flasher, the test shall be made with the specific ampere design load connected. For a variable load flasher, the test shall be made with both the minimum and maximum ampere design loads. The test shall be made in an ambient temperature of 24 °C ± 5 °C. The start time shall be measured and recorded for three starts, each of which is separated by a cooling interval of at least 5 min.

4.2.2 VOLTAGE DROP—The voltage drop across the flasher shall be measured between the input and the load terminals at the flasher and during the "on" period of each cycle. After the flasher has been operating for five consecutive cycles, the lowest voltage drop observed during each of three consecutive cycles shall be measured and recorded. For a fixed load flasher, the test shall be conducted with the specific ampere design load connected. For a variable load flasher, the test shall be conducted with both the minimum and maximum ampere design loads. The test shall be conducted in an ambient temperature of 24 °C ± 5 °C.

4.2.3 FLASH RATE AND PERCENT CURRENT ON TIME—The flash rate and percent current on time shall be measured and recorded after the flasher has completed five consecutive cycles and shall be the average of at least three consecutive cycles at each of the following bulb voltages and ambient temperature conditions:

a. 12.8 V (or 6.4 V or 25.6 V) and 24 °C ± 5 °C
b. 11.0 V (or 5.5 V or 22.0 V) and -17 °C ± 3 °C
c. 13.0 V (or 6.5 V or 26.0 V) and -17 °C ± 3 °C
d. 11.0 V (or 5.5 V or 22.0 V) and 50 °C ± 3 °C
e. 13.0 V (or 6.5 V or 26.0 V) and 50 °C ± 3 °C

The flashers shall be temperature stabilized before each test. For a fixed load flasher, the test shall be conducted with the specific ampere design load connected. For a variable load flasher, the test shall be conducted with both the minimum and maximum design loads connected.

4.2.4 EXTREME TEMPERATURE—The flasher shall be subjected to ambient temperatures of 63 °C ± 3 °C and -32 °C ± 3 °C until stabilized. The start time and flash rate shall be measured and recorded at each extreme temperature. The flash rate measurement must be completed within the first minute of energization. Otherwise, the procedure shall be as specified in 4.2.1 and 4.2.3a.

4.2.5 DURABILITY—The durability test shall be conducted under the following conditions:

a. 24 °C ± 5 °C ambient temperature
b. 13.0 V (6.5 V or 25.6 V) applied to the input terminals of the test circuit
c. Maximum specified ampere design load
d. Continuous flasher operation for 36 h

5. Performance Requirements

5.1 Start Time—The average of the three start time measurements (4.2.1) shall not exceed 1.5 s. No single measurement shall exceed 2.0 s.

5.2 Voltage Drop—The average of the three voltage drop measurements (4.2.2) shall not exceed 0.5 V. No single measurement shall exceed 0.8 V.

5.3 Flash Rate and Percent Current On Time—The average flash rate and percent current on time shall fall within 60 to 120 flashes per minute and 30 to 75% on time, respectively, under all conditions of 4.2.3.

5.4 Extreme Temperature—At the extreme temperature conditions, start time shall not exceed 3 s and flash rate shall be 50 to 130 flashes per minute.

5.5 Durability—The flasher shall conform to 5.1, 5.2, and 5.3 (under test condition 4.2.3a only) at the start and conclusion of the test.

COMBINATION TURN SIGNAL HAZARD WARNING SIGNAL FLASHERS—SAE J2068 JAN90

SAE Recommended Practice

Report of the Auxiliary Devices Standards Committee approved January 1990. Rationale statement available.

1. Scope—This document defines the test conditions, procedures, and minimum design requirements, for nominal 6, 12, and 24 V flashers used for both turn signal and hazard warning signaling.

2. Definition—This flasher is a device installed in a vehicle lighting system, which has the primary functions of causing the turn signal lamps to flash when the turn signal switch is actuated and the hazard warning signal lamps to flash when the hazard warning switch is activated.

3. Flasher Identification Code—Flashers conforming to this document may be identified by the code SAE J590/J945 in accordance with SAE J759.

4. Tests

4.1 Test Equipment—The standard test equipment and circuitry for performing flasher tests shall conform with the specifications in SAE J823.

4.2 Test Procedures—All of the following tests shall be performed at 12.8 V (or 6.4 V and/or 25.6 V) at the bulbs unless otherwise specified. The test shall be conducted at an ambient temperature of 24°C ± 5. The start time shall be measured and recorded for three starts, each of which is separated by a cooling interval of at least 5 min.

4.2.1 START TIME—The start time of a normally closed type flasher is the time to open the circuit after the voltage is applied provided the closed circuit remains closed for a minimum of 0.10 s. If the closed circuit opens in less than 0.10 s, the flasher shall be considered a normally open type flasher for this test. The start time of a normally open type flasher is the time to complete one cycle (close the circuit then open the circuit) after voltage is applied.

4.2.1.1 *Turn Signal Test*—For a fixed-load flasher, the test shall be made with the specific ampere design load connected. For a variable-load flasher, the test shall be made with both the minimum and maximum ampere design load.

4.2.1.2 *Hazard Warning Test*—The test shall be made with both the minimum and maximum ampere design load.

4.2.2 VOLTAGE DROP—The lowest voltage drop across the flasher shall be measured between the input and the load terminals at the flasher and during the "on" period. The voltage drop shall be measured and recorded during any three cycles after the flasher has been operating for a minimum of five consecutive cycles but less than twenty cycles. The test shall be made in an ambient temperature of 24°C ± 5.

4.2.2.1 *Turn Signal Test*—For a fixed-load flasher, the voltage drop is measured with the specific ampere design load connected. For a variable-load flasher, the voltage drop shall be measured with the maximum ampere design load connected.

4.2.2.2 *Hazard Warning Test*—The voltage drop shall be measured with the maximum ampere design load connected.

4.2.3 FLASH RATE AND PERCENT CURRENT ON TIME—The flash rate and percent current on time shall be measured and recorded after the flasher has completed five consecutive cycles and shall be an average of at least three consecutive cycles at the specified bulb voltages and ambient temperature conditions. The flashers shall be temperature stabilized before each test.

4.2.3.1 *Turn Signal Tests*

a. 12.8 V (or 6.4 or 25.6 V) and 24°C ± 5

b. 12.0 V (or 6.0 or 24.0 V) and −17°C ± 3
c. 15.0 V (or 7.5 or 30.0 V) and −17°C ± 3
d. 11.0 V (or 5.5 or 22.0 V) and 50°C ± 3
e. 14.0 V (or 7.0 or 28.0 V) and 50°C ± 3

For a fixed-load flasher, the test shall be made with the specific ampere design load connected. For a variable-load flasher, the test shall be made with both the minimum and maximum ampere design load.

4.2.3.2 *Hazard Warning Tests*
a. 12.8 V (or 6.4 or 25.6 V) and 24°C ± 5
b. 11.0 V (or 5.5 or 22.0 V) and −17°C ± 3
c. 13.0 V (or 6.5 or 26.0 V) and −17°C ± 3
d. 11.0 V (or 5.5 or 22.0 V) and 50°C ± 3
e. 13.0 V (or 6.5 or 26.0 V) and 50°C ± 3

The test shall be made with both the minimum and maximum ampere design load.

4.2.4 EXTREME TEMPERATURE—The flasher shall be subjected to ambient temperatures of 63°C ± 3 and −32°C ± 3 until stabilized. The start time and flash rate shall be measured and recorded at each extreme temperature. The flash rate measurement must be completed within the first minute of energization. Otherwise the procedure shall be as specified in 4.2.1.1 and 4.2.3.1(a) for turn signal and 4.2.1.2 and 4.2.3.2(a) for hazard warning.

4.2.5 DURABILITY—The durability test shall be conducted under the following conditions:

4.2.5.1 *Turn Signal Test Conditions:*
a. 24°C ± 5 ambient temperature.
b. 14.0 V (7.0 or 28.0 V) applied to the input terminals of the test circuit.
c. Specific ampere design load for fixed-load flashers and maximum specified ampere design load for variable-load flashers.

4.2.5.2 *Hazard Warning Test Conditions:*
a. 24°C ± 5 ambient temperature.
b. 13.0 V (6.5 or 26 V) applied to the input terminals of the test circuit.
c. Maximum specified ampere design load.

4.2.5.3 *Test Cycle*
a. 50 h of intermittent flashing (15 s on, 15 s off) per 4.2.5.1.
b. 8 continuous hours per 4.2.5.2.
c. 50 h of intermittent flashing (15 s on, 15 s off) per 4.2.5.1.
d. 8 continuous hours per 4.2.5.2.
e. 50 continuous hours per 4.2.5.1.
f. 20 continuous hours per 4.2.5.2.

5. *Performance Requirements*

5.1 **Start Time**—The average and maximum of the three start time measurements (4.2.1, 4.2.1.1, 4.2.1.2) for the flasher shall not exceed the values shown in Table 1.

TABLE 1—START TIME, SECONDS

FLASHER TYPE	AVERAGE TIME	MAXIMUM TIME
Normally closed	1.3	2.0
Normally open	1.5	

5.2 **Voltage Drop**—The average of the three voltage drop measurements (4.2.2) for the flasher shall not exceed 0.5 V. No single measurement shall exceed 0.8 V.

5.3 **Flash Rate and Percent Current On Time**—The average flash rate and percent current on time shall fall within 60 to 120 flashes/min and 30 to 75% on under all conditions of 4.2.3.

5.4 **Extreme Temperature**—At the extreme temperature conditions, start time shall not exceed 3 s and flash rate shall be 50 to 130 flashes/min.

5.5 **Durability**—The flasher shall conform to 5.1, 5.2, and 5.3 (under test conditions 4.2.3.1(a) and 4.2.3.2(a) only) at the start and conclusion of the test.

6. *Guidelines*

6.1 Turn signal secondary functions may include the visible pilot indication (required by SAE J588), an audible signal to indicate flasher operation, and indication of turn signal lamp outage.

6.2 Hazard warning secondary functions may include the visible pilot indicator required in hazard warning signal systems and an audible signal to indicate flasher operation. When included, the pilot function must operate under all hazard warning test conditions.

(R) WARNING LAMP ALTERNATING FLASHERS—SAE J1054 OCT89

SAE Recommended Practice

Report of Lighting Committee approved September 1973. Editorial change January 1977. Completely revised by the Auxiliary Devices Standards Committee October 1989. Rationale statement available.

1. *Scope*—This document defines the test conditions, procedures, and minimum design requirements for nominal 6, 12, and 24 V warning lamp alternating flashers.

2. *Definition*—The flasher is a device installed in a vehicle lighting system which has the primary function of causing warning lamps to alternately flash when the system is activated. Secondary functions may include the visible pilot(s) indication for the warning system and an audible signal to indicate when the flasher is operating (recommended by SAE J887 and J595).

3. *Flasher Identification Code*—Flashers conforming to this document may be identified in accordance with SAE J759.

4. *Tests*

4.1 **Test Equipment**—The standard test equipment and circuitry for performing flasher tests shall conform with the specifications in SAE J823.

4.2 **Test Procedures**—All the following tests shall be performed at 12.8 V (or 6.4 V or 25.6 V) at the bulbs unless otherwise specified.

4.2.1 START TIME—The start time is the time to complete one cycle (both load circuits have been energized and de-energized) after voltage is applied to the flasher. For fixed-load flashers, the test shall be made with the specific ampere design loads connected. For variable-load flashers, the test shall be made with both the minimum and maximum ampere design loads connected. The test shall be made in an ambient temperature of 24°C ± 5. The start time shall be measured and recorded for three starts, each of which is separated by a cooling interval of at least 5 min at 24°C ± 5.

4.2.2 VOLTAGE DROP—The lowest voltage drop across the flasher shall be measured between the input and each load terminal at the flasher and during the "on" period. The test shall be made with the specific maximum ampere design load connected and in an ambient temperature of 24°C ± 5. The voltage drop shall be measured and recorded during any three cycles after the flasher has been operating for five consecutive cycles.

4.2.3 FLASH RATE AND PERCENT CURRENT ON TIME—The flash rate and percent current on time of each load terminal shall be measured and recorded after the flasher has completed five consecutive cycles and shall be an average of at least three consecutive cycles at each of the following bulb voltages and ambient temperature conditions:
a. 12.8 V (or 6.4 V or 25.6 V) and 24°C ± 5
b. 12.0 V (or 6.0 V or 24.0 V) and -17°C ± 3
c. 15.0 V (or 7.5 V or 30.0 V) and -17°C ± 3
d. 11.0 V (or 5.5 V or 22.0 V) and 50°C ± 3
e. 14.0 V (or 7.0 V or 28.0 V) and 50°C ± 3

The flashers shall be temperature stabilized before each test. The test shall be made with the specific ampere design load connected for each circuit.

4.2.4 EXTREME TEMPERATURE TESTS—The flasher shall be subjected to ambient temperatures of 63°C ± 3 and -32°C ± 3 until stabilized. The start time and flash rate shall be measured and recorded at each extreme temperature. The measurements must be completed within the first minute of energization, otherwise the procedure shall be as specified in 4.2.1 and 4.2.3a.

4.2.5 DURABILITY—The durability test shall be conducted under the following conditions:
a. 24°C ± 5 ambient temperature
b. 13.0 V (6.5 V for 6.0 V nominal system or 26.0 V for 26.0 V nominal system) applied to the input terminal of the test circuit.
c. Specific maximum ampere design load.
d. 100 h of intermittent flashing (15 s on, 15 s off) followed by 50 h of continuous flashing.

5. *Performance Requirements*

5.1 **Start Time**—The average and maximum of the three start time measurements (4.2.1) for the flasher shall not exceed 1.5 and 2.0 s respectively.

5.2 Voltage Drop—The average of the three voltage drop measurements (4.2.2) shall not exceed 0.5 V. No single measurement may exceed 0.8 V.

5.3 Flash Rate and Percent Current On Time—At each load terminal, the flash rate shall be a minimum of 60 and a maximum of 120 per minute and the percent current "on" time shall be a minimum of 30 and a maximum of 75. The total of the percent current "on" times for the two terminals shall be a minimum of 90 and a maximum of 110.

5.4 Extreme Temperature—At the extreme temperature conditions, start time shall not exceed 5 s and flash rate shall be not less than 30 nor more than 150 flashes per minute.

5.5 Durability—The flasher shall conform to 5.1, 5.2, and 5.3 (under test procedure 4.2.3a only) at the start and conclusion of test.

φSPOT LAMPS—SAE J591 MAY89   SAE Standard

Report of Lighting Committee approved October 1951 and completely revised by the Lighting Coordinating Committee May 1989. Rationale statement available.

1. Scope—This SAE Standard provides test procedures and performance requirements for spot lamps.

2. Definitions

2.1 Spot Lamps—Lamps which provide a substantially parallel beam of light and are capable of being aimed as desired by the user.

3. Lighting Identification Code—Spot lights for vehicles may be identified with "0" in accordance with SAE J759 DEC87, Lighting Identification Code.

4. Tests

4.1 SAE J575—The following test procedures in SAE J575 DEC88, Tests for Motor Vehicle Lighting Devices and Components, are part of this standard, with modifications as indicated:

4.1.1 VIBRATION TEST
4.1.2 MOISTURE TEST
4.1.3 DUST TEST—(The dust test shall not be required for sealed units.)
4.1.4 CORROSION TEST

4.2 Color Test—SAE J578, Color Specification is a part of this standard.

5. Requirements

5.1 Performance Requirements—A device, when tested in accordance with the test procedures specified in Section 4, shall meet the following requirements indicated in SAE J575:

5.1.1 VIBRATION
5.1.2 MOISTURE
5.1.3 DUST
5.1.4 CORROSION

5.2 Color—The color of the emitted light shall be white as defined in SAE J578.

5.3 Beam Pattern—The spot lamp beam pattern shall be well defined and generally round or oval in shape.

6. Appendix—As a matter of additional information, attention is called to SAE J567 NOV87, Lamp Bulb Retention System, for requirements and gages used in socket design.

PARKING LAMPS (FRONT POSITION LAMPS)—SAE J222 DEC91   SAE Standard

Report of the Lighting Committee approved October 1951, and revised March 1986. Rationale statement available. Revised by the SAE Signalling and Marking Devices Standards Committee and the SAE Lighting Coordinating Committee December 1991.

1. Scope—This SAE Standard provides test procedures, requirements, and guidelines for parking lamps (front position lamps).

(R) *2. References*

2.1 Applicable Documents—The following publications form a part of this specification to the extent specified herein. The latest issue of SAE publications shall apply.

2.1.1 SAE PUBLICATIONS—Available from SAE, 400 Commonwealth Drive, Warrendale, PA 15096-0001.

SAE J567—Lamp Bulb Retention System
SAE J575—Tests for Motor Vehicle Lighting Devices and Components
SAE J576—Plastic Materials for Use in Optical Parts Such as Lenses and Reflectors of Motor Vehicle Lighting Devices
SAE J578—Color Specification
SAE J759—Lighting Identification Code

2.2 Definitions

(R) 2.2.1 PARKING LAMPS—Whether separate or in combination with other lamps, parking lamps are located on both the front left and right of the vehicle which show to the front and are intended to mark the vehicle when parked. In addition, these front lamps serve as a reserve front position indicating system in the event of headlamp failure.

3. Lighting Identification Code—Parking lamps may be identified by the code "P" in accordance with SAE J759.

4. Tests

4.1 SAE J575 is a part of this report. The following tests are applicable with the modifications as indicated:

4.1.1 VIBRATION TEST
4.1.2 MOISTURE TEST
4.1.3 DUST TEST
4.1.4 CORROSION TEST

4.1.5 PHOTOMETRY TEST—In addition to the test procedures in SAE J575, the following apply:

4.1.5.1 Photometric measurements shall be made with the light source of the lamp at least 3 m from the photometer. The H-V axis shall be taken as parallel to the axis of reference of the lamp as mounted on the vehicle.

4.1.6 WARPAGE TEST FOR DEVICES WITH PLASTIC COMPONENTS

4.2 Color Test—SAE J578 is a part of this report.

5. Requirements

5.1 Performance Requirements—A device, when tested in accordance with the test procedures specified in Section 4, shall meet the following requirements:

5.1.1 VIBRATION—SAE J575.
5.1.2 MOISTURE—SAE J575.
5.1.3 DUST—SAE J575.
5.1.4 CORROSION—SAE J575.
5.1.5 PHOTOMETRY—In addition to the photometric requirements in SAE J575, the following apply:

5.1.5.1 The lamp under test shall meet the photometric performance requirements contained in Table 1. The summation of the luminous intensity measurements at the specified test points in a zone shall be at least the value shown.

5.1.5.2 When a parking lamp is combined with the turn signal lamp, the signal lamp shall not be less than three times the luminous intensity of the parking lamp at any test point on or above horizontal; except that at H-V, H-5L, H-5R, and 5U-V, the (turn signal) lamp shall not be less than five times the luminous intensity of the parking lamp.

5.1.6 WARPAGE—SAE J575.
5.1.7 COLOR—The color of the light from a parking lamp shall be white or yellow as specified in SAE J578.

5.2 Materials Requirements — Plastic materials used in optical parts shall meet the requirements of SAE J576.

5.3 Design Requirements

5.3.1 If a turn signal lamp is optically combined with the parking lamp and a two-filament replaceable bulb is used, the bulb shall have an indexing base and the socket shall be designed so that bulbs with non-indexing bases cannot be used. Removable sockets shall have an indexing feature so that they cannot be reinserted into lamp housings in random positions, unless the lamp will perform its intended function with random light source orientation.

6. Guidelines

(R)TABLE 1 – PHOTOMETRIC REQUIREMENTS[1,2]

Zone	Test Points (Degrees)	Minimum Luminous Intensity (Candela)
1	10U - 5L	2.4
	5U - 20L	2.4
	5D - 20L	2.4
	10D - 5L	2.4
2	5U - 10L	3.0
	H - 10L	3.0
	5D - 10L	3.0
3	5U - V	16.8
	H - 5L	16.8
	H - V	16.8
	H - 5R	16.8
	5D - V	16.8
4	5U - 10R	3.0
	H - 10R	3.0
	5D - 10R	3.0
5	10U - 5R	2.4
	5U - 20R	2.4
	5D - 20R	2.4
	10D - 5R	2.4

[1] The measured values at each test point shall not be less than 60% of the minimum values in Table 2.
[2] Ratio requirements of 5.1.5.2 apply.

6.1 Photometric Design Guidelines for parking lamps, when tested in accordance with 4.1.5 of this report, are contained in Table 2.

6.2 Installation Guidelines — The following guidelines apply to parking lamps as used on the vehicle and shall not be considered part of the requirements.

6.2.1 Parking lamps on the front of the vehicle should be spaced as far apart laterally as practicable so that the signal will be clearly visible and its intent clearly understood.

6.2.2 The luminous intensity of incandescent filament bulbs will vary with applied voltage. The electrical wiring in the vehicle should be adequate to supply design voltage to the lamp filament.

6.2.3 Performance of lamps may deteriorate significantly as a result of dirt, grime, and/or snow accumulation on their optical surfaces. Installation of lamps on vehicles should be considered to minimize the effect of these factors.

(R)TABLE 2 – PHOTOMETRIC DESIGN GUIDELINES[1]

Test Points (Degrees)	Test Points (Degrees)	Minimum Luminous Intensity (Candela)
10U, 10D	5L, 5R	0.8
5U, 5D	20L, 20R	0.4
5U, 5D	10L, 10R	0.8
5U, 5D	V	2.8
H	10L, 10R	1.4
H	5L, 5R	3.6
H	V	4.0

[1] Ratio requirements of 5.1.5.2 apply.

6.2.4 Where it is expected that lamps must perform in extremely severe environments, such as in off-highway, mining, fuel haulage or where it is expected that they will be totally immersed in water, the user should specify lamps specifically designed for such use.

APPENDIX A

A.1 As a matter of additional information, attention is called to SAE J567 for requirements and gages used in socket design.

DAYTIME RUNNING LAMPS FOR USE ON MOTOR VEHICLES — SAE J2087 AUG91

SAE Recommended Practice

Report of the SAE DRL Task Force of the SAE Lighting Coordinating Committee approved August 1991. Rationale statement available.

1. Scope — This SAE Recommended Practice provides test procedures, requirements, and guidelines for daytime running lamps that are mounted on the exterior of a vehicle. It is applicable to daytime running lamps that are combined with or use headlamps, parking lamps, turn signal lamps, fog lamps, or other lamps on the front of the vehicle, as well as to daytime running lamps that use dedicated lamps.

2. References

2.1 Applicable Documents — The following publications form a part of this specification to the extent specified herein. The latest issue of SAE publications shall apply.

2.1.1. SAE PUBLICATIONS — Available from SAE, 400 Commonwealth Drive, Warrendale, PA 15096-0001.

SAE J567 — Lamp Bulb Retention System
SAE J575 — Tests for Motor Vehicle Lighting Devices and Components
SAE J576 — Plastic Materials for Use in Optical Parts Such as Lenses and Reflectors of Motor Vehicle Lighting Devices
SAE J578 — Color Specifications for Electric Signal Lighting Devices
SAE J759 — Lighting Identification Code
SAE J1050 — Describing and Measuring the Driver's Field of View

2.2 Related Publications — The following publications are provided for information purposes only and are not a required part of this document.

SAE Lighting Committee DRL Test Reports, 1974-1989, nine separate reports.

CIE TC4.13 Report — Automobile Daytime Running Lights (DRL), Third Draft, July 1990.

Canadian Motor Vehicle Safety Standard 108 — Light Equipment

2.3 Definitions

2.3.1 DAYTIME RUNNING LAMPS (DRL) — Steady burning lamps that are used to improve the conspicuity of a vehicle from the front and front sides when the regular headlamps are not required for driving.

2.3.2 DAYTIME RUNNING LAMP TELLTALE — An indicator that provides a visual signal to advise the driver that only his daytime running lamps are on and he should switch on the regular headlamps.

3. Lighting Identification Code — Daytime running lamps meeting the performance requirements of Section 5 of this document may be identified by the code Y2 in accordance with SAE J759.

4. Tests

4.1 SAE J575 is a part of this document. The following tests, from that document, are applicable with the modifications as indicated.

4.1.1 VIBRATION TEST
4.1.2 MOISTURE TEST
4.1.3 DUST TEST
4.1.4 CORROSION TEST
4.1.5 PHOTOMETRY — In addition to the test procedures in SAE J575, the following applies:

Photometric measurements shall be made with the light source of the DRL at least 3 m from the photometer. If the DRL is optically combined with a headlamp or a fog lamp then the photometric measurements shall be made with the light source of the DRL at least 18.3 m from the photometer.

4.1.6 WARPAGE TEST ON DEVICES WITH PLASTIC COMPONENTS—The bulb operation for this test shall be steady burning.

4.1.7 COLOR TEST—SAE J578 is a part of this document.

5. Requirements

5.1 Performance Requirements—A DRL, when tested in accordance with the test procedures specified in Section 4, shall meet the following requirements:

5.1.1 VIBRATION—SAE J575
5.1.2 MOISTURE—SAE J575
5.1.3 DUST—SAE J575
5.1.4 CORROSION—SAE J575
5.1.5 PHOTOMETRY—SAE J575—The DRL under test shall meet the photometric requirements contained in Table 1.
5.1.6 WARPAGE—SAE J575
5.1.7 COLOR—SAE J578—The color of the light from a DRL shall be white, white to yellow, white to selective yellow, yellow, or selective yellow as specified in SAE J578.

5.2 Materials Requirements—Plastic materials used in optical parts shall meet the requirements of SAE J576.

5.3 System Requirements—A DRL system shall consist of at least two lamps.

5.3.1 LOCATION REQUIREMENTS—The DRLs shall be located on the front, at the same mounting height, and symmetrically placed laterally relative to the centerline of the vehicle.

5.3.2 AREA REQUIREMENTS—The DRLs shall have a minimum unobstructed effective projected luminous lens area of 40 cm². In addition the DRL must provide an unobstructed view of the outer lens surface of at least 10 cm² measured at 45 degrees to the longitudinal axis of the vehicle.

6. Guidelines

6.1 Photometric Design Guidelines for DRLs, when tested in accordance with 4.1.5 of this document, are contained in Table 2.

6.2 Luminous Flux Maintenance Guideline—The applicable luminous flux maintenance test cycle established to verify the performance of halogen bulbs used at reduced voltages in daytime running lamps is shown in Table 3. Testing is to be done at room temperature with a steady-state DC applied voltage as specified in Table 3. It is not required to use AC, pulse width modulators, or dropping resistors but they may be used if desired to obtain the voltage as specified in Table 3.

6.2.1 Season the filament(s) and measure the original luminous flux output of the filament(s) in a spherical photometer at 12.8 V DC. Test the light source mounted in a 100 × 165 mm sealed beam size enclosure through 24 cycles where a test cycle is as defined in Table 3 and is sequenced from the same starting point in each cycle. Measure the final luminous flux output of the filament(s) after testing.

After the testing (see equation 1):

$$\frac{LUBF}{LUBO} \geq 95\% \quad \frac{LLBF}{LLBO} \geq 95\% \quad \text{(Eq. 1)}$$

where:
LUBF = Lumens Upper Beam, Final
LUBO = Lumens Upper Beam, Original
LLBF = Lumens Lower Beam, Final
LLBO = Lumens Lower Beam, Original

6.3 Telltale Guidelines—A DRL Telltale, if provided, shall be located on the instrument panel or in the driver's forward field of view. If a light sensor is used to activate the telltale, the sensor should be upward pointing and activate the telltale when the ambient light level is

TABLE 1—PHOTOMETRIC REQUIREMENTS

Test Point	Minimum Candela	Maximum Candela
5U - 10L	80	
5U - V	280	
5U - 10R	80	
H - 20L	40	
H - 10L	280	
H - 5L	360	
H - V	400	7000
H - 5R	360	
H - 10R	280	
H - 20R	40	
5D - 10L	80	
5D - V	280	
5D - 10R	80	

TABLE 2—PHOTOMETRIC DESIGN GUIDELINES

Test Point	Minimum Candela	Maximum Candela
5U - 10L	100	
5U - V	350	
5U - 10R	100	
H - 20L	50	
H - 10L	350	
H - 5L	450	
H - V	500	7000
H - 5R	450	
H - 10R	350	
H - 20R	50	
5D - 10L	100	
5D - V	350	
5D - 10R	100	

TABLE 3—TEST CYCLE

Lighting Mode	Lighting Mode	Voltage (V)	H/L Using Upper Beam For DRL Double-Fil. Period (h)	H/L Using Upper Beam For DRL Single-Fil. Period (h)	H/L Using Lower Beam For DRL Double-Fil. Period (h)	H/L Using Lower Beam For DRL Single-Fil. Period (h)
Headlamp	Upper Beam	12.8	0.25	0.25	0.25	—
(Nighttime)	Lower Beam	12.8	1.0	—	1.0	1.0
DRL	Upper Beam	6.4	6.25	6.25	—	—
(Daytime)	Lower Beam	10.5	—	—	6.25	6.25
OFF	OFF	0	0.5	1.5	0.5	0.75
Total (1 Cycle)	Total (1 Cycle)		8.0	8.0	8.0	8.0

less than 1000 lux, indicating dusk, night, or other reduced light conditions. If a light sensor is not used the telltale should provide a visual signal when the ambient light level, as measured by an upward pointing sensor, is less than 1000 lux, indicating dusk, night, or other reduced light conditions. The telltale is used to indicate to the driver:
  a. The DRLs are still illuminated and the ambient light level indicates the headlamps should be illuminated or
  b. The headlamps should be turned on.

The telltale may deactivate when the headlamps are switched on.

6.3.1 The telltale may also function as a bulb failure indicator for any exterior lighting functions.

6.3.2 The telltale should emit yellow colored light and have a minimum projected illuminated area of 18 mm^2. The minimum required illuminated area of the telltale shall be visible according to the procedures contained in SAE J1050. The steering wheel shall be turned to a straight-ahead position and in the design location for an adjustable wheel or column.

6.3.3 A DRL telltale need not be installed on vehicles having the following lighting equipment:
  a. Lower beam headlamps operating at full voltage used as DRLs, with all exterior marking lamps activated by the DRL system or
  b. An automatic photocell system for switching between DRL and night exterior lighting modes, unless this device can be manually switched to an inoperative status.

6.4 Operating Guidelines—These guidelines apply to how DRLs are used on the vehicle and are not part of the requirements.

6.4.1 The DRLs are to be activated without any switching by the operator (apart from the ignition switch).

6.4.2 No other lights are required to be illuminated with the DRLs, unless full intensity low beam headlamps are used as the DRLs. In this case those exterior lamps that are required to be illuminated with the low beam headlamps are to be illuminated with the DRLs.

6.4.3 The DRLs are to be deactivated when the low beam or high beam headlamps are turned on.

6.5 Installation Guidelines—For DRLs to be most effective they should be spaced as far apart laterally as practicable, to maximize the field of view and to facilitate estimation of distances by drivers in approaching vehicles.

DRLs should be designed to be mounted on the vehicle so the centers of the lenses are not less than 380 mm (15 in) nor more than 1820 mm (72 in) above the road surface.

6.6 Color Guidelines—The color of the emitted light from all DRLs on a vehicle shall be designed to be the same (see 5.1.7).

Appendix A

A.1 For information on requirements and gages used in socket design, refer to SAE J567.

CLEARANCE, SIDE MARKER, AND IDENTIFICATION LAMPS—SAE J592 JUN92

SAE Information Report

Report of the Lighting Division, approved January 1937, completely revised by the Lighting Committee January 1984. Rationale statement available. Reaffirmed by the Lighting Coordinating Committee March 1990. Revised by the SAE Signalling and Marking Devices Standards Committee and the SAE Lighting Coordinating Committee June 1992.

1. Scope—This SAE Information Report provides test procedures, requirements, and guidelines for clearance, side marker, and identification lamps.

2. References

(R) **2.1 Applicable Documents**—The following publications form a part of this specification to the extent specified herein. The latest issue of SAE publications shall apply.

(R) 2.1.1 SAE PUBLICATIONS—Available from SAE, 400 Commonwealth Drive, Warrendale, PA 15096-0001.

    SAE J567—Lamp Bulb Retention System
    SAE J575—Tests for Motor Vehicle Lighting Devices and Components
    SAE J576—Plastic Materials for Use in Optical Parts such as Lenses and Reflectors of Motor Vehicle Lighting Devices
    SAE J578—Color Specification for Electric Signal Lighting Devices
    SAE J759—Lighting Identification Code

2.2 Definitions

2.2.1 CLEARANCE LAMPS—Lamps mounted on the permanent structure of the vehicle as near as practicable to the upper left and right extreme edges that provide light to the front or rear to indicate the overall width and height of the vehicle.

2.2.2 SIDE MARKER LAMPS—Lamps mounted on the permanent structure of the vehicle as near as practicable to the front and rear edges, that provide light to the side to indicate the overall length of the vehicle. Additional lamps may also be mounted at intermediate locations on the sides of the vehicle.

2.2.3 COMBINATION CLEARANCE AND SIDE MARKER LAMPS—Single lamps which simultaneously fulfill the performance requirements of clearance and side marker lamps.

2.2.4 IDENTIFICATION LAMPS—Lamps used in groups of three, in a horizontal row, that provide light to the front or rear or both, having lamp centers that are spaced not less than 150 mm nor more than 310 mm apart, mounted on the permanent structure as near as practicable to the vertical centerline and the top of the vehicle to identify vehicles 2032 mm or more in overall width.

3. Lighting Identification Code—Clearance, side marker, or identification lamps may be identified by the code "P2," and combination clearance and marker lamps may be identified with the code "PC," and in accordance with SAE J759.

4. Tests

4.1 SAE J575 is a part of this report. The following tests are applicable with the modifications as indicated.

4.1.1 VIBRATION TEST
4.1.2 MOISTURE TEST
4.1.3 DUST TEST
4.1.4 CORROSION TEST
4.1.5 PHOTOMETRY TEST

4.1.5.1 Photometric tests shall be made at a lamp distance of at least 3 m. The H-V axis of a clearance lamp shall be taken as parallel with the longitudinal axis of the vehicle. The H-V axis of a combination clearance and side marker lamp shall be taken as parallel with the longitudinal axis of the vehicle when measuring clearance lamp test points, and normal to this vehicle axis when measuring side marker test points. In all cases, the H-V axis shall be taken as parallel to the surface on which the vehicle stands.

4.1.6 WARPAGE TEST ON DEVICES WITH PLASTIC COMPONENTS

4.2 Color Test—SAE J578 is a part of this report.

5. Requirements

5.1 Performance Requirements—A device which, when tested in accordance with the test procedures specified in Section 4, shall meet the following requirements:

5.1.1 VIBRATION—SAE J575
5.1.2 MOISTURE—SAE J575
5.1.3 DUST—SAE J575
5.1.4 CORROSION—SAE J575
5.1.5 PHOTOMETRY—SAE J575

5.1.5.1 The lamp under test shall meet the photometric performance requirements contained in Table 1 and its footnotes. The summation of the luminous intensity measurements at the specified test points in a zone shall be at least the value shown.

5.1.6 Warpage—SAE J575

5.1.7 Color—The color of light from front clearance lamps, front and intermediate side marker lamps, and front identification lamps shall be yellow. The color of light from rear clearance, side marker, and identification lamps shall be red. Color shall be as specified in SAE J578.

5.2 Materials Requirements—Plastic materials used in optical parts shall meet the requirements of SAE J576.

5.3 Design Requirements

5.3.1 A clearance lamp and/or side marker lamp may be combined optically with a turn signal and/or a stop lamp. A clearance lamp may not be combined optically with a tail lamp or an identification lamp.

5.3.2 If a clearance lamp or a side marker lamp is optically combined with a turn signal lamp or a stop lamp and a two-light source (two filament) bulb is used, the bulb shall have an indexing base and the socket shall be designed so that bulbs with nonindexing bases cannot be inserted. In addition, removable sockets shall have an indexing feature so that they cannot be reinserted into lamp housings in random positions, unless the lamp will perform its intended function with random light source orientation.

6. Guidelines

6.1 Photometric Design Guidelines—Photometric design guidelines for clearance, side marker, and identification lamps, when tested in accordance with 4.1.5 of this report, are contained in Table 2 and its footnotes.

6.2 Installation Guidelines—The following guidelines apply to clearance, side marker, and identification lamps as used on the vehicle and shall not be considered part of the requirements.

6.2.1 The luminous intensity of incandescent filament bulbs will vary with applied voltage. The electrical wiring in the vehicle should be adequate to supply design voltage to the lamp filament.

6.2.2 Performance of lamps can deteriorate significantly as a result of dirt, grime, and/or snow accumulation on their optical surfaces. Installation of lamps on vehicles should be considered to minimize the effect of these factors.

6.2.3 Where it is expected that lamps must perform in extremely severe environments, such as in off-highway, mining, fuel haulage, or where it is expected that they will be totally immersed in water, the user should specify lamps specifically designed for such use.

APPENDIX A

A.1 As a matter of additional information, attention is called to SAE J567 for requirements and gages used in socket design.

(R) TABLE 1—PHOTOMETRIC REQUIREMENTS

Zone	Test Points[1,2] (degrees)	Minimum Luminous Intensity (cd) See Notes [3,4] Red	Minimum Luminous Intensity (cd) See Notes [3,4] Yellow
1	45L-10U 45L-H 45L-10D	0.75	1.86
2	V-10U V-H V-10D	0.75	1.86
3	45R-10U 45R-H 45R-10D	0.75	1.86

[1] Maximum luminous intensities of red clearance and identification lamps shall not exceed 18 cd within the solid cone angle 45L to 45R and 10U to 10D. When red clearance lamps are optically combined with stop or turn signal lamps, the maximum applies only on or above horizontal. The maximum luminous intensity shall not be exceeded over any area larger than that generated by a 0.5 degree radius within the solid cone angle prescribed by the test points.

[2] The requirements for side markers used on vehicles less than 2032 mm wide need only be met for inboard test points at a distance of 4.6 m from the vehicle on a vertical plane that is perpendicular to the longitudinal axis of the vehicle and located midway between the front and rear side marker lamps.

[3] When calculating zone totals, the measured value at each test point shall not be less than 60% of the minimum values in Table 2.

[4] Combination clearance and side marker lamps shall conform with both clearance and side marker photometric performance requirements.

TABLE 2—PHOTOMETRIC DESIGN GUIDELINES

	Test Points[1,2] (degrees)	Minimum Luminous Intensity (cd) See Note [2] Red[1]	Minimum Luminous Intensity (cd) See Note [2] Yellow
10U	45L V 45R	0.25 0.25 0.25	0.62 0.62 0.62
H	45L V 45R	0.25 0.25 0.25	0.62 0.62 0.62
10D	45L V 45R	0.25 0.25 0.25	0.62 0.62 0.62

[1] The maximum design value of a lamp intended for the rear of the vehicle should not exceed the listed design maximum over any area larger than that generated by 0.25 degree radius within the solid angle defined by the test points in Table 2.

[2] For combined clearance and side marker lamps, both the clearance and side marker photometric design values should apply.

CLEARANCE, SIDEMARKER, AND IDENTIFICATION LAMPS FOR USE ON MOTOR VEHICLES 2032 mm OR MORE IN OVERALL WIDTH—SAE J2042 JUN91

SAE Standard

Report of the SAE Lighting Coordinating Committee and the Heavy-Duty Lighting Standards Committee approved June 1991. Rationale statement available.

1. Scope—This SAE Standard provides test procedures, requirements, and guidelines for clearance, sidemarker, and identification lamps intended for use on vehicles 2032 mm or more in overall width. A clearance lamp, a sidemarker lamp, or an identification lamp conforming to the requirements of this document may be used on vehicles less than 2032 mm in overall width.

2. References

2.1 Applicable Documents—The following publications form a part of this specification to the extent specified herein. The latest issue of SAE publications shall apply.

2.1.1 SAE PUBLICATIONS—Available from SAE, 400 Commonwealth Drive, Warrendale, PA 15096-0001.

SAE J567—Lamp Bulb Retention System
SAE J576—Plastic Material for Use in Optical Parts Such as Lenses and Reflectors of Motor Vehicle Lighting Devices
SAE J578—Color Specification
SAE J592—Clearance, Sidemarker, and Identification Lamps
SAE J759—Lighting Identification Code
SAE J2139—Tests for Lighting Devices, Reflective Devices and Components Used on Vehicles 2032 mm or More in Overall Width

2.1.2 OTHER PUBLICATIONS—Attention is called to the following documents for additional information on lamp design and installation requirements.

FMVSS 108
FHWA 393 Subpart B
TTMA #RP-9
TMC #RP-702

2.2 Definitions

2.2.1 CLEARANCE LAMP—A clearance lamp provides light to the front or rear of a vehicle to indicate the overall width and height.

2.2.2 SIDEMARKER LAMP—A sidemarker lamp provides light to the side of a vehicle to indicate the overall length of the vehicle. Additional sidemarker lamps may also be mounted at intermediate locations on the side of the vehicle. The rear sidemarker lamp used on a trailer is also referred to as a tracking lamp.

2.2.3 IDENTIFICATION LAMP—An identification lamp is a group of three lamps in a horizontal row which provide light to the front or rear or both, having a light center spacing of not less than 150 mm nor more than 300 mm apart, to identify vehicles 2032 mm or more in overall width.

2.2.4 COMBINATION CLEARANCE AND SIDEMARKER LAMP—A combination clearance and sidemarker lamp is a single lamp which simultaneously meets the requirements of a clearance and a sidemarker lamp.

3. Lighting Identification Code—Clearance, sidemarker, or identification lamps may be identified by the code "P3"; combination clearance and sidemarker lamps may be identified by the code "PC2," in accordance with SAE J759.

4. Tests

4.1 SAE J2139 is a part of this document. The following tests are applicable with modification as indicated.

4.1.1 VIBRATION
4.1.2 MOISTURE
4.1.3 DUST
4.1.4 CORROSION
4.1.5 PHOTOMETRY

4.1.5.1 The photometric test shall be made at a device distance of at least 3 m from the photometer.

4.1.5.2 The H-V axis of a clearance or identification lamp shall be taken to be parallel to the longitudinal axis of the vehicle, when the device is mounted in its design position.

4.1.5.3 The H-V axis of a sidemarker lamp shall be taken to be perpendicular to a vertical plane passing through the longitudinal axis of the vehicle, when mounted in its design position.

4.1.5.4 The H-V axis of a combination clearance and sidemarker lamp shall be taken to be parallel to the longitudinal axis of the vehicle when testing the clearance lamp function, and perpendicular to a vertical plane passing through the longitudinal axis of the vehicle when testing the sidemarker lamp function, when the device is mounted in its design position.

4.1.6 WARPAGE TEST ON DEVICES WITH PLASTIC COMPONENTS

4.2 Color—SAE J578 is a part of this document.

4.3 Plastic Materials—SAE J576 is a part of this document.

5. Requirements

5.1 Performance Requirements—The device when tested in accordance with the test procedures of this document shall meet the requirements of SAE J2139 or as indicated.

5.1.1 VIBRATION
5.1.2 MOISTURE
5.1.3 DUST
5.1.4 CORROSION
5.1.5 PHOTOMETRY—The device tested shall meet the photometric performance requirements of Table 1 and its footnotes, except that front yellow clearance and identification lamps that are roof mounted on the vehicle need not meet the photometric requirements at 20 degrees down; 45 degrees left to 45 degrees right as specified in footnote 7.

The summation of the luminous intensity measurements at the specified test points in a zone shall be at least the value shown.

5.1.5.1 When a clearance lamp is combined with a stop lamp or a turn signal lamp, the stop lamp or turn signal lamp intensity shall be not less than three times the luminous intensity of the clearance lamp at any test point, except that at H-V, H-5L, H-5R, and 5U-V, the stop lamp or turn signal lamp intensity shall be not less than five times the luminous intensity of the clearance lamp.

5.1.6 WARPAGE

TABLE 1—CLEARANCE, SIDEMARKER, AND IDENTIFICATION LAMP PHOTOMETRIC PERFORMANCE REQUIREMENTS

Zone	Test Points Deg.	Zone Total Luminous Intensity, Candela, Red	Zone Total Luminous Intensity, Candela, Yellow
1	10U — 45L H — 45L 10D — 45L	0.75	1.86
2	10U — V H — V 10D — V	0.75	1.86
3	10U — 45R H — 45R 10D — 45R	0.75	1.86
Maximum Luminous Intensity, Candela, Red, rear only		18.0	

1. The maximum luminous intensity shall not be exceeded over any area larger than that generated by a 0.5 degree radius within the area defined by the test point pattern of Table 2.
2. Unless otherwise specified, the lamp shall be considered to have failed the photometric requirements of this document if the luminous intensity at any test point is less than 60% of the values specified in Table 2.
3. Unless otherwise specified, the lamp shall be considered to have failed the photometric requirements of this document if the minimum luminous intensity between test points is less than 60% of the lower design values of Table 2 for the closest adjacent test points on a horizontal and vertical line as defined by the test point pattern.
4. The summation of the luminous intensity measurements at the specified test points in a zone shall be at least the values shown.
5. Combination clearance and sidemarker lamps shall conform with both clearance and sidemarker lamp photometric performance requirements.
6. When a clearance lamp is combined with a stop lamp or turn signal lamp, see 5.1.5.1 of this document for luminous intensity ratio requirements.
7. Throughout the photometric pattern defined by the corner points of 20U-45L, 20U-45R, 20D-45R, and 20D-45L, the light intensity shall be not less than 0.1 candela in red or 0.2 candela in yellow.

5.2 Color—The color of the light from the front clearance lamps, the front identification lamps, and the front and intermediate sidemarker lamps shall be yellow.

The color of the light from the rear clearance lamps, rear identification lamps, and the rear sidemarker lamp (aka a tracking lamp on a trailer) shall be red.

The color shall meet the requirements of SAE J578.

5.3 Plastic Materials—The plastic materials used in the optical parts shall meet the requirements of SAE J576.

5.4 Design Requirements

5.4.1 A clearance lamp shall not be combined with a tail lamp.

5.4.2 A clearance lamp may be combined with a turn signal lamp or a stop lamp.

5.4.3 A sidemarker lamp may be combined with a side turn signal lamp.

5.4.4 A clearance lamp, identification lamp, or a sidemarker lamp may be combined with a reflex reflector.

5.4.5 If a clearance lamp is combined with a turn signal lamp or a stop lamp, or if a sidemarker lamp is combined with a side turn signal lamp, and a replaceable multiple light source is used, the light source retention system shall be designed with an indexing means so that the light source is properly indexed. Removable light source retention systems shall have an indexing feature so that they cannot be reinserted into the lamp housing in a random position, unless the lamp will perform its intended function with random light source orientation.

5.5 Installation Requirements

5.5.1 Clearance lamps shall be mounted on the permanent structure of the vehicle as near as practicable to the upper left and right extreme edges of the vehicle.

5.5.2 Sidemarker lamps shall be mounted on the permanent structure of the vehicle not less than 380 mm above the road surface measured from the center of the device at vehicle curb weight.

An intermediate yellow sidemarker lamp shall be mounted at the midpoint on vehicles 7.6 m or more in overall length.

The red, rear sidemarker lamp, used on trailers, also referred to as a tracking lamp, shall be mounted on the permanent structure of the trailer not less than 380 mm nor more than 1525 mm above the road surface measured from the center of the lamp at trailer curb weight.

5.5.3 Identification lamps shall be mounted on the permanent structure of the vehicle as near as practicable to the vehicle centerline and to the top of the vehicle.

5.5.4 When the rear identification lamps are mounted at the extreme height of the vehicle, rear clearance lamps need not meet the requirement that they be mounted as close as practicable to the top of the vehicle.

5.5.5 Visibility of each clearance lamp, sidemarker lamp, or identification lamp shall not be obstructed by any part of the vehicle throughout the photometric test pattern for the lamp unless the lamp is designed to comply with all photometric and visibility requirements considered, except that front yellow clearance and identification lamps that are roof mounted on the vehicle need not meet the photometric requirements at 20 degrees down, 45 degrees left to 45 degrees right, as specified in footnote 7, Table 1.

6. Guidelines

6.1 Design Guidelines

6.1.1 PHOTOMETRICS—Photometric design guidelines are contained in Table 2 and its footnotes.

TABLE 2—CLEARANCE, SIDEMARKER, AND IDENTIFICATION LAMP PHOTOMETRIC DESIGN GUIDELINES

Test Points Deg.		Luminous Intensity, Candela, Red	Luminous Intensity, Candela, Yellow
10U —	45L	0.25	0.62
	V	0.25	0.62
	45R	0.25	0.62
H —	45L	0.25	0.62
	V	0.25	0.62
	45R	0.25	0.62
10D —	45L	0.25	0.62
	V	0.25	0.62
	45R	0.25	0.62
Maximum Luminous Intensity, Candela, Red, rear only		18.0	

1. The maximum luminous intensity shall not be exceeded over any area larger than that generated by a 0.25 degree radius within the solid angle defined by the test points in Table 2.
2. For combined clearance and sidemarker lamps, both clearance and sidemarker photometric design guidelines shall apply.
3. When a clearance lamp is combined with a stop lamp or a turn signal lamp, see 5.1.5.1 of this document for luminous intensity ratio requirements.
4. Throughout the photometric pattern defined by the corner points of 20U-45L, 20U-45R, 20D-45R, and 20D-45L, the light intensity shall be not less than 0.10 candela in red or 0.20 candela in yellow.

6.2 Installation Guidelines

6.2.1 The following guidelines apply to devices used on the vehicle and shall not be considered part of the requirements.

6.2.2 Performance of the lamps may deteriorate significantly as a result of dirt, grime, snow, and ice accumulation on the optical surfaces. Installation of the device on vehicles should be considered to minimize the effect of these factors.

6.2.3 Where it is expected that the device must perform in extremely severe environments such as in off-highway, mining, fuel haulage, or where it is expected to be totally immersed in water, the user should specify devices specifically designed for such use.

7. Advance Requirements—This section of the document gives advance notice to manufacturers and users of the device of a pending change in the requirements for clearance lamps, sidemarker lamps, and identification lamps. The change in requirements shall be effective on devices marketed and used on or after January 1, 1996.

See Tables 3 and 4 and the footnotes.

TABLE 3—CLEARANCE, SIDEMARKER, AND IDENTIFICATION LAMP PHOTOMETRIC PERFORMANCE REQUIREMENTS—ADVANCE REQUIREMENTS

Zone	Test Points Deg.	Zone Total Luminous Intensity, Candela, Red	Zone Total Luminous Intensity, Candela, Yellow
1	20U — 45L 10U — 45L H — 45L 10D — 45L 20D — 45L	1.4	3.4
2	10U — V H — V 10D — V	1.2	3.0
3	20U — 45R 10U — 45R H — 45R 10D — 45R 20D — 45R	1.4	3.4
Maximum Luminous Intensity, Candela, Red, rear only		18.0	

1. The maximum luminous intensity shall not be exceeded over any area larger than that generated by a 0.5 degree radius within the area defined by the test point pattern of Table 4.
2. Unless otherwise specified, the lamp shall be considered to have failed the photometric requirements of this document if the luminous intensity at any test point is less than 60% of the values specified in Table 4.
3. Unless otherwise specified, the lamp shall be considered to have failed the photometric requirements of this document if the minimum luminous intensity between test points is less than 60% of the lower design values of Table 4 for the closest adjacent test points on a horizontal and vertical line as defined by the test point pattern.
4. Combination clearance and sidemarker lamps shall conform with both clearance and sidemarker lamp photometric performance requirements.
5. The summation of the luminous intensity measurements at the specified test points in a zone shall be at least the values shown.
6. When a clearance lamp is combined with a stop lamp or a turn signal lamp, see 5.1.5.1 of this document for luminous intensity ratio requirements.

TABLE 4—CLEARANCE, SIDEMARKER, AND IDENTIFICATION LAMP PHOTOMETRIC DESIGN GUIDELINES—ADVANCE REQUIREMENTS

Test Points Deg.	Luminous Intensity, Candela, Red	Luminous Intensity, Candela, Yellow
20U — 45L 45R	0.10 0.10	0.20 0.20
10U — 45L V 45R	0.40 0.40 0.40	1.0 1.0 1.0
H — 45L V 45R	0.40 0.40 0.40	1.0 1.0 1.0
10D — 45L V 45R	0.40 0.40 0.40	1.0 1.0 1.0
20D — 45L 45R	0.10 0.10	0.20 0.20
Maximum Luminous Intensity, Candela, Red, rear only	18.0	

1. The maximum luminous intensity shall not be exceeded over any area larger than that generated by a 0.25 degree radius within the solid angle defined by the test points in Table 4.
2. For combined clearance and sidemarker lamps, both clearance and sidemarker photometric design guidelines shall apply.
3. When a clearance lamp is combined with a stop lamp or a turn signal lamp, see 5.1.5.1 of this document for luminous intensity ratio requirements.

ɸ BACKUP LAMPS (REVERSING LAMPS)—SAE J593 JUN87 — SAE Standard

Report of the Lighting Committee, approved August 1947, completely revised June 1987. Rationale statement available.

1. *Scope*—This SAE Standard provides test procedures, requirements, and guidelines for backup lamps.

2. *Definitions*

   2.1 A backup lamp is a device used to provide illumination behind the vehicle and to provide a warning signal to pedestrians and other drivers when the vehicle is backing up or is about to back up.

   2.2 **Point of Visibility**—Any point on the lens surface which is within an area bounded by the intersection of the lens surface with a 25 mm diameter cylinder, the centerline of which passes through the light source center and is oriented horizontally and normal to the longitudinal axis of the vehicle.

3. *Lighting Identification Code*—Backup lamps may be identified by the code "R" in accordance with SAE J759—Lighting Identification Code.

4. *Tests*

   4.1 SAE J575, Tests for Motor Vehicle Lighting Devices and Components, is a part of this report. The following tests are applicable with the modifications as indicated.

   4.1.1 VIBRATION TEST
   4.1.2 MOISTURE TEST
   4.1.3 DUST TEST
   4.1.4 CORROSION TEST
   4.1.5 PHOTOMETRIC TEST

   4.1.5.1 Photometric tests shall be made with the photometer at a distance of at least 3 mm from the lamp. The H-V axis shall be taken as parallel with the longitudinal axis of the vehicle.

   4.1.6 WARPAGE TEST FOR DEVICES WITH PLASTIC COMPONENTS

   4.2 **Color Test**—SAE J578 is a part of this report.

5. *Requirements*

   5.1 **Performance Requirements**—A device, when tested in accordance with the test procedures specified in Section 4, shall meet the requirements indicated in following sections of the SAE J575:

   5.1.1 VIBRATION TEST
   5.1.2 MOISTURE TEST
   5.1.3 DUST TEST
   5.1.4 CORROSION TEST
   5.1.5 PHOTOMETRIC TEST—In addition to the requirements in SAE J575, the following apply:

   5.1.5.1 The light from a single lamp, when used in a two-lamp system, shall meet the photometric requirements shown in Table 1.

   5.1.5.2 When only one backup lamp is used on the vehicle, it shall meet twice the photometric requirements of Table 1.

   5.1.5.3 When two asymmetrical lamps of the same or symmetrically opposite design are used, the reading along the vertical axis and the averages of the readings for the same angles left and right of vertical for one lamp shall be used to determine compliance with the requirements of Table 1. If two lamps of differing designs are used, they shall be tested individually and the values added to determine that the combined units meet twice the candlepower requirements of Table 1.

   5.1.6 WARPAGE—SAE J575.

   5.1.7 COLOR—The color of the light from a backup lamp shall be white, in accordance with SAE J578. A backup lamp may project incidental red, yellow, or white light through reflectors or lenses that are adjacent, close

TABLE 1—PHOTOMETRIC REQUIREMENTS[1,2]
(MINIMUM LUMINOUS INTENSITY REQUIREMENTS FOR BACKUP LAMPS)

Zone	Test Points (deg)	Minimum Luminous Intensity (cd)
1	45L— 5U 45L— H 45L— 5D	45
2	30L— H 30L— 5D	50
3	10L—10U 10L— 5U V—10U V— 5U 10R—10U 10R— 5U	100
4	10L— H 10L— 5D V— H V— 5D 10R— H 10R— 5D	360
5	30R— H 30R— 5D	50
6	45R— 5U 45R— H 45R— 5D	45

[1] The measured value at each test point shall not be less than 60% of the minimum value specified for that test point in Table 2.
[2] The maximum per lamp at H and the above shall be 300 cd for a two lamp system, and 500 cd for a single lamp system.

TABLE 2—PHOTOMETRIC DESIGN GUIDELINES
(MINIMUM LUMINOUS INTENSITY (cd))

Test Points (deg)	45L	30L	10L	V	10R	30R	45R
10U	—	—	10	15	10	—	—
5U	15	—	20	25	20	—	15
H	15	25	50	80	50	25	15
5D	15	25	50	80	50	25	15

Note: The maximum per lamp at H and above shall be 300 cd for a two-lamp system and 500 cd for a single lamp system.

to, or a part of the lamp assembly. If a lamp has portions of its lens which project non-white light, that light shall be regarded as incidental if, quantitatively, it does not exceed 20% of the total device output at all specified test points; the lamp shall also meet the photometric requirements of this standard with white light alone.

5.2 Materials Requirements—Plastic materials used in the optical parts shall meet the requirements of SAE J576, Plastic Materials For Use in Optical Parts Such As Lenses and Reflectors of Motor Vehicle Lighting Devices.

5.3 Installation Requirements

5.3.1 Backup lamps shall be mounted on the rear so that the point of visibility of at least one of the lamps is visible from any eye point that is (a) 0.6-1.8 m above the horizontal plane on which the vehicle is standing and (b) rearward of a vertical plane perpendicular to the longitudinal axis of the vehicle, 0.9 m to the rear of the vehicle and extending 0.9 m beyond each side of the vehicle.

5.3.2 The backup lamp shall be lighted only when the ignition switch is energized and reverse gear is engaged.

6. Guidelines

6.1 Photometric Design Guidelines for backup lamps, when tested in accordance with Section 4.1.5 of this report, are contained in Table 2—Photometric Design Guidelines and its footnote.

6.2 Installation Guidelines—The following guidelines apply to tail lamps as used on the vehicle and shall not be considered part of the requirements:

6.2.1 The luminous intensity of incandescent filament bulbs will vary with applied voltage. The electrical wiring in the vehicle should be adequate to supply design voltage to the lamp filament.

6.2.2 Performance of lamp may deteriorate significantly as a result of dirt, grime and/or snow accumulation on their optical surfaces. Installation of lamps on vehicles should be considered to minimize the effect of these factors.

6.2.3 Where it is expected that lamps must perform in extremely severe environments, such as in off-highway, mining, fuel haulage or where it is expected that they will be totally immersed in water, the user should specify lamps specifically designed for such use.

7. Appendix—As a matter of information, attention is called to SAE J567, Lamp Bulb Retention System, for requirements and gauges used in socket design.

(R) BACKUP LAMP SWITCH—
SAE J1076 MAR90

SAE Standard

Report of Lighting Committee approved February 1974. Completely revised by the Auxiliary Devices Standards Committee March 1990. Rationale statement available.

1. Scope—This standard defines the test conditions, procedures and performance specification for 6, 12, and 24 V backup lamp switches which are intended for use in motor vehicles.

2. Definitions—The backup lamp switch is an operator activated device intended primarily to control the function of the backup lamps. There are three types:

2.1 Type "A"—A transmission mounted backup lamp switch is that device which is mounted in or on the transmission and actuated by a moving part within the transmission that energizes the backup lamps when the transmission is shifted into reverse.

2.2 Type "B"—A backup lamp switch performing the same function as Type "A", except that it is operated by a mechanism external of the transmission but not mounted in the passenger compartment.

2.3 Type "C"—A backup lamp switch performing the same function as Type "A" but mounted in the passenger compartment and actuated by movement of the shift mechanism or linkage.

3. Test

3.1 Test Equipment and Instrumentation:

3.1.1 POWER SUPPLY—The power supply shall comply with the following specifications:

    a. Output current–capable of supplying the continuous and inrush currents of the design load (see 3.2.1.1).

    b. Regulation:

    (1) Dynamic—The output voltage at the supply shall not deviate more than 1.0 V from zero to maximum load (including inrush current) and should recover 63% of its maximum excursion within 100 ms.

    (2) Static—The output voltage at the supply shall not deviate more than 2% with changes in static load from zero to maximum (not including inrush current), and means shall be provided to compensate for static input line variations.

    c. Ripple Voltage–maximum 300 mV peak to peak.

3.1.2 VOLTMETER—0–30 V maximum full scale deflection, accuracy ±1/2%.

NOTE: A digital meter having at least a 3-1/2 digit readout with an accuracy of ±1% plus 1 digit is recommended for mV readings.

3.1.3 AMMETER—Capable of carrying full system load current, accuracy ±3%.

3.2 Test Procedures— Environmental conditions have been selected for this document to help assure satisfactory operation under general customer use conditions. It is essential to duplicate specific environmental conditions under which the device is expected to function.

3.2.1 ELECTRICAL LOADS

3.2.1.1 The design load applied to the switch is the electrical load specified by the number and type of lamp(s) or other electrical load device(s) to be operated by each circuit of the switch. For example, the design load for the backup lamp circuit may be two 1156 bulbs.

3.2.1.2 The switch shall be operated at 6.4 V DC ± 0.2 for a 6 V system, 12.8 V DC ± 0.2 V for a 12 V system, or 25.6 V DC ± 0.2 V for a 24 V system. These voltages shall be the open circuit voltage measured at the input termination on the switch.

3.2.2 Temperature Test Procedure

3.2.2.1 Type "A" and "B"—The switch shall be exposed for 1 h without electrical load to each of these temperatures: 25°C ± 5; 107 (+0°, -3°C); -32 (+3°C, -0°C). After each of the one h temperature exposures, the switch shall be manually cycled for ten cycles at the design electrical load to insure basic electrical and mechanical function at these temperatures.

3.2.2.2 Type "C"—The temperature test shall be conducted the same as for Type "A" and "B" except the ambient temperatures shall be 25°C ± 5°C; 74 (+0, -3°C); -32 (+3, -0°C).

3.2.2.3 This same switch shall be used for the endurance test described in 3.2.3.

3.2.3 Endurance Test Procedure

3.2.3.1 The switch shall be electrically connected to operate its design load (both primary and secondary circuit function design electrical loads) at a temperature of 25°C ± 5°C.

3.2.3.2 The switch shall be operated for a minimum of 30 000[1] cycles. One complete cycle shall consist of sequencing through each position (with dwell in each position) and return without dwelling in each of the intermediate positions to the initial position.

[1] 30 000 cycles represents 8 cycles of backup lamp switch operation every day for approximately 10 years, or one cycle for each 3.3 miles driven for 100 000 miles.

The test equipment shall be arranged to provide the following switch operating time requirements:

Travel Time: 0.1-0.5 s (time from one position to the next)
Dwell Time: 0.5-2.0 s (time in each position)
Make & Break Rate: 130-150 mm per s

3.2.3.3 At the completion of the cycle testing, the switch shall be operated for 1 h in each detect position with the design load(s) connected.

3.2.4 Voltage Drop Test Procedure

3.2.4.1 The voltage drop from the input terminal(s) to the corresponding output terminal(s) shall be measured at design load before and after the completion of the endurance test and shall be the average of three consecutive readings. If wiring is an integral part of the switch, the voltage drop measurement shall be made including 75 mm ± 6 mm of wire on each side of the switch; otherwise the measurement shall be made at the switch terminals.

4. Performance Requirements

4.1 During and after each of the cycles described in 3.2.2 and 3.2.3, the switch shall be electrically and mechanically operable.

4.2 The voltage drop shall not exceed 0.3 V when measured as in 3.2.4, before and after completion of the tests described in 3.2.3.

HEADLAMP SWITCH—SAE J253 DEC89

SAE Standard

Report of the Lighting Committee, approved July 1971, completely revised April 1984. Rationale statement available. Reaffirmed by the Auxiliary Devices Standards Committee and the Lighting Coordinating Committee December 1989.

1. Scope—This document defines the test conditions, procedures, and performance specifications for 6-, 12-, and 24-volt manually actuated headlamp switches (circuit breaker(s) may be incorporated for circuit overload protection).

2. Definition—The headlamp switch is an operator-activated device intended primarily to control functioning of headlamps, parking lamps, tail lamps, and certain marking lamps. Secondarily, the device may control functioning of various accessory and instrument lights.

3. Test Requirements

3.1 Test Equipment and Instrumentation

3.1.1 Power Supply—The power supply shall comply with the following specifications:

a. Output Current—capable of supplying the continuous and in-rush currents of the design load (Reference 3.2.1.1).

b. Regulation

Dynamic—the output voltage at the supply shall not deviate more than 1.0 V from zero to maximum load (including in-rush current) and should recover 63% of its maximum excursion within 100 ms.

Static—the output voltage at the supply shall not deviate more than 2% with changes in static load from zero to maximum (not including in-rush current), and means shall be provided to compensate for static input line voltage variations.

c. Ripple Voltage—maximum 300 mV peak-to-peak.

3.1.2 Voltmeter—0 to 30 V maximum full scale deflection, accuracy ± ½%.

Note: A digital meter having at least a 3-½ digit readout with an accuracy of ± 1% plus one digit is recommended for millivolt readings.

3.1.3 Ammeter—Capable of carrying full system load current, accuracy ± 3%.

3.2 Test Procedures—Environmental conditions have been selected for this document to help assure satisfactory operation under general customer use conditions. It is essential to duplicate specific environmental conditions under which the device is expected to function.

3.2.1 Electrical Loads

3.2.1.1 The design load applied to the switch is the electrical load specified by the number and type of bulbs (or other electrical load devices) to be operated by each circuit of the switch. For example, the design load for the headlamp circuit may be four sealed beam headlamp units (2-4651 and 2-4652) and four-#194 bulbs.

3.2.1.2 The switch shall be operated at 6.4 V DC ± 0.2 for a 6-V system, 12.8 V DC ± 0.2 for a 12-V system, or 25.6 V DC ± 0.2 for a 24-V system. These voltages shall be the open circuit voltage measured at the input termination on the switch.

3.2.2 Temperature Test Procedures

3.2.2.1 The switch shall be exposed for 1 h without electrical load to each of the following temperatures: 25°C ± 5; 74 + 0°C, −3°C; −32 + 3°C, −0°C. The switch shall be manually cycled at each temperature for ten cycles at design load.

3.2.2.2 The same switch shall be used for the endurance test described in 3.2.3.

3.2.3 Endurance Test Procedure

3.2.3.1 The switch shall be electrically connected to operate its design load (both primary and secondary circuit function design electrical loads) at a temperature of 25°C ±5.

3.2.3.2 The switch shall be operated for a minimum of 11 000 cycles[1]. One complete cycle shall consist of sequencing through each position (with dwell in each position) and return without dwell in intermediate positions to the initial position.

The test equipment shall be arranged to provide the following switch operating time requirements:

Travel Time: 0.1-0.5 s (time from one position to the next)
Dwell Time: 1.0-2.0 s (time in each position)
Make and Break Rate: 130-150 mm/s

3.2.3.3 At the completion of the cycle testing, the switch shall be operated for 1 h in the headlamp position with the design load(s) connected.

3.2.4 Voltage Drop Test Procedure

3.2.4.1 The voltage drop from the input terminal(s) to the corresponding output terminal(s) shall be measured at design load before and after the completion of the endurance test and shall be the average of three consecutive readings. These voltage drop readings should exclude the voltage drop across the circuit breaker(s). If wiring is an integral part of the switch, the voltage drop measurement shall be made including 75 mm ± 6 of wire on each side of the switch; otherwise the measurement shall be made at the switch terminals.

4. Performance Requirements

4.1 During and after each of the cycles described in 3.2.2.1 and 3.2.3, the switch shall be electrically and mechanically operable.

4.2 The voltage drop shall not exceed 0.3 V when measured as in 3.2.4, before and after completion of the tests described in 3.2.3.

[1] 11 000 cycles represents three cycles of headlamp switch operation every day for approximately 10 years, or one cycle for each 4.5 miles driven for 100 000 miles with 50% night driving.

φREFLEX REFLECTORS—SAE J594 MAY89   SAE Standard

Report of the Lighting Division, approved January 1951, completely revised by the Lighting Committee May 1989. Rationale statement available.

1. Scope—This SAE technical report provides test procedures, requirements, and guidelines for reflex reflectors.

2. Definitions

2.1 Reflex reflectors are devices that are used on vehicles to give an indication of presence to an approaching driver by reflected light from the headlamps on the approaching vehicle.

2.2 The observation angle is the angle between a line from the observation point to the center of the reflector and a second line from the center of the reflector to the source of illumination.

2.3 The entrance angle is the angle between the axis of the reflex reflector and a line from the center of the reflector to the source of illumination.

3. Identification Code—Reflex reflectors may be identified by the Code "A" in accordance with SAE J759 DEC87, Lighting Identification Code.

4. Tests

4.1 SAE J575 DEC88, Tests for Motor Vehicle Lighting Devices and Components is a part of this report. The following tests are applicable with the modifications as indicated.

4.1.1 VIBRATION TEST
4.1.2 MOISTURE
4.1.3 DUST TEST
4.1.4 CORROSION TEST
4.1.5 PHOTOMETRY—In addition to the test procedures in SAE J575, the following apply:

4.1.5.1 Test Setup—Photometric measurement shall be made at a distance of at least 30 m with the reflex reflector setup for testing as shown in Fig. 1. The reflex reflector shall be mounted in a goniometer with the center of the reflex area at the center of rotation and at the same horizontal level as the source of illumination.

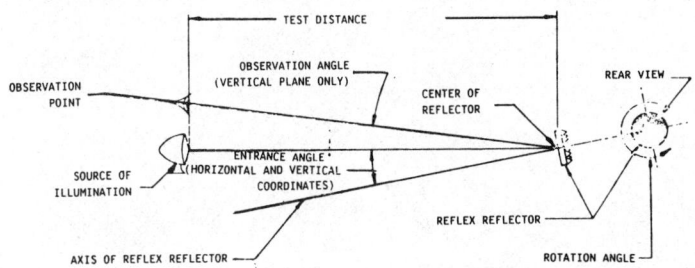

FIG. 1—SETUP FOR TESTING

4.1.5.2 Light Source and Sensor—The source of illumination shall be a projector with a 50 ± 5 mm effective diameter and a lamp filament operating at 2856 K (nominal) color temperature. In making photoelectric measurements, the opening to the photo cell shall not be more than 13 mm vertical by 25 mm horizontal with the observation point above (geometrically) the source of illumination.

4.1.5.3 Measurements—Reflex reflectors shall be photometered at the observation and entrance angles shown in Table 1. The entrance angle shall be designated left, right, up, and down in accordance with the position of the source of illumination with respect to the axis of the reflex reflector as viewed from behind the reflector. The H-V axis of reflex reflectors shall be taken parallel to the longitudinal axis of the vehicle for rear reflectors and perpendicular to a vertical plane parallel to the longitudinal axis of the vehicle for side reflectors.

Photometric measurements shall be made photoelectrically. The recorded value for each test point shall be the quotient of luminous intensity of the reflected light expressed as millicandela (candela[1]) divided by the illumination on the reflector measured in lux (foot candle). Also, the illumination on the reflex reflector from the source of illumination shall be measured in lux (foot candle). Reflex reflectors may have any linear or area dimension; but, for the photometric test, a maximum projected area of 7740 mm^2 contained within a 254 mm diameter circle shall be exposed.

4.1.5.4 Rotational Position—Reflex reflectors that do not have a fixed rotational position with respect to the vehicle shall be rotated 360 deg about their axis to find the minimum millicandela per incident lux (candela per incident foot candle), which shall be reported for each test point. If the output falls below the minimum requirement at any test point, the reflector shall be rotated ± 5 deg about its axis from the angle where the minimum output occurred; and the maximum millicandela per lux (candela per foot candle) within the angular range reported as a tolerance value.

Reflex reflectors that, by their design or construction, permit mounting on the vehicle in fixed rotational position shall be tested in this position. A visual locator, such as the word TOP, shall not be considered adequate to establish a fixed rotational position on the vehicle.

4.1.5.5 Uncolored Reflections—If uncolored reflections from the front surface interfere with photometric readings at any test point, the operator shall check 1 deg above, below, right, and left of the test point, and report the lowest reading and location. The latter must meet the minimum requirement for the test point.

4.2 Color Test—SAE J578, Color Specification is a part of this report. Additionally, the test sample may be either the reflex reflector or a disc of the same material, technique of fabrication, and dye formulation as the reflex reflector. If a disc is used for color determination by the transmission technique, the thickness should be twice the thickness of the reflector as measured from the face of the lens to the apexes of the reflecting elements. For either sample, a Source "A" illumination shall be used for color measurement.

5. Requirements

5.1 Performance Requirements—A reflex reflector, when tested in accordance with the test procedures specified in Section 4, shall meet the following requirements:

5.1.1 VIBRATION—SAE J575
5.1.2 MOISTURE—SAE J575, except that in the case of sealed units the alternate water submersion test (4.2.4) is required.
5.1.3 DUST—SAE J575
5.1.4 CORROSION—SAE J575
5.1.5 PHOTOMETRY—SAE J575
5.1.5.1 The reflex reflectors under test shall meet the photometric performance requirement contained in Table 1 or Table 1A.

TABLE 1—MINIMUM MILLICANDELAS PER INCIDENT LUX FOR A RED REFLEX REFLECTOR^a

Observation Angle (deg)	Entrance Angle (deg)				
	0 deg	10 deg Up	10 deg Down	20 deg Left	20 deg Right
0.2	420	280	280	140	140
1.5	6	5	5	3	3

5.1.6 COLOR—The color of the light from a reflex reflector shall be red, yellow, or white as defined in SAE J578.

5.2 Material Requirements—Plastic materials used in the optical portion of each reflex reflector unit shall meet the requirements of SAE J576 SEP86, Plastic Materials for Use in Optical Parts Such as Lenses and Reflectors of Motor Vehicle Lighting Devices.

5.3 Photometric Design Requirements:

5.3.1 If a reflex reflector is optically combined with signalling or marking bulb type devices, it shall be photometered independently by masking from the other functions and shall meet the performance values contained in Table 1 or 1A.

5.4 Installation Requirements—When installed on a vehicle, the visibility of the reflector to the front, side, or rear shall not be obstructed by any part of the vehicle throughout the photometric test angles of the device, unless designed to perform with the obstruction in place.

---

[1] "Candela" is used rather than "candlepower" as the preferred term in either metric or English units.

TABLE 1A—MINIMUM CANDLEPOWER PER INCIDENT FOOTCANDLE - RED REFLEX REFLECTOR[a]

Observation Angle (deg)	Entrance Angle (deg)				
	0 deg	10 deg Up	10 deg Down	20 deg Left	20 deg Right
0.2	4.5	3.0	3.0	1.5	1.5
1.5	0.07	0.05	0.05	0.03	0.03

[a]Yellow values shall be 2.5 times indicated red values and white values shall be 4 times indicated red values.

6. Guidelines

6.1 Photometric Design Guidelines—Reflex reflectors, when tested in accordance with paragraph 4.1.5, should be designed at least equal to the values contained in Table 1 or 1A.

6.2 Installation Guidelines—The following guidelines apply to reflex reflectors as used on the vehicle and shall not be considered a part of this report:

6.2.1 Reflex reflectors when used on the exterior of vehicles should be mounted to minimize the accumulation of dirt, grime, and/or snow so that adequate illumination is maintained from the low beam headlamps of approaching vehicles.

6.2.2 If reflex reflectors must perform in severe environments; such as, periodic total immersion in water, the user should specify reflex reflector designs suitable for such use.

REFLEX REFLECTORS FOR USE ON VEHICLES 2032 mm OR MORE IN OVERALL WIDTH—SAE J2041 JUN92

SAE Information Report

Report of the SAE Heavy-Duty Lighting Standards Committee approved June 1992. Rationale statement available.

1. Scope—This SAE Information Report provides test procedures, requirements, and guidelines for reflex reflectors used on vehicles 2032 mm or more in overall width and 7.6 m or more in length. Reflex reflectors conforming to these requirements may also be used on vehicles less than 2032 mm in overall width.

2. References

2.1 Applicable Documents—The following publications form a part of this specification to the extent specified herein. The latest issue of SAE publications shall apply.

2.1.1 SAE PUBLICATIONS—Available from SAE, 400 Commonwealth Drive, Warrendale, PA 15096-0001.

SAE J576—Plastic Material for Use in Optical Parts Such as Lenses and Reflectors of Motor Vehicle Lighting Devices

SAE J578—Color Specifications

SAE J594—Reflex Reflectors

SAE J759—Lighting Identification

SAE J2139—Test Methods and Equipment for Lighting Devices and Components for Use on Vehicles 2032 mm or More in Overall Width

2.2 Definitions

2.2.1 REFLEX REFLECTORS—Reflex reflectors are devices used on vehicles to alert an approaching driver by reflected light from the lamps on the approaching vehicle of a possible hazard.

2.2.1.1 *Type 1 Reflex Reflectors*—Type 1 reflex reflectors are devices having a photometric pattern ranging from 20 degrees left to 20 degrees right along the horizontal axis and from 10 degrees up to 10 degrees down along the vertical axis. (See SAE J594.)

2.2.1.2 *Type 2 Wide Angle Reflex Reflectors*—Type 2 wide angle reflex reflectors are devices having an expanded photometric pattern beyond that for a Type 1 reflex reflector. The photometric pattern covers the area defined by the entrance angle corner points of 45L-5U, 45R-5U, 45R-5D, 45L-5D and from 10 U and 10 D along the vertical axis.

Wide angle reflex reflectors may be directional and may require that they be mounted on the vehicle in a fixed orientation to be effective.

2.2.2 OBSERVATION ANGLE—The observation angle is the angle formed by a line from the observation point to the center of the reflective area and a second line from the center of the reflective area to the center of the source of illumination in the vertical plane only.

2.2.3 ENTRANCE ANGLE—The entrance angle is the angle between the axis of the reflex reflector and a line from the center of the reflective area to the center of the source of illumination, with both horizontal and vertical coordinates.

The entrance angle shall be designated left, right, up, and down in accordance with the position of the source of illumination with respect to the axis of the reflex reflector as viewed from behind the reflector.

3. Identification Code

3.1 Type 1 Reflex Reflectors may be identified by the code "A" in accordance with SAE J759.

3.2 Type 2 Wide Angle Reflex Reflectors may be identified by the code "A2" in accordance with SAE J759.

4. Tests

4.1 SAE J2139 is a part of this document. The following tests are applicable with the modifications as indicated.

4.1.1 VIBRATION—The device shall be mounted on the test fixture in accordance with the manufacturer's instruction and the horizontal mounting line marked.

The device shall be conditioned in a circulating air oven with the temperature controlled at 46 °C to 49 °C for 60 min. The device and test fixture shall be removed from the oven and without remounting (repressing of adhesive tape or tightening of mounting screws) placed on the vibration test machine and vibrated for 60 min.

4.1.2 MOISTURE

4.1.2.1 Either the Water Spray Moisture Test or Water Submersion Test may be used to test reflex reflectors.

4.1.2.2 Water Spray Moisture Test

4.1.2.2.1 If the reflex reflector is a separate unit and not combined with a lighting device, the light source on-off cycle during the water spray moisture test is not applicable.

4.1.2.2.2 Upon completion of the drain period, the interior of the device shall be observed for moisture accumulation that can be formed by tapping or tilting the device.

4.1.3 DUST

4.1.3.1 If the reflex reflector is a separate unit and not combined with a lighting device, the light source on-off cycle during the dust test is not applicable.

4.1.4 CORROSION

4.1.4.1 If the reflex reflector is a separate unit and not combined with a lighting device, the light source on-off cycle during the corrosion test is not applicable.

4.1.5 PHOTOMETRY

4.1.5.1 The reflex reflector shall be set up for testing as shown in Figure 1.

4.1.5.2 The reflex reflector shall be mounted on the goniometer in accordance with the manufacturer's instructions and with the center of the reflex area at the center of rotation and at the same horizontal level as the source of illumination.

4.1.5.3 The test distance shall be 30 m. The source of illumination shall be a lamp with a 50 mm ± 5 mm effective diameter and with a filament operating at nominal 2856 K color temperature.

4.1.5.4 The observation point shall be located directly above the source of illumination and the opening to the photocell shall not be more than 13 mm vertical by 25 mm horizontal.

4.1.5.5 The H-V axis of the reflex reflector shall be taken as being parallel to the longitudinal axis of the vehicle for front and rear reflectors and perpendicular to a vertical plane passing through the longitudinal axis of the vehicle for side reflectors.

4.1.5.6 Reflex reflectors may have any linear or area dimension but for the photometric test a maximum projected area of 7740 mm^2 contained within a 254 mm diameter circle shall be exposed.

4.1.5.7 If uncolored reflections from the front surface interfere with the photometric readings at any test point, the operator shall check 1 degree above, below, right, and left of the test point, and report the lowest reading and location. The lowest reading shall meet the minimum requirements for that test point.

4.1.5.8 Type 1 reflex reflectors, which do not require a fixed mounted position on the vehicle, shall be rotated about their axis through 360 degrees to find the millicandela per incident lux (minimum candela per incident footcandle) which shall be reported for each test point. If the output falls below the minimum requirements at any test point, the reflector shall be rotated ±5 degrees about its axis from the angle where the minimum output occurred and highest reading and location reported. The highest reading shall meet the minimum requirements for the test point.

4.1.5.9 Type 2, Wide Angle Reflex Reflectors, that maybe directional and may require a fixed orientation shall be mounted in their design position and in accordance with the manufacturer's instructions.

4.1.6 WARPAGE—The device mounted in its design position shall be placed in a circulating air oven with the temperature controlled at 46 °C to 49 °C for 60 min.

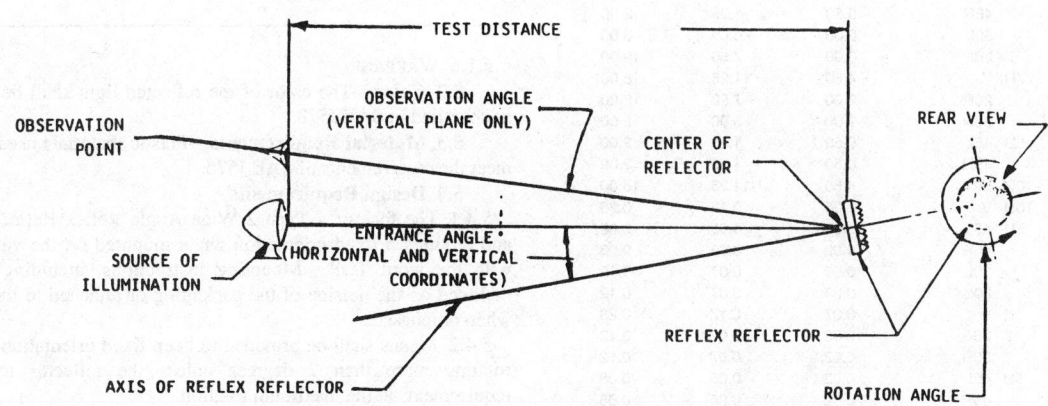

FIGURE 1—REFLEX REFLECTOR SETUP FOR TESTING

4.2 Color—The test sample may be either the reflex reflector or a disc of the same material, technique of fabrication and dye formulation as the reflex reflector.

If a disc is used, the thickness shall be twice the thickness of the reflector as measured from the face of the lens to the apexes of the reflecting elements.

4.3 Plastic Material—SAE J576.

5. Requirements

5.1 Performance Requirements—A device when tested in accordance with the test procedures of this document shall meet the requirements, of SAE J2139.

5.1.1 VIBRATION—Upon completion of the vibration test the reflector shall not have become separated from its mounting means. A Type 2 Wide Angle Reflex Reflector that is directional and requires a fixed orientation when mounted on the vehicle shall not rotate more than 2 degrees above or below the horizontal mounting line, unless the reflector meets all photometric requirements of Table 1 or Table 1A at the maximum rotation.

TABLE 1—WIDE ANGLE REFLEX REFLECTORS PHOTOMETRIC PERFORMANCE REQUIREMENTS
MILLICANDELA PER INCIDENT LUX

Obs. Angle (deg)	Ent. Angle (deg)		Red	Yellow	White
	10U	V	420	1050	1680
	5U	45L	50	125	200
		45R	50	125	200
		30L	185	465	740
		20L	280	700	1120
0.2	H	V	420	1050	1680
		20R	280	700	1120
		30R	185	465	740
	5D	45L	50	125	200
		45R	50	125	200
	10D	V	420	1050	1680
	10U	V	5	12	20
	5U	45L	2	5	8
		45R	2	5	8
		30L	3	7	12
		20L	3	7	12
1.5	H	V	6	15	24
		20R	3	7	12
		30R	3	7	12
	5D	45L	2	5	8
		45R	2	5	8
	10D	V	5	12	20

NOTE—Unless otherwise specified, the reflector shall be considered to have failed the photometric requirements of this document if the millicandela per incident lux between test points is less than the lowest values specified for the closest adjacent test points on a horizontal and vertical line defined by the test point pattern of Table 1A.

TABLE 1A—TYPE 2 WIDE ANGLE REFLEX REFLECTORS PHOTOMETRIC PERFORMANCE REQUIREMENTS
CANDELA PER INCIDENT FOOTCANDLE

Obs. Angle (deg)	Ent. Angle (deg)		Red	Yellow	White
	10U	V	4.50	11.25	18.00
	5U	45L	0.50	1.25	2.00
		45R	0.50	1.25	2.00
		30L	2.00	5.00	8.00
		20L	3.00	7.50	12.00
0.2	H	V	4.50	11.25	18.00
		20R	3.00	7.50	12.00
		30R	2.00	5.00	8.00
	5D	45L	0.50	1.25	2.00
		45R	0.50	1.25	2.00
	10D	V	4.50	11.25	18.00
	10U	V	0.05	0.12	0.20
	5U	45L	0.02	0.05	0.08
		45R	0.02	0.05	0.08
		30L	0.03	0.07	0.12
		20L	0.03	0.07	0.12
1.5	H	V	0.07	0.17	0.28
		20R	0.03	0.07	0.12
		30R	0.03	0.07	0.12
	5D	45L	0.02	0.05	0.08
		45R	0.02	0.05	0.08
	10D	V	0.05	0.12	0.20

NOTE—Unless otherwise specified, the reflector shall be considered to have failed the photometric requirements of this document if the candela per incident footcandle between test points is less than the lowest values specified for the closest adjacent test points on a horizontal and vertical line defined by the test point pattern of Table 1.

5.1.2 MOISTURE

5.1.2.1 Upon completion of the moisture test, the reflex reflector shall not have become separated from its mounting means.

5.1.3 DUST

5.1.3.1 Upon completion of the dust test, the reflex reflector shall not have become separated from its mounting means.

5.1.4 CORROSION

5.1.4.1 Upon completion of the corrosion test, the reflex reflector shall not have become separated from its mounting means.

5.1.5 PHOTOMETRY

5.1.5.1 Type 1, Reflex Reflectors shall meet the photometric performance requirements of Table 2 or Table 2A and the footnotes.

5.1.5.2 Type 2, Wide Angle Reflex Reflectors shall meet the photometric performance requirements of Table 1 or Table 1A and the footnotes.

TABLE 2—TYPE 1 REFLEX REFLECTORS PHOTOMETRIC PERFORMANCE REQUIREMENTS
MILLICANDELA PER INCIDENT LUX

Obs. Angle (deg)	Ent. Angle (deg)		Red	Yellow	White
	10U	V	280	700	1120
	20L		140	350	560
0.2	H	V	420	1050	1680
	20R		140	350	560
	10D	V	280	700	1120
	10U	V	5	12	20
	20L		3	7	12
1.5	H	V	6	15	24
	20R		3	7	12
	10D	V	5	12	20

TABLE 2A—TYPE 1 REFLEX REFLECTORS PHOTOMETRIC PERFORMANCE REQUIREMENTS
CANDELA PER INCIDENT FOOTCANDLE

Obs. Angle (deg)	Ent. Angle (deg)		Red	Yellow	White
	10U	V	3.0	7.5	12.00
	20L		1.5	3.75	6.00
0.2	H	V	4.5	11.25	18.00
	20R		1.5	3.75	6.00
	10D	V	3.0	7.5	12.00
	10U	V	0.05	0.12	0.20
	20L		0.03	0.07	0.12
1.5	H	V	0.07	0.17	0.28
	20R		0.03	0.07	0.12
	10D	V	0.05	0.12	0.20

5.1.6 WARPAGE

5.2 Color—The color of the reflected light shall be red, yellow, or white as specified in SAE J578.

5.3 Material Requirements—Plastic materials used in optical parts shall meet the requirements of SAE J576.

5.4 Design Requirements

5.4.1 The face of a Type 2 Wide Angle Reflex Reflector that is directional and requires a fixed orientation when mounted on the vehicle shall be marked with the word TOP. Mounting instructions, including a diagram, shall be included on the outside of the packaging or attached to the reflector and visible when purchased.

5.4.2 Means shall be provided to keep fixed orientation reflex reflectors from rotating more than 2 degrees unless the reflector meets all photometric requirements at the maximum rotation.

5.5 Installation Requirements

5.5.1 Type 2, Wide Angle Reflex Reflectors that are directional may require mounting on the vehicle in a fixed orientation to be effective and shall be mounted on the vehicle in accordance with the manufacturer's instructions.

5.5.2 Type 1 and Type 2 reflectors shall be mounted not less than 380 mm nor more than 1525 mm above the road surface as measured from the center of the reflector at vehicle curb weight.

5.5.3 Visibility of the reflected light shall not be obstructed by any part of the vehicle throughout the specified photometric test pattern unless the reflector is designed to comply with these obstructions in place.

5.5.4 The reflex reflector shall be mounted on the rigid structure of the vehicle.

6. Guidelines—The following guidelines apply to Type 1 and/or Type 2 Reflex Reflectors used on the vehicle and shall not be considered a part of the requirements.

6.1 Design Guidelines

6.1.1 A Type 1 or a Type 2 reflex reflector may be combined with a stop lamp, turn signal lamp, tail lamp, clearance lamp, side marker lamp, identification lamp, or side turn signal lamp.

6.2 Installation Guidelines

6.2.1 It is suggested that Type 2 wide angle reflex reflectors be mounted on the corners of the vehicle facing front and rear and to each side. Type 1 reflex reflectors may be used at other mounting locations as one method of improving passive vehicle delineation.

(R) EMERGENCY WARNING DEVICE (TRIANGULAR SHAPE)—SAE J774 DEC89

SAE Standard

Report of Lighting Committee approved June 1961 and last revised January 1971. Completely revised by the Heavy Duty Lighting Standards Committee December 1989. Rationale statement available.

1. Scope—This document provides test procedures, performance requirements and guidelines for emergency warning devices (triangular shape) that are designed to be carried in motor vehicles and intended for highway use.

2. References

SAE J575 DEC88, Tests for Motor Vehicle Lighting Devices and Components

SAE J576 SEP86, Plastic Materials for Use in Optical Parts Such as Lenses and Reflectors of Motor Vehicle Lighting Devices

SAE J578 MAY88, Color Specification

SAE J594 MAY89, Reflex Reflectors

SAE J759 DEC87, Lighting Identification Code

SAE J774c revised January 1971 for information on "TYPE 2, DOT over DOT" Flare design

Federal Motor Vehicle Safety Standard 125

Federal Highway Administration Parts and Accessories Necessary for Safe Operation Subpart "H", 393.95, Emergency Equipment

3. Definitions

3.1 Emergency Warning Device—A triangular shaped device placed on the highway to warn the driver of an approaching vehicle of a stationary hazard (disabled vehicle) by reflection of light from the headlamps of the approaching vehicle at night or by a fluorescent area in the daytime.

3.2 Fluorescent—The property of emitting visible light due to the absorption of radiation of a shorter wavelength which may be outside the visible spectrum.

4. Identification Code

4.1 Emergency warning devices (triangular shape) may be identified by the code W4 in accordance with SAE J759.

5. Tests

5.1 Emergency warning device (triangular shape) sample submitted for test shall be representative of the device as regularly manufactured and marketed.

5.2 SAE J575 is a part of this document. The following tests are applicable with the modifications as detailed. All tests shall be run on a single device in the order listed. At the conclusion of all tests, the device shall be photometered and shall meet the specified photometric values.

5.2.1 VIBRATION TEST—The complete device in its opaque container shall be tested in the stored position. If a means is not provided to attach the device securely to the vehicle, the device in its container shall be vibration tested in a metal box on the test equipment with a clearance of 25 mm (1 in) to the closest surface of the device when the device is at rest.

5.2.2 DUST TEST—The device shall be tested in its functional position. All units shall be subjected to this test whether sealed or not sealed.

5.2.3 MOISTURE TEST—The device shall be tested in its functional position.

5.2.4 CORROSION TEST—The device shall be tested in its functional position.

5.2.5 PHOTOMETRIC TEST

5.2.5.1 Submit the warning device to the following conditioning sequence, returning the device after each step in the sequence to ambient air at 20°C (68°F) for at least 2 h.

5.2.5.1.1 Low Temperature Test—The device in its functional position shall be conditioned at −40°C ± 3 (−40°F ± 5) for 16 h in a circulating air chamber using ambient air, which would have not less than 30% and not more than 70% relative humidity at 20°C (70°F).

5.2.5.1.2 High Temperature Test—The device in its functional position shall be conditioned at 65°C ± 3 (150°F ± 5) for 16 h in a circulating air chamber using ambient air, which would have not less than 30% and not more than 70% relative humidity at 20°C (70°F).

5.2.5.1.3 Humidity Test—The device in its functional position shall be conditioned at 38°C (100°F) and 90% relative humidity for 16 h.

5.2.5.1.4 Immersion Test—The device in its functional position shall be immersed for 2 h in water at a temperature of 38°C (100°F).

5.2.5.2 *Reflex Reflector Area*—Prevent the orange fluorescent material from affecting the photometric measurements of the reflectivity of the reflex reflector by masking.

The device shall be tested in its functional position in accordance with SAE J594, except that the candela return for each side and at each test point shall be not less than the values specified in Table 1. The total area for each side of the device shall be photometered either in whole or in parts with particular caution regarding beam uniformity.

TABLE 1—PHOTOMETRIC REQUIREMENTS FOR THE RED REFLEX REFLECTOR AREA, EACH SIDE OF THE EMERGENCY WARNING DEVICE

Obs. Angle	Ent. Angle	Minimum Candela Per Incident Footcandle	Minimum Milli-Candela Per Incident Lux
0.2	V-10U	80	7430
	H-30L	8	745
	H-20L	40	3715
	H-V	80	7430
	H-20R	40	3715
	H-30R	8	745
	V-10D	80	7430
1.5	V-10U	0.8	74
	H-30L	0.08	7
	H-20L	0.4	37
	H-V	0.8	74
	H-20R	0.4	37
	H-30L	0.08	7

5.2.5.3 *Fluorescent Area*—Prevent the red reflex reflective material from affecting the photometric measurement of the luminance of the orange fluorescent material by masking.

Using a 150 watt high pressure xenon compact arc lamp as the light source, illuminate the test sample at an angle of incidence of 45 deg and an angle of observation of 90 deg. Measure the luminance of the material at a perpendicular viewing angle with no ray of the viewing beam more than 5 deg from the perpendicular to the specimen.

Repeat the procedure for a flat magnesium oxide surface, and compute the quotient (percentage) of the luminance of the material relative to that of the magnesium oxide surface.

5.3 Stability Test (Wind Test)—The device in its functional position shall be placed on a horizontal brushed concrete surface both with and against the brush marks and subjected to a horizontal wind of 65 km/h (40 mph). The wind shall be directed for 3 min in each position; perpendicular to the device face, first on one side and then the other side and then at three intermediate positions.

5.4 Color

5.4.1 REFLEX REFLECTOR AREA—The test sample may be either the reflex reflector or a disc of the same material, technique of fabrication, and dye formulation as the reflex reflector. If a disc is used, the thickness shall be twice the thickness of the reflector as measured from the face of the reflector to the apexes of the reflecting elements.

5.4.2 FLUORESCENT MATERIAL AREA—A 150 watt high pressure xenon compact arc lamp shall illuminate the sample using the unmodified spectrum at an angle of incidence of 45 deg and an observation of 90 deg.

6. Requirements

6.1 Material Requirements—The plastic material used in optical parts shall meet the requirements of SAE J576.

6.2 Performance Requirements—The device when tested in accordance with the test procedures of SAE J575 and with the modifications detailed in this document shall meet the following requirements:

6.2.1 VIBRATION—The reflex reflector sections shall show no evidence of surface abrasion at the conclusion of the test.

6.2.2 MOISTURE—There shall be no visible moisture within the device at the conclusion of the test.

6.2.3 PHOTOMETRIC

6.2.3.1 *Reflex Reflector*—Both before and after the device has been conditioned the intensity for each side shall be not less than the values specified in Table 1.

6.2.3.2 *Fluorescent*—Both before and after the device has been conditioned the relative luminance shall not be less than 25% of a flat magnesium oxide surface and a minimum product of that relative luminance and width in inches of 44.

6.2.4 COLOR

6.2.4.1 *Reflex Reflector Area*—The color of the reflected light shall be red, as specified in SAE J578.

6.2.4.2 *Fluorescent Area*—The fluorescent material shall be orange and shall have the following characteristics when the source of illumination is a 150 watt high pressure xenon compact arc lamp, expressed in terms of the International Commission on Illumination (CIE) 1931 standard colorimetric observer system. The chromaticity coordinates of the orange fluorescent material shall lie within the region bounded by the spectrum locus and the lines on the diagram defined by the following:

YELLOW	$y = 0.49x + 0.17$
WHITE	$y = 0.93 - x$
RED	$y = 0.35$

6.2.5 STABILITY TEST (WIND TEST)—No part of the device shall slide more than 75 mm (3 in) from its initial position.

The triangular portion shall not tilt to a position that is more than 10 deg from vertical.

The device shall not turn through a horizontal angle of more than 10 deg in either direction from the initial position.

6.2.6 DURABILITY—After all testing has been completed, the device shall be functional and no part of the device shall be warped or separated from the rest of the device.

6.3 Design Requirements

6.3.1 The emergency warning device (triangular shape) shall form an equilateral triangle and each side shall display both a daytime and a nighttime warning area. The device shall stand in a plane not more than 10 deg from the vertical, with the lower base of the triangle horizontal and not less than 25 mm (1 in) above the road surface.

6.3.2 The daytime warning shall be an orange fluorescent area meeting the color and luminance requirements specified.

6.3.3 The nighttime warning shall be a red retroreflective area meeting the color and photometric requirements specified.

6.3.4 Each of the three legs of the triangular portion of the warning device shall not be less than 430 mm (17 in) and not more than 560 mm (22 in) and not more than 75 mm (3 in) wide. See Figure 1.

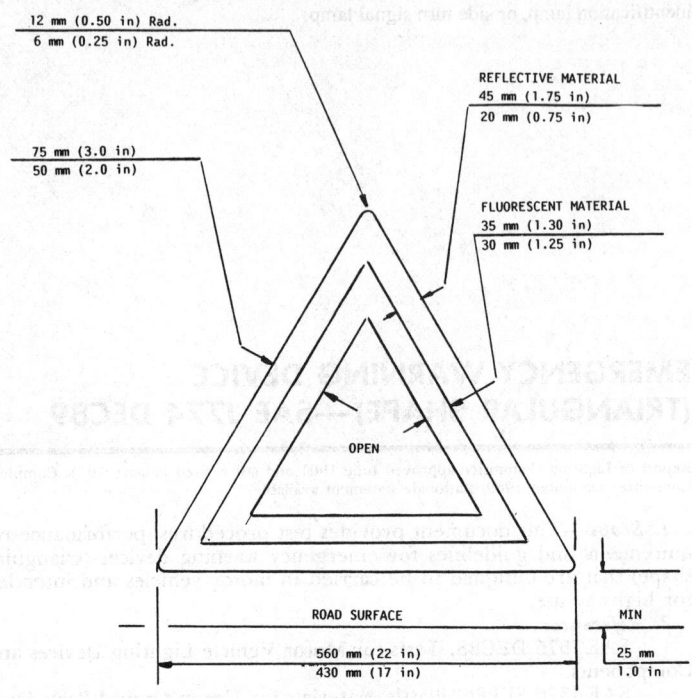

FIG. 1—DIMENSIONS OF THE EMERGENCY WARNING DEVICE (TRIANGULAR SHAPE)

6.3.5 Each face of the triangular portion of the warning device shall have an outer border of red reflex reflective material of uniform width not less than 20 mm (0.75 in) and not more than 45 mm (1.75 in) wide and an inner border of orange fluorescent material of uniform width not less than 30 mm (1.25 in) and not more than 35 mm (1.30 in) wide.

6.3.6 Each vertex of the triangular portion of the device shall have a radius of not less than 6 mm (0.25 in) and not more than 13 mm (0.5 in).

6.3.7 Each device shall have instructions for its erection and display. The instructions shall be either indelibly printed on the warning device or attached in such a manner that they cannot be easily removed.

6.3.8 The instructions shall include a recommendation that the driver activate the vehicular hazard warning signal lamps before leaving the vehicle to erect the warning devices.

6.3.9 Instructions shall include an illustration indicating the recommended positioning of the warning device on the highway. See Figure 2.

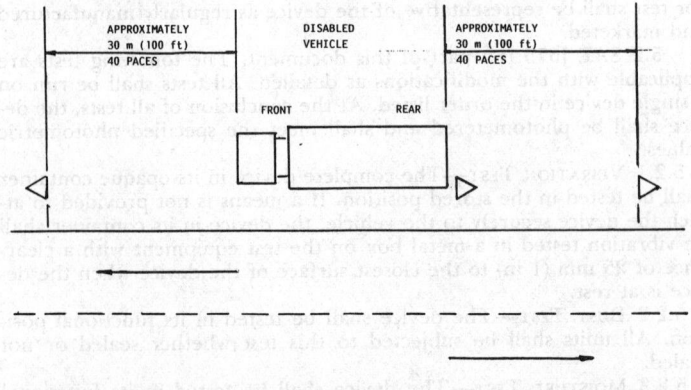

FIG. 2—RECOMMENDED WARNING DEVICE POSITIONING ON THE ROADWAY

FLASHING WARNING LAMPS FOR AUTHORIZED EMERGENCY, MAINTENANCE AND SERVICE VEHICLES—SAE J595 JAN90

SAE Recommended Practice

Report of the Lighting Committee, approved December 1948, completely revised August 1983. Rationale statement available. Reaffirmed by the Emergency Warning Devices Standards Committee January 1990.

1. Scope—This document provides design guidelines, test procedure references, and performance requirements for flashing incandescent warning lamps. It is intended to apply to, but is not limited to, surface land vehicles.

2. Purpose—The purpose of this document is to establish general requirements for flashing warning lamps for use on authorized emergency, maintenance, and service vehicles.

3. Definition—A flashing warning lamp is a lamp in which the light source is turned on and off by circuit interruption producing a repetitive flash of light which is directionally aimed and will project a flashing beam signal over a minimum area from 20 deg right to 20 deg left on a horizontal plane and from 10 deg up to 10 deg down on a vertical plane.

4. Test Procedures—The following sections of SAE J575, are a part of this document:

    Section 2.1—Lighting Devices
    Section 2.2—Bulbs
    Section 2.3—Test Fixture
    Section 3—Laboratory Facilities
    Section 4.1—Vibration Test
    Section 4.2—Moisture Test
    Section 4.3—Dust Test
    Section 4.4—Corrosion Test
    Section 4.5—Color Test (See SAE J578)
    Section 4.6—Photometry Test

Photometric measurements shall be made with the device mounted in its normal operating position and all luminous intensity measurements shall be made with the incandescent filament of the signal lamp 3 m or more from the photometer screen. The lamp shall be mounted so that the horizontal and vertical plane through the photometer axis also passes through the center of the test bulb filament. Photometry shall be done with the filament burning continuously.

    Section 4.8—Warpage Test on Devices with Plastic Components:

A sample device shall be mounted in its normal operating position, operating at a flash rate of 1.50 Hz ± 0.17 with a 50% ± 2 current, on time and at the voltage recommended by the bulb manufacturer in a circulating air oven for 1 h within a temperature range of 46 to 49°C.

5. Dimensional Requirements—The effective projected luminous area measured on a plane at right angles to the axis of the lamp shall be not less than 60 cm^2.

6. General Requirements

6.1 Photometric Design Guidelines—Design guidelines are listed in Table 1—Photometric Design Guidelines.

6.2 Lighting Identification Code—The lighting identification code should be "W" in accordance with SAE J759.

7. Performance Requirements

7.1 Lighting Devices—Sample devices submitted for laboratory tests shall be representative of devices as regularly manufactured and marketed.

7.2 Bulbs—Requirements are based on laboratories using accurately rated bulbs operated at their designed mean spherical luminous intensity. Sealed units shall be seasoned (lighted) at 12.8 V for 1% of their rated average laboratory life or 10 h maximum prior to photometry and then operated at their design voltage. (For units designed to operate on other than a 12-V circuit, check manufacturer for proper seasoning schedule.)

7.3 Vibration—Upon completion of the test, there shall be no observed rotation, displacement, cracking, or rupture of parts of the test device (except bulb(s) and sealed beam unit internal components) that would result in failure of any other tests contained in Section 4 of SAE J575. Cracking or rupture of parts of the device affecting its mounting shall also constitute a failure.

7.4 Moisture—Upon completion of the test, the moisture accumulation in the test device shall be 2 mL or less. For devices with an interior volume greater than 7000 mL, the maximum allowable moisture ac-

TABLE 1—PHOTOMETRIC DESIGN GUIDELINES

Test Points, deg	Luminous Intensity, Candela		
	White	Yellow	Red
5L	80	40	20
10U-V	200	100	50
5R	80	40	20
20L	80	40	20
10L	200	100	50
5L	400	200	100
5U-V	600	300	150
5R	400	200	100
10R	200	100	50
20R	80	40	20
20L	120	60	30
10L	300	150	75
5L	800	400	200
H-V	1200	600	300
5R	800	400	200
10R	300	150	75
20R	120	60	30
20L	80	40	20
10L	200	100	50
5L	400	200	100
5D-V	600	300	150
5R	400	200	100
10R	200	100	50
20R	80	40	20
5L	80	40	20
10D-V	200	100	50
5R	80	40	20

TABLE 2—PHOTOMETRIC REQUIREMENTS

Zones	Test Points deg	Luminous Intensity, Candela		
		White Zone Total	Yellow Zone Total	Red Zone Total
1	5U-10L 5U-20L H-20L 5D-20L 5D-10L	600	300	150
2	10U-5L 10U-V 10U-5R	320	160	80
3	5U-5L H-10L 5D-5L	1000	500	250
4	5U-V H-5L H-V H-5R 5D-V	3600	1800	900
5	5U-5R H-10R 5D-5R	1000	500	250
6	10D-5L 10D-V 10D-5R	320	160	80
7	5U-10R 5U-20R H-20R 5D-20R 5D-10R	600	300	150

cumulation in the test device shall be 0.03% of the total interior volume of the test device.

7.5 Dust—On completion of the test, the test device shall be considered to have met all the requirements of the dust test when complying with either of the following requirements:

7.5.1 No dust shall be found on the interior surface of the test device, or

7.5.2 The ratio of the maximum luminous intensities (exterior only cleaned to exterior and interior cleaned) shall be a minimum of 0.9.

7.6 Corrosion—On completion of the test, there shall be no observed corrosion that would result in the failure of any other test contained in Section 4 of SAE J575.

7.7 Color—The color of the light emitted from the flashing warning lamps shall be white, yellow, or red as specified in SAE J578.

7.8 Photometry—The lamp shall meet the zonal photometric requirements of Table 2.

7.8.1 For the device to comply with the photometric performance requirements, the summation of the luminous intensity values shall meet the values specified in Table 2.

7.8.2 The measured luminous intensity at each test point shall be not less than 60% of the values specified in Table 1.

7.8.3 An adjustment in lamp orientation from the design position may be made in determining compliance to the performance photometric requirements, provided such adjustment does not exceed 3 deg in any direction. All zone totals must comply after reaim.

7.9 Warpage—Upon completion of the test, there shall be no observed warpage of plastic components of the test device that would result in the failure of any other test contained in Section 4 of SAE J575.

8. Plastic Material—Plastic materials used in optical parts shall comply with the requirements of SAE J576.

9. General Installation Recommendations—These general recommendations apply to the device as used on the vehicle and are not part of the performance requirements.

9.1 Front and rear warning lamps should be mounted as high and as far apart as practicable. The location of front warning lamps should be such that they can be clearly distinguished when the headlamps are lighted on the lower beam.

9.2 Visibility of the warning lamps should be unobstructed by any part of the vehicle 10 deg above to 10 deg below the horizontal and from 45 deg to the right to 45 deg to the left of the centerline of the vehicle.

9.3 There should be a visible or audible means of giving a clear and unmistakable indication to the driver when the warning lamps are turned "on."

9.4 To improve the efficiency of the signal, it is recommended that, where practical, the area surrounding the lamp should be black.

9.5 Flash Rate—The flash rate when observed from a fixed position shall be between 1.0–2.0 Hz—when operated at the voltage recommended by the manufacturer. The "on" period of the lamp shall be as specified in SAE J945 and/or SAE J1054.

APPENDIX A

Appendix A contains additional information considered useful in application to this document.

Attention is called to SAE J567, for requirements and gages to be used in socket design.

HAZARD WARNING SIGNAL SWITCH—SAE J910 OCT88

SAE Standard

Report of Lighting Committee approved January 1965 and completely revised October 1988. Rationale statement available.

1. Scope—This standard defines the test conditions, procedures and performance specifications for 6, 12 and 24-V manually actuated hazard warning signal switch.

2. Definition—The hazard warning switch is an operator actuated device whose function is to cause at least one turn signal lamp on the left and right of the front, and left and right of the rear of the vehicle to flash simultaneously to indicate to the approaching driver the presence of a vehicular hazard.

2.1 Combination Turn Signal and Hazard Warning Signal

2.1.1 A combination switch is defined as a hazard warning switch combined in the same housing as the turn signal switch.

2.1.2 The operating motion of the hazard warning signal switch function shall differ from the actuating motion of the turn signal function.

3. Test Requirements

3.1 Test Equipment and Instrumentation

3.1.1 POWER SUPPLY—The power supply shall not generate any adverse transients not present in motor vehicles, and shall comply with the following specifications:

  a. Output Current—The power supply shall be capable of supplying the continuous current of the design electrical load and the in-rush current, as required by the bulb load complement.

  b. Output Regulation
Dynamic—The dynamic output voltage at the supply shall not deviate more than 1.0 V from zero to maximum load (including in-rush current) and shall recover 63% of its maximum excursion within 100 ms.
Static—The static output voltage at the supply shall not deviate more than 2% with changes in static load from zero to maximum (not including in-rush current), and means shall be provided to compensate for static input line voltage variations.

  c. Ripple Voltage—The ripple output voltage shall be a maximum of 300 mV peak-to-peak.

3.1.2 VOLTMETER—A voltmeter with a 0-30 V maximum full scale deflection and ±1/2% accuracy should be used. NOTE: A digital meter having at least a 3-1/2 digit readout with an accuracy of ±1% plus 1 digit is recommended for mV readings.

3.1.3 AMMETER—Capable of carrying full system load current, with an accuracy of ±3%.

3.2 Test Procedures—It is essential to duplicate specific environmental conditions under which the device is expected to function.

3.2.1 ELECTRICAL LOADS

3.2.1.1 The design load applied to the switch is the electrical load specified by the quantity and type of bulbs (or other electrical load devices) to be operated by each circuit of the hazard warning signal switch.

3.2.1.2 The switch shall be operated with the maximum design bulb load stated by the switch manufacturer with the flasher not included in the circuit unless the flasher is an integral part of the assembly.

3.2.1.3 The switch shall be operated at 6.4 ± 0.2 V DC for a 6-V system, 12.8 ± 0.2 V DC for a 12-V system, or 25.6 ± 0.2 V DC for a 24-V system. These voltages shall be the open circuit voltage measured at the input termination on the switch.

3.2.2 TEMPERATURE TEST PROCEDURE

3.2.2.1 The switch shall be manually cycled after a 1-h exposure with no electrical load at each of these temperatures: 25 ± 5°C; 74 + 0, −3°C; −32 +3, −0°C. The switch shall be manually cycled at each temperature for 10 cycles at the designed loads.

3.2.2.2 The same hazard warning signal switch shall be used for the described endurance test described in paragraph 3.2.3.

3.2.3 ENDURANCE TEST PROCEDURE

3.2.3.1 The switch shall be electrically connected to operate its design load for 7500 cycles at a temperature of 25 ± 5°C, followed by a 1-h "on" at a temperature of 25 ± 5°C.

3.2.3.2 One complete cycle shall consist of sequencing it through each position (with dwell in each position): Off, On, Off.

3.2.3.3 The test equipment shall be arranged to provide the following switch operating time requirements:
Travel Time - 0.1–0.5 s (time from one position to next)
Dwell Time - 1.0–2.0 s (time in each position)
Make and Break Rate - 130–150 mm/s

3.2.4 VOLTAGE DROP TEST PROCEDURE

3.2.4.1 The voltage drop from the input terminals to the corresponding output terminals shall be measured at the beginning of the test and immediately after the endurance test. Cycling the switch three times prior to taking readings is permitted and the reading should be taken right after cycling. A total of five readings can be taken and the average reading will prevail.

3.2.4.2 If wiring is an integral part of the switch, the voltage drop measurement shall be made including 75 ± 6 mm of wire on each side of the switch; otherwise, the measurement shall be made at the switch terminals.

4. Performance Requirements

4.1 During and after each of the tests described in paragraphs 3.2.2 and 3.2.3, the switch shall be electrically and mechanically operable.

4.2 Combination Turn Signal and Hazard Warning—The same combination switch shall be used for the test of each function. The hazard warning switch shall meet the requirements of this standard. The turn signal switch shall meet the requirements of SAE J589.

4.3 The voltage drop shall not exceed 0.30 V.

(R) 360 DEGREE WARNING DEVICES FOR AUTHORIZED EMERGENCY, MAINTENANCE, AND SERVICE VEHICLES—SAE J845 MAR92 — SAE Recommended Practice

Report of the Lighting Committee, approved January 1963, completely revised January 1984. Completely revised by the SAE Lighting Coordinating Committee and the SAE Emergency Warning Lamp and Devices Standards Committee March 1992. Rationale statement available.

1. Scope—This SAE Recommended Practice provides test procedures, requirements, and guidelines for single color, 360 degree warning devices.

2. References

2.1 Applicable Documents—The following publications form a part of this specification to the extent specified herein. The latest issue of SAE publications shall apply.

2.1.1 SAE PUBLICATIONS—Available from SAE, 400 Commonwealth Drive, Warrendale, PA 15096-0001.

SAE J575—Tests for Motor Vehicle Lighting Devices and Components
SAE J576—Materials for Use in Optical Parts Such as Lenses and Reflectors of Motor Vehicle Lighting Devices
SAE J578—Color Specification of Electric Signal Lighting Devices
SAE J590—Turn Signal Flashers

2.1.2 OTHER PUBLICATIONS

National Bureau of Standards Special Publication 480-16—Emergency Vehcle Warning Lights: State of the Arts

2.2 Definitions

2.2.1 360 DEGREE WARNING DEVICE—A device that projects light in a horizontal 360 degree arc. It will appear to project a regularly repeating pattern of flashes to an observer positioned at a fixed location. Its function is to inform other highway users to stop, yield right-of-way, or indicate the existence of a hazardous situation.

2.2.2 ROTATING SIGNAL DEVICE—A warning device in which the beam or beams rotate either because one or more lamps rotate around fixed axes or because one or more lenses, reflectors, or mirrors rotate around fixed 360 degree light sources projecting light through a 360 degree arc.

2.2.3 OSCILLATING SIGNAL DEVICE—A warning device in which the beam or beams oscillate (turn back and forth) through fixed angles, either because one or more lamps oscillate around fixed axes or because one or more lenses, reflectors, or mirrors oscillate around fixed 360 degree light sources projecting light though a 360 degree arc.

2.2.4 FLASHING SIGNAL DEVICE—A warning device in which the light source is turned on and off through circuit interruption producing repetitive flashes of light to all points on a 360 degree arc.

2.2.5 PRIMARY WARNING DEVICE—Devices or groups of devices that are intended to provide the primary visual warning signal as called out in each service class. Unless prohibited by law or regulation, a Class 1 device may be used in place of Class 2 device and a Class 1 or 2 device in place of Class 3 device.

2.2.6 SECONDARY WARNING DEVICES—Devices or groups of devices of lower performance that can be used to provide supplemental warning to that provided by the primary warning device or devices.

2.2.7 CLASS 1 WARNING DEVICES—Primary warning devices for use on authorized emergency vehicles responding to emergency situations. These devices are utilized to capture the attention of motorists and pedestrians and warn of a potentially hazardous activity or situation.

2.2.8 CLASS 2 WARNING DEVICES—Primary warning devices for use on authorized maintenance or service vehicles to warn of traffic hazards such as a lane blockage or slow moving vehicle.

2.2.9 CLASS 3 WARNING DEVICES—Primary warning devices for use on vehicles authorized to display a warning device for identification only.

3. Lighting Identification Code—360 Degree warning devices may be identified by the codes:
a. W3-1, Class 1
b. W3-2, Class 2
c. W3-3, Class 3
in accordance with SAE J759.

4. Tests

4.1 SAE J575 is a part of this report. The following tests are applicable with the modifications as indicated.

All tests are to be made at 12.8 V dc for devices intended for operation on 12.8 V systems and 25.6 V dc for 24 V systems.

4.1.1 VIBRATION TEST
4.1.2 MOISTURE TEST
4.1.3 DUST TEST
4.1.4 CORROSION TEST
4.1.5 PHOTOMETRY—In addition to the test procedures in SAE J575, the following apply:

4.1.5.1 *Flash Energy Method for Measuring Photometric Performance*—For 360 degree warning devices producing flashes of 0.125 s or less duration, photometric performance can be determined by measuring flash energy. The flash rate shall be measured and recorded at the end of the test.

4.1.5.1.1 *Alternate Method of Measuring Photometric Performance*—For 360 degree warning devices with a flash rate of 2 Hz or less, photometric performance can be determined by measuring the luminous intensity of the steady on light source (not flashing).

4.1.5.2 *Devices Flashed by Current Interruption*—Photometric measurements shall be made with the device mounted in its normal operating position and all measurements shall be made with the incandescent filament of the device at least 18 m from the photometer. The device shall be mounted so that the horizontal plane through the photometer axis passes through the center of the light source. The vertical axis through the center of the light source shall be perpendicular to this horizontal plane. The device shall be turned about its vertical axis until the photometer indicates minimum reading. This shall be the H-V point.

4.1.5.3 *Devices Flashed by Rotation or Oscillation* — Photometric measurements shall be made with the device mounted in its normal operating position and all measurements shall be made with the incandescent filament of the device at least 18 m from the photometer. The device shall be mounted so that the horizontal plane through the photometer axis passes through the center of the light source of the rotating element. The vertical axis through the center of rotation shall be perpendicular to this horizontal plane. The rotating element shall be turned about its vertical axis until the photometer indicates the maximum reading.

For a device with a symmetrical lens, filter, or lamp filament orientation, this shall be the H-V point.

For a device with an asymmetrical lens, filter, or lamp filament orientation, the H-V point shall be determined as previously stated after rotating the device about its vertical axis to determine the point of lowest photometric performance.

4.1.6 WARPAGE TEST FOR PLASTIC COMPONENTS
4.2 Color Test — SAE J578 is a part of this report.
4.3 Additional Tests
4.3.1 HIGH TEMPERATURE FLASH RATE TEST — The device shall be subjected to an ambient temperature of 50 °C ± 3 °C for a period of 6 h. The device shall be off (not operating) during the first hour and shall operate continuously for the next 5 h of the test. The flash range shall be measured before the test, not less than 3 min nor more than 4 min after the beginning of the second hour of the test, and not less than 3 min nor more than 4 min after the end of the test.

4.3.2 LOW TEMPERATURE FLASH RATE TEST — The device shall be subjected to an ambient temperature of -30 °C ± 3 °C for a period of 6 h. The device shall be off (not operating) during the first 5 h and shall operate continuously for the last hour of the test. The flash rate shall be measured before the test, not less than 3 min nor more than 4 min after the beginning of the last hour of the test, and not less than 3 min nor more than 4 min after the end of the test.

4.3.3 DURABILITY TEST — The device shall be operated continuously for 200 h at an ambient temperature of 25 °C ± 3 °C in cycles consisting of 50 min on and 10 min off. The flash rate shall be measured before the test and not more than 3 min after the last off period at the end of the test.

5. Requirements

5.1 Performance Requirements — A device, when tested in accordance with the test procedures specified in Section 4, shall meet the following requirements.

5.1.1 VIBRATION — SAE J575
5.1.2 MOISTURE — SAE J575
5.1.3 DUST — SAE J575
5.1.4 CORROSION — SAE J575
5.1.5 PHOTOMETRY — SAE J575

5.1.5.1 *Flash Energy* — The device shall meet the photometric requirements contained in Tables 1, 2, or 3, and Tables 4, 5, or 6, Photometric Requirements and their footnotes. The summation of the flash energy measurements at the specified test points in a zone shall be at least the value shown.

5.1.5.2 *Alternate Method* — The device shall meet the photometric requirements contained in Tables 7, 8, or 9 and Tables 10, 11, or 12, Photometric Requirements and their footnotes. The summation of the luminous intensity measurements at the specified test points in a zone shall be at least the value shown.

The steady-state totals for warning devices shown in the referenced Tables (7, 8, or 9, and 10, 11, or 12) apply to devices where two or more lamps, lenses, reflectors or mirrors rotate or oscillate around fixed axes.

5.1.5.3 For warning devices having only one lamp, lens, reflector or mirror, the steady-state zone totals and design guidelines shall be twice those shown in these Tables.

TABLE 1 — PHOTOMETRIC REQUIREMENTS — CLASS 1 WARNING LAMPS

Zone	Test Points Degrees	Minimum Flash Energy Candela Seconds White	Minimum Flash Energy Candela Seconds Yellow	Minimum Flash Energy Candela Seconds Red	Minimum Flash Energy Candela Seconds Signal Blue
1	5U-V 2.1/2U-V H-V 2.1/2D-V 5D-V	396	198	99	99

Notes:
1. A one-time adjustment in lamp orientation from design position may be made in determining compliance to Tables 1 and 4, provided each adjustment does not exceed 1 degree in any direction. The same shall comply after this one time, final reaim.
2. The measured value at each test point shall not be less than 60% of the minimum values in Table 4.

TABLE 2 — PHOTOMETRIC REQUIREMENTS — CLASS 2 WARNING LAMPS

Zone	Test Points Degrees	Minimum Flash Energy Candela Seconds White	Minimum Flash Energy Candela Seconds Yellow	Minimum Flash Energy Candela Seconds Red	Minimum Flash Energy Candela Seconds Signal Blue
1	5U-V 2.1/2U-V H-V 2.1/2D-V 5D-V	99	49.5	25	25

Notes:
1. A one-time adjustment in lamp orientation from design position may be made in determining compliance to Tables 2 and 5, provided such adjustment does not exceed 1 degree in any direction. The zone shall comply after this one time, final reaim.
2. The measured value at each test point shall not be less than 60% of the minimum values in Table 5.

5.1.6 WARPAGE—SAE J575
5.1.7 COLOR—The color of light emitted shall be white, yellow, red, or signal blue as specified in SAE J578.

5.2 Material Requirements—Plastic materials used in optical parts shall meet the requirements of SAE J576.

5.3 Additional Requirements

5.3.1 HIGH TEMPERATURE—There shall be no evidence of operating conditions which would result in failure to comply with Section 5 of this document. The flash rate at each of the required measurements shall be not less than 0.8 Hz nor more that 133% of the flash rate measured per 4.1.5.1.

However, if photometric performance was determined using the Alternate Method in 4.1.5.1.1, the flash rate at each of the required measurements shall be not less than 0.8 Hz nor more that 2.2 Hz.

5.3.2 LOW TEMPERATURE—There shall be no evidence of operating conditions which would result in failure to comply with Section 5 of this document. The flash rate at each of the required measurements shall be not less than 0.8 Hz nor more than 133% of the flash rate measured per 4.1.5.1.

However, if photometric performance was determined using the Alternate Method in 4.1.5.1.1, the flash rate at each of the required measurements shall be not less than 0.8 Hz nor more than 2.2 Hz.

5.3.3 DURABILITY—There shall be no evidence of operating conditions which would result in failure to comply with Section 5 of this document. The flash rate at each of the required measurements shall be not less than 1 Hz nor more than 133% of the flash rate measured in 4.1.5.1 up to a maximum of 4 Hz.

However, if photometric performance was determined using the Alternate Method in 4.1.5.2, the flash rate at each of the required measurements shall be not less than 1 Hz nor more than 2 Hz.

6. Guidelines

6.1 Photometric

6.1.1 FLASH ENERGY—For devices tested in accordance with 4.1.5.1, the Photometric Design Guidelines are contained in Tables 4, 5, or 6.

6.1.2 ALTERNATE METHOD—For devices tested in accordance with 4.1.5.1.1, the Photometric Design Guidelines are contained in Tables 10, 11, or 12.

6.2 Installation Guidelines—The following guidelines apply to 360 degree warning lamps as used on the vehicle and shall not be considered part of the requirements.

6.2.1 MOUNTING—The vertical axis of the device shall be installed normal to the longitudinal axis of the vehicle.

6.2.2 VISIBILITY—Visibility of the 360 degree warning lamp should be unobstructed by any part of the vehicle 5 degrees above to 5 degrees below the horizontal and provide a flashing light throughout a 360 degree circle.

6.2.3 INDICATOR—There should be a visible or audible means of giving a clear and unmistakable indication to the driver when the warning lamps are turned on and functioning normally.

6.2.4 "ON" TIME—For current interrupted devices the on period of the lamp shall be as specified in SAE J590.

APPENDIX A

Following the procedure outlined in the National Bureau of Standards Special Publication 480-16 entitled, Emergency Vehicle Warning Lights: State of the Arts, it is possible to directly calculate the flash energy produced by the sealed beam lamp typically used in rotating warning lights and which is used as the basis for the steady-state photometry tables.

In section 10.6 NBS demonstrates that the effective intensity (I_e) of a 360 degree warning lamp utilizing two sealed beam lamps having a 4.5 degree wide by 11 degree tall beam spread producing 90 fpm (45 rpm) is equal to 4.4% of the peak steady-state intensity (I).

Current definition of effective intensity (I_e) is:

TABLE 3—PHOTOMETRIC REQUIREMENTS—CLASS 3 WARNING LAMPS

Zone	Test Points Degrees	Minimum Flash Energy Candela Seconds White	Minimum Flash Energy Candela Seconds Yellow	Minimum Flash Energy Candela Seconds Red	Minimum Flash Energy Candela Seconds Signal Blue
1	5U-V 2.1/2U-V H-V 2.1/2D-V 5D-V	40	20	10	10

Notes:
1. A one-time adjustment is lamp orientation from design position may be made in determining compliance to Tables 3 and 6, provided such adjustment does not exceed 1 degree in any direction. The zone shall comply after this one time, final reaim.
2. The measured value at each test point shall not be less than 60% of the minimum values in Table 6.

TABLE 4—PHOTOMETRIC DESIGN GUIDELINES—CLASS 1 WARNING LAMPS

Zone	Test Points Degrees	Minimum Flash Energy Candela Seconds White	Minimum Flash Energy Candela Seconds Yellow	Minimum Flash Energy Candela Seconds Red	Minimum Flash Energy Candela Seconds Signal Blue
1	5U-V	10	9	4.5	4.5
	2.1/2U-V	90	45	22.5	22.5
	H-V	180	90	45	45
	2.1/2D-V	90	45	22.5	22.5
	5D-V	10	9	4.5	4.5

TABLE 5 — PHOTOMETRIC DESIGN GUIDELINES — CLASS 2 WARNING LAMPS

Zone	Test Points Degrees	Minimum Flash Energy Candela Seconds White	Minimum Flash Energy Candela Seconds Yellow	Minimum Flash Energy Candela Seconds Red	Minimum Flash Energy Candela Seconds Signal Blue
1	5U-V	4.5	2	1	1
	2.1/2U-V	22.5	11.5	6	6
	H-V	45	22.5	11	11
	2.1/2D-V	22.5	11.5	6	6
	5D-V	4.5	2	1	1

TABLE 6 — PHOTOMETRIC DESIGN GUIDELINES — CLASS 3 WARNING LAMPS

Zone	Test Points Degrees	Minimum Flash Energy Candela Seconds White	Minimum Flash Energy Candela Seconds Yellow	Minimum Flash Energy Candela Seconds Red	Minimum Flash Energy Candela Seconds Signal Blue
1	5U-V	1	1	0.5	0.5
	2.1/2U-V	9	4.5	2	2
	H-V	18	9	5	5
	2.1/2D-V	9	4.5	2	2
	5D-V	1	1	0.5	0.5

TABLE 7 — PHOTOMETRIC REQUIREMENTS (ALTERNATE METHOD) — CLASS 1 WARNING LAMPS

Zone	Test Points Degrees	Minimum Luminous Intensity Zone Totals Candela White	Minimum Luminous Intensity Zone Totals Candela Yellow	Minimum Luminous Intensity Zone Totals Candela Red	Minimum Luminous Intensity Zone Totals Candela Signal Blue
1	5U-V 2.1/2U-V H-V 2.1/2D-V 5D-V	39 600	19 800	9 900	9 900

Notes:
1. A one-time adjustment in lamp orientation from design position may be made in determining compliance to Tables 7 and 10 provided such adjustment does not exceed 1 degree in any direction. The zone shall comply after this one time, final reaim.
2. The measured value of each test point shall not be less than 60% of the minimum values in Table 10.

TABLE 8 — PHOTOMETRIC REQUIREMENTS (ALTERNATE METHOD) — CLASS 2 WARNING LAMPS

Zone	Test Points Degrees	Minimum Luminous Intensity Zone Totals Candela White	Minimum Luminous Intensity Zone Totals Candela Yellow	Minimum Luminous Intensity Zone Totals Candela Red	Minimum Luminous Intensity Zone Totals Candela Signal Blue
1	5U-V 2.1/2U-V H-V 2.1/2D-V 5D-V	9 900	4 950	2 475	2 475

Notes:
1. A one-time adjustment in lamp orientation from design position may be made in determining compliance to Tables 8 and 11 provided such adjustment does not exceed 1 degree in any direction. The zone shall comply after this one time, final reaim.
2. The measured value at each test point shall not be less than 60% of the minimum values in Table 11.

TABLE 9 — PHOTOMETRIC REQUIREMENTS (ALTERNATE METHOD) — CLASS 3 WARNING LAMPS

Zone	Test Points Degrees	Minimum Luminous Intensity Zone Totals Candela White	Minimum Luminous Intensity Zone Totals Candela Yellow	Minimum Luminous Intensity Zone Totals Candela Red	Minimum Luminous Intensity Zone Totals Candela Signal Blue
1	5U-V 2.1/2U-V H-V 2.1/2D-V 5D-V	3 960	1 980	990	990

Notes:
1. A one-time adjustment in lamp orientation from design position may be made in determining compliance to Tables 9 and 12 provided such adjustment does not exceed 1 degree in any direction. The zone shall comply after this one time, final reaim.
2. The measured value of each test point shall not be less than 60% of the minimum values in Table 12.

TABLE 10 — PHOTOMETRIC DESIGN GUIDELINES (ALTERNATE METHOD) — CLASS 1 WARNING LAMPS

Zone	Test Points Degrees	Minimum Luminous Intensity Steady-State Beam Candela White	Minimum Luminous Intensity Steady-State Beam Candela Yellow	Minimum Luminous Intensity Steady-State Beam Candela Red	Minimum Luminous Intensity Steady-State Beam Candela Signal Blue
1	5U-V	1 000	900	450	450
	2.1/2U-V	9 000	4 500	2 250	2 250
	H-V	18 000	9 000	4 500	4 500
	2.1/2D-V	9 000	4 500	2 250	2 250
	5D-V	1 000	900	450	450

TABLE 11 — PHOTOMETRIC DESIGN GUIDELINES (ALTERNATE METHOD) — CLASS 2 WARNING LAMPS

Zone	Test Points Degrees	Minimum Luminous Intensity Steady-State Beam Candela White	Minimum Luminous Intensity Steady-State Beam Candela Yellow	Minimum Luminous Intensity Steady-State Beam Candela Red	Minimum Luminous Intensity Steady-State Beam Candela Signal Blue
1	5U-V	250	225	113	113
	2.1/2U-V	2 250	1 125	562	562
	H-V	4 500	2 250	1 125	1 125
	2.1/2D-V	2 250	1 125	562	562
	5D-V	250	225	113	113

TABLE 12 — PHOTOMETRIC DESIGN GUIDELINES (ALTERNATE METHOD) — CLASS 3 WARNING LAMPS

Zone	Test Points Degrees	Minimum Luminous Intensity Steady-State Beam Candela White	Minimum Luminous Intensity Steady-State Beam Candela Yellow	Minimum Luminous Intensity Steady-State Beam Candela Red	Minimum Luminous Intensity Steady-State Beam Candela Signal Blue
1	5U-V	180	90	45	45
	2.1/2U-V	900	450	225	225
	H-V	1 800	900	450	450
	2.1/2D-V	900	450	225	225
	5D-V	180	90	45	45

$$I_e = \max_{t_1, t_2} \frac{\int_{t_1}^{t_2} I(t)dt}{0.2 + (t_2 - t_1)} \quad \text{(Eq. A1)}$$

where:
  I_e = the effective intensity of the flashing light
  $I(t)$ = the instantaneous actual intensity of the light at the time (t) during the course of a single flash
  t_1 and t_2 = the beginning and ending times of the useful, higher intensity portion of the flash
  0.2 = constant
  max t_1, t_2 = the maximum value obtainable through variation of both t_1 and t_2

In this equation, $\int_{t_1}^{t_2} I(t)\,dt$ is the flash energy contained in the signal from one lamp, candela seconds.

Following the NBS sample calculation 4.5 degree beam spread lamp rotating at 45 rpm (90 fpm) has a flash interval of:

$$\frac{4.5 \text{ degrees} \times 60 \text{ s/min}}{360 \frac{\text{deg}}{\text{rev}} \times 45 \frac{\text{rev}}{\text{min}}} = 0.0167 \text{ s} \quad \text{(Eq. A2)}$$

Thus, for a 20 000 cd lamp having a beam spread of 4.5 degree wide and rotating at 45 rpm (90 fpm), the flash energy (candela seconds) can be calculated as follows in equation A3:

Flash Energy (candela second)

$$\text{Flash Energy} = \int_{t_1}^{t_2} I(t)\,dt \quad \text{(Eq. A3)}$$

$$= I_e \times [0.2 + (t_2 - t_1)]$$

$$= \frac{I_e}{I_o} \times I \times [0.2 + (t_2 - t_1)]$$

$$= (0.044)(20\,000)(0.2 + 0.0167) = 190.7 \text{ cd s}$$

Rounding this value to 200 cd s yields the relationship of flash energy being numerically equal to 1% of peak intensity for the typical sealed beam lamp traditionally used in 360 degree rotating warning lamps, having a beam spread of 4.5 degree horizontal by 11 degree vertical and flashing at 90 fpm.

GASEOUS DISCHARGE WARNING LAMP FOR AUTHORIZED EMERGENCY, MAINTENANCE, AND SERVICE VEHICLES—SAE J1318 APR86

SAE Recommended Practice

Report of the Lighting Committee approved April 1986. Rationale statement available.

1. Scope—This SAE Recommended Practice provides test procedures, requirements, and guidelines for single color gaseous discharge warning lamps.

2. Definitions

2.1 Light Pulse—A sudden emission of light of short duration and high intensity.

2.2 Light Flash—A single light pulse or a train of pulses. In order to be considered a flash all pulse peaks must occur within 100 ms.

2.3 Gaseous Discharge Warning Lamp—A device that produces a regularly repeating pattern of light flashes when electrical current is discharged periodically through an ionized gas.

2.4 360 Deg Warning Lamp—A lamp that projects a light in a horizontal 360 deg arc. It will appear to project a regularly repeating pattern of flashes to an observer positioned at a fixed location. Its function is to inform other highway users to stop, yield right-of-way, or to indicate the existence of a hazardous situation.

2.5 Directional Warning Lamp—A lamp that produces a repetitive flash of light which is directionally aimed and will project a flashing beam signal over a minimum area from 20 deg right to 20 deg left on a horizontal plane and from 10 deg up to 10 deg down on a vertical plane.

2.6 Primary Warning Lamps—Lamps or groups of lamps that are intended to provide the primary visual warning signal as called out in each service class.

2.7 Secondary Warning Lamps—Lamps or groups of lamps that can be used to provide a supplemental warning signal for each service class.

2.8 Class 1 Warning Lamps—Primary warning lamps for use on authorized emergency vehicles responding to emergency situations. These lamps are utilized to capture the attention of motorists and pedestrians and to warn of a potentially hazardous activity or situation.

2.9 Class 2 Warning Lamps—Primary warning lamps for use on authorized maintenance and service vehicles to warn of traffic hazards such as an accident, slow moving service truck, etc.

2.10 Class 3 Warning Lamps—Primary warning lamps for use on vehicles that are authorized to display flashing warning lamps for identification only.

2.11 Flash Energy—Flash energy is the total luminous energy per unit solid angle contained in the entire flash in candela seconds.

$$\text{candela-second} = \int_{t_1}^{t_2} I\,dt$$

where: I = Instantaneous intensity (candela)
t_1 = Time at start of flash (seconds)
t_2 = Time at end of flash (seconds)
$(t_2 - t_1)$ = Flash Duration (seconds)

2.12 Light Center—The light center of a gaseous discharge warning lamp is the geometric center of the light emitting element (arc or light source) of the lamp.

2.13 Flash Cycle—A sequence of light flashes and dark intervals which, with regular repetition, constitutes the complete output cycle of a flashing lamp. For a simple flashing lamp, the full cycle consists of a single on-off sequence.

3. Lighting Identification Code—Gaseous discharge warning lamps may be identified in accordance with SAE J759, Lighting Identification Code, by the codes:

360 deg gaseous discharge lamps

(W5-1)—Class 1
(W5-2)—Class 2
(W5-3)—Class 3

Directional gaseous discharge lamps

(W5)—

4. Tests

4.1 SAE J575, Tests for Motor Vehicle Lighting Devices and Components, is a part of this report. The following tests are applicable with modifications as indicated:

4.1.1 VIBRATION TEST
4.1.2 MOISTURE TEST
4.1.3 DUST TEST
4.1.4 CORROSION TEST
4.1.5 PHOTOMETRY—In addition to the photometric test procedures in SAE J575, the following apply:

4.1.5.1 The device shall be allowed to operate for 15 min prior to making photometric measurements. In all instances where a device is required to be operated during a test specified in this report, the voltage applied to the input wires or terminals of the device shall be 12.8 V for nominal 12 V electrical systems and 25.6 V for nominal 24 V electrical systems.

4.1.5.2 *Photometric Measurement for 360 Deg Gaseous Discharge Warning Lamps*—Photometric measurements shall be made with the device mounted in its normal operating position and all flash energy measurements shall be made with the light source of the signal lamp at least 18 m from the photometer sensor. The lamps shall be mounted so that the horizontal plane through the photometer axis passes through the center of the light source. The vertical axis through the center of light source shall be perpendicular to this horizontal plane.

The lamp shall be turned about its vertical axis until the photometer indicates minimum flash energy. This shall be the H-V point.

4.1.5.3 *Photometric Measurement for Directional Gaseous Discharge Warning Lamps*—Photometric measurements shall be made with the device mounted in its normal operating position and all flash energy measurements shall be made with the light source of the warning lamp at least 18 m from the photometer sensor. The lamps shall be mounted so that the horizontal plane through the photometer sensor axis passes through the center of the light source. The vertical axis through the center of the light source shall be perpendicular to this horizontal plane.

4.1.5.4 Photometric luminous intensity measurements (candela seconds) shall be taken as the average of ten consecutive flash cycles. There shall be an off time before each flash of at least 50% of the total flash cycle time.

4.1.6 WARPAGE TEST ON DEVICE WITH PLASTIC COMPONENTS—The test described in paragraph 4.8.3.3 of SAE J575 shall be omitted and the following test conducted:

4.1.6.1 *Flash Tube Operation*—Unless otherwise specified, the gaseous discharge device shall be operated at design voltage and in a steady on, flashing operation.

4.2 Color Test—SAE J578, Color Specification for Electric Signal Lighting Devices, is a part of this report. Devices shall be tested with the light source normally supplied with the lamp. When it is not feasible to make measurements with this light source, a steady burning CIE Illuminant C (6774 K) light source shall be substituted.

4.3 Additional Tests

4.3.1 HIGH TEMPERATURE FLASH RATE TEST—The device shall be subjected to an ambient temperature of 50 ± 3°C for a period of 6 h. The device shall be off during the first hour and shall operate continuously for the next 5 h at 12.8 V for a nominal 12 V system and 25.6 V for a nominal 24 V system. The flash rate shall be measured before the test, not less than 3 min nor more than 4 min after the beginning of the second hour of the test, and not less than 3 min nor more than 4 min after the end of the test.

4.3.2 LOW TEMPERATURE FLASH RATE TEST—The device shall be subjected to an ambient temperature of −30 ± 3°C for a period of 6 h. The device shall be off during the first 5 h and shall operate continuously for the last hour of the test at 12.8 V for a nominal 12 V system and 25.6 V for a nominal 24 V system.

The flash rate shall be measured before the test, not less than 3 min nor more than 4 min after the beginning of the last hour of the test, and not less than 3 min nor more than 4 min after the end of the test.

4.3.3 DURABILITY TEST—The device shall be operated continuously for 200 h at an ambient temperature of 25 ± 3°C in cycles of 50 min on and 10 min off at 12.8 V for a nominal 12 V system and 25.6 V for a nominal 24 V system. The flash rate shall be measured before the test and not more than 3 min after the last off period at the end of the test.

5. Requirements

5.1 Performance Requirements—A device when tested in accordance with the test procedures specified in Section 4 shall meet the following requirements in SAE J575:

5.1.1 VIBRATION
5.1.2 MOISTURE
5.1.3 DUST
5.1.4 CORROSION
5.1.5 PHOTOMETRY—The lamp under test shall meet the photometric performance requirements contained in Tables 1, 2, 3, and 4. The summation of the flash energy measurements at the specified test points in a zone shall be at least the value shown.
5.1.6 WARPAGE—Shall meet the requirements of paragraph 4.8.4 of SAE J575.
5.1.7 COLOR—The color of the light emitted shall be white, yellow, red, or signal blue as specified in SAE J578.

5.2 Material Requirements—Plastic materials used in optical parts shall meet the requirements of SAE J576, Plastic Material for Use in Optical Parts such as Lenses and Reflectors of Motor Vehicle Lighting Devices.

5.3 Additional Requirements

5.3.1 HIGH TEMPERATURE—There shall be no evidence of operating conditions which would result in failure to comply with Section 5. After the unit has been allowed to operate for 3 min after the high temperature test, the flash rate shall not be less than 0.80 Hz nor more than 2.2 Hz.

5.3.2 LOW TEMPERATURE—There shall be no evidence of operating conditions which would result in failure to comply with Section 5. The lamp must flash and continue to flash within 20 s after the current is turned on or it is considered a failure. After the unit has been allowed to operate for 3 min after the low temperature test, the flash rate shall not be less than 0.80 Hz nor more than 2.2 Hz.

5.3.3 DURABILITY—There shall be no evidence of operating conditions which would result in failure to comply with Section 5. The flash rate shall be measured before the test and not more than 3 min after the last off period at the end of the test. The flash rate shall be not less than 1 Hz nor more than 2 Hz.

TABLE 1
PHOTOMETRIC REQUIREMENTS CLASS 1
360 DEG GASEOUS DISCHARGE WARNING LAMPS
Minimum Flash Energy Requirements
Zone Totals (Candela-Seconds)

Zone	Test Point Degree	Flash Energy—Candela Seconds			
		White	Yellow	Red	Signal Blue
#1	5U–V 2.5U–V H–V 2.5D–V 5D–V	396	198	99	*

* Not Recommended.

NOTES:
a. A one time adjustment in lamp orientation from design position may be made in determining compliance to Tables 5, 6, 7, and 8 provided such adjustment does not exceed 1 deg in any direction. The zone shall comply after this one time, final reaim.
b. When calculating zone totals, the measured value at each test point shall not be less than 60% of the minimum values in Tables 5, 6, 7, and 8.

TABLE 2
PHOTOMETRIC REQUIREMENTS CLASS 2
360 DEG GASEOUS DISCHARGE WARNING LAMPS
Minimum Flash Energy Requirements
Zone Totals (Candela-Seconds)

Zone	Test Point Degree	Flash Energy—Candela Seconds			
		White	Yellow	Red	Signal Blue
#1	5U–V 2.5U–V H–V 2.5D–V 5D–V	99	49.5	25	12.5

NOTES:
a. A one time adjustment in lamp orientation from design position may be made in determining compliance to Tables 5, 6, 7, and 8 provided such adjustment does not exceed 1 deg in any direction. The zone shall comply after this one time, final reaim.
b. When calculating zone totals, the measured value at each test point shall not be less than 60% of the minimum values in Tables 5, 6, 7, and 8.

TABLE 3
PHOTOMETRIC REQUIREMENTS CLASS 3
360 DEG GASEOUS DISCHARGE WARNING LAMPS
Minimum Flash Energy Requirements
Zone Totals (Candela-Seconds)

Zone	Test Point Degree	Flash Energy—Candela Seconds			
		White	Yellow	Red	Signal Blue
#1	5U–V 2.5U–V H–V 2.5D–V 5D–V	40	20	10	5

NOTES:
a. A one time adjustment in lamp orientation from design position may be made in determining compliance to Tables 5, 6, 7, and 8 provided such adjustment does not exceed 1 deg in any direction. The zone shall comply after this one time, final reaim.
b. When calculating zone totals, the measured value at each test point shall not be less than 60% of the minimum values in Tables 5, 6, 7, and 8.

TABLE 4

**PHOTOMETRIC REQUIREMENTS CLASS 1
DIRECTIONAL, GASEOUS DISCHARGE WARNING LAMPS**

Minimum Flash Energy Requirements
Zone Totals (Candela-Seconds)

Zone	Test Point Degree	Flash Energy—Candela Seconds			
		White	Yellow	Red	Signal Blue
#1	5U–10L 5U–20L H–20L 5D–20L 5D–10L	108	54	27	*
#2	10U–5L 10U–V 10U–5R	56	28	14	*
#3	5U–5L H–10L 5D–5L	184	92	46	*
#4	5U–V H–5L H–V H–5R 5D–V	664	332	116	*
#5	5U–5R H–10R 5D–5R	184	92	46	*
#6	10D–5L 10D–V 10D–5R	56	28	14	*
#7	5U–10R 5U–20R H–20R 5D–20R 5D–10R	108	54	27	*

* Not Recommended.

NOTES:
  a. A one time adjustment in lamp orientation from design position may be made in determining compliance to Tables 5, 6, 7, and 8 provided such adjustment does not exceed 1 deg in any direction. The zone shall comply after this one time, final reaim.
  b. When calculating zone totals, the measured value at each test point shall not be less than 60% of the minimum values in Tables 5, 6, 7, and 8.

TABLE 5

**PHOTOMETRIC DESIGN GUIDELINES
360 DEG GASEOUS DISCHARGE WARNING LAMPS**

Minimum Design Flash Energy Guidelines
Class 1 Warning Lamps

Test Point Degree	Flash Energy—Candela-Seconds			
	White	Yellow	Red	Signal Blue
5U–V	18	9	4.5	*
2.5U–V	90	45	22.5	*
H–V	180	90	45	*
2.5D–V	90	45	22.5	*
5D–V	18	9	4.5	*

* Not Recommended.

TABLE 6

**PHOTOMETRIC DESIGN GUIDELINES
360 DEG GASEOUS DISCHARGE WARNING LAMPS**

Minimum Design Flash Energy Guidelines
Class 2 Warning Lamps

Test Point Degree	Flash Energy—Candela-Seconds			
	White	Yellow	Red	Signal Blue
5U–V	4.5	2	1	0.5
2.5U–V	22.5	11.5	6	3
H–V	45	22.5	11	5.5
2.5D–V	22.5	11.5	6	3
5D–V	4.5	2	1	0.5

TABLE 7

**PHOTOMETRIC DESIGN GUIDELINES
360 DEG GASEOUS DISCHARGE WARNING LAMPS**

Minimum Design Flash Energy Guidelines
Class 3 Warning Lamps

Test Point Degree	Flash Energy—Candela-Seconds			
	White	Yellow	Red	Signal Blue
5U–V	2	1	0.5	0.25
2.5U–V	9	4.5	2	1
H–V	18	9	5	2.5
2.5D–V	9	4.5	2	1
5D–V	2	1	0.5	0.25

TABLE 8

**PHOTOMETRIC DESIGN GUIDELINES
DIRECTIONAL, GASEOUS DISCHARGE WARNING LAMPS**

Minimum Flash Energy Guidelines
Warning Lamps

Test Point Degree	Flash Energy—Candela-Seconds			
	White	Yellow	Red	Signal Blue
10U–5L	12	6	3	*
10U–V	32	16	8	*
10U–5R	12	6	3	*
5U–20L	12	6	3	*
5U–10L	32	16	8	*
5U–5L	68	34	17	*
5U–V	100	50	25	*
5U–5R	68	34	17	*
5U–10R	32	16	8	*
5U–20R	12	6	3	*
H–20L	20	10	5	*
H–10L	48	24	12	*
H–5L	132	66	33	*
H–V	200	100	50	*
H–5R	132	66	33	*
H–10R	48	24	12	*
H–20R	20	10	5	*
5D–20L	12	6	3	*
5D–10L	32	16	8	*
5D–5L	68	34	17	*
5D–V	100	50	25	*
5D–5R	68	34	17	*
5D–10R	32	16	8	*
5D–20R	12	6	3	*
10D–5L	12	6	3	*
10D–V	32	16	8	*
10D–5R	12	6	3	*

* Not Recommended.

6. Guidelines

6.1 Photometric Guidelines—Photometric design guidelines for 360 deg and directional gaseous discharge warning lamps, when tested in accordance with Section 4.1.5 of this report, are contained in Tables 5, 6, 7, and 8.

6.2 Installation Guidelines—The following guidelines apply to 360 deg and directional gaseous discharge warning lamps as used on the vehicle and shall not be considered part of the requirements:

6.2.1 Mounting—The vertical axis of the lamp should be installed normal to the longitudinal axis of the vehicle.

6.2.2 Visibility—Visibility of the 360 deg warning lamp should be unobstructed by any part of the vehicle 5 deg above to 5 deg below the horizontal and provide a flashing light throughout a 360 deg circle. Additional primary warning lamps may be used whenever vehicle size or design prevents a single primary warning lamp from projecting 360 deg of a full strength warning signal. These additional warning lamps shall be mounted so that the 360 deg of full strength signal is obtained around the vehicle.

Directional warning lamps should be mounted as high as practical and if mounted in pairs, as far apart as practical. Visibility to the front and to the rear of the vehicle should be unobstructed by any part of the vehicle from 10 deg up to 10 deg below horizontal and from 45 deg left to 45 deg right of the centerline of the vehicle. To improve the efficiency of the signal it is recommended that when practical, the area surrounding the lamps should be black.

6.2.3 INDICATOR—There should be a visible or audible means of giving a clear and unmistakable indication to the driver when the warning lamps are turned on and functioning normally.

7. Test Equipment Guidelines—The following guidelines apply to photometric test equipment and are not part of the technical requirements:

7.1 A pulse integrating photometer or other accepted means of measuring pulsed light signals shall have the following:

Response Time—1 µs or less

Sensor Response—Sensor shall be corrected to that of the 1931 C.I.E. standard observer (2 deg) photopic response curve. Sensor shall be calibrated for the color of the light being measured.

Range Linearity—Linearity of the sensor and photometer system shall be verified over the range of the luminous intensities being tested. Linearity deviation shall not deviate more than 2.5% from the calibration level to the extreme luminous intensity values measured.

7.2 The regulated D.C. power supply shall have the following minimum requirements:

Line regulations	±0.1%
Load regulation	±0.1%
Ripple voltage	±0.4%
Stability	±0.1% during test

References

ISO 4148, Road Vehicles—Special Warning Lights—Dimensions (1978).
NBS No. 480-3, Sirens and Emergency Warning Lights (June 1977).
NBS No. 480-36, Some Psychophysical Tests of the Conspicuities of Emergency Vehicle Warning Lights (July 1979).
NBS No. 480-37, Emergency Vehicle Warning Systems (May 1981).
SAE J595 AUG83.
SAE J845 JAN84.

φSCHOOL BUS WARNING LAMPS—SAE J887 AUG87

SAE Standard

Report of the Lighting Committeed, approved July 1964 and completely revised August 1987. Rationale statement available.

1. Scope—This SAE technical report provides test procedures, requirements and guidelines for red and yellow school bus warning lamps.

2. Definitions

2.1 School bus red warning lamps are lights alternately flashing at 1-2 Hz per lamp, mounted horizontally both front and rear, intended to inform other users of the highway that such vehicle is stopped on highway to take on or discharge school children.

2.2 School bus yellow warning lamps are lights alternately flashing at 1-2 to Hz per lamp, mounted horizontally both front and rear, intended to inform other users of the highway that such vehicle is about to stop to take on or discharge school children.

3. Lighting Identification Code—Lamps conforming to this technical report may be identified with the code W2 in accordance with SAE J759.

4. Tests

4.1 SAE J575, Tests for Motor Vehicle Lighting Devices and Components, is part of this report. The following tests are applicable with the modifications as indicated:

4.1.1 VIBRATION TEST
4.1.2 MOISTURE TEST
4.1.3 DUST TEST
4.1.4 CORROSION TEST
4.1.5 PHOTOMETRY TEST

4.1.5.1 All photometric measurements shall be made with the filament of the lamp at a distance of at least 3 m from the photometric screen. The lamp axis shall be taken as the horizontal line through the light source parallel to what would be the longitudinal axis of the vehicle, if the lamp were mounted in its normal position on the vehicle.

4.1.5.2 The school bus warning lamp shall be operated at design voltage.

4.1.5.3 An optional alternate measure of photometric performance can be made using flash energy.

4.1.5.3.1 The device shall be allowed to operate for 15 min prior to making photometric measurements. In all instances where a device is required to be operated during a test specified in this report, the voltage applied to the input wires or terminals of the device shall be 12.8 V for nominal 12 V electrical systems and 25.6 V for nominal 24 V electrical systems.

4.1.5.3.2 Photometric luminous intensity measurements (candela seconds) shall be taken as the average of ten consecutive flash cycles. There shall be an off time before each flash of at least 50% of the total flash cycle time.

4.1.6 Warpage test on device with plastic components.

4.2 SAE J578—Color Specification is a part of this report.

4.3 Sealed Units as described in SAE J571, Dimensional Specifications for Sealed Beam Headlamp Units, and SAE J760, Dimensional Specifications for General Service Sealed Lighting Units, and SAE J1132, 142 mm X 200 mm Sealeld Beam Headlamp Unit, designed for use as school bus warning lamps, when tested without the other parts of the lamp assembly, need only be tested to paragraphs 4.1.1 and 4.1.5.

5. Requirements

5.1 **Performance Requirements**—A device, when tested in accordance with the test procedures specified in Section 4, shall meet the following requirements. Sealed units, as described in SAE J571, SAE J760, and SAE J1132, when tested without the other parts of the lamp assembly, need only comply with paragraphs 5.1.11, 5.1.5, and 5.1.6.

5.1.1 VIBRATION—SAE J575.
5.1.2 MOISTURE—SAE J575.
5.1.3 DUST—SAE J575.
5.1.4 CORROSION—SAE J575.
5.1.5 PHOTOMETRY—SAE J575.

5.1.5.1 The lamp under test shall meet the photometric performance requirements contained in Table 1, Photometric Performance Requirements and its footnotes. The summation of the luminous intensity measurements at the specified test points in a zone shall be at least the value shown.

5.1.5.2 Alternate Method—The lamp under test shall meet the photometric performance requirements contained in Table 3, Photometric Performance Requirements and its footnotes. The summation of the flash energy measurements at the specified test points in a zone shall be at least the value shown.

5.1.6 WARPAGE—SAE J575.

5.1.7 COLOR—The lamp shall comply with the red or yellow requirements specified in SAE J578.

5.2 **Material Requirements**—Any plastic materials used in optical parts shall comply with the requirements in SAE J576, Plastic Materials for Use in Optical Parts Such as Lenses and Reflectors of Motor Vehicle Lighting Devices.

5.3 **Design Requirements**

5.3.1 The functional lighted lens area of a school bus warning lamp shall not be less than 120 cm².

5.3.2 Sealed units if used shall comply dimensionally with SAE J571, or SAE J1132 for sealed beam lamps.

5.3.3 AIMING PROVISIONS—The lamp shall be equipped with aiming pads, as described in SAE J571, SAE J760, or SAE J1132, on the lens face suitable for use with mechanical headlamp aimers as described in SAE J602. The lamp shall be designed so that with the aiming plane normal to the photometric axis, the beam shall meet the photometric specifications of Table 1.

6. Guidelines—The mounting and use of school bus warning lamps are specified by various legal agencies. The following guidelines, if followed, will enhance performance of the system and uniformity in use throughout the various jurisdictional agencies. They are not part of the test provisions, specifications, requirements, or procedures.

6.1 Photometric Design Guidelines for School Bus Warning Lamps, when tested in accordance with Section 4.1.5 of this report, are contained in Tables 2 and 4.

6.2 The yellow lamps should be automatically deactivated and the red lamps activated when the vehicle is stopped to take on or discharge school children.

6.3 For circuit interrupted incandescent filament devices, see SAE J1104, Service Performance Requirements for Warning Lamp Alternating Flashers. The 'on' period of the flasher should be long enough to permit a bulb filament to approach full brightness.

6.4 There should be a visible or an audible means of giving a clear and unmistakable indication to the driver when the warning lamps are activated.

6.5 Front and rear warning lamps should be spaced as far apart laterally as practical with the yellow lamps mounted inboard of the red lamps. In no case should the spacing between the inboard lamps be less than 1000 mm, as measured from the nearest edge of the lens.

6.6 The warning lamps should be mounted on the same horizontal centerline as high as practical at the front above the windshield and on the same horizontal centerline as high as practical at the rear so that the lower edge of the lenses is not lower than the top line of the side window openings.

6.7 The visibility of the front warning lamps to the front and of the rear warning lamps to the rear should be unobstructed by any part of the vehicle from 10 deg above to 10 deg below horizontal and from 30 deg to the right to 30 deg to the left of the centerline of the lamps.

6.8 To improve the effectiveness of the signal, the area of the vehicle immediately surrounding the warning lamp extending outward approximately 70 mm should be painted black.

6.9 The lamps should be mounted on the school bus with their aiming plane vertical and normal to the vehicle longitudinal axis. If lamps are aimed or inspected with a mechanical headlamp aimer, the graduation settings for aim should be 0 down and 0 sideways. The limits for inspection should be from 5 up to 5 down and from 10 right to 10 left.

7. Test Equipment Guidelines—The following guidelines apply to photometric test equipment and are not part of the technical requirements for the lamps:

7.1 A pulse integrating photometer or other accepted means of measuring pulsed light signals should have the following:

Response Time—1µs or less.

Sensor Response—Sensor should be corrected to that of the 1931 C.I.E. standard observer (2 deg) photopic response curve. Sensor should be calibrated for the color of the light being measured.

Range Linearity—Linearity of the sensor and photometer system should be verified over the range of the luminous intensities being tested. Linearity deviation should not deviate more than 2.5% from the calibration level to the extreme luminous intensity values measured.

7.2 The regulated D.C. power supply should have the following minimum requirement:

Line regulations ±0.1%
Load regulation ±0.1%
Ripple voltage ±1.4%
Stability ±0.1% during test

TABLE 1—PHOTOMETRIC PERFORMANCE REQUIREMENTS SCHOOL BUS WARNING LAMPS

Zone	Test Point Degree		Total Zonal Luminous Intensity (Candela)	
			Red	Yellow
1	H	30L	590	1475
	5D	30L		
	5U	20L		
	H	20L		
	5D	20L		
2	10U	5L	90	225
	10U	V		
	10U	5R		
3	5U	10L	1500	3750
	5U	5L		
	5U	V		
	5U	5R		
	5U	10R		
4	H	10L	2400	6000
	H	5L		
	H	V		
	H	5R		
	H	10R		
5	5D	10L	1950	4875
	5D	5L		
	5D	V		
	5D	5R		
	5D	10R		
6	10D	5L	120	300
	10D	V		
	10D	5R		
7	5U	20R	590	1475
	H	20R		
	5D	20R		
	H	30R		
	5D	30R		

Notes:
a. For the lamp to conform to the photometric zonal performance requirements, the summation of the candela measurements at the specific test points in a zone shall meet or exceed the value specified for that zone in Table 1.
b. The measured candela at each test point shall not be less than 60% of the requirements specified in Table 2.
c. An adjustment in lamp aim from design position may be made, provided that such adjustment does not exceed 3 deg. All zones shall comply after final re-aim.
d. See Fig. 1 for a graphical description of the zonal boundaries.

TABLE 2—PHOTOMETRIC DESIGN GUIDELINES SCHOOL BUS WARNING LAMPS

Test Point Degree		Luminous Intensity (Candela)	
		Red	Yellow
10U	5L	20	50
	V	50	125
	5R	20	50
5U	20L	150	375
	10L	300	750
	5L	300	750
	V	300	750
	5R	300	750
	10R	300	750
	20R	150	375
H	30L	30	75
	20L	180	450
	10L	400	1000
	5L	500	1250
	V	600	1500
	5R	500	1250
	10R	400	1000
	20R	180	450
	30R	30	75
5D	30L	30	75
	20L	200	500
	10L	300	750
	5L	450	1125
	V	450	1125
	5R	450	1125
	10R	300	750
	20R	200	500
	30R	30	75
10D	5L	40	100
	V	40	100
	5R	40	100

Notes:
a. An adjustment in lamp aim from design position may be made provided that such adjustment does not exceed 3 deg. The lamp should meet or exceed the values specified in Table 2 after final re-aim.
b. See Fig. 2 for a graphical description of photometric design guidelines.

TABLE 3—PHOTOMETRIC PERFORMANCE REQUIREMENTS (ALTERNATE METHOD) SCHOOL BUS WARNING LAMPS

Zone	Test Point		Total Zonal Flash Energy (Candela Seconds)	
	Degree		Red	Yellow
1	H	30L	141	351
	5D	30L		
	5U	20L		
	H	20L		
	5D	20L		
2	10U	5L	22	54
	10U	V		
	10U	5R		
3	5U	10L	360	890
	5U	5L		
	5U	V		
	5U	5R		
	5U	10R		
4	H	10L	571	1426
	H	5L		
	H	V		
	H	5R		
	H	10R		
5	5D	10L	465	1157
	5D	5L		
	5D	V		
	5D	5R		
	5D	10R		
6	10D	5L	30	72
	10D	V		
	10D	5R		
7	5U	20R	141	351
	H	20R		
	5D	20R		
	H	30R		
	5D	30R		

Notes:
a. For the lamp to conform to the photometric zonal performance requirements, the summation of the flash energy measurements at the specific test points in a zone shall meet or exceed the value specified for that zone in Table 3.
b. The measured flash energy at each test point shall not be less than 60% of the requirements specified in Table 4.
c. An adjustment in lamp aim from design position may be made, provided that such adjustment does not exceed 3 deg. All zones shall comply after final re-aim.
d. See Fig. 1 for a graphical description of the zonal boundaries.

TABLE 4—PHOTOMETRIC DESIGN GUIDELINES SCHOOL BUS WARNING LAMPS

Test Point		Flash Energy (Candela Seconds)	
Degree		Red	Yellow
10U	5L	5	12
	V	12	30
	5R	5	12
5U	20L	36	89
	10L	72	178
	5L	72	178
	V	72	178
	5R	72	178
	10R	72	178
	20R	36	89
H	30L	7	18
	20L	43	107
	10L	95	238
	5L	119	297
	V	143	356
	5R	119	297
	10R	95	238
	20R	43	107
	30R	7	18
5D	30L	7	18
	20L	48	119
	10L	72	178
	5L	107	267
	V	107	267
	5R	107	267
	10R	72	178
	20R	48	119
	30R	7	18
10D	5L	10	24
	V	10	24
	5R	10	24

Notes:
a. An adjustment in lamp aim from design position may be made provided that such adjustment does not exceed 3 deg. The lamp should meet or exceed the values specified in Table 2 after final re-aim.
b. See Fig. 2 for a graphical description of photometric design guidelines.

24.195

See Table 1 or 3 for Zone Values

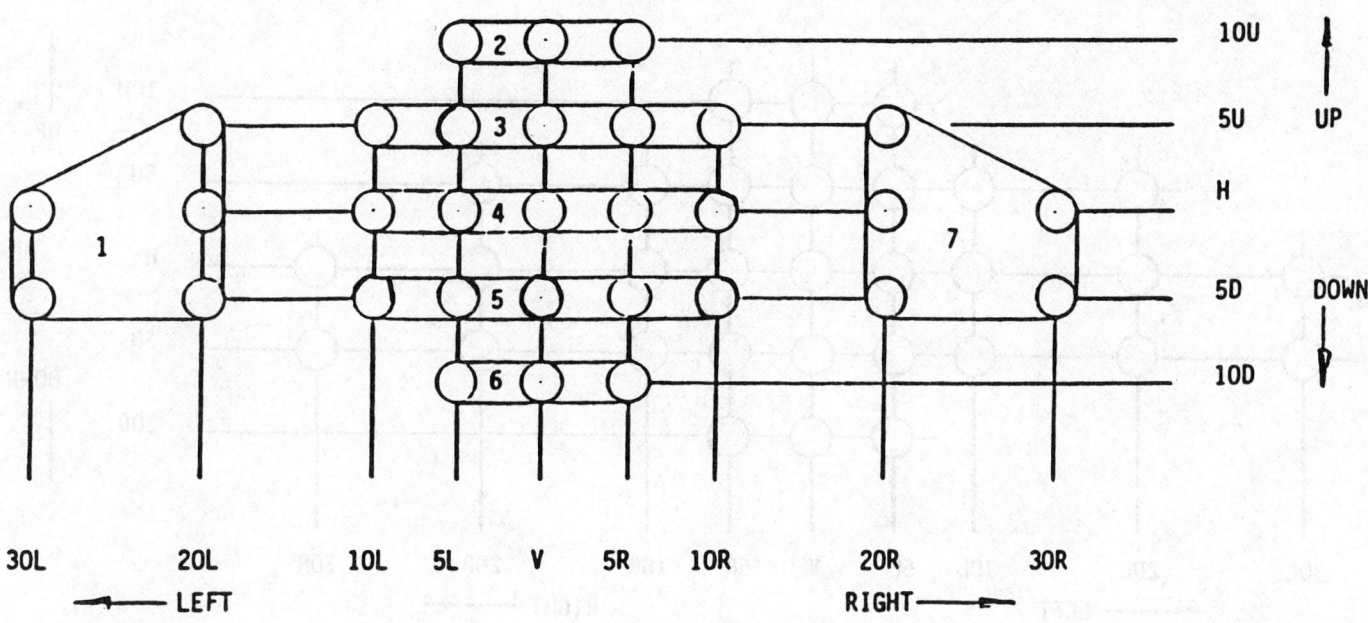

The line formed by the intersection of a vertical plane through the light source of the device and normal to the test screen is designated V. The line formed by the intersection of a horizontal plane through the light source and normal to the test screen is designated H. The point of intersection of these two lines is designated H-V. The other points on the test screen are measured in terms of degree from these two lines. Degrees to the right (R) and to the left (L) are regarded as being to the right and left of the vertical line when the observer stands behind the lighting device and looks in the direction of the emanating light beam when the device is properly aimed for photometry with respect to the H-V point. Similarly, the upward angles designated as U and the downward angles designated D, refer to light emanating at angles above and below the horizontal line, respectively.

FIG. 1—GRAPHICAL DESCRIPTION OF THE ZONAL BOUNDARIES

The Circles Indicate the Test Points for Distribution of Light. See Table 2 or 4 for Photometric Values

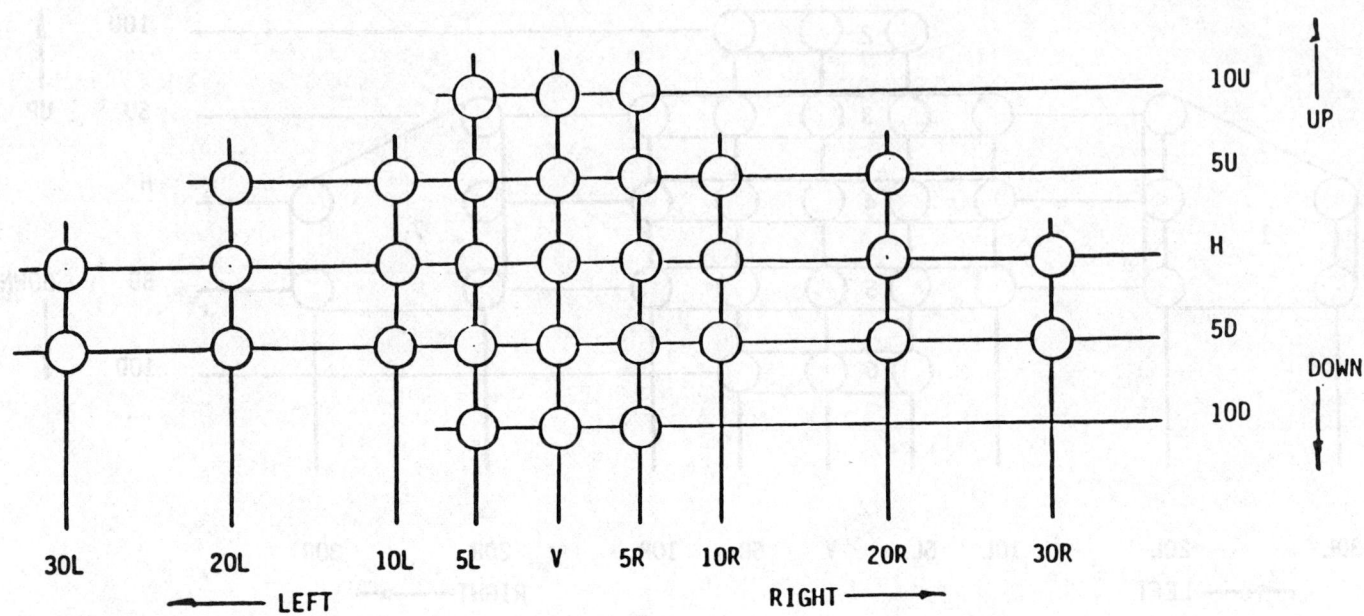

The line formed by the intersection of a vertical plane through the light source of the device and normal to the test screen is designated V. The line formed by the intersection of a horizontal plane through the light source and normal to the test screen is designated H. The point of intersection of these two lines is designated H-V. The other points on the test screen are measured in terms of degree from these two lines. Degrees to the right (R) and to the left (L) are regarded as being to the right and left of the vertical line when the observer stands behind the lighting device and looks in the direction of the emanating light beam when the device is properly aimed for photometry with respect to the H-V point. Similarly, the upward angles designated as U and the downward angles designated D, refer to light emanating at angles above and below the horizontal line, respectively.

FIG. 2—GRAPHICAL DESCRIPTION OF THE PHOTOMETRIC GUIDELINES

(R) SCHOOL BUS STOP ARM—SAE J1133 JUL89

SAE Recommended Practice

Report of the Lighting Committee, approved April 1976, completely revised April 1984. Rationale statement available. Completely revised by the Lighting Coordinating Committee July 1989. Rationale statement available.

1. Scope—This document provides test procedures, requirements, and guidelines for school bus stop arms.

2. Definition—A school bus stop arm is an auxiliary device used to signal that a school bus has stopped to load or discharge passengers. It supplements devices specified by SAE J887.

3. Lamps for use on school bus stop arms may be identified by the code "W6" in accordance with SAE J759.

4. Tests

4.1 SAE J575 is a part of this document. The following tests are applicable, with the modifications indicated:

4.1.1 VIBRATION TEST
4.1.2 MOISTURE TEST
4.1.3 DUST TEST
4.1.4 CORROSION TEST
4.1.5 PHOTOMETRY—In addition to the test procedures in SAE J575, the following apply:

4.1.5.1 Photometric measurements shall be made with the light source(s) of the lamp(s) at least 18 m from the photometer. The H-V axis shall be taken as parallel to the longitudinal axis of the vehicle.

4.1.5.2 Photometric measurements shall be made with the bulb filament steadily burning.

4.1.5.3 An optional alternate measure of photometric performance can be made using flash energy.

4.1.5.3.1 Photometric measurements shall be made with the device in its normal operating position and all flash energy measurements shall be made with the light source at least 18 m from the photometer sensor. The H-V axis shall be taken as parallel to the longitudinal axis of the vehicle.

4.1.5.3.2 The voltage applied to the input wires or terminals of the device shall be 12.8 V for nominal 12 V electrical systems and 25.6 V for nominal 24 V electrical systems.

4.1.5.3.3 Photometric luminous intensity measurements (candela seconds) shall be taken as the average of ten consecutive flash cycles.

4.1.6 WARPAGE TEST FOR DEVICES WITH PLASTIC COMPONENTS

4.2 Color Test—SAE J578 is a part of this document.

4.3 Durability—The device shall be subjected to a test of 45 000 cycles at a rate not to exceed 0.2 Hz and at a temperature of 25°C ± 3. A cycle shall consist of movement from the parked or retracted position to the fully extended position and return to the parked position.

5. Requirements

5.1 Performance Requirements—A device, when tested in accordance with the test procedures specified in Section 4, shall meet the following requirements:

5.1.1 VIBRATION—SAE J575
5.1.2 MOISTURE—SAE J575
5.1.3 DUST—SAE J575
5.1.4 CORROSION—SAE J575
5.1.5 PHOTOMETRY—In addition to the requirements of SAE J575, the school bus stop arm lamps shall meet the following photometric performance requirements:

5.1.5.1 The summation of the luminous intensity readings of the specific test points in a zone shall meet the values in Table 1.

5.1.5.2 When calculating the zone total, the measured luminous intensity for a test point shall not be less than 60% of the value specified for that test point in Table 2.

5.1.5.3 *Alternate Method*—The lamp under test shall meet the photometric performance requirements contained in Table 3. The summation of the flash energy measurements at the specified test points in a zone shall be at least the value shown. When calculating the zone total, the measured flash energy for a test point shall not be less than 60% of the value specified for that test point in Table 4.

5.1.6 WARPAGE—SAE J575
5.1.7 COLOR—The color of light emitted from the school bus stop arm lamps shall be red as specified in SAE J578.
5.1.8 DURABILITY—Failure of the device to operate in the intended electrical or mechanical manner during or at the conclusion of the test shall constitute a failure. Internal bulb failure shall not be considered a failure of the device.
5.1.9 FLASH RATE

5.1.9.1 For circuit-interrupted incandescent devices, the two lamps on each face shall flash alternately with the rate and percent "on" time as required in SAE J1054.

5.1.9.2 For gaseous discharge lamps, the two lamps on each face shall flash alternately with the flash rate not less than 0.80 Hz nor more than 2.2 Hz. There shall be an off time before each flash of at least 50% of the total flash cycle time.

5.2 Material Requirements—Plastic materials used in the optical parts shall meet the requirements of SAE J576.

5.3 Design Requirements

5.3.1 A school bus stop arm shall have on both the front and rear the word "STOP" in letters which are at least 150 mm in height and have a stroke width of at least 20 mm.

5.3.2 School bus stop arms shall have a minimum of two lamps to the front and two lamps to the rear, or two double-faced lamps may be used.

TABLE 1—PHOTOMETRIC PERFORMANCE REQUIREMENTS

Zone	Test Points (deg)	Total Zonal Luminous Intensity (cd)
1	10U-5L 5U-20L 5D-20L 10D-5L	52
2	5U-10L H-10L 5D-10L	100
3	5U-V H-5L H-V H-5R 5D-V	380
4	5U-10R H-10R 5D-10R	100
5	10U-5R 5U-20R 5D-20R 10D-5R	52

Notes:
1. For the lamp to conform to the photometric zonal performance requirements, the summation of the candela measurements at the specific test points in a zone shall meet or exceed the values specified for that zone in Table 1.
2. When calculating the zone total, the measured candela for a test point shall not be less than 60% of the value specified for that test point in Table 2.
3. See Fig. 1 for a graphical description of the Zonal Boundaries.

TABLE 2—PHOTOMETRIC DESIGN GUIDELINES

Test Points (deg)				Luminous Intensity (cd)
10U,	10D	5L,	5R	16
5U,	5D	V 10L, 20L,	10R 20R	70 30 10
H		V 5L, 10L,	5R 10R	80 80 40

Notes:
1. Any photometric measurements that fall below 60% of the test point value given in Table 2 shall not be used in the calculation of zone totals.
2. The luminous intensity values (candela) specified in Table 2 have been established by empirical and field evaluation techniques for lighting devices to perform their intended function in field service.
3. See Fig. 2 for a graphical description of Photometric Design Guidelines.

GRAPHICAL DESCRIPTION OF THE ZONAL BOUNDARIES
See Table 1 for Zone Values.

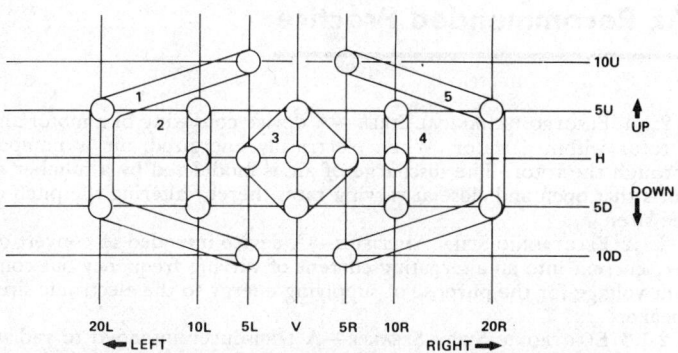

The line formed by the intersection of a vertical plane through the light source of the device and normal to the test screen is designated V. The line formed by the intersection of a horizontal plane through the light source and normal to the test screen is designated H. The point of intersection of these two lines is designated H-V. The other points on the test screen are measured in terms of degree from these two lines. Degrees to the right (R) and to the left (L) are regarded as being to the right and left of the vertical line when the observer stands behind the lighting device and looks in the direction of the emanating light beam when the device is properly aimed for photometry with respect to the H-V point. Similarly, the upward angles designated as U and the downward angles designated D, refer to light emanating at angles above and below the horizontal line, respectively.

FIG. 1

GRAPHICAL DESCRIPTION OF THE PHOTOMETRIC GUIDELINES
The Circles Indicate the Test Points for Distribution of Light. See Table 2 for Photometric Values.

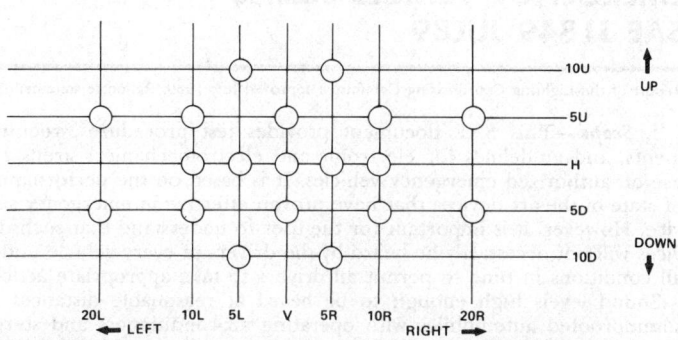

The line formed by the intersection of a vertical plane through the light source of the device and normal to the test screen is designated V. The line formed by the intersection of a horizontal plane through the light source and normal to the test screen is designated H. The point of intersection of these two lines is designated H-V. The other points on the test screen are measured in terms of degree from these two lines. Degrees to the right (R) and to the left (L) are regarded as being to the right and left of the vertical line when the observer stands behind the lighting device and looks in the direction of the emanating light beam when the device is properly aimed for photometry with respect to the H-V point. Similarly, the upward angles designated as U and the downward angles designated D, refer to light emanating at angles above and below the horizontal line, respectively.

FIG. 2—GRAPHICAL DESCRIPTION OF THE PHOTOMETRIC GUIDELINES

TABLE 3—PHOTOMETRIC PERFORMANCE REQUIREMENTS

Zone	Test Points (deg)	Total Zonal Flash Energy Candela-Second
1	10U-5L 5U-20L 5D-20L 10D-5L	14
2	5U-10L H-10L 5D-10L	26
3	5U-V H-5L H-V H-5R 5D-V	96
4	5U-10R H-10R 5D-10R	26
5	10U-5R 5U-20R 5D-20R 10D-5R	14

Notes:
1. For the lamp to conform to the photometric zonal performance requirements, the summation of the candela-second measurements at the specific test points in a zone shall meet or exceed the values specified for that zone in Table 3.
2. When calculating the zone total, the measured candela-second for a test point shall not be less than 60% of the value specified for that test point in Table 4.
3. See Fig. 1 for a graphical description of the Zonal Boundaries.

5.3.3 Lamps shall be activated at the commencement of the stop arm extension cycle and deactivated when the stop arm is retracted.

5.3.4 The functional lighted lens area shall not be less than 75 cm² (12 in²).

6. Guidelines

6.1 Photometric design guidelines for lamps used on school bus stop arms, when tested in accordance with 4.1.5, are contained in Table 2 or in Table 4.

6.2 Installation Guidelines—The following apply to school bus stop arms as used on the vehicle, and shall not be considered part of the requirements.

6.2.1 The school bus stop arm should be installed on the left outside of the bus body and be mounted so as to be seen readily by motorists approaching from either the front or rear of the bus.

6.2.2 If the device is operated by a manual switch, that switch shall be located so as to be easily accessible to the driver.

TABLE 4—PHOTOMETRIC DESIGN GUIDELINES

Test Points (deg)				Flash Energy Candela-Second
10U,	10D	5L,	5R	4
5U,	5D	V 10L, 20L,	10R 20R	18 8 3
H		V 5L, 10L,	5R 10R	20 20 10

Notes:
1. Any photometric measurements that fall below 60% of the test point value given in Table 4 shall not be used in the calculation of zone totals.
2. The flash energy values (candela-seconds) specified in Table 4 have been established by empirical and field evaluation techniques for lighting devices to perform their intended function in field service and calculations.
3. See Fig. 2 for a graphical description of Photometric Design Guidelines.

6.3 Design Guidelines

6.3.1 The lamps should be located in the extreme top and bottom portions of the stop arm, one above the other.

6.3.2 It is recommended that the word "STOP" be displayed as white letters against a red background, and that the stop arm have the shape of a regular octagon which is at least 450 × 450 mm. The octagon should have a white border at least 12 mm wide. The maximum extension should not exceed 560 mm beyond the left side of the vehicle. The school bus stop arm may also optionally be reflectorized.

6.3.3 The two lamps on each face should flash alternately with a flash rate of 1-2 Hz.

7. Test Equipment Guidelines—The following apply to photometric test equipment and are not part of the technical requirements:

7.1 A pulse integrating photometer or other accepted means of measuring pulsed light signals should have the following:
  a. Response Time – 1 μs or less.
  b. Sensor Response – Sensor should be corrected to that of the 1931 CIE standard observer (2 deg) photopic response curve. Sensor should be calibrated for the color of the light being measured.
  c. Range Linearity – Linearity of the sensor and photometer system should be verified over the range of the luminous intensities being tested. Linearity deviation should not deviate more than 2.5% from the calibration level to the extreme luminous intensity values measured.

7.2 The regulated DC power supply should have the following minimum requirements:
  a. Line Regulations – ±0.1%
  b. Load Regulation – ±0.1%
  c. Ripple Voltage – ±1.4%
  d. Stability – ±0.1% during test

EMERGENCY VEHICLE SIRENS—
SAE J1849 JUL89

SAE Recommended Practice

Report of the Lighting Coordinating Committee approved July 1989. Rationale statement available.

1. Scope—This SAE document provides test procedures, requirements, and guidelines for electronic and electromechanical sirens for use on authorized emergency vehicles. It is based on the performance of state-of-the-art devices that have proven effective in emergency service. However, it is important for the user to understand that such devices will not necessarily be heard by the drivers of every vehicle under all conditions in time to permit all drivers to take appropriate action.

Sound levels high enough to be heard at reasonable distances in soundproofed automobiles with operating air-conditioners and stereo sound systems are so high as to be environmentally unacceptable. The siren is a useful warning device for calling for the right-of-way by an emergency vehicle but must always be used in conjunction with effective visual warning devices and operated only by properly trained personnel.

2. Definitions

2.1 Siren—A device for producing standardized acoustical signals which have become recognized as the call for the right-of-way by an emergency vehicle.

2.1.1 ELECTROMECHANICAL SIREN—A device consisting of a motor and a rotor within a stator. When electrically energized, air is pumped through the rotor. The discharge of air is modulated by a number of ports that open and close at varying rates thereby altering the pitch of the siren.

2.1.2 ELECTRONIC SIREN AMPLIFIER—A device intended to convert direct current into an alternating current of varying frequency but constant voltage for the purpose of supplying energy to the electronic siren speaker.

2.1.3 ELECTRONIC SIREN SPEAKER—A transducer intended to radiate acoustical energy into the air with an acoustical waveform equivalent to the input electrical waveform. The electronic siren speaker shall include the electrical to mechanical transducer plus any and all mechanisms or housings required to couple and control transducer acoustical output.

2.1.4 ELECTRONIC SIREN SYSTEM—An assembly of matched devices including an electronic siren speaker, an electronic siren amplifier, and such controls as are necessary to operate the system. The number, wattage, and nominal impedance of electronic siren speakers are determined by the design of the amplifier and so specified on the nameplate of the electronic siren amplifier.

2.2 Siren Signals

2.2.1 WAIL—A tonal pattern of slow automatic increases and decreases in frequency at the rate of 10 to 30 cpm. This signal can be produced by both electronic and electromechanical sirens. The frequency of this signal shall not fall below 650 Hz for electronic siren systems nor rise above 2000 Hz for either system and shall encompass a range from high to low of at least 850 Hz.

2.2.2 YELP—A tonal pattern of rapid automatic increases and decreases in frequency at a rate of 150 to 250 cpm. This signal is usually produced only by electronic siren systems. The frequency of this signal shall not fall below 650 Hz for electronic siren systems nor rise above 2000 Hz for either system and shall encompass a range from high to low of at least 850 Hz.

2.2.3 MANUAL WAIL—Intermittently operated wail by a push button or other control means.

2.2.4 AUXILIARY SIGNALS—Any signal other than wail, yelp, or manual. While such signals may have legitimate uses in nonemergency situations and may be included as a part of an electronic siren system, it is recommended that only wail, yelp, or manual wail be used as a call for the right-of-way by an emergency vehicle.

2.3 Anechoic Chamber—An acoustical device testing room in which all six of the surfaces absorb at least 99% of the incident acoustic energy over the frequency range of interest. The chamber shall be in compliance with ANSI S1.13-1971 (3).

2.4 Sound Pressure Level—A quantity in decibels read from a sound level meter or other acoustical instrumentation system that fulfills the requirements of ANSI S1.4-1971-Type 1 (1) that is switched to the "Flat" or "Unweighted" network. The sound pressure level (Lp) is defined by Lp = 20 log 10 (p/po), where "p" is the sound pressure and "po" is the reference sound pressure.

2.5 "A" Weighted Sound Level—The setting on an acoustical instrumentation system that adds a filter network to modify the reading of sound pressure level. The "A" weighting network discriminates against the lower frequencies according to a relationship approximately equivalent to the auditory sensitivity of the human ear at moderate sound levels.

2.6 Electronic Siren Output Voltage—A potential expressed in V AC rms that is required to produce the sound pressure levels as defined by this document when applied to an electronic siren speaker.

2.7 Nominal Impedance—The impedance expressed in Ohms of an electronic siren speaker over the range of frequencies of interest. This nominal impedance for the purposes of this document shall be that of the speaker at 1000 Hz.

3. Identification Code and Marking

3.1 Identification Code

3.1.1 Devices conforming to this document which are intended to be mounted in the interior of the vehicle including the trunk or any other dry compartment designed for mounting equipment shall be identified with the code EVS1 in accordance with SAE J759.

3.1.2 Devices conforming to this document which are intended to be mounted outside the vehicle or in any other wet location (except under hood) shall be identified with the code EVS2 in accordance with SAE J759.

3.1.3 Devices conforming to this document which are intended to be mounted under hood shall be identified with the code EVS3 in accordance with SAE J759.

3.2 Markings—All markings shall be 3.0 mm or greater in height, permanently affixed to the device.

3.2.1 ELECTROMECHANICAL SIRENS—The name of the manufacturer, the model number, the nominal input voltage, operating amperage, mounting orientation, and the SAE identification code indicating mounting location shall be shown.

3.2.2 ELECTRONIC SIREN SPEAKERS—The name of the manufacturer, the model number, the nominal impedance, operating wattage, mounting orientation, and the SAE identification code indicating mounting location shall be shown. If two or more speakers are used as an array, the exact geometry in which the array was tested must be specified. Approval of an array of speakers is valid only when mounted on a vehicle in exact conformance with the specified geometry.

3.2.3 ELECTRONIC SIREN AMPLIFIERS AND ASSOCIATED CONTROLS—The name of the manufacturer, the model number, the nominal impedance, operating wattage and number of speaker(s) to which the device is intended to be connected, the intended input voltage, output voltage, and the SAE identification code indicating mounting location shall be shown.

If the amplifier and controls are separate components, each component shall be identified as to the system of which it was intended to be a part and the SAE identification code indicating mounting location shall be shown.

The controls for all functions shall be clearly identified and visible during operation both day and night.

3.2.4 SYSTEMS WHICH HAVE BEEN TESTED AS A WHOLE WITHOUT THE USE OF THE STANDARD SIGNAL GENERATOR/AMPLIFIER OR STANDARD RESISTIVE LOAD—The components of such a system shall be marked to indicate that they are approved only when used with each other and should not be interchanged with components or other systems.

4. Tests

4.1 General Information—Because siren systems can be composed of a single device or multiple, interconnected devices mounted in various locations inside or outside the vehicle, the tests required for any specific device or component are determined by its type, function and intended mounting location. To permit assembly of electronic siren systems using devices manufactured by more than one manufacturer, electronic siren speakers and electronic siren amplifiers may be independently certified in accordance to the requirements of this document.

4.1.1 ELECTRONIC SIREN SPEAKERS—In conformance with the requirements and procedures of this document, all tests of electronic siren speakers may be made using a standard signal source. For typical electronic siren speakers having 11 ohms impedance referenced at 1000 Hz, the input test voltage "E" is defined as:

$$E = \sqrt{11 \text{ ohms} \times \text{rated wattage of the siren speaker}}$$

Thus, the input voltage for testing an 11 ohm/100 watt speaker shall be 33.0 V rms ± 0.5.

Alternatively, the input test voltage for speakers of any impedance or wattage may be determined by applying a constant frequency signal of 1000 Hz to the speaker through a wattmeter and increasing the voltage until the nameplate wattage of the speaker is reached. This voltage shall then be the test voltage.

When multiple siren speakers are tested connected in parallel, the input test voltage shall be 90% of voltage determined by either of the previously mentioned methods. This reduced voltage is necessary to approximate the internal losses within a typical electronic siren amplifier and the external wiring when additional electronic siren speakers are connected.

4.1.2 ELECTRONIC SIREN AMPLIFIERS—In conformance with the requirements and procedures of this document, all tests of electronic siren amplifiers may be made using a standard, noninductive resistive load per 4.2.6. This load shall have a resistance of 11.0 ohms ± 0.2 (equivalent to the impedance of the typical electronic siren speaker referenced at 1000 Hz).

To test the performance of the amplifier if two such speakers are connected in parallel, the resistance shall be 5.5 ohms ± 0.1.

Electronic siren amplifiers designed to operate with nontypical electronic siren speakers (other than 11 ohm nominal impedance) may be tested using a noninductive, resistive load of the size matching the nominal impedance of the speaker intended for use.

4.1.3 FAMILY APPROVALS—Manufacturers of electronic siren amplifiers who market multiple variations of these amplifiers may elect to group such devices into product families. Representative models may be tested to obtain family wide approval as long as:

  a. Model numbers or identification codes are arranged to define such families and,

  b. The amplification and signal generation circuitry are identical in the family and,

  c. The test report shall list each specific model variation and the manufacturer shall have on record the details of each variation, and state that the performance of the tested and untested members of the family are the same.

4.1.4 ELECTRONIC SIREN SYSTEMS—While it is permissible to test an electronic siren system as a whole (siren amplifier connected to a siren speaker) without the use of the standard signal source and the standard resistive load, the devices comprising a system so tested, cannot be interchanged with any other devices or used as a part of any other system. The testing of a siren system as a whole shall be in conformance with the requirements and procedures of this document.

4.1.5 ELECTROMECHANICAL SIRENS—In conformance with the requirements and procedures of this document, all tests of electromechanical sirens shall be made using a timer to automatically operate a relay of sufficient current carrying capacity to energize and de-energize the siren motor at a rate of 10 to 30 times per minute.

Electromechanical sirens whose acoustical performance substantially exceeds the requirements of this document (5.1.4) are exempted from this requirement and may be tested by automatic operation at any convenient rate.

4.2 Test Equipment and Instrumentation

4.2.1 TEST VOLTAGE—Unless otherwise specified, the test voltage shall be 13.6 V DC ± 0.2 for devices intended for operation on 12 V systems and 27.2 V DC ± 0.2 for devices intended for operation on 24 V systems.

4.2.2 ACOUSTICAL TEST FIXTURE—A test base approximately 30 x 30 x 2.5 cm shall be used to mount the device. The support shall be capable of being rotated by a turntable from 0 to ±50 deg and shall position the acoustic axis of the device(s) at a height of at least 1.5 m. The axial positions shall contain no reflective surfaces with dimensions greater than 2.5 cm within 1.5 m of the siren or speaker(s) unless the design of the system incorporates external surfaces for proper operation. Such requirements, if any, shall be fully defined in the reports.

4.2.3 DIRECT CURRENT POWER SUPPLY—The power supply shall be regulated to ±0.1 V with a maximum ripple of 75 mV peak to peak. Output current capacity must be at least 1.5 times the rated current for the device under test. The output voltage shall be adjustable to provide the voltage required for specific tests.

Electromechanical sirens may be tested using an automotive battery of the correct nominal voltage for the siren provided the battery shall have a minimum cranking rate at -18°C of 450 amps for 30 s and a reserve capacity of 100 min at a discharge rate of 25 amps. A power supply as defined may be used during the test to maintain the battery charge.

4.2.4 INTEGRATING WATTMETER—The integrating wattmeter used for measuring the electronic siren speaker input power shall be of adequate capacity to record peak wattage and have a range from 100 Hz to 20 kHz with ±1% accuracy.

4.2.5 SOUND METERING SYSTEM—The sound metering system shall meet the requirements of SAE J184.

4.2.6 STANDARD NONINDUCTIVE, RESISTIVE LOAD—The noninductive resistor shall be of adequate size to dissipate the heat produced during the test. The resistor shall have a value of 11 ohms ± 0.2 if being used in place of a typical electronic siren speaker or 5.5 ohms ± 0.1 if used in place of two typical speakers connected in parallel. If the amplifier under test is designated by the manufacturer for use with nontypical siren speakers, the resistance values specified by the manufacturer shall be used.

4.2.7 STANDARD SIGNAL SOURCE—The signal generator/amplifier shall be capable of generating square wave signals at the required voltage from at least 500 to 2000 Hz into reactive loads with impedances as low as 4 ohms with rise and fall times of no more than 20 µs from 10 to 90% and 90 to 10% of the maximum waveform value. The amplifier shall remain stable regardless of the reactive nature of the load.

4.2.8 VOLTMETER FOR MEASURING ELECTRONIC SIREN AMPLIFIER OUTPUT—The voltmeter shall be capable of measuring the true AC rms value of nonsinusoidal waveforms with crest factors of up to 10 with a band width of at least 100 Hz to 20 kHz. Accuracy over this entire band width shall be ±1%.

4.3 Test Procedures

4.3.1 TEST SEQUENCE—The sequence of tests shall be:
 a. The tests under 4.3.2.1 in any order.
 b. The tests under 4.3.2.2 in any order.
 c. The test under 4.3.4 or 4.3.5 as required.
 d. The test under 4.3.3.

To reduce the time to test, two identical devices may be used. One device may be used for performing the entire test sequence except 4.3.3. The second device may be used to simultaneously run test 4.3.3.

NOTE: To determine whether or not an electromechanical siren is exempt from compliance with the cyclic rate requirement, it shall be necessary to run test 4.3.5 before the start of the test sequence. Regardless of the outcome, this test shall be performed again in the required sequence.

4.3.2 ENVIRONMENTAL TESTS

4.3.2.1 *Tests from SAE J575 with modifications as indicated*
 a. Vibration test
 b. Moisture test
 c. Dust test
 d. Corrosion test

Devices intended only for interior mounting in the vehicle do not need to be tested per b, c, and d.

4.3.2.2 *Additional Environmental Tests*

4.3.2.2.1 High Temperature Operating Test for All Devices Other Than Those Designed for Under Hood Mounting—The device shall be subjected to an ambient temperature of 50°C ± 3 for a period of 6 h. The device shall be connected to any other components necessary to form an operable system and shall be off during the first hour and shall then operate continuously for the next 5 h.

Other components may be mounted in the test chamber if they are to be tested or outside the chamber if they are not under test. The device shall operate before the start of the test in the normal fashion. If the siren produces more than one siren signal, the setting of the unit shall be changed periodically to reasonably test all siren signals.

The measurements shall be made before the test, not less than 3 min or more than 4 min after the beginning of the second hour of operation, and not less than 3 min or more than 4 min after the end of the test. At each required measurement:
 a. Electromechanical sirens shall have the maximum input amperage measured using an ammeter per 4.3.6.1 and the cyclic rate recorded.
 b. Electronic siren amplifiers shall have the output voltage determined per 4.3.6.2 and the cyclic rate (cpm) of wail and yelp (if present) shall be measured and recorded.
 c. Electronic siren speakers shall have their input wattage measured in accordance with 4.3.6.3.

4.3.2.2.2 High Temperature Test for Devices Designated by the Manufacturer as Permissible for Under Hood Mounting, i.e., In the Engine Compartment—Such devices shall be tested in accordance with 4.3.2.2.1 except at a temperature of 90°C ± 3.

4.3.2.2.3 Low Temperature Operating Test—All devices shall be subject to an ambient temperature of −30°C ± 3 for 6 h. The device shall be connected to any other components necessary to form an operable system and shall be turned off (not operating) during the first 5 h of the test and then turned on and operated during the last hour of the 6 h test.

Other components may be mounted in the test chamber if they are to be tested or outside the chamber if they are not under test. The device shall operate before the start of the test in the normal fashion. If the siren produces more than one siren signal, the setting of the unit shall be changed periodically to reasonably test all siren signals. At each required measurement:
 a. Electromechanical sirens shall have the maximum input amperage measured using an ammeter per 4.3.6.1 and the cyclic rate recorded.
 b. Electronic siren amplifiers shall have the output voltage determined per 4.3.6.2 and the cyclic rate (cpm) of wail and yelp (if present) shall be measured and recorded.
 c. Electronic siren speakers shall have their input wattage measured in accordance with 4.3.6.3.

The measurements shall be made before the test, not less than 3 min nor more than 4 min after the beginning of the last hour of operation, and not less than 3 min nor more than 4 min after the end of the test.

4.3.3 DURABILITY TEST—The device shall be connected to any other components necessary to form an operable system and the device shall be operated continuously for 200 h at an ambient temperature of 25°C ± 3 in cycles consisting of 30 min on and 30 min off. The device shall operate in the normal fashion before the start of the test. If the siren produces more than one siren signal, the setting of the unit shall be changed periodically to reasonably test all siren signals.

At each required measurement:
 a. Electromechanical sirens shall have the maximum input amperage measured using an ammeter per 4.3.6.1 and the cyclic rate recorded.
 b. Electronic siren amplifiers shall have the output voltage determined per 4.3.6.2 and the cyclic rate (cpm) of wail and yelp (if present) shall be measured and recorded.
 c. Electronic siren speakers shall have their input wattage measured in accordance with 4.3.6.3.

The measurements shall be made before the test, and not less than 3 min nor more than 4 min after the end of the test.

4.3.4 PERFORMANCE TEST FOR ELECTRONIC SIREN AMPLIFIERS—The electronic siren amplifier shall be connected to a DC power supply per 4.2.3 and to a load with a voltmeter conforming to 4.2.8 connected between the amplifier and the load.

The amplifier shall be allowed to operate for 5 min and output voltage readings taken per 4.3.6.2 for all siren signals. The frequency range and cyclic rate will also be determined for the siren tones of wail and yelp (if present).

The input voltage shall then be increased to 15.0 V DC for devices intended to operate on 12 V systems or 30 V for devices intended to operate on 24 V systems, and the procedure mentioned repeated.

4.3.5 ACOUSTICAL PERFORMANCE TEST—Electromechanical sirens and electronic siren speakers shall be tested in conformance with the following procedures:

4.3.5.1 *General*—Acoustical tests shall be conducted at a temperature of 25°C ±3. With the device under test mounted on the acoustical test fixture per 4.2.2 in an anechoic chamber, the microphone shall be positioned in line with the device at a distance of 3.0 m ± 0.01 from the edge of the device. The microphone shall be mounted at normal incidence to the device axis, in line with the device axis and at the same height as the device axis.

The device shall be located as far away from the walls of the anechoic chamber as possible. There shall be no significant reflecting surfaces within 1.5 m of the microphone used to measure the generated signal or within 1.5 m of the path between the device and the microphone. Position the device at 0 deg and connect the microphone to the instrumentation system.

Without changing the setup, adjust the measuring instrumentation to measure the "A weighted" sound level. Adjust the response characteristic of the system to provide a time constant of 0.02 s. Operate the device for 1 min and record the maximum and minimum "A weighted" sound pressure level at the 0 deg point. At the 0 deg point, determine the minimum and maximum fundamental frequency of sound produced by the device. The cyclic rate of wail and yelp (if present) shall also be measured and recorded.

Repeat the procedure by indexing the turntable until readings are obtained from 50 deg left to 50 deg right at 10 deg ± 0.5 intervals. Sound pressure readings should be recorded to the nearest 0.5 dB. Frequency and cyclic rate readings need be taken only at the 0 deg point.

During the test, technicians and observers shall remain outside the chamber. The instrumentation shall be calibrated using a pistonphone or calibrator before and after each period of use and at intervals not exceeding 2 h when the measuring instrumentation is used for a period longer than 2 h. The air temperature and pressure in the chamber shall be recorded during the test.

A record shall be kept of the calibration during the test period and any changes in calibration noted. The test shall be considered invalid if changes in calibration exceed 1 dB. All calibration equipment used during the performance of the measurements shall be operating correctly and traceable to the National Bureau of Standards.

4.3.5.2 *Electronic Siren Speaker(s)*—Siren speakers shall be securely mounted to the test fixture and connected to the signal generator/amplifier per 4.2.7. A signal of the appropriate test voltage varying in frequency from 650 to 2000 Hz at a rate of 10 to 30 cpm (approximating wail) shall be applied to the terminals of the device and the acoustical measurements made according to the general procedure.

The tests shall then be repeated using the same signal but increasing the rate to 150 to 250 cpm (approximating yelp).

If testing a complete system per 4.1.4, the electronic siren amplifier is used as the signal source and operated in the wail and yelp modes.

NOTE: If multiple siren speakers are tested, the exact geometry of the array must be reported and the test results applied only to mountings which conform to this geometry.

4.3.5.3 *Electromechanical Sirens*—The siren shall be securely mounted on the test fixture and the device connected to a power supply through a relay of adequate size. A timer shall be used to energize the relay to achieve a rate of 10 to 30 cpm. The periods of on and off need not be the same and can be set to maximize the performance of the siren.

Acoustical measurements are then made in accordance with the general procedure and the cyclic rate recorded.

4.3.6 SECONDARY TEST PROCEDURES USED IN PERFORMING THE ENVIRONMENTAL, PERFORMANCE, AND DURABILITY TESTS

4.3.6.1 *Input Amperage Test*—An ammeter of appropriate size shall be connected to the electromechanical siren between the device and the power supply. The maximum input amperage of the device shall be determined at an operating voltage of 13.6 V ± 0.2 at the terminals of the siren (27.2 V ± 0.2 if intended for 24 V operation).

4.3.6.2 *Output Voltage Test*—Amplifiers for electronic sirens shall be connected to any other components necessary to form an operable system and a voltmeter per 4.7.8 installed between the output terminals of the amplifier and the load. The output voltage of the amplifier shall be determined at 13.6 V ± 0.2 input voltage to the amplifier (27.2 V ± 0.2 if intended for 24 V operation) or such other voltage as called for in the test.

4.3.6.3 *Input Wattage Test*—An integrating wattmeter per 4.2.4 shall be connected between the output terminals of the signal generator/amplifier and the input terminals of the electronic siren speaker.

5. **Requirements**—These requirements apply to new, undamaged production units selected at random. The device shall be tested in accordance with the test procedures in Section 4.

5.1 Performance Requirements

5.1.1 When tested in conformance with 4.3.2, the device shall meet the requirements of SAE J575, with modifications as indicated.

5.1.1.1 Vibration
5.1.1.2 Moisture
5.1.1.3 Dust
5.1.1.4 Corrosion

Any device which when tested fails the test or shows signs of change that would cause failure under any of the tests is to be considered a failure.

5.1.2 ADDITIONAL ENVIRONMENTAL AND DURABILITY REQUIREMENTS—When tested according to 4.3.2.2 and 4.3.3, any device or component which fails to operate at any of the required measurements shall be considered a failure.

Also considered failures are:

a. Electromechanical sirens that require more than 125% of nameplate amperage at any required measurement or that were tested at a cyclic rate other than from 10 to 30 cpm. However, electromechanical sirens producing sound pressure levels 2 dB(A) or greater than the required values at all required measurements in 5.1.4, can be tested at any cyclic rate.

b. Electronic siren amplifiers that produce at any required measurement:

At 13.6 V (or 27.2 V) input, less than 92% or more than 108% of the nameplate voltage.

A wail signal at a cyclic rate other than from 10 to 30 cpm.

A yelp signal (if present) at a cyclic rate other than from 150 to 250 cpm.

c. Electronic siren speakers that when operated at the test voltage draw at any required measurement:

More than 108% of the nameplate wattage.

5.1.3 When tested in conformance to 4.3.4, an electronic siren amplifier shall be considered a failure if when operated at an input voltage of 13.6 V ± 0.2 (or 27.2 V ± 0.2), at any required measurement:

The output voltage is less than 92% or more than 108% of the nameplate output voltage.

A wail signal is measured at a cyclic rate other than from 10 to 30 cpm or at a frequency of less than 650 Hz or more than 2000 Hz or had a range of high to low of less than 850 Hz.

A yelp signal (if present) is measured at a cyclic rate other than from 150 to 250 cpm or at a frequency of less than 650 Hz or more than 2000 Hz or had a range of high to low of less than 850 Hz.

The amplifier is also considered a failure if when operated at an input voltage of 15.0 V (or 30 V), at a required measurement:

The output voltage is more than 125% of the nameplate voltage.

A wail signal is measured at a cyclic rate other than from 10 to 30 cpm or at a frequency of less than 650 Hz or more than 2000 Hz or had a range of high to low of less than 850 Hz.

A yelp signal (if present) is measured at a cyclic rate other than from 150 to 250 cpm or at a frequency of less than 650 Hz or more than 2000 Hz or had a range of high to low of less than 850 Hz.

5.1.4 ACOUSTICAL PERFORMANCE—When measured in accordance with 4.3.5, the wail signal and the yelp signal (if present in the system) when operating within the required cycle rates shall produce at least the following sound pressure levels:

Location Degrees	0	±10	±20	±30	±40	±50
Line A dB(A) SPL	118	117	116	115	113	111
Line B dB(A) SPL	111	110	109	108	107	106

The sound pressure at each point must equal or exceed the value shown in Line A during some portion of each wail and yelp cycle (if yelp is present) except that up to three points may fall 1 dB(A) below the value shown so long as the total of all points exceeds 1262.

The sound pressure at each point must not fall below the value shown in Line B during any portion of each wail and yelp cycle (if yelp is present) except that up to three points may fall 1 dB(A) below the value shown so long as the total of all points exceeds 1191.

The exemption for electromechanical sirens applies only if the measured sound pressure level exceeds the table of values by 2 or more dB(A) at all points in both Line A and Line B.

5.2 Effect of Failure of a Given Test—The failure of any device or component at any point in the required test sequence shall constitute a failure of the entire sequence. The entire test sequence including the tests that were passed shall be considered void and the entire test sequence shall be repeated using a new device or component or the original device or component after repair.

Failure of a component which is included in order to have an operable system but which is not itself under test shall have no effect on the test sequence. Such a component may be replaced with a similar component and the test sequence continued.

5.3 Matching of Components in Electronic Siren Systems—In accordance with their respective nameplates, only properly matched electronic siren speaker(s) and electronic siren amplifiers shall be interconnected. The number, wattage, and nominal impedance of connected electronic siren speakers shall be in conformance with the requirements presented on the nameplate of the electronic siren amplifier.

6. Guidelines

6.1 General—Proper installation is vital to the performance of the siren and the safe operation of the emergency vehicle. It is important to recognize that the operator of the emergency vehicle is under psychological and physiological stress caused by the emergency situation. The siren system should be installed in such a manner as to:

a. Not reduce the acoustical performance of the system.

b. Limit as much as practical the noise level in the passenger compartment of the vehicle.

c. Place the controls within convenient reach of the operator so that he can operate the system without losing eye contact with the roadway.

6.2 Mounting and Wiring—All devices should be mounted in accordance with the manufacturers instructions and securely fastened to vehicle elements of sufficient strength to withstand the forces applied to the device.

All wiring should conform to the minimum wire size and other recommendations of the manufacturer and be protected from moving parts and hot surfaces. Looms, grommets, cable ties, and similar installation hardware should be used to anchor and protect all wiring.

Fuses or circuit breakers should be located as close to the power take-off points as possible and properly sized to protect the wiring and devices.

Particular attention should be paid to the location and method of making electrical connections and splices to protect these points from corrosion and loss of conductivity. Ground terminations should be only made to substantial chassis components.

6.3 Placement of Electromechanical Sirens and Electronic Siren Speakers—The sound projecting opening should be pointed forward, parallel to the ground, and not obstructed or muffled by structural components of the vehicle. Concealed or under hood mounting will result in a dramatic reduction in performance and is not recommended.

To minimize potential hearing loss, it is recommended that a maximum sound pressure level of 85 dB(A) be present in the passenger area when the vehicle is in motion with the siren operating. Manufacturers and users of emergency vehicles should measure the actual sound pressure levels per SAE J336 at each riding position.

Electromechanical sirens and electronic siren speakers should be mounted as far from the occupants as possible using acoustically insulated compartments and isolation mountings to minimize the transmission of sound into the vehicle. It may be helpful to mount the device on the front bumper, engine cowl or fender, heavily insulate the passenger compartment, and operate the siren only with the windows closed.

Each of these approaches may cause significant operational problems including loss of siren performance from road slush, increased likelihood of damage to the siren in minor collisions, and the inability to hear the sirens on other emergency vehicles. Appropriate training of vehicle operators is recommended to alert them to these problems and minimize the effect of these problems on operations.

6.4 Electronic Amplifiers and Siren Controls—Devices should be mounted only in locations that conform to their SAE identification code. Devices for interior mounting should not be placed under hood, etc.

Controls should be placed within the convenient reach of the driver or if intended for two man operation the driver and/or the passenger. In some vehicles, multiple control switches may be necessary for convenient operation from two positions.

Convenient reach is defined as the ability of the operator of the siren system to manipulate the controls from his normal driving/riding position without excessive movement away from the seat back or loss of eye contact with the roadway.

LIGHTING INSPECTION CODE—SAE J599 MAY81   SAE Standard

Report of the Lighting Division, approved January 1937, last revised by the Lighting Committee December 1974, editorial change May 1981. Rationale statement available.

This code is intended only for the inspection and maintenance of lighting equipment on motor vehicles that are in use.

The original SAE code, adopted in 1937, was drafted for use in preparing Interstate Commerce Commission regulations for trucks and buses in interstate operation under the 1935 Motor-Carrier Act. Subsequently, the SAE code served as a basis for Section 2, Lighting Systems, of the American National Standard Code for Inspection Requirements for Motor Vehicles, ANSI D7-1939. The ANSI inspection requirements for lighting systems were adopted by the Society as the SAE Recommended Practice in January 1940.

1. Definitions

1.1 Sealed Beam Unit—An integral and indivisible optical assembly with the name "Sealed Beam" molded in the lens.

1.2 Upper Beam—A beam intended primarily for distant illumination and for use on the open highway when not meeting other vehicles.

1.3 Lower Beam—A beam intended to illuminate the road ahead of the vehicle without causing undue glare to other drivers.

1.4 7 in. (178 mm) Sealed Beam System—A system employing two 7 in. (178 mm) Sealed Beam units.

1.5 7 in. (178 mm) Type 2 Sealed Beam Unit—A 7 in. (178 mm) diameter unit (with a numeral 2 molded in the lens), which provides an upper and a lower beam. These units are mechanically aimable. NOTE: Original 7 in. (178 mm) Sealed Beam units which can be identified by the absence of "2" on the lens shall be aimed visually on the upper beam.

1.6 5¾ in. (146 mm) Sealed Beam System—A system employing four 5¾ in. (146 mm) Sealed Beam units: two Type 1 and two Type 2.

1.7 5¾ in. (146 mm) Type 1 Sealed Beam Unit—A 5¾ in. (146 mm) diameter unit having a single filament and used in a four-lamp system to provide the principal portion of the upper beam.

1.8 5¾ in. (146 mm) Type 2 Sealed Beam Unit—A 5¾ in. (146 mm) diameter unit having two filaments and used in a four-lamp system to provide the lower beam and a secondary portion of the upper beam.

1.9 4 x 6½ in. (100 x 165 mm) Sealed Beam System—A system employing four 4 x 6½ in. (100 x 165 mm) sealed beam units: two Type 1A and two Type 2A.

1.10 4 x 6½ in. (100 x 165 mm) Type 1A Sealed Beam Unit—A 4 x 6½ in. (100 x 165 mm) rectangular unit having a single filament and used in a four-lamp system to provide the principal portion of the upper beam.

1.11 4 x 6½ in. (100 x 165 mm) Type 2A Sealed Beam Unit—A 4 x 6½ in. (100 x 165 mm) rectangular unit having two filaments and used in a four-lamp system to provide the lower beam and a secondary portion of the upper beam.

1.12 Mechanically Aimable Sealed Beam Unit—A unit having three pads on the face of the lens forming a plane which is intended to be used to adjust and inspect the aim of the unit when installed on the vehicle.

1.13 Symmetrical Beam—A beam in which both sides are symmetrical with respect to the median vertical plane of the beam.

1.14 Asymmetrical Beam—A beam in which both sides are not symmetrical with respect to the median vertical plane of the beam. All lower beams are asymmetrical. NOTE: The inspector should see that the driver understands how to use multiple beam headlamps so as to obtain the best road lighting with minimum glare to other users of the highway.

2. Equipment—It is recommended that mechanically aimable headlamps be aimed and inspected for aim by mechanical aimers. Another aiming and inspection method is by visual means on a screen at a distance of 25 ft (7.6 m) ahead of the headlamps or on the screen of a headlamp testing machine.

2.1 The mechanical aimer used shall conform to the requirements of SAE J602. The device shall be in good repair, calibrated and used according to the manufacturer's instructions.

2.2 If a screen is used, it should be of adequate size with a matte-white surface well shaded from extraneous light and properly adjusted to the floor area on which the vehicle stands. Provision should be made for moving the screen or its vertical centerline so that it can be aligned with the vehicle axis. In addition to the vertical centerline, the screen should be provided with four laterally adjustable vertical tapes and two vertically adjustable horizontal tapes. The four movable vertical tapes should be located on the screen at the left and right limits called for in the specification with reference to centerlines ahead of each headlamp unit. The headlamp centerlines shall be spaced either side of the fixed centerline on the screen by the amount the headlamp units are to the left and right. The horizontal tapes should be located on the screen at the upper and lower limits called for in the specifications with reference to the height of lamp centers and the plane on which the vehicle rests, not the floor on which the screen rests. See Fig. 1.

2.3 The Headlamp Testing Machine used shall conform to the requirements of SAE J600. The device shall be in good repair, calibrated and used according to the manufacturer's instructions.

3. Preparation for Headlamp Aim or Inspection—Before checking beam aim, the inspector shall:

3.1 Remove ice or mud from under fenders.
3.2 See that no tire is noticeably deflated.
3.3 Check car springs for sag or broken leaves.
3.4 See that there is no load in the vehicle other than the driver.
3.5 Check functioning of any "level-ride" control.
3.6 Clean lenses and aiming pads.
3.7 Check for bulb burnout, broken mechanical aiming pads, and proper beam switching.
3.8 Stabilize suspension by rocking vehicle sideways.

4. Headlamp Aim Adjustment for Service Facilities

4.1 The following aim adjustment requirements should apply to dealers, service stations, and others who do headlamp adjusting.

4.2 It is recommended that mechanically aimable headlamps be aimed using mechanical aimers (paragraph 2.1). The aimers shall be calibrated for accuracy and shall be compensated for the level of the floor in the aiming area.

4.3 Mechanical Aiming

4.3.1 The correct mechanical aim for both Type 1 and Type 2 units is 0-0.
4.3.2 If a headlamp being serviced is not so aimed, the aim shall be corrected to 0-0.

4.4 Visual Aiming

4.4.1 The correct visual aim for Type 1 units is with the center of the high intensity zone at horizontal and straight ahead vertically. (See Fig. 2.)

4.4.2 The correct visual aim for Type 2 units is with the top edge of the high intensity zone of the lower beam horizontal and the left edge at vertical. (See Fig. 3.)

4.4.3 If the headlamp being serviced is not so aimed, it should be corrected to the above aim.

5. Headlamp Aim Inspection Limits for Vehicle Inspection Facilities

5.1 The following inspection limits should apply to stations that conduct mandatory inspection of vehicles.

5.2 It is recommended that mechanically aimable lamps be inspected using mechanical aimers (paragraph 2.1). The aimers shall be calibrated for accuracy and shall be compensated for the level of the floor in the inspection area.

5.3 Mechanical Aim Inspection

5.3.1 The mechanical inspection limits for both Type 1 and Type 2 units shall be 4 (100 mm) up to 4 (100 mm) down and 4 (100 mm) left to 4 (100 mm) right.

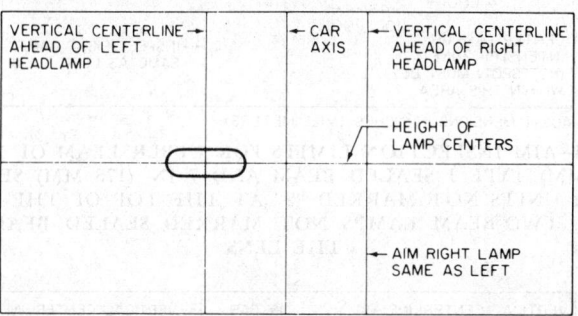

FIG. 2—HOW PROPERLY AIMED UPPER BEAM OF 5¾ IN (146 MM) TYPE 1 AND 7 IN (178 MM) SEALED BEAM (NOT MARKED "2" ON LENS) WILL APPEAR ON THE AIMING SCREEN 25 FT (7.6 M) IN FRONT OF VEHICLE. (SHADED AREA INDICATES HIGH INTENSITY ZONE)

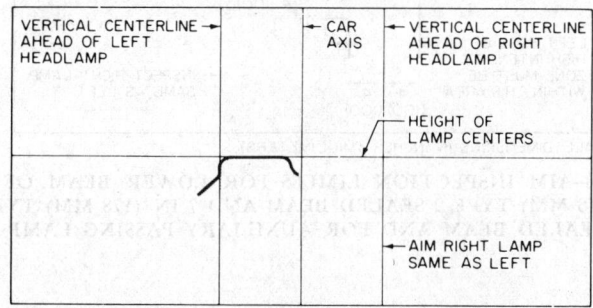

FIG. 3—HOW PROPERLY AIMED LOWER BEAM OF 5¾ IN (146 MM) AND 7 IN (178 MM) TYPE 2 SEALED BEAM WILL APPEAR ON THE AIMING SCREEN 25 FT (7.6 M) IN FRONT OF VEHICLE. (SHADED AREA INDICATES HIGH INTENSITY ZONE)

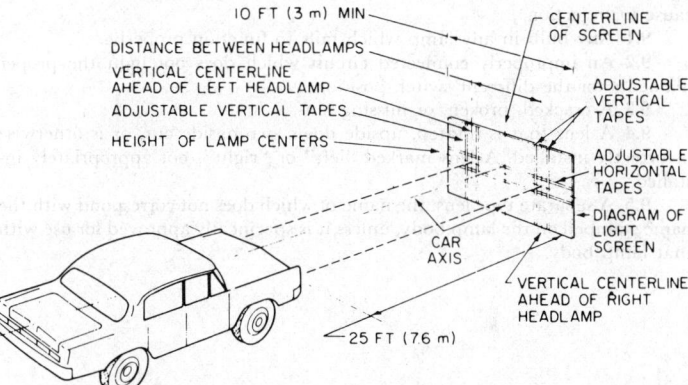

FIG. 1—ALIGNMENT OF HEADLAMP AIMING SCREEN

5.3.2 Failure to meet these limits shall be cause for rejection.

5.4 Visual Aiming

5.4.1 The visual inspection limits for Type 1 units shall be with the center of the high intensity zone from 4 (100 mm) up to 4 (100 mm) down and from 4 (100 mm) left to 4 (100 mm) right based in inches (millimeters) on a screen at 25 ft (7.6 m). (See Fig. 4.)

5.4.2 The visual inspection limits for Type 2 units shall be with the top edge of the high intensity zone from 4 (100 mm) up to 4 (100 mm) down and the left edge of the high intensity zone from 4 (100 mm) left to 4 (100 mm) right based in inches (millimeters) on a screen at 25 ft (7.6 m). (See Fig. 5.)

5.4.3 Failure to meet these limits shall be cause for rejection.

6. Fog Lamps (Symmetrical Beams) Aim Adjustment for Service Facilities

6.1 The following aim adjustment requirements should apply to dealers, service stations, and others who do headlamp adjusting.

6.2 The correct visual aim for fog lamps (symmetrical beams) is with the top edge of the high intensity zone 4 (100) below horizontal and the center of the high intensity zone straight ahead vertically based in inches (millimeters) on a screen at 25 ft (7.6m) (See Fig. 6.)

7. Fog Lamps (Symmetrical Beam) Aim Inspection Limits for Vehicle Inspection Facilities

7.1 The following inspection limits should apply to stations that conduct mandatory inspection of vehicles.

ed. 7.2 The visual inspection limits for fog lamps (symmetrical beam), which are installed with universal mounting applications so as to fit many different vehicle models, shall be with the top edge of the high intensity zone at horizontal or below and with the center of the high intensity zone from 4 (100) left to 4 (100) right based in inches (millimeters) on a screen at 25 ft (7.6m) (See Fig. 7.)

ed. 7.3 The visual inspection limits for fog lamps (symmetrical beam), which are designed to be integrated into one specific vehicle, shall be with the top edge of the high intensity zone at horizontal or below on a screen at 25 ft (7.6m). (See Fig. 8.)

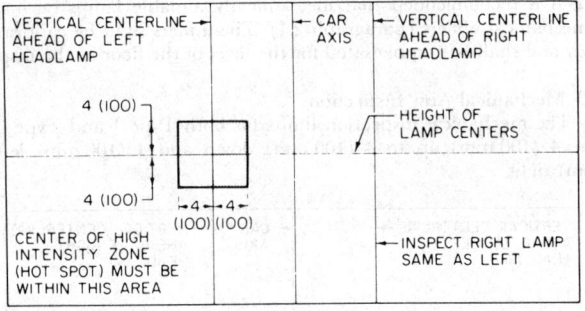

FIG. 4—AIM INSPECTION LIMITS FOR UPPER BEAM OF 5¾ IN (146 MM) TYPE 1 SEALED BEAM AND 7 IN (178 MM) SEALED BEAM UNITS NOT MARKED "2" AT THE TOP OF THE LENS. ALSO, TWO-BEAM LAMPS NOT MARKED SEALED BEAM ON THE LENS

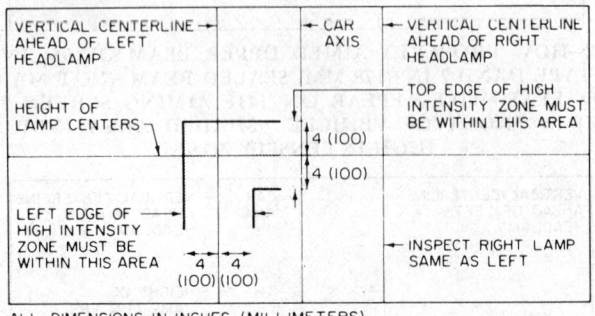

FIG. 5—AIM INSPECTION LIMITS FOR LOWER BEAM OF 5¾ IN (146 MM) TYPE 2 SEALED BEAM AND 7 IN (178 MM) TYPE 2 SEALED BEAM AND FOR AUXILIARY PASSING LAMP

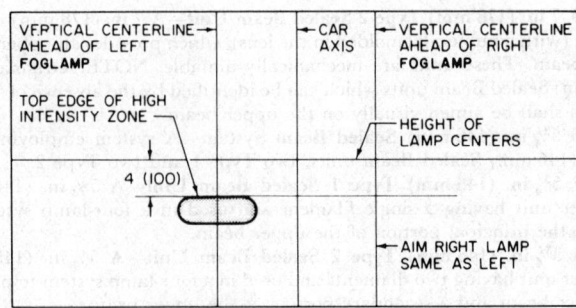

FIG. 6—HOW PROPERLY AIMED FOG LAMP (SYMMETRICAL ed. BEAM) WILL APPEAR ON THE AIMING SCREEN 25 FT (7.6 M) IN FRONT OF VEHICLE. (SHADED AREA INDICATES HIGH INTENSITY ZONE)

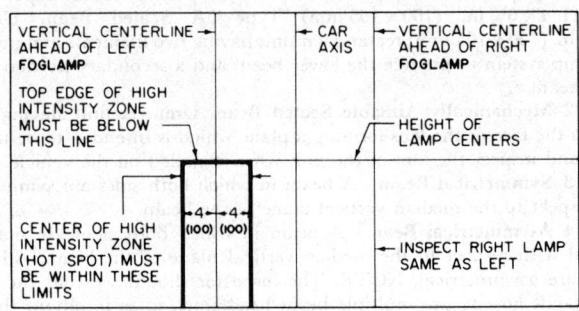

ed. FIG. 7—AIM INSPECTION LIMITS FOR FOG LAMPS (SYMMETRICAL BEAM) (UNIVERSAL APPLICATION)

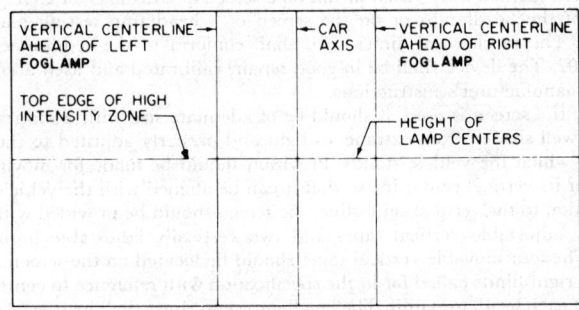

ed. FIG. 8—AIM INSPECTION LIMITS FOR FOG LAMPS (SYMMETRICAL BEAM) (INTEGRATED TYPE)

8. Fog Lamps (Asymmetrical Beam) and Passing Lamps Aim Adjustment and Inspection Limits

8.1 Lamp aim adjustment and inspection is the same as for Type 2 Sealed Beam headlamp units. See paragraph 4.4.2 for adjustment and 5.4.2 for inspection.

9. General Lamp Inspection Other Than Headlamp Aim Inspection—This includes the following types of lamps: head, tail, stop, license, clearance, signal, marker, reflex reflector, and fog. Any of the following defects shall be cause for rejection.

9.1 Any bulb in any lamp which fails to function properly.

9.2 An improperly connected circuit which does not light the proper filaments for the different switch positions.

9.3 A cracked, broken, or missing lens.

9.4 A lens that is rotated, upside down, wrongside out, or is otherwise incorrectly installed. A lens marked "left" or "right", not appropriately installed.

9.5 A separate type lens, the name of which does not correspond with the name stamped on the lamp body, unless it is specifically approved for use with that lamp body.

(R) HEADLAMP AIM TEST MACHINES —SAE J600 FEB93

SAE Recommended Practice

Report of the Lighting Committee approved December 1952, revised November 1963, editorial change May 1981. Rationale statement available. Completely revised by the SAE Road Illumination Devices Standards Committee February 1993. Rationale statement available.

1. **Scope**—This SAE Recommended Practice provides laboratory test procedures for testing headlamp aim test machines to determine their ability to aim or to check the aim of headlamps, fog lamps, and auxiliary driving and lower beam lamps. This report does not apply to aiming devices of the kind covered by SAE J602.

2. **References**

   2.1 **Applicable Documents**—The following publications form a part of this specification to the extent specified herein. The latest issue of SAE publications shall apply.

   2.1.1 SAE PUBLICATIONS—Available from SAE, 400 Commonwealth Drive, Warrendale, PA 15096-0001.

   SAE J575—Tests for Motor Vehicle Lighting Devices and Components
   SAE J599—Lighting Inspection Code
   SAE J602—Headlamp Aiming Device for Mechanically Aimable Sealed Beam Headlamp Units
   SAE J1383—Performance Requirements for Motor Vehicle Headlamps

   2.2 **Definitions**

   2.2.1 A headlamp aim test machine is an optical or photoelectric device used to aim or check the aim of forward lighting devices.

   2.2.2 The H and V readings are located relative to the H-V axis as defined in paragraph 2.7 in SAE J1383. The H reading is located at H-2R and the V reading at V-1D.

3. **Tests**

   3.1 **Samples for Test**

   3.1.1 Headlamp aim test machines submitted for laboratory tests should be representative of the device as regularly manufactured and marketed, except that in the case of a machine using a track, an abbreviated section of track may be supplied for the test. Each sample shall include all accessory equipment for the device and necessary to its service operation and calibration. Full assembly and operating instructions shall be provided, including information on how to check accuracy and maintain the device in calibration.

   3.1.2 Sample lamps for test shall include a group of upper (1A) and lower (2B) beam headlamp units which meet SAE specifications. Representative groups of fog lamps and auxiliary driving and lower beam lamps will be obtained in the same manner.

   3.2 **Laboratory Facilities**—The laboratory shall be equipped with facilities to make physical and optical tests required in this report, in accordance with established laboratory practice. It will include a goniometer, a test screen located at 7.6 m from the goniometer and calibrated photocells or equivalent light-detecting devices.

   3.3 **Test Procedures**

   3.3.1 SYMMETRICAL BEAM—UPPER BEAM HEADLAMPS AND AUXILIARY DRIVING LAMPS

   3.3.1.1 The sample lamp shall be mounted on a goniometer which meets the requirement of the Photometry Test in SAE J575.

   3.3.1.2 The sample lamp should be operated at rated voltage and then kept burning through the end of test.

   3.3.1.3 The headlamp aim test machine shall be positioned in front of the headlamp according to the manufacturer's instructions.

   3.3.1.4 The headlamp shall be aimed according to the aim test machine manufacturer's instructions, using the goniometer to position the lamp.

   3.3.1.5 Goniometer initial readings (Hi, Vi) shall be recorded.

   3.3.1.6 The aim test machine shall be removed from its position in the front of the headlamp.

   3.3.1.7 The headlamp upper beam shall then be photoelectrically balanced on the goniometer in accordance with the Beam Pattern Location Test for the upper beam in SAE J1383.

   3.3.1.8 The final goniometer reading (Hf, Vf) shall be recorded.

   3.3.2 ASYMMETRICAL BEAM—LOWER BEAM HEADLAMPS AND AUXILIARY LOWER BEAM LAMPS

   3.3.2.1 The lower beam headlamps and auxiliary lower beam lamps shall be used to test the headlamp aim test machine in accordance with 3.3.1.1 through 3.3.1.8, substituting lower beam for upper beam.

   3.3.3 FOG LAMPS

   3.3.3.1 Fog lamps shall be used to test the aim test machine in accordance with 3.3.1.1 through 3.3.1.8, substituting fog lamp for upper beam.

4. **Requirements**

   4.1 **For Testing With Upper Beam Headlamps and Symmetrical Beam Auxiliary Driving Lamps**

   4.1.1 The headlamp aim test machine shall permit determination of the vertical and horizontal aim of the geometric center of the high intensity zone on the upper beam within specified limits.

   4.1.2 The final goniometer aim reading (Hf, Vf), when compared to the initial aim reading (Hi, Vi) shall not vary from the initial reading in the vertical direction by more than 0.3 degrees (37 mm at 7.6 m) and in the horizontal direction by more than 0.6 degrees (75 mm at 7.6 m).

   4.2 **For Testing With Asymmetrical Beam Lower Beam Headlamps and Auxiliary Lower Beam Lamps**

   4.2.1 The headlamp aim test machine shall permit determination of the aim of the top and left edge cutoffs of the high intensity zone of the lower beam of headlamps and auxiliary lower beam lamps.

   4.2.2 The final goniometer aim reading (Hf, Vf), when compared to the initial aim reading (Hi, Vi) shall not vary from the initial reading in the vertical direction by more than 0.3 degrees (37 mm at 7.6 m) and in the horizontal direction by more than 0.6 degrees (75 mm at 7.6 m).

   4.3 **For Testing With Fog Lamps**

   4.3.1 The headlamp aim test machine shall permit determination of the vertical aim of the top cutoff of the high intensity zone and the horizontal aim of the geometric center of the high intensity zone.

   4.3.2 The final goniometer aim reading (Hf, Vf), when compared to the initial aim reading (Hi, Vi) shall not vary from the initial reading in the vertical direction by more than 0.3 degrees (37 mm at 7.6 m) and in the horizontal direction by more than 0.6 degrees (75 mm at 7.6 m).

   4.4 **General Requirements**

   4.4.1 The headlamp aim test machine shall incorporate a fixed track or equivalent for positioning the aiming device in front of the headlamps.

   4.4.1.1 Means shall be provided in the device for compensating within ±0.05 for variation in the floor slope and the adjustment method shall be clearly explained in the operating instructions.

   4.4.1.2 Means shall be provided in the device for lateral alignment within 0.1 degree with respect to the longitudinal axis of the vehicle.

   4.4.2 Aim test machines using a photoelectric means to determine aim shall also have a visual screen or equivalent upon which the beam pattern is projected proportional to its appearance and aim on a screen at 7.6 m. Such visual screen or equivalent shall be plainly visible to the operator and should have horizontal and vertical reference lines to permit visual appraisal of the aim of the lamp.

   4.4.3 Design of the headlamp aim test machine shall permit checking the aim of the lamps mounted at heights from 25 to 140 cm and spaced up to 130 cm from the center of the motor vehicle.

   4.4.4 The device and/or the instructions shall provide a means for calibration and/or verification of calibration.

   4.4.5 The spirit level or other means provided for indicating vertical aim shall be capable of indicating a deviation of at least 2.5 mm with a 0.2 degree (25 mm at 7.6 m) change in level.

   4.4.6 A vertical aim scale shall be provided with numerical gradations in steps, each of which represents 25 mm at 7.6 m to provide for variations in aim at least 100 mm above level to 250 mm below level.

   4.4.7 A lateral aim scale shall be provided with gradations in steps of not more than 50 mm at 7.6 m from straight ahead to at least 150 mm left and right.

   4.4.8 The instructions covering use of the device shall include those items in the Preparation for Aiming Section of SAE J599.

   4.4.9 Instructions furnished by the manufacturer for aiming lamps shall be such that the beam patterns when viewed on the screen will fall within the limits set in SAE J599.

   4.4.10 Headlamp Aim Test Machines must be capable of operating accurately in service garages with a temperature range of 0 to 40 °C.

(R) HEADLAMP AIMING DEVICE FOR MECHANICALLY AIMABLE HEADLAMP UNITS—SAE J602 DEC89

SAE Standard

Report of the Lighting Committee approved October 1957, revised October 1980. Completely revised by the Road Illumination Devices Standards Committee December 1989.

1. Scope—This document applies to the requirements of a device used in the field and inspection stations to aim and check aim of mechanically aimable headlamp units.

The purpose of this document is to provide a laboratory test procedure to determine whether the devices under test are capable of accurately positioning headlamp units from their aiming pads and maintaining their accuracy in service within the tolerances designated in this document.

2. Definitions

2.1 Headlamp Aiming Device—A device used to adjust and inspect the aim of mechanically aimable headlamp units consisting of one or more fixtures designed to seat against the three aiming pads (aiming plane) on mechanically aimable headlamp units installed on a vehicle to facilitate accurate aiming of such units, vertically and laterally.

2.2 Mechanically Aimable Headlamp Units—A unit having three pads on the face of the lens forming a mechanical aiming plane used to adjust and inspect the aim of the unit when installed on a vehicle.

2.3 Aiming Plane—A plane which is perpendicular to the longitudinal axis of the vehicle and tangent to the forward most aiming pad on the headlamp.

3. Samples For Test—Sample devices submitted for laboratory tests shall be representative of the devices as regularly manufactured and marketed. Each sample shall include all accessory equipment peculiar to the device. Full assembly and operating instructions shall be provided, including information on how to check accuracy and maintain the device in calibration.

4. Laboratory Facilities—The laboratory shall be equipped with all facilities necessary to make the tests in this document.

NOTE 1: All tests are to be made in air ambient temperature of 75°F ± 5 (24°C ± 3).

NOTE 2: If a vertical indication means other than a spirit level is used, equivalent accuracy shall be maintained.

5. Requirements

5.1 Design Requirements

5.1.1 The device shall be of such design that the seating portion will register only on the three aiming pads on the headlamp units as covered by SAE J1383.

5.1.2 No part of the device, except those parts (strings, sighting devices, scales, etc.) required for referencing lateral alignment between devices, shall extend beyond the dimensional limits of the headlamp aiming device locating plate (Figure 1, dimension C and Figures 2, 3, and 4, dimension 4.05 in (102.9 mm) maximum diameter).

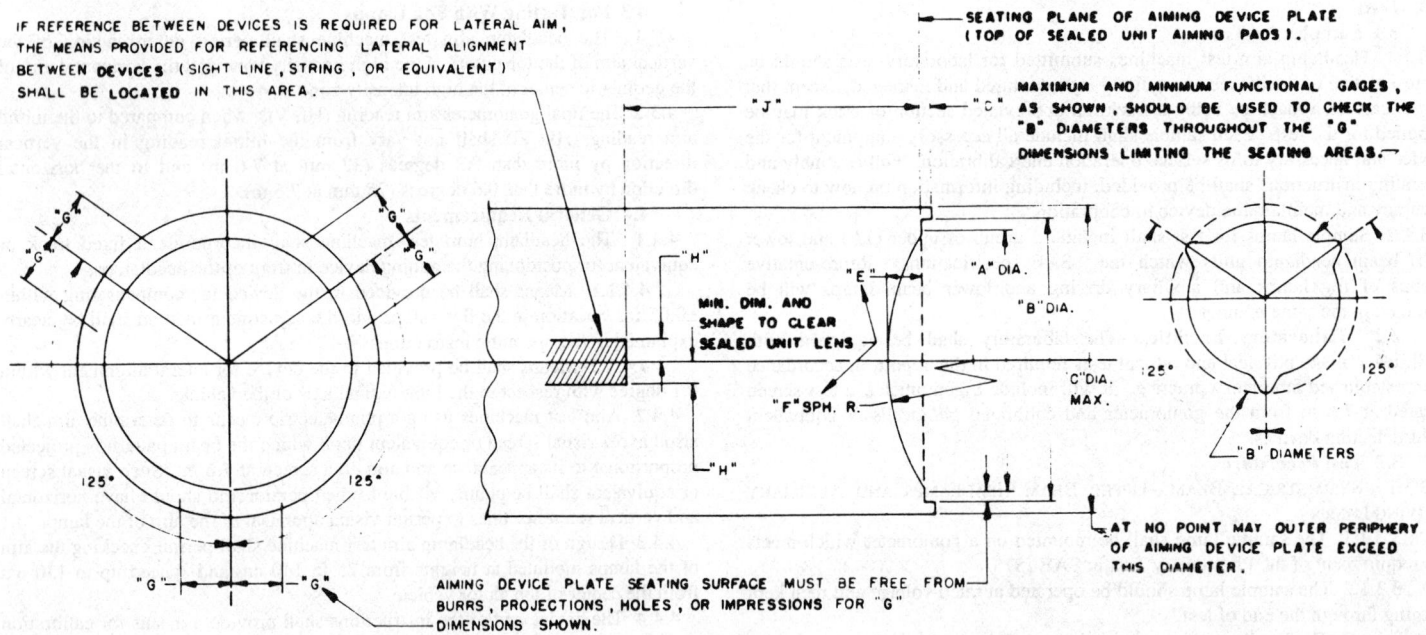

NOTE: There shall be no projections, tangs, lugs, etc. on this locating plate, which will permit locating the aiming device on any part of the headlamp other than the aiming pads on the mechanical headlamp unit.

Locating Plate	Unit of Measure	Dimensions											
		A		B		C	D		E	F	G	H	J
		Max	Min	Max	Min	Max	Max	Min	Ref	Ref	Min	Max	Min
5-3/4 in	in	4.830	4.770	5.375	5.345	5.700	0.165	0.145	0.70	4.40	0.70	1.00	9.50
(146 mm)	mm	122.7	121.2	136.5	135.8	144.8	4.19	3.68	17.8	111.8	17.8	25.4	241.3
7 in	in	6.140	6.080	6.710	6.680	7.031	0.180	0.160	0.96	5.60	0.70	1.00	10.25
(178 mm)	mm	156.0	154.4	170.4	169.7	178.6	4.57	4.06	24.4	142.2	17.8	25.4	260.4

FIG. 1—DIMENSIONAL SPECIFICATIONS FOR HEADLAMP AIMING DEVICE LOCATING PLATE

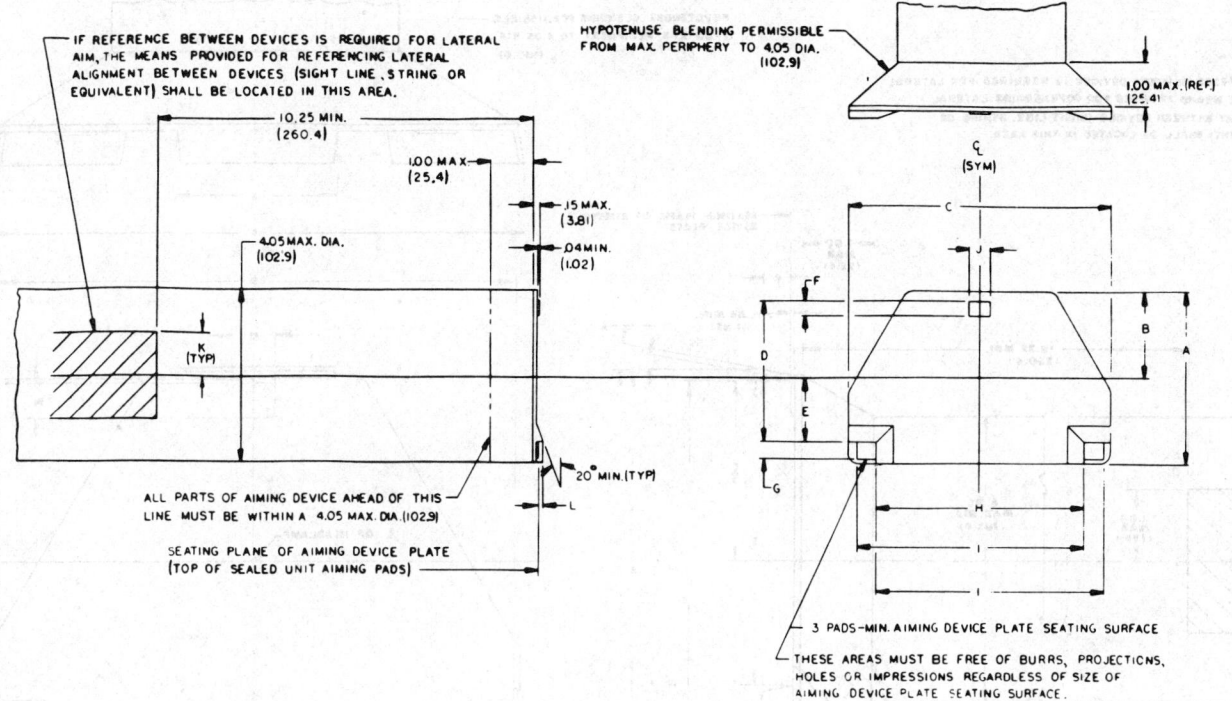

NOTE: There shall be no projections, tangs, lugs, etc. on this locating plate, which will permit locating the aiming device on any part of the headlamp other than the aiming pads on the mechanical headlamp unit.

ALL DIMENSIONS ENCLOSED (X.XX) ARE IN MM.

Locating Plate	Unit of Measure	Dimensions																						
		A		B		C		D		E		F		G		H		I		J		K		L	
		Max	Min	Max	Min	Max	Min	Max	Min	Max	Min	Max	Min	Max	Min	Max	Min	Max	Min	Max	Min	Max	Min	Max	Min
4 × 6½	in	3.935	3.925	1.975	1.953	6.001	5.991	3.320	3.300	1.550	1.540	0.370	0.350	0.330	0.310	5.088	5.018	—	5.421	—	0.400	1.000	—	0.080	0.060
100 × 165	mm	99.95	99.60	50.17	49.66	152.42	152.17	84.33	83.82	39.37	39.12	9.40	8.89	8.38	7.87	129.24	128.98	—	137.20	—	10.16	25.40	—	2.03	1.52

FIG. 2—DIMENSIONAL SPECIFICATIONS FOR HEADLAMP AIMING DEVICE LOCATING PLATE (100 × 165)

5.1.3 A device that uses adapters to fit more than one size headlamp unit shall meet all the requirements of this document with and without adapters.

5.1.4 The seating plane of the device shall meet the dimensions shown in Figures 1, 2, 3, and 4.

5.1.5 When aiming headlamp units spaced 90 in (2300 mm) apart, the torque exerted by the device at the aiming plane shall not exceed 18 lbf in (2.0 N·m) vertically and 12 lbf in (1.4 N·m) laterally.

5.1.6 The means of securing the device to the headlamp unit shall retain the device against the three aiming pads when an axially centered tensile force of 4.0 lb/ft (17.8 N·m) minimum is applied to the device.

5.1.7 The device shall be capable of being calibrated and shall have available for immediate use an independent calibration fixture and/or instructions to immediately recalibrate the device.

5.1.8 If a suction cup is used to retain the device to the headlamp unit, the effective diameter for 5-3/4 in (146 mm) and 7 in (178 mm) headlamps shall not exceed 3.5 in (90 mm) and the effective diameter for 4 × 6-1/2 in (100 × 165 mm) and 5 × 8 in (142 × 200 mm) headlamps shall not exceed 2.8 in (71 mm) when installed.

5.1.9 Means shall be provided in the device for compensating within ±0.1 deg through a slope range of ±1.5 deg from horizontal. The method for device compensation shall be clearly explained in the operating instructions.

5.1.10 If the horizontal aim is to be accomplished by reference between devices on opposite sides of the vehicle, the means provided for referencing lateral alignment between devices (sight line, string, or equivalent) shall be located as shown in Figures 1, 2, 3, and 4.

5.1.11 The spirit level or other means provided for indicating vertical aim shall be capable of showing at least a 0.1 in (2.5 mm) deviation with a 1 in (25 mm[1]) change in level.

5.1.12 A horizontal aim scale shall be provided with graduations in steps of not more than 2 in (51 mm) from straight ahead to at least 8 in (203 mm) left and right.

5.1.13 The instructions covering the use of the device shall include those items shown in Section 3 of SAE J599.

5.1.14 The vertical aim scale shall be marked 0 with the aiming plane vertical.

5.1.15 The vertical aim scale shall be provided with numerical graduations in steps, each of which represents 1 in (25 mm) to provide for variations in vertical aim from at least 8 in (203 mm) below 0.

5.2 Test Procedure—Assuming that the devices comply with the general requirements, they shall be considered acceptable if they comply with additional test requirements as follows:

5.2.1 With the aiming plane vertical and with the vertical scale on the device set at 0, the angle through which the aiming plane must be rotated vertically to center the bubble in the spirit level, or equivalent, shall not exceed 0.5 in (13 mm).

5.2.2 With the aiming planes in the same vertical plane and with the means provided for adjusting lateral aim in use, the angle through which the aiming plane must be rotated laterally to indicate straight ahead shall not exceed ±1 in (25 mm)[2] with the lamps 24 and 90 in (610 and 2300 mm) apart.

[1] Represents inches (millimeters) at 25 ft (7.6 m).
[2] Represents inches (millimeters) at 25 ft (7.6 m).

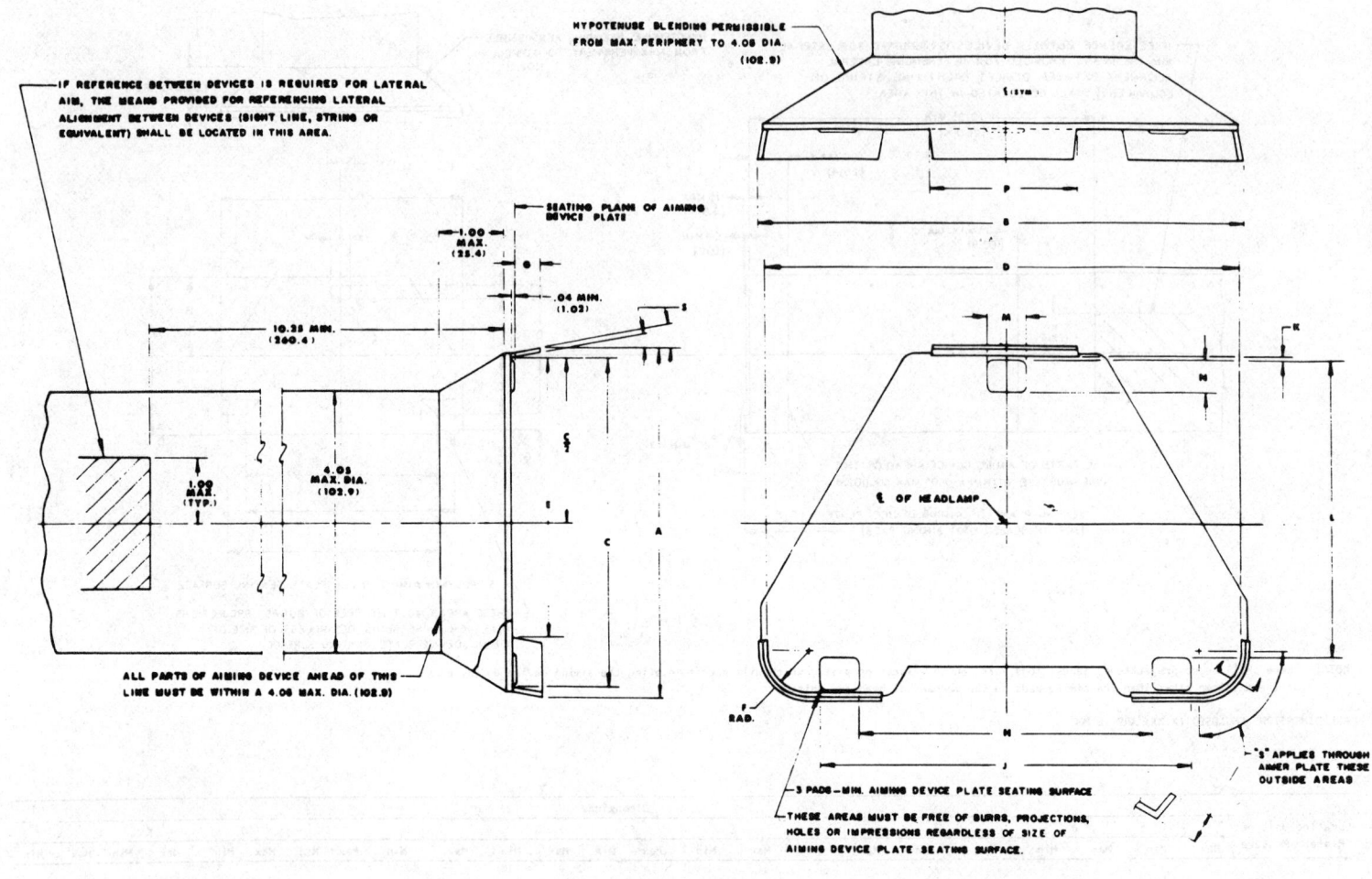

NOTE: There shall be no projections, tangs, lugs, etc. on this locating plate, which will permit locating the aiming device on any part of the headlamp other than the aiming pads on the mechanical headlamp unit.

ALL DIMENSIONS ENCLOSED (X.XX) ARE IN MM.

Unit of Measure	Dimensions															
	A ±0.010	B ±0.010	C ±0.005	D ±0.005	E ±0.010	F ±0.010	G ±0.010	H ±0.005	J ±0.005	K ±0.005	L ±0.005	M ±0.010	N ±0.010	P ±0.010	±0°	S 30 ft
in	5.275	7.340	4.995	7.030	4.285	0.625	0.395	4.450	5.663	0.005	4.487	0.610	0.485	2.125	10°	0 ft
mm	133.98	186.43	126.87	179.07	108.83	15.87	10.03	113.03	143.84	0.12	113.97	15.49	12.31	53.97		

FIG. 3—DIMENSIONAL SPECIFICATIONS FOR HEADLAMP AIMING DEVICE LOCATING PLATE (142 × 200)

5.2.3 With the aiming planes initially in the same vertical plane and subsequently toed inward and outward 6 in (152 mm) and with the means provided for checking lateral aim in use, the error in reading shall not exceed ±1 in (25 mm) with the lamps 60 in (1520 mm) apart.

5.2.4 With the aiming plane vertical and with the vertical scale on the device set at 0, the level on the aimer shall be adjusted prior to each of the following tests to center the bubble in the spirit level or equivalent.

5.2.4.1 Each step on the vertical aim scale shall be checked and in no case shall the variation from the correct aim exceed ±0.5 in (13 mm).

5.2.4.2 A pair of devices shall be stabilized at 20°F ± 5 (−7°C ± 3) and then installed on a pair of unlighted headlamp units spaced 60 in (1520 mm) apart at the 20°F (−7°C) ambient temperature. After a period of 30 min, the seating portion of the device shall continue to register against the three headlamp unit aiming pads, and the variation from correct vertical aim shall not exceed ±0.5 in (13 mm) and the variation from correct lateral aim shall not exceed ±1 in (25 mm)[3].

5.2.4.3 They shall then be installed on the pair of unlighted headlamp units spaced 60 in (1520 mm) apart and the variation from

[3] Represents inches (millimeters) at 25 ft (7.6 m).

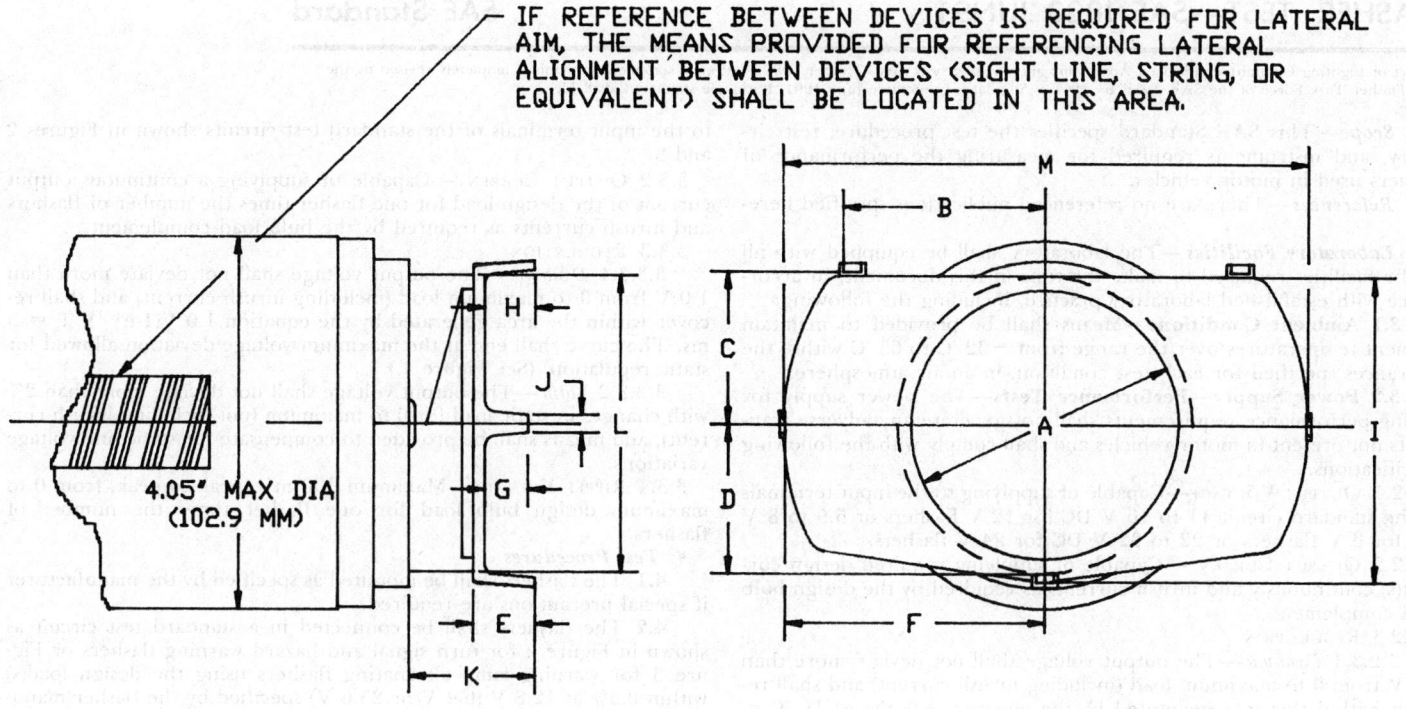

Locating Plate	Unit of Measure	Dimensions												
		A	B	C	D	E	F	G	H	I	J	K	L	M
	Tolerance inches (mm)	±0.010 (±0.254)	±0.005 (±0.127)	±0.010 (±0.254)	±0.010 (±0.254)	±0.010 (±0.254)	±0.005 (±0.127)	±0.010 (±0.254)	±0.001 (±0.025)	±0.010 (±0.254)	±0.010 (±0.254)	±0.005 (±0.127)	±0.010 (±0.254)	±0.010 (±0.254)
92 × 150	inches (mm)	2.925 (74.295)	2.264 (57.506)	1.575 (40.005)	1.634 (41.504)	0.676 (17.170)	2.853 (72.466)	0.472 (11.989)	0.551 (13.995)	3.300 (83.820)	0.197 (5.004)	1.375 (34.925)	3.225 (81.915)	5.720 (145.288)

FIG. 4 — DIMENSIONAL SPECIFICATIONS FOR HEADLAMP AIMING DEVICE LOCATING PLATE (92 × 150)

correct vertical aim shall not exceed ±0.5 in (13 mm) and the variation from correct lateral aim shall not exceed ±1 in (25 mm).[4]

5.2.4.4 A sample device shall be exposed to 35°F ± 5 (1.7°C ± 3) for 1 h and then immediately allowed to free fall onto a concrete floor three times from its normal operating position on a headlamp unit at a height of 40 in (1020 mm), after which it shall show no damage that would interfere with the proper calibration of the device. It shall then be installed in combination with its companion device on a pair of unlighted headlamp units spaced 60 in (1520 mm) apart and the variation from correct vertical aim shall not exceed 1 in (25 mm)[5] and the variation from the correct lateral aim shall not exceed 1 in (25 mm)[6]. (This test applies only to devices that are supported by the headlamp unit.)

5.2.4.5 Using the calibration fixture and/or instructions required by 5.1.7, the device shall be calibrated and checked for compliance with 5.2.1 and 5.2.2.

[4] Represents inches (millimeters) at 25 ft (7.6 m).

[5] Represents inches (millimeters) at 25 ft (7.6 m).

[6] Represents inches (millimeters) at 25 ft (7.6 m).

(R) FLASHER TEST—SAE J823 JUN91 — SAE Standard

Report of Lighting Committee approved April 1962 and completely revised October 1987. Rationale statement available. Completely revised by the SAE Flasher Task Force of the SAE Auxiliary Devices Standards Committee June 1991. Rationale statement available.

1. Scope—This SAE Standard specifies the test procedure, test circuitry, and instruments required for measuring the performance of flashers used in motor vehicles.

2. References—There are no referenced publications specified herein.

3. Laboratory Facilities—The laboratory shall be equipped with all of the facilities required to make the tests in this document, in accordance with established laboratory practice, including the following:

3.1 Ambient Conditions—Means shall be provided to maintain ambient temperatures over the range from $-32\,°C$ to $63\,°C$ within the tolerances specified for each test condition, in an air atmosphere.

3.2 Power Supply—Performance Tests—The power supply for testing performance requirements shall not generate any adverse transients not present in motor vehicles and shall comply with the following specifications:

3.2.1 OUTPUT VOLTAGE—Capable of supplying to the input terminals of the standard circuit 11 to 16 V DC for 12 V flashers or 5.5 to 8 V DC for 6 V flashers or 22 to 32 V DC for 24 V flashers.

3.2.2 OUTPUT CURRENT—Capable of supplying required design current(s) continuously and inrush currents as required by the design bulb load complement.

3.2.3 REGULATION

3.2.3.1 *Dynamic*—The output voltage shall not deviate more than 1.0 V from 0 to maximum load (including inrush current) and shall recover within the area generated by the equation $1.0\,V(1-e^{-t/T})$, $T = 100\,\mu s$. The curve shall end at the maximum voltage deviation allowed for static regulation. (See Figure 1.)

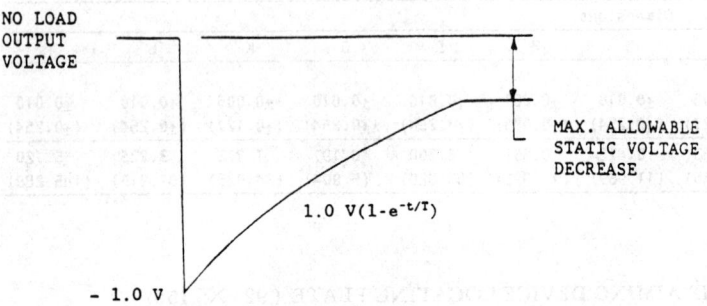

FIGURE 1—POWER SUPPLY—MAXIMUM DYNAMIC REGULATION CURVE

3.2.3.2 *Static*—The output voltage shall not deviate more than 2% with changes in static load from 0 to maximum (not including inrush current) nor for static line voltage variations.

3.2.4 RIPPLE VOLTAGE—Maximum 75 mV, peak to peak, from 0 to maximum design bulb load.

3.3 Power Supply—Durability Tests—The power supply for the durability test requirements shall not generate any adverse transients not present in motor vehicles and shall comply with the following specifications:

3.3.1 OUTPUT VOLTAGE—Capable of supplying, as required, 14 and 13 V (7 and 6.5 V DC or 28 and 26 V), according to the flasher rating, to the input terminals of the standard test circuits shown in Figures 2 and 3.

3.3.2 OUTPUT CURRENT—Capable of supplying a continuous output current of the design load for one flasher times the number of flashers and inrush currents as required by the bulb load complement.

3.3.3 REGULATION

3.3.3.1 *Dynamic*—The output voltage shall not deviate more than 1.0 V from 0 to maximum load (including inrush current) and shall recover within the area generated by the equation $1.0\,V(1-e^{-t/T})$, $T = 5$ ms. The curve shall end at the maximum voltage deviation allowed for static regulation. (See Figure 1.)

3.3.3.2 *Static*—The output voltage shall not deviate more than 2% with changes in static load for 0 to maximum (not including inrush current), and means shall be provided to compensate for static line voltage variations.

3.3.4 RIPPLE VOLTAGE—Maximum 300 mV, peak to peak, from 0 to maximum design bulb load for one flasher times the number of flashers.

4. Test Procedures

4.1 The flashers shall be mounted as specified by the manufacturer if special precautions are required.

4.2 The flashers shall be connected in a standard test circuit as shown in Figure 2 for turn signal and hazard warning flashers or Figure 3 for warning lamp alternating flashers using the design load(s) within 0.5% at 12.8 V (6.4 V or 25.6 V) specified by the flasher manufacturer.

4.3 A suitable high impedance measuring device connected to points X-Y in Figure 2, or to points X-Y_1 and to points X-Y_2 in Figure 3 shall be used for measuring flash rate, percent current "on" time, starting time, and voltage drop across the flasher. The measurement of these quantities shall not affect the circuit.

4.4 The resistance at A-B for each load circuit in Figure 2 or Figure 3 shall be measured with flasher and bulb loads each shorted out with removable shunt resistances not to exceed 0.005 Ω each.

The effective series resistance in the total circuit (Figure 2) or in each of the parallel circuits (Figure 3) between the power supply and bulb sockets (excluding the flasher and bulb loads by using the removable shunt resistances) shall be 0.10 Ω ± 0.01.

4.5 Adjust the voltage at the bulbs to 12.8 V (6.4 V or 25.6 V) as required for testing at C-D in Figure 2 or C-D and E-F in Figure 3 with the flasher shorted out by an effective shunt resistance not to exceed 0.005 Ω. The load current shall be held to the rated value for the total flasher design load(s) within 0.5% at 12.8 V (6.4 V or 25.6 V) by simultaneously adjusting trimmer resistors, R.

4.6 For testing fixed load flashers at other required voltages, adjust the power supply to provide required voltages at required temperatures at C-D in Figure 2 or C-D and E-F in Figure 3 without readjustment of trimming resistors, R.

4.7 For testing variable load flashers, the circuit shall be first adjusted at 12.8 V (6.4 V or 25.6 V) at C-D in Figure 2 or C-D and E-F in Figure 3 with a minimum required bulb load and the power supply shall be adjusted to provide other required test voltages at required temperatures at C-D in Figure 2 or C-D and E-F in Figure 3 without readjustment of trimming resistors, R (each required test voltage shall be set with a minimum bulb load in place). The required voltage tests with a maximum bulb load shall be conducted without readjusting each corresponding power supply voltage previous set with minimum bulb load.

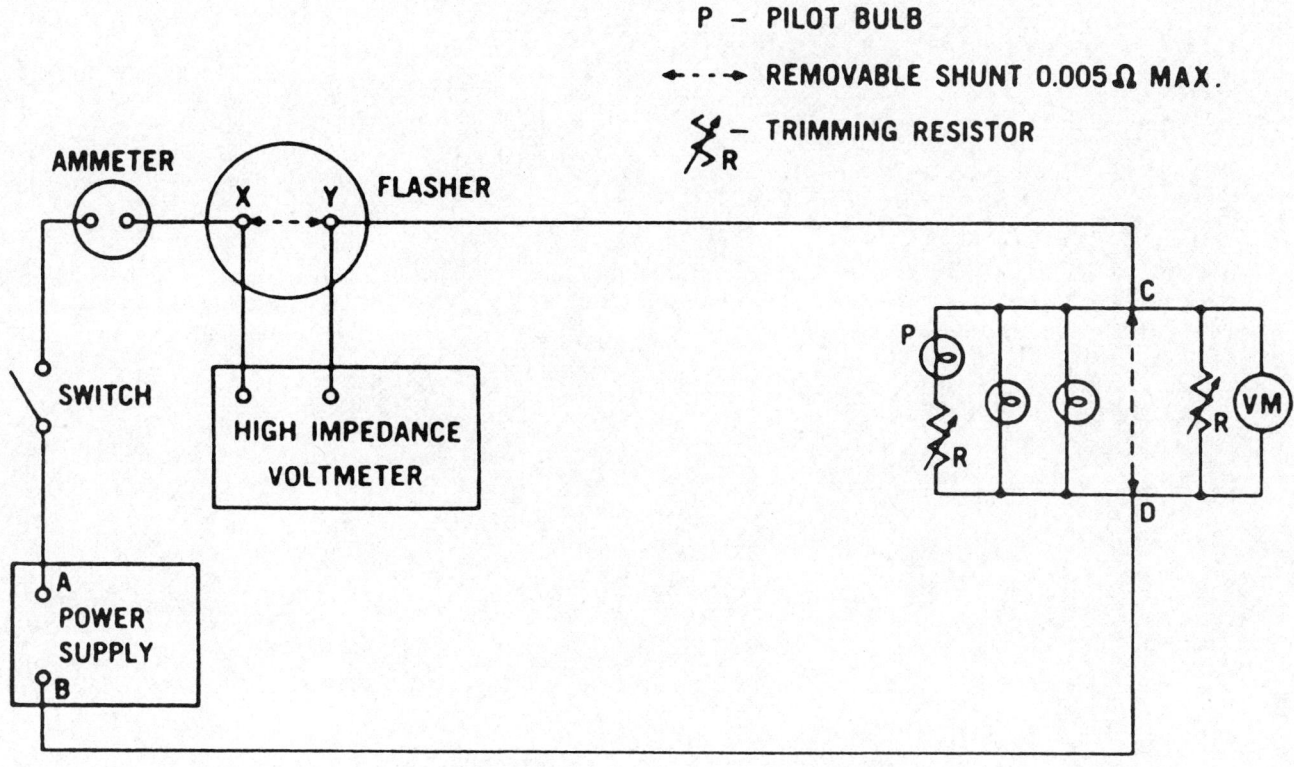

FIGURE 2—STANDARD TEST CIRCUIT—TURN SIGNAL AND HAZARD WARNING FLASHERS

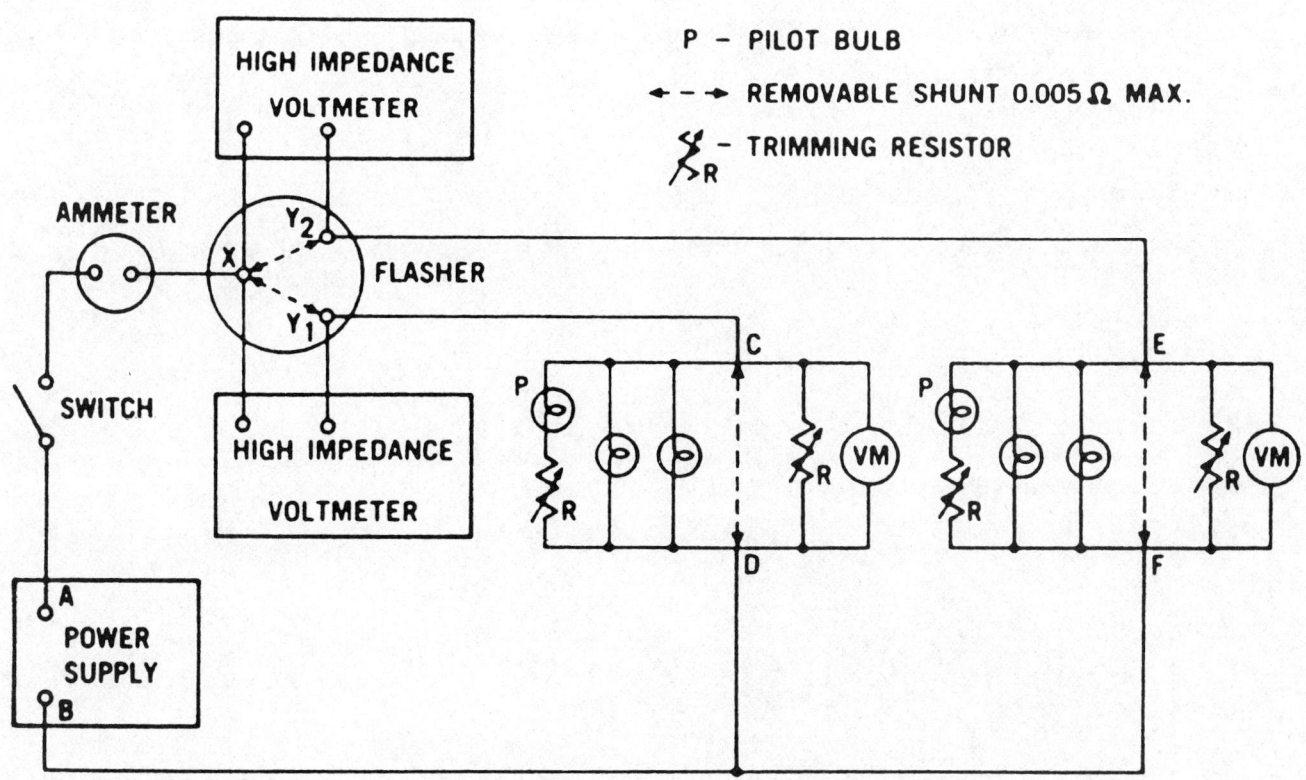

FIGURE 3—STANDARD TEST CIRCUIT—WARNING LAMP ALTERNATING FLASHERS

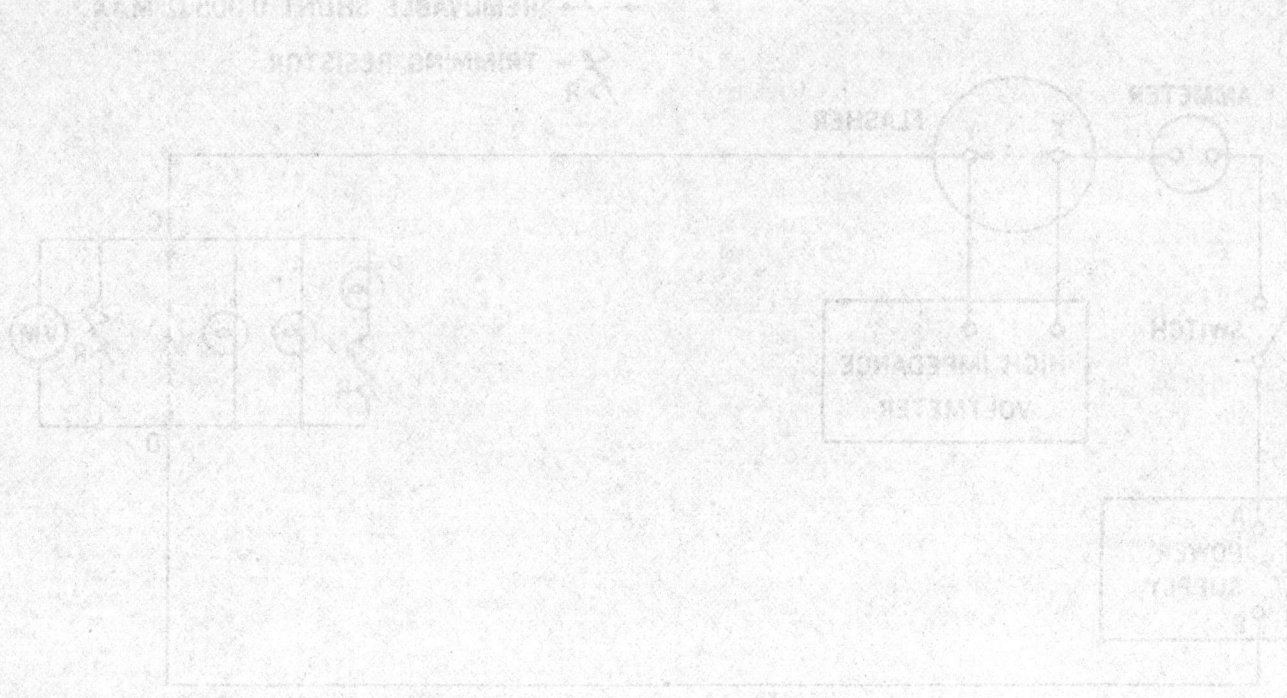
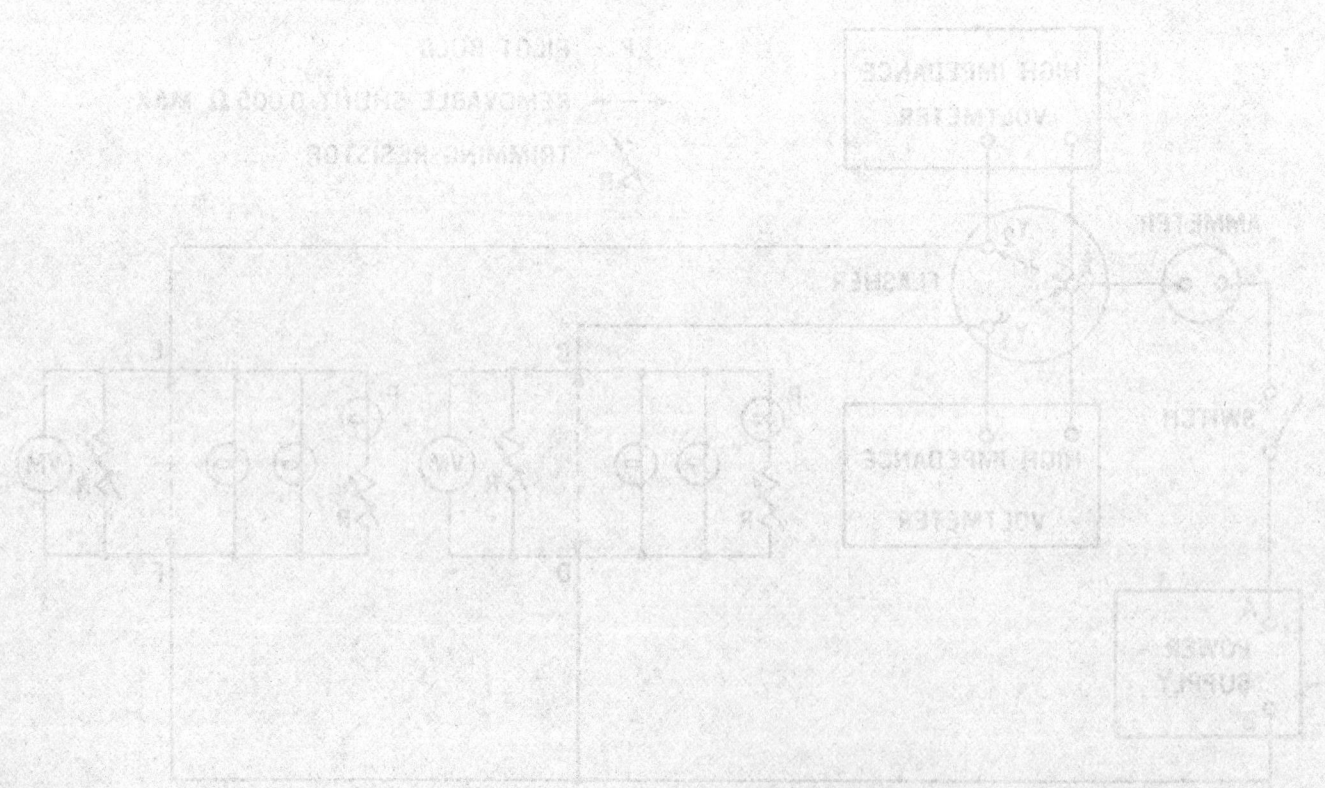

BRAKES

25 Brakes

R) ROAD VEHICLE—HYDRAULIC BRAKE HOSE ASSEMBLIES FOR USE WITH NONPETROLEUM-BASE HYDRAULIC FLUIDS—SAE J1401 JUN93

SAE Standard

Report of the Motorcoach and Motor Truck Division, approved January 1942. Revised by the Hydraulic Brake Systems Actuating Committee June 1985. Completely revised by the Automotive Brake and Steering Hose Standards Committee June 1990, and again June 1993. Rationale statement available.

1. Scope—This SAE Standard specifies the performance tests and requirements for hydraulic brake hose assemblies used in the hydraulic braking system of a road vehicle. It also specifies the methods used for identification of the hose manufacturer.

This document applies to brake hose assemblies made of a hose fabricated from yarn and natural or synthetic elastomers and assembled with metal end fittings for use with nonpetroleum-base brake fluids as specified in SAE J1703 and SAE J1705.

The nominal internal diameter of the brake hose shall be one of the following values:
 a. 3.2 mm (1/8 in) or,
 b. 4.8 mm (3/16 in)

2. References

2.1 Applicable Documents—The following publications form a part of this specification to the extent specified herein. The latest issue of SAE publications shall apply.

2.1.1 SAE PUBLICATIONS—Available from SAE, 400 Commonwealth Drive, Warrendale, PA 15096-0001.

 SAE J1703—Motor Vehicle Brake Fluid
 SAE J1705—Low Water Tolerant Brake Fluids

2.1.2 ASTM PUBLICATION—Available from ASTM, 1916 Race Street, Philadelphia, PA 19103-1187.

 ASTM B 117—Method of Salt Spray (Fog) Testing

2.1.3 ISO PUBLICATION—Available from ANSI, 11 West 42nd Street, New York, NY 10036-8002.

 ISO R147—Load calibration of testing machines for tensile testing of steel

2.2 Related Publications—The following publications are provided for information purposes only and are not a required part of this document.

2.2.1 SAE PUBLICATIONS—Available from SAE, 400 Commonwealth Drive, Warrendale, PA 15096-0001.

 SAE J1288—Packaging, Storage, and Shelf Life of Hydraulic Brake Hose Assemblies.
 SAE J1406—Application of Hydraulic Brake Hose to Motor Vehicles

2.2.2 ISO PUBLICATION—Available from ANSI, 11 West 42nd Street, New York, NY 10036-8002.

 ISO 3996—Hydraulic brake hose assemblies—Non-Petroleum hose hydraulic fluid standard

2.3 Definitions

2.3.1 BRAKE HOSE ASSEMBLY—A brake hose equipped with end fittings for use in a brake system.

2.3.2 BRAKE HOSE—A flexible conduit manufactured for use in a brake system to transmit and contain the fluid pressure medium used to apply force to the vehicle's brakes.

2.3.3 BRAKE HOSE END FITTING—A coupling, other than a clamp, designed for attachment to the end of a brake hose.

2.3.4 PERMANENTLY ATTACHED END FITTING—A coupling designed for permanent attachment to the ends of a brake hose by crimping or swaging.

2.3.5 FREE LENGTH—The linear measurement of brake hose exposed between the end fittings of a brake hose assembly while maintained in a straight position.

2.3.6 LEAKS, BURST—The loss of test fluid from the brake hose assembly other than by the designed inlet(s) and outlet(s).

2.3.7 CRACKING—The interruption of a surface due to environment and/or stress.

3. Performance Tests—Performance tests for hydraulic brake hose assemblies include all of the tests listed in Table 1. These tests shall be conducted on each I.D. size and type[1] from each hose manufacturer. A change in hose construction, that is, a change in material or a change in the manufacturing method, shall require a complete performance test. Accordingly, each coupler shall conduct the performance test on each coupling crimp design for each hose construction. A change of coupling crimp design shall require a complete performance test. Variations that do not influence the integrity of the hose coupling joint, such as variation in thread size, port dimensions, hex size, and the like, shall not be considered new design.

TABLE 1—HYDRAULIC BRAKE HOSE ASSEMBLY PERFORMANCE TEST SUMMARY[1]

Sample Size	Performance Test	Test Procedure (paragraph)	Performance Requirement (paragraph)
All	100% Pressure Test	3.2.1	4.1
All[2]	Constriction	3.2.2	4.2
4	Volumetric Expansion	3.2.3	4.3
	Followed by Burst	3.2.4	4.4
4	Brake Fluid Compatibility	3.2.5	4.5
4	Whip	3.2.6	4.6
4	Tensile	3.2.7	4.7
1	Cold Bend	3.2.8	4.8
1	Ozone	3.2.9	4.9
1	Salt Spray	3.2.10	4.10
	Water Absorption	3.2.11	4.11
4	Burst		
4	Whip		
4	Tensile		
4	Hot Impulse	3.2.12	4.12
4	Dynamic Ozone	3.2.13	4.13
39	Total Samples		

[1] When the hose assembly configurations make it impractical to conduct tests such as tensile, whip, and constriction, hose assemblies produced from equivalent type end fittings, production type equipment, and processes must be used to make the substitute brake hose assemblies.

[2] Four brake hose assemblies may be used if assemblies must be cut to conduct constriction tests.

3.1 Test Conditions—The assemblies for each performance test shall be new and unused and shall be at least 24 h old. The last 4 h prior to testing shall be at a temperature of 15 to 32 °C (60 to 90 °F). Prior to installation of the hose

---

[1] Various reinforcing cord(s) and/or elastomer(s).

assembly on a whip or cold bend test, all external appendages such as mounting brackets, spring guards, and metal collars shall be removed or long tubes shortened, or both. The temperature of the testing room shall be between 15 and 32 °C (60 and 90 °F) for all tests except brake fluid compatibility, cold bend, ozone, salt spray, and water absorption.

Samples for the dynamic ozone test are to be hose only without fittings or appendages.

3.2 Test Procedures

3.2.1 100% PRESSURE TEST—The hose assembly shall be subject to a pressure test, using inert gas, air, water, or brake fluid, and it shall meet SAE J1703/J1705 as the pressure medium. The test pressure shall be 10.3 MPa (1500 psi) minimum, 14.5 MPa (2100 psi) maximum for inert gas and air and 20.7 MPa (3000 psi) minimum, 24.8 MPa (3600 psi) maximum for water and nonpetroleum-base hydraulic brake fluid. Special care should be taken when gas or air is used. Under the pressure specified, gas or air is explosive if a failure should occur in the hose or hose assembly. The pressure shall be held for not less than 10 nor more than 25 s.

3.2.2 CONSTRICTION TEST—The constriction of the hose assemblies shall be measured with a gage plug as shown in Figure 1 and whose "A" dimension shall be as listed in Table 2.

Hold the assembly vertically at the fitting and insert the "A" diameter portion of the constriction gage into the end of the fitting for the full length of the probe. Repeat this step for the other end of the brake hose assembly.

Some hose assemblies have a fitting, which is designed to make it impossible to insert the gage plug externally. For these assemblies, insert a special elongated gage plug into the opposite fitting and pass the probe through the hose, into and through the crimped area of the fitting being tested. If the gage plug becomes misaligned at the entrance to the second fitting, it may be necessary to align the hose to allow the plug gage to pass through. The special gage plug shall meet all the requirements of Figure 1, with the exception of the 76 mm (3 in) length, which must be increased appropriately so that its tip will extend past the hose opening.

Some brake hose assemblies have fittings on both ends which cannot be entered with a gage plug. Cut these assemblies 50 mm ± 3 mm (2 in ± 0.1 in) from the end of the fitting and then test with the plug gage. (Reference Table 1, footnote 2).

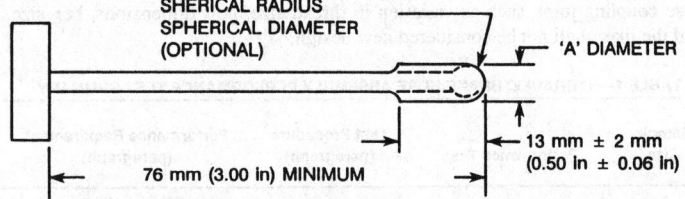

FIGURE 1—GAGE PLUG FOR TESTING CONSTRICTION OF BORE OF HOSE ASSEMBLY

TABLE 2—CONSTRICTION GAGE DIMENSIONS

Hose I.D. mm (in)	Constriction Gage "A" Dimension Min Diameter mm (in)
3.2 (1/8)	2.03 (0.080)
4.8 (3/16)	3.05 (0.120)

3.2.3 VOLUMETRIC EXPANSION TEST—The expansion test is designed to measure, by fluid displacement, the volumetric expansion of the free length of assembled hydraulic brake hose when subjected to specified internal pressures. The free length is the length measured between the fittings.

3.2.3.1 If the specimen used in this test has been subjected to a pressure above 10.3 MPa (1500 psi) using any medium prior to this test, allow it to recover for 15 min.

3.2.3.2 Carefully thread the hose assembly into the adapters designed to seal in the same manner as in actual use. Do not twist. Maintain the hose in a vertical, straight position, without tension, while under pressure.

3.2.3.3 Bleed all the air from the system by allowing approximately 0.25 L (0.5 pt) of water to flow from the reservoir tank through the hose assembly and into the buret. Removal of air bubbles may be facilitated by moving the hose back and forth. Close the valve to the buret and apply 10.3 MPa +0, -0.14 MPa (1500 psi +0, -20 psi) to the hose assembly. Within 10 s, inspect the hose assembly for leaks at the connections and then release the pressure completely in the hose. Adjust the water level in the buret to zero.

With the valve to the buret closed, apply 6.9 MPa +0, -0.14 MPa (1000 psi +0, -20 psi) to hose assembly and seal this pressure in the hose within 5 s ± 3 s. Within 3 s, open the valve to the buret for 10 s +3, -0 s and allow the water in the expanded hose to rise in the buret. The liquid level in the buret should be constant within that time period.

3.2.3.4 Repeat the preceding step two times, so the amount of water in the buret will be the total of the three expansions. Measure this buret reading to the nearest 0.05 cm^3.

3.2.3.5 The volumetric expansion is calculated by dividing the buret reading by three and subtracting the calibration factor. This figure divided by the free length in meters (feet) will give the volumetric expansion per meter (feet) of hose.

3.2.3.6 Readjust the water level in the buret to zero as previously stated and repeat the procedure to obtain the expansion at a pressure of 10.3 MPa +0, -0.14 MPa (1500 psi +0, -20 psi). If the pressure in the hose should inadvertently be raised to a value above that specified, but not above 24 MPa (3500 psi), completely release the pressure and allow the hose to recover for at least 15 min and then repeat the test. If the hose was subjected to a pressure above 24 MPa (3500 psi), repeat the test using a new brake hose. If at any time during the test an air bubble flows out of the hose, repeat the test after allowing at least 3 min for the hose to recover.

3.2.3.7 *Test Apparatus*—Test apparatus shall consist essentially of the following:

 a. A source for required fluid pressures, test fluid consisting of water without any additives and free of air or gas bubbles.
 b. A reservoir for water pressure gages, fittings where the hose assembly may be mounted vertically for application of pressure under controlled conditions.
 c. A graduated buret with 0.05 cm^3 increments for measuring the volume of liquid corresponding to the expansion of the hose under pressure.
 d. Plumbing hardware as required.

All piping and connections shall be smooth bore without recesses or offsets so all air may be freely removed from the system before running each test. Valves shall be capable of withstanding pressures involved without leakage. See Figure 2.

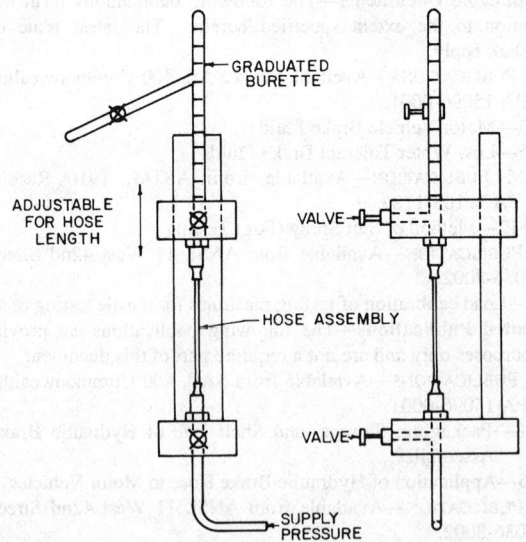

FIGURE 2—EXPANSION TEST APPARATUS

3.2.3.8 *Calibration of Apparatus*—The apparatus shall be tested prior to use to determine its calibration correction factors established at pressures of 6.9 and 10.3 MPa (1000 and 1500 psi) using a simulated hose assembly, which shall consist of 1.52 mm (0.060 in) minimum wall, hydraulic steel tubing with a free length of 305 mm ± 6 mm (12 in ± 0.2 in), and 6.3 mm (0.25 in) outside diameter. All fittings and adaptors used in testing of the hose assembly shall be in this system. This may require the attachment of the tubing to the brake hose fittings in the case of special end configurations. The calibration correction factors shall be subtracted from the expansion readings obtained on the test specimens. The maximum permissible calibration correction factor shall be 0.08 cm^3 at 10.3 MPa (1500 psi).

3.2.4 BURST STRENGTH TEST—Connect the specimen to the pressure system and fill completely with water or brake fluid, allowing all air to escape. Removal of air bubbles may be facilitated by moving the hose back and forth. Apply 27.6 MPa +0, -1.4 MPa (4000 psi +0, -200 psi) pressure at the rate specified in 3.2.4.1 and hold for 2 min +0, -10 s. At the expiration of this hold period, increase the pressure at 172.5 MPa/min ± 69 MPa/min (25 000 psi/min ± 10 000 psi/min) until the hose bursts. Read the maximum pressure obtained on the calibrated gage to the nearest 1 MPa (100 psi) and record as the bursting strength of the hose assembly.

3.2.4.1 Test Apparatus—The apparatus shall consist of a suitable pressure system where hose is connected so that controlled and measured fluid pressure may be applied internally. The pressure shall be obtained by means of a hand- or power-driven pump or an accumulator system and it shall be measured with a calibrated gage. Provision shall be made for filling the hose with water or brake fluid and allowing all air to escape through a relief valve prior to application of pressure. This is important as a safety measure. The hold and burst pressures shall be applied at a rate increase of 172.5 MPa/min ± 69 MPa/min (25 000 psi/min ± 10 000 psi/min). Since this type of hose withstands a minimum bursting pressure of 49 MPa (7000 psi) for 3.2 mm (1/8 in) and 34.5 MPa (5000 psi) for 4.8 mm (3/16 in), care must be taken that all piping, valves, and fittings are sufficiently rugged and adapted to high-pressure work. The apparatus described for the expansion test may be used when it conforms to these requirements.

3.2.5 BRAKE FLUID COMPATIBILITY, CONSTRICTION, AND BURST STRENGTH TEST

3.2.5.1 Attach a hose assembly or a manifold to which multiple hose assemblies may be attached, below a 0.5 L (1 pt) can reservoir filled with 100 mL of SAE Compatibility Brake Fluid, SAE RM 66-03 (see Figure 3).

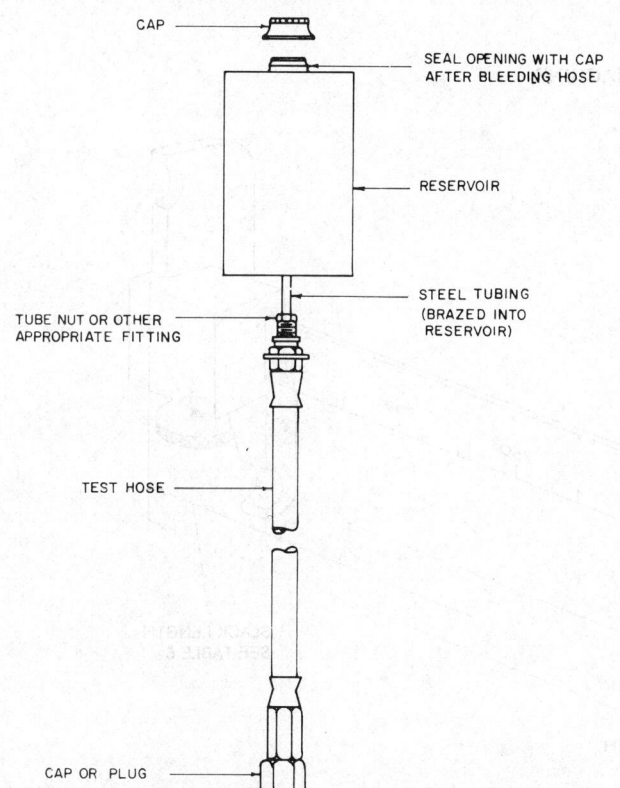

FIGURE 3—BRAKE FLUID COMPATIBILITY APPARATUS

3.2.5.2 Fill the hose assembly with SAE Compatibility Brake Fluid, seal the lower end, and place the test assembly in a vertical position in an oven.

3.2.5.3 Condition the hose assembly at 120 °C +5, -0 °C (248 °F +9, -0 °F) for 70 to 72 h.

3.2.5.4 After completion of the heat aging period, remove the hose assembly and cool at room temperature for 30 min ± 5 min.

3.2.5.5 Drain the brake hose assembly, and within 10 min, determine, per 3.2.2, that every applicable diameter of the hose assembly is not less than shown in Table 2.

3.2.5.6 The brake hose assembly shall be burst within 3 h using the test specified in 3.2.4.

3.2.6 WHIP TEST

3.2.6.1 Measure the free length of the hose assembly with the assembly in a vertical position with a mass of 567 g ± 3 g (20 oz ± 0.1 oz) attached to one end. Use a vernier caliper scale or equivalent and report the length between fittings to within a tolerance of 0.5 mm (0.02 in).

3.2.6.2 Equip the nonrotating header to permit attachment of each assembly with individual adjustment for length. When mounted in the whip test machine (see Figure 4), the projected length of the hose assembly shall be less than the free length by the amount indicated as slack in Table 3 (see Figure 5).

TABLE 3—WHIP TEST SLACK SETTING

Internal Diameter mm (in)	Free Length mm (in)	Slack Length mm (in)
3.2 (1/8)	200 to 400 (8 to 15-1/2) incl	44.45 ± 0.40 (1.750 ± 0.015)
	Over 400 to 480 (15-1/2 to 19) incl	31.75 ± 0.40 (1.250 ± 0.015)
	Over 480 to 600 (19 to 24) incl	19.05 ± 0.40 (0.750 ± 0.015)
4.8 (3/16)	250 to 400 (10 to 15-1/2) incl	25.40 ± 0.40 (1.000 ± 0.015)

3.2.6.3 Since the whip test results are very sensitive to error in setting this length, the projected length on the machine shall be within the limits specified. Take the projected length parallel to the axis of the rotating head.

3.2.6.4 Install the test specimen assemblies in the apparatus without any twist. Apply the water pressure and bleed all hose and passages to eliminate air pockets or bubbles. Start the motor rotating the movable head. Periodically check the rpm. Failure of the specimen by water leakage and subsequent loss of pressure terminates the test. Note the elapsed time of the test prior to termination.

3.2.6.5 Test Apparatus—The test apparatus shall provide the same motion to the specimens as the following: a movable header consisting of a horizontal bar mounted at each end on vertically rotating disks through bearings with centers placed 101.6 mm (4 in) from the disk centers, and an adjustable stationary header parallel to the movable header in the same horizontal plane as the centers of the disks. Each header is provided with end connections in which the hose assemblies are mounted in a parallel manner. The disks are revolved at a speed of 800 rpm ± 10 rpm, whereby the hose ends fastened to the moving header are rotated at this speed through a circle 203.20 mm ± 0.25 mm (8.000 in ± 0.010 in) in diameter, while the opposite hose ends remain stationary. The end connections on the movable header are tightly capped, while those on the stationary header are open to a manifold through which water pressure is supplied by a suitable means. The hose assemblies are subjected during testing to a constant water pressure, which shall be maintained between 1.55 and 1.72 MPa (225 and 250 psi). A limit switch shall be used to stop the machine when the water pressure drops, as in the case of hose failure, since it is essential that the machine stop if the pressure drops. An elapsed time indicator shall be provided.

3.2.7 TENSILE TEST—Apply an increasing tension load such that the moving head of the testing machine travels at a speed as indicated in Table 4 until the hose assembly fails. Record the total load at the time of failure, the type of failure, and the separation rate.

TABLE 4—TENSILE SEPARATION RATE AND MINIMUM LOAD

Internal Diameter mm (in)	Separation Rate mm/min (in/min)	Minimum Load N (lb)
3.2 (1/8), 4.8 (3/16)	25 ± 3 (1 ± 0.1)	1446 (325)
3.2 (1/8), 4.8 (3/16)	50 ± 3 (2 ± 0.1)	1646 (370)

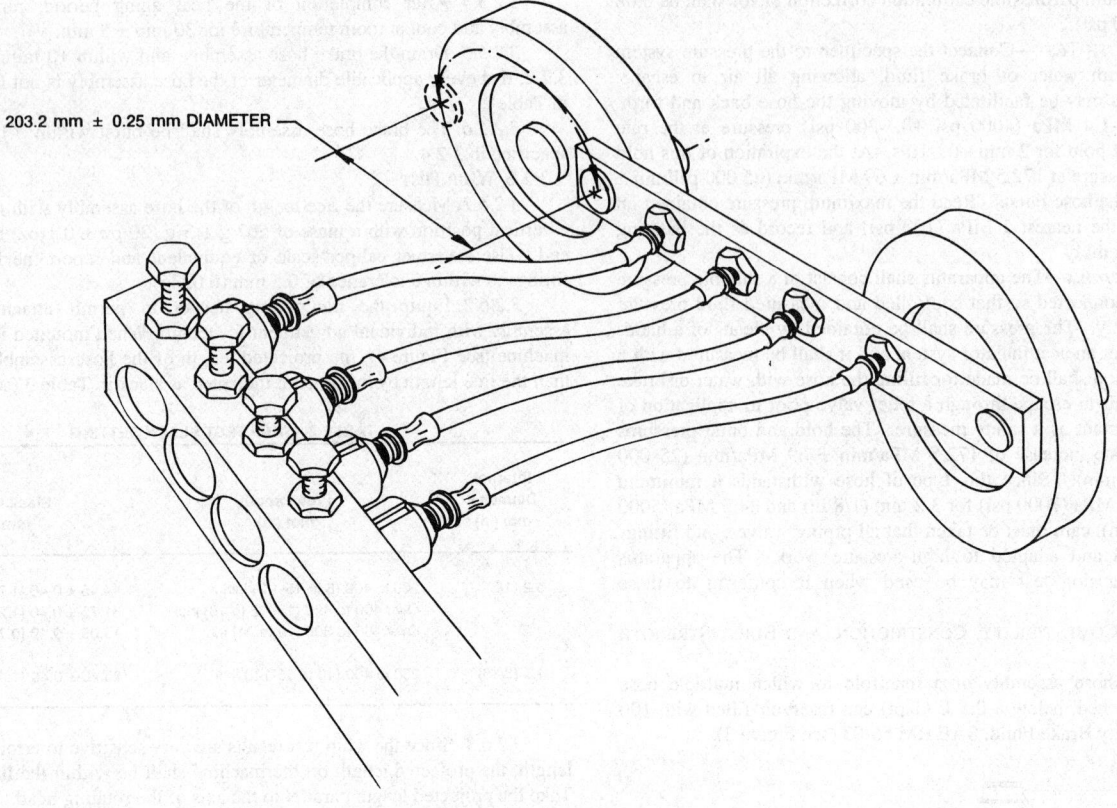

FIGURE 4—WHIP TEST MACHINE

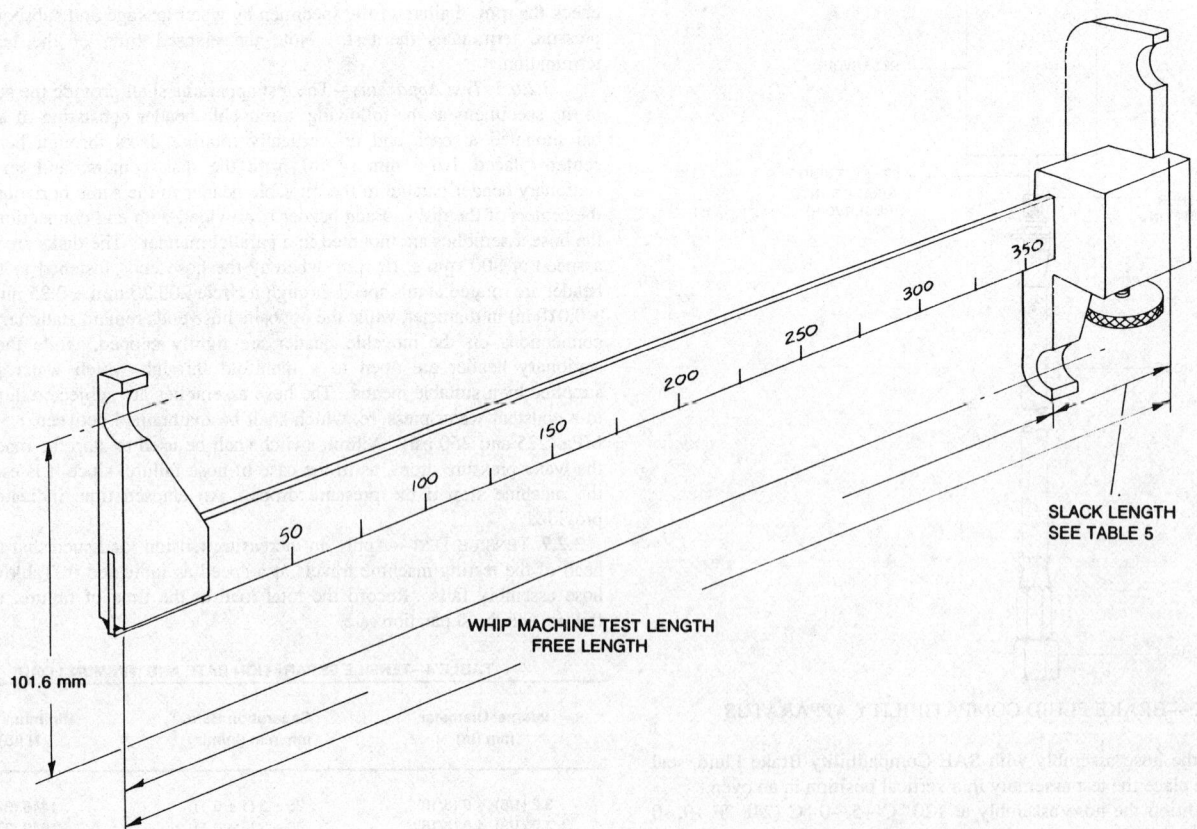

FIGURE 5—TYPICAL WHIP TEST SLACK SETTING FIXTURE

3.2.7.1 Test Apparatus—A tension testing machine conforming to the requirements of ISO R147 shall be used for the tensile test of the hose assembly. The machine shall be provided with a recording device to give the total pull in Newtons (pounds) at the conclusion of the test. A machine of 4.5 kN (1000 lbf) will be found suitable. The specimen shall be held so that the hose fittings have a straight centerline corresponding to the direction of the machine pull.

3.2.8 COLD BEND TEST—Condition the hose (in a straight position) and a mandrel of the diameter specified as follows, in air at -45 to -48° C (-50 to -55° F) for 70 to 72 h. Then while still at this temperature, bend the hose at least 180 degrees around the mandrel at a steady rate in a period of 3 to 5 s.

3.2.8.1 Examine the cover of the brake hose with a naked eye for cracks or breaks.

3.2.8.2 Test Apparatus—The mandrel diameter shall be 76.2 mm +1, -0 mm (3 in +0.04, -0 in) for 3.2 mm (1/8 in) hose and 88.9 mm +1, -0 mm (3.50 in +0.04, -0 in) for 4.8 mm (3/16 in) hose.

3.2.9 OZONE TEST

3.2.9.1 Bend a brake hose around a cylinder, the diameter of which shall be eight times the nominal outside diameter of the brake hose, and bind the ends. The cylinder and binding shall be made of metal or materials that prevent the consumption of ozone. If the hose collapses when bent around the cylinder, provide for internal support of the hose.

3.2.9.2 Condition the hose on the cylinder for 24 h ± 0.5 h room temperature, and then place it in an exposure chamber containing air mixed with ozone at the ozone partial pressure of 100 mPa ± 5 mPa (100 parts of ozone/100 million parts of air by volume ± 5 parts of ozone/100 million parts of air by volume) for 70 to 72 h. Ambient air temperature in chamber during test shall be 40 °C ± 3 °C (104 °F ± 5 °F).

3.2.9.3 Examine the cover of the hose for cracks under 7X magnification, ignoring the areas immediately adjacent to or within the area covered by the binding.

3.2.10 SALT SPRAY TEST

3.2.10.1 Test Apparatus—Utilize the apparatus described ASTM B 117 Appendix B. Construct the salt spray chamber so that:

3.2.10.1.1 The construction material does not affect the corrosiveness of the fog.

3.2.10.1.2 The hose assembly is supported or suspended between 15 and 30 degrees from the vertical and within the principal plane of the horizontal flow of fog through the chamber.

3.2.10.1.3 The hose assembly does not contact any metallic material or any material capable of acting as a wick.

3.2.10.1.4 Condensation, which falls from the assembly, does not return to the solution reservoir for respraying.

3.2.10.1.5 Condensation from any source does not fall on the brake hose assemblies or the solution collectors.

3.2.10.1.6 Spray from the nozzles is not directed onto the hose assembly.

3.2.10.2 Test Preparation

3.2.10.2.1 Mix a salt solution 5 parts ± 1 part by weight of sodium chloride to 95 parts of distilled water, using sodium chloride substantially free of nickel and copper, and containing on a dry basis not more than 0.1% of sodium iodide and not more than 0.3% total impurities. Ensure that the solution is free of suspended solids before the solution is atomized.

3.2.10.2.2 After atomization at 35 °C +1, -2 °C (95 °F +1.8, -3.6 °F) ensure that the collected solution is in the pH range of 6.5 to 7.2. Make the pH measurements at 25 °C ± 3 °C (77 °F ± 5 °F).

3.2.10.2.3 Maintain a compressed air supply to the nozzle free of oil and dirt and between 68.9 and 172.4 kPa (10 and 25 psi).

3.2.10.3 Plug each end of the hose assembly.

3.2.10.4 Subject the brake hose assembly to the salt spray continually for 24 h +0.2, -0 h.

3.2.10.5 Regulate the mixture so that each collector will collect from 1 to 2 mL (0.06 to 0.12 in^3) of solution per hour for each 80 cm^2 (12.4 in^2) of horizontal collecting area.

3.2.10.6 Maintain exposure zone temperature at 35 °C ± 2 °C (95 °F ± 4 °F).

3.2.10.7 Upon completion, remove the salt deposit from the surface of the hoses by washing gently or dipping in clean running water, not warmer than 37 °C (98.6 °F) and then drying with air within 2 min.

3.2.10.8 Examine the brake hose end fitting for base metal corrosion and record results.

3.2.11 WATER ABSORPTION TESTS

3.2.11.1 Immerse brake hose assemblies in water heated at 85 °C ± 2 °C (185 °F ± 3.6 °F) for a period of 70 to 72 h.

3.2.11.2 Within 10 min of removal from the water, start the burst and tensile tests.

3.2.11.3 The whip test shall be started within 10 to 30 min after removal from the water.

3.2.12 HOT IMPULSE TEST

3.2.12.1 Test Equipment

a. Pressure Cycling Apparatus—The pressure cycling apparatus shall be capable of applying a pressure of 11 MPa (1600 psi). It shall have automatic control of the time for the pressure apply/release cycle.

b. Circulating Air Oven—An insulated circulating air oven with a suitable thermostatically controlled heating system is required to maintain a temperature of 143 °C ± 3 °C (295 °F ± 5 °F).

c. Pressure Hold and Burst Strength Test Apparatus—An apparatus conforming to the requirements described in 3.2.4.1.

3.2.12.2 Connect the hose assemblies to the pressure cycling apparatus.

3.2.12.3 Fill the pressure cycling apparatus and hose assemblies with brake fluid, and bleed free of air.

3.2.12.4 Place the assemblies in the circulating air oven, and within 30 min attain an oven temperature of 143 °C ± 3 °C (295 °F ± 5 °F).

3.2.12.5 Subject the assemblies to a cycling internal pressure of 11 MPa +0.5, -0 MPa (1600 psi +75, -0 psi) for 1 min ± 0.1 min and 0 pressure for 1 min ± 0.1 min; pressures to be attained within 2 s.

3.2.12.6 Pressure cycle assemblies for 150 cycles minimum.

3.2.12.7 Remove the assemblies from the oven. Disconnect the assemblies from the impulse apparatus, and drain the fluid.

3.2.12.8 Cool the assemblies in air at room temperature for 45 min minimum.

3.2.12.9 Subject the assemblies to the burst test in 3.2.4.

3.2.13 DYNAMIC OZONE TEST

3.2.13.1 Test Apparatus—Brake hose cut lengths of 218 mm ± 3 mm (8.6 in ± 0.1 in), SAE dynamic ozone test apparatus that will flex the brake hose as shown in Figure 6 and the ozone test chamber.

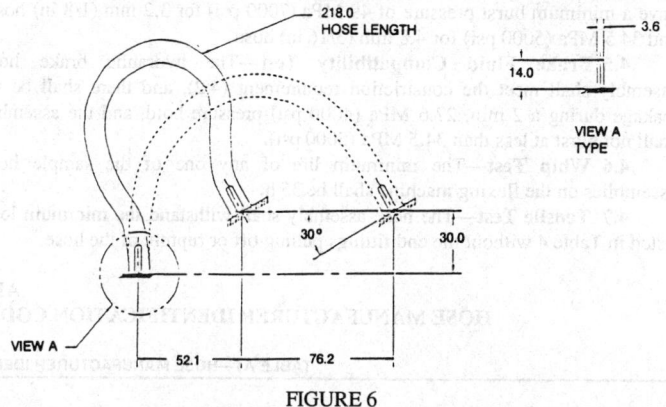

FIGURE 6

3.2.13.2 Precondition all the brake hose samples in a nonstressed condition at 27 °C ± 6 °C (80 °F ± 10°F) for at least 24 h prior to the start of the test.

3.2.13.3 Assemble the brake hose samples on the SAE dynamic ozone test apparatus so that they meet the relative position and flex parameters as shown in Figure 6. Install the brake hose over the fixture pins until the hose has bottomed out. Use band clamps to securely retain the brake hose on the pin. Install the test apparatus and assembled hoses in a stabilized ozone chamber. The chamber shall contain air mixed with ozone at the ozone partial pressure of 100 mPa ± 10 mPa (100 parts of ozone per 100 million parts of air by volume ± 10 parts of ozone per 100 million parts of air by volume). The air temperature in the chamber shall be 40 °C ±3 °C (104 °F ± 5 °F).

3.2.13.4 Start cycling when the chamber reaches the specified ozone concentration but no later than 1 h after putting the test apparatus in the ozone chamber. The flex rate shall be 0.30 Hz ± 0.05 Hz. The stroke shall be 76.2 mm ± 2.5 mm (3.0 in ± 0.1 in).

3.2.13.5 Examine the hoses for ozone cracks every 24 h ± 1 h. Remove the fixture from the cabinet and close the door immediately. Inspect for ozone cracks visible to the eye without magnification at the worst stress condition ignoring the areas immediately adjacent to or within the area covered by the band clamps. Do not remove the hoses from the fixture. Reinstall the fixture in the cabinet within 15 min of its removal. The test shall be run continuously except for the daily inspection periods. Inspection periods may be eliminated on non-work days if determined to be not critical.

3.2.13.6 Stop test when ozone cracks are observed. Record the hose identification and the number of days that elapsed until the first visible crack was observed.

4. Performance Requirements

4.1 100% Pressure Test—Hose assemblies showing leaks under this test shall be rejected and destroyed.

4.2 Constriction Test—Hose assemblies not allowing passage of the plug gage shall be rejected and destroyed. The constriction requirement does not apply to that part of the brake hose end fittings which does not contain hose.

4.3 Volumetric Expansion Test—The maximum expansion of any of the hose assemblies tested shall not exceed the values in Table 5.

TABLE 5—MAXIMUM EXPANSION OF FREE LENGTH HOSE

	6.9 MPa (1000 psi)		6.9 MPa (1000 psi)	
Hose I.D. mm (in)	Reg Exp Hose cm^3/m	(cm^3/ft)	Low Exp Hose cm^3/m	(cm^3/ft)
3.2 (1/8)	2.17	(0.66)	1.08	(0.33)
4.8 (3/16)	2.82	(0.86)	1.81	(0.55)

	10.3 MPa (1500 psi)		10.3 MPa (1500 psi)	10.3 MPa (1500 psi)
Hose I.D. mm (in)	Reg Exp Hose cm^3/m	(cm^3/ft)	Low Exp Hose cm^3/m	(cm^3/ft)
3.2 (1/8)	2.59	(0.79)	1.38	(0.42)
4.8 (3/16)	3.35	(1.02)	2.36	(0.72)

4.4 Burst Test—When tested under hydraulic pressure, each sample of hose shall withstand a 2 min pressure hold at 27.6 MPa (4000 psi), and shall have a minimum burst pressure of 49 MPa (7000 psi) for 3.2 mm (1/8 in) hose, and 34.5 MPa (5000 psi) for 4.8 mm (3/16 in) hose.

4.5 Brake Fluid Compatibility Test—The hydraulic brake hose assembly shall meet the constriction requirement (4.2), and there shall be no leakage during a 2 min, 27.6 MPa (4000 psi) pressure hold, and the assembly shall not burst at less than 34.5 MPa (5000 psi).

4.6 Whip Test—The minimum life of any one of the sample hose assemblies on the flexing machine shall be 35 h.

4.7 Tensile Test—The hose assembly shall withstand the minimum load listed in Table 4 without the end fittings pulling off or rupture of the hose.

4.8 Cold Bend Test—The hose cover shall not crack (visible without magnification) or break.

4.9 Ozone Test—The outer cover of the hose shall show no cracking when examined under 7X magnification.

4.10 Salt Spray Test—The hose assembly end connections shall have no base metal corrosion. The area of the fitting where crimping or the application of labeling information has caused the displacement of the protective coating is exempt from the corrosion requirements. Brass fittings have adequate corrosion resistance; therefore, salt spray testing of brass fittings is not required.

4.11 Water Absorption Tests—Water conditioned hose assemblies shall pass all burst (4.4), whip (4.6), and tensile (4.7) requirements as outlined for nonaged brake hose assemblies.

4.12 Hot Impulse Test

4.12.1 The hose assemblies shall withstand impulsing for 150 cycles without leakage.

4.12.2 There shall be no leakage during a 2 min, 27.6 MPa (4000 psi) pressure hold.

4.12.3 The assembly shall not burst at less than 34.5 MPa (5000 psi).

4.13 Dynamic Ozone Test—The hose shall not crack after testing a minimum of 48 h.

5. Construction

5.1 Hose—The hose shall consist of an elastomeric inner tube, two or more layers of reinforcing cord imbedded in and/or bonded to the elastomeric inner tube and outer cover. The cover must be a black stock, free from sulfur bloom, which will not crack when subjected to long periods of weather aging. The inner tube of this hose must be a stock which will effectively resist deterioration by nonpetroleum-base hydraulic brake fluids as designated in Section 1.

5.2 Hose Assembly—Each hydraulic brake hose assembly shall have permanently attached brake hose end fittings.

6. Hose Identification
—The brake hose of each manufacturer shall be identified by one or more colored yarns incorporated into the construction. Embossed or imprinted (3-dimensional) marking on the brake hose cover may be used.

NOTE—The R.M.A.[2] approved marker yarn color and the name trademark on cover designations for each brake hose manufacturer shall be registered with SAE.

---

[2] Rubber Manufacturers Association, 1400 K Street, N.W., Washington, DC 20005.

APPENDIX A
HOSE MANUFACTURER IDENTIFICATION CODE-COLORED YARN ASSIGNMENTS AS OF JANUARY 1989

TABLE A1—HOSE MANUFACTURER IDENTIFICATION CODE-COLORED YARN ASSIGNMENTS

Line Number New	Line Number Old	Yarn Color Code	Year Assigned	Assignments
1	1	Yellow	*	The BFGoodrich Co., Akron, Ohio
2	2	Green	*	The Goodyear Tire & Rubber Co., Akron, Ohio
3	3 (C)	Red	*	Uniroyal, Inc., Middlebury, Connecticut - Terminated 1984
4	4	Black	*	General Motors Corp., Inland Division, Dayton, Ohio
5	5	Blue	1978	Goodall Rubber Co., Trenton, New Jersey
6	6	Brown	*	Aeroquip Corp., Norwood, North Carolina
7	7	Violet	1959	Firestone Tire & Rubber Co., Akron, Ohio
8	8	Orange	1959	Dayco Products, Inc., Dayton, Ohio
9	9	Yellow Green	1972	Compagnie des Produits Industriels de l'Ouest, Nantes, France
10	14	Yellow Red	1959	Amerace Corp., Swan Hose Div., Worthington, Ohio - Terminated 1985 (see line 43)
11	15	Yellow Black	1959	Continental Gummi-Werke A.G., Hannover, F.R. Germany
12	16	Yellow Blue	1950	Hewitt-Robins, Ltd., Orangeburg, NY - Terminated 1979
13	17	Yellow Brown	1959	Plumley Rubber Co., Paris, Tennessee
14	-	Yellow Violet	1979	Avon Industrial Polymers, Trowbridge, Wilshire, England
15	-	Yellow Orange	1981	Buckeye Rubber Products, Inc., Lima, Ohio
16	10	Green Red	1950	Thermoid, Inc., Div. of H.K. Porter Co., Pittsburgh, Pennsylvania
17	11	Green Black	1950	The Gates Rubber Co., Denver, Colorado
18	12	Green Blue	1950	Crown Products Co., Ralston, Nebraska
19	13	Green Brown	1968	Toyoda Gosei Co., Ltd., Aichi Pref., Japan
20	-	Green Violet	1981	Garrett Flexible Products Inc., Garrett, Indiana

Continued

TABLE A1—HOSE MANUFACTURER IDENTIFICATION CODE-COLORED YARN ASSIGNMENTS (CONTINUED)

Line Number New	Line Number Old	Yarn Color Code	Year Assigned	Assignments
21	-	Green Orange	1983	Citla, S.A., Mexico City, Mexico
22	18	Red Black	*	Uniroyal, Ltd., Montreal, Canada (since 1966)
23	19	Red Blue	1950	Goodall Rubber Co., Trenton, New Jersey - Terminated 1978
24	20	Red Brown	1958	Continental Rubber Works, Erie, Pennsylvania - Terminated 1982
25	-	Red Violet	1983	Epton Industries, Inc., Kitchener, Ontario, Canada
26	-	Red Orange	1983	Stratoflex, Inc., Ft. Worth, Texas
27	21	Black Blue	1950	Electric Hose & Rubber Co. - Terminated 1981
28	22	Black Brown	1959	American Biltrite, Inc. - Terminated 1981
29	- (T)	Black Violet	1982	Association of Automotive Engineers & Technicians (AITA), Argentina
30	-	Black Orange	1981	Dana Corp./Boston Industrial Products, Hohenwald, Tennessee
31	23	Blue Brown	1958	C.F.W. Division Simrit, Montrond Les Bains, France
32	-	Blue Violet	1982	Flexigom, S.A., Federal Capital, Argentina
33	-	Blue Orange	1982	ICEMAP Argentina S.R.L., Buenos Aires, Argentina
34	-	Brown Violet	1982	Farloc Argentina S.A., Buenos Aires, Argentina
35	-	Brown Orange	1985	Pirelli Tecnica S.A.I.C., Buenos Aires, Argentina
36	-	Violet Orange	1981	Indomax, C.A., Caracas, Venezuela
37	24	Yellow Green Red	1959	Acme-Hamilton Mfg. Corp., Trenton, New Jersey - Terminated 1978
38	25	Yellow Green Black	1963	The Polymer Corporation, Reading, Pennsylvania
39	26	Yellow Green Blue	1971	Lectron Corp., American Hose Div., Winchester, Indiana - Terminated 1978
40	27	Yellow Green Brown	1973	Nephi Rubber Products, Inc., Nephi, Utah
41	- (T)	Yellow Green Violet	1987	Insulated Duct & Cable Co., Armstrong Hose Div., Trenton, NJ
42	-	Yellow Green Orange		
43	34	Yellow Red Black	1985	Anchor Swan, Division of Harvard Industries, Worthington, Ohio
44	35	Yellow Red Blue		
45	36	Yellow Red Brown	1977	Productos Pirelli, S.A., Barcelona, Spain
46	-	Yellow Red Violet		
47	-	Yellow Red Orange		
48	37	Yellow Black Blue	1976	Manifattura Gomma Cazzaniga, Dormelleto, Italy
49	38	Yellow Black Brown	1976	Getoflex Metzeler Industria, Sao Paulo, Brazil
50	-	Yellow Black Violet		
51	- (C)	Yellow Black Orange	1987	H.S. Parker Co., Ltd., Kyong-Nam, Korea
52	39	Yellow Blue Brown	1975	Shibami Industry Co., Ltd., Osaka, Japan
53	-	Yellow Blue Violet		
54	- (T)	Yellow Blue Orange	1988	Morenci Engineered Rubber Products, Morenci, Michigan
55	-	Yellow Brown Violet		
56	-	Yellow Brown Orange		
57	-	Yellow Violet Orange		
58	28	Green Red Black	1973	Parker-Hannifin Corp., Wickliffe, Ohio
59	29	Green Red Blue	1973	General Tire & Rubber Co., Wabash, Indiana
60	30	Green Red Brown	1974	Cooper Industrial Products, Inc., Auburn, Indiana
61	-	Green Red Violet		
62	-	Green Red Orange		
63	31	Green Black Blue	1978	Hadbar, Division of Purosil, Inc., Alhambra, California
64	32	Green Black Brown	1979	Republic Hose Manufacturing Corp., Youngstown, Ohio
65	-	Green Black Violet		
66	-	Green Black Orange		
67	33	Green Blue Brown	1977	Stratoflex, Ft. Worth, Texas - Withdrawn, not used (1983)
68	-	Green Blue Violet		
69	-	Green Blue Orange		
70	-	Green Brown Violet		
71	-	Green Brown Orange		
72	-	Green Violet Orange		
73	40	Red Black Blue	1972	Dunlop Australia, Ltd., Victoria, Australia
74	41	Red Black Brown	1972	Societa Meridianale Accessori Elastomerici, Salerno, Italy
75	-	Red Black Violet		
76	-	Red Black Orange		
77	42	Red Blue Brown	1971	Durkee-Atwood Co., Minneapolis, Minnesota
78	-	Red Blue Violet		
79	-	Red Blue Orange		
80	-	Red Brown Violet		
81	-	Red Brown Orange		
82	-	Red Violet Orange		
83	43	Black Blue Brown	1979	Semperit, Vienna, Austria
84	-	Black Blue Violet		
85	-	Black Blue Orange		

Continued

TABLE A1—HOSE MANUFACTURER IDENTIFICATION CODE-COLORED YARN ASSIGNMENTS (CONTINUED)

Line Number New	Line Number Old	Yarn Color Code	Year Assigned	Assignments
86	-	Black Brown Violet		
87	-	Black Brown Orange		
88	-	Black Violet Orange		
89	-	Blue Brown Violet		
90	-	Blue Brown Orange		
91	-	Blue Violet Orange		
92	- (T)	Brown Violet Orange	1985	Tong Yang Chemical Co., Ltd., Kyoung Nam, Korea
93	44	Yellow Yellow Green	1972	SAIAG, Torino, Italy
94	55	Yellow Yellow Red	1968	Ages & Co., Torino, Italy
95	57	Yellow Yellow Black	1967	Meiji Rubber & Chemical Co., Ltd., Tokyo, Japan
96	59	Yellow Yellow Blue	1967	Kugelfischer Georg Schafer & Co., Ebern, F.R. Germany
97	61	Yellow Yellow Brown	1962	Reserve - Tecalemit Ltd., Plymouth, England
98	-	Yellow Yellow Violet		
99	-	Yellow Yellow Orange		
100	45	Green Green Yellow	1971	Dunlop Industrial Products (Pty) Ltd., Benoni, Transvaal, Africa
101	47	Green Green Red	1971	Moldeados Industriales, S.A., Naucalpan, Mexico
102	49	Green Green Black	1978	Codan Gummi A/S, Koge, Denmark
103	51	Green Green Blue	1970	Vincke S.A., Palamos, Spain
104	-	Green Green Brown		
105	53	Green Green Violet	1969	P.B. Cow (Special Products) Ltd., Bucks, England
106	-	Green Green Orange		
107	54	Red Red Yellow	1968	STOP, Freins Hydrauliques, Saint-Ouen, France
108	46	Red Red Green	1971	BTR Hose Ltd., Farington, Leyland, Lancs., England
109	63	Red Red Black	1967	The Dunlop Co., Ltd. Polymer Engineering Div., Leicester, England
110	65	Red Red Blue	1962	Tecalemit Pty. Ltd., Finsbury, S. Australia
111	67	Red Red Brown	1962	Tecalon Brasileria de Autopecas, Sao Paulo, Brazil
112	- (C)	Red Red Violet	1987	Tubex Australia Pty. Ltd., Bayswater, Victoria, Australia
113	-	Red Red Orange	1985	First Trust Industrial Corp., Taipei, Taiwan, ROC
114	56	Blak Black Yellow	1967	Gates Canada Inc. - Terminated 1981 and combined w/Line 17
115	48	Black Black Green	1969	Dunlop India Ltd., Calcutta, India
116	62	Black Black Red	1962	Deutsche Tecalemit G.M.B.H., Bielefeld, Germany
117	69	Black Black Blue	1962	Tecalemit S.A., Paris, France
118	71	Black Black Brown	1966	Alfred Roberts & Sons Ltd., Birmingham, England - Terminated 1974
119	-	Black Black Violet		
120	-	Black Black Orange		
121	58	Blue Blue Yellow	1967	Nichirin Rubber Industrial Comapny, Ltd., Kobe, Japan
122	50	Blue Blue Green	1970	Nelson Stokes Ltd., Camelford, Cornwall, England
123	64	Blue Blue Red	1962	Tecamec Limited, Plymouth, England
124	68	Blue Blue Black	1961	Imperial Eastman Div., Clevite Industries Inc., Chicago, Illinois
125	73	Blue Blue Brown	1959	Mundener Gummiwerk GmbH, Hann. Munden, F.R. Germany
126	-	Blue Blue Violet	1988	Dunlop Metaloflex Industrial Ltda., Sao Paulo, Brazil
127	-	Blue Blue Orange	1985	Companhia Saad do Brazil, Sao Paulo, Brazil
128	60	Brown Brown Yellow	1962	Reserve - Tecalemit Ltd., Plymouth, England
129	52	Brown Brown Green	1970	Hitachi Cable Ltd., Tokyo, Japan
130	66	Brown Brown Red	1962	Tecalemit (India) Ltd., Calcutta, India
131	70	Brown Brown Black	1962	Tecalemit Italia, Turin, Italy
132	72	Brown Brown Blue	1987	Pirelli Sistemi Antivibranti SpA, Settimo Torinese, Italy
133	-	Brown Brown Violet		
134	-	Brown Brown Orange		
135	-	Violet Violet Yellow		
136	-	Violet Violet Green		
137	-	Violet Violet Red		
138	-	Violet Violet Black		
139	-	Violet Violet Blue		
140	-	Violet Violet Brown		
141	-	Violet Violet Orange		
142	-	Orange Orange Yellow		
143	-	Orange Orange Green		
144	-	Orange Orange Red		
145	-	Orange Orange Black		
146	-	Orange Orange Blue		
147	-	Orange Orange Brown		
148	-	Orange Orange Violet		

* Exact assignment date unknown, prior to 1950
(C) Change in entry since last edition
(T) Tentative assignment

NOTE— The information shown in this publication is based on the best information available to the RMA. No claims are made as to the completeness or currency of the information shown.

MOISTURE TRANSMISSION TEST PROCEDURE—HYDRAULIC BRAKE HOSE ASSEMBLIES—SAE J1873 JUN93

SAE Recommended Practice

Report of the Hydraulic Brake Systems Actuating Committee approved March 1988. Rationale statement available. Revised by the SAE Automotive Brake and Steering Hose Standards Committee June 1993.

(R) **1. Scope**—This SAE Recommended Practice is intended for all vehicle hydraulic brake hoses. It is an accelerated test which is intended to provide the user with a method of comparing the ability of hydraulic brake hose designs to retard the ingress of moisture into brake fluid.

This document does not specify a performance requirement. Interlaboratory reproducibility and correlation of data have not been defined, nor has correlation been established between field vehicle brake fluid moisture content and data obtained by this document.

1.1 Purpose—The purpose of this document is to simulate in the laboratory the transmission of moisture into a brake hose.

(R) **2. References**—There are no referenced publications specified herein.

3. Note—This procedure is extremely sensitive and care must be taken to minimize the amount of unwanted moisture introduced into the system due to experimental error.

4. Test Apparatus

4.1 Brake Hose Assemblies—305 mm ± 5 mm (12.0 in ± 0.2 in) free length with female threaded fittings no longer than 40 mm (1-1/2 in) on each end.

4.2 Threaded end plugs tapered to match fitting eyelet.
4.3 Sample glass vials with appropriate caps to insure seal.
4.4 Glass water bath container.
4.5 Karl Fischer test apparatus.
4.6 Squeeze bottle with "J" tube.
4.7 Deionized or distilled water.
4.8 Desiccator.
4.9 Unopened can of brake fluid.
4.10 Standard room environmental conditions of 23.0 °C ± 2.0 °C (73.4 °F ± 3.6 °F), 50% RH ± 5% RH.

5. Test Preparation

5.1 Precondition all the brake hose assemblies at 100 °C ± 2 °C (212 °F ± 3.6 °F) for 24 h ± 2 h. Precondition all glass sample vials and the "J" tube squeeze bottle at 70 °C ± 2 °C (158 °F ± 3.6 °F) for a minimum of 1 h. This will assure that the hoses, vials, and bottle initially will be relatively dry.

5.2 After removing the previous items from the oven, place them in a desiccator at standard room environmental conditions for a minimum of 1 h or until the items reach room temperature. Keep these items in the desiccator until required to assure they stay in a dry state.

5.3 All phases of this test, where brake fluid is added to or removed from the brake hoses, shall be done at standard room environmental conditions as described in 4.11.

6. Test Procedure

6.1 Remove the preconditioned brake hose assemblies from the desiccator and start the fill procedure. Open an unopened can of brake fluid and fill the "J" tube squeeze bottle. Fill the appropriate vials and brake hoses described as follows within 1 h. This will assure that the test will be started with "dry" brake hoses and "unused" brake fluid, and keep the exposure to air and moisture at a minimum.

6.1.1 Fill one conditioned glass vial with brake fluid using the conditioned "J" tube squeeze bottle. The vial shall be totally filled to minimize air entrapment. Cap the vial securely within 30 s of filling to minimize exposure to air and moisture. This vial will be used as a control.

6.1.2 Fill three brake hose assemblies with brake fluid using the "J" tube squeeze bottle, and then plug each end tightly within 30 s to minimize exposure to air and moisture. This is easily accomplished by bending the hose into a "U" shape with both ends level and injecting brake fluid into one of the ends. The "J" tube should be small enough to fit inside the brake hose fitting eyelet. Plug the ends of the hose with a threaded plug tapered at the tip to seal properly against the brake hose fitting eyelet. Make sure no air is entrapped inside the hose when the ends are plugged.

6.2 Place the brake hoses in a desiccator at standard room environmental conditions until ready to start the immersion test.

6.3 Remove the brake hoses from the desiccator and immerse them in a glass water bath container. Place the container and hoses in an oven maintained at 70 °C ± 2 °C (158 °F ± 3.6 °F) for 72 h ± 1 h.

6.3.1 The hose assemblies shall be held in a U-shape such that their centerlines are 76 mm ± 13 mm (3.0 in ±0.5 in) apart. This can be accomplished by designing the walls of the glass bath container to accommodate this or by using a wire to tie the ends together. Position the hose assemblies in the glass water bath container so that they are totally immersed in the water and not contacting each other.

6.3.2 Use deionized or distilled water. Maintain the volume of water to a minimum of 490 mL (30 in^3) per hose. Keep the complete hose assembly below the water line at all times.

6.3.3 Refill the water bath as required to compensate for evaporation and assure compliance with the previous conditions. A lid for the water bath container will minimize evaporation and generally eliminate the need to add water during the test. If additional water is necessary, use water at 70 °C ± 5 °C (158 °F ± 9 °F).

6.4 After the required exposure interval, remove the three brake hose assemblies from the water bath and empty the brake fluid from each hose into separate conditioned glass vials.

6.4.1 Dry the outside of the brake hose thoroughly. Let cool to room temperature at standard room environmental conditions for a minimum of 30 min.

6.4.2 Wipe the hose assemblies just prior to cutting them using a cloth wetted with isopropyl alcohol.

6.4.3 Cut through the hose section within 13 mm (0.5 in) from one of the end fittings. While maintaining the longer hose section of the brake hose assembly in a vertical position with the end fitting up, cut the hose within 13 mm (0.5 in) from the remaining end fitting allowing the brake fluid to empty into a clean dry sample vial. Size the vial so that it will be totally filled. No attempt shall be made to recover or include the brake fluid remaining in either end fitting.

6.4.4 Cap the vial within 30 s after filling the brake fluid.

6.4.5 Repeat the previous procedure for each hose.

6.5 Determine the percentage water content of the brake fluid in the brake hose assemblies and control vial to the nearest 0.01% using the Karl Fischer test method.

6.5.1 Run the brake fluid moisture content test on each brake hose assembly sample vial and control sample vial.

6.5.2 Calculate the percentage moisture transmission by subtracting the control vial sample percent water from the percent of water averaged from the three hose samples.

6.5.3 Report individual hose sample data points, three sample average, control point, and net moisture transmission.

DYNAMIC OZONE TEST PROCEDURE—HYDRAULIC BRAKE HOSE—SAE J1914 MAR88

SAE Recommended Practice

Report of the Hydraulic Brake Systems Actuating Committee approved March 1988. Rationale statement available.

1. Purpose—The purpose of this recommended practice is to evaluate in the laboratory the effect of flexing on a brake hose when exposed to a high ozone concentration environment.

2. Scope—This recommended practice is intended for all vehicle hydraulic brake hoses. It is an accelerated test which subjects the hose to dynamic ozone exposure.

3. Test Apparatus

3.1 Brake hose cut lengths of 218 ± 3 mm (8.6 ± 0.1 in).

3.2 SAE dynamic ozone test apparatus that will flex the brake hose as shown in Fig. 1.

3.3 Ozone test chamber.

4. Test Preparation

4.1 Pre-condition all the brake hose samples in a non-stressed condition at 27 ± 6°C (80 ± 10°F) for at least 24 h prior to the start of the test.

5. Test Procedure

5.1 Assemble the brake hose samples on the SAE dynamic ozone test apparatus so that they meet the relative position and flex parameters as shown in Fig. 1.

5.1.1 Install the brake hose over the fixture pins until the hose has bottomed out.

5.1.2 Use band clamps to securely retain the brake hose on the pin.

5.2 Install the test apparatus and assembled hoses in a stabilized ozone chamber.

5.2.1 The chamber shall contain air mixed with ozone at the ozone partial pressure of 100 ± 10 MPa (100 ± 10 parts of ozone per 100 million parts of air by volume).

5.2.2 The air temperature in the chamber shall be 40 ± 3°C (104 ± 5°F).

5.3 Start the test.

5.3.1 Start cycling when the chamber reaches the specified ozone concentration but no later than 1 h after putting the test apparatus in the ozone chamber.

5.3.2 The flex rate shall be 0.30 ± 0.05 Hz.

5.3.3 The stroke shall be 76.2 ± 2.5 mm (3.0 ± 0.1 in).

5.4 Examine the hoses for ozone cracks every 24 ± 1 h.

5.4.1 Remove the fixture from the cabinet and close the door immediately.

5.4.2 Inspect for ozone cracks visible to the eye without magnification at the worst stress condition ignoring the areas immediately adjacent to or within the area covered by the band clamps. Do not remove the hoses from the fixture.

5.4.3 Reinstall the fixture in the cabinet within 15 min of its removal.

5.4.4 The test shall be run continuously except for the daily inspection periods. Inspection periods may be eliminated on non-work days if determined to be not critical.

5.5 End of test.

5.5.1 Stop the test when ozone cracks are observed.

5.5.2 Record the hose identification and the number of days that elapsed until the first visible crack was observed.

6. Performance Requirement

6.1 The hose shall not crack after testing a minimum of 48 h.

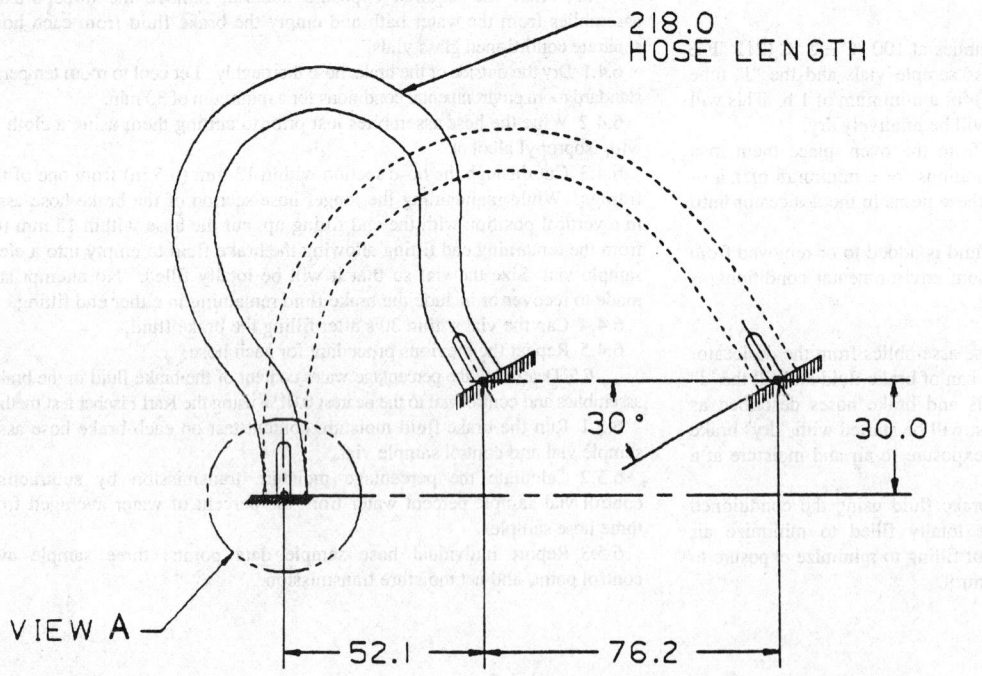
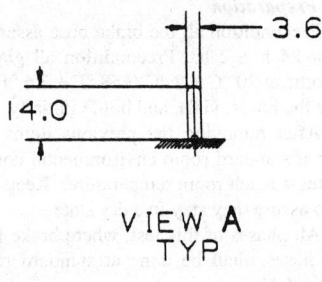

FIG. 1

AUTOMOTIVE AIR BRAKE HOSE AND HOSE ASSEMBLIES—SAE J1402 JUN85

SAE Recommended Practice

Report of the Motorcoach and Motor Truck Division, approved January 1942, last revised, Brake Committee, June 1985. This material was formerly designated SAE 40Rz.

1. *Scope*—This recommended practice covers minimum requirements for air brake hose assemblies made from reinforced elastomeric hose and suitable fittings for use in automotive air brake systems including flexible connections from frame to axle, tractor to trailer, trailer to trailer and other unshielded air lines that are exposed to potential pull or impact. This hose is not to be used where temperatures, external or internal, fall outside the range of −40 to +200°F (−40 to +93°C).

2. *Hose Dimensions*

    2.1 For Permanently Attached Fittings—When the hose is assembled with permanently attached fittings, the hose portion of the hose assembly shall conform to the dimensional requirements of Table A.

    2.2 For Reusable Fittings—When the hose is assembled with reusable fittings, the hose portion of the hose assembly shall conform to the dimensional requirements of Table AI or AII, with the exceptions of 3/8 in, 7/16 in and 1/2 in SP from Table A.

    2.3 Minimum Bend Radius

    2.3.1 Table 1 contains the minimum bend radii recommended for vehicle installations; however, smaller radii have been successfully used in the flexure test and some installations.

3. *Identification*

    3.1 Hose—Each hose manufacturer shall incorporate into the hose construction an identification yarn as assigned by the RMA and as shown in Appendix B of SAE J1401.

    Each air brake hose shall also be labeled in a color contrasting to that of the hose and labeling shall be repeated every 15 in (381 mm) or less along the entire length of hose in legible block capital letters at least 1/8 in (3.2 mm) high with the following minimum information in the order listed:

    (a) The hose manufacturer's identification XXX.
    (b) The words "Air Brake" to identify specific hose application.
    (c) The nominal hose inside diameter in fractions of an inch such as 3/8.
    (d) SAE J1402

    EXAMPLE: XXX Air Brake 3/8 SAE J1402

    In addition, each air brake hose shall be labeled with either an A, AI, or AII identifying whether the hose has been manufactured to the dimensions of Table A, AI, or AII. This additional labeling need not appear on the same layline as the above (a), (b), (c), and (d) information, but shall have the same minimum requirements of color contrast, spacing, and letter height.

    3.2 Fittings—Each reusable air brake hose fitting shall be permanently etched, embossed, or stamped in legible block capital letters at least 1/16 in (1.6 mm) high with the coupling manufacturer's identification and fractional hose size.

    3.3 Assemblies—Each air brake hose assembly shall be identified by means of a band around the hose. The band may move freely along the length of the assembly as long as it is retained by the end fitting. The band shall be permanently etched, embossed, or stamped in legible block capital letters at least 1/8 in (3.2 mm) high with the following information:

    (a) The month, day, and year or the month and year the assembly was made expressed in numerals. For example, 3/1/75 means March 1, 1975 or 3/75 means March 1975.

    (b) The assembler's identification and additional information as required.

4. *Performance*

    NOTE 1: In the interest of safety, all samples subjected to one or more performance tests other than Proof Pressure and Length Change shall be destroyed and discarded after completion of the tests and their analysis.

    NOTE 2: Unless otherwise specified, all tests are to be performed in accordance with ASTM D 622.

    4.1 Acceptance Performance—Hose or hose assemblies at the time of manufacture shall conform to the following:

    4.1.1 PROOF PRESSURE—Assemblies subjected to a pressure test using 300 psi ± 10 psi (2.07 MPa ± 0.07 MPa) air or nitrogen under water for a minimum of 30 s shall show no leaks.

    4.1.2 BURST STRENGTH—There shall be no hose burst or end fitting separation below 900 psi (6.21 MPa) when hose or hose assemblies are subjected to a hydrostatic burst test.

    4.1.3 ASSEMBLY TENSILE STRENGTH—The hose assembly complete with couplings shall be subjected to a tensile test until separation of the hose from the couplings or rupture of the hose occurs. Failure of the 1/4 in (6.4 mm) and smaller nominal I.D. size shall occur at no less than 250 lb (1.112 kN) and larger sizes at no less than 325 lb (1.446 kN).

TABLE A—INSIDE AND OUTSIDE DIAMETER OF HOSE

in (mm)	1/4 (6.4)	5/16 (7.9)	3/8[a] (9.5)	7/16[a] (11.1)	1/2 SP[a] (12.7)	5/8 (15.9)
Min I.D.	0.227 (5.8)	0.289 (7.3)	0.352 (8.9)	0.407 (10.3)	0.469 (11.9)	0.594 (15.1)
Max I.D.	0.273 (6.9)	0.335 (8.5)	0.398 (10.1)	0.469 (11.9)	0.531 (13.5)	0.656 (16.7)
Min O.D.	0.594 (15.1)	0.656 (16.7)	0.719 (18.3)	0.781 (19.8)	0.844 (21.4)	1.031 (26.2)
Max O.D.	0.656 (16.7)	0.719 (18.3)	0.781 (19.8)	0.843 (21.4)	0.906 (23.0)	1.094 (27.8)

[a] The sizes 3/8 in, 7/16 in, and 1/2 in SP can be assembled with reusable fittings if desired.

TABLE AI—INSIDE AND OUTSIDE DIAMETER OF HOSE

in (mm)	3/16 (4.8)	1/4 (6.4)	5/16 (7.9)	13/32 (10.3)	1/2 (12.7)	5/8 (15.9)
Min I.D.	0.188 (4.8)	0.250 (6.4)	0.312 (7.9)	0.406 (10.3)	0.500 (12.7)	0.625 (15.9)
Max I.D.	0.214 (5.4)	0.281 (7.1)	0.343 (8.7)	0.437 (11.1)	0.539 (13.7)	0.667 (16.9)
Min O.D.	0.472 (12.0)	0.535 (13.6)	0.598 (15.1)	0.714 (18.1)	0.808 (20.5)	0.933 (23.7)
Max O.D.	0.510 (13.0)	0.573 (14.6)	0.636 (16.2)	0.760 (19.3)	0.854 (21.7)	0.979 (24.9)

TABLE AII—INSIDE AND OUTSIDE DIAMETER OF HOSE

in (mm)	3/16 (4.8)	1/4 (6.4)	5/16 (7.9)	13/32 (10.3)	1/2 (12.7)	5/8 (15.9)
Min I.D.	0.188 (4.8)	0.250 (6.4)	0.312 (7.9)	0.406 (10.3)	0.500 (12.7)	0.625 (15.9)
Max I.D.	0.214 (5.4)	0.281 (7.1)	0.343 (8.7)	0.437 (11.1)	0.539 (13.7)	0.667 (16.9)
Min O.D.	0.500 (12.7)	0.562 (14.3)	0.656 (16.7)	0.742 (18.8)	0.898 (22.8)	1.054 (26.8)
Max O.D.	0.539 (13.7)	0.602 (15.3)	0.695 (17.7)	0.789 (20.1)	0.945 (24.0)	1.101 (27.9)

TABLE 1—RECOMMENDED MINIMUM BEND RADIUS

Nominal Hose I.D.		Minimum Bend Radius (To Inside of Bend)	
in	mm	in	mm
3/16	4.8	2	51
1/4	6.4	2-1/2	64
5/16	7.9	3	76
3/8	9.5	3-1/2	89
13/32	10.3	3-1/2	89
7/16	11.1	4	102
1/2	12.7	4	102
5/8	15.9	4-1/2	114

4.1.4 LENGTH CHANGE—Test for length change shall be conducted in accordance with ASTM D 622 except that the original measurement shall be made at 10 psi (0.07 MPa). The change in length shall be determined at 200 psi (1.38 MPa) and shall be from +5% to −7%.

4.1.5 ADHESION—Test for adhesion in all hose types, provided they are not wire reinforced hoses, shall be conducted in accordance with ASTM D 413 Machine Method, and the average load required to separate any adjacent layers shall be 8 lb/in of width (14 N/cm) minimum. Test for adhesion in wire reinforced hoses shall be conducted per paragraph 4.1.5.1. The adhesion test is to be made only on the original unaged specimen.

4.1.5.1 *Adhesion of Wire Reinforced Hose*—The requirements and method of testing cover adhesion shall be the same as 4.1.5 for all other hose types. The integrity of the inner tube adhesion shall be tested by subjecting a length of hose not less than 15 in (381 mm) long to the following requirements.

Place a steel ball (of the size specified in Table 2) in the bore of the hose. One end shall then be attached to a vacuum source and the other end plugged. A vacuum of 25 in (635 mm) of mercury shall be applied for a period of five (5) min while the hose is in an essentially straight position. At the conclusion of this period and while still under vacuum, the hose shall be bent 180 deg to the minimum bend radius in 2.3.1 in each of two directions 180 deg apart. After bending and returning to an essentially straight position and while still under vacuum, the ball shall be rolled from end to end of the hose. Failure of the ball to pass freely from end to end shall be indication of separation of the tube from the reinforcement layer and shall constitute failure.

4.2 Qualification Performance—For initial qualification under this specification all of the requirements under Acceptance Performance, Qualification Performance, and Flexure Test shall be met. Minimum sampling shall be per Table 3, including the specified sequential test procedure.

4.2.1 TEMPERATURE RESISTANCE

4.2.1.1[1] *High Temperature Resistance*—The hose shall show no cracks, charring, or disintegration externally or internally when straightened after being bent over a form for a period of 70 ± 2 h while in an air oven at 212 ± 3.6°F (100 ± 2.0°C). The radius of the test form shall be in accordance with Table 4.

4.2.1.2[1] *Low Temperature Resistance*—The hose shall show no cracks internally or externally when bent 180 deg over a form having the radius shown in paragraph 2.3.1 after hose and form have been exposed for a period of 70 ± 2 h in an air circulating chamber at −40 ± 3.6°F (−40 ± 2°C) and while still at this temperature. The hose and form shall be supported by a non-metallic surface during the entire period. The bend shall be completed in a period of 3–5 s.

4.2.2 RESISTANCE TO ENVIRONMENT

4.2.2.1 *Oil*—Specimens prepared from the inner tube and the cover shall show a volume increase when measured after removal from ASTM #3 Oil in which it has been immersed for 70 ± 2 h at 212 ± 3.6°F (100 ± 2°C) of not more than 100%.

4.2.2.2 *Water*—Condition hose assembly by immersion in tap water at room temperature for a period of 168 ± 2 h while bent over a form having the minimum bend radius shown in 2.3.1. Ends shall be completely capped during immersion. See Table 3 for next step.

4.2.2.3[1] *Ozone*—After being exposed for 70 ± 2 h in an ozone cabinet containing 50 pphm by volume of ozone at a temperature of 104 ± 3.6°F (40 ± 2°C) and while bent over a form having the radius shown in 2.3.1, the hose shall show no cracking under 7X magnification.

4.2.2.4 *Salt Spray Test*—Hose assembly end fittings while assembled on hose shall withstand 24 ± 1 h exposure to salt spray when tested in accordance with ASTM B 117 Method of Salt Spray (Fog) Testing. After this exposure, fittings shall show no base metal corrosion except red rust is acceptable in areas of identification stamping and crimp distortions. White corrosion products are acceptable.

5. End Fittings—End fittings shall be such as to permit conformance to all portions of this recommended practice. After assembly of the end fitting to the hose, the minimum I.D. of the end fitting or the hose shall not be less than 66% of the nominal hose I.D.

6. Flexure Test

6.1 Preparation of Test Samples

6.1.1 Prior to cutting the hose, apply a layline (of a color distinguishable from that of the hose cover) along the length of the hose (following the natural hose curvature which results from the hose being coiled in a roll—see Fig. 1).

6.1.2 Cut the hose to provide a hose assembly sample with a free hose length as shown in Fig. 2. Free hose length is the outside exposed hose length between the fittings in the finished hose assembly.

6.1.3 Fittings are to be assembled on the hose in accordance with the manufacturer's instructions.

6.2 Preconditioning—Subject each sample hose assembly to the preconditioning specified in paragraph 6.2.1, followed by the preconditioning specified by paragraph 6.2.2.

6.2.1 SALT SPRAY CONDITIONING—With the ends plugged, subject the hose assembly samples to 24 ± 1 h exposure to salt spray testing in accordance with ASTM B 117, Method of Salt Spray (Fog) Testing.

6.2.1.1 Allow no more than 168 h elapsed time between completion of salt spray conditioning and the starting of high temperature aging per paragraph 6.2.2.

6.2.2 HIGH TEMPERATURE AGING—With each hose assembly sample in a straight position, age the samples in air at a temperature of 212 ± 1.8°F (100 ± 1°C) for a period of 70 ± 2 h. During the entire conditioning period, the hose bore of each sample is to be exposed to the air within the oven.

6.2.2.1 Allow no more than 168 h elapsed time between completion of high temperature aging and starting of flex tests per paragraphs 6.3 and 6.4.

---

[1] The external surface of fabric covered hoses shall be exempt from inspection for cracks after 4.2.1.1, 4.2.1.2, and 4.2.2.3 as visual inspection is not practical.

TABLE 4—RADIUS FOR HIGH TEMPERATURE RESISTANCE TEST

Nominal Hose I.D.		Radius of Test Form	
in	mm	in	mm
3/16	4.8	1	25
1/4	6.4	1-1/2	38
5/16	7.9	1-3/4	45
3/8	9.5	1-3/4	45
13/32	10.3	1-7/8	48
7/16	11.1	2	51
1/2	12.7	2	51
5/8	15.9	2-1/2	64

TABLE 2—BALL SIZE FOR TESTING ADHESION OF WIRE REINFORCED HOSE

Hose I.D.	in (mm)	3/16 (4.7)	1/4 (6.4)	5/16 (7.9)	3/8 (9.5)	13/32 (10.3)	7/16 (11.1)	1/2 (12.7)	5/8 (15.9)
Ball Size	in (mm)	9/64 (3.6)	3/16 (4.8)	15/64 (6.0)	9/32 (7.1)	19/64 (7.5)	21/64 (8.3)	3/8 (9.5)	15/32 (11.9)

TABLE 3—MINIMUM SAMPLING AND SEQUENTIAL TEST PROCEDURE

Sample No.	Subjected to	Followed by
1	5.[a]	4.2.1.1 then 4.1.1
2	5.[a]	4.2.1.2 then 4.1.1
3	4.2.2.1	—
4	5.[a]	4.2.2.2 then 4.1.3
5	4.2.2.3	—
6	5.[a]	4.2.2.4 then 4.1.2
7	5.[a]	4.1.4 then 4.1.1 and 4.1.2
8	4.1.5	—
9	6.[a]	—

[a] Couple hose before starting tests or aging.

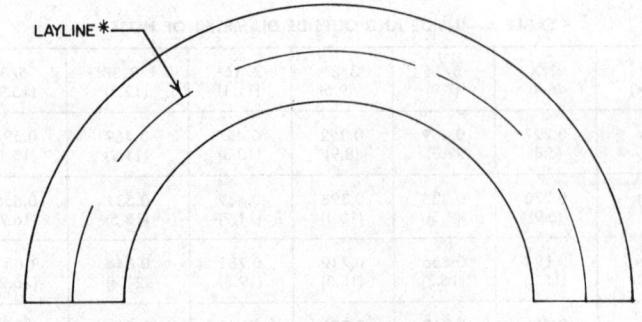

* MARKED WHEN PREPARING SAMPLES FOR TEST PER 6.1.1. NOT MANUFACTURER'S LAYLINE.

FIG. 1

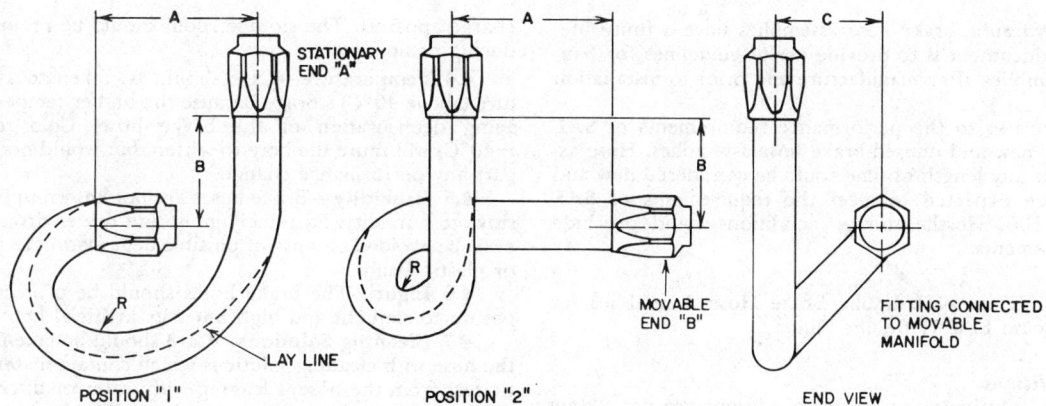

POSITION "1" ILLUSTRATES THE "LEFT" EXTREME OF TRAVEL.
POSITION "2" ILLUSTRATES THE "RIGHT" EXTREME OF TRAVEL.

Free Hose Length ±1/16 (1.6 mm)	Hose I.D., in (mm)	Dimensions							
		Position "1"				Position "2"			
		"A"	"B"	"C"	"R"a	"A"	"B"	"C"	"R"a
10 in (254 mm)	3/16 1/4 (4.8) (6.3)	3 in (76 mm)	2.75 in (70 mm)	3.75 in (95 mm)	1.4 in (34 mm)	3 in (76 mm)	2.75 in (70 mm)	3.75 in (95 mm)	1.2 in (30 mm)
11 in (279 mm)	5/16 3/8 13/32 (7.9) (9.7) (10.4)	3 in (76 mm)	3.5 in (89 mm)	4.5 in (114 mm)	1.7 in (43 mm)	3 in (76 mm)	3.5 in (89 mm)	4.5 in (114 mm)	1.3 in (33 mm)
14 in (355 mm)	7/16 1/2 5/8 (11.2) (12.7) (16.0)	3 in (76 mm)	4 in (102 mm)	5 in (127 mm)	2.2 in (56 mm)	3 in (76 mm)	4 in (102 mm)	5 in (127 mm)	1.8 in (46 mm)

a This is an approximate average radius.

FIG. 2

6.3 Installation of Samples in the Test Setup—After the samples have been preconditioned, they are to be installed in the test setup in the configuration specified in Fig. 2. The installation procedure is as follows:

6.3.1 With the movable manifold of the flex test machine at the center of its stroke, fitting "B" is connected to the movable manifold in such a manner that the layline is located at the top-center position (see Fig. 2).

6.3.2 After fitting "B" has been coupled with the movable manifold, the "A" fitting is then connected to the stationary manifold without imparting any twist to the hose, but allowing the hose to seek its natural curvature.

6.4 Test Procedure—Flex the samples by moving fitting "B" from the center stroke position to 3 in (76 mm) either side of center ("A" dimension in Fig. 2) alternating between Position "1" and "2" while simultaneously cycling the air pressure on for 1 min and off for 1 min.

6.4.1 FLEXURE/PRESSURE CYCLING TEST PARAMETERS

6.4.1.1 *Total Flexure Stroke*—6 in (152 mm) Tolerance: ±1/16 in (1.6 mm).

6.4.1.2 *Flexure Stroke Frequency*—100 ± 5 cpm.

6.4.1.3 *Ambient Temperature*—70–80°F (21–27°C).

6.4.1.4 *Internal Air Pressure on Test Samples*—150 psi ± 10 psi (1.03 ± 0.07 MPa). The air pressure shall be alternately fully "on" for 1 min ± 5 s and fully "off" for 1 min ± 5 s.

6.5 The failure point (number of flex stroke cycles) shall be determined by loss of air pressure through the failed sample. As the air pressure is alternately fully "on" for 1 min and fully "off" for 1 min, pressure loss shall be further described as failure of the system to be repressurized to 150 psi ± 10 psi (1.03 MPa ± 0.07 MPa) through a 0.062 + 0.001/ −0.000 in (1.6 + 0.03/ −0.00 mm) diameter orifice within 2 min. Failure shall not occur before completing one million flex cycles.

PACKAGING, STORAGE, AND SHELF LIFE OF HYDRAULIC BRAKE HOSE ASSEMBLIES— SAE J1288 MAR90

SAE Information Report

Report of the Hydraulic Brake Systems Actuating Committee, approved June 1985. Reaffirmed by the Automotive Brake and Steering Hose Standards Committee March 1990.

Foreword—This reaffirmed document has been changed to reflect the new SAE Technical Board format.

1. Scope—This SAE Information Report is the listing of recommendations for the proper packaging, storage, and shelf life limitations of new and unused hydraulic brake hose assemblies. The document embodies the testing, analysis, and experience of many users and manufacturers. Where specific manufacturer's recommendations are made, those recommendations shall supersede the recommendations of this document.

This document describes the successful procedures and practices associated with brake hose assemblies usage by a wide cross section of manufacturers and users over several years. The practices are expected to be applicable to all brake hose assemblies which qualify under SAE J1401.

1.1 Purpose—Hydraulic brake hose assemblies have a finite life. The purpose of this document is to provide useful guidelines for handling brake hose assemblies after manufacture and prior to installation onto a vehicle.

Another purpose relates to the performance requirements of SAE J1401 which apply to new and unused brake hose assemblies. Hose assemblies in storage for any length of time could be considered new and unused, and therefore expected to meet the requirements of SAE J1401 and FMVSS 106. Hostile storage conditions could preclude meeting those requirements.

2. References

SAE J1401, Road Vehicle—Hydraulic Brake Hose Assemblies for Use with Non-Petroleum Base Hydraulic Fluids
FMVSS 106

3. Packaging Conditions

3.1 Background—Optimum packaging conditions are not always feasible because of the design of the brake hose assembly, and the available storage environment. The objective of packaging is to protect the hose from harmful materials in the environment, identify the brake hose assembly, and provide marketing value. It is the responsibility of the brake hose assembly supplier, in conjunction with the distributor, to insure that the individual package is acceptable. Optimum conditions are recommended, and exceptions should be carefully evaluated.

3.2 Positioning of Brake Hose Assemblies—Brake hose assemblies should be stored in a straight position with no external forces on the hose. A straight position is essential to the proper installation of the brake hose assembly to the vehicle. Curved storage positions result in residual deformation which makes correct installation difficult, and could result in the brake hose contacting the tire or suspension after installation. Any curvature to the hose adds stress to the cover. The result of the stress is an acceleration of the normal aging effect of the polymers.

3.3 Compatibility of Brake Hose with Packaging Materials—Some customary packaging materials contain chemicals which cause a reaction with the brake hose. Among those materials are certain plastics, adhesives, and paper treatments. Care should be taken to insure that the packaging materials are inert to the hose assembly.

4. Storage Conditions

4.1 Background—All polymeric materials undergo changes in physical properties over time. Detrimental changes can be minimized by controlling the storage environment. Oils, solvents, lubricants, ozone, ultraviolet light, heat, adhesives, water, salts, and humidity are some of the natural forces which can act to degrade a brake hose assembly.

4.2 Cleanliness—It is imperative that the brake hose assembly be kept clean. Even foreign materials which do not affect the brake hose assembly can be detrimental to the brake system. The bases, bores, and sealing surfaces are particularly sensitive to foreign material contaminants.

4.3 Ozone and Oxidation Protection—Ozone is an especially active form of oxygen which causes deterioration of rubber products. Ozone is present in the natural form, but it can be generated by such equipment as electric motors, high intensity lamps, and voltage discharge apparati. The storage room should be remote from ozone producing equipment.

4.4 Temperature—Care should be taken to avoid high temperature (above 40°C) storage because the higher temperature increases the aging deterioration of the brake hose. Cold temperature (below −40°C) will cause the hose to stiffen, but would not be expected to impart any performance change.

4.5 Humidity—Brake hoses are not known to be performance sensitive to humidity. Extremely moist and dry environments should, however, be avoided because of possible degradation to the supporting yarn or plastic armor.

4.6 Light—The brake hoses should be protected from direct exposure to sunlight and high intensity artificial light.

4.7 Cleaning Solutions—Care should be taken to avoid contact of the hose with cleaning solutions which contain materials that leach plasticizers from the hose. Cleaning with water or discretionary use of isopropyl alcohol will not harm the hose. The hose should be dry before use.

4.8 Ammonia and Ammonia Derivatives—Many brake hose end fittings are made from brass which is susceptible to degradation due to ammonia contact. The source of degradation can be from animal waste, cleaning materials, fertilizers, or other available forms. Degradation of brass due to attack from ammonia compounds can be rapid, therefore, extreme care should be taken to avoid hose assembly contact with these types of materials.

4.9 Oils, Solvents, and Special Fluids—Oils, solvents, and other fluids can cause the rubber material to soften and swell. Further, certain brake hose protective sleeving is especially susceptible to performance loss due to attack by such materials. Therefore, the hose should be protected against contact with these materials.

4.10 Salts—The presence of salts can cause corrosion of metal components and degradation of rubbers. Such salts can be found in households, fertilizer, or in natural forms. Brake hose assemblies should be protected from all forms of salts.

5. Shelf Life of Brake Hose Assemblies

5.1 Background—Different hose constructions have different initial performance levels and different degradation rates relative to specific adverse packaging, handling, and storage conditions.

5.2 Inspection—The brake hose assembly should be inspected prior to installation on the vehicle. The inspection should include items such as straightness, flexibility, cleanliness, cracks, swell, and corrosion.

5.3 Qualified Shelf Life—Under optimum packaging and storage conditions, shelf life of brake hose assemblies is unrestricted. Shelf life will be reduced commensurate with the degree of departure from optimum conditions. Based upon the inspection, 5.2, the hose assembly can be put into service if it is not degraded or has not been mishandled.

5.3.1 SPECIAL CONDITIONING OF BRAKE HOSE PRIOR TO INSTALLATION—If brake hoses are deformed during packaging or storage, it is recommended that the hose be removed from the package and placed in a straight position and allowed to stabilize to a temperature of 15 to 40°C.

After temperature stabilization, the brake hose should be manually flexed to remove any residual deformation and inspected for cracks before attempting to install the brake hose to the vehicle.

(R) VACUUM BRAKE HOSE—SAE J1403 JUL89 — SAE Standard

Report of the Motorcoach and Motor Truck Division, approved January 1942, last revised, Nonmetallic Materials Committee, June 1985. This hose was formerly designated SAE 40R3. Completely revised by the Automotive Brake and Steering Hose Standards Committee July 1989.

1. Scope—The vacuum brake hose is intended for use in the power braking systems of vehicles or as connections on transmission lines in combinations of vehicles or systems thereof. For the purposes of clearly identifying hose classification and for specification simplification, vacuum brake hose is divided into two types: heavy-wall Type H, and light-wall Type L.

2. Performance Tests

2.1 Fuel Resistance Test

2.1.1 Fill a specimen of vacuum brake hose 300 mm ± 6 (11.8 in ± 0.2) long with ASTM Reference Fuel B.

2.1.2 Maintain the reference fuel in the hose at atmospheric pressure and room temperature for 48 h ± 1.0.

2.1.3 Drain the hose and within 5 min determine that every inside diameter of any section of hose is not less than 75% of the nominal inside diameter of the hose for heavy-wall and not less than 70% of the nominal inside diameter of the hose for light-wall hose. This determination can be performed by passage, end-to-end, of a ball having a diameter equal to or greater than the 75%, 70% of nominal I.D.

2.1.4 Within 10 min of fuel removal, subject the specimen to a vacuum of 88 + 0, −7 kPa (26 + 0, −2 in Hg) for 10 min ± 1.

2.1.5 Conduct the adhesion test for reinforced hose per 2.5.

2.2 Pressure Test—Subject a 450 mm ± 6 (18 in ± 0.2) length of hose to a hydrostatic pressure of 2.41 MPa ± 0.07 (350 psi ± 10) for 1 min + 10, −0 s.

2.3 High Temperature Exposure Test

2.3.1 Using calipers, measure the outside diameter of a 300 mm ± 6 (11.8 in ± 0.2) straight length of hose.

2.3.2 Subject the hose to an internal vacuum of 88 + 0, −7 kPa (26 + 0, −2 in Hg) at a temperature of 125°C ± 2 (257°F ± 3.6) for 96 h ± 2.0.

2.3.3 Within 5 min after completion of the hot vacuum aging period, measure the outside diameter at the point of greatest collapse using calipers.

2.3.4 Cool the hose for 4 h ± 0.25, and then condition the hose for 30 min ± 5 at room temperature.

2.3.5 Bend the hose around a mandrel having a diameter five times the nominal outside diameter of the hose.

2.3.6 Examine the hose for evidence of external and internal embrittlement or other evidence of degradation.

2.3.7 Subject the hose to a proof test pressure of 1.21 MPa ± 0.07 (175 psi ± 10) for 1 min + 10, −0 s.

2.4 Bend Test

2.4.1 Using calipers, measure the outside diameter of a specimen of vacuum brake hose of the length described in Table 1 at the middle section in the plane of the centerline.

TABLE 1—DIMENSIONS OF BEND TEST SPECIMEN AND MAXIMUM COLLAPSE OF O.D. OF VACUUM BRAKE HOSE

Inside Diameter of Hose		Length of Specimen				Maximum Collapse of O.D. (% of O.D.)	
		Heavy-Wall		Light-Wall		Heavy-Wall	Light-Wall
mm	in	mm	in	mm	in		
5.56	7/32	—	—	180	7	—	40%
6.35	1/4	205	8	—	—	20%	—
8.73	11/32	—	—	280	11	—	30%
9.53	3/8	305	12	—	—	20%	—
11.91	15/32	—	—	355	14	—	30%
12.70	1/2	405	16	—	—	20%	—
15.88	5/8	560	22	—	—	20%	—
19.05	3/4	710	28	—	—	20%	—
25.4	1	915	36	—	—	20%	—

2.4.2 Bend the specimen in the direction of its normal curvature until its ends just touch as shown in Fig. 1.

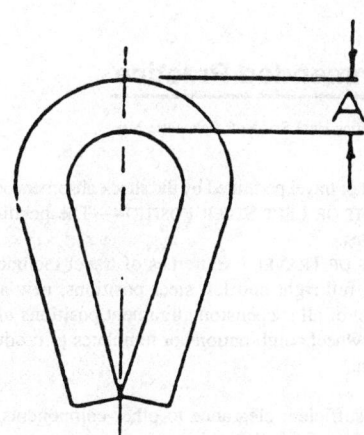

FIG. 1—BEND TEST OF VACUUM BRAKE HOSE

2.4.3 When the specimen is in bent configuration, again measure the outside diameter of the hose at the same middle section "A" in the plane of the centerline.

2.4.4 The difference between the two measurements shall be considered the collapse of the hose outside diameter on bending.

2.5 Adhesion Test

2.5.1 SPECIMEN PREPARATION

2.5.1.1 From the hose to be tested, cut a test specimen of length 25.0 + 3.0, −0 mm (1.00 + 0.12, −0 in).

2.5.1.2 Cut the layer to be tested of the test specimen longitudinally along its entire length of contact with the adjacent layer.

2.5.1.3 Peel the layer to be tested from the adjacent layer to create a flap large enough to permit attachment of a clamp.

2.5.1.4 Mount the test specimen on a freely rotating form with the separated layer attached to a clamp.

2.5.2 TEST APPARATUS—Utilize an appropriate tensile testing machine such that:

2.5.2.1 The recording head includes a freely rotating form with an outside diameter substantially the same as the inside diameter of the hose specimen to be placed on it.

2.5.2.2 The freely rotating form is mounted so that its axis of rotation is in the plane of the ply being separated from the specimen, and so that the applied force is perpendicular to the tangent of the specimen circumference at the line of separation.

2.5.2.3 The preferred rate of travel of the grip is 25 mm/min ± 3 (1.0 in/min ± 0.1), and the capacity of the machine is such that the maximum applied tension during the test is not more than 85% nor less than 15% of the machine's rated capacity.

2.5.2.4 The machine operates with no device for maintaining maximum load indication, and in a pendulum type machine, the weight lever swings as a free pendulum without engagement of pawls.

2.5.2.5 The machine produces a chart with length separation as one coordinate and applied tension as the other. The preferred chart speed is the same as the crosshead speed.

2.5.2.6 The adhesion value shall be the minimum force recorded on the portion of the chart corresponding to the actual separation of the part being tested.

2.6 Low Temperature Test

2.6.1 Condition the hose in a cold box in a straight position at −40°C ± 1 (−40°F ± 1.8) for 70 to 72 h. After conditioning, and while still at this temperature, bend the hose at least 180 deg around the mandrel at a steady rate in a period of 3 to 5 s.

2.6.2 Examine the cover of the hose with naked eye for cracks or breaks.

2.6.3 To qualify the fact that no breaks occurred in the tube, the hose shall be allowed to return to room temperature, after which it shall show no leaks when subjected to a proof pressure of 1.21 MPa ± 0.07 (175 psi ± 10) for 1 min + 10, −0 s.

2.6.4 TEST APPARATUS—The mandrel diameter shall be that shown in Table 2.

2.7 Ozone Test

2.7.1 Bend around the full circumference of the mandrel a specimen of hose approximately 250 mm (10 in) longer than the circumference of the required mandrel. Bind with tape or twine where the ends cross one another. If collapse of the hose occurs when bent around the mandrel, provide for internal support of the hose.

2.7.2 Condition the hose, on the mandrel, for 24.0 h ± 0.5 at room temperature.

2.7.3 While still on the mandrel, place the specimen in an exposure chamber containing air mixed with ozone at the ozone partial pressure of 100 mPa ± 5 (100 parts of ozone/100 million parts of air by volume ± 5) for 70 to 72 h. Ambient air temperature in chamber during test shall be 40°C ± 3 (104°F ± 5).

TABLE 2—DIMENSIONS OF LOW TEMPERATURE RESISTANCE TEST SPECIMEN LENGTH AND MANDREL DIAMETER

Nominal I.D. of Hose		Length of Specimen		Mandrel Diameter	
mm	in	mm	in	mm	in
5.56	7/32	444.5	17-1/2	76.2	3
6.35	1/4	444.5	17-1/2	76.2	3
8.73	11/32	482.6	19	88.9	3-1/2
9.53	3/8	482.6	19	88.9	3-1/2
11.91	15/32	520.7	20-1/2	101.6	4
12.70	1/2	520.7	20-1/2	101.6	4
15.88	5/8	558.8	22	114.3	4-1/2
19.05	3/4	609.6	24	127.0	5
25.4	1	723.9	28-1/2	165.1	6-1/2

2.7.4 Examine the outside surface of the specimen for cracks under 7X magnification, ignoring the areas immediately adjacent to or within the area covered by the binding.

2.7.5 TEST APPARATUS—The mandrel diameter shall be eight times the nominal outside diameter of the hose being tested.

3. Performance Requirements

3.1 Fuel Resistance Test—The inside diameter of any section of hose shall not be less than 75% of the nominal I.D. for heavy-wall hose and not less than 70% of the nominal I.D. for light-wall hose. The hose shall withstand a 10 min vacuum application without leakage or separation of inner tube from the fabric, if present. The minimum load required to separate the tube from the plies and the cover from the plies shall be 10.5 N/cm (6 lb/in) of width.

3.2 Pressure Test—There shall be no leakage or burst of the hose during a 1 min pressure hold.

3.3 High Temperature Exposure—The collapse of the outside diameter shall not exceed 10% of original O.D. for heavy-wall hose and 15% of original O.D. for light-wall hose. There shall be no external nor internal embrittlement or degradation. There shall be no leakage during a 1 min pressure hold.

3.4 Bend Test—The collapse of the hose outside diameter on bending shall not exceed the values given in Table 1. This requirement does not apply to preformed hoses molded to configurations that fit specific applications without further bending.

3.5 Adhesion Test—The minimum load required to separate the tube from the plies and the cover from the plies shall be 14 N/cm (8 lb/in) of width.

3.6 Low Temperature Test—The hose tube and cover shall not crack or break.

3.7 Ozone Test—The outside surface of the hose shall show no cracking when examined under 7X magnification.

4. Hose Identification—Each vacuum brake hose shall be labeled at intervals of not more than 152.4 mm (6 in) in block capital letters and numerals at least 3.2 mm (1/8 in) high. The manufacturer's designation (trademark, identifiable code letters, or number), the date code (consisting of month, day and year, or the month and year expressed in numerals), the nominal inside diameter, and the type designation shall appear on the outer surface of the hose. The wording "Vacuum Brake" may appear on the opposite side of the hose parallel to its longitudinal axis.

Type designation shall be indicated "VH" for heavy-wall or "VL" for light-wall hose.

Heavy-wall and light-wall vacuum brake hose dimensions and tolerances tabulations are shown in Appendix A for engineering reference only.

5. References
ASTM D 471, Rubber Property–Effect of Liquids
ASTM D 622, Rubber Hose for Automotive Air and Vacuum Brake System
ASTM D 1149, Rubber Deterioration–Surface Ozone Cracking in a Chamber (Flat Specimen)

APPENDIX A
ENGINEERING REFERENCE FOR DIMENSIONS AND TOLERANCES

HEAVY-WALL VACUUM BRAKE HOSE

Dimension	Hose Size, in					
	1/4	3/8	1/2	5/8	3/4	1
Inside diameter						
in	0.25	0.38	0.50	0.62	0.75	1.00
mm	6.35	9.53	12.70	15.88	19.05	25.40
Tolerance						
± in	0.03	0.03	0.03	0.03	0.03	0.06
± mm	0.76	0.76	0.76	0.76	0.76	1.52
Outside diameter						
in	0.56	0.81	0.94	1.06	1.19	1.47
mm	14.22	20.57	23.88	26.92	30.23	37.34
Tolerance						
± in	0.03	0.03	0.03	0.03	0.03	0.06
± mm	0.76	0.76	0.76	0.76	0.76	1.52

LIGHT-WALL VACUUM BRAKE HOSE

Dimension	Hose Size, in		
	7/32	11/32	15/32
Inside Diameter			
in	0.22	0.34	0.47
mm	5.56	8.73	11.91
Tolerance			
± in	0.03	0.03	0.03
± mm	0.76	0.76	0.76
Outside diameter			
in	0.44	0.69	0.81
mm	11.18	17.53	20.57
Tolerance			
± in	0.03	0.03	0.03
± mm	0.76	0.76	0.76

(R) APPLICATION OF HYDRAULIC BRAKE HOSE TO MOTOR VEHICLES—SAE J1406 JUN93

SAE Recommended Practice

Report of the Hydraulic Brake Systems Actuating Committee approved March 1981. Completely revised by the Hydraulic Brake and Steering Hose Standards Committee May 1989. Completely revised by the Automotive Brake and Steering Hose Standards Committee June 1993. Rationale statement available.

Foreword—The performance requirements in this SAE Recommended Practice represent the accumulation of the best information available from vehicle, brake hose, and brake hose assembly manufacturers. Since this document is subject to frequent change in order to keep pace with experience and technical advances, inclusion in any regulations where flexibility of revision is impractical is not recommended.

1. Scope—This SAE Recommended Practice covers the application of hydraulic brake hose (as defined by current issue of SAE J1401) as used to provide a flexible hydraulic connection between brake system components on motor vehicles.

1.1 Purpose—The purpose of this document is to outline design, operating, and service factors in routing a hydraulic brake hose assembly to a vehicle. It is intended to serve as a recommended practice for original equipment manufacturers. Vehicle design circumstances may exist that prevent strict adherence to this document. Any deviations should have the concurrence of all engineering functions involved.

2. References

2.1 Applicable Document—The following publication forms a part of this specification to the extent specified herein. The latest issue of SAE publications shall apply.

2.1.1 SAE PUBLICATION—Available from SAE, 400 Commonwealth Drive, Warrendale, PA 15096-0001.

SAE J1401—Road Vehicle—Hydraulic Brake Hose Assemblies for Use With Nonpetroleum-Base Hydraulic Fluids

2.2 Definitions

2.2.1 FULL JOUNCE POSITION—The attitude of the front or rear suspension compressed to metal-to-metal contact, with all rubber components of the suspension removed.

2.2.2 FULL REBOUND POSITION—The attitude of the front or rear suspension extended to the limit of travel permitted by the shock absorbers or other restrictions.

2.2.3 FULL RIGHT OR LEFT STEER POSITION—The position at metal-to-metal steering system stops.

2.2.4 EXTREMES OF TRAVEL—Extremes of travel include the full jounce and rebound positions, full right and left steer positions, new and fully worn brake pads, and full effect of all suspension alignment positions of caster, camber, and tow using tire and wheel combinations or templates to produce the most adverse clearance condition.

3. Objectives

3.1 Provide sufficient clearance to other components; i.e., tire chains, all tire/wheel sizes and configurations, drive shaft, etc.

3.2 Protect from exposure to adverse conditions such as heat, petroleum products, battery fluid, etc.

3.3 Minimize tension, severe bends, twist, and length.

3.4 Minimize exposure to mechanical damage due to operating the vehicle over underbrush, bumps, loose gravel, sand, mud, snow, ice, ruts, etc.

3.5 Minimize possibility of damage to hose during brake and other component assembly to vehicle.

3.6 Minimize potential for improper installation.

4. Design Factors

4.1 Tire/Wheel Clearance—Because of the potential for failure caused by wear through, it is extremely important not only to route the hose for adequate tire, tire chain, suspension and chassis component clearances, but also to assure that dynamic factors do not induce interference.

To reduce the likelihood of unanticipated interference, it is recommended that the designer establish the most "natural" path from caliper (or wheel cylinder) or intermediate support point to the bracket on frame or body. This will tend to avoid configurations which may cause the hose to suddenly take a new position when influenced by suspension motion and torsional effects. The sequence of motion or the rate of change may be the dominant factor, e.g., if the left-hand wheel moves from rebound to jounce in left steer and then to right steer, a twist in the hose might produce interference with the tire, whereas the opposite sequence (steer first, then elevate to jounce) would not cause hose-to-tire interference. Hose routings that produce a tight "S" curve are especially susceptible to a change in hose relative position, viz, the unsupported portion of the hose is capable of maintaining more than one stable position (generally "looped") depending on the sequence of suspension/steering motions just preceding these positions.

4.2 Dynamic Effects—Although adequate clearance may be obtained under static conditions, operating the vehicle over rough or "washboard" surfaces may produce hose contact with wheel/tire, suspension or chassis components, especially where long lengths of unsupported hose or looped sections are involved. If contact can occur, mechanical protection as described in Section 5 should be added. Road simulator equipment may be useful to investigate hose resonant effects by observing hose motion at varying frequencies.

4.3 Tension—Tension may be inherent in brake hose routings and should be minimized. The "fixed" frame or body attachment point for the hose is usually selected to balance full jounce position and full rebound position tensions, if other conditions permit.

4.4 Severe Bends—Severe bends may be inherent in brake hose routings and should be minimized. In some routings, for example, where an intermediate attachment of the hose is made between end connections, a section of the hose may be in compression. Aside from inconvenience in attaching the hose assembly to the vehicle, compressive force is not detrimental to the hose.

4.5 Twist—Twist may be inherent in hose attachment. Steering and/or suspension movements may accent the twist characteristic. Depending upon the design and type of suspension, a change in twist may also be minimized by vertical orientation of the fixed end connection.

4.6 Length—Length is determined by extremes of relative movement of the hose assembly and connection points. The two connection points are located by consideration of the following factors: clearance, manufacturing feasibility (vehicle assembly), vehicle servicing, vertical and rotational (steering) movement of the wheel, and location of tie rod, suspension spring, shock absorber (strut or damper), suspension control arm, etc. Length may need to be adjusted to limit the minimum bend radius and/or minimize tension.

4.7 Trapped Fluid Systems (Sustained High Pressure)—Vehicle manufacturer's recommendations should be considered in any hydraulic brake hose application in which brake fluid can be trapped and result in sustained pressure, as in hydraulic parking brake, antitheft, or antilock brake systems.

4.8 Heat—Deterioration of brake hose is generally a direct function of temperature over time. Within the space limitations of the routing envelope, exposure to heat sources (engine compartment, exhaust manifold, catalytic converter, muffler, and pipe system) may be minimized by hose location or shielding. Consideration should be given to prevent the exposure of the brake hose to heat due to an exhaust system failure.

4.9 Cold—Little can be done by means of routing to protect the hose from the effects of extreme cold beyond avoiding tight bends.

4.10 Installation—Fittings, brackets, and mating components should be designed to optimize correct installation to the vehicle and avoid operator sensitive conditions.

5. Design Verification—Static design verification should be performed on vehicles manufactured to the extremes of the tolerance range. Since it is impractical for the vehicle manufacturer to build vehicles to the extremes of the vehicle tolerance range, it may be desirable to verify prototype, pilot build, and early production vehicles to consider the effects of vehicle design tolerance ranges. In addition, it is advisable to perform durability tests to verify no adverse effects due to vehicle dynamics.

The static design verification should be performed for all combinations of the extremes of travel using hose assemblies that encompass the extremes of the assembly length and end fitting orientation tolerance ranges.

Recommended guidelines for static design verification are:
a. Minimum hose outside diameter bend radius: 25 mm
b. Minimum hose clearance to rotating components: 19 mm
c. Minimum hose clearance to nonrotating components: 13 mm
d. Where contact between hose and other components is positively prevented by stop, minimum clearance: 5 mm
e. Minimum hose clearance to heat sources: 25 mm
f. Hose should not be taut in any position as demonstrated by the ability to physically rotate the center section of the hose about a 25 mm swing radius.

Mechanical means may sometimes be added to the brake hose exterior to avoid abrasion when touch conditions are expected to occur during extremes of travel, or to direct the hose to improve clearances. These devices may include, but not be limited to rings, helixes, or sleeves.

6. Operating Factors—Some factors to which the hose assembly may be exposed in operation are:
a. Mud—If in splash path
b. Water—Rain water and salt water
c. Slush—Ice and snow with or without salt and/or cinders and sand
d. Road oil
e. Sun and ultraviolet exposure
f. Ozone
g. Detergent, degreaser, wax, hot water, and steam—Vehicle wash operations
h. Rustproofing and undercoating materials—Depending on proximity to the surfaces being treated or ease of protecting during undercoating
i. Road debris—If practicable, the brake hose should be routed to be protected by less vulnerable chassis components. For off-road vehicles this would include ruts, underbrush, tree stumps, tree branches, rocks, sand, gravel, barbed wire, etc. This guideline may be satisfied by keeping the lowest part of the hose above the wheel axis or skid plate.

The routing should be evaluated for effects of foreign material accumulating on the hose assembly.

7. Vehicle Assembly and Servicing Factors—A brake hose routing may provide adequate clearances and satisfactory operating conditions and yet fail because of vulnerability to damage in assembling or servicing the vehicle. Exercise caution to avoid:

7.1 Hooking or Pinching—The hose may be hooked by towing or lifting equipment or pinched between a chassis component and service equipment.

7.2 Stretching or Twisting—The hose may be overstressed if allowed to support the axle or caliper weight during spring and shock absorber servicing or brake pad replacement. The potential exists that the hose may be twisted 360 degrees (and routing clearances negated) if the caliper is rotated upon reinstallation.

7.3 Cutting or Abrasion—Sharp-edged tools, files, drills, grinding wheels, wire brushes, etc., should be used carefully when near brake hoses. The hose may also be vulnerable to damage by sharp-edged sheet metal forced into contact with the hose in order to gain access to some other vehicle part.

7.4 Misrouting—A satisfactory hose routing may be negated by inadequate strength of hose attachment brackets, which may be susceptible to bending during servicing of the hose or brake components, thereby adversely changing brake hose clearance.

7.5 Heat—Welding or cutting torches used carelessly near brake hoses may impair hose performance.

7.6 Degreasing Compounds, Penetrating Oil, etc.—Petroleum products and other chemicals used in vehicle service or repair to clean, remove rust, remove paint, or loosen bolts and nuts may adversely affect brake hose durability.

8. Related and Interfacing Components—The design of related and interacting components should be restricted to prevent failure of these components from causing a hose failure.

HOT IMPULSE TEST FOR HYDRAULIC BRAKE HOSE ASSEMBLIES—SAE J1833 NOV88

SAE Recommended Practice

Report of the Hydraulic Brake Systems Actuating Committee approved November 1988. Rationale statement available.

1. Purpose—This SAE Recommended Practice presents an accelerated test to verify the structural integrity of hydraulic brake hose and the hose-to-fitting seal. It is intended to simulate the effects of environmental aging and braking pressurization on brake hose assemblies. The test is a guide to assist hose designers and/or users in determining brake hose assembly performance characteristics under conditions of heat and pressure.

2. Scope—This recommended practice describes the equipment, test procedure, and performance requirements for high temperature impulsing of automotive brake hose assemblies with hydraulic brake fluid.

3. Test Procedure

3.1 Test Equipment

3.1.1 PRESSURE CYCLING APPARATUS—The pressure cycling apparatus shall be capable of applying a pressure of 11 MPa (1600 psi). It shall have automatic control of the time for the pressure apply/release cycle.

3.1.2 CIRCULATING AIR OVEN—An insulated circulating air oven with a suitable thermostatically-controlled heating system is required to maintain a temperature of 143 ± 3°C (295 ± 5°F).

3.1.3 PRESSURE HOLD AND BURST STRENGTH TEST APPARATUS—An apparatus conforming to the requirements described in SAE J1401.

3.2 Connect the hose assemblies to a pressure cycling apparatus capable of producing a pressure of 0 - 11 MPa (0 - 1600 psi).

3.3 Fill the pressure cycling apparatus and hose assemblies with brake fluid, and bleed free of air.

3.4 Place the assemblies in a circulating air oven, and within 30 min attain an oven temperature of 143 ± 3°C (295 ± 5°F).

3.5 Subject the assemblies to a cycling internal pressure of 11 + 0.5, -0 MPa (1600 + 75, -0 psi) for 1 ± 0.1 min and 0 pressure for 1 ± 0.1 min; pressures to be attained within 2 seconds.

3.6 Pressure cycle assemblies for 150 cycles minimum.

3.7 Remove the assemblies from the oven. Disconnect the assemblies from the impulse apparatus, and drain the fluid.

3.8 Cool the assemblies in air at room temperature for 45 min minimum.

3.9 Fill the assemblies with water or brake fluid, allowing air to escape. Apply 27.6 + 0, -1.4 MPa (4000 + 0, -200 psi) of brake fluid pressure, and hold for 2 min + 0, -10 seconds.

3.10 Subject the assemblies to the burst test per SAE J1401.

4. Performance Requirements

4.1 The assemblies shall withstand impulsing for 150 cycles without leakage.

4.2 There shall be no leakage during a 2–min, 27.6 MPa (4000 psi) pressure hold.

4.3 The assembly shall not burst at less than 34.5 MPa (5000 psi).

MATERIALS FOR PLASTIC PISTONS FOR HYDRAULIC DISC BRAKE CYLINDERS—SAE J1568 JUN93

SAE Standard

Report of the SAE Hydraulic Brake Plastics Standards Committee approved June 1993.

1. Scope—The materials defined by this SAE Standard are glass-fiber-reinforced, mineral-filled phenolic molding compounds suitable for compression molding. Preforms may be radio frequency preheated or screw preheated slugs. Compound for use in hydraulic disc brake caliper pistons.

2. References

2.1 Applicable Documents—The following publications form a part of this specification to the extent specified herein.

2.1.1 ASTM PUBLICATIONS—Available from ASTM, 1916 Race Street, Philadelphia, PA 19103-5585.

ASTM D 543—Standard Test Method for Resistance of Plastics to Chemical Reagents
ASTM D 570—Standard Test Method for Water Absorption of Plastics
ASTM D 638—Test Method for Tensile Properties of Plastics
ASTM D 648—Standard Test Method for Deflection Temperature of Plastics Under Flexural Load
ASTM D 695—Standard Test Method for Comprehensive Properties of Rigid Plastics
ASTM D 696—Standard Test Method for Coefficient of Linear Thermal Expansion of Plastics Between -30 Degrees C and 30 Degrees C
ASTM D 785—Test Method for Rockwell Hardness of Plastics and Electrical Insulating Materials
ASTM D 790—Standard Test Method for Flexural Properties of Unreinforced and Reinforced Plastics and Electrical Insulating Materials
ASTM D 794—Practice for Determining Permanent Effect of Heat on Plastics

2.1.2 ISO PUBLICATIONS—Available from ANSI, 11 West 42nd Street, New York, NY 10036-8002.

ISO 62—Plastics—Determination of water absorption
ISO 175—Plastics—Determination of the effects of liquid chemicals, including water
ISO 178—Plastics—Determination of flexural properties of rigid plastics
ISO/R 527—Plastics—Determination of tensile properties
ISO 2039/2—Plastics—Determination of hardness—Rockwell hardness

3. General Material Requirements

3.1 Conditioning—All specimens shall be compression-molded and postcured at 205 °C ± 5 °C for 16 h in an air-circulating oven prior to testing.

4. Test Requirements

4.1 Hardness—The material when tested by the procedure specified in 5.1 shall have a Rockwell Hardness of 90 min.

4.2 Compressive Strength—The material when tested by the procedure specified in 5.2 shall have a Compressive Strength of not less than 187 MPa.

4.3 Tensile Strength—The material when tested by the procedure specified in 5.3 shall have a Tensile Strength of not less than 41 MPa.

4.4 Modulus of Elasticity—The material when tested by the procedure specified in 5.4 shall have a Tangent Modulus of Elasticity (Flexural Modulus) of not less than 19 GPa.

4.5 Flexural Strength—The material when tested by the procedure specified in 5.5 shall have a Flexural Strength of not less than 76 MPa.

4.6 Coefficient of Linear Thermal Expansion—The material when tested by the procedure specified in 5.6 shall have a Coefficient of Linear Thermal Expansion of not more than 20×10^{-6} mm/mm/°C.

4.7 Heat Aging—The material when tested by the procedure specified in 5.7 shall not crack or blister, shall have a volume change of not more than -0.6% and shall have a hardness change of not less than 0 nor more than +6.

4.8 Fluid Aging—The material when tested by the procedure specified in 5.8 shall have a weight change of not more than -0.5%, a volume change of not more than -0.3%, and a hardness change of not less than 0 nor more than +6.

4.9 Water Aging—The material when tested by the procedure specified in 5.9 shall have a weight change of not more than 0.15% and a volume change of not more than 0.20%.

4.10 Deflection Temperature—The material when tested by the procedure specified in 5.10 shall have a Deflection Temperature of not less than 250 °C.

5. Test Procedures

5.1 Hardness—Determine the Rockwell Hardness of the material by ASTM D 785 using scale E (ISO 2039/2).

5.2 Compressive Strength—Determine the Compressive Strength of the material by ASTM D 695 (ISO N/A).

5.3 Tensile Strength—Determine the Tensile Strength of the material by ASTM D 638 using a Type I specimen (ISO/R 527 Type 1 specimen).

5.4 Modulus of Elasticity—Determine the Tangent Modulus of Elasticity (Flexural Modulus) of the material by ASTM D 790 using Method I and Procedure A (ISO 178).

5.5 Flexural Strength—Determine the Flexural Strength of the material by ASTM D 790 using Method I and Procedure A (ISO 178).

5.6 Coefficient of Linear Thermal Expansion—Using a temperature range of 25 to 177 °C, determine the coefficient of linear thermal expansion by ASTM D 696 (ISO N/A).

5.7 Heat Aging—Determine the heat aging of the material by exposing a 102 mm diameter and 6.4 mm thick test specimen to a temperature of 177 °C ± 2 °C for 130 h in accordance with ASTM D 794 (ISO N/A).

5.8 Fluid Aging—Determine the fluid aging of the material by immersing a 51 mm diameter and 6.4 mm thick test specimen in SAE RM 66-03 brake fluid for 168 h at a temperature of 149 °C ± 2 °C in accordance with ASTM D 543 (ISO 175).

5.9 Water Aging—Determine the water aging of the material by immersing a 51 mm diameter and 6.4 mm thick test specimen in distilled water for 24 h at a temperature of 23 °C ± 1 °C in accordance with ASTM D 570 (ISO 62).

5.10 Deflection Temperature—Determine the Deflection Temperature of ASTM D 648 (ISO 75) using a test specimen that is 127 × 12.7 × 12.7 mm and a load of 1820 kPa.

(R) RUBBER SEALS FOR HYDRAULIC DISC BRAKE CYLINDERS—SAE J1603 JUN90

SAE Standard

Report of the Hydraulic Brake Systems Actuating Committee, approved April 1974, reaffirmed without change March 1985. Formerly SAE j62. Currently under revision by Committee. Completely revised by the Hydraulic Brake Elastomeric Standards Committee June 1990.

1. Scope—This SAE Standard describes the performance and part requirements for elastomeric seals used in highway vehicle disc brake calipers. Seals covered by this specification may be the solid section type (square, rectangular, O-ring, etc.) mounted stationary in the cylinder bore or on the movable piston. The specification contains the following major sections:

1. Resistance to Fluid at Elevated Temperature–Physical Stability [loose parts in 120°C ± 2 (248°F ± 3.6) brake fluid for 70 h].
2. Resistance to Fluid at Elevated Temperature–Precipitation Characteristics [loose parts in 120°C ± 2 (248°F ± 3.6) brake fluid for 70 h].
3. Resistance to Elevated Temperatures in Dry Air [loose Parts In 175°C ± 2 (347°F ± 3.6) air for 22 h].
4. Ambient Temperature Stroking Test [tested in brake assembly for 500 000 cycles to 7 MPa ± 0.3 (1000 psi ± 50)].
5. High Temperature Stroking Test [tested in brake assembly for 70 h (70 000 strokes) at 120°C ± 2 (248°F ± 3.6) to 7 MPa ± 0.3 (1000 psi ± 50)].
6. Low Temperature Leakage Test [tested in brake assembly for 120 h at -40 to 42.8°C (-40 to 45°F)].
7. Cycling Humidity Storage Corrosion Test [tested in brake assembly for 14 days at 95% humidity cycling between 21 to 46°C (69.8 to 114.8°F)].

2. References
  2.1 Applicable Documents—ASTM D 573

3. Parts Requirements
  3.1 Parts shall conform to the pertinent drawing in all respects.
  3.2 Manufacturer's identification, when used, shall be that on record at the Rubber Manufacturers Association.
  3.3 All parts to be tested shall be cleaned prior to testing by rinsing in isopropyl alcohol and blown dry or wiped dry with a lint-free cloth. Seals shall not remain in alcohol for more than 30s.

4. Brake Test Fluid—Fluid for all test phases except 7 (see 7.7) shall be SAE compatibility fluid, or a mutually agreed upon commercial fluid which subscribes to SAE J1703. Fluid used in test 7 shall be SAE compatibility fluid or a mutually agreed upon commercial preservative type fluid.

NOTE: RM66-03 is obtainable from SAE.

5. Test Equipment—Equipment used for testing shall be as shown in Figures 1 and 2.

6. Test Requirements
  6.1 After test 1, Resistance to Fluid at Elevated Temperature–Physical Stability, the parts must subscribe to the following:
  6.1.1 Volume change must be within 0 and +15%.
  6.1.2 IRHD hardness change must be within 0 and -15 points.
  6.2 After test 2, Resistance to Fluid at Elevated Temperature Precipitation Characteristics, the parts must subscribe to the following:
  6.2.1 Not more than 0.05% vol sediment shall be formed.
  6.3 After test 3, Resistance to Elevated Temperature in Dry Air, the parts must subscribe to the following:
  6.3.1 Change in IRHD hardness must be within 0 and +15 points.

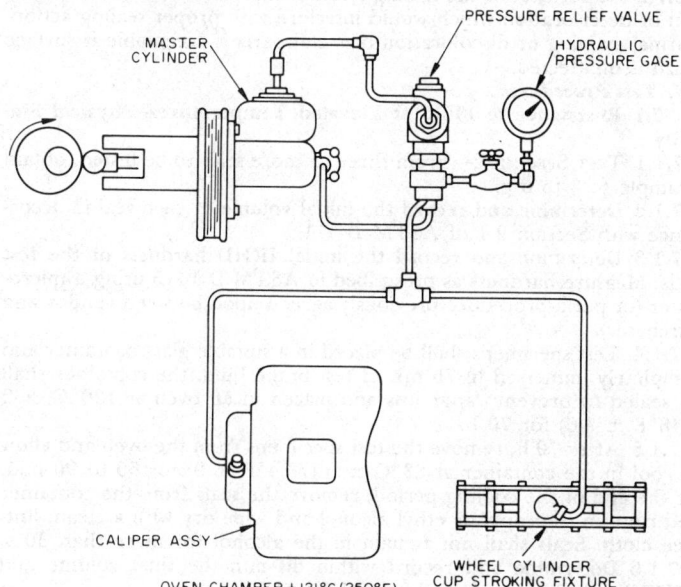

FIG. 1—HIGH TEMPERATURE STROKING TEST

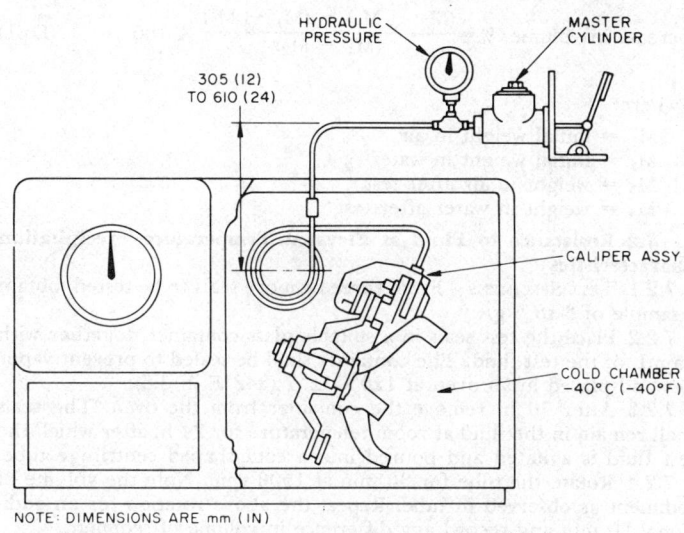

NOTE: DIMENSIONS ARE mm (IN)

FIG. 2—LOW TEMPERATURE LEAKAGE TEST

6.3.2 Seal Condition—Test parts shall show no evidence of tackiness, blistering, cracking, or change in shape from original appearance.

6.4 After test 4, Ambient Temperature Stroking Test, parts and assembly must subscribe to the following:

6.4.1 No leakage beyond normal wetting of the bore(s) is allowed during the stroking test.

6.4.2 No leakage beyond normal wetting of the bore(s) is allowed during static leak test (see 7.4.8).

6.5 After test 5, High Temperature Stroking Test, the parts must subscribe to the following:

6.5.1 No leakage beyond normal wetting of the bore(s) is allowed during the stroking test.

6.5.2 No leakage beyond normal wetting of the bore(s) is allowed during static leak test (see 7.5.11).

6.6 After test 6, Low Temperature Leakage Test, the parts and assemblies must subscribe to:

6.6.1 No leakage beyond normal wetting of the bore(s) is allowed during the test period or pressure applications.

6.6.2 The seal shall not crack and shall show evidence of rubber-like qualities during the flexibility bend test (see 7.6.10).

6.7 After test 7, Cycling Humidity Storage Corrosion Test, the parts and assembly must subscribe to the following:

6.7.1 No evidence of rubber adhesion of the test seal(s) is allowed during disassembly of the test brake.

6.7.2 No surfaces of the sealing systems shall show evidence of corrosion or deterioration which would interfere with proper sealing action. Normal staining or discoloration of metal parts is acceptable if surface finish is unaffected.

7. Test Procedures

7.1 Resistance to Fluid at Elevated Temperatures–Physical Stability

7.1.1 Test Specimens—From three or more seals to be tested, obtain a sample of 3 to 5 g.

7.1.2 Determine and record the initial volume of each seal in accordance with Section 9.1 of ASTM D 471.

7.1.3 Determine and record the initial IRHD hardness of the test seals. Measure hardness as prescribed in ASTM D 1415 using a microtester (or per a procedure previously agreed upon between vendor and purchaser).

7.1.4 Test specimens shall be placed in a suitable glass container and completely immersed in 75 mL of test brake fluid; the container shall be sealed to prevent vapor loss and placed in an oven at 120°C ± 2 (248°F ± 3.6) for 70 h.

7.1.5 After 70 h, remove the test specimens from the oven and allow to cool in the container at 23°C ± 5 (73.4°F ± 9) for 60 to 90 min. At the end of the cooling period, remove the seals from the container and rinse in isopropyl or ethyl alcohol and wipe dry with a clean, lint-free cloth. Seals shall not remain in the alcohol for more than 30 s.

7.1.6 Determine and record within 60 min the final volume and IRHD hardness of each seal per 7.1.2 and 7.1.3.

NOTE: The change in volume shall be reported at a percentage of the original volume. The calculations shall be made as follows:

$$\text{Increase in volume, \%} = \frac{(M_3 - M_4) - (M_1 - M_2)}{(M_1 - M_2)} \times 100 \quad \text{(Eq.1)}$$

where:

M_1 = initial weight in air
M_2 = initial weight in water
M_3 = weight in air after test
M_4 = weight in water after test

7.2 Resistance to Fluid at Elevated Temperature–Precipitation Characteristics

7.2.1 Test Specimens—From three or more seals to be tested, obtain a sample of 3 to 5 g.

7.2.2 Place the test seals in a suitable glass container, together with 75 mL of the test fluid. The container shall be sealed to prevent vapor loss and placed in an oven at 120°C ± 2 (248°F ± 3.6).

7.2.3 After 70 h, remove the container from the oven. The seals shall remain in the fluid at room temperature for 24 h, after which the test fluid is agitated and poured into a cone-shaped centrifuge tube.

7.2.4 Rotate the tube for 30 min at 1500 rpm. Note the volume of sediment as observed in tube. Repeat the above rotation for an additional 30 min and record any difference in volume of sediment.

7.2.5 Record amount of sediment per 6.2.1.

7.3 Resistance to Elevated Temperatures in Dry Air

7.3.1 Test Specimens—Two or more seals shall be used.

7.3.2 Measure and record the IRHD hardness of each seal per 7.1.3.

7.3.3 The test seals shall be placed in a circulating air oven, as prescribed in ASTM D 573 and held for 22 h at 175°C ± 2 (347°F ± 3.6).

7.3.4 At the termination of the heating period, the seals shall be removed from the oven and allowed to cool for 30 min to room temperature.

7.3.5 After cooling, measure and record the IRHD hardness per 7.1.3 and note any visual change such as cracking, blistering, distortion, etc.

7.4 Ambient Temperature Stroking Test

7.4.1 Test Specimens—Adequate test seals for at least one complete caliper shall be prepared.

7.4.2 Measure and record the IRHD hardness of the seal per 7.1.3.

7.4.3 Wet seals and caliper bores with the test brake fluid. Install test seals in caliper.

7.4.4 Complete the test caliper assembly, placing the piston to simulate half-worn lining position.

7.4.5 Assemble the test caliper assembly to a production spindle and disc assembly or equivalent simulating fixture.

7.4.6 Connect the test fixture to the pressure source. It may be necessary or desirable to include a fluid accumulator (such as the standard SAE J60 fixture).

7.4.7 Test Parameters

7.4.7.1 Temperature—18.3 to 32.2°C (65 to 90°F).

7.4.7.2 Pressure—Pressure will be applied to external means at a maximum rate-of-pressure rise of 21 MPa/s ± 1.4 (3000 psi/s ± 200) from 0 to 7 MPa ± 0.3 (0 to 1000 psi ± 40).

7.4.7.3 Cycles Required—500 000 total.

7.4.7.4 Cycle Rate—3600/h maximum.

7.4.8 Leakage Test—Observe leakage during and after the stroking test. After completion of the stroking test, run high and low pressure leak tests.

7.4.8.1 High Pressure Leak Test—Apply 0.7 MPa (100 psi) hydraulic pressure for 5 min and observe and record leakage, if any.

7.4.8.2 Low Pressure Leak Test—Remove the caliper from the test stand and connect the test caliper to a pressure source of 10 kPa ± 1.75 (1.5 psi ± 0.25) for 24 h. Observe leakage, if any.

NOTE: The low pressure source may be a static column of fluid. A 1200 mm (48 in) column will provide 10 kPa (1.5 psi).

7.4.9 Disassemble caliper and inspect seal. Record visual condition of seal, bore, and piston.

7.4.10 Wash seal per 3.3.

7.4.11 Measure and record IRHD hardness of the seal per 7.1.3.

7.5 High Temperature Stroking Test

7.5.1 Adequate test seals for at least one complete caliper shall be prepared.

7.5.2 Measure and record the IRHD hardness of seal per 7.1.3.

7.5.3 Wet seals and caliper bores with the test brake fluid. Install test seals in caliper.

7.5.4 Complete the test caliper assembly, placing the piston to simulate half-worn lining position.

7.5.5 Assemble the test caliper assembly to a production spindle and disc assembly or equivalent simulating fixture.

7.5.6 Place complete test fixture in an oven conforming to Section 4 of ASTM D 573. (Also, see Figure 1.)

7.5.7 Connect to the actuating pressure device. Device may be composed of a mechanically, pneumatically, or hydraulically actuated automotive type master cylinder whose rate of operation shall be controlled at 1000 strokes/h ± 100. The test fixture shall be connected to the actuating pressure device in conjunction with a suitable pressure relief valve (if mechanically actuated) and arranged in such a manner as to yield a maximum rate-of-pressure rise of 7 MPa/s (1000 psi/s) and a minimum dwell period below 0.18 MPa (25 psi) of 0.25 s. (It may be found necessary to install a fluid accumulator, such as a standard wheel cylinder in an SAE 60R2 fixture, to meet the prescribed curve.)

7.5.8 Test Parameters

7.5.8.1 Temperature—120°C ± 2 (248°F ± 3.6).

7.5.8.2 Pressure—7 MPa ± 0.3 (100 psi ± 50) at a rate-of-pressure rise of 7 MPa/s (1000 psi/s) maximum.

7.5.8.3 Elapsed Time—70 h.

7.5.8.4 Cycles Required—70 000 ± 5000.

7.5.9 After 70 h, discontinue stroking, shut off heat, open oven door, release hydraulic pressures in system, and allow oven to cool for 60 min. The circulating fan may be left on to aid in cooling.

7.5.10 After the 60 min cooling period, remove the test assembly and allow to complete cooling in open air for 25 h ± 5.

7.5.11 Leakage Test—Observe leakage during and after the 70 h stroking test. After completion of the 25 h cooling period, run high and low pressure leak test.

7.5.11.1 High Pressure Leak Test—Apply 0.7 MPa (100 psi) hydraulic pressure for 5 min and observe and record leakage, if any.

7.5.11.2 Low Pressure Leak Test—Remove the caliper from the test stand and connect the test caliper to a pressure source of 10 kPa ± 3.3 (1.5 psi ± 0.5) for 24 h. Observe leakage, if any.

NOTE: The low pressure source may be a static column of fluid. A 1200 mm (48 in) column will provide 10 kPa (1.5 psi).

7.5.12 Wash seals per 3.3.

7.5.13 Measure and record IRHD hardness of the seal per 7.1.3.

7.6 Low Temperature Leakage Test

7.6.1 Adequate test seals for at least one complete caliper shall be prepared.

7.6.2 Wet the seals and caliper bores with the test brake fluid. Install test seals in caliper.

7.6.3 Complete the test caliper assembly, placing the piston to simulate new lining position. Accommodations must be made to change piston position during the cold test to simulate new, half, two-thirds, and full-worn lining positions.

7.6.4 Assemble the test caliper assembly to a production spindle and disc assembly or equivalent simulating fixture.

7.6.5 Place the test fixture in a -40 to 43°C (-40 to 45.4°F) cold chamber and connect to a pressure source as shown in Figure 2. Pressure source shall be located to provide a static reservoir head of 300 to 600 mm (12 to 24 in).

7.6.6 Allow the caliper to soak for 72 h with the piston at new lining position.

7.6.7 After 72 h, stroke the actuating mechanism 6 times at 1 MPa ± 0.07 (150 psi ± 10) followed by 6 times at 4.2 MPa ± 0.35 (600 psi ± 50). Stroke shall be held for approximately 5 s and applied approximately 60 s apart. Immediately after stroking, remove the first shims and by means of the stroking mechanism, move the pistons into the half-worn lining position using minimum line pressure to establish the new location for all pistons. Observe and record leakage, if any, 30 min after new position is established. Allow the test caliper to continue to soak for 24 h.

7.6.8 After 96 h, repeat 7.6.7, except progress to two-thirds worn lining piston position.

7.6.9 After 120 h, repeat 7.6.7, except progress to full-worn lining piston position and discontinue test 30 min after establishing final piston position.

7.6.10 Bend Test Procedure

7.6.10.1 Place one seal in a test chamber at -40 to 43°C (-40 to 45.4°F).

7.6.10.2 After 22 h, the seal shall be folded back upon itself between the thumb and finger and released within 2 to 5 s. The cold seal shall be folded while in the cold chamber and shall be handled with cold gloves to prevent heating by fingers.

7.7 Cycling Humidity Storage Corrosion Test

7.7.1 Adequate test seals for at least one complete caliper shall be prepared.

7.7.2 Wet seals and caliper with test brake fluid. Install test seals in caliper.

7.7.3 Complete the test caliper assembly, placing the piston to simulate half-worn lining position. Caliper assembly need not be assembled to spindle or test fixture as long as provisions are made to hold the pistons in their proper positions and the boots are properly installed.

7.7.4 Place the test caliper in a humidity chamber capable of 95% relative humidity and a temperature range of 21 to 46°C (70 to 115°F). Caliper should be placed with inlet port open and facing down.

7.7.5 Hold the caliper at 43 to 46°C (110 to 115°F) and 95% humidity for 16 h.

7.7.6 Change temperature to 18 to 21°C (65 to 70°F) while maintaining 95% relative humidity, and hold for 8 h.

7.7.7 Continue above 24 h cycle for 12 days. When interrupted during incidence of one or more nonworking days, hold per 7.7.6 until temperature cycling can be resumed.

7.7.8 At the conclusion of the 12 day test, remove the test caliper for disassembly and inspection. Do not rotate caliper, and where possible, disassemble while holding in the test position.

7.7.9 Inspect and note all components for corrosion, pitting, adhesion, and other deleterious factors resulting from corrosion and/or interaction between the materials involved.

APPENDIX A—SAE RM-66-03 COMPATIBILITY FLUID[1]

A1. This fluid is a blend of four proprietary polyglycol brake fluids of fixed composition, in equal parts by volume. The four fluids selected comprise three factory-fill and one aftermarket fluid, as follows:
 a. DOW HD50-4
 b. Delco Supreme II
 c. DOW 455
 d. Olin HDS-79

---

[1] Obtainable from the Society of Automotive Engineers, Inc. 400 Commonwealth Drive, Warrendale, PA 15096.

(R) BRAKE MASTER CYLINDER RESERVOIR DIAPHRAGM GASKET—J1605 MAR92

SAE Standard

Report of the Hydraulic Brake Systems Actuating Committee, approved April 1974, reaffirmed without change March 1985. Currently under revision by Committee. Formerly SAE J66. Completely revised by the SAE Hydraulic Brake Elastomeric Standards Committee March 1992.

1. Scope—This SAE Standard covers performance requirements and methods of test for master cylinder reservoir diaphragm gaskets that will provide a functional seal and protection from outside dirt and water.

2. References

2.1 Applicable Documents—The following publications form a part of this specification to the extent specified herein. The latest issue of SAE publications shall apply.

2.1.1 SAE PUBLICATIONS—Available from SAE, 400 Commonwealth Drive, Warrendale, PA 15096-0001.

SAE J1601—Rubber Cups for Hydraulic Actuating Cylinders
SAE J1703—Motor Vehicle Brake Fluid

2.1.2 ASTM PUBLICATIONS—Available from ASTM, 1916 Race Street, Philadelphia, PA 19103.

ASTM D 395—Test Methods for Rubber Property—Compression Set
ASTM D 412—Test Methods for Rubber Properties in Tension
ASTM D 471—Test Method for Rubber Property—Effect of Liquids
ASTM D 573—Test Method for Rubber—Deterioration in an Air Oven
ASTM D 1149—Test Method for Rubber Deterioration—Surface Ozone Cracking in a Chamber (Flat Specimens)
ASTM D 1415—Test Method for Rubber Property—International Hardness

3. General Material Requirements

3.1 Composition—The materials used in the diaphragm gaskets shall be a rubber elastomer or combinations of elastomers and moisture barrier materials suitable for use with motor vehicle brake fluids of the nonpetroleum type conforming to SAE J1703.

3.2 Workmanship and Finish—The diaphragm gaskets shall be free from blisters, pin holes, cracks, embedded foreign material, or other physical defects, and shall conform to the dimensions specified on the drawings.

3.3 Marking—The identification mark of the manufacturer, as designated by the Rubber Manufacturers Association, and other details as specified on the drawing shall be molded into each diaphragm gasket.

4. Inspection and Rejection

4.1 All tests and inspections shall be made at the place of manufacture prior to shipment, unless otherwise specified. The manufacturer shall afford the inspector all reasonable facilities for the tests and specimens.

4.2 The purchaser may make tests and inspections to govern the acceptance or rejection of the diaphragm gaskets at a laboratory of choice. Such tests and inspections shall be made not later than 60 days after receipt of the material.

4.3 Any lot which fails to conform to one or more of the test requirements on the first sampling may be retested. For this purpose, two additional tests shall be made for the requirement in which failure occurred. Failure of either of the retests shall be cause for final rejection.

4.4 All rejected diaphragm gaskets shall be destroyed by the manufacturer to insure that no substandard diaphragms will be used.

5. Classification of Tests

5.1 Qualification Tests—The qualification tests shall include all tests specified herein.

5.2 Lot Acceptance Tests—Quality control tests for production lot acceptance shall include tests specified in 6.1.2, 6.3, and 6.4.1.

5.3 Sampling and Test Frequency—The quantity of parts and the frequency of qualification tests and lot acceptance tests used to control production shall be agreed upon by the supplier and purchaser.

6. Physical Properties Requirements

6.1 Rubber Hardness

6.1.1 QUALIFICATION—When tested as specified in 7.1, the rubber hardness shall be within the limits of 45 to 67 points. Hardness determinations from any one lot shall not vary more than ±5 points.

6.1.2 LOT ACCEPTANCE—When tested as specified in 7.1, the rubber hardness shall be equal to the qualifying value within ±5 points providing it is within the limits of 45 to 67 points.

6.2 Rubber Tension and Tear

6.2.1 QUALIFICATION—ORIGINAL PROPERTIES—When tested per ASTM D 412 and ASTM D 624, the rubber materials shall meet the following requirements:

a. Tensile Strength: 6.9 MPa (1000 psi) min
b. Ultimate Elongation: 250% min
c. Tear, Die C: 21.9 kN/m (125 lb) min

6.2.2 QUALIFICATION—ACCELERATED AGING AT 100 °C (212 °F)—The change from the original properties and compression set after dry heat aging per 7.2 shall conform to the following requirements:

a. Change in Hardness (IRHD or Durometer): 0 to +10 points
b. Change in Tensile Strength: -20% max
c. Change in Ultimate Elongation: -30% max
d. Compression Set: 25% max

6.3 Low Temperature Bendability at -40 °C (-40 °F)—After the low temperature test as described in 7.3, the diaphragm shall not crack and shall return to its approximate original shape within 1 min.

NOTE—Residual kinking of some convolutions shall not be cause for rejection.

6.4 Resistance to Fluids at Elevated Temperature 120 °C (248 °F)—After being subjected to the test for resistance to fluids at elevated temperature per 7.4, the diaphragm materials shall conform to the following requirements:

6.4.1 LOT ACCEPTANCE
a. Change in Hardness (IRHD or Durometer): -10 to +5 points
b. Change in Volume: -10 to +20%
c. Precipitation: Not more than light

6.4.2 QUALIFICATION
a. Change in Tensile Strength: -30% max
b. Change in Ultimate Elongation: -55% max

6.5 Ozone Resistance—After being tested per 7.5, the surface of the diaphragm shall evidence no cracking, rupture, or other deterioration when examined under 2X magnification.

6.6 Heat Pressure Stroking—After being subjected to the heat pressure stroking test prescribed in 7.6, there shall be no fluid dampness on the top surface of the diaphragm except where there is contact with metal.

6.7 Functional Design Test—After being subjected to the functional design test prescribed in 7.7, the convolutions of the diaphragm shall be fully extended, or the brake fluid level in the master cylinder reservoir shall not be greater than 3.2 mm (1/8 in) above the master cylinder porting.

7. Test Procedures

7.1 Rubber Hardness—The referee method of determining rubber hardness shall be as described in ASTM D 1415. An alternate procedure, as agreed upon between vendor and purchaser, may be used. Test each specimen submitted for test; record the range of IRHD (degrees).

Sample diaphragms, segments thereof, or specimens mutually acceptable to supplier and purchaser shall be plied together as necessary to provide the test thickness. The same operator shall make all hardness determinations for any one test.

7.2 Accelerated Aging at 100 °C (212 °F)

7.2.1 TEST METHOD—Rubber Hardness, Tensile Strength, and Elongation Charge

7.2.1.1 *Apparatus*—Circulating air oven as specified by ASTM D 573.

7.2.1.2 *Procedure*—Sample diaphragms, specimen sections thereof, or accepted specimens shall be rinsed in isopropyl alcohol or equivalent and wiped dry with a lint-free cloth to remove dirt and packing debris. The specimen shall not remain in the alcohol for more than 30 s.

The rubber hardness of unaged test specimens shall be determined as specified in 7.1 and recorded. The unaged tensile strength and ultimate elongation shall be determined and recorded in accordance with ASTM D 412.

New test specimens shall be placed in a circulating air oven as described in ASTM D 573 and held for 70 h at 100 °C ± 2 °C (212 °F ± 3.6 °F). The specimen shall then be removed from the oven, placed on a clean dry table top and allowed to cool for 16 to 96 h. The specimens shall then be retested for hardness as specified in 7.1, and tensile strength and ultimate elongation as specified in ASTM D 412. The tensile strength and ultimate elongation change shall be calculated per ASTM D 573.

7.2.2 TEST METHOD—Compression Set—Sample diaphragms, plied sections thereof, or accepted specimens, mutually agreed upon between supplier and purchaser, shall be rinsed in isopropyl alcohol or equivalent and wiped dry with a lint-free cloth to remove dirt and packing debris. The specimens shall not remain in the alcohol for more than 30 s.

The compression set of the specimens shall then be determined as specified in ASTM D 395, Method B with a heat treatment period of 22 h at 100 °C ± 2 °C (212 °F ± 3.6 °F).

7.3 Low Temperature Bendability

Sample diaphragms or accepted specimens shall be rinsed in isopropyl alcohol or its equivalent and wiped dry with a lint-free cloth. The specimen shall not remain in the alcohol for more than 30 s. The specimen shall then be placed in a suitable cold chamber and exposed for 22 h at -40 °C to -43 °C (-40 °F to 45 °F). It shall then be bent 180 degrees around a 6.35 mm (1/4 in) mandrel, conditioned at the test temperature, and released immediately. The time required for the part to return to its approximate original shape shall be noted. (The cold specimen shall be bent while in the cold chamber with cold gloves to prevent heating by the fingers.)

7.4 Resistance to Fluids at Elevated Temperature

7.4.1 APPARATUS—Circulating air oven as specified in ASTM D 573. Use tightly sealed glass containers of suitable size.

7.4.2 SPECIMEN—Sample diaphragms or acceptable specimens providing a suitable sample size weighing approximately 3 to 5 g shall be used for the rubber hardness, volume change, and precipitation test. Suitably prepared specimens per ASTM D 412 shall be used for the tensile strength and elongation change tests.

The specimens shall be stabilized at room temperature and then rinsed in isopropyl alcohol or its equivalent and wiped dry with a lint-free cloth to remove dirt and packing debris. The specimens shall not remain in the alcohol for more than 30 s.

7.4.3 TEST METHODS—Rubber Hardness and Volume Change—Select two test specimens for test. Establish the initial rubber hardness per 7.1 and record.

Establish the volume of the same two specimens per ASTM D 471, using the water displacement method and record. The specimens shall then be quickly dipped in alcohol to remove the water and dried with a lint-free cloth. Immediately after drying, each specimen shall be placed in a container and completely immersed in 75 mL of the specified test fluid RM-66-03[1]. Containers shall be sealed to prevent vapor loss, placed in the oven, and held at 120 °C ± 2 °C (248 °F ± 3.6 °F) for 70 h. At the end of the heating period, remove the specimen and container from the oven and allow to cool at 23 °C ± 5 °C (73 °F ± 3.6 °F) for 60 to 90 min. The specimens shall then be removed from their containers, rinsed in isopropyl alcohol, and wiped dry with a lint-free cloth. The final volume and hardness shall be determined within 60 min after rinsing in alcohol.

The weighings shall be the last operation before and the first operation after the immersion in the brake fluid and followed by a rubber hardness determination.

The volume change shall be calculated per ASTM D 471 as follows in Equation 1 to determine conformance to 6.4.1.

$$\text{Change in volume, \%} = \frac{(m_3 - m_4) - (m_1 - m_2)}{(m_1 - m_2)} \times 100 \quad \text{(Eq. 1)}$$

where:
- m_1 = initial mass in air
- m_2 = initial mass in water
- m_3 = final mass in air
- m_4 = final mass in water

7.4.4 TEST METHOD—PRECIPITATION—Two specimens approximately 3 to 5 g each shall be placed in a suitable glass container and completely immersed in 75 mL of the specified test fluid RM-66-03[2]. The container shall be sealed to prevent vapor loss and placed in the oven at 120 °C ± 2 °C (248 °F ± 3.6 °F) for 70 h. At the end of the heating period, remove the container from the oven and with the diaphragm specimens still in the fluid, allow it to stand at room temperature for 22 h. The resulting precipitate shall be classified as none, trace, light, medium, or heavy. (It is recommended that a blank be run on the fluid with each series of tests.)

a. A trace precipitate shall be flocculent and remain in suspension.
b. A light precipitate shall be flocculent, may settle, but possess no crystalline formations.
c. A medium precipitate shall be primarily flocculent material with a few small crystals.
d. A heavy precipitate shall be large in volume and contain both flocculent and crystalline material.

7.4.5 TEST METHOD—TENSILE STRENGTH AND ELONGATION—Sample diaphragms, sections thereof, or accepted specimens as specified in ASTM D 412 for tensile strength and ultimate elongation shall be immersed in the specified test fluid RM-66-03[2] for 70 h at 120 °C ± 2 °C (248 °F ± 3.6 °F). At the end of the heating period, remove the specimens and container from the oven and allow to cool at 23 °C ± 5 °C (73 °F ± 9 °F) for 60 to 90 min. The specimens shall then be removed from their container, rinsed in isopropyl alcohol, wiped dry with a lint-free cloth, and the tensile strength and ultimate elongation determined within 30 min after removal from the test fluid container.

The change in tensile strength and elongation shall be expressed as a percentage of change from the original value as determined in 6.2.1.

7.5 Ozone Resistance

7.5.1 APPARATUS—The apparatus shall consist of an ozone test chamber as described in ASTM D 1149 and capable of maintaining an ozone concentration of 50 mPa by volume.

7.5.2 SPECIMEN—The diaphragm test specimens shall be segments cut from the gasket area of the diaphragm or any substantially flat area.

7.5.3 TEST METHOD—The specimens shall be mounted flat and clamped at each end to provide a 15% stretch. They shall be allowed to rest in this position for 22 h at room temperature and then exposed in the test chamber to an ozone concentration of 50 mPa ± 5 mPa by volume at 40 °C ± 2 °C (104 °F ± 3.6 °F) for a period of 70 h.

7.6 Heat Pressure Stroking

7.6.1 APPARATUS

7.6.1.1 *Oven*—A well-designed uniformly heated standard dry air oven conforming to the requirements of ASTM D 573. The oven shall suitably contain the master cylinder as mounted on the stroking fixture.

7.6.1.2 *Stroking Fixture*—A pressure stroking device for actuating a master cylinder containing the specimen, at a rate of 1000 strokes/h ± 50 strokes/h and provide a master cylinder piston movement of 90% of the total available master cylinder stroke or 50.8 mm (2 in) maximum. It shall conform in general to the stroking test apparatus for SAE J1601, except that provision shall be made to return fluid from the bypass valve to the master cylinder reservoir below the diaphragm.

7.6.2 TEST PROCEDURE—A master cylinder designed for use with the test diaphragm shall be assembled on the stroking fixture and connected to a displacement apparatus which allows the master cylinder primary cup(s) to pass over the compensating port(s) at a pressure not exceeding 1.03 MPa (150 psi). Provision shall be made to return the fluid from the bypass valve to the master cylinder below the diaphragm.

The master cylinder reservoir shall then be half filled with the specified test fluid RM-66-03[3] and the test diaphragm and cap properly installed. The system shall then be actuated 70 h with the fluid temperature in the master cylinder maintained at 120 °C ± 2 °C (248 °F ± 3.6 °F). At the termination of the 70 h stroking period, the heat shall be shut off with the master cylinder in the off position. After a 1 h cooling period, the fluid line may be disconnected from the master cylinder and the cylinder removed from the oven. After a 21 h additional cooling period, the master cylinder reservoir cover shall be disassembled and the top of the diaphragm inspected for dampness.

7.7 Functional Design Test

7.7.1 APPARATUS—A brake master cylinder of the proper configuration to accept the test diaphragm.

7.7.2 TEST PROCEDURE—The master cylinder reservoir and cylinder bore shall be filled with the specified brake fluid RM-66-03[3] and the test diaphragm assembled. The master cylinder outlet port shall be connected to an open reservoir 305 mm (1 ft) below the master cylinder. The master cylinder shall then be actuated at room temperature at a rate not exceeding 200 strokes/h until the hydraulic brake fluid is no longer expelled from the master cylinder outlet(s). The cover shall then be removed from the master cylinder and the deflection of the diaphragm and the fluid level in the reservoir noted.

NOTE: An inspection hole may be made through the master cylinder cover to facilitate inspection of the diaphragm convolutions and deflection.

---

[1] SAE RM-66-03 is obtainable from SAE, 400 Commonwealth Drive, Warrendale, PA 15096-0001.

[2] SAE RM-66-03 is obtainable from SAE, 400 Commonwealth Drive, Warrendale, PA 15096-0001.

[3] SAE RM-66-03 is obtainable from SAE, 400 Commonwealth Drive, Warrendale, PA 15096-0001.

(R) RUBBER BOOTS FOR DRUM-TYPE HYDRAULIC BRAKE WHEEL CYLINDERS—SAE J1604 OCT89

SAE Standard

Report of the Hydraulic Brake Systems Actuating Committee, approved June 1976, reaffirmed without change March 1985. Currently under revision by Committee. Completely revised by the Hydraulic Brake Elastomeric Standards Committee October 1989.

1. Scope—This document covers molded rubber boots used as end closures on drum-type wheel brake actuating cylinders to prevent the entrance of dirt and moisture, which could cause corrosion and otherwise impair wheel brake operation.

The document includes performance tests of brake cylinder boots of both plain and insert types under specified conditions and does not include requirements relating to chemical composition, tensile strength, or elongation of the rubber compound. Further, it does not cover the strength of the adhesion of rubber to the insert material where an insert is used.

The rubber material used in these boots is classified as suitable for operation in a temperature range of -40 to $+120 \pm 2°C$ (-40 to $+248 \pm 3.6°F$).

2. General Requirements

2.1 Workmanship and Finish—Boots shall be free from blisters, pinholes, cracks, protuberances, embedded foreign material, or other physical defects, which can be detected by thorough inspection, and shall conform to the dimensions specified on the drawings.

2.2 Marking—The identification mark of the manufacturer as recorded by the Rubber Manufacturers Association, and other details, as specified on drawings, shall be molded into each boot.

2.3 Packaging—Boots shall be packaged to meet requirements specified by the purchaser.

2.4 Sampling—The minimum lot on which complete specification tests shall be conducted for quality control testing, or the frequency of any specific type test used to control production, shall be agreed on by the manufacturer and the purchaser.

3. Test Requirements

3.1 Resistance to Fluids at Elevated Temperature—After being subjected to the test for resistance to fluids at elevated temperatures, as prescribed in 4.4, a boot shall conform to the following requirements:

Change in volume: -10 to $+15\%$
Change in hardness: -10 to $+10$ points

3.2 Heat Stroking Test—After stroking, as detailed in 4.5, a boot shall be free of flex cracks which extend through the wall thickness, and shall fit tightly around the cylinder and push rod.

3.3 Low Temperature Stroking Test—During stroking, as detailed in 4.6, a boot shall not crack or separate from its assembled position on the cylinder or become loose on the push rod.

3.4 Tension Set Test—After being subjected to the tension set test prescribed in 4.7, a boot shall show no more than 75% tension set.

3.5 Heat Resistance Test (Static)—After the heat resistance test, as detailed in 4.8, a boot shall conform to the following requirements:

a. No cracking shall occur when flexed similarly to service conditions.
b. Change in hardness: -5 to $+10$ points

3.6 Ozone Resistance Test—At the end of the 70 h exposure period, as detailed in 4.9, test specimens shall be removed from the ozone chamber and examined under 2X magnification. Surface of the test specimens shall show no evidence of cracking, rupture, or other deterioration.

4. Test Procedures

4.1 Test Specimens—Specimens prepared for a particular test shall be cut from the same approximate location on different sample boots. Hardness test specimens shall be prepared to present a flat molded surface to the indentor.

4.2 Test Fluids—The brake fluid used for testing shall be RM 66-30.[1]

4.3 Hardness—The method of determining rubber hardness shall be as described in ASTM D 1415-83, Standard Test Method for Rubber Property—International Hardness. If agreed on by vendor and purchaser, an alternate procedure, ASTM D 2240-86, Standard Test Method for Rubber Property—Durometer Hardness, may be used.

4.4 Resistance to Fluids at Elevated Temperature

4.4.1 APPARATUS—Circulating air oven as specified in Section 5 of ASTM D 573-88, Standard Test Method for Rubber—Deterioration in an Air Oven.

Use a screw top, straight sided, round glass jar[2], having a capacity of approximately 250 cm^3 and inner dimensions of approximately 125 mm in height and 50 mm in diameter, and a tinned steel lid (no insert or organic coating).

4.4.2 TEST SPECIMENS—A section weighing approximately 3 to 5 g shall be cut from each of two boots.

4.4.3 PROCEDURE—The specimens shall be rinsed in isopropyl or ethyl alcohol and wiped dry with a clean, lint-free cloth to remove dirt and packing debris. Specimens shall not be left in the alcohol for more than 30 s.

Determine and record the initial hardness of the test specimens. Refer to 4.3.

The volume of each specimen shall be determined in the following manner:

Weigh each specimen in air (M_1) to the nearest milligram and then weigh the specimen immersed in distilled water[3] at room temperature (M_2). (A Jolly spring balance, adequately shielded from air currents, may be used for making these determinations.) Quickly dip each specimen in alcohol and then wipe dry with a clean lint-free cloth.

Two specimens shall be completely immersed in 75 cm^3 of the test fluid in the glass jar and tightly capped.

The jar shall be placed in an oven at $120°C \pm 2$ ($248°F \pm 3.6$) for a period of 70 h $\pm$ 2. At the end of the heating period, remove the specimens from the oven and allow to cool in the jar at $23°C \pm 5$ ($73.4°F \pm 9$) for 60 to 90 min. At the end of the cooling period, remove the specimens from the jar and rinse in isopropyl or ethyl alcohol and wipe dry with a clean, lint-free cloth. Specimens shall not remain in the alcohol for more than 30 s.

After removal from the alcohol and drying, place each specimen in a separate, tared, stoppered weighing bottle and weigh (M_3). Remove each specimen from its weighing bottle and weigh it immersed in distilled water[3] (M_4) to determine water displacement after hot fluid immersion.

The final volume and hardness of each specimen shall be determined within 60 min after rinsing alcohol.

4.4.4 CALCULATION AND REPORT

4.4.4.1 Volume change shall be reported as a percentage of the original volume. The calculation shall be made as follows:

$$\% \text{ change in volume} = \frac{(M_3 - M_4) - (M_1 - M_2)}{M_1 - M_2} \times 100$$

where:
M_1 = initial mass in air
M_2 = initial mass in water
M_3 = mass in air after immersion in test fluid
M_4 = mass in water after immersion in test fluid

4.4.4.2 Change in hardness shall be determined and recorded.

4.4.4.3 Examine the specimens for disintegration as evidenced by blisters or sloughing.

4.5 Heat Stroking Test

4.5.1 APPARATUS—Circulating air oven as specified in Section 5 of ASTM D 573.

Stroking fixtures as shown in Fig. 1 of SAE J1601 or equivalent.

4.5.2 TEST SPECIMENS—Two boots shall be used as test specimens.

4.5.3 PROCEDURE—Install two sample wheel cylinder boots on the cylinder for which they are designed, or equivalent. Then mount the cylinder into the actuator assembly, operating at 1000 strokes/h with a stroke length of 3.8 mm $\pm$ 1.7 (0.15 in $\pm$ 0.07).

Then place the cylinder assembly in the oven and actuate for 22 h $\pm$ 1 at $120°C \pm 2$ ($248°F \pm 3.6$). After the actuation of the assembly, it shall be removed from the oven, allowed to cool to room temperature, and the boots examined for flex cracks and general appearance.

---

[1] Obtainable from the Society of Automotive Engineers, Inc., 400 Commonwealth Drive, Warrendale, PA 15096.

[2] Suitable effect on rubber test jars (SAE part no. RM-51) and tinned lids (SAE part no. RM-52) can be obtained from Society of Automotive Engineers, Inc., 400 Commonwealth Drive, Warrendale, PA 15096.

[3] A trace of a suitable wetting agent not large enough to significantly affect the specific weight of the water should be added to the distilled water to eliminate small air bubbles trapped on the rubber surface during the weighing process.

4.6 Low Temperature Stroking Test

4.6.1 APPARATUS—The cold chamber, in which the test specimens are exposed to the low temperature, shall be of sufficient size to contain the apparatus assembled with test specimens and so arranged as to permit the operator to check and operate it without removal from the chamber. It shall be capable of maintaining a uniform atmosphere of cold dry air within the specified temperature range of −40°C to −43 (−40°F to −45.4).

Stroking fixture shall be as shown in Fig. 4 of SAE J1601 or equivalent.

4.6.2 TEST SPECIMENS—Two boots shall be used as test specimens.

4.6.3 PROCEDURE—Install the sample wheel cylinder boots on the cylinder, for which they are designed, or equivalent. Place the test boots and test apparatus in a cold chamber and expose to a temperature of −40°C to −43 (−40°F to −45.4) for 70 h ± 2. Then stroke the boots with the stroking apparatus for six strokes, 30 s apart, without removal from the cold chamber.

4.7 Tension Set Test

4.7.1 APPARATUS—Circular stretching mandrel having a diameter which will expand by 15% one or the other of the sealing ends attached to the wheel cylinder or to the actuating rod. The mandrel diameter is calculated as 115% of the molded diameter of the chosen boot end. The molded diameter shall be calculated from the average of two measurements made at right angles to each other. The mandrel shall be provided with a smooth lead-in chamfer to prevent cutting of the rubber and shall itself have a polished machine finish (0.40 μm [16 μin] AA maximum).

Circulating air oven as specified in Section 5 of ASTM D 573.

4.7.2 PROCEDURE

a. Measure to the nearest 0.02 mm (0.001 in) the inside diameters of the ends of three specimen boots and record. Assemble on stretching mandrels. Place assemblies in the oven and age for 70 h ± 2 at 120°C ± 2 (248°F ± 3.6). Remove the assemblies and cool at room temperature for 1 h. Remove the boots. Allow to recover between ½ and 1 h. Again measure the diameter and record.

b. Calculate the tension set as a percentage of the original stretch deflection:

$$\text{Tension set} = \frac{D_2 - D_1}{D_3 - D_1} \times 100$$

where:
D_1 = unaged inside diameter of the boot
D_2 = aged inside diameter of the boot
D_3 = diameter of the stretching mandrel

4.7.3 No less than three specimens shall be tested.

4.8 Heat Resistance Test—Static

4.8.1 APPARATUS—Circulating air oven as specified in Section 5 of ASTM D 573.

4.8.2 PROCEDURE—Select two sample boots for the heat resistance test. Determine the initial hardness of the boots as detailed in 4.3. Suspend the test specimens in the oven for 22 h ± 1 at 120°C ± 2 (248°F ± 3.6). Remove them from oven, allow to cool at room temperature for 16 to 96 h, then check for hardness and flexibility.

4.9 Ozone Resistance Test

4.9.1 APPARATUS—The apparatus shall consist of an ozone chamber as described in ASTM D 1149-86, Standard Test Method for Rubber Deterioration—Surface Ozone Cracking in a Chamber, and shall be capable of maintaining an ozone concentration of 50 mPa ± 5 (50 parts of ozone/100 million parts of air by volume ± 5).

Stretching mandrel (see 4.7.1).

4.9.2 TEST SPECIMENS—Test specimens shall be two boots.

4.9.3 PROCEDURE—Assemble the boots on the stretching mandrels (which will provide 15 + 0, −3% stretch in the boot bead section) and allow to rest for 22 h ± 1 at room temperature, then subject the boots installed on the mandrels to an ozone concentration of 50 mPa ± 5 by volume at 40°C ± 2 (104°F ± 3.6) for 70 h ± 2.

(R) HYDRAULIC WHEEL CYLINDERS FOR AUTOMOTIVE DRUM BRAKES— SAE J101 DEC89

SAE Standard

Report of the Hydraulic Brake Actuating Systems Committee, approved March 1973, editorial change March 1980, reaffirmed without change March 1985. Completely revised by the Hydraulic Brake Actuating Components Standards Committee December 1989.

1. Scope—This document specifies minimum performance and durability requirements for satisfactory vehicle usage, and it is applicable to wheel cylinder assemblies from commercial production, after production shipment, shelf storage, and remanufacture (factory rebuild).

2. Type—This document applies to wheel cylinder assemblies used on hydraulically operated, drum-type brakes of highway vehicles. It covers such cylinders where they are employed in passenger car, truck, bus, and like brake systems utilizing motor vehicle brake fluids that conform to SAE J1703.

3. Requirements—A wheel cylinder assembly shall, when tested in accordance with the procedures of Section 6, meet the following requirements:

3.1 Unrestricted Apply and Release—Per 6.1.2, the piston(s) must move smoothly throughout full stroke after starting, and must be completely returned to original position on the fifth stroke within 30 s by the force of the piston return spring, or, in the absence of a return spring, by 5 psi (34 kPa) maximum air pressure at inlet port.

3.2 Ozone Resistance—Per 6.2, the boot(s) shall not be perforated or cracked through in any areas. (6.2 is a conditioning test of the wheel cylinder for other tests which follow, and the cylinder should be rejected if deterioration of the boot obviously precludes meeting subsequent test requirements.)

3.3 Applied Leakage

3.3.1 Per 6.3.1, there shall be no drop in pressure in excess of 1 psi (7 kPa) in a 30 s interval.

3.3.2 Per 6.3.2, there shall be no drop in pressure in excess of 50 psi (345 kPa) in a 30 s interval.

3.4 Physical Strength—Per 6.4.1, the gage shall show no abrupt pressure drop and the cylinder shall show no signs of mechanical failure.

3.5 Humidity Operation

3.5.1 Per 6.5.1, the wheel cylinder piston(s) must fully apply the load fixture(s) to its bottom or stop and allow it to return to the full release stop within a stroke cycle of the master cylinder.

3.5.2 Per 6.5.3, see requirements 3.3.1 and 3.3.2.

3.6 High Temperature Durability

3.6.1 Per 6.6.1, see requirement 3.5.1.

3.6.2 Per 6.6.2, there shall be no visible leakage at the wheel cylinder bleed screw or hydraulic connector(s) and leakage at each boot of the wheel cylinder shall not be measurable.

3.6.3 Per 6.6.3, leakage at each boot shall not exceed five drops.

3.6.4 Per 6.6.4, see requirements 3.3.1 and 3.3.2.

3.7 Cold Temperature Operation

3.7.1 Per 6.7.1, see requirement 3.5.1.

3.7.2 Per 6.7.2, there shall be no visible leakage at the wheel cylinder bleed screw or hydraulic connector(s), and leakage at each boot of the wheel cylinder shall not exceed five drops.

3.7.3 Per 6.7.3, see requirements 3.3.1 and 3.3.2.

3.8 Storage Corrosion Resistance

3.8.1 Per 6.8.1, there shall be no visible leakage at the wheel cylinder bleed screw or hydraulic connector(s), and leakage at each boot shall not exceed five drops.

3.8.2 Per 6.8.2, piston(s) must start to move at 40 psi (276 kPa) maximum pressure.

3.8.3 Per 6.8.3, the wheel cylinder piston(s) must fully apply the load fixture to its bottom or stop and return it to full release stop.

3.8.4 Per 6.8.4, see requirements 3.3.1 and 3.3.2.

3.9 Static Leakage—Per 6.9.1, there shall be no visible fluid leakage in the trap(s).

3.10 Examination

3.10.1 Per 6.10.1, wheel cylinders up to and including 2 in (50.8 mm) bore diameter shall have 0.080 in (2.03 mm) minimum diameter at the smallest opening of the hydraulic inlet(s).

3.10.2 Because these specifications spell out only minimum requirements, none are called for per 6.10.3.

4. Test Apparatus—The basic apparatus shall be that shown and as arranged in Figure 1 or equivalent. All hydraulic lines and fittings shall be of sufficient size as to permit unrestricted fluid flow to and from the test wheel cylinder(s). The apparatus shall operate per the following description and as called for in Section 6.

4.1 Master Cylinder Assembly—The master cylinder should be one commercially representative of the brake system(s) in which the test wheel cylinder(s) is(are) used. Its bore size and stroke will depend on the numbers and bore sizes of wheel cylinders to be stroke tested simultaneously.

The referee master cylinder of SAE J1703 is one commercially representative.[1] A shutoff valve shall be provided at the hydraulic outlet of the master cylinder.

4.2 Pressure Actuating Mechanism—The pressure actuating mechanism shall apply an axial force to the master cylinder push rod without side thrust, and it shall allow the pressure in the master cylinder to return to 0 psi (0 kPa) when it is in the released position.

Means must be provided for the actuating mechanism to stroke the master cylinder both singly and cyclically. For single stroke operation, the means must be capable of generating pressures in the master cylinder up to 3000 psi (20.7 MPa) and it must have adjustment such that pressures of 20, 1000, and 3000 psi (0.14, 6.9, and 20.7 MPa) can be held statically after they are achieved. For cyclic operation, the pressure actuating mechanism must be capable of generating pressures in the master cylinder up to 1000 psi (6.9 MPa) and have adjustments such that pressure can be peaked out at both 500 psi (3.4 MPa) and 1000 psi (6.9 MPa). Further, it must build up both of these pressures uniformly in 1.6 to 2.0 s, and be capable of doing so at any stroke of the master cylinder up to 90% of its total stroke. The pressure actuating mechanism, when releasing, must permit the full retraction of the master cylinder push rod. The means for cycling the pressure actuating mechanism shall permit adjustment of uniform apply/release strokes at rates of both 500 and 1000 cycles/h.

4.3 Load Fixture—The load fixture, such as that shown in Figure 2, must stroke 0.12 + 0.0 − 0.5 in (3.05 + 0.0 − 0.5 mm) in each load cell at a hydraulic pressure buildup in the wheel cylinder within the limits shown by the curves of Figure 3. It must accommodate both single- and double-ended wheel cylinders and mount them as they would be mounted on their brakes; it must accommodate both insert and socket-type wheel cylinder pistons; it must be capable of placing 4 deg ± 0.5 angularity on connecting link(s) with respect to the wheel cylinder bore longitudinal axis; and it must accommodate wheel cylinders of 5/8 to 2 in (15.9 to 50.8 mm) diameter. (This is a composite description of load fixture requirements. At least two load fixtures with appropriate differences in design would be necessary for testing all types and sizes of wheel cylinders.)

Ends of connecting links engaging wheel cylinder pistons must be per those links actually used on brakes by the vehicle manufacturer from the standpoint of their piston and boot fits. Locks must be provided for the connecting links at their load cell ends.

NOTE: Actual brake assemblies, such as shown in SAE J1703 or equivalent fixtures, such as shown in SAE J1603 may be used in place of the above fixture providing they meet its requirements and those of Section 6. Single-end wheel cylinders, such as those for two leading, floating shoe brakes, and whose reactive loads are taken by brake shoes as well as cylinder mountings, should be tested in pairs as part of a complete brake and drum assembly.

4.4 Instrumentation—Two hydraulic pressure measuring devices shall be employed, each equipped with a shutoff valve. One shall have a range of 0 to 30 psi (0 to 207 kPa) and the other shall have a range of 0 to 5000 psi (0 to 34.5 MPa). Both shall be of a type that requires negligible hydraulic displacement and is equipped with a bleeder.

A mechanical or electrical counter shall be used to record the number of strokes during cyclic operation.

A leak trap shall be provided for each wheel cylinder boot. It shall have minimum exposed area in order to minimize evaporation.

A 24 to 48 in (610 to 1220 mm) height standpipe shall be provided as shown in Figure 1.

4.5 Environmental Equipment

4.5.1 OZONE CHAMBER—An ozone chamber as described in ASTM D 1149-81, Standard Test Method for Rubber Deterioration - Surface Ozone Cracking in a Chamber (Flat Specimens). It must maintain 50 pphm by volume at 37.7°C ± 3 (100°F ± 5) ozone concentration.

4.5.2 HEATED AIR BATH CABINET—An insulated oven or cabinet having sufficient capacity to house the load fixtures of the test apparatus. A suitable thermostatically controlled heating system is required to maintain a temperature of 100°C ± 5 (212°F ± 9). Heaters shall be shielded to prevent direct radiation to the wheel cylinders.

4.5.3 COLD CHAMBER—A cold chamber shall be provided having sufficient capacity to house the load fixtures of the test apparatus. It shall be such that the apparatus can be checked and operated without removal, and it shall be capable of maintaining a uniform atmosphere of cold dry air at −40 + 2, −5°C (−40 + 3.6, −9°F) temperature.

4.5.4 HUMIDITY CABINET—A humidity cabinet having sufficient capacity to house the load fixtures of the test apparatus. It shall be capable of maintaining a relative humidity of 95% ± 3 at temperatures of 21°C ± 3 (70°F ± 5) and 46.1°C ± 3 (115°F ± 5).

4.6 Test Fittings and Material—Test hydraulic fluid shall be per SAE J1703. The compatibility fluid of SAE J1703 is recommended.[2]

Wheel cylinder bolts and their lock washers shall be of types used by the vehicle manufacturer. The hydraulic connector to the wheel cylinder, whether tubing or hose, shall also be of the type used by the vehicle manufacturer.

5. Test Samples—Wheel cylinders for test shall not have been used after manufacture or rebuild, and they shall not be disassembled prior to testing.

6. Test Setup and Procedures—Tests shall be conducted in the sequence shown and at room temperature except where otherwise specified. Wheel cylinders shall not be disassembled until after all tests are completed or unless testing is discontinued.

6.1 Unrestricted Apply and Release

6.1.1 Remove the shipping plug from the wheel cylinder hydraulic inlet port(s).

6.1.2 Fully stroke the cylinder five times by hand and allow it to return under the piston return spring load. In the absence of a piston return spring, apply 5 psi (34 kPa) air pressure to the inlet port until the cylinder returns to its original position, five times. Record time for return on fifth stroke.

NOTE: Avoid stroking the wheel cylinder cups into the bleed screw and/or hydraulic inlet openings during this test.

6.2 Ozone Resistance—Seal the hydraulic port(s) from the atmosphere, install connecting link(s) where required to seal the small diameter of the boot, and place the cylinder in the ozone chamber. Subject the cylinder to an ozone concentration of 50 pphm ± 5 by volume at 37.7°C ± 3 (100°F ± 5) for 50 h.

Remove the cylinder from the ozone chamber and visually inspect the boot(s) without disassembling.

6.3 Applied Leakage—Install the wheel cylinder on its load fixture and tighten the mounting bolts by hand. (See Note of 4.3 for single-end wheel cylinders.) Making certain that the piston(s) is(are) not beyond the brake release position and that the connecting link(s) has(have) 4 deg ± 0.5 angularity with the wheel cylinder bore longitudinal axis, tighten the mounting bolts to the nominal torque specified by the vehicle manufacturer. Adjust the connecting link(s) so that the piston(s) is(are) in the release position.

Assemble the hydraulic connector of the test apparatus to the wheel cylinder and tighten to the nominal torque specified by the vehicle manufacturer. Fill the test setup with new hydraulic fluid and bleed at all points in the system as necessary to remove air. Tighten the bleed screw to the nominal torque specified by the vehicle manufacturer.

6.3.1 Drop the connecting link lock(s) in place per Figure 2 and open the valve to the low pressure gage. Apply the master cylinder to build up 20 psi ± 1 (138 kPa ± 6.9) pressure in the system, shut off the valve to the master cylinder, and release the master cylinder.

Allow the pressure to the wheel cylinder to stabilize for 15 to 20 s, and then record the pressure at the beginning and end of a 30 s ± 1 interval.

6.3.2 Open the valve to the master cylinder, close the valve to the low pressure gage, and open the valve to the high-pressure gage. Apply the master cylinder to build up 1000 psi ± 100 (6.9 MPa ± 0.7) pressure in the system, shut off the valve to the master cylinder, and release

---

[1] Referee master cylinders, of 1-1/8 in (28.6 mm) diameter bore and 1-7/16 in (36.5 mm) total stroke, may be obtained from Society of Automotive Engineers, Inc., 400 Commonwealth Drive, Warrendale, PA 15096.

[2] Compatibility fluid may be obtained from Society of Automotive Engineers, Inc., 400 Commonwealth Drive, Warrendale, PA 15096.

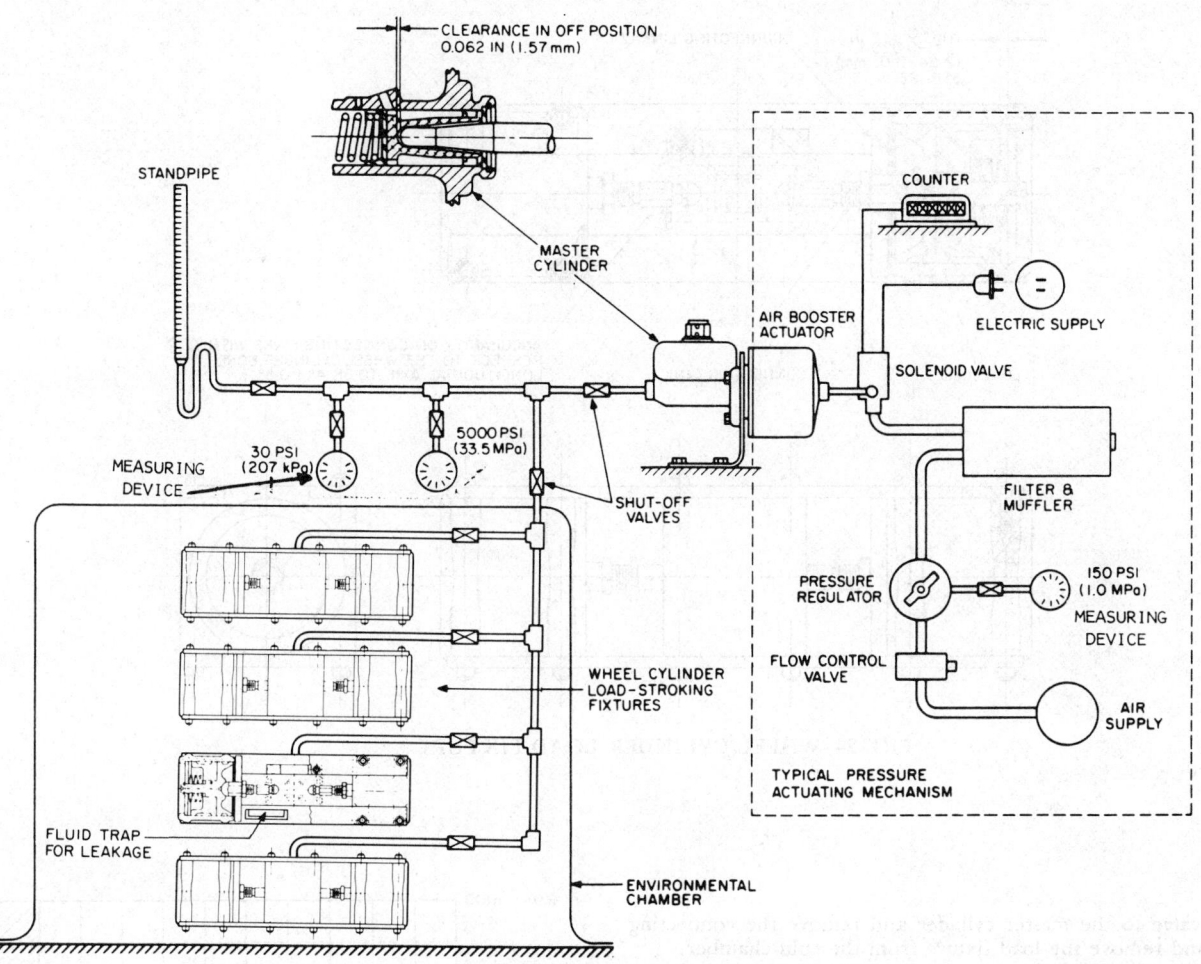

FIG. 1—TEST APPARATUS

the master cylinder. Allow the pressure to the wheel cylinder to stabilize for 15 to 20 s, and then record the pressure at the beginning and end of a 30 s ± 1 interval.

6.4 Physical Strength—Open the valve to the master cylinder and apply the master cylinder to build up 3000 psi ± 300 (20.7 MPa ± 2) pressure in the system. Hold the pressure for 15 s ± 5 and then release the master cylinder.

6.4.1 Observe the pressure gage during the test, and visually inspect the wheel cylinder and its mounting afterward for signs of leaks or structural failure.

Remove the connecting link lock(s).

6.5 Humidity Operation—Place the load fixture(s) with the wheel cylinder(s) in the humidity cabinet. Set the pressure actuating mechanism for cyclic operation at 1000 cycles/h ± 100 (3.27 s/cycle to 4.00 of apply and release stroke), and adjust it to build up a master cylinder output pressure of 500 psi ± 50 (3.4 MPa ± 0.3). Stroke 8 h at 46.1°C ± 3 (115°F ± 5) temperature and 95% ± 3 relative humidity; and then cease stroking for 16 h while at 21°C ± 3 (70°F ± 5) temperature and resultant relative humidity. Repeat this sequence.

6.5.1 Periodically observe the wheel cylinder action during stroking.

6.5.2 Remove the load fixture(s) with the wheel cylinder(s) from the humidity cabinet at the end of the second day (16 000 cycles stroking and 32 h static).

DO NOT DISTURB WHEEL CYLINDER BOOT(S).

6.5.3 Repeat Procedures 6.3.1 and 6.3.2.

Open the valve to the master cylinder and remove the connecting link lock(s).

6.6 High Temperature Durability—Set the pressure actuating mechanism for cyclic operation at 1000 cycles/h ± 100 (3.27 to 4.00 s/cycle of apply and release stroke). Empty the leak trap(s) and place the load fixture(s) with the cylinder(s) in the heat cabinet. Adjust the pressure actuating mechanism to build up a master cylinder output pressure of 1000 psi ± 100 (6.9 MPa ± 0.7). Place the leak trap under each wheel cylinder boot and commence stroking while raising the temperature of the cabinet to 100°C ± 5 (212°F ± 9) within 6 h.

6.6.1 Periodically observe the wheel cylinder action during stroking.

6.6.2 Discontinue stroking at the end of 100 000 cycles. Inspect the wheel cylinder for external leakage. Measure and record the fluid in each leak trap.

6.6.3 Empty the leak trap(s), shut off the valve to the master cylinder, and open the valve from the wheel cylinder(s) to the standpipe. Place the leak trap under each wheel cylinder boot and let the system stand idle for 12 to 18 h, during which the cabinet shall be allowed to cool to room temperature. Measure and record leakage.

6.6.4 Shut off the valve to the standpipe, open the valve to the master cylinder, and repeat 6.3.1 and 6.3.2.

Open the valve to the master cylinder, remove the connecting link lock(s), and remove the load fixture(s) from the heat cabinet.

6.7 Cold Temperature Operation—Empty the leak trap(s) and place the load fixture(s) with the wheel cylinder(s) in the cold chamber. Set the pressure actuating mechanism for cyclic operation at 500 cycles/h ± 50 (6.55 to 8.00 s/cycle of apply and release stroke), and adjust it to build up a master cylinder output pressure of 500 psi ± 50 (3.4 MPa ± 0.3). Place the leak trap under each wheel cylinder boot and lower the temperature of the chamber to −40 + 2, −5°C (−40 + 3.6, −9°F) within 18 h. Commence stroking after a minimum of 4 h soak at the test temperature.

6.7.1 Observe the wheel cylinder action during stroking.

6.7.2 Discontinue stroking at the end of 20 cycles and allow the load fixture(s) to come to room temperature. Inspect the wheel cylinder for external leakage. Measure and record the fluid in each leak trap.

6.7.3 Repeat 6.3.1 and 6.3.2.

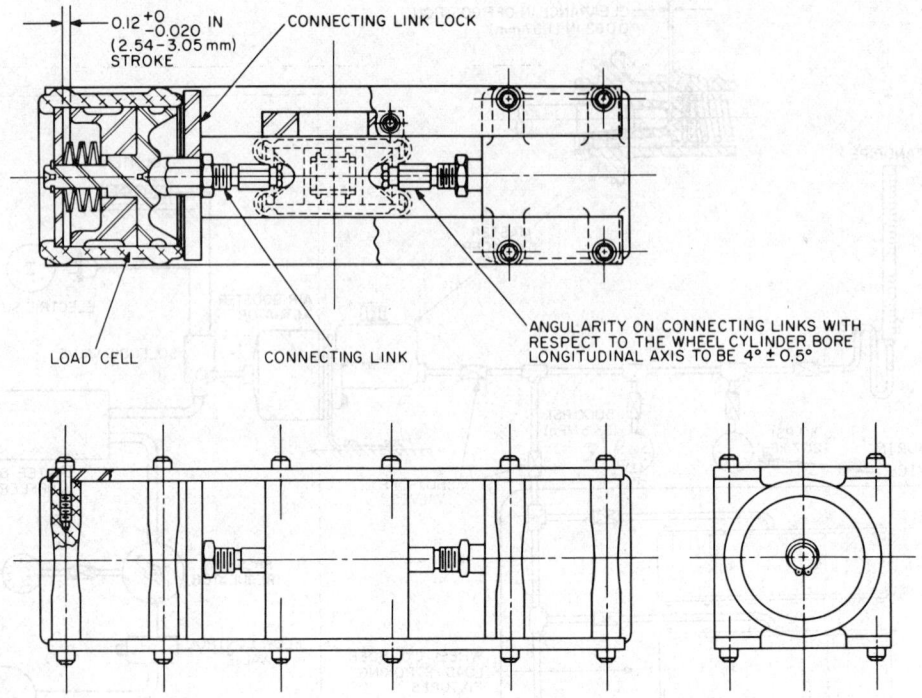

FIG. 2—WHEEL CYLINDER LOAD FIXTURE

Open the valve to the master cylinder and remove the connecting link lock(s), and remove the load fixture from the cold chamber.

6.8 Storage Corrosion Resistance—Remove the wheel cylinder connector at its juncture with the line to the master cylinder, and then, taking care not to empty fluid from the wheel cylinder or its connecting tubing/hose, install a vented plug in the open end of connecting tubing/hose. With the wheel cylinder on its load fixture or a like device that holds the piston(s) in release position, place empty leak trap(s) under the cylinder boot(s) and store the cylinder for seven days at room temperature.

6.8.1 At the end of seven days, examine the cylinder for visible leakage. Measure the amount of fluid in the leak trap(s).

6.8.2 Remove the piston clamps, if used, and remount the cylinder on the load fixture or like device allowing equivalent piston stroke. Reattach the cylinder to the test apparatus or equivalent and build up hydraulic pressure gradually until the piston(s) starts to move, measure and record this pressure.

6.8.3 Continue the pressure buildup until 500 psi ± 50 (3.4 MPa ± 0.3) is achieved and release it, meanwhile observing the wheel cylinder action.

6.8.4 Repeat 6.3.1 and 6.3.2.

Open the valve to the master cylinder and remove the connecting link lock(s).

6.9 Static Leakage—Remove the wheel cylinder boot(s) and drain any fluid that might be present. Shut off the valve to the master cylinder and open the valve from the wheel cylinder to the standpipe. Fill the standpipe with new fluid and let the system stand idle for 12 to 18 h with leak traps under each open end of the cylinder.

6.9.1 At the end of 12 to 18 h, measure any fluid leakage.

6.10 Examination

6.10.1 Tighten the hydraulic connector to the maximum torque specified by the vehicle manufacturer.

6.10.2 Remove the wheel cylinder from the load fixture and test apparatus, and carefully disassemble it. Measure the smallest diameter of port opening(s).

6.10.3 Examine parts and fluid for evidence which would indicate imminent failure of the cylinder on its continued usage in a vehicle.

The examination provides the tester with an indication of how far the test wheel cylinder would surpass minimum performance and durability requirements for satisfactory vehicle usage.

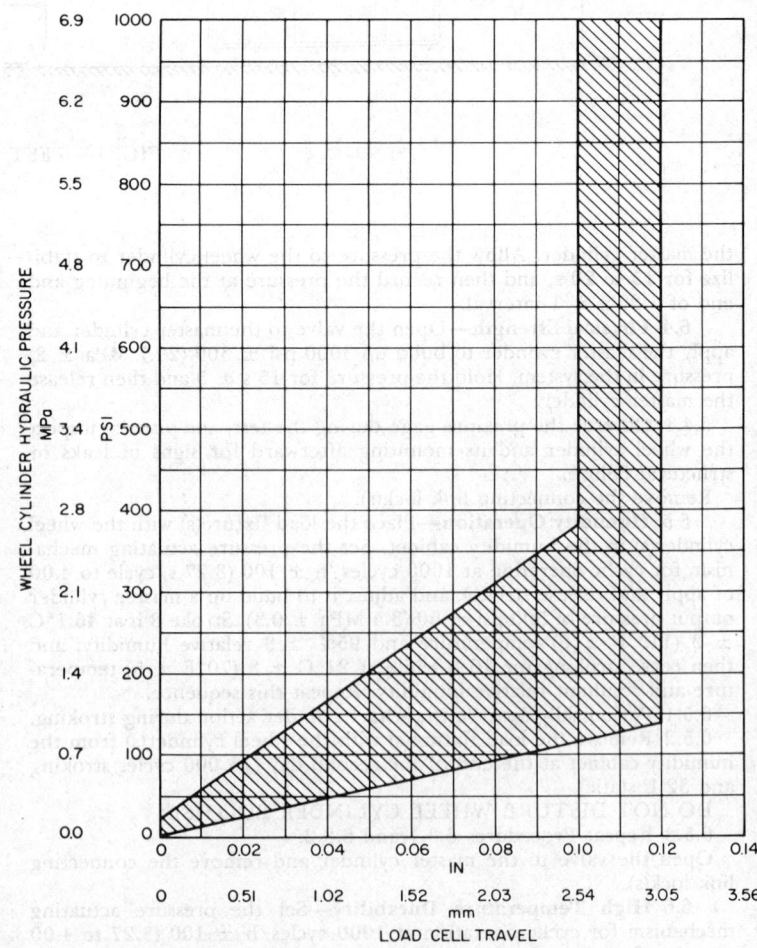

FIG. 3—LOAD FIXTURE PERFORMANCE CURVES

HYDRAULIC MASTER CYLINDERS FOR MOTOR VEHICLE BRAKES— TEST PROCEDURE—SAE J1153 JUN91

SAE Standard

Report of the Hydraulic Brake Systems Actuating Committee, approved July 1976, reaffirmed without change March 1985. Revised by the SAE Hydraulic Brake Actuating Components Standards Committee June 1991.

(R) **1. Scope**—This SAE Standard specifies the test procedure to determine minimum performance and durability characteristics for master cylinder assemblies of current established designs, components of which conform to SAE Standards. It is applicable to new assemblies from commercial production and remanufacture (factory rebuild).

The minimum performance and durability requirements are specified in SAE J1154.

(R) **1.1 Type**—This document applies to both single and dual output master cylinder assemblies used in hydraulically operated brake systems of highway vehicles. It covers such cylinders where they are employed in passenger car, truck, bus, and like brake systems utilizing motor vehicle brake fluids which conform to SAE J1703.

(R) **2. References**

2.1 Applicable Documents—The following publications form a part of this specification to the extent specified herein. The latest issue of SAE publications shall apply.

2.1.1 SAE PUBLICATIONS—Available from SAE, 400 Comonwealth Drive, Warrendale, PA 15096-0001.

SAE J1154—Hydraulic Master Cylinders for Motor Vehicle Brakes—Performance Requirements

3. Test Apparatus—The basic apparatus shall be that shown and as arranged in Figure 1 or equivalent. All hydraulic lines and fittings shall be of sufficient size as to permit unrestricted flow to and from test master cylinder(s). The apparatus shall operate per the following description and as called for in Section 5. It is desirable to have the test apparatus portable to facilitate cold, hot, and room temperature bench testing.

3.1 Displacement Mechanism—The displacement mechanism(s) connected to the master cylinder outlet(s) shall restrict the master cylinder(s) output performance to the shaded area of Figure 2. In addition, the heel of the pressure cup(s) on the piston(s) shall be past the vent port(s) before a pressure of 345 kPa (50.0 psi) is attained. The master cylinder outlet pressure(s) shall rise smoothly to a maximum of 6900 kPa ± 690 (1000 psi ± 100) within 60 to 80% of total piston travel.

3.2 Stroking Mechanism—The stroking mechanism shall contain a mounting plate to which the master cylinder can be attached. The actuating push rod shall be compatible with the master cylinder piston socket and shall operate coaxially within 2 degrees of the master cylinder bore longitudinal axis. The fixture shall be constructed such that full release of the master cylinder piston is obtained. The stroking mechanism may accommodate multiple master cylinders, if desired.

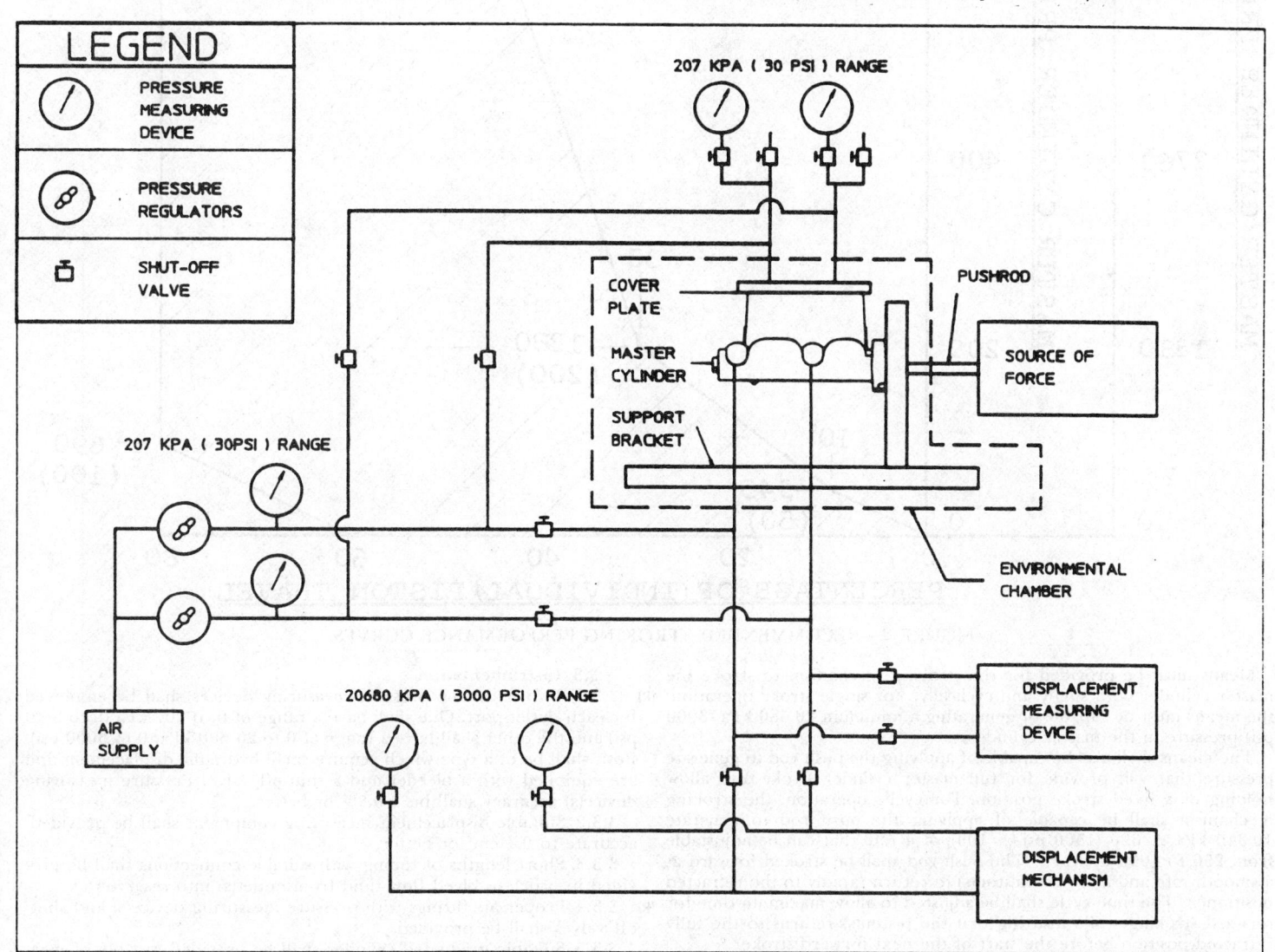

FIGURE 1—TEST APPARATUS

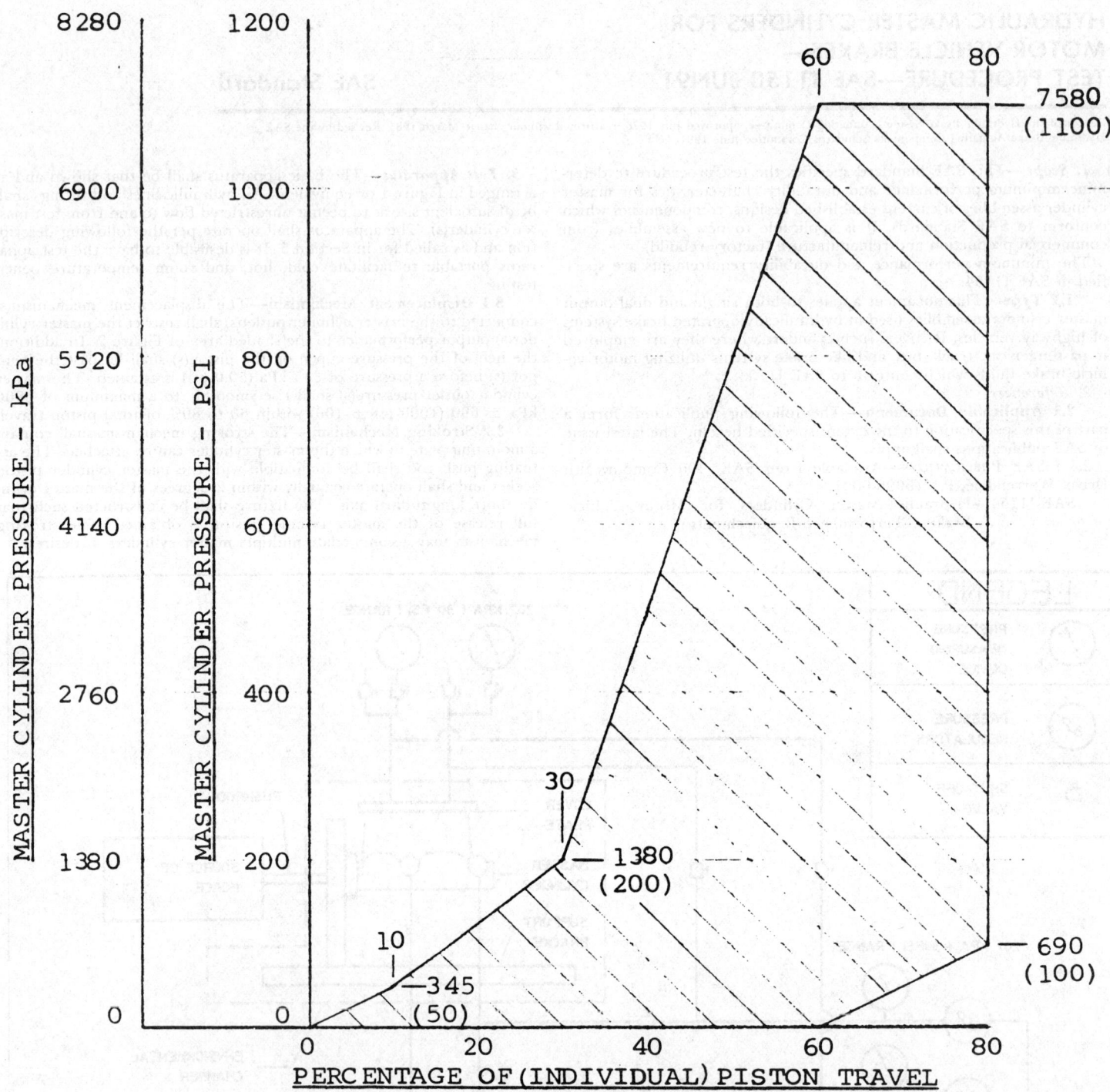

FIGURE 2—RECOMMENDED STROKING PERFORMANCE CURVES

(R) Means must be provided for the stroking mechanism to stroke the master cylinder both singly and cyclically. For single stroke operation, the means must be capable of generating a minimum 20 680 kPa (3000 psi) pressure in the master cylinder.

The means shall also be capable of applying the push rod to generate pressure that will provide for full master cylinder stroke and allow holding of a fixed stroke position. For cyclic operation, the stroking mechanism shall be capable of applying the push rod to generate 10 340 kPa ± 690 (1500 psi ± 100), at a rate that can be adjustable from 250 to 1000 cycles/h. The push rod shall be stroked forward at a smooth rate and allow the piston(s) to return rapidly to the retracted position(s). The time cycle shall be adjusted to allow maximum time for forward stroking while insuring that the piston(s) returns to the fully retracted position before the start of the next forward stroke.

Means must also be provided for a 207 kPa ± 7 (30 psi ± 1) air pressure source to be applied to the outlet port(s).

3.3 Instrumentation

(R) 3.3.1 Two hydraulic pressure measuring devices shall be employed for each outlet port. One shall have a range of 0 to 207 kPa (0 to 30.0 psi) and the other shall have a range of 0 to 20 680 kPa (0 to 3000 psi). Both shall be of a type which require small hydraulic displacement and are equipped with a bleeder and a shut-off valve. Pressure measuring device(s) accuracy shall be ±0.5% or better.

3.3.2 Suitable displacement measuring equipment shall be provided, accurate to 0.1 cm^3 or better.

3.3.3 Short lengths of tubing with suitable connections shall be provided in order to bleed flow fluid from outlet(s) into reservoir(s).

(R) 3.3.4 Proper air fittings with pressure measuring device(s) and shut-off valves shall be provided.

3.3.5 Suitable graduated cylinder shall be provided capable of measuring fluid volumes, accurate to 1 cm^3 or better.

3.4 Environmental Equipment

3.4.1 HEATED AIR BATH CABINET—An insulated oven or cabinet shall be provided having sufficient capacity to house test apparatus fixtures. A suitable thermostatically controlled heating system is required to maintain a temperature of 120 °C ± 3 (248 °F ± 5). Heaters shall be shielded to prevent direct radiation to master cylinder.

3.4.2 COLD CHAMBER—A cold chamber shall be provided having sufficient capacity to house test apparatus. It shall be capable of maintaining a uniform atmosphere of cold dry air at −40 to −42.8 °C (−40 to −45 °F).

3.4.3 HUMIDITY CABINET—A humidity cabinet shall be provided having sufficient capacity to house test apparatus fixtures. It shall be capable of maintaining a relative humidity between 80 and 90% at 21 °C ± 3 (70 °F ± 5) and 46.1 °C ± 3 (115 °F ± 5).

(R) **3.5 Test Fittings and Material**—Test hydraulic fluid shall conform to SAE J1703. The compatibility fluid of SAE J1703 is recommended.[1] The hydraulic connector to the master cylinder shall be of the type used by the vehicle manufacturer.

4. Test Sample—The master cylinder shall come from one of the sources described in Section 2. It shall not have been used after manufacture or rebuild, and it shall not be disassembled prior to testing.

5. Test Setup and Procedure—Tests shall be conducted in the sequence shown and at room temperature except where otherwise specified. The master cylinder shall not be disassembled until after all tests are completed or unless testing is discontinued.

NOTE:
1. When outlet ports are pressurized on dual master cylinders, both ports must be pressurized simultaneously.
2. When fully stroking master cylinders which do not incorporate internal stroke limiting means, care shall be exercised to avoid damage to spring(s), retainer(s), cup(s), etc.

5.1 Unrestricted Apply and Release—Remove shipping plug(s) from master cylinder outlet port(s) and stroke the piston through full design stroke five times and allow it to return by the return spring load.

5.2 Venting—Install the master cylinder on the mounting plate and tighten mounting bolts. Make certain that the push rod is properly aligned with the cylinder bore longitudinal axis within 2 degrees. Adjust push rod to allow piston(s) to return to normal release position.

5.2.1 Connect a 207 kPa ± 7 (30.0 psi ± 1) air supply to the outlet port(s). Bore venting will be indicated by air flow from the vent port(s).

5.2.2 Without changing the setup in procedure 5.2.1, stroke the input push rod a minimum of 5.1 mm (0.200 in). Apply 207 kPa ± 7 (30.0 psi ± 1) air pressure to the outlet port(s) and observe that no air is flowing from the vent port(s).

5.3 Residual Pressure Valve—For master cylinders with residual check valve(s) only. Release the push rod to allow piston(s) to return to the normal release position.

5.3.1 Cap (both) reservoir(s), apply 207 kPa ± 7 (30.0 psi ± 1) maximum air through cap with outlet port(s) open.

5.3.2 Open (both) reservoir(s) and apply air pressure to outlet port(s) as specified in manufacturer's data for residual valve operation check.

(R) **5.4 Applied Leakage**—Stroke piston(s) a minimum of 5.1 mm (0.200 in) such that the vent port(s) is (are) closed and mechanically restrain piston(s) from releasing. Apply 207 kPa ± 7 (30.0 psi ± 1) constant air pressure to the outlet port(s). Cap reservoir(s) with pressure measuring device(s) mounted through cap.

(R) 5.4.1 Stroke the pistons once to full design stroke at no more than 6.35 mm/s (0.250 in/s) and allow to return to starting restrained position. Record reservoir pressure measuring device(s) pressure after 30 ± 1 s stabilization period.

(R) 5.4.2 Remove piston restraint and disconnect air pressure source from outlet port(s). Connect low pressure hydraulic pressure measuring device(s). Fill the test setup with new clean brake fluid to the manufacturer's recommended level and bleed air from master cylinder and pressure measuring device(s) by stroking until the exiting fluid stream is free of bubbles. If bleed screws are present, open for required bleeding then tighten to nominal torque specified by the manufacturer.

NOTE: During the following procedures, 5.4.3 and 5.4.4, allow 15 to 20 s such that the pressure shall stabilize and then record pressure at the beginning and end of a 30 s ± 1 interval. If the specified pressure cannot be obtained simultaneously in both pressure chambers on any one application of a dual master cylinder, repeat the procedure to obtain the specified pressure for each individual chamber.

5.4.3 Apply master cylinder to build up 138 kPa ± 14 (20.0 psi ± 2) pressure. Hold push rod in applied position(s) and observe pressure measuring device(s) for pressure drop.

(R) 5.4.4 Replace low pressure measuring device(s) with high pressure measuring device(s) and bleed. Stroke piston(s) sufficiently for seal(s) to pass vent holes at approximately atmospheric pressure and restrain piston(s). Repeat procedure 5.4.3 for high pressure test of 6900 kPa ± 690 (1000 psi ± 100).

5.5 Fluid Displacement—Suitable fluid displacement measuring equipment shall be connected to cylinder outlet port(s) with shut-off valve(s) between measuring equipment and outlet(s). The cylinder and equipment shall be bled of air before starting test measurements. The cylinder shall be stroked to its full design stroke for five full applications at 2.5 mm/s (0.1 in/s) maximum velocity with a minimum of 5 s interval between strokes. Close shut-off valve(s) at end of each application and while cylinder is being returned to release position. Make-up fluid may be added to the reservoir(s). The fluid volume discharge from the outlet(s) at the end of each stroke shall be recorded. Calculate and record the average of all trials.

5.6 Replenishing—From the results obtained in procedure 5.5, calculate the variation between each application and the average obtained in procedure 5.5 and determine percentage.

(R) **5.7 Physical Strength**—Connect high pressure measuring device(s) to outlet port(s). Apply push rod force to develop 20 680 kPa ± 1030 (3000 psi ± 150) pressure for 15 s ± 5.

Observe pressure measuring device(s) for an abrupt decline in pressure and master cylinder for fluid leakage.

5.8 Humidity Operation—Place stroking mechanism with master cylinder mounted and filled with brake fluid to the manufacturer's recommended level into the temperature-humidity cabinet. Connect the displacement mechanism(s) to the outlet port(s) of the cylinder. The system shall be bled and carefully dried of fluid. Set the stroking mechanism to cycle at 1000 apply/release cycles/h ± 100 (3.27 to 4.00 s/cycle). Adjust the input force to the master cylinder and/or adjust the displacement mechanism(s) to stroke (each) master cylinder piston 60 to 80% of its full stroke at output pressures of 6900 kPa ± 690 (1000 psi ± 100). The rate of pressure rise versus travel shall fall within the shaded limits of Figure 2. Stroke 8 h at 46.1 °C ± 3 (115 °F ± 5) temperature and 80 to 90% relative humidity. Cease stroking for 16 h while at room temperature and resultant relative humidity. Repeat this sequence.

5.8.1 Periodically observe master cylinder for fluid disturbance in the reservoir(s) as an indication of venting.

NOTE: On dual output cylinders, often only one chamber will give fluid disturbance during venting.

5.8.2 Remove master cylinder from humidity cabinet at the end of the second day (16 000 apply/release cycles and 32 h static).

5.8.3 Remove the pressure line(s) from the cylinder outlet port(s) to the displacement mechanism(s), and attach bleeder loop(s) from the cylinder outlet port(s) into the reservoir(s). Stroke piston(s) five times to full design stroke and allow to return, observing smoothness of stroke and returnability.

5.8.4 Remove bleeder loop(s) and reinstall pressure line(s) from the cylinder outlet port(s) to the displacement mechanism(s). Repeat procedure 5.4.3.

5.8.5 Repeat procedure 5.4.4.

5.9 High Temperature Durability—Place stroking mechanism with master cylinder mounted and filled with brake fluid to the manufacturer's recommended level into the heated air bath cabinet. Connect the displacement mechanism(s) to the outlet port(s) of the cylinder.

The system shall be bled and carefully dried of fluid. Set the stroking mechanism to cycle at 1000 apply/release cycles/h ± 100 (3.27 to 4.00 s/cycle). Adjust the input force to the master cylinder and/or adjust the displacement mechanism(s) to stroke (each) master cylinder piston 60 to 80% of its full stroke at output pressure(s) of 6900 kPa ± 690 (1000 psi ± 100). The rate of pressure rise versus travel shall fall within the shaded limits of Figure 2. Place leak trap(s) under the entrance to the master cylinder bore(s) and commence stroking while raising the temperature of the cabinet to 120 °C ± 3 (248 °F ± 5) within 2 to 6 h.

5.9.1 Periodically observe master cylinder for fluid disturbance in the reservoir(s) as an indication of bore venting. See Note in 5.8.1.

5.9.2 Discontinue stroking at the end of 70 h continuous apply/release cycles. Inspect master cylinder for external leakage.

5.9.3 Remove master cylinder from heated air bath chamber and immediately repeat procedure 5.4.3.

5.9.4 Immediately following procedure 5.9.3, repeat procedure 5.4.4.

---

[1] Compatibility fluid may be obtained from Society of Automotive Engineers, Inc., 400 Commonwealth Drive, Warrendale, PA 15096.

5.10 Static Leakage

5.10.1 Immediately following procedure 5.9 (while master cylinder remains hot), disconnect displacement mechanism and plug outlet(s). Cylinder body and area(s) around outlet(s) and seal(s) shall be dry before starting test. Place master cylinder in design position, filled with brake fluid to manufacturer's recommended level for a minimum of 12 h. Observe and measure any fluid leakage.

(R) 5.10.2 MASTER CYLINDER IN INVERTED POSITION—Master cylinder with reservoir(s) sealed to the atmosphere shall be tested for reservoir seal leakage by mounting the master cylinder in an inverted position with the reservoir cover(s) on the bottom. Cylinder with vented reservoir cover(s) shall be tested by suitably plugging all cover vent(s). Mounting shall be such that the weight of the assembly or external means shall not aid the reservoir sealing. Cylinder body and area(s) around outlet(s) and seal(s) shall be dry before starting test. Allow master cylinder to remain in inverted position filled with brake fluid for a minimum of 20 min. Observe and measure any fluid leakage.

(R) **5.11 Cold Temperature Operation**—Place stroking mechanism with the master cylinder mounted and filled with brake fluid to the manufacturer's recommended level into the cold chamber. Connect the displacement mechanism(s) to the outlet port(s) of the cylinder. The system shall be bled and carefully dried of fluid. Set the stroking mechanism to cycle at 250 apply/release cycles/h ± 25 (13.1 to 16.0 s/cycle). Adjust the input force to the master cylinder and/or adjust the displacement mechanism(s) to stroke (each) master cylinder piston 60 to 80% of its full stroke at output pressure(s) of 6900 kPa ± 690 (1000 psi ± 100). The rate of pressure rise versus travel shall fall within the shaded limits of Figure 2 at ambient temperature. Place empty leak traps under the entrance to the master cylinder bore(s), and lower the temperature of the chamber to −40 to −42.8 °C (−40 to −45 °F) within 18 h. Commence stroking after a minimum of 4 h soak at the test temperature.

5.11.1 Observe master cylinder for fluid disturbance in the reservoir(s) as an indication of venting. See Note in 5.8.1.

5.11.2 Discontinue stroking at the end of 20 apply/release cycles. Inspect master cylinder for external leakage.

5.11.3 Remove master cylinder from cold chamber and immediately repeat procedure 5.8.3.

5.11.4 Immediately following paragraph 5.11.3, repeat procedure 5.4.3.

5.11.5 Immediately following paragraph 5.11.4, repeat procedure 5.4.4.

5.11.6 Allow master cylinder to return to room temperature.

5.12 Storage Corrosion Resistance—Disconnect master cylinder at its outlet(s) and plug. With the master cylinder on its mounting plate and the piston(s) in release position, place empty leak trap(s) under entrance to cylinder bore(s) and store cylinder for 7 days at room temperature.

5.12.1 At end of 7 days, examine cylinder for visible leakage. Measure amount of fluid in leak trap(s).

5.12.2 Remove outlet plug(s) and install bleeder loop(s). Gradually increase force on input rod(s) until piston(s) starts to move. Measure and record this force.

5.12.3 Stroke the piston to full design stroke five times and allow it to return by the return spring load.

5.12.4 Repeat procedure 5.4.3.

5.12.5 Repeat procedure 5.4.4.

5.13 Reservoir Capacity—With the master cylinder located in design position, plug the outlet port(s) and remove the reservoir cover(s). Syphon, syringe, or otherwise remove all fluid from the reservoir(s) down to the level of the vent port opening(s). Retain the removed fluid in a suitable clean container for inspection purposes. Using a graduated cylinder, refill reservoir(s) to minimum design level using new clean brake fluid and measure the amount(s) required.

5.14 Reservoir Fluid Depletion—Mount master cylinder in the design position filled with brake fluid to recommended level. Attach line(s) to cylinder outlet port(s) which contain sufficient restriction to pump all usable fluid out of the reservoir(s). Immerse open end(s) of line(s) in a shallow level of clean new brake fluid in a clean container of sufficient capacity. Stroke cylinder until all usable reservoir fluid is pumped into container, saving fluid for inspection purposes. On master cylinder so equipped, remove cylinder cover(s) and gasket(s) and examine cover gasket diaphragm(s) for distention.

5.15 Push Rod Retention—For manual brake type master cylinders with integral push rod only. With master cylinder mounted in suitable fixture, apply a direct (non-angular, non-twisting) tension load to the push rod. This load shall be the minimum load specified by the manufacturer.

5.16 Examination—Remove master cylinder from test fixture, and carefully disassemble it.

(R) 5.16.1 Tighten tube nut(s) to the maximum as specified by the manufacturer. Measure and record minimum resulting diameter at smallest opening of hydraulic outlet(s).

5.16.2 Examine parts and fluid for evidence which would indicate imminent failure of the cylinder upon its continued usage in the vehicle. (This examination provides the tester with an indication of how far the test master cylinder would surpass minimum performance and durability requirements for satisfactory vehicle usage.)

25.33

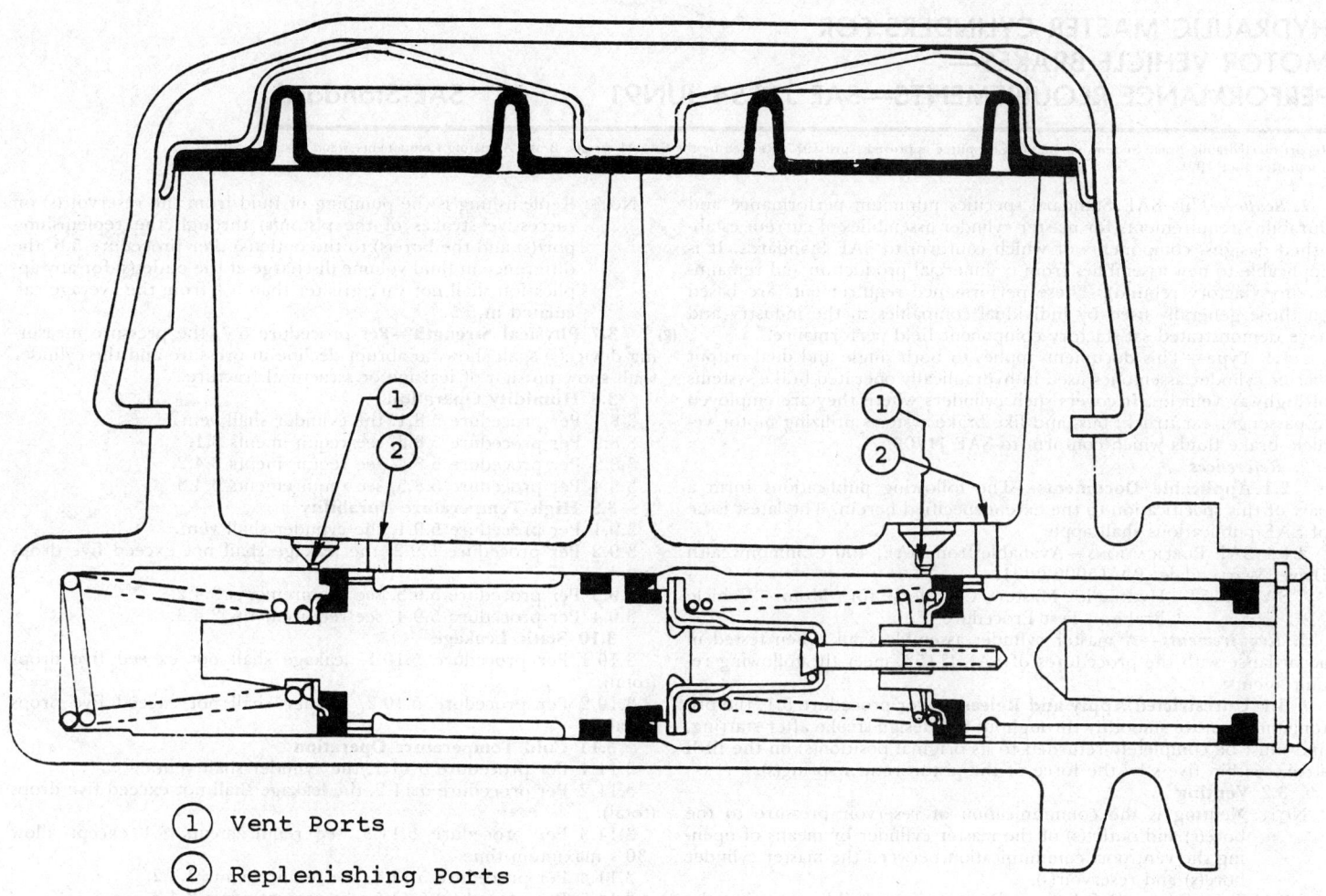

① Vent Ports
② Replenishing Ports

FIGURE 3—MASTER CYLINDER NOMENCLATURE

HYDRAULIC MASTER CYLINDERS FOR MOTOR VEHICLE BRAKES—PERFORMANCE REQUIREMENTS—SAE J1154 JUN91

SAE Standard

Report of Hydraulic Brake Systems Actuating Committee approved April 1977. Revised by the SAE Hydraulic Brake Actuating Components Standards Committee June 1991.

(R) **1. Scope**—This SAE Standard specifies minimum performance and durability requirements for master cylinder assemblies of current established designs, components of which conform to SAE Standards. It is applicable to new assemblies from commercial production and remanufacture (factory rebuild). These performance requirements are based on those generally used by individual companies in the industry and have demonstrated satisfactory component field performance.

(R) **1.1 Type**—This document applies to both single and dual output master cylinder assemblies used in hydraulically operated brake systems of highway vehicles. It covers such cylinders where they are employed in passenger car, truck, bus, and like brake systems utilizing motor vehicle brake fluids which conform to SAE J1703.

2. References

2.1 Applicable Documents—The following publications form a part of this specification to the extent specified herein. The latest issue of SAE publications shall apply.

2.1.1 SAE PUBLICATIONS—Avaliable from SAE, 400 Commonwealth Drive, Warrendale, PA 15096-0001.

SAE J1153—Hydraulic Master Cylinders for Motor Vehicle Brakes—Test Procedure

3. Requirements—A master cylinder assembly shall, when tested in accordance with the procedures of SAE J1153, meet the following requirements:

3.1 Unrestricted Apply and Release—Per procedure 5.1, the piston(s) must move smoothly throughout full design stroke after starting, and must be completely returned to its original position(s) on the fifth stroke within five s by the force of the piston return spring(s).

3.2 Venting

NOTE: Venting is the communication of reservoir pressure to the bore(s) and outlet(s) of the master cylinder by means of opening the vent port communication between the master cylinder bore(s) and reservoir(s).

3.2.1 Per procedure 5.2.1, the cylinder bore(s) shall be vented to the reservoir(s).

3.2.2 Per procedure 5.2.2, the cylinder bore(s) shall not be vented to the reservoir(s).

3.3 Residual Pressure Valve

NOTE: The requirements of 3.3 are applicable only to master cylinders which contain a residual check valve(s).

3.3.1 Per procedure 5.3.1, air shall flow through outlet port(s).

3.3.2 Per procedure 5.3.2, air at pressure below manufacturer's minimum residual valve specification shall not flow through outlet port(s) into reservoir(s), and above manufacturer's maximum specification air shall flow through outlet port(s) into reservoir(s).

3.4 Applied Leakage

(R) 3.4.1 Per procedure 5.4.1, the reservoir pressure measuring device(s) pressure(s) shall not exceed 7 kPa (1.0 psi) increase above atmospheric after 30 s ± 1 interval.

3.4.2 Per procedure 5.4.3, there shall be no drop in pressure in excess of 7 kPa (1.0 psi) after 30 s ± 1 interval.

3.4.3 Per procedure 5.4.4, there shall be no drop in pressure in excess of 345 kPa (50.0 psi) after 30 s ± 1 interval.

3.5 Fluid Displacement—Per procedure 5.5, the average fluid volume discharge at the outlet port(s) per stroke shall be as indicated by design specifications.

3.6 Replenishing

NOTE: Replenishing is the pumping of fluid from the reservoir(s) on successive strokes of the piston(s) through the replenishing port(s) and the bore(s) to the outlet(s). Per procedure 5.6, the difference in fluid volume discharge at the outlet(s) for any application shall not vary greater than 5% from the average calculated in 5.5.

(R) **3.7 Physical Strength**—Per procedure 5.7, the pressure measuring device(s) shall show no abrupt decline in pressure and the cylinder shall show no sign of leakage or structural fracture.

3.8 Humidity Operation

3.8.1 Per procedure 5.8.1, the cylinder shall vent.

3.8.2 Per procedure 5.8.3, see requirements 3.1.

3.8.3 Per procedure 5.8.4, see requirements 3.4.2.

3.8.4 Per procedure 5.8.5, see requirements 3.4.3.

3.9 High Temperature Durability

3.9.1 Per procedure 5.9.1, the cylinder shall vent.

3.9.2 Per procedure 5.9.2, the leakage shall not exceed five drops (total).

3.9.3 Per procedure 5.9.3, see requirements 3.4.2.

3.9.4 Per procedure 5.9.4, see requirements 3.4.3.

3.10 Static Leakage

3.10.1 Per procedure 5.10.1, leakage shall not exceed five drops (total).

3.10.2 Per procedure 5.10.2, leakage shall not exceed five drops (total).

3.11 Cold Temperature Operation

3.11.1 Per procedure 5.11.1, the cylinder shall vent.

3.11.2 Per procedure 5.11.2, the leakage shall not exceed five drops (total).

3.11.3 Per procedure 5.11.3, see requirements 3.1 except allow 30 s maximum time.

3.11.4 Per procedure 5.11.4, see requirements 3.4.2.

3.11.5 Per procedure 5.11.5, see requirements 3.4.3.

3.12 Storage Corrosion Resistance

3.12.1 Per procedure 5.12.1, leakage at entrance to bore(s) shall not exceed five drops (total).

3.12.2 Per procedure 5.12.2, piston(s) must start to move at 222 N (50.0 lb) maximum force.

3.12.3 Per procedure 5.12.3, see requirements 3.1.

3.12.4 Per procedure 5.12.4, see requirements 3.4.2.

3.12.5 Per procedure 5.12.5, see requirements 3.4.3.

3.13 Reservoir Capacity—Per procedure 5.13, the fluid volume(s) required to fill the reservoir(s) shall be no less than the design specification for reservoir fluid capacity(s).

3.14 Reservoir Fluid Depletion—Per procedure 5.14, the master cylinder usable fluid of the reservoir(s) shall be depleted. On master cylinders incorporating a cover diaphragm(s) the convolution(s) shall be distended.

3.15 Push Rod Retention—Per procedure 5.15, the push rod shall remain intact in the piston or master cylinder (when applicable).

3.16 Examination

3.16.1 Per procedure 5.16.1, master cylinders up to and including 51 mm (2.0 in) bore diameter shall have 2.03 mm (0.080 in) minimum diameter at smallest opening of hydraulic outlet(s).

3.16.2 Because these specifications spell out only minimum requirements, none are specified for procedure 5.16.2.

SHELF STORAGE OF HYDRAULIC BRAKE COMPONENTS—SAE J1825 APR88

SAE Information Report

Report of the Hydraulic Brake Systems Actuating Committee approved April 1988.

1. Scope—This report is the listing of recommendations for shelf storage for hydraulic brake components. Included in brake components are wheel cylinders, master cylinders, combination valves, and disc brake assemblies. This report is not a specification. This report embodies the analyses and experiences of many users and manufacturers. Where specific manufacturers' recommendations are made, those recommendations shall supersede the recommendations of this report. This report lists the successful procedures and practices associated with brake components based on long experience of a wide cross section of manufacturers and users. The practices are expected to be applied to all brake components where SAE standards are applicable.

2. Background—Hydraulic brake components in storage are not filled with brake fluid and are subject to environmental conditions from which they are protected after vehicle installation and when filled with fluid. All polymeric materials such as rubber and plastic parts undergo degradation in physical properties when stored over an extended period of time in an incompatible environment. Metal parts are also subject to corrosion and physical damage due to adverse storage environment. Detrimental changes can be minimized by controlling this storage environment.

3. Storage Conditions

3.1 Cleanliness—It is imperative that the brake component assembly be kept clean. Even foreign materials that do not affect the brake component assembly can be detrimental to the brake system. The bores and sealing surfaces are particularly sensitive to foreign material contaminants.

3.2 Ozone and Oxidation Protection—Ozone is an especially active form of oxygen that causes rapid deterioration of rubber, plastic and metal products. Ozone is present in the natural form, but it can be generated by such equipment as electric motors, high intensity lamps, and voltage discharge apparatuses. The storage area should be remote from ozone producing equipment.

3.3 Temperature—Care should be taken to avoid exposing brake components to storage temperatures above 40°C (104°F) because the higher temperature increases the aging deterioration of the rubber and plastic parts. Cold temperatures do not impart any performance change.

3.4 Humidity—Metal brake components should not be stored in high humidity areas since they are susceptible to corrosion. This corrosion can be internal, causing damage to the part. In some plastic parts, humidity will cause changes in physical properties.

3.5 Light—Brake component parts should be protected from direct exposure to sunlight and high intensity artificial light due to the damaging effects of ultraviolet rays.

3.6 Cleaning Solutions—Certain cleaning solutions contain materials that attack metal, plastic and rubber parts. Care should be taken to avoid contact of brake components with these types of materials.

3.7 Oils, Solvents, and Special Fluids—Oils, solvents, and other fluids can cause the rubber and plastic to deteriorate. Polymeric parts should be protected against contact with solutions containing petroleum, petroleum distillates or solvents and similar types of fluid.

3.8 Salts—The presence of salts can cause corrosion of metal components and degradation of rubber parts. The salts are available in household, fertilizer, or natural forms. Brake component assemblies should be protected from all forms of salt.

3.9 Packaging—Brake components that are intended for ocean shipment or for long term shelf storage (for example, service parts) should be protectively packaged to prevent the above identified environmental contamination.

R) HYDRAULIC VALVES FOR MOTOR VEHICLE BRAKE SYSTEMS TEST PROCEDURE—SAE J1118 JUN93

SAE Recommended Practice

Report of Hydraulic Brake Systems Actuating Committee approved May 1977. Completely revised by the SAE Hydraulic Brake Components Standards Committee June 1993.

NOTE—The "Requirements" (SAE J1137) for this updated procedure are under development.

1. Scope—The SAE Recommended Practice specifies the test procedure to assure valve assemblies which are satisfactory for vehicle usage, and it is applicable to new valve assemblies for commercial production. It covers such valves where they are employed in passenger car and light truck brake systems utilizing motor vehicle hydraulic brake fluids. This procedure and requirements (SAE J1137) was developed for brake fluids conforming to SAE J1703 and FMVSS 116 (DOT 3); however, it may be utilized for valves which use DOT 4 or DOT 5 brake fluid.

These procedure specifications were developed for base brake operation and do not consider the effects of ABS (anti-lock brake systems) or traction control systems which may have a significant effect on the valve. Careful analysis of the particular type ABS and/or traction control (if included in the system) should be made and additional tests are required which are not included in this document. Provisions for ABS and traction control will be incorporated in a future revision or covered in a separate document.

The minimum performance requirements are specified in SAE J1137.

1.1 Purpose—This document applies to valve assemblies used in hydraulically operated brake systems of highway vehicles. It is applicable for differential warning, metering hold-off, bypass function, or proportioning type valves or any combination thereof.

2. References

2.1 Applicable Documents—The following publications form a part of this specification to the extent specified herein. The latest issue of SAE publications shall apply.

2.1.1 SAE PUBLICATIONS—Available from SAE, 400 Commonwealth Drive, Warrendale, PA 15096-0001.

SAE J1137—Hydraulic Valves for Motor Vehicle Brake Systems—Performance Requirements

SAE J1703—Motor Vehicle Brake Fluid

2.1.2 FMVSS PUBLICATION—Available from the Superintendent of Documents, U.S. Government Printing Office, Washington, DC 20402.

FMVSS 116—Motor Vehicle Brake Fluids

3. Test Apparatus—The basic apparatus shall be that shown and as arranged in Figure 1. All hydraulic lines and fittings shall be of actual size and design similar to, or equivalent to, the representative vehicle installation.

3.1 Hydraulic Pressure Actuating Mechanism—The pressure actuating mechanism should be a dual master cylinder capable of a rate of pressure application and retraction to the valve of 3448 kPa/s ± 345 kPa/s (500 psi/s ± 50 psi/s), unless otherwise specified.

3.1.1 For single stroke operation, the actuating mechanism must be capable of generating pressures to the valve from 138 to 20 685 kPa (20 to 3000 psi) at a uniform apply and release rate and it must have adjustments such that the pressures can be held statically after they are achieved.

3.1.2 For cyclic operation the mechanism must be capable of generating pressures up to 6900 kPa (1000 psi) at a uniform apply and release rate. The actuating mechanisms, when releasing, must permit the full release of hydraulic pressure.

3.2 Load Fixture (Displacement for Foundation Brake Simulation)—The displacement devices need to simulate the pressure versus displacement curves for the actual vehicle within ±10%. If it is not known, use Figure 3 for total rear brake system and connect to the valve outlet to the rear brakes. Use Figure 4 for each front brake corner and connect to each valve outlet to front brakes.

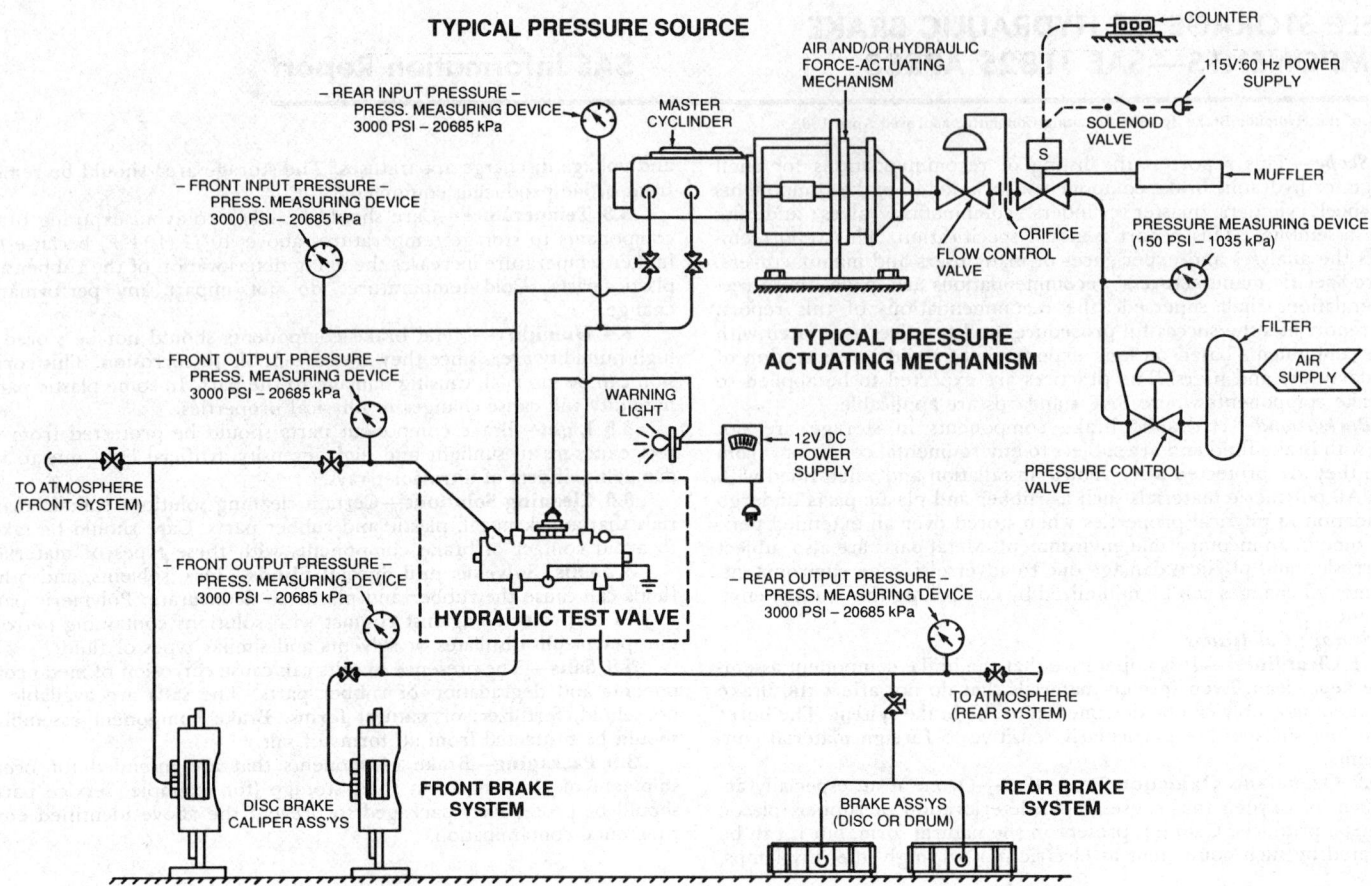

FIGURE 1—TEST APPARATUS

NOTE—It is recommended that the largest size actual brake assemblies for the manufacturers' intended valve usage be used in place of fixtures. In this case, it is not necessary to measure or meet the requirements of Figure 3 or Figure 4, but need to be plumbed as required for the particular vehicle for intended usage. Use large enough m/cylinder (bore size and stroke) to provide at least the required amount of brake fluid supply to each brake system.

3.3 Instrumentation—Hydraulic pressure measuring devices with a range to 20 685 kPa (3000 psi) are required and shall have accuracy of ±0.5%. An X-Y-Y plotter or equivalent (for functional testing) and a counter shall be used to record number of strokes during cyclic operation. A 12 VDC power source is required and connected to the warning lamp switch to indicate contact.

A leak trap shall be provided for each sealing point on the valve assembly as required for the various tests that require measuring leakage including the connections. They shall have minimum exposed area to minimize evaporation.

3.4 Environmental Equipment

3.4.1 OZONE CHAMBER—An ozone chamber as described in ASTM D 1149. It must maintain 50 pphm by volume at 37.7 °C ± 3 °C (100 °F ± 5 °F) ozone concentration.

3.4.2 HEATED AIR CHAMBER—An insulated oven or cabinet having sufficient capacity to house test valve. A suitable thermostatically controlled heating system is required to maintain a uniform and forced air circulation temperature of 120 °C ± 3 °C (248 °F ± 5 °F). It shall be such that the valves can be checked and operated without removal. Heaters shall be shielded to prevent direct radiation to valve assembly.

3.4.3 COLD CHAMBER—A cold chamber shall be provided having sufficient capacity to house test valve. It shall be such that valves can be checked and operated without removal, and it shall be capable of maintaining a uniform atmosphere of cold air at -40 to -42.8 °C (-40 to -45 °F) temperature.

3.5 Fluid and Test Fittings—Use brake fluid conforming to SAE J1703 and FMVSS 116 (DOT 3) or the recommended DOT brake fluid. The hydraulic tubing connections to the valve assembly shall be of the type intended for the particular vehicle intended usage and shall be torqued to recommended specifications.

4. Test SetUp and Procedures—All tests shall be conducted with the valve and fluid at room temperature 15 to 32 °C (60 to 90 °F) except where otherwise specified. Valve assemblies may be disassembled after the tests are completed unless otherwise specified or testing is discontinued.

4.1 Functional Characteristics—Connect hydraulic lines per Figure 1 and tighten all connections to nominal torque specified by the vehicle manufacturer. Fill the test setup with new brake fluid and bleed the system as necessary to remove all air. Using a X-Y-Y plotter or equivalent, set X as the front inlet pressure (when measuring fronts) or as rear inlet pressure (when measuring rears), and Y1 the front brake outlet pressure (either front outlet port) and Y2 the rear brake outlet pressure, and conduct the performance test.

4.1.1 PERFORMANCE TEST—Apply pressure to the inlet port at the rate described in 3.1 to 6900 kPa +345, -0 kPa (1000 psi +50, -0 psi). Plot the performance of apply and release (reference Figure 2). Analyzing the input-output curves will determine the metering performance (holdoff and blendback), proportioner performance (knee and percent slope), and hysteresis (difference of curves between apply and release).

4.1.1.1 Record holdoff of front metering pressure (point A, Figure 2 on the apply curve where the inlet pressure increases with no increase in output pressure).

4.1.1.2 Record the point where the metering pressure starts to increase again after the holdoff (point B, Figure 2).

4.1.1.3 Record blendback of metering pressures (point C, Figure 2 on the apply curve where the outlet and inlet pressures have returned to within 138 kPa [20 psi]).

4.1.1.4 Record the metering outlet pressure on apply and on release at 6900 kPa (1000 psi).

4.1.1.5 Record knee pressure of the proportioner (point on the apply curve where the rear outlet pressure departs from the inlet pressure curve).

4.1.1.6 Calculate percent slope of proportioner performance (divide the total increase of outlet pressure after the knee point by the total increase of inlet pressure after the knee point).

4.1.2 PROPORTIONER LEAKUP CHECK—Reapply to 6900 kPa (1000 psi) and hold for 30 s at the static pressure.

4.1.2.1 Record initial and final pressure (5 s stabilization time permitted).

TYPICAL VALVE PERFORMANCE

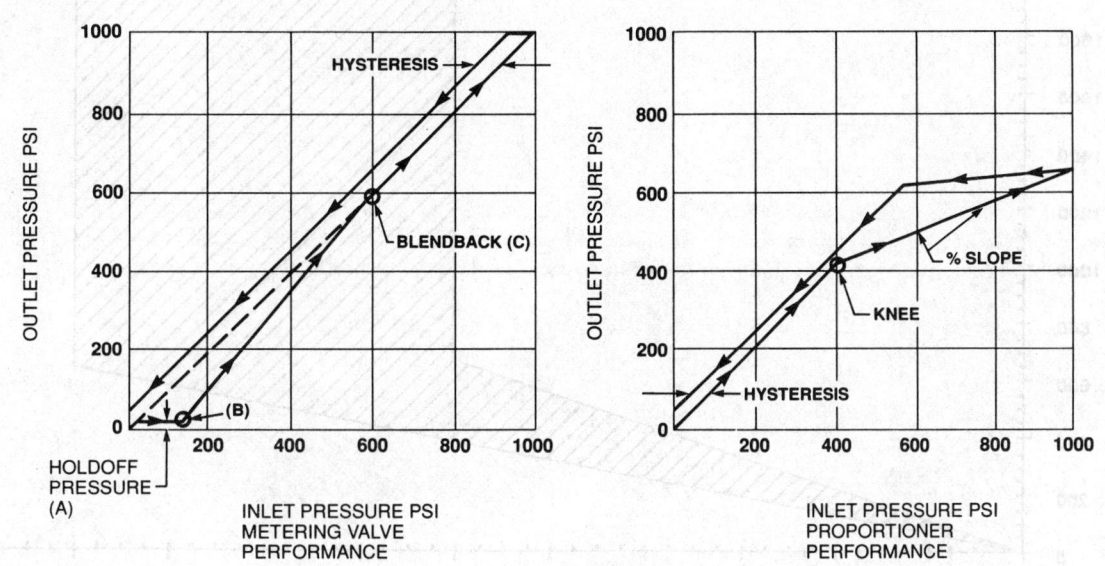

FIGURE 2—TYPICAL VALVE PERFORMANCE

4.2 Differential Warning Activation—Open front system outlet to atmosphere. The 12 to 14 VDC electrical circuit shall be connected to the warning lamp switch to indicate switch contact (reference Figure 1). Switch shall be open before applying pressure. The X of the plotter is to be set for measuring rear inlet pressure (change to front inlet when measuring front inlet pressure). Apply pressure at the prescribed rate to the rear system inlet port to 6895 kPa +345, -0 kPa (1000 psi +50, -0 psi) and hold for 30 s (5 s stabilization time permitted).

4.2.1 Record rear inlet pressure to activate warning lamp.

4.2.2 Record rear outlet pressure at 6900 kPa (1000 psi) inlet pressure after the differential piston has shuttled and the warning lamp is activated (bypass feature check).

4.2.3 Record the rear initial and final inlet and outlet pressures after 30 s.

4.2.4 Release pressure, close front outlet to atmosphere, and rebleed if air ingestion is suspected, being careful not to recenter the shuttle piston.

4.2.5 Reapply pressure to both valve inlet ports at the prescribed rate and record the pressure to recenter the differential piston by noting the pressure at which the warning lamp breaks contact.

4.2.6 Repeat this entire section (4.2) except open rear outlet system to atmosphere, change X on plotter to the front inlet, and apply pressure to the front inlet port.

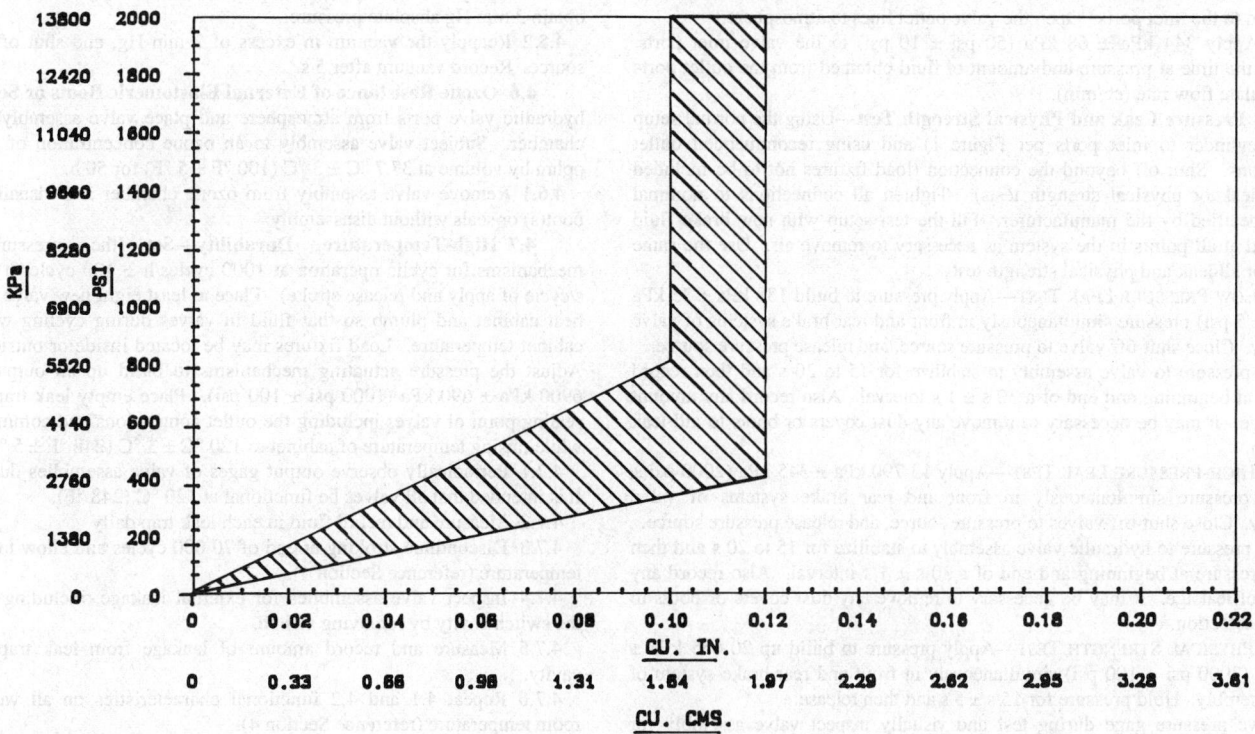

FIGURE 3—REAR BRAKES

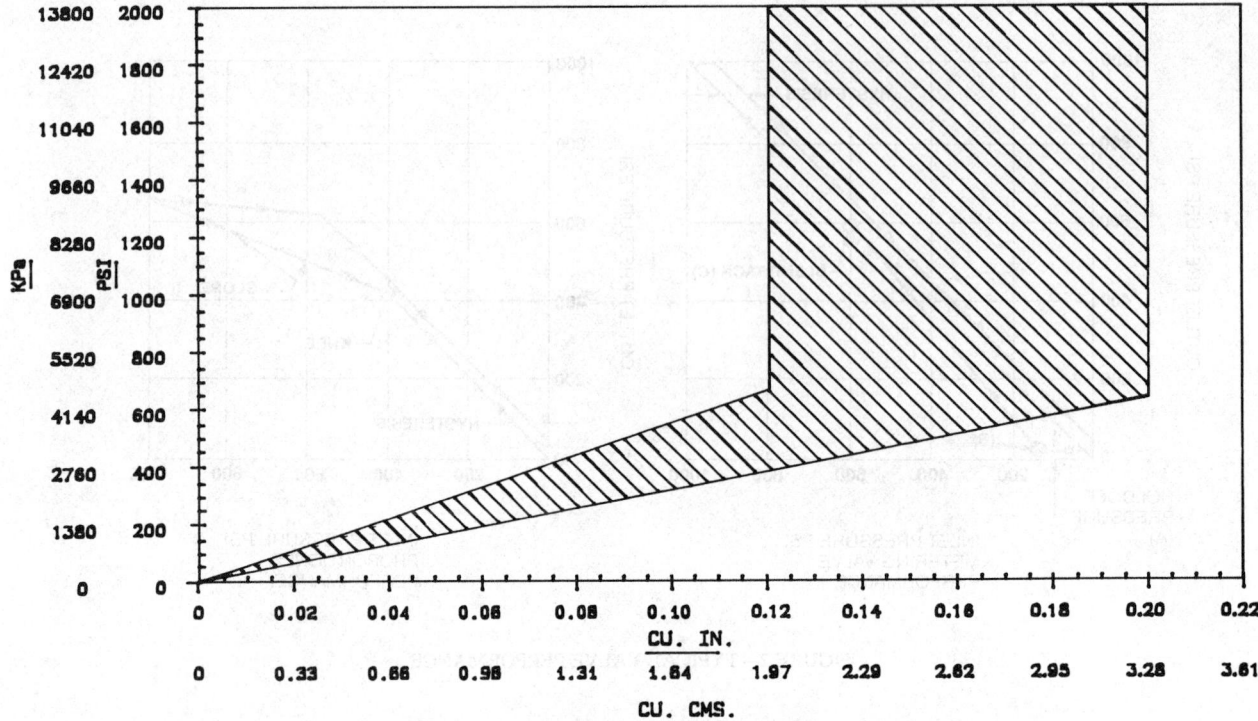

FIGURE 4—EACH FRONT BRAKE

4.3 Reverse Flow Tests—Reverse the valve connections so that the valve inlet lines are open to atmosphere and the master cylinder is supplying fluid to the valve outlets (check each front separately). Place a measuring beaker under each valve inlet port to measure the amount of fluid discharged. Apply 344 kPa ± 68 kPa (50 psi ± 10 psi) pressure to the outlet ports.

4.3.1 Measure the time at 344 kPa ± 68 kPa (50 psi ± 10 psi) and amount of fluid obtained from the inlet ports and calculate flow rate (cc/min).

4.3.2 Connect back to normal Figure 1 setup with the master cylinder connected to the inlet ports. Open the valve outlet lines to atmosphere.

4.3.3 Apply 344 kPa ± 68 kPa (50 psi ± 10 psi) to the valve inlet ports. Measure the time at pressure and amount of fluid obtained from the outlet ports and calculate flow rate (cc/min).

4.4 Pressure Leak and Physical Strength Test—Using the normal setup (master cylinder to inlet ports per Figure 1) and using recommended outlet connections. Shut off beyond the connection (load fixtures not to be included for any leak or physical strength tests). Tighten all connections to nominal torque specified by the manufacturer. Fill the test setup with new brake fluid and bleed at all points in the system as necessary to remove air. Use the same valves for all leak and physical strength tests.

4.4.1 LOW-PRESSURE LEAK TEST—Apply pressure to build 138 kPa ± 35 kPa (20 psi ± 5 psi) pressure simultaneously in front and rear brake systems of valve assembly. Close shut-off valve to pressure source, and release pressure source.

Allow pressure to valve assembly to stabilize for 15 to 20 s and then record pressure at beginning and end of a 30 s ± 1 s interval. Also record any amount of leakage. It may be necessary to remove any dust covers or boots to aid leak detection.

4.4.2 HIGH-PRESSURE LEAK TEST—Apply 13 790 kPa ± 345 kPa (2000 psi ± 50 psi) pressure simultaneously in front and rear brake systems of valve assembly. Close shut-off valves to pressure source, and release pressure source.

Allow pressure to hydraulic valve assembly to stabilize for 15 to 20 s and then record pressure at beginning and end of a 30 s ± 1 s interval. Also record any amount of leakage. It may be necessary to remove any dust covers or boots to aid leak detection.

4.4.3 PHYSICAL STRENGTH TEST—Apply pressure to build up 20 685 kPa ± 690 kPa (3000 psi ± 100 psi) simultaneously in front and rear brake system of valve assembly. Hold pressure for 15 s ± 5 s and then release.

Observe pressure gage during test and visually inspect valve assembly for signs of leakage or structural failure.

4.4.4 Repeat 4.1 and 4.2 of functional characteristics on all valves tested for leak and physical strength.

4.5 Vacuum Capability—Using new dry valves and connection tubes, setup to a vacuum source on the valve inlets. Install vacuum measuring devices beyond the outlet connections. Outlets are not to be connected to the load fixtures and are to be shut off after the vacuum measuring devices. Tighten all valve connections to the minimum recommended torque.

4.5.1 Apply vacuum to the inlet ports simultaneously, recording the time to obtain 2 mm Hg absolute pressure.

4.5.2 Reapply the vacuum in excess of 2 mm Hg, and shut off the vacuum source. Record vacuum after 5 s.

4.6 Ozone Resistance of External Elastomeric Boots or Seals—Seal the hydraulic valve ports from atmosphere and place valve assembly in the ozone chamber. Subject valve assembly to an ozone concentration of 50 pphm ± 5 pphm by volume at 37.7 °C ± 3 °C (100 °F ± 5 °F) for 50 h.

4.6.1 Remove valve assembly from ozone chamber and visually inspect the boot(s) or seals without disassembly.

4.7 High-Temperature Durability—Set the pressure actuating mechanisms for cyclic operation at 1000 cycles/h ± 100 cycles/h (3.27 to 4.00 s/cycle of apply and release stroke). Place at least eight new valve assemblies in heat cabinet and plumb so that fluid in valves during cycling will be at heat cabinet temperature. Load fixtures may be located inside or outside of cabinet. Adjust the pressure actuating mechanisms to build up an output pressure of 6900 kPa ± 690 kPa (1000 psi ± 100 psi). Place empty leak traps under each sealing point of valves including the outlet connections and commence stroking while raising temperature of cabinet to 120 °C ± 3 °C (248 °F ± 5 °F) within 6 h.

4.7.1 Periodically observe output gages of valve assemblies during stroking. It is intended that all valves be functional at 120 °C (248 °F).

4.7.2 Measure and record fluid in each leak trap daily.

4.7.3 Discontinue stroking at end of 70 000 cycles and allow to cool to room temperature (reference Section 4).

4.7.4 Inspect valve assemblies for external leakage, including inspection of the switch cavity by removing switch.

4.7.5 Measure and record amount of leakage from leak traps and switch cavity.

4.7.6 Repeat 4.1 and 4.2 functional characteristics on all valves tested at room temperature (reference Section 4).

4.8 Cold-Temperature Operation—Place new valve assemblies in cold chamber and bleed system. Place leak traps under each sealing point of valves including outlet connections. Set the pressure actuating mechanism for cyclic operation at 250 cycles/h ± 50 cycles/h (13.1 to 16.0 s/cycle). Adjust the displacement mechanism(s). Adjust pressure actuating mechanism to an output pressure of 6900 kPa ± 690 kPa (1000 psi ± 100 psi) and at an apply rate of 6900 kPa/s ± 690 kPa/s (1000 psi/s ± 100 psi/s). Adjust temperature to -40 to -42.8 °C (-40 to -45 °F) and commence stroking.

4.8.1 Periodically observe output gages of valve assemblies during stroking. It is intended that all valves be functional at -40 °C (-40 °F).

4.8.2 Discontinue stroking at the end of 20 apply/release cycles and allow to return to room temperature.

4.8.3 Repeat 4.1 and 4.2 functional characteristics on all valves tested at room temperature (reference Section 4).

4.8.4 Inspect and measure amount of leakage in the leak traps and in the switch cavity.

4.9 Examination—Examine parts and fluid for evidence which would indicate imminent failure of valve assembly upon its continued usage in a vehicle (this examination provides the tester with an indication of how far the test valve assembly would surpass minimum performance and durability requirements for satisfactory vehicle usage).

5. See SAE J1137 for requirements.

HYDRAULIC VALVES FOR MOTOR VEHICLE BRAKE SYSTEMS—PERFORMANCE REQUIREMENTS—SAE J1137 MAR85

SAE Recommended Practice

Report of the Hydraulic Brake Systems Actuating Committee, approved May 1977, reaffirmed without change, March 1985. Currently under revision by Committee.

1. General—This recommended practice applies to valve assemblies used in hydraulically operated brake systems of highway vehicles. It covers such valves where they are employed in passenger car and light truck brake systems utilizing motor vehicle non-petroleum hydraulic brake fluids per SAE J1702 and J1703. This recommended practice is applicable for differential warning, hold-off, or proportioning type valves or any combination thereof.

2. Scope—The recommended practice specifies minimum performance and durability requirements for satisfactory vehicle usage, and it is applicable to new valve assemblies from commercial production and re-manufacture (factory rebuild).

3. Requirements—Valve assemblies shall, when tested in accordance with the test procedures outlined in SAE J1118, Hydraulic Valves for Motor Vehicle Brake Systems—Test Procedure, meet the following requirements:

3.1 Ozone Resistance of External Elastomeric Boots or Seals

3.1.1 Per test procedure 5.1.1 of SAE J1118, the boot(s) or seals shall not be perforated or cracked through in any areas.

3.2 Compensation and Reverse Flow for Both Front and Rear Brake Systems

3.2.1 Per test procedure 5.2.1, 5.2.2, 5.2.3, and 5.2.4 of SAE J1118, the minimum flow rate shall not be less than 3 cm³/min.

3.2.2 Per test procedure 5.2.5 and 5.2.6 of SAE J1118, the minimum flow rate shall not be less than 300 cm³/min.

3.3 Pressure Leak Test

3.3.1 Low Pressure Leak Test—Front and Rear Brake Systems—Per test procedure 5.3.1 of SAE J1118, there shall be no drop in pressure in excess of 14 kPa (2 psi) in 30 s interval.

3.3.2 High Pressure Leak Test—Front and Rear Brake Systems—Per test procedure 5.3.2 of SAE J1118, there shall be no drop in pressure in excess of 345 kPa (50 psi) in 30 s interval.

3.4 Functional Characteristics

3.4.1 Differential Warning Actuation

3.4.1.1 Per test procedure 5.4.1.1 of SAE J1118, there shall be no drop in pressure in excess of 690 kPa (100 psi) in 30 s interval. Resistance of switch terminal to body must exceed 10 000 Ω.

3.4.1.2 Per test procedure 5.4.1.2 of SAE J1118, pressure to shuttle differential piston shall be within requirements specified by the vehicle manufacturer. Switch contact shall be established by an indicating light.

3.4.1.3 Per test procedure 5.4.1.3 of SAE J1118, pressure to re-center differential piston when switch turns off indicating light, where applicable, shall be within requirements specified by the vehicle manufacturer.

3.4.1.4 Per test procedure 5.4.1.4 of SAE J1118, see requirements 3.4.1.2.

3.4.1.5 Per test procedure 5.4.1.5 of SAE J1118, see requirements 3.4.1.3.

3.4.2 Hold-Off Valve Test—Per test procedure 5.4.2 of SAE J1118, hold-off section shall functionally comply with requirements specified by the vehicle manufacturer.

3.4.3 Proportioning Valve Test—Per test procedure 5.4.3 of SAE J1118, proportioning section and applicable by-pass feature shall functionally comply with requirements specified by the vehicle manufacturer.

3.5 Physical Strength Test—Per test procedure 5.5 of SAE J1118, the gage shall show no abrupt pressure drop and no leakage of brake fluid in excess of one drop per sealing point, nor shall the valve assembly show any signs of mechanical failure.

3.6 High Temperature Durability

3.6.1 Per test procedure 5.6.1 of SAE J1118, output pressure shall indicate the valve components are functioning. Shuttling of the differential piston during stroking tests, which results in the switch closing (contact), shall constitute a failure.

3.6.2 Per test procedure 5.6.2 of SAE J1118, leakage of brake fluid at each sealing point shall not exceed two drops.

3.6.3 Per test procedure 5.6.3 of SAE J1118, there shall be no leakage of brake fluid in excess of two drops per sealing point.

3.6.4 Per test procedure 5.6.4 of SAE J1118, see requirements 3.3.1 and 3.3.2.

3.7 Cold Temperature Operation

3.7.1 Per test procedure 5.7.1 of SAE J1118, there shall be no leakage of brake fluid in excess of two drops per sealing point.

3.7.2 Per test procedure 5.7.2 and 5.7.3 of SAE J1118, pressure drop must not exceed 69 kPa (10 psi) on 5.7.2 and 345 kPa (50 psi) on test procedure 5.7.3 of SAE J1118 and/or two drops per sealing point. Shuttling of the differential piston which results in the switch closing (contact) shall constitute a failure.

3.7.3 Per test procedure 5.7.4 and 5.7.8 of SAE J1118, pressure to shuttle differential piston shall not vary more than 100% from the normal original room temperature requirements as specified by the vehicle manufacturer. Switch contact must be established.

3.7.4 Per test procedure 5.7.5 and 5.7.6 of SAE J1118, see requirements 3.7.2.

3.7.5 Per test procedure 5.7.7 of SAE J1118, piston must re-center where applicable and light go out.

3.7.6 Per test procedure 5.7.9 of SAE J1118, see requirements 3.7.3 and 3.7.5.

3.8 Static Leak Test

3.8.1 Per test procedure 5.8.1 of SAE J1118, there shall be no leakage of brake fluid in excess of one drop per sealing point.

3.8.2 Per test procedure 5.8.2 of SAE J1118, ±25% of functional requirements of 3.4.

3.9 Examination—Because these specifications spell out only minimum requirements, none are specified for procedure 5.9 of SAE J1118.

MATERIALS FOR PLASTIC CHECK VALVES FOR VACUUM BOOSTER SYSTEMS—SAE J1875 JUN93

SAE Standard

Report of the Hydraulic Brake Plastics Standards Committee approved June 1993.

1. Scope—The materials defined by this SAE Standard are unreinforced thermoplastic acetal and thermoplastic 6/6 heat stabilized nylon, as both materials will function in this application.

The specific material chosen will depend on the final application's surroundings and heat requirements. They are for use in vacuum booster check valves for hydraulic brake systems.

2. References

2.1 Applicable Documents—The following publications form a part of this specification to the extent specified herein.

2.1.1 ANSI PUBLICATIONS—Available from ANSI, 11 West 42nd Street, New York, NY 10036-8002.

ISO 75—Plastics and ebonite—Determination of temperature of deflection under load

ISO 175—Plastics—Determination of the effects of liquid chemicals, including water

ISO 178—Plastics—Determination of flexural properties of rigid plastics

ISO 180—Plastics—Determination of Izod impact strength of rigid materials

ISO/R 527—Plastics—Determination of tensile properties

ISO 1183—Plastics—Methods for determining the density and relative density of non-cellular plastics

ISO 2039/2—Plastics—Determination of hardness—Rockwell hardness

2.1.2 ASTM PUBLICATIONS—Available from ASTM, 1916 Race Street, Philadelphia, PA 19103-1187.

ASTM D 256—Test Methods for Impact Resistance of Plastics and Electrical Insulating Materials

ASTM D 543—Standard Test Method for Resistance of Plastics to Chemical Reagents

ASTM D 638—Standard Test Method for Tensile Properties of Plastics

ASTM D 648—Standard Test Method for Deflection Temperature of Plastics Under Flexural Load

ASTM D 785—Standard Test Method for Rockwell Hardness of Plastics and Electrical Insulating Materials

ASTM D 790—Standard Test Methods for Flexural Properties of Unreinforced and Reinforced Plastics and Electrical Insulating Materials

ASTM D 792—Standard Test Methods for Density and Specific Gravity (Relative Density) of Plastics by Displacement

3. General Material Requirements

3.1 Conditioning—All test values are based on materials dry as molded or conditioned in a controlled atmosphere of 23 °C ± 2 °C and 50% ± 5% relative humidity for 24 h prior to testing and tested under the same conditions.

4. Test Requirements—See Table 1.

TABLE 1—TEST REQUIREMENTS

Property Tested	Units	Thermoplastic Acetal	Type 6/6 Nylon Heat Stabilized
Specific Gravity	—	1.41 ± 0.02	1.14 ± 0.02
Hardness, Rockwell	R	115 min	118 min[1]
Tensile Strength	MPa	62 min	55 min[1]
			44 min[2]
Flexural Modulus	MPa	2400 min	2550 min[1]
			1100 min[2]
Izod Impact	J/M	62 min	28 min[1]
			96 min[2]
Deflection Temperature			
0.5 kPa	°C	160 min	225 min
1.8 kPa	°C	125 min	85 min

[1] dry as molded
[2] 50% relative humidity

4.1 Fluid Aging—The material when tested by the producer specified in 5.7 shall have a weight gain of not more than 2.0%.

5. Test Procedures

5.1 Specific Gravity—Determine the Specific Gravity of the material by ASTM D 792 (ISO R 1183).

5.2 Hardness—Determine the Rockwell Hardness of the material by ASTM D 785 using scale R (ISO 2039/2).

5.3 Tensile Strength—Determine the Tensile Strength of the material by ASTM D 638 (ISO R 527). Type I sample size.

5.4 Modulus of Elasticity—Determine the Tangent of Modulus of Elasticity (Flexural Modulus) of the material by ASTM D 790 (ISO 178).

5.5 Impact Strength—Determine the notched Izod Strength of the material by ASTM D 256 (ISO 180). Samples to be 3.18 mm thick.

5.6 Deflection Temperature—Determine the Deflection Temperature by ASTM D 648, using a test specimen that is 127 × 12.7 × 12.7 mm and a load of 1.820 MPa (ISO 75).

5.7 Fluid Aging—Determine the Fluid Aging of the material by immersing a test specimen that is 51 mm in diameter and 6.4 mm thick in SAE RM-66-04 brake fluid for 168 h at a temperature of 70 °C in accordance with ASTM D 543 (ISO 175). Also test material in SAE 10W30 oil, antifreeze, and unleaded gasoline. Temperature of testing to be in accordance with ASTM specification.

MOTOR VEHICLE BRAKE FLUID—SAE J1703 JUN91 — SAE Standard

Report of the Nonmetalic Materials Committee, approved December 1946 and revised by the Hydraulic Brake Systems Actuating Committee, October 1988. Rationale statement available. Completely revised by the Motor Vehicle Brake Fluids Standards Committee June 1990. Rationale statement available. This document is similar to ISO 4925. Revised by the SAE Motor Vehicle Brake Fluids Standards Committee June 1991.

1. Scope—This SAE Standard covers motor vehicle brake fluids of the nonpetroleum type for use in the braking system of any motor vehicle such as a passenger car, truck, bus, or trailer. These fluids are not intended for use under arctic conditions. These fluids are designed for use in braking systems fitted with rubber cups and seals made from natural rubber (NR), styrene-butadiene rubber (SBR), or a terpolymer of ethylene, propylene, and a diene (EPDM).

2. References

(R) **2.1 Applicable Documents**—The following publications form a part of this specification to the extent specified herein. The latest issue of SAE publications shall apply.

2.1.1 SAE PUBLICATIONS—Available from SAE, 400 Commonwealth Drive, Warrendale, PA 15096-0001.

SAE J527—Brazed Double Wall Low Carbon Steel Tubing

2.1.2 ASTM PUBLICATIONS—Available from ASTM, 1916 Race Street, Philadelphia, PA 19103-1187.

ASTM D 91—Test Method for Precipitation Number of Lubricating Oils

ASTM D 344—Method of Test for Relative Dry Hiding Power of Paints

ASTM D 445—Test Method for Kinematic Viscosity of Transparent and Opaque Liquids (and the Calculation of Dynamic Viscosity)

ASTM D 664—Test Method for Neutralization Number of Potentiometric Titration

ASTM D 1120—Method of Test for Boiling Point of Engine Antifreezes

ASTM D 1415—Method of Test for International Hardness of Vulcanized Natural Rubber and Synthetic Rubbers

ASTM D 2240—Method of Test for Indentation Hardness of Rubber and Plastics by Means of a Durometer

ASTM D 3182—Recommended Practice for Rubber-Materials, Equipment, and Procedures for Mixing Standard Compounds and Preparing Standard Vulcanized Sheets

ASTM D 3185—Methods for Rubber-Evaluation of SBR (Styrene-Butadiene Rubber) including Mixtures with Oil

ASTM E 1—Specification for ASTM Thermometers

ASTM E 260-73—Standard Recommended Practice for General Gas Chromatography Procedure

3. Material—The quality of the materials used shall be such that the resulting product will conform to the requirements of this standard and insure uniformity of performance.

4. Requirements

4.1 Equilibrium Reflux Boiling Point—Brake fluid when tested by the procedure specified in 5.1 shall have an equilbrium reflux boiling point not less than 205 °C (401 °F).

4.2 Viscosity—Brake fluid when tested by the procedure specified in 5.5 shall have the following kinematic viscosities:

a. At −40 °C (−40 °F)—Not more than 1800 cSt (1800 mm^2/s).
b. At 100 °C (212 °F)—Not less than 1.5 cSt (1.5 mm^2/s).

4.3 pH Value—Brake fluid, when tested by the procedure specified in 5.6, shall have a pH value not less than 7, nor more than 11.5.

4.4 Fluid Stability

a. High Temperature Stability—When tested by the procedure specified in 5.7(a), the equilibrium reflux boiling point of the brake fluid shall not change by more than 5 °C (9 °F) increase or decrease.
b. Chemical Stability—When tested by the procedure specified in 5.7(b), the test fluid mixture shall show no chemical reversion as evidenced by a change in recorded temperature of more than 5 °C (9 °F) increase or decrease.

4.5 Corrosion—Brake fluid, when tested by the procedure specified in 5.8, shall not cause corrosion exceeding the limits shown in Table 1. The metal strips outside of the area where the strips are in contact shall neither be pitted nor roughened to an extent discernible to the naked eye, but staining or discoloration is permitted.

The fluid-water mixtures at end of test shall show no jelling at 23 °C ± 5 (73 °F ± 9). No crystalline type deposit shall form and adhere to either the glass jar walls or the surface of metal strips. The fluid-water mixture shall contain no more than 0.10% sediment by volume. The

(R) **TABLE 1—CORROSION TEST STRIPS AND WEIGHT CHANGES (SEE ALSO APPENDIX A)**

Test Strip[1]	RM No.	Max Permissible Weight Change, mg/cm^2 of Surface
Tinned Iron	6	0.2
Steel	7	0.2
Aluminum	8	0.1
Cast Iron	9	0.2
Brass	10	0.4
Copper	11	0.4
Zinc	ISO-2	0.4

[1] Obtainable from the Society of Automotive Engineers, Inc., 400 Commonwealth Drive, Warrendale, PA 15096.

fluid-water mixture shall have a pH value of not less than 7, nor more than 11.5.

The rubber cup at end of test shall show no disintegration, as evidenced by blisters or sloughing indicated by carbon black separation on the surface of the rubber cup. The hardness of the rubber cup shall not decrease by more than 15 degrees and the base diameter shall not increase by more than 1.4 mm (0.055 in).

4.6 Fluidity and Appearance at Low Temperatures

a. At −40 °C (−40 °F)—When brake fluid is tested by the procedure specified in 5.9(a), the fluid shall show no stratification, sedimentation, or crystallization. Upon inversion of sample bottle, the air bubble shall travel to the top of the fluid in not more than 10 s. Cloudiness is permissible, but on warming to room temperature 23 °C ± 5 (73 °F ± 9), the fluid shall regain its original uniformity, appearance, and clarity.
b. At −50 °C (−58 °F)—When brake fluid is tested by the procedure specified in 5.9(b), the fluid shall show no stratification, sedimentation, or crystallization. Upon inversion of sample bottle, the air bubble shall travel to the top of the fluid in not more than 35 s. Cloudiness is permissible, but on warming to room temperature 23 °C ± 5 (73 °F ± 9), the fluid shall regain its original uniformity, appearance, and clarity.

4.7 Evaporation—When brake fluid is tested by the procedure specified in 5.10, loss by evaporation shall not exceed 80% by weight. Residue from the brake fluid after evaporation shall contain no precipitate that remains gritty or abrasive when rubbed with the fingertip. Residue shall have a pour point below −5 °C (+23 °F).

4.8 Water Tolerance

a. At −40 °C (−40 °F)—When brake fluid is tested by the procedure specified in 5.11(a), the black contrast lines on a hiding power chart shall be discernible when viewed through the fluid in the centrifuge tube. The fluid shall show no stratification or sedimentation. Upon inversion of the centrifuge tube, the air bubble shall travel to the top of the fluid in not more than 10 s.
b. At 60 °C (140 °F)—When brake fluid is tested by the procedure specified in 5.11(b), the fluid shall show no stratification, and sedimentation shall not exceed 0.05% by volume after centrifuging when fluid is tested for qualification, or shall not exceed 0.15% by volume for a commercial packaged fluid.

4.9 Compatibility

a. At −40 °C (−40 °F)—When brake fluid is tested by the procedure specified in 5.12(a), the black contrast lines on a hiding power chart shall be discernible when viewed through the fluid in the centrifuge tube. The fluid shall show no stratification or sedimentation.
b. At 60 °C (140 °F)—When brake fluid is tested by the procedure specified in 5.12(b), the fluid shall show no stratification, and sedimentation shall not exceed 0.05% by volume after centrifuging.

4.10 Resistance to Oxidation—Brake fluid, when tested by the procedure specified in 5.13, shall not cause the metal strips outside the areas in contact with the tinfoil to be pitted or roughened to an extent discernible to the naked eye, but staining or discoloration is permitted. No more than a trace of gum shall be deposited on the test strips outside of the areas in contact with the tinfoil. The aluminum strips shall not decrease in weight by more than 0.05 mg/cm^2 and the cast iron strips shall not decrease in weight by more than 0.3 mg/cm^2.

4.11 Effect on Rubber

(R) 4.11.1 Rubber brake cups subjected to brake fluid, as specified in 5.14.1, shall show no increase in hardness, shall not decrease in hardness by more than 10 points, and shall show no disintegration as evidenced by blisters or sloughing indicated by carbon black separation on the surface of the rubber cup. The increase in the diameter of the base of the cups shall not be less than 0.15 mm (0.006 in), nor more than 1.4 mm (0.055 in).

(R) 4.11.2 Rubber brake cups subjected to brake fluid, as specified in 5.14.2, shall show no increase in hardness, shall not decrease in hardness by more than 15 points, and shall show no disintegration as evidenced by blisters or sloughing indicated by carbon black separation on the surface of the rubber cup. The increase in the diameter of the base of the cups shall not be less than 0.15 mm (0.006 in), nor more than 1.4 mm (0.055 in).

(R) 4.11.3 Rubber slab stock subjected to brake fluid, as specified in 5.14.3, shall show no increase in hardness, shall not decrease in hardness by more than 10 points, and shall show no disintegration as evidenced by blisters or sloughing indicated by carbon black separation on the surface of the test specimens. The test specimens shall not decrease in volume and the increase in volume shall not exceed 10%.

(R) 4.11.4 Rubber slab stock subjected to brake fluid, as specified in 5.14.4, shall show no increase in hardness, shall not decrease in hardness by more than 15 points, and shall show no disintegration as evidenced by blisters or sloughing indicated by carbon separation on the surface of the test specimens. The test specimens shall not decrease in volume and the increase in volume shall not exceed 10%.

4.12 Stroking Test Procedure
Brake fluid, when tested by the procedure specified in 5.15, shall meet the following performance requirements:

a. Metal parts shall not show corrosion as evidenced by pitting to an extent discernible to the naked eye, but staining or discoloration shall be permitted.
b. The initial diameter of any cylinder or piston shall not change by more than 0.13 mm (0.005 in) during test.
c. Rubber cups shall not decrease in hardness by more than 15 degrees and shall not be in an unsatisfactory operating condition as evidenced by excessive amounts of scoring, scuffing, blistering, cracking, chipping (heel abrasion), or change in shape from original appearance.
d. The base diameter of the rubber cups shall not increase by more than 0.9 mm (0.035 in).
e. The average lip diameter interference set of the rubber cups shall not be greater than 65%.
f. During any period of 24 000 strokes, the volume loss of fluid shall be not more than 36 mL.
g. The cylinder pistons shall not freeze nor function improperly throughout the test.
h. The volume loss of fluid during the 100 strokes at the end of the test shall not be more than 36 mL.
i. The fluid at the end of the test shall not be in an unsatisfactory operating condition as evidenced by sludging, jelling, or abrasive grittiness, and sedimentation shall not exceed 1.5% by volume after centrifuging.
j. No more than a trace of gum shall be deposited on brake cylinder walls or other metal parts during test. Brake cylinder shall be free of deposits which are abrasive or which cannot be removed when rubbed with a cloth wetted with isopropanol.

4.13 Wet Boiling Point
Brake fluid, when tested by the procedure specified in 5.17, shall have a wet boiling point of not less than 140 °C (284 °F).

5. Test Procedures

5.1 Equilibrium Reflux Boiling Point
Determine the equilibrium reflux boiling point of the fluid by ASTM D 1120 with the following exceptions:

5.2 Apparatus

5.2.1 3(d) THERMOMETER—ASTM E 1, 76 mm immersion, calibrated. Use ASTM 3C or 3F thermometer. For fluids boiling below 300 °C (572 °F), ASTM 2C or 2F thermometer may be used.

5.2.2 3(e) HEAT SOURCE—Use a suitable variac-controlled 100 mL heating mantle designed to fit the flask, capable of supplying the heat required to conform to the specified heating and reflux rates. (Supplier: GLAS COL Apparatus Co., Terre Haute, IN. Serial number: 135464. 230 W, 135 V [max]).

5.3 Preparation of Apparatus
5(d) Thoroughly clean and dry all glassware before use. Attach the flask to the condenser. Place the mantle under the flask and support it with a suitable ring clamp and laboratory type stand, holding the whole assembly in place by a clamp.

NOTE—Place the whole assembly in an area free from drafts or other types of sudden temperature changes.

5.4 Procedure
6(a) When everything is in readiness, turn on the condenser water and apply heat to the flask at such a rate that the fluid is refluxing in 10 min ± 2 at a rate in excess of 1 drop/s. Immediately adjust heat input to obtain a specified equilibrium reflux rate of 1 to 2 drops/s over the next 5 min ± 2 period. Maintain a timed and constant equilibrium reflux rate of 1 to 2 drops/s for an additional 2 min; record the average value of four temperature readings taken at 30 s intervals as the equilibrium reflux boiling point.

5.4.1 205 AND 232 °C (401 AND 450 °F) FLUIDS—Report the boiling point to the nearest degree Celsius (Fahrenheit). Duplicate runs which agree within 1 °C (2 °F) are acceptable for averaging (95% confidence level).

5.4.2 REPEATABILITY (SINGLE ANALYST)—The standard deviation of results (each the average of duplicates), obtained by the same analyst on different days, has been estimated to be 0.4 °C (0.88 °F) at 72 degrees of freedom. Two such values should be considered suspect (95% confidence level) if they differ by more than 1.5 °C (2.5 °F).

5.4.3 REPRODUCIBILITY (MULTILABORATORY)—The standard deviation of results (each the average of duplicates), obtained by analysts in different laboratories, has been estimated to be 1.8 °C (3.02 °F) at 17 degrees of freedom. Two such values should be considered suspect (95% confidence level) if they differ by more than 5 °C (9 °F).

5.4.4 288 °C (550 °F) FLUID—Report the boiling point to the nearest degree Celsius (Fahrenheit). Duplicate runs which agree within 3 °C (5 °F) are acceptable for averaging (95% confidence level)

5.4.5 REPEATABILITY (SINGLE ANALYST)—The standard deviation of results (each the average of duplicates), obtained by one analyst on different days, has been estimated to be 1.3 °C (2.38 °F) at 34 degrees of freedom. Two such values should be considered suspect (95% confidence level) if they differ by more than 4 °C (7 °F).

5.4.6 REPRODUCIBILITY (MULTILABORATORY)—The standard deviation of results (each the average of duplicates), obtained by analysts in different laboratories, has been estimated to be 3.5 °C (6.44 °F) at 15 degrees of freedom. Two such values should be considered suspect (95% confidence level) if they differ by more than 10.5 °C (19 °F).

5.5 Viscosity
Determine the kinematic viscosity of the fluid by ASTM D 445.

5.5.1 Report the viscosity to the nearest centistoke (mm^2/s). Duplicate runs which agree within 1.2% relative are acceptable for averaging (95% confidence level).

Repeatability (Single Analyst)—The coefficient of variation of results (each the average of duplicates), obtained by the same analyst on different days has been estimated to be 0.4% at 47 degrees of freedom. Two such values should be considered suspect (95% confidence level) if they differ by more than 1.2%.

Reproducibility (Multilaboratory)—The coefficient of variation of results (each the average of duplicates), obtained by analysts in different laboratories, has been estimated to be 1% at 15 degrees of freedom. Two such values should be considered suspect (95% confidence level) if they differ by more than 3%.

(R) **5.6 pH Value**—Mix the fluid with an equal volume of a mixture of 80% ethanol and 20% distilled water neutralized to a pH of 7. Determine the pH of the resulting solution electrometrically at 23 °C ± 5 (73.4 °F ± 9), using a pH meter equipped with a calibrated full range (0 to 14) glass electrode and a calomel reference electrode, as specified in ASTM D 664.

5.7 Fluid Stability

a. High Temperature Stability—Heat a new sample of the original test brake fluid to a temperature of 185 °C ± 2 (365 °F ± 3.6) by the procedure specified in 5.1 and maintain at that temperature for 2 h. Then determine the boiling point of this brake fluid as specified in 5.1. The difference between this observed boiling point and that previously determined in 5.1 shall be considered as the change in boiling point of the brake fluid.

b. Chemical Stability—Mix 30 mL of fluid with 30 mL of SAE Compatibility Fluid described in Appendix B (RM-6603). Determine the equilibrium reflux boiling point of this fluid mixture by use of the test apparatus specified in 5.1, applying heat to the flask at such a rate that the fluid is refluxing in 10 min ± 2 at a rate in excess of 1 drop/s. The reflux rate shall not exceed 5 drops/s. Record the maximum fluid temperature observed during the first minute after the fluid begins refluxing at a rate in excess of 1 drop/s. Over the next 15 min ± 1, adjust and maintain the rate of reflux to 1 to 2 drops/s. Maintain a timed and constant equilibrium reflux rate of 1 to 2 drops/s for an additional 2 min; record the average value of four temperature readings taken at 30 s intervals as the final equilibrium reflux boiling point. Chemical reversion is evidenced by the decrease in temperature between the maximum fluid temperature recorded and the final equilibrium reflux boiling point.

5.8 Corrosion—Prepare two sets of strips from each of the metals listed in Table 1, each strip having a surface area of 25 cm^2 ± 5 (approximately 8 cm long, 1.3 cm wide, and not more than 0.6 cm thick). Drill a hole between 4 and 5 mm in diameter and about 6 mm from one end of each strip. With the exception of the tinned iron strips, clean the strips by abrading them on all surface areas with 320A waterproof carborundum paper (RM-29) and isopropanol until all surface scratches, cuts, and pits are removed from the strips, using a new piece of carborundum paper for each different type of metal. Wash the strips, including the tinned iron, with isopropanol and dry the strips with a clean lint-free cloth and place strips in a desiccator containing desiccant maintained at 23 °C ± 5 (73.4 °F ± 9) for at least 1 h. Handle the strips with clean forceps after polishing to avoid fingerprint contamination.

Weigh each strip to the nearest 0.1 mg and assemble each set of strips on an uncoated steel bolt (RM-61) in the order tinned iron, steel, aluminum, cast iron, brass, copper, and zinc, so that the strips are in electrolytic contact. Bend the strips, other than cast iron, so that there is a separation of at least 3 mm between adjacent strips for a distance of about 6 cm from the free end of the strips. Immerse strip assemblies in isopropanol to eliminate fingerprints and then handle only with clean forceps.

(R) Measure the base diameter of two standard SBR cups (RM-3a) described in Appendix C, using an optical comparator or micrometer to the nearest 0.02 mm (0.001 in) along the centerline of the SAE and rubber type identifications and at right angles to this centerline. Take the measurements within 0.4 mm (0.015 in) of the bottom edge and parallel to the base of the cup. Discard any cup if the two measured diameters differ by more than 0.08 mm (0.003 in). Average the two readings of each cup. Support the rubber cup on a rubber anvil or cylinder having a flat circular top surface at least 19 mm in diameter, a thickness of at least 9 mm, and a hardness within 5 IRHD of the hardness of the rubber test cup. Determine the hardness of each cup thus supported by the procedure specified in ASTM D 1415 using the Standard Tester.

NOTE—ASTM D 2240 may be used for quality control and routine tests when a type A durometer is equipped with a fixture for keeping the plane of the pressure foot on the durometer parallel to the plane of the cup face during measurement.

Obtain two straight-sided round glass jars[1] (RM-49) having a capacity of approximately 475 mL and inner dimensions of approximately 100 mm in height and 75 mm in diameter.

To the RM-49 corrosion test jar, apply four wrappings of ¾ in Teflon tape around the jar threads allowing a ⅛ in height above the top of the jar. Place one rubber cup with lip edge facing up, in each of the two glass jars. Use only tinned steel lids vented with a hole 0.8 mm ± 0.1 in diameter (RM-64).

Insert a metal strip assembly inside each cup with the bolted end in contact with the concavity of the cup and the free end extending upward in the jar. Mix 760 mL of fluid with 40 mL of distilled water. Add 400 mL of the mixture to cover the metal strip assembly in each jar to a depth of approximately 10 mm above the tops of the strips. Tighten the lids and place the jars in an oven maintained at 100 °C ± 2 (212 °F ± 3.6) for 120 h ± 2. Allow the jars to cool at 23 °C ± 5 (73.4 °F ± 9) for 60 to 90 min. Immediately following the cooling period, remove the metal strips from the jars by use of a forceps, removing loose adhering sediment by agitation of the metal strip assembly in the fluid in jar. Examine test strips and test jars for adhering crystalline deposit, disassemble the metal strips, removing adhering fluid by flushing with water, and clean individual strips by wiping with a cloth wetted with isopropanol. Examine the strips for evidence of corrosion and pitting. Place strips in a desiccator containing a desiccant maintained at 23 °C ± 5 (73.4 °F ± 9) for at least 1 h. Weigh each strip to the nearest 0.1 mg. Determine the difference in weight of each metal strip and divide the difference by the total surface area of the metal strip measured in square centimeters. Average the measured quantities of the duplicates. In the event of a marginal pass on inspection, or of a failure in only one of the duplicates, another set of duplicate test samples shall be run. Both repeat samples must meet all the requirements of 4.5.

Immediately following the cooling period, remove the rubber cups from the jars by use of a forceps, removing loose adhering sediment by agitation of the cup in the fluid in jar. Rinse cups in isopropanol and air dry cups. Examine the cups for evidence of sloughing, blisters, and other forms of disintegration. Measure the base diameter and hardness of each cup within 15 min after removal from the fluid.

Examine the fluid-water mixture in the jars for jelling. Agitate the fluid in jars to suspend and uniformly disperse sediment and transfer a 100 mL portion of this fluid to an ASTM cone-shaped centrifuge tube and determine percent sediment as described in 5(b) of ASTM D 91. Measure the pH value of the corrosion test fluid by the procedure specified in 5.6.

5.9 Fluidity and Appearance at Low Temperatures

a. At −40 °C (−40 °F)—Place 100 mL of the test fluid in a glass sample bottle[2] (RM-59a) having a capacity of approximately 125 mL, an outside diameter of 37 mm ± 0.5, and an overall height of 165 mm ± 3. Stopper or cap the bottle tightly and place in a cold bath maintained at −40 °C ± 2 (−40 °F ± 3.6) for 144 h ± 4. Remove the bottle from the bath, quickly wipe the bottle with a clean lint-free cloth saturated with isopropyl alcohol, and examine the fluid for evidence of stratification, sediment, or crystals. Invert the bottle and determine the number of seconds required for the air bubble to travel to the top of the fluid. Allow the fluid to warm to room temperature 23 °C ± 5 (73 °F ± 9); if necessary, allow to stand for as long as 4 h. Examine the fluid for clarity and appearance by comparing it to an original sample of the test fluid in an identical container.

b. At −50 °C (−58 °F)—Place 100 mL of fluid in a glass sample bottle (same as in −40 °F test above.) Stopper or cap the bottle tightly and place in a cold bath maintained at −50 °C ± 2 (−58 °F ± 3.6) for 6 h ± 0.2. Remove the bottle from the bath, quickly wipe the bottle with a clean lint-free cloth saturated with isopropyl alcohol, and examine the fluid for evidence of stratification, sediment, or crystals. Invert the bottle and determine the number of seconds required for the air bubble to travel to the top of the fluid. Allow the fluid to warm to room temperature 23 °C ± 5 (73 °F ± 9); if necessary, allow to stand for as long as 4 h. Examine the fluid for clarity and appearance by comparing it to a sample of the original test fluid in an identical container.

5.10 Evaporation—Obtain the tare weight of four covered Petri dishes of approximately 100 mm in diameter and 15 mm high, weighing with cover in place to the nearest 0.01 g. Place approximately 25 mL of fluid in each of the four tared Petri dishes, replace proper covers, and reweigh to the nearest 0.01 g. Determine the weight of fluid from the difference in weights of filled and empty dishes.

Place the dishes inside the inverted covers in a top vented gravity convection oven at 100 °C ± 2 (212 °F ± 3.6) and maintain this temperature for a total of 168 h ± 2.

Remove the dishes from the oven. Allow to cool to 23 °C ± 5 (73.4 °F ± 9) with covers on and weigh each dish. Calculate the percentage of fluid evaporated from each dish. Average the percentage evaporated from all four dishes to determine the loss by evaporation.

Examine the residue in the dishes at the end of 1 h at 23 °C ± 5 (73.4 °F ± 9). Rub any sediment with the fingertip to determine grittiness or abrasiveness.

Combine the residue from the four dishes in an oil sample bottle (RM-59a), store in a vertical position at −5 °C ± 1 (23 °F ± 1.8) for 60 min ± 10, then remove quickly and turn to the horizontal. The residue must flow at least 5 mm (0.2 in) along tube wall within 5 s.

5.11 Water Tolerance

a. At −40 °C (−40 °F)—Pour 100 mL of fluid which has been humidified according to 5.16 into an ASTM cone-shaped centri-

---

[1] Obtainable from the Society of Automotive Engineers, Inc., 400 Commonwealth Drive, Warrendale, PA 15096-0001.

[2] Obtainable from the Society of Automotive Engineers, Inc., 400 Commonwealth Drive, Warrendale, PA 15096-0001.

fuge tube described in 3(a) in ASTM D 91. Stopper the tube with a cork and place in a cold bath maintained at −40 °C ± 2 (−40 °F ± 3.6) for 22 h ± 2. Remove the centrifuge tube from the bath, quickly wipe the tube with a clean lint-free cloth saturated with isopropanol, determine the transparency of the fluid by placing the tube against a hiding power test chart[3] (RM-28) and observing the clarity of the contrast lines on the chart when viewed through the fluid. Examine the fluid for evidence of stratification and sedimentation. Invert the tube and determine the number of seconds required for the air bubble to travel to the top of the fluid. (The air bubble shall be considered to have reached the top of the fluid when the top of the bubble reaches the 2 mL graduation of the centrifuge tube.)

b. At 60 °C (140 °F)—Place the centrifuge tube from 5.11(a) in an oven maintained at 60 °C ± 2 (140 °F ± 3.6) for 22 h ± 2. Remove the tube from the oven and immediately examine the contents for evidence of stratification. Determine percent sediment by volume as described in 5(b) of ASTM D 91.

(R) **5.12 Compatibility**

a. At −40 °C (−40 °F)—Mix 50 mL of fluid with 50 mL of SAE Compatibility Fluid described in Appendix B (RM-6603) and pour this mixture into an ASTM cone-shaped centrifuge tube described in 3(a) in ASTM D 91 and stopper with a cork. Place centrifuge tube for 22 h ± 2 in a bath maintained at −40 °C ± 2 (−40 °F ± 3.6). Remove the centrifuge tube from the bath, quickly wipe the tube with a clean lint-free cloth saturated with isopropanol, determine the transparency of the fluid by placing the tube against a hiding power test chart[3] (RM-28) and observing the clarity of the contrast lines on the chart when viewed through the fluid. Examine the fluid for stratification and sedimentation.

b. At 60 °C (140 °F)—Place the centrifuge tube mentioned in 5.12(a) in an oven maintained at 60 °C ± 2 (140 °F ± 3.6) for 22 h ± 2. Remove the tube from the oven and immediately examine the contents for evidence of stratification. Determine percent sediment by volume as described in 5(b) of ASTM D 91.

5.13 Resistance to Oxidation—Prepare two sets of aluminum and cast iron test strips (as listed in Table 1) by the procedure specified in 5.8. Weigh each strip to the nearest 0.1 mg and assemble a strip of each metal on an uncoated steel bolt (RM-62), separating the strips at each end with a piece of tinfoil[4] (RM-27) (99.5% tin, 0.20% lead, max) approximately 12 mm square and between 0.02 and 0.06 mm in thickness.

Place 30 mL ± 1 of fluid in a small glass bottle approximately 120 mL in capacity. Add 60 mg ± 2 of reagent grade benzoyl peroxide and 1.5 mL ± 0.05 distilled water to bottle. Stopper the bottle and shake the contents, avoiding getting the solution on the stopper. Place bottle in an oven at 70 °C ± 2 (158 °F ± 3.6) for 120 min ± 10, shaking every 15 min to effect solution of the peroxide. Remove the bottle from the oven, do not disturb the stopper, and cool in air at room temperature (23 °C ± 5) for 2 h.

(R) Place approximately 1/8 section of a standard SBR cup described in Appendix C (RM-3a) in the bottom of each of two test tubes about 22 mm in diameter and 175 mm in length. Add 10 mL of prepared test fluid to each test tube. Place a metal-strip assembly in each tube with the end of the strips resting on the rubber, the solution covering about one-half the length of the strips, and the bolted end remaining out of the solution. Stopper the tubes with corks and store upright for 22 h ± 2 at 23 °C ± 5 (73.4 °F ± 9). Loosen the stoppers and place the tubes for 168 h ± 2 in an oven maintained at 70 °C ± 2 (158 °F ± 3.6). After the heating period, remove and disassemble the metal strips. Examine the strips for gum deposits. Wipe the strips with a cloth wet with isopropanol and examine for pitting or roughening of surface. Place strips in a desiccator containing a desiccant maintained at 23 °C ± 5 (73.4 °F ± 9) for at least 1 h. Weigh each strip to the nearest 0.1 mg.

Determine corrosion loss by dividing the difference in weight of each metal strip by the total surface area of each metal strip measured in square centimeters. Average the measured quantities of the duplicates. In the event of a marginal pass on inspection, or of a failure in only one of the duplicates, another set of duplicate test samples shall be run. Both repeat samples must meet all the requirements of 4.10.

(R) **5.14 Effect on Rubber**—For test procedures 5.14.1 and 5.14.2, use standard SBR cups described in Appendix C (RM-3a). Measure the base diameter of all cups and hardness of all specimens as described in 5.8, discarding any cups whose diameters differ by more than 0.08 mm (0.003 in).

For test procedures 5.14.3 and 5.14.4, cut 1 × 1 in test specimens from standard EPDM slab stock, as described in Appendix D (RM-69). Determine the volume of each specimen in the following manner:

Weigh the specimen in air (M_1) to the nearest milligram and then weigh the specimens immersed in room temperature distilled water (M_2) containing no more than 0.2% of a suitable wetting agent. Pluronic L-61 (BASF Wyandotte) or equivalent has been found to be acceptable.

5.14.1 Test at 70 °C (158 °F)—Place two standard SBR cups in a straight sided round glass jar[5] (RM-51), having a capacity of approximately 250 mL and inner dimensions of approximately 125 mm in height and 50 mm in diameter, and a tinned steel lid (RM-52a). Add 75 mL of fluid to the jar and heat for 70 h ± 2 at 70 °C ± 2 (158 °F ± 3.6). Allow the jar to cool at 23 °C ± 5 (73.4 °F ± 9) for 60 to 90 min. Remove the cups from the jar, wash quickly with isopropanol, and air dry cups. Examine the cups for disintegration as evidenced by blisters or sloughing. Measure the base diameter and hardness of each cup within 15 min after removal from the fluid.

5.14.2 Test at 120 °C (248 °F)—Place two standard SBR cups (RM-3a) in a straight sided round glass jar[5] (RM-51), having a capacity of approximately 250 mL and inner dimensions of approximately 125 mm in height and 50 mm in diameter, and a tinned steel lid (RM-52a). Add 75 mL of fluid to the jar and heat for 70 h ± 2 at 120 °C ± 2 (248 °F ± 3.6). Allow the jar to cool at 23 °C ± 5 (73.4 °F ± 9) for 60 to 90 min. Remove the cups from the jar, wash quickly with isopropanol, and air dry cups. Examine the cups for disintegration as evidenced by blisters or sloughing. Measure the base diameter and hardness of each cup within 15 min after removal from the fluid.

5.14.3 Test at 70 °C (158 °F)—Place two 1 × 1 in standard test specimens (RM-69) in a straight sided round glass jar[6], having a capacity of approximately 250 mL and inner dimensions of approximately 125 mm in height and 50 mm in diameter, and a tinned steel lid. Add 75 mL of fluid to the jar. Heat the prepared glass jar for 70 h ± 2 at 70 °C ± 2 (158 °F ± 3.6). Allow the jar to cool at 23 °C ± 5 (73.4 °F ± 9) for 60 to 90 min. Remove the specimens from the jar, wash quickly with isopropanol, and air dry. Examine the specimens for disintegration as evidenced by blisters or sloughing. Weigh each specimen in air (M_3), again to the nearest milligram, then reweigh immersed in room temperature distilled water (M_4), to determine the volume after hot fluid immersion. Measure the hardness of each specimen. All weighings must be completed within 60 min after removal from the test fluid. Volume changes shall be reported as a percentage of the original volume, calculated as follows:

$$\text{Percent volume change} = \frac{(M_3 - M_4) - (M_1 - M_2) \times 100}{(M_1 - M_2)} \quad \text{(Eq.1)}$$

where:

M_1 = Initial mass in air
M_2 = Initial mass in water
M_3 = Final mass in air
M_4 = Final mass in water

5.14.4 Test at 120 °C (248 °F)—Place two 1 × 1 in standard test specimens (RM-69) in a straight sided round glass jar[6], having a capacity of approximately 250 mL and inner dimensions of approximately

---

[3] A suitable hiding power chart as described in ASTM D 344, Method of Test for Relative Dry Hiding Power of Paints, published by the American Society for Testing and Materials, or in Method 4112 of Federal Test Method Standard No. 141, is obtainable from the Society of Automotive Engineers, Inc., 400 Commonwealth Drive, Warrendale, PA 15096-0001.

[4] Obtainable from the Society of Automotive Engineers, Inc., 400 Commonwealth Drive, Warrendale, PA 15096-0001.

[5] Obtainable from the Society of Automotive Engineers, Inc., 400 Commonwealth Drive, Warrendale, PA 15096-0001.

[6] Obtainable from the Society of Automotive Engineers, Inc., 400 Commonwealth Drive, Warrendale, PA 15096-0001.

125 mm in height and 50 mm in diameter, and tinned steel lid. Add 75 mL of fluid to the jar. Heat the prepared glass jar for 70 h ± 2 at 120 °C ± 2 (248 °F ± 3.6). Allow the jar to cool to 23 °C ± 5 (73.4 °F ± 9) for 60 to 90 min. Remove the specimens from the jar, wash quickly with isopropanol, and air dry. Examine the specimens for disintegration as evidenced by blisters or sloughing. Determine the volume change as in 5.14.3. Measure the hardness of each specimen.

5.14.5 Report the rubber swell to the nearest 0.03 mm (0.001 in). Duplicate results which agree within 0.10 mm (0.004 in) are acceptable for averaging (95% confidence level).

Repeatability (Single Analyst)—The standard deviation of results (each the average of duplicate determinations) obtained by the same analyst on different days has been estimated to be 0.51 mm (0.002 in) at 46 degrees of freedom. Two such values should be considered suspect (95% confidence level) if they differ by more than 0.13 mm (0.005 in).

Reproducibility (Multilaboratory)—The standard deviation of results (each the average of duplicates) obtained by analysts in different laboratories has been estimated to be 0.08 mm (0.003 in) at 7 degrees of freedom. Two such values should be considered suspect (95% confidence level) if they differ by more than 0.20 mm (0.008 in).

5.15 Stroking Test Procedure—Use the following procedure to evaluate the lubrication quality of the brake fluid.

(R) 5.15.1 TEST APPARATUS AND MATERIAL[7]—Use the Figure 1 stroking fixture type apparatus with the following components arranged as shown in Figure 2. The drum and shoe apparatus as described in SAE J1703c may be used as an alternative test system.

a. Master Cylinder Assembly—One cast iron housing hydraulic brake master cylinder having a diameter of approximately 28 mm (1⅛ in) and fitted with an uncoated steel standpipe. Master cylinder used is SAE RM-15b 28 mm (1⅛ in) diameter or equivalent.

b. Brake Assemblies—Three cast iron housing straight bore hydraulic brake wheel cylinder assemblies having a diameter approximately 28 mm (1⅛ in). Wheel cylinder used is SAE RM-14b or equivalent with stroking fixture apparatus. Three fixture units are required, including appropriate adapter mounting plates to hold the brake wheel cylinder assemblies as shown in Figure 1.

c. Braking Pressure Actuating Mechanism—A suitable actuating mechanism for applying a force to the master cylinder push rod without side thrust.

The amount of force applied by the actuating mechanism shall be adjustable and capable of supplying sufficient stroke and thrust to the master cylinder to create a pressure of at least 70 kg/cm² (1000 lbf/in²) in the simulated brake system. A hydraulic gauge and pressure recorder capable of establishing the pressure curve of the system and monitoring the pressure developed shall be installed on a hydraulic line extending from the master cylinder to the outside of the oven. This line shall be provided with a shut-off valve and a bleeding valve for removing air from the connecting tubing.

The actuating mechanism shall be designed to provide a stroking rate of approximately 1000 strokes/h. The pressure buildup rate versus cylinder stroke and time shall correspond to Figure 3.

d. Heated Air Bath Cabinet—An insulated cabinet or oven having sufficient capacity to house the three wheel cylinder fixture assemblies, master cylinder, and necessary connections. A suitable thermostatically-controlled heating system is required to maintain a brake fluid temperature of 120 °C ± 5 (248 °F ± 9). Heaters shall be shielded to prevent direct radiation to wheel or master cylinders. Fluid temperature shall be monitored at random intervals during the test at the master cylinder reservoir, using a temperature recording device.

(R) 5.15.2 PREPARATION OF TEST APPARATUS

a. Wheel Cylinder Assemblies—Use new wheel cylinder assemblies SAE RM-14b or equivalent having diameters as specified in 5.15.1(b). Pistons (SAE RM-12 or equivalent) shall be made from unanodized SAE AA 2024 aluminum alloy. Disassemble cylinders and discard rubber cups. Clean all metal parts with isopropanol and dry with clean compressed air. Inspect the working surfaces of all metal parts for scoring, galling, or pitting and cylinder bore roughness, and discard all defective parts. Remove any stains on cylinder walls with crocus cloth and isopropanol. If stains cannot be removed, discard the cylinder. Measure the internal diameter of each cylinder at locations approximately 19 mm (0.75 in) from each end of the cylinder bore, taking measurements in line with the hydraulic inlet opening and at right angles to the centerline. Discard the cylinder if any of these four readings exceeds maximum or minimum limits of 28.66 to 28.60 mm (1.1285 to 1.126 in). Measure the outside diameter of each piston at two points approximately 90 degrees apart. Discard any piston if either reading exceeds maximum or minimum limits of 28.55 to 28.52 mm (1.124 to 1.123 in). Select parts to insure that the clearance between each piston and mating cylinder is within 0.08 to 0.13 mm (0.003 to 0.005 in). Use new standard SAE RM-3a SBR cups as specified in Figure 4 and Appendix C that are free of lint and dirt. Discard any cups showing imperfections such as cuts, tooling marks, molding flaws, or blisters. Measure the lip and base diameters of all test cups with an optical comparator or a micrometer to the nearest 0.025 mm (0.001 in) along the centerline of SAE and rubber type identifications and at right angles to this centerline. Determine base diameter measurements within 0.8 mm (0.032 in) of the bottom edge and parallel to the base of the cup. Discard any cups if the two measured lip or base diameters differ by more than 0.08 mm (0.003 in). Average the lip and base diameters of each cup. Determine the hardness of all cups by the procedure specified in 5.8. Clean rubber parts with isopropanol and a lint-free cloth. Dry with clean compressed air. Dip the rubber and metal parts of the wheel cylinders, except housings, in the fluid to be tested and install them in accordance with manufacturer's instructions. Rubber boots may be retained on the cylinders if a small section is removed on the bottom to observe leakage. Manually stroke the cylinders to insure that they operate easily. Install cylinders in the simulated brake system.

b. Master Cylinder Assembly—Use a new SAE RM-15b master cylinder or equivalent having an SAE RM-13 aluminum alloy piston or equivalent and new standard SAE RM-4a and RM-5a SBR cups as specified in Figures 5 and 6 and in Appendix C. Inspect and clean all parts as specified in 5.15.2(a). Measure each end of the master cylinder piston at two points approximately 90 degrees apart. Discard the piston if any of these readings exceed maximum or minimum limits of 28.55 to 28.52 mm (1.124 to 1.123 in). Dip the secondary cup in the test brake fluid, assemble on the piston, and maintain the assembly in a vertical position at 23 °C ± 5 (73.4 °F ± 9) for at least 2 h. Determine the lip and base diameter of the secondary cup as installed on the piston and the primary cup at locations shown in Figure 5. Inspect the relief and supply ports of the master cylinder and discard the cylinder if these ports have burrs or wire edges. Measure the internal diameter of the cylinder at two locations: approximately midway between the relief and

---

[7] Obtainable from the Society of Automotive Engineers, Inc., 400 Commonwealth Drive, Warrendale, PA 15096-0001.

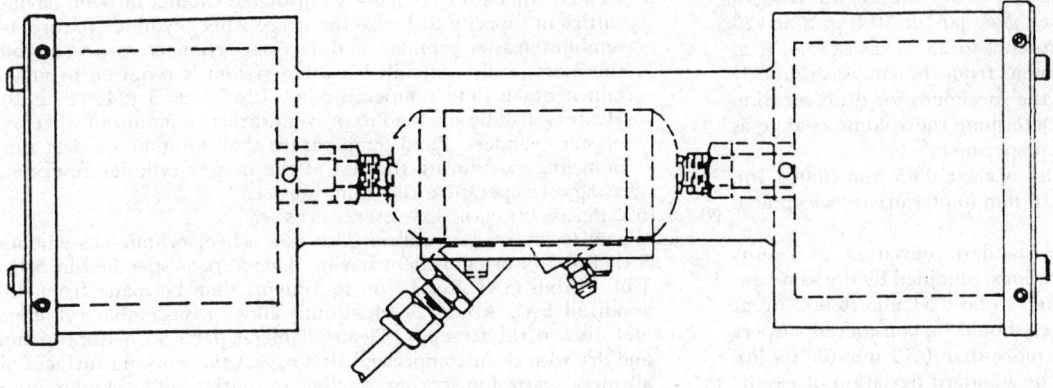

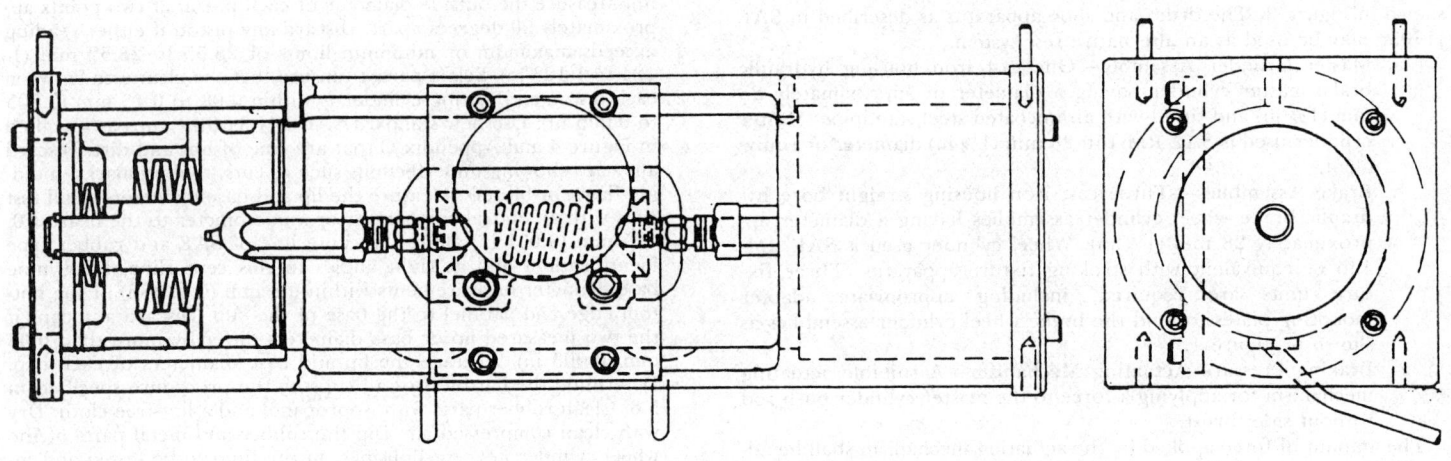

(R) FIGURE 1—STROKING FIXTURE APPARATUS

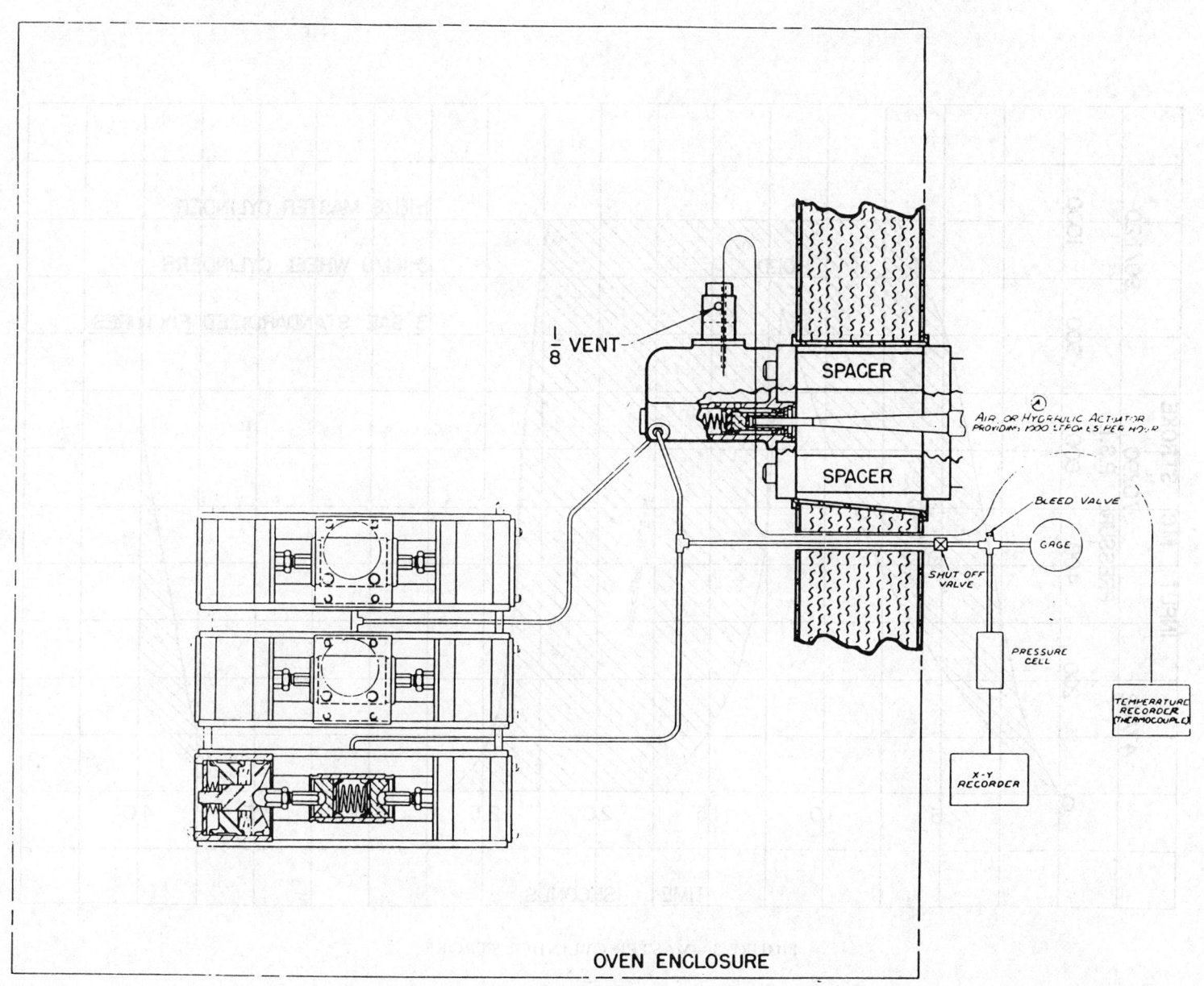

(R) FIGURE 2—STROKING TEST APPARATUS

25.48

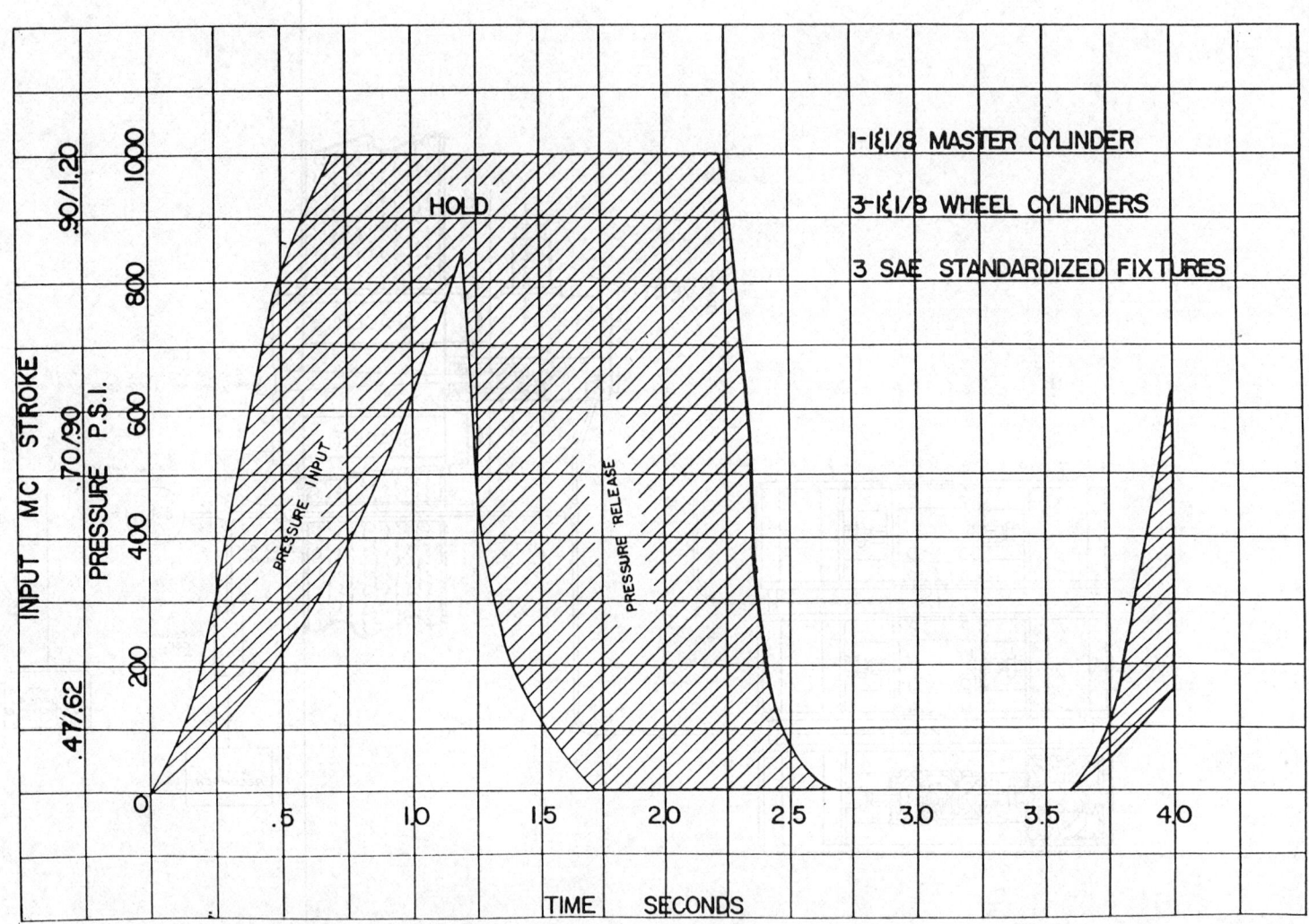

FIGURE 3—MASTER CYLINDER STROKE

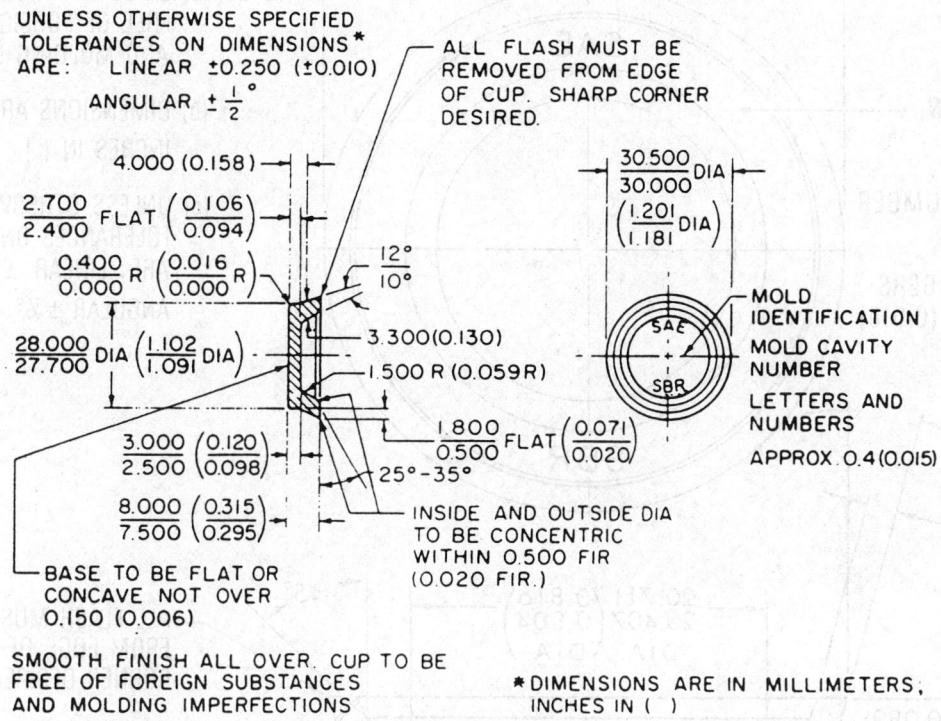

FIGURE 4—SAE TEST CUP WHEEL CYLINDER (RM-3a)

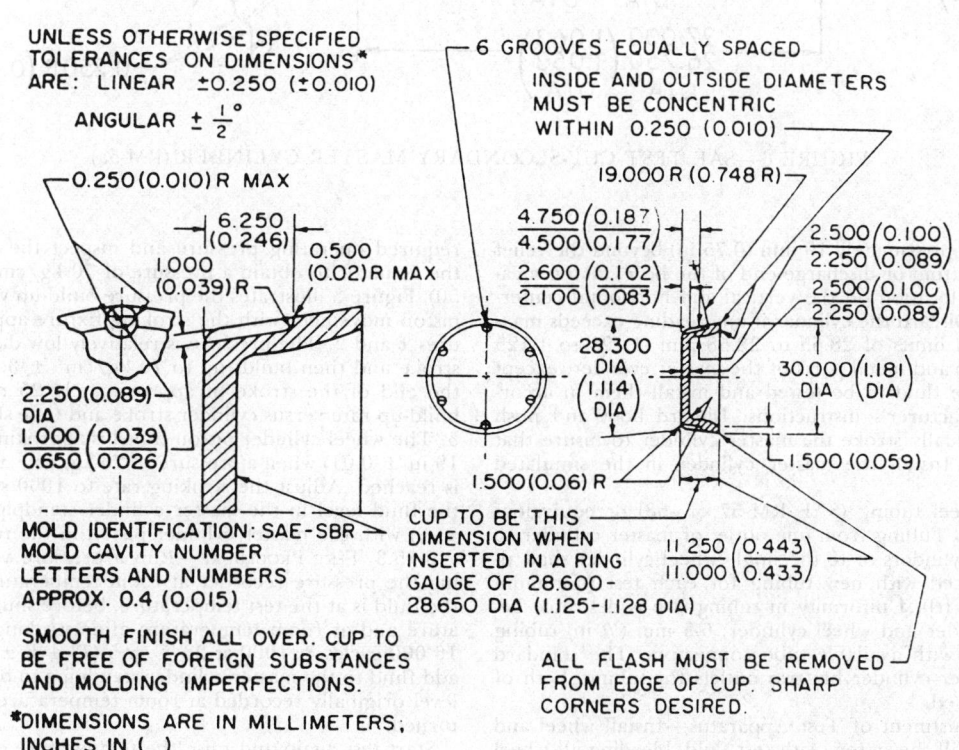

FIGURE 5—SAE TEST CUP-PRIMARY MASTER CYLINDER (RM-4a)

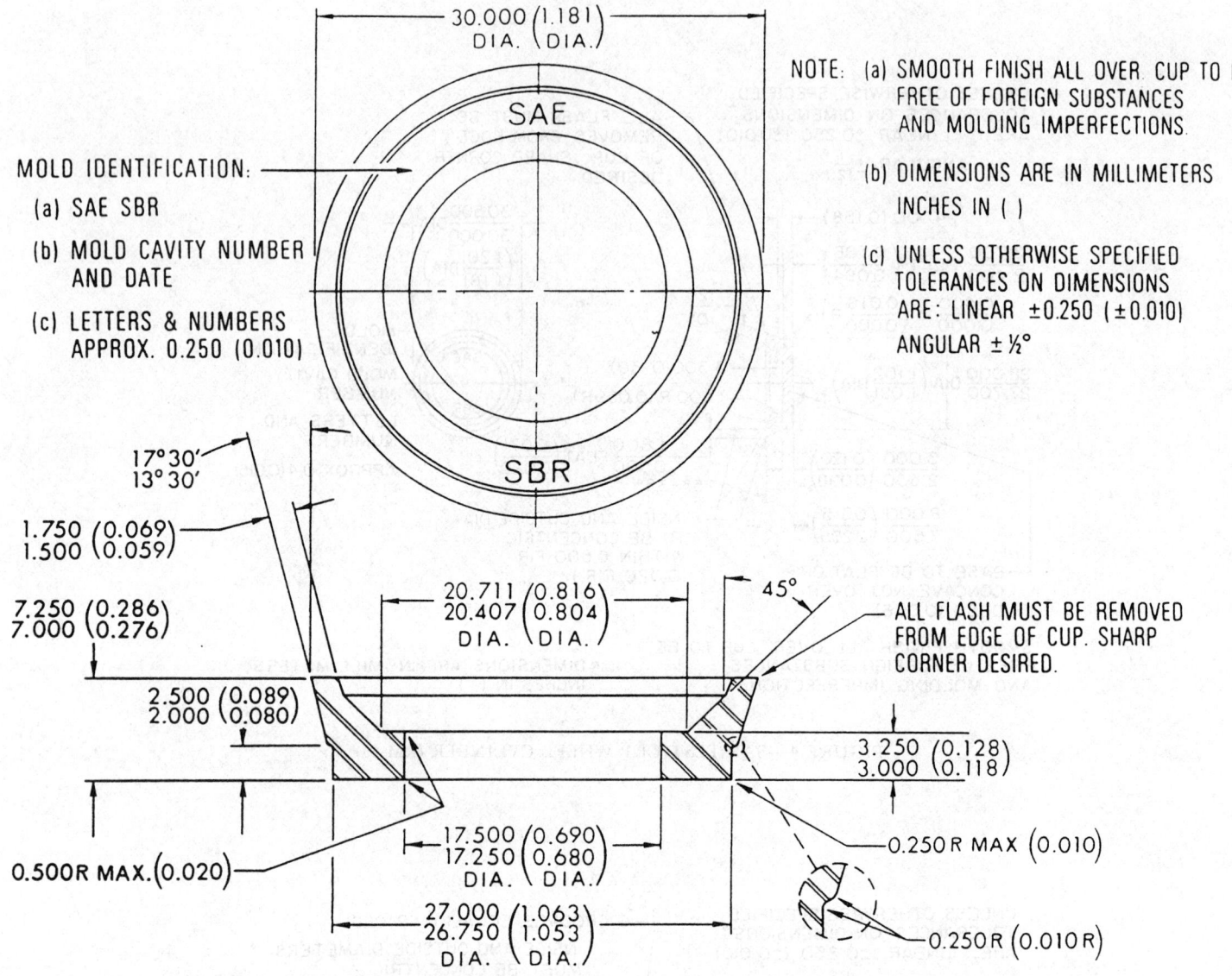

FIGURE 6—SAE TEST CUP-SECONDARY MASTER CYLINDER (RM-5a)

supply ports and approximately 19 mm (0.75 in) beyond the relief port toward the bottom or discharge end of the bore, taking measurements at each location on the vertical and horizontal centerlines of the bore. Discard the cylinder if any reading exceeds maximum or minimum limits of 28.65 to 28.58 mm (1.128 to 1.125 in). Dip the rubber and metal parts of the master cylinder, except the housing, in the fluid to be tested and install them in accordance with manufacturer's instructions. Discard boot and push rod assembly. Manually stroke the master cylinder to insure that it operates easily. Install the master cylinder in the simulated brake system.

c. Use double-wall steel tubing (SAE RM-57 or -58) or equivalent meeting SAE J527. Tubing from one outlet of master cylinder to the pair of wheel cylinders or to the single wheel cylinder shall alternately be replaced with new tubing for each test (minimum length 915 mm [3 ft]). Uniformity in tubing size is desirable between master cylinder and wheel cylinder; 6.3 mm (¼ in) tubing is more adaptable with available tube connectors. The standard SAE RM-15b master cylinder has two outlets for tubing, both of which should be used.

d. Assembly and Adjustment of Test Apparatus—Install wheel and master cylinders. Fill the system with test fluid, bleeding all wheel cylinders and the pressure equipment and gauges to remove entrapped air from the system.

Operate the actuator manually to apply a pressure of more than the required operating pressure and inspect the system for leaks. Adjust the actuator to obtain a pressure of 70 kg/cm^2 ± 3.5 (1000 lbf/in^2 ± 50). Figure 3 illustrates the pressure build-up versus the master cylinder piston movement with the stroking fixture apparatus illustrated in Figures 1 and 2. The pressure is relatively low during the first part of the stroke and then builds up to 70 kg/cm^2 ± 3.5 (1000 lbf/in^2 ± 50) at the end of the stroke of approximately 25 mm (1 in). The pressure build-up rate versus cylinder stroke and time shall correspond to Figure 3. The wheel cylinder piston travel is approximately 4.8 mm ± 0.25 (0.19 in ± 0.01) when a pressure of 70 kg/cm^2 ± 3.5 (1000 lbf/in^2 ± 50) is reached. Adjust the stroking rate to 1000 strokes/h ± 100. Record the fluid level in the master cylinder standpipe at 23 °C ± 5 (73.4 °F ± 9) with the master cylinder piston in the fully returned position.

5.15.3 TEST PROCEDURE—Run a pressure versus stroke curve utilizing the pressure recorder at room temperature before stroking, after the fluid is at the test temperature, before shutdown at the test temperature and at room temperature after stroking. Operate the system of 16 000 cycles ± 1000 at 23 °C ± 5 (73.4 °F ± 9). Repair any leaks and add fluid to the master cylinder standpipe to bring the fluid level to the level originally recorded at room temperature with the piston fully returned.

Start test again and raise the temperature of the fluid in the master cylinder within 6 h ± 2 to 120 °C ± 5 (248 °F ± 9). During test, observe operation of the master cylinder for complete piston return and wheel cylinders for proper operation. Observe fluid level in relation to the room temperature level at random intervals. Continue the test to 85 000 total recorded strokes which shall include the number of strokes during operation at 23 °C ± 5 (73.4 °F ± 9), the number of strokes required to bring the system to the operating temperature of 120 °C ± 5 (248 °F ± 9), plus the number of strokes at this operating temperature. Stop the test, and with the master cylinder piston in the fully returned position to relieve retained pressure in the system, allow the equipment to cool to room temperature.

Record the amount of fluid required to replenish any loss of fluid to the 23 °C ± 5 (73.4 °F ± 9) level originally recorded. Stroke the assembly an additional 100 strokes at 23 °C ± 5 (73.4 °F ± 9) and 70 kg/cm² ± 3.5 (1000 lbf/in² ± 50), examine wheel cylinders for leakage, and add and record volume of fluid required to bring the fluid level to the 23 °C ± 5 (73.4 °F ± 9) original level.

Within 16 h, remove the master and wheel cylinders from the system, retaining the fluid in the cylinders by immediately capping or plugging the ports. Disassemble the cylinders, collecting the fluid from the master cylinder and wheel cylinders in a glass jar. Record any sludge, jell, or abrasive grit present in the test fluid. When collecting the stroked fluid, all the residue which has deposited on the rubber and metal internal parts should be removed by rinsing and agitating such parts in the stroked fluid and using a soft brush to assure that all loose adhering sediment is collected.

Clean rubber cups in isopropanol and dry with clean, compressed air. Inspect cups for tackiness, scoring, scuffing, blistering, cracking, chipping (heel abrasions), and change in shape from original appearance. Within 1 h after disassembly, measure the lip and base diameter of each cylinder cup by the procedure specified in 5.15.2(b) with the exception that the lip or base diameters of cups may differ by more than 0.08 mm (0.003 in). Determine the hardness of each cup by the procedure specified in 5.8.

Within 1 h after draining cylinders, agitate fluid in glass jar to suspend and uniformly disperse sediment and transfer a 100 mL portion of this fluid to an ASTM cone-shaped centrifuge tube and determine percent sediment as described in 5(b) of ASTM D 91. Inspect cylinder parts, recording any gum deposits. Rub any deposits adhering to cylinder walls with a cloth wetted with isopropanol to determine abrasiveness and removability. Clean cylinder parts in isopropanol to determine abrasiveness and removability. Measure and record diameters of pistons and cylinders by the procedures specified in 5.15.2(a) and 5.15.2(b).

Calculate lip diameter interference set by the following formula:

$$\frac{D_1 - D_2}{D_1 - D_3} \times 100 = \% \text{ Lip Diameter Interference Set} \quad \text{(Eq.1)}$$

where:
D_1 = Original lip diameter
D_2 = Final lip diameter
D_3 = Original cylinder bore diameter

Repeat the test if mechanical failure occurs that may affect the evaluation of the test fluid.

5.16 Humidification Procedure—Lubricate the ground-glass joint of a 250 mm (9.89 in) I.D. bowl-form desiccator having matched tubulated glass cover and fitted with a No. 8 rubber stopper. Pour 450 mL ± 10 (15.22 oz ± 0.34) of distilled water into the desiccator and insert a perforated porcelain plate (Coors No. 60456 or equivalent). Immediately place one open RM-49 corrosion test jar containing 350 mL ± 5 of the test brake fluid into the desiccator. Place a second open RM-49 corrosion test jar containing 350 mL ± 5 of TEGME (triethylene glycol monomethyl ether, brake fluid grade-Appendix E) (RM-71) into the same desiccator. The water content of the TEGME control fluid at the start of exposure shall have been adjusted to 0.50% ± 0.05 by weight (Karl Fischer analysis or equivalent). Replace desiccator cover and insert at once into an ASTM E 145, Type II A, forced ventilation oven set at 50 °C ± 1 (122 °F ± 1.8).

Periodically, during oven humidification, remove the rubber stopper from the desiccator and, using a long needle hypodermic syringe, quickly sample the control fluid and determine its water content. When the water content of the control fluid has reached 3.70% ± 0.05 by weight, remove the desiccator from the oven and seal the test jar promptly using a screw-cap lid (RM-63). Allow the sealed jar to cool for 60 to 90 min at 23 °C ± 5 (73.4 °F ± 9).

5.17 Wet Boiling Point—Humidify the fluid as described in 5.16 and determine the boiling point as described in 5.1.

(R) APPENDIX A
STANDARD CORROSION TEST STRIPS[8]

Corrosion Test Strip	Material Specification	General Material Data	Dimensions	Surface Requirements
Tinned iron RM-6	ASTM A 624, Federal Specification QQ-T-425A	SR tin plate electrolytic, bright: No. 25, type MR Temper 3, base weight 85 lb Ferrostand and DOS oil	Approx. 8 cm long; 1.3 cm wide Thickness: As purchased Surface area: 25 cm² ± 5	As sheared. Clean and uniform tinning.
Steel RM-7	SAE 1018	Low carbon sheet Cold rolled Hardness: 40 to 72 RB	Approx. 8 cm long; 1.3 cm wide Thickness: Approx. 0.2 cm Surface area: 25 cm² ± 5	Edges machined to remove shearing marks. Clean uniform surfaces.
Aluminum RM-8	SAE AA2024	Wrought aluminum alloy Temper T3 Hardness: 75 RB typical	Approx. 8 cm long; 1.3 cm wide Thickness: Approx. 0.2 cm Surface area: 25 cm² ± 5	Edges machined to remove shearing marks. Clean uniform surfaces.
Cast iron RM-9	SAE G3000	Soft automotive cast iron. Must be free from shrinkage cavities, porosity, or any other defects detrimental to specification use of the material. Hardness: 86 to 98 RB	Approx. 8 cm long; 1.3 cm wide Thickness: Approx. 0.4 cm Surface area: 25 cm² ± 5	Surface grind 4 sides to dimension using a well-dressed No. 80 Alundum wheel. Clean uniform surfaces.
Brass RM-10	SAE CA260	Wrought alloy—yellow brass Rolled sheet or strip; half hard temper. Hardness: 57 to 74 RB	Approx. 8 cm long; 1.3 cm wide Thickness: Approx. 0.2 cm Surface area: 25 cm² + 5	Edges machined to remove shearing marks. Clean uniform surfaces.
Copper RM-11	SAE CA114	Cold rolled copper sheet or strip Half-hard temper Hardness: 35 to 56 RB	Approx. 8 cm long; 1.3 cm wide Thickness: Approx. 0.2 cm Surface area: 25 cm² ± 5	Edges machined to remove shearing marks. Clean uniform surfaces.
Zinc ISO-2	2N AL4 CU1 ISO/R 301	Die casting alloy strips. Hardness: 85 to 105 HB	Approx. 8 cm long; 1.3 cm wide Thickness: Approx. 0.2 cm Surface area: 25 cm² ± 5	Edges machined to remove shearing marks. Clean uniform surfaces.

NOTES: Drill hole between 4 and 5 mm in a diameter and approximately 6 mm from one end of each strip. Holes to be clean and free from burrs. Hardness ranges are commercial for the designated metals. Hardness is not specified for the tinned iron because it is not considered a practical requirement.

---

[8] Obtainable from the Society of Automotive Engineers, Inc., 400 Commonwealth Drive, Warrendale, PA 15096-0001.

(R) APPENDIX B
SAE RM-66-03 COMPATIBILITY FLUID[9]

This fluid is a blend of four proprietary polyglycol brake fluids of fixed composition, in equal parts by volume. The four fluids selected comprise three factory-fill and one aftermarket fluid, as follows:
  a. DOW HD50-4
  b. Delco Supreme II
  c. DOW 455
  d. Olin HDS-79

(R) APPENDIX C
STANDARD STYRENE-BUTADIENE RUBBER (SBR) BRAKE CUPS FOR TESTING SAE MOTOR VEHICLE BRAKE FLUIDS

C1. Formulation of Rubber Compound—(See Table C1)

TABLE C1—FORMULATION OF RUBBER COMPOUND

Ingredient	Parts by Weight
SBR type 1503[1]	100
Oil furnace black (NBS 378)	40
Zinc oxide (NBS 370)	5
Sulfur (NBS 371)	0.25
Stearic acid (NBS 372)	1
n-tertiary butyl-2-benzothiazole sulfenamide (NBS 384)	1
Symmetrical-dibetanophthyl-p-phenylene diamine	1.5
Dicumyl peroxide (40% on precipitated $CaCO_3$)[2]	4.5
Total	153.25

NOTE: The ingredients labeled (NBS_____) must have properties identical with those supplied by the National Bureau of Standards.

[1] Philprene 1503 has been found suitable.
[2] Use only within 90 d of manufacture and store at temperature below 27 °C (80 °F).

C2. Procedure for Mixing Rubber Compound—The rubber compound shall be mixed in accordance with the procedure given in ASTM D 3185 for Formula 2B.

C3. Properties of Rubber Compound—Vulcanizates cured for 12 min at 180 °C (356 °F) by the procedure described in ASTM D 3182 shall meet the requirements in Table C2:

TABLE C2—PROPERTIES OF RUBBER COMPOUND

Property	Requirement	ASTM Method
Hardness	63 ± 3	D 1415 or D 2240
Tensile strength	17.5 MPa (2500 lbf/in^2, min)	D 412
Ultimate elongation	350%, min	D 412
Tensile strength after 70 h at 125 °C (257 °F)	30% decrease, max	D 865
Ultimate elongation after 70 h at 125 °C (257 °F)	50% decrease, max	D 865
Hardness after 70 h at 125 °C (257 °F)	0 to 10 increase	D 865
Compression set after 22 h at 125 °C (257 °F)	15 to 20%	D 395 (Method B)
Brittleness temperature	−40 °C (−40 °F), max	D 746

C4. Brake Cups Prepared from Rubber Compound—Brake cups[10] shall be prepared from the rubber compound by vulcanization under the conditions required to obtain the properties given in Section C3. The dimensions of the cups shall be suitable for the brake cylinders used to determine stroking test procedure in 5.15. Cups may be used for testing brake fluids within 36 months from date of manufacture when stored in the dark at ambient temperatures not exceeding 38 °C (100 °F) and adequately protected from atmospheric and other contaminants. After removal of cups from storage, they shall be conditioned base down on a flat surface for at least 12 h at room temperature in order to allow cups to reach their true configuration before measurement.

APPENDIX D
STANDARD ETHYLENE, PROPYLENE, AND DIENE (EPDM) TERPOLYMER RUBBER SLABSTOCK (RM-69)

D1. Formulation of Rubber Compound—(See Table D1)

TABLE D1—FORMULATION OF RUBBER COMPOUND

Ingredient	Parts by Weight
EPDM type (nordel 1320)[1]	100
Zinc oxide (NBS 370)	5
Oil furnace black (NBS 378)	43
Polymerized 1,2-dihydro-2,2,4-trimethylquinoline	2
Dicumyl peroxide (40% on precipitated $CaCO_3$)[2]	10
Total	160

NOTE: The ingredients labeled (NBS_____) must have properties identical with those supplied by the National Bureau of Standards.

[1] E. I. DuPont Nordel EPDM 1320.
[2] Use only within 90 d of manufacture and store at a temperature below 27 °C (80 °F).

D2. Procedure for Mixing Rubber Compound—The rubber compound shall be mixed in accordance with the procedures given in ASTM D 3182.

D3. Properties of Rubber Compound—Vulcanizates cured for 25 min at 175 °C (347 °F) by the procedure described in ASTM D 3182 shall meet the requirements as in Table D2:

TABLE D2—PROPERTIES OF RUBBER COMPOUND

Property	Requirement	ASTM Method
Hardness, IRHD	70 ± 3	D 1415
Tensile strength, min	13.8 MPa, (2000 lbf/in^2)	D 412
Ultimate elongation, min	225%	D 412
Tensile strength, decrease after 22 h at 175 °C (347 °F), max	15%	D 865
Ultimate elongation, decrease after 22 h at 175 °C (347 °F), max	30%	D 865
Hardness, increase after 22 h at 175 °C (347 °F)	0 to 10	D 865
Compression set after 22 h at 175 °C (347 °F)	20% max	D 395 (Method B)
Brittleness temperature, max	−65 °F	D 746

D4. Slabstock Prepared from Rubber Compound—Test slabs approximately 150 × 150 × 1.9 mm = (6 × 6 × 0.075 in) shall be prepared from the rubber compound by vulcanization under the conditions stated in Section D3. These slabs may be used in testing brake fluids within 36 months from their date of manufacture, when stored in the dark at ambient temperatures not exceeding 38 °C (100 °F) and adequately protected from atmospheric or other contaminants.

When stored at other than 23 °C ± 5 (73.4 °F ± 9), the material shall be allowed to stabilize at laboratory temperature prior to measurements.

APPENDIX E
TRIETHYLENE GLYCOL MONOMETHYL ETHER (TEGME) BRAKE FLUID GRADE[11] (RM-71)

TABLE E1

Property	Requirement	Method
Assay	94 area %, min. Further, neither the material preceding nor that following TEGME through the column shall exceed 4 area %	Gas Chromatographic (GC) analysis (see below)
Water content	0.3% by weight, max	ASTM D 1364
Acidity	0.02% by weight, max, as acetic acid	ASTM D 1613
Suspended matter	Substantially free	
Appearance	Clear liquid; 100 APHA units, max	ASTM D 1209
ERBP	240 °C (464 °F), min	Paragraph 5.1 of SAE Standard J1703f

---

[9] Obtainable from the Society of Automotive Engineers, Inc., 400 Commonwealth Drive, Warrendale, PA 15096-0001.

[10] Obtainable from the Society of Automotive Engineers, Inc., 400 Commonwealth Drive, Warrendale, PA 15096-0001.

[11] Stabilized by addition of ¼% by weight of 4.4' isopropylidene diphenol.

E1. Gas Chromatographic Analysis—Analyze a representative sample using a Bendix Model 2200 dual column, programmed temperature gas chromatograph, or equivalent instrument, with a thermal conductivity detector and two 10 ft × 1/8 in Type 304 stainless steel columns packed with 10% CARBOWAX 20M-terephthalic acid on Chromosorb T, 40 to 60 mesh, as follows:

E1.1 Column Preparation—Use precleaned tubing or obtain two 10 ft lengths of 1/8 in 304 stainless steel tubing (0.02 in wall thickness) and clean as follows:

a. Rinse the tubing with 30 mL of concentrated nitric acid.
CAUTION—NITRIC ACID CAUSES SEVERE BURNS IF IT COMES IN CONTACT WITH ANY PART OF THE BODY.

b. Drain and rinse the tubing with distilled water; drain and rinse with acetone; dry with nitrogen.

E1.1.1 Weigh 5 g of CARBOWAX 20M-terephthalic acid into a 400 mL breaker. Add 200 mL of methylene chloride and stir with a magnetic stirrer until dissolved. Approximately 30 min will be required.

E1.1.2 Weigh 45 g of Chromosorb T, 40 to 60 mesh into a tared breaker and transfer to a 500 mL rotary evaporating flask.

E1.1.3 Add the CARBOWAX 20M-TPA solution to the flask and mix by gently swirling. If necessary, add additional methylene chloride to form a set slurry.

E1.1.4 Allow the slurry to stand for 10 min.

E1.1.5 Attach the flask to a rotary evaporator and apply vacuum slowly while degassing. Set the pressure at approximately 100 mm Hg. Use dry ice-acetone traps to protect the vacuum source.

E1.1.6 Rotate the flask at 10 rpm.

E1.1.7 Protect the contents of the flask from extreme cold or heat by means of a hot air gun (hair dryer at 50 to 60 °C).

E1.1.8 When all of the solvent has been removed, stop the evaporator and allow the contents of the flask to return to room temperature.

NOTE—If a rotary evaporator is not available, satisfactory packing may be prepared using the evaporating dish technique.

E1.1.9 Transfer the dried packing to a bottle having a volume about twice the volume of the packing.

E1.1.10 Add 0.5% by weight of powdered graphite and mix thoroughly until the mixture flows freely.

E1.1.11 Sieve the mixture using a combination of 30- and 60-mesh screens. Retain the portion that passes through the 30-mesh and is retained on the 60-mesh screen.

E1.1.12 Using a funnel, pack the columns with approximately 7 g of packing by gently tapping the side of the column with a suitable metal rod. Do not add large quantities of packing to the funnel at one time.

E1.1.13 Condition the columns by programming from ambient temperature to 200 °C at 2 °C/min and hold at 200 °C for at least 4 h. Repeatedly inject a sample until a good baseline is obtained.

E1.2 Operating Parameters
a. Recorder: 1mV
b. Chart speed: 0.5 in/min
c. Temperatures column: 150-225 °C programmed at 6 °C/min; hold at 225 °C for 30 min
d. Detector: 260 °C
e. Injection port: 230 °C
f. Carrier gas: Helium at 20 cc/min
g. Sample size: 1 µL
h. Total elution time: 45 min

E1.3 Procedure—Inject the sample into the chromatograph and obtain the chromatogram using the parameters outlined in E1.2.

E1.3.1 Measure the areas of all component peaks using an electronic integrator or a planimeter.

E1.3.2 Calculate and report the area percent of methoxytriglycol. The methoxytriglycol elutes at about 15 min.

E1.3.3 CALCULATION

$$\frac{AT \times 100}{D} = \text{area percent methoxytriglycol} \qquad (Eq.\ E1)$$

where:
A = peak area for methoxytriglycol
D = total area, sum of all areas corrected for attenuation
T = attenuation for component peak

(R) SERVICE MAINTENANCE OF SAE J1703 BRAKE FLUIDS IN MOTOR VEHICLE BRAKE SYSTEMS—SAE J1707 NOV91

SAE Recommended Practice

Report of the Hydraulic Brake Systems Actuating Committee, approved November 1985. Completely revised by the Motor Vehicle Brake Fluids Standards Committee November 1991.

1. Scope—This SAE Recommended Practice provides basic recommendations for dispensing and handling of J1703 Brake Fluids by Service Maintenance Personnel to assure their safe and effective performance when installed in or added to motor vehicle hydraulic brake actuating systems.

This document is concerned only with brake fluid and those system parts in contact with it. It describes general maintenance procedures that constitute good practice and that should be employed to help assure a properly functioning brake system. Recommendations that promote safety are emphasized. Specific step-by-step service instructions for brake maintenance on individual makes or models are neither intended nor implied. For these, one should consult the vehicle manufacturer's service brake maintenance procedures for the particular vehicle. Vehicle manufacturer's recommendations should always be followed.

2. References

2.1 Applicable Documents—The following publications form a part of this specification to the extent specified herein. The latest issue of SAE publications shall apply.

2.1.1 SAE PUBLICATIONS—Available from SAE, 400 Commonwealth Drive, Warrendale, PA 15096-0001.

SAE J1703—Motor Vehicle Brake Fluid

3. Motor Vehicle Brake Fluid Requirements

3.1 SAE Motor Vehicle Brake Fluid—SAE J1703—This specification covers motor vehicle brake fluids of the nonpetroleum type for use in the braking system of any motor vehicle, such as a passenger car, truck, bus or trailer. These fluids are not intended for use under arctic conditions.

3.2 Performance Requirements of Motor Vehicle Brake Fluids—While minimum anticipated atmospheric temperatures can generally be estimated, maximum operational brake fluid temperatures are a function of brake system design, service maintenance practices, variations in driving habits, and other factors which are difficult to evaluate. The motor vehicle manufacturer is best qualified to recommend the brake fluid type required in a specific motor vehicle brake actuating system. Whenever the vehicle manufacturer clearly specifies or otherwise indicates the brake fluid required, service maintenance personnel should use the brake fluid recommended by the vehicle manufacturer.

3.3 Effect of Contaminants Upon the Performance of Motor Vehicle Brake Fluids—Commercial brake fluids are susceptible to various types of contamination which can be detrimental to the performance and safety of brake actuating systems.

3.3.1 CONTAMINATION WITH PETROLEUM PRODUCTS—Due to the color and visual appearance of motor vehicle brake fluids, accidental or inadvertent contamination with petroleum products can occur, unless specific precautions are taken to prevent this type of contamination. Usually engine or transmission oils will partially separate from brake fluid when the percentage of such oil contamination is in excess of 1%. Petroleum distillates, cleaning solvents, or fabric cleaners are miscible with brake fluids in higher percentages before any visual separation occurs. Petroleum type contaminants will normally cause a visual cloudiness or opaque appearance when present in brake fluids.

"Hydraulic brake fluid" can be confused with service products identified as "hydraulic fluid," which may be a petroleum or a synthetic fluid having different chemical and physical properties and which are not intended for use in a motor vehicle brake system.

NOTE—Never use any hydraulic fluid either for motor vehicle brake fluid or in any part of the hydraulic brake system.

Petroleum products are rapidly and selectively absorbed by brake system rubber parts, resulting in a high degree of softening, dimensional swelling, and general deterioration of the functional properties of these rubber parts. This type of brake fluid contamination will result in unsafe braking action, and may be the direct cause of complete brake failure.

Do not use a brake fluid that is cloudy, opaque, an emulsion, separated into layers, or that contains drops of liquid. If petroleum contamination is suspected, dispose of fluid in accordance with all applicable laws and regulations.

3.3.2 CONTAMINATION WITH MISCELLANEOUS AUTOMOTIVE PRODUCTS—Many types of liquid automotive specialty products, solvents, and cleaners are used in service stations, garages, and other establishments where brake service or maintenance is performed. Contamination of motor vehicle brake fluid with any such products will result in a deterioration of many essential performance requirements of the brake fluids, causing improper brake action or actual brake failure.

3.3.3 WATER CONTAMINATION—J1703 motor vehicle brake fluids are hygroscopic and absorb moisture when exposed to the atmosphere and in service. Water contamination from any source including mechanical or accidental additions of free water, will appreciably lower the original boiling point of the brake fluid and increase its viscosity at low ambient temperatures. Water contamination may cause corrosion of brake cylinder bores and pistons, and may seriously affect the braking efficiency and safety of the brake actuating system.

3.3.4 CONTAMINATION WITH DIRT OR DUST—Dirt, moisture, and petroleum contaminants can enter the brake system from failure to clear the master cylinder cap before removing it to check the fluid level, or from careless handling of brake fluid or brake system parts in performing other maintenance or service operations. Dust, as well as moisture, can enter the brake system through damaged or improper cylinder boots or gasket seals.

Dirt and dust are abrasive and will score or scratch wheel cylinder seals and bores, resulting in fluid leakage or other operational problems which affect the braking efficiency and the safety of the motor vehicle brake system.

3.3.5 MIXING OF SAE J1703 MOTOR VEHICLE BRAKE FLUIDS—The mixing of different SAE J1703 brake fluids will not adversely affect the brake actuating system, but it may adversely affect the resulting properties of the mixture.

4. Recommended Service Instructions for Handling and Dispensing Brake Fluid

4.1 Storage of Motor Vehicle Brake Fluid—Store brake fluid in a dry place at or below room temperature and separate this storage area from similar storage of petroleum products, automotive specialty products, or any fluid materials used for shop maintenance purposes.

Large capacity drum type containers require transfer of brake fluid to some intermediate dispensing container. This type of handling tends to promote or cause contamination. It is recommended that brake fluid be purchased and stocked in container sizes that permit direct transfer of brake fluid from the container to the brake system. It is strongly recommended that containers do not exceed 1 gal in capacity. Brake fluid stock should be rotated to prevent extended storage periods. For servicing the motor vehicle brake system, use SAE specification brake fluid that conforms to the SAE grade recommended by the vehicle manufacturer.

4.2 Dispensing Equipment—Due to the increased possibility of contamination of the brake fluid, the use of auxiliary dispensing equipment should be avoided. When commercial dispensing equipment is used for adding brake fluid to the motor vehicle brake system such equipment must be assigned for use only with brake fluids, must be clean, and must be mechanically designed to eliminate any possibility of contamination of the brake fluid with dirt, petroleum lubricants, or moisture from condensation or exposure to the atmosphere. Any new dispensing equipment should be flushed with the brake fluid and the fluid examined for contamination before placing the unit in service. When a pressure bleeder tank is used, it must be designed to prevent any possible aeration of the brake fluid. Normally, air pressure is applied to the fluid by a mechanism which employs a rubber diaphragm. It is important that the compressed air supply be clean and dry. Many rubber compounds are quite permeable to both air and moisture. Butyl rubber diaphragms have been found superior for this application and should be used.

4.3 Essential Safety Precautions—In the service maintenance of brake systems, the handling and dispensing of motor vehicle brake fluids must be controlled and regulated to avoid any accidental or inadvertent contamination of the brake fluid added to the brake system. The possible toxicity and environmental effects of motor vehicle brake fluid must be considered in terms of any applicable laws and regulations. Obtain Material Safety Data Sheet from manufacturer before handling and disposing of brake fluid. The following are essential specific precautions for handling and dispensing brake fluid:

4.3.1 Store brake fluid only in its original container and keep the container tightly closed. Do not puncture the container to provide a "breather hole."

4.3.2 Before opening a brake fluid container, remove any dirt or other contamination from the top and other surfaces of the container.

4.3.3 When the brake fluid container is empty, dispose of the container in accordance with all local, state, and federal regulations. Do not reuse container for other liquids.

4.3.4 It is the best practice to pour the brake fluid directly from the original container to the brake system fill point. If the brake fluid is transferred from the original container to a dispenser, such dispensing equipment must be clean and dry and any unused brake fluid must not be poured back into the original brake fluid container. Discard fluid in accordance with local, state, and federal regulations.

4.3.5 Do not transfer brake fluid to a container or dispenser that has been used for oil, kerosene, gasoline, antifreeze, water cleaners, or any other liquids or chemicals. Do not transfer any material or product from a container or dispenser back into a brake fluid container.

4.3.6 In dispensing or storing brake fluids, use equipment that is specifically assigned for brake fluid and preferably used only in work areas assigned for servicing brake systems.

4.3.7 Do not reuse brake fluid from bleeding operations. Discard it in accordance with local, state, and federal regulations.

4.3.8 Do not spill brake fluid on brake linings or car finish. If spilled on brake lining, the best practice is to replace the lining. If spilled on car finish, do not wipe or rub, but immediately flush area with cold water.

4.3.9 Use only new brake fluid for flushing a brake system and then only in compliance with the vehicle manufacturer's service recommendations. Do not use flushing fluids, alcohol, or other solvents.

4.3.10 Whenever possible, clean new brake fluid should be used to clean brake parts. The use of alcohol for cleaning should be restricted to disassembled brake units and parts that can be visually inspected to determine freedom from alcohol contamination. Do not use alcohol for flushing assembled brake components or systems. Never use gasoline, kerosene, petroleum products, chlorinated or other type solvents to clean any brake system parts. Use only clean wipers or shop towels. Handle parts with clean and dry hands. Place brake parts on clean paper or clean lint-free cloths.

4.3.11 A clean and dry work area is essential in handling and dispensing brake fluid, in cleaning or handling brake parts, and for work on the brake system proper. Work should not be performed in the presence of air-borne dust, dirt, or water. Whenever possible, it is desirable to separate the area used for servicing or repairing the brake system from areas used for vehicle lubrication or other service or maintenance operations where brake fluid and brake system parts may be likely to become contaminated.

5. Recommended Service Procedures for Filling Brake Master Cylinders

5.1 Safety Precautions—The following are essential specific recommendations and procedures that are based upon acceptable service maintenance practices:

5.1.1 Before opening the master cylinder, clean the filler plug or cap and surrounding area to prevent dirt or grease from entering master cylinder reservoir. If necessary, use a wire brush and wipe area clean.

5.1.2 Do not open the master cylinder outdoors when it is raining or snowing. Do not open the master cylinder in an area where there is airborne dust or dirt that may deposit in the exposed reservoir.

5.1.3 Remove the filler plug or cap. Inspect the vent hole and clean if necessary. If the gasket shows evidence of deterioration or excessive leakage, replace it. If a reservoir sealing diaphragm is used, carefully remove and inspect the diaphragm. If the diaphragm shows evidence of deterioration or excessive leakage, replace the diaphragm with the proper part specified by the vehicle manufacturer. If diaphragm is extended, reset it.

5.1.4 Check the fluid level. Fill the master cylinder reservoir to the fluid level recommended by the vehicle manufacturer. Use brake fluid that conforms to the recommendation by the vehicle manufacturers. If the brake fluid level is below the master cylinder reservoir ports, air has been taken into the system. Frequently, the master cylinder reservoir can be refilled before the air has been forced into the brake lines and this air can be purged from the master cylinder by a few strokes on the brake pedal. When this procedure fails to restore brake pedal pressure and proper braking action, the brake system must be bled to remove trapped air as specified in Section 6.

5.1.5 Seal the master cylinder with the gasket and filler plug or cap. If the master cylinder is equipped with a diaphragm, carefully install it with clean hands or a clean dull tool, then replace the cover and tighten or otherwise seal completely.

5.1.6 Unless otherwise specified by the vehicle manufacturer, check the master cylinder fluid level at least every 6 months. Some vehicles are equipped with brake master cylinders that have a brake fluid level indicator or warning device and should be filled when the addition of brake fluid is indicated.

Fluid level checks should be so conducted that no possible brake fluid contamination can occur. The addition of contaminated brake fluid to both reservoirs of a dual or tandem-type master cylinder will cause the failure of the braking system units.

6. Recommended Service Procedures for Bleeding Brake System

6.1 Objectives and Provisions for Bleeding—The pressure applied to a fluid or liquid at one point in a closed system will be transferred to any point in the system that the fluid reaches. By means of the motor vehicle hydraulic brake system, the driver's foot pressure is amplified into a hydraulic pressure which is transferred to the wheel cylinders and converted into a frictional braking force on the wheels. Effective pressure transmission is obtained because brake fluid is relatively incompressible. However, air or other gases can be compressed. Air that is dispersed in brake fluid, or otherwise entrapped in the brake system, is compressible and will cause erratic and unsafe braking action due to loss of brake pedal pressure. The brake system must be bled free of air for the proper functioning of the system.

The presence of air in the brake system is indicated by a "spongy" feeling of the brake pedal when the brakes are applied. The air must be removed from the system by a process called "brake bleeding" before the brakes will function efficiently and safely. The vehicle manufacturer has provided for bleeding through strategically located bleeder screws.

6.2 Bleeding the Motor Vehicle Hydraulic Brake System—Whenever a brake line, brake hose, or other component part of the hydraulic brake system has been disconnected or replaced, when the brake fluid in the system is replaced, or whenever the brake pedal action indicates the presence of air in the system, the brake system must be bled. There are two basic methods of supplying brake fluid for bleeding the brake system, the manual bleeding method and the pressure bleeding method. Both methods require opening and closing of bleeder screws for removal of air. In using either method, the bled fluid is discharged into a glass jar containing new brake fluid. This is done by means of a tube or hose attached to the bleeder screw and immersed in the brake fluid in the jar. The bled fluid is observed for the presence of air bubbles. Bleeding is completed when the expelled brake fluid is free of air bubbles.

The manual bleeding method is normally a two-man operation. It requires one operator to slowly press the brake pedal, actuating the master cylinder to supply brake fluid to the system. Simultaneously, the other operator bleeds the hydraulic brake system.

The pressure bleeding method requires one operator. A brake pressure bleeder is connected to the master cylinder to supply brake fluid under pressure to the system automatically as the operator bleeds the system.

No standardized service procedures for brake bleeding can cover all types of commercial brake systems. The car manufacturer's service instructions for the individual vehicle make and model must be specifically followed. For example, special bleeding sequences are essential for variations in brake design, type of master cylinder, and optional equipment such as power brakes, hydraulic clutch, hill holder, etc. The use of complex pressure control valves and sensing mechanisms in motor vehicle brake systems may require specific or special procedures for brake bleeding. To insure proper bleeding of any brake system, service personnel should follow the vehicle manufacturer's instructions. Failure to follow the vehicle manufacturer's instructions could result in a brake system malfunction.

6.3 General Safety Precautions for Bleeding the Motor Vehicle Brake System—These recommendations are generally applicable to all motor vehicles:

6.3.1 Clean all dirt and grease from and around the bleeder screws before opening.

6.3.2 Follow recommendations and procedures specified in 5.1 for additions of brake fluid to brake master cylinder.

6.3.3 Follow recommendations and applicable precautions, as specified in 4.3 for handling and dispensing brake fluid used in brake bleeding operations.

6.3.4 Check brake hose assembly for evidence of leakage, cracking, abrasions, cuts, or tears in the outer covering before bleeding the system. Replace any brake hose when inspection indicates hose damage.

6.3.5 End of bleeder hose must be completely immersed in brake fluid in the glass jar to permit observation of air bubbles and to prevent air from being returned to the system when the pressure is released.

6.3.6 If the brake fluid drained from the system in bleeding operations shows evidence of abrasive material, cloudiness or severe discoloration, excessive precipitates, or sedimentation, carefully inspect the entire brake system for excessive wear, corrosion, and seal damage. Follow the vehicle manufacturer's service instructions.

6.3.7 After bleeding the system, check for satisfactory brake pedal action.

6.3.8 Do not reuse brake fluid from bleeding operations. Discard it in accordance with local, state, and federal regulations.

7. General Maintenance and Service Recommendations for Motor Vehicle Brake System—Brake fluid is only one factor affecting the operational performance and safety of the brake system. Other components or materials in the brake system can adversely affect the performance of the brake fluid. The following general maintenance and service instructions are minimum recommendations for providing safe and efficient functional performance of the brake fluid. When either brake component parts or brake fluid require replacement in the system, recommendations and specific applicable precautions specified in 4.3, 5.1, and 6.3 should be followed.

7.1 Regular or Persistent Loss of Brake Fluid—Whenever the master cylinder requires frequent or abnormal additions of brake fluid, the brake system should be completely inspected for evidence of fluid leakage. If leakage cannot be completely stopped, the defective component part should immediately be removed and properly repaired or replaced.

7.2 Replacement of Brake Fluid in the Braking System—Brake fluids in the motor vehicle braking system can become contaminated. (See 3.3.) Whenever wheel cylinders and/or calipers are removed for inspection, reconditioning or replacement, or when contamination is suspected, it is strongly recommended that the system be flushed and refilled. Only clean dry brake fluid of the grade recommended by the vehicle manufacturer should be used for both of these procedures. Follow the vehicle manufacturer's recommendations for service brake maintenance and fluid changes. Periodic changes of fluid in aging vehicles are not recommended unless wheel cylinders and calipers are disconnected to prevent any dirt, sludge, or abrasive materials in the system from being flushed into them. Otherwise they may cause scoring or scuffing of pistons, bores, cups, or seals, with possible leakage and system failure. Whenever a change of brake fluid in the system is indicated in the interest of preventive maintenance or safety, and the vehicle manufacturer has not recommended a procedure for changing the brake fluid, the following is recommended:

7.2.1 With all wheel cylinders and calipers disconnected, and using new brake fluid of the type and grade recommended by the vehicle manufacturer, flush all brake lines and hoses.

7.2.2 The vehicle master cylinder, wheel cylinders, and calipers should be carefully disassembled, cleaned, (see 4.3.10) and examined.

Corrosion of cylinder bores or pistons, scoring or scuffing of parts, or deterioration of boots, cups, and seals to an extent which may affect performance of the braking system, indicate that such parts should be replaced.

7.2.3 Examine brake hoses carefully for any scuffing, exterior cover cracks, or evidence of leakage. Replace with new hoses where necessary.

7.2.4 Use only clean dry brake fluid for flushing and bleeding the system. Do not use alcohols or any other fluids. They cannot be completely removed from the system and will contaminate the brake fluid subsequently added to the system.

7.2.5 Do not reuse brake fluid from bleeding, flushing, or washing operations. Discard it in accordance with local, state, and federal regulations.

7.3 Replacement of Defective Brake Component Parts or Assemblies—Follow the applicable recommendations in 7.2. These general service and maintenance instructions must be supplemented by any specific step-by-step service instructions for brake maintenance of individual makes or models of motor vehicle as recommended by the vehicle manufacturer.

7.4 Service Inspection of Brake System

7.4.1 BRAKE HOSE—Inspect for leaks, cracking, abrasions, cuts, or tears in the outer covering twice a year or as recommended by the vehicle manufacturer.

7.4.2 BRAKE LININGS—Inspect for wear every 20 000 miles or as recommended by the vehicle manufacturer. Whenever replacement of brake lining is indicated, the vehicle manufacturer's service instructions and replacement recommendations for the individual vehicle make and model should be specifically followed.

7.4.3 BRAKE ADJUSTMENTS—Variations in brake design and modifications require specific instructions and precautions which are not covered by the scope of this SAE document.

LOW WATER TOLERANT BRAKE FLUIDS—SAE J1705 OCT88

SAE Recommended Practice

Report of the Hydraulic Brake Systems Actuating Committee, approved March 1985 and revised October 1988.

1. Scope—This recommended practice was prepared by the Motor Vehicle Brake Fluids Subcommittee of the SAE Hydraulic Brake Systems Actuating Committee to provide engineers, designers, and manufacturers of motor vehicles with a set of minimum performance standards in order to assess the suitability of silicone and other low water tolerant type brake fluids (LWTF) for use in motor vehicle brake systems. These fluids are designed for use in braking systems fitted with rubber cups and seals made from natural rubber (NR), styrene-butadiene rubber (SBR), or a terpolymer of ethylene, propylene, and a diene (EPDM).

In the development of the recommended requirements and test procedures contained herein, it is concluded that the LWTF's must be functionally compatible with existing motor vehicle brake fluids conforming to SAE Standard J1703 and with braking systems designed for such fluids. To utilize LWTF's to the fullest advantage, they should not be mixed with other brake fluids. Inadvertent mixtures of LWTF's with fluids meeting SAE J1703 are not known to have any adverse effects on performance, but all combinations have not been tested. Vehicle manufacturers' recommendations should be followed where indicated. These fluids are not necessarily suitable for use in central hydraulic or pumped systems and are not intended for use below temperatures of −50°C (−58°F). Brake fluids covered under this recommended practice are not required to tolerate water and extreme caution should be exercised to prevent accidental entry of water which might lead to brake failure. Other performance characteristics of these LWTF's not covered in this recommended practice are discussed in the Appendix.

2. Requirements

2.1 Equilibrium Reflux Boiling Point (ERBP)

(a) When tested by the procedure specified in paragraph 3.1, the low water tolerant brake fluid shall have an ERBP of not less than 260°C (500°F).

(b) When tested by the procedure specified in paragraph 3.1.1, the low water tolerant brake fluid shall have a wet ERBP of not less than 180°C (356°F).

2.2 Viscosity—When tested by the procedure specified in paragraph 3.2, LWTF's shall have kinematic viscosities as follows:

(a) At −40°C (−40°F): not more than 900 mm^2/s (900 cSt).

(b) At 25°C (77°F): not more than 50 mm^2/s (50 cSt).

(c) At 100°C (212°F): not less than 1.5 mm^2/s (1.5 cSt).

2.3 Corrosion—LWTF's, when tested by the procedure specified in paragraph 3.3, shall not cause corrosion exceeding the limits shown in Table 1. The metal strips outside the area of contact shall be neither pitted nor roughened to an extent discernible by the eye without magnification, but staining or discoloration is permitted. Roughening caused during assembly and disassembly shall be disregarded.

The fluid at the end of the test shall show no gelling at 23 ± 5°C (73 ± 9°F). No crystalline type deposit shall form and adhere to either the glass jar walls or the surface of metal strips. The fluid shall contain no more than 0.10% sediment by volume.

The rubber cup at the end of the test shall show no appreciable disintegration as evidenced by blisters or sloughing as indicated by carbon black separation on the surface of the rubber cup. The hardness of the rubber cup shall not decrease by more than 15 International Rubber Hardness degrees (IRHD) and the base diameter shall not increase by more than 1.40 mm (0.055 in).

2.4 Fluidity and Appearance at Low Temperature

(a) At −40°C (−40°F)—When LWTF's are tested by the procedure specified in paragraph 3.4(a), the fluid shall show no stratification, sediment, or crystals. Upon inversion of the sample bottle, the air bubble shall travel to the top of the fluid in not more than 10 s. Cloudiness is permissible, but on warming to room temperature 23 ± 5°C (73 ± 9°F), the fluid shall regain its original uniformity, appearance, and clarity.

(b) At −50°C (−58°F)—When LWTF's are tested by the procedure specified in paragraph 3.4(b), the fluid shall show no stratification, sediment, or crystals. Upon inversion of the sample bottle, the air bubble shall travel to the top of the fluid in not more than 35 s. Cloudiness is permissible, but on warming to room temperature 23 ± 5°C (73 ± 9°F), the fluid shall regain its original uniformity, appearance, and clarity.

2.5 Tolerance to High Humidity

(a) At −40°C (−40°F)—When LWTF's are tested by the procedure specified in paragraph 3.5, the fluid shall show no stratification,

TABLE 1—CORROSION TEST STRIPS AND MASS CHANGES

Test Specimens[a]	Max Permissible Mass Change mg/cm^2 of Surface
Tinned Iron	0.1
Steel	0.1
Aluminum	0.1
Cast Iron	0.1
Brass	0.2
Copper	0.2

[a]Obtainable from the Society of Automotive Engineers, 400 Commonwealth Drive, Warrendale, PA 15096.

sediment, or crystals. Upon inversion of the centrifuge tube, the air bubble shall travel to top of the fluid in not more than 10 s. Cloudiness is permissible, but on warming to room temperature 23 ± 5°C (73 ± 9°F), the fluid shall regain its original uniformity, appearance, and clarity.

(b) At 60°C (140°F)—When LWTF's are tested by the procedure specified in paragraph 3.5, the fluid shall show no stratification; sediment shall not exceed 0.05% by volume after centrifuging.

2.6 Effect on Rubber—Rubber specimens, when tested as specified in paragraph 3.6, shall show no disintegration as evidenced by blisters or sloughing. Hardness changes, base diameter changes, and volume swell shall lie within the ranges given in Table 2.

2.7 Compatibility

(a) At −40°C (−40°F)—When LWTF's are tested by the procedure specified in paragraph 3.7(a), the fluid shall show no sedimentation or crystallization. Cloudiness is permissible, but on warming to room temperature 23 ± 5°C (73 ± 9°F), the mixture shall be no more cloudy than the original fluid under test.

(b) At 60°C (140°F)—When LWTF's are tested by the procedure specified in paragraph 3.7(b), sedimentation shall not exceed 0.05% by volume after centrifuging.

2.8 Fluid Stability—When the brake fluid is tested according to paragraph 3.8, the ERBP shall not be less than 260°C (500°F).

2.9 Stroking Test—An LWTF, when tested by the procedure specified in paragraph 3.9, shall meet the following performance requirements:

(a) Metal parts shall not show corrosion as evidenced by pitting to an extent discernible to the naked eye, but staining or discoloration shall be permitted.

(b) The initial diameter of any cylinder or piston shall not change by more than 0.13 mm (0.005 in) during test.

(c) Rubber cups shall not decrease in hardness by more than 15 IRHD and shall not be in an unsatisfactory operating condition as evidenced by excessive amounts of scoring, scuffing, blistering, cracking, or change in shape from original appearance.

(d) The base diameter of the rubber cups shall not increase by more than 0.9 mm (0.035 in).

(e) The average lip diameter interference set of the rubber cups shall not be greater than 65%.

(f) During any period of 24 000 strokes, the volume loss of fluid shall not be greater than 36 mL.

(g) The cylinder pistons shall not seize or function improperly throughout the test.

(h) The volume loss of fluid during the 100 strokes at the end of the test shall not be more than 36 mL.

(i) The fluid at the end of the test shall not be in an unsatisfactory operating condition as evidenced by sludging, gelling, or abrasive grittiness, and sediment in either the master cylinder or the wheel cylinders shall not exceed 1.5% by volume after allowing the fluid to stand 24 h at room temperature and then centrifuging.

(j) Brake cylinder walls and other metal parts shall be free of deposits which are abrasive or which cannot be removed when rubbed with a cloth wetted with isopropyl alcohol.

3. Test Procedure

3.1 Equilibrium Reflux Boiling Point—Determine the equilibrium reflux boiling point of the fluid by ASTM D1120, Method of Test for Boiling Point of Engine Antifreezes, with the following exceptions:

APPARATUS

3(d) Thermometer—ASTM E1, 76 mm immersion, calibrated. Use ASTM 3C or 3F thermometer. For fluids boiling below 300°C (572°F), ASTM 2C or 2F thermometer may be used.

3(e) Heat Source—Use a suitable Variac-controlled 100 mL heating mantle designed to fit the flask, capable of supplying the heat required to conform to the specified heating and reflux rates. [Supplier: GLAS COL Apparatus Co., Terre Haute, IN. Serial number: 135464. 230 W, 135 V (max).]

PREPARATION OF APPARATUS

5(d)—Thoroughly clean and dry all glassware before use. Attach the flask to the condenser. Place the mantle under the flask and support it with a suitable ring clamp and laboratory type stand, holding the whole assembly in place by a clamp.

NOTE: Place the whole assembly in an area free from drafts or other types of sudden temperature changes.

PROCEDURE

6(c)—When everything is in readiness, turn on the condenser water and apply heat to the flask at such a rate that the fluid is refluxing in 10 ± 2 min at a rate in excess of 1 drop/s.

The reflux rate shall not exceed 5 drop/s. Immediately adjust heat input to obtain a specified equilibrium reflux rate of 1–2 drop/s over the next 5 ± 2 min period. Maintain a timed and constant equilibrium reflux rate of 1–2 drop/s for an additional 2 min; record the average value of four temperature readings taken at 30 s intervals at the equilibrium reflux boiling point.

If the temperature exceeds 260°C (500°F), the test shall be stopped and the temperature recorded as exceeding 260°C (500°F).

3.1.1 WET ERBP—Humidify the fluid as described in paragraph 3.5 and determine the boiling point as described in paragraph 3.1.

3.2 Viscosity—Determine the kinematic viscosity of the fluid by ASTM D445.

3.2.1 Report the viscosity to the nearest centistoke (mm²/s).

3.3 Corrosion—Prepare two sets of strips from each of the metals listed in Table 1. Drill a hole between 4 and 5 mm (0.16 and 0.20 in) in diameter and about 6 mm (¼ in) from one end of each strip. With the exception of the tinned iron strips, clean the strips by abrading them on all surface areas with 320A and 400A waterproof carborundum papers and isopropyl alcohol until all surface scratches, cuts, and pits are removed from the strips, using a new piece of carborundum paper for each different type of metal. Wash the strips, including the tinned iron, with isopropyl alcohol and dry the strips with a clean lint-free cloth; then place them in a desiccator containing desiccant and maintained at 23 ± 5°C (73 ± 9°F) for at least 1 h before weighing. Handle the strips with clean forceps after polishing, to avoid fingerprint contamination.

Weigh each strip to the nearest 0.1 mg and assemble each set of strips on an uncoated steel bolt in the order tinned iron, steel, aluminum, cast iron, brass, and copper, so that the strips are in electrolytic contact. Bend the strips, other than cast iron, so that there is a separation of at least 3 mm between adjacent strips for a distance of at least 6 cm from the free ends of the strips. Immerse strip assemblies in 90% ethanol to eliminate fingerprints; dry thoroughly with clean com-

Φ**TABLE 2—EFFECT ON RUBBER**

Type	Test Specimen	Volume Swell %	Base Diameter Change	Change in Hardness	Test Temperature °C	Test Temperature °F	Time Hours
SBR	SAE RM-3a	(+5 to +20)	0.15 to 1.4 mm (0.006 to 0.055 in)	(0 to −10)	25 ± 2	77 ± 4	168 ± 2
	SAE RM-3a	(+5 to +20)	0.15 to 1.4 mm (0.006 to 0.055 in)	(0 to −10)	70 ± 2	158 ± 4	70 ± 2
	SAE RM-3a	(+5 to +20)	0.15 to 1.4 mm (0.006 to 0.055 in)	(0 to −15)	120 ± 2	248 ± 4	70 ± 2
Polychloroprene	SAE RM-68	(−5 to +10)	—	(+3 to −10)	25 ± 2	77 ± 4	168 ± 2
	SAE RM-68	(−5 to +10)	—	(+3 to −10)	70 ± 2	158 ± 4	70 ± 2
	SAE RM-68	(−5 to +10)	—	(+3 to −10)	100 ± 2	212 ± 4	70 ± 2
EPR	SAE RM-69	(0 to +10)	—	(0 to −10)	25 ± 2	77 ± 4	168 ± 2
	SAE RM-69	(0 to +10)	—	(0 to −10)	70 ± 2	158 ± 4	70 ± 2
	SAE RM-69	(0 to +10)	—	(0 to −10)	120 ± 2	248 ± 4	70 ± 2
Natural	SAE NR-X	(+5 to +20)	0.15 to 1.4 mm (0.006 to 0.055 in)	(0 to −10)	25 ± 2	77 ± 4	168 ± 2
	SAE NR-X	(+5 to +20)	0.15 to 1.4 mm (0.006 to 0.055 in)	(0 to −10)	70 ± 2	158 ± 4	70 ± 2

pressed air; handle only with clean, dry forceps thereafter.

Measure the base diameter of two standard SAE RM-3A SBR cups using an optical comparator or micrometer, to the nearest 0.02 mm (0.001 in) along the centerline of the SAE and rubber type identifications and at right angles to this centerline. Take the measurements within 0.4 mm (0.016 in) of the bottom edge and parallel to the base of the cup. Discard any cup if the two measured diameters differ by more than 0.08 mm (0.003 in). Average the two readings of each cup. Support the rubber cup on a rubber anvil or cylinder having a flat circular top surface of at least 19 mm diameter, a thickness of at least 9 mm, and hardness within 5 IRHD of the hardness of the rubber test cup. Determine the hardness of each cup thus supported, using the standard instrument and procedure specified in ASTM D1415, Method of Test for International Hardness of Vulcanized Natural and Synthetic Rubbers. (NOTE: ASTM D2240, Method of Test for Indentation Hardness of Rubber and Plastics by Means of a Durometer, may be used for quality control and routine tests if the Type A durometer is equipped with a fixture for keeping the plane of the pressure foot on the durometer parallel to the plane of the cup face during measurement.) Obtain two straight-sided, round, glass jars, having a capacity of approximately 475 mL and inner dimensions of approximately 100 mm in height and 75 mm in diameter (SAE RM-49). Grind the lip of each jar flat with the use of 400A waterproof carborundum paper and a flat surface as required to insure proper sealing. Place one rubber cup with lip edge facing up, in each of the two glass jars. Use only tinned steel lids vented with a hole 0.8 ± 0.1 mm in diameter. Place a Teflon disc seal with a hole diameter slightly larger than the hole in the tinned steel lid in each lid. Insert a metal strip assembly inside each cup with the bolted end in contact with the concavity of the cup and the free end extending upward in the jar.

Add test fluid, humidified in accordance with paragraph 3.5 in sufficient amount to cover the metal strip assembly to a depth of approximately 10 mm (0.39 in). Tighten the lids and place the two jars in a gravity convection oven conforming to ASTM E145 Type 1A and maintained at 100 ± 2°C (212 ± 4°F), for 120 ± 2 h. Allow the jars to cool at 23 ± 5°C (73 ± 9°F) for 60–90 min.

Immediately following the cooling period, take the rubber cups from the jars with forceps, removing loose adhering sediment by agitating the cup in the fluid. Rinse both cups in isopropyl alcohol and air dry. Examine them for evidence of sloughing, blisters, and other forms of disintegration. Measure the base diameter and hardness of each cup within 15 min after removal from the fluid.

Next, take the metal strips from the jars with forceps, again removing loose adhering sediment by agitation of the metal strip assembly in the fluid. Examine assemblies and jars for adhering crystalline deposits; disassemble the metal strips, removing adhering fluid by flushing with isopropyl alcohol. Clean individual strips by wiping with a cloth wetted with isopropyl alcohol. Examine the strips for evidence of corrosion and pitting. Place strips in a desiccator [containing desiccant and maintained at 23 ± 5°C (73 ± 9°F)] for at least 1 h. Weigh each strip to the nearest 0.1 mg. Determine the difference in mass of each metal strip and divide by its total surface area in square centimeters. Average the calculated values for the duplicates and report to the nearest 0.01 mg/cm^2. In the event of a marginal pass on inspection, or of a failure in only one of the duplicates, another set of duplicate test samples shall be run. Both repeat samples must meet all the requirements of paragraph 2.3. Examine the test fluid in the jars for gelling. Agitate the fluid in jars to suspend and uniformly disperse sediment, then transfer a 100 mL portion of this fluid to an ASTM cone-shaped centrifuge tube. Determine percent sediment as described in paragraph 5(b) of ASTM D91.

3.4 Fluidity and Appearance of Low Temperatures

(a) At −40°C (−40°F)—Place 100 mL of the test fluid in a screw-top glass sample bottle having a capacity of approximately 125 mL, an outside diameter of about 37 mm, and overall height of about 155 mm. Cap the bottle tightly and place in a cold bath maintained at −40 ± 2°C (−40 ± 4°F) for 144 ± 4 h. Remove the bottle from the bath, quickly wipe the bottle with a clean lint-free cloth saturated with isopropyl alcohol, and examine the fluid for evidence of stratification, sediment, or crystals. Invert the bottle and determine the number of seconds required for the air bubble to travel to the top of the fluid. Allow the fluid to warm to room temperature 23 ± 5°C (73 ± 9°F); if necessary allow to stand for as long as 4 h. Examine the fluid for clarity and appearance by comparing it to an original sample of the test fluid in an identical container.

(b) At −50°C (−58°F)—Place 100 mL of fluid in a glass sample bottle (same as in the −40°F test above). Cap the bottle tightly and place in a cold bath maintained at −50 ± 2°C (−58 ± 4°F) for 6 ± 0.2 h. Remove the bottle from the bath, quickly wipe the bottle with a clean lint-free cloth saturated with isopropyl alcohol, and examine the fluid for evidence of stratification, sediment, and crystals. Invert the bottle and determine the number of seconds required for the air bubble to travel to the top of the fluid. Allow the fluid to warm to room temperature 23 ± 5°C (73 ± 9°F); if necessary allow to stand for as long as 4 h. Examine the fluid for clarity and appearance by comparing it to a sample of the original test fluid in an identical container.

3.5 Tolerance for High Humidity

3.5.1 APPARATUS FOR HUMIDIFICATION (SEE FIG. 7)—Test apparatus shall consist of:

(a) Glass Jars and Lids—Four SAE RM-49 corrosion test jars. Each shall have a capacity of about 475 mL and approximate inner dimensions of 100 mm in height by 85 mm in diameter. Use matching lids (RM-63) having new, clean inserts, unaffected by the test fluid and providing water-vapor proof seals.

(b) Desiccator and Cover—Bowl-form glass desiccators, about 250 mm inside diameter, having matching tubulated covers fitted with No. 8 rubber stoppers. Four are required.

(c) Desiccator Plate—Porcelain desiccator plates perforated with 5 mm (0.20 in) diameter holes approximately 15 mm on centers. Approximately 230 mm diameter, without feet, glazed one side (Coors 60003 or equal). Four are required.

NOTE: Minor variations in the humidification apparatus will not affect the end results, provided that identical set ups are used for both the reference and test fluids.

3.5.2 REAGENTS AND MATERIALS
(a) Ammonium sulfate $(NH_4)_2SO_4$, Reagent or A.C.S. grade.
(b) Distilled water.
(c) SAE RM-71 reference fluid.

3.5.3 PREPARATION OF APPARATUS—Lubricate ground-glass joint of desiccator with suitable stopcock grease. Load each desiccator with 450 ± 25 g of ammonium sulfate and add 125 ± 10 mL of distilled water. Surface of the salt slurry shall lie within 45 ± 7 mm of the top surface of the desiccator plate. Place the desiccators in an area with the temperature controlled at 23 ± 5°C (73 ± 9°F) throughout the humidification procedure. Load the desiccators with the slurry and allow to condition with the desiccator cover on, test jar inside, and stoppers in place, at least 12 h before use. Use a fresh charge of salt slurry for each test.

3.5.4 HUMIDIFICATION PROCEDURE—Pipette 100 ± 1 mL of the brake fluid into a test jar through the desiccator cover and replace the rubber stopper. Prepare a duplicate test sample and two duplicate specimens of the RM-71 reference fluid. Adjust water contents of the SAE RM-71 fluid to 0.50 ± 0.05% by weight at start of test. At intervals, remove rubber stopper in the top of each desiccator containing SAE RM-71 fluid. Using a long needled hypodermic syringe, take a sample of not more than 2 mL from each jar and determine its water content by the Karl Fischer procedure (ASTM E 203) or equivalent. Remove no more than 10 mL of fluid from each SAE RM-71 sample during humidification. When the water content of the SAE fluid reaches 3.70 ± 0.05% by weight (average of the duplicates), remove the two test fluid specimens (jars) from their desiccators and promptly cap each tightly. Fill a cone-shaped centrifuge tube (as described in paragraph 3.1 of ASTM D91) with 100 mL of humidified fluid (remainder is available for other tests as required).

3.5.5 ALTERNATE (BULK) HUMIDIFICATION PROCEDURE (SEE NOTES FOLLOWING)—Lubricate the ground-glass joint of a 250 mm ID bowl-form desiccator having matching tubulated glass cover and fitted with a No. 8 rubber stopper. Pour 450 ± 10 mL of distilled water into the desiccator and insert a perforated porcelain plate (Coors No. 60003 or equal). Immediately place two open RM-49 corrosion test jars each containing 350 ± 5 mL of the test brake fluid, into the desiccator, replace desiccator cover, and insert at once into an ASTM E145, Type IIA, forced-ventilation oven set at 50 ± 1°C (122 ± 2°F). Place an identical desiccator set up having two jars containing Reference Fluid RM-71 in the oven at the same time. The water content of the RM-71 fluid at the start of exposure shall have been adjusted to 0.50 ± 0.05% by weight (Karl Fischer analysis or equivalent).

Periodically during oven humidification, remove the rubber stopper from the desiccator containing the control fluid. Using a hypodermic syringe with a long needle, quickly sample each jar and determine its individual water content. When the average water content of the control fluid has reached 3.70 ± 0.05% by weight, remove desiccator containing the test fluid at once from the oven and seal the test jars promptly, using screw-cap jar lids with suitable vapor-proof liners (RM-63). Allow the sealed jars to cool for 60–90 min at 23 ± 5°C (73 ± 9°F).

Notes on Alternate Procedure

(a) The bulk procedure may be run with either one, two, three, or four jars per desiccator, and jars may be filled from 150 to 350 mL as desired.

(b) Minor variations in equipment used are not critical. However, it is very important that all desiccator set ups in a given run be identical, and that an ASTM Type IIB oven of ample size be used.

3.5.6 PROCEDURE FOR HUMIDITY TOLERANCE DETERMINATION

(a) At $-40°C$ ($-40°F$)—Pour 100 mL of the humidified fluid into an ASTM coneshaped centrifuge tube described in paragraph 3.1 of ASTM D91. Seal the tube and place in a cold bath maintained at $-40 \pm 2°C$ ($-40 \pm 4°F$) for 144 ± 4 h. Remove the centrifuge tube from the bath, quickly wipe with a clean, lint-free cloth saturated with isopropyl alcohol or acetone, and examine the fluid for evidence of stratification, sediment, or crystals. Invert the tube and determine the number of seconds required for the air bubble to travel to the top of the fluid. (The air bubble shall be considered to have reached the top of the fluid when the top of the bubble reaches the 2 mL graduation of the centrifuge tube.) Allow the fluid to warm to room temperature $23 \pm 5°C$ ($73 \pm 9°F$); if necessary allow to stand for as long as 4 h. Examine the fluid for appearance and clarity by comparing it to an as-received sample of the test fluid in an identical container.

(b) At $60°C$ ($140°F$)—Place the centrifuge tube from paragraph 3.5.6 in a gravity convection oven conforming to ASTM E145 Type 1A and maintained at $60 \pm 2°C$ ($140 \pm 4°F$) for 22 ± 2 h. Then remove tube from oven and immediately examine the fluid for evidence of stratification. Determine percent sediment by volume as described in paragraph 5.2 of ASTM D91.

3.6 Effect on Rubber—Use test specimens for test procedures in paragraphs 3.6.1 and 3.6.2 (either standard cups or 1×1 in test specimens) as described in Table 2. The specimens shall be rinsed in isopropyl alcohol and quickly air-dried to remove any dirt and packing debris. The specimens shall not remain in the alcohol for more than 30 s. Allow the rubber specimens to stabilize at $23 \pm 5°C$ ($73 \pm 9°F$) prior to measuring volume and hardness, also base diameter of cups as described in paragraph 3.3. Weigh the specimen in air (M_1) to the nearest milligram and then weigh the specimens immersed in room temperature distilled water (M_2) containing no more than 0.2% of a suitable wetting agent. Pluronic L-61 (BASF Wyandotte) or equivalent has been found to be acceptable.

3.6.1 EXPOSURE AT ROOM TEMPERATURE—Place two rubber specimens in a straight-sided, screw-top, round glass jar having a capacity of approximately 250 mL, inner dimensions of approximately 125 mm in height and 50 mm in diameter, and containing sufficient glass beads (4–6 mm diameter) to cover approximately two-thirds of the jar bottom. Immerse the two specimens in 75 mL of test fluid. Promptly seal the jar, using a steel lid provided with a foil or polytetrafluoroethylene faced cork or pulp insert. Store at $25 \pm 2°C$ for 168 ± 2 h. Remove rubber specimens from fluid, rinse in isopropyl alcohol, and quickly air dry. Specimens shall not remain in alcohol for more than 30 s.

After drying, first weigh each specimen in air (M_3) to the nearest milligram, then weigh in distilled water (M_4) at $23 \pm 5°C$ ($73 \pm 9°F$). Next measure base diameter (on cups) and hardness. All measurements must be completed within 60 min after removal from the test fluid.

3.6.2 EXPOSURE AT ELEVATED TEMPERATURE—Place two rubber specimens in a straight-sided, screw-top, round glass jar having a capacity of approximately 250 mL, inner dimensions of approximately 125 mm in height and 50 mm in diameter, and containing sufficient glass beads (4–6 mm diameter) to cover approximately two-thirds of the jar bottom. Immerse the two specimens in 75 mL of test fluid. Promptly seal the jar, using a steel lid provided with a foil or polytetrafluoroethylene faced cork or pulp insert. Place in a forced ventilation oven (Type IIA of ASTM E145) and store at the specified test temperature (see Table 2). After completion of the heating period, remove from oven and allow to cool 60–90 min at $23 \pm 5°C$ ($73 \pm 9°F$). Remove rubber specimens from fluid, rinse in isopropyl alcohol, and quickly air dry. Specimens shall not remain in alcohol for more than 30 s.

After drying, first weigh each specimen in air (M_3) to the nearest milligram, then weigh in distilled water (M_4) at $23 \pm 5°C$ ($73 \pm 9°F$). Next measure base diameter (on cups) and hardness. All measurements must be completed within 60 min after removal from the test fluid.

3.6.3 CALCULATION AND REPORTING—Volume change shall be as a percentage of the original volume, calculated as follows:

$$\text{Percent volume change} = \frac{(M_3 - M_4) - (M_1 - M_2)}{(M_1 - M_2)} \times 100$$

Where: M_1 = Initial mass in air
M_2 = Initial mass in water
M_3 = Final mass in air
M_4 = Final mass in water

Calculate and report the change in volume and hardness after exposure. Measure and report the change in base diameter for cups. Examine the exposure rubber specimens for blistering or sloughing. Determine conformance to requirements as set forth in Table 2.

3.7 Compatibility

(a) At $-40°C$ ($-40°F$)—Mix 50 mL of LWTF with 50 mL of SAE RM-70-03 silicone base compatibility fluid and pour this mixture into an ASTM cone-shaped centrifuge tube described in paragraph 3.1 of ASTM D91, seal. Place the centrifuge tube for 144 ± 4 h in a cold bath maintained at $-40 \pm 2°C$ ($40 \pm 4°F$). Remove tube from bath, quickly wipe with a clean, lint-free cloth saturated with isopropyl alcohol or acetone, and examine fluid for sedimentation and crystallization. Allow fluid to warm to room temperature $23 \pm 5°C$ ($73 \pm 9°F$). If necessary, allow to stand for as long as 4 h. Examine the fluid for appearance and clarity by comparing it to an as-received sample of the test fluid in an identical container. Repeat the above procedure using RM-66-03.

(b) At $60°C$ ($140°F$)—Place the centrifuge tubes of paragraph 2.7 in a gravity convection oven conforming to ASTM E145 Type 1A and maintained at $60 \pm 2°C$ ($140 \pm 4°F$) for 22 ± 2 h. Remove tube from oven and determine percent sediment by volume as described in paragraph 5.2 of ASTM D91.

3.8 Fluid Stability—Heat a new sample of the original test brake fluid to a temperature $185 \pm 2°C$ ($365 \pm 4°F$) by the procedure specified in paragraph 3.1 and maintain at that temperature for 2 h. Then determine the boiling point of this fluid as specified in paragraph 3.1. If the temperature exceeds $260°C$ ($500°F$), the test can be stopped and the value reported as exceeding $260°C$ ($500°F$).

3.9 Stroking Test Procedure[1]—Use the following procedure to evaluate the lubrication quality of the brake fluid.

3.9.1 TEST APPARATUS AND MATERIAL—Use the Fig. 2 stroking fixture type apparatus with the following components arranged as shown in Fig. 1. The drum and shoe apparatus as described in SAE J1703 may be used as an alternative test system.

(a) Master Cylinder Assembly—One cast iron housing hydraulic brake master cylinder having a diameter of approximately 28 mm ($1\frac{1}{8}$ in) and fitted with an uncoated steel standpipe. Master cylinder used is SAE RM-15a 28 mm ($1\frac{1}{8}$ in) diameter or equivalent.

(b) Brake Assemblies—Three cast iron housing straight bore hydraulic brake wheel cylinder assemblies having a diameter approximately 28 mm ($1\frac{1}{8}$ in). Wheel cylinder used is SAE RM-14a with stroking adapter mounting plates to hold the brake wheel cylinder assemblies as shown in Fig. 2.

(c) Braking Pressure Actuating Mechanism—A suitable actuating mechanism for applying a force to the master cylinder push rod without side thrust.

The amount of force applied by the actuating mechanism shall be adjustable and capable of supplying sufficient stroke and thrust to the master cylinder to create a pressure of at least 70 kg/cm² (1000 psi) in the simulated brake system. A hydraulic gauge and pressure recorder, capable of establishing the pressure curve of the system and monitoring the pressure developed, shall be installed on a hydraulic line extending from the master cylinder to the outside of the oven. This line shall be provided with a shut-off valve and a bleeding valve for removing air from the connection tubing.

The actuating mechanism shall be designed to provide a stroking rate of approximately 1000 strokes/h. The pressure buildup rate versus cylinder stroke and time shall correspond to Fig. 3.

(d) Heated Air Bath Cabinet—An insulated cabinet or oven having sufficient capacity to house the three wheel cylinder fixture assemblies, master cylinder, and necessary connection. A suitable thermostatically-controlled heating system is required to maintain a brake fluid temperature of $120 \pm 5°C$ ($248 \pm 9°F$). Heaters shall be shielded to prevent direct radiation to wheel or master cylinders. Fluid temperature shall be monitored at random intervals during the test at the master cylinder reservoir, using a temperature recording device.

3.9.2 PREPARATION OF TEST APPARATUS

(a) Wheel Cylinder Assemblies—Use new wheel cylinder assemblies, SAE RM-14a or equivalent having diameters as specified in paragraph 3.9.1 (b). Pistons (SAE RM-12) shall be made from unanodized SAE AA2024 aluminum alloy. Disassemble cylinders and discard rubber cups. Clean all metal parts with isopropanol and dry with clean compressed air. Inspect the working surfaces of all metal parts for scoring, galling, or pitting, and cylinder bore roughness, and discard all defective parts. Remove any stains on cylinder walls with crocus cloth and

---

[1] This procedure is identical to the one found in SAE J1703 JAN80.

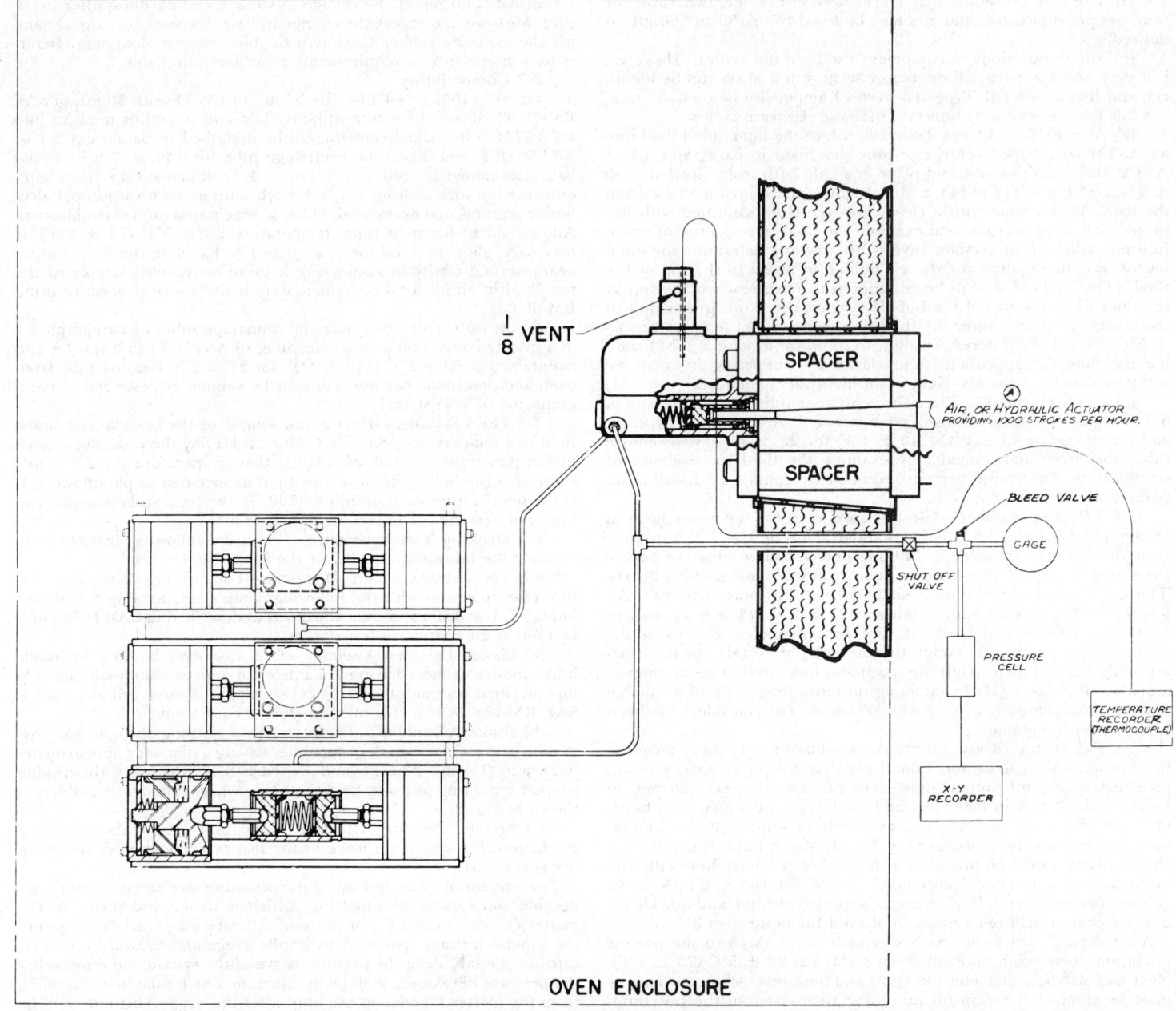

FIG. 1—STROKING TEST APPARATUS

isopropanol. If stains cannot be removed, discard the cylinder. Measure the internal diameter of each cylinder at locations approximately 19 mm (0.75 in) from each end of the cylinder bore, taking measurements in line with the hydraulic inlet opening and at right angles to the center line. Discard the cylinder if any of these four readings exceeds maximum or minimum limits of 28.66 − 28.60 mm (1.1285 − 1.126 in). Measure the outside diameter of each piston at two points approximately 90 deg apart. Discard any piston if either reading exceeds maximum or minimum limits of 28.55 − 28.52 mm (1.124 − 1.123 in). Select parts to insure that the clearance between each piston and mating cylinder is within 0.08 − 0.13 mm (0.003 − 0.005 in). Use new standard SAE RM-3 SBR cups as specified in Fig. 4 that are free of lint and dirt. Discard any cups showing imperfections such as cuts, tooling marks, molding flaws, or blisters. Measure the lip and base diameters of all test cups with an optical comparator or a micrometer to the nearest 0.025 mm (0.001 in) along the center line of SAE and rubber type identifications and right angles to this center line. Determine base diameter measurements within 0.8 mm (0.032 in) of the bottom edge and parallel to the base of the cup. Discard any cups if the two measured lip or base diameters differ by more than 0.08 mm (0.003 in). Average the lip and base diameters of each cup. Determine the hardness of all cups by the procedure specified in paragraph 3.3. Clean rubber parts with isopropanol and a lint-free cloth. Dry with clean compressed air. Dip the rubber and metal parts of the wheel cylinders, except housings, in the fluid to be tested and install them in accordance with manufacturer's instructions. Rubber boots may be retained on the cylinders if a small section is removed on the bottom to observe leakage. Manually stroke the cylinders to insure that they operate easily. Install cylinder in the simulated brake system.

(b) Master Cylinder Assembly—Use a new SAE RM-15b master cylinder having an SAE RM-13-02 aluminum alloy piston and new standard SAE RM-4a and RM-5a SBR cups as specified in Figs. 5 and 6. Inspect and clean all parts as specified in paragraph 3.9.2 (a). Measure each land of the master cylinder piston at two points approximately 90 deg apart. Discard the piston if any of these readings exceeds maximum or minimum limits of 28.55 − 28.52 mm (1.124 − 1.123 in). Dip the secondary cup in the test brake fluid, assemble on the piston, and maintain the assembly in a vertical position at 23 ± 5°C (73 ± 9°F) for at least 2 h. Determine the lip and base diameter of the secondary cup as installed on the piston and the primary cup at locations shown in Fig. 5. Inspect the relief and supply ports of the master cylinder and discard the cylinder if these ports have burrs or wire edges. Measure the internal diameter of the cylinder at two locations: approximately mid-way between the relief and supply ports and approximately 19 mm (0.75 in)

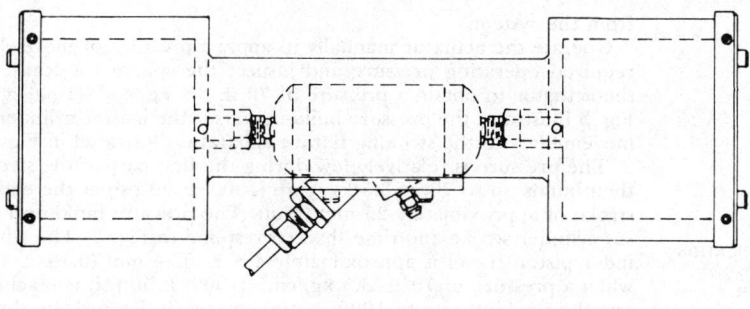
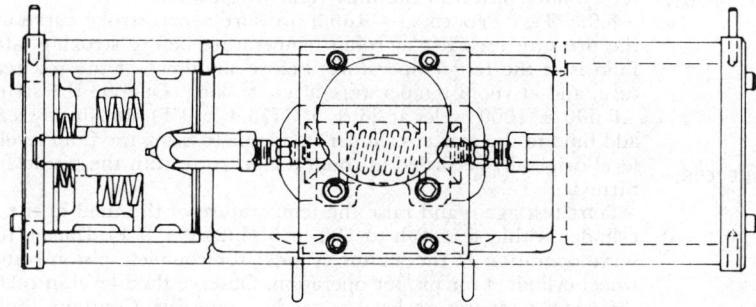
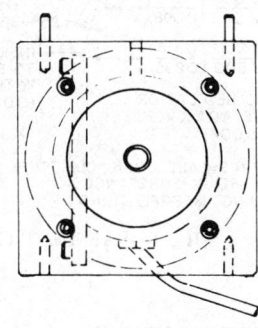

FIG. 2—STROKING FIXTURE APPARATUS

beyond the relief port toward the bottom or discharge end of the bore, taking measurements at each location in the vertical and horizontal centerlines of the bore. Discard the cylinder if any reading exceeds maximum or minimum limits of 28.65 – 28.58 mm (1.128 – 1.125 in). Dip the rubber and metal parts of the master cylinder, except the housing, in the fluid to be tested and install them in accordance with manufacturer's instructions. Discard boot and push rod assembly. Manually stroke the master cylinder to insure that it operates easily. Install the master cylinder in the simulated brake system.

(c) Use double-wall steel tubing (SAE RM-57 or -58) or equivalent, meeting SAE J527. Tubing from one outlet of master cylinder to the pair of wheel cylinders or to the single wheel cylinder, shall alternately be replaced with new tubing for each test [minimum length 915 mm (3 ft)]. Uniformity in tubing size is desirable between master cylinder and wheel cylinder; 6.3 mm (1/4 in) tubing is more adaptable with available tube connectors. The standard SAE RM-15a master cylinder has two outlets for tubing, both of which should be used.

(d) *Assembly and Adjustment of Test Apparatus*—Install wheel and master cylinders. Fill the system with test fluid, bleeding all wheel cylinders and the pressure equipment and gauges to remove entrapped air

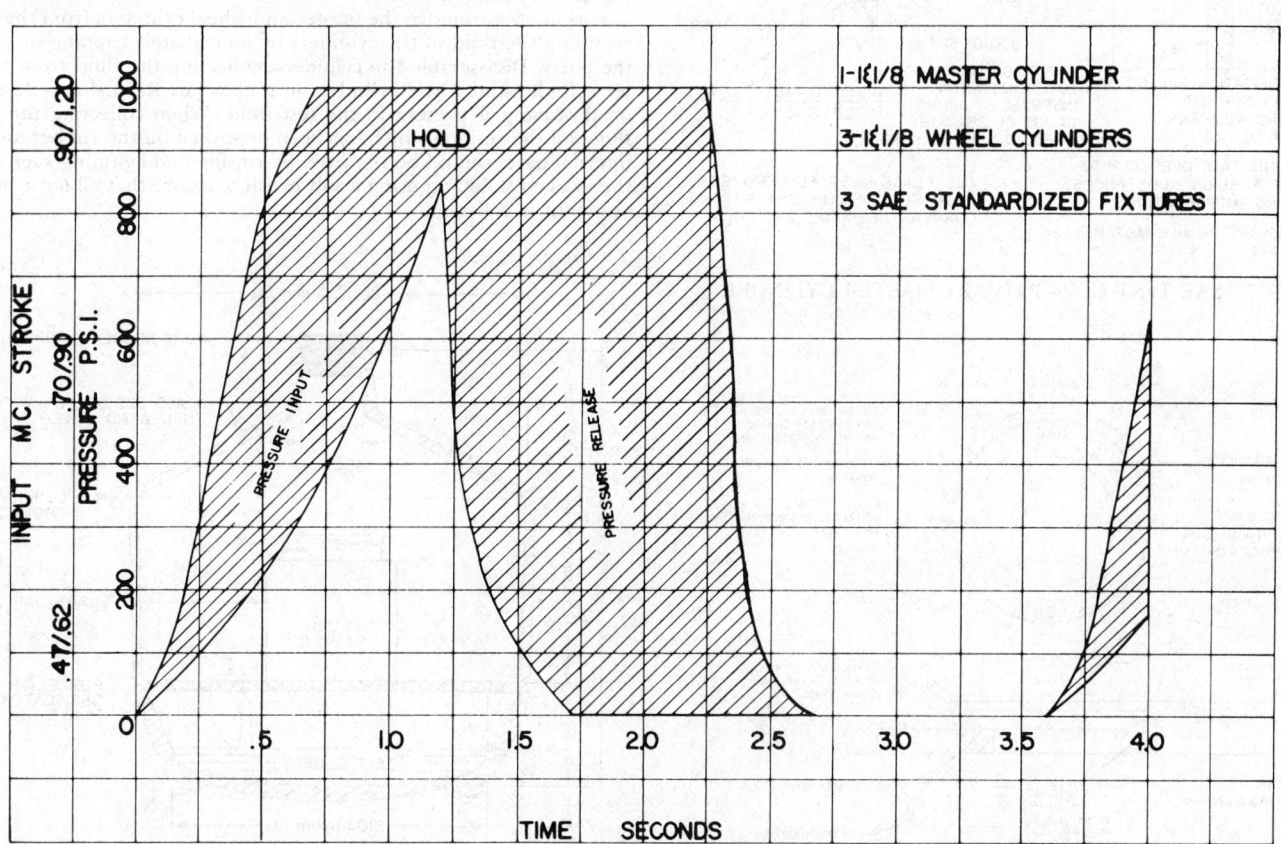

FIG. 3—MASTER CYLINDER STROKE

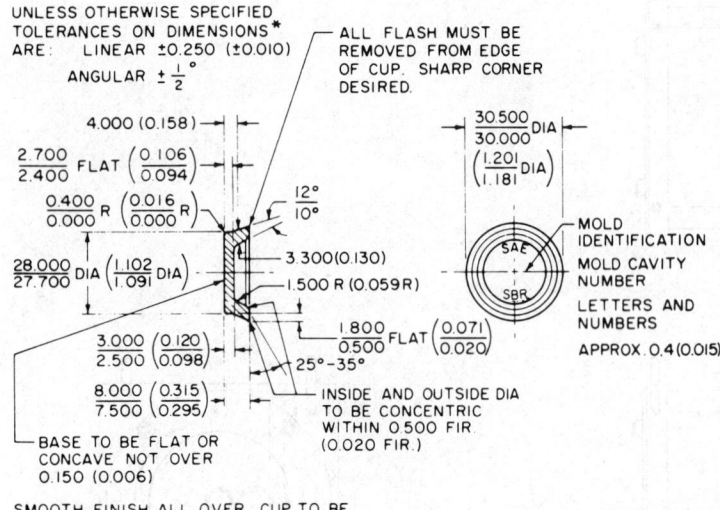

FIG. 4—SAE TEST CUP WHEEL CYLINDER

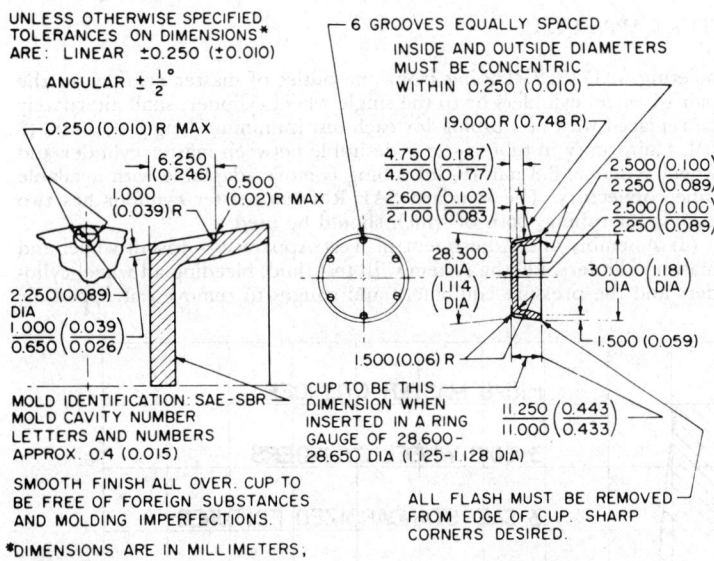

FIG. 5—SAE TEST CUP—PRIMARY MASTER CYLINDER

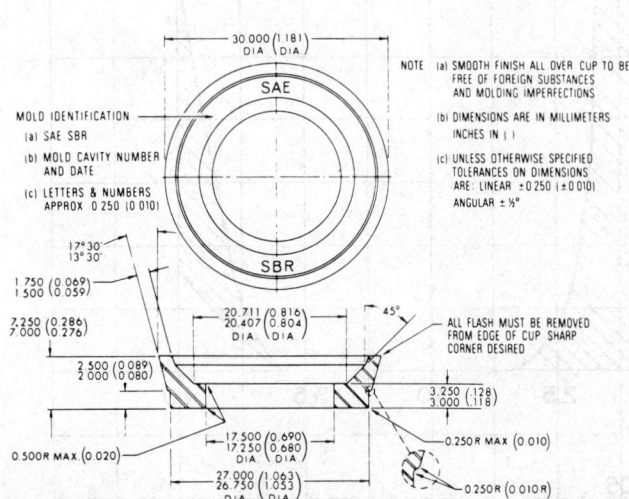

FIG. 6—SAE TEST CUP—SECONDARY MASTER CYLINDER

from the system.

Operate the actuator manually to apply a pressure of more than the required operating pressure and inspect the system for leaks. Adjust the actuator to obtain a pressure of 70 ± 3.5 kg/cm^2 (1000 ± 5 psi). Fig. 3 illustrates the pressure build-up versus the master cylinder piston movement with the stroking fixture apparatus illustrated in Figs. 1 and 2. The pressure is relatively low during the first part of the stroke and then builds up to 70 ± 3.5 kg/cm^2 (1000 ± 50 psi) at the end of the stroke of approximately 25 mm (1 in). The pressure build-up rate versus cylinder stroke and time shall correspond to Fig. 3. The wheel cylinder piston travel is approximately 4.8 ± 0.24 mm (0.19 ± 0.01 in) when a pressure of 70 ± 3.5 kg/cm^2 (1000 ± 50 psi) is reached. Adjust the stroking rate to 1000 ± 100 strokes/h. Record the fluid level in the master cylinder standpipe at $23 \pm 5°C$ ($73 \pm 9°F$) with the master cylinder piston in the fully returned position.

3.9.3 TEST PROCEDURE—Run a pressure versus stroke curve utilizing the pressure recorder at room temperature before stroking, after the fluid is at the test temperature, before shutdown at the test temperature, and at room temperature after stroking. Operate the system of 16 000 ± 1000 cycles at $23 \pm 5°C$ ($73.4 \pm 9°F$). Repair any leaks and add fluid to the master cylinder standpipe to bring the fluid level to the level originally recorded at room temperature with the piston fully returned.

Start test again and raise the temperature of the fluid in the master cylinder within 6 ± 2 h to $120 \pm 5°C$ ($248 \pm 9°F$). During test, observe operation of the master cylinder for complete piston return and wheel cylinders for proper operation. Observe fluid level in relation to the room temperature level at random intervals. Continue the test to 85 000 total recorded strokes which shall include the number of strokes during operation at $23 \pm 5°C$ ($73.4 \pm 9°F$), the number of strokes required to bring the system to the operating temperature of $120 \pm 5°C$ ($248 \pm 9°F$), plus the number of strokes at this operating temperature. Stop the test, and with the master cylinder piston in the fully returned position to relieve retained pressure in the system, allow the equipment to cool to room temperature.

Record the amount of fluid required to replenish any loss of fluid to the $23 \pm 5°C$ ($73 \pm 9°F$) level originally recorded. Stroke the assembly an additional 100 strokes $23 \pm 5°C$ ($73 \pm 9°F$) and 70 ± 3.5 kg/cm^2 (1000 ± 50 psi), examine wheel cylinders for leakage, and add and record volume of fluid required to bring the fluid level to the $23 \pm 5°C$ ($73 \pm 9°F$) original level.

Within 16 h, remove the master and wheel cylinders from the system, retaining the fluid in the cylinders by immediately capping or plugging the ports. Disassemble the cylinders, collecting the fluid from the master cylinder and wheel cylinders in a glass jar. Record any sludge, gel, or abrasive grit present in the test fluid. When collecting the stroked fluid, all the residue which has been deposited on the rubber and metal internal parts should be removed by rinsing and agitating such parts in the stroked fluid and using a soft brush to assure that all loose adhering sediment is collected.

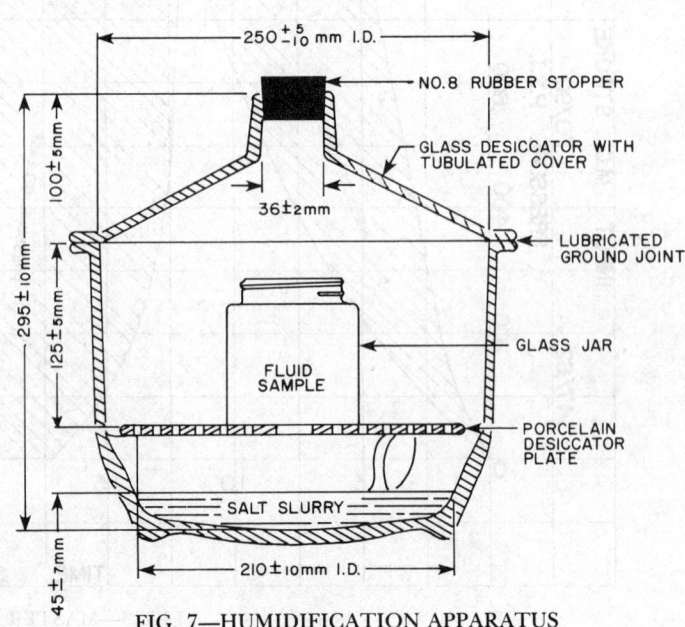

FIG. 7—HUMIDIFICATION APPARATUS

Clean rubber cups in isopropanol, and dry with clean, compressed air. Inspect cups for tackiness, scoring, scuffing, blistering, cracking, chipping (heel abrasions), and change in shape from original appearance. Within 12 h after disassembly, measure the lip and base diameter of each cylinder cup by the procedure specified in paragraph 3.9.2 (a) with the exception that the lip or base diameters of cups may differ by more than 0.08 mm (0.003 in). Determine the hardness of each cup by the procedure specified in paragraph 3.3.

Within 1 h after draining cylinders, agitate fluid in glass jar to suspend and uniformly disperse sediment and transfer a 100 mL portion of this fluid to an ASTM cone-shaped centrifuge tube and determine percent sediment as described in paragraph 5 (b) of ASTM D91. Inspect cylinder parts, recording any gum deposits, and rub any deposits adhering to cylinder walls with a cloth wetted with isopropanol to determine abrasiveness and removability. Clean cylinder parts in isopropanol, dry with compressed air, and inspect for pitting and scoring on pistons and cylinder walls. Measure and record diameters of pistons and cylinders by the procedures specified in paragraph 3.9.2 (a).

Calculate lip diameter interference by the following formula:

$$\frac{D_1 - D_2}{D_1 - D_3} \times 100 = \% \text{ Lip Diameter Interference Set}$$

Where: D_1 = Original lip diameter
D_2 = Final lip diameter
D_3 = Original cylinder bore diameter

Repeat the test if mechanical failure occurs that may affect the evaluation of the test fluid.

APPENDIX
PERFORMANCE CHARACTERISTICS OF SILICONE BASED BRAKE FLUIDS (SBBF)

To date, the only low water tolerant type brake fluids (LWTF) known to be commercially available are silicone based brake fluids. Other types of LWTF's are known to be in advanced stages of development. The following performance characteristics discussed in this Appendix relate only to SBBF's and may not necessarily be valid for other types of LWTF's.

Since 1974, SBBF's meeting the Department of Transportation Motor Vehicle Safety Standard No. 116 (MVSS 116) requirements for specification DOT-5 have been commercially available. In March of 1978, the U. S. Army issued an SBBF specification, MIL-B-46176. To date, four SBBF's have been qualified for use in military hydraulic brake systems. This report is the product of an intensive ten year study of SBBF's by the SAE Non-Conventional Brake Fluid Task Group and is based upon extensive laboratory work, numerous inter-laboratory studies, and two two-year field service trials.

CHARACTERISTIC DIFFERENCES BETWEEN SBBF AND SAE J1703 BRAKE FLUIDS

SBBF and SAE J1703 brake fluids are chemically different. SBBF's are essentially non-hygroscopic and non-water tolerant and are immiscible with fluids meeting SAE J1703.

Basic Fluid Performance Characteristics—Silicone based brake fluids meeting DOT-5 are designed for use in systems fitted with rubber seals made from natural rubber (NR), styrene butadiene rubber (SBR), and ethylene propylene rubber (EPR), and with hoses made from polychloroprene rubber (CR) compounds of the type used in conventional motor vehicle hydraulic brake systems.

SBBF must be functionally compatible with SAE J1703 motor vehicle brake fluids and with braking systems designed for such fluids. The term "functionally compatible" means that the braking system will perform satisfactorily should the two types of fluids be mixed, and that no undesirable interaction will occur. This does not imply complete miscibility since SBBF's are not completely miscible with present SAE J1703 fluids.

To utilize SBBF's to the fullest advantage, they should not be mixed with other brake fluids. Inadvertent mixtures of SBBF with fluids meeting SAE J1703 are not known to have any adverse effects, but all combinations have not been tested. Vehicle manufacturers' recommendations should be followed where indicated.

Fluid Performance Concerns

Water Tolerance—Because SBBF's are essentially non-water tolerant and have been found to absorb little water during in-service use, special emphasis was placed on the performance properties obtained after the fluid was first exposed to humid atmospheres. A brief discussion of each of the critical performance areas follows:

Water intolerant SBBF may absorb up to 0.1% (1000 ppm) of water when exposed to high humidity per MVSS 116 specification. With some of these fluids, it has been found that on chilling $-40°C$ ($-40°F$) or lower, fine crystals appear either in the vapor space or in the fluid. On re-warming to room temperature, these crystals may collect as fine droplets and are not readily redispersed. Such behavior has always been considered unacceptable in a brake fluid. SBBF's are now available that do not exhibit this behavior.

Vapor Lock—It has been demonstrated that the presence of free water in the non-water tolerant silicone based brake fluids will produce a vapor lock temperature at or about 100°C (212°F) in laboratory instruments.

SBBF's are not designed to accommodate water and caution should be exercised to prevent accidental entry of water which may lead to brake failure.

Lubricity—It has been generally accepted that the present SAE stroking test procedure (S4.12 SAE J1703 JAN80) does not assure adequate rubber-to-metal and metal-to-metal lubricity to determine service performance in the vacuum over hydraulic boosters used in vehicles of 4536 kg (10 000 lb) GVWR or larger, and possibly in hydraulic clutches and dual master cylinders, as well.

The NCBF Task Group of the Motor Vehicle Brake Fluids Subcommittee has recommended the development and use of a dual master cylinder stroking test in place of the obsolete single system master cylinder (RM-15a) presently used in the SAE stroke test (S4.12 of the SAE J1703 JAN80 Standard). Such substitutions will be more representative for determining the lubricity properties of brake fluids in current braking systems.

There is general agreement that an SBBF which meets both the Bendix Hydrovac stroke test and the Delco Moraine 701a dual master cylinder stroke test will have sufficient lubricity to be used wherever SAE J1703 brake fluids are now utilized. Whether an SBBF needs to pass both of these tests to perform satisfactorily in service has not yet been determined. SBBF's are now available that have been validated as meeting both the Bendix Hydrovac and Delco Moraine 701a performance test requirements. However, sufficient field test data are not yet available to demonstrate their functional capabilities in these applications.

These fluids are not necessarily suitable for use in central hydraulic or pumped systems.

Effect on Rubber—There is also general agreement that SBBF's affect braking system elastomeric components in quite a different manner than do fluids meeting SAE J1703. Legitimate questions have been raised concerning interpretation of the effect on rubber test for MVBF (S3.11 and 4.11 of SAE J1703 JAN80), particularly with respect to silicone-based fluids. More specifically, doubts have been raised as to (1) whether the present short-term test at elevated temperature provides an adequate measure of actual rubber swell in service, and (2) whether evaluations of only rubber swell and hardness change provide sufficient information to assure satisfactory performance. The task group has requested assistance from both the Hydraulic Brake Actuating Elastomeric and the Automotive Brake and Steering Hose Subcommittees on these questions.

In addition, test data on various mixtures of an SBBF with an SAE J1703 brake fluid have been accumulated which can be interpreted as indicating that a "mixed fluid" rubber swell test for SBBF is desirable. SBBF's are essentially immiscible with SAE J1703 brake fluids and when the two types are mixed, two liquid phases are formed. The test data indicates that migration of rubber swell agents across the liquid interface can occur. A mixed fluid rubber swell test, wherein an SBBF is mixed with the SAE RM-66-03 compatibility test fluid, has been considered which would provide some assurance that mixing of the two types of fluids in service will not lead to excessive swelling and/or softening of critical rubber components.

Of particular concern is the establishment of realistic rubber swell minimum-maximum values which relate to actual in-service performance.

Mixed Fluid Corrosion—Concerns have been expressed over the possibility of corrosion occurring due to the migration of additives when SBBF and SAE J1703 fluids are mixed. In 1973, the Non-Conventional Brake Fluid Task Force of the Motor Vehicle Brake Fluids Subcommittee carried out round-robin mixed fluid corrosion tests. These studies showed that mixtures tested met all the corrosion requirements of the SAE J1703 Standard. However, all possible combinations have not been studied. Attempts should be made to prevent mixing of the fluids and specific combinations should be studied before deliberately mixing the fluids in a motor vehicle braking system.

Field Test Data—Results of one- and two-year field service tests by the U. S. Army indicate that SBBF provides corrosion protection for the braking system that is significantly better than that provided by conventional brake fluids conforming to SAE Standard J1703

(USAMERDC Report 2132, AD A-12849 dated February 1975, and Report 21264, AD-02618 dated January 1976).

Vehicle tests by Automotive Research Associates, Inc. (ARA) for SAE demonstrated that SBBF's offer potentially higher temperature capabilities (resistance to vapor locking) than do conventional hygroscopic brake fluids. However, this can only be achieved by excluding conventional fluids as well as any free water from the braking system.

Viscosity—SBBF's provide a superior viscosity index when compared to SAE J1703 fluids. However, there is some concern that the ambient viscosity of SBBF at 25°C (which can be more than twice that of SAE J1703 type brake fluids), could result in additional assembly line filling and bleeding time. Vehicle manufacturers have expressed concern that viscosities over 25 mm^2/s may create fill problems both at the assembly plant and in servicing. As a result, specifications being issued by major vehicle producers may specify a lower maximum viscosity of 25 mm^2/s at 25°C.

ΦAir Solubility—It has been reported that dimethyl polysiloxane fluid, which is a major component of silicone based low water tolerant type brake fluids (SAE J1705), can typically contain dissolved air at a level of 16% ± 3% by volume at standard temperature and pressure. This compares with a typical level of 5% ± 2% by volume of dissolved air for glycol ether based SAE J1703 type brake fluids. An increase in brake pedal travel may be experienced under severe operating conditions, especially at higher altitudes and high temperature conditions.

The term "dissolved air" (air absorbed from the atmosphere) should not be confused with the term "entrapped" or "free air" since their effects on brake system performance can be entirely different. Air that has been absorbed from the atmosphere does not result in an increase in fluid or system volume, whereas, entrapped air or free air does occupy system volume and can be easily compressed when force is applied to the system.

Compressibility—Silicone based brake fluids are more compressible than conventional brake fluids and the difference is magnified at higher temperatures. The compressibility of SBBF's may be calculated at any combination of temperature and pressure [see J. A. Tichy and W. O. Winer, "A Correlation of Bulk Moduli and P-V-T Data for Silicone Fluids at Pressure up to 500,000 PSIF." ASLE Transactions 11, 338 − 344 (1968)].

The effect of brake system performance, specifically pedal travel, may also be calculated by the following formula:

$$\text{MC Piston Travel} = \frac{\text{Compressibility} \times \text{Fluid Volume}^2}{\text{MC Piston Area}}$$

$$\text{Pedal Travel} = \text{MC Piston Travel} \times \text{Pedal Ratio}$$

Development and improvement of SBBF is continuing with special attention being given to the requirements of individual manufacturers. The performance properties of SBBF will be more fully defined through field tests and laboratory test data demonstrating satisfactory SBBF functional capability in all types of motor vehicle braking systems. Other low water tolerant type brake fluids, although not presently available, may not necessarily exhibit similar physical property characteristics to the SBBF's and may require separate recommended practices.

[2] Vehicle testing has indicated that the most precise correlation with actual vehicle performance is obtained by using only the volume of fluid in the wheel cylinders and/or calipers with the pistons fully extended.

EUROPEAN BRAKE FLUID TECHNOLOGY
—SAE J1709 JUN93

SAE Information Report

Report of the SAE Borate Ester Specification Work Group of the SAE Motor Vehicle Brake Fluids Standards Committee approved June 1993. Rationale statement available.

Foreword—The SAE has established engineering guidelines which describe the tests and minimum performance properties for motor vehicle brake fluids. Such information is contained within SAE J1703 (Motor Vehicle Brake Fluid), proposed SAE J1704 (DOT 4 Type Brake Fluid), and SAE J1705 (Low Water Tolerant Brake Fluids). Brake fluids which meet the criteria set forth by these documents are non-mineral oil based. Common compositions may include synthetic glycol ethers, borate esters, or silicone.

The requirements for brake fluids are further described within the Federal Motor Vehicle Safety Standard 116, prepared by the National Highway Traffic Safety Administration. Within this Standard, motor vehicle brake fluids are classified as DOT 3, DOT 4, DOT 5, or DOT 5.1 based upon minimum wet and dry boiling points, viscosity characteristics, and in the DOT 5/5.1 case, base fluid chemistry.

Developments in Europe have resulted in the introduction of brake fluids which offer enhanced wet and dry boiling point performance. Employing both conventional and novel chemistry, these fluids are becoming an increasingly important factor in the European marketplace. The purpose of this document is to provide an overview of the performance and characterization of these fluids and the general trend toward the utilization of such technology. In utilizing the information provided, it should be recognized that field experience and engineering knowledge is greatest for DOT 4 followed by Super DOT 4. DOT 5.1 fluids have recently been introduced and more application experience continues to be gained. Additionally, the silicon ester technology described has well-established experience in racing and continues to be evaluated for use in passenger cars.

1. Scope—This SAE Information Report provides an overview of brake fluid technology developed and marketed in Europe which offers enhanced wet and dry boiling point performance. The information contained within this document applies to hydraulic brake fluids utilized in automotive braking systems which are designed to be compatible with SAE J1703 (DOT 3) fluids. The report reflects details received as of September 1991.

2. References

2.1 Applicable Documents—The following publications form a part of this specification to the extent specified herein. The latest issue of SAE publications shall apply.

2.1.1 SAE PUBLICATIONS—Available from SAE, 400 Commonwealth Drive, Warrendale, PA 15096-0001.

SAE J1703—Motor Vehicle Brake Fluid
Proposed SAE J1704—DOT 4 Type Brake Fluids
SAE J1705—Low Water Tolerant Brake Fluids

2.1.2 FMVSS PUBLICATION—Available from the Superintendent of Documents, U.S. Government Printing Office, Washington, DC 20402.

FMVSS No. 116—Motor Vehicle Brake Fluids

2.2 Related Publications—The following publications are provided for information purposes only and are not a required part of this document.

2.2.1 SAE PUBLICATION—Available from SAE, 400 Commonwealth Drive, Warrendale, PA 15096-0001.

SAE 740126—Brake Fluid Temperatures Obtained in Alpine Vehicle Trials

2.2.2 ISO PUBLICATION—Available from ANSI, 11 West 42nd Street, New York, NY 10036-8002.

ISO 4925—Road vehicle—Non petroleum based brake fluids

3. Background—Borate ester based DOT 4 brake fluids developed between 1960 to 1970 have been commonplace in Europe, and to a lesser extent North America, for many years. The requirements established within the FMVSS 116 dictate that these fluids possess a higher dry and wet equilibrium reflux boiling point than fluids classified as DOT 3. Introduction of DOT 4 fluids in Europe was designed to better accommodate the engineering trends and service requirements of European vehicles which are operated more often at high-speed conditions and are subject to regular safety inspections.

Since 1970, there has been an extension of the borate ester technology such that a wide range of fluids is now available offering higher wet and dry boiling point performance compared to DOT 4. Brake fluids are now being manufactured and marketed in Europe with properties which bridge the gap between DOT 4 and DOT 5 fluids. These so called Super DOT 4 fluids are gaining increased acceptance and are now utilized by some original equipment manufacturers in Europe for initial and service fill applications.

In the 1980's, brake fluids based upon silicon esters were introduced which offer further improvement to the boiling point properties beyond Super DOT 4 fluids. Such fluids possess very high wet and dry boiling point performance and have been utilized successfully in racing applications.

Also in the 1980's, fluids were developed in Europe which meet the performance requirements of DOT 5, but do not utilize silicone based chemistry. These fluids are classified by the recently adopted DOT 5.1 category. Such fluids employ borate ester bases resulting in formulations which are miscible with SAE J1703 fluids and meet the performance criteria of all established DOT categories. Non-silicone based DOT 5.1 brake fluids are being offered in the aftermarket in small but increasing quantities.

4. European Market Trends—It is estimated that over three-quarters of the European market for brake fluid is represented by DOT 4 product. In Germany close to 100% of the brake fluid market is covered by DOT 4, Super DOT 4, and DOT 5.1 fluids. DOT 4 fluids are recommended by most European vehicle manufacturers. DOT 3 fluids represent a minor percentage and are utilized in commercial applications and in certain passenger car vehicles.

Very little DOT 5 or 5.1 fluid is sold in Europe partly because they have only recently been introduced. Additionally, the low-temperature performance properties of DOT 3 or DOT 4 fluids are generally viewed as satisfactory for existing applications.

The largest growth area for brake fluid performance in Europe is the Super DOT 4 category. This type of fluid is used by some original equipment manufacturers as an initial fill and service fill fluid. Super DOT 4 fluids are considered by certain manufacturers as being necessary to cope with the high temperatures attained in braking systems under European driving conditions, in cars of modern design and to extend service intervals.

5. Brake Fluid Characteristics—Table 1 presents the respective performance and characterization of brake fluid technology in Europe versus established DOT 3, DOT 4, and DOT 5 benchmarks. The primary points of differentiation are dry and wet boiling point and low-temperature viscosity.

5.1 Super DOT 4 Fluids—Super DOT 4 brake fluids are commonly based upon mixtures of glycol ether and borate ester chemistry. Although an official specification has not been established to date (and is not planned), it is commonly accepted that such fluids should attain a minimum dry boiling point of 260 °C and a minimum wet boiling point of 180 °C as required by the DOT 5 (silicone brake fluid) or DOT 5.1 (non-silicone brake fluid) category. Low-temperature viscosity requirements are that of DOT 4. Super DOT 4 fluids in the European market are typically higher in dry and wet boiling point (280 °C ERBP/183 °C wet ERBP) with low-temperature viscosity meeting DOT 3 criteria. Additionally, Super DOT 4 fluids have recently been introduced which achieve 200 °C wet ERBP and DOT 3 low-temperature viscosity characteristics.

5.2 Borate Ester DOT 5.1 Fluids—Revisions to FMVSS 116 have been published to accommodate non-silicone based DOT 5 fluids. The fluids are categorized as DOT 5.1 and, unlike DOT 5 fluids which are dyed purple, can vary from clear to amber in color. DOT 5.1 brake fluids are based predominantly on borate ester chemistry. Such fluids typically possess wet/dry boiling points and low-temperature viscometric properties within the DOT 5 specification limits. Borate ester chemistry allows for SAE J1703 DOT 3 and DOT 4 fluid miscibility.

5.3 Silicon Ester Fluids—Silicon ester based brake fluids typically achieve boiling point characteristics above the requirements of the DOT 5 specification. Silicon esters do not have the same chemical structure as silicones and differ in physical properties. Silicon esters are similar in structure to borate esters. Figure 1 reflects the respective compositions of silicone and borate/silicon ester based fluids.

Silicon esters are miscible with glycol ethers and borate ester type fluids and achieve all performance criteria established by SAE J1703 and FMVSS 116 for DOT 3 and DOT 4 fluids. At present, these fluids are not utilized by manufacturers but are employed in racing.

TABLE 1—BRAKE FLUID PARAMETERS

Property	DOT 3 Requirement	DOT 4 Requirement	Super DOT 4 Accepted Limit	Super DOT 4 (Typical)	DOT 5 Requirement	Silicone[1] DOT 5 (Typical)	DOT 5.1 Requirement	Borate Ester DOT 5.1 (Typical)	Silicon Ester (Typical)
Dry ERBP (°C)	>205	>230	>260	270	>260	>310	>260	265	310
Wet ERBP (°C)	>140	>155	>180	185	>180	>310	>180	185	260
Kinematic Viscosity, mm^2/s -40 °C	<1500	<1800	<1800	1350	<900	250	<900	850	1350

[1] Silicone fluids do not possess a true boiling point.

```
CHEMISTRY                STRUCTURE

                          \   O       O       O   /
                           \ / \     / \     / \ /
Silicone                    Si   Si    Si    Si
                           / \  / \   / \   / \
                         CH3 CH3 CH3 CH3 CH3 CH3 CH3 CH3

                              OR'
                              |
Borate Ester              B - OR'
                              |
                              OR'

                            R                 R
                            |                 |
Silicon Ester           R - Si - OR'  or  R'O - Si - OR'
                            |                 |
                            R                 R

    R:  alkyl group
    R': glycol or glycol ether group
```

FIGURE 1—RESPECTIVE COMPOSITIONS OF SILICONE AND BORATE/SILICON ESTER BASED FLUIDS

6. Performance and Testing—Borate Ester Super DOT 4 and DOT 5.1 brake fluid manufactured in Europe all meet the minimum performance requirements set forth within SAE J1703 and FMVSS 116 for the respective DOT categories. Along with boiling point and viscometric requirements, this includes evaluations for chemical stability, high- and low-temperature stability, evaporation, water tolerance, fluid and elastomer compatibility, oxidation resistance, corrosion protection, and stroking properties.

In addition to the evaluation previously described, testing beyond SAE and Federal requirements has been conducted within Europe to support advanced brake fluid technology. This includes the following:
   a. Testing to motor manufacturers' specifications
   b. Alpine Trials
   c. Fleet Trials
   d. Water ingress into different hose/fluid combinations
   e. Hose Burst Tests
   f. Additional corrosion testing

Evaluation of fluid performance versus motor vehicle manufacturers' specifications will generally impose additional requirements beyond those established by SAE J1703 and FMVSS 116. These can include additional stroking tests and corrosion protection performance to provide improved safety margin during actual field use. Vehicle manufacturer specifications which include performance limits and test procedures are proprietary to equipment and fluid manufacturers.

Alpine trials are commonly conducted within Europe to monitor brake fluid temperatures and brake system protection under arduous mountain descent conditions. Such conditions have been found to result in extremely high brake fluid temperatures in passenger cars and supported the movement to brake fluid formulations possessing improved boiling point performance.

The effect of fluid chemistry on brake hoses is an important performance concern and an integral part of the brake fluid qualification process for fluid manufacturers. Burst strength studies have been conducted in Europe on hoses which are both temperature and water ingress aged. The chemistries of both the fluid and the hose materials play important roles in achieving compatibility. Careful selection of the fluid/hose combination becomes essential to ensure system durability, performance, and protection.

φMOTOR VEHICLE BRAKE FLUID CONTAINER COMPATIBILITY—SAE J75 OCT88

SAE Information Report

Report of Hydraulic Brake Systems Actuating Committee approved February 1965 and completely revised October 1988.

1. Purpose—Motor vehicle brake fluid must conform to the requirements of SAE J1703, not only when manufactured, but also after extended storage in any commercial packaging container. The purpose of this report is to generate an awareness of the major problems involved in the storage of brake fluids and, to some extent, provide means of circumventing them. It is also the purpose of this report to relate to experience and to test data accumulated and to list certain conclusions which should aid in the proper selection of containers for brake fluid.

2. Background—A problem in selecting containers for brake fluids is that, in the past, many containers have not been capable of preserving some of the brake fluids in their original state. For instance, SAE J1703 requires that no more than 0.05% by volume of sediment may be found in the fluid at the time of manufacture when tested by the water tolerance test at 140°F. Some commercially packaged brake fluids known to meet the SAE standards when manufactured have been found to exceed the 0.15% sediment permitted for packaged fluid by as much as ten fold, due to contamination from the container.

The reaction of certain inhibitors and other components commonly used in brake fluids with tin plate, soldered seams of metal cans, organic coated steel, and plastic containers, may create a storage problem because of the formation of precipitates. These precipitates may or may not be soluble in the brake fluid but are often precipitated under conditions of the water tolerance test and cause the brake fluid to fail this specification. Other properties of the brake fluid such as boiling point, corrosion and stability may be affected adversely by storage in certain containers.

3. Experimental Data: Metal Cans—Extensive tests have shown that storage of many brake fluids in soldered metal cans, as judged by the quantity of precipitate formed, may be improved by limiting the lead content of the solder and preferably having the solder seam on the outside of the can. The least reactive solders, and therefore the best for this use, would be 100% tin. However, there is no assurance that any solder will be suitable with every brake fluid. When metal cans are used, welded seams are preferred.

4. Polyethylene Containers—Corrosion tests and water tolerance tests have been run on brake fluids stored in high density polyethylene containers for three years. There was no increase in precipitate in the water tolerance test and the corrosion test was satisfactory.

Any moisture pickup during storage will cause a reduction in boiling point. The moisture pickup is minimized if the wall thickness of a high density polyethylene container is at least 0.03 in. For increased container strength which is needed for handling and shipping, 1 gal high density polyethylene containers should have a minimum wall thickness of 0.04 in.

5. Recommendations—Proper container selection is critical with respect to preserving the fluid in a satisfactory condition conforming to SAE J1703 standards. It is recommended that the following SAE J1703 test procedures be used to evaluate containers.
   1.) Boiling Point, paragraphs 3.1 and 4.1—for evaluation of moisture pickups, package sealing efficiency and permeability.
   2.) Corrosion, paragraphs 3.5 and 4.5—for evaluation of possible depletion of brake fluid inhibitor systems under storage conditions.
   3.) Water tolerance, paragraphs 3.8 and 4.8—for evaluation of precipitated, dispersed, or hydrolyzed precipitates resulting from the chemical activity between the brake fluid and container materials under storage conditions.
   4.) Resistance to oxidation, paragraphs 3.10 and 4.10—for evaluation of overall stability of motor vehicle brake fluids under storage conditions.

Valuable storage information can be obtained by subjecting containers of brake fluid to accelerated storage tests at 120-140°F. Tests performed on fluid samples withdrawn after 10-30 days will permit the selection of the most suitable container. The moisture pickups in high density polyethylene containers stored at 100% relative humidity and 70-75°F for three months is equivalent to about one year of storage under warehouse conditions where humidity is not controlled.

PLASTIC DUST SHIELD FOR HYDRAULIC DISC BRAKES—SAE J1876 MAR90

SAE Standard

Report of the Hydraulic Brake Plastics Standards Committee approved March 1990.

Foreword—This proposed SAE Standard, reviewed initially at the November 18, 1987 subcommittee meeting in New Orleans, Louisiana, was developed to characterize the material requirements for a dust shield for hydraulic disc brakes. Plastic dust shields are currently in production on some domestic passenger vehicles. The use of plastic in this application offers advantages in cost, corrosion resistance, and design flexibility relative to a conventional stamping.

1. Scope—The material defined by this SAE document is an impact modified, heat stabilized, 66 nylon reinforced with glass fibers. This material is for use in dust shields for hydraulic disc brakes.

NOTE: The applicability of a plastic dust shield must be evaluated for each individual brake system. Its use with solid rotors and/or high performance brake systems is not recommended.

2. References
ISO 178 (ASTM D 790)
ISO R527 (ASTM D 638)
ISO 180 (ASTM D 256)
ISO 75 (ASTM D 648)
ISO 1218 (ASTM D 789)

3. General Material Requirements

3.1 Conditioning—All test values indicated herein are based on materials conditioned in a controlled atmosphere of 23°C ± 2 and 50% ± 5 relative humidity for 24 h prior to testing and tested under the same controlled conditions.

4. Test Requirements

4.1 Tensile Strength—The material when tested by the procedure specified in 5.1 shall have a minimum tensile strength of 62 MPa.

4.2 Ultimate Elongation—The material when tested by the procedure specified in 5.1 shall have a minimum ultimate elongation of 4%.

4.3 Flexural Modulus—The material when tested by the procedure specified in 5.2 shall have a minimum flexural modulus (tangent modulus of elasticity) of 2000 MPa.

4.4 Flexural Strength—The material when tested by the procedure specified in 5.2 shall have a minimum flexural strength of 106 MPa.

4.5 Impact Strength—The material when tested by the procedure specified in 5.3 shall have the following minimum notched izod impact strengths:
a. Notched specimen at 23°C   85 J/m
b. Notched specimen at -40°C   28 J/m

4.6 Deflection Temperature—The material when tested by the procedure specified in 5.4 shall have the following minimum deflection temperatures:
a. At 455 kPa   240°C
b. At 1820 kPa   230°C

4.7 Melting Point—The material when tested by the procedure specified in 5.5 shall have a melting point of 256°C ± 6.

5. Test Procedures

5.1 Tensile Strength and Ultimate Elongation—Determine the tensile strength at maximum load and the elongation of the material at maximum load by ASTM D 638 (ISO R527), using 5 mm/min ± 1.25 testing speed.

5.2 Flexural Modulus and Flexural Strength—Determine the flexural modulus and the flexural strength of the material by ASTM D 790 (ISO 178).

5.3 Impact Strength—Determine the izod impact strength of the material by ASTM D 256 (ISO 180).

5.4 Deflection Temperature—Determine the deflection temperature of the material by ASTM D 648 (ISO 75). Test specimens shall be annealed prior to test by immersion in hot oil at 28°C ± 2 below the melting point of the material for 30 min. Allow the specimens to cool in air away from drafts to ambient temperature prior to test.

5.5 Melting Point—Determine the melting point of the material by ASTM D 789 (ISO 1218).

PRODUCTION, HANDLING AND DISPENSING OF SAE J1703 MOTOR VEHICLE BRAKE FLUIDS—SAE J1706 OCT88

SAE Recommended Practice

Report of the Hydraulic Brake Systems Actuating Committee approved October 1988.

1. Scope—This SAE Recommended Practice is intended to provide basic recommendations to aid in the development and use of safe and efficient practices for all operations involving the production, handling, and dispensing of SAE J1703 Motor Vehicle Brake Fluids.

2. References—
SAE J75 OCT88, Motor Vehicle Brake Fluid Container Compatibility
SAE J1703 APR88, Motor Vehicle Brake Fluid
SAE J1707 NOV85, Service Maintenance of SAE J1703 Brake Fluids in Motor Vehicle Systems

3. Contamination—SAE J1703 brake fluids are susceptible to various types of contamination which can be detrimental to the safe and efficient performance of brake actuating systems.

3.1 Water Contamination—SAE J1703 brake fluids are hygroscopic and absorb moisture when exposed to the atmosphere. The degree of moisture absorption is dependent upon a number of variables. Among these are the relative hygroscopicity of the brake fluid, the area of the surface exposed to the atmosphere, the time of exposure, and the temperature and relative humidity of the atmosphere. Water contamination can also occur from condensation resulting from atmospheric temperature changes or the mechanical entrance of the free water. This water contamination will lower the boiling point and increase the low temperature viscosity (see Fig. 1).

3.2 Contamination with Petroleum Products—Since motor vehicle brake fluids are handled and dispensed under conditions where contamination with petroleum products can occur, specific precautions must be taken to prevent contamination. Motor vehicle brake fluids contaminated by petroleum products will cause excessive swelling and softening of brake system rubber parts, resulting in brake systems failure.

3.3 Particle Contamination—Specific precautions must be taken to prevent contamination of motor vehicle brake fluids with particles or solid material of any sort which could interfere with the proper operation of the braking system components. In particular, abrasive particles will cause scuffing and wear of the pistons and cylinders and can damage the elastomer seal materials.

4. Recommendations—The following recommendations are applicable to most procedures.

4.1 Material Control—Material specifications should be established for all ingredients of the brake fluid formulation. Chemical and physical tests should be required to assure that such specifications are met.

4.2 Processing—Processing equipment used in the manufacture of components should be so controlled as to insure the degree of quality and uniformity required by paragraph 4.1. Use of dedicated storage tanks, pumps, and lines is suggested for each fluid component. Materials of construction should be compatible with all fluids that they contact. All tanks, pumps, pipes, etc. should be completely isolated from other production processes to reduce the chances of accidental contamination. Since brake fluid and its components are hygroscopic, the manufacturing operations must minimize the water content of the finished brake fluid.

4.3 Material Handling and Cleaning Processes—Normally, blending tanks, tank cars, and tank wagons will be used for handling a number of different motor vehicle brake fluid formulations. It is important to avoid cross-contamination.

a) Blending tanks should be checked before starting each batch to insure that they are empty and clean.
b) Pipelines should be drained and blown out with dry air or nitrogen. When such lines are cleaned or purged with brake fluid, the fluid should be discarded properly.
c) Whenever possible, all bulk containers for shipping and storage should be restricted to brake fluid. Bulk containers should be drained, purged with dry air or nitrogen, and inspected before filling.
d) New clean drums should be used for brake fluid, and the inner surface should be compatible with the particular fluid formulation. Drums should be visually inspected for foreign material. Lines to drum fillers should be prepared as in paragraph b).
e) Brake fluid should be filtered through a suitable filter of 10 μm or less before any filling operation.
f) Extreme care must be exercised in transferring fluid from its original container.
g) The possible toxicity hazard and environmental effects of motor vehicle brake fluids must be considered in terms of any applicable laws or regulations.

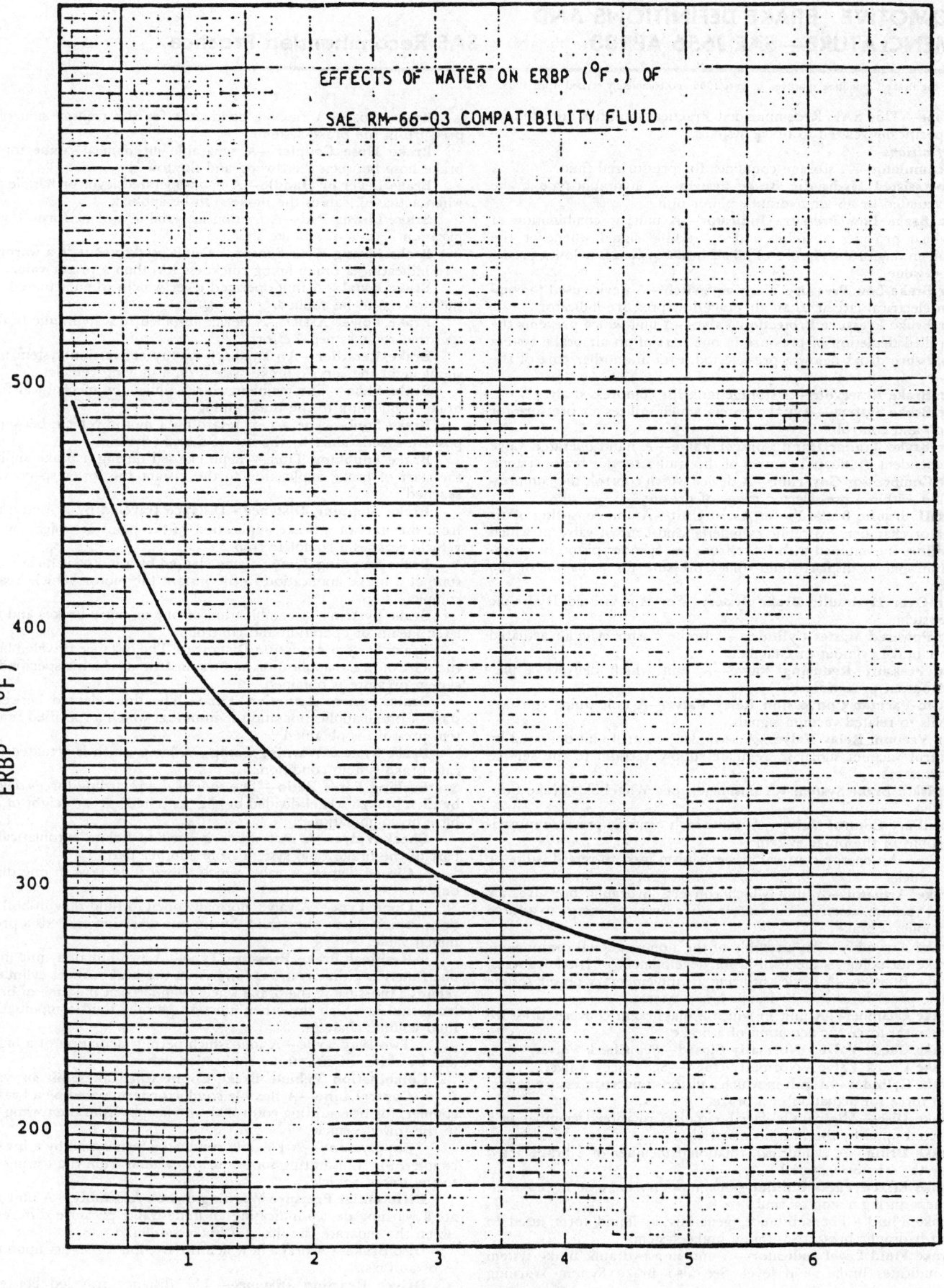

FIG. 1

AUTOMOTIVE BRAKE DEFINITIONS AND NOMENCLATURE—SAE J656 APR88

SAE Recommended Practice

Report of the Brake Committee approved January 1952 and completely revised April 1988.

1. Scope—This SAE Recommended Practice identifies and defines terms specifically related to brake systems.

2. Definitions

Accumulator—A storage container for pressurized fluid.

Air-Assisted Hydraulic Brake System—A hydraulic-type brake system actuated by an air hydraulic power unit.

Air Brake Low Pressure Indicator—A unit or combination of units which provides a visible and/or audible signal whenever the stored air in one of the circuits of an air brake system is below a predetermined value.

Air Brake Low Pressure Warning Switch—A device used to complete an electrical circuit to an air brake low pressure indicator.

Air Brake Pressure Protection Valve—A unit which prevents the uncontrolled depletion of pressure in one part of an air brake system when pressure drops below a pre-selected level in another part of the system.

Air Brake Reservoir—A storage tank for compressed air.

Air Brake System—A brake system which utilizes air pressure for operation and control.

Air Brake Trailer Hand Control Valve—A hand operated valve for independent graduated control of the trailer service brake system.

Air Compressor Governor—A device which controls the compression of air within a pre-selected range of pressure.

Air Hydraulic Brake Power Assist Unit—A unit consisting of air cylinder or chamber, hydraulic cylinder(s) and control valve in which driver effort is combined with force from the cylinder piston or chamber diaphragm to displace fluid under pressure for actuation of the brake(s).

Air Over Hydraulic Brake System—See Air-Assisted Hydraulic Brake System.

Air-Powered Master Cylinder—A brake master cylinder actuated by an air brake cylinder or chamber.

Air Pressure Reducing Valve—A unit which delivers a pre-selected output pressure.

Air-to-Vacuum Conversion Relay Valve—A unit which converts air signals to related vacuum signals.

Air Vacuum Relay Valve—A secondary control unit operated by the control vacuum signal to regulate supply vacuum to the service brake actuators.

Antiskid Brake System—A misnomer. See Wheel Slip Brake Control System.

Auxiliary Control Valve—A unit which controls pressure in various portions of the brake system.

Brake—An energy conversion mechanism used to retard, stop, or hold a vehicle.

Brake Actuator—A unit which converts hydraulic pressure, air pressure, vacuum, electrical current or other forms of energy to a force which applies a brake.

Brake Assembly—An assembly of the non-rotational components of a brake including its mechanism for development of friction forces.

Brake Booster—A device utilizing a supplementary power source to reduce pedal force in a hydraulic brake system.

Brake Chamber—A unit in which a diaphragm converts pressure to mechanical force for actuation of a brake.

Brake Check Valve - Normally Closed—See Check Valve.

Brake Check Valve - Normally Open—See Check Valve.

Brake Cylinder—A unit in which a piston converts pressure to mechanical force for actuation of a brake.

Brake Disc—The parallel-faced, circular, rotational member of a brake acted upon by the friction material.

Brake Drum—A cylindrical, rotational member of a brake acted upon by the friction material.

Brake Effectiveness Buildup—A temporary increase in brake effectiveness during a stop or snub.

Brake Fluid—The substance, generally in liquid form, used to transmit hydraulic pressure within a brake system.

Brake Fluid Level Indicator—A unit in a hydraulic brake system which indicates brake fluid level. See also Brake System Warning Lamp.

Brake Fluid Level Warning Switch Assembly—A unit used to actuate a warning device indicating a reservoir fluid level lower than a preset value.

Brake Hose—A flexible conductor for the transmission of fluid pressure in the brake system.

Brake Hose Coupler—A separable mechanical connector for a brake hose between the towing and the towed vehicle.

Brake Lever or Handle—A hand-operated lever or handle which, when actuated, causes the brake(s) to be applied.

Brake Lining Pad—A friction material in a plane form. Usual application is on a disc brake shoe.

Brake Lining Wear Sensor—A unit used to actuate a warning device indicating a brake lining thickness less than a preset value.

Brake Pedal—A foot-operated lever which, when actuated, causes the brake(s) to be applied.

Brake Power Assist—A device installed in a hydraulic brake system that reduces pedal effort.

Brake Retarder—An auxiliary energy conversion system used to supplement the service brake system on a moving vehicle.

Brake Shoe—The member which applies the mechanical force of brake application to the brake lining.

Brake Snub—The act of retarding a motor vehicle between two positive speed values by the use of a brake system.

Brake Snubbing Time—Time elapsed during a brake snub from the start of brake application to the instant the lower speed value is reached.

Brake Stopping Distance—Distance traveled by a motor vehicle from the start of a brake application to the point at which the motor vehicle reaches a complete stop.

Brake Stopping Time—Time elapsed by a motor vehicle from the start of a brake application to the instant the motor vehicle reaches a complete stop.

Brake System—A combination of one or more brakes and the related means of operation and control.

Brake System Actuation Distance—The distance traveled between the start of brake application and the instant at which a specified brake system pressure is obtained.

Brake System Actuation Time—The time elapsed between the start of brake application and the instant at which a specified brake system pressure is obtained.

Brake System Warning Lamp—A lamp which is actuated to indicate brake system condition.

Braking Force Ratio—The sum of the retarding forces developed by each braked wheel divided by the "as tested" gross weight of the vehicle or combination.

Check Valve—A unit which is used to isolate automatically one part of the brake fluid system from another part.

Closed Type—A valve which allows fluid flow in one direction only.

Open Type—A valve, normally open to fluid flow in both directions, which closes when fluid flow in one direction exceeds a predetermined value.

Residual Brake Pressure Type—A two-function unit in which one function either restricts fluid from the brake wheel cylinder(s) or retains a pressure in the brake wheel cylinder(s) at the time of brake release, and in which the other function permits fluid compensation for fluid volume changes.

Two-Way Type—A unit which permits actuation of a brake system by either of two brake application valves.

Combination Vehicle Brake Connecting Lines (air or vacuum)

Control Line—A flexible conductor terminated by a brake hose coupler, for transmitting control air or vacuum from the towing vehicle to the towed vehicle.

Supply Line—A Flexible conductor terminated by a brake hose coupler, for transmitting supply air or vacuum from the towing vehicle to the towed vehicle.

Differential Pressure Warning Switch Assembly—A unit to actuate a warning device indicating an undesirable pressure difference between the separate circuits of a brake system.

Disc Brake—A brake in which the friction forces act upon the faces of a disc(s).

Driver Reaction Distance—The distance traveled between the point at which the driver perceives a demand for braking and the start of brake application.

Driver Reaction Time—The time elapsed between the instant the driver perceives a demand for braking and the start of brake application.

Drum Brake—A brake in which the frictional forces act on the cylindrical surface(s) of a brake drum.

Dual Master Cylinder—A master cylinder for a split brake system having two fluid displacement sections.

Electric Brake

Armature—The rotating part of the brake electric actuating mechanism to which the magnet is attracted.

Controller—A variable resistance for graduated control of electric brake(s).

Load Compensating Resistor—A resistor unit in series with an electric brake control which can be preset to limit current to achieve brake balance.

Magnet—The part of the electric actuating mechanism, which when energized is attracted to the armature, creating a controlled force to apply the brake(s).

Emergency Brake System—A brake system used for stopping a vehicle in the event of a malfunction in the means of operation and control of the service brake system.

Emergency Valve—A unit under the control of the driver which, when actuated, will activate the emergency brake system.

Equivalent Braking Force—See Braking Force Ratio.

Fade

Heat Fade—Temporary reduction of brake effectiveness due to a loss of friction between braking surfaces, resulting from heat.

Water Fade—Temporary reduction of brake effectiveness due to a loss of friction between braking surfaces, resulting from water.

Follow-Up-Type Valve—A unit which responds to fluid displacement or mechanical linkage movement, to modulate pressure in a cylinder or chamber.

Forward Shoe—See Leading Shoe.

Foundation Brake Assembly—See Brake Assembly.

Glad Hand—An air brake hose coupler.

Hold-Off Valve—A unit which permits free fluid flow in either direction when the brakes are not applied, but prevents pressure buildup in one part of the brake system until pressure in the other part reaches a predetermined value.

Hydraulic Brake System—A brake system in which brake operation and control utilizes hydraulic fluid.

Initial Stopping Speed—Speed of the motor vehicle at the start of brake application.

Leading Shoe—A show in a non-servo brake which pivots around a fulcrum in the direction of normal drum rotation.

Load Proportional Brake Control—A system or device which regulates the input force to the brakes on an axle in proportion to the load on that axle.

Logic Controller—A part of the wheel slip brake control system which interprets input signals from the sensor(s) and transmits the controlling output signals to the modulator(s).

Manual Hydraulic Brake System—A hydraulic-type brake system which utilizes unassisted driver effort.

Master Cylinder—The primary unit for displacing hydraulic fluid under pressure in the brake system.

Metallic Friction Material—A sintered friction material formulated with metallic or metallic-ceramic materials.

Modulator—The unit(s) in a wheel slip brake control system which adjust(s) brake actuating force in response to input signals.

Moisture Ejection Valve—A unit which expels fluid contaminants from an air system.

Non-Asbestos Friction Material—A friction material containing no asbestos.

Non-Servo Brake—A drum brake in which all brake shoes are independently actuated.

Organic Friction Material—A friction material, having organic binders, substantially formulated with non-metallic fibers.

Parking Brake Control Valve—A valve that controls the parking brake system.

Parking Brake System—A brake system, intended to hold a vehicle stationary, in which one or more brakes may be held applied without continued application of force to the control.

Parking Brake Warning Switch Assembly—A unit used to actuate a warning device indicating the parking brake application mechanism is not in the fully released position.

Proportioning Valve (Pressure)—A unit in the brake system used to modify the pressure to the brakes on one or more axles.

Quick Release Valve—A control unit which accelerates the release of air pressure from various portions of brake systems.

Reactionary Type Valve—A unit which responds to fluid displacement and pressure, or mechanical linkage movement and force, to modulate pressure in a brake cylinder or chamber.

Relay Emergency Valve—A relay valve which also provides for automatic application of the trailer brakes in case of a breakaway or loss of pressure in the trailer supply (emergency) line.

Relay Valve—A control unit used to accelerate the air application and release of air pressure or vacuum in a part of the air system or vacuum brake system.

Reverse Shoe—See Trailing Shoe.

Safety Valve—A pressure release unit used to protect the air system against excessive pressure.

Semi-Metallic Friction Material—A friction material having organic binders formulated with metallic fibers and/or metal powders.

Service Brake System—The brake system generally used for retarding or stopping a vehicle.

Service Brake Valve—A foot-operated unit which is used for graduated control of all the brakes in the service brake system.

Servo Brake—A drum brake in which brake shoes are linked, such that the braking force of one shoe amplifies the input of the other shoe(s).

Slack—Brake actuator stroke prior to effective application of force.

Slack Adjuster—An adjustable member which transmits brake application force and permits compensation for lining wear.

Spitter Valve—See Moisture Ejection Valve.

Split Brake System—A service brake system having two or more separate fluid, electrical, mechanical or other circuits which upon failure in any circuit retains full or partial brake actuating capability.

Spring Type Brake Actuator—A unit which utilizes the stored energy in a spring(s) to actuate the brake(s).

Start of Brake Application—The instant at which the brake control system is actuated as determined by initial brake control movement.

Stop Light Switch—A switch which completes the electrical circuit to the stop lamp(s) when the brake(s) is (are) applied.

Straight Air Brake System—See Air Brake System.

Supplemental Brake System—An additional brake system used to assist the service brake system in retarding a vehicle.

Tractor Protection Control Valve—A manually operated valve that actuates the tractor protection valve.

Tractor Protection Valve—A unit which is part of a towing vehicle air brake system and which:

(a) Permits driver control of the opening and closing of the air brake lines to the towed vehicle whenever the air pressure in the towing vehicle exceeds a predetermined value.

(b) Closes the air brake lines automatically when the tractor brake system pressure is less than the predetermined value.

(c) Vents the trailer supply (emergency) line when closed either manually or automatically.

Trailing Shoe—A shoe in a non-servo brake which pivots around a fulcrum in a direction opposite to normal drum rotation.

Vacuum Assisted Hydraulic Brake System—A hydraulic-type brake system actuated by a vacuum hydraulic power unit.

Vacuum Brake Supply Line—The conduit for transmitting supply vacuum from a vacuum source to the vacuum reservoirs.

Vacuum Brake System—A brake system which utilizes vacuum means for operation and control.

Vacuum Hydraulic Power Unit—A unit consisting of a vacuum brake cylinder or chamber, hydraulic cylinder(s) and control valve, in which driver effort is combined with force from the cylinder piston or chamber diaphragm to displace fluid under pressure for actuation of the brake(s).

Vacuum Over Hydraulic Brake System—A hydraulic-type brake system actuated by a vacuum-powered master cylinder.

Vacuum Powered Master Cylinder—A brake master cylinder actuated by a vacuum cylinder or chamber.

Vacuum Pump—A device which creates vacuum to actuate the brakes.

Vacuum Reservoir—A storage container for air at pressures less than atmospheric.

Wheel Cylinder—A unit for converting hydraulic fluid pressure to mechanical force for actuation of a brake.

Wheel Lockup—A condition of 100% wheel slip.

Wheel Slip Brake Control System—A system which automatically controls rotational wheel slip during braking.

Wheel Slip Sensor—When used in combination with the wheel slip brake controls system, a unit which senses the rate of angular rotation of the wheel(s) and transmits signals to the logic controller.

ANTILOCK BRAKE SYSTEM REVIEW—SAE J2246 JUN92

SAE Standard

Report of the SAE Brake Standards Committee 10 ABS approved June 1992.

Foreword—The application of Antilock Brake Systems (ABS) to passenger cars and light trucks has grown in recent years. This has been fueled by advances in automotive electronics, competitive trends, and consumer safety awareness. Although technical literature exists regarding specific systems, hardware, and applications, little exists that addresses the topic from the viewpoint of the industry as a whole. Recognizing this need, the Antilock Brake Standards Committee was formed and began its work by compiling this document.

TABLE OF CONTENTS

1. Scope
2. References
2.1 Applicable Documents
2.2 Definitions
2.3 Nomenclature
3. Introduction
4. Historical Review
4.1 The Evolution of Passenger Car and Light Truck Antilock Braking Systems
4.2 ABS Installation Rates for Passenger Cars and Light Trucks
5. Basic ABS Theory
5.1 Goals of ABS Application
5.2 Inherent Limitations and Compromises
5.3 Overview of ABS
5.4 Braking Dynamics—Single Wheel Model
5.4.1 Tire-to-Road Interface Description
5.4.2 Braking Without ABS
5.4.3 Road Surface Friction Utilization
5.4.3.1 Longitudinal Force Utilization
5.4.3.2 Lateral vs. Longitudinal Force Utilization
5.4.3.2.1 Nondeformable Surface
5.4.3.2.2 Deformable Surface
5.4.4 ABS Objective
5.4.5 Simple ABS Modulation Logic
5.4.6 Example Run on Simulation
5.5 Braking Dynamics—Four Wheel Model
5.5.1 Straight Line Braking
5.5.2 Stability and Controllability in Response to Steering Inputs
5.5.3 Braking in a Turn
5.5.4 Split Coefficient Braking
5.5.5 Performance Tradeoffs
5.6 Control System Block Diagram
6. ABS Current Production System Profile

1. Scope—This SAE Information Report provides information applicable to production Original Equipment Manufacturer antilock braking systems found on some past and current passenger cars and light trucks. It is intended for readers with a technical background.

It does not include information about aftermarket devices or future antilock brake systems.

Information in this document reflects that which was available to the committee at the time of publication.

2. References

2.1 Applicable Documents—The following publications form a part of this specification to the extent specified herein. The latest issue of SAE Publications shall apply.

1. Zellner, John W., "An Analytical Approach to Antilock Brake System Design," SAE Paper 840249, 1984.
2. Leiber, Heinz, Czinczel, Armin, and Anlauf, Juergen, "Antiskid System (ABS) for Passenger Cars," Bosch Technische Berichte (English Special Edition), Feb. 1982.
3. Ehlbeck, Jim, Moore, Tony, and Young, Warren, "Antilock Brake Systems for the North America Truck Market," SAE Paper 901174.
4. Ellis, J. R., "Vehicle Dynamics," London Business Books, 1969.
5. Flaim, T. A., "Vehicle Brake Balance Using Objective Brake Factors," SAE Paper 890804, February 1989.
6. Rowell, J. Martin, Gritt, Paul S., editors, "Antilock Braking Systems for Passenger Cars and Light Trucks—A Preview (PT-29)," SAE, 1987.
7. Ward Automotive Reports; Ward's Communications
   January 25, 1988
   January 16, 1989
   February 5, 1990
   January 28, 1991
   February 25, 1991
   July 8, 1991
   August 19, 1991
8. Lowery, Joseph, "Jensen—Ferguson: 2+2 = 4;" February, 1966; p.u.
9. "Ward's Automotive Antilock Braking Systems In The 1990s," p. 2; Lamm, Michael, Ward's Communications, 1990.
10. ISO 611-1980 "Road vehicles—Braking of automotive vehicles and their trailers—Vocabulary"
11. ECE R13.05, Annex 13 "Requirements Applicable to Tests for Vehicles Equipped with Anti-Lock Devices"
12. SAE J670—Vehicle Dynamics Terminology

2.2 Definitions

2.2.1 ABS (ACRONYM FOR ANTILOCK BRAKE SYSTEMS)—A device which automatically controls the level of slip in the direction of rotation of the wheel on one or more wheels during braking (see ISO 611).

2.2.2 ACCUMULATOR—LOW PRESSURE (SUMP)—A low pressure brake fluid storage device not intended as an energy source.

2.2.3 ACCUMULATOR—HIGH PRESSURE—An energy storage device using pressurized brake fluid as the storage medium.

2.2.4 ACTUATION PRINCIPLE—PUMP BACK—An ABS system configuration where during modulation control, low pressure brake fluid is restored to high pressure by a pump and made available for a subsequent build cycle. The total amount of fluid available for modulation control for a given stop is limited to the amount of fluid provided by the master cylinder for that particular stop.

2.2.5 ACTUATION PRINCIPLE—REPLENISHMENT—An ABS system configuration employing an external source of high pressure fluid in addition to displaced master cylinder fluid for modulation control. This type of system has virtually an unlimited supply of high pressure fluid available during modulation control.

2.2.6 ADD-ON ABS SYSTEM—An ABS configuration in which both the ABS power supply and modulation control functions are independent from the base brake actuation system. The components of this system may be packaged together or separately.

2.2.7 CONTROL CHANNEL—A portion of the hydraulic brake circuit which can be operated independently from other portions of the hydraulic brake circuit. In ABS braking, it is a hydraulic brake circuit that controls a wheel or wheels independently of other wheels.

2.2.8 CONTROLLER—A component of the antilock braking system which interprets input signals from the sensor(s) and transmits the controlling output signals to the modulator(s) (see SAE J670e).

2.2.9 DIAGONAL SPLIT BRAKE SYSTEM—A brake system in which separate hydraulic circuits actuate the service brakes for one front wheel and one rear wheel on the opposite side.

2.2.10 DIRECTLY CONTROLLED WHEEL—A wheel whose braking force is modulated according to data provided at least by its own sensor (see ECE Regulation 13).

2.2.11 G-SWITCH/ACCELEROMETER G-SENSOR/(LATERAL and LONGITUDINAL)—A device by which acceleration or a change in acceleration of the vehicle is detected or confirmed.

2.2.12 INDIRECTLY CONTROLLED WHEEL—A wheel whose braking force is modulated according to data provided by the sensor(s) of other wheel(s) (see ECE Regulation 13).

2.2.13 INTEGRATED ABS SYSTEM—An ABS configuration in which some ABS and base brake actuation functions are shared. Most commonly, both systems may share a hydraulic power supply.

2.2.14 LATERAL FORCE COEFFICIENT—The ratio of the lateral force to the vertical load (see SAE J670e).

2.2.15 LONGITUDINAL FORCE COEFFICIENT—The ratio of the longitudinal force to the vertical load (see SAE J670e).

2.2.16 MODULATION CONTROL—The systematic regulation of braking force resulting from the build, decay, and/or hold of pressure to a given control channel.

2.2.17 MODULATOR—The component responsible for modulating the force developed by the brake actuators as a function of the order received from the controller (see ISO 611).

2.2.18 NONUNIFORM/NONHOMOGENEOUS COEFFICIENT OF FRICTION—A braking tractive surface in which variable surface conditions exist.

2.2.19 PEDAL FEEDBACK—A tactile sensation felt by the driver's foot on the brake pedal during modulation control.

2.2.20 PUMP MOTOR—A mechanical pump driven by an electric motor used to pressurize or move brake fluid.

2.2.21 SELECT HIGH—Multi-wheel control where the signal of that wheel which is the last to tend to lock controls the system for all the wheels of the group (see ISO 611).

2.2.22 SELECT LOW—Multi-wheel control where the signal of that wheel which is the first to tend to lock controls the system for all the wheels of the group (see ISO 611).

2.2.23 SLIP—The difference between the angular velocity of a freely rolling wheel (ω) and the angular velocity of the braked wheel (ω_B) divided by the angular velocity of the freely rolling wheel (ω), expressed as a percentage.

$$\% \text{ SLIP} = \frac{\omega - \omega_B}{\omega} \qquad \text{(Eq. 1)}$$

2.2.24 SLIP ANGLE—The angle between the wheel plane and the direction of travel of the center of the tire contact (see SAE J670e).

2.2.25 SOLENOID—An electromagnetic device in which an electrically energized magnet moves an armature to open or close a hydraulic flow path (see ISO 611).

2.2.26 SPLIT COEFFICIENT (SPLIT μ)—A braking tractive surface in which two significantly differing coefficients of friction exist at the left and right side of the vehicle.

2.2.27 TRANSITION COEFFICIENT (TRANSITION μ)—A braking tractive surface in which two significantly differing coefficients of friction exist in the direction of travel of the vehicle.

2.2.28 UNIFORM/HOMOGENEOUS COEFFICIENT OF FRICTION—A braking tractive surface in which no significantly differing coefficient of friction exist throughout the surface.

2.2.29 VERTICAL SPLIT BRAKE SYSTEM—A brake system in which separate hydraulic circuits actuate the service brakes, one for both front wheels and one for both rear wheels.

2.2.30 WHEEL SPEED SENSOR—The component responsible for sensing the condition of rotation of the wheel(s) and for transmitting this information to the controller.

2.2.31 YAW RATE (r)—The angular velocity about the (vehicle's) vertical axis (see Figure 1).

2.3 Nomenclature

- a_x — Acceleration along the x (longitudinal) axis of the vehicle
- e — Base of Napierian logarithmic system, (2.7182....)
- G — Brake specific torque
- g — Acceleration due to gravity
- J — Wheel inertia
- M — Vehicle mass
- P — Brake pressure
- P_0 — Initial brake pressure
- P_1 — Rate of change of brake pressure
- R — Radius of tire
- S — Stopping distance
- s — Slip
- s_0 — Initial slip
- T — Torque
- t — Continuous time
- u — Vehicle velocity along its x (longitudinal) axis
- V — Peripheral velocity of free straight rolling tire
- X — Longitudinal force
- Z — Vertical force
- μ — Longitudinal force coefficient
- α — Slip angle
- γ — Slope of longitudinal force coefficient curve
- τ — Lumped time constant
- ω — Angular velocity

2.4 Subscripts

- B Brake
- H High limit
- i Initial
- L Low limit
- p Peak
- R Road
- s Slide

3. Introduction—ABS may represent the single greatest advancement in automotive braking since the development of hydraulic brakes. Given the significance, this document has been written to provide the reader

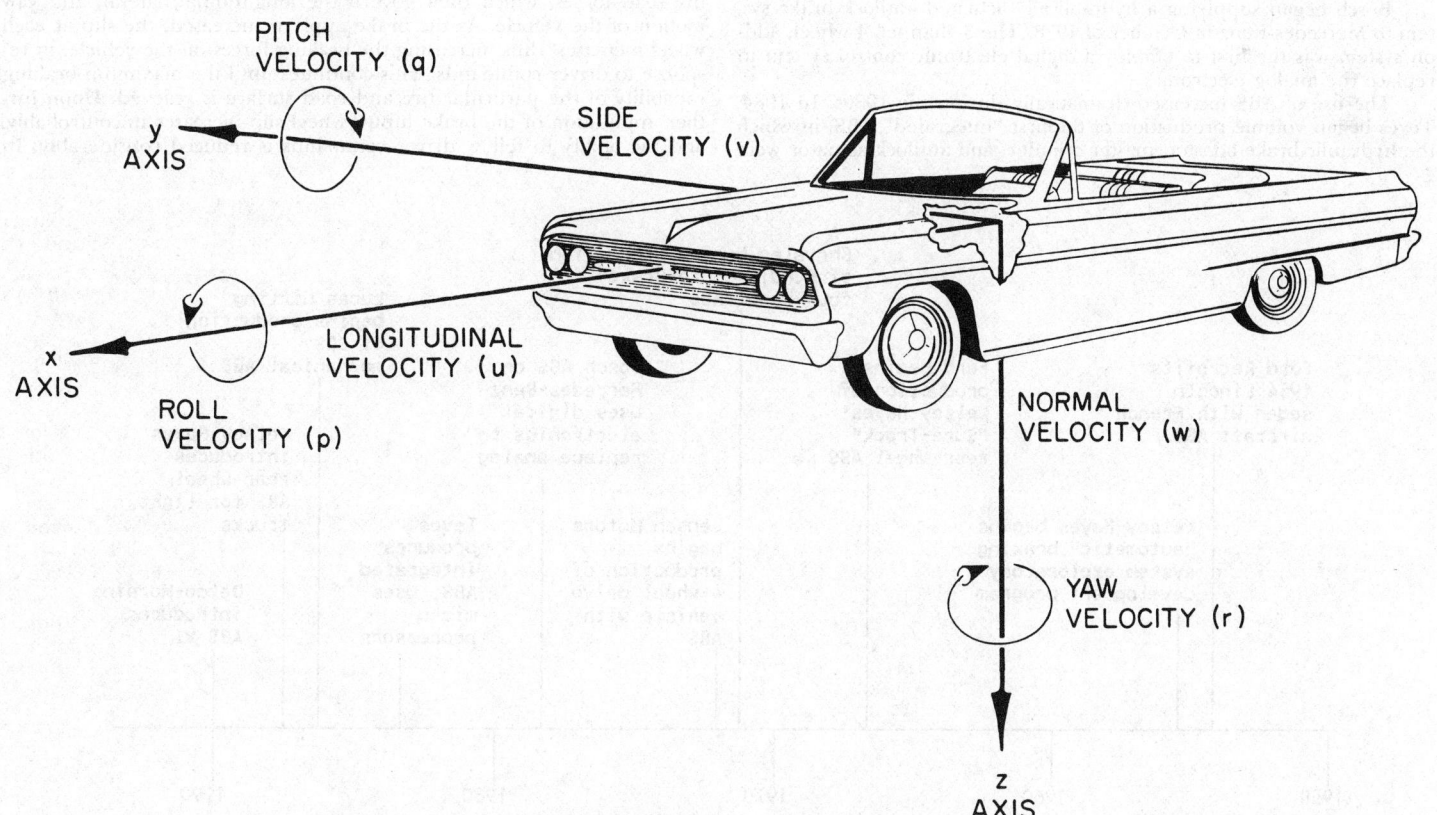

FIGURE 1—DIRECTIONAL CONTROL AXIS SYSTEM

with the following:
a. Historical Review of ABS
b. Basic ABS Theory
c. Profile of Current Antilock Brake Systems

This information is intended to provide the reader with an understanding of the fundamentals of ABS and its development. With this knowledge, the reader should have a better understanding of the present, and may have the tools to help understand future ABS developments.

4. Historical Review

4.1 The Evolution of Passenger Car and Light Truck Antilock Brake Systems—The current hydraulic antilock brake systems were conceived from systems developed for trains in the early 1900s. The development of passenger car and light truck ABS appears to have started around 1936, when Bosch received its first patent for an antilock brake system using electromagnetic wheel speed sensors. When the sensors detected a locked wheel, an electric motor controlled orifice at each brake line was activated, thus regulating the brake pressure.

Several ABS development projects began in the 1950s. The first project began in 1954 at Ford when a Lincoln sedan was fitted with an antilock brake system from a French aircraft. In 1957 Kelsey-Hayes began an "automatic" braking system exploratory development program. The program concluded that the system should prevent the loss of vehicle control and reduce the vehicle's stopping distance. 1957 also saw Chrysler begin research on a "skid control" brake system, however, it was not until 1966 that Chrysler began developing antilock brake systems that were intended for production.

The late 1960s saw the first antilock brake system enter production. Kelsey-Hayes completed development of a rear wheel ABS in 1968. The single channel vacuum powered system was first offered by Ford on its 1969 Thunderbird and Lincoln Continental Mark III, under the trade name "Sure-Track."

Chrysler introduced four-wheel ABS on the 1971 Imperial. The system, developed with Bendix, was a 3-channel, 4-wheel vacuum actuated system marketed under the trade name "Sure-Brake."

Jensen Motors became the first automobile manufacturer to offer ABS in conjunction with a viscous coupled 4-wheel drive system. In 1972 the Jensen Interceptor was made available with the Dunlop "Maxaret" antilock brake system. The system used a prop-shaft mounted speed sensor to operate a solenoid, which in turn operated air valves to reduce the brake vacuum servo output force.

Bosch began supplying a hydraulically actuated antilock brake system to Mercedes-Benz in October of 1978. The 3-channel, 4-wheel, add-on system was the first to employ a digital electronic control system to replace the analog electronics.

The use of ABS increased dramatically during the 1980s. In 1984, Teves began volume production of the first "integrated" ABS, in which the hydraulic brake booster, master cylinder, and antilock actuator were combined into a single component. The system, designated MK II, was also the first microprocessor based ABS and was first used on the 1985 Lincoln Mark VII.

Lucas Girling began supplying Ford of Europe with a mechanical ABS called Stop Control System (SCS) in 1986, a derivative of one developed in the early 1980s for motorcycles. The system used two mechanical wheel speed sensors on the front wheels. The rear wheels were valved to prevent lock up.

In 1986 Kelsey-Hayes introduced a single channel rear-wheel ABS on light trucks. The system saw widespread usage in the late 1980s, beginning with Ford in the 1987 model year.

Delco Moraine NDH began production of its ABS VI system in 1990. The system is unique in that the pistons used to control brake pressures are driven by electric motors via gear boxes and ball screws.

A summary of the previous information is shown in Figure 2.

4.2 ABS Installation Rates for Passenger Cars and Light Trucks—Figure 3 illustrates ABS installations for passenger cars and light trucks as a percent of production. The information shown represents both rear wheel and four wheel ABS.

5. Basic ABS Theory

5.1 Goals of ABS Application—The application of ABS to a vehicle can provide improvements in the vehicle performance under braking compared to a conventional brake system. Improvement is typically sought in the areas of stability, steerability, and stopping distance. In the following sections, some basic approaches to achieving these goals are examined.

5.2 Inherent Limitations and Compromises—It must be remembered that the addition of ABS to a vehicle does not release it from compliance to the basic laws of physics. The interface between the road and tire still defines the maximum braking force that can be applied, the vehicle chassis will still determine steer responses and load transfer patterns, and the vehicle geometric and inertial properties will still be a major factor in these responses. In addition, the vehicle brake system still must provide for brake application up to the time that ABS begins to control the braking. Even then, the ABS must work through the foundation brakes.

5.3 Overview of ABS—When a driver applies the brake pedal on a vehicle moving initially at a uniform rate of speed, the wheels tend to slow down relative to the ground, causing slip at the wheels. This slip between the tire and the road results in the generation of horizontal tire-road forces, which then govern the longitudinal, lateral, and yaw motion of the vehicle. As the brake apply is increased, the slip at each wheel increases, thus increasing the braking forces on the vehicle, in response to driver commands. This continues until the maximum braking capability of the particular tire and road surface is reached. Upon further application of the brake input, wheel slip increases uncontrollably, and the ability to follow driver commands is reduced considerably. In

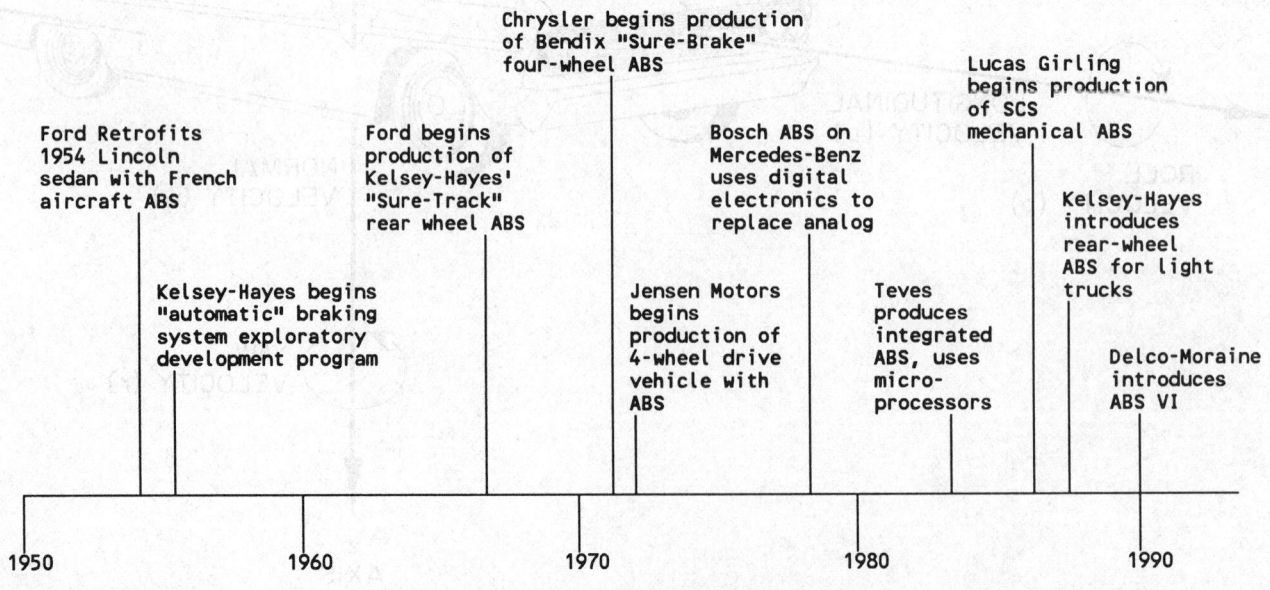

FIGURE 2—ANTILOCK BRAKE SYSTEM SIGNIFICANT EVENTS

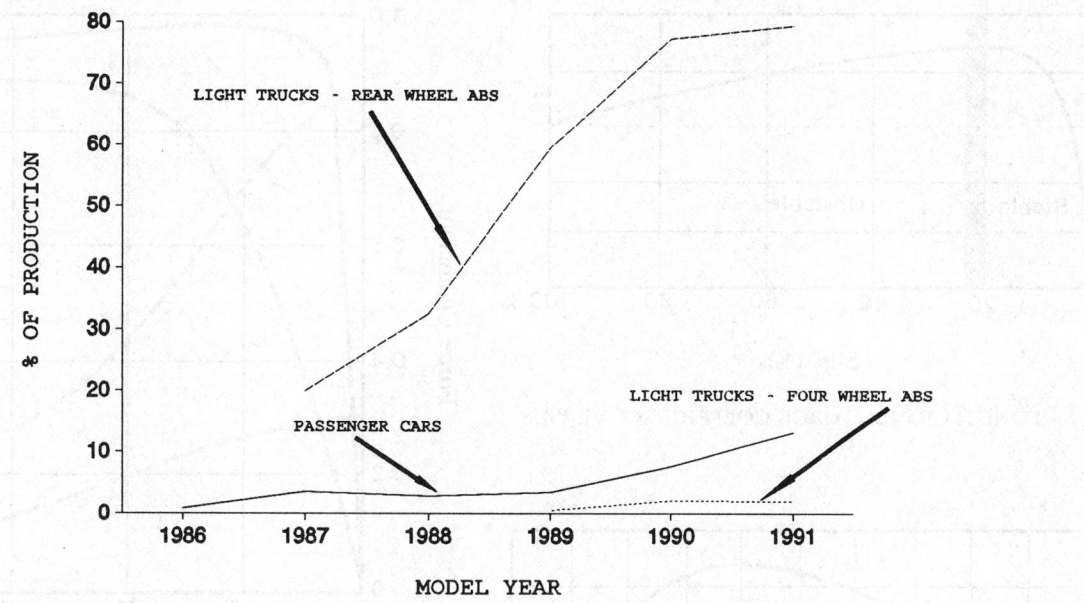

NOTE—ABS installation rates are bsed on passenger cars and light trucks assembled in the U.S., Canada, and Mexico, for the U.S. market. 1991 model year figures are estimates derived from production figures through March 31, 1991. Ward's Automotive Reports were used as a source for this data.

FIGURE 3—NORTH AMERICAN ABS INSTALLATIONS

particular, the ability to steer the vehicle or counteract disturbing lateral forces is diminished.

ABS is a feedback control system that attempts to maintain controlled braking under all operating conditions. This is accomplished by controlling the slip at each wheel so as to obtain optimum forces within the limits of the tire-road combination.

In the following sections, the various aspects of braking dynamics are explained. The first part of the discussion, which centers on the control of the slip at a given wheel, is based on a single wheel model of a vehicle. Some of the important characteristics including capabilities and limitations are discussed. Subsequent discussion covers the coordination of slip at the four wheels in an automobile. A four wheel model is used for this purpose. The effect of ABS control on the stability, steerability, and stopping ability of the vehicle are outlined. The characteristics of some of the more common wheel lock control strategies are examined.

5.4 Braking Dynamics—Single Wheel Model—The various forces and torques acting on a single wheel vehicle are shown in Figure 4. The vehicle is modeled to have only longitudinal motion. In a hydraulic brake, torque is developed at the wheel brake by means of brake pressure. The brake torque per unit pressure is called the specific torque. The normal load, Z, corresponds to the weight of the vehicle. The longitudinal road force, X, is the product of the longitudinal force coefficient and the normal load. This force retards the motion of the vehicle. The rotational motion of the wheel is governed by the torques resulting from the brake pressure (brake torque) and the road force (road torque).

The equations of motion can then be written as:

$$Z = M * g \qquad \text{(Eq. 2)}$$
$$X = \mu * Z \qquad \text{(Eq. 3)}$$
$$T_R = R * X \qquad \text{(Eq. 4)}$$
$$T_B = G * P \qquad \text{(Eq. 5)}$$
$$J * (d\omega/dt) = T_R - T_B \qquad \text{(Eq. 6)}$$
$$M * (du/dt) = -(X) \qquad \text{(Eq. 7)}$$

5.4.1 TIRE-TO-ROAD INTERFACE DESCRIPTION—The horizontal force generated at the tire-road interface has two components. The longitudinal force is along the length of the tire, while the lateral force is perpendicular to it. The normalized longitudinal force for a rubber tire is primarily a function of the longitudinal slip between the tire and the road. Wheel slip is defined as the difference of the angular velocity of a freely rolling wheel and that of the braked wheel, divided by the former (Equation 8). The road forces result from deformation of the tire and are transmitted via the contact patch at the tire-road interface.

$$s = \frac{u - \omega R}{u} \qquad \text{(Eq. 8)}$$

A typical relationship between the longitudinal force coefficient and the wheel slip is shown in Figure 5. Note that the coefficient increases with small slip values. At high slips, values decrease as slip increases. The maximum coefficient, μ_p, is the peak traction capability of the tire-road interface. The coefficient of friction for a locked wheel (slip = 100%) is called the sliding coefficient of friction.

The lateral force coefficient for an unbraked wheel is primarily a function of the slip angle. A typical curve is shown in Figure 6.

The force coefficient is dependent on a number of parameters. These include road surface condition (dry, wet, ice), tire construction, tire wear, surface roughness, normal force and tire pressure. Typical characteristics for different surfaces are shown in Figure 7. The curves for deformable surfaces, such as unpacked snow and loose gravel, are of particular interest.

On these surfaces, the longitudinal force increases continuously with increasing slip, until wheel lock occurs. This is typical of a deformable surface, where the ploughing action affects the braking significantly.

The friction characteristics are also a function of the operating conditions, including longitudinal and lateral slip, as shown in Figure 8. Vehicle velocity can also be an important parameter, especially on wet surfaces where hydroplaning can occur.

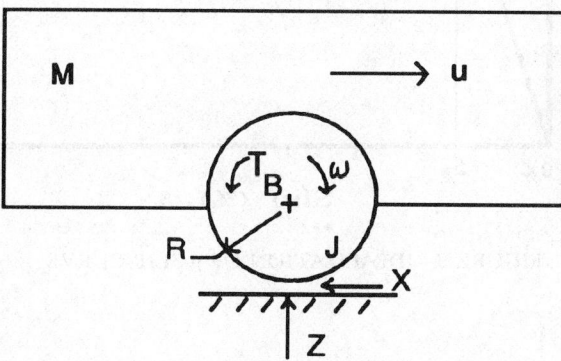

FIGURE 4—SINGLE WHEEL VEHICLE MODEL

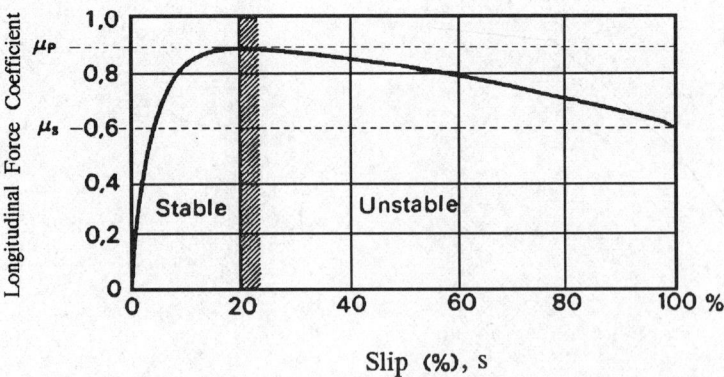

FIGURE 5 – LONGITUDINAL FORCE COEFFICIENT VERSUS SLIP

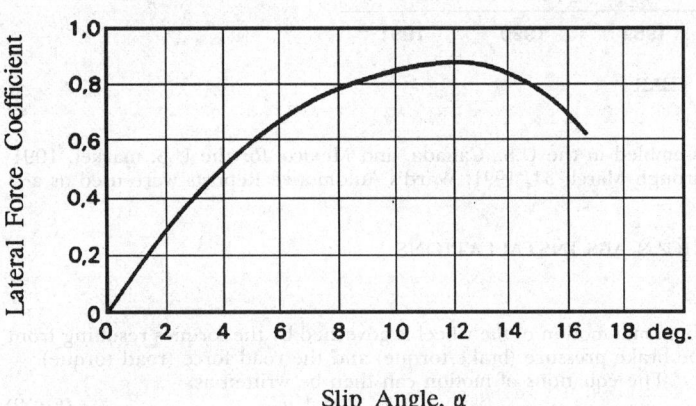

FIGURE 6 – LATERAL FORCE COEFFICIENT VERSUS SLIP ANGLE FOR AN UNBRAKED WHEEL

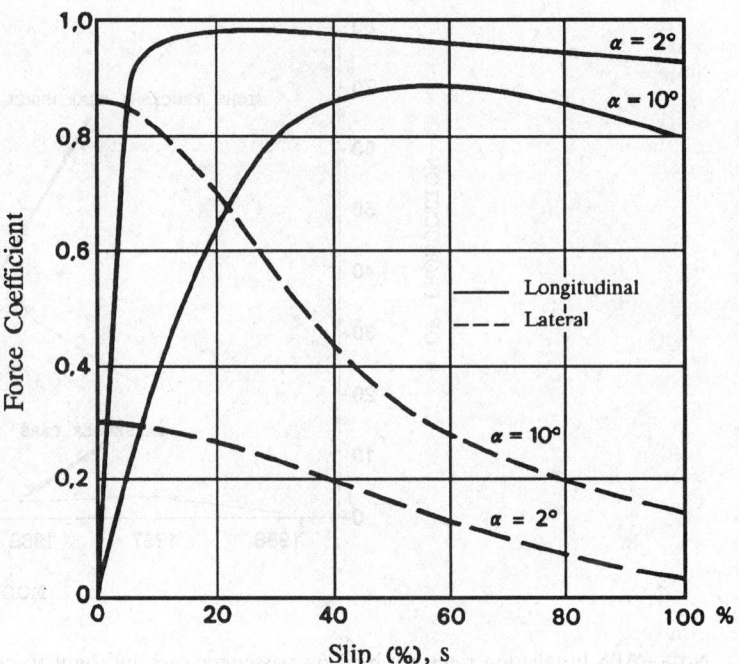

FIGURE 8 – FRICTION COEFFICIENT AS A FUNCTION OF LONGITUDINAL AND LATERAL SLIP

In order to develop the equations for longitudinal motion for the single wheel vehicle, the curve of Figure 5 will be idealized to that shown in Figure 9. Hereafter, unless otherwise mentioned, the term friction coefficient will refer to the longitudinal value. Hence,

$$\mu = \begin{cases} \mu_p * (s/s_p) & s \leq s_p \\ \mu_p - (\mu_p - \mu_s)\dfrac{(s - s_p)}{1 - s_p} \end{cases} \quad \text{(Eq. 9)}$$

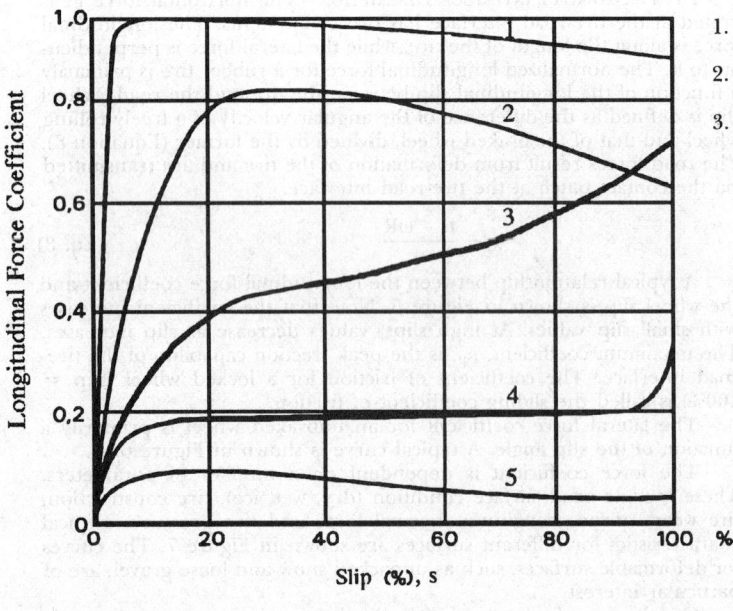

1. Dry Asphalt 4. Unpacked Snow
2. Wet Asphalt 5. Ice
3. Gravel

FIGURE 7 – TYPICAL LONGITUDINAL FORCE COEFFICIENT VERSUS SLIP

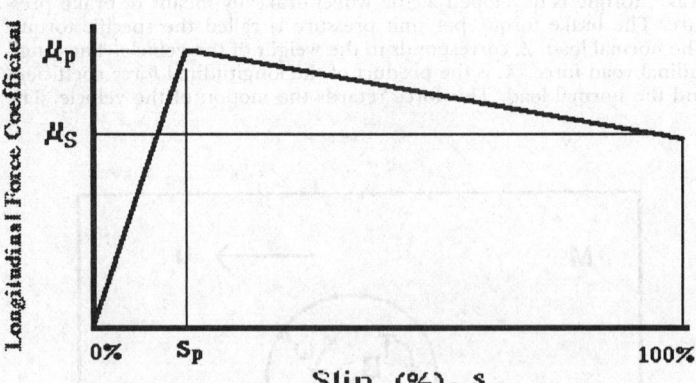

FIGURE 9 – IDEALIZATION OF μ-SLIP CURVE

5.4.2 Braking Without ABS—Let us consider an application of brakes without ABS action, where the brake pressure is ramped up continuously from an initial pressure, as in Equation 10.

$$P = P_0 + P_1 t \quad \text{(Eq. 10)}$$

For $s \leq s_p$;
Further, let the vehicle velocity be a constant. Then, Equations 2 through 10 can be solved to give the time trajectory for the slip as:

$$s(t) = s_o e^{-t/\tau} + \frac{s_p}{\mu_p} \left\{ \frac{KP_1}{ZR} [t + \tau(1-e^{-t/\tau})] - \frac{(KP_o)}{ZR}(1-e^{-t/\tau}) \right\} \quad \text{(Eq. 11)}$$

where:

$$\tau = \frac{JVs_p}{ZR\mu_p}$$

For $s > s_p$;

$$s(t) = (s_o - s_p)e^{-t/\tau} + \frac{1}{\gamma} \left\{ \frac{KP_1}{ZR}[-t + \tau(1-e^{-t/\tau})] - \frac{(KP_o)}{ZR} - \mu_p(1-e^{-t/\tau}) \right\} \quad \text{(Eq. 12)}$$

where:

$$\gamma = \frac{\mu_p - \mu_s}{1 - S_p}$$

and

$$\tau = \frac{-JV}{ZR\gamma}$$

The previous equations show all the parameters that govern the transient of slip. Of particular interest is the time constant in the two cases. When the slip is lower than the peak slip, the time constant is positive. This implies that the exponential terms in Equation 11 decay to zero, implying a stable condition. However, when the slip exceeds the peak slip, the time constant is negative and the exponential terms increase without bound, resulting in an unstable condition.

The time responses for various terminated ramp inputs of brake pressure are shown in Figure 10. For inputs 1, 2, and 3, the wheel stabilizes at a constant slip on the stable side of the μ-slip curve. With input 4, the wheel passes gradually into the unstable side, and then proceeds to a locked state although the pressure is held constant for the latter part of the curve. With input 5, the wheel approaches lock very rapidly because of the excessive brake pressure.

5.4.3 Road Surface Friction Utilization

5.4.3.1 *Longitudinal Force Utilization*—To obtain the maximum deceleration possible during braking, the force coefficient of friction between the road and the wheel should be at its peak value, μ_p. This is obtained when the longitudinal wheel slip is maintained constant at s_p and results in the minimum stopping distance. Then,

$$\text{vehicle deceleration} = a_x = \mu_p * g \quad \text{(Eq. 13)}$$
$$\text{stopping distance} = S = u_i^2/2\mu_p g \quad \text{(Eq. 14)}$$

The relationship between the peak force coefficient of friction and the stopping distance is shown graphically in Figure 11.

It must be noted that the physics of the tire-to-road interface dictates the minimum stopping distance. Further, if the wheel slip strays from the optimum value, s_p, vehicle deceleration and stopping distance will be degraded somewhat. Theoretical values for an initial speed of 60 km/h for different surfaces are included in Figure 12.

5.4.3.2 *Lateral vs. Longitudinal Force Utilization*

5.4.3.2.1 Nondeformable Surface—Let us first consider a typical nondeformable surface. In reference to the curves of Figures 5 through 8, maximum braking is obtained when the wheel slip is maintained to reach the peak coefficient.

However, in order to obtain the maximum lateral force, the wheel slip should be maintained at zero, or in a free rolling condition. It is not possible to simultaneously obtain both maximum longitudinal force and maximum lateral force.

5.4.3.2.2 Deformable Surface—To obtain maximum longitudinal friction on a deformable surface, the slip will have to be controlled to 100% (locked wheel). From Figure 9, the lateral force that is available in this situation will be very small. Once again, if maximum lateral force is desired, the longitudinal force capability will have to be severely compromised.

5.4.4 ABS Objective—ABS attempts to regulate the tire-road forces during braking to follow the driver's steering and braking commands within the constraints of the tire-road traction capability. This is accomplished by controlling the wheel slips to obtain a suitable balance between the longitudinal and lateral tire-road forces.

5.4.5 Simple ABS Modulation Logic—A driver normally brakes a vehicle by modulating the wheel brake pressure through the brake pedal to obtain the desired deceleration. If the braking capability of the tire and road surface is exceeded, the wheels tend to lock. It is at this time that the antilock brake system's control logic takes over the pressure regulation at the wheel in order to obtain optimum braking.

An example of a simple control logic to control the braking is shown in Figure 13a. To illustrate the concept, the system is greatly simplified. It is assumed that the brake pressure at the wheel can be regulated directly by the control logic. Further, this regulation can be through one of two modes: (1) pressure increase or "build" and (2) pressure decrease or "decay" modes. The desired operating slip region, from s_L to s_H, has been determined by the various considerations discussed earlier.

5.4.5 The control logic shown in Figure 13b is implemented when optimum braking is desired. As mentioned previously, this occurs only af-

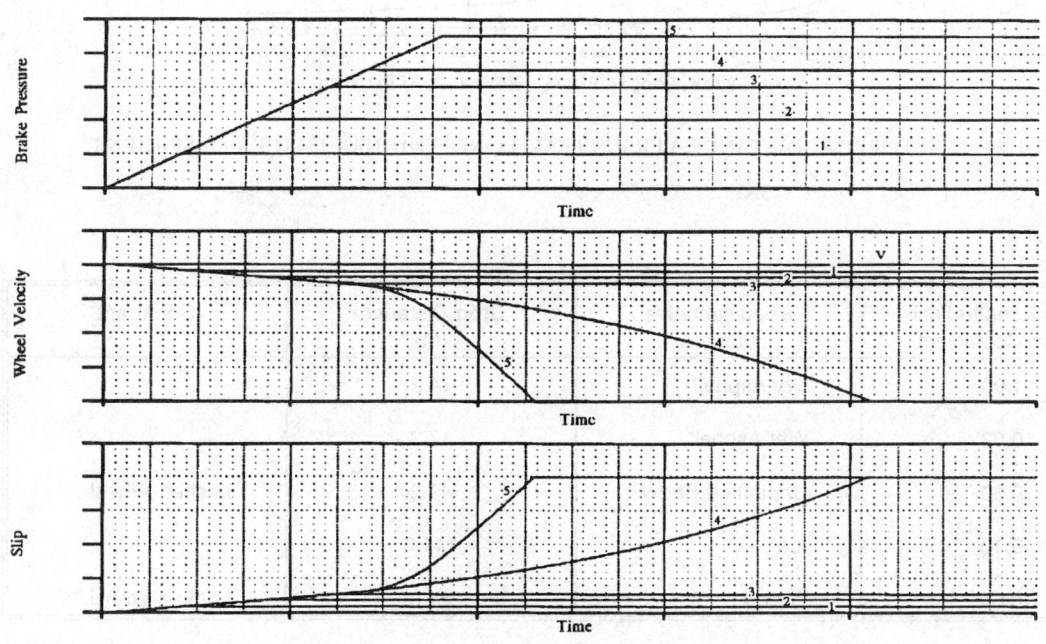

FIGURE 10—SYSTEM RESPONSE TO TERMINATED RAMP INPUTS

ter impending wheel lock is first sensed. The control mode is initially set to "build." In the "build" mode, when the slip exceeds the higher limit of the desired operating slip band, the control mode is changed to "decay," so that the slip may return to the desired range. When the slip drops below the lower threshold, the control mode is changed to "build." This allows the brake pressure to increase, thereby building slip once again. The cycle is repeated continuously, until the vehicle stops or the driver takes his foot off the brake.

5.4.6 EXAMPLE RUN ON SIMULATION—The control logic described previously, results in frequent switching between the "build" and "decay" modes to regulate the braking. This results in continuous "cycling" of the wheel slip, the brake pressure, and the road force, all varying in nominal bands around their operating values.

Per the idealized μ-slip characteristics of Figure 9, as the pressure changes the transient response of the wheel is governed by Equations 11 and 12. Hence, the actual range of the wheel slip and the brake pressure will depend on the various terms in these equations, including wheel inertia, the pressure rates in the "build" and "decay" modes, the specific torque, and so on.

The time responses of a typical run, on a surface with a peak longitudinal force coefficient of 1.0, are shown in Figure 14.

5.5 Braking Dynamics—Four Wheel Model—Several simplifications were made in the development of the single wheel model. A more complete description will include the generation of tire-road forces at each of the four wheels. Further, both longitudinal and lateral forces and motions need to be considered. In addition, suspension and drivetrain interactions and the longitudinal and lateral load transfer also affect the response of the vehicle.

Without going into a rigorous development of the directional response of a vehicle undergoing braking, an overview of the application of ABS to a vehicle is presented here. Three aspects of vehicle performance under braking conditions are considered; stability, controllability, and stopping distance. Straight line braking is considered first, the influence of ABS control on stopping distance and stability to external disturbances will be the focus. Controllability and stability in response to steering commands will then be considered. Braking in a turn and split coefficient braking issues are also examined.

The analysis presented here makes several simplifying assumptions to make the problem tractable. The quasi static response of the vehicle is considered to adequately characterize the more complex response of the vehicle. The vehicle is analyzed at trim conditions defined by lateral and longitudinal accelerations with wheel loads corresponding to the steady-state values dictated by load transfer. Additionally the influence of varying speed is neglected and results are presented for a range of speeds to demonstrate changes with speed. The influence of the ABS cycling is neglected, variations in slip level are assumed to average out to a value determined by a slip setpoint or operating level. The potentially nonlinear behavior of the tires for conditions near the limit of adhesion

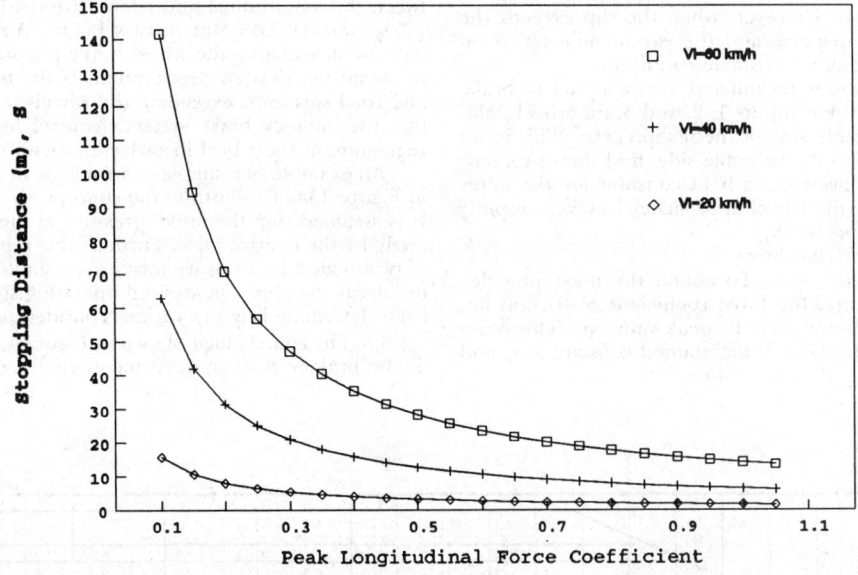

FIGURE 11—STOPPING DISTANCE VERSUS PEAK LONGITUDINAL FORCE COEFFICIENT

Peak Longitudinal Force Coefficient	Representative Surface	Stop Distance (m)	Comment
1.00	Dry Asphalt	14.2	
0.82	Wet Asphalt	17.3	
0.30	Unpacked Snow	47.2	Locked Wheel
0.10	Ice	141.6	
0.65	Gravel	21.8	Locked Wheel

FIGURE 12—THEORETICAL MINIMUM ACHIEVABLE STOPPING DISTANCE FOR REPRESENTATIVE SURFACES AT 60 km/h

```
IF    "APPLY"  THEN
   IF SLIP  ≥  S_H  THEN
      MODE  =  "RELEASE"    ;  SWITCH TO RELEASE
   ELSE
      MODE  =  "APPLY       ;  REMAIN IN APPLY
   ENDIF
ELSE   ("RELEASE")
   IF SLIP  ≤  S_L  THEN
      MODE  =  "APPLY"      ;  SWITCH TO APPLY
   ELSE
      MODE  =  "RELEASE"    ;  REMAIN IN RELEASE
   ENDIF
ENDIF
```

FIGURE 13a – SIMPLE ABS MODULATION LOGIC

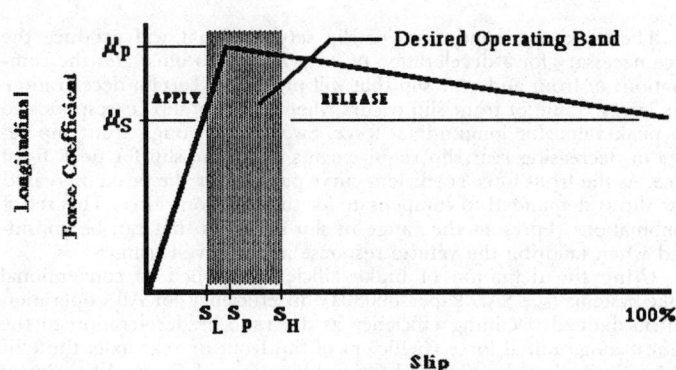

FIGURE 13b – OPERATING MODES FOR ABS MODULATION CONTROL

has been avoided by linearizing the tire properties about the trim condition. This linear representation is not intended to provide a complete description of the vehicle dynamic behavior at these conditions, but rather to illustrate the influence of ABS on vehicle directional response.

The focus of this discussion is on systems providing control over all four vehicle wheels. The control of slip level is considered paramount and the transients involved in arriving at the selected slip will be neglected. Features of single axle systems (rear axle only) are noted separately when significantly different performance is achieved. Various two channel systems, select high / select low, and diagonal control, for instance, will not be dealt with in detail.

5.5.1 STRAIGHT LINE BRAKING – A vehicle undergoing deceleration must develop retarding forces at the tire-road interface through the development of relative longitudinal slip between the tire and road surface as discussed previously. This deceleration generates a longitudinal load transfer from the rear axle to the front axle. This load transfer causes the maximum level of braking force available at each axle to vary as a function of deceleration. Utilization of this available force is a fundamental aspect of brake system design.

In a vehicle equipped with a conventional brake system, tailoring of the front and rear brake balance as a function of deceleration is accomplished through the relative magnitudes of front and rear specific torques, and the characteristics of proportioning valves. With ABS cycling, the balance of the front and rear braking forces is accomplished through slip regulation. To achieve minimum stopping distance, maintaining both front and rear slip at levels that correspond to the peak force production, is desirable. With this strategy, the load transfer effect on longitudinal force availability is compensated for and the vehicle braking efficiency could theoretically approach 100%. In practice this is not accomplished due to the system transients, the need to search for the peak longitudinal force and the uncertainties of peak force and slip values.

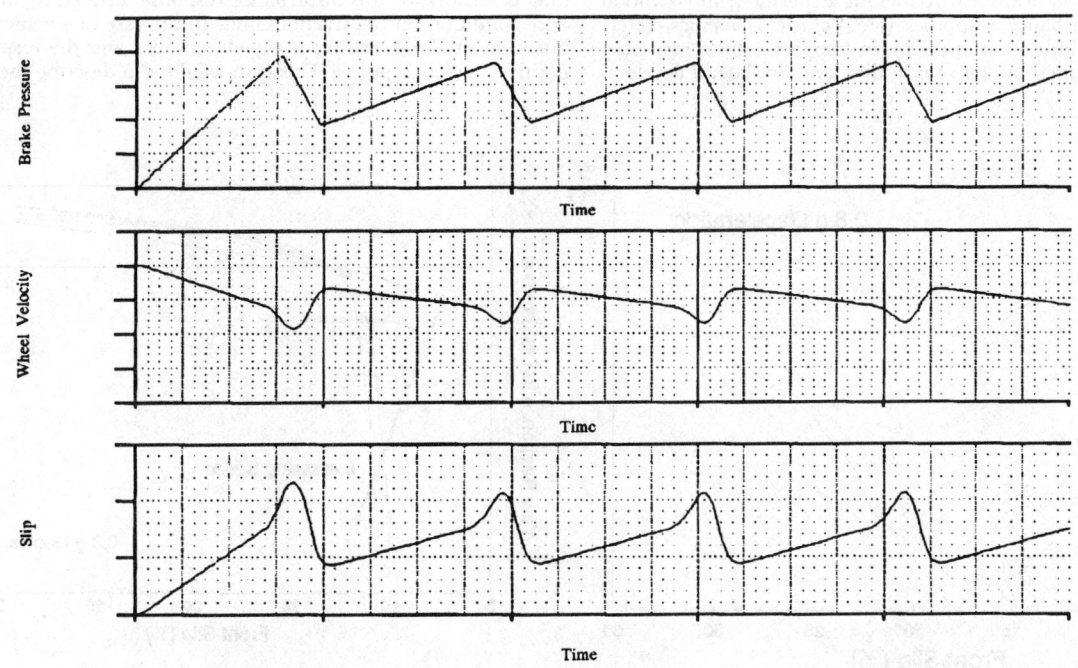

FIGURE 14 – ABS TRANSIENT RESPONSES FOR A SINGLE WHEEL SYSTEM

There exists a continuum of slip setpoints that will produce the force necessary for a deceleration of 0.8g. Figure 15 illustrates the combinations of front and rear slip that will produce a certain deceleration. The lowest value of front slip occurs when the rear slip corresponds to the peak value for longitudinal force. Sweeping through front slip results in decreasing rear slip requirements up to the slip for peak front force. As the front force coefficient curve passes over the peak, increased rear slip is demanded to compensate for the lost front force. This set of combinations represent the range of slip setpoints that can be considered when tailoring the vehicle response at this given trim.

Using the definition of brake efficiency applied to conventional brake systems (see SAE Paper 890804), an efficiency for ABS operation can be defined. Defining efficiency as the ratio of deceleration to the highest longitudinal force coefficient of the front or rear axle, the efficiency ascribed to the potential slip combinations of Figure 14 is shown in Figure 16. As mentioned previously, this measure of efficiency does not take into account any effects other than slip setpoint.

The determination of true vehicle velocity is one of the major obstacles to controlling slip at the individual wheels. This must be known accurately to establish the slip level. A reference velocity must be generated by the ABS to estimate the vehicle velocity from the available inputs. The exact algorithm used to calculate a reference velocity is generally a closely guarded secret. Often the wheel speed data collected from an undriven axle, or a diagonal pair of wheels, are used as initial indications of the vehicle velocity. Once braking is underway, the wheel speeds are monitored, typically to provide a lower limit on the reference velocity, and maximum allowable levels of deceleration are imposed to arrive at reference velocities. Many variations in vehicle velocity calculations are used. Several levels of maximum deceleration can be used to more accurately estimate speed on different surfaces. Switching between levels is controlled by the recovery rates of the wheels in some algorithms. Auxiliary transducers are sometimes employed, such as accelerometers, to determine the actual vehicle deceleration and appropriately adjust the reference velocity. The accuracy of reference velocity estimation is fundamental to achieving good control with the ABS.

In addition to minimizing stopping distance, vehicle stability is another aspect to straight line braking that must be considered. For the case of straight line braking with the steering held fixed in the straight ahead position, the vehicle should brake in a straight line in the presence of external disturbances. One example of such a disturbance is a lateral force applied at the vehicle center of gravity. This is chosen as it accurately represents the disturbance caused by road crossgrade, and approximates the response to a crosswind. The latter results in a force applied somewhere other than the vehicle center of gravity, but the general trends are the same.

To describe the vehicle response to this external disturbance the vehicle equations of motion are linearized about a specific trim condition and classic linear disturbance responses (see Vehicle Dynamics, reference 4) are used as an indication of the more complex vehicle response. The specific trim is defined by the vehicle deceleration, lateral acceleration (zero for this case), vehicle speed, and the selected front and rear slips. The yaw response to an external disturbance is shown for a range of vehicle speeds in Figure 17. Yaw responses with a positive sign indicate that the vehicle is turning away from the disturbing force, always a stable response. Negative responses indicate the existence of a speed above which the vehicle could have an unstable or increasing response to the input.

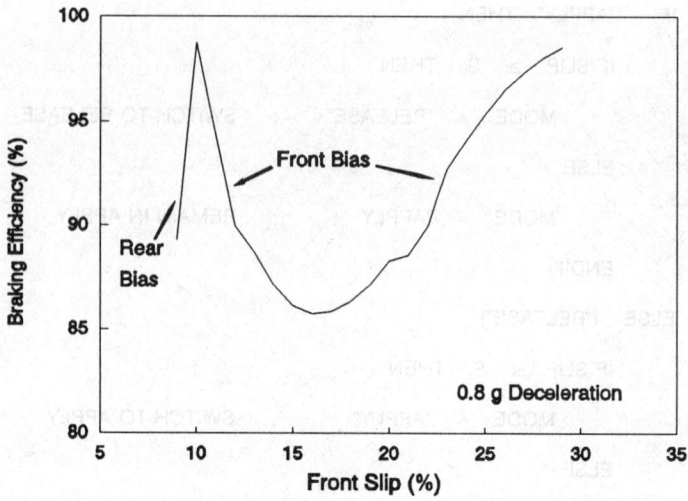

FIGURE 16 – BRAKING EFFICIENCY AS A FUNCTION OF FRONT SLIP SETPOINT

Figure 17 shows that the vehicle response becomes and remains positive with front slips above some value. This is due to the reduction in cornering stiffness of the tire with increasing slip as illustrated in Figure 8. Decreases in front cornering stiffness and increases in rear cornering stiffness cause the vehicle to have a more stable response to the disturbance. For high values of front slip, those in excess of the slip at peak force generation, the rear slip has to be increased to compensate for the loss in front longitudinal force. As a result, both front and rear cornering stiffnesses are reduced and the change in response is not as dramatic as for the lower slip levels.

For the vehicle not equipped with ABS there are three possible limiting behaviors for this disturbance response dictated by three brake balance conditions at the traction limit. In the case of a vehicle that is front biased at the limit, the front wheels will lock and the response will be a positive, stable response. This case would also describe the limiting con-

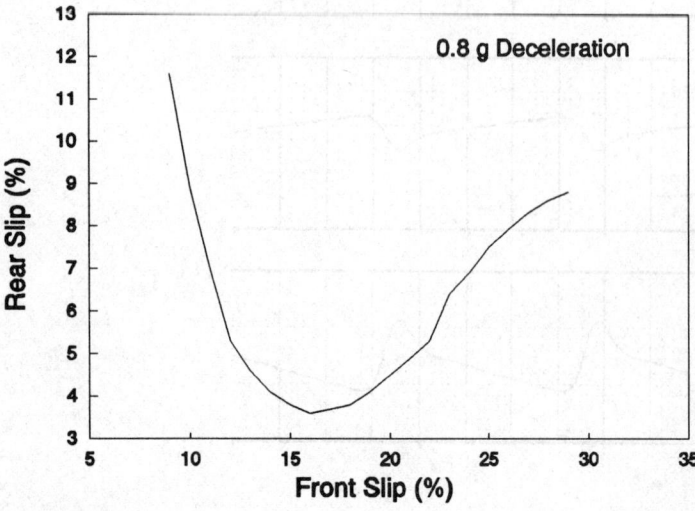

FIGURE 15 – SLIP COMBINATIONS FOR 0.8 g DECELERATION

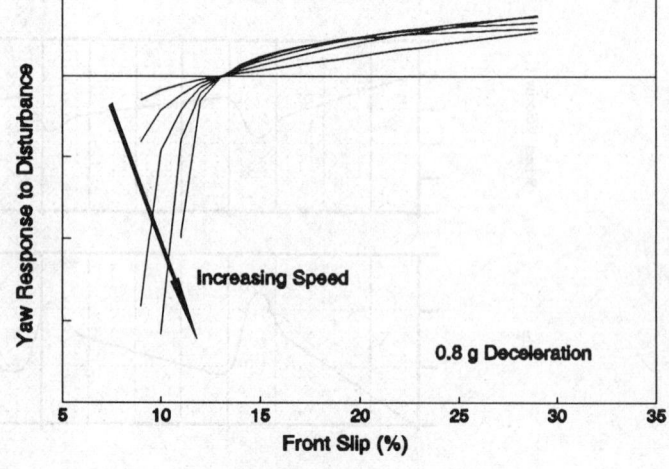

FIGURE 17 – DISTURBANCE RESPONSE AS A FUNCTION OF FRONT SLIP SETPOINT

dition for a vehicle equipped with a rear wheel only ABS. If the vehicle is rear biased at the limit, the response will be a large negative response and the vehicle response will be unstable for any nonzero speed. Neutral balance will result in a vehicle that develops a lateral velocity but no yaw velocity, a neutrally stable system.

5.5.2 STABILITY AND CONTROLLABILITY IN RESPONSE TO STEERING INPUTS—Preservation of steering control and stability during braking are prime goals of the application of ABS to vehicles. The stability of the straight-running vehicle in response to disturbance inputs has been discussed. This section will address the stability and response of the vehicle to steering inputs. Again the linearization of the vehicle equations of motion about the straight running, braking condition is used to demonstrate the influence of ABS control on vehicle performance.

The yaw velocity response to steering is used as a measure of both stability and steerability. The linearized yaw response to steering inputs is shown in Figure 18 for a range of vehicle speeds as a function of front slip.

With low levels of front slip, the yaw response to steering inputs is very high, particularly at high speed. For the two higher speed curves the response is unstable for the very low front slip cases. Increases in front slip reduce the yaw response observed at higher speeds, and the response is stable for all speeds. Figure 19 translates the responses of Figure 18 into a path radius response. The range of path radii achievable for a given steering input with changes in the front slip setpoint is large at high speeds. Variations in control sensitivity can present the operator with a considerable driving challenge.

Vehicles not equipped with a four wheel ABS are subject to the same limiting conditions as mentioned previously: front lock, rear lock, and four wheel lock. In terms of steering response, these vehicles represent the extremes. For the front lock case, the vehicle will not respond to steering input, the locked wheel always generating its force in a direction opposing its motion. This response, or lack of it, is also attributable to the four wheel lock and rear wheel ABS only cases. In the case of rear wheel lock the response to steering input is unstable, the rear tires providing virtually zero resistance to the buildup of yaw velocity.

5.5.3 BRAKING IN A TURN—This aspect of performance addresses the response of a vehicle negotiating a steady-state turn subjected to a braking input. In this maneuver the vehicle has already established a constant turn and the associated slip angles and loads have been developed at the tires. The presence of a lateral acceleration causes load to be transferred from the inside wheels to the outside. The distribution of this load transfer is controlled mainly by the suspension roll stiffness distribution.

The introduction of a deceleration to the vehicle will cause a longitudinal load transfer to be imposed upon the lateral load transfer. This will cause the front axle to be loaded and the rear to be unloaded. This load transfer is a destabilizing influence, increasing the front lateral force production and decreasing the rear. This causes the vehicle to increase its yaw velocity and decrease its turn radius. This scenario is true for maneuvers that do not challenge the traction available at any of the tires.

The lateral load transfer associated with the steady turn will cause the first wheel to lock to be on the inside of the turn. When this lockup occurs on a vehicle without ABS, the vehicle will tend to increase its turn radius with front wheel lock, and decrease its turn radius with rear wheel lock. With a limit defined by the point (deceleration) that both wheels on a given axle lock, the front and rear lock scenarios diverge widely. In the case of front axle lock, the vehicle will leave its curved path and proceed in a straight, tangential path. Rear axle lock causes the vehicle to increase its yaw velocity, decrease its turn radius and become unstable. These responses are illustrated in Figure 20.

ABS control of the slip at the front and rear axles eliminates the lockup scenarios described previously. Additionally, the control of slip level instead of torque causes a moment to be generated opposing that created by the load transfer effect. This occurs only at deceleration levels that challenge the traction of the inside wheel of an axle. Once the ABS takes control of the applied brake pressure, it attempts to control the slip operating point. With a heavily loaded outside wheel and a lightly loaded inside wheel operating at similar slips, the imbalance in forces will create a rigid body moment that tends to increase the turn radius.

Additional steps are often taken to preserve the rear cornering force available using a select low logic. With this type of logic, the brake pressure command to the two rear wheel brakes is controlled by the wheel most disposed to lock. In this manner, the cornering force of the inside tire is maintained at some level, and the outside wheel is operated at a slip well below its peak capability, leaving it with substantial cornering force available.

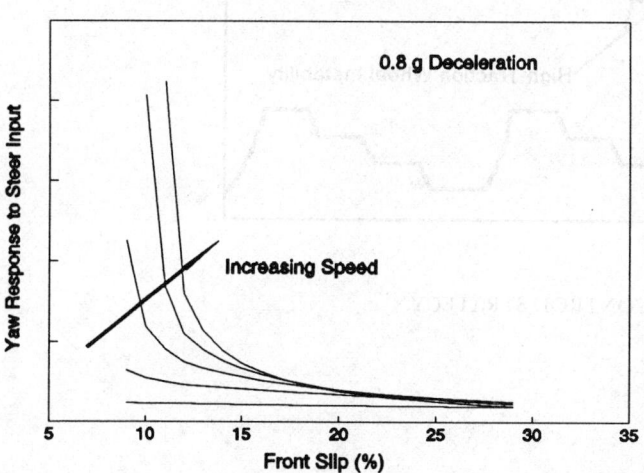

FIGURE 18—RESPONSE TO STEERING AS A FUNCTION OF FRONT SLIP SETPOINT

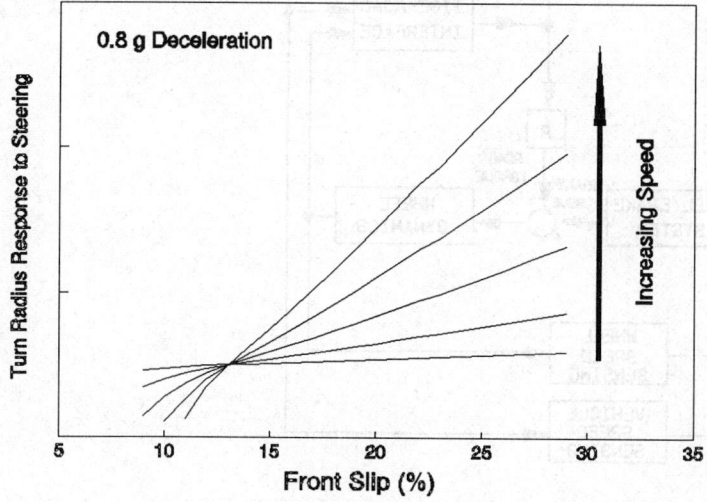

FIGURE 19—TURN RADIUS RESPONSE TO STEERING INPUT

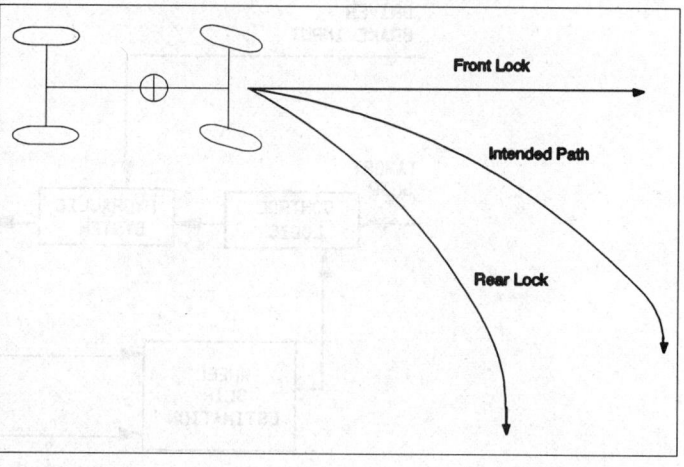

FIGURE 20—EXTREMES OF PATH DEVIATIONS WITH WHEEL LOCK

The slip setpoints of the front and rear axle can be used in a similar fashion to the previous discussion to modify the response of the vehicle once it is under ABS control. Increasing the slip level at the front will decrease the tendency for the vehicle to tighten its radius, while higher rear slip levels will increase this tendency.

With rear-wheel-only ABS the rear wheel lockup scenario is avoided, but front lock is still a possibility. This implies that unstable yaw response is avoided, but with front wheel lock, the curved path cannot be maintained.

5.5.4 SPLIT COEFFICIENT BRAKING — It is not uncommon to encounter a road surface with differing coefficients under the left and right tires. Brake applications that exceed the friction available on the low coefficient side cause imbalanced, side-to-side, longitudinal forces. The resultant rigid body moment tends to steer the vehicle to the higher coefficient surface.

A vehicle not equipped with ABS is prone to spin once the wheels on the low coefficient side have locked up. The rigid body moment will tend to turn the vehicle in such a way that the front tire on the high coefficient will develop forces sympathetic to the spin, and the load transfer will reduce the load on the rear axle and reduce its stabilizing influence. It is not unheard of for a non-ABS vehicle to spin along the split coefficient surface, while its center of gravity travels in a straight line along the split.

Again, ABS can assist in stabilizing the vehicle in this situation. Select low logic at the rear will preserve some level of cornering stiffness at the rear of the vehicle, aiding in stabilization of the yaw response. Another form of control, known colloquially as "yaw control," is also helpful. The main source of disturbance in this maneuver is the imbalance of front brake forces. Using select low logic on the front axle would eliminate this imbalance and the associated disturbance. However, this is not practical, as stopping distances may become unacceptably long. As a compromise, the yaw control logic recognizes the locking of the low coefficient wheel, and as it acts to recover that wheel, it also modifies the pressure to the high coefficient wheel. This can be accomplished by holding or releasing the pressure to the high coefficient wheel. The system then allows the high coefficient wheel to increase its pressure at a reduced rate until it reaches its limit. The onset of the disturbing moment and its subsequent increase is delayed, thus giving the driver time to react. The subsequent increase in brake force on this wheel allows the stopping distance to be decreased over that possible with a select low

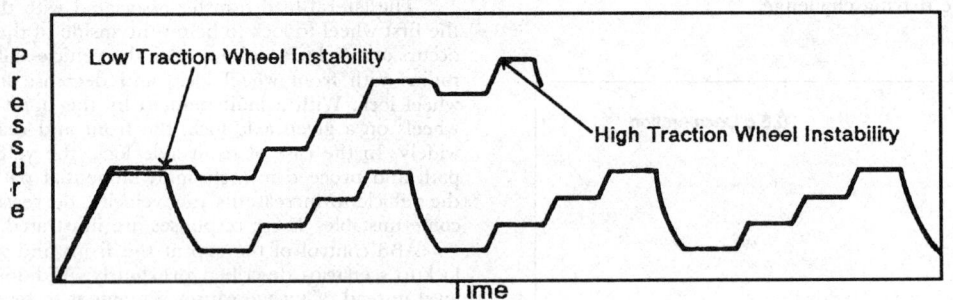

FIGURE 21 — EXAMPLE YAW CONTROL STRATEGY

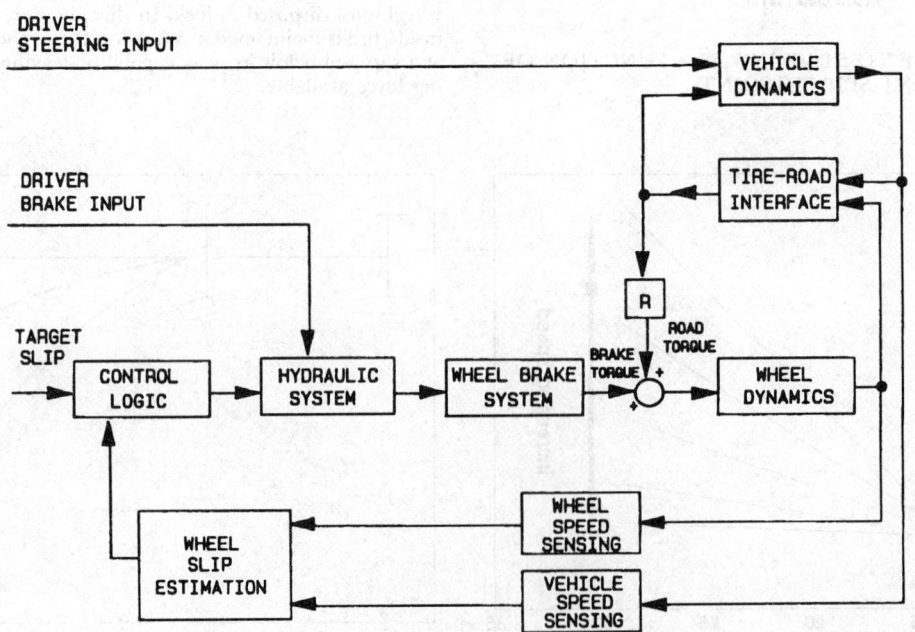

FIGURE 22 — CONTROL SYSTEM BLOCK DIAGRAM

logic. This control is illustrated in Figure 21.

A vehicle equipped with rear wheel only ABS will provide a higher degree of stability on a split coefficient surface than a vehicle with no ABS. The stability enhancement comes from the rear wheel control and attendant rear cornering stiffness preservation. These systems do not allow the opportunity to mitigate the disturbance arising from front braking force imbalance.

5.5.5 Performance Tradeoffs—There are many different aspects of performance for a vehicle that must be considered when applying ABS. Trading off split coefficient stopping distance and yaw stability is a prime example. Additionally, compromises must be struck between vehicle response to steering and its response to external disturbances. Most of the decisions made in the course of development of an ABS equipped vehicle will also involve more of the vehicle than just ABS, or even brakes. At a minimum, other chassis characteristics, suspension compliance and kinematics, and tire properties must be considered in design and development. The problem is complex with many competing factors to be considered. It is not reasonable to assume that one optimum design could be reached given a set of vehicle and system parameters.

5.6 Control System Block Diagram—A block diagram of the entire control system is shown in Figure 22.

This Figure also shows the major elements needed to implement an antilock brake control system with the control logic described earlier. These include:

a. Sensors to determine the wheel speed and the vehicle speed
b. Control logic to process the sensor signals and determine the desired regulation of the brake pressure
c. A means to implement the control logic
d. A means to regulate the brake pressure as dictated by the control logic

These generic requirements have been implemented in different ways by various manufacturers. A list of the current systems in the market is included later in this document.

In a typical application, variable reluctance sensors are used for wheel speed sensing. The vehicle speed is estimated from the wheel speeds, eliminating the need for a separate vehicle speed sensor. The control logic is implemented via microprocessor software in an electronic controller. The control software may also perform other functions, such as driver information and diagnostics. A wiring harness links the various sensors, the displays, the controller, the vehicle electrical system, and the modulator. The brake pressure regulation is typically done with the modulator employing solenoids that close or open different fluid paths to build or decay the brake pressure at the wheels.

6. ABS Current System Profile—Figure 23 lists current ABS manufacturers and their system profile that was available to the committee.

Figures 24, 25, and 26 show different generic antilock systems.

Manufacturer	Acronym	Base System		Actutation Principle			Number of Channels	Number of Speed Sensors	Modulation Method			Current Applications
		Add On	Integrated	Return to Reservoir	Return to M/C	Other			Build	Decay	Other	
Aisin Seiki	R-AL	x				x	1	1	x	x		R
Akebono	ABS 1	x			x		2	4	x	x	x	4
Allied Bendix	PLC-1		x	x			3	4	x	x	x	R/4
	PLC-3		x	x			3	4	x	x	x	F/A
	LC-6	x			x		3	4	x	x		F
Atsugi Unisia	AR3-1	x			x		3	3	x	x	x	R
Bosch	ABS2E/2U/2S	x				x	3 or 4	3 or 4	x	x	x	F/R/4/A
	ABS3		x	x			3 or 4	3 or 4	x	x	x	F
Delco Moraine	Power Master 3		x			x	3	4	x	x		F/R
	ABS VI	x				x	3	4			x	F
Honda	None				x		3 or 4	4	x	x		F/R
Kelsey-Hayes	RWAL	x				x	1	1	x	x	x	R/4/A
	4WAL	x			x		3	4	x	x	x	R/4/A
Lucas Automotive	ABS2/2	x				x	2	2	x	x		F
	ABS4/4	x		x			4	4	x	x	x	F/R
	SCS	x				x	2	2	x	x		F/4
Nippon ABS, Ltd.	ABS 2S/2E/2M/2L	x			x		2, 3, or 4	3 or 4	x	x	x	F/R/4/A
Nippondenso	Unknown	x					3	3 or 4	x	x	x	F/R/4/A
Nisshinbo	ABS MK II		x	x			3 or 4	4		x	x	F/R/4/A
Sumitomo Electric	ABS1	x			x	x	4	4	x	x	x	F/R
	Compact ABS	x			x		4	4	x		x	F/R
Teves	MKII		x	x			3 or 4	3 or 4		x	x	F/R
	MKIV	x		x			3 or 4	3 or 4		x	x	F/R
Wabco	Unknown		x	x			4	4	x	x	x	R/4/A

All systems use an Electronic Controller, (except the Lucas Automotive SCS system).
F = Front Wheel Drive   R = Rear Wheel Drive   A = All Wheel Drive   4 = Four Wheel Drive

FIGURE 23—ABS CURRENT PASSENGER CAR AND LIGHT TRUCK PRODUCTION SYSTEM PROFILE, JULY 1991

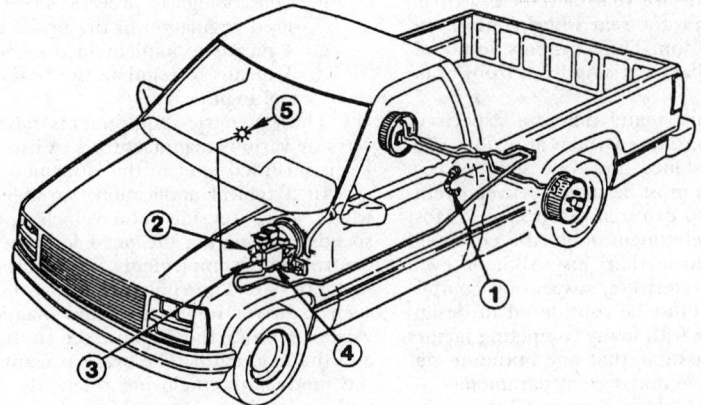

1. Wheel Speed Sensor
2. Booster/Master Cylinder
3. Modulator
4. Controller
5. Indicator Light(s)

Hydraulic Circuit = = = = =
Electronic Circuit ———

FIGURE 24—GENERIC REAR WHEEL ANTILOCK SYSTEM

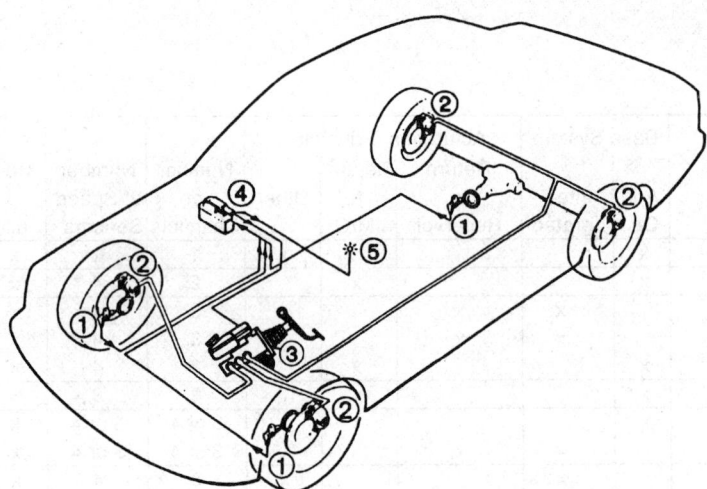

1. Wheel Speed Sensor
2. Wheel Brake
3. Booster/Master Cylinder/Modulator
4. Controller
5. Indicator Light(s)

Hydraulic Circuit = = = = =
Electronic Circuit ———

FIGURE 25—GENERIC INTEGRATED 3-CHANNEL ANTILOCK SYSTEM

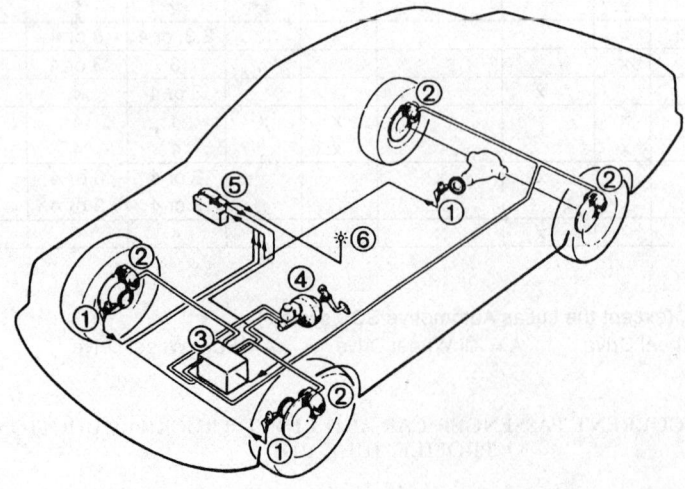

1. Wheel Speed Sensor
2. Wheel Brake
3. Modulator
4. Booster/Master Cylinder
5. Controller
6. Indicator Light(s)

Hydraulic Circuit = = = = =
Electronic Circuit ———

FIGURE 26—GENERIC ADD ON 3-CHANNEL ANTILOCK SYSTEM

25.85

IN-SERVICE BRAKE PERFORMANCE TEST PROCEDURE PASSENGER CAR AND LIGHT-DUTY TRUCK—SAE J201 MAY89

SAE Recommended Practice

Report of Brake Committee approved April 1976 and reaffirmed by the Brake Standards Committee 7 - Road Test Procedures May 1989.

1. Scope—This SAE Recommended Practice establishes a uniform procedure for testing the brake systems (service and parking) of all passenger cars, light-duty trucks, and multi-purpose passenger vehicles up to and including 10 000 lb (4500 kg) GVWR.

2. Purpose—The purpose of the test code is to evaluate brake system performance of vehicles in service for compliance with regulations.

 2.1 The test code is expected to be utilized as a basis for a brake evaluation conducted by State or Federal officials engaged in highway safety programs.

 2.2 The primary consideration is that this test requires a minimum of instrumentation time, driver skill, and cost to conduct.

3. Instrumentation
 3.1 Decelerometer (U-tube or equivalent)
 3.2 Pedal force indicator
 3.3 Pedal travel indicator
 3.4 Speedometer
 3.5 Stop watch

4. Installation Details
 4.1 Install and adjust decelerometer to vehicle.
 4.2 Install pedal force and travel indicator to manufacturer's procedure and calibrate.
 4.3 Brake system is to be tested as received without any changes or adjustments.

5. Vehicle Test Weight
 5.1 The vehicle is to be tested at the "as received" load excluding passengers. Official observer permissible.
 5.2 The load may be repositioned or removed if a hazardous condition exists.

6. Test Facility
 6.1 Selected test area shall be a paved 12 ft (3.7 m) lane of adequate length, dry, clean, straight, essentially level, and not heavily traveled.
 6.2 Provide a 15% grade of sufficient length and skid resistance to support the entire test vehicle.

7. Test Procedure
 7.1 Test Notes (to be recorded on data sheet—Fig. 1)
  7.1.1 Vehicle make, model name, year, and serial number.
  7.1.2 Vehicle odometer reading.
  7.1.3 Condition of each tire.
  7.1.4 Type of brake (disc, drum) (power, manual).
  7.1.5 Brake warning lamp operation.
  7.1.6 Brake stop lamps operation.
  7.1.7 Any change in pedal height when held for 10 s at 20 lb (89 N) pedal force.
  7.1.8 Any change in pedal height when held for 10 s at 150 lb (667 N) pedal force and if pedal reached its limit of travel.
  7.1.9 If parking brake application will lock the wheels with the vehicle on the 15% grade.
  7.1.10 Any unusual brake or vehicle noises.
  7.1.11 Any unusual brake action such as pulls, roughness, etc.
  7.1.12 If wheel slide occurred during the brake snubs and designate which wheel.
  7.1.13 Maximum sustained deceleration attained.
  7.1.14 Maximum pedal force required.
 7.2 Static Check
  7.2.1 Verify warning lamp operation as indicated by vehicle manufacturer.
  7.2.2 If vehicle is so equipped, observe brake warning lamp indicator during test.
  7.2.3 With vehicle stopped and the engine running, apply 20 lb (89 N) force to the brake pedal and hold for 10 s.
  7.2.4 Note if there is any change in pedal height during the 10 s.
  7.2.5 With vehicle stopped and the engine running, apply 150 lb (667 N) force to the brake pedal and hold for 10 s.
  7.2.6 Note if there is any change in pedal height during the 10 s or if the pedal reaches the limit of its travel.
 7.3 Parking Brake
  7.3.1 Drive vehicle up a 15% grade, apply the service brake to stop, and hold the vehicle.
  7.3.2 Place transmission selector in neutral.

  7.3.3 Apply the parking brake up to the regulatory load, but do not exceed 200 lb (890 N) force for a foot operated mechanism or 100 lb (445 N) for a hand operated mechanism. Remove foot or hand from the parking brake apply mechanism.
  7.3.4 Release the service brake.
  7.3.5 Observe if wheels remain locked.
 7.4 Preliminary Snub Test to Acquaint Driver with Vehicle [to be conducted within 12 ft (3.7 m) wide test lane]
  7.4.1 Snubs required—1.
   7.4.1.1 Snub speed—30-10 mph (48-16 km/h).
   7.4.1.2 Snub deceleration (sustained)—10 ft/s^2 (3 m/s^2).

Vehicle:
Make_____ Model Name_____ Year_____
Odometer Reading_____ Serial Number_____
Tire Condition:

	LF	RF	LR	RR
GOOD				
AVG.				
POOR				

Type of Brakes: Front-Disc_____ Drum_____
                Rear-Disc_____ Drum_____
Type of Actuation: Power_____ Manual_____
Driver_____ Observer (Official)_____

Performance Data:
1. Brake Warning Lamp Operable—YES_____ NO_____
2. Brake Stop Lamps Operable— YES_____ NO_____
3. Static Check
   Record—Pedal Travel

	20 (lb) 89 (N)	150 (lb) 667 (N)
Initial		
After 10 s		
Change		

   Limit of pedal travel reached YES_____ NO_____
4. Parking Brake Test—Hold vehicle on 15% grade with regulatory test force.
   Force Applied_____
   Locked Wheels—YES_____ NO_____
5. Preliminary snub a) 30-10 mph at 10 ft/s^2 (48-16 km/h at 3 m/s^2)
                    b) 40-20 mph at 16 ft/s^2 (64-32 km/h at 5 m/s^2)
   Record—Pedal Force_____
          Deceleration_____
          Brake Action(Wheel slide, pulls, roughness, light or heavy pedal, etc. and deceleration attained if less than specified.)
          _____
          _____
6. Highway Stopping Test—(1) stop from 50 to 60 mph (80 to 97 km/h) not to exceed 20 ft/s^2 (6 m/s^2)
   Record—Pedal Force_____
          Deceleration_____
          Brake Action (Wheel slide, pulls, roughness, light or heavy pedal, etc. and deceleration attained if less than specified.)
          _____
          _____

7. General Comments on Braking Performance
   _____
   _____
   _____
   _____
   _____

FIG. 1—REPORT FORM
BRAKE TEST—VEHICLE IN-SERVICE

7.4.1.3 Moderate apply rate [do not exceed 150 lb (667N)].
7.4.1.4 Abort snub if wheel slide occurs and discontinue test.
7.4.2 Snubs required—1.
7.4.2.1 Snub speed—40-20 mph (64-32 km/h).
7.4.2.2 Snub deceleration (sustained)—16 ft/s^2 (5 m/s^2).
7.4.2.3 Moderate apply rate [do not exceed 150 lb (667N)].
7.4.2.4 Abort snub if wheel slide occurs and discontinue test.
7.5 Highway Stopping Test [to be conducted within the 12 ft (3.7 m) wide test lane].
7.5.1 Stops required—1.
7.5.1.1 Initial speed—50-60 mph (80-97 km/h) or maximum practical speed attainable within the test area if less than 50 mph (80 km/h).
7.5.1.2 Sustained deceleration attainable—not to exceed 20 ft/s^2 (6m/s^2).
7.5.1.3 Pedal force—150 lb (667 N) maximum.
7.5.1.4 Brake apply rate—Maximum rate possible when maintaining deceleration control (not a spike) up to 20 ft/s^2 (6m/s^2).
7.6 Repeat Testing
7.6.1 If the vehicle brake systems exhibit marginal performance with respect to the regulatory requirements, tests for paragraphs 7.2 or 7.5 may be repeated.
8. Report Form—General Data and Report Form, Fig. 1.

IN-SERVICE BRAKE PERFORMANCE TEST PROCEDURE—VEHICLES OVER 4500 kg (10 000 lb)—SAE J1250 NOV92

SAE Recommended Practice

Report of the Brake Committee approved February 1980 and revised by the Truck and Bus Brake Committee December 1987. Rationale statement available. Reaffirmed by the SAE Truck and Bus Brake Systems Subcommittee of the SAE Truck and Bus Brake Committee November 1992.

Foreword—This reaffirmed document has been changed only to reflect the new SAE Technical Standards Board format.

1. Scope—This SAE Recommended Practice establishes a uniform practical series of subprocedures for level road testing of the brake performance of vehicles with gross vehicle weight ratings over 4500 kg (10 000 lb).

1.1 Purpose—The purpose of this practice is to establish a uniform method for use by operators and law enforcement agencies, to evaluate the condition of the brake systems of vehicles with GVWRs and GCWRs over 4500 kg (10 000 lb) under any condition of loading. By following the test procedures set forth, the operator and/or law enforcement agencies can ascertain if the vehicle meets the service brake and emergency brake stopping distance requirements of applicable State and Federal regulations.

2. References

2.1 Applicable Documents—The following publications form a part of this specification to the extent specified herein. The latest version of SAE publications shall apply.

2.1.1 SAE PUBLICATIONS—Available from SAE, 400 Commonwealth Drive, Warrendale, PA 15096-0001.

SAE J229—Service Brake Structural Integrity Test Procedure—Passenger Car

3. General—This practice is written as a quick assessment procedure to uncover the most common or gross performance inadequacies in the braking system; however, not all performance and maintenance problems may be uncovered. Good vehicle maintenance and preventative maintenance programs are of utmost importance for vehicle safety. Because of the maximum speed limitations of the typical inspection site and the safety of the tester, speed and temperature fade problems and worn components may not be uncovered.

4. Equipment and Location

4.1 Instrumentation and Equipment—The tests shall be run using a bumper or frame clamping fifth wheel capable of displaying vehicle speed and distance to stop, triggered by initial brake control movement or force sensor.

4.2 Test Area—The test area shall be substantially straight, level (not to exceed 1% grade), dry, smooth, hard surface roadway of portland cement, concrete, or equivalent, that is free from loose material and approximately 60 m (200 ft) in length with an access adequate to permit a truck to enter at 32 km/h (20 mph). The test surface shall be marked with a 3.7 m (12 ft) wide lane by marking the test surface or using pylons.

It is recommended that the desired stopping distance be identified by surface markings or pylons as a guide for the driver.

5. System Leak Check—The following checks are to be made after the engine has been run a sufficient time to build up normal air pressure, boost pressure, or boost vacuum (1 min minimum).

5.1 Air and Air Assist Hydraulic Systems

5.1.1 With engine off and brakes unapplied, note for sounds or other evidence of air leakage.

5.1.2 With engine off, make a full pressure application and hold for 1 min. Record the drop in reservoir pressure(s) after initial application, and note any sound or other evidence of leakage.

5.2 Straight Hydraulic and Power Assisted (Vacuum or Hydraulic) Systems—Turn engine off and depress the brake pedal with a light pressure for 10 s and then press hard for 10 s. Note any change in pedal height while being held and sound or evidence of leakage.

6. Stopping Ability Test

6.1 Pretest Check Out—The tester shall briefly examine the vehicle, the load and its retention for conditions that might prove unsafe during the test such as load shift, poor steering, and brake pedal response, excessive brake system leakage, and that brakes are functioning on all wheels. If the vehicle and/or load is judged unsafe, the test shall be delayed until the condition is corrected.

6.2 Procedure—The driver shall enter the test area as near as possible to 32 km/h (20 mph) and maintain the speed until the prescribed location, at which time he shall apply the service brakes as rapidly as possible, without locking the wheels, attempting to bring the vehicle to a complete stop within a 3.7 m (12 ft) wide lane. The initial speed and distance the vehicle travels from start of brake application to stop shall be noted and recorded. Record any brake pull or instability, and whether or not stop was made within 3.7 m (12 ft) wide lane.

7. Emergency Brake System Check

7.1 Procedure—From 8 km/h (5 mph) apply the emergency brakes. Record distance to stop.

8. Distance Correction Formula for Small Initial Stopping Speed Errors—If the initial speed is within ±3.2 km/h (±2 mph), stopping distances shall be corrected per SAE J229. If the initial stopping speed variation is greater, the test shall be run over.

9. Report Form—General Data and Report Form, Figure 1.

25.87

TEST DATA SHEET
IN-SERVICE BRAKE PERFORMANCE TEST
(FOR VEHICLES OVER 4500 kg (10 000 lb)

DATE_____
TIME_____

VEHICLE DESCRIPTION_____
_____ODOMETER READING_____
OWNER/OPERATOR_____
TEST LOCATION_____
EQUIPMENT USED_____

SYSTEM LEAK CHECK AIR OR AIR BOOST HYDRAULIC

WITH ENGINE OFF AND BRAKE UNAPPLIED—NOTE   NO LEAKS NOTED_____
FOR EVIDENCE OF LEAKAGE   LEAKS NOTED_____

WITH ENGINE OFF MAKE FULL APPLICATION FOR 1 MIN.

A. INITIAL PRESSURE_____
B. FINAL PRESSURE_____
C. PRESSURE DROP (A-B)_____

SYSTEM LEAK CHECK HYDRAULIC AND POWER ASSIST HYDRAULIC

WITH ENGINE OFF, PRESS PEDAL—NOTE MOVEMENT

PEDAL FIRM_____   PEDAL MOVED TOWARDS FLOOR_____

TEST SAFETY CHECK

BRIEFLY CHECK VEHICLE AND LOAD FOR CONDITIONS UNSAFE FOR TESTING.

VEHICLE AND LOAD   OK_____   UNSAFE_____   IF UNSAFE,
DESCRIBE_____

STOP TEST

MAKE 32 km/h (20 mph) COMPLETE STOP.
ACTUAL INITIAL SPEED_____   STOP DISTANCE_____
CORRECTED STOPPING DISTANCE_____
STAYED WITHIN 3.7 m (12 ft) LANE   YES _____   NO _____
VEHICLE PULL PROBLEM   YES _____   NO _____
IF YES, DESCRIBE_____

EMERGENCY BRAKE SYSTEM CHECK

MAKE 8 km/h (5 mph) STOP BY ACTUATING EMERGENCY BRAKE SYSTEM.
   ACTUAL INITIAL SPEED_____   STOP DISTANCE_____
   CORRECTED STOPPING DISTANCE_____

TESTER_____

FIGURE 1—GENERAL DATA AND REPORT FORM

GOGAN HARDNESS OF BRAKE LINING—SAE J379a   SAE Recommended Practice

Report of Brake Committee and Automotive Safety Committee approved January 1969 and last revised May 1972.

1. *Purpose*—To establish a uniform procedure for determining the Gogan hardness of brake lining.

2. *Scope*—Gogan hardness, a nondestructive (a penetrator causes shallow surface deformation) method of measuring compressibility, is used as a quality control check of the consistency of formulation and processing of brake lining. Gogan hardness alone shows nothing about a lining's ability to develop friction or to resist fade when used as a friction element in brakes. Gogan hardness varies with formulation, contour, and thickness of the lining.

The Gogan hardness and the range of Gogan hardness are peculiar to each formulation, thickness, and contour and, therefore, the acceptable values or range must be established for each formulation and part configuration by the manufacturer.

2.1 The hardness of sintered powder metal lining is usually determined with Rockwell Superficial hardness equipment. Reference ASTM B 347[1] (latest revision), Standard Method of Test for Hardness of Sintered Metal Friction Materials.

3. *Equipment*—A commercially available Gogan Model 911 (or equivalent) direct reading hardness testing machine is required. In this machine the flat end of a cylindrical penetrator is forced against the lining supported by a matching anvil. The Gogan hardness number is the distance, in units of 0.00025 in. (0.0064 mm) the penetrator advances into the lining, while the force on the penetrator increases from the initial or minor load to the final or major load. Two systems are used: 1500 kg or 3000 kg (14.71 or 29.42 kN) major loads with nominal 500 kg and 1500 kg (4.90 and 14.71 kN) minor loads, respectively.

3.1 The test cycle is initiated by closing an electric switch, causing the penetrator to approach and contact the lining sample and start to apply the test load. When the minor load is reached, the hardness indicator is engaged to the penetrator and the two timers that control the actual test cycle are activated. At 0.75 s (ET_1), the hardness indicator is disengaged and arrested. The major load is reached prior to the ET_1 setting, the time depending upon the hardness of the lining penetrator travel, and the force buildup in the hydraulic system. At 1.75 s (ET_2), the cycle is terminated and the machine returns to its idle position and is reset for the next test. The 1 s interval between ET_1 and ET_2 provides sufficient dwell for the operator to read the Gogan hardness dial indicator. Four test combinations are possible utilizing 1 or 0.75 in. (25.39 or 19.05 mm) diameter penetrators with either 1500 or 3000 kg (14.71 or 29.43 kN) major loads. This requires that Gogan hardness numbers be prefixed with a scale symbol representing the load and penetrator as listed in Table 1. The combination of the load and penetrator is selected to provide the greatest sensitivity with the least damage to the lining. Usually, this is accomplished when the Gogan readings fall within the ranges shown, although the desirable range will differ with formulation and the configuration of the lining, particularly its thickness and curvature.

4. *Test Machine Specifications*
4.1 Minor load—Nominal 500 or 1500 kg (4.90 or 14.71 kN).
4.2 Major load—1500 or 3000 kg (14.71 or 29.42 kN).
4.3 Penetrator diameter—1 or 0.75 in. (25.39 or 19.05 mm).
4.4 Split penetrator—For grooved lining, has a flat end face consisting of two semicircles, either 1 or 0.75 in. (25.39 or 19.05 mm) in diameter, spaced apart a minimum of the groove width plus 0.25 in. (6.4 mm) as shown in Fig. 1.
4.5 Timers—Reference, Gogan Machine Co., Wiring Diagram XE-1473. ET_1—0.75 s, ET_2—1.75 s.
4.6 Anvil for curved lining—Curved to minimum inside radius specified for lining. Cord length, 2 in. (50.8 mm); minimum width, 2 in. (50.8 mm). Anvil for flat lining—Minimum 1¾ in. (44.4 mm) diameter flat.
4.7 Penetrator travel to upper surface of lining—½ ± ⅛ in. (13 ± 3 mm).

5. *Operating Procedure*
5.1 Position anvil in socket of anvil, adjusting screw after making sure seating surfaces are clean.
5.2 Position brake lining on anvil. Adjust backstop so that penetrator is no closer than ⅛ in. (3 mm) to edge of lining.
NOTE: The penetrator must be completely supported by brake lining with a minimum of ⅛ in. (3 mm) to any edge of the lining or groove for a valid Gogan hardness test.
A split penetrator will be used when the dimensions of the lining grooves on either side of the braking or shoe surface make it impossible to provide ⅛ in. (3 mm) clearance between the OD of the penetrator and the edge of the lining or groove.

In positioning grooved brake lining on the anvil, adjust the back stop so the groove in the lining is centered with the penetrator and rotate the anvil so the groove in the lining and the groove in the penetrator are parallel. Gogan hardness numbers taken on grooved lining with a split penetrator are prefixed with an additional symbol, "S," as in Table 1; thus, GAS, GBS, GCS, and GDS.

5.3 With square bar gage, adjust space between penetrator and top of lining to ½ ± ⅛ in. (13 ± 3 mm). Tighten adjusting screw clamping nut.
5.4 Start machine pump. Allow 2 min minimum warmup. Operate machine a few times to seat anvil and eliminate backlash of anvil screw.
5.5 Test the lining at the desired locations and note the Gogan hardness from the dial indicator.

6. *Calibration Procedure*
6.1 Calibrate the Gogan hardness tester in accordance with the following:
(a) Timers set as follows: ET_1—0.75 s, ET_2—1.75 s.
(b) Penetrator size—10 mm ball. Anvil size—8 in. (203.2 mm) spherical radius.
(c) Check hardness tester with standard Brinell test block.
1500 kg (14.71 kN) load:
2.50–2.55* test block   25.1–25.5 hardness reading
2.55–2.60* test block   25.5–26.2 hardness reading
3000 kg (29.42 kN) load:
3.20–3.25* test block   32.0–32.5 hardness reading
*Brinell indentation diameter.

6.2 Major calibration and resetting of Gogan hardness tester:
(a) If machine is out of calibration, check the major load with either a load cell or proving ring, or by Brinelling a polished steel block of known hardness and reading the impression diameter with an accurate glass. Adjust major load pressure regulating valve to obtain correct load.
(b) If major load is correct and calibration is off, adjust minor load pressure switch until indicator shows correct Gogan reading for standard Brinell test block.

TABLE 1—GOGAN HARDNESS SCALES

Scale Symbol	Major Load		Nominal Minor Load		Penetrator Diameter		Recommended Range of Gogan (G) Numbers
	kg	kN	kg	kN	in	mm	
A	1500	14.71	500	4.90	1	25.39	GA10–GA80
B	3000	29.42	1500	14.71	1	25.39	GB10–GB80
C	1500	14.71	500	4.90	3/4	19.05	GC10–GC80
D	3000	29.42	1500	14.71	3/4	19.05	GD10–GD80

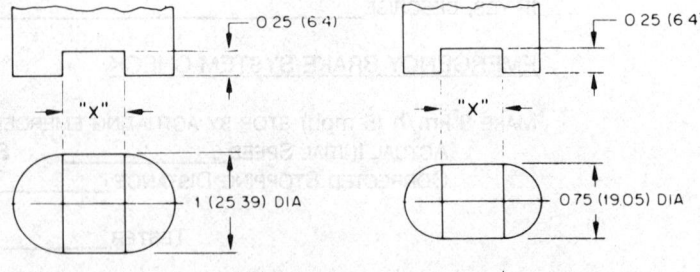

FIG. 1—SPLIT PENETRATOR END

[1] Published by the American Society for Testing and Materials, 1916 Race Street, Philadelphia, Pennsylvania 19103.

SPECIFIC GRAVITY OF BRAKE LINING—SAE J380 FEB93

SAE Recommended Practice

Report of the Brake Committee and Automotive Safety Committee approved February 1969. Editorial change August 1971. Reaffirmed by the SAE Brake Standards Committee 2—Brake Linings February 1993.

Foreword—This reaffirmed document has been changed only to reflect the new SAE Technical Standards Board format.

1. Scope—Specific gravity is a nondestructive test used as a quality control check of the consistency of formulation and processing of brake lining. Specific gravity alone shows nothing about a lining's ability to develop friction or to resist fade when used as a friction element in brakes. Specific gravity varies with the formulation of the lining.

The specific gravity of sintered metal powder linings, particularly those which have steel backing members, is usually determined somewhat differently. Reference ASTM B 376. The specific gravity and the range of specific gravity are peculiar to each formulation and, therefore, the acceptable values or range must be established for each formulation by the manufacturer.

1.1 Purpose—To establish a uniform procedure for determining the specific gravity of brake lining.

2. References

2.1 Applicable Documents—The following publications form a part of this specification to the extent specified herein.

2.1.1 ASTM PUBLICATIONS—Available from ASTM, 1916 Race Street, Philadelphia, PA 19103-1187.

ASTM B 376—Density of Sintered Metal Friction Material (Latest Version)

3. Equipment—(See Figure 1.)

3.1 A scale or balance that will weigh to 0.1 g.

3.2 A support for the scale or balance.

3.3 A container of distilled water 21.1 to 29.4 °C (70 to 85 °F) large enough to hold a completely submerged brake lining without contacting the inside surfaces of the container.

3.4 A monofilament cord and tray fastened to the weighing mechanism from which a brake lining can be suspended and be completely immersed in the water.

4. Procedure

a. Adjust the scale or balance to zero with the empty tray immersed in the water.

b. Place the lining on the scale or balance and determine the "weight in air." Record the weight in air to 0.1 g.

c. Place the lining on the tray and completely immerse the lining in the water. Record the "weight in water" of the brake lining to 0.1 g.

d. Subtract the "weight in water" from the "weight in air" and divide the "weight in air" by the difference.

The resultant figure, to the closest two decimal places, is the specific gravity of the brake lining.

EXAMPLE:

$$\text{Specific gravity} = \frac{A}{A-B} \quad \text{(Eq.1)}$$

where:

A = weight in air, g
B = weight in water, g

CAUTION—The weighing in water should be performed as rapidly as possible (within 15 s) to minimize the absorption of the liquid in the brake lining. Also, the addition of approximately 1 part of a wetting agent to 1000 parts of water will help to eliminate air bubbles from the specimen and will improve the accuracy of the test results.

It is recommended to test undrilled linings only, because there is a tendency for air bubbles to become trapped in the drilled holes.

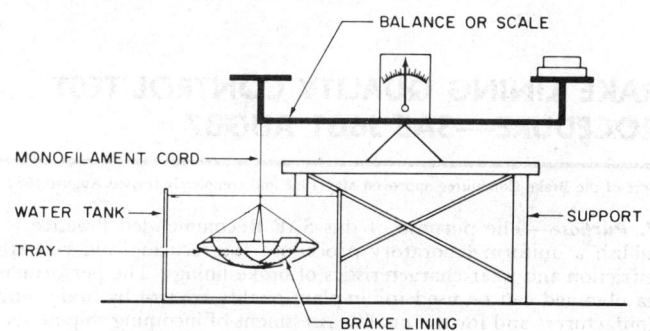

FIGURE 1—EQUIPMENT ARRANGEMENT

(R) FRICTION COEFFICIENT IDENTIFICATION SYSTEM FOR BRAKE LININGS—SAE J866 NOV90

SAE Recommended Practice

Report of the Brake Committee, approved March 1964, completely revised March 1985. Completely revised by the Brake Standards Committee 2—Brake Linings November 1990.

1. Scope—This SAE Recommended Practice is intended to provide a uniform means of identification which may be used to classify the friction coefficient of brake linings, based on data obtained from tests conducted in accordance with SAE J661.

NOTE: It is emphasized that this document does not establish friction requirements for brake linings, nor does it designate significant characteristics of brake linings which must be considered in overall brake performance. Due to other factors that include brake system design and operating environment, the friction coefficients obtained from this document cannot reliably be used to predict brake system performance.

2. References

2.1 Applicable Documents

2.1.1 SAE PUBLICATIONS—Available from SAE, 400 Commonwealth Drive, Warrendale, PA 15096-0001.

SAE J661—Brake Lining Quality Control Test Procedure

3. Coding—The code will consist of two letters reflecting the friction coefficients as in Table 1.

3.1 The first letter will represent normal friction coefficient and the second will represent hot friction coefficient.

3.2 Normal Friction Coefficient—Normal friction coefficient is defined as the average of four points on the second fade curve, located at 93°C (200°F), 121°C (250°F), 149°C (300°F), and 204°C (400°F).

3.3 Hot Friction Coefficient—Hot friction coefficient is defined as the average of 10 points located at 204°C (400°F) and 149°C (300°F)

TABLE 1—FRICTION COEFFICIENT

Code Letter	Friction Coefficient
C	Not over 0.15
D	Over 0.15 but not over 0.25
E	Over 0.25 but not over 0.35
F	Over 0.35 but not over 0.45
G	Over 0.45 but not over 0.55
H	Over 0.55
Z	Unclassified

on the first recovery; 232°C (450°F), 260°C (500°F), 288°C (550°F), 316°C (600°F), and 343°C (650°F) on the second fade; and 260°C (500°F), 204°C (400°F), and 149°C (300°F) on the second recovery.

NOTE: If any temperature point or points required to calculate friction coefficients are not reached in the prescribed time limit, the coefficient of friction value at 10 min shall be used to give the full number of points required.

3.4 Example—A lining having a normal friction coefficient of 0.29 and a hot friction coefficient of 0.40 would be coded "EF."

3.5 Location of Code—The appropriate code designation will be marked on an external noncontacting surface in letters not less than 2.8 mm (7/64 in) in height where a brake lining is 3.2 mm (1/8 in) or greater in thickness, or no more than 0.4 mm (1/64 in) less than the thickness where the brake lining thickness is less than 3.2 mm (1/8 in).

SWELL, GROWTH, AND DIMENSIONAL STABILITY OF BRAKE LININGS—SAE J160 JUN80

SAE Recommended Practice

Report of the Brake Committee, approved July 1970, editorial change January 1980, reaffirmed without change June 1980.

1. Scope—These tests are designed to check the thermal dimensional stability of brake linings in the laboratory under controlled conditions.

2. Swell and Growth

2.1 Obtain initial room temperature thickness readings, to the nearest 0.001 in (0.025 mm), measuring the specimen at not less than six points located approximately $\frac{1}{2}$–$\frac{3}{4}$ in (12.7–19.5 mm) in from the lining edge.

2.2 Place the unconfined specimen in an oven at room temperature. Increase oven temperature to 400 ± 5°F (204 ± 2.8°C). Rise time to 400 ± 5°F (204 ± 2.8°C) shall be $\frac{1}{2}$–1 h. Allow specimen to remain in oven for 30–40 min at 400 ± 5°F (204 ± 2.8°C).

2.3 Remove the specimen from the oven and while still hot measure thickness at the same points used for obtaining initial thickness. Record increase in thickness as *swell*.

2.4 Allow specimen to cool completely, in still air, to room temperature and again measure thickness. Record increase in thickness as *growth*.

NOTE: The dimensional stability test may be performed at the same time and on the same specimen as the swell and growth test, if desired.

3. Dimensional Stability Test (Arcuate Materials)

3.1 Place the arcuate specimen, at room temperature, on a piece of paper and trace its outline. Record the outside arc length of the specimen.

3.2 Place the unconfined specimen, on one of its arcuate edges, in an oven at room temperature. Increase oven temperature to 400 ± 5°F (204 ± 2.8°C). Rise time to 400 ± 5°F (204 ± 2.8°C) shall be $\frac{1}{2}$–1 h. Allow specimen to remain in oven for 30–40 min at 400 ± 5°F (204 ± 2.8°C).

3.3 Allow the specimen to cool completely, in still air, to room temperature and again place it on the paper so that the center of the specimen coincides with that of the original trace at the center and shows approximately equal deviation from the original trace at each end. Again trace specimen outline and compare it to the original.

4. Procedural Notes

4.1 Blisters and surface abnormalities at measured reference points and elsewhere shall be recorded.

4.2 Care should be taken that transfer to heat to the measuring device does not affect the accuracy of the measurements.

φBRAKE LINING QUALITY CONTROL TEST PROCEDURE—SAE J661 AUG87

SAE Recommended Practice

Report of the Brake Committee approved May 1958 and completely revised August 1987.

1. Purpose—The purpose of this SAE Recommended Practice is to establish a uniform laboratory procedure for securing and reporting the friction and wear characteristics of brake linings. The performance data obtained can be used for in-plant quality control by brake lining manufacturers and for the quality assessment of incoming shipments by the purchasers of brake linings.

2. Equipment—A typical, commercially available, machine as used in the preparation of this test procedure and known as a Friction Materials Test Machine is shown in Figs. 1 and 2. The Friction Materials Test Machine shall be equipped with suitable means for:

1. Measuring the drum temperature.
2. Heating the drum.
3. Controlling the drum heating rate.
4. Cooling the drum from the back side only.
5. Controlling the drum cooling rate.
6. Measuring friction force.
7. Measuring drum rotational speed.

Means shall be provided for measuring specimen thickness and mass.

The temperature measuring means shall incorporate a welded thermocouple, coin silver slip rings, silver-graphite brushes, and an indicator and/or recorder having a high input impedance.

The drum heating means shall be adjusted as follows and remain so during the test: with the drum rotating at 411 rpm, cool from 300°F (149°C) to 200°F (93°C) with cooling air on. Then cool to 180°F (82°C) with cooling air off. Turn on heaters at 180°F and start timing. Heat for 10 min. Drum temperature shall be 430 ± 25°F (221 ± 14°C) at 10 min.

The drum cooling means shall be adjusted as follows: with the drum rotating at 411 rpm, and after having heated the drum with the heater elements to 700°F (371°C), turn off heaters and turn on cooling air. Cool to 650°F (343°C) and start timing. Cool for 10 min. Drum temperature shall be 200 ± 25°F (93 ± 14°C) at 10 min.

The temperature measuring system shall have ±2% full scale accuracy.

The friction force measuring system shall have ±2% full scale accuracy.

The drum speed measuring system shall have ±2% full scale accuracy.

The drum shall be used only between the inside diameter limits of 277.0 - 280.0 mm and have three thermocouple locations, one each at depths of 2.55 (stamped number 1), 3.05 (stamped number 2) and 3.55 mm (stamped number 3) from the new drum surface diameter of 277.0 mm.

The thermocouple should be mounted in the position indicated:

Drum Inside Diameter	Location Stamp on Drum
277 - 278	1
278 - 279	2
279 - 280	3

3. Test Conditions—Actual tests for performance shall be started when preparations have been completed in accordance with Section 4.

3.1 Conduct of Test—All testing shall proceed without interruption.

3.2 Drum Speeds—All drum speeds (rpm) are based on a nominal 278.5 mm diameter drum with load applied to the specimen.

4. Procedure

4.1 Preparation of Test Specimen—The test specimen shall be taken from the center of the friction material approximately equidistant from each end.

The test specimen shall be 25.7/25.6 mm square (660 mm²), flat on the bottom, and the radius of the working surface shall conform to the radius of the test drum. On pre-ground linings, remove at least 0.3 mm, but not more than 0.5 mm from the working surface of the specimen. For un-ground linings (directly from molds), remove 1.0/1.2 mm to be certain that the resin impregnated surface is totally removed. Specimen thickness (or specimen plus shim) should be approximately 6 mm measured in the center of specimen. Excess of material must be removed from the side opposite the working surface of the specimen. In cases where nominal lining thickness is less than 5 mm, remove minimum amount of material from the side opposite the working surface to produce flatness.

The working surface of the specimen shall not be handled and shall be kept free from foreign material.

4.2 Preparation of Test Drum Surface

4.2.1 NEW OR RESURFACED DRUM—After grinding the drum surface on the test machine, remove all grinder marks by polishing with abrasive paper or cloth. Final polishing shall be with 320 grit. Remove dust from drum with clean dry cheesecloth, white paper toweling, or equivalent. Complete the surface preparation by running a reference specimen continuously at 440 N, 411 rpm and not over 200°F (93°C) until the friction coefficient has stabilized.

4.2.2 PRIOR TO EACH TEST—Polish the drum surface with abrasive paper or cloth. Final polishing shall be with 320 grit. Remove dust from the drum with clean cheesecloth, white paper toweling, or equivalent.

4.3 Conditioning of Test Specimen—The specimen is burnished at 308 rpm, 440 N and a maximum temperature of 200°F (93°C), for a minimum of 20 min, to obtain at least 95% contact.

4.4 Initial Thickness and Mass Measurements—Specimen thickness measurement is taken in three places along the axis parallel to the drum axis (open, center, and closed edges) and recorded. Weigh, to nearest milligram, and record. Reseat specimen by running continuously for 5 min at 220 N and 205 rpm. Initial clearance between specimen and drum should be 0.3-0.4 mm in the "OFF" position.

4.5 Initial Wear Measurement—With drum stationary and its temperature between 190°F (88°C) and 210°F (99°C) with 660 N on specimen, obtain indicator reading of height of specimen holder and record.

5. Test Runs

5.1 Baseline Run—Run 10 s "ON" (load applied) and 20 s "OFF" (load removed) at 660 N and 411 rpm for 20 applications.

Start run at a drum temperature of 180-200°F (82-93°C) and maintain the maximum and minimum temperature during each successive application between 180-220°F (82-104°C) with the use of cooling air. Turn cooling air off on 20th load application.

5.2 First Fade Run—Allow drum to cool with drum rotating and heating and cooling means off. At 180°F (82°C) apply specimen and energize heating elements. Run continuous drag at 660 N and 411 rpm. Run for either 10 min or until 550°F (288°C) is attained, whichever occurs first. Take readings of friction force at intervals of 50°F (28°C), starting at 200°F (93°C). Record time required to reach 550°F (288°C).

5.3 First Recovery Run—Immediately following completion of First Fade Run (paragraph 5.2), turn off heater and turn on cooling means and make a 10 s application at 660 N and 411 rpm at 500°F (260°C), 400°F (204°C), 300°F (149°C) and 200°F (93°C) during cooling.

5.4 Second Wear Measurement—Repeat Initial Wear Measurement (paragraph 4.5).

5.5 Wear Run—Run 20 s "ON," 10 s "OFF," at 660 N and 411 rpm for 100 applications. Start run at a drum temperature of 380-400°F (193-204°C) and maintain maximum and minimum during each application between 380-420°F (193-216°C) with use of cooling air.

5.6 Third Wear Measurement—Immediately upon completion of Wear Run (paragraph 5.5), cool to 190-210°F (88-99°C) and repeat Initial Wear Measurement (paragraph 4.5).

5.7 Second Fade Run—Upon completion of Third Wear Measurement, allow drum to cool with drum rotating and heating and cooling means off. At 180°F (82°C) apply specimen and energize heating elements. Run continuous drag at 660 N and 411 rpm. Run for either 10 min or until 650°F (343°C) is attained, whichever occurs first. Take readings of friction force at intervals of 50°F (28°C), starting at 200°F (93°C). Record time required to reach 650°F (343°C).

5.8 Second Recovery Run—Immediately upon completion of Second Fade Run (paragraph 5.7), turn off heater and turn on cooling means and make a 10 s application at 660 N and 411 rpm at 600°F (316°C), 500°F (260°C), 400°F (204°C), 300°F (149°C) and 200°F (93°C) during cooling.

5.9 Baseline Rerun—Repeat Baseline Run (paragraph 5.1).

5.10 Final Wear Measurement—Repeat Initial Wear Measurement (paragraph 4.5).

5.11 Final Thickness and Mass Measurements—Measure and weigh as described in Initial Thickness and Mass Measurements (paragraph 4.4).

6. Selection of Plot Point for Friction Coefficient Value—During intermittent application runs, the friction coefficient values are taken at the end of the application.

7. Presentation of Test Data

7.1 Data should be presented on Master Form Log Sheet.

7.2 Data should be plotted on Master Form Plot Sheet.

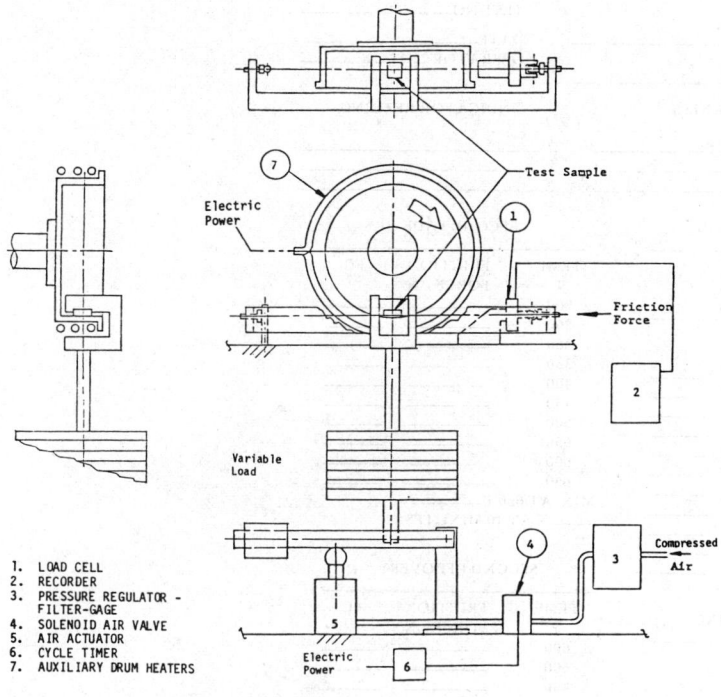

FIG. 1—SCHEMATIC DIAGRAM OF FRICTION MATERIALS TEST MACHINE

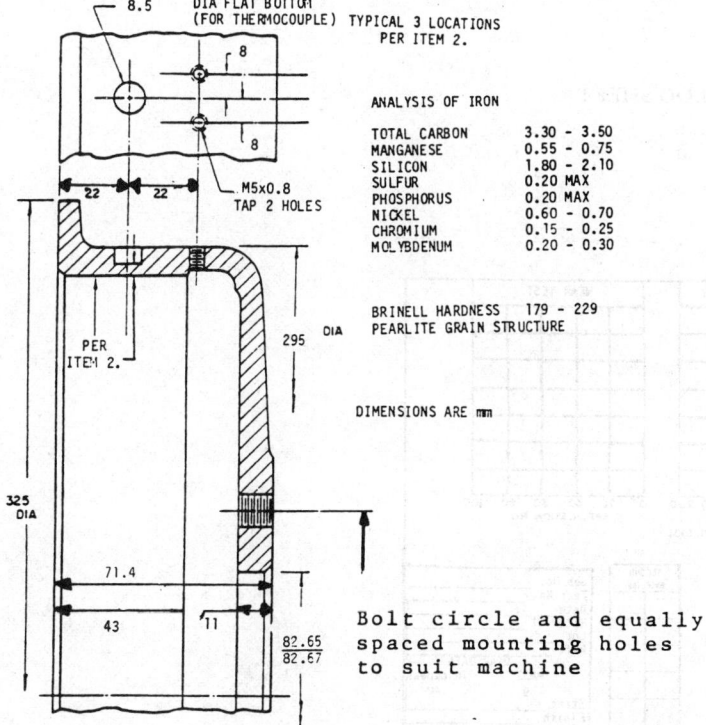

FIG. 2—FRICTION MATERIALS TEST MACHINE DRUM

MATERIAL_____ JOB NO._____
                                  TEST NO._____
LOT_____   DATE_____
                                  OPERATOR_____
REFERENCE_____

	MASS	THICKNESS	INDICATOR READING
INITIAL	_____	_____	_____
FINAL	_____	_____	_____
LOSS	_____	_____	_____

FIRST BASELINE

APPL NO	FRICTION FORCE N	FC*
1	_____	_____
5	_____	_____
10	_____	_____
15	_____	_____
20	_____	_____

FIRST FADE

TEMP F°	FRICTION FORCE N	FC
200	_____	_____
250	_____	_____
300	_____	_____
350	_____	_____
400	_____	_____
450	_____	_____
500	_____	_____
550	_____	_____

MIN. AT 550°F _____
_____ F AT 10 MINUTES

FIRST RECOVERY

TEMP F°	FRICTION FORCE N	FC
500	_____	_____
400	_____	_____
300	_____	_____
200	_____	_____

REMARKS: _____

*FC - FRICTION COEFFICIENT

INDICATOR READING

WEAR TEST

APPL NO	FRICTION FORCE N	FC
1	_____	_____
10	_____	_____
20	_____	_____
30	_____	_____
40	_____	_____
50	_____	_____
60	_____	_____
70	_____	_____
80	_____	_____
90	_____	_____
100	_____	_____

INDICATOR READING

SECOND FADE

TEMP °F	FRICTION FORCE N	FC
200	_____	_____
250	_____	_____
300	_____	_____
350	_____	_____
400	_____	_____
450	_____	_____
500	_____	_____
550	_____	_____
600	_____	_____
650	_____	_____

MIN. AT 650°F _____
_____ F AT 10 MINUTES

SECOND RECOVERY

TEMP °F	FRICTION FORCE N	FC
600	_____	_____
500	_____	_____
400	_____	_____
300	_____	_____
200	_____	_____

SECOND BASELINE

APPL NO	FRICTION FORCE N	FC
1	_____	_____
5	_____	_____
10	_____	_____
15	_____	_____
20	_____	_____

FIG. 3—MASTER FORM LOG SHEET

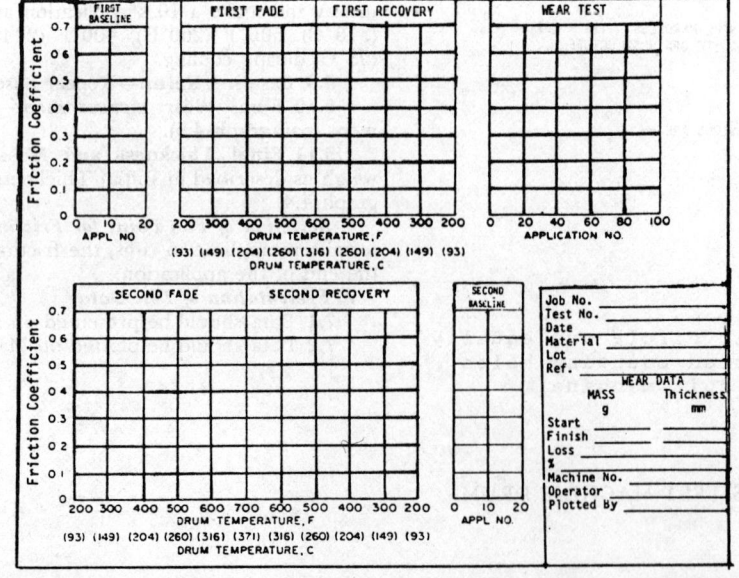

FIG. 4—MASTER FORM PLOT SHEET

(R) BRAKE BLOCK CHAMFER—SAE J662 NOV90 — SAE Standard

Report of the Brake Committee approved September 1953. Reaffirmed without change January 1968. Completely revised by the Brake Standards Committee 2—Brake Linings November 1990.

1. **Scope**—The scope of this SAE Standard is to provide a uniform chamfer on brake blocks.

2. **References**—There are no referenced publications specified herein.

3. **Chamfer Formation**

   3.1 Where the arc length of the block is 70 degrees or less, the chamfer is formed by cutting off the ends of the block parallel to the centerline of the block length. These cuts are to intersect the radial ends at the centerline of thickness as shown in Figure 1.

   3.2 Where the arc length of the block is more than 70 degrees, the ends shall be cut off parallel to a radius 35 degrees inside the ends. This line is to intersect the radial ends at a point one-half the thickness of the block, as shown in Figure 2.

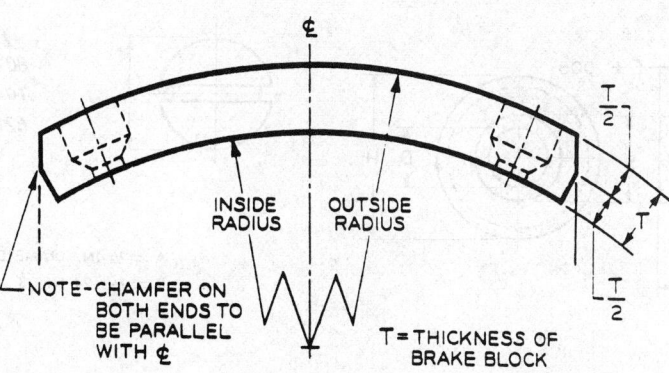

FIG. 1

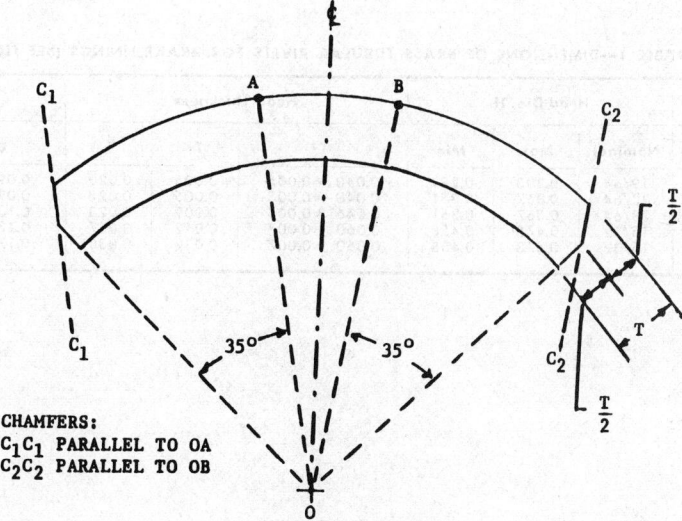

FIG. 2

RIVETS FOR BRAKE LININGS AND BOLTS FOR BRAKE BLOCKS—SAE J663b

SAE Standard

Report of Brake Committee approved February 1948 and last revised July 1968.

Brass Tubular Rivets for Brake Linings—Table 1 and Fig. 1 give dimensions for brass tubular rivets used for brake linings.

The eccentricity between rivet head and shank must be no greater than 0.010 in., and the eccentricity between the hole in the shank and the shank must be no greater than 0.010 in.

Bolts for Brake Blocks—The standard bolt for blocks shall be a brass slotted flat head capscrew with a 3/8-in. diameter body, a 3/4-in. diameter head having an inclusive angle of countersink of 82 deg, and conforming to Table 23, SAE J478.

The general bolt dimensions given with Fig. 2 are recommended. For other dimensions, refer to Table 23, SAE J478.

NOTE: For drill and countersink dimensions for rivet and bolt holes, see SAE J660.

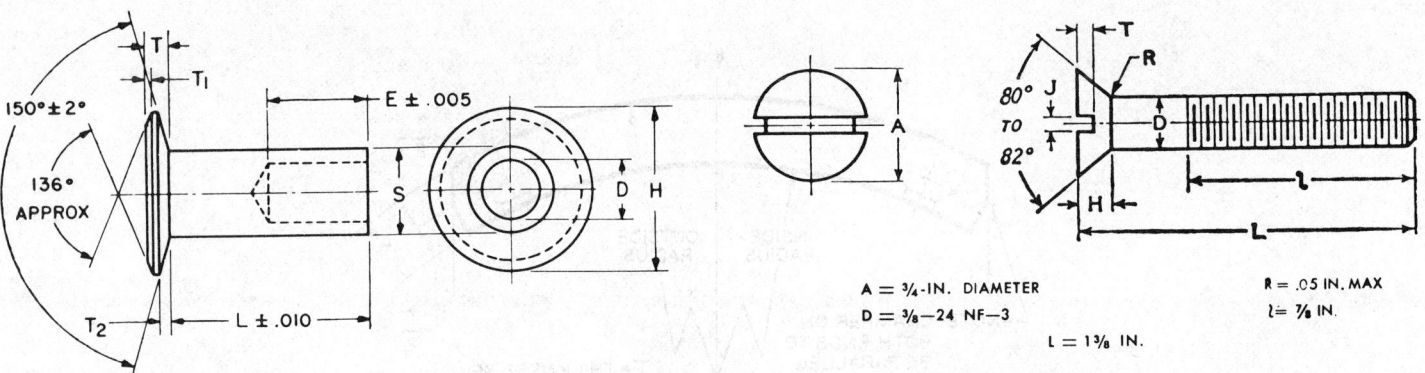

FIG. 1—BRASS TUBULAR RIVET FOR BRAKE LININGS

FIG. 2—SLOTTED FLAT HEAD CAPSCREW FOR BRAKE BLOCKS

TABLE 1—DIMENSIONS OF BRASS TUBULAR RIVETS FOR BRAKE LININGS (SEE FIG. 1)

Rivet Size No.	Body Dia, S			Head Dia, H			Head Thickness			Hole		Length, L
	Nominal	Max	Min	Nominal	Max	Min	T	T_1	T_2	Dia, D	Depth, E	
4	9/64	0.146	0.141	19/64	0.303	0.289	0.040 ±0.005	0.005	0.020	0.099–0.104	9/64	As specified in increments of 1/16 in.
5	9/64	0.146	0.141	23/64	0.367	0.351	0.040 ±0.005	0.005	0.028	0.099–0.104	9/64	
7	3/16	0.188	0.182	23/64	0.367	0.351	0.046 ±0.005	0.009	0.023	0.133–0.139	3/16	
8	3/16	0.188	0.182	15/32	0.478	0.458	0.060 ±0.007	0.012	0.037	0.133–0.139	3/16	
10	1/4	0.252	0.244	15/32	0.478	0.458	0.060 ±0.007	0.012	0.030	0.173–0.183	1/4	

TEST PROCEDURES FOR BRAKE SHOE AND LINING ADHESIVES AND BONDS—SAE J840 AUG82

SAE Recommended Practice

Report of the Brake Committee, approved November 1962, last revised August 1982.

1. Scope—This SAE Recommended Practice covers equipment and procedures for qualification of bonded brake shoe and lining assemblies and for quality control on materials and processes used in their manufacture.

2. Qualification Tests

2.1 Scope—The following tests cover equipment and procedures used to verify the structural integrity of the brake shoe, adhesive, and brake lining assembly. The Bond Plane Shear Test and either the Dynamometer Test or the Vehicle Abuse Test are used for qualification.

2.2 Bond Plane Shear Test

ϕ 2.2.1 PURPOSE—The purpose of this test is to provide values of lining-to-brake-shoe shear strength by measuring the load required to cause shear failure on complete shoe and lining assemblies, under both ambient and elevated temperature conditions.

2.2.2 EQUIPMENT—The equipment for performing this test consists of a compression test machine of sufficient capacity to shear the lining from the shoe, a fixture which shall provide means to hold the shoe firmly, and a movable ram through which the shear load is applied to the lining. Additional fixture requirements are:

2.2.2.1 *Drum Brake Shoes*—Fixture (Fig. 1) shall be so designed that the ram contacts the edge of the lining for its full length and thickness to within 0.005–0.020 in (0.13–0.51 mm) of the shoe table or rim. Load application on the ram shall be in a direction perpendicular to the plane of the shoe web and the shoe shall be supported to maintain uniform loading along the length of the lining.

2.2.2.2 *Disc Brake Shoes*—Fixture (Fig. 2) shall be so designed that the ram contacts the edge of the lining within 0.005–0.020 in (0.13–0.51 mm) of the shoe and conforms adequately to the lining edge contour to avoid crushing of the lining edge prior to failure. Normally, the ram shall contact the edge parallel to the long axis of the lining; the edge parallel to the short axis may be used if premature crushing of the lining is not incurred

2.2.3 PROCEDURE

2.2.3.1 *Ambient Destructive Shear Test*—The brake shoe and lining assembly shall be placed in the shear test fixture and the load shall be applied at a rate of 1000 ± 100 lb (453 ± 48 kg) per second, or 0.40 ± 0.04 in (10 ± 1 mm) per minute after the ram is in contact with the lining edge. Loading shall be continued until failure has occurred. The load at which observable lining movement or complete shear occurs and the shear pattern (paragraph 2.2.5) shall be recorded. Also, a check shall be made for state of cure using the Cotton Tack Test (paragraph 3.2.5).

2.2.3.2 *Hot Destructive Shear Test*—The brake shoe and lining assembly shall be placed in a heating fixture or oven that will bring the temperature up to the specified value uniformly throughout the bond line within ±10°F (±5.5°C). This specified temperature must be reached within 30 min. When the temperature is reached, the shoe assembly should be placed in the shear test fixture and tested as in paragraph 2.2.3.1. The load at which observable lining movement or complete shear occurs shall be recorded.

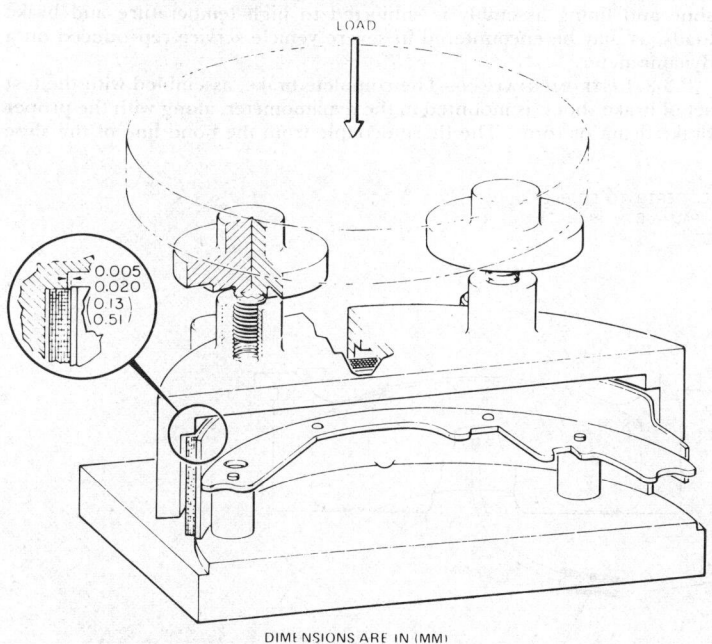

ϕ FIG. 1—BOND PLANE SHEAR TEST—DRUM BRAKE

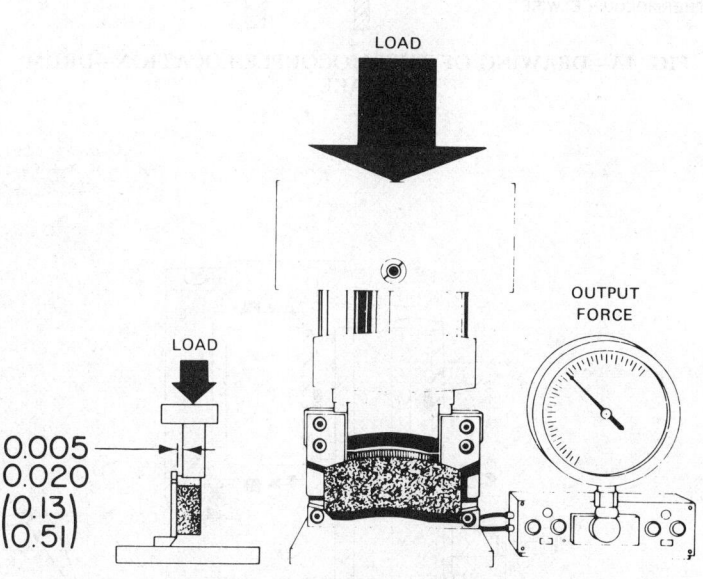

ϕ FIG. 2—DISC BRAKE SHEAR TEST FIXTURE

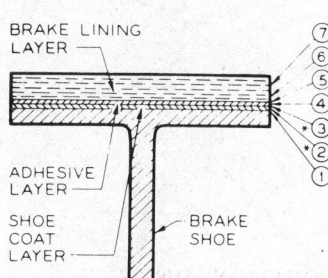

⑦ DEEP FAILURE WITHIN THE LINING
⑥ SHALLOW FAILURE WITHIN THE LINING (LESS THAN 0.02 IN (0.51 MM) OF LINING REMAINING
⑤ ADHESION FAILURE BETWEEN LINING AND ADHESIVE
④ COHESION FAILURE WITHIN THE ADHESIVE LAYER
*③ ADHESION FAILURE BETWEEN ADHESIVE AND SHOE COAT
*② COHESION FAILURE WITHIN THE SHOE COAT
① ADHESION FAILURE (BARE METAL) BETWEEN BRAKE SHOE METAL AND SHOE COAT OR BETWEEN BRAKE SHOE METAL AND ADHESIVE, WHEN SHOE COAT IS NOT USED

*ELIMINATE WHEN SHOE COAT IS NOT USED.
TO REPORT FRACTURE PATTERN, EXAMINE THE DESTROYED BOND TO DETERMINE EXACTLY WHERE THE FRACTURE TOOK PLACE, IE, BETWEEN THE ADHESIVE AND THE SHOE COAT (3), OR BETWEEN THE LINING AND THE ADHESIVE (5). SHOULD THE EXAMINATION SHOW MORE THAN ONE TYPE OF FRACTURE, REPORT, IN DECREASING ORDER, ALL OF THE DIFFERENT TYPES THAT ARE PRESENT, AND INDICATE THIER APPROXIMATE PERCENTAGE OF THE TOTAL AREA, IE 60 6, 40 2, OR 50 1, 30 3, 20 4.

FIG. 3—STANDARD METHODS OF BOND FRACTURE

NOTE: The heating fixture may be incorporated in the shear test fixture or external to it. If external to it, not more than 15 s should elapse between removal of the shoe assembly from the heating fixture and failure. Temperature of the bond line shall be observed by means of the bond line thermocouple shown in Figs. 4A and 4B.

2.2.3.3 *Resistance to Fluids Test*—This procedure is designed for testing adhesives for resistance to fluids encountered in service, and provides for reporting loss in shear strength after immersion in the test fluids. Individual specimens shall be totally immersed in each test liquid (paragraph 2.2.3.3(a)) in a separate container for 7 days at room temperature. The liquid shall be agitated every 24 h by moderate manual rotation of the container. The individual specimens shall be removed from the containers, blown off or wiped with a clean dry cloth, and tested immediately at room temperature; or, in the event of adverse effect on the lining, longer drying periods may be used. The brake shoe and lining assembly shall be placed in the shear fixture and loaded to destruction at the prescribed loading rate as per paragraph 2.2.3.1.

(a) Immersion Fluids:

(1) Reference fuels A and B as specified in ASTM D 471,[1] Tentative Method of Test for Change in Properties of Elastomeric Vulcanizates Resulting from Immersion in Liquids.

[1] Published by American Society for Testing and Materials, 1916 Race St., Philadelphia, Pennsylvania 19103.

(2) ASTM Oil No. 1, as specified in ASTM D 471.
(3) ASTM Oil No. 3, as specified in ASTM D 471.
(4) Calcium chloride 20% solution.
(5) Hypoid oil.
(6) Butyl cellosolve (brake fluid grade).
(7) Tap water.

2.2.4 REPORT

2.2.4.1 *Ambient Destructive Shear Test*

2.2.4.1.1 Record load at which observable movement or complete shear fracture of the lining relative to the shoe occurs.

2.2.4.1.2 Establish and record type of shear fracture pattern. (NOTE: Refer to paragraph 2.2.5.)

2.2.4.1.3 Record results of Cotton Tack Test (paragraph 3.2.5).

2.2.4.2 *Hot Destructive Shear Test*—Record load at which observable movement or complete shear of the lining relative to the shoe occurs.

2.2.4.3 *Resistance to Fluids Test*

2.2.4.3.1 Record load at which observable movement or complete fracture of the lining relative to the shoe occurs.

2.2.4.3.2 Establish and record type of shear fracture.

2.2.4.3.3 Report any visible adverse effects on the adhesive from the immersion fluid.

2.2.5 REPORT, STANDARD METHOD OF REPORTING BOND FRACTURE—Fig. 3 shows the seven possible planes of fracture between the brake lining and the brake shoe. Each of these planes has been assigned a number from 1 to 7. The report should include the type or types of fractures encountered, by indicating the appropriate number from 1 to 7, together with the relative areas of each fracture type expressed as a percentage of the total area in decreasing order. (Example: 60 No. 4, 30 No. 6, 10 No. 7—Note that the percent mark is not used.)

2.3 Dynamometer Test—High Temperature Bond Abuse

2.3.1 PURPOSE—The purpose of this test is to determine the effectiveness of the bond of a combination of lining and adhesive when the brake shoe and lining assembly is subjected to high temperature and brake loads, as may be encountered in severe vehicle service reproduced on a dynamometer.

2.3.2 INSTRUMENTATION—The complete brake, assembled with the test set of brake shoes, is mounted in the dynamometer, along with the proper brake drum or rotor. The thermocouple from the bond line of the shoe

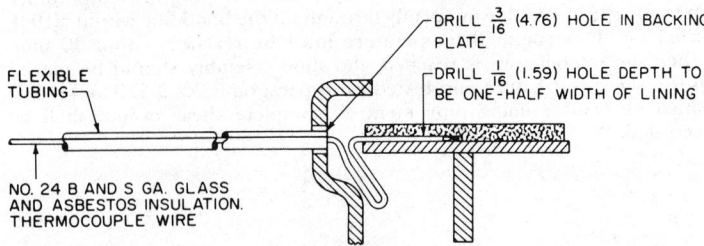

FIG. 4A—DRAWING OF THERMOCOUPLE LOCATION—DRUM BRAKE

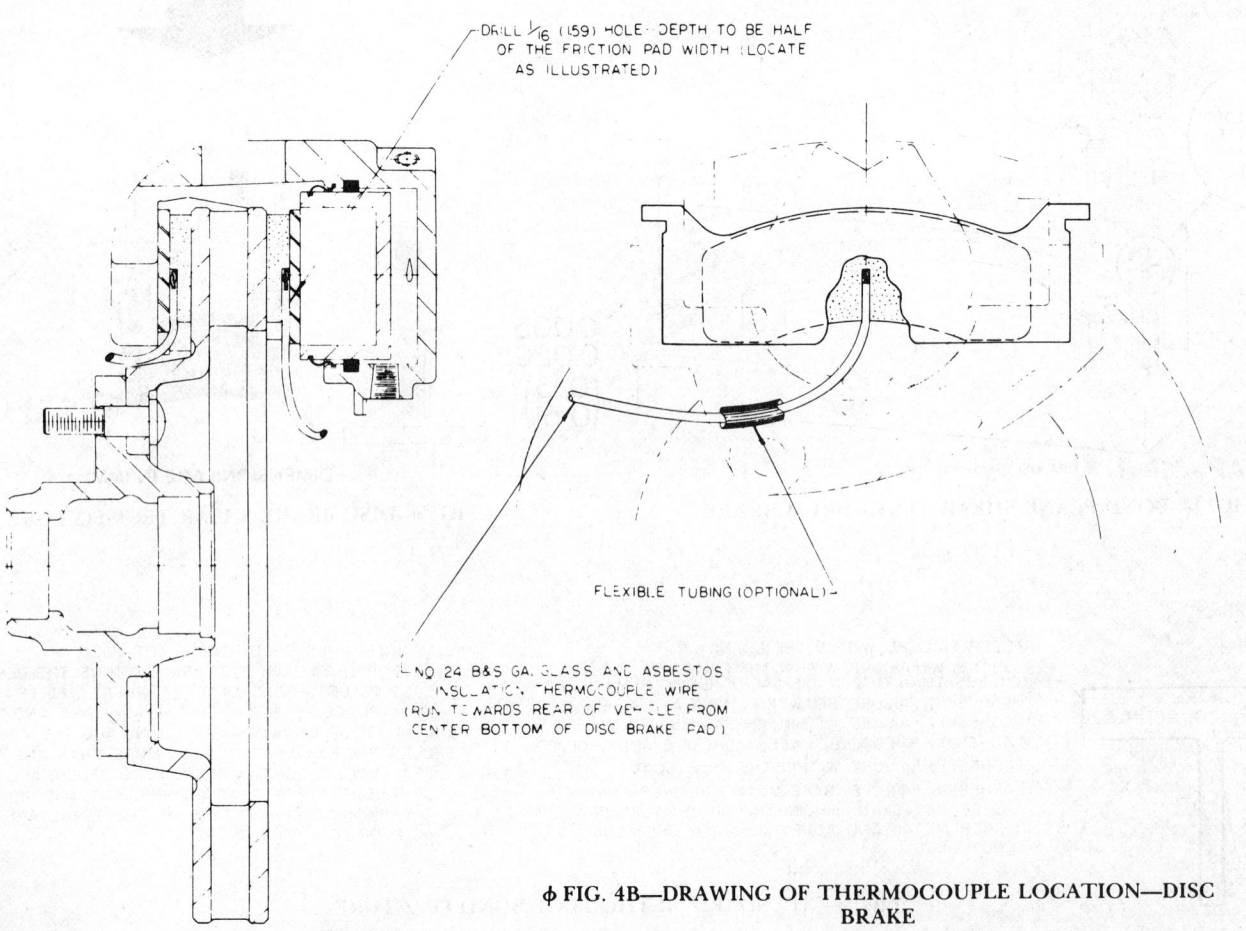

φ FIG. 4B—DRAWING OF THERMOCOUPLE LOCATION—DISC BRAKE

is connected to the temperature recording instrument. The flywheel loading shall correspond with the work load imposed on the brake as it is used in the specific vehicle. Adjust brake shoe to drum clearance or brake shoe to rotor clearance as recommended by the manufacturer.

2.3.3 PROCEDURE

2.3.3.1 *Preparation of Specimens*—Sufficient shoes for dynamometer tests shall be prepared. Test shoes and lining assemblies shall be processed over regular production equipment. For the purpose of recording the bond line temperature, drill a $\frac{1}{16}$ in (1.59 mm) diameter hole edgewise in the brake lining and at the bond line to a depth of approximately one-half the lining width, as shown in Fig. 4A for the drum brake or Fig. 4B for the disc brake. The thermocouple must be imbedded in the adhesive layer. When testing a drum brake shoe and lining assembly, locate the hole at approximately the high pressure point of the lining on the shoe which is producing the maximum brake effectiveness and install the thermocouple to the bottom of the hole. For a disc brake shoe and lining assembly, the thermocouple should be located in the bond line at the center of the inboard shoe.

2.3.3.2 *Burnish*—Turn blower on. Make consecutive stops from a flywheel speed corresponding to 60 mph (96.5 km/h) at an average rate of deceleration of 10 ft/s/s (3.66 m/s/s) allowing the temperature to drop to approximately 200°F (93.5°C) between applications until linings show at least 80% contact.

2.3.3.3 *Bond Test, Constant Temperature, 650°F (343.3°C)*—Turn blower on at 100°F (37.8°C) below test temperature. Make stops from a flywheel speed corresponding to 70 mph (112.66 km/h) at an average rate of deceleration of 15 ft/s/s (4.57 m/s/s) until the bond line temperature reaches 650°F (343.3°C). Make as many stops as possible, a maximum of 50 or until the lining wears out, at 650°F (343.3°C) at a deceleration rate of 15 ft/s/s. Make applications at time intervals that will result in maintaining the predetermined average temperature throughout the test. If the 15 ft/s/s deceleration rate cannot be maintained, continue the ϕ test at the maximum line pressure encountered during the previous 15 ft/s/s stops.

2.3.3.4 *Bond Test—Ultimate Temperature*—Repeat burnish and bond test above except at increased temperature levels in increments of 50°F (27.7°C). Run three tests, each with new samples, at each temperature level until a failure occurs. Bond line temperature at which consistent failures occur in less than 50 stops is considered the ultimate temperature resistance of the adhesive being investigated.

NOTE: This test is generally used as a research evaluation method for adhesive, and is recommended to be used in testing new adhesive formulations to gain comparative values only.

2.3.4 REPORT—Record pertinent data and test results on the form as shown in Fig. 5.

2.4 Vehicle Abuse Test

2.4.1 PURPOSE—The purpose of this test is to determine the effectiveness of the bond for a combination of lining and adhesive when the brake shoes have been subjected to the stresses of shock loading and heat, as may be encountered in severe vehicle service.

2.4.2 EQUIPMENT—The equipment for performing this test shall consist of an appropriate test vehicle modified to receive a thermocouple or thermocouples at the bond line of the brake shoe and lining assembly, a thermocouple or thermocouples, and a pyrometer.

2.4.3 PROCEDURE

2.4.3.1 Select and prepare brake shoe and lining assemblies as in paragraph 2.3.3.1 except that in testing a disc brake shoe and lining assembly, the lining should be ground to 0.100 in (2.54 mm) in thickness before installation to accelerate the heat transfer to the bond line.

2.4.3.2 A complete test for bonded brake shoe and lining assembly shall consist of:

2.4.3.2.1 Shock Test—Cold
2.4.3.2.2 Heat Test
2.4.3.2.3 Repeat Shock Test—Intermediate
2.4.3.2.4 Continued Heat Test
2.4.3.2.5 Repeat Shock Test—Hot

2.4.3.3 *Test Procedure*—Install a set of shoes and adjust brakes according to manufacturer's recommendations.

2.4.3.3.1 Shock Test—Cold—Make one forward and one reverse stop from approximately 15–20 mph (24.1–32.2 km/h) at maximum rate of deceleration and minimum time interval. Repeat two more times. Be alert for wheel drag. A heavy drag or rubbing noise may indicate lining slippage due to bond failure.

2.4.3.3.2 Heat Test at 250–300°F (121–149°C)—In any appropriate gear, drive the vehicle while dragging the service brakes. Continue until a bond line temperature at 250–300°F (121–149°C) is reached within 3–5 min (suggested speed 20–40 mph (32.2–64.4 km/h)).

2.4.3.3.3 Repeat Shock Test—Intermediate—When the bond line tem-

---

DYNAMOMETER TEST OF BONDED SHOE AND LINING ASSEMBLIES

GENERAL
TEST NUMBER _____ PURPOSE _____ DATE _____
BRAKE TYPE _____ SIZE _____ WHEEL CYLINDER DIAMETER _____

ADHESIVE
SOURCE _____ COMPOUND _____ BATCH NO. _____

LINING
SOURCE—PRIMARY OR FORWARD DRUM SHOE OR OUTER DISC SHOE COMPOUND IDENT. _____
(strike out one)
SOURCE—SECONDARY OR REVERSE DRUM SHOE OR INNER DISC SHOE COMPOUND IDENT. _____
(strike out one)

DRUM OR DISC (strike out one)
SOURCE _____ TYPE _____ PART NO. _____

SAMPLE PREPARATION
ADHESIVE PATTERN _____ TYPE _____ PART NO. _____
ROOM TEMP. DRY TIME _____ FORCE DRY TIME _____ °F (____°C)
BOND CURE _____ MINUTES AT _____ °F (____°C) AT APPROX.
_____ PSI (_____ N/M²)

DYNAMOMETER FLYWHEEL EQUIVALENT OF ____ FT-LB K.E. AT ____ MPH
(____ N/M² AT ____ KM/H)

TEST NO.	BURNISH				BOND TEST					REMARKS
	SPEED	DECEL.	TEMP, MIN	STOPS	SPEED	DECEL.	TEST TEMP, AVERAGE	STOPS	
1									
2									
3									
4									
5									
6									
7									
8									
9									
10									
11									
12									
13									
14									
15									
16									
17									
18									
19									
20									
21									
22									

ϕ FIG. 5—DYNAMOMETER TEST REPORT

peratures of the shoes are above 250°F (121°C), but not over 300°F (149°C), immediately repeat Shock Test as in paragraph 2.4.3.3.1.

2.4.3.3.4 *Continued Heat Test*—Immediately continue, dragging the service brakes as in paragraph 2.4.3.3.2 for 30 min. Record bond line temperatures in 2 min intervals and plot on the chart shown in Fig. 6A. During the test, the bond line temperature curve must stay within the limits of the envelope of the two curves shown or the test is not to be considered valid.

NOTE: It may be possible to stay within this envelope only on either the two front wheels or the two rear wheels but not both (suggested speed 20–40 mph (32.2–64.4 km/h)).

2.4.3.3.5 *Repeat Shock Test—Hot*—Immediately after completion of Continued Heat Test in paragraph 2.4.3.3.4, repeat paragraph 2.4.3.3.1.

2.4.3.3.6 *Lining Test*—Shear or chisel lining from test shoes (see paragraphs 3.2.3 and 3.2.4), record failing loads and/or fracture pattern.

2.4.4 Report results on the form shown in Fig. 6B.

3. **Quality Control Tests**
   3.1 **Material Tests**
   3.1.1 SCOPE—These tests are conducted on the adhesive product being used to determine its consistency within the limits established on a qualified product.
   3.1.2 VISCOSITY TEST
   3.1.2.1 *Purpose*—This is a quality control test on the bonding agent to determine its viscosity. The viscosity of an adhesive is defined as the internal friction resistance to flow. This viscosity characteristic is important in the development of application techniques.
   3.1.2.2 *Equipment*
   3.1.2.2.1 *Viscometer*—Variable speed, spindle type Synchro-electric viscometer.
   3.1.2.2.2 *Container*—One quart (0.94 L) round, friction topped can, 4.5 in (114.5 mm) in diameter and 4.875 in (124 mm) high, with a 3.25 in (82.5 mm) opening.
   3.1.2.2.3 *Thermometer*—Accurate thermometer to read in the 77 ± 1°F (25 ± 0.5°C) range.
   3.1.2.2.4 *Mixer*—High-speed agitator to stir adhesive before testing.

3.1.2.3 *Procedure*
3.1.2.3.1 Agitate sample before testing. Amount and type of mixing to be determined and specified by adhesive supplier.
3.1.2.3.2 Adjust sample to 77 ± 2°F (25 ± 0.5°C).
3.1.2.3.3 Immerse spindle to proper level in adhesive and start motor.
3.1.2.3.4 The reading should be taken when the viscosity reading has stabilized.
3.1.2.4 *Report*—Make and model of the viscometer, the spindle number, spindle speed, temperature, viscosity reading, and scale factor.
3.1.3 SOLIDS CONTENT TEST
3.1.3.1 *Purpose*—This is a quality control test to determine the nonvolatile content of the adhesive. Dry film coverage is directly proportional to the solids content. The test is performed by evaporating the solvent from a sample of known weight and weighing the residue.
3.1.3.2 *Equipment*
3.1.3.2.1 Circulating hot air oven or vacuum oven equipped with a thermometer.
3.1.3.2.2 A 3.0 oz (85 g) ointment tin with cover 2⅜ in (60.5 mm) in diameter.
3.1.3.3 *Procedure*
3.1.3.3.1 Mix sample thoroughly.
3.1.3.3.2 Weigh empty container and cover to 0.01 g.
3.1.3.3.3 As rapidly as possible, pour approximately 5 g of adhesive into the container. Replace cover on ointment tin at once. Weigh accurately to 0.01 g.
3.1.3.3.4 Remove cover, heat sample in the oven at 220 ± 5°F (105 ± 3°C) for 30 min.

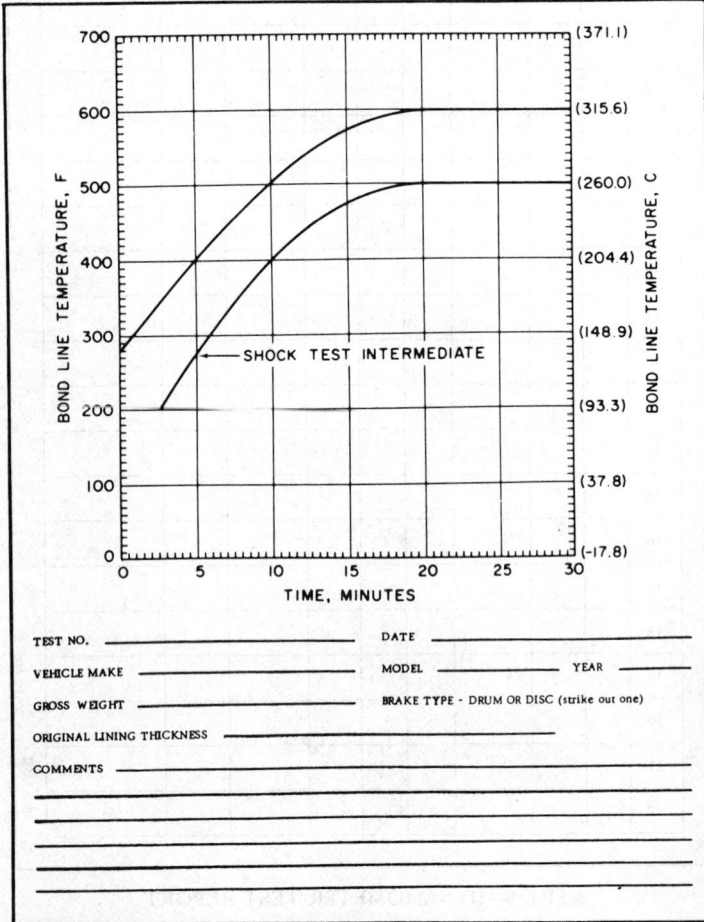

FIG. 6A—ENVELOPE VEHICLE ABUSE TEST

φ FIG. 6B—REPORT FORM FOR VEHICLE ABUSE TEST

3.1.3.3.5 Reweigh sample, container, and cover to 0.01 g.
3.1.3.3.6 Calculations:

$$\frac{\text{Weight sample after heating}}{\text{Weight sample before heating}} \times 100 = \text{Total solids, \%}$$

3.1.3.4 *Report*—Total solids in percent, time of heating, temperature of heating, and type of oven.

3.1.4 FLOW TEST

3.1.4.1 *Purpose*—This test is used to determine the flow properties of the bonding agent. The flow of the dried adhesive film under bonding or curing conditions indicates its ability to wet the surfaces.

3.1.4.2 *Equipment*

3.1.4.2.1 Steel discs 0.250 in (6.3 mm) thick, 1.125 in (28.6 mm) in diameter, SAE 1010.

3.1.4.2.2 Steel strip 0.250 × 1.250 × 4.250 in (6.3 × 31.8 × 110.8 mm), SAE 1010.

3.1.4.2.3 SAE disc shear bonding press.

3.1.4.2.4 Alternate equipment for bonding—use spring loaded fixtures and circulating air oven.

3.1.4.2.5 Micrometer capable of measuring 0.0001 in (0.0025 mm).

3.1.4.2.6 Doctor blade (Fig. 7).

3.1.4.3 *Procedure*

3.1.4.3.1 If the adhesive is a liquid, a dry film must be cast; if a tape, use as supplied.

3.1.4.3.2 Pour a portion of the liquid adhesive on a clean glass or metal plate covered with a polyethylene film or directly onto a polytetrafluoroethylene coated plate.

3.1.4.3.3 Draw the doctor blade (Fig. 7) across the adhesive, casting a sufficiently thick wet film to give a dry film 0.008–0.010 in (0.20–0.25 mm) thick.

3.1.4.3.4 In lieu of a specific recommendation, dry 3 h minimum at room temperature or as required to obtain a smooth film. Follow by heating for 20 min at an oven air temperature of 175 ± 5°F (80 ± 3°C).

3.1.4.3.5 Using a circular die, cut a 0.75 in (19 mm) diameter circle from the dried film. Remove the polyethylene film. Measure the film thickness to 0.0001 in (0.0025 mm) with the spring micrometer. Use an average of five readings.

3.1.4.3.6 Place a 1.125 in (28.6 mm) diameter circle of heat resistant cellophane over the center of the 0.250 × 1.125 × 4.250 in (6.3 × 28.6 × 108.0 mm) steel bar used in the Disc Shear Test. Mount the film specimen in the center of the cellophane. Cover the specimen with a second circle of cellophane 1.125 in (28.6 mm) in diameter. Use the 1.125 in (28.6 mm) steel shear disc to complete the assembly.

3.1.4.3.7 Place the assembly in the standard SAE disc bonding press (Fig. 8) or in a spring loaded fixture if an oven is to be used for heating. Load to 100 psi (7.0 kg/cm²). Heat at the recommended temperature and time cycle. In lieu of a specific recommendation, use rate as shown in Fig. 9. As a supplemental test, the pressure and rate of heating could be the same as that used in paragraph 2.2.

3.1.4.3.8 Remove the assembly from the press or oven and cool to room temperature. Remove the adhesive disc from the plate.

3.1.4.3.9 Soak the disc in water to remove the cellophane.

3.1.4.3.10 Measure the new film thickness with a micrometer to the nearest 0.0001 in (0.0025 mm). Use average of five readings. As an alternate method, measure the area using a planimeter.

3.1.4.3.11 Calculate flow by:

$$\text{Flow, \%} = \frac{\text{Original thickness} - \text{New thickness}}{\text{Original thickness}} \times 100$$

3.1.4.3.12 Alternate calculation:

$$\text{Flow, \%} = \frac{\text{New area} - \text{Original area}}{\text{Original area}} \times 100$$

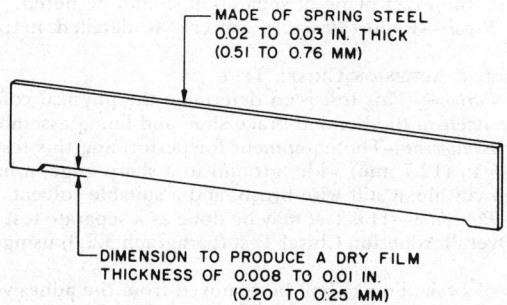

FIG. 7—DOCTOR BLADE

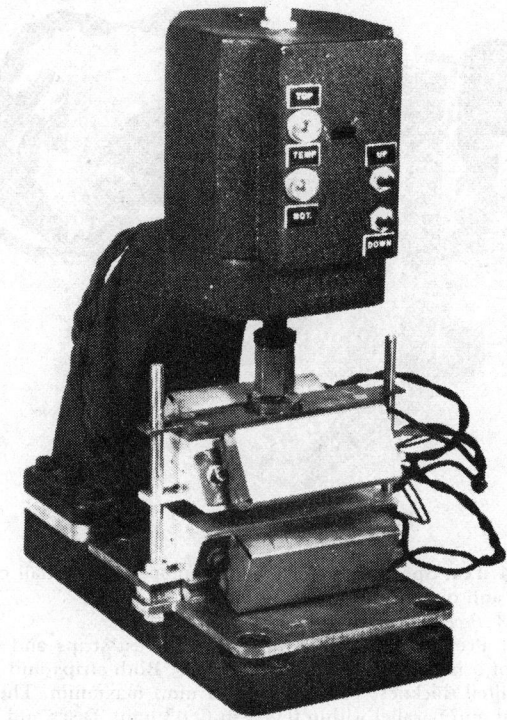

FIG. 8—DISC SHEAR BONDING PRESS

3.1.4.4 *Report*—Percent flow and method of measurement and calculation (based on thickness or area change).

3.1.5 DISC SHEAR TEST

3.1.5.1 *Purpose*—This test is used to determine the shear strength of an adhesive intended for bonding brake linings to brake shoes. Disc shear specimens are tested at room temperature and at elevated temperature.

3.1.5.2 *Equipment*

3.1.5.2.1 Testing Machine—The testing machine shall be capable of compression loading and shall be so selected that the breaking load of the specimens falls between 15 and 85% of the full-scale capacity. The testing machine shall be capable of maintaining a uniform rate of loading of 1200 psi (85 kg/cm²) per minute. This rate of loading will be approximately obtained by a free crosshead speed of 0.05 in (1.27 mm) per minute.

3.1.5.2.2 Disc Shear Fixture—The shear fixture (Fig. 10) consists of a semicircular anvil and a rectangular opening to receive the bonded test disc and strip.

3.1.5.2.3 SAE Bonding Press—A suitable press for bonding the disc shear specimens is shown in Fig. 8. It consists of upper and lower heated platens and an air cylinder for applying pressure during the bonding cycle. The bonding of the disc shear specimens may be done in any suitable manner which conforms to the bonding conditions specified.

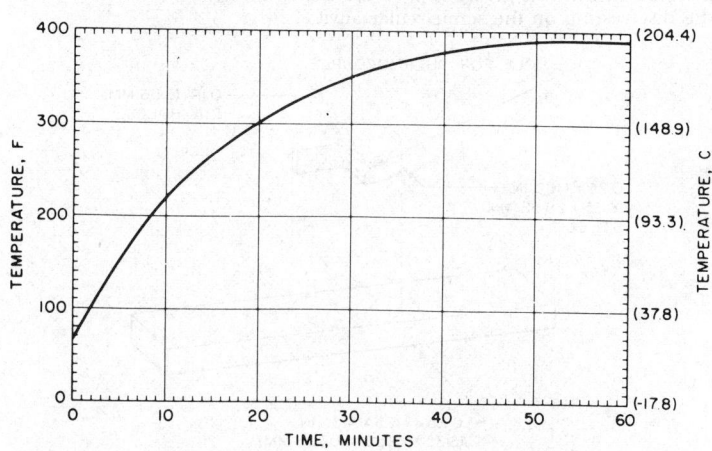

FIG. 9—RATE OF HEATING CURVE

FIG. 10—DISC SHEAR FIXTURE

3.1.5.2.4 *Test Specimens*—Disc and strip specimens shall conform to the shape and dimensions shown in Fig. 11.

3.1.5.3 *Procedure*

3.1.5.3.1 *Preparation of Test Specimens*—Test strips and discs shall be made of a mild steel, such as SAE 1010. Both strips and discs shall have an initial thickness of 0.252 in (6.4 mm) maximum. They shall be ground flat and parallel within 0.001 in (0.03 mm). Discs and strips may be reused by removing the old adhesive and refinishing the surfaces as described, but the thickness shall not be reduced to less than 0.240 in (6.10 mm).

The bonding surfaces of the strips and discs shall be prepared as follows:

(1) Clean with a hot degreasing solvent such as trichlorethylene.

(2) The surfaces to be bonded shall be finished with 180 grit aluminum oxide cloth or grit blasted. (G 40 grit has been found satisfactory.)

(3) Follow with a methyl-ethyl-ketone rinse.

(4) Apply adhesive to the prepared surface immediately.

The adhesive shall be applied as follows:

(1) If a tape adhesive is used, a 1¼ in (31.75 mm) diameter disc of adhesive shall be cut and placed between the disc and the strip. Record dry film thickness.

(2) If a liquid adhesive is used, the adhesive shall be spread on the surface of the disc and the appropriate portion of the strip. The wet film thickness shall be sufficient to produce a dry film thickness of 0.002–0.003 in (0.05–0.08 mm) on each surface. The adhesive on the disc and strip shall be air dried for 3 h minimum. Follow by heating for 20 min at an oven air temperature of 175 ± 5°F (80.4 ± 3°C). The mating surfaces shall then be placed together.

The bonding procedure shall be as follows. The specimens shall be bonded at the recommended temperature, pressure, and bonding time. They shall be allowed to cool to room temperature before shear testing.

3.1.5.3.2 *Testing of Specimens*

(1) Room Temperature Shear—The bonded test specimen shall be inserted in the slot at the top of the shear fixture with the bottom of the disc resting on the semicircular anvil.

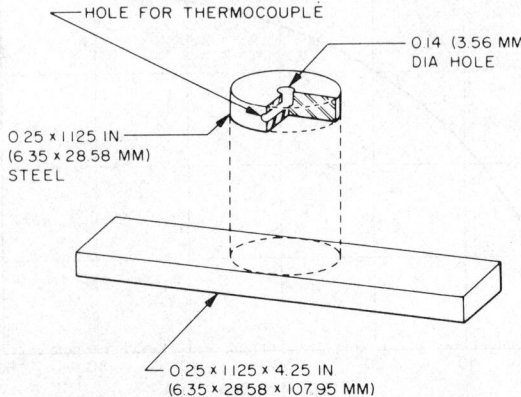

FIG. 11—STEEL DISC AND STRIP

(2) Elevated Temperature Shear—The shear fixture shall be heated to the specified temperature before the specimen is inserted; then the specimen shall be inserted, and when the bond line reaches the specified temperature, the load is applied. A thermocouple inserted in the disc as shown in Fig. 11 is used to check temperature.

3.1.5.4 *Report*—The report shall include the following:

3.1.5.4.1 Complete identification of adhesive tested, source including type, manufacturer's code number, date of manufacture, and test date.

3.1.5.4.2 Bonding apparatus used.

3.1.5.4.3 Metal preparation.

3.1.5.4.4 Air or oven drying time and temperature.

3.1.5.4.5 Dry film thickness for each specimen to nearest 0.001 in (0.03 mm).

3.1.5.4.6 Curing time, temperature, and pressure.

3.1.5.4.7 Rate of curing (time-temperature curve).

3.1.5.4.8 Temperature at which shear test was performed.

3.1.5.4.9 Shear stress at failure for each specimen.

3.1.5.4.10 Nature of failure, including the average estimated percentage of failure in cohesion and adhesion.

3.2 Process Tests

3.2.1 SCOPE—These tests are conducted on the brake shoe and lining assembly to determine the consistency of the previously qualified production process.

3.2.2 BOND PLANE SHEAR TEST

3.2.2.1 *Purpose*—The Bond Plane Shear Test is included in this paragraph on Process Tests, as well as in paragraph 2 on Qualification Tests, to denote that it is to be used as a test of the bonding process on production brake shoe and lining assemblies to determine any variance from a previously approved process.

3.2.2.2 *Equipment*—See paragraph 2.2.2.

3.2.2.3 *Procedure*

3.2.2.3.1 Nondestructive Proof Load—If production assemblies are to be proof tested, select a proper temperature with an appropriate tolerance and apply the proof load up to the specified value after the ram is in contact with the lining. Maintain the proof load for 1½ s minimum.

3.2.2.3.2 For the Ambient Destructive Shear Test on an appropriate statistical sampling procedure, see paragraph 2.2.3.1.

3.2.2.4 *Report*

3.2.2.4.1 Nondestructive Proof Load

(1) Report length, width, and drum diameter of shoe and lining assembly.

(2) Report type of lining, bonding process, and time of sampling.

(3) Report proof load in pounds (kilograms) and testing temperature.

(4) Report any movement of lining relative to the brake shoe table or rim.

3.2.2.4.2 Ambient Destructive Shear Test on Statistical Sample

(1) Report length, width, and drum diameter of shoe and lining assembly.

(2) Report type of lining, bonding process, and type of sampling.

(3) Report failing load in pounds (kilograms).

(4) Report fracture pattern (see paragraph 2.2.5).

(5) Report statistical frequency of test as a percentage of production.

3.2.3 OVERALL ADHESION CHISEL TEST

3.2.3.1 *Purpose*—This test is used to determine the overall quality of lining to brake shoe adhesion with simple hand tools.

3.2.3.2 *Equipment*—The equipment for performing this test consists of a chisel as wide or wider than the lining and ground to a sharp edge, a hammer, and a vise.

3.2.3.3 *Procedure*—The brake shoe and lining assembly is held in the vise and the lining removed from the shoe with the hammer and the wide chisel, starting from the end of the lining with the point of the chisel at the adhesive layer. It is important that the lining be removed from the brake shoe as close as possible to the adhesive line. During the operation, the exact plane of separation should be noted.

3.2.3.4 *Report*—Report fracture pattern as detailed in paragraph 2.2.5.

3.2.4 SPECIFIC ADHESION CHISEL TEST

3.2.4.1 *Purpose*—This test is to determine the physical condition of the adhesive itself in the bonded brake shoe and lining assembly.

3.2.4.2 *Equipment*—The equipment for performing this test consists of a chisel ½ in (12.7 mm) wide, ground to a sharp edge, a hammer, a vise, a rough cut file, a stiff wire brush, and a suitable solvent.

3.2.4.3 *Procedure*—This test may be done as a separate test or it may follow the Overall Adhesion Chisel Test (paragraph 3.2.3) using the same sample.

All traces of brake lining shall be removed from the adhesive layer in a lateral strip, approximately 1 in (25.4 mm) wide, using the chisel, the file, and finally the wire brush. The condition of the adhesive layer should be noted. It is suggested that the Cotton Tack Test (paragraph 3.2.5) be utilized; any tackiness of the adhesive layer should be reported.

3.2.4.4 *Report*—The physical condition of the layer should be reported as to its amount of flow, its wetting to the metal, its continuity or sponginess, and its reaction to the Cotton Tack Test.

3.2.5 COTTON TACK TEST

3.2.5.1 *Purpose*—This test is to determine by simple means the approximate state of cure of some adhesives on a fractured shoe and lining assembly, following Specific Adhesion Chisel Test (paragraph 3.2.4) or other test.

3.2.5.2 *Material*—Long fiber absorbent cotton, an eye dropper, and a suitable solvent are required.

3.2.5.3 *Procedure*

3.2.5.3.1 Expose the adhesive layer as indicated in paragraph 3.2.4.3.

3.2.5.3.2 With a dropper, apply 2 or 3 drops of suitable solvent.

3.2.5.3.3 While the solvent is evaporating, use a small wad of long fiber absorbent cotton to dab (not wipe) the moistened surface repeatedly at a rate of approximately two dabs per second, until the surface is completely dry.

3.2.5.3.4 Blow lightly on the surface to remove any stray cotton fibers which may have simply fallen onto the surface.

3.2.5.3.5 Examine the surface to determine whether any cotton remains stuck to the surface.

3.2.5.4 *Report*

3.2.5.4.1 If there are no cotton fibers stuck to the surface, report "no tack."

3.2.5.4.2 If there are cotton fibers stuck to the surface, report "tack."

3.2.5.5 *Interpretation of Results*

3.2.5.5.1 Tack may be an indication that the adhesive is not completely cured and results should be verified by performing the State of Cure Test (paragraph 3.2.6). NOTE: This test may not properly reflect state of cure on all types of adhesive. It is recommended that the suitability of this test, as well as the proper solvent, be determined.

3.2.6 STATE OF CURE TEST

3.2.6.1 *Purpose*—This test is used to determine the state of cure of the adhesive of a bonded shoe and lining assembly.

3.2.6.2 *Equipment*—The equipment for performing this test consists of a vise, hack saw, state of cure fixture shown in Fig. 12, a small "C" clamp, a pyrometer with a thermocouple, and a gas burner or gas torch.

3.2.6.3 *Procedure*—The brake shoe shall be mounted in a vise and a 1 in² (6.5 cm²) segment of lining is isolated along one edge with a hack saw, making sure the saw reaches the bare metal of the shoe rim. Drill a hole for the thermocouple at the bond line adjacent to the saw cut. Insert the thermocouple wire. With the spring in the loaded position, the state of cure fixture is mounted on the shoe as illustrated in Fig. 12. The sliding punch is brought to bear on the cut lining segment and secured by tightening the nuts. The fixture is then unclamped. This brings the spring load onto the edge of the lining segment. The compressed spring length should be measured and agree with the previously calibrated length for 100 lb (45.4 kg). The "C" clamp is mounted on the spring side of the toggle clamp support and the shoe rim. (NOTE: It is not clamped tightly and is used to prevent the fixture cocking from the shoe.) Heat

FIG. 12—STATE OF CURE FIXTURE

is now applied to the assembly at a uniform rate to permit the bond line to reach the test temperature of 400–420°F (204.4–205.6°C) within 45–60 s. A dwell at the test temperature is required for 2–3 min, thereby assuring the entire area is at temperature.

3.2.6.4 *Results*—If no failure occurs, the adhesive is considered to be properly cured. Incomplete cure is evidenced by any movement of the lining with respect to the shoe rim, providing the cause can be established as an adhesive failure rather than a lining failure.

3.2.6.5 *Special Note*—A soft lining might compress under the 100 lb (45.4 kg) load. In this case, reduce the ratio of the area in shear to the bearing area at the punch by one-half, that is, cut a ½ in² (3.2 cm²) lining segment, 1 in (25.4 mm) wide by ½ in (12.7 mm) deep and reduce the spring load to 50 lb (22.7 kg).

φTRUCK AND BUS GRADE PARKING PERFORMANCE TEST PROCEDURE—SAE J360 OCT88

SAE Recommended Practice

Report of Truck and Bus Brake Committee approved September 1968 and completely revised October 1988. Rationale Statement available.

1. Purpose—This SAE Recommended Practice establishes a uniform procedure for determining the parking performance on a grade by the parking brake systems of new trucks and buses over 10 000 lb (4500 kg) gross vehicle weight rating intended for roadway use.

2. Scope—This practice establishes methods to determine grade parking performance with respect to:

2.1 Ability of the parking brake system to lock the braked wheels.

2.2 The vehicle holding or sliding on the grade, fully loaded or unloaded.

2.3 Applied manual effort.

2.4 Unburnished or burnished brake lining friction conditions.

2.5 Down and up grade directions.

3. Introduction—The ability to hold a vehicle stationary on a grade involves two performance factors: 1) overcoming the downhill grade force with the parking brake system by preventing rotation of the braked wheels, and 2) having sufficient weight on the braked wheels to prevent the vehicle from sliding on the roadway. By the use of this procedure, the manually applied input effort required to prevent braked wheel rotation can be measured and the stability of a vehicle parked on a grade can be observed.

4. Instrumentation

4.1 Force measuring device - 0 - 200 lb (0 - 890 N).
4.2 Decelerometer (0 - 1 G).
4.3 Temperature measuring device - 0 - 1000°F (0 - 540°C).
4.4 Stopwatch.

5. Test Preparations

5.1 On the brakes applied by the parking brake system, use new lining and drums or discs of original equipment material and install in accordance with the vehicle manufacturer's specifications.

5.2 Parking brake assemblies and actuation systems are to be installed, lubricated, adjusted and inspected in accordance with the vehicle manufacturer's specifications.

5.3 The vehicle is to be tested in both the fully loaded and unloaded condition. For the purposes of this test procedure fully loaded shall be the manufacturer's gross vehicle weight rating (GVWR) distributed proportionately to individual axle weight ratings (GAWR's).

Unloaded shall be the fully serviced vehicle weight with no payload, with driver and observer on the front seat, and with test equipment. The sequence of testing may be arranged to suit the desired order of vehicle loading and unloading.

5.4 Tires are to be of the largest diameter specified for the vehicle, new or not more than 20% worn, and inflated to pressures specified by the vehicle manufacturer.

6. Test Notes

6.1 Conduct the test on a dry, smooth Portland cement concrete surface (or other surface of equivalent coefficient of surface friction) that is free from loose materials and has a grade equal to or greater than any specified grade requirement for the test vehicle, as designated in SAE J293.

6.2 Brake drums or discs are to be within a temperature range of 40 - 150°F (4 - 66°C) during the test.

6.3 With variable input systems, conduct the test to establish the applied manual effort or pressure required to lock the braked wheels. With pressure-applied/mechanically locked actuators, make observations after releasing the application pressure.

6.4 With fixed input systems, determine the manual effort required to actuate the system control and observe whether the braked wheels lock or roll.

6.5 For vehicles having any equipment (such as driver-controlled interaxle differentials or multi-speed axles), which can be placed in or out of engagement by the driver so as to vary either the number of axles braked or the amount of torque imposed on the parking brake or brakes, conduct the test to determine parking performance under the condition that requires highest parking brake torque and under the condition that requires highest tire-to-road tractive force.

6.6 Parking brake systems employing service brakes or those employing non-service brakes for which the manufacturer provides the purchaser with a published burnish procedure need not be tested before burnishing because of the difficulty in obtaining reliable and repeatable preburnish data.

6.7 Vehicles equipped with parking brake systems employing cable or rod actuation of service brakes shall be allowed to roll off the grade after each data reading. This will improve repeatability of results by allowing realignment between axle and actuation parts and by removing system hysteresis.

6.8 Method of Conducting On-Grade Testing

6.8.1 Place driver selected equipment (see paragraph 6.5) in desired condition.

6.8.2 Drive the test vehicle up the test grade, stop, and hold with the service brakes.

6.8.3 With vehicle declutched or transmission in neutral, allow the vehicle to creep very slowly under control of the service brakes. Slowly apply the parking brake and, as the parking brake energizes, slowly release the service brakes.

6.8.4 For variable input systems, record the minimum applied effort or pressure to lock the braked wheels. For fixed input systems, record the applied effort or pressure and observe if the braked wheels are rolling or locked. Observe vehicle stability and record whether vehicle holds for at least 5 min or slides on the grade.

6.8.5 Apply service brakes and release the parking brake and repeat test to obtain a minimum of three consistent readings. If the system is a cable or rod actuated service brake, remove vehicle from grade after each reading.

6.8.6 Repeat all steps in the down-grade position.

6.8.7 Repeat all steps for each test condition of driver selected equipment.

6.9 Data sheets should provide for recording the following data: loaded vehicle weight and items making up the weight, unloaded weight, braked axle weight, identification of equipment affecting braked weight, percent grade or grade angle, indication of whether or not brakes were burnished, direction of vehicle on the grade, condition of driver selected equipment, applied input effort, and observation of wheel roll or lock and vehicle hold or slide.

7. Test Procedure

7.1 For vehicles with a parking brake system that does not utilize the service brakes:

7.1.1 Burnish parking brake in accordance with manufacturer's instructions to purchaser. If no instructions are provided, no burnish is to be performed.

7.1.2 Adjust the parking brake and its actuation system in accordance with the vehicle manufacturer's specifications.

7.1.3 Conduct unloaded vehicle test in accordance with Section 6.8.

7.1.4 Conduct loaded vehicle test in accordance with Section 6.8.

7.2 For vehicles with a parking brake system that utilizes friction elements of the service brake system:

7.2.1 Burnish the loaded vehicle in accordance with SAE J880.

7.2.2 Adjust the parking brake and its actuation system in accordance with the vehicle manufacturer's specifications.

7.2.3 Conduct loaded vehicle test in accordance with Section 6.8.

7.2.4 Conduct unloaded vehicle test in accordance with Section 6.8.

8. Reporting Of Performance

8.1 Vehicle parking performance shall be expressed as the amount and type of effort or pressure applied to the parking brake system required to lock the braked wheels and whether the vehicle holds or slides on the grade for: a vehicle weight, a percent grade or grade angle, and the condition of driver-selected equipment. The applied effort or pressure shall be the higher value with respect to up and down grade directions. If pertinent to the parking performance, the items making up vehicle weight and their location should be reported.

φVEHICLE GRADE PARKING PERFORMANCE REQUIREMENTS—SAE J293 OCT88

SAE Recommended Practice

Report of Brake Committee approved June 1972 and completely revised by the Truck and Bus Brake Committee October 1988. Rationale Statement available.

1. Purpose—This SAE Recommended Practice presents on-grade parking performance requirements for new motor vehicles and trailers intended for roadway use.

2. Scope—This practice establishes minimum performance requirements for trucks, buses, truck-tractors, full trailers and semitrailers with gross vehicle weight ratings greater than 10 000 lb (4500 kg) with re-

gard to:
- **2.1** Vehicle classification.
- **2.2** Vehicle load.
- **2.3** Percent grade.
- **2.4** Application force.

3. General Notes—Performance requirements established for this recommended practice are based on test procedures specified in SAE J360 or SAE J1452 for motor vehicles and trailers, respectively.

4. Requirements—With maximum application forces of 125 lb (556 N) for hand-operated or 150 lb (667 N) for foot-operated systems:

4.1 Truck, bus, trailer, and semitrailer parking brake systems shall be capable of holding the vehicle stationary (to the limit of traction on the braked wheels) for 5 min in both up and down grade directions on a 20% grade.

4.2 Truck-tractor parking brake systems shall be capable of holding the vehicle stationary (to the limit of traction of the braked wheels) for 5 min in both up and down grade directions on either a 20% grade (required for 2-axle tractors) or a 15% grade (required for tractors with more than two axles).

4.3 While the parking brake system is applied, the parking brake or brakes shall be held in the applied position with the required effectiveness by energy other than fluid pressure, air pressure or electric energy. Release of the system shall not be possible from the driver's seat unless energy is available for immediate reapplication to the required effectiveness level.

BRAKE TEST CODE— INERTIA DYNAMOMETER—SAE J667

SAE Recommended Practice

Report of Brake Committee approved April 1952 and last revised June 1961.

Purpose—This SAE Recommended Practice establishes a uniform laboratory procedure for evaluating performance and wear of automotive brakes and brake drums by an inertial dynamometer simulation of vehicle test and operating conditions.

Scope—The code is applicable to hydraulic, air, or electrically actuated brakes. For electric brakes, "ampere" values are substituted for "line pressure" where specified in the procedure or on data and curve sheets.

Tests—The procedure includes the following tests:
1. Torque output and deceleration (various speeds and brake input).
2. Evaluation of lining life (wear tests).
3. Brake characteristics (fade and recovery tests, noise, chatter, grabbiness).
4. Drum evaluation (wear, heat checking, scoring).

Graphs and Charts—The following graphs are plotted and charts compiled on standard forms:
1. Torque output versus brake input and deceleration versus brake input graphs (effectiveness tests). See Figs. 3 and 4.
2. Characteristic torque graphs from effectiveness tests at maximum line pressures (individual stops). See Fig. 5.
3. Lining fade and recovery graphs (deceleration and temperature versus stop number). See Fig. 6.
4. Lining wear tabulation—wear per test phase and total test. See Fig. 7.
5. Drum—statement of condition (and wear, if any). See Fig. 7.

Test Conditions—Actual tests for performance shall be started when preliminary preparations and lining "break in" have been completed. See Procedure.

Dynamometer inertia shall correspond with the work load imposed on the particular brake as it is to work in the specific vehicle.

Instrumentation—The dynamometer shall be capable of speed control within ±2% of designated speeds (see Procedure) and inertia loading within ±5% of that designated under Test Conditions.

The dynamometer shall be equipped with accurate means for determining torque output of the brake. These shall include both direct torque recording instruments and devices for indicating the average flywheel deceleration which can be translated into brake torque output. Means shall be provided for accurate control and recording of input pressure (line pressure) to the brake wheel cylinder.

The dynamometer shall be equipped with a blower for control of drum temperature within the limits specified and a pyrometer which records drum temperature as described under Procedure. A direct reading pyrometer may be used for observation of brake lining temperature.

It is desirable that the dynamometer be equipped with a means to measure the volumetric fluid requirements of hydraulic brakes.

It is also desirable that the dynamometer be equipped with a torque balancing device or mechanism to maintain specified deceleration rates during wear tests.

Procedure

1. Preliminary Preparations—Before the brake unit is assembled to the dynamometer for test, the lining should be attached and ground per manufacturer's specifications.

The brake shoe and lining assembly thickness shall be measured accurately at established toe, center and heel points on each side of each shoe. Record measurements, see Fig. 7.

New drums are recommended for each test, surface finish in accordance with manufacturer's recommendations, with careful attention to insure uniform surface finish from test to test. Make and record necessary measurements for drum wear determination.

The brake drum shall be washed thoroughly on the braking surface with a suitable solvent to remove contaminants.

Mount the brake and drum on the dynamometer, center, and record runout.

The drum thermocouple junction shall be located 0.090 in. from the braking surface, and in the center of the brake track.

Complete Master Forms 1 and 2 (Figs. 1 and 2). This requires calculation of flywheel load and speeds.

2. Initial Lining Break In—Adjust brake shoes to drum clearance recom-

FIG. 1—MASTER SHEET NO. 1

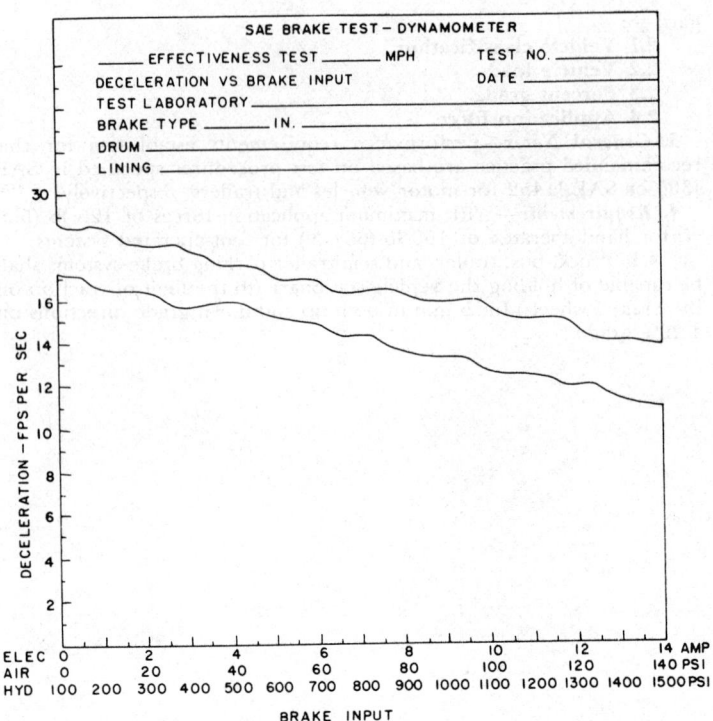

FIG. 3—DECELERATION VERSUS BRAKE INPUT

mended by the manufacturer. Record clearance. Regulate line pressure to maintain a deceleration of approximately 11 feet per second per second (fpsps). Regulate the blower air velocity and stop cycle to maintain a starting drum temperature of 125-150 F (200 F for truck brakes). Make a sufficient number of stops from 40 mph to obtain a minimum of 80% lining contact. After completion of break in, remove shoes and repeat lining measurements. Record results and general condition of lining. Reassemble shoes, adjust and record clearance.

3. First Effectiveness Test—Run effectiveness tests at 30, 50, and 70 mph (10 mph increments optional) on passenger car brakes, and 30, 50, and 60 mph (10 mph increments optional) on truck brakes. Make one stop at each speed and pressure. Use line pressures in increments of 100 psi or multiples thereof, using a minimum of 5 points through the range to determine the characteristics. Blower shall be regulated to provide air velocity of 2200 fpm (25 mph) at the brake, distributed uniformly and continuously over the projected area during test, using air at room temperature. With drum rotating, cool the brake drum to 125-150 F between successive stops. Inspect brake linings and drum surface, and record observations.

4. First Fade and Recovery Test (50 mph)—Establish line pressures to give a deceleration of 15 fpsps at both 30 and 50 mph prior to start of fade test. Cool drums to 125 F between trial stops. Blower shall be regulated to provide air velocity of 2200 fpm (25 mph) at the brake, distributed uniformly and continuously over the projected area during test, using air at room temperature. Make ten 50 mph stops using established line pressures at a cycle time of 1 minute, accelerating to 50 mph test speed at the end of each stop. Reset line pressure to 30 mph value (previously established) and, after 3 minutes cooling time with drum rotating at 30 mph equivalent, make ten

FIG. 2—MASTER SHEET NO. 2

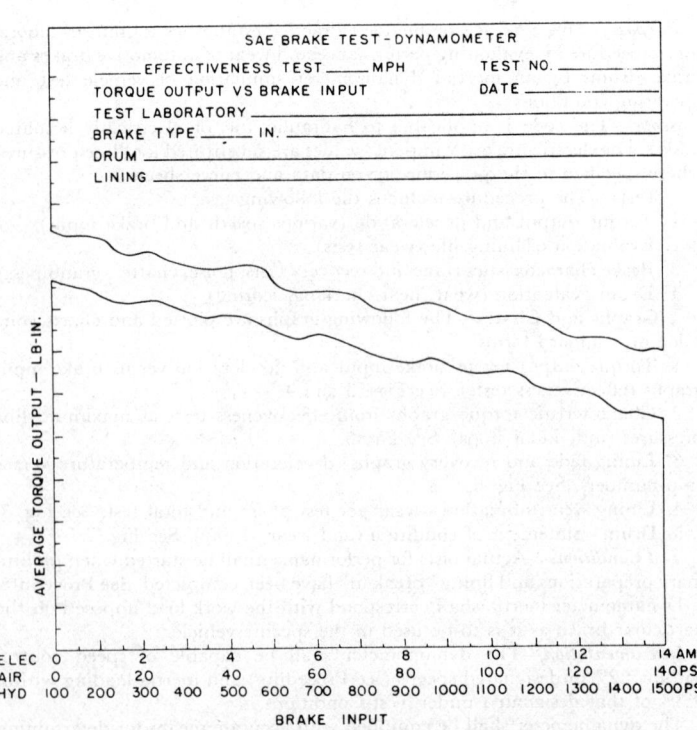

FIG. 4—TORQUE OUTPUT VERSUS BRAKE INPUT

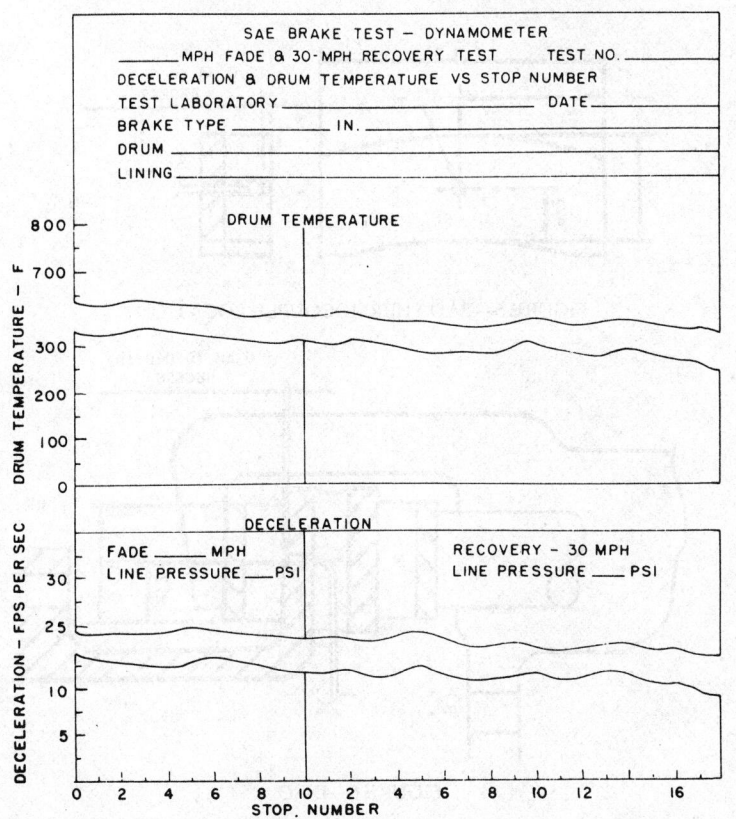

FIG. 5—TORQUE CHARACTERISTIC CURVES

FIG. 6—DECELERATION AND DRUM TEMPERATURE VERSUS STOP NUMBER

FIG. 7—LINING AND DRUM WEAR DATA

30 mph stops at a cycle time of 1 minute, accelerating to 30 mph at the completion of each stop. If lining does not show definite recovery in 10 stops at 1 minute intervals, continue to 15 stops. Record deceleration for each stop. Inspect brake linings and drum surface, and record observations.

5. Second Fade and Recovery Test (70 mph-Passenger Car, 60 mph-Truck)—Use same procedure as in Test 4 establishing line pressures for applicable test speed. Make 5 fade stops from test speed. Recovery is same as in Test 4. Inspect brake linings and drum surface and record observations.

6. Second Effectiveness Test—Use same procedure as in Test 3. Inspect brake lining and drum surface, and record observations.

7. Lining Wear Test, Glazing (30 mph)—Measure and record lining thickness at exposed edges of shoes at beginning and end of test. (See Fig. 7). Make 500 stops at 30 mph. Adjust line pressure or regulate torque balancing mechanism to maintain a deceleration of approximately 11 fpsps. Obtain line pressure graph and record drum temperature every 10 stops. Use blower air velocity and stop cycle to maintain cycle starting temperature of 200 F.

8. Lining Wear Test (40 mph)—Use procedure as in Test 7, except that there shall be 300 stops at 40 mph and the cycle starting temperature shall be 300 F. Measure and record lining thickness at exposed edges of shoes at end of each test. See Fig. 7.

9. Third Effectiveness Test—Use procedure as in Test 3. Inspect brake linings and drum surface, and record observations. See Fig. 7.

10. Third Fade and Recovery Test (50 mph)—Repeat fade and recovery test as in Test 4.

11. Fourth Fade and Recovery Test (70 mph-Passenger Car, 60 mph-Truck)—Repeat fade and recovery test as in Test 5.

12. Lining Wear Test (60 mph)—Use procedure as in Test 7 (including measuring and recording of lining thickness at exposed edges of shoes at beginning and end of test), except that there shall be 100 stops at 60 mph and the cycle starting temperature shall be 350 F.

13. Lining Wear Evaluation—Complete test. Remove shoes from brake assembly. Measure linings at all designated points. Subtract from initial measurements, and record total wear.

14. Drum Evaluation—Measure drum to determine surface wear and record. Comments on drum condition requested on Lining and Drum Wear Data Sheet (Fig. 7) should indicate surface change resulting from various test phases as noted at measurement and observation points specified in the procedure.

BRAKE PERFORMANCE AND WEAR TEST CODE COMMERCIAL VEHICLE INERTIA DYNAMOMETER—SAE J2115 JUN93

SAE Standard

Report of the Truck and Bus Foundation Brake Subcommittee of the SAE Truck and Bus Brake Committee approved June 1993. Rationale statement available.

1. **Scope**—This SAE Standard provides test procedures for air, air/hydraulic, and hydraulic drum and disc brakes used on highway commercial vehicles over 4536 kg (10 000 lb) GVWR.

   1.1 **Purpose**—The purpose of this code is to provide a procedure for evaluating performance and wear of a brake and drum or disc when tested on an inertia-type brake dynamometer.

2. **References**—There are no referenced publications specified herein.

3. **Test**—The procedures include the following:
   a. Effectiveness tests at various conditions and speeds
   b. Fade and recovery
   c. Wear and effectiveness at various temperatures

4. **Instrumentation, Equipment, and Test Conditions**

   4.1 The dynamometer inertia should be equivalent to the maximum loading conditions to which the brakes are normally subjected. Rotation speeds should be established based on revolutions per kilometer (mile) for the tires normally used to carry such wheel loads. Dynamometer inertia is to be based on the maximum static radius and half the GAWR.

   4.2 **Thermocouples**—Install thermocouples in drum or disc and in lining (optional) as shown in Figures 1 to 6.

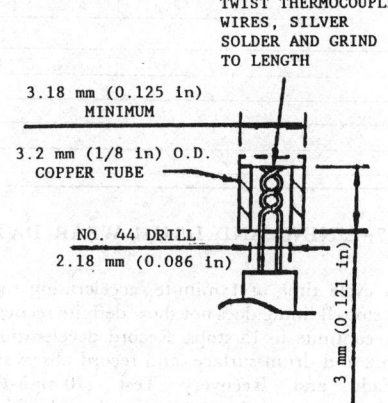

FIGURE 1—THERMOCOUPLE CONSTRUCTION

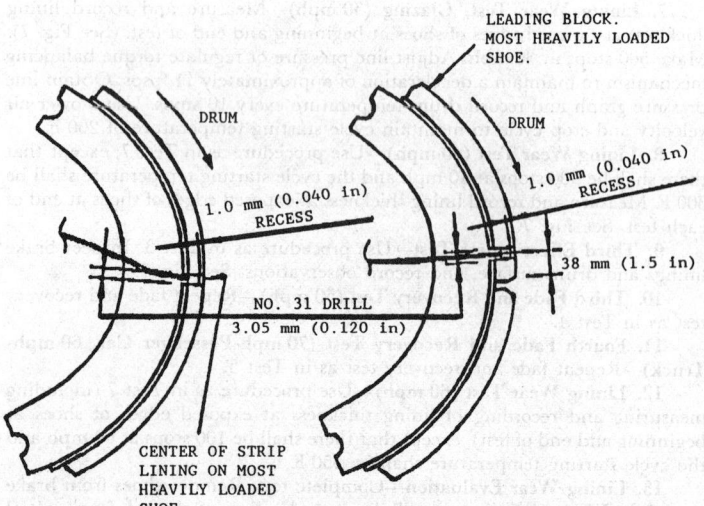

FIGURE 2—BRAKE SHOE THERMOCOUPLE INSTALLATION

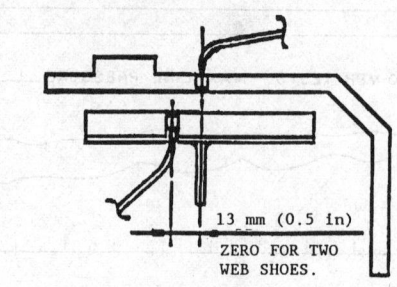

FIGURE 3—DRUM THERMOCOUPLE AND SINGLE WEB SHOE THERMOCOUPLE LOCATION

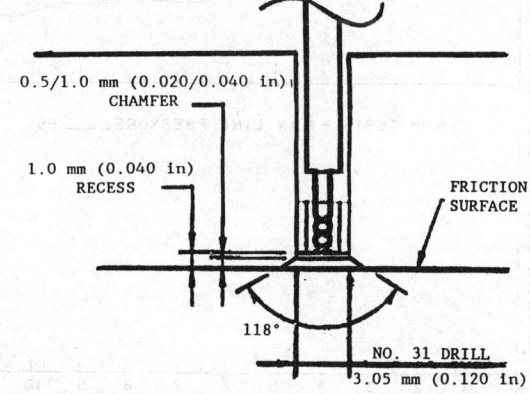

FIGURE 4—DRUM OR DISC THERMOCOUPLE INSTALLATION

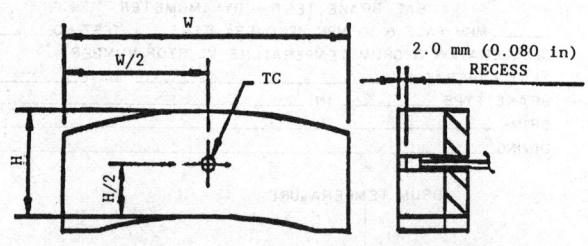

FIGURE 5 PAD THERMOCOUPLE LOCATION

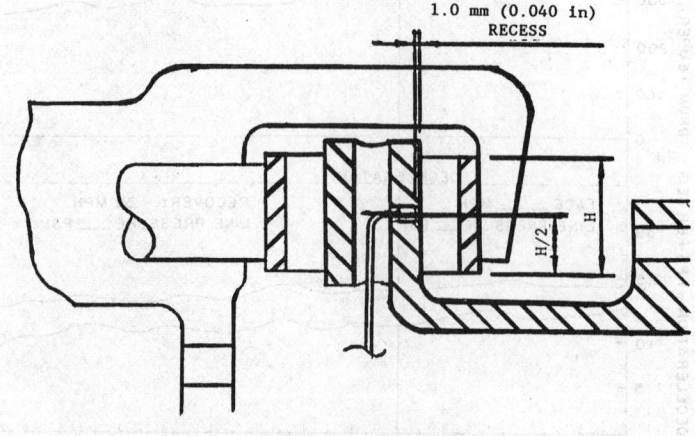

FIGURE 6—DISC

4.3 Warmup—If required, make a series of 48 to 0 km/h (30 to 0 mph) stops at 3 mpsps (10 fpsps) to obtain the initial drum or disc temperature of 93 to 121 °C (200 to 250 °F).

4.4 Cooling Speed—Unless otherwise specified, cooling speed is at the speed of the next stop.

4.5 Air Flow—Unless otherwise specified, air at ambient temperature is directed uniformly and continuously over the brake drum or disc at a velocity of 670 m/min (2200 ft/min).

4.6 Ambient temperature is to be between 15 and 40 °C (59 and 104 °F).

4.7 Chamber air pressure rise rate is to be 1.5 MPa/s ± 0.3 MPa/s (220 psi/s ± 45 psi/s).

4.8 Tests described herein are for air or air/hydraulic brakes. For hydraulic brakes, all effectiveness tests should be run from 1.40 to 12.4 MPa (200 to 1800 psi) at 1.4 MPa (200 psi) increment.

5. Performance Procedure

5.1 The lining should be attached and ground per manufacturer specifications. New drums or discs are recommended for each test. Surface finish is to be in accordance with manufacturer recommendations with careful attention to insure uniform surface finish from test to test. Measure swept surface hardness at 3 points, 120 degrees apart. Mount the brake and drum or disc assembly on the dynamometer, center, and record runout. Maximum runout is 0.254 mm (0.010 in) T.I.R. at open end of drum wear surface or 0.254 mm (0.010 in) at outer radius of disc wear surface. Measure lining and drum or disc as in 6.1.

5.2 Preburnish Effectiveness—Adjust brake to manufacturer's specifications. Warm brake to initial drum or disc temperature of 93 to 121 °C (200 to 250 °F). Make an 80 km/h (50 mph) stop at 69 kPa (10 psi). Cool brake at 48 km/h (30 mph). Make successive stops at 138, 276, 414, 552, and 690 kPa (20, 40, 60, 80, and 100 psi) with an initial drum or disc temperature of 93 to 121 °C (200 to 250 °F).

5.3 Preburnish Static Torque—With drum or disc temperature 24 to 52 °C (75 to 125 °F), measure static breakaway torque and actuator stroke in forward and reverse direction with calibrated actuator. Repeat three times. Measure static breakaway torque in forward and reverse directions at 276, 414, 552, and 690 kPa (40, 60, 80, and 100 psi).

5.4 Burnish—Make 200 stops from 64 km/h (40 mph) at 3.0 mpsps (10 fpsps) with an initial drum or disc temperature of 177 to 204 °C (350 to 400 °F). Make 200 additional stops at 260 to 288 °C (500 to 550 °F). Inspect lining and drum or disc and record percentage of lining area contact of each shoe and pad. Make sketch or photo of contact.

5.5 Postburnish Static Torque—Adjust brakes to manufacturer specifications. Repeat 5.3.

5.6 80 km/h (50 mph) First Effectiveness—Warm brake to initial drum or disc temperature of 93 to 121 °C (200 to 250 °F). Make an 80 km/h (50 mph) stop at 69 kPa (10 psi). Cool brake. Make successive stops at 103, 138, 276, 414, 552, and 690 kPa (15, 20, 40, 60, 80, and 100 psi) with an initial drum or disc temperature of 93 to 121 °C (200 to 250 °F).

5.7 32 km/h (20 mph) First Effectiveness—Repeat 5.6 except at 32 km/h (20 mph).

5.8 96 km/h (60 mph) First Effectiveness—Repeat 5.6 except at 96 km/h (60 mph).

5.9 Recovery Baseline—Make three 48 km/h (30 mph) stops at 3.6 mpsps (12 fpsps) from drum or disc temperature of 93 to 121 °C (200 to 250 °F).

5.10 Fade—Warm brake to initial drum or disc temperature of 93 to 121 °C (200 to 250 °F). Make 10 snubs from 80 to 24 km/h (50 to 15 mph) at 2.7 mpsps (9 fpsps) with 72 s intervals between start of each snub with pressure limited to 690 kPa (100 psi).

5.11 Hot Stop—One minute after end of last fade snub deceleration make a 32 km/h (20 mph) stop attempting to maintain 4.3 mpsps (14 fpsps) with pressure limited to 690 kPa (100 psi).

5.12 Recovery—Two minutes after end of hot stop decelerations, make 20 stops from 48 km/h (30 mph) at a deceleration rate of 3.6 mpsps (12 fpsps) at 1 min intervals between the start of each stop.

5.13 Reburnish—Make 50 stops from 64 km/h (40 mph) at 3.0 mpsps (10 fpsps) with an initial drum or disc temperature of 260 to 288 °C (500 to 550 °F).

5.14 80 km/h (50 mph) Second Effectiveness—Repeat 5.6.

5.15 32 km/h (20 mph) Second Effectiveness—Repeat 5.7.

5.16 96 km/h (60 mph) Second Effectiveness—Repeat 5.8.

6. Wear and Third Effectiveness Test

6.1 Preparation—(Use lining from previous test.) The brake shoe and lining assembly thickness shall be measured at established toe, center, and heel points on each side of each shoe. Weigh the shoe and lining assembly. Measure maximum and minimum diameter or thickness at the center of swept width of drum or disc. Location of previous measurements must be permanently marked so that future measurements will be taken at the same location.

Measure swept surface hardness at 3 points, 120 degrees apart. Mount the brake and drum assembly on the dynamometer, center, and record runout. Maximum runout is 0.254 mm (0.010 in) T.I.R. at open end of drum wear surface or 0.254 mm (0.010 in) at outer radius of disc wear surface.

Record line pressure at every 100th snub. Measure lining and drum or disc at conclusion of snub cycle. Record lining and drum or disc condition including comments regarding noise, chatter, grabbiness, heat checks, and scoring. Adjust brakes to manufacturer specifications.

6.2 Wear and Effectiveness Tests—Run 64 to 24 km/h (40 to 15 mph) at 2.1 mpsps (7 fpsps). Run snub cycles as listed in Table 1:

Before demounting assembly for measurements and after each snub cycle:
 a. Let brake cool to ambient.
 b. Adjust brake to manufacturer specifications.
 c. Warm up brake to snub cycle temperature using the wear test procedure.
 d. Repeat 5.14, 5.15, and 5.16 except run at the temperature specified in 6.2c.

TABLE 1—SNUB CYCLE REQUIREMENTS

Drum or Disc Temperature	Snub
121 °C (250 °F)	1000
177 °C (350 °F)	1000
246 °C (475 °F)	1000
316 °C (600 °F)	1000
399 °C (750 °F)	500[1]
121 °C (250 °F)	1000

[1] For disc brake only

BRAKE SYSTEM ROAD TEST CODE—PASSENGER CAR AND LIGHT-DUTY TRUCK—SAE J843 NOV90

SAE Recommended Practice

Report of Brake Committee approved January 1963 and last revised May 1971. Editorial change March 1973. Reaffirmed by the SAE Brake Standards Committee 2 November 1990.

Foreword—This reaffirmed document has been changed only to reflect the new SAE Technical Standards Board format.

1. Scope—This SAE Recommended Practice establishes a uniform procedure for the level road test of the brake systems of new light-duty trucks and new multipurpose passenger vehicles[1] up to and including 2700 kg (6000 lb) gvw and all classes of new passenger cars.

1.1 Purpose—The purpose of the test code is to establish brake system capabilities with regard to:

1.1.1 Deceleration versus input, as affected by vehicle speed, brake temperature, and usage
1.1.2 Brake system integrity
1.1.3 Stopping ability during emergency or inoperative power assist conditions
1.1.4 Water recovery characteristics

2. References—There are no referenced publications specified herein.

3. Instrumentation
  3.1 Line pressure or pedal force gage
  3.2 Decelerometer
  3.3 Direct reading temperature instrument
  3.4 Speedometer (calibrated) or fifth wheel pousometer
  3.5 Tire pressure gage
  3.6 Odometer (calibrated)
  3.7 Thermometer—ambient (or ambient sensitive thermocouple)
  3.8 Stopmeter (fifth wheel, distance only)
  3.9 Optional instrumentation
    a. Pedal travel gage
    b. Solenoid stop counter
    c. Stopwatch

4. Installation Details

4.1 Friction Material Preparation—Attach and finish friction material per vehicle manufacturer's specifications.

4.2 Thermocouples—Install plug-type thermocouple in each brake. See Figure 1.

4.3 Brake Drum (or Rotor) and Hub Assembly—New drums (or rotors) recommended. Surface finish, dimensional characteristics, with special emphasis on runout of rubbing surface, shall be in accordance with vehicle manufacturer's specifications.

4.4 Brake Assembly—Brakes shall be prepared in accordance with vehicle manufacturer's specification with special attention to required load characteristics on all brake springs. New linings are recommended on all brakes. Adjust brakes to manufacturer's specifications.

4.5 Vehicle Test Weight—Vehicle manufacturer's recommended axle test loading[2] shall be maintained throughout full test procedure except during the minimum load test (5.8).

5. Test Procedure

5.1 Test Notes

5.1.1 Effectiveness, fade, and recovery test stops shall be conducted on a substantially level (not to exceed a ±1% grade), dry, smooth, hardsurfaced roadway of Portland cement concrete (or other surface with equivalent coefficient of surface friction) that is free from loose materials.

5.1.2 During all phases of this procedure, any unusual performance such as wrap-up or noise characteristics are to be noted and recorded. Note any uncontrollable braking action causing the vehicle to pull or swerve out of a 3.7 m (12 ft) wide roadway lane.

5.1.3 "Initial brake temperature" is defined as 0.3 km (0.2 mile) before stop (average temperature of brakes on hottest axle), brakes off.

5.1.4 If brakes require warming to prescribed temperature, use burnish procedure and shorten interval if necessary.

5.1.5 Because variations in ambient temperature have a significant effect on test results, fade and recovery tests must be conducted within a range of ambient temperature of 4.4 to 32.2 °C (40 to 90 °F).

5.1.6 Decelerations used in the various fade, recovery, or warmup procedures refer to values at which the decelerometer is held approximately constant during the stop by varying the input pressure.

5.1.7 Deceleration and line pressure (pedal force) readings shall not be taken below 8 km/h (5 mph).

5.1.8 On vehicles with manual transmissions, disengage clutch below 16 km/h (10 mph).

5.2 Preburnish Check—In order to allow for a general check of instrumentation, brakes, and vehicle function, the following stops are to be run: 10 stops, 48-0 km/h (30-0 mph), 3 m/s^2 (10 ft/s^2), 1.6 km (1 mile) interval, 64 km/h (40 mph) cooling speed in normal driving gear.

Record—Maximum line pressure (pedal force).

NOTE: Assuming instrumentation, brakes, and vehicle are functioning satisfactorily, proceed immediately with First Effectiveness Test.

5.3 First (Preburnish) Effectiveness Test—Initial brake temperature, 93.3 °C (200 °F) before each application.

5.3.1 STOP SPEED—48 and 97 km/h (30 and 60 mph) (full stops in neutral).

5.3.2 INCREMENTS—Curve to be defined to point of incipient skid by adequate number of points.

5.3.3 RECORD—Deceleration and line pressure (pedal force) and method of brake application (that is, machine or manual). When using manual method, full stops shall be defined by maximum line pressure (pedal force) and minimum deceleration. Also note, at the appropriate stop, which wheel or wheels skidded.

5.4 Burnish

5.4.1 STOP SPEED—64 to 0 km/h (40 to 0 mph).
5.4.2 STOP DECELERATION—3.7 m/s^2 (12 ft/s^2) (in normal driving gear).
5.4.3 STOP INTERVAL—As required to achieve 121 °C (250 °F) "initial brake temperature"[3] or a maximum of 1.6 km (1 mile).

---

[1] Multipurpose passenger vehicle means a motor vehicle with power, except a trailer, designed to carry 10 persons or less, which is constructed either on a truck chassis or with special features for occasional off-road operations.
[2] For light truck—Manufacturer's gvw rating not to be exceeded.
For passenger car—Normally curb weight plus 270 kg (+600 lb) for vehicles four or more passengers.

[3] See Test Notes, 5.1.3.

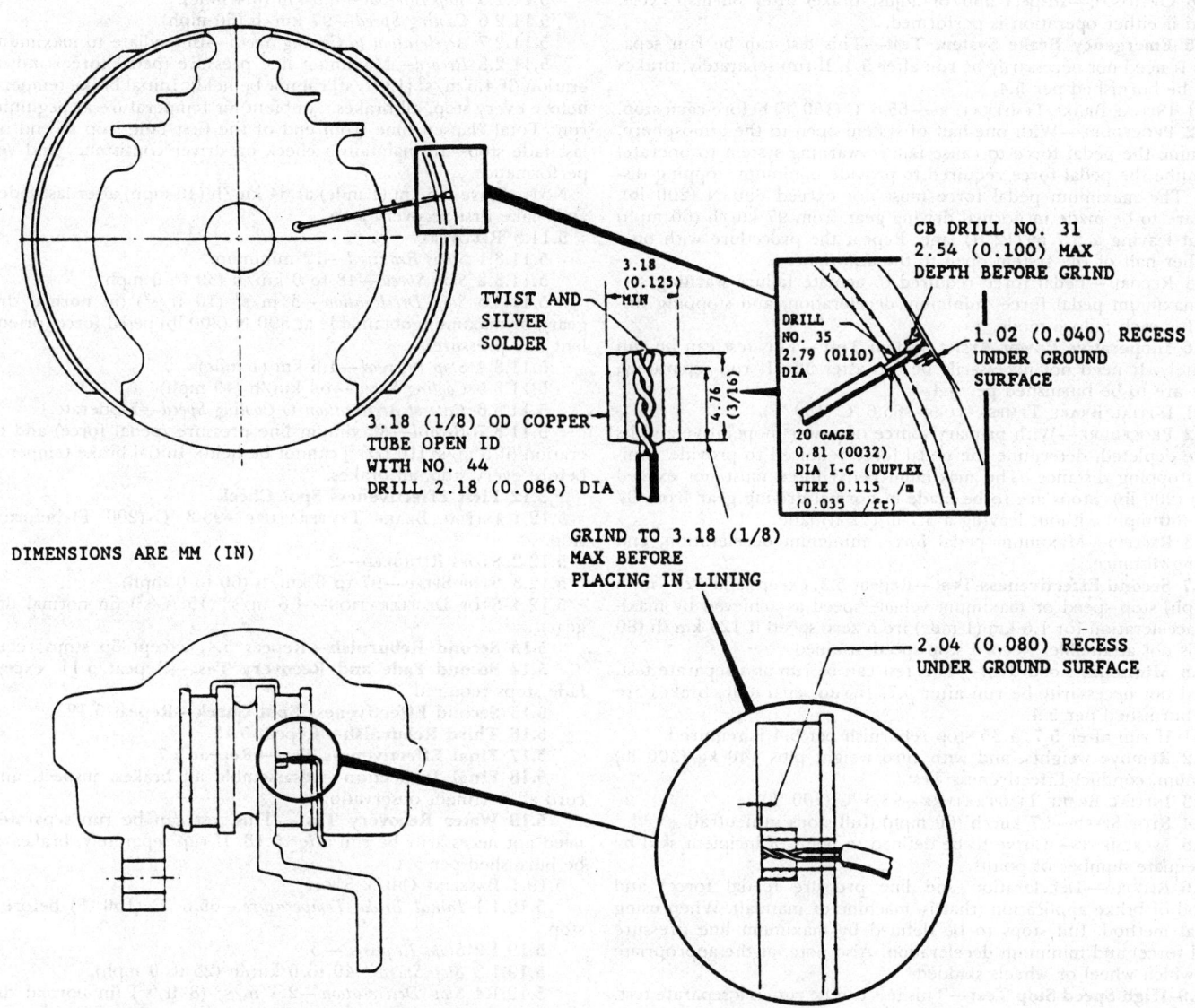

FIGURE 1—TYPICAL PLUG THERMOCOUPLE INSTALLATIONS

NOTE: The 1.6 km (1 mile) maximum must be observed even though the initial temperature exceeds 121 °C (250 °F).

5.4.4 COOLING SPEED—64 km/h (40 mph) (moderate acceleration to cooling speed).

5.4.5 STOPS REQUIRED—200.

5.4.6 OPTIONAL—Inspect and/or adjust brakes after burnish cycle. Record if either operation is performed.

5.5 Emergency Brake System Test—This test can be run separately. It need not necessarily be run after 5.4. If run separately, brakes are to be burnished per 5.4.

5.5.1 INITIAL BRAKE TEMPERATURE—65.6 °C (150 °F) before each stop.

5.5.2 PROCEDURE—With one-half of system open to the atmosphere, determine the pedal force to cause failure warning system to operate. Determine the pedal force required to provide minimum stopping distance. The maximum pedal force must not exceed 890 N (200 lb). Stops are to be made in normal driving gear from 97 km/h (60 mph) without leaving a 3.7 m (12 ft) lane. Repeat the procedure with only the other half of the system open to the atmosphere.

5.5.3 RECORD—Pedal force required to actuate failure warning system, maximum pedal force, minimum deceleration, and stopping distance for each failure mode.

5.6 Inoperative Power Assist System Test—This test can be run separately. It need not necessarily be run after 5.5. If run separately, brakes are to be burnished per 5.4.

5.6.1 INITIAL BRAKE TEMPERATURE—65.6 °C (150 °F).

5.6.2 PROCEDURE—With primary source of power inoperative and its reserve depleted, determine the pedal force required to provide minimum stopping distance. The maximum pedal force must not exceed 890 N (200 lb). Stops are to be made in normal driving gear from 97 km/h (60 mph) without leaving a 3.7 m (12 ft) lane.

5.6.3 RECORD—Maximum pedal force, minimum deceleration, and stopping distance.

5.7 Second Effectiveness Test—Repeat 5.3, except add 129 km/h (80 mph) stop speed or maximum vehicle speed as achieved by maximum acceleration for 1.6 km (1 mile) from zero speed if 129 km/h (80 mph) is not attainable. Record stop speed attained.

5.8 Minimum Load Test—This test can be run as a separate test. It need not necessarily be run after 5.7. If run separately, brakes are to be burnished per 5.4.

5.8.1 If run after 5.7, a 35 stop reburnish per 5.4 is required.

5.8.2 Remove weights, and with curb weight plus 140 kg (300 lb) maximum, conduct Effectiveness Test.

5.8.3 INITIAL BRAKE TEMPERATURE—93.3 °C (200 °F).

5.8.4 STOP SPEED—97 km/h (60 mph) (full stops in neutral).

5.8.5 INCREMENTS—Curve to be defined to point of incipient skid by an adequate number of points.

5.8.6 RECORD—Deceleration and line pressure (pedal force) and method of brake application (that is, machine or manual). When using manual method, full stops to be defined by maximum line pressure (pedal force) and minimum deceleration. Also note, at the appropriate stop, which wheel or wheels skidded.

5.9 High Speed Stop Test—This test can be run as a separate test. It need not necessarily be run after 5.8. If run separately, brakes are to be burnished per 5.4.

5.9.1 Conduct at original test weight per 4.5.

5.9.2 INITIAL BRAKE TEMPERATURE—65.6 °C (150 °F).

5.9.3 STOPS REQUIRED—1.

5.9.4 STOP SPEED—As achieved by maximum obtainable acceleration for 1.6 km (1 mile) from zero speed but not to exceed 100 mph (161 km/h).

5.9.5 STOP DECELERATION—4.6 m/s² (15 ft/s²) in normal driving gear or maximum attainable at 890 N (200 lb) pedal force.

5.9.6 RECORD—Stop speed, maximum line pressure (pedal force), and deceleration (if 4.6 m/s² [15 ft/s²] cannot be held).

5.10 First Reburnish—Repeat 5.4, except 35 stops required.

5.11 First Fade and Recovery Test

5.11.1 BASELINE CHECK STOPS

5.11.1.1 *Initial Brake Temperature*—65.6 °C (150 °F) before each stop.

5.11.1.2 *Stops Required*—3.

5.11.1.3 *Stop Speed*—48 to 0 km/h (30 to 0 mph).

5.11.1.4 *Stop Deceleration*—3 m/s² (10 ft/s²) (in normal driving gear).

5.11.1.5 *Record*—Maximum line pressure (pedal force).

5.11.2 FADE

5.11.2.1 *Initial Brake Temperature*—65.6 °C (150 °F) before first stop.

5.11.2.2 *Stops Required*—10.

5.11.2.3 *Stop Speed*—97 to 0 km/h (60 to 0 mph).

5.11.2.4 *Stop Deceleration*—4.6 m/s² (15 ft/s²) (in normal driving gear) or maximum obtainable at 890 N (200 lb) pedal force (or equivalent line pressure).

5.11.2.5 *Stop Interval*—0.6 km (0.4 mile).

5.11.2.6 *Cooling Speed*—97 km/h (60 mph).

5.11.2.7 *Acceleration to Cooling Speed*—Immediate to maximum.

5.11.2.8 *Record*—Maximum line pressure (pedal force) and deceleration (if 4.6 m/s² [15 ft/s²] cannot be held). Initial brake temperature before every stop, all brakes. Ambient air temperature at beginning of run. Total elapsed time from end of the first fade stop to end of the last fade stop—to maintain a check on driver consistency and vehicle performance.

NOTE: Drive 1.6 km (1 mile) at 64 km/h (40 mph) after last fade stop and make first recovery stop.

5.11.3 RECOVERY

5.11.3.1 *Stops Required*—12 minimum.

5.11.3.2 *Stop Speed*—48 to 0 km/h (30 to 0 mph).

5.11.3.3 *Stop Deceleration*—3 m/s² (10 ft/s²) (in normal driving gear), or maximum obtainable at 890 N (200 lb) pedal force (or equivalent line pressure).

5.11.3.4 *Stop Interval*—1.6 km (1 mile).

5.11.3.5 *Cooling Speed*—64 km/h (40 mph).

5.11.3.6 *Rate of Acceleration to Cooling Speed*—Moderate.

5.11.3.7 *Record*—Maximum line pressure (pedal force) and deceleration (if 3 m/s² [10 ft/s²] cannot be held). Initial brake temperatures before every stop, all brakes.

5.12 First Effectiveness Spot Check

5.12.1 INITIAL BRAKE TEMPERATURE—93.3 °C (200 °F) before each stop.

5.12.2 STOPS REQUIRED—2.

5.12.3 STOP SPEED—97 to 0 km/h (60 to 0 mph).

5.12.4 STOP DECELERATION—4.6 m/s² (15 ft/s²) (in normal driving gear).

5.13 Second Reburnish—Repeat 5.4, except 35 stops required.

5.14 Second Fade and Recovery Test—Repeat 5.11, except 15 fade stops required.

5.15 Second Effectiveness Spot Check—Repeat 5.12.

5.16 Third Reburnish—Repeat 5.13.

5.17 Final Effectiveness Test—Repeat 5.7.

5.18 Final Inspection—Disassemble all brakes, inspect, and record all pertinent observations.

5.19 Water Recovery Test—This test can be run separately. It need not necessarily be run after 5.18. If run separately, brakes are to be burnished per 5.4.

5.19.1 BASELINE CHECK STOPS

5.19.1.1 *Initial Brake Temperature*—65.6 °C (150 °F) before each stop.

5.19.1.2 *Stops Required*—3.

5.19.1.3 *Stop Speed*—40 to 0 km/h (25 to 0 mph).

5.19.1.4 *Stop Deceleration*—2.4 m/s² (8 ft/s²) (in normal driving gear).

5.19.1.5 *Record*—Maximum line pressure (pedal force) for each stop.

5.19.2 WETTING OF BRAKES

5.19.2.1 *Wetting Time*—2 min minimum.

5.19.2.2 *Wetting Procedure*—With the brakes fully released, wet all brakes thoroughly by slowly driving through a trough of suitable depth or equivalent method. Start recovery stops not more than 1 min after wetting brakes. Do not exceed 40 km/h (25 mph) prior to recovery stops.

5.19.3 WATER RECOVERY STOPS

5.19.3.1 *Stop Speed*—40 to 0 km/h (25 to 0 mph).

5.19.3.2 *Speed Between Stops*—40 km/h (25 mph).

5.19.3.3 *Stop Deceleration*—2.4 m/s² (8 ft/s²) (in normal driving gear) or maximum obtainable at 890 N (200 lb) pedal force (or equivalent line pressure).

5.19.3.4 *Stop Interval*—0.5 mile.

5.19.3.5 *Stops Required*—15.

5.19.3.6 *Record*—Maximum line pressure (pedal force) for each stop and deceleration (if 2.4 m/s² [8 ft/s²] cannot be held).

6. Report Forms and Graph Sheets

6.1 General Data and Summary Sheet, Figure 2.

6.2 Initial Effectiveness, Emergency Brake, and Inoperative Power System Data Sheet, Figure 3.

6.3 Second Effectiveness and First Fade Data Sheet, Figure 4.

6.4 Second Fade, Recovery and Final Effectiveness Data Sheet, Figure 5.

6.5 Final Inspection and Water Recovery Data Sheet, Figure 6.

6.6 Sample of Layout of Effectiveness Test Graph Coordinates, Figure 7.

6.7 Sample of Layout of Fade and Recovery Test Graph Coordinates, Figure 8.

VEHICLE: MAKE ______ MODEL ______ YEAR ______
ENGINE ______ TRANSMISSION ______ AXLE ______
WEIGHT ______ kg (lb) FRONT WEIGHT ______ kg (lb) REAR WEIGHT ______ kg (lb)
BALLAST ______ kg (lb) TEST WEIGHT
EMPTY WEIGHT ______ GENERAL DATA ______
TIRE SIZE: FRONT ______ REAR ______ MAKE: FRONT ______ REAR ______
MANUFACTURER'S DESIGNATION: FRONT ______ REAR ______

BRAKES: FRONT-SIZE ______ TYPE ______ CYL DIA ______
REAR-SIZE ______ TYPE ______ CYL DIA ______
LINING-FRONT ______ REAR ______
DRUM (ROTOR) TYPE-FRONT ______ REAR ______
MASTER CYL DIA ______ PEDAL RATIO ______ POWER BRAKE: YES ____ NO ____ TYPE ______

TEST INFORMATION: THERMOCOUPLE INSTALLATION METHOD: ______
SPECIAL EQUIPMENT ______
TESTED BY ______ LOCATION ______ DATE ______

TEST PHASE	TEST RESULTS
PREBURNISH CHECK	PF LB
EFFECTIVENESS TESTS	1st 2nd 3rd
48 km/h (30 mph) at ____ m/s² (ft/s²)	____ ____ ____ N (lb) PF
97 km/h (60 mph) at ____ m/s² (ft/s²)	____ ____ ____ N (lb) PF
129 km/h (80 mph) at ____ m/s² (ft/s²)	____ ____ ____ N (lb) PF
EMERGENCY BRAKE TEST	
WARNING SYSTEM ACTUATION	TYPE: POWER MANUAL
	FRT ____ RR ____ N (lb) PF
97 km/h (60 mph) STOPPING DISTANCE	FRT ____ m (ft) ____ N (lb) PF
	RR ____ m (ft) ____ N (lb) PF
INOPERATIVE POWER ASSIST SYSTEM TEST	
97 km/h (60 mph) STOPPING DISTANCE	____ m (ft) ____ N (lb) PF
MINIMUM LOAD TEST	____ m/s² (ft/s²) ____ N (lb) PF
HIGH SPEED STOP TEST	____ m/s² (ft/s²) ____ N (lb) PF
FIRST FADE AND RECOVERY TEST	
FADE STOPS 1–4	____ . ____ . ____ . ____ N (lb)
RECOVERY STOPS 1–5	____ m/s² (ft/s²) BY ____ N (lb)
RECOVERY STOPS 6–12	____ N (lb) PF MAX
SECOND FADE AND RECOVERY TEST	
FADE STOPS 1–8	____ . ____ . ____ . ____
	____ . ____ . ____ . N (lb) PF
RECOVERY STOPS 1–5	____ m/s² (ft/s²) BY ____ N (lb)
RECOVERY STOPS 6–12	____ N (lb) PF MAX
STABILITY DURING EFFECTIVENESS TESTS	CONTROLLABLE BRAKING
	BELOW ____ m/s² (ft/s²)
	YES ____ NO ____
FINAL INSPECTION	
LINING INTEGRITY	YES ____ NO ____
MECHANICAL INTEGRITY	YES ____ NO ____
HYDRAULIC INTEGRITY	YES ____ NO ____
WATER RECOVERY TEST	
AVERAGE BASELINE PF	AVG BASE ____ N (lb) PF
RECOVERY STOPS 4–6	____ N (lb) PF MAX
RECOVERY STOPS 7–14	____ N (lb) PF MAX
RECOVERY STOP 15	____ N (lb) PF FROM BASE

COMMENTS ______

REPORTED BY ______ DATE ______

FIGURE 2—GENERAL DATA AND SUMMARY SHEET

25.112

VEHICLE _____

TESTED BY _____

DATE _____

INITIAL EFFECTIVENESS, EMERGENCY BRAKE, AND INOPERATIVE POWER SYSTEM DATA

INPUT CORRELATION ENGINE IDLING IN NEUTRAL	
PEDAL FORCE - PF, N (LB)	LINE PRESSURE - LP N/m^2 (PSI)

BURNISH 64-0 km/h (40-0 mph), 3.7 m/s^2 (12 ft/s^2) IN GEAR, 121°C (250°F) IBT EACH STOP BUT 1.6 km (1 mile) MAX INTERVAL		
STOP	PF--LP	COMMENTS
1		
20		
40		
60		
80		
100		
120		
140		
160		
180		
200		

PREBURNISH CHECK 43-0 km/h (30-0 mph) 3 m/s^2 (10 ft/s^2) IN GEAR, 1.6 km (1 mile) INTERVAL		
STOP	PF--LP	COMMENTS
1		
2		
3		
4		
5		
6		
7		
8		
9		
10		

EMERGENCY BRAKE SYSTEM 97 km/h (60-0 mph) IN GEAR 65.6°C (150°F) IBT EACH APPLICATION		
MODE	PF--LP TO ACTUATE WARNING SYSTEM	
MODE	PF--LP	m (ft)

FIRST (PREBURNISH) EFFECTIVENESS 93.3°C (200°F) IBT EACH APPLICATION			
48 km/h (30 mph) IN NEUTRAL		97 km/h (60 mph) IN NEUTRAL	
PF--LP	m/s^2 (ft/s^2)	PF--LP	m/s^2 (ft/s^2)
SKID		SKID	

INOPERATIVE POWER SYSTEM 97-0 km/h (60-0 mph) IN GEAR 65.6°C (150°F) IBT EACH APPLICATION		
PF--LP	m (ft)	COMMENTS

FIGURE 3—INITIAL EFFECTIVENESS, EMERGENCY BRAKE, AND INOPERATIVE POWER SYSTEM DATA SHEET

25.113

VEHICLE _____

TESTED BY _____

DATE _____

SECOND EFFECTIVENESS AND FIRST FADE DATA

SECOND EFFECTIVENESS
93.3°C (200°F) IBT EACH APPLICATION

48 km/h (30 mph) IN NEUTRAL

PF--LP	m/s² (ft/s²)
SKID	

97 km/h (60 mph) IN NEUTRAL

PF--LP	m/s² (ft/s²)
SKID	

129 km/h (80 mph) IN NEUTRAL

PF--LP	m/s² (ft/s²)	PF--LP	m/s² (ft/s²)
SKID		SKID	

FIRST BASELINE CHECK
48-0 km/h (30-0 mph) 3 m/s² (10 ft/s²) IN GEAR, 1.6 km (1 mile) INTERVAL

STOP	PF--LP	COMMENTS
1		
2		
3		

FIRST FADE
97-0 km/h (60-0 mph), 4.6 m/s² (15 ft/s²) IN GEAR, 0.6 km (0.4 mile) INTERVAL 65.6°C (150°F) IBT FIRST STOP

AMBIENT _____ °C (°F)   TIME FOR 10 STOPS _____ S

STOP	FRONT IBT	REAR IBT	PF--LP	COMMENTS
1				
2				
3				
4				
5				
6				
7				
8				
9				
10				

FIRST RECOVERY
48-0 km/h (30-0 mph) 3 m/s² (10 ft/s²) IN GEAR, 1.6 km (1 mile) INTERVAL AT 64 km/h (40 mph)

STOP	FRONT IBT	REAR IBT	PF--LP	COMMENTS
1				
2				
3				
4				
5				
6				
7				
8				
9				
10				
11				
12				

FIRST EFFECTIVENESS SPOTCHECK
97-0 km/h (60-0 mph) 4.6 m/s² (15 ft/s²) IN GEAR, 93.3°C (200°F) IBT

STOP	PF--LP	COMMENTS
1		
2		

FIRST REBURNISH
64 km/h (40-0 mph), 3.7 m/s² (12 ft/s²) IN GEAR, 121°C (250°F) IBT EACH STOP BUT 1.6 km (1 mile) MAX INTERVAL

STOP	PF--LP	COMMENTS
1		
10		
25		
35		

SECOND BASELINE CHECK
48-0 km/h (30-0 mph), 3 m/s² (10 ft/s²) IN GEAR, 1.6 km (1 mile) INTERVAL

STOP	PF--LP	COMMENTS
1		
2		
3		

FIGURE 4—SECOND EFFECTIVENESS AND FIRST FADE DATA SHEET

25.114

VEHICLE _____

TESTED BY _____

DATE _____

SECOND FADE, RECOVERY, AND FINAL EFFECTIVENESS DATA

SECOND FADE

97-0 km/h (60-0 mph)
4.6 m/s² (15 ft/s²) IN GEAR,
0.6 km (0.4 mile) INTERVAL
65.6°C (150°F) IBT FIRST STOP

AMBIENT ____°C (°F) TIME FOR 15 STOPS ____S

STOP	FRONT IBT	REAR IBT	PF--LP	COMMENTS
1				
2				
3				
4				
5				
6				
7				
8				
9				
10				
11				
12				
13				
14				
15				

SECOND RECOVERY

48-0 km/h (30-0 mph)
3 m/s² (10 ft/s²) IN GEAR,
1.6 km (1 mile) INTERVAL AT 64 km/h (40 mph)

STOP	FRONT IBT	REAR IBT	PF--LP	COMMENTS
1				
2				
3				
4				
5				
6				
7				
8				
9				
10				
11				
12				

SECOND EFFECTIVENESS SPOTCHECK

97-0 km/h (60-0 mph)
4.6 m/s² (15 ft/s²) IN GEAR,
93.3°C (200°F) IBT

STOP	PF--LP	COMMENTS
1		
2		

SECOND REBURNISH

64 km/h (40-0 mph), 3.7 m/s² (12 ft/s²) IN GEAR,
121°C (250°F) IBT EACH STOP
BUT 1.6 km (1 mile) MAX INTERVAL

STOP	PF--LP	COMMENTS
1		
10		
25		
35		

FINAL EFFECTIVENESS

93.3°C (200°F) IBT EACH APPLICATION

48 km/h (30 mph) IN NEUTRAL		129 km/h (80 mph) IN NEUTRAL	
PF--LP	m/s² (ft/s²)	PF--LP	m/s² (ft/s²)
SKID			
97 km/h (60 mph) IN NEUTRAL			
PF--LP	m/s² (ft/s²)		
SKID		SKID	

FIGURE 5—SECOND FADE, RECOVERY AND FINAL EFFECTIVENESS DATA SHEET

25.115

VEHICLE _____
TESTED BY _____
DATE _____

FINAL INSPECTION

FRICTION MATERIAL CONDITION:	
LF	
RF	
LR	
RR	
DRUM (OR ROTOR) CONDITION:	
LF	
RF	
LR	
RR	
MECHANICAL COMPONENT CONDITION:	
LF	
RF	
LR	
RR	
BRAKE PEDAL	
POWER BRAKE	
STOPLIGHTS	
HYDRAULIC COMPONENT CONDITION:	
LF	
RF	
LR	
RR	
MASTER CYLINDER	

INSPECTION COMMENTS: _____

FIGURE 6—FINAL INSPECTION DATA SHEET

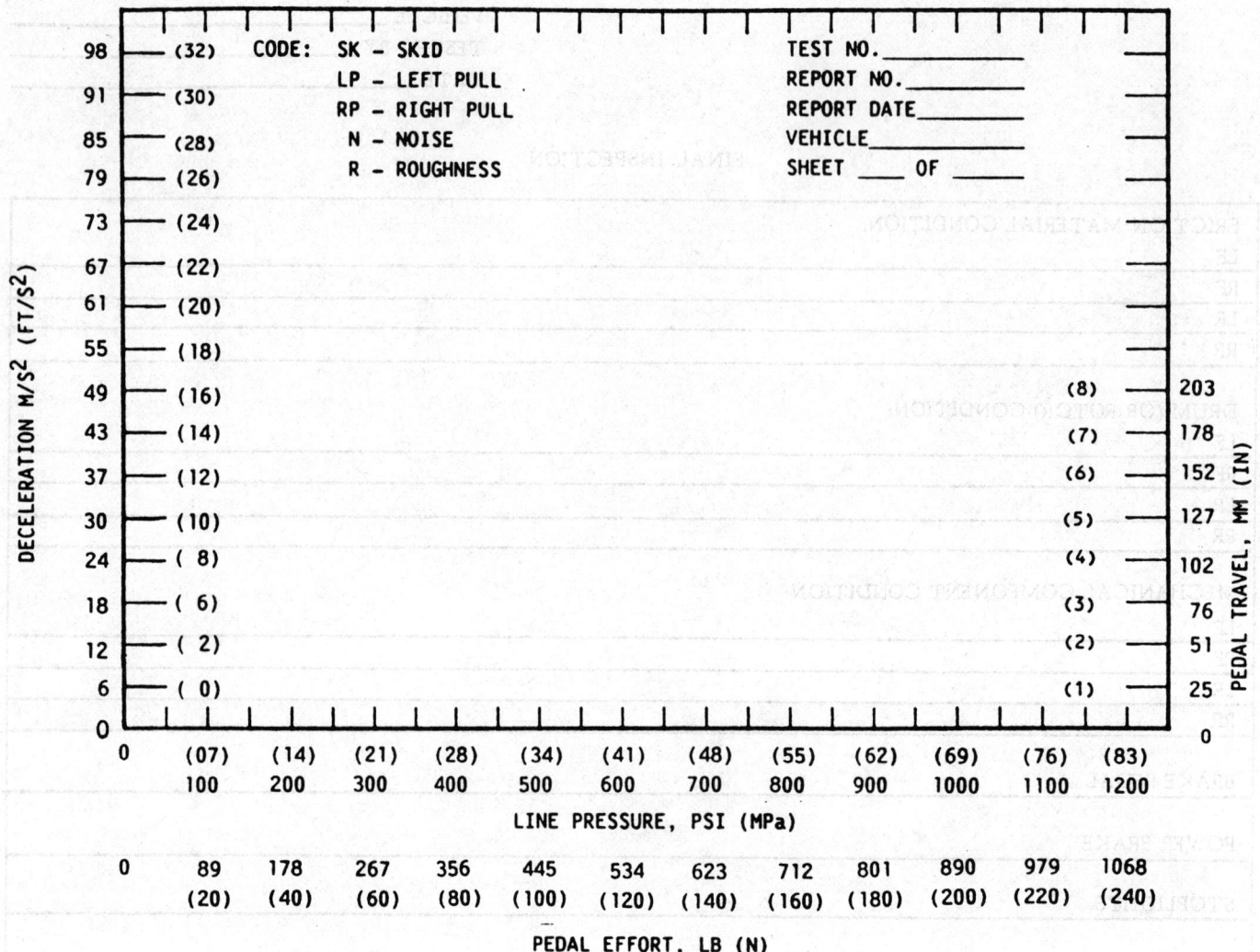

FIGURE 7—SAMPLE OF LAYOUT OF EFFECTIVENESS TEST GRAPH COORDINATES

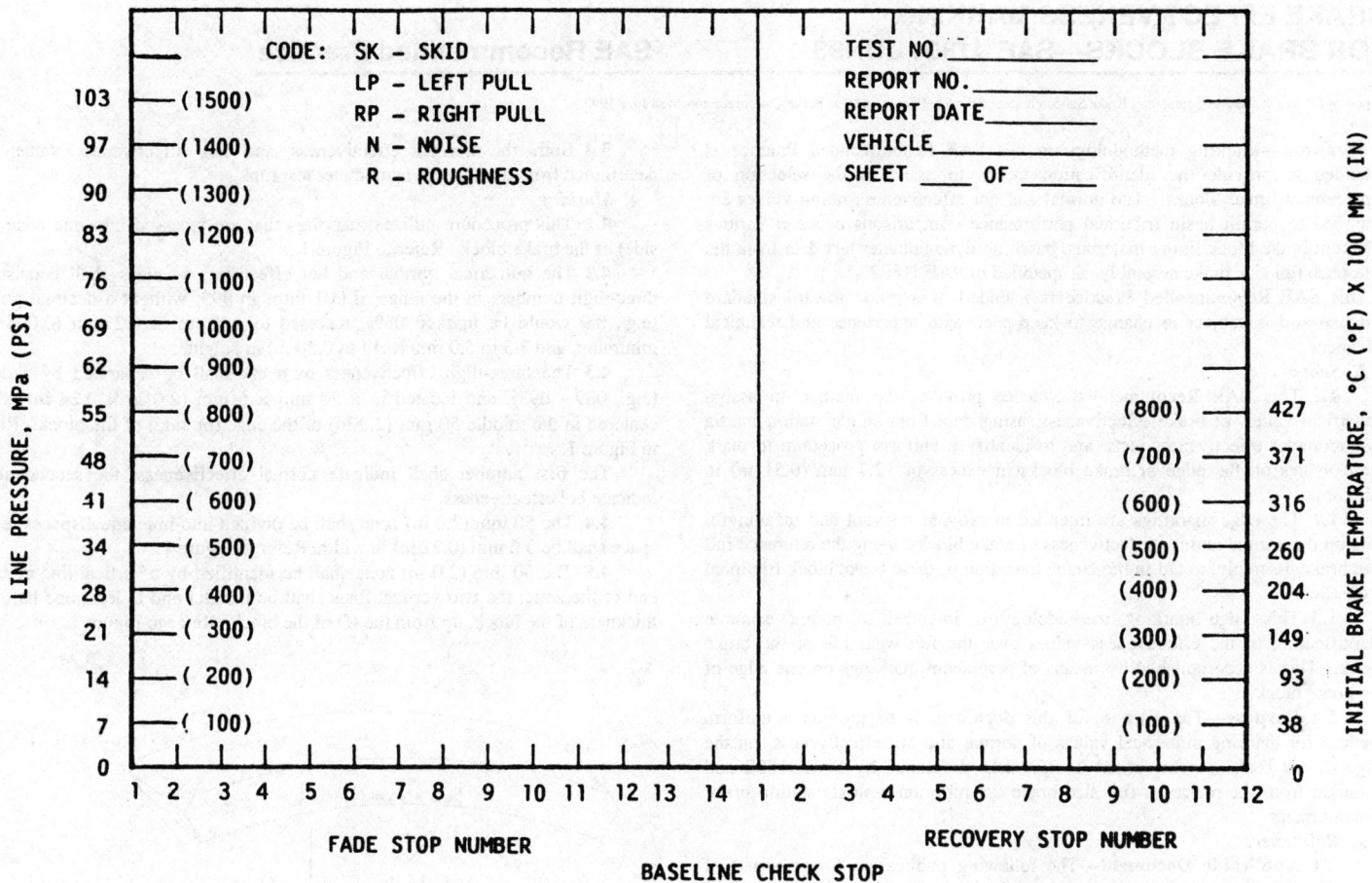

FIGURE 8—SAMPLE OF LAYOUT OF FADE AND RECOVERY TEST GRAPH COORDINATES

BRAKE EFFECTIVENESS MARKING FOR BRAKE BLOCKS—SAE J1801 JUN93

SAE Recommended Practice

Report of the Truck and Bus Foundation Brake Subcommittee of the SAE Truck and Bus Brake Committee approved June 1993.

Foreword—Marking methodology in this SAE Recommended Practice is intended to provide the identification means to assist in the selection of replacement brake blocks. The normal and hot effectiveness rating values are intended to permit basic frictional performance comparisons between various different brake block lining materials, based on dynamometer test data from the reference full size brake assembly, as specified in SAE J1802.

This SAE Recommended Practice is intended as a guide toward standard practice and is subject to change to keep pace with experience and technical advances.

1. Scope

1.1 This SAE Recommended Practice provides the method to assign numerical values of brake effectiveness, using data from single station inertia dynamometer effectiveness tests, and to identify a uniform procedure to mark these values on the edge of brake blocks in excess of 12.7 mm (0.51 in) in thickness.

1.2 The edge markings are intended to provide relevant and meaningful data on the normal and hot effectiveness of brake blocks, using the reference full size brake assembly, to aid in the characterization of these brake block frictional properties.

1.3 This edge marking methodology is intended to permit accurate identification of the effectiveness values over the full wear life of the brake block. This is accomplished by means of permanent markings on one edge of the brake block.

1.4 Purpose—The purpose of this document is to provide a uniform method for marking numerical values of normal and hot effectiveness, on the edge of the brake blocks, based on test data developed by SAE J1802 and obtained from the reference full size brake assembly on a single station brake dynamometer.

2. References

2.1 Applicable Document—The following publication forms a part of this specification to the extent specified herein. The latest issue of SAE publications shall apply.

2.1.1 SAE PUBLICATION—Available from SAE, 400 Commonwealth Drive, Warrendale, PA 15096-0001.

SAE J1802—Brake Block Effectiveness Rating

3. Effectiveness Determination

3.1 The brake effectiveness, for this document, is a nondimensional value that is the average calculated from a minimum of three inertia dynamometer brake tests. Brake effectiveness is the ratio of the brake frictional output to brake force input as determined by the SAE J1802 test procedure.

3.2 The normal effectiveness values from a minimum of three dynamometer tests are arithmetically averaged to obtain the normal effectiveness rating.

3.3 Similarly, the hot effectiveness values from a minimum of three dynamometer tests are arithmetically averaged to obtain the hot effectiveness rating.

3.4 Both the normal effectiveness and hot effectiveness value are determined from the same dynamometer test run.

4. Marking

4.1 This procedure utilizes markings that are recessed into one edge (or side) of the brake block. Refer to Figure 1.

4.2 The numerical normal and hot effectiveness values shall consist of three-digit numbers in the range of 001 through 999, without a decimal point, (e.g., 8.9 would be marked 089), recessed to a depth of 0.2 mm (0.008 in) minimum, and 3.5 to 5.0 mm (0.14 to 0.20 in) in height.

4.3 The three-digit effectiveness numbers shall be separated by a dash (e.g., 089 - 093), and located in a 50 mm × 6 mm (2.0 in × 0.24 in) zone, centered in the middle 50 mm (2.0 in) of the edge (or side) of the block. Refer to Figure 1.

The first number shall indicate normal effectiveness; the second shall indicate hot effectiveness.

4.4 The 50 mm (2.0 in) zone shall be divided into nine equal spaces; each space shall be 5.6 mm (0.22 in) in width. Refer to Figure 1.

4.5 The 50 mm (2.0 in) zone shall be identified by a vertical line at each end of the zone; the two vertical lines shall be parallel and at least one half the thickness of the block, up from the ID of the block. Refer to Figure 1.

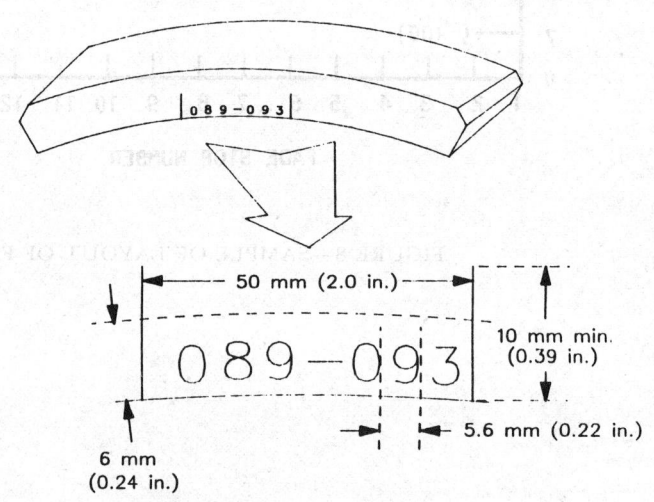

FIGURE 1—BRAKE BLOCK MARKING PROCEDURE

BRAKE BLOCK EFFECTIVENESS RATING
—SAE J1802 JUN93

SAE Recommended Practice

Report of the SAE Truck and Bus Foundation Brake Subcommittee of the SAE Truck and Bus Brake Committee approved June 1993.

Foreword—The brake block test methodologies in this SAE Recommended Practice provide the techniques and procedures required to determine brake effectiveness for the entire range of brake blocks, whether asbestos-based, nonasbestos organic, semimetallic, or full metallic in composition. Brake effectiveness is the nondimensional measurement of braking performance. This document is intended to supersede that of SAE J661 for brake blocks. An inertia dynamometer with one reference brake assembly is utilized to determine brake effectiveness under both normal temperature and high temperature conditions. This full-sized reference brake assembly is used to characterize the effectiveness performance of complete brake assemblies, thereby eliminating the shortcomings and uncertainties inherent in the use of small brake lining specimens.

1. Scope—This SAE Recommended Practice provides the test procedures and methods to calculate the effectiveness of brake blocks, using an inertia dynamometer. To minimize testing variability, and to optimize standardization and correlation, a single, high volume size of brake block is specified (FMSI No. 4515E) and evaluated in a reference S-cam brake assembly of 419 mm × 178 mm (16.5 in × 7.0 in) size, using a specified brake drum.

1.1 Purpose—The purpose of this document is to establish a uniform procedure for the determination and classification of brake effectiveness for commercial vehicle brake blocks. This document will permit comparison of basic frictional properties for brake blocks. Service usage classification ratings and applicable axle load ratings are not included in this document.

2. References

2.1 Applicable Documents—The following publication forms a part of this specification to the extent specified herein.

2.1.1 FMSI PUBLICATION—Available from Friction Materials Standards Institute Inc., Monroe, CT.

FMSI No. 4515E

3. Test Procedure—This test procedure applies to a reference air brake assembly of 419 mm × 178 mm (16.5 in × 7.0 in) size that utilizes S-cam actuation, with cam rotation same as brake drum rotation. Reference Appendix A and Figure A1.

Specifications for this brake assembly and the associated component parts, including installation/assembly requirements and maintenance procedures, are specified in Appendix B and C.

Dimensional, chemical, metallurgical, and surface finish specifications and requirements for the reference brake drums are included in Appendix D.

Brake lining block grinding requirements are shown in Appendix E.

Installation and mounting requirements for the brake assembly to the dynamometer stub axle fixture can be found in SAE J1802/1 (Draft). SAE Draft Technical Report J1802/1 can be obtained from SAE.

A complete parts list, which includes all components of the reference brake assembly, can be found in SAE J1802/1 (Draft). SAE Draft Technical Report J1802/1 can be obtained from SAE.

3.1 Dynamometer Test Conditions

3.1.1 The dynamometer inertia is to be 1134 kg-m^2 (837 slug-ft^2).

3.1.1.1 Reference

Nominal wheel load - 4536 kg (10 000 lb)

Effective radius - 0.5 m (19.7 in)

3.1.2 Ambient temperature for the dynamometer cooling air is to be between 25 °C and 40 °C (77 °F and 104 °F).

3.1.3 Air at ambient temperature is to be directed uniformly and continuously over the brake drum during the test procedures at a velocity that when cooled from 250 °C (482 °F) and rotating at a uniform 60 rpm ± 5 rpm with no brake drag, the cooling time from 200 °C (392 °F) to 100 °C (212 °F) will be 10.0 min ± 1.0 min 30 s. If cooling characteristics are not achieved, abort test.

3.1.4 The brake drum temperature is to be measured with one thermocouple, installed at the center of the braking surface. The thermocouple hole is to be positioned within 3 to 4 mm (0.12 to 0.16 in) of the rubbing surface. Installation is per Appendix F.

3.1.5 Brake temperatures are to be within ±5 °C (±9 °F) of those temperatures as specified.

3.1.6 Brake apply pressures are to be maintained within -0 to +15 kPa (-0 to +2.0 psi) of those pressures as specified.

3.1.7 Chamber air pressure rise rate is to be 1.50 mPa/s ± 0.3 mPa/s (220 psi/s ± 45 psi/s).

3.1.8 The maximum air pressure in the chamber is not to exceed 700 kPa (100 psi). If a specified deceleration requires pressures above the pressure limit, continue at the limit pressure and resulting deceleration until the deceleration returns or the test section is completed.

3.1.9 The brake drum temperature is increased to a specified level by conducting one or more stops from 320 rpm at a deceleration of 3.0 m/s^2 ± 0.15 m/s^2 (9.8 ft/s^2 ± 0.5 ft/s^2) and an interval of not less than 60 s, unless otherwise specified.

3.1.10 The brake drum temperature is decreased to a specified level by rotating the drum at a constant 240 rpm. Air flow rates are not to be varied during the test sequence, except during periods of measurement and inspection.

3.1.11 ADJUSTMENT—With brake fully released and the brake drum temperature less than 70 °C (158 °F), set clevis position such that the center of the pin to chamber mounting face is 69.85 mm ± 3.18 mm (2.75 in ± 0.125 in) with the brake drum temperature less then 70 °C (158 °F) adjust slack adjuster to 12.7 to 25.4 mm (0.50 to 1.0 in) of air chamber stroke at 70 kPa (10 psi), with peak brake drag less than 11.3 N·m (100 in-lb) with the brake released.

3.1.12 BRAKE ASSEMBLY GEOMETRY CHECK—With brake, drum, and stub axle assembled, actuate brake with 35 kPa (5 psi) chamber pressure; using a 0.025 mm (0.001 in) feeler gage, confirm shoe to drum contact near shoe centers. This check is to be performed after the brake lining blocks have been ground per Appendix E.

3.1.13 All rotation speeds referenced are to be within ±10 rpm.

3.2 Brake Burnish Procedure

3.2.1 BRAKES ARE BURNISHED BEFORE TESTING AS FOLLOWS—Place the reference brake assembly, with new test linings (ground per the requirements of Appendix E) and a new drum on the inertia dynamometer and set up the brake as specified in Appendix A and 3.1. Make 200 stops from 320 rpm at a deceleration rate of 3.0 m/s^2 ± 0.15 m/s^2 (9.8 ft/s^2 ± 0.5 ft/s^2) with an initial brake drum temperature on each stop of 200 °C (392 °F).

3.2.2 NORMAL TEMPERATURE TEST FOR BRAKE EFFECTIVENESS

3.2.2.1 With an initial brake drum temperature of 100 °C (212 °F), conduct a stop from 400 rpm maintaining brake chamber air pressure at a constant 70 kPa (10 psi). Record the brake output torque, and air pressure and stroke or chamber force exerted by the brake from the time the specified air pressure is reached until the brake stops. Perform this procedure at 70 kPa (10 psi), 103 kPa (15 psi), 138 kPa (20 psi), 172 kPa (25 psi), 207 kPa (30 psi), 241 kPa (35 psi), 276 kPa (40 psi), 310 kPa (45 psi), up to a maximum of 345 kPa (50 psi).

3.2.2.2 Normal Temperature Inspection—For the test to be acceptable, 90% minimum lining contact is required on all four lining blocks. If specified contact pattern is not achieved, the test is invalid.

3.2.3 Warm the brake by conducting stops from 320 rpm, at a deceleration of 3.0 m/s^2 ± 0.15 m/s^2 (9.8 ft/s^2 ± 0.5 ft/s^2), at 45 s intervals until the drum reaches 315 °C (599 °F).

3.2.4 Make 200 additional stops from 320 rpm at a deceleration of 3.0 m/s^2 ± 0.15 m/s^2 (9.8 ft/s^2 ± 0.5 ft/s^2) with an initial brake drum temperature on each stop of 300 °C (572 °F).

3.2.5 Evaluate brake drum cooling characteristics, as specified in 3.1.3, ±1 min 30 sec at the completion of the burnish procedure. If cooling time from 200 °C (392 °F) to 100 °C (212 °F) is not 10 min ± 1 min 30 s, the test is to be aborted.

3.3 Hot Temperature Test for Brake Effectiveness

3.3.1 BRAKE ADJUSTMENT—Cool brake to a drum temperature below 70 °C (158 °F). Adjust the brake as specified in 3.1.11.

3.3.2 Warm the brake by conducting stops from 320 rpm, at a deceleration of 3.0 m/s² ± 0.15 m/s² (9.8 ft/s² ± 0.5 ft/s²), at 45 s intervals until the drum reaches 315 °C (599 °F); make five additional stops from 300 °C (572 °F) initial brake temperature at 320 rpm and 3.0 m/s² ± 0.15 m/s² (9.8 ft/s² ± 0.5 ft/s²) deceleration.

3.3.3 With an initial brake drum temperature of 300 °C (572 °F), conduct a stop from 400 rpm, maintaining brake chamber air pressure at a constant 70 kPa (10 psi). Record the brake output torque, brake chamber air pressure, and stroke or chamber force exerted by the brake chamber from the time the specified air pressure is reached until the brake stops.

Perform this procedure at 70 kPa (10 psi), 103 kPa (15 psi), 138 kPa (20 psi), 172 kPa (25 psi), 207 kPa (30 psi), 241 kPa (35 psi), 276 kPa (40 psi), 310 kPa (45 psi), up to a maximum of 345 kPa (50 psi).

3.3.4 FINAL INSPECTION—For the effectiveness tests to be acceptable, 90% minimum lining contact is required on all four lining blocks. If specified contact pattern is not achieved, the tests are invalid.

3.4 Effectiveness Determination

3.4.1 The brake effectiveness for this document is the nondimensional value that is calculated from the measured average brake torque output and the average torque input, using a distance based averaging method. The average torque input is the product of average chamber force and nominal slack arm length—all incompatible measurement units.

Chamber force can be obtained by using chamber calibration curves with input pressure and stroke, or measured directly using force transducers.

3.4.2 Brake effectiveness is defined as the slope of the linear regression line which describes the relationship between average output torque, as determined by this procedure, and the input average torque.

To calculate brake effectiveness, the following equation is to be used:

$$\text{Brake Effectiveness} = \frac{\sum XY - \frac{\sum X \sum Y}{N}}{\sum X^2 - \frac{(\sum X)^2}{N}} \quad \text{(Eq.1)}$$

where:
X = Average Input Torque
Y = Average Output Torque
N = Number of Stops (9 for this procedure)

3.4.3 The same interval is to be used for measuring pressure, input torque, and output torque.

3.4.4 The start of the interval is defined as the point at which the air pressure measured at the input point to the air chamber reaches the specified value.

3.4.5 The end of the interval is defined as the point at which the brake is released or 0 rpm, whichever is higher. The speed at which the brake is released shall not exceed 10 rpm.

NOTE—Some "rockback" is to be expected during 3.3.

3.5 Normal Effectiveness

3.5.1 Normal effectiveness is determined from the data developed in 3.2.2.1 and the brake effectiveness formula defined in 3.4.2, using the average brake torque values and the corresponding effective input torques.

3.6 Hot Effectiveness

3.6.1 Hot effectiveness is determined from the data developed in 3.3.3 and the brake effectiveness formula defined in 3.4.2, using the average brake torque values and the corresponding effective input torques.

APPENDIX A
REFERENCE BRAKE ASSEMBLY

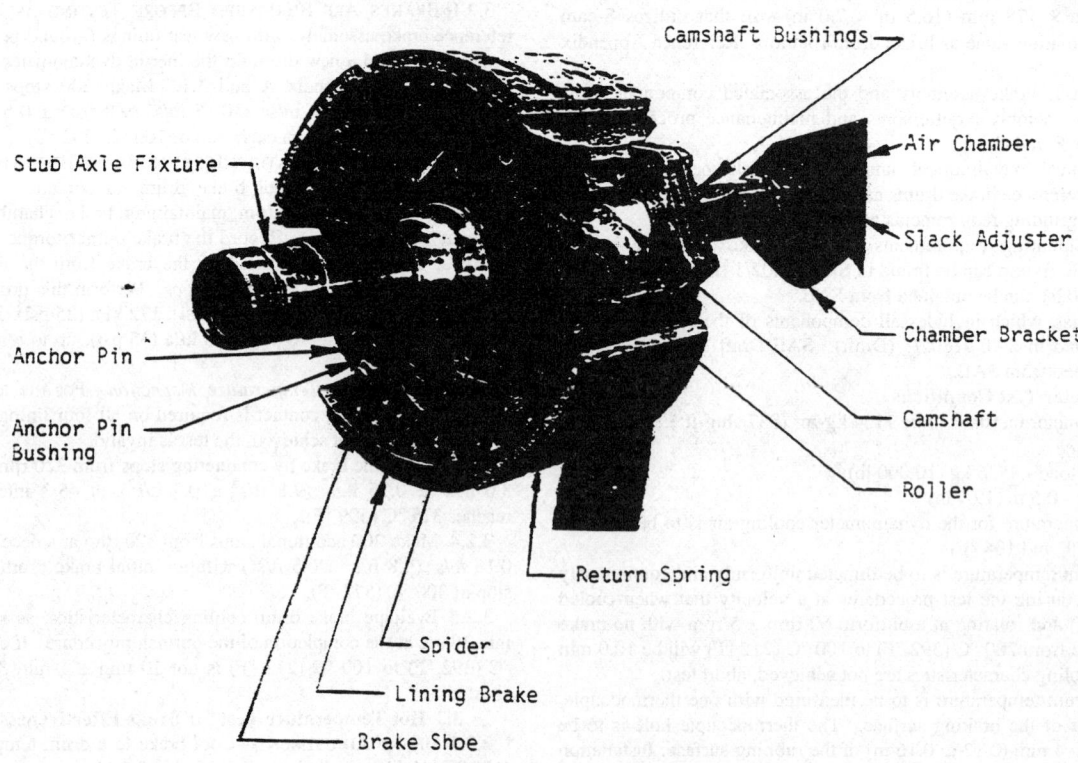

FIGURE A1—REFERENCE BRAKE ASSEMBLY

APPENDIX B
BRAKE ASSEMBLY SPECIFICATIONS

B.1 Design
Single sourced, dimensionally certified
Only certified components to be used
Test component identification must be recorded

B.2 Spider
Specified spindle alignment and axle mounting/assembly
Refer to Appendix A (SAE J1802/1)

B.3 Camshaft
Prescribed rise rate
Hardened and ground surfaces
Minimal, restricted journal clearance
Hardnesses specified (surface, case, and core)
Strain gaging optional (for calibration checking purposes only)

B.4 Air Chamber
Rotochamber type, calibrated

B.5 Shoes
Precision, pivot locations
Controlled shoe table dimensions
Bolted block attachment
Prescribed block grind as specified in Appendix E
Lining retention bolt torque: 9 to 11 N·m (80 to 100 in-lb)

B.6 Slack Adjuster
137.50 mm (5.50 in) effective length
Manual design

B.7 Chamber Bracket
68.75 mm (2.75 in) clevis pin to chamber mounting face

B.8 Return Springs
Specified spring rate
Specified installed load

B.9 Anchor Pins
Hardened journals

B.10 Rollers
Hardened trunnions

APPENDIX C
MAINTENANCE GUIDELINES

C.1 Camshaft Axial/Radial Play

C.1.1 Axial play of camshaft should be adjusted by adding/subtracting spacing washers between slack adjuster and snap ring. Maximum axial play: 0.75 mm (0.030 in).

C.1.2 Radial play of camshaft within the bushing should not exceed 0.37 mm (0.015 in) total reading. Replace and rebore bushings as necessary.

C.2 Lubrication—Lubricate camshaft bushings and roller ID prior to each test with an appropriate high temperature NGLI #1 grease. DO NOT apply excessive grease. Cam head and roller surface should not be lubricated. Grease shall not run onto lining during test.

C.3 Brake Drum—Brake drum runout should not exceed 0.2 mm (0.008 in) total indicator reading 25.4 mm (1.0 in) from the open end of the brake drum.

C.4 Linings—Use a new set of linings for each individual test. Linings must be ground on shoe as indicated in Appendix E.

C.5 Anchor Pins/Rollers—Check anchor pins and rollers for flat spots. Replace as necessary.

C.6 Return Spring—Replace return spring if installed load is greater/less than 2.3 kg (5.0 lb) from that specified.

C.7 Air Chamber—Check air chamber calibration every 12 tests. Replace if output force varies more than 1% from previous calibration.

C.8 Slack Adjuster—Position slack so that clevis pin is located at 68.75 mm ± 1.55 mm (2.75 in ± 0.062 in) from air chamber mounting face. If clevis hole elongates, allowing greater than 1.50 mm (0.060 in) movement, replace components as necessary.

APPENDIX D
BRAKE DRUM SPECIFICATIONS

D.1 Composition—Specified chemistry, microstructure, hardness, graphite size, and type.

D.2 Design—Single sourced, specified configuration.

D.3 Dimension
Maximum runout: 0.2 mm (0.008 in)
Width: 175.0 mm (7.0 in) minimum
Diameter: 412.75/412.25 mm (16.510/16.490 in)

D.4 Braking Surface Finish—1.5 to 2.3 µm (60 to 90 µin)

D.5 Balance—0.14 N·m (20 oz-in)

D.6 Mounting
Hub piloted
10 hole bolt pattern
281.25 (11.25 in) bolt circle

D.7 Usage—New drum required for each test.

APPENDIX E
BRAKE LINING BLOCK GRINDING REQUIREMENTS

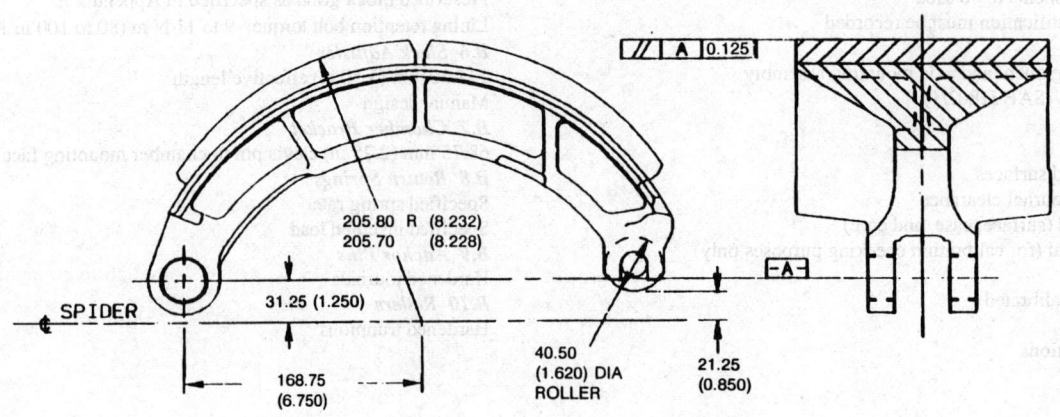

NOTES – SHOE/LINING GROUND TO INDICATED RADIUS AFTER ASSEMBLY

LINING MUST FIT TO SHOE SURFACE SUCH THAT 0.15 mm
(0.006 in) FEELER GAGE WILL NOT FIT BETWEEN COMPONENTS

TORQUE LINING RETENTION BOLTS TO 9 TO 11 N·m (80 TO 100 in-lb)

ALL DIMENSIONS IN MILLIMETERS (INCHES)

FIGURE E1—BRAKE LINING BLOCK GRINDING REQUIREMENTS

APPENDIX F
DRUM OR DISC THERMOCOUPLE INSTALLATION

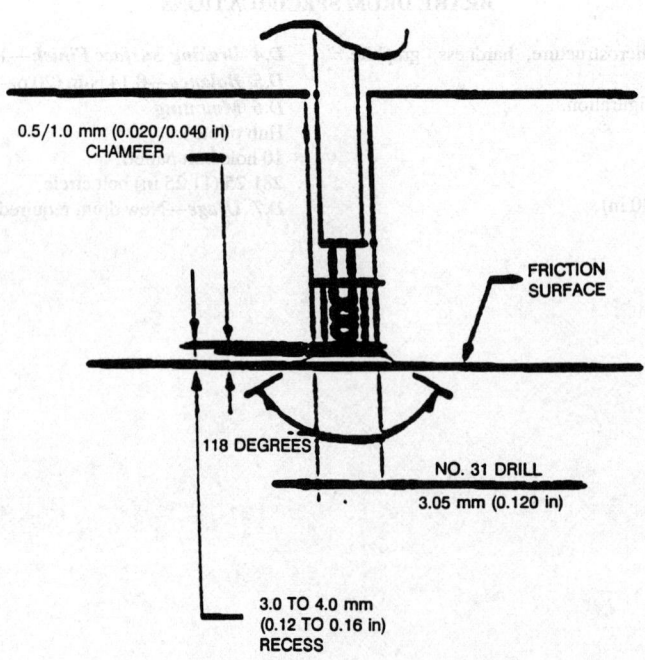

FIGURE F1—DRUM OR DISC THERMOCOUPLE INSTALLATION

SIMULATED MOUNTAIN BRAKE PERFORMANCE TEST PROCEDURE—SAE J1247 APR80

SAE Recommended Practice

Report of the Brake Committee, approved April 1980.

1. Introduction—This procedure was developed in response to requests for a flat road test procedure that would simulate the power and energy duty cycles occurring during mountain driving. A number of different mountain roads were investigated. In general, the mountain roads in the Eastern United States tend to have steeper grades, sharper curves, and shorter descent distances as compared to many mountain roads in the Western United States. This procedure simulates the longer descents found in the Western United States and complements the fade schedules in SAE J843d (March, 1973) which requires a higher average power input to the brakes, but for a shorter time. A general correlation with actual mountain descents has been established by analysis of actual power, energy, and temperature measurements. This procedure is intended as a development test. Line pressure usage rather than pedal force is recommended to provide a more precise measure of the input to the brakes.

2. Purpose—The purpose of this test code is to establish brake system capabilities with regard to fade resistance, balance, stability, recovery, and maximum brake fluid temperatures under simulated mountain driving conditions.

3. Scope—This SAE Recommended Practice establishes a uniform procedure for a flat road simulation of a mountain fade test of the brake systems of light-duty trucks and multipurpose passenger vehicles up to and including 10 000 lb (4500 kg) GVW and all classes of passenger cars.

4. Instrumentation
 4.1 Line pressure or pedal force gauge.
 4.2 Decelerometer.
 4.3 Speedometer (calibrated vehicle unit or fifth wheel type).
 4.4 Stop watch or cycle timer.
 4.5 Thermometer or ambient sensitive thermocouple.
 4.6 Odometer (calibrated).
 4.7 Direct reading temperature instrument.
 4.8 Brake lining and brake fluid thermocouples.
 4.9 Optional instrumentation:
  (a) Pedal travel gauge.
  (b) Solenoid stop counter.
  (c) Brake torque transducers.
  (d) Brake drum (or rotor) thermocouples.
  (e) Slip rings (for brake drum or rotor temperatures).

5. Installation Details
 5.1 Thermocouples
 5.1.1 LINING—Install plug-type thermocouples in each brake per current SAE J843d (March, 1973).
 5.1.2 BRAKE FLUID—Install fluid thermocouples in each brake per current SAE J291 (June, 1972).
 5.1.3 BRAKE DRUM (OR ROTOR) (OPTIONAL)—Install disc/drum thermocouples in each brake per current SAE J79 (August, 1972).
 5.2 Friction Material Preparation—Attach and finish friction material per vehicle manufacturer's specification.
 5.3 Brake Drum (or Rotor) and Hub Assembly—New drums or rotors are recommended. Surface finish and dimensional characteristics, with special emphasis on runout of rubbing surface, shall be in accordance with vehicle manufacturer's specifications.
 5.4 Brake Assembly—Brakes shall be prepared in accordance with vehicle manufacturer's specifications. New linings and springs are recommended on all brakes. Adjust brakes to manufacturer's specifications.
 5.5 Uniformity of Brake Components (Optional)—The objectives for each test shall determine the selection of uniformity requirements. Since braking system balance during the high temperature portion of this test may be affected by uniformity of components, care should be exercised in selecting components such that unbalance is controlled (that is, if lining stability is to be evaluated, other components such as drums or rotors should be matched or mismatched left to right with respect to surface finish, mass, runout, etc.).
 NOTE: Specifications for the brake components and the installation may not be available for experimental development tests. In such cases, the selection will have to be made by the testing agency.
 5.6 Vehicle Test Weight—The test weight should be selected by the user based on the test objectives. Normal practice in evaluating production vehicles is to select the manufacturer's maximum recommended weight.
 5.7 Vehicle Power Train—The power train should be selected by the user based on the test objectives. Normal practice in evaluating production vehicles is to select a power train providing the least engine braking.
 NOTE: The variation in engine braking among power trains will normally have only a small effect on test results but may have a significant effect on brake performance obtained during actual mountain descents. The reason is that drag affects the power absorbed by the power train only when the vehicle is coasting or when the brakes are applied. This is a small part of the total time for the test but may be a large part of the total time for actual mountain descents. Caution should be exercised in interpreting comparative results between vehicles having widely differing drag characteristics.
 Since vehicle drag affects braking severity during actual descents, drag should be measured and recorded with the data.
 The final speed for the snub may be adjusted as a function of vehicle drag. At this date, there is insufficient information to define the adjustment to the snub final speed as a function of vehicle drag, or if such a correction will simulate braking duty as a function of vehicle drag. The snub final speed of 17 mph (27 km/h) stated in paragraph 6.5 is based on best simulation of mountain descents for full-size domestic vehicles equipped with automatic transmissions and standard axle ratios.

6. Test Procedure
 6.1 Test Notes
 6.1.1 With the exception of the burnish portion of this test, all test stops and snubs shall be conducted on a substantially level (not to exceed ±1% grade), dry, smooth, hard surfaced roadway of Portland cement concrete (or other surface with equivalent coefficient of surface friction) that is free from loose materials.
 6.1.2 During all phases of this procedure, any unusual performance such as wrap-up or noise characteristics are to be noted and recorded. Note any uncontrollable braking action causing the vehicle to pull or swerve out of a 12 ft (3.66 m) wide roadway lane.
 6.1.3 *Initial Brake Temperature* is the lining temperature 0.2 mile (0.3 km) before stop or snub (average temperature of brakes on hottest axle), brakes off.
 6.1.4 Because variations in ambient temperature have a significant effect on test results, this test should be conducted within a range of ambient temperature of 40–90°F (4–32°C). Record ambient temperature as indicated on data sheets. Wind velocity should be less than 10 mph (16 km/h) and should be recorded.
 6.1.5 Decelerations used in test stops and snubs refer to values at which the decelerometer is held approximately constant during the stop or snub by varying the input pressure.
 6.1.6 Deceleration and line pressure readings shall not be taken below 5 mph (8 km/h).
 6.1.7 All stops and snubs should normally be made with vehicle in normal driving gear.
 6.1.8 Paragraphs 6.3–6.10 of this procedure shall be run continuously without interruption; 102 min, 30 s are required to complete these paragraphs.
 6.1.9 To permit a check of consistency in performing the test, each segment of the test (except paragraphs 6.2 and 6.11) should be timed, and vehicle mileage covered should be recorded. The average speed provides the parameter for measuring consistency.
 6.1.10 The test should be terminated if the pedal force reaches 200 lb (900 N) (or a line pressure equivalent to 200 lb (900 N) pedal force) and/or, if the pedal travel becomes excessive, or at any time the vehicle is considered to be unsafe.
 6.2 Burnish—Condition the braking system as follows:
  (a) Stop speed—40–0 mph (64–0 km/h).
  (b) Stop deceleration—12 ft/s² (3.66 m/s²).
  (c) Stop interval—As required to achieve 250°F (120°C) *initial brake temperature* or a maximum of 1 mile (1.6 km).
  NOTE: The 1 mile (1.6 km) maximum must be observed even though the initial temperature exceeds 250°F (120°C).
  (d) Cooling speed—40 mph (64 km/h) (moderate acceleration to cooling speed).
  (e) Stops required—200 (first 10 stops shall be used as a general check of instrumentation, brakes, and vehicle function).
  (f) Optional—Inspect and adjust brakes after burnish cycle. Record if either operation is performed.
  (g) Record odometer and time at end of first stop, maximum line pressure, stop interval, and initial brake temperature for the first and every twentieth stop, and odometer and time at end of last stop.
 6.3 Initial Brake Temperature—Warm the brakes to 150–200°F (65–95°C) initial brake temperature using the burnish procedure and shorten the interval if necessary. Record odometer at the last stop used to warm the brakes. Proceed immediately to paragraph 6.4.

NOTE: If only a limited range of ambient temperature exists, the brakes can be warmed by making 10 stops from 40 mph (64 km/h) at a 60 s interval. This procedure specifies the braking duty and would result in lower initial brake temperatures for systems that have good heat dissipation and, therefore, more nearly reflects real-world conditions. The disadvantage is that the initial brake temperature becomes dependent upon the ambient temperature. This procedure is, therefore, only recommended for a test series in which small changes in ambient temperature are expected.

6.4 First (Cold) Effectiveness Check—Immediately after the previous stop, moderately accelerate to 60 mph (97 km/h) and drive at 60 mph (97 km/h) for 200 s including acceleration time. Make check stops as follows:

(a) Stop speed—60-0 mph (97-0 km/h).
(b) Stop deceleration—15 ft/s^2 (4.6 m/s^2).
(c) Stops required—3.
(d) Cooling speed—60 mph (97 km/h) (moderate acceleration at cooling speed).
(e) Stop interval—Begin stop every 200 s.
(f) Record—Maximum line pressure for each stop, brake lining and fluid temperatures immediately before each stop, and odometer at the end of the third stop.

6.5 First Simulated Mountain Descent—Immediately after effectiveness check stop 3, moderately accelerate to 35 mph (56 km/h) and drive at 35 mph (56 km/h) for 15 s including acceleration time. Begin the simulated descent as follows:

(a) Snub speeds—35-17 mph (56-27 km/h).
(b) Snub deceleration—8 ft/s^2 (2.4 m/s^2).
(c) Snubs required—80.
(d) Snub interval—Begin snub every 15 s.
(e) Speed between snubs—35 mph (56 km/h) with moderate acceleration.
(f) Record—Maximum line pressure every fourth snub, brake lining and fluid temperatures immediately before every eighth snub, and odometer at the end of the last snub.

6.6 Second (Hot) Effectiveness Check—Immediately after the last snub in the simulated descent, moderately accelerate to 60 mph (97 km/h) and drive at 60 mph (97 km/h) for 50 s including the acceleration time. Make check stops as follows:

(a) Stop speed—60-0 mph (97-0 km/h).
(b) Stop deceleration—15 ft/s^2 (4.6 m/s^2).
(c) Stops required—3.
(d) Cooling speed—60 mph (97 km/h) (moderate acceleration to cooling speed).
(e) Stop interval—Begin stop every 50 s.
(f) Record—Maximum line pressure for each stop, brake lining and fluid temperatures immediately before each stop, and odometer at the end of the third stop.

6.7 Recovery—Immediately after the third stop in the second effectiveness check, moderately accelerate to 40 mph (64 km/h) and drive at 40 mph (64 km/h) for 120 s including the acceleration time. Make recovery stops as follows:

(a) Stop speed—40-0 mph (64-0 km/h).
(b) Stop deceleration—12 ft/s^2 (3.7 m/s^2).
(c) Stops required—10.
(d) Cooling speed—40 mph (64 km/h) (moderate acceleration cooling speed).
(e) Stop interval—Begin stop every 120 s.
(f) Record—Maximum line pressure for each stop, brake lining and fluid temperatures immediately before each stop, and odometer at the end of the last stop.

6.8 Second Simulated Mountain Descent—Immediately after the tenth stop in the first recovery, moderately accelerate to 35 mph (56 km/h) and drive at 35 mph (56 km/h) for 15 s including acceleration time. Repeat the simulated descent as stated in paragraph 6.5.

6.9 Soak—Immediately after the final snub in the second mountain descent, drive the vehicle the minimum distance possible to allow the vehicle to be parked in a windless area with the engine running at idle, and soak for 20 min. At 1 min intervals during the soak, apply the brakes to the average line pressure measured in paragraph 6.4 and hold for approximately 5 s before releasing. Note if pedal is spongy or goes to floor and if brake warning indicator illuminates during the applications. Record brake lining and fluid temperatures immediately before each fourth application. Do not proceed to paragraph 6.10 if the pedal travel is excessive during the final check at the end of the soak.

6.10 Third (Cold) Effectiveness Check—Immediately after the soak, repeat paragraph 6.4.

6.11 Reburnish—Repeat paragraph 6.2 except 35 stops required. Record odometer and time at end of the first stop, maximum line pressure, stop interval, and initial brake temperature for the first and every fifth stop, and odometer and time at the end of the stop.

6.12 Fourth (Cold) Effectiveness Check—Immediately after the last burnish stop, repeat paragraph 6.4.

6.13 Final Inspection—Disassemble all brakes, inspect, and record all pertinent observations.

7. Vehicle Drag Measurement (Optional)—Make 10 closed throttle coast-downs with the transmission in the gear being simulated (usually top gear, but could be a lower gear). Ambient wind velocity should be less than 5 mph (8 km/h). The direction should be alternated to reduce the effects of ambient wind. The coast-downs should be started at 40 mph (64 km/h).

8. Report Forms and Data Sheets

GENERAL DATA
SIMULATED MOUNTAIN BRAKE PERFORMANCE TEST

Vehicle

Make _____ Model _____ Year _____

Engine _____ Transmission _____ Axle Ratio _____

Test Weight _____ Curb Weight _____

Tire Type and Size _____ Other Information _____

Brake System

Front: Size _____ Type _____ Cyl Dia _____

Rear: Size _____ Type _____ Cyl Dia _____

Front Lining _____

Master Cylinder Dia _____ Power System _____

Other Information _____

Test Information

Location _____ Surface _____

Start Date _____ Completion Date _____

Mileage Travelled _____ Amb Temp Range _____

Wind Velocity _____ Direction _____

Driver _____ Observer _____

Instrumentation Description _____

Comments

Abbreviations

IBT   Average temperature of linings on hottest axle, 0.2 mile (0.3 km) before stop.
LP   Maximum line pressure during stop.
LFLT   Temperature of left front lining immediately before stop.
RFLT   Temperature of right front lining immediately before stop.
LRLT   Temperature of left rear lining immediately before stop.
RRLT   Temperature of right rear lining immediately before stop.
LFFT   Temperature of left front fluid immediately before stop.
RFFT   Temperature of right front fluid immediately before stop.
LRFT   Temperature of left rear fluid immediately before stop.
RRFT   Temperature of right rear fluid immediately before stop.

25.125

Test No. _____
Vehicle _____
Tested by _____
Date _____

Odometer at end of last stop to warm brakes _____
Time at end of last stop used to warm brakes (time zero) _____
Ambient temperature _____

Burnish (Paragraph 6.2)—200 stops, 40–0 mph (64–0 km/h), 12 ft/s² (3.7 m/s²), 250°F (120°C) initial brake temperature for each stop (but 1 mile (1.6 km) maximum interval).

Odometer at end of first stop _____
Time at end of first stop _____

Stop	Time	LP	LFLT	RFLT	LRLT	RRLT	LFFT	RFFT	LRFT	RRFT
1	3:20				Comments:					
2	6:40				Comments:					
3	10:00				Comments:					

Odometer at end of stop 3 _____ Average speed _____

Stop	Dist	LP	LFLT	RFLT	LRLT	RRLT	LFFT	RFFT	LRFT	RRFT
1					Comments:					
20					Comments:					
40					Comments:					
60					Comments:					
80					Comments:					
100					Comments:					
120					Comments:					
140					Comments:					
160					Comments:					
180					Comments:					
200					Comments:					

First Simulated Mountain Descent (Paragraph 6.5)—80 snubs, 35–17 mph (56–27 km/h), 8 ft/s² (2.4 m/s²), 15 s interval, first snub at 10 min 15 s.

Snub	Time	LP	LFLT	RFLT	LRLT	RRLT	LFFT	RFFT	LRFT	RRFT
4	11:00									
8	12:00				Comments:					
12	13:00									
16	14:00				Comments:					
20	15:00									
24	16:00				Comments:					
28	17:00									
32	18:00				Comments:					
36	19:00									
40	20:00				Comments:					
44	21:00									
48	22:00				Comments:					

Odometer at end of last stop _____
Time at end of last stop _____ Average speed _____
Ambient temperature range _____

Initial Brake Temperature (Paragraph 6.3)—Make burnish stops as required to warm brakes to 150–200°F (65–95°C) initial brake temperature.

Odometer at end of last stop used to warm brakes _____
Time at end of last stop used to warm brakes (time zero) _____
Ambient temperature _____

First (Cold) Effectiveness Check (Paragraph 6.4)—3 stops, 60–0 mph (97–0 km/h), 15 ft/s² (4.6 m/s²), 200 s interval, 150–200°F initial brake temperature for the first stop, first stop at 30 min 20 s.

Snub	Time	LP	LFLT	RFLT	LRLT	RRLT	LFFT	RFFT	LRFT	RRFT
52	23:00									
56	24:00		Comments:							
60	25:00									
64	26:00		Comments:							
68	27:00									
72	28:00		Comments:							
76	29:00									
80	30:00		Comments:							

Odometer at end of snub 80 _____ Average speed _____
Additional comments: _____

Second (Hot) Effectiveness Check (Paragraph 6.6)—3 stops, 60–0 mph (97–0 km/h), 15 ft/s² (4.6 m/s²), 50 s interval, first stop at 30 min 50 s.

Stop	Time	LP	LFLT	RFLT	LRLT	RRLT	LFFT	RFFT	LRFT	RRFT
1	30:50		Comments:							
2	31:40		Comments:							
3	32:30		Comments:							

Odometer at end of stop 3 _____ Average speed _____
Comments: _____

Recovery (Paragraph 6.7)—10 stops, 40–0 mph (64–0 km/h), 12 ft/s² (3.7 m/s²), 120 s interval, first stop at 34 min 30 s.

Stop	Time	LP	LFLT	RFLT	LRLT	RRLT	LFFT	RFFT	LRFT	RRFT
1	34:30		Comments:							
2	36:30		Comments:							
3	38:30		Comments:							
4	40:30		Comments:							
5	42:30		Comments:							
6	44:30		Comments:							
7	46:30		Comments:							
8	48:30		Comments:							
9	50:30		Comments:							
10	52:30		Comments:							

Odometer at end of stop 10 _____ Average speed _____
Comments: _____

Second Simulated Mountain Descent (Paragraph 6.8)—80 snubs, 35–17 mph (56–27 km/h), 8 ft/s² (2.4 m/s²), 15 s interval, first snub at 52 min 45 s.

Snub	Time	LP	LFLT	RFLT	LRLT	RRLT	LFFT	RFFT	LRFT	RRFT
4	53:30									
8	54:30		Comments:							
12	55:30									

25.127

Snub	Time	LP	LFLT	RFLT	LRLT	RRLT	LFFT	RFFT	LRFT	RRFT
16	56:30		Comments:							
20	57:30									
24	58:30		Comments:							
28	59:30									
32	60:30		Comments:							
36	61:30									
40	62:30		Comments:							
44	63:30									
48	64:30		Comments:							
52	65:30									
56	66:30		Comments:							
60	67:30									
64	68:30		Comments:							
68	69:30									
72	70:30		Comments:							
76	71:30									
80	72:30		Comments:							

Odometer at end of snub 80 _____ Average speed _____
Additional comments: _____

Soak (Paragraph 6.9)—20 brake applications to average line pressure measured in the first effectiveness check at 1 min interval, first application at 73:30.

Time	Apply	LP	LFLT	RFLT	LRLT	RRLT	LFFT	RFFT	LRFT	RRFT
76:30	4		Comments:							
80:30	8		Comments:							
84:30	12		Comments:							
88:30	16		Comments:							
92:30	20		Comments:							

Odometer at application 1 _____
Additional comments: _____

Third (Cold) Effectiveness Check (Paragraph 6.10)—3 stops, 60–0 mph (97–0 km/h), 15 ft/s² (4.6 m/s²), 200 s interval, first stop at 95 min 50 s.

Stop	Time	LP	LFLT	RFLT	LRLT	RRLT	LFFT	RFFT	LRFT	RRFT
1	95:50		Comments:							
2	99:10		Comments:							
3	102:30		Comments:							

Odometer at end of stop 3 _____ Average speed _____
Comments: _____

25.128

Reburnish (Paragraph 6.11)—35 stops, 40–0 mph (64–0 km/h), 12 ft/s² (3.7 m/s²), 250°F initial brake temperature for each stop (but 1 mile (1.6 km) maximum interval).

Odometer at end of first stop _____

Time at end of first stop _____

Stop	Dist	LP	LFLT	RFLT	LRLT	RRLT	LFFT	RFFT	LRFT	RRFT
1			Comments:							
5			Comments:							
15			Comments:							
25			Comments:							
35			Comments:							

Odometer at end of last stop _____

Time at end of last stop _____   Average speed _____

Ambient temperature range _____

Fourth (Cold) Effectiveness Check (Paragraph 6.12)—3 stops, 60–0 mph (97–0 km/h), 15 ft/s² (4.6 m/s²), 200 s interval, 150–200°F initial brake temperature for the first stop, first stop at 3 min 20 s.

Odometer at end of last stop to warm brakes _____

Time at end of last stop used to warm brakes (time zero) _____

Ambient temperature _____

Stop	Time	LP	LFLT	RFLT	LRLT	RRLT	LFFT	RFFT	LRFT	RRFT
1	3:20		Comments:							
2	6:40		Comments:							
3	10:00		Comments:							

Odometer at end of stop 3 _____   Average speed _____

Final Inspection (Paragraph 6.13)

Observations: _____
_____
_____
_____
_____
_____
_____

Vehicle Drag Measurement (Optional) (Paragraph 7)—10 coast-downs, 40–20 mph (64–32 km/h), alternate directions, record 35–20 mph (56–32 km/h) coast-down times.

Coast	Direction	Coast-Down Time
1		
2		
3		
4		
5		
6		
7		
8		
9		
10		

Average coast-down time _____   Transmission gear position _____

Ambient wind velocity _____   Ambient wind direction _____

BRAKE SYSTEM DYNAMOMETER TEST PROCEDURE— PASSENGER CAR—SAE J212 JUN80

SAE Recommended Practice

Report of the Brake Committee, approved July 1971, editorial change March 1973, reaffirmed without change June 1980.

1. Introduction—This SAE Recommended Practice is based upon SAE J843 and is intended to provide a laboratory simulation of vehicle brake system performance (based on simultaneous testing of one front and one rear brake). Certain details of this dynamometer procedure have been purposely left flexible because of varying equipment, and results should not be construed as providing absolute correlation with road tests.

2. Scope—This procedure establishes a uniform laboratory dynamometer method of testing all classes of passenger car brake systems.

3. Purpose—The purpose of the practice is to establish brake system capabilities with regard to:

3.1 Deceleration versus input, as affected by speed, brake temperature, and usage.

3.2 Brake system integrity.

3.3 Stopping ability during emergency or inoperative power assist conditions.

4. Equipment and Instrumentation
 4.1 Equipment
  4.1.1 An inertia type dual brake dynamometer.
  4.1.2 Means for varying brake cooling.
  4.1.3 Means for simulating partial brake system failure (half of system open to atmosphere).
  4.1.4 Means for applying brake system pressure at a specified rate.
 4.2 Instrumentation
  4.2.1 REQUIRED
   4.2.1.1 Means for recording hydraulic line pressures.
   4.2.1.2 Means for recording brake torques.
   4.2.1.3 Means for recording brake lining temperatures.
   4.2.1.4 Means for recording shaft speed.
   4.2.1.5 Cooling air temperature indicators.
   4.2.1.6 Revolutions to stop indicator for measurement of equivalent stopping distance.
  4.2.2 OPTIONAL INSTRUMENTATION
   4.2.2.1 Cooling air velocity indicators.
   4.2.2.2 Drum or disc temperature indication and/or recording equipment.
   4.2.2.3 Fluid displacement indicators.
   4.2.2.4 Stopping time indicator.
 4.3 System Accuracy
  4.3.1 ACCURACY OF INSTRUMENTATION—The overall system accuracy for all recording or indicating instruments shall be ±2% of full-scale or better.
  4.3.2 CONTROL PARAMETER ACCURACY
   4.3.2.1 Line pressures, torques, and temperatures shall be maintained within ±5% of the desired value.
   4.3.2.2 Speed shall be maintained within ±2% of the desired value.
   4.3.2.3 Test moment of inertia shall be within ±1.5 ft-lb-s² of value calculated from paragraph 5.7

5. Test Preparation and Installation Details

5.1 Friction Material Preparation—Attach and finish friction material per manufacturer's specifications, unless otherwise noted.

5.2 Thermocouples—Install plug type thermocouples in each brake as shown in Fig. 1. All thermocouples are to be located in the approximate center of the most heavily loaded shoe, one per brake. Indicate location on data sheet.

5.3 Brake Drum or Disc Assembly—New drums or discs should be used for each test. Surface finish, dimensional characteristics (with special emphasis on thickness variation and runout of rubbing surface), and material properties shall be in accordance with manufacturer's specifications.

5.4 Brakes shall be prepared in accordance with manufacturer's specifications. Adjust brakes to manufacturer's specifications where applicable.

5.5 Brake Mounting—Shall be mounted essentially as in service.

5.6 Hydraulic System—Shall incorporate pressure proportioning valve and/or hold-off valve if used on the vehicle being simulated.

5.7 Test Moment of Inertia—Calculate the moment of inertia required as follows:

$$I = \frac{Wr^2}{2g}$$

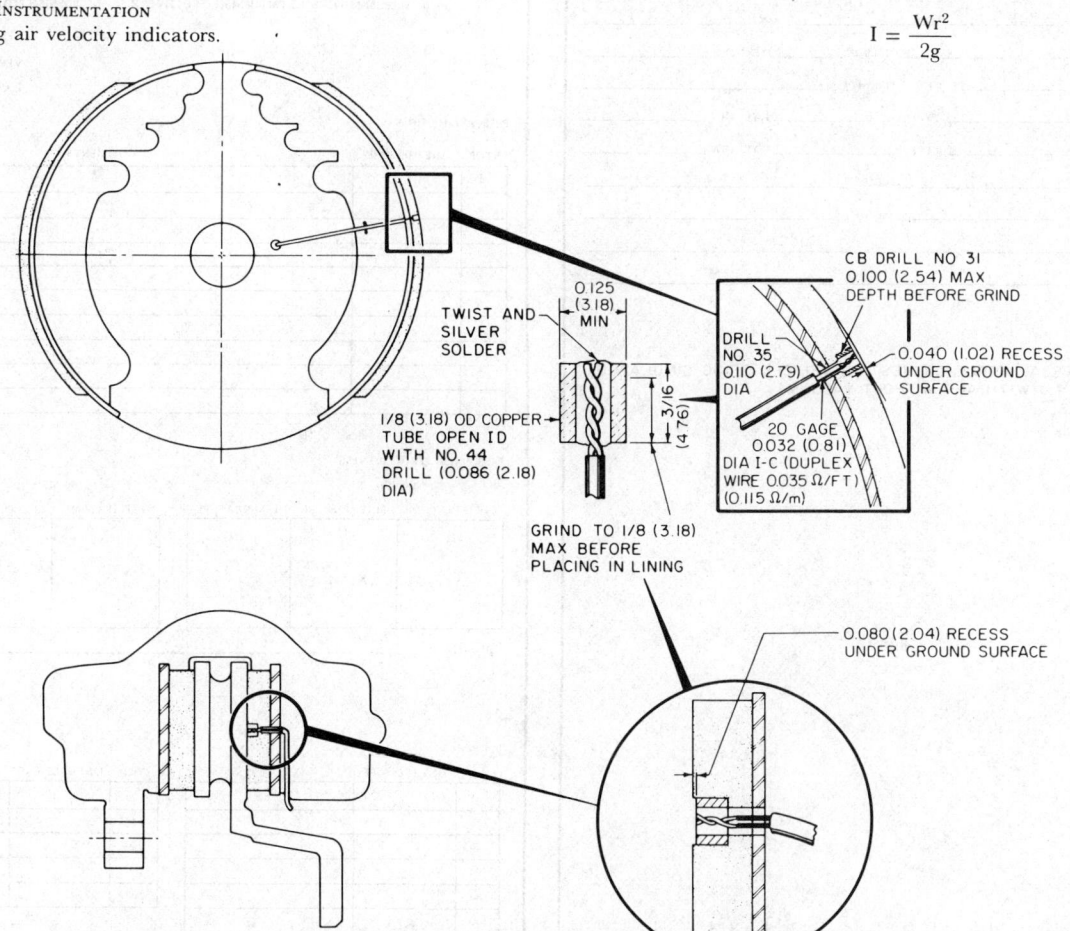

FIG. 1—TYPICAL PLUG THERMOCOUPLE INSTALLATIONS

where: I = moment of inertia required, ft-lb-s²
W = car test weight, lb × 0.86 (0.86 = correction for parasitic losses). Car test weight is normally curb + 600 lb
r = effective radius of tire, ft

$$= \frac{5280}{2\pi \times \text{wheel rpm}}$$

g = 32.2 fpsps

5.8 Test rpm—As required to simulate specified test speeds. Calculate rpm as follows:

$$\text{rpm} = \frac{14.02 \times \text{mph}}{r}, \text{ or rpm} = \frac{\text{Tire rpm} \times \text{mph}}{60}$$

where: r = effective tire radius, ft.

5.9 Test Deceleration—All control decelerations shall be converted to torque for the dynamometer settings by the following formula:

$$T = \frac{W \times r \times a}{2g}$$

where: T = torque required, ft-lb
a = control deceleration, fpsps
W,r,g = from paragraph 5.7

6. Test Procedure—Performance

6.1 Test Notes

6.1.1 During all phases of this procedure, any unusual performance characteristics such as noise, roughness, etc., are to be noted and recorded.

6.1.2 Initial brake temperature is defined as the lining temperature at which the brake application is initiated.

6.1.3 If the brakes require warming to prescribed initial temperature, use the burnish procedure, except cycle time is not to be less than 45 s.

6.1.4 "Sustained" torque or line pressure normally is interpreted herein to mean that torque or line pressure at which a leveling off occurs during a stop. If no leveling off occurs, "sustained" values should be recorded as that indicated at one-half the stopping time.

6.1.5 "Final" torque or line pressure readings shall be taken at an rpm equivalent to 5 mph.

6.1.6 The cooling speed (speed at which the rotating portion of the brake is moving between successive brake applications) shall be set at approximately stop speed for all phases. Stop speed is that at which the brake is applied.

6.1.7 During all phases of this procedure, cooling air speeds for each brake must be controlled to produce the brake temperatures normally experienced on the particular vehicle or vehicle brake system being simulated. In other words, for each brake system, a particular set of cooling air control settings must be worked out on a "baseline" test for which comparable vehicle test data are available. It is especially important that proper temperatures be attained during burnish, fades, and recoveries.

6.1.8 Rate of pressure rise during all phases of the test shall be 1000–2000 psi/s.

6.2 Preburnish Check—In order to allow for a general check of instrumentation, brakes, and dynamometer function, run the following stops: 10 stops, 30 mph, 10 fpsps, 90 s cycle.

6.3 First (Preburnish) Effectiveness Test

6.3.1 Initial brake temperature (each stop)—200 F (hottest brake).

6.3.2 Stop speed—30 and 60 mph.

6.3.3 Test methods—Curve to be defined at each speed by adequate number of points. Optional methods are as follows (specify which used):

6.3.3.1 A minimum of five consecutive stops at constant line pressure increments. If this method is used, use line pressure increments of not over 100 psi at 30 mph to 30 fpsps; of not over 150 psi at 60 mph to 30 fpsps; and of not over 200 psi at 80 mph to 30 fpsps (on second and final effectiveness tests only).

6.3.3.2 A series of consecutive stops at constant deceleration increments. If this method is used, make checks at 5, 10, 15, 20, 25, and 30 fpsps at each speed.

BRAKE DYNAMOMETER TEST DATA

VEHICLE SIMULATED ______ DATE ______

TEST WEIGHT ______ LB ______ FT-LB-S² ______ TEST NO. ______

EFFECTIVE TIRE RADIUS ______ FT TIRE SIZE ______

BRAKES - FRONT: SIZE ______ TYPE ______ CYL DIA ______

REAR: SIZE ______ TYPE ______ CYL DIA ______

BRAKE LININGS - FRONT ______ REAR ______

ROTOR DRUM TYPE - FRONT ______ REAR ______

SPECIAL EQUIPMENT ______

GENERAL INFORMATION ______

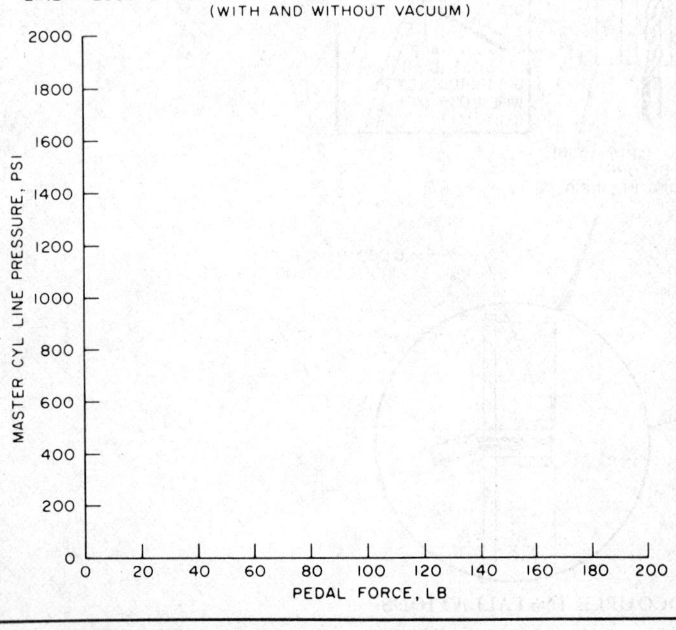

FIG. 2—BRAKE DYNAMOMETER TEST DATA

PREBURNISH CHECK, PREBURNISH EFFECTIVENESS, AND BURNISH REPORT FORM

TEST NO. ______

VEHICLE ______

DATE ______

PREBURNISH CHECK

STOP	LINE PRESSURE	COMMENTS
1		
2		
3		
4		
5		
6		
7		
8		
9		
10		

PREBURNISH EFFECTIVENESS

30 MPH			60 MPH		
DECELERATION	LINE PRESSURE	TEMPERATURE	DECELERATION	LINE PRESSURE	TEMPERATURE

COMMENTS: ______

BURNISH

STOP	LINE PRESSURE	COMMENTS	STOP	LINE PRESSURE	COMMENTS
1			120		
20			140		
40			160		
60			180		
80			200		
100					

FIG. 3—PREBURNISH CHECK, PREBURNISH EFFECTIVENESS, AND BURNISH REPORT FORM

6.3.4 Report—Maximum line pressure on constant deceleration stops, and minimum torque on constant pressure stops.

6.4 Burnish
6.4.1 Stop speed—40 mph.
6.4.2 Stops required—200.
6.4.3 Stop deceleration—12 fpsps.
6.4.4 Stop cycle—As required to maintain 250 F initial brake temperature on hottest brake, or a maximum of 90 s.
6.4.5 Report—Maximum line pressure every 20 stops.

6.5 Second (Burnished) Effectiveness Test—Repeat paragraph 6.3, except add 80 mph stop speed.

6.6 High Speed Stop Test
6.6.1 Stop speed—As achieved by maximum attainable acceleration for 1 mile from zero speed but not to exceed 100 mph (to be determined with actual vehicle being simulated).
6.6.2 Stops required—1.
6.6.3 Stop deceleration—15 fpsps.
6.6.4 Initial brake temperature—150 F (hottest brake).
6.6.5 Report—Maximum line pressure and deceleration if 15 fpsps cannot be held.

6.7 First Reburnish—Repeat paragraph 6.4, except 35 stops.

6.8 First Fade and Recovery Test
6.8.1 Baseline Check Stops
6.8.1.1 Stop speed—30 mph.
6.8.1.2 Stops required—3.
6.8.1.3 Stop deceleration—10 fpsps.
6.8.1.4 Initial brake temperature—150 F hottest brake each stop.
6.8.1.5 Report—Maximum line pressures.
6.8.2 Fade
6.8.2.1 Stop speed—60 mph.
6.8.2.2 Stops required—10.
6.8.2.3 Stop deceleration—15 fpsps.
6.8.2.4 Initial brake temperature—150 F hottest brake for first stop.
6.8.2.5 Stop cycle—35 s.
6.8.2.6 Report—Maximum line pressure, initial brake temperature, cooling air temperature, deceleration values.
6.8.3 Note—Run 90 s at 30 mph after last fade stop and make first recovery stop.
6.8.4 Recovery
6.8.4.1 Stop speed—30 mph.
6.8.4.2 Stops required—12.
6.8.4.3 Stop deceleration—10 fpsps.
6.8.4.4 Stop cycle—2 min.
6.8.4.5 Report—Same as for fade run (paragraph 6.8.2.6).

6.9 First Effectiveness Spot Check
6.9.1 Stop speed—60 mph.
6.9.2 Stops required—2.
6.9.3 Stop deceleration—15 fpsps.
6.9.4 Initial brake temperature—200 F hottest brake each stop.
6.9.5 Report—Maximum line pressure.

6.10 Second Reburnish—Repeat paragraph 6.4, except 35 stops.
6.11 Second Fade and Recovery Test—Repeat paragraph 6.8, except 15 fade stops in paragraph 6.8.2.2.
6.12 Second Effectiveness Spot Check—Repeat paragraph 6.9.
6.13 Third reburnish—Repeat paragraph 6.4, except 35 stops.
6.14 Final Effectiveness Test—Repeat paragraph 6.5.
6.15 Fourth Reburnish—Repeat paragraph 6.4, except 35 stops.
6.16 Emergency Brake System and Inoperative Power System Test
6.16.1 Test Notes
6.16.1.1 Calculate wheel revolutions equivalent to 600 ft for the vehicle being simulated.
6.16.1.2 Obtain "with vacuum" and "no vacuum" pedal forceline pressure calibration curves for the vehicle being simulated.
6.16.1.3 All stops in this section are to be made at constant line pressure from 60 mph at an initial brake temperature of 150 F (hottest brake).
6.16.2 Stopping Test with Failed Front System
6.16.2.1 Determine the constant *master cylinder* hydraulic pressure to stop in the equivalent of 500 + 0, − 60 ft, using the rear brake only (front system open to atmosphere).

FIG. 4—BURNISHED EFFECTIVENESS, HIGH SPEED STOP, FIRST REBURNISH, AND FIRST BASELINE CHECK REPORT FORM

FIG. 5—FIRST FADE, FIRST RECOVERY, AND FIRST EFFECTIVENESS SPOT CHECK REPORT FORM

6.16.2.2 Report—Constant master cylinder pressure to stop with rear brake only (P_r), the "with vacuum" pedal force to produce P_r (F_r), the actual revolutions to stop (RTS_r).

6.16.3 Stopping Test with Failed Rear System

6.16.3.1 Determine the constant *master cylinder* hydraulic pressures to stop in the equivalent of 600 + 0, −60 ft using the front brake only (rear system open to atmosphere).

6.16.3.2 Report—Constant master cylinder pressure to stop with front brake only (P_f), the "with vacuum" pedal force to produce P_f (F_f), the actual revolutions to stop (RTS_f).

6.16.4 Stopping Test with Inoperative Power System

6.16.4.1 From the "no vacuum" pedal force-line pressure calibration curve, determine the master cylinder hydraulic pressure (p_{nv}) produced by a pedal force of 200 lb.

6.16.4.2 Determine the revolutions to stop (RTS_{nv}) for a master cylinder hydraulic line pressure of P_{nv} using both front and rear brakes.

6.16.4.3 Report—Revolutions to stop (RTS_{nv}).

6.17 Final Inspection—Disassemble brakes, inspect, and record all pertinent observations.

*7. **Report Forms***

7.1 Brake Dynamometer Test Data, Fig. 2.

7.2 Preburnish Check, Preburnish Effectiveness, and Burnish Report Form, Fig. 3.

7.3 Burnished Effectiveness, High Speed Stop, First Reburnish, and First Baseline Check Report Form, Fig. 4.

7.4 First Fade, First Recovery, and First Effectiveness Spot Check Report Form, Fig. 5.

7.5 Second Reburnish, Second Baseline Check, and Second Fade Report Form, Fig. 6.

7.6 Second Recovery, Second Effectiveness Spot Check, Third Reburnish, and Final Effectiveness Report Form, Fig. 7.

7.7 Fourth Reburnish, and Emergency Brake System and Inoperative Power System Test Report Form, Fig. 8.

SECOND REBURNISH, SECOND BASELINE CHECK, AND SECOND FADE REPORT FORM

TEST NO. ____________

VEHICLE ____________

DATE ____________

SECOND REBURNISH

STOP	LINE PRESSURE	COMMENTS
1		
10		
25		
35		

SECOND BASELINE CHECK

STOP	LINE PRESSURE	COMMENTS
1		
2		
3		

SECOND FADE

STOP	DECELERATION	LINE PRESSURE	INITIAL TEMPERATURES	COMMENTS
1				
2				
3				
4				
5				
6				
7				
8				
9				
10				
11				
12				
13				
14				
15				

COOLING AIR TEMPERATURE:

FIG. 6—SECOND REBURNISH, SECOND BASELINE CHECK, AND SECOND FADE REPORT FORM

SECOND RECOVERY, SECOND EFFECTIVENESS SPOT CHECK, THIRD REBURNISH, AND FINAL EFFECTIVENESS REPORT FORM

TEST NO. ____________

VEHICLE ____________

DATE ____________

SECOND RECOVERY

STOP	DECELERATION	LINE PRESSURE	INITIAL TEMPERATURES	COMMENTS
1				
2				
3				
4				
5				
6				
7				
8				
9				
10				
11				
12				

SECOND EFFECTIVENESS SPOT CHECK

STOP	DECELERATION	LINE PRESSURE	INITIAL TEMPERATURES	COMMENTS
1				
2				

THIRD REBURNISH

STOP	LINE PRESSURE	COMMENTS
1		
10		
25		
35		

FINAL EFFECTIVENESS

30 MPH			60 MPH			80 MPH		
DECELERATION	LINE PRESSURE	TEMPERATURES	DECELERATION	LINE PRESSURE	TEMPERATURES	DECELERATION	LINE PRESSURE	TEMPERATURES

FIG. 7—SECOND RECOVERY, SECOND EFFECTIVENESS SPOT CHECK, THIRD REBURNISH, AND FINAL EFFECTIVENESS REPORT FORM

FOURTH REBURNISH, AND EMERGENCY BRAKE SYSTEM AND INOPERATIVE POWER SYSTEM TEST REPORT FORM

TEST NO. ____________

VEHICLE ____________

DATE ____________

FOURTH REBURNISH

STOP	LINE PRESSURE	COMMENTS
1		
10		
25		
35		

EMERGENCY BRAKE SYSTEM AND INOPERATIVE POWER SYSTEM TEST

WHEEL REVOLUTIONS EQUIVALENT TO 600 FT ____________

FAILED FRONT SYSTEM

CONSTANT MASTER CYLINDER PRESSURE TO STOP IN 600 FT WITH REAR ONLY (P_r) ____________

"WITH VACUUM" PEDAL FORCE TO PRODUCE P_r (F_r) ____________

ACTUAL REVOLUTIONS TO STOP (RTS_r) ____________

FAILED REAR SYSTEM

CONSTANT MASTER CYLINDER PRESSURE TO STOP IN 600 FT WITH FRONT ONLY (P_f) ____________

"WITH VACUUM" PEDAL FORCE TO PRODUCE P_f (F_f) ____________

ACTUAL REVOLUTIONS TO STOP (RTS_f) ____________

INOPERATIVE POWER SYSTEM

"NO VACUUM" MASTER CYLINDER PRESSURE PRODUCED BY 200 LB PEDAL FORCE (Pnv) ____________

REVOLUTIONS TO STOP USING Pnv ____________

FIG. 8—FOURTH REBURNISH, AND EMERGENCY BRAKE SYSTEM AND INOPERATIVE POWER SYSTEM TEST REPORT FORM

SERVICE BRAKE STRUCTURAL INTEGRITY TEST PROCEDURE—PASSENGER CAR—SAE J229 JUN80

SAE Recommended Practice

Report of the Brake Committee and Automotive Safety Committee, approved March 1971, reaffirmed without change June 1980.

1. Scope—This SAE Recommended Practice establishes a method of evaluating the structural integrity of the entire brake system of all passenger cars under extreme braking conditions.

2. Purpose—The main purpose of this recommended practice is to evaluate the structural integrity of a vehicle's braking system. However, other areas, such as the steering or suspension system, may also be evaluated during the test, providing that the criteria and procedure detailed below are not modified in any way. For repeatability, it is recommended that a brake apply device be utilized whenever possible, since it will eliminate the variations in application times and efforts of different operators.

3. Equipment

3.1 Brake apply device (optional, but recommended).

3.2 Calibrated speedometer and odometer (a fifth wheel pousometer may be used as an alternative).

3.3 Pedal force transducer (30 Hz minimum response if used for spike stops.

3.4 Decelerometer.

3.5 Ambient temperature gage.

3.6 Recording equipment (pedal force versus time) 30 Hz minimum response.

3.7 Tire pressure gage.

3.8 Wheel alignment equipment.

3.9 Torque wrench.

3.10 Brake lining temperature measuring instrumentation.

4. Test Preparation

4.1 Calibrate instrumentation as required and note calibration on data sheet.

4.2 Install new brake drums and/or rotors and brake assemblies to the manufacturer's specifications with special attention to the torque specifications of all brake fasteners.

4.3 Adjust brakes per manufacturer's specifications.

4.4 Check front and rear alignment, adjust to manufacturer's mean specifications, and record. Rear wheel toe is not required on a vehicle with a solid axle.

4.5 Vehicle test weight shall be curb weight, plus additional weights of accessories or optional equipment over 5 lb which are offered but not installed, plus a four-passenger load of 600 lb. On vehicles designed for less than four passengers, use 150 lb per passenger. All test equipment shall be part of this weight.

4.6 Install the tires and wheels offered for the vehicle by the manufacturer which produce the largest moment of inertia. Tires must be in good condition. Set tire pressure per manufacturer's specifications for vehicle test weight specified in this recommended practice.

4.7 Install plug type thermocouples in each brake (reference SAE J843). All thermocouples to be located in the approximate center of the most heavily loaded shoe, one per brake.

5. General Notes

5.1 All tests shall be conducted on a substantially level (not to exceed a ±1% grade), dry, smooth, hard surfaced roadway of Portland cement concrete (or other surface with equivalent coefficient of surface friction) that is free from loose materials.

5.2 A spike brake application is accomplished by applying a pedal force of 200 lb (100 lb overshoot permitted) while recording pedal force versus time. Rate of apply shall be 2500 lb/s. To achieve this rate, instantaneous rates can vary from 1000 to 4000 lb/s.[1] At least 160 lb of the 200 lb force shall be within this tolerance. See Fig. 1 for a typical spike brake curve. Maintain pedal force until car has stopped.

5.3 In any series of spike brake applications, the initial brake temperature for the first stop shall be 200°F. Initial brake temperature is defined as 0.2 mile before stop (average temperature of brakes on hottest axle), brakes off. If brakes require warming, use burnishing procedure (paragraph 6.3.1).

5.4 Spike brake applications are to be made in normal driving gear. On cars with manual transmission, disengage the clutch.

5.5 Driving speed between spike brake applications to be at the subsequent test speed. Acceleration to all test speeds is to be moderate.

5.6 Vehicle function stops (paragraphs 6.2.2, 6.4.2, etc.) may be made as soon as convenient after spike brake applications and need not be made at any specific initial brake temperature.

5.7 During all phases of this procedure, note and record any unusual braking or handling characteristics of the vehicle.

6. Procedure

6.1 *Preburnish Check*—In order to allow for a general check of instrumentation, brakes, and vehicle function, the following stops are to be run, noting pedal force for each stop:

Number of Stops—10.
Speed—30-0 mph.
Deceleration—10 fpsps.
Interval—1 mile.
Cooling Speed—40 mph, normal driving gear.

6.2 *Preburnish Spike Brake Application*

6.2.1 Make one spike brake application from 20 mph and immediately drive 0.5 mile and make one spike brake application from 40 mph.

6.2.2 Check brakes and vehicle function by making a stop from 30 mph at 10 fpsps. Note pedal force.

6.3 *Burnish*

6.3.1 Make 200 burnishing stops from 40 mph at 12 fpsps in normal driving gear. Stop interval shall be as required to achieve 250 F "initial brake temperature" or a maximum of 1.0 mile.

NOTE: The 1.0 mile maximum must be observed even though the initial temperature exceeds 250 F. Cooling speed shall be 40 mph with moderate acceleration to cooling speed.

6.3.2 Adjust brakes after burnishing per manufacturer's specifications.

6.4 *Post Burnish Spike Brake Applications*

6.4.1 Make one spike brake application at each of the following speeds at 1.0 mile intervals: 10, 20, and 40 mph.

6.4.2 Check brakes and vehicle function by making a stop from 30 mph at 10 fpsps. Note pedal force.

6.4.3 Make two consecutive spike brake applications in reverse from 10 mph (estimated).

6.4.4 Check brakes and vehicle function by making a stop from 30 mph at 10 fpsps. Note pedal force.

6.4.5 Make two spike brake applications from 10 mph at 0.25 mile intervals.

6.4.6 Check brakes and vehicle function by making a stop from 30 mph at 10 fpsps. Note pedal force.

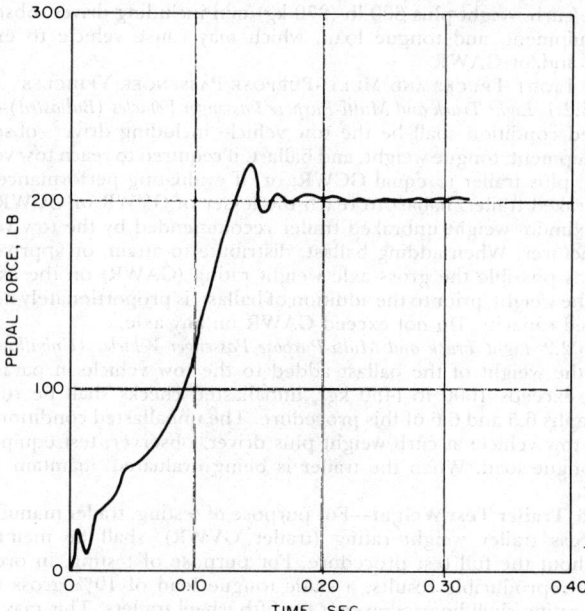

FIG. 1—TYPICAL SPIKE BRAKE APPLICATION

---

[1] Although tolerances of 1000-4000 lb/s may appear too broad, actual test curves, both manual and machine applied, show that instantaneous slopes do vary considerably. Furthermore, 40 lb of the 200 lb pedal force has been exempted from this rate in order to accommodate the typical dip in the curve, as well as miscellaneous short duration spikes. Thus, it is intended that the basic slope of the curve should be 2500 lb/s for 160 lb of the 200 lb force. However, slight deviations in the curve must be tolerated in order to achieve this rate.

6.4.7 Make three spike brake applications from 40 mph at 1.0 mile intervals.

6.4.8 Check brakes and vehicle function by making a stop from 30 mph at 10 fpsps. Note pedal force.

6.5 Inspection

6.5.1 Check and record front end alignment and rear wheel toe. (Rear wheel toe is not required on vehicles with solid rear axle.)

6.5.2 Inspect all components of the brake system.

ϕ BRAKE SYSTEM ROAD TEST CODE—PASSENGER CAR AND LIGHT DUTY TRUCK-TRAILER COMBINATIONS—SAE J134 MAY85

SAE Recommended Practice

Report of the Brake Committee, approved December 1970, completely revised June 1979, reaffirmed without change May 1985.

1. Introduction—This recommended practice, in conjunction with SAE J135a (June, 1979),[1] is intended for use primarily by:

(a) Tow vehicle manufacturers' testing with unbraked trailers to determine the maximum unbraked trailer weight which can be towed;

(b) Tow vehicle manufacturers' testing with braked trailers to evaluate tow vehicle braking performance for vehicle combinations;

(c) Trailer or brake system suppliers to evaluate trailer brake and actuation system performance.

This procedure assumes a tow vehicle complying with existing applicable legal requirements. It is recommended that tow vehicles incorporate that manufacturer's trailering package. Tow vehicle manufacturer's recommendations regarding hitch type shall be followed. Trailer loading shall be in accordance with trailer manufacturer's recommendations except as modified in this procedure. Tires shall be inflated to vehicle manufacturer's recommendations.

2. Scope—This SAE Recommended Practice establishes a uniform procedure for the level road test of the brake systems of all combinations of new multi-purpose passenger vehicles, new light-duty trucks up to and including 10 000 lb (4500 kg) GVW and new passenger cars when coupled with new trailers (braked or unbraked).

3. Purpose—The purpose of the test code is to establish a uniform test procedure to determine capabilities with regard to:

3.1 Deceleration versus input, as affected by vehicle speed, brake temperature and usage.

3.2 Brake system integrity within the limits of this test.

3.3 Stopping ability during:

3.3.1 Emergency (partial brake) conditions; and,

3.3.2 Inoperative power assist conditions.

4. Instrumentation

4.1 Tow vehicle line pressure and/or pedal force gage.

4.2 Decelerometer (U-tube or equivalent).

4.3 Direct reading temperature instrument.

4.4 Speedometer (calibrated vehicle unit or fifth wheel type).

4.5 Tire pressure gage.

4.6 Odometer (calibrated).

4.7 Thermometer-ambient (or ambient sensitive thermocouple).

4.8 Stopmeter (fifth wheel, distance only).

4.9 Voltmeter and ammeter (where applicable).

4.10 Stop watch.

4.11 Trailer line pressure gage (where applicable).

4.12 Optional instrumentation.

4.12.1 Pedal travel gage.

4.12.2 Stop counter.

4.12.3 Strain gage ball or equivalent and required equipment to record fore and aft loads imposed on the tow vehicle.

5. Installation Details

5.1 **Friction Material Preparation**—Attach and finish friction material per vehicle manufacturer's specifications.

5.2 **Thermocouples**—Install thermocouples in each tow vehicle and trailer brake per current SAE J843d (March, 1973), Brake System Road Test Code—Passenger Car.

5.3 **Brake Drum (or Rotor) and Hub Assembly**—New drums (or rotors) recommended for each complete test (Section 6). Surface finish and dimensional characteristics including runout of rubbing surface to be in accordance with manufacturer's specifications.

5.4 **Brake Assembly**

5.4.1 Tow Vehicle—Brakes to be prepared in accordance with manufacturer's specifications. New springs and linings recommended on all brakes for each complete test (Section 6). Adjust brakes to manufacturer's specifications.

5.4.2 Trailer—Applicable only when evaluating trailers or combination brake systems. For all other tests, trailers should be unbraked. Brakes are to be prepared in accordance with manufacturer's specifications. New springs, linings, magnets, and armatures where applicable are recommended on all brakes for each complete test (Section 6). Adjust brakes to manufacturer's specifications.

5.5 **Tow Vehicle Test Weight**

5.5.1 Passenger Cars

5.5.1.1 *Passenger Car (Rated for less than four passengers)*—Test loading shall be curb weight plus 300 lb (135 kg) min including driver, observer, test equipment, and ballast, if necessary. Tongue load is additional, which may cause vehicle to exceed GVWR and/or GAWR.

5.5.1.2 *Passenger Car (Rated for four or more passengers)*—Test loading shall be curb weight plus 600 lb (270 kg) min including driver, observer, test equipment, and tongue load, which may cause vehicle to exceed GVWR and/or GAWR.

5.5.2 Light Trucks and Multi-Purpose Passenger Vehicles

5.5.2.1 *Light Truck and Multi-Purpose Passenger Vehicles (Ballasted)*—The ballasted condition shall be the tow vehicle including driver, observer, test equipment, tongue weight, and ballast, if required to reach tow vehicle GVWR, plus trailer to equal GCWR; or, if evaluating performance with an unbraked trailer, ballast to reach the lesser of GVWR or GCWR with the maximum weight unbraked trailer recommended by the tow vehicle manufacturer. When adding ballast, distribute to attain, or approach as nearly as possible the gross axle weight rating (GAWR) on the axle on which the weight, prior to the addition of ballast, is proportionately nearest the rated capacity. Do not exceed GAWR on any axle.

5.5.2.2 *Light Truck and Multi-Purpose Passenger Vehicles (Unballasted)*—When the weight of the ballast added to the tow vehicle in paragraph 5.5.2.1 exceeds 1000 lb (450 kg), unballasted checks shall be run per paragraphs 6.5 and 6.6 of this procedure. The unballasted condition shall be the tow vehicle at curb weight plus driver, observer, test equipment, plus tongue load. When the trailer is being evaluated, maintain trailer GVWR.

5.6 **Trailer Test Weight**—For purpose of testing, trailer manufacturer's gross trailer weight rating (trailer GVWR), shall be maintained throughout the full test procedure. For purpose of testing, in order to achieve reproducible results, a static tongue load of 10% gross trailer weight rating shall be used except for fifth wheel trailers. This may cause trailer axle loading in excess of manufacturer's ratings. Fifth wheel trailers shall follow manufacturer's recommendations.

---

[1] For passenger cars only. Light truck requirements to be determined.

5.6.1 The tow vehicle manufacturer's maximum recommended gross trailer weight shall be maintained when the tow vehicle is being evaluated except when paragraph 5.5.2.1 is applicable.

5.6.2 The trailer manufacturer's gross trailer weight rating (trailer GVWR) shall be maintained when the trailer is being evaluated.

5.7 Weight Distributing Hitch Adjustment—When a weight distributing hitch is used, the hitch shall be adjusted as follows:

The hitch ball height on the tow vehicle prior to hook-up, at curb weight, shall be set so that the trailer is level. Check the height from hitch to ground on the tow vehicle. Connect trailer and adjust the hitch so that the hitch point checked is as high as but not more than an inch higher than before the trailer was connected.

6. Test Procedure

6.1 Test Notes

6.1.1 Effectiveness, stopping distance, fade, and recovery test stops shall be conducted on a substantially level (not to exceed a ±1% grade), dry, smooth, hard-surfaced roadway of Portland cement concrete (or other surface with equivalent coefficient of surface friction) that is free from loose materials.

6.1.2 During all phases of this procedure, any unusual performance such as wrap-up or noise characteristics are to be noted and recorded. Also note, at the appropriate stops, which wheel or wheels of the tow vehicle or trailer skidded. Note any uncontrollable braking action causing either of the vehicles to pull or swerve out of a 12 ft (3.7 m) wide roadway lane.

6.1.3 Initial brake temperature is considered to be the average temperature of brakes on the hottest axle with brakes off 0.2 mile (0.3 km) before stop.

6.1.4 If brakes require warming to a prescribed temperature, use burnish procedure and shorten interval if necessary.

6.1.5 Because variations in ambient temperature have a significant effect on test results, fade and recovery tests must be conducted within a range of ambient temperature of 40–90°F (4.4–32.2°C).

6.1.6 Decelerations used in the various fade, recovery, or warm-up procedures refer to values at which the decelerometer is held approximately constant during the stop by varying the input pressure.

6.1.7 Deceleration and line pressure (pedal force) readings shall not be taken below 5 mph (8 km/h).

6.1.8 Vehicles with manual transmissions should be declutched below 10 mph when stops are made in normal driving gear.

6.2 Preburnish Check
—In order to allow for a general check of instrumentation, brakes, and vehicle function, the following stops are to be run: 10 stops, 30–0 mph (48–0 km/h), 10 fpsps (3 m/s²), 1 mile (1.6 km) interval, 40 mph (64 km/h) cooling speed in normal driving gear. Record tow vehicle line pressure (pedal force) and trailer brake input.

Note: Assuming instrumentation, brakes, and vehicle are functioning satisfactorily, proceed immediately with First Effectiveness Test.

6.3 First (Preburnish) Effectiveness Test
—Initial brake temperature, 200°F (93.3°C) before each application.

Stop Speed—30 and 60 mph (48 and 97 km/h) (full stops in neutral).

Increments—Curve to be defined to point of loss of lateral control or 16 fpsps (4.9 m/s²) by adequate number of points (wheel slide permitted).

Record—Deceleration, tow vehicle line pressure (pedal force), trailer brake input, and method of brake application (that is, machine or manual). When using manual method, full stops are to be made at each deceleration level and maximum line pressure (pedal force) recorded. Optional—Record fore and aft load at ball.

6.4 Burnish

Stop Speed—40–0 mph (64–0 km/h).

Stop Deceleration—12 fpsps (3.7 m/s²) (in normal gear).

Stop Interval—As required to achieve 250°F (121°C) *initial brake temperature*[2] or a maximum of 1 mile (1.6 km).

Note: The 1 mile (1.6 km) maximum must be observed even though the initial temperature exceeds 250°F (121°C).

Cooling Speed—40 mph (64 km/h) (moderate acceleration to cooling speed).

Stops Required—200. Record tow vehicle line pressure (pedal force), trailer brake input and brake temperature for stops 1, 20, and each succeeding 20th stop.

Optional—Record fore and aft load at ball.

After Burnish Cycle:
  (a) Inspect and adjust trailer brakes.
  (b) Inspect and adjust towing vehicle brakes.
  (c) Record any operations performed.

[2] See test notes paragraph 6.1.3.

6.5 Second Effectiveness Test
—Repeat paragraph 6.3. Also, if tow vehicle additional payload capacity (ballast) with trailer coupled exceeds 1000 lb (450 kg), repeat this test in the unballasted condition. (See paragraph 5.5.2.2.) In this case, leave tow vehicle unballasted for the next test sequence.

6.6 Emergency System Test

6.6.1 Initial Brake Temperature—150°F (65.6°C) before each stop.

6.6.2 Procedure—With one subsystem of the tow vehicle brake system open to atmosphere, determine the shortest stopping distances, (a) with 150 lb (667 N) maximum allowable pedal force and, if no more than one wheel slides, (b) with 200 lb (890 N) maximum allowable pedal force. Stops are to be made in normal driving gear from 60 mph (97 km/h) without any portion of the vehicles leaving a 12 ft (3.7 m) lane. Repeat the procedure for each other subsystem of the tow vehicle brake system open to atmosphere. Three stops are to be made at each test condition and the average of the three recorded in the summary sheet.

6.6.3 Record—Pedal forces (maximum) and stopping for each failure mode. If first portion of this test was run at the unballasted condition (paragraph 5.5.2.2) following paragraph 6.5, reballast to paragraph 5.5.2.1 and repeat this section.

6.7 Inoperative Power System Test

6.7.1 Initial Brake Temperature—150°F (65.6°C).

6.7.2 Procedure—With the tow vehicle brake system's primary source of power assist inoperative and its reserve depleted, determine the shortest stopping distances, (a) with 150 lb (667 N) maximum allowable pedal force and, if no more than one wheel slides, (b) with 200 lb (890 N) maximum allowable force. Stops are to be made in normal driving gear from 60 mph (97 km/h) without leaving a 12 ft (3.7 m) lane. Three stops are to be made at each test condition and the average of the three recorded in the summary sheet.

6.7.3 Record—Pedal forces (maximum) and stopping distances.

6.8 First Fade and Recovery Test

6.8.1 Baseline Check Stops

  Initial Brake Temperature—150°F (65.6°C) before each stop.
  Stops Required—3.
  Stop Speed—30–0 mph (48.0 km/h).
  Stop Deceleration—10 fpsps (3 m/s²) (in normal driving gear).
  Record—Tow vehicle line pressure (pedal force) and trailer brake input.

6.8.2 Fade

  Initial Brake Temperature—150°F (65.6°C) before first stop.
  Stops Required—10.
  Stop Speed—60–0 mph (97–0 km/h).
  Stop Deceleration—15 fpsps (4.6 m/s²) (in normal driving gear) or maximum obtainable at 200 lb (890 N) pedal force (or equivalent line pressure).
  Stop Interval—0.8 miles (1.2 km).
  Cooling Speed—60 mph (97 km/h).
  Acceleration to Cooling Speed—Intermediate at a moderate rate.
  Record—Maximum tow vehicle line pressure (pedal force) and deceleration [if 15 fpsps (4.6 m/s²) cannot be held] and trailer brake input. Brake temperatures 0.2 mile (0.3 km) before every stop, all brakes. Ambient air temperature at beginning of run. Total elapsed time from end of the first fade stop to end of last fade stop—to maintain a check on driver consistency and car performance.
  Optional—Record fore and aft load at ball.

Note: Drive 1 mile at 40 mph (1.6 km at 64 km/h) immediately after last fade stop and make first recovery stop.

6.8.3 Recovery

  Stops Required—12.
  Stop Speed—30–0 mph (48–0 km/h).
  Stop Deceleration—10 fpsps (3 m/s²) (in normal driving gear), or maximum obtainable at 200 lb (890 N) pedal force (or equivalent line pressure).
  Stop Interval—1 mile (1.6 km).
  Cooling Speed—40 mph (64 km/h).
  Rate of Acceleration to Cooling Speed—Moderate.
  Record—Maximum tow vehicle line pressure (pedal force), deceleration [if 10 fpsps (3 m/s²) cannot be held] and trailer brake input. Initial brake temperatures before every stop, all brakes. Optional—Record fore and aft load at ball.

6.9 First Effectiveness Spot Check

  Initial Brake Temperature—200°F (93.3°C) before each stop.
  Stops Required—2.
  Stop Speed—60–0 mph (97–0 km/h).
  Stop Deceleration—15 fpsps (4.6 m/s²) (in normal driving gear).
  Record—Maximum tow vehicle line pressure (pedal force) and trailer brake input. Optional—Record fore and aft load at ball.

6.10 First Reburnish—Repeat paragraph 6.4, except 35 stops required.

6.11 Second Fade and Recovery Test—Repeat paragraph 6.8, except 15 fade stops required.

6.12 Second Effectiveness Spot Check—Repeat paragraph 6.9.

6.13 Second Reburnish—Repeat paragraph 6.10.

6.14 Third Effectiveness Test—Repeat paragraph 6.3.

6.15 Final Inspection—Disassemble all brakes, inspect and record all pertinent observations.

7. Report Forms—The recommended report forms listed provide space for the data required for this road test code as well as non-mandatory data.

7.1 General Data, Fig. 1.
7.2 Summary Sheet, Fig. 2.
7.3 Input Correlation and Preburnish Check Data Sheet, Fig. 3.
7.4 First (Preburnish) Effectiveness Data Sheet, Fig. 4.
7.5 Burnish and Inoperative Power System Test Data Sheet, Fig. 5.
7.6 Emergency System Test Data Sheet, Fig. 6.
7.7 Second Effectiveness Test Data Sheet, Fig. 7.
7.8 First Baseline Check and First Fade Data Sheet, Fig. 8.
7.9 First Recovery, First Effectiveness Spot Check and First Reburnish Data Sheet, Fig. 9.
7.10 Second Baseline Check and Second Fade Test Data Sheet, Fig. 10.
7.11 Second Recovery, Second Effectiveness Spot Check and Second Reburnish Data Sheet, Fig. 11.
7.12 Third Effectiveness Test Data Sheet, Fig. 12.
7.13 Final Inspection Data Sheet, Fig. 13.

TRAILER
  Trailer Make _____ Model _____ Year _____
  Number of Axles _____ Number of Brakes _____ Tire Size _____
  Tire Mfg. and Type _____ Tire Pressure _____
  Weight _____ lb (kg) + _____ lb (kg) Ballast = _____ lb (kg) (Uncoupled)
  Tongue Load at Coupling _____ Percent of Total _____ %
  Type of Hitch _____
  Trailer Axle(s) Weight (Coupled):
  Front _____ lb (kg) Rear _____ lb (kg) Total _____ lb (kg)
  Brakes
  Size _____ Type _____ Cyl Dia _____
  Lining _____
  Type of Actuation _____

TOWING VEHICLE
  Make _____ Model _____ Year _____
  Engine _____ Transmission _____ Axle Ratio _____
  Curb Weight: Front _____ lb (kg) Rear _____ lb (kg) Total _____ lb (kg)
  Test Weight: Front _____ lb (kg) Rear _____ lb (kg) Total _____ lb (kg)
    (Trailer Coupled)
  GVWR or GCWR Weights _____ _____ _____ _____
  Minimum Weights _____ _____ _____ _____
  Tire Mfgr _____ Size _____ Pressure: F _____ R _____
  Brakes
  Front Size _____ Description Type _____ Cyl Dia _____
  Rear Size _____ Description Type _____ Cyl Dia _____
  Lining Front _____ Rear _____
  Drum (Disc) Type: Front _____ Rear _____
  Master Cyl Dia _____ Stroke _____ Split: Front _____ % Rear _____ %
  Pedal: Pedal Ratio _____ Available Travel _____
  Power Brake: Yes _____ No _____ Type _____
  Hydraulic System   Front Metering   Rear Proportioning   Other
         _____ psi (kPa) Split _____ psi (kPa)
                        Slope _____

TEST INFORMATION
  Thermocouple Installation Method _____
  Tested by _____ Location _____
  Date: Test Started _____ Test Completed _____
  Ambient Temperature Range: High _____ °C (°F) Low _____ °C (°F)

FIG. 1—GENERAL DATA SHEET

Test No. _____
Test Phase   Actual
Preburnish Check   _____ Min _____ Max lb (N) Pedal Force
Effectiveness Tests   1st   2nd   3rd
  30 mph (48 km/h)
  at 16 ft/s² (5.2 m/s²)   _____ _____ _____ lb (N) Pedal Force
  60 mph (97 km/h)
  at 16 ft/s² (5.2 m/s²)   _____ _____ _____ lb (N) Pedal Force
Emergency System Test
  Warning System Actuation   Type: Power _____ Manual _____
  60 mph (97 km/h) Stopping Distance
                    GVWR or GCWR
  Front Operating _____ ft (m) _____ lb (N) Pedal Force
  Front Operating _____ ft (m) _____ lb (N) Pedal Force
  Rear Operating _____ ft (m) _____ lb (N) Pedal Force
  Rear Operating _____ ft (m) _____ lb (N) Pedal Force
                Minimum Tow Vehicle Weight
  Front Operating _____ ft (m) _____ lb (N) Pedal Force
  Front Operating _____ ft (m) _____ lb (N) Pedal Force
  Rear Operating _____ ft (m) _____ lb (N) Pedal Force
  Rear Operating _____ ft (m) _____ lb (N) Pedal Force
Inoperative Power System Test
  60 mph (97 km/h) Stopping Distance
  _____ ft (m) _____ lb (N) Pedal Force
  _____ ft (m) _____ lb (N) Pedal Force
First Fade and Recovery Test
  Fade Stops 1-4
  _____ _____ _____ lb (N) Pedal Force (or Min Decel)
  Recovery Stops 1-5
  _____ ft/s² (m/s²) at _____ lb (N) Max Pedal Force
  Recovery Stops 6-12
  _____ lb (N) Max Pedal Force
Second Fade and Recovery Test
  Fade Stops 1-8
  _____ _____ _____ _____ _____
  _____ _____ lb (N) Pedal Force (or Min Decel)
  Recovery Stops 1-5
  _____ ft/s² (m/s²) at _____ lb (N) Max Pedal Force
  Recovery Stops 6-12
  _____ lb (N) Max Pedal Force
Stability During   Controllable Braking Through
Effectiveness Tests   16 ft/s² (5.2 m/s²)
                                    Yes _____ No _____
Final Inspection
  Lining Integrity   Yes _____ No _____
  Mechanical Integrity   Yes _____ No _____
  Hydraulic Integrity   Yes _____ No _____
Comments _____
_____
_____

Reported By: _____ Date _____

FIG. 2—SUMMARY SHEET

25.137

Test No. _____ Odometer Finish _____
Weather Condition _____
Date _____ Start _____
Road Condition _____
Driver _____ Total _____
Observer _____
Ambient Temperature: High _____ °C (°F)
Low _____ °C (°F)

INPUT CORRELATION

Engine Idling in Neutral _____ in Hg (kPa) Vacuum
(Power Equipped Tow Vehicle Only)

Line Pressure, PST (Pa)	1st Recording	Pedal Force, lb (N)			Average
		2nd Recording	3rd Recording		

PREBURNISH CHECK

30 mph (48 km/h) 10 ft/s² (3 m/s²) in Gear, 1 mile (1.6 km) Interval

Stop No.	Tow Veh. Input (Line Pressure or Pedal Force)	Trailer Input	Pedal Travel, in (mm)
1			
2			
3			
4			
5			
6			
7			
8			
9			
10			

Final Temperatures After 19th Stop
Tow Vehicle
L.F. _____ °C (°F)   R.F. _____ °C (°F)
L.R. _____ °C (°F)   R.R. _____ °C (°F)
Trailer:
L.F. _____ °C (°F)   R.F. _____ °C (°F)
L.R. _____ °C (°F)   R.R. _____ °C (°F)
Summary of Performance: _____

FIG. 3—INPUT CORRELATION AND PREBURNISH CHECK DATA SHEET

Test No. _____ Weather Condition _____
Date _____ Road and Track Condition _____
Driver _____ Ambient Temperature: High _____ °C (°F)
Low _____ °C (°F)
Observer _____ Method of Application _____
Odometer: Finish _____ Start _____ Total miles (km) _____

FIRST (PREBURNISH) EFFECTIVENESS TEST

30 mph (48 km/h) in Neutral, 93.3°C (200°F) IBT Each Application

Tow Veh. Input LP or PF	Trailer Input	Deceleration ft/s² (m/s²)	*Pedal Travel, in (mm)	Tow Veh. Temp. °C (°F)				Trailer Temp. °C (°F)				Remarks
				L.F.	R.F.	L.R.	R.R.	L.F.	R.F.	L.R.	R.R.	

60 mph (97 km/h) in Neutral, 93.3°C (200°F) IBT Each Application

Tow Veh. Input LP or PF	Trailer Input	Deceleration ft/s² (m/s²)	*Pedal Travel, in (mm)	Tow Veh. Temp. °C (°F)				Trailer Temp. °C (°F)				Remarks
				L.F.	R.F.	L.R.	R.R.	L.F.	R.F.	L.R.	R.R.	

Summary of Performance: _____

*Optional

FIG. 4—FIRST (PREBURNISH) EFFECTIVENESS DATA SHEET

Test No. _____ Weather Condition _____
Date _____ Road and Track Condition _____
Driver _____ Ambient Temperature: High _____ °C (°F)
Observer _____ Low _____ °C (°F)
Odometer: Finish _____ Start _____ Total miles (km) _____

BURNISH

40–0 mph (64–0 km/h) 12 ft/s² (3.7 m/s²) in Gear,
121°C (250°F) IBT Each Stop But 1 Mile (1.6 km) Max Interval

Stop No.	Tow Veh. Input LP or PF	Trailer Input	*Pedal Travel in (mm)	Tow Veh. Temp. °C (°F)				Trailer Temp. °C (°F)				Remarks
				L.F.	R.F.	L.R.	R.R.	L.F.	R.F.	L.R.	R.R.	
1												
20												
40												
60												
80												
100												
120												
140												
160												
180												
200												

Record Any Operations Performed _____

_____

*Optional

Date _____ Weather Condition _____
Driver _____ Road and Track Condition _____
Observer _____ Ambient Temperature: High _____ °C (°F)
Odometer: Finish _____ Start _____ Low _____ °C (°F)
Total miles (km) _____

INOPERATIVE POWER SYSTEM TEST

60–0 mph (97–0 km/h) in Gear, 65.6°C (150°F) IBT Each Application,
150 and 200 lb (667 and 890 N) Maximum Pedal Force

Tow Veh. Input LP or PF	Trailer Input	Stopping Distance ft (m)	Tow Veh. Temp. °C (°F)				Trailer Temp. °C (°F)				Remarks
			L.F.	R.F.	L.R.	R.R.	L.F.	R.F.	L.R.	R.R.	
Average												
Average												

Summary of Performance _____

FIG. 5

Test No. _____ Weather Condition _____
Date _____ Road and Track Condition _____
Driver _____ Ambient Temperature: High _____ °C (°F)
Observer _____ Low _____ °C (°F)
Odometer: Finish _____ Start _____ Total miles (km) _____

EMERGENCY SYSTEM TEST

60–0 mph (97–0 km/h) in Gear, 65.6°C (150°F) IBT Each Application,
150 and 200 lb (667 and 890 N) Maximum Pedal Force

	System Operating	Tow Veh. Input LP or PF	Trailer Input	Stopping Distance ft (m)	Warning Light on		Tow Veh. Temp. °C (°F)	Trailer Temp. °C (°F)				Remarks
					Yes	No	L R	L.F.	R.F.	L.R.	R.R.	
GVWR or GCWR	Front											
	Front											
	Front											
	Average of Front											
	Front											
	Front											
	Front											
	Average of Front											
	Rear											
	Rear											
	Rear											
	Average of Rear											
	Rear											
	Rear											
	Rear											
	Average of Rear											

60–0 mph (97–0 km/h) in Gear, 65.6°C (150°F) IBT Each Application,
150 and 200 lb (667 and 890 N) Maximum Pedal Force

	System Operating	Tow Veh. Input LP or PF	Trailer Input	Stopping Distance ft (m)	Warning Light on		Tow Veh. Temp. °C (°F)	Trailer Temp. °C (°F)				Remarks
					Yes	No	L R	L.F.	R.F.	L.R.	R.R.	
Minimum Tow Vehicle Weight	Front											
	Front											
	Front											
	Average of Front											
	Front											
	Front											
	Front											
	Average of Front											
	Rear											
	Rear											
	Rear											
	Average of Rear											
	Rear											
	Rear											
	Rear											
	Average of Rear											

Summary of Performance: _____

FIG. 6

25.139

Test No. _____ Weather Condition _____
Date _____ Road and Track Condition _____
Driver _____ Ambient Temperature: High _____ °C (°F)
                                  Low _____ °C (°F)
Observer _____ Method of Application _____
Odometer: Finish _____ Start _____ Total miles (km) _____

SECOND EFFECTIVENESS TEST

30 mph (48 km/h) in Neutral, 93.3°C (200°F) IBT Each Application

Tow Veh. Input LP or PF	Trailer Input	Deceleration ft/s² (m/s²)	*Pedal Travel, in (mm)	Tow Veh. Temp. °C (°F)				Trailer Temp. °C (°F)				Remarks
				L.F.	R.F.	L.R.	R.R.	L.F.	R.F.	L.R.	R.R.	

60 mph (97 km/h) in Neutral, 93.3°C (200°F) IBT Each Application

Tow Veh. Input LP or PF	Trailer Input	Deceleration ft/s² (m/s²)	*Pedal Travel, in (mm)	Tow Veh. Temp. °C (°F)				Trailer Temp. °C (°F)				Remarks
				L.F.	R.F.	L.R.	R.R.	L.F.	R.F.	L.R.	R.R.	

Summary of Performance: _____

*Optional

FIG. 7—SECOND EFFECTIVENESS TEST DATA SHEET

Test No. _____ Weather Condition _____
Date _____ Road and Track Condition _____
Driver _____ Ambient Temperature: High _____ °C (°F)
Observer _____ Low _____ °C (°F)
Odometer: Finish _____ Start _____ Total miles (km) _____
Check if truck tow vehicle unballasted

FIRST BASELINE CHECK

30 mph (43 km/h), 10 ft/s² (3 m/s²) in Gear, 65.6°C (150°F) IBT First Stop

Stop No.	Tow Veh. Input LP or PF	Trailer Input	*Pedal Travel in (mm)	Tow Veh. Temp. °C (°F)				Trailer Temp. °C (°F)				Remarks
				L.F.	R.F.	L.R.	R.R.	L.F.	R.F.	L.R.	R.R.	
1												
2												
3												

Start Time _____ Finish Time _____ Lapsed Time _____

FIRST FADE TEST

60 mph (97 km/h) 15 ft/s² (4.6 m/s²) in Gear, 0.4 mile (0.6 km) Interval, 65.6°C (150°F) IBT First Stop

Stop No.	Tow Veh. Input LP or PF	Trailer Input	*Pedal Travel in (mm)	Tow Veh. Temp. °C (°F)				Trailer Temp. °C (°F)				Remarks
				L.F.	R.F.	L.R.	R.R.	L.F.	R.F.	L.R.	R.R.	
1												
2												
3												
4												
5												
6												
7												
8												
9												
10												

Summary of Performance: _____

*Optional

FIG. 8—FIRST BASELINE CHECK AND FIRST FADE TEST DATA SHEET

Test No. _____ Weather Condition _____
Date _____ Road and Track Condition _____
Driver _____ Ambient Temperature: High _____ °C (°F)
Observer _____ Low _____ °C (°F)
Odometer: Finish _____ Start _____ Total miles (km) _____

FIRST RECOVERY

30 mph (48 km/h), 10 ft/s² (3 m/s²), in Gear,
1 mile (1.6 km) Interval at 40 mph (64 km/h)

Stop No.	Tow Veh. Input LP or PF	Trailer Input	*Pedal Travel in (mm)	Tow Veh. Temp. °C (°F)				Trailer Temp. °C (°F)				Remarks
				L.F.	R.F.	L.R.	R.R.	L.F.	R.F.	L.R.	R.R.	
1												
2												
3												
4												
5												
6												
7												
8												
9												
10												
11												
12												

Summary of Performance: _____

FIRST EFFECTIVENESS SPOT CHECK

60 mph (97 km/h), 15 ft/s² (4.6 m/s²) in Gear, 93.3°C (200°F) IBT

Stop No.	Tow Veh. Input LP or PF	Trailer Input	*Pedal Travel in (mm)	Tow Veh. Temp. °C (°F)				Trailer Temp. °C (°F)				Remarks
				L.F.	R.F.	L.R.	R.R.	L.F.	R.F.	L.R.	R.R.	
1												
2												

FIRST REBURNISH

40-0 mph (64-0 km/h), 12 ft/s² (3.7 m/s²) in Gear,
121°C (250°F) IBT Each Stop, But 1 mile (1.6 km) Max Interval

Stop No.	Tow Veh. Input LP or PF	Trailer Input	*Pedal Travel in (mm)	Tow Veh. Temp. °C (°F)				Trailer Temp. °C (°F)				Remarks
				L.F.	R.F.	L.R.	R.R.	L.F.	R.F.	L.R.	R.R.	
1												
10												
25												
35												

Summary of Performance: _____

*Optional

FIG. 9—FIRST RECOVERY, FIRST EFFECTIVENESS SPOT CHECK, AND FIRST REBURNISH DATA SHEET

Test No. _____ Weather Condition _____
Date _____ Road and Track Condition _____
Driver _____ Ambient Temperature: High _____ °C (°F)
Observer _____ Low _____ °C (°F)
Odometer: Finish _____ Start _____ Total miles (km) _____

SECOND BASELINE CHECK

30 mph (48 km/h), 10 ft/s² (3 m/s²), in Gear, 65.6°C (150°F) IBT

Stop No.	Tow Veh. Input LP or PF	Trailer Input	*Pedal Travel in (mm)	Tow Veh. Temp. °C (°F)				Trailer Temp. °C (°F)				Remarks
				L.F.	R.F.	L.R.	R.R.	L.F.	R.F.	L.R.	R.R.	
1												
2												
3												

Start Time _____ Finish Time _____ Lapsed Time _____

SECOND FADE TEST

60 mph (97 km/h), 15 ft/s² (4.6 m/s²) in Gear,
0.4 mile (0.6 m) Interval, 65.6°C (150°F) IBT First Stop

Stop No.	Tow Veh. Input LP or PF	Trailer Input	*Pedal Travel in (mm)	Tow Veh. Temp. °C (°F)				Trailer Temp. °C (°F)				Remarks
				L.F.	R.F.	L.R.	R.R.	L.F.	R.F.	L.R.	R.R.	
1												
2												
3												
4												
5												
6												
7												
8												
9												
10												
11												
12												
13												
14												
15												

Summary of Performance: _____

*Optional

FIG. 10—SECOND BASELINE CHECK AND SECOND FADE TEST DATA SHEET

Test No. _____ Weather Condition _____
Date _____ Road and Track Condition _____
Driver _____ Ambient Temperature: High _____ °C (°F)
Observer _____ Low _____ °C (°F)
Odometer: Finish _____ Start _____ Total miles (km) _____

SECOND RECOVERY

30 mph (48 km/h), 10 ft/s² (3 m/s²), in Gear,
1 mile (1.6 km) Interval at 40 mph (64 km/h)

Stop No.	Tow Veh. Input LP or PF	Trailer Input	*Pedal Travel in (mm)	Tow Veh. Temp. °C (°F)				Trailer Temp. °C (°F)				Remarks
				L.F.	R.F.	L.R.	R.R.	L.F.	R.F.	L.R.	R.R.	
1												
2												
3												
4												
5												
6												
7												
8												
9												
10												
11												
12												

Summary of Performance: _____

SECOND EFFECTIVENESS SPOT CHECK

60-0 mph (97.0 km/h), 15 ft/s² (4.6 m/s²) in Gear, 93.3°C (200°F) IBT

Stop No.	Tow Veh. Input LP or PF	Trailer Input	*Pedal Travel in (mm)	Tow Veh. Temp. °C (°F)				Trailer Temp. °C (°F)				Remarks
				L.F.	R.F.	L.R.	R.R.	L.F.	R.F.	L.R.	R.R.	
1												
2												

SECOND REBURNISH

40-0 mph (64-0 km/h), 12 ft/s² (4.6 m/s²) in Gear,
121°C (250°F) IBT Each Stop, But 1 mile (1.6 km) Max Interval

Stop No.	Tow Veh. Input LP or PF	Trailer Input	*Pedal Travel in (mm)	Tow Veh. Temp. °C (°F)				Trailer Temp. °C (°F)				Remarks
				L.F.	R.F.	L.R.	R.R.	L.F.	R.F.	L.R.	R.R.	
1												
10												
25												
35												

Summary of Performance: _____

*Optional

FIG. 11—SECOND RECOVERY, SECOND EFFECTIVENESS SPOT CHECK, SECOND REBURNISH DATA SHEET

Test No. _____ Weather Condition _____
Date _____ Road and Track Condition _____
Driver _____ Ambient Temperature: High _____ °C (°F)
Low _____ °C (°F)
Observer _____ Method of Application _____
Odometer: Finish _____ Start _____ Total miles (km) _____

THIRD EFFECTIVENESS TEST

30 mph (48 km/h) in Neutral, 93.3°C (200°F) IBT Each Application

Tow Veh. Input LP or PF	Trailer Input	Deceleration ft/s² (m/s²)	*Pedal Travel, in (mm)	Tow Veh. Temp. °C (°F)				Trailer Temp. °C (°F)				Remarks
				L.F.	R.F.	L.R.	R.R.	L.F.	R.F.	L.R.	R.R.	

60 mph (97 km/h) in Neutral, 93.3°C (200°F) IBT Each Application

Tow Veh. Input LP or PF	Trailer Input	Deceleration ft/s² (m/s²)	*Pedal Travel, in (mm)	Tow Veh. Temp. °C (°F)				Trailer Temp. °C (°F)				Remarks
				L.F.	R.F.	L.R.	R.R.	L.F.	R.F.	L.R.	R.R.	

Summary of Performance: _____

*Optional

FIG. 12—THIRD EFFECTIVENESS TEST DATA SHEET

Test No. _____ Inspected By _____ Date _____

FRICTION MATERIAL CONDITION

Tow Vehicle:
L.F. _____
R.F. _____
L.R. _____
R.R. _____

Trailer:
L.F. _____
R.F. _____
L.R. _____
R.R. _____

DRUM (DISC) CONDITION

Tow Vehicle:
L.F. _____
R.F. _____
L.R. _____
R.R. _____

Trailer:
L.F. _____
R.F. _____
L.R. _____
R.R. _____

MECHANICAL COMPONENT CONDITION

Tow Vehicle:
L.F. _____
R.F. _____
L.R. _____
R.R. _____

Trailer:
L.F. _____
R.F. _____
L.R. _____
R.R. _____

HYDRAULIC/OTHER COMPONENT CONDITION

Tow Vehicle:
L.F. _____
R.F. _____
L.R. _____
R.R. _____

Trailer:
L.F. _____
R.F. _____
L.R. _____
R.R. _____

ACTUATOR SYSTEM CONDITION

Brake Pedal _____
Power Brake _____
Master Cylinder _____
Stop Lights _____
_____

Inspection Comments: _____
_____
_____
_____
_____

FIG. 13—FINAL INSPECTION DATA SHEET

SERVICE BRAKE SYSTEM PERFORMANCE REQUIREMENTS—PASSENGER CAR—TRAILER COMBINATIONS—SAE J135 MAY85

SAE Recommended Practice

Report of the Brake Committee, approved March 1973, last revised June 1979, reaffirmed without change May 1985.

1. **Introduction**—The performance requirements in this SAE Recommended Practice are a modification of SAE J937b and represent the accumulation of the best information available from investigations of the service brake system performance of combinations of new passenger cars and new trailers (braked or unbraked) designed for roadway use. They also represent the minimum performance recognized as acceptable by vehicle, brake system, and component manufacturers. The towing vehicles must comply with SAE J937b (March, 1978). This recommended practice may be used to determine the maximum weight of unbraked trailers the towing vehicle is recommended to pull.

2. **Scope**—This SAE Recommended Practice presents service brake performance requirements for brake systems of all combinations of new passenger cars and new trailers (braked or unbraked) intended for roadway use (excluding special-purpose vehicles such as ambulances, hearses, etc.). Acceptable performance requirements are based on data obtained from SAE J134c (June, 1979).

3. **Purpose**—The purpose of this practice is to establish the minimum service brake system performance requirements with regard to:

   3.1 Stopping Ability
   3.1.1 Of cold brakes as affected by vehicle speed.
   3.1.2 Of hot brakes as affected by vehicle speed and duty cycles.
   3.1.3 Of cold brakes during emergency or inoperative power assist conditions.

   3.2 Pedal Force—Maximum and/or minimum force allowable.
   3.3 Brake Stability
   3.4 Brake system Integrity

4. **Instrumentation**—See SAE J134c, Section 4.
5. **Installation Details**—See SAE J134c, Section 5.
6. **Test Procedure**—See SAE J134c, Section 6.
7. **Acceptable Performance Requirements**
   7.1 Preburnish Check—See SAE J134c, paragraph 6.2.
   7.1.1 Pedal force shall be between 10 and 55 lb (45 and 245 N) inclusive, for 10 ft/s² (3 m/s²) stops from 30 mph (48 km/h).
   7.2 Effectiveness Test—See SAE J134c, paragraphs 6.3, 6.5, and 6.14.
   7.2.1 30 mph (48 km/h)—Pedal force shall be between 15 and 100 lb (67 and 445 N) inclusive, for 16 ft/s² (4.9 m/s²).
   7.2.2 60 mph (96 km/h)—Pedal force shall be between 15 and 120 lb (67 and 534 N) inclusive, for 16 ft/s² (4.9 m/s²).
   7.3 Emergency Brake System Test—See SAE J134c, paragraph 6.6.
   7.3.1 Maximum stopping distance of 1000 ft (305 m) with a maximum pedal force of 200 lb (890 N) without causing any tire of either vehicle to leave a 12 ft (3.7 m) lane.
   7.4 Inoperative Power System Test—See SAE J134c, paragraph 6.7.
   7.4.1 Maximum stopping distance of 600 ft (183 m) with a maximum pedal force of 200 lb (890 N) without leaving a 12 ft (3.7 m) lane.

GENERAL DATA AND SUMMARY REPORT FORM
SERVICE BRAKE SYSTEM PERFORMANCE REQUIREMENTS: PASSENGER CAR-TRAILER COMBINATIONS

TEST PHASE	REQUIRED		ACTUAL
PREBURNISH CHECK	10–55 lb pf	44–245 N pf	min ___ max ___ lb (N) pf
EFFECTIVENESS TESTS			1st 2nd Final
30 mph at 16 ft/s² (48 km/h at 4.9 m/s²)	15–100 lb pf	67–445 N pf	___ ___ ___ lb (N) pf
60 mph at 16 ft/s² (97 km/h at 4.9 m/s²)	15–120 lb pf	67–534 N pf	___ ___ ___ lb (N) pf
EMERGENCY BRAKE TEST 60 mph (97 km/h) Stopping Distance	1000 ft and 200 lb max	305 m and 890 N max pf	Front ___ ft (m) ___ lb (N) pf Rear ___ ft (m) ___ lb (N) pf
INOPERATIVE POWER SYSTEM TEST 60 mph (97 km/h) Stopping Distance	Manual ___ Power ___ 600 ft and 200 lb max pf	183 m and 890 N max pf	___ ft (m) ___ lb (N) pf
FIRST FADE AND RECOVERY TEST Fade Stops 1–4 Recovery Stops 1–5 Recovery Stops 6–12	120, 147, 173, 200 lb pf 5 ft/s² at 200 lb max pf 10 ft/s² at 150 lb pf	534, 654, 770, 890 N pf 1.5 m/s² at 890 N max pf 3 m/s² at 667 N pf	___, ___, ___, ___ lb (N) pf ___ ft/s² (m/s²) at ___ lb (N) max pf ___ lb (N) max pf
SECOND FADE AND RECOVERY TEST Fade stops 1–8 Recovery stops 1–5 Recovery stops 6–12	120, 132, 143, 155, 166, 177, 189, 200 lb pf 5 ft/s² at 200 lb max pf 10 ft/s² at 150 lb pf	534, 587, 636, 689, 738, 787, 841, 890 N 1.5 m/s² at 890 N max pf 3 m/s² at 667 N pf	___, ___, ___, ___, ___, ___, ___, ___ lb (N) pf ___ ft/s² (m/s²) at ___ lb (N) max pf ___ lb (N) max pf
STABILITY DURING Effectiveness Tests	No Uncontrollable Braking Causing Car to Leave 12 ft Lane Below 16 ft/s²	No Uncontrollable Braking Causing Car to leave 3.7 m Lane Below 4.9 m/s²	Controllable Braking Below 16 ft/s² (4.9 m/s²) Yes ___ No ___
FINAL INSPECTION Lining Integrity Mechanical Integrity Hydraulic Integrity Electrical Integrity	 Intact and No Cracks Intact and Functional Leakfree Intact and Functional	 Intact and No Cracks Intact and Functional Leakfree Intact and Functional	 Yes ___ No ___ Yes ___ No ___ Yes ___ No ___ Yes ___ No ___

Comments: _____

Reported By: _____ Date: _____

FIG. 1—GENERAL DATA AND SUMMARY REPORT FORM

7.5 First Fade and Recovery Test—See SAE J134c, paragraph 6.8.

7.5.1 FADE—Pedal force for first four 15 ft/s² (4.6 m/s²) stops shall not exceed 120, 147, 173, and 200 lb (534, 654, 770, and 890 N), respectively.

7.5.2 RECOVERY—A minimum of 5 ft/s² (1.5 m/s²) shall be maintained at a maximum pedal force of 200 lb (890 N) for the first five recovery stops, and the pedal force shall be below 150 lb (667 N) at 10 ft/s² (3 m/s²) by stop 6.

7.6 Second Fade and Recovery Test—See SAE J134c, paragraph 6.11.

7.6.1 FADE—Pedal force for first eight at 15 ft/s² (4.6 m/s²) stops shall not exceed 120, 132, 143, 155, 166, 177, 189, and 200 lb (534, 587, 636, 689, 738, 787, 841, and 890 N), respectively.

7.6.2 RECOVERY—Same as First Recovery Requirement, paragraph 7.5.

7.7 Stability Requirements—See SAE J134c, paragraphs 6.3, 6.5, and 6.14.

7.7.1 No uncontrollable braking action causing any tire of either vehicle to leave a 12 ft (3.7 m) wide roadway lane is permissible below 16 ft/s² (4.9 m/s²). (Wheel slide permitted.)

7.8 Final Inspection—See SAE J134c, paragraph 6.15.

7.8.1 LINING—Shall be firmly attached and intact on shoes. (Minor cracks that do not impair attachment are acceptable.)

7.8.2 MECHANICAL—All components of the brake system shall be intact and functional.

7.8.3 HYDRAULIC—All hydraulic components of the brake system shall be leakfree.

7.8.4 ELECTRICAL—All electrical components of the trailer brake system, excluding stop light circuits, shall be intact and functional.

8. Report Form—General Data and Summary Report Form, Fig. 1.

BRAKE SYSTEM ROAD TEST CODE—MOTORCYCLES—SAE J108 MAR87

SAE Recommended Practice

Report of the Motorcycle Committee and Brake Committee approved July 1969, editorial change March 1973, and reaffirmed by the Motorcycle Committee March 1987.

1. Scope—This SAE Recommended Practice establishes a uniform procedure for the level road test of the brake systems of all classes of motorcycles intended for highway use.

2. Purpose—The purpose of the recommended practice is to establish brake system capabilities with regard to:
 1. Deceleration or stopping distance versus input, as affected by vehicle speed, brake temperature, and usage.
 2. Lining characteristics.
 3. Drum characteristics.

2.1 Section A—Instrumentation
Line pressure of pedal and lever force gauges.
Decelerometer.
Direct reading temperature instrument.
Speedometer (calibrated).
Odometer (calibrated).
Thermometer—ambient (or ambient sensitive thermocouple).
Optional Instrumentation
 Pedal travel gauge.
 Stopmeter ("fifth wheel," distance only or shot marker).
 Solenoid stop counter.
 Stopwatch.

2.2 Section B—Installation Details

1. *Friction Material Preparation*—Attach and finish friction material per manufacturer's specifications.

2. *Thermocouples*—Install the desired type of thermocouples in each brake. Any one of the following installations may be used:
 (a) Plug type. See Fig. 1.
 (b) Web-rim junction type, welded or otherwise, in intimate contact with the brake shoe near the web-rim junction.
 (c) Thermocouple inserted in a hole drilled from the lining edge, approximately one-half the width of the lining in depth and as close to the shoe rim as possible.

All thermocouples are to be located in the approximate center of the most heavily loaded shoe, one per brake (two per brake on double-leading shoe brake).

3. *Brake Drum or Rotor*—Surface finish and dimensional characteristics (with special emphasis on runout of rubbing surface) of brake drums or rotor shall be in accordance with manufacturer's specifications for each test.

4. *Brake Assembly*—Brakes shall be prepared in accordance with manufacturer's specifications with special attention to required load characteristics on all brake springs. Adjust brakes to manufacturer's specifications.

5. Vehicle Test Weight—Vehicle test weight (W) in pounds shall be determined by the following formula for all tests:

$$W = C + 150\,S + P$$

For the Effectiveness Test, the test weight for two-wheeled machines designed to carry more than one person may be determined by the following formula:

$$W = C + 200\,\text{lb}$$

where: C = curb weight which is the weight in pounds of the vehicle with standard equipment, including maximum capacity of fuel and oil
 S = vehicle's designed seating capacity
 P = manufacturer's specified payload for three-wheeled motorcycle only

6. Test Conditions

(a) *Test Course*—Effectiveness, fade and recovery test stops shall be conducted on a substantially level (not to exceed ±1% grade), dry, smooth, hard-surfaced roadway of Portland cement concrete (or other surface with equivalent coefficient of surface friction) that is free from loose materials. Also, guides to indicate a 12 ft wide roadway lane shall be provided on the test course.

(b) *Ambient Air Temperature*—Ambient air temperature at fade and recovery tests shall be between 40 and 90° F.

(c) *Applied Point and Direction of Lever Force and Pedal Force*—As shown in Fig. 2, the point of application of the lever force shall be one 1.2 in from the end of the brake lever grip and the direction of the force shall be perpendicular, on the plane along which the brake lever rotates, to the handle grip. The point of application of the pedal force shall be the center of the foot contact pad of the brake pedal and the direction of the force shall be perpendicular, on the plane along which the brake pedal rotates, to the foot contact pad.

(d) *Test Speed*—Vehicles shall be tested at the specified speed for each test. Those vehicles which cannot obtain the specified speed shall be driven at not less than 4 mph nor more than 8 mph below the practical maximum speed with the second number in the speed figure being either 5 or 0. (For example: if 69 mph is the maximum speed, the test speed for the 70 mph requirement would be 65 mph.)

7. Test Procedure

(a) During all phases of this procedure, any unusual performance such as grab, noise characteristics, or wheel skid are to be noted and recorded. Note any uncontrollable braking action causing the vehicle to lose stability,

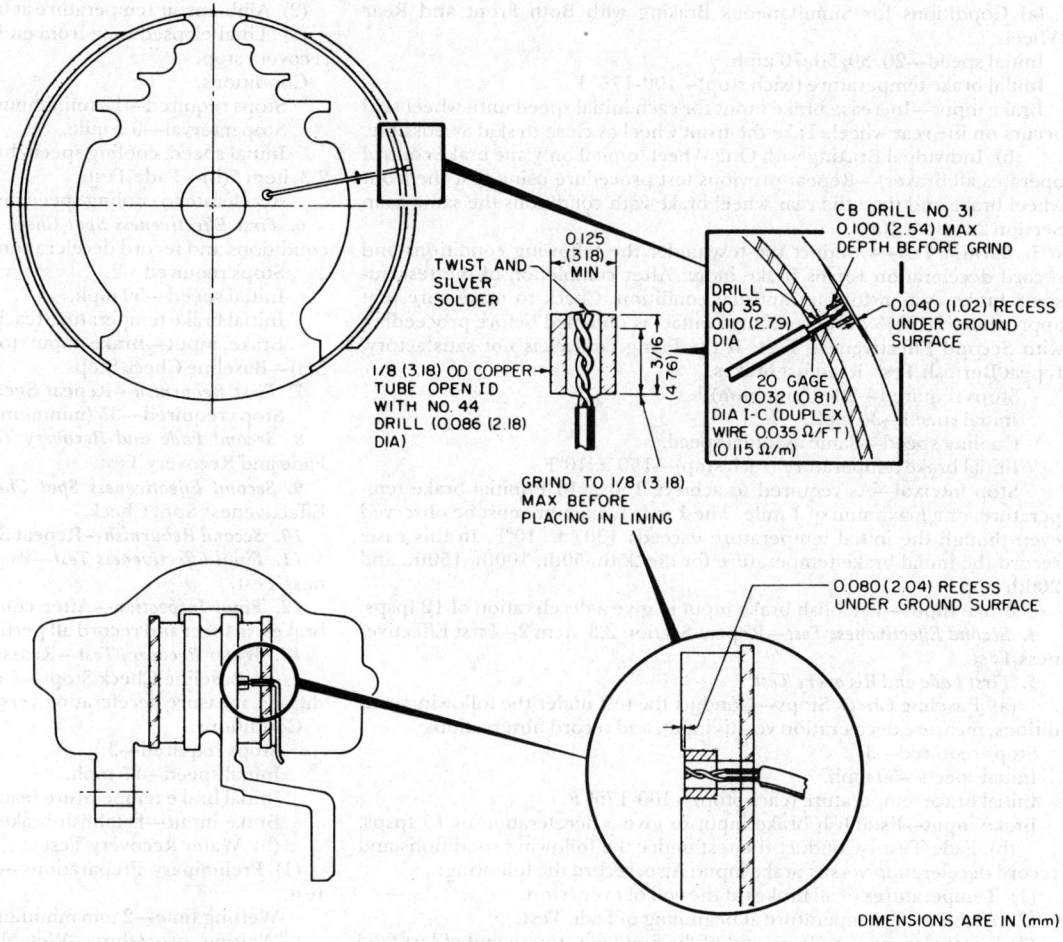

FIG. 1—TYPICAL PLUG TYPE THERMOCOUPLE INSTALLATION

to overturn, or to swerve out of a 12 ft wide roadway lane.
(b) If brakes require warming to prescribed temperature, use the burnish procedure and shorten interval if necessary.
(c) All stops shall be made with the clutch disengaged, or in neutral.
(d) Brake input on the tests other than preburnish check and effectiveness test shall conform to the prescribed value and be held constant during the brake operation.
(e) When a recording decelerometer is used, deceleration shall be determined by the average of four measurements, which are made at the four points dividing the braking time into five equal portions.
(f) When stopping distance is used, deceleration may be determined by the following formula:

$$a = \frac{1.075 \, V^2}{S}$$

where: V = velocity, mph
S = stopping distance, ft

(g) Initial brake temperature is defined as that of the highest reading thermocouple taken 0.1 mile before stop with the brakes released.

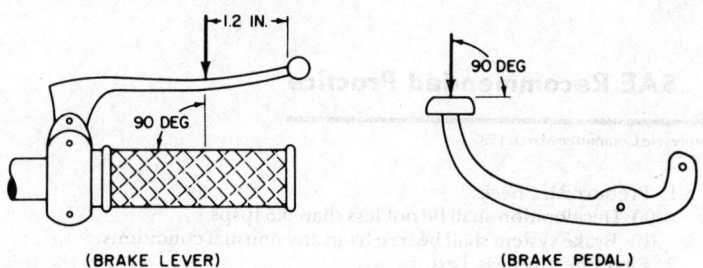

FIG. 2

2.3 Section C—Test Procedure—The test procedure shall conform to the following:

1. Preburnish Check
(a) *Brake Operational Check*—In order to allow for a general check of brakes, instrumentation, and vehicle operation, the following steps are to be run:

Conditions:
Stops required—10.
Stop interval—0.5 mile
Initial speed—30 mph.
Cooling speed—30 mph.
Brake input—Establish brake input to give a deceleration of 10 fpsps.

(b) *Establishing Brake Input Proportioning*—Assuming the above-mentioned test is performed satisfactorily, establish brake input proportioning to be used in the succeeding tests, except effectiveness tests, according to the following procedures. (If one pedal or lever operates all brakes, this step may be omitted.)

(1) Conduct stopping tests at 50 mph (or maximum vehicle speed if 50 mph is not obtainable) with a brake temperature for each brake between 100 and 175° F before each stop.

(2) The stopping tests should be conducted with the rear brake alone and then the front brake alone. The lever or pedal inputs shall be 5 lb for the first stop and increased at increments not exceeding 10 lb on succeeding stops until sufficient plotting data are obtained. Care should be taken to avoid wheel skid. The average or steady deceleration or total stopping distance shall be measured for each stop.

(3) This information shall then be plotted as deceleration versus pedal and hand lever pressure.

(4) This plot shall then be used to determine the appropriate brake control input force to achieve a 60/40 front to rear retardation proportioning.

2. First Effectiveness Test—Conduct the test and record actual brake input and deceleration at each initial speed. Stay alert for any hazardous condition which may cause the vehicle to lose stability due to wheel skid, to overturn, or to swerve out of the specified width roadway lane.

(a) Conditions for Simultaneous Braking with Both Front and Rear Wheels

Initial speed—20, 30, 50, 70 mph.
Initial brake temperature (each stop)—100-175° F.
Brake input—Increase brake input for each initial speed until wheel skid occurs on the rear wheel. Take the front wheel as close to skid as possible.

(b) Individual Braking with One Wheel (omit if only one brake control operates all brakes)—Repeat previous test procedure using first the front wheel brake and then the rear wheel brake with conditions the same as in Section 2.3, item 2(a).

3. *Burnish Test*—Conduct the test under the following conditions and record deceleration versus brake input. After completion of the test, inspect brake system for any unusual condition. Check to make sure that approximately 80% or more lining contact is obtained before proceeding with Second Effectiveness Test. If the lining contact is not satisfactory, repeat Burnish Test. Readjust brakes.

Stops required—200 (minimum).
Initial speed—30 mph.
Cooling speed—Same as initial speed.
Initial brake temperature (each stop)—150 ± 10°F.
Stop interval—As required to achieve 150 ± 10°F initial brake temperature, or a maximum of 1 mile. The 1 mile maximum must be observed even though the initial temperature exceeds 150 ± 10°F. In this case, record the initial brake temperature for the 25th, 50th, 100th, 150th, and 200th stop.
Brake Input—Establish brake input to give a deceleration of 12 fpsps.

4. *Second Effectiveness Test*—Repeat Section 2.3, item 2—First Effectiveness Test.

5. *First Fade and Recovery Test*

(a) Baseline Check Stops—Conduct the test under the following conditions, measure deceleration versus input, and record observations:
Stops required—3.
Initial speed—50 mph.
Initial brake temperature (each stop)—100-175° F.
Brake input—Establish brake input to give a deceleration of 15 fpsps.

(b) Fade Test—Conduct the test under the following conditions and record deceleration versus brake input. Also, record the following:
(1) Temperatures of all brakes at the end of every stop.
(2) Ambient air temperature at beginning of Fade Test.
(3) Total elapsed time from end of the first fade stop to end of last fade stop.
Conditions:
Stops required—10.
Stop interval—0.25 mile.
Initial speed—50 mph.
Cooling speed—Same as initial speed.
Initial brake temperature—First stop only, 150 ± 10°F.
Brake input—Establish brake input to give a deceleration of 15 fpsps for first stop. Use these input forces on all 10 fade stops.
Accelerate to cooling speed moderately.
(4) Drive the vehicle for 1 mile at cooling speed after last fade stop and then perform the Recovery Test.

(c) Recovery Test—Conduct the test under the following conditions and record deceleration versus brake input. Record, at the same time, observations on the following items:
(1) Temperatures of all brakes at the end of each stop.
(2) Ambient air temperature at beginning of Recovery Test.
(3) Total elapsed time from end of the first recovery stop to end of last recovery stop.
Conditions:
Stops required—12 (minimum).
Stop interval—0.5 mile.
Initial speed, cooling speed, brake input—Same as specified in Section 2.3, item 5(b)—Fade Test.
Accelerate to cooling speed moderately.

6. *First Effectiveness Spot Check*—Conduct the test under the following conditions and record deceleration versus brake input:
Stops required—2.
Initial speed—50 mph.
Initial brake temperature (each stop)—100-175°F.
Brake input—Brake input to be the same as in Section 2.3, item 5(a)—Baseline Check Stops.

7. *First Reburnish*—Repeat Section 2.3, item 3—Burnish Test, except:
Stops required—35 (minimum).

8. *Second Fade and Recovery Test*—Repeat Section 2.3, item 5—First Fade and Recovery Test.

9. *Second Effectiveness Spot Check*—Repeat Section 2.3, item 6—First Effectiveness Spot Check.

10. *Second Reburnish*—Repeat Section 2.3, item 7—First Reburnish.

11. *Final Effectiveness Test*—Repeat Section 2.3, item 2—First Effectiveness Test.

12. *Final Inspection*—After completion of tests 1—11, disassemble all brakes; inspect and record all pertinent observations.

13. *Water Recovery Test*—Reassemble brakes (Final Inspection).

(a) Baseline Check Stops—Conduct the test under the following conditions, measure deceleration versus brake input, and record observations.
Conditions:
Stops required—3.
Initial speed—25 mph.
Initial brake temperature (each stop)—100-175° F.
Brake input—Establish brake input to give a deceleration of 8 fpsps.

(b) Water Recovery Test
(1) Preliminary Preparations—Perform wetting of brakes prior to the test.
Wetting time—2 min minimum.
Wetting procedure—Wet all brakes thoroughly by slowly rolling through a trough sufficiently deep to wet the brakes.
Start recovery stops not more than 1 min after wetting brakes. Do not exceed initial speed prior to recovery stop.
(2) Tests—Conduct the test under the following conditions and inspect deceleration versus brake input. Record, at the same time, observations on the following items:
(a) Temperatures of all brakes at the end of each stop.
(b) Ambient air temperature at beginning of run.
(c) Total elapsed time from end of the first stop to end of last stop.
Conditions:
Stops required—15.
Stop interval—0.25 mile.
Initial speed—25 mph.
Speed between stops—Same as initial speed.
Brake input—Establish brake input to be the same as in Section 2.3, item 13(a)—Brake Input.

SERVICE BRAKE SYSTEM PERFORMANCE REQUIREMENTS—MOTORCYCLES AND MOTOR-DRIVEN CYCLES—SAE J109 MAR87

SAE Recommended Practice

Report of the Motorcycle Committee and Brake Committee approved July 1969 and reaffirmed by the Motorcycle Committee March 1987.

1. *Scope*—This SAE Recommended Practice establishes performance requirements for the service brake systems of all classes of motorcycles intended for highway use.

2. *Performance Requirements*—When subjected to the test procedures in SAE J108, by a skilled rider, the following requirements must be satisfied by the service brake system:

1. Preburnish Check
(a) Deceleration shall be not less than 9.5 fpsps.
(b) Brake system shall be free from any unusual conditions.

2. First Effectiveness Test
(a) Shall meet the requirements specified in Tables 1 and 2 when equipped with independent brake controls for front and rear.

(b) During all phases of the test, no condition shall be permitted to cause the vehicle to lose stability due to wheel skid, to overturn, or to pull or swerve out of the specified width roadway lane.

3. *Burnish Test*—No unusual dispersion in deceleration values versus brake input at any stop shall be permissible.

4. *Second Effectiveness Test*—Same as in item 2—First Effectiveness Test.

5. First Fade and Recovery Test

(a) *First Fade Test*—First eight stops shall be achieved with a deceleration of not less than 11 fpsps, and not less than 9.5 fpsps on the ninth and tenth stops.

(b) *First Recovery Test*—First five stops shall be achieved with deceleration not less than 9.5 fpsps and on sixth and after, not less than 11 fpsps.

6. *First Effectiveness Spot Check*—No unusual deceleration value versus brake input shall be permissible.

7. First Reburnish

(a) No unusual dispersion in deceleration values versus brake input shall be permissible.

(b) No unusual condition shall be permissible on the brake system.

8. Second Fade and Recovery Test

(a) *Second Fade Test*—First eight stops shall be achieved with a deceleration of not less than 11 fpsps, and 9.5 fpsps on ninth and tenth stops.

(b) *Second Recovery Test*—First five stops shall be achieved with deceleration not less than 9.5 fpsps and on sixth and after, not less than 10 fpsps.

9. *Second Effectiveness Spot Check*—No unusual deceleration value versus brake input shall be permissible.

10. *Second Reburnish*—Same as in item 7, First Reburnish.

11. *Final Effectiveness Test*—Same as in item 2, First Effectiveness Test, and item 4, Second Effectiveness Test.

12. Final Inspection

(a) Lining shall be firmly attached and intact on shoes. (Minor cracks that do not impair attachment are acceptable.)

(b) No unusual wear that may interfere with braking function shall be permissible on lining.

(c) Scores or cracks that may interfere with brake action shall not develop on drums.

(d) All components of the brake system shall be intact and functional.

13. *Water Recovery Test*—Deceleration at all stops shall be within 35-120% of the average of deceleration values of the baseline check stops. Also, brakes shall recover a minimum of 80% of the average of deceleration values of the baseline check stops before the final stop.

TABLE 1—TOTAL BRAKE SYSTEM PERFORMANCE

Initial Speed, mph	Brake Input, lb		Deceleration, fpsps (more than)	Braking Distance, ft
	Lever Force	Pedal Force		
20	5-55	10-90	20	20
30	5-55	10-90	20	50
50	5-55	10-90	20	135
70	5-55	10-90	20	265

TABLE 2—SINGLE BRAKE PERFORMANCE

Initial Speed, mph	Brake Input, lb		Deceleration, fpsps (more than)	Braking Distance, ft
	Lever Force	Pedal Force		
20	5-55	10-90	6.0	70
30	5-55	10-90	6.0	160
50	5-55	10-90	6.0	450
70	5-55	10-90	6.0	880

BRAKE FORCE DISTRIBUTION TEST CODE—COMMERCIAL VEHICLES—SAE J1505 MAY85

SAE Recommended Practice

Report of the Truck and Bus Brake Committee, approved May 1985.

1. *Purpose*—This code provides a method to determine the brake force distribution for commercial vehicles.

2. *Scope*—This code provides the test procedure and instructions for air braked combination vehicles. It also provides recommendations for:

2.1 Instrumentation and equipment.

2.2 Vehicle preparation.

2.3 Calculating distribution of brake force.

3. Instrumentation and Equipment

3.1 Each combination vehicle should be equipped with:

3.1.1 A speed measuring device capable of measuring vehicle speed to within ± 0.2 mph (±0.32 km/h).

3.1.2 A means of measuring time to within ± 0.1 s between two speeds such as a strip chart recorder or digital timer that receives a signal from the speed measuring device.

3.1.3 Two pressure transducers and a recording device accurate to within ± 0.5 psi (3.5 kPa).

3.1.4 Brake lining thermocouples as per SAE J786a and a suitable temperature readout device accurate to within ± 20°F (±11°C). If brake lining thermocouples are not used, a contact pyrometer or other method may be used to ensure that adequate cooling time is provided between tests to ensure that brake heating and fade do not contaminate results.

3.1.5 Valving to allow brakes on each axle or axle set to be operated and evaluated independently. Valving must not disturb the inherent pressure differentials in the vehicle brake system. Provide driver override capability.

3.1.6 An external application system (see Fig. 1) that can apply the braking system with a constant input is recommended. (Use of the driver's foot to apply the brakes may increase variation in the test results.) This system must be designed, installed, and utilized in such a way as to take into account the inherent pressure differentials in the vehicle brake system.

4. Vehicle Information and Data

4.1 Vehicle information sheet (Fig. 2) to be filled in prior to starting test.

4.2 Test data sheet (Fig. 3) to be used as a work sheet during testing.

4.3 Vehicle information and test data sheets may be used for or extended to single or multiple combination vehicles.

5. *Vehicle Condition*—Brake force distribution may be determined for any lining condition new or in service; however, to ensure the accuracy and validity of results, the following conditions should be met:

5.1 Adjust brakes to manufacturer's specifications.

5.2 Linings should be in good condition, free from oil, grease, glazing, or exposure to excessive heat. Drums should be clean and free from cracks.

5.3 All components between the air compressor and each of the brake chambers must be functioning properly.

5.4 Initial testing each day should be preceded by a series of brake snubs which will warm the brakes to 200–300°F (93–149°C).

5.5 Testing of a new vehicle for compliance to a recommended practice currently under development shall be preceded by a burnish procedure as specified in Federal Motor Vehicle Safety Standard No. 121, Section 6.1.8.1.

6. Braking Distribution Tests

6.1 Install and secure instrumentation and test equipment in the test vehicle. One of the pressure transducers is to be installed at the

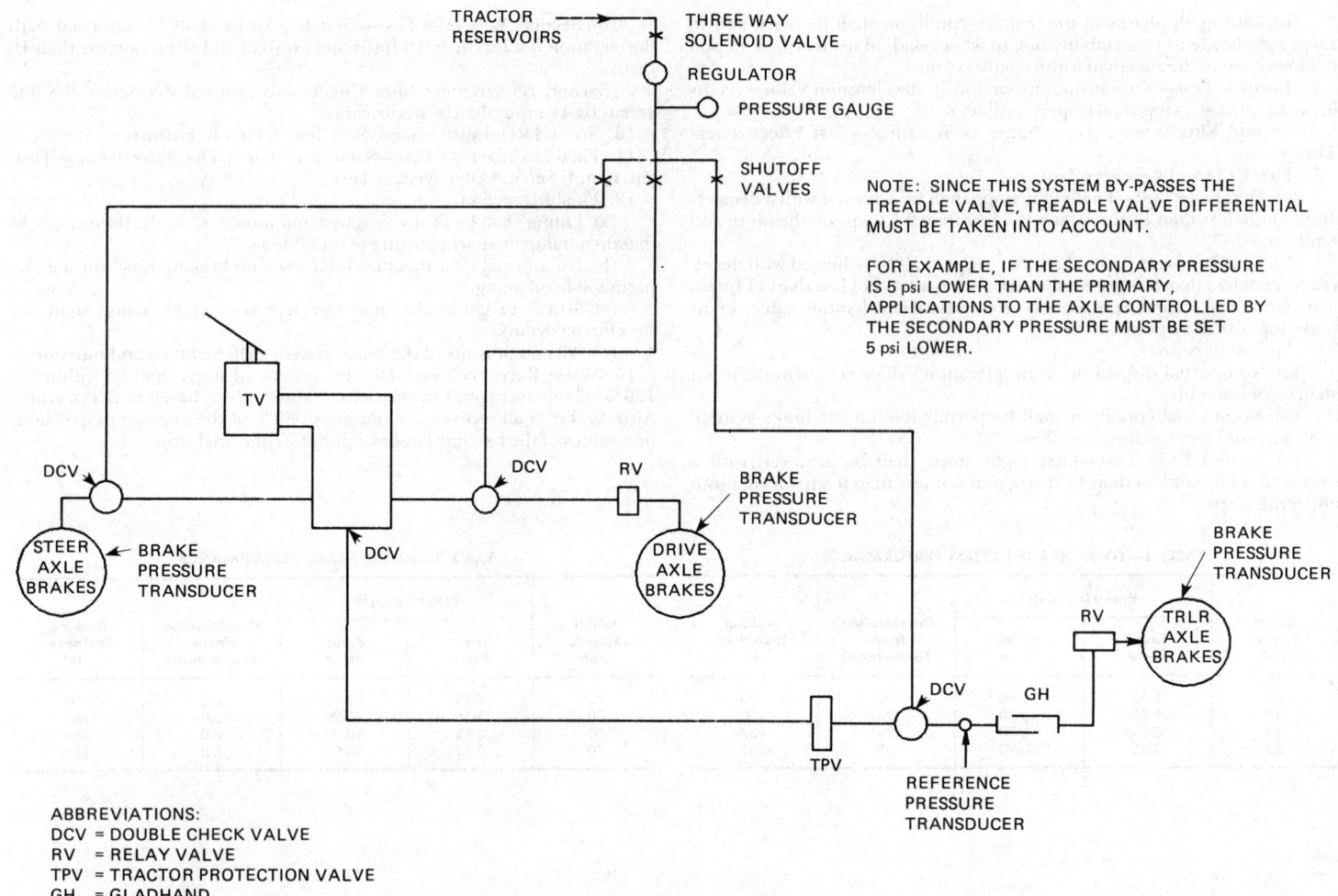

FIG. 1—EXTERNAL SYSTEM FOR BRAKE APPLICATION

axle under test, down stream of any relay or proportioning valves; the other is to be installed at the tractor to trailer control gladhand as the reference transducer.

The axle transducer is moved from axle to axle as necessary and is used to ensure that brakes are fully activated (i.e. at steady state) before data collection occurs. The reference transducer at the gladhand is used to ensure that the desired input level is achieved. All data is related back to the reference level for braking force distribution calculations. The axle transducer may be omitted if the relationship between the reference pressure and the brake pressure is known and it is certain that the brakes are fully applied when the data collection is initiated.

6.2 Weigh the test vehicle in the as-tested configuration on an axle-by-axle or axle-set basis. Weight should be distributed approximately in accordance with GAWR's and be sufficient to prevent wheel lock-up of one axle with respect to the others.

6.3 Determine the gladhand (reference) pressure level at which braking starts to occur at each brake by raising the vehicle and rotating the wheel by hand while gradually increasing input to the brake system. Record the gladhand (reference) pressure level at which brake torque is first evident on Fig. 3. Continue to increase pressure to approximately 40 psi (275 kPa) and then slowly decrease it until the point at which the brake is fully released is determined. Record the gladhand (reference) pressure level at brake release on Fig. 3. The average of these two recorded values is defined as the brake push-out pressure. Calculate the push-out pressure for each brake and average these values for each axle or tandem set.

6.4 Test the braking on each axle (or axle set) of the combination vehicle independently using the following procedure.

6.4.1 Bring the brakes to the desired initial temperature [200–300°F (93–149°C) hottest brake is recommended] by making snubs or by cooling as necessary.

6.4.2 Accelerate the vehicle to 45 mph (73 km/h), shift to neutral or de-clutch and immediately apply the brakes to the desired gladhand (reference) pressure level. Pressure overshoot is not permitted. Ensure that a constant pressure is achieved at the test brakes before the vehicle decelerates to 40.0 mph (64.37 km/h). Continue to decelerate until the vehicle reaches 35.0 mph (56.33 km/h) and immediately release the brakes. Measure the time between 40.0 and 35.0 mph (64.37 and 56.33 km/h), i.e. delta V = 5 mph (8.047 km/h), average V = 37.5 mph (60.35 km/h) and record on Fig. 3. Lower speed levels can be used if desired; however, the speed change measured should be maintained at 5 mph (8.047 km/h). Speeds above those shown should not be used as significant in-stop fade can occur due to high energy input to the brakes.

NOTE: The speeds indicated have been determined to work well on vehicles up to 80 000 lbs (36 300 kg) GCW.

For vehicles greater than 80 000 lbs (36 300 kg) GCW, speeds should be reduced to limit energy input to the braking system in accordance with the following table:

GCW		Test Starting Speed		Timed Speeds			
				Initial		Final	
lbs	kg	mph	km/h	mph	km/h	mph	km/h
100 000	45 400	37	60	32	51.50	27	43.45
120 000	54 400	33	53	28	45.06	23	37.01
140 000	63 500	29	47	24	38.62	19	30.58

6.4.3 Repeat 6.4.1 and 6.4.2 at the same input level (at the control gladhand) for a total of 4 runs, 2 in each direction on the same area of the roadway, to allow grade effects to be averaged out. If an external

| Test No. _____ | Test Date _____ |

Test Facility and Location _____
Tractor Year, Make, and Model _____
Trailer Year, Make, and Model _____
Tractor V.I.N. _____ Trailer V.I.N. _____
G.A.W.R.: Tractor: Front _____ Rear _____ Trailer: _____
Weight Distribution:

Tractor: Front _____ Rear _____ Total _____
Trailor Axle(s) _____
Gross Total _____

Brakes: Type[1]   Size   Make   Lining
Tractor: Front   _____   _____   _____   _____
         Rear   _____   _____   _____   _____
Trailer: _____   _____   _____   _____

Brake Drum/Rotor: Make   Type[2]   Part No.
Tractor: Front   _____   _____   _____
         Rear   _____   _____   _____
Trailer: _____   _____   _____

Actuation Details:

	Air Chamber Make/Model/Size	Slk. Adj. Lgth. or Wedge Angle	Adjustment Type[3]	Stroke @ 80 psi (550 kPa)	Cam[4] Rotation
Tractor:					
Front	_____	_____	_____	_____	_____
Rear-fwd.	_____	_____	_____	_____	_____
Rear-rear	_____	_____	_____	_____	_____
Trailer:					
Front	_____	_____	_____	_____	_____
Rear	_____	_____	_____	_____	_____

	Tire(s) Type[5]	Size	Press.	Measured Static Rolling Radius
Tractor: Front	_____	_____	_____	_____
Rear-fwd.	_____	_____	_____	_____
Rear-rear	_____	_____	_____	_____
Trailer: Front	_____	_____	_____	_____
Rear	_____	_____	_____	_____

Special conditions which might affect brake performance: _____

[1] Cam, disc, wedge, etc.
[2] Cast or composite drum, vented or non-vented rotor, etc.
[3] Automatic, manual, etc.
[4] With or opposite drum, etc.
[5] Single or dual, radial or bias-ply, etc.

FIG. 2—VEHICLE INFORMATION SHEET

application system is not used, variation resulting from use of the driver's foot may require additional runs to be made.

6.4.4 Repeat 6.4.1 to 6.4.3 for a range of gladhand (reference) pressure levels to cover the desired operating spectrum. As a minimum, pressure levels of 15, 20, and 40 psi (103.4, 137.9, and 275.8 kPa) are recommended. Terminate test run if wheel lock occurs. Because all axles must be run over the same range to calculate distribution, it is desirable to test the axle which is expected to lock at the lowest pressure first to establish the range for the other axles.

6.4.5 Repeat steps 6.4.1 to 6.4.4 for each axle (or axle set).

6.4.6 Repeat steps 6.4.1 to 6.4.4 with all brakes operational (to serve as a data check).

6.4.7 Repeat steps 6.4.2 to 6.4.3 with no braking (coast down test) to determine parasitic drag.

7. **Analysis of Data (Fig. 4)**

   7.1 **Definition of Terms**

   t = time in seconds.
   d = deceleration in ft/s² (m/s²).
   d_i = average deceleration of an individual axle or axle set (corrected for drag).
   1 = tractor front axle.
   2 = tractor rear axle or axle set.
   3 = trailer axle or axle set.
   d_a = deceleration with all brakes operational (corrected for drag).
   d_c = average deceleration with no brakes applied—coast down deceleration.
   d_{avg} = the average of the individual deceleration runs for each test condition.
   d_n = the sum of the individual average decelerations for each axle or axle set.
   % Error = the sum of the individual axle average decelerations compared to the average deceleration with all brakes applied.
   %B_i = the percentage braking contributed by each axle or axle set of brakes.
   F_i = the braking force contributed by each axle or axle set of brakes.
   w = total combination vehicle weight in pounds (kilograms).
   g = 32.2 ft/s² (9.81 m/s²).
   ΣF_n = the sum of all axle/axle set/brake forces.

   7.2 Calculate the deceleration for each of the coast down runs using:

   English $\quad$ Metric

   $$d = \frac{7.33}{t} \quad \left(d = \frac{2.234}{t}\right) \quad \text{(Equation 1)}$$

   Average the four coast down decelerations to obtain d_c.

   NOTE: Do *not* average times first, or grade effects will not be removed from the data correctly.

   7.3 Use Equation 1 to calculate deceleration for each of the runs with braking. Average the four decelerations for each axle or axle set at each pressure level. Subtract the average coast down deceleration (d_c) from the average braking deceleration (d_{avg}) to obtain (d_i).

   7.4 At each reference pressure evaluated, add the individual axle average braking decelerations together and perform a data check by comparing this sum (d_n) to the deceleration with all brakes operational (d_a) using:

   $$\% \text{ Error} = \frac{d_n - d_a}{d_a} \times 100 \quad \text{(Equation 2)}$$

   A negative error (greater than 20%) at all input levels indicates possible brake fade due to the greater energy inputs per brake during the individual axle tests. This is not usually a problem at the test speeds specified in this procedure. If it is, lower test speeds may be necessary. A large random error indicates lack of repeatability in the brake application pressure level and more than four runs should be made.

   7.5 Calculate brake force for each axle or axle set using:

   $$F_i = \frac{w}{g} \times d_i \quad \text{(Equation 3)}$$

   Calculate total brake force (F_n) by adding the individual axle or axle set of brake forces.

   Plot brake force (F_i) for each axle vs. gladhand reference pressure level on a common plot to show actual braking force distribution (See example, Fig. 5.) Each curve has its origin ($F_i = 0$) at the average push-out pressure determined in paragraph 6.3. Straight line segments should be used to connect the individual data points.

   Plot total tractor brake force for each data point. The origin of the curve will be at the lowest average tractor push-out pressure determined

25.150

Tractor: _____ Trailer: _____ Total Weight: _____ lb (kg)
Measured Deceleration Times, seconds:

Operational Brakes	Run #	CONTROL LINE PRESSURE psi (kPa)					
		15(103)	20(138)	40(276)			
Front	1						
	2						
	3						
	4						

Control Gladhand Pressure at Front Brake Push-Out
 Increasing Pressure: LF _____ RF _____
 Decreasing Pressure: LF _____ RF _____ Overall Average _____
 Average: LF _____ RF _____

Rear	1						
	2						
	3						
	4						

Control Gladhand Pressure at Rear Brake Push-Out
 Increasing Pressure: LI _____ RI _____ LR _____ RR _____
 Decreasing Pressure: LI _____ RI _____ LR _____ RR _____ Overall Average _____
 Average: LI _____ RI _____ LR _____ RR _____

Trailer	1						
	2						
	3						
	4						

Control Gladhand Pressure at Trailer Brake Push-Out
 Increasing Pressure: LF _____ RF _____ LR _____ RR _____
 Decreasing Pressure: LF _____ RF _____ LR _____ RR _____ Overall Average _____
 Average: LF _____ RF _____ LR _____ RR _____

All	1						
	2						
	3						
	4						

Coastdown Deceleration Times: (seconds)
1- _____ 2- _____ 3- _____ 4- _____

Start: Odo- _____ Date: _____ Time: _____
Amb. Temp: _____ °F Wind Vel: _____ Wind Dir: _____
Finish: Odo- _____ Date: _____ Time: _____
Amb. Temp: _____ °F Wind Vel: _____ Wind Dir: _____
Driver: _____ Observer: _____

FIG. 3—BRAKE FORCE DISTRIBUTION TEST DATA SHEET

in paragraph 6.3. Note that variation in push-out pressure for individual axles will change the slope of the curve as each axle begins braking.

7.6 Calculate the percent of total vehicle braking at each gladhand reference pressure for each axle by using the braking forces from the force distribution described in paragraph 7.5 and the following equation:

$$\% \, B_i = \frac{F_i}{\Sigma F_n} \times 100 \quad \text{(Equation 4)}$$

Increments no larger than 1 psi (7 kPa) up to 15 psi (105 kPa) gladhand (reference) pressure, 2 psi (14 kPa) up to 20 psi (138 kPa) and 5 psi (35 kPa) above 20 psi (138 kPa) are recommended.

Plot the braking distribution for each axle or axle set of brakes over the range 0–100% and 0–40 psi (276 kPa). (See example, Fig. 6.)

7.7 If the vehicle weight is not known, brake force cannot be determined, but braking distribution may be calculated for each of the reference pressure level data points as follows:

$$\% \, B_i = \frac{d_i}{\Sigma d_n} \times 100 \quad \text{(Equation 5)}$$

Braking distribution may be calculated for intermediate points by assuming a vehicle weight and calculating braking force as in paragraph 7.5. Braking distribution may then be calculated as in paragraph 7.6.

25.151

TRACTOR: _____ TRAILER: _____
COMBINED AXLE WEIGHT (W): _____

Operational Brakes	Run #	Deceleration (d)		
		15 (103)	20 (138)	40 (276)
Front Axle	1			
	2			
	3			
	4			
Total				
Average				
$d_1 = d_{AVE} - d_C$				
Rear Axle(s)	1			
	2			
	3			
	4			
Total				
Average				
$d_2 = d_{AVE} - d_C$				
Trailer Axle(s)	1			
	2			
	3			
	4			
Total				
Average				
$d_3 = d_{AVE} - d_C$				
All Axles	1			
	2			
	3			
	4			
Total				
Average $d_A = d_{AVE} - d_C$				
$\Sigma d_N = d_1 + d_2 + d_3$				
$\Sigma d_N - d_A$				
% Error				

Prepared By: _____
Date: _____

Coastdown (d_c)	
Run #	d
1	
2	
3	
4	
Total	
Average	

Braking Distribution and Force			
	15 (103)	20 (138)	40 (276)
% B_1			
F_1			
% B_2			
F_2			
% B_3			
F_3			
ΣF_N			

Equations:

(1) $d = \dfrac{7.33}{t}$ — English

$\left(d = \dfrac{2.234}{t}\right)$ — SI

(2) % Error $= \dfrac{\Sigma d_N - d_A}{d_A} \times 100$

(3) $F_I = \dfrac{W}{g} \times d_I$

(4) % $B_I = \dfrac{F_I}{\Sigma F_N} \times 100$

(5) % $B_I = \dfrac{d_I}{\Sigma d_N} \times 100$

FIG. 4—CALCULATION SHEET

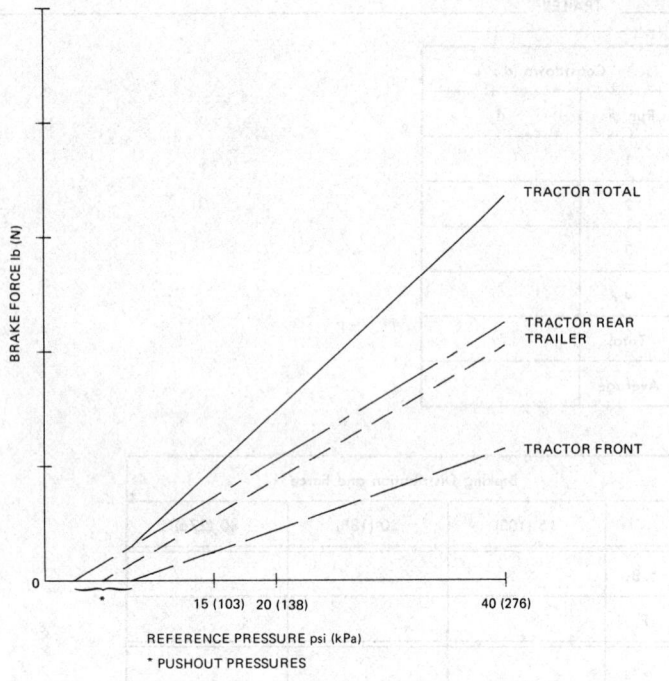

FIG. 5—BRAKE FORCE VS. REFERENCE PRESSURE, EXAMPLE

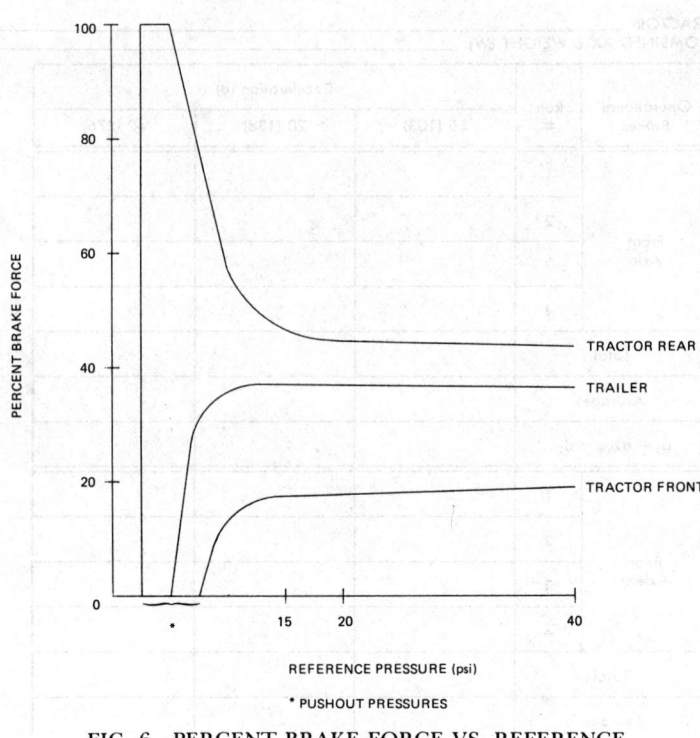

FIG. 6—PERCENT BRAKE FORCE VS. REFERENCE PRESSURE, EXAMPLE

(R) BRAKE POWER RATING TEST CODE— COMMERCIAL VEHICLE INERTIA DYNAMOMETER—SAE J971 JUN91

SAE Recommended Practice

Report of Brake Committee approved September 1966. Editorial change June 1967. Completely revised by the SAE Truck and Bus Foundation Brake Subcommittee of the SAE Truck and Bus Brake Committee June 1991. Rationale statement available.

1. Scope—The code provides test procedures and methods of calculating a brake rating from the data obtained for brakes used in highway commercial vehicles over 4.5 T (10 000 lbs) GVWR air and hydraulic. Some general correlation may be expected between brake ratings established by this means and those obtained from vehicle tests such as outlined in SAE J880. The brake rating power, kW (hp) calculated by conduct of this code is an arbitrary index of performance of the brake and drum when tested by this procedure and may be appreciably different from the values obtained by other techniques.

1.1 Purpose—The purpose of this SAE Recommended Practice is to provide a method for determining a brake rating based on the energy absorption and dissipation capacity of the brake and drum when tested on an inertia-type brake dynamometer.

2. References

2.1 Applicable Documents—The following publications form a part of this specification to the extent specified herein. The latest issue of SAE publications shall apply.

2.1.1 SAE PUBLICATIONS—Available from SAE, 400 Commonwealth Drive, Warrendale, PA 15096-0001.

SAE J880—Brake System Rating Test Code—Commercial Vehicles

3. Instrumentation, Equipment, and Test Conditions

3.1 The dynamometer inertia should be equivalent to the maximum loading conditions to which brakes tested are normally subjected. Rotation speeds should be established based on revolutions per kilometer (mile) for the tires normally used to carry such wheel loads. Dynamometer inertia is to be based on the maximum static rolling radius and half the GAWR.

3.2 Thermocouples—Install thermocouples in drum or rotor and in lining (optional) as shown in Figures 1 to 6.

3.3 Warmup—If required, make a series of 48 to 0 km/h (30 to 0 mph) stops at 3 mpsps (10 fpsps) to obtain the initial drum or disc temperature of 93 to 121 °C (200 to 250 °F).

3.4 Cooling Speed—Unless otherwise specified, cooling speed is at the speed of the next stop.

3.5 Air Flow—Unless otherwise specified, air at ambient temperature is directed uniformally and continuously over the brake drum or disc at a velocity of 670 m/min (2200 ft/min).

3.6 Ambient temperature is to be between 15 and 40 °C (59 and 104 °F).

3.7 Chamber air pressure rise rate is to be 1.50 MPa/s ± 0.3 (220 psi/s ± 45).

3.8 Tests described herein are for air brakes. For hydraulic brakes, all test pressure should be limited to 12.4 MPa (1800 psi) at 1.4 MPa (200 psi) increment.

4. Performance Procedure

4.1 The lining should be attached and ground per manufacturer specifications. New drums or discs are recommended for each test. Surface finish is to be in accordance with manufacturer recommendations with careful attention to insure uniform surface finish from test to test. Mount the brake and drum or disc assembly on the dynamometer, center, and record runout. Maximum runout is 0.254 mm (0.010 in) T.I.R. at open end of drum wear surface or 0.254 mm (0.010 in) at outer radius of disc wear surface.

4.2 Burnish—Make 200 stops from 64 km/h (40 mph) at 3.0 mpsps (10 fpsps) with an initial drum or disc temperature of 177 to 204 °C (350 to 400 °F). Make 200 additional stops at 260 to 288 °C (500 to

550 °F). Inspect lining and drum or disc and record percentage of lining area contact of each shoe and pad. If 80% drum or disc contact is not obtained on each block segment of each shoe or pad, repeat the test with new lining until 80% is obtained. Make a sketch or photo of contact.

4.3 80 km/h (50 mph) Effectiveness—Warm brake to initial drum or disc temperature of 93 to 121 °C (200 to 250 °F). Make an 80 km/h (50 mph) stop at 69 kPa (10 psi). Cool, brake. Make successive stops at 103, 138, 276, 414, 552, and 690 KPa (15, 20, 40, 60, 80, and 100 psi) with an initial drum or disc temperature of 93 to 121 °C (200 to 250 °F).

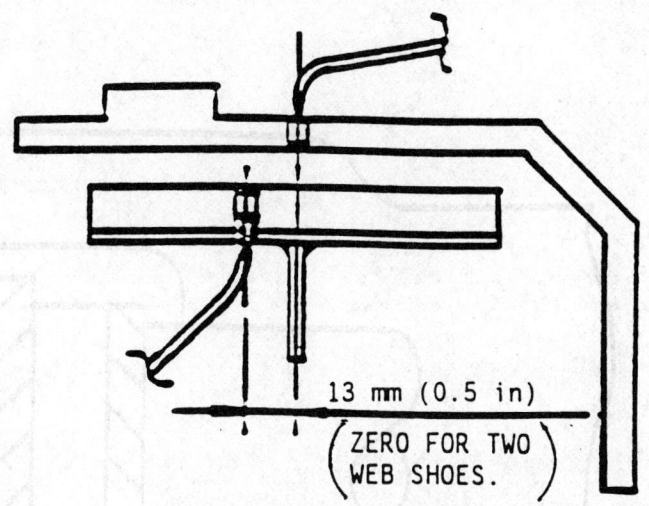

FIGURE 3—DRUM THERMOCOUPLE AND SINGLE WEB SHOE THERMOCOUPLE LOCATION

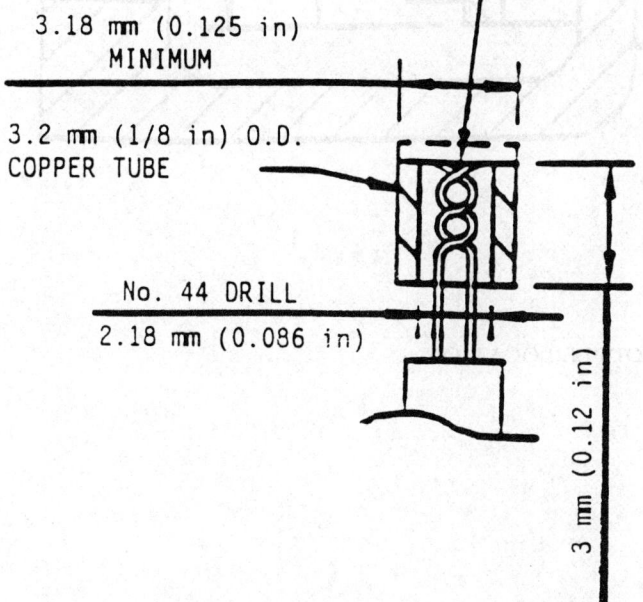

FIGURE 1—THERMOCOUPLE CONSTRUCTION

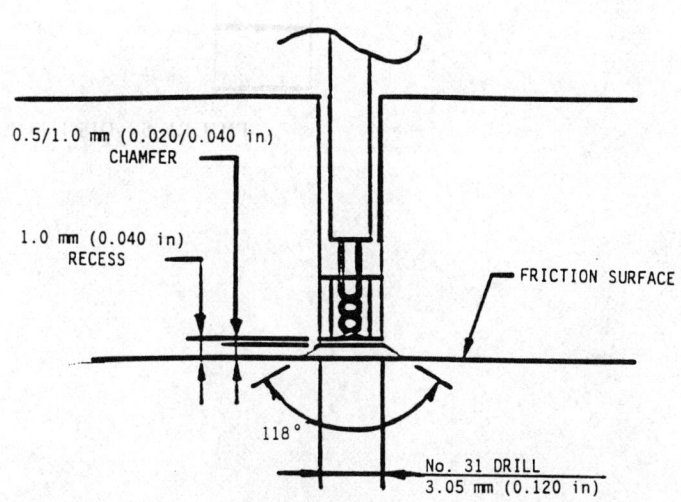

FIGURE 4—DRUM OR DISC THERMOCOUPLE INSTALLATION

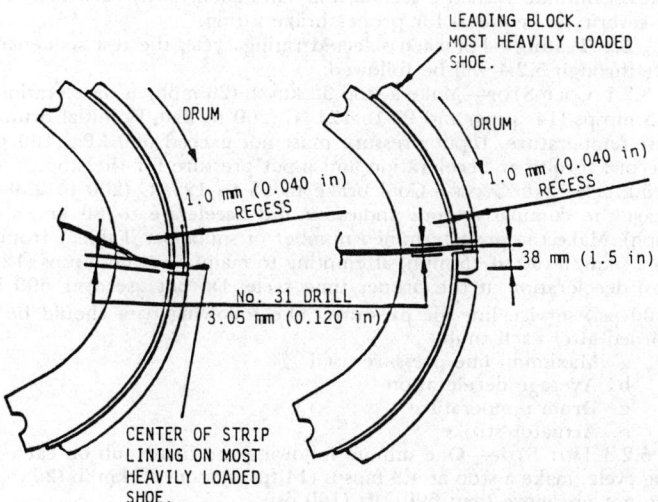

FIGURE 2—BRAKE SHOE THERMOCOUPLE INSTALLATION

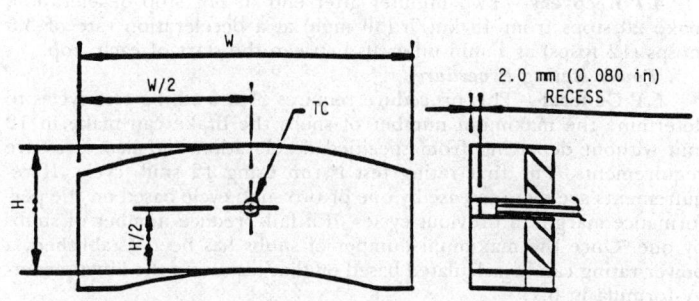

FIGURE 5—PAD THERMOCOUPLE LOCATION

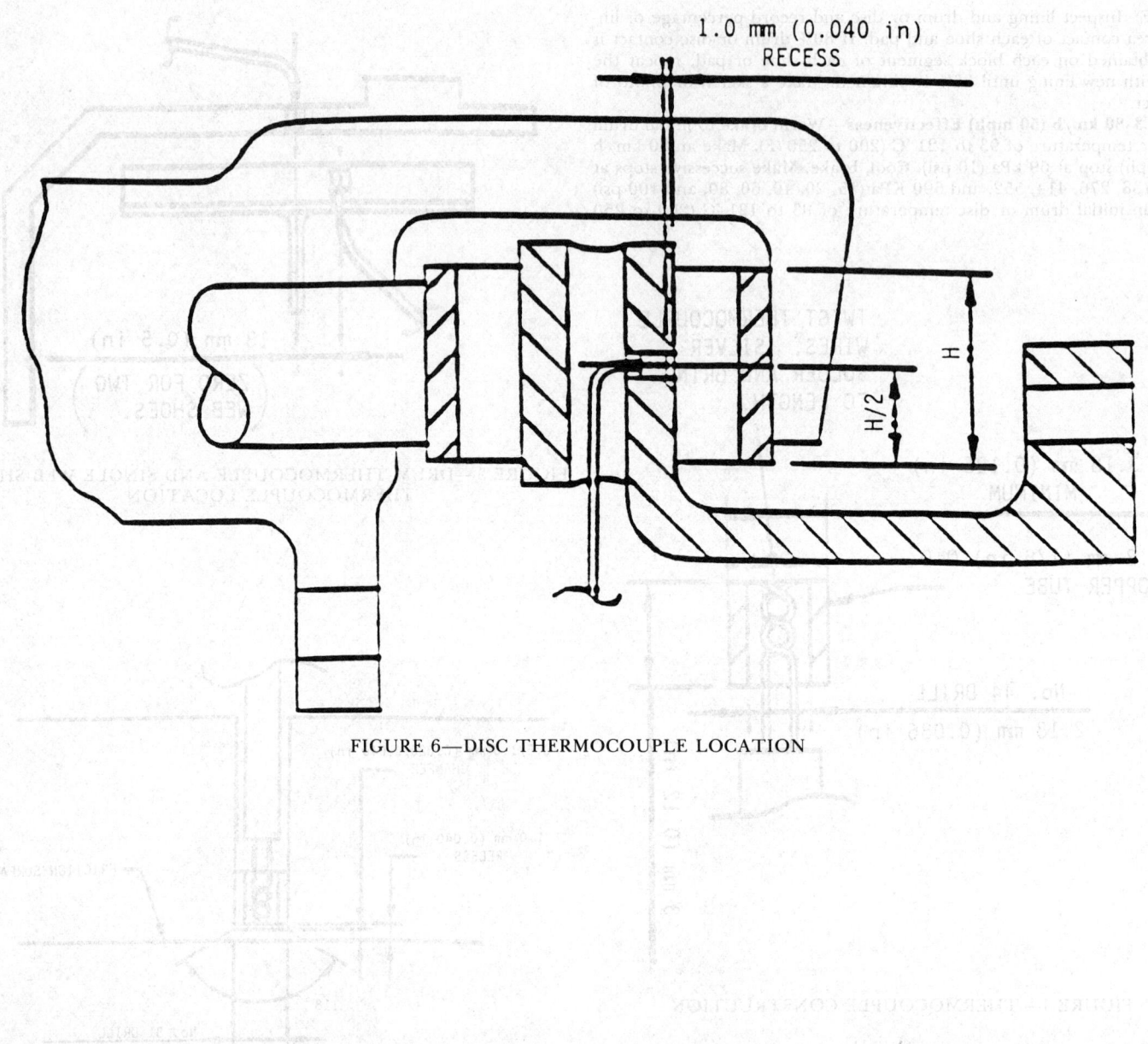

FIGURE 6—DISC THERMOCOUPLE LOCATION

4.4 Recovery Baseline—Make three 48 km/h (30 mph) stops at 3.6 mpsps (12 fpsps) from drum or disc temperature of 93 to 121 °C (200 to 250 °F).

4.5 Fade—Warm brake to initial drum or disc temperature of 93 to 121 °C (200 to 250 °F). Make 10 snubs from 80 to 24 km/h (50 to 15 mph) at 2.7 mpsps (9 fpsps) with 72 s intervals between start of each snub with the pressure limited to 690 kPa (100 psi).

4.6 Hot Stop—One minute after end of last fade snub deceleration, make a 32 km/h (20 mph) stop attempting to maintain 4.3 mpsps (14 fpsps) with pressure limited to 690 kPa (100 psi).

4.7 Recovery—Two minutes after end of hot stop deceleration, make 20 stops from 48 km/h (30 mph) at a deceleration rate of 3.6 mpsps (12 fpsps) at 1 min intervals between the start of each stop.

5. Power Rating Procedure

5.1 General—This procedure requires 3 to 5 rating test cycles to determine the maximum number of snubs the brake can make in 12 min without departing from specified deceleration and max pressure requirements. The first rating test is run using 12 snub cycle. If requirements are met increase by one or two snub cycle based on the performance margin of previous cycles. If it fails, reduce number of snubs by one. Once the maximum number of snubs has been established, a power rating can be calculated based on that test using the kinetic energy formula in 6.1.

Snubs should be made in accordance with the brake apply time shown in Table 1, thus continually compensating for fractions of seconds. Continue with the second and subsequent tests, each increasing in severity as required for proper brake rating.

5.2 Testing—For each selected rating cycle, the test sequence 5.2.1 through 5.2.4 will be followed:

5.2.1 COLD STOP—Make a stop 32 km/h (20 mph) at deceleration of 4.3 mpsps (14 fpsps) and 93 to 121 °C (200 to 250 °F) initial drum or disc temperature. Input pressure must not exceed 690 kPa (100 psi). Record torque or deceleration and input pressure for the stop.

5.2.2 RATING CYCLE—Cool brake to 93 to 121 °C (200 to 250 °F). Start the cumulative time indicator and accelerate to 80 km/h (50 mph). Make the predetermined number of snubs per Table 1 from 80 to 24 km/h (50 to 15 mph) attempting to maintain a 3.6 mpsps (12 fpsps) deceleration at the proper time cycle. Do not use over 690 kPa (100 psi) service line air pressure. The following data should be recorded after each snub:

    a. Maximum line pressure used
    b. Average deceleration
    c. Drum temperature
    d. Actuator stroke

5.2.3 HOT STOP—One minute following the final snub on each rating cycle, make a stop at 4.3 mpsps (14 fpsps) from 32 km/h (20 mph). Do not use more than 690 kPa (100 psi).

5.2.4 RECOVERY—Repeat 4.7. Do not use more than 690 kPa (100 psi).

5.3 Performance Requirements—In order to have completed the rating test satisfactorily, the brake must have:

5.3.1 Max air pressure of 690 kPa (100 psi) during the rating cycle.

5.3.2 A cold stop of not less than an average deceleration equivalent to 4.3 mpsps (14 fpsps) and max of 690 kPa (100 psi).

5.3.3 A hot stop of not less than the average deceleration equivalent to 3.3 mpsps (11 fpsps) and max of 690 kPa (100 psi).

5.3.4 The brake must give no evidence of unusual characteristic or performance during the recovery test (see 5.2.4). No excessive over recovery, as indicated by a reduction in line pressure to 50% or less of the line pressure obtained in 4.4 and must attain an average of 3.6 mpsps (12 fpsps) during all recovery stops.

6. Horsepower Rating Calculation

6.1 The brake rating power, kw (brhp), can be determined from the formula:

$$kW = \frac{N(V_1^2 - V_2^2)n}{2gt \times 1000} \quad \text{(Eq.1)}$$

where:

- N = Wheel weight, N (lb)
- V_1 = Initial speed, mps (fps)—80 km/h equivalent
- V_2 = Final speed, mps (fps)—24 km/h equivalent
- n = Number of snubs made in test cycle
- t = Cycle time, seconds
- g = Acceleration due to gravity, mpss (fpss)

The preceding formula for the standard 80 to 24 km/h (50 to 15 mph) rating cycle may be simplified to:

$$kW = 3.22 \times 10^5 \, Nn \quad \text{(Eq.2)}$$
$$kW = 1.34 \, hp$$

TABLE 1—BRAKE APPLY TIME SCHEDULE FOR VARIOUS TOTAL NUMBER OF RATING TEST BRAKE SNUBS[1]—TOTAL NUMBER OF BRAKE SNUBS (CYCLE TIME, SECONDS)

Snub No.	8 (90)	9 (80)	10 (72)	11 (68.5)	12 (60)	13 (55.5)	14 (51.5)	15 (48)	16 (45)	17 (42.4)	18 (40)	19 (37.9)	20 (36)	21 (34.3)	22 (32.7)
1	1-25	1-15	1-07	1-0.5	0-55	0-50.5	0-46.5	0-43	0-40	0-37.4	0-35	0-32.9	0-31	0-29.3	0-27.7
2	2-55	2-35	2-19	2-06	1-55	1-46	1-38	1-31	1-25	1-19.7	1-15	1-10.8	1-07	1-03.6	1-0.4
3	4-25	3-55	3-31	3-11.5	2-55	2-41.5	2-29.5	2-19	2-10	2-2.0	1-55	1-48.6	1-43	1-37.9	1-33.2
4	5-55	5-15	4-43	4-17	3-55	3-37	3-21	3-07	2-55	2-45.4	2-35	2-26.5	2-19	2-12.1	2-05.9
5	7-25	6-35	5-55	5-22.5	4-55	4-32.5	4-12.5	3-55	3-40	3-26.8	3-15	3-04.4	2-55	2-46.4	3-38.6
6	8-55	7-55	7-07	6-28	5-55	5-28	5-04	4-43	4-25	4-09.1	3-55	3-42.3	3-31	3-20.7	3-11.3
7	10-25	9-15	8-19	7-33.5	6-55	6-23.5	5-55.5	5-31	5-10	4-51.5	4-35	4-20.2	4-07	3-55.0	3-44.0
8	11-55	10-35	9-31	8-39	7-55	7-19	6-47	6-19	5-55	5-33.8	5-15	4-58.0	4-43	4-29.3	4-16.8
9		11-55	10-43	9-44.5	8-55	8-14.5	7-38.5	7-07	6-40	6-16.2	5-55	5-35.9	5-19	5-03.6	4-49.5
10			11-55	10-50	9-55	9-10	8-30	7-55	7-25	6-58.5	6-35	6-13.8	5-55	5-37.9	5-22.2
11				11-55.5	10-55	10-05.5	9-21.5	8-43	8-10	7-40.9	7-15	6-51.7	6-31	6-12.1	5-54.9
12					11-55	11-01	10-13	9-31	8-55	8-23.2	7-55	7-29.6	7-07	6-46.4	6-27.6
13						11-56.5	11-04.5	10-19	9-40	9-05.6	8-35	8-07.5	7-43	7-20.7	7-0.4
14							11-56	11-07	10-25	9-47.9	9-15	8-45.4	8-19	7-55.0	7-33.1
15								11-55	11-10	10-30.3	9-55	9-23.3	8-55	8-29.3	8-05.8
16									11-55	11-12.6	10-35	10-11	9-3.1	9-03.6	8-38.5
17										11-55	11-15	10-39.1	10-07	9-37.8	9-11.2
18											11-55	11-16.9	10-43	10-12.1	9-44.0
19												11-54.8	11-19	10-46.4	10-16.7
20													11-55	11-20.7	10-49.4
21														11-55.0	11-22.1
22															11-54.8

[1] Brake apply time is shown as minutes-seconds elapsed. Time interval includes 5s to allow for brake-on time.

φTRAILER GRADE PARKING PERFORMANCE TEST PROCEDURE—SAE J1452 OCT88

SAE Recommended Practice

Report of the Brake Committee, approved June 1985 and completely revised by the Truck and Bus Brake Committee October 1988. Rationale statement available.

1. Purpose—This SAE Recommended Practice establishes a uniform procedure for determining the parking performance on a grade of any new trailer with manufacturer's maximum weight rating of more than 10 000 lb (4500 kg) intended for roadway use.

2. Scope—This practice establishes methods to determine grade parking performance with respect to:

2.1 Ability of the parking brake system to lock the braked wheels.

2.2 The trailer holding or sliding on the grade, fully loaded or unloaded.

2.3 Applied manual effort.

2.4 Unburnished or burnished brake lining friction conditions.

2.5 Down and up grade directions.

3. Introduction—The ability to hold a trailer stationary on a grade involves two performance factors: 1) overcoming the downhill grade force with the parking brake system by preventing rotation of the braked wheels, and 2) having sufficient weight on the braked wheels to prevent the trailer from sliding on the roadway. By the use of this procedure, the manually applied input effort required to prevent braked wheel rotation can be measured and the stability of a trailer parked on a grade can be observed.

4. Instrumentation

4.1 Force measuring device — 0 - 200 lb (0 - 890 N)

4.2 Decelerometer (0 - 1 G)

4.3 Temperature measuring device — 0 - 1000°F (0 - 540°C)

4.4 Stopwatch

5. Test Preparations

5.1 On brakes applied by the parking brake system, use new lining and drums or discs of original equipment material installed in accordance with the trailer manufacturer's specifications.

5.2 Parking brake assemblies and actuation systems are to be installed, lubricated, adjusted, and inspected in accordance with the trailer manufacturer's specifications.

5.3 All trailers are to be tested in both the fully loaded and unloaded condition.

5.3.1 The fully loaded tests shall be conducted on full trailers with the trailer loaded to the manufacturer's maximum rated weight (GVWR if applicable) with the load distributed proportionately to the individual axle GAWRs; the fully loaded weight shall include the weight of any test equipment.

5.3.2 The fully loaded tests shall be conducted on semitrailers with the front end of the semitrailer supported by a dolly and the semitrailer/dolly combination loaded to a weight (including weight of test equipment) which is equivalent to the sum of the GAWRs of the semitrailer axle(s); the load on the dolly axle(s) shall not exceed 20% of the loaded combination test weight.

5.3.3 The unloaded tests shall be conducted on full trailers with no payload but with test equipment.

5.3.4 The unloaded tests shall be conducted on semitrailers, utilizing a dolly to support the front end of the semitrailer, with no payload but with test equipment.

5.4 Tires are to be of the largest diameter specified for the trailer, new or not more than 20% worn, and inflated to pressures specified by the trailer manufacturer.

6. Test Notes

6.1 Conduct the test on a dry, smooth Portland cement concrete surface (or other surface of equivalent coefficient of surface friction) that is free from loose materials and has a grade equal to or greater than any specified grade requirement for the test vehicle, as designated in SAE J293.

6.2 Parking brake system components are to be within a temperature range of 40 - 150°F (4 - 66°C) during the test.

6.3 Parking brake systems employing service brakes shall be tested after burnishing because of the difficulty of obtaining reliable and repeatable preburnish data. The burnish schedule is specified in SAE J880. Parking brake systems which employ a friction brake that is not a part of the service brake system shall be tested after being burnished per the published procedure provided to the purchaser by the trailer manufacturer. If no such procedure is provided, test without burnish.

6.4 Trailer shall be positioned on a test grade either by a powered unit in a manner consistent with normal usage, or by other mechanical means (example—block and tackle).

6.5 Data sheets should provide for recording the following data—gross vehicle test weight (including test equipment) and axle weights for the fully loaded and unloaded condition, percent grade or grade angle, identification of parking brake system, direction of trailer on the grade, applied input effort, and observation of wheel roll or lock and trailer hold or slide.

7. Test Procedure

7.1 Connect towing equipment to full trailer or dolly that supports semitrailer in a normally connected attitude.

7.2 Ascend 20% grade until trailer is fully on grade.

7.3 With trailer held on grade by towing equipment, apply trailer parking brake system to a force not exceeding 125 lb (556 N) for hand-operated or 150 lb (667 N) for foot-operated system.

7.4 Render trailer brake system independent of towing equipment brake system.

7.5 Disengage towing equipment so that no retarding force is supplied to the trailer. Deactivate any brakes on a dolly supporting a semitrailer so that they do not retard the trailer. Observe parking performance for at least 5 min.

NOTE: Suitable precautions must be taken to stop trailer in case of breakaway on test grade.

7.6 In case of trailer creep, note whether wheels roll or slide.

7.7 Reconnect towing equipment. Release parking brakes and repeat steps 7.2 through 7.6 facing the opposite direction on the grade.

8. Reporting Of Performance
Trailer parking performance shall be expressed as described in paragraph 6.5.

STOPPING DISTANCE TEST PROCEDURE—SAE J299 AUG87

SAE Recommended Practice

Report of the Brake Committee and Automotive Safety Committee, approved August 1972, completely revised by the Brake Committee January 1980, and reaffirmed by the Brake Committee August 1987.

1. Scope
This test procedure provides a method for determining stopping distances of all motor vehicles with any type of brake system.

2. Purpose
This code provides the test procedure and instructions to determine motor vehicle stopping distances on any level road surface from any desired initial vehicle speed. It allows the user to impose test conditions specified by any source and designates the preferred instrumentation and techniques for achieving the accuracy that is practical with current equipment.

3. Definitions

3.1 Motor Vehicle—Every device which is self-propelled and equipped with driver controls in, upon, or by which any person or property is or may be transported or drawn upon a highway or upon natural terrain, excepting devices moved by human or animal power or used exclusively upon stationary rails.

3.2 Start of Brake Application—The initial movement of the brake system control.

3.3 Stopping Distance—The distance traveled by a motor vehicle from the start of a brake application to the point at which the motor vehicle stops.

3.4 Initial Stopping Speed—The speed of the motor vehicle at the start of brake application.

3.5 Instrumentation System Delay—The time between the start of brake application and the start of stopping distance readout.

4. Instrumentation and Equipment
All instrumentation and equipment used in this test procedure must maintain required accuracy throughout the test period.

4.1 Speed Indicator—Fifth wheel type device that not only monitors vehicle speed, but also makes an instrumented recording of actual initial stopping speed. Error must not exceed ±0.5 mph (0.8 km/h) or ±0.5% of the actual speed, whichever is greater.

4.2 Stopping Distance Measuring Instrumentation—Fifth wheel type distance meter triggered by contact or travel switch which detects movement within the first 0.125 in (3.2 mm) of travel of the center of the brake pedal pad, the tip of the brake treadle, or the tip of the break control handle (initial movement). Total instrumentation system delay shall not exceed 0.020 s. Error of distance measuring instrumentation shall not exceed ±0.50 ft (±0.15 m) or ±1% of actual distance, whichever is greater.

5. Motor Vehicle Preparation

5.1 Perform motor vehicle and brake system preparation required to conform to specific desired test conditions. Record these conditions and operations.

5.2 Install and calibrate instrumentation. Record pertinent instrumentation information.

6. Test Procedure
The following test sequence shall be conducted at the test site:

6.1 Attain a speed sufficiently above the desired initial stopping speed to allow the driver to perform operations in paragraphs 6.2 and 6.3 and still comply with the requirements of paragraph 6.4. However, this speed shall not exceed desired initial stopping speed by more than 5 mph (8 km/h).

6.2 Release throttle.

6.3 If the stop is to be made in neutral or with clutch disengaged, perform the desired operation(s).

6.4 At the desired initial stopping speed, apply the brake control at the desired rate to any required limit(s) and maintain braking at the desired limit(s) until the motor vehicle reaches a full stop. The limit(s) shall be determined by the specific desired conditions and may be wheel skid, pedal force, deceleration, pressure, brake control movement, vehicle control, lane boundaries, or a combination of these.

NOTE: For vehicles with standard transmission, if stop is made in gear, the clutch should be disengaged when vehicle speed is reduced to below 10 mph (16 km/h) or as engine nears idle speed, whichever is the greater vehicle speed.

6.5 Record measured stopping distance and actual initial vehicle stopping speed plus wind velocity, wind direction, road grade (if other than level), vehicle direction, road surface data, vehicle data, test conditions, etc.

6.6 Repeat paragraphs 6.1, 6.2, 6.3, 6.4, and 6.5 as many times as specified for each set of conditions.

7. Distance Correction Formula for Small Initial Stopping Speed Errors
This correction can only be made if actual initial vehicle stop-

ping speed is visibly recorded as recommended in paragraph 4.1. Stopping distance corrections for initial speed errors greater than ±2 mph (3.2 km/h) are invalid due to inaccuracy.

$$S_c = S_m \frac{V_d^2}{V_a^2}$$

where: V_d = desired initial vehicle stopping speed, mph (km/h)
V_a = actual initial vehicle stopping speed, mph (km/h)
S_m = measured stopping distance, ft (m)
S_c = calculated stopping distance from V_d, ft (m)

BRAKE DISC AND DRUM THERMOCOUPLE INSTALLATION—SAE J79

SAE Recommended Practice

Report of Brake Committee and Automotive Safety Committee approved August 1972.

1. *Purpose*—To provide a uniform method for installing thermocouples in automotive brake discs and drums.
2. *Scope*—This SAE Recommended Practice describes two methods for installing thermocouples in automotive brake discs or drums for the purpose of indicating bulk temperature during inertia dynamometer and vehicle road tests.
3. *Equipment*
   Method A—Resistance welder
   Method B—General shop tools
4. *Procedure*
   Method A—Use a resistance welder to attach each wire of the thermocouple to the disc or drum in the locations shown in Fig. 1.

Method B—Drill two holes in the brake disc or drum in the locations shown in Fig. 2. Stake the thermocouple wires in the holes with a center punch.

When thermocouples are installed in a ribbed or finned drum the location shall be between the ribs or fins.

Thermocouple wires shall be 20 gage, 0.032 in. (0.8 mm) dia, with glass braid insulation.

In all installations the wires shall be protected from water, dirt, salt, and oxidation and against breakage from fatigue by surrounding the junction with high temperature silicone adhesive sealant, or equivalent.

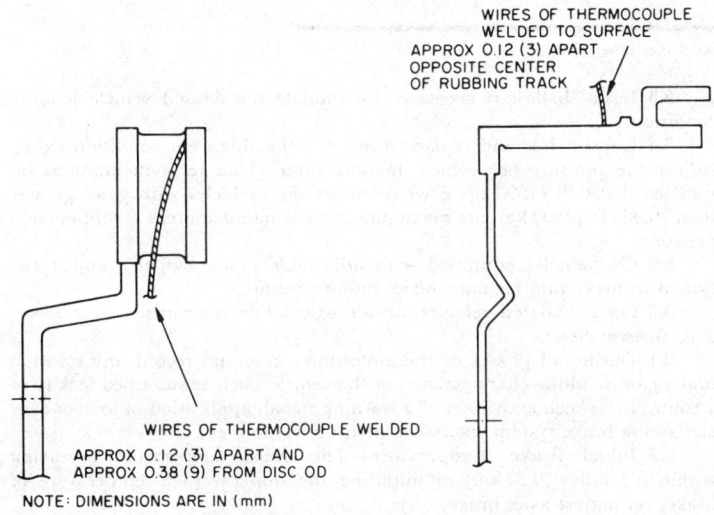

FIG. 1—THERMOCOUPLE INSTALLATION, METHOD A

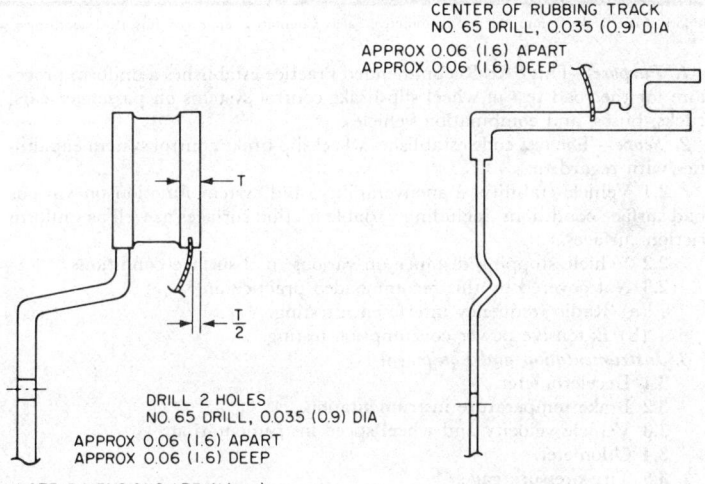

FIG. 2—THERMOCOUPLE INSTALLATION, METHOD B

DETERMINATION OF BRAKE FLUID TEMPERATURE—SAE J291 JUN80

SAE Recommended Practice

Report of the Brake Committee, Automotive Safety Committee, and Hydraulic Brake Systems Actuating Committee, approved June 1972, reaffirmed without change June 1980.

1. *Purpose*—To establish a uniform procedure for obtaining and determining a maximum temperature of brake fluid in the automotive braking system of vehicles equipped with disc brakes. This procedure is a uniform means of heating the brakes and the brake fluid and is not a simulation of any other test procedure.
2. *Scope*—This code provides a test procedure for obtaining and determining extremely high brake fluid temperature encountered in the brake system of a vehicle that is equipped with disc brakes. Vehicles in normal operation may or may not produce brake fluid temperatures that are obtained in this procedure.
3. *Equipment*
   3.1 Thermocouples

3.2 Line pressure gages, 0–300 psi (0–2.07 MPa) range
3.3 Temperature instrumentation, 650 F (616 K) capability
3.4 Stopwatch
3.5 Calibrated speedometer and odometer
3.6 Ambient temperature thermometer

4. *Procedure*[1]

4.1 Preliminary Preparations

4.1.1 Install thermocouples in brake bleed screws as shown in Fig. 1. Install bare brake shoes (no linings) in disc brakes. If vehicle is equipped with rear drum brakes, install production shoe and lining assemblies in rear brakes.

4.1.2 Instrument vehicle with line pressure gage(s) on axle(s) with disc brakes. Install temperature instrumentation and connect to thermocouples with proper lead wires.

4.1.3 Check all instruments for operational accuracy.

4.2 Testing Instructions

4.2.1 Test not to be conducted if ambient temperature is under 50 F (283 K).

4.2.2 Record ambient temperature.

4.2.3 Apply brake pedal until 150 psi (1.03 MPa) line pressure is achieved at front brakes. While maintaining this pressure, accelerate the vehicle to 30 mph (38.3 km/h) and maintain this speed for 4 miles (6.4 km).

4.2.4 Read and record temperatures at 1 min intervals.

4.2.5 Park vehicle with engine running and foot off brake pedal. Record temperature rise at 1 min intervals until temperatures start to decline.

[1]Caution—This test may be destructive to the brake system and may affect the handling or stopping ability of the vehicle. It is suggested that the test be conducted by skilled personnel under carefully controlled conditions.

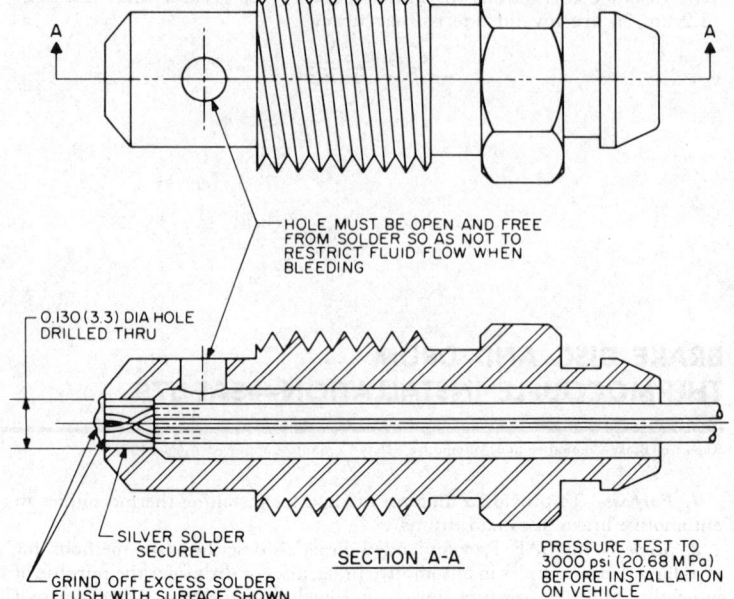

FIG. 1

WHEEL SLIP BRAKE CONTROL SYSTEM ROAD TEST CODE—SAE J46 JUN80

SAE Recommended Practice

Report of the Brake Committee and Automotive Safety Committee, approved July 1973, reaffirmed without change June 1980.

1. *Purpose*—This SAE Recommended Practice establishes a uniform procedure for the road test of wheel slip brake control systems on passenger cars, trucks, buses, and combination vehicles.

2. *Scope*—The test code establishes wheel slip brake control system capabilities with regard to:

2.1 Vehicle stability, maneuverability, and system function on various road surface conditions, including variable friction surfaces as well as uniform friction surfaces.

2.2 Vehicle stopping distance on various road surface conditions.

2.3 Not covered by this recommended practice are:
(a) Radio frequency interference testing.
(b) Extensive power consumption testing.

3. *Instrumentation and Equipment*

3.1 Decelerometer.
3.2 Brake temperature instrumentation.
3.3 Vehicle velocity and wheel speed instrumentation.
3.4 Odometer.
3.5 Tire pressure gage.
3.6 Stopping distance instrumentation.
3.7 Vehicle yaw measuring device.
3.8 Articulation restraints for trailers (optional where applicable).
3.9 Means for disabling wheel slip control system.
3.10 System pressure instrumentation (optional where applicable).
3.11 Vehicle stabilizer (optional when needed for vehicle stability).
3.12 Means to designate the point at which the brakes are applied (such as a detonator).

4. *Facilities*—Twelve foot (3.7 m) wide road surfaces of various friction levels are required with sufficient space on all sides for approach, spin-out, and recovery conditions. (See Figs. 2–4.)

4.1 Road surface description and suggested (guidelines only) lengths of uniform surface facilities assuming that all the brakes are working normally and the maximum speed is as indicated in Table 1.

4.2 Pylons as required.

5. *Vehicle Preparation*

5.1 Inspect the brake friction elements and replace if over 25% are worn or if any abnormal condition exists. Severity of tests sequences may require frequent checks to avoid overadjustment of the brakes.

5.2 Install and calibrate equipment. See Fig. 1 for brake thermocouple installation method.

5.3 Install ballast if necessary to simulate the desired vehicle loading condition.

5.4 Inspect tires and replace if an objectionable wear condition exists. Adjust tire pressure per vehicle manufacturer's load recommendations on vehicles 10,000 lb (4500 kg) gvwr or under. For vehicles with gvwr greater than 10,000 lb (4500 kg), use maximum vehicle manufacturer's recommended pressure.

5.5 On vehicles equipped with adjustable power systems, adjust the system to maximum recommended cutout pressure.

5.6 On articulated vehicles, install articulation restraints.

6. *General Notes*

6.1 During all phases of this procedure, note and record any unusual braking or handling characteristics of the vehicle, such as sustained lockup of a controlled wheel, activation of a warning signal, application of a secondary or parking brake system, excessive lateral deviation, etc.

6.2 *Initial Brake Temperature*—The brake temperature occurring within 0.2 miles (0.32 km) of initiating the stop (average temperature of brakes on hottest axle, brakes off).

6.2.1 When not otherwise specified, initial brake temperature shall be 150–200°F (66–93°C).

6.3 On vehicles so equipped, charge the system power supply pressure (vacuum) to maximum before each test stop and note the system pressure (vacuum) at the end of the stop.

TABLE 1

Surface	Suggested Length		Assumed Max[a] Speed	
	ft	m	mph	km/h
Very low friction—smooth ice or equiv.	400	122	20	32
Low friction—wet jennite or equiv.	400	122	30	48
Medium friction—wet asphalt or wet concrete	300	91	40	64
High friction—dry asphalt or dry concrete	400	122	60	97
Special—graded loose gravel	250	76	30	48

[a]Recommend moving up to speed in steps.

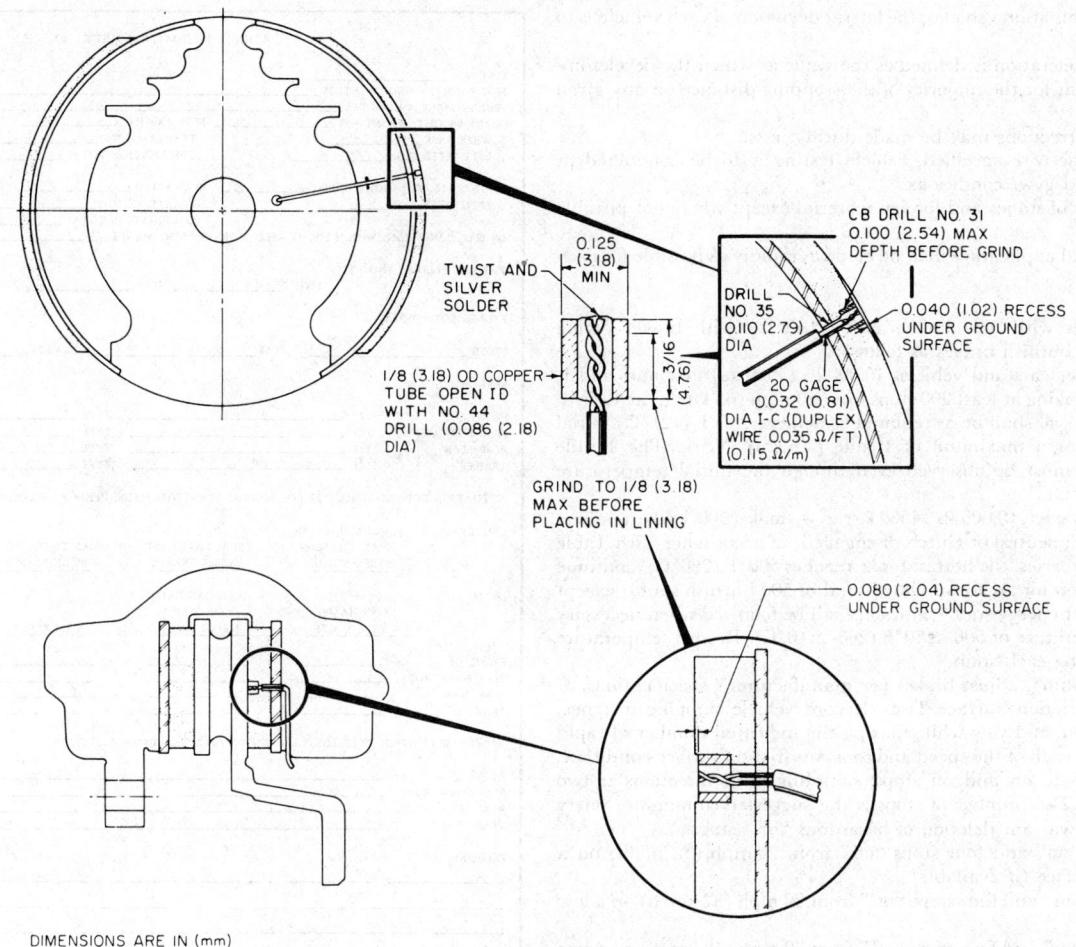

FIG. 1—TYPICAL PLUG THERMOCOUPLE INSTALLATIONS

6.4 Brake applications are to be made in normal driving gear. On vehicles with manual transmission, disengage the clutch.

6.5 Stopping Distance is defined as the distance traveled between the point at which the driver starts to move the braking control and the point at which the vehicle comes to rest. Since stopping distance on a given road surface varies approximately as the square of initial speed, comparative checks will require that the vehicle speed at which the brake control is initially moved be within 1% of the nominal value for each test condition.

Alternatively, speed variation up to a suggested maximum of ±5% can be tolerated by multiplying each measured stopping distance by the square of the initial speed ratio (nominal speed/actual speed).

6.6 Vehicle "Yaw" is defined as the vehicle's angular deviation between the point at which the brake control is actuated and the point at which the vehicle comes to rest. (That is, one complete revolution in clockwise direction would be a yaw of +360 deg, while a quarter of a revolution in the counter-clockwise direction would be −90 deg.) For combination vehicles, the yaw angle of each vehicle is to be noted.

6.7 Vehicle lateral deviation is defined as the greater of the distance between a reference point on the front and the rear of the vehicle at the longitudinal centerline and the centerline of the lane in which a stop is to be

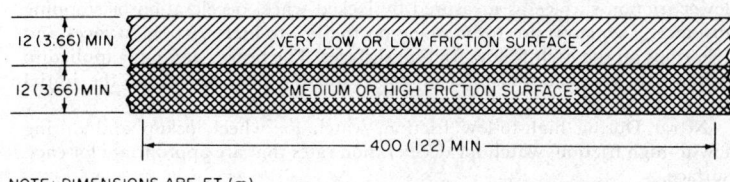

FIG. 2—SPLIT FRICTION SURFACE TEST FACILITY

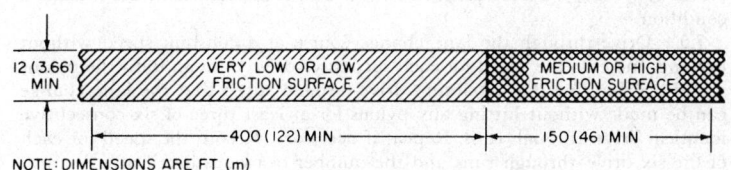

FIG. 3—CHANGING FRICTION SURFACE TEST FACILITY

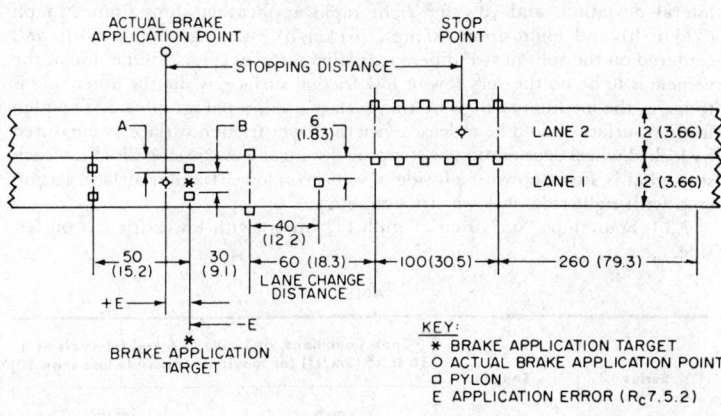

FIG. 4—LANE CHANGE TEST FACILITY (LOW FRICTION SURFACE)

completed. For combination vehicles, the lateral deviation of each vehicle is to be noted.

6.8 Vehicle deceleration is defined as the value at which the decelerometer is nearly constant for the majority of the stopping distance on any given surface condition.

6.9 Steering corrections may be made during tests.

6.10 Unless otherwise specified, vehicle testing is to be conducted at unloaded and loaded gvwr conditions.

6.11 Clear tires of stones and foreign material except where not possible for special surfaces.

6.12 Brake pedal application shall be made as rapidly as possible for each test.

7. Procedure

7.1 For vehicles with new linings and drums or with brakes giving inconsistent results, burnish brakes as follows:

7.1.1 For passenger cars and vehicles 10,000 lb (4500 kg) gvw and under, burnish brakes by making at least 200 stops from 40 mph (64 km/h) at 12 ft/s^2 (3.7 m/s^2). Stop interval shall be as required to achieve 250°F (121°C) initial brake temperature or a maximum of 1 mile (1.6 km). NOTE: The 1 mile (1.6 km) maximum must be observed even though the initial temperature exceeds 250°F (121°C).

7.1.2 For vehicles over 10,000 lb (4500 kg) gvw, make 500 brake applications (transmission in neutral or clutch disengaged) in accordance with Table 2. When, during any series, the hottest brake reaches 500°F (260°C), continue at that snub condition for an accumulated total of 500 burnish snubs, except that a higher or lower energy snub condition shall be followed when necessary to maintain a temperature of 500 ±50°F (260 ±10°C). Record temperature immediately following each snub.

7.1.3 After burnishing, adjust brakes per manufacturer's specifications.

7.2 **Constant Friction Surface Test**—Record vehicle stopping distance, final lateral deviation, and yaw while making the indicated number of rapid application stops for each of the speed and constant friction surface conditions listed below. Alternate on and off stops, sampling both directions if two directions are used. The number of stops is the suggested minimum. Safety considerations may warrant deletion of hazardous "off" stops.

7.2.1 Four stops "on" and four stops "off" from 20 mph (32 km/h) on a very low friction surface (if available).

7.2.2 Four stops "on" and four stops "off" from 20 mph (32 km/h) on a low friction surface.

7.2.3 Four stops "on" and four stops "off" from 30 mph (48 km/h) on a low friction surface.

7.2.4 Four stops "on" and two stops "off" from 20 mph (32 km/h) on a medium friction surface.

7.2.5 Four stops "on" and two stops "off" from 40 mph (64 km/h) on a medium friction surface.

7.2.6 Four stops "on" and four stops "off" from 30 mph (48 km/h) on a special friction surface.

7.2.7 Four stops "on" and two stops "off" from 20 mph (32 km/h) on a high friction surface.

7.2.8 Four stops "on" and two stops "off" from 40 mph (64 km/h) on a high friction surface. Optional for passenger and vehicles under 10,000 lb (4500 kg).

7.2.9 Four stops "on" from 60 mph (97 km/h) on a high friction surface.

7.3 **Split Friction Surface Test**—Record vehicle stopping distance, final lateral deviation, and yaw for eight rapid application stops from 20 mph (32 km/h) and eight from 40 mph (64 km/h), starting in line with and centered on the split of two different friction surfaces (Fig. 2). One side of the vehicle is to be on the very low or low friction surface, while the other side is to be on the medium or high friction surface used in paragraph 7.2. The high friction surface should be at least twice the lower friction surface as measured by locked wheel deceleration or stopping distance. Reverse stop direction each stop (that is, first stop with left side of vehicle on lower friction surface, second stop with right side on lower friction, etc.).

7.3.1 Four stops "on" from 20 mph (32 km/h) with lower friction on left side.

TABLE 2

Series	Snubs	Snub Conditions, at 1 mile (1.6 km) intervals at 10 ft/s^2 (3m/s^2) (or maximum possible less than 10)	
		mph	km/h
1	175	40 to 20	64 to 32
2	25	45 to 20	72 to 32
3	25	50 to 20	80 to 32
4	25	55 to 20	88 to 32
5	25	60 to 20	97 to 32

FIG. 5—VEHICLE INFORMATION SHEET

7.3.2 Four stops "on" from 20 mph (32 km/h) with lower friction on right side.

7.3.3 Four stops "on" from 40 mph (64 km/h) with lower friction on left side.

7.3.4 Four stops "on" from 40 mph (64 km/h) with lower friction on right side.

7.4 **Changing Friction Surface Test**—Record vehicle deceleration, final lateral deviation, and yaw for four rapid application stops from 40 mph (64 km/h) while traveling from one surface friction condition to another (Fig. 3). One surface is to be very low or low friction, while the other is to be medium or high friction. The high friction surface should be at least twice the lower friction surface as measured by locked wheel deceleration or stopping distance. Two stops are to be made on the low-to-high friction surfaces and two stops on the high-to-low friction surfaces. The brakes are to be applied to achieve the friction transition at approximately three-fourths of the initial speed at the axle that is being tested.

NOTE: During high-to-low friction, watch for wheel lockup and during low-to-high friction, watch for deceleration rates that are appropriate for each surface.

7.4.1 Two stops from 40 mph (64 km/h) on lower-to-high friction surfaces.

7.4.2 Two stops from 40 mph (64 km/h) on high-to-lower friction surfaces.

7.5 **Lane Change Test, If Applicable**—Make the lane change maneuver on the low friction course detailed in Fig. 4 with the vehicle in the unloaded condition.

7.5.1 Drive through the lane change course at a constant speed without braking. Increase the speed for each successive drive through until the pylons are hit. Determine the maximum drive-through speed that the lane change can be made without hitting any pylons for at least three of six consecutive identical drive-through runs. Repeat if necessary. Record the speed for each of the six drive-through runs and the number of pylons hit, if any.

7.5.2 Drive to the brake application point of the lane change course at a constant speed, make a rapid application stop, and steer through the course (Fig. 4). Increase the initial braking speed for each successive braking run

WHEEL SLIP CONTROL PERFORMANCE TEST
DATA SHEET

☐ CONSTANT FRICTION SURFACE TEST
☐ SPLIT FRICTION SURFACE TEST
☐ CHANGING FRICTION SURFACE TEST
SURFACE DESCRIPTION ____
SURFACE CONDITION ____
VEHICLE LOADING ____

DATE ____
TEST NO. ____
VEHICLE(S) ____
DRIVER ____
OBSERVER ____

STOP NO.	SPEED		SYSTEM ON/OFF	STOPPING DISTANCE		VEHICLE DECELERATION		APPLICATION PRESSURE	LATERAL DEVIATION	YAW, DEG.	COMMENTS
	MPH	km/h		FT	m	FT/S^2	m/s^2				

COMMENTS: ____

FIG. 6—WHEEL SLIP CONTROL PERFORMANCE TEST DATA SHEET

WHEEL SLIP CONTROL LANE CHANGE TEST
DATA SHEET

SURFACE DESCRIPTION ____
SURFACE CONDITION ____
VEHICLE LOADING ____

DATE ____
TEST NO. ____
VEHICLE(S) ____
DRIVER ____
OBSERVER ____

RUN NO.	DRIVE-THROUGH SPEED, V_d		NO. PYLONS HIT	COMMENTS
	MPH	km/h		

STOP NO.	BRAKING SPEED, V_d		NO. PYLONS HIT	APPLICATION ERROR - E		STOPPING DISTANCE		LATERAL DEVIATION	YAW, DEG.	COMMENTS
	MPH	km/h		FT	m	FT	m			

LANE CHANGE PERFORMANCE FACTOR $P = \frac{V_b \text{ MAX.}}{V_d \text{ MAX.}} =$ ____

FIG. 7—WHEEL SLIP CONTROL LANE CHANGE TEST DATA SHEET

until the pylons are hit. Determine the maximum initial braking speed that the lane change can be made without hitting any pylons for at least three of six consecutive identical braking runs. Repeat if necessary. Record the initial braking speed for each braking run and the number of pylons hit, if any. Also record application error, stopping distance, final vehicle lateral deviation, and final vehicle yaw. The brake application error, E, must be within ±5 ft (1.5 m) to be used in determining maximum braking speed.

7.5.3 Lane Change Performance—The lane change performance is the ratio of the maximum braking speed, V_b, to the maximum drive through speed, V_d:

$$P = \frac{V_b}{V_d}$$

8. *Report Forms*

8.1 Vehicle Information Sheet, Fig. 5.
8.2 Performance Test Data Sheet, Fig. 6.
8.3 Lane Change Test Data Sheet, Fig. 7.

25.162

MINIMUM REQUIREMENTS FOR WHEEL SLIP BRAKE CONTROL SYSTEM MALFUNCTION SIGNALS—SAE J1230 OCT79

SAE Recommended Practice

Report of the Brake Committee, approved October 1979.

1. Purpose—This SAE Recommended Practice establishes a uniform minimum set of functional areas to be monitored in the process of detecting malfunctions in discrete wheel slip brake control systems for motor vehicles, and establishes a minimum reaction to those detected malfunctions.

2. Scope—It is recognized that a malfunction in any one of the specified areas can degrade intended performance, but that levels of malfunction or combinations thereof must be considered by the vehicle designer in determining the point at which a failure indication is warranted. Consequently, the minimum reaction recommended by this document consists of making available a malfunction signal.

2.1 A discrete wheel slip brake control system is a wheel slip brake control system designed to control a single wheel, pair of wheels on the same axle, or other combination of wheels somehow coupled typically by a suspension, (such as a Tandem axle suspension).

2.2 A malfunction signal is the output or lack thereof given by a discrete wheel slip brake control system indicating the existence of a functional degradation as specified by Section 3 of this document. The malfunction signal may be indicated by visual, audio, electrical, or other appropriate means. When a visual malfunction indication involves a color, it shall be amber.

2.3 A failure warning signal is the signal given to the vehicle operator or inspector when certain malfunction signals or combinations thereof exist which result in unacceptable vehicle performance. The failure warning signal may be indicated by visual, audio, electrical, or other appropriate means. When a visual failure warning indication involves a color, it shall be red.

2.4 In motor vehicles, excluding trailers, the malfunction *or* failure warning indication shall be continuously available when the signal exists, providing that the vehicle electrical power system is intact and energized. As a minimum, failure warning must be visual.

2.5 In towed vehicles, or other vehicles without independent electrical power sources, the indication can be either active or nonactive.

2.5.1 If active, satisfactory system performance shall be indicated when the stoplight circuit is energized. Absence of this indication when the stoplight circuit is energized indicates existence of either a malfunction or a failure.

2.5.2 If nonactive, system condition shall be determinable with a minimum amount of effort on the part of the inspector.

3. Functional Monitoring

3.1 Electrical Power—Any failure within a system which causes detrimental loss of electrical power to any electronic control module within the discrete wheel slip brake control system shall result in a malfunction signal.

3.2 Loss of continuity of any speed sensor or its cable shall result in a malfunction signal. If the system contains more than one sensor, loss of electrical output from any one sensor at constant speeds above 20 mph may be used in place of the continuity criterion.

3.3 Solenoid—Loss of solenoid continuity shall result in a malfunction signal.

3.4 Brakes—Single electrical or electronic defects within a discrete wheel slip control system which totally deny braking action shall result in a malfunction signal, and the discrete wheel slip brake control system shall be capable of restoring braking action to the affected brakes.

3.5 General—Any cause, intentional by design, for disabling some or all of the discrete wheel slip brake control system shall result in a malfunction signal.

TUBING—MOTOR VEHICLE BRAKE SYSTEM HYDRAULIC—SAE J1047 JUN90

SAE Recommended Practice

Report of Hydraulic Brake Systems Actuating Committee approved October 1974. Revised by the Hydraulic Brake Components Standards Committee June 1990.

Foreword—This is a performance requirement qualification specification intended to give reasonable flexibility in the initial engineering selection of materials when fabricated by currently accepted manufacturing processes. It is not intended to be used for quality control purposes.

1. Scope—This SAE Recommended Practice covers the tubing intended primarily for use as hydraulic brake lines on motor vehicles. It covers materials, manufacturing processes, and general properties required to meet the wide range of service encountered in automotive applications. To meet this need, it must be formed, assembled, handled, and installed in accordance with sound engineering and manufacturing practices. This specification covers only the basic tubing and does not include attachments such as protective armor or end fittings. This document is not intended to be used for quality control purposes. Design guidelines are shown in Appendix A.

2. References
  2.1 Applicable Documents
  SAE J533, Flares for Tubing
  SAE J1703, Motor Vehicle Brake Fluid
  ASTM B 117
  ASTM Bulletin #187 of January 1953

3. Materials—The material(s) must be metallic and compatible with the type of tube construction.

4. Construction—The construction must be limited to either of the following types:
  4.1 Multiple Ply Tubing—Wherein the plies are continuously bonded by a metallurgical bond, and if made from separate strips, the joints in adjacent plies, which occur as a result of the forming operation, are separated by at least 120 degrees (2.1 rad).
  4.2 Seamless Tubing

5. Requirements—The tubing must meet the following requirements:

5.1 Bursting Strength—When tested with hydraulic pressure, sections of tubing—with a minimum expanded length of 18 in (45.7 cm)— must be capable of withstanding an internal pressure of 8000 psi (5.52 MN/m^2).

5.2 Bending Properties

5.2.1 Using suitable bending fixtures, an adequate length of tubing must withstand bending around a mandrel whose diameter is equal to five times the nominal diameter of the tubing, through 360 degrees (6.28 rad) without kinking, cracking, or developing other flaws.

5.2.2 The reduction in tubing outside diameter must not exceed 20%.

5.2.3 After the bending test, the tubing must be capable of meeting the corrosion resistance requirements (see 5.6).

5.3 Flaring Test

5.3.1 The tubing must be capable of being expanded over a tapered mandrel, having a slope (based on radius) of one in ten, until the outside diameter at the expanded end is increased 20% without cracking or developing other flaws. Prior to the expansion test, the tubing must be cut off square, edge crowned, and deburred. It must be held firmly and squarely in the die, and the punch must be guided on the axis of the tubing.

5.3.2 The tubing must be capable of being flared with double 45 degree (0.79 rad) flares in accordance with SAE J533. There must not be cracks on the sealing surface nor other imperfections which would prevent sealing. (R)

5.4 Fatigue Resistance—Straight lengths of new tubing, when tested as outlined in 6.1, must have a minimum fatigue limit of 24 000 psi (16.55 MN/m²) at 10^7 cycles for steel, or an equivalent degree of fatigue resistance if made of other materials.

5.5 Heat Resistance

5.5.1 After soaking at a temperature of 425°F ± 25 (218°C ± 14) for 30 min ± 5, the tubing must be capable of meeting the bursting strength requirements (see 5.1).

5.5.2 After being subjected to a temperature of 425°F ± 25 (218°C ± 14) for 30 min ± 5, the tubing must be capable of the corrosion resistance requirements (see 5.6).

5.6 Corrosion Resistance—Either of the following tests, both of which are described in 6.2, are satisfactory to assure compliance.

5.6.1 SALT SPRAY TEST—After 60 days of exposure as described in 6.2.1, the tubing must be capable of meeting the bursting strength requirements (see 5.1) at a reduced internal pressure of 2000 psi (1.38 MN/m²).

5.6.2 CYCLIC HUMIDITY TEST—After 170 cycles of exposure as described in 6.2.2, the tubing must be capable of meeting the bursting strength requirements (see 5.1) at a reduced internal pressure of 2000 psi (1.38 MN/m²).

5.7 Impact Resistance—The tubing must withstand an impact load of 1.5 ft-lb (0.21 kg-m) in the transverse plane by a 60 degree (1 rad) included angle hardened steel knife edge. After impact, the tubing must be capable of meeting the burst strength requirements (see 5.1).

5.8 Brake Fluid Compatibility—The tubing and SAE J1703 RM-1 compatibility fluid shall, when tested in accordance with the procedures outlined in 6.3, meet the following requirements.

5.8.1 WEIGHT LOSS—The tubing must not experience a weight loss greater than 0.000007 oz per in² (0.02 mg/cm²) of internal area.

5.8.2 SEDIMENT—The test brake fluid must not contain more than 0.010 percent sediment by volume.

5.8.3 pH—The pH value of the fluid must not be less than 7 nor more than 11.5.

5.8.4 BOILING POINT—The boiling point of the fluid must not have changed by more than 5°F (3°C) plus 0.09°F (0.050°C) for each 1.8°F (1°C) that the boiling point exceeds 437°F (235°C).

6. Test Procedure

6.1 Fatigue Resistance—A standard rotating beam fatigue testing machine (for wire or tubing) should be used for testing tubing specimens by loading them as bent, pin-ended columns.

A series of identical samples of a given kind of tubing shall be tested in air at room temperature at various calculated levels of maximum bending stress until failure by fracture occurs or until an acceptably large number of cycles for the material tested has been reached without failure of the specimen. For each sample tested, the value of applied stress shall be plotted versus the number of cycles of stress applications to produce a typical S-N diagram for the kind of tubing tested. The fatigue limit, the maximum stress that a tube will withstand for the minimum specified number of cycles, is determined by inspection of the S-N diagram obtained.

6.2 Corrosion Resistance—Prepare the specimen by installing tube nuts, flaring both ends with double 45 degree (0.79 rad) flares per SAE J533 and sealing both ends with appropriate fittings. Clean the assemblies by simple immersion in trisodium phosphate at 160 to 180°F (71 to 82°C). Coat both ends for a distance of 2 in (5.08 cm) with wax or equivalent to protect fittings against corrosion.

6.2.1 SALT SPRAY (FOG) TEST—Run this test as outlined in ASTM B 117.

6.2.2 CYCLIC HUMIDITY TEST—Run this test as outlined in ASTM Bulletin #187 of January 1953 (except as noted below). (see Appendix B for condensation of this bulletin.)

a. The dipping solution no longer contains 0.1% sulphuric acid by weight.

b. The test sample must be electrically insulated from the cabinet and from each other.

6.3 Brake Fluid Compatibility—Form each piece of tubing (that is, eight for $^3/_{16}$ OD—4.75 mm—or smaller or four if the diameter is larger) into a "U" shape about a 2 in ± ⅛ (50.8 mm ± 3.2) mandrel. Each piece of tubing is to be 36 in ± 1 (91 cm ± 2.5) long before bending. Measure and record the length and internal diameter of each tube. Weigh the tubes to the nearest 0.0000035 oz (0.1 mg). Support the formed tubes in an appropriate fixture so that the straight sections of the tubes are vertical. Fill the tubes to within 2 in (5.1 cm) of the open ends of the tubes with brake fluid conforming to SAE J1703 RM-1. Place the fluid tubes in an oven maintained at 212°F ± 3.6 (100°C ± 2) for 120 h ± 2.

After the test, collect the test brake fluid in a clean beaker. Remove loose adhering sediment from the interior of the tubes by flushing with water. Dry the tubes in an oven at 212°F ± 3.6 (100°C ± 2) for 30 min. Weigh the tubes to the nearest 0.0000035 oz (0.1 mg) and calculate the weight loss per in² (cm²) of internal area.

Determine the percent sediment by volume, the pH value and the boiling point of the test brake fluid per procedures specified in SAE J1703.

APPENDIX A

A1. Design Guidelines—The best tubing will be unsatisfactory unless it is used properly. The following should be considered:

a. Since tubing may suffer damage and/or loss of corrosion resistance as a result of gravel impact, it should be adequately protected in areas of potential damage.

b. Tubing should be adequately protected against hoist or towing fixture damage.

c. Tubing should be routed or otherwise protected so that under no condition can the tubing or its protective conduit come in contact with any vibrating or moving component (that is, if the tubing is attached to the frame, the underbody is one item considered to be a "vibrating component"). The tubing should never cross under (or over) an exhaust pipe, muffler, or catalytic converter unless it is adequately protected against excessive movement of the pipe, muffler, or catalytic converter such as may occur if a hanger failed.

d. Tubing should be so routed that its stress limits will not be exceeded during flexing.

e. Tubing should be so routed that it will not be in, or form, a pocket which will trap salt or other de-icing chemicals.

f. Tubing should avoid, or be protected from, exhaust systems or other areas of extreme heat.

g. The design engineer should take into account possible electrolytic corrosion resulting from contact between dissimilar metals (that is, brake pipes and protective conduit, clips, fittings, and mounting surfaces).

h. The design engineer should determine the minimum tubing inside diameter based on brake system actuation time. Other factors affecting actuation time are: (R)

(1) Brake fluid viscosity
(2) Operating temperature
(3) Pipe length
(4) Fluid flow rate as determined by brake system displacement requirements

A2. Cyclic Humidity Test—Figure A1 shows four of the five main parts of the apparatus: the humidifying tower, the heating cabinet, the corrosion chest, and the dip mechanism. Not shown is a drying train which is used on part of the cycle.

A cyclic variation of humidity is obtained basically by the variation of the temperature of the water in the humidifying tower. The temperature of the water is cycled thermostatically between limits such that the relative humidity of air bubbling through the water will vary between about 50 and 100% when the air is brought to 125°F (52°C). Extension of the range of relative humidity is accomplished by adding a drying period to the basic humidity cycle described above. The relay circuit is so arranged that the drying period is switched on when the relative humidity has descended to about 50%. The length of the drying period has been 3 h; the entire cycle takes about 8 h. The lowest relative humidity regularly obtained is 8 to 10% and the highest is 100%. Figure A2 portrays the cycle schematically.

The drying train, through which the air is switched during the drying period, has a concentrated sulfuric acid bubbler for primary dehumidification. The secondary stage is a desiccating tower containing anhydrous calcium sulfate.

The block diagram of air distribution is shown in Figure A3. The humidifying tower and the drying train are in parallel, with the solenoid valves switching the air flow through one or the other.

The corrosion chest is a coated stainless-steel box within which the specimens are exposed. It is surrounded by a heating cabinet which is thermostatically maintained at 125°F ± 2 (52°C ± 1). This 125°F (52°C) temperature is an accentuated variable leading to the acceleration of the test; it has been held constant during the test procedure. The air leaving the chest passes through a wet and dry bulb hygrometer used to show gross variations in the relative humidity.

The dipping device is essentially a leveling bottle operating through a drain in the bottom of the corrosion chest. By such a method very few extraneous variables are introduced into the test by the dip. The

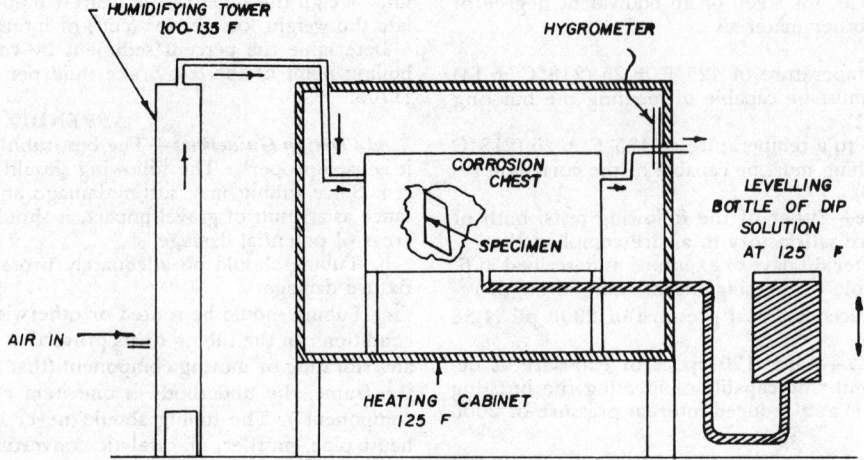

FIG. A1

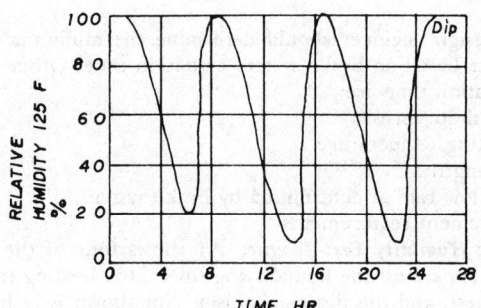

FIG. A2

dip solution is heated to 125°F (52°C) to prevent cooling the chest. The concentration of solution is 1% sodium chloride, and 1% calcium chloride. For the first dip, the solution is siphoned into the corrosion chest and allowed to stand for about 25 min and then drained. On each succeeding workday during the test, the daily dip is accomplished at that point in the humidity cycle where the relative humidity is close to 100% (the short horizontal line marked "dip" at about 25 h on Figure 3) to decrease the perturbation of the cycle due to the dip. After the solution has been siphoned in and has remained in the chest for 5 min, almost all of it is drained and replaced three times to provide a washing action. Finally the solution is drained. This operation takes about 20 to 25 min/d.

The proper conditions for the start of the test are such that the test cycle will not be disturbed greatly; for example, the cabinet should be at 125°F (52°C) and the relative humidity within the corrosion chest should be about 100%. At this time the rack of specimens previously prepared is placed in the corrosion chest and the dip solution is siphoned into the chest. The solution is later drained and the humidity variations then control the test until it is completed.

The chest is preferably kept shut for the remainder of the test to prevent the introduction of extraneous variables. The variation of humidity and the dipping on each workday continue until 170 cycles (57 workdays) have been completed. At that time the specimens are removed from the cabinet and subjected to the Bursting Strength Test (see 5.1).

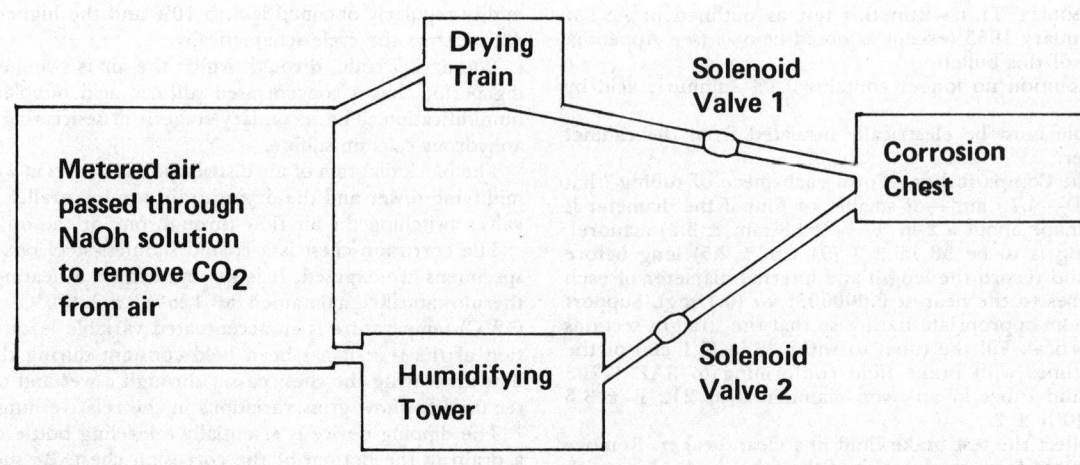

FIG. A3

25.165

φAUTOMOTIVE HYDRAULIC BRAKE SYSTEM—METRIC TUBE CONNECTIONS—SAE J1290 MAY89

SAE Standard

Report of the Hydraulic Brake Systems Actuating Committee, approved July 1980 and completely revised May 1989.

1. Scope—This standard documents dimensional metric specifications for hydraulic brake system tubing with flared ends, threaded ports, and male tube nuts for the interconnection of major components in automotive hydraulic brake systems.

The purpose of this document is to recommend preferred metrically dimensioned components (including alternative choices), that are intended to be functionally compatible with International Organization for Standardization Specification, ISO 4038-1977 (E). Some applications may require sizes of forms other than those shown herein, and this document does not preclude such other details when they are required.

2. Tubing and Flares—Tubing and tubing end flares should be dimensioned as shown in Fig. 1 and Table 1.

3. Threaded Ports (Tube Nuts)—Threaded ports for disc brake calipers, drum brake wheel cylinders, combination valves, pressure switches, metering valves, proportioning valves, master cylinders, brake hose end fittings, tubing fittings, and other brake circuit components should be dimensioned as shown in Fig. 2 and Table 2.

4. Male Tube Nuts (Fittings)—Male tube nuts should be dimensioned as shown in Fig. 3 and Table 3.

NOTE: This standard supersedes SAE J1258 (Cancelled 1980). Threaded ports for banjo bolts, and banjo bolts are covered by SAE Recommended Practice J1291 MAR85.

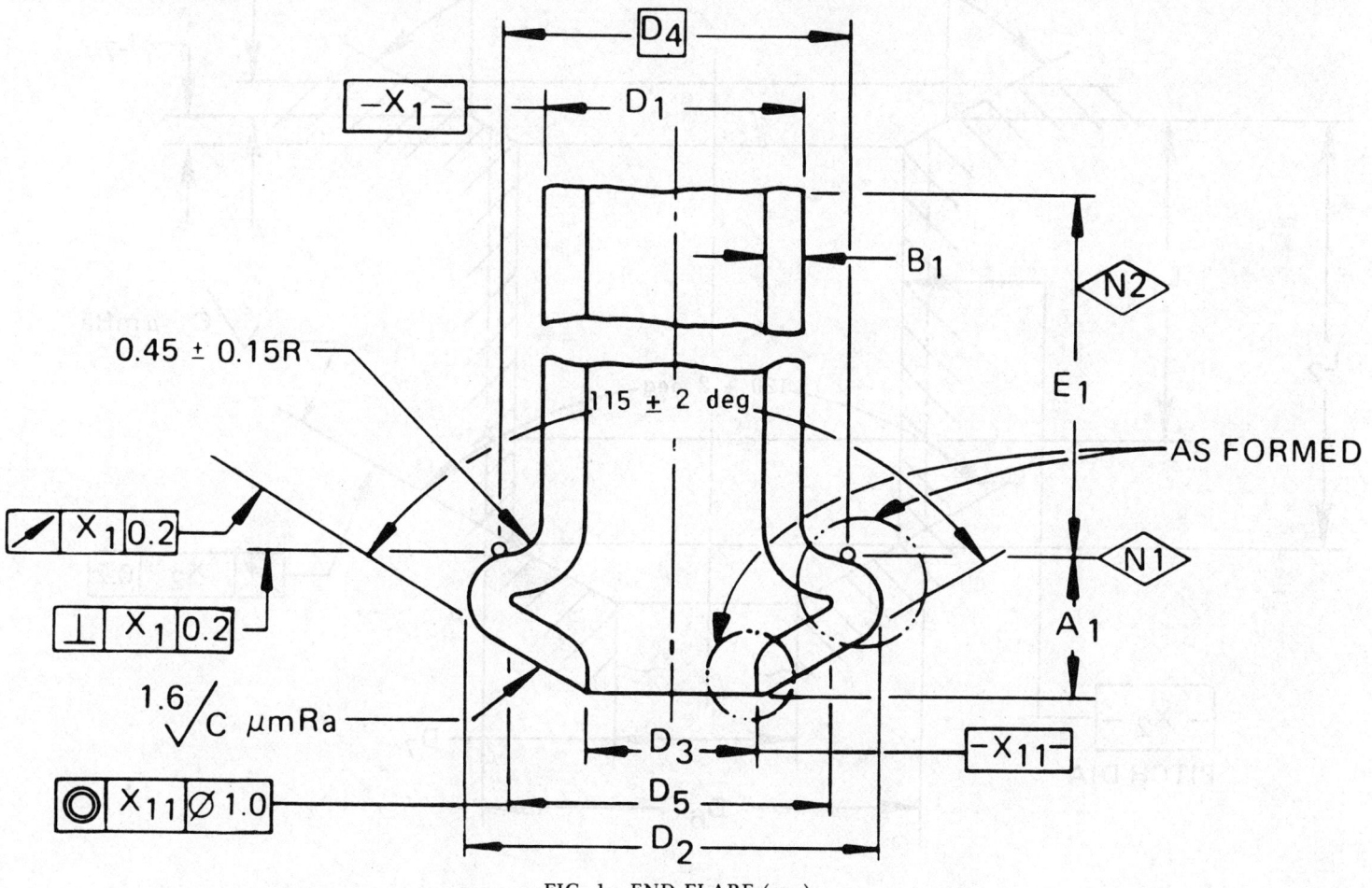

FIG. 1—END FLARE (mm)

TABLE 1—DIMENSIONS AND TOLERANCES FOR TUBING AND END FLARE (mm)

D_1		D_2	D_3	D_4	D_5	A_1	B_1	E_1
Bare Tube ±0.07	Coated Tube Φ max	±0.18	+0.3 −0.2		min	±0.3	±0.07	min
4.75	4.87	7.1	3.2	6.0	5.5	2.5	0.70	16
6.00	6.12	8.4	4.5	7.3	6.8	2.5	0.70	18
8.00	8.12	10.7	6.5	9.3	8.8	2.7	0.70	24
10.00	10.12	12.7	8.5	11.3	10.8	3.0	0.70	28

N1 Datum line
N2 Squareness and runout applies about diameter D_1 over length E_1
For information relative to brazed double-wall low-carbon steel tubing, refer to SAE J527 JAN83 and SAE J1047 OCT74. For additional information on flares, refer to SAE J533 JAN72 and ISO 4038-1977.

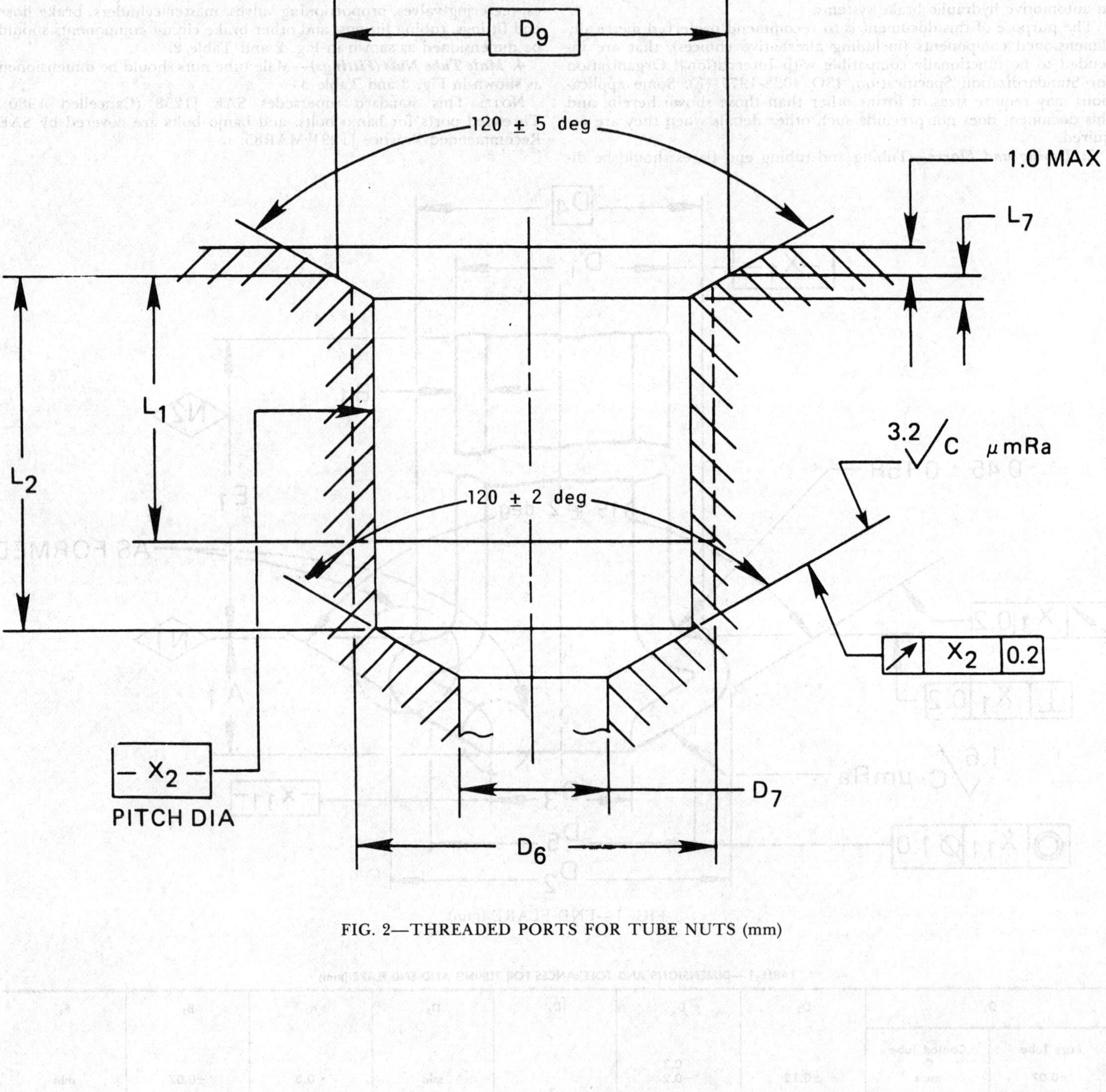

FIG. 2—THREADED PORTS FOR TUBE NUTS (mm)

TABLE 2—DIMENSIONS AND TOLERANCES FOR THREADED PORTS FOR TUBE NUTS (mm)

D_1 Nom Tube OD	D_6 Straight Thread						Selection Preference[b]	D_7 +0.0 −0.4	D_9	L_1 Full Thread min	L_2 +0.0 −0.5	L_7	
	Nom Size	Pitch	Pitch 6H		Minor 6H							min	max
			max	min[a]	max	min							
4.75	M10	1	9.500	9.350	9.153	8.917	1	3.3	10.5	7.0	10.0	0.35	0.50
	M12	1	11.510	11.350	11.153	10.917	2	3.3	12.5	7.0	10.0	0.35	0.50
	M12	1.5	11.216	11.026	10.676	10.376		3.3	12.5	6.0	10.0	0.47	0.63
	M14	1.5	13.216	13.026	12.676	12.376		3.3	14.5	6.0	10.0	0.47	0.63
	M11	1	10.500	10.350	10.153	9.917	3	3.3	11.5	7.0	10.0	0.35	0.50
	M11	1.5	10.206	10.026	9.676	9.376		3.3	11.5	6.0	10.0	0.47	0.63
	M13	1	12.510	12.350	12.153	11.917		3.3	13.5	7.0	10.0	0.35	0.50
	M13	1.5	12.216	12.026	11.676	11.376		3.3	13.5	6.0	10.0	0.47	0.63
	M15	1.5	14.216	14.026	13.676	13.376		3.3	15.5	6.0	10.0	0.47	0.63
6.00	M12	1	11.510	11.350	11.153	10.917	1	4.6	12.5	9.0	12.0	0.35	0.50
	M12	1.5	11.216	11.026	10.676	10.376	2	4.6	12.5	8.0	12.0	0.47	0.63
	M14	1.5	13.216	13.026	12.676	12.376		4.6	14.5	8.0	12.0	0.47	0.63
	M16	1.5	15.216	15.026	14.676	14.376		4.6	16.5	8.0	12.0	0.47	0.63
	M13	1	12.510	12.350	12.153	11.917	3	4.6	13.5	9.0	12.0	0.35	0.50
	M13	1.5	12.216	12.026	11.676	11.376		4.6	13.5	8.0	12.0	0.47	0.63
	M15	1.5	14.216	14.026	13.676	13.376		4.6	15.5	8.0	12.0	0.47	0.63
8.00	M14	1.5	13.216	13.026	12.676	12.376	1	6.6	14.5	12.5	16.5	0.47	0.63
	M16	1.5	15.216	15.026	14.676	14.376	2	6.6	16.5	12.5	16.5	0.47	0.63
	M15	1.5	14.216	14.026	13.676	13.376	3	6.6	15.5	12.5	16.5	0.47	0.63
	M17	1.5	16.216	16.026	15.676	15.376		6.6	17.5	12.5	16.5	0.47	0.63
10.00	M16	1.5	15.216	15.026	14.676	14.376	1	8.6	16.5	13.5	17.5	0.47	0.63
	M18	1.5	17.216	17.026	16.676	16.376	2	8.6	18.5	13.5	17.5	0.47	0.63
	M17	1.5	16.216	16.026	15.676	15.376	3	8.6	17.5	13.5	17.5	0.47	0.63

[a]These values are also the basic pitch diameter.
[b]Ports recognized as standard for respective tube diameters are listed as preference 1. To avoid proliferation where ports having other thread diameter pitch combinations must be used to satisfy design or installation requirements, it is recommended they be selected from sizes listed under preferences 2 and 3, respectively.
For information relative to tube nut ports, refer to SAE Standard J512 OCT80 and International Standard, Road Vehicles - Hydraulic Braking Systems - Pipes, Tapped Holes and Male Fittings - ISO 4038-1977.

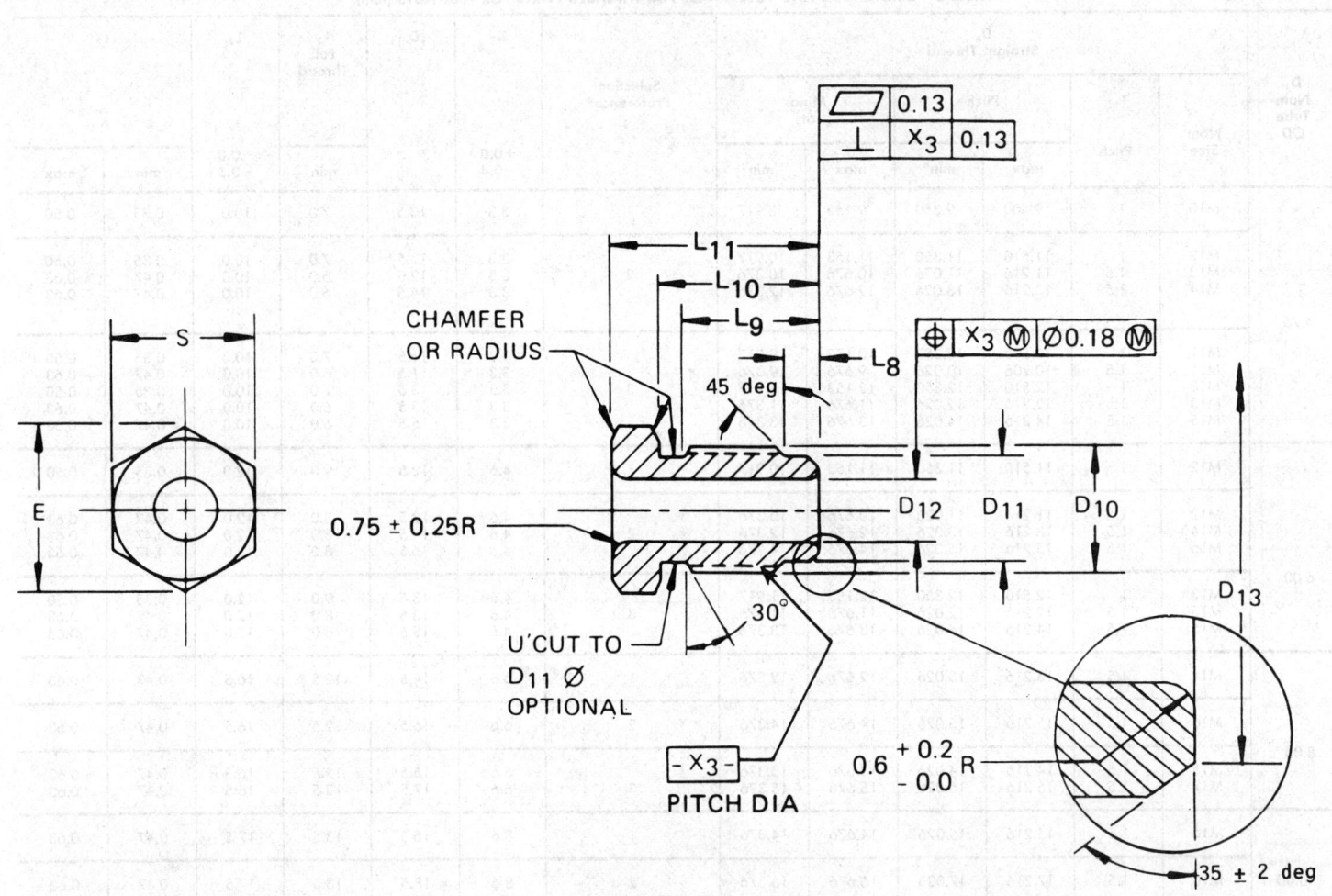

FIG. 3—MALE TUBE NUT (mm)

TABLE 3—DIMENSIONS AND TOLERANCES FOR MALE TUBE NUTS (mm)

D_1 Nom Tube OD	D_{10} Straight Thread				Selection Preference[b]	D_{11} +0.0 −0.2	D_{12} +0.13 −0.00	D_{13} +0.5 −0.0	E min	S[c] +0.00 −0.13	L_8 +0.5 −0.0	L_9 Full Thread min	L_{10} +0.25 −0.00	L_{11} +0.50 −0.00
	Nom Size	Pitch	Pitch[a] 6g max	Pitch[a] 6g min										
4.75	M10	1	9.324	9.212	1	8.4	4.95	7.3	12.16	11.00	2.3	10.10	12.60	16.85
	M12	1	11.324	11.206	2	10.4	4.95	7.3	14.47	13.00	2.6	10.70	13.20	17.45
	M12	1.5	10.994	10.854		9.7	4.95	7.3	14.47	13.00	3.1	10.95	13.45	17.70
	M14	1.5	12.994	12.854		11.7	4.95	7.3	16.78	15.00	3.7	11.55	14.05	18.30
	M11	1	10.324	10.212	3	9.4	4.95	7.3	13.32	12.00	2.3	10.40	12.90	17.15
	M11	1.5	9.994	9.862		8.7	4.95	7.3	13.32	12.00	2.8	10.60	13.10	17.35
	M13	1	12.324	12.206		11.4	4.95	7.3	15.63	14.00	2.9	11.00	13.50	17.75
	M13	1.5	11.994	11.854		10.7	4.95	7.3	15.63	14.00	3.4	11.20	13.70	17.95
	M15	1.5	13.994	13.854		12.7	4.95	7.3	17.94	16.00	4.0	11.85	14.35	18.60
6.00	M12	1	11.324	11.206	1	10.4	6.20	8.6	14.47	13.00	2.3	12.35	14.85	20.10
	M12	1.5	10.994	10.854	2	9.7	6.20	8.6	14.47	13.00	3.1	12.60	15.10	20.35
	M14	1.5	12.994	12.854		11.7	6.20	8.6	16.78	15.00	3.3	13.15	15.65	20.90
	M16	1.5	14.994	14.854		13.7	6.20	8.6	19.09	17.00	3.9	13.75	16.25	21.50
	M13	1	12.324	12.206	3	11.4	6.20	8.6	15.63	14.00	2.5	12.65	15.15	20.40
	M13	1.5	11.994	11.854		10.7	6.20	8.6	15.63	14.00	3.0	12.90	15.40	20.65
	M15	1.5	13.994	13.854		12.7	6.20	8.6	17.94	16.00	3.6	13.50	16.00	21.25
8.00	M14	1.5	12.994	12.854	1	11.7	8.20	10.9	16.78	15.00	3.3	17.45	19.95	25.20
	M16	1.5	14.994	14.854	2	13.7	8.20	10.9	19.09	17.00	3.3	18.05	20.55	25.80
	M15	1.5	13.994	13.854	3	12.7	8.20	10.9	17.94	16.00	3.3	17.75	20.25	25.50
	M17	1.5	15.994	15.854		14.7	8.20	10.9	20.25	18.00	3.5	18.30	20.80	26.05
10.00	M16	1.5	14.994	14.854	1	13.7	10.20	12.9	19.09	17.00	3.3	18.40	20.90	26.15
	M18	1.5	16.994	16.854	2	15.7	10.20	12.9	21.41	19.00	3.3	19.00	21.50	26.75
	M17	1.5	15.994	15.854	3	14.7	10.20	12.9	20.25	18.00	3.3	18.75	21.25	26.50

[a]Pitch diameter specified shall apply to plain (unplated or uncoated) nuts before plating or coating. The basic pitch diameter (minimum pitch diameter of internal thread shown in Table 2), shall apply to plated or coated nuts after plating or coating.

[b]Nuts recognized as standard for respective tube diameters are listed as preference 1. To avoid proliferation where nuts having other thread diameter pitch combinations must be used to satisfy design or installation requirements, it is recommended they be selected from sizes listed under preferences 2 and 3, respectively.

[c]For hexagon sizes refer to DIN 176.

For information relative to materials, finishes, and workmanship, refer to SAE Standard J512 OCT80. For thread sizes, refer to International Standard, Road Vehicles - Hydraulic Braking Systems - Pipes, Tapped Holes and Male Fittings - ISO 4038-1977.

⌀AUTOMOTIVE HYDRAULIC BRAKE SYSTEM—
METRIC BANJO BOLT CONNECTIONS—
SAE J1291 MAY89

SAE Standard

Report of the Hydraulic Brake Systems Actuating Committee, approved July 1980 and completely revised May 1989.

1. Scope—This standard documents dimensional metric specifications for hydraulic brake system threaded ports and banjo bolts for the interconnection of major components in automotive hydraulic brake systems. Banjo blocks are not covered by this standard.

The purpose of this document is to recommend preferred metrically dimensioned components (including alternative choices). Some applications may require sizes or forms other than those shown herein, and this document does not preclude such other details when they are required.

2. Threaded Ports (Banjo Bolts)—Threaded ports for banjo bolts should be dimensioned as shown in Fig. 1 and Table 1.

3. Banjo Bolts—Banjo bolts should be dimensioned as shown in Fig. 2 and Table 2.

NOTE: This standard supersedes SAE J1258 (Cancelled 1980). Tubing, end flares, threaded ports for tube nuts, and male tube nuts are covered by SAE Recommended Practice J1290 MAR85.

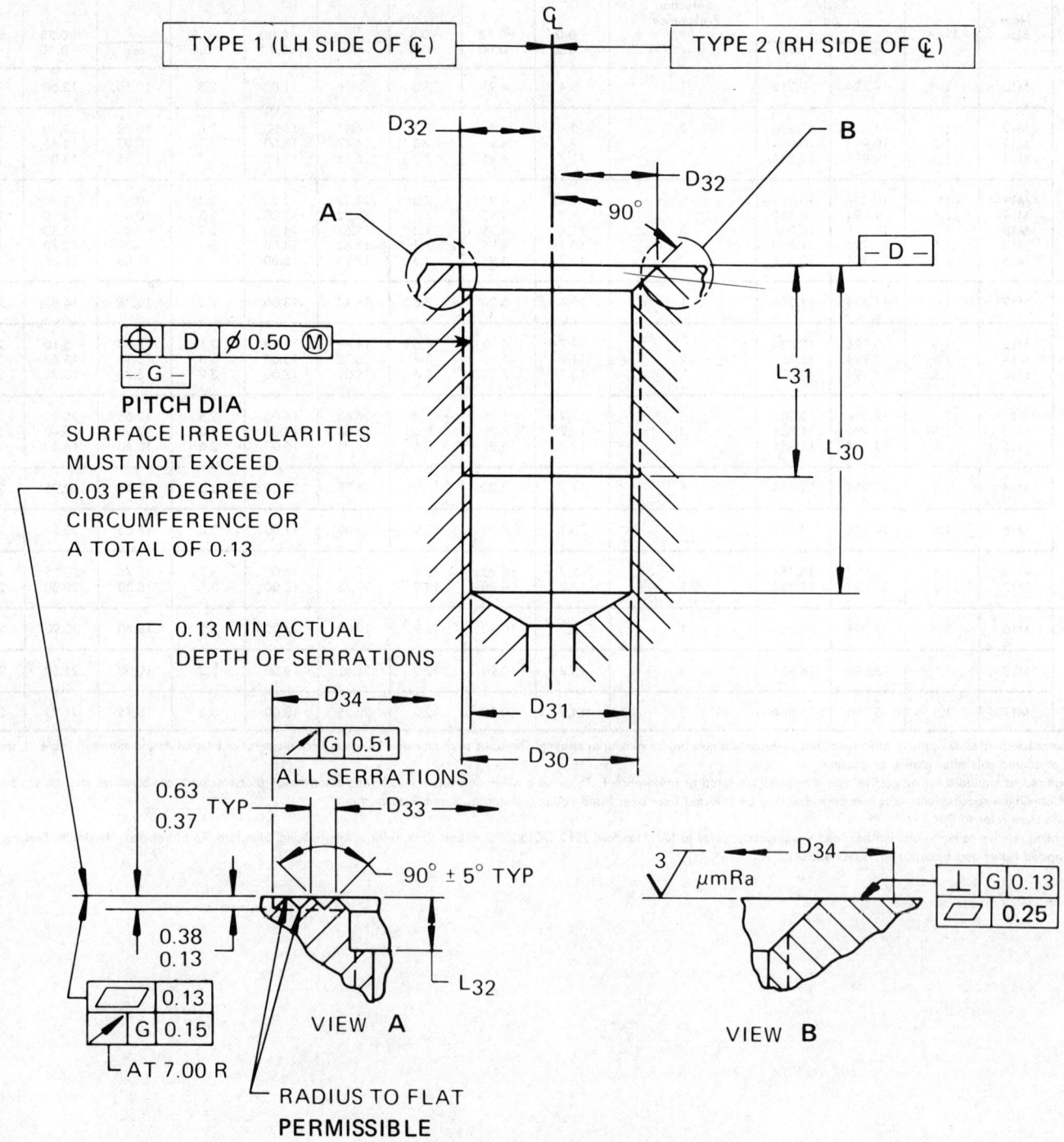

FIG. 1—THREADED PORTS FOR BANJO BOLTS (mm)

TABLE 1—DIMENSIONS AND TOLERANCES FOR THREADED PORTS FOR BANJO BOLTS (mm)

D_{30} Thread 6H Preference 1	2	Pitch	Pitch max	Pitch min	Minor max	Minor min	D_{31} +0.0 −0.3	D_{32} +0.0 −0.2 Type 1	D_{32} +0.0 −0.2 Type 2	D_{33} +0.0 −0.5	D_{34} +0.0 −0.5 Type 1	D_{34} +0.0 −0.5 Type 2	L_{30} +0.0 −1.0	L_{31} min	L_{32} +0.0 −1.0
M10		1.0	9.500	9.350	9.153	8.917	8.50	10.25	11.6	11.25	18.25	48.3	18.5	12.5	1.80
M10		1.5	9.206	9.026	8.676	8.376	8.70	10.25	11.6	11.25	18.25	48.3	18.5	12.5	1.80
	M11	1.0	10.500	10.350	10.153	9.917	9.40	11.25	12.7	12.25	19.25	48.3	18.5	12.5	1.80
	M11	1.5	10.206	10.026	9.676	9.376	9.70	11.25	12.7	12.25	19.25	48.3	18.5	12.5	1.80
M12		1.0	11.510	11.350	11.153	10.917	10.30	12.25	13.8	13.25	20.25	48.3	18.5	12.5	1.80
M12		1.5	11.216	11.026	10.676	10.376	10.70	12.25	13.8	13.25	20.25	48.3	18.5	12.5	1.80

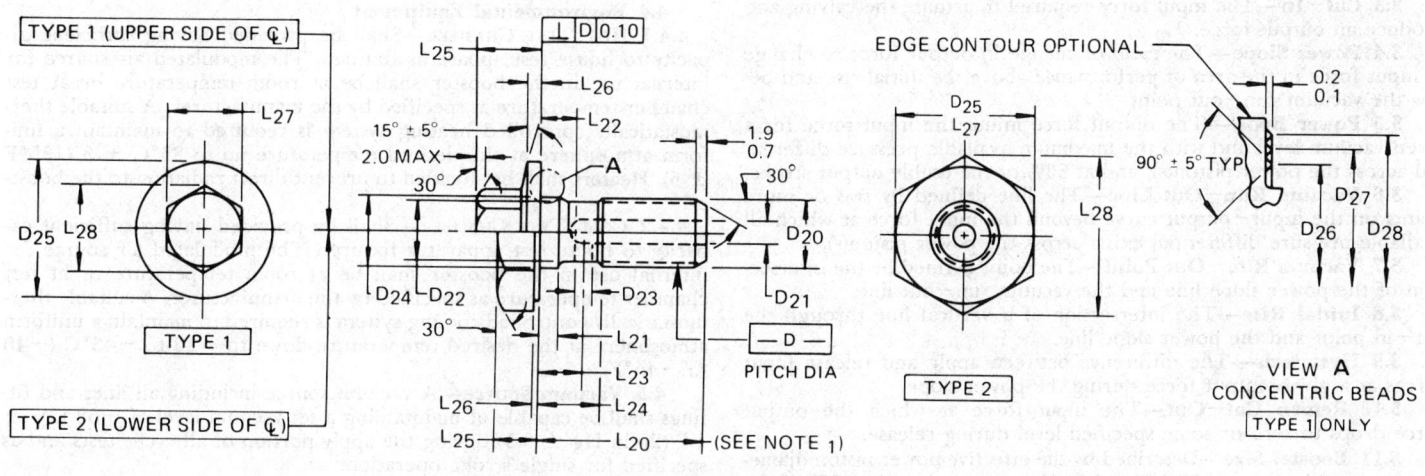

FIG. 2—BANJO BOLTS (mm)

TABLE 2—DIMENSIONS AND TOLERANCES FOR BANJO BOLTS (mm)

D_{20} Thread 6g Pref 1	2	Pitch	Pitch min	Pitch max	D_{21} +0.0 −0.3	D_{22} +0.0 −0.1	D_{23} +0.0 −0.3	D_{24} +0.0 −0.1	D_{25} Type 1 +0.0 −1.0	D_{25} Type 2	D_{26} +0.0 −0.5	D_{27} +0.0 −0.5	D_{28} +0.0 −0.5	L_{20} +0.0 −0.8	L_{21} +0.0 −0.6	L_{22} +0.0 −1.0	L_{23} +0.0 −0.6	L_{24} +0.0 −0.6	L_{25} Type 1 +0.0 −1.1	L_{25} Type 2 +0.0 −0.5	L_{26} Type 1 min	L_{26} Type 2 +0.0 −0.3	L_{27} Type 1 +0.0 −0.3	L_{27} Type 2 +0.0 −0.3	L_{28} Type 1 +0.0 −0.6	L_{28} Type 2 +0.0 −0.6
M10		1.0	9.212	9.324	3.4	8.99	2.6	10.0	18.0		13.0	14.3	15.6	3.0	4.0	6.5	8.6	9.6	6.6	1.5	0.6	11.0	15.0	12.7	17.3
M10		1.5	8.862	8.994	3.4	8.99	2.6	10.0	18.0		13.0	14.3	15.6	3.0	4.0	6.5	8.6	9.6	6.6	1.5	0.6	11.0	15.0	12.7	17.3
	M11	1.0	10.212	10.324	3.4	9.99	2.6	11.0	19.0	L_{27}	14.0	15.3	16.6	See Note	3.0	4.0	6.5	8.6	10.6	7.2	1.8	0.6	12.0	16.0	13.9	18.5
	M11	1.5	9.862	9.994	3.4	9.99	2.6	11.0	19.0		14.0	15.3	16.6		3.0	4.0	6.5	8.6	10.6	7.2	1.8	0.6	12.0	16.0	13.9	18.5
M12		1.0	11.206	11.324	3.4	10.99	2.6	12.0	20.0		15.0	16.3	17.6	3.0	4.0	6.5	8.6	11.6	7.8	2.0	0.6	13.0	17.0	15.0	19.6
M12		1.5	10.854	10.994	3.4	10.99	2.6	12.0	20.0		15.0	16.3	17.6	3.0	4.0	6.5	8.6	11.6	7.8	2.0	0.6	13.0	17.0	15.0	19.6

NOTE: Dimension dependent on thickness of banjo block, which is not covered by this standard. Pitch diameters shown above apply to plain (unplated or uncoated) bolts. Basic pitch diameters (minimum pitch diameters for internal threads in Table 1) shall apply to plated or coated bolts after plating or coating. For information relative to materials, finishes, and workmanship, refer to SAE J512 OCT80. For thread sizes, refer to DIN 13. For hex sizes, refer to DIN 176.

VACUUM POWER ASSIST BRAKE BOOSTER TEST PROCEDURE—
SAE J1808 OCT89

SAE Standard

Report of the Hydraulic Brake Actuating Components Standards Committee approved October 1989.

1. Scope—This document applies to direct acting vacuum power assist brake boosters only, exclusive of the master cylinder or other brake system prime mover devices for passenger cars and light trucks [4500 kg GVW (10 000 lb)]. It specifies the test procedure to determine minimum performance and durability characteristics which are specified in SAE J1902 (under development).

2. Purpose—This document specifies standards for direct acting vacuum power assist brake boosters of current established designs. It is applicable to assemblies from commercial production, after production shipment, and remanufacture.

3. Definition Of Terms—In order to establish and maintain a continuity of discussion throughout this document, the following definition of terms will be used. NOTE: See Fig. 1 for illustration of terms as applied to a typical input−output force curve.

3.1 Released—The unapplied, fully returned state with no force on the input push rod.

3.2 Poise—The condition of placing the control valve in a steady state of force equilibrium so that the vacuum power assist brake booster is neither fully applied, nor fully released.

3.3 Cut−In—The input force required to actuate the valving and produce an output force.

3.4 Power Slope—The ratio of change in output force to change in input force in the area of performance above the initial rise and below the vacuum run−out point.

3.5 Power Boost—The output force minus the input force for a given vacuum level and with the maximum available pressure differential across the power piston(s), and at 80% of the usable output stroke.

3.6 Vacuum Run−Out Line—The line defined by two or more points on the input−output curve beyond the input force at which all available pressure differential exists across the power piston(s).

3.7 Vacuum Run−Out Point—The point defined by the intersection of the power slope line and the vacuum run−out line.

3.8 Initial Rise—The intersection of a vertical line through the cut−in point and the power slope line.

3.9 Hysteresis—The difference between apply and release input forces at a given output force during the power slope.

3.10 Return Cut−Out—The input force at which the output force drops to zero or some specified level during release.

3.11 Booster Size—Described by the effective power piston diameter(s), power boost at −68 kPa (20 in Hg) vacuum, and at 80% of the usable output stroke, and whether single or tandem.

4. Test Apparatus—The basic apparatus shall be that shown and as arranged in Fig. 2 or equivalent. The apparatus shall operate per Section 4 and as called for in Section 6. It is desirable to have the test apparatus portable to facilitate cold, hot, and room temperature testing.

4.1 Force Absorbing Mechanism—The force absorbing mechanism shall be connected to the vacuum power assist brake booster front housing. This mechanism shall be capable of absorbing a minimum of 150% of booster vacuum run−out point output force, or 9000 N (2000 lb) output force, whichever is greater. In addition, this mechanism shall be capable of restricting the output force (stroke relationship to the shaded area of Fig. 3).

4.2 Stroking Mechanism—The stroking mechanism shall contain a rigid mounting plate to which the booster can be attached. The actuator shall be compatible with the input push rod(s) of the booster and shall operate coaxially within 3 deg of the longitudinal axis of the booster. The fixture shall be constructed such that full release of the booster is obtained. The stroking mechanism may accommodate multiple boosters if desired, and shall be designed so that it does not apply tensile force to input push rod. Means must be provided for the stroking mechanism to stroke the booster both singly and cyclically.

4.2.1 For single stroke operation, the mechanism shall be capable of generating input push rod forces up to 4500 N (1000 lb) at its maximum travel and holding this position.

4.2.2 For cyclic operation, the stroking mechanism shall be capable of applying the input push rod to generate 80% ± 10 of the vacuum run−out point output force at a rate that can be adjusted from 250 to 1000 apply/release cycles per hour. The input push rod shall be stroked forward at a smooth rate and allowed to return rapidly to its fully released position. The time cycle shall be adjusted to allow maximum time for forward stroking while insuring that the input push rod returns to the fully released position before the start of the next forward stroke.

4.2.3 In 6.6 structural test, utilize a suitable compression test apparatus for performing structural test.

4.3 Instrumentation

4.3.1 Two−channel X−Y electronic plotter (with real time capability) or equivalent that is compatible with 4.3.2 and 4.3.3.

4.3.2 Force transducers or equivalent measuring devices to measure input and output push rod forces of 0 to 4500 N (0 to 1000 lb) and 0 to 9000 N (0 to 2000 lb) respectively. Overall accuracy ±1/2% of full scale.

4.3.3 Linear transducers or equivalent measuring devices must be provided to measure

4.3.3.1 Input push rod travel of 0 to 50 mm (0 to 2 in).

4.3.3.2 Output push rod travel of 0 to 50 mm (0 to 2 in).

Overall accuracy of ±1/2% of full scale.

4.3.4 Vacuum gage or equivalent capable of measuring 0 to −100 kPa ±1/2% full scale (0 to 30 in Hg).

4.4 Environmental Equipment

4.4.1 HOT TEST CHAMBER—Shall be provided having sufficient capacity to house test apparatus fixtures. The modulated air source for internal use in the booster shall be at room temperature or at test chamber temperature as specified by the manufacturer. A suitable thermostatically controlled heating system is required to maintain a uniform atmosphere at the desired temperature up to 85°C ± 3 (185°F ± 5). Heaters shall be shielded to prevent direct radiation to the booster.

4.4.2 COLD TEST CHAMBER—Shall be provided having sufficient capacity to house test apparatus fixtures. The modulated air source for internal use in the booster shall be at room temperature or at test chamber temperature as specified by the manufacturer. A suitable thermostatically controlled cooling system is required to maintain a uniform atmosphere at the desired temperature down to −40 to −43°C (−40 to −45°F).

4.5 Vacuum Source—A vacuum source including all lines and fittings shall be capable of maintaining a set vacuum level of −68 kPa ± 1.7 (20 in Hg ± 0.5) during the apply portion of all cyclic tests and as specified for single stroke operation.

5. Test Sample—The booster shall come from one of the sources described in Section 1. It shall be new or not used after rebuild, and it shall not be disassembled prior to testing. Separate boosters may be used for the various tests.

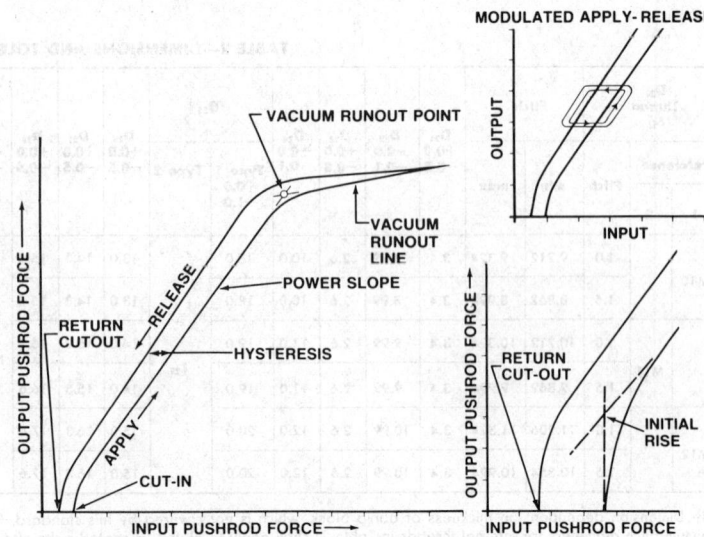

FIG. 1—TYPICAL INPUT−OUTPUT (X−Y) CURVE

25.173

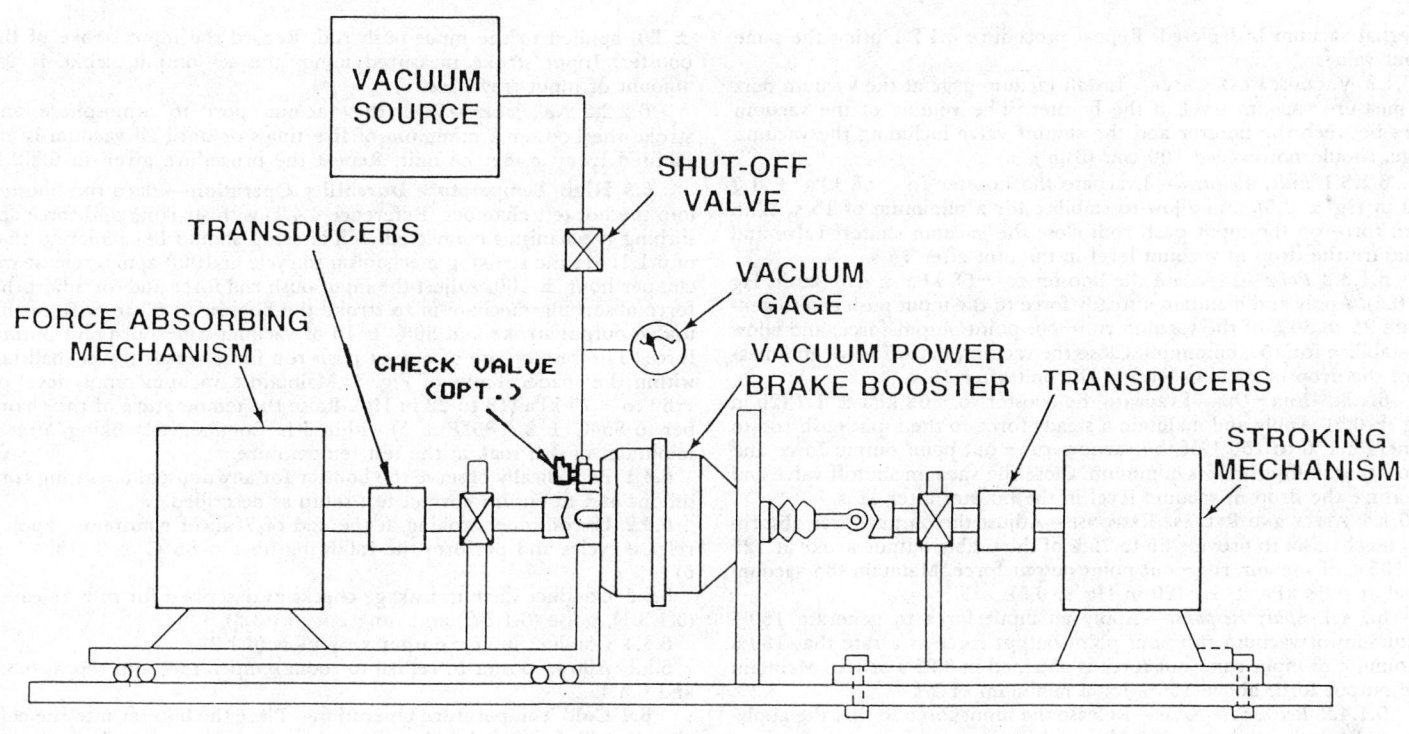

FIG. 2—SCHEMATIC OF RECOMMENDED SETUP

6. Test Setup And Procedure—Tests shall be conducted in the sequence shown and at room temperature except where otherwise specified. The booster shall not be disassembled until after all tests are completed or unless testing is discontinued.

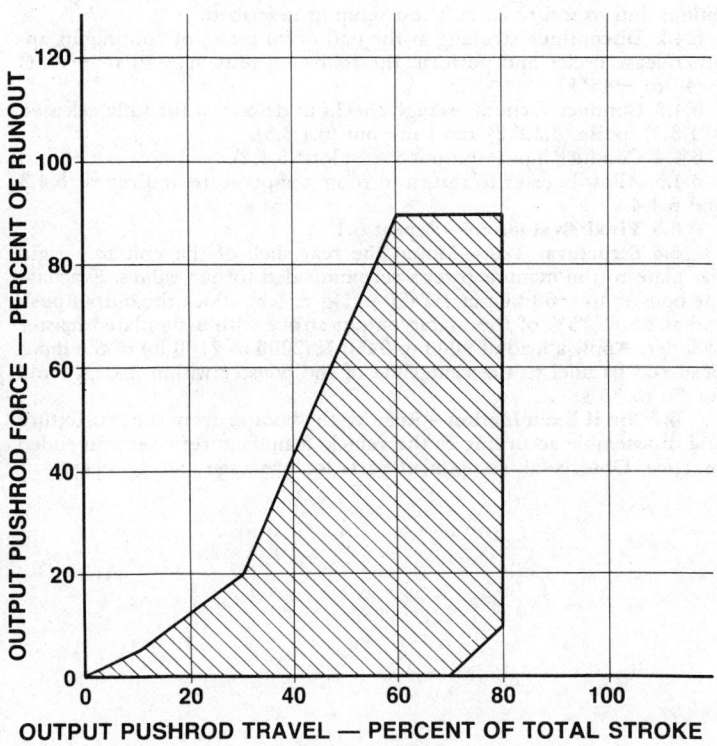

FIG. 3—RECOMMENDED STROKING PERFORMANCE CURVES FOR FORCE ABSORBING MECHANISM

6.1 Initial Evaluation—This phase of the test is for the purpose of obtaining a performance profile of the booster without disrupting or otherwise altering its parts or integrity.

6.1.1 INITIAL SETUP—Install the booster on the mounting plate, as shown in Fig. 2, and tighten mounting fasteners to the recommended torque. Make certain that the input push rod is properly aligned with the longitudinal axis of the booster within 3 deg. Adjust the apply actuator and/or input push rod to allow the booster to return to a fully released position with zero preload on the input push rod. Attach the force absorbing mechanism to the front housing of the booster and adjust according to Fig. 3. Install vacuum line and gage per 6.1.3 from the vacuum port to the vacuum source.

In cases where the booster is supplied with an integrally mounted check valve at the booster vacuum port or an in—line check valve, the vacuum supply line must be connected between the check valve and the vacuum port. In this case, the check valve should be vented to atmosphere.

6.1.2 RECORDER—Connect the input push rod force transducer to the X—axis and the output push rod force transducer to the Y—axis of the X—Y plotter. Assure that the output push rod is not preloaded.

6.1.2.1 *Apply—Release*—Evacuate the booster to −68 kPa ± 1.7 (20 in Hg ± 0.5). Apply force to the input push rod at a rate of 135 N/s ± 90 (30 lb/s ± 20). After the apply force generates 110 to 150% of the vacuum run—out point output force, release the force to the input push rod at the same rate and allow the booster to return to a fully released condition.

6.1.2.2 *Modulated Apply—Release*—Repeat the procedure given in 6.1.2.1 except modulate the apply force to generate output force increments of 450 N ± 90 (100 lb ± 20) from 25 to 110% of the vacuum run—out point output force. During the modulated apply mode, reduce the output push rod force 450 N ± 90 (100 lb ± 20), then reapply. Do not reduce the output push rod force to zero before reapply. When complete, return to the fully released condition.

6.1.2.3 *Reserve Apply—Release Capability*—Evacuate the booster to −68 kPa ± 1.7 (20 in Hg ± 0.5) and disconnect the vacuum supply. Apply force to the input push rod at a rate of 135 N/s ± 90 (30 lb/s ± 20) to generate 25 to 30% of the vacuum run—out point output force and 25 to 35% of the usable output stroke. Hold for 15 s, and return the input push rod force to zero. Repeat this apply—release cycle and record, to the nearest cycle, the number of cycles until power assist is reduced to zero.

6.1.2.4 *"No Power" Apply—Release*—With vacuum supply disconnected and vacuum port vented to atmosphere, stroke booster until all internal vacuum is depleted. Repeat procedure 6.1.2.1 using the same input values.

6.1.3 VACUUM LEAK CHECK—Install vacuum gage at the vacuum port to measure vacuum level in the booster. The volume of the vacuum lines between the booster and the shutoff valve including the vacuum gage, should not exceed 100 cm^3 (6 in^3).

6.1.3.1 *Fully Released*—Evacuate the booster to -68 kPa $\pm$ 1.7 (20 in Hg $\pm$ 0.5), and allow to stabilize for a minimum of 15 s. With zero force on the input push rod, close the vacuum shutoff valve and measure the drop in vacuum level in the unit after 15 s.

6.1.3.2 *Poise*—Evacuate the booster to -68 kPa $\pm$ 1.7 (20 in Hg $\pm$ 0.5). Apply and maintain a steady force to the input push rod to generate 25 to 30% of the vacuum run−out point output force, and allow to stabilize for 15 s minimum. Close the vacuum shutoff valve and measure the drop in vacuum level in the unit after 15 s.

6.1.3.3 *Run−Out*—Evacuate the booster to -68 kPa $\pm$ 1.7 (20 in Hg $\pm$ 0.5). Apply and maintain a steady force to the input push rod to generate 110 to 150% of the vacuum run−out point output force and allow to stabilize for 15 s minimum. Close the vacuum shutoff valve and measure the drop in vacuum level in the booster after 15 s.

6.1.4 APPLY AND RELEASE RESPONSE—Adjust the output force absorbing mechanism to provide 65 to 75% of the usable output stroke at 125 to 135% of vacuum run−out point output force. Maintain the vacuum level at -68 kPa $\pm$ 1.7 (20 in Hg $\pm$ 0.5).

6.1.4.1 *Apply Response*—Apply an input force to generate 150% minimum of vacuum run−out point output force at a rate that 130% minimum of input run−out force is obtained in 0.25 s or less. Maintain the output force above 150% for a minimum of 2 s.

6.1.4.2 *Release Response*—Release the input force so that the applying device retracts faster than the input rod.

6.1.4.3 Record the input and output forces versus time for the respective apply and release tests.

6.2 Usable Output Stroke and Input Travel Loss

6.2.1 USABLE OUTPUT STROKE—It is the intent of this phase of the test to measure the usable output stroke of the booster. It is the lesser of 6.2.1.1 or 6.2.1.2 results.

6.2.1.1 *Power*—Evacuate the booster to -68 kPa $\pm$ 1.7 (20 in Hg $\pm$ 0.5). With linear transducers connected to measure the output push rod travel relative to the front housing mounting surface, remove the force absorbing mechanism and stroke the booster its full usable output stroke with 670 N $\pm$ 90 (150 lb $\pm$ 20) applied to the input push rod. Record the output stroke of the booster. CAUTION: If excessive force is applied to obtain full stroke, internal damage to the booster could result.

6.2.1.2 *No Power*—Vent the vacuum port to atmosphere and stroke the booster a minimum of five times or until all vacuum is exhausted from inside the unit. Repeat the procedure given in 6.2.1.1.

6.2.2 INPUT TRAVEL LOSS—It is the intent of this phase of the test to measure the input push rod travel loss and, thereby, determine the required input stroke of the booster.

The input travel loss is the greater of either 6.2.2.1 or 6.2.2.2.

The amount of input push rod stroke required for a given output stroke is the sum of the input travel loss and the given output stroke.

6.2.2.1 *Power*—Evacuate the booster to -68 kPa $\pm$ 1.7 (20 in Hg $\pm$ 0.5). Set, block, and record the output push rod at 65 to 75% of the usable output stroke (6.2.1) with a rigid fixture simulating a master cylinder mounted to the front shell of the booster so that the axial motion of the front housing is not restricted.

With a linear transducer connected to measure the input push rod travel relative to the rear housing mounting surface, stroke the booster from 65 to 75% of its usable output stroke with 670 N $\pm$ 90 (150 lb $\pm$ 20) applied to the input push rod. Record the input stroke of the booster. Input stroke measured minus the set output stroke is the amount of input travel loss.

6.2.2.2 *No Power*—Vent the vacuum port to atmosphere and stroke the booster a minimum of five times or until all vacuum is exhausted from inside the unit. Repeat the procedure given in 6.2.2.1.

6.3 High Temperature Durability Operation—Place the booster into the hot test chamber (Reference 4.4.1) with stroking and force absorbing mechanisms connected. Initial setup should be similar to that of 6.1.1. Set the stroking mechanism to cycle at 1000 apply/release cycles per hour $\pm$ 100. Adjust the input push rod force and/or adjust the force absorbing mechanism to stroke the booster at 60 to 80% of the usable output stroke and 80% $\pm$ 10 of vacuum run−out point output force. The rate of rise of output push rod force versus travel shall fall within the shaded limits of Fig. 3. Maintain a vacuum supply level of -60 to -75 kPa (18 to 22 in Hg). Raise the temperature of the chamber to 85°C $\pm$ 3 (185°F $\pm$ 5) within 6 h. Commence stroking, after a minimum of 4 h soak at the test temperature.

6.3.1 Periodically observe the booster for any unusual operating conditions and to ensure correct test setup as described.

6.3.2 Discontinue stroking at the end of 70 h of continuous apply/release cycles and perform the following tests at 85°C $\pm$ 3 (185°F $\pm$ 5)

6.3.3 Conduct vacuum leakage checks as described for fully released (6.1.3.1), poise (6.1.3.2) and run−out (6.1.3.3).

6.3.4 Conduct input−output x−y plots (6.1.2).

6.3.5 Allow booster to return to room temperature and repeat 6.3.3 and 6.3.4.

6.4 Cold Temperature Operation—Place the booster into the cold chamber (Reference 4.4.2) with stroking and force absorbing mechanisms connected. Initial setup should be similar to that of 6.1.1. Set the stroking mechanism to cycle at 250 apply/release cycles per hour $\pm$ 25. Adjust the input push rod force and/or adjust the force absorbing mechanism to stroke the booster(s) at 60 to 80% of the usable output stroke and 80% $\pm$ 10 of vacuum run−out point output force. The rate of rise of output push rod force versus travel shall fall within the shaded limits of Fig. 3. Maintain a vacuum supply level of -60 to -75 kPa (18 to 22 in Hg). Lower the temperature of the chamber from -40 to -43°C (-40 to -45°F) within 18 h. Commence stroking, after a minimum of 4 h soak at the test temperature.

6.4.1 Periodically observe the booster for any unusual operating conditions and to insure correct test setup as described.

6.4.2 Discontinue stroking at the end of 20 cycles of continuous apply/release cycles and perform the following tests at -40 to -43°C (-40 to -45°F):

6.4.3 Conduct vacuum leakage checks as described for fully released (6.1.3.1), poise (6.1.3.2) and run−out (6.1.3.3).

6.4.4 Conduct input−output x−y plots (6.1.2).

6.4.5 Allow booster to return to room temperature and repeat 6.4.3 and 6.4.4.

6.5 Final Evaluation—Repeat 6.1.

6.6 Structural Test—Mount the rear shell of the unit to a rigid flat plate to the manufacturer's recommended torque values. Evacuate the booster to -68 kPa $\pm$ 1.7 (20 in Hg $\pm$ 0.5). Block the output push rod at 65 to 75% of the usable output stroke with a simulated master cylinder. Apply a load of 9000 to 9450 N (2000 to 2100 lb) to the input push rod parallel to the centerline of the booster within 3 deg. Hold for 30 to 35 s.

6.7 Final Examination—Remove the booster from the test fixture and disassemble according to the vehicle manufacturer's recommended practice. Observe all component parts for breakage and/or wear.

RUBBER CUPS FOR HYDRAULIC ACTUATING CYLINDERS—SAE J1601 NOV90

SAE Standard

Report of the Hydraulic Brake Systems Actuating Committee, approved May 1975, reaffirmed without change March 1985. Currently under revision by Committee. SAE J60a has been discontinued and replaced with this report. Revised by the Hydraulic Brake Actuating Elastomeric Standards Committee November 1990. Rationale statement available.

(R) **1. Scope**—These specifications cover molded cups 51 mm (2 in) in diameter and under, compounded from high temperature resistant rubber for use in hydraulic actuating cylinders employing motor vehicle brake fluid conforming to the requirements specified in SAE J1703 and SAE J1705.

These specifications cover the performance tests of hydraulic brake cups under specified conditions and do not include requirements relating to chemical composition, tensile strength, and elongation of the rubber compound.

Disc brake seals are not covered by this document.

(R) **2. References**
SAE J1703—Motor Vehicle Brake Fluid
SAE J1705—Low Water Tolerant Brake Fluids
ASTM D 91—Test for Precipitation Number of Lubricating Oils
ASTM D 1415—Test Method for Rubber Property—International Hardness
ASTM D 2240—Test Method for Indention Hardness of Rubber and Plastics by Means of a Durometer
ASTM E 145—Specifications for Gravity—Convection and Forced—Ventilation Ovens

3. General Material Requirements

3.1 Workmanship and Finish—Cups shall be free from blisters, pinholes, cracks, protuberances, embedded foreign material, or other physical defects which can be detected by thorough inspection, and shall conform to the dimensions specified on the drawings.

3.2 Marking—The identification mark of the manufacturer as recorded by the Rubber Manufacturers Association and other details as specified on drawings shall be molded into each cup.

3.3 Packaging—Cups shall be packaged to meet requirements specified by the purchaser.

3.4 Sampling—The minimum lot on which complete specification tests shall be conducted for quality control testing, or the frequency of any specific type test used to control production, shall be agreed upon by the manufacturer and the purchaser.

4. Test Requirements

4.1 Resistance to Fluid at Elevated Temperature—After being subjected to the test for resistance to fluid at elevated temperature as prescribed in 5.1, the cups shall conform to the requirements specified in Table 1. The cups shall show no excessive disintegration as evidenced by blisters or sloughing.

4.2 Precipitation—Not more than 0.3% sediment by volume shall be formed in the centrifuge tube after the cups have been tested as specified in 5.2.

4.3 Wheel Cylinder Heat Pressure Stroking—Wheel cylinder cups when tested by the procedure specified in 5.3 shall meet the following performance requirements.

4.3.1 Lip Diameter Change—The minimum lip diameter of wheel cylinder cups after the stroking test shall be greater than the wheel cylinder bore by the minimum dimensions specified in Table 2.

4.3.2 Leakage—Constant dampness past the cups or fluid discoloration of the filter paper on two or more inspections shall be cause for rejection.

4.3.3 Corrosion—Pistons and cylinder bore shall not show corrosion as evidenced by pitting to an extent discernible to the naked eye, but stain or discoloration shall be permitted.

4.3.4 Change in Hardness—Rubber cups shall not decrease in hardness by more than 15 degrees when tested in accordance with the procedure as specified in 5.7.

4.3.5 Condition of Test Cup—Wheel cylinder cups shall not show excessive deterioration such as scoring, scuffing, blistering, cracking, chipping (heel abrasion), or change in shape from original appearance.

4.4 Master Cylinder Heat Pressure Stroking—Master cylinder cups when tested by the procedure specified in 5.4, shall meet the following performance requirements:

4.4.1 Lip Diameter Change—The minimum lip diameter of master cylinder cups after the stroking test shall be greater than the master cylinder bore by the minimum dimensions specified in Table 3.

4.4.2 Leakage—Constant dampness past the secondary cup or fluid discoloration of the filter paper on two or more inspections shall be cause for rejection.

TABLE 1—FLUID RESISTANCE AT ELEVATED TEMPERATURES

Change in	
Volume	+5 to +20%
Outside diameter, lip	0 to +5.75%
Outside diameter, base	0 to +5.75%
Hardness, ASTM D 1415, degrees	−15 to 0

TABLE 2—WHEEL CYLINDER CUPS—EXCESS OVER BORE

Diameter, mm (in)	mm (in) min
Through 25.4 (1)	0.508 (0.020)
Over 25.4 (1) through 38.1 (1-1/2)	0.635 (0.025)
Over 38.1 (1-1/2) through 50.8 (2)	0.762 (0.030)

TABLE 3—MASTER CYLINDER CUPS—EXCESS OVER BORE

Diameter, mm (in)	mm (in) min
Through 25.4 (1)	0.381 (0.015)
Over 25.4 (1) through 38.1 (1-1/2)	0.508 (0.020)
Over 38.1 (1-1/2) through 50.8 (2)	0.635 (0.025)

4.4.3 Corrosion—Piston and cylinder bore shall not show corrosion as evidenced by pitting to an extent discernible to the naked eye, but staining or discoloration shall be permitted.

4.4.4 Change in Hardness—The hardness of the primary and secondary master cylinder test cups shall not decrease in hardness by more than 15 degree when tested according to the procedure specified in 5.7.

4.4.5 Condition of Test Cups—The primary and secondary cups shall not show excessive deterioration such as scoring, scuffing, blistering, cracking, chipping (heel abrasion), or change in shape from original appearance.

4.5 Low Temperature Performance

4.5.1 Leakage—No leakage of fluid shall occur when cylinder cups are tested according to the procedure specified in 5.5.1.

4.5.2 Bend Test—The cylinder cup shall not crack and shall return to its approximate original shape within 1 min when tested according to the procedure specified in 5.5.2.

4.6 Oven Aging—Cylinder cups when tested according to the procedure specified in 5.6 shall meet the following requirements:

4.6.1 Change in Hardness—The change in hardness shall be within the limits of −5 to +5 degrees.

4.6.2 Condition of Test Cups—The cups shall show no evidence of deterioration, or change in shape from original appearance.

4.7 Corrosion Resistance—Cups when tested by the procedure specified in 5.8 shall not cause corrosion exceeding the limits shown in Table 4. The metal strips outside of the area where the strips are in contact shall neither be pitted nor roughened to an extent discernible to the naked eye, but staining or discoloration is permitted.

The fluid water mixtures at end of test shall show no jelling at 23 °C ± 5 (73.4 °F ± 9). No crystalline type deposits shall form and adhere to either the glass jar walls or the surface of metal strips. The fluid-water mixture shall contain no more than 0.20% sediment by volume.

4.8 Storage Corrosion—After 12 cycles in the humidity cabinet when run according to the procedure specified in 5.9, there shall be no evidence of corrosion adhering to or penetrating the wall of the cylinder bore which was in contact with the test cup. Slight discoloration (staining) or any corrosion or spots away from the contact surface of the test cups shall not be cause for rejection.

5. Test Procedures

5.1 Resistance to Fluid at Elevated Temperature—Dimensional test.

5.1.1 Apparatus—Micrometer, shadowgraph, or other suitable apparatus to measure accurately to 0.02 mm (0.001 in) and glass containers[1] of approximately 250 cm³ (½ pt) capacity which can be tightly sealed (RM-51 and RM-52).

(R)TABLE 4—CORROSION TEST STRIPS AND WEIGHT CHANGES

Test strip[a]	RM No.	Max Permissible Weight Change, mg/cm² of Surface
Tinned Iron	6	0.2
Steel	7	0.2
Aluminum	8	0.1
Cast Iron	9	0.2
Brass	10	0.4
Copper	11	0.4

[a]Test strips may be obtained from Society of Automotive Engineers, Inc., 400 Commonwealth Drive, Warrendale, PA 15096.

(R) 5.1.2 TEST SPECIMENS—Four cups shall be used for testing at 120°C (248°F).

(R) 5.1.3 PROCEDURE—The cups shall be rinsed in isopropyl alcohol and wiped dry with a clean, lint-free cloth to remove dirt and packing debris. Cups shall not be left in the alcohol for more than 30 s. The lip and base diameters shall be measured to the nearest 0.02 mm (0.001 in) taking the average of two readings at right angles to each other. Care shall be taken when measuring the diameters before and after aging that the measurements be taken in the same manner and at the same locations.

Determine and record the initial hardness of the test cups. Refer to 5.7 and Figure 5.

The volume of each cup shall be determined in the following manner: Weigh the cups in air (M_1) to the nearest milligram and then weigh the cup immersed in distilled water at room temperature (M_2)[2] Quickly dip each specimen in isopropyl alcohol and then blot dry with filter paper free of lint and foreign material.

Two cups shall be completely immersed in 75 cm³ of each of the specified test fluids, as outlined in Appendices A and B, in suitable glass containers, and the containers shall be sealed to prevent vapor loss. The containers shall be placed in an oven at 120°C ± 2 (248°F ± 3.6) for a period of 70 h ± 2. At the end of the heating period, remove the cups from the oven and allow to cool in the containers at 23°C ± 5 (73.4°F ± 9) for 60 to 90 min. At the end of the cooling period, remove the cups from the containers and rinse in isopropyl alcohol and wipe dry with a clean, lint-free cloth. Cups shall not remain in the alcohol for more than 30 s.

After removal from the alcohol and drying, place each cup in a separate, tared, stoppered weighing bottle and weigh (M_3). Remove each cup from its weighing bottle and weigh immersed in distilled water (M_4) to determine water displacement after hot fluid immersion.

The final volume, dimensions, and hardness of each cup shall be determined within 30 to 60 min after rinsing in alcohol.

(R) 5.1.4 CALCULATION AND REPORT

5.1.4.1 *Volume Change*—Shall be reported as a percentage of the original volume. The calculation shall be made as follows:

$$\% \text{ increase in volume} = \frac{(M_3 - M_4) - (M_1 - M_2)}{(M_1 - M_2)} \times 100 \quad \text{(Eq.1)}$$

where:
- M_1 = initial mass in air
- M_2 = initial mass in water
- M_3 = mass in air after immersion in test fluid
- M_4 = mass in water after test

5.1.4.2 *Dimensional Changes*—The original measurements shall be subtracted from the measurements taken after the test and the difference reported as a percentage of the original diameters.

5.1.4.3 Change in hardness shall be determined and recorded.

5.1.4.4 Examine the cups for disintegration as evidenced by blisters or sloughing.

(R) **5.2 Precipitation Test**

5.2.1 APPARATUS—Glass containers[3] having a capacity of approximately 250 cm³ (½ pt) and inner dimensions of approximately 125 mm (5 in) in height and 50 mm (2 in) in diameter which can be tightly sealed, and a cone-shaped centrifuge tube of 100 cm³ capacity.

(R) 5.2.2 TEST SPECIMENS—Four cups shall be used.

5.2.3 PROCEDURE—To determine the precipitation characteristics of the test cups, place two cups in a suitable glass container containing 75 cm³ of each of the test fluids specified in Appendices A and B. The container shall be sealed to prevent vapor loss and placed in an oven at 120°C ± 2 (248°F ± 3.6) for 70 h. At the end of the heating period, remove the containers from the oven and allow to cool at room temperature for 24 h after which the cups are removed. The contents of the jar shall be thoroughly agitated and transferred to a cone-shaped centrifuge of 100 cm³ capacity and the sediment determined as described in 5(b) of ASTM D 91.[4]

(R) **5.3 Wheel Cylinder Heat Pressure Stroking**

5.3.1 APPARATUS

5.3.1.1 *Oven*—A well-designed, uniformly heated, standard dry air oven conforming to the requirements for Type 11B oven in ASTM E 145.

5.3.1.2 *Actuating Heat Pressure Stroking Fixture for Wheel Cylinder Cups*—The actuating heat pressure stroking fixture shall be designed to provide a 3.8 mm ± 1.7 (0.15 in) ± 0.07 movement of each piston. During the total movement of the piston, the pressure shall increase to 7.0 MPa ± 0.3 (1000 psi ± 50). The rate of operation shall be held at a uniform reciprocating motion of 1000 strokes/h ± 100. Figure 2 illustrates a recommended pressure MPa (psi) versus wheel cylinder piston movement curve for wheel cylinders within 12.7 to 50.8 mm (½ to 2 in) diameter.

NOTE: A new wheel cylinder assembly must be used for each test.

5.3.2 TEST SPECIMENS—Two wheel cylinder cups shall be used as test specimens.

(R) 5.3.3 PROCEDURE—The wheel cylinder cups shall be rinsed in isopropyl alcohol and wiped dry with a clean, lint-free cloth to remove dirt and packing debris. Cups shall not remain in the alcohol for more than 30 s.

The lip diameter measurement shall be determined to the nearest 0.02 mm (0.001 in), taking the average of two readings at right angles to each other. In the case of double lip cups, these measurements shall be taken after the cup has been assembled on the piston. Determine and record the initial hardness of the test cups. The internal parts, which may include among other things cups, piston springs, expanders, etc., shall be installed in a wheel cylinder of known diameter using the test fluid specified in Appendix A as a lubricant. (Boots shall not be used.) Install the wheel cylinder assembly on the stroking fixture. The stroking fixture assembly shall be placed in an oven and actuated for 70 h at 120°C ± 2 (248°F ± 3.6). After 1 h minimum operation, place a sheet of filter paper under each end of the wheel cylinder to catch and determine leakage. Inspect filter paper at least twice for discoloration at not less than 24 h intervals. Filter paper is to be changed if discolored from fluid leakage. The actuating means and the oven heater shall be shut off at the termination of the 70 h stroking period with the master cylinder piston in the "off" position to relieve retained pressure in the system. After 1 h cooling period with the oven door open and a ventilating fan on, disconnect the fluid line at the wheel cylinder inlet. Remove the entire stroking test fixture containing the test wheel cylinder from the oven and allow to cool for 22 h ± 2 at room temperature. Immediately after completion of the 22 h cooling period, careful inspection shall be made to check for fluid leaks past the cups and results recorded, the fluid shall be drained from the system, and the cups shall be removed from the wheel cylinder. Double lip cups shall be measured before removal from the pistons. The cups shall be rinsed in isopropyl alcohol and dried with compressed air. Cups shall not remain in the alcohol for more than 30 s. Inspect cups for scoring, scuffing, blistering,

---

[1] Suitable test jars and tinned steel lids can be obtained from Society of Automotive Engineers, Inc., 400 Commonwealth Drive, Warrendale, PA 15096.

[2] A trace of a suitable wetting agent not large enough to significantly affect the specific weight of the water should be added to the distilled water to eliminate small air bubbles from being trapped on the rubber surface during the weighing process.

[3] See Footnote 1.

[4] Published by the American Society for Testing and Materials, 1916 Race Street, Philadelphia, PA 19103.

cracking, chipping (heel abrasion), and change in shape from original appearance.

Inspect cylinder parts, recording any pitting on pistons and cylinder walls.

Determine and record the change in hardness.

The lip diameter of each cylinder cup shall be measured within 30 to 60 min after removal from the wheel cylinder and the difference between the actual cylinder bore and the lip diameter after the test shall be reported (excess over bore—Table 2).

5.4 Master Cylinder Heat Pressure Stroking

5.4.1 APPARATUS

(R) 5.4.1.1 A well-designed, uniformly heated, standard dry air oven conforming to the requirements for Type 11B oven prescribed in ASTM E 145.

5.4.1.2 *Actuating Heat Pressure Stroking Machine for Master Cylinder Cups*—The stroking machine shall consist of a suitable means for actuating the master cylinder containing the test specimens at the rate of 1000 strokes/h ± 100. The total piston movement shall be sufficient to cover approximately 90% of the total available stroke. On all master cylinders having a total stroke of 63 mm (2½ in) or more, they shall be heat, pressure, and stroke tested at 90% of the 63 mm (2½ in) stroke, or 57 mm (2¼ in). The rate of stroke shall be 800 strokes/h ± 80. Full pressure 7 MPa (1000 psi) shall be attained and maintained for 3 mm (⅛ in) of the stroke or 1 s maximum.

Figure 1 illustrates a typical master cylinder cup stroking apparatus. Figure 3 illustrates typical pressures in MPa (psi) versus the master cylinder piston movement obtained with three wheel cylinders of approximately 22 mm (⅞ in) diameter mounted in the three stroking fixtures as shown in Figure 1 actuated by a 25 mm (1 in) diameter master cylinder. The total stroke of the master cylinder shall be 25 mm (1 in).

The initial movement of approximately 14 to 15 mm (9/16 to ⅝ in) shall be at a rate providing a gradual buildup of pressure not to exceed 1 MPa (150 psi). This shall permit the primary cup to pass over the compensating port at a low pressure. The balance of the stroke shall provide a gradual buildup of pressure to 7.0 MPa ± 0.3 (1000 psi ± 50) during the last 1.6 to 3.20 mm (1/16 to ⅛ in) of stroke. This remaining stroke at 7.0 MPa ± 0.3 1000 psi ± 50) shall be held constant by an adjustable relief valve.

The master cylinder shall be located in a uniformly heated, dry air oven and the fluid temperature in the master cylinder reservoir shall be maintained at 120°C ± 2 (248°F ± 3.6).

NOTE: A new master cylinder must be used for each test. It is recommended that at least 0.05 to 0.13 mm (0.002 to 0.005 in) clearance be allowed between the master cylinder piston and the master cylinder bore when conducting a master cylinder stroking test.

5.4.2 TEST SPECIMENS—One primary and one secondary cup shall be used for test specimens.

(R) 5.4.3 PROCEDURE—The cups shall be rinsed in isopropyl alcohol and wiped dry with a clean, lint-free cloth to remove dirt and packing debris. Cups shall not remain in the alcohol for more than 30 s.

Determine and record the initial hardness of the test cups. The lip diameter of the primary and secondary cups shall be measured and recorded to the nearest 0.02 mm (0.001 in), taking the average of two readings at right angles to each other. The lip diameter of the secondary cup shall be measured after the cup has been assembled on the piston.

The cups and master cylinder internal parts shall be dipped in the test fluid specified in Appendix A and the cylinder walls coated with the specified test fluid before assembly.

The master cylinder assembly, after installation in an oven, shall be operated for 70 h at the rate of 1000 strokes/h ± 100 at a temperature of 120°C ± 2 (248°F ± 3.6) as described in 5.4.1. After approximately 1 h of stroking to allow for evaporation of excess lubricant used at assembly, place a sheet of filter paper under the secondary cup of the master cylinder to catch and determine leakage past the secondary cup. The heat and actuating means shall be shut off at the termination of the 70 h stroking period with the master cylinder in the "off" position to relieve retained pressure in the master cylinder. After 1 h cooling period with the oven door open and the ventilating fan on, disconnect the fluid line at the master cylinder outlet. Remove the master cylinder from the oven and allow to cool for 22 h ± 2 at room temperature. Immediately after completion of the 22 h cooling period, careful inspection shall be made to check for fluid leakage past the master cylinder secondary cup. The fluid shall be drained from the master cylinder. The primary cup shall be removed from the cylinder, rinsed with isopropyl alcohol, and dried with compressed air. The secondary cup on the piston shall be rinsed in isopropyl alcohol, dried with compressed air, and the lip diameter measured before removal from the piston. Cups shall not remain in the alcohol for more than 30 s. Inspect cups for deterioration such as scoring, scuffing, blistering, cracking, chipping (heel abrasion), and change in shape from original appearance. Inspect cylinder parts, recording any pitting on piston or cylinder walls. The lip diameter of the primary cup shall be measured within 30 to 60 min after removal from the cylinder and the difference between the actual cylinder bore and the lip diameter after the test shall be determined and recorded for both primary and secondary cups.

Determine and record the change in hardness within 30 to 60 min after removal from the cylinder.

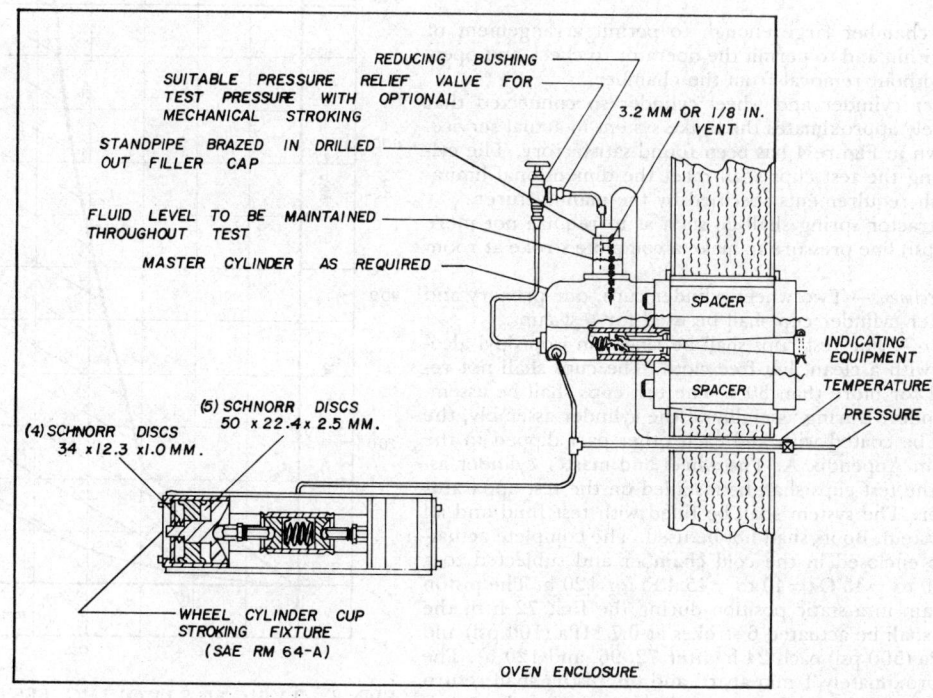

FIG. 1

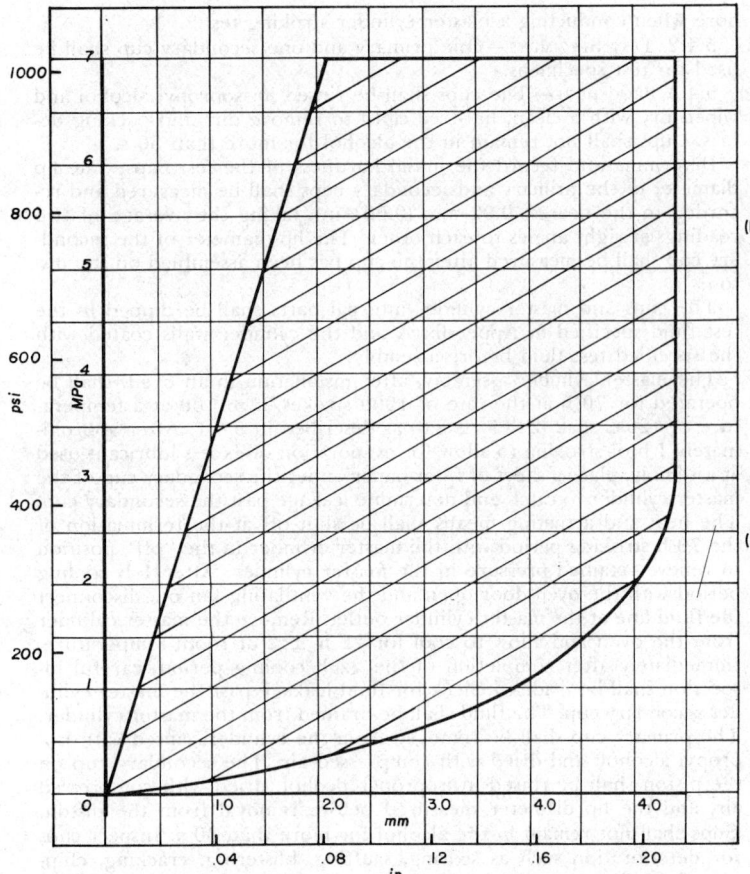

FIG. 2—TYPICAL WHEEL CYLINDER CUP MOVEMENT VERSUS PRESSURE—12.7 to 50.8 mm (½ to 2 in) DIAMETER

5.5 Low Temperature Performance

5.5.1 LEAKAGE

5.5.1.1 *Apparatus*—The leakage test apparatus shall include the following:

5.5.1.1.1 A cold chamber large enough to permit arrangement of the test apparatus within and to permit the operator to check and operate the apparatus without removal from the chamber.

5.5.1.1.2 A master cylinder and wheel cylinder so connected that their operation closely approximates the brake system in actual service. The apparatus shown in Figure 4 has been found satisfactory. The cylinder bore containing the test cups shall meet the dimensional limitations and bore finish requirements specified by the manufacturer.

5.5.1.1.3 The retractor spring shall be such as to require not more than 0.35 MPa (50 psi) line pressure to make a complete stroke at room temperature.

5.5.1.2 *Test Specimens*—Two wheel cylinder cups, one primary and one secondary master cylinder cup shall be used for test cups.

(R) 5.5.1.3 *Procedure*—The test cups shall be rinsed in isopropyl alcohol and wiped dry with a clean, lint-free cloth. The cups shall not remain in the alcohol for more than 30 s. The test cups shall be assembled in the test cylinder. During assembly of the cylinder assembly, the cylinder walls shall be coated with and each other part dipped in the test fluid specified in Appendix A. The wheel and master cylinder assembly containing the test cups shall be installed on the test apparatus in the cold chamber. The system shall be filled with test fluid and all air bled from the system. Boots shall not be used. The complete actuating system shall be enclosed in the cold chamber and subjected to a temperature of −40 to −43°C (−40 to −45.4°F) for 120 h. The piston and cups shall remain in a static position during the first 72 h of the test and thereafter shall be actuated 6 strokes at 0.7 MPa (100 psi) and 6 strokes at 3.5 MPa (500 psi) each 24 h (after 72, 96, and 120 h). The strokes shall be approximately 1 min apart, and the piston shall return to the stop after each stroke. No leakage shall occur during the 120 h test period.

5.5.2 BEND TEST

5.5.2.1 *Test Specimen*—One cup shall be used.

5.5.2.2 *Procedure*—The test cup after being subjected to 22 h at −40 to −43°C (−40 to −45.4°F) shall be bent through an angle of approximately 90 degree and immediately released. (The cold cup shall be bent while in the cold chamber and shall be handled to prevent warming). Within 1 min, examine test cup for cracking and change in shape from original appearance.

5.6 Oven Aging

(R) 5.6.1 Two test cups shall be rinsed in isopropyl alcohol and wiped dry with a clean, lint-free cloth to remove dirt and packing debris. Cups shall not remain in the alcohol for more than 30 s. The hardness of the cups shall be determined and recorded. The two test cups shall be placed in a Type 11B oven, as prescribed in ASTM E 145, and subjected to hot air heating at 100°C ± 2 (212°F ± 3.6) for 70 h. At the termination of the 70 h heating period, the cups shall be removed from the oven and allowed to cool for 16 to 96 h at room temperature.

The cups shall be inspected for blistering, or change in shape from original appearance. The hardness after aging shall be determined and recorded.

5.7 Hardness Determination

5.7.1 APPARATUS

(R) 5.7.1.1 International rubber hardness tester as described in ASTM D 1415.

NOTE: A type A durometer as described in ASTM D 2240.

5.7.1.2 Rubber anvil or cylinder having a flat circular top surface at least 19 mm (0.65 in) in diameter, a thickness of at least 9 mm (0.35 in), and a hardness within 5 IRHD of the hardness of the rubber test cup. See Figure 5 for one design of anvil.

5.7.2 PROCEDURE—The rubber cup shall be placed on a rubber anvil or cylinder as shown in Figure 5. The hardness of the cups is then measured by the procedure specified in ASTM D 1415 (ASTM D 2240 if a type A durometer is used).

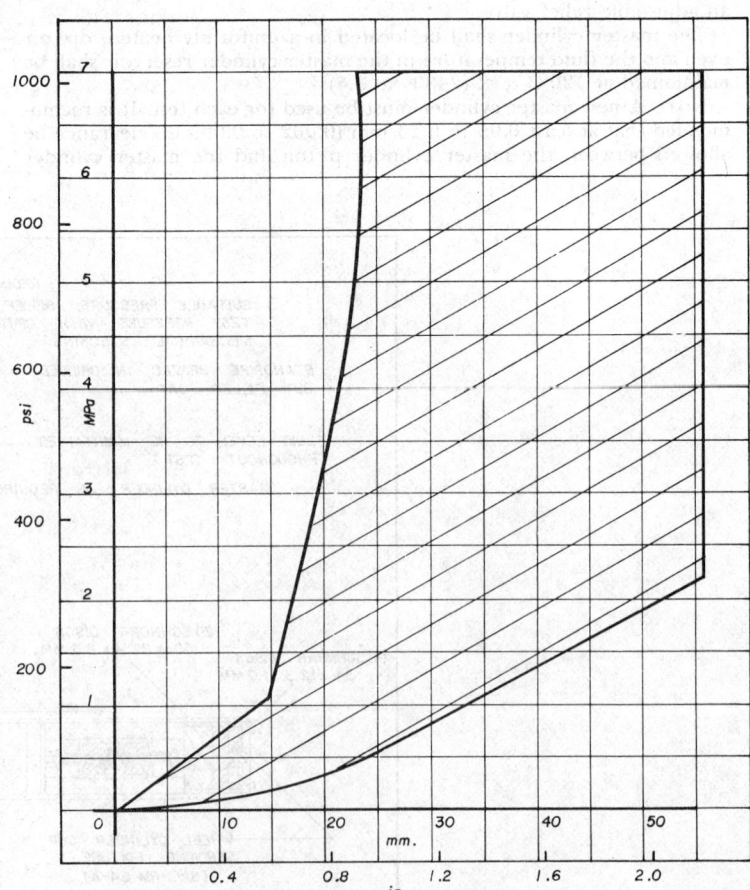

FIG. 3—TYPICAL STROKING TEST FOR MASTER CYLINDER CUP
(25 mm or 1 in)

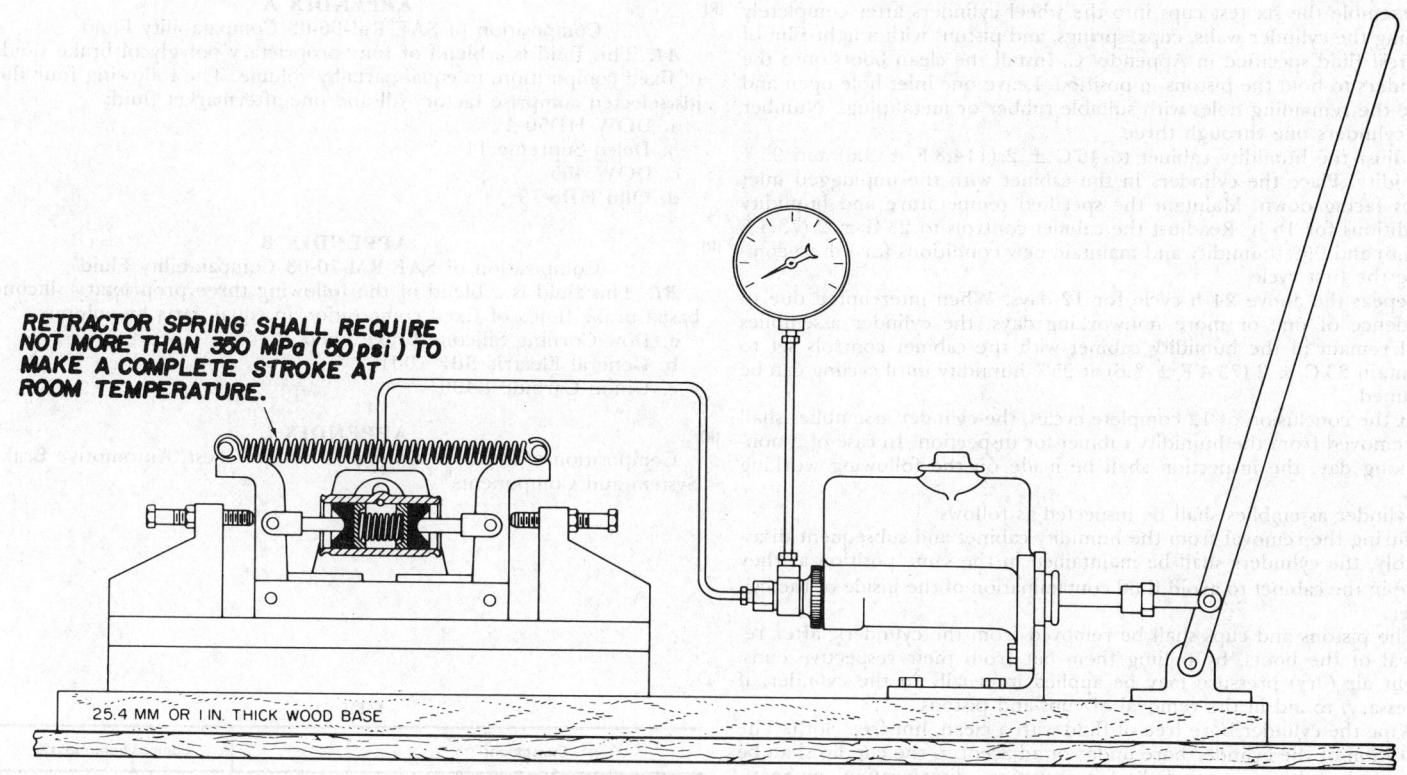

FIG. 4—LOW TEMPERATURE LEAKAGE TEST APPARATUS

(R) **5.8 Corrosion**—Prepare two sets of strips from each of the metals listed in Table 4, each strip having a surface area of 25 cm^2 ± 5 (3.875 in^2 ± 0.775), (approximately 8 cm (3.15 in) long, 1.3 cm (0.50 in) wide, and not more than 0.6 cm (0.24 in) thick. Drill a hole between 4 and 5 mm (0.16 and 0.20 in) in diameter and about 6 mm (0.24 in) from one end of each strip. With the exception of the tinned iron strips, clean the strips by abrading them on all surface areas with 320A waterproof carborundum paper and isopropyl alcohol until all surface scratches, cuts, and pits are removed from the strips, using a new piece of carborundum paper for each different type of metal. Wash the strips, including the tinned iron, with isopropyl alcohol and dry the strips with a clean, lint-free cloth and place the strips in a desiccator containing desiccant maintained at 23°C ± 5 (73.4°F ± 9) for at least 1 h. Handle the strips with clean forceps after polishing to avoid fingerprint contamination.

Weigh each strip to the nearest 0.1 mg and assemble each set of strips on an uncoated steel bolt in the order tinned iron, steel, aluminum, cast iron, brass, and copper, so that the strips are in electrolytic contact. Bend the strips, other than cast iron, so that there is a separation of at least 3 mm (0.125 in) between adjacent strips for a distance of about 6 cm (2.36 in) from the free end of the strips. Immerse strip assemblies in isopropyl alcohol to eliminate fingerprints and then handle only with clean forceps. Place one rubber cup with lip edges facing up, in each of two straight sided round glass jars[5] having a capacity of approximately 475 mL and inner dimensions of approximately 100 mm (4 in) in height and 75 mm (3 in) in diameter. Use only tinned steel lids vented with a hole 0.8 mm ± 0.1 (0.032 ± 0.003 in) in diameter. Insert a metal strip assembly inside one cup in each jar with the pinned end in contact with the concavity of the cup and the free end extending upward in the jar. Mix 760 mL of SAE RM66-03 compatibility fluid with 40 mL of distilled water.

Add a sufficient amount of the mixture to cover the metal strip assembly in each jar to a depth of approximately 10 mm (0.40 in) above the tops of the strips listed in Table 4. Tighten the lids and place the jars in an oven maintained at 100°C ± 2 (212°F ± 3.6) for 120 h ± 2. Allow the jars to cool at 23°C ± 5 (73.4°F ± 9) for 60 to 90 min. Immediately following the cooling period, remove the metal strips from the jars by use of forceps, removing loose adhering sediment by agitation of the metal strip assembly in the fluid in the jar. Examine test strips and test jars for adhering crystalline deposit, disassemble the metal strips, remove adhering fluid by flushing with water, and clean individual strips by wiping with a cloth wetted with isopropyl alcohol. Examine the strips for evidence of corrosion and pitting. Place strips in a desiccator containing a desiccant maintained at 23°C ± 5 (73.4°F ± 9) for at least 1 h. Weigh each strip to the nearest 0.1 mg. Determine the difference in weight of each metal strip and divide the difference by the total surface area of the metal strip measured in square centimeters. Average the results for the two strips of each type of metal.

Examine the fluid-water mixture in the jars for jelling. Agitate the fluid in jars to suspend and uniformly disperse sediment and transfer a 100 mL portion of this fluid to an ASTM cone-shaped centrifuge tube and determine percent sediment as described in 5(b) of ASTM D 91.

5.9 Storage Corrosion Test

5.9.1 APPARATUS—The apparatus shall include a humidity cabinet capable of maintaining a temperature of 21 to 46°C (70 to 115°F) at 95% humidity, three wheel cylinder assemblies of proper size for the cups being tested.

(R) 5.9.2 PROCEDURE—Disassemble the three cylinder assemblies and using a clean, lint-free cloth, wipe all fluids from the cylinders, pistons, boots, and springs. Cylinder or parts showing light stains or corrosion shall be discarded.

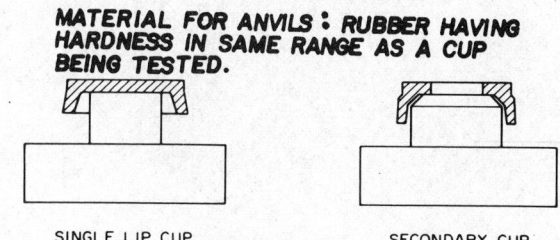

FIG. 5—ANVIL FOR MEASURING HARDNESS

[5] See Footnote 1.

Assemble the six test cups into the wheel cylinders after completely coating the cylinder walls, cups, springs, and pistons with a light film of the test fluid specified in Appendix C. Install the clean boots onto the cylinders to hold the pistons in position. Leave one inlet hole open and close the remaining holes with suitable rubber or metal plugs. Number the cylinders one through three.

Adjust the humidity cabinet to 46°C ± 2 (114.8°F ± 3.6) and 95% humidity. Place the cylinders in the cabinet with the unplugged inlet holes facing down. Maintain the specified temperature and humidity conditions for 16 h. Readjust the cabinet controls to 23°C ± 2 (73.4°F ± 3.6) and 95% humidity and maintain new conditions for 8 h to complete the first cycle.

Repeat the above 24 h cycle for 12 days. When interrupted due to incidence of one or more nonworking days, the cylinder assemblies shall remain in the humidity cabinet with the cabinet controls set to maintain 23°C ± 2 (73.4°F ± 3.6) at 95% humidity until cycling can be resumed.

At the conclusion of 12 complete cycles, the cylinder assemblies shall be removed from the humidity cabinet for inspection. In case of a nonworking day, the inspection shall be made on the following working day.

Cylinder assemblies shall be inspected as follows:

During the removal from the humidity cabinet and subsequent disassembly, the cylinders shall be maintained in the same position as they were in the cabinet to avoid fluid contamination of the inside of the cylinder.

The pistons and cups shall be removed from the cylinders, after removal of the boots, by pulling them out from their respective ends. Slight air (dry) pressure may be applied internally in the cylinder, if necessary, to aid in the removal of cups and pistons.

Wipe the cylinder bore free of fluid with a clean, lint-free cloth. The condition of the cylinder bore under or adjacent to the cup lip shall be inspected under a strong light for corrosion, discoloration, or spots, noting particularly the area of the ring left by the lip of the cup during its exposure in the humidity cabinet. Any corrosion or spots away from the contact surface of the cups shall be disregarded.

(R) APPENDIX A
Composition of SAE RM-66-03 Compatability Fluid[6]

A1. This fluid is a blend of four proprietary polyglycol brake fluids of fixed composition, in equal parts by volume. The following four fluids selected comprise factory-fill and one aftermarket fluid:
a. DOW HD50-4
b. Delco Supreme 11
c. DOW 455
d. Olin HDS-79

(R) APPENDIX B
Composition of SAE RM-70-03 Compatability Fluid[7]

B1. This fluid is a blend of the following three proprietary silicone based brake fluids of fixed composition in equal parts by volume:
a. Dow Corning Silicone Brake Fluid
b. General Electric SBF 1001
c. Union Carbide L490

(R) APPENDIX C
Composition of Fluid for Storage Corrosion Test, Automotive Brake System and Components[8]

TABLE C1

Constituent		Composition, mass%
Castor oil		38.7 ± 1.0
Diethylene glycol monomethyl ether[a]		31.4 ± 0.5
Ethylene glycol monobutyl ether[a]		27.3 ± 0.5
Borax-glycol condensate[b]		2.1 ± 0.2
Di-t-butyl-p-Cresol[a]		0.475 ± 0.050
M-Cresol		0.025 ± 0.005
	Total	100.000

[a]The diethylene glycol monomethyl ether, ethylene glycol monobutyl ether, and di-t-butyl-p-Cresol shall be technical grade.
[b]The borax-glycol condensate shall be 30 parts by mass of sodium borate decahydrate ($Na_2B_4O_7 \cdot 10H_2O$) combined with 100 parts by mass of 1, 2-propylene glycol.

C1. The physical and chemical requirements of the preservative fluid shall be as follows:

TABLE C2

Boiling point, @ 760 mm Hg	167 to 186°C (333 to 367°F)
pH value	9.9 to 10.7
Specific gravity at 20/20°C	0.9620 to 0.9680

---

[6] See Footnote 1.
[7] See Footnote 1.
[8] See Footnote 1.

(R) RUBBER DUST BOOTS FOR THE HYDRAULIC DISK BRAKE PISTON—SAE J1570 SEP91

SAE Standard

Report of the Hydraulic Brake Actuating Elastomeric Committee approved October 1988. Completely revised by the SAE Hydraulic Brake Actuating Elastomeric Committee September 1991. Rationale statement available.

1. Scope
This SAE Standard covers molded rubber boots used on disc brake pistons to prevent the entrance of dirt and moisture which could cause corrosion and otherwise impair caliper operation.

The specification includes performance tests of brake piston boots of both plain and insert types under specified conditions and does not include requirements relating to chemical composition or physical properties of the rubber compound. Further, it does not cover the strength of the adhesion of rubber to the insert material where an insert is used.

The rubber material used in these boots is classified as suitable for operating in a temperature range of –40 to +120 °C ± 2 °C (–40 to +248 °F ± 3.6 °F).

2. References
2.1 Applicable Documents—The following publications form a part of this specification to the extent specified herein. The latest issue of SAE publications shall apply.

2.1.1 SAE PUBLICATIONS—Available from SAE, 400 Commonwealth Drive, Warrendale, PA 15096-0001.

SAE J1601—Rubber Cups for Hydraulic Actuating Cylinders

2.1.2 ASTM PUBLICATIONS—Available from ASTM, 1916 Race Street, Philadelphia, PA 19103.

ASTM D 471—Test Method for Rubber Property-Effect of Liquids
ASTM D 573—Test Method for Rubber-Deterioration in an Air Oven
ASTM D 1149—Test Method for Rubber Deterioration-Surface Ozone Cracking in a Chamber
ASTM D 1415—Test Method for Rubber Property-International Hardness
ASTM D 2240—Test Method for Rubber Property-Durometer Hardness

3. General Requirements
3.1 Workmanship and Finish—Boots shall be free from blisters, pinholes, cracks, protuberance, embedded foreign material, or other physical defects which can be detected by thorough inspection, and shall conform to the dimensions specified on the drawings.

3.2 Marking—The identification mark of the manufacturer as recorded by the Rubber Manufacturers Association and other details as specified on drawings shall be molded into each boot.

3.3 Packaging—Boots shall be packaged to meet requirements specified by the purchaser.

3.4 Sampling—The minimum lot on which complete specification tests shall be conducted for quality control testing, or the frequency of any specific type test used to control production, shall be agreed upon by the manufacturer and the purchaser.

4. Test Requirements
4.1 Resistance to Fluid at Elevated Temperature—After being subjected to the test for resistance to fluid at elevated temperatures as prescribed in 5.4, boots shall conform to the following requirements:
  a. Change in volume: –5 to +20%
  b. Change in hardness: –15 to +5 points
  c. IRHD (Shore A)

4.2 Heat Stroking Test—After stroking as detailed in 5.5, boots will have passed the requirements if the following criteria are met:
  a. Both beads must remain in installed position during stroking.
  b. IRHD hardness change must not exceed –10 to +15 points.
  c. Boots must not be cracked through (surface cracking allowed) after completion of the stroking and removal from the caliper or fixture.
  d. Clearance must not occur at the interference points measured per 5.5.2.

4.3 Low Temperature Stroking Test—During stroking as detailed in 5.6, boots shall be considered to have passed the test if the following criteria are met:
  a. Boot beads shall remain in the installed position during stroking.
  b. No cracks caused by flexing at low temperature shall be in evidence.

4.4 Heat Resistance Test (Static)—After the heat resistance test, as described in 5.7, boots shall conform to the following requirements:
  a. The boots must not be cracked through (surface cracking allowed) after installation, flexing, and removal.
  b. Change in IRHD (Shore A) hardness: –5 to +20 points

4.5 Ozone Resistance Test—At the end of the 168 h exposure period, as detailed in 5.8, test specimens shall be removed from the ozone chamber and examined under 2X magnification while in the installed position. Surface of the test specimens shall show no evidence of cracking, rupture, or other deterioration.

5. Test Procedures
5.1 Test Specimens—Specimens prepared for a particular test shall be cut from the same approximate location on different sample boots. Hardness test specimens shall be prepared to present a flat molded surface to the indentor.

5.2 Test Fluid—The brake fluid used for testing shall be that specified in Appendix A.

5.3 Hardness—The method of determining rubber hardness shall be as described in ASTM D 1415. If agreed upon by vendor and purchaser, an alternate procedure, ASTM D 2240 may be used.

5.4 Resistance to Fluids at Elevated Temperatures

5.4.1 APPARATUS
  a. Oven—Circulating air oven as specified in Section 5 of ASTM D 573.
  b. Container—Use a screw top, straight-sided, round glass jar, having a capacity of approximately 250 cm^3 and inner dimensions of approximately 125 mm in height and 50 mm in diameter, and tinned steel lid (no insert or organic coating).

5.4.2 TEST SPECIMENS—Two new test segments weighing 0.5 to 5 g shall be cut from complete boots.

5.4.3 TEST PROCEDURE
  a. The segments shall be rinsed in isopropyl alcohol, and blown dry or wiped dry with a lint-free cloth to remove dirt and packaging debris. The segments shall not remain in the alcohol for more than 30 s.
  b. Determine and record the volume of the segments in accordance with Section 10 of ASTM D 471, except use isopropyl alcohol where acetone is specified.
  c. Determine and record the initial IRHD hardness of test segments.
  d. The segments shall be placed in a suitable glass container and completely immersed in 75 mL of the approved brake fluid, and the container shall be sealed to prevent vapor loss and placed in an oven at 120 °C ± 2 °C (248 °F ± 3.6 °F) for 70 h followed by removal per ASTM D 471.
  e. Determine and record the final volume and IRHD (Shore A) hardness of the segments.

5.4.4 CALCULATIONS
  a. Volume Change—The change in volume shall be reported as percentage of the original volume. The calculations shall be made as follows:

$$\% \text{ Increase in Volume} = \frac{(M_3 - M_4) - (M_1 - M_2)}{(M_1 - M_2)} \times 100 \quad (Eq.1)$$

where:
  M_1 = initial mass in air
  M_2 = initial mass in water
  M_3 = mass in air after test
  M_4 = mass in water after test

  b. Hardness Change—Calculate and record the change in IRHD (Shore A) hardness.

5.5 Heat Stroking Test

5.5.1 APPARATUS
  a. Oven—Circulating air oven as specified in Section 5 of ASTM D 573.
  b. Actuated Stroking Fixture—This fixture shall be composed of a production caliper assembly or simulating fixture capable of being stroked 0.020 in at the one-half worn lining piston position.
  c. Actuating Stroking Fixture—This fixture shall be composed of a mechanically, pneumatically or hydraulically actuated device whose rate of operation shall be controlled at 1000 strokes per hour ± 100 strokes per hour.
  d. Drawing—A drawing of a typical actuated and actuating system is shown in SAE J1601, Figure 1.

5.5.2 TEST SPECIMEN PREPARATION
  a. Two new boots shall be rinsed in isopropyl alcohol and blown dry or wiped dry with a lint-free cloth to remove dirt and packaging debris.

b. The boots shall not remain in the alcohol for more than 30 s. Measure and record IRHD (Shore A) hardness of boots.
c. Measure, record, and calculate boot-to-piston and boot-to-caliper groove interferences (diametral and/or side interference at sealing bead, as designed).
d. Install the dry boots on the caliper assembly or fixture.

5.5.3 TEST PROCEDURE
a. Place the caliper assemblies or fixtures (adjusted to simulate the one-half worn lining piston position) in the oven and connect as required.
b. Stroke for 70 h at 1000 strokes per hour ± 100 strokes per hour at 120 °C ± 2 °C (248 °F ± 3.6 °F). If simulating fixtures are used, a stroke of 0.020 in shall be maintained.
c. After 70 h, discontinue stroking, shut off heat and allow oven to cool for 1 h with door open. Remove boots for inspection within 30 h.
d. Note and record general condition of boots relative to cracks, retention of shape, and tackiness.
e. Remeasure and record interference per 5.5.2.
f. Remeasure and record IRHD (Shore A) hardness of boots.

5.6 Low Temperature Stroking Test

5.6.1 APPARATUS
a. Cold Chamber—A cold chamber large enough to permit installation of the test caliper (or fixture) and capable of continuous −40 °C (−40 °F) operation.
b. Caliper—A test caliper (or fixture) per SAE J1601 Figure 1, except modified to allow for stroking from unworn lining piston position to full worn lining piston position.
c. Actuator—An actuator capable of actuating the boot test caliper or fixture.

5.6.2 TEST SPECIMEN PREPARATION
a. Two new boots shall be rinsed in isopropyl alcohol and blown dry or wiped dry with a lint-free cloth. The test boot shall not remain in alcohol for more than 30 s.
b. Install boots on calipers or fixtures and place in cold chamber.

5.6.3 TEST PROCEDURE
a. The test caliper or fixture shall be subjected to a temperature of −40 °C ± 2 °C (−40 °F ± 3.6 °F). After a total of 70 h, stroke the calipers (or fixtures) from full lining piston position to full worn lining piston position for six strokes, 10 s apart, without removal from the cold chamber.
b. After return to ambient temperature, inspect for boot retention and remove boots from fixtures and note and record general condition of boots relative to cracks.

5.7 Heat Resistance Test (Static)

5.7.1 APPARATUS—Circulating air oven as specified in Section 5 of ASTM D 573.

5.7.2 TEST SPECIMEN PREPARATION
a. Two new boots shall be rinsed in isopropyl alcohol and blown dry or wiped dry with a lint-free cloth to remove dirt and packing debris. The boots shall not remain in the alcohol for more than 30 s.
b. Measure and record IRHD (Shore A) hardness of the boots per 5.3.

5.7.3 TEST PROCEDURE
a. The test boots shall be placed in the circulating air oven and held for 22 h at 175 °C ± 2 °C (347 °F ± 3.6 °F).
b. At the termination of the heating period, the boots shall be removed from the oven and allowed to cool for 30 min at room temperature.
c. The boots shall be visually inspected and the IRHD hardness, after aging, determined and recorded. Also, record any visual change in the boot.
d. Install boots in normal manner, flex 10 times from new to full worn lining position and remove.

5.8 Ozone Resistance Test

5.8.1 APPARATUS—The apparatus shall consist of an ozone chamber as described in ASTM D 1149, and shall be capable of maintaining an ozone partial pressure of 100 MPa ± 10 MPa (100 ± 10 parts of ozone/100 million parts of air by volume).

5.8.2 TEST SPECIMEN PREPARATION—Two new test boots shall be installed on a caliper assembly or fixture which shall simulate the normal actual installation stretch at the piston and caliper assembly groove beads and the full worn lining piston position.

5.8.3 TEST PROCEDURE
a. Assemble two new test boots on the calipers or fixtures.
b. Immediately after assembly, the calipers or fixtures shall be placed in the chamber and exposed to an ozone partial pressure of 100 MPa ± 10 MPa (100 ± 10 parts of ozone/100 million parts of air by volume) at a temperature of 40 °C ± 1 °C (104 °F ± 1.8 °F) for 7 days.
c. After the 7-day period, remove the caliper or fixture and examine the boots for cracks under a 2X magnification while in the installed position.

APPENDIX A—SAE RM-66-03 COMPATIBILITY FLUID[1]

A.1 This fluid is a blend of four proprietary polyglycol brake fluids of fixed composition, in equal parts by volume. The four fluids selected comprise three factory-fill and one aftermarket fluid, as follows:

DOW HD50-4
DOW 455
Delco Supreme II
Olin HDS-79

---

[1] Obtainable from the Society of Automotive Engineers, Inc., 400 Commonwealth Drive, Warrendale, PA 15096-0001.

POWERPLANT COMPONENTS AND ACCESSORIES

26 Powerplant Components and Accessories

ENGINE TERMINOLOGY AND NOMENCLATURE—GENERAL—SAE J604 JAN86

SAE Recommended Practice

Report of the Engine Committee, approved May 1958, reaffirmed January, 1986.

1. This SAE Recommended Practice is applicable to all types of reciprocating engines including two-stroke cycle and free piston engines, and was prepared to facilitate clear understanding and promote uniformity in nomenclature.

Modifying adjectives in some cases were omitted for simplicity. However, it is good practice to use adjectives when they add to clarity and understanding.

2. Geometry Terminology

  2.1 *Compression Ratio* = $\dfrac{\text{Maximum cylinder volume}}{\text{Minimum cylinder volume}}$

  2.2 **Valve or Port Areas**—Full open areas measured immediately adjacent to the cylinder.

Example: For poppet valves

$$\text{Area} = (\pi) \times (\text{head outer diameter}) \times (\text{full lift})$$

Example: For rectangular port in the cylinder wall

Area = (height at cylinder surface) × (width, developed at cylinder surface)

  2.3 **Valve or Port Timing**—Geometric crankshaft positions at which ports or valves open or close.

  2.4 **Top Center**—The geometric crankshaft position at which piston motion reverses direction and the cylinder volume is at, or near, a minimum.

  2.5 **Bottom Center**—The geometric crankshaft position at which piston motion reverses direction and the cylinder volume is at, or near, a maximum.

  2.6 **Combustion Chamber Surface-to-Volume Ratio**[1]—Area of

[1] A major source of unburned hydrocarbons in the exhaust gas of spark ignition engines is the quenching of the flame by the relatively cold combustion chamber walls. A useful way to compare different engine designs as to their potential for low exhaust emission values is to compare their combustion chamber surface-to-volume ratios.

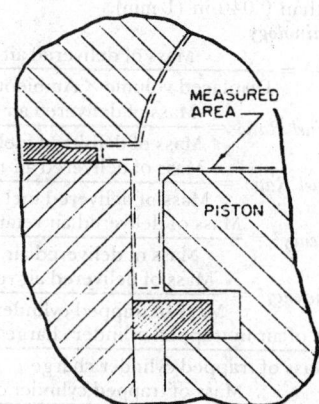

FIG. 2—HEAD GASKET AND TOP RING LAND AREA

chamber divided by volume at top center. Fig. 1 illustrates the surface area and the volume of a typical combustion chamber. Figs. 2–4 and the following list define the chamber area in detail:

Include:
1. Head cavity area.
2. Head flat or quench area within head gasket outline.
3. Cylinder block top surface area within head gasket outline.
4. Side area of head gasket outline.
5. Valve side areas, including cylindrical side of valve head and that part of the face projecting into the chamber.
6. Valve head surface area.
7. Piston top surface area.
8. Piston top ring land area.
9. Area of top surface of piston ring exposed between top land diameter and cylinder bore diameter.

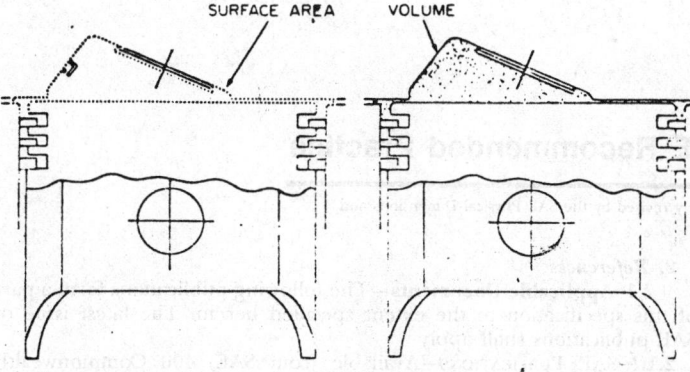

FIG. 1—TYPICAL COMBUSTION CHAMBER

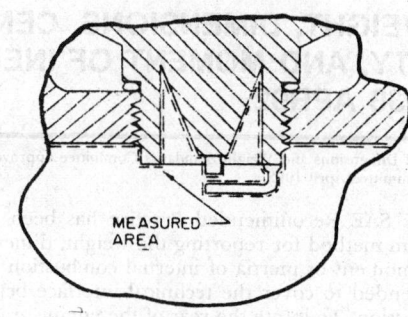

FIG. 3—SPARK PLUG AREA

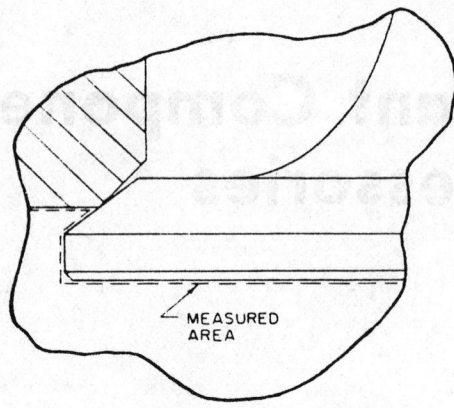

FIG. 4—VALUE AREA

10. Cylinder bore surface area above top ring.
11. Spark plug cavity area.

Exclude:
1. Area behind top ring.
2. Gasket area inside first bead.
3. Chamber less than 0.040 in (1 mm).

3. **Performance Terminology**

3.1 *Delivery Ratio*[2,3] $= \dfrac{\text{Mass of delivered air}}{\text{Displaced volume} \times \text{Ambient density}}$

3.2 *Delivered Air-Fuel Ratio* $= \dfrac{\text{Mass of delivered air}}{\text{Mass of delivered fuel}}$

3.3 *Trapped Air-Fuel Ratio* $= \dfrac{\text{Mass of delivered air retained}}{\text{Mass of delivered fuel retained}}$

3.4 *Trapping Efficiency*[2] $= \dfrac{\text{Mass of delivered air retained}}{\text{Mass of delivered air}}$

3.5 *Scavenging Efficiency*[2] $= \dfrac{\text{Mass of delivered air retained}}{\text{Mass of trapped cylinder charge}}$

3.6 *Purity* $= \dfrac{\text{Mass of air in trapped cylinder charge}}{\text{Mass of trapped cylinder charge}}$

3.7 *Relative Charge*[3] $= \dfrac{\text{Mass of trapped cylinder charge}}{\text{Displaced volume} \times \text{Ambient density}}$

3.8 *Charging Efficiency*[2,3] $= \dfrac{\text{Mass of delivered air retained}}{\text{Displaced volume} \times \text{Ambient density}}$

3.9 *Excess Air Factor* $= \dfrac{\text{Trapped air-fuel ratio}}{\text{Stoichiometric ratio}}$

4. **Nomenclature—Multiple Expansion Piston Engines**

4.1 **Compound Engine**—An engine in which the output power is delivered by both reciprocating and rotating expanders.

---

[2] If scavenging is done with air-fuel mixture (example given, carburetor engine) "mixture" is to be substituted for "air" and "Mixture density at ambient pressure and temperature" is to be substituted for "Ambient density."

[3] When ambient density is unknown, the density of dry air at SAE standard reference atmospheric conditions (0.0719 lbm/ft³) (1.1517 kg/m³) is to be used.

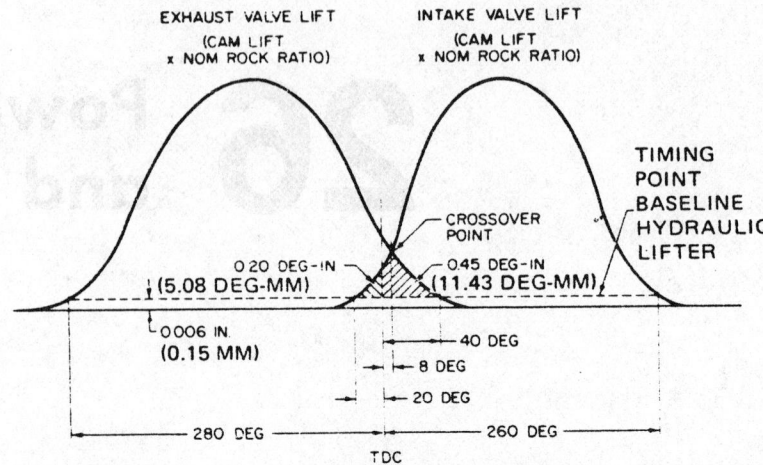

FIG. 5—TYPICAL VALVE TIMING AND OVERLAP DESIGNATION (POPPET VALVES)

4.2 **Reheater**—A combustor or heat exchanger wherein heat is added to the working fluid between stages of expansion.

4.3 **Afterburner**—A combustor in which heat is added to the working fluid after the last expansion stage.

5. **Valve Timing and Valve Overlap (Poppet Valves)**—Valve timing events are the valve opening and closing points in the operating cycle, while valve overlap describes that part of the cycle in which both the intake and exhaust valves are open. These are illustrated in Fig. 5 and further defined as follows:

5.1 Timing Events are stated in crankshaft degrees from piston top dead center, rounded to the nearest whole degree. They are based on reference valve lift points at a timing point baseline reference as follows[4]:

(a) Hydraulic Lifters—Timing point baseline is at 0.006 in (0.15 mm) valve lift.

(b) Mechanical Lifters—Timing point baseline is at a valve lift of 0.006 in (0.15 mm) plus the specified lash for each valve.

5.2 Valve Overlap Area is specified as two separate overlap areas expressed as deg-in (deg-mm) rounded to the nearest hundredth. Overlap areas are the areas bounded by the exhaust closing valve lift curve, the intake valve lift curve and the timing point baseline from TDC to the respective opening or closing event as defined in paragraph 5.1.

5.3 The Crossover Point is defined as the angular crankshaft position at which exhaust closing valve lift and intake opening valve lift are equal.

---

[4] For the purpose of defining valve events, overlap, and crossover points, the valve lift curves are obtained by multiplication of the cam lift values by the nominal valve mechanism lift ratio (for example, rocker arm ratio). In the case of mechanical lifters, the specified lash is subtracted from each valve lift value.

(R) ENGINE WEIGHT, DIMENSIONS, CENTER OF GRAVITY, AND MOMENT OF INERTIA —SAE J2038 APR92

SAE Recommended Practice

Report of the Physical Dimensions and Weight Standards Committee approved April 1990. Completely revised by the SAE Physical Dimensions and Weight Standards Committee April 1992.

1. Scope—This SAE Recommended Practice has been developed to provide a uniform method for reporting the weight, dimensions, center of gravity, and moment of inertia of internal combustion engines. SAE J2038 is not intended to cover the technical interface between the engine and transmission. To locate the rear of the engine crankshaft in relationship to the rear of the flywheel housing, refer to SAE J617.

2. **References**

2.1 **Applicable Documents**—The following publications form a part of this specification to the extent specified herein. The latest issue of SAE publications shall apply.

2.1.1 SAE PUBLICATIONS—Available from SAE, 400 Commonwealth Drive, Warrendale, PA 15096-0001.

SAE J617—Engine Flywheel Housings
SAE J824—Engine Rotation and Cylinder Numbering
2.1.2 ISO PUBLICATIONS—Available from ANSI, 11 West 42nd Street, New York, NY 10036-8002.

ISO/DIS 1204—Reciprocating internal combustion engines—Designation of the direction of rotation and of cylinders and valves in cylinder heads, and definition of right-hand and left-hand in-line engines and locations of an engine.

2.2 Terminology

2.2.1 BASIC ENGINE—The "basic engine" is defined as a runnable engine equipped with built-in accessories, such as fuel, oil, and coolant pumps, emission control equipment, and includes all standard equipment as clearly defined by each engine manufacturer. Standard equipment includes a defined or no specified SAE standard flywheel, flywheel housing, and auxiliary drives. The dry engine weight will be reported.

2.2.2 AS SHIPPED ENGINE—The "as shipped engine" is the "basic engine" equipped with accessories required for users' application. A variety of options, which may include lube oil and coolant, will be included in the weight. The "as shipped engine" corresponds to the user's specifications for a specified part number. Shipping stands and other items required for shipment are not included.

2.2.3 FULLY EQUIPPED ENGINE—A "fully equipped engine" is an "as shipped engine" equipped with accessories necessary to perform its intended service. This includes, but is not restricted to, intake air system, cooling system, alternator, starting motor, noise control equipment, power steering pump system, air compressor and vacuum pumps.

2.2.4 The manufacturer should clearly define what is included in each engine arrangement.

3. Weight Measurement Procedure

3.1 Engine weights should be established within ±3%. Various methods (actual weighing, computer analysis) may be used to determine these weights.

3.2 Dry weight is defined as the weight of an engine that has been run and then had fluids drained. (Filters are not emptied.)

3.3 Wet weight is defined as the dry weight, plus the weight of fluids as recommended by engine manufacturer (fuel, oil, coolant). This weight may be established by weighing wet engines or by calculating weight of fluids and adding to dry weight.

3.4 The manufacturer should clearly define what equipment is included in each engine weight, as defined in 2.2.

4. Dimension Measurement Procedure

4.1 Forward length of the engine is defined as the horizontal distance from the rear face of the engine cylinder block to the front most point of the engine.

4.2 Rearward length of the engine is defined as the horizontal distance from the rear face of the engine cylinder block to the rearmost point of the engine.

4.3 Overall length of the engine is the sum of 4.1 and 4.2.

4.4 Upper height of the engine is defined as the vertical distance from the centerline of the crankshaft upward to the highest point on the engine.

4.5 Lower height of the engine is defined as the vertical distance from the centerline of the crankshaft downward to the lowest point on the engine.

4.6 Overall height of the engine is the sum of 4.4 and 4.5.

4.7 Left side width is defined as the horizontal distance from the centerline of the crankshaft to the widest point on the left side of the engine. Refer to SAE J824 or ISO/DIS 1204 for definition of right and left side of engine.

4.8 Right side width is defined as the horizontal distance from the centerline of the crankshaft to the widest point on the right side of the engine. Refer to SAE J824 or ISO/DIS 1204 for definition of right and left side of engine.

4.9 Overall width of the engine is the sum of 5.7 and 5.8.

5. Center of Gravity Location and Dimensioning Procedure

5.1 The center of gravity of a basic engine, as shipped engine, or fully equipped engine is the point at which all the weight of the engine or engine package can be considered to be concentrated.

5.2 The center of gravity of an engine or engine package is located from three reference planes.

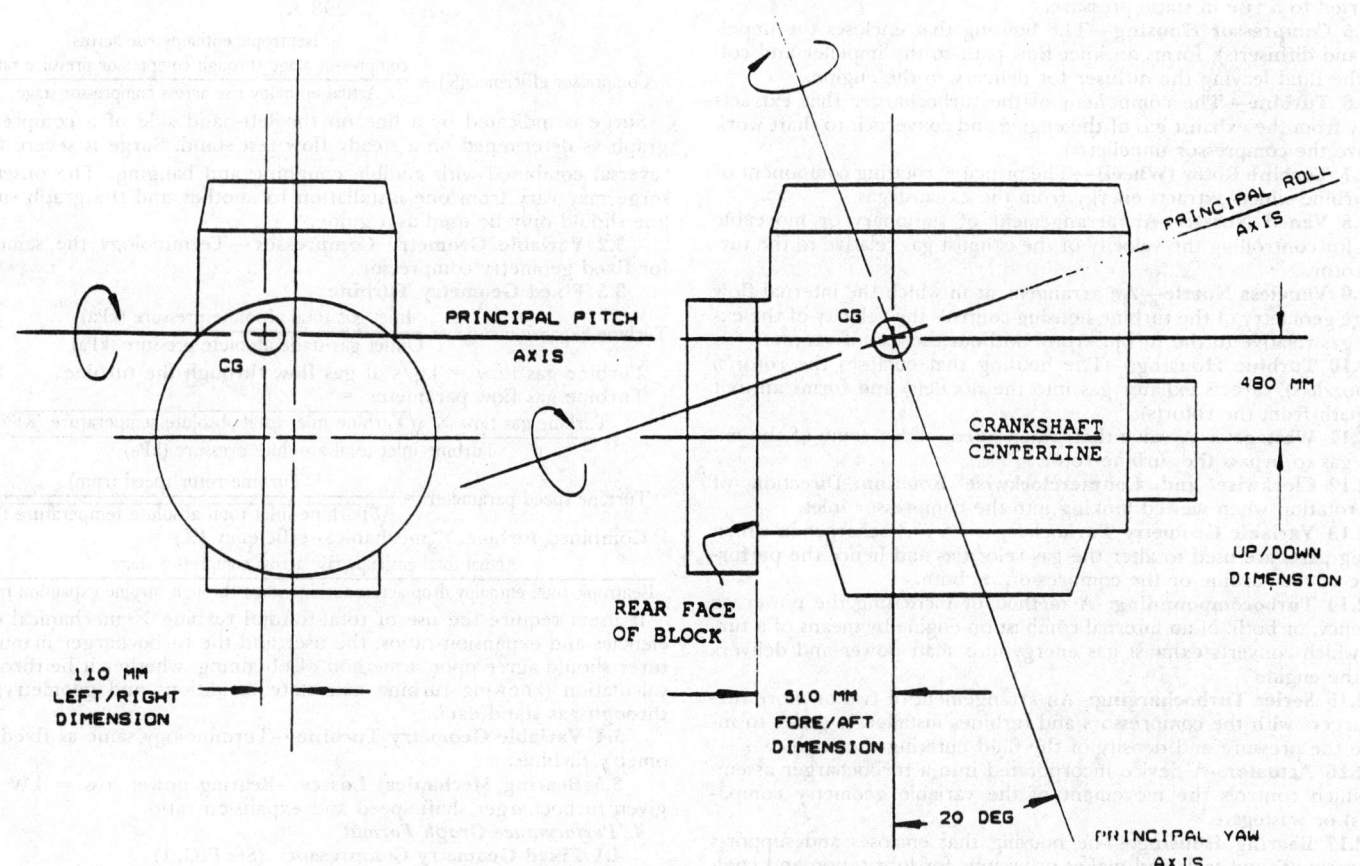

FIGURE 1—SAMPLE ILLUSTRATION OF CENTER OF GRAVITY AND MOMENT OF INERTIA

5.3 The fore-aft location is defined as the horizontal distance from the rear face of the engine cylinder block to the center of gravity.

5.4 The vertical location is defined as the distance above or below the centerline of the crankshaft to the center of gravity.

5.5 The lateral location is defined as the right or left horizontal distance from the centerline of the engine crankshaft to the center of gravity. For definition of engine right and left side refer to SAE J824 or ISO/DIS 1204.

6. Moment of Inertia Location and Dimensioning Procedure

6.1 The moment of inertia values of an engine or engine package acts about three principal axes that pass through the engine center of gravity.

6.2 The principal roll or fore-aft axis generally slopes downward toward the rear of the engine.

6.3 The principal yaw or near vertical axis lies in the same vertical plane as the roll axis, 90 degrees adjacent.

6.4 The principal pitch or lateral axis lies in the same near-horizontal plane as the roll axis, 90 degrees adjacent.

6.5 Figure 1 shows a sample illustration of a center of gravity and moment of inertial specification.

φTURBOCHARGER NOMENCLATURE AND TERMINOLOGY—SAE J922 JUL88

SAE Recommended Practice

Report of the Engine Committee, approved July 1965, and completely revised July 1988. Rationale statement available.

1. Scope—This recommended practice applies to nomenclature of turbocharger parts and terminology of performance.

2. Nomenclature

2.1 Turbocharger—A device used for increasing the pressure and density of the fluid entering an internal combustion engine using a compressor driven by a turbine which extracts energy from the exhaust gas.

2.2 Compressor—The component of the turbocharger that raises the pressure and density of the inlet fluid.

2.3 Compressor Impeller (Rotor, Wheel)—The principal rotating component of the compressor which imparts energy to the fluid.

2.4 Compressor Diffuser—A component of the compressor in which the kinetic energy of the fluid leaving the impeller is partially converted to a rise in static pressure.

2.5 Compressor Housing—The housing that encloses the impeller(s) and diffuser(s), forms an inlet flow path to the impeller and collects the fluid leaving the diffuser for delivery to the engine.

2.6 Turbine—The component of the turbocharger that extracts energy from the exhaust gas of the engine and converts it to shaft work to drive the compressor impeller(s).

2.7 Turbine Rotor (Wheel)—The principal rotating component of the turbine which extracts energy from the exhaust gas.

2.8 Vaned Nozzle—An arrangement of stationary or moveable vanes for controlling the velocity of the exhaust gas relative to the turbine rotor.

2.9 Vaneless Nozzle—An arrangement in which the internal flow passage geometry of the turbine housing controls the velocity of the exhaust gas relative to the turbine rotor without the use of vanes.

2.10 Turbine Housing:—The housing that encloses the rotor(s) and nozzle(s), directs exhaust gas into the nozzle(s) and forms an exit flow path from the rotor(s).

2.11 Wastegate:—A valve that, when open, allows some of the exhaust gas to bypass the turbine rotor.

2.12 Clockwise and Counterclockwise Rotation:—Direction of shaft rotation when viewed looking into the compressor inlet.

2.13 Variable Geometry Turbocharger:—A turbocharger in which moving parts are used to alter the gas velocities and hence the performance of the turbine or the compressor, or both.

2.14 Turbocompounding:—A method of increasing the power or efficiency, or both, of an internal combustion engine by means of a turbine which converts exhaust gas energy into shaft power and delivers it to the engine.

2.15 Series Turbocharging:—An arrangement of two or more turbochargers with the compressors and turbines installed in series to increase the pressure and density of the fluid entering the engine.

2.16 Actuator:—A device incorporated into a turbocharger assembly which controls the movement of the variable geometry component(s) or wastegate.

2.17 Bearing Housing:—The housing that encloses and supports the bearing(s) and seals and makes provisions for lubrication and cooling.

3. Performance Terminology

3.1 Fixed Geometry Compressor

$$\text{Compressor pressure ratio} = \frac{\text{Outlet air static absolute pressure (kPa)}}{\text{Inlet air total absolute pressure (kPa)}}$$

Compressor air mass flow = kg/s of air mass flow through the compressor

$$\text{Corrected compressor air mass flow} = \frac{\text{Compressor air mass flow} \times \sqrt{\frac{\text{Compressor inlet total absolute temperature (K)}}{298 \text{ K}}}}{\text{Compressor inlet total absolute pressure (kPa)}/100 \text{ kPa}}$$

$$\text{Corrected compressor speed} = \frac{\text{Compressor impeller speed (rpm)}}{\sqrt{\frac{\text{Compressor inlet total absolute temperature (K)}}{298 \text{ K}}}}$$

$$\text{Compressor efficiency (\%)} = \frac{\text{Isentropic enthalpy rise across compressor stage through compressor pressure ratio}}{\text{Actual enthalpy rise across compressor stage}}$$

Surge is indicated by a line on the left-hand side of a compressor graph as determined on a steady flow test stand. Surge is severe flow reversal combined with audible coughing and banging. The onset of surge may vary from one installation to another and the graph surge line should only be used as a guide.

3.2 Variable Geometry Compressor—Terminology the same as for fixed geometry compressor.

3.3 Fixed Geometry Turbine

$$\text{Turbine expansion ratio} = \frac{\text{Inlet gas total absolute pressure (kPa)}}{\text{Outlet gas static absolute pressure (kPa)}}$$

Turbine gas flow = kg/s of gas flow through the turbine

$$\text{Turbine gas flow parameter} = \frac{\text{Turbine gas flow} \times \sqrt{\text{Turbine inlet total absolute temperature (K)}}}{\text{Turbine inlet total absolute pressure (kPa)}}$$

$$\text{Turbine speed parameter} = \frac{\text{Turbine rotor speed (rpm)}}{\sqrt{\text{Turbine inlet total absolute temperature (K)}}}$$

$$\text{Combined turbine} \times \text{mechanical efficiency (\%)} = \frac{\text{Actual total enthalpy rise across compressor stage}}{\text{Isentropic total enthalpy drop across turbine stage through turbine expansion ratio}}$$

If users require the use of total-to-total turbine × mechanical efficiencies and expansion ratios, the user and the turbocharger manufacturer should agree upon a method of obtaining, whether it be through calculation (knowing turbine exit state conditions and geometry) or through gas stand data.

3.4 Variable Geometry Turbine—Terminology same as fixed geometry turbine.

3.5 Bearing Mechanical Losses—Bearing power loss = kW at a given turbocharger shaft speed and expansion ratio.

4. Performance Graph Format

4.1 Fixed Geometry Compressor (See FIG. 1)

4.2 Variable Geometry Compressor—The performance graph format would be the same as for the fixed geometry compressor. Un-

```
INLET DIA. _____ (mm)*     * = AT PRESSURE
OUTLET DIA. _____ (mm)*         MEASURING STATIONS
INLET TYPE: _____
OUTLET TYPE: _____
IMPELLER INERTIA: _____ (N-m-sec²)
```

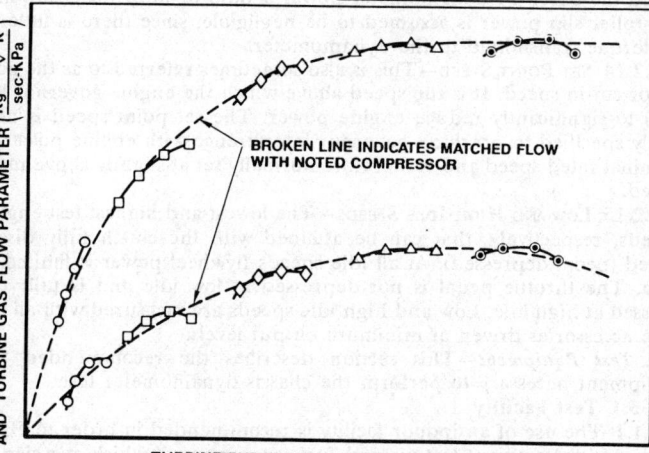

FIG. 1—TYPICAL TURBOCHARGER COMPRESSOR PERFORMANCE GRAPH

```
COMPRESSOR DESCRIPTION: _____
TURBINE HOUSING TYPE: _____
IF LIQUID COOLED:
    LIQUID DESCRIPTION: _____
    LIQUID TEMP. INTO HSG: _____ (°C)
    LIQUID PRESS. INTO HSG: _____ (kPa)
    DISCHARGE CONNECTION DESCRIPTION: _____
T₁T/T₁C = _____ OR T₁T _____ (°C)  T₁T = 600 ± 20°C FOR DIESEL ENG.
                                         T₁T = 900 ± 20°C FOR S.I. ENG.
OIL TYPE: _____       T₁C = COMPRESSOR INLET TEMP. (°C)
OIL SUPPLY TEMP. = _____ (°C) 100°C RECOMMENDED
ROTOR AND SHAFT INERTIA = _____ (N-m-sec²)
D₁/D₂ = _____ } TURBOCHARGER
L/D₂  = _____   REF. TEST CODE
```

FIG. 2 — TYPICAL TURBOCHARGER TURBINE PERFORMANCE GRAPH

less otherwise agreed to by the user and the turbocharger manufacturer, three graphs should typically be supplied as follows:
   a) Optimized for minimum surge.
   b) Optimized for peak efficiency.
   c) Optimized for maximum flow.

4.3 Fixed Geometry Turbine (See FIG. 2)

4.4 Variable Geometry Turbine—The performance graph format would be the same as for the fixed geometry turbine. The number of graphs to be supplied to the user to adequately describe the performance range should be agreed to with the turbocharger manufacturer. Three graphs should typically be supplied as follows:
   a) Optimized for minimum flow.
   b) Optimized for mid-point flow.
   c) Optimized for maximum flow.

4.5 Wastegate—If users require the use of wastegate data (for example, flow versus lift and lift versus pressure), the user and the manufacturer should agree on a format. (See FIG. 3)

4.6 Oil Flow (See FIG. 4)

4.7 Liquid Flow—If users require the use of liquid flow data (for example, pressure drop, temperature rise, surface temperature, etc.) for cooled bearing or turbine housings, the user and the manufacturer should agree upon a format.

5. References—
SAE J916 JUN82.
SAE J1349 JUN83.
"Principles of Turbomachinery," D. G. Shepherd, Macmillan, 1956
"Thermodynamics of Turbomachinery," S. L. Dixon, Pergamon, 1978

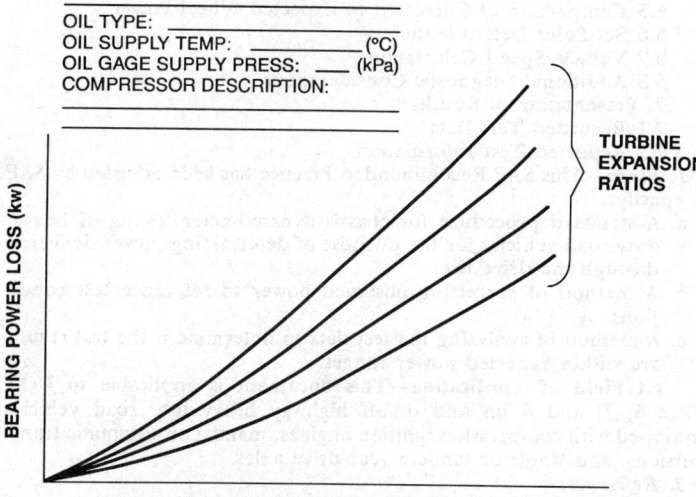

FIG. 3 — BEARING POWER LOSS GRAPH

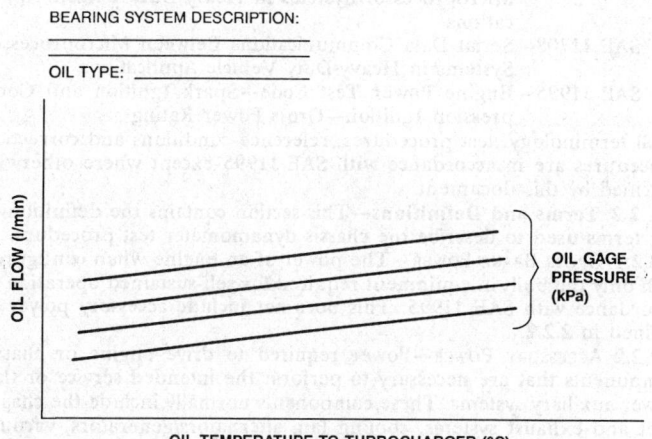

FIG. 4 — TURBOCHARGER OIL FLOW GRAPH

CHASSIS DYNAMOMETER TEST PROCEDURE— HEAVY-DUTY ROAD VEHICLES— SAE J2177 APR92

SAE Recommended Practice

Report of the SAE Power Test Code Standards Committee 4 approved April 1992.

TABLE OF CONTENTS

1. Scope
   1.1 Field of Application
2. References
   2.1 Applicable Documents
   2.2 Terms and Definitions
3. Test Equipment
   3.1 Test Facility
   3.2 Chassis Dynamometer
   3.3 Instrumentation
4. Test Procedure
   4.1 Vehicle Preparation
   4.2 Operating Conditions
   4.3 Test Measurements
   4.4 Instrumentation and Equipment Removal
5. Data Correction
   5.1 Reference Test Conditions
   5.2 Power Corrections
   5.3 Correction Factor Calculations
6. Data Analysis
   6.1 Additional Units and Symbols
   6.2 Determination of Observed Wheel Power
   6.3 Calculation of Corrected Wheel Power
   6.4 Calculation of Expected Wheel Power
   6.5 Comparison of Corrected to Expected Wheel Power
   6.6 Set Point Determination
   6.7 Vehicle Speed Calculation
   6.8 Additional Diagnostic Considerations
7. Presentation of Results
   7.1 Recorded Test Data
   7.2 Reported Test Information

1. Scope—This SAE Recommended Practice has been adopted by SAE to specify:
   a. A standard procedure for chassis dynamometer testing of heavy-duty road vehicles for the purpose of determining power delivered through the drive tires.
   b. A method of correcting observed power to reference test conditions.
   c. A method of analyzing the test data to determine if the test results are within expected power ranges.

1.1 Field of Application—This document is applicable to DOT Class 6, 7, and 8 on and on/off highway heavy-duty road vehicles equipped with compression ignition engines, manual or automatic transmissions, and single or tandem rear drive axles.

2. References

2.1 Applicable Documents—The following publications form a part of this specification to the extent specified herein. The latest issue of SAE publications shall apply.

SAE J1587—Joint SAE/TMC Electronic Data Interchange Between Microprocessor Systems in Heavy-Duty Vehicle Applications

SAE J1708—Serial Data Communications Between Microprocessor Systems in Heavy-Duty Vehicle Applications

SAE J1995—Engine Power Test Code—Spark Ignition and Compression Ignition—Gross Power Rating

All terminology, test procedures, reference conditions and correction procedures are in accordance with SAE J1995 except where otherwise specified by this document.

2.2 Terms and Definitions—This section contains the definition of key terms used to describe the chassis dynamometer test procedure.

2.2.1 GROSS BRAKE POWER—The power of an engine when configured with only the built-in equipment required for self-sustained operation in accordance with SAE J1995. This does not include accessory power as defined in 2.2.2.

2.2.2 ACCESSORY POWER—Power required to drive engine or chassis components that are necessary to perform the intended service or that power auxiliary systems. These components normally include the chassis inlet and exhaust systems, cooling fan, alternators/generators, vacuum pumps, power steering pumps, and air and freon compressors, operating as described in Sections 3 and 4. All engine or chassis parasitic power losses incurred during the chassis dynamometer test that are not included in the gross brake power shall for the purpose of this procedure be considered accessory power.

2.2.3 FLYWHEEL POWER—The power of an engine as installed in the chassis and measured at the flywheel. Flywheel power is the net difference between gross brake power and accessory power.

2.2.4 DRIVE TRAIN EFFICIENCY—The ratio of the sum total of the power delivered to all drive wheels to the flywheel power. This includes power losses generated by the transmission, driveshaft, carriers, and axles. It does not include tire rolling resistance losses. Drive train efficiency may vary with the speed of the vehicle as well as the amount of torque transmitted. It is normally supplied by the vehicle component manufacturer.

2.2.5 ROLLING RESISTANCE POWER—The power required to overcome the rolling resistance of all drive tires. This power normally varies with the type of tires, tread pattern, inflation pressure, vehicle weight on the drive tires, vehicle speed, and/or dynamometer roller surface.

2.2.6 TIRE-ROLLER SLIP POWER—The net difference between the power delivered through the drive tires and that transmitted to the dynamometer rollers. This power loss is due to tire to roller relative slip and is normally dissipated as heat. Tire-roller slip power can be significant and may vary with the type of tire, tread pattern, inflation pressure, roller surface, roller diameter, as well as the test speed and torque application.

2.2.7 OBSERVED DYNAMOMETER POWER—The observed vehicle power with no corrections made as measured by the chassis dynamometer using the equipment and procedures defined in Sections 3 and 4.

2.2.8 DYNAMOMETER MACHINE LOSSES—The net difference between the power delivered to the dynamometer rollers and that measured by the dynamometer load cell. These include windage or bearing power losses as well as any unaccounted for load cell calibration error.

2.2.9 OBSERVED WHEEL POWER—The observed dynamometer power adjusted only for dynamometer machine losses, if any. If these losses are not known and cannot be accounted for, then observed wheel power is assumed equivalent to observed dynamometer power.

2.2.10 CORRECTED WHEEL POWER—The observed wheel power corrected to reference test conditions.

2.2.11 EXPECTED WHEEL POWER—The vehicle power that would normally be expected to be transmitted to the dynamometer if the test were run at reference test conditions. This is calculated by subtracting estimated accessory, drive train, and rolling resistance power losses from the engine gross brake power.

2.2.12 REFERENCE TEST CONDITIONS—The standard or reference test conditions to which all power corrections are made per 5.1. These corrections are made to reference engine inlet air and fuel conditions as well as to account for tire-roller slip.

2.2.13 REFERENCE TEST SPEED—The highest engine speed that can be attained in direct drive gear with the throttle fully depressed and the dynamometer rollers turning, but fully unloaded. At the reference test speed the observed dynamometer power should be at or near zero and tire-roller slip power is assumed to be negligible, since there is little or no torque transmitted to the dynamometer.

2.2.14 SET POINT SPEED—This is also sometimes referred to as the governor cut-in speed. It is the speed above which the engine governor begins to significantly reduce engine power. The set point speed is normally specified to preclude governor interference with engine power at nominal rated speed and is therefore normally set at slightly above rated speed.

2.2.15 LOW AND HIGH IDLE SPEEDS—The lowest and highest test engine speeds, respectively, that can be attained with the clutch fully disengaged (pedal depressed). At all idle speeds flywheel power is limited to zero. The throttle pedal is not depressed at low idle and is fully depressed at high idle. Low and high idle speeds are measured with all engine accessories driven at minimum output levels.

3. Test Equipment—This section describes the recommended test equipment necessary to perform the chassis dynamometer test.

3.1 Test Facility

3.1.1 The use of an indoor facility is recommended in order to effectively control ambient test air and fuel temperatures, which can significantly affect vehicle power. The test facility should be able to control ambient air temperature to within 25 °C ± 15 °C.

3.1.2 The test facility should be equipped with a fresh air ventilation system of sufficient capacity to maintain an adequate air supply to the test vehicle. Ambient air pressure must be maintained at a constant level in the test facility throughout the chassis dynamometer test.

3.1.3 The facility should be equipped with an exhaust ventilator system that permits engine exhaust gas to be removed directly from the test room. This ventilator tube is normally vented to atmosphere, or to a separate reservoir, either of which must have a total pressure of within 0.75 kPa of the ambient air pressure in the test facility.

3.1.4 The test facility should be equipped with safety chains and mounts for the test vehicle rear axle as well as tie-down straps for the front wheels. Chocks are acceptable for the front wheels only if the drive tires cradle or are cradled by the dynamometer rollers. The safety chain mounts should be positioned so as to minimize the downward force on the test vehicle when the chains are installed, resulting in an undesirable increase in rolling resistance. Ideally the safety chain mounts should allow the chains to be positioned as close to horizontal as possible, provided they prevent lateral movement of the test vehicle during operation.

3.1.5 For vehicles having engines equipped with chassis mounted air-to-air charge air cooling systems, it is recommended that a portable auxiliary cooling fan be available. This cooling fan should be placed directly in front of the radiator on any chassis dyno tests where the inlet manifold air temperature exceeds the recommended test limits as described in 4.2.6.

3.1.6 The test facility must also be equipped with other safety equipment as required by the dynamometer manufacturer, facility owner, or appropriate federal, state, and local codes. This would normally include, but not be limited to, a fire extinguisher, safety glasses, ear protection, etc.

3.2 Chassis Dynamometer

3.2.1 A computer-based dynamometer automatic control and data acquisition system is recommended, but not required. This type of dynamometer can be used to automatically control test cycles and acquire, analyze, plot, and report data. Such dynamometer control systems can generally improve test uniformity, repeatability, and overall accuracy levels as well as reduce operator error and fatigue.

If the test dynamometer does not have the capability to analyze the test results in accordance with Sections 5 and 6, it is recommended that a PC-based computer program be used to analyze the test results. Refer to the engine or vehicle manufacturer for this program. In such cases observed test data would manually be entered into the program in order to calculate corrected and expected wheel power levels.

3.2.2 The chassis dynamometer should be designated by the manufacturer for use in this type of testing. The dynamometer must be of the non-portable variety designed to measure whole vehicle horsepower at the drive tires on single and tandem axle heavy-duty road vehicles. The expected wheel power should not exceed the dynamometer manufacturer's stated power absorption capabilities.

3.2.3 On tandem drive axle vehicles, the chassis dynamometer must be capable of adjustment as required to maintain proper tire-roller contact on all drive tires of the test vehicle.

3.2.4 Whenever possible, large diameter dynamometer rollers are recommended in order to increase tire-roller contact area and reduce the possibility of large tire-roller slippage power losses and tire overheating. A roller diameter of 0.9 m (36 in) or greater is recommended, although a minimum roller size of 0.5 m (20 in) diameter is permitted. On any test where excessive tire-roller slippage or tire overheating is apparent as described in 4.2.8, a set of auxiliary or "slave" test drive tires is also recommended.

3.2.5 The test dynamometer should be calibrated by a method approved by the manufacturer at periodic intervals. On many dynamometers this is a dead weight load cell calibration check only. In such cases the calibration period is not to exceed every 6 months. Prior to conducting each chassis dynamometer test, check to insure that this calibration period has not expired.

The dynamometer calibration should also include accounting for bearing, windage, or other machine losses. On certain dynamometers these losses can be measured and automatically included in the observed dynamometer power reading. In such cases observed dynamometer power will be equal to observed wheel power. If these losses are not automatically included in the test dynamometer calibration, then they should be accounted for in accordance with 6.2.

3.3 Instrumentation

3.3.1 The following minimum test measurements and instrumentation accuracy is recommended for the chassis dynamometer test at each test point:
 a. Engine Speed — ±0.2% of measured value
 b. Dynamometer Speed — ±0.2% of measured value
 c. Observed Dynamometer Power — ±1.0% of measured value
 d. Fuel Flow — ±1.0% of measured value
 e. Inlet Fuel Temperature — ±2.0 °C
 f. Inlet Air Temperature — ±2.0 °C
 g. Inlet Manifold Air Temperature — ±2.0 °C
 h. Inlet Fuel Pressure — ±2 kPa
 i. Inlet Manifold Pressure — ±2 kPa

3.3.2 On vehicles equipped with tandem drive axles, where each axle drives a separate set of rollers (either single or paired), dynamometer speed measurements must be obtained for both axle-roller sets. On dynamometers so equipped, observed dynamometer power should also be recorded for each set of rollers. Observed dynamometer power is then corrected individually for each drive axle in accordance with 6.3.

3.3.3 The following measurements and instrumentation accuracy should be taken only once at some convenient point immediately before, during, or after the dynamometer test:
 a. Fuel Density — ±0.001 kg/ℓ (0.2 API)
 b. Fuel Viscosity (optional) — ±0.1 mm²/s
 c. Barometric Pressure — ±0.1 kPa
 d. Vapor Pressure (optional) — ±0.1 kPa

3.3.4 Fuel viscosity and ambient air vapor pressure measurements are optional. If measured, they will improve the accuracy of the correction factors used in Section 5 to calculate corrected wheel power. Note, however, that fuel viscosity is only critical on pump-line-nozzle (P-L-N) injection systems; therefore it is not necessary to measure on test engines equipped with unit injectors.

3.3.5 The use of the standard vehicle tachometer to obtain engine speed is generally not recommended because the accuracy level required per 3.3.1 cannot normally be obtained. Engine and dynamometer speed values must be extremely accurate since they are used to calculate tire-roller slip power. A digital tachometer is recommended to measure engine speed and should be part of the basic facility instrumentation. Since the dynamometer test is conducted in direct gear, the speed pickup may be placed at any convenient engine or driveshaft location.

3.3.6 The dynamometer or facility fuel flow measurement system should be capable of measuring fuel density with the required accuracy and correcting observed values to the 15 °C ASTM reference standard. If not, a hydrometer having a minimum 0.825 to 0.875 kg/ℓ (30 to 40 API) range and the previously referenced accuracy should be used. The temperature of the fuel sample must also be recorded with a thermometer accurate to within ±1 °C in order to correct the observed hydrometer reading to the reference temperature.

3.3.7 Kinematic viscosity should be measured with a viscometer having a minimum 2.0 to 3.2 mm²/s range and the required accuracy. If the viscometer or viscosity measurement system does not correct the observed sample readings to the 40 °C ASTM reference temperature standard, then a thermometer having ±1 °C accuracy should be used for this purpose.

3.3.8 Inlet air temperature shall be measured in the center of the airstream at a location downstream of the air cleaner and as close as possible to the entrance to the turbocharger or inlet manifold on naturally aspirated engines.

3.3.9 Inlet manifold air temperature and pressure shall be measured as static values in a section of the manifold which is common to several cylinders. In such cases, dynamic pressure is assumed to be negligible.

3.3.10 Inlet fuel temperature and pressure taps shall be located immediately prior to the entrance to the high pressure injection pump on engines equipped with pump-line-nozzle injection systems and in the high pressure common rail on engines equipped with unit injectors. The inlet fuel pressure is a static reading.

3.3.11 When testing electronically controlled engines, any or all of the required measurements may be obtained directly through the vehicle's serial communications port, so long as the information is available, sensor locations meet the previous specifications, and the required accuracies can be maintained. In such cases it is recommended that the data transmission be in accordance with SAE standards J1708 and J1587. A compatible connector and data acquisition and display system is also required, either as part of the dynamometer control system or as separate instrumentation.

4. Test Procedure—This section details the basic chassis dynamometer test procedure.

4.1 Vehicle Preparation

4.1.1 Check to ensure that the vehicle is in a fully operational condition. This includes checking all engine, transmission, and carrier fluid levels. Also check to insure all wheel lugs are tight and that all drive tires are the same nominal diameter and have been set to the tire or vehicle manufacturer's recommended inflation pressure. Remove stones, salt pellets, or any other debris embedded in the tire tread.

4.1.2 Connect the rear axle safety chain(s) as described in 3.1.4 to the rear axle, frame, or frame cross member. Install the front wheel tie-down straps/chocks. Remove the chain slack per the dynamometer manufacturer's recommended procedure.

CAUTION—The chain(s) should not be tensioned and should not exert a downward force on the drive tires.

4.1.3 Lock the engine cooling fan in the fully "on" or engaged position. The required method will depend upon the type of cooling fan used on the test vehicle. Most vehicles are equipped with one of the following fan drives:

a. If equipped with a direct coupled fan, it should automatically be in the "on" position.
b. If equipped with an air pressure activated fan drive, engage the fan by removing or supplying air as required.
c. On viscous fan equipped engines a locking method approved by the manufacturer must be used. This normally requires either use of an external locking collar tool or removal of the temperature sensing element.
d. If equipped with an electric fan, the fan should be switched to the "on" position.

4.1.4 Install the digital tachometer as described in 3.3.5. Install the inlet air, manifold and fuel temperature, and pressure instrumentation as described in 3.3.8, 3.3.9, and 3.3.10.

CAUTION—Use extreme care in making all fuel oil instrumentation connections in order to minimize fire hazards. All fuel hoses should be routed away from the exhaust system or any moving parts. High temperature hose material and/or hose insulation with a noncombustible material is also recommended.

4.1.5 Install the necessary fuel flow measurement equipment as required by the dynamometer, facility, or engine manufacturer.

4.1.6 Turn off all vehicle electrical devices, including lights, wipers, air conditioner, etc. All power take-offs (PTO) should be disengaged. If applicable, the fuel heater should also be turned off. On vehicles equipped with tandem drive axles, the power divider should be run in accordance with the dynamometer and vehicle manufacturer's recommended procedures. If specific instructions are not available, the power divider is normally unlocked. Disconnect ABS or ATC on vehicles so equipped.

4.2 Operating Conditions

4.2.1 For safety considerations, all nonessential personnel should leave the dynamometer test cell prior to starting the vehicle and/or dynamometer.

4.2.2 Start and fully warm the vehicle and dynamometer. The recommended warm-up procedure is to idle the engine at 1200 rpm for a period of not less than 5 min. The dynamometer is then engaged and engine load is gradually increased until full load is achieved and the engine and drive train have come fully up to normal operating temperatures as defined by the engine or vehicle manufacturer.

4.2.3 Following warm-up, all engine accessories except the cooling fan should be operating at minimum power output levels. Brake system air pressure should be fully up to normal operating levels.

4.2.4 At no time during the chassis dynamometer testing should vehicle coolant, lubricant, or exhaust gas temperatures exceed the engine manufacturer's published limits.

4.2.5 At no time during the chassis dynamometer testing should engine inlet air temperature exceed the range limitations for ambient air of 25 °C ± 15 °C as defined in 3.1.1.

4.2.6 On test vehicles with air-to-air chassis mounted charge air cooling, if the inlet manifold temperature exceeds the manufacturer's stated limits or if it exceeds the manufacturer's nominal test temperature by more than 15 °C, it is recommended that an auxiliary cooling fan be installed per 3.1.5.

4.2.7 With either manual or automatic transmissions, place or lock the transmission in direct drive. This will not necessarily be top gear since many transmissions are overdrive units. Failure to place the transmission in direct drive can affect test results due to reduced drive train efficiency as well as rolling resistance power effects.

On certain automatic transmission equipped vehicles where the torque converter cannot be locked, the test should be run as close to direct drive as is possible. In such cases driveshaft speed should be measured and used in lieu of engine speed for the slip correction factor calculations per 6.3.3.

4.2.8 After warm-up and periodically during test point stabilization and data acquisition, check to insure the drive tires are not overheating. Methods to diagnose overheating include tires that are abnormally hot to the touch, excessive tire to roller rubber displacement or excessive tire-roller slip per 5.2.2. If overheating is apparent the test should be terminated and a set of auxiliary test drive tires should be installed.

On tandem drive axle vehicles where each axle drives a separate set of rollers, check to insure that the dynamometer speed or load differential between the roller sets does not exceed 10%. Speed or load differentials in excess of this amount can cause interaxle power divider damage. It is normally the result of excessive tire-roller slip, but may also indicate improperly matched axle ratios, a malfunctioning power divider, or other problems.

4.2.9 If excessive tire-roller slip is noted during the dynamometer test at a given engine rpm, and this slip can be significantly reduced by changing vehicle speed, then it is permissible to conduct the test in a gear other than direct. However, if this is done the effect on drive train efficiency and rolling resistance must be accounted for when calculating expected wheel power, and the deviation from standard test practices must be noted in accordance with 7.2.

4.3 Test Measurements

4.3.1 Each test point shall be fully stabilized prior to data acquisition. A test point will be considered fully stabilized when observed dynamometer power and dynamometer and engine speed have been maintained within ±1% during a 1 min period. In order to meet this criterion a total stabilization time at each test point of 2 min or more would normally be expected. This stabilization time can be minimized on computer-controlled dynamometers that have been programmed to determine when each test point is fully stabilized. Each test speed shall not deviate from the nominal target engine speed by more than ±1%.

4.3.2 Test measurements as defined in 3.3.1 should be obtained at each target test speed with the throttle pedal fully depressed. The following test speeds are recommended as determined by the engine manufacturer:

a. Reference test speed
b. Set point speed
c. Rated or governed speed
d. Intermediate speeds
e. Peak torque speed
f. Optional speeds

Intermediate speeds are generally defined by the engine manufacturer, but if not defined should be obtained at approximately 200 rpm increments from rated speed to peak torque. Optional speeds are any other speeds as defined by the engine manufacturer necessary to adequately define the shape of the power curve.

4.3.3 On vehicles equipped with road speed limiting controls, it may not be possible to obtain the required test engine speeds, particularly if direct gear is the top gear. In such cases it is recommended that the road speed controls be deactivated or reset in order to conduct the chassis dynamometer test. If this is not possible, then it may be necessary to conduct the test in a lower gear than direct. In such cases the deviation from standard test practices should be noted in accordance with 7.2.

4.3.4 During the chassis dynamometer test the set point speed must be determined. This is the test speed above which observed dynamometer power begins to rapidly drop off. Normally the engine speed range immediately above rated speed (0 to 100 rpm) must be traversed in order to determine the set point speed. The rate of power drop-off used to determine the set point speed should be supplied by the engine or vehicle manufacturer.

4.3.5 All test data required as well as additional data as defined in Section 7 should be recorded on a test log sheet or otherwise appropriate format.

4.3.6 After the running dynamometer portion of the testing has been completed, disengage the engine fan (if applicable). With the transmission clutch disengaged (pedal depressed) and the engine still fully warmed and the cooling fan operating at minimum output levels, record the low idle (no throttle) and high idle (full throttle) speeds.

4.3.7 If fuel density or viscosity analysis is not automatically obtained by the fuel measurement system, extract a sample of fuel from the vehicle fuel tanks. Measure and record the sample fuel density, temperature, and viscosity (P-L-N systems only) in accordance with 3.3.6 and 3.3.7. Correct fuel density and viscosity to the reference temperatures using ASTM reference charts.

4.3.8 Unless automatically obtained by the dynamometer or test facility, record ambient air (barometric) pressure. If available, also record ambient air vapor pressure.

4.4 Instrumentation and Equipment Removal—After the chassis dynamometer test has been completed, remove all instrumentation that was required for the test. Disconnect the safety chains and remove the wheel tie-down straps or chocks. Disconnect the exhaust ventilator. Reinstall the original drive tires, if applicable.

5. Data Correction—This section contains the general formula for correcting observed wheel power to reference test conditions.

5.1 Reference Test Conditions—In order for the chassis dynamometer test results to be meaningful, observed wheel power must be corrected to account for variations in ambient test air and fuel conditions as well as power losses due to tire-roller slippage.

5.1.1 The reference test condition for tire-roller slip is 0% slippage. Since tire-roller slip exists in nearly all chassis dynamometer tests, for each test point the power lost due to slippage must be calculated and added to the observed wheel power.

5.1.2 The following are reference test conditions for ambient fuel and air to which observed wheel power corrections are made:
 a. Dry ambient air supply pressure (absolute)—99 kPa
 b. Engine inlet air supply temperature—25 °C
 c. Engine inlet manifold temperature—Manufacturer's specification
 d. Fuel density @ 15 °C—0.850 kg/ℓ
 e. Fuel kinematic viscosity @ 40 °C—2.6 mm²/s
 f. Fuel temperature—40 °C

5.2 Power Corrections

5.2.1 On any vehicle where engine power output is automatically adjusted to compensate for changes induced by one or more of the listed inlet air or fuel supply conditions, no power correction for that test parameter shall be made.

5.2.2 The magnitude of the power corrections shall not exceed 5% for the air or fuel correction factor or 10% for the slip correction factor. If any correction factor exceeds these values, it shall be noted in accordance with 7.2.

5.3 Correction Factor Calculations

5.3.1 UNITS AND SYMBOLS—Units and symbols used for correction factor calculations are shown in Table 1.

TABLE 1—UNITS AND SYMBOLS

Symbol	Term	Units
CFS	Slip Correction Factor	Nondimensional
CFA	Air Correction Factor	Nondimensional
CFF	Fuel Correction Factor	Nondimensional
R	Speed Ratio	See 5.3.3
fa	Atmosphere Factor	Nondimensional
fm	Engine Factor	Nondimensional
fd	Fuel Density Factor	Nondimensional
fv	Fuel Viscosity Factor	Nondimensional
ft	Fuel Temperature Factor	Nondimensional
α	Ambient Pressure Exponent	Nondimensional
β	Inlet Air Temperature Exponent	Nondimensional
ε	Manifold Temperature Exponent	Nondimensional
S	Viscosity Sensitivity Coefficient	Nondimensional
D	Engine Displacement	ℓ
B	Barometric Pressure	kPa
t	Inlet Air Supply Temperature	°C
m	Inlet Manifold Temperature	°C
f	Inlet Fuel Temperature	°C
P	Inlet Manifold Absolute Pressure	kPa
r	Pressure Ratio	Nondimensional
q	Fuel Delivery Parameter	m/ℓ cycle
n	Engine Speed	min⁻¹
N	Dynamometer Speed	As specified
F	Fuel Flow	g/s
SG	Fuel Density @ 15 °C	kg/ℓ
V	Fuel Viscosity @ 40 °C (optional)	mm²/s

5.3.2 SUBSCRIPTS
 o Refers to data observed or measured at actual test conditions
 c Refers to data measured and corrected to reference test conditions
 d Refers to the dry portion of the barometric pressure (subtract the vapor pressure from the observed barometric pressure to obtain the observed dry barometric)
 r Refers to reference test conditions per 5.1
 ref Refers to the reference test speed per 2.2.13
 e Refers to the expected or nominal wheel power levels

5.3.3 CALCULATION OF SLIP CORRECTION FACTOR—A correction factor for tire-roller slippage power losses can be determined by comparing the engine to dynamometer speed ratio at each test point to the reference test speed ratio. The reference ratio is obtained at the maximum attainable engine speed with the dynamometer engaged but completely unloaded per 2.2.13. At this reference point tire-roller slip is assumed negligible since there is no torque transmittal. A correction factor for tire-roller slip must be calculated for each test point (R_{ref} is calculated only once for each roller set) and for each roller set on tandem drive axle vehicles, using Equation 1:

$$CFS = \frac{R}{R_{ref}} \qquad (Eq.\ 1)$$

where:

$$R = \frac{n}{N} \qquad (Eq.\ 2)$$

$$R_{ref} = \frac{n_{ref}}{N_{ref}} \qquad (Eq.\ 3)$$

Depending upon the units on the dynamometer readout, the speed ratio, R, may not be a dimensionless quantity. However, for the purpose of these calculations it may be treated as dimensionless so long as R_{ref} is calculated using the same units.

5.3.4 CALCULATION OF AIR CORRECTION FACTOR—The air correction factor corrects observed wheel power to that which would be obtained at reference test air supply and inlet manifold conditions as described in 5.1.

$$CFA = (fa)^{fm} \qquad (Eq.\ 4)$$

where:

$$fa = \left(\frac{B_{dr}}{B_{do}}\right)^{\alpha} \left(\frac{t_0 + 273}{t_r + 273}\right)^{\beta} \left(\frac{m_0}{m_r}\right)^{\varepsilon} \qquad (Eq.\ 5)$$

or:

$$fa = \left(\frac{99}{B_{do}}\right)^{\alpha} \left(\frac{t_0 + 273}{298}\right)^{\beta} \left(\frac{m_0}{m_r}\right)^{\varepsilon} \qquad (Eq.\ 6)$$

and values of α, β, and ε are listed in Table 2.

TABLE 2—VALUES OF α, β, AND ε

Pressure Charging System	Charge Air Cooling System	α	β	ε
Naturally aspirated	None	1.0	0.7	0
Turbocharged	None	0.7	1.2	0
Turbocharged	Jacket Water	0.7	0.7	0
Turbocharged	Air-to-Air	0.7	0.4	0.6

The value of fm is given as shown in Table 3.

TABLE 3—VALUE OF fm

q/r	fm
Less than 37.2	0.2
Between 37.2 and 65	0.036 (FA) − 1.14
More than 65	1.2

where for 4-stroke engines:

$$q = \frac{120\ 000\ (F_0)}{D(n)} \qquad (Eq.\ 7)$$

and:

$$r = P_0/B_0 \quad (r = 1 \text{ if naturally aspirated}) \qquad (Eq.\ 8)$$

NOTE—For 2-stroke engines the value of q/r as previously stated must be reduced by a factor of two.

NOTE—If water vapor pressure was not measured, it shall be assumed 1 kPa. In such cases $B_{do} = B_0 - 1$.

NOTE—The previous air correction factor calculations are in accordance with SAE J1995, except that the manifold temperature term has been added for air-to-air charge air cooled engines. This allows the inlet air and manifold air temperature terms to be corrected independently when conducting the dynamometer test.

5.3.5 CALCULATION OF FUEL CORRECTION FACTOR—The fuel correction factor corrects observed wheel power to that which would be obtained at reference fuel density, viscosity, and temperature levels as described in 5.1. Observed fuel density levels must first be corrected to 15 °C. Observed fuel viscosity levels must first be corrected to 40 °C.

$$CFF = fd\ (fv)\ ft \qquad (Eq.\ 9)$$

where:

$$fd = 1 + 0.70\ \frac{SG_r - SG_0}{SG_0} = 1 + 0.70\ \frac{0.850 - SG_0}{SG_0} \qquad (Eq.\ 10)$$

and:

$$fv = \frac{1 + S/V_0}{1 + S/V_r} = \frac{1 + S/V_0}{1 + S/2.6} \qquad (Eq.\ 11)$$

and:

$$ft = 1 + 0.00079\ (f_0\text{-}f_r) + 0.0073\ (S)\ (f_0\text{-}f_r)$$
$$= 1 + 0.00079\ (f_0\text{-}40) + 0.0073\ (S)\ (f_0\text{-}40) \qquad (Eq.\ 12)$$

Values of S shall be determined by the engine manufacturer. If no values are available, the following shall be used:
 Pump-line-nozzle systems (P-L-N) - S = 0.15
 Unit injector systems - S = 0.0

NOTE—For the purpose of correcting observed mass fuel flow rates to reference test conditions, Equation 13 should be used:

$$F_c = F_o ((SGr/SG_o) \, fv \, (ft)) \quad \text{(Eq. 13)}$$

NOTE—The previous fuel correction factor calculations are in accordance with SAE J1995 except that the fuel temperature factor (ft) has been added. This is because J1995 requires fuel inlet temperature to be regulated to 40 °C, whereas in a chassis dynamometer test such regulation would not be practical. This correction accounts for both the effect of temperature on fuel density and the effect of temperature on fuel viscosity on P-L-N systems.

6. Data Analysis—This section contains the methods for interpreting the chassis dynamometer test results.

6.1 The additional units and symbols shown in Table 4 are needed to analyze the test results, using definitions as described in 2.2.

TABLE 4 — ADDITIONAL UNITS AND SYMBOLS

Symbol	Term	Units
dp	Dynamometer Power	kW
DML	Dynamometer Machine Losses	kW
wp	Wheel Power	kW
bp	Engine Gross Brake Power	kW
ap	Accessory Power	kW
rp	Rolling Resistance Power	kW
η	Drive Train Efficiency	Nondimensional

6.2 Determination of Observed Wheel Power—The observed dynamometer power must first be adjusted to account for dynamometer machine losses, if any. On certain electric dynamometers these losses may be accounted for automatically and included in the observed dynamometer power. These losses should be accounted for on all chassis dynamometer testing in accordance with the dynamometer manufacturer's recommended test correction procedures. However, if this procedure, or the appropriate corrections, are not known, the correction shall be assumed negligible and in such cases observed dynamometer power will equal observed wheel power.

$$wp_o = dp_o + DML \quad \text{(Eq. 14)}$$

6.3 Calculation of Corrected Wheel Power—Following the chassis dynamometer test the observed wheel power must be corrected for ambient test conditions and tire-roller slippage as defined in Section 5.

$$wp_c = wp_o \, (CFS) \, (CFA) \, (CFF) + rp \, ((CFA) \, (CFF)-1) + ap \, (\eta) \, ((CFA)(CFF)-1) \quad \text{(Eq. 15)}$$

Equation 15 is recommended for general use, but if required can be further reduced with very little loss of accuracy to Equation 16:

$$wp_c = wp_o \, (CFS) \, (CFA) \, (CFF) \quad \text{(Eq. 16)}$$

On tandem drive axle vehicles where each axle drives a separate set of rollers, it is necessary to calculate the slip correction factor (CFS) at each test point for each roller set. Corrected wheel power is then calculated for each drive axle and summed to obtain the whole vehicle corrected wheel power per Equation 16. If observed dynamometer power is not measured independently for each roller set, consult the dynamometer manufacturer to obtain the recommended procedure for estimating these values. Otherwise, it must be assumed that power is split evenly between drive axles.

6.4 Calculation of Expected Wheel Power—Expected wheel power is the power that would be expected to be delivered to the drive tires for a given test vehicle. Expected power is based on the engine manufacturer's advertised gross power per SAE J1995 after subtracting power losses due to accessories, drive train efficiency and rolling resistance. Expected wheel power is given by Equation 17:

$$wp_e = \eta \, (bp - ap) - rp \quad \text{(Eq. 17)}$$

6.4.1 If it is supplied by the engine manufacturer or otherwise known, the actual engine power as tested on an engine-only dynamometer may be substituted for the advertised gross brake power.

6.4.2 Accessory power and drive train efficiency should be supplied by the vehicle manufacturer based on test conditions existing during the chassis dynamometer test. All engine accessories are normally operated at minimum power output levels except the cooling fan, which operates at full output per 4.1.3.

6.4.3 Rolling resistance power should be supplied by the tire manufacturer as a function of rotational speed and weight (per tire). In the absence of this data Equation 18 may be used:

$$rp = (Ro11 \, (KPH) + Ro12 \, (KPH)^2) \, W \times 10^{-5} \quad \text{(Eq. 18)}$$

where:
W is vehicle weight on the drive tires in kilograms (kg)
KPH is vehicle speed in kilometers per hour
Ro11 and Ro12 are tire constants defined as follows in Table 5:

NOTE—Vehicle weight on the drive tires should be supplied by the vehicle manufacturer. This is particularly important on vehicles where the "tractor" is rigidly connected to the vehicle body (fire trucks, dump trucks, garbage trucks, etc.) and the required weight may not commonly be known. If the weight on the drive tires is not known, it should be measured on an appropriate scale.

TABLE 5 — Ro11 AND Ro12

Tire Type	Ro11	Ro12
Bias	1.69	0.0066
Radial	0.73	0.0102

6.5 Comparison of Corrected to Expected Wheel Power—Under normal circumstances if the vehicle is operating properly and the chassis dynamometer test procedure has been properly followed, corrected wheel power (wp_c) should approximately equal expected wheel power (wp_e). It is not the intent of this document to address vehicle power anomalies. The allowable deviation between corrected and expected wheel power beyond which may indicate a vehicle power problem must be determined by the vehicle manufacturer or appropriate service personnel.

6.6 Set Point Determination—Check to insure the measured set point speed is within the manufacturer's published specifications. If it is not, check with the engine or vehicle manufacturer regarding proper adjustment procedures. Check to make sure engine low and high idle speeds are within the engine manufacturer's specifications.

6.7 Vehicle Speed Calculation—In addition to the speedometer reading, vehicle speed can also be calculated for each test point using Equation 19:

$$\text{Vehicle Speed} = CFS(N) \quad \text{(Eq. 19)}$$

It may be necessary to convert vehicle speed to the appropriate units in Equation 19, depending upon the units used for dynamometer speed output. However, in most cases this conversion factor is equal to one. On tandem drive axle vehicles where each axle drives a separate set of rollers, the vehicle speed must be calculated for each roller set at each test point, then averaged to obtain the correct vehicle speed.

6.8 Additional Diagnostic Considerations—In addition to comparing corrected to expected wheel power levels, several other parameters measured during the chassis dyno test can be independently used to diagnose vehicle or engine power problems.

6.8.1 Check to insure that inlet fuel pressure and corrected fuel flow are within the manufacturer's specifications.

6.8.2 Check to insure that inlet manifold pressures and temperatures measured during the chassis dynamometer test are within the engine manufacturer's specifications. If not, this could be an indication of other engine or charge air cooler problems.

6.8.3 Although the effects of inlet fuel temperature, density, and viscosity are accounted for in the corrected wheel power calculations, significant variation of these parameters from expected levels can be a source of over-the-road power complaints. Fuel density at 15 °C for Type 2-D diesel fuel would normally be expected to be within 0.825 to 0.875 kg (30 to 40 API), kinematic viscosity at 40 °C would normally range from 2.0 to 3.2 mm²/s. Fuel inlet temperature levels during the dynamometer test should be at reasonable or expected levels considering the ambient test conditions. If not, this could be an indication of chassis fuel system plumbing problems.

6.8.4 Inlet air temperature at the engine or turbocharger inlet should not significantly exceed ambient air temperature. If so, this could be an indication of a chassis air supply problem. Consult the engine or vehicle manufacturer regarding any specifications in this area.

7. Presentation of Results—This section contains recommended procedures for recording and reporting the chassis dynamometer test results.

7.1 Recorded Test Data

7.1.1 The following minimum test information should be recorded prior to conducting the dynamometer test:
a. Date of test
b. Name and location of test facility
c. Dynamometer manufacturer and model
d. Vehicle model and serial number
e. Engine and transmission model
f. Nominal rated engine power and rated speed
g. Axle ratio and number of drive axles
h. Tire manufacturer, model, designation size, and type

7.1.2 The following information should be recorded once during the chassis dynamometer test:
a. All measurements listed in 3.3.3
b. Engine set point speed
c. Engine high idle speed
d. Engine low idle speed

7.1.3 The following test information should be recorded at each speed
  a. All measurements listed in 3.3.1
  b. Slip correction factor
  c. Air correction factor
  d. Fuel correction factor
  e. Corrected wheel power
  f. Expected wheel power
  g. Vehicle speed per 6.7

7.1.4 In addition to the tabulated information, corrected and expected wheel power for each test point should be plotted versus engine speed.

7.2 Reported Test Information—All chassis dynamometer tests conducted in accordance with this test procedure should carry the notation: "Results obtained and corrected in accordance with SAE J2177." In such cases any deviation from this document shall be noted. All reported chassis dynamometer test results bearing the SAE J2177 notation shall include a minimum of the following information at each test point:
  a. Engine speed
  b. Corrected wheel power

(R) ENGINE POWER TEST CODE—SPARK IGNITION AND COMPRESSION IGNITION—GROSS POWER RATING—SAE J1995 JUN90

SAE Standard

Report of the Power Test Code Standards Committee approved January 1990. Completely revised by the Power Test Code Standards Committee June 1990.

Table of Contents

1. SCOPE AND FIELD OF APPLICATION
2. REFERENCES
3. TERMS AND DEFINITIONS
4. REFERENCE TEST CONDITIONS AND CORRECTIONS
5. LABORATORY AND ENGINE EQUPMENT
6. TEST PROCEDURES
7. PRESENTATION OF RESULTS
8. CORRECTION FORMULAS
TABLE 1 Reference Atmospheric Conditions
TABLE 2 Reference SI Gasoline Specifications
TABLE 3 Reference CI Fuel Specifications
TABLE 4 Engine Equipment
TABLE 5 Atmospheric Correction Factor Exponents

1. Scope and Field of Application

1.1 Scope—This document has been adopted by SAE to specify:
a. A basis for gross engine power rating.
b. Reference inlet air and fuel supply test conditions.
c. A method for correcting observed power to reference conditions.
d. A method for determining gross full load engine power with a dynamometer.

1.2 Field of Application—This test code document is applicable to both four stroke and two stroke spark ignition (SI) and compression ignition (CI) engines, naturally aspirated and pressure charged, with and without charge air cooling. This document does not apply to aircraft or marine engines.

2. References

2.1 This test code supersedes those portions of SAE J1349 dealing with gross power rating.

2.2 Standard CI diesel fuel specifications are range mean values for Type 2-D EPA test fuel per Title 40, Code of Federal Regulations, Part 86.1313-87.

2.3 The corresponding test code for net power rating is SAE J1349.

2.4 The document for mapping engine performance is SAE J1312.

2.5 Relationship to ISO 2534–ISO 2534 (1972) differs from SAE J1995 in several areas, among which the most important are:
  a. This document is not limited to road vehicles.
  b. This document requires inlet fuel temperature be controlled to 40°C on CI engines.
  c. This document includes a reference fuel specification and requires that engine power be corrected to that specification on all CI and certain SI engines.
  d. This document includes a different procedure for testing engines with a laboratory charge air cooler (ISO method optional).
  e. This document includes a different procedure for correcting power to reference atmospheric conditions on turbocharged CI engines.

2.6 Complete correlation has not been established with ISO 3046. It is expected that this power test code will eventually align with ISO 1585 and ISO 2534.

3. Terms and Definitions
This section contains the definitions of key terms used to describe the gross power test.

3.1 Gross Brake Power— The power of an engine when configured as a "basic" engine as defined in 3.4 and 5.2, and tested and corrected in accordance with this document.

3.2 Rated Gross Power—Engine gross power as declared by the manufacturer at "rated speed".

3.3 Rated Speed—The speed determined by the manufacturer at which the engine power is rated.

3.4 Basic Engine—A "basic" engine is an engine configured with only the built in equipment required for self-sustained operation. A basic engine does not include accessories that are necessary only to per-

TABLE 1—REFERENCE ATMOSPHERIC CONDITIONS

	Standard Condition	Test Range Limits
Inlet Air Supply Pressure (absolute)	100 kPa	—
Dry Air Pressure (absolute)	99 kPa	90–105 kPa
Inlet Air Supply Temperature	25°C	15–40°C

TABLE 2—REFERENCE SI GASOLINE SPECIFICATIONS

	Regular Fuel	Premium Fuel
Research Octane No.:	92 ± 0.5	97 ± 0.5
Motor Octane No.:	83 ± 0.5	87 ± 0.5
Lower Heating Value:	43.3 MJ/kg ± 0.1	43.1 MJ/kg ± 0.1

TABLE 3—REFERENCE CI FUEL SPECIFICATIONS

	Standard Condition	Test Range Limits
Fuel Density @ 15°C	0.850 kg/l	0.840–0.860 kg/l
Fuel Kinematic Viscosity @ 40°C	2.6 mm²/s	2.0–3.2 mm²/s
Fuel Inlet Temperature	40°C	39–41°C (pump/line/nozzles) or 37–43°C (unit injectors)

form its intended service or that power auxiliary systems. If these accessories are integral with the engine or for any reason are included on the test engine, the power absorbed may be determined and added to the gross brake power. Common "basic" engine accessory examples are listed in Table 4.

3.5 Reference Test Conditions—The standard or reference engine inlet air supply (atmospheric) and inlet fuel conditions to which all power corrections are made.

3.6 Friction Power—The power required to drive the engine alone as equipped for the power test. Friction power may be established by one of the following methods (the value is needed for power correction of spark ignition engines):
 a. Assume 85% mechanical efficiency.
 b. Hot Motoring Friction—Record friction torque at wide open throttle at each test speed run on the power test. All readings are to be taken at the same coolant and oil temperature as observed on the power test points ±3°C.

3.7 Indicated Power—The power developed in the cylinders. It is defined as the sum of the brake power and friction power for the purpose of this document.

4. Reference Test Conditions and Corrections—This section contains reference air and fuel supply test conditions and specifications, recommended test ranges, and applicability of the correction procedures.

4.1 Reference Atmospheric Conditions—Table 1 is reference atmospheric conditions and test ranges for which the correction procedures are valid.

4.2 Reference SI Gasoline Specifications—Reference gasoline research and motor octane numbers in Table 2 have been determined corresponding to "regular" and "premium" test fuels. Reference gasoline is required for all SI engines equipped with knock sensors or other devices that control spark advance as a function of spark knock. Other SI engines may use any gasoline with an octane number sufficient to prevent knock.

4.3 Reference CI Fuel Specifications—Reference fuel specifications are per Title 40, Code of Federal Regulations, Part 86.1313-87, and represent range mean values for Type 2-D diesel fuel. The reference fuel characteristics in Table 3 have been determined to affect engine test power, and are listed with the applicable test ranges for which the correction procedures are valid:

Observed engine power is also corrected for variations in lower heating value (LHV) based on an empirical relationship between LHV and fuel density per 8.4.2.

4.4 Alternate Fuels—Reference values for alternate SI and CI fuels, both liquid and gaseous, are not presented in this document.

Therefore, when alternate fuels are used for the gross power engine test, no corrections to reference fuel conditions shall be made.

4.5 Power Corrections—The performance of SI and CI engines is affected by the density of the inlet combustion air as well as by the characteristics of the test fuel. Therefore, in order to provide a common basis of comparison, it may be necessary to apply correction factors to the observed gross power to account for differences between reference air and fuel conditions and those at which the test data were acquired.

4.5.1 All power correction procedures for atmospheric air are based on the conditions of the engine inlet air supply immediately prior to the entrance into the engine inlet system. This may be ambient (atmospheric) air or a laboratory air plenum that maintains air supply conditions within the range limits defined per 4.1.

4.5.2 On any engine where the power output is automatically controlled to compensate for changes in one or more of the listed inlet air and fuel supply test conditions, no correction for that test parameter shall be made.

4.5.3 The magnitude of the power correction should not exceed 5% for inlet air or 3% for inlet fuel corrections. If the correction factor exceeds these values, it shall be noted in accordance with 7.1.

4.6 Correction Formulas—The applicable correction formulas for spark ignition and compression ignition engines are listed in Section 8. These correction formulas are designed for correction of gross brake power at full throttle operation; however, for CI engines the formulas may also be used to correct partial load power for the purpose of determining specific fuel consumption. These correction formulas are not intended for altitude derating.

5. Laboratory and Engine Equipment—This section contains a list of laboratory and engine equipment used in the gross power test.

5.1 Laboratory Equipment—The following standard laboratory test equipment is required for the gross power test.

5.1.1 INLET SYSTEM—Any laboratory system that provides a supply of air to the basic engine. The inlet system begins at the point where air enters from the supply source (atmosphere or lab plenum) and ends at the entrance to the throttle body, inlet manifold, or turbocharger inlet, on engines as appropriate. Restriction induced by the inlet system may be at minimum levels.

5.1.2 EXHAUST SYSTEM—Any laboratory system that vents exhaust gas from the outlet of the basic engine. The exhaust system begins at the exhaust manifold outlet or at the turbine outlet on engines so equipped. Restriction induced by the exhaust system may be at minimum levels.

5.1.3 FUEL SUPPLY SYSTEM—Any laboratory system that provides a supply of fuel to the fuel inlet of the basic engine. The fuel supply system must be capable of controlling fuel supply temperature to within the ranges specified in 4.3 for CI engines. The fuel supply system shall not exceed the manufacturer's maximum permissible restriction requirements, if applicable.

5.1.4 CHARGE AIR COOLER—For charge cooled engines a laboratory auxiliary cooler may be employed for test purposes. If used, one of the following test methods is required and the appropriate correction procedure is applied per Section 8:
 a. Standard Method: This is the preferred test method. The laboratory unit is set to simulate intended in-service charge air cooler restriction and inlet manifold temperatures as if the ambient and inlet supply air temperatures were 25°C.

 b. Optional Method: The laboratory unit is set to duplicate the charge air cooler restriction and inlet manifold temperatures that would be obtained during intended service operation at the observed inlet air test conditions.

5.1.5 AUXILIARY POWER SUPPLY—Electrically driven engine components determined to be part of the basic engine may be operated via an external power supply. In such cases, the power required must be determined and subtracted from the corrected gross brake power.

5.2 Engine Equipment—A basic engine, as defined in 3.4, is used for the gross power test. Table 4 lists basic engine accessories and control settings required for the gross power test.

6. Test Procedures—This section contains the required test proce-

TABLE 4—ENGINE EQUIPMENT

System	Required	Comments
1. Inlet Air System	Optional	See 5.1.1.
Air Ducting	Optional	
Air Cleaner	Optional	
Air Preheat	No	
2. Pressure Charging System	Yes	
Boost Control Settings	Manufacturer's Specification	For all engines equipped with variable boost as a function of other engine parameters (speed/load/fuel octane, etc.), the boost pressure controls must be set to reflect intended in-service operation.
3. Charge Air Cooling System	Yes	If applicable.
Charge Air Cooler	Yes	See 5.1.4 for auxiliary cooler options.
Cooling Pump or Fan	Conditional	Not required if it can be shown to be functioning less than 20% of running time during intended in-service operation at reference test conditions.
4. Electrical System	Yes	See 5.1.5.
Ignition System	Yes	
Starter	No	
Generator/Alternator	Conditional	Required only if needed to operate the basic engine in a self sustained continuous manner and an external power supply is not used. In this case, the generator shall operate at a load level only sufficient to power the required components (i.e., fuel injectors, electric fuel pump).
Ignition and Timing Control Settings	Manufacturer's Specification	For any engine equipped with electronic controls and/or knock sensors, the spark or timing advance must be adjusted to reflect intended in-service operation.
5. Emissions Control System	Optional	If used, all control settings or adjustments must be set to reflect intended in-service operation.
6. RFI/EMI Controls (radio frequency or electromagnetic interference)	Manufacturer's Specification	Control settings must reflect intended in-service operation.

TABLE 4 (CONTINUED)

System	Required	Comments
7. Fuel Supply System	Yes	
Fuel Filters/Prefilters	Optional	See 5.1.3.
Fuel Supply Pump	Yes	Or equivalent electrical load if applicable.
Injection Pump/Carburetor or Fuel Metering Control Settings	Manufacturer's Specification	Control settings must reflect intended in-service operation.
8. Engine Cooling System (liquid)	Yes	
Cooling Pump	Yes	
Radiator	Optional	Functionally equivalent laboratory system recommended.
Thermostat	Optional	If not used, then coolant temperature and flow shall be regulated to intended in-service levels.
Cooling Fan	No	If used, power absorbed should be calculated and added to the gross brake power.
Engine Cooling System (Air)	Yes	
Blower	Conditional	Required if not disconnectable. On variable speed units the fan can be disconnected if it can be shown to be functioning less than 20% of engine running time during intended in-service operation at reference test conditions.
9. Lubrication System	Yes	The basic engine closed loop lubrication system is used. Oil fill shall be at manufacturer's full level. Oil temperatures shall reflect in-service levels at reference test conditions.
10. Exhaust System	Optional	See 5.1.2.
11. Auxiliary Drives		
Power Steering Pump	No	
Freon Compressor	No	
Vacuum Pumps	Conditional	Required only if needed to drive other required systems listed and it functions in that capacity more than 20% of engine running time during intended in-service operation.
Air Compressors	Conditional	See above comments - same as vacuum pumps.

dures for determining gross engine power.

6.1 Instrumentation Accuracy—The following minimum test instrumentation accuracy is required:
  a. Torque: ±0.5% of measured value
  b. Speed: ±0.2% of measured value
  c. Fuel Flow: ±1% of measured value
  d. Temperatures: ±2°C
  e. Air Supply Pressure: ±0.1 kPa
  f. Other Gas Pressures: ±0.5 kPa

6.2 Adjustments and Run-in

6.2.1 Adjustments shall be made before the test in accordance with the manufacturer's instructions. No changes or adjustments shall be made during the test.

6.2.2 The engine shall be run-in according to the manufacturer's recommendation. If no such recommendation is available, the engine shall be run-in until corrected brake power is repeatable within 1% over an 8 h period.

6.3 Pressure and Temperature Measurement

6.3.1 Pressure and temperature of the inlet air supply, used for the purpose of engine power corrections, shall be measured in a manner to obtain the total (stagnation) condition at the entrance to the engine inlet system. On those tests where the engine air supply is ambient air, this pressure is the barometric pressure; on those tests where the air supply is test cell ambient air, this pressure is the cell barometric pressure.

6.3.2 Inlet manifold pressure and temperature shall be measured as static values with probes located in a section common to several cylinders. In such installations dynamic pressure is assumed zero.

6.3.3 On charge air cooled engines in which a laboratory cooler is employed for testing, precooler charge air pressure must also be measured for the purpose of setting in-service restrictions per 5.1.4. Precooler pressure must be measured upstream of the auxiliary unit in a manner to obtain the total (stagnation) value. Auxiliary cooler restriction is the difference between the precooler and inlet manifold pressures.

6.3.4 Coolant temperatures in liquid cooled engines shall be measured at the inlet and outlet of the engine, in air cooled engines at points specified by the manufacturer.

6.3.5 Oil pressure and temperature shall be measured at the entrance to the main oil gallery.

6.3.6 Fuel temperature shall be measured at the inlet to the carburetor or fuel injector rail for SI engines, and at the inlet to the high pressure injection pump or unit injector rail for CI engines, and at the outlet of the volumetric flow meter for gaseous fueled engines.

6.4 Test Operating Conditions

6.4.1 The engine must be started and warmed up in accordance with manufacturer's specifications. No data shall be taken until torque and speed have been maintained within 1% and temperatures have been maintained within ±2°C for at least 1 min.

6.4.2 Engine speed shall not deviate from the nominal speed by more than ±1% or ±10 min^{-1}, whichever is greater.

6.4.3 Coolant outlet temperature for a liquid cooled engine shall be controlled to within ±3°C of the nominal thermostat value specified by the manufacturer. Coolant inlet air temperature for an air cooled engine is regulated to 35°C ± 5.

6.4.4 Fuel inlet temperature for diesel fuel injection shall be controlled to 40°C ± 3 for unit injector systems, and 40°C ± 1 for pump/line/nozzle systems. Test fuel temperature control is not required on SI engine power tests.

6.4.5 The exhaust gas must be vented to a reservoir having a total pressure within 0.75 kPa of the inlet air supply pressure.

6.5 Test Points—Record full throttle data for at least 5 approximately evenly spaced operating points to define the power curve between 600 rpm (or the lowest stable speed) and the maximum engine speed recommended by the manufacturer. One of the operating speeds shall be the rated speed, one shall be the peak torque speed.

7. Presentation of Results—This section contains a listing of test data to be recorded and procedures for presenting results.

7.1 Reporting Requirements—All reported engine test data shall carry the notation: "Performance obtained and corrected in accordance with SAE J1995". Any deviation from this document, its procedures, or limits, shall be noted. All reported or advertised test data bearing the SAE J1995 notation shall include a minimum of the following information at each test point:
  a. Engine speed
  b. Corrected gross brake power (or torque)

7.2 Recorded Test Conditions—Record the following ambient air, fuel, and lubricating oil test conditions and specifications.

7.2.1 INLET AIR SUPPLY CONDITIONS:
  a. Air supply pressure
  b. Air supply vapor pressure
  c. Air supply temperature

7.2.2 SPARK IGNITION ENGINE FUEL–LIQUID:
  a. Fuel type and/or blend
  b. Research and motor octane numbers
  c. Lower heating value

7.2.3 SPARK IGNITION ENGINE FUEL–GASEOUS:
  a. Fuel type or grade
  b. Composition
  c. Density @ 15°C and 101 kPa
  d. Lower heating value

7.2.4 DIESEL FUELS:
  a. ASTM or other fuel grade
  b. Density @ 15°
  c. Viscosity @ 40°
  d. Lower heating value (optional)

7.2.5 LUBRICATING OIL:
  a. API engine service classification
  b. SAE–Viscosity grade
  c. Manufacturer and brand name

7.3 Recorded Test Data—Record the following minimum information at each data test point:
  a. Brake torque
  b. Friction torque (if measured)
  c. Engine speed
  d. Fuel flow rate
  e. Fuel supply pressure and temperature
  f. Ignition and/or injection timing
  g. Oil pressure and temperature
  h. Coolant temperature
  i. Inlet manifold air temperature and pressure
  j. Total pressure drop across the auxiliary cooler (if applicable)
  k. Smoke (optional–CI engines only)

7.4 Engine Equipment—Record all engine equipment listed per 5.2. Additionally, record engine manufacturer, displacement, bore and stroke, number and configuration of cylinders, carburetion or injection system type, plus type of pressure charging system, if applicable. If a laboratory charge air cooler is used, record the test method per 5.1.4.

For SI engines equipped with knock sensors, the engine should be designated as a "regular" or "premium" fuel engine. For those SI engines without knock sensors, the minimum octane number for which knock does not occur shall be recorded as stated by the engine manufacturer.

7.5 Additional Recorded Information—Record any other pertinent test data as determined by the manufacturer. This may include, but is not limited to: test date, engine serial number, test number, test location, etc.

8. Correction Formulas—This section includes all formulas necessary to correct observed engine power performance for deviations in inlet air and fuel supply conditions.

8.1 Symbols and Units

SYMBOLS	TERM	UNITS
CA	Air correction factor	
CF	Fuel correction factor	
fa	Atmospheric factor	
fm	Engine factor	
fd	Fuel density factor	
fv	Fuel viscosity factor	
α	Pressure sensitivity exponent	
β	Temperature sensitivity exponent	
S	Viscosity sensitivity coefficient	
D	Engine displacement	l
B	Inlet air supply total pressure	kPa
t	Inlet air supply temperature	°C
P	Inlet manifold total pressure	kPa
r	Pressure ratio	
q	Fuel delivery	mg/L cycle
bp	Brake power	kW
fp	Friction power	kW
ip	Indicated power	kW
n	Engine speed	min^{-1}
F	Fuel flow	g/s
SG	Fuel density @ 15°C	kg/l
V	Fuel viscosity @ 40°C	mm^2/s

8.2 Subscripts

c = Refers to data corrected to reference inlet air and fuel supply conditions.
o = Refers to data observed at the actual test conditions.
d = Refers to the dry air portion of the total inlet air supply pressure.
r = Refers to the reference test conditions per Section 4.

8.3 Spark Ignition Correction Formulas—These spark ignition engine correction formulas are only applicable at full (WOT) throttle positions.

$$bp_c = CA \times bp_o \quad \text{(Eq. 1)}$$

Calculation of atmospheric correction factor, CA. If 85% mechanical efficiency is assumed:

$$CA = 1.18 \left[\left(\frac{99}{B_{do}}\right) \left(\frac{t_o + 273}{298}\right) \right]^{.5} - 0.18 \quad \text{(Eq. 2)}$$

If friction power is measured:

$$bp_c = ip_c \cdot fp_o \quad \text{(Eq. 3)}$$

where:

$$ip_c = ip_o \left(\frac{99}{B_{do}}\right) \left(\frac{t + 273}{298}\right)^{.5}$$

and:

$$ip_o = fp_o + bp_o$$

NOTE: If a lab auxiliary charge air cooler is used in conjunction with the standard test method per 5.1.4, no inlet air temperature corrections shall be made. In this case, the temperature correction exponent becomes zero. Otherwise use the above formula.

8.4 Compression Ignition Engine Correction Formulas—These CI engine correction formulas are applicable at all speed and load levels.

$$bp_c = (CA \times CF) bp_o \quad \text{(Eq. 4)}$$

8.4.1 CALCULATION OF ATMOSPHERIC CORRECTION FACTOR, CA:

$$CA = (fa)^{fm} \quad \text{(Eq. 5)}$$

where:

$$fa = \left(\frac{B_{dr}}{B_{do}}\right)^{\alpha} \left(\frac{t_o + 273}{t_r + 273}\right)^{\beta} = \left(\frac{99}{B_{do}}\right)^{\alpha} \left(\frac{t_o + 273}{298}\right)^{\beta}$$

and values for α and β, are summarized in Table 5:

TABLE 5—ATMOSPHERIC CORRECTION FACTOR EXPONENTS

Pressure Charging System	Charge Air Cooling System	α	β
Naturally Aspirated	None	1.0	0.7
Mechanically Supercharged	All	1.0	0.7
Turbocharged	None	0.7	1.2
Turbocharged	Air-to-Air	0.7	1.2
Turbocharged	Jacket Water	0.7	0.7
Turbocharged	Lab Auxiliary (Standard)	0.7	0.4
Turbocharged	Lab Auxiliary (Optional)	0.7	1.2

Where "standard" and "optional", refer to the lab auxiliary cooler test method described in 5.1.4.

The value of fm is given as:

q/r	fm
Less than 37.2	0.2
Between 37.2 and 65	(0.036 × q/r) − 1.14
More than 65	1.2

(Eq. 6)

where:

q = 120 000 F/Dn for four stroke engines
q = 60 000 F/Dn for two stroke engines
r = P_o/B_o for all engines (r = 1 if naturally aspirated)

8.4.2 CALCULATION OF FUEL CORRECTION FACTOR, CF:

$$CF = fd \times fv$$

where:

$$fd = 1 + 0.70 \left(\frac{SG_r - SG_o}{SG_o}\right) = 1 + 0.70 \left(\frac{0.850 - SG_o}{SG_o}\right) \quad \text{(Eq. 7)}$$

and:

$$fv = \frac{1 + S/V_o}{1 + S/V_r} = \frac{1 + S/V_o}{1 + S/2.6}$$

NOTE: The above formulas correct observed power to reference fuel density and viscosity levels. A correction coefficient of 0.70 in the above density factor equation is added to account for typical changes in lower heating value at differing density levels, based on an empirical LHV-SG relationship.

Values of S shall be determined by the engine manufacturer. If no values are available, the following shall be used:

Pump/Line/Nozzle Systems 0.15
Unit Injectors 0.0

NOTE: If used for the purpose of determining specific fuel consumption the corrected fuel flow is given by the following:

$$F_c = (SG_r/SG_o \times fv) F_o \quad \text{(Eq. 8)}$$

(R) ENGINE POWER TEST CODE—SPARK IGNITION AND COMPRESSION IGNITION— NET POWER RATING—SAE J1349 JUN90

SAE Standard

Report of the Engine Committee approved December 1980, completely revised June 1985. Completely revised by the Power Test Code Standards Committee January 1990 and again in June 1990.

Table of Contents

1. SCOPE AND FIELD OF APPLICATION
2. REFERENCES
3. TERMS AND DEFINITIONS
4. REFERENCE TEST CONDITIONS AND CORRECTIONS
5. LABORATORY AND ENGINE EQUIPMENT
6. TEST PROCEDURES
7. PRESENTATION OF RESULTS
8. CORRECTION FORMULAS

TABLE 1 Reference Atmospheric Conditions
TABLE 2 Reference SI Gasoline Specifications
TABLE 3 Reference CI Fuel Specifications
TABLE 4 Engine Equipment
TABLE 5 Atmospheric Correction Factor Exponents

1. Scope and Field of Application

1.1 Scope—This document has been adopted by SAE to specify:
a. A basis for net engine power rating.
b. Reference inlet air and fuel supply test conditions.

c. A method for correcting observed power to reference conditions.
d. A method for determining net full load engine power with a dynamometer.

1.2 Field of Application—This test code document is applicable to both four stroke and two stroke spark ignition (SI) and compression ignition (CI) engines, naturally aspirated and pressure charged, with and without charge air cooling. This document does not apply to aircraft or marine engines.

2. References

2.1 This test code supersedes those portions of SAE J1349 JUN85 dealing with net power rating.

2.2 Standard CI diesel fuel specifications are range mean values for Type 2-D EPA test fuel per Title 40, Code of Federal Regulations, Part 86.1313-87.

2.3 The corresponding test code for gross power rating is SAE J1995 JAN90.

2.4 The document for mapping engine performance is SAE J1312.

2.5 Relationship to ISO 1585–ISO 1585 (DIS in 1989) differs from SAE J1349 in several areas, among which the most important are:
a. This document is not limited to road vehicles.
b. This document requires inlet fuel temperature be controlled to 40°C on CI engines.
c. This document includes a reference fuel specification and requires that engine power be corrected to that specification on all CI and certain SI engines.
d. This document includes a different procedure for testing engines with a laboratory charge air cooler (ISO method optional).
e. This document stipulates a 20% duty cycle limit on variable speed cooling fans in order to qualify for testing at the minimum power loss settings.

2.6 Complete correlation has not been established with ISO 3046, ISO 2288, ISO 9249, or with ISO 4106. It is expected that these power test codes will eventually align with ISO 1585.

3. Terms and Definitions
—This section contains the definitions of key terms used to describe the net power test.

3.1 Net Brake Power— The power of an engine when configured as a "fully equipped" engine as defined in 3.4 and 5.2, and tested and corrected in accordance with this document.

3.2 Rated Net Power—Engine net power as declared by the manufacturer at "rated speed".

3.3 Rated Speed—The speed determined by the manufacturer at which the engine power is rated.

3.4 Fully Equipped Engine—A "fully equipped" engine is an engine equipped with only those accessories necessary to perform its intended service. A fully equipped engine does not include components that are used to power auxiliary systems. If these components are integral with the engine or for any reason are included on the test engine, the power absorbed may be determined and added to the net brake power. Common "fully equipped" engine accessory examples are listed in Table 4.

3.5 Reference Test Conditions—The standard or reference engine inlet air supply (atmospheric) and inlet fuel conditions to which all power corrections are made.

3.6 Friction Power—The power required to drive the engine alone as equipped for the power test. Friction power may be established by one of the following methods (the value is needed for power correction of spark ignition engines):
a. Assume 85% mechanical efficiency.
b. Hot Motoring Friction—Record friction torque at wide open throttle at each test speed run on the power test. All readings are to be taken at the same coolant and oil temperature as observed on the power test points ±3°C.

3.7 Indicated Power—The power developed in the cylinders. It is defined as the sum of the brake power and friction power for the purpose of this document.

4. Reference Test Conditions and Corrections
—This section contains reference air and fuel supply test conditions and specifications, recommended test ranges, and applicability of the correction procedures.

4.1 Reference Atmospheric Conditions—Table 1 is reference atmospheric conditions and test ranges for which the correction procedures are valid.

TABLE 1—REFERENCE ATMOSPHERIC CONDITIONS

	Standard Condition	Recommended Test Range Limits
Inlet Air Supply Pressure (absolute)	100 kPa	—
Dry Air Pressure (absolute)	99 kPa	90-105 kPa
Inlet Air Supply Temperature	25°C	15-40°C

4.2 Reference SI Gasoline Specifications—Reference gasoline research and motor octane numbers in Table 2 have been determined corresponding to "regular" and "premium" test fuels. Reference gasoline is required for all SI engines equipped with knock sensors or other devices that control spark advance as a function of spark knock. Other SI engines may use any gasoline with an octane number sufficient to prevent knock.

TABLE 2—REFERENCE SI GASOLINE SPECIFICATIONS

	Regular Fuel	Premium Fuel
Research Octane No.:	92 ± 0.5	97 ± 0.5
Motor Octane No.:	83 ± 0.5	87 ± 0.5
Lower Heating Value:	43.3 MJ/kg ± 0.1	43.1 MJ/kg ± 0.1

4.3 Reference CI Fuel Specifications—Reference fuel specifications are per Title 40, Code of Federal Regulations, Part 86.1313-87, and represent range mean values for Type 2-D diesel fuel. The reference fuel characteristics in Table 3 have been determined to affect engine test power, and are listed with the applicable test ranges for which the correction procedures are valid.

TABLE 3—REFERENCE CI FUEL SPECIFICATIONS

	Standard Condition	Test Range Limits
Fuel Density at 15°C	0.850 kg/l	0.840–0.860 kg/l
Fuel Kinematic Viscosity at 40°C	2.6 mm²/s	2.0–3.2 mm²/s
Fuel Inlet Temperature	40°C	39–41°C (pump/line/nozzles) or 37–43°C (unit injectors)

Observed engine power is also corrected for variations in lower heating value (LHV) based on an empirical relationship between LHV and fuel density per 8.4.2.

4.4 Alternate Fuels—Reference values for alternate SI and CI fuels, both liquid and gaseous, are not presented in this document. Therefore, when alternate fuels are used for the net power engine test, no corrections to reference fuel conditions shall be made.

4.5 Power Corrections—The performance of SI and CI engines is affected by the density of the inlet combustion air as well as by the characteristics of the test fuel. Therefore, in order to provide a common basis of comparison, it may be necessary to apply correction factors to the observed net power to account for differences between reference air and fuel conditions and those at which the test data were acquired.

4.5.1 All power correction procedures for atmospheric air are based on the conditions of the engine inlet air supply immediately prior to the entrance into the engine inlet system. This may be ambient (atmospheric) air or a laboratory air plenum that maintains air supply conditions within the range limits defined per 4.1.

4.5.2 On any engine where the power output is automatically controlled to compensate for changes in one or more of the listed inlet air and fuel supply test conditions, no correction for that test parameter shall be made.

4.5.3 The magnitude of the power correction should not exceed 5% for inlet air or 3% for inlet fuel corrections. If the correction factor exceeds these values, it shall be noted in accordance with 7.1.

4.6 Correction Formulas—The applicable correction formulas for spark ignition and compression ignition engines are listed in Section 8. These correction formulas are designed for correction of net brake power at full throttle operation; however, for CI engines the formulas may also be used to correct partial load power for the purpose of determining specific fuel consumption. These correction formulas are not intended for altitude derating.

5. Laboratory and Engine Equipment—This section contains a list of laboratory and engine equipment used in the net power test.

5.1 Laboratory Equipment—The following standard laboratory test equipment is required for the net power test.

5.1.1 INLET SYSTEM—The intended service inlet system or any laboratory system that provides equivalent restriction at all speeds and loads. The inlet system begins at the point where air enters from the supply source (atmosphere or lab plenum) and ends at the entrance to the throttle body, inlet manifold, or turbocharger inlet, on engines as appropriate.

5.1.2 EXHAUST SYSTEM—A complete intended service exhaust system (including mufflers, catalytic converters, resonators, etc.) or any laboratory system that provides equivalent restriction at all speeds and loads. The exhaust system begins at the exhaust manifold outlet or at the turbine outlet on engines so equipped.

5.1.3 FUEL SUPPLY SYSTEM—Any laboratory system that provides a supply of fuel to the fuel inlet of the fully equipped engine. The fuel supply system must be capable of controlling fuel supply temperature to within the ranges specified in 4.3 for CI engines. The fuel supply system shall not exceed the manufacturer's maximum permissible restric-

TABLE 4—ENGINE EQUIPMENT

System	Required	Comments
1. Inlet Air System	Yes	See 5.1.1.
Air Ducting	Yes	
Air Cleaner	Yes	
Air Preheat	No	
2. Pressure Charging System	Yes	
Boost Control Settings	Manufacturer's Specification	For all engines equipped with variable boost as a function of other engine parameters (speed/load/fuel octane, etc.), the boost pressure controls must be set to reflect intended in-service operation.
3. Charge Air Cooling System	Yes	If applicable.
Charge Air Cooler	Yes	See 5.1.4 for auxiliary cooler options.
Cooling Pump or Fan	Conditional	Not required if it can be shown to be functioning less than 20% of running time during intended in-service operation at reference test conditions.
4. Electrical System	Yes	See 5.1.5.
Ignition System	Yes	
Starter	No	
Generator/Alternator	Conditional	Required only if needed to operate the fully equipped engine in a self sustained manner and an external power supply is not used. In this case, the generator shall operate at a load level only sufficient to power the required components (i.e., fuel injectors, electric fuel pump).
Ignition and Timing Control Settings	Manufacturer's Specification	For any engine equipped with electronic controls and/or knock sensors, the spark or timing advance must be adjusted to reflect intended in-service operation.
5. Emissions Control System	Yes	All control settings or adjustments must be set to reflect intended in-service operation.
6. RFI/EMI Controls (radio frequency or electromagnetic interference)	Manufacturer's Specification	Control settings must reflect intended in-service operation.

TABLE 4 (CONTINUED)

System	Required	Comments
7. Fuel Supply System	Yes	
Fuel Filters/Prefilters	Optional	See 5.1.3.
Fuel Supply Pump	Yes	Or equivalent electrical load if applicable.
Injection Pump/Carburetor or Fuel Metering Control Settings	Manufacturer's Specification	Control settings must reflect intended in-service operation.
8. Engine Cooling System (liquid)	Yes	
Cooling Pump	Yes	
Radiator	Optional	Functionally equivalent laboratory system recommended.
Thermostat	Optional	If not used, then coolant temperature and flow shall be regulated to intended in-service levels.
Cooling Fan	Yes	On variable speed units the fan may be run at minimum power consumption levels if it can be shown to be functioning less than 20% of engine running time during intended in-service operation at reference test conditions. NOTE: If for any reason the fan is omitted, the minimum allowable fan power should be determined and subtracted from the net brake power. If run at full output, the fan power absorbed should be calculated and the difference between it and the minimum allowable fan power shall be added to the net brake power.
Engine Cooling System (Air)	Yes	
Blower	Yes	See above comments – same as liquid cooling fan.
9. Lubrication System	Yes	The fully equipped engine closed loop lubrication system is used. Oil fill shall be at manufacturer's full level. Oil temperatures shall reflect in-service levels at reference test conditions.
10. Exhaust System	Yes	See 5.1.2.
11. Auxiliary Drives		
Power Steering Pump	No	
Freon Compressor	No	
Vacuum Pumps	Conditional	Required only if needed to drive other required systems listed, and it functions in that capacity more than 20% of engine running time during intended in-service operation.
Air Compressors	Conditional	See above comments – same as vacuum pumps.

tion requirements, if applicable.

5.1.4 CHARGE AIR COOLER—For charge cooled engines a laboratory auxiliary cooler may be employed for test purposes. If used, one of the following test methods is required and the appropriate correction procedure is applied per Section 8:

a. Standard Method: This is the preferred test method. The laboratory unit is set to simulate intended in-service charge air cooler restriction and inlet manifold temperatures as if the ambient and inlet supply air temperatures were 25°C.
b. Optional Method: The laboratory unit is set to duplicate the charge air cooler restriction and inlet manifold temperatures that would be obtained during intended service operation at the observed inlet air test conditions.

5.1.5 AUXILIARY POWER SUPPLY—Electrically driven engine components determined to be part of the basic engine may be operated via an external power supply. In such cases, the power required must be determined and subtracted from the corrected net brake power.

5.2 Engine Equipment—A fully equipped engine, as defined in 3.4, is used for the net power test. Table 4 lists fully equipped engine accessories and control settings required for the net power test.

6. Test Procedures—This section contains the required test procedures for determining net engine power.

6.1 Instrumentation Accuracy—The following minimum test instrumentation accuracy is required:

a. Torque: ±0.5% of measured value
b. Speed: ±0.2% of measured value
c. Fuel Flow: ±1% of measured value
d. Temperatures: ±2°C
e. Air Supply, Inlet and Exhaust Pressures: ±0.1 kPa
f. Other Gas Pressures: ±0.5 kPa

6.2 Adjustments and Run-in

6.2.1 Adjustments shall be made before the test in accordance with the manufacturer's instructions. No changes or adjustments shall be made during the test.

6.2.2 The engine shall be run-in according to the manufacturer's recommendation. If no such recommendation is available, the engine shall be run-in until corrected brake power is repeatable within 1% over an 8 h period.

6.3 Pressure and Temperature Measurement

6.3.1 Pressure and temperature of the inlet air supply, used for the purpose of engine power corrections, shall be measured in a manner to obtain the total (stagnation) condition at the entrance to the engine inlet system. On those tests where the engine air supply is ambient air, this pressure is the barometric pressure; on those tests where the air supply is test cell ambient air, this pressure is the cell barometric pressure.

6.3.2 Inlet air pressure, used for the purpose of determining inlet system restriction, shall be measured in a manner to obtain the total (stagnation) pressure immediately prior to the end of the inlet system as defined in 5.1.1.

6.3.3 Inlet manifold pressure and temperature shall be measured as static values with probes located in a section common to several cylinders. In such installations dynamic pressure is assumed zero.

6.3.4 On charge air cooled engines in which a laboratory cooler is employed for testing, precooler charge air pressure must also be measured for the purpose of setting in-service restrictions per 5.1.4. Precooler pressure must be measured upstream of the auxiliary unit in a manner to obtain the total (stagnation) value. Auxiliary cooler restriction is the difference between the precooler and inlet manifold pressures.

6.3.5 Coolant temperatures in liquid cooled engines shall be measured at the inlet and outlet of the engine, in air cooled engines at points specified by the manufacturer.

6.3.6 Oil pressure and temperature shall be measured at the entrance to the main oil gallery.

6.3.7 Fuel temperature shall be measured at the inlet to the carburetor or fuel injector rail for SI engines, and at the inlet to the high pressure injection pump or unit injector rail for CI engines, and at the outlet of the volumetric flow meter for gaseous fueled engines.

6.3.8 Exhaust pressure shall be measured in a manner to obtain the total (stagnation) pressure in a straight section of piping not less than three nor more than six diameters downstream of the entrance to the exhaust system as defined in 5.1.2.

6.4 Test Operating Conditions

6.4.1 The engine must be started and warmed up in accordance with manufacturer's specifications. No data shall be taken until torque and speed have been maintained within 1% and temperatures have been maintained within ±2°C for at least 1 min.

6.4.2 Engine speed shall not deviate from the nominal speed by more than ±1% or ±10 min^{-1}, whichever is greater.

6.4.3 Coolant outlet temperature for a liquid cooled engine shall be controlled to within ±3°C of the nominal thermostat value specified by the manufacturer. Coolant inlet air temperature for an air cooled engine is regulated to 35°C ± 5.

6.4.4 Fuel inlet temperature for diesel fuel injection shall be controlled to 40°C ± 3 for unit injector systems, and 40°C ± 1 for pump/line/nozzle systems. Test fuel temperature control is not required on SI engine power tests.

6.4.5 The exhaust gas must be vented to a reservoir having a total pressure within 0.75 kPa of the inlet air supply pressure.

6.5 Test Points—Record full throttle data for at least five approximately evenly spaced operating points to define the power curve between 600 rpm (or the lowest stable speed) and the maximum engine speed recommended by the manufacturer. One of the operating speeds shall be the rated speed, one shall be the peak torque speed.

7. Presentation of Results—This section contains a listing of test data to be recorded and procedures for presenting results.

7.1 Reporting Requirements—All reported engine test data shall carry the notation: "Performance obtained and corrected in accordance with SAE J1349". Any deviation from this document, its procedures, or limits shall be noted. All reported or advertised test data bearing the SAE J1349 notation shall include a minimum of the following information at each test point:

a. Engine speed
b. Corrected net brake power (or torque)

7.2 Recorded Test Conditions—Record the following ambient air, fuel, and lubricating oil test conditions and specifications.

7.2.1 INLET AIR SUPPLY CONDITIONS:
a. Air supply pressure
b. Air supply vapor pressure
c. Air supply temperature

7.2.2 SPARK IGNITION ENGINE FUEL–LIQUID:
a. Fuel type and/or blend
b. Research and motor octane numbers
c. Lower heating value

7.2.3 SPARK IGNITION ENGINE FUEL–GASEOUS:
a. Fuel type or grade
b. Composition
c. Density at 15°C and 101 kPa
d. Lower heating value

7.2.4 DIESEL FUELS:
a. ASTM or other fuel grade
b. Density at 15°C
c. Viscosity at 40°C
d. Lower heating value (optional)

7.2.5 LUBRICATING OIL:
a. API engine service classification
b. SAE–viscosity grade
c. Manufacturer and brand name

7.3 Recorded Test Data—Record the following minimum information at each data test point:

a. Brake torque
b. Friction torque (if measured)
c. Engine speed
d. Fuel flow rate
e. Fuel supply pressure and temperature
f. Ignition and/or injection timing
g. Oil pressure and temperature
h. Coolant temperature
i. Inlet manifold air temperature and pressure
j. Total pressure drop across the inlet air system
k. Total pressure drop across the auxiliary cooler (if applicable)
l. Total pressure drop across the exhaust system
m. Smoke (optional–CI engines only)

7.4 Engine Equipment—Record all engine equipment listed per 5.2. Additionally, record engine manufacturer, displacement, bore and stroke, number and configuration of cylinders, carburetion or injection system type, plus type of pressure charging system, if applicable. If a laboratory charge air cooler is used, record the test method per 5.1.4.

For SI engines equipped with knock sensors, the engine should be designated as a "regular" or "premium" fuel engine. For those SI engines without knock sensors, the minimum octane number for which knock does not occur shall be recorded as stated by the engine manufacturer.

7.5 Additional Recorded Information—Record any other pertinent test data as determined by the manufacturer. This may include, but is not limited to: test date, engine serial number, test number, test location, etc.

8. Correction Formulas—This section includes all formulas necessary to correct observed engine power performance for deviations in inlet air and fuel supply conditions.

8.1 Symbols and Units

SYMBOLS	TERM	UNITS
CA	Air correction factor	
CF	Fuel correction factor	
fa	Atmospheric factor	
fm	Engine factor	
fd	Fuel density factor	
fv	Fuel viscosity factor	
α	Pressure sensitivity exponent	
β	Temperature sensitivity exponent	
S	Viscosity sensitivity coefficient	
D	Engine displacement	l
B	Inlet air supply total pressure	kPa
t	Inlet air supply temperature	°C
P	Inlet manifold total pressure	kPa
r	Pressure ratio	
q	Fuel delivery	mg/L cycle
bp	Brake power	kW
fp	Friction power	kW
ip	Indicated power	kW
n	Engine speed	min^{-1}
F	Fuel flow	g/s
SG	Fuel density at 15°C	kg/l
V	Fuel viscosity at 40°C	mm^2/s

8.2 Subscripts:

c = Refers to data corrected to reference inlet air and fuel supply conditions.
o = Refers to data observed at the actual test conditions.
d = Refers to the dry air portion of the total inlet air supply pressure.
r = Refers to the reference test conditions per Section 4.

8.3 Spark Ignition Correction Formulas—These spark ignition engine correction formulas are only applicable at full (WOT) throttle positions.

$$bp_c = CA \times bp_o \quad (Eq.1)$$

Calculation of atmospheric correction factor, CA. If 85% mechanical efficiency is assumed:

$$CA = 1.18 \left[\left(\frac{99}{B_{do}} \right) \left(\frac{t_o + 273}{298} \right)^{.5} \right] - 0.18 \quad (Eq.2)$$

If friction power is measured:

$$bp_c = ip_c - fp_o \quad (Eq.3)$$

where:

$$ip_c = ip_o \left(\frac{99}{B_{do}} \right) \left(\frac{t + 273}{298} \right)^{.5}$$

and:

$$ip_o = fp_o + bp_o$$

NOTE: If a lab auxiliary charge air cooler is used in conjunction with the standard test method per 5.1.4, no inlet air temperature corrections shall be made. In this case, the temperature correction exponent becomes zero. Otherwise use the above formula.

8.4 Compression Ignition Engine Correction Formulas—These CI engine correction formulas are applicable at all speed and load levels.

$$bp_c = (CA \times CF) bp_o \quad (Eq.4)$$

8.4.1 CALCULATION OF ATMOSPHERIC CORRECTION FACTOR, CA:

$$CA = (fa)^{fm} \quad (Eq.5)$$

where:

$$fa = \left(\frac{B_{dr}}{B_{do}} \right)^\alpha \left(\frac{t_o + 273}{t_r + 273} \right)^\beta = \left(\frac{99}{B_{do}} \right)^\alpha \left(\frac{t_o + 273}{298} \right)^\beta$$

and values for α and β, are summarized in Table 5:

TABLE 5—ATMOSPHERIC CORRECTION FACTOR EXPONENTS

Pressure Charging System	Charge Air Cooling System	α	β
Naturally Aspirated	None	1.0	0.7
Mechanically Supercharged	All	1.0	0.7
Turbocharged	None	0.7	1.2
Turbocharged	Air-to-Air	0.7	1.2
Turbocharged	Jacket Water	0.7	0.7
Turbocharged	Lab Auxiliary (Standard)	0.7	0.4
Turbocharged	Lab Auxiliary (Optional)	0.7	1.2

Where "standard" and "optional", refer to the lab auxiliary cooler test method described in 5.1.4.

The value of fm is given as:

q/r	fm	
Less than 37.2	0.2	(Eq.6)
Between 37.2 and 65	(0.036 x q/r) − 1.14	
More than 65	1.2	

where:

q = 120 000 F/Dn for four stroke engines
q = 60 000 F/Dn for two stroke engines
r = P$_o$/B$_o$ for all engines (r = 1 if naturally aspirated)

8.4.2 CALCULATION OF FUEL CORRECTION FACTOR, CF:

$$CF = fd \times fv$$

where:

$$fd = 1 + 0.70 \left(\frac{SG_r - SG_o}{SG_o} \right) = 1 + 0.70 \left(\frac{0.850 - SG_o}{SG_o} \right) \quad (Eq.7)$$

and:

$$fv = \frac{1 + S/V_o}{1 + S/V_r} = \frac{1 + S/V_o}{1 + S/2.6}$$

NOTE: The above formulas correct observed power to reference fuel density and viscosity levels. A correction coefficient of 0.70 in the above density factor equation is added to account for typical changes in lower heating value at differing density levels, based on an empirical LHV-SG relationship.

Values of S shall be determined by the engine manufacturer. If no values are available, the following shall be used:

a. Pump/Line/Nozzle Systems   0.15
b. Unit Injectors   0.0

NOTE: If used for the purpose of determining specific fuel consumption, the corrected fuel flow is given by the following:

$$F_c = (SG_r/SG_o \times fv) F_o \quad (Eq.8)$$

26.22

(R) PROCEDURE FOR MAPPING ENGINE PERFORMANCE—SPARK IGNITION AND COMPRESSION IGNITION ENGINES—SAE J1312 JAN90

SAE Standard

Report of the SAE/DOT Advisory Committee, approved June 1980 and completely revised by the Engine Committee May 1987. Completely revised by the Power Test Code Standards Committee January 1990.

1. Scope—The purpose of this SAE code is to provide a standardized test procedure for generating engine performance maps. An engine performance map is a listing of engine fuel flow rates versus torque or power obtained at specific engine speeds and loads. Engine performance maps as specified by this code can be used in fuel economy simulation programs.

This document is applicable to both four-stroke spark ignition (SI) and compression ignition (CI) engines, naturally aspirated and pressure charged, with or without charge air cooling.

2. Reference—SAE J1349 JAN90, Engine Power Test Code—Spark Ignition and Compression Ignition—Net Power Rating

SAE J1349 is the sole reference document for this document. All engine test procedures, including engine and test equipment, correction procedures, instrumentation, reference conditions, terminology, etc., shall be in accordance with SAE J1349 except where otherwise specified by this document.

3. Engine Equipment—A "fully equipped" engine is used for this test, in accordance with SAE J1349, Sections 4.4 and 6. Please refer to SAE J1349 for specific equipment details.

4. Power Corrections—Measured engine torque or power levels and fuel flow rates must be corrected to standard reference air and fuel supply test conditions. The corrections are made in accordance with SAE J1349, Sections 5 and 9 with the following exceptions:

4.1 For CI engines, the correction procedure shall be used for both full and part load operation.

4.2 Since no adequate part load correction procedure is documented for SI engines, no corrections shall be made to the measured data. Therefore, inlet air must be controlled to reference test conditions.

5. Test Procedures—The test data for performance maps must be generated at specific engine speeds and loads. All other test procedures are conducted in accordance with SAE J1349, Section 7.

5.1 Test data must be obtained at each of the following speeds:
a. Rated or maximum speed
b. Peak torque speed
c. Idle speed
d. Intermediate speeds
e. Optional speeds.

5.1.1 Rated (or maximum), peak torque and idle speeds are defined by the manufacturer.

5.1.2 Intermediate speeds should be obtained at equal 10% increments of rated (or maximum) speed, (for example, 90% of rated, 80% of rated, etc.). Intermediate speeds may be rounded to the nearest 100 rpm. Intermediate speeds falling within 100 rpm of any other required test speeds may be omitted.

5.1.3 Optional speeds are any additional speeds deemed by the manufacturer as necessary to adequately define the shape of the power curve or to simulate vehicle operational requirements.

5.2 At each test speed, the engine shall be run at no less than five loads, one of which shall be full load. Specific test load levels shall be determined by the engine manufacturer, based on the engine's intended service operating characteristics. This may include motoring (negative) load levels. If intended service operating characteristics are not known, the engine shall be run at equally spaced load intervals (example 0%, 25%, 50%, 75%, 100%).

5.2.1 Data acquisition at certain speeds and loads may be subject to the capabilities of the test dynamometer. In such cases, these test points may be omitted from the map, at the manufacturer's discretion, if it can be shown that vehicle operational requirements do not utilize these data.

5.3 If engine performance is affected by vehicle related hardware, it may be necessary to generate additional maps to simulate specific operating conditions. Examples include individual vehicle restriction effects, or engine control devices based on torque converter position, gear ratio, vehicle speed, etc. When additional maps are run, each shall be clearly labeled with applicable test conditions or hardware.

6. Presentation Of Results

6.1 Data shall be recorded for test documentation in accordance with SAE J1349, Section 8. Additional data may be recorded at the manufacturer's discretion.

6.2 Engine data obtained and required specifically for the performance map shall be tabulated in a format acceptable for use by the vehicle simulation program. The following sample format may be used:

Engine Speed	% Load	Reference Condition Torque or Power	Reference Fuel Flow Rate
Rated:	X	Highest ↓ Lowest	X ↓ X	X ↓ X
1st Intermediate	X	Highest ↓ Lowest	X ↓ X	X ↓ X
2nd Intermediate:	X	etc.	etc.	etc.
etc.			

SMALL SPARK IGNITION ENGINE TEST CODE—SAE J607 AUG88

SAE Standard

Report of the Engine Committee, approved January 1956, completely revised June 1984, and reaffirmed August 1988.

1. Note—SAE J607 was established to provide a uniform means of comparing small gasoline engines. The document is now considered obsolete, and it is recommended that J1349 be used for rating these small engines formerly covered by SAE J607. SAE J607 was last published in the 1984 Handbook.

2. Scope—The purpose of this code is to prove a uniform means of comparing small gasoline engines of 0.82 L (50 in^3) or less in displacement and 15.1 kW (20 hp) or less in power output. It is not intended as a laboratory manual. Governor characteristics are considered part of the end item specification.

SMALL ENGINE POWER RATING PROCEDURE—SAE J1940 JUN89

SAE Standard

Report of the Specialized Vehicle Engine Committee approved June 1989. Rationale statement available.

1. Scope—This standard is applicable to small spark ignition and diesel engines, commonly operated outdoors, powering lawn and garden, construction, and general utility equipment. It is not intended to cover engines powering boats, chainsaws or weedeaters.

2. References
  2.1 SAE J1349

3. Test Code
  3.1 Test the engine in accordance with the applicable parts of SAE J1349.

4. Engine Rating
  4.1 When supplying any written quantitative SAE rated data concerning engine power, the engine manufacturer shall provide the following information:
  a. Rated in accordance with SAE J1940.
  b. Power Rating—either corrected gross or corrected net brake power.
  c. The rpm at which the power is rated (typically 3600 rpm). The rating speed shall not exceed the engine manufacturer's maximum operating speed.

  4.2 Production engines, after adjustment and run in, shall develop not less than 95% of the corrected gross or net rated brake power at rated rpm.

(R) PROCEDURE FOR EVALUATING TRANSIENT RESPONSE OF SMALL ENGINE DRIVEN GENERATOR SETS—SAE J1444 JUN91

SAE Recommended Practice

Report of the Engine Committee, approved December 1983. Completely revised by the SAE Small Engine and Powered Equipment Committee June 1991.

This document is equivalent to ISO 8528 part 2.

1. Scope—The scope of this SAE Recommended Practice is to provide a uniform practice for the testing of small engine powered alternating current generator sets. The small engines addressed are the reciprocating piston type of 14.9 kw (20 BHP) or less.

2. References
  2.1 Applicable Documents—The following publications form a part of this specification to the extent specified herein. The latest issue of SAE publicatons shall apply.
  2.1.1 SAE PUBLICATIONS—Available from SAE, 400 Commonwealth Drive, Warrendale, PA 15096-0001.
  SAE J1349—Engine Power Test Code—Spark Ignition and Diesel
  SAE J1940—Small Engine Power Rating Procedure

3. Test Set-Up—The engine and generator set shall be assembled as a complete unit and connected to a 100% resistive load bank with a maximum of 5% light bulb load. The engine fuel and oil shall be consistent with the recommendations of the engine manufacturer. The type of fuel used shall be recorded on the test log. The lubricating oil should also be identified in the test log by manufacturer, brand name, API duty rating, and SAE viscosity rating.

4. Instruments—The principal instruments for measuring the performance of an engine generator set are: a recording frequency meter, a volt meter, and either an ammeter or watt meter, or both. All electrical meters should be laboratory quality RMS reading instruments with accuracy within ±0.75% of full scale value.

5. Engine Run-In—If the engine is new, it must be run-in and stabilized according to the engine manufacturer's recommendation prior to the test run.

6. Correction Factor—SAE J1349 and J1940 contain instructions for instrumentation and calculations for the ambient condition correction factor (see 7.2). The engine manufacturer will specify which of the documents should be referred to for each specific engine.

7. Governor Performance Test
  7.1 Stabilization—Start and operate the generator set and allow the set to stabilize at rated load, rated voltage, and rated frequency. During this period, operate the recording frequency meter at a chart speed of not less than 152 mm/h and record all instrument readings including thermal instrumentation at maximum intervals of 10 min. Adjustments to the load voltage and frequency must be made to maintain corrected rated load at rated voltage and frequency. Adjustments to the voltage and frequency shall be limited to those adjustments available to the operator, specifically adjustments to the voltage or frequency adjust devices. On sets with a droop type speed control system as the prime speed control, the speed and droop portions of the control may be adjusted. Adjustments to load, voltage, or frequency controls shall be recorded on the recording meter chart at the time of adjustment. The engine generator set is considered stabilized when run continuously for a minimum of 1/2 h at rated load.

  7.2 After 20 min of the stabilization period has elapsed, the correction factor parameters will be measured and recorded on the data sheet (see Figure 1). The correction factor can then be calculated per SAE J1349 sections 8.1 and 8.2. The generator load will then be adjusted from rated load to corrected rated load. This corrected rated load and fractions of it will then prevail for the remainder of the test.

$$\text{Corrected Rated Load} = \frac{\text{Rated Load}}{\text{Correction Factor}} \quad \text{(Eq.1)}$$

  7.3 Test—The recording meter chart speed shall be not less than 2.54 mm/s (6 in/min) during this period.

After the generator set is stabilized, operate the set at each of the load conditions listed as follows in the sequence indicated. Running time at each load condition shall be the recovery time plus at least 10 s.

The loading sequence is:

1. Corrected Rated Load
2. No Load
3. Corrected Rated Load
4. No Load
5. 3/4 Corrected Rated Load
6. No Load
7. 3/4 Corrected Rated Load
8. No Load
9. 1/2 Corrected Rated Load
10. No Load
11. 1/2 Corrected Rated Load
12. No Load
13. 1/4 Corrected Rated Load
14. No Load
15. 1/4 Corrected Rated Load
16. No Load
17. Corrected Rated Load
18. No Load
19. 1/4 Corrected Rated Load
20. 1/2 Corrected Rated Load
21. 3/4 Corrected Rated Load
22. Corrected Rated Load
23. 3/4 Corrected Rated Load
24. 1/2 Corrected Rated Load
25. 1/4 Corrected Rated Load
26. No Load

8. Data Reduction

8.1 Prepare the recording instrument chart according to Figures 1 and 2. Record all 26 steps in the loading sequence on the data sheet, Figure 1.

9. Prescribed Standards—(See Table 1)

TABLE 1—PRESCRIBED STANDARDS

	Governor Performance Class "A"	Governor Performance Class "B"
A-Steady-State Speed Regulation Droop	3 Hz	4 Hz
B1-Recovery Time, Increasing Load	4 s	8 s
B2-Recovery Time, Decreasing Load	4 s	8 s
C-Overshoot	3 Hz	5 Hz
D-Undershoot	3 Hz	5 Hz
E-Steady-State Speed Band	0.6 Hz	0.8 Hz

NOTE: See Figure 2 for definition of terms.

FREQUENCY REGULATION DATA

Seq.	Obs. Watts	Corr. Watts	Volts	G Hz	A Hz	B1, 2 Sec.	C, D Hz	E Hz	Remarks
1									
2									
3									
4									
5									
6									
7									
8									
9									
10									
11									
12									
13									
14									
15									
16									
17									
18									
19									
20									
21									
22									
23									
24									
25									
26									

Test Date ______
Tester Name ______
Test Number ______
Sheet Number ______
Generator ______
Engine ______
Rated Load ______ kW
Rated Speed ______ rpm
Type of Fuel ______
Oil ______
Duty ______ Visc. ______
Corrected Barometer ______
Obs. ______ Temp. ______ True ______
Wet Bulb ______ Dry Bulb ______
Air Temp.—Intake ______ Amb. ______
Correction Factor ______

A—Frequency Regulation
B—Recovery Time
C—Overshoot or Undershoot
E—Steady State Frequency Band
G—Median of Frequency Band

FIGURE 1—FREQUENCY REGULATION FORM EXAMPLE

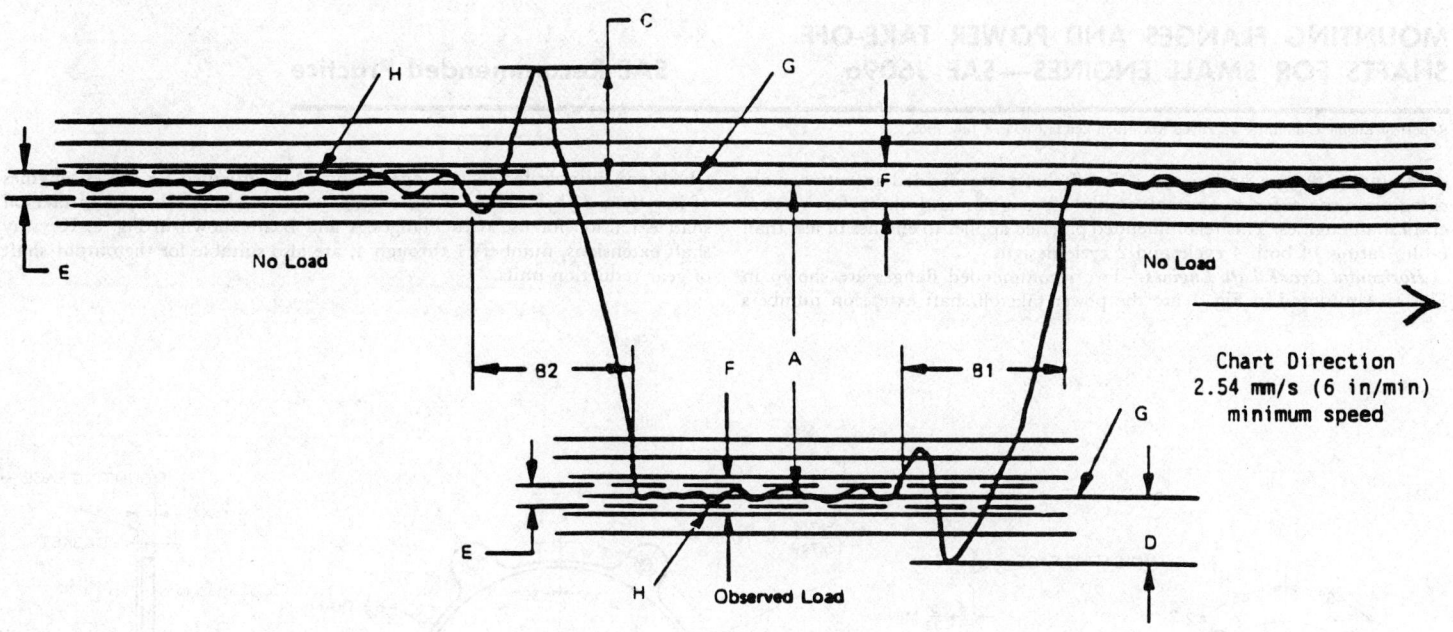

Diagram Illustrating Definition of Terms

A - Steady State Speed Regulation (Droop)
B1 - Recovery Time, Increasing Load
B2 - Recovery Time, Decreasing Load
C - Overshoot or Momentary Overspeed
D - Undershoot or Momentary Underspeed
E - Steady State Speedband - Observed
F - Allowable Steady State Speedband (refer 9)
G - Median of Observed Speedband
H - Actual Instrument Trace of Function

FIGURE 2—DIAGRAM ILLUSTRATING DEFINITION OF TERMS

MOUNTING FLANGES AND POWER TAKE-OFF SHAFTS FOR SMALL ENGINES—SAE J609a

SAE Recommended Practice

Report of Engine Committee approved May 1958 and last revised July 1965.

Mounting flanges and power take-off shafts are divided into two main categories: those for horizontal crankshaft engines and those for vertical crankshaft engines. This recommended practice applies to engines of less than 6-bhp rating, of both 4 cycle and 2 cycle designs.

Horizontal Crankshaft Engines—Two recommended flanges are shown in Fig. 1. Also noted in Fig. 1 are the power take-off shaft extension numbers which are suitable for each flange. Flange A is intended for use on engines of less than 3 bhp, Flange B on engines of over 3 bhp. The power take-off shaft extension for use with Flanges A and B are shown in Fig. 2. Keyway shaft extensions, numbers 1 through 4, are also suitable for the output shaft of gear reduction units.

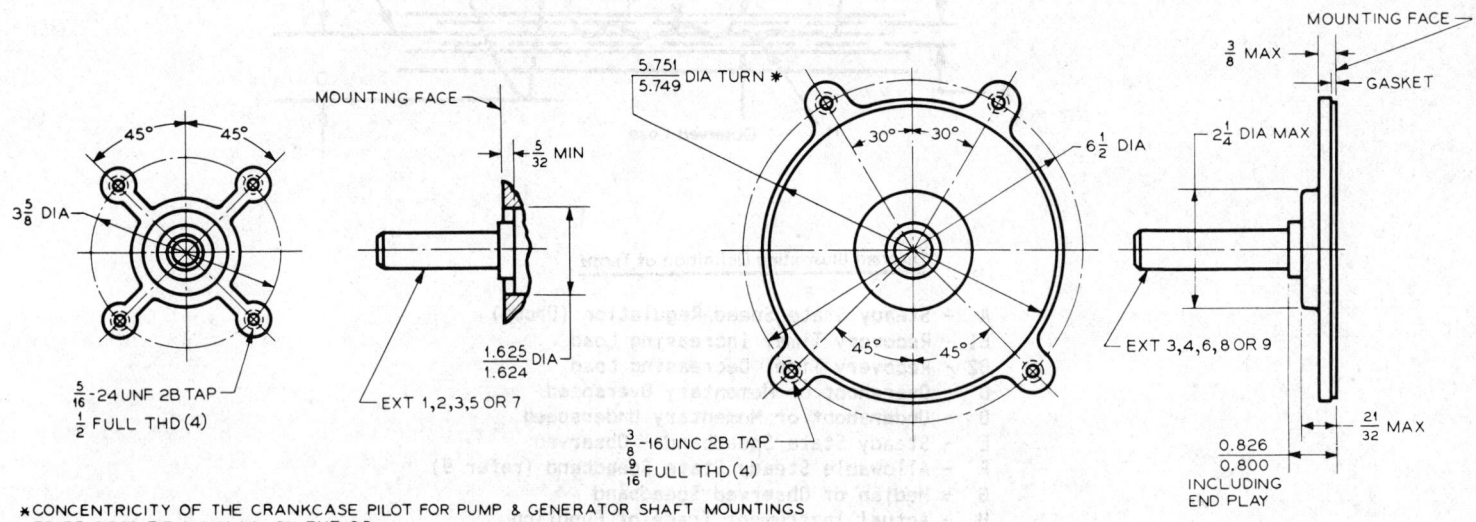

*CONCENTRICITY OF THE CRANKCASE PILOT FOR PUMP & GENERATOR SHAFT MOUNTINGS TO BE 0.005 TIR MAXIMUM ON THE O.D.

FOR PUMP AND GENERATOR MOUNTINGS SQUARENESS OF THE CRANKSHAFT TO THE MOUNTING FACE TO BE 0.0015 TIR PER INCH OF DIAMETER.

MOUNTING HOLES TO BE WITHIN 0.015 IN. OF TRUE POSITION.

Flange A   Flange B

FIG. 1—MOUNTING FLANGES FOR HORIZONTAL CRANKSHAFT ENGINES

26.27

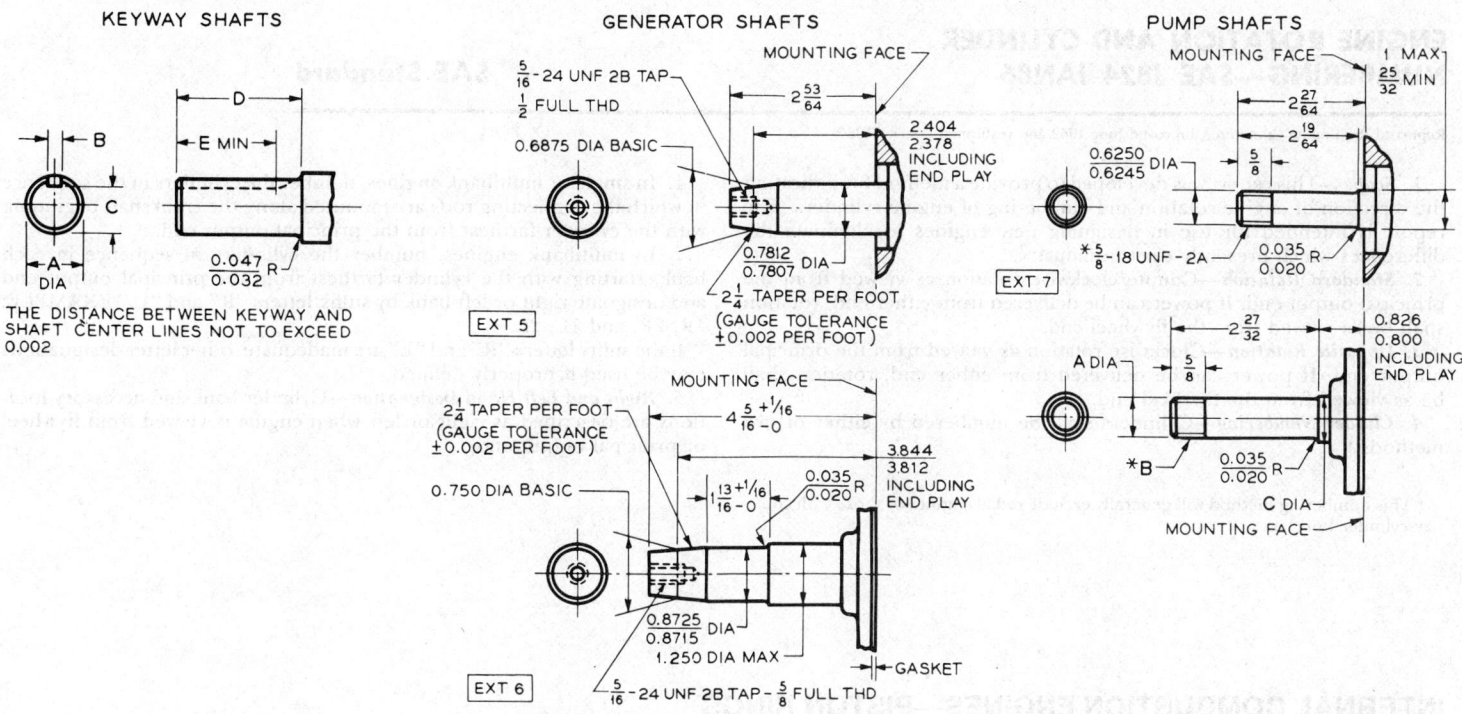

FIG. 2—POWER TAKE-OFF SHAFT EXTENSIONS FOR HORIZONTAL CRANKSHAFT ENGINES

Vertical Crankshaft Engine—Two recommended flanges and suitable crankshaft power take-off extensions are shown in Fig. 3.

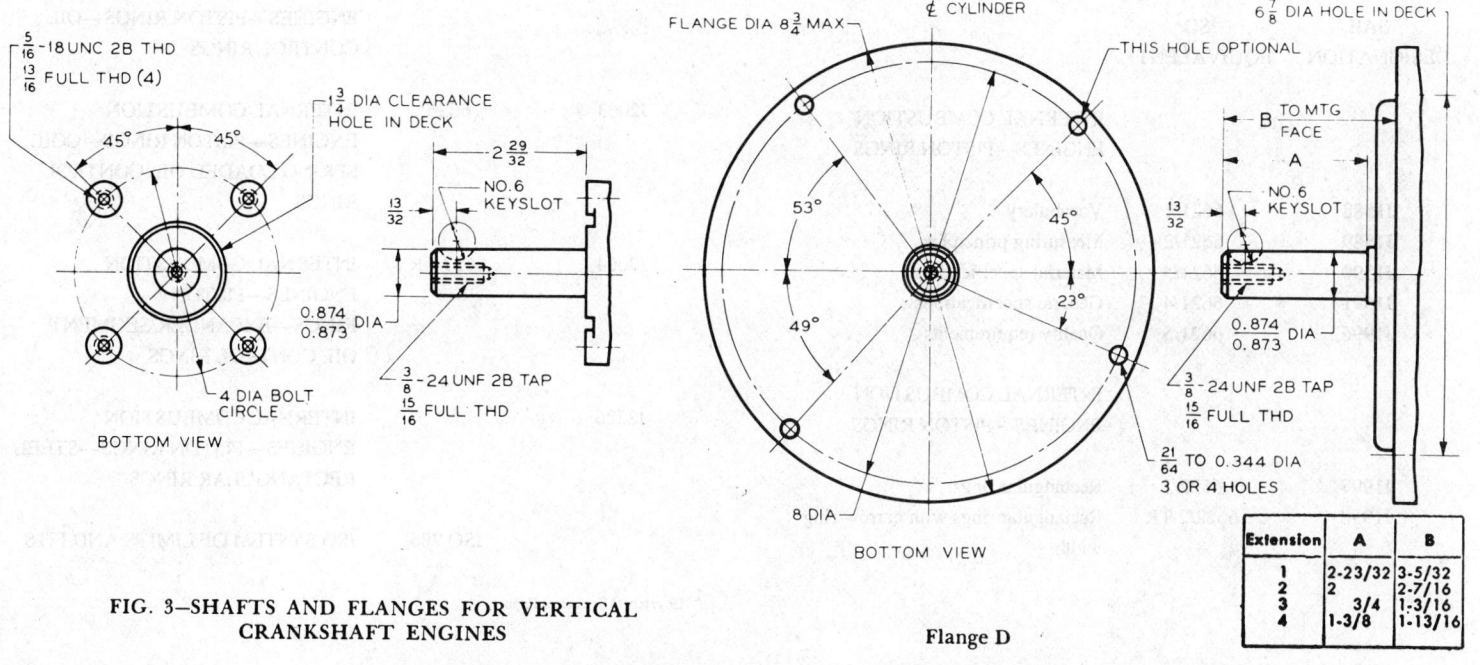

FIG. 3—SHAFTS AND FLANGES FOR VERTICAL CRANKSHAFT ENGINES

Flange D

ENGINE ROTATION AND CYLINDER NUMBERING—SAE J824 JAN86

SAE Standard

Report of the Engine Committee approved June 1962 and reaffirmed January 1986.

1. Scope—This report was developed to provide a method for indicating the direction of engine rotation and numbering of engine cylinders. The report is intended for use in designing new engines to eliminate the differences which presently exist in industry.

2. Standard Rotation—Counterclockwise rotation as viewed from the principal output end. If power can be delivered from either end, rotation shall be as viewed from the flywheel end.

3. Opposite Rotation—Clockwise rotation as viewed from the principal output end. If power can be delivered from either end, rotation shall be as viewed from the flywheel end.

4. Clinder Numbering—Cylinders shall be numbered by either of two methods:[1]

1. In single or multibank engines, number the cylinders in the sequence in which the connecting rods are mounted along the crankshaft beginning with the cylinder farthest from the principal output end.

2. In multibank engines, number the cylinders in sequence in each bank, starting with the cylinder farthest from the principal output end and designate right or left bank by suffix letters "R" and "L." EXAMPLE: 1R, 2R, and 1L, 2L.

If the suffix letters "R" and "L" are inadequate, other letter designations may be used if properly defined.

5. Right and Left Hand Designation—Cylinder bank and accessory locations are described as right or left when engine is viewed from flywheel or principal output end.

[1] This numbering method will generally exclude radial engines or those with coplanar cylinder bore axes.

INTERNAL COMBUSTION ENGINES—PISTON RINGS—VOCABULARY—SAE J1588 OCT92

SAE Standard

Report of the Piston Ring Standard Committee approved September 1990. Revised by the SAE Piston Ring Standards Committee 7 October 1992.

This SAE Standard is equivalent to ISO Standard 6621/1.

(R) 1. Scope and Field of Application—Differences, where they exist, are shown in Appendix A with associated rationale.

This SAE Standard defines the most commonly used terms for piston rings. These terms designate either types of piston rings or certain characteristics and phenomena of piston rings.

The terms and definitions apply to piston rings for reciprocating internal combustion engines and compressors working under analogous conditions.

Appendix B is included which lists equivalent terms in English, French, Russian, German, Spanish, Portuguese, Italian, and Japanese.

(R) 2. References

SAE DESIGNATION	ISO[1] EQUIVALENT	
		INTERNAL COMBUSTION ENGINES—PISTON RINGS
J1588	6621/1	Vocabulary
J1589	6621/2	Measuring principles
J1590	6621/3	Material specifications
J1591	6621/4	General specifications
J1996	6621/5	Quality requirements
		INTERNAL COMBUSTION ENGINES—PISTON RINGS
J1997	6622/1	Rectangular rings
J1998	6622/2 TR	Rectangular rings with narrow ring width
J1999	6623	INTERNAL COMBUSTION ENGINES—PISTON RINGS—SCRAPER RINGS
		INTERNAL COMBUSTION ENGINES—PISTON RINGS
J2000	6624/1	Keystone rings
J2001	6624/2 TR	Half keystone rings
J2002	6625	INTERNAL COMBUSTION ENGINES—PISTON RINGS—OIL CONTROL RINGS
J2003	6626	INTERNAL COMBUSTION ENGINES—PISTON RINGS—COIL SPRING LOADED OIL CONTROL RINGS
J2004	6627 TR	INTERNAL COMBUSTION ENGINES—PISTON RINGS—EXPANDER/SEGMENT OIL CONTROL RINGS
J2226		INTERNAL COMBUSTION ENGINES—PISTON RINGS—STEEL RECTANGULAR RINGS
	ISO 286	ISO SYSTEM OF LIMITS AND FITS

[1] TR refers to Technical Report

3. Piston Ring Classification

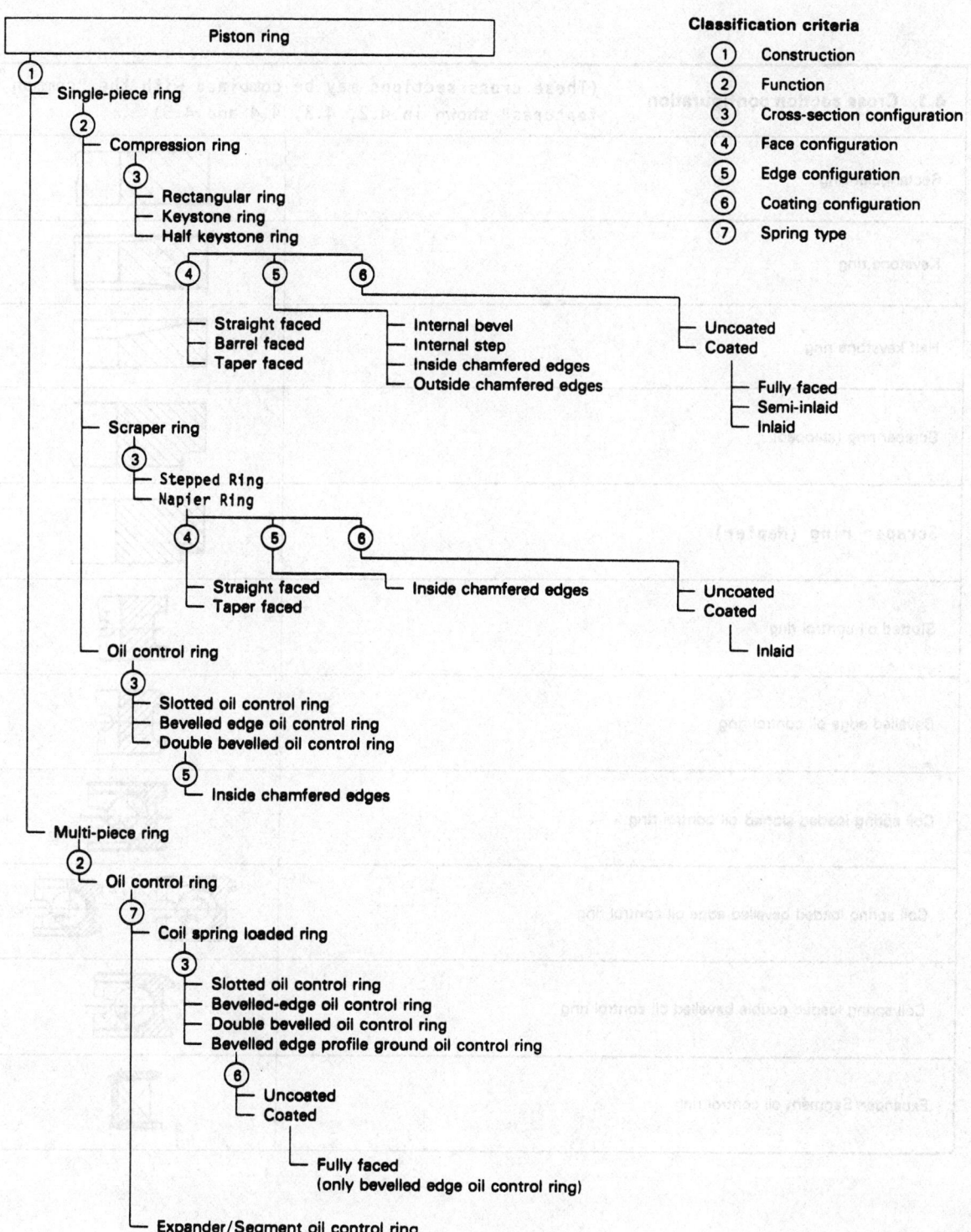

FIGURE 1—PISTON RING CLASSIFICATION

4. Piston Ring Types

4.1 Cross section configuration	(These cross sections may be combined with the "common features" shown in 4.2, 4.3, 4.4 and 4.5)
Rectangular ring	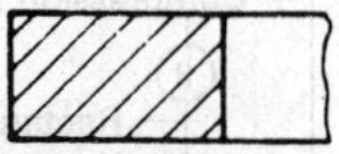
Keystone ring	
Half keystone ring	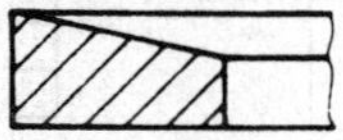
Scraper ring (stepped)	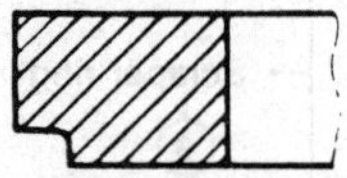
Scraper ring (Napier)	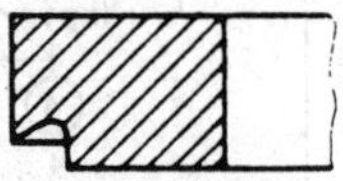
Slotted oil control ring	
Bevelled edge oil control ring	
Coil spring loaded slotted oil control ring	
Coil spring loaded bevelled edge oil control ring	
Coil spring loaded double bevelled oil control ring	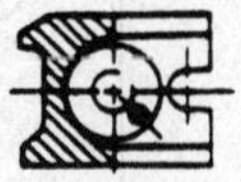
Expander/Segment oil control ring	

FIGURE 2—PISTON RING TYPES

4.2 Face configuration	
Straight faced	
Barrel faced	
Taper faced	

4.3 Edge configuration	
Internal bevel top (positive twist type)	
Internal step top (positive twist type)	
Internal bevel bottom (negative twist type)	
Internal step bottom (negative twist type)	
Inside edges chamfered	
Outside edges chamfered	
Inside and outside edges chamfered	

FIGURE 2 (CONTINUED)

4.4 Coating configuration	
Uncoated	
Coated	
— Fully faced	
— Semi-inlaid	
— Inlaid	
4.5 Joint configuration	
Joint with side notch	
Joint with internal notch	

FIGURE 2 (CONTINUED)

5. Piston Ring Nomenclature

5.1 Free (Unstressed) Ring

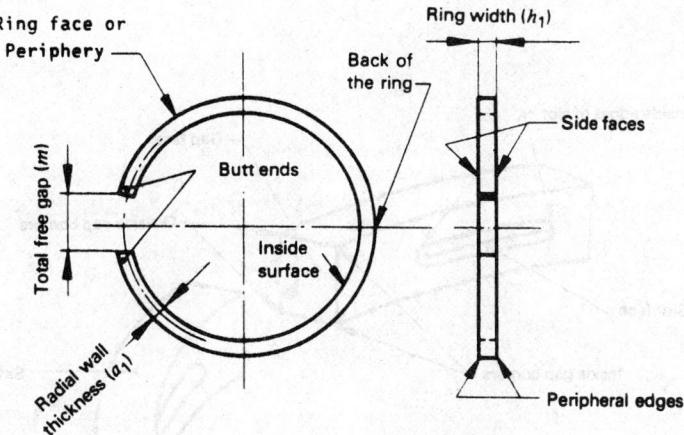

FIGURE 3—FREE (UNSTRESSED) RING

5.2 Closed Ring

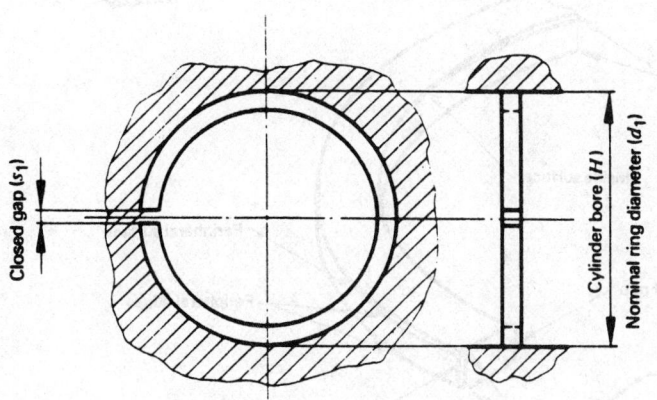

FIGURE 4—CLOSED RING

5.3 Ring Clearances

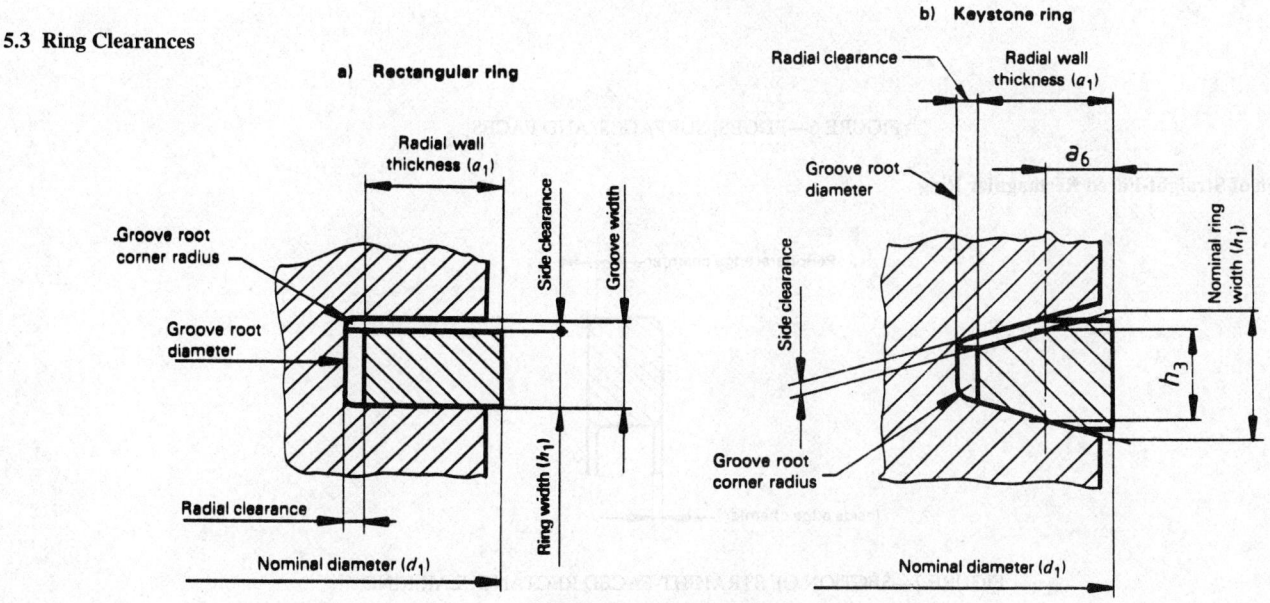

Method A: a_6 ref., h_3 measured
Method B: h_3 ref., a_6 measured

FIGURE 5—RING CLEARANCES

5.4 Edges, Surfaces, and Faces

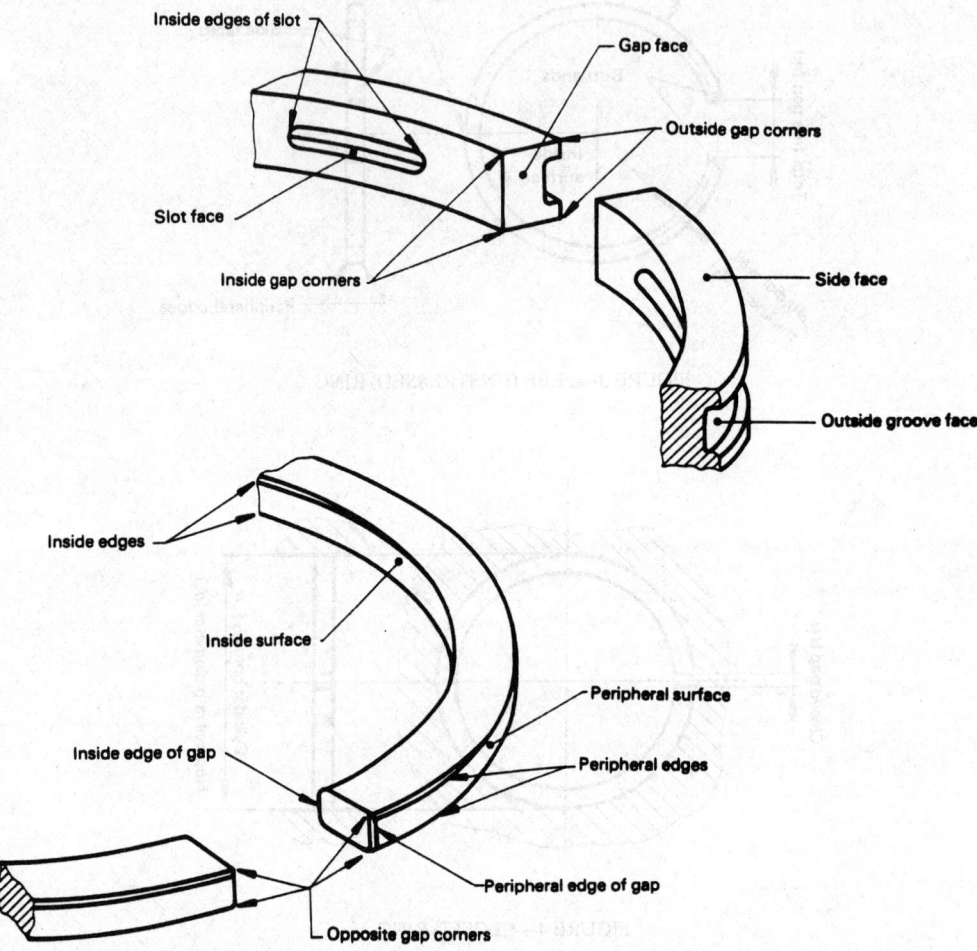

FIGURE 6—EDGES, SURFACES, AND FACES

5.5 Section of Straight-Faced Rectangular Ring

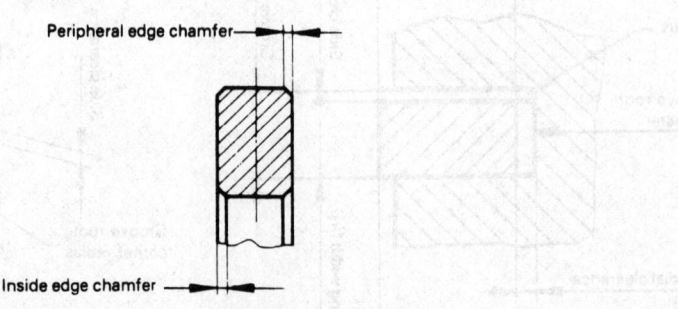

FIGURE 7—SECTION OF STRAIGHT-FACED RECTANGULAR RING

5.6 Section of Scraper Ring

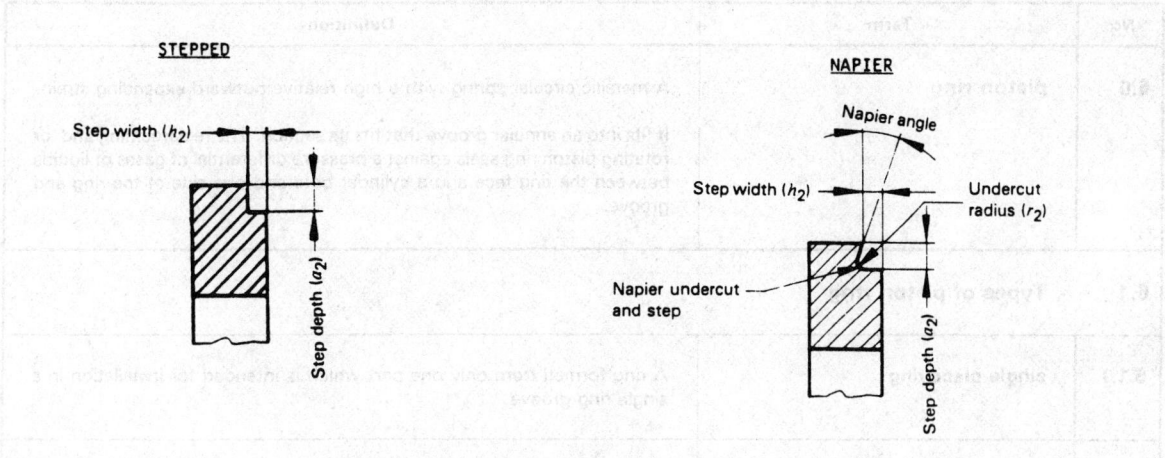

FIGURE 8—SECTION OF SCRAPER RING

5.7 Slotted Oil Control Ring

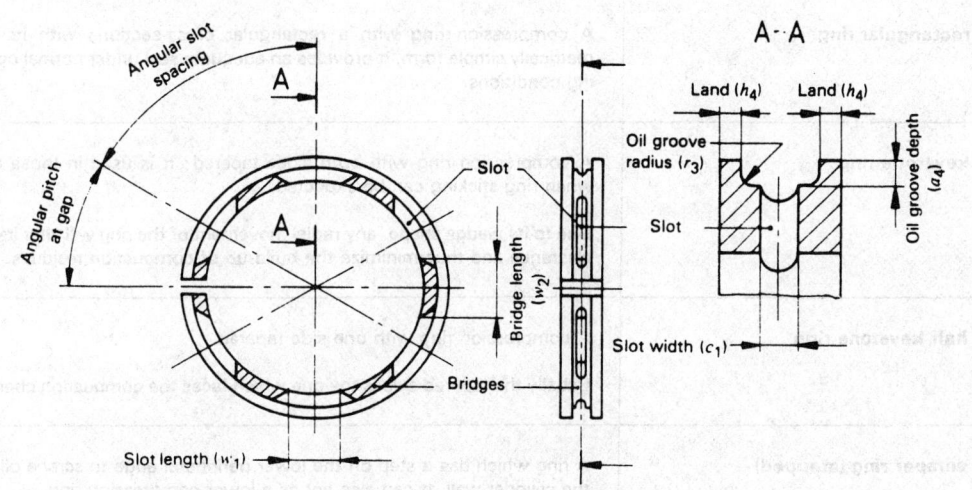

FIGURE 9—SLOTTED OIL CONTROL RING

6. Terms and Definitions

No.	Term	Definition
6.0	piston ring	A metallic circular spring with a high relative outward expanding strain. It fits into an annular groove that fits its section. The reciprocating and/or rotating piston ring seals against a pressure differential of gases or liquids between the ring face and a cylinder bore and one side of the ring and groove.
6.1	**Types of piston ring**	
6.1.1	single-piece ring	A ring formed from only one part which is intended for installation in a single ring groove.
6.1.2	multi-piece ring	A ring comprising two or more component parts which are intended for installation in a single ring groove.
6.1.3	compression ring	A ring intended primarily to prevent the leakage of gas past the piston.
6.1.4	oil control ring	A ring having oil return slots or an equivalent and intended to scrape oil from the cylinder wall.
6.1.5	rectangular ring	A compression ring with a rectangular cross-section; with its geometrically simple form, it provides an adequate seal under normal operating conditions.
6.1.6	keystone ring	A compression ring with both sides tapered; it is used in those cases when ring sticking can be expected. Due to its wedge shape, any radial movement of the ring will alter its axial clearance and thus minimize the build-up of combustion residues.
6.1.7	half keystone ring	A compression ring with one side tapered. Usually the tapered side is the one which faces the combustion chamber.
6.1.8	scraper ring (stepped)	A ring which has a step on the lower peripheral edge to scrape oil from the cylinder wall. It can also act as a lower compression ring.
6.1.9	scraper ring (Napier)	A scraper ring with an undercut step.
6.1.10	slotted oil control ring	A slotted oil control ring with parallel sides and two contact lands. Due to the narrow lands of this type of ring, a high unit pressure is achieved.

No.	Term	Definition
6.1.11	bevelled-edge oil control ring	A slotted oil control ring with parallel sides and two lands. The peripheral edges of both lands are chamfered, in order to achieve a further increase in unit pressure and thereby a better oil scraping effect.
6.1.12	double-bevelled oil control ring	A ring similar to type 6.1.11 except that both lands are chamfered on their upward facing edges. By chamfering the edges of both lands in the same direction, the oil scraping effect is even further improved.
6.1.13	coil spring loaded slotted oil control ring	A ring similar to type 6.1.10, the radial pressure of which is increased by means of a cylindrical coil spring. This spring acts equally in all directions against the inside of the ring.
6.1.14	coil spring loaded bevelled-edge oil control ring	A ring similar to type 6.1.11 but coil spring loaded with both lands chamfered on their outer edges.
6.1.15	coil spring loaded double-bevelled oil control ring	A ring similar to type 6.1.12 but coil spring loaded with both lands chamfered in the same direction on their upward facing edges.
6.1.16	coil spring loaded bevelled-edge chromium plated oil control ring	A ring similar to type 6.1.14 but with both lands chromium plated and chamfered on their inner and outer edges. May or may not be profile ground.
6.1.17	expander/segment oil control ring	A three-piece oil control ring comprised of an expander-spacer and two segments. Expander-spacer design will vary with manufacturer.
6.2	Physical characteristics of rings	
6.2.1	nominal ring diameter (symbol d_1)	The nominal diameter is identical to the nominal cylinder bore (H).
6.2.2	witness line	A narrow continuous line of contact lapped on the periphery of the ring, which can be seen around the circumference with normal vision.
6.2.3	joint	The joint at the butt ends of the ring.
6.2.4	butting	An effect which occurs when the gap faces of the ring touch.
6.2.5	effective free gap	The total free gap, m (see figure in 5.1), minus the measured closed gap s_1 (see figure in 5.2); it is the free gap used in the formulae for the calculations of E value, tangential and diametral forces and stresses.
6.2.6	pressure pattern	The radial pressure distribution around the circumference of the ring when closed in its nominal cylinder bore.

No.	Term	Definition
6.2.7	contact pressure	Pressure, in newtons per square millimetre, which a ring exerts radially against the cylinder wall.
6.2.8	pin point or burry light	Intermittent pin points of bright light or hazy light, but not bright direct light, observed in the test for light-tightness.
6.3	**Piston part**	
6.3.1	ring groove	The groove in the piston in which the piston ring is fitted.
6.4	**Measuring devices**	
6.4.1	ring gauge	A solid annular gauge having an inside diameter of nominal (H) cylinder bore.
6.4.2	datum surface	The plane on which the ring lies for measurements, except where otherwise specified.

APPENDIX A

A.1 This SAE Standard has been established to harmonize the ISO and SAE piston ring standards. The U.S. Technical Advisory Group, with the support of the National Engine Parts Manufacturers Association, has worked for the last 10 years with other national organizations on this worldwide standard. Some of the wording and phrasing may differ slightly from U.S. terminology for translation purposes.

In preparing this SAE Standard, the Scope and Field of Application and Reference sections of the ISO 6621/1 have been editorially revised and reorganized.

Paragraph numbering has been changed to reflect this reorganization.

In section 5.1 the label "Periphery" has been changed to "Ring Face or Periphery" to conform with wording elsewhere in the document.

Section 5.6 "Section of Scraper Ring" (original ISO section 6.6 "Section of Napier Ring") has been modified to include the scraper cross section which is predominant in North America.

APPENDIX B
LIST OF EQUIVALENT TERMS IN ENGLISH, FRENCH, RUSSIAN, GERMAN, SPANISH, PORTUGUESE, ITALIAN, AND JAPANESE

English	French	Russian	German	Spanish	Portuguese	Italian	Japanese
barrel faced	portée bombée	бочкообразная рабочая поверхность	ballige Lauffläche	bombeado	face de contacto bombeada	bombatura della superficie periferica	バレル・フェース
barrel on periphery	bombée sur la périphérie	бочкообразность по окружности	Balligkeit	periferia bombeada	periferia bombeada	bombatura sul diametro esterno	バレル面
bevelled edge oil control ring	segment racleur régulateur d'huile chanfreiné symétrique	маслосъёмное кольцо со встречными фасками	Dachfasenring	segmento de engrase con patines biselados simétricos	segmento de óleo de chanfros simétricos	anello raschiaolio a pattini smussati convergenti	ベベル型オイル・リング
butting	arc bout-à-bout	смыкание замка	Berührung der Stoßflächen	contacto de las puntas	contacto das pontas	contatto delle estremità dell'anello	突合せ
cam turned	tourné en forme	обточенный по копиру	formgedreht	torneado de forma	torneado de forma	tornito con camma	仕上げカム
closed gap	jeu à la coupe	тепловой зазор	Stoßspiel	ajuste de puntas	folga entre pontas	gioco delle estremità dell'anello	合口すきま
coating layer thickness	épaisseur de revêtement/ incrustation	толщина покрытия	Schichtdicke	espesor de recubrimiento/inserto	espessura do revestimento/enchimento	spessore rivestimento/rivestimento centrale	被覆層厚度
coil spring loaded oil control ring	segment racleur régulateur d'huile mis en charge par ressort hélicoïdal	маслосъёмное кольцо с витым расширителем	Schlauchfederring	segmento de engrase con expansor helicoidal	segmento de óleo de mola helicoidal	anello raschiaolio caricato con molla elicoidale	背位スプリング付オイル・リング
compression ring	segment de compression	компрессионное кольцо	Verdichtungsring	segmento de compresión	segmento de compressão	anello di compressione	圧力リング
datum surface	surface de référence	базовая плоскость (для измерений)	Meßebene	superficie de referencia	superfície de referência	piano di riferimento	基準面
diametral force	tare diamétrale	диаметральная сила	Diametralkraft	carga diametral	força diametral	forza diametrale	直径力
double-bevelled oil control ring	segment racleur d'huile chanfreiné parallèle	маслосъёмное кольцо с параллельными фасками	Gleichfasenring	segmento de engrase con patines biselados asimétricos	segmento de óleo de chanfros paralelos com mola helicoidal	anello raschiaolio a pattini smussati con doppio smussi paralleli	両面ベベル型オイル・リング
effective free gap	ouverture libre efficace	раствор замка минус тепловой зазор	effektive Maulweite	abertura libre efectiva	abertura livre efectiva	gioco effettivo delle estremità ad anello libero	有効自由線間
free flatness	planéité dans un état libre, sans contrainte	отклонение от плоскости кольца в свободном состоянии	Ebenheit, in unbelastetem Zustand	planicidad en estado libre	planeza no estado livre	planarità dell'anello libero	自由平面度

APPENDIX B (CONTINUED)

English	French	Russian	German	Spanish	Portuguese	Italian	Japanese
fully-faced	portée totale revêtue	с покрытием рабочей поверхности	Laufflächen beschichtung	totalmente recubierto	face de contacto totalmente revestida	superficie periferica interamente rivestita	全面成形
half keystone ring	segment demi trapézoidal	трапециевидное кольцо одностороннее	einseitiger Trapezring	segmento trapezial simple	segmento semitrapezoidal	anello semi trapezoidale	ハーフ・キーストン・リング
heat-formed	mise en forme thermique	с горячей формовкой	thermisch gespannt	abertura térmica	conformado térmicamente	sagomato a caldo	熱成形
inlaid	incrusté	с заполненной канавкой	Füllung	inserto	incrustado	rivestimento centrale sulla superficie periferica	填層
inside edges chamfered	arêtes intérieures chanfreinées	с внутренними фасками	Innenkantenbruch	cantos interiores biselados	arestas interiores chanfradas	spigoli interni smussati	内部角面取り
inside and peripheral edges chamfered	arêtes intérieures et extérieures chanfreinées	с внутренними и наружными фасками	Innen- und Außen kantenbruch	cantos interiores y exteriores biselados	arestas interiores e exteriores chanfradas	spigoli interni e esterni smussati	内面周辺修正（面取り）
internal bevel bottom (negative twist type)	chanfrein intérieur de torsion bas (torsion négative)	кольцо с внутренней нижней фаской (обратного скручивания)	Innenfase unten (negative Vertwistung)	chaflán interior inferior (tipo torsional negativo)	chanfro de torção interior inferior (torção negativa)	smusso sul diametro interno inferiore (tipo a torsione negativa)	インナー・ベベル・ボトム（負ねじり型）
internal bevel top (positive twist type)	chanfrein intérieur de torsion haut (torsion positive)	кольцо с внутренней верхней фаской (прямого скручивания)	Innenfase oben (positive Vertwistung)	chaflán interior superior (tipo torsional positivo)	chanfro de torção interior superior (torção positiva)	smusso sul diametro interno superiore (tipo a torsione positiva)	インナー・ベベル・トップ（正ねじり型）
internal step bottom (negative twist type)	épaulement intérieur bas (torsion négative)	кольцо с внутренней нижней выточкой (обратного скручивания)	Innenwinkel unten (negative Vertwistung)	escalón interior inferior (tipo torsional negativo)	rebaixo interior inferior (torção negativa)	spallamento sul diametro interno inferiore (tipo a torsione negativa)	内階底（負ねじり型）
internal step top (positive twist type)	épaulement intérieur haut (torsion positive)	кольцо с внутренней верхней выточкой (прямого скручивания)	Innenwinkel oben (positive Vertwistung)	escalón interior superior (tipo torsional positivo)	rebaixo interior superior (torção positiva)	spallamento sul diametro interno superiore (tipo a torsione positiva)	内階上面（正ねじり型）
joint	coupe	замок	Stoß	corte	corte	estremità	継手
joint with internal notch	coupe avec encoche intérieure	замок с торцевой фиксацией	Stoß mit Innensicherung	entalla para fijo interior	corte com entalhe interna	estremità con tacca interna	内切欠き継手
joint with side notch	coupe avec encoche frontale	замок с внутренней фиксацией	Stoß mit Flankensicherung	entalla para fijo lateral	corte com entalhe lateral	estremità con tacca frontale	側切欠き継手
keystone angle	angle du trapèze	угол трапеции	Trapezwinkel	angulo trapecial	ângulo do trapézio	angolo del trapezio	keystone角
keystone ring	segment trapézoidal	кольцо трапециевидное двустороннее	Trapezring	segmento trapecial	segmento trapezoidal	anello trapezoidale	keystoneリング
land offset	dépôt des lèvres	смещение перемычки	Laufstegversatz	desplazamiento cordón	desencontro das faces	disassamento pattino	ランド・オフセット

APPENDIX B (CONTINUED)

English	French	Russian	German	Spanish	Portuguese	Italian	Japanese
land width	largeur des lèvres	высота перемычки	Laufsteghöhe	espesor cordón	espessura dos cordões	altezza pattino	ランド幅
light tightness	étanchéité à la lumière	плотность прилегания	Lichtspaltdichtheit	estanqueidad	vedação à luz	tenuta alla (prova di) luce	光案度
modulus of elasticity	module d'élasticité	модуль упругости	Elastizitätsmodul	módulo de elasticidad	módulo de elasticidade	modulo di elasticità	弾性計数
multi piece ring	segment multipièce	составное кольцо (многоэлементное)	mehrteiliger Kolbenring	segmento múltiple	segmento de multiplas peças	anello multipezzo	コンバインド・リング
Napier ring	segment bec d'aigle	скребковое кольцо с подрезанием	скребковое кольцо с поднутрением канавки	segmento rascador de uña	segmento raspador "Napier"	anello napier	ナピーア・リング
Napier ring, taper faced	segment bec d'aigle à portée conique	скребковое кольцо с поднутрением и с конической рабочей поверхностью	Nasenminutenring, hinterstochen	segmento rascador de uña con periferia cónica	segmento raspador "Napier" com face de contacto cónica	anello napier con superficie periferica conica	ナピーア・リング斜面成形
nominal ring diameter	diamètre nominal du segment	номинальный диаметр кольца	Nenndurchmesser	diámetro nominal del segmento	diâmetro nominal do segmento	diametro nominale dell'anello	リング呼び径
obliqueness	obliquité	коробление	Schieflage	inclinación	inclinação	obliquità	斜面度
oil control ring	segment racleur régulateur d'huile	маслосъемное кольцо	Ölabstreifring	segmento control de aceite	segmento de óleo	anello raschiaolio	オイル・リング
ovality or circularity	ovalisation ou circularité	овальность	Ovalität	ovalidad	ovalização	ovalità	楕円度或円型
peripheral edges chamfered	arêtes extérieures chanfreinées	с наружными фасками	Außenkantenbruch	cantos exteriores biselados	arestas exteriores chanfradas	spigoli esterni smussati	面縁修正
pin point or burry light	pointe d'épingle ou lumière irisée	мерцающий просвет	leichter unterbrochener Lichtschimmer	luz difusa	passagem de luz difusa	puntinatura o diffusione di luce	ピン継手
piston ring	segment de piston	поршневое кольцо	Kolbenring	segmento de pistón	segmento do êmbolo	anello	ピストン・リング
point deflection	point de flexion	прогиб стыкования	Stoßeinfall	punto de flexión	ponto de flexão	inflessione delle punte	点偏差
pressure pattern	diagramme de pression	эпюра распределения давлений	Druckverteilung	distribución de la presión	distribuição da pressão	distribuzione della pressione	圧力パターン
radial wall thickness	épaisseur radiale	радиальная толщина	radiale Wanddicke	espesor radial	espessura radial	spessore radiale	T寸
rectangular ring	segment rectangulaire	прямоугольное кольцо	Rechteckring	segmento rectangular	segmento rectangular	anello rettangolare	長方型リング
ring gauge	bague étalon	кольцевой калибр	Kontrollring	calibre para segmento	calibre do segmento	calibro ad anello	リング・ゲージ
ring groove	gorge du piston	канавка поршня	Ringnut	ranura de pistón	caixa do êmbolo	gola dell'anello	リング溝
ring width	hauteur du segment	высота кольца	Ringhöhe	altura de segmento	altura do segmento	altezza dell'anello	リング幅

APPENDIX B (CONTINUED)

English	French	Russian	German	Spanish	Portuguese	Italian	Japanese
scraper ring (stepped)	segment racleur mixte (épaulé droit)	скребковое кольцо	Nasenring	segmento rascador	segmento raspador	anello di tenuta con funzioni di raschiaolio	油かきリング (階型)
scraper ring (stepped), taper-faced	segment racleur mixte (épaulé droit) à portée conique	скребковое кольцо с конической поверхностью	Nasenminutenring	segmento cónico rascador	segmento raspador com face de contacto cónica	anello di tenuta con funzioni di raschiaolio a superficie periferica conica	油かきリング (階型) テーパ型
semi-inlaid	semi-incrusté	с полузаполненной канавкой	einseitige Füllung	semi inserto	semi incrustado	riporto seminterno	部分埋層
single piece ring	segment monopièce	одноэлементное кольцо	einteiliger Kolbenring	segmento de una sola pieza	segmento de uma só peça	anello monopezzo	単体リング
slotted oil control ring	segment racleur régulateur d'huile à fentes	маслосъёмное кольцо с прорезями	Ölschlitzring	segmento de engrase con ventanas	segmento de óleo com rasgos	anello raschiaolio con feritoie scarico olio	すり割り付 油リング
straight faced	portée cylindrique	кольцо с цилиндрической рабочей поверхностью	zylindrische Lauffläche	cara recta	face de contacto plana	superficie periferica rettilinea	直面成形
tangential force	tare tangentielle	тангенциальная сила	Tangentialkraft	carga tangencial	força tangencial	carico tangenziale	接線力
taper-faced ring	segment à portée conique	кольцо с конической рабочей поверхностью	Minutenring	segmento cónico	segmento com face de contacto cónica	anello con superficie periferica conica	斜面形テーパ・リング
taper faced keystone ring	segment trapézoïdal à portée conique	трапециевидное кольцо с конической рабочей поверхностью	Trapezminutenring	segmento cónico trapecial	segmento trapezoidal com face de contacto cónica	anello trapezoidale a superficie periferica conica	斜面形 keystone リング
taper on periphery	conicité de la périphérie	конусность рабочей поверхности	Winkligkeit (Zylindrizität, Konizität)	perfil cónico	conicidade da face de contacto	conicità sulla periferia	斜傾表面
total free gap	ouverture libre totale	размер замка кольца в свободном состоянии	Maulweite	abertura libre	abertura livre total	gioco totale delle estremità ad anello libero	全自由隙
twist	torsion	скручивание	Verwistung	torsión/torsional	torção	torsione	ねじり
uncoated ring	segment non revêtu	кольцо без покрытия	unbeschichteter Ring	segmento sin recubrimiento	segmento sem revestimento	anello non rivestito	無被覆リング
unevenness	inégalité	шероховатость	Unebenheit	ondulación	ondulação	ondulazione	非平坦
wind	vrillage	смещение стыковании	axialer Stoßversatz	giro	desvio das pontas (empeno)	spostamento assiale (delle estremità dell'anello)	巻上げ
witness line	ligne témoin	линия контакта	Tragspiegel	línea de testimonio	linha testemunha	linea di riferimento o di testimonio	証示線

INTERNAL COMBUSTION ENGINES—PISTON RINGS—RECTANGULAR RINGS WITH NARROW RING WIDTH—SAE J1998 OCT92

SAE Standard

Report of the Piston Ring Standards Committee 7 approved June 1990. Revised by the SAE Piston Ring Standards Committee 7 October 1992.

This SAE Standard is equivalent to ISO Standard 6622/2 TR.

1. Scope and Field of Application—Differences, where they exist, are shown in Appendix A.

This SAE Standard specifies the essential dimensional features of R, B, and M rectangular piston ring types with narrow ring width.

Dimensional Tables 8 and 9 allow for the use of cast iron (Table 8) or steel (Table 9). Since the modulus of elasticity of steel rings is higher than that of cast iron rings, the fluctuation in the surface pressure will become greater if the free gap is set as the reference for force. Therefore, forces are set using the surface pressure as the reference, in order to minimize the effect of the fluctuation.

The requirements of this document apply to rectangular rings for reciprocating internal combustion engines up to and including 90 mm diameter for cast iron rings and up to and including 100 mm diameter for steel. They may also be used for piston rings of compressors working under similar conditions.

2. References

SAE DESIGNATION	ISO[1] EQUIVALENT	
		INTERNAL COMBUSTION ENGINES—PISTON RINGS
J1588	6621/1	Vocabulary
J1589	6621/2	Measuring principles
J1590	6621/3	Material specifications
J1591	6621/4	General specifications
J1996	6621/5	Quality requirements
		INTERNAL COMBUSTION ENGINES—PISTON RINGS
J1997	6622/1	Rectangular rings
J1998	6622/2 TR	Rectangular rings with narrow ring width
J1999	6623	INTERNAL COMBUSTION ENGINES—PISTON RINGS—SCRAPER RINGS
		INTERNAL COMBUSTION ENGINES—PISTON RINGS
J2000	6624/1	Keystone rings
J2001	6624/2 TR	Half keystone rings
J2002	6625	INTERNAL COMBUSTION ENGINES—PISTON RINGS—OIL CONTROL RINGS
J2003	6626	INTERNAL COMBUSTION ENGINES—PISTON RINGS—COIL SPRING LOADED OIL CONTROL RINGS
J2004	6627 TR	INTERNAL COMBUSTION ENGINES—PISTON RINGS—EXPANDER/SEGMENT OIL CONTROL RINGS
J2226		INTERNAL COMBUSTION ENGINES—PISTON RINGS—STEEL RECTANGULAR RINGS
	1101	TECHNICAL DRAWINGS—GEOMETRICAL TOLERANCING—TOLERANCING OF FORM, ORIENTATION, LOCATION AND RUN-OUT—GENERALITIES, DEFINITIONS, SYMBOLS INDICATIONS ON DRAWINGS

[1] TR refers to Technical Report

3. Ring Types and Designation Examples
3.1 Type R—Straight Faced Rectangular Ring

3.1.1 GENERAL FEATURES
NOTE—See Table 8 or 9 for dimensions and forces.

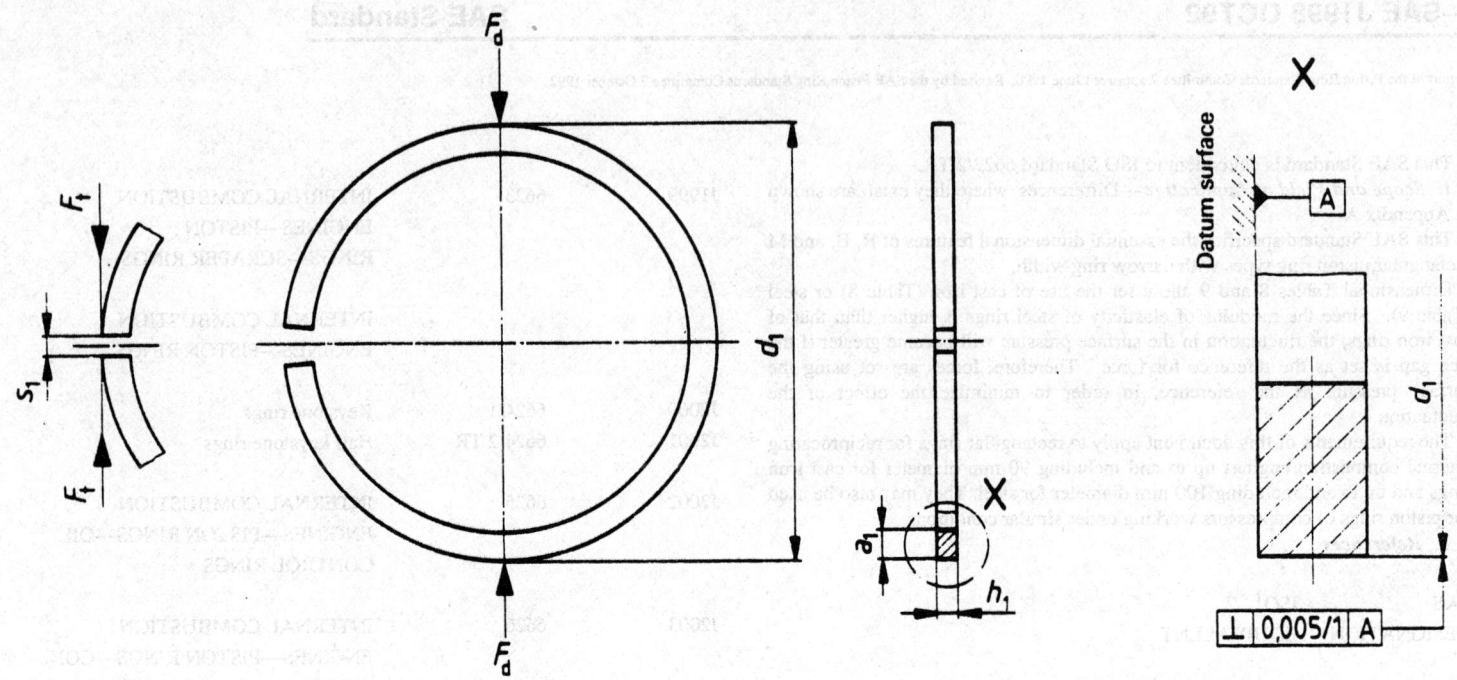

FIGURE 1—TYPE R

3.1.2 DESIGNATION EXAMPLE—Designation of a straight faced rectangular ring with narrow ring width of $d_1 = 60$ mm nominal diameter, $h_1 = 1.2$ mm ring width made of spheroidal graphite cast iron heat-treated (material subclass 51), general features as shown in Figure 1, and periphery chromium coated fully faced design 0.1 mm minimum thickness.

a. Piston ring SAE J1998 R-60 × 1.2-MC51/CR2

3.2 Type B—Barrel Faced Rectangular Ring
3.2.1 GENERAL FEATURES
NOTE—See Table 8 or 9 for dimensions and forces.

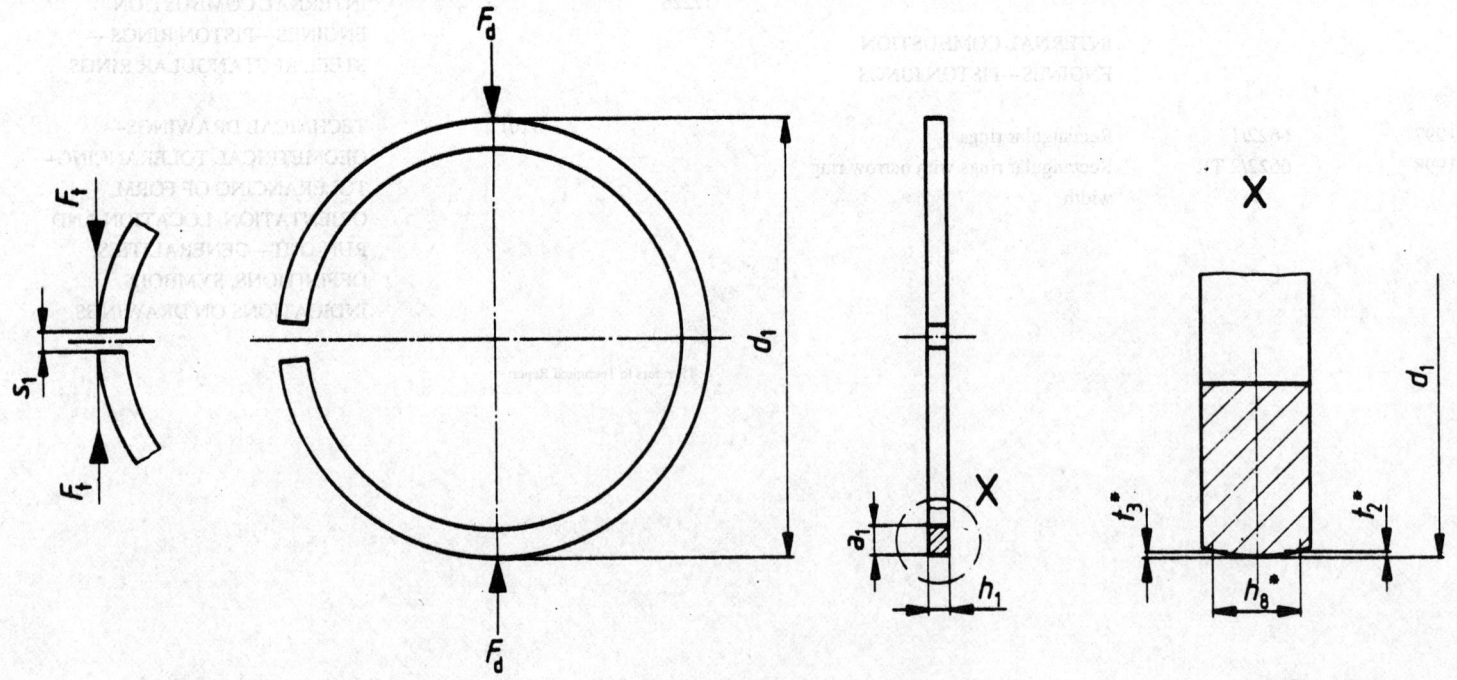

*See Table 1.

FIGURE 2—TYPE B

TABLE 1—GAUGE WIDTH (h_8) AND BARREL DIMENSIONS

Dimensions in millimeters

h_1	h_8	t_2, t_3	t_2, t_3	Maximum Peak Off Center
1.2	0.6	0.001	0.013	0.2
1.5	0.8	0.002	0.016	0.25

3.2.2 DESIGNATION EXAMPLE—Designation of a barrel faced rectangular ring with narrow ring width of $d_1 = 60$ mm nominal diameter, $h_1 = 1.2$ mm ring width made of steel (material subclass 62), general features as shown in Figure 2, and periphery chromium coated fully faced design 0.1 mm minimum thickness:
  a. Piston ring SAE J1988 B-60 × 1.2-MC62/CR2

3.3 Type M—Taper Faced Rectangular Ring
3.3.1 GENERAL FEATURES
NOTE—See Table 8 or 9 for dimensions and forces.

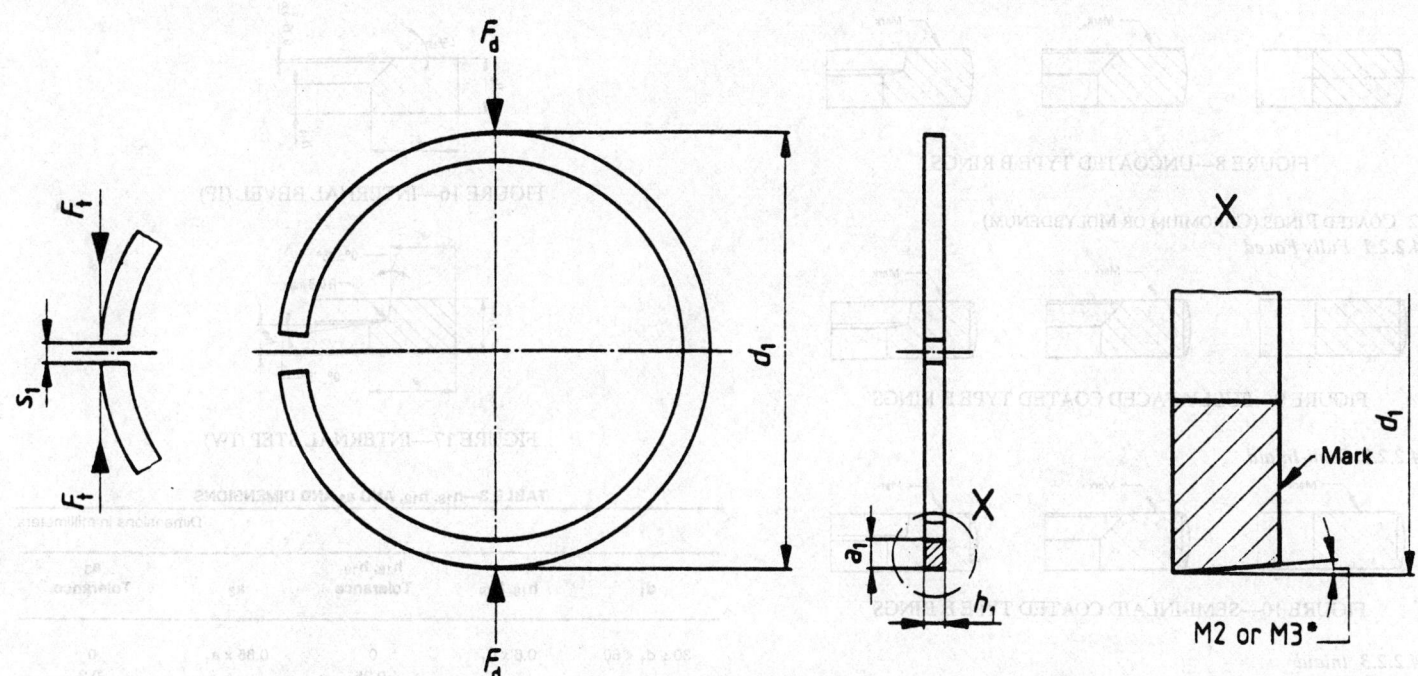

FIGURE 3—TYPE M

TABLE 2—TAPER UNCOATED AND COATED RINGS (MOLYBDENUM OR CHROMIUM WITH TAPERED PERIPHERY GROUND)

Taper	Tolerance[2]	Tolerance[2]	With IF or IW[1] (Top Side) Tolerance[2]	With IF or IW[1] (Top Side) Tolerance[2]
M2	30'	+50' 0	30'	+60' 0
M3	60'		60'	

[1] IF and IW are explained in Figures 16 and 17.
[2] For coated rings with tapered periphery not ground, the tolerance shall be increased by 10' (e.g., M3 = 60': +60' 0 for M-rings or -70' 0 for M-rings with IF or IWI)

3.3.2 DESIGNATION EXAMPLE—Designation of a taper faced rectangular ring with narrow ring width of $d_1 = 60$ mm nominal diameter, $h_1 = 1.2$ mm ring width made of spheroidal graphite cast iron, heat-treated (material subclass 51), general features as shown in Figure 3 with taper M3 = 60', and periphery chromium coated fully faced design 0.1 mm minimum thickness:
  a. Piston ring SAE J1998 M3-60 × 1.2-MC51/CR2

4. Common Features
4.1 Type R—Straight Faced Rectangular Ring
4.1.1 UNCOATED RINGS

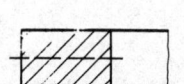 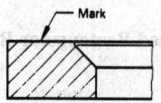 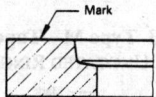

FIGURE 4—UNCOATED TYPE R RINGS

4.1.2 COATED RINGS (CHROMIUM OR MOLYBDENUM)
4.1.2.1 Fully Faced

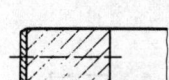 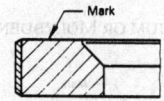 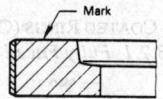

FIGURE 5—FULLY FACED COATED TYPE R RINGS

4.1.2.2 Semi-Inlaid

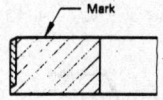 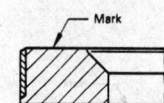 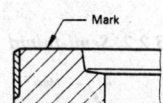

FIGURE 6—SEMI-INLAID COATED TYPE R RINGS

4.1.2.3 Inlaid

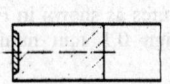 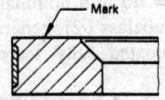 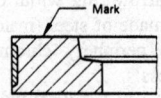

FIGURE 7—INLAID COATED TYPE R RINGS

4.2 Type B—Barrel Faced Rectangular Ring
4.2.1 UNCOATED RINGS

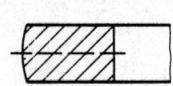 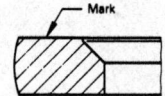 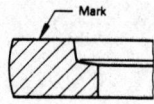

FIGURE 8—UNCOATED TYPE B RINGS

4.2.2 COATED RINGS (CHROMIUM OR MOLYBDENUM)
4.2.2.1 Fully Faced

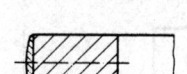 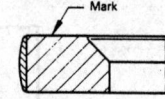 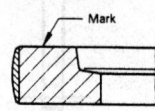

FIGURE 9—FULLY FACED COATED TYPE B RINGS

4.2.2.2 Semi-Inlaid

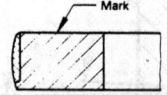 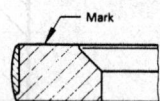 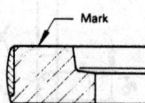

FIGURE 10—SEMI-INLAID COATED TYPE B RINGS

4.2.2.3 Inlaid

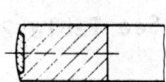 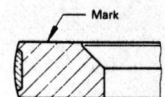 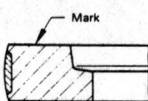

FIGURE 11—INLAID COATED TYPE B RINGS

4.3 Type M—Taper Faced Rectangular Rings
4.3.1 UNCOATED RINGS

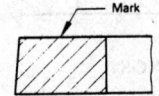 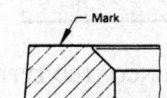 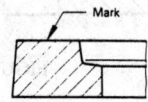

FIGURE 12—UNCOATED TYPE M RINGS

4.3.2 COATED RINGS (CHROMIUM OR MOLYBDENUM)
4.3.2.1 Fully Faced

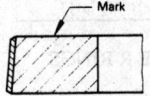 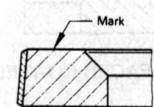 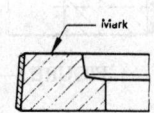

FIGURE 13—FULLY FACED COATED TYPE M RINGS

4.3.2.2 Semi-Inlaid

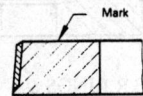 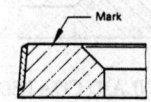 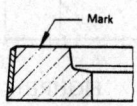

FIGURE 14—SEMI-INLAID COATED TYPE M RINGS

4.3.2.3 Inlaid

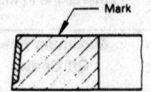 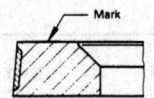 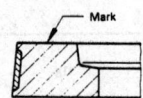

FIGURE 15—INLAID COATED TYPE M RINGS

4.4 R, B, and M Rings (Positive Twist Type)—Internal Bevel (Top Side) or Internal Step (Top Side)

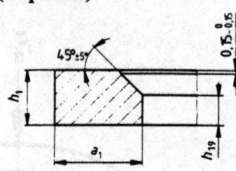

FIGURE 16—INTERNAL BEVEL (IF)

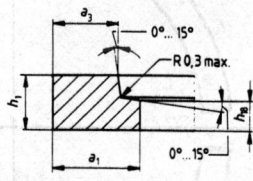

FIGURE 17—INTERNAL STEP (IW)

TABLE 3—h_{18}, h_{19}, AND a_3 AND DIMENSIONS

Dimensions in millimeters

d_1	h_{18}, h_{19}	h_{18}, h_{19} Tolerance	a_3	a_3 Tolerance
$30 \leq d_1 < 60$	$0.6 \times h_1$	0 -0.25	$0.85 \times a_1$	0 -0.2
$60 \leq d_1 \leq 100$	$0.6 \times h_1$	0 -0.25	$0.9 \times a_1$	0 -0.3

4.5 Chamfered Edges (Cast Iron Rings)
4.5.1 R AND B RINGS—OUTSIDE CHAMFERED EDGES (KA)

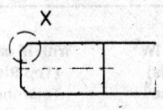 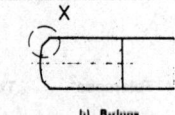

FIGURE 18—OUTSIDE CHAMFERED EDGES (KA)

4.5.2 R, B, AND M RINGS—INSIDE CHAMFERED EDGES (KI)

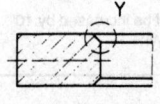 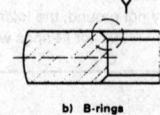 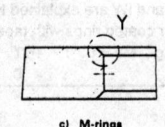

FIGURE 19—INSIDE CHAMFERED EDGES (KI)

4.5.3 R AND B RINGS—OUTSIDE AND INSIDE CHAMFERED EDGES (KA + KI)
NOTE: See Table 4 for dimensions.

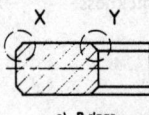 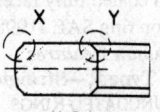

FIGURE 20—OUTSIDE AND INSIDE CHAMFERED EDGES (KA + KI)

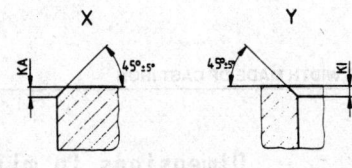

FIGURE 21—DETAILS OF FIGURES 18, 19, AND 20

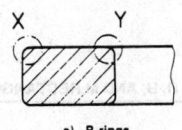

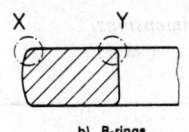

FIGURE 22—OUTSIDE AND INSIDE ROUNDED EDGES

TABLE 4—KA AND KI DIMENSIONS

KA	KI
0.15 ± 0.1	max 0.2

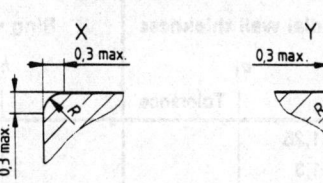

FIGURE 23—DETAILS OF FIGURE 22

4.7 R, B, and M Rings (Fully Faced, Semi-Inlaid, and Inlaid)—Layer Thickness

NOTE—See Table 5 for dimensions.

4.6 R and B Steel Rings—Outside and Inside Rounded Edges

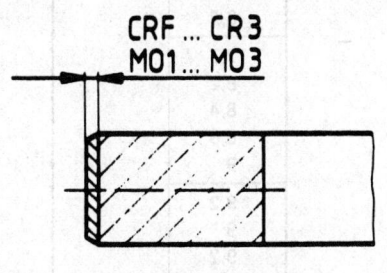

a) Fully faced

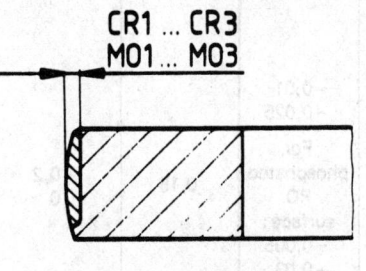

b) Semi-inlaid

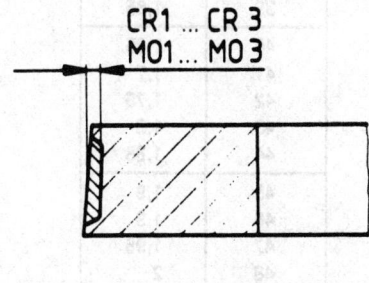

c) Inlaid

FIGURE 24—LAYER THICKNESS

TABLE 5—LAYER THICKNESS

Dimensions in millimeters

Chromium	Molybdenum	Thickness min
CRF	-	0.005
CR1	MO1	0.05
CR2	MO2	0.1
CR3[1]	MO3[1]	0.15

[1] CR3 and MO3 apply to rings with nominal diameters of 50 mm or greater.

5. Force Factors—The tangential and diametral forces given in Tables 8 shall be corrected, when additional features and/or materials other than grey cast iron with a modulus of elasticity of 100 000 MPa are being used.

For common features, multiplier correction factors given in Tables 6 and 7 and the force correction factors given in SAE J1591 shall be used.

Also the tangential and diametral force given in Table 9 shall be corrected when additional features are being used.

TABLE 6—FORCE CORRECTION FACTORS FOR R, B, AND M RINGS WITH FEATURES KA, KI, IF, IW, AND TAPER

Factor KA	Factor KI	Factor Taper M3	Factor IF	Factor IW $d_1 < 60$	Factor IW $d_1 > 60$	Factor KA and KI	Factor KA and Taper M3	Factor KA and IF	Factor KA and IW $d_1 < 60$	Factor KA and IW $d_1 > 60$
0.98	0.98	0.98	0.87	0.77	0.82	0.96	0.96	0.85	0.75	0.8

TABLE 7—FORCE CORRECTION FACTORS FOR COATED R, B, AND M RINGS (FULLY FACED, SEMI-INLAID, AND INLAID TYPE)

d_1 mm	Factor CRF	Factor CR1	Factor CR2/MO1	Factor CR3	Factor MO2	Factor MO3
$30 \leq d_1 < 50$	1	0.81	0.7	-	0.64	-
$50 \leq d_1 \leq 100$	1	0.9	0.85	0.81	0.81	0.75

6. Dimensions

TABLE 8—DIMENSIONS OF R, B, AND M RECTANGULAR RINGS WITH NARROW RING WIDTH MADE OF CAST IRON

Dimensions in millimeters

Nominal diameter d_1	Radial wall thickness a_1	Tolerance	Ring width h_1	Tolerance	Closed gap s_1	Tolerance	Tangential force F_t, N	Tolerance	Diametral force F_d, N	Tolerance
30	1,25								6	
31	1,3								6,2	
32	1,35							—		6,7	
33	1,4								6,9	
34	1,4								6,5	
35	1,45								6,9	
36	1,5								7,1	
37	1,55								7,5	
38	1,6								7,7	
39	1,65								8,2	
40	1,65								7,7	
41	1,7				−0,01				8,2	
42	1,75				−0,025			—	8,4	
43	1,8				For				8,8	
44	1,85				phosphated				9	
45	1,9				PO	0,15	+0,2			9,2	
46	1,9				surface:		0			9	
47	1,95				−0,005			—	9,2	
48	2				−0,03				9,7	
49	2,05								9,9	
50	2,1							4,8		10,3	
51	2,15							4,9		10,5	
52	2,15							4,7		10,1	
53	2,2	± 0,15						4,9		10,5	
54	2,25	Within						5		10,8	
55	2,3	a ring:	1,2					5,2		11,2	± 30 %
56	2,35	0,15						5,4		11,6	
57	2,4	max.						5,5		11,8	
58	2,4							5,3		11,4	
59	2,45							5,5		11,8	
60	2,5							5,6		12	
61	2,55							5,7		12,3	
62	2,6							5,9		12,7	
63	2,65							6,1		13,1	
64	2,65							5,9	± 30 %	12,7	
65	2,7							6		12,9	
66	2,75							6,2		13,3	
67	2,8				−0,01	0,2	+0,2	6,3		13,3	
68	2,85				−0,03		0	6,5		14	
69	2,9				For			6,6		14,2	
70	2,9				phosphated			6,4		13,8	
71	2,95				PO			6,6		14,2	
72	3				surface:			6,7		14,4	
73	3,05				−0,005			6,9		14,8	
74	3,1				−0,03			7,1		15,3	
75	3,15							7,2		15,5	
76	3,15							7		15,1	
77	3,2					0,25	+0,25	7,1		15,3	
78	3,25						0	7,3		15,7	
79	3,3							7,4		15,9	

This table is shown in ISO format. Commas represent decimal points.

TABLE 8—DIMENSIONS OF R, B, AND M RECTANGULAR RINGS WITH NARROW RING WIDTH MADE OF CAST IRON (CONTINUED)

Dimensions in millimeters

Nominal diameter d_1	Radial wall thickness a_1	Tolerance	Ring width h_1	Tolerance	Closed gap s_1	Tolerance	Tangential force F_t, N	Tolerance	Diametral force F_d, N	Tolerance
80	3,35						7,6		16,3	
81	3,4						7,8		16,8	
82	3,4			−0,01 −0,03			7,6		16,3	
83	3,45	± 0,15 Within a ring: 0,15 max.		For phosphated PO surface: −0,005 −0,03		+0,25 0	7,7	± 30 %	16,6	± 30 %
84	3,5		1,2		0,25		7,9		17	
85	3,55						8		17,2	
86	3,6						8,2		17,6	
87	3,65						8,3		17,8	
88	3,65						8,1		17,4	
99	3,7						8,3		17,8	
90	3,75						8,5		18,2	

NOTES

[1] For intermediate sizes (for example repair sizes), the radial wall thickness of the next smaller nominal diameter should be applied.

[2] The values for F_t and F_d, given in Table 8, apply to as cast grey cast iron with a typical modulus of elasticity (E_n) of 100 000 MPa. Multiplying factors for materials having a different modulus (E_n) are given in SAE J1591.

Mean forces are calculated for nominal radial wall thickness (a_t) and mean ring width (h_t).

[3] For the sole purpose of this document, the assumed average ratio F_d/F_t is 2.15. However, for rings up to 50 mm, the ratio F_d/F_t shall be determined between the manufacturer and client.

This table is shown in ISO format. Commas represent decimal points.

TABLE 9—DIMENSIONS OF R, B, AND M RECTANGULAR RINGS WITH NARROW RING WIDTH MADE OF STEEL

Dimensions in millimeters

Nominal diameter d_1	Radial wall thickness a_1	Tolerance	Ring width h_1 Column 1	Ring width h_1 Column 2	Tolerance	Closed gap s_1	Tolerance	Tangential force F_t, N — For h_1 shown in column 1	Tangential force F_t, N — For h_1 shown in column 2	Tolerance	Diametral force F_d, N — For h_1 shown in column 1	Diametral force F_d, N — For h_1 shown in column 2	Tolerance
30	1,1							—	—		6,8	8,5	
31											7	8,8	
32											7,3	9,1	
33											7,5	9,4	
34	1,3										8,2	10,5	
35								—	—		8,4	10,8	
36											8,6	11,1	
37											8,9	11,4	
38											9,2	11,8	
39	1,5										9,5	12	
40								—	—		9,7	12,3	
41											9,9	12,5	
42											10,1	12,9	
43											10,3	13,1	
44	1,7				−0,01	0,15	+0,2 / 0				10,5	13,3	
45								—	—		11	13,5	
46											11,2	14	
47											11,4	14,2	
48											11,6	14,6	
49	1,9										11,8	14,8	
50					−0,025			5,6	7		12	15,1	
51								5,7	7,2		12,3	15,5	
52		± 0,15 Within a ring: 0,15 max.	1,2	1,5				5,8	7,3		12,5	15,7	± 30 % if F_d < 21,5 N; ± 20 % if F_d > 21,5 N
53								6	7,5		12,9	16,1	
54								6,1	7,6		13,1	16,3	
55	2,1							6,2	7,7		13,3	16,6	
56								6,3	7,9		13,5	17	
57								6,4	8		13,8	17,2	
58								6,5	8,2		14	17,6	
59								6,6	8,3		14,2	17,8	
60	2,3							6,7	8,5		14,4	18,3	
61								6,9	8,6		14,8	18,5	
62								7	8,7		15,1	18,7	
63								7,1	8,9	± 30 % if F_t < 10 N; ± 20 % if F_t > 10 N	15,3	19,1	
64								7,2	9		15,5	19,4	
65	2,5							7,3	9,2		15,7	19,8	
66						0,2	+0,2 / 0	7,4	9,3		15,9	20	
67								7,5	9,4		16,1	20,2	
68								7,6	9,6		16,3	20,6	
69					−0,01 / −0,03			7,8	9,7		16,8	20,9	
70	2,7							7,9	9,9		12	15,1	
71								8	10		17,2	21,5	
72							+0,2 / 0	8,1	10,1		17,4	21,7	
73								8,2	10,3		17,6	22,1	
74								8,3	10,4		17,8	22,4	
75	2,9							8,4	10,6		18,1	22,8	
76								8,5	10,7		18,3	23	
77								8,7	10,8		18,7	23,2	
78								8,8	11		18,9	23,7	
79	3,1							8,9	11,1		19,1	23,9	

This table is shown in ISO format. Commas represent decimal points.

TABLE 9—DIMENSIONS OF R, S, AND M RECTANGULAR RINGS WITH NARROW RING WIDTH MADE OF STEEL (CONTINUED)

Dimensions in millimeters

Nominal diameter d_1	Radial wall thickness a_1	Tolerance	Ring width h_1 Column 1	Ring width h_1 Column 2	Tolerance	Closed gap s_1	Tolerance	Tangential force F_t, N For h_1 shown in column 1	For h_1 shown in column 2	Tolerance	Diametral force F_d, N For h_1 shown in column 1	For h_1 shown in column 2	Tolerance
80								9	11,3		19,4	24,3	
81								9,1	11,4		19,6	24,5	
82	3,1							9,2	11,6		19,8	24,9	
83								9,3	11,7		20	25,2	
84						0,25	+0,25 0	9,4	11,8		20,2	25,4	
85								9,6	12		20,6	25,8	
86	3,3							9,7	12,1		20,9	26	
87		± 0,15						9,8	12,3		21,1	26,4	
88		Within a ring:			−0,01			9,9	12,4		21,3	26,7	
89		0,15 max.	1,2	1,5	−0,03			10	12,5	± 30 % if F_t < 10 N	21,5	26,9	± 30 % if F_d < 21,5 N
90								10,1	12,7	± 20 % if F_t > 10 N	21,7	27,3	± 20 % if F_d > 21,5 N
91	3,5							10,2	12,8		21,9	27,5	
92								10,3	13		22,1	28	
93								10,5	13,1		22,6	28,2	
94								10,6	13,2		22,8	28,4	
95						0,3	+0,25 0	10,7	13,4		23	28,8	
96	3,7							10,8	13,5		23,2	29	
97								10,9	13,7		23,4	29,5	
98								11	13,8		23,7	29,7	
99								11,1	13,9		23,9	29,9	
100	3,9							11,2	14		24,1	30,1	

NOTES

1. For intermediate sizes (for example repair sizes), the radial wall thickness of the next smaller nominal diameter should be applied.

2. The values for F_t and F_d, given in Table 9, apply to steel with a typical modulus of elasticity (E_n) of 200 000 MPa.

   Mean forces are calculated for nominal radial wall thickness (a_1), mean ring width (h_1), and no coating.

3. For the sole purpose of this document, the assumed average ratio F_d/F_t is 2.15. However, for rings up to 50 mm, the ratio F_d/F_t shall be determined between the manufacturer and client.

This table is shown in ISO format. Commas represent decimal points.

APPENDIX A

A.1 This SAE Standard has been established to harmonize the ISO and SAE piston ring Standards. The U.S. Technical Advisory Group, with the support of the National Engine Parts Manufacturers Association, has worked with other national organizations on this worldwide Standard. Some of the wording and phrasing may differ slightly from U.S. terminology for translation purposes.

In preparing this SAE Standard, the Scope and Field of Application, and Reference sections of the ISO 6622/2 have been editorially revised and reorganized.

The tolerances specified in this document represent a six sigma quality level.

INTERNAL COMBUSTION ENGINES—PISTON RINGS—INSPECTION MEASURING PRINCIPLES—SAE J1589 OCT92 — SAE Standard

Report of the Piston Ring Standards Committee 7 approved January 1990. Revised by the SAE Piston Ring Standards Committee 7 October 1992.

This SAE Standard is equivalent to ISO Standard 6621/2.

(R) **1. Scope and Field of Application**—Differences, where they exist, are shown in Appendix A.

This SAE Standard defines the measuring principles to be used for measuring piston rings. It applies to piston rings up to and including 200 mm diameter for reciprocating combustion engines.

These inspection measuring principles may also be used for piston rings for compressors working under analogous conditions.

(R) **2. References**

SAE DESIGNATION	ISO[1] EQUIVALENT	
		INTERNAL COMBUSTION ENGINES—PISTON RINGS
J1588	6621/1	Vocabulary
J1589	6621/2	Measuring principles
J1590	6621/3	Material specifications
J1591	6621/4	General specifications
J1996	6621/5	Quality requirements
		INTERNAL COMBUSTION ENGINES—PISTON RINGS
J1997	6622/1	Rectangular rings
J1998	6622/2 TR	Rectangular rings with narrow ring width
J1999	6623	INTERNAL COMBUSTION ENGINES—PISTON RINGS—SCRAPER RINGS
		INTERNAL COMBUSTION ENGINES—PISTON RINGS
J2000	6624/1	Keystone rings
J2001	6624/2 TR	Half keystone rings
J2002	6625	INTERNAL COMBUSTION ENGINES—PISTON RINGS—OIL CONTROL RINGS
J2003	6626	INTERNAL COMBUSTION ENGINES—PISTON RINGS—COIL SPRING LOADED OIL CONTROL RINGS
J2004	6627 TR	INTERNAL COMBUSTION ENGINES—PISTON RINGS—EXPANDER/SEGMENT OIL CONTROL RINGS
J2226		INTERNAL COMBUSTION ENGINES—PISTON RINGS—STEEL RECTANGULAR RINGS
	ISO 468	SURFACE ROUGHNESS
	ISO 1302	TECHNICAL DRAWINGS

[1] TR refers to Technical Report

3. Measuring Principles

3.1 General Measuring Conditions—The following general notes are applicable to all measuring principles unless otherwise specified:

a. The ring shall rest on the datum surface in the free or open condition. No additional force shall be applied to load the ring on the datum surface.

b. Certain measurements are made with the ring in the closed condition in a gauge of nominal cylinder bore diameter. When orientated rings are measured in this way, they shall be so placed that the top is towards the datum surface.

c. Measurements shall be made using instruments with a resolution not to exceed 10% of the tolerance of the dimension being measured.

3.2 Characteristics and Measuring Principles

Sub-clause	Characteristics of the ring	Symbol
	Principal characteristics of the ring	
3.2.1	Ring width	
	a) parallel sided rings	h_1
	b) keystone rings	h_3, a_6
3.2.2	Radial wall thickness	a_1
3.2.3	Total free gap	m, p
3.2.4	Closed gap	s_1
3.2.5	Tangential force	F_t
3.2.6	Diametral force	F_d
	Characteristics of ring shape	
3.2.7	Ovality or circularity	U
3.2.8	Point deflection	W
3.2.9	Light tightness	—
	Associated with peripheral surface	
3.2.10	Taper on periphery	—
3.2.11	Barrel on periphery	t_2, t_3, h_8
3.2.12	Land width	h_4, h_5
3.2.13	Land offset	—
3.2.14	Coating/inlay thickness	—
	Associated with sides	
3.2.15	Keystone angle	—
3.2.16	Obliqueness	—
3.2.17	Twist	—
3.2.18	Unevenness	Te_f, Te_u
	Other	
3.2.19	Helix (axial displacement of butt ends)	—
3.2.20	Free flatness	—
3.2.21	Surface roughness	R_a, R_z

Term	Definition	Measuring principles	Illustration of measuring principles
3.2.1 Ring width (in millimeters) a) Parallel sided rings, h_1	The distance between the sides, at any particular point perpendicular to the datum surface (see figures 1 and 2).	Measure with spherical measuring probes each of radius $1,5 \pm 0,05$ mm, exerting a measuring force of approximately 1 N (see figure 3). In the case of slotted oil rings, the measurement shall be made between the slots and not across them (see figure 2).	Figure 1 Figure 2 Figure 3

This table is shown in ISO format. Commas represent decimal points.

Term	Definition	Measuring principles	Illustration of measuring principles
b) Keystone rings, h_3	The distance between the sides at a specified distance a_6 from the peripheral surface (see figure 4).	a) Method A This method determines h_3 (see figure 4) for a specified value of a_6. Measure with spherical measuring probes each of radius $1,5 \pm 0,05$ mm exerting a measuring force of approximately 1 N (see figure 5). If the measuring equipment is set up with parallel gauges instead of keystone gauges the use of spherical measuring probes will give rise to an error as follows: for 6° keystone angle : 0,004 mm for 15° keystone angle : 0,026 mm. To obtain the correct measured width of the keystone ring the above values shall be deducted from the measured values. Values of a_6 are given in ISO 6624/1. b) Method B This method determines a_6 for a specified width h_3 (see figure 4). Measure with a flat face probe exerting a measuring force of approximately 1 N. The ring shall be placed between two sharp edged circular discs which are spaced apart at the specified gauge width h_3 (see figure 6). Values of h_3 are given in ISO 6624/1.	Figure 4 Figure 5 Figure 6

This table is shown in ISO format. Commas represent decimal points.

Term	Definition	Measuring principles	Illustration of measuring principles
3.2.2 Radial wall thickness, a_1 (in millimeters)	The radial distance between the periphery and the inside surface of the ring (see figure 7).	a) Measure radially between a flat measuring surface on the periphery and a spherical measuring surface of radius approximately 4 mm on the inside surface, and using a measuring force of 3 to 10 N (see figure 8). b) Measure radially between cylindrical inserts or rollers of radius approximately 4 mm and with a measuring force of 3 to 10 N. The length of the rollers shall be greater than the ring width (see figure 9).	Figure 7 Figure 8 Figure 9

This table is shown in ISO format. Commas represent decimal points.

Term	Definition	Measuring principles	Illustration of measuring principles
3.2.3 Total free gap m, p (in millimeters)	The chordal distance between the butt ends of the ring in a free, unstressed state, measured at the centreline of the radial wall thickness (see figure 10). For rings with an internal notch for a peg, the total free gap is defined by the chordal distance marked as p in figure 11.	Measure with a steel rule to the nearest 0,25 mm	Figure 10 Figure 11
3.2.4 Closed gap, s_1 (in millimeters)	The gap at the butt ends of the ring, measured at the narrowest point of the gap, which the ring would have when fitted in a gauge of nominal cylinder bore size (see figure 12). The closed gap s_1 is related to the nominal diameter d_1.	Measure in a bore gauge of nominal diameter using a wedge gauge or feeler gauges and using a measuring force of approximately 1 N (see figure 12). The diameter of the bore gauge shall comply with the following deviations from the nominal ring diameter: Tolerance : $^{+\,0,001}_{0} d_1$ Correction shall be made for any deviation of the bore gauge from the nominal ring diameter.	Figure 12

This table is shown in ISO format. Commas represent decimal points.

Term	Definition	Measuring principles	Illustration of measuring principles
3.2.5 Tangential force, F_t (in Newtons) a) For single-piece rings	The force necessary to maintain the ring at the closed gap condition by means of a tangential pull on the ends of a circumferential metal tape or hoop (see figure 13).	a) Tape method (see figure 14) The encircling steel tape of thickness 0,08 to 0,10 mm is carried round 10 mm diameter rollers set 20 mm apart (see figure 14). In tightening the tape, the ring is closed to the point where the butt ends touch and then opened to the closed gap dimension previously measured. The ring force is then read off from the precision measuring scale. The gap of the ring shall be symmetrically disposed between the rollers. b) Hoop method (see figure 15) The ring is placed in a correctly sized hoop with its gap aligned to the gap of the hoop. The hoop is then closed in a precision loading machine until the loading pins are at a predetermined distance apart at which point the hoop is precisely at the cylinder bore diameter appropriate to the ring (see figure 15). The force is then read off from the display.	Figure 13 — Closed gap Figure 14 — Diameter of rollers 10 mm Figure 15 — Loading pin spacing to suit machine

This table is shown in ISO format. Commas represent decimal points.

Term	Definition	Measuring principles	Illustration of measuring principles
b) For multi-piece rings	The force which is necessary to maintain the ring at the closed gap condition by means of a tangential pull on the ends of a circumferential metal tape or hoop whilst vibrating the butt ends of the ring [see figure 13a)].	For the measurement of coil spring loaded rings or similar rings where the spring is supported in the inside grooved surface of the ring, the gap of the spring shall be positioned at 180° to the gap of the cast iron part. For the measurement of multi-piece steel rail oil control rings, the ring assembly shall be mounted in a carrier simulating the ring groove. The gap of the spring element is placed at 180° to the gap of the rails, both of which shall be in line. For the measurement of a ring provided with a wavy spring, or other spring which is groove root supported, the ring assembly shall be mounted in a carrier simulating the groove, the root diameter of which is equal to the mean diameter of the piston ring groove in which the ring will be used. Tolerance on carrier root diameter ± 0,02 mm. The gap of the wavy spring shall be at 180° to the gap of the cast iron part. a) Tape method Identical procedures are used as for single piece rings but an appropriate vibration shall be applied to the tape loading mechanism to relieve forces of friction [see figure 14a)]. A suitable level is 40 to 50 Hz at an amplitude of 0,15 mm.	Figure 13a) Diameter of rollers 10 mm Figure 14a)

This table is shown in ISO format. Commas represent decimal points.

Term	Definition	Measuring principles	Illustration of measuring principles
		b) **Hoop method** Identical procedures are used as for single-piece rings but an appropriate vibration shall be applied to the hoop loading mechanism to relieve all forces of friction (see figure 15a). NOTES 1 Before tangential force measurements are made, rings must be degreased and lightly coated with thin machine oil. 2 It is recommended that closed gap measurements be made immediately prior to measuring tangential force. 3 In order to improve consistency of measurement and particularly with coil spring loaded rings which have been oxided or phosphated it is permissible to rotate the spring forwards and backwards to smooth the surface before carrying out measurements. 4 The reproducibility of tangential force measurements has not been high in the past but current machines using tape and hoop methods give an overall reproducibility of the order of 6,5 %. It is recommended that customer and supplier agree on a suitable factor to take account of different machines, different locations and different operators.	(diagram showing ring with d_1, Vibration, Loading pin spacing to suit machine, Measuring scale, F_t) **Figure 15a)**

This table is shown in ISO format. Commas represent decimal points.

Term	Definition	Measuring principles	Illustration of measuring principles
3.2.6 Diametral force, F_d (in Newtons) NOTE — This method is only applicable to single-piece rings.	The force, acting diametrically at 90° to the gap, necessary to maintain the ring at the nominal diameter condition measured in the direction of the force (see figure 16).	Diametral force is measured in purpose built machines which incorporate flat plates for closing the rings (see figure 16).	F_d, d_1 Figure 16
3.2.7 Ovality or circularity, U (in millimeters) NOTE — This method is only applicable to single-piece rings.	The difference between the mutually perpendicular diameters d_3 and d_4 when the ring is drawn to closed gap within a flexible tape. It may be either positive ($d_3 > d_4$) or negative ($d_3 < d_4$) (see figure 17).	Measure, with the ring drawn to true closed gap in a flexible steel tension tape or band of thickness 0,08 to 0,10 mm using a diametral measuring device with a measuring force of not more than 1 N (see figure 17). With the ring closed within the tape it is an acceptable alternative to clamp it between plates and then remove the tape prior to measuring the diameters d_3 and d_4. NOTE — Clamping of the ring between plates is not applicable to oil control rings with slots.	Closed gap, d_4, d_3 Figure 17

This table is shown in ISO format. Commas represent decimal points.

Term	Definition	Measuring principles	Illustration of measuring principles
3.2.8 Point deflection, W (in millimeters)	The deviation of the butt ends from the true circle when restrained in a gauge of nominal cylinder bore diameter (see figure 18).	Measure with a probe of spherical radius of 1,5 mm ± 0,05 mm with measuring force of approximately 1 N, the ring being mounted in a gauge of nominal cylinder bore diameter relieved over the gauge angle, θ, (see figure 18). The gauge angle, θ, shall be agreed between manufacturer and customer. It normally relates to port angle. The following gauge tolerances apply for this test: Angle θ : ± 1° Diameter : $^{+0,001}_{0}\, d_1$ Circularity : 0,000 1 d_1 max.	Figure 18
3.2.9 Light tightness (percentage of ring circumference)	The ability of the periphery of a ring when mounted in a gauge of nominal cylinder bore diameter to exclude the passage of light (see figure 19). A ring showing only pin-point, burry or fuzzy light shall be considered as light tight.	Measure in a gauge equipped with a suitable light source and determine the percentage of the ring circumference which will allow light to pass (see figure 19). It is permissible to rotate the ring in the gauge to remove any slight surface roughness on the periphery. Unless otherwise specified examination and measurement should be made without magnification and with normal eyesight. It is important to avoid errors of parallax and to protect the viewer against stray light penetration. Illuminance behind the ring to be 400 to 1 500 lux above the ambient conditions. The following gauge tolerances apply for this test: Diameter : $^{+0,001}_{0}\, d_1$ Circularity : 0,000 1 d_1 max.	Figure 19

¹This table is shown in ISO format. Commas represent decimal points.

Term	Definition	Measuring principles	Illustration of measuring principles
3.2.10 Taper on periphery (in micrometers or degrees)	Taper is the intentional deviation of the periphery from a line perpendicular to the datum surface (see figure 20).	a) Method A Measure at the back of the ring perpendicular to the datum surface using flat faced probes exerting a force of approximately 1 N (see figure 21). The measurement recorded is the difference in radial dimension of the ring peripheral surface between two points, near the top and near the bottom, distance H apart. The dimension H shall be approximately two-thirds of the total width of the ring and the recorded measurement may be converted to the taper angle in degrees or minutes. b) Method B The ring shall be mounted on a datum surface and the peripheral surface of the back of the ring graphed perpendicular to the datum surface using a profile recorder. Magnification used shall be clearly indicated. NOTE — The same methods may be used to determine the unintentional taper which may be present on, for example, a nominally straight faced rectangular ring.	Figure 20 Figure 21

This table is shown in ISO format. Commas represent decimal points.

Term	Definition	Measuring principles	Illustration of measuring principles
3.2.11 Barrel on periphery, t_2, t_3, h_8 (in millimeters)	The barrel is the intentional convex deviation of the peripheral surface from a line perpendicular to the datum surface (see figure 22).	a) Method A Measure at the back of the ring perpendicular to the ring datum surface using flat ended probes exerting a force of approximately 1 N (see figure 23). The measurement recorded is the difference in radial dimension of the ring peripheral surface between two points, one at the peak of the barrel (at or near the center line of the ring) and the second at half gauge width $\frac{h_8}{2}$, from the peak. b) Method B The ring shall be mounted on a datum surface and the peripheral surface of the back of the ring graphed, perpendicular to the datum surface, using a profile recorder. Magnification used shall be clearly indicated. (Recommended ratio between vertical and horizontal magnifications 10 or 25.) NOTE — The same methods may be used to determine the unintentional barrel which may be present on, for example, a nominally straight faced rectangular ring.	Figure 22 Figure 23

This table is shown in ISO format. Commas represent decimal points.

Term	Definition	Measuring principles	Illustration of measuring principles
3.2.12 Land width, h_4, h_5 (in millimeters)	The width of the land which theoretically should be in contact with the cylinder bore (see figure 24).	**a) Method A** For all forms of land (sharp edge, chamfered or radiused) measure with a measuring microscope or on a projector. The measurement shall be made only on the periphery of the lands (see figure 25). **b) Method B** For all forms, the ring shall be mounted on a datum surface and the lands shall be graphed on a profile recorder. Magnification used shall be clearly stated. NOTE — Land offset, (see 3.2.13), can be included and obtained from this measurement at the back of the ring.	Figure 24 Figure 25

This table is shown in ISO format. Commas represent decimal points.

Term	Definition	Measuring principles	Illustration of measuring principles
3.2.13 Land offset (in millimeters)	The displacement of the two peripheral surfaces of a slotted or drilled oil control ring in relation to each other in a radial direction (see figure 26).	a) Method A Measure at the back of the ring from a line perpendicular to the datum surface (see figure 26) using flat measuring probes exerting a force of approximately 1 N. The ring shall be loaded against the measuring instrument in the direction of and in the position of the force F (see figure 27). Value of force F to be in the range of 3 to 5 N. b) Method B See Method B in 3.2.12.	Figure 26 Figure 27
3.2.14 Coating/inlay thickness (in millimeters)	The distance between the outer surface of the coating/inlay and the surface of the base ring material (see figure 28).	Measure non-destructively in the middle of the width of coating using a calibrated inductive thickness measuring instrument. The calibration shall be made using a master ring of equal dimension and material to the ring being tested. Specific points for measuring shall be at the back of the ring and at 15 mm from each butt end.	Figure 28

This table is shown in ISO format. Commas represent decimal points.

Term	Definition	Measuring principles	Illustration of measuring principles
3.2.15 Keystone angle (in degrees)	The angle enclosed by the two sides of the ring (see figure 29). Alternatively, the sum of both side face angles, i.e. included angle.	**a) Method A** Measure, in a true radial direction at the back of the ring, the difference in ring width at two points of known distance apart, using spherical probes each of radius $1,5 \pm 0,05$ mm exerting a force of approximately 1 N. The keystone angle can then be calculated as the sum of both side angles (see figure 30). **b) Method B** Measure, in a true radial direction at the back of the ring, the difference in ring width, using two probe systems formed as knife edges, each probe exerting a force of approximately 1 N. The keystone angle can then be calculated as the sum of both side angles (see figure 31). **c) Methods C and D** Both methods involve the use of a probe which traverses in a true radial direction, a known distance across the side face at the back of the ring. The probe has a spherical radius of $1,5 \pm 0,05$ mm and exerts a force of approximately 1 N. The datum surface plate on which the ring rests for measurement is provided with a location to ensure that the line of measurement is truly radial (see figure 33). **1) Method C** The datum surface plate is inclined at an angle equal to the nominal side face angle of the ring and hence the contact surface of the probe traverses a path nominally parallel to the axis of motion of the probe. The probe measures any deviation of the side face from parallel and allows the actual angular deviation to be calculated. Hence the actual side face angle can be determined.	Included angle **Figure 29** **Figure 30** **Figure 31**

This table is shown in ISO format. Commas represent decimal points.

Term	Definition	Measuring principles	Illustration of measuring principles
		The ring is measured on both sides and the sum of the side face angles gives the keystone angle (see figure 32). 2) Method D The datum surface plate is parallel to the axis of motion of the probe and the ring side face lies at an angle to the datum surface equal to the side face angle of the ring. The contact surface of the probe in traversing the side face describes the full movement equivalent to the side face angle : the latter can then be calculated directly. The ring is measured on both sides and the sum of the side face angles gives the keystone angle.	**Figure 32** — Side face horizontal; Nominal angle of side face **Figure 33** — Line of measurement; Datum surface plate
3.2.16 Obliqueness (in degrees)	The unintentional deviation of the bisector of the keystone included angle from parallelism with the datum surface (see figure 34). Not applicable to rings with designed twist.	The measuring principles are identical with those given for keystone angle, see 3.2.15. When each side face angle is available, obliqueness is one half the difference between the two side face angles, for example, with a 15° included angle ring where one side is 7° 40' and the other is 7° 20', the obliqueness is 10'.	**Figure 34** — Obliqueness

This table is shown in ISO format. Commas represent decimal points.

Term	Definition	Measuring principles	Illustration of measuring principles
3.2.17 Twist (in millimeters)	The intentional torsional deviation of the section of the ring from the datum surface when the ring is restricted to nominal diameter (as in the case of asymmetrical rings such as those internally or externally stepped or bevelled) (see figure 35).	Measure, over a true radial gauge length, the deviation of the ring side from a plane parallel to the datum surface, the ring being closed to nominal diameter in a bore gauge. For nonkeystone rings, the measurement shall be made at the back of the ring on the side opposite the bevel or step, using a spherical probe of radius 1,5 ± 0,05 mm exerting a force of approximately 1 N (see figures 36 and 37). Twist for keystone rings is the difference in bottom side angle values measured free and confined at the back of the ring. NOTE — Twist is measured as the linear deviation ir distance from the datum surface over a gauge length of 2 mm or a minimum of 60 % of the available radial thickness. See 3.2.15 for measuring keystone ring angles.	**Figure 35** — Positive twist / Negative twist **Figure 36** — Positive reading equals positive twist **Figure 37** — Positive reading equals negative twist

This table is shown in ISO format. Commas represent decimal points.

Term	Definition	Measuring principles	Illustration of measuring principles
3.2.18 Unevenness, Te_r, Te_u (in millimeters)	Unintentional deviation of the sides of the ring from parallelism to the datum surface, i.e. twisted or dished rings (see figures 38 and 40). NOTE — Not applicable to rings with designed twist, as covered by term 3.2.17.	a) In the radial direction Measure with a probe of spherical radius 1,5 ± 0,05 mm exerting a force of approximately 1 N on the upper side of the ring (see figures 39 and 42), centrally between the loaded points. The largest of the four readings is taken as the measure of unevenness.	Figure 38 Figure 39 Figure 40

This table is shown in ISO format. Commas represent decimal points.

Term	Definition	Measuring principles	Illustration of measuring principles
		b) In the circumferential direction Measure with probe of spherical radius $1,5 \pm 0,05$ mm exerting a force of approximately 1 N on the upper side of the ring (see figures 41 and 42), in the middle of the ring wall and centrally between the loaded points. The difference between the greatest and the least deflection is taken as the measure of the unevenness. Loading of rings : Rings shall be loaded prior to measurement using 5 load points, one each side of the gap, one at 90°, one at 180° and one at 270°. In the case of slotted oil control rings, load points and measurement areas shall be on the nearest available bridge and not over slot areas. The following loads shall be used on each load point : For rings < 80 mm diameter : 2,5 N; > 80 mm diameter : 5,0 N.	**Figure 41** **Figure 42** • Load points + Measurement points
3.2.19 Helix (axial displacement of butt ends) (in millimeters)	The displacement of the butt ends perpendicular to the datum surface (see figure 43).	The butt end of the ring already in contact with the datum surface shall be loaded or clamped with a force, F, approximately 10 N. Measure the displacement of the adjacent butt end with a measuring microscope or magnifier. The loading device or clamp shall be confined to within 15° of arc from the appropriate butt end.	**Figure 43**

This table is shown in ISO format. Commas represent decimal points.

Term	Definition	Measuring principles	Illustration of measuring principles

3.2.20 Free flatness	The relationship between the ring in the free state and a plane parallel to its datum surface.	The clean and dry ring shall drop freely of its own weight between vertical plates (see figure 44).
		
		The distance apart of the plates shall be equal to the maximum ring width plus the out-of-plane allowance shown below.
		
		Ring diameter **Out-of-plane allowance**
		Below 100 0,050
		100 to 125 0,075
		125 to above 0,100
		
		NOTE — For rings 1,5 mm wide or less, add 0,025 mm to the out-of-plane allowance.
		
		Surface plates :
		   Size : equal to or greater than the largest free diameter of the ring
		   Flatness : ± 0,002 5 mm
		   Roughness : $R_a = 0{,}25$ μm
		   Tolerance on spacing of plates to be $^{+0,01}_{0}$ mm
			Figure 44

3.2.21 Surface roughness, R_a, R_z (in micrometers)	In accordance with ISO 468.	Measure in accordance with ISO 468 using any suitable profile measuring machine. NOTE — For indications on drawings, see ISO 1302.	

This table is shown in ISO format. Commas represent decimal points.

APPENDIX A

A.1 This SAE Standard has been established to harmonize the ISO and SAE piston ring standards. The U.S. Technical Advisory Group, with the support of the National Engine Parts manufacturers Association, has worked with other national organizations on this worldwide standard. Some of the wording and phrasing may differ slightly from U.S. terminology for translation purposes.

In preparing this SAE document, the Introduction, Scope and Field of Application, and Reference sections of the ISO 6621/2 have been editorially revised and reorganized.

INTERNAL COMBUSTION ENGINES—PISTON RINGS—MATERIAL SPECIFICATIONS—SAE J1590 OCT92

SAE Standard

Report of the Piston Ring Standards Committee 7 approved October 1989. Revised by the SAE Piston Ring Standards Committee 7 October 1992.

This SAE Standard is equivalent to ISO Standard 6621/3.

1. Scope and Field of Application
Differences, where they exist, are shown in Appendix A.

This SAE Standard establishes a classification of materials intended for the manufacture of piston rings based on mechanical properties and the stresses that these materials are capable of withstanding.

This document applies to the manufacture of piston rings up to and including 200 mm diameter for reciprocating internal combustion engines. It also applies to piston rings for compressors working under similar conditions.

(R) ### 2. References

SAE DESIGNATION	ISO[1] EQUIVALENT	
		INTERNAL COMBUSTION ENGINES—PISTON RINGS
J1588	6621/1	Vocabulary
J1589	6621/2	Measuring principles
J1590	6621/3	Material specifications
J1591	6621/4	General specifications
J1996	6621/5	Quality requirements
		INTERNAL COMBUSTION ENGINE—PISTON RINGS
J1997	6622/1	Rectangular rings
J1998	6622/2 TR	Rectangular rings with narrow ring width
J1999	6623	INTERNAL COMBUSTION ENGINES—PISTON RINGS—SCRAPER RINGS
J2000	6624/1	Keystone rings
J2001	6624/2 TR	Half keystone rings
J2002	6625	INTERNAL COMBUSTION ENGINES—PISTON RINGS—OIL CONTROL RINGS
J2003	6626	INTERNAL COMBUSTION ENGINES—PISTON RINGS—COIL SPRING LOADED OIL CONTROL RINGS
J2004	6627 TR	INTERNAL COMBUSTION ENGINES—PISTON RINGS—EXPANDER/SEGMENT OIL CONTROL RINGS
J2226		INTERNAL COMBUSTION ENGINES—PISTON RINGS—STEEL RECTANGULAR RINGS
	ISO/R 80	ROCKWELL HARDNESS TEST (B AND C SCALES) FOR STEEL

[1] TR refers to Technical Report

3. Mechanical Properties
See Table 1.

TABLE 1—MECHANICAL PROPERTIES

Class	Mechanical Properties MPa or N/mm² Typical Modulus of Elasticity	Mechanical Properties MPa or N/mm² Minimum Bending Strength	Materials Meeting the Required Mechanical Properties Type of Material	Materials Meeting the Required Mechanical Properties Minimum Hardness	Materials Meeting the Required Mechanical Properties Specific Details	Materials Meeting the Required Mechanical Properties Subclass
10	90 000	300	Grey cast iron	93 HRB	Nonheat-treated	11
	100 000	350		95 HRB		12
20		450		23 HRC		21
		450		28 HRC		22
	115 000	450	Grey cast iron	40 HRC	Heat-treated	23
		500		32 HRC		24
	130 000	650		37 HRC		25
30	145 000	550	Carbidic cast iron	25 HRC	Heat-treated pearlitic	31
		500		30 HRC	Heat-treated martensitic	32
		600		95 HRB	Heat-treated pearlitic	41
		600		22 HRC	Heat-treated martensitic	42
40	160 000	600	Malleable cast iron	30 HRC	Heat-treated martensitic	43
		1000		27 HRC	Heat-treated carbidic	44
		1100		23 HRC	Heat-treated martensitic	51
		1300		23 HRC	Heat-treated martensitic	52
50	160 000	1300	Spheroidal graphite cast iron	28 HRC	Heat-treated martensitic	53
		1300		95 HRB	Pearlitic	54
		--		97 HRB	Ferritic	55
				38 HRC	CrMoV - alloyed	61
60	200 000	--	Steel	40 HRC	CrSi - alloyed	62
				48 HRC	CrSi - alloyed	63

NOTE—The hardness values are averages from three measurements on one ring, one being at the gap and one each 90 and 180 degrees around from the gap. HRB and HRC hardness testing is in accordance with ISO/R 80.

The application of the hardness measuring methods HRB and HRC is restricted, due to the geometry and the material of piston rings. The hardness values stated are used only for grouping the materials into the individual subclasses. Other hardness measuring methods and their equivalent values shall be agreed upon between customer and manufacturer.

APPENDIX A

A.1 This SAE Standard has been established to harmonize the ISO and SAE piston ring standards. The U.S. Technical Advisory Group, with the support of the National Engine Parts Manufacturers Association, has worked with other national organizations on this worldwide standard. Some of the wording and phrasing may differ slightly from U.S. terminology for translation purposes.

In preparing this SAE Standard, the Scope and Field of Application and Reference sections of the ISO 6621/3 have been editorially revised and reorganized.

INTERNAL COMBUSTION ENGINES—PISTON RINGS—QUALITY REQUIREMENTS—SAE J1996 OCT92

SAE Standard

Report of the Piston Ring Standards Committee approved September 1990. Revised by the Piston Ring Standards Committee 7 October 1992.

This SAE Standard is equivalent to ISO Standard 6621/5.

(R) 1. Scope and Field of Application—Differences, where they exist, are shown in Appendix A.

This SAE Standard specifies the quality aspects that are capable of definition but not normally found on a drawing specification.

The difficulty of trying to define in absolute terms the quality attainable in normal commercial manufacture of piston rings is well known. In this document the commonly encountered aspects of quality in terms of casting defects and other departures from ideal are quantified. Many minor defects are clearly quite acceptable; other defects because of size or numbers are inadmissible.

This document covers the following:
  a. Single piece piston rings of grey, carbidic, malleable, spheroidal graphite cast iron or steel
  b. Multipiece piston rings (oil control rings) consisting of cast iron parts and spring components
  c. Single piece and multipiece rings of steel, i.e., oil control rings in the form of strip steel components or steel segments (rails) with spring expander components

In addition to specifying certain limits of acceptance relating to inspection measuring principles (covered by SAE J1589) this document also covers those features for which no recognized quantitative measurement procedures exist and which are only checked visually with normal eyesight and without magnification. Such features (superficial defects) are additional to the standard tolerances of ring width, radial wall thickness, and closed gap.

The requirements of this document apply to all rings up to and including 200 mm diameter covered by the above classification for both reciprocating internal combustion engines and compressors.

(R) 2. References

SAE DESIGNATION	ISO[1] EQUIVALENT	
		INTERNAL COMBUSTION ENGINES—PISTON RINGS
J1588	6621/1	Vocabulary
J1589	6621/2	Measuring principles
J1590	6621/3	Material specifications
J1591	6621/4	General specifications
J1996	6621/5	Quality requirements
		INTERNAL COMBUSTION ENGINES—PISTON RINGS
J1997	6622/1	Rectangular rings
J1998	6622/2 TR	Rectangular rings with narrow ring width
J1999	6623	INTERNAL COMBUSTION ENGINES—PISTON RINGS—SCRAPER RINGS
		INTERNAL COMBUSTION ENGINES—PISTON RINGS
J2000	6624/1	Keystone rings
J2001	6624/2 TR	Half keystone rings
J2002	6625	INTERNAL COMBUSTION ENGINES—PISTON RINGS—OIL CONTROL RINGS
J2003	6626	INTERNAL COMBUSTION ENGINES—PISTON RINGS—COIL SPRING LOADED OIL CONTROL RINGS
J2004	6627 TR	INTERNAL COMBUSTION ENGINES—PISTON RINGS—EXPANDER/SEGMENT OIL CONTROL RINGS
J2226		INTERNAL COMBUSTION ENGINES—PISTON RINGS—STEEL RECTANGULAR RINGS
	2859	SAMPLING PROCEDURES AND TABLES FOR INSPECTION BY ATTRIBUTES

[1] TR refers to Technical Report

3. Terminology—The terminology used in this document is as given in SAE J1588.

4. Visible Defects

4.1 General—Visible defects are divided into two principle classes as described in 4.2 to 4.5.

The first class covers those defects frequently found in castings and includes such defects as porosity, sand inclusions, cavities, etc.

The second class of defects covers mechanical abrasions which may occur during machining or handling of rings, and includes scratches, dents, chipping, burns, and cracks.

Inspection of piston rings for such defects is generally carried out visually, without magnification, by inspectors having normal eyesight, corrected if necessary.

It is not intended that every ring be rigorously inspected for size and distribution of defects but rather that the values given in the tables and text be used as a general guide. However, in case of doubt, the values given should be used as the means of judging the quality of the rings.

4.2 Pores, Cavities, and Sand Inclusions—Such defects are permissible on uncoated surfaces and edges provided that the values given in Table 1 for size, number, and spacing are not exceeded.

NOTE—The depth of porosity cannot be checked visually and, therefore, no limiting values are given.

4.3 Scratches, Indentations, Depressions, and Cracks

4.3.1 SCRATCHES—Isolated scratches are permissible provided that:
  a. No burrs are produced exceeding the permissible values given in 4.4.1.1
  b. On the periphery with turned surface they are not deeper than the tool marks or, for peripheries without a turned surface, not deeper than 0.004 mm
  c. On the side faces they are not deeper than 0.01 mm
  d. On other surfaces they are not deeper than 0.06 mm

4.3.2 INDENTATIONS AND DEPRESSIONS—Indentations and Depressions are permissible provided that:
  a. The values given in Table 1 for number and spacing of defects are met
  b. No burrs are produced exceeding the permissible values given in 4.4.1.1
  c. They do not exceed the values for size and depth given in Table 2

Rings of a coated/inlaid type shall not have indentations or depressions on the periphery.

NOTE—Indentations arising from hardness measurements on the side faces are acceptable provided that they do not exceed the limits given in Tables 2 and 11.

TABLE 1—PERMISSIBLE VALUES OF SIZE, NUMBER AND SPACING OF PORES, CAVITIES AND SAND INCLUSIONS

Dimensions are millimeters

Nominal diameter d_1	Defect size max on periphery[1]	Defect size max on other surfaces[1]	Defect size max on peripheral edges	Defect size max on other edges[2]	Number per ring max	Spacing[3] min
$30 \leq d_1 < 60$	0.1	0.3	0.1	0.1	2	4
$60 \leq d_1 < 100$	0.15	0.5	0.1	0.2	4	4
$100 \leq d_1 < 150$	0.2	0.5	0.1	0.3	6	8
$150 \leq d_1 \leq 200$	0.2	0.8	0.1	0.4	6	8

[1] The defects should not be closer to an edge than one-half of the maximum permissible size of the defect, with a minimum of 0.2 mm.
[2] Not on inside gap edges of piston rings with internal notch.
[3] Spacing includes defects on adjacent or opposite surfaces.

TABLE 2—PERMISSIBLE SIZE OF INDENTATIONS AND DEPRESSIONS

Dimensions are millimeters

Nominal diameter d_1	Defect size max on periphery	Defect size max on side face	Depth max
$30 \leq d_1 < 100$	0.3	0.6	10% of corresponding max defect size surface
$100 \leq d_1 < 200$	0.5	1	

4.3.3 CRACKS—No cracks are permissible.
See also 4.5.4 for chromium-plated peripheries.

4.4 Edges

4.4.1 EDGE CONFIGURATION—All edges of the piston ring shall be sharp; ideally they should be free from burrs and from ragged edges whether arising from crumbling of material or from deburring. Such conditions are almost impossible to achieve regularly in volume production and, hence, both burrs and removal of edge material is permitted up to the maximum sizes given in 4.4.1.1 and 4.4.1.2.

4.4.1.1 *Burrs*—Burrs are permitted up to the maximum values given in Table 3. The orientation and direction of burrs shall relate to the functional surfaces of the piston ring, that is to say any burr present should point in the direction of sliding motion of the ring and not normal to the direction of sliding.

Any burrs remaining on the edges of rings should be firmly adherent, forming an integral part of the edge.

4.4.1.2 *Edge Material Removal*—To eliminate protruding burrs in any direction, it is permissible to remove material from the edges to the values given in Table 4.

4.4.2 CHIPPING AND SIMILAR DEFECTS ON PERIPHERAL EDGES, PERIPHERAL EDGES AT THE GAP, OUTSIDE GAP CORNERS AND ON PERIPHERAL CHAMFERS

4.4.2.1 Chipping and similar defects are permitted at these points provided that:
a. They are free of loosely adhering particles
b. No burrs are produced exceeding the values permitted in 4.4.1.1
c. They do not exceed half the width of any witness land on, for example, taper-faced rings
d. They do not exceed the values given in the following tables:
  (1) Table 5 for uncoated rings
  (2) Table 6 for plated, coating rings
  (3) Table 7 for spray-coated rings
  (4) Table 8 for chamfers on all rings

Typical defects are illustrated in Figures 1 to 6.
(See 4.4.2.2 for explanations of F and K.)

TABLE 3—MAXIMUM SIZE OF BURRS FOR ALL SIZES OF RING

Dimensions are millimeters

Location of burr	Size of burr[1] max
On edges adjacent to	
the peripheral surface and side faces	0.006
the butt ends (gap surfaces)	0.04
the outside groove face (oil rings)	0.2
the inside surface and the ends of the slots (oil rings)	0.5
all other surfaces	0.1

[1] Maximum values of burrs on steel rings are to be agreed on between manufacturer and purchaser.

TABLE 4—EDGE MATERIAL REMOVAL IN DEBURR OPERATIONS

Dimensions are millimeters

Location of edge	Removal of material max
On peripheral edges	0.08
On peripheral edges of the gap[1]	0.15
On other edges	0.25

[1] Does not apply to rings coated on the periphery that have gap edge chamfers.

TABLE 5—PERMISSIBLE SIZE OF CHIPPING AND DEFECTS ON UNCOATED RINGS ON PERIPHERAL EDGES, PERIPHERAL EDGES OF THE GAP AND OUTSIDE GAP CORNERS[1]

Dimensions are millimeters

Ring width h_1	Land width h_4 or h_5	Defect in direction normal to periphery[2] on peripheral edge of gap F_1, F_3	Defect in direction normal to periphery[2] on peripheral edge F_2, F_3	Defect in direction along edge[2] on peripheral edge of gap[3] K_1, K_3	Defect in direction along edge[2] on peripheral edge K_2, K_3
$h_1 < 2$	-	0.2	0.2	0.5	0.5
$2 \leq h_1 < 4$	-	0.2	0.2	0.6	0.6
$4 \leq h_1 \leq 6$	-	0.3	0.3	0.8	0.8
-	$h_4, h_5 < 0.5$	0.1	0.1	0.1	0.6
-	$h_4, h_5 \geq 0.5$	0.2	0.2	0.2	0.8

[1] Number and spacing of defects to be in accordance with Table 1.
[2] See Figures 1, 2, 3, 4, and 5.
[3] Subject to a maximum of one-third of peripheral width of ring or land.

TABLE 6—PERMISSIBLE SIZE OF CHIPPING AND DEFECTS ON COATED RINGS WITH CHROMIUM PLATED PERIPHERY, ON PERIPHERAL EDGES, PERIPHERAL EDGES OF THE GAP AND OUTSIDE GAP CORNERS[1]

Dimensions are millimeters

Ring width h_1	Land width h_5	Defect in direction normal to periphery[2] on peripheral edge of gap F_1, F_3	Defect in direction normal to periphery[2] on peripheral edge F_2, F_3	Defect in direction along edge[2] on peripheral edge of gap[3] K_1, K_3	Defect in direction along edge[2] on peripheral edge K_2, K_3
$h_1 < 2$	–	0.2	0.2	0.3	0.3
$2 \leq h_1 < 4$	–	0.2	0.2	0.4	0.3
$4 \leq h_1 \leq 6$	–	0.3	0.3	0.4	0.4
–	$h_5 < 0.5$	0.1	0.1	0.1	0.6
–	$h_5 \geq 0.5$	0.2	0.1	0.2	0.6

[1] Number and spacing of defects to be in accordance with Table 1.
[2] See Figures 1, 2, 3, 4, and 5.
[3] Subject to a maximum of one-third of peripheral width of ring or land.

TABLE 7—PERMISSIBLE SIZE OF CHIPPING AND DEFECTS ON SPRAY COATED RINGS ON PERIPHERAL EDGES, PERIPHERAL EDGES OF THE GAP AND OUTSIDE GAP CORNERS[1]

Dimensions are millimeters

Ring width h_1	Defect in direction normal to periphery[2] on peripheral edges of the gap F_1, F_3	Defect in direction normal to periphery[2] on peripheral edges[3] F_2, F_3	Defect in direction along edge[2] on peripheral edge of the gap[4] K_1, K_3	Defect in direction along edge[2] on peripheral edge[3] K_2, K_3
$h_1 < 2$	0.3	0.3	0.5	0.5
$2 \leq h_1 < 4$	0.3	0.3	0.6	0.6
$4 \leq h_1 \leq 6$	0.4	0.4	0.8	0.8

[1] Number and spacing of defects to be in accordance with Table 1.
[2] See Figures 1, 2, 3, 4, and 5.
[3] Only for fully faced and semi-inlaid design.
[4] Subject to a maximum of one-third of peripheral width of ring or coating.

TABLE 8—PERMISSIBLE SIZE OF CHIPPING AND DEFECTS ON CHAMFERS AT THE PERIPHERAL EDGE AND PERIPHERAL EDGE OF THE GAP[1]

Dimensions are millimeters

Ring width h_1	Size of defect L_1/L_2 max
$h_1 < 2$	0.5
$2 \leq h_1 < 4$	0.8
$4 \leq h_1 \leq 6$	1.2

[1] Number and spacing of defects to be in accordance with Table 1.

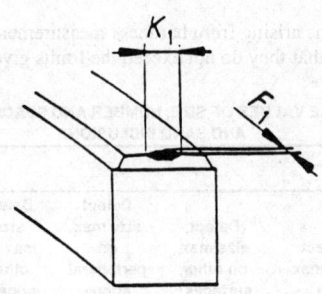

FIGURE 1—CHIPPING ON PERIPHERAL EDGES OF THE GAP

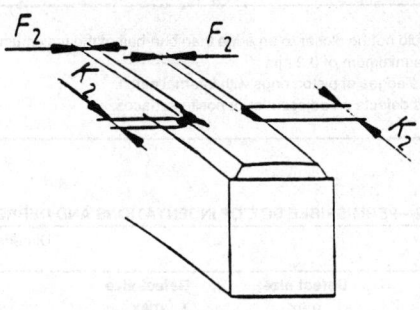

FIGURE 2—CHIPPING ON PERIPHERAL EDGES

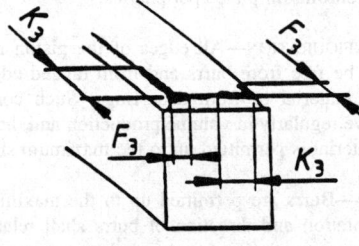

FIGURE 3—CHIPPING ON OUTSIDE GAP CORNERS

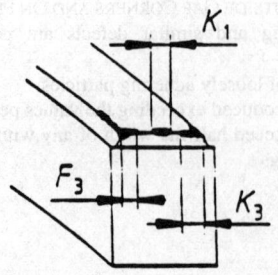

FIGURE 4—COMBINATION OF FIGURES 1 AND 3

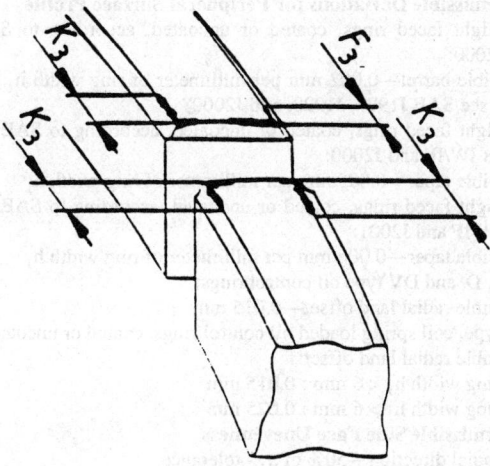

FIGURE 5—CHIPPING ON OPPOSITE GAP CORNERS

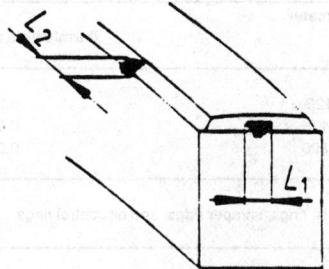

FIGURE 6—CHIPPING ON CHAMFERS

4.4.2.2 K_1, K_2, K_3, F_1, F_2, and F_3 Dimensions (See Figures 1 to 6)—K_1, K_2, K_3 are always the dimensions of the defect measured along the edge cut by the defect.

F_1, F_2, F_3 are always the dimensions of the defect measured normal to the edge cut by the defect.

However, when chipping or other defects occur on outside gap corners, i.e., when the defect crosses the intersecting edges of the peripheral edge and the peripheral edge of the gap, a convention is required.

The defect is taken as appropriate to the edge that contains the larger amount of the defect. For example in Figure 3, most of the left side defect is on the peripheral edge and, therefore, the defect is appropriate to that edge. Hence, the K value lies along the peripheral edge and is denoted K_3 while the F value, although it lies along the peripheral edge of the gap, is taken as the dimension measured normal to the peripheral edge and is denoted F_3.

In the case of the defect on the right side corner, most of the defect lies along the peripheral edge of the gap and the defect is, therefore, appropriate to this edge. The measurement K_3 in this case, therefore, is measured along the peripheral edge of the gap and F_3 is its dimension normal to the edge of the gap.

4.4.2.3 The limitations for chipping and similar defects on peripheral edges, peripheral edges of the gap and opposite gap corners are given in a to c.

 a. Peripheral edges—Defects to be included in the assessment of the peripheral edge are all values of F_2 and K_2 as well as the F_3, K_3 values of outside gap corner defects if these are appropriate to the peripheral edges, e.g., the left hand defect illustrated in Figure 3.
 Maximum sizes are given in Tables 5, 6, and 7 (peripheral edge column).
 b. Peripheral edges of the gap—Defects to be included in the assessment of the peripheral edges of the gap are all values of F_1 and K_1 as well as the F_3, K_3 values of outside gap corner defects if these are appropriate to the peripheral edges of the gap, e.g., the right hand defect illustrated in Figure 3.
 Maximum sizes are given in Tables 5, 6, and 7 (peripheral edge of gap column).
 However, an additional limitation is that the sum of the defect sizes measured in the axial direction, i.e., along the peripheral edge of the gap, shall not exceed the values given in Tables 5, 6, and 7.
 The defects to be added taken from the examples in Figure 4 are K_3 (right hand corner) + K_1 + F_3 (left hand corner).
 c. Opposite gap corners—Defects at outside gap corners are accounted for in the assessments shown in a and b either as peripheral edge defects or as peripheral edge of the gap defects.
 However, an additional limitation is that the sum of the defects measured circumferentially on opposite corners shall not exceed the values given in Tables 5, 6, and 7.
 The defects to be added in Figure 5 are the K_3 value of the left hand corner plus the K_3 value on the opposite corner and the F_3 value of the right hand corner plus the K_3 value of the opposite corner.

4.4.2.4 The limitations for chipping and similar defects on the chamfers at the peripheral edge and at the peripheral edge of the gap are as follows:

This type of defect is illustrated in Figure 6 and is more likely to occur on chromium plated chamfers (machined or unmachined), on machined chamfers on metal sprayed rings (fully coated), and on machined chamfers on grey iron rings. The maximum values of the defects allowable are given in Table 8 and are the same for all rings with chamfers on peripheral edge and peripheral edge of the gap.

Defects counted as on the chamfers shall not cut peripheral edges or peripheral edges of the gap but may just cut side faces or gap faces.

4.4.3 CHIPPING AND DEFECTS ON INNER EDGES AND OTHER EDGES—Chipping and defects on inner edges and other edges are permissible provided that:

 a. No burrs are produced exceeding the values given in 4.4.1.1
 b. They do not exceed the maximum established values given in Table 1 for pores, cavities, and sand inclusions

4.4.4 CHIPPING AND DEFECTS ON INSIDE GAP CORNERS—Chipping and defects on inside gap corners are permitted provided that:

 a. No burrs are produced exceeding the values given in 4.4.1.1
 b. The rings do not have an internal notch
 c. They do not exceed 0.3 mm in the radial direction and 0.5 mm in the circumferential and axial directions for coil-spring-loaded oil control rings
 d. They do not exceed the values given in Table 9 for remaining ring designs

TABLE 9—PERMISSIBLE SIZE OF CHIPPING AND DEFECTS ON INSIDE GAP CORNERS

Dimensions are millimeters

Nominal diameter d_1	Size of defect measured axially[1]	Size of defect measured radially[1]	Size of defect measured circumferentially
$30 \leq d_1 < 100$	0.6	0.8	1
$100 \leq d_1 \leq 200$	0.8	1	1.5

[1] Subject to a maximum of one-third of the ring width or radial wall thickness.

4.5 Other Characteristics Subject to Visual Inspection Only

4.5.1 DISCOLORING OR STAINING OF SURFACE—Discoloring or staining spread evenly or unevenly over the ring surfaces is permissible. This does not include rust.

4.5.2 CASTING SKIN AND DEPOSITS ON INSIDE SURFACE—The following defects are permitted:

 a. Unmachined (NCU) areas within 5 degrees of the gap ends
 b. Firmly adherent deposits arising from processing of the ring

4.5.3 CHIPPING ON UNCOATED SURFACES—This is permissible provided the chip sizes do not exceed the maximum values established in Table 1 for pores, cavities, and sand inclusions.

4.5.4 CHROMIUM PLATED PERIPHERY—The chromium plating shall be fully coherent, i.e., there shall be no visible macro cracks, pores, blisters, chromium beads (undercut bulge in the surface), or pin holes.

Exceptions with regard to pin holes may be agreed on between manufacturer and purchaser.

4.5.5 SPRAY COATINGS—Spray coatings are not homogeneous. The acceptance conditions may be agreed on between manufacturer and purchaser; otherwise, manufacturer's specifications apply.

5. Material

5.1 Specifications—The basic material specifications are given in SAE J1590.

The detail specifications and acceptance conditions may be agreed on between manufacturer and purchaser; otherwise, the manufacturer's specifications apply.

5.2 Loss of Tangential Force Under Temperature Effects—Some loss of tangential force at engine operating conditions is acceptable. For the purposes of establishing quality, test conditions, and loss of tangential force with the ring closed to nominal diameter are given in Table 10.

TABLE 10—TEST CONDITIONS TO MEASURE TANGENTIAL FORCE LOSS

SAE reference	Material class	Loss of tangential force max %	Test conditions (ring closed to nominal diameter) Temperature °C	Test conditions (ring closed to nominal diameter) Time h
J1997, J1998, J1999, J2000, J2001, J2002, and J2003	10, 20, 30	12	300	3
	40, 50, 60	8	300	3
Rings with cast iron parts WF[1]	10, 20, 30, 40, 50	25	250	5
	10, 20, 30, 40, 50	12	250	5
Single and multipiece steel oil control rings WF[1]	60	30	220	5
	-	15	220	5

[1] WF = reduced heat set.

6. Raised Material Caused by Marking of Rings—Raised material is permitted subject to the values given in Table 11.

TABLE 11—MAXIMUM PERMITTED RAISED MATERIAL

Dimensions in millimeters

Nominal diameter d_1	Value above surface max
$30 \leq d_1 < 100$	0.008
$100 \leq d_1 \leq 200$	0.01

7. Machining of Periphery and Sides—Unintentional Departure From Ideal Profile and Flatness—Machining operations are not perfect and the peripheral and side faces cannot be machined precisely to profiles and dimensions given or implied in the general specifications (see Section 3).

7.1 Permissible Deviations for Peripheral Surface Profile

7.1.1 Straight faced rings, coated or uncoated, according to SAE J1997, J1998, and J2000:

a. Permissible barrel—0.002 mm per millimeter of ring width h_1 (measuring points: see SAE J1997, J1998, and J2000)

7.1.2 Straight faced rings, coated or uncoated, according to SAE J1997 and J1998 without IW/IF and J2000:

a. Permissible taper—0.005 mm per millimeter of ring width h_1

7.1.3 Straight faced rings, coated or uncoated, according to SAE J1997 and J1998 with IW/IF and J2001:

a. Permissible taper—0.006 mm per millimeter of ring width h_1

7.1.4 S, G, D, and DV type oil control rings:

a. Permissible radial land offset—0.015 mm

7.1.5 SF type, coil spring loaded oil control rings, coated or uncoated:

a. Permissible radial land offset:
  (1) Ring width $h_1 < 6$ mm : 0.015 mm
  (2) Ring width $h_1 > 6$ mm : 0.025 mm

7.2 Permissible Side Face Unevenness

In the radial direction—50% of h_1 - tolerance.

NOTE—This does not apply to twist rings, scraper rings, half keystone rings, and keystone rings.

In the circumferential direction, see Table 12.

TABLE 12—PERMISSIBLE SIDE FACE UNEVENNESS IN THE CIRCUMFERENTIAL DIRECTION[1]

Dimensions in millimeters

Nominal diameter d_1	Permissible unevenness
$30 \leq d_1 < 125$	0.02
$125 \leq d_1 \leq 175$	0.03
$175 \leq d_1 \leq 200$	0.04

[1] This does not apply to twist rings, scraper rings, and oil control rings.

7.3 Permissible Helix—See Table 13.

TABLE 13—PERMISSIBLE HELIX

Dimensions in millimeters

Nominal diameter d_1	Permissible helix	Ring width h_1
$30 \leq d_1 < 80$	0.5	$h_1 \leq 1.5$
	0.3	$h_1 > 1.5$
$80 \leq d_1 < 125$	0.5	
$125 \leq d_1 < 175$	0.7	
$175 \leq d_1 \leq 200$	1	

INTERNAL COMBUSTION ENGINES—PISTON RINGS—RECTANGULAR RINGS—SAE J1997 OCT92

SAE Standard

Report of the Piston Ring Standards Committee 7 approved October 1989. Revised by the SAE Piston Ring Standards Committee 7 October 1992.

This SAE Standard is equivalent to ISO Standard 6622/1.

(R) 1. Scope and Field of Application—Differences, where they exist, are shown in Appendix A.

This SAE Standard specifies the essential dimensional features of R, B, and M rectangular piston ring types.

Dimensional Tables 8 and 9 offer the choice of two radial wall thicknesses:
  a. Radial wall thickness "regular" (Table 8)
  b. Radial wall thickness "D/22" (Table 9)

The requirements of this document apply to rectangular rings for reciprocating internal combustion piston engines up to and including 200 mm diameter. They may also be used for piston rings of compressors working under similar conditions.

(R) 2. References

SAE DESIGNATION	ISO[1] EQUIVALENT	
		INTERNAL COMBUSTION ENGINES—PISTON RINGS
J1588	6621/1	Vocabulary
J1589	6621/2	Measuring principles
J1590	6621/3	Material specifications
J1591	6621/4	General specifications
J1996	6621/5	Quality requirements
		INTERNAL COMBUSTION ENGINES—PISTON RINGS
J1997	6622/1	Rectangular rings
J1998	6622/2 TR	Rectangular rings with narrow ring width
J1999	6623	INTERNAL COMBUSTION ENGINES—PISTON RINGS—SCRAPER RINGS
		INTERNAL COMBUSTION ENGINES—PISTON RINGS
J2000	6624/1	Keystone rings
J2001	6624/2 TR	Half keystone rings
J2002	6625	INTERNAL COMBUSTION ENGINES—PISTON RINGS—OIL CONTROL RINGS
J2003	6626	INTERNAL COMBUSTION ENGINES—PISTON RINGS—COIL SPRING LOADED OIL CONTROL RINGS
J2004	6627 TR	INTERNAL COMBUSTION ENGINES—PISTON RINGS—EXPANDER/SEGMENT OIL CONTROL RINGS
J2226		INTERNAL COMBUSTION ENGINES—PISTON RINGS—STEEL RECTANGULAR RINGS
	1101	TECHNICAL DRAWINGS—GEOMETRICAL TOLERANCING—TOLERANCING OF FORM, ORIENTATION, LOCATION AND RUN-OUT—GENERALITIES, DEFINITIONS, SYMBOLS INDICATIONS ON DRAWINGS

[1] TR refers to Technical Report

3. Ring Types and Designation Examples

3.1 Type R—Straight Faced Rectangular Ring

3.1.1 GENERAL FEATURES

NOTE—See Table 8 or 9 for dimensions and forces.

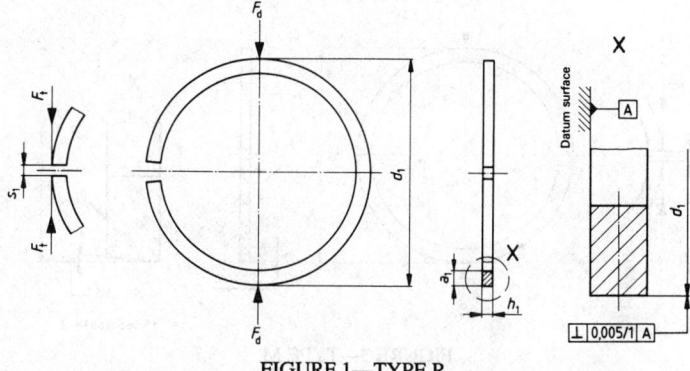

FIGURE 1—TYPE R

3.1.2 DESIGNATION EXAMPLE—Designation of a straight faced rectangular ring of $d_1 = 90$ mm nominal diameter, radial wall thickness "regular", $h_1 = 2.5$ mm ring width, made of grey cast iron, nonheat-treated (material subclass 12), general features as shown in Figure 1, and phosphated all over.

3.2 Type B—Barrel Faced Rectangular Ring

3.2.1 GENERAL FEATURES

NOTE—See Table 8 or 9 for dimensions and forces.

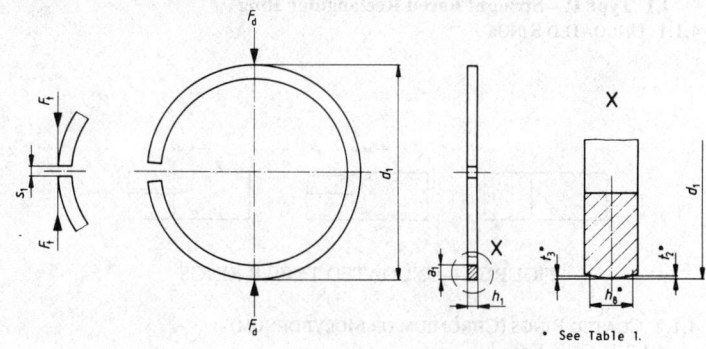

FIGURE 2—TYPE B

TABLE 1—GAUGE WIDTH (h_8) AND BARREL DIMENSIONS

Dimensions in millimeters

h_1	h_8	t_2, t_3	Maximum Peak Off Center
1.5	0.8		0.25
1.75	1.0		
2.0	1.2	0.002/0.016	0.3
2.5	1.6		0.4
3.0	2.0		
3.5	2.4	0.005/0.020	0.5
4.0	2.8		
4.5	3.2	0.005/0.023	0.6

3.2.2 DESIGNATION EXAMPLE—Designation of a barrel faced rectangular ring of d_1 = 90 mm nominal diameter, radial wall thickness "regular", h_1 = 2.5 mm ring width, made of spheroidal graphite cast iron (material subclass 51), general features as shown in Figure 2, and periphery chromium coated fully faced design, 0.15 mm minimum thickness.

3.3 Type M—Taper Faced Rectangular Ring

3.3.1 GENERAL FEATURES

NOTE—See Table 8 or 9 for dimensions and forces.

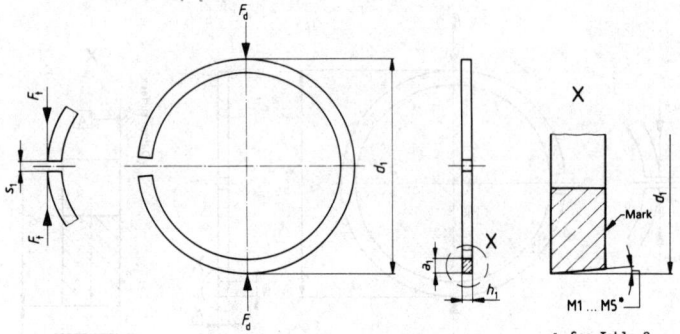

FIGURE 3—TYPE M

3.3.2 DESIGNATION EXAMPLE—Designation of a taper faced rectangular ring of d_1 = 90 mm nominal diameter, radial wall thickness "regular", h_1 = 2.5 mm ring width, made of grey cast iron, heat-treated (material subclass 23), general features as shown in Figure 3, with taper M1 = 10′, and periphery molybdenum coated inlaid design, 0.10 mm minimum thickness.

4. Common Features

4.1 Type R—Straight Faced Rectangular Ring

4.1.1 UNCOATED RINGS

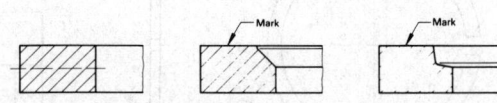

FIGURE 4—UNCOATED TYPE R RINGS

4.1.2 COATED RINGS (CHROMIUM OR MOLYBDENUM)

4.1.2.1 Fully Faced

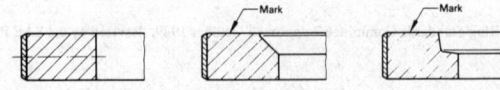

FIGURE 5—FULLY FACED COATED TYPE R RINGS

4.1.2.2 Semi-Inlaid

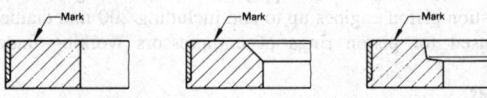

FIGURE 6—SEMI-INLAID COATED TYPE R RINGS

4.1.2.3 Inlaid

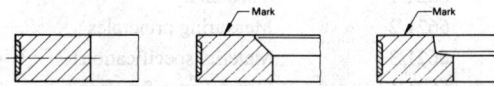

FIGURE 7—INLAID COATED TYPE R RINGS

4.2 Type B—Barrel Faced Rectangular Ring

4.2.1 UNCOATED RINGS

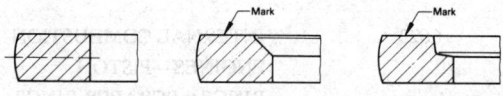

FIGURE 8—UNCOATED TYPE B RINGS

4.2.2 COATED RINGS (CHROMIUM OR MOLYBDENUM)

4.2.2.1 Fully Faced

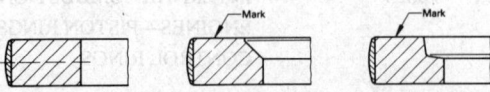

FIGURE 9—FULLY FACED COATED TYPE B RINGS

26.81

4.2.2.2 Semi-Inlaid   4.2.2.3 Inlaid

FIGURE 10—SEMI-INLAID COATED TYPE B RINGS   FIGURE 11—INLAID COATED TYPE B RINGS

TABLE 2—TAPER

Taper	Uncoated and Coated Rings (Molybdenum or Chrome With Tapered Periphery Ground)	Uncoated and Coated Rings (Molybdenum or Chrome With Tapered Periphery Ground) Tolerance[3]	Uncoated and Coated Rings (Molybdenum or Chrome With Tapered Periphery Ground) With IF or IW[1] (Top Side)	Uncoated and Coated Rings (Molybdenum or Chrome With Tapered Periphery Ground) With IF or IW[1] (Top Side) Tolerance[3]	Uncoated and Coated Rings (Molybdenum or Chrome With Tapered Periphery Ground) With IFU or IWU[1] (Bottom Side)[2]	Uncoated and Coated Rings (Molybdenum or Chrome With Tapered Periphery Ground) With IFU or IWU[1] (Bottom Side)[2] Tolerance
M1	10'	+40' / 0	10'		-	-
M2	30'		30'		-	-
M3	60'		60'		60'	
M4	90'	+50' / 0	90'	+60' / 0	90'	+60' / 0
M5	120'		120'		120'	

[1] IF, IW, IFU, and IWU are explained in Figures 16 to 19.
[2] For M-rings (negative twist type) M3, M4, and M5, the twist angle should not exceed 90% of the minimum taper angle.
[3] For coated rings with tapered periphery not ground, the tolerance shall be increased by 10' (for example, M3 = 60': +60' / 0 for M-rings or +70' / 0 for M-rings with IF or IW).

4.3 Type M—Taper Faced Rectangular Ring
4.3.1 Uncoated Rings

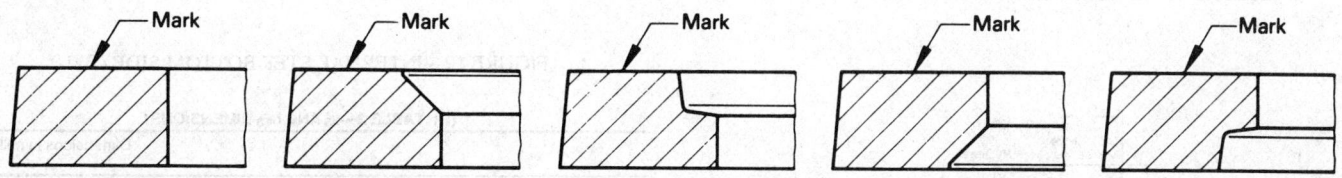

FIGURE 12—UNCOATED TYPE M RINGS

4.3.2 Coated Rings (Chromium or Molybdenum)
4.3.2.1 Fully Faced

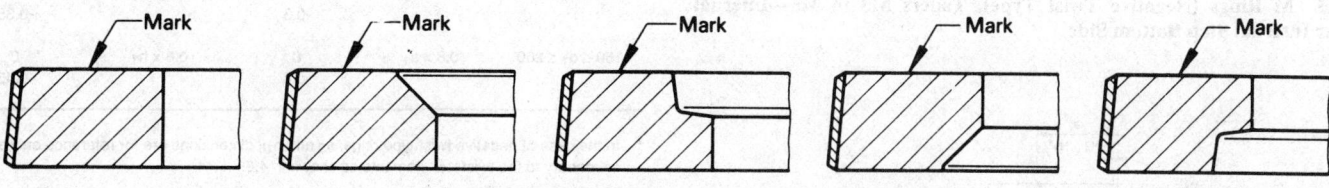

FIGURE 13—FULLY FACED COATED TYPE M RINGS

4.3.2.2 Semi-Inlaid

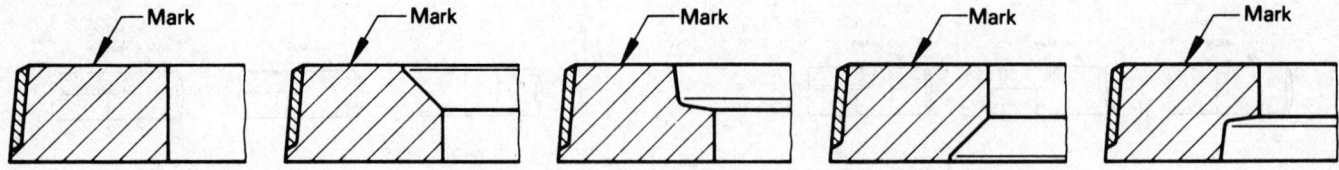

FIGURE 14—SEMI-INLAID COATED TYPE M RINGS

4.3.2.3 Inlaid

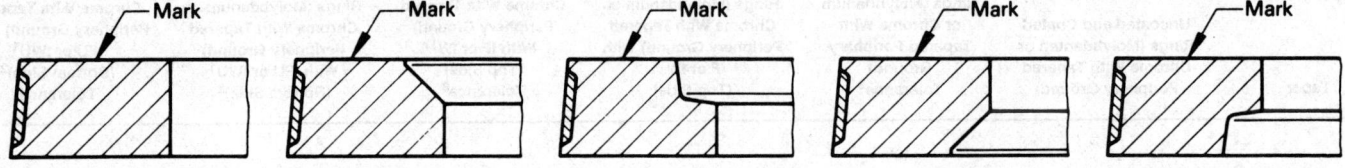

FIGURE 15—INLAID COATED TYPE M RINGS

4.4 R, B, and M Rings (Positive Twist Type)—Internal Bevel or Internal Step Top Side

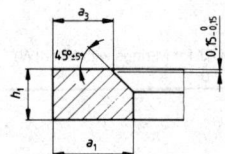

FIGURE 16—INTERNAL BEVEL TOP SIDE (IF)

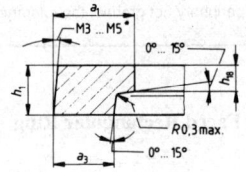

FIGURE 19—INTERNAL STEP BOTTOM SIDE (IWU)

FIGURE 17—INTERNAL STEP TOP SIDE (IW)

4.5 M Rings (Negative Twist Type), Tapers M3 to M5—Internal Bevel or Internal Step Bottom Side

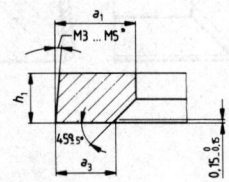

FIGURE 18—INTERNAL BEVEL BOTTOM SIDE (IFU)

(R) TABLE 3—a_3 AND h_{18} DIMENSIONS[1]

Dimensions in millimeters

d_1	a_3	a_3 Tolerance	h_{18}	h_{18} Tolerance
$30 \leq d_1 < 80$	$0.8 \times a_1$	0 -0.2	$0.6 \times h_1$	0 -0.25
$80 \leq d_1 < 100$	$0.8 \times a_1$	0 -0.3	$0.6 \times h_1$	0 -0.25
$100 \leq d_1 < 150$	$0.8 \times a_1$	0 -0.3	$0.6 \times h_1$	0 -0.35
$150 \leq d_1 \leq 200$	$0.8 \times a_1$	0 -0.4	$0.6 \times h_1$	0 -0.45

[1] In the case of negative twist type rings, a_3 and h_{18} dimensions are for reference only and are secondary to the twist requirements as shown in 4.6.

4.6 R, B, and M Rings (Positive Twist Type) and M-Rings (Negative Twist Type)—Variable Internal Bevel—When the standard twist of 0.01/0.05 for rings ≤2 mm axial width and 0.01/0.04 mm for rings >2 mm axial width per 2 mm of radial ring thickness is specified, the dimension a_3, the angle φ and the width of the bevel are at the discretion of the manufacturer. In such cases, the design should correspond to the design shown in Figure 20, a) or b).

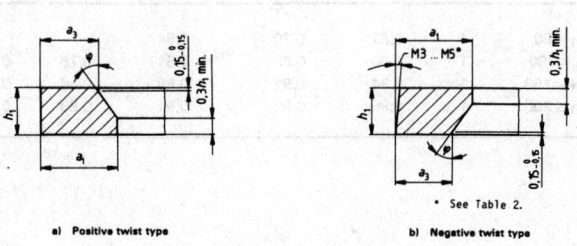

FIGURE 20—VARIABLE INTERNAL BEVEL

4.7 R and B Rings—Outside Chamfered Edges (KA)

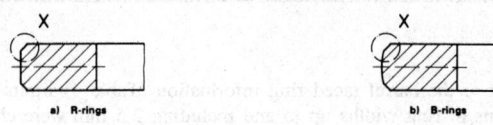

FIGURE 21—OUTSIDE CHAMFERED EDGES (KA)

4.8 R, B, and M Rings—Inside Chamfered Edges (KI)

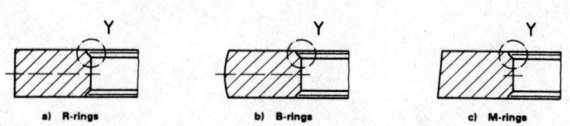

FIGURE 22—INSIDE CHAMFERED EDGES (KI)

4.9 R and B Rings—Outside and Inside Chamfered Edges (KA + KI) (KA Applies to Uncoated Rings Only)

FIGURE 23—OUTSIDE AND INSIDE CHAMFERED EDGES (KA + KI)

FIGURE 24—DETAILS OF FIGURES 21, 22, AND 23

(R) TABLE 4—KA AND KI DIMENSIONS

Dimensions in millimeters

d_1	KA	KI
$30 \leq d_1 < 50$	0.2 max	0.2 max
$50 \leq d_1 < 125$	0.3 ± 0.1	0.3 ± 0.15
$125 \leq d_1 < 175$	0.4 ± 0.1	0.4 ± 0.15
$175 \leq d_1 \leq 200$	0.5 ± 0.1	0.6 ± 0.2

4.10 R, B, and M Rings (Fully Faced, Semi-Inlaid, and Inlaid)—Layer Thickness

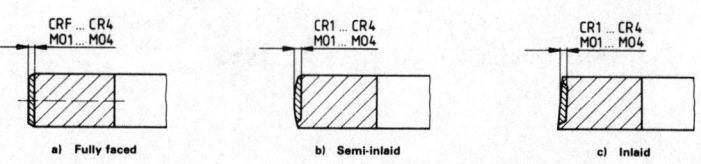

FIGURE 25—LAYER THICKNESS (SEE TABLE 5 FOR DIMENSIONS)

TABLE 5—LAYER THICKNESS

Dimensions in millimeters

Chromium	Molybdenum	Thickness min
CRF	-	0.005
CR1	MO1	0.05
CR2	MO2	0.10
CR3	MO3	0.15
CR4	MO4	0.20

5. Force Factors—The tangential and diametral forces given in Tables 8 and 9 shall be corrected, when additional features and/or materials other than grey cast iron with a modulus of elasticity of 100 000 MPa are being used.

For common features, the multiplier correction factors given in Tables 6 and 7 and the force correction factors given in ISO 6621/4 shall be used.

The factors of Table 7 have been calculated with mean coating thickness.

(R) TABLE 7—FORCE CORRECTION FACTORS FOR COATED R, B, AND M RINGS (FULLY FACED, SEMI-INLAID, AND INLAID TYPE)

d_1 mm	CRF	CR1	CR2/MO1	CR3/MO2	MO3	MO4
$30 \leq d_1 < 50$	1	0.81	0.70	0.64	-	-
$50 \leq d_1 < 100$	1	0.90	0.85	0.81	0.75	0.71
$100 \leq d_1 < 150$	1	0.94	0.91	0.88	0.86	0.83
$150 \leq d_1 \leq 200$	1	0.96	0.93	0.91	0.89	0.87

(R) TABLE 6—FORCE CORRECTION FACTORS FOR R, B, AND M RINGS WITH FEATURES KA, KI, IF, AND IW

d_1 mm	KA	KI	Taper M2 or M3	Taper M4 or M5	IF	IW	KA and KI	KI and taper M2 or M3	KI and taper M4 or M5	KA and IF	KA and IW
$30 \leq d_1 < 50$	1	1	0.97	0.93	0.88	0.75	1	0.97	0.93	0.88	0.75
$50 \leq d_1 \leq 200$	0.97	0.97	0.98	0.96	0.88	0.78	0.94	0.95	0.93	0.85	0.76

APPENDIX A

A.1 This SAE Standard has been established to harmonize the ISO and SAE piston ring Standards. The U.S. Technical Advisory Group, with the support of the National Engine Parts Manufacturers Association, has worked with other national organizations on this worldwide Standard. Some of the wording and phrasing may differ slightly from U.S. terminology for translation purposes.

In preparing this SAE Standard, the Scope and Field of Application, and Reference sections of the ISO 6622/1 have been editorially revised and reorganized.

With respect to the barrel faced ring information (Table 1) limits of barrel drop dimensions of ring widths up to and including 2.5 mm were changed to conform with U.S. standard practice.

In the case of negative twist type rings, a_3 and h_{18} dimensions are for reference only and are secondary to the twist requirements as shown in 4.6.

Certain of the closed gap and ring width dimensions were changed to conform with established U.S. standard practice (Tables 8 and 9).

The tolerances specified in this document represent a six sigma quality level.

6. Dimensions

TABLE 8—DIMENSIONS OF R, B, AND M RECTANGULAR RINGS (RADIAL WALL THICKNESS "REGULAR")

Dimensions in millimeters

Nominal diameter d_1	Radial wall thickness "regular" a_1	Tolerance	Ring width h_1 Column 1	2	3	4	Tolerance	Closed gap s_1	Tolerance	Tangential force F_t, N — For h_1 shown in column 1	2	3	4	Tolerance	Diametral force F_d, N — For h_1 shown in column 1	2	3	4	Tolerance
30	1,25														7,5	8,6	9,9	12,5	
31	1,3														8,0	9,2	10,5	13,1	
32	1,35									—	—	—	—		8,2	9,7	11	13,8	
33	1,4														8,6	10,1	11,6	14,6	
34	1,4														8,2	9,5	11	13,8	
35	1,45														8,6	10,1	11,4	14,4	
36	1,5														9	10,5	12	15,1	
37	1,55									—	—	—	—		9,5	11	12,7	15,7	
38	1,6														9,9	11,4	13,1	16,6	
39	1,65														10,3	12	13,8	17,2	
40	1,65							−0,010								9,7	11,4	13,1	16,3	
41	1,7							−0,030 For phosphated PC surface: −0,005 −0,035								10,1	11,8	13,5	17	
42	1,75								0,15	+0,2 / 0	—	—	—	—		10,5	12,3	14,2	17,6	
43	1,8															11	12,9	14,6	18,3	
44	1,85															11,4	13,3	15,3	19,1	
45	1,9															11,8	13,8	15,7	19,6	
46	1,9															11,2	13,1	15,1	18,7	
47	1,95										—	—	—	—		11,6	13,5	15,5	19,4	
48	2															12	14	16,1	20,2	
49	2,05															12,5	14,6	16,6	20,9	
50	2,1										6	7	8	10		12,9	15,1	17,2	21,5	
51	2,15										6,2	7,2	8,3	10,3		13,3	15,5	17,8	22,1	
52	2,15	±0,15 Within a ring: 0,15 max.								5,9	6,9	7,9	9,9		12,7	14,8	17	21,3	±30 % if $F_d < 21{,}5$N ±20 % if $F_d \geq 21{,}5$N
53	2,2										6,1	7,2	8,2	10,3		13,1	15,5	17,6	22,1	
54	2,25		1,5	1,75	2	2,5				6,3	7,4	8,5	10,6		13,5	15,9	18,3	22,8	
55	2,3										6,5	7,6	8,7	10,9		14	16,3	18,7	23,4	
56	2,35										6,7	7,8	9	11,2		14,4	16,8	19,4	24,1	
57	2,4										6,9	8,1	9,2	11,6		14,8	17,4	19,8	24,9	
58	2,4										6,7	7,8	8,9	11,2		14,4	16,8	19,1	24,1	
59	2,45										6,9	8	9,2	11,5		14,8	17,2	19,8	24,7	
60	2,5										7	8,2	9,4	11,7		15,1	17,6	20,2	25,2	
61	2,55										7,2	8,4	9,6	12,1		15,5	18,1	20,6	26	
62	2,6										7,4	8,6	9,9	12,4		15,9	18,5	21,3	26,7	
63	2,65										7,6	8,9	10,1	12,7	±30 % if $F_t < 10$N ±20 % if $F_t \geq 10$N	16,3	19,1	21,7	27,3	
64	2,65										7,3	8,6	9,8	12,3		15,7	18,5	21,1	26,4	
65	2,7										7,5	8,8	10,1	12,6		16,1	18,9	21,7	27,1	
66	2,75										7,7	9	10,3	12,9		16,6	19,4	22,1	27,7	
67	2,8								0,2	+0,2 / 0	7,9	9,3	10,6	13,3		17	20	22,8	28,6	
68	2,85										8,1	9,5	10,9	13,6		17,4	20,4	23,4	29,2	
69	2,9										8,3	9,7	11,1	13,9		17,8	20,9	23,9	29,9	
70	2,9										8,1	9,4	10,8	13,5		17,4	20,2	23,2	29	
71	2,95										8,3	9,7	11,1	13,8		17,8	20,9	23,9	29,7	
72	3										8,5	9,9	11,3	14,2		18,3	21,3	24,3	30,5	
73	3,05										8,6	10,1	11,6	14,5		18,5	21,7	24,9	31,2	
74	3,1										8,8	10,3	11,8	14,8		18,9	22,1	25,4	31,8	
75	3,15										9	10,5	12	15,1		19,4	22,6	25,8	32,5	
76	3,15										8,8	10,2	11,7	14,7		18,9	21,9	25,2	31,6	
77	3,2								0,25	+0,25 / 0	8,9	10,5	12	15		19,1	22,6	25,8	32,3	
78	3,25										9,1	10,7	12,2	15,3		19,6	23	26,2	32,9	
79	3,3										9,3	10,9	12,5	15,6		20	23,4	26,9	33,5	

This table is shown in ISO format. Commas represent decimal points.

TABLE 8—DIMENSIONS OF R, B, AND M RECTANGULAR RINGS (RADIAL WALL THICKNESS "REGULAR") (CONTINUED)

Dimensions in millimeters

Nominal diameter d_1	Radial wall thickness "regular" a_1	Tolerance	Ring width h_1 Column 1	2	3	4	Tolerance	Closed gap s_1	Tolerance	Tangential force F_t, N For h_1 shown in column 1	2	3	4	Tolerance	Diametral force F_d, N For h_1 shown in column 1	2	3	4	Tolerance
80	3,35									9,5	11,1	12,7	16		20,4	23,9	27,3	34,4	
81	3,4									9,7	11,4	13	16,3		20,9	24,5	28	35	
82	3,4									9,5	11,1	12,7	15,9		20,4	23,9	27,3	34,2	
83	3,45									9,7	11,3	12,9	16,2		20,9	24,3	27,7	34,8	
84	3,5		1,5	1,75	2	2,5		0,25	+0,25 / 0	9,9	11,5	13,2	16,5		21,3	24,7	28,4	35,5	
85	3,55						−0,010 / −0,030			10,1	11,8	13,5	16,8		21,7	25,4	29	36,1	
86	3,6						For phosphated PO surface: −0,005 / −0,035			10,3	12	13,7	17,2		22,1	25,8	29,5	37	
87	3,65	±0,15 Within a ring: 0,15 max.								10,4	12,2	14	17,5		22,4	26,2	30,1	37,6	
88	3,65									10,2	11,9	13,6	17,1		21,9	25,6	29,2	36,8	
89	3,7									10,4	12,2	13,9	17,4		22,4	26,2	29,9	37,4	
90	3,75									12,3	14,1	17,6	21,2		26,4	30,3	37,8	45,6	
91	3,8									12,5	14,3	18	21,6		26,9	30,7	38,7	46,4	
92	3,85									12,8	14,6	18,3	22		27,5	31,4	39,3	47,3	
93	3,9									13	14,9	18,6	22,4		28	32	40	48,2	
94	3,9		1,75	2	2,5	3				12,7	14,5	18,2	21,9		27,3	31,2	39,1	47,1	
95	3,95									12,9	14,8	18,5	22,3		27,7	31,8	39,8	47,9	
96	4									13,2	15,1	18,8	22,6		28,4	32,5	40,4	48,6	
97	4,05									13,4	15,3	19,2	23		28,8	32,9	41,3	49,5	
98	4,1									13,6	15,6	19,5	23,4		29,2	33,5	41,9	50,3	
99	4,15							0,3	+0,25 / 0	13,8	15,8	19,8	23,8		29,7	34	42,6	51,2	
100	4,15									15,5	19,4	23,3			33,3	41,7	50,1		
101	4,2									15,7	19,7	23,7			33,8	42,4	51		
102	4,25									—	16	20	24		—	34,4	43	51,6	
103	4,25									16,2	20,3	24,4			34,8	43,6	52,5		
104	4,3									15,9	19,9	23,9			34,2	42,8	51,4		
105	4,35		—	2	2,5	3				16,1	20,1	24,2		±30% if F_t < 10N	34,6	43,2	52		±30% if F_d < 21,5N
106	4,4									16,3	20,4	24,6			35	43,9	52,9		
107	4,4									—	16	20	24,1	±20% if F_t > 10N	—	34,4	43	51,8	±20% if F_d > 21,5N
108	4,45									16,2	20,3	24,4			34,8	43,6	52,5		
109	4,5									16,4	20,6	24,8			35,3	44,3	53,3		
110	4,55									20,8	25	29,2			44,7	53,8	62,8		
111	4,55									20,4	24,5	28,6			43,9	52,7	61,5		
112	4,6									—	20,7	24,9	29		—	44,5	53,5	62,4	
113	4,65									21	25,2	29,4			45,2	54,2	63,2		
114	4,7									21,3	25,6	29,8			45,8	55	64,1		
115	4,7	±0,2 Within a ring: 0,20 max.								20,9	25,1	29,3			44,9	54	63		
116	4,75									21,1	25,4	29,7			45,4	54,6	63,9		
117	4,8									—	21,4	25,8	30,1		—	46	55,5	64,7	
118	4,85									21	25,3	29,5			45,2	54,4	63,4		
119	4,85							0,35	+0,30 / 0	21,3	25,6	29,9			45,8	55	64,3		
120	4,9									21,6	25,9	30,3			46,4	55,7	65,1		
121	4,95									21,9	26,3	30,7			47,1	56,5	66		
122	4,95		—	2,5	3	3,5				—	21,5	25,8	30,1		—	46,2	55,5	64,7	
123	5									21,8	26,1	30,5			46,9	56,1	65,6		
124	5,05									22	26,5	30,9			47,3	57	66,4		
125	5,05						−0,010 / −0,040 For phosphated PO surface: −0,005 / −0,045			21,6	26	30,4			46,4	55,9	65,4		
126	5,1									21,9	26,3	30,7			47,1	56,5	66		
127	5,15									—	22,2	26,7	31,1		—	47,7	57,4	66,9	
128	5,2									22,5	27	31,5			48,4	58,1	67,7		
129	5,2									22,1	26,5	31			47,5	57	66,7		
130	5,25									22,3	26,8	31,3			47,9	57,6	67,3		
131	5,3									22,6	27,1	31,6			48,6	58,3	67,9		
132	5,3							0,4	+0,35 / 0	—	22,2	26,6	31,1		—	47,7	57,2	66,9	
133	5,35									22,4	27	31,5			48,2	58,1	67,7		
134	5,4									22,7	27,3	31,9			48,8	58,7	68,6		

This table is shown in ISO format. Commas represent decimal points.

TABLE 8—DIMENSIONS OF R, B, AND M RECTANGULAR RINGS (RADIAL WALL THICKNESS "REGULAR") (CONCLUDED)

Dimensions in millimeters

Nominal diameter d_1	Radial wall thickness "regular" a_1	Tolerance	Ring width h_1 Column 1	2	3	4	Tolerance	Closed gap s_1	Tolerance	Tangential force F_t, N For h_1 shown in column 1	2	3	4	Tolerance	Diametral force F_d, N For h_1 shown in column 1	2	3	4	Tolerance
135	5,4									—	22,3	26,8	31,3		—	47,9	57,6	67,3	
136	5,45									—	22,6	27,2	31,7		—	48,6	58,5	68,2	
137	5,5		—	2,5	3	3,5				—	22,9	27,5	32,1		—	49,2	59,1	69	
138	5,5						−0,010 −0,040			—	22,5	27	31,6		—	48,4	58,1	67,9	
139	5,55						For phosphated PO surface: −0,005 −0,045			—	22,8	27,3	31,9		—	49	58,7	68,6	
140	5,6							0,4	+0,35 0	—	27,7	32,3	36,9		—	59,6	69,4	79,3	
141	5,65									—	28	32,7	37,4		—	60,2	70,3	80,4	
142	5,65									—	27,5	32,2	36,8		—	59,1	69,2	79,1	
143	5,7									—	27,8	32,5	37,2		—	59,8	69,9	80	
144	5,75									—	28,2	32,9	37,6		—	60,6	70,7	80,8	
145	5,75									—	27,7	32,4	37		—	59,6	69,7	79,6	
146	5,8									—	28	32,7	37,4		—	60,2	70,3	80,4	
147	5,85									—	28,3	33,1	37,9		—	60,8	71,2	81,5	
148	5,85									—	27,9	32,6	37,3		—	60	70,1	80,2	
149	5,9									—	28,2	33	37,7		—	60,6	71	81,1	
150	5,95									—	28,3	33,1	37,8		—	60,8	71,2	81,3	
152	6									—	28,2	32,9	37,7		—	60,6	70,7	81,1	
154	6,05		—	3	3,5	4				—	28,1	32,8	37,5		—	60,4	70,5	80,6	
155	6,1									—	28,4	33,2	37,9		—	61,1	71,4	81,5	
156	6,15									—	28,7	33,5	38,3		—	61,7	72	82,3	
158	6,2									—	28,6	33,4	38,2		—	61,5	71,8	82,1	
160	6,25	±0,2 Within a ring: 0,20 max.						0,5	+0,4 0	—	28,5	33,2	38	±30 % if F_t < 10 N ±20 % if F_t > 10 N	—	61,3	71,4	81,7	±30 % if F_d < 21,5 N ±20 % if F_d > 21,5 N
162	6,35									—	29	33,9	38,8		—	62,4	72,9	83,4	
164	6,4									—	28,9	33,8	38,7		—	62,1	72,7	83,2	
165	6,4									—	28,5	33,3	38,1		—	61,3	71,6	81,9	
166	6,45									—	28,8	33,7	38,5		—	61,9	72,5	82,8	
168	6,5									—	28,7	33,5	38,4		—	61,7	72	82,6	
170	6,6									—	29,3	34,2	39,1		—	63	73,5	84,1	
172	6,65									—	29,2	34,1	39		—	62,8	73,3	83,9	
174	6,7									—	29,1	34	38,8		—	62,6	73,1	83,4	
175	6,75									—	34,1	39	44		—	73,3	83,9	94,6	
176	6,8									—	34,5	39,4	44,4		—	74,2	84,7	95,5	
178	6,85									—	34,3	39,3	44,2		—	73,7	84,5	95	
180	6,9									—	34,2	39,1	44,1		—	73,5	84,1	94,8	
182	6,95									—	34,1	39	43,9		—	73,3	83,9	94,4	
184	7,05									—	34,7	39,7	44,7		—	74,6	85,4	96,1	
185	7,05									—	34,3	39,2	44,2		—	73,7	84,3	95	
186	7,1									—	34,6	39,6	44,6		—	74,4	85,1	95,9	
188	7,15		—	3,5	4	4,5		0,6	+0,45 0	—	34,5	39,5	44,4		—	74,2	84,9	95,5	
190	7,2									—	34,4	39,3	44,3		—	74	84,5	95,2	
192	7,25									—	34,3	39,2	44,2		—	73,7	84,3	95	
194	7,35									—	34,9	39,9	44,9		—	75	85,8	96,5	
195	7,35									—	34,5	39,5	44,4		—	74,2	84,9	95,5	
196	7,4									—	34,8	39,8	44,8		—	74,8	85,6	96,3	
198	7,45									—	34,7	39,7	44,7		—	74,6	85,4	96,1	
200	7,5									—	34,6	39,6	44,5		—	74,4	85,1	95,7	

NOTES

[1] For intermediate sizes (for example repair sizes), the radial wall thickness of the next smaller nominal diameter should be applied.

[2] The values for F_t and F_d, given in Table 8, apply to as cast grey cast iron with a typical modulus of elasticity (E_n) of 100 000 MPa. Multiplying factors for materials having a different modulus (E_n) are given in SAE J1591.

Mean forces are calculated for nominal radial wall thickness (a_t) and mean ring width (h_t).

[3] For the sole purpose of this document, the assumed average ratio F_d/F_t is 2.15. However, for rings up to 50 mm, the ratio F_d/F_t shall be determined between the manufacturer and client.

This table is shown in ISO format. Commas represent decimal points.

TABLE 9—DIMENSIONS OF R, B, AND M RECTANGULAR RINGS (RADIAL WALL THICKNESS "D/22")

Dimensions in millimeters

Nominal diameter d_1	Radial wall thickness "D/22" a_1	Tolerance	Ring width h_1 Column 1	2	3	4	Tolerance	Closed gap s_1	Tolerance	Tangential force F_t, N For h_1 shown in column 1	2	3	4	Tolerance	Diametral force F_d, N For h_1 shown in column 1	2	3	4	Tolerance
50	2,25						−0,010 −0,030 For phosphated PO surface: −0,005 −0,035	0,15	+0,2 0	7,4	8,7	9,9	12,4		15,9	18,7	21,3	26,7	
51	2,3									7,6	8,9	10,2	12,7		16,3	19,1	21,9	27,3	
52	2,35									7,8	9,1	10,4	13,1		16,8	19,6	22,4	28,2	
53	2,4									8	9,4	10,7	13,4		17,2	20,2	23	28,8	
54	2,45									8,2	9,6	11	13,8		17,6	20,6	23,7	29,7	
55	2,5									8,4	9,9	11,3	14,1		18,1	21,3	24,3	30,3	
56	2,55									8,6	10,1	11,5	14,5		18,5	21,7	24,7	31,2	
57	2,6									8,8	10,3	11,8	14,8		18,9	22,1	25,4	31,8	
58	2,65									9	10,6	12,1	15,2		19,4	22,8	26	32,7	
59	2,7									9,3	10,8	12,4	15,5		20	23,2	26,7	33,3	
60	2,75									9,4	11	12,6	15,7		20,2	23,7	27,1	33,8	
61	2,75									9,1	10,6	12,2	15,2		19,6	22,8	26,2	32,7	
62	2,8									9,3	10,9	12,4	15,6		20	23,4	26,7	33,5	
63	2,85									9,5	11,1	12,7	15,9		20,4	23,9	27,3	34,2	
64	2,9									9,7	11,3	13	16,3		20,9	24,3	28	35	
65	2,95							0,2	+0,2 0	9,9	11,6	13,3	16,6		21,3	24,9	28,6	35,7	
66	3									10,1	11,8	13,5	16,9		21,7	25,4	29	36,3	
67	3,05									10,3	12,1	13,8	17,3		22,1	26	29,7	37,2	
68	3,1									10,5	12,3	14,1	17,6		22,6	26,4	30,3	37,8	
69	3,15		1,5	1,75	2	2,5				10,7	12,5	14,4	18		23	26,9	31	38,7	
70	3,2									10,9	12,8	14,6	18,3		23,4	27,5	31,4	39,3	
71	3,25									11,1	13	14,9	18,7		23,9	28	32	40,2	
72	3,25									10,8	12,7	14,5	18,1		23,2	27,3	31,2	38,9	
73	3,3	±0,15 Within a ring: 0,15 max.								11	12,9	14,8	18,5	±30% if F_t < 10N ±20% if F_t > 10N	23,7	27,7	31,8	39,8	±30% if F_d < 21,5N ±20% if F_d ≥ 21,5N
74	3,35									11,2	13,1	15	18,8		24,1	28,2	32,3	40,4	
75	3,4									11,4	13,3	15,2	19,1		24,5	28,6	32,7	41,1	
76	3,45									11,6	13,6	15,5	19,4		24,9	29,2	33,3	41,7	
77	3,5									11,8	13,8	15,8	19,8		25,4	29,7	34	42,6	
78	3,55									12	14	16,1	20,1		25,8	30,1	34,6	43,2	
79	3,6									12,2	14,3	16,3	20,5		26,2	30,7	35	44,1	
80	3,65									12,4	14,5	16,6	20,8		26,7	31,2	35,7	44,7	
81	3,7							0,25	+0,25 0	12,6	14,8	16,9	21,1		27,1	31,8	36,3	45,4	
82	3,75									12,8	15	17,2	21,5		27,5	32,3	37	46,2	
83	3,75									12,5	14,6	16,7	21		26,9	31,4	35,9	45,2	
84	3,8									12,7	14,9	17	21,3		27,3	32	36,6	45,8	
85	3,85									12,9	15,1	17,3	21,6		27,7	32,5	37,2	46,4	
86	3,9									13,1	15,3	17,6	22		28,2	32,9	37,8	47,3	
87	3,95									13,3	15,6	17,8	22,3		28,6	33,5	38,3	47,9	
88	4									13,5	15,8	18,1	22,7		29	34	38,9	48,8	
89	4,05									13,7	16,1	18,4	23		29,5	34,6	39,6	49,5	
90	4,1									16,2	18,6	23,2	27,9		34,8	40	49,9	60	
91	4,15									16,5	18,8	23,6	28,3		35,5	40,4	50,7	60,8	
92	4,2									16,7	19,1	23,9	28,7		35,9	41,1	51,4	61,7	
93	4,25									16,9	19,4	24,3	29,2		36,3	41,7	52,2	62,8	
94	4,25		1,75	2	2,5	3				16,6	19	23,7	28,5		35,7	40,9	51	61,3	
95	4,3									16,8	19,2	24,1	28,9		36,1	41,3	51,8	62,1	
96	4,35									17	19,5	24,4	29,3		36,6	41,9	52,5	63	
97	4,4							0,3	+0,25 0	17,3	19,8	24,8	29,8		37,2	42,6	53,3	64,1	
98	4,45									17,5	20,1	25,1	30,2		37,6	43,2	54	64,9	
99	4,5									17,8	20,3	25,5	30,6		38,3	43,6	54,8	65,8	
100	4,55	±0,2 Within a ring: 0,20 max.									20,6	25,8	31			44,3	55,5	66,7	
101	4,6										20,8	26,1	31,3			44,7	56,1	67,3	
102	4,65		—	2	2,5	3				—	21,1	26,4	31,7		—	45,4	56,8	68,2	
103	4,7										21,3	26,7	32,1			45,8	57,4	69	
104	4,75										21,6	27	32,4			46,4	58,1	69,7	

This table is shown in ISO format. Commas represent decimal points.

TABLE 9—DIMENSIONS OF R, B, AND M RECTANGULAR RINGS (RADIAL WALL THICKNESS "D/22") (CONCLUDED)

Dimensions in millimeters

Nominal diameter d_1	Radial wall thickness "D/22" a_1	Tolerance	Ring width h_1 Column 1	2	3	4	Tolerance	Closed gap s_1	Tolerance	Tangential force F_t, N For h_1 shown in column 1	2	3	4	Tolerance	Diametral force F_d, N For h_1 shown in column 1	2	3	4	Tolerance
105	4,75		—	2	2,5	3		0,35	+0,30 / 0	—	21,1	26,4	31,8		—	45,4	56,8	68,4	
106	4,8										21,4	26,7	32,1			46	57,4	69	
107	4,85										21,6	27,1	32,5			46,4	58,3	69,9	
108	4,9										21,8	27,4	32,9			46,9	58,9	70,7	
109	4,95										22,1	27,7	33,2			47,5	59,6	71,4	
110	5										27,9	33,5	39,1			60	72	84,1	
111	5,05										28,2	33,8	39,5			60,6	72,7	84,9	
112	5,1										28,5	34,2	40			61,3	73,5	86	
113	5,15										28,8	34,6	40,4			61,9	74,4	86,9	
114	5,2										29,1	34,9	40,8			62,6	75	87,7	
115	5,25										29,4	35,3	41,2			63,2	75,9	88,6	
116	5,25										28,8	34,6	40,4			61,9	74,4	86,9	
117	5,3										29,1	35	40,8			62,6	75,3	87,7	
118	5,35										29,4	35,3	41,3			63,2	75,9	88,8	
119	5,4										29,7	35,7	41,7			63,9	76,8	89,7	
120	5,45										30	36	42,1			64,5	77,4	90,5	
121	5,5										30,3	36,4	42,5			65,1	78,3	91,4	
122	5,55										30,6	36,7	42,9			65,8	78,9	92,2	
123	5,6										30,9	37,1	43,3			66,4	79,8	93,1	
124	5,65										31,2	37,4	43,7			67,1	80,4	94	
125	5,7	±0,2 Within a ring: 0,20 max.	—	2,5	3	3,5	−0,010 −0,040 For phosphated PO surface −0,005 −0,045			—	31,4	37,8	44,1	±30% if F_t < 10N ±20% if F_t > 10N	—	67,5	81,3	94,8	±30% if F_d < 21,5N ±20% if F_d > 21,5N
126	5,75										31,7	38,1	44,5			68,2	81,9	95,7	
127	5,75										31,2	37,5	43,8			67,1	80,6	94,2	
128	5,8										31,5	37,8	44,2			67,7	81,3	95	
129	5,85										31,8	38,2	44,6			68,4	82,1	95,9	
130	5,9										32	38,4	44,8			68,8	82,6	96,3	
131	5,95										32,2	38,7	45,2			69,2	83,2	97,2	
132	6										32,5	39,1	45,6			69,9	84,1	98	
133	6,05										32,8	39,4	46			70,5	84,7	98,9	
134	6,1										33,1	39,8	46,4			71,2	85,6	99,8	
135	6,15										33,4	40,1	46,8			71,8	86,2	100,6	
136	6,2								0,4	+0,35 / 0		33,7	40,4	47,2			72,5	86,9	101,5	
137	6,25										33,9	40,8	47,6			72,9	87,7	102,3	
138	6,25										33,4	40,1	46,8			71,8	86,2	100,6	
139	6,3										33,7	40,4	47,2			72,5	86,9	101,5	
140	6,35										40,8	47,6	54,5			87,7	102,3	117,2	
141	6,4										41,1	48	54,9			88,4	103,2	118	
142	6,45										41,4	48,4	55,4			89	104,1	119,1	
143	6,5										41,8	48,8	55,8			89,9	104,9	120	
144	6,55										42,1	49,2	56,2			90,5	105,8	120,8	
145	6,6		—	3	3,5	4					42,4	49,6	56,7			91,2	106,6	121,9	
146	6,65										42,8	49,9	57,1			92	107,3	122,8	
147	6,7										43,1	50,3	57,6			92,7	108,1	123,8	
148	6,75										43,4	50,7	58			93,3	109	124,7	
149	6,75										42,8	49,9	57,1			92	107,3	122,8	
150	6,8							0,5	+0,40 / 0		42,8	50	57,1			92	107,5	122,8	

NOTES

1. For intermediate sizes (for example repair sizes), the radial wall thickness of the next smaller nominal diameter should be applied.

2. The values for F_t and F_d, given in Table 9, apply to as cast grey cast iron with a typical modulus of elasticity (E_n) of 100 000 MPa. Multiplying factors for materials having a different modulus (E_n) are given in SAE J1591.

   Mean forces are calculated for nominal radial wall thickness (a_t) and mean ring width (h_t).

3. For the sole purpose of this document, the assumed average ratio F_d/F_t is 2.15.

This table is shown in ISO format. Commas represent decimal points.

INTERNAL COMBUSTION ENGINES—PISTON RINGS—SCRAPER RINGS—SAE J1999 OCT92

SAE Standard

Report of the Piston Ring Standards Committee 7 approved October 1989. Revised by the SAE Piston Ring Standards Committee 7 October 1992.

This SAE Standard is equivalent to ISO Standard 6623.

(R) **1. Scope and Field of Application**—Differences, where they exist, are shown in Appendix A.

This SAE Standard specifies the essential dimensional features of N, NM, E, and EM scraper piston ring types.

Dimensional Tables 7 and 8 offer the choice of two radial wall thicknesses:
a. Radial wall thickness "regular" (Table 7)
b. Radial wall thickness "D/22" (Table 8)

The requirements of this document apply to scraper rings for reciprocating internal combustion piston engines up to and including 200 mm diameter. They may also be used for piston rings of compressors working under similar conditions.

(R) **2. References**

SAE DESIGNATION	ISO[1] EQUIVALENT	
		INTERNAL COMBUSTION ENGINES—PISTON RINGS
J1588	6621/1	Vocabulary
J1589	6621/2	Measuring principles
J1590	6621/3	Material specifications
J1591	6621/4	General specifications
J1996	6621/5	Quality requirements
		INTERNAL COMBUSTION ENGINES—PISTON RINGS
J1997	6622/1	Rectangular rings
J1998	6622/2 TR	Rectangular rings with narrow ring width
J1999	6623	INTERNAL COMBUSTION ENGINES—PISTON RINGS—SCRAPER RINGS
		INTERNAL COMBUSTION ENGINES—PISTON RINGS
J2000	6624/1	Keystone rings
J2001	6624/2 TR	Half keystone rings
J2002	6625	INTERNAL COMBUSTION ENGINES—PISTON RINGS—OIL CONTROL RINGS
J2003	6626	INTERNAL COMBUSTION ENGINES—PISTON RINGS—COIL SPRING LOADED OIL CONTROL RINGS
J2004	6627 TR	INTERNAL COMBUSTION ENGINES—PISTON RINGS—EXPANDER/SEGMENT OIL CONTROL RINGS
J2226		INTERNAL COMBUSTION ENGINES—PISTON RINGS—STEEL RECTANGULAR RINGS

[1] TR refers to Technical Report

3. Ring Types and Designation Examples

3.1 Types N, NM, E, and EM—Scraper Rings—Common General Features

NOTE—See Table 7 or 8 for dimensions and forces.

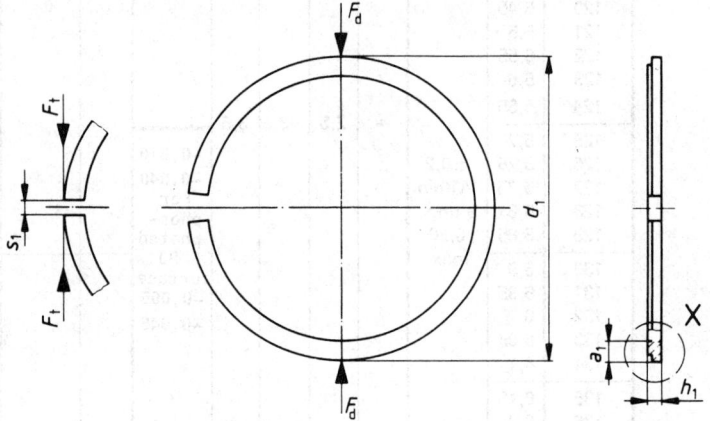

FIGURE 1—TYPES N, NM, E, AND EM

3.2 Type N—Scraper Ring (Napier)
3.2.1 GENERAL FEATURES

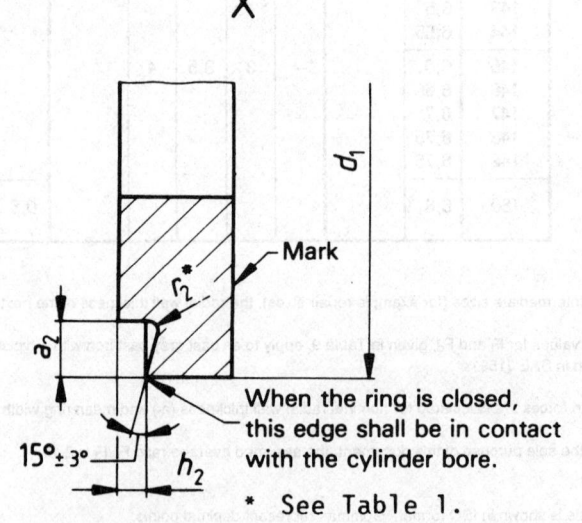

When the ring is closed, this edge shall be in contact with the cylinder bore.

* See Table 1.

FIGURE 2—TYPE N—DETAIL OF FIGURE 1

(R) TABLE 1—r_2 DIMENSIONS

Dimensions in millimeters

d_1	r_2 max
$30 \leq d_1 < 175$	0.3
$175 \leq d_1 \leq 200$	0.7

3.2.2 DESIGNATION EXAMPLE—Designation of a Napier ring of d_1 = 90 mm nominal diameter, radial wall thickness "regular", h_1 = 2.5 mm ring width, made of grey cast iron, nonheat-treated (material subclass 12), general features as shown in Figures 1 and 2, and inside chamfered edges.

3.3 Type NM—Scraper Ring (Napier), Taper Faced
3.3.1 GENERAL FEATURES

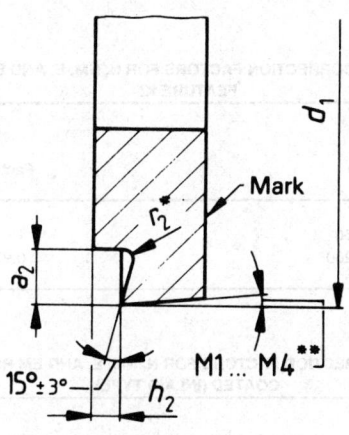

* See Table 1.
** See Table 2.

FIGURE 3—TYPE NM—DETAIL OF FIGURE 1

TABLE 2—TAPER

Taper	Uncoated and coated rings (molybdenum)	Uncoated and coated rings (molybdenum) Tolerance
M1	10'	+60' / 0
M2	30'	+60' / 0
M3	60'	+60' / 0
M4	90'	+60' / 0

3.3.2 DESIGNATION EXAMPLE—Designation of a Napier ring, taper faced M4 = 90', of d_1 = 90 mm nominal diameter, radial wall thickness "regular", h_1 = 2.5 mm ring width, made of grey cast iron, heat-treated (material subclass 21), general features as shown in Figures 1 and 3, and phosphate all over:
Piston ring SAE J1999 or ISO 6623 - NM4 - 90 x 2.5 - MC21 PO

3.4 Type E—Scraper Ring (Stepped)
3.4.1 GENERAL FEATURES

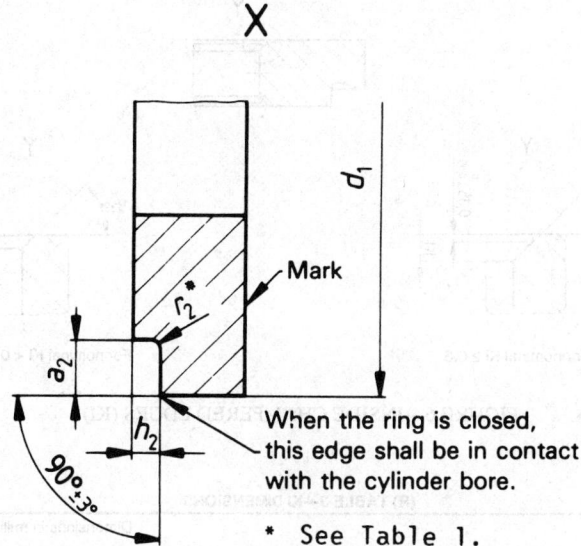

When the ring is closed, this edge shall be in contact with the cylinder bore.

* See Table 1.

FIGURE 4—TYPE E—DETAIL OF FIGURE 1

3.4.2 DESIGNATION EXAMPLE—Designation of a scraper ring of d_1 = 90 mm nominal diameter, radial wall thickness "regular", h_1 = 2.5 mm ring width, made of grey cast iron, nonheat-treated (material subclass 12), general features as shown in Figures 1 and 4, and periphery molybdenum coated inlaid design, 0.10 mm minimum thickness.

3.5 Type EM—Scraper Ring (Stepped) Taper Faced
3.5.1 GENERAL FEATURES

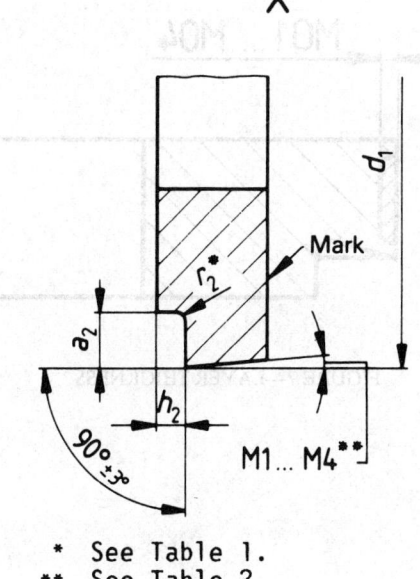

* See Table 1.
** See Table 2.

FIGURE 5—TYPE EM—DETAIL OF FIGURE 1

3.5.2 DESIGNATION EXAMPLE—Designation of a scraper ring taper faced M2 = 30', of d_1 = 90 mm nominal diameter, radial wall thickness "regular", h_1 = 2.5 mm ring width, made of grey cast iron, heat-treated (material subclass 22), general features as shown in Figures 1 and 5, and inside chamfered edges.

4. Common Features
4.1 N, NM, E, and EM Rings—Inside Chamfered Edges (KI)
NOTE—See Table 3 for dimensions.

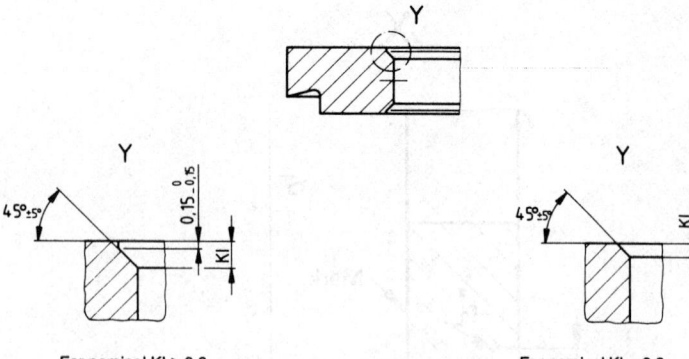

FIGURE 6—INSIDE CHAMFERED EDGES (KI)

(R) TABLE 3—KI DIMENSIONS

Dimensions in millimeters

d_1	KI
30 ≤ d_1 < 50	0.2 max
50 ≤ d_1 < 125	0.3 ± 0.15
125 ≤ d_1 < 175	0.4 ± 0.15
175 ≤ d_1 ≤ 200	0.6 ± 0.2

4.2 N, NM, E, and EM Rings, Coated (Molybdenum Inlaid)—Layer Thickness

NOTE—See Table 4 for dimensions.

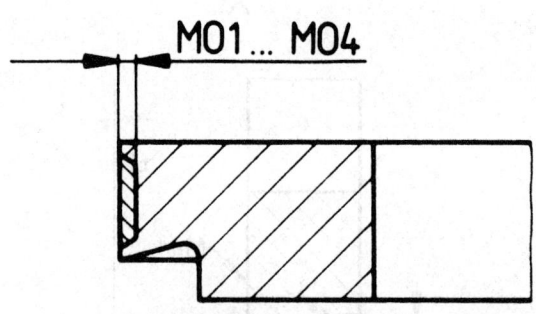

FIGURE 7—LAYER THICKNESS

TABLE 4—LAYER THICKNESS

Dimensions in millimeters

Molybdenum	Thickness min
MO1	0.05
MO2	0.10
MO3	0.15
MO4	0.20

5. Force Factors—The tangential and diametral forces given in Tables 7 and 8 shall be corrected when additional features and/or materials other than grey cast iron with a modulus of elasticity of 100 000 MPa are being used.

For common features, the multiplier correction factors given in Tables 5 and 6 and the force correction factors given in SAE J1591 shall be used.

The factors of Table 6 have been calculated with mean coating thickness.

TABLE 5—FORCE CORRECTION FACTORS FOR N, NM, E, AND EM RINGS, WITH FEATURE KI

d_1 mm	Factor
30 ≤ d_1 < 50	1
50 ≤ d_1 ≤ 200	0.97

TABLE 6—FORCE CORRECTION FACTORS FOR N, NM, E, AND EM RINGS, MOLYBDENUM COATED (INLAID TYPE)

d_1 mm	Factor MO1	Factor MO2	Factor MO3	Factor MO4
30 ≤ d_1 < 50	0.81	0.75	-	-
50 ≤ d_1 < 100	0.90	0.86	0.83	0.80
100 ≤ d_1 < 150	0.94	0.91	0.89	0.87
150 ≤ d_1 ≤ 200	0.95	0.94	0.92	0.90

6. Dimensions—See Tables 7 and 8.

APPENDIX A

A.1 This SAE Standard has been established to harmonize the ISO and SAE piston ring documents. The U.S. Technical Advisory Group, with the support of the National Engine Parts Manufacturers Association, has worked with other national organizations on this worldwide document. Some of the wording and phrasing may differ slightly from U.S. terminology for translation purposes.

In preparing this SAE document, the Scope and Field of Application, and Reference Sections of ISO 6623 have been editorially revised and reorganized.

Certain closed gap and ring width dimensions were changed to conform with established U.S. standard practice (Tables 7 and 8).

The tolerances specified in this document represent a six sigma quality level.

TABLE 7—DIMENSIONS FOR N, NM, E, AND EM SCRAPER RINGS (RADIAL WALL THICKNESS ["REGULAR"])

Dimensions in millimeters

Nominal diameter d_1	Radial wall thickness "regular" a_1	Tolerance	Ring width h_1 Column 1	2	3	4	Tolerance	Closed gap s_1	Tolerance	Axial width of step h_2 For h_1 shown in column 1	2	3	4	Radial depth of step a_2	Tangential force F_t, N For h_1 shown in column 1	2	3	4	Tolerance	Diametral force F_d, N For h_1 shown in column 1	2	3	4	Tolerance
30	1,25													0,3 ±0,15	—	—	—	—		6,2	7,5	8,6	10,8	
31	1,3																			6,7	8	9	11,4	
32	1,35																			7,1	8,4	9,7	12	
33	1,4																			7,1	8,6	9,9	12,5	
34	1,4																			6,7	8	9,2	11,6	
35	1,45													0,4 ±0,15	—	—	—	—		7,1	8,4	9,7	12,3	
36	1,5																			7,5	8,8	10,3	12,9	
37	1,55																			8	9,2	10,8	13,5	
38	1,6																			8,2	9,7	11,2	14,2	
39	1,65																			8,6	10,1	11,8	14,8	
40	1,65														—	—	—	—		8,2	9,7	11,2	14,2	
41	1,7																			8,6	10,1	11,6	14,8	
42	1,75							0,15	−0,010 −0,030 For phosphated PO surface: −0,005 −0,035											9	10,5	12,3	15,5	
43	1,8																			9	10,8	12,5	15,7	
44	1,85									+0,2 0										9,5	11,2	12,9	16,3	
45	1,9	±0,15 Within a ring: 0,15 max	1,5	1,75	2	2,5				0,45 ±0,15	0,5 ±0,15	0,5 ±0,15	0,6 ±0,15	0,5 ±0,15	—	—	—	—		9,7	11,6	13,3	16,8	
46	1,9																			9,2	11	12,7	16,1	
47	1,95																			9,7	11,4	13,1	16,8	±30% if $F_d < 21,5$ N
48	2																			10,1	11,8	13,8	17,4	±20% if $F_d ≥ 21,5$ N
49	2,05																			10,5	12,3	14,2	18,1	
50	2,1															5	5,9	6,9	8,7		10,8	12,7	14,8	18,7	
51	2,15															5,2	6,2	7,1	9		11,2	13,3	15,3	19,4	
52	2,15															5	5,9	6,8	8,6		10,8	12,7	14,6	18,5	
53	2,2															5	6	6,9	8,7		10,8	12,9	14,8	18,7	
54	2,25															5,2	6,2	7,1	9		11,2	13,3	15,3	19,4	
55	2,3														0,6 ±0,15	5,4	6,4	7,4	9,3		11,6	13,8	15,9	20	
56	2,35															5,6	6,6	7,6	9,6		12	14,2	16,3	20,6	
57	2,4															5,7	6,8	7,8	9,9		21,3	14,6	16,8	21,3	
58	2,4															5,5	6,6	7,6	9,6		11,8	14,2	16,3	20,6	
59	2,45															5,7	6,8	7,8	9,9	±30% if $F_t < 10$ N ±20% if $F_t ≥ 10$ N	12,3	14,6	16,8	21,3	
60	2,5															5,9	6,9	8	10,1		12,7	14,8	17,2	21,7	
61	2,55															6	7,1	8,2	10,4		12,9	15,3	17,6	22,4	
62	2,6															6,1	7,2	8,3	10,5		13,1	15,5	17,8	22,6	
63	2,65								0,2	+0,2 0						6,2	7,4	8,5	10,8		13,3	15,9	18,3	23,2	
64	2,65															6	7,1	8,3	10,5		12,9	15,3	17,8	22,6	
65	2,7														0,7 ±0,15	6,2	7,3	8,5	10,8		13,3	15,7	18,3	23,2	
66	2,75															6,4	7,6	8,7	11,1		13,8	16,3	18,7	23,9	
67	2,8															6,6	7,8	9	11,4		14,2	16,8	19,4	24,5	
68	2,85															6,7	8	9,2	11,7		14,4	17,2	19,8	25,2	
69	2,9															6,9	8,2	9,4	12		14,8	17,6	20,2	25,8	

This table is shown in ISO format. Commas represent decimal points.

TABLE 7—DIMENSIONS FOR N, NM, E, AND EM SCRAPER RINGS (RADIAL WALL THICKNESS ["REGULAR"]) (CONTINUED)

Dimensions in millimeters

Nominal diameter d_1	Radial wall thickness "regular" a_1	Tolerance	Ring width h_1 Column 1	2	3	4	Tolerance	Closed gap s_1	Tolerance	Axial width of step h_2 For h_1 shown in column 1	2	3	4	Radial depth of step a_2	Tangential force F_t, N For h_1 shown in column 1	2	3	4	Tolerance	Diametral force F_d, N For h_1 shown in column 1	2	3	4	Tolerance
70	2,9													0,7 ±0,15	6,7	7,9	9,2	11,6		14,4	17	19,8	24,9	
71	2,95														6,9	8,1	9,4	11,9		14,8	17,4	20,2	25,6	
72	3							0,2	+0,2 0						6,9	8,2	9,5	12		14,8	17,6	20,4	25,8	
73	3,05														7,1	8,4	9,7	12,3		15,3	18,1	20,9	26,4	
74	3,1														7,2	8,6	9,9	12,6		15,5	18,5	21,3	27,1	
75	3,15													0,8 ±0,15	7,4	8,8	10,1	12,8		15,9	18,9	21,7	27,5	
76	3,15														7,2	8,5	9,8	12,5		15,5	18,3	21,1	26,9	
77	3,2														7,4	8,7	10,1	12,8		15,9	18,7	21,7	27,5	
78	3,25														7,5	8,9	10,3	13,1		16,1	19,1	22,1	28,2	
79	3,3									0,4 ±0,15	0,45 ±0,15	0,5 ±0,15	0,6 ±0,15		7,7	9,1	10,5	13,4		16,6	19,6	22,6	28,8	
80	3,35		1,5	1,75	2	2,5	−0,010 −0,030								7,9	9,3	10,8	13,7		17	20	23,2	29,5	
81	3,4	±0,15 Within a ring: 0,15 max.					For phosphated PO surface: −0,005 −0,035	0,25	+0,25 0						7,9	9,4	10,8	13,8		17	20,2	23,2	29,7	
82	3,4														7,7	9,1	10,6	13,4		16,6	19,6	22,8	28,8	
83	3,45														7,9	9,3	10,8	13,7		17	20	23,2	29,5	
84	3,5														8	9,5	11	14		17,2	20,4	23,7	30,1	
85	3,55													0,9 ±0,15	8,2	9,7	11,3	14,3		17,6	20,9	24,3	30,7	
86	3,6														8,4	9,9	11,5	14,6		18,1	21,3	24,7	31,4	
87	3,65														8,6	10,2	11,7	14,9		18,5	21,9	25,2	32	
88	3,65														8,4	9,9	11,5	14,6	±30% if F_t < 10N	18,1	21,3	24,7	31,4	±30% if F_d < 21,5N
89	3,7														8,5	10,1	11,7	14,9		18,3	21,7	25,2	32	
90	3,75														10,3	11,9	15,1	18,1	±20% if F_t > 10N	22,1	25,6	32,5	38,9	±20% if F_d > 21,5N
91	3,8														10,3	11,9	15,2	18,2		22,1	25,6	32,7	39,1	
92	3,85														10,5	12,1	15,4	18,5		22,6	26	33,1	39,8	
93	3,9														10,7	12,4	15,7	18,9		23	26,7	33,8	40,6	
94	3,9		1,75	2	2,5	3				0,45 ±0,15	0,5 ±0,15	0,6 ±0,15	0,75 ±0,15		10,5	12,1	15,4	18,5		22,6	26	33,1	39,8	
95	3,95													1 ±0,15	10,7	12,3	15,7	18,8		23	26,4	33,8	40,4	
96	4														10,9	12,6	16	19,2		23,4	27,1	34,4	41,3	
97	4,05														11,1	12,8	16,3	19,5		23,9	27,5	35	41,9	
98	4,1														11,3	13	16,6	19,9		24,3	28	35,7	42,8	
99	4,15							0,3	+0,25 0						11,5	13,3	16,9	20,2		24,7	28,6	36,3	43,4	
100	4,15															16,5	19,8	23,1			35,5	42,6	49,7	
101	4,2															16,6	19,9	23,2			35,7	42,8	49,9	
102	4,25	±0,2 Within a ring: 0,20 max.													—	16,8	20,2	23,6		—	36,1	43,4	50,7	
103	4,3															17,1	20,5	24			36,8	44,1	51,6	
104	4,3															16,8	20,1	23,5			36,1	43,2	50,5	
105	4,35		—	2,5	3	3,5				—	0,6 ±0,15	0,75 ±0,15	0,9 ±0,15	1,1 ±0,15		17	20,4	23,8			36,6	43,9	51,2	
106	4,4															17,3	20,7	24,2			37,2	44,5	52	
107	4,4														—	16,9	20,3	23,7		—	36,3	43,6	51	
108	4,45															17,2	20,6	24,1			37	44,3	51,8	
109	4,5															17,5	21	24,5			37,6	45,2	52,7	

This table is shown in ISO format. Commas represent decimal points.

TABLE 7—DIMENSIONS FOR N, NM, E, AND EM SCRAPER RINGS (RADIAL WALL THICKNESS ["REGULAR"]) (CONTINUED)

Dimensions in millimeters

Nominal diameter d_1	Radial wall thickness "regular" a_1	Tolerance	Ring width h_1 Column 1	2	3	4	Tolerance	Closed gap s_1	Tolerance	Axial width of step h_2 For h_1 shown in column 1	2	3	4	Radial depth of step a_2	Tangential force F_t, N For h_1 shown in column 1	2	3	4	Tolerance	Diametral force F_d, N For h_1 shown in column 1	2	3	4	Tolerance
110	4,55													1,1 ±0,15		21,2	24,8	28,5			45,6	53,3	61,3	
111	4,55															20,8	24,3	28			44,7	52,2	60,2	
112	4,6														—	20,9	24,3	28,1		—	44,9	52,2	60,4	
113	4,65															21,2	24,7	28,5			45,6	53,1	61,3	
114	4,7															21,5	25,1	28,9			46,2	54	62,1	
115	4,7														1,2 ±0,15		21,1	24,6	28,4			45,4	52,9	61,1	
116	4,75															21,4	25	28,8			46	53,8	61,9	
117	4,8						−0,010 −0,030 For phos- phated PO surface: −0,005 −0,035	0,35	+0,30 0						—	21,7	25,3	29,2		—	46,7	54,4	62,8	
118	4,8															21,3	24,9	28,7			45,8	53,5	61,7	
119	4,85															21,6	25,2	29,1			46,4	54,2	62,6	
120	4,9														1,2 ±0,15		21,9	25,6	29,5			47,1	55	63,4	
121	4,95															22,2	25,9	29,9			47,7	55,7	64,3	
122	4,95		—	3	3,5	4				—	0,75 ±0,15	0,9 ±0,15	1 ±0,15		—	21,8	25,5	29,4		—	46,9	54,8	63,2	
123	5															21,9	25,5	29,5			47,1	54,8	63,4	
124	5,05															22,2	25,9	29,9			47,7	55,7	64,3	
125	5,05	±0,2 Within a ring: 0,20 max.												1,3 ±0,15		21,8	25,4	29,3	±30% if $F_t < 10$ N ±20% if $F_t ≥ 10$ N		46,9	54,6	63	±30% if $F_d < 21,5$ N ±20% if $F_d ≥ 21,5$ N
126	5,1															22,1	25,8	29,7			47,5	55,5	63,9	
127	5,15						−0,010 −0,040 For phos- phated PO surface: −0,005 −0,045								—	22,4	26,1	30,1		—	48,2	56,1	64,7	
128	5,2															22,7	26,5	30,6			48,8	57	65,8	
129	5,2															22,3	26	30			47,9	55,9	64,5	
130	5,25															22,5	26,3	30,3			48,4	56,5	65,1	
131	5,3															22,8	26,6	30,7			49	57,2	66	
132	5,3															—	22,4	26,2	30,2		—	48,2	56,3	64,9
133	5,35															22,7	26,6	30,6			48,8	57	65,8	
134	5,4															22,8	26,6	30,7			49	57,2	66	
135	5,4															22,4	26,2	30,2			48,2	56,3	64,9	
136	5,45															22,7	26,5	30,6			48,8	57	65,8	
137	5,5															—	23	26,8	31		—	49,5	57,6	66,7
138	5,5															22,6	26,4	30,5			48,6	56,8	65,6	
139	5,55							0,4	+0,35 0					1,4 ±0,15		22,9	26,7	30,9			49,2	57,4	66,4	
140	5,6															27,1	31,3				58,3	67,3		
141	5,65															27,4	31,6				58,9	67,9		
142	5,65														—	27	31,1	—		—	58,1	66,9	—	
143	5,7															27,3	31,5				58,7	67,7		
144	5,75		—	3,5	4	—				—	0,9 ±0,15	1 ±0,15	—			27,7	31,9				59,6	68,6		
145	5,75															27,2	31,4				58,5	67,5		
146	5,8															27,3	31,5				58,7	67,7		
147	5,85														1,5 ±0,2	—	27,6	31,9	—		—	59,3	68,6	—
148	5,85															27,2	31,4				58,5	67,5		
149	5,9															27,5	31,8				59,1	68,4		

This table is shown in ISO format. Commas represent decimal points.

TABLE 7—DIMENSIONS FOR N, NM, E, AND EM SCRAPER RINGS (RADIAL WALL THICKNESS ["REGULAR"]) (CONCLUDED)

Dimensions in millimeters

Nominal diameter d_1	Radial wall thickness "regular" a_1	Tolerance	Ring width h_1 Column 1	2	3	4	Tolerance	Closed gap s_1	Tolerance	Axial width of step h_2 For h_1 shown in column 1	2	3	4	Radial depth of step a_2	Tangential force F_t, N For h_1 shown in column 1	2	3	4	Tolerance	Diametral force F_d, N For h_1 shown in column 1	2	3	4	Tolerance
150	5,95														—	27,6	31,9	—		—	59,3	68,6	—	
152	6													1,5 ±0,2	—	27,6	31,8	—		—	59,3	68,4	—	
154	6,05														—	27,5	31,7	—		—	59,1	68,2	—	
155	6,1														—	27,8	32,1	—		—	59,8	69	—	
156	6,15														—	28,1	32,4	—		—	60,4	69,7	—	
158	6,2														—	27,8	32,1	—		—	59,8	69	—	
160	6,25														—	27,7	32	—		—	59,6	68,8	—	
162	6,35							0,5	+0,4 / 0					1,6 ±0,2	—	28,3	32,7	—		—	60,8	70,3	—	
164	6,4														—	28,2	32,6	—		—	60,6	70,1	—	
165	6,4														—	27,8	32,1	—		—	59,8	69	—	
166	6,45														—	28,1	32,5	—		—	60,4	69,9	—	
168	6,5														—	28	32,4	—		—	60,2	69,7	—	
170	6,6						−0,010 / −0,040 For phosphated PO surface: −0,005 / −0,045								—	28,4	32,8	—		—	61,1	70,5	—	
172	6,65	±0,2 Within a ring: 0,20 max.	—	3,5	4	—									—	28,3	32,7	—	±30% if $F_t < 10$ N ±20% if $F_t > 10$ N	—	60,8	70,3	—	±30% if $F_d < 21,5$ N ±20% if $F_d > 21,5$ N
174	6,7														—	28,2	32,6	—		—	60,6	70,1	—	
175	6,75									0,9 ±0,15	1 ±0,15			1,7 ±0,2	—	28,4	32,8	—		—	61,1	70,5	—	
176	6,8														—	28,7	33,2	—		—	61,7	71,4	—	
178	6,85														—	28,6	33,1	—		—	61,5	71,2	—	
180	6,9														—	28,6	33	—		—	61,5	71	—	
182	6,95														—	28,5	32,9	—		—	61,3	70,7	—	
184	7,05														—	28,8	33,3	—		—	61,9	71,6	—	
185	7,05														—	28,4	32,9	—		—	61,1	70,7	—	
186	7,1													1,8 ±0,2	—	28,7	33,2	—		—	61,7	71,4	—	
188	7,15							0,6	+0,45 / 0						—	28,7	33,1	—		—	61,7	71,2	—	
190	7,2														—	28,6	33	—		—	61,5	71	—	
192	7,25														—	28,5	33	—		—	61,3	71	—	
194	7,35														—	29,1	33,6	—		—	62,6	72,2	—	
195	7,35														—	28,7	33,2	—		—	61,7	71,4	—	
196	7,4													1,9 ±0,2	—	28,8	33,3	—		—	61,9	71,6	—	
198	7,45														—	28,7	33,2	—		—	61,7	71,4	—	
200	7,5														—	28,6	33,1	—		—	61,5	71,2	—	

NOTES

[1] For intermediate sizes (for example repair sizes), the radial wall thickness of the next smaller nominal diameter should be applied.

[2] The values for F_t and F_d, given in Table 6, apply to as cast grey cast iron with a typical modulus of elasticity (E_n) of 100 000 MPa. Multiplying factors for materials having a different modulus (E_n) are given in SAE J1591.

Mean forces are calculated for nominal radial wall thickness (a_t) and mean ring width (h_t).

[3] For the sole purpose of this document, the assumed average ratio F_d/F_t is 2.15. However, for rings up to 50 mm, the ratio F_d/F_t shall be determined between the manufacturer and client.

This table is shown in ISO format. Commas represent decimal points.

TABLE 8—DIMENSIONS FOR N, NM, E, AND EM SCRAPER RINGS (RADIAL WALL THICKNESS "D/22")

Dimensions in millimeters

Nominal diameter d_1	Radial wall thickness "D/22"	Tolerance	Ring width h_1 Column 1	2	3	4	Tolerance	Closed gap s_1	Tolerance	Axial width of step h_2 For h_1 shown in column 1	2	3	4	Radial depth of step a_2	Tangential force F_t, N For h_1 shown in column 1	2	3	4	Tolerance	Diametral force F_d, N For h_1 shown in column 1	2	3	4	Tolerance
50	2,25							0,15	+0,2 / 0					0,6 ±0,15	6,1	7,2	8,3	10,6		13,1	15,5	17,8	22,8	
51	2,3														6,3	7,4	8,6	10,9		13,5	15,9	18,5	23,4	
52	2,35														6,5	7,7	8,8	11,2		14	16,6	18,9	24,1	
53	2,4														6,7	7,9	9,1	11,5		14,4	17	19,6	24,7	
54	2,45														6,9	8,1	9,4	11,9		14,8	17,4	20,2	25,6	
55	2,5														7	8,3	9,6	12,2		15,1	17,8	20,6	26,2	
56	2,55														7,2	8,5	9,9	12,5		15,5	18,3	21,3	26,9	
57	2,6														7,2	8,6	9,9	12,6		15,5	18,5	21,3	27,1	
58	2,65														7,4	8,8	10,2	12,9		15,9	18,9	21,9	27,7	
59	2,7														7,6	9	10,4	13,2		16,3	19,4	22,4	28,4	
60	2,75														0,7 ±0,15	7,8	9,2	10,6	13,5		16,8	19,8	22,8	29	
61	2,75															7,5	8,9	10,3	13		16,1	19,1	22,1	28	
62	2,8															7,7	9,1	10,5	13,3		16,6	19,6	22,6	28,6	
63	2,85															7,9	9,3	10,8	13,6		17	20	23,2	29,2	
64	2,9															8,1	9,5	11	14		17,4	20,4	23,7	30,1	
65	2,95	±0,15 Within a ring: 0,15 max.	1,5	1,75	2	2,5	−0,010 / −0,030 For phosphated PO surface: −0,005 / −0,035	0,2	+0,2 / 0	0,4 ±0,15	0,45 ±0,15	0,5 ±0,15	0,6 ±0,15	0,8 ±0,15	8,3	9,8	11,3	14,3	±30% if F_t < 10N ±20% if F_t > 10N	17,8	21,1	24,3	30,7	±30% if F_d < 21,5N ±20% if F_d > 21,5N
66	3															8,3	9,8	11,3	14,4		17,8	21,1	24,3	31	
67	3,05															8,4	10	11,6	14,7		18,1	21,5	24,9	31,6	
68	3,1															8,6	10,2	11,8	15		18,5	21,9	25,4	32,3	
69	3,15															8,8	10,4	12,1	15,3		18,9	22,4	26	32,9	
70	3,2															9	10,7	12,3	15,6		19,4	23	26,4	33,5	
71	3,25															9,2	10,9	12,6	16		19,8	23,4	27,1	34,4	
72	3,25															8,9	10,6	12,2	15,5		19,1	22,8	26,2	33,3	
73	3,3															9,1	10,8	12,5	15,8		19,6	23,2	26,9	34	
74	3,35															9,3	11	12,7	16,1		20	23,7	27,3	34,6	
75	3,4														0,9 ±0,15	9,3	11	12,7	16,1		20	23,7	27,3	34,6	
76	3,45															9,4	11,2	12,9	16,5		20,2	24,1	27,7	35,5	
77	3,5															9,6	11,4	13,2	16,8		20,6	24,5	28,4	36,1	
78	3,55															9,8	11,6	13,5	17,1		21,1	24,9	29	36,8	
79	3,6															10	11,9	13,7	17,4		21,5	25,6	29,5	37,4	
80	3,65							0,25	+0,25 / 0						10,2	12,1	14	17,7		21,9	26	30,1	38,1	
81	3,7															10,4	12,3	14,2	18		22,4	26,4	30,5	38,7	
82	3,75															10,6	12,5	14,5	18,4		22,8	26,9	31,2	39,6	
83	3,75															10,3	12,2	14,1	17,9		22,1	26,2	30,3	38,5	
84	3,8															10,3	12,2	14,1	18		22,1	26,2	30,3	38,7	
85	3,85														1 ±0,15	10,5	12,4	14,4	18,3		22,6	26,7	31	39,3	
86	3,9															10,7	12,6	14,6	18,6		23	27,1	31,4	40	
87	3,95															10,8	12,9	14,9	18,9		23,2	27,7	32	40,6	
88	4															11	13,1	15,1	19,2		23,7	28,2	32,5	41,3	
89	4,05															11,2	13,3	15,4	19,6		24,1	28,6	33,1	42,1	

This table is shown in ISO format. Commas represent decimal points.

TABLE 8—DIMENSIONS FOR N, NM, E, AND EM SCRAPER RINGS (RADIAL WALL THICKNESS "D/22") (CONTINUED)

Dimensions in millimeters

Nominal diameter d_1	Radial wall thickness "D/22" a_1	Tolerance	Ring width h_1 Column 1	2	3	4	Tolerance	Closed gap s_1	Tolerance	Axial width of step h_2 For h_1 shown in column 1	2	3	4	Tolerance	Radial depth of step a_2	Tangential force F_t, N For h_1 shown in column 1	2	3	4	Tolerance	Diametral force F_d, N For h_1 shown in column 1	2	3	4	Tolerance
90	4,1														1 ±0,15	13,4	15,5	19,8	23,7		28,8	33,3	42,6	51	
91	4,15															13,7	15,8	20,1	24,1		29,5	34	43,2	51,8	
92	4,2	±0,15 Within a ring: 0,15 max.	1,75	2	2,5	3				0,45 ±0,15	0,5 ±0,15	0,6 ±0,15	0,75 ±0,15			13,7	15,8	20,1	24,2		29,5	34	43,2	52	
93	4,25															13,9	16,1	20,5	24,5		29,9	34,6	44,1	52,7	
94	4,25															13,6	15,7	20	24		29,2	33,8	43	51,6	
95	4,3														1,1 ±0,15	13,8	16	20,3	24,4		29,7	34,4	43,6	52,5	
96	4,35															14	16,2	20,6	24,8		30,1	34,8	44,3	53,3	
97	4,4															14,2	16,5	20,9	25,1		30,5	35,5	44,9	54	
98	4,45															14,4	16,7	21,3	25,5		31	35,9	45,8	54,8	
99	4,5							0,3	+0,25 0							14,7	17	21,6	25,9		31,6	36,6	46,4	55,7	
100	4,55																21,9	26,3	30,6			47,1	56,5	65,8	
101	4,6																21,9	26,3	30,7			47,1	56,5	66	
102	4,65									—	0,6 ±0,15	0,75 ±0,15	0,9 ±0,15		1,2 ±0,15	—	22,2	26,6	31,1		—	47,7	57,2	66,9	
103	4,7			2,5	3	3,5											22,5	27	31,5			48,4	58,1	67,7	
104	4,75																22,7	27,3	31,9			48,8	58,7	68,6	
105	4,75																22,3	26,7	31,2			47,9	57,4	67,1	
106	4,8																22,6	27,1	31,6			48,6	58,3	67,9	
107	4,85															—	22,8	27,4	32	±30% if F_t < 10N	—	49	58,9	68,8	±30% if F_d < 21,5N
108	4,9																23,1	27,8	32,4			49,7	59,8	69,7	
109	4,95																23,4	28,1	32,8			50,3	60,4	70,5	
110	5																28	32,7	37,8	±20% if F_t > 10N		60,2	70,3	81,3	±20% if F_d > 21,5N
111	5,05																28,4	33,1	38,2			61,1	71,2	82,1	
112	5,1	±0,2 Within a ring: 0,20 max.					For phos- phated PO surface: −0,005 −0,035								1,3 ±0,15	—	28,7	33,5	38,7		—	61,7	72	83,2	
113	5,15																29	33,9	39,1			62,4	72,9	84,1	
114	5,2																29,4	34,3	39,5			63,2	73,7	84,5	
115	5,25																29,7	34,7	40			63,9	74,6	86	
116	5,25																29,1	34	39,2			62,6	73,1	84,3	
117	5,3										—	0,75 ±0,15	0,9 ±0,15	1 ±0,15			—	29,5	34,4	39,7		—	63,4	74	85,4	
118	5,35			3	3,5	4											29,8	34,8	40,1			64,1	74,8	86,2	
119	5,4							0,35	+0,30 0								29,8	34,8	40,2			64,1	74,8	86,4	
120	5,45																30,1	35,2	40,6			64,7	75,7	87,3	
121	5,5																30,4	35,6	41			65,4	76,5	88,2	
122	5,55														1,4 ±0,15	—	30,8	35,9	41,5		—	66,2	77,2	89,2	
123	5,6																31,1	36,3	41,9			66,9	78	90,1	
124	5,65							−0,010 −0,040 For phos- phated PO surface: −0,005 −0,045									31,4	36,7	42,3			67,5	78,9	90,9	
125	5,7																31,7	37,1	42,8			68,2	79,8	92	
126	5,75																32,1	37,5	43,2			69	80,6	92,9	
127	5,75														1,5 ±0,15		31,5	36,8	42,5			67,7	79,1	91,4	
128	5,8																31,5	36,8	42,5			67,7	79,1	91,4	
129	5,85																31,8	37,2	42,9			68,4	80	92,2	

This table is shown in ISO format. Commas represent decimal points.

TABLE 8—DIMENSIONS FOR N, NM, E, AND EM SCRAPER RINGS (RADIAL WALL THICKNESS "D/22") (CONCLUDED)

Dimensions in millimeters

Nominal diameter d_1	Radial wall thickness "D/22" a_1	Tolerance	Ring width h_1 Column 1	2	3	4	Tolerance	Closed gap s_1	Tolerance	Axial width of step h_2 For h_1 shown in column 1	2	3	4	Tolerance	Radial depth of step a_2	Tangential force F_t, N For h_1 shown in column 1	2	3	4	Tolerance	Diametral force F_d, N For h_1 shown in column 1	2	3	4	Tolerance
130	5,9																32	37,4	43,2			68,8	80,4	92,9	
131	5,95																32,4	37,8	43,6			69,7	81,3	93,7	
132	6														1,5 ±0,15		32,7	38,2	44,1			70,3	82,1	94,8	
133	6,05			3	3,5	4					0,75 ±0,15	0,9 ±0,15	1 ±0,15				33	38,5	44,5			71	82,8	95,7	
134	6,1																33,3	38,9	44,9			71,6	83,6	96,5	
135	6,15																33,6	39,3	45,3			72,2	84,5	97,4	
136	6,2																33,6	39,3	45,4			72,2	84,5	97,6	
137	6,25						−0,010 −0,040 For phosphated PO surface: −0,005 −0,045	0,4	+0,35 0					1,6 ±0,15		33,9	39,6	45,8		±30% if F_t < 10 N ±20% if F_t > 10 N		72,9	85,1	98,5		±30% if F_d < 21,5 N ±20% if F_d > 21,5 N
138	6,25	±0,2 Within a ring: 0,20 max.															33,4	39	45			71,8	83,9	96,8	
139	6,3																33,7	39,4	45,4			72,5	84,7	97,6	
140	6,35																39,7	45,9				85,4	98,7		
141	6,4																40,1	46,3				86,2	99,5		
142	6,45																40,4	46,7				86,9	100,4		
143	6,5																40,8	47,1				87,7	101,3		
144	6,55										0,9 ±0,15	1 ±0,15					41,2	47,5				88,6	102,1		
145	6,6			3,5	4												41,1	47,5				88,4	102,1		
146	6,65																41,5	47,9				89,2	103		
147	6,7														1,7 ±0,2		41,8	48,3				89,9	103,8		
148	6,75																42,2	48,7				90,7	104,7		
149	6,75																41,6	48				89,4	103,2		
150	6,8							0,5	+0,4 0								41,6	48,1				89,4	103,4		

NOTES

1. For intermediate sizes (for example repair sizes), the radial wall thickness of the next smaller nominal diameter should be applied.

2. The values for F_t and F_d, given in Table 7, apply to as cast grey cast iron with a typical modulus of elasticity (E_n) of 100 000 MPa. Multiplying factors for materials having a different modulus (E_n) are given in SAE J1591.

   Mean forces are calculated for nominal radial wall thickness (a_t) and mean ring width (h_t).

3. For the sole purpose of this document, the assumed average ratio F_d/F_t is 2.15.

This table is shown in ISO format. Commas represent decimal points.

INTERNAL COMBUSTION ENGINES—PISTON RINGS—KEYSTONE RINGS—SAE J2000 OCT92

SAE Standard

Report of the Piston Ring Standards Committee 7 approved October 1989. Revised by the SAE Piston Ring Standards Committee 7 October 1992.

This SAE Standard is equivalent to ISO Standard 6624/1.

1. Scope and Field of Application—Differences, where they exist, are shown in Appendix A.

This SAE Standard specifies the essential dimensional features of T, TB, TM, K, KB, and KM keystone piston ring types.

The requirements of this document apply to compression rings for reciprocating internal combustion engines up to and including 200 mm diameter.

(R) **2. References**

SAE DESIGNATION	ISO[1] EQUIVALENT	
		INTERNAL COMBUSTION ENGINES—PISTON RINGS
J1588	6621/1	Vocabulary
J1589	6621/2	Measuring principles
J1590	6621/3	Material specifications
J1591	6621/4	General specifications
J1996	6621/5	Quality requirements
		INTERNAL COMBUSTION ENGINES—PISTON RINGS
J1997	6622/1	Rectangular rings
J1998	6622/2 TR	Rectangular rings with narrow ring width
J1999	6623	INTERNAL COMBUSTION ENGINES—PISTON RINGS—SCRAPER RINGS
		INTERNAL COMBUSTION ENGINES—PISTON RINGS
J2000	6624/1	Keystone rings
J2001	6624/2 TR	Half keystone rings
J2002	6625	INTERNAL COMBUSTION ENGINES—PISTON RINGS—OIL CONTROL RINGS
J2003	6626	INTERNAL COMBUSTION ENGINES—PISTON RINGS—COIL SPRING LOADED OIL CONTROL RINGS
J2004	6627 TR	INTERNAL COMBUSTION ENGINES—PISTON RINGS—EXPANDER/SEGMENT OIL CONTROL RINGS
J2226		INTERNAL COMBUSTION ENGINES—PISTON RINGS—STEEL RECTANGULAR RINGS

[1] TR refers to Technical Report

3. Ring Types and Designation Examples

3.1 Type T—Straight Faced Keystone Ring 6°

3.1.1 GENERAL FEATURES

NOTE—See Table 7 for dimensions and forces.

3.1.2 DESIGNATION EXAMPLE—Designation of a straight faced keystone ring 6°, of d_1 = 90 mm nominal diameter, h_1 = 2.5 mm ring width, made of grey cast iron, nonheat-treated (material subclass 12), general features as shown in Figure 1, and periphery coated fully faced with chromium, 0.10 mm minimum thickness.

3.2 Type TB—Barrel Faced Keystone Ring 6°

3.2.1 GENERAL FEATURES

NOTE—See Table 7 for dimensions and forces.

TABLE 1—GAUGE WIDTH (h_8) AND BARREL DIMENSIONS

Dimensions in millimeters

(h_1)	h_8	t_2, t_3	Maximum Peak Off Center
2.0	1.2	0.002/0.016	0.30
2.5	1.6		0.40
3.0	2.0	0.005/0.020	0.50
3.5	2.4		
4.0	2.8	0.005/0.023	0.60
4.5	3.2		

3.2.2 DESIGNATION EXAMPLE—Designation of a barrel faced keystone ring 6°, of d_1 = 90 mm nominal diameter, h_1 = 2.5 mm ring width, made of spheroidal graphite cast iron, heat-treated martensitic (material subclass 53), general features as shown in Figure 2, and periphery semi-inlaid coated with molybdenum, 0.2 mm minimum thickness.

3.3 Type TM—Taper Faced Keystone Ring 6°

3.3.1 GENERAL FEATURES

NOTE—See Table 7 for dimensions and forces.

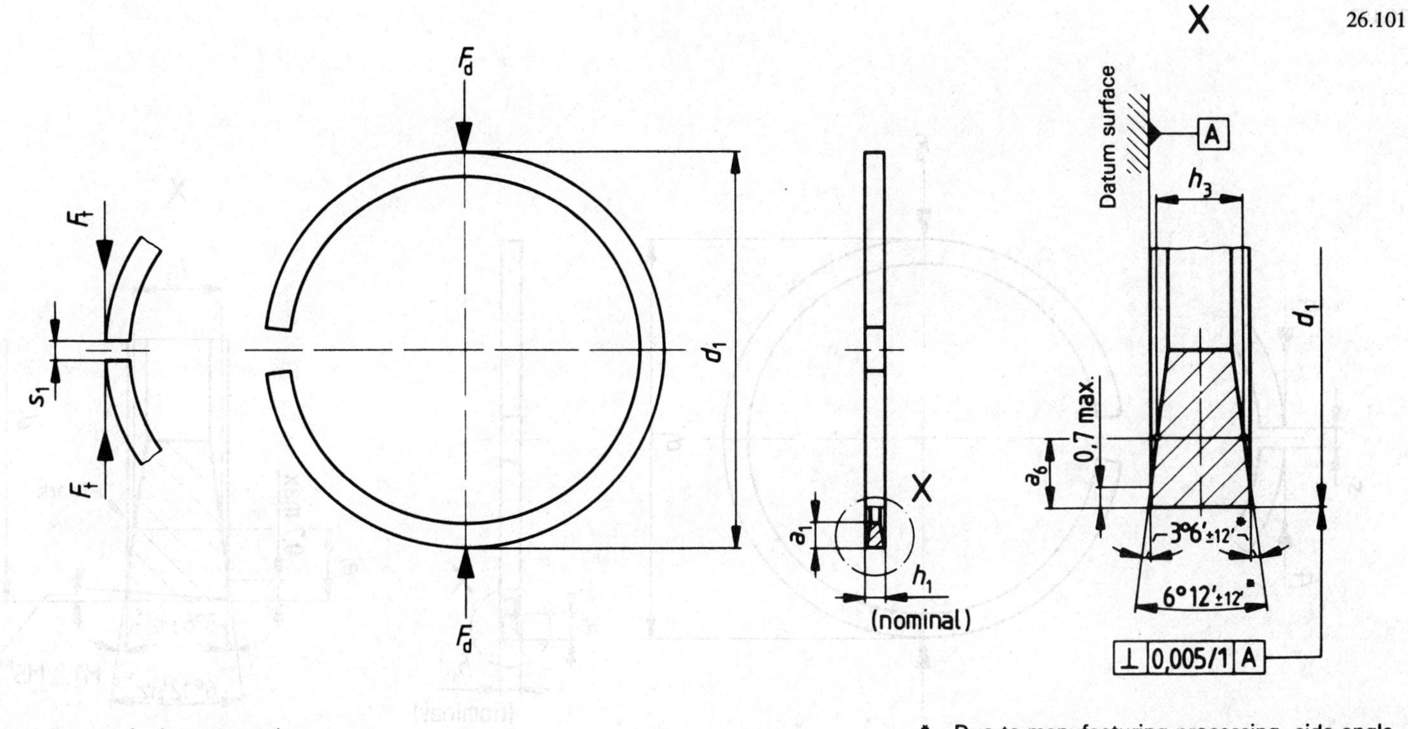

26.101

Method A: a_6 ref., h_3 measured
Method B: h_3 ref., a_6 measured

* Due to manufacturing processing, side angle tolerances are not cumulative.

FIGURE 1—TYPE T

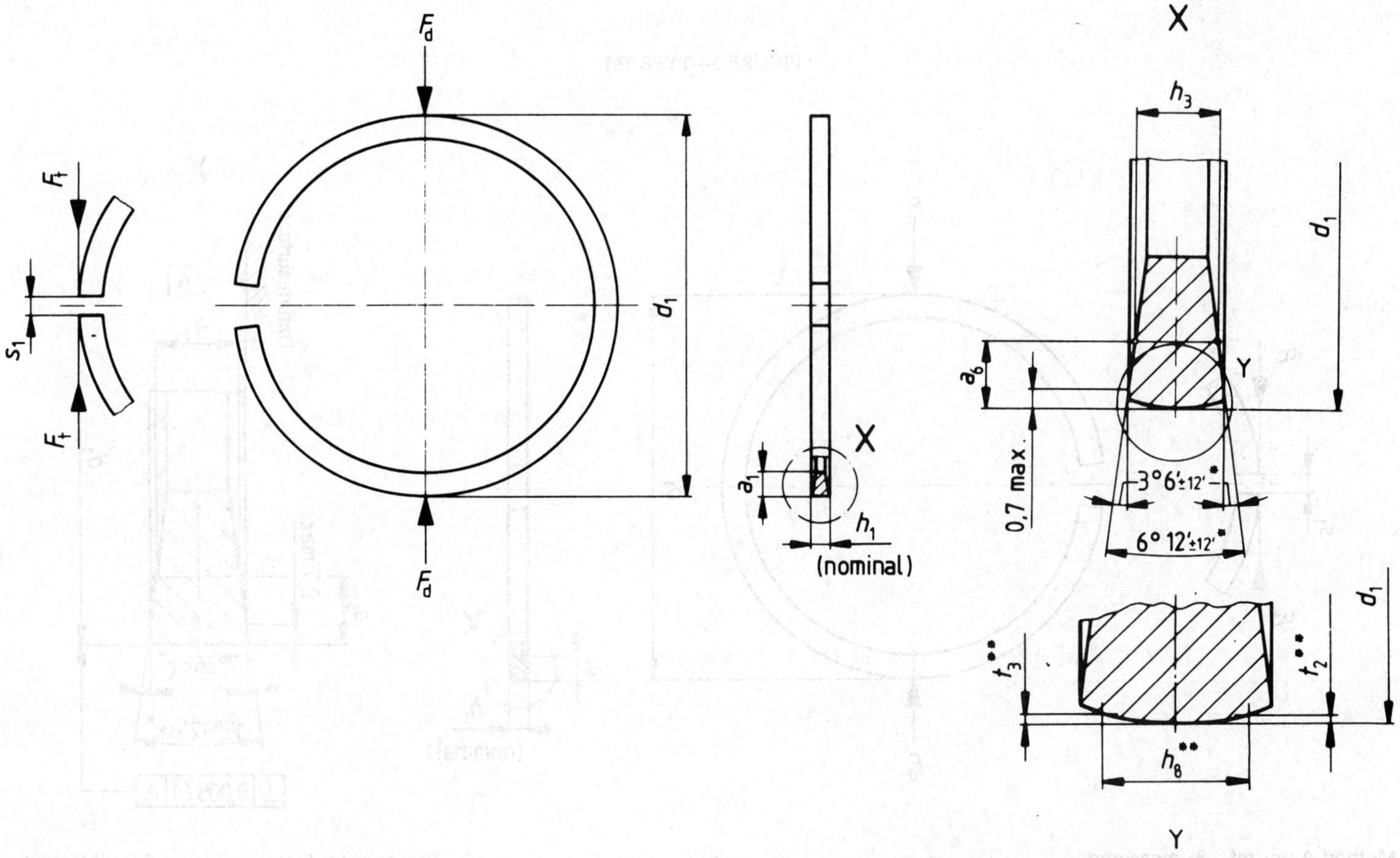

Method A: a_6 ref., h_3 measured
Method B: h_3 ref., a_6 measured

* Due to manufacturing processing, side angle tolerances are not cumulative.

** See Table 1.

FIGURE 2—TYPE TB

26.102

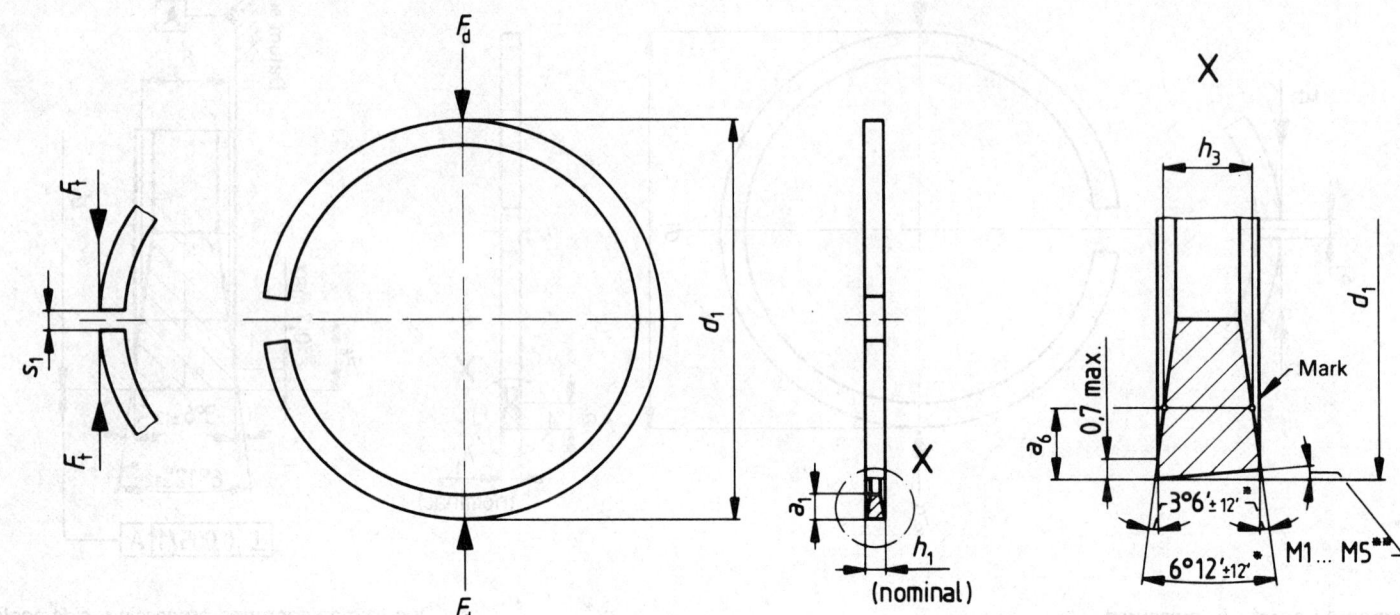

Method A: a_6 ref., h_3 measured
Method B: h_3 ref., a_6 measured

* Due to manufacturing processing, side angle tolerances are not cumulative.

** See Table 2.

FIGURE 3—TYPE TM

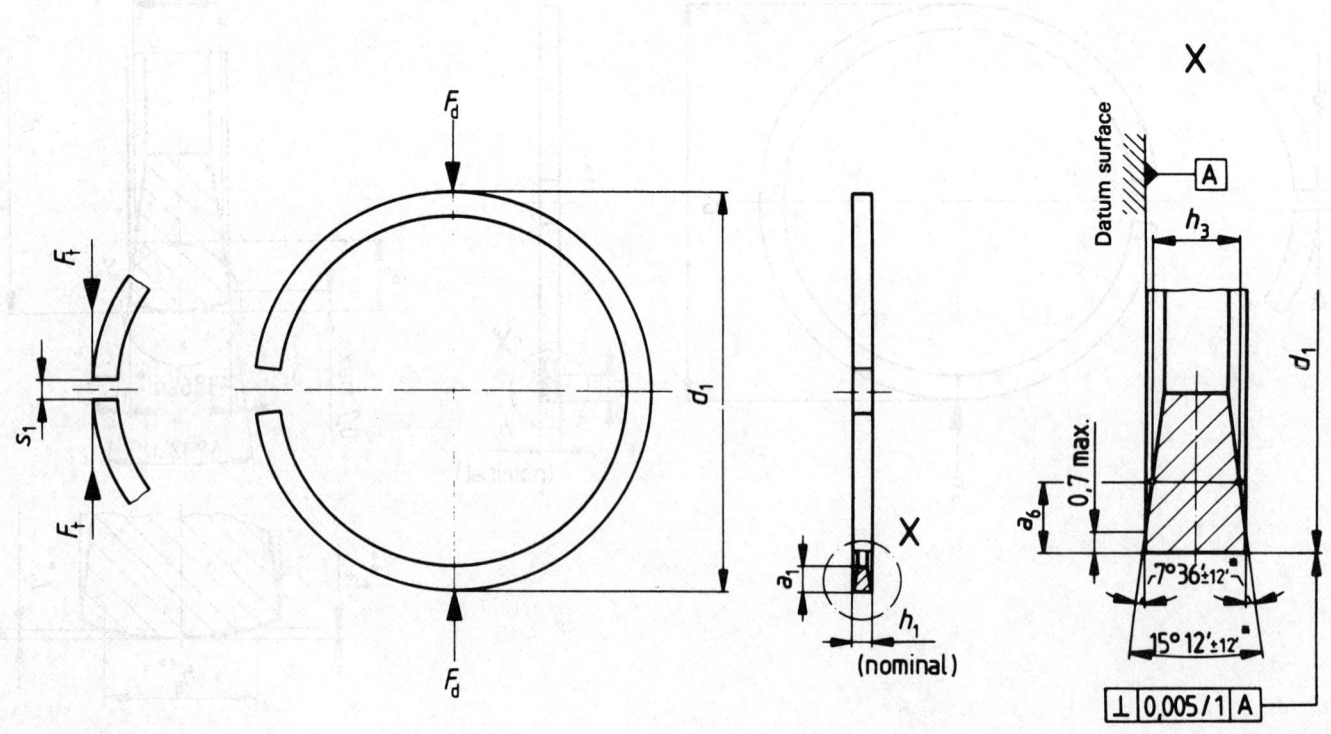

Method A: a_6 ref., h_3 measured
Method B: h_3 ref., a_6 measured

* Due to manufacturing processing, side angle tolerances are not cumulative.

FIGURE 4—TYPE K

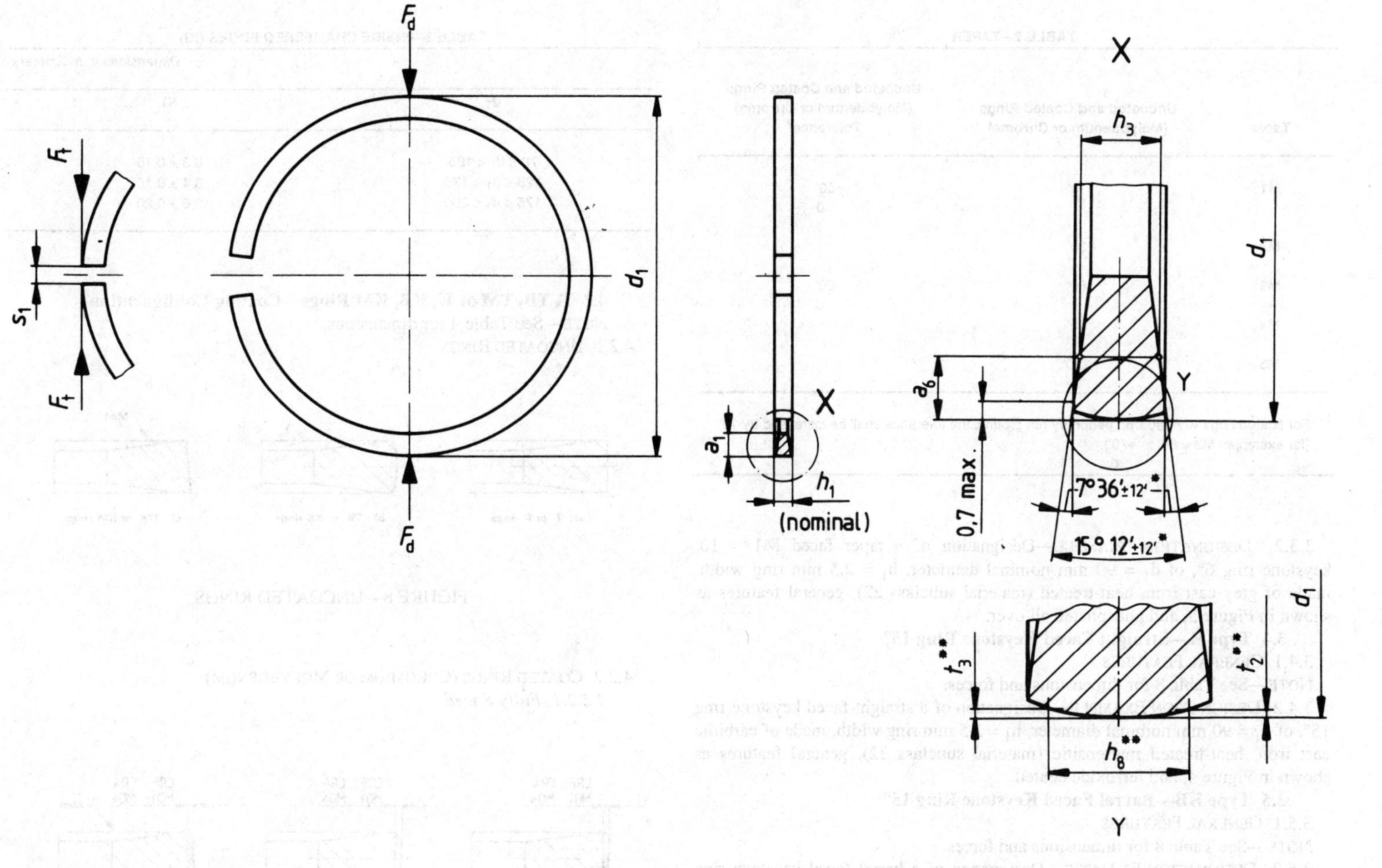

26.103

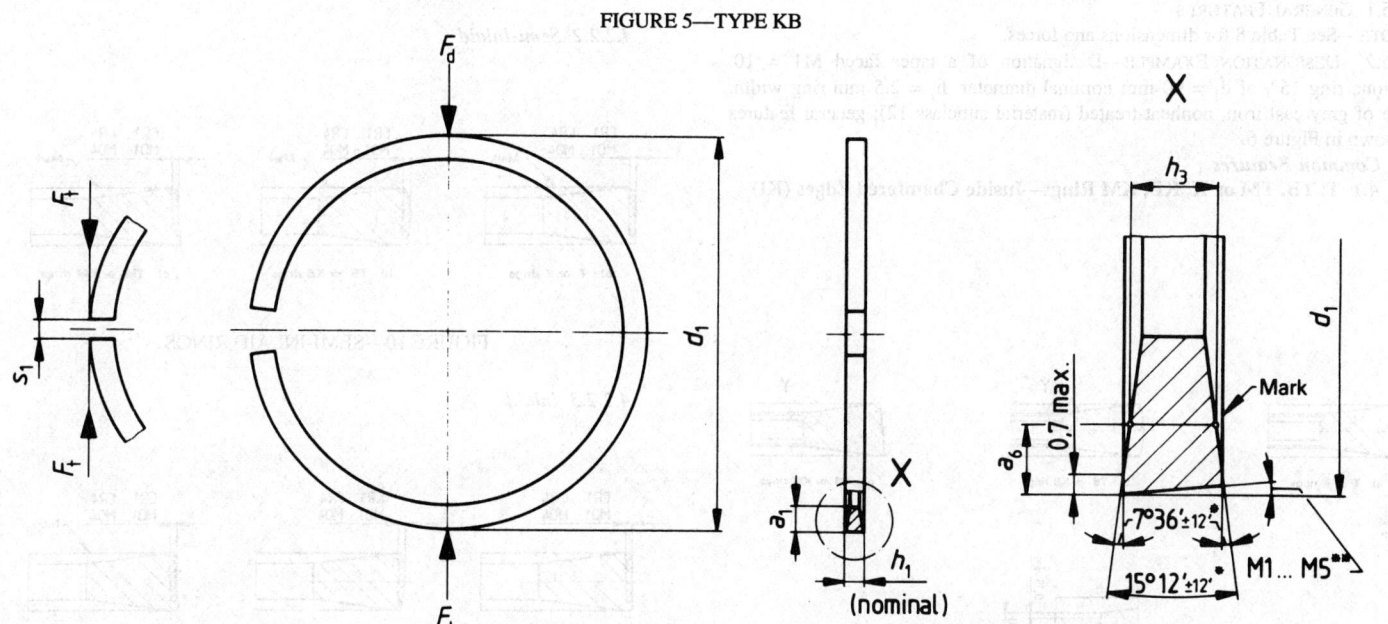

Method A: a_6 ref., h_3 measured
Method B: h_3 ref., a_6 measured

* Due to manufacturing processing, side angle tolerances are not cumulative.

** See Table 1.

FIGURE 5—TYPE KB

Method A: a_6 ref., h_3 measured
Method B: h_3 ref., a_6 measured

* Due to manufacturing processing, side angle tolerances are not cumulative.

** See Table 2.

FIGURE 6—TYPE KM

TABLE 2—TAPER

Taper	Uncoated and Coated Rings (Molybdenum or Chrome)	Uncoated and Coated Rings (Molybdenum or Chrome) Tolerance[1]
M1	10'	+50' / 0
M2	30'	
M3	60'	+60' / 0
M4	90'	
M5	120'	

[1] For coated rings with tapered periphery not ground, the tolerance shall be increased by 10' (for example: M3 = 60' : +70' / 0).

3.3.2 DESIGNATION EXAMPLE—Designation of a taper faced M1 = 10' keystone ring 6°, of d_1 = 90 mm nominal diameter, h_1 = 2.5 mm ring width, made of grey cast iron, heat-treated (material subclass 22), general features as shown in Figure 3, and phosphated all over.

3.4 Type K—Straight Faced Keystone Ring 15°
3.4.1 GENERAL FEATURES
NOTE—See Table 8 for dimensions and forces.
3.4.2 DESIGNATION EXAMPLE—Designation of a straight faced keystone ring 15°, of d_1 = 90 mm nominal diameter, h_1 = 2.5 mm ring width, made of carbidic cast iron, heat-treated martensitic (material subclass 32), general features as shown in Figure 4, and ferroxide coated.

3.5 Type KB—Barrel Faced Keystone Ring 15°
3.5.1 GENERAL FEATURES
NOTE—See Table 8 for dimensions and forces.
3.5.2 DESIGNATION EXAMPLE—Designation of a barrel faced keystone ring 15°, of d_1 = 90 mm nominal diameter, h_1 = 2.5 mm ring width, made of malleable cast iron, heat-treated pearlitic (material subclass 41), general features as shown in Figure 5, and periphery fully faced, coated with molybdenum, 0.2 mm minimum thickness.

3.6 Type KM—Taper Faced Keystone Ring 15°
3.6.1 GENERAL FEATURES
NOTE—See Table 8 for dimensions and forces.
3.6.2 DESIGNATION EXAMPLE—Designation of a taper faced M1 = 10' keystone ring 15°, of d_1 = 90 mm nominal diameter, h_1 = 2.5 mm ring width, made of grey cast iron, nonheat-treated (material subclass 12); general features as shown in Figure 6.

4. Common Features
4.1 T, TB, TM or K, KB, KM Rings—Inside Chamfered Edges (KI)

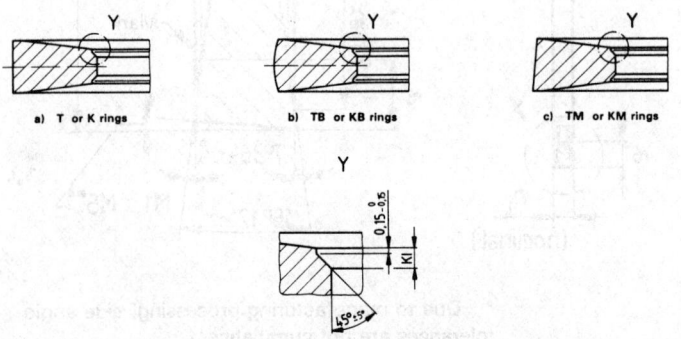

FIGURE 7—INSIDE CHAMFERED EDGES (KI)

TABLE 3—INSIDE CHAMFERED EDGES (KI)

Dimensions in millimeters

d_1	KI
70 ≤ d_1 < 125	0.3 ± 0.15
125 ≤ d_1 < 175	0.4 ± 0.15
175 ≤ d_1 ≤ 200	0.6 ± 0.20

4.2 T, TB, TM or K, KB, KM Rings—Coating Configuration
NOTE—See Table 4 for dimensions.
4.2.1 UNCOATED RINGS

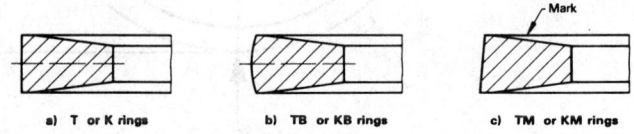

FIGURE 8—UNCOATED RINGS

4.2.2 COATED RINGS (CHROMIUM OR MOLYBDENUM)
4.2.2.1 Fully Faced

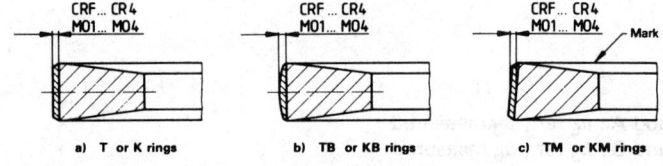

FIGURE 9—FULLY FACED RINGS

4.2.2.2 Semi-Inlaid

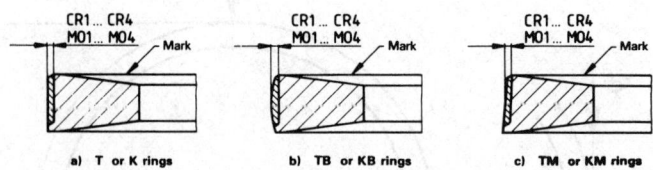

FIGURE 10—SEMI-INLAID RINGS

4.2.2.3 Inlaid

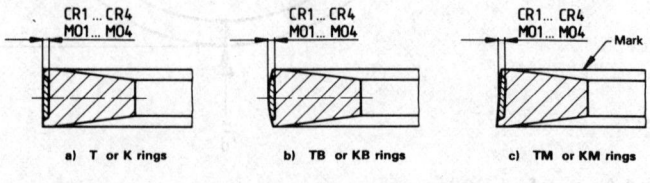

FIGURE 11—INLAID RINGS

TABLE 4—LAYER THICKNESS

Dimensions in millimeters

Chromium	Molybdenum	Thickness Minimum
CRF	-	0.005
CR1	MO1	0.05
CR2	MO2	0.10
CR3	MO3	0.15
CR4	MO4	0.20

5. Force Factors—The tangential and diametral forces given in Tables 7 and 8 shall be corrected when additional features and/or materials other than grey cast iron with a modulus of elasticity of 100 000 MPa are being used.

For common features, the multiplier correction factors given in Tables 5 and 6 and the force correction factors given in SAE J1591 shall be used.

The factors of Table 6 have been calculated with mean coating thickness.

TABLE 5—FORCE CORRECTION FACTORS FOR T, TB, TM, K, KB, AND KM RINGS WITH FEATURE KI

KI	Taper M2 or M3	Taper M4 or M5	KI and Taper M2 or M3	KI and Taper M4 or M5
0.96	0.98	0.96	0.94	0.92

TABLE 6—FORCE CORRECTION FACTORS FOR COATED T, TB, TM, K, KB, AND KM RINGS (FULLY FACED, SEMI-INLAID, AND INLAID TYPES)

CRF	CR1	CR2/MO1	CR3/MO2	CR4/MO3	MO4
1	0.94	0.91	0.88	0.85	0.83

APPENDIX A

A.1 This SAE Standard has been established to harmonize the ISO and SAE piston ring documents. The U.S. Technical Advisory Group, with the support of the National Engine Parts Manufacturers Association, has worked with other national organizations on this worldwide document. Some of the wording and phrasing may differ slightly from U.S. terminology for translation purposes.

In preparing this SAE document, the Scope and Field of Application, and Reference sections of the ISO 6624/1 have been editorially revised and reorganized.

Certain barrel drop (Table 1) and closed gap (Tables 7 and 8) dimensions were changed to conform with established U.S. standard practice.

The tolerances specified in this document represent a six sigma quality level.

6. Dimensions

TABLE 7—DIMENSIONS OF T, TB, TM KEYSTONE RINGS 6°

Dimensions in millimeters

Nominal diameter d_1	Radial wall thickness "regular" a_1	Tolerance	Nominal value of ring width (h_1) Column 1	Column 2	a_6 (ref.)	Method A Measured value h_3 Column 1	Column 2	Method B h_3 (ref.) Column 1	Column 2	Measured value a_6	Closed gap s_1	Tolerance	Tangential force F_t, N For h_1 shown in column 1	Column 2	Tolerance	Diametral force F_d, N For h_1 shown in column 1	Column 2	Tolerance
70	2,90										0,20	+0,20 / 0	9,9	12,6		21,3	27,1	
71	2,95												10,1	12,9		21,7	27,7	
72	3,00												10,3	13,2		22,1	28,4	
73	3,05												10,5	13,4		22,6	28,8	
74	3,10												10,7	13,7		23,0	29,5	
75	3,15												10,9	13,9		23,4	29,9	
76	3,15					1,832	2,332			1,61			10,6	13,6		22,8	29,2	
77	3,20					0	0			0			10,8	13,9		23,2	29,9	
78	3,25					−0,024	−0,024			−0,22			11,0	14,1		23,7	30,3	
79	3,30			2,0	2,5	1,5	For phosphated PO surface: +0,01 / −0,024	For phosphated PO surface: +0,01 / −0,024	1,82	2,32	For phosphated PO surface: +0,09 / −0,22			11,3	14,4		24,3	31,0	
80	3,35	±0,15 Within a ring: 0,15 max.									0,25		11,5	14,7		24,7	31,6	
81	3,40												11,7	15,0		25,2	32,3	
82	3,40												11,4	14,6		24,5	31,4	
83	3,45												11,6	14,9		24,9	32,0	
84	3,50												11,8	15,2		25,4	32,7	
85	3,55												12,0	15,4		25,8	33,1	
86	3,60												12,2	15,7		26,2	33,8	
87	3,65												12,5	16,0		26,9	34,4	
88	3,65												12,2	15,6		26,2	33,5	
89	3,70												12,4	15,9		26,7	34,2	
90	3,75											+0,25 / 0	16,1	19,6		34,6	42,1	
91	3,80												16,3	20,0		35,0	43,0	
92	3,85												16,6	20,3		35,7	43,6	
93	3,90												16,9	20,6	±30% if F_t < 10N	36,3	44,3	±30% if F_d < 21,5N
94	3,90												16,5	20,2		35,5	43,4	
95	3,95												16,8	20,5	±20% if F_t > 10N	36,1	44,1	±20% if F_d > 21,5N
96	4,00												17,1	20,9		36,8	44,9	
97	4,05												17,3	21,2		37,2	45,6	
98	4,10												17,6	21,5		37,8	46,2	
99	4,15										0,30		17,9	21,9		38,5	47,1	
100	4,15						2,278	2,778			2,08			17,5	21,4		37,6	46,0	
101	4,20						0	0			0			17,7	21,7		38,1	46,7	
102	4,25						−0,024	−0,024			−0,22			18,0	22,0		38,7	47,3	
103	4,30						For phosphated PO surface: +0,01 / −0,024	For phosphated PO surface: +0,01 / −0,024			For phosphated PO surface: +0,09 / −0,22			18,2	22,3		39,1	47,9	
104	4,30		2,5	3,0	2,0			2,27	2,77				17,9	21,9		38,5	47,1	
105	4,35												18,1	22,2		38,9	47,7	
106	4,40												18,3	22,5		39,3	48,4	
107	4,40												18,0	22,0		38,7	47,3	
108	4,45	±0,20 Within a ring: 0,20 max.											18,2	22,3		39,1	47,9	
109	4,50												18,4	22,6		39,6	48,6	
110	4,55												18,6	22,8		40,0	49,0	
111	4,55												18,2	22,4		39,1	48,2	
112	4,60												18,5	22,7		39,8	48,8	
113	4,65												18,7	22,9		40,2	49,2	
114	4,70												18,9	23,2		40,6	49,9	
115	4,70										0,35	+0,30 / 0	18,6	22,8		40,0	49,0	
116	4,75												18,8	23,1		40,4	49,7	
117	4,80												19,0	23,4		40,9	50,3	
118	4,80												18,7	22,9		40,2	49,2	
119	4,85												18,9	23,2		40,6	49,9	

This table is shown in ISO format. Commas represent decimal points.

TABLE 7 (CONTINUED)

Dimensions in millimeters

Nominal diameter d_1	Radial wall thickness "regular" a_1 Tolerance	Nominal value of ring width (h_1) Column 1	Nominal value of ring width (h_1) Column 2	a_6 (ref.)	Method A Measured value h_3 Column 1	Method A Measured value h_3 Column 2	Method B h_3 (ref.) Column 1	Method B h_3 (ref.) Column 2	Method B Measured value a_6	Closed gap s_1	Closed gap Tolerance	Tangential force F_t, N For h_1 shown in column 1	Tangential force F_t, N For h_1 shown in column 2	Tolerance	Diametral force F_d, N For h_1 shown in column 1	Diametral force F_d, N For h_1 shown in column 2	Tolerance
120	4,90				2,278 0 −0,024 For phosphated PO surface: +0,01 −0,024	2,778 0 −0,024 For phosphated PO surface: +0,01 −0,024			2,08 0 −0,22 for phosphated PO surface: +0,09 −0,22	0,35	+0,30 0	19,1	23,5		41,1	50,5	
121	4,95											19,3	23,8		41,5	51,2	
122	4,95	2,5	3,0	2,0			2,27	2,77				19,0	23,3		40,9	50,1	
123	5,00											19,2	23,6		41,3	50,7	
124	5,05											19,4	23,9		41,7	51,4	
125	5,05											23,4	27,8		50,3	59,8	
126	5,10											23,7	28,1		51,0	60,4	
127	5,15											24,0	28,5		51,6	61,3	
128	5,20											24,2	28,8		52,0	61,9	
129	5,20											23,8	28,3		51,2	60,8	
130	5,25											24,0	28,5		51,6	61,3	
131	5,30											24,3	28,9		52,2	62,1	
132	5,30											23,9	28,4		51,4	61,1	
133	5,35				2,724 0 −0,024	3,224 0 −0,024			2,63 0 −0,22			24,1	28,7		51,8	61,7	
134	5,40											24,4	29,0		52,5	62,4	
135	5,40				For phosphated PO surface: +0,01 −0,024	For phosphated PO surface: +0,01 −0,024			For phosphated PO surface: +0,09 −0,22			24,0	28,5		51,6	61,3	
136	5,45											24,3	28,8		52,2	61,9	
137	5,50											24,5	29,1		52,7	62,6	
138	5,50											24,1	28,7		51,8	61,7	
139	5,55									0,40	+0,35 0	24,4	29,0		52,5	62,4	
140	5,60	±0,20 Within a ring: 0,20 max.										24,6	29,3	±30% if $F_t < 10$N ±20% if $F_t > 10$N	52,9	63,0	±30% if $F_d < 21,5$N ±20% if $F_d > 21,5$N
141	5,65											24,9	29,6		53,5	63,6	
142	5,65											24,5	29,1		52,7	62,6	
143	5,70											24,7	29,4		53,1	63,2	
144	5,75	3,0	3,5	2,5			2,71	3,21				25,0	29,7		53,8	63,9	
145	5,75											24,6	29,3		52,9	63,0	
146	5,80											24,9	29,6		53,5	63,6	
147	5,85											25,1	29,9		54,0	64,3	
148	5,85											24,7	29,4		53,1	63,2	
149	5,90											25,0	29,7		53,8	63,9	
150	5,95											25,0	29,8		53,8	64,1	
152	6,00											24,9	29,7		53,5	63,9	
154	6,05				2,724 0 −0,029	3,224 0 −0,029			2,63 0 −0,26			24,8	29,5		53,3	63,4	
155	6,10											25,0	29,8		53,8	64,1	
156	6,15				For phosphated PO surface: +0,01 −0,029	For phosphated PO surface: +0,01 −0,029			For phosphated PO surface: +0,09 −0,27			25,2	30,1		54,2	64,7	
158	6,20											25,1	29,9		54,0	64,3	
160	6,25											25,0	29,8		53,8	64,1	
162	6,35											25,4	30,3		54,6	65,1	
164	6,40										0,50	+0,40 0	25,3	30,2		54,4	64,9	
165	6,40											25,0	29,8		53,8	64,1	
166	6,45											25,2	30,0		54,2	64,5	
168	6,50											25,1	29,9		54,0	64,3	
170	6,60				3,172 0 −0,029	3,672 0 −0,029			3,20 0 −0,27			30,4	35,4		65,4	76,1	
172	6,65											30,3	35,2		65,1	75,7	
174	6,70	3,5	4,0	3,0	For phosphated PO surface: +0,01 −0,029	For phosphated PO surface: +0,01 −0,029	3,15	3,65	For phosphated PO surface: +0,09 −0,27			30,2	35,1		64,9	75,5	
175	6,75									0,60	+0,45 0	30,3	35,2		65,1	75,7	
176	6,80											30,5	35,5		65,6	76,3	
178	6,85											30,4	35,4		65,4	76,1	

This table is shown in ISO format. Commas represent decimal points.

TABLE 7 (CONCLUDED)

Dimensions in millimeters

Nominal diameter d_1	Radial wall thickness "regular" a_1		Nominal value of ring width (h_1)		a_6 (ref.)	Method A — Measured value h_3 Column		Method B h_3 (ref.) Column		Measured value a_6	Closed gap s_1		Tangential force F_t, N — For h_1 shown in column			Diametral force F_d, N — For h_1 shown in column		
		Tolerance	Column 1	2		1	2	1	2			Tolerance	1	2	Tolerance	1	2	Tolerance
180	6,90												30,3	35,2		65,1	75,7	
182	6,95												30,1	35,1		64,7	75,5	
184	7,05												30,6	35,7		65,8	76,8	
185	7,05					3,172	3,672			3,20			30,3	35,2		65,1	75,7	
186	7,10	±0,20				0	0			0			30,5	35,5		65,6	76,3	
188	7,15	Within a ring: 0,20 max.	3,5	4,0	3,0	−0,029 For phosphated PO surface: +0,01 −0,029	−0,029 For phosphated PO surface: +0,01 −0,029	3,15	3,65	−0,27 For phosphated PO surface: +0,09 −0,27	0,60	+0,45 0	30,4	35,4	±30 % if $F_t < 10$ N; ±20 % if $F_t > 10$ N	65,4	76,1	±30 % if $F_d < 21,5$ N; ±20 % if $F_d > 21,5$ N
190	7,20												30,3	35,2		65,1	75,7	
192	7,25												30,1	35,1		64,7	75,5	
194	7,35												30,6	35,7		65,8	76,8	
195	7,35												30,2	35,2		64,9	75,7	
196	7,40												30,5	35,5		65,6	76,3	
198	7,45												30,4	35,4		65,4	76,1	
200	7,50												30,2	35,2		64,9	75,7	

NOTES

[1] For intermediate sizes (for example repair sizes), the radial wall thickness of the next smaller nominal diameter should be applied.

[2] For values for F_t and F_d, given in Table 7, apply to as cast grey cast iron with a typical modulus of elasticity (E_n) of 100 000 MPa. Multiplying factors for materials having a different modulus (E_n) are given in SAE J1591.

Mean forces are calculated for nominal radial wall thickness (a_1) and mean ring width (h_1).

[3] For the sole purpose of this document, the assumed average ratio F_d/F_t is 2.15.

This table is shown in ISO format. Commas represent decimal points.

TABLE 8—DIMENSIONS OF K, KB, KM KEYSTONE RINGS 15°

Dimensions in millimeters

Nominal diameter d_1	Radial wall thickness "regular" a_1	Tolerance	Nominal value of ring width (h_1) Column 1	Nominal value of ring width (h_1) Column 2	a_6 (ref.)	Method A Measured value h_3 Column 1	Method A Measured value h_3 Column 2	Method B h_3 (ref.) Column 1	Method B h_3 (ref.) Column 2	Method B Measured value a_6	Closed gap s_1	Closed gap Tolerance	Tangential force F_t, N For h_1 shown in column 1	Tangential force F_t, N For h_1 shown in column 2	Tolerance	Diametral force F_d, N For h_1 shown in column 1	Diametral force F_d, N For h_1 shown in column 2	Tolerance
80	3,35										0,25		12,8	16,0		27,5	34,4	
81	3,40												13,0	16,3		28,0	35,0	
82	3,40												12,7	15,9		27,3	34,2	
83	3,45												12,9	16,2		27,7	34,8	
84	3,50												13,1	16,5		28,2	35,5	
85	3,55					2,097 0 −0,029 For phosphated PO surface: +0,01 −0,029	2,597 0 −0,029 For phosphated PO surface: +0,01 −0,029			1,49 0 −0,11 For phosphated PO surface: +0,04 −0,11			13,3	16,7		28,6	35,9	
86	3,60												13,5	17,0		29,0	36,6	
87	3,65	±0,15											13,7	17,3		29,5	37,2	
88	3,65	Within a ring: 0,15 max.	2,5	3,0	1,5			2,10	2,60				13,4	16,9		28,8	36,3	
89	3,70												13,6	17,1		29,2	36,8	
90	3,75												13,7	17,3		29,5	37,2	
91	3,80											+0,25 0	13,9	17,6		29,9	37,8	
92	3,85												14,1	17,8		30,3	38,3	
93	3,90												14,3	18,1		30,7	38,9	
94	3,90												14,0	17,7		30,1	38,1	
95	3,95												14,1	17,9		30,3	38,5	
96	4,00												14,3	18,2		30,7	39,1	
97	4,05												14,5	18,5		31,2	39,8	
98	4,10												14,7	18,7		31,6	40,2	
99	4,15												14,9	19,0		32,0	40,9	
100	4,15										0,30		18,5	22,5		39,8	48,4	
101	4,20												18,8	22,8		40,4	49,0	
102	4,25												19,0	23,1		40,9	49,7	
103	4,30												19,2	23,3		41,3	50,1	
104	4,30												18,8	22,9		40,4	49,2	
105	4,35												19,0	23,1	±30% if $F_t < 10$ N ±20% if $F_t > 10$ N	40,9	49,7	±30% if $F_d < 21,5$ N ±20% if $F_d > 21,5$ N
106	4,40												19,2	23,4		41,3	50,3	
107	4,40					2,463 0 −0,034 For phosphated PO surface: +0,01 −0,034	2,963 0 −0,034 For phosphated PO surface: +0,01 −0,034			2,05 0 −0,13 For phosphated PO surface: +0,04 −0,13			18,8	22,9		40,4	49,2	
108	4,45												19,0	23,2		40,9	49,9	
109	4,50												19,2	23,5		41,3	50,5	
110	4,55												19,4	23,6		41,7	50,7	
111	4,55												19,0	23,2		40,9	49,9	
112	4,60		3,0	3,5	2,0			2,45	2,95				19,2	23,4		41,3	50,3	
113	4,65												19,4	23,7		41,7	51,0	
114	4,70												19,6	24,0		42,1	51,6	
115	4,70	±0,20 Within a ring: 0,20 max.											19,2	23,5		41,3	50,5	
116	4,75												19,4	23,7		41,7	51,0	
117	4,80												19,6	24,0		42,1	51,6	
118	4,80												19,2	23,5		41,3	50,5	
119	4,85											+0,30 0	19,4	23,8		41,7	51,2	
120	4,90										0,35		19,6	24,0		42,1	51,6	
121	4,95												19,8	24,3		42,6	52,2	
122	4,95												19,4	23,8		41,7	51,2	
123	5,00												19,6	24,1		42,1	51,8	
124	5,05												19,8	24,3		42,6	52,2	
125	5,05					2,830 0 −0,034 For phosphated PO surface: +0,01 −0,034	3,330 0 −0,034 For phosphated PO surface: +0,01 −0,034			2,61 0 −0,13 For phosphated PO surface: +0,04 −0,13			23,9	28,3		51,4	60,8	
126	5,10												24,1	28,6		51,8	61,5	
127	5,15												24,3	28,9		52,2	62,1	
128	5,20												24,6	29,2		52,9	62,8	
129	5,20												24,1	28,7		51,8	61,7	
130	5,25		3,5	4,0	2,5			2,80	3,30				24,3	28,9		52,2	62,1	
131	5,30												24,5	29,2		52,7	62,8	
132	5,30										0,40	+0,35 0	24,1	28,7		51,8	61,7	
133	5,35												24,3	28,9		52,2	62,1	
134	5,40												24,5	29,2		52,7	62,8	

This table is shown in ISO format. Commas represent decimal points.

TABLE 8 (CONCLUDED)

Dimensions in millimeters

Nominal diameter d_1	Radial wall thickness "regular" a_1 Tolerance	Nominal value of ring width (h_1) Column 1	Nominal value of ring width (h_1) Column 2	a_6 (ref.)	Method A Measured value h_3 Column 1	Method A Measured value h_3 Column 2	Method B h_3 (ref.) Column 1	Method B h_3 (ref.) Column 2	Method B Measured value a_6	Closed gap s_1	Closed gap Tolerance	Tangential force F_t, N For h_1 shown in column 1	Tangential force F_t, N For h_1 shown in column 2	Tangential Tolerance	Diametral force F_d, N For h_1 shown in column 1	Diametral force F_d, N For h_1 shown in column 2	Diametral Tolerance
135	5,40											24,1	28,7		51,8	61,7	
136	5,45											24,4	29,0		52,5	62,4	
137	5,50											24,6	29,3		52,9	63,0	
138	5,50											24,2	28,8		52,0	61,9	
139	5,55											24,4	29,1		52,5	62,6	
140	5,60											24,6	29,3		52,9	63,0	
141	5,65											24,8	29,6		53,3	63,6	
142	5,65										0,40	+0,35 / 0	24,4	29,1		52,5	62,6	
143	5,70												24,6	29,4		52,9	63,2	
144	5,75												24,8	29,6		53,3	63,6	
145	5,75					2,830 0 −0,034 For phosphated PO surface: +0,01 −0,034	3,330 0 −0,034 For phosphated PO surface: +0,01 −0,034			2,61 0 −0,13 For phosphated PO surface: +0,04 −0,13			24,4	29,2		52,5	62,8	
146	5,80												24,6	29,4		52,9	63,2	
147	5,85	3,5	4,0	2,5			2,80	3,30				24,8	29,7		53,3	63,9	
148	5,85												24,4	29,2		52,5	62,8	
149	5,90												24,6	29,5		52,9	63,4	
150	5,95												24,7	29,5		53,1	63,4	
152	6,00												24,5	29,3		52,7	63,0	
154	6,05												24,3	29,1		52,2	62,6	
155	6,10												24,5	29,4		52,7	63,2	
156	6,15												24,7	29,6		53,1	63,6	
158	6,20												24,5	29,4		52,7	63,2	
160	6,25	±0,20 Within a ring: 0,20 max.											24,3	29,2	±30% if $F_t < 10$ N ±20% if $F_t > 10$ N	52,2	62,8	±30% if $F_d < 21,5$ N ±20% if $F_d > 21,5$ N
162	6,35										0,50	+0,40 / 0	24,6	29,7		52,9	63,9	
164	6,40												24,5	29,5		52,7	63,4	
165	6,40												24,1	29,1		51,8	62,6	
166	6,45												24,3	29,3		52,2	63,0	
168	6,50												24,1	29,1		51,8	62,6	
170	6,60												29,5	34,5		63,4	74,2	
172	6,65												29,3	34,3		63,0	73,7	
174	6,70												29,1	34,1		62,6	73,3	
175	6,75												29,2	34,2		62,8	73,5	
176	6,80												29,4	34,4		63,2	74,0	
178	6,85					3,191 0 −0,039 For phosphated PO surface: +0,01 −0,039	3,691 0 −0,039 For phosphated PO surface: +0,01 −0,039			2,98 0 −0,15 For phosphated PO surface: +0,04 −0,15			29,2	34,2		62,8	73,5	
180	6,90												29,0	34,0		62,4	73,1	
182	6,95												28,8	33,9		61,9	72,9	
184	7,05	4,0	4,5	3,0			3,20	3,70		0,60	+0,45 / 0	29,2	34,3		62,8	73,7	
185	7,05												28,8	33,9		61,9	72,9	
186	7,10												29,0	34,1		62,4	73,3	
188	7,15												28,8	33,9		61,9	72,9	
190	7,20												28,7	33,7		61,7	72,5	
192	7,25												28,5	33,6		61,3	72,2	
194	7,35												28,8	34,0		61,9	73,1	
195	7,35												28,5	33,6		61,3	72,2	
196	7,40												28,7	33,8		61,7	72,7	
198	7,45												28,5	33,6		61,3	72,2	
200	7,50												28,3	33,4		60,8	71,8	

NOTES

[1] For intermediate sizes (for example repair sizes), the radial wall thickness of the next smaller nominal diameter should be applied.

[2] For values for F_t and F_d, given in Table 8, apply to as cast grey cast iron with a typical modulus of elasticity (E_n) of 100 000 MPa. Multiplying factors for materials having a different modulus (E_n) are given in SAE J1591.

Mean forces are calculated for nominal radial wall thickness (a_1) and mean ring width (h_1).

[3] For the sole purpose of this document, the assumed average ratio F_d/F_t is 2.15.

This table is shown in ISO format. Commas represent decimal points.

26.111

INTERNAL COMBUSTION ENGINES—PISTON RINGS—HALF KEYSTONE RINGS—SAE J2001 OCT92

SAE Standard

Report of the Piston Ring Standards Committee approved September 1990. Revised by the SAE Piston Ring Standards Committee 7 October 1992.

This SAE Standard is equivalent to ISO Standard 6624/2 TR.

1. Scope and Field of Application—Differences, where they exist, are shown in Appendix A.

This SAE Standard specifies the essential dimensional features of HK- and HKB-half keystone rings with narrow ring width types.

Dimensional Tables 6 and 7 allow for the use of cast iron (Table 6) or steel (Table 7). Since the modulus of elasticity of steel rings is higher than that of cast iron rings, the fluctuation in the surface pressure will become greater if the free gap is set as the reference for forces. Therefore, forces are set using the surface pressure as the reference in order to minimize the effect of the fluctuation.

The requirements of this document apply to half keystone rings of reciprocating internal combustion engines up to and including 70 mm diameter for cast iron rings and up to and including 100 mm diameter for steel rings.

(R) 2. References

SAE DESIGNATION	ISO[1] EQUIVALENT	
		INTERNAL COMBUSTION ENGINES—PISTON RINGS
J1588	6621/1	Vocabulary
J1589	6621/2	Measuring principles
J1590	6621/3	Material specifications
J1591	6621/4	General specifications
J1996	6621/5	Quality requirements
		INTERNAL COMBUSTION ENGINES—PISTON RINGS
J1997	6622/1	Rectangular rings
J1998	6622/2 TR	Rectangular rings with narrow ring width
J1999	6623	INTERNAL COMBUSTION ENGINES—PISTON RINGS—SCRAPER RINGS
		INTERNAL COMBUSTION ENGINES—PISTON RINGS
J2000	6624/1	Keystone rings
J2001	6624/2 TR	Half keystone rings
J2002	6625	INTERNAL COMBUSTION ENGINES—PISTON RINGS—OIL CONTROL RINGS
J2003	6626	INTERNAL COMBUSTION ENGINES—PISTON RINGS—COIL SPRING LOADED OIL CONTROL RINGS
J2004	6627 TR	INTERNAL COMBUSTION ENGINES—PISTON RINGS—EXPANDER/SEGMENT OIL CONTROL RINGS
J2226		INTERNAL COMBUSTION ENGINES—PISTON RINGS—STEEL RECTANGULAR RINGS
	1101	TECHNICAL DRAWINGS—GEOMETRICAL TOLERANCING—TOLERANCING OF FORM, ORIENTATION, LOCATION AND RUN-OUT—GENERALITIES, DEFINITIONS, SYMBOLS INDICATIONS ON DRAWINGS

[1] TR refers to Technical Report

3. Ring Types and Designation Examples

NOTE—For the angle of half keystone rings, the same definition and measurement apply as for keystone rings (see SAE J1589).

3.1 Type HK—Straight Faced Half Keystone Ring 7 Degrees

3.1.1 GENERAL FEATURES

NOTE—See Table 6 or 7 for dimensions and forces.

3.1.2 DESIGNATION EXAMPLE—Designation of a straight faced half keystone ring 7 degrees of $d_1 = 60$ mm nominal diameter, $h_1 = 1.25$ mm ring width, made of steel (material subclass 62), general features as shown in Figure 1, and periphery chromium coated fully faced design 0.1 mm minimum thickness:

a. Piston ring SAE J2001 HK-60 × 1.25-MC62/CR2

3.2 Type HKB—Barrel Faced Half Keystone Ring 7 Degrees

3.2.1 GENERAL FEATURES

NOTE—See Table 6 or 7 for dimensions and forces.

3.2.2 DESIGNATION EXAMPLE—Designation of a barrel faced half keystone ring 7 degrees of $d_1 = 60$ mm nominal diameter, $h_1 = 1.25$ mm ring width, made of steel (material subclass 62), general features as shown in Figure 2 and Table 1, and periphery molybdenum coated inlaid design 0.1 mm minimum thickness:

a. Piston ring SAE J2001 HKB-60 × 1.25-MC62/MO2

TABLE 1—GAUGE WIDTH (h_a) AND BARREL DIMENSIONS

Dimensions in millimeters

h_1	h_a	t_2, t_3	t_2, t_3	Maximum Peak Off Center
1.25	0.6	0.001	0.013	0.2
1.55	0.8	0.02	0.016	0.25

Dimensions in millimeters.

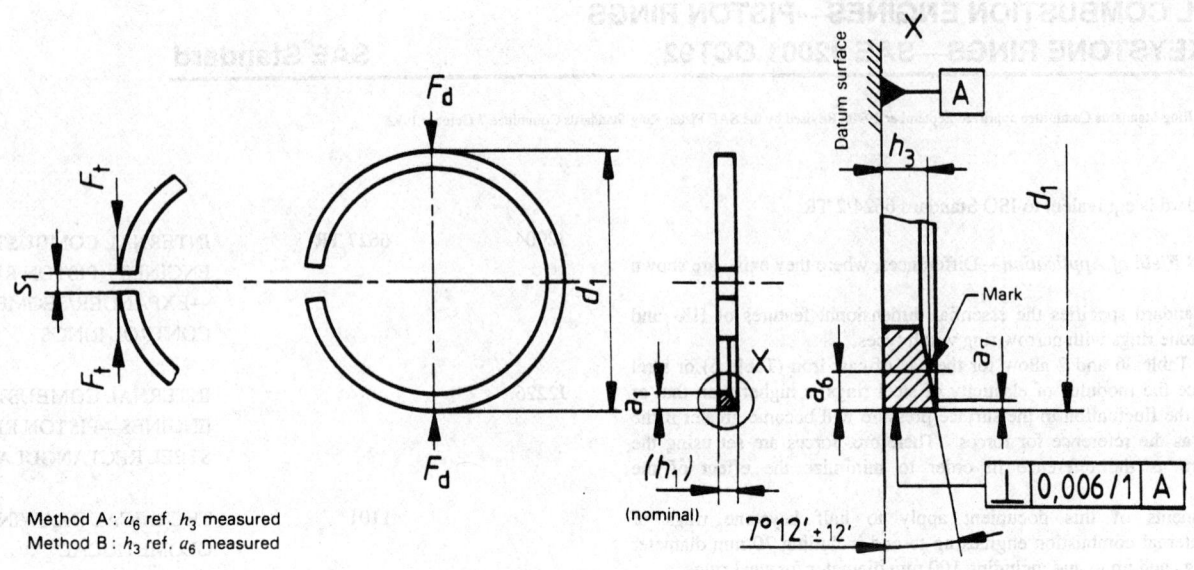

Method A: a_6 ref. h_3 measured
Method B: h_3 ref. a_6 measured

FIGURE 1—TYPE HK

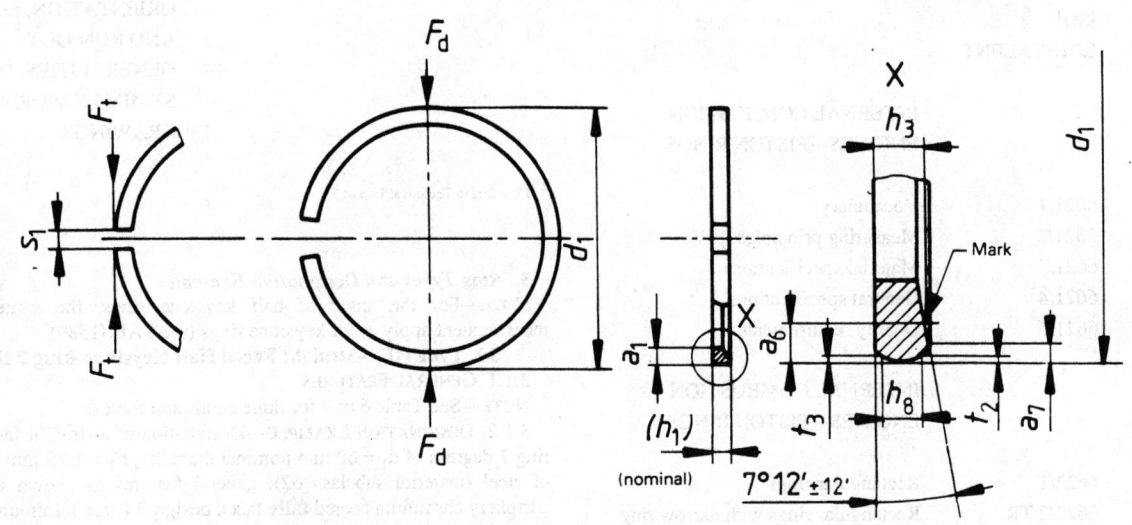

FIGURE 2—TYPE HKB

4. Common Features
4.1 Types HK and HKB—Half Keystone Ring—(See Figures 3 to 6.)
4.1.1 UNCOATED RINGS

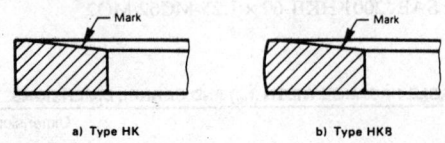

FIGURE 3—UNCOATED RINGS

4.1.2 COATED RINGS (CHROMIUM OR MOLYBDENUM)
4.1.2.1 Fully Faced

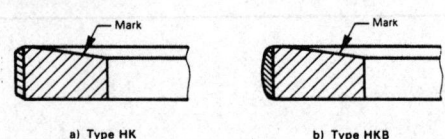

FIGURE 4—FULLY FACED COATED RINGS

4.1.2.2 Semi-Inlaid

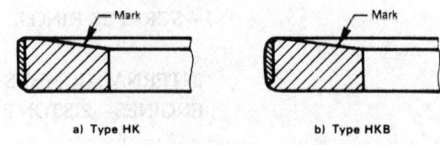

FIGURE 5—SEMI-INLAID COATED RINGS

4.1.2.3 Inlaid

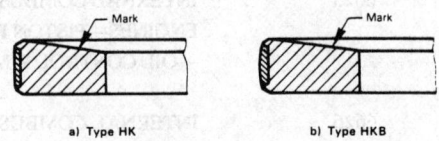

FIGURE 6—INLAID COATED RINGS

4.2 Chamfered Edges (Cast Iron Rings)
4.2.1 HK AND HKB RINGS—OUTSIDE CHAMFERED EDGES (KA) (see Figures 7 to 9)—See Table 2 for dimensions.

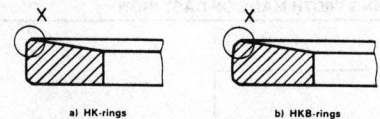

FIGURE 7—OUTSIDE CHAMFERED EDGES (KA)

4.2.2 HK AND HKB RINGS—INSIDE CHAMFERED EDGES (KI)

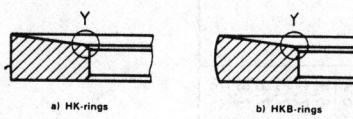

FIGURE 8—INSIDE CHAMFERED EDGES (KI)

4.2.3 HK AND HKB RINGS—OUTSIDE AND INSIDE CHAMFERED EDGES (KA + KI)

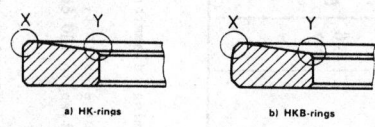

FIGURE 9—OUTSIDE AND INSIDE CHAMFERED EDGES (KA + KI)

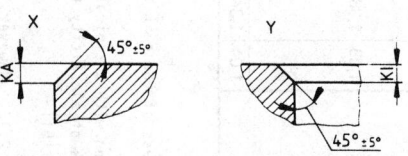

FIGURE 10—DETAILS OF FIGURES 7, 8, AND 9

TABLE 2—KA AND KI DIMENSIONS

Dimensions in millimeters

KA	KI
0.15 ± 0.1	max 0.2

4.3 HK and HKB Steel Rings—Outside and Inside Rounded Edges—(See Figure 11.)

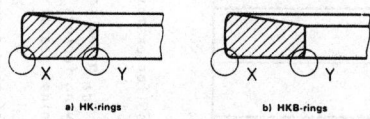

FIGURE 11—OUTSIDE AND INSIDE ROUNDED EDGES

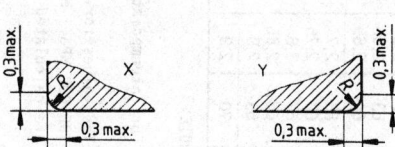

FIGURE 12—DETAILS OF FIGURE 11

4.4 HK and HKB Rings (Fully Faced, Semi-Inlaid, and Inlaid)—Layer Thickness (see Figure 13)—See Table 3 for dimensions.

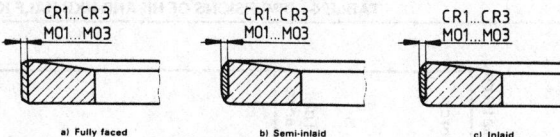

FIGURE 13—LAYER THICKNESS

TABLE 3—LAYER THICKNESS

Dimensions in millimeters

Chromium	Molybdenum	Thickness min
CR1	MO1	0.05
CR2	MO2	0.1
CR3[1]	MO3[1]	0.15

[1] CR3 and MO3 apply to rings with nominal diameters of 50 mm or greater.

5. Force Factors—The tangential and diametral forces given in Table 6 shall be corrected when additional features and/or materials other than grey cast iron with a modulus of elasticity of 100 000 MPa are being used.

For common features, multiplier correction factors given in Tables 4 and 5 and the force correction factors given in SAE J1591 shall be used. Also, the tangential and diametral forces given in Table 7 shall be corrected when additional features are being used.

TABLE 4—FORCE CORRECTION FACTORS FOR HK AND HKB RINGS WITH FEATURES KA AND KI

	Factor	
KA	KI	KA and KI
0.98	0.98	0.96

(R) TABLE 5—FORCE CORRECTION FACTORS FOR COATED HK AND HKB RINGS (FULLY FACED, SEMI-INLAID, AND INLAID TYPES)

d_1 (Reference Coating) mm	CR1	CR2/MO1	Factor CR3	Factor MO2	MO3
38 ≤ d_1 < 50	0.81	0.7	-	0.64	-
50 ≤ d_1 ≤ 100	0.9	0.85	0.81	0.81	0.75

6. Dimensions—See Tables 6 and 7.

APPENDIX A

A.1 This SAE Standard has been established to harmonize the ISO and SAE piston ring standards. The U.S. Technical Advisory Group, with the support of the National Engine Parts Manufacturers Association, has worked with other national organizations on this worldwide standard. Some of the wording and phrasing may differ slightly from U.S. terminology for translation purposes.

In preparing this SAE Standard, the Scope and Field of Application, and Reference sections of the ISO 6624/2 TR have been editorially revised and reorganized.

Certain barrel drop (Table 1) dimensions were changed to comply with established U.S. standard practice.

The tolerances specified in this SAE Standard represent a six sigma quality level.

TABLE 6—DIMENSIONS OF HK AND HKB HALF KEYSTONE RINGS WITH NARROW RING WIDTH MADE OF CAST IRON

Dimensions in millimeters.

Nominal diameter d_1	Radial wall thickness a_1	Tolerance	Ring width nominal value For h_1 shown in column 1	Ring width nominal value For h_1 shown in column 2	a_6 (ref.)	a_7	Method A Measured value For h_3 shown in column 1	Tolerance	Method A Measured value For h_3 shown in column 2	Tolerance	Method B Measured value h_3 (ref.) Column 1	Method B Measured value h_3 (ref.) Column 2	Tolerance a_6	Closed gap s_1	Tolerance	Tangential force F_t, N For h_1 shown in column 1	Tangential force F_t, N For h_1 shown in column 2	Tolerance	Diametral force F_d, N For h_1 shown in column 1	Diametral force F_d, N For h_1 shown in column 2	Tolerance
38	1,6	±0,15 Within a ring: 0,15 max.	1,25	1,55	0,8	0,5 max.	1,143	0 −0,024 For phosphated PO surface: +0,01 −0,024	1,418	0 −0,024 For phosphated PO surface: +0,01 −0,024	1,13		0 −0,19 For phosphated PO surface: +0,08 −0,19		+0,2 / 0	4,4	5,6	±30 %	7,2		±30 %
39	1,65															4,5	5,8		7,6		
40	1,65				1		1,118				1,11					4,3	5,5		7,2		
41	1,7															4,5	5,7		7,6		
42	1,75															4,4	5,7		7,8		
43	1,8												0,91			4,6	5,9		8,2	10,5	
44	1,85																		8,3	10,8	
45	1,9						1,093		1,393		1,08	1,38	1,06			4,7	6		8,3	11,2	
46	1,9															4,9	6,2		8,5	10,6	
47	1,95															4,9	6,4		8,5	10,9	
48	2												1,3			4,7	6,2		8,9	11,4	
49	2,05															4,9	6,4		9	11,8	
50	2,1				1,2	0,6 max.	1,09		1,39		1,08	1,38	1,28		0,15	5	6,5		9,4	12,1	
51	2,15															5,1	6,7		9,6	12,5	
52	2,15															5,3	6,9		9,2	11,9	
53	2,2															5,4	7		9,6	12,3	
54	2,25															5,3	6,7		9,7	12,7	
55	2,3															5,3	6,9		10,1	13	
56	2,35															5,5	7,1		10,4	13,4	
57	2,4															5,3	7		10,5	13,8	
58	2,4															5,5	7,3		10,1	13,4	
59	2,45															5,7	7,5		10,5	13,8	
60	2,5															5,3	7,1		10,7	14	
61	2,55															5,5	6,9		10,9	14,4	
62	2,6															5,3	7		11,3	14,8	
63	2,65															5,3	6,7		11,7	15	
64	2,65															5,5	6,9		11,3	14,4	
65	2,7				1,5		1,053		1,353		1,04	1,34	1,6		0,2	5,5	6,9		11,5	14,8	
66	2,75															5,5	7,1		11,7	15,3	
67	2,8															5,5	7,3		11,9	15,6	
68	2,85															5,7	7,5		12,3	16	
69	2,9															5,7	7,6		12,4	16,2	
70	2,9															5,6	7,3		12	15,7	

NOTES

[1] For intermediate sizes (for example repair sizes), the radial wall thickness of the next smaller nominal diameter shall be applied.

[2] The values for F_t and F_d, given in Table 6, apply to as-cast grey cast iron with a typical modulus of elasticity (E_n) of 100 000 MPa. Multiplying factors for materials having a different modulus of elasticity (E_n) are given in ISO 6621/4. Mean forces are calculated for nominal radial wall thickness (a_1) and mean ring width (h_1).

[3] For the sole purpose of this document, the assumed average ratio F_d/F_t is 2.15. However, for rings up to 50 mm the ratio F_d/F_t shall be determined between manufacturer and client.

This table is shown in ISO format. Commas represent decimal points.

TABLE 7—DIMENSIONS OF HK AND HKB HALF KEYSTONE RINGS WITH NARROW RING WIDTH MADE OF STEEL

Dimensions in millimeters.

Nominal diameter d_1	Radial wall thickness a_1	Tolerance	Ring width nominal value For h_1 shown in column 1	Ring width nominal value For h_1 shown in column 2	a_6 (ref.)	a_7	Method A Measured value For h_3 shown in column 1	Tolerance	Method A Measured value For h_3 shown in column 2	Tolerance	h_3 (ref.) Column 1	h_3 (ref.) Column 2	Method B Measured value a_6	Tolerance	a_7	Closed gap s_1	Tolerance	Tangential force F_t, N For h_1 shown in column 1	Tangential force F_t, N For h_1 shown in column 2	Tolerance	Diametral force F_d, N For h_1 shown in column 1	Diametral force F_d, N For h_1 shown in column 2	Tolerance
38	1,5	±0,15 Within a ring: 0,15 max.	1,25	1,55	0,8	0,5 max.	1,143	0 / −0,024	1,418	0 / −0,024	1,13		0,91	0 / −0,19	0,5 max.		+0,2 / 0			±30% if F_t <10N / ±20% if F_t >10N			±30% if F_d <21,5N / ±20% if F_d >21,5N
39	1,5																					8,8	12,6	
40	1,7																					9,1	12,8	
41	1,7																					9,3	13	
42	1,7																					9,5	13,4	
43	1,7																					9,7	13,6	
44	1,7																					9,8	13,9	
45	1,9				1		1,118		1,41		1,11	1,41	1,06					5,3	6,7		10	14,1	
46	1,9																		5,4	6,8		10,5	14,3	
47	1,9																		5,5	6,9		10,6	14,7	
48	1,9																		5,6	7,1		10,8	14,9	
49	1,9																		5,7	7,1		10,9	15,1	
50	2,1						1,093		1,393		1,08	1,38	1,3					5,8	7,2		11,1	15,3	
51	2,1																0,15		5,9	7,4		11,3	15,6	
52	2,1				1,2														6	7,5		11,6	16	
53	2,1																		6	7,6		11,8	16,2	
54	2,1																		6,1	7,7		12	16,4	
55	2,1																					12,2	16,6	
56	2,1																					12,4	17,2	
57	2,1																					12,6	17,4	
58	2,1																					12,8	17,6	
59	2,1																					12,9	17,8	
60	2,3						1,09		1,39		1,08	1,38	1,28		0,6 max.			6,2	8		13,1	18	
61	2,3																		6,3	8,1		13,2	18,4	
62	2,3																		6,4	8,2		13,6	18,6	
63	2,3																		6,5	8,3		13,9	18,8	
64	2,3																		6,6	8,4		13,9	19	
65	2,5																0,2		6,6	8,6		14,1	18	
66	2,5																		6,7	8,6		14,3	18,4	
67	2,5																		6,8	8,7		14,5	18,6	
68	2,5																		6,8	8,8		14,7	19	
69	2,5																		7	8,9		15,1	19,2	
70	2,7				1,5		1,053		1,353		1,04	1,34	1,6					7,1	9,1		15,3	19,6	
71	2,7																		7,2	9,2		15,5	19,8	
72	2,7																		7,3	9,3		15,7	20	
73	2,7																		7,3	9,4		15,7	20,1	
74	2,9																		7,4	9,5		15,8	20,4	

This table is shown in ISO format. Commas represent decimal points.

TABLE 7—CONCLUDED

Dimensions in millimeters.

Nominal diameter d_1	Radial wall thickness a_1		Ring width nominal value For h_1 shown in column		a_6 (ref.)	a_7	Method A Measured value For h_3 shown in column				Method B				Closed gap s_1		Tangential force F_t, N			Diametral force F_d, N		
		Tolerance	1	2			1	Tolerance	2	Tolerance	h_3 (ref.) Column 1	2	a_6 Measured value	Tolerance	a_7		Tolerance	For h_1 shown in column 1	2	Tolerance	For h_1 shown in column 1	2	Tolerance
75	2,9																	7,5	9,6		16,1	20,7	
76																		7,6	9,7		16,3	20,9	
77			1,25	1,55	1,5		1,053	0 −0,024	1,353	0 −0,024	1,04	1,34	1,6	0 −0,19	0,6 max.	0,25	+0,25 0	7,7	9,8	±30% if F_t <10N ±20% if F_t >10N	16,6	21,1	±30% if F_d <21,5N ±20% if F_d >21,5N
78																		7,7	10		16,6	21,6	
79		±0,15 Within a ring: 0,15 max.																7,8	10,1		16,8	21,7	
80	3,1																	7,9	10,3		17,1	22,1	
81																		8	10,4		17,2	22,3	
82																		8,1	10,6		17,4	22,7	
83																		8,1	10,5		17,4	22,7	
84																		8,2	10,6		17,6	22,9	
85	3,3																	8,4	10,8		17,9	23,2	
86																		8,4	10,9		18,2	23,4	
87																		8,5	11,1		18,4	23,8	
88																		8,5	11		18,3	23,8	
89																		8,6	11,1		18,5	23,9	
90	3,5				2	0,6 max.			1,289				2,07			0,3		11,3			24,3		
91													1,28						11,4			24,5		
92																			11,6			24,9		
93																			11,5			24,8		
94																			11,6			25		
95	3,7																		11,8			25,3		
96																			11,9			25,5		
97																			12,1			26		
98																			12,1			26,1		
99	3,9																		12,2			26,3		
100																			12,4			26,7		

NOTES

[1] For intermediate sizes (for example repair sizes), the radial wall thickness of the next smaller nominal diameter shall be applied.

[2] The values for F_t and F_d, given in Table 7, apply to steel with a typical modulus of elasticity (E_n) of 200 000 MPa. Mean forces are calculated for nominal radial wall thickness (a_1) and mean ring width (h_1).

[3] For the sole purpose of this document, the assumed average ratio F_d/F_t is 2.15. However, for rings up to 50 mm the ratio F_d/F_t shall be determined between manufacturer and client.

This table is shown in ISO format. Commas represent decimal points.

INTERNAL COMBUSTION ENGINES—PISTON RINGS—OIL CONTROL RINGS—SAE J2002 OCT92

SAE Standard

Report of the Piston Ring Standards Committee approved October 1989. Revised by the SAE Piston Ring Standards Committee 7 October 1992.

This SAE Standard is equivalent to ISO Standard 6625.

1. Scope and Field of Application—Differences, where they exist, are shown in Appendix A.

This SAE Standard specifies the dimensional features of S, G, D, and DV oil control piston ring types.

The normal range for the axial width of oil control rings (2.5 to 8 mm inclusive) is divided into 0.5 or 1.0 increments. In Table 7, dimensions in inch units are given for oil control rings with axial width 4.75 mm (equal to 3/16 in) for existing applications.

The requirements of this document apply to oil control rings for reciprocating internal combustion piston engines up to and including 200 mm in diameter. They may also be used for piston rings of compressors working under similar conditions.

(R) **2. References**

SAE DESIGNATION	ISO[1] EQUIVALENT	
		INTERNAL COMBUSTION ENGINES—PISTON RINGS
J1588	6621/1	Vocabulary
J1589	6621/2	Measuring principles
J1590	6621/3	Material specifications
J1591	6621/4	General specifications
J1996	6621/5	Quality requirements
		INTERNAL COMBUSTION ENGINES—PISTON RINGS
J1997	6622/1	Rectangular rings
J1998	6622/2 TR	Rectangular rings with narrow ring width
J1999	6623	INTERNAL COMBUSTION ENGINES—PISTON RINGS—SCRAPER RINGS
		INTERNAL COMBUSTION ENGINES—PISTON RINGS
J2000	6624/1	Keystone rings
J2001	6624/2 TR	Half keystone rings
J2002	6625	INTERNAL COMBUSTION ENGINES—PISTON RINGS—OIL CONTROL RINGS
J2003	6626	INTERNAL COMBUSTION ENGINES—PISTON RINGS—COIL SPRING LOADED OIL CONTROL RINGS
J2004	6627 TR	INTERNAL COMBUSTION ENGINES—PISTON RINGS—EXPANDER/SEGMENT OIL CONTROL RINGS
J2226		INTERNAL COMBUSTION ENGINES—PISTON RINGS—STEEL RECTANGULAR RINGS

[1] TR refers to Technical Report

3. Ring Types and Designation Examples
   3.1 Type S—Slotted Oil Control Ring
3.1.1 GENERAL FEATURES
NOTE—See Table 5 or 7 for dimensions and forces.

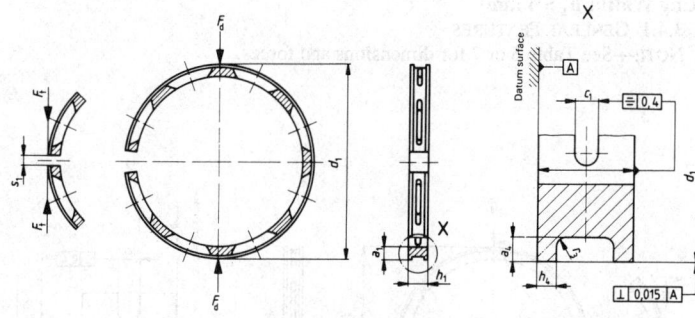

FIGURE 1—TYPE S

3.1.2 DESIGNATION EXAMPLE—Designation of a slotted oil control ring of d_1 = 90 mm nominal diameter, h_1 = 4 mm ring width, made of grey cast iron, nonheat-treated (material subclass 12), general features as shown in Figure 1, and inside chamfered edges.

   3.2 Type G—Double Bevelled Oil Control Ring
3.2.1 GENERAL FEATURES
NOTE—See Table 6 or 7 for dimensions and forces.

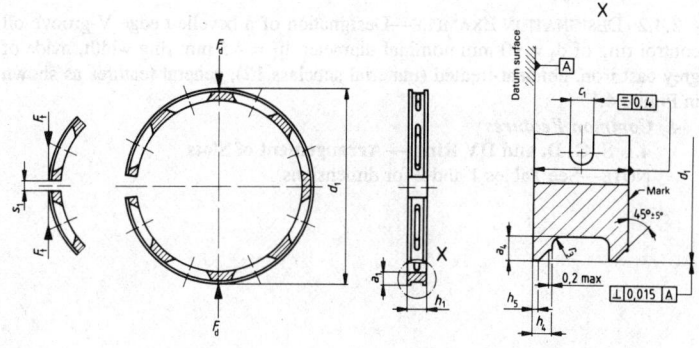

FIGURE 2—TYPE G

3.2.2 DESIGNATION EXAMPLE—Designation of a double bevelled oil control ring of $d_1 = 90$ mm nominal diameter, $h_1 = 4$ mm ring width, made of grey cast iron, nonheat-treated (material subclass 12), general features as shown in Figure 2, and phosphate coated.

3.3 Type D—Bevelled Edge Oil Control Ring
3.3.1 GENERAL FEATURES
NOTE—See Table 6 or 7 for dimensions and forces.

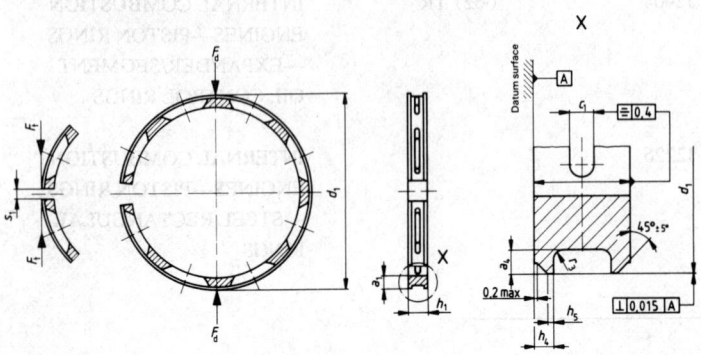

FIGURE 3—TYPE D

3.3.2 DESIGNATION EXAMPLE—Designation of a bevelled edge oil control ring of $d_1 = 90$ mm nominal diameter, $h_1 = 4$ mm ring width, made of grey cast iron, nonheat-treated (material subclass 12), general features as shown in Figure 3.

3.4 Type DV—Bevelled Edge V-Groove Oil Control Ring (Only for Ring Widths $h_1 > 4$ mm)
3.4.1 GENERAL FEATURES
NOTE—See Table 6 or 7 for dimensions and forces.

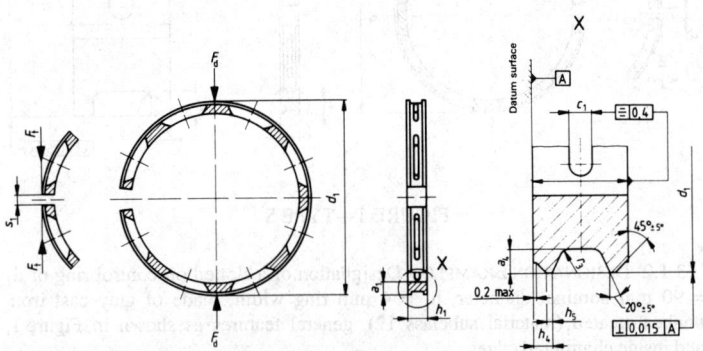

FIGURE 4—TYPE DV

3.4.2 DESIGNATION EXAMPLE—Designation of a bevelled edge V-groove oil control ring of $d_1 = 90$ mm nominal diameter, $h_1 = 4.5$ mm ring width, made of grey cast iron, nonheat-treated (material subclass 12), general features as shown in Figure 4.

4. Common Features
4.1 S, G, D, and DV Rings—Arrangement of Slots
NOTE—See Tables 1 and 2 for dimensions.

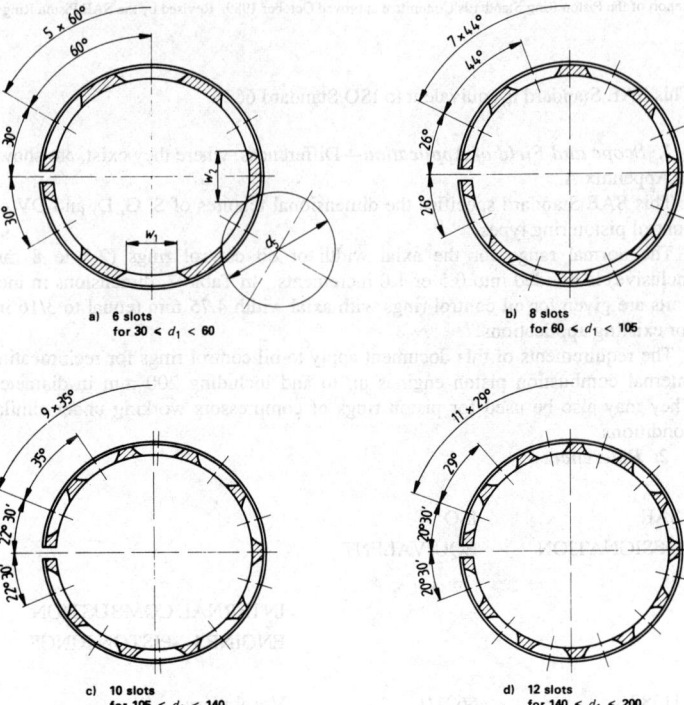

a) 6 slots for $30 < d_1 < 60$
b) 8 slots for $60 < d_1 < 105$
c) 10 slots for $105 < d_1 < 140$
d) 12 slots for $140 < d_1 < 200$

FIGURE 5—ARRANGEMENT OF SLOTS

TABLE 1—CUTTER DIAMETER

Dimensions in millimeters

d_1	Cutter Diameter d_5 Max
$30 \leq d_1 < 50$	55
$50 \leq d_1 < 170$	60
$170 \leq d_1 \leq 200$	75

TABLE 2—SLOT LENGTH

Dimensions in millimeters

d_1	Slot Length w_1	Slot Length Tolerance	Permissible Difference Between w_1 and w_2
$30 \leq d_1 < 36$	5	±2	-
$36 \leq d_1 < 40$	6	±2	-
$40 \leq d_1 < 50$	8	±2	-
$50 \leq d_1 < 170$	$w_1 = w_2$	-	2
$170 \leq d_1 \leq 200$	$w_1 = w_2$	-	4

4.2 S, G, D, and DV Rings—Inside Chamfered Edges (KI)
NOTE—See Table 3 for dimensions.

TABLE 3—KI DIMENSIONS

Dimensions in millimeters

d_1	KI
$30 \leq d_1 < 125$	0.3 ± 0.15
$125 \leq d_1 < 175$	0.4 ± 0.15
$175 \leq d_1 \leq 200$	0.6 ± 0.2

5. Force Factors—The tangential and diametral forces, given in Tables 5, 6, and 7 shall be corrected when additional features and/or materials other than grey cast iron with a modulus of elasticity of 100 000 MPa are being used.

For common features, the multiplier correction factors given in Table 4 and the force correction factors given in SAE J1591 shall be used.

TABLE 4—FORCE CORRECTION FACTORS FOR S, G, D, AND DV RINGS WITH KI FEATURE

d_1 mm	Factor
$30 \leq d_1 < 50$	1
$50 \leq d_1 < 100$	0.98
$100 \leq d_1 < 150$	0.98
$150 \leq d_1 \leq 200$	0.97

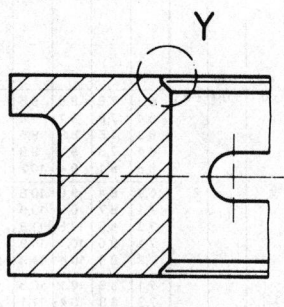

a) S rings

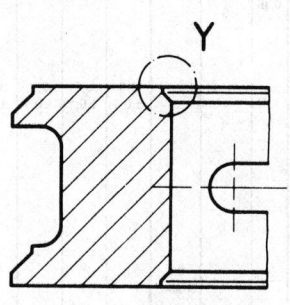

b) G rings

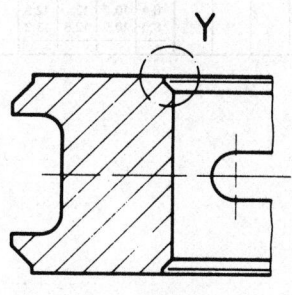

c) D rings

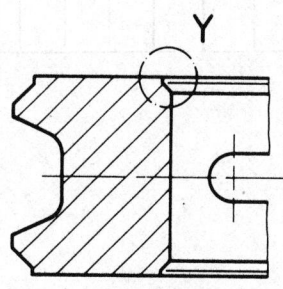

d) DV rings

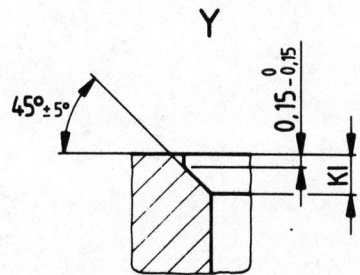

Nominal KI > 0,3

FIGURE 6—INSIDE CHAMFERED EDGES

TABLE 5—DIMENSIONS OF S OIL CONTROL RINGS

Dimensions in millimeters

Nominal diameter d_1	Radial wall thickness "regular" a_1	Tolerance	Ring width h_1 Column 1	2	3	4	Tolerance	Closed gap s_1	Tolerance	Radius r_3	Land width h_4 For h_1 shown in column 1	2	3	4	Groove depth a_4	Number of slots	Slot width c_1 For h_1 shown in column 1	2	3	4	Tangential force F_t, N For h_1 shown in column 1	2	3	4	Tolerance	Diametral force F_d, N For h_1 shown in column 1	2	3	4	Tolerance
30	1,25														0,4 ±0,1						—	—	—	—		7,5	9,2	10,8	11,5	
31	1,3																									8	9,9	11,4	12,3	
32	1,35																									8,6	10,3	12,3	13,1	
33	1,4																									9	11	12,9	13,8	
34	1,4																									8,6	10,3	12	13,1	
35	1,45																									9	11	12,9	13,8	
36	1,5																				—	—	—	—		9,5	11,6	13,5	14,6	
37	1,55																									9,9	12,3	14,2	15,5	
38	1,6																									10,5	12,9	15,1	16,1	
39	1,65							0,2 max																		11	13,5	15,7	17	
40	1,65																									9,9	12,3	14,2	15,3	
41	1,7																									10,5	12,9	14,8	15,9	
42	1,75														0,5 ±0,1						—	—	—	—		11	13,3	15,7	16,8	
43	1,8																									11,4	14	16,3	17,6	
44	1,85																									12	14,6	17	18,3	
45	1,9	±0,15 Within a ring: 0,15 max.	2,5	3	3,5	4	−0,010 −0,030 For phosphated P0 surface +0,005 −0,025	0,15	+0,2 0							6										12,3	15,1	17,4	18,9	
46	1,9																					—	—	—	—		11,6	14,4	16,8	18,1	
47	1,95																									12,3	14,8	17,4	18,9	
48	2																								±30% if F_d < 21,5 N	12,7	15,5	18,1	19,6	
49	2,05											0,5 ±0,1	0,6 ±0,1	0,7 ±0,1	0,7 ±0,1			0,7 ±0,1	0,7 ±0,1	0,8 ±0,1	1 ±0,1					13,1	16,1	18,9	20,4	
50	2,1																					6,2	7,5	8,8	9,5	±20% if F_d > 21,5 N	13,3	16,1	18,9	20,4	
51	2,15																					6,4	7,8	9,1	9,8		13,8	16,8	19,6	21,1	
52	2,15																					6,2	7,5	8,8	9,5		13,3	16,1	18,9	20,4	
53	2,2																					6,4	7,8	9,1	9,8		13,8	16,8	19,6	21,1	
54	2,25														0,6 ±0,1							6,6	8,1	9,4	10,2		14,2	17,4	20,2	21,9	
55	2,3																					6,9	8,4	9,8	10,6		14,8	18,1	21,1	22,8	
56	2,35																					7,1	8,7	10,1	10,9		15,3	18,7	21,7	23,4	
57	2,4																					7,3	8,9	10,5	11,3	±30% if F_t < 10 N	15,7	19,1	22,6	24,3	
58	2,4																					7,1	8,6	10,1	10,9		15,3	18,5	21,7	23,4	
59	2,45								0,3 max													7,3	8,9	10,4	11,3		15,7	19,1	22,4	24,3	
60	2,5																					7,1	8,6	10,1	10,8	±20% if F_t > 10 N	15,3	18,5	21,7	23,2	
61	2,55																					7,3	8,9	10,4	11,1		15,7	19,1	22,4	23,9	
62	2,6																					7,5	9,2	10,7	11,5		16,1	19,8	23	24,7	
63	2,65																					7,7	9,4	11	11,8		16,6	20,2	23,7	25,4	
64	2,65								0,2	+0,2 0						0,8 ±0,1	8					7,5	9,1	10,7	11,4		16,1	19,6	23	24,5	
65	2,7																					7,7	9,4	11	11		16,6	20,2	23,7	25,4	
66	2,75																					7,9	9,7	11,3	12,2		17	20,9	24,3	26,2	
67	2,8																					8,2	10	11,6	12,5		17,6	21,5	24,9	26,9	
68	2,85																					8,4	10,2	12	12,9		18,1	21,9	25,8	27,7	
69	2,9																					8,6	10,5	12,3	13,3		18,5	22,6	26,4	28,6	

This table is shown in ISO format. Commas represent decimal points.

TABLE 5—(CONTINUED)

Dimensions in millimeters

Nominal diameter d_1	Radial wall thickness "regular" a_1		Ring width h_1					Closed gap s_1		Radius r_3	Land width h_4					Groove depth a_4	Number of slots	Slot width c_1					Tangential force F_t, N					Diametral force F_d, N				
		Tolerance	Column 1	2	3	4	Tolerance		Tolerance		For h_1 shown in column 1	2	3	4			For h_1 shown in column 1	2	3	4	For h_1 shown in column 1	2	3	4	Tolerance	For h_1 shown in column 1	2	3	4	Tolerance
70	2,9																					8,4	10,2	11,9	12,9		18,1	21,9	25,6	27,7	
71	2,95																					8,6	10,5	12,3	13,2		18,5	22,6	26,4	28,4	
72	3							0,2	+0,2 0													8,8	10,8	12,6	13,6		18,9	23,2	27,1	29,2	
73	3,05																					9,1	11,1	12,9	14		19,6	23,9	27,7	30,1	
74	3,1		2,5	3	3,5	4				0,3 max	0,5 ±0,1	0,6 ±0,1	0,7 ±0,1	0,7 ±0,1	0,8 ±0,1			0,7 ±0,1	0,7 ±0,1	0,8 ±0,1	1 ±0,1	9,3	11,4	13,3	14,3		20	24,5	28,6	30,7	
75	3,15																					9,5	11,6	13,5	14,6		20,4	24,9	29	31,4	
76	3,15																					9,2	11,3	13,2	14,2		19,8	24,3	28,4	30,5	
77	3,2																					9,5	11,6	13,5	14,6		20,4	24,9	29	31,4	
78	3,25																					9,7	11,8	13,8	15		20,9	25,4	29,7	32,3	
79	3,3																					9,9	12,1	14,2	15,4		21,3	26	30,5	33,1	
80	3,35						−0,010 −0,030 For phosphated PO surface: +0,005 −0,025															11,9	13,9	14,9	16,8		25,6	29,9	32	36,1	
81	3,4																					12,2	14,3	15,3	17,2		26,2	30,7	32,9	37	
82	3,4	±0,15 Within a ring: 0,15 max						0,25	+0,25 0													11,9	13,9	14,9	16,8		25,6	29,9	32	36,1	
83	3,45																					12,2	14,2	15,3	17,2		26,2	30,5	32,9	37	
84	3,5																					12,5	14,6	15,7	17,6		26,9	31,4	33,8	37,8	
85	3,55																8					12,8	14,9	16	18		27,5	32	34,4	38,7	
86	3,6																					13	15,2	16,4	18,4		28	32,7	35,3	39,6	
87	3,65																					13,3	15,6	16,8	18,8		28,6	33,5	36,1	40,4	
88	3,65										0,6 ±0,1	0,7 ±0,1	0,7 ±0,1	0,8 ±0,1	1 ±0,1			0,7 ±0,1	0,8 ±0,1	1 ±0,1	1,2 ±0,1	13	15,2	16,4	18,4		28	32,7	35,3	39,6	
89	3,7		3	3,5	4	4,5																13,3	15,5	16,7	18,8		28,6	33,3	35,9	40,4	
90	3,75																					13,5	15,8	17	19,1	±30 % if $F_t < 10N$ ±20 % if $F_t > 10N$	29	34	36,6	41,1	±30 % if $F_d < 21,5N$ ±20 % if $F_d > 21,5N$
91	3,8																					13,8	16,1	17,4	19,6		29,7	34,6	37,4	42,1	
92	3,85																					14,1	16,5	17,8	20		30,3	35,3	38,3	43	
93	3,9																					14,4	16,8	18,2	20,4		31	36,1	39,1	43,9	
94	3,9																					14,1	16,4	17,8	20		30,3	35,3	38,3	43	
95	3,95									0,5 max												14,4	16,8	18,1	20,4		31	36,1	38,9	43,9	
96	4																					14,7	17,1	18,5	20,8		31,6	36,8	39,8	44,7	
97	4,05																					15	17,5	18,9	21,2		32,3	37,6	40,6	45,6	
98	4,1																					15,3	17,8	19,3	21,6		32,9	38,3	41,5	46,4	
99	4,15							0,3	+0,25 0													15,6	18,2	19,7	22,1		33,5	39,1	42,4	47,5	
100	4,15																					17,1	18,3	20,6	23,2		36,8	39,3	44,3	49,9	
101	4,2																					17,4	18,7	21	23,7		37,4	40,2	45,2	51	
102	4,25																					17,7	19	21,4	24,1		38,1	40,9	46	51,8	
103	4,25																					18	19,3	21,8	24,5		38,7	41,5	46,9	52,7	
104	4,3																					17,6	18,9	21,3	24		37,8	40,6	45,8	51,6	
105	4,35	±0,2 Within a ring: 0,20 max																				17,9	19,3	21,7	24,4		38,5	41,5	46,7	52,5	
106	4,4										0,7 ±0,1	0,7 ±0,1	0,8 ±0,1	0,9 ±0,1	1,2 ±0,1		10	0,8 ±0,1	1 ±0,1	1,2 ±0,1	1,2 ±0,1	18,2	19,6	22,1	24,8		39,1	42,1	47,5	53,3	
107	4,4		3,5	4	4,5	5																17,8	19,2	21,6	24,3		38,3	41,3	46,4	52,2	
108	4,45																					18,1	19,5	22	24,8		38,9	41,9	47,3	53,3	
109	4,5																					18,4	19,9	22,3	25,4		39,6	42,8	47,9	54,2	
110	4,55																					18,7	20,2	22,7	25,5		40,2	43,4	48,8	54,8	
111	4,55																					18,3	19,8	22,2	25		39,3	42,6	47,7	53,8	
112	4,6							0,35	+0,30 0													18,6	20,1	22,6	25,4		40	43,2	48,6	54,6	
113	4,65																					18,9	20,4	23	25,9		40,6	43,9	49,5	55,7	
114	4,7																					19,2	20,8	23,3	26,3		41,3	44,7	50,1	56,5	

This table is shown in ISO format. Commas represent decimal points.

TABLE 5—(CONTINUED)

Dimensions in millimeters

Nominal diameter d_1	Radial wall thickness "regular" a_1	Ring width h_1 Column 1	2	3	4	Tolerance	Closed gap s_1	Tolerance	Radius r_3	Land width h_4 For h_1 shown in column 1	2	3	4	Tolerance	Groove depth a_4	Number of slots	Slot width c_1 For h_1 shown in column 1	2	3	4	Tolerance	Tangential force F_t, N For h_1 shown in column 1	2	3	4	Tolerance	Diametral force F_d, N For h_1 shown in column 1	2	3	4	Tolerance
115	4,7																					18,8	20,4	22,9	25,8		40,4	43,9	49,2	55,5	
116	4,75									0,7	0,7	0,8	0,9									19,1	20,7	23,3	26,2		41,1	44,5	50,1	56,3	
117	4,8									±0,1	±0,1	±0,1	±0,1									19,4	21	23,6	26,6		41,7	45,2	50,7	57,2	
118	4,85																					19,1	20,6	23,2	26,1		41,1	44,3	49,9	56,1	
119	4,85	3,5	4	4,5	5				0,5 max						1,2 ±0,1		0,8	1	1,2	1,2		19,4	21	23,6	26,5		41,7	45,2	50,7	57	
120	4,9																	±0,1	±0,1	±0,1	±0,1		19,7	21,3	23,9	27		42,4	45,8	51,4	58,1	
121	4,95																						20	21,6	24,3	27,4		43	46,4	52,2	58,9	
122	4,95						0,35	+0,30 0														19,6	21,2	23,9	26,9		42,1	45,6	51,4	57,8	
123	5																					19,9	21,6	24,2	27,3		42,8	46,4	52	58,7	
124	5,05																					20,2	21,9	24,6	27,7		43,4	47,1	52,9	59,6	
125	5,05																					20,4	23	25,9	31,3		43,9	49,5	55,7	67,3	
126	5,1																					20,7	23,3	26,2	31,8		44,5	50,1	56,3	68,4	
127	5,15																10						21	23,7	26,6	32,2		45,2	51	57,2	69,2	
128	5,2																						21,4	24	27	32,7		46	51,6	58,1	70,3	
129	5,2							−0,010 −0,040														21	23,6	26,6	32,1		45,2	50,7	57,2	69	
130	5,25							For phosphated P0 surface: −0,005 −0,035															21,2	23,9	26,9	32,5		45,6	51,4	57,8	69,9	
131	5,3			±0,2 Within a ring: 0,20 max																			21,6	24,2	27,3	33		46,4	52	58,7	71	
132	5,3								0,7 max														21,2	23,8	26,8	32,4	±30% if F_t < 10N	45,6	51,2	57,6	69,7	±30% if F_d < 21,5N
133	5,35																						21,5	24,2	27,2	32,9	±20% if F_t > 10N	46,2	52	58,5	70,7	±20% if F_d ≥ 21,5N
134	5,4									0,7 ±0,1	0,8 ±0,1	0,9 ±0,1	1,1 ±0,1										21,8	24,5	27,6	33,4		46,9	52,7	59,3	71,8	
135	5,4															1,5 ±0,1		1 ±0,1	1,2 ±0,1	1,2 ±0,1	1,4 ±0,1		21,4	24,1	27,1	32,8		46	51,8	58,3	70,5	
136	5,45																						21,8	24,4	27,5	33,3		46,9	52,5	59,1	71,6	
137	5,5	4	4,5	5	6																		22,1	24,8	27,9	33,8		47,5	53,3	60	72,7	
138	5,5																						21,7	24,4	27,5	33,2		46,7	52,5	59,1	71,4	
139	5,55																						22	24,7	27,8	33,7		47,3	53,1	59,8	72,5	
140	5,6																						22,3	25,1	28,2	34,1		47,9	54	60,6	73,3	
141	5,65							+0,35															22,6	25,4	28,6	34,6		48,6	54,6	61,5	74,4	
142	5,65						0,4	0															22,3	25	28,2	34,1		47,9	53,8	60,6	73,3	
143	5,7																						22,6	25,3	28,5	34,5		48,6	54,4	61,3	74,2	
144	5,75																						22,9	25,7	28,9	35		49,2	55,3	62,1	75,3	
145	5,75																						22,5	25,3	28,5	34,4		48,4	54,4	61,3	74	
146	5,8																						22,8	25,6	28,9	34,9		49	55	62,1	75	
147	5,85																12						23,1	26	29,2	35,4		49,7	55,9	62,8	76,1	
148	5,85																						22,8	25,6	28,8	34,8		49	55	61,9	74,8	
149	5,9																						23,1	25,9	29,2	35,3		49,7	55,7	62,8	75,9	
150	5,95																						24,9	28	33,9	39,8		53,5	60,2	72,9	85,6	
152	6																						24,8	27,9	33,8	39,7		53,3	60	72,7	85,4	
154	6,05	4,5	5	6	7		0,5	+0,4 0		0,8 ±0,1	0,9 ±0,1	1,1 ±0,1	1,3 ±0,15		1,8 ±0,15			1,2 ±0,1	1,2 ±0,1	1,4 ±0,1	1,6 ±0,1		24,8	27,9	33,8	39,6		53,3	60	72,7	85,1	
155	6,1																						25,1	28,2	34,2	40,1		54	60,6	73,5	86,2	
156	6,15																						25,4	28,6	34,6	40,6		54,6	61,5	74,4	87,3	
158	6,2																						25,3	28,5	34,5	40,5		54,4	61,3	74,2	87,1	

This table is shown in ISO format. Commas represent decimal points.

TABLE 5—(CONCLUDED)

Dimensions in millimeters

Nominal diameter d_1	Radial wall thickness "regular" a_1	Ring width h_1 Column				Ring width Tolerance	Closed gap s_1	Closed gap Tolerance	Radius r_3	Land width h_4 For h_1 shown in column				Groove depth a_4	Number of slots	Slot width c_1 For h_1 shown in column				Tangential force F_t, N For h_1 shown in column				Tangential force Tolerance	Diametral force F_d, N For h_1 shown in column				Diametral force Tolerance
		1	2	3	4					1	2	3	4			1	2	3	4	1	2	3	4		1	2	3	4	
160	6,25																			25,3	28,5	34,5	40,5		54,4	61,3	74,2	87,1	
162	6,35																			25,9	29,2	35,3	41,5		55,7	62,8	75,9	89,2	
164	6,4																			25,9	29,1	35,2	41,4		55,7	62,6	75,7	89	
165	6,4	4,5	5	6	7		0,5	+0,4 / 0		0,8 ±0,1	0,9 ±0,1	1,1 ±0,1	1,3 ±0,15	1,8 ±0,15		1,2 ±0,1	1,2 ±0,1	1,4 ±0,1	1,6 ±0,1	25,5	28,7	34,8	40,8		54,8	61,7	74,8	87,7	
166	6,45																			25,8	29,1	35,2	41,3		55,5	62,6	75,7	88,8	
168	6,5																			25,8	29	35,1	41,2		55,5	62,4	75,5	88,6	
170	6,6																			26,6	29,7	36	42,2		56,8	63,9	77,4	90,7	
172	6,65																			26,3	29,7	35,9	42,1		56,5	63,9	77,2	90,5	
174	6,7																			26,3	29,6	35,8	42		56,5	63,6	77	90,3	
175	6,75																			29	35,1	41,3	48,7	±30% if $F_t < 10$N	62,4	75,5	88,8	104,7	±30% if $F_d < 21,5$N
176	6,8							-0,010 -0,040 For phosphated PO surface: +0,005 -0,035												29,3	35,5	41,7	49,2		63	76,3	89,7	105,8	
178	6,85	±0,2 Within a ring: 0,20 max													12					29,3	35,5	41,6	49,1		63	76,3	89,4	105,6	
180	6,9								0,7 max.											29,2	35,4	41,6	49	±20% if $F_t > 10$N	62,8	76,1	89,4	105,4	±20% if $F_d > 21,5$N
182	6,95																			29,2	35,3	41,5	48,9		62,8	75,9	89,2	105,1	
184	7,05																			29,9	36,1	42,4	50		64,3	77,6	91,2	107,5	
185	7,05																			29,5	35,7	41,9	49,4		63,4	76,8	90,1	106,2	
186	7,1	5	6	7	8		0,6	+0,45 / 0		0,9 ±0,1	1,1 ±0,1	1,3 ±0,15	1,6 ±0,15	2 ±0,15		1,2 ±0,1	1,4 ±0,1	1,6 ±0,1	1,8 ±0,1	29,8	36,1	42,3	49,9		64,1	77,6	90,9	107,3	
188	7,15																			29,8	36	42,3	49,8		64,1	77,4	90,9	107,1	
190	7,2																			29,7	36	42,2	49,7		63,9	77,4	90,7	106,9	
192	7,25																			29,7	35,9	42,1	49,6		63,9	77,2	90,5	106,6	
194	7,35																			30,3	36,7	43,1	50,7		65,1	78,9	92,7	109	
195	7,35																			29,9	36,2	42,5	50,1		64,3	77,8	91,4	107,7	
196	7,4																			30,3	36,6	43	50,6		65,1	78,7	92,5	108,8	
198	7,45																			30,2	36,6	42,9	50,5		64,9	78,7	92,2	108,6	
200	7,5																			30,2	36,5	42,9	50,4		64,9	78,5	92,2	108,4	

NOTES

1. For intermediate sizes (for example repair sizes), the radial wall thickness of the next smaller nominal diameter should be applied.

2. For values for F_t and F_d, given in Table 5, apply to as cast grey cast iron with a typical modulus of elasticity (E_n) of 100 000 MPa. Multiplying factors for materials having a different modulus (E_n) are given in SAE J1591.

   Mean forces are calculated for nominal radial wall thickness (a_1) and mean ring width (h_1).

3. For the sole purpose of this document, the assumed average ratio F_d/F_t is 2.15. However, for rings up to 50 mm, the ratio F_d/F_t shall be determined between the manufacturer and client.

This table is shown in ISO format. Commas represent decimal points.

TABLE 6—DIMENSIONS FOR G, D, AND DV OIL CONTROL RINGS

Dimensions in millimeters

Nominal diameter d_1	Radial wall thickness "regular" a_1	Ring width h_1 Column				Tolerance	Closed gap s_1	Tolerance	Radius r_3	Land width h_4 For h_1 shown in column				Land width h_5 For h_1 shown in column				Groove depth a_4	Number of slots	Slot width c_1 For h_1 shown in column				Tolerance	Tangential force F_t, N For h_1 shown in column				Diametral force F_d, N For h_1 shown in column				Tolerance
		1	2	3	4					1	2	3	4	1	2	3	4			1	2	3	4		1	2	3	4	1	2	3	4	
30	1,25					Tolerance ±0,15 Within a ring: 0,15 max													0,4 ±0,1											7,1	8,2	8,8	9,5	
31	1,3																													7,5	8,6	9,5	10,3	
32	1,35																													8	9,2	10,1	11	
33	1,4																													8,4	9,9	11	11,8	
34	1,4																													8	9,2	10,3	11,2	
35	1,45																													8,4	9,9	11	11,8	
36	1,5																													9	10,3	11,6	12,7	
37	1,55								0,2 max																	—	—	—	—	9,5	11	12,3	13,3	
38	1,6																													9,9	11,6	12,9	14,2	
39	1,65																													10,5	12,3	13,8	14,8	
40	1,65																													9,5	11	12,3	13,1	
41	1,7																													9,9	11,6	12,9	14	
42	1,75					0,15		+0,2 0											0,5 ±0,1	6										10,3	12	13,5	14,6	
43	1,8																													11	12,7	14,2	15,5	
44	1,85																													11,4	13,3	14,8	16,1	
45	1,9																													11,6	13,8	15,5	16,8	±30% if F_d < 21,5 N
46	1,9																													11,2	13,1	14,6	15,9	
47	1,95					–0,010 –0,030 For phosphated PO surface: +0,005 –0,025																				—	—	—	—	11,6	13,8	15,5	16,8	
48	2																													12,3	14,4	16,1	17,4	
49	2,05	2,5	3	3,5	4																									12,7	14,8	16,8	18,3	
50	2,1									0,5 ±0,1	0,6 ±0,1	0,7 ±0,1	0,7 ±0,1	0,28 ±0,08	0,28 ±0,08	0,28 ±0,08	0,28 ±0,08			0,7 ±0,1	0,7 ±0,1	0,8 ±0,1	1 ±0,1		5,9	6,9	7,8	8,4	12,7	14,8	16,8	18,1	±20% if F_d > 21,5 N
51	2,15																									6,2	7,2	8,1	8,8	13,3	15,5	17,4	18,9	
52	2,15																		0,6 ±0,1							5,9	6,9	7,8	8,4	12,7	14,8	16,8	18,1	
53	2,2																									6,1	7,2	8,1	8,8	13,1	15,5	17,4	18,9	
54	2,25																									6,4	7,5	8,4	9,2	13,8	16,1	18,1	19,8	
55	2,3																									6,6	7,8	8,8	9,5	14,2	16,8	18,9	20,4	
56	2,35																									6,8	8,1	9,1	9,9	14,6	17,4	19,6	21,3	±30% if F_t < 10N
57	2,4																									7,1	8,4	9,4	10,3	15,3	18,1	20,2	22,2	
58	2,4																									6,8	8,1	9,1	9,9	14,6	17,4	19,6	21,3	
59	2,45								0,3 max																	7,1	8,4	9,4	10,3	15,3	18,1	20,2	22,1	
60	2,5						0,2	+0,2 0																		6,8	8	9	9,7	14,6	17,2	19,4	20,9	±20% if F_t > 10N
61	2,55																									7	8,3	9,3	10	15,1	17,8	20	21,5	
62	2,6																									7,2	8,5	9,6	10,4	15,5	18,3	20,6	22,4	
63	2,65																									7,5	8,8	10	10,7	16,1	18,9	21,5	23	
64	2,65																		0,8 ±0,1	8						7,2	8,5	9,6	10,4	15,5	18,3	20,6	22,4	
65	2,7																									7,5	8,8	10	10,7	16,1	18,9	21,5	23	
66	2,75																									7,7	9,1	10,3	11,1	16,6	19,6	22,1	23,9	
67	2,8																									7,9	9,4	10,6	11,4	17	20,2	22,8	24,5	
68	2,85																									8,1	9,6	10,9	11,8	17,4	20,6	23,4	25,4	
69	2,9																									8,4	9,9	11,2	12,2	18,1	21,3	24,1	26,2	

This table is shown in ISO format. Commas represent decimal points.

TABLE 6—(CONTINUED)

Dimensions in millimeters

Nominal diameter d_1	Radial wall thickness "regular" a_1	Ring width h_1 Column 1	2	3	4	Tolerance	Closed gap s_1	Tolerance	Radius r_3	Land width h_4 For h_1 shown in column 1	2	3	4	Tolerance	Land width h_5 For h_1 shown in column 1	2	3	4	Tolerance	Groove depth a_4	Number of slots	Slot width c_1 For h_1 shown in column 1	2	3	4	Tolerance	Tangential force F_t, N For h_1 shown in column 1	2	3	4	Tolerance	Diametral force F_d, N For h_1 shown in column 1	2	3	4	Tolerance
70	2,9						0,2	+0,2 / 0	0,3 max	0,5	0,6	0,7	0,7	±0,1	0,28	0,28	0,28	0,28	±0,08	0,8		0,7	0,7	0,8	1	±0,1	8,1	9,6	10,9	11,8		17,4	20,6	23,4	25,4	
71	2,95																			±0,1		±0,1	±0,1	±0,1	±0,1		8,4	9,9	11,2	12,2		18,1	21,3	24,1	26,2	
72	3	2,5	3	3,5	4																						8,6	10,2	11,6	12,5		18,5	21,9	24,9	26,9	
73	3,05																										8,8	10,5	11,9	12,9		18,9	22,6	25,6	27,7	
74	3,1																										9,1	10,8	12,2	13,3		19,6	26,2	26,2	28,6	
75	3,15																										9,2	11	12,5	13,5		19,8	23,7	26,9	29	
76	3,15																										9	10,7	12,1	13,2		19,4	23	26	28,4	
77	3,2																										9,2	11	12,5	13,5		19,8	23,7	26,9	29	
78	3,25																										9,4	11,2	12,8	13,9		20,2	24,1	27,5	29,9	
79	3,3																										9,7	11,5	13,1	14,3		20,9	24,7	28,2	30,7	
80	3,35					−0,010 / −0,030 For phosphated PO surface: +0,005 / −0,025	0,25	+0,25 / 0																			11,3	12,8	13,8	15,1		24,3	27,5	29,7	32,5	
81	3,4																										11,6	13,2	14,2	15,5		24,9	28,4	30,5	33,3	
82	3,4	±0,15 Within a ring: 0,15 max																									11,3	12,8	13,8	15,1		24,3	27,5	29,7	32,5	
83	3,45																										11,6	13,2	14,2	15,5		24,9	28,4	30,5	33,3	
84	3,5																										11,9	13,5	14,6	15,9		25,6	29	31,4	34,2	
85	3,55																					8					12,1	13,8	14,9	16,3		26	29,7	32	35	
86	3,6																										12,4	14,1	15,3	16,7		26,7	30,3	32,9	35,9	
87	3,65																										12,7	14,5	15,6	17,1		27,3	31,2	33,5	36,8	
88	3,65																										12,4	14,1	15,3	16,8		26,7	30,3	32,9	36,1	
89	3,7										0,6	0,7	0,7	0,8	±0,1	0,28	0,28	0,28	0,28	±0,08	1 ±0,1		0,7	0,8	1	1,2	±0,1	12,7	14,5	15,6	17,2	±30% if F_t < 10N	27,3	31,2	33,5	37	±30% if F_d < 21,5N
90	3,75	3	3,5	4	4,5																		±0,1	±0,1	±0,1	±0,1		12,9	14,7	15,9	17,5		27,7	31,6	34,2	37,6	
91	3,8																											13,2	15,1	16,3	17,9		28,4	32,5	35	38,5	
92	3,85																											13,5	15,4	16,7	18,3	±20% if F_t > 10N	29	33,1	35,9	39,3	±20% if F_d > 21,5N
93	3,9																											13,8	15,7	17,1	18,7		29,7	33,8	36,8	40,2	
94	3,9																											13,5	15,4	16,7	18,3		29	33,1	35,9	39,3	
95	3,95							0,5 max																			13,8	15,7	17,1	18,7		29,7	33,8	36,8	40,2	
96	4																										14,1	16,1	17,4	19,2		30,3	34,6	37,4	41,4	
97	4,05																										14,4	16,4	17,8	19,6		31	35,3	38,3	42,1	
98	4,1																										14,6	16,7	18,2	20		31,4	35,9	39,1	43	
99	4,15						0,3	+0,25 / 0																			14,9	17,1	18,6	20,4		32	36,8	40	43,9	
100	4,15																											16	17,2	18,9	20,6		34,4	37	40,6	44,7	
101	4,2																											16,3	17,6	19,3	21,3		35	37,8	41,5	45,8	
102	4,25																											16,6	17,9	19,7	21,7		35,7	38,5	42,4	46,7	
103	4,25																											16,9	18,2	20	22,1		36,3	39,1	43	47,5	
104	4,3																											16,5	17,8	19,6	21,6		35,5	38,3	42,1	46,4	
105	4,35																											16,8	18,2	20	22,1		36,1	39,1	43	47,5	
106	4,4	±0,2 Within a ring: 0,20 max	3,5	4	4,5	5					0,7	0,7	0,8	0,9	±0,1	0,28	0,28	0,28	0,28	±0,08	1,2 ±0,1	10	0,8	1	1,2	1,2	±0,1	17,1	18,5	20,4	22,5		36,8	39,8	43,9	48,4	
107	4,4																							±0,1	±0,1	±0,1	±0,1		16,8	18,1	20	22		36,1	38,9	43	47,3	
108	4,45																											17,1	18,5	20,3	22,4		36,8	39,8	43,6	48,2	
109	4,5																											17,4	18,8	20,7	22,9		37,4	40,4	44,5	49,2	
110	4,55																											17,6	19,1	21	23,2		37,8	41,1	45,2	49,9	
111	4,55																											17,3	18,7	20,6	22,7		37,2	40,2	44,3	48,8	
112	4,6						0,35	+0,30 / 0																			17,6	19	21	23,2		37,8	40,9	45,2	49,9	
113	4,65																											17,9	19,4	21,3	23,6		38,5	41,7	45,8	50,7	
114	4,7																											18,2	19,7	21,7	24		39,1	42,4	46,7	51,6	

This table is shown in ISO format. Commas represent decimal points.

TABLE 6—(CONTINUED)

Dimensions in millimeters

Nominal diameter d_1	Radial wall thickness "regular" a_1	Tolerance	Ring width h_1 Column 1	2	3	4	Tolerance	Closed gap s_1	Tolerance	Radius r_3	Land width h_4 For h_1 shown in column 1	2	3	4	Land width h_5 For h_1 shown in column 1	2	3	4	Groove depth a_4	Number of slots	Slot width c_1 For h_1 shown in column 1	2	3	4	Tangential force F_t, N For h_1 shown in column 1	2	3	4	Tolerance	Diametral force F_d, N For h_1 shown in column 1	2	3	4	Tolerance
115	4,7																								17,8	19,3	21,3	23,5		38,3	41,5	45,8	50,5	
116	4,75																								18,1	19,6	21,7	24		38,9	42,1	46,7	51,6	
117	4,8																								18,4	20	22	24,4		39,6	43	47,3	52,5	
118	4,85										0,5 max														18,1	19,6	21,6	23,9		38,9	42,1	46,4	51,4	
119	4,85		3,5	4	4,5	5					0,7 ±0,1	0,7 ±0,1	0,8 ±0,1	0,9 ±0,1	0,28 ±0,08	0,28 ±0,08	0,28 ±0,08	0,28 ±0,08	1,2 ±0,1		0,8 ±0,1	1 ±0,1	1,2 ±0,1	1,2 ±0,1	18,4	19,9	22	24,3		39,6	42,8	47,3	52,2	
120	4,9																								18,7	20,3	22,3	24,7		40,2	43,6	47,9	53,1	
121	4,95							0,35	+0,30 0																19	20,6	22,7	25,2		40,9	44,3	48,8	54,2	
122	4,95																								18,6	20,2	22,3	24,7		40	43,4	47,9	53,1	
123	5																								18,9	20,6	22,7	25,1		40,6	44,3	48,8	54	
124	5,05																								19,2	20,9	23	25,5		41,3	44,9	49,5	54,8	
125	5,05																								19,4	21,4	23,6	28,1		41,7	46	50,7	60,4	
126	5,1																				10					19,7	21,7	24	28,5		42,4	46,7	51,6	61,3	
127	5,15						−0,010																			20	22,1	24,4	29		43	47,5	52,5	62,4	
128	5,2						−0,040																			20,3	22,4	24,8	29,5		43,6	48,2	53,3	63,4	
129	5,2						For phosphated PO surface:																			20	22	24,3	29		43	47,3	52,2	62,4	
130	5,25																									20,2	22,3	24,7	29,3		43,4	47,9	53,1	63	
131	5,3						+0,005																			20,5	22,6	25,1	29,8		44,1	48,6	54	64,1	±30% if F_d < 21,5N
132	5,3	±0,2 Within a ring: 0,20 max					−0,035																			20,2	22,3	24,6	29,3	±30% if F_t < 10N	43,4	47,9	52,9	63	
133	5,35																									20,5	22,6	25	29,8		44,1	48,6	53,8	64,1	
134	5,4																									20,8	23	25,4	30,2		44,7	49,5	54,6	64,9	±20% if F_d > 21,5N
135	5,4																									20,4	22,6	25	29,7		43,9	48,6	53,8	63,9	
136	5,45		4	4,5	5	6					0,7 ±0,1	0,8 ±0,1	0,9 ±0,1	1,1 ±0,1	0,28 ±0,08	0,28 ±0,08	0,28 ±0,08	0,35 ±0,1	1,5 ±0,1		1 ±0,1	1,2 ±0,1	1,2 ±0,1	1,4 ±0,1	20,8	22,9	25,4	30,2	±20% if F_t > 10N	44,7	49,2	54,6	64,9	
137	5,5																									21,1	23,3	25,7	30,6		45,4	50,1	55,3	65,8	
138	5,5							0,4	+0,35 0	0,7 max															20,7	22,9	25,3	30,2		44,5	49,2	54,4	64,9	
139	5,55																									21	23,2	25,7	30,6		45,2	49,9	55,3	65,8	
140	5,6																									21,3	23,6	26,1	31,1		45,8	50,7	56,1	66,9	
141	5,65																									21,6	23,9	26,5	31,5		46,4	51,4	57	67,7	
142	5,65																									21,3	23,5	26,1	31		45,8	50,5	56,1	66,7	
143	5,7																									21,6	23,9	26,4	31,5		46,4	51,4	56,8	67,7	
144	5,75																									21,9	24,2	26,8	31,9		47,1	52	57,6	68,6	
145	5,75																									21,6	23,8	26,4	31,4		46,4	51,2	56,8	67,5	
146	5,8																									21,9	24,2	26,8	31,9		47,1	52	57,6	68,6	
147	5,85																				12					22,2	24,5	27,2	32,4		47,7	52,7	58,5	69,7	
148	5,85																									21,8	24,1	26,7	31,9		46,9	51,8	57,4	68,6	
149	5,9																									22,1	24,5	27,1	32,3		47,5	52,7	58,3	69,4	
150	5,95																									23,4	25,9	30,8	35,5		50,3	55,7	66,2	76,3	
152	6																									23,3	25,8	30,8	35,4		50,1	55,5	66,2	76,1	
154	6,05		4,5	5	6	7		0,5	+0,4 0		0,8 ±0,1	0,9 ±0,1	1,1 ±0,1	1,3 ±0,15	0,35 ±0,1	0,35 ±0,1	0,35 ±0,1	0,4 ±0,1	1,8 ±0,15		1,2 ±0,1	1,2 ±0,1	1,4 ±0,1	1,6 ±0,1	23,3	25,8	30,8	35,4		50,1	55,5	66,2	76,1	
155	6,1																									23,6	26,2	31,2	35,9		50,7	56,3	67,1	77,2	
156	6,15																									23,9	26,5	31,6	36,4		51,4	57	67,9	78,3	
158	6,2																									23,9	26,5	31,6	36,4		51,4	57	67,9	78,3	

This table is shown in ISO format. Commas represent decimal points.

26.127

TABLE 6—(CONCLUDED)

Dimensions in millimeters

Nominal diameter d_1	Radial wall thickness "regular" a_1		Ring width h_1					Closed gap s_1		Radius r_3	Land width h_4 For h_1 shown in column				Land width h_5 For h_1 shown in column				Groove depth a_4	Number of slots	Slot width c_1 For h_1 shown in column				Tangential force F_t, N For h_1 shown in column					Diametral force F_d, N For h_1 shown in column				
		Tolerance	Column 1	2	3	4	Tolerance		Tolerance		1	2	3	4	1	2	3	4			1	2	3	4	1	2	3	4	Tolerance	1	2	3	4	Tolerance
160	6,25																								23,9	26,5	31,5	36,3		51,4	57	67,7	78	
162	6,35																								24,5	27,1	32,4	37,3		52,7	58,3	69,7	80,2	
164	6,4																								24,4	27,1	32,4	37,3		52,5	58,3	69,7	80,2	
165	6,4							0,5	+0,4 / 0			0,8 ±0,1	0,9 ±0,1	1,1 ±0,1	1,3 ±0,15	0,35 ±0,1	0,35 ±0,1	0,35 ±0,1	0,4 ±0,1	1,8 ±0,15		1,2 ±0,1	1,2 ±0,1	1,4 ±0,1	1,6 ±0,1	24,1	26,7	31,9	36,8		51,8	57,4	68,6	79,1	
166	6,45		4,5	5	6	7																			24,4	27,1	32,3	37,3		52,5	58,3	69,4	80,2	
168	6,5																								24,4	27,1	32,3	37,2		52,5	58,3	69,4	80	
170	6,6																								25	27,7	33,1	38,2		53,8	59,6	71,2	82,1	
172	6,65																								25	27,7	33,1	38,2		53,8	59,6	71,2	82,1	
174	6,7																								24,9	27,7	33,1	38,1		53,5	59,6	71,2	81,9	
175	6,75	±0,2 Within a ring: 0,20 max							−0,010 −0,040 For phosphated PO surface: −0,005 −0,035		0,7 max										12					27,1	32,3	37,3	42,9	±30% if F_t < 10N	58,3	69,4	80,2	92,2	±30% if F_d < 21,5N
176	6,8																								27,4	32,7	37,7	43,5		58,9	70,3	81,1	93,5	
178	6,85																								27,4	32,7	37,7	43,4		58,9	70,3	81,1	93,3	
180	6,9																								27,3	32,6	37,7	43,4	±20% if F_t > 10N	58,7	70,1	81,1	93,3	±20% if F_d > 21,5N
182	6,95																								27,3	32,6	37,6	43,4		58,7	70,1	80,8	93,3	
184	7,05																								28	33,4	38,6	44,4		60,2	71,8	83	95,5	
185	7,05							0,6	+0,45 / 0			0,9 ±0,1	1,1 ±0,1	1,3 ±0,15	1,6 ±0,15	0,35 ±0,1	0,35 ±0,1	0,4 ±0,1	0,5 ±0,1	2 ±0,15		1,2 ±0,1	1,4 ±0,1	1,6 ±0,1	1,8 ±0,1	27,6	33	38,1	43,9		59,3	71	81,9	94,4	
186	7,1		5	6	7	8																			27,9	33,4	38,5	44,4		60	71,8	82,8	95,5	
188	7,15																								27,9	33,3	38,5	44,4		60	71,6	82,8	95,5	
190	7,2																								27,9	33,3	38,5	44,3		60	71,6	82,8	95,2	
192	7,25																								27,9	33,3	38,4	44,3		60	71,6	82,6	95,2	
194	7,35																								28,5	34,1	39,3	45,3		61,3	73,3	84,5	97,4	
195	7,35																								28,1	33,6	38,9	44,8		60,4	72,2	83,6	96,3	
196	7,4																								28,5	34	39,3	45,3		61,3	73,1	84,5	97,4	
198	7,45																								28,4	34	39,3	45,3		61,1	73,1	84,5	97,4	
200	7,5																								28,4	34	39,3	45,2		61,1	73,1	84,5	97,2	

NOTES

1 For intermediate sizes (for example repair sizes), the radial wall thickness of the next smaller nominal diameter should be applied.

2 For values for F_t and F_d, given in Table 6, apply to as cast grey cast iron with a typical modulus of elasticity (E_n) of 100 000 MPa. Multiplying factors for materials having a different modulus (E_n) are given in SAE J1591.

Mean forces are calculated for nominal radial wall thickness (a_1) and mean ring width (h_1).

3 For the sole purpose of this document, the assumed average ratio F_d/F_t is 2.15. However, for rings up to 50 mm, the ratio F_d/F_t shall be determined between the manufacturer and client.

This table is shown in ISO format. Commas represent decimal points.

TABLE 7—DIMENSIONS FOR S, G, D, AND DV OIL CONTROL RINGS WITH SPECIAL RING WIDTH h_1 = 4.75 mm (3/16 in)

Dimensions in millimeters

Nominal diameter d_1	Radial wall thickness "regular" a_1 Tolerance	Ring width h_1 Tolerance	Closed gap s_1 Tolerance	Radius r_3	Land width h_4	Land width h_5	Groove depth a_4	Number of slots	Slot width c_1	Type S Tangential force F_t, N Tolerance	Type S Diametral force F_d, N Tolerance	Types G and D Tangential force F_t, N Tolerance	Types G and D Diametral force F_d, N Tolerance
50	2,1									11,1	23,9	9,5	20,4
51	2,15									11,5	24,7	9,9	21,3
52	2,15									11	23,7	9,5	20,4
53	2,2			0,15	+0,2 / 0			0,6 ±0,1	6	11,5	24,7	10	21,5
54	2,25									11,9	25,6	10,4	22,4
55	2,3									12,3	26,4	10,8	23,2
56	2,35									12,8	27,5	11,2	24,1
57	2,4									13,2	28,4	11,7	25,2
58	2,4									12,8	27,5	11,3	24,3
59	2,45									13,2	28,4	11,7	25,2
60	2,5									12,6	27,1	10,9	23,4
61	2,55									13	28	11,3	24,3
62	2,6									13,4	28,8	11,7	25,2
63	2,65									13,8	29,7	12,1	26
64	2,65				0,2 max					13,4	28,8	11,8	25,4
65	2,7									13,8	29,7	12,2	26,2
66	2,75		−0,010 / −0,030 For phosphated PO surface: +0,005 / −0,025							14,2	30,5	12,6	27,1
67	2,8	±0,15 Within a ring: 0,15 max		0,2	+0,2 / 0					14,6	31,4	13	28
68	2,85									15,1 ±30% if F_t < 10N	32,5 ±30% if F_d < 21,5N	13,4 ±30% if F_t < 10N	28,8 ±30% if F_d < 21,5N
69	2,9		4,75			0,8 ±0,1	0,28 ±0,08	0,8 ±0,1	1,2 ±0,1	14,6	31,4	13,8	29,7
70	2,9									15 ±20% if F_t > 10N	32,3 ±20% if F_d > 21,5N	13,4 ±20% if F_t > 10N	28,8 ±20% if F_d > 21,5N
71	2,95								8	15,5	33,3	13,9	29,9
72	3									15,9	34,2	14,3	30,7
73	3,05									16,3	35	14,7	31,6
74	3,1									16,8	36,1	15,1	32,5
75	3,15									17,1	36,8	15,5	33,3
76	3,15									16,7	35,9	15,1	32,5
77	3,2									17,1	36,8	15,5	33,3
78	3,25									17,5	37,6	15,9	34,2
79	3,3									18	38,7	16,4	35,3
80	3,35									17,4	37,4	15,7	33,8
81	3,4									17,9	38,5	16,2	34,8
82	3,4			0,25	+0,25 / 0					17,4	37,4	15,8	34
83	3,45									17,9	38,5	16,2	34,8
84	3,5				0,5 max			1 ±0,1		18,3	39,3	16,6	35,7
85	3,55									18,7	40,2	17	36,6
86	3,6									19,1	41,1	17,4	37,4
87	3,65									19,6	42,1	17,9	38,5
88	3,65									19,1	41,1	17,5	37,6
89	3,7									19,6	42,1	17,9	38,5

This table is shown in ISO format. Commas represent decimal points.

TABLE 7—(CONTINUED)

Dimensions in millimeters

Nominal diameter d_1	Radial wall thickness "regular" a_1	Tolerance	Ring width h_1	Tolerance	Closed gap s_1	Tolerance	Radius r_3	Land width h_4	Land width h_5	Groove depth a_4	Number of slots	Slot width c_1	Type S Tangential force F_t, N	Tolerance	Type S Diametral force F_d, N	Tolerance	Types G and D Tangential force F_t, N	Tolerance	Types G and D Diametral force F_d, N	Tolerance
90	3,75												19,9		42,8		18,2		39,1	
91	3,8												20,4		43,9		18,7		40,2	
92	3,85	±0,15 Within a ring: 0,15 max								1 ±0,1	8		20,8		44,7		19,1		41,1	
93	3,9												21,2		45,6		19,5		41,9	
94	3,9		4,75										20,8		44,7		19,1		41,1	
95	3,95												21,2		45,6		19,6		42,1	
96	4												21,7		46,7		20		43	
97	4,05												22,1		47,5		20,4		43,9	
98	4,1												22,6		48,6		20,9		44,9	
99	4,15				0,3	+0,25 / 0							23		49,5		21,3		45,8	
100	4,15												21,4		46		19,7		42,4	
101	4,2												21,8		46,9		20,1		43,2	
102	4,25												22,2		47,7		20,5		44,1	
103	4,3												22,6		48,6		20,9		44,9	
104	4,3												22,1		47,5		20,4		43,9	
105	4,35												22,5		48,4		20,8		44,7	
106	4,4						0,5 max						22,9		49,2		21,2		45,6	
107	4,4												22,5		48,4		20,8		44,7	
108	4,45							0,8 ±0,1	0,28 ±0,08			1,2 ±0,1	22,9	±30 % if F_t < 10N	49,2	±30 % if F_d < 21,5N	21,2	±30 % if F_t < 10N	45,6	±30 % if F_d < 21,5N
109	4,5												23,2		49,9		21,6		46,4	
110	4,55			-0,010 / -0,030 For phosphated PO surface: +0,005 / -0,025									23,6	±20 % if F_t ≥ 10N	50,7	±20 % if F_d ≥ 21,5N	21,9	±20 % if F_t ≥ 10N	47,1	±20 % if F_d ≥ 21,5N
111	4,55									1,2 ±0,1			23,1		49,7		21,5		46,2	
112	4,6	±0,2 Within a ring: 0,20 max											23,5		50,5		21,9		47,1	
113	4,65												23,9		51,4		22,3		47,9	
114	4,7		4,75								10		24,3		52,2		22,6		48,6	
115	4,7												23,8		51,2		22,2		47,7	
116	4,75												24,2		52		22,6		48,6	
117	4,8												24,6		52,9		23		49,5	
118	4,8												24,2		52		22,6		48,6	
119	4,85				0,35	+0,30 / 0							24,5		52,7		23		49,5	
120	4,9												24,9		53,5		23,3		50,1	
121	4,95												25,3		54,4		23,7		51	
122	4,95												24,9		53,5		23,3		50,1	
123	5												25,3		54,4		23,7		51	
124	5,05												25,6		55		24,1		51,8	
125	5,05												23,8		51,2		22,2		47,7	
126	5,1						0,7 max			1,5 ±0,1			24,2		52		22,6		48,6	
127	5,15												24,6		52,9		22,9		49,2	
128	5,2												24,9		53,5		23,3		50,1	
129	5,2												24,5		52,7		22,9		49,2	

This table is shown in ISO format. Commas represent decimal points.

TABLE 7—(CONCLUDED)

Dimensions in millimeters

Nominal diameter d_1	Radial wall thickness "regular" a_1	Tolerance	Ring width h_1	Tolerance	Closed gap s_1	Tolerance	Radius r_3	Land width h_4	Land width h_5	Groove depth a_4	Number of slots	Slot width c_1	Type S Tangential force F_t, N	Tolerance	Type S Diametral force F_d, N	Tolerance	Types G and D Tangential force F_t, N	Tolerance	Types G and D Diametral force F_d, N	Tolerance
130	5,25												24,8		53,3		23,2		49,9	
131	5,3												25,2		54,2		23,6		50,7	
132	5,3												24,7		53,1		23,2		49,9	
133	5,35												25,1		54		23,5		50,5	
134	5,4										10		25,5		54,8		23,9		51,4	
135	5,4												25		53,8		23,5		50,5	
136	5,45												25,4		54,6		23,9		51,4	
137	5,5			−0,010									25,8		55,5		24,2		52	
138	5,5	±0,2		−0,040									25,3		54,4		23,8		51,2	
139	5,55	Within a ring: 0,20 max		For phosphated PO surface: +0,005 −0,035	0,4	+0,35 0	0,7 max	0,8 ±0,1	0,28 ±0,08	1,5 ±0,1		1,2 ±0,1	25,7	±30 % if $F_t < 10$ N	55,3	±30 % if $F_d < 21,5$ N	24,2	±30 % if $F_t < 10$ N	52	±30 % if $F_d < 21,5$ N
140	5,6		4,75										26,1	±20 % if $F_t > 10$ N	56,1	±20 % if $F_d > 21,5$ N	24,5	±20 % if $F_t > 10$ N	52,7	±20 % if $F_d > 21,5$ N
141	5,65												26,4		56,8		24,9		53,5	
142	5,65												26		55,9		24,5		52,7	
143	5,7												26,4		56,8		24,9		53,5	
144	5,75										12		26,7		57,4		25,2		54,2	
145	5,75												26,3		56,5		24,8		53,3	
146	5,8												26,7		57,4		25,2		54,2	
147	5,85												27		58,1		25,5		54,8	
148	5,85												26,6		57,2		25,1		54	
149	5,9												27		58,1		25,5		54,8	
150	5,95				0,5	+0,4 0				1,8 ±0,1			25,8		55,5		24,3		52,2	

NOTES

1. For intermediate sizes (for example repair sizes), the radial wall thickness of the next smaller nominal diameter should be applied.

2. For values for F_t and F_d, given in Table 7, apply to as cast grey cast iron with a typical modulus of elasticity (E_n) of 100 000 MPa. Multiplying factors for materials having a different modulus (E_n) are given in SAE J1591.

   Mean forces are calculated for nominal radial wall thickness (a_1) and mean ring width (h_1).

3. For the sole purpose of this document, the assumed average ratio F_d/F_t is 2.15.

This table is shown in ISO format. Commas represent decimal points.

APPENDIX A

A.1 This SAE Standard has been established to harmonize the ISO and SAE piston ring documents. The U.S. Technical Advisory Group, with the support of the National Engine Parts Manufacturers Association, has worked with other national organizations on this worldwide document. Some of the wording and phrasing may differ slightly from U.S. terminology for translation purposes.

In preparing this SAE document, the Scope and Field of Application, and Reference sections of the ISO 6625 have been editorially revised and reorganized.

Certain closed gap and ring width dimensions were changed to conform with established U.S. standard practice (Tables 5, 6, and 7).

The tolerances specified in this document represent a six sigma quality level.

INTERNAL COMBUSTION ENGINES—PISTON RINGS—COIL SPRING LOADED OIL CONTROL RINGS—SAE J2003 OCT92

SAE Standard

Report of the Piston Ring Standards Committee approved September 1990. Revised by the SAE Piston Ring Standards Committee 7 October 1992.

This SAE Standard is equivalent to ISO Standard 6626.

1. Scope and Field of Application—This SAE Standard is equivalent to ISO Standard 6626.

Differences, where they exist, are shown in the appendix with associated rationale.

This document specifies the essential dimensions of piston ring types DSF-C, DSF-CNP, SSF, GSF, DSF, DSF-NG, and SSF-L coil spring loaded oil control rings.

For the cast iron part the recommended material is class 10 according to SAE J1590. For special applications material classes 20 to 50 may be used.

Variation in face design and spring groove from these may be used, as recommended by individual manufacturers, in plain or chromed versions.

The tangential forces of coil spring loaded oil control rings can be varied over a wide range. Explanations and recommendations are given in Section 6.

The normal range for axial width of coil spring loaded oil control rings (3 to 8 mm inclusive) is divided into 0.5 or 1.0 mm steps. In Tables 15 to 20 dimensions are given for coil spring loaded oil control rings within an axial width of 4.75 mm (i.e., 3/16 in) for existing applications in inch units.

This document applies to coil spring loaded oil control rings up to 200 mm inclusive for reciprocating internal combustion engines. It may also be used for piston rings of compressors working under similar conditions.

2. References

SAE DESIGNATION	ISO[1] EQUIVALENT	
		INTERNAL COMBUSTION ENGINES—PISTON RINGS
J1588	6621/1	Vocabulary
J1589	6621/2	Measuring principles
J1590	6621/3	Material specifications
J1591	6621/4	General specifications
J1996	6621/5	Quality requirements
		INTERNAL COMBUSTION ENGINES—PISTON RINGS
J1997	6622/1	Rectangular rings
J1998	6622/2 TR	Rectangular rings with narrow ring width
J1999	6623	INTERNAL COMBUSTION ENGINES—PISTON RINGS —SCRAPER RINGS
J2000	6624/1	INTERNAL COMBUSTION ENGINES—PISTON RINGS Keystone rings
J2001	6624/2 TR	Half keystone rings
J2002	6625	INTERNAL COMBUSTION ENGINES—PISTON RINGS —OIL CONTROL RINGS
J2003	6626	INTERNAL COMBUSTION ENGINES—PISTON RINGS —COIL SPRING LOADED OIL CONTROL RINGS
J2004	6627 TR	INTERNAL COMBUSTION ENGINES—PISTON RINGS —EXPANDER/SEGMENT OIL CONTROL RINGS
J2226		INTERNAL COMBUSTION ENGINES—PISTON RINGS —STEEL RECTANGULAR RINGS
	ISO 1101	TECHNICAL DRAWINGS —Tolerancing of form, orientation, location and run-out—Generalities, definitions, symbols indications on drawings

[1] TR refers to Technical Report

3. Piston Ring Types and Designation Examples

3.1 Type DSF-C—Coil Spring Loaded Bevelled Edge Oil Control Ring, Chromium-Plated and Profile Ground

3.1.1 GENERAL FEATURES

NOTE—See Tables 9 and 15 for dimensions and forces.

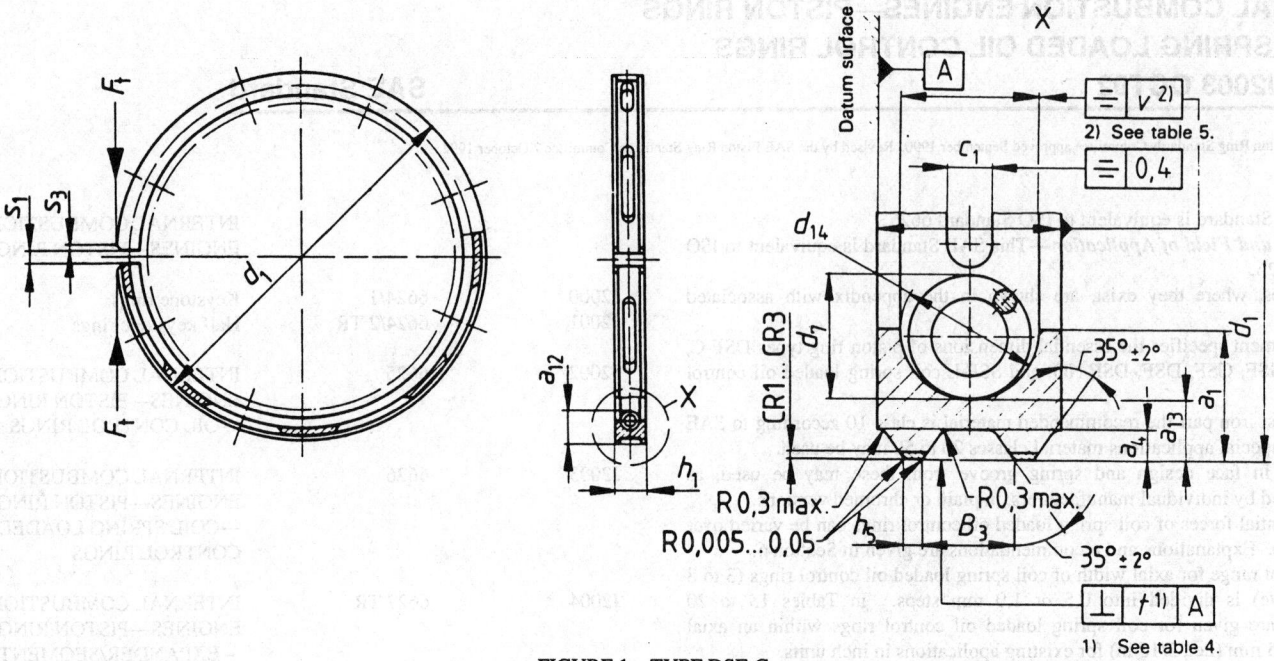

FIGURE 1—TYPE DSF-C

3.1.2 DESIGNATION EXAMPLE—Designation of a piston ring complying with the requirements of SAE J2003 being a coil spring loaded bevelled edge oil control ring, chromium-plated and profile ground (DSF-C), of nominal diameter $d_1 = 125$ mm (125), a nominal ring width $h_1 = 5$ mm (5), made of grey cast iron, nonheat-treated, material subclass 11 (MC 11). A selected closed gap of 0.2 mm (S02), a chromium layer thickness on the lands of 0.15 mm min (CR3) phosphated on all cast iron surfaces to depth of 0.002 mm min (PO), reduced slot length (WK), coil spring with reduced heat set (WF), and variable pitch with coil diameter d_7 ground (CSE), tangential force F_t according to the medium nominal contact pressure class (PNM) and the ring marked with manufacturer's mark (MM).

3.2 Type DSF-CNP—Coil Spring Loaded Bevelled Edge Oil Control Ring, Chromium-Plated Not Profile Ground

3.2.1 GENERAL FEATURES

NOTE—See Tables 10 and 16 for dimensions and forces.

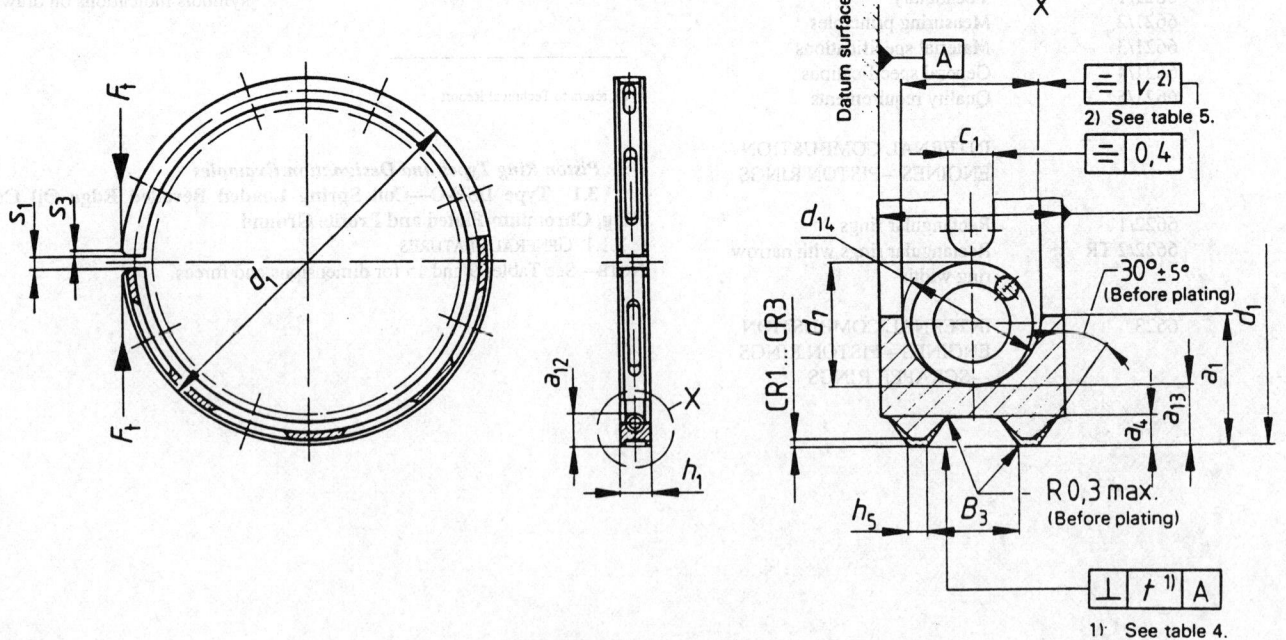

FIGURE 2—TYPE DSF-CNP

3.2.2 DESIGNATION EXAMPLE—Designation of a piston ring complying with the requirements of SAE J2003 being a coil spring loaded bevelled edge oil control ring, chromium-plated not profile ground (DSF-CNP), of nominal diameter $d_1 = 180$ mm (180) and nominal ring width $h_1 = 8$ mm (8), made of grey cast iron, nonheat-treated, material subclass 12 (MC12), a chromium layer thickness on the lands of 0.05 mm min (CR1), constant pitch spring (CSN) and tangential force F_t according to the high nominal contact pressure class (PNH).

3.3 Type SSF—Coil Spring Loaded Slotted Oil Control Ring

3.3.1 GENERAL FEATURES

NOTE—See Tables 11 and 17 for dimensions and forces.

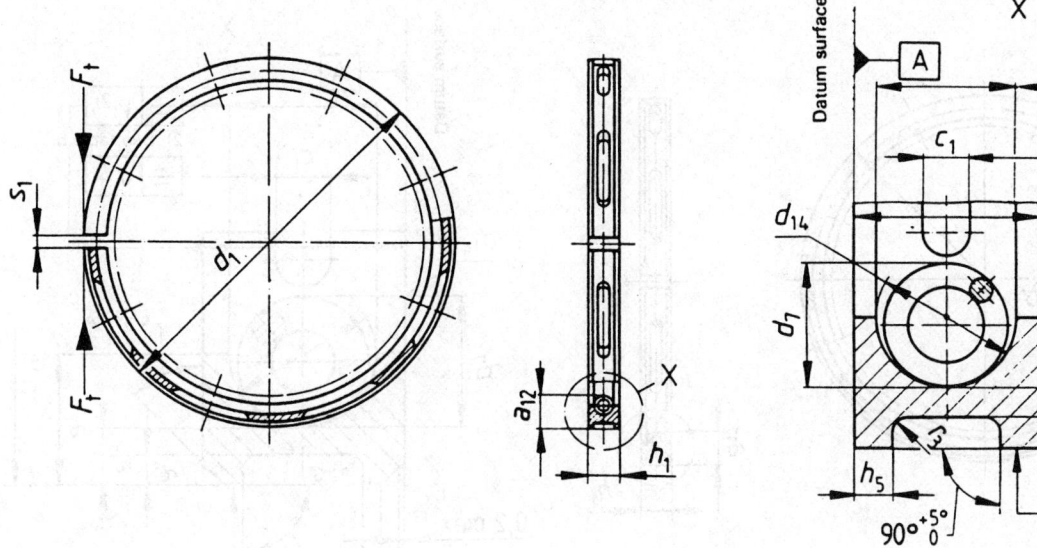

FIGURE 3—TYPE SSF

3.3.2 DESIGNATION EXAMPLE—Designation of a piston ring complying with the requirements of SAE J2003 being a coil spring loaded slotted oil control ring (SSF), of nominal diameter d_1 = 80 mm (80) and nominal ring width h_1 = 4 mm (4), made of grey cast iron, nonheat-treated, material subclass 12 (MC12), constant pitch spring (CSN) and tangential force F_t according to the low nominal contact pressure class (PNL).

3.4 Type GSF—Coil Spring Loaded Double Bevelled Oil Control Ring

3.4.1 GENERAL FEATURES
NOTE—See Tables 12 and 18 for dimensions and forces.

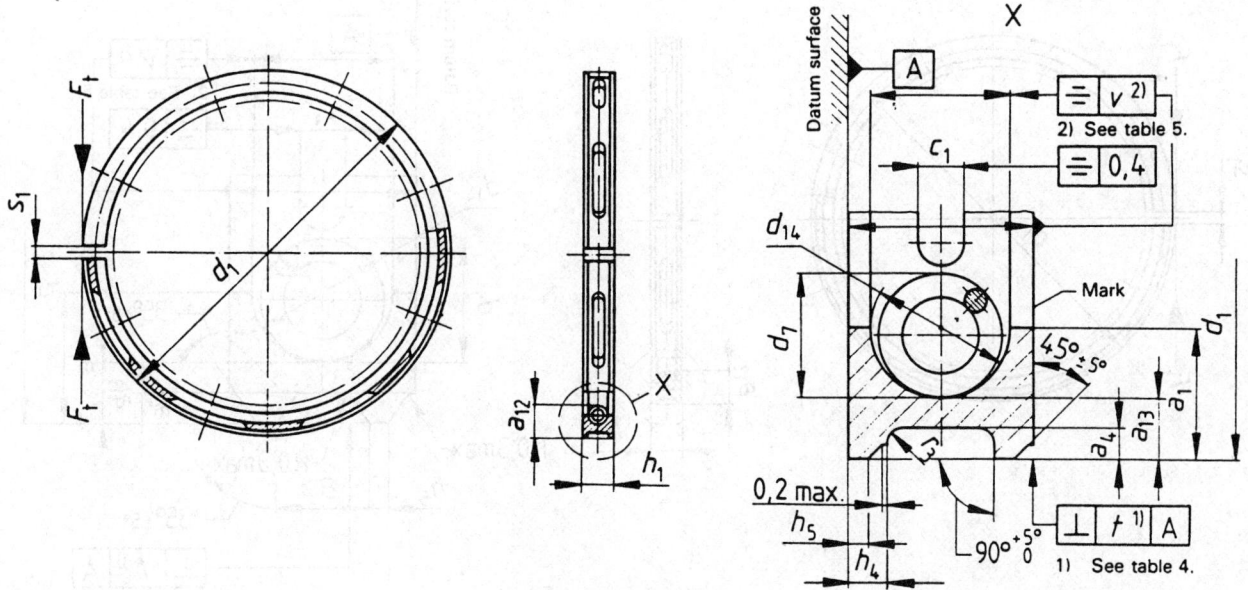

FIGURE 4—TYPE GSF

3.4.2 DESIGNATION EXAMPLE—Designation of a piston ring complying with the requirements of SAE J2003 being a coil spring loaded double bevelled oil control ring (GSF), of nominal diameter d_1 = 75 mm (75), a nominal ring width h_1 = 3 mm (3), made of grey cast iron, nonheat-treated, material subclass 12 (MC12), constant pitch spring (CSN) and a tangential force F_t according to the low nominal contact pressure class (PNL).

3.5 Type DSF—Coil Spring Loaded Bevelled Edge Oil Control Ring

3.5.1 GENERAL FEATURES
NOTE—See Tables 12 and 18 for dimensions and forces.

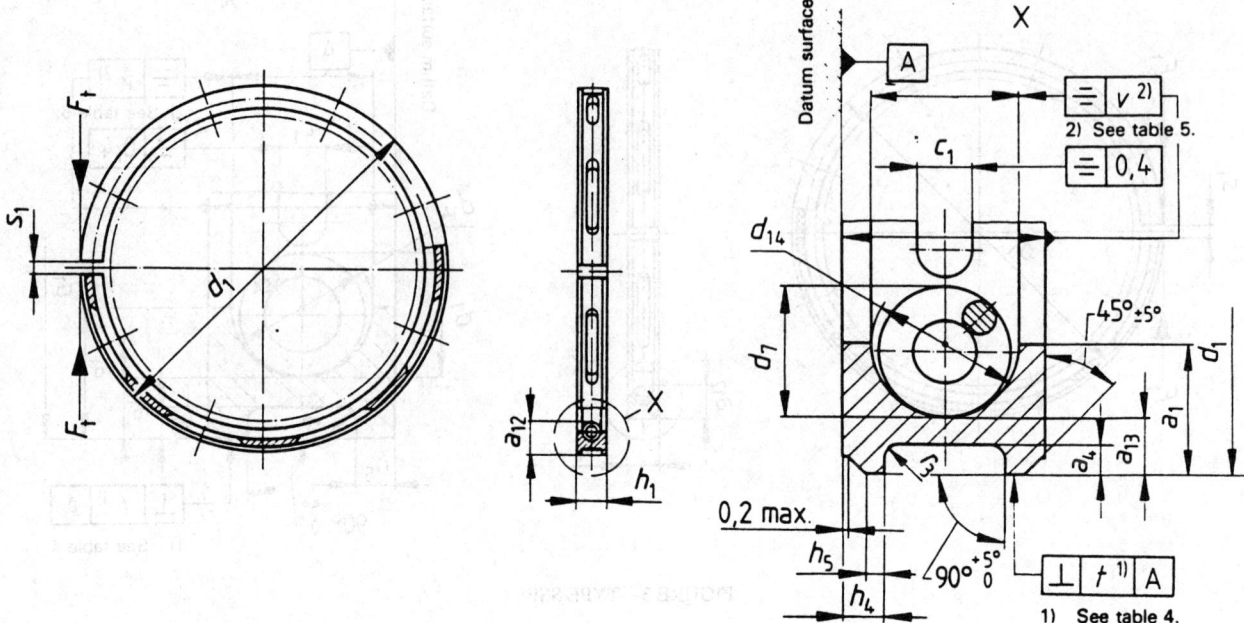

FIGURE 5—TYPE DSF

3.5.2 DESIGNATION EXAMPLE—Designation of a piston ring complying with the requirements of SAE J2003 being a coil spring loaded bevelled edge oil control ring (DSF), of nominal diameter d_1 = 90 mm (90) and nominal ring width h_1 = 3.5 mm (3.5), made of grey cat iron, nonheat-treated, material subclass 12 (MC12), constant pitch spring (CSN) and tangential force F_t according to the reduced nominal contact pressure class (PNR).

3.6 Type DSF-NG—Coil Spring Loaded Bevelled Edge Oil Control Ring (Face Geometry Similar to Type DSF-C or Type DSF-CNP)

3.6.1 GENERAL FEATURES

NOTE—See Tables 13 and 19 for dimensions and forces.

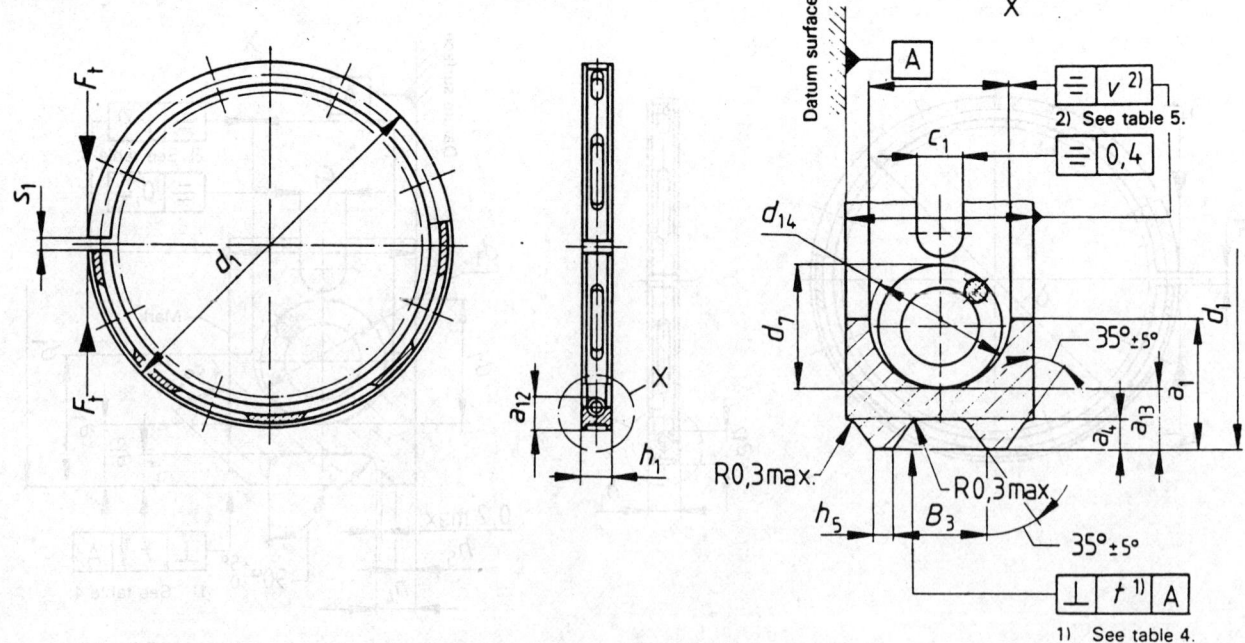

FIGURE 6—TYPE DSF-NG

3.6.2 DESIGNATION EXAMPLE—Designation of a piston ring complying with the requirements of SAE J2003 being a coil spring loaded bevelled edge oil control ring (DSF-NG) of nominal diameter d_1 = 140 mm (140) and nominal ring width h_1 = 6 mm (6), made of grey cast iron, nonheat-treated, material subclass 12 (MC12), constant pitch spring (CSN) and a tangential force F_t according to the medium nominal contact pressure class (PNM).

3.7 Type SSF-L—Coil Spring Loaded Slotted Oil Control Ring With 0.6 mm Nominal Land Width

3.7.1 GENERAL FEATURES

NOTE—See Tables 14 and 20 for dimensions and forces.

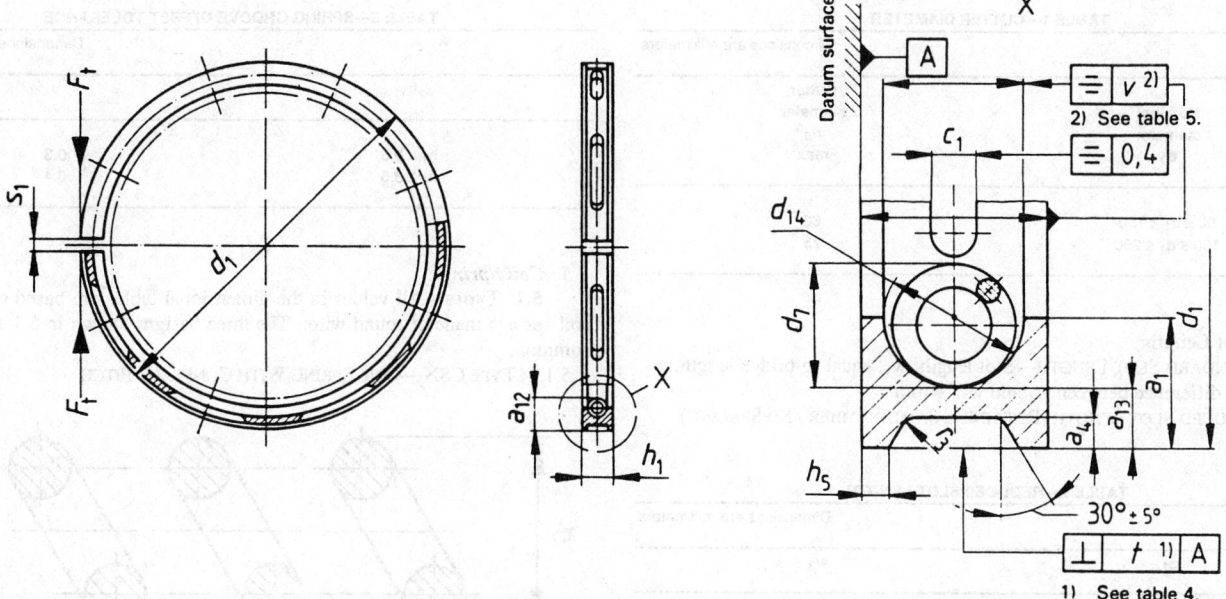

FIGURE 7—TYPE SSF-L

3.7.2 DESIGNATION EXAMPLE—Designation of a piston ring complying with the requirements of SAE J2003 being a coil spring loaded slotted oil control ring with 0.6 mm nominal land width (SSF-L), of nominal diameter $d_1 = 100$ mm (100) and nominal ring width $h_1 = 4.5$ mm (4.5), made of grey cast iron, nonheat-treated, material subclass 12 (MC12), constant pitch spring (CSN) and tangential force F_t according to the reduced nominal contact pressure class (PNR).

4. Common Features
4.1 Arrangement of Slots

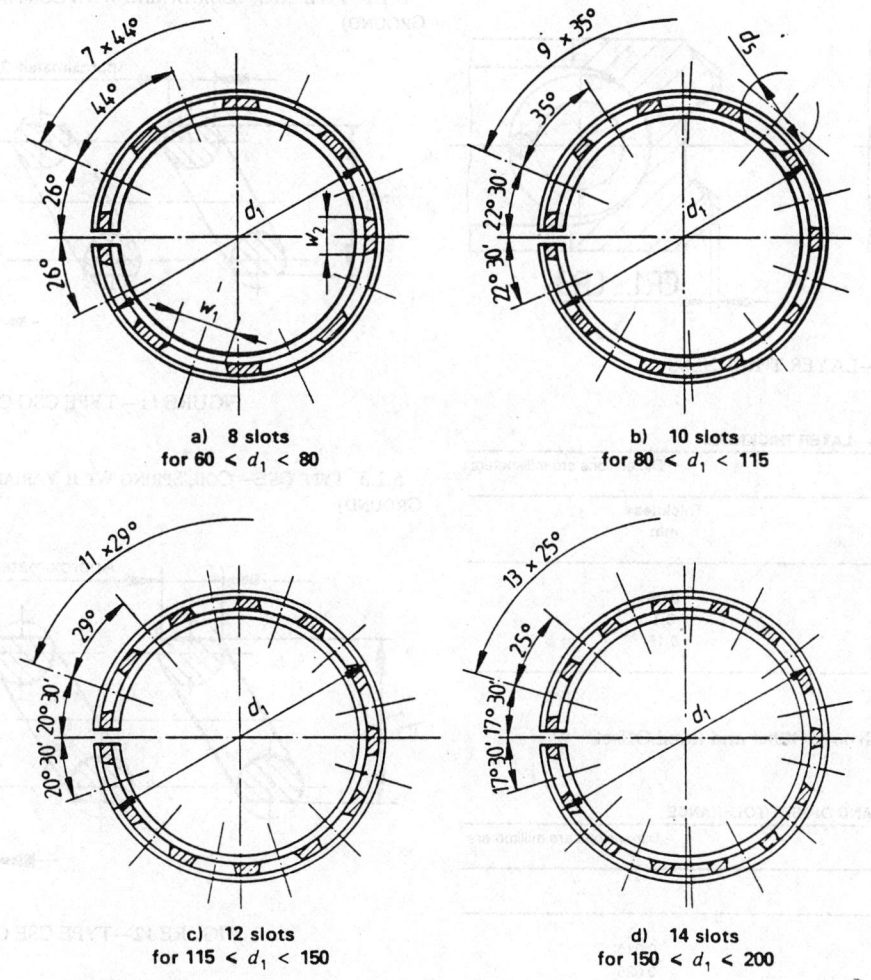

a) 8 slots for $60 < d_1 < 80$

b) 10 slots for $80 < d_1 < 115$

c) 12 slots for $115 < d_1 < 150$

d) 14 slots for $150 < d_1 < 200$

FIGURE 8—ARRANGEMENT OF SLOTS

TABLE 1—CUTTER DIAMETER

Dimensions are millimeters

Nominal diameter d_1	Cutter diameter d_5 max
$60 \leq d_1 < 150$	60
$150 \leq d_1 \leq 200$	75

4.2 Slot Length

4.2.1 STANDARD SLOT LENGTH—Slot length, w_1, equal to bridge length, w_2. Tolerance on difference between w_1 and w_2: 4 mm.

4.2.2 REDUCED SLOT LENGTH (RETAINING SAME NUMBER AND SPACING)

TABLE 2—REDUCED SLOT LENGTH

Dimensions are millimeters

d_1	w_1
$60 \leq d_1 < 80$	8.5 ± 2.5
$80 \leq d_1 < 115$	10.5 ± 2.5
$115 \leq d_1 < 150$	12.5 ± 2.5
$150 \leq d_1 \leq 200$	15.0 ± 3.0

4.3 DSF-C and DSF-CNP—Layer Thickness

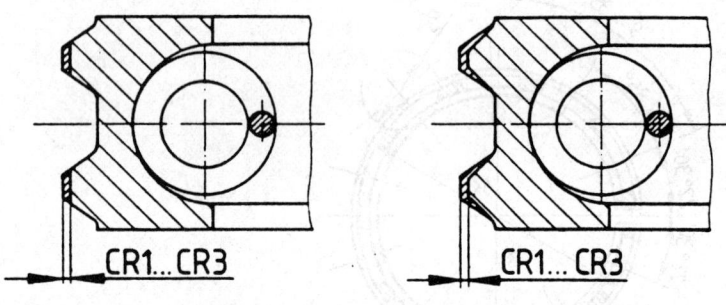

FIGURE 9—LAYER THICKNESS

TABLE 3—LAYER THICKNESS

Dimensions are millimeters

Chromium	Thickness min
CR1	0.05
CR2	0.1
CR3	0.15

4.4 Tolerances of Spring Groove Offset and Land Offset

TABLE 4—LAND OFFSET TOLERANCE

Dimensions are millimeters

h_1	t
$3 \leq h_1 < 5$	0.015
$5 \leq h_1 \leq 8$	0.025

TABLE 5—SPRING GROOVE OFFSET TOLERANCE

Dimensions are millimeters

h_1	v
$h_1 < 3.5$	0.3
$h_1 \geq 3.5$	0.4

5. Coil Spring

5.1 Types—All values in the dimensional tables are based on cylindrical coil springs made of round wire. The three designs shown in 5.1.1 to 5.1.3 are common.

5.1.1 TYPE CSN—COIL SPRING WITH CONSTANT PITCH

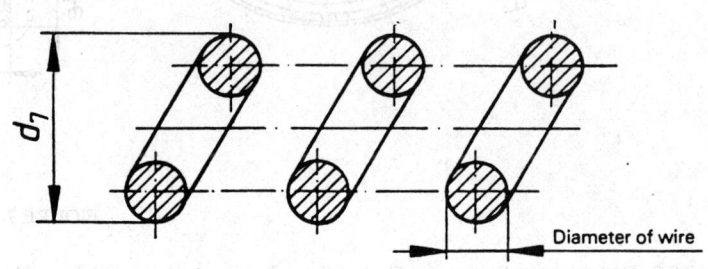

FIGURE 10—TYPE CSN COIL SPRING

5.1.2 TYPE CSG—COIL SPRING WITH CONSTANT PITCH (COIL DIAMETER, D_7, GROUND)

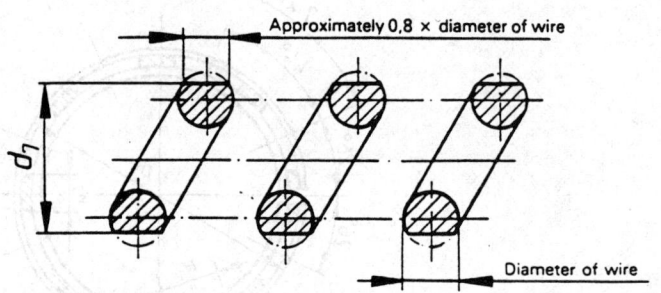

FIGURE 11—TYPE CSG COIL SPRING

5.1.3 TYPE CSE—COIL SPRING WITH VARIABLE PITCH (COIL DIAMETER, D_7, GROUND)

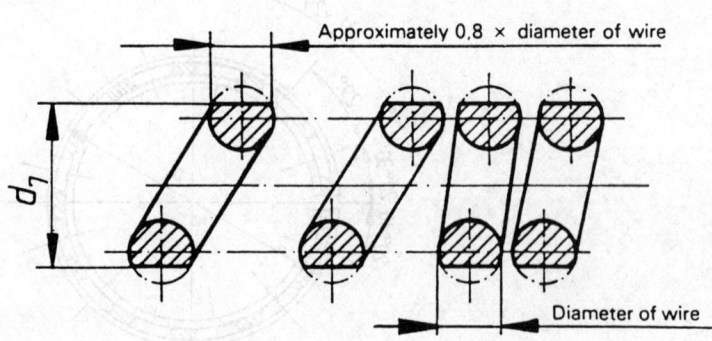

FIGURE 12—TYPE CSE COIL SPRING

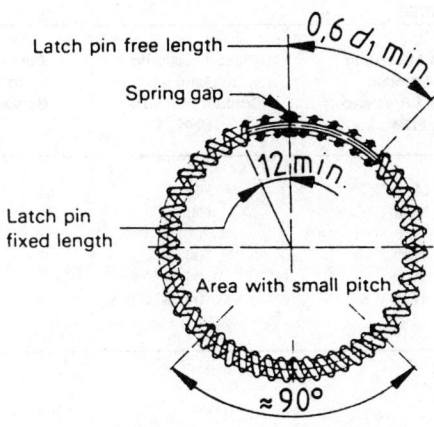

FIGURE 13—POSITION OF AREA WITH SMALL PITCH

NOTE—The use of different spring designs can be agreed between manufacturer and client. Changed spring groove configurations and dimensions may then be necessary.

5.2 Excursion—Coil spring excursion is the distance between the ends of the ring gap with unstressed ring, measured in the middle of the spring groove (see Figure 14).

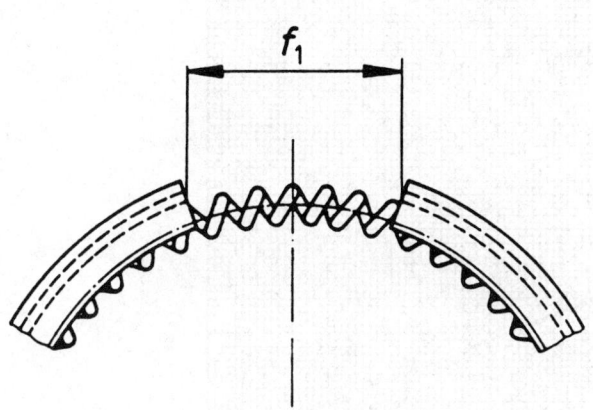

FIGURE 14—COIL SPRING EXCURSION

TABLE 6—COIL SPRING EXCURSION

Dimensions are millimeters

d_1	f_1 max
$60 \leq d_1 < 125$	$0.13 \times d_1$
$125 \leq d_1 \leq 200$	$0.12 \times d_1$

5.3 Position of Coil Spring Gap and Fixing—The spring gap shall be approximately 180° from the ring gap and the spring gap ends fixed with a connecting or latch pin.

5.4 Material—Coil springs are made of valve spring wire, oil heat-treated. A suitable material for coil spring expanders is subclass 63 (see SAE J1590).

Springs are available with two different heat set resistance levels (loss of tangential force under load and temperature):

a. Standard heat resistance
b. Reduced heat set, code WF

The test conditions and the permissible loss of tangential forces are given in SAE J1996, Table 10.

6. Tangential Force and Nominal Contact Pressure

6.1 Tangential Force—The tangential force of coil spring loaded oil control rings is mainly determined by the force of the spring. The cast iron part itself has a very small tangential force due to its low radial wall thickness and the decreased ratio "total free gap/nominal diameter".

The tangential force measurement only can be used because of the flexible design of the cast iron part of coil spring loaded oil control rings.

6.2 Force Factors—Because of the small contribution of the cast iron part in the tangential force, force factors are not necessary when additional features and/or materials other than subclass 12 are being used (see SAE J1590).

6.3 Tangential Force, F_t—The tangential force, F_t, of a spring loaded oil control ring is determined by the following:

a. Nominal diameter, d_1, in millimeters
b. Land width, h_t, in millimeters
c. Required nominal contact pressure, p_o, in newtons per square millimeter and can be calculated from the following equation:

$$F_t = \frac{1}{2} \cdot d_1 \cdot 2h_5 \cdot p_o \qquad \text{(Eq.1)}$$

The land width, h_5, depends on ring type, nominal diameter and ring width. The nominal contact pressure, p_o, can be selected over a wide range to suit the application and the required oil scraping effect.

6.3.1 SPECIFIC TANGENTIAL FORCE, F_{tc}—F_{tc} is the specific tangential force required to maintain a spring loaded oil control ring at a unit contact pressure p_{ou}, of 1 N/mm^2:

$$F_{tc} = \frac{1}{2} \cdot d_1 \cdot 2h_5 \qquad \text{(Eq.2)}$$

In Section 7, F_{tc} is tabulated for every ring type.

6.3.2 ACTUAL TANGENTIAL FORCE, F_t, AND TOLERANCE—The actual tangential force of a spring loaded oil control ring can be calculated with the F_{tc} value and the required nominal contact pressure from the following equation:

$$F_t = \frac{p_o}{p_{ou}} \cdot F_{tc} \qquad \text{(Eq.3)}$$

The tolerance on F_t is the actual value $F_t \pm 20\%$.

6.4 Classes of Nominal Contact Pressure—The range of the nominal contact pressure has been subdivided into six classes, in accordance with Table 7.

TABLE 7—CLASSES OF NOMINAL CONTACT PRESSURE

Class	Meaning
PNE	very low nominal contact pressure
PNL	low nominal contact pressure
PNR	reduced nominal contact pressure
PNM	medium nominal contact pressure
PNH	high nominal contact pressure
PNV	very high nominal contact pressure

The nominal contact pressure, p_o, normally decreases with increasing nominal diameter. Figure 15 shows characteristic values of p_o depending on d_1.

The range possible for the nominal contact pressure, p_o, varies according to the ring type, ring dimensions, and features of the contact lands (if plain or chromium-plated).

The classification given in Table 8 is taken from the dimensional tables in Section 7, where p_o values are given in the following three categories for every nominal diameter:

a. "Low" category—Low friction design oil control rings
b. "Mean" category—Normal design oil control rings
c. "High" category—High contact pressure oil control rings for high oil scraping effect

TABLE 8—OIL CONTROL RING PRESSURE CLASSES

Ring Type	Class According to Nominal Contact Pressure PNE	Class According to Nominal Contact Pressure PNL	Class According to Nominal Contact Pressure PNR	Class According to Nominal Contact Pressure PNM	Class According to Nominal Contact Pressure PNH	Class According to Nominal Contact Pressure PNV
DSF-C	(-)	Low	Mean	Mean	High	(x)
DSF-CNP	(-)	Low	Mean	Mean	High	(x)
SSF	Low	Mean	High	(x)	(-)	(-)
GSF	(x)	Low	Mean	High	(x)	(-)
DSF	(x)	Low	Mean	High	(x)	(-)
DSF-NG	(x)	Low	Mean	High	(x)	(-)
SSF-L	Low	Mean	High	(-)	(-)	(-)

NOTES:

(x) For special applications.
(-) This pressure class is not used with this ring type.

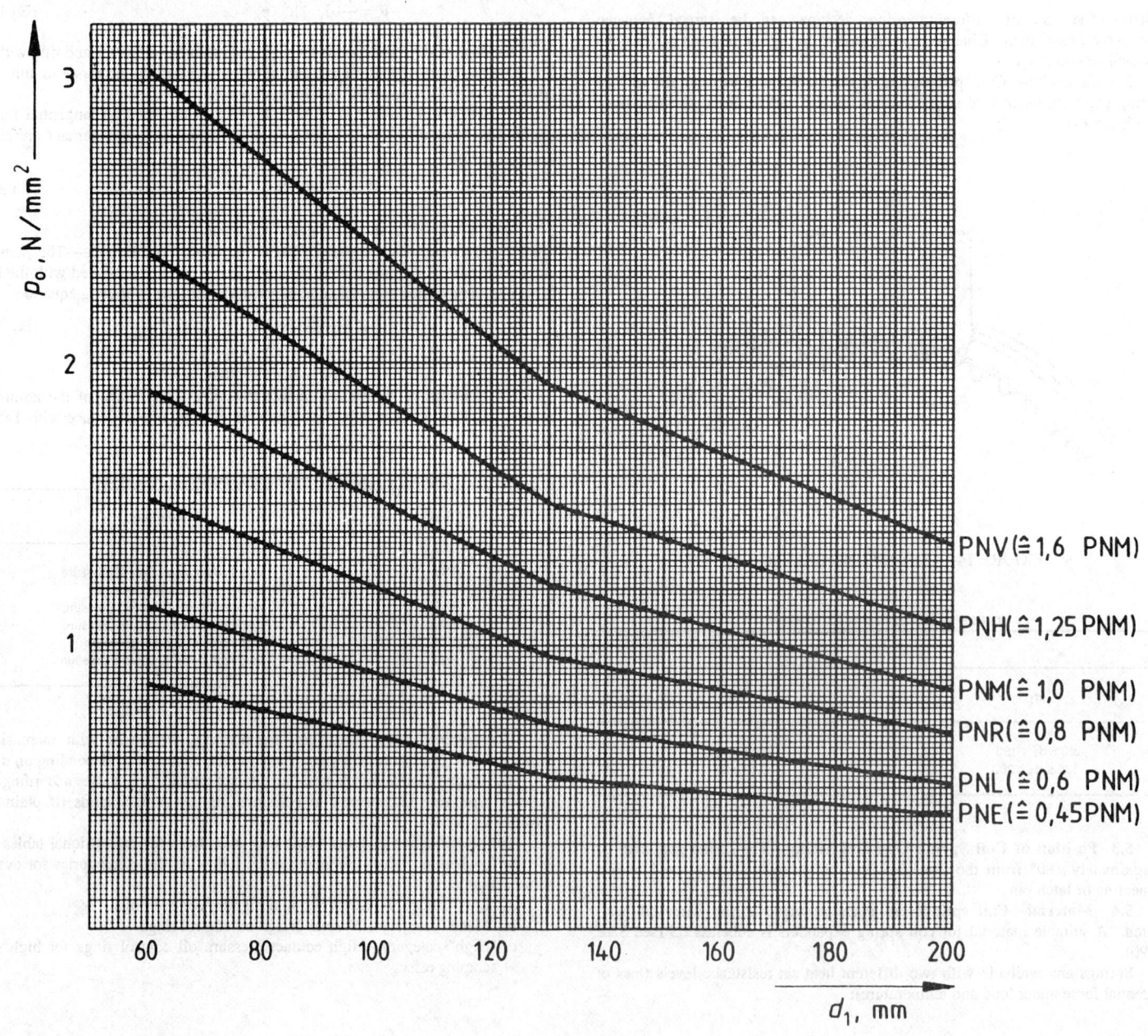

FIGURE 15—CHARACTERISTIC VALUES OF p_o DEPENDING ON d_1

7. Dimensions

TABLE 9—DIMENSIONS FOR DSF-C COIL SPRING LOADED OIL CONTROL RINGS

Dimensions are millimeters

Nominal diameter d_1	Radial thickness over coil spring a_{12} for h_1 shown in column				Ring width h_1 Column					Closed gap s_1	Radial wall thickness a_1 for h_1 shown in column					Land width h_5 for h_1 shown in column				Land spacing $\approx B_3$ for h_1 shown in column				Groove depth a_4 for h_1 shown in column			
	1	2	3	4	1	2	3	4	Tolerance		1	2	3	4	Tolerance	1	2	3	4	1	2	3	4	1	2	3	4
60 61 62 63 64	—	3,5 0 -0,25	3,5 0 -0,25						-0,010 -0,030 For phosphated PO surface: +0,005 -0,025	0,2 +0,2 0	—	2,4	2,4		±0,15 within a ring 0,15 max.												
65 66 67 68 69	—	3,6 0 -0,25	3,7 0 -0,25	3,7 0 -0,25							—	2,45	2,5	2,5													
70 71 72 73 74	—	—	3,8 0 -0,25	3,9 0 -0,25	—	3	3,5	4			—	2,55	2,6	2,6		—	0,35 ±0,07	0,35 ±0,07	0,4 ±0,07	—	1,55	1,85	2,3	—	0,45 -0,1	0,45 -0,1	0,5 -0,1
75 76 77 78 79	—	3,7 0 -0,25	3,9 0 -0,25	4 0 -0,25							—	2,6	2,7	2,7													
80 81 82 83 84	3,8 0 -0,25	4 0 -0,25	4,1 0 -0,25	4,1 0 -0,25						0,25 +0,25 0	2,7	2,7	2,8	2,8													
85 86 87 88 89	3,9 0 -0,25										2,8	2,8	2,9	2,9													
90 91 92 93 94		4,1 0 -0,25	4,2 0 -0,25	4,2 0 -0,25	3	3,5	4	4,5			2,85	2,9	2,95	3		0,35 ±0,07	0,35 ±0,07	0,4 ±0,07	0,4 ±0,07	1,55	1,85	2,3	2,3	0,45 -0,1	0,45 -0,1	0,5 -0,1	0,5 -0,1
95 96 97 98 99	4 0 -0,25										2,95	3	3,05	3,1													
100 101 102 103 104	4,2 0 -0,25	4,4 0 -0,25	4,5 0 -0,25	4,7 0 -0,25						0,3 +0,25 0	3,05	3,1	3,15	3,2													
105 106 107 108 109											3,1	3,15	3,2	3,3													
110 111 112 113 114	4,3 0 -0,25	4,5 0 -0,25	4,6 0 -0,25	4,8 0 -0,25	3,5	4	4,5	5			3,2	3,25	3,3	3,4	±0,20 within a ring 0,20 max	0,35 ±0,07	0,4 ±0,07	0,4 ±0,07	0,4 ±0,07	1,85	2,3	2,8	2,9	0,45 -0,1	0,5 -0,1	0,5 -0,1	0,7 -0,1
115 116 117 118 119	4,4 0 -0,25	4,6 0 -0,25	4,7 0 -0,25	4,9 0 -0,25						0,35 +0,30 0	3,3	3,35	3,4	3,5													
120 121 122 123 124			4,8 0 -0,25	5 0 -0,25							3,4	3,45	3,5	3,6													

This table is shown in ISO format. Commas represent decimal points.

TABLE 9—DIMENSIONS FOR DSF-C COIL SPRING LOADED OIL CONTROL RINGS (CONTINUED)

Dimensions are millimeters

Groove depth and bridge a_{13} for h_1 shown in column				Number of slots	Slot width c_1 for h_1 shown in column				Coil spring groove diameter d_{14} for h_1 shown in column				Coil spring diameter d_7 for h_1 shown in column				Tangential force F_{tc}, N for unit pressure $p_{ou} = 1$ N/mm² for h_1 shown in column				Recommended class of nominal contact pressure N/mm²		
1	2	3	4		1	2	3	4	1	2	3	4	1	2	3	4	1	2	3	4	PNL Low	PNM Mean	PNH High
—	1,5 0 −0,15			8	—	0,7 ±0,1	0,8 ±0,1		—	2,1 +0,1 0	2,3 +0,1 0		—	2 0 −0,1	2,2 0 −0,1		—	21 21,4 21,7 22,1 22,4	21 21,4 21,7 22,1 22,4	24 24,4 24,8 25,2 25,6	1,14 1,13 1,13 1,12 1,12	1,9 1,89 1,88 1,87 1,86	2,38 2,36 2,35 2,34 2,33
		1,5 0 −0,15	1,5 0 −0,15															22,8 23,1 23,5 23,8 24,2	22,8 23,1 23,5 23,8 24,2	26 26,4 26,8 27,2 27,6	1,11 1,1 1,1 1,09 1,09	1,85 1,84 1,83 1,82 1,81	2,31 2,3 2,29 2,28 2,26
	1,6 0 −0,15																	24,5 24,9 25,2 25,6 25,9	24,5 24,9 25,2 25,6 25,9	28 28,4 28,8 29,2 29,6	1,08 1,07 1,07 1,06 1,06	1,8 1,79 1,78 1,77 1,76	2,25 2,24 2,23 2,21 2,2
		1,6 0 −0,15	1,6 0 −0,15					1 +0,1				2,5 +0,1 0				2,4 0 −0,1		26,3 26,6 27 27,3 27,7	26,3 26,6 27 27,3 27,7	30 30,4 30,8 31,2 31,6	1,05 1,04 1,04 1,03 1,03	1,75 1,74 1,73 1,72 1,71	2,19 2,18 2,16 2,15 2,14
1,8 0 −0,15	1,7 0 −0,15			10													28 28,4 28,7 29,1 29,4	28 28,4 28,7 29,1 29,4	32 32,4 32,8 33,2 33,6	32 32,4 32,8 33,2 33,6	1,02 1,01 1,01 1 1	1,7 1,69 1,68 1,67 1,66	2,13 2,11 2,1 2,09 2,08
	1,8 0 −0,15		1,7 0 −0,15														29,8 30,1 30,5 30,8 31,2	29,8 30,1 30,5 30,8 31,2	34 34,4 34,8 35,2 35,6	34 34,4 34,8 35,2 35,6	0,99 0,98 0,98 0,97 0,97	1,65 1,64 1,63 1,62 1,61	2,06 2,05 2,04 2,03 2,01
1,9 0 −0,15					0,7 ±0,1	0,8 ±0,1	1 ±0,1	1,2 ±0,1	2,1 +0,1 0	2,3 +0,1 0	2,5 +0,1 0	2,5 +0,1 0	2 0 −0,1	2,2 0 −0,1	2,4 0 −0,1	2,4 0 −0,1	31,5 31,9 32,2 32,6 32,9	31,5 31,9 32,2 32,6 32,9	36 36,4 36,8 37,2 37,6	36 36,4 36,8 37,2 37,6	0,96 0,95 0,95 0,94 0,94	1,6 1,59 1,58 1,57 1,56	2 1,99 1,98 1,96 1,95
	1,9 0 −0,15	1,8 0 −0,15	1,8 0 −0,15														33,3 33,6 34 34,3 34,7	33,3 33,6 34 34,3 34,7	38 38,4 38,8 39,2 39,6	38 38,4 38,8 39,2 39,6	0,93 0,92 0,92 0,91 0,91	1,55 1,54 1,53 1,52 1,51	1,94 1,93 1,91 1,9 1,89
2 0 −0,15																	35 35,4 35,7 36,1 36,4	40 40,4 40,8 41,2 41,6	40 40,4 40,8 41,2 41,6	40 40,4 40,8 41,2 41,6	0,9 0,89 0,89 0,88 0,88	1,5 1,49 1,48 1,47 1,46	1,88 1,86 1,85 1,84 1,83
	2 0 −0,2	1,9 0 −0,2	1,9 0 −0,2														36,8 37,1 37,5 37,8 38,2	42 42,4 42,8 43,2 43,6	42 42,4 42,8 43,2 43,6	42 42,4 42,8 43,2 43,6	0,87 0,86 0,86 0,85 0,85	1,45 1,44 1,43 1,42 1,41	1,81 1,8 1,79 1,78 1,76
2,1 0 −0,2	2,1 0 −0,2	2 0 −0,2	2 0 −0,2	12	0,8 ±0,1	1 ±0,1	1,2 ±0,1	1,2 ±0,1	2,3 −0,1 0	2,5 +0,1 0	2,7 +0,1 0	2,9 +0,1 0	2,2 0 −0,1	2,4 0 −0,1	2,6 0 −0,1	2,8 0 −0,1	38,5 38,9 39,2 39,6 39,9	44 44,4 44,8 45,2 45,6	44 44,4 44,8 45,2 45,6	44 44,4 44,8 45,2 45,6	0,84 0,83 0,83 0,82 0,82	1,4 1,39 1,38 1,37 1,36	1,75 1,74 1,73 1,71 1,7
		2,1 0 −0,2	2,1 0 −0,2														40,3 40,6 41 41,3 41,7	46 46,4 46,8 47,2 47,6	46 46,4 46,8 47,2 47,6	46 46,4 46,8 47,2 47,6	0,81 0,8 0,8 0,79 0,79	1,35 1,34 1,33 1,32 1,31	1,69 1,68 1,66 1,65 1,64
2,2 0 −0,2	2,2 0 −0,2																42 42,4 42,7 43,1 43,4	48 48,4 48,8 49,2 49,6	48 48,4 48,8 49,2 49,6	48 48,4 48,8 49,2 49,6	0,78 0,77 0,77 0,76 0,76	1,3 1,29 1,28 1,27 1,26	1,63 1,61 1,6 1,59 1,58
		2,2 0 −0,2	2,2 0 −0,2																				

This table is shown in ISO format. Commas represent decimal points.

TABLE 9—DIMENSIONS FOR DSF-C COIL SPRING LOADED OIL CONTROL RINGS (CONTINUED)

Dimensions are millimeters

Nominal diameter d_1	Radial thickness over coil spring a_{12} for h_1 shown in column				Ring width h_1 Column					Closed gap s_1	Radial wall thickness a_1 for h_1 shown in column					Land width h_5 for h_1 shown in column				Land spacing $\approx B_3$ for h_1 shown in column				Groove depth a_4 for h_1 shown in column			
	1	2	3	4	1	2	3	4	Tolerance		1	2	3	4	Tolerance	1	2	3	4	1	2	3	4	1	2	3	4
125 126 127 128 129	4,7 0 −0,25	4,9 0 −0,25	5 0 −0,25	5,2 0 −0,25						0,35 +0,30 0	3,5	3,6	3,6	3,7													
130 131 132 133 134			5,1 0 −0,25	5,3 0 −0,25							3,5	3,6	3,7	3,8													
135 136 137 138 139					4	4,5	5	6	−0,010 −0,040 For phosphated surface: +0,005 −0,035	0,4 +0,35 0	3,6	3,7	3,8	3,9		0,4 ±0,07	0,4 ±0,07	0,4 ±0,07	0,5 ±0,07	2,3	2,8	2,9	3,5	0,5 ±0,1	0,7 ±0,1	0,7 ±0,1	0,9 ±0,1
140 141 142 143 144	4,9 0 −0,25	5,1 0 −0,25	5,3 0 −0,25	5,5 0 −0,25							3,7	3,8	3,9	4													
145 146 147 148 149											3,8	3,9	4	4,1													
150 152 154 155 156 158	5,4 0 −0,25	5,4 0 −0,25	5,5 0 −0,25	5,5 0 −0,25							3,9	4	4,1	4,2	±0,20 within a ring 0,20 max.												
160 162 164					4,5	5	6	7		0,45 +0,35 0						0,4 ±0,07	0,4 ±0,07	0,5 ±0,07	0,6 ±0,09	2,8	2,9	3,5	3,9	0,7 ±0,1	0,7 ±0,1	0,9 ±0,1	1,2 ±0,1
165 166 168	5,4 0 −0,25	5,6 0 −0,25	5,8 0 −0,25	6 0 −0,25							4	4,2	4,3	4,4													
170 172 174																											
175 176 178 180 182 184	5,8 0 −0,35	6 0 −0,35	6,3 0 −0,35	6,7 0 −0,35							4,6	4,7	4,8	5													
185 186 188					5	6	7	8		0,55 +0,40 0						0,4 ±0,07	0,5 ±0,07	0,6 ±0,09	0,6 ±0,09	2,9	3,5	3,9	4,3	0,7 ±0,1	0,9 ±0,1	1,2 ±0,15	1,5 ±0,15
190 192 194	6,2 0 −0,4	6,5 0 −0,4	6,7 0 −0,4	7,1 0 −0,4							4,9	5	5,1	5,3													
195 196 198 200																											

NOTES:

[1] For intermediate sizes (e.g. repair sizes), the radial thickness of the next smaller nominal diameter applies.

[2] Values of specific tangential force F_{tc} are calculated with mean land width (h_5).

This table is shown in ISO format. Commas represent decimal points.

TABLE 9—DIMENSIONS FOR DSF-C COIL SPRING LOADED OIL CONTROL RINGS (CONCLUDED)

Dimensions are millimeters

Groove depth and bridge a_{13} for h_1 shown in column				Number of slots	Slot width c_1 for h_1 shown in column				Coil spring groove diameter d_{14} for h_1 shown in column				Coil spring diameter d_7 for h_1 shown in column				Tangential force F_{tc}, N for unit contact pressure $p_{ou} = 1$ N/mm² for h_1 shown in column				Recommended class of nominal contact pressure N/mm²		
1	2	3	4		1	2	3	4	1	2	3	4	1	2	3	4	1	2	3	4	PNL Low	PNM Mean	PNH High
		2.2 0 −0.2	2.2 0 −0.2						2.5 +0.1 0	2.7 +0.1 0	2.9 +0.1 0	3.1 +0.1 0	2.4 0 −0.1	2.6 0 −0.1	2.8 0 −0.1	3 0 −0.1	50 50.4 50.8 51.2 51.6	50 50.4 50.8 51.2 51.6	50 50.4 50.8 51.2 51.6	62,5 63 63,5 64 64,5	0,75 0,74 0,74 0,73 0,73	1,25 1,24 1,23 1,22 1,21	1,56 1,55 1,54 1,53 1,51
2.3 0 −0.2	2.3 0 −0.2	2.3 0 −0.2	2.3 0 −0.2	12	1 ±0.1	1.2 ±0.1	1.2 ±0.1	1.4 ±0.1									52 52.4 52.8 53.2 53.6	52 52.4 52.8 53.2 53.6	52 52.4 52.8 53.2 53.6	65 65,5 66 66,5 67	0,72 0,72 0,71 0,71 0,71	1,2 1,2 1,19 1,19 1,18	1,5 1,49 1,49 1,48 1,48
																	54 54.4 54.8 55.2 55.6	54 54.4 54.8 55.2 55.6	54 54.4 54.8 55.2 55.6	67,5 68 68,5 69 69,5	0,71 0,7 0,7 0,7 0,69	1,18 1,17 1,17 1,16 1,16	1,47 1,46 1,46 1,45 1,44
									2.7 +0.1 0	2.9 +0.1 0	3.1 +0.1 0	3.3 −0.15 0	2.6 0 −0.1	2.8 0 −0.1	3 0 −0.1	3.2 0 −0.1	56 56.4 56.8 57.2 57.6	56 56.4 56.8 57.2 57.6	56 56.4 56.8 57.2 57.6	70 70,5 71 71,5 72	0,69 0,69 0,68 0,68 0,68	1,15 1,15 1,14 1,14 1,13	1,44 1,43 1,43 1,42 1,41
																	58 58.4 58.8 59.2 59.6	58 58.4 58.8 59.2 59.6	58 58.4 58.8 59.2 59.6	72,5 73 73,5 74 74,5	0,68 0,67 0,67 0,67 0,66	1,13 1,12 1,12 1,11 1,11	1,41 1,4 1,39 1,39 1,38
	2.4 0 −0.2	2.3 0 −0.2	2.3 0 −0.2								3.3 +0.15 0				2.8 0 −0.1	3.2 0 −0.1	60 60.8 61.6	60 60.8 61.6	75 76 77	90 91,2 92,4	0,66 0,65 0,65	1,1 1,09 1,08	1,38 1,36 1,35
																	62 62.4 63.2	62 62.4 63.2	77,5 78 79	93 93,6 94,8	0,65 0,64 0,64	1,08 1,07 1,06	1,34 1,34 1,33
2.6 0 −0.2					1.2 ±0.1	1.2 ±0.1	1.4 ±0.1	1.6 ±0.1	2.9 −0.1 0	3.1 +0.1 0	3.3 −0.15 0		2.8 0 −0.1	3 0 −0.1	3.2 0 −0.1		64 64.8 65.6	64 64.8 65.6	80 81 82	96 97,2 98,4	0,63 0,63 0,62	1,05 1,04 1,03	1,31 1,3 1,29
	2.6 0 −0.2	2.6 0 −0.2	2.4 0 −0.2									3.7 −0.15 0				3.6 0 −0.1	66 66.4 67.2	66 66.4 67.2	82,5 83 84	99 99,6 100,8	0,62 0,61 0,61	1,03 1,02 1,01	1,28 1,28 1,26
				14													68 68.8 69.6	68 68.8 69.6	85 86 87	102 103,2 104,4	0,6 0,59 0,59	1 0,99 0,98	1,25 1,24 1,23
2.8 0 −0.2	2.8 0 −0.2								3.1 −0.1 0	3.3 −0.15 0	3.7 −0.15 0	4.15 −0.15 0	3 0 −0.1	3.2 0 −0.1	3.6 0 −0.1	4 0 −0.12	70 70.4 71.2	87,5 88 89	105 105,6 106,8	105 105,6 106,8	0,59 0,58 0,58	0,98 0,97 0,96	1,22 1,21 1,2
																	72 72.8 73.6	90 91 92	108 109,2 110,4	108 109,2 110,4	0,57 0,56 0,56	0,95 0,94 0,93	1,19 1,18 1,16
		2.7 0 −0.2	2.7 0 −0.2		1.2 ±0.1	1.4 ±0.1	1.6 ±0.1	1.8 ±0.15									74 74.4 75.2	92,5 93 94	111 111,6 112,8	111 111,6 112,8	0,56 0,55 0,55	0,93 0,92 0,91	1,16 1,15 1,14
3 0 −0.2	2.9 0 −0.2								3.3 −0.15 0	3.7 +0.15 0	4.15 −0.15 0	4.55 −0.15 0	3.2 0 −0.1	3.6 0 −0.1	4 0 −0.12	4.4 0 −0.12	76 76.8 77.6	95 96 97	114 115,2 116,4	114 115,2 116,4	0,54 0,53 0,53	0,9 0,89 0,88	1,13 1,11 1,1
																	78 78.4 79.2	97,5 98 99	117 117,6 118,8	117 117,6 118,8	0,53 0,52 0,52	0,88 0,87 0,86	1,09 1,09 1,08
																	80	100	120	120	0,51	0,85	1,06

This table is shown in ISO format. Commas represent decimal points.

TABLE 10—DIMENSIONS FOR DSF-CNP COIL SPRING LOADED OIL CONTROL RINGS

Dimensions are millimeters

Nominal diameter d_1	Radial thickness over coil spring a_{12} for h_1 shown in column				Ring width h_1 Column					Closed gap s_1	Radial wall thickness a_1 for h_1 shown in column					Land width h_5 for h_1 shown in column				Land spacing $\approx B_3$ for h_1 shown in column				Groove depth a_4 for h_1 shown in column			
	1	2	3	4	1	2	3	4	Tolerance		1	2	3	4	Tolerance	1	2	3	4	1	2	3	4	1	2	3	4
60 61 62 63 64	—	3,5 0 −0,25	3,5 0 −0,25	3,7 0 −0,25		3	3,5	4	−0,010 −0,030 For phosphated PO surface: +0,005 −0,025	0,2 +0,2 0	—	2,4	2,4	2,5	±0,15 within a ring 0,15 max.	—	0,4 ±0,12	0,4 ±0,12	0,4 ±0,12	—	1,5	1,8	2,3	—	0,5 −0,1	0,5 ±0,1	0,5 ±0,1
65 66 67 68 69	—	3,6 0 −0,25	3,7 0 −0,25								—	2,45	2,5														
70 71 72 73 74	—		3,8 0 −0,25	3,9 0 −0,25							—	2,55	2,6	2,6													
75 76 77 78 79	—	3,7 0 −0,25	3,9 0 −0,25	4 0 −0,25						0,25 +0,25 0	—	2,6	2,7	2,7													
80 81 82 83 84	3,8 0 −0,25	4 0 −0,25	4,1 0 −0,25	4,1 0 −0,25	3	3,5	4	4,5			2,7	2,7	2,8	2,8		0,4 ±0,12	0,4 ±0,12	0,4 ±0,12	0,4 ±0,12	1,5	1,8	2,3	2,65	0,5 ±0,1	0,5 ±0,1	0,5 ±0,1	0,5 ±0,1
85 86 87 88 89											2,8	2,8	2,9	2,9													
90 91 92 93 94	3,9 0 −0,25	4,1 0 −0,25	4,2 0 −0,25	4,2 0 −0,25							2,85	2,9	2,95	3													
95 96 97 98 99	4 0 −0,25									0,3 +0,25 0	2,95	3	3,05	3,1													
100 101 102 103 104	4,2 0 −0,25	4,4 0 −0,25	4,5 0 −0,25	4,7 0 −0,25							3,05	3,1	3,15	3,2													
105 106 107 108 109											3,1	3,15	3,2	3,3													
110 111 112 113 114	4,3 0 −0,25	4,5 0 −0,25	4,6 0 −0,25	4,8 0 −0,25	3,5	4	4,5	5			3,2	3,25	3,3	3,4	±0,20 within a ring 0,20 max.	0,4 ±0,12	0,4 ±0,12	0,4 ±0,12	0,4 ±0,12	1,8	2,3	2,65	2,9	0,5 ±0,1	0,5 ±0,1	0,5 ±0,1	0,5 ±0,1
115 116 117 118 119	4,4 0 −0,25	4,6 0 −0,25	4,7 0 −0,25	4,9 0 −0,25						0,35 +0,30 0	3,3	3,35	3,4	3,5													
120 121 122 123 124			4,8 0 −0,25	5 0 −0,25							3,4	3,45	3,5	3,6													

This table is shown in ISO format. Commas represent decimal points.

TABLE 10—DIMENSIONS FOR DSF-CNP COIL SPRING LOADED OIL CONTROL RINGS (CONTINUED)

Dimensions are millimeters

Groove depth and bridge a_{13} for h_1 shown in column				Number of slots	Slot width c_1 for h_1 shown in column				Coil spring groove diameter d_{14} for h_1 shown in column				Coil spring diameter d_7 for h_1 shown in column				Tangential force F_{tc}, N for unit contact pressure $p_{ou} = 1$ N/mm² for h_1 shown in column				Recommended class of nominal contact pressure N/mm²		
1	2	3	4		1	2	3	4	1	2	3	4	1	2	3	4	1	2	3	4	PNL Low	PNM Mean	PNH High
—	1,5 0 −0,15	1,5 0 −0,15	1,5 0 −0,15	8	—	0,7 ±0,1	0,7 ±0,1	0,8 ±0,1	—	2,1 +0,1 0	2,1 +0,1 0	2,3 −0,1 0	—	2 0 −0,1	2 0 −0,1	2,2 0 −0,1	—	24 24,4 24,8 25,2 25,6	24 24,4 24,8 25,2 25,6	24 24,4 24,8 25,2 25,6	1,14 1,13 1,13 1,12 1,12	1,9 1,89 1,88 1,87 1,86	2,38 2,36 2,35 2,34 2,33
—	1,6 0 −0,15	1,6 0 −0,15	1,5 0 −0,15	8	—	0,7 ±0,1	0,8 ±0,1		—	2,1 +0,1 0	2,3 +0,1 0	2,3 −0,1 0	—	2 0 −0,1	2,2 0 −0,1	2,2 0 −0,1	—	26 26,4 26,8 27,2 27,6	26 26,4 26,8 27,2 27,6	26 26,4 26,8 27,2 27,6	1,11 1,1 1,1 1,09 1,09	1,85 1,84 1,83 1,82 1,81	2,31 2,3 2,29 2,28 2,26
—	1,6 0 −0,15	1,6 0 −0,15	1,6 0 −0,15	8	—	0,7 ±0,1	0,8 ±0,1	1 +0,1	—	2,1 +0,1 0	2,3 +0,1 0	2,5 +0,1 0	—	2 0 −0,1	2,2 0 −0,1	2,4 0 −0,1	—	28 28,4 28,8 29,2 29,6	28 28,4 28,8 29,2 29,6	28 28,4 28,8 29,2 29,6	1,08 1,07 1,07 1,06 1,06	1,8 1,79 1,78 1,77 1,76	2,25 2,24 2,23 2,21 2,2
—	1,7 0 −0,15	1,6 0 −0,15	1,6 0 −0,15	8	—				—				—				—	30 30,4 30,8 31,2 31,6	30 30,4 30,8 31,2 31,6	30 30,4 30,8 31,2 31,6	1,05 1,04 1,04 1,03 1,03	1,75 1,74 1,73 1,72 1,71	2,19 2,18 2,16 2,15 2,14
1,8 0 −0,15	1,8 0 −0,15	1,7 0 −0,15	1,7 0 −0,15	10	0,7 ±0,1	0,8 ±0,1	1 ±0,1	1,2 ±0,1	2,1 +0,1 0	2,3 +0,1 0	2,5 +0,1 0	2,5 +0,1 0	2 0 −0,1	2,2 0 −0,1	2,4 0 −0,1	2,4 0 −0,1	32 32,4 32,8 33,2 33,6	32 32,4 32,8 33,2 33,6	32 32,4 32,8 33,2 33,6	32 32,4 32,8 33,2 33,6	1,02 1,01 1,01 1 1	1,7 1,69 1,68 1,67 1,66	2,13 2,11 2,1 2,09 2,08
1,9 0 −0,15	1,9 0 −0,15	1,8 0 −0,15	1,8 0 −0,15	10													34 34,4 34,8 35,2 35,6	34 34,4 34,8 35,2 35,6	34 34,4 34,8 35,2 35,6	34 34,4 34,8 35,2 35,6	0,99 0,98 0,98 0,97 0,97	1,65 1,64 1,63 1,62 1,61	2,06 2,05 2,04 2,03 2,01
				10													36 36,4 36,8 37,2 37,6	36 36,4 36,8 37,2 37,6	36 36,4 36,8 37,2 37,6	36 36,4 36,8 37,2 37,6	0,96 0,95 0,95 0,94 0,94	1,6 1,59 1,58 1,57 1,56	2 1,99 1,98 1,96 1,95
				10													38 38,4 38,8 39,2 39,6	38 38,4 38,8 39,2 39,6	38 38,4 38,8 39,2 39,6	38 38,4 38,8 39,2 39,6	0,93 0,92 0,92 0,91 0,91	1,55 1,54 1,53 1,52 1,51	1,94 1,93 1,91 1,9 1,89
2 0 −0,15	2 0 −0,2	1,9 0 −0,2	1,9 0 −0,2	10													40 40,4 40,8 41,2 41,6	40 40,4 40,8 41,2 41,6	40 40,4 40,8 41,2 41,6	40 40,4 40,8 41,2 41,6	0,9 0,89 0,89 0,88 0,88	1,5 1,49 1,48 1,47 1,46	1,88 1,86 1,85 1,84 1,83
				10													42 42,4 42,8 43,2 43,6	42 42,4 42,8 43,2 43,6	42 42,4 42,8 43,2 43,6	42 42,4 42,8 43,2 43,6	0,87 0,86 0,86 0,85 0,85	1,45 1,44 1,43 1,42 1,41	1,81 1,8 1,79 1,78 1,76
2,1 0 −0,2	2,1 0 −0,2	2 0 −0,2	2 0 −0,2	12	0,8 ±0,1	1 ±0,1	1,2 ±0,1	1,2 ±0,1	2,3 +0,1 0	2,5 +0,1 0	2,7 +0,1 0	2,9 +0,1 0	2,2 0 −0,1	2,4 0 −0,1	2,6 0 −0,1	2,8 0 −0,1	44 44,4 44,8 45,2 45,6	44 44,4 44,8 45,2 45,6	44 44,4 44,8 45,2 45,6	44 44,4 44,8 45,2 45,6	0,84 0,83 0,83 0,82 0,82	1,4 1,39 1,38 1,37 1,36	1,75 1,74 1,73 1,71 1,7
		2,1 0 −0,2	2,1 0 −0,2	12													46 46,4 46,8 47,2 47,6	46 46,4 46,8 47,2 47,6	46 46,4 46,8 47,2 47,6	46 46,4 46,8 47,2 47,6	0,81 0,8 0,8 0,79 0,79	1,35 1,34 1,33 1,32 1,31	1,69 1,68 1,66 1,65 1,64
2,2 0 −0,2	2,2 0 −0,2	2,2 0 −0,2	2,2 0 −0,2	12													48 48,4 48,8 49,2 49,6	48 48,4 48,8 49,2 49,6	48 48,4 48,8 49,2 49,6	48 48,4 48,8 49,2 49,6	0,78 0,77 0,77 0,76 0,76	1,3 1,29 1,28 1,27 1,26	1,63 1,61 1,6 1,59 1,58

This table is shown in ISO format. Commas represent decimal points.

TABLE 10—DIMENSIONS FOR DSF-CNP COIL SPRING LOADED OIL CONTROL RINGS (CONTINUED)

Dimensions are millimeters

Nominal Diameter d_1	Radial thickness over coil spring a_{12} for h_1 shown in column				Ring width h_1 Column					Closed gap s_1	Radial wall thickness a_1 for h_1 shown in column					Land width h_5 for h_1 shown in column				Land spacing $\approx B_3$ for h_1 shown in column				Groove depth a_4 for h_1 shown in column			
	1	2	3	4	1	2	3	4	Tolerance		1	2	3	4	Tolerance	1	2	3	4	1	2	3	4	1	2	3	4
125 126 127 128 129	4,7 0 -0,25	4,9 0 -0,25	5 0 -0,25	5,2 0 -0,25						0,35 +0,30 0	3,5	3,6	3,6	3,7													
130 131 132 133 134	4,7 0 -0,25	4,9 0 -0,25	5,1 0 -0,25	5,3 0 -0,25							3,5	3,6	3,7	3,8													
135 136 137 138 139					4	4,5	5	6	-0,010 -0,040 For phosphated PO surface: +0,005 -0,035	0,4 -0,35 0	3,6	3,7	3,8	3,9		0,4 ±0,12	0,4 ±0,12	0,4 ±0,12	0,5 ±0,12	2,3	2,65	2,9	3,5	0,5 ±0,1	0,5 ±0,1	0,7 ±0,1	0,9 ±0,1
140 141 142 143 144	4,9 0 -0,25	5,1 0 -0,25	5,3 0 -0,25	5,5 0 -0,25							3,7	3,8	3,9	4													
145 146 147 148 149											3,8	3,9	4	4,1													
150 152 154 155 156 158	5,4 0 -0,25	5,4 0 -0,25	5,5 0 -0,25	5,5 0 -0,25							3,9	4	4,1	4,2	±0,20 within a ring 0,20 max.												
160 162 164					4,5	5	6	7		0,45 -0,35 0						0,4 ±0,12	0,4 ±0,12	0,5 ±0,12	0,6 ±0,12	2,65	2,9	3,5	3,9	0,5 ±0,1	0,7 ±0,1	0,9 ±0,1	1,2 ±0,15
165 166 168 170 172 174	5,4 0 -0,25	5,6 0 -0,25	5,8 0 -0,25	6 0 -0,25							4	4,2	4,3	4,4													
175 176 178 180 182 184	5,8 0 -0,35	6 0 -0,35	6,3 0 -0,35	6,7 0 -0,35							4,6	4,7	4,8	5													
185 186 188					5	6	7	8		0,55 -0,40 0						0,4 ±0,12	0,5 ±0,12	0,6 ±0,12	0,6 ±0,12	2,9	3,5	3,9	4,3	0,7 ±0,1	0,9 ±0,1	1,2 ±0,15	1,5 ±0,15
190 192 194 195 196 198 200	6,2 0 -0,4	6,5 0 -0,4	6,7 0 -0,4	7,1 0 -0,4							4,9	5	5,1	5,3													

NOTES:

[1] For intermediate sizes (e.g. repair sizes), the radial thickness of the next smaller nominal diameter applies.

[2] Values of specific tangential force F_{tc} are calculated with mean land width (h_5).

This table is shown in ISO format. Commas represent decimal points.

TABLE 10—DIMENSIONS FOR DSF-CNP COIL SPRING LOADED OIL CONTROL RINGS (CONCLUDED)

Dimensions are millimeters

Groove depth and bridge a_{13} for h_1 shown in column				Number of slots	Slot width c_1 for h_1 shown in column				Coil spring groove diameter d_{14} for h_1 shown in column				Coil spring diameter d_7 for h_1 shown in column				Tangential force F_{tc}, N for unit contact pressure $p_{ou} = 1$ N/mm² for h_1 shown in column				Recommended class of nominal contact pressure N/mm²		
1	2	3	4		1	2	3	4	1	2	3	4	1	2	3	4	1	2	3	4	PNL Low	PNM Mean	PNH High
		2,2 0 −0,2	2,2 0 −0,2	12	1 ±0,1	1,2 ±0,1	1,2 ±0,1	1,4 ±0,1	2,5 +0,1 0	2,7 +0,1 0	2,9 +0,1 0	3,1 +0,1 0	2,4 0 −0,1	2,6 0 −0,1	2,8 0 −0,1	3 0 −0,1	50 50,4 50,8 51,2 51,6	50 50,4 50,8 51,2 51,6	50 50,4 50,8 51,2 51,6	62,5 63 63,5 64 64,5	0,75 0,74 0,74 0,73 0,73	1,25 1,24 1,23 1,22 1,21	1,56 1,55 1,54 1,53 1,51
2,3 0 −0,2	2,3 0 −0,2																52 52,4 52,8 53,2 53,6	52 52,4 52,8 53,2 53,6	52 52,4 52,8 53,2 53,6	65 65,5 66 66,5 67	0,72 0,72 0,71 0,71 0,71	1,2 1,2 1,19 1,19 1,18	1,5 1,49 1,49 1,48 1,48
		2,3 0 −0,2	2,3 0 −0,2														54 54,4 54,8 55,2 55,6	54 54,4 54,8 55,2 55,6	54 54,4 54,8 55,2 55,6	67,5 68 68,5 69 69,5	0,71 0,7 0,7 0,7 0,69	1,18 1,17 1,17 1,16 1,16	1,47 1,46 1,46 1,45 1,44
									2,7 +0,1 0	2,9 +0,1 0	3,1 +0,1 0	3,3 −0,15 0	2,6 0 −0,1	2,8 0 −0,1	3 0 −0,1	3,2 0 −0,1	56 56,4 56,8 57,2 57,6	56 56,4 56,8 57,2 57,6	56 56,4 56,8 57,2 57,6	70 70,5 71 71,5 72	0,69 0,69 0,68 0,68 0,68	1,15 1,15 1,14 1,14 1,13	1,44 1,43 1,43 1,42 1,41
																	58 58,4 58,8 59,2 59,6	58 58,4 58,8 59,2 59,6	58 58,4 58,8 59,2 59,6	72,5 73 73,5 74 74,5	0,68 0,67 0,67 0,67 0,66	1,13 1,12 1,12 1,11 1,11	1,41 1,4 1,39 1,39 1,38
	2,4 0 −0,2	2,3 0 −0,2	2,3 0 −0,2		1,2 ±0,1	1,2 ±0,1	1,4 ±0,1	1,6 ±0,1				3,3 −0,15 0				3,2 0 −0,1	60 60,8 61,6	60 60,8 61,6	75 76 77	90 91,2 92,4	0,66 0,65 0,65	1,1 1,09 1,08	1,38 1,36 1,35
2,6 0 −0,2									2,9 +0,1 0	3,1 +0,1 0	3,3 −0,15 0		2,8 0 −0,1	3 0 −0,1	3,2 0 −0,1		62 62,4 63,2	62 62,4 63,2	77,5 78 79	93 93,6 94,8	0,65 0,64 0,64	1,08 1,07 1,06	1,34 1,34 1,33
																	64 64,8 65,6	64 64,8 65,6	80 81 82	96 97,2 98,4	0,63 0,63 0,62	1,05 1,04 1,03	1,31 1,3 1,29
	2,6 0 −0,2	2,6 0 −0,2	2,4 0 −0,2									3,7 −0,15 0				3,6 0 −0,1	66 66,4 67,2	66 66,4 67,2	82,5 83 84	99 99,6 100,8	0,62 0,61 0,61	1,03 1,02 1,01	1,28 1,28 1,26
																	68 68,8 69,6	68 68,8 69,6	85 86 87	102 103,2 104,4	0,6 0,59 0,59	1 0,99 0,98	1,25 1,24 1,23
				14													70 70,4 71,2	87,5 88 89	105 105,6 106,8	105 105,6 106,8	0,59 0,58 0,58	0,98 0,97 0,96	1,22 1,21 1,2
2,8 0 −0,2	2,8 0 −0,2								3,1 −0,1 0	3,3 −0,15 0	3,7 −0,15 0	4,15 −0,15 0	3 0 −0,1	3,2 0 −0,1	3,6 0 −0,1	4 0 −0,12	72 72,8 73,6	90 91 92	108 109,2 110,4	108 109,2 110,4	0,57 0,56 0,56	0,95 0,94 0,93	1,19 1,18 1,16
																	74 74,4 75,2	92,5 93 94	111 111,6 112,8	111 111,6 112,8	0,56 0,55 0,55	0,93 0,92 0,91	1,16 1,15 1,14
		2,7 0 −0,2	2,7 0 −0,2		1,2 ±0,1	1,4 ±0,1	1,6 ±0,1	1,8 −0,15									76 76,8 77,2	95 96 97	114 115,2 116,4	114 115,2 116,4	0,54 0,53 0,53	0,9 0,89 0,88	1,13 1,11 1,1
3 0 −0,2	2,9 0 −0,2								3,3 +0,15 0	3,7 −0,15 0	4,15 −0,15 0	4,55 −0,15 0	3,2 0 −0,1	3,6 0 −0,12	4 0 −0,12	4,4 0 −0,12	78 78,4 79,2	97,5 98 99	117 117,6 118,8	117 117,6 118,8	0,53 0,52 0,52	0,88 0,87 0,86	1,09 1,09 1,08
																	80	100	120	120	0,51	0,85	1,06

This table is shown in ISO format. Commas represent decimal points.

TABLE 11—DIMENSIONS FOR SSF COIL SPRING LOADED OIL CONTROL RINGS

Dimensions are millimeters

Nominal diameter d_1	Radial thickness over coil spring a_{12} for h_1 shown in column				Ring width h_1 Column				Closed gap s_1	Radial wall thickness a_1 for h_1 shown in column					Land width h_5 for h_1 shown in column				Radius r_3	Groove depth a_4 for h_1 shown in column			
	1	2	3	4	1	2	3	4	Tolerance	1	2	3	4	Tolerance	1	2	3	4		1	2	3	4
60 61 62 63 64	—	3,5 0 -0,25	3,5 0 -0,25	3,7 0 -0,25		3	3,5	4		—	2,4	2,4	2,5		—	0,6 ±0,1	0,7 ±0,1	0,7 ±0,1		—	0,5 ±0,1	0,5 ±0,1	0,5 ±0,1
65 66 67 68 69	—	3,6 0 -0,25	3,7 0 -0,25	3,7 0 -0,25					0,2 +0,2 0	—	2,45	2,5											
70 71 72 73 74			3,8 0 -0,25	3,9 0 -0,25						—	2,55	2,6	2,6										
75 76 77 78 79	—	3,7 0 -0,25	3,9 0 -0,25	4 0 -0,25					-0,010 -0,030 For phosphated PO surface: +0,005 -0,025	—	2,6	2,7	2,7	±0,15 within a ring 0,15 max.									
80 81 82 83 84	3,8 0 -0,25	4 0 -0,25	4,1 0 -0,25	4,1 0 -0,25					0,25 +0,25 0	2,7	2,7	2,8	2,8										
85 86 87 88 89	3,9 0 -0,25									2,8	2,8	2,9	2,9		0,6 ±0,1	0,7 ±0,1	0,7 -0,1	0,8 -0,1	0,2 max.	0,5 ±0,1	0,5 ±0,1	0,5 ±0,1	0,7 ±0,1
90 91 92 93 94		4,1 0 -0,25	4,2 0 -0,25	4,2 0 -0,25	3	3,5	4	4,5		2,85	2,9	2,95	3										
95 96 97 98 99	4 0 -0,25								0,3	2,95	3	3,05	3,1										
100 101 102 103 104	4,2 0 -0,25	4,4 0 -0,25	4,5 0 -0,25	4,7 0 -0,25					+0,25 0	3,05	3,1	3,15	3,2										
105 106 107 108 109										3,1	3,15	3,2	3,3										
110 111 112 113 114	4,3 0 -0,25	4,5 0 -0,25	4,6 0 -0,25	4,8 0 -0,25	3,5	4	4,5	5		3,2	3,25	3,3	3,4	±0,20 within a ring 0,20 max.	0,7 ±0,1	0,7 ±0,1	0,8 -0,1	0,9 -0,1		0,5 ±0,1	0,5 ±0,1	0,7 ±0,1	0,7 ±0,1
115 116 117 118 119	4,4 0 -0,25	4,6 0 -0,25	4,7 0 -0,25	4,9 0 -0,25					0,35 +0,30 0	3,3	3,35	3,4	3,5										
120 121 122 123 124			4,8 0 -0,25	5 0 -0,25						3,4	3,45	3,5	3,6										

This table is shown in ISO format. Commas represent decimal points.

TABLE 11—DIMENSIONS FOR SSF COIL SPRING LOADED OIL CONTROL RINGS (CONTINUED)

Dimensions are millimeters

Groove depth and bridge a_{13} for h_1 shown in column				Number of slots	Slot width c_1 for h_1 shown in column				Coil spring groove diameter d_{14} for h_1 shown in column				Coil spring diameter d_7 for h_1 shown in column				Tangential force F_{tc}, N for unit contact pressure $p_{ou} = 1\ N/mm^2$ for h_1 shown in column				Recommended class of nominal contact pressure N/mm^2		
1	2	3	4		1	2	3	4	1	2	3	4	1	2	3	4	1	2	3	4	PNE Low	PNL Mean	PNR High
—	1,5 0 −0,15	1,5 0 −0,15	1,5 0 −0,15	8	—	0,7 ±0,1	0,7 ±0,1	0,8 ±0,1	—	2,1 +0,1 0	2,1 +0,1 0	2,3 +0,1 0	—	2 0 −0,1	2 0 −0,1	2,2 0 −0,1	—	36 36,6 37,2 37,8 38,4	42 42,7 43,4 44,1 44,8	42 42,7 43,4 44,1 44,8	0,86 0,85 0,85 0,84 0,84	1,14 1,13 1,13 1,12 1,12	1,52 1,51 1,5 1,5 1,49
—	1,6 0 −0,15	1,6 0 −0,15	1,5 0 −0,15	8	—	0,7 ±0,1	0,8 ±0,1		—	2,1 +0,1 0	2,3 +0,1 0		—	2 0 −0,1	2,2 0 −0,1		—	39 39,6 40,2 40,8 41,4	45,5 46,2 46,9 47,6 48,3	45,5 46,2 46,9 47,6 48,3	0,83 0,83 0,82 0,82 0,81	1,11 1,1 1,1 1,09 1,09	1,48 1,47 1,46 1,46 1,45
—			1,6 0 −0,15	8				1 +0,1				2,5 +0,1 0				2,4 0 −0,1	—	42 42,6 43,2 43,8 44,4	49 49,7 50,4 51,1 51,8	49 49,7 50,4 51,1 51,8	0,81 0,81 0,8 0,8 0,79	1,08 1,07 1,07 1,06 1,06	1,44 1,43 1,42 1,41 1,41
—	1,7 0 −0,15	1,6 0 −0,15		8													—	45 45,6 46,2 46,8 47,4	52,5 53,2 53,9 54,6 55,3	52,5 53,2 53,9 54,6 55,3	0,79 0,78 0,78 0,77 0,77	1,05 1,04 1,04 1,03 1,03	1,4 1,39 1,38 1,38 1,37
1,8 0 −0,15	1,7 0 −0,15			10	0,7 ±0,1	0,8 ±0,1	1 ±0,1	1,2 ±0,1	2,1 +0,1 0	2,3 +0,1 0	2,5 +0,1 0	2,5 +0,1 0	2 0 −0,1	2,2 0 −0,1	2,4 0 −0,1	2,4 0 −0,1	48 48,6 49,2 49,8 50,4	56 56,7 57,4 58,1 58,8	56 56,7 57,4 58,1 58,8	64 64,8 65,6 66,4 67,2	0,77 0,76 0,76 0,75 0,75	1,02 1,01 1,01 1 1	1,36 1,35 1,34 1,34 1,33
	1,8 0 −0,15	1,7 0 −0,15		10													51 51,6 52,2 52,8 53,4	59,5 60,2 60,9 61,6 62,3	59,5 60,2 60,9 61,6 62,3	68 68,8 69,6 70,4 71,2	0,74 0,74 0,74 0,73 0,73	0,99 0,98 0,98 0,97 0,97	1,32 1,31 1,3 1,3 1,29
1,9 0 −0,15				10													54 54,6 55,2 55,8 56,4	63 63,7 64,4 65,1 65,8	63 63,7 64,4 65,1 65,8	72 72,8 73,6 74,4 75,2	0,72 0,72 0,71 0,71 0,7	0,96 0,95 0,95 0,94 0,94	1,28 1,27 1,26 1,26 1,25
	1,9 0 −0,15	1,8 0 −0,15	1,8 0 −0,15	10													57 57,6 58,2 58,8 59,4	66,5 67,2 67,9 68,6 69,3	66,5 67,2 67,9 68,6 69,3	76 76,8 77,6 78,4 79,2	0,7 0,69 0,69 0,68 0,68	0,93 0,92 0,92 0,91 0,91	1,24 1,23 1,22 1,22 1,21
2 0 −0,15				10													70 70,7 71,4 72,1 72,8	70 70,7 71,4 72,1 72,8	80 80,8 81,6 82,4 83,2	90 90,9 91,8 92,7 93,6	0,68 0,67 0,67 0,66 0,66	0,9 0,89 0,89 0,88 0,88	1,2 1,19 1,18 1,18 1,17
		2 0 −0,2	1,9 0 −0,2	10													73,5 74,2 74,9 75,6 76,3	73,5 74,2 74,9 75,6 76,3	84 84,8 85,6 86,4 87,2	94,5 95,4 96,3 97,2 98,1	0,65 0,65 0,64 0,64 0,63	0,87 0,86 0,86 0,85 0,85	1,16 1,15 1,14 1,14 1,13
2,1 0 −0,2	2,1 0 −0,2	2 0 −0,2	2 0 −0,2	12	0,8 ±0,1	1 ±0,1	1,2 ±0,1	1,2 ±0,1	2,3 −0,1 0	2,5 +0,1 0	2,7 +0,1 0	2,9 +0,1 0	2,2 0 −0,1	2,4 0 −0,1	2,6 0 −0,1	2,8 0 −0,1	77 77,7 78,4 79,1 79,8	77 77,7 78,4 79,1 79,8	88 88,8 89,6 90,4 91,2	99 99,9 100,8 101,7 102,6	0,63 0,63 0,62 0,62 0,61	0,84 0,83 0,83 0,82 0,82	1,12 1,11 1,1 1,1 1,09
		2,1 0 −0,2	2,1 0 −0,2	12													80,5 81,2 81,9 82,6 83,3	80,5 81,2 81,9 82,6 83,3	92 92,8 93,6 94,4 95,2	103,5 104,4 105,3 106,2 107,1	0,61 0,6 0,6 0,59 0,59	0,81 0,8 0,8 0,79 0,79	1,08 1,07 1,06 1,06 1,05
2,2 0 −0,2	2,2 0 −0,2	2,2 0 −0,2	2,2 0 −0,2	12													84 84,7 85,4 86,1 86,8	84 84,7 85,4 86,1 86,8	96 96,8 97,6 98,4 99,2	108 108,9 109,8 110,7 111,6	0,59 0,58 0,58 0,57 0,57	0,78 0,77 0,77 0,76 0,76	1,04 1,03 1,02 1,02 1,01

This table is shown in ISO format. Commas represent decimal points.

TABLE 11—DIMENSIONS FOR SSF COIL SPRING LOADED OIL CONTROL RINGS (CONTINUED)

Dimensions are millimeters

Nominal diameter d_1	Radial thickness over coil spring a_{12} for h_1 shown in column				Ring width h_1 Column				Closed gap s_1	Radial wall thickness a_1 for h_1 shown in column					Land width h_5 for h_1 shown in column				Radius r_3	Groove depth a_4 for h_1 shown in column			
	1	2	3	4	1	2	3	4	Tolerance	1	2	3	4	Tolerance	1	2	3	4		1	2	3	4
125 126 127 128 129	4,7 0 −0,25	4,9 0 −0,25	5 0 −0,25	5,2 0 −0,25					0,35 +0,30 0	3,5	3,6	3,6	3,7										
130 131 132 133 134			5,1 0 −0,25	5,3 0 −0,25						3,5	3,6	3,7	3,8										
135 136 137 138 139					4	4,5	5	6	0,4 +0,35 0 −0,010 −0,040 For phosphated PO surface: +0,005 −0,035	3,6	3,7	3,8	3,9	±0,20 within a ring 0,20 max.	0,7 ±0,1	0,8 ±0,1	0,9 ±0,1	1,1 ±0,1	0,2 max.	0,5 ±0,1	0,7 ±0,1	0,7 ±0,1	0,9 ±0,1
140 141 142 143 144	4,9 0 −0,25	5,1 0 −0,25	5,3 0 −0,25	5,5 0 −0,25						3,7	3,8	3,9	4										
145 146 147 148 149										3,8	3,9	4	4,1										
150 152 154	5,4 0 −0,25	5,4 0 −0,25	5,5 0 −0,25	5,5 0 −0,25						3,9	4	4,1	4,2										
155 156 158																							
160 162 164					4,5	5	6	7	0,45 −0,35 0						0,8 ±0,1	0,9 ±0,1	1,1 ±0,1	1,3 ±0,15		0,7 ±0,1	0,7 ±0,1	0,9 ±0,1	1,2 ±0,15
165 166 168	5,4 0 −0,25	5,6 0 −0,25	5,8 0 −0,25	6 0 −0,25						4	4,2	4,3	4,4										
170 172 174																							
175 176 178	5,8 0 −0,35	6 0 −0,35	6,3 0 −0,35	6,7 0 −0,35						4,6	4,7	4,8	5										
180 182 184																							
185 186 188					5	6	7	8	0,55 −0,40 0						0,9 ±0,1	1,1 ±0,1	1,3 ±0,15	1,6 ±0,15	0,5 max.	0,7 ±0,1	0,9 ±0,1	1,2 ±0,15	1,5 ±0,15
190 192 194	6,2 0 −0,4	6,5 0 −0,4	6,7 0 −0,4	7,1 0 −0,4						4,9	5	5,1	5,3										
195 196 198 200																							

NOTES:

[1] For intermediate sizes (e.g. repair sizes), the radial thickness of the next smaller nominal diameter applies.

[2] Values of specific tangential force F_{tc} are calculated with mean land width (h_5).

This table is shown in ISO format. Commas represent decimal points.

TABLE 11—DIMENSIONS FOR SSF COIL SPRING LOADED OIL CONTROL RINGS (CONCLUDED)

Dimensions are millimeters

Groove depth and bridge a_{13} for h_1 shown in column				Number of slots	Slot width c_1 for h_1 shown in column				Coil spring groove diameter d_{14} for h_1 shown in column				Coil spring diameter d_7 for h_1 shown in column				Tangential force F_{tc}, N for unit contact pressure $p_{ou}=1\ N/mm^2$ for h_1 shown in column				Recommended class of nominal contact pressure N/mm^2		
1	2	3	4		1	2	3	4	1	2	3	4	1	2	3	4	1	2	3	4	PNE Low	PNL Mean	PNR High
		2,2 0 −0,2	2,2 0 −0,2	12					2,5 +0,1 0	2,7 +0,1 0	2,9 +0,1 0	3,1 +0,1 0	2,4 0 −0,1	2,6 0 −0,1	2,8 0 −0,1	3 0 −0,1	87,5 88,2 88,9 89,6 90,3	100 100,8 101,6 102,4 103,2	112,5 113,4 114,3 115,2 116,1	137,5 138,6 139,7 140,8 141,9	0,56 0,56 0,55 0,55 0,54	0,75 0,74 0,74 0,73 0,73	1 0,99 0,98 0,98 0,97
2,3 0 −0,2	2,3 0 −0,2				1 ±0,1	1,2 ±0,1	1,2 ±0,1	1,4 ±0,1									91 91,7 92,4 93,1 93,8	104 104,8 105,6 106,4 107,2	117 117,9 118,8 119,7 120,6	143 144,1 145,2 146,3 147,4	0,54 0,54 0,54 0,53 0,53	0,72 0,72 0,71 0,71 0,71	0,96 0,96 0,95 0,95 0,94
		2,3 0 −0,2	2,3 0 −0,2														94,5 95,2 95,9 96,6 97,3	108 108,8 109,6 110,4 111,2	121,5 122,4 123,3 124,2 125,1	148,5 149,6 150,7 151,8 152,9	0,53 0,53 0,52 0,52 0,52	0,71 0,7 0,7 0,7 0,69	0,94 0,94 0,93 0,93 0,92
									2,7 +0,1 0	2,9 +0,1 0	3,1 +0,1 0	3,3 +0,15 0	2,6 0 −0,1	2,8 0 −0,1	3 0 −0,1	3,2 0 −0,1	98 98,7 99,4 100,1 100,8	112 112,8 113,6 114,4 115,2	126 126,9 127,8 128,7 129,6	154 155,1 156,2 157,3 158,4	0,52 0,52 0,51 0,51 0,51	0,69 0,69 0,68 0,68 0,68	0,92 0,92 0,91 0,91 0,9
																	101,5 102,2 102,9 103,6 104,3	116 116,8 117,6 118,4 119,2	130,5 131,4 132,3 133,2 134,1	159,5 160,6 161,7 162,8 163,9	0,51 0,5 0,5 0,5 0,5	0,68 0,67 0,67 0,67 0,66	0,9 0,9 0,89 0,89 0,88
	2,4 0 −0,2	2,3 0 −0,2	2,3 0 −0,2								3,3 −0,15 0				3,2 0 −0,1		120 121,6 123,2	135 136,8 138,6	165 167,2 169,4	195 197,6 200,2	0,5 0,49 0,49	0,66 0,65 0,65	0,88 0,87 0,86
2,6 0 −0,2				14	1,2 ±0,1	1,2 ±0,1	1,4 ±0,1	1,6 ±0,1	2,9 −0,1 0	3,1 +0,1 0	3,3 +0,15 0		2,8 0 −0,1	3 0 −0,1	3,2 0 −0,1		124 124,8 126,4	139,5 140,4 142,2	170,5 171,6 173,8	201,5 202,8 205,4	0,48 0,48 0,48	0,65 0,64 0,64	0,86 0,86 0,85
																	128 129,6 131,2	144 145,8 147,6	176 178,2 180,4	208 210,6 213,2	0,47 0,47 0,46	0,63 0,62 0,62	0,84 0,83 0,82
		2,6 0 −0,2	2,6 0 −0,2									3,7 −0,15 0				3,6 0 −0,1	132 132,8 134,4	148,5 149,4 151,2	181,5 182,6 184,8	214,5 215,8 218,4	0,46 0,46 0,45	0,62 0,61 0,61	0,82 0,82 0,81
			2,4 0 −0,2														136 137,6 139,2	153 154,8 156,6	187 189,2 191,4	221 223,6 226,2	0,45 0,45 0,44	0,6 0,6 0,59	0,8 0,79 0,78
	2,8 0 −0,2	2,8 0 −0,2							3,1 −0,1 0	3,3 +0,15 0	3,7 +0,15 0	4,15 −0,15 0	3 0 −0,1	3,2 0 −0,1	3,6 0 −0,1	4 0 −0,12	157,5 158,4 160,2	192,5 193,6 195,8	227,5 228,8 231,4	280 281,6 284,8	0,44 0,44 0,43	0,59 0,58 0,58	0,78 0,78 0,77
																	162 163,8 165,6	198 200,2 202,4	234 236,6 239,2	288 291,2 294,4	0,43 0,42 0,42	0,57 0,56 0,56	0,76 0,75 0,74
		2,7 0 −0,2	2,7 0 −0,2		1,2 ±0,1	1,4 ±0,1	1,6 −0,1 0	1,8 ±0,15									166,5 167,4 169,2	203,5 204,6 206,8	240,5 241,8 244,4	296 297,6 300,8	0,42 0,41 0,41	0,56 0,55 0,55	0,74 0,74 0,73
																	171 172,8 174,6	209 211,2 213,4	247 249,6 252,2	304 307,2 310,4	0,41 0,4 0,4	0,54 0,53 0,53	0,72 0,71 0,7
3 0 −0,2	2,9 0 −0,2								3,3 +0,15 0	3,7 +0,15 0	4,15 −0,15 0	4,55 −0,15 0	3,2 0 −0,1	3,6 0 −0,12	4 0 −0,12	4,4 0 −0,12	175,5 176,4 178,2	214,5 215,6 217,8	253,5 254,8 257,4	312 313,6 316,8	0,4 0,39 0,39	0,53 0,52 0,52	0,7 0,7 0,69
																	180	220	260	320	0,38	0,51	0,68

This table is shown in ISO format. Commas represent decimal points.

TABLE 12—DIMENSIONS FOR GSF AND DSF COIL SPRING LOADED OIL CONTROL RINGS

Dimensions are millimeters

Nominal diameter d_1	Radial thickness over coil spring a_{12} for h_1 shown in column				Ring width h_1 Column					Closed gap s_1	Radial wall thickness a_1 for h_1 shown in column					Land width h_4 for h_1 shown in column				Land width h_5 for h_1 shown in column				Radius r_3
	1	2	3	4	1	2	3	4	Tolerance		1	2	3	4	Tolerance	1	2	3	4	1	2	3	4	
60 61 62 63 64	—	3,5 0 −0,25	3,5 0 −0,25								—	2,4	2,4											
65 66 67 68 69	—	3,6 0 −0,25	3,7 0 −0,25	3,7 0 −0,25	—	3	3,5	4		0,2 +0,2 0	—	2,45	2,5	2,5		—	0,6 ±0,1	0,7 ±0,1	0,7 ±0,1	—	0,28 ±0,08	0,35 ±0,1	0,4 ±0,1	
70 71 72 73 74	—		3,8 0 −0,25	3,9 0 −0,25							—	2,55	2,6	2,6										
75 76 77 78 79	—	3,7 0 −0,25	3,9 0 −0,25	4 0 −0,25					−0,010 −0,030 For phosphated PO surface: +0,005 −0,025		—	2,6	2,7	2,7	±0,15 within a ring 0,15 max.									
80 81 82 83 84	3,8 0 −0,25	4 0 −0,25	4,1 0 −0,25	4,1 0 −0,25						0,25 +0,25 0	2,7	2,7	2,8	2,8										
85 86 87 88 89	3,9 0 −0,25				3	3,5	4	4,5			2,8	2,8	2,9	2,9		0,6 ±0,1	0,7 ±0,1	0,7 ±0,1	0,8 ±0,1	0,28 ±0,08	0,35 ±0,1	0,4 ±0,1	0,4 ±0,1	0,2 max.
90 91 92 93 94		4,1 0 −0,25	4,2 0 −0,25	4,2 0 −0,25							2,85	2,9	2,95	3										
95 96 97 98 99	4 0 −0,25									0,3 +0,25 0	2,95	3	3,05	3,1										
100 101 102 103 104	4,2 0 −0,25	4,4 0 −0,25	4,5 0 −0,25	4,7 0 −0,25							3,05	3,1	3,15	3,2										
105 106 107 108 109											3,1	3,15	3,2	3,3										
110 111 112 113 114	4,3 0 −0,25	4,5 0 −0,25	4,6 0 −0,25	4,8 0 −0,25	3,5	4	4,5	5			3,2	3,25	3,3	3,4	±0,20 within a ring 0,20 max.	0,7 ±0,1	0,7 ±0,1	0,8 ±0,1	0,9 ±0,1	0,35 ±0,1	0,4 ±0,1	0,4 ±0,1	0,4 ±0,1	
115 116 117 118 119	4,4 0 −0,25	4,6 0 −0,25	4,7 0 −0,25	4,9 0 −0,25						0,35 +0,30 0	3,3	3,35	3,4	3,5										
120 121 122 123 124			4,8 0 −0,25	5 0 −0,25							3,4	3,45	3,5	3,6										

This table is shown in ISO format. Commas represent decimal points.

TABLE 12—DIMENSIONS FOR GSF AND DSF COIL SPRING LOADED OIL CONTROL RINGS (CONTINUED)

Dimensions are millimeters

Goove depth a_4 for h_1 shown in column				Groove depth and bridge a_{13} for h_1 shown in column				Number of slots	Slot width c_1 for h_1 shown in column				Coil spring groove diameter d_{14} for h_1 shown in column				Coil spring diameter d_7 for h_1 shown in column				Tangential force F_{tc}, N for unit contact pressure $P_{ou} = 1$ N/mm² for h_1 shown in column				Recommended class of nominal contact pressure N/mm²		
1	2	3	4	1	2	3	4		1	2	3	4	1	2	3	4	1	2	3	4	1	2	3	4	PNL Low	PNR Mean	PNM High
—	0,5 ±0,1	0,5 ±0,1	0,5 0	—	1,5 0 −0,15	1,5 0 −0,15	1,5 0 −0,15	8	—	0,7 ±0,1	0,7 ±0,1	0,8 ±0,1	—	2,1 +0,1 0	2,1 +0,1 0	2,3 +0,1 0	—	2 0 −0,1	2 0 −0,1	2,2 0 −0,1	—	16,8 17,1 17,4 17,6 17,9	21 21,4 21,7 22,1 22,4	24 24,4 24,8 25,2 25,6	1,14 1,13 1,13 1,12 1,12	1,52 1,51 1,5 1,5 1,49	1,9 1,89 1,88 1,87 1,86
					1,6 0 −0,15	1,5 0 −0,15					0,8 ±0,1			2,1 +0,1 0	2,3 +0,1 0			2 0 −0,1	2,2 0 −0,1		18,2 18,5 18,8 19 19,3	22,8 23,1 23,5 23,8 24,2	26 26,4 26,8 27,2 27,6	1,11 1,1 1,1 1,09 1,09	1,48 1,47 1,46 1,46 1,45	1,85 1,84 1,83 1,82 1,81
					1,6 0 −0,15							1 ±0,1			2,5 +0,1 0				2,4 0 −0,1			19,6 19,9 20,2 20,4 20,7	24,5 24,9 25,2 25,6 25,9	28 28,4 28,8 29,2 29,6	1,08 1,07 1,07 1,06 1,06	1,44 1,43 1,42 1,42 1,41	1,8 1,79 1,78 1,77 1,76
					1,7 0 −0,15	1,6 0 −0,15															21 21,3 21,6 21,8 22,1	26,3 26,6 27 27,3 27,7	30 30,4 30,8 31,2 31,6	1,05 1,04 1,04 1,03 1,03	1,4 1,39 1,38 1,38 1,37	1,75 1,74 1,73 1,72 1,71
0,5 ±0,1	0,5 ±0,1	0,5 ±0,1	0,7 ±0,1	1,8 0 −0,15	1,7 0 −0,15	1,7 0 −0,15		10	0,7 ±0,1	0,8 ±0,1	1 ±0,1	1,2 ±0,1	2,1 +0,1 0	2,3 +0,1 0	2,5 +0,1 0	2,5 +0,1 0	2 0 −0,1	2,2 0 −0,1	2,4 0 −0,1	2,4 0 −0,1	22,4 22,7 23 23,2 23,5	28 28,4 28,7 29,1 29,4	32 32,4 32,8 33,2 33,6	32 32,4 32,8 33,2 33,6	1,02 1,01 1,01 1 1	1,36 1,35 1,34 1,34 1,33	1,7 1,69 1,68 1,67 1,66
				1,9 0 −0,15	1,8 0 −0,15	1,7 0 −0,15															23,8 24,1 24,4 24,6 24,9	29,8 30,1 30,5 30,8 31,2	34 34,4 34,8 35,2 35,6	34 34,4 34,8 35,2 35,6	0,99 0,98 0,98 0,97 0,97	1,32 1,31 1,3 1,3 1,29	1,65 1,64 1,63 1,62 1,61
																					25,2 25,5 25,8 26 26,3	31,5 31,9 32,2 32,6 32,9	36 36,4 36,8 37,2 37,6	36 36,4 36,8 37,2 37,6	0,96 0,95 0,95 0,94 0,94	1,28 1,27 1,26 1,26 1,25	1,6 1,59 1,58 1,57 1,56
					1,9 0 −0,15	1,8 0 −0,15	1,8 0 −0,15														26,6 26,9 27,2 27,4 27,7	33,3 33,6 34 34,3 34,7	38 38,4 38,8 39,2 39,6	38 38,4 38,8 39,2 39,6	0,93 0,92 0,92 0,91 0,91	1,24 1,23 1,22 1,22 1,21	1,55 1,54 1,53 1,52 1,51
0,5 ±0,1	0,5 ±0,1	0,7 ±0,1	0,7 ±0,1	2 0 −0,15	2 0 −0,2	1,9 0 −0,2	1,9 0 −0,2	12	0,8 ±0,1	1 ±0,1	1,2 ±0,1	1,2 ±0,1	2,3 −0,1 0	2,5 +0,1 0	2,7 +0,1 0	2,9 −0,1 0	2,2 0 −0,1	2,4 0 −0,1	2,6 0 −0,1	2,8 0 −0,1	35 35,4 35,7 36,1 36,4	40 40,4 40,8 41,2 41,6	40 40,4 40,8 41,2 41,6	40 40,4 40,8 41,2 41,6	0,9 0,89 0,89 0,88 0,88	1,2 1,19 1,18 1,18 1,17	1,5 1,49 1,48 1,47 1,46
																					36,8 37,1 37,5 37,8 38,2	42 42,4 42,8 43,2 43,6	42 42,4 42,8 43,2 43,6	42 42,4 42,8 43,2 43,6	0,87 0,86 0,86 0,85 0,85	1,16 1,15 1,14 1,14 1,13	1,45 1,44 1,43 1,42 1,41
				2,1 0 −0,2	2,1 0 −0,2	2 0 −0,2	2 0 −0,2														38,5 38,9 39,2 39,6 39,9	44 44,4 44,8 45,2 45,6	44 44,4 44,8 45,2 45,6	44 44,4 44,8 45,2 45,6	0,84 0,83 0,83 0,82 0,82	1,12 1,11 1,1 1,1 1,09	1,4 1,39 1,38 1,37 1,36
						2,1 0 −0,2	2,1 0 −0,2														40,3 40,6 41 41,3 41,7	46 46,4 46,8 47,2 47,6	46 46,4 46,8 47,2 47,6	46 46,4 46,8 47,2 47,6	0,81 0,8 0,8 0,79 0,79	1,08 1,07 1,06 1,06 1,05	1,35 1,34 1,33 1,32 1,31
				2,2 0 −0,2	2,2 0 −0,2	2,2 0 −0,2	2,2 0 −0,2														42 42,4 42,7 43,1 43,4	48 48,4 48,8 49,2 49,6	48 48,4 48,8 49,2 49,6	48 48,4 48,8 49,2 49,6	0,78 0,77 0,77 0,76 0,76	1,04 1,03 1,02 1,02 1,01	1,3 1,29 1,28 1,27 1,26

This table is shown in ISO format. Commas represent decimal points.

TABLE 12—DIMENSIONS FOR GSF AND DSF COIL SPRING LOADED OIL CONTROL RINGS (CONTINUED)

Dimensions are millimeters

Nominal diameter d_1	Radial thickness over coil spring a_{12} for h_1 shown in column				Ring width h_1 Column				Tolerance	Closed gap s_1	Radial wall thickness a_1 for h_1 shown in column				Tolerance	Land width h_4 for h_1 shown in column				Land width h_5 for h_1 shown in column				Radius r_3
	1	2	3	4	1	2	3	4			1	2	3	4		1	2	3	4	1	2	3	4	
125 126 127 128 129	4.7 0 −0.25	4.9 0 −0.25	5 0 −0.25	5.2 0 −0.25						0.35 +0.30 0	3.5	3.6	3.6	3.7										
130 131 132 133 134			5.1 0 −0.25	5.3 0 −0.25							3.5	3.6	3.7	3.8										
135 136 137 138 139					4	4.5	5	6		0.4 −0.35 0	3.6	3.7	3.8	3.9		0.7 ±0.1	0.8 ±0.1	0.9 ±0.1	1.1 ±0.1	0.4 ±0.1	0.4 ±0.1	0.4 ±0.1	0.5 ±0.1	
140 141 142 143 144	4.9 0 −0.25	5.1 0 −0.25	5.3 0 −0.25	5.5 0 −0.25					−0.010 −0.040 For phosphated PO surface: +0.005 −0.035		3.7	3.8	3.9	4										0.2 max
145 146 147 148 149											3.8	3.9	4	4.1										
150 152 154	5.4 0 −0.25	5.4 0 −0.25	5.5 0 −0.25	5.5 0 −0.25							3.9	4	4.1	4.2	±0.20 within a ring 0.20 max									
155 156 158																								
160 162 164					4.5	5	6	7		0.45 −0.35 0						0.8 ±0.1	0.9 ±0.1	1.1 ±0.1	1.3 ±0.15	0.4 ±0.1	0.4 ±0.1	0.5 ±0.1	0.6 ±0.1	
165 166 168	5.4 0 −0.25	5.6 0 −0.25	5.8 0 −0.25	6 0 −0.25							4	4.2	4.3	4.4										
170 172 174																								
175 176 178 180 182 184	5.8 0 −0.35	6 0 −0.35	6.3 0 −0.35	6.7 0 −0.35							4.6	4.7	4.8	5										
185 186 188					5	6	7	8		0.55 −0.40 0						0.9 ±0.1	1.1 ±0.1	1.3 ±0.15	1.6 ±0.15	0.4 ±0.1	0.5 ±0.1	0.6 ±0.1	0.6 ±0.1	0.5 max
190 192 194	6.2 0 −0.4	6.5 0 −0.4	6.7 0 −0.4	7.1 0 −0.4							4.9	5	5.1	5.3										
195 196 198 200																								

NOTES:

[1] For intermediate sizes (e.g. repair sizes), the radial thickness of the next smaller nominal diameter applies.

[2] Values of specific tangential force F_{tc} are calculated with mean land width (h_5).

This table is shown in ISO format. Commas represent decimal points.

TABLE 12—DIMENSIONS FOR GSF AND DSF COIL SPRING LOADED OIL CONTROL RINGS (CONCLUDED)

Dimensions are millimeters

Groove depth a_4 for h_1 shown in column				Groove depth and bridge a_{13} for h_1 shown in column				Number of slots	Slot width c_1 for h_1 shown in column				Coil spring groove diameter d_{14} for h_1 shown in column				Coil spring diameter d_7 for h_1 shown in column				Tangential force F_{tc}, N for unit contact pressure $P_{ou} = 1\ N/mm^2$ for h_1 shown in column				Recommended class of nominal contact pressure N/mm^2		
1	2	3	4	1	2	3	4		1	2	3	4	1	2	3	4	1	2	3	4	1	2	3	4	PNL Low	PNR Mean	PNM High
						2,2 0 −0,2	2,2 0 −0,2						2,5 +0,1 0	2,7 +0,1 0	2,9 +0,1 0	3,1 +0,1 0	2,4 0 −0,1	2,6 0 −0,1	2,8 0 −0,1	3 0 −0,1	50 50,4 50,8 51,2 51,6	50 50,4 50,8 51,2 51,6	50 50,4 50,8 51,2 51,6	62,5 63 63,5 64 64,5	0,75 0,74 0,74 0,73 0,73	1 0,99 0,98 0,98 0,97	1,25 1,24 1,23 1,22 1,21
0,5 ±0,1	0,7 ±0,1	0,7 ±0,1	0,9 ±0,1	2,3 0 −0,2	2,3 0 −0,2	2,3 0 −0,2	2,3 0 −0,2	12	1 ±0,1	1,2 ±0,1	1,2 ±0,1	1,4 ±0,1	2,7 +0,1 0	2,9 +0,1 0	3,1 +0,1 0	3,3 +0,15 0	2,6 0 −0,1	2,8 0 −0,1	3 0 −0,1	3,2 0 −0,1	52 52,4 52,8 53,2 53,6 54 54,4 54,8 55,2 55,6 56 56,4 56,8 57,2 57,6 58 58,4 58,8 59,2 59,6	52 52,4 52,8 53,2 53,6 54 54,4 54,8 55,2 55,6 56 56,4 56,8 57,2 57,6 58 58,4 58,8 59,2 59,6	52 52,4 52,8 53,2 53,6 54 54,4 54,8 55,2 55,6 56 56,4 56,8 57,2 57,6 58 58,4 58,8 59,2 59,6	65 65,5 66 66,5 67 67,5 68 68,5 69 69,5 70 70,5 71 71,5 72 72,5 73 73,5 74 74,5	0,72 0,72 0,71 0,71 0,71 0,71 0,7 0,7 0,7 0,69 0,69 0,69 0,68 0,68 0,68 0,68 0,67 0,67 0,67 0,66	0,96 0,96 0,95 0,95 0,94 0,94 0,94 0,93 0,93 0,92 0,92 0,92 0,91 0,91 0,9 0,9 0,9 0,89 0,89 0,88	1,2 1,2 1,19 1,19 1,18 1,18 1,17 1,17 1,16 1,16 1,15 1,15 1,14 1,14 1,13 1,13 1,12 1,12 1,11 1,11
0,7 ±0,1	0,7 ±0,1	0,9 ±0,1	1,2 ±0,15	2,6 0 −0,2	2,4 0 −0,2	2,3 0 −0,2	2,3 0 −0,2	14	1,2 ±0,1	1,2 ±0,1	1,4 ±0,1	1,6 ±0,1	2,9 +0,1 0	3,1 +0,1 0	3,3 +0,15 0	3,3 +0,15 0	2,8 0 −0,1	3 0 −0,1	3,2 0 −0,1	3,2 0 −0,1	60 60,8 61,6 62 62,4 63,2 64 64,8 65,6	60 60,8 61,6 62 62,4 63,2 64 64,8 65,6	75 76 77 77,5 78 79 80 81 82	90 91,2 92,4 93 93,6 94,8 96 97,2 98,4	0,66 0,65 0,65 0,65 0,64 0,64 0,63 0,62 0,62	0,88 0,87 0,86 0,86 0,86 0,85 0,84 0,83 0,82	1,1 1,09 1,08 1,08 1,07 1,06 1,05 1,04 1,03
					2,6 0 −0,2	2,6 0 −0,2	2,4 0 −0,2									3,7 +0,15 0				3,6 0 −0,1	66 66,4 67,2 68 68,8 69,6	66 66,4 67,2 68 68,8 69,6	82,5 83 84 85 86 87	99 99,6 100,8 102 103,2 104,4	0,62 0,61 0,61 0,6 0,59 0,59	0,82 0,82 0,81 0,8 0,79 0,78	1,03 1,02 1,01 1 0,99 0,98
0,7 ±0,1	0,9 ±0,1	1,2 ±0,15	1,5 ±0,15	2,8 0 −0,2	2,8 0 −0,2	2,7 0 −0,2	2,7 0 −0,2		1,2 ±0,1	1,4 ±0,1	1,6 ±0,1	1,8 ±0,15	3,1 −0,1 0	3,3 +0,15 0	3,7 +0,15 0	4,15 +0,15 0	3 0 −0,1	3,2 0 −0,1	3,6 0 −0,1	4 0 −0,12	70 70,4 71,2 72 72,8 73,6 74 74,4 75,2	87,5 88 89 90 91 92 92,5 93 94	105 105,6 106,8 108 109,2 110,4 111 111,6 112,8	105 105,6 106,8 108 109,2 110,4 111 111,6 112,8	0,59 0,58 0,58 0,57 0,56 0,56 0,56 0,55 0,55	0,78 0,78 0,77 0,76 0,75 0,75 0,74 0,74 0,73	0,98 0,97 0,96 0,95 0,94 0,93 0,93 0,92 0,91
				3 0 −0,2	2,9 0 −0,2								3,3 +0,15 0	3,7 +0,15 0	4,15 +0,15 0	4,55 +0,15 0	3,2 0 −0,1	3,6 0 −0,1	4 0 −0,12	4,4 0 −0,12	76 76,8 77,6 78 78,4 79,2 80	95 96 97 97,5 98 99 100	114 115,2 116,4 117 117,6 118,8 200	114 115,2 116,4 117 117,6 118,8 200	0,54 0,53 0,53 0,53 0,52 0,52 0,51	0,72 0,71 0,7 0,7 0,7 0,69 0,68	0,9 0,89 0,88 0,88 0,87 0,86 0,85

This table is shown in ISO format. Commas represent decimal points.

TABLE 13—DIMENSIONS FOR DSF-NG COIL SPRING LOADED OIL CONTROL RINGS

Dimensions are millimeters

Nominal diameter d_1	Radial thickness over coil spring a_{12} for h_1 shown in column				Ring width h_1 Column					Closed gap s_1	Radial wall thickness a_1 for h_1 shown in column					Land width h_5 for h_1 shown in column				Land spacing $\approx B_3$ for h_1 shown in column				Groove depth a_4 for h_1 shown in column			
	1	2	3	4	1	2	3	4	Tolerance		1	2	3	4	Tolerance	1	2	3	4	1	2	3	4	1	2	3	4
60 61 62 63 64	—	3,5 0 −0,25	3,5 0 −0,25							0,2 +0,2 0	—	2,4	2,4														
65 66 67 68 69	—	3,6 0 −0,25	3,7 0 −0,25	3,7 0 −0,25	—	3	3,5	4			—	2,45	2,5	2,5		—	0,35 ±0,1	0,35 ±0,1	0,4 ±0,1	—	1,55	1,85	2,3		0,5 ±0,1	0,5 ±0,1	0,5 ±0,1
70 71 72 73 74			3,8 0 −0,25	3,9 0 −0,25							—	2,55	2,6	2,6													
75 76 77 78 79	—	3,7 0 −0,25	3,9 0 −0,25	4 0 −0,25					−0,010 −0,030 For phosphated PO surface: +0,005 −0,025		—	2,6	2,7	2,7	±0,15 within a ring 0,15 max.												
80 81 82 83 84	3,8 0 −0,25	4 0 −0,25	4,1 0 −0,25	4,1 0 −0,25						0,25 +0,25 0	2,7	2,7	2,8	2,8													
85 86 87 88 89					3	3,5	4	4,5			2,8	2,8	2,9	2,9		0,35 ±0,1	0,35 ±0,1	0,4 ±0,1	0,4 ±0,1	1,55	1,85	2,3	2,65	0,5 ±0,1	0,5 ±0,1	0,5 ±0,1	0,7 ±0,1
90 91 92 93 94	3,9 0 −0,25	4,1 0 −0,25	4,2 0 −0,25	4,2 0 −0,25							2,85	2,9	2,95	3													
95 96 97 98 99	4 0 −0,25									0,3	2,95	3	3,05	3,1													
100 101 102 103 104	4,2 0 −0,25	4,4 0 −0,25	4,5 0 −0,25	4,7 0 −0,25						+0,25 0	3,05	3,1	3,15	3,2													
105 106 107 108 109											3,1	3,15	3,2	3,3													
110 111 112 113 114	4,3 0 −0,25	4,5 0 −0,25	4,6 0 −0,25	4,8 0 −0,25	3,5	4	4,5	5			3,2	3,25	3,3	3,4	±0,20 within a ring 0,20 max	0,35 ±0,1	0,4 ±0,1	0,4 ±0,1	0,4 ±0,1	1,85	2,3	2,65	2,9	0,5 ±0,1	0,5 ±0,1	0,7 ±0,1	0,7 ±0,1
115 116 117 118 119	4,4 0 −0,25	4,6 0 −0,25	4,7 0 −0,25	4,9 0 −0,25						0,35 +0,30 0	3,3	3,35	3,4	3,5													
120 121 122 123 124			4,8 0 −0,25	5 0 −0,25							3,4	3,45	3,5	3,6													

This table is shown in ISO format. Commas represent decimal points.

TABLE 13—DIMENSIONS FOR DSF-NG COIL SPRING LOADED OIL CONTROL RINGS (CONTINUED)

Dimensions are millimeters

Groove depth and ring a_{13} for h_1 shown in column				Number of slots	Slot width c_1 for h_1 shown in column				Coil spring groove diameter d_{14} for h_1 shown in column				Coil spring diameter d_7 for h_1 shown in column				Tangential force F_{tc}, N for unit contact pressure $p_{ou} = 1$ N/mm² for h_1 shown in column				Recommended class of nominal contact pressure N/mm²		
1	2	3	4		1	2	3	4	1	2	3	4	1	2	3	4	1	2	3	4	PNL Low	PNR Mean	PNM High
—	1,5 0 −0,15	1,5 0 −0,15	1,5 0 −0,15	8	—	0,7 ±0,1	0,8 ±0,1	1 +0,1	—	2,1 +0,1 0	2,3 +0,1 0	2,5 +0,1 0	—	2 0 −0,1	2,2 0 −0,1	2,4 0 −0,1	—	21 21,4 21,7 22,1 22,4	21 21,4 21,7 22,1 22,4	24 24,4 24,8 25,2 25,6	1,14 1,13 1,13 1,12 1,12	1,52 1,51 1,5 1,5 1,49	1,9 1,89 1,88 1,87 1,86
—	1,6 0 −0,15	1,6 0 −0,15															—	22,8 23,1 23,5 23,8 24,2	22,8 23,1 23,5 23,8 24,2	26 26,4 26,8 27,2 27,6	1,11 1,1 1,1 1,09 1,09	1,48 1,47 1,46 1,46 1,45	1,85 1,84 1,83 1,82 1,81
																	—	24,5 24,9 25,2 25,6 25,9	24,5 24,9 25,2 25,6 25,9	28 28,4 28,8 29,2 29,6	1,08 1,07 1,07 1,06 1,06	1,44 1,43 1,42 1,42 1,41	1,8 1,79 1,78 1,77 1,76
	1,7 0 −0,15	1,6 0 −0,15															—	26,3 26,6 27 27,3 27,7	26,3 26,6 27 27,3 27,7	30 30,4 30,8 31,2 31,6	1,05 1,04 1,04 1,03 1,03	1,4 1,39 1,38 1,38 1,37	1,75 1,74 1,73 1,72 1,71
1,8 0 −0,15	1,7 0 −0,15			10													28 28,4 28,7 29,1 29,4	28 28,4 28,7 29,1 29,4	32 32,4 32,8 33,2 33,6	32 32,4 32,8 33,2 33,6	1,02 1,01 1,01 1 1	1,36 1,35 1,34 1,34 1,33	1,7 1,69 1,68 1,67 1,66
	1,8 0 −0,15	1,7 0 −0,15			0,7 ±0,1	0,8 ±0,1	1 ±0,1	1,2 ±0,1	2,1 +0,1 0	2,3 +0,1 0	2,5 +0,1 0	2,5 +0,1 0	2 0 −0,1	2,2 0 −0,1	2,4 0 −0,1	2,4 0 −0,1	29,8 30,1 30,5 30,8 31,2	29,8 30,1 30,5 30,8 31,2	34 34,4 34,8 35,2 35,6	34 34,4 34,8 35,2 35,6	0,99 0,98 0,98 0,97 0,97	1,32 1,31 1,3 1,3 1,29	1,65 1,64 1,63 1,62 1,61
1,9 0 −0,15																	31,5 31,9 32,2 32,6 32,9	31,5 31,9 32,2 32,6 32,9	36 36,4 36,8 37,2 37,6	36 36,4 36,8 37,2 37,6	0,96 0,95 0,95 0,94 0,94	1,28 1,27 1,26 1,26 1,25	1,6 1,59 1,58 1,57 1,56
	1,9 0 −0,15	1,8 0 −0,15	1,8 0 −0,15														33,3 33,6 34 34,3 34,7	33,3 33,6 34 34,3 34,7	38 38,4 38,8 39,2 39,6	38 38,4 38,8 39,2 39,6	0,93 0,92 0,92 0,91 0,91	1,24 1,23 1,22 1,22 1,21	1,55 1,54 1,53 1,52 1,51
2 0 −0,15																	35 35,4 35,7 36,1 36,4	40 40,4 40,8 41,2 41,6	40 40,4 40,8 41,2 41,6	40 40,4 40,8 41,2 41,6	0,9 0,89 0,89 0,88 0,88	1,2 1,19 1,18 1,18 1,17	1,5 1,49 1,48 1,47 1,46
	2 0 −0,15	1,3 0 −0,2	1,9 0 −0,2														36,8 37,1 37,5 37,8 38,2	42 42,4 42,8 43,2 43,6	42 42,4 42,8 43,2 43,6	42 42,4 42,8 43,2 43,6	0,87 0,86 0,86 0,85 0,85	1,16 1,15 1,14 1,14 1,13	1,45 1,44 1,43 1,42 1,41
2,1 0 −0,2	2,1 0 −0,2	2 0 −0,2	2 0 −0,2		0,8 ±0,1	1 ±0,1	1,2 ±0,1	1,2 ±0,1	2,3 −0,1 0	2,5 +0,1 0	2,7 +0,1 0	2,9 −0,1 0	2,2 0 −0,1	2,4 0 −0,1	2,6 0 −0,1	2,8 0 −0,1	38,5 38,9 39,2 39,6 39,9	44 44,4 44,8 45,2 45,6	44 44,4 44,8 45,2 45,6	44 44,4 44,8 45,2 45,6	0,84 0,83 0,83 0,82 0,82	1,12 1,11 1,1 1,1 1,09	1,4 1,39 1,38 1,37 1,36
		2,1 0 −0,2	2,1 0 −0,2														40,3 40,6 41 41,3 41,7	46 46,4 46,8 47,2 47,6	46 46,4 46,8 47,2 47,6	46 46,4 46,8 47,2 47,6	0,81 0,8 0,8 0,79 0,79	1,08 1,07 1,06 1,06 1,05	1,35 1,34 1,33 1,32 1,31
2,2 0 −0,2	2,2 0 −0,2	2,2 0 −0,2	2,2 0 −0,2	12													42 42,4 42,7 43,1 43,4	48 48,4 48,8 49,2 49,6	48 48,4 48,8 49,2 49,6	48 48,4 48,8 49,2 49,6	0,78 0,77 0,77 0,76 0,76	1,04 1,03 1,02 1,02 1,01	1,3 1,29 1,28 1,27 1,26

This table is shown in ISO format. Commas represent decimal points.

TABLE 13—DIMENSIONS FOR DSF-NG COIL SPRING LOADED OIL CONTROL RINGS (CONTINUED)

Dimensions are millimeters

Nominal diameter d_1	Radial thickness over coil spring a_{12} for h_1 shown in column				Ring width h_1 Column					Closed gap s_1	Radial wall thickness a_1 for h_1 shown in column					Land width h_5 for h_1 shown in column				Land spacing $\approx B_3$ for h_1 shown in column				Groove depth a_4 for h_1 shown in column			
	1	2	3	4	1	2	3	4	Tolerance		1	2	3	4	Tolerance	1	2	3	4	1	2	3	4	1	2	3	4
125 126 127 128 129	4,7 0 −0,25	4,9 0 −0,25	5 0 −0,25	5,2 0 −0,25						0,35 +0,30 0	3,5	3,6	3,6	3,7													
130 131 132 133 134			5,1 0 −0,25	5,3 0 −0,25							3,5	3,6	3,7	3,8													
135 136 137 138 139					4	4,5	5	6			3,6	3,7	3,8	3,9		0,4 ±0,1	0,4 ±0,1	0,4 ±0,1	0,5 ±0,1	2,3	2,65	2,9	3,5	0,5 ±0,1	0,7 ±0,1	0,7 ±0,1	0,9 ±0,1
140 141 142 143 144	4,9 0 −0,25	5,1 0 −0,25	5,3 0 −0,25	5,5 0 −0,25					−0,010 −0,040 For phosphated PO surface: +0,005 −0,035	0,4 +0,35 0	3,7	3,8	3,9	4													
145 146 147 148 149											3,8	3,9	4	4,1													
150 152 154 155 156 158	5,4 0 −0,25	5,4 0 −0,25	5,5 0 −0,25	5,5 0 −0,25							3,9	4	4,1	4,2	±0,20 within a ring 0,20 max.												
160 162 164					4,5	5	6	7		0,45 +0,35 0						0,4 ±0,1	0,4 ±0,1	0,5 ±0,1	0,6 ±0,12	2,65	2,9	3,5	3,9	0,7 ±0,1	0,7 ±0,1	0,9 ±0,1	1,2 ±0,15
165 166 168	5,4 0 −0,25	5,6 0 −0,25	5,8 0 −0,25	6 0 −0,25							4	4,2	4,3	4,4													
170 172 174																											
175 176 178 180 182 184	5,8 0 −0,35	6 0 −0,35	6,3 0 −0,35	6,7 0 −0,35							4,6	4,7	4,8	5													
185 186 188 190					5	6	7	8		0,55 −0,40 0						0,4 ±0,1	0,5 ±0,1	0,6 ±0,12	0,6 ±0,12	2,9	3,5	3,9	4,3	0,7 −0,1	0,9 ±0,1	1,2 ±0,15	1,5 ±0,15
192 194 195 196 198 200	6,2 0 −0,4	6,5 0 −0,4	6,7 0 −0,4	7,1 0 −0,4							4,9	5	5,1	5,3													

NOTES:

[1] For intermediate sizes (e.g. repair sizes), the radial thickness of the next smaller nominal diameter applies.

[2] Values of specific tangential force F_{tc} are calculated with mean land width (h_5).

This table is shown in ISO format. Commas represent decimal points.

TABLE 13—DIMENSIONS FOR DSF-NG COIL SPRING LOADED OIL CONTROL RINGS (CONCLUDED)

Dimensions are millimeters

Groove depth and bridge a_{13} for h_1 shown in column				Number of slots	Slot width c_1 for h_1 shown in column				Coil spring groove diameter d_{14} for h_1 shown in column				Coil spring diameter d_7 for h_1 shown in column				Tangential force F_{tc}, N for unit contact pressure $p_{ou} = 1\ N/mm^2$ for h_1 shown in column				Recommended class of nominal contact pressure N/mm^2		
1	2	3	4		1	2	3	4	1	2	3	4	1	2	3	4	1	2	3	4	PNL Low	PNR Mean	PNM High
		2.2 0 −0.2	2.2 0 −0.2	12	1 ±0.1	1.2 ±0.1	1.2 ±0.1	1.4 ±0.1	2.5 +0.1 0	2.7 +0.1 0	2.9 +0.1 0	3.1 +0.1 0	2.4 0 −0.1	2.6 0 −0.1	2.8 0 −0.1	3 0 −0.1	50 50.4 50.8 51.2 51.6	50 50.4 50.8 51.2 51.6	50 50.4 50.8 51.2 51.6	62.5 63 63.5 64 64.5	0,75 0,74 0,74 0,73 0,73	1, 0,99 0,98 0,98 0,97	1,25 1,24 1,23 1,22 1,21
2.3 0 −0.2	2.3 0 −0.2	2.3 0 −0.2	2.3 0 −0.2														52 52.4 52.8 53.2 53.6	52 52.4 52.8 53.2 53.6	52 52.4 52.8 53.2 53.6	65 65.5 66 66.5 67	0,72 0,72 0,71 0,71 0,71	0,96 0,96 0,95 0,95 0,94	1,2 1,2 1,19 1,19 1,18
																	54 54.4 54.8 55.2 55.6	54 54.4 54.8 55.2 55.6	54 54.4 54.8 55.2 55.6	67.5 68 68.5 69 69.5	0,71 0,7 0,7 0,7 0,69	0,94 0,94 0,93 0,93 0,92	1,18 1,17 1,17 1,16 1,16
									2.7 +0.1 0	2.9 +0.1 0	3.1 +0.1 0	3.3 +0.15 0	2.6 0 −0.1	2.8 0 −0.1	3 0 −0.1	3.2 0 −0.1	56 56.4 56.8 57.2 57.6	56 56.4 56.8 57.2 57.6	56 56.4 56.8 57.2 57.6	70 70.5 71 71.5 72	0,69 0,69 0,68 0,68 0,68	0,92 0,92 0,91 0,91 0,9	1,15 1,15 1,14 1,14 1,13
																	58 58.4 58.8 59.2 59.6	58 58.4 58.8 59.2 59.6	58 58.4 58.8 59.2 59.6	72.5 73 73.5 74 74.5	0,68 0,67 0,67 0,67 0,66	0,9 0,9 0,89 0,89 0,88	1,13 1,12 1,12 1,11 1,11
	2.4 0 −0.2	2.3 0 −0.2	2.3 0 −0.2								3.3 +0.15 0					3.2 0 −0.1	60 60.8 61.6	60 60.8 61.6	75 76 77	90 91.2 92.4	0,66 0,65 0,65	0,88 0,87 0,86	1,1 1,09 1,08
																	62 62.4 63.2	62 62.4 63.2	77,5 78 79	93 93.6 94.8	0,65 0,64 0,64	0,86 0,86 0,85	1,08 1,07 1,06
2.6 0 −0.2					1.2 ±0.1	1.2 ±0.1	1.4 ±0.1	1.6 ±0.1	2.9 +0.1 0	3.1 +0.1 0	3.3 +0.15 0		2.8 0 −0.1	3 0 −0.1	3.2 0 −0.1		64 64.8 65.6	64 64.8 65.6	80 81 82	96 97.2 98.4	0,63 0,62 0,62	0,84 0,83 0,82	1,05 1,04 1,03
		2.6 0 −0.2	2.6 0 −0.2	2.4 0 −0.2								3.7 −0.15 0				3.6 0 −0.1	66 66.4 67.2	66 66.4 67.2	82.5 83 84	99 99.6 100.8	0,62 0,61 0,61	0,82 0,82 0,81	1,03 1,02 1,01
				14													68 68.8 69.6	68 68.8 69.6	85 86 87	102 103.2 104.4	0,6 0,59 0,59	0,8 0,79 0,78	1 0,99 0,98
2.8 0 −0.2	2.8 0 −0.2								3.1 +0.1 0	3.3 +0.15 0	3.7 −0.15 0	4.15 −0.15 0	3 0 −0.1	3.2 0 −0.1	3.6 0 −0.1	4 0 −0.12	70 70.4 71.2	87.5 88 89	105 105.6 106.8	105 105.6 106.8	0,59 0,58 0,58	0,78 0,78 0,77	0,98 0,97 0,96
																	72 72.8 73.6	90 91 92	108 109.2 110.4	108 109.2 110.4	0,57 0,56 0,56	0,76 0,75 0,75	0,95 0,94 0,93
		2.7 0 −0.2	2.7 0 −0.2		1.2 ±0.1	1.4 ±0.1	1.6 ±0.1	1.8 ±0.15									74 74.4 75.2	92.5 93 94	111 111.6 112.8	111 111.6 112.8	0,56 0,55 0,55	0,74 0,74 0,73	0,93 0,92 0,91
																	76 76.8 77.6	95 96 97	114 115.2 116.4	114 115.2 116.4	0,54 0,53 0,53	0,72 0,71 0,7	0,9 0,89 0,88
3 0 −0.2	2.9 0 −0.2								3.3 +0.15 0	3.7 +0.15 0	4.15 +0.15 0	4.55 −0.15 0	3.2 0 −0.1	3.6 0 −0.1	4 0 −0.12	4.4 0 −0.12	78 78.4 79.2	97,5 98 99	117 117.6 118.8	117 117.6 118.8	0,53 0,52 0,52	0,7 0,7 0,69	0,88 0,87 0,86
																	80	100	120	120	0,51	0,68	0,85

This table is shown in ISO format. Commas represent decimal points.

TABLE 14—DIMENSIONS FOR SSF-L COIL SPRING LOADED OIL CONTROL RINGS

Dimensions are millimeters

Nominal diameter d_1	Radial thickness over coil spring a_{12} for h_1 shown in column				Ring width h_1 Column					Closed gap s_1	Radial wall thickness a_1 for h_1 shown in column					Land width h_5	Radius r_3	Groove depth a_4 for h_1 shown in column			
	1	2	3	4	1	2	3	4	Tolerance		1	2	3	4	Tolerance			1	2	3	4
60																					
61		3,5 0 −0,25	3,5 0 −0,25								−	2,4	2,4								
62	−																				
63																					
64				3,7 0 −0,25										2,5								
65			3,7 0 −0,25							0,2 +0,2 0	−	2,45	2,5								
66																					
67		3,6 0 −0,25																			
68	−																				
69																					
70			3,8 0 −0,25	3,9 0 −0,25	−	3	3,5	4			−	2,55	2,6	2,6					−	0,5 ±0,1	0,5 ±0,1	0,5 ±0,1
71																					
72	−																				
73																					
74																					
75									−0,010 −0,030 For phosphated PO surface: +0,005 −0,025												
76		3,7 0 −0,25	3,9 0 −0,25	4 0 −0,25							−	2,6	2,7	2,7							
77	−																				
78																					
79																±0,15 within a ring 0,15 max.					
80	3,8 0 −0,25									0,25 +0,25 0											
81																					
82											2,7	2,7	2,8	2,8							
83																					
84		4 0 −0,25	4,1 0 −0,25	4,1 0 −0,25																	
85																					
86											2,8	2,8	2,9	2,9							
87																					
88																					
89	3,9 0 −0,25				3	3,5	4	4,5									0,6 ±0,1	0,2 max	0,5 ±0,1	0,5 ±0,1	0,5 ±0,1	0,7 ±0,1
90											2,85	2,9	2,95	3							
91																					
92																					
93																					
94		4,1 0 −0,25	4,2 0 −0,25	4,2 0 −0,25																	
95																					
96	4 0 −0,25										2,95	3	3,05	3,1							
97																					
98										0,3 +0,25 0											
99																					
100																					
101											3,05	3,1	3,15	3,2							
102																					
103																					
104	4,2 0 −0,25	4,4 0 −0,25	4,5 0 −0,25	4,7 0 −0,25																	
105																					
106											3,1	3,15	3,2	3,3							
107																					
108																					
109																					
110																±0,20 within a ring 0,20 max					
111	4,3 0 −0,25	4,5 0 −0,25	4,6 0 −0,25	4,8 0 −0,25	3,5	4	4,5	5			3,2	3,25	3,3	3,4					0,5 ±0,1	0,5 ±0,1	0,7 ±0,1	0,7 ±0,1
112																					
113																					
114																					
115										0,35 +0,30 0											
116			4,7 0 −0,25	4,9 0 −0,25							3,3	3,35	3,4	3,5							
117																					
118																					
119	4,4 0 −0,25	4,6 0 −0,25																			
120																					
121			4,8 0 −0,25	5 0 −0,25							3,4	3,45	3,5	3,6							
122																					
123																					
124																					

This table is shown in ISO format. Commas represent decimal points.

TABLE 14—DIMENSIONS FOR SSF-L COIL SPRING LOADED OIL CONTROL RINGS (CONTINUED)

Dimensions are millimeters

Groove depth and bridge a_{13} for h_1 shown in column				Number of slots	Slot width c_1 for h_1 shown in column				Coil spring groove diameter d_{14} for h_1 shown in column				Coil spring diameter d_7 for h_1 shown in column				Tangential force F_{tc}, N for unit contact pressure $p_{ou}=1$ N/mm²	Recommended class of nominal contact pressure N/mm²		
1	2	3	4		1	2	3	4	1	2	3	4	1	2	3	4		PNE Low	PNL Mean	PNR High
—	1,5 0 −0,15			8	—	0,7 ±0,1	0,7 ±0,1		—	2,1 +0,1 0	2,1 +0,1 0		—	2 0 −0,1	2 0 −0,1	2,2 0 −0,1	36	0,86	1,14	1,52
		1,5 0 −0,15															36,6	0,85	1,13	1,51
			1,5 0 −0,15				0,8 ±0,1				2,3 +0,1 0						37,2	0,85	1,13	1,5
	1,6 0 −0,15																37,8	0,84	1,12	1,5
																	38,4	0,84	1,12	1,49
																	39	0,83	1,11	1,48
																	39,6	0,83	1,1	1,47
																	40,2	0,82	1,1	1,46
																	40,8	0,82	1,09	1,46
																	41,4	0,81	1,09	1,45
		1,6 0 −0,15								2,1 +0,1 0				2 0 −0,1			42	0,81	1,08	1,44
											2,3 +0,1 0				2,2 0 −0,1		42,6	0,81	1,07	1,43
																	43,2	0,8	1,07	1,42
																	43,8	0,8	1,06	1,41
								1 +0,1				2,5 +0,1 0				2,4 0 −0,1	44,4	0,79	1,06	1,41
	1,7 0 −0,15		1,6 0 −0,15														45	0,79	1,05	1,4
																	45,6	0,78	1,04	1,39
																	46,2	0,78	1,04	1,38
																	46,8	0,77	1,03	1,38
																	47,4	0,77	1,03	1,37
1,8 0 −0,15	1,7 0 −0,15			10													48	0,77	1,02	1,36
																	48,6	0,76	1,01	1,35
	1,8 0 −0,15		1,7 0 −0,15														49,2	0,76	1,01	1,34
																	49,8	0,75	1	1,34
																	50,4	0,75	1	1,33
1,9 0 −0,15					0,7 ±0,1	0,8 ±0,1	1 ±0,1	1,2 ±0,1	2,1 +0,1 0	2,3 +0,1 0	2,5 +0,1 0	2,5 +0,1 0	2 0 −0,1	2,2 0 −0,1	2,4 0 −0,1	2,4 0 −0,1	51	0,74	0,99	1,32
																	51,6	0,74	0,98	1,31
																	52,2	0,74	0,98	1,3
																	52,8	0,73	0,97	1,3
																	53,4	0,73	0,97	1,29
	1,9 0 −0,15	1,8 0 −0,15	1,8 0 −0,15														54	0,72	0,96	1,28
																	54,6	0,72	0,95	1,27
																	55,2	0,71	0,95	1,26
																	55,8	0,71	0,94	1,26
																	56,4	0,7	0,94	1,25
																	57	0,7	0,93	1,24
																	57,6	0,69	0,92	1,23
																	58,2	0,69	0,92	1,22
																	58,8	0,68	0,91	1,22
																	59,4	0,68	0,91	1,21
2 0 −0,15																	60	0,68	0,9	1,2
																	60,6	0,67	0,89	1,19
	2 0 −0,2	1,9 0 −0,2	1,9 0 −0,2														61,2	0,67	0,89	1,18
																	61,8	0,66	0,88	1,18
																	62,4	0,66	0,88	1,17
																	63	0,65	0,87	1,16
																	63,6	0,65	0,86	1,15
																	64,2	0,64	0,86	1,14
																	64,8	0,64	0,85	1,14
																	65,4	0,63	0,85	1,13
2,1 0 −0,2	2,1 0 −0,2	2 0 −0,2	2 0 −0,2	12	0,8 ±0,1	1 ±0,1	1,2 ±0,1	1,2 ±0,1	2,3 +0,1 0	2,5 +0,1 0	2,7 +0,1 0	2,9 +0,1 0	2,2 0 −0,1	2,4 0 −0,1	2,6 0 −0,1	2,9 0 −0,1	66	0,63	0,84	1,12
																	66,6	0,63	0,83	1,11
																	67,2	0,62	0,83	1,1
																	67,8	0,62	0,82	1,1
																	68,4	0,61	0,82	1,09
		2,1 0 −0,2	2,1 0 −0,2														69	0,61	0,81	1,08
																	69,6	0,6	0,8	1,07
																	70,2	0,6	0,8	1,06
																	70,8	0,59	0,79	1,06
																	71,4	0,59	0,79	1,05
2,2 0 −0,2	2,2 0 −0,2																72	0,59	0,78	1,04
		2,2 0 −0,2	2,2 0 −0,2														72,6	0,58	0,77	1,03
																	73,2	0,58	0,77	1,02
																	73,8	0,57	0,76	1,02
																	74,4	0,57	0,76	1,01

This table is shown in ISO format. Commas represent decimal points.

TABLE 14—DIMENSIONS FOR SSF-L COIL SPRING LOADED OIL CONTROL RINGS (CONTINUED)

Dimensions are millimeters

Nominal diameter d_1	Radial thickness over coil spring a_{12} for h_1 shown in column				Ring width h_1 Column					Closed gap s_1	Radial wall thickness a_1 for h_1 shown in column					Land width h_5	Radius r_3	Groove depth a_4 for h_1 shown in column			
	1	2	3	4	1	2	3	4	Tolerance		1	2	3	4	Tolerance			1	2	3	4
125 126 127 128 129	4,7 0 -0,25	4,9 0 -0,25	5 0 -0,25	5,2 0 -0,25						0,35 +0,30 0	3,5	3,6	3,6	3,7							
130 131 132 133 134			5,1 0 -0,25	5,3 0 -0,25							3,5	3,6	3,7	3,8							
135 136 137 138 139					4	4,5	5	6		0,4 +0,35 0	3,6	3,7	3,8	3,9			0,2 max.	0.5 ±0.1	0.7 ±0.1	0.7 ±0.1	0.9 ±0.1
140 141 142 143 144	4,9 0 -0,25	5,1 0 -0,25	5,3 0 -0,25	5,5 0 -0,25					-0,010 -0,040 For phosphated PO surface: +0,005 -0,035		3,7	3,8	3,9	4							
145 146 147 148 149											3,8	3,9	4	4,1							
150 152 154 155 156 158	5,4 0 -0,25	5,4 0 -0,25	5,5 0 -0,25	5,5 0 -0,25							3,9	4	4,1	4,2	±0,20 within a ring 0,20 max.	0,6 ±0,1						
160 162 164					4,5	5	6	7		0,45 +0,35 0									0.7 ±0.1	0.7 ±0.1	0.9 ±0.1	1.2 ±0.15
165 166 168	5,4 0 -0,25	5,6 0 -0,25	5,8 0 -0,25	6 0 -0,25							4	4,2	4,3	4,4							
170 172 174																					
175 176 178 180 182 184	5,8 0 -0,35	6 0 -0,35	6,3 0 -0,35	6,7 0 -0,35							4,6	4,7	4,8	5							
185 186 188					5	6	7	8		0,55 -0,40 0							0,5 max	0.7 ±0.1	0.9 ±0.1	1.2 ±0.15	1.5 ±0.15
190 192 194	6,2 0 -0,4	6,5 0 -0,4	6,7 0 -0,4	7,1 0 -0,4							4,9	5	5,1	5,3							
195 196 198 200																					

NOTES:

[1] For intermediate sizes (e.g. repair sizes), the radial thickness of the next smaller nominal diameter applies.

[2] Values of specific tangential force F_{tc} are calculated with mean land width (h_5).

This table is shown in ISO format. Commas represent decimal points.

TABLE 14—DIMENSIONS FOR SSF-L COIL SPRING LOADED OIL CONTROL RINGS (CONCLUDED)

Dimensions are millimeters

Groove depth and bridge a_{13} for h_1 shown in column				Number of slots	Slot width c_1 for h_1 shown in column				Coil spring groove diameter d_{14} for h_1 shown in column				Coil spring diameter d_7 for h_1 shown in column				Tangential force F_{tc}, N for unit contact pressure $p_{ou} = 1\ N/mm^2$	Recommended class of nominal contact pressure N/mm^2		
1	2	3	4		1	2	3	4	1	2	3	4	1	2	3	4		PNE Low	PNL Mean	PNR High
2,3 0 −0,2	2,3 0 −0,2	2,2 0 −0,2 2,3 0 −0,2	2,2 0 −0,2 2,3 0 −0,2	12	1 ±0,1	1,2 ±0,1	1,2 ±0,1	1,4 ±0,1	2,5 +0,1 0	2,7 +0,1 0	2,9 +0,1 0	3,1 +0,1 0	2,4 0 −0,1	2,6 0 −0,1	2,8 0 −0,1	3 0 −0,1	75 75,6 76,2 76,8 77,4	0,56 0,56 0,55 0,55 0,54	0,75 0,74 0,74 0,73 0,73	1 0,99 0,98 0,98 0,97
																	78 78,6 79,2 79,8 80,4	0,54 0,54 0,54 0,53 0,53	0,72 0,72 0,71 0,71 0,71	0,96 0,96 0,95 0,95 0,94
									2,7 −0,1 0	2,9 +0,1 0	3,1 +0,1 0	3,3 +0,15 0	2,6 0 −0,1	2,8 0 −0,1	3 0 −0,1	3,2 0 −0,1	81 81,6 82,2 82,8 83,4	0,53 0,53 0,52 0,52 0,52	0,71 0,7 0,7 0,7 0,69	0,94 0,94 0,93 0,93 0,92
																	84 84,6 85,2 85,8 86,4	0,52 0,52 0,51 0,51 0,51	0,69 0,69 0,68 0,68 0,68	0,92 0,92 0,91 0,91 0,9
																	87 87,6 88,2 88,8 89,4	0,51 0,5 0,5 0,5 0,5	0,68 0,67 0,67 0,67 0,66	0,9 0,9 0,89 0,89 0,88
2,6 0 −0,2	2,4 0 −0,2 2,6 0 −0,2	2,3 0 −0,2 2,6 0 −0,2	2,3 0 −0,2 2,4 0 −0,2	14	1,2 ±0,1	1,2 ±0,1	1,4 ±0,1	1,6 ±0,1	2,9 −0,1 0	3,1 +0,1 0	3,3 +0,15 0	3,3 +0,15 0	2,8 0 −0,1	3 0 −0,1	3,2 0 −0,1	3,2 0 −0,1	90 91,2 92,4	0,5 0,49 0,49	0,66 0,65 0,65	0,88 0,87 0,86
																	93 93,6 94,8	0,48 0,48 0,48	0,65 0,64 0,64	0,86 0,86 0,85
																	96 97,2 98,4	0,47 0,47 0,46	0,63 0,62 0,62	0,84 0,83 0,82
												3,7 +0,15 0				3,6 0 −0,1	99 99,6 100,8	0,46 0,46 0,45	0,62 0,61 0,61	0,82 0,82 0,81
																	102 103,2 104,4	0,45 0,45 0,44	0,6 0,6 0,59	0,8 0,79 0,78
2,8 0 −0,2	2,8 0 −0,2	2,7 0 −0,2	2,7 0 −0,2		1,2 ±0,1	1,4 ±0,1	1,6 ±0,1	1,8 ±0,15	3,1 −0,1 0	3,3 −0,15 0	3,7 −0,15 0	4,15 +0,15 0	3 0 −0,1	3,2 0 −0,1	3,6 0 −0,1	4 0 −0,12	105 105,6 106,8	0,44 0,44 0,43	0,59 0,58 0,58	0,78 0,78 0,77
																	108 109,2 110,4	0,43 0,42 0,42	0,57 0,56 0,56	0,76 0,75 0,74
3 0 −0,2	2,9 0 −0,2																111 111,6 112,8	0,42 0,41 0,41	0,56 0,55 0,55	0,74 0,74 0,73
									3,3 −0,15 0	3,7 −0,15 0	4,15 −0,15 0	4,55 +0,15 0	3,2 0 −0,1	3,6 0 −0,1	4 0 −0,12	4,4 0 −0,12	114 115,2 116,4	0,41 0,4 0,4	0,54 0,53 0,53	0,72 0,71 0,7
																	117 117,6 118,8	0,4 0,39 0,39	0,53 0,52 0,52	0,7 0,7 0,69
																	120	0,38	0,51	0,68

This table is shown in ISO format. Commas represent decimal points.

TABLE 15—DIMENSIONS FOR DSF-C COIL SPRING LOADED OIL CONTROL RINGS WITH SPECIAL RING WIDTH h_1 = 4.75 mm (3/16 in)

Dimensions are millimeters

Nominal diameter d_1	Radial thickness over coil spring a_{12}	Ring width h_1	Closed gap s_1	Radial wall thickness a_1	Land width h_5	Land spacing $\approx B_3$	Groove depth a_4	Groove depth and bridge a_{13}	Number of slots	Slot width c_1	Coil spring groove diameter d_{14}	Coil spring diameter d_7	Tangential force F_{tc}, N for unit contact pressure p_{ou} = 1 N/mm²	Recommended class of nominal contact pressure N/mm²		
				Tolerance										PNL Low	PNM Mean	PNH High
60													24	1,14	1,9	2,38
61													24,4	1,13	1,89	2,36
62													24,8	1,13	1,88	2,35
63													25,2	1,12	1,87	2,34
64	3,7 / 0 / -0,25			2,5							2,3 +0,1 0	2,2 0 -0,1	25,6	1,12	1,86	2,33
65			0,2					1,5 0 -0,15					26	1,11	1,85	2,31
66													26,4	1,1	1,84	2,3
67			+0,2 0										26,8	1,1	1,83	2,29
68													27,2	1,09	1,82	2,28
69									8				27,6	1,09	1,81	2,26
70													28	1,08	1,8	2,25
71	3,9 / 0 / -0,25			2,6									28,4	1,07	1,79	2,24
72													28,8	1,07	1,78	2,23
73													29,2	1,06	1,77	2,21
74													29,6	1,06	1,76	2,2
75								1,6 0 -0,15					30	1,05	1,75	2,19
76	4 / 0 / -0,25			2,7	±0,15 within a ring 0,15 max.								30,4	1,04	1,74	2,18
77													30,8	1,04	1,73	2,16
78													31,2	1,03	1,72	2,15
79							0,5 ±0,1						31,6	1,03	1,71	2,14
80			0,25										32	1,02	1,7	2,13
81				2,8									32,4	1,01	1,69	2,11
82			+0,25 0								2,5 +0,1 0	2,4 0 -0,1	32,8	1,01	1,68	2,1
83													33,2	1	1,67	2,09
84	4,1 / 0 / -0,25							1,7 0 -0,15					33,6	1	1,66	2,08
85													34	0,99	1,65	2,06
86													34,4	0,98	1,64	2,05
87			-0,010 -0,030 For phos- pnated PO surface +0,005 -0,025	2,9									34,8	0,98	1,63	2,04
88		4,75								1,2 ±0,1			35,2	0,97	1,62	2,03
89													35,6	0,97	1,61	2,01
90													36	0,96	1,6	2
91													36,4	0,95	1,59	1,99
92				3,0		2,65	0,4 ±0,07						36,8	0,95	1,58	1,98
93													37,2	0,94	1,57	1,96
94	4,2 / 0 / -0,25							1,8 0 -0,15					37,6	0,94	1,56	1,95
95													38	0,93	1,55	1,94
96				3,1									38,4	0,92	1,54	1,93
97									10				38,8	0,92	1,53	1,91
98													39,2	0,91	1,52	1,9
99			0,3										39,6	0,91	1,51	1,89
100													40	0,9	1,5	1,88
101			+0,25 0										40,4	0,89	1,49	1,86
102				3,15									40,8	0,89	1,48	1,85
103													41,2	0,88	1,47	1,84
104	4,5 / 0 / -0,25							1,9 0 -0,2					41,6	0,88	1,46	1,83
105													42	0,87	1,45	1,81
106													42,4	0,86	1,44	1,8
107				3,2									42,8	0,86	1,43	1,79
108													43,2	0,85	1,42	1,78
109													43,6	0,85	1,41	1,76
110													44	0,84	1,4	1,75
111	4,6 / 0 / -0,25			3,3	±0,20 within a ring 0,20 max.		0,7 ±0,1	2,0 0 -0,2			2,7 +0,1 0	2,6 0 -0,1	44,4	0,83	1,39	1,74
112													44,8	0,83	1,38	1,73
113													45,2	0,82	1,37	1,71
114													45,6	0,82	1,36	1,7
115			0,35										46	0,81	1,35	1,69
116	4,7 / 0 / -0,25		-0,30 0					2,1 0 -0,2					46,4	0,8	1,34	1,68
117				3,4									46,8	0,8	1,33	1,66
118													47,2	0,79	1,32	1,65
119													47,6	0,79	1,31	1,64
120									12				48	0,78	1,3	1,63
121	4,8 / 0 / -0,25							2,2 0 -0,2					48,4	0,77	1,29	1,61
122				3,5									48,8	0,77	1,28	1,6
123													49,2	0,76	1,27	1,59
124													49,6	0,76	1,26	1,58

This table is shown in ISO format. Commas represent decimal points.

TABLE 15—DIMENSIONS FOR DSF-C COIL SPRING LOADED OIL CONTROL RINGS WITH SPECIAL RING WIDTH h_1 = 4.75 mm (3/16 in) (CONCLUDED)

Dimensions are millimeters

Nominal diameter d_1	Radial thickness over coil-spring a_{12}	Ring width h_1	Closed gap s_1	Radial wall thickness a_1	Land width h_5	Land spacing ≈ B_3	Groove depth a_4	Groove depth and bridge a_{13}	Number of slots	Slot width c_1	Coil-spring groove diameter d_{14}	Coil-spring diameter d_7	Tangential force F_{tc}, N for unit contact pressure p_{ou} = 1 N/mm²	Recommended class of nominal contact pressure N/mm² PNL Low	PNM Mean	PNH High
			Tolerance		Tolerance											
125			0,35										50	0,75	1,25	1,56
126													50,4	0,74	1,24	1,55
127			+0,30										50,8	0,74	1,23	1,54
128			0										51,2	0,73	1,22	1,53
129	4,9							2,3					51,6	0,73	1,21	1,51
130	0			3,6				0					52	0,72	1,2	1,5
131	−0,25							−0,2					52,4	0,72	1,2	1,49
132													52,8	0,71	1,19	1,49
133													53,2	0,71	1,19	1,48
134													53,6	0,71	1,18	1,48
135												2,7 2,6	54	0,71	1,18	1,47
136												+0,1 0	54,4	0,7	1,17	1,46
137				3,7					12			0 −0,1	54,8	0,7	1,17	1,46
138			0,4										55,2	0,7	1,16	1,45
139													55,6	0,69	1,16	1,44
140		−0,010											56	0,69	1,15	1,44
141		−0,040	+0,35					2,5					56,4	0,69	1,15	1,43
142	5,1	For phos-	0	3,8				0					56,8	0,68	1,14	1,43
143	0	phated						−0,2					57,2	0,68	1,14	1,42
144	−0,25	PO											57,6	0,68	1,13	1,41
145		surface:											58	0,68	1,13	1,41
146		+0,005											58,4	0,67	1,12	1,4
147		−0,035											58,8	0,67	1,12	1,39
148													59,2	0,67	1,11	1,39
149													59,6	0,66	1,11	1,38
150				3,9									60	0,66	1,1	1,38
152					±0,20								60,8	0,65	1,09	1,36
154					within a ring								61,6	0,65	1,08	1,35
155		4,75			0,20	0,4	2,65	0,7			1,2		62	0,65	1,08	1,34
156					max.	±0,07		±0,1			±0,1		62,4	0,64	1,07	1,34
158													63,2	0,64	1,06	1,33
160	5,4		0,45					2,6				2,9 2,8	64	0,63	1,05	1,31
162	0		+0,35					0				+0,1 0	64,8	0,63	1,04	1,3
164	−0,25		0					−0,2				0 −0,1	65,6	0,62	1,03	1,29
165													66	0,62	1,03	1,28
166				4									66,4	0,61	1,02	1,28
168													67,2	0,61	1,01	1,26
170													68	0,6	1	1,25
172													68,8	0,59	0,99	1,24
174									14				69,6	0,59	0,98	1,23
175													70	0,59	0,98	1,22
176	5,8							2,8				3,1 3	70,4	0,58	0,97	1,21
178	0							0				−0,1 0	71,2	0,58	0,96	1,2
180	−0,35			4,6				−0,2				0 −0,1	72	0,57	0,95	1,19
182													72,8	0,56	0,94	1,18
184													73,6	0,56	0,93	1,16
185			0,55										74	0,56	0,93	1,16
186			+0,40										74,4	0,55	0,92	1,15
188			0										75,2	0,55	0,91	1,14
190													76	0,54	0,9	1,13
192	6,2							3				3,3 3,2	76,8	0,53	0,89	1,11
194	0			4,9				0				+0,15 0	77,6	0,53	0,88	1,1
195	−0,4							−0,2				0 −0,1	78	0,53	0,88	1,09
196													78,4	0,52	0,87	1,09
198													79,2	0,52	0,86	1,08
200													80	0,51	0,85	1,06

NOTES:

[1] For intermediate sizes (e.g. repair sizes), the radial thickness of the next smaller nominal diameter applies.

[2] Values of specific tangential force F_{tc} are calculated with mean land width (h_5).

This table is shown in ISO format. Commas represent decimal points.

TABLE 16—DIMENSIONS FOR DSF-CNP COIL SPRING LOADED OIL CONTROL RINGS WITH SPECIAL RING WIDTH h_1 = 4.75 mm (3/16 in)

Dimensions are millimeters

Nominal diameter d_1	Radial thickness over coil spring a_{12}	Ring width h_1 Tolerance	Closed gap s_1	Radial wall thickness a_1 Tolerance	Land width h_5	Land spacing $\approx B_3$	Groove depth a_4	Groove depth and bridge a_{13}	Number of slots	Slot width c_1	Coil spring groove diameter d_{14}	Coil spring diameter d_7	Tangential force F_{tc}, N for unit contact pressure p_{ou} = 1 N/mm²	Recommended class of nominal contact pressure N/mm² PNE Low	PNL Mean	PNR High
60													24	1,14	1,9	2,38
61													24,4	1,13	1,89	2,36
62													24,8	1,13	1,88	2,35
63													25,2	1,12	1,87	2,34
64	3,7 0 −0,25			2,5							2,3 +0,1 0	2,2 0 −0,1	25,6	1,12	1,86	2,33
65			0,2										26	1,11	1,85	2,32
66			+0,2 0					1,5 0 −0,15					26,4	1,1	1,84	2,3
67													26,8	1,1	1,83	2,29
68													27,2	1,09	1,82	2,28
69													27,6	1,09	1,81	2,26
70									8				28	1,08	1,8	2,25
71													28,4	1,07	1,79	2,24
72	3,9 0 −0,25			2,6									28,8	1,07	1,78	2,23
73													29,2	1,06	1,77	2,21
74													29,6	1,06	1,76	2,2
75													30	1,05	1,75	2,19
76								1,6 0 −0,15					30,4	1,04	1,74	2,18
77	4 0 −0,25			2,7	±0,15 within a ring 0,15 max.								30,8	1,04	1,73	2,16
78													31,2	1,03	1,72	2,15
79													31,6	1,03	1,71	2,14
80			0,25				0,5 ±0,1						32	1,02	1,7	2,13
81			+0,25 0										32,4	1,01	1,69	2,11
82				2,8									32,8	1,01	1,68	2,1
83								1,7 0 −0,15					33,2	1	1,67	2,09
84	4,1 0 −0,25	−0,010 −0,030 For phosphated PO surface +0,005 −0,025									2,5 +0,1 0	2,4 0 −0,1	33,6	1	1,66	2,08
85													34	0,99	1,65	2,06
86													34,4	0,98	1,64	2,05
87				2,9									34,8	0,98	1,63	2,04
88													35,2	0,97	1,62	2,03
89										1,2 ±0,1			35,6	0,97	1,61	2,01
90		4,75											36	0,96	1,6	2
91						2,65							36,4	0,95	1,59	1,99
92				3,0	0,4 ±0,12								36,8	0,95	1,58	1,98
93													37,2	0,94	1,57	1,96
94	4,2 0 −0,25							1,8 0 −0,15					37,6	0,94	1,56	1,95
95													38	0,93	1,55	1,94
96													38,4	0,92	1,54	1,93
97				3,1					10				38,8	0,92	1,53	1,91
98													39,2	0,91	1,52	1,9
99			0,3										39,6	0,91	1,51	1,89
100			+0,25 0										40	0,9	1,5	1,88
101													40,4	0,89	1,49	1,86
102				3,15									40,8	0,89	1,48	1,85
103													41,2	0,88	1,47	1,84
104	4,5 0 −0,25							1,9 0 −0,2					41,6	0,88	1,46	1,83
105													42	0,87	1,45	1,81
106													42,4	0,86	1,44	1,8
107				3,2									42,8	0,86	1,43	1,79
108													43,2	0,85	1,42	1,78
109													43,6	0,85	1,41	1,76
110													44	0,84	1,4	1,75
111													44,4	0,83	1,39	1,74
112	4,6 0 −0,25			3,3	±0,20 within a ring 0,20 max.		0,7 ±0,1	2,0 0 −0,2			2,7 −0,1 0	2,6 0 −0,1	44,8	0,83	1,38	1,73
113													45,2	0,82	1,37	1,71
114													45,6	0,82	1,36	1,7
115													46	0,81	1,35	1,69
116			0,35					2,1 0 −0,2					46,4	0,8	1,34	1,68
117	4,7 0 −0,25		+0,30 0	3,4									46,8	0,8	1,33	1,66
118													47,2	0,79	1,32	1,65
119									12				47,6	0,79	1,31	1,64
120													48	0,78	1,3	1,63
121													48,4	0,77	1,29	1,61
122	4,8 0 −0,25			3,5				2,2 0 −0,2					48,8	0,77	1,28	1,6
123													49,2	0,76	1,27	1,59
124													49,6	0,76	1,26	1,58

This table is shown in ISO format. Commas represent decimal points.

TABLE 16—DIMENSIONS FOR DSF-CNP COIL SPRING LOADED OIL CONTROL RINGS WITH SPECIAL RING WIDTH h_1 = 4.75 mm (3/16 in) (CONCLUDED)

Dimensions are millimeters

Nominal diameter d_1	Radial thickness over coil spring a_{12}	Ring width h_1 Tolerance	Closed gap s_1	Radial wall thickness a_1 Tolerance	Land width h_5	Land spacing $\approx B_3$	Groove depth a_4	Groove depth and bridge a_{13}	Number of slots	Slot width c_1	Coil spring groove diameter d_{14}	Coil spring diameter d_7	Tangential force F_{tc}, N for unit contact pressure p_{ou} = 1 N/mm²	Recommended class of nominal contact pressure N/mm² PNL Low	PNM Mean	PNH High
125			0,35										50	0,75	1,25	1,56
126			+0,30										50,4	0,74	1,24	1,55
127			0										50,8	0,74	1,23	1,54
128													51,2	0,73	1,22	1,53
129	4,9							2,3					51,6	0,73	1,21	1,51
130	0		3,6					0					52	0,72	1,2	1,5
131	−0,25							−0,2					52,4	0,72	1,2	1,49
132													52,8	0,71	1,19	1,49
133													53,2	0,71	1,19	1,48
134													53,6	0,71	1,18	1,48
135											2,7	2,6	54	0,71	1,18	1,47
136				3,7							+0,1	0	54,4	0,7	1,17	1,46
137									12		0	−0,1	54,8	0,7	1,17	1,46
138			0,4										55,2	0,7	1,16	1,45
139													55,6	0,69	1,16	1,44
140			+0,35					2,5					56	0,69	1,15	1,44
141	5,1		0					0					56,4	0,69	1,15	1,43
142	0			3,8				−0,2					56,8	0,68	1,14	1,43
143	−0,25	−0,010											57,2	0,68	1,14	1,42
144		−0,040											57,6	0,68	1,13	1,41
145		For											58	0,68	1,13	1,41
146		phos-											58,4	0,67	1,12	1,4
147		phated											58,8	0,67	1,12	1,39
148		PO											59,2	0,67	1,11	1,39
149		surface											59,6	0,66	1,11	1,38
150		+0,005		3,9	±0,20								60	0,66	1,1	1,38
152		−0,035			within								60,8	0,65	1,09	1,36
154					a ring								61,6	0,65	1,08	1,35
155		4,75			0,20	0,4	2,65	0,7			1,2		62	0,65	1,08	1,34
156					max.	±0,12		±0,1			±0,1		62,4	0,64	1,07	1,34
158													63,2	0,64	1,06	1,33
160			0,45					2,6			2,9	2,8	64	0,63	1,05	1,31
162	5,4		+0,35					0			+0,1	0	64,8	0,63	1,04	1,3
164	0		0					−0,2			0	−0,1	65,6	0,62	1,03	1,29
165	−0,25												66	0,62	1,03	1,28
166				4									66,4	0,61	1,02	1,28
168													67,2	0,61	1,01	1,26
170													68	0,6	1	1,25
172													68,8	0,59	0,99	1,24
174									14				69,6	0,59	0,98	1,23
175													70	0,59	0,98	1,22
176								2,8			3,1	3	70,4	0,58	0,97	1,21
178	5,8			4,6				0			+0,1	0	71,2	0,58	0,96	1,2
180	0							−0,2			0	−0,1	72	0,57	0,95	1,19
182	−0,35												72,8	0,56	0,94	1,18
184													73,6	0,56	0,93	1,16
185			0,55										74	0,56	0,93	1,16
186			+0,40										74,4	0,55	0,92	1,15
188			0										75,2	0,55	0,91	1,14
190													76	0,54	0,9	1,13
192	6,2							3			3,3	3,2	76,8	0,53	0,89	1,11
194	0			4,9				0			−0,15	0	77,6	0,53	0,88	1,1
195	−0,4							−0,2			0	−0,1	78	0,53	0,88	1,09
196													78,4	0,52	0,87	1,09
198													79,2	0,52	0,86	1,08
200													80	0,51	0,85	1,06

NOTES:

[1] For intermediate sizes (e.g. repair sizes), the radial thickness of the next smaller nominal diameter applies.

[2] Values of specific tangential force F_{tc} are calculated with mean land width (h_5).

This table is shown in ISO format. Commas represent decimal points.

TABLE 17—DIMENSIONS FOR SSF COIL SPRING LOADED OIL CONTROL RINGS WITH SPECIAL RING WIDTH h_1 = 4.75 mm (3/16 in)

Dimensions are millimeters

Nominal diameter d_1	Radial thickness over coil spring a_{12}	Ring width h_1 Tolerance	Closed gap s_1	Radial wall thickness a_1 Tolerance	Land width h_5	Radius r_3	Groove depth a_4	Groove depth and bridge a_{13}	Number of slots	Slot width c_1	Coil spring groove diameter d_{14}	Coil spring diameter d_7	Tangential force F_{tc}, N for unit contact pressure p_{ou} = 1 N/mm²	Recommended class of nominal contact pressure N/mm² PNE Low / PNL Mean / PNR High
60													48	0,86 / 1,14 / 1,52
61													48,8	0,85 / 1,13 / 1,51
62													49,6	0,85 / 1,13 / 1,5
63													50,4	0,84 / 1,12 / 1,5
64	3,7 / 0 / -0,25			2,5							2,3 +0,1 / 0	2,2 / 0 / -0,1	51,2	0,84 / 1,12 / 1,49
65			0,2				1,5 / 0 / -0,15						52	0,83 / 1,11 / 1,48
66			+0,2										52,8	0,83 / 1,1 / 1,47
67			0										53,6	0,82 / 1,1 / 1,46
68													54,4	0,82 / 1,09 / 1,46
69									8				55,2	0,81 / 1,09 / 1,45
70													56	0,81 / 1,08 / 1,44
71	3,9 / 0 / -0,25			2,6									56,8	0,81 / 1,07 / 1,43
72													57,6	0,8 / 1,07 / 1,42
73													58,4	0,8 / 1,06 / 1,41
74													59,2	0,79 / 1,06 / 1,41
75													60	0,79 / 1,05 / 1,4
76	4 / 0 / -0,25			2,7	±0,15 within a ring 0,15 max.		1,6 / 0 / -0,15						60,8	0,78 / 1,04 / 1,39
77													61,6	0,78 / 1,04 / 1,38
78													62,4	0,77 / 1,03 / 1,38
79													63,2	0,77 / 1,03 / 1,37
80			0,25					0,5 ±0,1					64	0,77 / 1,02 / 1,36
81			+0,25										64,8	0,76 / 1,01 / 1,35
82			0	2,8									65,6	0,76 / 1,01 / 1,34
83													66,4	0,75 / 1 / 1,34
84	4,1 / 0 / -0,25						1,7 / 0 / -0,15				2,5 +0,1 / 0	2,4 / 0 / -0,1	67,2	0,75 / 1 / 1,33
85		-0,010 -0,030 for phosphated PO surface +0,005 -0,025											68	0,74 / 0,99 / 1,32
86				2,9									68,8	0,74 / 0,98 / 1,31
87													69,6	0,74 / 0,98 / 1,3
88													70,4	0,73 / 0,97 / 1,3
89										1:2 ±0,1			71,2	0,73 / 0,97 / 1,29
90		4,75		3,0		0,8 ±0,1	0,2 max.						72	0,72 / 0,96 / 1,28
91													72,8	0,72 / 0,95 / 1,27
92													73,6	0,71 / 0,95 / 1,26
93													74,4	0,71 / 0,94 / 1,26
94	4,2 / 0 / -0,25						1,8 / 0 / -0,15						75,2	0,7 / 0,94 / 1,25
95													76	0,7 / 0,93 / 1,24
96				3,1					10				76,8	0,69 / 0,92 / 1,23
97													77,6	0,69 / 0,92 / 1,22
98													78,4	0,68 / 0,91 / 1,22
99			0,3										79,2	0,68 / 0,91 / 1,21
100			+0,25										80	0,68 / 0,9 / 1,2
101			0										80,8	0,67 / 0,89 / 1,19
102				3,15									81,6	0,67 / 0,89 / 1,18
103													82,4	0,66 / 0,88 / 1,18
104	4,5 / 0 / -0,25						1,9 / 0 / -0,2						83,2	0,66 / 0,88 / 1,17
105													84	0,65 / 0,87 / 1,16
106													84,8	0,65 / 0,86 / 1,15
107				3,2									85,6	0,64 / 0,86 / 1,14
108													86,4	0,64 / 0,85 / 1,14
109													87,2	0,63 / 0,85 / 1,13
110													88	0,63 / 0,84 / 1,12
111	4,6 / 0 / -0,25					0,7 ±0,1	2,0 / 0 / -0,2				2,7 -0,1 / 0	2,6 / 0 / -0,1	88,8	0,63 / 0,83 / 1,11
112				3,3	±0,20 within a ring 0,20 max.								89,6	0,62 / 0,83 / 1,1
113													90,4	0,62 / 0,82 / 1,1
114													91,2	0,61 / 0,82 / 1,09
115			0,35										92	0,61 / 0,81 / 1,08
116			+0,30										92,8	0,6 / 0,8 / 1,07
117	4,7 / 0 / -0,25		0	3,4			2,1 / 0 / -0,2						93,6	0,6 / 0,8 / 1,06
118													94,4	0,59 / 0,79 / 1,06
119									12				95,2	0,59 / 0,79 / 1,05
120													96	0,59 / 0,78 / 1,04
121	4,8 / 0 / -0,25						2,2 / 0 / -0,2						96,8	0,58 / 0,77 / 1,03
122				3,5									97,6	0,58 / 0,77 / 1,02
123													98,4	0,57 / 0,76 / 1,02
124													99,2	0,57 / 0,76 / 1,01

This table is shown in ISO format. Commas represent decimal points.

TABLE 17—DIMENSIONS FOR SSF COIL SPRING LOADED OIL CONTROL RINGS WITH SPECIAL RING WIDTH h_1 = 4.75 mm (3/16 in) (CONCLUDED)

Dimensions are millimeters

Nominal diameter	Radial thickness over coil spring	Ring width	Closed gap	Radial wall thickness	Land width	Radius	Groove depth	Groove depth and bridge	Number of slots	Slot width	Coil spring groove diameter	Coil spring diameter	Tangential force F_{tc}, N for unit contact pressure	Recommended class of nominal contact pressure N/mm²		
d_1	a_{12}	h_1 Tolerance	s_1	a_1 Tolerance	h_5	r_3	a_4	a_{13}		c_1	d_{14}	d_7	p_{ou} = 1 N/mm²	PNE Low	PNL Mean	PNR High
125			0,35										100	0,56	0,75	1
126													100,8	0,56	0,74	0,99
127			+0,30										101,6	0,55	0,74	0,98
128			0										102,4	0,55	0,73	0,98
129	4,9			3,6				2,3					103,2	0,54	0,73	0,97
130	0							0					104	0,54	0,72	0,96
131	−0,25							−0,2					104,8	0,54	0,72	0,96
132													105,6	0,54	0,71	0,95
133													106,4	0,53	0,71	0,95
134													107,2	0,53	0,71	0,94
135													108	0,53	0,71	0,94
136				3,7							2,7	2,6	108,8	0,53	0,7	0,94
137									12		+0,1	0	109,6	0,52	0,7	0,93
138			0,4								0	−0,1	110,4	0,52	0,7	0,93
139													111,2	0,52	0,69	0,92
140		−0,010	+0,35										112	0,52	0,69	0,92
141	5,1	−0,040	0	3,8				2,5					112,8	0,52	0,69	0,92
142	0	For						0					113,6	0,51	0,68	0,91
143	−0,25	phos-						−0,2					114,4	0,51	0,68	0,91
144		phated			0,2 max.								115,2	0,51	0,68	0,9
145		PO											116	0,51	0,68	0,9
146		surface											116,8	0,5	0,67	0,9
147		+0,005											117,6	0,5	0,67	0,89
148		−0,035											118,4	0,5	0,67	0,89
149													119,2	0,5	0,66	0,88
150				3,9									120	0,5	0,66	0,88
152					±0,20								121,6	0,49	0,65	0,87
154					within a ring		0,7			1,2			123,2	0,49	0,65	0,86
155		4,75			0,20 max.	0,8	±0,1			±0,1			124	0,48	0,65	0,86
156													124,8	0,48	0,64	0,86
158													126,4	0,48	0,64	0,85
160	5,4		0,45					2,6			2,9	2,8	128	0,47	0,63	0,84
162	0							0			+0,1	0	129,6	0,47	0,62	0,83
164	−0,25		+0,35					−0,2			0	−0,1	131,2	0,46	0,62	0,82
165			0										132	0,46	0,62	0,82
166				4									132,8	0,46	0,61	0,82
168													134,4	0,45	0,61	0,81
170													136	0,45	0,6	0,8
172													137,6	0,45	0,6	0,79
174									14				139,2	0,44	0,59	0,78
175													140	0,44	0,59	0,78
176	5,8							2,8			3,1	3	140,8	0,44	0,58	0,78
178	0			4,6				0			+0,1	0	142,4	0,43	0,58	0,77
180	−0,35							−0,2			0	−0,1	144	0,43	0,57	0,76
182													145,6	0,42	0,56	0,75
184													147,2	0,42	0,56	0,74
185			0,55										148	0,42	0,56	0,74
186					0,5 max.								148,8	0,41	0,55	0,74
188			+0,40										150,4	0,41	0,55	0,73
190			0										152	0,41	0,54	0,72
192	6,2			4,9				3			3,3	3,2	153,6	0,4	0,53	0,71
194	0							0			−0,15	0	155,2	0,4	0,53	0,7
195	−0,4							−0,2			0	−0,1	156	0,4	0,53	0,7
196													156,8	0,39	0,52	0,7
198													158,4	0,39	0,52	0,69
200													160	0,38	0,51	0,68

NOTES:

[1] For intermediate sizes (e.g. repair sizes), the radial thickness of the next smaller nominal diameter applies.

[2] Values of specific tangential force F_{tc} are calculated with mean land width (h_5).

This table is shown in ISO format. Commas represent decimal points.

26.169

TABLE 18—DIMENSIONS FOR GSF AND DSF COIL SPRING LOADED OIL CONTROL RINGS WITH SPECIAL RING WIDTH h_1 = 4.75 mm (3/16 in)

Dimensions are millimeters

Nominal diameter d_1	Radial thickness over coil spring a_{12}	Ring width h_1 Tolerance	Closed gap s_1	Radial wall thickness a_1 Tolerance	Land width h_4	Land width h_5	Radius r_3	Groove depth a_4	Groove depth and bridge a_{13}	Number of slots	Slot width c_1	Coil spring groove diameter d_{14}	Coil spring diameter d_7	Tangential force F_{tc}, N for unit contact pressure p_{ou} = 1 N/mm²	Recommended class of nominal contact pressure N/mm²		
															PNL Low	PNR Mean	PNM High
60														24	1,14	1,52	1,9
61														24,4	1,13	1,51	1,89
62	3,7											2,3	2,2	24,8	1,13	1,5	1,88
63	0			2,5								+0,1	0	25,2	1,12	1,5	1,87
64	-0,25		0,2									0	-0,1	25,6	1,12	1,49	1,86
65									1,5					26	1,11	1,48	1,85
66									0					26,4	1,1	1,47	1,84
67			+0,2						-0,15					26,8	1,1	1,46	1,83
68			0											27,2	1,09	1,46	1,82
69										8				27,6	1,09	1,45	1,81
70														28	1,08	1,44	1,8
71	3,9													28,4	1,07	1,43	1,79
72	0			2,6										28,8	1,07	1,42	1,78
73	-0,25													29,2	1,06	1,42	1,77
74														29,6	1,06	1,41	1,76
75									1,6					30	1,05	1,45	1,75
76	4								0					30,4	1,04	1,39	1,74
77	0			2,7	±0,15				-0,15					30,8	1,04	1,38	1,73
78	-0,25				within a ring									31,2	1,03	1,38	1,72
79					0,15 max.		0,5							31,6	1,03	1,37	1,71
80							±0,1							32	1,02	1,36	1,7
81			0,25											32,4	1,01	1,35	1,69
82				2,8										32,8	1,01	1,34	1,68
83			+0,25											33,2	1	1,34	1,67
84	4,1		0						1,7			2,5	2,4	33,6	1	1,33	1,66
85	0	-0,010							0			+0,1	0	34	0,99	1,32	1,65
86	-0,25	-0,030		2,9					-0,15			0	-0,1	34,4	0,98	1,31	1,64
87		For phos-												34,8	0,98	1,3	1,63
88		phated												35,2	0,97	1,3	1,62
89		P0												35,6	0,97	1,29	1,61
90		surface												36	0,96	1,28	1,6
91	4,75	+0,005												36,4	0,95	1,27	1,59
92		-0,025		3,0		0,8	0,4	0,2			1,2			36,8	0,95	1,26	1,58
93						±0,1	±0,1	max.			±0,1			37,2	0,94	1,26	1,57
94	4,2								1,8					37,6	0,94	1,25	1,56
95	0								0					38	0,93	1,24	1,55
96	-0,25								-0,15					38,4	0,92	1,23	1,54
97				3,1						10				38,8	0,92	1,22	1,53
98														39,2	0,91	1,22	1,52
99			0,3											39,6	0,91	1,21	1,51
100														40	0,9	1,2	1,5
101			+0,25											40,4	0,89	1,19	1,49
102			0	3,15										40,8	0,89	1,18	1,48
103														41,2	0,88	1,18	1,47
104	4,5								1,9					41,6	0,88	1,17	1,46
105	0								0					42	0,87	1,16	1,45
106	-0,25								-0,2					42,4	0,86	1,15	1,44
107				3,2										42,8	0,86	1,14	1,43
108														43,2	0,85	1,14	1,42
109														43,6	0,85	1,13	1,41
110														44	0,84	1,12	1,4
111	4,6							0,7	2,0			2,7	2,6	44,4	0,83	1,11	1,39
112	0			3,3	±0,20			±0,1	0			+0,1	0	44,8	0,83	1,1	1,38
113	-0,25				within a ring				-0,2			0	-0,1	45,2	0,82	1,1	1,37
114					0,20 max.									45,6	0,82	1,09	1,36
115			0,35						2,1					46	0,81	1,08	1,35
116	4,7								0					46,4	0,8	1,07	1,34
117	0		+0,30	3,4					-0,2					46,8	0,8	1,06	1,33
118	-0,25		0											47,2	0,79	1,06	1,32
119														47,6	0,79	1,05	1,31
120										12				48	0,78	1,04	1,3
121	4,8								2,2					48,4	0,77	1,03	1,29
122	0			3,5					0					48,8	0,77	1,02	1,28
123	-0,25								-0,2					49,2	0,76	1,02	1,27
124														49,6	0,76	1,01	1,26

This table is shown in ISO format. Commas represent decimal points.

TABLE 18—DIMENSIONS FOR GSF AND DSF COIL SPRING LOADED OIL CONTROL RINGS WITH SPECIAL RING WIDTH h_1 = 4.75 mm (3/16 in) (CONCLUDED)

Dimensions are millimeters

Nominal diameter	Radial thickness over coil spring	Ring width		Closed gap	Radial wall thickness		Land width	Land width	Radius	Groove depth	Groove depth and bridge	Number of slots	Slot width	Coil spring groove diameter	Coil spring diameter	Tangential force F_{tc}, N for unit contact pressure	Recommended class of nominal contact pressure N/mm²		
d_1	a_{12}	h_1	Tolerance	s_1	a_1	Tolerance	h_4	h_5	r_3	a_4	a_{13}		c_1	d_{14}	d_7	p_{ou} = 1 N/mm²	PNL Low	PNR Mean	PNM High
125				0,35												50	0,75	1	1,25
126																50,4	0,74	0,99	1,24
127				+0,30												50,8	0,74	0,98	1,23
128				0												51,2	0,73	0,98	1,22
129	4,9 0 −0,25				3,6						2,3 0 −0,2					51,6	0,73	0,97	1,21
130																52	0,72	0,96	1,2
131																52,4	0,72	0,96	1,2
132																52,8	0,71	0,95	1,19
133																53,2	0,71	0,95	1,19
134																53,6	0,71	0,94	1,18
135																54	0,71	0,94	1,18
136																54,4	0,7	0,94	1,17
137				0,4	3,7							12		2,7 +0,1 0	2,6 0 −0,1	54,8	0,7	0,93	1,17
138																55,2	0,7	0,93	1,16
139																55,6	0,69	0,92	1,16
140			−0,010	+0,35												56	0,69	0,92	1,15
141			−0,040	0							2,5					56,4	0,69	0,92	1,15
142	5,1 0 −0,25		For phos-		3,8						0 −0,2					56,8	0,68	0,91	1,14
143			phated													57,2	0,68	0,91	1,14
144			PO							0,2 max.						57,6	0,68	0,9	1,13
145			surface													58	0,68	0,9	1,13
146			+0,005													58,4	0,67	0,9	1,12
147			−0,035													58,8	0,67	0,89	1,12
148																59,2	0,67	0,89	1,11
149																59,6	0,66	0,88	1,11
150																60	0,66	0,88	1,1
152					3,9											60,8	0,65	0,87	1,09
154						±0,20 within a ring 0,20 max.	0,8 ±0,1	0,4 ±0,1		0,7 ±0,1			1,2 ±0,1			61,6	0,65	0,86	1,08
155		4,75														62	0,65	0,86	1,08
156																62,4	0,64	0,86	1,07
158																63,2	0,64	0,85	1,06
160				0,45							2,6			2,9	2,8	64	0,63	0,84	1,05
162	5,4 0 −0,25			+0,35							0 −0,2			+0,1 0	0 −0,1	64,8	0,62	0,83	1,04
164				0												65,6	0,62	0,82	1,03
165																66	0,62	0,82	1,03
166					4											66,4	0,61	0,82	1,02
168																67,2	0,61	0,81	1,01
170																68	0,6	0,8	1
172																68,8	0,59	0,79	0,99
174												14				69,6	0,59	0,78	0,98
175																70	0,59	0,78	0,98
176											2,8			3,1	3	70,4	0,58	0,78	0,97
178	5,8 0 −0,35				4,6						0 −0,2			+0,1 0	0 −0,1	71,2	0,58	0,77	0,96
180																72	0,57	0,76	0,95
182																72,8	0,56	0,75	0,94
184																73,6	0,56	0,75	0,93
185				0,55												74	0,56	0,74	0,93
186				+0,40					0,5 max							74,4	0,55	0,74	0,92
188				0												75,2	0,55	0,73	0,91
190																76	0,54	0,72	0,9
192											3			3,3	3,2	76,8	0,53	0,71	0,89
194	6,2 0 −0,4				4,9						0 −0,2			−0,15 0	0 −0,1	77,6	0,53	0,7	0,88
195																78	0,53	0,7	0,88
196																78,4	0,52	0,7	0,87
198																79,2	0,52	0,69	0,86
200																80	0,51	0,68	0,85

NOTES:

[1] For intermediate sizes (e.g. repair sizes), the radial thickness of the next smaller nominal diameter applies.
[2] Values of specific tangential force F_{tc} are calculated with mean land width (h_5).

This table is shown in ISO format. Commas represent decimal points.

TABLE 19—DIMENSIONS FOR DSF-NG COIL SPRING LOADED OIL CONTROL RINGS WITH SPECIAL RING WIDTH h_1 = 4.75 mm (3/16 in)

Dimensions are millimeters

Nominal diameter d_1	Radial thickness over coil spring a_{12}	Ring width h_1 Tolerance	Closed gap s_1	Radial wall thickness a_1 Tolerance	Land width h_5	Land spacing ≈ B_3	Groove depth a_4	Groove depth and bridge a_{13}	Number of slots	Slot width c_1	Coil spring groove diameter d_{14}	Coil spring diameter d_7	Tangential force F_{tc}, N for unit contact pressure p_{ou} = 1 N/mm²	Recommended class of nominal contact pressure N/mm² PNL Low	PNR Mean	PNM High
60													24	1,14	1,52	1,9
61													24,4	1,13	1,51	1,89
62													24,8	1,13	1,5	1,88
63													25,2	1,12	1,5	1,87
64	3,7 0 −0,25			2,5							2,3 +0,1 0	2,2 0 −0,1	25,6	1,12	1,49	1,86
65			0,2					1,5 0 −0,15					26	1,11	1,48	1,85
66													26,4	1,1	1,47	1,84
67			+0,2 0										26,8	1,1	1,46	1,83
68													27,2	1,09	1,46	1,82
69													27,6	1,09	1,45	1,81
70									8				28	1,08	1,44	1,8
71	3,9 0 −0,25			2,6									28,4	1,07	1,43	1,79
72													28,8	1,07	1,42	1,78
73													29,2	1,06	1,42	1,77
74													29,6	1,06	1,41	1,76
75													30	1,05	1,4	1,75
76	4 0 −0,25			2,7	±0,15 within a ring 0,15 max.			1,6 0 −0,15					30,4	1,04	1,39	1,74
77													30,8	1,04	1,38	1,73
78													31,2	1,03	1,38	1,72
79							0,5 ±0,1						31,6	1,03	1,37	1,71
80			0,25										32	1,02	1,36	1,7
81				2,8									32,4	1,01	1,35	1,69
82			+0,25 0										32,8	1,01	1,34	1,68
83													33,2		1,34	1,67
84	4,1 0 −0,25							1,7 0 −0,15			2,5 +0,1 0	2,4 0 −0,1	33,6	1	1,33	1,66
85													34	0,99	1,32	1,65
86				2,9									34,4	0,98	1,31	1,64
87													34,8	0,98	1,3	1,63
88		−0,010 −0,030 For phos- phated PO surface +0,005 −0,025								1,2 ±0,1			35,2	0,97	1,3	1,62
89													35,6	0,97	1,29	1,61
90													36	0,96	1,28	1,6
91		4,75											36,4	0,95	1,27	1,59
92				3,0	0,4 ±0,1	2,65							36,8	0,95	1,26	1,58
93													37,2	0,94	1,26	1,57
94	4,2 0 −0,25							1,8 0 −0,15					37,6	0,94	1,25	1,56
95													38	0,93	1,24	1,55
96													38,4	0,92	1,23	1,54
97				3,1					10				38,8	0,92	1,22	1,53
98													39,2	0,91	1,22	1,52
99			0,3										39,6	0,91	1,21	1,51
100													40	0,9	1,2	1,5
101			+0,25 0										40,4	0,89	1,19	1,49
102				3,15									40,8	0,89	1,18	1,48
103													41,2	0,88	1,18	1,47
104	4,5 0 −0,25							1,9 0 −0,2					41,6	0,88	1,17	1,46
105													42	0,87	1,16	1,45
106													42,4	0,86	1,15	1,44
107				3,2									42,8	0,86	1,14	1,43
108													43,2	0,85	1,14	1,42
109													43,6	0,85	1,13	1,41
110													44	0,84	1,12	1,4
111	4,6 0 −0,25							2,0 0 −0,2					44,4	0,83	1,11	1,39
112				3,3	±0,20 within a ring 0,20 max.		0,7 ±0,1				2,7 −0,1 0	2,6 0 −0,2	44,8	0,83	1,1	1,38
113													45,2	0,82	1,1	1,37
114													45,6	0,82	1,09	1,36
115			0,35										46	0,81	1,08	1,35
116	4,7 0 −0,25							2,1 0 −0,2					46,4	0,8	1,07	1,34
117				3,4									46,8	0,8	1,06	1,33
118			+0,30 0										47,2	0,79	1,06	1,32
119													47,6	0,79	1,05	1,31
120									12				48	0,78	1,04	1,3
121	4,8 0 −0,25							2,2 0 −0,2					48,4	0,77	1,03	1,29
122				3,5									48,8	0,77	1,02	1,28
123													49,2	0,76	1,02	1,27
124													49,6	0,76	1,01	1,26

This table is shown in ISO format. Commas represent decimal points.

TABLE 19—DIMENSIONS FOR DSF-NG COIL SPRING LOADED OIL CONTROL RINGS WITH SPECIAL RING WIDTH h_1 = 4.75 mm (3/16 in) (CONCLUDED)

Dimensions are millimeters

Nominal diameter d_1	Radial thickness over coil spring a_{12}	Ring width h_1	Closed gap s_1	Radial wall thickness a_1	Land width h_5	Land spacing $\approx B_3$	Groove depth a_4	Groove depth and bridge a_{13}	Number of slots	Slot width c_1	Coil spring groove diameter d_{14}	Coil spring diameter d_7	Tangential force F_{tc}, N for unit contact pressure p_{ou} = 1 N/mm²	Recommended class of nominal contact pressure N/mm² PNL Low	PNR Mean	PNM High
			Tolerance		Tolerance											
125			0,35										50	0,75	1	1,25
126			+0,30										50,4	0,74	0,99	0,24
127			0										50,8	0,74	0,98	1,23
128													51,2	0,73	1,98	1,22
129	4,9							2,3					51,6	0,73	0,97	1,21
130	0			3,6				0					52	0,72	0,96	1,2
131	-0,25							-0,2					52,4	0,72	0,96	1,2
132													52,8	0,71	0,95	1,19
133													53,2	0,71	0,95	1,19
134													53,6	0,71	0,94	1,18
135													54	0,71	0,94	1,18
136				3,7							2,7	2,6	54,4	0,7	0,94	1,17
137									12		+0,1	0	54,8	0,7	0,93	1,17
138											0	-0,1	55,2	0,7	0,93	1,16
139			0,4										55,6	0,69	0,92	1,16
140		-0,010	+0,35										56	0,69	0,92	1,15
141	5,1	-0,040	0					2,5					56,4	0,69	0,92	1,15
142	0	For		3,8				0					56,8	0,68	0,91	1,14
143	-0,25	phos-						-0,2					57,2	0,68	0,91	1,14
144		phated											57,6	0,68	0,9	1,13
145		PO											58	0,68	0,9	1,13
146		surface											58,4	0,67	0,9	1,12
147		+0,005											58,8	0,67	0,89	1,12
148		-0,035											59,2	0,67	0,89	1,11
149													59,6	0,66	0,88	1,11
150				3,9									60	0,66	0,88	1,1
152					±0,20								60,8	0,65	0,87	1,09
154		4,75			within a ring	0,4	2,65	0,7			1,2		61,6	0,65	0,86	1,08
155					0,20	±0,1		±0,1			±0,1		62	0,65	0,86	1,08
156					max.								62,4	0,64	0,86	1,07
158													63,2	0,64	0,85	1,06
160	5,4		0,45					2,6			2,9	2,8	64	0,63	0,84	1,05
162	0		+0,35					0			+0,1	0	64,8	0,62	0,83	1,04
164	-0,25		0					-0,2			0	-0,1	65,6	0,62	0,82	1,03
165				4									66	0,62	0,82	1,03
166													66,4	0,61	0,82	1,02
168													67,2	0,61	0,81	1,01
170													68	0,6	0,8	1
172													68,8	0,59	0,79	0,99
174									14				69,6	0,59	0,78	0,98
175													70	0,59	0,78	0,98
176	5,8							2,3			3,1	3	70,4	0,58	0,78	0,97
178	0			4,6				0			-0,1	0	71,2	0,58	0,77	0,96
180	-0,35							-0,2			0	-0,1	72	0,57	0,76	0,95
182													72,8	0,56	0,75	0,94
184													73,6	0,56	0,75	0,93
185			0,55										74	0,56	0,74	0,93
186			+0,40										74,4	0,55	0,74	0,92
188			0										75,2	0,55	0,73	0,91
190													76	0,54	0,72	0,9
192	6,2			4,9				3			3,3	3,2	76,8	0,53	0,71	0,89
194	0							0			-0,15	0	77,6	0,53	0,7	0,88
195	-0,4							-0,2			0	-0,1	78	0,53	0,7	0,88
196													78,4	0,52	0,7	0,87
198													79,2	0,52	0,69	0,86
200													80	0,51	0,68	0,85

NOTES:

[1] For intermediate sizes (e.g. repair sizes), the radial thickness of the next smaller nominal diameter applies.

[2] Values of specific tangential force F_{tc} are calculated with mean land width (h_5).

This table is shown in ISO format. Commas represent decimal points.

TABLE 20—DIMENSIONS FOR SSF-L COIL SPRING LOADED OIL CONTROL RINGS WITH SPECIAL RING WIDTH h_1 = 4.75 mm (3/16 in)

Dimensions are millimeters

Nominal diameter d_1	Radial thickness over coil spring a_{12}	Ring width h_1 Tolerance	Closed gap s_1	Radial wall thickness a_1 Tolerance	Land width h_5	Radius r_3	Groove depth a_4	Groove depth and bridge a_{13}	Number of slots	Slot width c_1	Coil spring groove diameter d_{14}	Coil spring diameter d_7	Tangential force F_{tc}, N for unit contact pressure p_{ou} = 1 N/mm²	Recommended class of nominal contact pressure N/mm² PNE Low	PNL Mean	PNR High
60													36	0,86	1,14	1,52
61													36,6	0,85	1,13	1,51
62													37,2	0,85	1,13	1,5
63													37,8	0,84	1,12	1,5
64	3,7 0 −0,25										2,3 +0,1 0	2,2 0 −0,1	38,4	0,84	1,12	1,49
65			0,2	2,5									39	0,83	1,11	1,48
66								1,5					39,6	0,83	1,1	1,47
67			+0,2					0					40,2	0,82	1,1	1,46
68			0					−0,15					40,8	0,82	1,09	1,46
69									8				41,4	0,81	1,09	1,45
70													42	0,81	1,08	1,44
71	3,9 0 −0,25			2,6									42,6	0,81	1,07	1,43
72													43,2	0,8	1,07	1,42
73													43,8	0,8	1,06	1,41
74													44,4	0,79	1,06	1,41
75													45	0,79	1,05	1,4
76								1,6					45,6	0,78	1,04	1,39
77	4 0 −0,25			2,7	±0,15 within a ring 0,15 max.			0 −0,15					46,2	0,78	1,04	1,38
78													46,8	0,77	1,03	1,38
79													47,4	0,77	1,03	1,37
80													48	0,77	1,02	1,36
81			0,25					0,5 ±0,1					48,6	0,76	1,01	1,35
82			+0,25	2,8									49,2	0,76	1,01	1,34
83			0										49,8	0,75	1	1,34
84	4,1 0 −0,25	4,75 −0,010 −0,030 For phosphated PO surface +0,005 −0,025						1,7 0 −0,15			2,5 +0,1 0	2,4 0 −0,1	50,4	0,75	1	1,33
85													51	0,74	0,99	1,32
86													51,6	0,74	0,98	1,31
87				2,9									52,2	0,74	0,98	1,3
88													52,8	0,73	0,97	1,3
89										1,2 ±0,1			53,4	0,73	0,97	1,29
90													54	0,72	0,96	1,28
91					0,6 ±0,1	0,2 max.							54,6	0,72	0,95	1,27
92				3,0									55,2	0,71	0,95	1,26
93								1,8 0 −0,15					55,8	0,71	0,94	1,26
94	4,2 0 −0,25												56,4	0,7	0,94	1,25
95													57	0,7	0,93	1,24
96													57,6	0,69	0,92	1,23
97				3,1					10				58,2	0,69	0,92	1,22
98													58,8	0,68	0,91	1,22
99			0,3										59,4	0,68	0,91	1,21
100													60	0,68	0,9	1,2
101			+0,25 0										60,6	0,67	0,89	1,19
102				3,15									61,2	0,67	0,89	1,18
103													61,8	0,66	0,88	1,18
104	4,5 0 −0,25							1,9 0 −0,2					62,4	0,66	0,88	1,17
105													63	0,65	0,87	1,16
106													63,6	0,65	0,86	1,15
107				3,2									64,2	0,64	0,86	1,14
108													64,8	0,64	0,85	1,14
109													65,4	0,63	0,85	1,13
110													66	0,63	0,84	1,12
111								2,0					66,6	0,63	0,83	1,11
112	4,6 0 −0,25			3,3	±0,20 within a ring 0,20 max.		0,7 ±0,1	0 −0,2			2,7 −0,1 0	2,6 0 −0,1	67,2	0,62	0,83	1,1
113													67,8	0,62	0,82	1,1
114													68,4	0,61	0,82	1,09
115													69	0,61	0,81	1,08
116			0,35					2,1					69,6	0,6	0,8	1,07
117	4,7 0 −0,25			3,4				0 −0,2					70,2	0,6	0,8	1,06
118			+0,30 0										70,8	0,59	0,79	1,06
119													71,4	0,59	0,79	1,05
120									12				72	0,59	0,78	1,04
121								2,2					72,6	0,58	0,77	1,03
122	4,8 0 −0,25			3,5				0 −0,2					73,2	0,58	0,77	1,02
123													73,8	0,57	0,76	1,02
124													74,4	0,57	0,76	1,01

This table is shown in ISO format. Commas represent decimal points.

TABLE 20—DIMENSIONS FOR SSF-L COIL SPRING LOADED OIL CONTROL RINGS WITH SPECIAL RING WIDTH h_1 = 4.75 mm (3/16 in) (CONCLUDED)

Dimensions are millimeters

Nominal diameter d_1	Radial thickness over coil spring a_{12}	Ring width h_1 Tolerance	Closed gap s_1	Radial wall thickness a_1 Tolerance	Land width h_5	Radius r_3	Groove depth a_4	Groove depth and bridge a_{13}	Number of slots	Slot width c_1	Coil spring groove diameter d_{14}	Coil spring diameter d_7	Tangential force F_{tc}, N for unit contact pressure p_{ou} = 1 N/mm²	Recommended class of nominal contact pressure N/mm² PNE Low / PNL Mean / PNR High
125			0.35										75	0,56 / 0,75 / 1
126													75,6	0,56 / 0,74 / 0,99
127			+0.30										76,2	0,55 / 0,74 / 0,98
128			0										76,8	0,55 / 0,73 / 0,98
129	4,9							2,3					77,4	0,54 / 0,73 / 0,97
130	0			3,6				0					78	0,54 / 0,72 / 0,96
131	−0,25							−0.2					78,6	0,54 / 0,72 / 0,96
132													79,2	0,54 / 0,71 / 0,95
133													79,8	0,53 / 0,71 / 0,95
134													80,4	0,53 / 0,71 / 0,94
135											2,7	2,6	81	0,53 / 0,71 / 0,94
136											+0.1	0	81,6	0,53 / 0,7 / 0,94
137				3,7					12		0	−0.1	82,2	0,52 / 0,7 / 0,93
138			0,4										82,8	0,52 / 0,7 / 0,93
139													83,4	0,52 / 0,69 / 0,92
140		−0.010	+0,35										84	0,52 / 0,69 / 0,92
141	5,1	−0.040	0					2,5					84,6	0,52 / 0,69 / 0,92
142	0	For		3,8				0					85,2	0,51 / 0,68 / 0,91
143	−0,25	phos-						−0.2					85,8	0,51 / 0,68 / 0,91
144		phated			0.2								86,4	0,51 / 0,68 / 0,9
145		PO			max.								87	0,51 / 0,68 / 0,9
146		surface											87,6	0,5 / 0,67 / 0,9
147		+0.005											88,2	0,5 / 0,67 / 0,89
148		−0.035											88,8	0,5 / 0,67 / 0,89
149													89,4	0,5 / 0,66 / 0,88
150				3,9									90	0,5 / 0,66 / 0,88
152					±0.20								91,2	0,49 / 0,65 / 0,87
154					within a ring								92,4	0,49 / 0,65 / 0,86
155		,4,75			0.20 max.	0.6 ±0.1	0,7 ±0.1			1 2 ±0.1			93	0,48 / 0,65 / 0,86
156													93,6	0,48 / 0,64 / 0,86
158													94,8	0,48 / 0,64 / 0,85
160	5,4		0,45					2,6			2,9	2,8	96	0,47 / 0,63 / 0,84
162	0							0			+0.1	0	97,2	0,47 / 0,62 / 0,83
164	−0,25		+0,35					−0.2			0	−0.1	98,4	0,46 / 0,62 / 0,82
165			0										99	0,46 / 0,62 / 0,82
166				4									99,6	0,46 / 0,61 / 0,82
168													100,8	0,45 / 0,61 / 0,81
170													102	0,45 / 0,6 / 0,8
172													103,2	0,45 / 0,6 / 0,79
174													104,4	0,44 / 0,59 / 0,78
175									14				105	0,44 / 0,59 / 0,78
176													105,6	0,44 / 0,58 / 0,78
178	5,8							2,8			3 1	3	106,8	0,43 / 0,58 / 0,77
180	0			4,6				0			−0.1	0	108	0,43 / 0,57 / 0,76
182	−0,35							−0.2			0	−0.1	109,2	0,42 / 0,56 / 0,75
184													110,4	0,42 / 0,56 / 0,74
185			0,55										111	0,42 / 0,56 / 0,74
186					0.5 max.								111,6	0,41 / 0,55 / 0,74
188			+0,40										112,8	0,41 / 0,55 / 0,73
190			0										114	0,41 / 0,54 / 0,72
192	6,2							3			3,3	3,2	115,2	0,4 / 0,53 / 0,71
194	0			4,9				0			−0.15	0	116,4	0,4 / 0,53 / 0,7
195	−0,4							−0.2				−0.1	117	0,4 / 0,53 / 0,7
196													117,6	0,39 / 0,52 / 0,7
198													118,8	0,39 / 0,52 / 0,69
200													120	0,38 / 0,51 / 0,68

NOTES:

[1] For intermediate sizes (e.g. repair sizes), the radial thickness of the next smaller nominal diameter applies.

[2] Values of specific tangential force F_{tc} are calculated with mean land width (h_5).

This table is shown in ISO format. Commas represent decimal points.

APPENDIX A

A.1 This SAE Standard has been established to harmonize the ISO and SAE piston ring standards. The U.S. Technical Advisory Group, with the support of the National Engine Parts Manufacturers Association, has worked with other organizations on this worldwide Standard. Some of the wording and phrasing may differ slightly for translation purposes.

In preparing this SAE Standard, the Introduction, Scope, Field of Application, and Reference sections of the ISO 6626 Standard have been editorially revised and reorganized.

The tolerances specified in this SAE Standard represent a six sigma quality level.

INTERNAL COMBUSTION ENGINES—PISTON RINGS—EXPANDER/SEGMENT OIL CONTROL RINGS—SAE J2004 OCT92

SAE Standard

Report of the Piston Ring Standards Committee approved September 1990. Revised by the SAE Piston Ring Standards Committee 7 October 1992.

This SAE Standard is equivalent to ISO Standard 6627 TR.

1. Scope and Field of Application—Differences, where they exist, are shown in Appendix A with associated rationale.

This SAE Standard specifies the dimensional features of commonly used oil control rings having two steel segments (rails) separated and expanded by one steel expander/spacer.

The segments vary in width from 0.4 to 0.6 mm. The assembly width ranges from 2.5 to 4.75 mm. The 4.75 mm width is equivalent to existing 3/16 in applications. Expander design will vary considerably with piston ring manufacturer.

The total circumferential deflection and the piston groove depth should be considered when designing these oil rings to optimize the fit of the ring assembly into the piston groove.

This document applies to oil control rings up through 125 mm for reciprocating internal combustion engines. It may also be used for piston rings of compressors working under similar conditions.

R) **2. References**

SAE DESIGNATION	ISO[1] EQUIVALENT	
		INTERNAL COMBUSTION ENGINES—PISTON RINGS
J1588	6621/1	Vocabulary
J1589	6621/2	Measuring principles
J1590	6621/3	Material specifications
J1591	6621/4	General specifications
J1996	6621/5	Quality requirements
		INTERNAL COMBUSTION ENGINES—PISTON RINGS
J1997	6622/1	Rectangular rings
J1998	6622/2 TR	Rectangular rings with narrow ring width
J1999	6623	INTERNAL COMBUSTION ENGINES—PISTON RINGS—SCRAPER RINGS
		INTERNAL COMBUSTION ENGINES—PISTON RINGS
J2000	6624/1	Keystone rings
J2001	6624/2 TR	Half keystone rings
J2002	6625	INTERNAL COMBUSTION ENGINES—PISTON RINGS—OIL CONTROL RINGS
J2003	6626	INTERNAL COMBUSTION ENGINES—PISTON RINGS—COIL SPRING LOADED OIL CONTROL RINGS
J2004	6627 TR	INTERNAL COMBUSTION ENGINES—PISTON RINGS—EXPANDER/SEGMENT OIL CONTROL RINGS
J2226		INTERNAL COMBUSTION ENGINES—PISTON RINGS—STEEL RECTANGULAR RINGS
	1101	TECHNICAL DRAWINGS—Tolerancing of form, orientation, location and run-out—Generalities, definitions, symbols, indications on drawings

[1] TR refers to Technical Report

26.176

3. Expander/Segment Assembly—When assembled in the engine the segments and expander shall be positioned as shown in Figure 1.

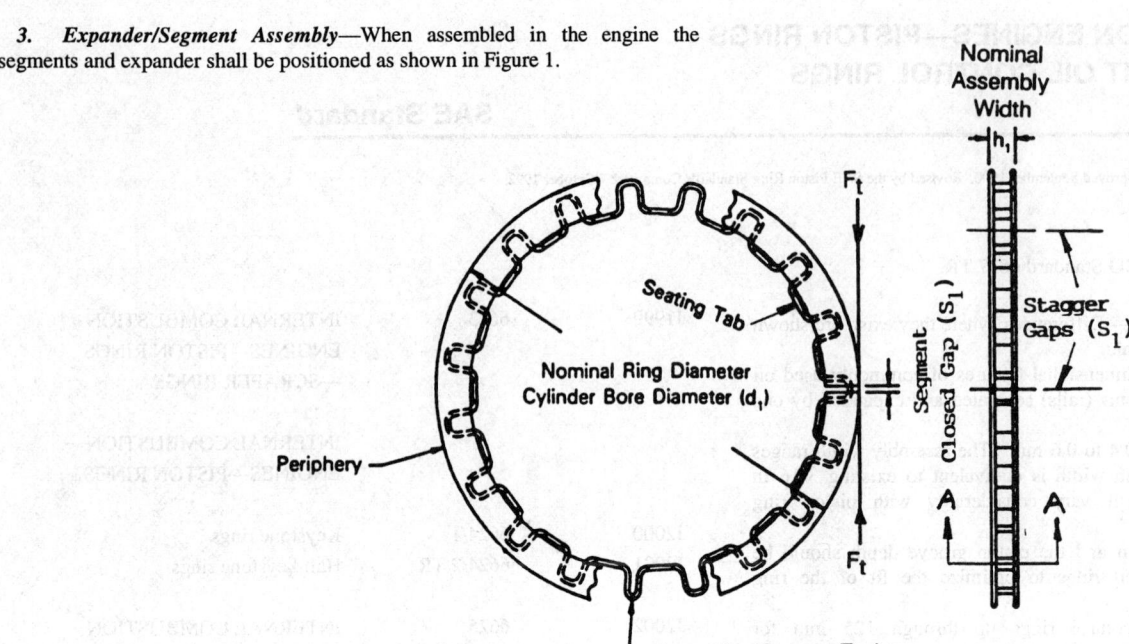

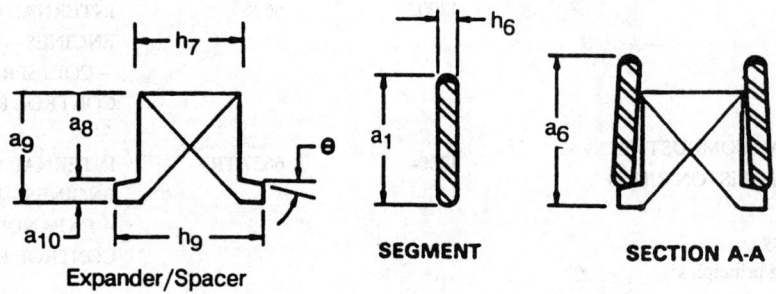

a_1 Segment Radial Wall Thickness	h_1 Nominal Assembly Width	d_1 Nominal Assembly Diameter
a_8 Spacer Radial Height	h_7 Spacer Width	θ Seating Tab Angle
a_{10} Seating Tab Height	h_9 Expander Width	F_t Tangential Force
a_6 Assembly Radial Thickness	h_6 Segment Width	
a_9 Expander Radial Height	s_1 Segment Closed Gap	

NOTE—For measuring purposes only, segment gaps shall be in line and expander/spacer ends shall be at the back of the segments (see SAE J1589 3.2.5b).

FIGURE 1—EXPANDER/SEGMENT ASSEMBLY

3.1 Ring Types—There are any number of possible oil ring expander designs. The more common designs in use today are designated in Figures 2 through 6.

3.2 Cross Section Configuration

3.2.1 ES-1 Type

3.2.2 ES-2 Type

FIGURE 2—ES-1 TYPE

FIGURE 3—ES-2 TYPE

3.2.3 ES-3 Type

FIGURE 4—ES-3 TYPE

3.2.4 ES-4 Type

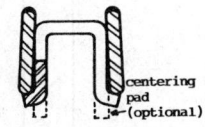

FIGURE 5—ES-4 TYPE

3.2.5 ES-5 Type

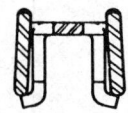

FIGURE 6—ES-5 TYPE

4. Common Features

4.1 Expander Seating Tab Angle—(See Figure 7.)

The expander may be designed with the seating tabs at a slight angle resulting in side sealing between the segment and the side of the piston groove.

FIGURE 7—EXPANDER SEATING TAB ANGLE

4.2 Segment Chromium Thickness—(See Figure 8.)
(Common feature of segment is chrome plated periphery.)

FIGURE 8—SEGMENT CHROMIUM THICKNESS

TABLE 1—LAYER THICKNESS

Dimensions in millimeters

Chromium	Thickness min
CR1	0.05
CR2	0.10
CR3	0.15

4.3 Segment Width (h_6)—(See Figure 9.)

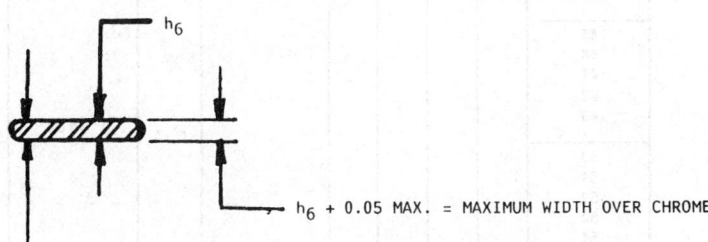

$h_6 + 0.05$ MAX. = MAXIMUM WIDTH OVER CHROME

$h_6 + 0.02$ MAX. = MAXIMUM WIDTH AT I.D. AFTER COILING

FIGURE 9—SEGMENT WIDTH (h_6)

5. Materials

5.1 Expander
Austenitic Stainless Steel (SAE J1590)
Carbon Steel (as agreed upon between client and manufacturer)

5.2 Segment—Carbon Steel

6. Force and Nominal Contact Pressure

6.1 Tangential Force (F_t)—The force of expander/segment oil control rings is determined by the force of the expander and can be calculated with the following equation:

$$F_t = 1/2\, P_o \cdot d_1 \cdot 2h_6 \qquad (Eq.1)$$

P_o = Required Nominal Contact Pressure

The force exerted by the segments is negligible.

In Table 3 of Section 7, Dimensions, the F_t has been tabulated at a unit pressure of $P_o = 1$ N/mm^2.

6.2 Nominal Contact Pressure P_o—F_t for selected unit pressures is calculated by use of the multiplying factors in Table 2:

TABLE 2—NOMINAL CONTACT PRESSURE P_o

Multiplying Factor × F_t (From Table 3)	Unit Pressure P_o
1.60[1]	1.60 N/mm^2
1.25	1.25 N/mm^2
1.00	1.00 N/mm^2
0.80	0.80 N/mm^2
0.60[2]	0.60 N/mm^2

[1] Recommended for repair sizes only.
[2] Not recommended <d_1 = 65 mm.

26.178

7. Dimensions

TABLE 3—DIMENSIONS

Dimensions in millimeters

Nominal diameter d_1	Segment width h_6 Column				Tolerance	Segment closed gap s_1 Nom.	Tolerance	F_t ($P_o = 1\ N/mm^2$) Column				Tolerance	Nominal assembly width h_1 (using available h_6) Column				
	1	2	3	4				1	2	3	4		1-3	1-3	1-4	1-4	4
40								16	18	20							
41								16,4	18,4	20,5							
42								16,8	18,9	21							
43								17,2	19,3	21,5							
44								17,6	19,8	22							
45								18	20,2	22,5							
46								18,4	20,7	23							
47								18,8	21,1	23,5							
48								19,2	21,6	24							
49								19,6	22	24,5			0,15	+0,50 / 0				
50								20	22,5	25							
51								20,4	22,9	25,5							
52								20,8	23,4	26							
53								21,2	23,8	26,5							
54								21,6	24,3	27							
55								22	24,7	27,5							
56								22,4	25,2	28							
57								22,8	25,6	28,5							
58								23,2	26,1	29							
59								23,6	26,5	29,5				2,5				
60								24	27	30	36						
61								24,4	27,4	30,5	36,6						
62								24,8	27,9	31	37,2						
63								25,2	28,3	31,5	37,8	±30 % if $F_t < 30\ N$					
64	0,4	0,45	0,5	0,6	+0,025 / 0			25,6	28,8	32	38,4			3	3,5	4	4,75
65								26	29,2	32,5	39	±20 % if $F_t \geq 30\ N$					
66								26,4	29,7	33	39,6						
67					0,2			26,8	30,1	33,5	40,2						
68								27,2	30,6	34	40,8						
69								27,6	31	34,5	41,4						
70								28	31,5	36	42						
71								28,4	31,9	36,5	42,6						
72								28,8	32,4	36	43,2						
73								29,2	32,8	36,5	43,8						
74								29,6	33,3	37	44						
75							+0,75 / 0	30	33,7	37,5	45						
76								30,4	34,2	38	45,6						
77								30,8	34,6	38,5	46,2						
78								31,7	35,1	39	46,8						
79								31,6	35,5	39,5	47,4						
80					0,25			32	36	40	48						
81								32,4	36,4	40,5	48,6						
82								32,8	36,9	41	49,2						
83								33,2	37,3	41,5	49,8						
84								33,6	37,8	42	50,4						
85								34	38,2	42,5	51						
86								34,4	38,7	43	51,6						
87								34,8	39,1	43,5	52,2						
88								35,2	39,6	44	52,3						
89								35,6	40	44,5	53,4						

TABLE 3—DIMENSIONS (CONTINUED)

Nominal diameter d_1	Segment width h_6 Column 1	2	3	4	Tolerance	Segment closed gap s_1 Nom.	Tolerance	F_t ($P_o = 1\ N/mm^2$) Column 1	2	3	4	Tolerance	Nominal assembly width h_1 (using available h_6) Column 1-3	1-3	1-4	1-4	4
90								36	40,5	45	54						
91								36,4	40,9	45,5	54,6						
92								36,8	41,4	46	55,2						
93								37,2	41,8	45,5	55,8						
94								37,6	42,3	46	56,4						
95								38	42,7	47,5	57						
96								38,4	43,2	48	57,6						
97								38,8	43,6	48,5	58,2						
98								39,2	44,1	49	58,8						
99						0,3		39,6	44,5	49,5	59,4						
100								40	45	50	60						
101								40,4	45,4	50,5	60,6						
102								40,8	45,9	51	61,2						
103								41,2	46,3	51,5	61,8						
104								41,6	46,8	52	62,4						
105								42	47,2	52,5	63	± 30 % if					
106								42,4	47,7	53	63,6	$F_t < 30\ N$					
107	0,4	0,45	0,5	0,6	+0,025 / 0		+0,75 / 0	42	48,1	53,5	64,2		3	3,5	4		4,75
108								43,2	48,6	54	64,8	± 20 % if					
109								43,6	49	54,5	65,4	$F_t \geq 30\ N$					
110								44	49,5	55	66						
111								44,4	49,9	55,5	66,6						
112								44,8	50,4	56	67,2						
113								45,2	50,8	56,5	67,8						
114								45,6	51,3	57	68,4						
115								46	51,7	57,5	69						
116								46,4	52,5	58	69,6						
117						0,35		46,8	52,6	58,5	70,2						
118								47,2	53,1	59	70,8						
119								47,6	53,5	59,5	71,4						
120								48	54	60	72						
121								48,4	54,4	60,5	72,6						
122								48,8	54,9	61	73,2						
123								49,2	55,3	61,5	73,8						
124								49,6	55,8	62	74,4						
125								50	56,2	62,5	75						

NOTES

The radial wall thickness of the segment will vary with the design of the expander. Common ratios d_1/a_1 are approximately between 40 and 20.

The maximum radial thickness of the selected assembly depends on its design (expander and segments) and has to be defined by the manufacturer.

APPENDIX A

A.1 This SAE Standard has been established to harmonize the ISO and SAE piston ring standards. The U.S. Technical Advisory Group, with the support of the National Engine Parts Manufacturers Association, has worked with other national organizations on this worldwide standard. Some of the wording and phrasing may differ slightly from U.S. terminology for translation purposes.

In preparing this SAE Standard, the Scope and Field of Application and Reference sections of the ISO 6627 TR have been editorially revised and reorganized.

The tolerances specified in this document represent a six sigma quality level.

INTERNAL COMBUSTION ENGINES—PISTON RINGS—STEEL RECTANGULAR RINGS—SAE J2226 OCT92

SAE Standard

Report of the SAE Piston Ring Standards Committee 7 approved October 1992.

1. Scope and Field of Application
There is no standard equivalent to this SAE Standard. However, SAE J2226 is similar to and represents an extension of SAE J1998 (equivalent to ISO Standard 6622/2 TR).

This SAE Standard specifies the essential dimensional features of Type B barrel faced steel rectangular piston rings. Only fully faced and inlaid coated rings are included, consistent with current U.S. practice.

The requirements of this SAE Standard apply to steel piston rings of reciprocating internal combustion engines, up to and including 200 mm diameter and 4.5 mm width.

Tolerances specified in this document represent a six sigma quality level.

2. References

SAE DESIGNATION	ISO[1] EQUIVALENT	
		INTERNAL COMBUSTION ENGINES—PISTON RINGS
J1588	6621/1	Vocabulary
J1589	6621/2	Measuring principles
J1590	6621/3	Material specifications
J1591	6621/4	General specifications
J1996	6621/5	Quality requirements
		INTERNAL COMBUSTION ENGINES—PISTON RINGS
J1997	6622/1	Rectangular rings
J1998	6622/2 TR	Rectangular rings with narrow ring width
J1999	6623	INTERNAL COMBUSTION ENGINES—PISTON RINGS—SCRAPER RINGS
		INTERNAL COMBUSTION ENGINES—PISTON RINGS
J2000	6624/1	Keystone rings
J2001	6624/2 TR	Half keystone rings
J2002	6625	INTERNAL COMBUSTION ENGINES—PISTON RINGS—OIL CONTROL RINGS
J2003	6626	INTERNAL COMBUSTION ENGINES—PISTON RINGS—COIL SPRING LOADED OIL CONTROL RINGS
J2004	6627 TR	INTERNAL COMBUSTION ENGINES—PISTON RINGS—EXPANDER/SEGMENT OIL CONTROL RINGS
	6507/3	METALLIC MATERIALS—HARDNESS TEST—VICKERS TEST—PART 3: LESS THAN HV 0.2

[1] TR refers to Technical Report

3. Ring Types and Designation Examples
3.1 Type B—Barrel Faced Rectangular Ring
3.1.1 GENERAL FEATURES
NOTE—See Table 4 for dimensions and forces.

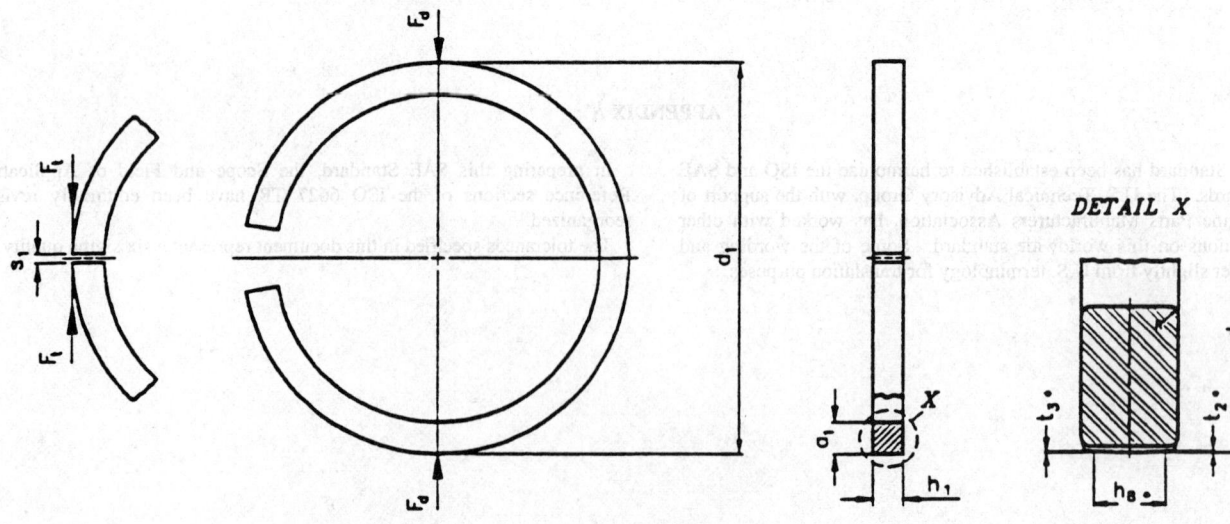

FIGURE 1—TYPE B

*See Table 1

TABLE 1—GAUGE WIDTH (h_R) AND BARREL DIMENSIONS

Dimensions in millimeters

h_1	h_R	t_2, t_3	t_2, t_3	Maximum peak off center
1.2	0.6	0.001	0.013	0.2
1.5	0.8	0.002	0.016	0.25
2.0	1.2	0.002	0.016	0.4
2.5	1.6	0.002	0.016	0.4
3.0	2.0	0.005	0.020	0.5
3.5	2.4	0.005	0.020	0.5
4.0	2.8	0.005	0.023	0.6
4.5	3.2	0.005	0.023	0.6

3.1.2 DESIGNATION EXAMPLE—Designation of a barrel faced rectangular ring with narrow ring width of $d_1 = 60$ mm nominal diameter, $h_1 = 1.2$ mm ring width made of steel (material subclass 62), general features as shown in Figure 2, and periphery chromium coated fully faced design 0.1 mm minimum thickness:

Piston ring SAE J2226 B-60 x 1.2-MC62/CR2

4. Common Features

4.1 Type B Steel Rings—Inside Rounded Edges—(See Figures 2 and 3.)

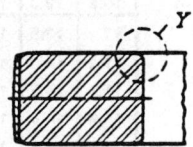

Type B-rings

FIGURE 2—INSIDE ROUNDED EDGES

Dimensions "A" and "B" to be determined between customer and supplier

FIGURE 3—DETAIL OF FIGURE 2

4.2 Coated Rings (Fully Faced and Inlaid)—Layer Thickness
NOTE—See Table 2 for dimensions.

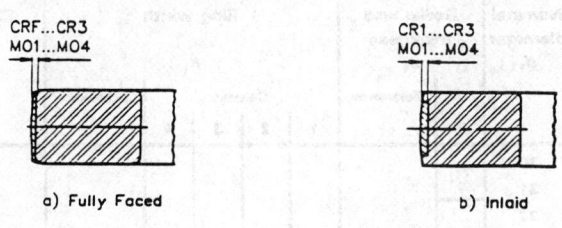

a) Fully Faced   b) Inlaid

FIGURE 4—LAYER THICKNESS

TABLE 2—LAYER THICKNESS

Dimensions in millimeters

Chromium	Molybdenum	Thickness min
CRF	-	0.005
CR1	MO1	0.05
CR2	MO2	0.1
CR3[1]	MO3[1]	0.15
-	MO4[1]	0.20

[1] CR3 and MO3/MO4 apply to rings with nominal diameters of 50 mm or greater.

5. Force Factors—For common features, multiplier correction factors given in Table 3 and the force correction factors given in SAE J1591 shall be used.

Additionally, the tangential and diametral force given in Table 4 shall be corrected when additional features are being used.

TABLE 3—FORCE CORRECTION FACTORS FOR COATED TYPE B RINGS

d_1 mm	Factor CRF	Factor CR1	Factor CR2/MO1	Factor CR3	Factor MO2	Factor MO3	Factor MO4
$30 \leq d_1 < 50$	1	0.81	0.70	-	0.64	-	-
$50 \leq d_1 < 100$	1	0.90	0.85	0.81	0.81	0.75	0.71
$100 \leq d_1 < 150$	1	0.94	0.91	0.88	0.88	0.86	0.83
$150 \leq d_1 \leq 200$	1	0.96	0.93	0.91	0.91	0.89	0.87

TABLE 4—DIMENSIONS OF TYPE B RECTANGULAR RINGS MADE OF STEEL

Dimensions in millimeters

Nominal diameter d_1	Radial wall thickness a_1	Tolerance	Ring width h_1 Column 1	2	3	4	Tolerance	Closed gap s_1	Tolerance	Tangential force F_t N For h_1 shown in column 1	2	3	4	Tolerance	Diametral force F_d N For h_1 shown in column 1	2	3	4	Tolerance
30	1.1									–	–	–	–		6.8	8.5	11.3	14.2	
31										–	–	–	–		7	8.8	11.7	14.6	
32										–	–	–	–		7.3	9.1	12.2	15.2	
33										–	–	–	–		7.5	9.4	12.5	15.6	
34	1.3									–	–	–	–		8.2	10.5	13.7	17.1	
35										–	–	–	–		8.4	10.8	14	17.5	
36										–	–	–	–		8.6	11.1	14.3	17.9	
37										–	–	–	–		8.9	11.4	14.8	18.5	
38										–	–	–	–		9.2	11.7	15.3	19.2	
39	1.5									–	–	–	–		9.5	12	15.8	19.8	
40										–	–	–	–		9.7	12.3	16.2	20.2	
41										–	–	–	–		9.9	12.5	16.5	20.6	
42								−0.010	0.15	+0.2	–	–	–	–	–	10.1	12.9	16.8	21	
43								−0.025		0	–	–	–	–		10.3	13.1	17.2	21.5	
44	1.7										–	–	–	–		10.5	13.3	17.5	21.9	
45								For			–	–	–	–		11	13.5	18.3	22.9	
46								Phosphated			–	–	–	–		11.2	14	18.7	23.3	
47								PO			–	–	–	–		11.4	14.2	19	23.7	
48								Surfaces			–	–	–	–		11.6	14.6	19.3	24.2	
49	1.9										–	–	–	–		11.8	14.8	19.7	24.6	
50											5.6	7	9.3	11.6		12	15.1	20	25	
51								−0.005			5.7	7.2	9.5	11.9		12.3	15.5	20.5	25.6	
52								−0.030			5.8	7.3	9.7	12.1		12.5	15.7	20.8	26	
53		± 0.15									6	7.5	10	12.5		12.9	16.1	21.5	26.9	± 30% if F_d < 21.5 N
54		Within									6.1	7.6	10.2	12.7		13.1	16.3	21.8	27.3	
55	2.1	a ring:	1.2	1.5	2.0	2.6					6.2	7.7	10.3	12.9		13.3	16.6	22.2	27.7	± 20% if F_d ≥ 21.5 N
56		0.15 max.									6.3	7.9	10.5	13.1		13.5	17	22.5	28.1	
57											6.4	8	10.7	13.4		13.8	17.2	23	28.7	
58											6.5	8.2	10.9	13.6		14	17.6	23.3	29.2	
59											6.6	8.3	11	13.8		14.2	17.8	23.7	29.6	
60	2.3										6.7	8.5	11.2	14		14.4	18.3	24	30	
61											6.9	8.6	11.5	14.3		14.8	18.5	24.7	30.8	
62											7	8.7	11.7	14.6		15.1	18.7	25.2	31.5	
63											7.1	8.9	11.9	14.8	± 30% if F_t < 10 N	15.3	19.1	25.5	31.9	
64											7.2	9	12	15		15.5	19.4	25.8	32.3	
65	2.5										7.3	9.2	12.2	15.2	± 20% if F_t ≥ 10 N	15.7	19.8	26.2	32.7	
66											7.4	9.3	12.3	15.4		15.9	20	26.5	33.1	
67									0.2	+0.2	7.5	9.4	12.5	15.6		16.1	20.2	26.8	33.5	
68										0	7.6	9.6	12.6	15.8		16.3	20.6	27.2	33.9	
69								−0.010			7.8	9.7	13	16.3		16.8	20.9	28	35	
70	2.7							−0.030			7.9	9.9	13.2	16.5		17	21.3	28.3	35.4	
71								For			8	10	13.3	16.7		17.2	21.5	28.7	35.8	
72								Phosphated			8.1	10.1	13.5	16.9		17.4	21.7	29	36.2	
73								PO			8.2	10.3	13.6	17.1		17.6	22.1	29.3	36.7	
74								Surface			8.3	10.4	13.8	17.2		17.8	22.4	29.7	37.1	
75	2.9										8.4	10.6	14	17.5		18.1	22.8	30.2	37.7	
76								−0.005			8.5	10.7	14.2	17.7		18.3	23	30.5	38.1	
77								−0.035			8.7	10.8	14.5	18.1		18.7	23.2	31.2	38.9	
78											8.8	11	14.6	18.3		18.9	23.7	31.5	39.4	
79	3.1										8.9	11.1	14.8	18.5		19.1	23.9	31.8	39.8	

TABLE 4—DIMENSIONS OF TYPE B RECTANGULAR RINGS MADE OF STEEL (CONTINUED)

Dimensions in millimeters

Nominal diameter d_1	Radial wall thickness a_1	Tolerance	Ring width h_1 Column 1	2	3	4	Tolerance	Closed gap s_1	Tolerance	Tangential force F_t, N For h_1 shown in column 1	2	3	4	Tolerance	Diametral force F_d, N For h_1 shown in column 1	2	3	4	Tolerance
80	3.1	±0.15 Within a ring: 0.15 max.						0.25	+0.25 0	9	11.3	15	18.8		19.4	24.3	32.3	40.4	
81										9.1	11.4	15.2	19		19.6	24.5	32.7	40.8	
82										9.2	11.6	15.3	19.2		19.8	24.9	33	41.2	
83										9.3	11.7	15.5	19.4		20	25.2	33.3	41.7	
84										9.4	11.8	15.7	19.6		20.2	25.4	33.7	42.1	
85	3.3									9.6	12	16	20		20.6	25.8	34.3	42.9	
86										9.7	12.1	16.2	20.2		20.9	26	34.8	43.5	
87										9.8	12.3	16.4	20.4		21.1	26.4	35.2	43.9	
88										9.9	12.4	16.5	20.6		21.3	26.7	35.5	44.4	
89										10	12.5	16.7	20.8		21.5	26.9	35.8	44.8	
90	3.5		1.2	1.5	2.0	2.5				10.1	12.7	16.8	21		21.7	27.3	36.2	45.2	
91										10.2	12.8	17	21.2		21.9	27.5	36.5	46.2	
92										10.3	13	17.1	21.4		22.1	28	36.8	46	
93										10.5	13.1	17.5	21.9		22.6	28.2	37.7	47.1	
94										10.6	13.2	17.7	22.1		22.8	28.4	38	47.5	
95	3.7									10.7	13.4	17.8	22.3		23	28.8	38.3	47.9	
96										10.8	13.5	18	22.5		23.2	29	38.7	48.3	
97										10.9	13.7	18.1	22.7		23.4	29.5	39	48.7	
98										11	13.8	18.4	23		23.7	29.7	39.5	49.4	
99								0.3		11.1	13.9	18.5	23.2		23.9	29.9	39.8	49.8	
100	3.9		1.5	2	2.5	3	-0.010 -0.030			14.2	18.9	23.6	28.4		30.5	40.7	50.8	61	
101										13.9	18.5	23.1	27.8		29.9	39.9	49.8	59.7	
102										13.6	18.1	22.7	27.2		29.2	39	48.7	58.5	
103										13.3	17.8	22.2	26.6		28.6	38.2	47.7	57.3	
104										13.1	17.4	21.8	26.1		28.1	37.4	46.8	56.1	
105								For Phosphated PO Surface		12.8	17.1	21.3	25.6	±30% if F_t < 10 N ±20% if F_t ≥ 10 N	27.5	36.7	45.8	55	±30% if F_d < 21.5 N ±20% if F_d ≥ 21.5 N
106										17	22.7	28.3	34		36.6	48.7	60.9	73.1	
107										16.7	22.2	27.8	33.3		35.8	47.8	59.7	71.7	
108										16.3	21.8	27.2	32.7		35.1	46.8	58.6	70.3	
109										16	21.4	26.7	32		34.5	45.9	57.4	68.9	
110	4.3		2.0	2.5	3.0	3.5	-0.006 -0.035			21	26.2	31.4	36.7		45.1	56.3	67.6	78.9	
111										20.6	25.7	30.8	36		44.2	55.2	66.3	77.3	
112										20.2	25.2	30.3	35.3		43.4	54.2	65	75.9	
113										19.8	24.7	29.7	34.6		42.6	53.2	63.8	74.5	
114										19.4	24.3	29.1	34		48.8	52.2	62.7	73.1	
115		+0.2 -0.2 Within a ring: 0.2 max.								19.1	23.8	28.6	33.4		41	51.3	61.5	71.8	
116										24.7	30.9	37.1	43.3		53.1	66.4	79.7	93	
117										24.3	30.3	36.4	42.5		52.2	65.2	78.3	91.3	
118										23.8	29.8	35.8	41.7		51.2	64.1	76.9	89.7	
119										23.4	29.3	35.1	41		50.3	62.9	75.5	88.1	
120	4.7							0.35	+0.30 0	23	28.7	34.5	40.2		49.4	61.8	74.2	86.5	
121										22.6	28.2	33.9	39.5		48.6	60.7	72.9	85	
122										22.2	27.8	33.3	38.9		47.7	59.7	71.6	83.6	
123										21.8	27.3	32.7	38.2		46.9	58.7	70.4	82.1	
124										21.5	26.8	32.2	37.5		46.1	57.7	69.2	80.7	
125										21.1	26.4	31.6	36.9		45.3	56.7	68	79.4	
126										26.8	33.5	40.1	46.8		57.5	71.9	86.3	100.7	
127										26.3	32.9	39.5	46.1		56.6	70.7	84.9	99	
128										25.9	32.4	38.8	45.3		55.6	69.6	83.5	97.4	
129										25.5	31.8	38.2	44.6		54.7	68.4	82.1	95.8	
130	5.1							0.4	+0.35 0	25	31.3	37.6	43.8		53.8	67.3	80.8	94.2	
131										24.6	30.8	37	43.1		53	66.2	79.5	92.7	
132										24.2	30.3	36.4	42.4		52.1	65.2	78.2	91.2	
133										23.9	29.8	35.8	41.8		51.3	64.1	76.9	89.8	
134										23.5	29.4	35.2	41.1		50.5	63.1	75.7	88.4	

TABLE 4—DIMENSIONS OF TYPE B RECTANGULAR RINGS MADE OF STEEL (CONTINUED)

Dimensions in millimeters

Nominal diameter d_1	Radial wall thickness s_1	Tolerance	Ring width h_1 Column 1	2	3	4	Tolerance	Closed gap s_1	Tolerance	Tangential force F_t N For h, shown in column 1	2	3	4	Tolerance	Diametral force F_d N For h, shown in column 1	2	3	4	Tolerance
135	5.1		2.0	2.5	3.0	3.5	−0.010 −0.030 For Phosphated PO Surface −0.005 −0.035	0.4	+0.35 0	23.1	28.9	34.7	40.5		49.7	62.1	74.6	87	
136										28.8	36	43.2	50.4		61.9	77.4	92.9	108.4	
137										28.4	35.6	42.5	49.6		61	76.2	91.5	106.7	
138										27.9	34.9	41.9	48.9		60	75.1	90.1	105.1	
139										27.5	34.4	41.3	48.1		59.1	73.9	88.7	103.5	
140										33.9	40.8	47.4	54.2		72.8	87.4	101.9	116.5	
141										33.4	40	46.7	53.4		71.7	86	100.4	114.7	
142	5.5									32.9	39.4	46	52.6		70.6	84.8	98.9	113	
143										32.4	38.8	45.3	51.8		69.6	83.5	97.4	111.4	
144			2.5	3.0	3.5	4.0				31.9	38.3	44.7	51		68.6	82.3	96	109.7	
145										31.4	37.7	44	50.3		67.6	81.1	94.6	108.1	
146										38.6	46.3	54	61.7		82.9	99.5	116.1	132.7	
147										38	45.6	53.2	60.8		81.7	98.1	114.4	130.8	
148										37.5	45	52.5	60		80.6	96.7	112.8	128.9	
149										36.9	44.3	51.7	59.1		79.4	95.3	111.2	127.1	
150										36.4	43.7	51	58.3		78.3	94	109.6	125.3	
151										35.9	43.1	50.3	57.4		77.2	92.6	108.1	123.5	
152	5.9									35.4	42.5	49.6	56.7		76.1	91.3	106.6	121.8	
153										34.9	41.9	48.9	55.9		75.1	90.1	105.1	120.1	
154										34.4	41.3	48.2	55.1		74	88.9	103.7	118.5	
155										34	40.8	47.6	54.4		73	87.6	102.2	116.9	
156										41.1	49.4	57.6	65.9		88.4	106.1	123.8	141.5	
157										40.6	48.7	56.8	64.9		87.2	104.7	122.1	139.6	
158							−0.010 −0.040			40	48	56	64		86.1	103.3	120.5	137.7	
159							For Phosphated PO Surface −0.005 −0.045			39.5	47.4	55.3	63.2		84.9	101.9	118.9	135.9	
160										39	46.8	54.6	62.4		83.8	100.5	117.3	134.1	
161		+0.2 −0.2							+0.4 0	38.5	46.2	53.9	61.5	± 30% if $F_t < 10$ N ± 20% if $F_t ≥ 10$ N	82.7	99.2	115.8	132.3	± 30% if $F_d < 21.5$ N ± 20% if $F_d ≥ 21.5$ N
162	6.3							0.5		38	45.5	53.1	60.7		81.6	97.9	114.3	130.6	
163		Within a ring: 0.2 max.								37.5	45	52.5	59.9		80.6	96.7	112.8	128.9	
164										37	44.4	51.8	59.2		79.5	95.4	111.3	127.2	
165										36.5	43.8	51.1	58.4		78.5	94.2	109.9	125.6	
166										43.7	52.4	61.2	69.9		93.9	112.7	131.5	150.3	
167										43.1	51.8	60.4	69		92.7	111.3	129.8	148.4	
168										42.6	51.1	59.6	68.1		91.6	109.9	128.2	146.5	
169										42.1	50.5	58.9	67.3		90.4	108.5	126.6	144.7	
170										41.5	49.8	58.1	66.4		89.3	107.1	125	142.9	
171										41	49.2	57.4	65.6		88.2	105.8	123.5	141.1	
172	6.7									40.5	48.6	56.7	64.8		87.1	104.5	121.9	139.4	
173										40	48	56	64		86	103.2	120.4	137.7	
174										39.5	47.4	55.3	63.2		85	102	119	136	
175										46.9	54.7	62.5	70.3		100.8	117.5	134.3	151.1	
176										55.5	64.7	74	83.2		119.3	139.2	159.1	179	
177										54.8	64	73.1	82.2		117.9	137.5	157.2	176.8	
178			3.0	3.5	4.0	4.5				54.2	63.2	72.2	81.3		116.5	135.9	155.3	174.7	
179										53.6	62.5	71.4	80.3		115.1	134.3	153.5	172.6	
180										52.9	61.7	70.5	79.4		113.7	132.7	151.7	170.6	
181										52.3	61	69.7	78.4		112.4	131.1	149.9	168.6	
182	7.1							0.6	+0.45 0	51.7	60.3	68.9	77.5		111.1	129.6	148.1	166.7	
183										51.1	59.6	68.1	76.6		109.8	128.1	146.4	164.7	
184										50.5	58.9	67.3	75.7		108.6	126.7	144.7	162.8	
185										49.9	58.2	66.6	74.9		107.3	125.2	143.1	161	
186										58.6	68.3	78.1	87.8		125.9	146.9	167.9	188.9	
187										57.9	67.5	77.2	86.8		124.5	145.2	166	186.7	
188	7.5									57.2	66.8	76.3	85.9		123.1	143.6	164.1	184.6	
189										56.6	66	75.5	84.9		121.7	142	162.3	182.5	

TABLE 4—DIMENSIONS OF TYPE B RECTANGULAR RINGS MADE OF STEEL (CONCLUDED)

Dimensions in millimeters

Nominal diameter d_1	Radial wall thickness a_1	Tolerance	Ring width h_1 Column 1	2	3	4	Tolerance	Closed gap s_1	Tolerance	Tangential force F_t, N For h, shown in column 1	2	3	4	Tolerance	Diametral force F_d, N For h, shown in column 1	2	3	4	Tolerance
190	7.5									56	65.3	74.6	84		120.3	140.4	160.5	180.5	
191							−0.010			55.4	64.6	73.8	83		119	138.8	158.7	178.5	
192							−0.040			54.7	63.9	73	82.1		117.7	137.3	156.9	176.5	
193										54.1	63.2	72.2	81.2	± 30% if	116.4	135.8	155.2	174.6	± 30% if
194		+0.2					For Phosphated			53.6	62.5	71.4	80.3	$F_t < 10$ N	115.1	134.3	153.5	172.7	$F_d < 21.5$ N
195		−0.2 Within	3.0	3.5	4.0	4.5	PO Surface	0.6	+0.45 0	53	61.8	70.6	79.5	± 20% if	113.8	132.9	151.8	170.8	± 20% if
196		a ring: 0.2								61.6	71.9	82.2	92.4	$F_t \geq 10$ N	132.5	154.6	176.7	198.8	$F_d \geq 21.5$ N
197		max.								61	71.1	81.3	91.4		131.1	152.9	174.8	196.6	
198	7.9						−0.005			60.3	70.4	80.4	90.5		129.7	151.3	172.9	194.5	
199							−0.045			59.7	69.6	79.6	89.5		128.3	149.7	171.1	192.4	
200										59	68.9	78.7	88.6		126.9	148.1	169.2	190.4	

NOTES

1 For intermediate sizes (for example repair sizes), the radial wall thickness of the next smaller nominal diameter should be applied.

2 The values for F_t and F_d, given in Table 4, apply to as-cast steel with a typical modulus of elasticity (E_n) of 200 000 MPa.

Mean forces are calculated for nominal radial wall thickness (a_1) and mean ring width (h_1).

3 For the sole purpose of this Technical Report, the assumed average ratio F_d/F_t is 2.15. However, for rings up to 50 mm the ratio F_d/F_t shall be determined between manufacturer and client.

INTERNAL COMBUSTION ENGINES—PISTON RINGS—GENERAL SPECIFICATIONS—SAE J1591 OCT92

SAE Standard

Report of the Piston Ring Standards Committee approved September 1990. Revised by the SAE Piston Ring Standards Committee 7 October 1992.

This SAE Standard is equivalent to ISO Standard 6621/4.

1. Scope and Field of Application—Differences, where they exist, are shown in Appendix A.

This SAE Standard specifies the general characteristics of piston rings and individual dimensional criteria, which are specified as appropriate in the documents in Section 2.

This document also provides a system for coding, designation, and marking of piston rings.

This document applies to the manufacture of piston rings up to and including 200 mm diameter for reciprocating internal combustion engines. It also applies to piston rings for compressors working under similar conditions.

(R) 2. References

SAE DESIGNATION	ISO[1] EQUIVALENT	
		INTERNAL COMBUSTION ENGINES—PISTON RINGS
J1588	6621/1	Vocabulary
J1589	6621/2	Measuring principles
J1590	6621/3	Material specifications
J1591	6621/4	General specifications
J1996	6621/5	Quality requirements
		INTERNAL COMBUSTION ENGINES—PISTON RINGS
J1997	6622/1	Rectangular rings
J1998	6622/2 TR	Rectangular rings with narrow ring width
J1999	6623	INTERNAL COMBUSTION ENGINES—PISTON RINGS—SCRAPER RINGS
		INTERNAL COMBUSTION ENGINES—PISTON RINGS
J2000	6624/1	Keystone rings
J2001	6624/2 TR	Half keystone rings
J2002	6625	INTERNAL COMBUSTION ENGINES—PISTON RINGS—OIL CONTROL RINGS
J2003	6626	INTERNAL COMBUSTION ENGINES—PISTON RINGS—COIL SPRING LOADED OIL CONTROL RINGS
J2004	6627 TR	INTERNAL COMBUSTION ENGINES—PISTON RINGS—EXPANDER/SEGMENT OIL CONTROL RINGS
J2226		INTERNAL COMBUSTION ENGINES—PISTON RINGS—STEEL RECTANGULAR RINGS
	6507/3	METALLIC MATERIALS—HARDNESS TEST—VICKERS TEST—PART 3: LESS THAN HV 0.2

[1] TR refers to Technical Report

3. Terminology—The terminology used in this document is as given in SAE J1588.

4. Piston Ring Codes—Codes used for piston rings shall be as given in Table 1 with their explanatory descriptions.

5. Designation of Piston Rings

5.1 Designation Elements and Order—To designate piston rings complying with the relevant International Standards and the SAE documents, the following details shall be given in the order shown.

The codes given in Table 1 shall be used.

5.1.1 MANDATORY ELEMENTS—The following mandatory elements shall constitute the designation of a piston ring:
    a. Designation, i.e., piston ring
    b. Number of SAE Standards
    c. Type of piston ring, e.g., R
    d. Hyphen
    e. Size of piston ring, $d_1 \times h_1$
    f. Code D22 if the selected wall thickness, in accordance with SAE J1997 and J1999, is D/22
        (1) Hyphen
        (2) Material code, e.g., MC11

5.1.2 ADDITIONAL ELEMENTS—The following optional elements may be added to the designation of a piston ring; in this case, they shall be added on the second line beneath the mandatory elements, or separated by a slant (/):
    a. Code for reduced ratio m/d_1, MR
    b. Code for ring shape, e.g., MZ
    c. Code for the selected nominal closed gap if it differs from the closed gap specified in the dimensional tables, e.g., S05
    d. Code for the selected coating, e.g., CR3
    e. Code for uncoated rings with fully lapped periphery or taper faced rings with partial lapped periphery, e.g., LF or LP
    f. Code for the selected surface treatment, e.g., PO
    g. Code for the selected peripheral edge feature, e.g., KA
    h. Code for the selected inside edge feature, e.g., KI
    i. Code for the inside step or bevel, e.g., IWU
    j. Code for the selected notch to prevent ring rotation, e.g., NH1
    k. Code if reduced slot length is required, WK
    l. Code if the coil spring with reduced heat set is required, e.g., WF
    m. Code for the selected type of coil spring, e.g., CSG
    n. Code for the selected pressure class, e.g., PNM

5.1.3 ELEMENTS FOR ADDITIONAL MARKING—Any additional marking shall follow the additional elements specified in 5.1.2:
    a. Code if manufacturer's mark is required, MM
    b. Code for marking of required ring shape, e.g., MZ
    c. Code for material, MX (see Table 1, footnote 1)
    d. Code for any other marking, MU (see Table 1, footnote 2)

TABLE 1—CODES AND DESCRIPTIONS

Code	Description	Relevant SAE Standard
R	Straight faced rectangular ring	J1997/J1998
B	Barrel faced rectangular ring	J1997/J1998
M1 to M5	Taper faced rectangular ring	J1997/J1998
N	Scraper ring (Napier)	J1999
NM1 to NM4	Taper faced scraper ring (Napier)	J1999
E	Scraper ring (stepped)	J1999
EM1 to EM4	Scraper ring (stepped), taper faced	J1999
T	Straight faced keystone ring 6°	J2000
TB	Barrel faced keystone ring 6°	J2000
TM1 to TM5	Taper faced keystone ring 6°	J2000
K	Straight faced keystone ring 15°	J2000
KB	Barrel faced keystone ring 15°	J2000
KM1 to KM5	Taper faced keystone ring 15°	J2000
HK	Straight faced half keystone ring 7°	J2001
HKB	Barrel faced half keystone ring 7°	J2001
S	Slotted oil control ring	J2002
G	Double bevelled oil control ring	J2002
D	Bevelled edge oil control ring	J2002
DV	Bevelled edge V-groove oil control ring	J2002
DSF-C	Coil spring loaded bevelled edge oil control ring, chromium plated and profile ground	J2003
DSF-CNP	Coil spring loaded bevelled edge oil control ring, chromium plated not profile ground	J2003
SSF	Coil spring loaded slotted oil control ring	J2003
GSF	Coil spring loaded double bevelled oil control ring	J2003
DSF	Coil spring loaded bevelled edge oil control ring	J2003
DSF-NG	Coil spring loaded bevelled edge oil control ring (face geometry similar type DSF-C or DSF-CNP)	J2003
SSF-L	Coil spring loaded slotted oil control ring with 0.6 mm nominal land width	J2003
D22	Radial wall thickness $a_1 = d_1/22$	J1997/J1999
MC11 to MC63	Material subclasses	J1590
MR	Ratio m/d_1 reduced	J1591, 7.4
Z	Ring shape round	
Y	Ring shape negative ovality	J1591, 7.1
S00 to S10	Closed gap (minimum values)	J1591, 7.3
CRF to CR4	Periphery chromium coated fully faced design	
CR1E to CR4E	Periphery chromium coated semi-inlaid design	J1591, 9.3
CR1F to CR4F	Periphery chromium coated inlaid design	
MO1 to MO4	Periphery molybdenum coated fully faced design	
MO1E to MO4E	Periphery molybdenum coated semi-inlaid design	J1591, 9.3
MO1F to MO4F	Periphery molybdenum coated inlaid design	
LF	Uncoated ring periphery or uncoated land periphery, fully lapped	J1591, 7.2
LP	Taper faced piston ring with lapped land over the whole circumference but not over the whole width of the periphery	J1591, 7.2
FE	Ferrox coated	J1591, 10.1
PO	Phosphate coated (0.002 mm min)	
PR	Phosphate coated (0.002 mm max)	J1591, 10.2 & 10.3
KA	Peripheral edges chamfered	
KI	Inside edges chamfered	J1997/J1998
IF	Internal bevel (top side)	
IFU	Internal bevel (bottom side)	
IW	Internal step (top side)	J1997/J1998
IWU	Internal step (bottom side)	
IFV	Variable internal bevel (top side)	
IFVU	Variable internal bevel (bottom side)	

TABLE 1—CODES AND DESCRIPTIONS (CONT.)

Code	Description	Relevant SAE Standard
NE1 to NE3	Ring joint with lateral stop	
NH1 to NH3	Ring joint with internal stop	J1591, 8.1
WK	Reduced slot length	J2003
WF	Reduced heat set	J1996, J2003
CSN, CSG, CSE	Type of coil spring	J2003
PNE, PNL, PNR, PNM, PNH, PNV	Contact pressure class	J2003
	Additional marking:	
MM	manufacturer's mark	
MZ	mark for required ring shape "round"	
MY	mark for required ring shape "negative ovality"	J1591, 6.2
MX	material mark[1]	
MU	any other additional mark[2]	

[1] Material mark (for alternative materials) at the discretion of the manufacturer.
[2] Any other additional marking on purchaser's request shall be quoted clearly in the order; it shall be agreed on between manufacturer and purchaser.

5.2 Designation Examples

5.2.1 Designation of a piston ring complying with the requirements of SAE J1997,
 a. A straight faced rectangular ring (R)
 b. Of nominal diameter $d_1 = 90$ mm (90)
 c. And a nominal ring width $h_1 = 2.5$ mm (2.5)
 d. Made of grey cast iron, nonheat-treated, material subclass 11 (MC11):
  (1) Piston ring SAE J1997 R - 90 × 2.5 - MC11

5.2.2 Designation of a piston ring complying with the requirements of SAE J2000,
 a. A keystone ring 6°, taper faced (TM1)
 b. Of nominal diameter $d_1 = 105$ mm (105)
 c. And nominal ring width $h_1 = 2.5$ mm (2.5)
 d. Made of spheroidal graphite cast iron, martensitic type, material subclass 51 (MC51)
 e. Ring shape round (MZ)
 f. With selected closed gap of 0.3 mm (S03)
 g. Inside edges chamfered (KI)
 h. Periphery chromium plated, with layer thickness 0.1 mm min (CR2):
  (1) Piston ring SAE J2000 TM1 - 105 × 2.5 - MC51
  (2) Z S03 KI CR2

5.2.3 Designation of a piston ring complying with the requirements of SAE J2003,
 a. A coil spring loaded, bevelled edge oil control ring, chromium plated and profile ground (DSF-C)
 b. Of nominal diameter $d_1 = 125$ mm (125)
 c. And nominal ring width $h_1 = 5$ mm (5)
 d. Made of grey cast iron, nonheat-treated, material subclass 11 (MC11)
 e. A selected closed gap of 0.2 mm (S02)
 f. A chromium layer thickness on the lands of 0.15 mm min (CR3)
 g. Phosphated on all cast iron surfaces to a depth of 0.002 mm min (PO)
 h. Reduced slot length (WK)
 i. Coil spring with reduced heat set (WF)
 j. And variable pitch with coil diameter d_1 ground (CSE)
 k. Tangential force F_t according to the medium nominal contact pressure class (PNM)
 l. And the ring marked with manufacturer's mark (MM):
  (1) Piston ring SAE J2003 DSF-C-125 × 5 - MC11
  (2) S02 CR3 PO WK WF CSE PNM MM

6. *Marking of Piston Rings*—The requirements and recommendations for piston ring marking in 6.1 and 6.2 apply to piston rings of 1.6 mm radial wall

thickness and above. Marking of piston rings below 1.6 mm is at the choice of the manufacturer.

6.1 Mandatory Top Side Marking—All rings requiring orientation shall be marked to indicate the top side only, i.e., the side nearest to the combustion chamber.

In the absence of any other mark agreed on between manufacturer and purchaser, the mark TOP should be used.

Marking of the top side applies to the following types of ring:
a. All taper faced rings
b. All internally bevelled or stepped rings
c. All semi-inlaid rings
d. All scraper rings
e. All half keystone rings
f. All directional oil control rings

All such rings requiring marking are shown in the "common features" clause of the appropriate SAE Standards: J1997, J1998, J1999, J2000, J2001, J2002, and J2003.

6.2 Additional Marking—Additional marking of piston rings is optional or at the purchaser's request.

Such additional marking comprises the following:
a. Manufacturer's mark
b. Mark for required ring shape
c. Material mark (for alternative materials)
d. Any other additional mark agreed on between manufacturer and purchaser

7. General Characteristics

7.1 Ring Shape—Degrees of ovality only apply to rectangular rings (SAE J1997, J1998), scraper rings (SAE J1999), and keystone rings (SAE J2000, J2001). The forms of ovality are:
a. Positive ovality, i.e., standard without a code
b. Round, code MZ
c. Negative ovality, code MY

Values are given in Table 2.

TABLE 2—OVALITY

Dimensions in millimeters

Nominal Diameter d_1	Positive Ovality	Round[1]	Negative Ovality[2]
$30 \leq d_1 < 60$	0 to +0.60	-0.30 to +0.30	-0.60 to 0
$60 \leq d_1 < 100$	+0.05 to +0.85	-0.35 to +0.35	-0.70 to 0
$100 \leq d_1 < 150$	+0.10 to +1.10	-0.45 to +0.45	-0.95 to -0.05
$150 \leq d_1 \leq 200$	+0.15 to +1.35	-0.50 to +0.50	-1.10 to -0.10

[1] For taper faced coated and uncoated rings with lapped land, the recommended ring shape is round.
[2] Not applicable for material class 10 of SAE J1590.

7.2 Light Tightness—At least 90% of the piston ring periphery shall be light tight.

Taper faced rings with coated and ground periphery without lapped land shall be at least 95% light tight.

The following piston ring designs shall be 100% light tight.
a. Uncoated piston rings with periphery fully lapped
b. Taper faced piston rings uncoated or coated with lapped land over the whole circumference but not over the whole width of the periphery
c. Piston rings with periphery chromium plated or molybdenum coated (fully faced, semi-inlaid, or inlaid design) with lapped land over the whole width of periphery

NOTE—In the case of piston rings with treated surface, the light tightness is normally measured prior to surface treatment. When it is checked after treatment, rotation of the ring in the gauge will be required.

In the case of rings with negative point deflection, visible light is permitted at the butt ends but should be confined to the angle defined in SAE J1589.

7.3 Closed Gap—Whenever the selected closed gap differs from that given in the dimensional tables of the specific International Standards, the codes in Table 3 apply and the tolerances remain the same.

TABLE 3—CLOSED GAP

Dimensions in millimeters

Code	Closed Gap
S00	0.05
S01	0.1
S02	0.2
S03	0.3
S04	0.4
S05	0.5
S06	0.6
S07	0.7
S08	0.8
S09	0.9
S10	1.0

7.4 Tangential Force, F_t, and Diametral Force, F_d, of Single Piece Piston Rings—The individual types of piston rings are given in SAE J1997, J1998, J1999, J2000, J2001, and J2002. The definitions of F_t and F_d are given in SAE J1589.

7.4.1 CALCULATION OF F_t AND F_d VALUES IN DIMENSION TABLES OF DIMENSIONAL STANDARDS—The tangential and diametral forces of piston rings are tabulated in the dimension tables of the dimensional documents.

The values are calculated for the following:
a. The basic feature of each piston ring type
b. Nominal radial wall thickness, a_1, and mean ring width, h_1 or h_3
c. Piston ring material with a modulus of elasticity of 100 000 MPa
d. A ratio of total free gap to nominal diameter (m/d_1) according to Table 4

NOTE—The calculation of the tangential forces and diametral forces of the following:
e. Rectangular rings with narrow ring width made of steel (see SAE J1998)
f. Half keystone rings made of steel (see SAE J2001)

is based on a theoretical contact pressure of approximately 0.19 MPa.

The ratio m/d_1 is quite different to the values given in Table 4 and depends on the nominal diameter and the special radial wall thickness. This radial wall thickness is not in a constant ratio to nominal diameter because there are steps of wall thickness which belong to a range of nominal diameters (e.g., $a_1 = 2.1$ mm for $d_1 = 52$ to 57 mm).

TABLE 4—REGULAR RATIO OF TOTAL FREE GAP TO NOMINAL DIAMETER

d_1	m/d
$30 \leq d_1 < 100$	$15 \dfrac{1}{100}$
$100 \leq d_1 \leq 200$	$17 - \dfrac{2d_1}{100} \dfrac{1}{100}$

7.4.2 CORRECTION OF F_t AND F_d VALUES—The F_t and F_d values shall be corrected when:
a. Additional features such as rings with:
   (1) Coated periphery and/or
   (2) Inside chamfered edges and/or
   (3) Outside chamfered edges and/or
   (4) Taper and/or
   (5) Internal step or
   (6) Internal bevel
b. Piston ring materials with a modulus of elasticity other than 100 000 MPa
c. A ratio of total free gap to nominal diameter (m/d_1) other than that given in Table 4

The formula for the regular ratio of total free gap to nominal diameter (m/d_1 regular) is given in Table 4.

7.4.2.1 Multiplier Factors for Common Features—For common features, the necessary multiplier correction factors are tabulated in the dimensional documents SAE J1997, J1998, J1999, J2000, J2001, and J2002 under clause 5 "force factors".

7.4.2.2 Multiplier Force Correction Factors for Materials—For materials specified in SAE J1590, the force correction factors given in Table 5 are recommended.

7.4.2.3 Multiplier Force Correction Factors for Ratio m/d_1—Piston rings made of materials in classes 30 to 60 increase the tangential force and diametral force in relationship to the modulus of elasticity (see Table 5) when ratio m/d_1 regular is used.

For limitation of such increased forces, it is common to use reduced values of m/d_1. In Table 6 the recommended correction factors for m/d_1 regular and m/d_1 reduced are given.

For calculation of real values of ratio m/d_1 reduced, the factors given in Table 6 apply. Therefore, the values of m/d_1 calculated with the formula given in Table 4 have to be corrected with the correction factors in Table 6.

TABLE 5—MATERIAL FORCE CORRECTION FACTORS

Material Class	Material Force Correction Factor
10	0.9 to 1[1]
20	1.1 to 1.3[1]
30	1.45
40	1.6
50	1.6
60	2.0

[1] Force correction factors for materials depend on the modulus of elasticity in the manufacturer's material specification. See (Eq.1):

$$\text{Correction factor} = \frac{\text{Typical modulus of elasticity in MPa}}{100\ 000\ \text{MPa}} \quad \text{(Eq.1)}$$

TABLE 6—FORCE CORRECTION FACTORS FOR RATIO m/d_1

Material Class	Factor for m/d_1 Regular	Factor for m/d_1 Reduced[1]
10	1	-
20	1	-
30	1	0.825
40	1	0.75
50	1	0.75
60	-	0.75

[1] Ratio m/d_1 reduced is given the code MR.

7.4.3 Examples for Correction of F_t and F_d

7.4.3.1 Designation of Piston Ring
a. SAE J1997 B - 95 × 2.5 - MC53
b. MR CR2 IW

7.4.3.1.1 Multiplying Factors
a. 1.6 for material subclass 53
b. 0.75 for ratio m/d_1 reduced
c. 0.85 for periphery chromium plated CR2
d. 0.78 for internal step

7.4.3.1.2 Calculation
Total force correction factor:
$$1.6 \times 0.75 \times 0.85 \times 0.78 = 0.796 \quad \text{(Eq.2)}$$
Basic values F_t and F_d according to J1997:
$$F_t = 18.5\ \text{N and } F_d = 39.8\ \text{N} \quad \text{(Eq.3)}$$
Corrected values:
$F_t = 0.796 \times 18.5\ \text{N} \pm 20\%$
$F_t = 14.7\ \text{N} \pm 20\%$
$F_t = 11.8$ to 17.6 N
and
$F_d = 0.796 \times 39.8\ \text{N} \pm 20\%$
$F_d = 31.7\ \text{N} \pm 20\%$
$F_d = 25.6$ to 38 N

7.4.3.2 Designation of Piston Ring
a. SAE J1999 N - 70 × 2 D22 - MC24/MO2

7.4.3.2.1 Multiplying Factors
1.15 for material subclass 24
0.86 for periphery molybdenum coated MO2F (inlaid type)

7.4.3.2.2 Calculation
Total force correction factor:
$$1.15 \times 0.86 = 0.989 \quad \text{(Eq.4)}$$
Basic values F_t and F_d according to SAE J2000:
$$F_t = 9.2\ \text{N and } F_d = 19.8\ \text{N} \quad \text{(Eq.5)}$$
Corrected values:
$F_t = 0.989 \times 9.2\ \text{N} \pm 30\%$
$F_t = 9.1\ \text{N} \pm 30\%$
$F_t = 6.4$ to 11.8 N
and
$F_d = 0.989 \times 19.8\ \text{N} \pm 30\%$
$F_d = 19.6\ \text{N} \pm 30\%$
$F_d = 13.7$ to 25.5 N

7.4.3.3 Designation of Piston Ring
a. SAE J2000 KB - 140 × 4 - MC42/MO4 KI

7.4.3.3.1 Multiplying Factors
a. 1.6 for material subclass 42
b. 0.83 for periphery molybdenum coated MO4 (fully faced type)
c. 0.96 for inside chamfered edges KI

7.4.3.3.2 Calculation
Total force correction factor:
$$1.6 \times 0.83 \times 0.96 = 1.275 \quad \text{(Eq.6)}$$
Basic values F_t and F_d according to SAE J2000:
$$F_t = 29.3\ \text{N and } F_d = 63\ \text{N} \quad \text{(Eq.7)}$$
Corrected values:
$F_t = 1.275 \times 29.3\ \text{N} \pm 20\%$
$F_t = 37.4\ \text{N} \pm 20\%$
$F_t = 29.9$ to 44.9 N
and
$F_d = 1.275 \times 63\ \text{N} \pm 20\%$
$F_d = 80.3\ \text{N} \pm 20\%$
$F_d = 64.2$ to 96.4 N

7.4.3.4 Designation of Piston Ring
a. SAE J2002 G - 120 × 5 - MC11/KI

7.4.3.4.1 Multiplying Factors
a. 0.9 for material subclass 11
b. 0.98 for inside chamfered edges KI

7.4.3.4.2 Calculation
Total force correction factor:
$$0.9 \times 0.98 = 0.882 \quad \text{(Eq.8)}$$
Basic values F_t and F_d according to SAE J2002:
$$F_t = 24.7\ \text{N and } F_d = 53.1\ \text{N} \quad \text{(Eq.9)}$$
Corrected values:
$F_t = 0.882 \times 24.7\ \text{N} \pm 20\%$
$F_t = 21.8\ \text{N} \pm 20\%$
$F_t = 17.4$ to 26.2 N
and
$F_d = 0.882 \times 53.1\ \text{N} \pm 20\%$
$F_d = 46.8\ \text{N} \pm 20\%$
$F_d = 37.4$ to 56.2 N

7.5 Tangential Force F_t of Multipiece Oil Control Rings as Specified in SAE J2003—The tangential force of a coil spring loaded oil control ring depends on:

a. Piston ring type
b. Class of nominal contact pressure
c. Specific tangential force F_{tc} for unit contact pressure of 1 MPa

The values for nominal contact pressure and specific tangential forces are tabulated in SAE J2003.

The formula for calculating the actual tangential force is also given in SAE J2003.

7.5.1 EXAMPLES FOR CALCULATING THE TANGENTIAL FORCE F_t

7.5.1.1 Designation of Piston Ring
SAE J2003 DSF-C - 100 × 4 - MC11/CR1 CSG PNM

7.5.1.1.1 PRESSURE CLASS AND SPECIFIC TANGENTIAL FORCE

a. Class of nominal contact pressure:
 (1) PNM = 1.49 MPa
b. Specific tangential force for unit contact pressure of 1 MPa:
 (1) F_{tc} = 40.4 N

7.5.1.1.2 CALCULATION

Tangential force, $\quad F_t = 1.49 \times 40.4 \text{ N} \pm 20\%$ (Eq.10)
$\quad F_t = 60.2 \text{ N} \pm 20\%$
$\quad F_t = 48 \text{ to } 72 \text{ N}^2$

7.5.1.2 Designation of Piston Ring
a. SAE J2003 SSF - 175 × 6 MC11/CSG PNE

7.5.1.2.1 PRESSURE CLASS AND SPECIFIC TANGENTIAL FORCE

a. Class of nominal contact pressure:
 (1) PNE = 0.59 MPa
b. Specific tangential force for unit contact pressure of 1 MPa:
 (1) F_{tc} = 192.5 N

7.5.1.2.2 CALCULATION

Tangential force, $\quad F_t = 0.59 \times 192.5 \text{ N} \pm 20\%$ (Eq.11)
$\quad F_t = 113.6 \text{ N} \pm 20\%$
$\quad F_t = 91 \text{ to } 136 \text{ N}^3$

8. Notches to Prevent Ring Rotation

8.1 Ring Joint With Internal Notch (only for compression rings of types R, B, M, T, K, and HK) (see Figure 1)—See Tables 7 and 8 for dimensions.

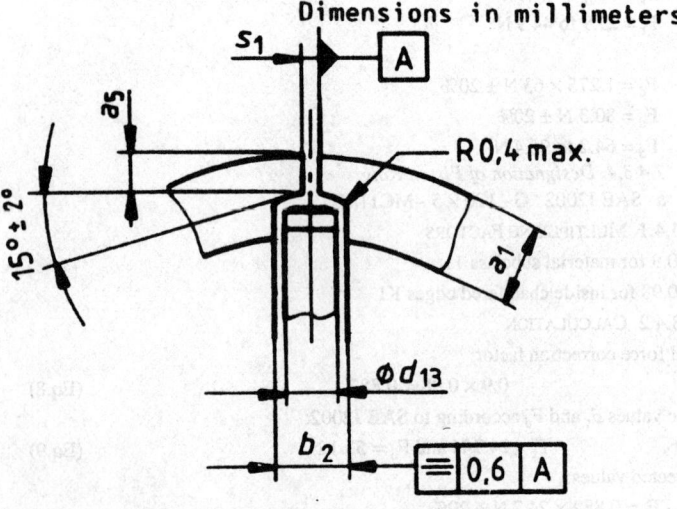

FIGURE 1—INTERNAL NOTCH

[2,3] Actual values of tangential force should be rounded up or down as follows:

$F_t < 50$ N: to the nearest 0.5 N

$F_t \geq 50$ N: to the nearest 1 N, where 0.5 N is to be rounded up

TABLE 7—DIMENSIONS OF INTERNAL NOTCH

Dimensions in millimeters

Code	Pin Diameter d_{13}	Notch b_2	Notch Tol.	Notch r_6	Notch Tol.
NH1	1.5	2		0.8	
NH2	2	2.5	+0.2 −0.1	0.9	±0.1
NH3	2.5	3		1	

$b_2 - d_{13} > s_1$ nom.

NOTE—r_6 applies only to notch design according to Figure 2.

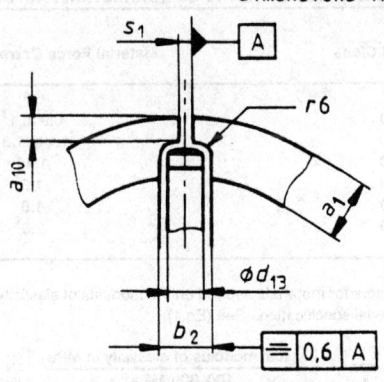

FIGURE 2—INTERNAL NOTCH—OPTION FOR PISTON RINGS WITH RADIAL WALL THICKNESS $a_1 > 2.1$

TABLE 8—WIDTH OF OVERLAP a_5 AND OPTIONAL a_{10} FOR INTERNAL NOTCH

Dimensions in millimeters

Radial Wall Thickness a_1	Overlap a_5	Overlap Tol.	Overlap a_{10}	Overlap Tol.
$1.5 \leq a_1 < 2.1$	0.6		-	
$2.1 \leq a_1 < 2.7$	0.7		0.6	
$2.7 \leq a_1 < 3.1$	1	±0.1	0.7	±0.1
$3.1 \leq a_1 < 3.5$	1.2		0.8	
$3.5 \leq a_1 < 3.9$	1.4		0.9	
$3.9 \leq a_1 \leq 4.1$	1.6		1	

8.2 Ring Joint With Side Notch (see Figure 3)—See Table 9 for dimensions.

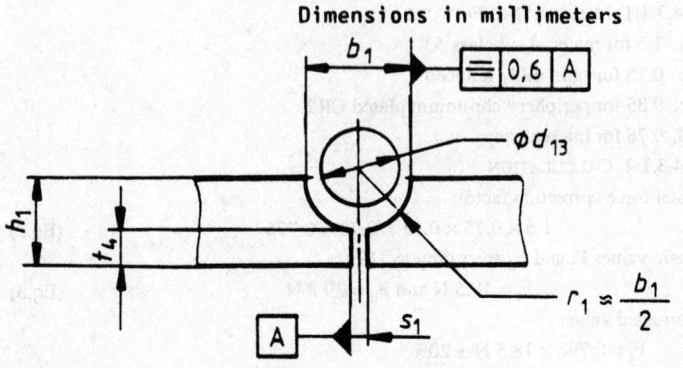

FIGURE 3—SIDE NOTCH

TABLE 9—DIMENSIONS OF SIDE NOTCH

Dimensions in millimeters

Code	Ring Width h_1	Pin Diameter d_{13}	Notch[1] b_1	Notch[1] Tol.	Notch[1] t_4	Notch[1] Tol.
NE1	1.2[2]				0.5	
	1.5				0.7	
	1.75	1.5	2		0.95	
	2				1.2	
	2.5				1.7	
NE2	2	2	2.5	-0.2 / -0.1	0.9	0 / -0.15
	2.5				1.4	
NE3	2.5	2.5	3		1	
	3				1.5	

[1] $b_1 - d_{13} > s_1$ nom.
[2] Not applicable for material class 10 in SAE J1590.

9. Machining and Coating Surfaces

9.1 Machining of Periphery—Where normal standards of machining are required, no code is needed for identification.

These normal standards are:
a. All uncoated rings—periphery fine turned
b. Rectangular or keystone rings with coated periphery, straight or barrel faced and chromium coated oil control rings and chromium plated segments of expander/segment oil control rings—machine (ground or lapped) over full face
c. Rectangular or keystone rings with coated periphery, taper faced—witness machining on part of width only

NOTE—Roughness values and measurement method may be agreed on between manufacturer and purchaser as there is no standard method available that is applicable in all cases.

9.2 Machining of Side Faces—The standard method of machining is by grinding of side faces; no code is required.

The standard side face finish is R_z 4 or R_a 0.8.

In the case of piston rings with treated surfaces (FE, PO, or PR), the roughness shall be measured before surface treatment.

9.3 Coatings on Periphery—Codes are required when chromium or molybdenum coatings are used as specified in the dimensional documents.

9.3.1 CHROMIUM COATING—Layer Thickness
a. CRF—Thickness 0.005 mm min
b. CR1—Thickness 0.05 mm min
c. CR2—Thickness 0.1 mm min
d. CR3—Thickness 0.15 mm min
e. CR4—Thickness 0.2 mm min

9.3.1.1 Chromium Coated Rings of Fully Faced Design—(See Figure 4.)
a. Code CRF to CR4

Piston rings with chromium coated periphery are normally designed fully faced.

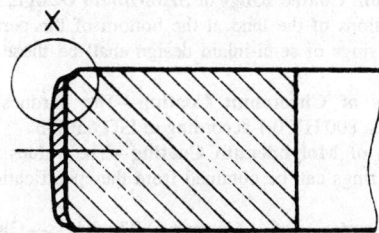

FIGURE 4—CHROMIUM COATED RING OF FULLY FACED DESIGN, CODE CRF TO CR4

9.3.1.2 Chromium Coated Rings of Semi-Inlaid Design—(See Figures 5 and 6.)

a. Code CR1E to CR4E
NOTE—This ring design needs a minimum chromium thickness CR1.

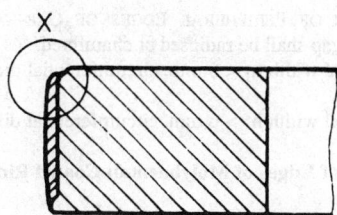

FIGURE 5—CHROMIUM COATED RING OF SEMI-INLAID DESIGN, CODE CR1E TO CR4E

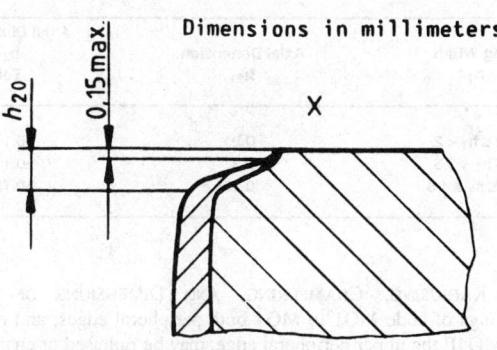

FIGURE 6—DESIGN OF PERIPHERAL EDGE (DETAIL OF FIGURES 4 AND 5)

9.3.1.3 Chromium Coated Rings of Inlaid Design
a. Code CR1F to CR4F
NOTE—This ring design needs a minimum chromium thickness CR1.

9.3.2 MOLYBDENUM COATING—Layer Thickness
a. MO1—Thickness 0.05 mm min
b. MO2—Thickness 0.1 mm min
c. MO3—Thickness 0.15 mm min
d. MO4—Thickness 0.2 mm min

9.3.2.1 Molybdenum Coated Rings of Fully Faced Design
a. Code MO1 to MO4

9.3.2.2 Molybdenum Coated Rings of Semi-Inlaid Design
a. Code MO1E to MO4E

9.3.2.3 Molybdenum Coated Rings of Inlaid Design
a. Code MO1F to MO4F

9.4 Designation of Edges of Chromium Coated Rings—See Table 10 for dimensions.

TABLE 10—AXIAL DIMENSIONS, h_{20}, OF PERIPHERAL EDGES OF CHROMIUM PLATED RINGS, FULLY FACED AND UPPER PERIPHERAL EDGE OF CHROMIUM PLATED RINGS, SEMI-INLAID[1]

Dimensions in millimeters

Ring Width h_1	Axial Dimension h_{20} max
$1.2 \leq h_1 \leq 3.5$	0.3
$3.5 < h_1 \leq 4.5$	0.4

[1] Values do not apply to chromium plated oil control rings.

9.4.1 Radiusing, Chamfering, and Dimensions of Peripheral Edges—Rings of code CRF to CR4 both peripheral edges, and rings of code CR1E to CR4E the upper peripheral edge, may be radiused or chamfered before plating.

9.4.2 Designation of Peripheral Edges of Gap—After plating, the peripheral edges at the gap shall be radiused or chamfered.
  a. For rings with axial width $h_1 < 6$ mm, circumferential dimension—0.4 mm max
  b. For rings with axial width $h_1 > 6$ mm, circumferential dimension—0.6 mm max

9.5 Designation of Edges of Molybdenum Coated Rings—See Table 11 for dimensions.

TABLE 11—AXIAL DIMENSIONS, h_{21}, OF PERIPHERAL EDGES OF MOLYBDENUM COATED RINGS, FULLY FACED AND UPPER PERIPHERAL EDGE OF MOLYBDENUM COATED RINGS, SEMI-INLAID

Dimensions in millimeters

Ring Width h_1	Axial Dimension h_{21}	Axial Dimension h_{21} Tol.
$1.2 \leq h_1 < 2$	0.2	±0.1
$2 \leq h_1 < 3.5$	0.3	±0.1
$3.5 \leq h_1 \leq 4.5$	0.35	±0.15

9.5.1 Radiusing, Chamfering, and Dimensions of Peripheral Edges—Rings of code MO1 to MO4 both peripheral edges, and rings of code MO1E to MO4E the upper peripheral edge, may be radiused or chamfered at the choice of the manufacturer. (See Figures 7 to 9.)

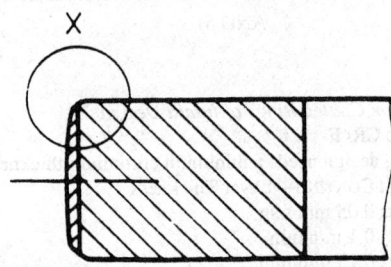

FIGURE 7—MOLYBDENUM COATED RING OF FULLY FACED DESIGN, CODE MO1 TO MO4

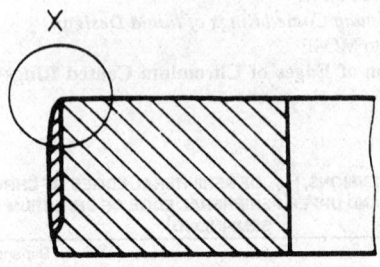

FIGURE 8—MOLYBDENUM COATED RING OF SEMI-INLAID DESIGN, CODE MO1E TO MO4E

9.5.2 Designation of Peripheral Edges of Gap—The peripheral edges at the gap may be radiused or chamfered.
  a. Circumferential dimension—0.4 mm max

9.6 Molybdenum Coated Rings of Inlaid Design, Code MO1F to MO4F—(See Figure 10.)

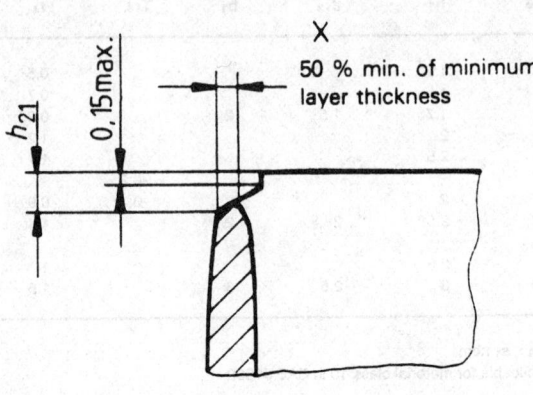

FIGURE 9—DESIGN OF PERIPHERAL EDGE (DETAIL OF FIGURES 7 AND 8)

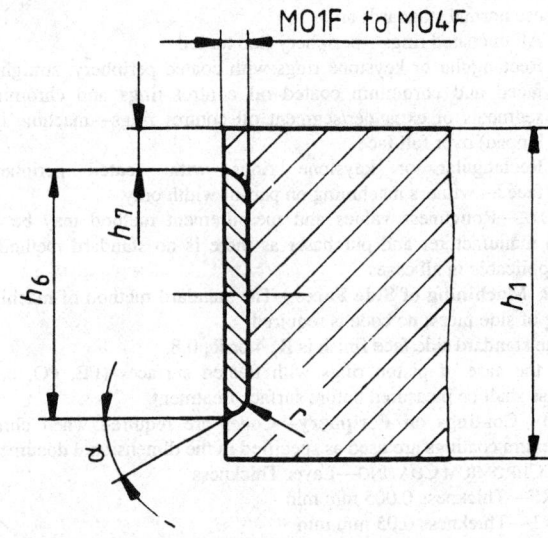

NOTE — Dimension r and angle α at the manufacturer's choice.

FIGURE 10—MO1F TO MO4F RINGS

9.7 Molybdenum Coated Rings of Semi-Inlaid Design, Code MO1E to MO4E—The dimensions of the land at the bottom of the peripheral edge for molybdenum coated rings of semi-inlaid design shall be those given for h_1 in Table 12.

9.8 Hardness of Chromium Coating—The hardness of chromium plating shall be at least 800 HV 0.1 according to ISO 6507-3.

9.9 Hardness of Molybdenum Coating—The values for hardness of molybdenum coated rings can be obtained from the specification of the piston ring manufacturer.

9.10 Tolerance of Coating Thickness—Chromium and Molybdenum—Usually a minimum specification does not call for tolerances. If a tolerance is required on the coating thickness, it shall be agreed on between manufacturer and client.

The tolerances given in Table 13 are suggested as a guideline.

TABLE 12—DIMENSIONS OF GROOVE AND LAND OF PERIPHERAL EDGES FOR MOLYBDENUM COATED RINGS

Dimensions in millimeters

Ring Width h_1	Groove Dimension h_6 min	Land Dimension h_7 MC10, 20, 30	Land Dimension h_7 40, 50, 60
1.2	0.6		
1.5	0.8	0.15 to 0.45	0.1 to 0.4
1.75	1.05		
2	1.3		
2.5	1.7		
3	2.2	0.15 to 0.5	0.1 to 0.45
3.5	2.5		
4	3	0.15 to 0.55	0.1 to 0.5
4.5	3.5		

TABLE 13—TOLERANCE OF LAYER THICKNESS

Dimensions in millimeters

Layer Thickness	Tolerance $d_1 < 150$	Tolerance $150 \leq d_1 \leq 200$
CR1 to CR4	+0.15 / 0	-0.2 / 0
MO1 to MO4	-0.2 / 0	-0.25 / 0

10. Treated Surfaces

10.1 Ferroix Coated, Code FE, Coating Thickness 0.003 mm min—Only uncoated rings (i.e., without chromium or molybdenum coated periphery) can be ferroix coated.

10.2 Phosphate Coated, Code PO, Coating Thickness 0.002 mm min—This phosphating applies only to piston rings made of cast iron materials.

10.3 Phosphate Coated, Code PR, Coating Thickness 0.002 mm max—This phosphating is for rust protection purposes only, and it applies only to piston rings made of cast iron materials.

11. Packing—To accommodate automatic mounting, the packing unit shall contain only one type of piston ring per packing unit.

If orientation is required, all piston rings shall be oriented in the same direction. In this case a mark for the orientation shall be shown on the package.

APPENDIX A

A.1 This document has been established to harmonize the ISO and SAE piston ring documents. The U.S. Technical Advisory Group, with the support of the National Engine Parts Manufacturers Association, has worked with other national organizations on this worldwide document. Some of the wording and phrasing may differ slightly from U.S. terminology for translation purposes.

In preparing this SAE Standard, the Scope and Field of Application, and Reference sections of the ISO 6621/4 document have been editorially revised and reorganized.

The tolerances specified in this document represent a six sigma quality level.

SLEEVE TYPE HALF BEARINGS—SAE J506b

SAE Standard

Report of Engine Committee approved June 1951 and completely revised November 1978.

1. Scope—This standard defines the normal dimensions, dimensioning practice, tolerances, specialized measurement techniques, and glossary of terms for bearing inserts commonly used in reciprocating machinery.

The standard sizes cover a range which permits a designer to employ, in proper proportion, the durability and lubrication requirements of each application, while utilizing the forming and machining practices common in manufacture of sleeve type half bearings.

Not included are considerations of hydrodynamic lubrication analysis or mechanical stress factors of associated machine structural parts which determine the nominal sizes to be used, selection of bearing material as related to load carrying capacity, and economics of manufacture. For information concerning materials, see SAE J459 (July, 1974) and SAE J460e (October, 1974).

These suggested sizes provide guidelines which may result in minimal costs of tooling but do not necessarily represent items which can be ordered from stock.

2. Applications and General Considerations—Sleeve type half bearings, sometimes called thinwall bearings, are most commonly found in the connecting rod and main bearing positions of gasoline and diesel engines used in the automotive, construction equipment, and farm equipment industries. Normally, they are lubricated with oil which is supplied under pressure.

Much theoretical and experimental work has been done with respect to the operation of sleeve type bearings, and both bearing manufacturers and large volume users are familiar with the theories of hydrodynamic lubrication so they can perform the associated calculations with relative ease. The theory involves the development of pressures within an oil film as a result of a journal rotating inside the bearing. Shearing of the oil film generates heat, however, oil is supplied in sufficient quantity to control temperatures and viscosities. For a given set of operating conditions, calculations can be made for oil film thickness and expected operating temperatures. Sometimes, such calculations highlight potential problems and point the way for modification to the original design. Such changes can involve oil viscosity, oil supply pressure, geometry of the bearing, clearances, grooving, and other factors.

Mechanically, a thinwall bearing should be flexible and elastic in comparison with the housing. When the housing is made round and true in relation to mating parts, the designer can be confident that the bearing, with its relative flexibility, will conform to the housing, thereby ensuring desired shape and alignment. Also, this flexible characteristic permits the bearings to be economically made from material which generally is prepared as a continuous strip from which the bearings are blanked, formed, and precisely machined, providing a finished bearing of high quality and reliability.

3. Glossary of Terms (See Fig. 1.)

3.1 **Annular Oil Groove**—A groove, uniform in cross section, through the entire 180 deg arc of the half shell, installed for the purpose of promoting oil flow from the center to the edges of the bearing and also for the creation of a constant and uniform oil supply from main journals to connecting rod journals by means of drilled passages in the crankshaft. Sometimes an annular oil groove is used near the end of a straight shell bearing for the purpose of draining oil away from the seal.

3.2 **Back**—The steel OD of half shell bearings.

3.3 **Bimetal**—A type of bearing construction in which a single layer of bearing material is bonded to a steel backing. A common example of this is a babbitt bearing. (See Trimetal.)

3.4 **Bore or Housing ID**—The diametral size of the housing into which the bearing is assembled. The housing ID which supports and retains the bearing in assembly. (See Gage.)

3.5 **Bore Distortion**—The elastic deformation of the housing which occurs because of the stress imposed by interference fit between housing ID and assembled bearing OD. Magnitude of bore distortion is normally a small value, but because housings generally have non-uniform cross sections, adjustments to the concentricity of the bearing walls are made to accommodate various degrees of non-uniform outward displacement of the housing. These adjustments to the bearing wall are normally achieved by controlling bearing eccentricity. Also, frequently considered are distortions produced by external or inertial loads. (See Eccentricity.)

3.6 **Centerline Wall**—The bearing wall thickness at a location 90 deg from the parting lines. Sometimes called vertical centerline.

3.7 **Crush**—The amount by which circumferential length of a half shell exceeds one half the circumference of the housing ID. This excess length ensures the interference fit which holds the mating half shells in place. (See Parting Line Height.)

3.8 **Eccentricity**—The gradual reduction in bearing wall thickness, normally from centerline wall to the parting line relief, which tends to create additional diametral clearance between the bearing and journal near the parting lines. The magnitude of eccentricity may vary as dictated by studies of bore distortion characteristics. Except in rare instances, eccentricity is positive, meaning that the wall thickness near the parting line is less than the wall thickness at the vertical centerline.

3.9 **Ends**—The surfaces or faces which determine the two planes that define the bearing length.

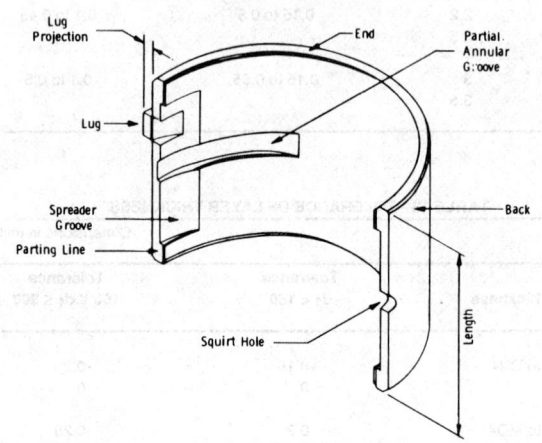

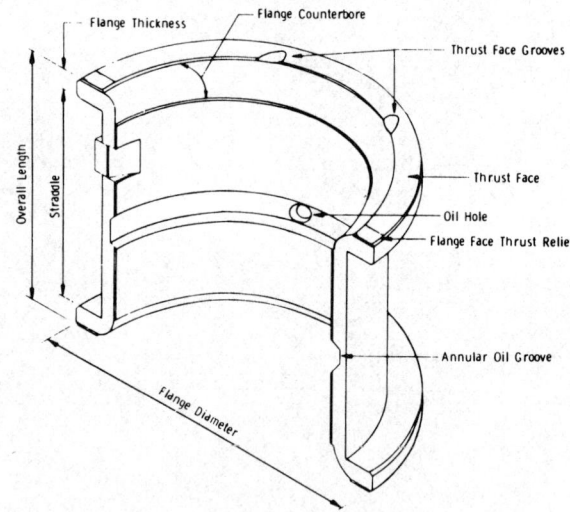

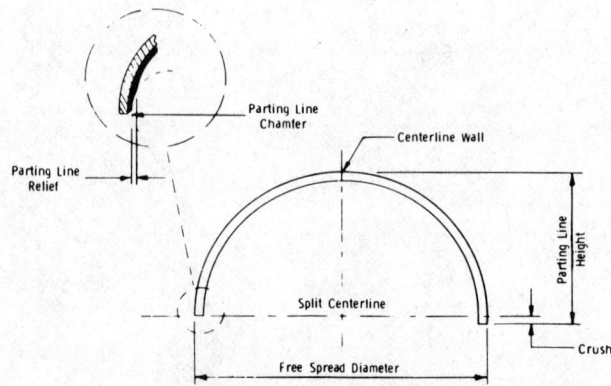

FIG. 1—ILLUSTRATION OF COMMON BEARING CHARACTERISTICS AND TERMINOLOGY

3.10 Flange Counterbore—Machined radius to aid in lubricant flow and clearance with crankshaft fillet.

3.11 Flange Diameter—The OD measurement of flanges in the assembled state. The maximum flange OD should not exceed 1.3 times the maximum housing ID if forming difficulties are to be avoided.

3.12 Flange Thickness—The thickness of the flange on a flange bearing.

3.13 Free Spread Diameter—The diametral dimension of the half shell bearing in its free state. Normally, this dimension will exceed the maximum housing ID by at least 0.5 mm (0.020 in) for straight shells, and by about one-tenth this amount for flange bearings. This deliberate increased diameter aids assembly by ensuring that each half shell will have sufficient friction within its intended housing to remain in place during engine assembly operations. Its exact values are not critical.

3.14 Gage Diameter—The numerical bore size which is equal to the high limit dimension of the housing ID.

3.15 Length—The overall axial dimension of the half bearing.

3.16 Lining—The bearing material which is bonded to the steel back.

3.17 Lug—The projection from the OD of the bearing half shell provided on straight shell bearings to ensure proper axial location of the half shell in the housing. It is sometimes referred to as a tang, notch, or nick. Lugs are not intended to secure the bearing against rotation within the housing. Crush does that. Lugs should be on one parting line only. Commonly, lugs on both half bearings are assembled on the same side of the housing.

3.18 Lug Projection—The dimension from the bearing back to the outside surface of the lug at the parting line.

3.19 Oil Hole—A hole through the bearing shell which is used to index with drilled oil passages in the bearing's housing. (See Squirt Hole.)

3.20 Overall Length—The dimension between thrust faces on a common flange bearing and equal to the maximum axial dimension of the bearing.

3.21 Overlay—A thin surface layer of soft bearing material on a harder lining material which, in turn, becomes an intermediate layer of high load capacity. Normally, overlays are deposited by electroplating, and have a nominal thickness of 0.025 mm (0.001 in) or less. The result is then a trimetal bearing. (See Trimetal.)

3.22 Partial Groove—This is a groove similar in nature and cross section to the full annular groove, generally for the purpose of extending the full annular groove into a mating half bearing but preventing its extension into the most heavily loaded portion of a main bearing.

3.23 Parting Line—The face or surface of the half shell which butts against a mating surface of another half shell to form a full round bearing.

3.24 Parting Line Chamfer—A small chamfer added to the inside surface of the parting line along the entire length of the bearing to eliminate sharp disturbances to the oil flow which could otherwise result from minor conditions of cap shift or misalignment.

3.25 Parting Line Height—A measurement of half shell circumference normally made with the parting lines of the bearing loaded in compression and with the back of the bearing seated and conforming to the ID of a precisely made inspection block which is normally equal in size to the gage diameter. The measured difference between parting line height and the inspection block radius equals the measured indicator crush. The load under which the bearing is measured is specified to be large enough to ensure adequate seating and reproducible measurements without causing permanent deformation of the bearing.

3.26 Parting Line Relief—The removal of bearing material near the parting line to aid smoothness of oil flow where the parting lines of the half shells butt against each other.

3.27 Spreader Groove—A cross groove, generally in a normally unloaded area, used to promote oil flow without increasing clearances.

3.28 Squirt Hole—A small cutout at parting line surface, sometimes used in rod bearings to provide a squirt of oil onto cylinder walls during operation.

3.29 Straddle—The dimension between the inside surfaces of the flanges on a flange bearing.

3.30 Thrust Face—That exterior portion of a flange bearing which runs against a mating thrust face of a shaft to control axial shaft movement and load.

3.31 Thrust Face Groove—A small groove incorporated into the thrust face for the purpose of promoting oil flow.

3.32 Trimetal—A type of bearing construction in which a heavy duty bearing material (lining) is bonded to a steel back and then a thin layer of softer bearing material is applied to the ID of the high strength bearing material. Normally, this surface layer is obtained by electroplating and is referred to as the overlay, or overlay plate. It is thin enough that the high strength of the intermediate layer determines the ultimate bearing strength from a fatigue standpoint. This type of bearing is normally used in heavy duty applications. (See Bimetal.)

3.33 Wall—The total thickness of the bearing half shell which is the sum of the steel back thickness, lining thickness and, when applicable, overlay thickness. When given without other qualifications, it is normally assumed to be centerline wall.

4. Wall Size Recommendations—Wall sizes for various shaft sizes are presented graphically in Fig. 2. The recommended wall sizes are based on the most common bearing strip preparations.

Corresponding housing sizes can be easily determined by adding to nominal shaft size, two times the bearing wall plus the diametral clearance.

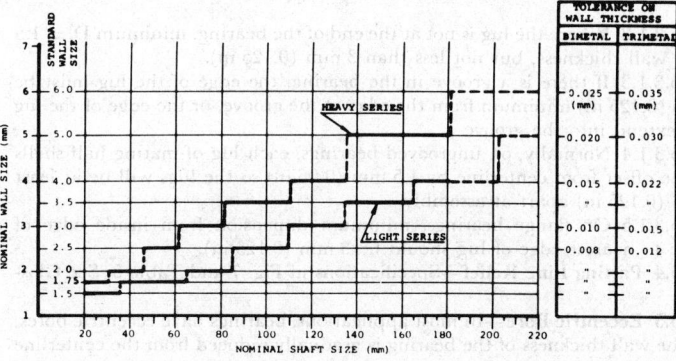

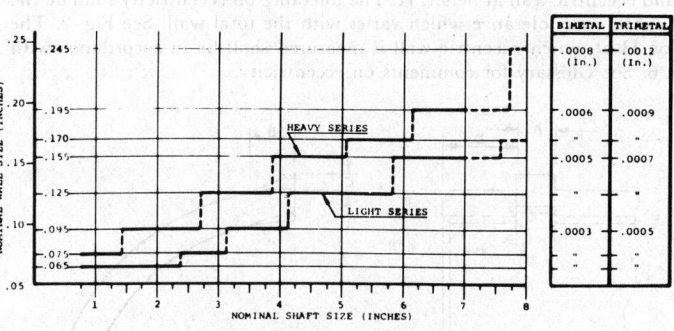

FIG. 2—BEARING WALL THICKNESS RECOMMENDATIONS

FIG. 3—STANDARD TOLERANCES OF STRAIGHT SHELL BEARINGS

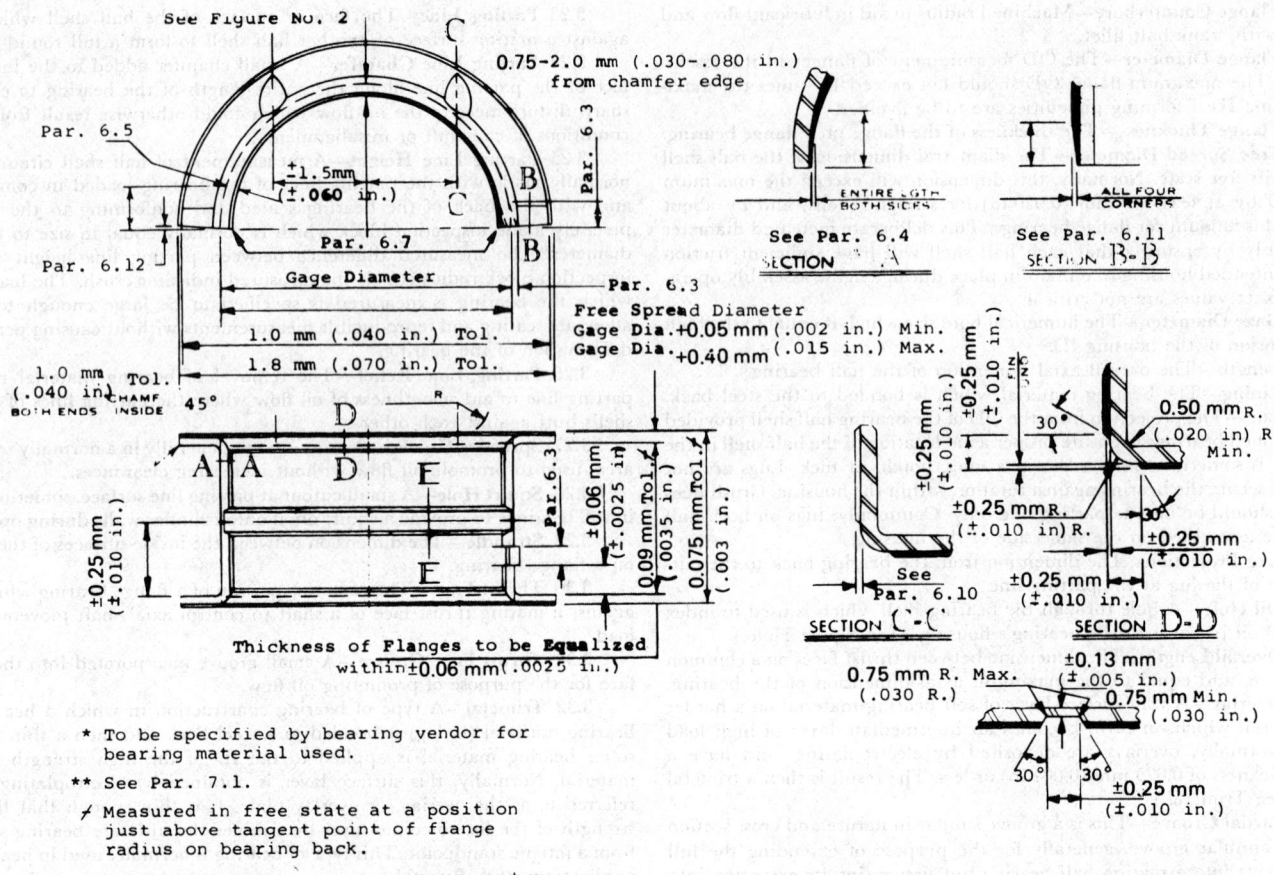

FIG. 4—STANDARD TOLERANCES OF FLANGED BEARINGS

Regarding the selection of light or heavy series wall sizes, the following can be used as a guide:

SAE Light Series— Connecting rod bearings for passenger car and similar light duty engines.

SAE Heavy Series— Main bearings for passenger car and similar light duty engines.
—Connecting rod and main bearings for heavy duty engines.

SAE Extra-Heavy Series—Main bearings for heavy duty engines sometimes use the next heavier wall size for a given shaft size than SAE Heavy Series.

5. Standard Features and Tolerances—The following standards generally are referred to from Figs. 2, 3, and 4.

5.1 Wall Thickness—The tolerance on wall thickness will depend upon whether the bearing is bimetal or trimetal. Recommended values are shown in Fig. 2.

5.2 Length—Bearing lengths will be determined by the application. Tolerances for various sizes are shown in Table 1.

5.3 Locating Lugs and Lug Slots—Dimensions of the locating lug and the notch in the housing should be as shown in Figs. 5 and 6, and Tables 2, 3, and 4.

5.3.1 DIMENSIONING

5.3.1.1 The lug may be produced at the end of the bearing, in which case D = 0.

5.3.1.2 Where the lug is not at the end of the bearing, minimum D = 1.5 times wall thickness, but not less than 3 mm (0.125 in).

5.3.1.3 If there is a groove in the bearing, the edge of the lug must be 3 mm (0.125 in) minimum from the edge of the groove, or the edge of the lug may extend into the groove.

5.3.1.4 Normally, on ungrooved bearings, each lug of mating half shells will be offset from centerline by 1.5 mm (0.06 in) so the lugs will be at least 3 mm (0.125 in) apart at assembly.

5.3.1.5 On flange bearings, minimum dimension from inside edge of flange to nearest edge of lug should be 3 mm (0.125 in).

5.4 Parting Line Relief—Specifications in Fig. 7 and Table 5. See Glossary.

5.5 Eccentric Bore—In most applications, bearings have eccentric bores, i.e., the wall thickness of the bearing is gradually reduced from the centerline wall to the parting lines. Eccentricity equals the difference between centerline wall and eccentric wall at height H. The tolerance on eccentricity shall be the same as the wall tolerance which varies with the total wall. See Fig. 2. The position H at which eccentric wall is measured shall be in accordance with Table 6. See Glossary for comments on eccentricity.

TABLE 1—BEARING LENGTH TOLERANCES

Values in Millimeters		Values in Inches	
Shaft Dia.	Limits	Shaft Dia.	Limits
20–120	+0 / −0.25	0.75–5	+0 / −0.010
125–260	+0 / −0.50	5–10	+0 / −0.020

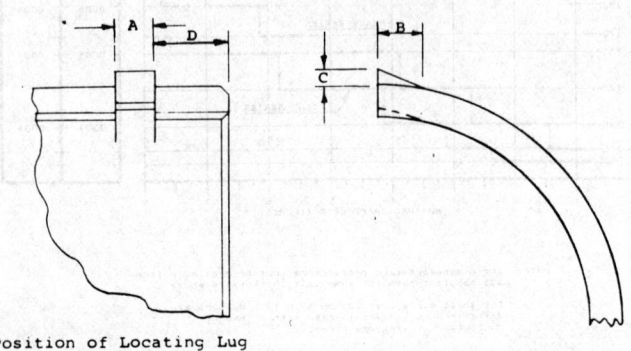

Position of Locating Lug

FIG. 5—BEARING LUG DIMENSIONING

TABLE 2—RECOMMENDED LUG DIMENSIONS

Values in Millimeters				Values in Inches			
Shaft Dia.	A	B	C	Shaft Dia.	A	B	C
20–40	4.45/4.57	3.0/4.0	0.8/1.1	0.75–1.50	0.175/0.180	0.125/0.155	0.031/0.041
40–65	4.45/4.57	5.0/6.0	1.0/1.3	1.50–2.50	0.175/0.180	0.190/0.220	0.031/0.041
65–85	6.00/6.20	5.0/6.0	1.2/1.5	2.50–3.50	0.238/0.243	0.190/0.220	0.045/0.055
85–200	9.20/9.35	8.0/9.0	1.5/1.65	3.50–7.50	0.363/0.368	0.310/0.340	0.055/0.065
200–260	12.00/12.50	8.0/9.0	1.5/1.65	7.50–10.00	0.488/0.493	0.310/0.340	0.055/0.065

TABLE 3—TOLERANCE ON LUG LOCATION

Values in Millimeters		Values in Inches	
Shaft Dia.	Limits on D	Shaft Dia.	Limits on D
20–120	+0.15 / −0	0.75–5.0	+0.005 / −0
125–260	+0.20 / −0	5.0–10.0	+0.008 / −0

TABLE 4—TOLERANCES AND SIZES OF HOUSING LUG SLOTS

Values in Millimeters				Values in Inches			
Shaft Dia.	E	F	G	Shaft Dia.	E	F	G
20–40	4.70–4.82	5.60	1.57	0.75–1.50	0.185–0.190	0.219	0.062
40–65	4.70–4.82	7.15	1.78	1.50–2.50	0.185–0.190	0.281	0.070
65–85	6.30–6.42	7.15	1.98	2.50–3.50	0.248–0.253	0.281	0.078
85–200	9.48–9.60	10.30	2.36	3.50–7.50	0.373–0.378	0.406	0.093
200–260	12.65–12.78	10.30	2.36	7.50–10.00	0.498–0.503	0.406	0.093

TABLE 5—SIZES OF PARTING LINE RELIEF

Values in Millimeters			Values in Inches		
Shaft Dia.	H (Nom.)	J	Shaft Dia.	H (Nom.)	J
20–25	4.0	0.012–0.038	0.75–1.00	0.16	0.0005–0.0015
25–41	4.0	0.025–0.050	1.00–1.63	0.16	0.001–0.002
41–50	5.0	0.025–0.050	1.63–2.00	0.19	0.001–0.002
50–65	5.5	0.025–0.050	2.00–2.50	0.22	0.001–0.002
65–75	6.0	0.025–0.050	2.50–2.90	0.25	0.001–0.002
75–86	7.8	0.025–0.050	2.90–3.40	0.31	0.001–0.002
86–105	9.7	0.025–0.050	3.40–4.15	0.38	0.001–0.002
105–125	11.0	0.025–0.050	4.15–5.00	0.44	0.001–0.002
125–145	23.0	0.050–0.080	5.00–5.75	0.90	0.002–0.003
145–170	26.0	0.050–0.080	5.75–6.75	1.02	0.002–0.003
170–200	30.5	0.050–0.080	6.75–7.75	1.20	0.002–0.003
200–260	35.0	0.050–0.080	7.75–10.00	1.38	0.002–0.003

TABLE 6—LOCATION OF ECCENTRIC WALL MEASUREMENT

Values in Millimeters		Values in Inches	
Shaft Dia.	Height H (Nom.)	Shaft Dia.	Height H (Nom.)
20–25	4.8	0.75–1.00	0.19
25–41	6.4	1.00–1.63	0.25
41–86	9.7	1.63–3.40	0.38
86–145	16.0	3.40–5.75	0.62
145–260	25.4	5.75–10.00	1.00

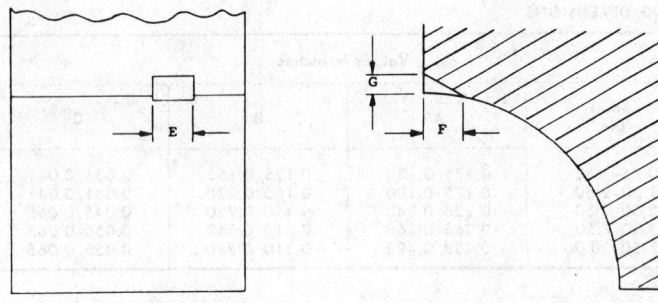

FIG. 6—HOUSING LUG SLOT

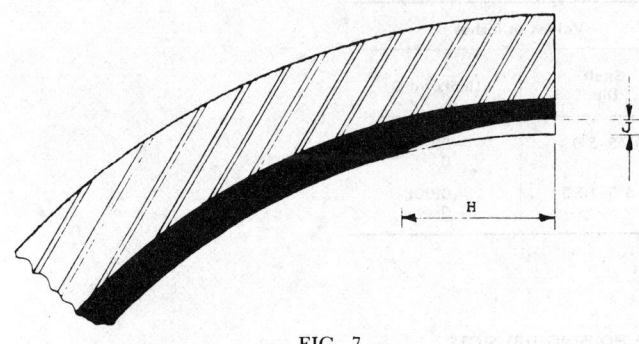

FIG. 7

5.6 Groove Forms—Groove sizes are determined by functional requirements. The following points should be considered when specifying grooves:

5.6.1 Preferred groove form is shown in Fig. 9. Table 8 shows suggested sizes.

5.6.2 For central annular grooves, the position should be specified as indicated in Fig. 9.

5.6.3 The minimum wall thickness at the back of the groove should be the larger of the following two values:

0.7 mm	0.028 in
or	or
0.35 × bearing wall thickness (mm)	0.35 × bearing wall thickness (in)

The following limits, shown in Table 7, will apply to the size so determined. See Fig. 9.

TABLE 7—TOLERANCE ON WALL MEASUREMENT AT BASE OF GROOVE

Values in Millimeters		Values in Inches	
Shaft Dia.	Limits on K	Shaft Dia.	Limits on K
20–120	+0.20 / –0	0.75–5.0	+0.008 / –0
125–260	+0.35 / –0	5.00–10.0	+0.014 / –0

5.7 Chamfers—All sharp edges should be deburred. If machined chamfers are required, these are generally at 45 deg with tolerance of ±0.3 mm (±0.010 in).

Parting line chamfers are generally 0.1/0.4 mm (0.005/0.015 in).

5.8 Oil Holes—Oil holes may be drilled or pierced before or after forming and are usually centrally located in the oil groove where applicable. All sharp edges should be removed.

Tolerance on the location of oil holes should be ±0.25 mm (±0.010 in) from the end of the bearing. Dimension the oil hole from the same end of the bearing that is used to dimension the lug.

5.9 Parting Line Height—Many factors must be considered in establishing recommended values for parting line height. For most applications, using cast iron or steel housings, Table 9 is satisfactory as a guideline.

After assembly of the bearings into the housing, then the actual total crush is the sum of: (1) indicator crush, (2) crush due to the load applied in the parting line height gage, (3) crush due to the housing bore being smaller than its maximum blueprint value.

See Section 7 for measurement procedures.

5.10 Thrust Face Grooves—Suggested specifications are shown in Fig. 10 and Table 10.

5.11 Flange Face Thrust Relief

5.12 Flange Parting Line Relief—The amount of relief at all four corners, measured at flange OD from the plane of the parting line, is nominally 0.13–0.64 mm (0.005–0.025 in).

5.13 Flange Counterbore—See Figs. 12 and 1.

5.14 Flange—Overall Length and Straddle—Recommended tolerances are:

Overall Length: 0.080 mm (0.003 in) tolerance
Straddle: 0.090 mm (0.0035 in) tolerance

6. Measurement Procedures—Parting Line Height—See Glossary.

Checking loads for this measurement will normally be agreed upon between bearing manufacturer and user. Often, bearings over 150 mm (6.0 in) gage are inspected in a fixture that has no fixed stop as illustrated, and load is applied at both sides.

The indicator gage in the above diagram is set with the use of a master. Table 9 shows typical indicator crush and its tolerances.

7. Surface Finishes

7.1 Half Shell ID—Since the bearing surface is relatively soft in comparison with journal surfaces, bearing finish is less critical than journal finish. ID surface finish of half shell bearings should not exceed 0.9 μm (35 μin).

7.2 Journals—Surface finish of journals can be extremely critical, and often affects durability of a given application. Generally, the table below shows values which will be acceptable.

Type Journal	μm R_a	μin R_a
Plain Steel Journal Ground and Polished	0.50	20
Cast Iron Journal Ground and Polished	0.40	15
Hardened Steel Journal	0.25	10

Often, the character of the finish is more critical than the surface finish reading, and very fine finishes can have a character of sharp peaks or edges which may cause premature failure.

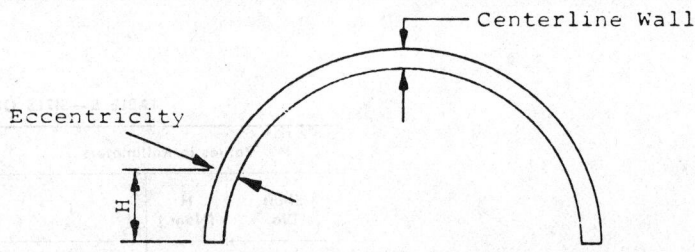

FIG. 8—ECCENTRICITY

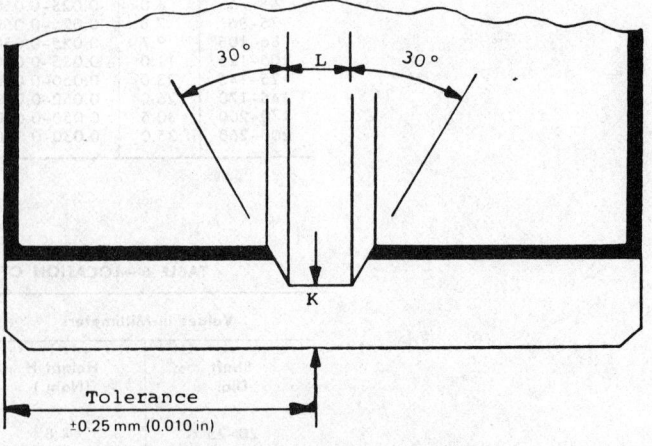

L = Groove Width
K = Groove Wall Thickness

FIG. 9—GROOVE DIMENSIONS

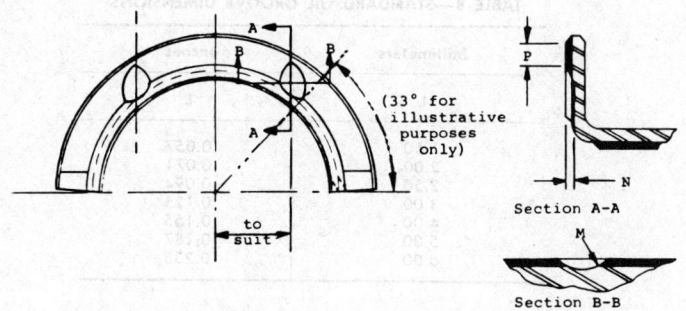

FIG. 10—TYPICAL THRUST FACE GROOVING

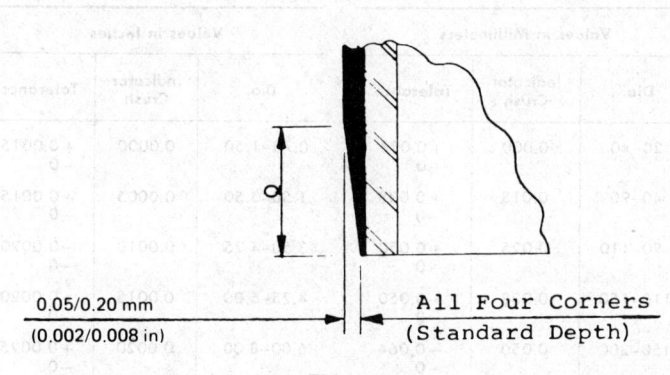

FIG. 11

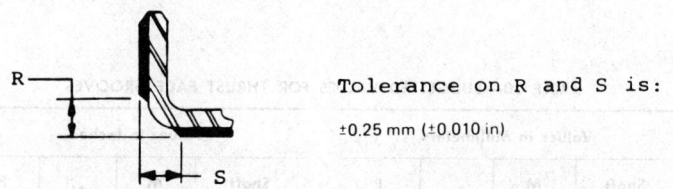

FIG. 12—FLANGE COUNTERBORE TOLERANCES

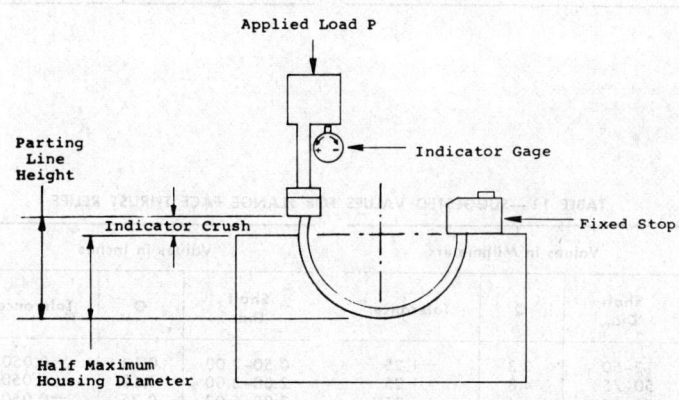

FIG. 13—SCHEMATIC OF A TYPICAL CHECKING GAGE

TABLE 8—STANDARD OIL GROOVE DIMENSIONS

Millimeters	Inches
L	L
1.50	0.057
2.00	0.071
2.50	0.094
3.00	0.125
4.00	0.155
5.00	0.187
6.00	0.235

TABLE 9—TYPICAL INDICATOR CRUSH

Values in Millimeters			Values in Inches		
Dia.	Indicator Crush	Tolerance	Dia.	Indicator Crush	Tolerance
20–40	0.000	+0.038 / −0	0.75–1.50	0.0000	+0.0015 / −0
40–90	0.013	+0.038 / −0	1.50–3.50	0.0005	+0.0015 / −0
90–110	0.025	+0.050 / −0	3.50–4.25	0.0010	+0.0020 / −0
110–150	0.038	+0.050 / −0	4.25–6.00	0.0015	+0.0020 / −0
150–200	0.050	+0.064 / −0	6.00–8.00	0.0020	+0.0025 / −0
200–260	0.076	+0.076 / −0	8.00–10.00	0.0030	+0.0030 / −0

TABLE 10—SUGGESTED VALUES FOR THRUST FACE GROOVES

Values in Millimeters				Values in Inches			
Shaft Dia.	M Nom.	N	P Nom.	Shaft Dia.	M Nom.	N	P Nom.
20–50	4.8	0.25 / 0.38	0.76	0.75–2.00	0.188	0.010 / 0.015	0.03
50–90	7.1	0.50 / 0.64	1.52	2.00–3.50	0.281	0.020 / 0.025	0.06
90–130	9.5	0.64 / 0.89	2.29	3.50–5.00	0.375	0.025 / 0.035	0.09

TABLE 11—SUGGESTED VALUES FOR FLANGE FACE THRUST RELIEF

Values in Millimeters			Values in Inches		
Shaft Dia.	Q	Tolerance	Shaft Dia.	Q	Tolerance
12–50	3.3	±1.25	0.50–2.00	0.13	±0.050
50–75	4.8	±1.25	2.00–3.00	0.19	±0.050
75–125	6.4	±1.25	3.00–5.00	0.25	±0.050

SPLIT TYPE BUSHINGS—
DESIGN AND APPLICATION—SAE J835 DEC81

SAE Standard

Report of the Engine Committee, approved June 1962, completely revised December 1981.

1. **Scope**—This SAE Standard presents the standard sizes, important dimensions, specialized measurement techniques and tolerances for split type bushings. Both SI and inch sizes are shown; their dimensions are not exact equivalents. New designs shall use SI units.

Unless specifically stated as ±, all tolerances are total.

2. **Split Bushings (Finished in Place)**
   2.1 Outside Diameter Tolerances[a]

A, O.D. Range	As Formed O.D.[a]	Ground or Sized O.D.[a]	As Formed O.D. With Graphite I.D.[a]
mm	mm	mm	mm
Up to 50	0.040	0.025	0.050
50– 80	0.045	0.035	0.055
80–120	0.050	0.040	0.065
120–152	0.085	0.070	0.100
in	in	in	in
Up to 1.95	0.0015	0.0010	0.0020
1.95–3.15	0.0015	0.0015	0.0020
3.15–4.70	0.0020	0.0015	0.0025
4.70–6.00	0.0035	0.0030	0.0040

[a] Refer to O.D. checking procedures in paragraph 2.14 of this standard.

2.2 Wall Thickness

Shaft Dia.	B, Recommended Nominal Wall
mm	mm
Up to 25	0.75[a]
15– 75	1.50
40–115	2.50
50–145	3.00
80 and Over	3.50
in	in
Up to 1.00	0.031[a]
0.50–3.00	0.062
1.50–4.50	0.094
2.00–5.70	0.125
3.15 and Over	0.156

[a] All 0.75 mm (0.031 in) wall bushings are furnished by manufacturer in solid bronze only.

2.3 **Wall Thickness Tolerance**—Furnished by manufacturer, as formed, for:

B, Wall	As Formed Wall Tolerance for Finished-in-Place Bushings
mm	mm
0.75	0.08
1.50	0.08
2.50	0.10
3.00	0.14
3.50	0.18
in	in
0.031	0.003
0.062	0.003
0.094	0.004
0.125	0.005
0.156	0.007

NOTE: Choosing sizes in this table can expedite samples and small production runs, as partial tooling may be available.

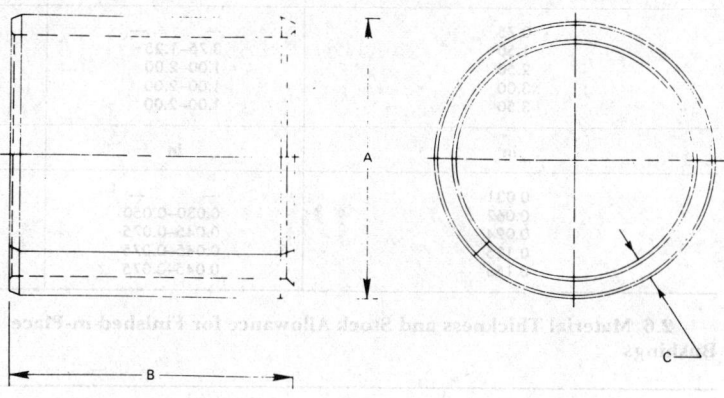

FIG. 1

2.4 **Bushing Length Tolerance** (See Fig. 1, Dimension C)
±0.25 mm (±0.010 in) for machined chamfer
±0.50 mm (±0.020 in) for die formed chamfer (plain)
±0.65 mm (±0.025 in) for die formed chamfer (with holes)

2.5 **Chamfers**
2.5.1 HOUSING BORE CHAMFERS
0.8 mm (0.03 in) × 40/60 deg included angle for up through 25 mm (1.00 in) dia. hole
1.5 mm (0.06 in) × 40/60 deg included angle for over 25 mm (1.00 in) dia. hole
2.5.2 BUSHING (ENDFACE CHAMFERS)
Machined chamfers not recommended for 0.75 mm (0.031 in) wall.

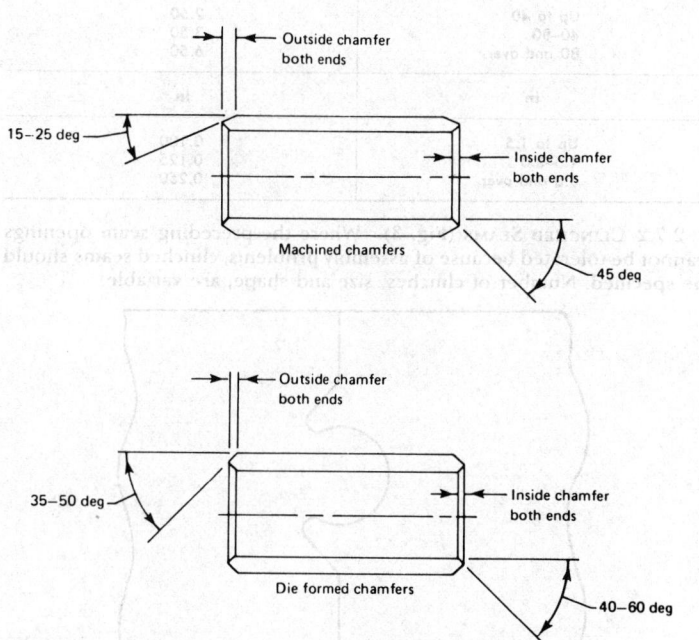

FIG. 2

Nominal Bushing Wall, B	Machined Chamfers		Die Formed Chamfers	
	Outside	Inside	Outside	Inside
mm	mm	mm	mm	mm
0.75	—	—	0.15–0.40	0.15–0.65
1.50	0.75–1.25	0.25–0.75	0.40–1.00	0.25–0.75
2.50	1.00–2.00	0.40–1.20	0.75–1.50	0.25–1.00
3.00	1.00–2.00	0.50–1.30	1.00–2.00	0.50–1.30
3.50	1.00–2.00	0.50–1.30	—	—
in	in	in	in	in
0.031	—	—	0.005–0.015	0.005–0.025
0.062	0.030–0.050	0.010–0.030	0.015–0.035	0.010–0.030
0.094	0.045–0.075	0.015–0.045	0.030–0.060	0.010–0.040
0.125	0.045–0.075	0.020–0.050	0.045–0.075	0.020–0.050
0.156	0.045–0.075	0.020–0.050	—	—

2.6 Material Thickness and Stock Allowance for Finished-in-Place Bushings

Nominal Wall	Actual Wall As Supplied, B	Min. Lining Thickness as Supplied	Min. Stock Allowance Per Side, As Supplied		
			Bored	Reamed	Burnished
mm	mm	mm	mm	mm	mm
0.75	0.86–0.94	—	0.13	0.08	0.025
1.50	1.66–1.74	0.40	0.13	0.08	0.025
2.50	2.65–2.75	0.45	0.15	0.10	0.025
3.00	3.23–3.37	0.50	0.25	0.13	0.025
3.50	3.81–3.99	0.55	0.30	0.15	0.025
in	in	in	in	in	in
0.031	0.036–0.039	—	0.005	0.003	0.001
0.062	0.068–0.071	0.015	0.005	0.003	0.001
0.094	0.100–0.104	0.017	0.006	0.004	0.001
0.125	0.135–0.140	0.020	0.010	0.005	0.001
0.156	0.168–0.175	0.022	0.012	0.006	0.001

2.7 Seams
2.7.1 Straight Seams

Bushing O.D., A	Maximum Seam Opening (In Free State)
mm	mm
Up to 40	2.50
40–80	3.50
80 and over	6.50
in	in
Up to 1.5	0.100
1.5–3.0	0.125
3.0 and over	0.250

2.7.2 Clinched Seams (Fig. 3)—Where the preceding seam openings cannot be tolerated because of assembly problems, clinched seams should be specified. Number of clinches, size and shape, are variable.

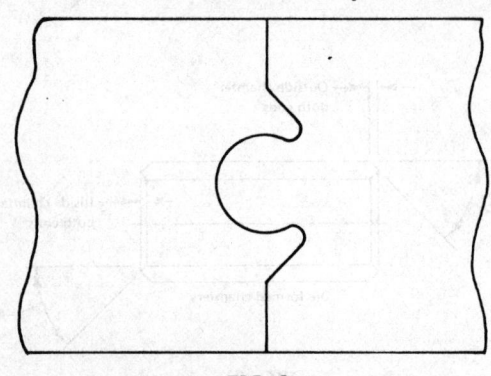

FIG. 3

2.8 Grooves (Fig. 4)
2.8.1 Width—Width of groove will vary depending on the lubrication requirements of the specific application.

For an economical method of inspection, it is customary to dimension remaining wall from back of bushing to bottom of groove. Tolerance on this dimension to be 0.25 mm (0.010 in).

Groove cross sections

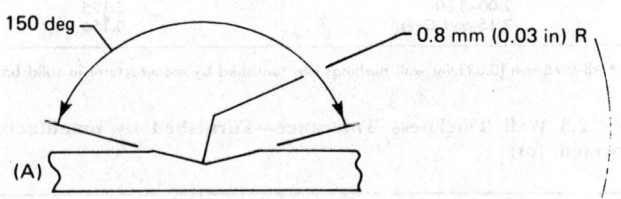

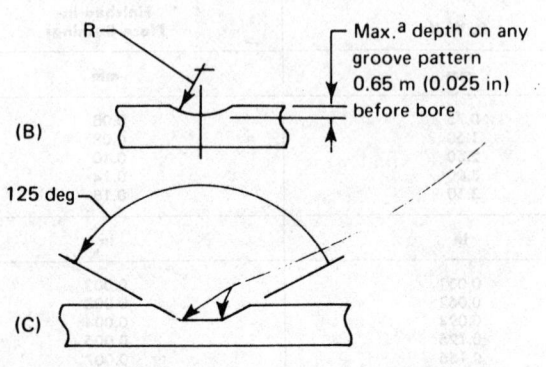

[a]Grooves not recommended for walls less than 1.5 mm (0.062 in)

FIG. 4

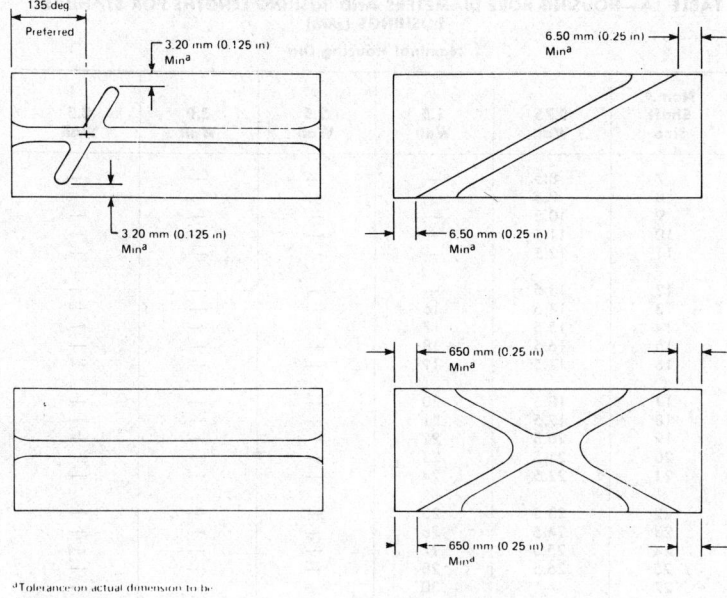

FIG. 5

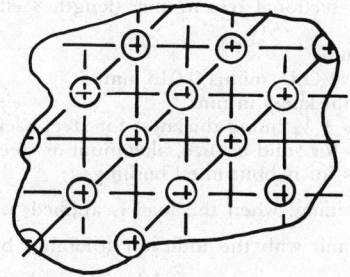

FIG. 7

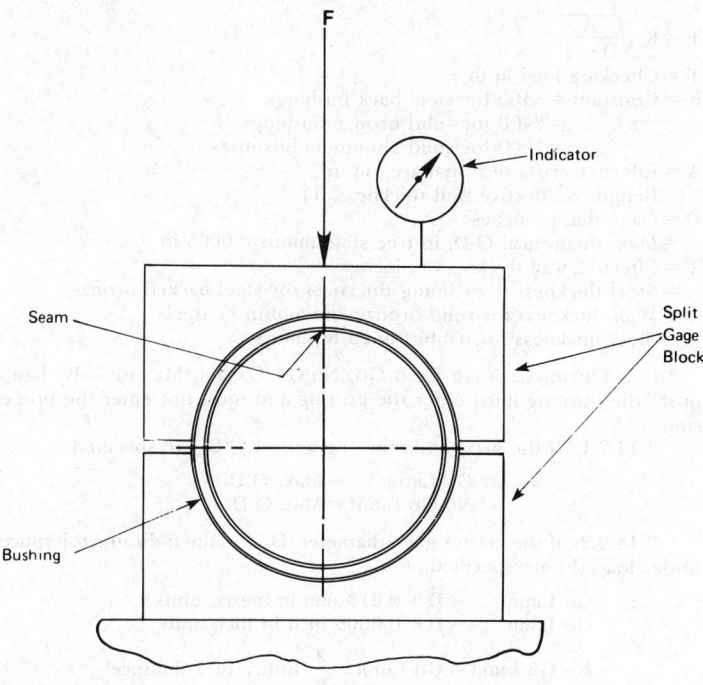

FIG. 8

To have a common understanding between the inspection departments of both vendor and customer, the following note must appear on the bushing drawing: *Slight cracks in groove allowable. No loose bearing material permissible.*

Because of groove distortion in forming, groove dimensions are to be checked in the blank.

On bushings with coined grooves, the following note should appear: "Wall thickness build-up within 10 mm (0.40 in) of groove not to exceed 0.10 mm (0.004 in) (babitt and aluminum alloy) and 0.05 mm (0.002 in) (leaded bronze) over maximum wall." The note also applies to reliefs stamped around holes and slots.

Grooving Patterns (Fig. 5)—Since numerous means of lubrication are employed, grooving design is quite varied. The following figures illustrate some of the more popular groove designs now in use.

2.9 Groove and Hole Location Tolerance (Fig. 6)

2.9.1 In Blank: ±0.80 mm (±0.030 in)

2.9.2 In Formed Bushing (Machined Chamfer): ±0.50 mm (±0.020 in) from end

2.10 Hole Size—Holes pierced in blank must have a minimum dimension of 3.0 mm (0.12 in). Avoid holes at 90 deg from either end of the seam split.

Distortion of holes due to forming is permissible as long as the minimum dimension is maintained.

Some special applications require odd-shaped lubricant cutouts; there is no limitation to the variations of these cutouts as long as manufacturing, gaging, and assembly considerations are satisfied.

2.11 Ball Indentations (Fig. 7)—Ball indentations on the lining surface of the bushing provide small reservoirs which retain the lubricant and are very beneficial in grease-lubricated applications, ensuring good lubrication during engine starting.

This indenting of the bushing lining surface may be used to complement the lubricating grooves discussed in paragraph 2.8 of this standard.

Size and location of indents will vary slightly depending on the manufacturer.

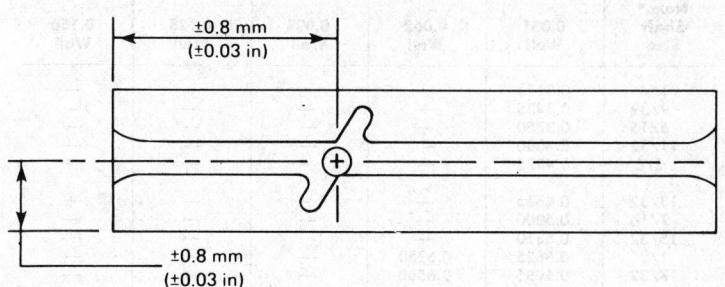

FIG. 6

2.12 Surface Finishes

2.12.1 Housing Holes—Minimum: 1.5 μm (60 μin) R_a
Maximum: 3.0 μm (120 μin) R_a

2.12.2. Bushing Inside Diameter—The inside diameter surface finish of prebored bushings and those finished in place *SHOULD NOT EXCEED* 1.0 μm (40 μin) R_a.

2.12.3. Finishes on Shafts *SHOULD NOT EXCEED* 0.4 μm (15 μin) R_a; a *MAXIMUM OF* 0.25 μm (10 μin) R_a *IS PREFERRED*. On very severe applications, such as piston pins, a finer finish is required.

Note: The service life realized with any bushing material application depends very greatly on the finish condition of the shaft and bushing inside diameter.

2.13 Bushing Materials—For a complete discussion on the methods of manufacture, characteristics, and application of bushing materials, see SAE Standard, Bearing and Bushing Alloys—SAE J460 (July 1974).

The steel backing generally specified for bushings is SAE 1010 (February 1973).

2.14 O.D. Checking Procedures

2.14.1 Split Block Method

The outside diameter is measured in a split gage block with a load F applied to the block.

Metric units:

$$F = K \sqrt{\frac{A}{D-T}}$$

F^1 = checking load in kN
K = Constant = 3.0 for steel back bushings
2.1 for solid bronze bushings
1.7 for solid aluminum bushings

[1] The basic method of checking outside diameter is the same as ISO standards. However, the checking load is calculated according to long established U.S.A. practice and is not the same as the ISO load.

A = Effective cross-sectional area in mm² (length × effective wall thickness)
D = Gage dia. in mm
  = Max. theoretical O.D. minus 0.013 mm
T = Effective wall thickness in mm
  = Steel thickness + ½ lining thickness for steel backed bronze
  = Wall thickness for solid bronze, aluminum or steel
  = Steel thickness for babbitt-lined bushings.

The indicator reading, when the load is applied, will be taken from zero as the high limit with the indicator tolerance being $\frac{\pi}{2}$ times the O.D. tolerance (Refer to paragraph 1.1). The indicator is set at zero using a master gage of "D" diameter at an applied load of "F."

Inch Units:

$$F = K \sqrt{\frac{A}{D-T}}$$

F = Checking load in lb
K = Constant = 3400 for steel back bushings
  = 2400 for solid bronze bushings
  = 1900 for solid aluminum bushings
A = Effective cross-sectional area in in²
  (length × effective wall thickness, T)
D = Gage dia. in inches
  = Max. theoretical O.D. in free state minus 0.0005 in
T = Effective wall thickness in inches
  = Steel thickness + ½ lining thickness for steel backed bronze
  = Wall thickness for solid bronze, aluminum or steel
  = Steel thickness for babbitt lined bushings.

2.14.2 OPTIONAL SOLID RING GO, NO-GO GAGING METHOD—By hand push, the bushing must enter the go ring and must not enter the no-go ring.

2.14.2.1. If the max. and min. theoretical O.D. are specified:

   Go Limit   = Max. O.D.
   No-Go Limit = Min. O.D.

2.14.2.2. If the master gage diameter (D) and the indicator tolerance under load (F) are specified:

   Go Limit = D + 0.013 mm in metric units
   Go Limit = D + 0.0005 inch in inch units
   No Go Limit = Go Limit − $\frac{2}{\pi}$ (Indicator Tolerance)

3. Split Bushings (Precision—Prebored by Supplier)

3.1 Precision bushings are prebored by the bushing manufacturer to the following tolerances. It must be recognized that due to the accumulation of bushing and housing tolerances, this type of bushing does not give as close I.D. tolerance and alignment control as realized with finished-in-place bushings.

Except for the two variations noted as follows, all other previously discussed standards for finished-in-place bushings also apply for precision prebored bushings.

Wall Thickness Tolerance:

  0.025 mm (0.001 in) up to 76 mm (3 in) dia and less than 50 mm (2 in) length
  0.040 mm (0.0015 in) over 76 mm (3 in) dia and/or 50 mm (2 in) length

Lining Material Thickness

Nominal Wall Size, B	Min. Lining Material Thickness
mm	mm
1.50	0.20
2.50	0.35
3.00	0.40
3.50	0.45
in	in
0.062	0.008
0.094	0.015
0.125	0.017
0.156	0.018

TABLE 1A—HOUSING BORE DIAMETERS AND BUSHING LENGTHS FOR STANDARD BUSHINGS (MM)

Nom.[a] Shaft Size	Nominal Housing Dia				
	0.75 Wall	1.5 Wall	2.5 Wall	3.0 Wall	3.5 Wall
7	8.5	—	—	—	—
8	9.5	—	—	—	—
9	10.5	—	—	—	—
10	11.5	—	—	—	—
11	12.5	—	—	—	—
12	13.5	—	—	—	—
13	14.5	16	—	—	—
14	15.5	17	—	—	—
15	16.5	18	—	—	—
16	17.5	19	—	—	—
17	18.5	20	—	—	—
18	19.5	21	—	—	—
19	20.5	22	—	—	—
20	21.5	23	—	—	—
21	22.5	24	—	—	—
22	23.5	25	—	—	—
23	24.5	26	—	—	—
24	25.5	27	—	—	—
25	26.5	28	—	—	—
27	—	30	—	—	—
29	—	32	—	—	—
31	—	34	—	—	—
33	—	36	—	—	—
35	—	38	—	—	—
37	—	40	—	—	—
39	—	42	44	—	—
41	—	44	46	—	—
43	—	46	48	—	—
45	—	48	50	—	—
47	—	50	52	—	—
49	—	52	54	—	—
51	—	54	56	57	—
55	—	58	60	61	—
59	—	62	64	65	—
63	—	66	68	69	—
67	—	70	72	73	—
71	—	74	76	77	—
75	—	78	80	81	—
79	—	—	84	85	86
83	—	—	88	89	90
87	—	—	92	93	94
91	—	—	96	97	98
95	—	—	100	101	102
99	—	—	104	105	106
103	—	—	108	109	110
107	—	—	112	113	114
111	—	—	116	117	118
115	—	—	—	121	122
119	—	—	—	125	126
123	—	—	—	129	130
127	—	—	—	133	134
131	—	—	—	137	138
135	—	—	—	141	142
140	—	—	—	146	147
145	—	—	—	151	152

TABLE 1B—HOUSING BORE DIAMETERS AND BUSHING LENGTHS FOR STANDARD BUSHINGS (IN)

Nom.[a] Shaft Size	Nominal Housing Dia.				
	0.031 Wall	0.062 Wall	0.094 Wall	0.125 Wall	0.156 Wall
1/4	0.3125	—	—	—	—
9/32	0.3435	—	—	—	—
5/16	0.3750	—	—	—	—
11/32	0.4060	—	—	—	—
3/8	0.4375	—	—	—	—
13/32	0.4685	—	—	—	—
7/16	0.5000	—	—	—	—
15/32	0.5310	—	—	—	—
1/2	0.5625	0.6250	—	—	—
17/32	0.5935	0.6560	—	—	—

(Table continued on next page)

TABLE 1B—HOUSING BORE DIAMETERS AND BUSHING LENGTHS FOR STANDARD BUSHINGS (IN) (CONTINUED)

Nominal Housing Dia.

Nom.[a] Shaft Size	0.031 Wall	0.062 Wall	0.094 Wall	0.125 Wall	0.156 Wall
9/16	0.6250	0.6875	—	—	—
19/32	0.6560	0.7185	—	—	—
5/8	0.6874	0.7500	—	—	—
21/32	0.7185	0.7810	—	—	—
11/16	0.7500	0.8125	—	—	—
23/32	0.7810	0.8435	—	—	—
3/4	0.8125	0.8750	—	—	—
25/32	0.8435	0.9060	—	—	—
13/16	0.8750	0.9375	—	—	—
27/32	0.9060	0.9685	—	—	—
7/8	0.9375	1.0000	—	—	—
29/32	0.9685	1.0310	—	—	—
15/16	1.0000	1.0625	—	—	—
31/32	1.0310	1.0935	—	—	—
1	1.0625	1.1250	—	—	—
1- 1/16	—	1.1875	—	—	—
1- 1/8	—	1.2500	—	—	—
1- 3/16	—	1.3125	—	—	—
1- 1/4	—	1.3750	—	—	—
1- 5/16	—	1.4375	—	—	—
1- 3/8	—	1.5000	—	—	—
1- 7/16	—	1.5625	—	—	—
1- 1/2	—	1.6250	1.6875	—	—
1- 9/16	—	1.6875	1.7500	—	—
1- 5/8	—	1.7500	1.8125	—	—
1-11/16	—	1.8125	1.8750	—	—
1- 3/4	—	1.8750	1.9375	—	—
1-13/16	—	1.9375	2.0000	—	—
1- 7/8	—	2.0000	2.0625	—	—
1-15/16	—	2.0625	2.1250	—	—
2	—	2.1250	2.1875	2.2500	—
2- 1/8	—	2.2500	2.3125	2.3750	—
2- 1/4	—	2.3750	2.4375	2.5000	—
2- 3/8	—	2.5000	2.5625	2.6250	—
2- 1/2	—	2.6250	2.6875	2.7500	—
2- 5/8	—	2.7500	2.8125	2.8750	—
2- 3/4	—	2.8750	2.9375	3.0000	—
2- 7/8	—	3.0000	3.0625	3.1250	—
3	—	3.1250	3.1875	3.2500	—
3- 1/8	—	—	3.3125	3.3750	3.437
3- 1/4	—	—	3.4375	3.5000	3.562
3- 3/8	—	—	3.5625	3.6250	3.687
3- 1/2	—	—	3.6875	3.7500	3.812
3- 5/8	—	—	3.8125	3.8750	3.937
3- 3/4	—	—	3.9375	4.0000	4.062
3- 7/8	—	—	4.0625	4.1250	4.187
4	—	—	4.1875	4.2500	4.312
4- 1/8	—	—	—	4.3750	4.437
4- 1/4	—	—	—	4.5000	4.562
4- 3/8	—	—	—	4.6250	4.687

TABLE 1B—HOUSING BORE DIAMETERS AND BUSHING LENGTHS FOR STANDARD BUSHINGS (IN) (CONTINUED)

Nominal Housing Dia.

Nom.[a] Shaft Size	0.031 Wall	0.062 Wall	0.094 Wall	0.125 Wall	0.156 Wall
4- 1/2	—	—	—	4.7500	4.812
4- 5/8	—	—	—	4.8750	4.937
4- 3/4	—	—	—	5.0000	5.062
4- 7/8	—	—	—	5.1250	5.187
5	—	—	—	5.2500	5.312
5- 1/4	—	—	—	5.5000	5.562
5- 1/2	—	—	—	5.7500	5.812
5- 3/4	—	—	—	6.0000	6.062
6	—	—	—	6.2500	6.312

[a] This SAE Standard uses nominal shaft diameters as a base. ISO Standards use nominal housing bores as a base. The two standards are in agreement as to dimensions and tolerances for bushings of the same size.

TABLE 2A—SUGGESTED MINIMUM PRESS FIT FOR NORMAL STEEL OR CAST IRON HOUSING[a] (MM)

Nominal O.D.	Short Length (Equal to the Dia or Less)		Medium Length (Over Dia Including 2 × Dia)		Long Length (Over 2 × Dia)	
	Solid Bronze	Bi-Metal[b]	Solid Bronze	Bi-Metal[b]	Solid Bronze	Bi-Metal[b]
7	0.025	0.025	0.025	0.025	0.025	0.013
10	0.025	0.025	0.025	0.025	0.025	0.013
13	0.040	0.040	0.040	0.040	0.025	0.025
16	0.050	0.040	0.050	0.040	0.040	0.025
19	0.065	0.050	0.050	0.040	0.040	0.025
22	0.075	0.065	0.065	0.050	0.050	0.040
25	0.090	0.065	0.075	0.050	0.050	0.040
28	0.100	0.075	0.090	0.065	0.065	0.040
31	0.100	0.075	0.090	0.065	0.065	0.040
35	0.115	0.090	0.100	0.075	0.075	0.050
39	0.115	0.090	0.100	0.075	0.075	0.050
45	0.125	0.100	0.100	0.090	0.075	0.050
51	0.125	0.100	0.100	0.090	0.075	0.050
57	0.140	0.115	0.115	0.090	—	—
64	0.140	0.115	0.115	0.090	—	—
70	0.150	0.125	0.125	0.100	—	—
76	0.150	0.125	0.125	0.100	—	—
89	0.180	0.150	—	—	—	—
102	0.200	0.180	—	—	—	—

[a] NOTE: Bushing applications, in which any of the following conditions exist, should be given special considerations: 1. Aluminum housings. 2. Heavily loaded applications. 3. Light section housings.
[b] Steel-backed.

TABLE 2B—SUGGESTED MINIMUM PRESS FIT FOR NORMAL STEEL OR CAST IRON HOUSING[a] (IN)

Nominal O.D.	Short Length (Equal to the Dia or Less)		Medium Length (Over Dia Including 2 × Dia)		Long Length (Over 2 × Dia)		Nominal O.D.	Short Length (Equal to the Dia or Less)		Medium Length (Over Dia Including 2 × Dia)		Long Length (Over 2 × Dia)	
	Solid Bronze	Bi-Metal[b]	Solid Bronze	Bi-Metal[b]	Solid Bronze	Bi-Metal[b]		Solid Bronze	Bi-Metal[b]	Solid Bronze	Bi-Metal[b]	Solid Bronze	Bi-Metal[b]
1/4	0.001	0.001	0.001	0.001	0.001	0.0005	1-1/2	0.0045	0.0035	0.004	0.003	0.003	0.002
3/8	0.001	0.001	0.001	0.001	0.001	0.0005	1-3/4	0.005	0.004	0.004	0.0035	0.003	0.002
1/2	0.0015	0.0015	0.0015	0.0015	0.001	0.001							
5/8	0.002	0.0015	0.002	0.0015	0.0015	0.001	2	0.005	0.004	0.004	0.0035	0.003	0.002
3/4	0.0025	0.002	0.002	0.0015	0.0015	0.001	2-1/4	0.0055	0.0045	0.0045	0.0035	—	—
7/8	0.003	0.0025	0.0025	0.002	0.002	0.0015	2-1/2	0.0055	0.0045	0.0045	0.0035	—	—
							2-3/4	0.006	0.005	0.005	0.004	—	—
1	0.0035	0.0025	0.003	0.002	0.002	0.0015							
1-1/8	0.004	0.003	0.0035	0.0025	0.0025	0.0015	3	0.006	0.005	0.005	0.004	—	—
1-1/4	0.004	0.003	0.0035	0.0025	0.0025	0.0015	3-1/2	0.007	0.006	—	—	—	—
1-3/8	0.0045	0.0035	0.004	0.003	0.003	0.002	4	0.008	0.007	—	—	—	—

[a] NOTE: Bushing applications, in which any of the following conditions exist, should be given special considerations: 1. Aluminum housings. 2. Heavily loaded applications. 3. Light section housings.
[b] Steel-backed.

THRUST WASHERS—DESIGN AND APPLICATION—SAE J924 JAN81

SAE Standard

Report of the Engine Committee, approved July 1965, completely revised January 1981.

1. Scope—This SAE Standard presents the basic size and tolerance information for the design and manufacture of thrust washers. In most cases the standard employs nominal figures in both metric and inch-pound units and, therefore, does not necessarily provide exact equivalents.

2. Full Round Washers

2.1 Outside Diameter and Inside Diameter Tolerances—Stamped

Thickness	Outside Diameter	Type of Grooves	I.D. and O.D. Tolerance
mm	mm		mm
0.75 thru 3	up to 75	none	±0.13
0.75 thru 3	up to 75	machined	±0.13
0.75 thru 3	up to 75	coined	±0.25
0.75 thru 3	over 75	any	±0.25
over 3	all	any	±0.25
in	in		in
0.03 thru 0.12	up to 3	none	±0.005
0.03 thru 0.12	up to 3	machined	±0.005
0.03 thru 0.12	up to 3	coined	±0.010
0.03 thru 0.12	over 3	any	±0.010
over 0.12	all	any	±0.010

Washers requiring closer outside or inside diameter tolerances than shown require machine finishing.

2.2 Thickness and Parallel Tolerances—Stamped

Thickness	Bimetal Roll Finish		Solid Bronze Roll Finish		Bimetal or Solid Bronze Ground Finish	
	Thickness Tolerance	Parallel Tolerance	Thickness Tolerance	Parallel Tolerance	Thickness Tolerance	Parallel Tolerance
mm	mm	mm	mm	mm	mm	mm
0.75	—	—	±0.025	0.03	—	—
1.5	±0.025	0.03	±0.025	0.03	—	—
2.5	±0.025	0.04	±0.025	0.04	—	—
3.0	±0.04	0.05	±0.04	0.08	±0.02	0.03
4.0	±0.05	0.08	±0.05	0.10	±0.02	0.03
5.0	—	—	±0.05	0.10	±0.025	0.03
6.5	—	—	±0.08	0.15	±0.025	0.05
in	in	in	in	in	in	in
0.03	—	—	±0.001	0.001	—	—
0.06	±0.001	0.001	±0.001	0.001	—	—
0.09	±0.001	0.002	±0.001	0.002	—	—
0.12	±0.002	0.002	±0.002	0.003	±0.001	0.001
0.16	±0.002	0.003	±0.002	0.004	±0.001	0.001
0.19	—	—	±0.002	0.004	±0.001	0.001
0.25	—	—	±0.003	0.006	±0.001	0.002

Washers are generally used as fabricated from strip stock. For applications requiring closer tolerances than shown, a ground finish is necessary.

2.3 Flatness Tolerances—A thrust load, when exerted against a thrust surface solidly backed up, will flatten or straighten the washer surface to a degree dependent upon the relative strength of the two. Washers of a thickness through 4 mm (0.16 in) are, therefore, used as stamped. Washers of 5 and 6.5 mm (0.19 and 0.25 in) thickness are supplied with a ground finish. Following are standard commercial washer flatness tolerances:

Description	Thickness	Flatness Tolerance
	mm	mm
Plain (no grooves)	0.75	0.20 up to 55 mm O.D. 0.23 up to 75 mm O.D. 0.25 over 75 mm O.D.
Plain or with machined grooves	1.5 and 2.5	0.07 + 0.03 for each 25 mm of O.D. or portion over 75
	3 and 4	0.07 + 0.015 for each 25 mm of O.D. or portion over 75
	5 and 6.5 (ground finish only)	0.03 + 0.015 for each 25 mm of O.D. or portion over 75
With bent lugs and/or coined grooves	1.5 and 2.5	0.12 + 0.03 for each 25 mm of O.D. or portion over 75
	3 and 4	0.12 + 0.015 for each 25 mm of O.D. or portion over 75
	5 and 6.5 (ground finish only)	0.03 + 0.015 for each 25 mm of O.D. or portion over 75
	in	in
Plain (no grooves)	0.03	0.008 up to 2 in O.D. 0.009 up to 3 in O.D. 0.010 over 3 in O.D.
Plain or with machined grooves	0.06 and 0.09	0.003 + 0.001 per in of O.D. or portion over 3
	0.12 and 0.16	0.003 + 0.0005 per in of O.D. or portion over 3
	0.19 and 0.25 (ground finish only)	0.001 + 0.0005 per in of O.D. or portion over 3
With bent lugs and/or coined grooves	0.006 and 0.09	0.005 + 0.001 per in of O.D. or portion over 3
	0.12 and 0.16	0.005 + 0.0005 per in of O.D. or portion over 3
	0.19 and 0.25 (ground finish only)	0.001 + 0.0005 per in of O.D. or portion over 3

2.4 Flatness Check—Parallel Plate Method—Pieces must drop by their own weight between vertical parallel plates spaced a distance apart equal to the tolerance shown in paragraph 2.3 plus the maximum washer thickness.

2.5 Surface Finish

2.5.1 NON-GROUND—Surfaces may be rolled, matted, or specially prepared as required, and should be free of nicks and gouges with raised metal and other imperfections which would adversely affect performance. The degree of surface imperfections allowable should be agreed to between vendor and customer.

2.5.2 GROUND—When grinding is required, faces will be ground to 0.8 μm (30 μin) R_a on the thrust face and 1.8 μm (72 μin) R_a on the back face. Mating thrust faces should have 0.2–0.3 μm (8–12 μin) R_a surface finish ground and polished on a circumferential pattern.

2.6 Lining Thickness—Bimetal Washers—It is advisable to keep lining thickness low for ease of manufacture and economy. A lining thickness of 0.20–0.40 mm (0.008–0.016 in) is recommended in both cast and sintered materials.

2.7 Thrust Washer Materials—A complete discussion of the materials commonly used in thrust washer applications can be found in the SAE Standard "Bearing and Bushing Alloys," SAE J460.

2.8 Steel Backing—The steel backing commonly specified for bimetal thrust washers is SAE 1008 or 1010.

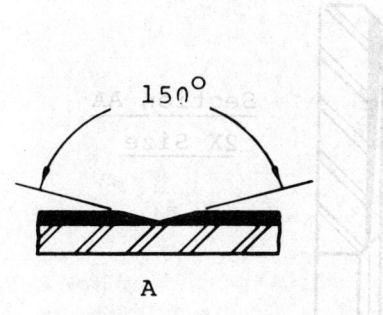

0.40 mm (0.015 in.) max. depth on any groove pattern   ±0.07 mm (±0.0025 in.) groove wall

2.9 Chamfers—Normally washers are not machine chamfered, only tumbled to remove burrs and sharp edges. Specific chamfers require additional machining and add to the cost of the part.

2.10 Grooves—The major portion of thrust washer applications can be accommodated by simple type grooving (see paragraphs 2.11, 2.12, and 2.13). In cases of marginal and difficult lubrication or high loading, special grooves can be utilized to improve washer life (for example see paragraph 2.14). Typical groove types are listed below in order of increasing effectiveness and cost with approximate load carrying capacities. The load capacities shown were developed under specific test conditions and are not intended to be absolute maximums, but rather show relative differences at real magnitudes. Applications with load values approaching or exceeding the values shown should be undertaken only after thorough bearing analysis.

(a) Straight through "V" type and plain tear drops coined—1400 kPa (200 psi) maximum.
(b) Blended tear drops—machined—2100 kPa (300 psi) maximum.
(c) Tapered land type—machined—3500 kPa (500 psi) maximum.

2.11 Standard Groove Cross Sections—The figure above illustrates the typical cross sections of coined oil grooves preferred for all thrust washer materials. Only groove sections "A" and "B" should be used in high lead alloys.

To have a common understanding between the inspection departments of both vendor and customer, the following note must appear on bimetal washer drawings: "Cracks in the lining material in the grooved area allowable. No loose material permissible."

2.12 Groove Depth—The maximum depth for coined grooves is 0.40 mm (0.015 in). If greater depth is required, a machining operation is necessary. Tolerance on groove depth is ±0.07 mm (±0.0025 in).

From the standpoint of maintaining flatness in a washer with coined grooves, it is advisable to keep the number of grooves and groove depth to a minimum.

2.13 Groove Patterns—Grooves can be made either stopped off or run through as shown. If stopped off, the tolerances shown for location should be used for most economical manufacture.

Most grooves may be furnished on one or both sides and may vary in number depending upon material and specific job requirements.

Frequently, it is desirable to provide more than ordinary means of oil flow to a thrust surface. To do this, scallops can be used. These can be punched on either the inside diameter, outside diameter, or both, with a tolerance of ±0.25 mm (±0.010 in) on the radius and depth.

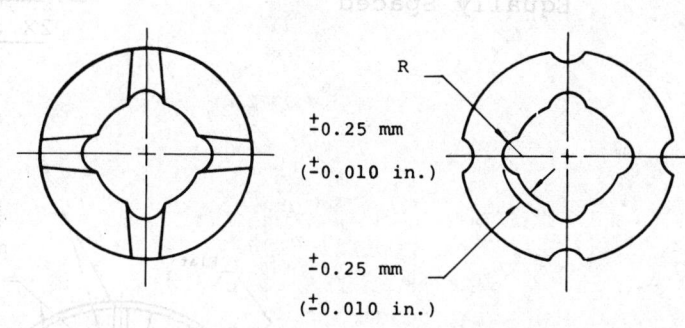

2.14 Special Grooving and Thrust Face Contour—For applications having higher loads, washers employing blended tear drop grooves or tapered lands can be used. Applications of this type should be referred to a bearing designer for analysis.

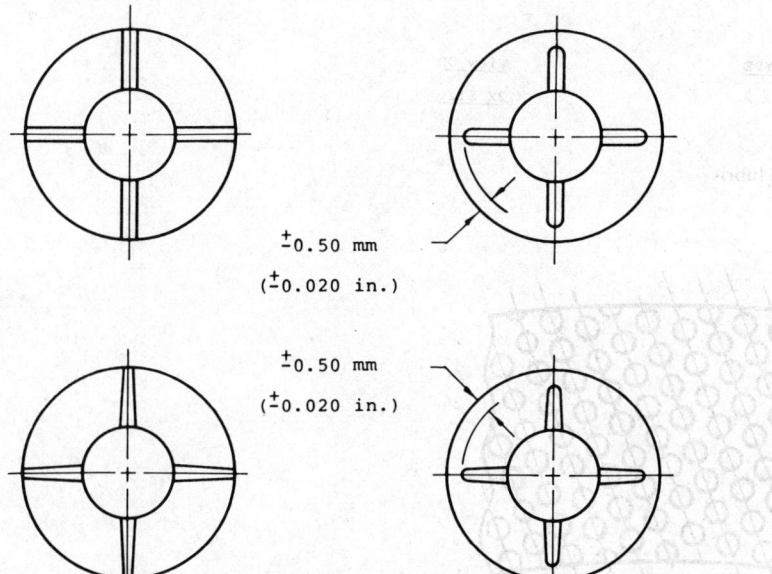

26.208

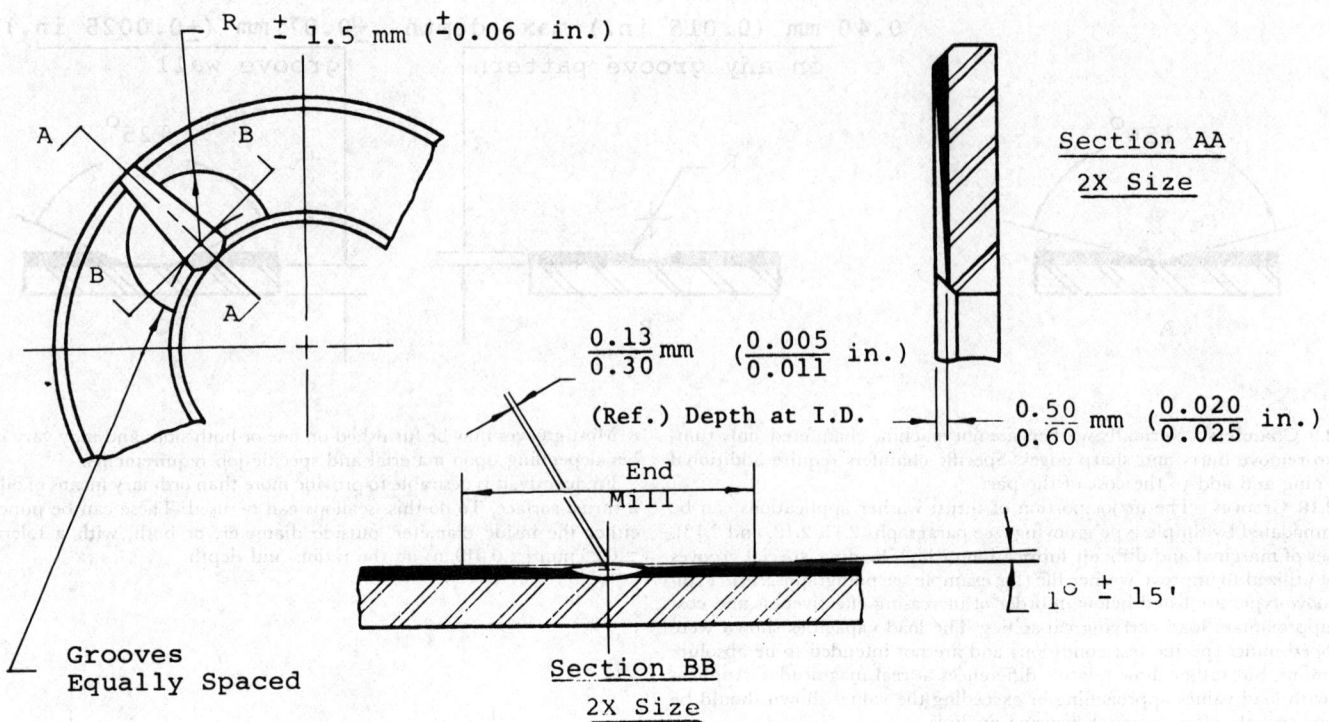

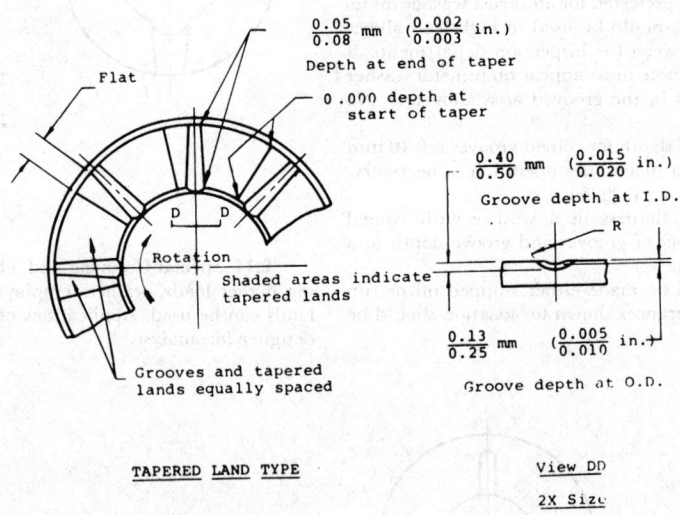

TAPERED LAND TYPE

For applications having marginal lubrication, it is possible to provide lubricant retention by using ball indented or graphited stock.

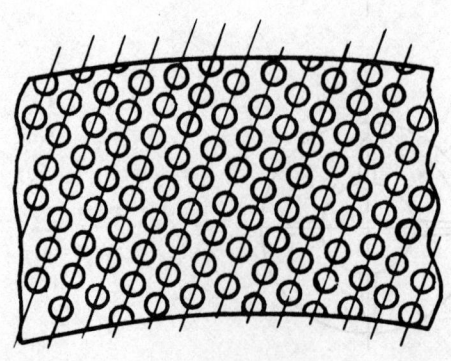

Size and location of indents will vary slightly depending on the manufacturer.

2.15 Locating and Retention Features—Locating or locking features, used except in the case of floating washers, can be supplied in the form of notches or tabs with standard manufacturing tolerances as shown.

Thickness	Thickness Tolerance
mm	mm
0.75—steel or solid bronze only	±0.025
1.5	±0.025
2.5	±0.025
3.0	±0.04
4.0	±0.05
in	in
0.03—steel or solid bronze only	±0.001
0.06	±0.001
0.09	±0.001
0.12	±0.002
0.16	±0.002

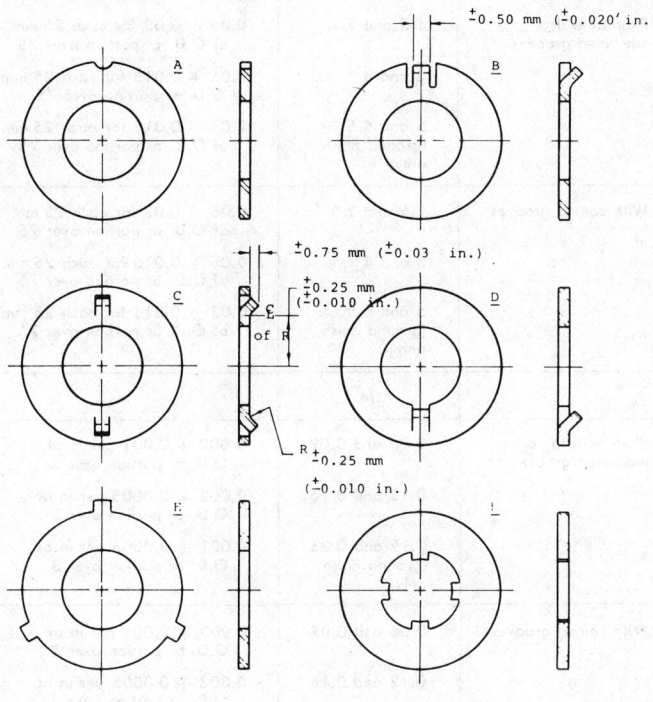

Outside Diameter	I.D. and O.D. Tolerance	Spherical Radius Tolerance
mm	mm	mm
up to 75	±0.25	0.08
over 75	±0.40	0.13
in	in	in
up to 3	±0.010	0.003
over 3	±0.015	0.005

Design A shows a notched outside diameter which receives a dowel pin to prevent rotation.

Designs B, C, and D show formed tabs which can be located either on the inside diameter, outside diameter, or on the face.

Designs E and F show flat inside and outside diameter tabs which fit into slots, notches, keyways, or splines which offer positive locking with a minimum of cost.

Design G shows a 90 deg formed tab which can be located either on the inside or outside diameter.

Cracks at bend of lug which are free of loose and foreign material and do not significantly affect strength are permissable.

2.16 Spherical washers can be produced from plain steel, solid bronze, or steel backed bronze. All materials may be ball indented for added oil retention and steel washers may be case hardened.

Tolerances for spherical washers are as follows:

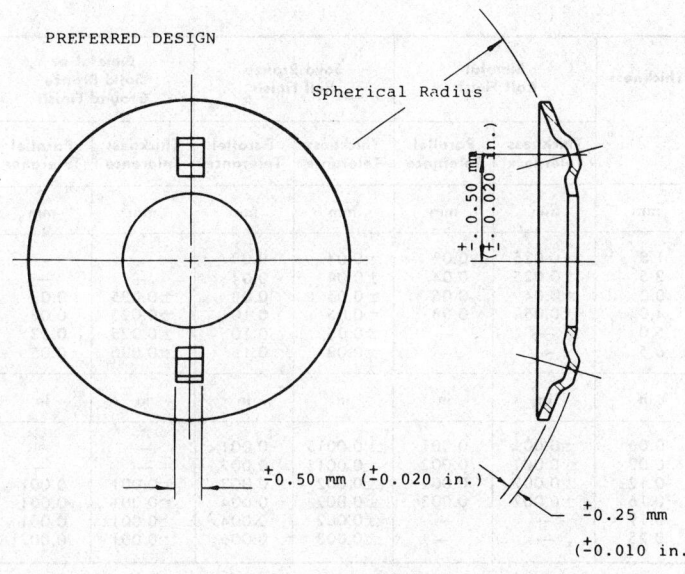

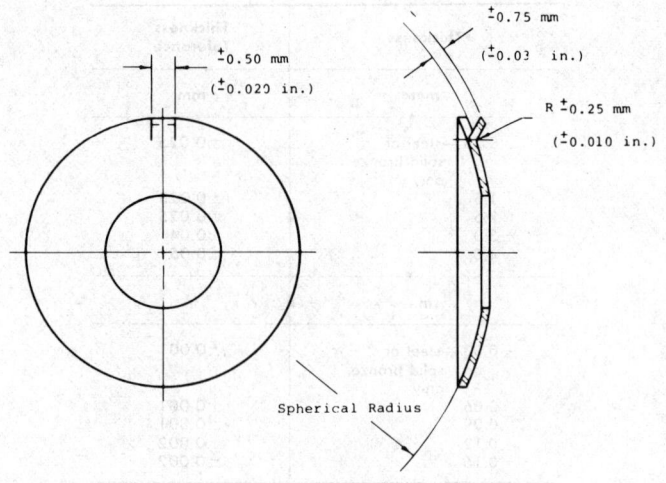

3. Half Washers
3.1 Outside Diameter and Inside Diameter Tolerances—Stamped

Thickness	Outside Diameter	Type of Grooves	I.D. and O.D. Tolerance
mm	mm		mm
0.75 thru 3	up to 100	none	±0.13
0.75 thru 3	up to 100	machined	±0.13
0.75 thru 3	up to 100	coined	±0.25
0.75 thru 3	over 100	any	±0.25
over 3	all	any	±0.25
in	in		in
0.03 thru 0.12	up to 4	none	±0.005
0.03 thru 0.12	up to 4	machined	±0.005
0.03 thru 0.12	up to 4	coined	±0.010
0.03 thru 0.12	over 4	any	±0.010
over 0.12	all	any	±0.010

Half washers requiring closer outside or inside diameter tolerances than shown require machine finishing.

3.2 Thickness and Parallel Tolerances—Stamped

Thickness	Bimetal Roll Finish		Solid Bronze Roll Finish		Bimetal or Solid Bronze Ground Finish	
	Thickness Tolerance	Parallel Tolerance	Thickness Tolerance	Parallel Tolerance	Thickness Tolerance	Parallel Tolerance
mm	mm	mm	mm	mm	mm	mm
1.5	±0.025	0.03	±0.04	0.03	—	—
2.5	±0.025	0.04	±0.04	0.04	—	—
3.0	±0.04	0.05	±0.05	0.08	±0.025	0.03
4.0	±0.05	0.08	±0.05	0.10	±0.025	0.03
5.0	—	—	±0.05	0.10	±0.025	0.03
6.5	—	—	±0.08	0.15	±0.025	0.05
in	in	in	in	in	in	in
0.06	±0.001	0.001	±0.0015	0.001	—	—
0.09	±0.001	0.002	±0.0015	0.002	—	—
0.12	±0.002	0.002	±0.002	0.003	±0.001	0.001
0.16	±0.002	0.003	±0.002	0.004	±0.001	0.001
0.19	—	—	±0.002	0.004	±0.001	0.001
0.25	—	—	±0.003	0.006	±0.001	0.002

3.3 Flatness Tolerances

Description	Thickness	Flatness Tolerance
	mm	mm
Plain or with machined grooves	1.5 and 2.5	0.05 + 0.03 for each 25 mm of O.D. or portion over 75
	3 and 4	0.05 + 0.015 for each 25 mm of O.D. or portion over 75
	5 and 6.5 (ground finish only)	0.03 + 0.015 for each 25 mm of O.D. or portion over 75
With coined grooves	1.5 and 2.5	0.08 + 0.03 for each 25 mm of O.D. or portion over 75
	3 and 4	0.08 + 0.015 for each 25 mm of O.D. or portion over 75
	5 and 6.5 (ground finish only)	0.03 + 0.015 for each 25 mm of O.D. or portion over 75
	in	in
Plain or with machined grooves	0.06 and 0.09	0.002 + 0.001 per in of O.D. or portion over 3
	0.12 and 0.16	0.002 + 0.0005 per in of O.D. or portion over 3
	0.19 and 0.25 (ground finish only)	0.001 + 0.0005 per in of O.D. or portion over 3
With coined grooves	0.06 and 0.09	0.003 + 0.001 per in of O.D. or portion over 3
	0.12 and 0.16	0.003 + 0.0005 per in of O.D. or portion over 3
	0.19 and 0.25 (ground finish only)	0.001 + 0.0005 per in of O.D. or portion over 3

3.4 Flatness Check—Parallel Plate Method—See paragraph 2.4.
3.5 Surface Finish—Ground—See paragraph 2.5.
3.6 Lining Thickness—Bimetal

Lining Material	Minimum Lining Thickness	
	mm	in
Babbitt	0.25	0.010
Aluminum	0.25	0.010
Bronze	0.20	0.008

3.7 See paragraph 2.7.
3.8 See paragraph 2.8.
3.9 See paragraph 2.9.
3.10 See paragraph 2.10.
3.11 See paragraph 2.11.
3.12 See paragraph 2.12.

3.13 Groove Patterns—Grooves can be made either stopped off or run through as shown. If stopped off, the tolerances shown should be used for most economical manufacture.

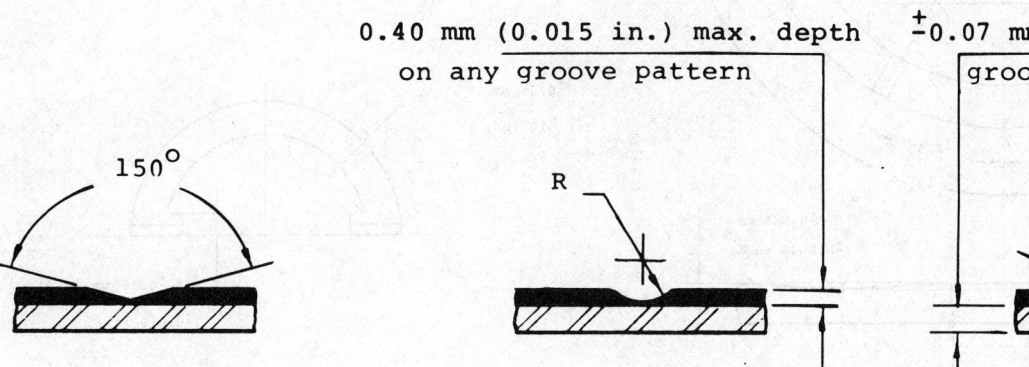

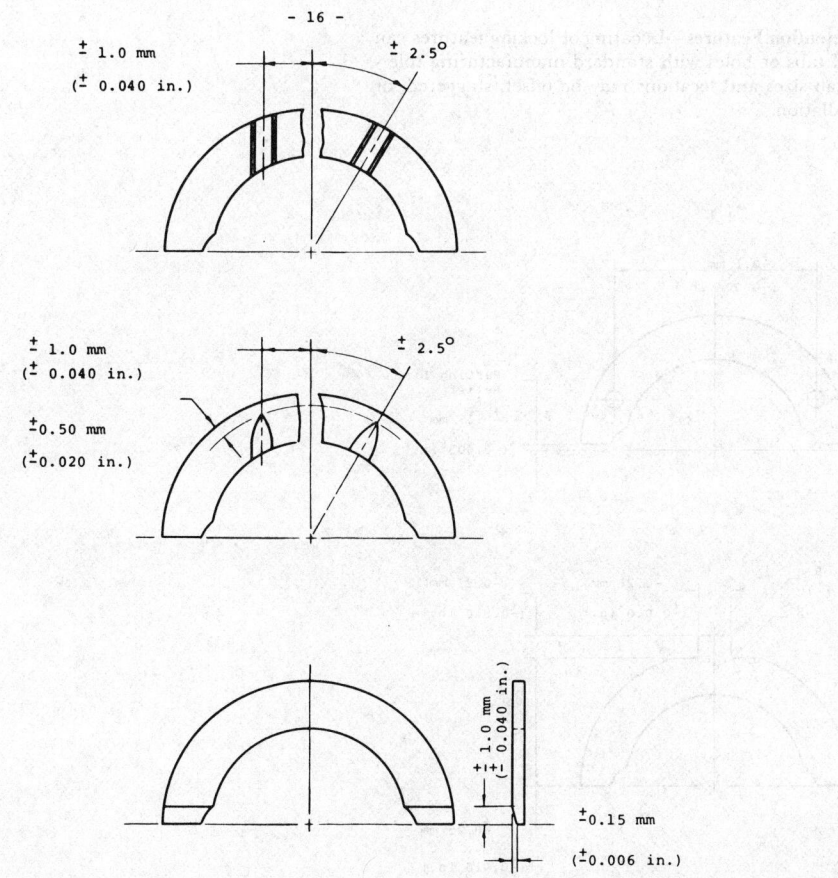

3.14 Special Thrust Face Contour—Washers having special contours can be produced to provide greater load capacity. Applications of this type should be referred to the vendor for design analysis.

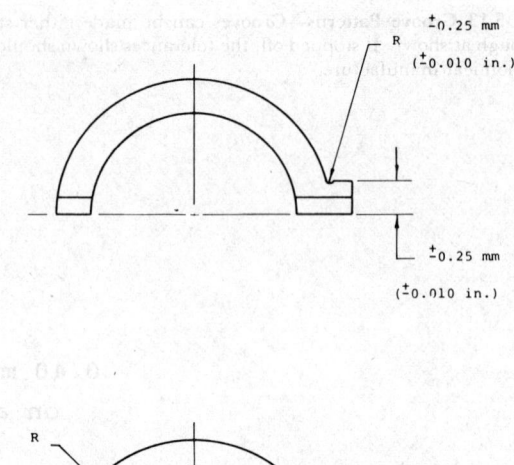

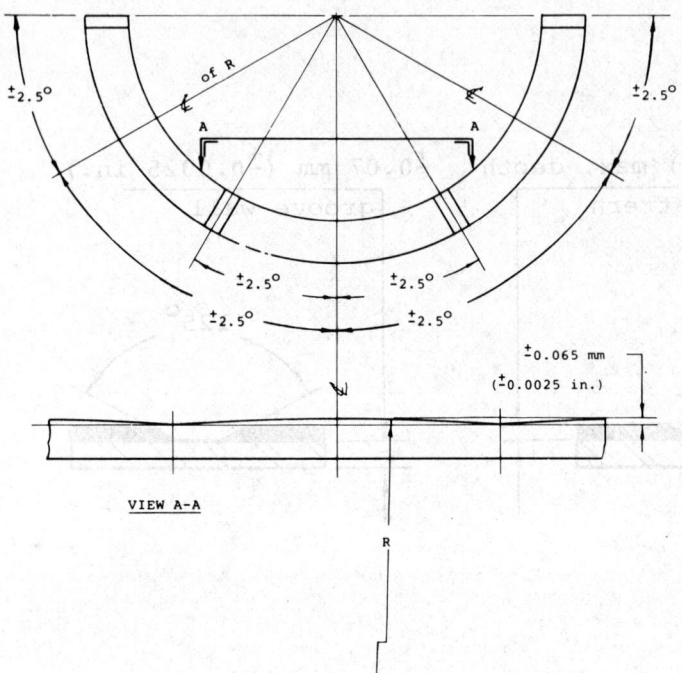

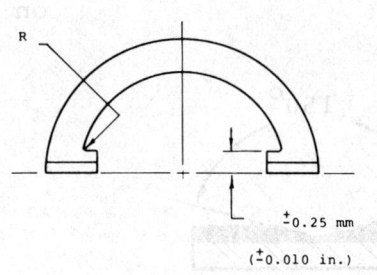

3.16 Parting Line Relief Tolerances—Parts made to less than a full half circle should have the amount of parting line relief dimensioned from the theoretical centerline.

3.15 Locating and Retention Features—Locating or locking features can be supplied in the form of tabs or holes with standard manufacturing tolerances as shown. Hole or tab sizes and locations may be offset, staggered, or varied to prevent misinstallation.

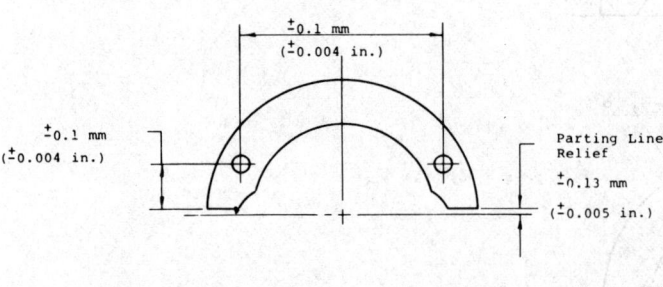

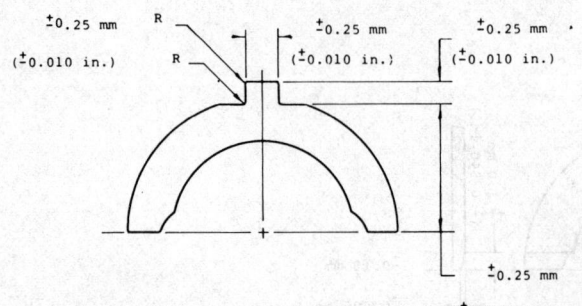

AIR CLEANER ELEMENTS—SAE J1141 JUL87

SAE Information Report

Report of the Engine Committee approved June 1977 and completely revised July 1987.

Over the past several years many differently sized dry-type air cleaner elements have been manufactured to service the numerous makes and models of United States domestic passenger cars and light trucks. This has led to inventory and stock control inefficiencies that affect the motoring public, the automobile manufacturers, and the air cleaner producers. Currently, there are hundreds of different air cleaner elements required to service these vehicles.

The SAE Air Cleaner Test Code Subcommittee, recognizing the need for standardization of air cleaners has compiled a list of the 25 most used round type and the most used panel type elements. Automotive engineers responsible for the design of engine air induction systems on new applications are encouraged to select a dry-type air cleaner element from these lists. It is expected that by adhering to this recommendation that the currently most popular elements will be chosen for new applications to eliminate further proliferation.

A periodic review will be made by the Air Cleaner Test Code Subcommittee to revise this listing, based on changes in sales volume rating for these and other models used on United States domestic production vehicles.

Underhood space limitations, available air cleaner housings and engine induction air requirements dictate the choice of the proper element to the automotive engineer. After these general parameters are established refer to one of the tables.

1. Selection Procedure—The air cleaner elements have been arranged in order of descending sales volume popularity in the following listings. The table includes the information on nominal height, inside diameter and outside diameter for the round elements and the nominal length, width, and height for the panel elements. The exact dimensions, tolerances, seal configurations, general construction and filtration characteristics information on these elements may be obtained from the air cleaner manufacturers.

(1) By referring to either listing, choose an appropriate size element for the application.

(2) If more than one element appears to fulfill the requirements, then choose the most popular.

(3) Get all necessary information and assistance to complete the necessary design details by contacting the air cleaner element manufacturers.

NOTE: The higher popularity ratings are based on physical dimensions and should take precedence only after it has been determined that the construction, materials, and functional characteristics satisfy the performance requirements.

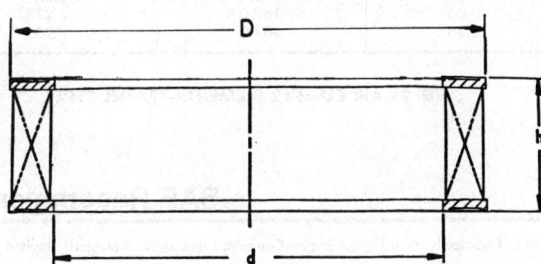

Volume Popularity Rating	Height-h		Inside Dia-d		Outside Dia-D	
	mm	in	mm	in	mm	in
1	88.0	3.46	249.0	9.82	306.0	12.04
2	66.5	2.62	230.0	9.04	286.0	11.26
3	70.0	2.75	191.5	7.54	248.0(T)	9.76(T)
					262.0(B)	10.30(B)
4	62.5	2.46	249.0	9.82	306.0	12.04
5	90.0	3.53	208.0	8.20	247.0	9.74
6	77.0	3.04	295.0	11.62	351.5	13.84
7	49.0	1.92	208.0	8.20	247.0	9.74
8	55.0	2.17	227.0	8.95	266.5	10.49
9	58.5	2.30	219.0	8.62	258.0	10.16
10	71.0	2.78	282.5	11.12	339.0	13.34
11	54.0	2.12	244.0	9.62	283.5	11.16
12	57.5	2.26	202.0	7.96	241.0	9.50
13	77.0	3.04	164.5	6.48	204.0	8.02
14	102.0	4.03	295.0	11.61	351.5	13.84
15	69.5	2.74	280.5	11.04	319.5	12.58
16	82.0	3.22	168.5	6.63	207.5	8.17
17	72.0	2.82	208.0	8.20	247.0	9.47
18	139.0	5.46	249.0	9.82	306.0	12.04
19	49.0	1.92	226.5	8.92	266.0	10.46
		(OVAL)	146.0	5.75	185.0	7.28
20	54.0	2.12	231.0	9.09	270.0	10.63
		(OVAL)	194.0	7.64	233.0	9.17
21	60.5	2.38	202.0	7.94	262.0	10.31
22	63.0	2.49	176.0	6.93	216.0	8.50
23	63.5	2.50	287.5	11.32	326.0	12.85
24	66.0	2.62	275.0	10.84	314.5	12.38
25	44.5	1.75	229.0	9.02	292.0	11.51

(T)—TOP (B)—BOTTOM

FIG. 1—AIR CLEANER ELEMENTS—ROUND TYPE

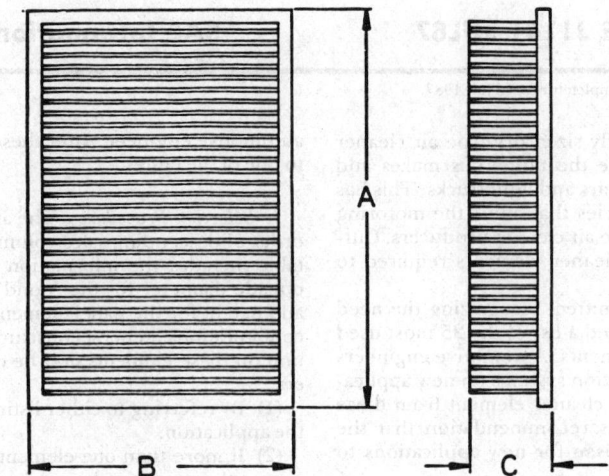

Volume Popularity Rating	Length-A		Width-B		Height-C	
	mm	in	mm	in	mm	in
1	205.0	8.07	133.5	5.25	41.0	1.62
2	334.0	13.15	133.5	5.25	41.0	1.62
3	273.0	10.75	133.5	5.25	41.0	1.62
4	271.0	10.65	159.0	6.27	36.0	1.42
5	212.0	8.36	201.0	7.90	57.0	2.27

FIG. 2—AIR CLEANER ELEMENTS—PANEL TYPE

(R) AIR CLEANER TEST CODE—SAE J726 JUN93

SAE Recommended Practice

Report of the Tractor and Equipment Division, approved January 1941. Completely revised by the Engine Committee June 1987. Completely revised by the SAE Air Cleaner Test Code Standards Committee June 1993.

Foreword—The basic performance characteristics of greatest interest are airflow restriction or pressure drop, dust collection efficiency, dust capacity, and air cleaner structural integrity. This test code, therefore, addresses itself to the measurement of these parameters.

The objective of the test code is to provide a uniform means for evaluating the performance characteristics of air cleaners on bench test equipment. The data collected by this test code can be used to establish standards of performance for air cleaners tested in this manner. The actual field operating conditions (humidity, vibration, contaminant, etc.) are difficult to duplicate. However, by use of these standard test methods, the test conditions are controlled, and comparisons of lab performance of filters may be made with a high degree of confidence.

The equipment specified in these standard tests assures that all particles of test dust are evenly dispersed, eliminating agglomerates, to a degree not possible in prior standard tests. For this reason, apparent filtration efficiency and dust capacity by these tests can be significantly lower than if the same filter were tested by the older methods. These results are more repeatable and reliable than the earlier test methods and correlations among the laboratories using these methods are now possible.

To simplify and improve the clarity of this test code, Section 1, covers general information and definitions applicable to all air cleaner testing covered in this test code. Section 3 covers the testing of dry type air cleaners normally used on automobile internal combustion engines. Section 4 covers the testing of industrial dry type air cleaners and precleaners for mobile and stationary internal combustion engines. Section 5 covers the testing of oil bath air cleaners used for mobile and stationary internal combustion engines.

1. Scope—The air cleaner test code has been established to cover dry type and oil bath air cleaners used on internal combustion engines and to present a uniform method of determining and reporting air cleaner performance.

1.1 Purpose—The purpose of this test code is to establish and specify uniform testing procedures, conditions, equipment, and a performance report to permit the direct laboratory performance comparison of dry type and oil bath type air cleaners, respectively.

1.2 Measurement Accuracy

1.2.1 Measure the airflow rate within 2% of the actual value.

1.2.2 Measure the pressure drop and restriction within 0.25 mbar (0.025 kPa) of the actual value.

1.2.3 Measure the temperature within 0.5 °C of the actual value.

1.2.4 Measure the weight within 1% of the actual value except where noted.

1.2.4.1 Weigh the absolute filter(s) to ±0.01 g.

1.2.5 Measure the relative humidity with an accuracy of 2% R.H.

1.2.6 Measure the barometric pressure within 3 mbar (0.3 kPa).

1.2.7 The measurement equipment shall be calibrated at regular intervals to ensure the required accuracy.

1.3 Test Conditions and Material

1.3.1 The test dust contaminant shall be standardized and shall be of two grades labeled FINE and COARSE. The following chemical analysis (Table 1) is typical:

TABLE 1—CHEMICAL ANALYSIS OF TEST DUST

Chemical	% of Weight
SiO_2	65 - 76
Al_2O_3	11 - 17
Fe_2O_3	2.5 - 5.0
Na_2O	2 - 4
CaO	3 - 6
MgO	0.5 - 1.5
TiO_2	0.5 - 1.0
V_2O_3	0.10
ZrO	0.10
BaO	0.10
Loss on Ignition	2 - 4

1.3.1.1 Before using the test dust, a quantity sufficient to cover test requirements shall be mixed in a sealed container for a minimum of 15 min. This test dust shall then be dried to a constant mass at a temperature of 105 °C ± 5 °C. The test dust shall then be allowed to become acclimatized to a constant mass under the prevailing test conditions.

NOTE—To ensure a constant rate of dust feed with some dust feeders, it may be found necessary to heat the dust prior to being fed to the injector.

1.3.2 The particle size distribution by volume as measured with an L & N Microtrac Analyzer shall be as follows in Table 2:

TABLE 2—PARTICLE SIZE DISTRIBUTION BY VOLUME, %

Size, Microns	Fine Grade (% less than)	Coarse Grade (% less than)
5.5	38 ± 3	13 ± 3
11	54 ± 3	24 ± 3
22	71 ± 3	37 ± 3
44	89 ± 3	56 ± 3
88	97 ± 3	84 ± 3
176	100	100

1.3.2.1 (Reference Only)—The particle size distribution by mass as measured by the Andreason method is given in Table 3:

TABLE 3—PARTICLE SIZE DISTRIBUTION, ANDREASON METHOD PERCENTAGE BY MASS

Size, µm	Fine Grade, %	Coarse Grade, %
<125	----	98.5 ± 1.5
<75	98 ± 2	84.5 ± 5.5
<40	84 ± 3	51 ± 2
<20	67 ± 3	32 ± 2
<10	49 ± 3	19.5 ± 1.5
<5	35 ± 3	10 ± 1
<2	17.5 ± 2.5	----

1.3.2.2 (Reference Only)—The particle size distribution as measured by a Roller analyzer and described in % weight is given in Table 4.

TABLE 4—PARTICLE SIZE DISTRIBUTION BY WEIGHT, %

Size, Microns	Fine Grade	Coarse Grade
0 - 5	39 ± 2	12 ± 2
5 - 10	18 ± 3	12 ± 3
10 - 20	16 ± 3	14 ± 3
20 - 40	18 ± 3	23 ± 3
40 - 80	9 ± 3	30 ± 3
80 - 200	----	9 ± 3

1.3.2.3 Test Dust (Coarse and Fine) is manufactured by Powder Technology Inc. Dust capacity differences may occur between different dust batches. Therefore, it is recommended that comparison testing of filters be performed using a single batch of dust per test program whenever possible.

1.3.3 ABSOLUTE FILTER MATERIALS—The absolute filter shall consist of fiberglass media with a minimum thickness of 12.7 mm and a minimum density of 9.5 kg/m³. The fiber diameter shall be 0.76 to 1.27 µm and the moisture absorption shall be less than 1% by weight after exposure of 49 °C and 95% relative humidity for 96 h. The filter shall be installed with nap side facing upstream in an airtight holder that adequately supports the media. The face velocity shall not exceed 50 m/min to maintain media integrity.

To reduce any subsequent errors in the measurements caused by loss of fibers or materials, the absolute filter shall be subject to a flow of at least 110% of the rated flow of ambient air for 15 min before the test weighings.

1.3.4 ABSOLUTE FILTER WEIGHT—The absolute filter shall be weighed to the nearest 0.01 g after the weight has stabilized and while in a ventilated oven at 105 °C ± 5 °C.

NOTE—If stabilization cannot be determined, a minimum time of 4 h is required.

1.3.5 TEMPERATURE AND HUMIDITY—All tests shall be conducted with air entering the air cleaner at a temperature of 23 °C ± 5 °C and a relative humidity of 55% ± 15%, the permissible variation at each weighing stage throughout each single test being ±2% (relative humidity).

NOTE—Since atmospheric conditions affect the test results, when comparing the performance of the filters designed for the same application, the tests should be conducted within the narrowest range of temperature and humidity possible.

1.3.6 AIR DRYER—To prevent the dust from caking and to prevent icing of the injector nozzle, an effective air dryer of sufficient size should be installed in the air supply line.

1.3.7 See Appendix A for:

a. Test material supplier list
b. Test equipment supplier list

2. References

2.1 Related Publications—The following publications are provided for information purposes only and are not a required part of this document.

2.1.1 SAE PUBLICATIONS—Available from SAE, 400 Commonwealth Drive, Warrendale, PA 15096-0001. The latest issue of SAE publications shall apply.

SAE J1124—Glossary of Terms Related to Fluid Filters and Filter Testing
SAE TSB 003—Rules for SAE Use of SI (Metric) Units

2.1.2 ASME PUBLICATIONS—Available from ASME, 22 Law Drive, Fairfield, NJ 07007.

Fluid Meters: Their Theory and Application—6th Edition 1971

2.1.3 ISO PUBLICATIONS—Available from ANSI, 11 West 42nd Street, New York, NY 10036-8002.

ISO 5011—Inlet air cleaning equipment for internal combustion engines and compressors—Performance testing

2.2 Definitions

2.2.1 AIR CLEANER—For the purpose of this test code, the air cleaners shall be of the dry type or oil bath type and may consist of one or more stages of filtration which remove particles suspended in the fresh charge as it is drawn into the engine.

2.2.1.1 Dry Type Air Cleaner—An air cleaning device consisting of one or more layers of filter media that may or may not have a wettant added to the media. Filtration is accomplished by removal of the contaminant as air passes through the filter media.

2.2.1.2 Oil Bath Air Cleaner—An air cleaning device in which dust laden air is directed on a reservoir of oil where some of the dust is collected. This air, carrying a mist of oil from the reservoir, then passes upward through the filter element for further filtering. The oil returning to the reservoir carries with it dust that was collected in the filter element.

2.2.2 UNIT UNDER TEST—Either a single air cleaner element, a precleaner, or a complete air cleaner assembly.

2.2.3 SINGLE-STAGE AIR CLEANER—An air cleaner that does not incorporate a separate precleaner.

2.2.4 MULTISTAGE AIR CLEANER—An air cleaner consisting of two or more stages, the first usually being a precleaner followed by one or more filter elements. If two elements are employed, the first shall be called the primary element, and the second one shall be called the secondary element.

2.2.5 PRECLEANER—A device usually employing inertial or centrifugal means to remove a portion of the dust prior to reaching the filter element. Precleaners may be integral with the air cleaner housing, attached to the housing or separate. They may employ a scavenge airflow, an unloader valve, atmospheric discharge, or other means to remove or store separated dust.

2.2.6 TEST AIRFLOW—A measure of the quantity of air drawn through the air cleaner outlet per unit time. The flow rate shall be expressed in cubic meters per minute corrected to standard conditions.

2.2.7 RATED AIRFLOW—The flow rate specified by the user or supplier and may be used as the test airflow. When rated airflow is not known, for test purposes, it can be approximated using the formula described in Appendix B.

2.2.8 SCAVENGE AIRFLOW—A measure of the quantity of air employed to remove the collected dust from a precleaner, expressed as a percentage of the test airflow, typically 10%.

2.2.9 PRESSURE DROP—(ΔP_d) A measure, in mbar (kPa), of the difference in static pressure measured immediately upstream and downstream of the unit under test.

2.2.9.1 Refer to Appendix C for correcting the recorded pressure drop values to the standard conditions.

2.2.9.2 Static Pressure—Pressure in a duct at the observed airflow rate, measured by connecting a pressure gage to a hole or holes drilled in the wall of the duct.

In the tests specified herein a static pressure is measured by a manometer (usually a liquid manometer) as a negative pressure difference against the atmospheric pressure and in the formula is treated as a positive value.

2.2.9.3 Pressure Loss—Measure of the loss of energy caused by an air cleaner at the observed airflow rate, expressed as the pressure drop corrected for any difference in the dynamic head at the measuring points.

When pressure drop across an air cleaner has been measured ($P_2 - P_1$ in Table 5), any difference in the cross-sectional area of the ducts at the upstream and downstream pressure tapping points must be taken into account in determining the pressure loss across the air cleaner. The pressure loss across the cleaner is given by the equation:

$$\Delta P_l = \Delta P_d - \Delta P_c \qquad (Eq.1)$$

where: ΔP_d is measured pressure drop

$$\Delta P_c = \frac{\rho \cdot V_2^2}{2} - \frac{\rho \cdot V_1^2}{2} \qquad (Eq.2)$$

in which:

ρ is the density of the air

V_1 is the velocity of the air in the duct at the upstream pressure tapping point

V_2 is the velocity of the air in the duct at the downstream pressure tapping point

When the upstream pressure is equal to atmospheric and therefore only the static pressure in the downstream duct has been measured, the pressure loss across the air cleaner can be calculated from the dynamic head required to accelerate the air from rest to its velocity in the downstream duct. The pressure loss across the cleaner is then given by the equation:

$$\Delta P_l = P_r - P_{dynamic} \qquad (Eq.3)$$
$$= P_2 - \frac{\rho \cdot V_2^2}{2}$$

where:

$P_r = P_2$ is the restriction/static pressure at the downstream pressure tapping point

ρ is the density of the air

V_2 is the velocity of the air at the downstream pressure tapping point

TABLE 5—PRESSURE DROP

Term	Air Cleaner Drawing Air from the Atmostphere	Air Cleaner Drawing Air Through an Inlet Duct
Static pressure upstream of air cleaner	----	P_1
Restriction = Static pressure downstream of air cleaner	$\Delta P_r = P_2$	$\Delta P_r = P_2$
Pressure drop	----	$\Delta P_d = \Delta P_r - P_1 = P_2 - P_1$
Pressure loss	$\Delta P_l = \Delta P_r - P_{dynamic}$ $= P_2 - \frac{\rho \cdot V_2^2}{2}$	$\Delta P_l = \Delta P_d - \Delta P_c$ $= (P_2 - P_1) - \frac{\rho}{2}(V_2^2 - V_1^2)$

2.2.10 RESTRICTION—(ΔP_r) A measure, in mbar (kPa) gage, of the static pressure measured immediately downstream of the unit under test.

2.2.11 ASSEMBLY RESTRICTION/PRESSURE DROP—The airflow pressure resistance across the complete assembly (test shroud and/or housing and element).

2.2.12 TARE RESTRICTION/PRESSURE DROP—The airflow pressure resistance across the test shroud and/or housing only (no element).

2.2.13 ELEMENT RESTRICTION/PRESSURE DROP—The assembly restriction/pressure drop minus the tare restriction/pressure drop.

2.2.14 TERMINATING RESTRICTION/PRESSURE DROP—The airflow pressure resistance across the unit under test at which the capacity is measured.

2.2.15 ABSOLUTE FILTER—The filter downstream of the unit under test to retain the contaminant passed by the unit under test.

2.2.16 EFFICIENCY—The ability of the air cleaner or the unit under test to remove contaminant. This will be expressed by the following formulas:

2.2.16.1 *Automobile Air Cleaners*

$$\text{Efficiency, \%} = \frac{\text{Increase in weight of unit under test}}{\text{Increase in weight of unit under test + Increase in weight of the absolute filter}} \times 100 \qquad (Eq.4)$$

2.2.16.2 *Industrial Air Cleaners*

$$\text{Efficiency, \%} = 1 - \frac{\text{Increase in weight of absolute filter}}{\text{Weight of dust fed}} \times 100 \qquad (Eq.5)$$

2.2.17 CAPACITY—The quantity of contaminant removed and defined as follows:

2.2.17.1 *Automobile Air Cleaners*—The total weight GAIN in grams of the unit under test at the terminating restriction or pressure drop.

2.2.17.2 *Industrial Air Cleaners*—The total weight in grams of the test dust fed to the air cleaner to produce a specified terminating restriction or pressure drop.

2.2.17.3 *Precleaners (Collection Type)*—The total weight in grams of test dust collected when the dust collected reaches the recommended servicing level.

2.2.18 STANDARD CONDITION—All airflow measurements are to be corrected to a standard condition of 20 °C at 1013 mbar (101.3 kPa).

2.2.19 OIL CARRYOVER—The appearance of oil at the air cleaner outlet.

2.3 Units of Measurement—For the purpose of this test code, all airflow measurements will be expressed in cubic meters per minute (m³/min). All pressure measurements will be expressed in bar followed by kilopascals (kPa) and temperature measurements will be expressed in degrees Celsius.

2.3.1 USEFUL EQUIVALENT CONVERSIONS

1 kPa = 4.019 in H_2O
1 kPa = 0.2953 in Hg
1 kPa = 0.1450 psi
1 m³/h = 0.58858 cfm
1 m/min = 3.2808 ft/min
1 ℓ = 61.024 in³
1 in H_2O = 0.2488 kPa
1 in Hg = 3.3865 kPa
1 psig = 6.895 kPa
1 cfm = 1.699 m³/h
1 ft/min = 0.3048 m/min
1 cfm = 0.028 316 8S m³/min
1 bar = 100 kPa

3. Automotive Air Cleaner Test Procedures

3.1 General—The automotive air cleaner test code has been established to cover the dry type air cleaners generally used in automobiles. The air cleaners used in truck, off-highway, construction, and some industrial applications are classified as industrial types and are covered in Section 4. The oil bath air cleaner test code is covered in Section 5.

3.2 Test Equipment—To determine the resistance to airflow, contaminant holding capacity, contaminant removal characteristics, sealing characteristics, and rupture/collapse characteristics, use equipment described in Figures 1 to 6 (test setups).

3.2.1 Use a dust metering device, which when used with the dust injector with sufficient compressed air supply, is capable of metering dust over the range of delivery rates required. This dust feed system shall not change the primary particle size distribution of the contaminant. The average delivery rate shall be within 5% of the desired rate and the deviation in delivery rate from the average shall be no more than 5%. The recommended minimum supply pressure is 1 bar (100 kPa).

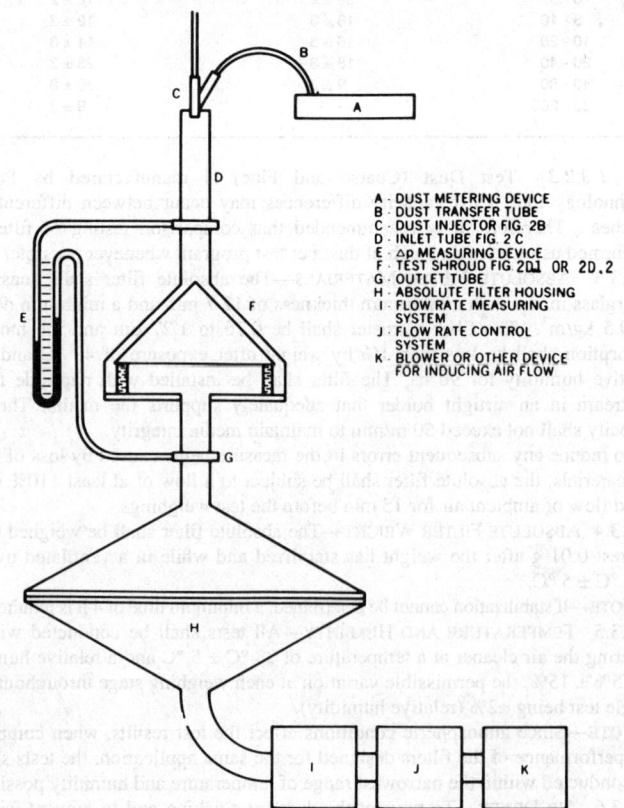

A: DUST METERING DEVICE
B: DUST TRANSFER TUBE
C: DUST INJECTOR FIG. 2B
D: INLET TUBE FIG. 2C
E: Δp MEASURING DEVICE
F: TEST SHROUD FIG 2D.1 OR 2D.2
G: OUTLET TUBE
H: ABSOLUTE FILTER HOUSING
I: FLOW RATE MEASURING SYSTEM
J: FLOW RATE CONTROL SYSTEM
K: BLOWER OR OTHER DEVICE FOR INDUCING AIR FLOW

FIGURE 1—EFFICIENCY/CAPACITY AIR FILTER ELEMENT TEST SETUP

3.2.1.1 Validation of the Dust Feed System
3.2.1.1.1 Charge the dust metering device with a preweighed amount of dust.
3.2.1.1.2 Simultaneously start the dust feed system and timer.
3.2.1.1.3 At 5 min intervals, determine the weight of the dust dispensed. Continue the weight determinations of the dust increments for 30 min.
3.2.1.1.4 Determine the average delivery rate and the maximum deviation in delivery rate.

3.2.1.2 Use a dust transfer tube sized to maintain a minimum velocity of 914 m/min for dust suspension.

3.2.1.3 Use the dust injector described in Figure 8. The specified injector has been shown satisfactorily to feed test dust at rates up to 40 g/min. The injector is operated with a minimum supply pressure to the nozzle of 1 bar (100 kPa). It should be noted that the design of the system feeding test dust to the injector may affect this maximum rate of dust feed. Where dust feed rates greater than 40 g/min are required, the dust injector shown in Figure 16 shall be used.

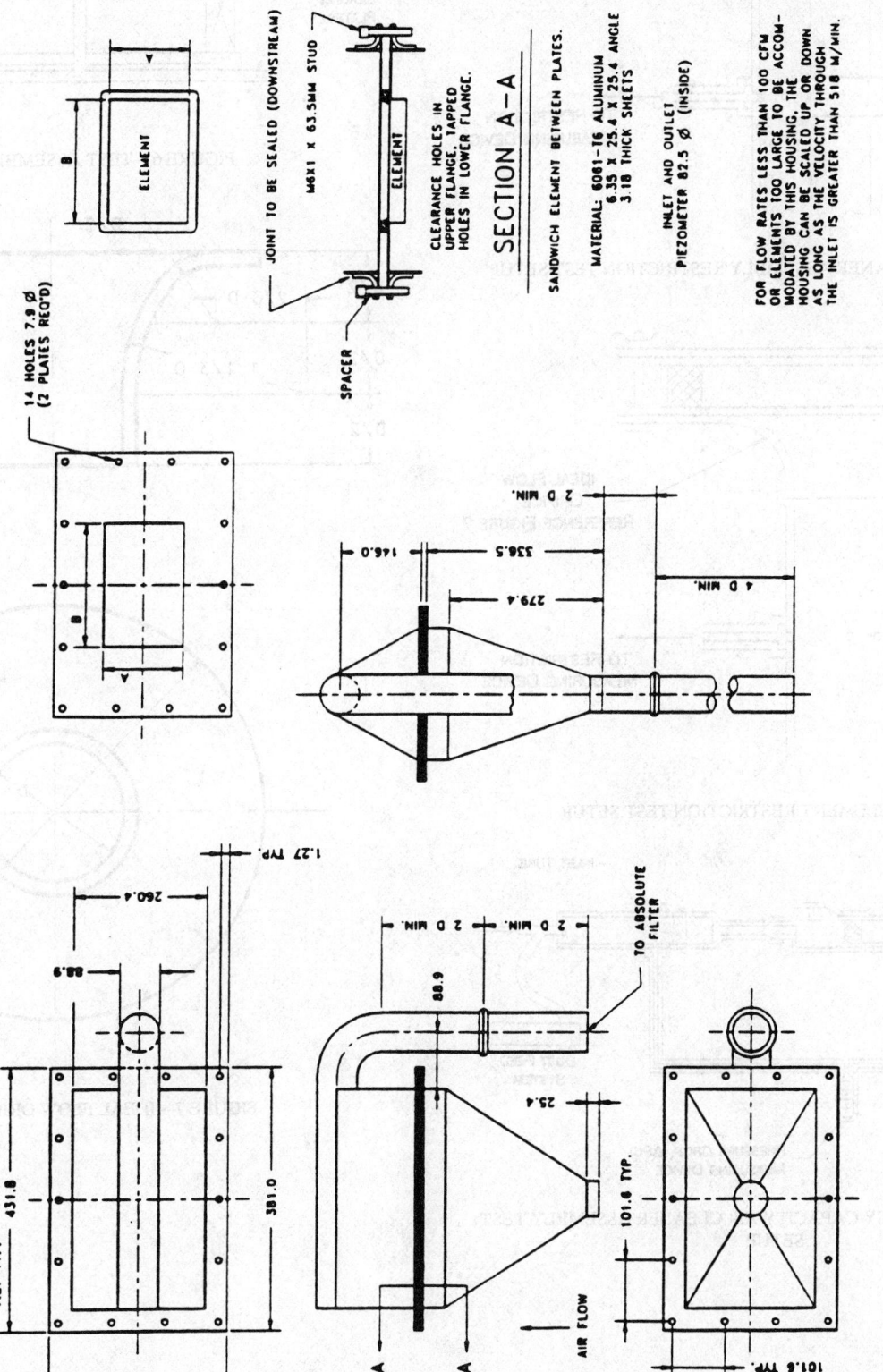

FIGURE 2—PANEL FILTER UNIVERSAL TEST HOUSING

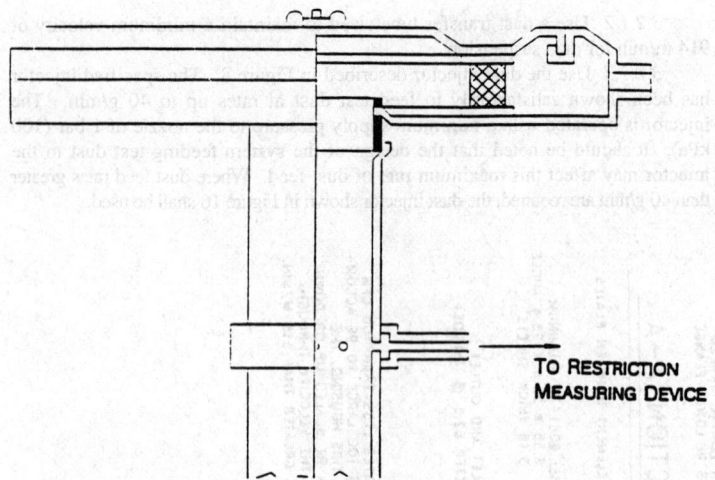

FIGURE 3—AIR CLEANER ASSEMBLY RESTRICTION TEST SETUP

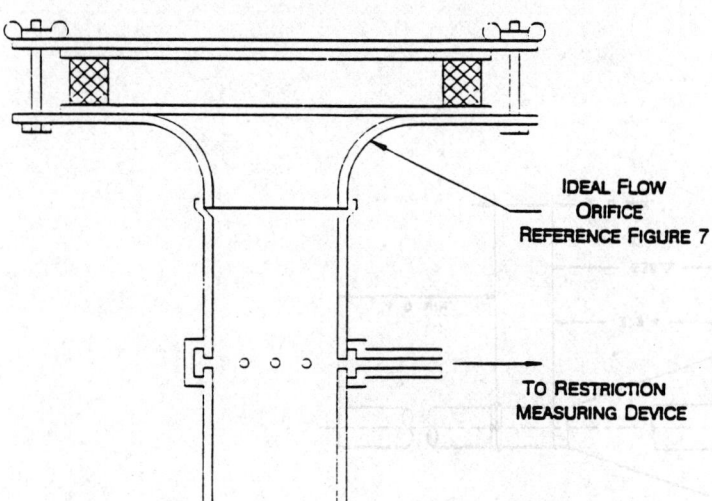

FIGURE 4—ELEMENT RESTRICTION TEST SETUP

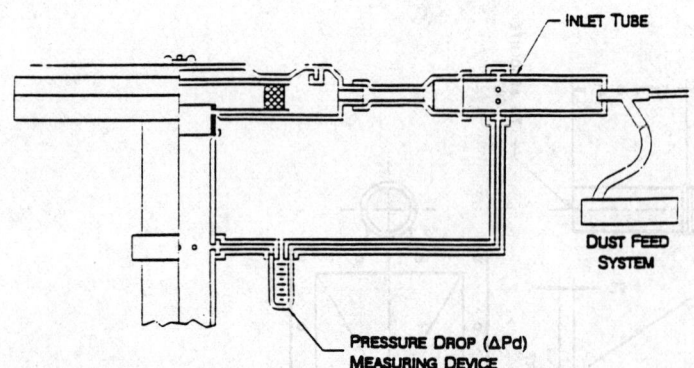

FIGURE 5—EFFICIENCY CAPACITY AIR CLEANER ASSEMBLY TEST SETUP

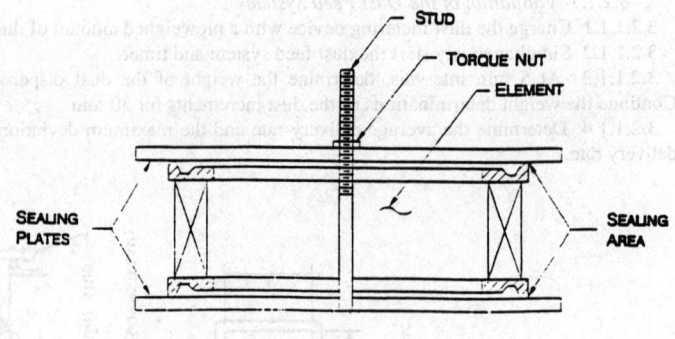

FIGURE 6—TEST ASSEMBLY

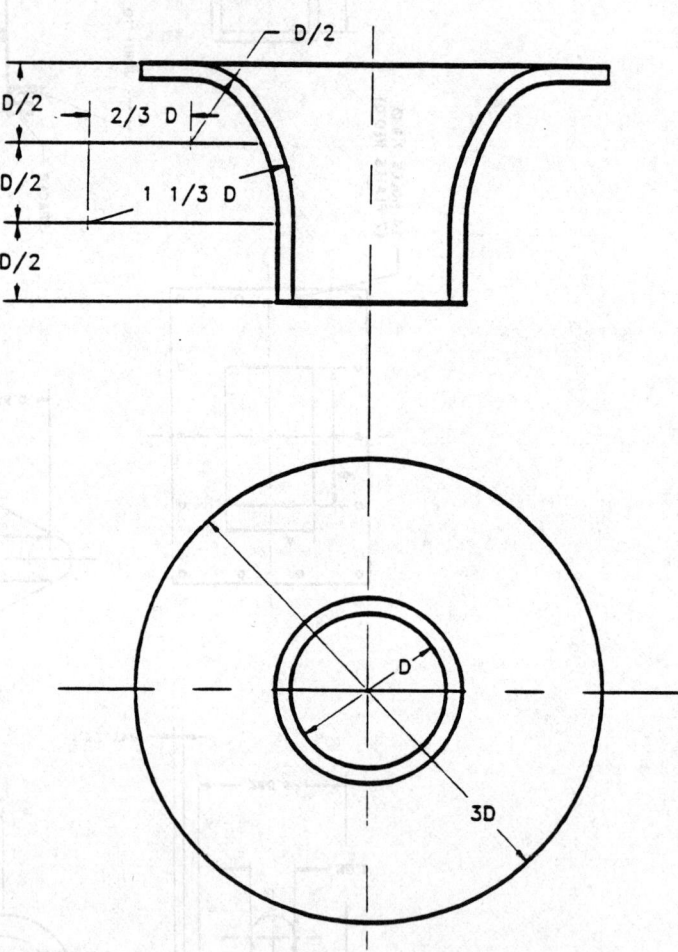

FIGURE 7—IDEAL FLOW ORIFICE

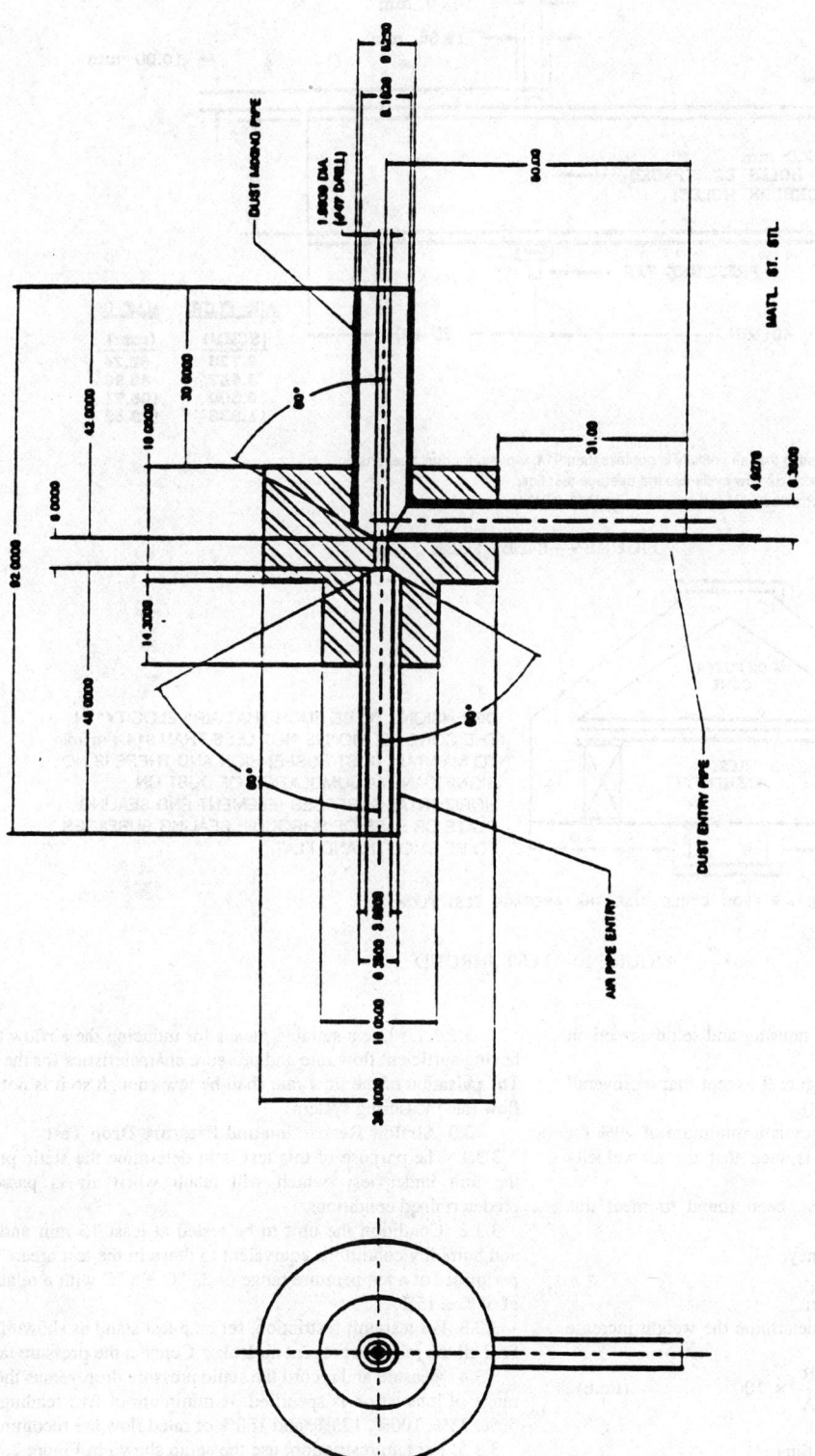

FIGURE 8—DUST INJECTOR FOR LIGHT-DUTY OR AUTOMOTIVE USE ONLY

DIMENSIONS in MILLIMETERS

3.2.1.4 Use an inlet tube conforming to Figure 9.

3.2.1.5 Use a manometer or other differential pressure measuring device with the accuracy described in 1.3.2.

3.2.1.6 For air filter element testing, use a test shroud conforming to Figure 2 or 10.

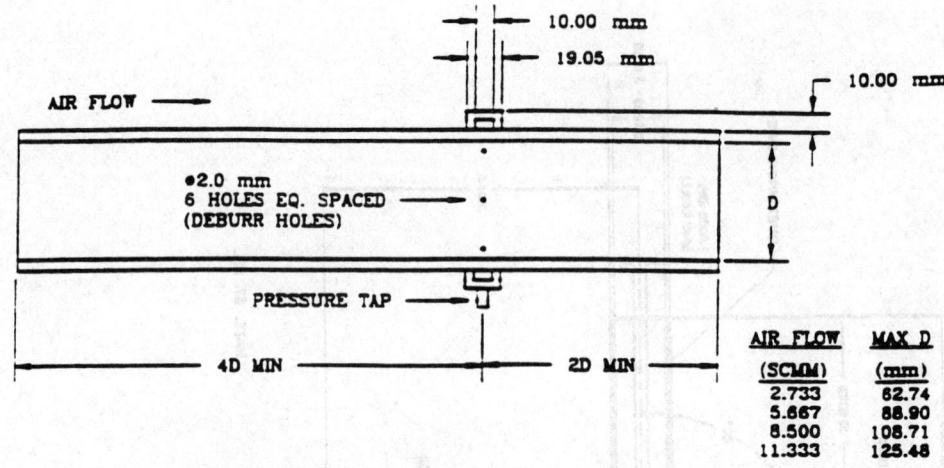

FIGURE 9—INLET TUBE

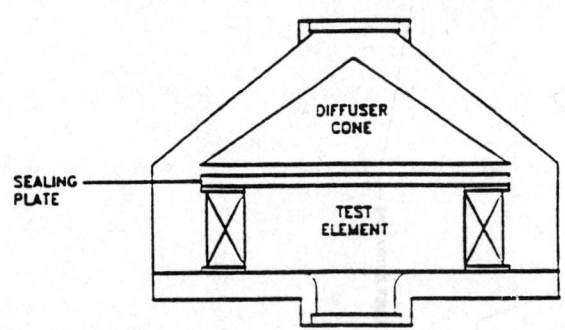

*FOR VARIABLE AIR FLOW CYCLE, USE THE AVERAGE TEST FLOW.

FIGURE 10—TEST SHROUD

3.2.1.6.1 For air cleaner assembly testing use a housing and setup agreed on by the supplier and user (Reference Figure 5).

3.2.1.7 Use an outlet tube conforming to Figure 9 except that the overall length is 4 diameters min (air velocity is not critical).

3.2.1.8 Use an absolute filter whose efficiency is a minimum of 99% for the contaminant presented to it and whose size is such that the air velocity through it does not exceed 50 m/min.

NOTE—Refer to 1.3 for materials that have been found to meet this validation requirement.

3.2.1.8.1 Validation of the absolute filter efficiency.
3.2.1.8.1.1 Weigh the absolute filters per 1.3.4.
3.2.1.8.1.2 Arrange two absolute filters in tandem.
3.2.1.8.1.3 Perform a filter efficiency test and determine the weight increase of each absolute filter.

$$\text{Absolute filter efficiency (\%)} = 1 - \frac{B}{A} \times 100 \quad (Eq.6)$$

where:

A = Weight increase of upstream absolute filter
B = Weight increase of downstream absolute filter

3.2.1.9 Use an airflow rate measuring system having the accuracy described in 1.3.1.

3.2.1.9.1 Validate the airflow rate measuring system according to ASME Fluid Meters, Sixth Edition. The airflow meter shall be an accepted design, such as a calibrated orifice, traceable, once removed to a standard calibrating source. Corrections shall be made for variations in absolute pressure and temperature at the meter inlet and the airflow rate shall be expressed in cubic meters/min corrected at 20 °C at 1.103 bar (101.3 kPa).

3.2.1.10 Use an airflow rate control system capable of maintaining the indicated flow rate within 1% of the selected value during both the steady-state and variable airflow operation.

3.2.1.11 Use a suitable means for inducing the airflow through the system having sufficient flow rate and pressure characteristics for the filters to be tested. The pulsation of the flow rate shall be low enough so it is not measurable by the flow rate measuring system.

3.3 Airflow Restriction and Pressure Drop Test

3.3.1 The purpose of this test is to determine the static pressure drop across the unit under test, which will result when air is passed through under predetermined conditions.

3.3.2 Condition the unit to be tested at least 15 min under the temperature and humidity conditions equivalent to those in the test area. The tests should be performed at a temperature range of 23 °C ± 5 °C with a relative humidity range of 55% ± 15%.

3.3.3 For test unit restriction, set up a test stand as shown in Figure 2, 3, or 4. Seal all the joints to prevent air leaks. Connect the pressure taps.

3.3.4 Measure and record the static pressure drop versus the flow rate over the range of interest or as specified. A minimum of five readings at approximately 50%, 75%, 100%, 125%, and 150% of rated flow are recommended.

3.3.5 For tare restriction, use the setup shown in Figure 2, 3, or 4 without the air cleaner. Repeat 3.3.4.

NOTE—It is possible, due to inertial effects, for this value to be greater than the assembly restriction.

3.3.6 Determine the test unit net restriction by subtracting the values obtained in 3.3.5 from the values obtained in 3.3.4.

3.3.7 Record the data.

3.4 Efficiency Test

3.4.1 The purpose of this test is to determine the retention capabilities of the unit under test. This test can be conducted with either constant or variable airflow and with coarse dust or fine dust contaminant. If desired, the efficiency tests can be performed concurrently with 3.5.

3.4.2 Three types of efficiency tests can be performed. These are:

3.4.2.1 Full life efficiency determined when terminating pressure drop (ΔP_d) is reached.

3.4.2.2 Incremental efficiency is determined at 10, 25, and 50% of the terminating pressure drop (ΔP_d) minus the initial pressure drop (ΔP_d).

3.4.2.3 The initial efficiency is determined after the addition of 20 g of contaminant or the number of grams numerically equivalent to 0.0973% of the test airflow in SCMH, whichever is greater.

3.4.3 The determination of efficiency at constant test airflow can be performed at the rated airflow or any percentage thereof as agreed on by the user and supplier.

3.4.3.1 Based on the test flow, calculate the dust contaminant feed rate using a dust concentration of 1.0 g/m^3 of air: in special cases (e.g., small filters) 0.25 and 0.50 g/m^3 may be allowed.

3.4.3.2 Condition the unit under test according to 3.3.2 and then weigh, in grams, to within 1% of the actual value.

3.4.3.3 Weigh the absolute filter per 1.3.4.

3.4.3.4 Set up the test stand as shown in Figure 1 or Figure 11 for air filter elements. Seal all of the joints to prevent air leaks.

NOTE—The panel filter elements can be tested with a setup similar to Figure 1 using the panel filter test housing from Figure 2.

3.4.3.5 Record the temperature and relative humidity.

3.4.3.6 Weigh out the specified test dust equal to 125% of the estimated capacity of the test unit. Record the weight to the nearest 0.1 g and place in the dust metering device.

3.4.3.7 Weigh and record the dust feed system with dust, in grams, to within 1% of the actual value.

3.4.3.8 Start the airflow through the stand and stabilize at the test flow. Record the pressure drop (ΔP_d).

3.4.3.9 Set the flow rate to that at which the test dust is to be added if this is different from the test flow rate used in 3.4.3.8.

3.4.3.10 Start the dust feeder and adjust the feed rate to inject dust at the concentration calculated in 3.4.3.1.

3.4.3.11 At prescribed time intervals (a minimum of five points is recommended), record the pressure drop (ΔP_d) at test flow and elapsed test time.

3.4.3.12 Continue the test until the terminating pressure drop is reached.

3.4.3.13 Repeat 3.4.3.5.

3.4.3.14 Repeat 3.4.3.7 and determine the difference in weight. This amount is the dust fed.

3.4.3.15 Carefully remove the unit under test without losing any dust. Note any evidence of seal leakage or any unusual conditions. Weigh, in grams, to within 1% of the actual value. The increase in weight of the unit under test is this weight minus the weight determined in 3.4.3.2.

3.4.3.16 Brush any observed dust on the downstream side of the test unit onto the absolute filter. Carefully remove the absolute filter. Repeat 3.4.3.3 and determine the difference in weight. This is the increase in weight of the absolute filter.

3.4.3.17 Calculate the material balance of the test dust. This value must be within the range of 0.98 to 1.02 to be a valid test.

$$\text{Material balance of test dust} = \frac{\text{Increased weight of absolute filter} + \text{Increased weight of unit under test}}{\text{Total weight of dust fed}}$$

(Eq.7)

3.4.3.18 Calculate the efficiency by the following method:

$$\text{Efficiency, \%} = \frac{\text{Increase in weight of unit under test}}{\text{Increase in weight of unit under test} + \text{Increase in weight of the absolute filter}} \times 100$$

(Eq.8)

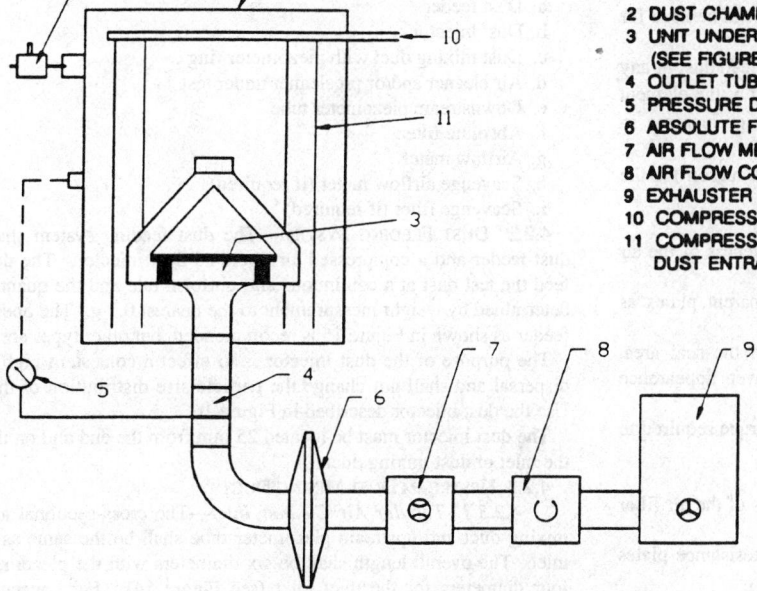

1 DUST INJECTOR
2 DUST CHAMBER
3 UNIT UNDER TEST WITH DIFFUSER CONE (SEE FIGURE 10)
4 OUTLET TUBE (SEE FIGURE 9)
5 PRESSURE DROP MEASURING DEVICE
6 ABSOLUTE FILTER
7 AIR FLOW METER
8 AIR FLOW CONTROL
9 EXHAUSTER
10 COMPRESSED AIR FEED
11 COMPRESSED AIR FLEXIBLE TUBES (FOR MAINTAINING DUST ENTRAINMENT)

NOTE: IN THIS FIGURE A SINGLE AIR CLEANER ELEMENT IS INSTALLED.

FIGURE 11—EFFICIENCY/CAPACITY TEST SETUP USING A DUST CHAMBER

3.4.4 Variable airflow efficiency can be determined by using a variable airflow cycle similar to Figure 12. In the case of large air cleaners (e.g., flow rate > 5 m^3/min), the duration of every partial flow section may be 5 min instead of 1 min.

3.4.4.1 Based on the average test flow for the cycle being used, calculate the dust feed rate as in 3.4.3.1. The dust feed rate should remain constant.

3.4.4.2 All pressure drop determinations are to be made at the maximum airflow.

3.4.4.3 Perform the test as in 3.4.3 using variable airflow in place of the constant airflow; however, with the following changes:

After the end of each cycle the pressure drop shall be determined at the maximum flow. The efficiency shall be determined at least after 3 cycles if the duration of partial flow section is 1 min, and after every cycle if the duration of partial flow section is 5 min, and after the end of test.

3.5 Capacity Test

3.5.1 The purpose of this test is to determine the total weight gain of the unit under test at the terminating pressure drop. This test can be conducted with either constant or variable airflow and with coarse or fine test dust contaminant. If desired, capacity determination can be performed concurrently with 3.4.

3.5.2 Perform the test as described in 3.4.3.1 through 3.4.3.15.

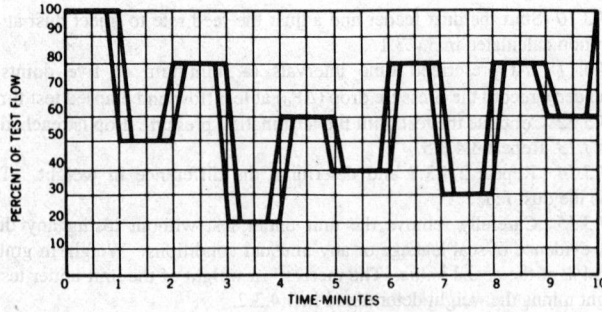

FIGURE 12—TYPICAL VARIABLE FLOW CYCLE AVERAGE FLOW 60%

3.5.3 Assuming a constant ratio of elapsed time versus weight gain of the test unit, plot curve of restriction versus weight gain. Refer to 3.4.3.11 for restriction and time interval data. Use the following formula to determine the weight gain values:

$$\text{Weight gain at end of each time interval} = \frac{\text{Total time to end of interval}}{\text{Total time to end of test}} \times \text{Total weight gain of unit under test}$$

(Eq.9)

3.5.4 The capacity of the test unit is the point at which the curve reaches the terminating restriction. This restriction does not include the restriction added by the dust metering device and test shroud.

3.6 Flow Pressure Collapse Test

3.6.1 The purpose of this test is to determine the ability of an air filter element to withstand a specified differential pressure and/or the differential pressure at which collapse occurs.

3.6.2 Set up the test stand to perform the basic dust capacity test as per Figure 1 or 5, or Figure 1 with modification for panel filter test housing (Figure 2). An element from a prior capacity or efficiency test or a new element can be used for this test.

3.6.3 Increase the airflow through the stand and, if necessary, feed dust at any convenient rate until the specified pressure drop (ΔP_d) is reached or the element collapse is evidenced by any decrease in pressure drop (ΔP_d) or increase in airflow.

3.6.4 Record the maximum pressure drop (ΔP_d) attained, the reason for the terminating test, and the condition of the element after the test.

3.7 Seal Effectiveness Test

3.7.1 The purpose of this test is to determine the seal effectiveness of the air filter elements.

3.7.1.1 Locate the filter element between the two transparent plates as shown in Figure 6 and torque to specified requirements.

3.7.1.2 Check visually for any irregularities or voids in the seal area. Proper sealing can normally be identified by a dark or wet appearance continuous through the sealing contact area.

3.7.1.3 Report and comment on the seal quality and the torque required to effect a seal.

3.8 Temperature Extreme Test

3.8.1 The purpose of this test is to determine the effectiveness of the air filter to withstand temperature extremes.

3.8.2 Locate the filter element between the two temperature resistance plates as shown in Figure 6 and torque to specified requirement.

3.8.3 Subject the assembly to the specified hot and cold cycle. In absence of the specific values, the following is recommended:

24 h at 121 °C ± 3 °C
24 h at -40 °C ± 3 °C
24 h at 121 °C ± 3 °C
24 h at -40 °C ± 3 °C

Allow the unit under test to adjust to room temperature between cycles.

3.8.4 After the test cycle, allow to adjust to room temperature. Remove the plates and inspect the element for conditions that will impair performance. If necessary, repeat 3.4.

3.8.5 Report all conditions of the test and visual observations.

3.9 Presentation of Data

3.9.1 The purpose is to standardize a test data reporting form. Refer to Figure 13.

4. Industrial Air Cleaner Test Procedures

4.1 General

4.1.1 This section of the air cleaner test code has been established to cover dry type air cleaners generally used in heavy trucks, construction equipment, agricultural tractors, and industrial applications. Air cleaners used in automobiles are classified as automotive and are covered in Section 3. The oil bath air cleaner test code is covered in Section 5.

4.1.2 The performance tests shall be performed on a precleaner or a complete air cleaner including precleaner, primary element, and secondary element, if normally provided. The tests shall consist of an airflow restriction/pressure drop test, an initial efficiency test, and a combined efficiency and dust capacity test.

4.1.3 TEST DUST AND CONCENTRATION—It is difficult, if not impossible, to select a test dust size distribution and concentration which will be representative of all service conditions. Therefore, based on primarily practical considerations, the different types of air cleaners have been classified as to their most probable service conditions, and the test dust grade and concentration selected accordingly. See Table 6.

TABLE 6—INDUSTRIAL AIR CLEANER RECOMMENDED DUST SELECTION AND CONCENTRATION

Air Cleaner Type	Test Dust	Concentration
Single Stage	Coarse or Fine	1 g/m³
Multistage	Coarse or Fine	2 g/m³
Precleaner	Coarse	2 g/m³

4.2 Test Equipment

4.2.1 TEST DUCT—The test duct shall consist of the following major components and be arranged as shown in Figure 14.
 a. Dust feeder
 b. Dust injector
 c. Dust mixing duct with piezometer ring
 d. Air cleaner and/or precleaner under test
 e. Downstream piezometer tube
 f. Absolute filter
 g. Airflow meter
 h. Scavenge airflow meter (if required)
 i. Scavenge filter (if required)

4.2.2 DUST FEEDING SYSTEM—The dust feeding system shall consist of a dust feeder and a compressed air operated dust injector. The dust feeder shall feed the test dust at a continuous and uniform rate and the quantity fed shall be determined by weight measurement to the nearest 0.1 g. The open tray type dust feeder as shown in Figure 15 is recommended, but other types are optional.

The purpose of the dust injector is to effect a consistent high degree of dust dispersal and shall not change the particle size distribution of the contaminant. Use the dust injector described in Figure 16.

The dust injector must be located 25 mm from the end and on the centerline of the inlet or dust mixing duct.

4.2.3 UPSTREAM DUST MIXING DUCT

4.2.3.1 Tubular Air Cleaner Inlet—The cross-sectional area of the dust mixing duct and upstream piezometer tube shall be the same as the air cleaner inlet. The overall length shall be six diameters with the piezometer ring placed four diameters for the duct inlet (see Figure 14). For concentrated inlets of noncircular area, use a transition to a circular area and treat as a tubular inlet.

4.2.3.2 Rectangular or Open Face Inlet—Same as 4.2.3.1, except the overall length and placement of the piezometer shall be 24 and 16 times the hydraulic radius, respectively. (Hydraulic radius = area ÷ perimeter.)

4.2.3.3 Peripheral Air Inlet or Stack Type Precleaners—The peripheral air inlet or stack type precleaners shall be tested in a chamber which ensures the even distribution and delivery of test dust to the inlet of the unit. Care should be taken in the design of the chamber to ensure that all the test dust is fed to the unit under test. If dust settling occurs, then compressed air jets may be used to re-entrain the test dust. Typical examples of chambers are shown in Figure 17.

When using compressed air for agitating dust, care should be taken not to eject dust out of the chamber. To ensure that no dust is ejected, a negative pressure should be maintained between the chamber interior and the atmosphere.

TEST UNIT DESCRIPTION _____	DATE _____
MANUFACTURED _____ TEST CONDUCTED BY _____	EQUIPMENT NO. _____

TEST CONDITIONS

RATED AIR FLOW _____ scmh   TEST AIR FLOW _____ scmh   STEADY _____   VARIABLE _____

TEST DUST: Fine ____ Coarse ____ BATCH NO. _____ TERMINATING RESTRICTION OR PRESSURE DROP _____

RELATIVE HUMIDITY BEFORE TEST _____ % AFTER TEST _____ %

BAROMETRIC PRESSURE BEFORE TEST _____ kPa   AFTER TEST _____ kPa

TEMPERATURE BEFORE TEST _____ °C   AFTER TEST _____ °C

TEST RESULTS

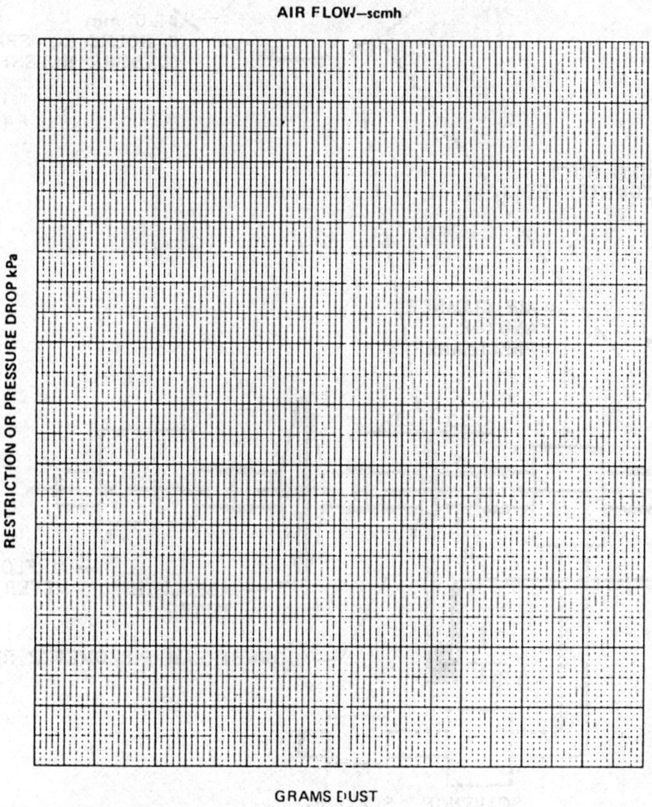

AIR FLOW—scmh

RESTRICTION OR PRESSURE DROP kPa

GRAMS DUST

GENERAL COMMENTS: _____

FULL LIFE EFFICIENCY _____ %

INITIAL EFFICIENCY
_____ % _____ g

INCREMENTAL EFFICIENCY AT PER CENT OF TERMINATION
@ 10% Eff. _____ %
@ 25% Eff. _____ %
@ 50% Eff. _____ %

INITIAL RESTRICTION
_____ kPa @ _____ scmh
(See Graph)

CAPACITY
g _____ to _____ kPa
(See Graph)

MATERIAL BALANCE FOR VALIDATION OF TEST _____

FLOW PRESSURE COLLAPSE REMARKS _____

SEAL TEST REMARKS _____

TEMPERATURE EXTREME TEST
Hours each _____
Hot _____ °C   Cold _____ °C
Times Repeated _____
Remarks _____

FIGURE 13—TEST DATA REPORTING FORM
SECTION 3—AUTOMOTIVE AIR CLEANERS

4.2.4 DOWNSTREAM PIEZOMETER—The inside diameter of the downstream piezometer shall be the same as the air cleaner outlet tube. (See Figure 14.)

4.2.5 ABSOLUTE FILTER—The absolute filter shall contain the material specified in 1.3.3.

4.2.6 AIRFLOW METER—The airflow meter shall be calibrated annually to assure a known accuracy of ±2%, by using a flow meter conforming to the construction set forth in ASME Fluid Meters, Sixth Edition.

4.3 Airflow Restriction and Pressure Drop Test

4.3.1 The purpose of this test is to determine restriction/pressure drop/pressure loss across the unit under test which results when air is passed through under predetermined conditions. Airflow restriction or pressure drop (see 2.2.9) is measured with a clean filter element, or elements, at five equally spaced airflows between 50 and 150% of rated airflow or as agreed upon between the user and the manufacturer.

4.3.2 AIRFLOW RESTRICTION (ΔP_r)

4.3.2.1 Set up the air cleaner per Figure 18.

4.3.2.2 Record the downstream static gage pressure at 50%, 75%, 100%, 125%, and 150% of rated airflow.

4.3.2.3 Record the ambient temperature, pressure, and relative humidity.

4.3.2.4 Correct the recorded restriction values to standard conditions per Appendix C and plot as shown in

Figure 19 indicating restriction on ordinate.

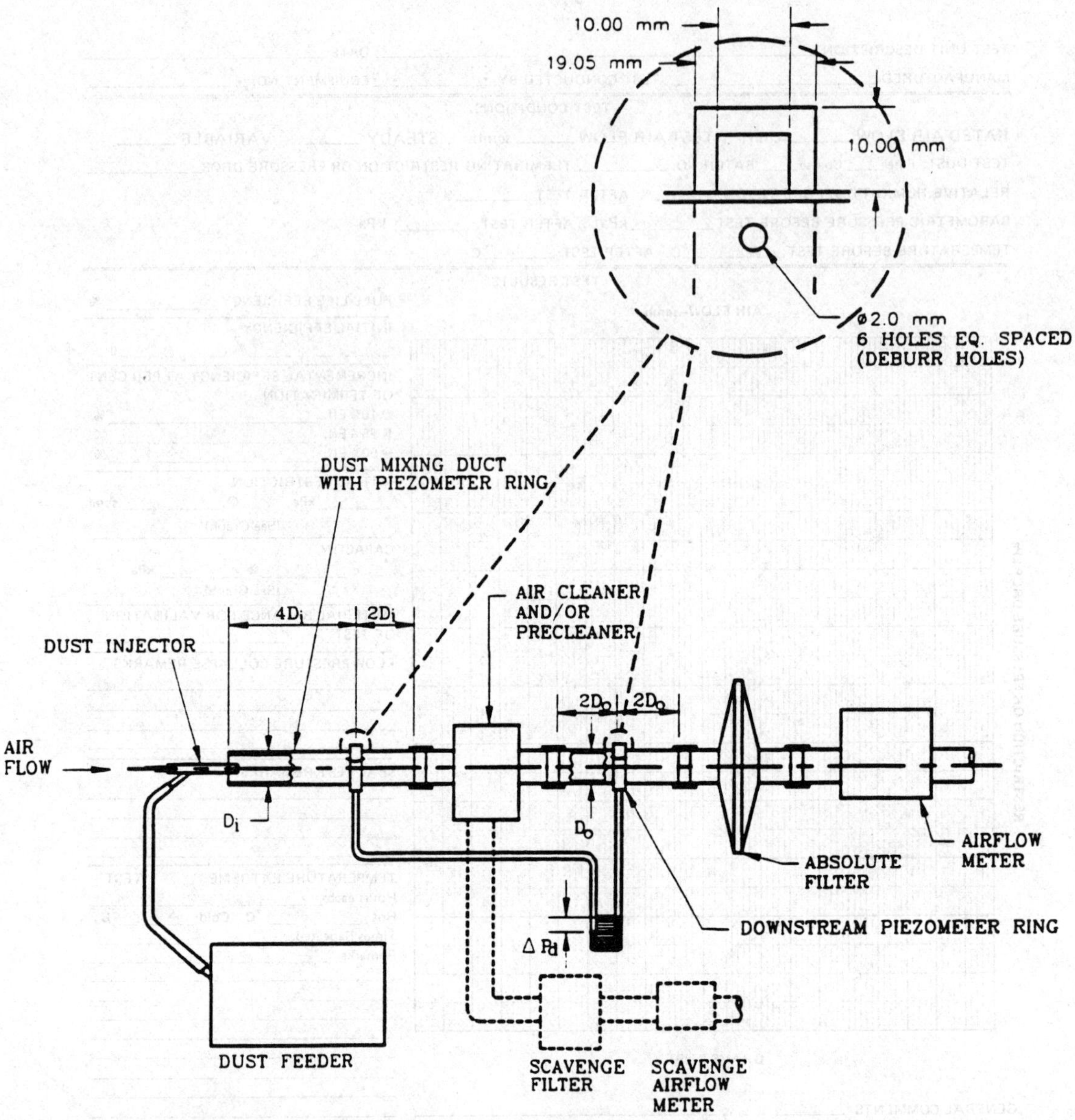

FIGURE 14—TUBULAR INLET AIR CLEANER TEST DUCT ASSEMBLY

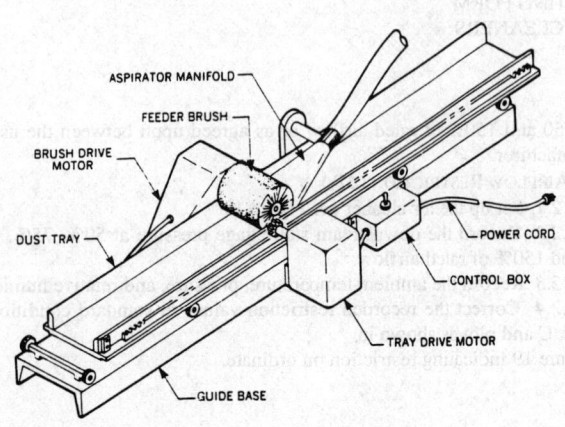

FIGURE 15—DUST FEEDER

4.3.3 PRESSURE DROP (ΔP_d)

4.3.3.1 Set up the air cleaner per Figure 20.

4.3.3.2 Record the difference in static pressure measured upstream and downstream of the air cleaner at 50%, 75%, 100%, 125%, and 150% of the rated airflow.

4.3.3.3 Record the ambient temperature, pressure, and relative humidity.

4.3.3.4 Correct the recorded pressure drop values to standard conditions, per Appendix C, and plot as shown in Figure 19 indicating the pressure drop on ordinate.

4.4 Initial Efficiency

4.4.1 The air cleaners covered within the scope of this test code exhibit a collection efficiency that is dependent on airflow, increases with dust load on the filter element, and decreases with dust particle size.

Two efficiency determinations are thus made: an initial efficiency test, and an accumulative efficiency conducted during the dust capacity test.

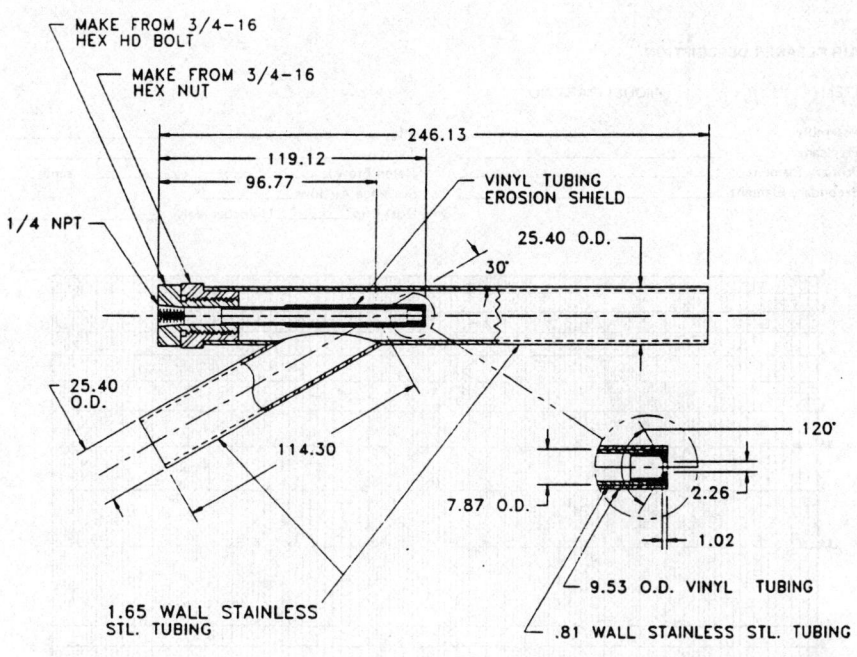

FIGURE 16—DUST INJECTOR FOR HEAVY-DUTY OR INDUSTRIAL USE ONLY

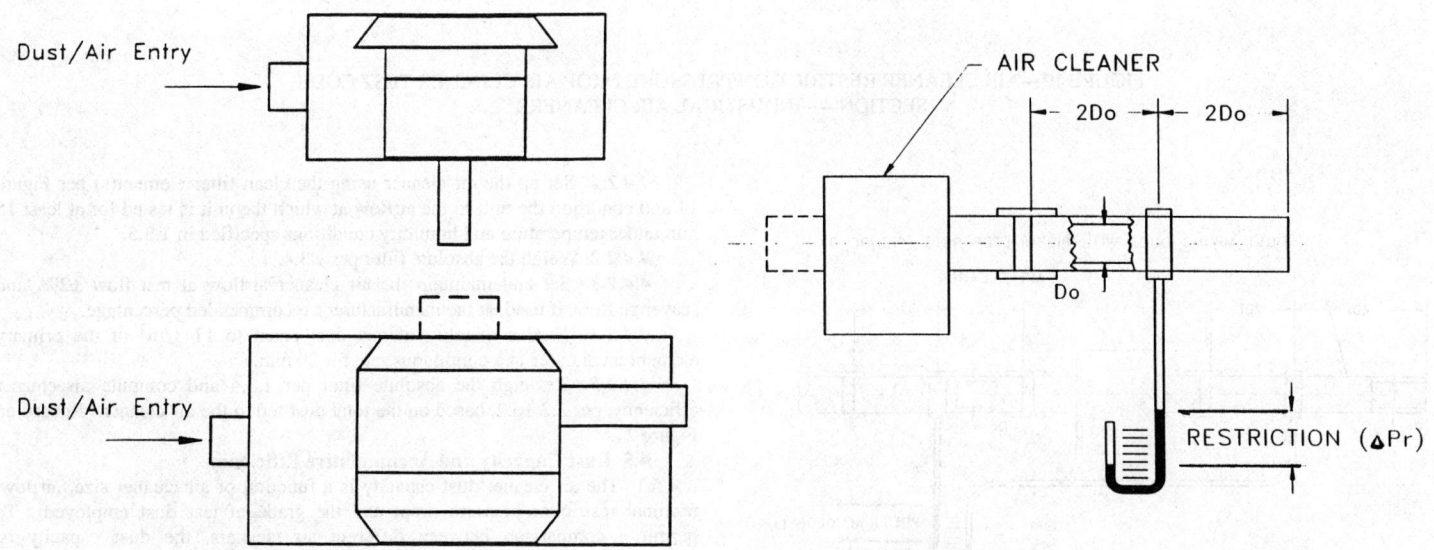

FIGURE 17—NONTUBULAR INLET AIR CLEANER TEST CHAMBER ASSEMBLIES

FIGURE 18—RESTRICTION TEST SETUP

AIR CLEANER DESCRIPTION

ITEM	MODEL/PART NO.
Assembly	_____
Precleaner	_____
Primary Element	_____
Secondary Element	_____

Mfg. _____
Type _____
Rated Flow _____ scmh
Scavenge Airflow _____ %
Dust Cup _____ Unloader Valve _____

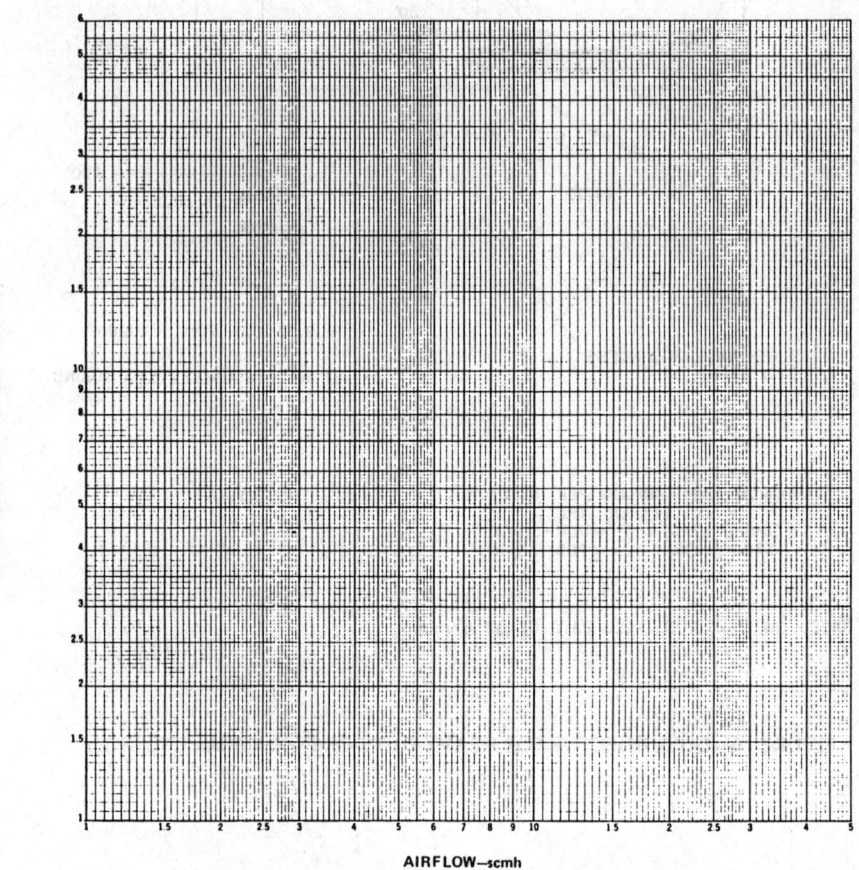

FIGURE 19—AIR CLEANER RESTRICTION/PRESSURE DROP AIR CLEANER TEST CODE
SECTION 4—INDUSTRIAL AIR CLEANERS

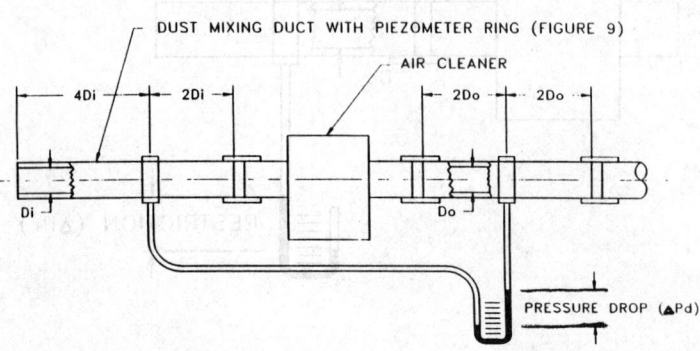

FIGURE 20—PRESSURE DROP TEST SETUP

4.4.2 PROCEDURE

4.4.2.1 Set up the air cleaner using the clean filter element(s) per Figure 14 and condition the unit to the airflow at which the unit is tested for at least 15 min under temperature and humidity conditions specified in 1.3.5.

4.4.2.2 Weigh the absolute filter per 1.3.4.

4.4.2.3 Set and maintain the air cleaner airflow at test flow ±2%, and scavenge flow, if used, at the manufacturer's recommended percentage.

4.4.2.4 Feed a quantity of test dust equal to 11 g/m^2 of the primary element media area at a continuous rate for 30 min.

4.4.2.5 Reweigh the absolute filter per 1.3.4 and compute air cleaner efficiency, per 2.2.16.2, based on the total dust fed to the air cleaner. Record on Figure 21.

4.5 Dust Capacity and Accumulative Efficiency

4.5.1 The air cleaner dust capacity is a function of air cleaner size, airflow, terminal restriction/pressure drop, and the grade of test dust employed. To permit a comparison between different air cleaners, the dust capacity is, therefore, determined at test airflow to a terminal restriction/pressure drop of 60 mbar (6 kPa), or as specified, with several intermediate points. This restriction does not include the restriction added by the dust metering device and test shroud. After correcting the observed restriction/pressure drop data to standard conditions, the values are shown plotted in curve form versus the weight of dust fed.

AIR CLEANER DESCRIPTION

ITEM	MODEL/PART NO.
Assembly	_____
Precleaner	_____
Primary Element	_____
Secondary Element	_____

Mfg. _____
Type _____
Test Flow _____ scmh
Scavenge Air Flow _____ %
Dust Cup _____ Unloader Valve _____

TEST CONDITIONS AND RESULTS

Initial Efficiency _____ %
Test Dust Fine Batch _____
Dust Fed _____ g
Initial Restr/Press. Drop _____ kPa
Final Restr/Press. Drop _____ kPa
Relative Humidity _____ %
Temperature _____ °C

Dust Capacity _____ g _____ scmh
Test Dust _____ Batch _____
Concentration _____ g/m^3
Accumulative Eff. _____ %
Precleaner Eff. _____ %
Dust cup serviced _____ times

[Graph: RESTRICTION OR PRESSURE DROP – kPa (ordinate) vs. DUST CAPACITY – g (abscissa)]

**FIGURE 21—AIR CLEANER DUST CAPACITY AND EFFICIENCY AIR CLEANER TEST CODE
SECTION 4—INDUSTRIAL AIR CLEANERS**

4.5.2 PROCEDURE

4.5.2.1 Set up the air cleaner using clean filter elements per Figure 14 or 17 and condition the unit to the airflow at which the unit is tested for at least 15 min under temperature and humidity conditions specified in 1.3.5.

4.5.2.2 Weigh the absolute filter per 1.3.4.

4.5.2.3 Set and maintain the air cleaner airflow at test flow ±2%, and scavenge flow, if used, at the manufacturer's recommended percentage.

4.5.2.4 Feed the test dust of and at the grade and concentration specified in 4.1.3, until the restriction/pressure drop reaches 60 m bar (6 kPa), or as specified.

4.5.2.5 Record several intermediate values of the weight of test dust fed to the air cleaner and the corresponding restriction/pressure drop at approximately uniform time intervals.

4.5.2.6 Correct the restriction/pressure drop values to stand conditions per Appendix C, and plot versus dust fed to the air cleaner, per Figure 21. Label ordinate either restriction or pressure drop, as appropriate.

4.5.2.7 Reweigh the absolute filter per 1.3.4 and compute the air cleaner efficiency, per 2.2.16.2, based on the total weight of dust fed to the air cleaner at the corrected restriction/pressure drop value of 60 mbar (6 kPa), or as specified.

NOTE—Certain types of air cleaner, e.g., cyclone air cleaners, have no limiting dust capacity. In such cases, the test shall not be stopped before the cleaner has been fed with a sufficient quantity of dust for its efficiency to be determined as accurately as required. The minimum quantity shall be 50 g of dust.

4.5.3 PRESENTATION OF DATA—Use Figure 19 to present data on restriction/pressure drop. Use Figure 21 to present data on efficiency and capacity.

4.6 Precleaner Performance Test

4.6.1 PRECLEANER DUST REMOVAL—When testing with precleaners that employ either a scavenge flow, an automatic dust unloading valve, or a dust cup, the following provisions are to be made:

4.6.1.1 Scavenge Flow—The scavenge flow recommended by the manufacturer shall be maintained during all the tests. The dust feed rate shall be based on the test airflow plus the scavenge airflow. (See 4.8.)

4.6.1.2 Automatic Unloader Valve—For test purposes, a sealed jar or container may be substituted for the unloader valve.

4.6.1.3 Dust Cup—The dust cup shall not be emptied during the dust capacity test until at least two-thirds full. Also, the number of servicings shall be noted in the performance report.

4.6.1.4 When the unit under test is a precleaner only (e.g., when tested as an add-on unit), the provisions of 4.1 through 4.3 and Figures 14, 17, 18, and 19 remain applicable.

4.6.1.5 Care should be taken to insure that the position and orientation of the dust injector does not significantly enhance or detract from the precleaner performance.

NOTE—The user should be aware that the previous provisions insure optimum air cleaner performance and is advised to consult the air cleaner manufacturer for specific instructions or test procedures for any given air cleaner installation.

4.6.2 PRECLEANER EFFICIENCY—The precleaner efficiency may be determined during the dust capacity test of the air cleaner system, or as a separate component. Efficiency is based on the total weight of dust fed to the air cleaner system or precleaner component and with the sum of the gain in weight of the primary and secondary elements and absolute filter, or the weight of dust removed by the precleaner. When testing an external precleaner, efficiency should be determined throughout its recommended airflow range.

4.7 Secondary Air Filter Test Procedure

4.7.1 GENERAL

4.7.1.1 This section of the air cleaner test code has been established to cover dry type secondary air filters generally used in applications as prescribed under 4.1.1. The intended purpose of a secondary filter element is to provide engine protection in the event of a catastrophic failure of the primary filter or its omission or improper installation. The secondary filter is not intended to provide improved efficiency or capacity.

4.7.1.2 Two different test setups are specified reflecting current industry practice.

One of these, hereafter referred to as the multistage test setup, includes a complete air cleaner with precleaner and secondary element installed. A failed primary element is simulated by removal of the primary filter media but with the end caps and guard sleeves retained.

The other setup, hereafter referred to as the single-stage test setup, includes only the secondary element in a special test shroud and sized to match the element.

A secondary element can be tested in either setup, regardless of its application. However, the two setups will each yield a different efficiency and dust capacity for the same element.

4.7.1.3 Test Dust and Concentration—(See Table 7.)

TABLE 7—TEST DUST AND CONCENTRATION

Test Setup	Test Dust	Concentration
Single Stage	Fine	0.1 g/m^3
Multistage	Coarse	1 g/m^3

NOTE—Dust concentration shall be reduced if necessary to provide a minimum test duration of 30 min.

4.7.2 TEST EQUIPMENT—For the multistage test setup, the test equipment is identical to that used for the complete air cleaner test (see 4.2).

For the single-stage test setup, the test equipment is also described by 4.2, except the "air cleaner under test" is replaced by the test shroud, as shown in Figures 22 and 23.

NOTE—Figures 22 and 23 assume the element is cylindrical. If it is not, a special shroud must be constructed and thoroughly described in the test report.

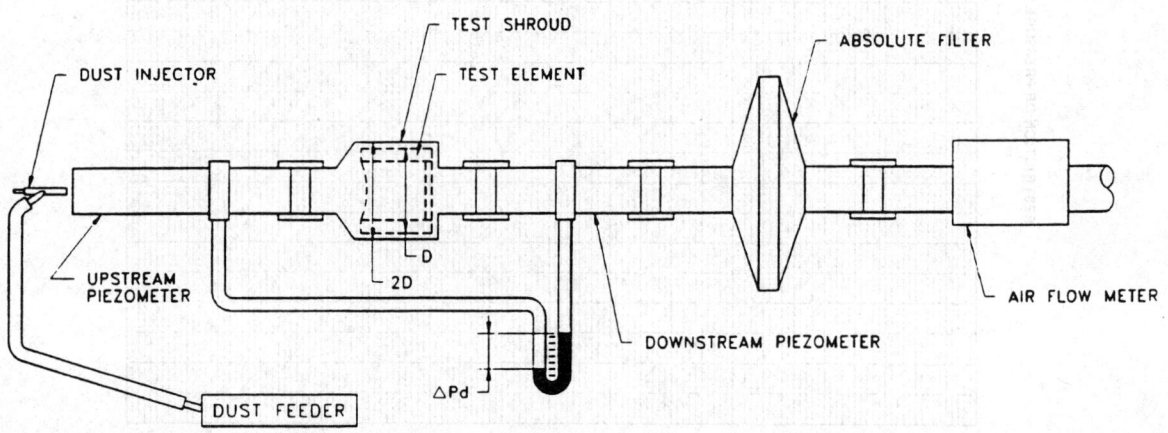

FIGURE 22—SINGLE-STAGE SETUP FOR SECONDARY ELEMENTS

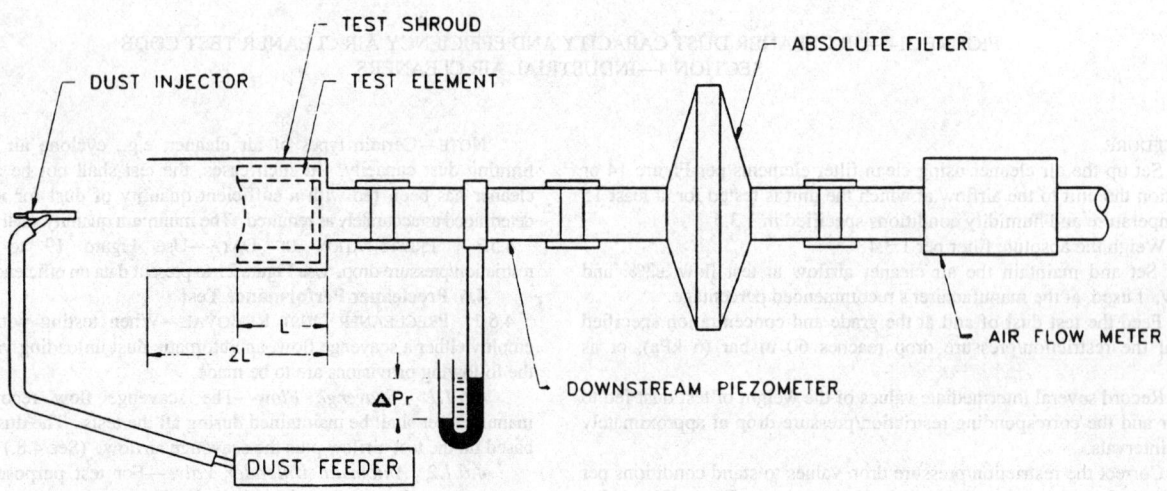

FIGURE 23—SINGLE-STAGE SETUP FOR SECONDARY ELEMENTS

4.7.3 AIRFLOW RESTRICTION AND PRESSURE DROP TEST—For the multistage test setup, repeat the tests specified in 4.3.2 or 4.3.3 with a normal primary element installed (not the simulated primary element described in 4.7.1.2). Do this both with and without a secondary element installed. The difference between pressure readings in the two tests is the restriction or pressure drop of the secondary element.

For the single-stage test setup, repeat the tests in 4.3.2 or 4.3.3 with Figure 19 replaced by Figure 22 and Figure 18 replaced by Figure 23.

4.7.4 EFFICIENCY AND DUST CAPACITY

4.7.4.1 Weigh the secondary element to the nearest 0.1 g.

4.7.4.2 For the multistage test setup, set up the air cleaner per Figure 14 or 17, using the clean secondary element and the simulated primary element per 4.7.1.2.

For the single-stage test setup, set up per Figure 22 or 23.

4.7.4.3 Weigh the absolute filter per 1.3.4.

4.7.4.4 Set and maintain the test airflow ±2%. For the multistage test setup, set and maintain any scavenge flow at the manufacturer's recommended value ±2%.

4.7.4.5 Feed a quantity of test dust 4.7.1.3 at a continuous rate until the restriction/pressure drop reaches 60 mbar (6 kPa), or as specified. The feed rate should be set to reach terminating restriction in a period of 30 min or longer.

4.7.4.6 Reweigh the absolute filter per 1.3.4.

4.7.4.7 Reweigh the secondary element to the nearest 0.1 g.

4.7.4.8 For the single-stage test setup, report the dust capacity as the weight gain of the element.

For the multistage test setup, report both the total dust fed and the weight gain of the element.

4.7.4.9 For the single-stage test setup, report efficiency per 2.2.16.1.

For the multipurpose test setup, report efficiency per 2.2.16.2.

4.7.4.10 Clearly indicate on the test report whether the multistage or single-stage test setup was used and whether the test setup was per Figure 14, 17, 22, or 23.

4.8 Scavenged Air Cleaner and/or Precleaner Performance Test

4.8.1 GENERAL

4.8.1.1 This clause describes those variations in the test procedures specified in this SAE Standard that are necessary for the testing of air cleaners that are scavenged in operation by a proportion of the air input that is bled off for that purpose.

4.8.1.2 The flow equation is as follows:

$$\dot{V}_B = \dot{V}_A - \dot{V}_C \qquad (Eq.10)$$

where:

$\dot{V}_A$ is the inlet airflow
$\dot{V}_B$ is the cleaned airflow
$\dot{V}_C$ is the scavenged airflow

4.8.2 ADDITIONAL EQUIPMENT—A typical test arrangement is shown in Figure 24.

4.8.2.1 Exhauster—An exhauster shall be provided to handle the scavenged flow and shall be capable of maintaining it at a steady state during the whole test.

4.8.2.2 Airflow Meter—An airflow meter shall be provided to measure the scavenged airflow rate within ±2%.

4.8.2.3 Pressure Tapping—The pressure tappings used shall conform to Figure 9.

4.8.2.4 Scavenged Air Filter—A filter shall be provided in the scavenged airflow of sufficient efficiency and capacity to protect the apparatus downstream from it against the effect of the dust in the scavenged airflow.

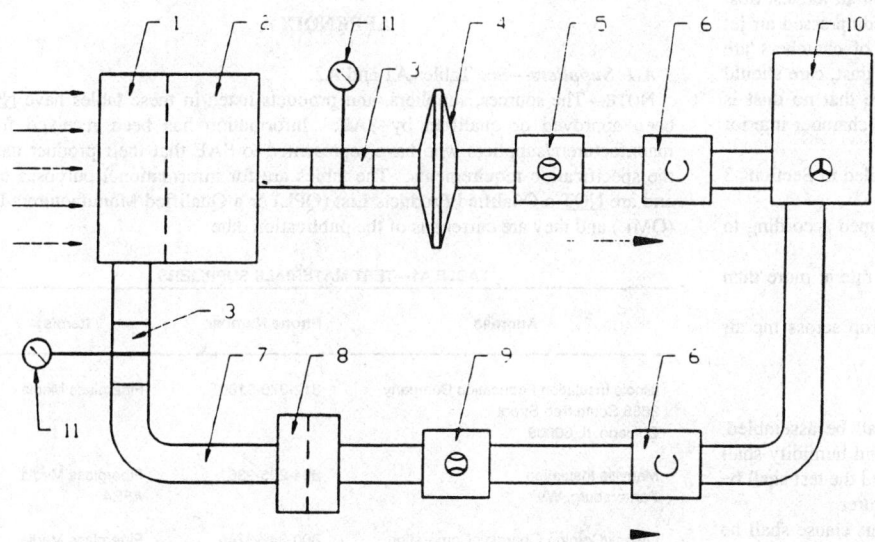

1 + 2 UNIT UNDER TEST
1 PRECLEANER, SCAVENGED
2 MAIN CLEANER
3 OUTLET TUBE (SEE FIGURE 9)
4 ABSOLUTE FILTER
5 AIR FLOW METER
6 AIR FLOW CONTROL
7 SCAVENGE AIR DUCT
8 SCAVENGE AIR DUCT FILTER
9 SCAVENGE AIR DUCT FLOW METER
10 EXHAUSTER
11 RESTRICTION MEASURING DEVICE

FIGURE 24—SCAVENGED AIR CLEANER EFFICIENCY/CAPACITY TEST SETUP

4.8.3 RESTRICTION AND PRESSURE DROP TEST—The test shall be conducted in accordance with 3.3 with the following changes:

4.8.3.1 The scavenged airflow shall be started before the cleaned airflow.

4.8.3.2 The scavenged airflow shall preferably be stopped at the same time as the cleaned airflow; it shall not be stopped before the cleaned airflow.

4.8.3.3 Measurements shall be made with the scavenged airflow adjusted to be a specified proportion of the cleaned airflow (interaction between the scavenged airflow and the cleaned airflow may require some re-adjustment to be made to maintain this proportion).

4.8.4 FULL LIFE EFFICIENCY AND CAPACITY TEST

4.8.4.1 Most of the air cleaners that are scavenged in operation by a proportion of the air input that is bled off for this purpose are comparatively large in size. The absolute filter test method is therefore recommended.

4.8.4.2 Unless otherwise specified, the scavenged airflow shall be maintained at a fixed proportion of the cleaned airflow, as agreed between the manufacturer and the user.

4.8.4.3 The test dust concentration shall be that in the inlet airflow.

4.8.4.4 The scavenged airflow shall be started before the cleaned airflow.

4.8.4.5 The scavenged airflow should preferably be stopped at the same time as the cleaned airflow; it shall not be stopped before the cleaned airflow.

4.8.4.6 Precleaner Efficiency—The precleaner efficiency shall be determined during the dust capacity test based on the total weight of dust fed to the air cleaner, and either the sum of the gain in weight of the primary and secondary elements and absolute filter, or the weight of dust removed by the precleaner.

4.8.4.7 The capacity of the unit shall be calculated as follows:

Capacity = Total mass of dust fed × $\dfrac{\text{Cleaned airflow}}{\text{Inlet airflow}}$ − Increase in mass of absolute filter

(Eq.11)

4.8.5 PRESENTATION OF DATA—For presentation of data, use Figures 18 or 19 or equivalent.

5. Oil Bath Air Cleaner Test Procedures

5.1 General

5.1.1 This section of the air cleaner test code has been specifically established for those internal combustion engine applications where oil bath air cleaners are used. Most commonly, these applications are military, stationary, or off-highway although automotive and truck applications may be included. The performance tests shall be performed on a complete oil bath air cleaner. The tests shall consist of a restriction/pressure drop test, of an oil carryover test, a combined capacity and efficiency test, and a recovery test.

5.1.2 The definitions, contaminants, procedures, and test equipment from Sections 3 and 4 shall apply unless specific variations are needed.

5.1.3 Test oil for oil bath types shall be of two grades: a. (light) SAE 10 (330 mm^2/s); and b. (heavy) SAE 30 (85 mm^2/s). The oil used shall be stabilized in the environmental temperature condition of the test area for at least 1 h prior to use.

5.1.4 All tests shall be carried out with the air cleaner in a level position unless otherwise specified by the user or the particular section of the test procedure. Before the test, the air cleaner shall be prepared in the following manner:

5.1.4.1 Thoroughly wash and dry the air cleaner.

5.1.4.2 Fill the oil cup/reservoir to the indicated level with the specified oil.

5.1.4.3 Flow air through the cleaner at rated airflow for 15 min.

5.1.4.4 Stop the airflow or allow a draining period of 15 min.

5.1.4.5 Refill the cup/reservoir with oil to the specified level for the particular test.

5.1.5 A typical arrangement for testing oil bath air cleaners of the tubular inlet type is shown in Figure 14.

5.1.6 The air cleaners of the peripheral inlet type shall be tested in a chamber that ensures the even distribution and delivery of test dust to the inlet of the unit. Care should be taken in the design of the chamber to ensure that all the test dust is fed to the unit under test. If dust settling occurs, then the compressed air jet may be used to re-entrain the test dust. Typical examples of chambers are shown in Figure 17. When using compressed air for agitating dust, care should be taken not to eject the dust out of the chamber. To ensure that no dust is ejected, a negative pressure should be maintained between the chamber interior and the atmosphere.

5.1.7 All tests shall be carried out under the conditions detailed in Sections 3 and 4.

5.2 Restriction and Pressure Drop Test—Tests performed according to 3.3 and 4.3 with the following changes:

5.2.1 Perform the restriction/pressure drop test versus flow rate at more than 100% rated airflow only as long as no oil carryover occurs.

5.2.2 The airflow shall be maintained until the pressure drop across the air cleaner has stabilized.

5.3 Oil Carryover Test

5.3.1 The dust shall not be fed to the cleaner during the test.

5.3.2 The cleaner, prepared in accordance with 5.1.4, shall be assembled, weighed, and attached to the test rig. The room temperature and humidity shall be recorded. The recommended oil shall be used for the test and the test shall be conducted at a temperature agreed on by the user and manufacturer.

5.3.3 Each oil bath air cleaner tested in accordance with this clause shall be tested in one of two ways, as agreed on by the manufacturer and the purchaser. The two ways are as follows:

5.3.3.1 A test at a single flow rate, above the rated flow, as agreed on by the manufacturer and the purchaser, to determine whether or not oil carryover occurs at that flow rate.

5.3.3.2 A test at increasing flow rates, starting at 80% of the rated flow and increasing in increments of 10% of the rated flow, to determine the airflow rate at which oil carryover occurs.

5.3.4 The test in 5.3.3.1 shall be conducted for a minimum of 60 min for each filter tested. The test in 5.3.3.2 shall be conducted for at least 10 min at each flow rate.

5.3.5 At the end of the test at each flow rate, the air cleaner outlet shall be examined for signs of oil carryover using an observation chamber with a target plate covered with chalk or a suitable paper, which turns transparent at the impact of oil droplets (see Figure 25).

5.3.6 At the end of the test described in 5.3.3, the air cleaner shall be removed and weighed again and the loss of oil by weight shall be recorded.

5.3.7 If an oil bath air cleaner is to be or may be operated in an inclined position, the tests described in 5.3.3.1 and 5.3.3.2 shall be repeated in full with the cleaner inclined at the angles and directions in which it may be required to operate with such additional margins as agreed on by the manufacturer and the purchaser.

5.4 Full Life Efficiency and Capacity Test—The dust capacity/efficiency characteristics of the oil bath air cleaners shall be assessed by the methods described in 3.4, 3.5, 4.4, and 4.5 with the exceptions detailed as follows. It is essential, when testing oil bath cleaners, to ensure that no oil carryover occurs at the rated test airflow. Significant oil losses of this kind will affect the weights recorded for the absolute filter and/or unit under test, which will influence the final test results. The tests can be conducted with either constant or variable airflow according to 3.4.4. The exceptions in test procedures are the following:

5.4.1 Condition the unit under test according to 5.1.4 and measure and record the weight.

5.4.2 Use the test dust at the concentration detailed in 3.4.3.1 or 4.1.3.

5.4.3 At the end, perform an oil carryover test according to 5.3.3.2.

5.5 Recovery Test—After the capacity test drain and the unit under test is cleaned to the precise instructions recommended by the cleaner manufacturer, resume the test without dust feed for 20 min noting the restriction/pressure drop at 5 min intervals during this period. The recovery capabilities of the test unit will be assessed by comparison of these results with those obtained for a new, unused test cleaner.

APPENDIX A

A.1 Suppliers—See Tables A1 and A2.

NOTE—The sources, suppliers, and products listed in these tables have <u>NOT</u> been approved or qualified by SAE. Information has been received from manufacturers/suppliers who have represented to SAE that their product meets the specification requirements. The tables are for informational purposes only and are <u>NOT</u> a Qualified Products List (QPL) or a Qualified Manufacturers List (QML) and they are current as of the publication date.

TABLE A1—TEST MATERIALS SUPPLIERS

Address	Phone Number	Item(s)
Illinois Insulation Fabrication Company 3636 South Iron Street Chicago, IL 60609	312-376-3100	Fiberglass Media
Manville Insulation Parkersburg, WV	304-295-9361	Fiberglass Media AFS4
Owens/Corning Fiberglas Corporation Fiberglas Tower Toledo, OH 43659 Attn: CTR Meeks	800-342-3745 419-248-8028	Fiberglass Media FM004
Powder Technology, Inc. P.O. Drawer 1464 Burnsville, MN 55337	612-894-8737	Fine & Coarse Test Dust Contaminant

NOTE—This list of suppliers reflects those known by members of the SAE Air Cleaner Test Code Committee. Other materials or manufacturers may be available.

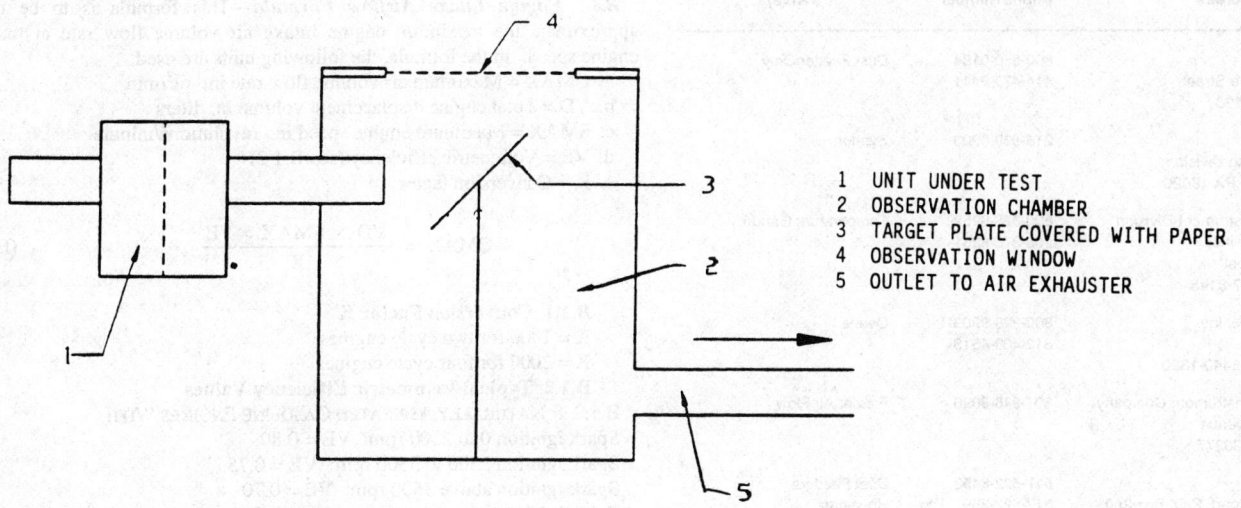

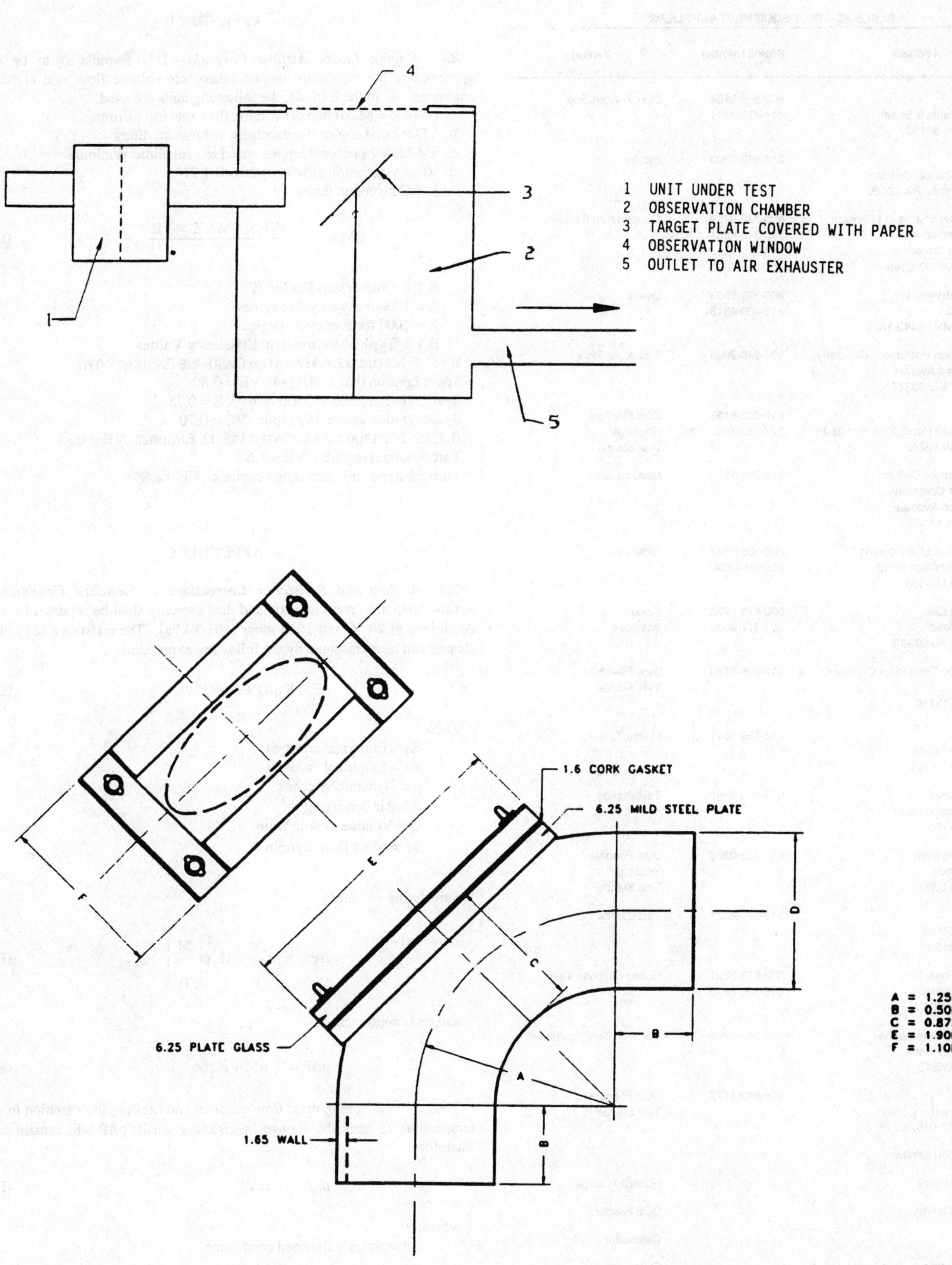

1 UNIT UNDER TEST
2 OBSERVATION CHAMBER
3 TARGET PLATE COVERED WITH PAPER
4 OBSERVATION WINDOW
5 OUTLET TO AIR EXHAUSTER

1.6 CORK GASKET
6.25 MILD STEEL PLATE
6.25 PLATE GLASS
1.65 WALL

A = 1.25D
B = 0.50D
C = 0.875D
E = 1.90D
F = 1.10D

FIGURE 25—OIL CARRYOVER TEST: OBSERVATION CHAMBER

TABLE A2—TEST EQUIPMENT SUPPLIERS

Address	Phone Number	Item(s)
AccuRate 746 East Milwaukee Street Whitewater, WI 53190	800-558-0184 414-473-2441	Dust Feeder Only
Ametek Schutte and Koering Division Cornwells Heights, PA 19020	215-639-0900	Syphon
Ashcroft Desser Instrument Division Dresser Industries, Inc. 250 East Main Street Stratford, CT 06497-5145	800-328-8258 203-378-8281	Temperature Gauge
Despatch Industries, Inc. P.O. Box 1320 Minneapolis, MN 55440-1320	800-828-9903 612-469-4513	Ovens
Frazier Precision Instrument Company 16761 Oakmont Avenue Gaithersburg, MD 20877	301-948-9026	Frazier Air Flow
IBR, Inc. 11599 Morrissey Road, P.O. Box 250 Grass Lake, MI 49240	517-522-8453 517-522-3695 Fax	Dust Feeders Housings Test Stands
Meriam Instrument Division Scott & Fetzer Company 10920 Madison Avenue Cleveland, OH 44102	216-281-1100	Manometers
Mettler Instrument Corporation Princeton-Hightstown Road Hightstown, NJ 08520	800-638-8537 609-448-3000	Balances
Ohaus Corporation 35 Hanover Road Florham Park, NJ 07932	800-672-7722 201-377-9000	Scale Balances
Penn-Anderson Equipment Company P.O. Box 305 Oakmont, PA 15139	412-828-6760	Dust Feeders Test Stands
B. F. Perkins 939 Chicopee Street Chicopee, MA 01013	413-536-1311	Mullen Tester
Scans Associates, Inc. 13034 Farmington Road Livonia, MI 48150	313-478-9687	Test Stands
Testand Corporation 38 Fortin Avenue Pawtucket, RI 02860	401-724-0306	Dust Feeders Housings Test Stands
Testek 12271 Globe Road Livonia, MI 48150	313-591-2271	Test Stands
Testing Machines, Inc. 398 Bayview Avenue Amityville, NY 11701	800-678-3221 516-842-5400	Gurley Stiffness Test
Unholtz-Dickie Corporation 3000 Whitney Avenue Hamden, CT 06518	203-265-3929	Vibration Test Equipment
Whitely 939C East Street Tewksbury, MA 01876	508-640-1177	Dust Feeders Test Stands
ADDRESS UNKNOWN:		
Airguide Instrument		Humidity Gauge
Auburn Engineering		Dust Feeder
Bell Art		Desiccator
Precision Scientific		Ovens
Sun Equipment		Test Stands

NOTE—This list of suppliers reflects those known by members of the SAE Air Cleaner Test Code Committee. Other materials or manufacturers may be available.

APPENDIX B

B.1 Engine Intake Airflow Formula—This formula is to be used to approximate the maximum engine intake air volume flow rate at maximum engine speed. In the formula, the following units are used:
 a. QMAX = Maximum air volume flow rate in: m^3/min
 b. VD = Total engine displacement volume in: liters
 c. SMAX = Maximum engine speed in: revolutions/minute
 d. VE = Volumetric efficiency (see B.1.2)
 e. K = Conversion factor

$$QMAX = \frac{VD \times SMAX \times VE}{K} \quad (Eq.B1)$$

B.1.1 Conversion Factor K
K = 1000 for two cycle engines
K = 2000 for four cycle engines

B.1.2 Typical Volumetric Efficiency Values
B.1.2.1 NATURALLY ASPIRATED GASOLINE ENGINES WITH
Spark ignition 0 to 2500 rpm: VE = 0.80
Spark ignition 2500 to 3500 rpm: VE = 0.75
Spark ignition above 3500 rpm: VE = 0.70
B.1.2.2 NATURALLY ASPIRATED DIESEL ENGINES: VE = 0.85
Turbocharged engines: VE = 1.60
Turbocharged and intercooled engines: VE = 1.85

APPENDIX C

C.1 Airflow and Resistance Corrections to Standard Conditions—The airflow restriction/pressure drop and dust capacity shall be reported for standard conditions of 20 °C and 1013 mbar (101.3 kPa). The resistance (ΔP) of the air cleaner can be represented by the following expression:

$$\Delta P = K_1 \mu Q + K_2 \rho Q^2 \quad (Eq.C1)$$

where:
 K_1 = Empirical constant
 K_2 = Empirical constant
 μ = Dynamic viscosity
 ρ = Air density kg/m^3
 Q = Volume flow m^3/min
 $\dot{M}$ = Mass flow, kg/min

Substituting $\frac{\dot{M}}{\rho}$ for Q

$$\Delta P = K_1 \mu \frac{\dot{M}}{\rho} + K_2 \rho \left(\frac{\dot{M}}{\rho}\right)^2 \quad (Eq.C2)$$

and rearranging terms:

$$\rho \Delta P = K_1 \mu \dot{M} + K_2 \dot{M}^2 \quad (Eq.C3)$$

Thus, by maintaining mass flow constant and limiting the variation in ambient temperature to keep the change in viscosity small, $\rho \Delta P$ will remain constant. Therefore:

$$\rho_O \Delta P_O = \rho \Delta P \quad (Eq.C4)$$

where:
 subscript $_O$ = standard conditions

$$\Delta P_O = \frac{\rho}{\rho_O} \Delta P \quad (Eq.C5)$$

The observed restriction/pressure drop values shall, therefore, be corrected to standard conditions by using the following equation:

$$\Delta P_O = \frac{P}{1013} \times \frac{293}{(t+273)} \times \Delta P_r \text{ or } \Delta P_d \qquad \text{(Eq.C6)}$$

where:

P and t are the observed ambient pressure and temperature, respectively, and ΔP_r or ΔP_d is the measured air cleaner restriction/pressure drop.

ENGINE AND TRANSMISSION DIPSTICK MARKING—SAE J614b

SAE Recommended Practice

Report of Engine and Transmission Committees approved May 1957 and completely revised by Engine Committee August 1977.

General Requirements

1. Lettering should be raised or depressed characters of the simple block type. One eighth in (3 mm) high minimum, but 3/16 in (5 mm) or higher is preferred. Depressed lettering may be filled with pigmented enamel to improve legibility.

2. Space permitting, lettering should not be positioned between FULL and first ADD lines.

3. The lower portion of the dipstick, below the FULL mark should be knurled, perforated, or coated to improve visibility of wetted level. Perforations of 0.090 in (2.29 mm) are ideal in a dipstick wide enough not to be seriously weakened by them. Level marks should not be stamped across the perforations. ADD lines and quantity required in quarts, pints, or litres may be included on the lower portion if the manufacturer so desires.

4. Additional information for checking procedure or identification may be added to the dipstick above the FULL mark or on an attached tag at the manufacturer's discretion. Detailed checking procedure should be described and illustrated in the owner's or service manual.

5. Locating dipsticks and filler spouts to make them accessible from the curb side of the vehicle, away from traffic is desirable. They should, however, be readily accessible regardless of that consideration and preferably both accessible from one position.

ENGINE MOUNTINGS—SAE J615 APR91

SAE Standard

Report of the Engine Division approved April 1919 and last revised July 1985, Rationale statement available. Revised by the SAE Clutch Flywheel and Flywheel Standards Committee April 1991.

1. Scope—This SAE Standard defines engine mounting dimensions for industry standardization and interchangeability. Table 1 and Figure 1 are dimensions for arm type mountings. Table 2 and Figure 2 are for side pad mountings. For engine housing SAE flange dimensions, see SAE J617. For engine foot type mountings (front and rear), see SAE J616.

2. References

2.1 Applicable Documents—The following publications form a part of this specification to the extent specified herein. The latest issue of SAE publications shall apply.

2.1.1 SAE PUBLICATIONS—Available from SAE, 400 Commonwealth Drive, Warrendale, PA 15096-0001.

SAE J616—Engine Foot Mounting (Front and Rear)
SAE J617—Engine Flywheel Housings

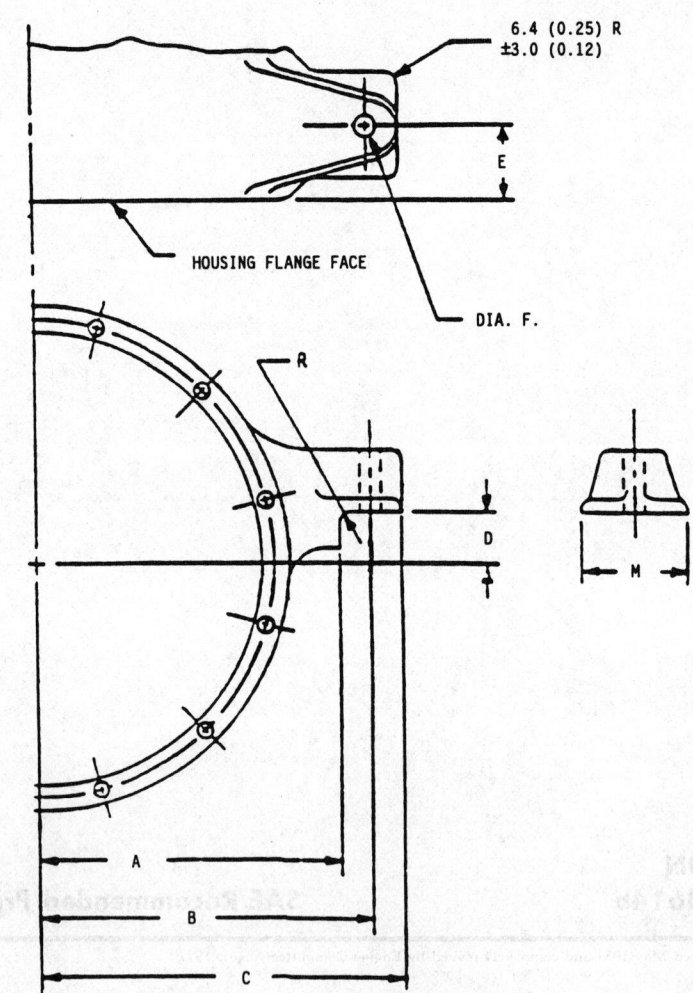

FIGURE 1—ARM TYPE MOUNTING DIMENSIONS

TABLE 1—ARM TYPE MOUNTING DIMENSIONS—REFER TO FIGURES 1 AND 2

SAE Arm No. Tolerance	A mm ±0.25	A in ±0.010	B mm ±0.25	B in ±0.010	C mm ±3.18	C in ±0.12	D mm ±0.25	D in ±0.010
1	298.45	11.750	339.73	13.375	365.13	14.375	50.80	2.000
3	285.75	11.250	311.15	12.250	327.03	12.875	63.50	2.500
4	285.75	11.250	311.15	12.250	327.03	12.875	63.50	2.500
5	285.75	11.250	311.15	12.250	327.03	12.875	63.50	2.500
2T	254.0	10.000	276.23	10.875	298.45	11.750	76.20	3.000

TABLE 1—(CONTINUED)

SAE Arm No. Tolerance	E mm ±0.25	E in ±0.010	F mm ±0.25	F in ±0.010	R mm Max	R in Max	M mm ±0.25	M in ±6.35
1	76.20	3.000	19.84	0.781	11.2	0.44	4.00	101.6
3	76.20	3.000	16.66	0.656	6.4	0.25	3.50	88.9
4	76.20	3.000	16.66	0.656	6.4	0.25	2.75	69.8
5	76.20	3.000	16.66	0.656	6.4	0.25	2.75	69.8
2T	76.20	3.000	16.66	0.656	6.4	0.25	4.00	101.6

(R) TABLE 2—SIDE PAD MOUNTING DIMENSIONS FLAT OR LOCATING LEDGE TYPE—REFER TO FIGURE 2

SAE Pad No[1] Tolerance	A mm ±0.25	A in ±0.010	B mm ±0.25	B in ±0.010	C mm ±0.25	C in ±0.010	D mm ±0.25	D in ±0.010
00	—	—	—	—	—	—	—	—
0	88.90	3.500	114.30	4.500	19.05	0.750	25.40	1.000
½	88.90	3.500	114.30	4.500	19.05	0.750	25.40	1.000
1	63.50	2.500	88.90	3.500	15.88	0.625	44.45	1.750
2	63.50	2.500	88.90	3.500	15.88	0.625	44.45	1.750
3	50.80	2.000	66.68	2.625	15.88	0.625	42.88	1.688
4	50.80	2.000	66.68	2.625	15.88	0.625	42.88	1.688
5	50.80	2.000	66.68	2.625	15.88	0.625	42.88	1.688
6	50.80	2.000	66.68	2.625	15.88	0.625	42.88	1.688

TABLE 2—(CONTINUED)

SAE Pad No[1] Tolerance	E mm ±0.25	E in ±0.010	Tapped Hole Size[2]	F mm ±0.25	F in ±0.010	G mm ±3.18	G in ±0.12	R mm Max	R in Max
00	—	—	—	—	—	—	—	—	—
0	368.30	14.500	⅝-11	19.05	0.750	381.30	15.01	1.3	0.05
½	330.20	13.000	⅝-11	19.05	0.750	343.20	13.51	1.3	0.05
1	285.75	11.250	⅝-11	19.05	0.750	298.75	11.76	1.3	0.05
2	260.35	10.250	½-13	15.87	0.625	269.85	10.63	1.3	0.05
3	260.35	10.250	½-13	15.87	0.625	269.85	10.63	0.8	0.03
4	212.72	8.375	½-13	15.87	0.625	222.22	8.88	0.8	0.03
5	212.72	8.375	½-13	15.87	0.625	222.22	8.88	0.8	0.03
6	187.32	7.375	½-13	15.87	0.625	196.82	7.78	0.8	0.03

[1] Pads of a given number can be matched with housings of the same SAE number or smaller size (larger SAE number).
[2] Tapped holes shall be threaded in accordance with UNC Class 2B tolerances of ANSI B1.1, threads. The minimum length of thread engagement shall be 1.5 times the nominal diameter for gray iron housings and 2 times the nominal diameter for aluminum housings.

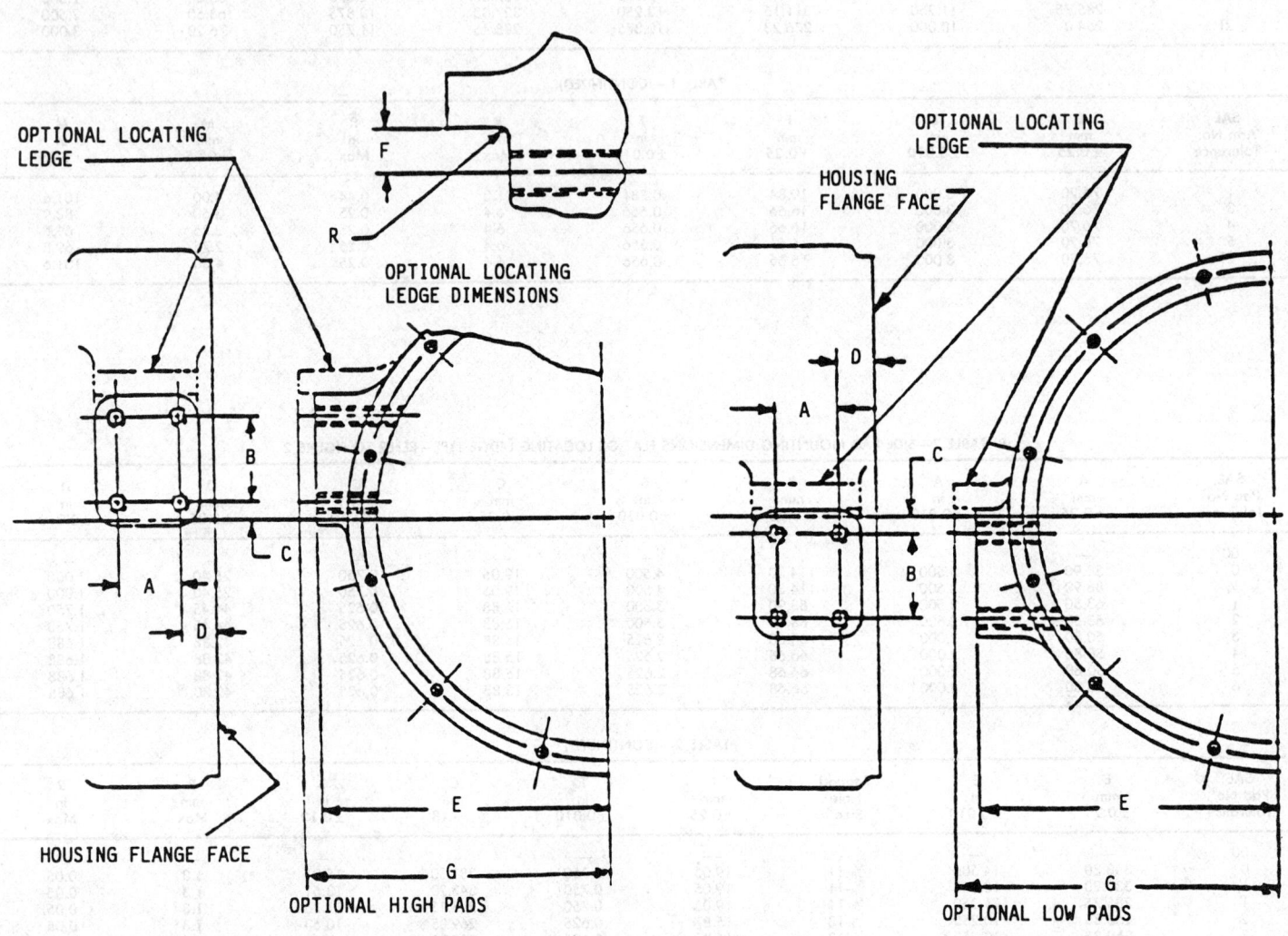

FIGURE 2—SIDE PAD MOUNTING DIMENSIONS

FLYWHEEL DIMENSIONS FOR TRUCK AND BUS APPLICATIONS—SAE J1857 MAY87

SAE Recommended Practice

Report of the Truck and Bus Powertrain Committee approved May 1987.

1. Purpose—This SAE Recommended Practice defines flywheel configuration to promote standardization of the 14 and 15.5 size clutch mounting.

2. Scope—Although not limited to, these clutches are normally used in Class 7 and 8 trucks.

FLYWHEEL DIMENSIONS FOR THE 14 & 15.5 CLUTCHES

14-2 (POT TYPE)

	in	mm
A.	7.1(2)	180.3
D.	14.725/14.755	374.0/374.78
E.	3.94	100.1
F.	15.5	393.7
G.	2.5	63.5
H.	14.750/14.753	374.65/374.73
J.	2.942/2.932	74.73/74.47
K.	0.75	19.1
L.	2.834(1)	72.00
M.	0.18(Min)	4.57
Holes(d)	3/8-16	
U.	0.75	19.1
V.	0.4990/0.5005	12.67/12.71
W.	12	—

15.5-2 (FLAT TYPE)

	in	mm
A.	8.76(2)	222.5
C.	16.625	422.3
E.	3.94	100.1
G.	2.5	63.5
H.	17.156/17.159	435.76/435.84
K.	0.75	19.1
L.	2.934(1)	72.00
M.	0.18(Max)	4.57
Holes(d)	7/16-14	

(1) Tolerance to be determined by engine manufacturer. Optional size of 2.441 for smaller bearing.
(2) Larger dimension required for some high capacity dampers. Check with clutch manufacturer.

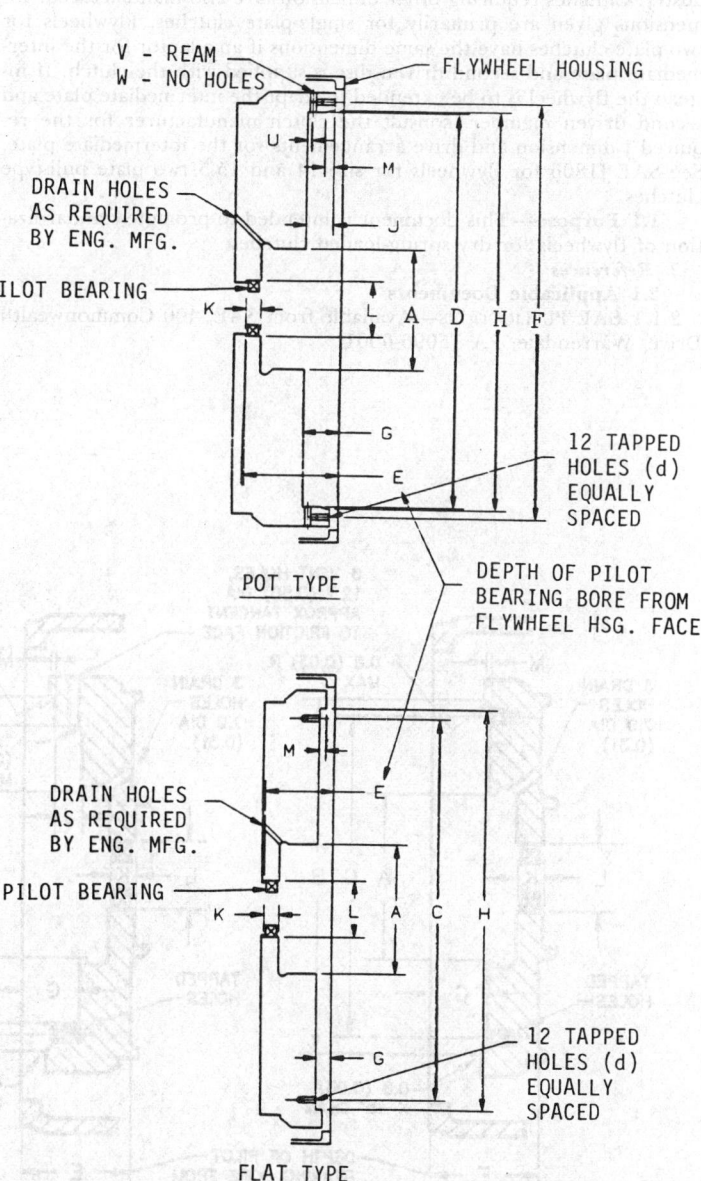

FIG. 1

(R) FLYWHEELS FOR SINGLE-PLATE SPRING-LOADED CLUTCHES—SAE J618 JAN91

SAE Recommended Practice

Report of the Construction and Industrial Machinery Technical Committee and Engine Committee, approved November 1957, last revised by the Engine Committee May 1974, reaffirmed without change June 1980. Completely revised by the Clutch, Flywheel and Housing Standards Committee January 1991.

1. **Scope**—This SAE Recommended Practice applies to flywheels for dry spring-loaded clutches used on internal combustion engines. Figure 1 and Tables 1, 2, and 3 report information currently used in the industry. Clutches requiring other dimensions are also manufactured. Dimensions given are primarily for single-plate clutches. Flywheels for two plate clutches have the same dimensions if an adaptor for the intermediate plate and second driven disc is supplied with the clutch. If instead the flywheel is to be extended to adapt the intermediate plate and second driven member, consult the clutch manufacturer for the required J dimension and drive arrangements for the intermediate plate. See SAE J1806 for flywheels for size 14 and 15.5 two plate pull-type clutches.

   1.1 **Purpose**—This document is intended to promote standardization of flywheels for dry spring-loaded clutches.

2. **References**

   2.1 **Applicable Documents**

   2.1.1 SAE PUBLICATIONS—Available from SAE, 400 Commonwealth Drive, Warrendale, PA 15096-0001.

   SAE J1033—Procedure for Measuring Bore and Face Runout of Flywheels, Flywheel Housings, and Flywheel Housings Adapters
   SAE J1806—Clutch Dimensions for Truck & Bus Applications

   2.1.2 ANSI AND ISO PUBLICATIONS—Available from ANSI, 1430 Broadway, New York, NY 10018.

   ANSI/ASME B.1-1989—Unified Inch Screw Threads (UN & UNR Thread Form)
   ISO 965-3:1980—General Purpose Metric Screw Threads—Tolerances—Part 3: Deviations for Constructional Threads

3. **Tolerances**—Maximum runout of pilot bearing bore or clutch piloting bore is 0.13 mm (0.005 in) FIM. Maximum face runout of clutch mounting surface and friction surface is 0.013 mm (0.0005 in) FIM per 25.4 mm (1 in) of measured diameter. The procedure for measuring bore and face runouts is described in SAE J1033.

4. **Tapped Holes**—Tapped holes shall be threaded in accordance with coarse pitch series thread class fit 6H per ISO 965111 or UNC class 2B thread per ANSI B.I.I. Minimum thread engagement shall be 1.5 times nominal bolt diameter for gray iron flywheels, 1.0 times nominal bolt diameter for ductile iron flywheels.

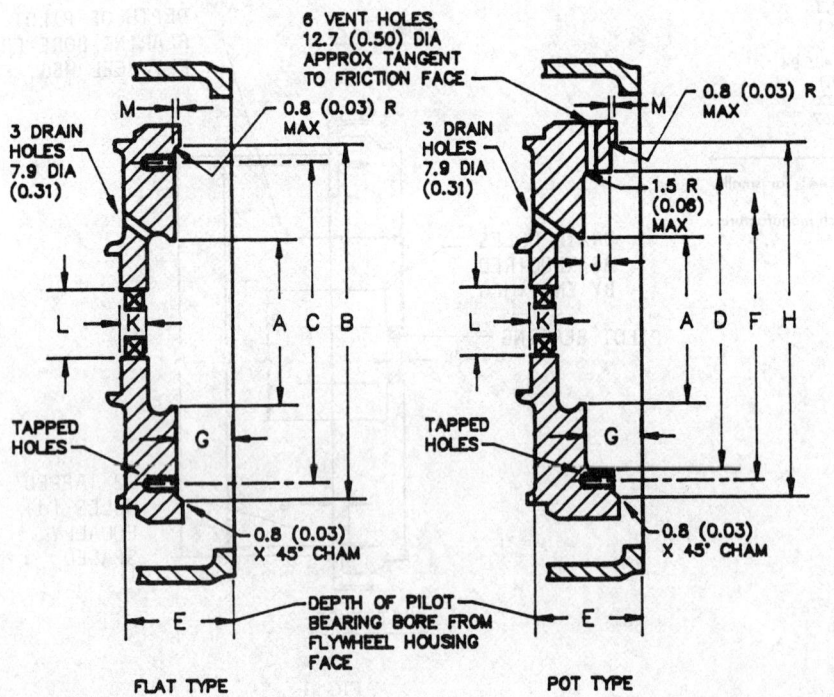
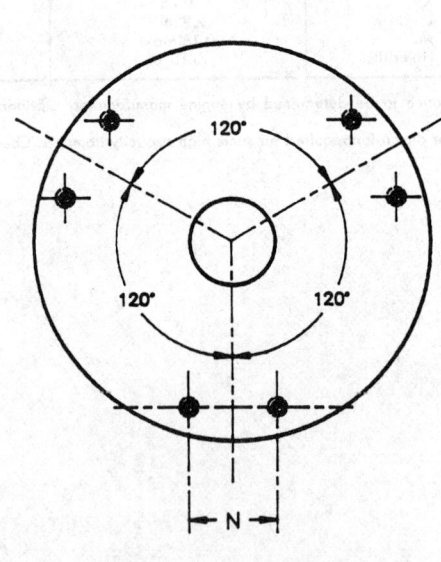

FIG. 1

TABLE 1—STANDARD DIMENSIONS FOR FLAT OR POT-TYPE FLYWHEELS mm (in)

Nominal Clutch Size	A	D[1]	E	G	K[2]	L[2]	M[3]
140 (5.5)	85.9 (3.38)	142.7 (5.62)	71.4 (2.81)	41.1 (1.62)	14.2 (0.56)	40.00 (1.5748)	3.0 (0.12)
150 (6.0)	91.9 (3.62)	155.4 (6.12)	71.4 (2.81)	41.1 (1.62)	14.2 (0.56)	40.00 (1.5748)	3.0 (0.12)
165 (6.5)	104.6 (4.12)	168.1 (6.62)	71.4 (2.81)	41.1 (1.62)	14.2 (0.56)	40.00 (1.5748)	3.0 (0.12)
215 (8.5)	127.0 (5.00)	218.9 (8.62)	71.4 (2.81)	41.1 (1.62)	14.2 (0.56)	40.00 (1.5748)	3.0 (0.12)
230 (9.0)	136.7 (5.38)	231.6 (9.12)	71.4 (2.81)	41.1 (1.62)	17.3 (0.68)	47.00 (1.8504)	3.0 (0.12)
255 (10.0)	149.4 (5.88)	257.0 (10.12)	100.1 (3.94)	66.5 (2.62)	17.3 (0.68)	47.00 (1.8504)	3.0 (0.12)
280 (11.0)	155.4 (6.12)	282.4 (11.12)	100.1 (3.94)	66.5 (2.62)	17.3 (0.68)	52.00 (2.0472)	3.0 (0.12)
305 (12.0)	165.1 (6.50)	307.8 (12.12)	100.1 (3.94)	66.5 (2.62)	17.3 (0.68)	52.00 (2.0472)	3.0 (0.12)
330 (13.0)	184.1 (7.25)	333.2 (13.12)	100.1 (3.94)	66.5 (2.62)	17.3 (0.68)	62.00 (2.4409)	4.6 (0.18)
355 (14.0)	187.4 (7.38)	358.6 (14.12)	100.1 (3.94)	66.5 (2.62)	17.3 (0.68)	62.00 (2.4409)	4.6 (0.18)
380 (15.0)	203.2 (8.00)	384.0 (15.12)	100.1 (3.94)	63.5 (2.50)	19.0 (0.75)	72.00 (2.8346)	4.6 (0.18)

[1] Pot-type clutches may not be commercially available in all sizes.
[2] K is length of bore for pilot bearing. L is nominal diameter of bearing. Diameter and fit are to suit installation.
[3] Most flat-type clutches are piloted by the mounting bolts and a piloting diameter is not used.

TABLE 2—CLUTCH MOUNTING DIMENSIONS OF FLAT-TYPE FLYWHEELS mm (in)

Nominal Clutch Size	B[1]	C	Tapped Holes No.	Tapped Holes Spacing	Tapped Holes N	Tapped Holes Size
140 (5.5)	185.4 (7.30)	163.07 (6.420)	6	Pairs	70.10 (2.760)	5/16-18
150 (6.0)	197.1 (7.76)	174.62 (6.875)	6	Pairs	70.10 (2.760)	5/16-18
165 (6.5)	219.1 (8.63)	203.20 (8.000)	6	Pairs	50.80 (2.000)	5/16-18
165 (6.5)	228.6 (9.00)	186.33 (7.336)	6	Pairs	70.10 (2.760)	5/16-18
215 (8.5)	266.7 (10.50)	244.46 (9.625)	6	Pairs	50.80 (2.000)	5/16-18
230 (9.0)	279.4 (11.00)	263.52 (10.375)	6	Pairs	72.64 (2.860)	5/16-18
230 (9.0)	288.3 (11.35)	268.00 (10.551)	6	Equal	-- --	5/16-18
255 (10.0)	304.8 (12.00)	288.92 (11.375)	6	Pairs	79.63 (3.135)	5/16-18
255 (10.0)	314.3 (12.38)	295.28 (11.625)	6	Equal	-- --	3/8 -16
280 (11.0)	330.2 (13.00)	314.32 (12.375)	6	Pairs	86.61 (3.410)	3/8 -16[2]
280 (11.0)	340.0 (13.43)	320.68 (12.625)	6	Equal	-- --	3/8 -16
305 (12.0)	373.1 (14.69)	342.90 (13.500)	6	Pairs	88.75 (3.494)	3/8 -16
330 (13.0)	390.5 (15.38)	358.78 (14.125)	12	Equal	-- --	3/8 -16
330 (13.0)	390.5 (15.38)	371.48 (14.625)	12	Equal	-- --	3/8 -16
355 (14.0)	417.8 (16.45)	393.70 (15.500)	12	Equal	-- --	3/8 -16
355 (14.0)	414.7 (16.32)	386.08 (15.200)	12	Equal	-- --	3/8 -16

[1] Most flat-type clutches are piloted by the mounting bolts and do not use a piloting diameter. Dimension given is the minimum diameter required for clutch clearance.
[2] 5/16-18 is also used.

NOTE: The first flywheel listed in any size is the preferred design for that size. Subsequent listings for the size are also in use.

TABLE 3—CLUTCH MOUNTING DIMENSIONS OF POT-TYPE FLYWHEELS mm (in)

Nominal Clutch Size	F	H[1]	J	Tapped Holes No.	Tapped Holes Spacing	Tapped Holes N	Tapped Holes Size
255 (10)	269.88 (10.625)	288.92 (11.375)	30.15 (1.187)[2]	6	Pairs	50.80 (2.000)	5/16-18
255 (10)	269.88 (10.625)	284.15 (11.187)	30.15 (1.187)	6	Equal	— —	5/16-18
280 (11)	298.45 (11.750)	315.90 (12.437)	36.58 (1.440)	6	Equal	— —	3/8-16
280 (11)	303.23 (11.938)	323.85 (12.750)	34.11 (1.343)	3	Equal	— —	5/16-18
305 (12)	330.20 (13.000)	355.60 (14.000)	35.81 (1.410)	6	Equal	— —	1/2-13
355 (14)	374.65 (14.750)	393.70 (15.500)	30.23 (1.190)	12	Equal	— —	3/8-16
355 (14)	374.65 (14.750)	393.70 (15.500)	51.82 (2.040)	12	Equal	— —	3/8-16
380 (15)	403.22 (15.875)	422.28 (16.625)	44.45 (1.750)	12	Equal	— —	3/8-16

[1] Diameter tolerance of clutch pilot bore is +0.08 mm (0.003 in) − 0.00.
[2] Also available with 37.59 (1.480) pot depth.

NOTE: Flat-type flywheels are generally preferred over pot-type flywheels.

FLYWHEELS FOR TWO-PLATE SPRING-LOADED CLUTCHES—SAE J619 JUN80

SAE Recommended Practice

Report of the Engine Committee, approved June 1960, last revised May 1974, reaffirmed without change June 1980.

Scope—This SAE Recommended Practice defines flywheel configuration to promote standardization of flywheels for dry spring-loaded clutches.

Clutches to fit flywheels with configurations per this recommended practice may not be commercially available. Availability should be ascertained prior to flywheel design.

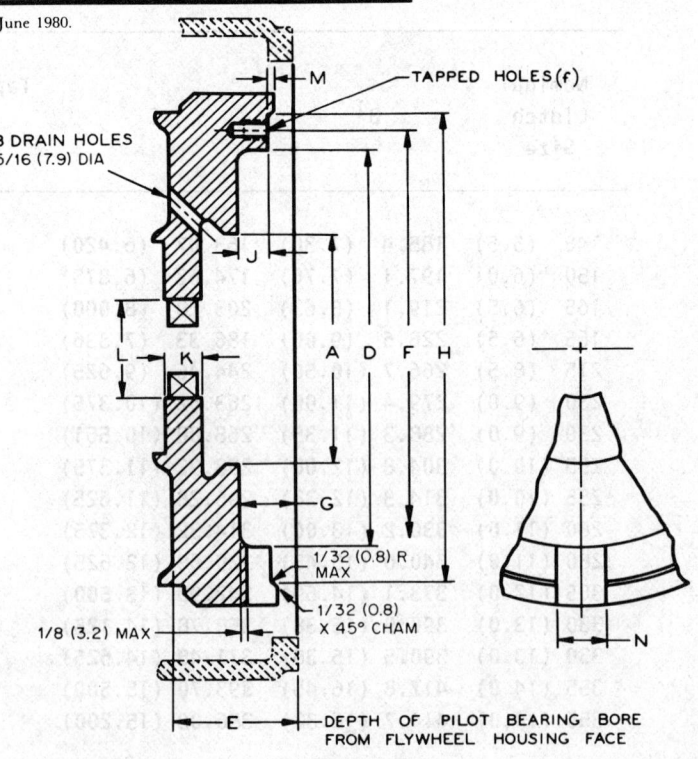

POT TYPE

NOTE: SEE "FLYWHEELS FOR SINGLE PLATE SPRING LOADED CLUTCHES" (FLAT TYPE) IF DRIVE FOR INTERMEDIATE PLATE IS SUPPLIED WITH TWO PLATE CLUTCH

TABLE 1—DIMENSIONS OF FLYWHEELS FOR TWO-PLATE SPRING-LOADED CLUTCHES

Nominal Clutch Size	A		D		E		F		G		H[a,c]		J	
	in	mm	in	mm	in	mm	in	mm	in	mm	in	mm	in	mm
6	3.62	91.9	6.25	158.8	2.81	71.4	7.375	187.32	1.62	41.1	8.125	206.38	0.750	19.05
7	4.38	111.3	7.25	184.2	2.81	71.4	8.375	212.72	1.62	41.1	9.125	231.78	0.750	19.05
8	5.00	127.0	8.25	209.6	2.81	71.4	9.375	238.12	1.62	41.1	10.125	257.18	0.750	19.05
9	5.62	142.7	9.25	235.0	2.81	71.4	10.375	263.52	1.62	41.1	11.125	282.58	0.938	23.83
10	5.88	149.4	10.25	260.4	3.94	100.1	11.375	288.92	2.62	66.5	12.125	307.98	1.062	26.97
11	6.12	155.4	11.25	285.8	3.94	100.1	12.500	317.50	2.62	66.5	13.375	339.72	1.062	26.97
12	6.88	174.8	12.25	311.2	3.94	100.1	13.500	342.90	2.62	66.5	14.375	365.12	1.062	26.97
13	7.50	190.5	13.25	336.6	3.94	100.1	14.625	371.48	2.62	66.5	15.500	393.70	1.062	26.97
14	8.00	203.2	14.25	362.0	3.94	100.1	15.625	396.88	2.62	66.5	16.500	419.10	1.375	34.92
15	8.50	215.9	15.25	387.4	3.94	100.1	16.625	422.28	2.62	66.5	17.500	444.50	1.375	34.92
16	9.25	235.0	16.25	412.8	3.94	100.1	17.750	450.85	2.62	66.5	18.625	473.08	1.562	39.67
17	9.88	251.0	17.25	438.2	3.94	100.1	18.875	479.42	2.62	66.5	19.750	501.65	1.562	39.67

Nominal Clutch Size	K[b]		L[b]		M		N, Slot			Tapped Holes[d]	
							Width		No.		
	in	mm	in	mm	in	mm	in	mm		No.	Size, in
6	0.56	14.2	1.5748	40.000	0.12	3.0	1.000	25.40	3	9	5/16–18
7	0.56	14.2	1.5748	40.000	0.12	3.0	1.000	25.40	3	9	5/16–18
8	0.56	14.2	1.5748	40.000	0.12	3.0	1.000	25.40	3	9	5/16–18
9	0.68	17.3	1.8504	47.000	0.12	3.0	1.500	38.10	3	9	5/16–18
10	0.68	17.3	1.8504	47.000	0.12	3.0	1.500	38.10	3	9	5/16–18
11	0.68	17.3	2.0472	52.000	0.12	3.0	1.500	38.10	3	12	3/8 –16
12	0.68	17.3	2.0472	52.000	0.12	3.0	1.750	44.45	3	12	3/8 –16
13	0.68	17.3	2.0472	52.000	0.18	4.6	1.750	44.45	3	12	3/8 –16
14	0.68	17.3	2.0472	52.000	0.18	4.6	2.000	50.80	3	12	3/8 –16
15	0.75	19.0	2.8346	72.000	0.18	4.6	2.000	50.80	3	12	3/8 –16
16	0.75	19.0	2.8346	72.000	0.18	4.6	2.000	50.80	4	12	3/8 –16
17	0.88	22.4	3.1496	80.000	0.18	4.6	2.000	50.80	3	12	3/8 –16

NOTE: Suggested tolerances and eccentricities to be measured on assembled engines. All holes and slots are equally spaced.

[a] Diameter tolerance of clutch pilot bore H and B is +0.003in (0.08mm) –0.000; maximum eccentricity is 0.005 in (0.13mm) TIR. For B, see SAE J618.

[b] K is length of bore for pilot bearing; L is nominal diameter of bearing. Diameter and fit are to suit installation. Maximum eccentricity is 0.005in (0.13mm) TIR.

[c] Face runout is 0.0005in (0.013mm) maximum TIR per inch (25.4mm) of diameter.

[d] Tapped holes should be threaded in accordance with UNC class 2B tolerances of ANSI B1.1, Screw Threads.

HOUSING INTERNAL DIMENSIONS FOR SINGLE- AND TWO-PLATE SPRING-LOADED CLUTCHES—SAE J373 APR93   SAE Recommended Practice

Report of the Engine Committee approved November 1968, completely revised September 1978, and reaffirmed by the Engine Committee August 1987. Reaffirmed by the SAE Clutch, Flywheel, and Housing Standards Committee April 1993.

Foreword—This reaffirmed document has been changed only to reflect the new SAE Technical Standards Board format.

1. **Scope**—This SAE Recommended Practice defines the minimum internal dimensions for clutch housings to provide adequate clearance for single- and two-plate spring-loaded clutches. (See Figure 1.)

Consult SAE J617 for housing flange dimensions. Consult SAE J618 and J619 for spring-loaded clutch flywheel dimensions F and G and other dimensional data.

Table 1 provides housing minimum internal dimensions for single- and two-plate spring-loaded clutches.

2. **References**

   2.1 **Applicable Documents**—The following publications form a part of this specification to the extent specified herein. The latest issue of SAE publications shall apply.

   2.1.1 SAE PUBLICATIONS—Available from SAE, 400 Commonwealth Drive, Warrendale, PA 15096-0001.

   SAE J617—Engine Flywheel Housing and Mating Transmission Housing Flanges
   SAE J618—Flywheels for Single-Plate Spring-Loaded Clutches
   SAE J619—Flywheels for Two-Plate Spring-Loaded Clutches

TABLE 1—HOUSING MINIMUM INTERNAL DIMENSIONS, mm (In)

Clutch Number	SAE Flywheel Housing Number[1]	A	B	C	D	E
6	6	235.0 (9.25)	190.5 (7.50)	22.4 (0.88)	41.1 (1.62)	12.7 (0.50)
7	6	235.0 (9.25)	190.5 (7.50)	22.4 (0.88)	41.1 (1.62)	12.7 (0.50)
8	5	241.3 (9.50)	206.2 (8.12)	35.1 (1.38)	53.8 (2.12)	12.7 (0.50)
9	5	279.4 (11.00)	241.3 (9.50)	38.1 (1.50)	62.0 (2.44)	12.7 (0.50)
10	4	292.1 (11.50)	254.0 (10.00)	15.7 (0.62)	42.9 (1.69)	12.7 (0.50)
11	4	317.5 (12.50)	266.7 (10.50)	30.2 (1.19)	49.3 (1.94)	12.7 (0.50)
12	3	342.9 (13.50)	292.1 (11.50)	31.8 (1.25)	55.6 (2.19)	25.4 (1.00)
13	3	374.6 (14.75)	311.2 (12.25)	36.6 (1.44)	58.7 (2.31)	25.4 (1.00)
14	2	381.0 (15.00)	323.8 (12.75)	39.6 (1.56)	68.3 (2.69)	25.4 (1.00)
15	1	425.4 (16.75)	349.2 (13.75)	41.1 (1.62)	68.3 (2.69)	25.4 (1.00)
16	1	469.9 (18.50)	387.4 (15.25)	60.5 (2.38)	95.2 (3.75)	50.8 (2.00)
17	1/2	520.7 (20.50)	419.1 (16.50)	60.5 (2.38)	95.2 (3.75)	50.8 (2.00)

[1] Column shows minimum SAE Flywheel Housing Size adaptable to clutch.

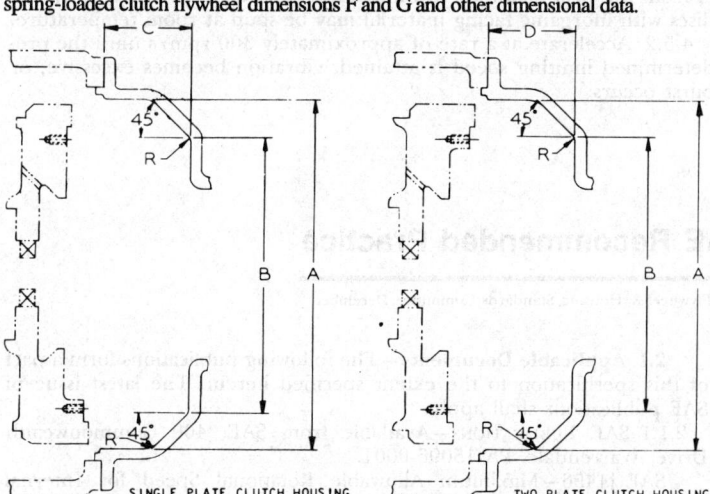

FIGURE 1—SINGLE- AND TWO-PLATE CLUTCH HOUSING

ENGINE FLYWHEEL HOUSINGS WITH SEALED FLANGES—SAE J1172 MAY93

SAE Recommended Practice

Report of the Engine Committee approved May 1977 and reaffirmed October 1988. Reaffirmed by the SAE Clutch, Flywheel, and Housing Standards Committee May 1993.

Foreword—This reaffirmed document has been changed only to reflect the new SAE Technical Standards Board format.

1. Scope—This SAE Recommended Practice defines flywheel housing flange configurations for applications requiring "O" ring sealing of the flange pilot bore.

Table 1 and Figure 1 show dimensions that are different from those in SAE J617. All other dimensions and tolerances of SAE J617 apply.

2. Reference

2.1 Applicable Document—The following publication forms a part of this specification to the extent specified herein. The latest issue of SAE publications shall apply.

2.1.1 SAE PUBLICATION—Available from SAE, 400 Commonwealth Drive, Warrendale, PA 15096-0001.

SAE J617—Engine Flywheel Housing and Mating Transmission Housing Flanges

TABLE 1—FLYWHEEL HOUSING FLANGE DIMENSIONS—mm (in)

SAE No.	F Min Pilot Bore Depth	G Min Radial Clearance for Mating Flange Pilot	H Chamfer ± 0.5 (0.02)
00	19.0 (0.75)	14.0 (0.55)	3.0 (0.12)
0	19.0 (0.75)	14.0 (0.55)	3.0 (0.12)
1/2	19.0 (0.75)	14.0 (0.55)	3.0 (0.12)
1	16.0 (0.63)	13.0 (0.51)	2.5 (0.10)
2	16.0 (0.63)	13.0 (0.51)	2.5 (0.10)
3	16.0 (0.63)	13.0 (0.51)	2.5 (0.10)
4	14.0 (0.55)	11.0 (0.43)	2.0 (0.08)
5	14.0 (0.55)	11.0 (0.43)	2.0 (0.08)
6	14.0 (0.55)	11.0 (0.43)	2.0 (0.08)

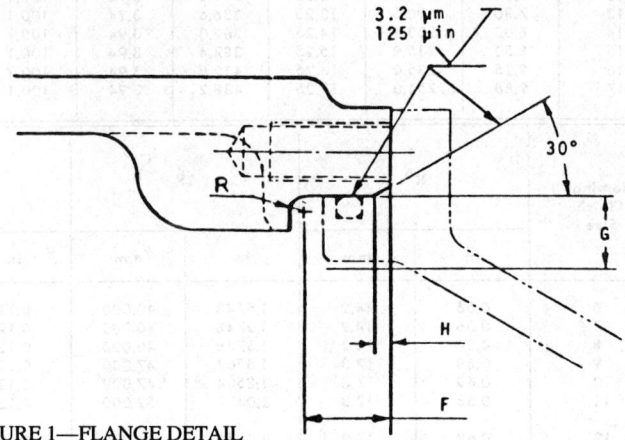

FIGURE 1—FLANGE DETAIL

(R) SPRING-LOADED CLUTCH SPIN TEST PROCEDURE—SAE J1073 JUN90

SAE Recommended Practice

Report of the Engine Committee, approved December 1973, reaffirmed without change June 1980. Completely revised by the Clutch, Flywheel, and Housing Standards Committee June 1990. Rationale statement available.

1. Scope—This SAE Recommended Practice applies to spring-loaded clutches such as are used with manual shift type transmissions.

1.1 Purpose—This document is intended to provide a uniform test procedure for spring-loaded clutches to determine rotative speeds at which they will either burst or withstand a specified limiting speed.

2. References

2.1 Applicable Documents

SAE J1240, Flywheel Spin Test Procedure

3. Test Equipment—Refer to SAE J1240 for a description of typical test equipment requirements.

4. Test Procedure

4.1 Prior to spin testing, the item to be tested is to be identified, checked for material and dimensional defects, and balanced to print specification.

4.2 Test should be performed in a minimum ambient temperature of 15°C (60°F). Record ambient temperature.

4.3 Mounting—Test sample is to be mounted on a spin arbor. The arbor may be operated in a vertical or horizontal position; the arbor design must be such that vibration will be minimized.

4.3.1 COVER ASSEMBLIES—Mount the cover assembly on a steel test flywheel in the position it would be when engaged with a new driven disc. To simulate the driven disc thickness, suitable spacers between the flywheel and pressure plate may be used, or the flywheel may be machined to effectively position the pressure plate in the new position.

4.3.2 PRESSURE PLATE OR INTERMEDIATE PLATE OF MULTIPLE-PLATE CLUTCHES—Mount the plate on a test fixture piloting by either the same method as used in the cover assembly or flywheel such as drive pins or by the I.D. Use enough clamping force to hold the test piece in place but not enough to counteract centrifugal force.

4.4 Cover Assemblies, Pressure Plates and Intermediate Plates

4.4.1 Spin test sample to 2000 rpm then accelerate at a rate not to exceed 80 rpm/s until the predetermined limiting speed is attained, vibration becomes excessive, or burst occurs.

4.5 Driven Disc Assemblies

4.5.1 Subject driven discs with organic facing material to 260°C ± 3 (500°F ± 5) for 15 min. The maximum elapsed time between removal from the heat chamber and initial acceleration shall be 25 s. Driven discs with inorganic facing material may be spun at room temperature.

4.5.2 Accelerate at a rate of approximately 300 rpm/s until the predetermined limiting speed is attained, vibration becomes excessive, or burst occurs.

(R) FLYWHEEL SPIN TEST PROCEDURE—SAE J1240 DEC91

SAE Recommended Practice

Report of the Engine Committee approved December 1978. Completely revised by the SAE Clutch Flywheel & Housing Standards Committee December 1991. Rationale statement available.

1. Scope—This SAE Recommended Practice applies to flywheels and flywheel-starter ring gear assemblies used with internal combustion engines of spark ignition and diesel type. Engine sizes are those capable of using SAE No. 6 through SAE No. 00 size flywheel housings.

1.1 Purpose—This document is intended to provide a uniform test procedure for flywheel assemblies to determine the rotative speeds at which they will either burst or withstand a specified limiting speed.

2. References

2.1 Applicable Documents—The following publications form a part of this specification to the extent specified herein. The latest issue of SAE publications shall apply.

2.1.1 SAE PUBLICATIONS—Available from SAE, 400 Commonwealth Drive, Warrendale, PA 15096-0001.

SAE J1456—Maximum Allowable Rotational Speed for Internal Combustion Engine Flywheels

3. Test Facility—The potential for serious injury to test personnel requires that greatest attention be given to safety. The spin chamber enclosing covers or doors must be properly designed and constructed to contain the flywheel and the failed material. All rotating parts of the drivetrain between the power source and the spin chamber must have protective guards. The controls area should be safely isolated from the spin pit and drivetrain areas. A ventilation system should be provided for the removal of iron dust to prevent injury to eyes and lungs by the flying iron dust when the spin chamber doors are opened. The test area must have the cautionary signs and warning bells, etc.

The following items are normally required for a test facility:

3.1 Burst test chamber and associated drive equipment.

3.2 Speed counter capable of recording burst speed or maximum speed attained. Recommend two speed counters be used, one as a backup. Speed counters should be calibrated periodically.

3.3 Expendable type adapter flange for attaching the flywheel to the spindle shaft, etc.

3.4 Flywheel to adapter mounting hardware.

3.5 Soft lining of pit (wood, cardboard, etc. — when retrieval of the failed flywheel pieces is necessary for inspection).

3.6 Dial indicator to check the concentricity and runout of the flywheel after it is mounted on the adapter flange.

3.7 Lifting crane.

3.8 Storage area and the necessary tools.

3.9 High-speed camera utilization capability, if desired.

4. Test Equipment and Drive System—The equipment may provide for testing of the flywheels in the horizontal or vertical axis position. The drive system adapters, etc., must be so designed that vibration is kept to a minimum. A disconnect clutch or equivalent cutoff system is recommended between the power source and the spindle to prevent extensive damage to the drive system components in case of sudden spindle seizure. If desired, spin chamber may be evacuated to control the flywheel test temperature or to reduce its uncontrolled vibration and prevent quill shaft failure, which may be caused by air turbulence when testing inside small test chambers.

5. Test Procedure

5.1 Flywheels should be spin tested with the ring gear installed unless the ring gear is not part of the flywheel assembly.

5.2 Flywheels without counterweight are usually spin tested one at a time. If the drive system is capable, flywheels with counterweight may be spin tested in pairs; heavy or light side of the flywheels positioned opposite in relation to each other. If tested one at a time, flywheels with counterweight should make a balanced assembly when mated with the adapter flange.

5.3 Prior to spin testing, the flywheel is to be identified, checked for material and dimensional defects, and balanced to print specification. Hardness readings should be taken on all critical areas of the flywheel and recorded. See SAE J1456 for recommended hardness checking areas for typical flywheel profiles.

5.4 Clean all clamping faces and piloting bores on the flywheel and adapter free of dust and dirt.

5.5 Assemble the adapter flange to the spindle shaft and the flywheel to adapter flange.

5.6 Check and record flywheel concentricity and runout as assembled on the spin test adapter.

5.7 Record ambient and flywheel test temperature prior to spin testing.

5.8 Spin each flywheel assembly to the predetermined speed limit and hold at that speed for 5 seconds or until burst occurs or vibration becomes excessive. Acceleration rate to suit flywheel size and the drive system capability.

5.9 Cracking or bursting of the flywheel assembly below the predetermined speed limit shall constitute failure.

6. Inspection of Test Parts after Spin Testing—When desired, chemical and physical data can be obtained after testing the flywheel. Material strength is to be established by test bars removed from the critical areas of the broken flywheel parts. Chamber walls must be cushion-lined to assure the reliability of the test bar data. If broken parts from the critical areas are too small and do not permit removal of test bar, then test bar may be removed from another area of the flywheel with similar hardness level. If this is also not possible, a flywheel with similar strength level may be cut for test bar removal. However, the flywheel must have similar hardness, porosity, and defect levels in all critical areas and must be:

a. From the same pour for cast type gray iron, nodular iron, and steel flywheels
b. From the same heat for forged type steel flywheels

7. Maximum Allowable Rotational Speed for Flywheels—For maximum allowable rotational speed for internal combustion engine flywheels, see J1456.

(R) MAXIMUM ALLOWABLE ROTATIONAL SPEED FOR INTERNAL COMBUSTION ENGINE FLYWHEELS—SAE J1456 JUN90 — SAE Recommended Practice

Report of the Engine Committee, approved December 1984. Rationale statement available. Completely revised by the Clutch, Flywheel, and Housing Standards Committee June 1990.

1. Scope—This SAE Recommended Practice applies to flywheels and flywheel-starter ring gear assemblies used with internal combustion engines of the spark ignition and diesel type equipped with a governor or speed limiting device. Engine sizes are those capable of using SAE No. 6 through SAE No. 00 flywheel housings.

This document applies to methods used to determine the rotational speed capability of flywheels for stresses imposed by centrifugal forces only.

1.1 Purpose—This document is intended to establish maximum allowable rotational speeds for flywheels under centrifugally imposed stresses.

1.2 Exclusions

1.2.1 GENERAL—This document does not encompass the selection of flywheel profile, materials, and manufacturing methods.

The influence of the following items which may reduce the speed capability of the flywheel below acceptable speed limit are not considered in this document and must be evaluated separately:

a. Material fatigue
b. Material yielding before the limiting speed or burst speed is reached
c. Clutch heat loading and cracks
d. Additional loading due to drive coupling members assembled to the flywheel or due to the flywheel drive system type used

1.2.2 APPLICATION SPEED RESTRICTIONS—This document does not apply to any application where the engine is not equipped with a governor or speed limiting device or when overspeed (No.) may exceed 50% above the rated engine speed (N_R); (No. $> 1.5\ N_R$).

1.2.3 FIELD USAGE AND REWORK RESTRICTIONS—This document does not apply when the following is done without the approval of the engine manufacturer:

1.2.3.1 The flywheel profile is reworked for another application.
1.2.3.2 The flywheel clutch friction face is refaced beyond the material removal limits recommended by the engine manufacturer.
1.2.3.3 The engine and flywheel package are used in an application other than the one for which it was originally designed.
1.2.3.4 The flywheel is removed from the original engine and installed on another engine where the rated speed or overspeed requirement is higher.

2. References

2.1 Applicable Documents

SAE J1240, Flywheel Spin Test Procedure
ASTM E 8, Standard Test Methods of Tension Testing of Metallic Materials

3. Rotational Speed Criteria

3.1 General—Flywheel limiting speed or burst speed capability should be established by spin test(s) and must be adjusted to minimum expected material tensile strength value.

Flywheel speed capability may also be established by an analytical method if the analytical method used closely correlates with the spin test results and spin test data is available on similar profile, size, and same material flywheels.

Flywheel material strength is to be established by test bars which are to be removed from the critical areas of the flywheel. Strength data obtained from separately poured test bars is not to be used for burst speed evaluation.

3.2 Predetermined Speed Limit or Minimum Burst Speed— Flywheel predetermined speed limit or minimum burst speed is to be based on the following equation:

$$N_B = 2.50 \times N_R \qquad \text{(Eq. 1)}$$

where:

N_R = Application full load, governed, rated engine speed (rpm)
N_B = Predetermined speed limit, actual or corrected minimum burst speed of flywheel (rpm)

3.2.1 Burst Speed Correction—When the spin test flywheel material has higher strength than the specified minimum tensile strength value, the measured test burst speed (N_T) shall be derated by Eq. 2 to obtain the corrected minimum burst speed (N_C):

$$N_C = N_T \sqrt{\frac{\text{Minimum Specified Tensile Strength}}{\text{Measured Tensile Strength of Test Specimen}}} \qquad \text{(Eq. 2)}$$

The corrected minimum burst speed must be equal to or greater than the minimum burst speed requirement ($N_C \geq N_B$).

4. General Instructions

4.1 Flywheel Drawing Specifications—Flywheel material expected minimum strength and hardness specification should be defined and critical areas or sections of the flywheel should be shown on the drawings. Unless otherwise indicated, the following areas shall be considered as critical areas or sections:

a. Flywheel hub
b. Thin section in hub area with abrupt profile and section thickness change

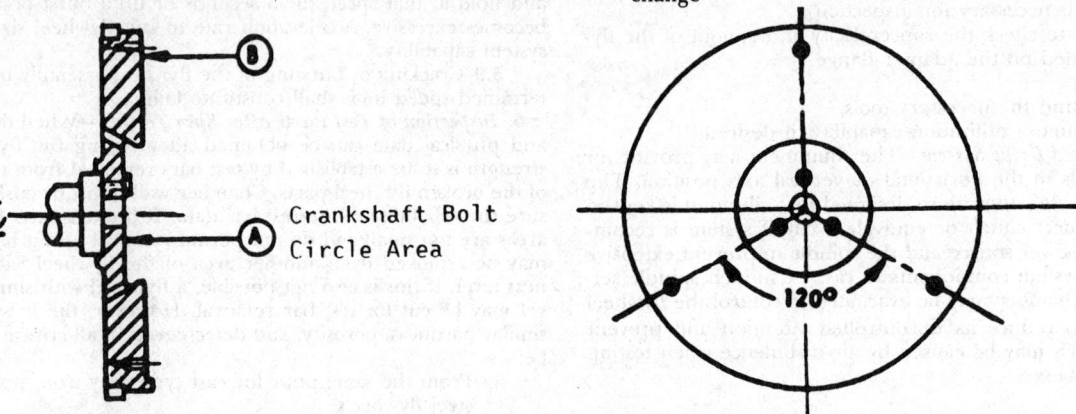

2. **Medium Pot Type Flywheels**
Some spring loaded clutch and most industrial driving ring type overcenter clutch flywheels.

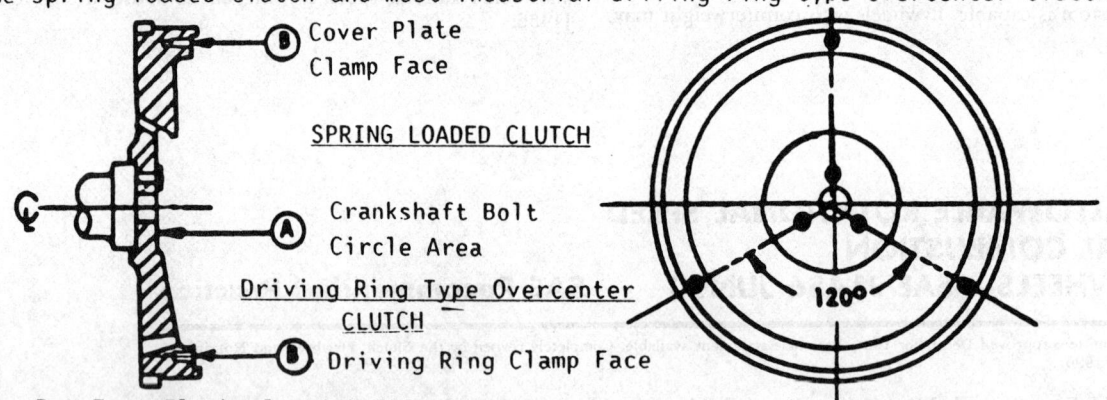

3. **Deep Pot Type Flywheels With Thin Rim**
Some multi-plate spring loaded clutch, hydraulic coupling, etc. flywheels.

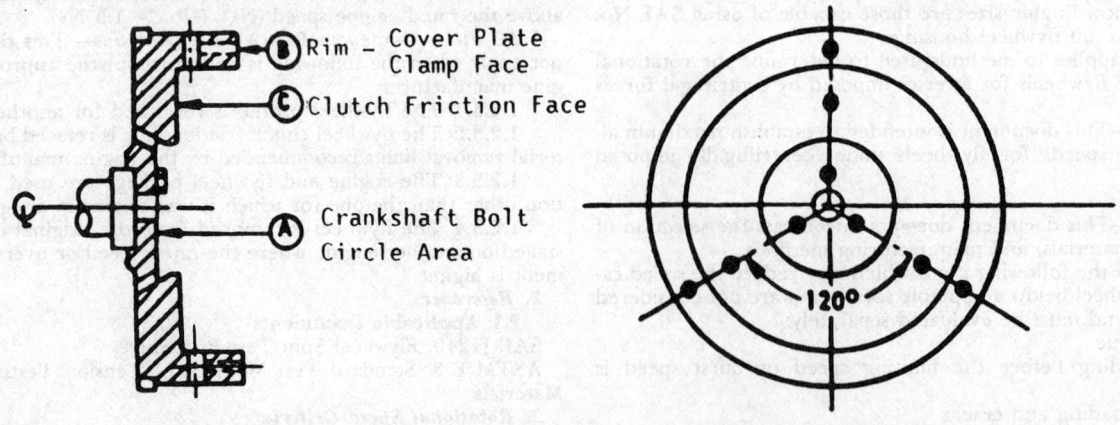

FIG. 1—TYPICAL FLYWHEEL PROFILES AND RECOMMENDED HARDNESS CHECKING AREAS

c. Thick heavy outer rims joining with thin body sections utilizing a small fillet radius

d. Thin rims of deep pot type flywheels where the rim has deep clutch or coupling mounting holes or balance holes

e. Rims of pot type flywheels which are interrupted by radially cut slots such as those used for driving the clutch intermediate pressure plate, etc.

4.2 Flywheel Redesign—Flywheel speed capability should be reconsidered when the flywheel profile or material specifications or heat treatment is altered, or when the casting process or supplier is changed.

4.3 Hardness Checks

4.3.1 Hardness readings can be used to indicate the strength and the uniformity of the flywheel material after hardness-to-strength relationship has been established. See Figure 1 for recommended hardness checking areas, and the number of checks for typical flywheel profiles.

CAUTION: Hardness readings are not a substitute for tensile test bar data. When estimating the tensile strength of gray iron from the measured hardness values, exercise caution; hardness to strength relationship can vary considerably.

4.3.2 Hardness readings are to be taken on surfaces ground flat to 3.2 µm (125 µin) AA or less on flywheels selected for spin testing.

4.4 Tensile Test Bars—Tensile test bars removed from the critical areas of the flywheels should be machined per ASTM E 8 (latest issue).

4.5 Spin Testing

4.5.1 The required number of flywheels for spin testing is to be selected by the engine manufacturer. It is recommended that three flywheels, randomly selected, be spin tested. If the flywheels burst, the lowest test burst speed is to be used for establishing the rotational speed capability of the flywheel. See 3.2.1 for correcting the test burst speed.

4.5.2 For flywheel spin testing procedure and inspection of test parts after spin testing, see SAE Recommended Practice, J1240.

TAPERS FOR SHAFT ENDS AND HUBS FOR FUEL INJECTION PUMPS—SAE J1419 FEB88   SAE Standard

Report of the Engine Committee approved February 1988. This report references ISO 6519.

1. Purpose and Description—This SAE Standard specifies the dimensional requirements necessary for the interchangeability of tapered shaft ends and hubs for fuel injection pumps of diesel engines.

2. Dimensions and Tolerances
2.1 Shaft Ends with Taper

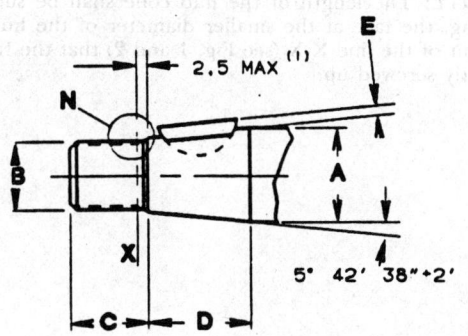

FIG. 1—SHAFT END, TYPE I

NOTE: The shaft ends may be made optionally according to type 1 or 2. However, it shall be possible to screw the go-gauge for the thread up to the chain line X-X.

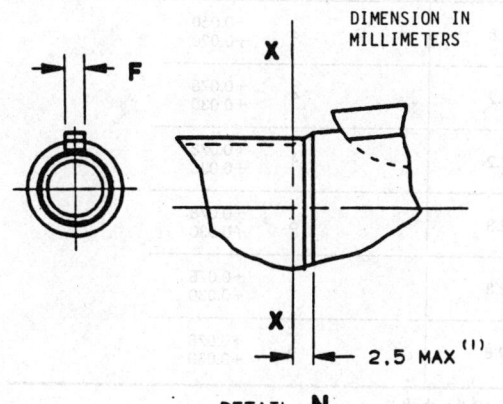

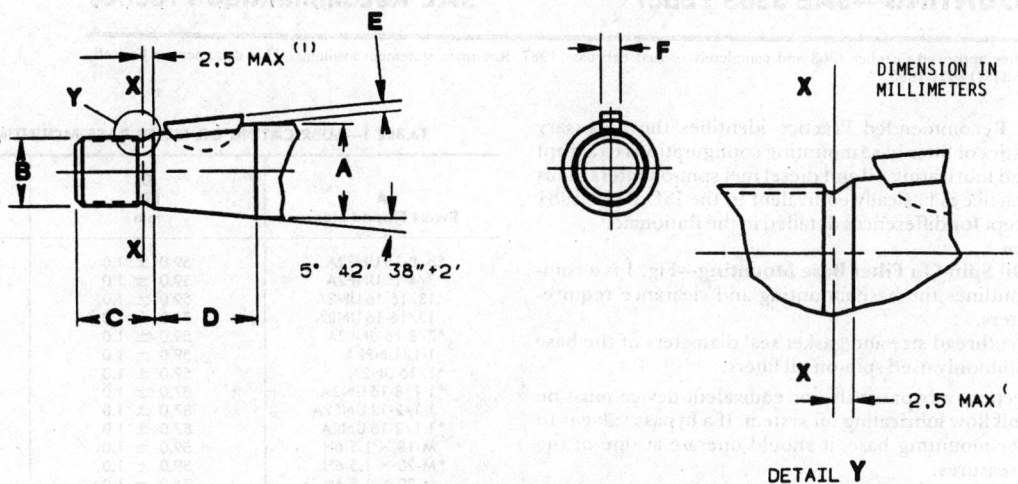

FIG. 2—SHAFT END, TYPE 2

NOTE: The shaft ends may be made optionally according to type 1 or 2. However, it shall be possible to screw the go-gauge for the thread up to the chain line X-X.

TABLE 1—SHAFT ENDS
Dimensions in millimetres

A[a] nom	B	C max	D -1	E₁ max	F₁ h9
17	M12	14.5	18	1.6	3 0 / −0.025
20	M14 × 1.5	16.5	20	2.0	4 0 / −0.03
22	M14 × 1.5	16.5	20	2.0	4 0 / −0.03
22	M16 × 1.5[b]	18.0	20	2.0	4 0 / −0.03
25	M18 × 1.5	20.0	25	2.6	5 0 / −0.03
30	M20 × 1.5	23.0	30	2.6	5 0 / −0.03
35	M24 × 1.5	27.0	35	2.6	5 0 / −0.03

[a] The tolerance for dimension A depends on the type of shaft bearing.
[b] The thread M16 × 1.5 is to be preferred for the shaft ends with 22 mm diameter.

TABLE 2—HUB
Dimensions in millimetres

A[a] nom	E₂ min	F₂ D10	
17	1.8	3	+0.060 / +0.020
20	2.2	4	+0.078 / +0.030
22	2.2	4	+0.078 / +0.030
25	2.8	5	+0.078 / +0.030
30	2.8	5	+0.078 / +0.030
35	2.8	5	+0.078 / +0.030

[a] A is the nominal diameter of the shaft.

2.2 Keyways of hub with taper

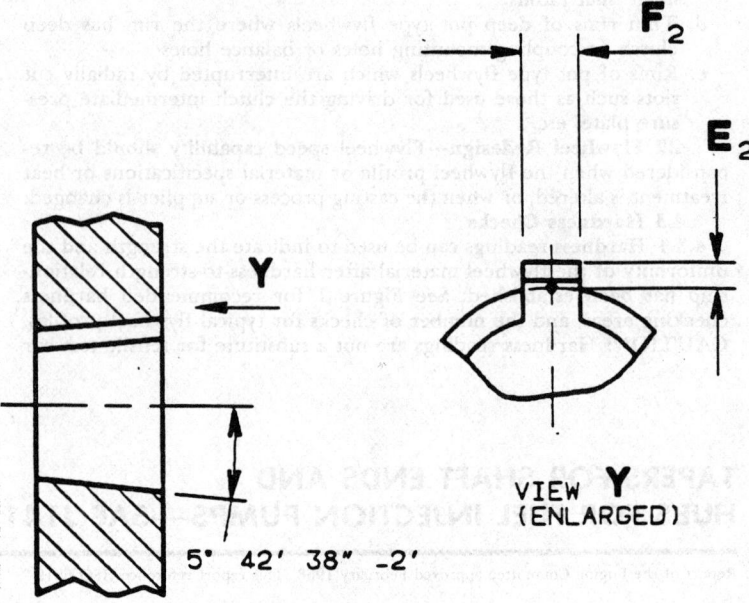

FIG. 3—HUB

NOTE: The length of the hub cone shall be such that, after assembling, the face at the smaller diameter of the hub cone lies so far in front of the line X-X (see Fig. 1 and 2) that the fixing nut can be correctly screwed up.

FILTER BASE MOUNTING—SAE J363 FEB87   SAE Recommended Practice

Report of the Engine Committee approved October 1968 and completely revised February 1987. Rationale statement available. This document is basically equivalent to ISO 6415-1981 and ISO 7654-1983.

1. Scope—This SAE Recommended Practice identifies the necessary dimensional characteristics of filter base mounting configurations to accept the more commonly used lubricating oil and diesel fuel spin-on filters. This SAE Recommended Practice is basically equivalent to the ISO 6415-1981 and ISO 7654-1983 except for differences detailed in the Rationale.

2. General Information

2.1 Lubricating Oil Spin-On Filter Base Mounting—Fig. 1 is a composite drawing which outlines the base mounting and clearance requirements for spin-on oil filters.

Table 1 lists the insert thread size and gasket seal diameters of the base mounts for the more commonly used spin-on oil filters.

For equipment protection, a bypass valve or equivalent device must be incorporated into the full flow lubricating oil system. If a bypass valve is to be located into the filter mounting base, it should operate at one of the following *recommended* pressures.

  60 ± 10 kPa
  90 ± 10 kPa
  140 ± 15 kPa
  200 ± 20 kPa
  250 ± 20 kPa

TABLE 1—LUBRICATING OIL FILTER BASE MOUNTING DIMENSIONS

A Insert Thread Sizes	B I.D. mm	C O.D. mm
*5/8-18 UNF2A	59.0 ± 1.0	79.0 ± 1.0
*3/4-16 UNF2A	59.0 ± 1.0	79.0 ± 1.0
*13/16-16 UN2A	59.0 ± 1.0	79.0 ± 1.0
13/16-16 UN2A	75.0 ± 1.0	93.0 ± 1.0
*7/8-16 UNF2A	59.0 ± 1.0	79.0 ± 1.0
1-12 UNF2A	59.0 ± 1.0	79.0 ± 1.0
*1/16 UN2A	59.0 ± 1.0	79.0 ± 1.0
*1 1/8-16 UN2A	87.0 ± 1.0	113.0 ± 1.0
1 1/2-12 UNF2A	87.0 ± 1.0	113.0 ± 1.0
*1 1/2-16 UN2A	87.0 ± 1.0	113.0 ± 1.0
M-18 × 1.5-6H	59.0 ± 1.0	79.0 ± 1.0
*M-20 × 1.5-6H	59.0 ± 1.0	79.0 ± 1.0
*M-20 × 1.5-6H	75.0 ± 1.0	93.0 ± 1.0

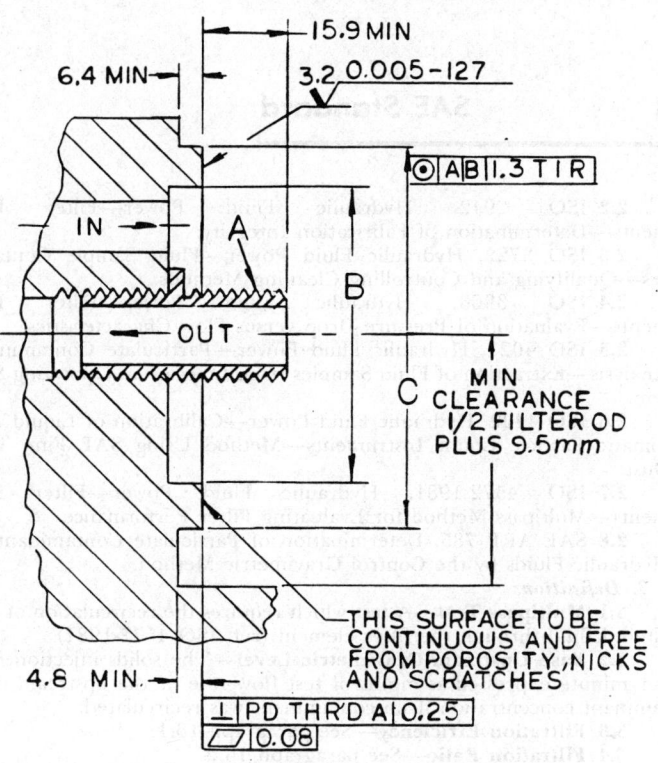

FIG. 1

TABLE 2—DIESEL FUEL FILTER BASE MOUNTING DIMENSIONS

A Insert Thread Sizes	B I.D. mm	C O.D. mm
5/8-11 UNC2A	59.0 ± 1.0	79.0 ± 1.0
13/16-12 UN2A	75.0 ± 1.0	93.0 ± 1.0
*13/16-18 UNS2A	59.0 ± 1.0	79.0 ± 1.0
*7/8-14 UNF2A	59.0 ± 1.0	79.0 ± 1.0
7/8-16 UN2A	59.0 ± 1.0	79.0 ± 1.0
*15/16-16 UN2A	59.0 ± 1.0	79.0 ± 1.0
1-12 UNF2A	75.0 ± 1.0	93.0 ± 1.0
*1-14 UNS2A	59.0 ± 1.0	79.0 ± 1.0
*1 1/16-16 UN2A	87.0 ± 1.0	113.0 ± 1.0
*1 1/4-12 UNF2A	87.0 ± 1.0	113.0 ± 1.0
*1 3/8-16 UNF2A	87.0 ± 1.0	113.0 ± 1.0
*M-14 × 1.5-6H	59.0 ± 1.0	79.0 ± 1.0
M-16 × 1.5-6H	59.0 ± 1.0	79.0 ± 1.0

2.2 Diesel Fuel Spin-On Filter Base Mounting—Fig. 2 is a composite drawing which outlines the base mounting and clearance requirements for spin-on fuel filters. Optional methods of preventing fuel leakage by the insert threads may be provided by means of a radial seal boss Type E or a compression seal member type D as shown respectively in the upper and lower halves of the drawing. The insert is to be threaded full distance F plus at least one full thread.

Table 2 lists the insert thread size and gasket seal diameters of the base mounts for the more commonly used spin-on fuel filters.

NOTE—The insert thread sizes listed in Tables 1 and 2 which are recommended for use are those designated by an asterisk. Others listed are available but not recommended as standard.

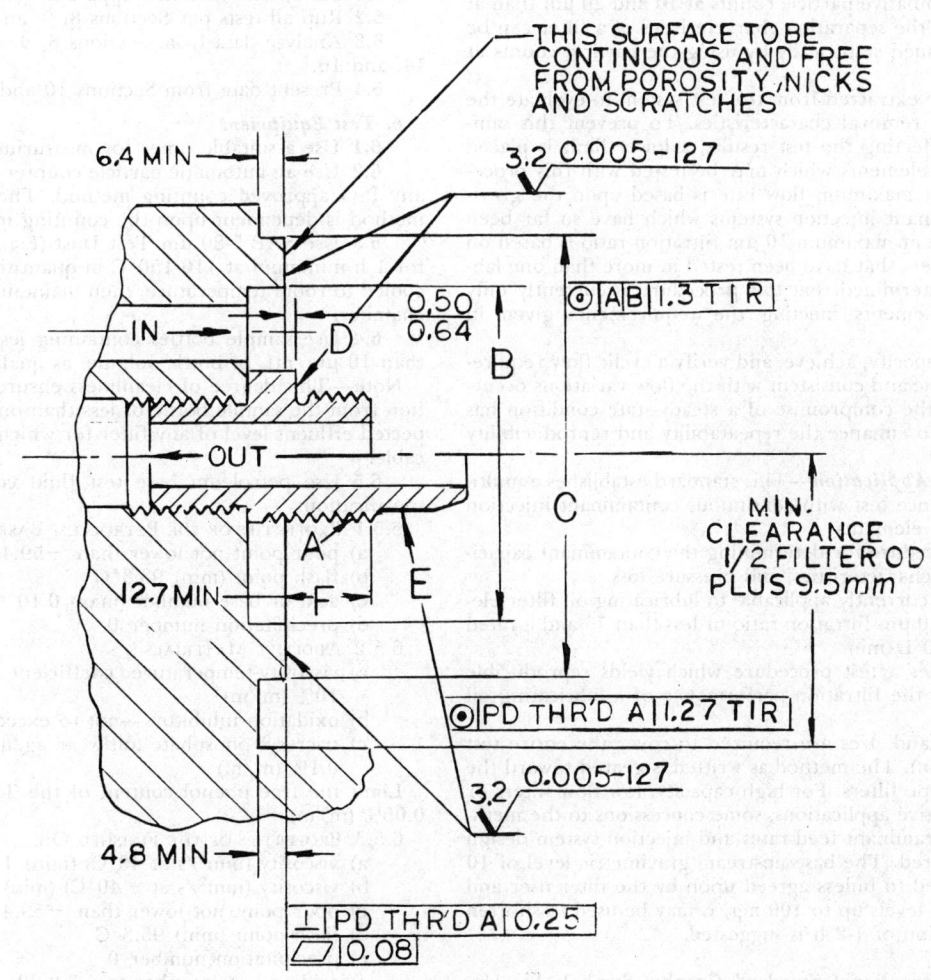

FIG. 2

FULL FLOW LUBRICATING OIL FILTERS— MULTIPASS METHOD FOR EVALUATING FILTRATION PERFORMANCE—SAE J1858 JUN88

SAE Standard

Report of the Engine Committee approved June 1988.

Introduction—The ideal filter for an engine lubricating oil system offers infinite restriction to the passage of particulate contaminants, exhibits zero resistance to the flow of fluid, and provides unlimited capacity for retained contaminant.

An actual filter cannot exhibit such phenomenal performance characteristics. Therefore, test procedures must be available to establish its degree of ideality (filter capacity and efficiency).

The performance characteristics of a filter are a function of the element (configuration and material) and the housing (general configuration and seal design).

In practice, a filter is subjected to a continuous flow of contaminant entrained in the oil until a specified terminal pressure drop (relief valve pressure) results.

Both the length of operating time (prior to reaching the terminal pressure drop) and the contaminant level at any point in the system are functions of the rate of contaminant addition (that is, the rates of contaminant ingression and generation) and the capability of the filter to remove contaminants.

Therefore, a realistic laboratory test which establishes filter capacity must provide the filter with a continuous supply of ingressed contaminant and allow the periodic monitoring of the performance characteristics of the filter.

The contaminant level of the fluid immediately downstream of a filter is directly related to the contaminant level of the upstream fluid. The contamination level of a fluid is given by the particle size distribution. This distribution can be accurately measured for particles greater than 10 μm using currently available automatic particle counters. However, particle size distributions associated with an operating system always exhibit higher cumulative particle counts at 10 and 20 μm than at larger sizes. Therefore, the separation characteristics of a filter can be most accurately determined statistically by using the particle counts at the lower μm sizes.

Fluid samples must be extracted from the test system to evaluate the filter element's particle removal characteristics. To prevent this sampling from adversely affecting the test results, a lower limit is placed upon the rated flow of elements which may be tested with this procedure. Thus, the current maximum flow rate is based upon the gravimetric level of contaminant injection systems which have so far been qualified while the current maximum 10 μm filtration ratio is based on the highest ratio for filters that have been tested in more than one laboratory. It has been determined that this procedure is currently only applicable for filter elements meeting the requirements given in section 1.

Since it is difficult to specify, achieve, and verify a cyclic flow requirement that is both realistic and consistent with the flow variations occurring in actual systems, the compromise of a steady-state condition has been used for this test to enhance the repeatability and reproducibility of results.

1. Scope and Field of Application—This standard establishes a multipass filtration performance test with continuous contaminant injection for lubricating oil filter elements.

It also includes a procedure for determining the contaminant capacity, particulate removal characteristics, and pressure loss.

It also includes a test currently applicable to lubricating oil filter elements which exhibit a 10 μm filtration ratio of less than 75 and a rated flow between 4 and 600 L/min.

This standard provides a test procedure which yields reproducible test data for evaluating the filtration performance of a lubricating oil filter element.

Note—Several test stand sizes are required to cover the entire test flow range (4-600 L/min). The method as written is geared toward the testing of heavy duty type filters. For high capacity, low flow filters or small filters for automotive applications, some concessions to the method in fluid volume, contaminant feed rates and injection system design may have to be considered. The base upstream gravimetric level of 10 mg/L should be adhered to unless agreed upon by the filter user and supplier. Base upstream levels up to 100 mg/L may be used to shorten test times. A test duration of 1-2 h is suggested.

2. References

2.1 ISO 1219, International Standard Graphic Symbols for Hydraulic and Pneumatic Equipment.

2.2 ISO 2942, Hydraulic Fluid Power—Filter Elements—Determination of Fabrication Integrity.

2.3 ISO 3722, Hydraulic Fluid Power—Fluid Sample Containers—Qualifying and Controlling Cleaning Methods.

2.4 ISO 3968, Hydraulic Fluid Power—Filter Elements—Evaluation of Pressure Drop versus Flow Characteristics.

2.5 ISO 4021, Hydraulic Fluid Power—Particulate Contaminant Analysis—Extraction of Fluid Samples from Lines of an Operating System.

2.6 ISO 4402, Hydraulic Fluid Power—Calibration of Liquid Automatic Particle Count Instruments—Method Using SAE Fine Test Dust.

2.7 ISO 4572-1981, Hydraulic Fluid Power—Filter Elements—Multipass Method for Evaluating Filter Performance.

2.8 SAE ARP 785, Determination of Particulate Contaminant in Hydraulic Fluids by the Control Gravimetric Method.

3. Definitions

3.1 *Multipass Test*—A test which requires the recirculation of unaltered fluid through the filter element (ref. ISO 4572-1981).

3.2 *Base Upstream Gravimetric Level*—The solids injection rate per minute expressed as mg/L of test flow rate or the upstream contaminant concentration if no contaminant was recirculated.

3.3 *Filtration Efficiency*—See paragraph 16.1.

3.4 *Filtration Ratio*—See paragraph 16.2.

4. Graphical Symbols

4.1 Graphical symbols used are in accordance with ISO 1219.

5. General Procedure

5.1 Set up and maintain apparatus per Sections 6 and 7.

5.2 Run all tests per Sections 8, 9, and 10.

5.3 Analyze data from Sections 8, 9, and 10 per Sections 11, 12, 14, and 16.

5.4 Present data from Sections 10 and 12 per Sections 13 and 15.

6. Test Equipment

6.1 Use a suitable timer for measuring minutes and seconds.

6.2 Use an automatic particle counter calibrated per ISO 4402, or any ISO approved counting method. The accuracy of this filter test method is dependent upon the counting method used.

6.3 Use SAE 5-80 μm Test Dust (Fig. 2) which has been predried for 1 h minimum at 110-150°C in quantities no larger than 200 g and cooled to room temperature, then maintained until use in a desiccating chamber.

6.4 Use sample bottles containing less than 1.5 particles greater than 10 μm/mL of bottle volume, as qualified per ISO 3722.

Note—This degree of cleanliness ensures a contamination contribution from the sample bottle of less than one-tenth of the minimum expected effluent level of any filter for which this test procedure is applicable.

6.5 Use petroleum base test fluid conforming to the following specifications

6.5.1 PROPERTIES OF THE PETROLEUM BASE STOCK
    a) pour point not lower than −59.4°C
    b) flash point (min) 93.3°C
    c) acid or base number (max) 0.10
    d) precipitation number 0

6.5.2 ADDITIVE MATERIALS
    a) viscosity-temperature coefficient improver—not to exceed 10% (m/m)
    b) oxidation inhibitors—not to exceed 2% (m/m)
    c) tricresyl phosphate antiwear agent—in the amount of 0.5 ± 0.1% (m/m)

Limit the free phenol content of the TCP agent to a maximum of 0.05% (m/m).

6.5.3 PROPERTIES OF THE FINISHED OIL
    a) viscosity (mm^2/s at 40°C) (min) 10.0
    b) viscosity (mm^2/s at −40°C) (min) 500
    c) pour point not lower than −59.4°C
    d) flash point (min) 93.3°C
    e) precipitation number 0
    f) acid or base number (max) 0.20

Note—1 mm^2/s = 1 cSt

6.5.4 COLOR OF FINISHED OIL—Use oil that is clear and transparent and which contains a red dye in a proportion not greater than one part of dye per 10 000 parts of oil (m/m) (used for identification only).

Note—The use of test fluid conforming to these specifications ensures greater reproducibility of results and is based upon current practices and other accepted filter standards. Fluid conforming to these specifications is available worldwide as aircraft hydraulic oil MIL-H-5606.

6.6 Use a filter performance test circuit comprising a "filter test system," and a "contaminant injection system." A typical layout is shown in Fig. 1.

6.6.1 The filter test system consists of:

6.6.1.1 A reservoir constructed with a conical bottom having an included angle of not more than 90 deg and where the oil entering is diffused below the fluid surface.

Note—This reservoir design avoids a horizontal bottom and thus minimizes contaminant settling while the subsurface diffusion reduces the entrainment of air.

6.6.1.2 An oil pump which is essentially insensitive to contaminant at the operating pressure.

Warning—Pumps exhibiting excessive flow pulses will cause erroneous results.

6.6.1.3 A system cleanup filter capable of providing an initial system contamination level of less than 15 particles greater than 10 μm/mL.

6.6.1.4 Pressure gauges, temperature indicator and controller, and flow meter as shown (Fig. 1).

6.6.1.5 Pressure taps in accordance with ISO 3968.

6.6.1.6 A means for turbulent sampling upstream and downstream of the test filter. Sample in accordance with ISO 4021.

6.6.1.7 Interconnecting lines which ensure that turbulent mixing conditions exist throughout the filter test system and that contaminant traps, silting areas, and combinations of cyclonic separation zones and quiescent chambers are avoided.

6.6.2 The contaminant injection system consists of:

6.6.2.1 A reservoir constructed with a conical bottom having an included angle of not more than 90 deg and where the oil entering is diffused below the fluid surface.

Note—This reservoir design avoids a horizontal bottom and thus minimizes contaminant settling while the subsurface diffusion reduces the entrainment of air.

6.6.2.2 A system cleanup filter capable of providing an initial system contamination level of less than 1000 particles greater than 10 μm/mL and a gravimetric level less than 2% of the calculated level at which the test is being conducted.

6.6.2.3 An oil pump (centrifugal or of another type) which does not fracture the contaminant particles.

6.6.2.4 A sampling means for the extraction of a small portion of the injection flow from a point in the contaminant injection system where active circulation of the fluid exists. Sample in accordance with ISO 4021.

6.6.2.5 Interconnecting lines which ensure that turbulent mixing conditions exist throughout the filter test system and that contaminant traps, silting areas, and combinations of cyclonic separation zones and quiescent chambers are not present. In particular, turbulent mixing conditions must exist throughout the length of the line carrying the injection fluid.

6.7 Use membranes and associated laboratory equipment suitable for carrying out the double membrane gravimetric method (SAE ARP 785).

7. Test Conditions Accuracy—Set up and maintain equipment accuracy within the limits given in Table 1.

TABLE 1—TEST CONDITIONS ACCURACY

Test Condition	Units	Maintain Within True Value
Flow	L/min	±2%
Pressure	Pascal	±2%
Temperature	°C	40° ±2°C
Volume	Liters	±2%

8. Filter Performance Test Circuit Validation Procedures

Note—These validation procedures reveal the effectiveness of the filter performance test circuit in maintaining contaminant entrainment and/or preventing contaminant size modification.

8.1 Validation of Filter Test System

8.1.1 Validate at the minimum flow rate at which the filter test system will be operated.

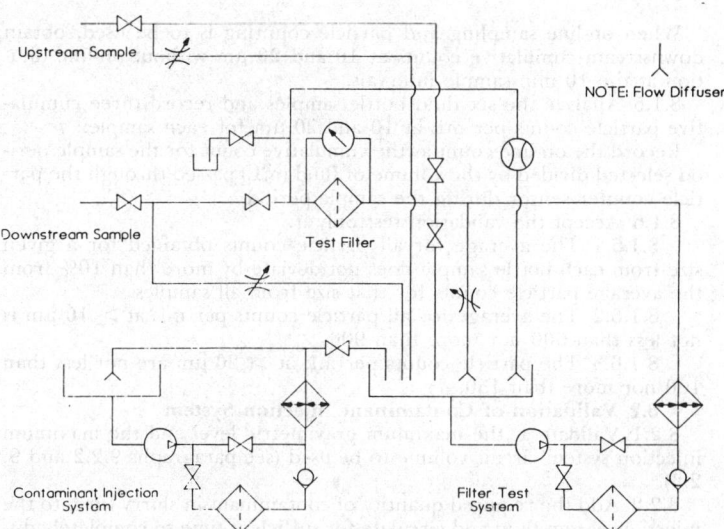

FIG. 1—TYPICAL FILTER PERFORMANCE TEST CIRCUIT WITH PROVISION FOR ON-LINE SAMPLING

Note—Install a conduit in place of a test filter during validation.

8.1.2 Adjust the total system volume to be numerically equal to one-fourth of the value of the minimum flow volume per minute through the filter *plus 4 liters*.

Note—This is the volume to flow ratio required for the filter test procedure (see paragraph 9.3.3).

8.1.3 Contaminate the system fluid to the calculated gravimetric level of 5 mg/L using SAE 5-80 μm Test Dust.

Note—This contamination level is below the saturation limitations of automatic particle counters.

8.1.4 Circulate the fluid in the test system for 1 h and extract fluid samples at 10 min intervals from the downstream sampling valve.

PRICE LIST - PP7B

SAE 5-80 MICROMETER TEST DUST

Container	Approx. Net. Wt. Per Jar	Cost Per Jar
1 Gallon Jar	8 pounds	$45.92 ($5.74/Lb.)

* Price is in U.S. Dollars/Pound.
* Each shipment will include Particle Size Distribution as analyzed with a Coulter Counter.
* f.o.b. - shipping point.
* Price is subject to change without notice.
* All orders less than $200 are subject to a $15.00 handling charge.

SPECIFIED PARTICLE SIZE DISTRIBUTION

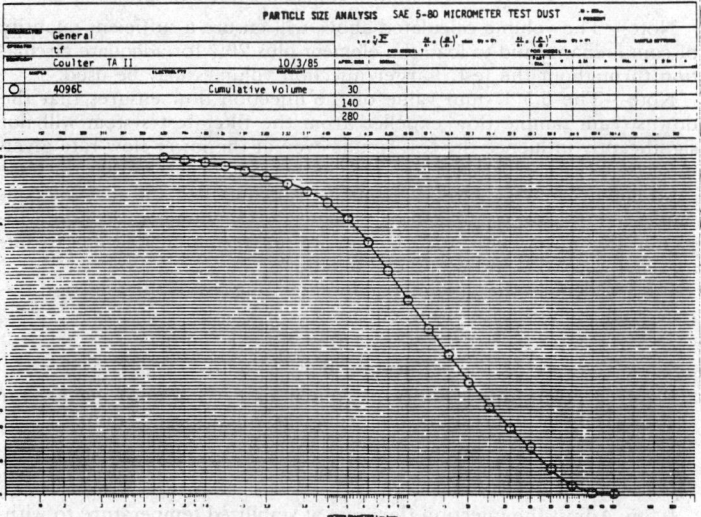

FIG. 2

When on-line sampling and particle counting is to be used, obtain downstream cumulative counts at 10 and 20 µm without on-line dilution at the 10 min sample intervals.

8.1.5 Analyze the six fluid bottle samples and record three cumulative particle counts per mL at 10 and 20 µm for each sample.

Record the on-line count as the cumulative count for the sample period selected divided by the volume of fluid (mL) passed through the particle counter sensor during the sample period.

8.1.6 Accept the validation test only if:

8.1.6.1 The average for all particle counts obtained for a given size from each bottle sample does not deviate by more than 10% from the average particle counts for that size from all samples.

8.1.6.2 The average for all particle counts per mL at > 10 µm is not less than 600 nor more than 900.

8.1.6.3 The particle counts per mL at > 20 µm are not less than 100 nor more than 150.

8.2 Validation of Contaminant Injection System

8.2.1 Validate at the maximum gravimetric level and the maximum injection system circuit volume to be used (see paragraphs 9.2.2 and 9.2.3).

8.2.2 Add the required quantity of contaminant in slurry form to the injection system fluid and circulate for sufficient time to completely disperse the contaminant.

Note—All systems may not disperse contaminant at the same rate.

8.2.3 Extract fluid samples at the point where the injection fluid is discharged into the filter test system at 30, 60, 90, and 120 min and analyze each sample gravimetrically. These samples should be taken at the intended test injection flow rate.

8.2.4 Accept the validation test only if the gravimetric level of each sample is within ±5% of the average of the four samples and within ±5% of the calculated gravimetric value.

9. Preliminary Preparation

9.1 Test Filter Assembly

9.1.1 Ensure that the test fluid cannot bypass the filter element to be evaluated.

9.1.2 Subject the test filter element to a fabrication integrity test in accordance with ISO 2942 using MIL-H-5606 fluid prior to the multipass test or following the multipass test if the element is not readily accessible as in the spin-on configuration.

9.1.3 Disqualify the element from further testing if it fails to meet the designated test pressure.

9.2 Contaminant Injection System

9.2.1 Using 10 mg/L as the base upstream gravimetric level, calculate the predicted test time (T') in minutes by the following equation:

$$T' = \frac{\text{(apparent capacity of filter element, mg)}}{(10 \text{ mg/L}) * \text{(test flow rate, L/min)}}$$

Note—A second element may be tested for capacity analysis if the value of the apparent capacity of the test element is not supplied by the filter manufacturer.

9.2.2 Calculate the minimum volume required for the operation of the injection system (V, liters) which is compatible with the above predicted test time (T') and a value for the injection flow (0.5 L/min) using the following equation:

$$V = 1.2 * (T', \text{min}) * (\text{injection flow, L/min})$$

Note—The volume calculated above will ensure a sufficient quantity of contaminant fluid to load the element plus 20% for adequate circulation throughout the test. Larger injection volumes may be used.

Note—The 0.5 L/min value of the injection flow ensures that the downstream sample flow expelled from the filter test system will not significantly influence the test results even at the lower flow rate given in section 1. *Lower injection flow rates may be used* provided that the base upstream gravimetric level of 10 mg/L is maintained and that the contaminant injection system can be validated at the intended flow rate. Injection flow rates below 0.25 L/min are not recommended due to silting characteristics and accuracy limitations.

9.2.3 Calculate the gravimetric level (Y', mg/L) of the injection system fluid using the following equation:

$$Y' = \frac{(10 \text{ mg/L}) * (\text{test flow, L/min})}{(\text{injection flow, L/min})}$$

9.2.4 Calculate the quantity of contaminant (w, grams) needed for the contaminant injection system using the following equation:

$$W = \frac{(Y', \text{mg/L}) * (\text{injection system volume, l})}{1000}$$

9.2.5 Adjust the injection flow rate at stabilized temperature to within ±5% of the value selected in paragraph 9.2.2 and maintain throughout the test.

9.2.6 Adjust the total volume of the contaminant injection system to the value determined in paragraph 9.2.2.

9.2.7 Circulate the fluid in the contaminant injection system through its system clean-up filter until a contamination level of less than 1,000 particles greater than 10 µm/mL and a gravimetric level of less than 2% of the value determined in paragraph 9.2.3 are attained.

9.2.8 Bypass the system clean-up filter after the required initial contamination has been achieved.

9.2.9 Add in slurry form the quantity of contaminant (grams) as determined in paragraph 9.2.4 to the injection system reservoir.

9.2.10 Add 0.25 mL of Shell ASA-3 antistatic agent, or equivalent, per 10 L of test oil to the charged injection system and circulate the fluid to thoroughly disperse the contaminant. Add the antistatic agent each time the system is charged.

Note—All systems may not disperse contaminant at the same rate.

9.3 Filter Test System

9.3.1 Install the filter housing (without the test element) in the filter test system.

9.3.2 Circulate the fluid in the filter test system at the rated flow and a stabilized test temperature of 40 ± 2°C and record the pressure drop of the empty filter housing.

9.3.3 Adjust the total fluid volume of the filter test system (exclusive of the system clean-up filter circuit) such that it is numerically equal to one-fourth of the value of the test volume flow per minute through the filter *plus 4 liters.*

Note—Repeatable results require that the system volume be held constant. The specified 1:4 volume to flow ratio plus 4 L minimizes the physical size of the system reservoir as well as the quantity of test fluid required and maximizes the mixing conditions in the reservoir. The additional 4 liters provides sufficient fluid volume to conduct tests with flows as low as 4 L/min.

9.3.4 Circulate the fluid in the filter test system through the clean-up filter until a contamination level of less than 15 particles greater than 10 m/mL is attained.

Note—The time required to achieve the contamination level is directly proportional to the particle separation capability of the clean-up filter used.

9.3.5 When using the bottle sampling method, select and install suitable lengths of capillary tubing restriction upstream and downstream of the test filter such that the initial upstream sample flow is 0.3 ± 0.05 L/min and the downstream sample flow is within 5% of the injection flow. Maintain uninterrupted flow from the two sampling points during the entire test.

When using on-line sampling and particle counting, also provide the bottle sampling capability just described. The on-line sample delivery lines consist of small bore tubing loops originating at the bottle sampling point, extending to the on-line dilution and particle counting system, with a return to the test system reservoir. Flow in the upstream and downstream sample line loops is to be maintained at 0.3 - 0.5 L/min. Silting in the sample lines is to be avoided by sizing the tubing bore to maintain line velocity (see Fig. 1.).

9.3.6 Return the bottle upstream sampling flow of the test filter directly to the reservoir when sampling is not in progress.

9.3.7 Collect the downstream bottle sampling flow outside of the filter test system to assist in maintaining a constant system volume which should be kept within 2% of the required system volume.

10. Filter Performance Test

10.1 Install the filter element in its housing and subject the assembly to the specified test conditions (test flow with test temperatures of 40 ± 2°C) and recheck the fluid level. Add 0.25 mL of Shell ASA-3 antistatic agent or equivalent per 10 L of test oil to the reservoir at the start of each test.

10.2 Measure and record the clean assembly pressure drop. Calculate and record the clean element pressure drop (clean assembly pressure drop minus the housing pressure drop measured in paragraph 9.3.2).

10.3 Calculate the pressure drop corresponding to an increase of 80% and 100% of the net pressure drop (terminal pressure drop minus the clean element pressure drop).

10.4 Obtain a sample upstream of the test filter element (paragraph 9.3.6) to determine the system initial contamination level.

Note—Take all samples in such a manner as to minimize the aeration of the fluid sample.

10.5 Obtain a fluid sample from the contaminant injection system.

10.6 Measure and record the injection flow rate.

10.7 Initiate the filter test as follows:

10.7.1 Bypass the system clean-up filter.

SAE J1858 MULTIPASS TEST - LUBRICATING OIL FILTERS

DATE: / /	TECHNICIAN:		TEST #:
AMBIENT TEMP:	REL HUMIDITY:		PROJECT #:
FILTER IDENTIFICATION:			
FABRICATION INTEGRITY TEST RESULTS:			
INJECTION SYSTEM FLOW RATE: L/MIN		INJECTION SYSTEM VOLUME:	L
CONTAMINANT TYPE/BATCH:		THEORETICAL GRAV LEVEL:	MG/L
CONTAMINANT ADDED TO INJECTION SYSTEM: G		ASA-3 ADDED:	CC
INJ SYSTEM INITIAL CLEANLINESS:	PARTICLES/ML > 10um		MG/L
TEST FLOW RATE: L/MIN	TEST VOLUME: L	ASA-3 ADDED:	CC
TEST SYSTEM INITIAL CLEANLINESS:	PARTICLES/ML > 10 um		
HOUSING PRESSURE DROP:		CLEAN ASSY PRESSURE DROP:	
CLEAN ELEMENT PRESSURE DROP:		TERMINAL PRESSURE DROP:	
NET PRESSURE DROP:		80% NET PRESSURE DROP:	

% NET DROP	-	-	-	-	-	-	80%	100%
ASSY DROP								
TIME (MIN)								

INJ FLOW RATE INITIAL:	L/MIN	FINAL:	L/MIN	AVERAGE:		L/MIN
INJ GRAV LEVEL INITIAL:	MG/L	FINAL:	MG/L	AVERAGE:		MG/L
ACTUAL BASE UPSTREAM GRAV LEVEL:			MG/L	FINAL SUMP GRAV:		MG/L

FILTRATION RATIO AND EFFICIENCY (FROM 3B)						
PARTICLE SIZE	> 10um	> 15um	> 20um	> 25um	> 30um	> 40um
AVG FILTRATION RATIO						
AVG EFFICIENCY %						

CAPACITY:	GRAMS	MINIMUM 10um RATIO:

COMMENTS:

FIG. 3A

SAE J1858 MULTIPASS TEST - LUBRICATING OIL FILTERS

DATE: / /	TECHNICIAN:		TEST #:
AMBIENT TEMP:	REL HUMIDITY:		PROJECT #:
FILTER IDENTIFICATION:			

PARTICLE DISTRIBUTION ANALYSIS (PARTICLES PER MILLILITER)
COUNTS CORRECTED FOR DILUTION (MEASURED COUNTS X DILUTION FACTOR)

PARTICLE SIZE	UPSTREAM COUNTS	DOWNSTREAM COUNTS	FILTRATION RATIO	EFFICIENCY (%)
SAMPLE TAKEN AT _____ MINUTES				
> 10um				
> 15um				
> 20um				
> 25um				
> 30um				
> 40um				
SAMPLE TAKEN AT _____ MINUTES				
> 10um				
> 15um				
> 20um				
> 25um				
> 30um				
> 40um				
SAMPLE TAKEN AT _____ MINUTES				
> 10um				
> 15um				
> 20um				
> 25um				
> 30um				
> 40um				

FIG. 3B

10.7.2 Allow the injection flow to enter the filter test system reservoir.

10.7.3 Start the timer.

10.7.4 Start the downstream sample flow.

10.8 Record the test time (minutes) required for the pressure drop across the filter assembly to increase by 80% and 100% of the net pressure drop.

10.9 Extract upstream and downstream samples simultaneously when using bottle sampling at 10 min from test initiation and at 10-min intervals to terminal pressure drop. Extract an upstream sample for gravimetric analysis when the pressure drop across the filter assembly has increased by 80% of the net pressure drop.

Note—The time of the upstream and downstream sample of each pair should not differ by more than 30 s when using sample bottles.

10.10 Divert into the on-line dilution and particle counting system fluid from the upstream and downstream sampling lines when using on-line sampling. Simultaneously, collect the flow from the downstream bottle sampling point outside the filter test system to assist in maintaining a constant system volume which should be kept within 2% of the required system volume.

10.11 Conclude the test by stopping the flow to the test filter.

10.12 Obtain a final fluid sample from the contaminant injection system.

10.13 Measure and record the final injection flow rate.

11. Data Accuracy—Select and maintain instrumentation so that data accuracy is within the limits of Table 2, unless otherwise specified.

TABLE 2—DATA ACCURACY

Quantity	Unit	Accuracy Within True Value
Injection Flow Rate	l/min	±5%
Base Upstream Gravimetric Level	mg/l	± mg/l

12. Calculations

12.1 Analyze the 10-min interval and final bottle samples extracted from the filter test system by determining the number of particles greater than 10, 15, 20, 25, 30, and 40 μm with an automatic particle counter calibrated per ISO 4402 or any ISO approved counting method.

Note—Care should be taken to dilute samples appropriately to avoid exceeding the saturation limit determined by the approved calibration procedure for the particular counting method used.

Note—When on-line sampling and particle counting is used, adjust the dilution at the start of the test to the anticipated maximum dilution required during the test to avoid particle counter saturation. Sample flow rates of approximately 25 mL per min into the dilution system are recommended. Obtain upstream and downstream cumulative particle counts at the six sizes indicated at 10-min sample intervals. Record the on-line count at the 10-min intervals as the cumulative count for the sample interval in minutes divided by the milliliters of fluid passed through the particle counter sensor in the same sample interval and multiplied by the dilution factor.

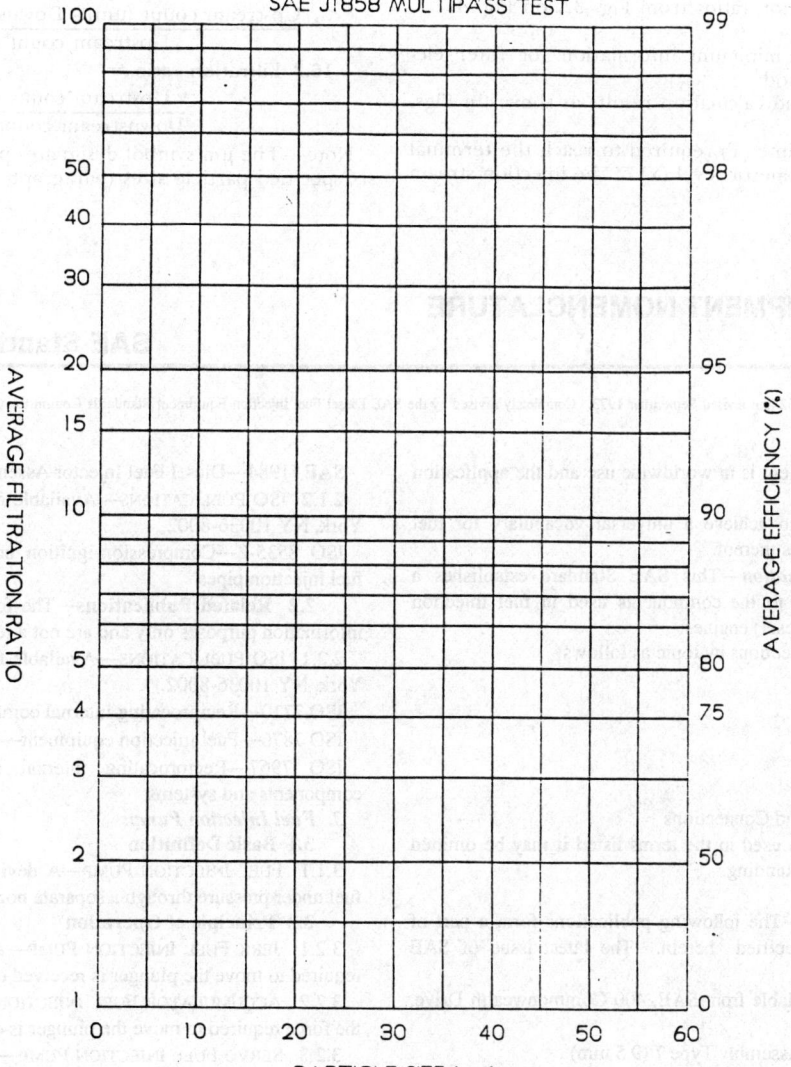

FIG. 4

12.2 Accept the test only if the number of particles greater than 10 μm/mL in the initial bottle sample from the filter test system is less than 15. For on-line sampling, start the test when the particle count per mL is less than eight after correcting for dilution.

12.3 Conduct a gravimetric analysis on the two samples extracted from the contaminant injection system and on the upstream sample extracted from the filter test system at the 80% sample point.

Note—The final sample for gravimetric analysis is taken at the 80% point because it often overlaps the 100% point.

12.3.1 Record the 80% gravimetric value as the final system gravimetric level.

12.3.2 Calculate the average (Y) of the gravimetric levels for the two samples from the contaminant injection system.

12.3.3 Accept the test only if the gravimetric level of each sample is within ±10% of this average.

12.4 Calculate and record the injection flow rate by averaging the measurements taken at the beginning and end of the test.

12.4.1 Accept the test only if this value is equal to the selected value ±5%.

12.5 Calculate and record the actual base upstream gravimetric level by multiplying the average injection gravimetric level (Y, mg/L) by the average injection flow rate (l/min) per paragraph 12.3 and dividing by the test flow rate (l/min).

12.5.1 Accept the test only if this value is equal to 10 ± 1 mg/L or as covered under the scope and field of application.

12.6 Calculate the filtration ratio and filtering efficiency as shown in Section 16.

12.6.1 Record these calculated ratios as shown in Figs. 3A and 3B.

12.6.2 Record these calculated efficiencies as shown in Figs. 3A and 3B.

12.6.3 Plot the average filtration ratios from Fig. 3A on Fig. 4.

13. Data Presentation

13.1 Record the following minimum information for filter elements evaluated using this method:

13.1.1 Present all test data and calculation results as shown in Figs. 3A and 3B.

13.2 Using the actual test time (T) required to reach the terminal pressure drop, the average gravimetric level (Y) of the injection stream and the average injection flow rate, calculate the filter element test dust capacity (a) using the following equation:

$$a, \text{grams} = \frac{(Y, \text{mg/L}) * (\text{injection flow, L/min}) * (T, \text{min})}{1000}$$

13.2.1 Record the test dust capacity as shown in Figs. 3A and 3B.

13.3 Report the values of the gravimetric levels obtained in paragraph 12.3.

13.4 Have available a record of the following minimum test data in test reports referencing this method:

a) All physical values pertaining to the test.
b) All additional provisions or modifications pertaining to the test.
c) Record the particle counting method used.

14. Criteria for Acceptance

14.1 Compare the average filtration ratio and filtration efficiency with the designated values.

14.2 Compare the filter element test dust capacity (a) with the designated value.

14.3 Check that there is no visual evidence of filter element damage as a result of performing this test.

15. Summary of Designated Information—The following designated information is needed when applying this standard to a particular application or use:

a) Fabrication integrity test pressure per ISO 2942.
b) Filter element test flow rate.
c) Terminal pressure drop.
d) The acceptable average filtration ratio and average efficiency.
e) The minimum acceptable filter element capacity (a).

16. Calculations

16.1 Filtering efficiency percent =

$$\frac{\text{Upstream count (μm)} - \text{Downstream count (μm)}}{\text{Upstream count (μm)}} * 100$$

16.2 Filtration ratio =

$$\frac{\text{Upstream count (μm)}}{\text{Downstream count (μm)}} * 100$$

Note—The μm symbol designates particle counts taken at one of the six specified particle sizes (paragraph 12.1).

(R) FUEL INJECTION EQUIPMENT NOMENCLATURE—SAE J830 OCT92

SAE Standard

Report of Engine Committee approved April 1962 and revised September 1973. Completely revised by the SAE Diesel Fuel Injection Equipment Standards Committee October 1992.

Foreword—Fuel injection equipment is in worldwide use and the application of such requires precise terminology.

This SAE Standard is intended to achieve a universal vocabulary for fuel injection systems and the components thereof.

1. Scope and Field of Application—This SAE Standard establishes a vocabulary and definitions relating to the components used in fuel injection systems for compression ignition (diesel) engines.

Definitions are separated into six sections by topic as follows:

Section 3—Fuel Injection Pumps
Section 4—Fuel Injectors
Section 5—Unit Injectors
Section 6—Governors
Section 7—Timing Devices
Section 8—High Pressure Pipes and Connections

NOTE - When the word "fuel" is used in the terms listed it may be omitted providing there can be no misunderstanding.

2. References

2.1 Applicable Documents—The following publications form a part of this specification to the extent specified herein. The latest issue of SAE publications shall apply.

2.1.1 SAE PUBLICATIONS—Available from SAE, 400 Commonwealth Drive, Warrendale, PA 15096-0001.

SAE J347—Diesel Fuel Injector Assembly Type 7 (9.5 mm)
SAE J1949—Diesel Engines—High Pressure Fuel Injection Pipe End Connections with 60 degree Female Cone
SAE J1958—Diesel Engines—Steel Tubes for High Pressure Fuel Injection Pipes (Tubing)
SAE J1984—Diesel Fuel Injector Assembly—Type 28 (9.5 mm)

2.1.2 ISO PUBLICATIONS—Available from ANSI, 11 West 42nd Street, New York, NY 10036-8002.

ISO 8535-2—Compression-ignition engines—Steel tubes for high-pressure fuel injection pipes

2.2 Related Publications—The following publications are provided for information purposes only and are not a required part of this document.

2.2.1 ISO PUBLICATIONS—Available from ANSI, 11 West 42nd Street, New York, NY 10036-8002.

ISO 2710—Reciprocating internal combustion engines—Vocabulary
ISO 7876—Fuel injection equipment—Vocabulary
ISO 7967—Reciprocating internal combustion engines—Vocabulary of components and systems

3. Fuel Injection Pumps

3.1 Basic Definition

3.1.1 FUEL INJECTION PUMP—A device which delivers metered volumes of fuel under pressure through a separate nozzle or nozzles.

3.2 Principle of Operation

3.2.1 JERK FUEL INJECTION PUMP—An injection pump in which the force required to move the plunger is received directly from the prime motivator.

3.2.2 ACCUMULATOR FUEL INJECTION PUMP—An injection pump in which the force required to move the plunger is drawn from an energy accumulator.

3.2.3 SERVO FUEL INJECTION PUMP—An injection pump in which the force required to move the plunger comes from an energy source external to the pump, with or without an intermediate intensifying device.

3.3 Form of Energy Input

3.3.1 MECHANICAL FUEL INJECTION PUMP—An injection pump which is actuated solely by mechanical means.

3.3.2 ELECTRICAL FUEL INJECTION PUMP—An injection pump which is actuated solely by electrical means.

3.3.3 HYDRAULIC FUEL INJECTION PUMP—An injection pump which is actuated solely by hydraulic means.

NOTE—An injection pump may be actuated by a combination of the above three energy inputs (e.g., hydromechanical, hydroelectrical, etc.)

3.4 Mode of Energy Application

3.4.1 RECIPROCATING FUEL INJECTION PUMP—A mechanical injection pump without an integral camshaft for reciprocating the plunger(s) of the pumping element(s).

3.4.2 ROLLER FUEL INJECTION PUMP (ROLLER PUMP)—A reciprocating injection pump with a roller tappet.

3.4.3 DRIVESHAFT FUEL INJECTION PUMP—A mechanical injection pump with an integral driveshaft or camshaft for actuating the plunger(s) of the pumping element(s).

Such a pump may consist also of a separate cambox having mounted on it a number of individual fuel injection pumps.

3.4.4 CAMSHAFT FUEL INJECTION PUMP—A driveshaft fuel injection pump with an integral camshaft for actuating the plunger(s) of the pumping element(s) and having any of the types of attachment defined herein.

3.5 Cylinder Arrangement

3.5.1 SINGLE CYLINDER FUEL INJECTION PUMP—An injection pump with only one pumping element and one outlet.

NOTE—A single cylinder pump serving only one engine cylinder may be termed "individual fuel injection pump."

3.5.2 IN-LINE FUEL INJECTION PUMP—An injection pump having the axes of its pumping elements arranged parallel to each other and in one plane.

3.5.3 CYLINDRICAL FUEL INJECTION PUMP—An injection pump in which the axes of its pumping elements are arranged cylindrically and concentrically with the driveshaft axis.

3.5.4 ROTARY FUEL INJECTION PUMP—A driveshaft injection pump in which the axis (axes) of the pumping element(s) rotate about a common axis to perform the working cycles.

3.6 Type of Distribution

3.6.1 MULTICYLINDER FUEL INJECTION PUMP—An injection pump having multiple pumping elements and an equal number of outlets.

NOTE—A multicylinder pump serving more than one engine cylinder may be termed "block type fuel injection pump."

3.6.2 DISTRIBUTOR FUEL INJECTION PUMP—An injection pump in which the fuel deliveries are directed to the appropriate nozzle(s) by at least one distributing device.

3.7 Type of Attachment

3.7.1 BASE MOUNTED FUEL INJECTION PUMP—A driveshaft injection pump having a flat mounting plane which is parallel to the driveshaft axis and at right angles to the axis(es) of the pumping elements.

3.7.2 BASE FLANGE MOUNTED FUEL INJECTION PUMP—A reciprocating injection pump with a mounting flange at right angles to the axis(es) of the pumping element(s) and with the fuel inlet, fuel control mechanism and fuel outlet above the flange.

3.7.3 HIGH FLANGE MOUNTED FUEL INJECTION PUMP—A reciprocating injection pump with a mounting flange at right angles to the axis(es) of the pumping element(s) and with the fuel inlet, fuel control mechanism below the flange.

3.7.4 SIDE MOUNTED FUEL INJECTION PUMP—An injection pump with a mounting face parallel to the pumping element(s) and parallel to the axis of the camshaft (whether included in the pump or not).

3.7.5 END FLANGE MOUNTED FUEL INJECTION PUMP—A driveshaft injection pump having a mounting flange arranged perpendicular to the driveshaft.

3.7.6 CRADLE MOUNTED FUEL INJECTION PUMP—A driveshaft injection pump having a cylindrical mounting surface coaxial with the driveshaft.

3.8 Fuel Metering Process

3.8.1 METERING—The process of establishing any required fuel delivery within the operating range of the fuel injection system using various means of control.

3.8.2 PORT AND HELIX METERING—A system of metering by means of one or more oblique grooves in the plunger and one or more ports in the barrel, or inversely.

3.8.3 SLEEVE METERING—A system of metering incorporating a movable sleeve by which port opening and/or closing is controlled.

3.8.4 INLET METERING—A system of metering by controlling the amount of fuel entering the pumping chamber during the filling or charging portion of the pump cycle.

3.8.5 VARIABLE STROKE METERING—A system of metering by controlling the plunger stroke.

3.8.6 VALVE METERING—A system of metering by cyclic operation of valve(s) to control the effective pumping stroke.

3.8.7 SHUTTLE (DISPLACEMENT) METERING—A system of metering by variable displacement of an auxiliary free piston.

3.8.8 CAM LIFT—The geometric difference between the base of the cam profile and the cam peak.

3.8.9 PLUNGER STROKE—The nominal distance through which a plunger moves between two successive reversals of its direction of motion.

3.8.10 PLUNGER LIFT TO CUT-OFF PORT CLOSING—That part of the plunger stroke from initial movement of the plunger to the position at which the cut-off port closes, determining the beginning of geometric delivery.

3.8.11 ANGLE TO CUT-OFF PORT CLOSING—The angular movement of the driveshaft equivalent to plunger lift to cut-off port closing.

3.8.12 PLUNGER LIFT TO SPILL PORT OPENING—That part of the plunger stroke from initial movement of the plunger to the position at which the spill port opens, determining the end of geometric delivery.

3.8.13 ANGLE OF SPILL PORT OPENING—The angular movement of the driveshaft equivalent to plunger lift to spill port opening.

3.8.14 CUT-OFF PORT—That port which is closed by the movement of the plunger at the beginning of geometric delivery.

3.8.15 SPILL PORT—The port which is opened by the movement of the plunger at the end of the geometric delivery.

3.8.16 INLET PORT—A port which admits fuel to the pumping chamber.

NOTE—The inlet port may or may not serve also as the cut-off and/or spill port(s).

3.8.17 GEOMETRIC FUEL DELIVERY STROKE—That part of the plunger stroke between the beginning and end of geometric fuel delivery.

3.8.18 RETRACTION VOLUME (UNLOADING VOLUME)—The volume by which the volume of the high pressure system is increased after the end of delivery.

3.8.19 GEOMETRIC RETRACTION VOLUME—The geometric volume by which the volume of the high pressure system is increased after the end of geometric fuel delivery.

3.8.20 RETRACTION STROKE—The calculated or real part of the plunger stroke corresponding to the retraction volume or the geometric retraction volume.

3.8.21 EFFECTIVE STROKE—The difference between the delivery stroke and the retraction stroke, geometric or otherwise, respectively.

3.8.22 REMAINDER STROKE—The part of the plunger stroke, if any, between the end of geometric fuel delivery stroke and the end of the plunger stroke.

3.8.23 HEAD CLEARANCE—The distance between the top surface of a fuel injection pump plunger (or plunger assembly) at the end of the plunger stroke and the nearest component that would limit its further movement.

3.8.24 FUEL DELIVERY—The metered volume of fuel delivered by a fuel injection system during one working cycle.

3.8.25 GEOMETRIC FUEL DELIVERY—The nominal volume of fuel displaced as defined by the geometric fuel delivery stroke.

3.9 Components and Assemblies

3.9.1 INJECTION PUMP ASSEMBLY—An assembly consisting of the injection pump proper, governor, supply or feed pump, and any other additional devices which together form a unit.

3.9.2 PUMPING ELEMENT—The combination in an injection pump of a pumping plunger and its barrel.

3.9.3 PUMPING ASSEMBLY—That combination of parts in an injection pump extending between the plunger foot and the high pressure pipe (or nozzle in a pump injector).

3.9.4 METERING SPILL VALVE—A controllable valve the operation of which controls the commencement and/or termination of the geometric delivery by permitting spill from the pumping chamber.

3.9.5 SPILL VALVE—A valve which by cyclic operation permits spill from the pumping chamber and therefore controls the commencement and/or termination of the geometric delivery.

3.9.6 INLET VALVE—An automatic valve which permits fuel to enter the pumping chamber.

3.9.7 METERING INLET VALVE—A device which performs inlet metering.

3.9.8 DELIVERY VALVE—A valve located at the outlet of the pumping chamber. Various types of delivery valves, by suitable design, can perform one or more of the various functions listed as follows:
 a. non-return
 b. constant volume unloading
 c. variable volume unloading
 d. pressure time volume unloading (including reverse flow damping)
 e. constant pressure unloading
 f. variable pressure unloading

3.9.9 DELIVERY VALVE HOLDER—A device which retains the delivery valve and associated components and sometimes forms a pump outlet.

3.9.10 INJECTION PUMP HOUSING—The housing which contains or holds the functional components of the injection pump proper including the driveshaft or camshaft when incorporated in the pump.

3.9.11 FUEL GALLERY—The passageway(s) of an injection pump through which fuel is transferred to and/or from the pumping elements.

3.9.12 CONTROL ROD (RACK)—The rod by means of which the fuel delivery is controlled.

3.9.13 CONTROL ARM (PINION)—An intermediate component or assembly which engages the metering device (plunger) and the control rod (rack).

3.9.14 METERING SLEEVE—A movable component by which sleeve metering is achieved.

3.9.15 MAXIMUM FUEL STOP (FULL LOAD STOP)—A device which limits the maximum fuel delivery of an injection pump for a given application.

3.9.16 HYDRAULIC HEAD ASSEMBLY—The assembly containing the pumping elements, and metering and distributing components and may include the delivery valve(s) for a distributor fuel injection pump.

3.10 Auxiliary Devices

3.10.1 EXCESS FUEL DEVICE—A device (automatic or manual) which permits, for engine starting only, a fuel delivery in excess of that controlled by the maximum fuel stop.

3.10.2 BOOST CONTROL (BOOST PRESSURE CONTROLLED MAXIMUM FUEL STOP)—A device which limits the maximum fuel delivery in relation to the engine boost (charge air) pressure.

3.10.3 ALTITUDE CONTROL (ALTITUDE CONTROLLED MAXIMUM FUEL STOP)—A device which limits the maximum fuel delivery in relation to the altitude pressure (barometric pressure) where the engine operates.

3.10.4 TORQUE CONTROL—A device which modifies the maximum fuel delivery at speeds below engine rated speed.

3.11 General Terms

3.11.1 DIRECTION OF ROTATION—The direction of rotation is designated "clockwise" or "counterclockwise" when viewing the driven end of the driveshaft.

3.11.2 FUEL INJECTION PUMP SPEED—The speed of the injection pump driveshaft when incorporated in the pump; alternatively, if the driveshaft is not incorporated in the pump, the frequency of delivery from an outlet.

3.11.3 INJECTION ORDER—The order in which the several pump outlets, with a specific direction of driveshaft rotation, deliver fuel. Where applicable the outlets shall be numbered in ascending sequence with number one nearest the driven end of the camshaft. In other cases the outlets shall be identified.

3.11.4 RESIDUAL PRESSURE—The mean pressure at the high pressure outlet of an injection pump before the start of any event related to the next working cycle.

3.11.5 PHASING—The geometric (commonly angular) relationship of the delivery periods between two or more fuel injection pump or system outlets.

3.11.6 STATIC PHASING (SPILL PHASING)—A method of phasing in which the beginning or end of delivery is established by observation of changes in spill flow.

3.11.7 DYNAMIC PHASING—A method of phasing in which a specific event in the injection cycle is established while the injection system is in operation.

4. Fuel Injectors

4.1 Main Definitions

4.1.1 FUEL INJECTOR—Assembly comprising a nozzle (2.2) and a nozzle holder (2.3) through which a metered volume of fuel under high pressure is injected into the combustion chamber.

Specific designs of injector may include the nozzle as an integral part of the nozzle holder (which is specified in SAE J347/J1984).

4.1.2 NOZZLE—Valve comprising two principal components, namely the nozzle body and the needle (valve needle), through which the fuel is atomized when open.

4.1.3 NOZZLE HOLDER—Assembly to which the nozzle is attached for location in the cylinder head. It comprises all parts of an injector except the nozzle.

4.2 Principle of Operation

4.2.1 ATOMIZATION—Conversion of liquid fuel under high pressure into a high velocity spray(s) of very small droplets.

4.3 Method of Operation

4.3.1 (CONVENTIONAL) FUEL INJECTOR—Injector actuated solely by the pressure of the metered fuel.

4.3.2 MECHANICAL FUEL INJECTOR—Injector actuated by external mechanical means.

4.3.3 ELECTRICAL FUEL INJECTOR—INJECTOR ACTUATED BY APPROPRIATE ELECTRICAL MEANS.

4.3.4 HYDRAULIC FUEL INJECTOR—Injector actuated by hydraulic means separate from the fuel pressure.

NOTE—An injector may be actuated by a combination of the four methods of operation given in this clause (e.g., hydromechanical, etc.).

4.4 Means of Angular Location

4.4.1 FIXED FLANGE-LOCATED INJECTOR—Injector where the angular location in the engine is determined by a fixed (integral) flange on the nozzle holder body.

4.4.2 FLATS-LOCATED INJECTOR—Injector where the angular location in the engine is determined by the flats on the nozzle holder body and the corresponding shape of the retaining flange or clamp.

4.4.3 BALL/DOWEL-LOCATED INJECTOR—Injector where the angular location in the engine is determined by a ball or dowel on the nozzle holder body and a corresponding slot in the mounting hole.

4.5 Types of Attachment

4.5.1 FLANGE-MOUNTED FUEL INJECTOR—Injector held in the engine by means of a loose or integral flange arranged perpendicular to the axis of the injector and secured with a minimum of two studs or bolt(s).

4.5.2 CLAMP-MOUNTED FUEL INJECTOR—Injector held in the engine by means of a single or double finger clamp and secured with stud(s) or bolt(s).

4.5.3 SCREW-MOUNTED FUEL INJECTOR—Injector held and secured in the engine by means of a gland nut.

4.5.4 SCREW-IN FUEL INJECTOR—Injector held and secured in the engine by means of an external thread on the nozzle holder body or nozzle retaining nut.

4.6 Nozzle Holder Types and Components

4.6.1 HIGH SPRING NOZZLE HOLDER—Design where the spring is remote from the pressure face.

4.6.2 LOW SPRING NOZZLE HOLDER—Design where the spring is near the pressure face.

4.6.3 TWO SPRING NOZZLE HOLDER—Design which provides two stage nozzle actuation.

4.6.4 COOLED NOZZLE HOLDER—Design which contains passages through which a coolant is passed.

4.6.5 NON-LEAK-OFF NOZZLE HOLDER—Design which does not require a back-leakage connection.

4.6.6 NOZZLE HOLDER BODY—Component having integral fuel passages and to which all other components are added in forming the nozzle holder.

4.6.7 ADAPTER PLATE—Component positioned between the nozzle and the nozzle holder body to provide a lift stop for the nozzle needle.

4.6.8 NOZZLE HOLDER CAP NUT—Component which covers and seals the adjustment feature for nozzle opening pressure.

4.6.9 SPINDLE—Component of some length positioned between the spring and the needle.

4.6.10 SPRING SEAT—Short spindle used in low spring injectors.

4.6.11 PRESSURE-ADJUSTING SCREW—Screw by means of which the spring force on the needle is adjusted.

4.6.12 PRESSURE-ADJUSTING SHIM—Shim by means of which the spring force on the needle is adjusted.

4.6.13 NEEDLE LIFT ADJUSTING SHIM—Shim by means of which the needle lift stop is adjusted.

4.6.14 SPRING CAP NUT—Component which houses the spring and may function as an adjusting screw.

4.6.15 FUEL INLET CONNECTION—That part of the nozzle holder to which the high pressure pipe is attached.

4.6.16 FUEL INLET CONNECTOR; INLET STUD—Adaptor attached to the nozzle holder body which functions as an inlet connection.

4.6.17 EDGE FILTER—Type of fuel inlet filter housed in the inlet connector or nozzle body.

4.6.18 BACK-LEAKAGE CONNECTION—Part of the nozzle holder to which a pipe is attached to remove back-leakage.

4.6.19 NOZZLE RETAINING NUT (NOZZLE CAP NUT)—Component which secures the nozzle and adaptor plate where applicable to the nozzle holder body.

4.6.20 GLAND NUT—Freely rotating threaded component assembled coaxially on the nozzle holder body, used to attach an injector to the engine.

4.6.21 HEAT SHIELD (SEAL)—Component used to reduce the heating effect on the nozzle from combustion.

4.7 Nozzle Types

4.7.1 PINTLE NOZZLE—Nozzle which has a needle with a profiled protrusion (the pintle) which extends through a coaxial hole in the body.

4.7.2 DELAY (THROTTLE) PINTLE NOZZLE—Pintle nozzle with a needle protrusion profile which throttles the fuel flow during the initial lift of the needle.

4.7.3 FLATTED PINTLE NOZZLE—Delay pintle nozzle with one or more flats on the needle protrusion profile which influences the fuel flow during the initial needle lift.

4.7.4 PINTAUX NOZZLE—Delay pintle nozzle with auxiliary hole or holes which permit fuel to bypass the throttling stage of the initial needle lift.

4.7.5 HOLE-TYPE NOZZLE—Nozzle with one or more injection/spray holes and a needle which does not affect hole area. These are commonly termed single- or multi-hole nozzles.

4.7.6 V.C.O. ("VALVE" NEEDLE COVERED ORIFICE) NOZZLE—Hole-type nozzle with injection/spray holes positioned in the body seat which are covered by the needle when closed.

4.7.7 POPPET NOZZLE—Nozzle with an outward opening mushroom-shaped needle.

4.7.8 COOLED NOZZLE—Nozzle with a body containing passages through which coolant is passed.

4.8 General Terms

4.8.1 INJECTOR SHANK DIAMETER—Diameter of the injector which determines co-axial location in the engine.

4.8.2 INJECTOR SHANK LENGTH—Distance from the primary sealing face of the nozzle retaining nut (with nozzle fitted) to a reference point on the nozzle holder determined by the particular type of injector attachment.

4.8.3 PRESSURE FACE—Faces on the nozzle body, nozzle holder body and adaptor plate where applicable which are held together in an assembled injector to form a fuel-tight seal.

4.8.4 BACK-LEAKAGE; LEAK-OFF—Fuel which leaks through the clearance between the needle and the nozzle body.

4.8.5 INJECTOR (NOZZLE) OPENING PRESSURE (NOP)—Lowest hydraulic pressure (applied at a slowly increasing rate) at which flow through the injector commences.

4.8.6 INJECTOR (NOZZLE) WORKING PRESSURE—Stabilized nozzle opening pressure for the correct function of the injector in the engine.

4.8.7 INJECTOR (NOZZLE) SETTING PRESSURE—Nozzle opening pressure at which an injector is initially set to ensure the correct working pressure after stabilization.

4.8.8 INJECTOR (NOZZLE) CLOSING PRESSURE—Highest hydraulic pressure at which the needle starts to close.

4.8.9 SEALING FACE—Face upon which the injector seats to make a gas-tight seal with the engine. Commonly this face is on the nozzle retaining nut.

4.8.10 DIFFERENTIAL ANGLE—Difference between the angles of the mating conical sealing surfaces of the needle and nozzle body.

4.8.11 DIFFERENTIAL RATIO—Ratio of the needle guide diameter to the needle seat diameter. This is expressed in terms of the diameter values in millimeters (e.g., 6×3).

4.8.12 SPRAY CONE ANGLE—Included angle of the cone embracing the axes of the injection/spray holes in a multi- hole nozzle. Particular nozzle designs may have more than one spray cone angle.

4.8.13 SPRAY DISPERSAL ANGLE—Included angle of the cone of fuel leaving a pintle or poppet nozzle or a single-hole injection/spray hole in a hole-type nozzle.

4.8.14 SPRAY CONE OFFSET ANGLE; SPRAY INCLINATION ANGLE—Angle between the axis of the spray cone angle and the axis of the nozzle.

4.8.15 OVERLAP—Length of the throttling (delay) portion of the pintle penetrating the orifice in a delay pintle nozzle, measured closed.

4.8.16 NOZZLE SAC; SAC HOLE—Chamber within a hole-type nozzle tip from which the fuel enters the injection/spray holes.

4.8.17 NOZZLE SAC VOLUME—Volume within a hole-type nozzle tip between the needle and the entry to the injection/spray holes determined with the nozzle closed.

4.8.18 NOZZLE SEAT—Contact line/surface between the needle and body of a closed nozzle which prevents fuel flow to injection/spray holes.

4.8.19 INJECTOR DEAD VOLUME—High pressure volume contained inside an injector between the needle when seated and the bottom of the cone of the inlet connection.

5. Unit Injectors

5.1 Main Definition

5.1.1 UNIT INJECTOR—An assembly which combines the features of a single cylinder pump and an injector in one unit through which a metered volume of fuel under high pressure is injected into the combustion chamber.

5.2 Types of Unit Injector—Unit injectors are categorized by three principal features. These are:

    a. Actuation (Energy Input)
    b. Timing Control
    c. Metering Control

Possible forms of these features may be mechanical, hydraulic, electrical, or electronic. A full description of a unit injector shall specify all the previously mentioned features.

Example—A mechanically actuated unit injector with electronic timing and metering control.

5.3 Components

5.3.1 TAPPET (FOLLOWER) ASSEMBLY—The component(s) that transmits linear movement to the plunger from external driving parts.

5.3.2 TAPPET HEAD—A separate part of the tappet assembly that contacts the external driving parts.

5.3.3 THRUST PAD—That part of the tappet assembly which is positioned between the tappet head and the plunger.

5.3.4 TAPPET BODY—That part of the tappet assembly which slides in contact with the tappet guide.

5.3.5 TAPPET GUIDE—The component or feature of the body that guides the movement of the tappet or tappet assembly.

5.3.6 TAPPET RETAINER—The component that retains the tappet assembly in its assembled configuration when the unit injector is not mounted in the engine.

5.3.7 RETURN SPRING—The component(s) that maintains contact of the tappet with the external driving parts and returns the plunger to the beginning of its stroke.

5.3.8 UNIT INJECTOR BODY—An assembly or sub-assembly which contains the pumping element.

5.3.9 NOZZLE SPRING HOUSING—The component which holds the nozzle spring and has fuel passages.

5.3.10 NOZZLE SPRING—The spring which applies a predetermined force to the needle of the nozzle.

5.3.11 SPRING HOUSING RETAINING NUT (SPRING HOUSING CAP NUT)—The component which secures the spring housing and intermediate parts where applicable, to the unit injector body assembly.

5.3.12 NOZZLE RETAINING NUT (NOZZLE CAP NUT)—The component which secures the nozzle and adapter plate where applicable, to the spring housing.

5.3.13 NOZZLE AND HOUSING RETAINING NUT (NOZZLE AND HOUSING CAP NUT)—The component which secures the nozzle, spring housing, and other components to the unit injector body assembly.

5.3.14 METERING DEVICE—An integral device which determines the quantity of fuel injected. This may utilize one of the metering processes defined in Section 3 of this document.

6. Governors

6.1 Basic Definition

6.1.1 (ENGINE SPEED) GOVERNOR—A device which compares the setting speed to actual speed and, under specific operating conditions, causes a modification to the fuel delivery into the engine in order to adjust the actual speed of the engine toward the setting speed. It may or may not be part of the injection pump.

6.2 Types

6.2.1 SINGLE SPEED GOVERNOR—A governor which regulates from one specified engine speed. In cases where the specified speed is the maximum permissible operating speed, this governor may be called a Maximum Speed Governor.

6.2.2 IDLE AND LIMITING SPEED (MIN-MAX) GOVERNOR—A governor which regulates at the idling speed and the limiting speed of the engine. Intermediate speeds are determined by the control lever position and engine load.

6.2.3 MULTIPLE-SPEED GOVERNOR—A governor which regulates at one of several predetermined engine speeds.

6.2.4 ALL-SPEED (VARIABLE-SPEED) GOVERNOR—A governor which regulates at any selected engine speed, between two predetermined limits.

6.2.5 COMBINATION GOVERNOR—A governor which has similar features to an idle and limiting speed governor but with an extended upper and/or lower controlled speed range.

6.3 Governor Action

NOTE—The dynamic behavior of the governor depends upon the relationship between the output signal and the speed error value. The relationship may have differing characteristics for different values of the setting speed signal.

6.3.1 PROPORTIONAL-ACTION GOVERNOR (P GOVERNOR)—A governor where the output signal is proportional to the speed error value. A change in load results in a change of steady-state speed.

6.3.2 PROPORTIONAL INTEGRAL GOVERNOR (PI GOVERNOR)—A governor where the output signal consists of a signal proportional to the speed error value modified by a signal which is proportional to the time integral of that speed error value. A change in load may or may not result in a change of steady-state speed.

6.3.3 PROPORTIONAL INTEGRAL DIFFERENTIAL GOVERNOR (PID GOVERNOR)—A proportional integral governor which additionally corrects the output signal proportionally to the rate of speed change.

6.4 Method of Operation

6.4.1 MECHANICAL GOVERNOR—A governor where the speed input signal is determined by the centrifugal action of a flyweight assembly and the output signal operates directly without any external amplification.

6.4.2 MECHANICAL-HYDRAULIC GOVERNOR—A mechanical governor with hydraulic amplification of the output signal.

6.4.3 MECHANICAL-PNEUMATIC GOVERNOR—A mechanical governor with pneumatic amplification of the output signal.

6.4.4 PNEUMATIC GOVERNOR—A governor where the input signal is determined by the inlet manifold pressure and the output signal operates directly.

6.4.5 HYDRAULIC GOVERNOR—A governor where the speed input signal is hydraulic pressure and the output signal operates directly.

6.4.6 ELECTRONIC/ELECTRIC GOVERNOR—A governor where the speed input signal(s) is determined by an input sensor(s) and the electrically/electronically amplified output signal operates directly.

6.4.7 ELECTROHYDRAULIC GOVERNOR—An electronic/electric governor with additional hydraulic amplification of the output signal.

6.4.8 ELECTROPNEUMATIC GOVERNOR—An electronic/electrical governor with additional pneumatic amplification of the output signal.

6.5 Characteristics/Terms

6.5.1 (GOVERNOR) INPUT SIGNAL—The signal supplied to the governor which is a measure of a system condition, e.g., speed, load, temperature.

6.5.2 (GOVERNOR) SPEED INPUT SIGNAL—The governor input signal which is a measure of the instantaneous engine speed.

6.5.3 GOVERNOR SETTING SPEED SIGNAL—The signal supplied to the governor which is a measure of the setting speed.

6.5.4 SETTING SPEED—A steady-state speed on a speed-power characteristic curve determined by the speed setting device, according to the required engine power. The setting speed may also be defined as the theoretical speed on the governor characteristic curve at which there would be zero fuel delivery.

6.5.5 SPEED-POWER CHARACTERISTIC CURVE—A curve of steady-state speeds over the engine power range for a given speed setting.

6.5.6 GOVERNOR CHARACTERISTIC (CONTROL ROD) CURVES—The curves which show the relationship between governor output signal and the steady-state speed of the pump or engine, for different given operating conditions.

6.5.7 SPEED ERROR VALUE—A measure of the instantaneous difference between the speed input signal and the current setting speed signal.

6.5.8 GOVERNOR OUTPUT SIGNAL—The signal delivered by the governor which is used to adjust the fuel delivery.

6.5.9 GOVERNOR GAIN (LEVER RATIO)—The ratio of the governor output signal to the speed error value. For mechanical governors the lever ratio is the ratio of the control rod travel to axial travel of the fly weight. This is equivalent to steady-state governor gain in such systems. The lever ratio may be constant or variable over the entire range of the output signal.

6.5.10 MINIMUM SENSITIVITY OF THE GOVERNOR (INSENSITIVITY)—The maximum variation of input signal which does not produce a change in output signal, expressed as a percentage of the full range of the input signal.

6.5.11 GOVERNOR REGULATION (SPEED DROOP)—The speed difference between the no-load speed and a specified speed, at a given power or load, expressed as a percentage of the specified speed, for a fixed speed setting. (Previously used terms are pull-off, run-out, and permanent droop.)

6.5.12 SPEED DROOP GOVERNING—Governing where for a specified setting speed the speed droop is greater than 0%.

6.5.13 ISOCHRONOUS GOVERNING—Governing where for a specified setting speed the governor maintains one steady-state speed over the load range. The speed droop is 0%.

6.5.14 OVERSHOOT—The transient excess engine speed due to sudden acceleration resulting from a sudden reduction in load.

6.5.15 UNDERSHOOT—The transient drop in engine speed due to engine deceleration or due to sudden load application.

6.5.16 LOAD CONTROL—Direct control of the injected fuel quantity by the engine or vehicle operator.

6.5.17 SPEED CONTROL—The governing principle which adjusts the injected fuel quantity in attempting to maintain a selected engine speed whilst subjected to changing load.

6.5.18 RESPONSE (RECOVERY TIME)—The time that it takes for the engine speed to attain a value within some specified tolerance of its final steady-state value when it has been disturbed by a specified sudden change of load or demand.

6.5.19 TORQUE CONTROL—The modification of the maximum fuel delivery at speeds below engine rated speed.

6.5.20 NEGATIVE TORQUE CONTROL—Where the torque control increases full load delivery with respect to an increase in speed.

6.5.21 POSITIVE TORQUE CONTROL—Where the torque control decreases full load delivery with respect to an increase in speed.

6.5.22 TORQUE CONTROL TRAVEL—The maximum change in control rod position as a result of the operation of the torque control device over its speed range.

6.5.23 LOAD BACK-UP (TORQUE BACK-UP)—Where the fuel delivery curve has the characteristic shape of increasing fuel delivery as speed decreases from the maximum rated speed. This can be either the natural fuel delivery curve or be achieved using a positive torque control device.

6.5.24 FAST IDLE SPEED—An increased low idle speed setting often used with cold engine start and during engine warm up. It may be either manual or automatic adjustment.

6.5.25 HIGH IDLE—The maximum no-load governed speed.

6.5.26 ANTI-STALL—The prevention of excessive undershoot of engine speed on deceleration.

6.5.27 GOVERNOR CONTROL LEVER (THROTTLE LEVER)—The device by means of which the set point or demand signal (e.g., speed, load, torque) is supplied to the governor.

6.5.28 SETTING STOP—A device or means by which the limits of travel of the governor control lever can be set.

6.5.29 WORK CAPACITY—The maximum work available from the governor as its output shaft or arm moves through its full available travel.

6.5.30 MAXIMUM FORCE—The maximum value of the force at the output arm at some specified position of the travel.

6.5.31 MAXIMUM TORQUE—The maximum value of the torque at the output shaft at some specified position of the travel.

6.5.32 GOVERNOR DRIVE TORQUE—The torque required to drive the speed sensing element of the governor. This normally only applies to independent governing devices.

6.5.33 GOVERNOR FORCE CURVES (HARTNELL DIAGRAM)—The curves which show the relationship between governor force and speed (pump or engine) for different governor weight positions.

6.5.34 LOAD SENSING—The direct measurement or sensing of engine load (or torque), e.g., in order to improve the transient response of a governor.

7. Timing Devices

7.1 Basic Definition

7.1.1 TIMING/ADVANCE DEVICE—A device which modifies the dynamic injection timing under specific operating conditions which may or may not be part of the pump. Its operation may be automatic or by manual control or both.

NOTE—Negative advance is commonly defined as retard and this word may be substituted for "advance" where appropriate.

7.2 Types

7.2.1 SPEED ADVANCE—A device which changes the advance angle in relation to engine speed.

7.2.2 LOAD ADVANCE—A device which changes the advance angle in relation to engine load.

7.2.3 SPEED & LOAD ADVANCE—A device which changes the advance angle in relation to both engine speed and load.

7.2.4 START ADVANCE—A device which modifies the advance angle at engine start conditions.

7.2.5 COLD IDLE ADVANCE—A device which modifies the advance angle at low idle speed dependent upon engine operating temperature.

7.2.6 TWO STAGE ADVANCE—A device which provides an advance curve with two distinct stages which have different advance characteristics.

7.3 Characteristics/Terms

7.3.1 DYNAMIC ENGINE INJECTION TIMING—The relationship between the commencement of the injection of fuel into the engine and the engine piston position, normally expressed in terms of degrees of engine crank angle rotation relative to a reference point, e.g., BDC or TDC.

7.3.2 DYNAMIC PUMP INJECTION TIMING—The relationship between the commencement of the injection of fuel and a datum point on the pump driveshaft, normally expressed in terms of degrees of pump driveshaft rotation.

7.3.3 START OF ADVANCE—The condition at which the advance angle starts to change, normally expressed in terms of speed, load, etc.

7.3.4 END OF ADVANCE—The condition at which the advance angle ceases to change and is expressed in terms of speed, load, etc.

7.3.5 ADVANCE ANGLE—The expression of movement within the timing device in degrees of pump driveshaft rotation. This angle is not necessarily equal to the change in dynamic injection timing of either the pump or the engine.

7.3.6 TOTAL ADVANCE ANGLE—The maximum advance angle of the device.

7.3.7 ADVANCE SPEED RANGE—The speed range between the start and end of advance determined by speed advance.

7.3.8 ADVANCE CURVE(S)—The curve(s) which illustrate the relationship between advance angle and speed for different operating conditions.

7.3.9 LOAD SENSITIVITY—The undesirable change of advance angle as a result of changes to pump fuel delivery.

8. High Pressure Pipes and Connections

8.1 Basic Definitions

8.1.1 HIGH PRESSURE FUEL INJECTION PIPE—A cut length of steel tube complying with SAE J1958.

8.1.2 HIGH PRESSURE FUEL INJECTION PIPE ASSEMBLY—A high pressure fuel injection pipe fitted with a connector nut at each end and with each pipe end formed to couple a female cone. The pipe may or may not have preformed bends for its intended use. Additional components may be included in this assembly on specific applications, e.g., low pressure seals.

8.1.3 ASSEMBLED PIPE SET—Two or more high pressure fuel injection pipe assemblies clamped together for placement on an engine. This may not include all the high pressure fuel injection pipe assemblies for the pump application.

8.1.4 END CONNECTION—The components and features which enable the high pressure fuel injection pipe assembly to be coupled to the fuel injection pump and injector.

8.2 Fuel Injection Tube

8.2.1 SEAMLESS TUBE—A seamless cold-drawn single-wall tube complying with SAE J1958.

8.2.2 COMPOSITE TUBE—A tube which does not have a single-wall and complies with ISO 8535-2. The internal bore may or may not have a seam. The structure can be produced in various forms.

8.2.3 SEAMED COMPOSITE (WRAPPED) TUBE—A composite tube with a seam on the internal bore and a cross-sectional structure in the form of a spiral.

8.2.4 SEAMLESS COMPOSITE TUBE—A composite tube with a seamless inner (liner) tube and an outer tube which may be seamless or wrapped.

8.2.5 BORE GRADE—A quantitative description of the number and description of imperfections allowed on the inner surface of the tube per tube cross section. This only applies to tubes where the surface of the inner bore is seamless. See SAE J1958.

8.3 End Connections—All end connections must comply with the requirements of SAE J1949.

8.3.1 SEALING FACE—The contact surface which forms the high pressure seal between the fuel injection pipe assembly and the female cone to which it is coupled.

8.3.2 FEMALE CONE—The fitting, conforming to SAE J1949, which couples to a fuel injection pipe assembly.

8.3.3 CONNECTION END—The fabricated end (conforming to SAE J1949) of a fuel injection pipe assembly which couples to a female cone. The exact structure and/or shape of the end connection may have various forms.

8.3.4 CONNECTOR (UNION) NUT—The component of a fuel injection pipe assembly which secures the connection end to the female cone.

8.3.5 CONNECTOR COLLAR—An optional component placed between the connector nut and the connection end of the high pressure pipe assembly used where necessary to improve coupling conditions.

8.3.6 REFERENCE DIAMETER—The basic diameter common to the female cone and the connection end (to which other dimensions refer) forming the initial contact line of the sealing face.

8.3.7 PIPE END ASSEMBLY—The components and features of the end connection which belong to the high pressure injection pipe assembly.

8.4 Pipe Assembly

8.4.1 BEND RADIUS—The radial dimension to the tube centerline in a formed bend.

8.4.2 ASSEMBLY CLAMP—A device which is used to position and hold one or more pipes to another pipe and/or to the engine.

8.4.3 PIPE INSIDE DIAMETER—The diameter of the circle with an area equal to the cross-sectional area of the high pressure fuel injection pipe bore.

DIESEL ENGINES — STEEL TUBES FOR HIGH PRESSURE FUEL INJECTION PIPES (TUBING)—SAE J1958 APR89

SAE Standard

Report of the Diesel Fuel Injection Equipment Standards Committee approved April 1989. Rationale statement available.

1. Scope—This SAE Standard specifies dimensions and requirements for single-wall steel tubing intended for use as high pressure fuel injection pipes on a wide range of engines (Class A), and for fuel injection pump testing (Class B, Reference SAE J1418). Tubing shall be cold drawn, annealed or normalized, seamless tubing suitable for cold swaging, cold upsetting and cold bending.

2. Dimensions and Tolerances

2.1 Sizes—The recommended outside diameter and inside diameter tubing sizes are shown on Table 1. Other sizes may be available by agreement.

2.2 Tolerances:

2.2.1 INSIDE DIAMETER[1]:

Standard: $\dfrac{\text{mm}}{\pm 0.06}$

[1] (Refer to SAE J1418 for Class B)

2.2.2 OUTSIDE DIAMETER:

Standard: $\dfrac{\text{mm}}{\pm 0.06}$

2.2.3 DEGREE OF ECCENTRICITY—The degree of eccentricity of the outside diameter of the tube relative to the inside diameter is proportioned to the wall thickness as shown in the following figure.

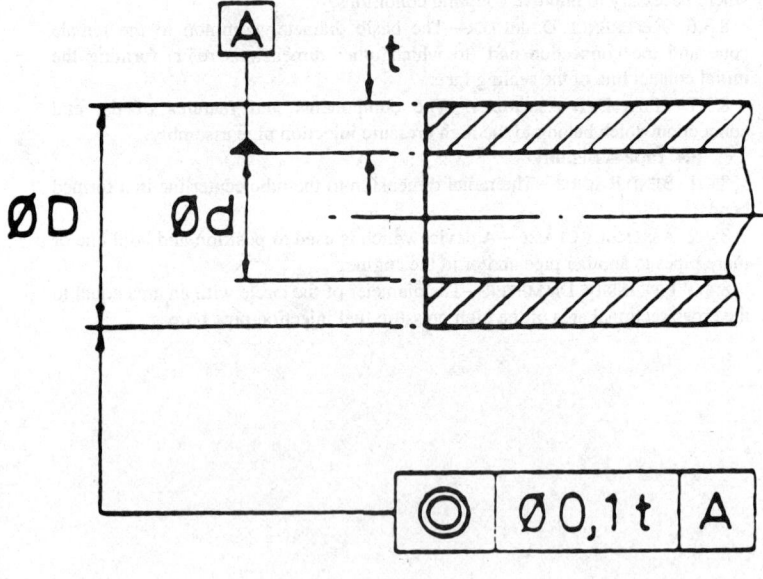

FIG. 1

2.2.4 LENGTH—Length and tolerance on length shall be by agreement between purchaser and supplier.

2.3 Straightness—Unless otherwise specified, the tubes shall be straight within a maximum error of 1 in 400 without any localized deformation.

3. Material Processing

3.1 Material—Tubing shall be manufactured from an unalloyed or low alloy steel produced by a steelmaking process providing a very homogenous structure. Upon request, the supplier shall state the method of the steel making and deoxydation process used.

3.2 Manufacturing of Tubes—The tubing shall be cold drawn from steel hollows processed in a manner that allows the manufacturer to obtain tubes in conformance with agreed upon portions of this standard. After forming, the tubing, if annealed, shall be processed in such a manner as to prevent formation of scale on the inside surface and produce a finished product which will meet all agreed to requirements of this standard.

3.3 Mechanical Properties—The furnished tubing shall have mechanical properties[2] as tabulated below:

	Yield Strength Min	Tensile Strength	Elongation in 2 in (51 mm) % Min	Hardness Rockwell B Scale Max
Grade One	205 N/mm² (29 730 psi)	310-379 N/mm² (44 960 - 54 965 psi)	30	65
Grade Two	220 N/mm² (31 905 psi)	360-480 N/mm² (52 210 - 69 615 psi)	23	80
Grade Three	355 N/mm² (51 485 psi)	490-630 N/mm² (71 065 - 91 370 psi)	22	87

[2] Per ASTM A 370, Supplement II, Section T-2

3.4 Surface Quality

3.4.1 GENERAL—The outside and inside surfaces of finished tubing shall be clean and free from scale, rust, seams, laps, laminations, deep pits, non-metallics or other injurious defects.

3.4.2 O.D. SURFACE:

3.4.2.1 *Condition*—The outside surface of annealed or normalized tubes may have a slight discoloration, but shall be free from loose scale and represent a smooth cold finished surface. Imperfections in any tube heat treated or not, deeper than 0.13 mm will be a cause for rejection. Scratches resulting from a mechanical finishing process such as polishing or grinding are prohibited.

3.4.2.2 *Coatings*—The outside surface of the tubes may be coated utilizing a product and method agreed upon by purchaser and manufacturer. The inside surface of the tube shall remain uncoated.

3.4.3 I.D. SURFACE—The inside of tubing shall be clean and free from any contaminations which will impair the processing or serviceability of the tubing and shall be finished to ensure a smooth bore of accurate size and shall conform to one of the following bore designations described in Table 2.

Bore designation, as described in Table 2, shall be specified on the face of the order.

TABLE 2

BORE DESIGNATION	PERMITTED IMPERFECTIONS	MAGNIFICATION
A	5 imperfections (max) between 0.08 - 0.13 mm deep in any cross-section and no imperfections deeper than 0.13 mm	100 X
B	5 imperfections (max) over 0.05 mm to 0.08 mm maximum deep	200 X
C	5 imperfections (max) over 0.02 mm to 0.05 mm maximum deep	200 X
D	All imperfections less than 0.03 mm deep. As an alternate an imperfection can be classified as having a depth that is at least twice the width.	200 X

3.4.3.1 *I.D. Bore Optional Requirement*—Inclusions occurring within 10% of the wall thickness as measured from the I.D. surface is to be considered as continuous to the I.D.

De-carb (C.F.D. = Carbon Free Depth) of 0.050 mm maximum on the O.D. and 0.020 mm maximum on the I.D. C.F.D. shall be determined at 200 X.

3.5 Microstructure

3.5.1 The microstructure shall be essentially ferrite with pearlite permitted to the extent necessary to fulfill the mechanical requirements.

3.5.2 Grain size shall be five or finer.

TABLE 1 - Recommended Inside and Outside Diameters in mm

INSIDE DIAMETER (d)	OUTSIDE DIAMETER (D)
	5 6 8
1.25	
1.4	
1.5	
1.6	
1.7	
1.8	
1.9	
2.0	
2.12	
2.24	
2.36	
2.50	
2.65	
2.80	
3.00	
3.15	
3.35	
3.55	
3.75	
4.00	

4. Testing

The finished tubing shall satisfactorily meet the following performance tests as well as the mechanical properties using standard sampling techniques for testing to determine compliance. Test specimens shall be taken from tubing which has not been subjected to cold working after final annealing/normalizing of the finished tubing.

4.1 Cold Upsetting Test—The tubing shall withstand cold upsetting from a length of 1/2 to 1/4 in (12.7 - 6.4 mm) without showing other than superficial outside surface ruptures.

4.2 Bend Test—Tubing shall be capable of being formed into bends and 360 deg circles without showing evidence of cracking, kinking, or other flaws rendering it useless for its intended end use when the following minimum bend radius is used: 3 x O.D.

4.3 Dimensional Testing—The dimensions of the tube shall comply with the dimensions and tolerances indicated under section 2 of this standard. All tubes in the lot shall be tested except that, by agreement, an acceptable statistical process of testing may be substituted.

4.4 Testing of Mechanical Properties—The tubes shall comply with properties stated under 3.3 of this standard. Tests shall be made according to ASTM A 370, Supplement II, with the average of results obtained from three samples per heat tested being reported on a test certificate, if requested.

4.5 Surface Quality Test—A visual check of the O.D. shall be conducted on all tubes to assure compliance with requirements of section 4.

4.5.1 If mutually agreed between the purchaser and supplier, the visual inspections may be replaced by a non-destructive testing procedure.

4.5.2 If a proof of a specified defect depth is required, an inspection by attributes per AQL shall be agreed upon when ordering, and the proof shall be made on the metallographic cross-section of the tube utilizing the specified magnification in Table 2.

4.5.3 Surface coatings on tubes with plated or treated surfaces shall satisfy tests on the coatings as agreed between purchaser and supplier.

4.6 Pressure Proof Test—Unless otherwise specified, tubing supplied under this standard shall have been tested hydrostatically, with no evidence of permanent internal deformation, at a pressure which will subject the material to a fiber stress of 75% of minimum yield strength. Test pressure shall be determined by the Lame formula:

$$P = S \frac{(D^2 - d^2)}{D^2 + d^2}$$

Where: D = nominal outside diameter of tubing, mm (in)
d = nominal inside diameter of tubing, mm (in)
P = hydrostatic pressure, MPa (psi)
S = allowable fiber stress (75% of minimum yield strength), N/mm^2 (psi)

4.7 Test Certificate—A Mill Test Certificate, if requested, shall be issued for each shipment, confirming that the supplied tubes meet the standard. Further, the certificate shall reflect the mechanical properties per 3.4, grade designation, heat number and carbon, manganese, phosphorus and sulfur contents of each heat of steel supplied. When agreed to, between purchaser and supplier, the records of a continuous production inspection procedure can be used to substantiate compliance of any clause(s) of this standard.

Other test certificates or letters of conformity may be agreed to between purchaser and supplier.

5. Markings—As agreed upon between supplier and purchaser.

6. Corrosion Protection—The inside and outside of the finished tubing shall be protected against corrosion during shipment and normal storage. If a corrosion preventive compound is applied, it shall be such that after normal storage periods it can readily be removed by cleaning agents normally used in manufacturing.

26.263

FUEL INJECTION PUMPS—HIGH PRESSURE PIPES (TUBING) FOR TESTING—SAE J1418 DEC87

SAE Standard

Report of the Engine Committee approved December 1987. This report references ISO 4093.

1. *Purpose*—This standard specifies the dimensional requirement of a range of high pressure pipes for use in the bench testing and setting of fuel injection pumps.

Only dimensions and requirements affecting the hydraulic characteristic of the pipes are defined. Other requirements, such as the type of end connections and shape of the pipes when bent, are not included. These depend on the connections provided at pump outlets and injector inlets, and on the design features of individual pumps and test benches.

2. *Description*—The range of pipes specified enables pump and engine manufacturers to choose suitable pipe sizes for pump deliveries up to 300 mm^3 per stroke per cylinder. The particular pipe to be used shall be identified by the pump manufacturer in the test specification for each individual pump type and application.

3. *Dimension*—The seven standardized sizes of pipes are shown in Table 1. Dimensions are in millimeters.

4. *General Requirements*

4.1 The pipes may be of ferrous material, usually cold-drawn mild steel, conforming to SAE J529 MAR85.

4.2 After end connections are made, any closing-in or reduction in opening of the pipe shall be removed to a depth of at least twice the length of the deformed end of the pipe. Any closing-in of the ends after extended use shall also be eliminated.

4.3 Pipes shall be cleaned internally after the ends are made and bent in order to remove extraneous matter.

4.4 During storage, the pipes should be protected internally against corrosion and contamination.

4.5 Flow specifications for straight lines are not applicable to bent lines; therefore, straight-line flow specifications are not provided.

TABLE 1

Item #	Internal Diameter	External Diameter min	Length	Minimum Central Line Bend Radius[a]
1	2.0 ± 0.025	6	600 ± 5	16
2	2.0 ± 0.025	6	845 ± 5	16
3	3.0 ± 0.025	6	600 ± 5	25
4	3.0 ± 0.025	6	1000 ± 5	25
5	3.0 ± 0.025	6	750 ± 5	25
6	1.6 ± 0.025	6	600 ± 5	16
7	2.0 ± 0.025	6	450 ± 5	16

[a]Bends may affect the pump fuel delivery. Pipes should be straight and uniform with as large as possible bending radii.

GASOLINE FUEL INJECTOR—SAE J1832 NOV89

SAE Recommended Practice

Report of the Fuel Injection Subcommittee approved November 1989. Rationale statement available.

1. *Scope*—This SAE Recommended Practice promotes uniformity in the evaluation and qualification tests conducted on fuel injectors used in gasoline engine applications. Its scope is limited to electronically actuated fuel injection devices used in automotive port or throttle body fuel injection systems where fuel supply pressure is below 500 kPa. It is further restricted to bench type tests. More specifically this document is intended for use as a guide to the following:

1.1 Identify and define those parameters that are used to measure fuel injector characteristics or performance. The parameters included in this document are listed along with their recommended symbol where appropriate:

Closing Time (CT)
Coil Inductance (L)
Coil Resistance (R)
Dynamic Flow (Q_d)
Dynamic Flow Calculated (Q_{dc})
Dynamic Flow Rate (Q)
Dynamic Minimum Operating Voltage (DMOV)
Dynamic Set Point (PW_{xx})
Dynamic Set Point Flow (Q_{sp})
External Leakage
Flow-Offset (Y)
Insulation Resistance (IR)
Linear Flow Range (LFR)
Linearity Deviation (LD)
Maximum Overload Voltage
Opening Time (OT)
Operating Voltage Range
Period (P)
Pulse Width (PW)
Pressure Drop Ratio (PDR)
Repeatability
Slope (m)
Slope Approximated (m_a)
Spray Pattern
Stability (S)
Static Drop-Out Current (I/S-OFF)
Static Flow Rate (Q_s)
Static Minimum Operating Voltage (SMOV)
Static Pull-In Current (I/S-ON)
Time-Offset (X)
Working Flow Range (WFR)

1.2 Establish test procedures, and recommend test equipment and methods to measure and quantify these parameters.

1.3 Standardize use of nomenclature specifically related to fuel injectors.

2. *Injector Types*

2.1 Fuel Flow Path—Injectors may be classified as top or bottom feed based on the fuel path. Fuel enters at the top of a "top feed" injector (Fig. 1a) and near the bottom or side of the "bottom feed" (Fig. 1b) injector. Metered fuel exits through the bottom in both type injectors.

A fuel filter is typically designed as an integral part of the injector at the fuel entrance. This filter is not normally designed to be serviceable or the only fuel system filtration device. Its main purpose is to prevent initial fuel line and rail contaminants from entering the injector. It also provides contamination protection during testing or servicing of the fuel system. An in-line serviceable filter functions as the primary fuel system filtration device. These filters usually have the capability of removing smaller contaminants than filters used on carbureted systems.

From the inlet filter section, the fuel moves to the fuel metering portion of the injector. The metering of fuel is accomplished by a valve and seat in conjunction with a metering orifice. On energization of the coil, magnetic force pulls the valve away from the seat allowing fuel to pass through the valve/seat and out through the metering orifice. The Qs of the injector depends on the fuel pressure and the hydraulic losses of the fuel circuit. By design, the major restriction to flow is in the metering section. The metering orifice usually presents the largest pressure drop; however, most designs also rely on a pressure drop across the valve and seat. As the stroke or lift of the valve is varied, the Qs is changed and can be used as a means for the manufacturer to obtain the specified Qs. A spring is incorporated to return the valve to the closed position when the coil is de-energized. This ensures that the injector is normally closed and fuel can only flow on energization of the coil. The spring also affects the time required for the valve to open and close, which changes the injector dynamic characteristics. Many designs, therefore, provide the manufacturer with the ability to adjust the preload of the spring to obtain the dynamic set point flow.

Most designs use highly finished metal-to-metal surfaces to provide a leak tight seal and minimize flow shift due to life cycling. Metal also helps minimize changes in stroke when operated at extreme environmental temperatures. Ball/seat, mating conical surfaces and flat surface designs are in use (Fig. 2).

From the metering section, fuel flows out through some type of spray pattern generating feature. The spray is produced by the metering ori-

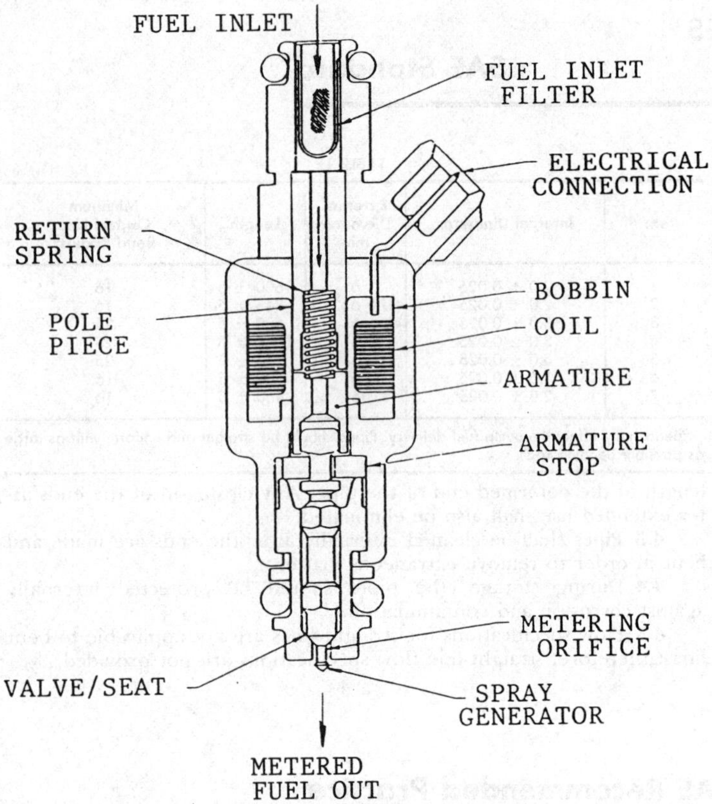

FIG. 1a—TOP FEED INJECTOR

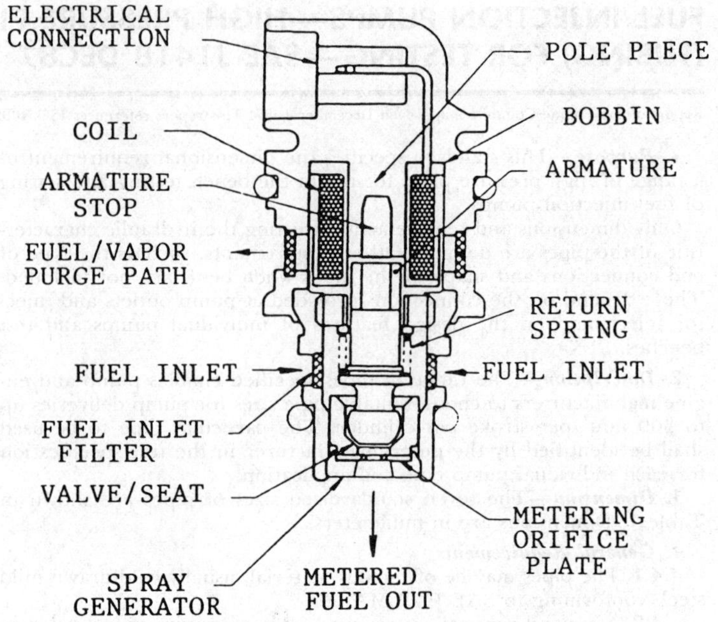

FIG. 1b—BOTTOM FEED INJECTOR

FIG. 1—TYPICAL FUEL INJECTOR DESIGNS

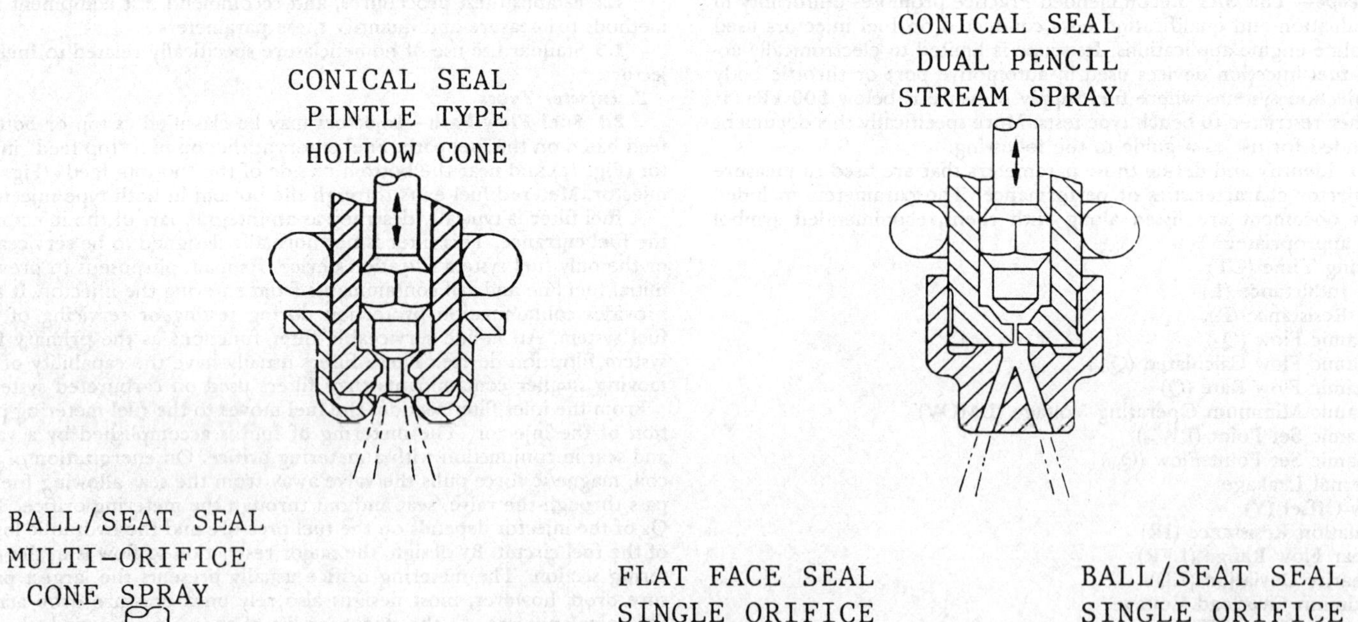

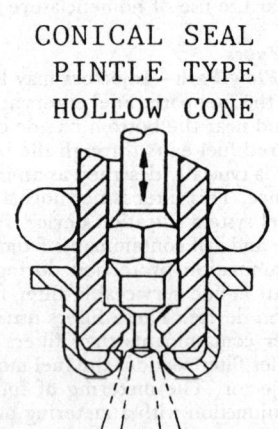

CONICAL SEAL
PINTLE TYPE
HOLLOW CONE

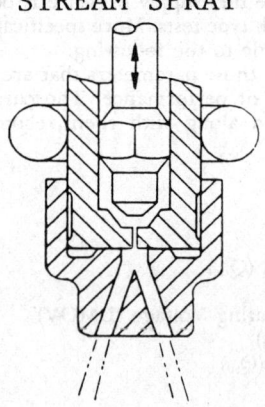

CONICAL SEAL
DUAL PENCIL
STREAM SPRAY

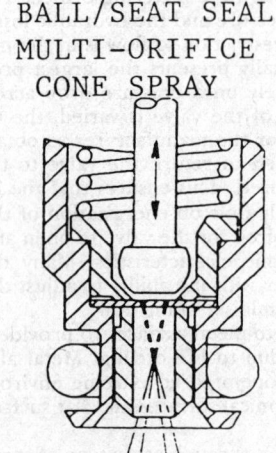

BALL/SEAT SEAL
MULTI-ORIFICE
CONE SPRAY

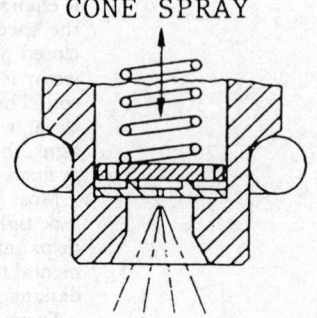

FLAT FACE SEAL
SINGLE ORIFICE
CONE SPRAY

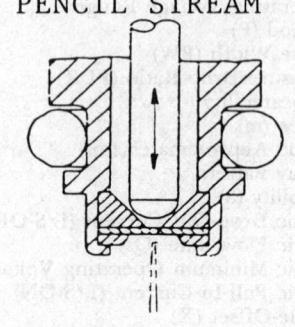

BALL/SEAT SEAL
SINGLE ORIFICE
PENCIL STREAM

FIG. 2—FUEL METERING AND SPRAY GENERATION

fice or by a separate spray generating part. Injectors used for throttle body injection generally use multiple orifices to generate a wide angle, hollow cone. The current port fuel injection designs are more diversified in spray pattern generation. Multiple orifice, single orifice, and pintle styles exist. These generate patterns ranging from a pencil stream to cones of varying angle distribution. With the introduction of multiple intake valve per cylinder engine designs, dual stream port fuel injectors are sometimes used. Multiple sprays can target fuel to each inlet valve to promote more uniform air fuel mixture.

Top feed injector designs have the fuel traveling axially through the center of the coil and down into the metering area. This design is readily adaptable to PFI, but vapors generated within the injector are not easily purged due to the opposing incoming fuel flow. To minimize vapor generation and provide adequate hot fuel handling, the fuel system pressure is maintained at a relatively high level of 250 to 500 kPa.

Bottom feed injectors have been primarily used on central fuel injection systems where one or more injectors are located in the throttle body. This design allows packaging of an injector in a fuel metering body such that the injector is centered above the throttle plates. An advantage of bottom feed injectors is the incorporation of a vapor purge path. Since fuel comes in near the bottom of the injector, the fuel vapors can rise above the metering area. By providing a direct passage out of the injector, vapors are more easily purged. This vapor purge passage allows the use of lower fuel pressure (100 kPa) while still maintaining adequate hot fuel handling performance. Although typically used on central fuel injection applications, there is interest in using the bottom feed design for port applications.

2.2 Magnetic Circuit—The solenoid assembly of an electromagnetic fuel injector supplies the force to actuate the fuel metering valve. Two commonly used solenoid designs are the plunger style armature, the most prevalent, and the flat face disk armature. The major components of the solenoid assembly include the coil assembly, solenoid body, pole piece, armature, and return spring.

The function of the coil assembly is to produce a magnetic field when energized. The coil consists of a specified number of turns of insulated wire wound around a bobbin. The material and size of the wire is chosen to provide a given number of turns to develop the magnetic force while achieving the required electrical resistance for the injector driver circuit. Copper is typically used for low resistance injectors and brass for the high resistance injectors. The total electrical resistance of an injector is essentially equal to that of the coil assembly, which is a function of the resistivity of the wire, length of wire, and the termination of the wire to an electrical connector. Inductance, on the other hand, is a function of the number of turns in the coil, materials used for the body, armature, and pole piece, geometric construction, air gaps, etc. Inductance, therefore, must be measured on the completed assembly.

The magnetic field produced by the coil generates flux, which travels in a closed loop around the coil assembly. It is the function of the components in the magnetic path to carry the flux efficiently. The solenoid body is the segment of the circuit that carries flux from the pole piece to the armature. The armature is the moving element of the solenoid, which controls the flow of fuel by opening and closing a valve. The pole piece is the nonmoving element that attracts the armature when the coil is energized. The predetermined clearance between the pole piece and the armature is called the working air gap. It is through this air gap that the force of attraction is generated. The working air gap consists of both a fixed and a variable air gap, the latter of which is dependent on the position or stroke of the armature. The fixed air gap prevents contact between the armature and pole piece to minimize the effect of residual magnetism. Since contact between these two parts would result in longer closing times, a stop or spacer made from nonmagnetic material is used to ensure that a fixed clearance is maintained when the valve is fully opened. The magnetic force of attraction is most important in the working air gap; consequently the magnetic properties of the materials used and the geometry of the parts are critical to the performance of the design. The remaining parts of the magnetic circuit have a lesser effect on the total reluctance and response of the solenoid.

The major difference between the two solenoid designs shown in Fig. 3 is the path of the flux into the armature from the solenoid body. In the plunger design, radial and working air gaps are used. The radial gap is necessary to permit the armature to move axially relative to the nonmoving components of the solenoid assembly. The working air gap accommodates the motion of the armature assembly. In the flat face disk design, an additional working air gap is required with no radial air gap between the armature and solenoid body.

2.3 Cold Start Injector—The cold start injector is a simplified version of an electromagnetic top feed fuel injector used to provide additional fuel during crank at cold ambients. It is utilized primarily on PFI systems and is located in the inlet air passage to distribute fuel to all cylinders. It is normally energized by a bimetallic switch that provides battery voltage to the injector coil independent of the electronic control module or injector driver. Dynamic performance is not critical since it is operated only fully "on" or "off". Q_s, atomization of the fuel, and minimum operating voltage are the important criteria.

3. Standard Test Conditions—Unless otherwise specified, the following test conditions are implied:

3.1 Test Fluid—Fuel injectors are designed to spray gasoline, and most development and testing involves verification of injector performance with that fuel. Gasoline, however, is volatile and variable in physical properties. A less volatile hydrocarbon liquid (normal Heptane) with known physical properties and a viscosity and density near gasoline is, therefore, recommended for measurement of injector performance characteristics discussed in Section 4. It is recognized that no one test fluid is ideal and will duplicate the performance of multihydrocarbon fuels found in the field. n-Heptane was selected as the best compromise on the basis of (1) worldwide availability, (2) stability if reused, and (3) fluid properties close to that of gasoline. Tests that are more application-related or conducted for quality control, durability, etc., may be performed with fluids specified by the user.

n-Heptane is one of the two common pure hydrocarbons used to define the octane number scale, and is stocked in all petroleum refinery control and analytical laboratories worldwide to known levels of purity and consistency. It sprays with gasoline-like qualities through the known fuel injector designs, and yields excellent reproducibility of spray form and flow rates. n-Heptane is not gasoline, however, and flows in the most common injector design of 1987 at about 3 to 4% lower rate than Indolene, the U.S. EPA emission test gasoline (see Table 1). For different injector designs, other values of this gasoline to n-Heptane difference will be found.

The most significant advantage of n-Heptane over gasoline is reusability. It is a pure hydrocarbon compound and not a mixture of naturally occurring compounds, as in gasoline. Repeated flow testing, resulting in partial evaporation of the test fluid, does not change the flow properties of the remaining fluid. Thus, repeated flow measurements are more consistent. When correlation work with gasoline is carried out, the gasoline should be tested "once-through" the injector, and fresh gasoline used for replicate tests. In this way, property change in the gasoline due to partial evaporation will be avoided.

Both gasoline and n-Heptane are light, volatile liquids, and have flash points well below normal room temperature. Vapor from both liquids, therefore, poses a fire hazard that must be continuously controlled in the laboratory areas.

Factory-floor environments seldom permit routine use of flammable liquids above their flash points. Manufacture of fuel injectors generally involves a flow measurement of the completed (or partially completed) injector, and a heavier-than-gasoline liquid is usually chosen for this purpose, so that the flash point of the liquid is higher than the factory working area temperature. One such liquid is a typical medium petroleum distillate usually called "mineral spirits" in the USA, which has a flash point over 43°C. Such liquids require a laboratory-to-factory floor calibration step in which a flow rate on the test fluid corresponding to the desired gasoline flow rate must be established for each individual injector design and test condition. No general relationship is appropriate, or possible, from this document. Middle distillates for factory floor use are normally procured from convenient local facilities near the makers' plants, and valid-for-the-batch calibrations are made to determine and maintain production quality of the finished injectors. Physical

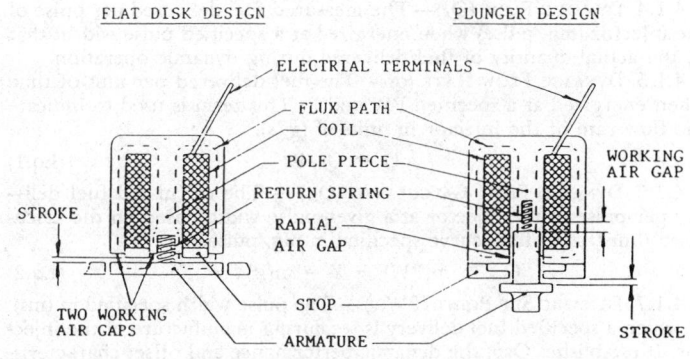

FIG. 3—MAGNETIC CIRCUIT

TABLE 1—INJECTOR FLOW RATES OF INDOLENE VERSUS n-HEPTANE

	Viscosity cSt @ 20°C	Specific Gravity	Static Flow g/s	Dynamic Flow mg/pulse
Indolene	0.619	0.75	1.993	3.63
n-Heptane	0.605	0.68	1.918	3.52
% Difference			−3.8	−3.0

property specifications for n-Heptane, Indolene, and mineral spirits are given in Table 2.

3.2 Fluid Temperature—Should be measured at the injector inlet and stabilized at 20°C ± 1.0 (Reference 6.1).

3.3 Injector Temperature—Should be stabilized at room temperature 22°C ± 3 prior to test.

3.4 Pressure—The differential pressure across the injector will be determined by the application and held to within ±0.10 kPa of this value throughout the test. It should be specified in (kPa) units and measured at the injector inlet (Reference 6.1).

3.5 Period (P)—The time elapsed between the beginning of one injection pulse to the beginning of the next pulse shall be 10 ms/pulse ± 0.001 for both port and throttle body injection applications whether the fuel delivery method is simultaneous single or double fire, sequential, etc.

3.6 Pulse Width (PW)—The increment of time (ms) that the injectors are commanded to deliver fuel shall be determined by the type of test being conducted and held within ±0.001 ms.

3.7 Injector Driver—Determined by application and of instrument 4.1 for typical driver types). Voltage supplied to the driver shall be 14.0 V DC ± 0.05.

3.8 Polarity—Maintained constant throughout all testing and same as that used in the application.

3.9 Test Apparatus—The type of flow fixture used will vary depending on the particular injector parameter being evaluated (Reference Section 6).

3.9.1 INSTRUMENTS—Stabilized per manufacturers' recommendations.

3.9.2 INJECTOR POSITION—The injector should be mounted vertically except when tested as an integral part of a fuel rail assembly.

3.9.3 PRECONDITIONING—Purge injectors and test apparatus with test fluid to remove all air, vapors, and shipping fluids. Flow injectors for 10 000 pulses at 5.0 ms PW and 10 ms P. Discard purge fluid.

3.9.4 FLOW MEASUREMENT—Flow rate may be measured by either volume or mass flow with the latter being the preferred method. Data should be reported in mass flow units (g/s or mg/pulse).

4. Basic Characteristics and Definitions

4.1 Functional Parameters—The following are parameters used in describing and/or measuring the basic functional characteristics of the injector.

4.1.1 PERIOD (P)—The reciprocal of the frequency of injection; that is, the time elapsed between the beginning of one injection to the beginning of the next injection, expressed in units of (ms/pulse).

4.1.2 PULSE WIDTH (PW)—Increment of time that the injectors are commanded to deliver fuel for a single injection event (ms).

4.1.3 STATIC FLOW RATE (Q_s)— The rate of fuel delivered (g/s) by an injector when energized in the fully opened position. It is the maximum flow rate of the injector and can be used to approximate the slope (m_a) of the injector flow curve (Reference 4.1.11).

4.1.4 DYNAMIC FLOW (Q_d)—The measured fuel delivered per pulse of the injector (mg/pulse) when energized at a specified pulse width: that is, the actual quantity of fuel delivered during dynamic operation.

4.1.5 DYNAMIC FLOW RATE (Q)— The fuel delivered per unit of time when energized at a specified PW and P. This term is used to indicate the flow rate of the injector in units of (g/s).

$$Q = Q_d/P \quad \text{(Eq.1)}$$

4.1.6 DYNAMIC FLOW CALCULATED (Q_{dc})—The calculated fuel delivery per pulse of the injector at a given pulse width based on the calculated (linearized) flow curve specified in mg/pulse.

$$Q_{dc} = m(PW) - Y = m(PW - X) \quad \text{(Eq.2)}$$

4.1.7 DYNAMIC SET POINT (PW_{xx})—The pulse width specified in (ms), at which a specified fuel delivery is set during manufacture of the injector. It establishes Qsp, the dynamic performance and offset characteristics of the injector. It represents a Q_d point or pulse width at which the flow variation of a population of injectors is minimized.

4.1.8 DYNAMIC SET POINT FLOW (Qsp)—The measured fuel delivered per pulse of the injector in mg/pulse when energized at the dynamic set point. This flow is used for incoming inspection and quality control.

4.1.9 LINEARITY DEVIATION (LD)— Ideally, the flow from an injector should be linear and directly proportional to pulse width over the full flow range of the injector. This is not the actual case, for significant deviation from linearity occurs at the extremities of the flow curve (Figs. 4 and 5). In order to measure the deviation from linearity, a least squares regression analysis is performed on five intermediate flow points at 3, 4, 5, 6, and 7 ms PW with a period of 10 ms/pulse. Qd and PW are, respectively, the dependent and independent variables. The resulting curve is referred to as the linearized flow curve. Deviation

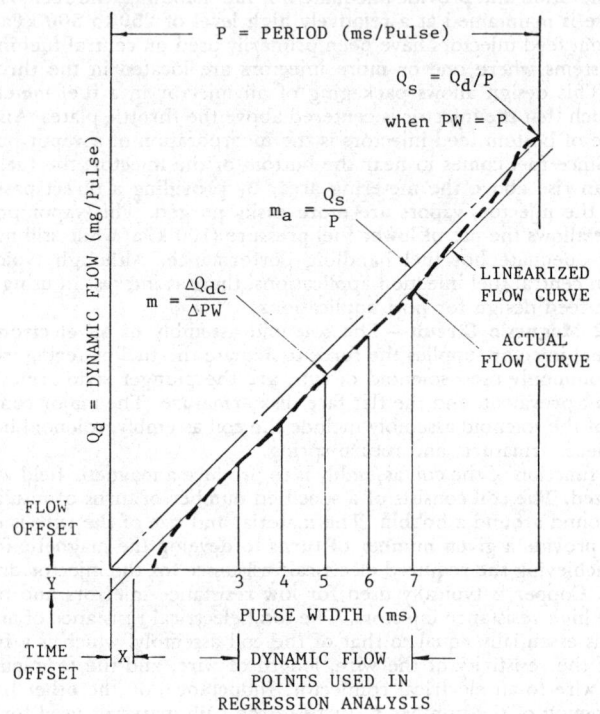

DYNAMIC FLOW CALCULATED (mg/Pulse) = Q_{dc} = m(PW) − Y = m(PW−X)

DYNAMIC FLOW RATE (g/s) = Q = Q_d/P

FIG. 4—CHARACTERISTIC INJECTOR FLOW CURVE

from linearity is then defined as the percent difference between the measured or actual flow (Q_d) and calculated flow (Q_{dc}) taken at a given pulse width divided by the calculated flow.

$$LD = \frac{Q_d - Q_{dc}}{Q_{dc}} \times 100 = \% \quad \text{(Eq.3)}$$

4.1.10 SLOPE (m)—The change in Q_d per unit of pulse width based on the calculated linear regression flow curve (mg/pulse/ms).

$$m = \frac{\Delta Q_{dc}}{\Delta PW} \quad \text{(Eq.4)}$$

4.1.11 SLOPE APPROXIMATED (m_a)— An approximation of the injector flow curve m using the Q_s expressed in units of (mg/pulse) when the P and PW are equal (see Fig. 4).

$$m_a = Q_s/P \quad \text{(Eq.5)}$$

When: Q_s = (mg/pulse) and P = PW (ms)

4.1.12 TIME-OFFSET (x)—The displacement of the calculated linear regression flow curve from the origin along the abscissa or pulse width axis (Fig. 4).

4.1.13 FLOW-OFFSET (Y)—The displacement of the calculated linear regression flow curve from the origin along the ordinate or Q_d axis (Fig. 4).

TABLE 2—PHYSICAL PROPERTY SPECIFICATIONS FOR TEST FLUIDS

PROPERTY	TEST FLUIDS		
Name of Fluid	Normal Heptane (CP Grade)	Indolene* (Clear)	Mineral Spirits
Purpose of Fluid	Primary reference fuel grade pure hydrocarbon liquid in gasoline boiling range, for injector flow tests in engineering and development.	Engineering and emission test gasoline for injector tests that require a full-boiling gasoline.	Middle distillate for production floor quality control tests.
Specific Gravity at 15.6°C (60°F) (ASTM D 1298)	0.681 to 0.685	0.739 to 0.749	0.775 to 0.785
Kinematic Viscosity at 20°C (68°F) (ASTM D 445)	0.60 to 0.64 cSt	0.59 to 0.65 cSt	1.195 to 1.205 cSt
Absolute Viscosity at 20°C (68°F) (ASTM D 445)	0.418 cps	0.418 to 0.468 cps	0.94 to 0.96 cps
Flash Point (typical) (ASTM D 56)	−1°C (30°F)	−40°C (−40°F)	43.3°C (110°F)
Distillation (ASTM D 86) Initial 10% Evaporated 50% Evaporated 90% Evaporated End Point	97.7°C (208°F) — 98.4°C (209°F) — 99.4°C (211°F)	24 to 35°C (75 to 95°F) 49 to 57°C (120 to 135°F) 93 to 110°C (200 to 230°F) 149 to 163°C (300 to 325°F) 213°C max (415°F)	150°C min (300°F) — — — 210°C max (410°F)
Pentane Insolubles (ASTM D 893) % by weight	0.001 max	0.001 max	0.01 max
Water Content (ASTM D 1744) % by weight	0.001 max	0.001 max	0.10 max

*Trademark of Amoco Oil Co.

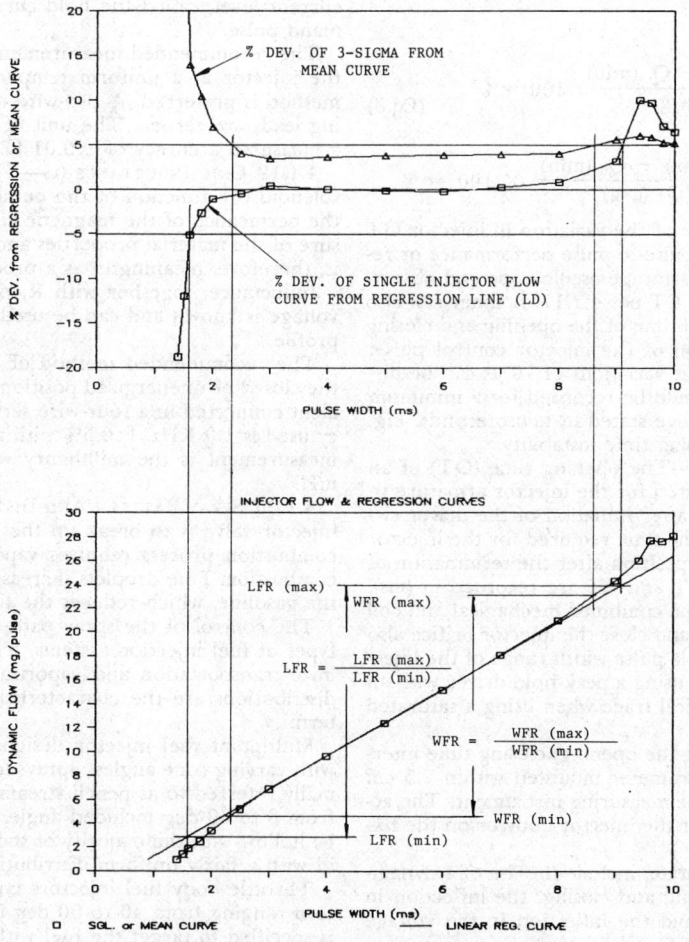

FIG. 5—CALCULATION OF FLOW RANGE

4.1.14 FLOW RANGE (LFR and WFR)—One of the more important characteristics of an injector is the usable minimum and maximum fuel flow. Flow range, specified as a ratio of the maximum to minimum dynamic flows, is used as a measure of this capability. Two methods are used for calculating the ratio: LFR and WFR.

4.1.14.1 *Linear Flow Range (LFR)*—A number based on the linearized flow curve of a single injector. It is used to compare the linear range between injectors of different design or manufacture. The number is defined as the maximum linearized flow point LFR (max) divided by the minimum linearized flow point LFR (min) at their respective pulse widths where the measured flows deviate ±5.0% from the linearized flow curve (see Fig. 5).

$$LFR = \frac{LFR\ (max)}{LFR\ (min)} \quad (Eq.6)$$

4.1.14.2 *Working Flow Range (WFR)*—A number based on the mean flow curve of a population (24 minimum) of injectors representative of a normal production distribution. The application engineer is interested in this ratio, since it provides a relative measure of production injector-to-injector variability; that is, the flow range where a population of injectors will be within a specified tolerance. It is defined as the maximum working flow point WFR (max) divided by the minimum working flow point WFR (min) where all injectors are within ±5.0% of the mean flow curve at three standard deviations (see Fig. 5).

$$WFR = \frac{WFR\ (max)}{WFR\ (min)} \quad (Eq.7)$$

4.1.15 REPEATABILITY—Ideally a measure of the pulse-to-pulse flow rate variation of the injector. It is an important element from the standpoint of engine control for emissions. The equipment required to accurately determine the mass flow for one injection pulse is expensive and resolution is limited. The following procedure is, therefore, recommended as a compromise:

The injector is tested at standard conditions at two points on the flow curve: (1) the Q_s, and (2) the 2.5 ms Q_d point. A minimum of 30 tests is conducted at each test point. Repeatability, given in percent, is then defined as:

$$\text{Static Repeatability} = \frac{Q_s\ (max) - Q_s\ (min)}{Q_s\ (avg)} \times 100 = \% \quad (Eq.8)$$

$$\text{Dynamic Repeatability} = \frac{Q_d\ (max) - Q_d\ (min)}{Q_d\ (avg)} \times 100 = \%$$

4.1.16 STABILITY (S)—S is a measure of the variation in injector OT and CT and is an indirect measure of pulse-to-pulse performance or repeatability. It is determined by using a storage oscilloscope and observing the trace representing the OT and CT per 4.1.17. A timer can also be used to directly measure the elapsed time of the opening and closing events by triggering it at the initiation of the injector control pulse. Resolution should be adjusted such that variations of 10 μs can be discerned. Opening and closing times should be recorded for a minimum of 1000 consecutive pulses and the range stated in microseconds. Fig. 6 shows a typical measurement of closing time instability.

4.1.17 OPENING AND CLOSING TIME—The opening time (OT) of an injector is a measure of the time required for the injector armature to first reach its fully opened positioned after initiation of the driver circuit pulse input. Closing time (CT) is the time required for the injector armature to first reach its fully closed position after the termination of the driver circuit pulse input. Both OT and CT are recorded in (ms). They provide a relative indication of the combined mechanical and coil response time. The total time to open and close the injector orifice also provides a relative measure of the usable pulse width range of the injector. Fig. 7a shows a typical trace when using a peak-hold driver with an injector of low R. Fig. 7b shows a typical trace when using a saturated driver and an injector with high R.

The preferred method of measuring the opening/closing time interval is by the use of a precision accelerometer mounted within 2.5 cm of the injector body and a suitable time measuring instrument. The accelerometer may be mounted on either the injector body or on the fixture as shown in Fig. 8.

Alternate methods, though less accurate, include the use of a dynamic pressure transducer for both opening and closing, the inflection in the current trace for opening only, and the inflection in the voltage trace for closing only (Reference Figs. 6 and 7).

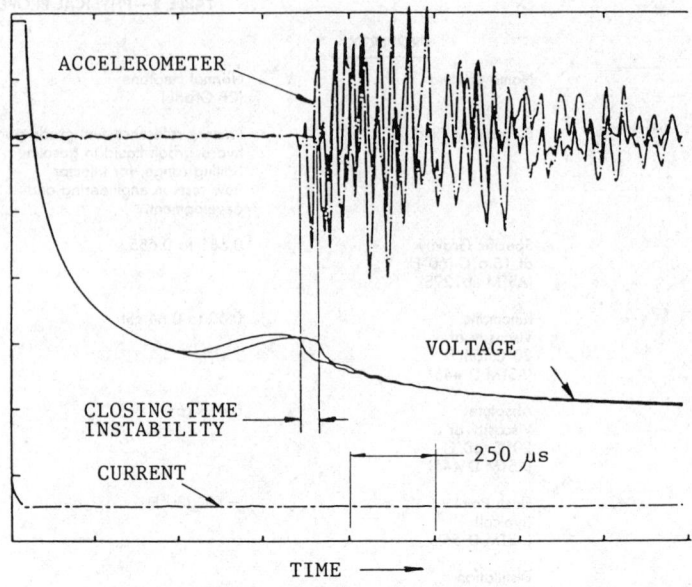

FIG. 6—MEASUREMENT OF CLOSING TIME INSTABILITY

4.1.18 COIL RESISTANCE (R)—Measurement of R permits the injector user to determine the current requirement for a given system voltage. Presently, injector solenoid designs consist of either high or low resistance coils. The high resistance design is typically 12 to 16 Ω and used with a saturated circuit driver. The driver turns battery voltage on and off allowing current to be controlled by the R. The low resistance design is typically 1 to 4 Ω and uses a current limiting driver. This system allows a peak current for rapid opening response followed by a lower current level, called the hold current, for the remainder of the command pulse.

The recommended measurement technique is by direct current with the injector at a uniform temperature of 20°C ± 1. The four-wire method is preferred. A two-wire measurement is acceptable if measuring leads are zeroed. The unit of measurement is to be the ohm with a measured accuracy of ±0.01 Ω.

4.1.19 COIL INDUCTANCE (L)—The inductance of an electromagnetic solenoid is a function of the number of turns in the solenoid coil and the permeance of the magnetic circuit. Inductance is an indirect measure of the material properties and geometry of the flux path. Its value is, therefore, meaningful as a production control parameter.

Inductance, together with R, provides the time constant when the voltage is known and can be used in predicting the initial current rise profile.

The recommended method of measurement is with the injector in the closed or unenergized position using a Wheatstone Bridge or equivalent connected in a four-wire series mode. The standard test frequency used is 1.0 KHz + 0.5% with a potential of 1.0 V rms. The unit of measurement is the millihenry with a reported resolution of ±0.01 mH.

4.1.20 SPRAY PATTERN AND DISTRIBUTION—One function of the fuel injector valve is to break up the fuel into fine droplets. The engine's combustion process requires vaporization of the gasoline for proper combustion. Fine droplets increase the surface area to volume ratio of the gasoline, which reduces the time for the gasoline to vaporize.

The control of the spray pattern characteristics is important for all types of fuel injection systems. The spray pattern targets fuel to optimize transportation and vaporization. The radial and segmental fuel distributions are the characteristics specified to define the spray pattern.

Multipoint fuel injector designs can have single or multiple sprays with varying cone angles. Sprays less than 8 deg included angle are normally referred to as pencil stream. The larger spray cone angles range from 8 to 30 deg included angle. The spray cone fuel distribution can be hollow with the majority of the fuel in the outer cone surface or solid with a fairly uniform distribution of fuel within the cone angle.

Throttle body fuel injectors typically have a hollow spray cone pattern ranging from 40 to 60 deg included angle. The spray cone angle is specified to target the fuel with respect to the throttle plate.

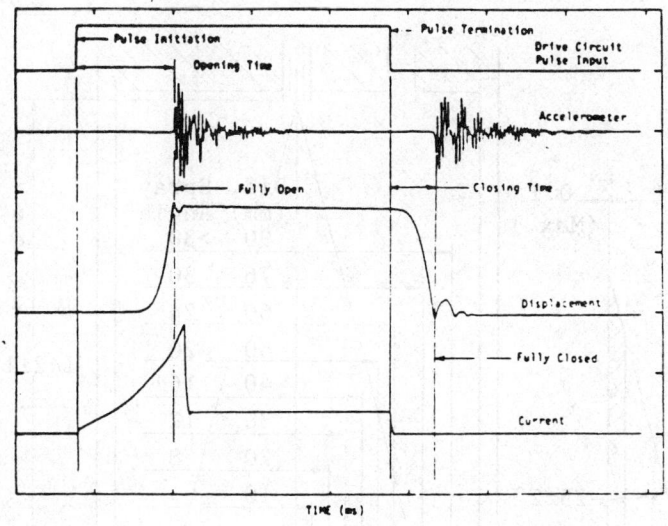

FIG. 7a—TYPICAL OPENING AND CLOSING TIME WITH PEAK AND HOLD DRIVER

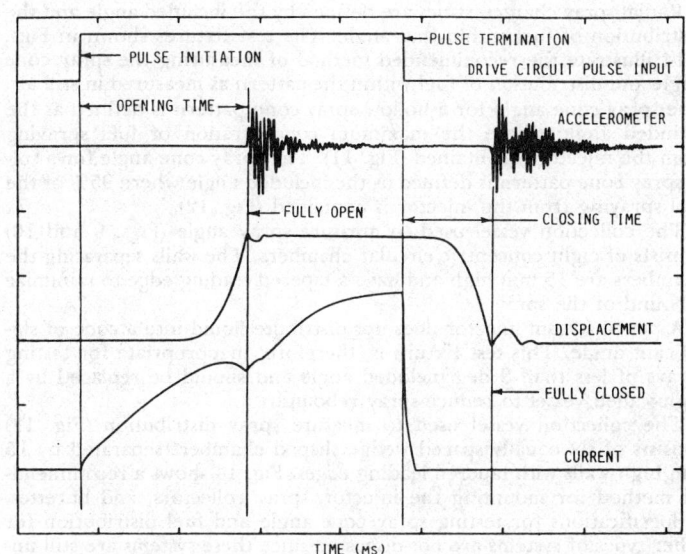

FIGURE 7b – Typical Opening and Closing Time With Saturated Driver

FIG. 7—OPENING AND CLOSING TIME

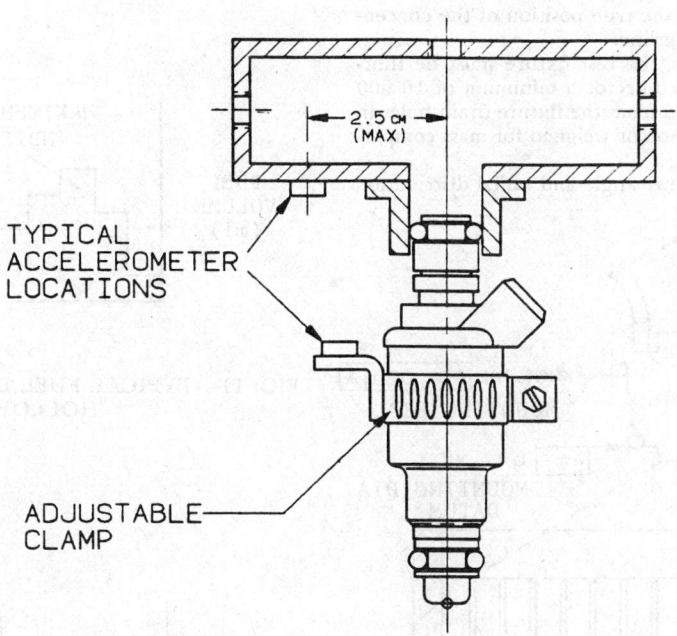

FIG. 8—ACCELEROMETER MOUNTING LOCATIONS

Radial spray characteristics are defined by the included angle and the distribution of fuel within that angle. The test fixtures shown in Figs. 9-14 illustrate the recommended method of measuring the spray cone angle and distribution of fuel within the pattern as measured in still air. The spray cone angle for a hollow spray cone pattern is defined as the included angle where the maximum concentration of fuel spraying from the injector is contained (Fig. 11). The spray cone angle for a solid spray cone pattern is defined as the included angle where 95% of the fuel spraying from the injector is contained (Fig. 12).

The collection vessel used to measure spray angle (Figs. 9 and 10) consists of eight concentric circular chambers. The walls separating the chambers are 15 mm high and have a tapered leading edge to minimize rebound of the spray.

A pencil stream injector does not distribute liquid into a cone of significant angle. This test fixture is, therefore, inappropriate for testing sprays of less than 8 deg included angle and should be replaced by a high-walled vessel to reduce spray rebound.

The collection vessel used to measure spray distribution (Fig. 13) consists of six equally spaced wedge shaped chambers separated by 15 mm high walls with tapered leading edges. Fig. 14 shows a recommended method for mounting the injector, spray collectors, and burettes.

Specifications for testing spray cone angle and fuel distribution for other types of systems are not discussed since these systems are still under development.

4.1.20.1 *Spray Measurement Procedure:*
 a. Tests to be conducted with n-Heptane as the control fluid.
 b. Tests to be conducted at pressure specified by application.
 c. Tests to be conducted both at the dynamic set point pulse width and Q_s.
 d. The collection vessel must be located below the point of origin of the injector's spray cone angle as indicated in Table 3 and Fig. 10. This distance is determined by the application. The point of origin of the spray cone angle must be specified from the injector's mounting surface datum.
 e. The fuel injector must be centered over the collection vessel. The centerline of the injector mounting diameter datum must be held to within 0.025 mm of the true position of the concentric ring collection vessel's centerline.
 f. To provide repeatable readings, the test fixture must be thoroughly wet down by pulsing the injector a minimum of 10 000 pulses. The fuel may be collected from the fixture drain holes in burettes for volumetric comparison or weighed for mass comparison.
 g. The distribution of fluid for spray angle and radial distribution shall be defined by the user.

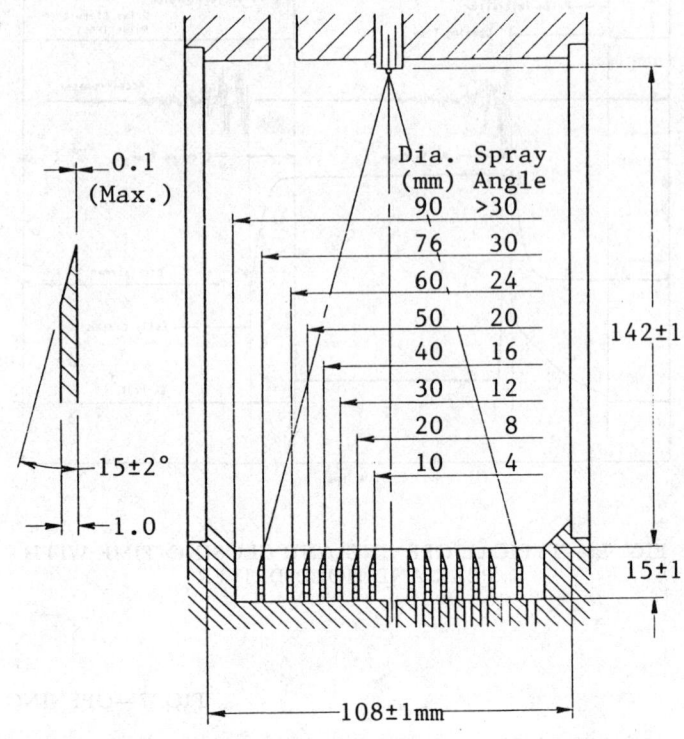

FIG. 10—TYPICAL SETUP FOR 30 DEG CONE SPRAY

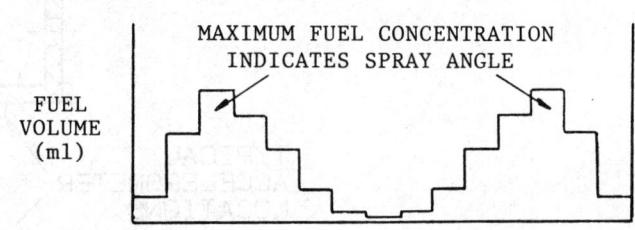

FIG. 11—TYPICAL FUEL DISTRIBUTION OF WIDE ANGLE HOLLOW CONE SPRAY

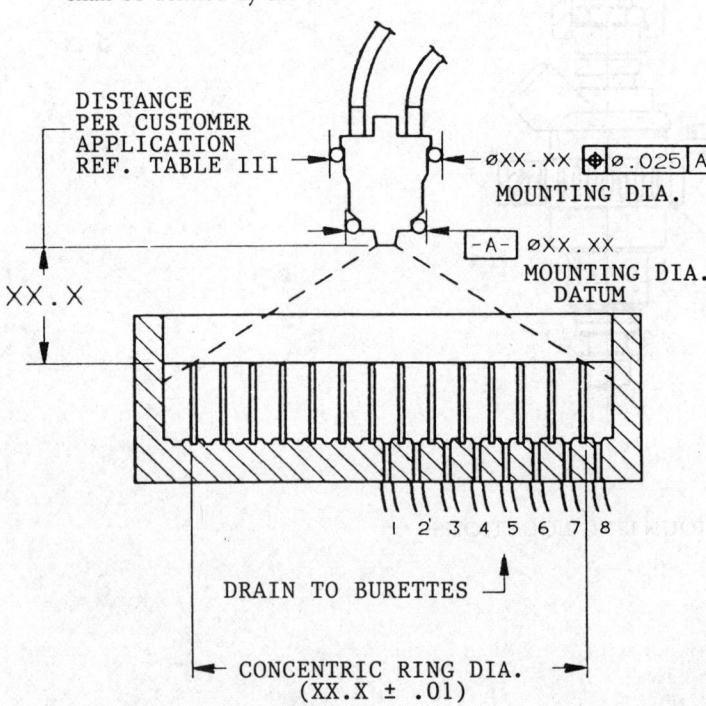

FIG. 9—RADIAL SPRAY COLLECTION VESSEL

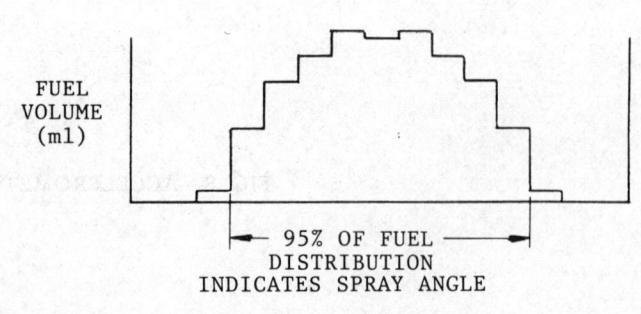

FIG. 12—TYPICAL FUEL DISTRIBUTION OF SOLID CONE SPRAY

TABLE 3—INJECTOR LOCATION AND CORRESPONDING SPRAY ANGLE

Ring No.	1	2	3	4	5	6	7	8
Diameter (mm)	10.0	20.0	30.0	40.0	50.0	60.0	76.0	90.0
Height (mm)				SPRAY CONE ANGLE (DEG)				
142.0	4.0	8.1	12.1	16.0	20.0	23.9	30.0	>30.0
91.7	6.2	12.4	18.6	24.6	30.5	36.2	45.0	>45.0
65.8	8.7	17.3	25.7	33.8	41.6	49.0	60.0	>60.0
38.0	15.0	29.5	43.1	55.5	66.7	76.6	90.0	>90.0

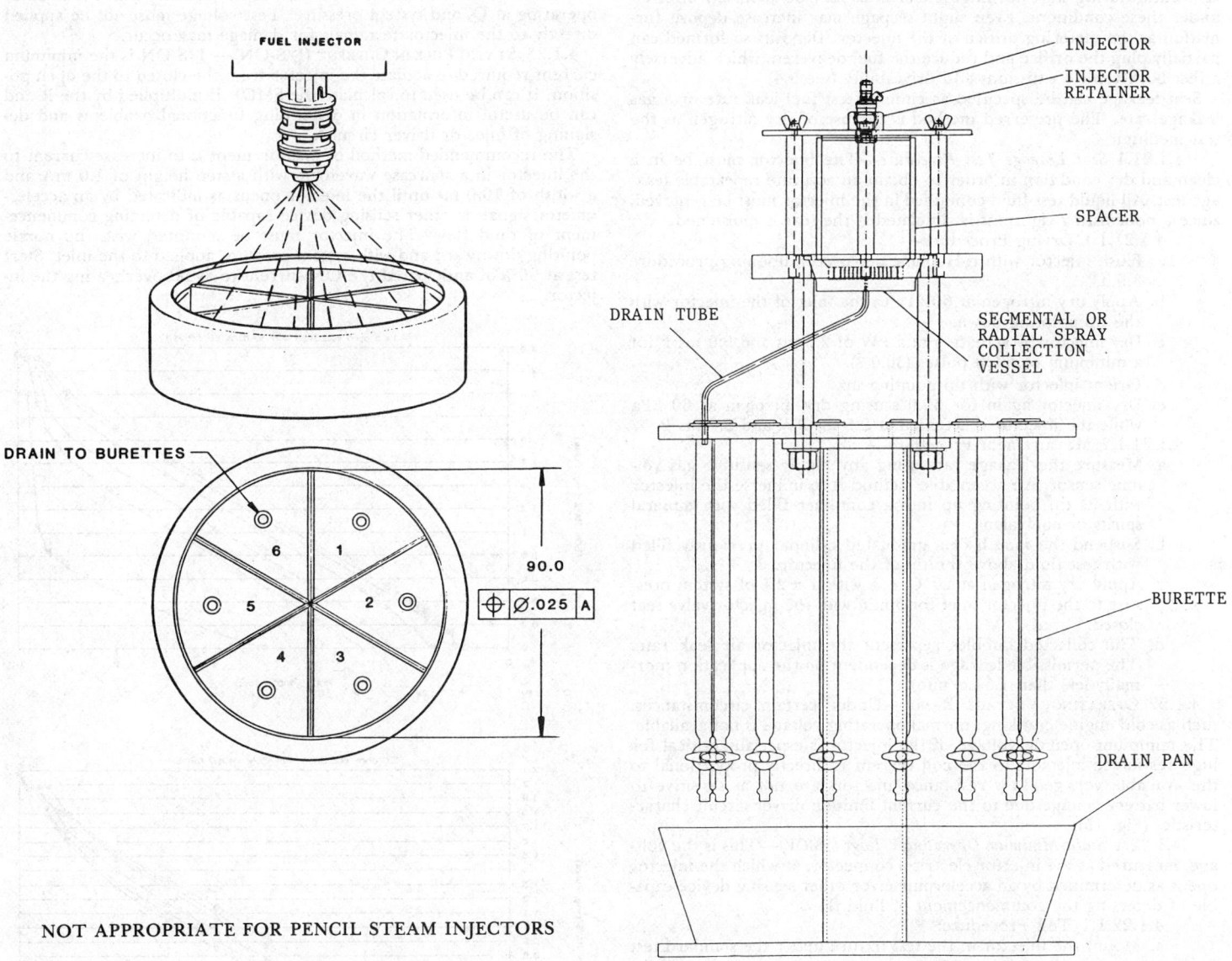

FIG. 13—SEGMENTED SPRAY COLLECTION VESSEL

NOT APPROPRIATE FOR PENCIL STEAM INJECTORS

FIG. 14—SPRAY DISTRIBUTION TEST FIXTURE

4.1.21 EXTERNAL LEAKAGE—No visible fluid leakage at working pressure is permitted in any area other than the seat. This includes the body, its seals, and the external O-rings.

The injector seat generally has metal-to-metal sealing surfaces for durability reasons. A leak tight interface is required between the mating surfaces of the injector valve seat components to maintain fuel pressure, prevent abnormal fuel accumulation in the intake manifold, and reduce deposit formation at the injector orifice. Seat leakage may result in a leaner or richer A/F ratio depending on such variables as leakage rate, temperature, and soak period.

Loss of pressure during extended periods of engine shutdown under hot ambient conditions may cause excessive fuel vaporization and incorrect fuel delivery during start-up. The increased time required to achieve desired rail pressure in combination with vapor entrained in the fuel delivered can result in increased cranking time, rough idles, stalls and/or a no-start condition.

Fuel leakage into the intake manifold may cause a richer than normal A/F ratio during start-up. Engine emissions may be adversely affected under these conditions. Even slight seepage may increase deposit formation at the metering orifice of the injector. Deposits so formed can partially plug the orifice and reduce the fuel delivered, which adversely affect both vehicle emissions and driveability (see 5.4).

Seat leakage can be specified as either a test fuel leak rate or a gas leakage rate. The preferred method is the use of dry nitrogen as the test medium.

4.1.21.1 *Seat Leakage Test Procedure*—The injector must be in a clean and dry condition in order to obtain an accurate repeatable leakage test. All liquid test fluid contained in the injector must be removed, since a nonexistent tightness is simulated if the seat is moistened.

4.1.21.1.1 Drying Procedure—
a. Flush injector with n-Heptane per preconditioning procedure 3.9.3.
b. Apply dry nitrogen at 60 kPa to the inlet of the injector with the tip pointing down.
c. Dry injector by actuating at a PW of 2.5 ms and 5.0 ms P for a minimum of 6000 pulses (30.0 s).
d. Orient injector with tip pointing up.
e. Dry injector again for 30.0 s using dry nitrogen at 60 kPa while the injector is actuated at 2.5 ms PW and 5.0 ms P.

4.1.21.1.2 Measurement Procedure—
a. Measure the leakage rate using any highly sensitive gas volume sensor. An acceptable method is to immerse the injector with its tip pointing up into a container filled with mineral spirits or equivalent.
b. Suspend the mouth of a graduated cylinder previously filled with test fluid above the tip of the injector.
c. Apply dry nitrogen at 20°C ± 1 within ±2% of system pressure to the injector inlet for 5 min with the injector valve seat closed.
d. The collected bubbles represent the injector air leak rate. The permissible leakage is dependent on the application (normally less than 1.5 cc/min).

4.1.22 OPERATING VOLTAGE RANGE—Under certain circumstances, such as cold engine cranking, normal operating voltage is not available. The minimum opening voltage of the injector is especially critical for high resistance injectors as the coil current is directly proportional to the available voltage. Low resistance injectors are not as sensitive to lower battery voltage due to the current limiting driver circuit characteristics (Fig. 15).

4.1.22.1 *Static Minimum Operating Voltage (SMOV)*—This is the voltage, measured at the injector electrical connector, at which the injector opens as determined by an accelerometer or other sensing device capable of detecting the commencement of fluid flow.

4.1.22.1.1 Test Procedure:
a. Mount the injector in the test fixture under the standard test conditions per Section 3.
b. While measuring the voltage at the electrical connector, raise the supply voltage in 0.10 V steps every 20.0 ms.
c. Record the voltage when the injector opens.

4.1.22.2 *Dynamic Minimum Operating Voltage (DMOV)*—This parameter is a measure of the fuel injection systems response to low voltage conditions. The driver circuit has a significant effect on the injector response characteristics; therefore, the driver circuit as used in the vehicle must be used.

4.1.22.2.1 Test Procedure:
a. Mount the injector in the test fixture under the standard test conditions per Section 3.
b. Using the driver circuit from the vehicle application, raise the supply voltage to the driver in steps of 0.10 V starting at the SMOV. The logic pulse to the driver is to be a period of 20.0 ms and a PW of 10.0 ms. A minimum of 1000 pulses are to be supplied at each step.
c. Measure the supply voltage and the injector flow rate at each step point.
d. The DMOV is that voltage at which the flow rate exceeds 50% of the flow at 14.0 V.

This test may also be modified to generate a voltage compensation curve for the system by continuing readings beyond the minimum opening point and using other PWs to create the family of curves similar to those in Fig. 15.

4.1.22.3 *Maximum Overload Voltage*—The injector must maintain calibration after a maximum of 24.0 V is applied to the driver for a period of 60 s. This test is intended to simulate an incorrect jump start condition.

The injector and driver are tested as an assembly with the injector operating at Q_s and system pressure. Test voltage must not be applied directly to the injector terminals for damage may occur.

4.1.23 STATIC PULL-IN CURRENT (I/S-ON)— I/S-ON is the minimum current required to actuate the injector from the closed to the open position. It can be used to calculate the SMOV if multiplied by the R and can be useful information in diagnosing functional problems and designing of injector driver circuits.

The recommended method of measurement is to increase current to the injector in a staircase waveform with a step height of 1.0 mA and a width of 10.0 ms until the injector opens as indicated by an accelerometer signal or other sensing device capable of detecting commencement of fluid flow. The injector must be mounted with the nozzle pointing downward and with system pressure applied to the inlet. Start test at 90% of anticipated I/S-ON current to avoid overheating the injector.

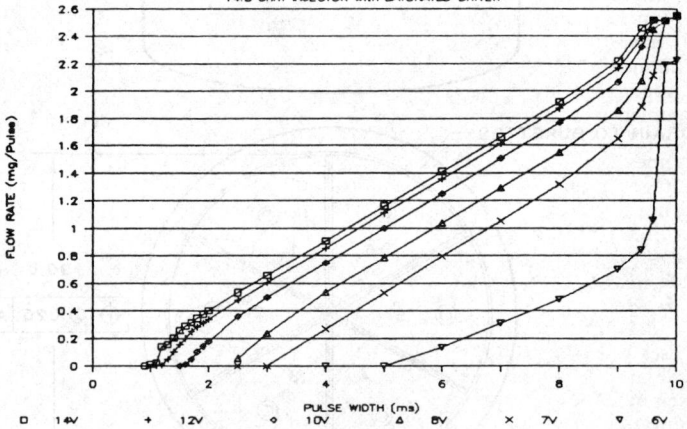

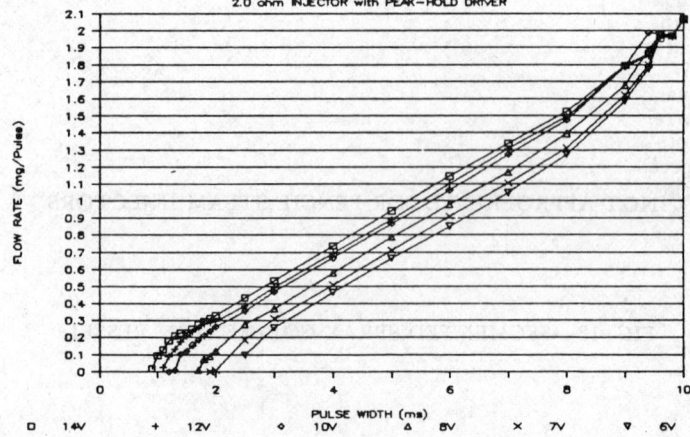

FIG. 15—EFFECT OF VOLTAGE ON PERFORMANCE

4.1.24 STATIC DROP-OUT CURRENT (I/S-OFF)—I/S-OFF is the minimum current required to hold the injector open after it has been energized. It is used to determine the holding current calibration of the peak-hold driver.

The recommended method of measurement is to apply sufficient current to assure that the injector is in the open position and then to decrease the current in a staircase waveform with a step height of 1.0 mA and a width of 10.0 ms until the injector closes as indicated by an accelerometer signal or other sensing device capable of detecting cessation of fluid flow. The injector must be mounted with the nozzle pointing downward and with system pressure applied to the inlet.

The I/S-OFF is always lower than the I/S-ON because the solenoid air gap is smaller by the amount of the stroke, which will yield higher magnetic forces for any given current level. The drop-out current is generally measured immediately following the pull-in current measurement.

4.1.25 PRESSURE DROP RATIO (PDR)—A ratio of the fuel pressure drop across the metering orifice of an injector compared to the fuel pressure drop across the valve seat.

The PDR is useful in evaluating the sensitivity of an injector design relative to the metering of hot fuel. Calculations can be made to determine if fuel exiting the metering area of the injector will vaporize if the PDR, fuel inlet pressure, fuel volatility, and maximum application temperature are known. This assumes that the valve/seat and the metering orifice are the primary restrictions to flow. If additional significant pressure drops occur within an injector design, they must be accounted for before any hot fuel handling predictions are made. Since an injector is designed to meter a liquid, fuel vapor cannot be accurately metered and steps to minimize this condition should be taken.

4.1.26 INSULATION RESISTANCE (IR)—This test is designed to check for a potential failure of the insulation between the coil assembly and the case of the injector. It is usually performed on completed injectors to ensure that the coil insulation has not been damaged during the assembly process and that the terminal to case clearance has been adequately maintained. IR is also measured after any mechanical integrity testing to ensure that the insulating abilities of the injector coil assembly have not been degraded.

In most existing applications, a positive battery voltage is continuously applied to one of the injector terminals and the unit is energized by completing the ground circuit. In a situation where the IR breaks down, the injector could become continuously energized and flood the engine. On systems that energize the unit by providing the positive voltage through the driver, breakdown can cause the injector to misfire or not fire at all depending on where the coil assembly is shorted.

4.1.26.1 *Test Procedure*—Connect a standard megaohmmeter tester, set to 750 V DC, between the injector case and a coil terminal post. The minimum allowable resistance reading shall be 340k Ω after 2 s.

5. Application Related Parameters and Tests

5.1 Injector Driver—An electronic circuit that supplies voltage pulses to an electromagnetic fuel injector for a precise increment of time and at a given repetition rate. The accuracy of these pulses and their repetition is normally ±0.001 ms. The peak-hold driver and the saturated driver are most commonly used by the industry for vehicle applications.

5.1.1 PEAK-HOLD DRIVER—A driver that uses two levels of current to operate the injector (Fig. 16). The driver circuit applies battery voltage to the injector until a predetermined current level is reached. The current is then reduced and held at a lower level for the duration of the PW. This type of driver is normally used with injectors having low resistance coils (typically around 2 Ω). The accuracy of the driver peak current level (I_p) and the hold current level (I_h) is held to ±0.50%. Note—In order to minimize OT instability, it is advisable to match the injector and driver characteristics such that the switch from peak-to-hold does not occur until after the injector is fully open.

ADVANTAGES: The high peak current minimizes OT response and the low hold current minimizes CT response. This method of control results in an increased linear range of injector operation.

DISADVANTAGES: Heat is primarily dissipated at the driver. Circuitry is more complex than that of the saturated driver.

5.1.2 SATURATED DRIVER—A power transistor driver that turns fully on for the entire duration of the injector PW (Fig. 17). This type of driver is used with injectors having high resistance coils (typically 12 to 16 Ω) or with injectors having low resistance coils in combination with a ballast resistor.

ADVANTAGES: Heat is primarily dissipated through the injector or ballast resistor and not at the driver circuit.

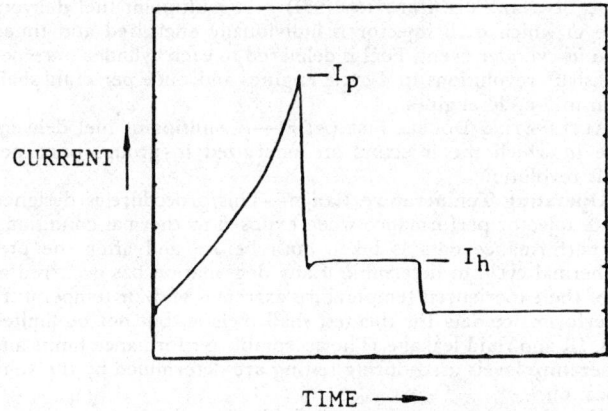

FIG. 16—TYPICAL INJECTOR CURRENT WITH PEAK-HOLD DRIVER

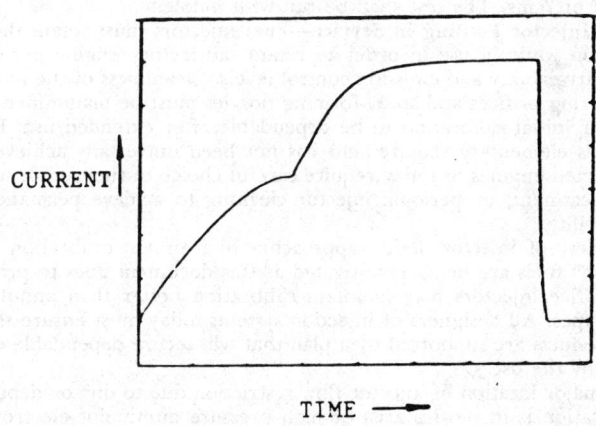

FIG. 17—TYPICAL INJECTOR CURRENT WITH SATURATED DRIVER

Circuitry is simplified compared to the peak-hold driver.

DISADVANTAGES: The inherently slower dynamic response of this system decreases the injector's usable flow range. The Q of an injector used with this type of circuit is more duty cycle sensitive due to heat dissipation considerations. This driver's inductive suppression, which may be resistance-capacitance or zener, significantly affects the injector's Q_d rates due to variations in the circuit's current decay rate. This decay results in a change of the injector's closing time.

5.2 Firing Strategies—Both the peak-hold and the saturated drivers are used in various firing strategies. Three commonly used schemes are simultaneous double fire, sequential single fire, and alternating double fire (see Fig. 18).

5.2.1 SIMULTANEOUS DOUBLE FIRE (SDF)—A multipoint fuel delivery technique in which all injectors in a 4-cycle engine are energized simultaneously once per crankshaft revolution.

5.2.2 SEQUENTIAL FUEL INJECTION (SFI)— A multipoint fuel delivery technique in which each injector is individually energized and timed relative to its cylinder event. Fuel is delivered to each cylinder once per two crankshaft revolutions in 4-cycle engines and once per crankshaft revolution in 2-cycle engines.

5.2.3 ALTERNATING DOUBLE FIRE (ADF)—A multipoint fuel delivery technique in which the injectors are energized in groups once per crankshaft revolution.

5.3 Operating Temperature Range—This procedure is designed to evaluate injector performance when exposed to thermal conditions. Injector performance data is taken both before and after the prescribed thermal cycle to determine if any degradation has occurred as a result of the experienced temperature extremes and/or temperature cycles. Performance data for this test shall include, but not be limited to, Q_s, Q, IR and fluid leakage. The acceptable performance limits and the temperature levels used during testing are determined by the vehicle application.

5.3.1 HOT STATIC SOAK—Purge all fluids from the injector and expose it to a stabilized ambient temperature of 150°C ± 2 for 16 h. The injector is not operated during this test.

5.3.2 COLD STATIC SOAK—Purge all fluids from the injector and expose it to a stabilized ambient of −40°C ± 2 for 16 h. The injector is usually not operated during this test.

5.3.3 CYCLIC SOAK—Expose the injector to the hot thermocycle shown in Fig. 19 for a total of 140 cycles. The injector is to be operational only during segment D-E of the cycle, using a PW of 2 ms and a period of 7 ms. The test shall be run with indolene.

5.4 Injector Fouling in Service—Fuel injectors must retain their calibration while in use in order to retain satisfactory engine performance, driveability and emission control levels. Cleanliness of the injector metering orifices and spray-forming nozzles must be maintained in order for initial calibration to be dependable after extended use. Because this elementary requirement has not been universally achieved, fuel injected engines of today require careful choice of fuel supply, additive treatment, or periodic injector cleaning to achieve permanent flow stability.

A variety of injector design approaches to maintain calibration on "problem" fuels are being investigated as this document goes to press. Open-orifice injectors may maintain calibration better than annular-orifice types. All designers of injection systems today must ensure that their products are supported by a plan that will secure dependable operation by the users.

The major location of injector flow restriction due to dirt or deposit accumulation is in the tip area of high pressure multipoint electronic injectors. Unstable gasolines can, in multistop city driving, build up a varnish-like deposit on the injector tip. These deposits reduce fuel flow through the annular orifice and cause lean engine operation and all the attendant driveability faults. The faulty operation persists until the injectors are cleaned or replaced.

Cleaning of injectors in service can be accomplished with in-tank additives to fuel, by solvent-cleaning of installed injectors, or by ultrasonic cleaning of removed injectors.

Fuel chemistry can stop or reverse deposit accumulation in the injector tip area. Gasolines with low olefin content (as produced from refineries with no cracking units) seem immune to the problem. Gasoline with substantial levels of cracked or polymer components require careful additive treatment with detergent-dispersant chemicals to maintain injector cleanliness. Higher levels of these detergent-dispersant additives can reverse deposit buildup in service and clean injectors, which were deposit-restricted on other fuels. Some gasoline marketers in the United States identify such injector-cleaning gasoline formulations at the point of purchase.

Fuels with high levels of both lead alkyl antiknock and sulfur (>2.0 gPb/usg and >0.15% S) may plug injectors regardless of base fuel formation. High additive levels may always be required in such fuels and an injector of fouling-resistant design may be required.

5.5 Salt Corrosion—This test is to be applied to gasoline injectors to ensure that they can withstand the effects of external salt corrosion, which is likely to be encountered on seaside marine applications and salty road conditions (Reference ASTM B 117-85).

5.5.1 TEST PROCEDURE:
a. The injector shall have its inlet and outlet suitably sealed. The injector is to have its electrical connector fitted; however, it is not to be energized.
b. Samples shall be supported or suspended between 15 and 30 deg from the vertical, with any significant surface parallel to the principal direction of fog flow.

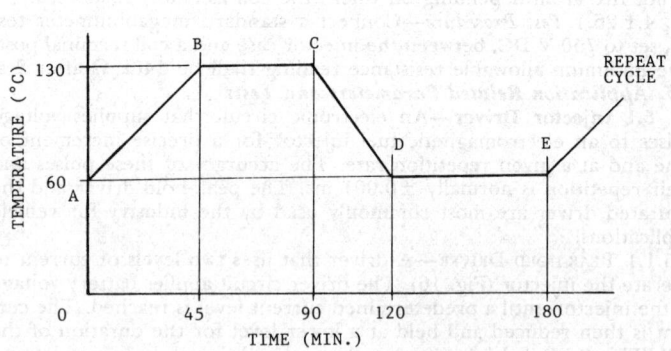

FIG. 18—PORT FUEL INJECTION FIRING STRATEGIES

FIG. 19—THERMOCYCLE TIME SCHEDULE

c. Samples shall be at least 30 mm from each other and 35 mm from a chamber wall.
d. Sample shall be so placed as to permit free settling of fog on all samples, but avoid salt solution from one sample dripping onto another.
e. Expose the samples in the fog chamber for a period of 96 h. Inspect the samples and operation of the chamber as required. Opening of the chamber shall be limited to the frequency necessary to carry out the operations. Inspections should not be made more than twice a day for a total of 15 min. Position of the samples may be varied during the inspection period.
f. Remove the samples, wash gently in water not warmer than 38°C, and dry by blotting with absorbent paper.

5.5.2 CORROSION ASSESSMENT—Each injector tested shall be examined immediately after testing. The degree of corrosion shall be visually assessed in accordance with 5.5.3.

Each injector tested shall meet the performance requirements detailed in 5.5.4 prior to and after the completion of testing.

5.5.3 SURFACE CORROSION AND BLISTERING—For the purpose of this document, a "significant surface" is defined as an area designated to be coated that can be touched by a 6.35 mm diameter ball.

Any spots of red rust on a significant surface shall be cause for rejection. Rust spots on surfaces, which cannot be touched by a 6.35 mm ball, shall be noted but will not be cause for rejection.

Any blistering of the coating shall be noted as to location on the sample piece, blistering size, and number of blisters. Blistering on significant surfaces shall be cause for rejection. An excess of three blisters of 1.0 mm diameter on surfaces, which cannot be touched by a 6.35 mm ball, shall be cause for rejection.

5.5.4 PERFORMANCE ASSESSMENT—Each injector shall be tested for Q_s (4.1.3), Q (4.1.4), IR (4.1.26), R (4.1.18), and leakage (4.1.21) prior to and after completion of corrosion tests. Q_s and Q deviation shall be reported as a percentage (Reference 5.9.3).

5.6 Fluid Compatibility (External)—This test is to be applied to the external surface of a gasoline injector, excluding external O-rings, to ensure the external surfaces can withstand contact with common automotive fluids.

5.6.1 TEST FLUIDS—It is not practical to evaluate all variations of test fluids, therefore, agreed reference fluids are employed in testing. The most commonly used fluids are listed as follows:

A reference fluid is defined as a material representative of its particular group, which is sufficiently well defined in all respects so that supplies from different sources are essentially identical in action for which the test is intended (Reference ASTM D 471-79).

ANTIFREEZE 50%—Shall be an ethylene glycol based material as described in SAE J814 mixed with an equal volume of water.

AUTOMATIC TRANSMISSION FLUID—Shall be Dexron II (General Motors Registered Trademark) or equivalent.

AXLE AND MANUAL TRANSMISSION LUBRICANTS—Gear Oil API-GL-5 as described in SAE J308. (Viscosity as defined in SAE J306 may be specified at the option of the test engineer.)

BATTERY ELECTROLYTE—Shall be reagent-grade sulfuric acid diluted with water to a specific gravity of 1.25 to 1.28.

BRAKE FLUID—Is a mixture of polyglycols and cellosolves conforming to DOT-3 of 49 CFR 571.116.

BUTYLCELLOSOLVE—Technical grade.

DIESEL FUEL (NO. 2)—As described by SAE J313. The fluid shall have an Aniline point of 60 to 70°C. It is preferred that emissions-grade diesel fuel conforming to 40 CFR 86.113-82 be used.

ENGINE OIL—Shall be ASTM Reference Oil No. 3.

INDOLENE—Shall be US EPA emission data fuel.

GASOLINE AND 20% METHANOL—Shall be 20% v/v reagent-grade methanol added to gasoline.

GREASE (LITHIUM SOAP BASED)—Shall be an extended-lubrication-interval grease as described in SAE J310. When tested to ASTM D 128, it shall contain not less than 4% by weight Lithium (12-Hydroxystearate type).

MINERAL SPIRITS—See Table 2.

1,1,1-TRICHLOROETHANE—Shall meet the requirements of Military Standard MIL-C-81533.

HIGH PRESSURE PULSED WATER—High pressure pulsed water, as found in commercial car washing machines, is to be sprayed over the exterior surfaces of the injector for 15 s. NOTE: Last step of 5.6.2 does not apply.

STEAM CLEANING—Steam cleaning, as found in commercial equipment, is to be sprayed over the exterior surface of the injector for 15 s. NOTE: Last step of 5.6.2 does not apply.

5.6.2 TEST PROCEDURE—Testing an injector with one fluid followed by another fluid is to be avoided.
  a. Each injector shall have its inlet and outlet suitably sealed. The injector is to have its electrical connector fitted; however, it is not to be energized.
  b. The injector shall have its exposed surfaces dampened either by spraying or brush application and allowed to stand for 24 h.
  c. On completion of the 24-h period, the injector shall be degreased using 1,1,1-Trichloroethane or equivalent.

5.6.3 ASSESSMENT OF RESULTS—Each injector tested shall be examined immediately after testing. The external surface condition shall be visually assessed in accordance with 5.5.3.

Each injector tested shall meet the performance requirements detailed in 5.5.4 prior to and after the completion of testing.

5.7 Fluid Compatibility (Internal)—This test is used to determine the compatibility of internal fuel injector components with fluids similar to those that may be found in the vehicle's fuel system.

5.7.1 TEST FLUIDS—It is not practical to evaluate all test fluids, therefore, agreed reference fluids are employed in this test. The most commonly used fluids are listed.

A reference fluid is defined as a material representative of its particular group, which is sufficiently well defined in all respects so that supplies from different sources are essentially identical in action for which the test is intended (Reference ASTM D 471-79).

WATER/GASOLINE SOLUTION—Mix by volume, 98% unleaded gasoline and 2% corrosive water. Corrosive water is a solution formed by dissolving the following amounts of anhydrous sodium salts in one liter of distilled water at 40°C to aid the mixing:

Sodium Sulfate	14.8 mg
Sodium Chloride	16.5 mg
Sodium Bicarbonate	13.5 mg

GASOLINE AND 20% METHANOL—Shall be 20% by volume reagent-grade methanol added to unleaded gasoline.

GASOHOL—Shall be 10% by volume reagent-grade ethanol added to unleaded gasoline.

BUTYLCELLOSOLVE—Shall be 5% by volume technical-grade Butylcellosolve added to unleaded gasoline.

OXIDIZED GASOLINE (Sour Gas)—Shall be mixed by the following procedure to achieve a peroxide number of 180 millimole/liter:
  a. Stock Fuel—Mix 70/30% by volume n-Heptane/Toluene
  b. Copper Ion Stock Solution—Due to the hazardous nature of these chemicals, the solution must be prepared sequentially in the following three steps:

NOTE: FAILURE TO FOLLOW PROPER PROCEDURES CAN RESULT IN FIRE. REFER TO OSHA MATERIAL SAFETY DATA SHEETS FOR ADDITIONAL INFORMATION:
  (1) Add 10 mL of 12% copper ion concentrate solution to 990 mL of Stock Fuel.
  (2) Add 100 mL of the solution from Step 1 to 1040 mL of Stock Fuel.
  (3) Dilute 10 mL of the solution from Step 2 with 990 mL of Stock Fuel.
  c. Peroxide Stock Solution—Dilute 335 mL of 90% by weight t-butyl hydroperoxide with 665 mL of n-Heptane.
  d. Test Solution—Dilute 60 mL of Peroxide Stock Solution with 10 mL of Copper Ion Stock Solution and 930 mL of Stock Fuel.

5.7.2 TEST PROCEDURE:
  a. A new injector must be used for each test fluid.
  b. Precondition per 3.9.3.
  c. Submerge the injector in the solution for 30 days at room temperature with the inlet port orientated in a manner to keep the injector full.

5.7.3 ASSESSMENT OF RESULTS—Each injector tested shall be evaluated under standard test conditions (per Section 3) and performance measured per 5.5.4 prior to and after completion of the tests.

5.8 Physical Tests—Fuel injectors are subjected to various mechanical loads. These loads occur during manufacturing, testing, handling, and installation in the vehicle fuel system. During actual vehicle service, the injector is also subjected to various shock and vibrational loads. The purpose of physical testing is to ensure that the injector's flow performance, electrical characteristics, and absence of leakage will remain within specification after being subjected to these types of forces.

5.8.1 TEST FIXTURE MOUNTING—Axial load, bending moment, and torsional loads are applied to the injector during installation and removal to overcome the resistance of O-rings or other elastomeric members used for the purpose of sealing, thermal and mechanical isolation, and compensation for part variations. They are also subjected to loads resulting from mounting the fuel rail and/or retaining the injector.

On most designs, the upper half of the injector contains the electrical connector and the lower half the fuel metering section. The loads are usually applied to the injector by pushing and twisting the upper half relative to the lower half. One fixture design will not accommodate all injectors because of the variety of injector designs and mounting methods in use. The fixture must take into account the specific injector design and application. The loads and fixtures used may, therefore, necessarily deviate from that recommended in the following sections.

5.8.2 ASSESSMENT OF PERFORMANCE— Reference 5.5.4.

5.8.3 AXIAL TEST PROCEDURE—The injector shall be mounted to a suitable fixture per the application requirement (Fig. 20) and subjected

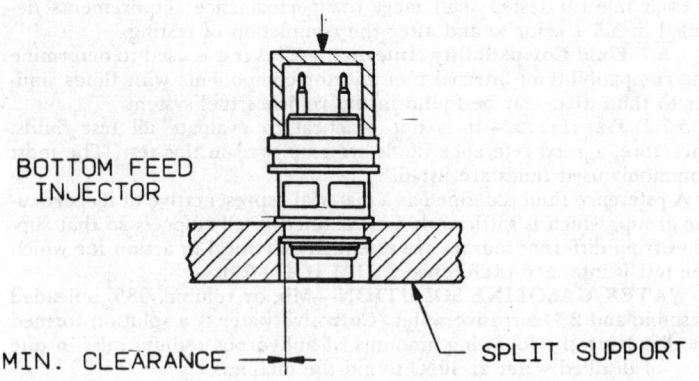

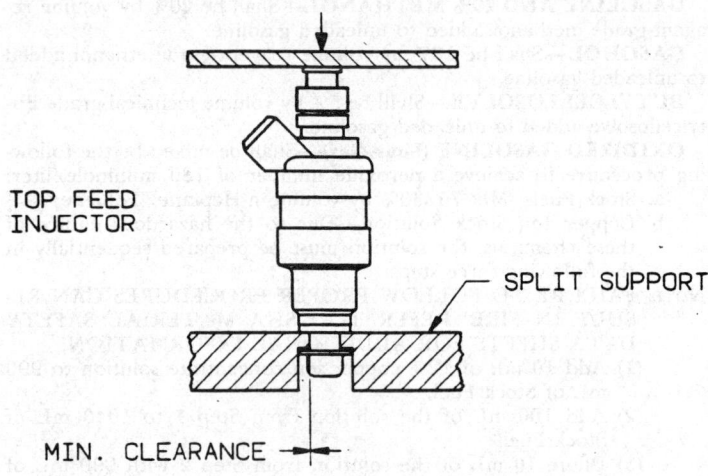

FIG. 20—AXIAL LOAD FIXTURE

to a force (typically 600 N) applied along the longitudinal axis of the injector.

5.8.4 TORSIONAL MOMENT TEST PROCEDURE—The injector shall be mounted in a suitable fixture with the body fixed (Fig. 21) and a torsional moment (typically 0.6 Nm) applied to both ends for 5 s about the longitudinal axis of the injector.

5.8.5 BENDING MOMENT—The injector shall be mounted in a suitable fixture (Fig. 22) and subjected to a bending moment (typically 6.0 Nm) for 30 s.

5.8.6 VIBRATION TEST PROCEDURE—Cap the injector inlet and outlet with the injector full of test fluid. Mount the injector in a vibration test fixture (Fig. 23) that duplicates the application mounting. Vibrate the injector in each of the three axes (longitudinal, lateral, and vertical). The vibration schedule is to be random with the frequency controlled over the range of 5 to 2000 Hz, based on the actual vehicle application. Test duration is to be per the customer requirements and may be accelerated per random vibration practices (Reference Military Standard 810-D). Accelerometers should be placed at the same location as per vehicle data acquisition. Vehicle data collection should encompass any conditions that may represent worst customer usage.

5.8.7 SHOCK TEST PROCEDURE—Cap the injector inlet and outlet with the injector full of test fluid. Mount the fuel injector onto a shock test fixture that duplicates the application mounting (Fig. 23). Perform impacts at 30 G's input load for a duration of 11 to 14 ms, six times with the injector in the vertical position (valve tip pointing downward) and six times with the injector in the horizontal position.

5.9 Bench Durability—This procedure defines a uniform test for evaluating the minimum acceptable durability level of a given gasoline fuel injector design.

5.9.1 FLUSHING PROCEDURE—All test samples are to be cycled with test fluid for 1 min at a PW of 2.5 ms and a period of 5.0 ms.

5.9.2 INJECTOR TEST CONDITIONS:
  a. Fuel Temperature Test Range—17 to 30°C
  b. Fluid—Indolene
  c. Fluid Condition—A fresh supply of fluid is to be used at the start of every test, and changed at every 100×10^6 cycles of test (approximately 139 h)
  d. Pressure—Determined by application
  e. PW—2.5 ms
  f. Period—(repetition rate)—5.0 ms
  g. Duration: Subject to application
        Typical PFI 600×10^6 pulses
        Typical CFI 900×10^6 pulses

5.9.3 PERFORMANCE CRITERIA—The test samples shall be evaluated for the following flow criteria:

  a. Percent Static Flow Shift

$$\frac{\text{End of Test Flow} - \text{Start of Test Flow}}{\text{Start of Test Flow}} \times 100 = \% \text{ Static Flow Shift}$$

  b. Percent Q_d Shift at 2.5 ms PW and 10 ms Period (calculated same as above)
  c. Leakage Specification—All test samples must be evaluated for leakage as specified in 4.1.21.1.

5.9.4 FREQUENCY OF PERFORMANCE CHECKS—The test samples shall be evaluated for the above criteria at least prior to the start and on completion of the test schedule.

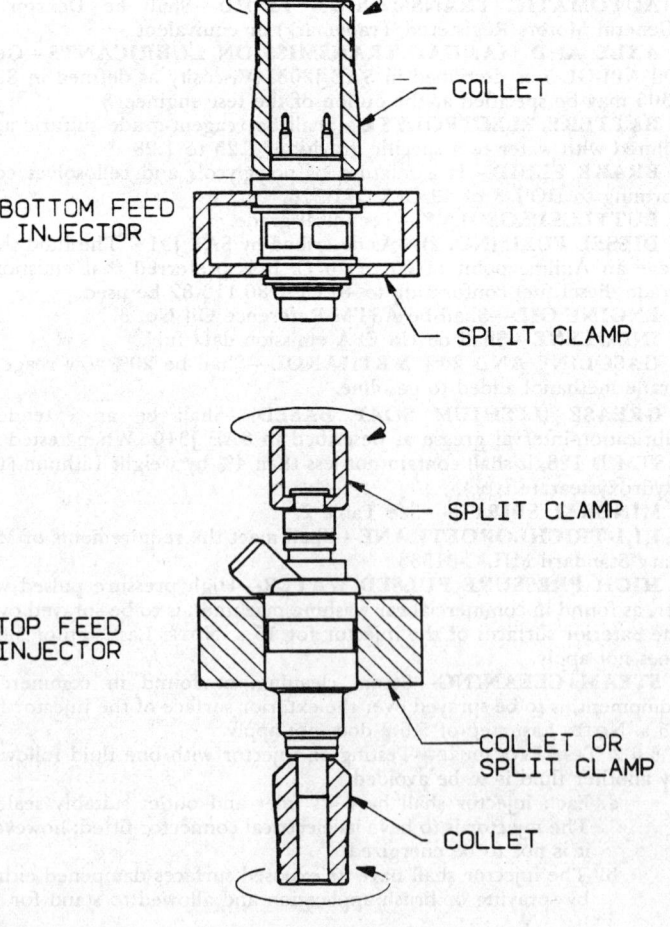

FIG. 21—TORSIONAL LOAD FIXTURE

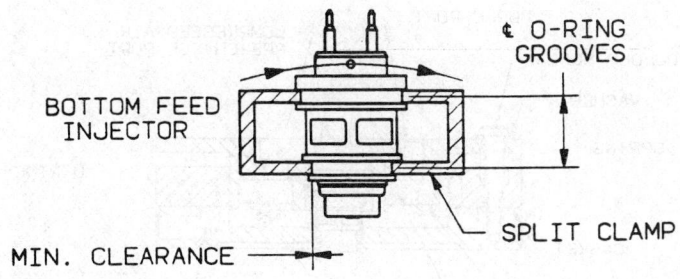

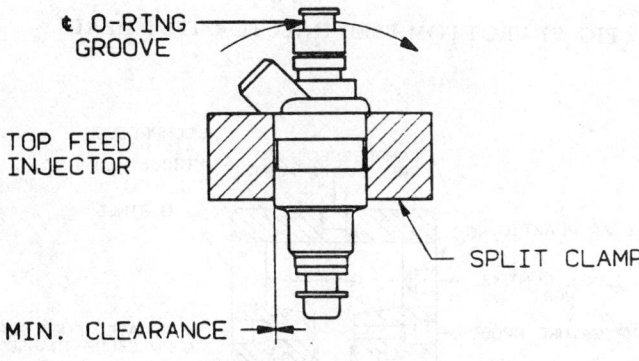

FIG. 22—BENDING MOMENT FIXTURE

USE SAME MOUNTING HARDWARE & CAVITY DIMENSIONS AS USED IN VEHICLE APPLICATION.

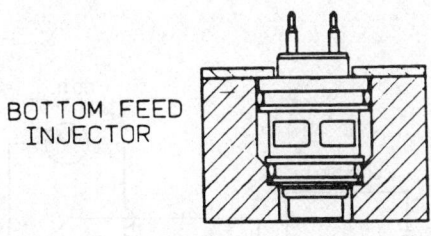

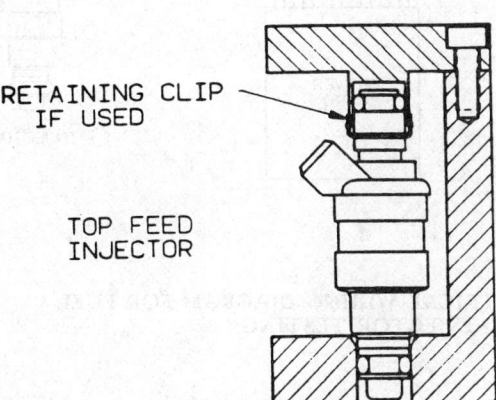

FIG. 23—SHOCK AND VIBRATION FIXTURE

New Designs/First Piece Samples will require additional flow test points during the accumulation of test cycles. See the recommended flow schedule for such cases as follows:
- 0 Cycles
- 25 x 10⁶ Cycles
- 50 x 10⁶ Cycles
- 100 x 10⁶ Cycles
- Continue at 100 x 10⁶ Cycle intervals until end of test

5.10 Hot Lean Shift—Fuel injectors exhibit varying degrees of "hot lean shift" when operated at high temperatures. An extreme condition will usually occur in hot weather after a running engine has been shut down for about 30 min and then restarted. Both the injector and fuel temperature will rise during the "soak" period, causing reduced flow rates after the restart. The injector flow rates will gradually recover to normal levels as temperatures drop to a stable operating range. The magnitude of the lean shift is related to the injector design, fuel system components, injector's environment and the fuel characteristics. The degree of lean shift and its effect on driveability are best evaluated in a running vehicle. Fig. 24 illustrates the observed hot lean shift and the gradual recovery to stable conditions for three injector designs.

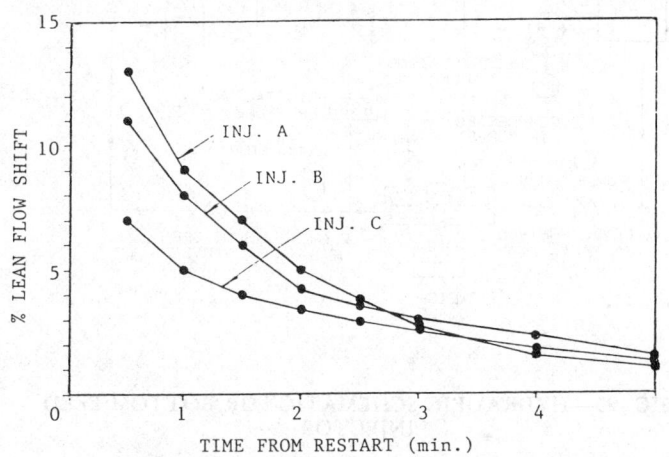

FIG. 24—HOT LEAN SHIFT

6. Test Equipment

The purpose of this section is to provide basic information on test equipment used in measuring fuel injector performance characteristics according to the specifications described in Sections 4 and 5. Recommendations are given for the design of the test flow head and hydraulic circuit for the bottom and top feed injectors, instrumentation resolution, and driver board connections.

The test equipment must be capable of precisely testing various electronic injector designs, preferably in mass flow units. The equipment and installations must be in compliance with all applicable fire and safety codes and regulations.

6.1 Hydraulic System—To assure that the test stand is versatile enough to evaluate the various injector design characteristics, the hydraulic system should be capable of operating with n-Heptane, Indolene, and mineral spirits as specified in 3.1. The recommended hydraulic system plumbing diagram is shown for a bottom feed injector in Fig. 25 and for a top feed injector in Fig. 26.

The test fluid pump must be capable of delivering 20 g/s of test fluid at a stable pressure of 700 kPa to the inlet of each injector. The test fluid filtering system must be 5 μm absolute or better. The test stand must be equipped with a heat exchanger to assure that the test fluid temperature is maintained constant within ±1°C throughout the test. To assure that the required accuracy to measure injector performance is met, the test fluid pressure regulation system must be capable of controlling the inlet test fluid pressure to the injector inlet with an accuracy of ±0.10 kPa in the range of 40 to 500 kPa. To eliminate any effect of the fluid head on injector inlet pressure, the pressure measuring device should be mounted in the same horizontal plane as the injector inlet. To minimize the effect of fluid pressure pulsations caused by the pump, the test stand hydraulic system must include an air dampened flow head. The trapped air test head for use with bottom feed injectors is shown in Fig. 27 and that for the top feed injectors in Fig. 28. The trapped air pocket must be precharged to maintain a minimum air volume of 30 cc at fuel system pressure.

Table 4 summarizes the test equipment resolution required to achieve the desired accuracy of the reported data and control variables.

6.2 Electrical—The wiring diagram for the injector, injector driver, and power supply is shown in Fig. 29. Because of the different injector drivers used in the automotive industry, it is recommended that the test stand be equipped with an interchangeable (plug-in) type driver board, 4-1/2 x 6-1/2 in with the numerics on the component side (see Fig. 30).

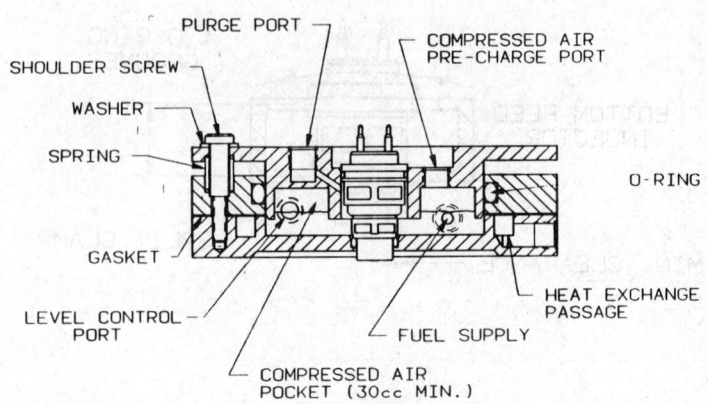

FIG. 27—BOTTOM FEED INJECTOR TEST HEAD

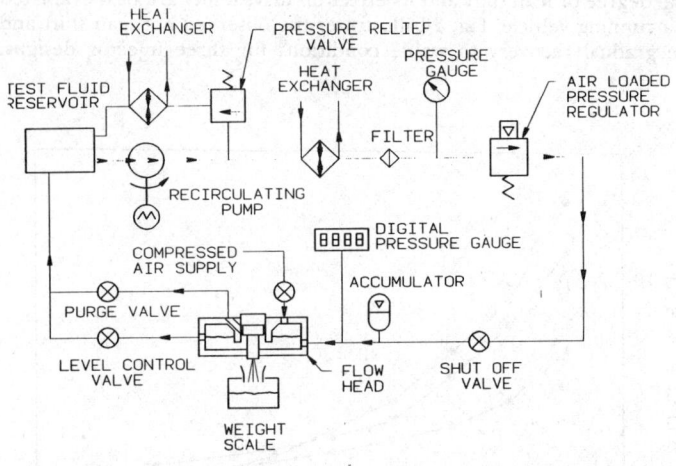

FIG. 25—HYDRAULIC SCHEMATIC FOR BOTTOM FEED INJECTOR

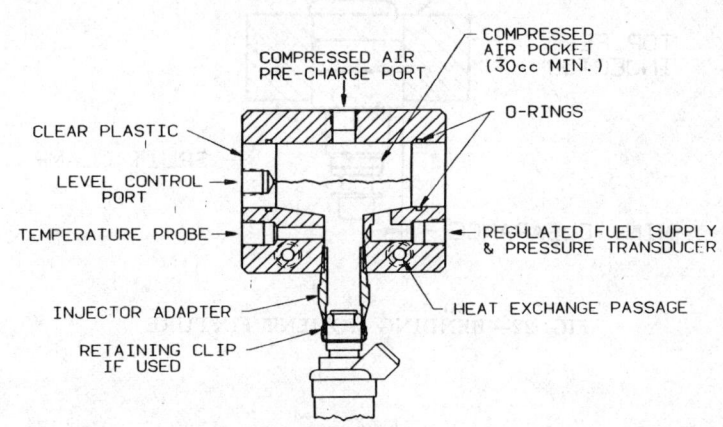

FIG. 28—TOP FEED INJECTOR TEST HEAD

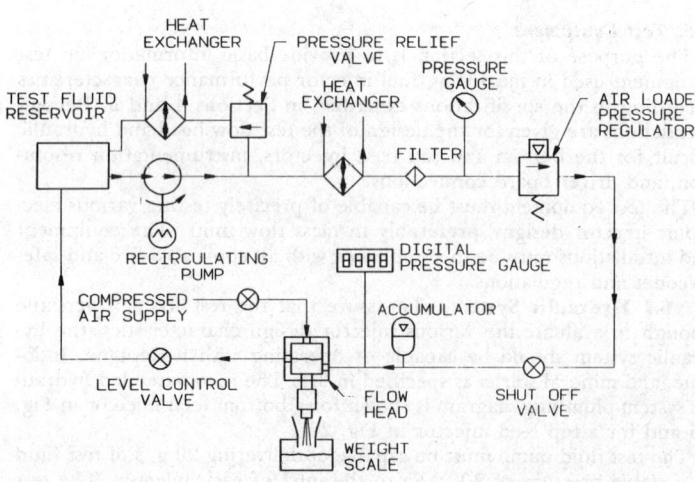

FIG. 26—HYDRAULIC SCHEMATIC FOR TOP FEED INJECTOR

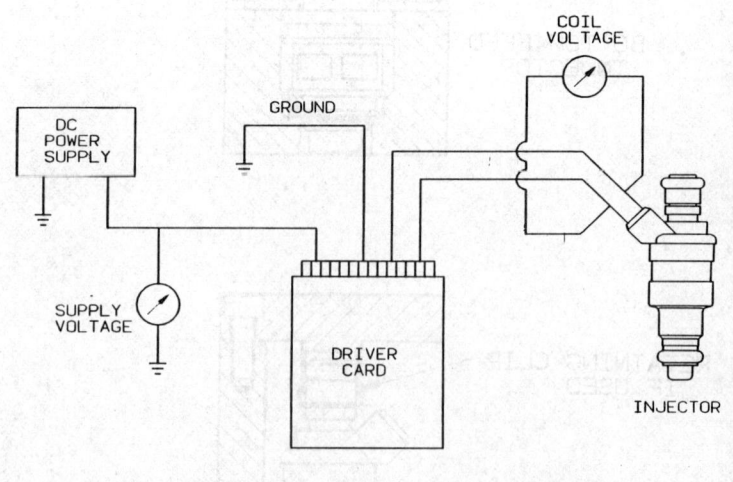

FIG. 29—ELECTRICAL WIRING DIAGRAM FOR FUEL INJECTOR TESTING

TABLE 4—ELECTRONIC FUEL INJECTOR TEST EQUIPMENT MEASUREMENT ACCURACY

CHARACTERISTIC	REPORTED RESOLUTION	EQUIPMENT RESOLUTION	CONTROL
Voltage Across Injector Terminal (volts)	XX.XX	±0.005	±0.01
Test Fluid:			
• Pressure (kPa)	XXX.X	±0.05	±0.10
• Temperature (°C)	XX.X	±0.05	±1.0
• Specific Gravity	X.XXX	±0.0005	
• Kinematic Viscosity (cSt)	X.XXX	±0.0005	
Static Pull-In Current (amps)	X.XX	±0.001	
Static Drop-Out Current (amps)	X.XX	±0.001	
Injector Coil Resistance (ohms)	XX.XX	±0.005	
Inductance (mH)	X.XX	±0.005	
Opening Time (ms)	X.XX	±0.005	
Closing Time (ms)	X.XX	±0.005	
Ambient Room Temperature (°C)	XX.X	±0.05	±3.0
Min Opening Voltage (volts)	X.XX	±0.005	
Pulse Width (ms)	XX.XX	±0.005	±0.001
(Range – 0.01 to 99.99 ms)			
Period (ms)	XX.XX	±0.005	±0.001
(Range – 5.0 to 99.99 ms)			
Weight of the Test Fluid (g)	X.XX	±0.005	
Dynamic Flow (mg/pulse)	XX.XX	Calculated	
Static Flow Rate (g/s)	XX.XXX	Calculated	

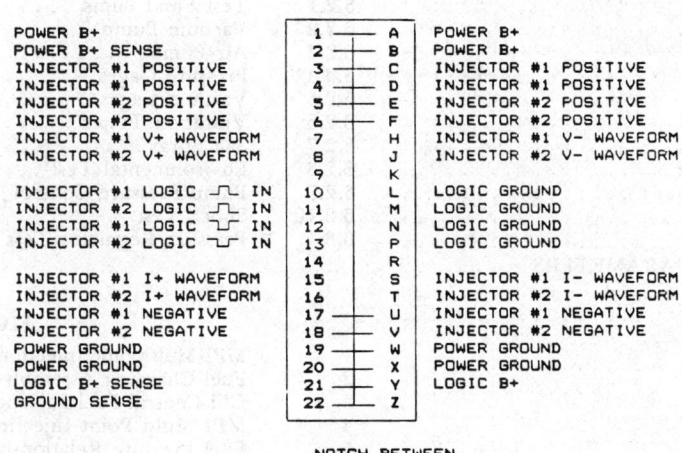

FIG. 30—INJECTOR DRIVER EDGE CONNECTIONS

FUEL INJECTION SYSTEM
FUEL PRESSURE REGULATOR AND
PRESSURE DAMPER—SAE J1862 FEB90

SAE Recommended Practice

Report of the Gasoline Fuel Injection Standards Committee approved February 1990.

Foreword—A fuel pressure regulator in an electronic fuel injection system is located downstream of the injectors and maintains a controlled differential pressure across the injectors for all engine operating conditions. Fuel delivered to the engine is metered by the length of time the injector is open and the controlled differential pressure. Typical fuel injection systems are shown in Figures 1 and 2.

The most commonly used pressure regulator is a hydromechanical bypass pressure relief device that is capable of sealing the fuel from returning to the fuel tank when closed. Typical fuel pressure regulators are shown in Figures 3 and 4. The regulator consists of two chambers separated by a diaphragm. The fuel chamber contains a fuel inlet, a fuel outlet, and a valve. The air chamber contains a spring and is usually referenced to the pressure condition of the discharge side of the injector(s). When the fuel pressure exceeds the set point pressure, the diaphragm raises the valve off the seat of the outlet passage and the excess pressure is relieved by permitting fuel to flow back to the fuel tank. The actual fuel system pressure can deviate from the desired value due to an accumulation of the following effects: slope, repeatability, hysteresis, response accuracy to a reference pressure signal, and production variability from regulator to regulator. Figure 5 illustrates the relationship of the fuel pressure to the manifold pressure.

Several other types of regulators have been developed but are generally used to correct a system problem such as the need for dual system pressure or the need to extend the dynamic range of the fuel system.

Also, bleed-down types are used in some CFI applications to allow pressure and vapor relief during soak periods.

A fuel pressure damper in an electronic fuel injection system is used to attenuate the pressure pulsations generated by the fuel pump and the opening and closing of the fuel injectors. The hydraulic pulsations in the fuel injection system can cause an objectionable audible noise, which is transmitted to the vehicle passenger compartment. The fuel pulsations can create standing pressure waves in the fuel supply system, which will affect the pressure differential across the fuel injectors and cause an uneven fuel delivery. Also, the high frequency pressure pulsations can adversely affect the pressure regulator performance and durability. A fuel pressure damper may be used to reduce the above problem.

The pressure damper commonly used in electronic fuel injection systems consists of two chambers, a fuel and an air chamber separated by a loaded diaphragm. Some pressure dampers use a mechanical spring, shown on Figure 6, and some pneumatic, shown on Figure 7. During engine operation, the fuel passes through the pressure damper, which reduces the pressure pulsations. Controlling fuel system hydraulic pulsation has been accomplished by various means, with remote mounted surge chambers and pulse dampers as the most common. A fuel rail assembly with internal damping features has also proven to be effective in attenuating injector induced pulsations.

TABLE OF CONTENTS

1.	SCOPE	
1.1	Purpose	
2.	REFERENCES	
3.	BASIC CHARACTERISTICS	
3.1	Pressure Regulator	
3.1.1	Pressure Set Point	
3.1.2	Slope	
3.1.3	Working Flow Range	
3.1.4	Repeatability	
3.1.5	Hysteresis	
3.1.6	Response to Reference Signal	
3.1.7	Internal Leakage (Leakdown)	
3.2	Pressure Damper Performance	
3.3	External Leakage	
3.3.1	External Leakage-Fuel Side	
3.3.2	External Leakage-Air Side	
4.	APPLICATION RELATED PARAMETERS	
4.1	Structural	
4.1.1	Vibration	
4.1.2	Shock	
4.1.3	Burst	
4.1.4	Physical Tests	
4.2	Durability-Life Cycling	
4.2.1	Flow Cycling	
4.2.2	Pressure Cycling	
4.2.3	Thermal Cycling	
4.2.4	Thermal Soak	
4.3	Material Integrity-External	
4.3.1	Salt Corrosion	
4.3.2	Resistance to Vehicle Fluids	
4.4	Material Integrity-Internal	
4.4.1	Reference Fluids	
4.4.2	Test Procedure	
4.4.3	Assessment of Results	
5.	STANDARD TEST CONDITIONS	
5.1	Test Fluid	
5.2	Ambient Test Temperature	
5.3	Test Fluid Temperature	
5.4	Return Line Pressure	
5.5	Mounting Configuration	
5.6	Preparation of Test Specimen	
6.	TEST APPARATUS & EQUIPMENT	
6.1	Introduction	
6.2	Test Equipment Description	
6.2.1	Test Fluid Pump	
6.2.2	Vacuum Pump	
6.2.3	Air Pump	
6.2.4	Pressure Gages	
6.2.5	Vacuum Gages	
6.2.6	Vibration Tester	
6.2.7	Salt Spray	
6.2.8	Environmental Test	
6.2.9	Pump Pressure Control	
6.2.10	Shock Test	
6.3	Pressure Regulator Test Equipment Measurement Accuracy	

FIGURES

1	MPI-Multipoint Injection
2	Fuel Charging Assembly-CFI-Central Fuel
3	CFI-Central Fuel Injection, Fuel Pressure
4	MPI Multi Point Injection System, Fuel Pressure
5	Fuel Pressure Relationship to Manifold Pressure
6	Fuel Pressure Damper-Mechanical Spring
7	Fuel Pressure Damper-Hydraulic Spring
8	Regulator Slope
9	Flow Range
10	Response to Reference Signal
11	Pressure Damper Overall Acceptance
12	Pressure Damper Frequency Domain
13	Flow Cycle Schedule
14	Pressure Cycle Schedule
15	Thermal Cycle Schedule
16	Pressure Regulator Performance Test Stand
17	Internal Leakage Test Stand
18	Pressure Damper Performance Test Stand

1. Scope—This document applies to electronic fuel injection systems, fuel pressure regulators and pressure dampers. It is intended as a guide toward a standard practice and is subject to change to keep pace with experience and technical advances. Any deviations from this test standard must be stated when referencing this document.

Many of the tests specify the use of mineral spirits for safety and fuel stability considerations. Since the properties of mineral spirits differ significantly from those of gasoline, it is recommended that vehicle verification development tests be conducted to determine the acceptability of a regulator and/or damper design for each vehicle application.

1.1 Purpose—The purpose of this document is to promote uniformity in the specifications for the qualification test procedures for gasoline fuel-injection pressure regulators and dampers.

More specifically this practice is intended as a guide to:

a. Identify and define those parameters that are required to measure pressure regulator and damper characteristics and performance.
b. Establish uniform, meaningful test procedures and recommend test equipment to evaluate and quantify these parameters.
c. Standardize the nomenclature as related to fuel injection systems.

2. References
SAE J308, Axle and Manual Transmission Lubricants
SAE J814, Engine Coolants
SAE J1541, Fuel Injection Nomenclature--Spark Ignition Engines
ASTM B 117, Salt Spray (Fog) Testing

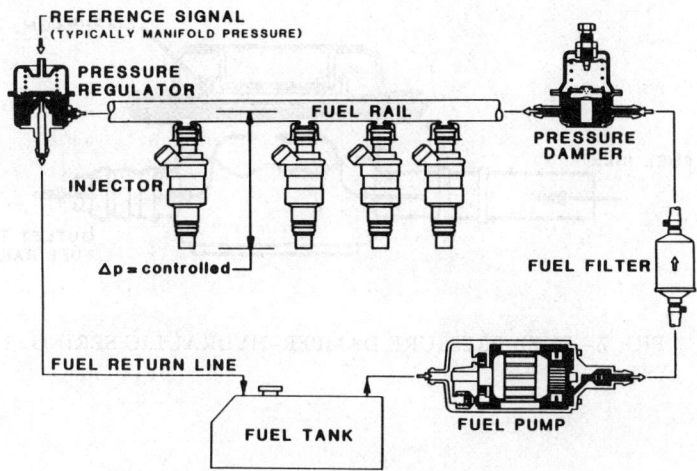

FIG. 1—MPI-MULTIPOINT INJECTION

ASTM D 56, Flash Point by Tag Closed Tester
ASTM D 86, Distillation of Petroleum Products
ASTM D 445, Test Method for Petroleum Products
ASTM D 893, Test for Insolubles in Lubricating Oils
ASTM D 1298, Test for Density and Specific Gravity
ASTM D 1744, Water in Liquid Petroleum Products

3. Basic Characteristics

3.1 Pressure Regulator

3.1.1 PRESSURE SET POINT

a. *Definition*—The test fluid pressure to which the regulator is set at a specified fluid flow rate. It is specified as pressure (in kPa) at a fluid flow rate (in g/s).

b. *Background*—The regulator pressure set point determines the nominal operating pressure for a fuel injection system. It is critical to the design of the system that this pressure be known and controlled. The specified regulator set pressure is meaningless without the corresponding flow rate since the pressure will vary as a function of flow rate. The set point specified for different applications may differ due to the differences in required injector flow rates.

c. *Objective*—To measure the regulator set point pressure at the specified test fluid flow rate of 10 g/s.

d. *Test Apparatus*—See 6.2.1, 6.2.4 and 6.2.9.

e. *Test Conditions*—Refer to 5.1 through 5.6.

f. *Test Method*:

(1) Slowly increase test fluid pressure until fluid just starts to flow through the regulator (approximately 1.0 g/s).

(2) Gradually increase the fluid flow until the flow rate of 10 g/s is achieved (within ±0.5 g/s). Do not allow the flow to exceed the specified flow rate; if so, restart the procedure.

(3) Allow the flow to stabilize for 15 to 30 s then record the regulator inlet pressure.

g. *Data Required*—Record the regulator inlet pressure, outlet pressure, and actual fluid flow rate. Report set point pressure as: xxx.x kPa at 10 g/s.

Reported population data for "pressure set point"

Measure the regulator set point for a random sample (24 minimum) of regulators representative of a normal production distribution. Determine the sample mean (X_s) and standard deviation (sigma). Report the population set point as the sample mean (X_s) ± four times the sample standard deviation (4 sigma): xxx.x ± xx.x kPa at 10 g/s. Also report the 4 sigma$_s$ as a percentage of X_s: ±xx.x%.

3.1.2 SLOPE

a. *Definition*—The change in regulator pressure over a linear flow range which results from a change in fluid flow rate through the regulator, all other parameters remaining constant. It is specified as pressure change per unit flow rate (i.e., kPa/(g/s)) between well defined lower and upper flow rates. A positive slope represents increasing pressure with increasing flow rate (see Figure 8).

b. *Background*—An ideal regulator maintains the desired fuel system pressure regardless of changes in fuel flow rate through the regulator, all other conditions remaining constant. However, with most production regulators the fuel system pressure varies from the desired pressure as a function of fuel flow rate through the regulator. This variation of pressure with respect to flow rate should be very close to

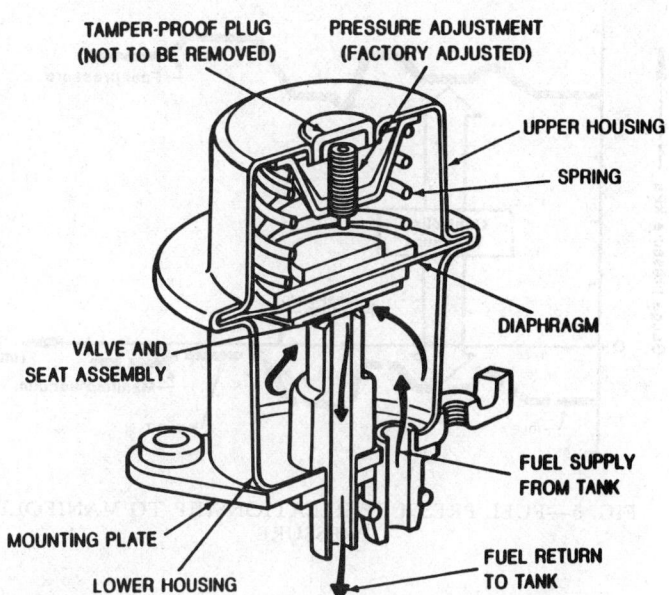

FIG. 3—CFI-CENTRAL FUEL INJECTION, FUEL PRESSURE

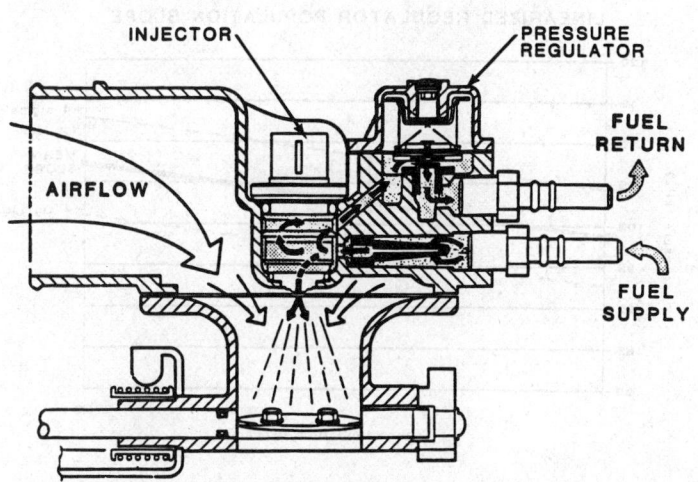

FIG. 2—FUEL CHARGING ASSEMBLY-CFI-CENTRAL FUEL

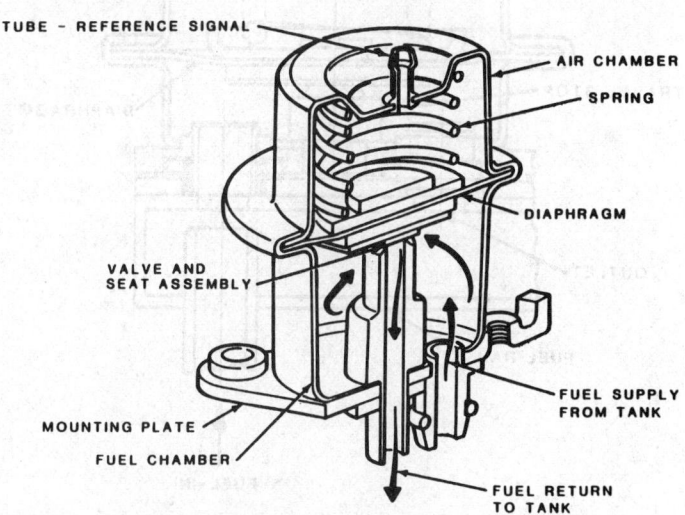

FIG. 4—MPI-MULTI POINT INJECTION SYSTEM, FUEL PRESSURE

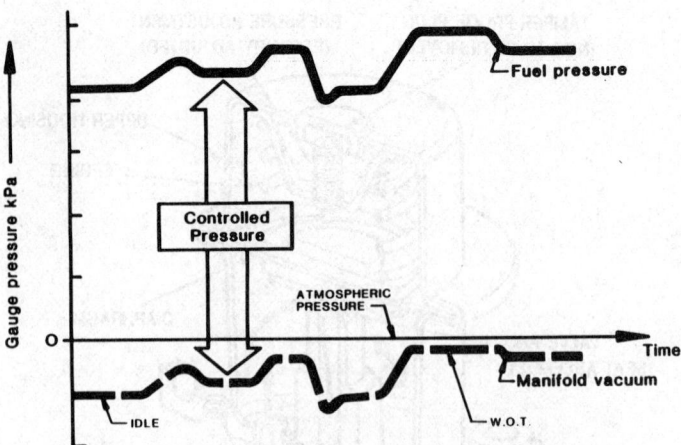

FIG. 5—FUEL PRESSURE RELATIONSHIP TO MANIFOLD PRESSURE

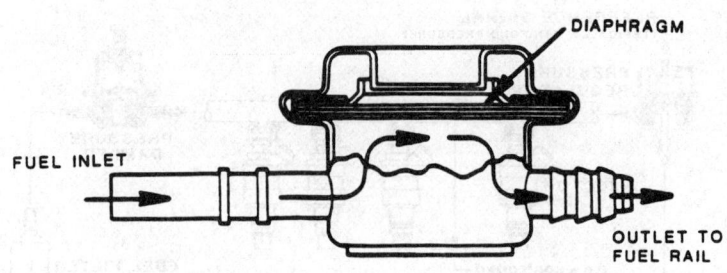

FIG. 7—FUEL PRESSURE DAMPER–HYDRAULIC SPRING

f. Test Method
  (1) From the set point flow rate, with the regulator air side at ambient pressure, reduce the flow rate to 1 to 2 g/s and allow the regulator to stabilize for 15 to 30 s.
  (2) Gradually increase the flow rate to the specified lower flow rate of 5.0 g/s ± 0.5, allow the regulator to stabilize for 15 to 30 s, and record the regulator inlet pressure. Do not allow the flow rate to exceed this flow value prior to making the pressure measurement. If it does, restart the procedure.
  (3) Continue with this same procedure, making regulator pressure measurements at flow rates of 10, 20, and 40 (±0.5) g/s.

linear over the operating flow range, and a low value for the slope is generally desired.

It is important when designing/specifying a fuel injection system to know the regulator slope. The fuel flow rate through the regulator will vary due to a variety of factors which can include engine load, rpm, battery voltage, and others. Subsequent changes in fuel system pressure, due to flow variation, will change injector flow, which has to be considered in the fuel injection system design.

  c. Objective—To measure the change in regulator pressure with change in fluid flow rate, from 5 to 40 g/s, through the regulator and to establish the regulator slope.
  d. Test Apparatus—See 6.2.1, 6.2.4, and 6.2.9.
  e. Test Conditions—See 5.1 through 5.6.

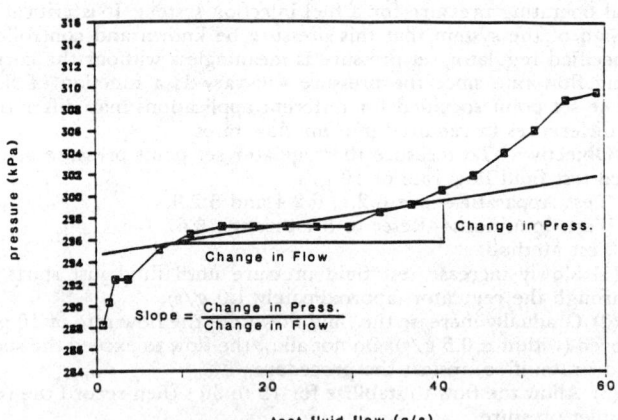

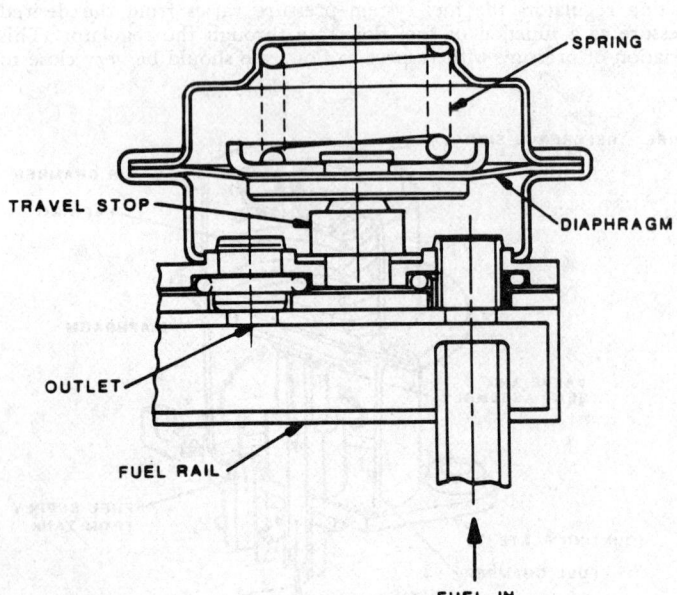

FIG. 6—FUEL PRESSURE DAMPER–MECHANICAL SPRING

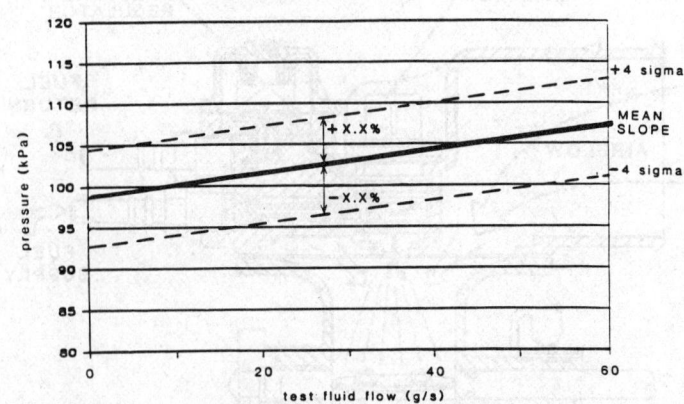

FIG. 8—REGULATOR SLOPE

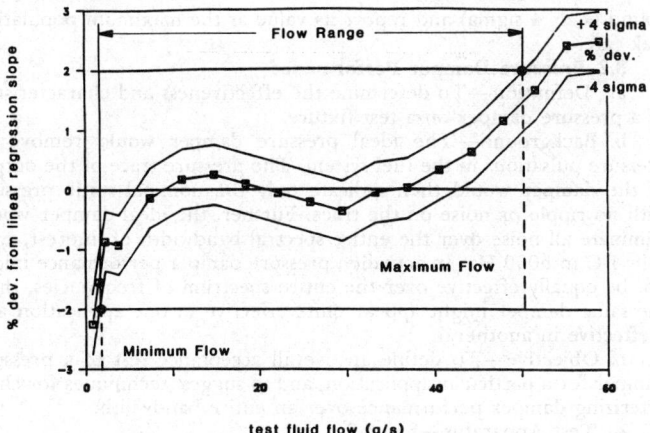

FIG. 9—FLOW RANGE

flow point (see Figure 9). The maximum value for the flow range is the flow value where the +4 sigma$_i$ pressure deviates by 2% from the linear regression slope, and the minimum value for the flow range is the flow value where the −4 sigma$_i$ pressure deviates by 2% from the linear regression slope. Report the population flow range as: xx.x g/s to xxx g/s minimum.

3.1.4 REPEATABILITY

a. Definition—The variability in the set pressure for a given regulator over a number of repeated pressure on-off cycles.

b. Background—An ideal regulator would produce the same set pressure every time it is cycled from zero pressure to the set point flow rate. Repeatability is important so that the fuel system pressure is predictable for consistent fuel system performance.

c. Objective—To determine the variability of the set pressure for a fuel pressure regulator.

d. Test Apparatus—See 6.2.1, 6.2.4, and 6.2.9.

e. Test Conditions—See 5.1 through 5.6.

f. Test Method:

(1) Use the test method defined in 3.1 to determine the regulator set point pressure.

(2) After taking the set point pressure measurement, gradually reduce pressure in the regulator to zero and allow the regulator to relax for 15 to 30 s.

(3) Repeat the above procedure to obtain 20 set point pressure measurements.

g. Data Required—Determine the sample mean and standard deviation for the above pressure measurements.

The repeatability is defined as four times the standard deviation as a percentage of the sample mean.

Report repeatability as: ±xx.x%

h. Reported Population Data—For a minimum random sample size of 24 regulators, determine the sample mean (X_i) and the individual regulator set pressure standard deviations (sigma). The population repeatability is defined as four times the sample mean standard deviation (4 sigma) as a percentage of the population set point (X_s). Report as: ±xx.x%.

3.1.5 HYSTERESIS

a. Definition—The change in regulator set point pressure as a result of approaching the set point flow from high flow versus approaching from low flow.

b. Background—An ideal regulator would maintain the same set point pressure regardless of whether the fluid flow through the regulator has exceeded or has been less than the set point flow. Variability in set pressure due to hysteresis has the same effect as variation due to repeatability.

c. Objective—To determine the hysteresis of a fuel pressure regulator by measuring the set point pressure when approaching the set point flow rate from both directions.

d. Test Apparatus—See 6.2.1, 6.2.4, and 6.2.9.

e. Test Conditions—See 5.1 through 5.6

f. Test Method:

(1) Use the method defined in 3.1 to determine the set point pressure as approached from a lower flow rate.

(2) After taking the set point pressure measurement, gradually increase the fluid flow rate to 40 g/s.

(3) Gradually reduce the fluid flow rate through the regulator to the set point flow rate and allow the regulator to stabilize for 15 to 30 s. Record the regulator inlet pressure. Do not allow the flow to fall below the set point flow rate prior to making this measurement. If so, restart the procedure.

g. Data Required—The hysteresis reported is the difference in increasing and decreasing flow set point pressure measurements as a percent of the set point pressure measured with increasing flow.

Report hysteresis as: xx.x%

h. Reported Population Data—Measure the regulator set point pressure with increasing and decreasing flow for a random sample (24 minimum) of regulators representative of a normal production distribution. Make 20 measurements of each regulator and determine the difference between the increasing flow and decreasing flow set pressures. Determine the maximum difference for each regulator. Next, determine the sample mean and standard deviation of the maximum differences. Report the population hysteresis as ±4 sigma as a percent of the population set point pressure: ±xx.x% ((4 sigma/set point pressure) x 100).

3.1.6 RESPONSE TO REFERENCE SIGNAL

a. Definition—The change in regulator pressure, which results from a change in the reference pressure signal being applied to the "air side" of the fuel pressure regulator. The response of the pressure regu-

g. Data Required—A minimum of the above four measurements should be made from the specified lower to upper flow rates inclusive. More intermediate points may be used if desired.

Determine the best fit slope for the data using least squares linear regression analysis techniques for data reduction. If data varies by more than ±2% from linear, the lower and/or upper specified limits may have to be reset or the larger error from linear must be noted with the slope.

Report slope as: xx.x kPa/(g/s) from 5 to 40 g/s. Deviations from the standard specifications must be stated when specifying the slope.

h. Reported Population Data—Determine the regulator slopes for a random sample (24 minimum) of regulators representative of a normal production distribution. Determine the sample mean (X_s) and standard deviation (sigma) of the slopes. Report the population slope as the sample mean (X_s) ± four times the sample standard deviation (4 sigma): xx.xx ± x.x kPa/(g/s) from 5 to 40 g/s. Also report the 4 sigma as a percentage of X_s: ±xx.x% (see Figure 8).

3.1.3 WORKING FLOW RANGE

a. Definition—The range of fluid flow rate over which the regulator pressure response is linear (to within 2%) with respect to flow rate (see Figure 9).

b. Background—The flow range should equal or exceed the operating flow range the regulator will be exposed to in an actual application.

c. Objective—To measure the minimum and maximum regulator fluid flow rates between which the regulator pressure response is linear with respect to flow rate.

d. Test Apparatus—See 6.2.1, 6.2.4, and 6.2.9.

e. Test Conditions—See 5.1 through 5.6.

f. Test Method:

(1) Slowly increase test fluid pressure until fluid just starts to flow through the regulator (approximately 1.0 g/s).

(2) Gradually increase flow rate, while measuring regulator pressure, until the pressure response becomes nonlinear or the test stand capability is reached. Allow regulator to stabilize at each measurement point.

g. Data Required—Flow rate measurement increments must be small enough to accurately determine the flow range, within ±10% at minimum flow rate, ±5% at maximum.

Using least squares linear regression analysis for data reduction, for the data points between 5 and 40 g/s, determine the minimum and maximum regulator fluid flow rates between which the regulator pressure response is linear with respect to flow rate to within ±2%. If the maximum flow rate is beyond the capability of the test stand (at least 75 g/s), specify the test stand maximum flow rate followed by a plus sign.

Report working flow range as: xx.x g/s to xxx g/s.

h. Reported Population Data—Measure the regulator pressures at a series of flow points as specified for a random sample (24 minimum) of regulators representative of a normal production distribution. Determine the sample mean pressure (X_i) and standard deviation (sigma$_i$) at each flow point. Determine the slope from 5 to 40 g/s using least squares linear regression of the mean pressure values (X_i). Plot the deviation from the determined slope and the ±4 sigma$_i$ values for each

lator to a change in the reference signal is both the degree of regulator pressure change as well as the time required to respond to changes of the reference signal.

b. Background—The fuel system pressure is controlled by the pressure regulator relative to a reference pressure signal applied to the "air side" of the regulator. The reference signal is usually air cleaner pressure (CFI) or intake manifold pressure (MPI). It will be assumed that the desired change in system pressure will equal the change in reference pressure unless otherwise specified.

c. Objective—To measure the response time and accuracy of regulated pressure change of a fuel pressure regulator relative to a change in the reference signal.

d. Test Apparatus—See 6.2.1 through 6.2.5, and 6.2.9.

e. Test Conditions—See 5.1 through 5.6.

f. Test Method:

(1) Start the test with the regulator operating at set point pressure and flow rate, and ambient pressure for the reference signal.

(2) Apply a sudden change in reference signal pressure (not greater than 50 ms to achieve 90% of the final value), maintain a constant flow rate, and trace the regulator pressure.

g. Data Required—Perform the above test with reference signal of 10 and 50 kPa vacuum and 10 and 50 kPa pressure (± 1 kPa). Additional reference signal values are optional.

Determine the stabilized regulator pressure with the new reference signal pressure. Express the change of regulator pressure as a percentage relative to the change in signal pressure.

Determine the time required for the regulated pressure to reach 90% of its final value as measured from the time that the reference pressure achieves 90% of its final value.

Report regulator response as: xx.x% response accuracy, xx.x ms response time, with xx kPa vacuum/pressure signal. Reference Figure 10.

h. Reported Population Data—Determine the response time and response accuracy for each regulator in a random sample (24 minimum) of regulators representative of a normal production distribution. Report the mean response time and mean response accuracy of the sample.

3.1.7 INTERNAL LEAKAGE (LEAKDOWN)

a. Definition—The fluid leakage from the inlet of the regulator to the lower pressure regulator outlet, at a pressure below the regulator set pressure.

b. Background—A regulator can be designed to either shut off fuel flow (checking) or allow fuel flow through the regulator (bleed-down) when the fuel system pump is de-energized. The purpose, respectively, is to maintain pressure in the system, or bleed-off pressure in the system, after the pump is de-energized.

c. Objective—To measure the internal leakage of a fuel pressure regulator.

d. Test Apparatus—See 6.2.1, 6.2.4, 6.2.9, and Figure 17.

e. Test Conditions—See 5.1 through 5.6.

f. Test Method-Fluid (for Shutoff or Bleed-Down Regulators):

(1) Start the test with the regulator operating at set point pressure and flow rate and ambient pressure for the reference signal.

(2) Reduce the regulator inlet pressure to 60% of set pressure, and maintain regulator pressure at that level with the regulator outlet at ambient pressure. Allow regulator to stabilize for 60 s minimum.

(3) Measure the mass of fluid that leaks out the outlet for a 60 s minimum time period.

g. Data Required-Fluid—Report regulator leakdown as: xx.x g/s test fluid at xxx kPa.

h. Test Method-Air (Optional Method for Shutoff Regulators): Regulator must be empty of all fluid.

(1) Apply dry, filtered air at the set pressure for 30 s minimum to dry the regulator interior and sealing surfaces. The airflow rate or pressure may need to be adjusted to prevent harmful vibration of internal components.

(2) Reduce the regulator pressure to 60% of set pressure to the regulator inlet with the regulator outlet at ambient pressure. Allow the regulator to stabilize for 60 s minimum.

(3) Measure the volume of air that leaks out the outlet for a 60 s minimum time period.

i. Data Required-Air—Convert measured volume of air over a measured time period to a standard airflow rate. Scc/min.

Report regulator leakdown as: xx.x Scc/min air at xxx kPa.

j. Reported Population Data—Determine the leak rate (either air or fluid) for each regulator in a random sample (24 minimum) representative of a normal production distribution. Using the mean leak rate and standard deviation for the sample, determine the 4 sigma upper bound (x + 4 sigma) and report its value as the maximum population leak rate.

3.2 Pressure Damper Performance

a. Definition—To determine the effectiveness and characteristics of a pressure damper on a test fixture.

b. Background—The ideal pressure damper would remove all pressure pulsations in the fuel system. The pressure trace of the output of the damper would then indicate only the nominal static pressure with no ripple or noise on the trace. Further, the ideal damper would eliminate all noise over the entire spectral bandwidth of interest, typically DC to 5000 Hz. In actuality, pressure damper performance might not be equally effective over the entire spectrum of frequencies, thus, the same damper might appear quite effective in one application and ineffective in another.

c. Objective—To define an overall acceptance test of a pressure damper for a particular application, and to suggest techniques for characterizing damper performance over an entire bandwidth.

d. Test Apparatus—See 6.2.

e. Test Conditions—See Section 5.

f. Test Method:

(1) Part 1-Overall Acceptance Test—For this test, a good high speed system digital voltmeter is necessary, which can store multiple readings (at least 1000), average those readings, find maximum/minimum, and read true root mean square voltages (rms).

The basic test procedure involves obtaining three pieces of data from the input pressure trace, and the output pressure trace, then comparing the measurements and expressing them in terms of percent differences. This test should be performed at the intended system nominal pressure and at points over the range of fuel system gauge pressures in steps of 10 kPa. (i.e., for a normally aspirated engine at 300 kPa, take readings at 200, 210, 220 ...300 kPa). The three readings are:

(a) Average Static Pressure Level—This is determined by taking 1000 readings of the DC equivalent (static) pressure level both at the inlet and outlet, and calculating the average of those readings (x_{in} and x_{out}). The ideal pressure damper has x_{in} and x_{out} equal, which means that no pressure drop has occurred through the device.

(b) Random Pressure Noise-Efficiency—This value is the peak-to-peak reading determined from the maximum and minimum values from the same 1000 readings taken in part 1. The value characterizes random pressure excursions present on the trace, and is to be reported as the difference between them (maximum-minimum) for the inlet and outlet traces. The final reported value is the ratio of the outlet to the inlet expressed in percent ((out/in) x 100).

(c) Root/Mean Square Pressure Ripple—With the pressure traces AC coupled to the digital voltmeter, the true rms value is obtained. This measurement characterizes periodic pressure disturbance with the difference between the in and out values indicating the device effectiveness. Report the ratio of the outlet to the inlet expressed in percent ((out/in) x 100).

g. Data Required:

(1) Test results of:
  a. Average static pressure level
  b. Random pressure noise-efficiency
  c. rms pressure ripple as shown on Figures 11 and 12.

(2) Specify the excitor characteristics and method.

h. Reported Population Data—Report the mean and 4 sigma values on a minimum random sample size of 24.

(1) Part 2-Characterizing Damper Performance—This test attempts to characterize a pressure damper for any application by measuring the performance characteristics in the frequency domain. A fast fourier transform (FFT) spectrum analyzer is required for this test, preferably a two channel condition would be one in which the input pressure trace contains random noise throughout the entire spectral bandwidth of DC to 5000 Hz.

The input and output pressure traces are AC coupled to the spectrum analyzer, and a transmissibility trace is obtained. This trace is a good qualitative indication of the performance of the damper over the entire bandwidth. The ideal pressure damper would demonstrate equal performance over the entire bandwidth. As an alternative to the transmissibility trace, a single channel analyzer may be used to display the power spectral density (PSD) of the input and output traces. A visual comparison of the two traces would yield the same basic information as the transmissibility trace but with less detail.

3.3 External Leakage

3.3.1 EXTERNAL LEAKAGE-FUEL SIDE

a. Definition—Leakage from the fuel side of the regulator or damper to the exterior and to the air side of the device.

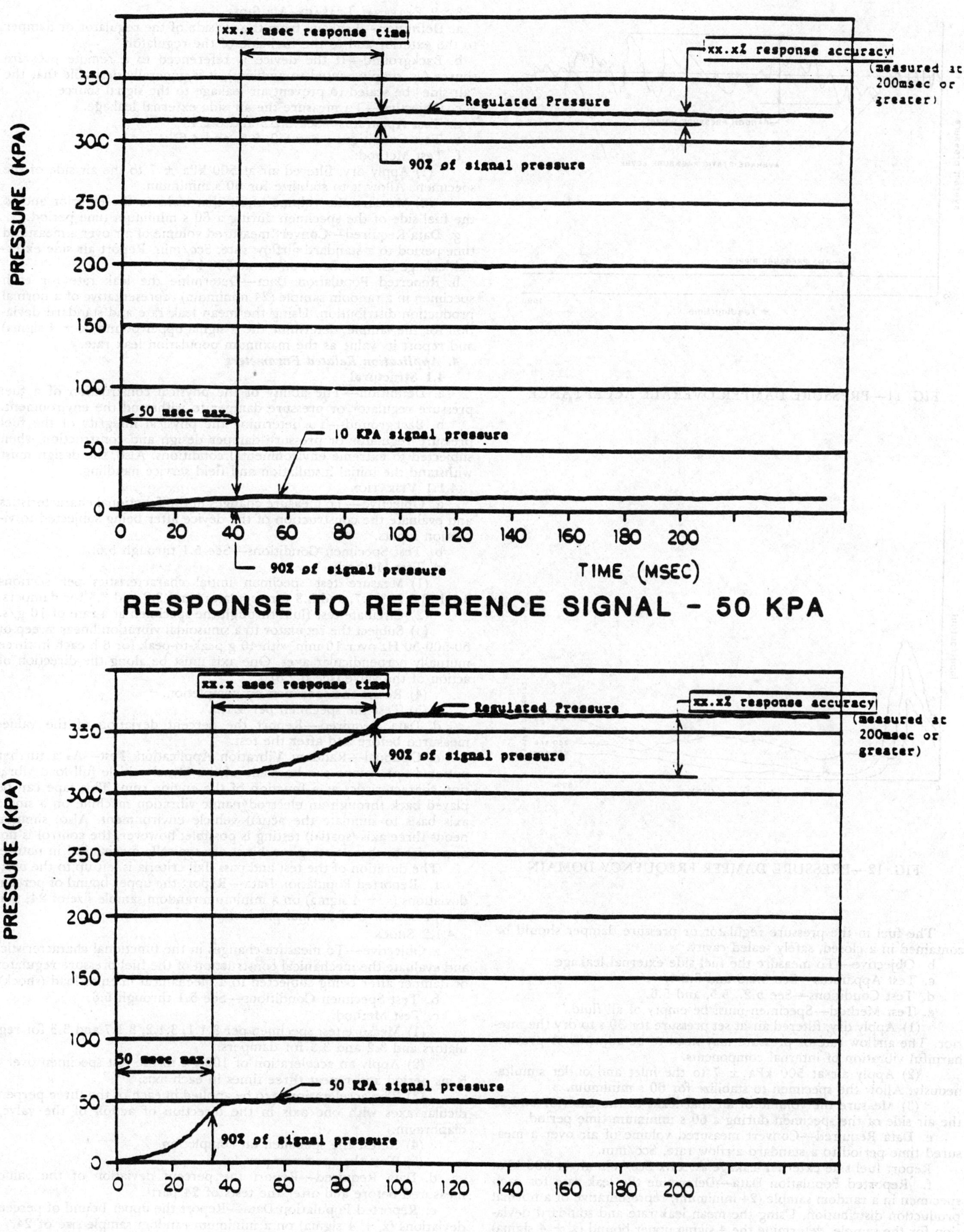

26.285

RESPONSE TO REFERENCE SIGNAL - 50 KPA

(Top chart labels: xx.x msec response time; xx.x% response accuracy (measured at 200msec or greater); Regulated Pressure; 90% of signal pressure; 50 msec max; 10 KPA signal pressure; 90% of signal pressure)

(Bottom chart labels: xx.x msec response time; xx.x% response accuracy (measured at 200msec or greater); Regulated Pressure; 90% of signal pressure; 50 msec max.; 50 KPA signal pressure; 90% of signal pressure)

Axes: PRESSURE (KPA) vs TIME (MSEC)

FIG. 10—RESPONSE TO REFERENCE SIGNAL

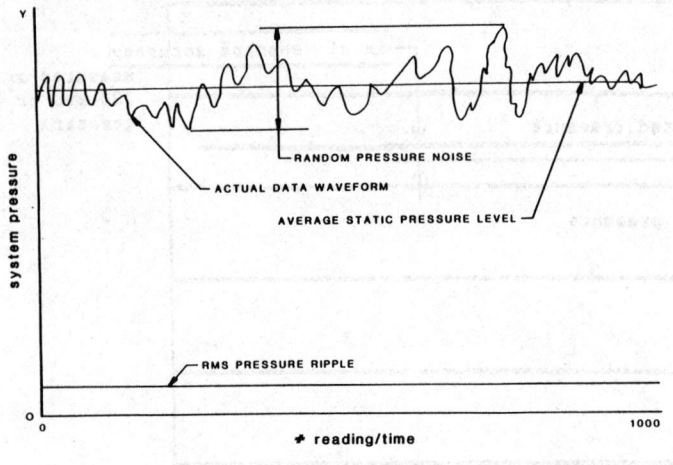

FIG. 11—PRESSURE DAMPER OVERALL ACCEPTANCE

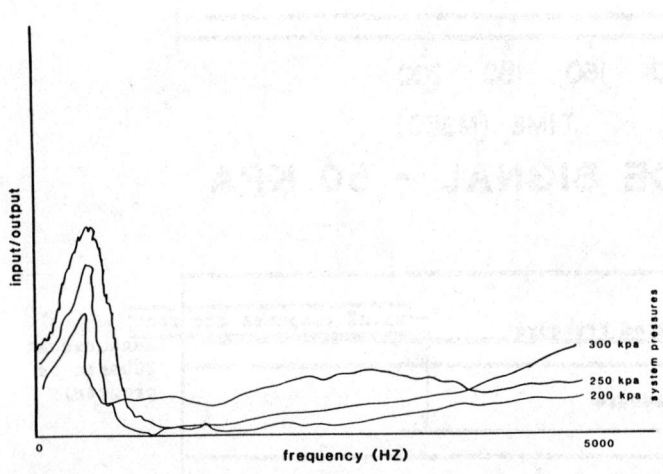

FIG. 12—PRESSURE DAMPER FREQUENCY DOMAIN

The fuel in the pressure regulator or pressure damper should be contained in a closed, safely sealed cavity.

   b. Objective—To measure the fuel side external leakage.
   c. Test Apparatus—See 6.2.3 and 6.2.4.
   d. Test Conditions—See 5.2., 5.5, and 5.6.
   e. Test Method—Specimen must be empty of all fluid.
     (1) Apply dry, filtered air at set pressure for 30 s to dry the interior. The airflow rate or pressure may need to be adjusted to prevent harmful vibration of internal components.
     (2) Apply air at 500 kPa ± 7 to the inlet and outlet simultaneously. Allow the specimen to stabilize for 60 s minimum.
     (3) Measure the volume of air that leaks to the exterior and to the air side of the specimen during a 60 s minimum time period.
   e. Data Required—Convert measured volume of air over a measured time period to a standard airflow rate, Scc/min.
Report fuel side external leakage as: xx.x Scc/min air at 500 kPa.
   f. Reported Population Data—Determine the leak rate for each specimen in a random sample (24 minimum) representative of a normal production distribution. Using the mean leak rate and standard deviation for the sample, determine the 4 sigma upper bound (x + 4 sigma) and report its value as the maximum population leak rate.

3.3.2 External Leakage–Air Side

   a. Definition—Leakage from the air side of the regulator or damper to the exterior and to the fuel side of the regulator.
   b. Background—If the device is referenced to a remote pressure source (vs. simply vented to ambient), it is generally desirable that the "air-side" be sealed to prevent air leakage to the signal source.
   c. Objective—To measure the air side external leakage.
   d. Test Apparatus—See 6.2.3 and 6.2.4.
   e. Test Conditions—See 5.2, 5.5, and 4.6.
   f. Test Method:
     (1) Apply dry, filtered air at 500 kPa ± 7 to the air side of the specimen. Allow it to stabilize for 60 s minimum.
     (2) Measure the volume of air that leaks to the exterior and to the fuel side of the specimen during a 60 s minimum time period.
   g. Data Required—Convert measured volume of air over a measured time period to a standard airflow rate, Scc/min. Report air side external leakage as: xx.x Scc/min air at 500 kPa.
   h. Reported Population Data—Determine the leak rate for each specimen in a random sample (24 minimum) representative of a normal production distribution. Using the mean leak rate and standard deviation for the sample, determine the 4 sigma upper bound (x + 4 sigma) and report its value as the maximum population leak rate.

4. Application Related Parameters

4.1 Structural

   a. Definition—The ability of the physical construction of a fuel pressure regulator or pressure damper to withstand the environment.
   b. Background—To determine the physical integrity of the fuel pressure regulator or pressure damper design and construction when subjected to extreme environmental conditions. Also, the design must withstand the initial installation and field service handling.

4.1.1 Vibration

   a. Objective—To measure changes in the functional characteristics and evaluate the construction of the device after being subjected to vibration inputs.
   b. Test Specimen Conditions—See 5.1 through 5.6.
   c. Test Method:
     (1) Measure test specimen initial characteristics per sections 3.1.1, 3.1.2, 3.1.7, and 3.3 for regulators and 3.2 and 3.3 for dampers.
     (2) Circulate test fluid through the specimen at a rate of 10 g/s.
     (3) Subject the regulator to a sinusoidal vibration linear sweep of 50-500-50 Hz over 10 min with 40 g peak-to-peak for 8 h each in three mutually perpendicular axes. One axis must be along the direction of action of the valve (see 6.2.6).
     (4) Retest the specimen on completion.
     (5) Test the specimen per 4.1.3.
   d. Data Required—Report the percent deviation of the values measured before and after the test.
   e. Optional—Random Vibration Application Test—As a further optional test, a tape can be made of the actual vehicle full load vibration characteristics as a function of the engine rpm. This tape can be played back through an electrodynamic vibration machine on a single axis basis to simulate the actual vehicle environment. Also, simultaneous three axis (spatial) testing is possible; however, the control is not as precise as these testing machines are typically mechanical in nature.

The duration of the test and pass/fail criteria is left up to the user.
   f. Reported Population Data—Report the upper bound of percent deviations (x + 4 sigma) on a minimum random sample size of 24, that is representative of normal production.

4.1.2 Shock

   a Objective—To measure changes in the functional characteristics and evaluate the mechanical construction of the fuel pressure regulator or damper after being subjected to a mechanical impulse load (shock).
   b. Test Specimen Conditions—See 5.1 through 5.6.
   c. Test Method:
     (1) Measure test specimen per 3.1.1, 3.1.2, 3.1.7 and 3.3 for regulators and 3.2 and 3.3 for dampers.
     (2) Apply an acceleration of 100 g's to the test specimen over a 5 ms period and repeat three times in each axis.
     (3) The acceleration is to be applied in each of the three perpendicular axes with one axis in the direction of action of the valve/diaphragm.
     (4) Retest the specimen on completion.
     (5) Test the specimen per 4.1.3.
   d. Data Required—Report the percent deviation of the values measured before and after the tests of 24 parts.
   e. Reported Population Data—Report the upper bound of percent deviations (x + 4 sigma) on a minimum random sample size of 24.

4.1.3 Burst

a. Objective—To determine the pressure at which the regulator or damper will develop an external leak and/or a fuel leak path to the air side of the diaphragm.

b. Test Apparatus—Hydraulic or air pump and fixture capable of performing the burst pressure test as described in the test method section. Hydraulic methods are preferred although air is acceptable. Proper safety precautions must be taken due to the high pressures and catastrophic nature of failures. Test is to be performed in an enclosure suitable to contain any failure of components.

c. Test Method:

(1) Gradually apply a pressure of 1000 kPa ± 20 to the fuel chamber of the specimen and maintain for a period of 60 s.

(2) Slowly increase pressure to test points in increments of 100 kPa holding at each point for 10 s until failure or 3500 kPa is reached.

d. Data Required—Record the pressure reading taken before the test point at which a sudden drop in pressure (greater than 200 kPa) takes place.

e. Reported Population Data—Report the lower bound of the sample population data (x −4 sigma) on a minimum random sample size of 24.

4.1.4 PHYSICAL TESTS

a. Connection Integrity—All fuel and vacuum connections must be tested according to the schedule in Table 1. The specific test requirements are dependent on the construction method used for the production of the regulator or damper assembly and usage in normal conditions in the vehicle application. Leakage is to be determined as in 3.3 both before and after the test.

TABLE 1—PHYSICAL TESTS

PULL-OUT TEST	TORQUE TEST(*)	BENDING TEST
Apply 600 N	Apply 20 N·m torque	Perform 30° bend

*Applicable for tubes with thread or hexagon feature at the base.

b. After performing the leak tests perform a burst test per 4.1.3.

4.2 Durability-Life-Cycling

a. Definition—The ability of the regulator or damper to function consistently over a given period of time, which would be representative of vehicle life.

b. Background—To determine the ability of the regulator or damper to fulfill performance requirements during and after a simulation of the parts life cycle under normal and extreme environmental conditions.

4.2.1 FLOW CYCLING

a. Objective—To measure the functional changes and leakage of the regulator or damper due to high speed/long duration flow cycling.

b. Test Apparatus—See 6.2.

c. Test Conditions—See Section 5.

d. Test Method:

(1) Measure the specimen per 3.1.1, 3.1.2, 3.1.7, and 3.3 for regulators and 3.2 and 3.3 for dampers.

(2) Flow cycle the specimen between 25 and 10 g/s. Use a square wave duty cycle as shown in Figure 13. The test duration is 100 million cycles at a frequency of 30 ms per cycle followed by 200 million cycles with a frequency of 10 ms per cycle.

(3) Retest the specimen on completion per step 1.

(4) Test the specimen per 4.1.3.

c. Data Required—Report the percent deviation between the before and after readings and burst test results.

d. Reported Population Data—Report the upper bound of percent deviations (x+4 sigma) on a minimum random sample size of 24 pieces.

4.2.2 PRESSURE CYCLING

a. Objective—To measure the functional changes and leakage of the regulator or damper due to pressure cycling.

b. Test Apparatus—See 6.2.

c. Test Conditions—See Section 5.

d. Test Method:

(1) Measure the specimen per 3.1.1, 3.1.2, 3.1.7, and 3.3 for regulators and 3.2 and 3.3 for dampers.

(2) Pressure cycle the specimen between the set point and 20% of that value. Use a square wave duty cycle consisting of a frequency of 10 s per test cycle. The test is performed with gasoline as a test fluid at a temperature of 15 to 25°C. The test duration is 40 000 cycles (to simulate 120 000 miles). Reference Figure 14.

(3) Retest the specimen on completion.

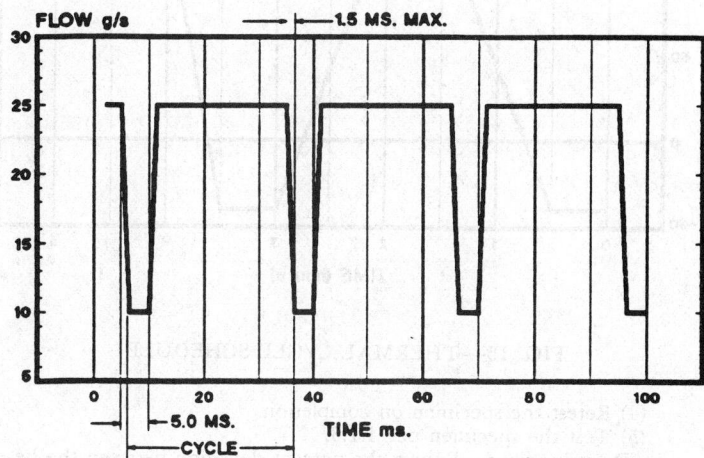

FIG. 13—FLOW CYCLE SCHEDULE

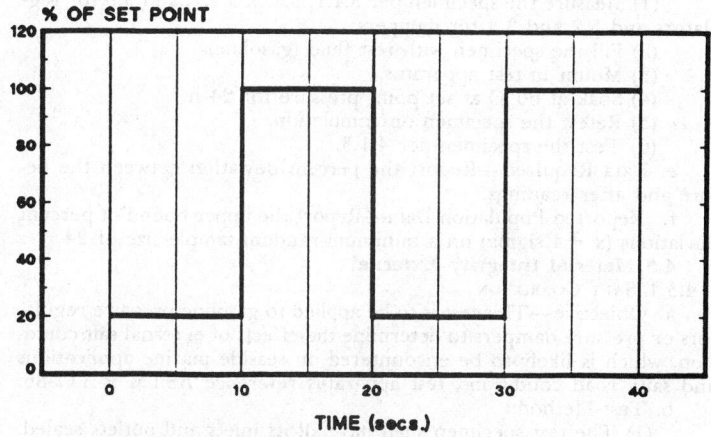

FIG. 14—PRESSURE CYCLE SCHEDULE

(4) Test the specimen per 4.1.3.

e. Data Required—Report the percent deviation between the before and after readings.

f. Reported Population Data—Report the upper bound of percent deviations (x +4 sigma) on a minimum random sample size of 24.

4.2.3 THERMAL CYCLING

a. Objective—To measure the functional changes and leakage of the regulator or damper due to thermal cycling.

b. Test Apparatus—See 6.5.

c. Test Conditions—See Section 5.

d. Test Method:

(1) Measure the specimen per 3.1.1, 3.1.2, 3.1.7, and 3.3 for regulators and 3.2 and 3.3 for dampers.

(2) The specimen must be empty of all fuel.

(3) Temperature cycle the regulator for 10 cycles between −40 and 125°C. Each cycle consists of a duty cycle of 3 h per cycle with 1 h transitions and 0.5 h holding periods (see Figure 15).

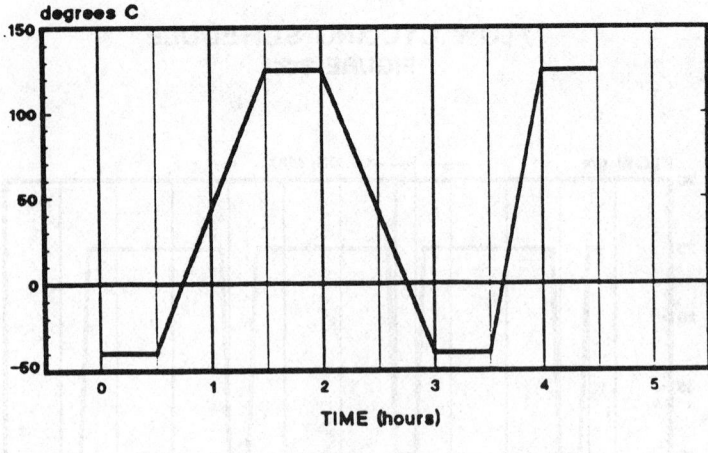

FIG. 15—THERMAL CYCLE SCHEDULE

(4) Retest the specimen on completion.
(5) Test the specimen per 4.1.3.

e. Data Required—Report the percent deviation between the before and after readings.

f. Reported Population Data—Report the upper bound of percent deviations (x +4 sigma) on a minimum random sample size of 24.

4.2.4 THERMAL SOAK

a. Objective—To measure the functional changes and leakage of the regulator or damper due to high temperature thermal soak.

b. Test Apparatus—See 6.5.

c. Test Conditions—See Section 5.

d. Test Method:

(1) Measure the specimen per 3.1.1, 3.1.2, 3.1.7, and 3.3 for regulators and 3.2 and 3.3 for dampers.
(2) Fill the specimen with test fluid (gasoline).
(3) Mount in test apparatus.
(4) Soak at 60°C at set point pressure for 24 h.
(5) Retest the specimen on completion.
(6) Test the specimen per 4.1.3.

e. Data Required—Report the percent deviation between the before and after readings.

f. Reported Population Data—Report the upper bound of percent deviations (x +4 sigma) on a minimum random sample size of 24.

4.3 Material Integrity-External

4.3.1 SALT CORROSION

a. Objective—This test is to be applied to gasoline pressure regulators or pressure dampers to determine the effects of external salt corrosion, which is likely to be encountered on seaside marine applications and salty road conditions, test apparatus reference ASTM B 117-85.

b. Test Method:

(1) The test specimen shall have all its inlets and outlets sealed.
(2) Samples shall be supported or suspended between 15 and 30 degrees from the vertical, with any significant surface parallel to the principal direction of fog flow.
(3) Samples shall be at least 30 mm from each other and 35 mm from a chamber wall.
(4) Sample shall be so placed as to permit free settling of fog on all samples, but avoid salt solution from one sample dripping onto another.
(5) Expose the samples in the fog chamber for a minimum period of 96 h. Inspect the samples and operation of the chamber as required. Opening of the chamber shall be limited to the frequency necessary to carry out the operations. Inspections should not be made more than twice a day for a total of 15 min. Position of the samples may be varied during the inspection period.
(6) Remove the samples, wash gently in water not warmer than 38°C, and dry by blotting with absorbent paper.

c. Data Required—Each specimen tested shall be examined immediately after testing. The degree of corrosion shall be visually assessed in accordance with 4.3.1.1 with the results and number of hours reported.

Each specimen tested shall be tested per 3.1.1, 3.1.2, 3.1.7, 3.2 and 3.3 as they apply prior to and after the completion of testing.

After completion of all other tests, perform the tests per 4.1.3 and 4.1.4. These tests are mutually exclusive and cannot be performed on the same specimen.

4.3.1.1 *Surface Corrosion & Blistering*—For the purpose of this specification, a "significant surface" is defined as an area designated to be coated that can be touched by a 6.35 mm diameter ball.

Any spots of red rust on a significant surface shall be cause for rejection. Rust spots on surfaces that cannot be touched by a 6.35 mm ball shall be noted but will not be cause for rejection.

Any blistering of the coating shall be noted as to location on the sample piece, blistering size, and number of blisters. Blistering on significant surfaces shall be cause for rejection. An excess of three blisters of 1.0 mm diameter on surfaces that cannot be touched by a 6.35 mm ball shall be cause for rejection.

4.3.2 RESISTANCE TO VEHICLE FLUIDS

a. Objective—This test is to be applied to the external surface of a gasoline regulator or damper excluding external 0-rings, to determine the ability of external surfaces to withstand contact with common automotive fluids.

b. Test Fluids—It is not practical to evaluate all variations of test fluids; therefore, agreed reference fluids are employed in testing. The most commonly used fluids are listed below.

A reference fluid is defined as a material representative of its particular group, which is sufficiently well defined in all respects so that supplies from different sources are essentially identical in action for which the test is intended (see ASTM D 471-79).

(1) Antifreeze 50%—Shall be an ethylene glycol based material as described in SAE J814 mixed with an equal volume of water.
(2) Automatic Transmission Fluid—Shall be Dexron II (General Motors Registered Trademark) or equivalent.
(3) Axle and Manual Transmission Lubricants—Gear Oil API-GP-5 as described in SAE J308. (Viscosity as defined in SAE J306 may be specified at the option of the test engineer).
(4) Battery Electrolyte—Shall be reagent grade sulfuric acid diluted with water to a specific gravity of 1.25 to 1.28.
(5) Brake Fluid—Is a mixture of polyglycols and cellosolves conforming to DOT-3 of 49 CFR 571.116.
(6) Butylcellosolve—Technical grade.
(7) Diesel Fuel No. 2—As described by SAE J313. The fluid shall have an aniline point of 60 to 70°C. It is preferred that emissions grade diesel fuel conforming to 40 CFR 86.113-82 be used.
(8) Engine Oil—Shall be ASTM Reference Oil No. 3.
(9) Indolene—Shall be U.S. EPA emission data fuel.
(10) Gasoline and 20% Methanol—Shall be 20% vol/vol reagent grade methanol added to unleaded gasoline.
(11) Grease (Lithium Soap Based)—Shall be an extended lubrication interval grease as described in SAE J310. When tested to ASTM D 128, it shall contain not less than 4% by weight lithium (12-hydroxystearate type).
(12) Mineral Spirits—See Section 5.1.
(13) Trichloroethane (1,1,1)—Shall meet the requirements of Military Standard MIL-C-81533.
(14) High Pressure Pulsed Water—High pressure pulsed water, as found in commercial car washing machines is to be sprayed over the exterior surfaces of the device for 15 s.

NOTE: The last step of in 4.3.2 c does not apply.

(15) Steam Cleaning—Steam cleaning, as found in commercial equipment, is to be sprayed over the exterior surface of the specimen for 15 s.

NOTE: The last step in 4.3.2 c does not apply.

c. Test Procedure:

(1) Testing with one fluid followed by another fluid is to be avoided.
(2) Each specimen shall have its inlet and outlet suitably sealed.
(3) The specimen shall have its exposed surfaces dampened either by spraying or brush application and allowed to stand for 24 h.
(4) On completion of the 24 h period, the specimen shall be degreased using trichloroethane (1,1,1) or equivalent.

d. Assessment of Results—Each device tested shall be examined immediately after testing. The external surface condition shall be visually assessed in accordance with 4.3.1.

The percent deviation from the initial test results is to be reported. Test per 3.1.1, 3.1.2, 3.1.7, 3.2, and 3.3 as they apply.

e. Reported Population Data—Report the upper bound of percent deviations (x + 4 sigma) on a minimum random sample size of 24.

4.4 Material Integrity-Internal

a. Objective—This test is to determine the compatibility of the components to the type of materials that may be present in the vehicle. It is not practical to evaluate all variations of test fluids; therefore, agreed reference fluids are employed in testing.

4.4.1 REFERENCE FLUIDS—A reference fluid is defined as representative of its particular group, which is sufficiently well defined in all respects so that supplies from different sources are essentially identical in performance for which the test is intended (Reference ASTM D 471-79).

a. Water/Gas Solution—Mix by volume 98% unleaded gasoline and 2% corrosive water. Corrosive water being a solution formed by dissolving the following amounts of anhydrous sodium salts in 1 L of distilled water at 40°C to aid the mixing:

(1) Sodium Sulfate   14.8 mg.
(2) Sodium Chloride   16.5 mg.
(3) Sodium Bicarbonate   13.5 mg.

b. Gasoline and 20% Methanol—Shall be 20% by volume reagent grade methanol added to unleaded gasoline.

c. Gasohol—Shall be 10% by volume reagent grade ethanol added to unleaded gasoline.

d. Butylcellosolve—Shall be 5% by volume technical grade butylcellosolve added to unleaded gasoline.

e. Oxidized Gasoline (Sour Gas)—Shall be mixed by the following procedure to achieve a peroxide number of 180 millimole/liter:

(1) Stock Fuel Mix—A 70:30% by volume of n-heptane:toluene.

(2) Copper Ion Stock Solution—Due to the hazardous nature of these chemicals, the solution must be prepared sequentially in the following three steps: (1) Add 10 ml of 12% copper ion concentrate solution to 990 ml stock fuel; (2) Add 100 ml of the solution from step 1 to 1040 ml of stock fuel, and (3) Dilute 10 ml of the solution from step 2 with 990 ml of stock fuel.

NOTE: FAILURE TO FOLLOW PROPER PROCEDURES CAN RESULT IN FIRE. FOR FURTHER HANDLING AND SAFETY INFORMATION REFER TO THE MATERIAL SAFETY DATA SHEETS (MSDS) REQUIRED BY O.S.H.A. FOR THE COMPOUNDS USED.

(3) Peroxide Stock Solution—Dilute 335 ml of 90% by weight t-butyl hydroperoxide with 665 ml of heptane.

f. Test Solution—Dilute 60 ml of peroxide stock solution with 10 ml of copper ion stock solution and 930 ml of stock fuel.

4.4.2 TEST PROCEDURE

a. Testing with one fluid followed by another fluid is to be avoided.

b. Each specimen shall have its vacuum reference port suitably sealed.

c. Flush the specimen with the solution for 1 min at an outlet port flow of 25 g/s minimum.

d. Submerge the specimen in the solution for 30 days at 60°C with the inlet port orientated up in a manner to keep the specimen full.

4.4.3 ASSESSMENT OF RESULTS—Each specimen tested shall be evaluated by the performance requirements detailed in 3.1.1, 3.1.2, 3.1.7, 3.2, and 3.3 as they apply prior to and after the completion of testing.

Report the percent deviation between the before and after readings.

a. Reported Population Data—Report the upper bound of percent deviations (x +4 sigma) on a minimum random sample size of 24.

5. Standard Test Conditions
Unless otherwise specified, the following test conditions are implied:

5.1 Test Fluid
a. Gasoline is a very volatile fluid and is very hard to maintain in a stable state; therefore, it is recommended that all the bench performance tests be conducted with a mineral spirits fluid of known composition.

It is recommended that the test fluid meet the following specifications:

(1) Specific Gravity at +15.6°C (60°F) ASTM D 1298: 0.775 to 0.790

(2) Kinematic Viscosity at 20°C (68°F) ASTM D 445: 1.195 to 1.205 cSt

(3) Absolute Viscosity at 20°C (68°F) ASTM D 445: 0.94 to 0.96 cSt

(4) Flash Point ASTM D 56—43.3°C (110°F) min

(5) Distillation—150°C (300°F) min

b. In order to predict the performance of the pressure regulator or damper under actual conditions, each design must be individually evaluated using mineral spirits for bench testing and gasoline (i.e., Indolene Clear HO III (AMOCO Trademark)) for bench durability (see 4.2). Correlation should be established between the test results obtained with test fluid and indolene.

NOTE: All tests must be confirmed in the vehicle application tests.

5.2 Ambient Test Temperature
All tests should be conducted at an ambient temperature of 22°C ± 3, unless otherwise specified.

5.3 Test Fluid Temperature
A stabilized test fluid temperature of 22°C ± 3 will be maintained and should be measured at the pressure regulator or pressure damper inlet port.

5.4 Return Line Pressure
The fluid return line pressure should be between 0 to 20% of the predetermined set pressure; this pressure range should not be exceeded throughout the test flow range and measured no more than 100 mm from the exit point of the regulator.

5.5 Mounting Configuration
Mounting configuration of the pressure regulator or damper will be such that the line of action of the valve/diaphragm is in the vertical position and is an additive component to the spring force.

It is noted that the performance of some regulators or dampers differ relative to the line of action of the valve/diaphragm and, therefore, should also be tested at the angle it was designed for. In doing so, the line of action of the valve/diaphragm should be included in the data submitted.

5.6 Preparation of Test Specimen
a. Flush for 10 s and fill the fuel chamber of the pressure regulator or pressure damper with test fluid.

b. Cap the fuel inlet and outlet ports.

c. Place the test specimen in a position such that the total volume of the fuel chamber is filled with test fluid.

d. At a stabilized temperature of 22°C ± 3, let the specimen soak in test fluid for a minimum of 30 min.

e. Install the pressure regulator or damper on its specified test fixture and cycle the valve (open/closed) six times.

f. Testing of the specimen may now commence.

6. Test Apparatus and Equipment

6.1 Introduction
The test equipment must provide a means to test fuel pressure regulator or damper performance characteristics according to the specifications described in Sections 3, 4, and 5.

The test equipment must be capable of precisely testing various design fuel pressure regulators or dampers, preferably in mass flow units. The equipment and installations must be in compliance with all the applicable fire and safety codes and regulations.

To assure that the test stand is versatile enough to evaluate the various design characteristics, the hydraulic system should be capable of operating with lead free gasoline (indolene), and mineral spirits specified by the supplier in his manufacturing process. The recommended hydraulic system plumbing diagrams are shown in Figures 16 and 17.

6.2 Test Equipment Description
The test equipment components should meet the following minimum functional requirements:

6.2.1 TEST FLUID PUMP—Capable of delivering up to 150 g/s at 700 kPa.

6.2.2 VACUUM PUMP—Must be capable of delivering 1.0 kPa absolute dead head pressure.

6.2.3 AIR PUMP—Must be capable of delivering 0 to 700 kPa filtered dry air.

6.2.4 PRESSURE GAGES—Capable of measuring 0 to 700 kPa gage for performance test and 0 to 3500 kPa for burst pressure.

6.2.5 VACUUM GAGES—Must be capable of measuring 0 to 100 kPa absolute.

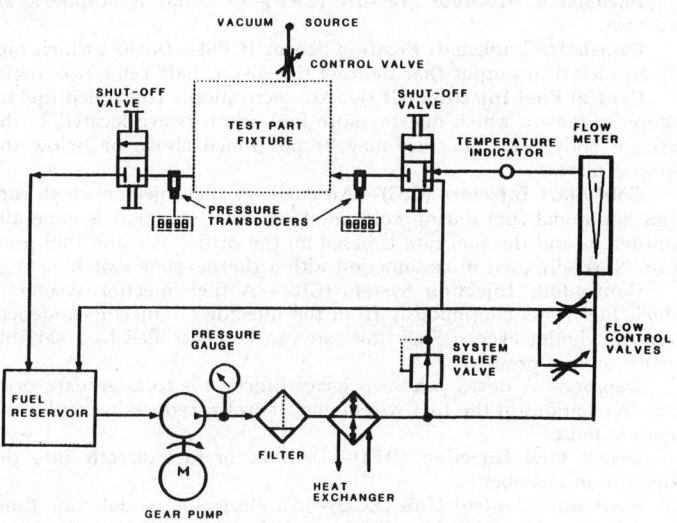

FIG. 16—PRESSURE REGULATOR PERFORMANCE TEST STAND

6.2.6 VIBRATION TESTER—Must be capable of sinusoidal cycling at 50 g amplitude.

6.2.7 SALT SPRAY—An enclosed chamber capable of generating a 5% salt concentration mist at 95°F (see ASTM B 117).

6.2.8 ENVIRONMENTAL TEST—A container capable of holding regulator or damper specimens submerged in corrosive liquids for extended time periods. Also, an enclosed chamber where test specimens can be cycled between temperature ranges of −30 to +130°C. Provisions must be present to pressure cycle the specimen at the test conditions.

6.2.9 PUMP PRESSURE CONTROL—Must be capable of adjusting test fluid pressure within ±0.5 kPa.

6.2.10 SHOCK TEST—Must be capable of performing 100 g over 5.0 ms.

6.3 Pressure Regulator Test Equipment Measurement Accuracy

TABLE 2—PRESSURE REGULATOR TEST EQUIPMENT MEASUREMENT ACCURACY

CHARACTERISTIC	REPORTED RESOLUTION	EQUIPMENT RESOLUTION	CONTROL
Test Fluid:			
Pressure (kPa)	XXX.	±0.5	±3.5
Temperature (xC)	XX.	±0.5	±1.0
Specific Gravity	X.XXX	±0.001	±0.002
Kinematic Viscosity (cSt)	X.XXX	±0.001	±0.015
Flow Rate (g/s)	XX.X	±0.01	±0.2
Air:			
Temperature (°C)	XX.	±0.50	±2.0
Pressure (kPa)	XXX.	±0.5	±1.5
Flow (Scc/min)	XX.X	±0.05	±0.10
Time (ms)	XX.X	±0.01	±0.2
Ambient Room Temperature (°C)	XX.X	±0.50	±3.0

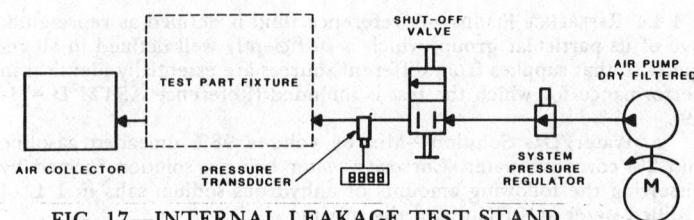

FIG. 17—INTERNAL LEAKAGE TEST STAND

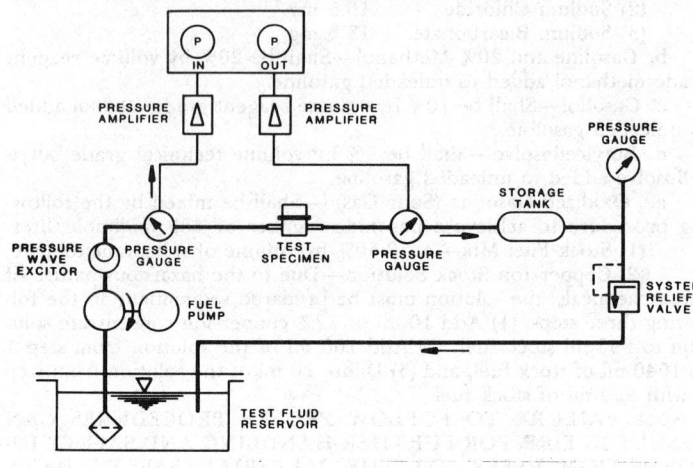

FIG. 18—PRESSURE DAMPER PERFORMANCE TEST STAND

FUEL INJECTION NOMENCLATURE—SPARK IGNITION ENGINES—SAE J1541 FEB88

SAE Recommended Practice

Report of the Engine Committee approved February 1988.

1. Scope—This recommended practice was developed to provide a common understanding and usage of the technical terms useful in fuel injection systems for spark ignition engines. Additional nomenclature related to specific components is covered in the individual component sections and SAE J1933.

Accumulator—A device to maintain fuel system pressure after the fuel pump is turned off. The device may also dampen pressure fluctuations and fuel pump noise during normal system operation.

Air Assisted Fuel Injection (AAFI)—A fuel injection system where auxiliary air is mixed with the metered fuel to aid in vaporization or transport, or both, of the fuel.

Airflow Sensor—A sensor that provides an electrical output proportional to the flow rate of the intake air to the engine.

Barometric Absolute Pressure (BAP)—Absolute atmospheric air pressure.

Camshaft/Crankshaft Position Sensor (CPS)—Devices which supply an electrical output that indicate the given shaft reference angle.

Central Fuel Injection (CFI)—An electronically controlled fuel injection system in which one or more fuel injectors are located in the throttle body. The injectors may be positioned above or below the throttle plates.

Cold Start Injectors (CSI)—An auxiliary fuel injector which supplies additional fuel during cold cranking. Fuel injection is generally continuous and the fuel rate is based on the orifice size and fuel pressure. Normally used in conjunction with a thermo-time switch.

Continuous Injection System (CIS)—A fuel injection system in which fuel flows continuously from the injector or injectors independent of cylinder events. Fuel flow rate can be controlled by a variable orifice or fuel pressure.

Damper—A device whose primary function is to attenuate pressure fluctuations in the fuel system and, thereby, reduce hydraulic and acoustic noise.

Direct Fuel Injection (DFI)—Delivery of fuel directly into the combustion chamber.

Electronic Control Unit (ECU)—An electronic module one function of which is to calculate a command signal for the injector driver circuit based on inputs from engine operating sensors.

Electronic Fuel Injection (EFI)—A general term referring to any fuel injection system in which fuel metering is controlled electronically. Quantity of fuel delivered is scheduled by an electronic control unit. Its output signal is based on information received from several sensors that monitor the operating conditions of the engine.

Engine Coolant Temperature Sensor (ECTS)—A sensor that provides an electrical output proportional to the engine coolant temperature.

Exhaust Gas Oxygen Sensor (EGOS)—A sensor located in the exhaust system that provides an electrical output which indicates oxygen content.

Fuel Injection (FI)—A general term referring to any type of fuel injection system, that is, CFI, CIS, EFI, MPI, PFI, SFI, SPI, TBI, etc. Fuel delivery may be mechanically or electronically controlled. Systems with fuel pressure > 150 kPa will be referred to as "high pressure" and those = < 150 kPa as "low pressure".

Fuel Injector (INJ)—An electromagnetic or mechanical device used to direct delivery or meter pressurized fuel or both.

Fuel-Injector Driver (FID)—An electronic circuit that converts the fuel control signal from the ECU into the proper voltage and current signal required. The two commonly used drivers in the industry are a peak-hold driver used with low resistance injectors as a "on-off" (saturated driver) used with high resistance injectors.

Fuel Pressure Regulator (REG)—A device to maintain a controlled fuel pressure at the fuel injector or a controlled differential pressure across the injector.

Fuel Pump (Pump)—A device that provides a specified fuel flow at the required system pressure.

Fuel Rail (Rail)—A fuel manifold that distributes fuel to the individual fuel injectors. It is used with multi-point injection systems.

Group Fuel Injection (GFI)—A multi-point delivery technique in which fuel is delivered in a predetermined sequence by groups of two or more injectors.

Idle Speed Control (ISC)—A general term used to indicate any device or system which provides programmed control of engine idle speed. Speed control is usually accomplished by adjusting the amount of air bypassing the throttle plate or by adjusting the position of the throttle plate.

Intake Air Temperature Sensor (IATS)—A sensor that provides an electrical output proportional to the intake air temperature. It is typically mounted within or ahead of any airflow measuring device.

Intake-Manifold Absolute Pressure Sensor (IMAPS)—A sensor which provides an electrical output proportional to the absolute pressure within the intake manifold downstream of the throttle plate.

Intake-Manifold Charge Temperature Sensor (IMCTS)—A sensor located in the intake manifold that provides an electrical output proportional to the temperature of the air or air/fuel mixture.

Multi-Point Injection (MPI)—A fuel delivery system in which each cylinder is fueled by at least one injector. The injector is normally located in the intake manifold or port close to the intake valve.

Port Fuel Injection (PFI): See Multi-Point Injection.

Sequential Fuel Injection (SFI)—A multi-point fuel delivery technique in which each injector is individually energized and timed relative to its cylinder event. Fuel is delivered to each cylinder once per two crankshaft revolutions in 4-cycle engines and once per crankshaft revolution in 2-cycle engines.

Simultaneous Double Fire (SDF)—A multi-point fuel delivery technique in which all injectors in a 4-cycle engine are energized simultaneously usually once per crankshaft revolution.

Single Point Injection (SPI)—A fuel delivery system in which the total fuel requirement of the engine is delivered at one specific location in the intake system. More than one injector may be used.

Speed Density (SD)—A system in which the mass air flow rate is calculated based on cylinder displacement and the measured intake manifold absolute pressure, engine speed, intake manifold air temperature and theoretical volumetric efficiency.

Thermo-Time Switch (TTS)—A bi-metal switch used to control the amount of fuel delivered by the cold start injector during cold cranking.

Throttle Body (TB)—A device used to vary the supply of intake air to the engine by means of one or more adjustable orifices and/or throttle plates.

Throttle Body Injection (TBI)—See Central Fuel Injection.

Throttle Position Sensor (TPS)—A device that provides an electrical output as a function of throttle plate position.

VALIDATION TESTING OF ELECTRIC FUEL PUMPS FOR GASOLINE FUEL INJECTION SYSTEMS— SAE J1537 JUN90

SAE Recommended Practice

Report of the Gasoline Fuel Injection Standards Committee approved June 1990.

TABLE OF CONTENTS

1	SCOPE
1.1	Purpose
2	REFERENCES
2.1	Applicable Documents
3	FUEL PUMP FLOW PERFORMANCE TESTING
3.1	Fuel Pump Flow Test Equipment
3.1.1	Adjustable DC Power Supply
3.1.2	DC Ammeter
3.1.3	DC Voltmeter
3.1.4	Pressure Gage (Transducer, Manometer, etc.)
3.1.5	Filter (Optional)
3.1.6	Adjustable Flow Restrictor or Pressure Regulator
3.1.7	Flowmeter
3.1.8	Tubing
3.1.9	Tank
3.1.10	Test Fluid for Flow Testing
3.2	Flow Stand Operating Procedure
3.2.1	Pump Installation
3.2.2	Adding the Test Fluid
3.2.3	Flow Testing the Pump
3.2.4	Flow Stand Shutdown
3.3	Presentation of Performance Data
4	VALIDATION TEST REQUIREMENTS
4.1	Vibration
4.1.1	Validation Requirements
4.2	Temperature Cycling
4.2.1	Validation Requirements
4.3	Compatibility with Fuels
4.3.1	Chemical Exposure
4.3.2	Validation Requirements
4.4	Hot Fuel Handling (Priming)
4.4.1	Validation Requirements
4.4.2	Definitions
4.4.3	Fuel
4.4.4	Fuel Volume
4.4.5	Heating the Fuel
4.4.6	Fuel Heating Rate
4.4.7	Pressure Settings
4.4.8	Operating Voltage
4.4.9	Test Sequence
4.5	Cold Magnet Knockdown
4.5.1	Validation Requirement
4.6	Load Dump Transient
4.6.1	Validation Requirement
4.7	Endurance Test
4.7.1	Test Set Up
4.7.2	Pump Operating Set Points
4.7.3	Test Cycle
4.7.4	Test Fuels
4.7.5	Validation Requirements

LIST OF FIGURES

Figure 1 Electric Fuel Pump Flow Performance Test Set Up
Figure 2 Electric Fuel Pump Parameters
Figure 3 Test Data Sheet
Figure 4 Pump Vibration Test Fixture Typical Station
Figure 5 Temperature Cycle Test Schedule
Figure 6 Hot Fuel Test Set-Up
Figure 7 Hot Fuel Handling Fuel Heating Rate

1. Scope—This SAE Recommended Practice defines the minimum design verification testing required to verify the suitability of in-tank mounted electric motor driven fuel pumps used for pumping gasoline or gasoline blend fuels.

Additional tests not specified in SAE J1537 will be required for frame mounted pump applications or pumps intended for use on aircraft, motorcycles, or marine equipment.

1.1 Purpose—J1537 is intended to provide a nucleus of basic validation tests which the users and suppliers of electric fuel pumps agree are required to qualify a particular pump for use with automotive gasoline fuel injection systems. Numerous tests found in the unique specifications of both users and suppliers have been omitted from SAE J1537 because common agreement as to their value could not be reached. The seven basic tests defined in Section 4 include the most expensive, most time-consuming tests in a validation program. By providing an industry wide standard, SAE J1537 reduces the cost of validation to the benefit of all concerned. However, it is to be expected that the unique applications and experience of both users and suppliers will dictate that additional tests be conducted to complete qualification for any specific application.

2. References

2.1 Applicable Documents—SAE J1113, Electromagnetic Susceptibility Measurement Procedures for Vehicle Component (Except Aircraft)

3. Fuel Pump Flow Performance Testing—This section describes the equipment, set up, and procedure for measuring fuel pump flow performance. All tests in Section 4 require that the pumps used in these tests be flowed and their performance recorded prior to the running of these tests. All new pumps must receive a break-in period of at least 30 min prior to measuring initial flow.

3.1 Fuel Pump Flow Test Equipment (See Figure 1)—This section describes the basic equipment and interconnection considerations required to produce repeatable flow performance data from different test set ups.

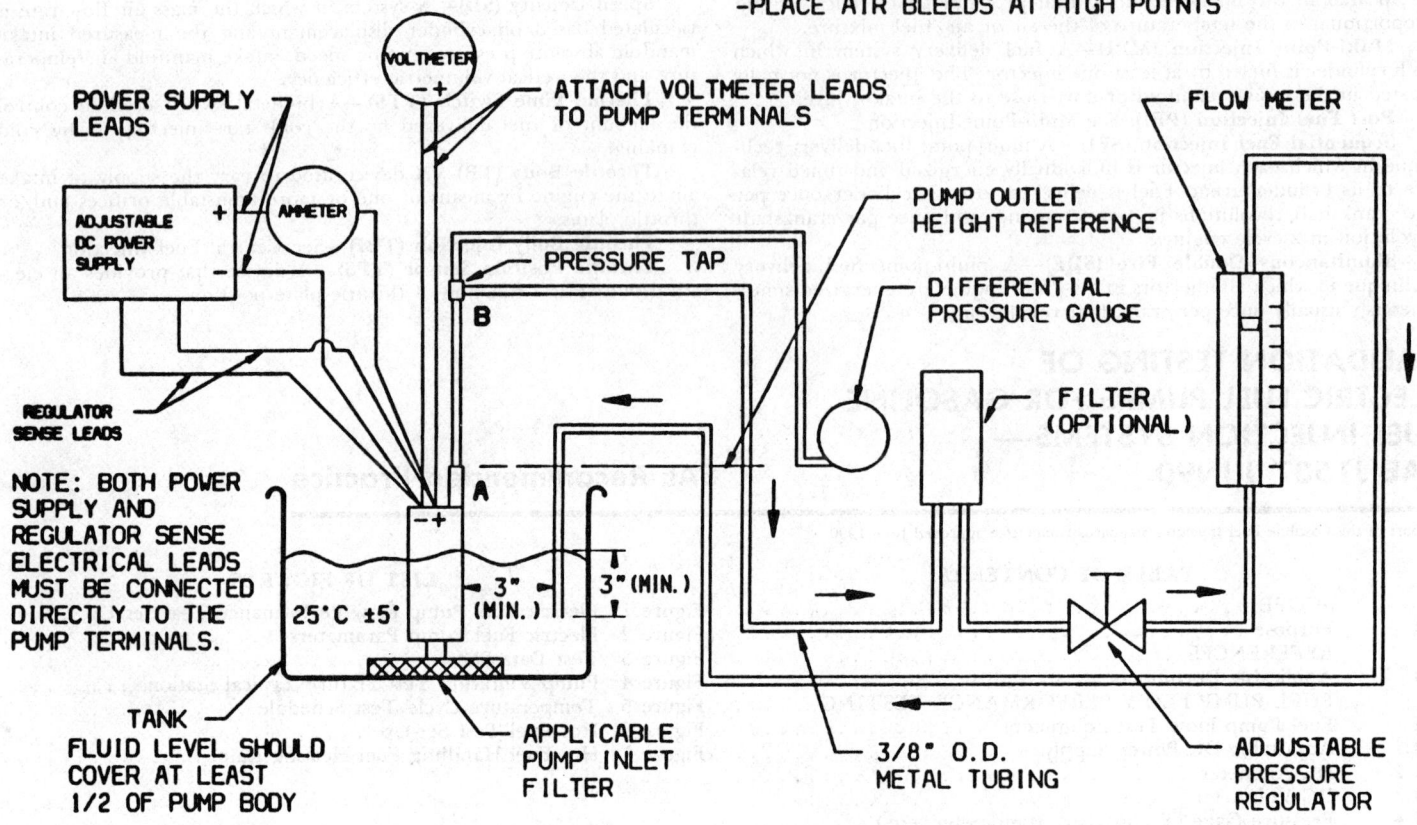

FIG. 1—ELECTRIC FUEL PUMP FLOW PERFORMANCE TEST
SET-UP

The basic test set-up must provide:
 a. Measurement of four electric fuel pump parameters as shown in Figure 2:

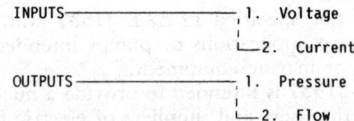

FIG. 2—ELECTRIC FUEL PUMP PARAMETERS

 b. Adjustment of two electric fuel pump parameters:
  (1) Voltage
  (2) Pressure
3.1.1 ADJUSTABLE DC POWER SUPPLY
 a. Voltage Range: 0 to 24 V DC
 b. Current Range: 0 to 20 Amps (Minimum)
 c. Stability: Must maintain set voltages at the pump terminals during flow measurement within ±0.1 V.

A power supply with external voltage sense capability is recommended. This allows the power supply to sense and stabilize the voltage at the pump terminals.
3.1.2 DC AMMETER
 a. Current Range: 0 to 20 Amps DC
 b. Accuracy: ±0.1 Amp
3.1.3 DC VOLTMETER
 a. Voltage Range: 0 to 24 V DC
 b. Accuracy: ±0.1 V
 c. Connection: Voltmeter leads are to be connected directly to pump terminals (see Figure 1).

3.1.4 PRESSURE GAGE (TRANSDUCER, MANOMETER, ETC.)
 a. Pressure Ranges:
  (1) 0 to 138 kPa (0 to 20 psi)
  (2) 0 to 690 kPa (0 to 100 psi)
  (3) Higher pressure ranges may be required depending on the pump type and application.
 b. Accuracy: ±1% of full scale
 c. Location—The differential gage must be referenced to the same level as the pump outlet (see Figure 1).
3.1.5 FILTER (OPTIONAL)—Maximum pressure differential at a flow of 40 g/s = 3 kPa.

The filter should contain an air bleed or have a configuration which eliminates trapped air.
3.1.6 ADJUSTABLE FLOW RESTRICTOR OR PRESSURE REGULATOR— Maximum pressure differential with restrictor wide open at flow of 40 g/s = 7 kPa. Maximum restriction: full closed.
3.1.7 FLOWMETER
 a. Type: Mass flow
 b. Range: 0 to 50 g/s
 c. Accuracy: ±0.5 g/s, with Stoddard solvent test fluid
3.1.8 TUBING
 a. Size: 3/8 in O.D. (minimum)
 b. Length: Short as possible
 c. Bends: None preferable
 d. Materials:
  (1) 1st choice–Metal tubing
  (2) 2nd choice–Steel braided rubber hose
  (3) Not recommended–Cloth, braided, or nonreinforced rubber hose
3.1.9 TANK
3.1.9.1 Size—Large enough for the test fluid to cover at least one-half of a vertically mounted fuel pump and to allow the fluid return tube to be at least 3 in away from the pump inlet.
3.1.10 TEST FLUID FOR FLOW TESTING—MIL-C-7024B (Type II) Laboratory Test Fluid, for example, Stoddard solvent.

The test fluid temperature should be maintained at 25°C ± 5 during performance testing. Test fluids should be changed periodically to prevent the accumulation of contaminants.

Alternate test fluids to Stoddard solvent are mineral spirits, Esso Varsol 130/180, Amosol Naphtha No. 395 HF and other fluids which have properties approximating the Stoddard solvent values specified in MIL-C-7024B.

NOTE: There is a difference between these test fluids and gasoline which varies in its effect on performance according to pump type and design. It is, therefore, the manufacturer's responsibility to develop the correlation factor between the fluid used and actual flow performance in gasoline for his pump.

3.2 Flow Stand Operating Procedure

3.2.1 PUMP INSTALLATION—Install the fuel pump with its appropriate inlet filter into the test tank maintaining the conditions prescribed in Figure 1. Note particularly the position of the pressure gage in relation to the pump outlet, the point of connection for the voltmeter leads, and the position of the return line in relation to the pump and the anticipated fluid height.

3.2.2 ADDING THE TEST FLUID—Add the test fluid (3.1.10) to the tank and make sure that at least one half of the pump body and at least 3 in of the return line is submerged in the fluid (see Figure 1). Test fluid temperature must be maintained between 20 and 30°C.

3.2.3 FLOW TESTING THE PUMP—A break-in period of at least 30 min should be given to new pumps before running a performance test.

Turn on the power supply and adjust the voltage applied to the pump terminals.

Adjust the test stand regulator to the specified operating pressure. Observe the flow meter for stable output. If stable flow readings cannot be obtained, check to be sure the return line is not too close to the pump inlet (see Figure 1), and that fuel temperature is within specified limits. If instability persists, cycle the regulator several times to increase flow and/or bleed the system to purge entrapped air.

With system reset at the desired operating pressure and voltage, record the pump flow. As flow data are taken, regulated pressure should be verified, adjusting as required.

3.2.4 FLOW STAND SHUTDOWN—After all the desired flow data have been obtained, the system should be shut down with pressure removed from the system first and electrical power to the pump shut off before disconnecting power and voltmeter lead wires. Pump fluid lines should be disconnected only after the verifying pressure in the system is at zero.

3.3 Presentation of Performance Data

The sample data sheet (see Figure 3) provides a format for recording and reporting validation test data. At the top of the sheet are spaces to fill in pump identification numbers, the regulated pressure at which the pump was tested, the test fluid used in the flow test, and the nominal pressure drop of the inlet filter used, if any.

The next data block provides a format for recording the results of the Initial Flow Test which is obtained after a minimum break-in of 30 min but prior to running any particular test from Section 4 for which the pump flow was chosen. It is intended that flow data be obtained in g/s mass flow units but a blank column is also provided for conversion to or from alternate units. If the pump performance is to be characterized as a function of pressure, as well as voltage, a new data sheet for each operating pressure should be used. A blank line is provided between 12 and 14 V for those who wish to measure pump performance at the charging system voltage.

The next section of the data sheet provides a space to identify the tests in Section 4 that the flow data are validating and to record any noteworthy observations about the test or deviations from the procedure that might aid in interpreting the data.

The Final Flow section is for recording the flow data obtained after the particular test in Section 4 has been completed.

The last section is for presenting the Change in Flow (initial flow minus final flow) which can be attributed to the wear and tear inflicted upon the pump by the particular test in Section 4.

4. Validation Test Requirements

The following seven tests comprise the SAE J1537 validation test requirements:

4.1 Vibration

The pump is to be vibrated in a mounting fixture as shown in Figure 4 for a total of 3 h, 1 h in each of the three major, mutually perpendicular planes of the pump as mounted. The test is to be run with the pump dry and no voltage applied to the pump terminals.

a. Vibration Displacement: 0.75 mm
b. Frequency Range: 10 Hz to 55 Hz
c. Max. Acceleration (reference): 4.5 g
d. Cycle time: 60 s (10-55-10 Hz)

4.1.1 VALIDATION REQUIREMENTS—The pump must meet or exceed the rated flow established for the pump.

4.2 Temperature Cycling (see Figure 5)

The pump shall be alternately heated and cooled from +65 to −30°C for 20 cycles. The test is to run per Figure 5. The pump is not to be run in fuel (that is, the test should be run with the pump dry). No voltage is to be applied to the pump terminals during the test.

4.2.1 VALIDATION REQUIREMENTS—The pump must meet or exceed the rated flow established for the pump.

4.3 Compatibility With Fuels

4.3.1 CHEMICAL EXPOSURE—Pumps are to be immersed in the solutions listed below (1/2 to 1 L/pump). The pumps are to be run a few seconds to fill them with solution. The container is then to be sealed and maintained for 30 d at 20°C ± 10.

 a. Commercial grade of unleaded gasoline.
 b. Unleaded gasoline with 0.05% thiophene added (contains sulphur).
 c. Unleaded gasoline with 250 ppm tertiary butyl hydroperoxide.
 d. Unleaded gasoline with 10% by volume, methanol.
 e. Unleaded gasoline with 10% by volume, ethanol.
 f. Unleaded gasoline with 10% by volume, methyl tertiary butyl ether.
 g. Unleaded gasoline with 10% by volume, isopropyl alcohol.
 h. Leaded gasoline (3 g/gal lead).

4.3.2 VALIDATION REQUIREMENTS—The pump must start unassisted in the test fluid with 10 V DC applied to the terminals. The pump must meet or exceed its rated flow at rated voltage within 5 min of operation.

4.4 Hot Fuel Handling (Priming)

WARNING: This test requires extreme caution. Both the test stand and room should be "explosion proof,": all electrical equipment should meet NEC Class I, Division I, Group D requirements.

The Hot Fuel Handling Test involves submerging the pump completely in fuel contained in a tank wherein a small regulated pressure is maintained. The fuel temperature is then increased from 20 to 65°C. The "time-to-prime" and the pump flow at 12 V DC are to be measured at every 5 degrees of temperature rise.

4.4.1 VALIDATION REQUIREMENTS
 a. Time-to-prime:
  (1) 3 s   20 to 50°C
  (2) 5 s   50 to 65°C
 b. Both flow and pressure are to be measured and recorded. Flow must be no less than 75% of rated flow within the time span specified in 4.4.9.

4.4.2 DEFINITIONS—Time-to-prime shall be the time between application of pump voltage and the attaining of 75% of the pump's rated pressure.

4.4.3 FUEL—Unleaded gasoline is to be used:
12 RVP ± 0.5 (initial measure)

4.4.4 FUEL VOLUME

 a. The fuel volume in the tank must be approximately 30 L and must be sufficient to completely cover the pump inlet throughout the test.
 b. The priming volume must be a minimum of 300 mL. The priming volume is the volume in the fuel line between the pump outlet and the pressure regulator including the flowmeter.

4.4.5 HEATING THE FUEL (REQUIRES EXTREME CAUTION!) (SEE FIGURE 6)—The fuel should be heated indirectly by circulating the fuel in the tank through an external heat exchanger. This process can be done safely only by placing the test tank inside an explosion proof thermal environmental chamber. Some experimentation will be necessary to determine the flow rates through the heat exchanger, the temperature feeding into the heat exchanger, and the temperature in the environmental chamber required to produce the required fuel heating rate.

4.4.6 FUEL HEATING RATE (SEE FIGURE 7)—The rate of temperature increase of the fuel shall be as shown in Figure 7:

a. 1°C/min from 20 to 50°C
b. 1/2°C/min from 50 to 65°C

4.4.7 PRESSURE SETTINGS—The tank pressure regulator shall be set not to exceed 15 kPa.

The pump pressure regulator shall be set at system-rated pressure with the pump operating at rated flow at 12 V DC.

4.4.8 OPERATING VOLTAGE—The applied pump voltage shall be 12 V DC ± 0.1 during the test.

4.4.9 TEST SEQUENCE—The pump prime shall be checked every 5 degrees rise in fuel temperature beginning with 20°C. The pump flow must be determined within 1 min after the priming determination has been made. The pump must then be off for a minimum of 4 min allowing the pressure in the line between the pump and regulator to bleed down through the by-pass orifice.

CONDITIONS:

Pump Serial No.: _____ Model: _____
Regulated Pressure: _____ kPa
Test Fluid: _____ at 25°C + 5
Inlet Filter (Optional): _____ ΔP, kPa

INITIAL FLOW TEST:

Operating Volts DC	Pump Flow Spec.	g/s	Current Draw-Amp
6			
8			
10			
12			
14			
16			
18			

TEST RESULTS SECTION: _____
_____
_____
_____
_____
_____
_____

FINAL FLOW (Same Conditions As Above):

Operating Volts DC	Pump Flow Spec.	g/s	Current Draw-Amp
6			
8			
10			
12			
14			
16			
18			

CHANGE IN FLOW (Initial Flow Minus Final Flow):

Operating Volts DC	Pump Flow Spec.	g/s	Current Draw-Amp
6			
8			
10			
12			
14			
16			
18			

Validation Testing of Electric Fuel Pumps for Gasoline Fuel Injection Systems

FIG. 3—TEST DATA SHEET

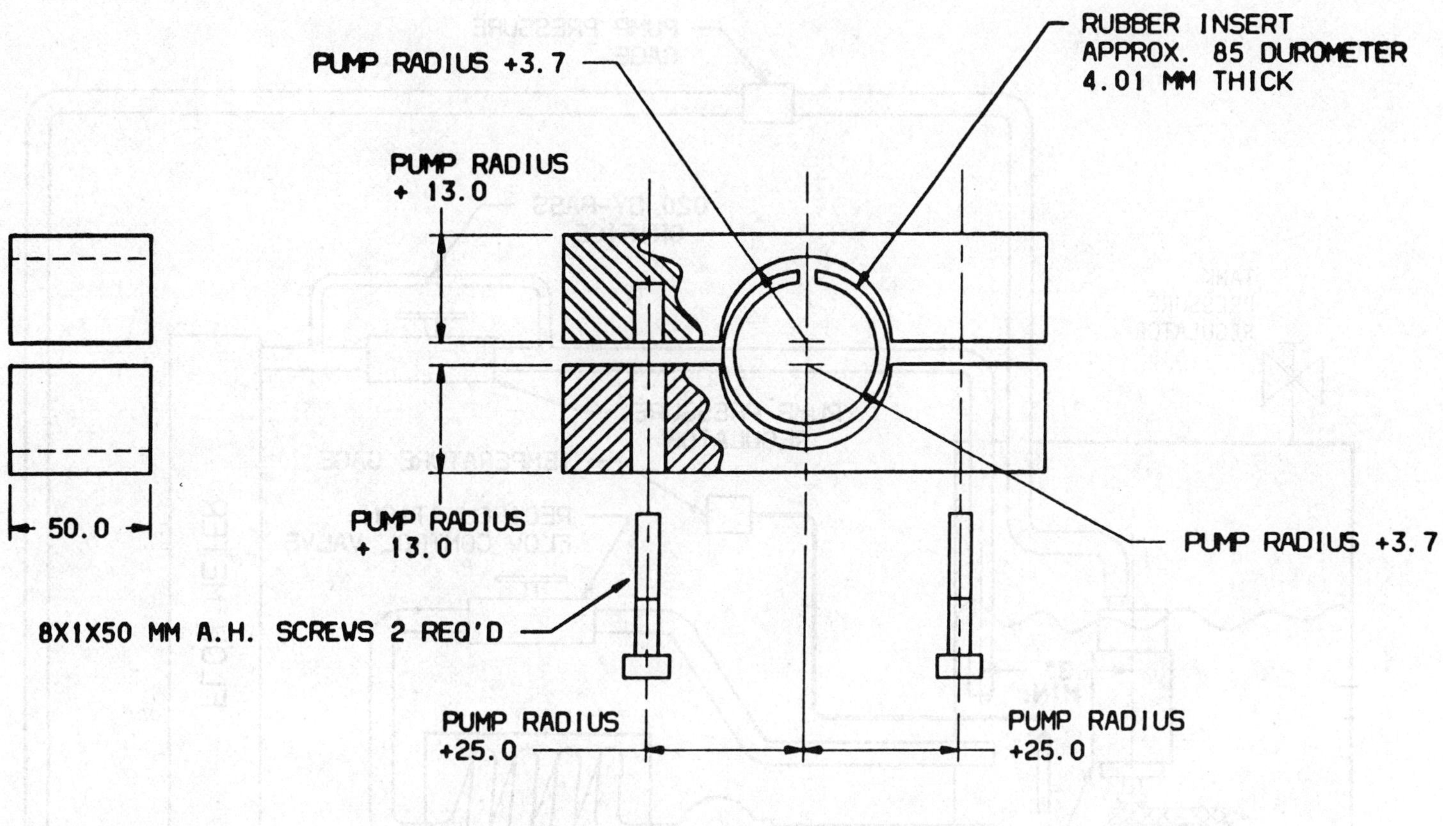

RUBBER INSERT
APPROX. 85 DUROMETER
4.01 MM THICK
PUMP RADIUS +3.7
PUMP RADIUS
+ 13.0
50.0
PUMP RADIUS
+ 13.0
8X1X50 MM A.H. SCREWS 2 REQ'D
PUMP RADIUS +3.7
PUMP RADIUS
+25.0
PUMP RADIUS
+25.0

FIG. 4—PUMP VIBRATION TEST FIXTURE TYPICAL STATION

70
60
50
40
30
20
10
0
-10
-20
-30
TEMPERATURE - CELSIUS
0
1
2
3
4
5
6
TIME - HOURS

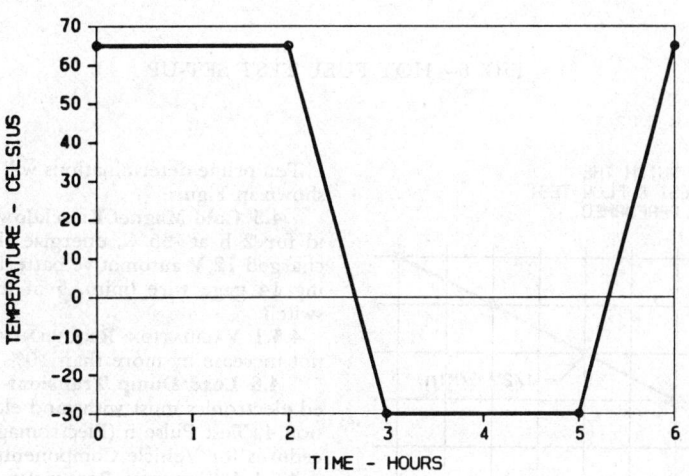

FIG. 5—TEMPERATURE CYCLE TEST SCHEDULE

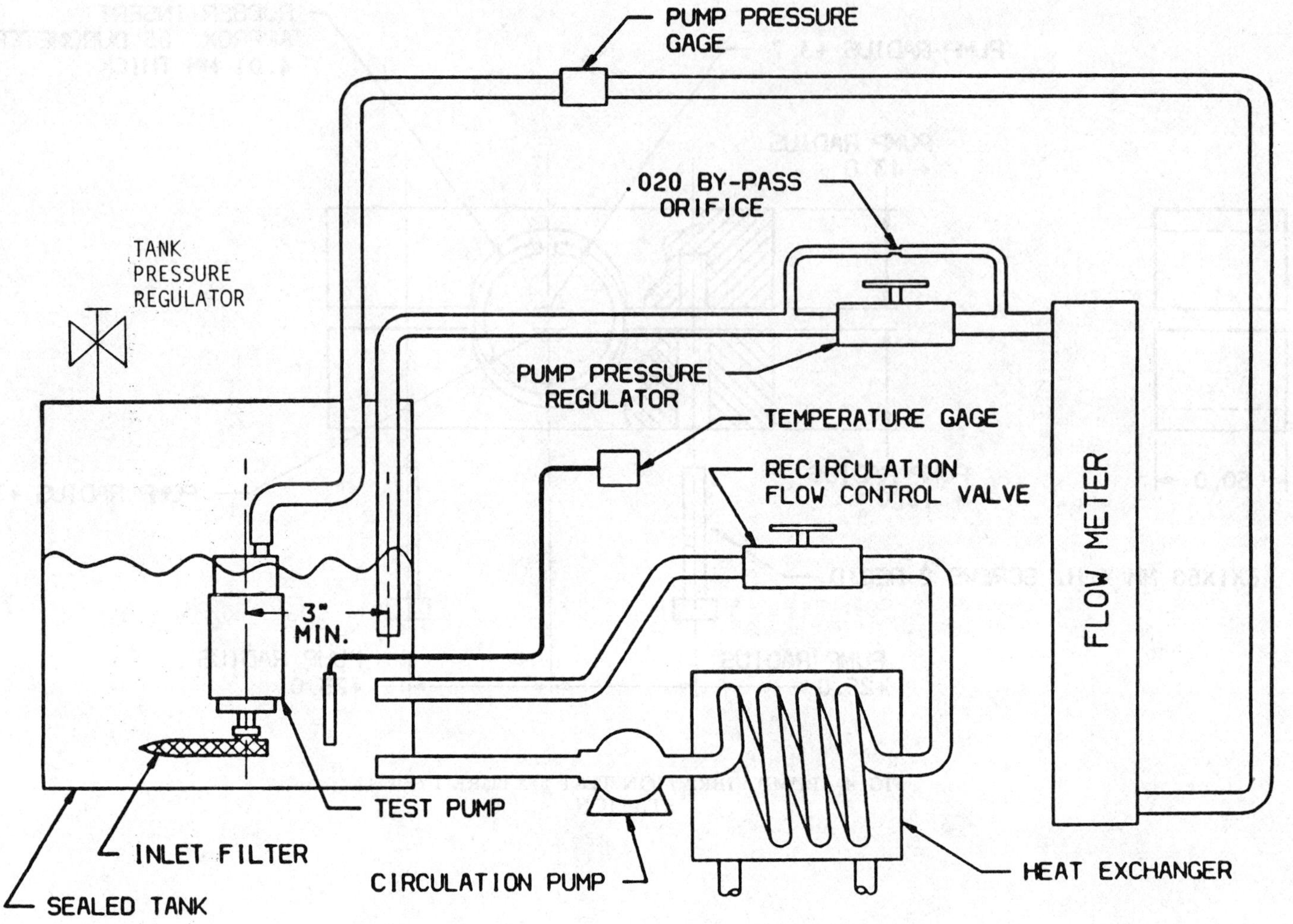

FIG. 6—HOT FUEL TEST SET-UP

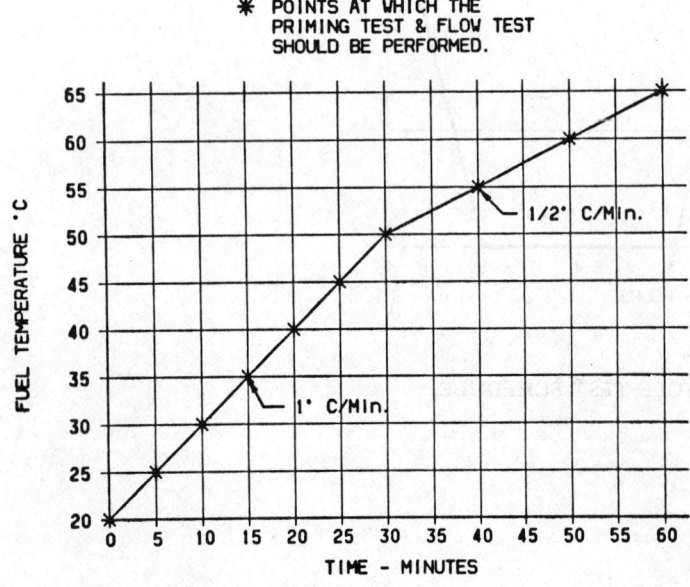

FIG. 7—HOT FUEL HANDLING FUEL HEATING RATE

Ten prime determinations will be made, requiring 1 h to complete as shown in Figure 7.

4.5 Cold Magnet Knockdown—After soaking the pump in test fluid for 2 h at -35°C, energize the unit twice for 5 s using two fully-charged 12 V automotive batteries connected to the pump in series using 14 gage wire (min), 5 m long (max) connected through a knife switch.

4.5.1 VALIDATION REQUIREMENT—Current draw of the pump must not increase by more than 10% at rated voltage.

4.6 Load Dump Transient—The electric fuel pump and associated electronics must withstand electrical transients per SAE J1113, Section 4, Test Pulse 5 (Electromagnetic Susceptibility Measurement Procedures for Vehicle Components Except Aircraft).

4.6.1 VALIDATION REQUIREMENT—The pump must meet or exceed the rated flow established for the pump.

4.7 Endurance Test

CAUTION: This procedure involves the use of flammable test fluids. Appropriate precautions must be taken.

The Endurance or Life Cycle Test involves running pumps for 3000 h in test fuel on a cycle which includes start and stop operation. A small portion of the total cycle time is to be conducted at temperature extremes. Three different test fuels are specified and pumps must be validated in each.

4.7.1 TEST SET UP

a. Tank Size: The tank, where the pumps are to be tested, must be of sufficient size to accommodate a minimum of 5 L of test fuel for each pump running in the tank, usually six or more for a representative statistical sample.

b. Fuel Change Interval: The fuel in the tank must be changed every 84 h ± 16 (twice a week).

c. Fuel Level: The test fuel must cover the pump inlet and at least 1/2 of the pump body.

d. Pump Mounting: The pump should be mounted in the tank oriented in the same attitude as it would be in the intended application. If the application is unknown, the pump should be mounted with the centerline of the pump in the vertical direction.

e. Filters (Inlet and Outlet):

(1) Inlet: The pump should be fitted with an inlet filter that meets the particle pass and pressure drop specification of the pump manufacturer.

(2) Outlet: Each pump outlet should be fed into an automotive in-line filter used in gasoline fuel injection applications (typically 10 to 20 m).

4.7.2 Pump Operating Set Points

a. Electrical Input: The pumps shall be operated at 13.5 V DC ± 0.5 measured at the pump terminals. A provision for automatic periodic start and stop should be provided.

b. Pump Output Pressure: The pumps must be operated at the rated output pressure ±10 kPa.

4.7.3 Test Cycle—The pump must be run for 3000 h at the nominal fuel temperature of +33°C ± 3. During the test, the pump shall be turned off every 20 min for 6 s ± 2. The pumps must be run for 20 h each at -33°C ± -3 and 62°C ± 3 at 1000 h intervals during the test (60 h total at each temperature extreme).

At the completion of the 3000 h test after the final flow performance has been determined, several pumps from each test fuel should be disassembled and the moving parts measured for wear.

One or more of the pumps tested should be run beyond the 3000 h mark to identify the point in time and the mode in which ultimate failure will occur.

4.7.4 Test Fuels—The test outlined in 4.7.3 must be performed with the following three test fuels:

a. Unleaded gasoline.

b. Methanol blend—Unleaded gasoline with 10% volume methanol and 2% tertiary butyl alcohol (TBA).

c. Ethanol blend—Unleaded gasoline with 10% volume ethanol (gasoline grade).

Note: Additional specific test fluids may be required by some users.

4.7.5 Validation Requirements—The pumps must meet or exceed the rated flow established for the pump at 500 h intervals throughout the test.

φDIESEL FUEL INJECTION—END MOUNTING FLANGES FOR FUEL INJECTION PUMPS—SAE J626 JUN88 — SAE Standard

Report of Engine Committee approved January 1953 and completely revised June 1988.

1. Scope and Field of Application—This SAE Standard specifies dimensional requirements for eight types of end-mounting flange for fuel injection pumps for use in diesel (compression ignition) engines.

2. Reference—SAE J1419, Tapers for shaft ends and hubs for fuel injection pumps. (Ref. ISO 6519.)

3. Dimensions and Tolerances

3.1 General—Engine manufacturers are encouraged to use the tolerance H8 for the female register diameter.

If functionally necessary, the tolerance g8 of the pump spigot diameter (ΦA in the figures) may be replaced by f7, and the tolerance H8 of the female register diameter may be replaced by H7, by mutual agreement between supplier and user.

3.2 Fuel Injection Pumps

3.2.1 Type 1 Flange—See Fig. 1 and Table 1.

3.2.2 Type 2 Flange—See Fig. 2 and Table 2.

3.2.3 Type 3 Flange—See Fig. 3 and Table 3.

TABLE 1
Dimensions in millimetres

A g8	B[a] nominal	C ±0.5	D max	E
50 or 68	17 or 20	12.5	11	13-16
		26	24.5	

[a] See SAE J1419.

TABLE 2
Dimensions in millimetres

A g8	B[a] nominal	C ±0.5	D max	E
50 or 68	17 or 20	12.5	11	13-16
		17.4	16	
		26	24.5	

[a] See SAE J1419.

TABLE 3
Dimensions in millimetres

A g8	B[a] nominal	C ±0.5	D max	E
50 or 68	17 or 20	9.5	8.2	13-16
		12.5	11	
		17.4	16	
		26	24.5	

[a] See SAE J1419.

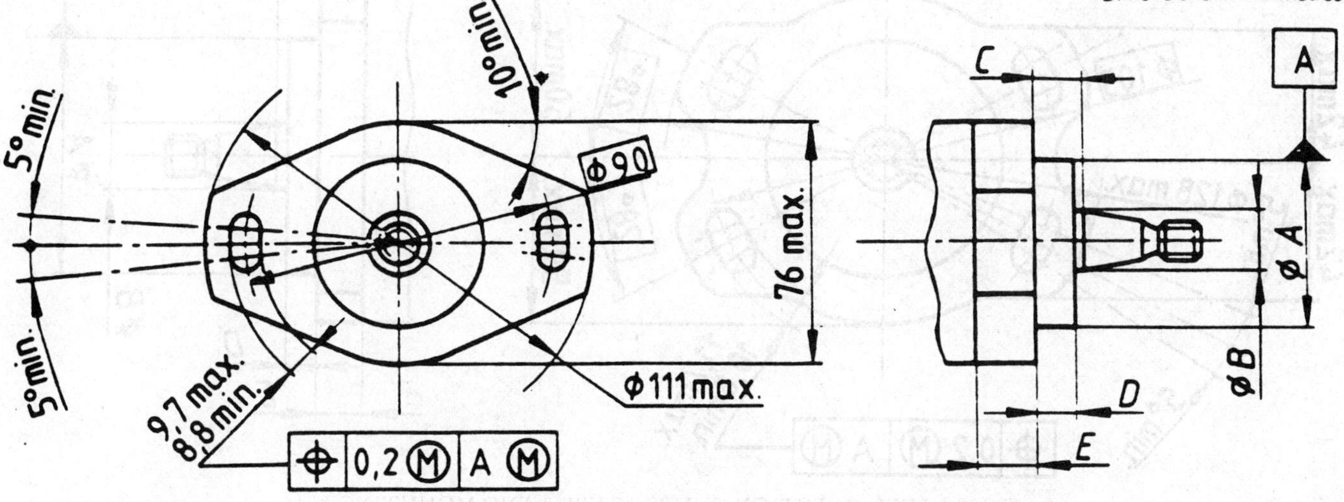

FIG. 1—FUEL INJECTION PUMPS—TYPE 1 END-MOUNTING FLANGE

26.298

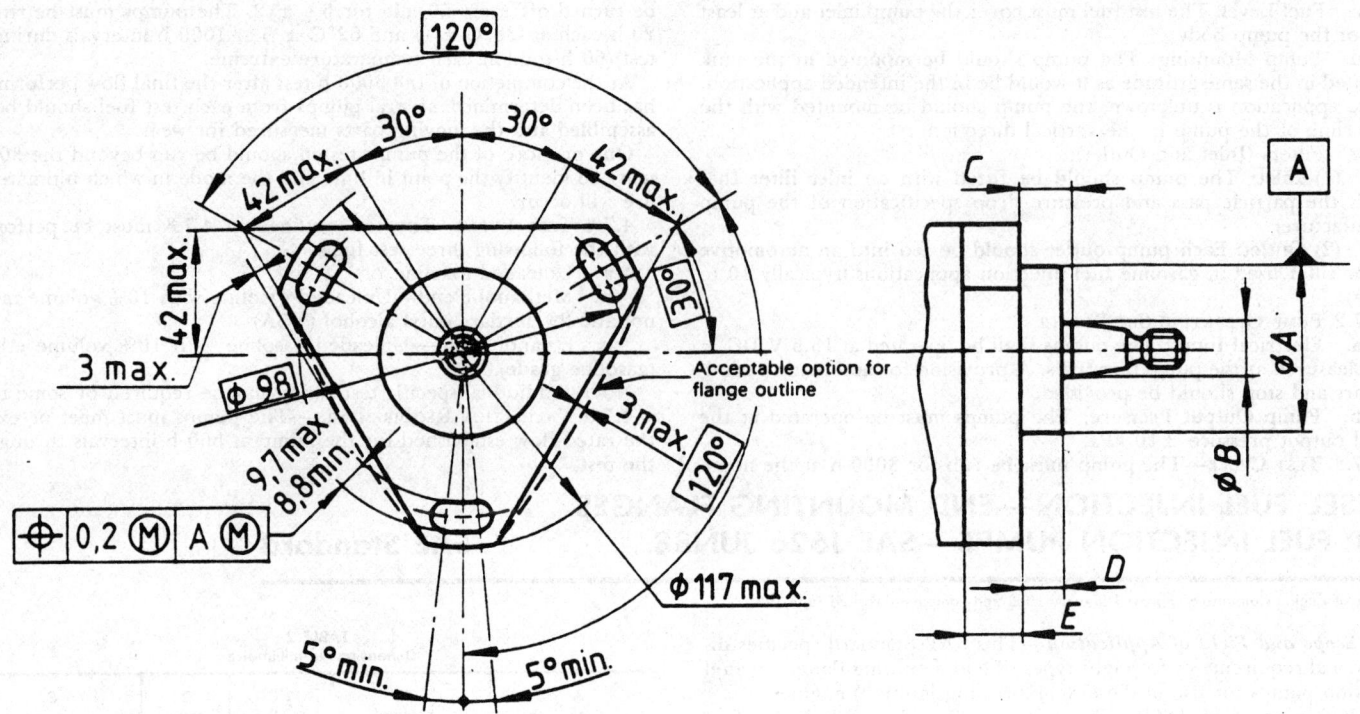

FIG. 2—FUEL INJECTION PUMPS—TYPE 2 END-MOUNTING FLANGE

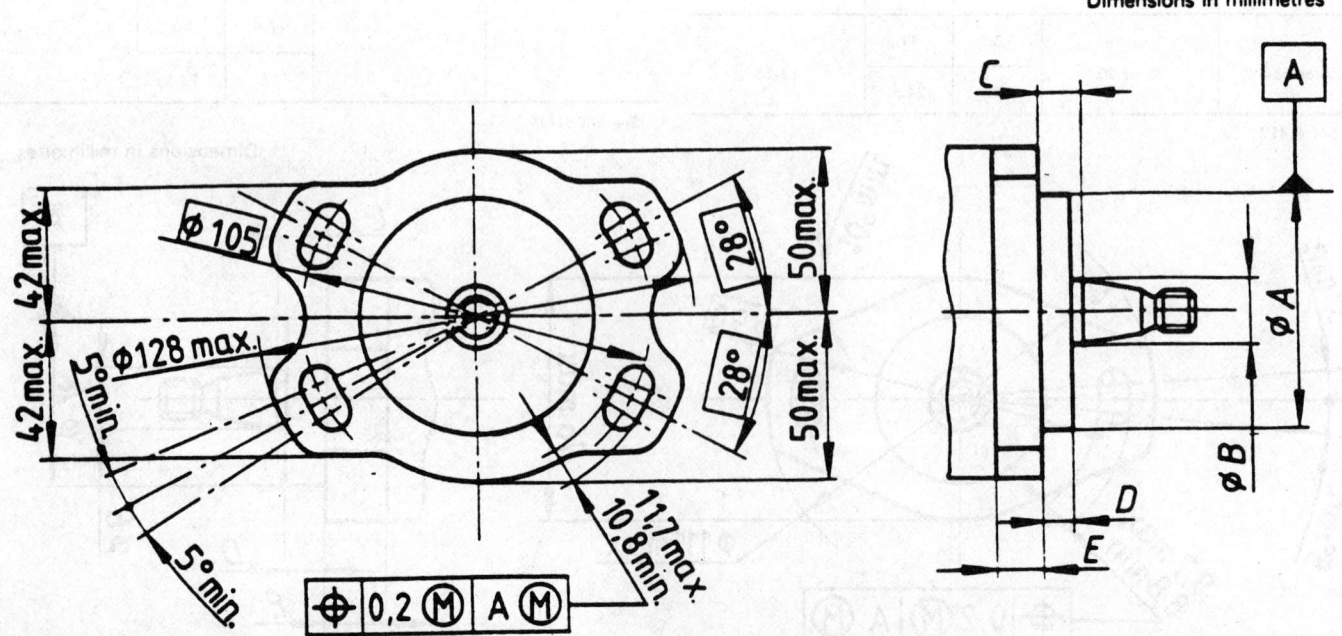

FIG. 3—FUEL INJECTION PUMPS—TYPE 3 END-MOUNTING FLANGE

3.2.4 Type 4 Flange—See Fig. 4 and Table 4.
3.2.5 Type 5 Flange—See Fig. 5 and Table 5.
3.2.6 Type 6 Flange—See Fig. 6 and Table 6.

3.2.7 Type 7 Flange—See Fig. 7 and Table 7.
3.2.8 Type 8 Flange—See Fig. 8 and Table 8.

TABLE 4
Dimensions in millimetres

A g8	B[a] nominal	C ±0.5	D max	E
68	17	9.5	8	10-16

[a] See SAE J1419.

TABLE 5
Dimensions in millimetres

A g8	B[a] nominal	C ±0.5	D max	E
74 or 76	17	9.5	8	8-10

[a] See SAE J1419.

TABLE 6
Dimensions in millimetres

A g8	B[a] nominal	C ±1	D max	E
68 or 97 or 112	20 or 22	4.5	7.5	17-18 or 24-26

[a] See SAE J1419.

TABLE 7
Dimensions in millimetres

A g8	B[a] nominal	C ±0.5	D max	E
68 or 85	17 or	4.5	8.2	15-18
	20 or	9.5	8.2	
	22 or 25	12.5	11	
		17.4	16	

[a] See SAE J1419.

TABLE 8
Dimensions in millimetres

A g8	B[a] nominal	C ±0.5	D max	E
95 or 107	20 or	9.5	8.2	17-20
	25 or 30	13.5	12	
	or 35	18.5	17	

[a] See SAE J1419.

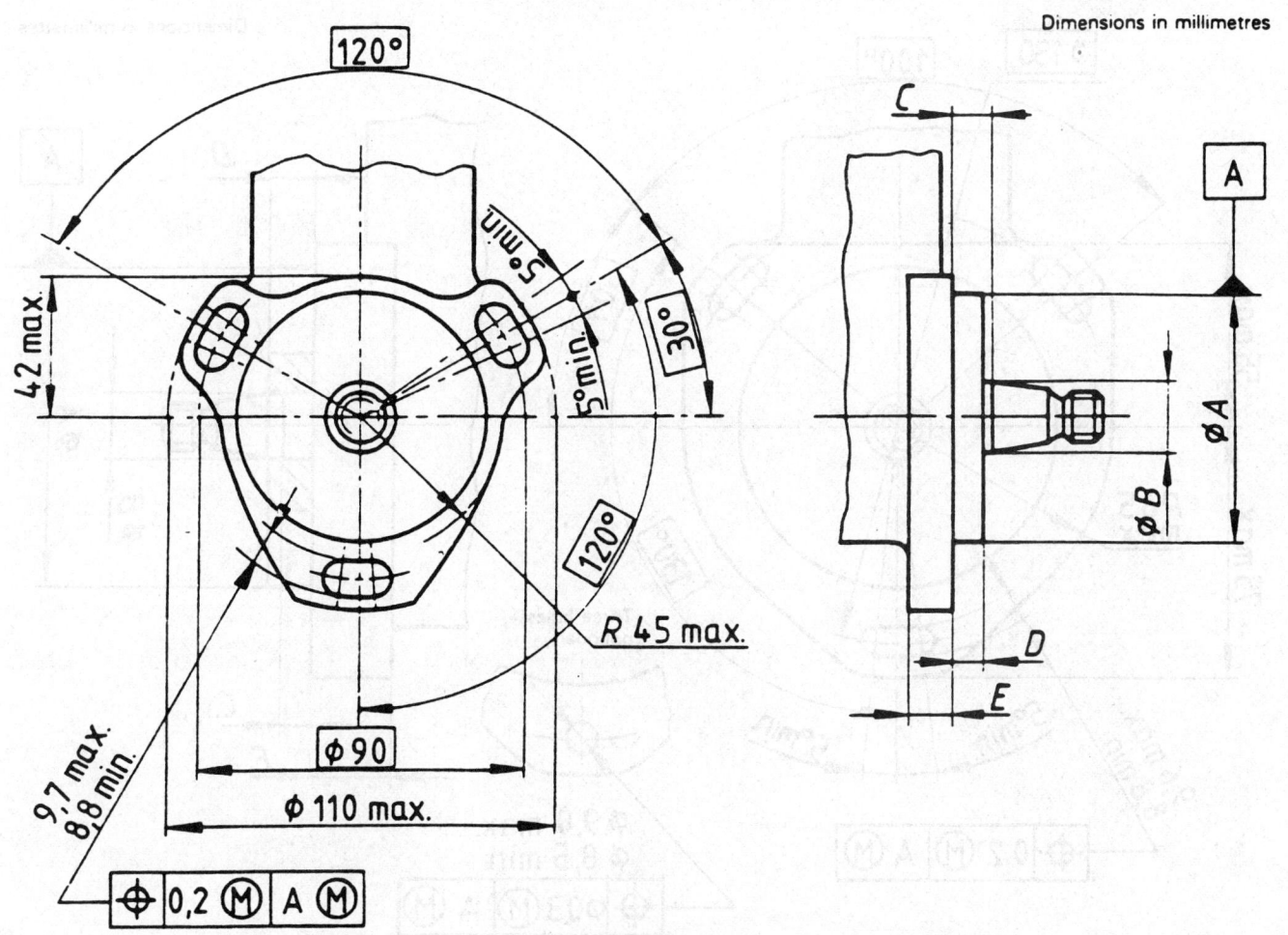

FIG. 4—FUEL INJECTION PUMPS—TYPE 4 END-MOUNTING FLANGE

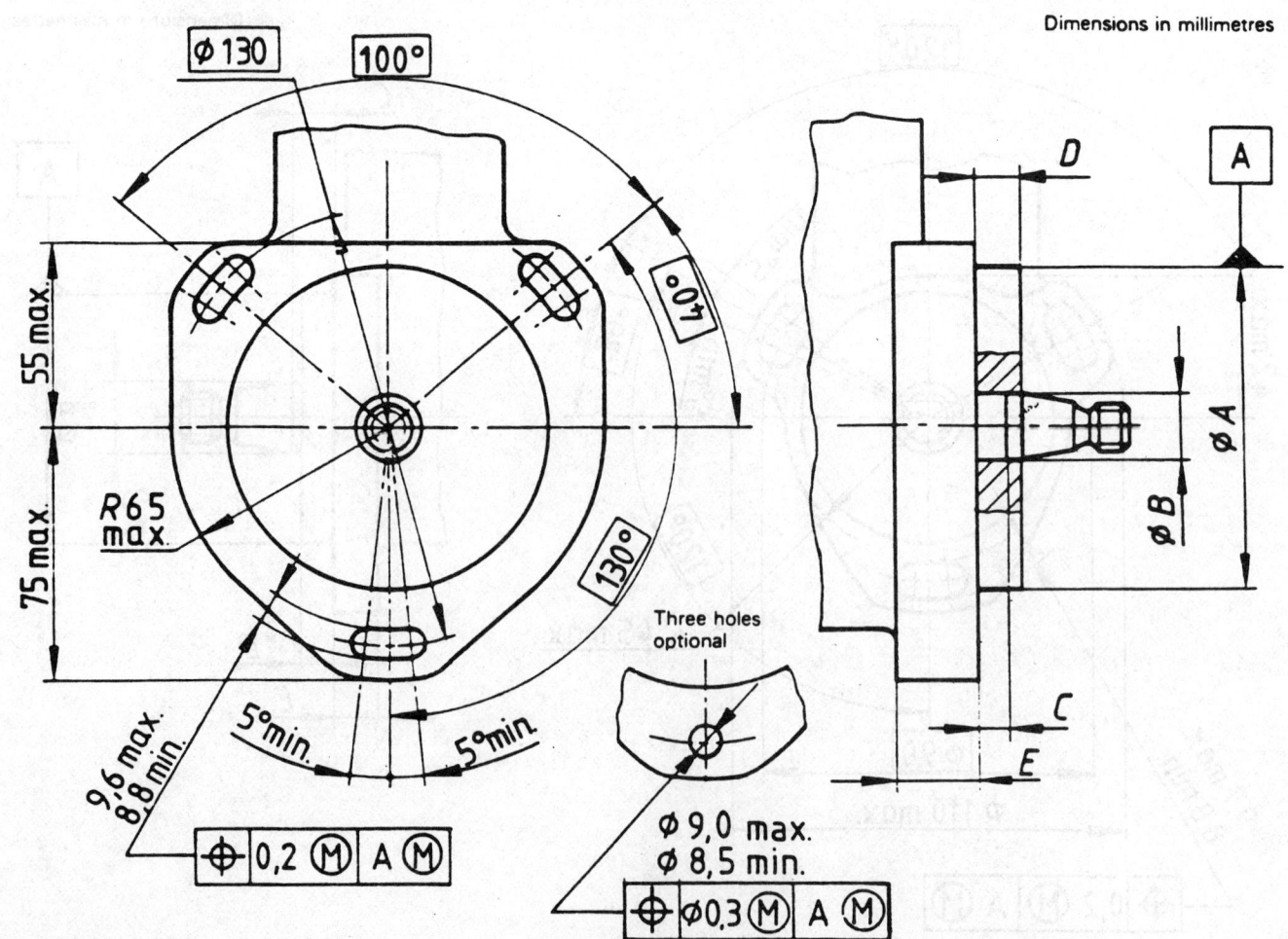

FIG. 5—FUEL INJECTION PUMPS—TYPE 5 END-MOUNTING FLANGE

FIG. 6—FUEL INJECTION PUMPS—TYPE 6 END-MOUNTING FLANGE

26.301

Dimensions in millimetres

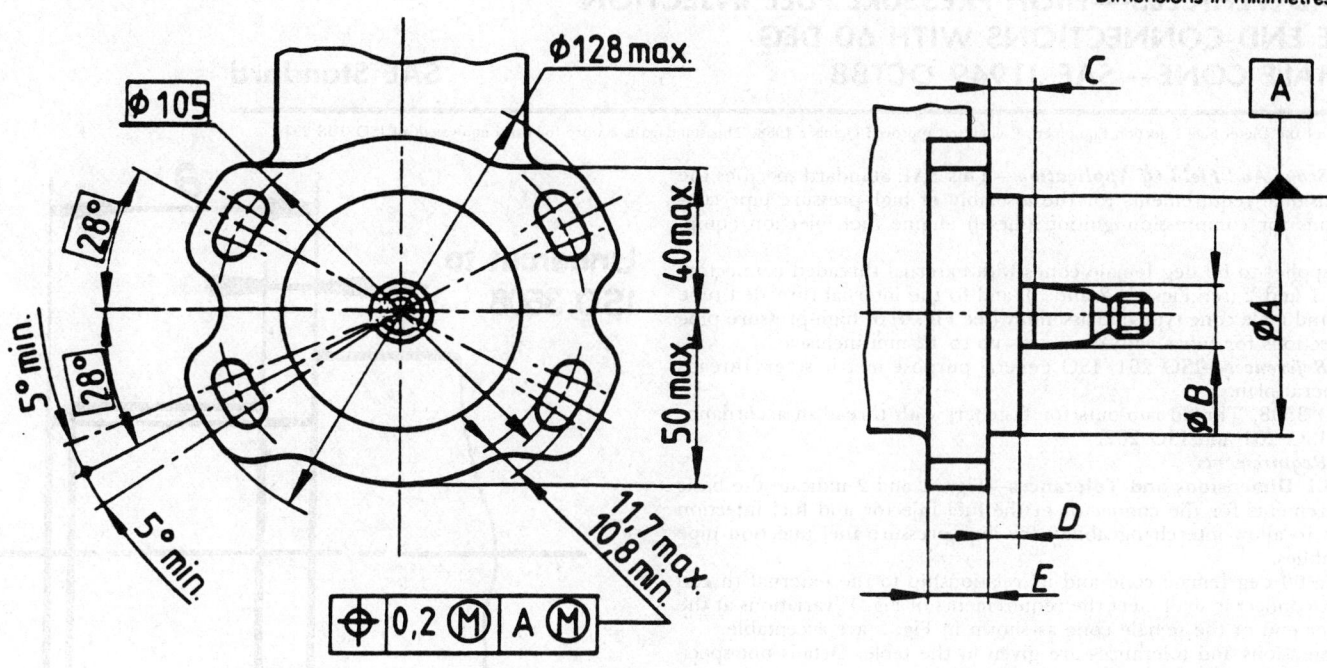

FIG. 7—FUEL INJECTION PUMPS—TYPE 7 END-MOUNTING FLANGE

Dimensions in millimetres

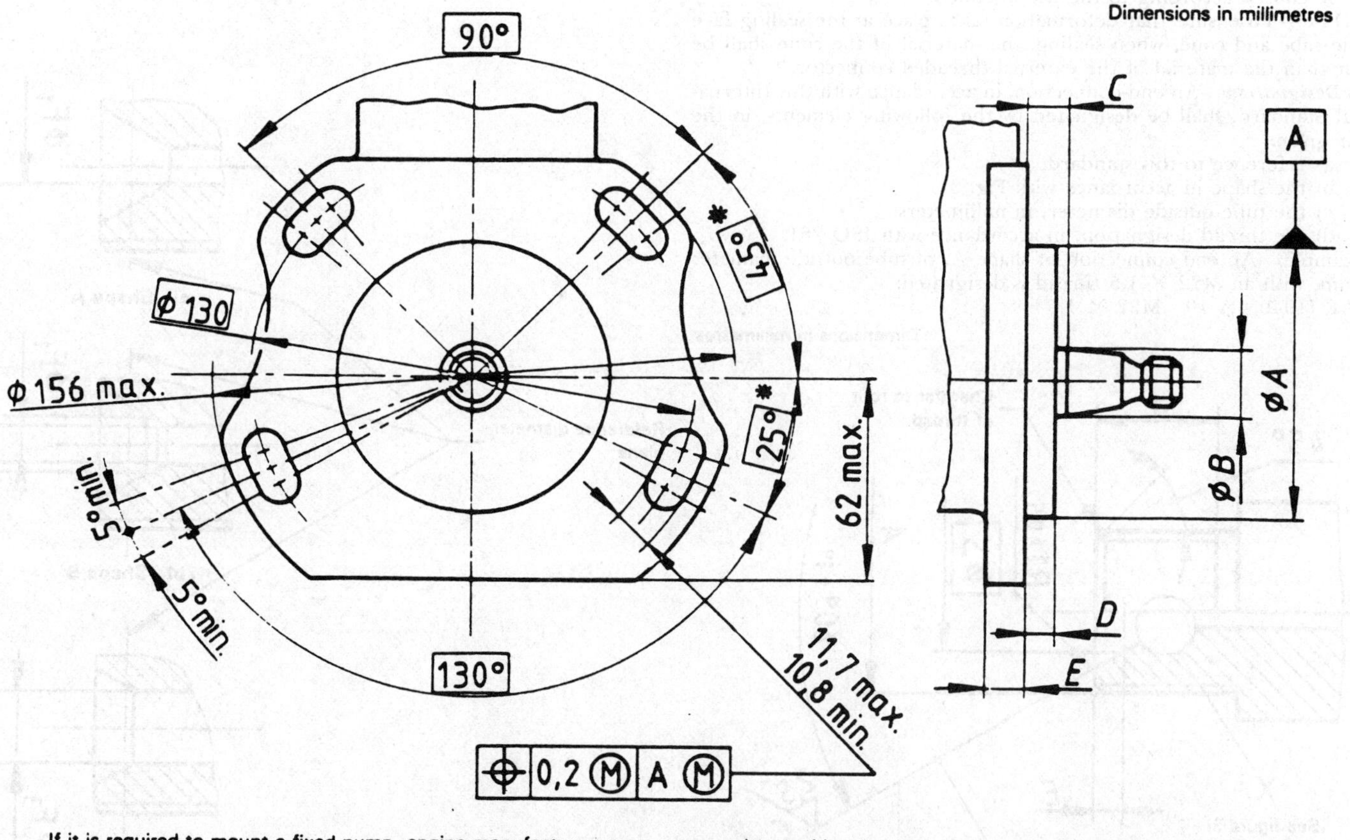

If it is required to mount a fixed pump, engine manufacturers are encouraged to position the engine studs at true (theoretically correct) angles of 40° above and 20° below the horizontal centreline.

FIG. 8—FUEL INJECTION PUMPS—TYPE 8 END-MOUNTING FLANGE

ROAD VEHICLES—HIGH PRESSURE FUEL INJECTION PIPE END-CONNECTIONS WITH 60 DEG FEMALE CONE—SAE J1949 OCT88

SAE Standard

Report of the Diesel Fuel Injection Equipment Committee approved October 1988. This standard is a word-for-word equivalent of ISO/DIS 2974.

1. Scope And Field Of Application—This SAE Standard specifies the dimensional requirements for the assembly of high-pressure pipe connections for compression-ignition (diesel) engine fuel injection equipment.

It applies to 60 deg female cones with external threaded connectors types 1 and 2 (see Figs. 1, 2 and 3), and to the internal threaded tube nuts and male cone type end assembly (see Fig. 4) of high-pressure pipe connections for tubes with diameters up to 12 mm inclusive.

2. References—ISO 261, ISO general purpose metric screw threads - General plan.

ISO 3508, Thread run-outs for fasteners with thread in accordance with ISO 261 and ISO 262.

3. Requirements

3.1 Dimensions and Tolerances—Figs. 1 and 2 indicate the basic requirements for the connector at the fuel injector and fuel injection pump to allow interchangeability for high-pressure fuel injection pipe assemblies.

The 60 deg female cone and its relationship to the external thread of the connector shall meet the requirements of Fig. 1: variations at the smaller end of the female cone as shown in Fig. 3 are acceptable.

Dimensions and tolerances are given in the table. Details not specified are left to the manufacturer's choice.

With reference to dimension a in Figs. 1 and 2, the external thread may be made optionally to type 1 or 2. However, it shall be possible to screw the GO-gauge for the thread up to the plane specified by dimension a for both types.

Fig. 4 identifies the nut and tube assembly dimensions which are important to sealing, normally on the leading edge of the cone on the high-pressure fuel injection pipe (see also the note in 3.2).

3.2 Materials—The specification of material and heat treatment shall be chosen according to the use intended.

NOTE—To ensure that deformation takes place at the sealing face of the tube and cone, when sealing, the material of the cone shall be softer than the material of the external threaded connector.

4. Designation—An end-connection, in accordance with this International Standard, shall be designated by the following elements, in the order given:

a) reference to this standard;
b) the shape in accordance with Fig. 3;
c) the tube outside diameter, in millimeters;
d) the thread designation, in accordance with ISO 261.

Example: An end connection of shape A, of tube outside diameter 10 mm, with an M22 × 1.5 thread is designated:

SAE J1949 - A 10 - M22 × 1.5

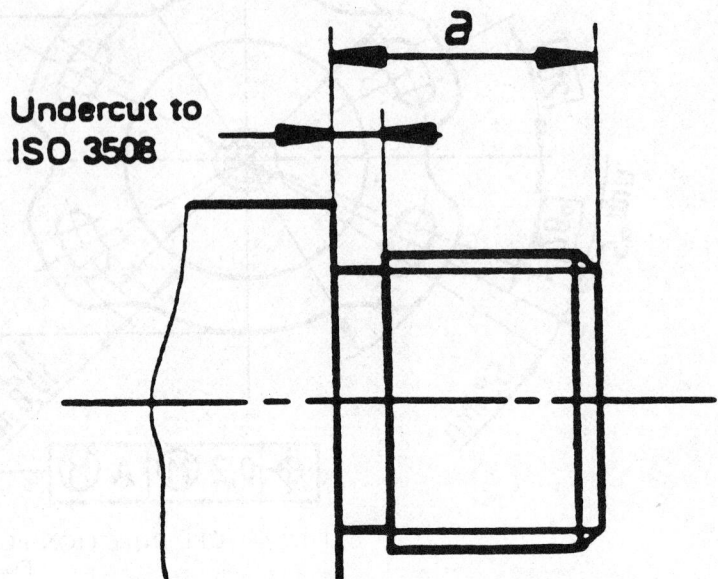

FIG. 2—EXTERNAL THREADED CONNECTOR, TYPE 2

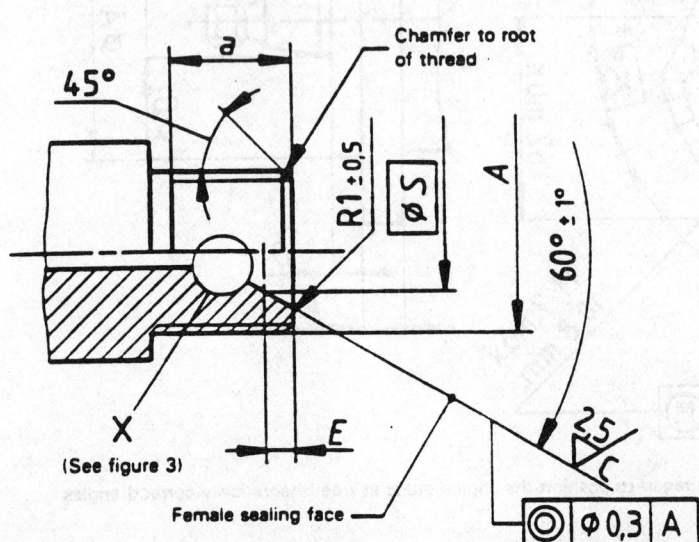

FIG. 1—EXTERNAL THREADED CONNECTOR, TYPE 1

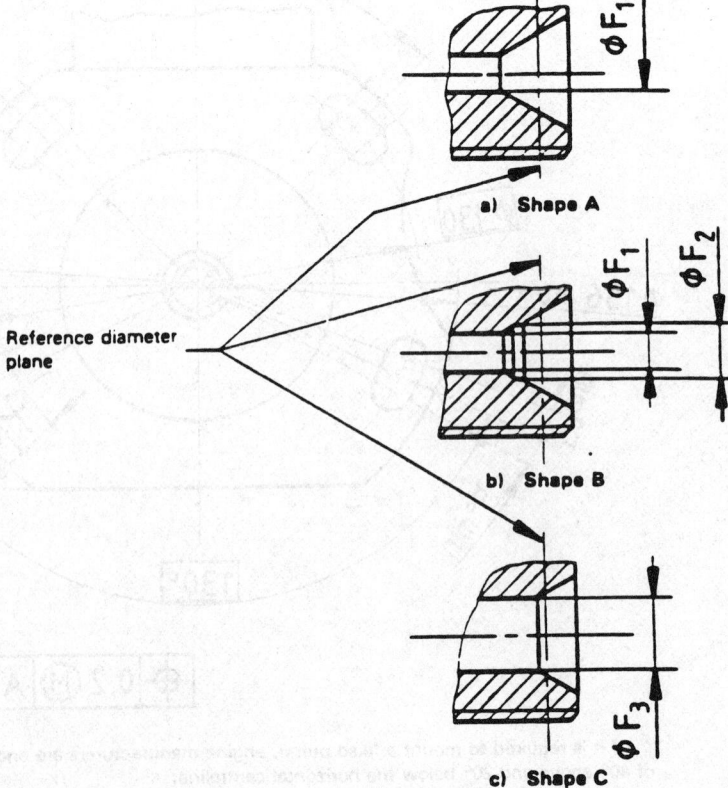

FIG. 3—SHAPES OF DETAIL X ON EXTERNAL THREADED CONNECTORS, TYPES 1 AND 2

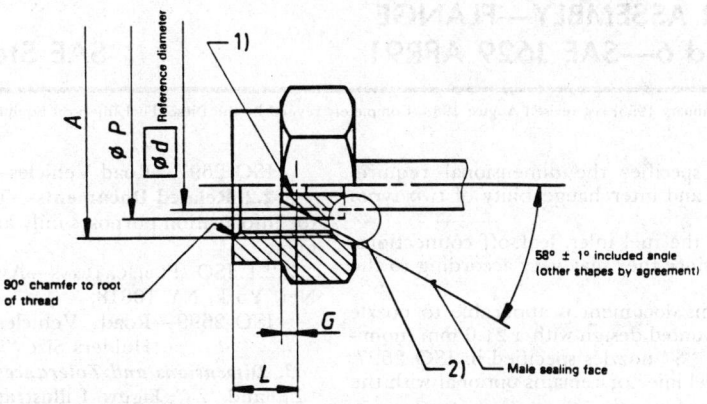

1) The tube end bore entrance configuration shall be so chosen that, after final assembly, the inside diameter of the tube is not reduced.
2) Design of the sealing cone shoulder and the cap nut shall be agreed between customer and manufacturer.

FIG. 4—NUT AND TUBE END ASSEMBLY

TABLE 1

Tube outside diameter	Thread[1] A	Reference diameter, d	F_1[2] ±0.1	F_2 +0.3 0	F_3[2] max.	E +0.3 0	P ±0.5	G +0.5 0	L max.	a min.
5	M10 × 1.25 M12 × 1.5	5.5	1.25 to 2.5	2.5		0.8	7.5	0.5	7	10
6	M12 × 1.5 M14 × 1.5	6.5	1.5 to 3	5	6.1	0.8	9.0	0.8	8	11
8	M16 × 1.5 M18 × 1.5 M22 × 1.5	8.5	2 to 4	4	7.3	2.6	11.5	0.9	11	16.5
10	M20 × 1.5 M22 × 1.5 M24 × 1.5	10.5	2.5 to 5	5	9.3	2.6	13.5	0.9	12.5	18
12	M22 × 1.5 M26 × 1.5	12.5	3 to 5	5	10.3	2.6	15.5	1.8	15.5	21

1) Tolerance classes of threads. 6g for external threaded connector: 6H for connector nuts.
2) Dimension F shall be adapted to the tube inner diameter for the sake of optimum flow conditions. If required, for instance for edge filters, application of dimension F_3 is allowed.

DIESEL FUEL INJECTOR ASSEMBLY—FLANGE MOUNTED TYPES 5 and 6—SAE J629 APR91

SAE Standard

Report of the Engine Committee, approved January 1953, last revised August 1983. Completely revised by the Diesel Fuel Injection Equipment Standards Committee 2 April 1991.

1. Scope—This SAE Standard specifies the dimensional requirements necessary for the mounting and interchangeability of two types of fuel injectors in diesel engines.

The location and dimensions of the fuel inlet, leak-off connections, and flange design are not defined since they may vary according to the particular application.

1.1 Field of Application—This document is applicable to nozzle holder types 5 and 6 of a flange mounted design with a 21.0 mm (nominal) shank diameter used with size "S" nozzles specified in ISO 2697. The internal construction of the fuel injector remains optional with the manufacturer.

2. References

2.1 Applicable Documents—The following publications form a part of this publication to the extent specified herein. The latest issue of SAE publications shall apply.

2.1.1 ISO Publications—Available from ANSI, 11 West 42nd Street, New York, NY 10036.

ISO 2697—Road Vehicles—Fuel Injection Nozzles—Size "S"

2.2 Related Documents—The following publications are provided for information purposes only and are not a required part of this document.

2.2.1 ISO Publications—Available from ANSI, 1430 Broadway, New York, NY 10018.

ISO 2699—Road Vehicles—Flange-Mounted Injection Nozzle Holders Size "S"—Types 2, 3, 4, 5, and 6

3. Dimensions and Tolerances—With the aid of detail enlargements "Z_1" and "Z_2", Figure 1 illustrates the length and diameters of the nozzle, sealing washer, nozzle retaining nut, and the nozzle holder as related to the interface between the injector and the bore in the engine.

Note that two basic nozzle types are shown: (A) hole-type and (B) pintle.

Dimensions and requirements not given in this document are left to the discretion of the manufacturer.

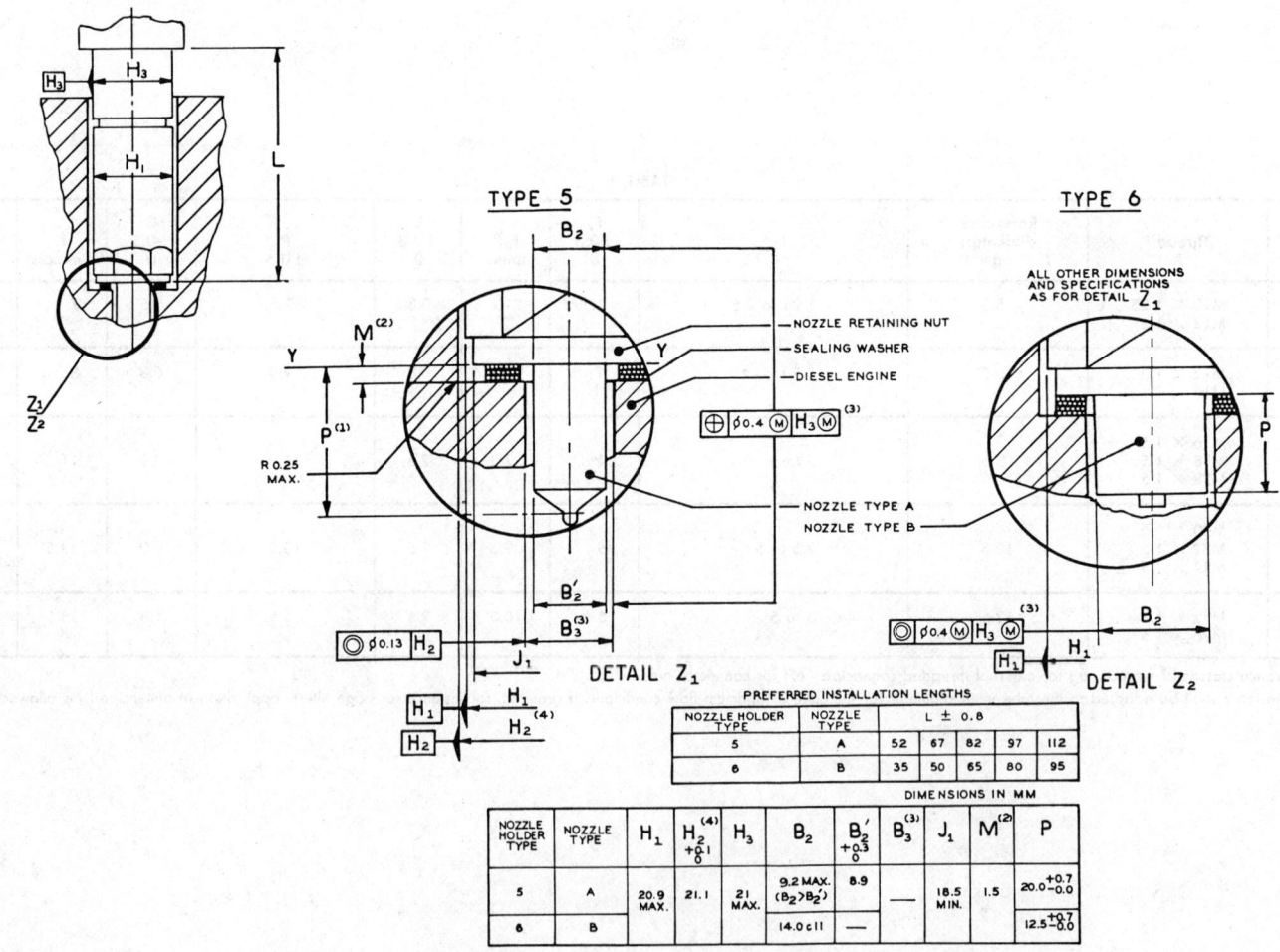

PREFERRED INSTALLATION LENGTHS

NOZZLE HOLDER TYPE	NOZZLE TYPE	L ± 0.8				
5	A	52	67	82	97	112
6	B	35	50	65	80	95

DIMENSIONS IN MM

NOZZLE HOLDER TYPE	NOZZLE TYPE	H_1	$H_2^{(4)}$ +0.1 / 0	H_3	B_2	B_2' +0.3 / 0	$B_3^{(3)}$	J_1	$M^{(2)}$	P
5	A	20.9 MAX.		21.1	21 MAX.	9.2 MAX. ($B_2 \geq B_2'$)	8.9	—	18.5 MIN.	1.5	20.0 +0.7/-0.0
6	B					14.0 c11	—				12.5 +0.7/-0.0

(1) Y-Y and the center of the nozzle tip radius on the nozzle axis which is generally the apex of orifice spray.

(2) With commercial tolerances (before compression).

(3) The determination of the diameter B_3 in the cylinder head is left to the manufacturer's choice. For that purpose the maximum value for the nozzle shank which is given as a result of the Maximum Material Principle Ⓜ and the maximum tolerance value of the cylinder head hole must be taken into account. The clearance shall be kept to a minimum to facilitate nozzle cooling.

(4) For type 5 and 6 nozzle holders without shanks, dimension H_2 should be reduced by 0.1mm. In this case the Maximum Material Principle Ⓜ in details Z_1 and Z_2 applies on diameter H_1, instead of diameter H_3.

FIGURE 1—FLANGE-MOUNTED NOZZLE HOLDER SIZE "S", TYPES 5 AND 6

DIESEL FUEL INJECTOR ASSEMBLY TYPE 7 (9.5 mm)—SAE J347 JUL88

SAE Standard

Report of the Engine Committee approved July 1968 and completely revised July 1988. This standard is equivalent to ISO 2698.

1. Scope—This standard specifies the dimensional requirements necessary for the mounting and interchangeability of a type of fuel injector used in diesel engines.

2. Field of Application—This standard is applicable to a 9.5 mm (nominal) shank diameter clamp mounted fuel injector and deals only with the interface between the injector and the engine. The internal construction of the fuel injector remains optional with the manufacturer.

3. Description—Fig. 1 illustrates the shank length "L". Dimensions of the fuel inlet, leak-off connections, and method of hold-down are optional and, therefore, are not shown.

Fig. 2 illustrates the engine bore details.

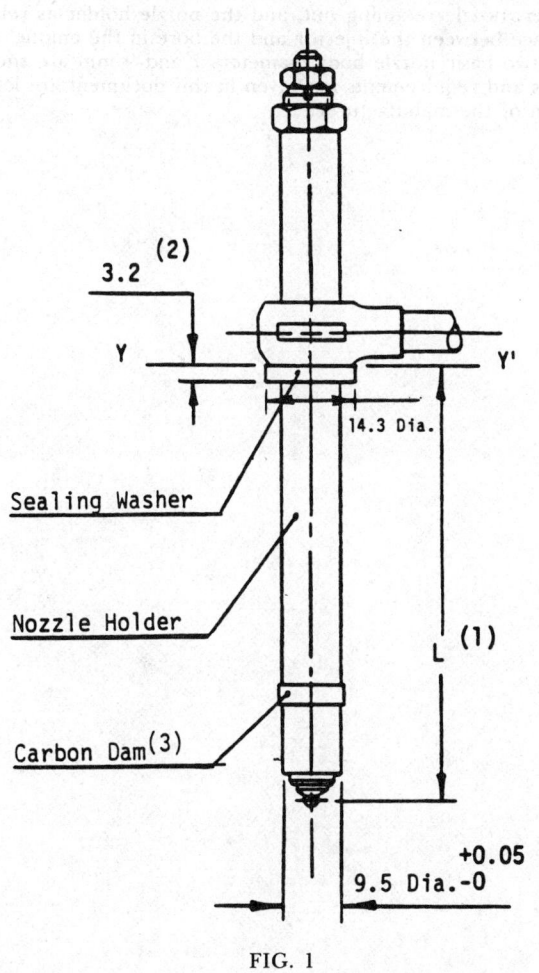

FIG. 1

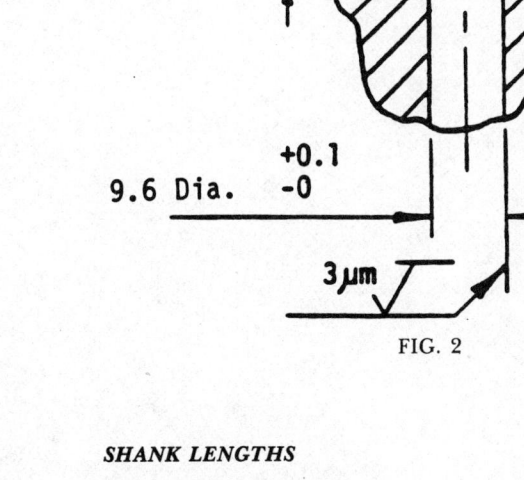

FIG. 2

1. This dimension determines the distance between the reference plane Y-Y' and the point of intersection of nozzle orifice with the fuel injector axis;
2. With commercial tolerances, before compression;
3. Carbon Dam O.D. to provide radial seal with engine bore;

SHANK LENGTHS

(1) L ± 0.3
Variable

(R) DIESEL FUEL INJECTOR ASSEMBLY —TYPES 8, 9, 10, AND 11 —SAE J265 APR91

SAE Standard

Report of the Engine Committee, approved September 1971, completely revised August 1983. Completely revised by the SAE Diesel Fuel Injection Equipment Standards Committee 2 April 1991.

1. Scope—This SAE Standard specifies the dimensional requirements necessary for the mounting and interchangeability of four types of fuel injectors in diesel engines. Two of the types specified are flats-located injectors.

The location and dimensions of the fuel inlet, leak-off connections, and type of attachment are not defined since they may vary according to the particular application.

1.1 Field of Application—This document is applicable to nozzle holder types 8 and 10 of an unspecified means of angular location and flats-located types 9 and 11 with a 17.0 mm (nominal) shank diameter. The internal construction of the fuel injector remains optional with the manufacturer.

2. References

2.1 Related Publications—The following publications are provided for information purposes only and are not a required part of this document.

ISO 3539—Road Vehicles—Injection Nozzle Holder with body, types 8 and 10, and injection nozzle holder with fixing flats, types 9 and 11

3. Dimensions and Tolerances—With the aid of detail enlargement "Z", Figures 1 and 2 illustrate the length and diameters of the nozzle, sealing washer, nozzle retaining nut, and the nozzle holder as related to the interface between the injector and the bore in the engine.

Note that two basic nozzle body diameters 7 and 9 mm are shown.

Dimensions and requirements not given in this document are left to the discretion of the manufacturer.

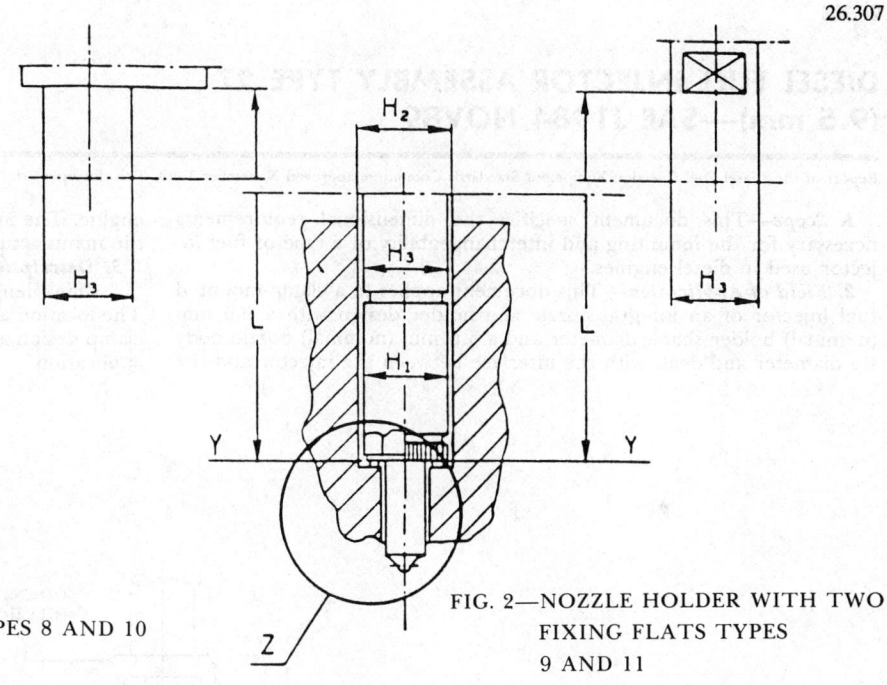

FIG. 2—NOZZLE HOLDER WITH TWO FIXING FLATS TYPES 9 AND 11

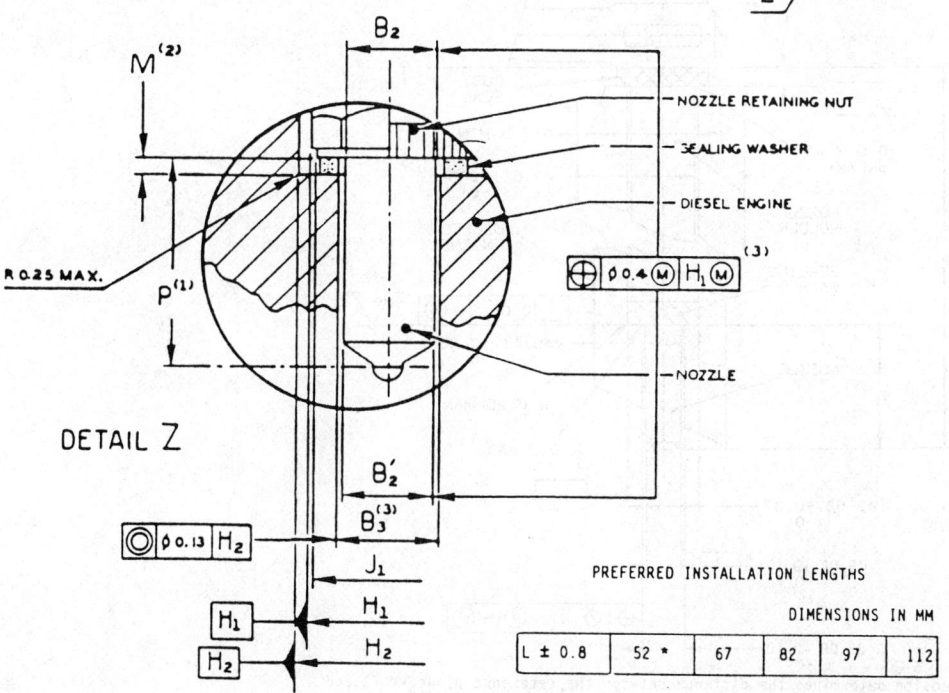

FIG. 1—NOZZLE HOLDER WITH BODY TYPES 8 AND 10

PREFERRED INSTALLATION LENGTHS

DIMENSIONS IN MM

| L ± 0.8 | 52 * | 67 | 82 | 97 | 112 |

* APPLIES TO TYPES 8 AND 10 ONLY

DIMENSIONS IN MM

NOZZLE HOLDER TYPES	H_1	H_2	H_3	B_2 ($B_2 > B_2'$)	B_2'	B_3 [3]	J_1	M [2]	P [1]
8 AND 9	17.0 MAX	17.1 $^{+0.1}_{0}$	16.9 MAX	9.2 MAX	8.9 $^{+0.3}_{0}$	---	14.5 MIN	1.5 NOM	20.0 $^{+0.7}_{0}$
10 AND 11				7.2 MAX	6.9 $^{+0.3}_{0}$				

[1] Y-Y and the center of the nozzle tip radius on the nozzle axis which is generally the apex of orifice spray.
[2] With commercial tolerances (before compression).
[3] The determination of the diameter B_3 in the cylinder head is left to the manufacturer's choice. For that purpose the maximum value for the nozzle shank which is given as a result of the Maximum Material Principle Ⓜ and the maximum tolerance value of the cylinder head hole must be taken into account. The clearance shall be kept to a minimum to facilitate nozzle cooling.

DIESEL FUEL INJECTOR ASSEMBLY TYPE 27 (9.5 mm)—SAE J1984 NOV89

SAE Standard

Report of the Diesel Fuel Injection Equipment Standards Committee approved November 1989. This document references ISO 2698.

1. Scope—This document specifies the dimensional requirements necessary for the mounting and interchangeability of a type of fuel injector used in diesel engines.

2. Field of Application—This document applies to a clamp-mounted fuel injector of an integral nozzle and holder design with a 9.5 mm (nominal) holder shank diameter and a 5.5 mm (nominal) nozzle body tip diameter and deals with the interface between the injector and the engine. The internal construction of the injector remains optional with the manufacturer.

3. Description—Fig. 1 illustrates interface dimensions and length "L" shank length plus the tip protrusion dimension from the holder. The location and dimensions of the fuel inlet, leak-off connections, and clamp design are not defined since they vary according to the particular application.

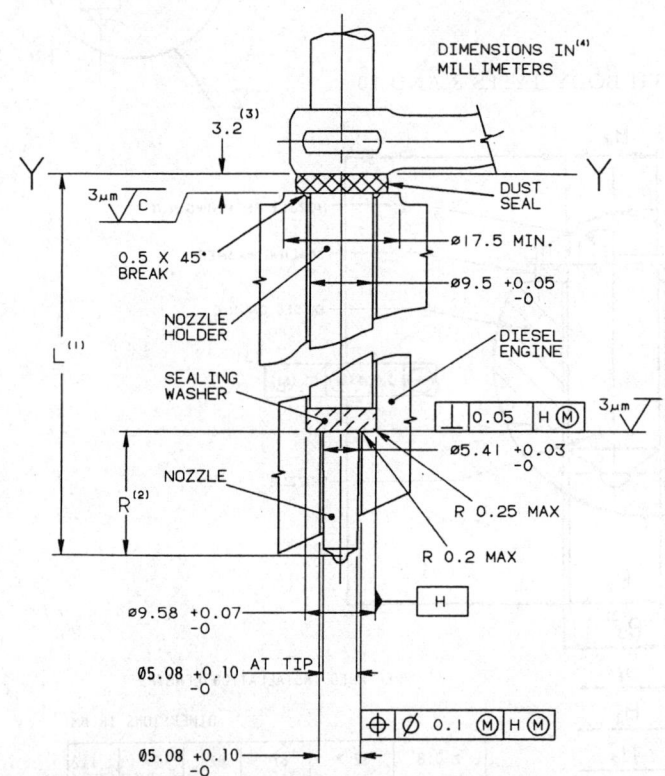

[1] This dimension determines the distance between the reference plane (Y-Y) and the point of intersection of the nozzle orifices with the fuel injector axis.

[2] This dimension determines the distance between the compressed sealing washer and the point of intersection of the nozzle orifices with the fuel injector axis and varies with nozzle sac design.

[3] With commercial tolerances.

[4] Dimensions are conversions from existing English units.

FIG. 1—DIMENSIONS AND TOLERANCES

(R) DIESEL INJECTION PUMP TESTING—PART 1: CALIBRATING NOZZLE AND HOLDER ASSEMBLIES—SAE J968/1 MAY91

SAE Recommended Practice

Report of the Engine Committee, approved August 1966, last revised November 1975, reaffirmed without change March 1985. Completely revised by the SAE Diesel Fuel Injection Equipment Standards Committee May 1991. This SAE Recommended Practice is similar to ISO 7440/1.

Foreword—This SAE Recommended Practice is similar to ISO 7440/1.

The calibrating nozzle and holder assemblies are intended to simulate closely the function of the nozzle and holder assemblies in the fuel injection system of a diesel engine. They are flow gauges and require careful handling and maintenance.

The manufacturer of the injection equipment and/or the manufacturer of the engine shall specify the type of calibrating nozzle and holder assembly, the appropriate single hole orifice plate size or pintle nozzle (as applicable), high pressure pipes, exact limits, etc. to be used.

1. Scope—This part of SAE J968 specifies two types of calibrating nozzle and holder assemblies intended for the testing and setting of diesel injection pumps on test benches.

It applies to:
a. A calibrating nozzle and holder assembly with a single hole orifice plate;
b. A calibrating nozzle and holder assembly with a delay pintle type nozzle.

The approximate range of the calibrating nozzle and holder assembly is up to:
a. 300 mm^3/stroke with the single hole orifice plate;
b. 150 mm^3/stroke with the delay pintle type nozzle.

Setting and maintenance requirements are specified in ISO 4008/3.

2. References

2.1 Applicable Documents—The following publications form a part of this specification to the extent specified herein. The latest issue of SAE publications shall apply.

2.1.1 SAE PUBLICATIONS—Available from SAE, 400 Commonwealth Dr., Warrendale, PA 15096-0001.

SAE J967—Calibration Fluid for Diesel Injection Equipment
SAE J968/2—Orifice Plate Flow Measurement

2.1.2 ISO PUBLICATIONS—Available from ANSI, 11 West 42nd St., New York, NY 10036.

ISO 2697 Road vehicles—Fuel injection nozzles—Size "S"
ISO 2699 Road vehicles—Flange-mounted injection nozzle holders size "S",—Types 2, 3, 4, 5 and 6
ISO 4008/3 Road vehicles—Fuel injection pump testing—Part 3: Application and test procedures
ISO 4010 Road vehicles—Calibrating nozzle, delay pintle type
ISO 4093 Road vehicles—Fuel injection pumps—High pressure pipes for testing
ISO 7440/1 Road vehicles—Fuel injection equipment testing—Part 1: Calibrating nozzle and holder assemblies

3. Required Characteristics—The functionally critical dimensions are shown in the following diagrams:

3.1 Calibrating Nozzle and Holder Assemblies
With single hole orifice plate, Figure 1.
With delay pintle type nozzle, Figure 2.

3.2 Components
Holder body with and without vent, Figure 3.
Spring, standard, and low rate, Figure 4.
Inlet stud with edge filter, Figure 5.
Suggested edge filter flow measuring system, Figure 6.
NOTE—The outlet of the fixture (within the dotted lines) shall be at ambient pressure.
Needle valve assembly, Figure 7.
Single hole orifice plate and flow ranges, Figure 8.
Delay pintle type nozzle is specified in ISO 4010.
Distance sleeve, Figure 9.
Retaining nuts, Figures 10 and 11.

4. Designation

4.1 General Example—There are many possible combinations of nozzle, nozzle holder, nozzle retaining nut, and nozzle opening pressure. To facilitate identification of calibrating nozzle and holder assemblies, they shall be designated in code form as given by the following example:

SAE J968-53-117L

4.2 Explanation of the Designation—The designation comprises the following three elements:

a. 1st element: SAE J968
This specifies the SAE number where the calibrating nozzle and holder assemblies are defined.

b. 2nd element: ...53...
This two-digit code is specified in Tables 1 and 2 and defines the components of the calibrating nozzle and holder assembly. In this example they are:

5—(Tens-digit code): Orifice plate No. : 5
—3 (Units-digit code): Vented holder body, as specified in Figure 3 (execution 1), not provided with a spray damper (see Figures 1 and 2).

c. 3rd element: ...117L
The first three digits define the nozzle opening pressure setting in 10^5 pascals (bar). 'L' indicates the low rate spring. No letter is used for the standard rate spring.

The preferred opening pressures are shown below:
With orifice plate and standard rate spring: 207 + 3 bar
With 0.5 mm orifice plate and low rate spring: 117 + 3 bar
With delay pintle type nozzle, either: 172 + 3 bar
or: 147 + 3 bar
or: 125 + 3 bar

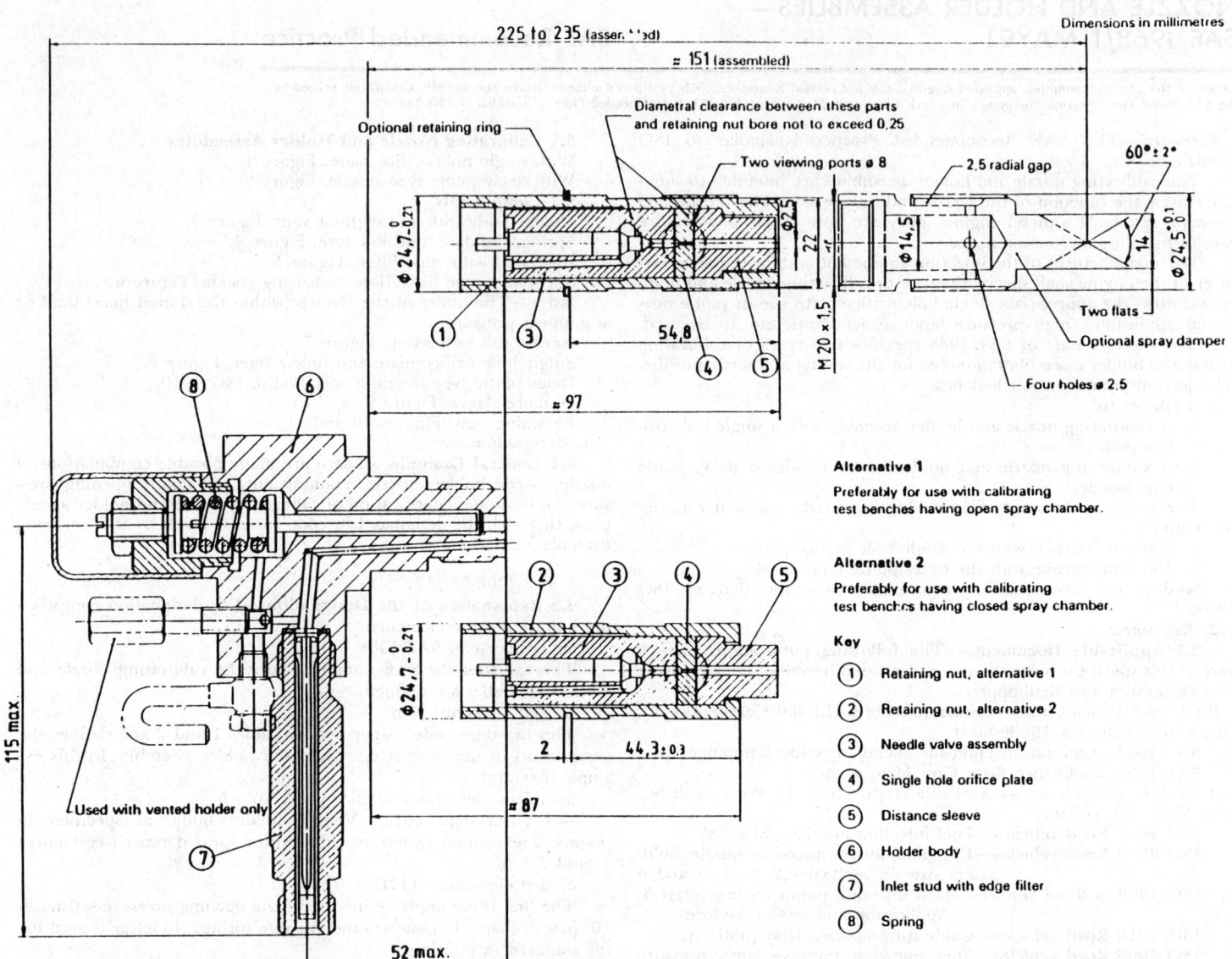

FIGURE 1—CALIBRATING NOZZLE AND HOLDER ASSEMBLY WITH SINGLE HOLE ORIFICE PLATE

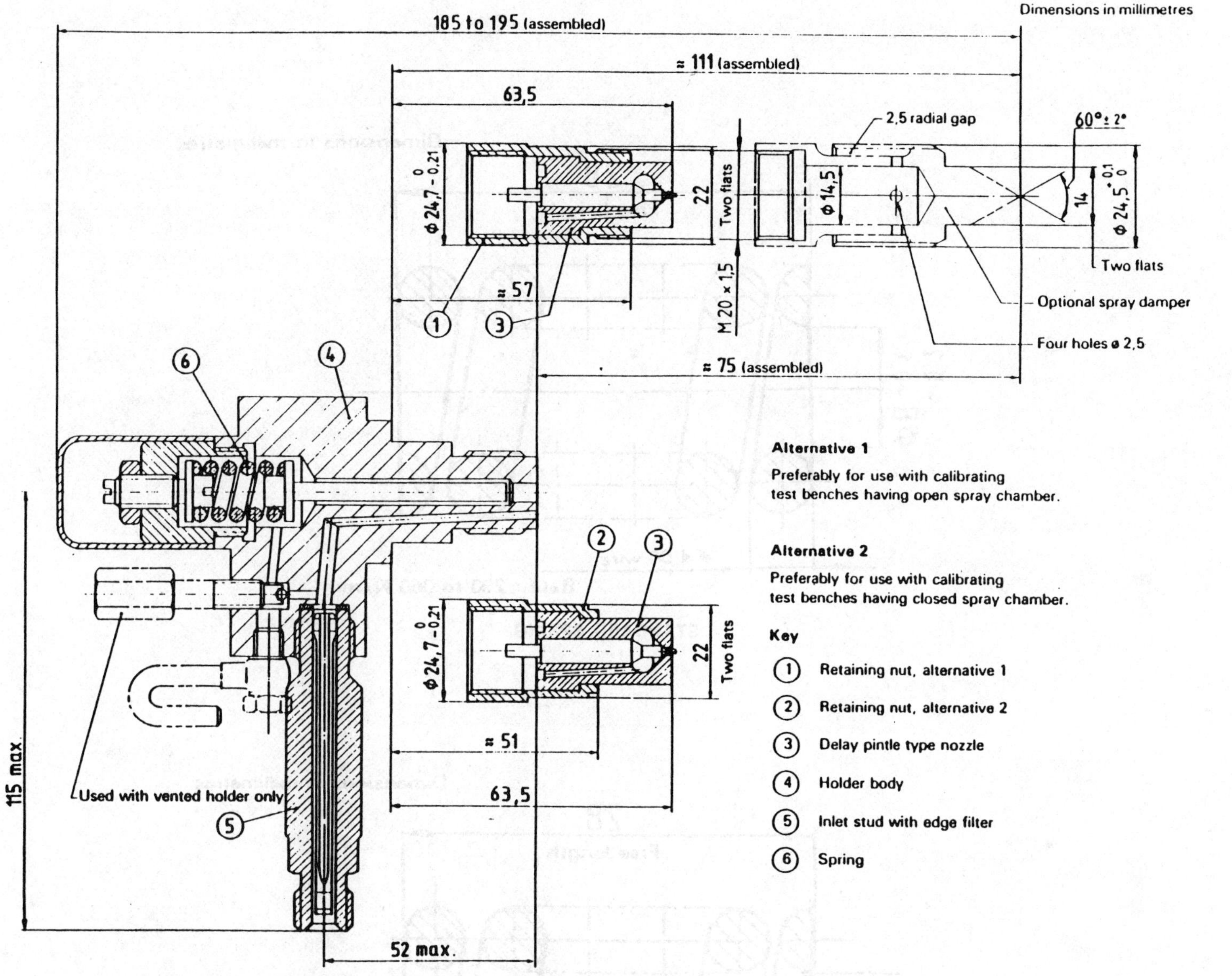

Dimensions in millimetres
185 to 195 (assembled)
≈ 111 (assembled)
63,5
2,5 radial gap
60° ± 2°
Two flats
Optional spray damper
Four holes ø 2,5
≈ 57
≈ 75 (assembled)
M 20 × 1,5
Alternative 1
Preferably for use with calibrating test benches having open spray chamber.
Alternative 2
Preferably for use with calibrating test benches having closed spray chamber.
Key
1 Retaining nut, alternative 1
2 Retaining nut, alternative 2
3 Delay pintle type nozzle
4 Holder body
5 Inlet stud with edge filter
6 Spring
≈ 51
63,5
115 max.
Used with vented holder only
52 max.

FIGURE 2—CALIBRATING NOZZLE AND HOLDER ASSEMBLY WITH DELAY PINTLE TYPE NOZZLE

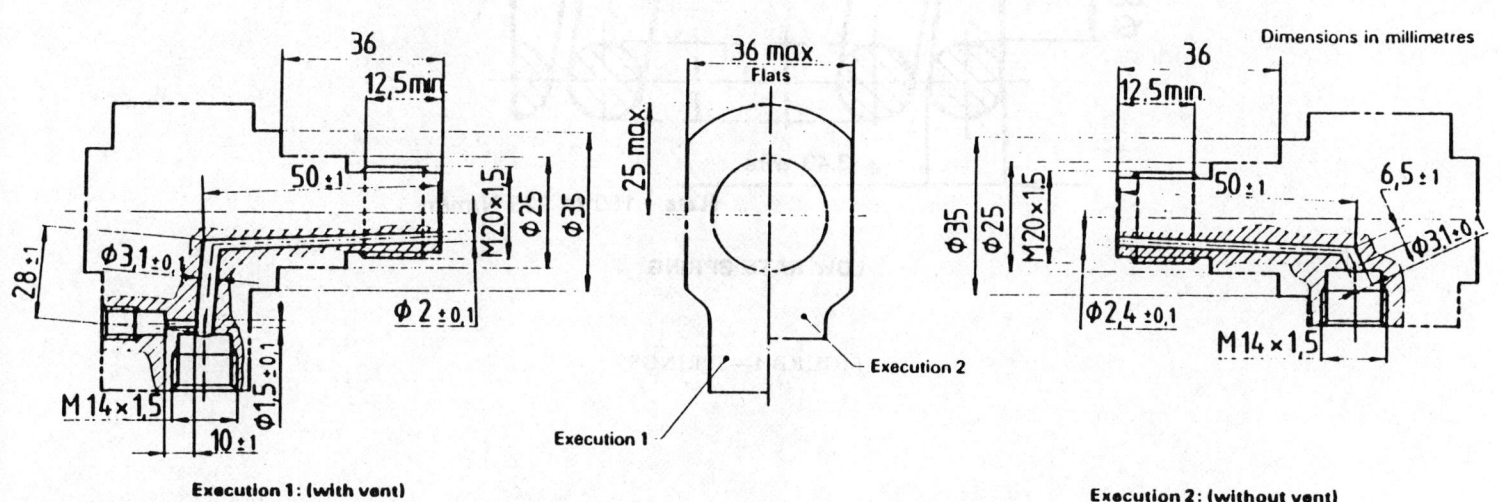

Dimensions in millimetres
36
12,5 min
50 ± 1
M20×1,5
Φ25
Φ35
Φ3,1 ± 0,1
28 ± 1
Φ 2 ± 0,1
Φ1,5 ± 0,1
M 14 × 1,5
10 ± 1
36 max
Flats
25 max.
Execution 1
Execution 2
6,5 ± 1
Φ31 ± 0,1
Φ2,4 ± 0,1
Execution 1: (with vent)
Execution 2: (without vent)

FIGURE 3—HOLDER BODY

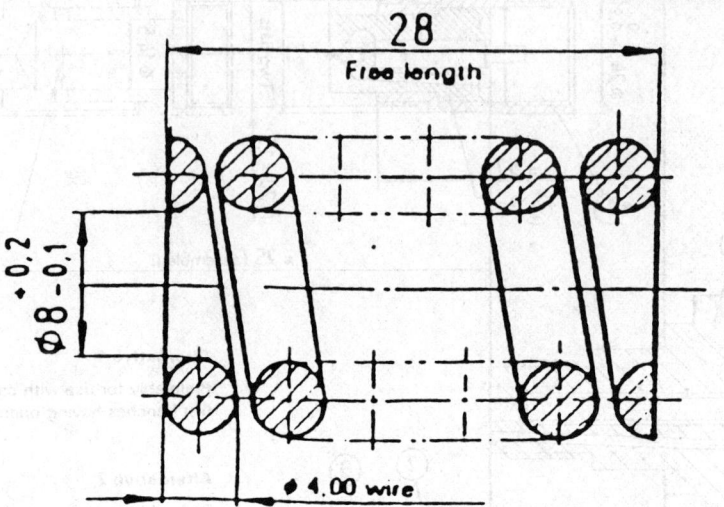

STANDARD SPRING

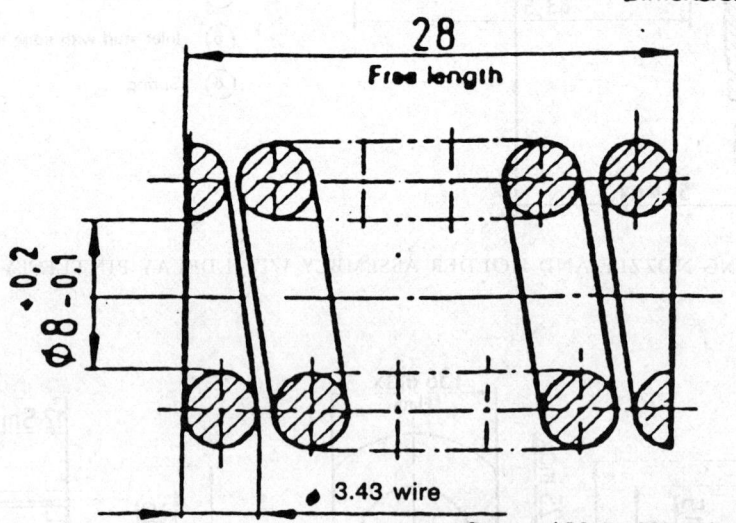

LOW RATE SPRING

FIGURE 4—SPRINGS

Dimensions in millimetres

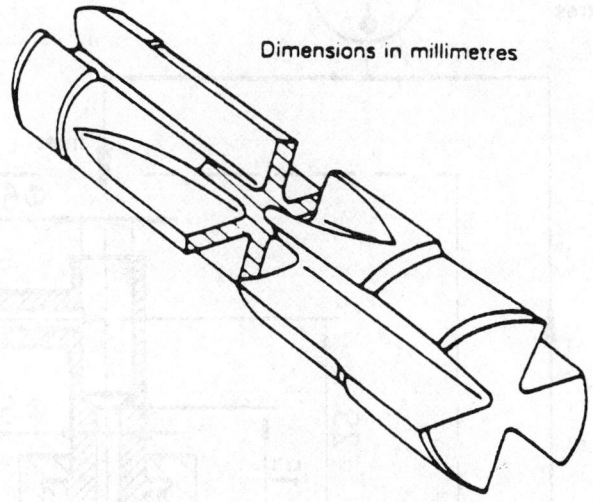

a) Edge filter with flutes (schematic)

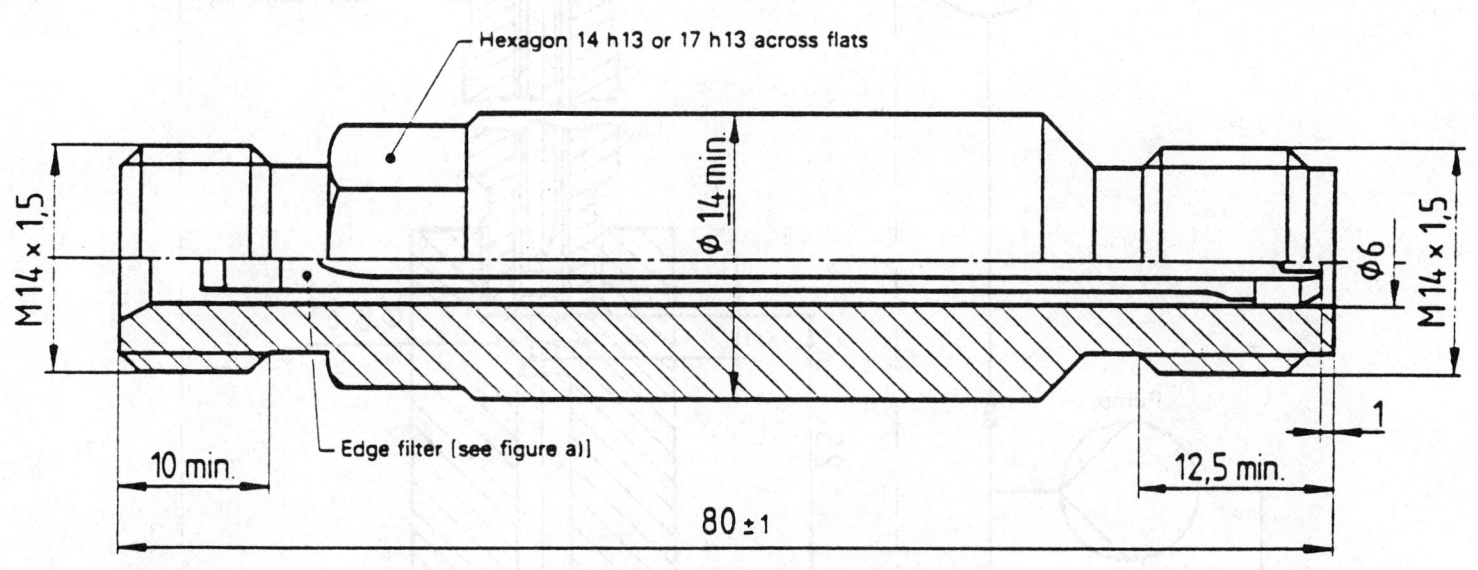

b) Inlet stud with edge filter

Characteristic	Value
Filter clearance (radial)	0,02 mm to 0,033 mm
Filter clearance area	7,3 mm² min.
Flow area for three flutes (see figure a)]	7,5 mm² min.
Internal volume	1 450 mm³ ± 40 mm³
Filter flow at 0,3 MPa (3 bar) pressure [1]	4 750 cm³/min to 6 500 cm³/min
Minimum flow of assembled holder body with inlet stud and edge filter at 0,3 MPa (3 bar) pressure [1]	2 000 cm³/min

[1] These tests shall be made using calibration fluid as specified in ISO 4113 at 40 °C ± 1 °C and the tolerance of the supply pressure shall be ± 0,03 MPa (0,3 bar).

FIGURE 5—CHARACTERISTIC DIMENSIONS AND VALUES OF INLET STUD WITH EDGE FILTER

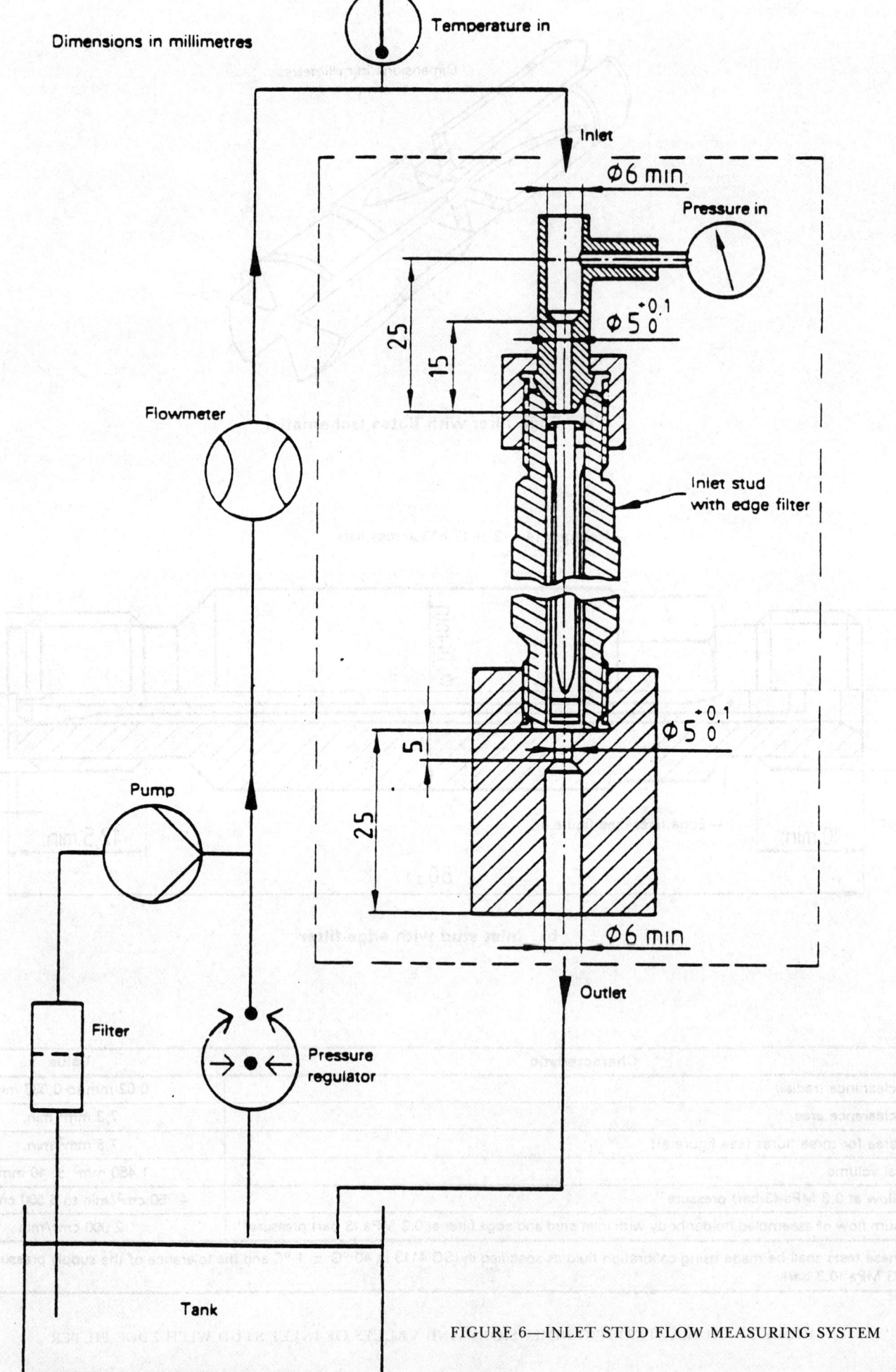

FIGURE 6—INLET STUD FLOW MEASURING SYSTEM

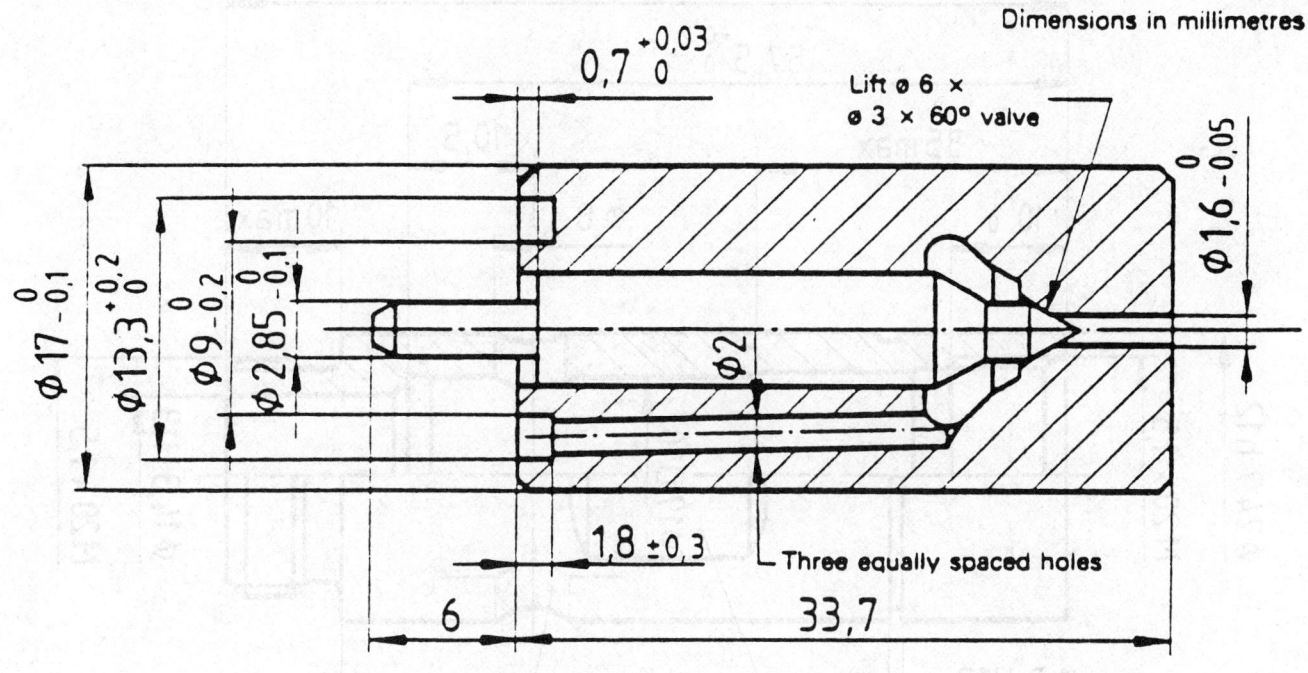

FIGURE 7—NEEDLE VALVE ASSEMBLY

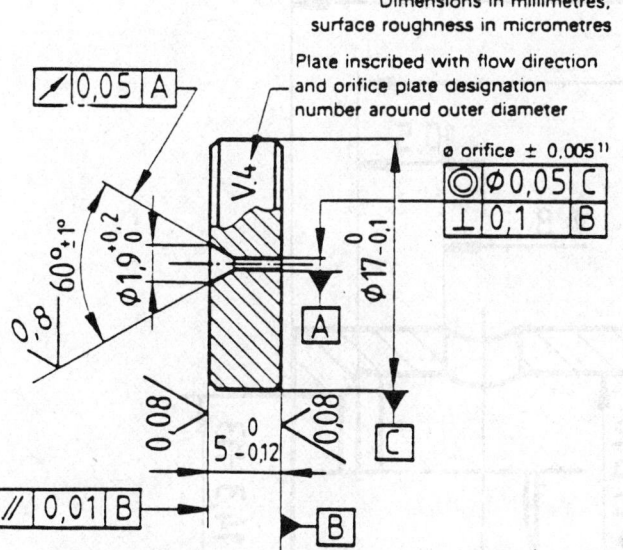

1) Tolerance of orifice hole is only a guide for manufacturing. Its purpose is to assist in meeting the final static flow limits specified in the table when flow-tested under the test conditions described in ISO 7440-2.

Orifice plate number		4	5	6	7	8
Orifice diameter, mm		0,4	0,5	0,6	0,7	0,8
Flow, cm³/min	min.	964	1 528	2 190	3 024	3 948
	max.	1 014	1 606	2 302	3 180	4 150

FIGURE 8—SINGLE HOLE ORIFICE PLATE

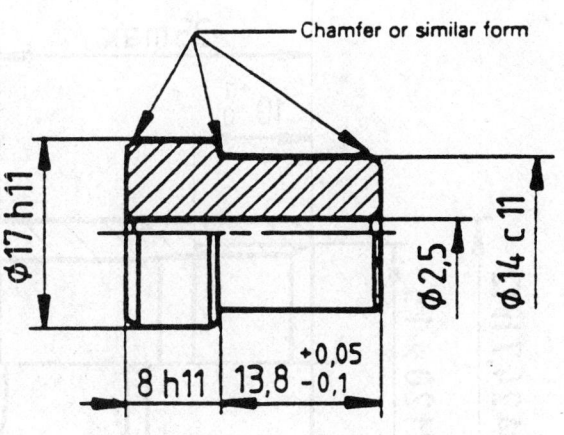

FIGURE 9—DISTANCE SLEEVE

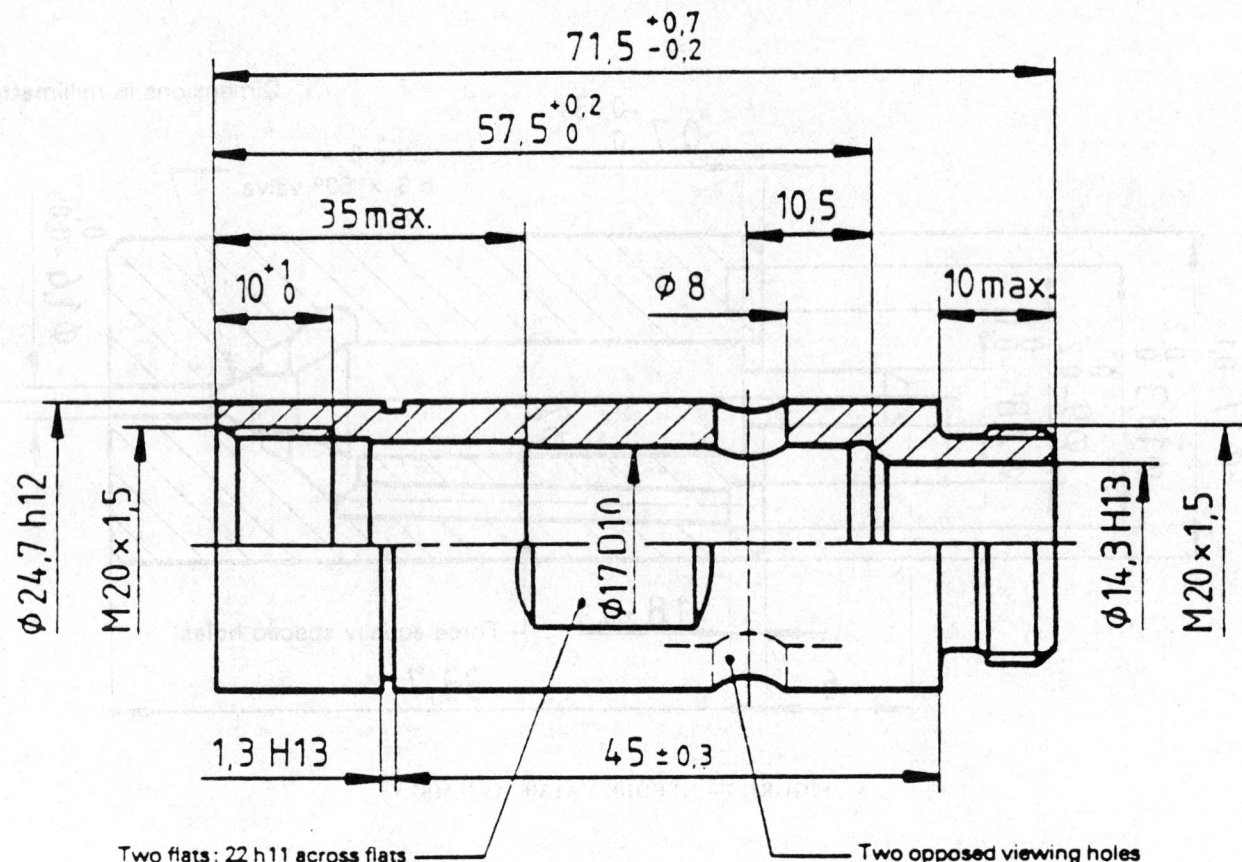

a) Alternative 1. (for use with open spray chamber)

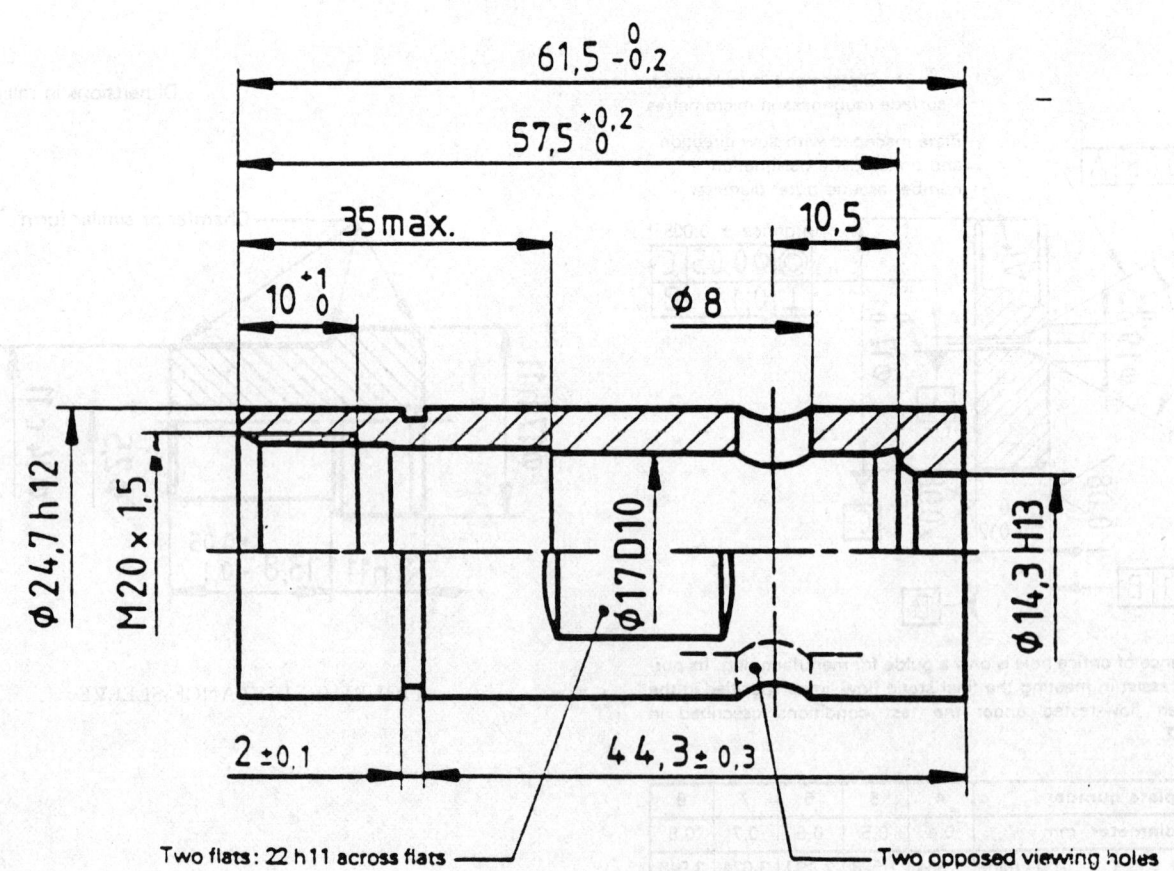

b) Alternative 2 (for use with closed spray chamber)

FIGURE 10—RETAINING NUTS (CAPNUTS) FOR CALIBRATING NOZZLE AND HOLDER ASSEMBLY WITH SINGLE HOLE ORIFICE PLATE

26.317

Dimensions in millimetres

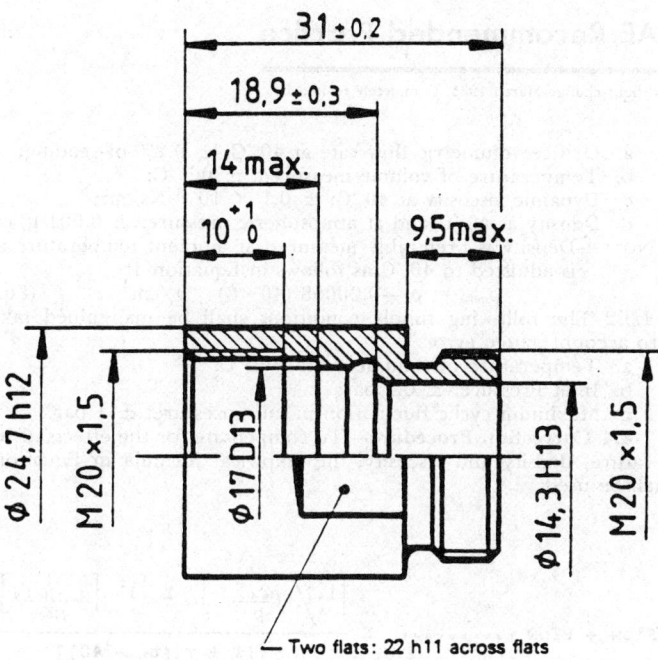

a) Alternative 1 (for use with open spray chamber)

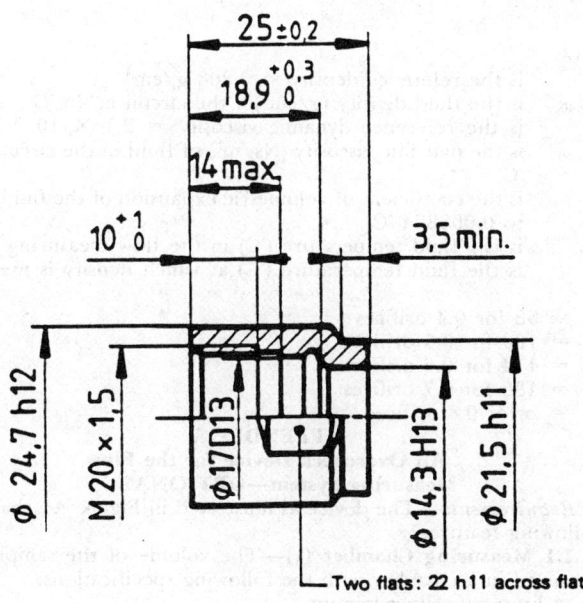

b) Alternative 2 (for use with closed spray chamber)[1]

FIGURE 11—RETAINING NUTS (CAPNUTS) FOR CALIBRATING NOZZLE AND HOLDER ASSEMBLY WITH DELAY PINTLE TYPE NOZZLE

[1] Retaining nut as specified in ISO 2699 (nozzle holder type 2) and in ISO 2697 (dimensions and tolerances of the nozzle capnut).

4.3 Structure of the 2nd Element of the Code:

180 to 225 N/mm

TABLE 1—TENS-DIGIT CODE

Code	Type of Nozzle or Orifice Plate
0-	none
1-	nozzle type: ISO 4010
2-	*
3-	*
4-	orifice plate No.: 4
5-	orifice plate No.: 5
6-	orifice plate No.: 6
7-	orifice plate No.: 7
8-	orifice plate No.: 8
9-	*

NOTES:
[1] The spaces marked with an asterisk (*) are available for future use.
[2] Code 0- (tens-digit) refers to a bare nozzle holder assembly suitable for accepting either type of nozzle assembly.

TABLE 2—UNITS-DIGIT CODE

Code	Type of Nozzle Holder Vent	Type of Nozzle Holder Damper
-0	none	none
-1	not specified	not specified
-2	yes	not specified
-3	yes	no
-4	yes	yes
-5	*	*
-6	no	not specified
-7	no	no
-8	no	yes
-9	*	*

NOTE:
[1] The spaces marked with an asterisk (*) are available for future use.

(R) DIESEL INJECTION PUMP TESTING—PART 2: ORIFICE PLATE FLOW MEASUREMENT—SAE J968/2 MAY91

SAE Recommended Practice

Report of the Engine Committee, approved August 1966, last revised November 1975, reaffirmed without change March 1985. Completely revised by the SAE Diesel Fuel Injection Equipment Standards Committee May 1991.

Foreword—This SAE Recommended Practice is similar to ISO 7440/2.

1. Scope—This part of SAE J968 specifies the flow measuring system, including the fixture, to be used for flow testing the single hole orifice plates used in an orifice plate type nozzle and holder assembly (described in SAE J968/1) which is intended for testing and setting diesel fuel injection pumps on test benches.

The flow measuring system and fixture ensure accurate flow testing of the entire range of orifices from 0.4 to 0.8 mm diameter as specified in SAE J968/1. It is intended primarily for use by the manufacturers of single hole orifice plates.

2. References

2.1 Applicable Documents—The following publications form a part of this specification to the extent specified herein. The latest issue of SAE publications shall apply.

2.1.1 SAE PUBLICATIONS—Available from SAE, 400 Commonwealth Dr., Warrendale, PA 15096-0001.

SAE J967—Calibration Fluid for Diesel Injection Equipment
SAE J968/1—Calibrating Nozzle and Holder Assemblies

2.1.2 ISO PUBLICATIONS—Available from ANSI, 11 West 42nd Street, New York, NY 10036.

ISO 7440/2—Road vehicles—Fuel injection equipment testing—Part 2: Orifice plate flow measurement

3. Orifice Plate Flow Measuring System

3.1 General Description—An orifice plate flow measuring system shall comprise:
a. Orifice Plate Holding Fixture (shown in Figure 1);
b. Fluid Supply Circuit and Flow Measuring Device (an example is shown in Figure 2);
c. Flow Overcheck Device—OPTIONAL. (This device is for periodic overchecking of the accuracy of the primary flow measurement device in (b) and is specified in Appendix A).

The system shall be designed and constructed to minimize fluid evaporation, contamination, aeration, and instability of pressure and temperature.

Flow values obtained by alternative systems shall require correlation with values determined using the above system.

3.2 Orifice Plate Holding Fixture—The functionally critical details and dimensions of the fixture are shown in Figure 1. Dimensions not specified and construction techniques are left to the discretion of the manufacturer of the fixture, but shall be such that the fixture has unrestricted and undisturbed flow.

4. Flow Measuring Procedure

4.1 General Requirements
a. The orifice plates shall be clamped in a fixture conforming to Figure 1.
b. The following monitoring points shall be provided and the values measured or verified:
 (1) orifice inlet temperature
 (2) orifice inlet pressure
 (3) orifice outlet pressure
c. The fluid shall enter the orifice through the entry cone.
d. The flow through each orifice plate shall be measured individually.
e. The fluid in the flow circuit shall conform to SAE J967. Viscosity and specific gravity shall be measured frequently and the fluid shall be discarded if it ceases to conform to SAE J967.

4.2 Reference Conditions—The orifice plate flow values stated in Part 1 of this document refer to volumetric flow rate measured (or adjusted to) 40 °C. The other reference conditions shall be as follows:
a. Inlet pressure minus outlet pressure = 100 bar
b. Outlet pressure = ambient pressure
c. Inlet temperature of fluid = 40 °C
d. Dynamic viscosity at 40 °C = 2.1×10^{-3} Ns/m^2
e. Density at 40 °C = 0.808 g/cm^3

NOTE—Dynamic viscosity at 40 °C = kinematic viscosity at 40 °C × density at 40 °C.

4.3 Accuracy of Measurement

4.3.1 Measurements shall be taken with the following accuracies

a. Orifice volumetric flow rate at 40 °C: ± 0.2% of reading;
b. Temperature of volume measured: ± 0.5 °C;
c. Dynamic viscosity at 40 °C: $\pm\ 0.1 \times 10^{-3}$ Ns/m^2;
d. Density at 40 °C and at atmospheric pressure: ± 0.001 g/cm^3;

NOTE—Density is preferably measured at ambient temperature and is adjusted to 40 °C as follows in Equation 1:

$$\rho_{m40} = \rho_a - 0.00068\ (40 - t_a)\ \text{g/cm}^3 \quad \text{(Eq. 1)}$$

4.3.2 The following supply conditions shall be maintained taking into account gauge error:
a. Temperature of fluid at inlet: ± 1 °C;
b. Inlet Pressure: ± 0.1 bar;
c. Maximum cyclic fluctuation in inlet pressure: ± 1 bar.

4.4 Correction Procedure—To compensate for the effects of temperature, density and viscosity, the empirical formula in Equation 2 shall be used:

$$\text{Flow} = \text{Flow}_{(measured)} \times \frac{\left[1.7\sqrt{\frac{\rho_{m40}}{\rho}}\right] \times \left[1 - \left[\frac{\mu - \mu_{m40}}{\mu K}\right]\right]}{[1 + \gamma\ (t_m - 40)]}$$

(Eq. 2)

where:

ρ	is the reference density = 0.808 g/cm^3
ρ_{m40}	is the fluid density (g/cm^3) in the circuit at 40 °C
μ	is the reference dynamic viscosity = 2.1×10^{-3} Ns/m^2
μ_{m40}	is the dynamic viscosity (Ns/m^2) of fluid in the circuit at 40 °C
γ	is the coefficient of volumetric expansion of the fluid, equal to 0.000824 °C^{-1}
t_m	is the fluid temperature (°C) in the flow measuring device
t_a	is the fluid temperature (°C) at which density is measured

and

K = 88 for 0.4 orifices
K = 100 for 0.5 orifices
K = 114 for 0.6 orifices
K = 135 for 0.7 orifices
K = ∞ for 0.8 orifices

APPENDIX A
An Overcheck Device for the Flow Measuring System—OPTIONAL

A.1 Requirements—The device as illustrated in Figure A1 shall have the following features:

A.1.1 Measuring Chamber (1)—The volume of the sample shall be measured in a chamber with the following specifications:
a. 4 liters capacity minimum
b. minimum internal wetted surface
c. conical upper and lower surfaces
d. vertical graduated tube of 25 mm diameter maximum extending upwardly from the top surface of the chamber
e. graduations in 1 mL increments and correct at 50 °C to ± 2 mL after draining for 1 min
f. chamber made from low thermal expansion material, or appropriate allowance for differential expansion to be made in the measured flow correction formula (see 4.4)
g. a 25 mm effective diameter drain valve in the center of the base
h. independent upwardly facing fluid inlet
i. the measuring chamber shall have a temperature sensor in a central position lying on the axis of the inlet connection

A.1.2 Level Sensor (2)—A level sensor looking through the graduate shall stop the sample collection and timer simultaneously.

A.1.3 Three-Port Valve (3)—A symmetrical power operated three-port valve with zero leakage (or two separate similar two-port

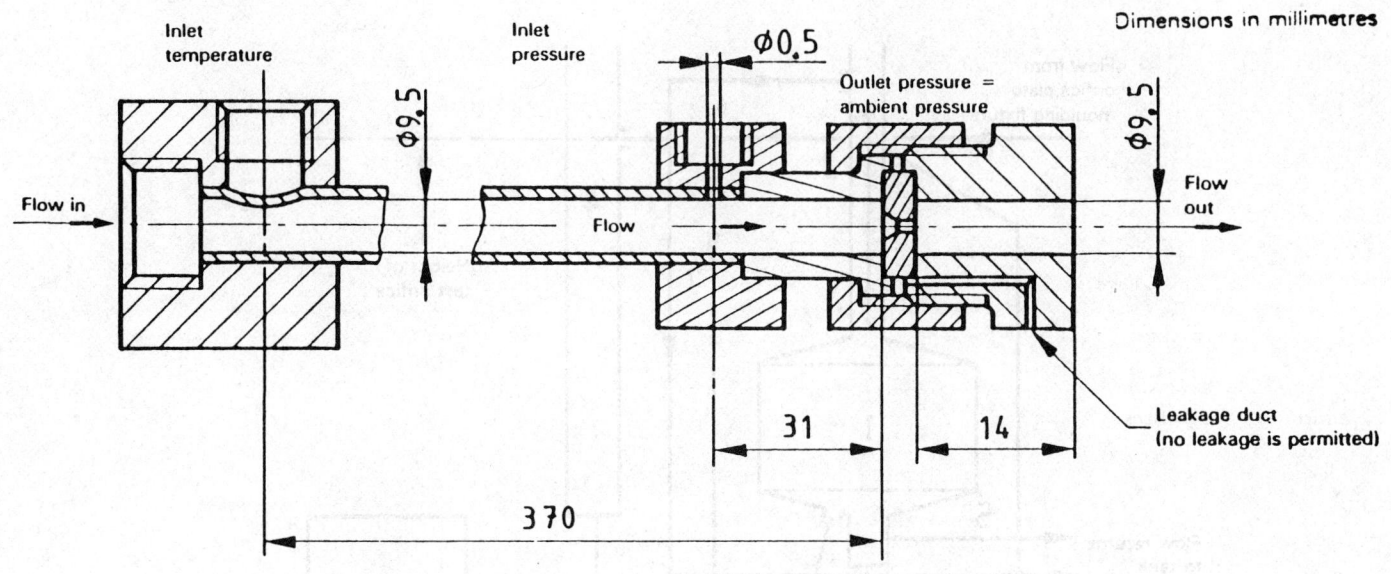

FIGURE 1—ORIFICE PLATE HOLDING FIXTURE FOR FLOW MEASUREMENT

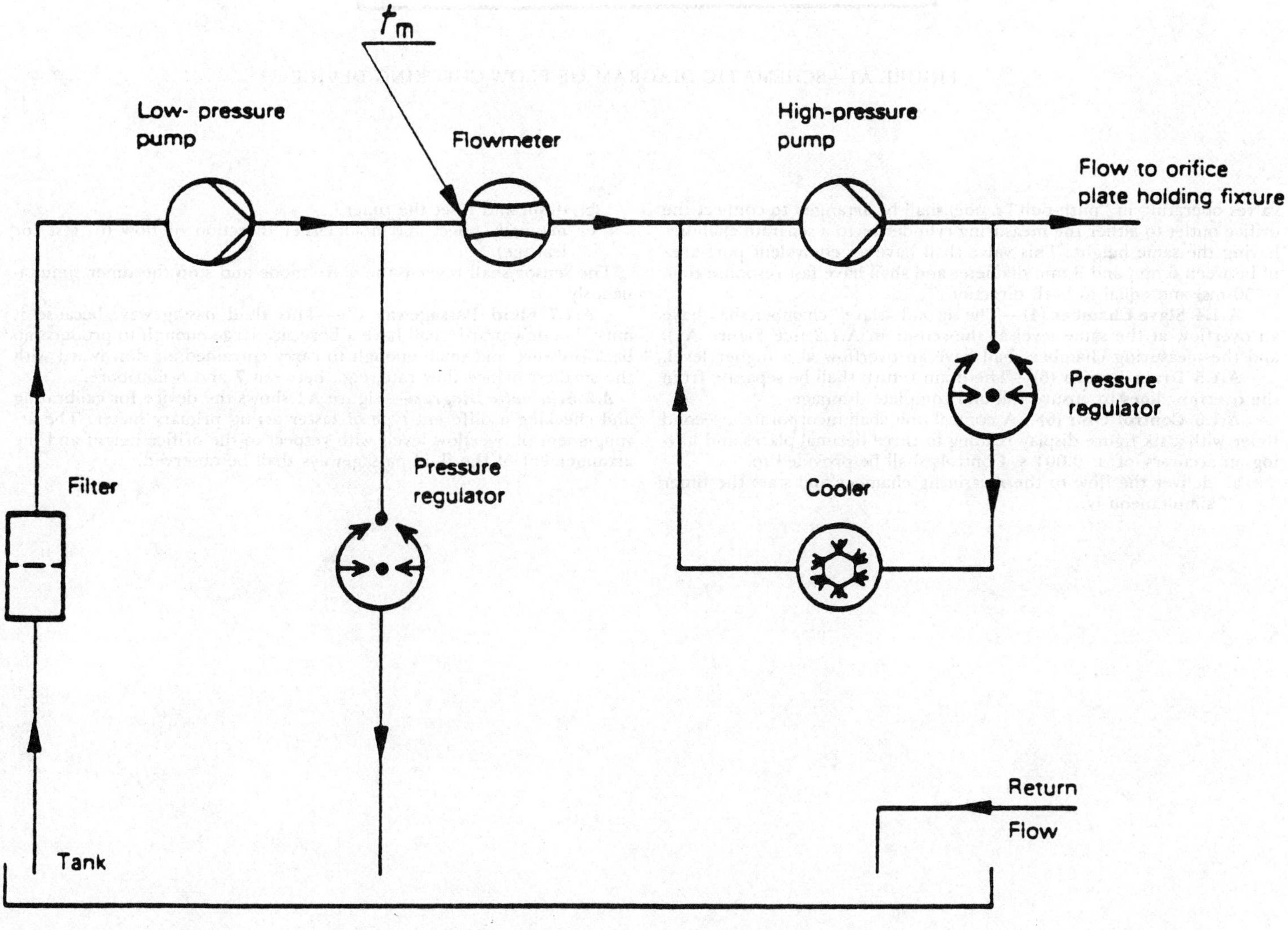

FIGURE 2—FLUID SUPPLY CIRCUIT WITH PRESSURE AND TEMPERATURE
CONTROL AND FLOW-MEASURING DEVICE

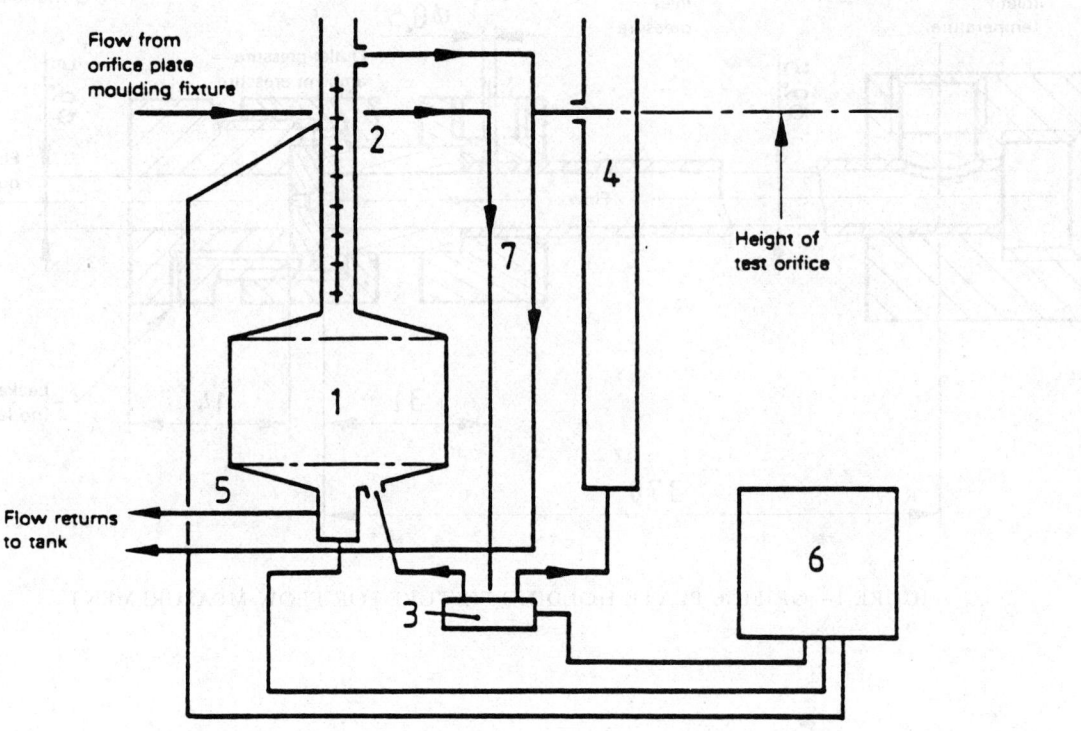

FIGURE A1—SCHEMATIC DIAGRAM OF FLOW-CHECKING DEVICE

valves operating in "push-pull" mode) shall be arranged to connect the orifice outlet to either the measuring cylinder or to a separate chamber having the same height. This valve shall have an equivalent port area of between 6 mm and 8 mm diameter and shall have fast response time (<50 ms) and equal in both directions.

A.1.4 Slave Chamber (4)—The second "slave" chamber shall have an overflow at the same level as the sensor in A.1.2 (see Figure A1), and the measuring chamber shall have an overflow at a higher level.

A.1.5 Drain Return (5)—The drain return shall be separate from the overflow lines to ensure full and complete drainage.

A.1.6 Control Unit (6)—A control unit shall incorporate a second timer with a six figure display reading to three decimal places and having an accuracy of ± 0.001 s. Controls shall be provided to:

a. deliver the flow to the measuring chamber and start the timer simultaneously.

b. drain and reset the timer.

c. manually select and hold either direction of flow (to test for leakage).

The sensor shall reverse the valve mode and stop the timer simultaneously.

A.1.7 Fluid Passageway (7)—This fluid passageway, because it must run downward, shall have a bore size large enough to produce no back pressure and small enough to carry entrained air downward with the smallest orifice flow rate, e.g., between 7 and 8 mm bore.

A.2 Schematic Diagram—Figure A1 shows the device for calibrating and checking a different type of faster acting primary meter. The arrangement of overflow levels with respect to the orifice height and the arrangement of the fluid passageways shall be observed.

DIESEL FUEL INJECTION PUMP—VALIDATION OF CALIBRATING NOZZLE HOLDER ASSEMBLIES—SAE J1549 APR88

SAE Recommended Practice

Report of the Engine Committee approved April 1988.

1. Purpose—The fuel injection pump is intended to validate the accuracy of calibrating nozzle and holder assemblies for applications using 0.4 − 0.8 mm diameter orifice plates and to assist in identifying problems in fuel injection pump test stands.

This SAE Recommended Practice is divided into two parts:

Part I—Design, Description and Specifications of the Fuel Injection Pump; and

Part II—Test Conditions and Procedures for Using the Fuel Injection Pump.

2. Part I—Design, Description and Specifications—Part I of this SAE Recommended Practice describes and establishes specifications and features of a fuel injection pump and its auxiliary equipment required to perform the above purpose.

2.1 Design and Description—This fuel injection pump shall be a single-cylinder, self-contained and cam-operated. Details as to the manufacturer's source are available from SAE. This pump shall provide fuel deliveries and injection pressures capable to test calibrating nozzle and holder assemblies as listed in Part II.

It shall incorporate positive control rack positioning and may have preset stops corresponding to a specified delivery for each orifice size of calibrating nozzle holders.

2.2 Pump Specifications—The fuel injection pump described here shall incorporate the following features:

2.2.1 DELIVERY VALVE	9 mm dia 75 mm^3 retraction volume Lower seated
2.2.2 DELIVERY VALVE SPRING	0.9 mm wire dia 4.8 mm spring O.D. Total number of coils — 18 Number of effective coils — 16 Free length — 26.2 mm Ends square and ground
2.2.3 DELIVERY VALVE HOLDER	14 × 1.5 mm thread, 60 deg cone seat for injection tubing
2.2.4 SUPPLY PUMP	Plunger dia 22 mm Stroke 10 mm Displacement 3800 mm^3/stroke Regulating pressure 2.4 bar (35 lbf/in^2) @ 60 rpm (full by-pass)
2.2.5 PLUNGER & BARREL ASSEMBLY	11 mm dia, left hand lower helix, 24 mm lead Lift to Port Closure: 3 mm Number of Ports: 2 Size of Ports: 3 mm
2.2.6 CAM TYPE, PROFILE	Tangential type Maximum velocity 1.94 m/s @ 1000 rpm, 10 mm lift
2.2.7 TAPPET ASSEMBLY	Roller dia 25 mm
2.2.8 OVERFLOW VALVE	For system operating pressures 1.2 - 1.3 bar (17 − 19 lbf/in^2) @ 1000 rpm
2.2.9 DRIVE COUPLING	Backlash free, to mate with test stand coupling and approved by the test stand manufacturer.
2.2.10 AUXILIARY EQUIPMENT:	
2.2.11 FUEL FILTER	Final stage to be plumbed between supply pump and pump gallery. Filter efficiency details can be obtained from SAE.
2.2.12 PRESSURE MEASUREMENTS	0 - 2 bar (0 - 30 lbf/in^2), ANSI B 40.1-1985 Grade A, measurement downstream of pump mounted final stage filter, adequately snubbed to prevent damage from severe pulsations.
2.2.13 CALIBRATION FLUID TEMPERATURE MEASUREMENTS	Must be capable of measuring calibration fluid temperature prior to entry into the gallery at 40°C with an accuracy of ±0.5°C.

3. Part II—Test Conditions and Procedure—Part II of this SAE Recommended Practice is intended to establish conditions and procedures for flow testing calibrating nozzle and holder assemblies in order to verify their flow accuracy.

For the purposes of these tests, two types of calibrating nozzle and holder assemblies are defined as follows:

"Working" type are those in continuous use for calibrating and testing pumps.

"Reference" type are those with deliveries within the mean flow band of the manufacturer's limits, with higher than average repeatability and used only for reference.

The objective of these tests is to compare deliveries of "working" types to "reference" calibrating nozzle and holder assemblies at pump preset deliveries under identical test conditions. Working and reference calibrating nozzle and holder assemblies must be of the same design and specifications.

Note: Before flow tests are initiated, make certain that "working" type calibrating nozzle and holder assembly leak tests and valve opening pressures meet SAE J968.

3.1 General—Checking and testing calibrating nozzle and holder assemblies is accomplished most effectively by dynamically flow testing them. The tests described below shall be accomplished on a test stand with the following features:

Power	3.7 KW (5 HP) min
Tachometer	Digital, accuracy ±1 digit, update 1 s max
Moment of Intertia	0.127 kgm^2 min[1]
Speed Stability	±1 rpm at 1000 rpm for 1 min duration, min.

In addition, the measuring system must be repeatable and checked prior to the start of these tests.

The calibrating nozzle and holder assemblies that are intended to be tests using this fuel injection pump are described in SAE J968.

3.2 Test Conditions—The following general guidelines and conditions shall be adhered to during the tests:

CALIBRATION FLUID	SAE J967
CALIBRATION FLUID TEMPERATURE	40 ± 0.5°C (104 ± 1°F) measured at fuel gallery inlet
CALIBRATION FLUID PRESSURE	1.2-1.3 bar (17 - 19lbf/in^2) at fuel gallery inlet
DRIVE COUPLING	Backlash free
HIGH PRESSURE TUBING	6.35 mm (1/4 in) O.D. × 1.6 mm (0.062 in) I.D. × 600 mm (23.6 in) long (SAE J1418)
SETTLING TIME[2]	30 s and consistent for each draw
DRAIN TIME[2]	20-30 s and consistent for each draw

3.3 Test Procedure—To test a set of "working" calibrating nozzle and holder assemblies (or a single assembly), proceed as follows:

3.3.1 Connect a "reference" calibrating nozzle and holder assembly to the injection pump described in Part I using high pressure tubing size as per this recommended practice.

3.3.2 Run pump at 1000 rpm for a minimum of 20 min to stabilize the system and achieve specified calibration fluid temperature.

3.3.3 Select control rack stop to achieve the approximate delivery per the table.

3.3.4 Take two draws, but do not record.

---

[1] Per ISO 4008/I assuming: Max delivery 260 mm^3/str; Peak Injection Pressure 641 bar (9300 lbf/in^2); Pump Speed 1000 rpm

[2] If graduates are used.

3.3.5 Take three consecutive draws with equal drain time. After settling time, record each reading. If readings are within 1 cc of each other, calculate the average and record.

Note: If readings are not repeatable, check test stand systems, pump and nozzle holder and correct problem before proceeding.

3.3.6 Remove "reference" calibrating nozzle and holder assembly and install a "working" unit of the same orifice size. DO NOT MOVE CONTROL RACK.

3.3.7 Repeat Step 2 (paragraph 3.3.2) for a minimum of 5 min to stabilize the temperature of the "working" calibrating nozzle and holder assembly.

3.3.8 Repeat Steps 4 and 5 (paragraphs 3.3.4 and 3.3.5).

3.3.9 Readings obtained with a "working" nozzle holder must be within ±3% of the reading of the "reference" calibrating nozzle and holder assembly and within a 2% band in a set.

3.3.10 If a calibrating nozzle and holder assembly is found to be defective, replace with new or repair according to manufacturer's instructions.

For each calibrating nozzle and holder assembly, set control rack to achieve approximate delivery per Table 1. The recommended number of strokes should be according to the test stand manufacturer's instructions. In the absence of specific instructions, fill graduates to a minimum of 50% of their capacity.[3]

TABLE 1

Orifice Diameter	Approximate Fuel Delivery cc/1000 str
0.4	70
0.5	110
0.6	160
0.7	215
0.8	250

4. References

SAE J968 MAR85, Calibrating Nozzle and Holder Assembly for Diesel Fuel Injection Systems.

SAE J1418 DEC87, Fuel Injection Pumps—High Pressure Pipes (Tubing) for Testing.

ISO 4008/1, Fuel Injection Pump Testing—Dynamic Conditions (1980-11-15).

ISO 7440, Calibrating Nozzle and Holder Assemblies.

[3] If graduates are used.

DIESEL ENGINES—FUEL INJECTION PUMP TESTING—SAE J1668 JUN93

SAE Standard

Report of the SAE Diesel Fuel Injection Equipment Standards Committee approved June 1993.

1. Scope—The correct setting and adjustment of fuel injection pumps requires standardized testing conditions.

This SAE Standard summarizes the design and operating parameters for test benches so that, using certain information supplied by the pump manufacturer, the pump test schedule, and certain information supplied by the test bench manufacturer, it can be determined whether a particular test bench is suitable for driving a particular injection pump.

This document is in most cases a summary of the ISO Standard 4008, Parts 1, 2, and 3 and is intended to provide its critical aspects. Standard ISO 4008 should be referred to for more details.

1.1 Field of Application—This document is primarily applicable to test benches suitable for the calibration of fuel injection pumps for diesel engines requiring a fuel delivery of up to 300 mm^3/st/cylinder at full load.

2. References

2.1 Applicable Documents—The following publications form a part of this specification to the extent specified herein. The latest issue of SAE publications shall apply.

2.1.1 SAE PUBLICATIONS—Available from SAE, 400 Commonwealth Drive, Warrendale, PA 15096-0001.

SAE J967—Calibration Fluid
SAE J968—Calibrating Nozzle and Holder Assemblies
SAE J1549—Diesel Fuel Injection Pump—Validation of Calibrating Nozzle Holder Assemblies

2.1.2 ANSI PUBLICATION—Available from ANSI, 11 West 42nd Street, New York, NY 10036-8002.

ANSI/ASME B40.1-1985

2.1.3 ISO PUBLICATIONS—Available from ANSI, 11 West 42nd Street, New York, NY 10036-8002.

ISO 4008—Part 1, Fuel injection pump testing—Dynamic conditions
ISO 4008—Part 2, Fuel injection pump testing—Static conditions
ISO 4008—Part 3, Fuel injection pump testing—Application and test procedures
ISO 4020/2—Fuel filters for automotive compression-ignition engines—Part 2, Test values and classification
ISO 8984—Testing of fuel injectors for compression-ignition engines

2.1.4 OTHER PUBLICATIONS

"Fuel Injection and Controls," 1962 by P. Burman and F. De Luca

3. Dynamic Timing

3.1 Speed Variation at Constant Load—The mean speed of the driveshaft under any steady admissible load shall not vary more than ±1/4% above 800 rpm and ±2 rpm below 800 rpm for at least 1 min.

3.2 Flywheel Inertia—The test bench flywheel shall ensure that the instantaneous speed drop of a pump during injection (i.e., cyclic speed variation) shall be less than 1% at any critical calibration point. Based on this criterion, the required flywheel inertia can be calculated from Equations 1 and 2:

$$I = \frac{Q_{max} \times p_{pa}}{1.31 \times n^2} \quad \text{(Eq.1)}$$

or

$$Q_{max} = \frac{I \times n^2 \times 1.31}{p_{pa}} \quad \text{(Eq.2)}$$

where:

I = Moment of inertia (kg·m^2)
Q_{max} = Pump delivery at calibration point (mm^3/st/outlet)
p_{pa} = Peak Line Pressure (bar)
n = Pump Test Speed (rpm)

To prevent torsional resonance with other mass in the transmission, the flywheel shall be as close to the coupling output as possible.

3.3 Driveshaft Stiffness—The maximum instantaneous angular driveshaft deflection under any operating conditions shall be 0.02 degrees.

The required driveshaft stiffness is calculated as shown in Equation 3:

$$S_d = \frac{Q_{max} \times p_{pa}}{22.4} \quad \text{(Eq.3)}$$

where:

S_d = Driveshaft Stiffness (Nm/°)
Q_{max} = Maximum pump delivery (mm^3/st/outlet)
p_{pa} = Peak Line Pressure (bar)

The driveshaft is considered to be the part connecting the flywheel to the test bench coupling. (On some test benches there is no driveshaft because the coupling is mounted directly to the flywheel.)

3.4 Coupling Stiffness—The maximum instantaneous angular deflection of the drive coupling under operating conditions shall be 0.1 degree.

The required coupling stiffness is given by Equation 4:

$$S_c = \frac{Q_{max} \times p_{pa}}{125.6} \quad \text{(Eq.4)}$$

where:

S_c = Coupling Stiffness (Nm/°)
Q_{max} = Maximum pump delivery (mm³/st/outlet)
p_{pa} = Peak Line Pressure (bar)

In Equations 1 to 4, it is assumed that the mean injection pressure is represented by a sine wave and the injection duration is 10 degrees of pump rotation.

3.5 Pump Mounting Stiffness and Alignment—The injection pump mounting shall be stiff enough to ensure that deflection of the pump body flange (or base bracket) about its driveshaft is less than 0.02 degree with respect to ground during injection. Special apparatus to measure pump mounting stiffness is fully described in ISO 4008 Part 1.

In addition, the pump mounting shall align the pump driveshaft with the test bench output within 0.13 mm radially and 0.05 degree angularly (or 0.25 mm over 300 mm length).

3.6 Backlash—There shall be no backlash between the flywheel and the pump drive coupling.

3.7 Angular Creep—There shall be no angular movement between connecting parts situated between the flywheel and the pump drive coupling when subject to torque reversals equal to twice the peak injection torque according to Equation 5:

$$T \geq Q_{max} \quad \text{(Eq.5)}$$

where:

T = Peak injection torque (N·m)
Q_{max} = Numeric value of pump max delivery

The formulae for driveshaft and coupling stiffness as well as limits for pump mounting stiffness and angular creep were derived during extensive testing.

3.8 Power Output—Test bench power output may be significantly less than the drive motor rated power and is not constant through the speed range. It depends on the type of transmission used. Therefore, output power shall be checked using a dynamometer connected to the test bench driveshaft, and results plotted on a graph as shown in Figure 1. Speed droop (no load to full load) if any, and speed variation at constant load over 1 min shall also be measured using a dynamometer.

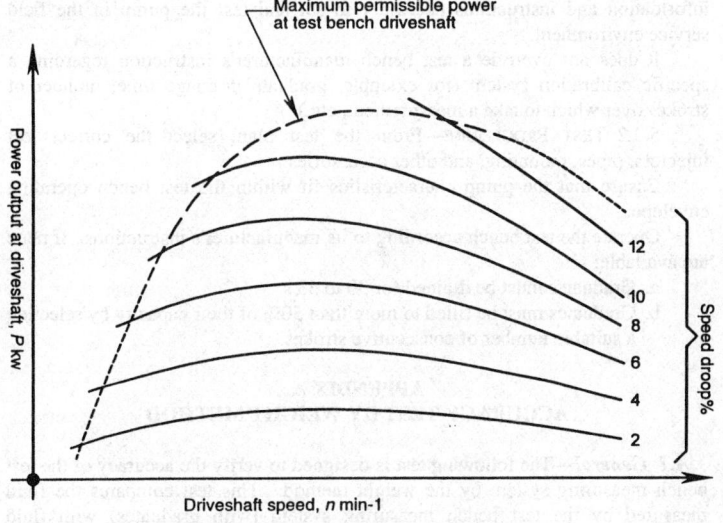

FIGURE 1—POWER CURVE SAMPLE

3.9 Conditions—In order to meet 3.2, 3.3, 3.4, 3.5, and 3.8 of this part of the Standard, the test bench manufacturer and the injection pump manufacturer must provide information to the user to enable him to determine the suitability of a certain test bench to test and calibrate a specific injection pump.

a. Test bench manufacturer shall publish
  (1) Continuous horsepower at the drive coupling available to drive the injection pump. This should be presented in the form of a graph as shown in Figure 1. If maximum horsepower is given, then maximum duration, ambient temperature, or other conditions and limitations must be stated.
  (2) Flywheel moment of inertia (kg·m²)
  (3) Coupling stiffness (Nm/°)
  (4) Driveshaft stiffness (Nm/°)
  (5) Flange and/or base bracket stiffness (Nm/°)

Furthermore, the test bench manufacturer shall provide a graph showing the permissible operating envelope based on formulae given in 3.2, 3.3, and 3.4. For simplicity, a peak injection pressure of 628 bar is assumed for this graph. If actual peak pressure is not 628 bar, then fuel delivery must be corrected as described in 3.10 in order to use the graph. An example of the graph is shown in Figure 2. (If the curve showing the stiffness of the flange or base brackets lies outside of the permissible operating envelope, it does not have to be shown in the graph.)

b. Injection pump manufacturer shall publish
  (1) Maximum horsepower that the injection pump absorbs
  (2) Fuel delivery (mm³/stroke)
  (3) Peak injection pressure (bar)

at each critical test point in the pump specification.

If horsepower is not specified, it can be approximated by Equation 6[1]:

$$HP = \frac{p_{pa} \times Q \times n \times 1.66 \times 10^{-9}}{\eta} \quad \text{(Eq.6)}$$

where:

HP = Horsepower (kW)
P_{pa} = Peak injection pressure (bar)
Q = Fuel delivery (including retraction volume) (mm³/st)
n = Number of injections per minute
η = 0.75 (accounting for pump efficiency and effect of retraction volume)

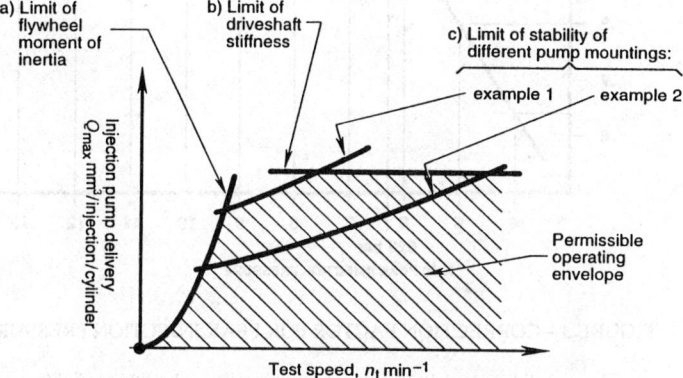

FIGURE 2—PERMISSIBLE OPERATING ENVELOPE

3.10 Application and Use of Given Information—Assume a 6-cylinder pump delivers 180 mm³/st at 600 rpm with a peak injection pressure of 800 bar.

a. Step 1—Determine Corrected Fuel Delivery—From Figure 3, correction factor at 800 bar is 1.27 (800/628). Therefore, the corrected delivery will be 1.27 × 180 = 228.6 mm³/st.

b. Step 2—Plot the point on the graph (Figure 2) where the corrected delivery (228.6 mm³/st) intersects the speed (600 rpm). This must be within the operating envelope of the test bench (shaded area in Figure 2).

c. Step 3—Calculate power requirement.

If not given by the pump manufacturer, calculate HP from Equation 6. In this case see Equation 7:

$$HP = \frac{800 \times 180 \times 3600 \times 1.66 \times 10^{-9}}{0.75} \quad \text{(Eq.7)}$$

Plot this point on graph (Figure 1). It must lie below the dotted curved line for the test stand to be capable of running this particular pump.

4. Static Requirements

4.1 Delivery Measurement System—The calibration fluid delivered from each test injector shall be routed to the device for measuring delivery and displaying it directly or by simple calculation as an average expressed in cubic millimeters per stroke per outlet.

---

[1] Reference: Derived from a formula given in "Fuel Injection and Controls."

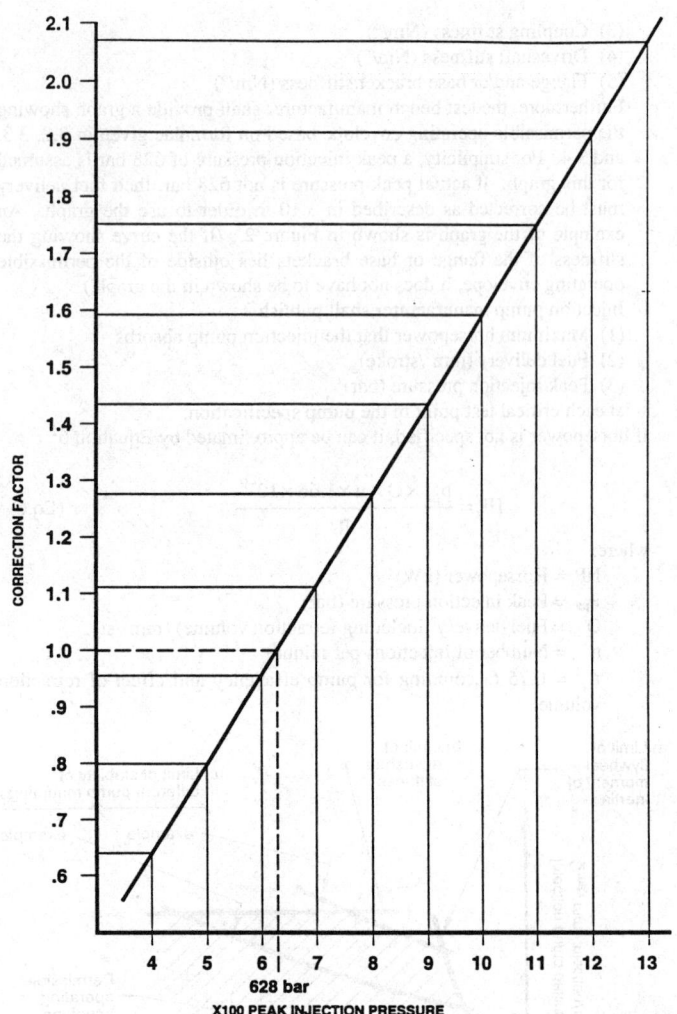

FIGURE 3—CORRECTION FACTOR FOR PEAK INJECTION PRESSURE

There shall be provision for observing visually, or by other suitable means, the moment when flow from any of the test injectors stops.

4.1.1 METHOD OF TESTING THE PUMP DELIVERY MEASUREMENT SYSTEM—Each test bench make and model shall pass the acceptance test. This test shall consist of:

a. Type Test—To be performed on at least one sample test bench and is aimed at proving the capability of a design to achieve the desired accuracy; a sophisticated apparatus is required.
Allowable error shall not exceed 1% if no specific instructions are given by the test stand manufacturer regarding graduate selection or stroke count; error shall not exceed 2% if instructions are given.

b. Field Test—To be performed periodically in the field on all test benches to ensure continued satisfactory operation; this test requires no special apparatus. Graduate zero error shall not exceed 0.5% of its maximum capacity. Count error shall not exceed one turn.

The complete test procedure for the Type and Field Tests are found in ISO 4008, Part 2, Annex A. An alternate and equally acceptable Type Test by weight method is described in Appendix A.

If graduates are used, the minimum pitch of graduations shall be at least 1 mm.

One graduation shall not be greater than 1% of the total volume. A certain number of graduations from zero upwards may be omitted.

If a digital display is used, then, at a minimum, fuel delivery shall be displayed in mm^3/st with a resolution of 0.1 mm^3/st.

4.2 Calibration Fluid System—The calibration fluid supply shall be at a flowrate of at least 2.5 times the delivery flow rate from the pump under test.

At any supply pressure over 0.3 bar, the maximum pressure fluctuation (from peak to peak) shall be no more than 5% of the reading with the supply connection capped off.

Calibration fluid temperature, as measured immediately behind the bulkhead supply connection, must be held to 40 °C ± 2 °C, irrespective of the pump demand.

The supply pipe shall be 1 m + 0.1 m long with an inside diameter of at least 9.5 mm.

a. Temperature measuring accuracy: ±0.5 °C @ 40 °C
b. Filtration: according to ISO 4020/2
c. Calibration fluid heater, if installed, shall be designed so as not to cause breakdown of the fluid

4.3 Other Requirements

4.3.1 Pressure and Vacuum Measuring Instruments
a. Accuracy: ANSI/ASME B40.1-1985 Grade A (2-1-2% FS) or better

4.3.2 Driveshaft Tachometer
a. Analog: Not acceptable
b. Digital: Resolution 1 rpm
c. Accuracy: ±1 digit

4.3.3 Output Shaft Angular Measuring Arrangement
a. Resolution of angular readout: 1 degree max
b. Error between any two divisions: 15 min max
c. Datum mark shall not produce parallax error

5. Application and Equipment Maintenance Recommendations

5.1 Test Benches—When a statement of ISO Test Condition is requested it shall conform with Annex E to ISO 4008, Part 3.

5.2 Test Injectors—All calibrating nozzle and holder assemblies conforming to the appropriate Standard must be accompanied by documentation (marking or labels in packaging) supplied by the manufacturer and vouching for their conformance to the document.

Other test injectors and nozzles may be specified in the appropriate pump test specifications.

For purposes of maintenance, a record shall be kept of the number of pumps tested with each test injector set.

For calibrating nozzle and holder assemblies and test injectors maintenance and servicing schedules, refer to Appendix B. It gives information on frequency of maintenance, testing procedure, recommended settings and limits for opening pressures, seat leakage, back leakage, and tightening torque values for nozzle retaining nuts and edge filter assembly (inlet studs).

5.3 Testing

5.3.1 TEST SCHEDULE—The test schedule shall contain all the necessary information and instructions necessary to set and test the pump in the field service environment.

It does not override a test bench manufacturer's instruction regarding a specific calibration system (for example, graduate drainage time, number of strokes over which to take a measurement, etc.).

5.3.2 TEST PROCEDURE—From the test plan, select the correct test injectors, pipes, mounting, and other accessories.

Ensure that the pump characteristics fit within the test bench operating envelope.

Operate the test bench according to its manufacturer's instructions. If none are available:

a. Graduates must be drained for 30 to 33 s.
b. Graduates must be filled to more than 50% of their capacity by selecting a suitable number of consecutive strokes.

APPENDIX A
ACCURACY TEST BY WEIGHT METHOD

A.1 General—The following test is designed to verify the accuracy of the test bench measuring system by the weight method. This test compares the fluid measured by the test bench measuring system (with graduates) with fluid measured by weight as delivered by a stable injection pump at controlled conditions. This method measures the weight of fluid delivered (in grams) over a large number of injection strokes. By knowing the fluid's specific gravity and temperature, the weight can be converted into volume and combined with the number of strokes, it can then be computed and expressed in cubic millimeters per stroke, which is the industry's basic unit for measurement of an injection pump delivery.

A.2 Equipment Required—See Table A1.

A.3 Preparation for Tests—Withdraw a sample of calibration fluid from the test stand reservoir after some stirring and prior to operating the stand.

a. Set glass cylinder in an area where it will not be disturbed and temperature is fairly stable.
b. Immerse glass thermometer (or temperature sensor) and hydrometer into the glass cylinder.

c. Measure and record specific gravity and temperature (SG_x and t_x respectively) after the calibration fluid has soaked at a stable ambient temperature for a minimum of 6 h. Record temperature to a resolution of 0.1 °C and specific gravity to a resolution of 0.0005.
d. Connect plumbing and electrical connections according to Figure A1.
e. Mount injection pump on the stand using accessories in accordance with the test stand manufacturer's instructions.
f. Connect cable of digital tachometer in parallel with the test stand's pickup. If stand does not have any pickup (magnetic, optical, or other) one must be installed in order to obtain a digital display of test stand rpm and stroke counts.
g. Connect plastic tubing from solenoid valve port B to test stand reservoir (return flow).
h. Connect timer and switch in the circuit as shown and connect to wall outlet.
i. With the pump running and discharging fluid to port A of the de-energized solenoid valve, the fluid is routed through port B to the reservoir.
j. Reset scale display to read zero (see scale manufacturer's instructions).
k. Place empty and dry container on scale.
l. Preselect number of strokes on the stroke counter.
m. Press reset button on electronic timer.

A.4 Test Schedule—Operate injection pump at speeds, delivery settings, and with Calibrating Nozzle Holders of orifice sizes at speed settings and deliveries as shown in SAE J1549. At each speed setting the test shall be conducted first by routing the fluid to the graduate in the test stand and then to the container for measuring its weight. If design permits, switch fluid routing from the test stand graduate to the fluid weighing apparatus while pump is operating. As a single-cylinder pump is used, each graduate must be tested separately.

A.5 Test Procedure—Operate the stand, in accordance with test stand manufacturer's instructions, running the pump at 500 rpm to stabilize calibration fluid temperature and to warm up the complete system.

At each speed and orifice size combination run pump for a minimum of 10 min and then take readings. Take always three readings of graduates and calculate average. If the three readings are not within 0.5 cc there must be a problem with one or more of the following: test stand, pump, injector, or high-pressure fuel line. Correct the problem before continuing. Allow equal time to drain graduates (30 to 35 s). Read graduates as soon as foam disappears, if any.

To operate the fluid weight measuring apparatus (mass flow measuring device) refer to Figure A1.

TABLE A1—EQUIPMENT REQUIRED

Name	Description
Electronic Balance	Digital readout: 0 to 999.99 Capacity: 0 to 500 g Resolution: 0.01 g Precision: ±0.007 g Error: 0.01 g
Glass Thermometer	Range: 15 to 40 °C (min) Resolution: 0.1 °C
Cal. Fluid Temperature Measurement	Accuracy of measurement: ±0.5 °C Calibrated @ 40 °C
Electronic Timer	Clock oscillator: 0.001 s/pulse Accuracy: 0.01%
Tachometer/Counter	With preset count switch Start/stop switch 0 to 999 rpm, 0 to 9999 counts Accuracy: ±1 digit
Hydrometer	Specific gravity range: 0.760 to 0.830 g/cc Resolution: 0.0005 (min)
Solenoid Valve	3-way, 3/32 in orifice (min)
Glass Cylinder	Capacity: 250 mL ungraduated
Injection Pump	Inline single cylinder per SAE J1549[1]
Calibrating Nozzle Holders	SAE J968[1]
High Pressure Tubing	(Dependent on pump used)
Calibration Fluid	SAE J967

[1] As an alternate an equally acceptable injection pump can be a fixed rack ("Reference Pump") and test injectors supplied for this purpose by the pump manufacturer.

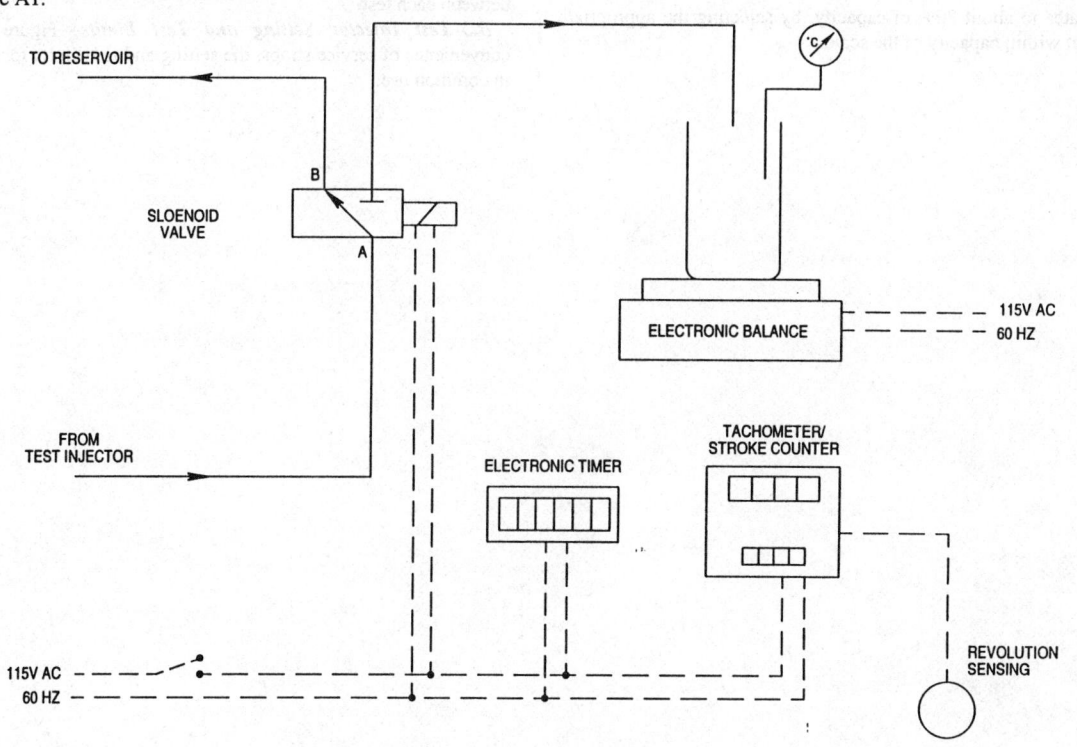

FIGURE A1—BLOCK DIAGRAM
TEST SET UP—ACCURACY BY WEIGHT METHOD

When ready to take a reading, press start switch. This will simultaneously energize solenoid valve, start routing fluid into container on scale, start stroke counter, and start electronic timer.

At the start of the test record temperature (t) of calibration fluid entering the container.

After the preselected number of strokes is achieved, solenoid valve will de-energize, fluid will be diverted to the reservoir, electronic timer will stop and scale display will indicate weight. Record weight, the final fluid temperature (if different from before) and reading on electronic timer.

The previous tests must be performed with each graduate.

To calculate true or "absolute" fuel delivery as measured by the scale, use Equation A1:

$$Q_a = \frac{W \times 1000}{SG \times N} \qquad (Eq.\,A1)$$

where:
Q_a = Delivery mm³/st
W = Weight in grams
SG = Specific gravity g/cc at temperature (t)
N = Number of strokes

NOTE—To find specific gravity (SG) at temperature (t) use either a graph showing specific gravity change with temperature (available from the supplier of calibration fluid) or use Equation A2:

$$SG = SG_x - 0.00037(t - t_x) \qquad (Eq.\,A2)$$

where:
SG = Specific gravity g/cc at temperature (t)
SG_x = Specific gravity g/cc at temperature (t_x)
t_x = Temperature at which SG_x was measured

0.00037 is the typical coefficient of cubic expansion for calibration fluid.

To verify rpm use Equation A3:

$$RPM_c = \frac{N \times 60}{T} \qquad (Eq.\,A3)$$

where:
RPM_c = Calculated rpm
N = Number of strokes
T = Time in seconds

NOTE—Fill graduates to about 80% of capacity, by selecting the appropriate number of strokes but within capacity of the scale.

APPENDIX B
TEST INJECTOR MAINTENANCE

B.1 This appendix provides recommended schedule and guidelines for the maintenance of test injectors. It also contains setting and test limits (see Figure B1).

B.2 Definitions—Test Injector is an injector designated by the manufacturer in the pump test specifications to be used in calibrating the pump. Calibrating Injector is a test injector that conforms to a Standard (SAE J968, ISO 7440, or other). (Therefore, all calibrating injectors are test injectors but all test injectors are not calibrating injectors.)

B.2.1 Maintenance Schedules—To facilitate scheduled maintenance of test injectors, a record shall be kept of the number of pumps tested with each injector. The maintenance schedule shown in Figure B2 is mandatory for all calibrating injectors.

B.2.2 Nozzle Opening Pressure Checking and Setting—Use Calibration Fluid SAE J967 and a hand-operated nozzle tester that meets ISO 8984-1.

CAUTION—The spray from a test injector can pierce human skin. To avoid injury keep clear of spray.

Nozzle opening pressures for calibrating injectors are given in Figure B1.
Where applicable, check for nozzle chatter.

B.2.3 Seat Leakage Check—For pintle nozzle injectors, with the nozzle facing down, wipe the tip and check that no drops separate from the tip after holding the pressure as stated in Figure B1 for 10 s.

For orifice type injectors, with the nozzle facing downward at 45 degrees, spray the nozzle a few times. Hold the pressure to within 5 bar of opening pressure until a drop falls from the tip. Now hold the pressure to that stated in Figure B1 and measure the time between the fall of the second and third drop. This time shall be less than 10 s.

If chatter is unsatisfactory or seat leakage is out of limits, repair or replace defective part(s).

B.2.4 Checking Back Leakage—After raising pressure to "A" (Figure B1), let go of pump handle and allow the pressure to fall slowly of its own accord.

Measure the calibration fluid temperature in the nozzle tester. Using a stopwatch, measure the time it takes for the pressure "B" to fall to pressure "C." If the time is below minimum back leakage t_{min}, as indicated by the curve, Figure B3, replace the nozzle. Repeat the test several times but flush the nozzle between each test.

B.3 Test Injector Setting and Test Limits—Figure B1 gives, for the convenience of service shops, the setting and test limits for test injectors that are in common use.

No.	Injector or nozzle type	Nozzle opening pressure[1] bar	Seat leakage pressure[1] bar	Cap nut torque[1] N-m	Edge filter assembly torque[1] N-m	Back leakage test pressures		
						A bar	B bar	C bar
		For all types						
1	Nozzle holder assemblies[1] with Delay pintle type nozzle							
	a. calibrating nozzle to ISO 4010	125^{+3}_{0}						
	b. calibrating nozzle to ISO 4010	147^{+3}_{0}						
	c. calibrating nozzle to ISO 4010	172^{+3}_{0}	20 bar below actual nozzle opening pressure	60^{+3}_{0}	45^{+20}_{0}	120	100	70
2	Single hole orifice plate[1]							
	a. No. .4 (φ0.4 mm orifice)	207^{+3}_{0}						
	b. No. .5 (φ0.5 mm orifice)	207^{+3}_{0}						
	c. No. .6 (φ0.6 mm orifice)	207^{+3}_{0}						
	d. No. .7 (φ0.7 mm orifice)	207^{+3}_{0}						
	e. No. .8 (φ0.8 mm orifice)	207^{+3}_{0}						
	f. No. .5 (φ0.5 mm orifice)	117^{+3}_{0}				110		

[1] As specified in SAE J968/1.

FIGURE B1—SETTING AND TEST LIMITS

Schedule No.	Frequency	Action
1	After injector maintenance	a. Run the injector, preferably with the highest output pump for which the injector is specified, at full speed, full load for 5 min. b. Check opening pressure and reset if necessary.
2	Weekly or after 100 pumps (max.), whichever occurs first	a. Check opening pressure and reset if necessary. b. Check back leakage. c. Check seat leakage.
3	a. Orifice plate injector assemblies: every 1000 pumps	a. Only for injectors with single hole orifice plate: fit new needle valve assembly and/or orifice plate or complete nozzle and holder assembly as necessary in accordance with manufacturer's instructions.
	b. Pintle nozzle injector assemblies: every 1000 pumps	b. Only for injectors with delay pintle type nozzle: fit new nozzle assembly.
4	Every 5000 pumps (max.)	Check flow-rate of inlet stud with edge filter.
5	Every 5000 pumps	Visually examine functional surfaces of injector components.

FIGURE B2—CALIBRATING INJECTOR MAINTENANCE SCHEDULE

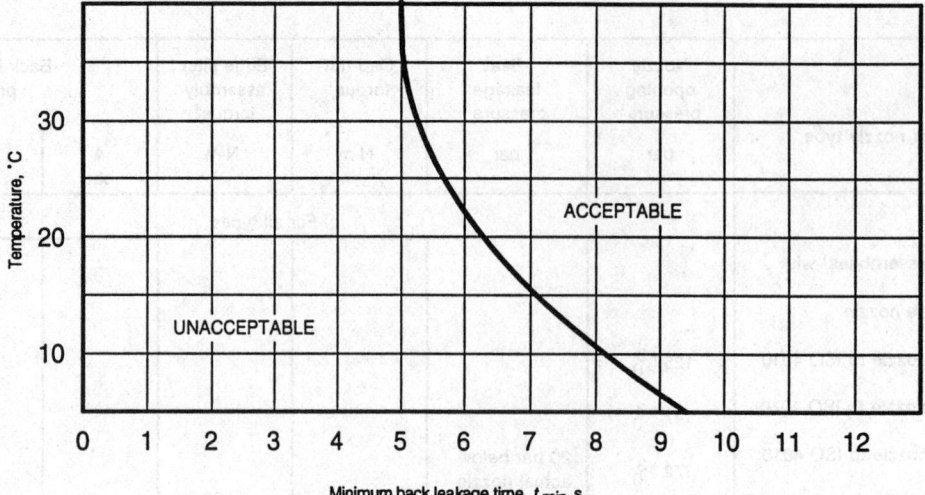

FIGURE B3—RELATIONSHIP BETWEEN TEMPERATURE AND MINIMUM BACK LEAKAGE TIME, t_{min}

φCALIBRATION FLUID FOR DIESEL INJECTION EQUIPMENT—SAE J967 SEP88   SAE Standard

Report of the Engine Committee, approved August 1966 and completely revised September 1988. This SAE Standard meets the specifications for ISO 4113 calibration fluid for diesel injection equipment.

1. Scope—This SAE Standard defines the requirements of a calibration fluid recommended for flow bench testing, calibration, and flushing of fuel injection equipment, in production facilities, in laboratories, and in service establishments.

2. International Standard—This SAE Standard meets the specifications for ISO 4113 calibration fluid for diesel injection equipment.

3. Property Requirements—The calibration fluid shall be formulated from refined and deodorized fuel stocks. It shall meet the specifications shown in Table 1. Anti-wear additives may be included.

The calibration fluid shall not contain components in such a concentration that it could irritate the normal skin.

The calibration fluid shall have good storage and thermal stability, and shall be such that without cleaning of the equipment after calibration, proper functioning of the equipment is ensured after being stored one year in normal conditions.

4. Certification—A supplier of this calibration fluid must supply, with each container or bulk shipment, certification that the fluid meets this standard, state the specific revision, the date of manufacture, and indicate the presence of anti-wear additives. The presence of anti-wear additives must also be included on the shipping container.

TABLE 1—CALIBRATION FLUID SPECIFICATIONS

Property	Specification Limit	Test
Density	0.820 - 0.830 g/ml at 15°C (59°F)	ISO 3675
Specific Gravity	0.819 - 0.829 at 15.5°C (60°F)	(ASTM D 1298)
Flash Point	75°C (167°F) min	ISO 2719 (ASTM D 56)
Viscosity	2.45 - 2.75 cSt (mm²/s) at 40°C (104°F)[a] 2.55 - 2.85 cSt (mm²/s) at 37.8°C (100°F)[a]	ISO 3104 (ASTM D 445)
Distillation	5% volume max at 210°C (410°F) 95% volume min at 360°C (680°F)	ISO 3405 (ASTM D 86)
Oxidation Stability (Catalyzed-48 h)		IP 306/82
Total Sludge	0.05% max by weight	
Total Acidity after oxidation	0.7 max mgKOH/g[b]	
Cloud Point	-10°C (14°F) max	ISO 3015 (ASTM D 2500)
Rust Protection (Polished Panels-50 h)	Pass 5 out of 6 faces of three panels	ASTM D 1748
Corrosion Tests		
- Ferrous Metal	Pass-24 h	ASTM D 665A
- Copper	Pass-classification 1 at 100°C (212°F) 3 h	ISO 2160 (ASTM D 130)
Galvanic Corrosion	Pass-10 days	FSTM 5322.1
Sulfur	0.4% weight, max	ASTM D 129
Trace Sediment (including water)	0.01% volume, max	ASTM D 2273
Aromatic components	12% C_A max	ASTM D 2140
Foaming Tendency (After 5 min blowing)	50 ml max at 24°C (75°F)	ASTM D 892
Foaming Stability (after 2 min settling)	0 ml max at 24°C (75°F)	ASTM D 892
Color		ISO 2049 (ASTM D 1500)
New Product	2 max	
After 6 months' storage	3 at 43°C (109.4°F) max	

NOTES:
[a] It is recommended that the Calibration Fluid be renewed when the viscosity increases above 3.0 cSt (mm²/s) at 40°C (104°F) or 3.1 cSt (mm²/s) at 37.8°C (100°F).
[b] Sum of volatile and soluble acidity.

LOCATION OF ATOMIZER OF ETHER SYSTEMS FOR DIESEL ENGINES—SAE J2079 MAY92

SAE Information Report

Report of the SAE Off-Road Machinery Technical Committee SC15—Cold Weather Operations of the SAE Off-Road Machinery Technical Committee approved May 1992. Rationale statement available.

1. Scope—This SAE Information Report recommends the requirements for the location of ether atomizers.

2. References—There are no referenced publications specified herein.

3. Rationale—Ether systems are used on diesel engines for cold weather starting below 5 °C. The location of the atomizer in the air intake system is of vital concern. Most air intake systems provide a tapped hole somewhere in the air intake system to accommodate an atomizer, however, some engines have the hole in the intake manifold where it is almost impossible to provide even distribution of ether to all cylinders.

4. Recommendation—The atomizer should be located above the intake manifold as close as possible to the inlet of the manifold preferably in the air horn leading to the intake manifold. On turbocharged diesels a desirable location is between the turbo and intake manifold. On air to air cooling systems the atomizer should be located after the air leaves the cooling radiator as close to the air intake manifold as possible. (See Figure 1.)

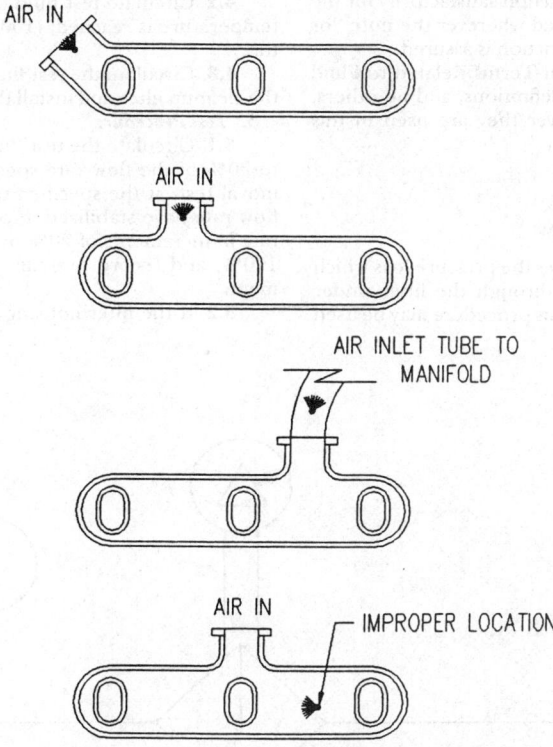

FIGURE 1—EXAMPLES OF ATOMIZER LOCATIONS IN AIR INTAKE SYSTEM

ϕ FUEL FILTER TEST METHODS—SAE J905 JAN87

SAE Recommended Practice

Report of the Engine Committee approved November 1964 and completely revised January 1987. This report supersedes HS J905.

The purpose of this fuel filter test method is to provide standardized methods for evaluating the performance characteristics of fuel filters by bench test methods. This, combined with data obtained from application tests, may be used to establish standards of performance for filters when tested by these standard methods. Many variations in requirements of filtration to protect fuel supply equipment on engines and variations in operating conditions make it difficult to specify meaningful "in-service" performance standards by which a filter may be judged. By the use of these standard test methods, test conditions are always the same, and comparisons of the performance of filters may be made with a high degree of confidence. Once the requirements of a particular application are known, performance standards for suitable filters may be established by these test methods, and adequacy of performance of filters for the job may be determined.

In order to achieve the highest degree of reliability of test results, the procedures and equipment must conform with those specified in this code. No minimum performance requirements for filters have been specified, since these are the responsibility of the user and manufacturer. Only the methods of determining, interpreting and reporting performance characteristics are the proper province of this document.

Separate chapters cover the test methods necessary to evaluate the several functional capabilities and mechanical properties of the filter. Each chapter is complete with recommended materials, apparatus and procedures for testing and evaluation. The chapters are:

1. Resistance to Flow
2. Filter Capacity and Contaminant Removal Characteristics
3. Media Migration Test
4. Collapse Test
5. Ability to Meet Environmental Conditions
6. Installation and Removal
7. Mechanical Tests

To simplify the chapters covering test methods, information of importance but not directly involved in test methods is covered in appendices as follows:

A. Test Contaminants and Fluids
B. Methods for Sample Analysis
C. Graphic Symbols
D. Source Information

In the various test methods, references are made to specific equipment, which is recommended because it is known to function satisfactorily for the intended purpose. Equivalents may be substituted wherever the note "or equivalent" is found, provided that equivalent function is assured.

Reference should be made to the "Glossary of Terms Related to Fluid Filters and Filter Testing", SAE J1124. These definitions, and no others, are to be applied to the terms defined wherever they are used in this document.

CHAPTER 1
RESISTANCE TO FLOW

1. Scope—The resistance to flow test determines the pressure loss which will result when standard test fluid is passed through the filter under standard conditions of flow and fluid viscosity. This procedure may be used to establish flow capacities, and is the method for measuring resistance to flow of sample filters against an established performance standard.

2. Test Materials

2.1 Test Fluid—Viscor Fuel Filter Test Fluid L-4264V, or as otherwise specified.

3. Test Apparatus

3.1 Fuel filter flow test stand (or equivalent). A pump, flowmeter, thermometer and manometers or differential pressure gauge are necessary as shown in Fig. 1. It is usually convenient to use the fuel filter test stand as described in Chapter 2 (Fig. 4).

3.2 Filter housing or mounting plate with pressure taps located to give pressure loss across the filter element (or across the entire filter unit, in the case of "spin-on" filters). See Figs. 2 and 3 for housing and mounting plate details.

3.3 If special housings are necessary or desirable, care must be exercised to locate inlet and outlet pressure taps to obtain true pressure values.

4. Test Preparation

4.1 Install a cleanup filter on the test stand and add 5 gal (20 L) of prefiltered test fluid.

4.2 Circulate test fluid through the bypass system until the specified temperature is reached; (100°F (38°C) is the recommended test temperature).

4.3 Circulate the test fluid through the cleanup filter for 1 h. Remove the cleanup filter and install the test filter.

5. Test Procedure

5.1 Circulate the test fluid through the test filter and adjust flow rate to 20% of the flow rate specified for the Capacity and Contaminant Removal test, at the specified test temperature. When the temperature and flow rate have stabilized, record the pressure differential. Adjust the flow rate in increments of 20% of the specified test flow rate, to a maximum of 120%, and record pressure differential after stabilization at each increment.

5.2 If the filter housing does not have pressure taps located to read

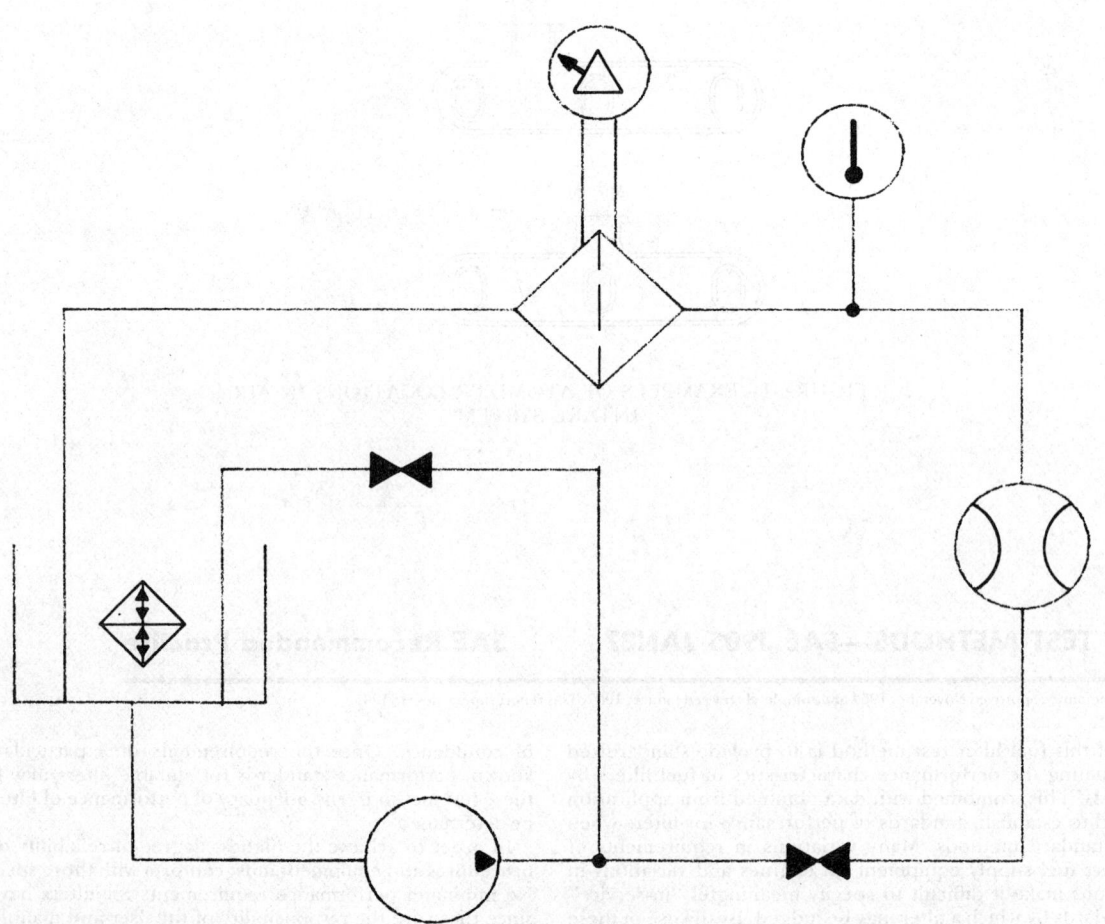

FIG. 1—FLOW TEST STAND

element pressure only, an empty housing pressure differential must be obtained. The difference between the pressure loss for the empty housing and for the housing with element is the pressure loss for the element only. For "spin-on" filters, pressure loss for the complete unit may be the only value of interest, in which case pressure loss for the element only will not be required.

6. Presentation of Data

6.1 Tabulate pressure differential at each required flow and temperature.

6.2 Plot a curve of flow rate versus pressure differential using the ordinate for pressure differential and the abscissa for flow rate.

CHAPTER 2

FILTER CAPACITY AND CONTAMINANT REMOVAL CHARACTERISTICS

1. General Information—This procedure has been developed to provide a standard method for evaluating filter performance in terms of contaminant holding capacity and ability to maintain effluent cleanliness within specified contamination limits. These characteristics are defined as follows:

Contaminant Holding Capacity—The amount of abrasive contaminant removed from a recirculating flow of test fluid and held by the filter before a specified pressure drop across the filter is reached. The limiting pressure drop is specified by the user and is related to the limitations of the fuel system served by the filter.

Fuel Cleanliness—The measure of the level of contaminant remaining in the filter effluent samples at specified time intervals during the test with abrasive contaminant. Cleanliness is specified in terms of mass of contaminant per unit volume of fluid, or mass of contaminant per unit mass of fluid.

Test fluids and contaminants, described in Appendix A, have been selected because they represent or accurately simulate the behavior of fuels and contaminants encountered in normal fuel filter operation. Their characteristics are closely controlled so that, used in accordance with standard procedures contained in this document, test results will be repeatable and comparable among laboratories using these test procedures. Producers of the test materials will not alter any characteristics unless they have first ascertained that the proposed change will have negligible effect on all tests in which they are used. No changes are made without consultation with the SAE Filter Test Methods Subcommittee.

CAUTIONARY NOTES—Prior standard tests incorporated in SAE J905 used a nondispersant fluid for the tests of filtration efficiency and dirt capacity. The test fluid specified for the tests in this chapter has significant dispersancy. For this reason, apparent filtration efficiency and capacity by this revision will be significantly lower than if the same filter was evaluated by the older method. These results are more realistic than earlier test results and are more repeatable and reliable.

This test is an evolution of the previously used test which was derived from years of experience in the testing of fuel filters. *However, the procedure is limited strictly to the laboratory comparison of filters. Test results are not directly relatable to the field performance of the filter.*

The differences between laboratory results and field experience are attributable to the types of contaminant used in the test procedure and that found in the field. General experience is that fuel filters in the field become restricted rapidly due to asphaltenic type materials in the fuel. Asphaltenic materials are micron or submicron particles of carbon with a resinous coating. The abrasive contaminant used in the test procedure provides a poor approximation of this type of clogging action. This is supported by the experience that plugged field filters have only a small fraction of the contaminant weight gain as one would expect from this test procedure.

The SAE Filter Test Methods Subcommittee has current programs to evaluate new test methods for the investigation of fuel filter capacity and efficiency. These programs address the contaminant problem and the relation of test results to field performance. In addition, new testing avenues such as particle counting are being examined in order to create a new procedure which can provide more information about the filter.

This test procedure has been included in the interim because there are

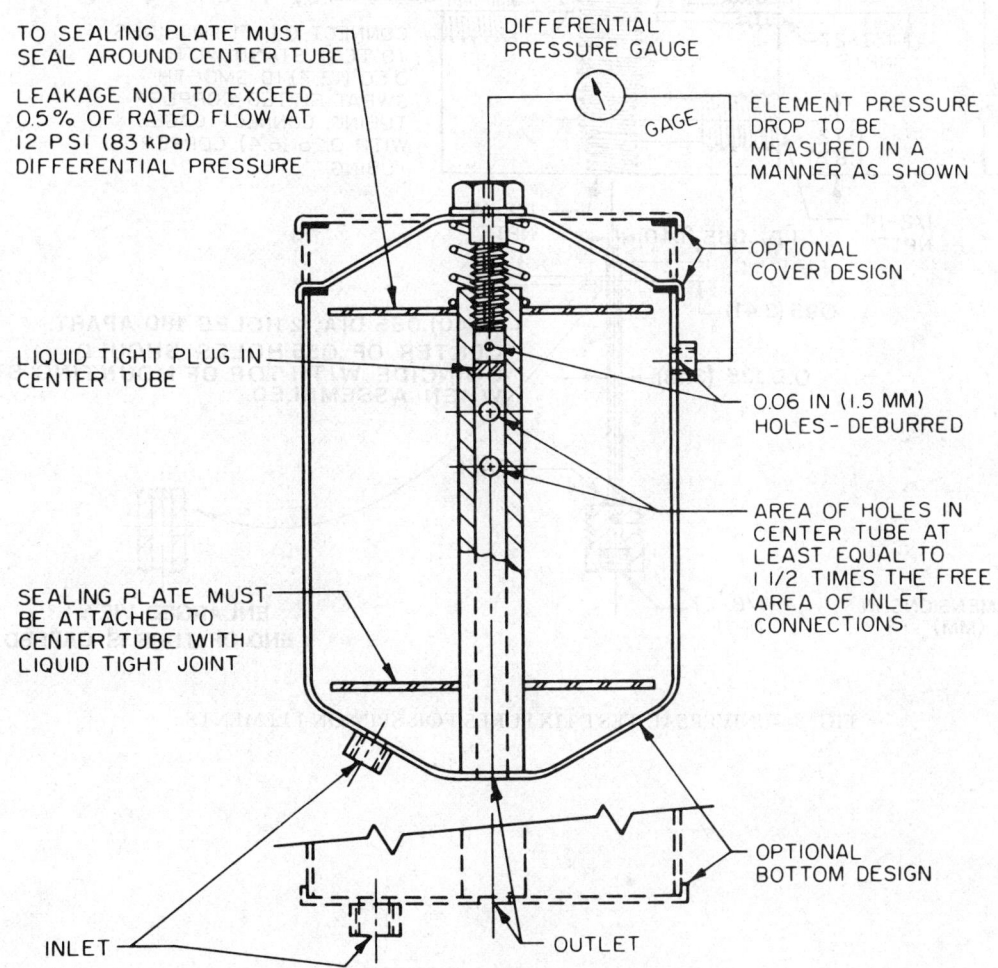

FIG. 2—TEST FILTER HOUSING FOR FULL-FLOW ELEMENTS

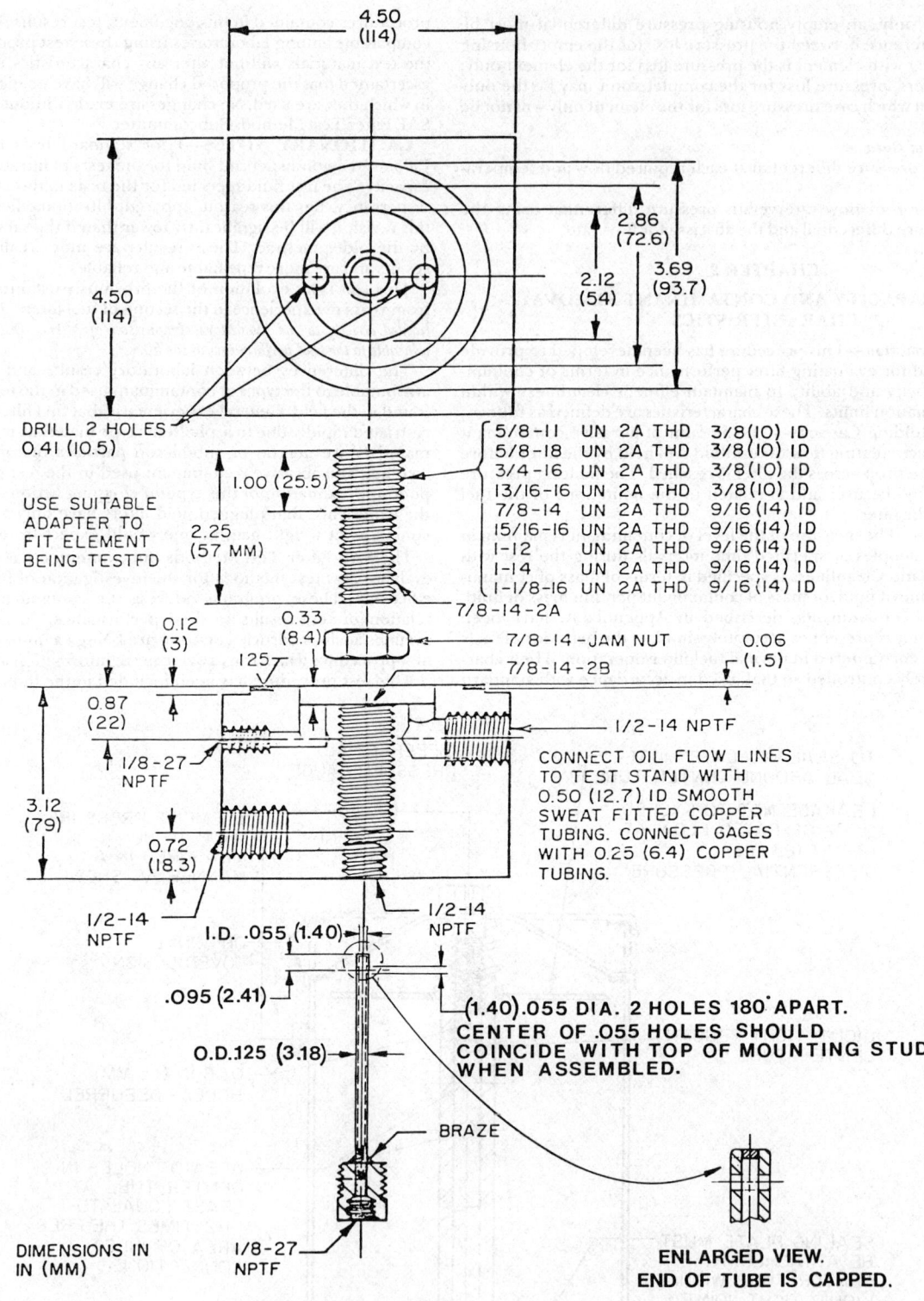

FIG. 3—UNIVERSAL TEST FIXTURES FOR SPIN-ON ELEMENTS

years of background and data which have been generated through its use. Most users have significant experience with the procedure and have developed special in-house methods to extend the test results to field applications. Although this is not the optimal situation, the FTMSC feels that this procedure will be useful in the interim, as long as users continue to exercise care in the interpretation of the test results.

Test results using ultrasonically dispersed test dust show slightly lower efficiency and dirt capacity than when the Waring blender is used. The difference is within the normal expected variation among laboratories, but is real. The Waring blender is specified because it will reliably disperse the test dust adequately. Other blenders have been tried, and results show that the dust was not well dispersed. Blenders which can be demonstrated to disperse contaminant as well as the Waring, verified by test results which correlate well with test results using a Waring blender, may be used. The Subcommittee has no evidence which would recommend other blenders, but some may be available.

2. Scope—This test determines the contaminant holding capacity and abrasive contaminant removal characteristics of fuel filters. While the test is generally applied to filters intended to handle fuel oils for diesel engines, it is equally applicable to filters for other classes of liquid fuel. The specified test fluid has viscosity characteristics similar to No. 2 diesel fuel.

In the construction of the test stand, care must be taken to insure that any contaminant added to the test sump is properly dispersed and fully presented to the test filter. This implies that the system must not have any traps or settling zones where contaminant could be lost, nor should the pumping system alter the contaminant in any way. The test apparatus recommendations are made with these points in mind.

3. Test Materials *(refer to Appendix A)*
   3.1 Test Fluid—Viscor Fuel Filter Test Fluid L-4264V
   3.2 Test Contaminant—AC Fine Test Dust

4. Test Apparatus
   4.1 Fuel Filter Test Stand, as shown in Fig. 4, with or without optional continuous contaminant feeder.
   4.1.1 PUMP—Progressive cavity type (Moyno pump or equivalent).
   4.1.2 HEAT EXCHANGER—Water jacketed length of system tubing.
   4.1.3 SUMP SAMPLE LINE—Positioned to deliver a constant flow from the pump outlet, returning to the sump.
   4.1.4 SYSTEM TUBING AND FITTINGS—½ in nominal tubing recommended; all fittings should be of such construction that they do not create any contaminant traps.
   4.2 Appropriate housing or standard mounting base for spin-on filters (Figs. 2 and 3).
   4.3 Ultrasonic Bath (5 w/in^2) with slurry beaker suspended ½ in above the tank bottom (see Fig. 5). Use the following or proven equivalent: Westinghouse MAGNATRAK portable generator Model G6C, 600w 115V and MAGNATRAK transducerized tank, high power density Model T6C4.3, 600w, 4.3 gal. capacity, 10 in × 10 in × 10 in inside dimensions of tank.

Alternately, a heavy duty laboratory Waring Blender may be used. Use the following or proven equivalent: Waring Blender Model 7011S, 2 speed, with Blender Container SS610, 1 L bakelite bottom with drawn one piece shell. Blender speed 21,000 rpm, no load. (NOTE: Blade wear may cause inconsistent dispersion; based on use, blades should be replaced at periodic intervals.)

Report the specific type and model of ACFTD dispersing device when reporting test results. Test results may not be consistent between the two recommended methods. The use of devices other than those recommended may compound the problem.

   4.4 Analytical apparatus for effluent sample analysis (see Appendix B).

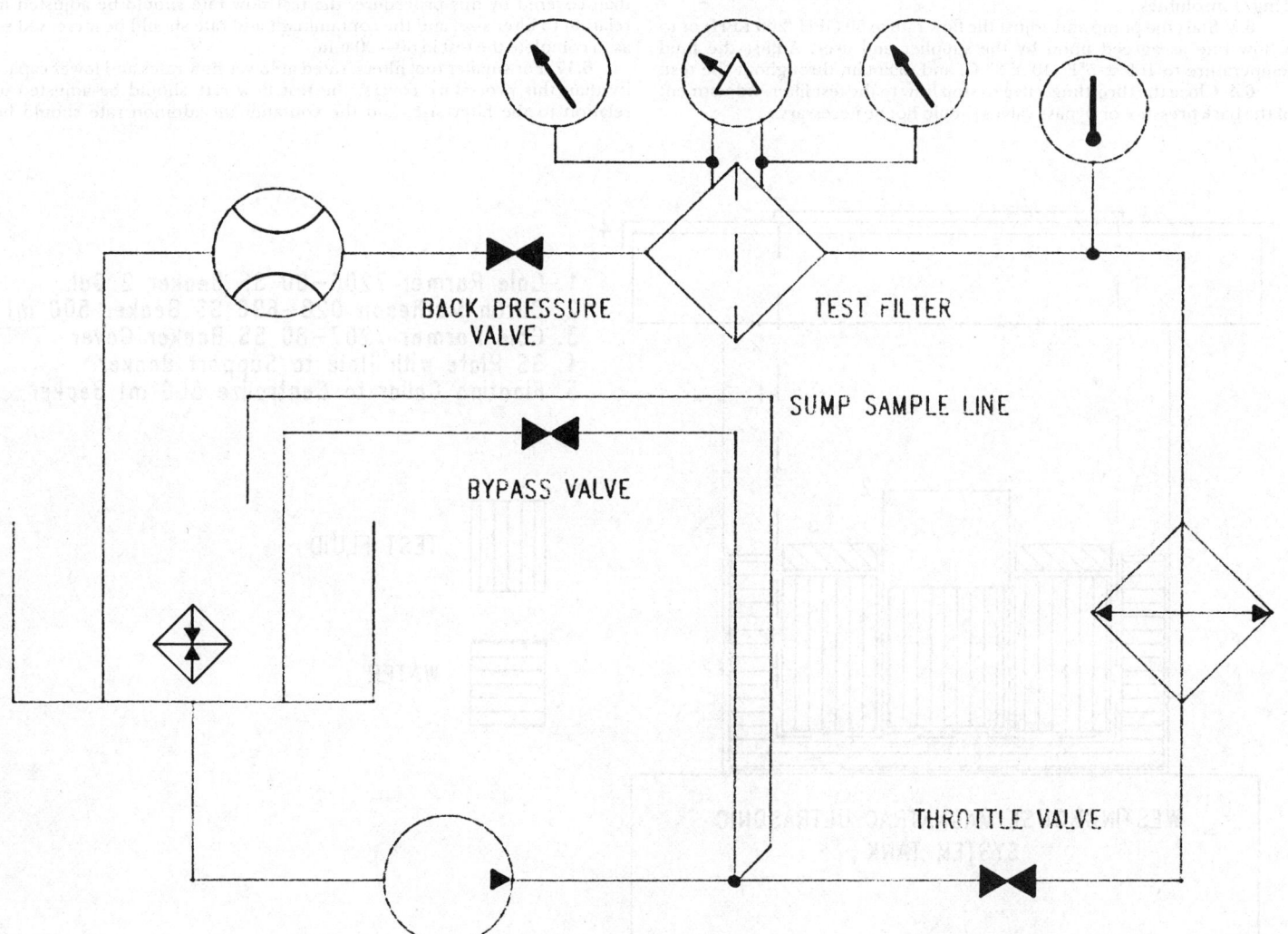

FIG. 4—FUEL FILTER TEST STAND

5. Preparation of ACFTD Contaminant Slurry

5.1 Prepare the AC Fine Test Dust slurry using either the ultrasonic or the alternate blender dispersion method. Equipment is as described in previous section.

NOTE: Dry the test dust in an open container in a convection oven at a temperature of 230—300°F (110—150°C) for 2 h. The depth of the test dust in the drying container should be less than 1 in (25.4 mm). Store the dried dust in a desiccator prior to use.

5.2 Ultrasonic Dispersion Method

5.2.1 Place the specified mass of dried ACFTD required for a 5 min incremental addition in the 500 ml beaker shown in Fig. 5.

5.2.2 Mix test dust with 50 ml of test fluid using a rubber policeman for 15 sec.

5.2.3 Dilute mixture to 350—400 ml total volume with test fluid. Test fluid used should be taken from the test stand in use for this test.

5.2.4 Place beaker with dust/fluid mixture in the ultrasonic bath and disperse for 2.5 ± 0.5 min. The preparation of the slurry should be performed in the 5 min interval between slurry additions to the test system.

5.3 Blender Dispersion Method

5.3.1 Place the specified mass of dried ACFTD required for a 5 min incremental addition in the blender one liter container.

5.3.2 Add approximately 700 ml of test fluid to the dust and blend at high speed (21,000 rpm, no load) for 2.5 ± 0.5 min. The preparation of the slurry should be performed in the 5 min interval between slurry additions to the test system.

6. Test Procedure

6.1 Fill the test stand sump with 5 gal (20 L) of test fluid. Install a cleanup filter in the test filter location and circulate the test fluid until a 100 ml sample has an insoluble concentration of less than 2 mg/l.

6.2 Remove the cleanup filter and install the test filter. Adjust the test fluid volume to 5 gal (20 L) using test fluid prefiltered to a level of less than 2 mg/l insolubles.

6.3 Start the pump and adjust the flow rate to 50 GPH (200 LPH) or to a flow rate as agreed upon by the supplier and user. Adjust the fluid temperature to $100 \pm 5°F$ ($40 \pm 3°C$) and maintain throughout the test.

6.4 Close the throttling valve to stop flow to the test filter. Adjustment of the back pressure or bypass valves should not be necessary.

6.5 Add to the test sump 2 g of ACFTD prepared as in paragraphs 5.2 or 5.3. Modified addition rates may be used as agreed upon by the user and supplier. Wash all the contaminant from the slurry container with test fluid from the sump sample line.

6.6 After 2 min of mixing, open the throttling valve. The test flow rate should re-establish itself if the bypass valve has not been adjusted. Maintain required flow rate throughout the test.

6.7 The instant of opening the valve to start flow through the test filter is time zero.

6.8 Take effluent samples of 100 ml each at times of 0.5, 1.0, 2.0 and 4.5 min from time zero.

6.9 Replenish the test system with 100 ml of prefiltered (> 2 mg/l) fluid immediately after removing each sample.

6.10 At time zero plus 5 min, add 1 g of ACFTD prepared as in paragraph 5.2 or 5.3. Rinse the container as in paragraph 6.5. Modified add rates may be used as agreed upon by the user and supplier.

6.11 Repeat each 5 min of the test until the terminating pressure drop is reached (10 psi (70 kPa) if no limit is specified).

6.12 Take effluent samples of 100 ml each 1 min before each slurry addition after the initial charge until the pressure drop reaches 25% of the terminating value, then at 50% and 75% of the limiting value. Replenish the test system with 100 ml of clean test fluid immediately after taking these samples.

6.13 Record the pressure drop across the filter just prior to removing each effluent sample.

6.14 Take a final 100 ml effluent sample and a 100 ml sump sample just prior to shutdown of the test when the terminating pressure drop has been reached.

6.15 Analyze the samples and present the data as specified in Sections 7 and 8.

6.16 For larger filters, rated at higher flow rates and greater capacities than covered by this procedure, the test flow rate should be adjusted in relation to filter size, and the contaminant add rate should be increased so as to complete the test in 60—90 min.

6.17 For smaller fuel filters, rated at lower flow rates and lower capacity than this procedure covers, the test flow rate should be adjusted in relation to the filter size, and the contaminant addition rate should be

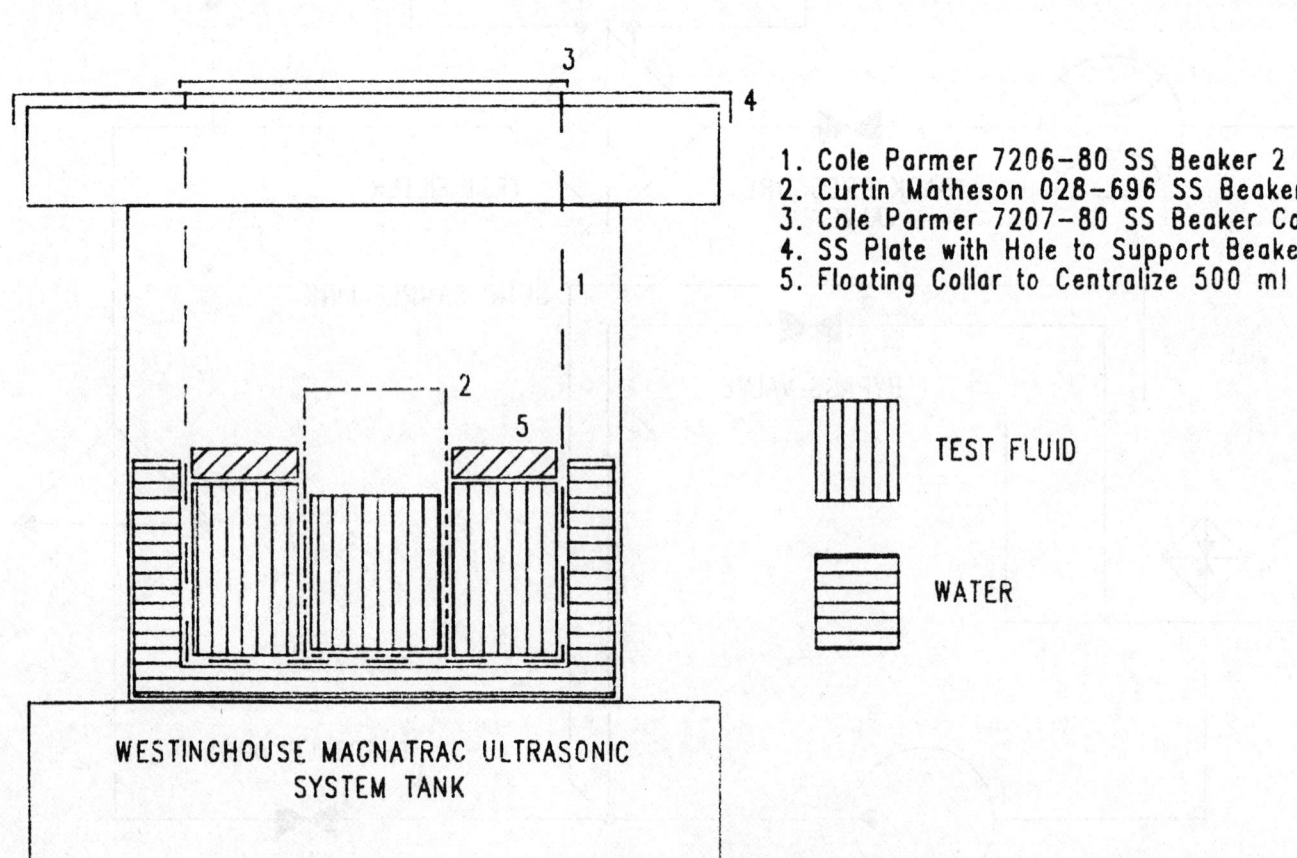

FIG. 5—ULTRASONIC SLURRY PREPARATION BATH

adjusted so as to complete the test in 30—60 min. For smaller filters, the following 5 min incremental ACFTD additions are suggested as options to meet the 30—60 min requirement:

1 ± 0.050 grams
0.5 ± 0.025 grams
0.25 ± 0.0125 grams
0.10 ± 0.005 grams

6.18 This test may be modified to provide for continuous feeding of contaminant. The contaminant batch is prepared by adding 10 ± 0.1 grams of ACFTD to the 500 ml beaker shown in Fig. 5. Disperse in the ultrasonic bath for 10 min. Dilute the slurry in the dispenser to a concentration of 10 g/l with prefiltered test fluid. Use proportionately higher concentrations for larger filters and lower concentrations for smaller filters. Prepare as much contaminant as is expected to be required to complete the test. Set the dispensing rate to 20 ml/min. (NOTE: Constant agitation of the slurry is needed to maintain homogeneity.) Time of feeding of the contaminant until test termination determines the quantity of solids added (i.e., 20 ml/min at 10 g/l = 1 g solids per 5 min). Start flow through the test filter (timed) after 2 g of contaminant have been dispensed into the test sump. For tests with altered ACFTD concentrations, dispense into the sump for 10 min before starting the test. Modified feed rates may be used as agreed upon by the user and supplier.

Conduct continuous feed tests in accordance with paragraphs 6.7, 6.8, 6.9 (mod), 6.12 (mod), 6.13, 6.14 and 6.15.

Modify paragraph 6.9 to read "replenish the system with 100 ml of clean (2 mg/l or less) fluid immediately after the removal of the 4.5 min sample."

Modify 6.12 to read "take effluent samples of 100 ml each, at 10 min from time zero and at 5 min intervals thereafter until the pressure drop reaches 25% of the terminating value, then at 50% and 75% of the limiting value. Replenish the sump with 100 ml of prefiltered fluid after each of these samples is taken."

7. Sample Analysis
7.1 Analyze all samples by the Millipore filtration method specified in Appendix B.

8. Data Presentation
8.1 Record all data on data reporting sheet.
8.2 Plot pressure drop versus contaminant added and mg/l or % insolubles versus contaminant added.
8.3 Report capacity, calculated as:

Capacity = total grams of contaminant added to the system minus (mg/l contaminant in final sump sample × liters of fluid in test system).

CHAPTER 3
MEDIA MIGRATION TEST

1. Scope—The purpose of this test is to determine if the filter introduces contaminant into the fuel system downstream of the filter. This contamination is referred to as media migration, although its source is not necessarily the filter media. Any of the filter components or the manufacturing techniques used in building the filter can also be a source of contamination.

The test method consists basically of flowing clean fluid through several filters and collecting the contaminant from the effluent on a wire cloth screen from which weight determinations and visual observations may be made.

2. Test Materials
2.1 Test Fluid—Viscor Fuel Filter Test Fluid L-4264V, prefiltered.
2.2 Four test filters
2.3 Solvent, analytical grade
2.3.1 n-Pentane
2.3.2 Petroleum Ether

3. Test Apparatus
3.1 Test Stand—See Fig. 6.
3.2 Media Migration Filter Screen
3.2.1 WIRE CLOTH SCREEN—41 m nominal.
3.3 Analytical Equipment—See Appendix B.

4. Test Preparation
4.1 Install wire cloth screen in holder in series with and downstream of the test filter location. Fill sump with test fluid. Install a cleanup filter in place of the test filter.
4.2 Circulate test fluid through the system at rated flow and 100°F (38°C) for 4 h and determine the system cleanliness by method described in paragraph 5.4. Repeat as often as necessary to achieve a cleanliness level of 2.0 mg maximum.

5. Test Procedure
5.1 After satisfying paragraph 4.2, install a test filter.
5.2 Install clean wire cloth screen in holder.
5.3 Flow test fluid through test filter at rated flow and temperature for 30 min. Repeat this cycle of 30 min circulation on a total of four test filters without a change or addition to the test sump. Discard any spillage that occurs in changing test filters. Do not replenish the sump with any make-up fluid.
5.4 Carefully remove the wire cloth screen from the holder and wash collected migration from the screen into a clean beaker with prefiltered n-pentane or petroleum ether. No less than 300 ml of solvent should be used for a thorough washing. Filter the washing through a tared $0.8\mu m$, 47 mm diameter analytical membrane filter disc following the techniques recommended in Appendix B. Determine the total weight of contaminant contained in the washing to the nearest 0.1 mg.
5.5 Analyze the contaminant to determine its nature.

6. Presentation of Data
6.1 The total weight of media migration (contaminant from washing) shall be divided by four to obtain the media migration per filter.
6.2 Report the nature of the contaminant from the test filters.

7. Determination of Inorganic Material of Migration
7.1 Test Apparatus
7.1.1 Evaporating dish or crucible made of platinum, silica, or porcelain of 90—120 ml capacity.
7.1.2 Electric muffle furnace capable of maintaining a temperature of 775°C ± 25°C.
7.2 Procedure
7.2.1 Heat empty evaporating dish or crucible at 700—800°C for 30 min. Cool for 30 min to room temperature and weigh to the nearest 0.1 mg. Repeat until no weight change occurs.
7.2.2 Weigh the evaporating dish or crucible with the membrane filter from the media migration determination to the nearest 0.1 mg. Heat in furnace at 775°C ± 25°C for 30 min. Cool container for 30 min to room temperature and reweigh. Repeat the heating and reweighing until consecutive weights are within 0.2 mg.
7.3 Calculation of Results
7.3.1 Report results as:

$$\text{Inorganic (\%)} = W_a/W_b \times 100$$

where W_a = mass of ash in grams and W_b = mass of contaminant on membrane filter.

CHAPTER 4
COLLAPSE TEST

1. Scope—The purpose of this test is to assure that a filter element will withstand the anticipated maximum differential pressure without bypassing due to breakage or collapse while filtering. The test method will determine element strength by means of pumping contaminated fluid through the element until collapse occurs or the anticipated maximum differential pressure is reached.

2. Test Materials
2.1 Test Fluid—SAE J1260, Standard Oil Filter Test Oil (RM99) or Viscor Fuel Filter Test Fluid L-4264V.
2.2 Contaminant—AC Fine Test Dust or other suitable choking type contaminant.

3. Test Apparatus
3.1 A suitable filter housing with pressure taps to sense filter element differential pressure.
3.2 A pump and motor capable of pumping the test fluid at minimum specified collapse pressure.
3.3 A reservoir of sufficient size.
3.4 Necessary piping, fittings, and valves to flow oil and control flow to the test filter (see Fig. 1).
3.5 A differential pressure gauge, with a maximum indicating pointer, capable of measuring at least 150 psig (1040 kPa).
3.6 A device for heating and controlling the temperature of the test fluid, if an elevated temperature is desired.
3.7 The fuel filter test stand (Fig. 4) may be used for this test.

4. Test Preparation
4.1 The filter element to be tested may be one previously subjected to a capacity and contaminant removal test or a new element. The test element should be examined for any apparent damage before the collapse test is performed.
4.2 Block or eliminate relief valve, if present.

5. Test Procedure
5.1 Place the test element in the housing on the test stand.
5.2 With the main valve closed and the bypass valve open, start the pump and circulate the fluid until the desired temperature is reached.
5.3 Open the main valve and close the bypass valve as needed to establish flow through the test element. All air must be bled from the housing at this time.
5.4 After the air has been bled from the housing, start adding contam-

inant to the test sump; this may be done manually or automatically.

5.5 When a differential pressure of approximately 25 psig (172 kPa) across the test element is reached, stop the contaminant addition and allow the pressure to increase. If the pressure drop increase stops, add more contaminant until the pressure drop starts to rise again. Adjust the bypass valve as needed to maintain flow through the test element.

5.6 Differential pressure should be increased until the element collapses as evidenced by a sudden drop in differential pressure, or until the specified minimum collapse pressure is reached.

6. Presentation of Data and Evaluation of Results

6.1 The report shall include the following information:

6.1.1 Pretreatment of filter element, i.e., capacity test, preflow, or new element.

6.1.2 Maximum differential pressure attained.

6.1.3 Reason for terminating test.

6.1.4 Condition of filter element after test.

CHAPTER 5
ABILITY TO MEET ENVIRONMENTAL CONDITIONS

1. Scope—This test method evaluates the effects of water in fuel upon the performance characteristics of fuel filters.

2. Test Method for the Effect of Water in Fuel on Filter Capacity

2.1 Test Materials

2.1.1 Distilled or de-ionized water

2.1.2 Viscor Fuel Filter Test Fluid L-4264V

2.2 Test Apparatus

2.2.1 Fuel filter test stand (Fig. 4)

2.2.2 Water addition system (Fig. 7)

2.3 Test Preparation

2.3.1 Prepare the test stand as described in Chapter 2, Section 4.

2.3.2 Install the water addition system in place above the sump of the test stand and fill water add system with distilled or de-ionized water.

2.4 Calibration and Adjustment of Water Addition System

2.4.1 Partially open needle valve "A" (see Fig. 7) so that water is deliv-

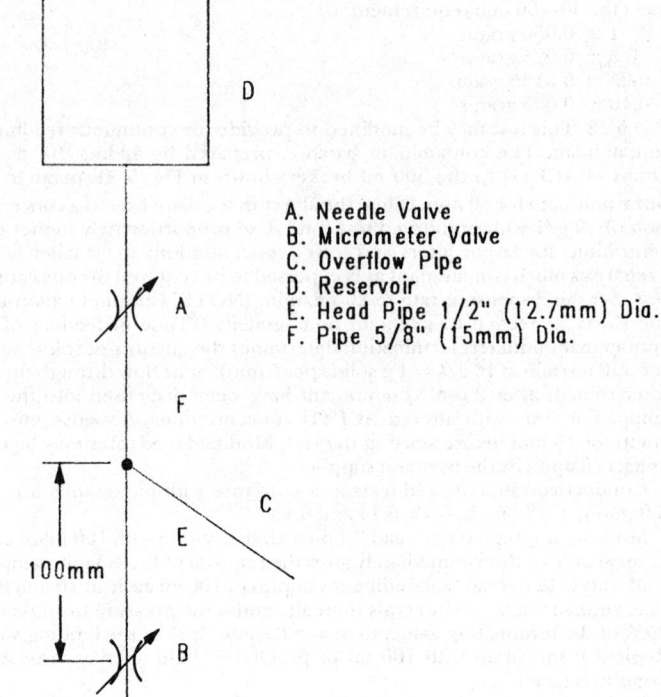

FIG. 7—WATER ADDITION SYSTEM

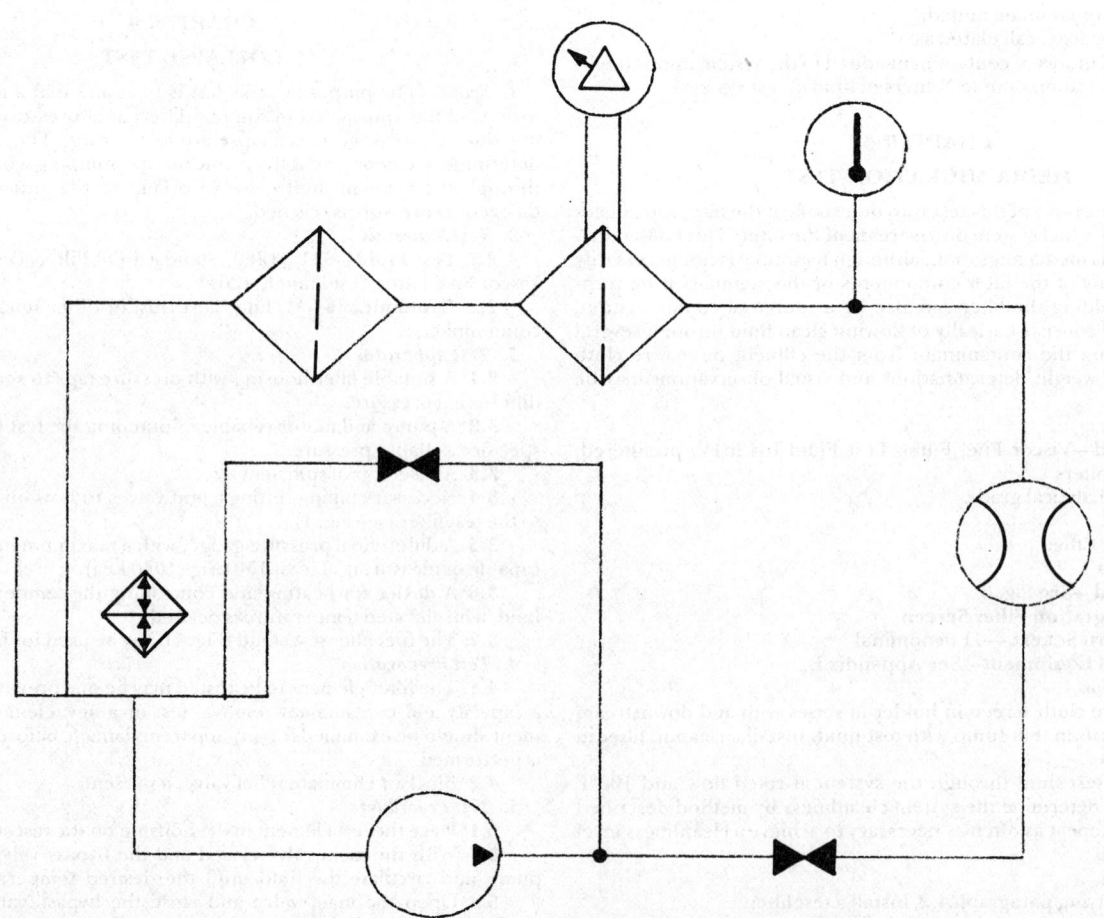

FIG. 6—MEDIA MIGRATION TEST STAND

ered to the head pipe and flow control valve. Divert flow from control valve "B" into a suitable container.

2.4.2 Open the micrometer flow control valve "B" so that the zero of the vernier is in line with the "1" on the scale.

2.4.3 Adjust the needle valve "A" so that there is a very small, constant flow from the overflow drain pipe "C". This provides a constant head pressure to the control valve.

2.4.4 Record the time required to deliver a predetermined volume of water through valve "B" and calculate the delivery rate in suitable units (i.e., L/h).

2.4.5 Repeat paragraphs 2.4.2 through 2.4.4 for other micrometer dial settings until the full range has been covered.

2.4.6 Plot micrometer setting versus delivery rate. From this plot, the approximate micrometer setting for any desired delivery rate can be obtained. Final adjustments can then be made to obtain the exact delivery rate.

2.5 Test Procedure

2.5.1 The test procedure is the same as the capacity and contaminant removal test described in Chapter 2, with the addition of the following procedure:

2.5.2 With needle valve "A" closed, empty any water from the head pipe and control valve. Set the micrometer to deliver the desired flow rate.

2.5.3 With effluent from the control valve diverted to a suitable container, open needle valve "A" and adjust flow so that there is a very small, constant flow from overflow drain pipe "C".

2.5.4 Divert the flow from control valve into test sump at test time zero.

2.5.5 Recommended water delivery rate is 0.5% of the test fluid flow rate to the test filter.

2.6 Presentation of Data

2.6.1 Results should be reported as in Chapter 2 with the additional notation of the amount of water added during the test.

CHAPTER 6
INSTALLATION AND REMOVAL

1. Seal Test

1.1 Scope—This test assures that the seal material is suitable for the intended application.

1.2 Submit three seals to tests as specified by SAE J200 MAY85 to assure that they meet the requirements of the specified rubber type.

2. Installation Sealing Torque

2.1 Scope—This test method relates to the static sealing characteristics of spin-on fuel filters. Results indicate the ability of the filter to seal at the sealing surface of the mounting head when installed with the specified torque.

2.2 Test Material—No. 2 Diesel Fuel or Viscor Fuel Filter Test Fluid L-4264V.

2.3 Test Apparatus (see Fig. 9).

2.3.1 Test apparatus used for generating and measuring pressure is the same as that specified in Chapter 7, Section 1.

2.3.2 Torque wrench readable to 5 in/lb (0.56 Nm).

2.4 Test Preparation

2.4.1 Check that gasket is bottomed out in filter gasket retaining groove.

2.4.2 Measure filter flange to determine dimensional relationship of gasket groove and threaded mounting hole of filter.

2.5 Test Procedure

2.5.1 Apply thin film of ASTM #3 oil to gasket surface and fill test filter with test fluid.

2.5.2 Screw filter onto mounting head until the gasket makes contact.

2.5.3 Tighten the filter to the test fixture with the application of 25 in/lb (2.82 Nm) of torque.

2.5.4 Position the plastic shield between the test filter and the operator.

2.5.5 Using the hydraulic pump, pressurize the assembly until the first

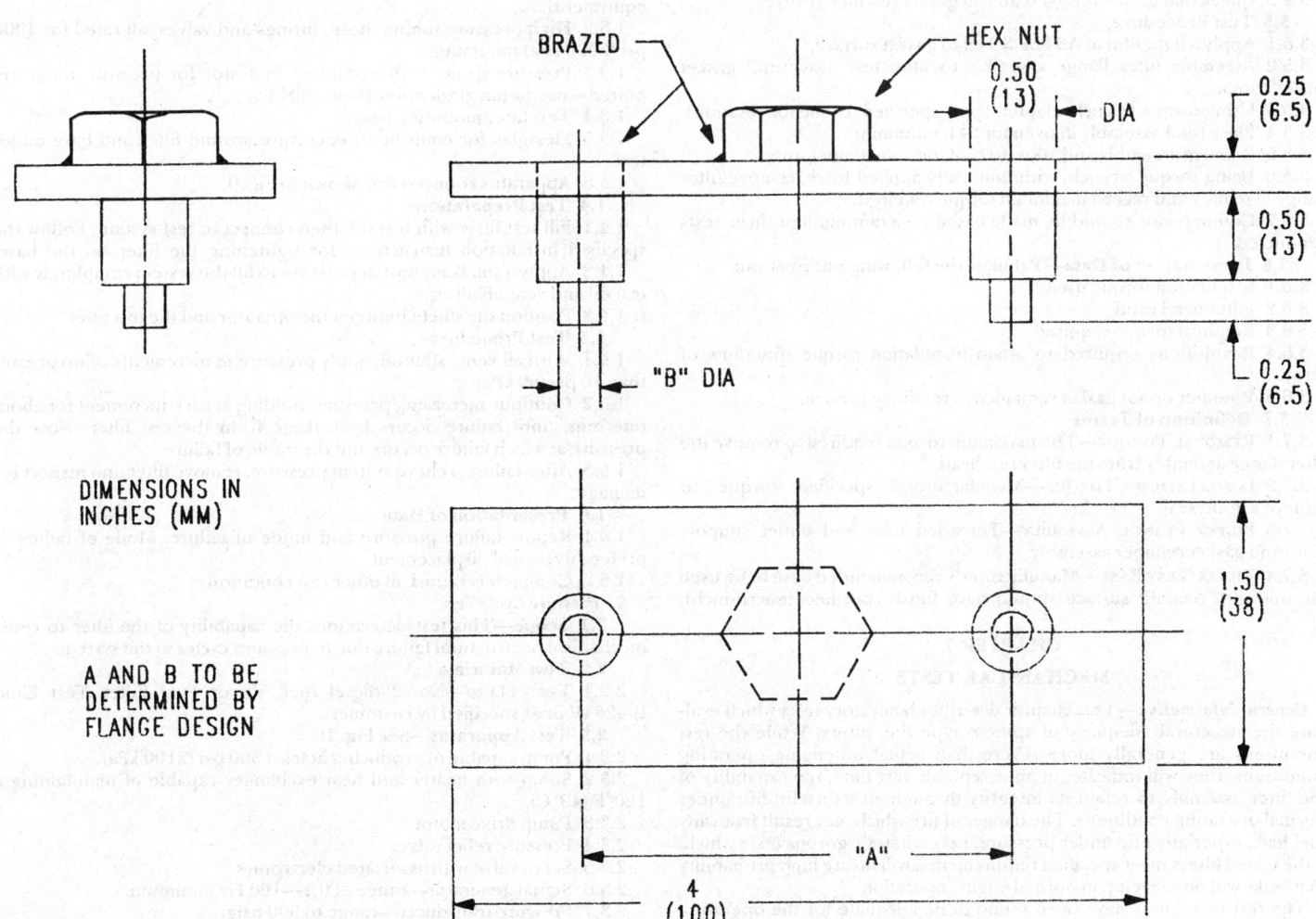

FIG. 8—TORQUE WRENCH ADAPTER

sign of leakage. Record leakage pressure.

2.5.6 Determine leakage pressure at increasing torque intervals of 25 in/lb (2.82 Nm) until the recommended torque or number of turns from gasket contact have been reached.

2.5.7 Remove filter from fixture and re-measure to determine if any permanent deformation of the filter flange has occurred.

2.6 Presentation of Data

2.6.1 Graph the data collected as follows: abscissa—torque; ordinate—pressure to produce leakage.

2.6.2 Report any deformation of the filter flange and the maximum pressure reached.

2.6.3 Note torque required to attain seal pressure desired.

3. Removal Torque Test Method

3.1 Scope—This test method relates to the torque required to remove spin-on fuel filters after exposure to specified installation torque pressure and simulated engine temperatures.

3.2 Test Material—Test fluid: ASTM #3 oil, or as otherwise specified.

3.3 Test Apparatus

3.3.1 Constant temperature oven, capable of maintaining 125 ± 2°F (52 ± 2°C).

3.3.2 Thermometer (immersion type) with a temperature range of 0—250°F (0—120°C).

3.3.3 Torque wrench with a torque range of 0—500 in/lb (0—56.5 Nm) readable to 5 in/lb (0.56 Nm).

3.3.4 Torque wrench adapter. Suggested design is shown in Fig. 9.

3.3.5 FILTER TEST BASE—Use manufacturer's specified base.

3.3.6 PRODUCT TO BE EVALUATED—Gasket and spin-on filter flange assembly (unit end plate assembly, nut plate assembly, etc.).

3.4 Test Preparation

3.4.1 Preheat oven to 125°F (52°C).

3.4.2 Secure filter base in vise; be sure that threads are clean and properly sized.

3.4.3 Check that gasket is fully seated in gasket retainer groove.

3.5 Test Procedure

3.5.1 Apply a light film of ASTM #3 oil to gasket surface.

3.5.2 Assemble filter flange assembly to filter test head until gasket makes contact.

3.5.3 Using torque wrench adapter, apply specified torque (or rotation).

3.5.4 Place total assembly in oven for 24 h minimum.

3.5.5 Remove assembly and allow to cool to room temperature.

3.5.6 Using torque wrench, with uniformly applied force, remove filter flange assembly and record maximum torque required.

3.5.7 Comparisons should be made based on a minimum of three tests conducted.

3.6 Presentation of Data—Tabulate the following information:

3.6.1 Installation torque used.

3.6.2 Filter head used.

3.6.3 Removal torque required.

3.6.4 Revolutions required to attain installation torque (fractions of turn).

3.6.5 Whether or not gasket remained in retaining groove.

3.7 Definition of Terms

3.7.1 REMOVAL TORQUE—The maximum torque required to remove the filter flange assembly from the filter test head.

3.7.2 INSTALLATION TORQUE—Manufacturer's specified torque to achieve a good seal.

3.7.3 FILTER FLANGE ASSEMBLY—Threaded inlet and outlet support plate and gasket retainer assembly.

3.7.4 FILTER TEST BASE—Manufacturer's recommended base to be used for this test. Sealing surface should have finish specified (micro-inch).

CHAPTER 7
MECHANICAL TESTS

General Information—This chapter describes laboratory tests which evaluate the structural adequacy of spin-on type fuel filters. While the test conditions are generally more severe than actual on-engine operating conditions, they will indicate, in an acceptable test time, the capability of the filter assembly to retain its integrity throughout its useful life under normal operating conditions. The danger of fire which may result from any fuel leak, especially one under pressure, necessitates rigorous tests which, if the tested filters meet specified requirements, will insure high probability that leaks will not develop in normal engine operation.

The test conditions have been found to be adequate for the operating conditions of most engines. However, if maximum reliability is to be assured, each engine application should be investigated to determine the particular pressure levels and pressure cycle range to which the filter will be subjected. Test conditions should be modified, if necessary, to account for operating conditions more severe than average.

Filter mounting brackets and other hardware must be capable of withstanding the same test conditions. Usually, it will be advantageous to include these items in the installations for pressure cycle testing and vibration testing unless previous tests have established their adequacy.

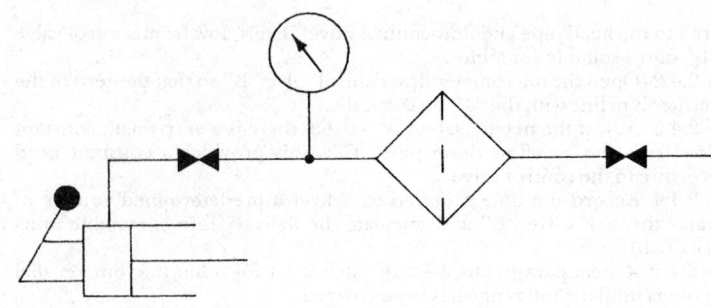

FIG. 9—BURST TEST STAND

1. Burst Test

1.1 Scope—This test determines the capability of the filter to withstand maximum fuel system surge pressure.

1.2 Test Materials

1.2.1 TEST FLUID—SAE 10 engine oil, hydraulic fluid or similar fluid compatible with the test apparatus and filter seals.

1.3 Test Apparatus

1.3.1 Hand operated high pressure hydraulic pump (Enerpac P80 or equivalent).

1.3.2 High pressure tubing, hose, fittings and valves, all rated for 1000 psi (6900 kPa) minimum.

1.3.3 Pressure gauge with maximum indicator for pressure range required—maximum graduation 10 psi (69 kPa).

1.3.4 Test filter mounting base.

1.3.5 Plexiglas (or equivalent) enclosure around filter and base under test.

1.3.6 Apparatus connected as shown in Fig. 9.

1.4 Test Preparation

1.4.1 Fill test filter with test oil, then connect to test system. Follow the specified installation instructions for tightening the filter on the base.

1.4.2 Apply a small amount of pressure to fill the system completely with test oil and vent off all air.

1.4.3 Position the shield between the operator and the test filter.

1.5 Test Procedure

1.5.1 With all vents shut off, apply pressure in increments of no greater than 10 psi (69 kPa).

1.5.2 Continue increasing pressure, holding at each increment for about one min, until failure occurs by leakage from the test filter. Note the pressure at which failure occurs and the mode of failure.

1.5.3 After failure, relieve system pressure, remove filter and inspect for damage.

1.6 Presentation of Data

1.6.1 Report failure pressure and mode of failure. Mode of failure is preferably by seal displacement.

1.6.2 Completely report all other test conditions.

2. Pressure Cycle Test

2.1 Scope—This test determines the capability of the filter to resist mechanical or structural failure due to pressure cycles in the system.

2.2 Test Materials

2.2.1 TEST FLUID—No. 2 diesel fuel, Viscor Fuel Filter Test Fluid L-4264V or as specified by customer.

2.3 Test Apparatus—See Fig. 10.

2.3.1 Pump capable of producing at least 300 psi (2100 kPa).

2.3.2 Sump with heater and heat exchanger capable of maintaining a 120°F (49°C).

2.3.3 Pump drive motor.

2.3.4 Pressure relief valve.

2.3.5 Servo valve with associated electronics.

2.3.6 Signal generator—range of 0.1—100 Hz minimum.

2.3.7 Pressure transducer—range to 300 psig.

2.3.8 Oscilloscope.

2.3.9 Filter mounting base(s).

2.3.10 Leakage catch sump(s) with level detector(s).

2.3.11 Piping, tubing, fittings, etc. to complete apparatus per Fig. 10.
2.4 Test Procedure
2.4.1 Mount a test filter on each test base. Fill each test filter with test fluid before installing.

2.4.2 Start pump and at low pressure bleed air from test filters through bleed valve. This valve may be left slightly open during test to vent any entrained air.

2.4.3 Using pressure hydraulic pressure control and signal generator electrical controls, adjust pressure pulse to the specified amplitude, offset, and frequency as monitored by the pressure transducer and oscilloscope.

2.4.4 When the specified pressure pulse is obtained, set cycle counters to zeros.

2.4.5 Terminate a filter test when that filter fails (leaks) or when the specified number of cycles is reached.

2.5 Test Conditions
Test conditions should be established to reflect the most severe operating conditions anticipated or measured for the filter in service. The following are suggested test conditions which may be applied when actual service conditions are not known. The specified waveform applies to all test conditions.

2.5.1 The pressure waveform should be approximately sinesoidal, with no pressure spikes.

2.5.2 Pressure cycle should be 0 psi to 1.5 × normal fuel system mean operating pressure, or 0 psi to the maximum fuel system pressure observed under any operating conditions, whichever is greater.

2.5.3 If cyclic pressure variations of the fuel system have been measured for the engine on which the filter is used, the pressure cycle may be the normal mean system pressure varied by ± 3 × the observed pressure variations.

2.6 Presentation of Data
2.6.1 If a failure occurs, report the number of cycles to failure, failure mode and location on the test filter. Report all test conditions.

2.6.2 If no failure occurs before specified number of cycles, report test conditions, number of cycles completed, and that no failure occurred.

3. Vibration Fatigue Test
3.1 Scope
This test determines the ability of the filter to withstand the vibratory forces that occur during engine operation.

3.2 Test Materials
Test fluid: SAE 10 oil or equivalent.

3.3 Test Apparatus
3.3.1 Vibration test machine with automatic variable frequency control.

3.3.2 Mounting fixtures, lines, valves, pressure source and gauges as required.

3.3.3 Ultraviolet light source.

3.4 Test Preparation
3.4.1 Fill filter with test fluid and mount to vibration table following manufacturer's instructions.

3.4.1.1 Test filter including the mounting base and any brackets must be rigid so that the vibration table movement is transmitted to the test filter undiminished.

3.4.1.2 Filter should be mounted so that the direction of vibration simulates as closely as possible the motion experienced on the engine.

3.4.2 Set the vibration machine controls to obtain a vibration amplitude of 0.030 in (0.76 mm) with a frequency varied from 10-50-10 Hz once per min, or as otherwise specified.

3.4.3 Connect pressure line to filter assembly and adjust to obtain 70 psi (483 kPa) or as specified.

3.4.4 If it is desired that the filter be tested at resonant frequency, the frequency cycling controls can be shut off and the test conducted continuously at the filter's resonance frequency as determined by a measurement of filter amplitude versus frequency.

3.5 Test Procedure
3.5.1 Pressurize filter and begin cycling at time zero.

3.5.2 Observe motion of filter using a stroboscopic light if required. The

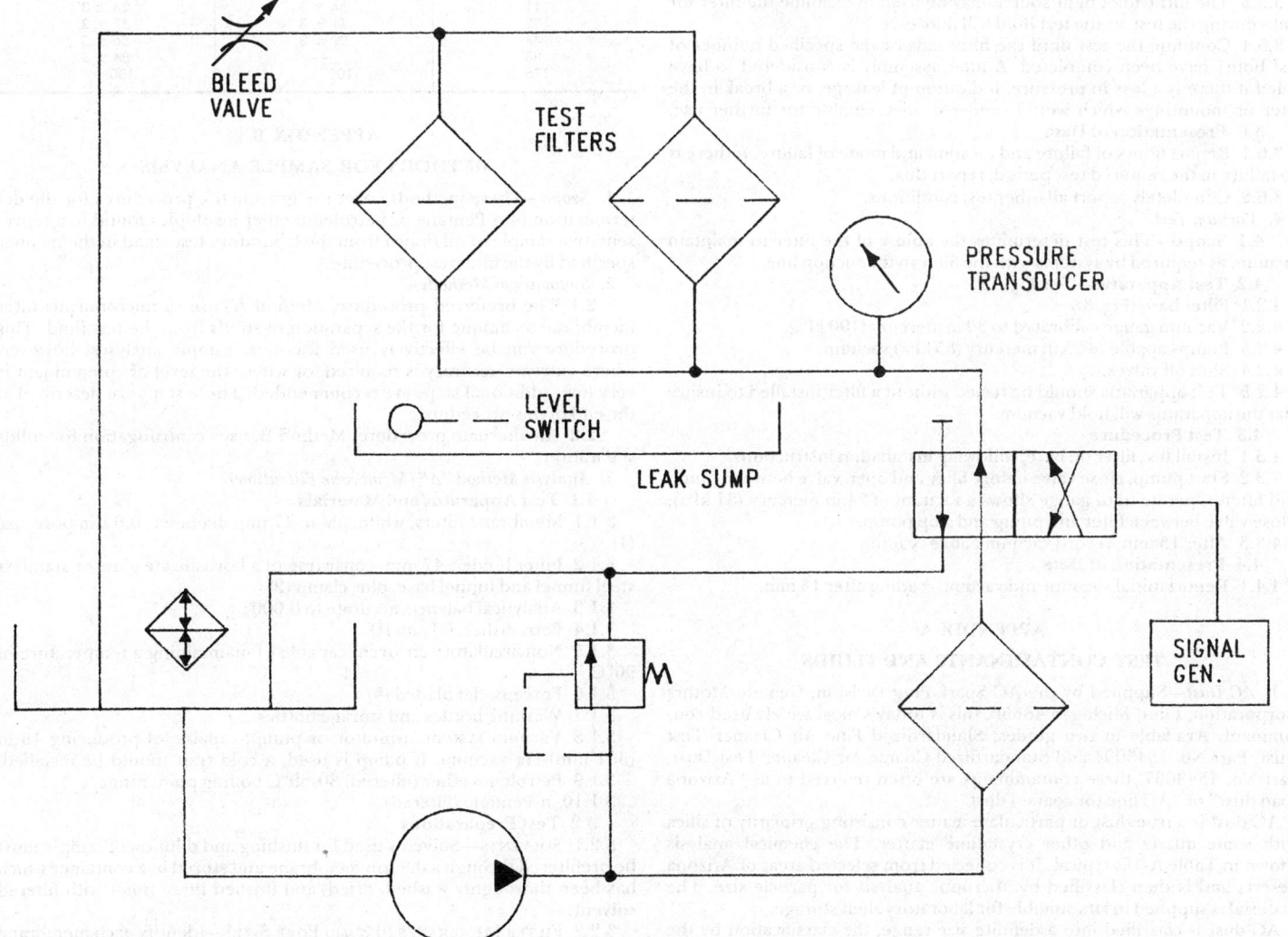

FIG. 10—PRESSURE CYCLE TEST STAND

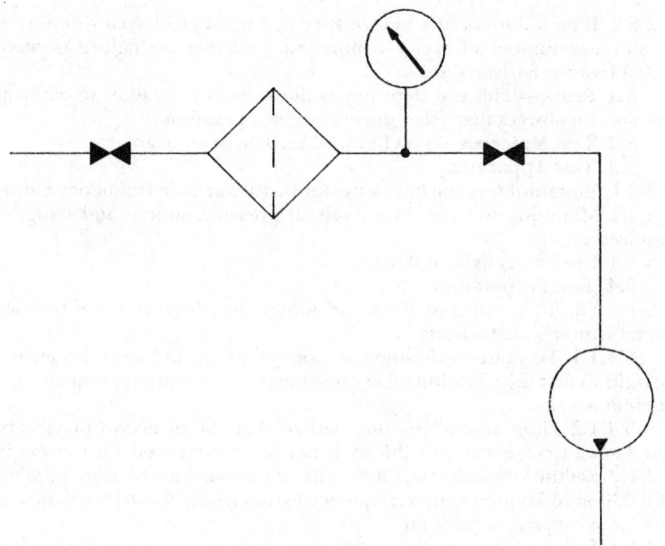

FIG. 11—VACUUM TEST APPARATUS

acceleration of the filter during test may be measured using an accelerometer coupled to a vibration meter.

3.5.3 The ultraviolet light source may be used to examine the filter for leaks during the test, as the test fluid will fluoresce.

3.5.4 Continue the test until the filter fails or the specified number of test hours have been completed. A filter assembly is considered to have failed if there is a loss in pressure, indication of leakage, or a break in the filter or mountings which would render it unacceptable for further use.

3.6 Presentation of Data

3.6.1 Report hours of failure and location and mode of failure. If there is no failure in the required test period, report this.

3.6.2 Completely report all other test conditions.

4. Vacuum Test

4.1 Scope—This test determines the ability of the filter to maintain vacuum, as required by systems with the filter in the suction line.

4.2 Test Apparatus—See Fig. 11.

4.2.1 Filter base (Fig. 3.)

4.2.2 Vacuum gauge calibrated to 30 in mercury (100 kPa).

4.2.3 Pump capable of 25 in mercury (85 kPa) vacuum.

4.2.4 Shut off valves.

4.2.5 Test apparatus should be tested without a filter installed to insure that the apparatus will hold vacuum.

4.3 Test Procedure

4.3.1 Install test filter on base, following installation instructions.

4.3.2 Start pump, close valve before filter and open valve between pump and filter. Operate until gauge shows a vacuum of 24 in mercury (81 kPa). Close valve between filter and pump and stop pump.

4.3.3 After 15 min, record vacuum gauge reading.

4.4 Presentation of Data

4.4.1 Report initial vacuum and vacuum reading after 15 min.

APPENDIX A

TEST CONTAMINANTS AND FLUIDS

1. AC Dust—Supplied by the AC Spark Plug Division, General Motors Corporation, Flint, Michigan 48556, this is today's most widely used contaminant. Available in two grades, Standardized Fine Air Cleaner Test Dust, Part No. 1543094 and Standardized Coarse Air Cleaner Test Dust, Part No. 1543637, these contaminants are often referred to as "Arizona road dust" or "AC fine (or coarse) dust."

AC dust is a true dust or particulate matter consisting primarily of silica with some quartz and other crystalline matter. The chemical analysis shown in Table A-1 is typical. It is collected from selected areas of Arizona deserts and is then classified by Microtrac analysis for particle size. The material is supplied in jars suitable for laboratory shelf storage.

AC dust is classified into a definite size range, the classification by the supplier is expressed in weight fractions. The comparative percentages for AC dust are shown in Table A-2.

Although each particle of dust is impervious to water and oils, care must be taken in handling this contaminant because it can form agglomerates which will give false performance data.

2. Test Fluid—The test fluid recommended for most tests in this document is Viscor Fuel Filter Test Fluid L-4264V, supplied by Viscosity Oil Company, 3200 South Western Avenue, Chicago, Illinois 60608. This product is specially blended for the recommended tests in terms of both physical properties and additive package. Both of these characteristics will affect performance data and are therefore closely controlled. The "clean test fluid" specified in several procedures is test fluid which has been prefiltered to an insoluble level of no more than 10 mg/l.

TABLE A-1—CHEMICAL ANALYSIS OF TYPICAL AC DUST

	Weight %
SiO_2	65-76
Al_2O_3	11-17
Fe_2O_3	2.5-5.0
Na_2O	2-4
CaO	3-6
MgO	0.5-1.5
TiO_2	0.5-1.0
V_2O_3	0.10
ZrO	0.10
BaO	0.10
Loss on Ignition	2-4

TABLE A-2—PARTICLE SIZE DISTRIBUTION STANDARDIZED AIR CLEANER TEST DUST

Micrometer	Weight % Fine	Smaller Than Coarse
5.5	38 ± 3	13 ± 3
11	54 ± 3	24 ± 3
22	71 ± 3	37 ± 3
44	89 ± 3	56 ± 3
88	—	84 ± 3
176	100	100

APPENDIX B

METHODS FOR SAMPLE ANALYSIS

1. Scope—These methods cover the gravimetric procedures for the determination of n-Pentane or petroleum ether insolubles found in a representative sample of oil drawn from the laboratory test stand in the manner specified by the filter test procedure.

2. Summary of Methods

2.1 The preferred procedure, Method A, uses a microporous filter membrane technique for the separation of solids from the test fluid. This procedure can be effectively used for most sample analyses; however, where extreme accuracy is required, or where the level of contaminant is very low, additional steps are recommended. These steps are described at the end of this procedure.

2.2 An alternate procedure, Method B, uses centrifugation for solids separation.

3. Analysis Method "A" (Membrane Filtration)

3.1 Test Apparatus and Materials

3.1.1 Membrane filters, white, plain, 47 mm diameter, 0.8 μm pore size (1).

3.1.2 Filter holder, 47 mm, consisting of a borosilicate glass or stainless steel funnel and funnel base, plus clamp (2).

3.1.3 Analytical balance, accurate to 0.0001 g.

3.1.4 Petri dishes, 60 mm ID.

3.1.5 Noncirculatory air oven, capable of maintaining a temperature of 90°C.

3.1.6 Forceps, flat bladed (3).

3.1.7 Washing bottles and storage bottles.

3.1.8 Vacuum system, aspirator or pump, capable of producing 15 in (381 mm)-Hg vacuum. If pump is used, a cold trap should be installed.

3.1.9 Petroleum ether (filtered) 30-60°C boiling point range.

3.1.10 n-Pentane (filtered).

3.2 Test Preparations

3.2.1 SOLVENTS—Solvents used for flushing and dilution of sample must be prefiltered through a 0.8 μm membrane and stored in a container which has been thoroughly washed, dried, and flushed three times with filtered solvent.

3.2.2 FILTER MEMBRANES (0.8 μm PORE SIZE)—Identify each membrane with a sample number by marking the rim (sealing edge) with a ball-point pen.

3.2.3 EQUIPMENT—The filter funnel, Petri dishes, and all other glassware used must be washed with detergent, rinsed thoroughly, oven dried, and flushed with filtered solvent before use.

3.3 Test Procedure

3.3.1 With clean forceps, place filter membrane in an open Petri dish and oven dry for 30 min minimum at 90°C.

3.3.2 Stand Petri dish, with the cover slightly ajar, in the balance room, or area near balance. Membrane must be protected from airborne contamination.

3.3.3 Allow 30 min (minimum) for the membrane to equilibrate with the ambient temperature and humidity before weighing.

3.3.4 Weigh membrane to the nearest 0.0001 g.

3.3.5 Place membrane centrally on funnel base, center funnel position, and clamp assembly securely. Place a large watchglass on top of funnel to protect the membrane until used.

NOTE: Static electricity is frequently generated during the filtration process. Since the solvents are highly flammable, it is recommended that the equipment and operator be safely grounded. In addition, the vacuum pump must be located in a well-ventilated area and/or the pump exhaust vented to a safe area.

3.3.6 If the liquid sample is stored in a container and the weight of the sample must be known, the outside surface of the container should be rinsed thoroughly with prefiltered solvent, and then the cap removed. Allow the container to reach room temperature again (about 5 min), then record weight of container and sample (no cap) to nearest 0.0001 g.

3.3.7 Pour 0.150 L of solvent into the filter funnel and add the sample, rinsing the sample bottle inner surface only. Put the clean sample bottle aside for reweighing. Apply 15 in (381 mm) Hg vacuum to the flask and maintain a liquid head in the funnel until filtration is completed. During this operation, rinse the inner surface of the funnel, using the washbottle. Repeat this washdown several times, using at least 0.150 L of solvent. About 0.400 L of solvent should be used in the entire analysis.

3.3.8 With the vacuum still applied, carefully remove the clamp and funnel. Wash the sealing rim of the membrane with solvent by directing a gentle stream from the washbottle. Direct the stream toward the center of the membrane, taking care not to wash off any of the contaminant. Also wash the sealing rim of the filter funnel onto the membrane.

3.3.9 Disconnect vacuum and carefully remove filter membrane and place into a covered Petri dish. Use clean forceps for handling.

3.3.10 Dry membrane and reweigh as described in paragraphs 3.3.1 and 3.3.4.

3.3.11 Reweigh the empty sample container and record the weight. Be sure that the outside of the container is free from oil before reweighing.

3.4 Evaluation of Results

The difference in weight between paragraphs 3.3.10 and 3.3.4 is the weight of contaminant solids collected. The weight of oil analyzed is the difference in weight between paragraphs 3.3.6 and 3.3.11, expressed in % w/w:

Percent n-Pentane (or P.E.) insolubles:

$$\frac{\text{Weight of contaminant solids collected}}{\text{Weight of oil sample analyzed}} \times 100$$

3.5 Additional Suggestions of Technique

3.5.1 During periods of high humidity, the cooling effect of the solvent evaporation will cause moisture condensation on the disc. This may retard the complete solvent washing and solid deposit; an intermediate air drying step followed by an additional wash with solvent may be necessary.

3.5.2 The analyst may insert between the filter membrane and the base a porous absorbent pad to insure a clean surface for the membrane and to minimize "freezing" of the filter to the base.

3.5.3 Visual examination of the membrane under ultraviolet light for absorbed oil residue will quickly determine if the rinsing operation was thorough. Traces of residual oil will show a fluorescent effect, whereas a clean membrane will not.

3.6 Additional Operations for Improved Accuracy of Analysis

3.6.1 CONDITIONING FILTER MEMBRANE

3.6.1.1 All filter membranes must be conditioned by a warm water soak to remove extraneous manufacturing material and to assure accurate and constant tare weight. Place the filter membrane in a clean beaker containing prefiltered distilled water at a temperature of 90–100°F (32–38°C). Cover the beaker and soak the filter membrane for 1 h.

3.6.1.2 Carefully remove the filter membrane with clean forceps. Allow all surface water to run off the membrane.

3.6.1.3 Place filter membrane in an open Petri dish. Place Petri dish, with cover slightly ajar, in the oven and dry for 60 min at 90°C minimum.

3.6.1.4 Remove the Petri dish, with membrane, from the oven and place in a desiccator, or constant temperature and humidity area, with cover slightly ajar. Allow 30 min (minimum) for the membrane to equilibrate with environmental conditions before weighing.

3.6.2 CONTROL FILTER MEMBRANE

3.6.2.1 A control filter membrane should be carried along for all analysis work. Subject this membrane to all preparation, handling, and weighing techniques of paragraphs 3.6.1.1-3.6.1.4.

3.6.2.2 Insert the control filter membrane directly beneath a test membrane in the filter holder in at least one instance during a single or group test run. Subject it to all handling and weighing techniques specified in paragraphs 3.3.9 and 3.3.10.

3.6.2.3 Apply the weight change of the control filter as a correction factor to test results, subtracting this factor when the control filter shows a weight increase or adding the factor when the control shows a weight decrease.

4. Analysis Method "B" (Centrifugation), Alternate

4.1 Test Apparatus and Materials

4.1.1 Air oven capable of maintaining 105°C.
4.1.2 Clamp—Curved type.
4.1.3 Centrifuge tube(s), 0.100 L size.
4.1.4 Centrifuge, capable of 98 km/s^2.
4.1.5 Desiccator (5).
4.1.6 Analytical balance.
4.1.7 Acetone—30-60°C boiling point range (or equivalent).
4.1.8 n-Pentane or equivalent.
4.1.9 Stiff wire—suggest stainless steel.
4.1.10 Washbottle.

4.2 Test Preparation

4.2.1 Solvents—See Section 3.
4.2.2 Wash forceps and centrifuge tubes in a detergent.
4.2.3 Rinse with water to remove suds.
4.2.4 Completely rinse inside and outside of centrifuge tube(s) with acetone. During this procedure, centrifuge tube should not be touched by hand.
4.2.5 Place centrifuge tube(s) in a 105 ± 3°C oven for 45 min.
4.2.6 Place centrifuge tube(s) in desiccator while cooling to room temperature.
4.2.7 Weigh centrifuge tube(s) and record weight.

4.3 Test Procedure

4.3.1 Heat sample to 180°F (82°C) or until sufficiently mobile to allow shaking of sample. The sample shall be shaken until the sediment is homogeneously suspended in the oil.

4.3.2 Weigh 10 g of sample into a pre-weighed centrifuge tube.

4.3.3 Fill centrifuge tube to the 0.100 L mark with n-Pentane and place stopper in centrifuge tube.

4.3.4 Shake centrifuge tube well so that contents are thoroughly mixed.

4.3.5 Centrifuge for 20 min at 98 km/s^2 in a high-speed centrifuge.

4.3.6 Decant all but 0.003 L from the centrifuge tube without disturbing the precipitate.

4.3.7 Add 0.010 L of n-Pentane to the centrifuge tube.

4.3.8 With a clean stiff wire, dislodge the precipitate and disperse it in the n-Pentane.

4.3.9 Wash all insolubles adhering to the wire back into the centrifuge tube using n-Pentane and fill the tube to the 0.050 L mark.

4.3.10 Shake tube until mixture is homogeneous.

4.3.11 Centrifuge for 20 min at 98 km/s^2.

4.3.12 Decant supernatant liquid being careful not to disturb the precipitate.

4.3.13 Rinse only the outside of the centrifuge tube with acetone (during and after rinse, centrifuge tube should not be touched by hands).

4.3.14 Place tube in an oven at 105 ± 3°C for 45 min.

4.3.15 Place tube in a desiccator until centrifuge tube returns to room temperature.

4.3.16 Weigh centrifuge tube to the nearest 0.0001 g and calculate weight of solids.

Supplies and Suppliers

1. Membrane filters, plain white, 47 mm diameter, 0.8 μm pore size—Millipore AAWP 047 00, Gelman 60110, Sartorius SM-11306, or other equivalent.

2. Filter holder, 47 m diameter—Millipore XX10 047 00 or XX10 047 30, Gelman 4200 or 4371, Sartorius 16306, or other equivalent.

3. Stainless steel forceps—Millipore XX62 000 06, Gelman 51147, or equivalent.

4. Centrifuge, 98 km/s^2—Fisher Scientific Co., VWR Scientific, Inc., DeLaval Separator Co., or other equivalent.

5. Desiccator—Fisher Scientific 08-594B, or other equivalent.

Suppliers' Addresses

1. Millipore Corp.
   Bedford, MA 01730

2. Gleman Instrument Co.
   600 South Wagner Road
   Ann Arbor, MI 48106
3. Sartorius-Membranfilter GmbH
   34 Gottingen, Germany
4. Fisher Scientific Co., Ltd.
   711 Forbes Avenue
   Pittsburgh, PA 15219
5. VWR Scientific, Inc.
   P.O. Box 999
   South Plainfield, NJ 07080
6. DeLaval Separator Co.
   Poughkeepsie, NY

APPENDIX C
GRAPHIC SYMBOLS

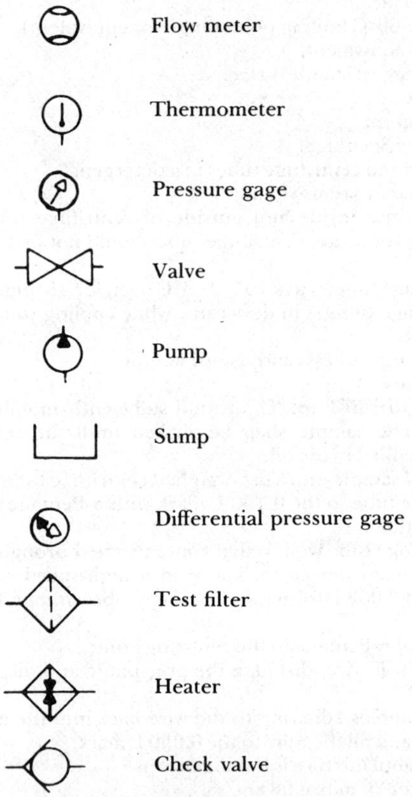

- Flow meter
- Thermometer
- Pressure gage
- Valve
- Pump
- Sump
- Differential pressure gage
- Test filter
- Heater
- Check valve

Reference:
ANS Y32.10-1967
NFPA S T3.10.68.4

FIG. C-1

APPENDIX D
SOURCE INFORMATION

Wire Cloth
41 micrometer—Catalog No. 1ST 50 x 700

Michigan Dynamics Div., AMBAC
32400 Ford Road
Garden City, MI 48135

Filter Holder
142 mm diameter—Catalog No. YY 30 142 36
293 mm diameter—Catalog No. YY 30 293 02

Millipore Corp.
Bedford, MA 01730

Waring Blender
Eberbach Corp.
P.O. Box 1024
Ann Arbor, MI 48106

Vibration Test Machine
LAB Corp.
999 Onondaga Street
Skaneateles, NY 13152

J1260 Test Oil, RM 99 (RFO-3-79)

Society of Automotive Engineers, Inc.
400 Commonwealth Drive
Warrendale, PA 15096

GLOSSARY OF TERMS RELATED TO FLUID FILTERS AND FILTER TESTING—SAE J1124 MAR87

SAE Recommended Practice

Report of the Engine Committee approved June 1976 and revised March 1987. Rationale statement available.

ɸ *Forward*—Over the years during which fluid filtration systems have been developing, many terms have come into use for descriptions of characteristics of filter media, filter assemblies, test methods, and test materials. Inevitably, some terms have been applied loosely, so that the same term may have different meaning to different people, or in different frames of reference.

Recognizing the need for clearly defined terms, which can have only one meaning for all persons in all circumstances, so that documents dealing with standard methods of evaluation of filters will have only one interpretation, the Filter Test Methods Subcommittee of the SAE Engine Committee has compiled this Glossary of related terms. No attempt has been made to produce an all-inclusive document, containing definitions of all terms related to all types of fluid filters. Instead, the Glossary is confined to the terms likely to be encountered in relation to filters for lubricating oil and fuels. At the same time, we have recognized that some terms are common to all types of fluid filters, and have been careful to avoid conflict with the definitions published by other standardizing groups. If not identical, the definitions of these terms are at least worded to convey an identical meaning, hopefully in fewer, simpler or more precise words.

We hope that this effort will be effective in helping to eliminate the ambiguities which have resulted from imprecise use of terminology and filtration.

This Glossary is referenced in the SAE filter test methods documents. Terms used in those documents are intended to have the definitions shown by this Glossary, and no other.

As new terms and their definitions become associated with the science of filtration and are relevant to the documents prepared by this subcommittee, revisions to the Glossary will be made, either by issuance of addenda or by revision and republication of the entire document.

1. General Terms

1.1 Anti-Drainback Valve—A check valve used in a filter housing in some spin-on filters to prevent oil drainage from the filter during engine shutdown.

1.2 Base—The foundation or support for the filter which may also contain one or more ports and a by-pass valve.

1.3 By-Pass—An alternate flow path around the filter element. Flow of fluid through the by-pass is usually blocked by a valve until a predetermined differential pressure across the element is reached.

1.4 Cleanable Element—A filter element, which when clogged, can be restored by a suitable cleaning process to an acceptable percentage of its original flow/pressure differential characteristics.

1.5 Clogging Indicator—An indicator which is activated when a predetermined pressure differential across the filter is reached.

1.6 Coalescer—Filter unit whose primary function is to cause finely dispersed droplets in an emulsion to form into larger droplets.

1.7 Combination Medium—A filter medium composed of two or more types, grades, or arrangements of filter media to provide properties which are not available in a single filter medium.

1.8 Crest—The outer fold of a pleat.

1.9 Depth Medium—A filter medium which primarily separates and retains contaminant within tortuous passages within the body of the medium.

1.10 Differential Pressure Indicator—A device which indicates continuously during operation the differential pressure across a filter element.

1.11 Disposable Element—A filter element which is discarded and replaced at the end of its service life.

1.12 Disposable Filter—A filter consisting of a filter element encased in a housing which is discarded and replaced in its entirety at the end of the service life of the element.

1.13 Dual Porosity Element—An element which contains two media of different porosity in parallel.

1.14 Dual Porosity Filter—A filter which contains two media of different porosity offering parallel flow paths to the fluid.

1.15 Edge Medium—A filter medium whose passages are formed by the adjacent surfaces of stacked discs, edge wound ribbon, or single layer filaments.

1.16 Effective Area—The area of a filter medium through which fluid flows.

1.17 Etched Medium—A filter medium having passages produced by chemical or electrolytic removal of unwanted material.

1.18 Filter—A device having a porous medium, whose primary function is the separation and retention of particulate contaminants from a fluid. The major filter components are the housing and the element.

1.19 Filter Drain—A port located near the lowest extremity of a filter through which fluid may be removed.

1.20 Filter Element—A sub-assembly of a filter which contains the filter medium or media.

1.21 Filter Housing—A ported enclosure which contains the filter element and directs fluid flow through it.

1.22 Filter Medium—The porous material which performs the process of particle separation and retention.

1.23 Filter Paper—A porous paper, matted or felted sheet of fibers (usually vegetable, but sometimes mineral, animal, or synthetic).

1.24 Final Filter—The last stage of a multi-stage filter system.

1.25 Full Flow Filter—A filter which is designed to filter all of the system fluid flow.

1.26 Inside-Out Flow Element—A filter element designed for flow outward from and perpendicular to its longitudinal axis.

1.27 Multi-Stage Filter—A filter system with two or more filters in series.

1.28 Non-Woven Medium—A filter medium composed of a mat of fibers. Filter paper is one type of non-woven medium.

1.29 Open Area—The total pore area of a filter medium, often expressed as a percent of total area of filter medium.

1.30 Outside-In Flow Element—A filter element designed for flow towards and perpendicular to its longitudinal axis.

1.31 Partial Flow Filter—A filter which filters only a part of the total system fluid.

1.32 Pleats—A series of folds in the filter medium used to increase effective filter area within a given space.

1.33 Pore Size Distribution—The ratio of the number of pores of given sizes to the total number of pores, per unit of area.

1.34 Porosity (Void Fraction)—The ratio of pore volume to total volume of a filter medium.

1.35 Precoat—A filter medium in loose powder or fiber form introduced into the upstream fluid to form a filter bed on the face of an element.

1.36 Primary Filter—The first stage of a multi-stage filter system.

1.37 Root—The inner fold of a pleat.

1.38 Secondary Filter—The second stage of a multi-stage filter system.

1.39 Sintered Medium—A metallic or non-metallic filter medium composed of particles of the medium processed to cause diffusion bonds at all contacting points.

1.40 Spin-On Filter—A disposable filter which mates to a permanent base and is attached by turning it onto a threaded base stud.

1.41 Surface Medium—A filter medium which primarily separates and retains contaminant on the influent surface face.

1.42 Total Area—The entire surface area of a porous medium, whether effective or not, in a filter element.

1.43 Two Stage Element—A filter element assembly composed of two filter media in series.

1.44 Wound Medium—A filter medium comprised of layers of crossed helical wraps of a continuous filament or strand of roving.

2. Contaminant Terms

2.1 Agglomerate (n)—A group of two or more particles combined, joined, or clustered, by any means.

2.2 Aggregate—A relatively stable assembly of dry particles formed under the influence of physical forces.

2.3 Automatic Particle Count—A particle count obtained by an electro-mechanical or electronic device as opposed to visual microscopic counting technique.

2.4 Background Contamination—The apparent contamination concentration determined by analysis of test fluid before addition of contaminant.

2.5 Classified Particle Contaminant—A test contaminant which has been separated and classified to a standard particle size distribution. These are usually hard, non-deformable particles, and are sometimes referred to as abrasive test contaminants.

2.6 Clean Element—A new or re-conditioned filter element which is essentially free of contamination introduced during manufacture, assembly, storage, installation, or use.

2.7 Cleanliness Level—The antonym of contamination level.

2.8 Clogged Element—A filter element which has collected a quantity of contaminant, such that it cannot maintain rated flow without excessive differential pressure.

2.9 Clogging Contaminant—A contaminant which is soft and more deformable than the classified particle contaminants, used to simulate, in the laboratory, the type of filter clogging experienced by filters in service.

2.10 Concentration—Quantity of solid, liquid, or gaseous material related to that of another material in which it is found in the form of a mixture, suspension, or solution.

2.11 Contaminant, Generated—Contaminant generated by the operation of a system or component.

2.12 Contaminant—Any material or substance which is unwanted in a system.

2.13 Contaminant, Built-In—Initial residual contaminant in a fluid system or component.

2.14 Contaminant, Environmental—Contaminant(s) present in the immediate surroundings, introduced into a fluid system or component.

2.15 Contamination Level—A quantitative term specifying the degree of contamination.

2.16 Dissolved Water—Water which is present in a fluid forming a single phase.

2.17 Emulsified Water—Small droplets of water in a stable suspension of water in fuel.

2.18 Feret's Diameter (of a particle)—The distance between two tangents, on opposite sides of the particle profile, that are parallel to some fixed direction. Remark: This dimension is used in visual microscope counting. The fixed direction is to be vertical, that is, from top to bottom, in the field of view that includes the image of the particle. Ferret, Assoc., Int pour l 'Essai des Mat 2 Group d, Zurich (1931).

2.19 Fiber—A fiber is a particle whose length is greater than $100 \mu m$ and ϕ

at least 10 times its width.

2.20 Free Air—Air present as a dispersed phase in a fluid.

2.21 Free Water—Water present in a fluid which may separate as a result of the difference in densities.

2.22 Gravimetric Value—The weight of suspended solids per unit volume of fluid.

2.23 Martin's Diameter (of a particle)—The length of the line that bisects the image of the particle; the line may be drawn in any direction which is to be maintained constant for all the image measurements. Remark: This dimension is sometimes used in connection with electron microscopy. The bisecting line is to be drawn horizontally, that is, from left to right, in the field of view that includes the image of the particle. Martin, Blyther, Tongue Trans Ceram Soc. England, 23, 6 (1924).

2.24 Media Migration—Migration of materials which make up the filter medium.

2.25 Micrometre—Unit of measurement one millionth of a metre long.

2.26 Migration—Movement of particles from the filter assembly into the effluent.

2.27 Non-Combustible Residue—Matter not changed to gaseous state when heated to 815°C in the presence of air.

2.28 Particle—A minute piece of matter with observable length, width, and thickness; usually measured in micrometres.

2.29 Particle Count Blank—An allowance for the determinable background contamination.

2.30 Particle Size Analysis—The whole of the operation by which a particle size distribution is determined.

2.31 Precipitate (n)—Particles separated from a fluid as a result of a difference in densities and the action of gravitational force.

2.32 Projected Area Diameter (of a particle) (Symbol: d_a)—The diameter of a circle having the same projected area as the particle when it is viewed in a direction perpendicular to a plane of stability. Remark: This dimension is the basis of measurement by the Vickers Projection Microscope and the HIAC automatic counter.

2.33 Raw Particle Count—The actual number of particles counted in each size of a specified size range or above a specified size or sizes.

2.34 Sieve Diameter (of a particle) (Symbol: d_Δ)—The width of the minimum square aperture through which the particle will pass.

2.35 Silt—Fine particulate matter, generally less than 5 μm in size.

2.36 Stoke's Diameter (of a particle) (Symbol: d_{st})—The diameter of a sphere having the same density as the particle and the same free-falling speed as the particle in a laminar flow region (Re 0.2) in a fluid of the same density and viscosity as the fluid concerned. Remark: This dimension is used as the basis for particle size measurement by liquid sedimentation methods of those fractions of powders that pass through a 75 μm sieve.

The particle size terms and definitions above are taken from "Particle Size Measurement," T. Allen, University of Bradford, 1968, Chapman & Hall, p. 17. The symbols have general agreement and are in conformity with publications by H. Heywood, Loughborough University.

2.37 Surface Diameter (of a particle) (Symbol: d_s)—The diameter of a sphere having the same surface area as the particle.

2.38 Test Contaminant—Particulate matter of controlled composition and particle size distribution or, water of controlled droplet size, used in the laboratory test of filters.

2.39 Unloading—The release of contaminant that was initially retained by the filter medium.

2.40 Visual Counting of Particles—Any optical counting procedure used to measure the number of particles of specified dimensions.

2.41 Volume Diameter (of a particle) (Symbol: d_v)—The diameter of a sphere having the same volume as the particle. Remark: This dimension is the basis of measurement by a Coulter Counter.

3. Filter Characteristics Terms

3.1 Filter Performance Test Terms

3.1.1 Apparent Capacity—The actual weight (grams) of contaminant injected into a laboratory filter test system before the terminal pressure drop is reached. Also referred to as life in grams.

3.1.2 Base Upstream Gravimetric Level—Theoretical upstream contamination level for a filter of 100% efficiency.

φ **3.1.3 Batch Contaminant Addition**—A filter test condition under which a specified contaminant is added in batches of specified size at specified time intervals.

φ **3.1.4 Combined Life and Particle Retention Test**—A filter Life Test in which specified amounts of inorganic classified contaminant are admitted to the influent stream at specified intervals during the test, to determine the particle separation efficiency of the filter at various stages of clogging.

An organic clogging contaminant may be used in this test, in combination with the inorganic classified contaminant.

3.1.5 Constant Rate of Fluid Flow—A filter performance test condition which specifies that test flow rate through the filter is held constant throughout the test, usually by varying the inlet pressure to compensate for increases of pressure drop across the filter.

3.1.6 Continuous Contaminant Addition—A filter test condition under which a specified contaminant is added to the test fluid at a continuous specified rate for the duration of the test.

3.1.7 Cumulative Filter Efficiency—A method of calculating filter efficiency based on the total cumulative amount of contaminant presented to the filter from the start of the test to any designated sampling time.

3.1.8 Effluent—The fluid leaving a component.

3.1.9 Filter Capacity for Contaminants—The weight of specified contaminant removed and held from the fluid by a filter at a specified termination point. The termination point is specified as a pressure differential, reduction in flow, filtration efficiency, or fluid contamination level.

3.1.10 Filter Efficiency—The ability, expressed as percent, of a filter to remove specified artificial contaminant from a specified fluid under specified test conditions.

3.1.11 Filter Life Test—A type of filter capacity test in which a clogging contaminant is added to the influent of a filter, under specified test conditions, to produce a given rise in pressure drop across the filter or until a specified reduction of flow is reached. Filter life may be expressed as test time required to reach terminal conditions, at a specified contaminant addition rate.

3.1.12 Filter (or Element) Pressure Differential—The drop in pressure due to flow across a filter or element at anytime. The term may be qualified by adding one of the words "initial," "final," or "mean."

3.1.13 Filter Rated Flow—The maximum flow rate, in litres/hour, of a fluid of specified viscosity, for which a filter is designed. The standard fluid viscosity for flow rating of lubricating oil filters is 20 cst (100 S.U.S.). Suitability of the filter for use at flows and/or viscosities beyond the rated flow is left to the discretion of the user.

3.1.14 Filtration Ratio—The ratio of the number of particles greater than a given size in the influent to the number of particles greater than the same size in the filter effluent.

3.1.15 Incremental Filter Efficiency—A method of calculating Filter Efficiency based on the total amount of contaminant presented to the filter during any specified segment of the test.

3.1.16 Influent—The fluid entering a component.

3.1.17 Initial Particle Retention Rating—A test conducted on new filter elements to establish the initial retention characteristics, of particles of a specified size range.

3.1.18 Instantaneous Filter Efficiency—A method of calculating filter efficiency based on the ratio of contaminant level of the effluent to that of the influent at an instant of time during the test.

3.1.19 Largest Particle Passed Test (Abs. Particle Retention Rating)—A form of Particle Retention Rating Test in which the largest hard particle that will pass through a filter under defined test conditions is determined. The term largest particle passed is intended to indicate the largest opening in a filter.

3.1.20 Multipass or Recirculation Test—Filter performance tests in which the contaminated fluid is allowed to recirculate through the filter for the duration of the test. Contaminant is usually added to the test fluid during the test.

3.1.21 Oil Contamination Level—The amount of contaminant present in the test fluid at any specified sampling time. The level is usually expressed as the weight of contaminant or particle count per unit of weight or volume of fluid.

3.1.22 Particle Retention Rating Test—A test which determines a filter's ability to separate and retain particles of a specified size. It is usually conducted as a single pass test with Classified Contaminant of a narrow size range, and results are expressed as percent efficiency.

3.1.23 Permeability—The relationship of flow per unit area to differential pressure across a filter medium.

3.1.24 Single Pass Test—Filter performance tests in which contaminant which passes through a test filter is not allowed to recirculate back to the test filter.

3.1.25 Variable Rate of Fluid Flow (Constant Pressure)—A filter performance test condition which specifies that pressure difference across the filter shall be held constant throughout the test, which results in decreasing flow rate as the filter becomes clogged.

3.1.26 Variable Rate of Fluid Flow (Variable Pressure)—A filter test specification which requires flow rate changes according to a specified time cycle.

3.2 Filter and Element Strength and Integrity Terms—Tests pertaining to strength, seal integrity, and temperature effects on the filters and filter elements.

3.2.1 Additive Removal Test—A test conducted by recirculating a specified additive type of oil through a test element for a specified period of time. The amount of ash type additive removed by the element is determined by oil analysis.

3.2.2 ANTI-DRAINBACK VALVE TEST—The test used to evaluate the performance of an anti-drainback valve in disposable spin-on type filters with respect to leakage rate under a static pressure head.

3.2.3 BY-PASS COMPONENT CHARACTERISTIC TEST—A test which determines the performance of By-Pass relief valve with respect to its leakage rate, opening and reseat pressures, and resistance to flow.

3.2.4 BY-PASS VALVE LEAKAGE TEST—A test to determine the flow rate, expressed as volume per unit of time, through the closed valve, at specified static pressure head on the valve.

3.2.5 BY-PASS VALVE OPENING PRESSURE TEST—A test to determine the pressure required to open the valve to permit flow at a specified minimum rate.

3.2.6 BY-PASS VALVE RESEAT PRESSURE TEST—A test to determine the pressure at which the valve closes and restricts the flow rate to a level below a specified minimum rate as the pressure across the valve decreases.

3.2.7 COMBINED PRESSURE PULSE AND VIBRATION FATIGUE TEST—A more sophisticated test method on which the laboratory test simulates simultaneously the service condition of impulse and fatigue.

3.2.8 Element End Load Proof Test—A non-destructive end load test in which the filter must remain free of visible damage and must not leak after being subjected to a specified end load for a specified time.

3.2.9 FILTER ELEMENT BURST PRESSURE TEST—A test procedure which determines the least pressure drop across a filter element arising from inside to outside flow that causes structural failure or medium rupture in a filter element.

3.2.10 FILTER ELEMENT COLLAPSE PRESSURE TEST—A test procedure which determines the least pressure drop across a filter element arising from outside to inside flow that causes structural failure or medium rupture in a filter element.

3.2.11 FILTER ELEMENT END LOAD FAILURE TEST—A test procedure which determines the least axial force applied to the end of a filter element that causes seal failure or failure due to permanent deformation of the element.

3.2.12 FILTER ELEMENT MEDIA FLOW FATIGUE TEST—A method of test to determine the ability of filter element media to withstand the flexing caused by cyclic differential pressure.

3.2.13 FILTER HOUSING DESIGN EVALUATION—Tests which are devised to reveal problems related to poor filter housing design, some examples of which are filter element unseating during flow surges, eroding of the media by high velocity influent flow, and excessive attaching thread leakage in disposable spin-on type filters.

3.2.14 FILTER RATED OPERATING PRESSURE—A maximum steady-state pressure at which a filter is designed to operate without structural damage to the filter housing.

3.2.15 FLEXIBLE LINE TESTS—Tests which measure burst strength, estimate rates of deterioration in service, and determine flexure fatigue and fluid compatibility.

3.2.16 HYDROSTATIC BURST-PRESSURE TEST—A destructive test which determines the minimum static pressure which a filter can sustain without leaking or rupture.

3.2.17 INSTALLATION SEALING TORQUE TEST—The test method which indicates the ability of the filter to be sealed at the junction of the filter with its base when installed in a specified manner.

3.2.18 LEAKAGE BUBBLE TEST—A method of test by which the fabrication integrity of an element may be determined. Air pressure is applied gradually to the inside of a filter element, which is submerged in a specified liquid, until the first positive stream of bubbles appears. The pressure is recorded and the point of origin of the bubbles is observed.

3.2.19 MECHANICAL TESTS—These test methods describe laboratory testing of filters to prove their mechanical integrity. Filters are tested to determine vibration fatigue life, pressure impulse fatigue life, and the hydrostatic burst pressure strength of the housing material and construction. The tests simulate conditions of engine vibrations and engine lubrication system cyclic pressure pulsation and maximum pressure surges. The test condition values should correspond with those actually measured on the intended engine applicable when such information is available.

3.2.20 MEDIA MIGRATION AND INITIAL FILTER CLEANLINESS TEST—The test used to determine the amount of particulate matter built into a filter or introduced during shipment and storage. It consists of flowing clean fluid through the test filter on a single pass basis and collecting all of the filter contamination downstream of the filter. The material so collected may be weighed and identified by microscopic means. Loose filtering media may be differentiated from other contaminants.

3.2.21 PRESSURE PULSE FATIGUE TEST—This laboratory test determines the ability of the filter to withstand lubricating system cyclic pressures encountered by the filter during engine operation.

3.2.22 PROOF PRESSURE TEST—A test pressure that a filter shall withstand without permanent deformation, external leakage, or other malfunction.

3.2.23 REMOVAL TORQUE TEST—The test method which measures φ torque required to remove spin-on filter after exposure to specified installation torque and engine operating temperatures.

3.2.24 SEAL TESTS—Tests which are devised to assess the sealing characteristics for high and low temperature sealing application, fluid compatibility, and shear resistance under laboratory simulated service conditions.

3.2.25 TEST FOR EFFECT OF HIGH TEMPERATURE ON ELEMENT COLLAPSE STRENGTH—A test similar to the Collapse Pressure Test except that the element shall have been subjected to hot oil immersion prior to the collapse test. This may be either a collapse pressure test or a proof test.

3.2.26 VIBRATION FATIGUE TEST—This laboratory test determines the ability of the filter to withstand forces due to vibration during engine operation.

(R) EMULSIFIED WATER/FUEL SEPARATION TEST PROCEDURE—SAE J1488 MAY90

SAE Recommended Practice

Report of the Engine Committee, approved June 1985. Completely revised by the Filter Test Methods Standards Committee May 1990.

Foreword—Water in fuels is one of the major causes of diesel engine maintenance problems. The effects of water in fuel are characterized by corrosion of fuel system parts, plugging of filters and orifices and, in some cases, failure of fuel injection equipment. Water in fuel often dissolves sulfur compounds, becomes acidic, and enhances corrosion in fuel injection systems as well as in the engine itself. The presence of water also encourages microbiological growth, which generates orifice and filter restricting sludge. Further, due to displacement of fuel lubrication in close tolerance injector parts, and rapid expansion of heated water at the fuel injector tip, galling, and more serious failure may also occur.

During transportation, transfer, and storage of fuel, water may become entrained in a variety of ways. The mode and timing of water entry in the handling sequence before use, as well as the chemistry of the fuel itself (additives and surfactants), will determine what form the contaminant takes. In systems where water and fuel pass through high shear pumps, fuel/water interfacial tension is relatively low, and settling time is minimized, fine emulsions may predominate. In systems where water enters before or after low shear pumps, or where there is a prolonged settling time in high interfacial tension fuel, larger water droplets may predominate. In some systems, both fine emulsions and large droplets may be present simultaneously. Generally, fine emulsions are more likely to predominate on the pressure side of high shear pumps, whereas larger water droplets are more likely to predominate on the suction side of pumps. (A water removal test procedure designed for applications where large water droplets predominate is also recommended. This procedure is given in SAE J1839.)

The following test procedure is relevant to finely dispersed or emulsified water separation devices whether applied on the suction or discharge side of engine fuel transfer pumps.[1] The procedure is well suited to lower flow rates, although it may be applied with due caution to higher flow rates [up to approximately 1.6×10^{-3} m^3/s (~25 gpm)]. It has been designed to approximate field conditions in a practical manner. A 3500 rpm centrifugal pump is used to disperse water in the fuel, simulating most fuel loading pumps (5 to 10 μm mean droplet size). The test fuel may be an actual fuel sample (with additives) that is to be used in the field, or it may be No. 2 fuel oil that has been clay treated (conditioned) so as to enable equal and reproducible laboratory comparisons of various test devices. Test fuel conditioning is recommended for laboratory comparisons only, as this treatment may yield water removal efficiency results, which are significantly different from those obtained using water separating devices in untreated fuel. Furthermore, testing unused "clean" water separators may provide water removal efficiencies which are far superior to those obtained from the same water separators after very short exposure to natural fuel and natural fuel contaminants.

1. Scope—To determine the ability of a fuel/water separator to separate emulsified or finely dispersed water from fuels.

2. References

SAE J1839, Fuel/Water Coarse Droplet Separation Test Procedure for Suction Side Applications

ASTM D 971

3. Test Apparatus—(A test stand meeting the following requirements is available from American Filtrona Company, P.O. Box 34668, Richmond, VA 23234.)

A test system,[2] as illustrated in Figure 1, is to include:

3.1 A flat bottom, corrosion resistant (for example, polyethylene) fuel container with a fuel outlet not less than 4 cm from the bottom of the container. The container must have a volume of at least five times the test flow rate per minute.

3.2 A 3500 rpm ± 100 centrifugal pump capable of delivering at least five times (but not greater than 20 times) the test flow rate at 1×10^5 Pa (15 psi). Note that magnetically coupled and multistaged centrifugal pumps should not be used.[3] The pump should be capable of producing the required pressure to overcome the system and fuel/water separator resistances.

3.3 Fuel flow meter capable of measuring with an accuracy of better than or equal to 5% of actual flow.

3.4 Temperature indicator with an accuracy of ±1.5°C.

3.5 A final water separator assembly such that not more than 30 ppm by volume of undissolved water are recycled on an average basis under test conditions.

3.6 A corrosion resistant (for example, polyethylene) water sump with approximate capacity of 2×10^{-2} m^3 (5 gal).

3.7 Water flow meter (0 to 8×10^{-7} m^3/s, or as required) with flow regulating valve, capable of measuring water at 0.25% of test flow rate, with an accuracy of 5% of actual flow.

3.8 Automatic Karl Fischer Titration Apparatus for water content analysis (for example, Seragen Diagnostics, Indiana, or equivalent).

3.9 In-line static mixer with at least three internal mixing units. The inner pipe diameters of the mixers should be as follows:

0 to 3.785×10^{-4} m^3/s (0 to 6 gpm): 1.27×10^{-2} (1/2 in) ID mixer (for example, Kenics Corp., Model No. 1/2-KMS-6 304 s s, or equivalent).

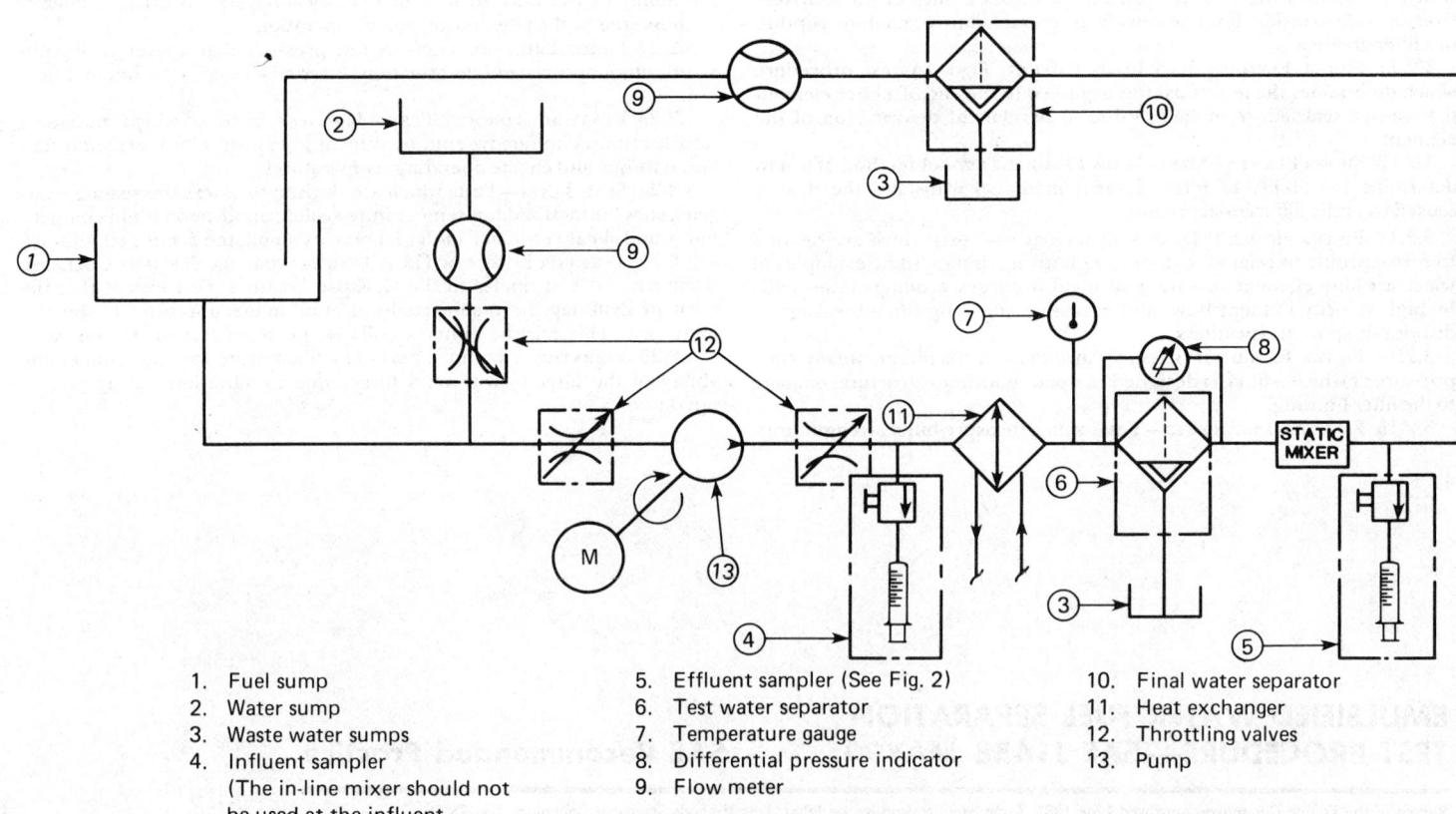

1. Fuel sump
2. Water sump
3. Waste water sumps
4. Influent sampler
   (The in-line mixer should not be used at the influent sampler)
5. Effluent sampler (See Fig. 2)
6. Test water separator
7. Temperature gauge
8. Differential pressure indicator
9. Flow meter
10. Final water separator
11. Heat exchanger
12. Throttling valves
13. Pump

FIG. 1

---

[1] This procedure recommends pressure side location of the test unit for ease and convenience of testing only. Modification of the procedure to place the test unit on the suction side of the pump should not alter test results as long as the water droplet size distribution remains unaltered.

[2] For test flow rates lower than approximately 4.73×10^{-5} m^3/s (0.75 gpm), the test system may be operated at a minimum design flow rate, with a small (desired flow rate) portion of this flow diverted into the test fuel/water separator. For this "slip stream method," generally two final (cleanup) fuel/water separators are required. Using this slip stream method, all the piping velocity and pump pressure flow rate requirements can be met.

[3] Magnetically coupled pumps have some slippage and this will affect the drop size distribution. Similarly multistaged centrifugal pumps will result in different drop size distribution.

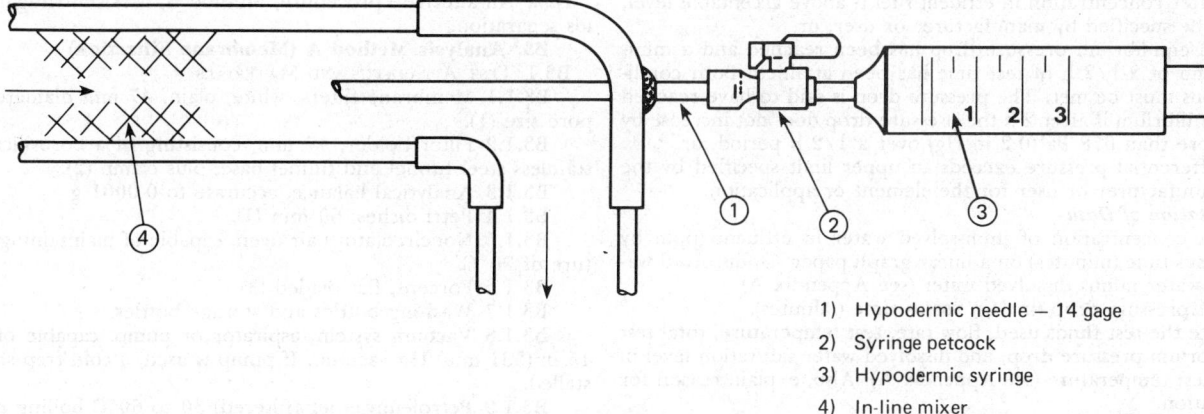

1) Hypodermic needle — 14 gage
2) Syringe petcock
3) Hypodermic syringe
4) In-line mixer

FIG. 2—EFFLUENT SAMPLER

For larger flow rates, the inner diameter should be such that the fuel velocity through an equivalent empty pipe is greater than 1.22 m/s (4 ft/s).

3.10 The fuel/water separation stand piping should have a velocity equal to or greater than 0.72 m/s. Any nonrusting and nonreacting pipe material may be used.

3.11 A differential pressure gauge with a readability of 338.64 Pa (0.1 in Hg) and an accuracy of at least 340 Pa (~0.1 in Hg).

3.12 Syringe sampler in accordance with Figure 2. Note[4] that the influent sampler does not have an in-line static mixer while the effluent sampler does.

3.13 Temperature control system capable of maintaining test temperature as specified in 5.2.

3.14 Double pipe heat exchanger (cooling and heating, if required) with fuel in the tube side. This should be a single pass, single tube, double pipe exchanger. If adequate cooling can be accomplished after the final (cleanup) fuel/water separator (see Figure 1), then any type of heat exchanger may be used.

3.15 Fuel/water interfacial tension measuring device. Preferably a platinum ring detachment method (ASTM D 971) should be used (for example, Cenco Instruments, or equivalent).

4. Test Materials

4.1 Test Fluid—Since fuel oil contains various constituents, the test oil type should be categorized and recorded as one of the following:

  a. A sample of the fluid used in the application.
  b. No. 2 diesel fuel, locally available.
  c. Specially treated fluid, per Appendix C.
  d. A standard reference fluid to be specified.

In all these cases, it should be understood that the results are relevant to this fuel and that some amount of variance in performance can be expected with different fuels, depending on the particular design of the test fuel/water separator.

4.2 Clean, distilled, or deionized water with a surface tension greater than 6×10^{-2} N/m (60 dynes/cm) at 20°C ± 1.5.

4.3 A 0.45 μm filter and associated equipment as listed in Appendix B.

5. Test Conditions

5.1 Volume of Fuel—Shall be five times the flow rate, per minute, with a minimum of 3.8×10^{-2} m³ (38 L).

5.2 Temperature—26.6°C ± 2.5 measured at the test separator inlet.

5.3 Flow Rate of Fuel—Rated flow of unit to be tested or as specified.

5.4 Water Flow Rate—0.25% of fuel flow rate.

5.5 Water must be injected at the suction side of the pump.

6. Test Procedure

NOTE: If Fuller's Earth treated fuel [4.1(c)] is selected as the fuel, then it may be reused after treating again. If Fuller's Earth treated fuel is not the selected fluid, use a fresh quantity of fluid.

6.1 For every fresh batch of fuel, determine the water saturation level in test fuel according to Appendix A.

6.2 Install cleanup filter (95% efficiency at 5 μm), in place of test filter; fill fuel tank; start circulation of fuel through the cleanup filter flow until a 500 mL sample of fluid has an insoluble contaminant level less than 5 mg/L (see Appendix B) and the total undissolved water concentration is less than 30 ppm.

6.3 Install test fuel/water separator or filter on the discharge side of the pump (see Figure 1); adjust fuel flow rate by throttling the discharge side valve; take the initial (fuel only) pressure drop reading at the rated flow.

6.4 Open the water valve and adjust water flow rate to be 0.25% of fuel flow. Start the clock at the same time water begins to flow. This point is zero test time.

NOTE: To insure the proper water flow rate, the water line from water sump to pump should be free of air and completely full of water. Further, the water line must feed into the fuel line, as close to the suction line fitting of the pump as is practical.

6.5 Periodically, drain the water from the water collection sump of the unit under test. (Obviously, this is not necessary if an automatic water sensing switch/drain is available.) Do not let water build up beyond the maximum recommended level of the water sump. DO NOT TAKE ANY SAMPLES WHEN ASSEMBLY IS BEING DRAINED.

6.6 After 10 min, insert a clean, water-free 5 mL syringe in the syringe petcock of the effluent sampler (see Figure 2). Flush syringe 3 to 5 times and slowly withdraw 2 to 3 mL of sample over a period of approximately 10 to 15 s. Analyze the sample immediately, using the automatic Karl Fischer Titration Apparatus. For flow rates greater than 3.875×10^{-4} m³/s (6 gpm), larger sample volumes should be withdrawn. Determine water concentration in ppm by volume.[5] Repeat this sampling procedure every 20 min thereafter until termination of the test.

6.7 Within the first 10 min of the test, take a sample (approximately 0.5 mL) of the fuel/water mixture entering the test separator and analyze the sample using the Karl Fischer Titration device, as described in 6.6. This is to confirm that approximately 2500 ppm ± 300 of water are being emulsified into the fuel stream. If this requirement is not met, adjust the water flow rate accordingly. Reconfirm this influent water concentration between every alternate effluent sample (for example, take influent samples at 10 min, 50 min, 90 min, etc; that is, every 40 min after the first 10 min sample). Record the influent concentrations with respect to time.

6.8 Record the differential pressure across the test fuel/water separator at each effluent sample interval.

6.9 Terminate test if one or more of the following conditions is met:

---

[4] Since the water at the influent section is emulsified, an in-line mixer is not needed prior to the influent sampler.

[5] Many Karl Fischer devices will determine micrograms. Convert this to ppm by volume.

$$\text{ppm by volume} = \frac{\text{Karl Fischer reading (μg)}}{\text{sample volume (mL)}}$$

$$\text{or} = \frac{\text{Karl Fischer reading (μg)} \times \text{fuel sp. gr.}}{\text{sample wt. (gms)}}$$

a. Water concentration in effluent fuel is above acceptable level, to be specified by manufacturer or user, or
b. An equilibrium pressure drop has been reached and a minimum of 2-1/2 h of test time has been attained. Both conditions must be met. The pressure drop is said to have reached equilibrium if after 2 h the pressure drop does not increase by more than 678 Pa (0.2 in Hg) over a 1/2 h period, or
c. Differential pressure exceeds an upper limit specified by the manufacturer or user for the element or application.

7. Presentation of Data

7.1 Plot concentration of undissolved water in effluent (ppm by volume) versus time (minutes) on a linear graph paper. Undissolved water = total water minus dissolved water (see Appendix A).

7.2 Plot pressure drop (in Hg) versus time (minutes).

7.3 Note the test fluids used, flow rate, test temperature, total test time, equilibrium pressure drop, and dissolved water saturation level in the fuel at test temperature (see Appendix A). Also, explain reason for test termination.

7.4 Calculate and report the time average undissolved effluent water level in ppm by volume.

$$E_{av} = \Sigma \, E_i \, (t_i - t_{i-1})/t_{total} \quad \text{(Eq. 1)}$$

where:

E_i is the effluent water concentration, ppm by volume at time t_i, min; t_{total} is the total test time, min.

7.5 Calculate and report average dispersed water separation efficiency.

$$\text{Average efficiency} = \frac{2500 - E_{av}}{2500} \times 100 \quad \text{(Eq. 2)}$$

Appendix A
Method for Determining Saturation Level of Dissolved Water in Fuel

A1. Wash a clean sample bottle (with a rubber diaphragm cap; minimum 100 mL capacity) with distilled water so as to remove traces of detergent; dry bottle in oven.

A2. Take about 150 mL of test fuel [4.1(a) or 4.1(b)] and filter fuel through a 0.45 μm membrane compatible with fuel oil.

A3. Determine total water concentration in the fuel by the Karl Fischer method in ppm by volume.

A4. If the water concentration is below or equal to 100 ppm, proceed to Section A5; if not, repeat Sections A2 and A3. If necessary, cool fluid to −4°C before filtering (Section A2).

A5. Fill the dried, clean sample bottle almost to the top with filtered fuel.

A6. Place the bottle on a magnetic stirrer. Using a hypodermic syringe with a long needle, gently insert 20 to 30 mL of clean distilled water in the bottom of the sample bottle; gently insert a magnetic mixer and gently fill the bottle up to the top with "dewatered" fuel; cap the bottle. DO NOT SHAKE OR MOVE THE BOTTLE after the water has been injected. Start the stirrer at the lowest possible speed. Make sure that the fluid/water interface is not strongly agitated and that no appreciable vortex develops as a result of mixing.

A7. After mixing for at least 18 h (overnight), gently remove from mixer and place bottle in a water bath at the test temperature for 2 h. Then, insert a clean, dried hypodermic syringe through the rubber diaphragm in the cap; gently withdraw 2 mL of the top fuel layer and analyze for water content using the Karl Fischer method. Take three readings. The average is the water saturation.

A8. Convert water saturation to ppm by volume; if the concentration is not between 100 to 150 ppm by volume, then repeat steps A6 through A8 to confirm. Report this as the dissolved water saturation level in fuel.

Appendix B
Methods for Oil Sample Analysis

B1. Scope—These methods cover the gravimetric procedures for the determination of n-Pentane or petroleum ether insolubles found in a representative sample of oil drawn from the laboratory test stand in the manner specified by the filter test procedure.

B2. Summary of Methods

B2.1 The preferred procedure, Method A, uses a microporous filter membrane technique for the separation of solids from the test fluid. This procedure can be effectively used for most sample analyses; however, where extreme accuracy is required or where the level of contaminant is very low, additional steps are recommended. These steps are described at the end of this procedure.

B2.2 An alternate procedure, Method B, uses centrifugation for solids separation.

B3. Analysis Method A (Membrane Filtration)

B3.1 TEST APPARATUS AND MATERIALS

B3.1.1 Membrane filters, white, plain, 47 mm diameter, 0.45 μm pore size (1).

B3.1.2 Filter holder, 47 mm, consisting of a borosilicate glass or stainless steel funnel and funnel base, plus clamp (2).

B3.1.3 Analytical balance, accurate to 0.0001 g.

B3.1.4 Petri dishes, 60 mm ID.

B3.1.5 Noncirculatory air oven, capable of maintaining a temperature of 90°C.

B3.1.6 Forceps, flat bladed (3).

B3.1.7 Washing bottles and storage bottles.

B3.1.8 Vacuum system, aspirator or pump, capable of producing 15 in (381 mm) Hg vacuum. If pump is used, a cold trap should be installed.

B3.1.9 Petroleum ether (filtered) 30 to 60°C boiling point range.

B3.1.10 n-Pentane (filtered).

B3.2 TEST PREPARATIONS

B3.2.1 Solvents—Solvents used for flushing and dilution of sample must be prefiltered through a 0.45 μm membrane and stored in a container, which has been thoroughly washed, dried, and flushed three times with filtered solvent.

B3.2.2 Filter Membranes (0.45 μm pore size)—Identify each membrane with a sample number by marking the rim (sealing edge) with a ballpoint pen.

B3.2.3 Equipment—The filter funnel, petri dishes, and all other glassware used must be washed with detergent, rinsed thoroughly, oven dried, and flushed with filtered solvent before use.

B3.3 TEST PROCEDURE

B3.3.1 With clean forceps, place filter membrane in an open petri dish and oven dry for 30 min minimum at 90°C.

B3.3.2 Stand petri dish, with the cover slightly ajar, in the balance room or area near balance. Membrane must be protected from airborne contamination.

B3.3.3 Allow 30 min (minimum) for the membrane to equilibrate with the ambient temperature and humidity before weighing.

B3.3.4 Weigh membrane to the nearest 0.0001 g.

B3.3.5 Place membrane centrally on funnel base, center funnel position, and clamp assembly securely. Place a large watchglass on top of funnel to protect the membrane until used.

NOTE: Static electricity is frequently generated during the filtration process. Since the solvents are highly flammable, it is recommended that the equipment and operator be safely grounded. In addition, the vacuum pump must be located in a well ventilated area and/or the pump exhaust vented to a safe area.

B3.3.6 If the liquid sample is stored in a container and the weight of the sample must be known, the outside surface of the container should be rinsed thoroughly with prefiltered solvent, and then the cap removed. Allow the container to reach room temperature again (about 5 min), then record weight of container and sample (no cap) to nearest 0.0001 g.

B3.3.7 Pour 0.150 L of solvent into the filter funnel and add the sample, rinsing the sample bottle inner surface only. Put the clean sample bottle aside for reweighing. Apply 15 in (381 mm) Hg vacuum to the flask and maintain a liquid head in the funnel until filtration is completed. During this operation, rinse the inner surface of the funnel, using the wash bottle. Repeat this wash down several times, using at least 0.150 L of solvent. About 0.400 L of solvent should be used in the entire analysis.

B3.3.8 With the vacuum still applied, carefully remove the clamp and funnel. Wash the sealing rim of the membrane with solvent by directing a gentle stream from the wash bottle. Direct the stream toward the center of the membrane, taking care not to wash off any of the contaminant. Also wash the sealing rim of the filter funnel onto the membrane.

B3.3.9 Disconnect vacuum and carefully remove filter membrane and place into a covered petri dish. Use clean forceps for handling.

B3.3.10 Dry membrane and reweigh as described in B3.3.1 and B3.3.4.

B3.3.11 Reweigh the empty sample container and record the weight. Be sure that the outside of the container is free from oil before reweighing.

B3.4 EVALUATION OF RESULTS—The difference in weight between B3.3.10 and B3.3.4 is the weight of contaminant solids collected. The weight of oil analyzed is the difference in weight between B3.3.6 and B3.3.11, expressed in % w/w:

Percent n-Pentane (or P.E.) insolubles:

$$\frac{\text{Weight of contaminant solids collected}}{\text{Weight of oil sample analyzed}} \times 100$$

B3.5 ADDITIONAL SUGGESTIONS OF TECHNIQUE

B3.5.1 During periods of high humidity, the cooling effect of the solvent evaporation will cause moisture condensation on the disc. This may retard the complete solvent washing and solid deposit; an intermediate air drying step followed by an additional wash with solvent may be necessary.

B3.5.2 The analyst may insert, between the filter membrane and the base, a porous absorbent pad to insure a clean surface for the membrane and to minimize "freezing" of the filter to the base.

B3.5.3 Visual examination of the membrane under ultraviolet light for absorbed oil residue will quickly determine if the rinsing operation was thorough. Traces of residual oil will show a fluorescent effect, whereas a clean membrane will not.

B3.6 ADDITIONAL OPERATIONS FOR IMPROVED ACCURACY OF ANALYSIS

B3.6.1 Conditioning Filter Membrane

B3.6.1.1 All filter membranes must be conditioned by a warm water soak to remove extraneous manufacturing material and to assure accurate and constant tare weight. Place the filter membrane in a clean beaker containing prefiltered distilled water at a temperature of 90 to 100°F (32 to 38°C). Cover the beaker and soak the filter membrane for 1 h.

B3.6.1.2 Carefully remove the filter membrane with clean forceps. Allow all surface water to run off the membrane.

B3.6.1.3 Place filter membrane in an open petri dish. Place petri dish, with cover slightly ajar, in the oven and dry for 60 min at 90°C minimum.

B3.6.1.4 Remove the petri dish, with membrane, from the oven and place in a dessicator, or constant temperature and humidity area, with cover slightly ajar. Allow 30 min (minimum) for the membrane to equilibrate with environmental conditions before weighing.

B3.6.2 Control Filter Membrane

B3.6.2.1 A control filter membrane should be carried along for all analysis work. Subject this membrane to all preparation, handling, and weighing techniques of B3.6.1.1 through B3.6.1.4.

B3.6.2.2 Insert the control filter membrane directly beneath a test membrane in the filter holder in at least one instance during a single or group test run. Subject it to all handling and weighing techniques specified in B3.3.9 and B3.3.10.

B3.6.2.3 Apply the weight change of the control filter as a correction factor to test results, subtracting this factor when the control filter shows a weight increase or adding the factor when the control shows a weight decrease.

B4. Analysis Method B (Centrifugation), Alternate

B4.1 TEST APPARATUS AND MATERIALS

B4.1.1 Air oven capable of maintaining 105°C.
B4.1.2 Clamp curved type.
B4.1.3 Centrifuge tube(s), 0.100 L size.
B4.1.4 Centrifuge, capable of 98 km/s^2.
B4.1.5 Desiccator (5).
B4.1.6 Analytical balance.
B4.1.7 Acetone–30 to 60°C boiling point range (or equivalent).
B4.1.8 n-Pentane or equivalent.
B4.1.9 Stiff wire–suggest stainless steel.
B4.1.10 Wash bottle.

B4.2 TEST PREPARATION

B4.2.1 Solvents–see Section B3.
B4.2.2 Wash forceps and centrifuge tubes in a detergent.
B4.2.3 Rinse with water to remove suds.
B4.2.4 Completely rinse inside and outside of centrifuge tube(s) with acetone. During this procedure, centrifuge tube should not be touched by hand.
B4.2.5 Place centrifuge tube(s) in a 105°C ± 3 oven for 45 min.
B4.2.6 Place centrifuge tube(s) in desiccator while cooling to room temperature.
B4.2.7 Weigh centrifuge tube(s) and record weight.

B4.3 TEST PROCEDURE

B4.3.1 Heat sample to 180°F (82°C) or until sufficiently mobile to allow shaking of sample. The sample shall be shaken until the sediment is homogeneously suspended in the oil.
B4.3.2 Weigh 10 g of sample into a preweighed centrifuge tube.
B4.3.3 Fill centrifuge tube to the 0.100 L mark with n-Pentane and place stopper in centrifuge tube.
B4.3.4 Shake centrifuge tube well so that contents are thoroughly mixed.
B4.3.5 Centrifuge for 20 min at 98 km/s^2 in a high speed centrifuge.
B4.3.6 Decant all but 0.003 L from the centrifuge tube without disturbing the precipitate.
B4.3.7 Add 0.010 L of n-Pentane to the centrifuge tube.
B4.3.8 With a clean, stiff wire, dislodge the precipitate and disperse it in the n-Pentane.
B4.3.9 Wash all insolubles adhering to the wire back into the centrifuge tube using n-Pentane and fill the tube to the 0.050 L mark.
B4.3.10 Shake tube until mixture is homogeneous.
B4.3.11 Centrifuge for 20 min at 98 km/s^2.
B4.3.12 Decant supernatant liquid being careful not to disturb the precipitate.
B4.3.13 Rinse only the outside of the centrifuge tube with acetone (during and after rinse, centrifuge tube should not be touched by hands).
B4.3.14 Place tube in an oven at 105°C ± 3 for 45 min.
B4.3.15 Place tube in a desiccator until centrifuge tube returns to room temperature.
B4.3.16 Weigh centrifuge tube to the nearest 0.0001 g and calculate weight of solids.

B4.4 SUPPLIES AND SUPPLIERS

1. Membrane filters, plain white, 47 mm diameter, 0.45 μm pore size–Millipore HAWP4700, Gelman GA-6, Sartorius SM-11306, or equivalent.
2. Filter holder, 47 mm diameter–Millipore XX100047 30 or XX15 047 00, Gelman 4370, Sartorius 16306, or equivalent.
3. Stainless steel forceps–Millipore XX62 000 06, Gelman 7050, or equivalent.
4. Centrifuge, 98 km/s^2–Fisher Scientific Co., Will Scientific, Inc., DeLaval Separator Co., or equivalent.
5. Desiccator–Fisher Scientific 8-594B, or equivalent.

B4.5 SUPPLIERS' ADDRESSES

1. Millipore Corp.
   Bedford, MA 01730
2. Gelman Instrument Co.
   P.O. Box 1448
   Ann Arbor, MI 48106
3. Sartorius-Membranfilter GmbH
   34 Gottingen
   Germany
4. Fisher Scientific Co., Ltd.
   711 Forbes Avenue
   Pittsburgh, PA 15219
5. Will Scientific, Inc.
   Box 23, Highbridge Station
   New York, NY 10452
6. DeLaval Separator Co.
   Poughkeepsie, NY

Appendix C
Fuel Treatment to Obtain Fluid as Specified in 4.1(c)

C1. Take the required volume of No. 2 fuel oil in the fuel sump and continuously recirculate the fuel by filtering the fuel through commercially available Fuller's Earth or clay cartridge filters in place of the cleanup filter.

C2. Periodically (about every 2 h), take a sample of the fuel in a beaker. Filter the sample through a 0.45 μm membrane and measure the interfacial tension (IFT) with distilled water at 20°C ± 1.5. The platinum ring detachment method (ASTM D 971) is recommended although other correlatable methods may be used. Stop the fuel treatment when the interfacial tension is between 30 to 35 dynes/cm (mN/m).

C3. Remove the Fuller's Earth filters from the test loop, or adjust valve positions such that the Fuller's Earth filter is isolated from the test loop.

C4. Add to the Fuller's Earth treated fuel 0.1% (1000 ppm) of cetane number improver Ethyl DII3,[6] or equivalent.

C5. Circulate the fuel with additive through the pump until there has been at least two complete turnovers of the fuel volume in the sump. The fuel treatment is now complete.

---

[6] Since DII3 is a commercially available additive and since it may have a finite shelf life, it is advisable to check the interfacial tension of this treated fuel (with DII3) against distilled water (ASTM D 971). If the fuel treatment is successful and the DII3 additive is acceptable, the interfacial tension within 1 h of DII3 addition should be between 23 and 28 dynes/cm.

STANDARD OIL FILTER TEST OIL—SAE J1260 APR89

SAE Standard

Report of the Engine Committee, Filter Test Methods Subcommittee, approved June 1983, and revised by the Filter Test Methods Standards Committee April 1989.

φ 1. **Scope**—This SAE Standard defines the requirements for an oil to be used in the SAE HS J806 JUN83 Oil Filter Test Procedures.

2. **Property Requirements**—The filter test oil shall be formulated from solvent refined petroleum stocks and blended with additives required to meet the specifications shown in Table 1. It shall not contain any viscosity index improver.

The filter test oil shall also meet engine oil performance and engine service classification SE/CC as outlined in SAE J183 MAR88.

3. **Performance Characteristics Approval**—The following oil performance characteristics must be approved by a panel of test oil users selected by the SAE Filter Test Methods Standards Committee:

3.1 **0.8 µm Membrane Filterability**—A panel judgment of slow filtration rates, relatable to additive removal or contamination, would constitute grounds for rejection.

3.2 **SAE HS J806 Filterability**—A panel judgment of filter performance, which deviates measurably from previously qualified batches of oil, would constitute grounds for rejection.

3.3 **Odor**—A panel judgment of objectionable odor, generated during any portion of SAE HS J806, would constitute grounds for rejection.

4. **Source Information**—Approved (February 1980) Filter Test Oil, RM 99, is available by contacting the Society of Automotive Engineers, Inc., 400 Commonwealth Drive, Warrendale, PA 15096. This oil has generally been called RFO-3-79 in the past.

TABLE 1—OIL FILTER TEST OIL SPECIFICATIONS

Property	Specification Limit	Test Designation
Viscosity	114–127 cSt (mm²/s) at 100°F (37.8°C) 100–112 cSt (mm²/s) at 40°C	ASTM D 445
	11.6–12.5 cSt (mm²/s) at 210°F (98.89°C) 11.2–12.2 cSt (mm²/s) at 100°C	ASTM D 445
Specific Gravity	0.888–0.876 at 60/60°F (15.5/15.5°C) 0.889–0.877 at 15/15°C	ASTM D 1298
Color	5.5 Max	ASTM D 1500
Viscosity Index	95 Min	ASTM D 2270
Flash Point	410°F (210°C) Min	ASTM D 92
Pour Point	0°F (−17.8°C) Max	ASTM D 97
Trace Sediment (including water)	0.05%, Volume, Max	ASTM D 2273
Foaming Tendency (after 5 min blowing)	Seq. I 10 mL Seq. II 50 mL Seq. III 10 mL	ASTM D 892
Foaming Stability (after 10 min setting)	Seq. I 0 mL Seq. II 0 mL Seq. III 0 mL	ASTM D 892
Additive Composition		
Sulfated Ash Wt., %	0.90–1.10	ASTM D 874
Calcium Wt., %	0.220–0.260	Atomic Absorption
Zinc Wt., %	0.110–0.130	Atomic Absorption
Phosphorus Wt., %	0.090–0.110	Atomic Absorption

(R) PRESSURE RELIEF FOR COOLING SYSTEM—SAE J151 JUN91

SAE Recommended Practice

Report of the Engine Committee, approved January 1970, reaffirmed without change, March 1985. Completely revised by the SAE Cooling Systems Standards Committee 3 June 1991.

1. **Scope**—This SAE Recommended Practice specifies requirements for pressure relief means and pressure relief rating identification for cooling systems of liquid cooled engines to reduce the possibility of injuries during opening of the cooling system.

2. **References**

2.1 **Applicable Documents**—The following publications form a part of this specification to the extent specified herein. The latest issue of SAE publications shall apply.

2.1.1 SAE PUBLICATIONS—Available from SAE, 400 Commonwealth Drive, Warrendale, PA 15096-0001.

SAE J164—Radiator Caps and Filler Necks

2.2 **Definitions**

2.2.1 COOLING SYSTEM CAP—Herein referred to as cap, means a device which closes the opening used to add coolant to the pressure section of the engine cooling system.

2.2.2 PRESSURE RELEASE MECHANISM—Means a device which, when activated, allows the pressure in a cooling system to be released.

2.2.3 LEGIBLE—Means readable under daylight lighting conditions by an observer having 20/20 vision (Snellen) whose eyepoint is located in proximity to the marked surface.

3. **Application**—This document applies to passenger cars, multipurpose passenger vehicles, trucks, chassis cabs, buses, construction vehicles, and to equipment produced for motor vehicles manufactured after the effective date of this document.

4. **Requirements**

4.1 Except as provided in 4.4, the cooling system of a liquid cooled engine shall have a pressure release mechanism conforming to the provisions of 4.2 or 4.3.

4.2 Pressure release mechanism incorporated in the cap and/or filler neck shall conform to one of the following:

4.2.1 If pressure relief can be accomplished by cap rotation, a mechanical stop shall be incorporated that prevents cap removal by rotation only. Removal requires the cap to be moved axially and further rotated. (Reference SAE J164.)

4.2.2 If pressure relief is accomplished through other than cap rotation, a means shall be provided to prevent removal of the cap before the pressure relief is actuated.

4.3 Pressure release mechanism not incorporated in the cap and/or filler neck shall have legibly and permanently marked on or adjacent to the cap instructions for relief of system pressure prior to cap removal.

4.4 "Sealed cooling system" which the manufacturer recommends servicing at stated intervals and which does not meet the requirements of 4.2 and 4.3 shall require use of a tool to accomplish cap removal. The words "Do Not Remove—See Manufacturer's Instructions" shall be legibly and durably marked on or immediately adjacent to the cap.

5. The cap or pressure relief valve assembly shall be legibly and durably marked with its normal pressure rating and notation "Warning Do Not Remove Hot." Other notations commonly used are: Caution-Remove Slowly, Caution-Pressure Remove Slowly, Never Open Hot, Do Not Open When Hot, Keep Tight-Do Not Open Hot, and Pressurized-Open Slowly.

(R) MULTIPOSITION SMALL ENGINE EXHAUST SYSTEM FIRE IGNITION SUPPRESSION—SAE J335 SEP90

SAE Recommended Practice

Report of Engine Committee approved April 1968 and completely revised September 1988. J335 MAY88 was not included in the SAE Handbook due to technical changes. Completely revised by the Spark Standards Committee September 1990.

1. Scope—This SAE Recommended Practice establishes equipment and test procedures for determining the performance of exhaust systems of multiposition engines used in hand-held portable applications. It is not applicable to spark arresters used in vehicles or stationary equipment.

1.1 Purpose—This document provides a method of testing to evaluate the fire ignition potential of exhaust systems of small multiposition portable engines.

1.2 Performance—Recommended performance criteria are given in Appendix A which are adequate for the severe fire hazard condition posed by heavy vegetative fuels. However, during periods of extreme fire danger, exhaust systems meeting this standard may not give absolute protection against fires. Additional control of operations may be necessary during such periods.

2. References

2.1 Applicable Documents

SAE J997—Spark Arrester Test Carbon.
SAE J1349—Engine Power Test Code—Spark Ignition and Diesel.
D. S. Stocksted, "Spontaneous and Piloted Ignition of Pine Needles." Research Note INT 194, USDA Forest Service, Northern Forest Fire Laboratory (1975).
D. S. Stocksted, "Spontaneous and Piloted Ignition of Cheatgrass." Prepublication Research Note, USDA Forest Service, Northern Forest Fire Laboratory (1976).
G. C. Kaminski, "Ignition Time versus Temperature for Selected Forest Fuels." Project Record, USDA Forest Service, San Dimas Equipment Development Center (1974).
R. T. Harrison, "Danger of Ignition of Ground Cover Fuels by Vehicle Exhaust Systems." ED & T Report 5100-15, USDA Forest Service, San Dimas Equipment Development Center (1970).
Anon, "Standard Test Procedure for General Spark Arresters." USDA Forest Service, San Dimas Equipment Development Center (1982).

2.2 Definitions

2.2.1 MULTIPOSITION SMALL ENGINE (MSE)—A hand-held power unit having an internal combustion engine operable in more than one position. MSE configurations include chain saws, weed trimmers, brushcutters, blowers, hedge trimmers, and cutoff saws.

2.2.2 POWER UNIT—An MSE unit exclusive of removable extensions such as chain saw bar and chain, brushcutter and trimmer shaft assemblies, hedge trimmer blade, cutoff saw blade extension, blower ducts, etc.

2.2.3 SPARK ARRESTER—An exhaust system having the ability to control the amount and size of particulate carbon or metal particles emitted into the atmosphere.

2.2.4 CONTACT PLANE—An imaginary flat surface defined by at least three points of contact on the surface of MSE power unit extremities.

2.2.5 EXPOSED SURFACES TEMPERATURE TEST—A test which measures exhaust system temperatures at the points where the engine exhaust system comes in contact with or intersects the contact plane surfaces established by the extremities of the MSE.

2.2.6 EXHAUST GAS TEMPERATURE TEST—A test which measures exhaust gas temperature at the points where the hottest gasses impinge a contact plane.

2.2.7 SHALL—Indicate a mandatory requirement exclusive of all other methods.

2.2.8 SHOULD—An advisory condition.

2.2.9 MAY—A permissive condition.

2.2.10 BEST POWER—Power at maximum torque achievable by a given test MSE at the maximum continuous corrected net brake power speed. (See SAE J1349.)

3. Instrumentation—The following instrumentation is required:

3.1 Calibrated Loading Device—An apparatus designed to test MSE's at given loads and speeds which display torque or power and is compatible with a multipoint recorder or data logger.

3.2 Tachometer—A device for determining the rotational speed of an MSE. It may be separate or integrated with the calibrated loading device or multipoint recorder or data logging device.

3.3 Thermocouple—Standard J or K type thermocouple wire with welded tip and grounded and shielded probe.

3.4 Graduated Cylinders—Calibrated containers for measuring amounts of fuel and lubricant for the fuel mix with an accuracy of ±2% of the amount actually contained.

3.5 Photographic Equipment—Cameras and adequate lighting, appropriate to record the test setup.

3.6 Wire Plug Gage—A round wire gage, with a diameter of 0.001 in greater than the Carbon Particle Retention Limit (see Appendix A) with precision ground squared ends. Examine ends with a 40× optical magnifier for any rounding of the corners. Replace if any rounding is noted.

4. Screen Test—The following screen test shall be made on all MSE screen-type spark arresters:

4.1 With the Screen Installed—With the screen installed, probe its periphery for any gaps in the mounting structure. Any penetration using the wire plug test gage fails the test. When gaging, do not exceed 2 oz (1730 dynes) of force.

4.2 Test with System Removed—Remove the exhaust system and randomly probe the screen at least 20 times. Especially pay attention to any bends, molding, or edges. Any penetration using the 0.024 in gage fails the test. Record results as GO for penetrations and NO GO for nonpenetration attempts.

4.3 Maximum Diameter Test—Using various gage sizes, find the largest diameter gage to penetrate the screen. Record results as "Max. gage size." To the "Max. gage size," add 0.001 in and record the results as "No-go wire gage size for screen openings."

5. Contact Plane Determination—The contact planes are determined by the following method:

5.1 General

5.1.1 Determine the configuration of the MSE to be tested in accordance with 5.2 through 5.7.

5.1.2 Slowly roll the test plate around the MSE exhaust system, or roll the MSE around the exhaust system on a flat surface, paying close attention to the area around the exhaust outlet. For each plane checked, mark at least three contact points to form a straight consistent line. (Refer to Figures 1 and 2.) When locating contact points, pay close attention to areas where the test plane touches the muffler assembly. Mark the contact points.

5.1.3 Use ⅛ to ¼ in pin stripe drafting tape or use a stiff, straight, ⅛ in (3.175 mm) thick hard wire or welding rod to construct the planes. These types of plane indicators may be fixed to the test unit for more accurate thermocouple probe alignment during the test.

5.2 Chain Saws

5.2.1 Assemble per manufacturer's specification, less bar and chain, fuel, and oil.

5.2.1.1 If spikes or bumper bars are supplied and required by the manufacturer, test the saw with spikes in place. Locate the test planes from the roots of the bumper spikes.

5.2.2 Fit a spacer, the same thickness as the bar, between the clutch cover and body and reinstall the bolts. The spacer shall not protrude beyond the body of the saw.

5.2.3 Place the chain brake in the "OFF" or disengaged position and deflect the hand guard by the weight of the saw into its most rearward position. Use tape to secure the chain brake and hand guard in this position.

5.2.4 Determine the planes per 5.1.

5.3 String Trimmers and Brushcutters—Use the power head and lower end shaft to determine the planes per 5.1.

5.4 Blowers—Use 5.1. Exception: Where the exhaust port is vented within the blower duct, thermocouples should be attached to the outlet surface of the exhaust outlet duct. Thermocouples should be attached to the outlet surface of the exhaust outlet.

5.5 Hedge Trimmers—Excluding the cutter bar, use bar mount to determine planes per 5.1.

5.6 Cutoff Saws—Remove blade and rotate blade housing. Establish plane as the closest possible plane to the muffler and exhaust outlet.

5.7 Other MSE Configurations—Determine planes for other configurations (that is, power drills, impact devices, and compactors) without attachments or detachable accessories.

6. Temperature Test—Perform the following temperature tests on all MSE's:

6.1 Exposed Surfaces Temperature Test—Measure exhaust system temperatures at the points where the engine exhaust system comes in contact with the plane surfaces established per 5.1.

6.1.1 Test Apparatus—Use thermocouples welded or brazed to the exhaust system to determine temperatures. For chain saws, a calibrated loading device is required (see Section 7).

Note: During all tests, a thermocouple should be mounted to the cylinder head or spark plug gasket to monitor the head temperature.

6.2 Exhaust Gas Temperature Test—Measure exhaust gas temperature at the points where the hottest gasses impinge a plane established per 5.1.

6.2.1 Test Apparatus—Use shielded and grounded thermocouples. For chain saws, a calibrated loading device is required.

Note: During all tests, a thermocouple should be mounted to cylinder head or spark plug gasket to monitor head temperature.

7. Detailed Temperature Test Methods—Use the following temperature test methods to test MSE exhaust systems:

7.1 Chain Saws

7.1.1 Preparation

7.1.1.1 Mount the test unit to the calibrated loading device with the bar clamp.

7.1.1.2 Weld or braze surface probes to all points found by the methods described in 5.1 and 5.2 where the test plane touches any part of the muffler surface. Use spark plug or cylinder head thermocouples for all tests.

7.1.1.3 Prepare a fresh fuel and oil mixture per the manufacturer's recommendations ±2%.

7.1.1.4 Establish contact planes by setting the narrow line of tape or 1/8 in welding rod sections between marked points.

7.1.1.5 Run engine until it reaches operating temperature. Set loading device for the best power speed stated by manufacturer. Adjust the high-speed mixture jet adjustment screw setting to maximum torque at best power, then back it out slightly to allow for proper lubrication. Adjust load to maintain head temperatures within 30°F (14.1°C) below manufacturer's maximum head temperature limit.

7.1.1.6 By carefully using a portable, hand-held, shielded probe, crisscross the plane surfaces to find the hottest spots. Mount shielded probes to monitor exhaust gas temperatures during the test at these locations.

7.1.2 Test

7.1.2.1 Run test engine at best power speed for at least 3 min, simultaneously recording time, temperatures, engine speed, torque and head temperature.

Note: If engine head temperature continues to rise after 3 min, recheck high-speed mixture setting and repeat.

7.1.2.2 Increase speed slowly by 1000 rpm above best power and run for at least 3 min, simultaneously recording time, temperatures, engine speed, torque and head temperature.

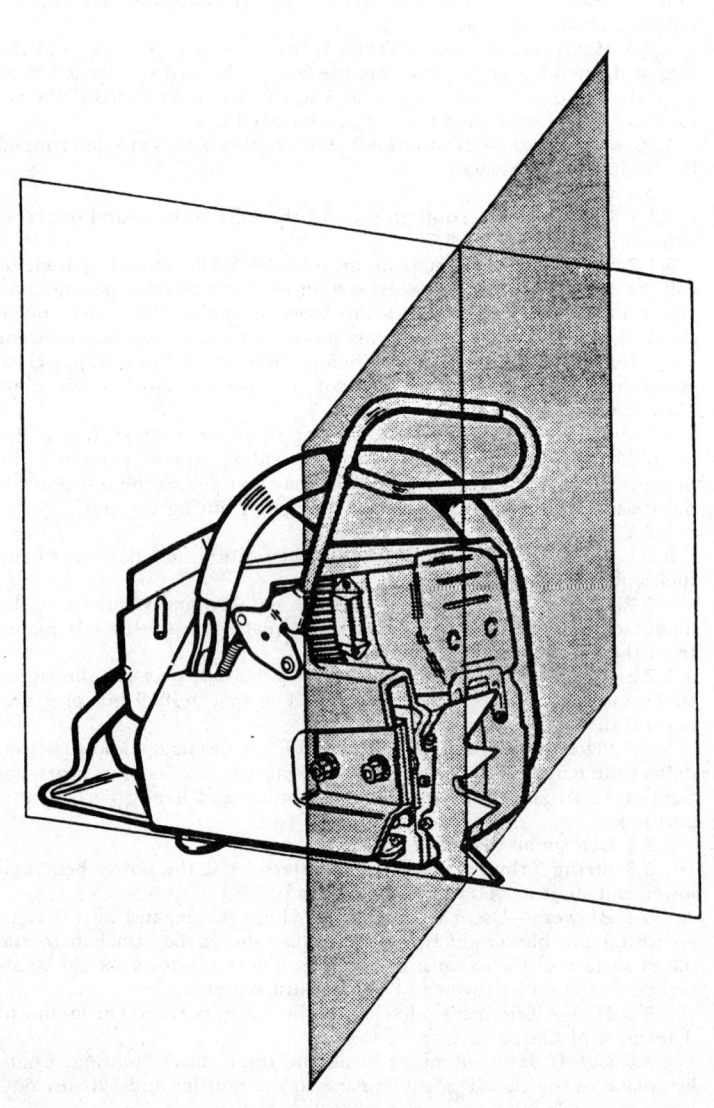

FIG. 1—CHAIN SAW WITH PLANES

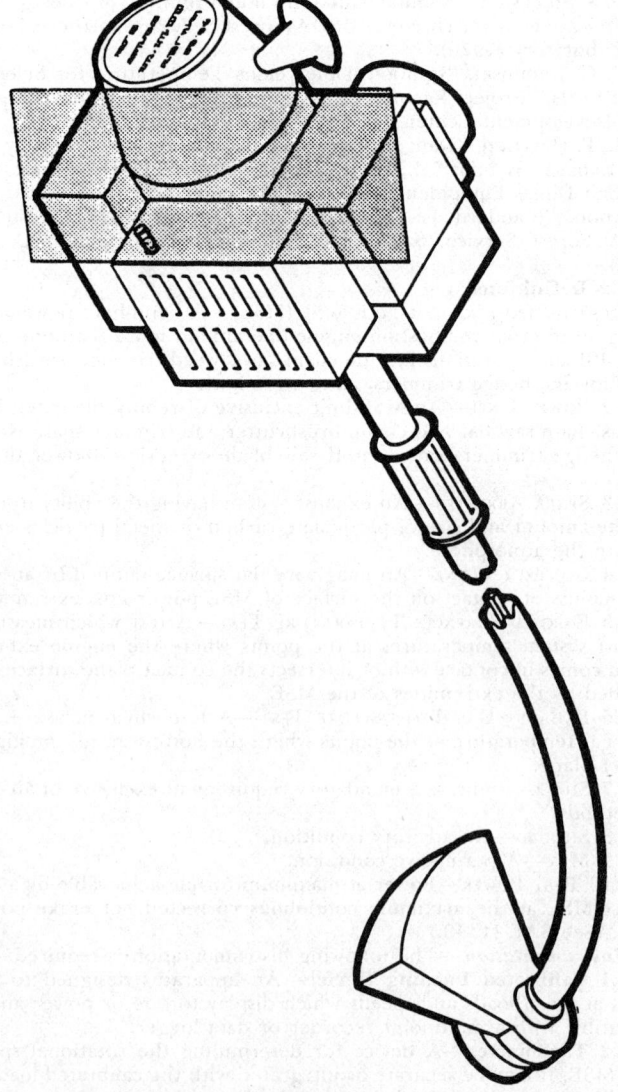

FIG. 2—BRUSHCUTTER WITH PLANES

7.1.2.3 Decrease speed slowly to 1000 rpm below best power in steps of 500 rpm. Record torque for each step. When the 2000 rpm speed reduction has been reached, run for 3 min simultaneously recording time, temperatures, engine speed, torque and head temperature.

EXAMPLE: If best power is 8000, then Test 7.1.2.2 would be conducted at 9000 rpm and Test 7.1.2.3 at 7000 rpm.

7.1.2.4 Complete documentation per Section 8.

7.2 String Trimmers and Brushcutters

7.2.1 PREPARATION

7.2.1.1 Instrument test unit to measure speed and temperatures per Section 6.

7.2.1.2 Run test unit with cutting head attachment until the head reaches operating temperature.

7.2.1.3 Establish operating engine speed by extending the line to the line limiter and adjusting the high-speed mixture jet for best power speed, or use the best power speed specified by the manufacturer. To achieve required speed, lengthen or shorten the cutting line. Record the speed and line length and diameter to document loading.

7.2.1.4 Adjust the high-speed mixture adjustment screw setting to maximum best power speed, then back out slightly to allow for proper lubrication at maximum power.

7.2.1.5 Do not exceed manufacturer supplied maximum engine head temperature during test.

7.2.2 TEST

7.2.2.1 Run engine at best power for at least 3 min. Record time, temperatures, engine speed, head temperature, and line length and diameter.

NOTE: If engine head temperature continues to rise after 3 min, recheck the high-speed mixture setting.

7.2.2.2 Increase speed to 1000 rpm above best power by cutting the length of the line. Run the unit for at least 3 min. Record time, temperature, speed, and head temperature, and line length.

7.2.2.3 Decrease speed to 1000 rpm below best power by lengthening the line. Run the unit for at least 3 min. Record time, temperatures, engine speed, head temperature and line length.

EXAMPLE: If the best power speed is 7000, Test 7.2.2.2 would be conducted at 8000 rpm and Test 7.2.2.3 at 6000 rpm.

7.2.2.4 Complete documentation per Section 8.

7.3 Blowers

7.3.1 PREPARATION

7.3.1.1 Instrument test unit to measure speed and temperatures per Section 6.

7.3.1.2 Run test unit until the engine is at operating temperature.

7.3.1.3 To check the operational speed, set the unit up without tubes or nozzles. (Blowers are constructed to run at full throttle with very little load on the engine.) Blowers usually run at a much higher speed than the best power speed.

7.3.1.4 Run unit for 3 min, adjust high-speed mixture adjustment screw for the best operational power speed, then back out slightly to allow for proper lubrication. If engine head temperature continues to rise, recheck the high-speed mixture setting and repeat after 3 min.

7.3.2 TEST

7.3.2.1 Run test unit at wide open throttle at least 3 min without tubes or nozzles. Record time, temperatures, speed, and head temperature.

7.3.2.2 Increase speed by 1000 rpm above the speed found in 7.3.2.1, by partially covering the blower intake or discharge ports. Run at least 3 min. Record time, temperatures, speed, head temperature and approximate percentage of the port area blocked.

EXAMPLE: If the wide open throttle speed is 7000 for step 7.3.2.1, then run Test 7.3.2.2 at 8000 rpm.

7.3.2.3 Complete documentation per Section 8.

7.4 Hedge Trimmers and Cutoff Saws

7.4.1 PREPARATION

7.4.1.1 Instrument test unit to measure speed and temperatures per Section 6.

7.4.1.2 Run the engine and let it warm up for 3 min.

7.4.2 TEST—Run the test with the throttle wide open. (Hedge trimmers and cutoff saws are constructed to run at full throttle with very little load on the engine. Hedge trimmers usually run at a much higher speed than the peak torque speed.)

7.4.3 Complete documentation per Section 8.

7.5 Other MSE Configurations—Test methods for configurations other than those listed above may be developed by agreement between the testing agency and the manufacturer with respect to the limits dictated by Appendix A of this document.

8. Documentation Procedure—Prepare at least the following documentation for all temperature tests, in addition to data required in Section 7:

8.1 Establish planes and photograph the procedure.

8.2 Photograph to show

  a. General test setup

  b. Thermocouple and probe locations with respect to the body of the test MSE unit

  c. Thermocouple and probe locations with respect to the planes and exhaust system

8.3 Draw a sketch to record the position of exhaust probes and thermocouples.

8.4 Record the hottest temperatures for each engine speed and torque. Record running time.

9. Spark Arrester Effectiveness Test

9.1 Introduction—This test method determines the proportion of carbon particles of two sizes retained by the spark arrester portion of the exhaust system. By definition, an arrester is 100% effective for the retention of carbon particles larger than the largest actual opening of its screens or baffles through which all gases pass. **No test is required for carbon particle retention of the sizes larger than the actual openings, thus no retention test is required for screen type arresters which are found to have a "go" size of 0.023 in or smaller per 4.3.**

9.2 Test Position of Spark Arrester—Test the spark arrester in the position where it is the least effective. This position is determined by attaching the spark arrester to a short length of hose from the air source. The hose I.D. should be equal to or larger than the spark arrester inlet.

Establish conditions described in 9.5.2 and 9.5.3 and then move spark arrester to different positions while watching for the emission of test carbon particles. The position at which the largest number of particles escape is the position of least effectiveness.

9.3 Test Apparatus—The test apparatus consists of a suitable blower with air directed through the testing apparatus, a flowmeter, a flow controlling valve or orifice, a back pressure manometer, a screened exhaust vent, a carbon injector, and a trap for collecting the particles. Figure 3 shows one acceptable arrangement with some optional features.

9.3.1 FLOWMETER—Use a flowmeter of an established design, such as a calibrated orifice. The indicated flow shall be within ±5% of actual.

9.3.2 INJECTING MECHANISM—Inject the test carbon with a feeder mechanism that does not crush or grind the material or affect the normal flow of air through the apparatus, located approximately as shown in Figure 3. Inject test carbon into the airstream at uniform rate over a period of 15 min ± 5.

9.4 Test Carbon—Make separate test runs using SAE J997.

9.4.1 Use 5 g ± 0.1 of carbon used for each test run.

9.4.2 Test carbon may be reused if the guidelines given in SAE J997 are followed.

9.5 Back Pressure and Flow

9.5.1 Make provision for measuring back pressure (differential pressure from intake to discharge of the arrester) and flow rate through the arrester.

9.5.2 Unless 9.5.3 applies, test the spark arrester at its assigned maximum flow capacity and at its assigned flow capacity with both coarse and fine test carbon:

9.5.2.1 Establishment of Assigned Flow Capacity and Assigned Maximum Flow Capacity: The assigned maximum flow capacity is the constant airflow rate resulting when a spark arrester is subjected to a pressure differential from intake to discharge of 1 lbf/in^2 gage (6.9 kPa) without test carbon being injected. The assigned flow capacity is the constant airflow rate resulting when a spark arrester is subjected to a pressure differential of 0.5 lbf/in^2 gage (3.45 kPa).

9.5.3 Some spark arresters may clog before the entire 5 g of test carbon is injected into the arrester. In these instances, use the following procedure:

9.5.3.1 *Condition 1*—Initially establish an airflow to cause a back pressure of 1 lbf/in^2 gage (6.89 kPa). Then allow the back pressure to build up to 1.3 lbf/in^2 gage (8.96 kPa). Once the 1.3 lbf/in^2 gage (8.96 kPa) is reached, maintain this pressure for the balance of the test.

9.5.3.2 *Condition 2*—Initially establish an airflow to cause a back pressure of 0.5 lbf/in^2 gage (3.45 kPa). Then allow the back pressure to build up to 0.8 lbf/in^2 gage (5.52 kPa). Once the 0.8 lbf/in^2 gage (5.52 kPa) is reached, maintain this pressure for the balance of the test.

9.6 Carbon Particle Collection—Carefully brush all carbon inside the test apparatus which has escaped through the spark arrester during each run into the trap and then hand-sieve lightly on a U.S. Standard No. 30 [0.023 in (0.584 mm) opening] screen. Weigh the carbon particles retained.

9.6.1 DETERMINATION OF ARRESTING EFFECTIVENESS—Use the following formula to determine arresting effectiveness for both SAE fine and SAE coarse carbon:

$$\% \text{ effectiveness fine or course} = \frac{\text{Weight used} - \text{Weight of Carbon Retained on No. 30 U.S. screen}}{\text{weight used}} \times 100 \quad \text{(Eq.1)}$$

Use the following formula to determine the arrester's combined effectiveness:

$$\% \text{ Combined effectiveness} = \frac{\% \text{ effectiveness with fine} + \% \text{ effectiveness with coarse}}{2} \quad \text{(Eq.2)}$$

APPENDIX A

A1. Recommended Performance Levels—The following exhaust system performance recommendations are based on the ignition characteristics of heavy vegetative fuels common to the Pacific Northwest. Maximum allowable surface and exhaust gas temperatures were determined from time versus ignition temperature studies conducted by the U.S.D.A Forest Service.

a. Effectiveness: The spark arrester used with the engine exhaust system should be designed to retain or destroy 90% or more of the carbon particles having a major diameter greater than 0.023 in (0.584 mm), as determined when testing in accordance with Section 9.

b. Exposed Surface Temperature: The exhaust system should be designed so that the exposed surface temperature does not exceed 550°F (288°C) as determined when tested in accordance with Section 6.

c. Exhaust Gas Temperature: The exhaust system should be designed so that the exhaust gas temperature does not exceed 475°F (246°C) as determined when tested in accordance with Section 6.

d. Debris Accumulation: The exhaust system should be designed so that there are no external pockets where flammable material could accumulate.

e. Durability: All parts of the exhaust system must have a service life of not less than 50 h. Cleaning of parts should not be required more frequently than once for each 8 h of operation.

f. Serviceability: The spark arrester should be so designed that it may be field inspected without a major disassembly of the power head. The spark arrester must be readily cleaned.

g. Marking: The exhaust system should be identified by manufacturer, trademark, or model number.

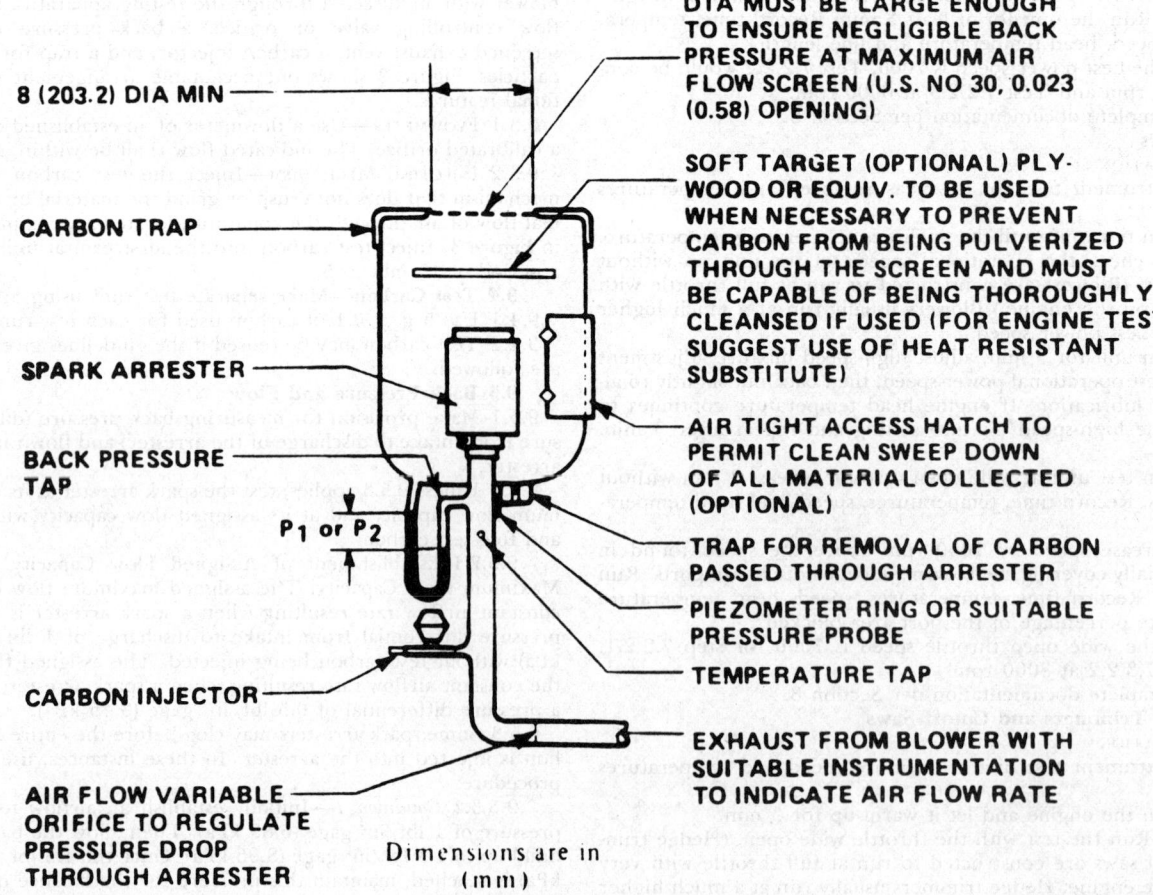

FIG. 3—SPARK ARRESTER TEST APPARATUS

TURBOCHARGER GAS STAND TEST CODE—SAE J1826 APR89

SAE Recommended Practice

Report of the Turbocharger Standards Committee approved April 1989.

1. Purpose—The purpose of this SAE Recommended Practice is to provide a recommended laboratory test procedure and presentation format for establishing the component performance for a turbocharger. It is intended that this test procedure be used to determine turbocharger compressor and turbine performance characteristics. The resulting data are intended for use in turbocharger component performance assessment and development and for engine/turbocharger matching.

2. Scope—The test procedures outlined in this recommended practice are applicable to single rotor turbochargers having either fixed- or variable-geometry with the following caveat:

At this stage in the development of variable-geometry (VG) turbochargers, it would be impractical to generate a detailed practice to cover all types of VG turbochargers which may evolve. However, there is a requirement to quote performance data within a stipulated degree of accuracy and to furnish comprehensive performance information. This will form a basis for this document where further refinements may be added as experience and necessity dictate.

3. Terminology—(See also SAE J922, Section 2.)

3.1 Turbocharger Hardware

3.1.1 FIXED-GEOMETRY TURBOCHARGER—Turbocharger having no moving parts in the aerodynamic flowpath other than the compressor impeller and turbine rotor.

3.1.2 VARIABLE-GEOMETRY TURBOCHARGER—Turbocharger incorporating moving parts such as, but not limited to, compressor inlet guide vanes, variable-geometry compressor diffusor, moveable turbine inlet nozzle vanes and/or a wastegate.

3.1.3 AUXILIARY EQUIPMENT—Equipment not directly associated with the compression or expansion processes, but necessary for proper turbocharger/engine system operation, such as:

a. actuator for variable-geometry device (example: wastegate actuator).

3.2 Turbocharger Performance—(See also SAE J922, Section 3.)

3.2.1 FLOW

$$\text{Compressor air mass flow} = \text{kg/s of air mass flow through the compressor}$$

$$\text{Corrected compressor air mass flow} = \text{Compressor air mass flow} \times \frac{\sqrt{\dfrac{\text{Compressor-inlet total absolute temperature (K)}}{298 \text{ K}}}}{(\text{Compressor inlet total absolute pressure (kPa)}/100 \text{ kPa})}$$

$$\text{Turbine gas flow} = \text{kg/s of gas flow through the turbine}$$

$$\text{Turbine gas flow parameter} = \frac{\text{Turbine gas flow} \times \sqrt{\text{Turbine-inlet total absolute temperature (K)}}}{\text{Turbine inlet total absolute pressure (kPa)}}$$

3.2.2 PRESSURE RATIO (EXPANSION RATIO)

$$\text{Compressor pressure ratio} = \frac{\text{Outlet-air static absolute pressure (kPa)}}{\text{Inlet-air total absolute pressure (kPa)}}$$

$$\text{Turbine expansion ratio} = \frac{\text{Inlet-gas total absolute pressure (kPa)}}{\text{Outlet-gas static absolute pressure (kPa)}}$$

3.2.3 EFFICIENCY

$$\text{Compressor efficiency} = \frac{\text{Isentropic enthalpy rise across compressor stage calculated using compressor pressure ratio}}{\text{Actual total enthalpy rise across compressor stage}}$$

$$\text{Combined turbine} \times \text{mechanical efficiency} = \frac{\text{Actual total enthalpy rise across compressor stage}}{\text{Isentropic enthalpy drop across turbine stage calculated using turbine expansion ratio}}$$

3.2.4 SPEED PARAMETER

$$\text{Corrected compressor speed} = \frac{\text{Compressor impeller speed (r/min)}}{\sqrt{\dfrac{\text{Compressor-inlet total absolute temperature (K)}}{298 \text{ K}}}}$$

$$\text{Turbine speed parameter} = \frac{\text{Turbine rotor speed (r/min)}}{\sqrt{\text{Turbine inlet total absolute temperature (K)}}}$$

3.2.5 SURGE—A line on the left-hand side of a compressor map as determined on a steady-flow test stand. Surge is the boundary of an area of severe flow reversal combined with audible coughing and banging.

4. Test Measurement and Accuracy—The test measurements below are required in turbocharger performance determination. The measurement accuracies specified do not include human or other probable errors in the reading.

4.1 Airflow ±1%
4.2 Pressure ±0.5%
4.3 Temperature ±0.5%
4.4 Speed ±0.5%
4.5 Duct diameters at static pressure measuring stations ±0.5%.

5. Apparatus/Test Stand

5.1 **2-Loop Hot Gas Stand**—The most commonly used test stand is the independent gas circuit (2-loop) hot gas stand as shown in Fig. 1.

5.2 **1-Loop (Bootstrap) Hot Gas Stand**—Less frequently used for component development and performance. Shown in Fig. 2. Useful for extended durability testing.

5.3 **Turbine Dynamometer**—Used for extended mapping of turbocharger turbine well outside the range of matched compressor and turbine flow and power. See Fig. 3.

6. Test Procedure

6.1 **Installation**—The test turbocharger shall be representative of the manufacturer's production units. Extraordinary nonproduction gaskets, sealants, etc. shall not be used.

6.1.1 LEAK CHECKING—In many production turbochargers, some small leakage is expected. However, leakage in the test stand and instrumentation shall be limited to that of good laboratory practice.

6.1.2 INSULATION—The compressor housing (cover) shall be insulated.

6.2 **Lubrication**—The turbocharger shall be supplied with SAE 30 lubricating oil at 350 kPa and 100°C unless otherwise specified by the manufacturer.

6.3 **Cooling Liquid**—Normally turbocharger housings are tested dry, with no insulation (except as noted in paragraph 6.1.2). However, if any turbocharger housings are ordinarily liquid cooled in the end application, testing can be performed with a supply of the appropriate coolant. In this event, the supply conditions (pressure, temperature and flow rate) shall be agreed upon by the tester and end user, and recorded on the performance maps.

6.4 **Operating Conditions**—Operating speeds and the number of data points per speedline vary with the turbocharger manufacturer. However, good practice dictates that point and line spacing be sufficiently close that undue interpolation is not necessary. Common practice is to begin testing at the lowest desired speed and move from high flow to low flow (surge) on the speed line.

6.4.1 STABILIZATION CRITERIA—Data shall not be taken until reasonable thermal stability is achieved (example: successive computations of compressor $\Delta T/T$ $((T_{out}-in)/T_{in})$ vary by less than 1/2%).

7. Data Acquisition and Computations

7.1 Minimum data to be recorded. The instrumentation list accompanying Fig. 3 can be used with appropriate test stand schematic (Figs. 1-3) to identify necessary data. In addition, the upper portions of Figs. 4 and 5 display additional data (held fixed during a given test) to be recorded.

7.2 Gas Properties—Most 2-loop gas stands use diesel fuel as the source of heat for the turbine inlet gas. The following are useful relationships for computation of the gas properties:

7.3 Nomenclature

T – Temperature, K
  K = 273.15 + °C
  °C = 5/9 (°F − 32)
  K = Degrees Kelvin
  °C = Degrees Celsius
  °F = Degrees Fahrenheit

γ – Ratio of specific heats
A/F – Gravimetric Air-to-Fuel Ratio
h – Enthalpy, kJ/kg
C_p – Specific Heat, kJ/(kg − K)
R – Gas Constant, kJ/(kg − K)

7.4 Air

$\gamma_a = 1.42592 - 8.03974 \times 10^{-5} T$
$h_a = 0.919848\, T^{1.01457}$
$R_a = 0.28699$
$Cp_a = R_a \dfrac{\gamma_a}{\gamma_a - 1}$

7.5 Combustion Products

$h_c = 2.3260\,(0.1284 + 2.3 \times 10^{-4} A/F)\,(1.8\,T)^A$
where $A = 1.0954 - 2.967 \times 10^{-4} A/F$
$R_c = 5.38038 \times 10^{-3}\,(53.341 - 0.4425/A/F)$
$C_{Pc} = 4.1868\,(0.0717 + A/F \times 10^{-4})\,(1.8T)^B$
where $B = 0.1883 - 2.903 \times 10^{-4} A/F$
$\gamma_c = C_{Pc} / (C_{Pc} - R_c)$

8. Data Presentation—(See also SAE J922, Section 4.)

8.1 Performance curves for compressor, turbine and bearing system losses are shown in Figs. 4 and 5. At present, for variable-geometry turbochargers, multiples of the figures shown, each at fixed-geometry increments, shall be used.

8.2 Formatting for electronic data transfer and computer simulation are recommended below:

8.2.1 COMPRESSOR DATA FORMAT

Line 1:	Description (supplier, model name, compressor nomenclature, reference test number) A15, A10, A20, A10
Line 2:	Inlet diameter (mm), outlet diameter (mm), inlet type, outlet type, impeller inertia (N-m-s²) F10, F10, A15, A15, F10
Line 3, 4, 5:	Additional comments (can be left blank) A80
Line 6 - N:	Corrected speed (r/min), corrected mass flow (kg/s), pressure ratio (T-S), efficiency (decimal) F10, F10, F10, F10

NOTE: Corrected mass flow rates are listed in ascending order. Speeds are also listed in ascending order.

Example:

SUPPLIER	ABC1		6789XY-C	87.0323C
76.2	50.0	HOSE	HOSE	0.0678
46250.00	0.032	1.2150	0.5250	
46250.00	0.050	1.2150	0.5800	
46250.00	0.056	1.2100	0.6300	
46250.00	0.065	1.2100	0.6520	
46250.00	0.081	1.2050	0.6550	
46250.00	0.097	1.1800	0.6180	
46250.00	0.116	1.1500	0.5200	
84100.00	0.084	1.8000	0.6850	
84100.00	0.094	1.8100	0.6950	
84100.00	0.116	1.8200	0.7210	
84100.00	0.138	1.8100	0.7200	
84100.00	0.166	1.7600	0.7130	
84100.00	0.191	1.6900	0.6650	
84100.00	0.216	1.5350	0.5370	
105850.00	0.129	2.4200	0.6920	
105850.00	0.143	2.4200	0.7000	
105850.00	0.161	2.4200	0.7110	
105850.00	0.189	2.3800	0.7180	
105850.00	0.210	2.3100	0.7000	
105850.00	0.234	2.1500	0.6380	
105850.00	0.247	1.8600	0.5370	
120300.00	0.154	2.9000	0.6540	
120300.00	0.175	2.9000	0.6700	
120300.00	0.193	2.8750	0.6850	
120300.00	0.218	2.8300	0.6850	
120300.00	0.238	2.6950	0.6520	
120300.00	0.256	2.5000	0.5850	
120300.00	0.261	2.1500	0.5000	

8.2.2 TURBINE DATA FORMAT

Line 1:	Description (supplier, model name, turbine nomenclature, reference test number) A15, A10, A20, A10
Line 2:	Test compressor, housing type, discharge connection description A20, A20, A20
Line 3:	Inlet gas temperature (°C) or turbine inlet-to-compressor discharge temperature ratio (K/K), oil type, oil temperature (°C), rotor/shaft inertia (N-m-s²) F10, A10, F10, F10
Line 4:	Cooling liquid description (if any), inlet temperature (°C), inlet pressure (kPa) A20, F10, F10
Line 5, 6, 7:	Additional comments (can be blank) A80
Line 8 - N:	Speed parameter (r/min − K), mass flow parameter (kg − K/s-kPa), expansion ratio (T-S), turbine × mechanical efficiency (decimal) F10, F10, F10, F10

NOTE: Expansion ratios are listed in ascending order. Speeds are also listed in ascending order.

Example:

SUPPLIER	ABC1		6789XY-T	87.0323T
ABC1	6789XY-C	HI SILICON		107.5 MM MARMAN
3.1	10W30	100.0	0.1442	
NONE	0	0		
2100.00	0.074	1.1390	0.5060	

Progression similar to compressor

| 4840.00 | 0.142 | 2.7580 | 0.6180 | |

9. References

SAE J244 JUN83, Measurement of Intake Air or Exhaust Gas Flow of Diesel Engines
SAE J916 MAY85, Rules for SAE Use of SI (Metric) Units
SAE J1349 JUN85, Engine Power Test Code—Spark Ignition and Diesel

Principles of Turbomachinery, D. G. Shepherd, MacMillan
Gas Properties, Keenan and Kaye, John Wiley & Sons
Experimental Methods for Engineers, J. P. Holman, McGraw-Hill

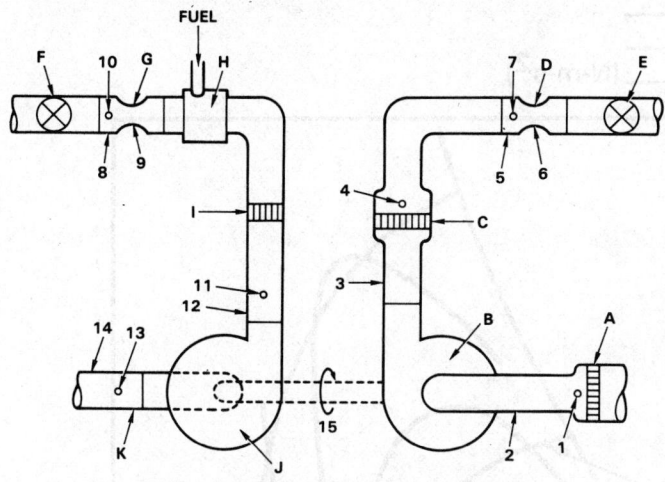

FIG. 1—2-LOOP HOT GAS STAND

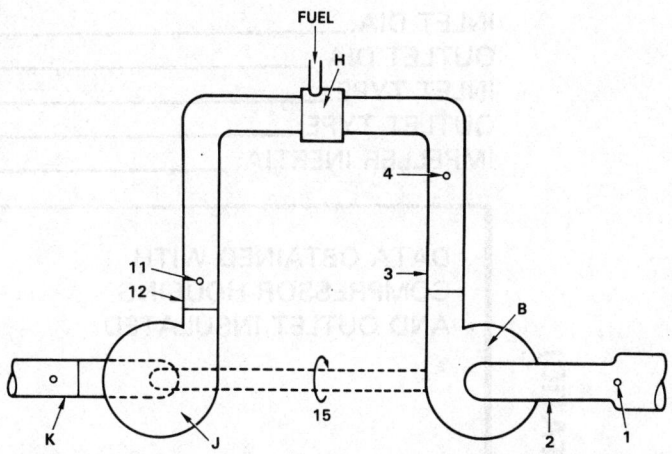

FIG. 2—1-LOOP (BOOTSTRAP) HOT GAS STAND

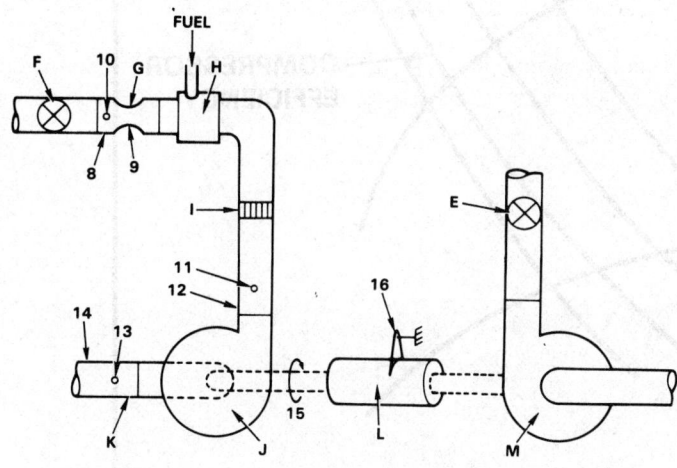

COMPONENTS

A COMPRESSOR INLET FLOW STRAIGHTENER
B COMPRESSOR STAGE
C COMPRESSOR DISHARGE FLOW STRAIGHTENER
D COMPRESSOR FLOW MEASURING SECTION
E COMPRESSOR THROTTLE VALVE
F TURBINE THROTTLE VALVE
G TURBINE FLOW MEASURING SECTION
H BURNER
I TURBINE INLET FLOW STRAIGHTENER
J TURBINE STAGE
K DISCHARGE DUCT
L DYNAMOMETER
M LOAD COMPRESSOR (OPTIONAL)

INSTRUMENTATION

1 COMPRESSOR INLET TOTAL TEMPERATURE
2 COMPRESSOR INLET STATIC PRESSURE
3 COMPRESSOR DISCHARGE STATIC PRESSURE (TOTAL OPTIONAL)
4 COMPRESSOR DISCHARGE
5 FLOW MEASURING SECTION INLET STATIC
6 PRESSURE DIFFERENTIAL
7 FLOW MEASURING SECTION TOTAL TEMPERATURE
8 FLOW MESURING SECTION INLET STATIC PRESSURE
9 PRESSURE DIFFERENTIAL
10 FLOW MEASURING SECTION TOTAL TEMPERATURE
11 TURBINE INLET TOTAL TEMPERATURE
12 TURBINE INLET TOTAL PRESSURE (STATIC OPTIONAL IF TOTAL CALUCULATIED)
13 TURBINE DISCHARGE TOTAL TEMPERATURE
14 TURBINE DISCHARGE STATIC PRESSURE
15 SHAFT SPEED
16 TORQUE

NOTE: Recommended Turbine Discharge Pipe Diameter (D_2) to be between 1.0 and 1.5 times the rotor exit diameter (D_1). Static Pressure taps to be placed 2 to 3 pipe diameters (L) Downstream of Rotor.

FIG. 3—TURBINE DYNAMOMETER

INLET DIA. _____ (mm)* 　　* = AT PRESSURE
OUTLET DIA. _____ (mm)* 　　　　MEASURING STATIONS
INLET TYPE: _____
OUTLET TYPE: _____
IMPELLER INERTIA: _____ (N-m-s^2)

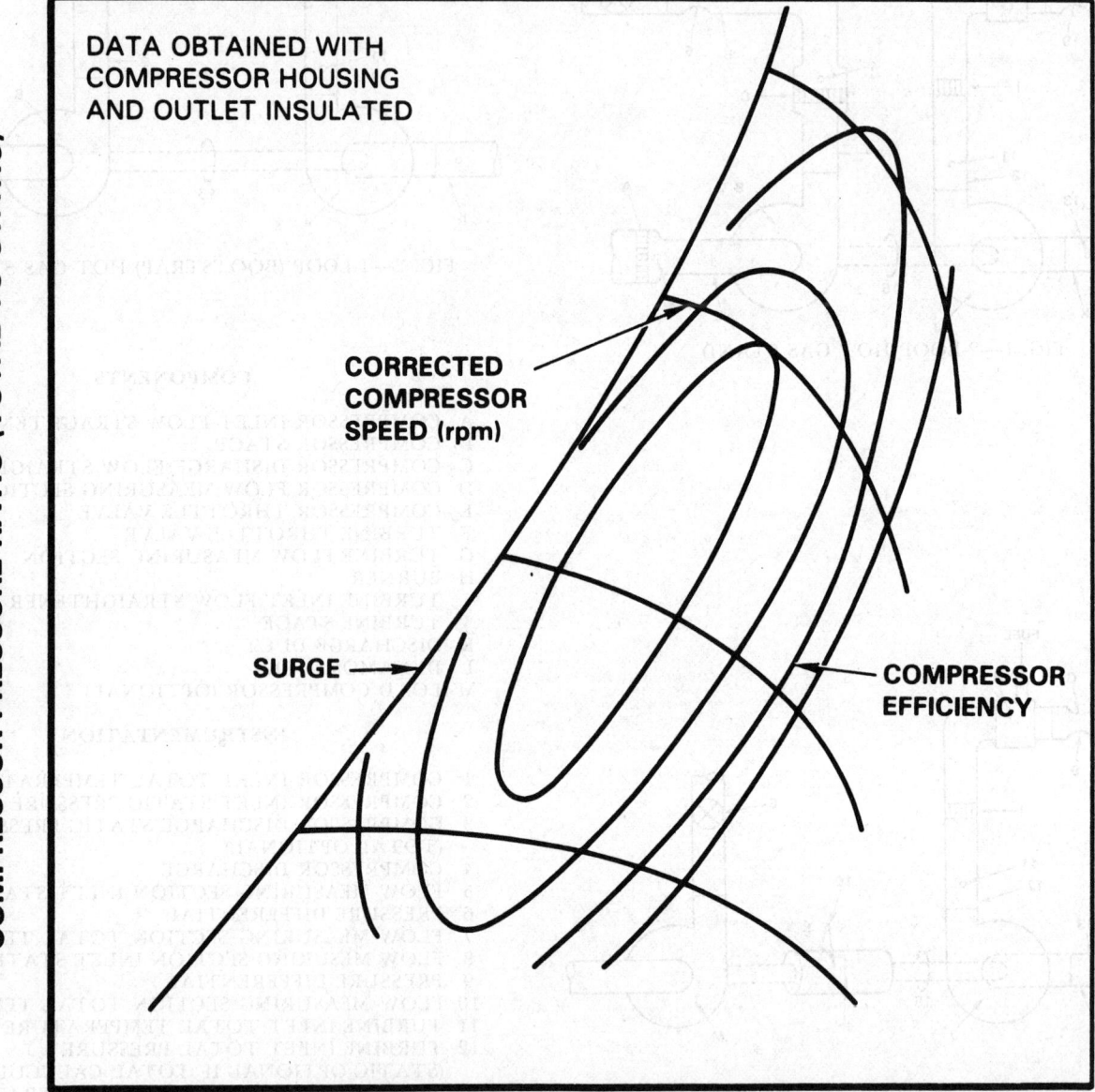

FIG. 4—TYPICAL TURBOCHARGER COMPRESSOR PERFORMANCE GRAPH

COMPRESSOR DESCRIPTION:_____
TURBINE HOUSING TYPE:_____
  IF LIQUID COOLED:
    LIQUID DESCRIPTION:_____
    LIQUID TEMP. INTO HSG:_____(°C)
    LIQUID PRESS. INTO HSG:_____(KPa)
    DISCHARGE CONNECTION DESCRIPTION:_____

$T_{1T}/T_{1C}(K/K)$ = _____ OR T_{1T} = _____ (°C) T_{1T} = 600 ± 20°C FOR DIESEL ENG.

T_{1T} = 900 ± 20°C FOR S.I. ENG.

OIL TYPE:_____

OIL SUPPLY TEMP. = _____ (°C) 100°C RECOMMENDED

ROTOR AND SHAFT INERTIA = _____ (N-m-s^2)

D_2/D_1 = _____   (SEE NOTE FIG. 3)

L/D_2 = _____

DATA APPLY ONLY WITH NOTED COMPRESSOR

SYMBOL	TURBINE SPEED PARAMETER
○	
□	
◇	
△	
⊙	

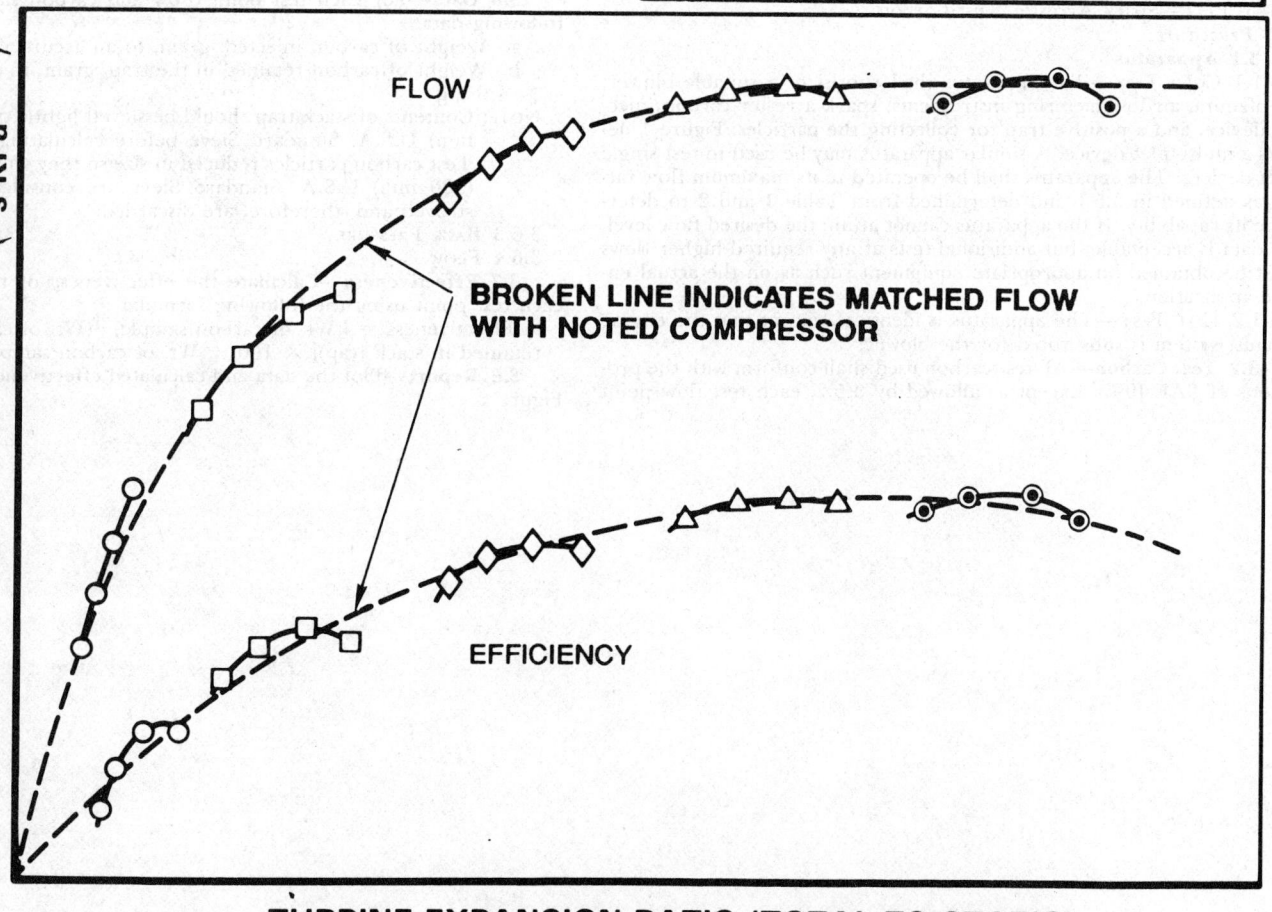

FIG. 5—TYPICAL TURBOCHARGER TURBINE
PERFORMANCE GRAPH

SPARK ARRESTER TEST PROCEDURE FOR LARGE SIZE ENGINES—
SAE J342 JAN91

SAE Recommended Practice

Report of the Engine Committee, approved June 1968, completely revised November 1980. Completely revised by the Spark Arrester Standards Committee January 1991. Rationale statement available.

1. Scope—This SAE Recommended Practice establishes equipment and procedures for the evaluation of the effectiveness and other performance characteristics of spark arresters or turbochargers used on the exhaust system of large engines normally used in a railroad locomotive, stationary power plant, and other similar applications. This document does not cover applications requiring flame arresting, exhaust gas cooling, or isolation from explosive gases. Two test methods are presented; a laboratory test using ambient air (cold test) and an engine test using exhaust gases (hot test). The hot test is preferred.

Arresters tested by the provisions of this document can be expected to perform as tested when tilted no more than 45 degrees from their normal position. Test results from a spark arrester or turbocharger evaluated by the hot test can be applied to different engines of similar design, provided the data shows it to be effective in the applicable flow ranges.

Certain design and performance characteristics, which represent current requirements by regulatory agencies for qualification and approval under this document, are listed in Appendix A.

1.1 Purpose—The purpose of this document is to provide a standard method of testing to evaluate spark arresters or turbochargers as spark arresters for use with large compression ignition internal combustion engines.

This document provides a method to evaluate the effectiveness of various spark arresters, but is not intended to establish the performance level required for adequate fire protection (see Appendix A).

2. References

2.1 Applicable Documents

SAE J997—Spark Arrester Test Carbon

3. Procedure

3.1 Apparatus

3.1.1 COLD TEST—The apparatus shall consist of a suitable blower, air plenum, airflow metering instruments, spark arrester carbon injector device, and a positive trap for collecting the particles. Figure 1 depicts a multi-inlet device. A similar apparatus may be used to test single inlet devices. The apparatus shall be operated to its maximum flow rating as defined in 3.5.1 and determined from Table 1 and 2 to determine its capability. If the apparatus cannot attain the desired flow level, the data is acceptable, but additional tests at any required higher flows must be obtained on appropriate equipment such as on the actual engine application.

3.1.2 HOT TEST—The apparatus is identical, except that the engine exhaust system is substituted for the blower.

3.2 Test Carbon—All test carbon used shall conform with the provisions of SAE J997. Except as allowed by 3.5.2, each test flow point shall be run using both fine and coarse test carbon.

Total test carbon injected, per Figure 1, shall be 100 g per engine cylinder. Single inlet devices shall have a minimum of 400 g of carbon injected.

3.3 Back Pressure—Back pressure shall be measured during each run at each flow, at the inlet of or in each manifold leg, during both hot and cold tests. A static pressure probe, such as illustrated in Figure 1, shall be used.

3.4 Injection Rate—Test carbon shall be injected into the inlet or each manifold leg as applicable for all flow rates tested at a uniform rate during a period of 15 min ± 5.

3.5 Flow Range

3.5.1 COLD TEST—The arrester shall be checked for effectiveness and back pressure at not less than five points between 30 and 100% of the rated flow of the arrester. Rated flow is defined as the calculated flow range for the maximum engine size application as determined per Table 1 and 2 or stated by the engine manufacturer. Note that maximum engine size application will be limited by maximum allowable back pressure requirements. One point shall be at 100% of rated flow, the remaining points shall be approximately evenly spaced relative to flow with the lowest point at approximately 30% of rated flow.

3.5.2 HOT TEST—The arrester shall be checked for effectiveness and back pressure at all numbered throttle positions. Where eight or more numbered throttle positions are involved, alternating carbon size (fine and coarse) may be used in lieu of testing each throttle position with both carbon sizes. Low effectiveness at any setting with one carbon size warrants further test with the other carbon size.

3.6 Data—For each test point (flow and carbon size), obtain the following data:

a. Weight of carbon injected, gram, to an accuracy of 0.1 g
b. Weight of carbon retained in the trap, gram, to an accuracy of 0.1 g

NOTE: Contents of stack trap should be sieved lightly on No. 30 (600 mm) U.S.A. Standard Sieve before calculating effectiveness. Test carbon particles reduced in size so they will pass a No. 30 (600 mm) U.S.A. Standard Sieve are considered to be destroyed and, therefore, are discarded.

3.6.3 BACK PRESSURE
3.6.4 FLOW

3.7 Effectiveness—Calculate the effectiveness of the arrester at each test point using the following formula:

$$\% \text{ Effectiveness} = [(\text{Wt. of carbon sample}) - (\text{Wt. of carbon retained in stack trap})] \times 100 \div \text{Wt. of carbon sample} \quad (\text{Eq.1})$$

3.8 Report—Plot the data and calculated effectiveness as shown in Figure 2.

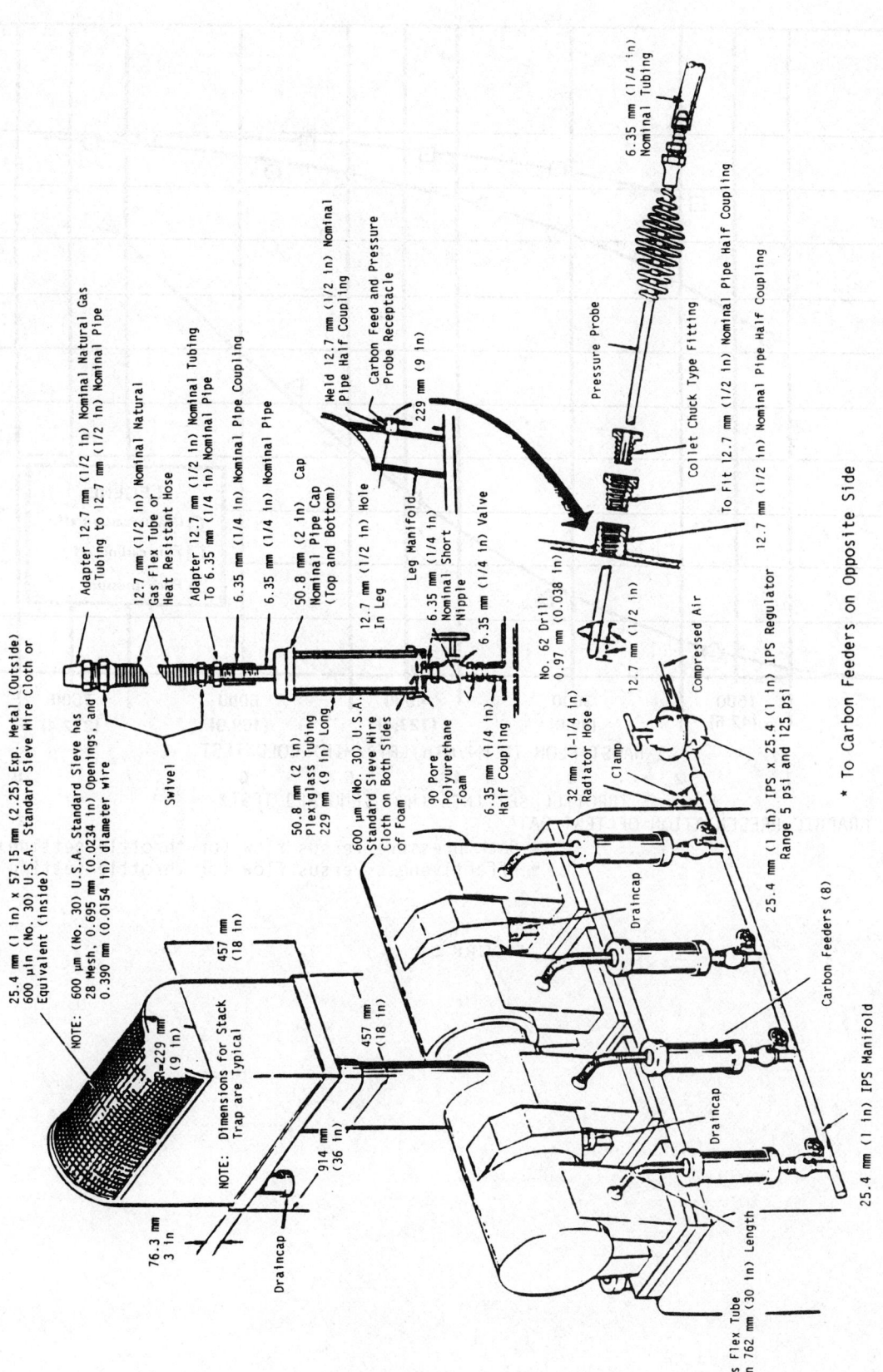

FIGURE 1

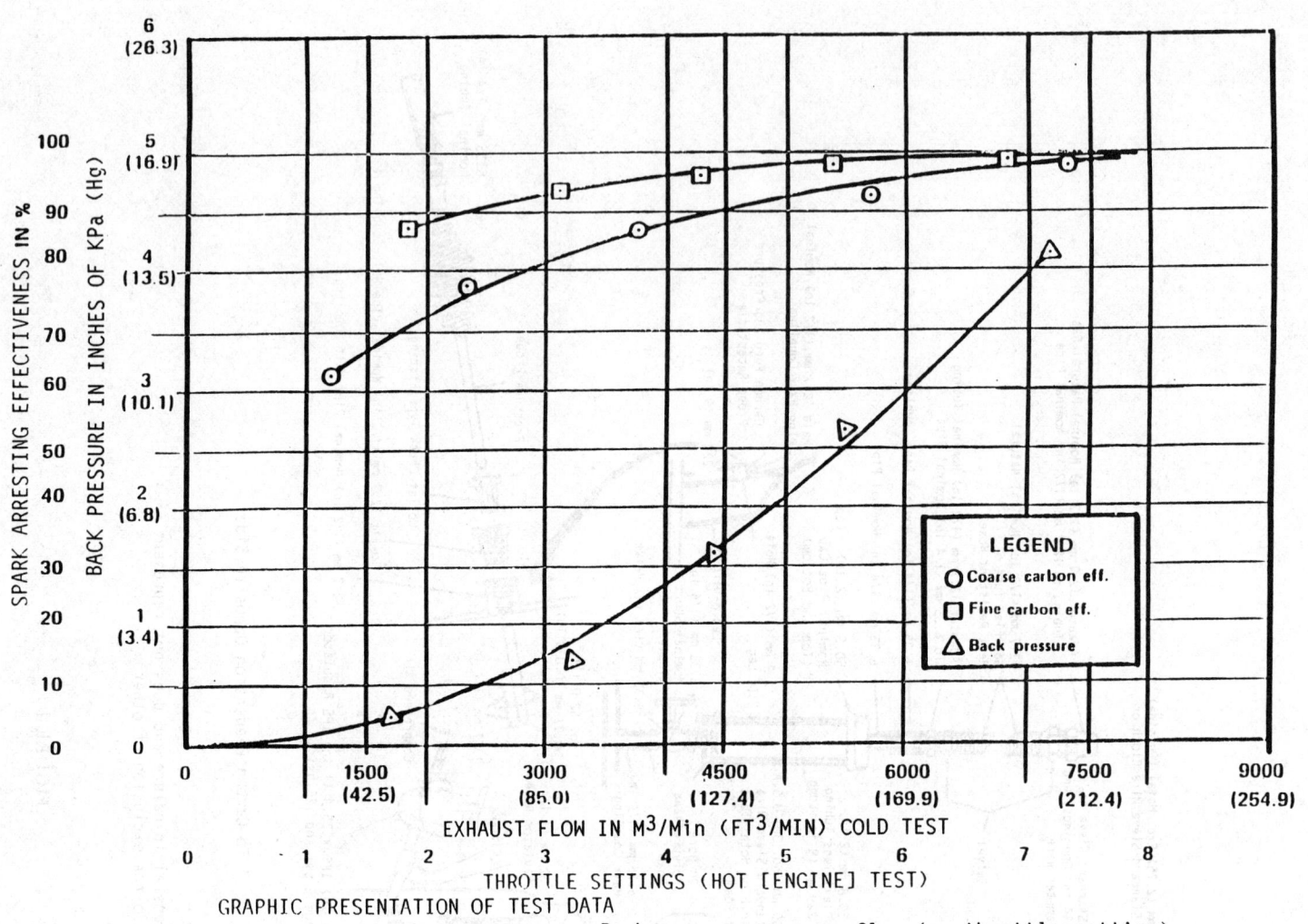

GRAPHIC PRESENTATION OF TEST DATA
• Back pressure versus flow (or throttle setting)
• Effectiveness versus flow (or throttle setting)

FIGURE 2

26.363

TABLE 1—EXHAUST FLOW IN CUBIC FEET PER MINUTE AT MAXIMUM RATINGS[1] FOR 4-CYCLE DIESEL ENGINES[2]

Engine Displacement (in^3)[3]	Rev Per Min 800	Rev Per Min 1000	Rev Per Min 1200	Rev Per Min 1400	Rev Per Min 1600	Rev Per Min 1800	Rev Per Min 2000	Rev Per Min 2200	Rev Per Min 2400	Rev Per Min 2600	Rev Per Min 2800	Rev Per Min 3000	rpm[4] Differential
1	0	1	1	1	1	1	1	1	1	1	2	2	0
2	1	1	1	2	2	2	2	2	3	3	3	3	0
3	1	2	2	2	3	3	3	4	4	4	5	5	0
4	2	2	3	3	4	4	4	5	5	6	6	7	0
5	2	3	3	4	4	5	6	6	7	7	8	8	1
6	3	3	4	5	5	6	7	7	8	9	9	10	1
7	3	4	5	5	6	7	8	9	9	10	11	12	1
8	4	4	5	6	7	8	9	10	11	12	13	13	1
9	4	5	6	7	8	9	10	11	12	13	14	15	1
10	4	6	7	8	9	10	11	12	13	15	16	17	1
20	9	11	13	16	18	20	22	25	27	29	31	34	2
30	13	17	20	24	27	30	34	37	40	44	47	51	3
40	18	22	27	31	36	40	45	49	54	58	63	67	4
50	22	28	34	39	45	51	56	62	67	73	79	84	6
60	27	34	40	47	54	61	67	74	81	88	94	101	7
70	31	39	47	55	63	71	79	87	94	102	110	118	8
80	36	45	54	63	72	81	90	99	108	117	126	135	9
90	40	51	61	71	81	91	101	111	121	131	142	152	10
100	45	56	67	79	90	101	112	124	135	146	157	169	11
200	90	112	135	157	180	202	225	247	270	292	314	337	22
300	135	168	202	236	270	303	337	371	404	438	472	506	34
400	180	225	270	315	360	404	449	494	539	584	629	674	45
500	224	281	337	393	449	505	562	618	674	732	786	843	56
600	269	337	405	472	539	606	674	742	809	876	944	1011	67
700	314	393	472	550	629	708	786	865	944	1023	1101	1180	79
800	359	449	540	629	719	809	899	989	1079	1169	1258	1348	90
900	404	506	607	708	809	910	1011	1113	1213	1315	1415	1517	101
1000	449	562	674	786	899	1011	1124	1236	1348	1461	1573	1685	112

[1] Volumetric effectiveness 80%; exhaust temperatures 900°F.
[2] For 2-cycle engines, multiply value by 3; for 2-cycle supercharged engines, multiply value by 4; for 4-cycle supercharged engines, multiply value by 1.25.
[3] 1 cm^3 = 0.061 in^3 (number of cubic centimeters x 0.061 = cubic inches).
[4] The "rpm Differential" column gives the difference between rpm columns for interpolation purposes. Entries are to the nearest whole number.

GENERAL: All chart values are proportional, so a flow can be calculated for a 1000 in^3 engine by doubling the flow figure for an identical engine but with 500 in^3 displacement. The same rule applies to rpm. Therefore, a flow rate can be calculated of a 3000 maximum rpm engine by doubling the flow value of an identical engine but with 1500 maximum rpm.

The chart may also be used like an interest table. If the engine has a cubic inch displacement of 438 at 2600 rpm, select readings from 400, 30, and 8 at 2600 rpm. The sum of these is the flow rate. When rpm falls between columns, that is, 1600 to 1800, 1800 to 2000, etc., take the next highest rpm column. To be more exact, interpolate by using the "rpm Differential" value.

EXAMPLE: A 4-cycle diesel engine has a total displacement of 633 in^3. If its maximum rpm is 2600, what is the maximum exhaust flow?

SOLUTION: 600 in^3 displacement at 2600 rpm = 876; 30 = 44; and 3 = 4; sum = 924.

TABLE 2—EXHAUST FLOW IN LITERS PER SECOND AT MAXIMUM RATINGS[1] FOR 4-CYCLE DIESEL ENGINES[2]

Engine Displacement (in³)[3]	Rev Per Min 800	Rev Per Min 1000	Rev Per Min 1200	Rev Per Min 1400	Rev Per Min 1600	Rev Per Min 1800	Rev Per Min 2000	Rev Per Min 2200	Rev Per Min 2400	Rev Per Min 2600	Rev Per Min 2800	Rev Per Min 3000	rpm[4] Differential
10	0.1	0.2	0.2	0.2	0.3	0.3	0.3	0.4	0.4	0.4	0.5	0.5	0
20	0.3	0.3	0.4	0.5	0.5	0.6	0.6	0.7	0.8	0.8	0.9	1.0	0.1
30	0.4	0.5	0.6	0.7	0.8	0.9	1.0	1.1	1.2	1.3	1.4	1.5	0.1
40	0.5	0.6	0.8	0.9	1.0	1.2	1.3	1.4	1.6	1.7	1.8	1.9	0.1
50	0.6	0.8	1.0	1.1	1.3	1.5	1.6	1.8	1.9	2.1	2.3	2.4	0.2
60	0.8	1.0	1.2	1.4	1.6	1.8	2.0	2.2	2.4	2.6	2.8	3.0	0.2
70	0.9	1.1	1.4	1.6	1.8	2.0	2.3	2.5	2.7	3.0	3.2	3.4	0.2
80	1.0	1.3	1.6	1.8	2.1	2.3	2.6	2.9	3.1	3.4	3.6	3.9	0.3
90	1.2	1.5	1.7	2.0	2.3	2.6	2.9	3.2	3.5	3.8	4.1	4.4	0.3
100	1.3	1.6	1.9	2.2	2.6	2.9	3.2	3.5	3.9	4.2	4.5	4.8	0.3
200	2.6	3.2	3.9	4.5	5.2	5.8	6.5	7.1	7.8	8.4	9.1	9.7	0.7
300	3.8	4.8	5.8	6.8	7.8	8.8	9.8	10.8	11.8	12.8	13.8	14.8	1.0
400	5.2	6.5	7.8	9.1	10.4	11.7	13.0	14.3	15.6	16.9	18.2	19.5	1.3
500	6.5	8.1	9.7	11.3	13.0	14.6	16.2	17.8	19.5	21.1	22.7	24.3	1.6
600	7.8	9.7	11.7	13.6	15.5	17.4	19.3	21.3	23.1	25.0	26.9	28.8	2.0
700	9.0	11.3	13.6	15.9	18.1	20.4	22.7	25.0	27.3	29.5	31.9	34.2	2.3
800	10.3	13.0	15.6	18.2	20.7	23.3	25.9	28.5	31.1	33.7	36.3	38.9	2.6
900	11.7	14.6	17.5	20.4	23.3	26.2	29.2	32.1	35.0	37.9	40.8	43.7	3.0
1000	12.9	16.2	19.4	22.7	26.0	29.3	32.6	35.9	39.2	42.5	45.8	49.1	3.3
2000	25.9	32.4	38.9	45.4	51.8	58.3	64.8	71.4	77.9	84.4	90.9	97.4	6.5
3000	38.8	48.6	58.3	68.0	77.7	87.4	97.1	106.8	116.5	126.2	135.9	145.6	9.7
4000	51.7	64.7	77.7	90.7	103.7	116.7	129.7	142.7	155.7	168.7	181.7	194.7	13.0
5000	64.8	81.0	97.2	113.4	129.6	145.9	162.1	178.4	194.6	210.8	227.0	243.2	16.2
6000	77.6	97.1	116.6	136.1	155.6	175.1	194.6	214.1	233.6	253.1	272.6	292.1	19.5
7000	90.7	113.4	136.0	158.7	181.4	204.2	226.9	249.8	272.5	295.3	318.0	340.8	22.7
8000	103.4	129.4	155.4	181.4	207.4	233.4	259.4	285.4	311.4	337.4	363.4	389.4	26.0
9000	116.1	145.4	174.7	204.0	233.2	262.5	291.8	321.1	350.4	379.7	409.0	438.3	29.3
10 000	129.3	161.9	194.3	226.7	259.3	291.7	324.3	356.8	389.3	421.8	454.3	486.8	32.5
20 000	258.6	323.6	388.6	453.6	518.6	583.6	648.6	713.6	778.6	843.6	908.6	973.6	65.0

[1] Volumetric effectiveness 80%; exhaust temperatures 482°C.
[2] For 2-cycle engines, multiply value by 3; for 2-cycle supercharged engines, multiply value by 4; for 4-cycle supercharged engines, multiply value by 2.
[3] 1 in³ = 16.39 cm³ (number of cubic inches × 16.39 = cubic centimeters).
[4] The "rpm Differential" column gives the difference between rpm columns for interpolation purposes. Entries are to the nearest tenth.

GENERAL: All chart values are proportional, so a flow can be calculated for a 1000 cm³ engine by doubling the flow figure for an identical engine but with 500 cm³ displacement. The same rule applies to rpm. Therefore, a flow can be calculated of a 3000 maximum rpm engine by doubling the flow value of an identical engine but with 1500 maximum rpm.

The chart may also be used like an interest table. If the engine has a cubic centimeter displacement of 4380 at 2600 rpm, select readings from 4000, 300, and 80 at 2600 rpm. The sum of these is the flow rate. When rpm falls between columns, that is, 1600 to 1800, 1800 to 2000, etc., take the next highest rpm column. To be more exact, interpolate by using the "rpm Differential" value.

EXAMPLE: A 4-cycle diesel engine has a total displacement of 6330 cm³. If its maximum rpm is 2600, what is the maximum exhaust flow?

SOLUTION: 6000 cm³ displacement at 2600 rpm = 253.1; 300 = 12.8; and 30 = 1.3; sum = 267.2.

APPENDIX A

A.1 Turbochargers—Turbochargers are generally very effective spark arresters.

A.2 Performance Requirements—When required performance levels are established, consideration should be given to the area in which the engine is operating. For example, in areas where daytime relative humidity of the atmosphere is below 30% for relatively long periods of time and there is considerable combustible material adjacent to the engine, the best spark arrester should be utilized. In areas of high humidity and little or no combustible material, spark arresters of lower ratings could be employed.

A.3 Hot Test—Cold Test Equivalence—Empirical data indicates that 80% effectiveness using the cold test procedure is approximately equivalent to 90% effectiveness using the hot test procedure.

A.4 Back Pressure Measurements—When selecting or testing spark arresters, the engine manufacturer's back pressure limits should be considered. If no such limits are stated, back pressure should not exceed 11.8 kPa (3.5 in Hg) average in all manifold inlet legs and should not exceed 13.5 kPa (4 in Hg) in any single leg. Single inlet devices, when installed on a manifold stack, should not cause manifold inlet legs to exceed the maximum allowable limits.

Most regulatory agencies require back pressure to be measured by the hot test method.

A.5 Performance Specifications—Most regulatory agencies in the U.S. that require spark arresters, require spark arresters to meet the following requirements:

A.5.1 Effectiveness—The arrester shall have effectiveness of 80% by the cold test or 90% by the hot test.

A.5.2 Cleaning Trap—The arrester, if of the retention type, shall have provisions for the easy disposal of the accumulated particles.

A.5.3 Spark Trap Capacity—The capacity of spark retention traps, if used, shall be not less than 2380 cm^3 (200 in^3) per trap, or not less than 490 cm^3 (30 in^3) per cylinder.

A.5.4 Marking and Labeling—Each arrester shall be permanently marked with manufacturer's model designing, serial number or date, and a trademark or other identification of manufacturer.

A.5.5 Warranty—The arrester manufacturer shall warrant that the arrester will maintain the required spark arresting effectiveness during the warranty period.

SPARK ARRESTER TEST PROCEDURE FOR MEDIUM SIZE ENGINES—SAE J350 JAN91

SAE Recommended Practice

Report of the Engine Committee, approved July 1968, and completely revised October 1988. Rationale statement available. Completely revised by the Spark Arrester Standards Committee January 1991.

1. Scope—This SAE Recommended Practice establishes equipment and procedures for testing spark arresters used on medium-size, single-position internal combustion engines, normally used in transportable, stationary, and vehicular applications, such as highway trucks, agricultural tractors, industrial tractors, other mobile equipment, and motorcycles. This document provides two methods of testing (laboratory testing and engine testing) which may be used to evaluate a spark arrester. It also includes special requirements for screen type devices and an endurance test procedure for screen type spark arresters.

1.1 Spark Arrester Application—Spark arresters qualified by an engine test should be used on only engines of the same design and size (or smaller) and which do not exceed the calculated flow rate of the test engine by more than 15%. Calculated flow rate shall be determined at maximum power engine speed or at maximum governed engine speed, whichever is lower.

This document has been established to cover most medium sizes and type of engines and engine applications. Arresters tested under this procedure should not be used above their rated flow unless tested and found to be adequately effective at the intended flow.

1.2 Purpose—This document provides a method of testing to evaluate spark arresters for compression ignition or spark ignition internal combustion engines. It is not intended to establish the performance level required for adequate fire protection, nor does it cover applications requiring flame arresting, exhaust gas cooling, or isolation from explosive gases.

2. References

2.1 Applicable Documents
SAE J726—Air Cleaner Test Code
SAE J997—Spark Arrester Test Carbon

3. Test Equipment Requirements

3.1 Laboratory Test

3.1.1 TEST EQUIPMENT—The test unit consists of a blower with air directed through the metering instruments, a carbon injector, the test spark arrester, a trap for collecting particles, and a method of measuring the gas flow through the spark arrester. The apparatus shall permit mounting of arresters in all positions in which they may be used and shall discharge into the positive trap in a manner similar to that shown in Figure 1.

3.1.2 TEST CARBON—Test carbon shall conform to SAE J997.

3.1.3 CARBON INJECTOR—The test carbon shall be injected by a feeder mechanism that does not crush or grind the carbon or affect the normal flow of air through the apparatus. It shall be located approximately as shown in Figure 1.

3.1.4 DIFFERENTIAL PRESSURE—The test equipment shall include a piezometer ring, probe, or other method for monitoring pressure at the inlet of the arrester. If a piezometer ring is used, it shall be as specified in SAE J726. The differential pressure shall be measured between the piezometer ring or probe and the pressure tap on the discharge side of the arrester or in the side of the positive trap chamber. (See Figure 1.)

A piezometer ring or probe calibration run shall be made over the anticipated airflow range, with no spark arrester in place. The indicated pressure recorded shall be plotted and the resulting calibration curve shall be used to correct data recorded during subsequent tests. At each test point, the flow rate through the arrester shall be held constant during the carbon feed cycle.

3.1.5 FLOW MEASUREMENTS—Any method which measures the gas flow through the spark arrester, with an accuracy of ±5% of the actual value, may be used.

3.2 Engine Test—When the spark arrester is to be tested on an engine, the equipment used shall be similar to those described in 3.1. Facilities for loading the engine shall be provided. The engine exhaust shall be used in place of the blower air supply. The measurement of gas exhaust flow is not required. Gas temperature shall be that normally produced by the engine after stabilization under conditions given in 4.2.4.

4. Test Procedure

4.1 Laboratory Test

4.1.1 SETUP AND PRE-TEST CHECKS—Test equipment shall be arranged in a general configuration similar to that shown in Figure 1. The arrester shall be mounted to the equipment in the position specified by the arrester manufacturer. The inlet duct shall be sized to fit the arrester as in a normal installation. The following checks of the test equipment shall be made with no spark arrester installed. A straight pipe equal in length and inlet inside diameter to that of the test arrester shall be installed.

4.1.1.1 Determine system back pressure, with air delivery flow from 10% of that flow which would occur at 1 lbf/in² gage with the spark arrester installed, to maximum expected. See 4.2.4.

4.1.1.2 Ensure that there is no pulverization of the test carbon caused by the injection apparatus or in the positive test trap by running a sample of each test carbon at the highest flow point.

4.1.1.3 Check that the outlet screen is of sufficient cross-sectional area to maintain minimum back pressure in the test apparatus discharge.

4.1.1.4 *Sample Run*—To establish the amount of test carbon to be used for each test, make a sample run. This consists of operating the test equipment with the spark arrester installed, at a back pressure of 1 lbf/in² gage carbon being injected, for a period of 12 min. At the end of 12 min, the airflow and temperature are recorded. From this information, the corrected air flow is calculated. See 5.2. The amount of test carbon to be used for each test run is the corrected flow at 1 lbf/in² gage times to 0.2 g. A minimum of 25 g and a maximum of 200 g shall be used.

4.1.2 TEST—Run the test as follows:

4.1.2.1 *Carbon*—Prepare sufficient samples of each size of test carbon. See 4.1.1.4.

4.1.2.2 *Duration of Test*—Inject test carbon into the air stream at a uniform rate over a period of 15 min ± 5.

4.1.2.3 *Flow Points*—Operate the test rig at 10, 32.5, 55, 77.5, and 100% of the flow at 1 lbf/in² gage differential pressure, with each fine and coarse carbon. Record data specified in 4.4.1. Calculate and plot the results as specified in Figure 2. Any points which do not fall on a smooth curve should be rejected and the data re-run. Data may be recorded at higher flows when potential arrester applications indicate a need for this additional data. Any effectiveness data within 2% of the minimum allowable is justification for additional intermediate data test points.

4.1.3 Data to be collected (for each flow rate)

a. Weight of test sample (W_s) for both SAE coarse and SAE fine carbon tests, grams;

NOTE: Under low flow rate conditions, stack air velocity may be insufficient to carry all of the test sample into the arrester. In these instances, W_s shall be determined by weighing the total of collected material found in the positive trap chamber and cleaned out of the arrester.

b. Weight of carbon in the positive trap chamber that is retained on a U.S. Standard No. 30 Sieve (W_t), gram; Record W_t (for coarse or fine carbon as applicable) for each individual run;

c. Air flow at test conditions, dm³s (ft³/m);

d. Air temperature at spark arrester inlet, °C (°F);

e. Pressure in the system without the spark arrester installed, at airflow rates per 4.2.4 (p1), (lbf/in² gage);

f. Pressure in the system with spark arrester installed, (P_2), lbf/in² gage;

NOTE: With some arresters, back pressure and flow will vary widely during the 15 min test period. In these instances, maintain constant flow during the test period, and record both minimum and maximum back pressure. In plotting the flow curve, the average differential pressure shall be used. See Figure 2.

g. Differential pressure, P_A, equal to P_2-P_1.

4.2 Engine Test—The engine test is to be conducted at a test facility utilizing a dynamometer for loading.

4.2.1 SETUP—Test equipment shall be arranged in a manner similar to that shown in Figure 1, with the exception that the engine exhaust shall be used as the air source rather than the blower. The arrester shall be mounted in the orientation that it is intended to operate on the engine and discharge into the positive trap.

4.2.2 QUANTITY OF CARBON—See 4.1.1.4.

4.2.3 DURATION OF TEST—See 4.1.2.2.

4.2.4 FLOW POINTS—Data shall be recorded separately for each size carbon, at the following operating conditions:

a. No load and governed high idle speed;
b. 100% rated load at rated speed;
c. 50% of attainable load at one-half rated speed;
d. Where applicable (such as for trucks), 80% of rated net load at rated speed.

4.2.5 DATA—Data to be obtained: (See Figure 3.)

a. Engine speed (rpm);
b. Engine load, hp (W);

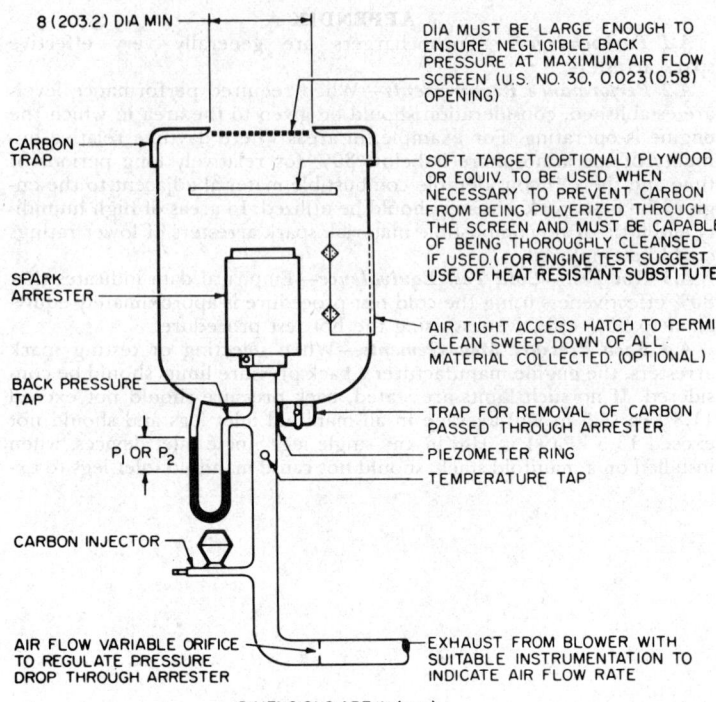

FIG. 1—SUGGESTED TEST APPARATUS

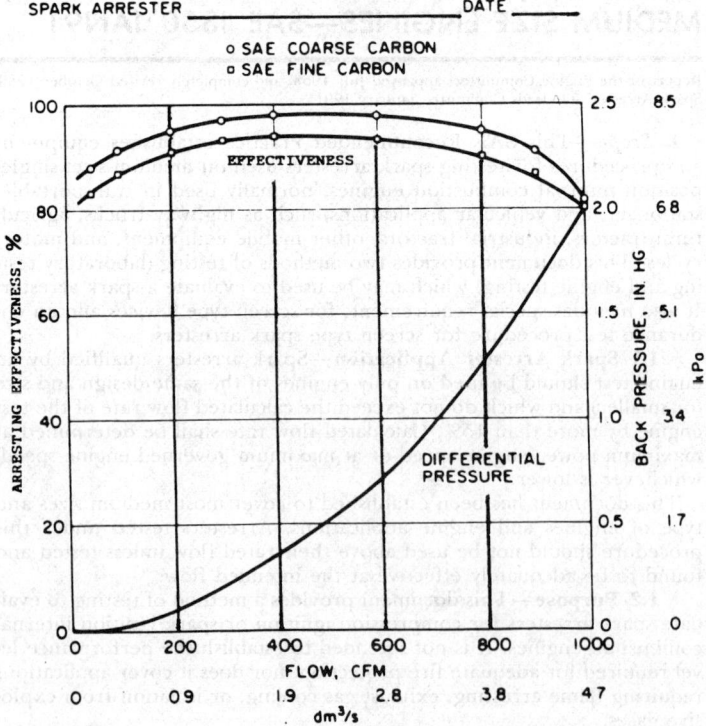

FIG. 2—DATA PRESENTATION

```
                    SPARK ARRESTER - ENGINE TEST RESULTS

                                        DATE _____

                                        DISPLACEMENT _____

ENGINE IDENTIFICATION _____

     RATING _____ BHP AT _____ RPM _____

SPARK ARRESTER _____

                              TEST DATA

| ENGINE RPM | ENGINE BHP | TEST CARBON SIZE | ARRESTER EFFECTIVENESS % | ARRESTER BACK PRESSURE, IN Hg (Pa) |
|------------|------------|------------------|--------------------------|------------------------------------|
|            |            |                  |                          |                                    |

                                        SIGNATURE _____

                                        ADDRESS _____
```

FIG. 3—SPARK ARRESTER ENGINE TEST RESULTS REPORT FORM

    c. Back pressure induced by the spark arrester, lbf/in² gage (Pa);
    d. Gas temperature at the inlet of the spark arrester, °C (°F);
    e. Weight of each size test carbon used, gram;
    f. Weight of each size test carbon retained in the trap that will not pass a U.S. No. 30 sieve or equivalent, gram.

4.3 Special Requirements for Screen Type Devices

4.3.1 Effective exhaust area of screen (total area of all screen openings) shall not be less than 200% of the total exhaust port area.

If it is necessary to clean the spark arrester screen, the time of such cleaning shall be recorded. (See 4.4.1.b.) No screen cleaning shall be allowed during the final 12.5 h of the 50-h period. Since the final examination is to determine qualification status of the arresting unit, it is imperative that the screen and other components not be cleaned or altered in any way following completion of the test.

4.3.2 Screen material shall be heat and corrosion resistant. The screen component shall provide a minimum of 50 h of service life when tested in accordance with the endurance test for screen type spark arresters. See 4.4.

4.4 Endurance Test for Screen Type Spark Arresters—The arrester shall be mounted as specified by the arrester manufacturer on an engine of the type and maximum size for which the arrester is to be used. Engine size shall be determined by the maximum calculated flow rate. If the arrester is intended for both two- and four-stroke cycle applications, separate tests shall be conducted for each type of engine.

The test shall be conducted for a minimum of 50 h, which need not be continuous. A 1-h cycling period, representative of actual operating conditions, including idle at no load through maximum operating engine speed at full load, shall be conducted continuously throughout the entire 50-h period. Manufacturer's recommended air-to-fuel ratios shall be observed. Engine manufacturer's recommended oil-to-fuel ratios shall be observed for all two-stroke cycle tests.

4.4.1 During the 50-h test, the following data shall be recorded at increments of not more than 2 h:
    a. Operating time;
    b. Engine speed;
    c. Temperature in screen area;

d. Engine load or output horsepower;
e. Exhaust back pressure;
f. Time and circumstance of any part failure or malfunction, including any required screen cleaning.

4.4.2 The following shall be recorded once for each arrester:
a. Complete information on the engine, including make, model, serial number, fuel used, etc.
b. A description of the test setup, including photographs

5. Calculations

5.1 Spark Arresting Effectiveness Formula

$$= \frac{(W_s - W_t) \times 100}{W_s} \quad \text{(Eq. 1)}$$

where:

  = spark arresting effectiveness, %
W_s = weight of the carbon sample grams
W_t = weight of the carbon found in the positive trap chamber that is retained on a U.S. Standard No. 30 sieve, grams

5.2 Exhaust Gas Flow Formula—(See 4.2.2)

$$\text{Flow in cfm} = W_m \times \frac{1}{V} = W_m \times \frac{T + 460}{0.6523 \times P} \quad \text{(Eq. 2)}$$

$$\text{Flow in dm}^3/\text{s} = W_m \times \frac{1}{V} = W_m \times \frac{T + 273}{3.49 \times P}$$

where:

W_m = airflow determined by standard laboratory instrumentation or by airflow as specified by the engine manufacturer, g/s (lb/min)
V = specific weight of flowing gas, g/m^3 (lb/ft^3)
P = gas pressure at the inlet of the spark arrester, lbf/in^2 gage (kPa)
T = gas temperature at the inlet of the spark arrester, °C (°F)

6. Presentation of Results

6.1 Laboratory Test—Test results shall be presented graphically as shown in Figure 2. In general, all tests shall be carried to a flow at which the differential pressure induced is 1 lbf/in^2 gage. Tests may be carried to greater flows when that information is required. Curves of effectiveness and differential pressure shall be presented as shown.

6.2 Engine Test—Engine test results shall be recorded in tabular form, as shown in Figure 3, including engine speed and load, spark arresting effectiveness with each carbon size, and differential pressure at each data point.

APPENDIX A

A1. Arrester Performance—When required performance levels are established, consideration should be given to the area in which the engine is operating. For example, in areas where daytime relative humidity of the atmosphere is below 30% for relatively long periods of time and there is considerable combustible material adjacent to the engine, the best spark arrester should be utilized. In areas of high humidity and little or no combustible material, spark arresters of lower ratings could be employed.

A2. Laboratory Test—Under normal operating conditions, taking into account the speed of engine and exhaust flow rate, a spark arrester having an effectiveness of at least 80% with both fine and coarse carbon as determined by the method of this standard is considered, by most authorities, adequate for most conditions and areas.

A3. Engine Test—Experience has shown that at flows causing back pressure up to 1 lbf/in^2 gage, 90% effectiveness by the engine test is equivalent to 80% by the laboratory test, with both carbon sizes.

In applications where back pressure above 1 lbf/in^2 gage is developed, it is essential that additional testing be completed at the higher flow rate necessary to develop the actual maximum back pressure.

Arresters meeting this standard should not be expected to arrest sparks adequately when titled more than 45 degrees from their normal operating position.

APPENDIX B

B1. Qualifications for Approval—A spark arrester is "qualified" under this document if:

B1.1 The arrester demonstrates arresting effectiveness equal to or greater than those shown in Appendix A of this document throughout the flow range for which qualification is sought.

B1.2 The arrester is easily cleaned (if cleaning is necessary).

B1.3 The arrester has a minimum service life of 1000 h, or a service life equal to that of the engine if a visual examination of the internal components is not possible, or, in the case of a screen type arrester, 100 h.

B1.4 Arrester is identified by appropriate name, trademark, or model number.

(R) SPARK ARRESTER TEST CARBON—SAE J997 SEP90 — SAE Standard

Report of the Engine Committee approved August 1967 and completely revised October 1988. J997 JUN88 was not included in the SAE Handbook due to technical changes. Completely revised by the Spark Arrester Standards Committee approved September 1990.

1. Scope—This SAE Standard establishes physical properties required of SAE Coarse Test Carbon and SAE Fine Test Carbon, and establishes test methods to ensure that these requirements are met.

1.1 Purpose—The purpose of this document is to establish specifications for test carbon to be used when performing the tests described in SAE J335, J342, and J350.

2. References

2.1 Applicable Documents

SAE J335—Multiposition Small Engine Exhaust System Fire Ignition Suppression
SAE J342—Spark Arrester Test Procedure for Large Size Engines
SAE J350—Spark Arrester Test Procedure for Medium Size Engines
ASTM E 11

3. Test Carbon Requirements

3.1 Type—The test carbon shall be petroleum coke.
3.2 Form—The test carbon shall be granular in form.
3.3 Size—The size distribution of particles of test carbon shall be as follows:

3.3.1 FINE CARBON
a. Pass through U.S. Standard No. 16 Sieve: 100% by mass
b. Retained on U.S. Standard No. 20 Sieve: 70% by mass
c. Retained on U.S. Standard No. 30 Sieve: 30% by mass

3.3.2 COARSE CARBON
a. Pass through U.S. Standard No. 8 Sieve: 100% by mass
b. Retained on U.S. Standard No. 12 Sieve: 60% by mass
c. Retained on U.S. Standard No. 16 Sieve: 40% by mass

3.3.3 MIXING—Before use, mix screened fractions by pouring from one container to another at least five times.

3.4 Activity—The test carbon shall conform to the commercial definition of "activated carbon."

3.5 Apparent Density—0.5 to 0.6 g/cm^3 when vibrated to minimum volume (see 4.1.)

3.6 Strength—75 to 90% (see 4.2.)

4. Test Methods

4.1 Apparent Density

4.1.1 PRINCIPLE—The apparent density is the mass of a unit volume of the sample, including the pores and voids between the particles. It is obtained by measuring the volume of a weighed sample in a graduate.

4.1.2 APPARATUS

4.1.2.1 A balance of 100 g capacity, accurate to 0.01 g.
4.1.2.2 A 100 cm^3 graduated cylinder, accurate to 0.01 cm^3.
4.1.2.3 A vibrating table, such as shown in Figure 1. The table shown in Figure 1 is a ¾ in (19 mm) air vibrator fastened to the vibrat-

ing table which is supported on rubber belting. Any shaker that provides adequate packing of the sample may be used.

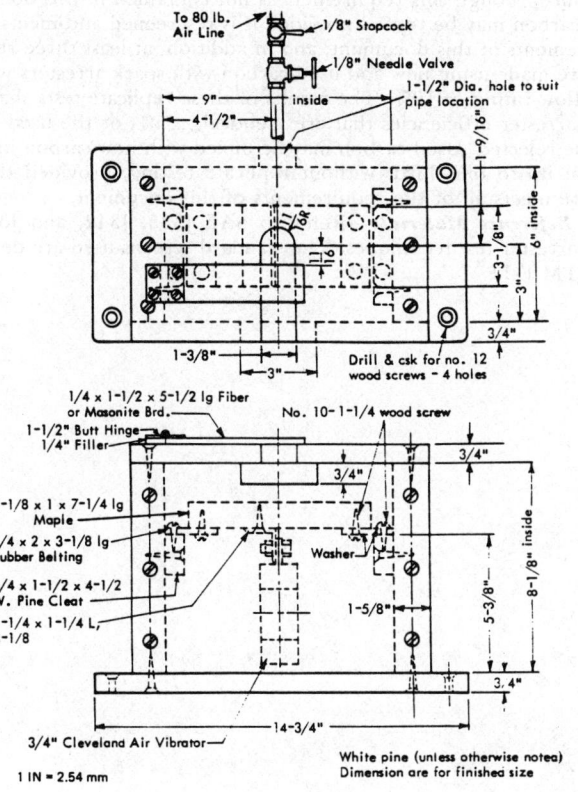

FIG. 1—APPARENT DENSITY OF ACTIVATED CARBON

4.1.3 PROCEDURE

4.1.3.1 Dry 50 g of the sample for at least 6 h at 150°F. Either rough screened or final screened carbon may be used. See A1. and A2.

4.1.3.2 Loosely support the graduated cylinder on the vibrating table and start the vibrator.

4.1.3.3 Weigh out a 39.5 to 40.5 g portion of the dried sample.

4.1.3.4 Pour the sample slowly into the graduated cylinder. The vibrator shall be in motion all during the time the sample is being poured into the graduate, and shall not be turned off until 60 s after the last of the sample enters the graduated cylinder.

4.1.4 CALCULATION—(See Eq. 1)

$$\text{Apparent Density} = \frac{\text{mass of dried sample, g}}{\text{volume of sample, cm}} \quad \text{(Eq.1)}$$

4.2 Strength

4.2.1 PRINCIPLE—The average particle size of a sample is determined. The sample is then shaken for a definite time interval with steel balls and the final average particle size again determined. The strength, or resistance to abrasion, is the ratio of the final average particle size to the original.

An alternative strength measurement method may be used, provided that the accuracy of the alternative method is established by at least three replicate tests using the alternative method and the method described in 4.2 of this document. No alternative method shall be used if it does not provide the same strength values for all three replicate samples, within 2% of the strength value obtained using the method described in 4.2.

4.2.2 APPARATUS

4.2.2.1 A W.S. Tyler Company Ro-Tap sieve shaking machine with sieves and special strength pan, or equivalent.

4.2.2.2 Six U.S. Standard Sieves, No. 4, 6, 8, 12, 16, and 20.

4.2.2.3 A special strength pan with a concave bottom. The pan may be constructed from a standard 8 in (203 mm) diameter Tyler pan (or equivalent) by replacing the bottom of the pan with a brass plate which has been turned 5/16 in (7.9 mm) thick at the circumference and tapers to 1/8 in (3.17 mm) thick in the center. The taper should be cut on a 43 in (1.09 m) radius and should extend from the center to the edge of the pan.

4.2.2.4 Twenty 1/2 in (12.7 mm) steel balls and ten 3/4 in (19 mm) steel balls.

4.2.2.5 A balance of 100 g capacity, accurate to 0.01 g.

4.2.3 PROCEDURE

4.2.3.1 Dry a 100 to 110 g sample for at least 6 h at 150°F. Either rough screened or final screened carbon may be used. See A1. and A2.

4.2.3.2 Place 100 g of the sample on the top sieve of the six sieves. Shake the sieves for 10 min on the sieve shaker. Weigh the material remaining on each sieve. Compute the average particle size of the sample from the weighted average of each sieve fraction as described in Table 1:

TABLE 1[1]

Retained On	Mass, g	Mean Sieve Opening	Weighted Average
No. 6 Sieve	22.8	4.01 mm	91.4
No. 8 Sieve	36.5	2.84 mm	103.7
No. 12 Sieve	24.1	2.00 mm	48.2
No. 16 Sieve	15.1	1.41 mm	21.3
No. 20 Sieve	0.9	1.00 mm	0.9
Pan	0.6	0.00 mm	0.0

[1] Mass × Mean Sieve Opening = Weight Average

4.2.3.3 After the original particle size of the sample has been determined, combine the screen fractions and shake the whole sample on the shaker for 20 min in the special strength pan with the 30 steel balls. At the end of the 20 min shaking interval, pour the sample on the top nested sieve, remove the balls, and shake the sieves for 10 min as before to separate the screen fractions.

4.2.3.4 Weigh the carbon on each sieve and determine the average particle size as in 4.2.3.2.

4.2.4 CALCULATION—(See Eq. 2)

$$\text{Strength} = \frac{\text{Final Average Particle Size}}{\text{Initial Average Particle Size}} \times 100 \quad \text{(Eq.2)}$$

APPENDIX A

A1. Carbon Size—Experience has shown that commercial suppliers cannot consistently supply carbon that meets this document. For this reason, it is suggested that the user perform the final sieving of the carbon to obtain samples for test use. The following commercially available "rough screened" carbon sieve analyses have proven to be satisfactory for final crushing and screening:

A1.1 For Coarse Carbon

TABLE A1—SUPPLIERS COARSE CARBON ROUGH SCREEN ANALYSIS SPECIFICATION

Retained on U.S. Standard Sieve No.	Mass %
8	0-10
12	40-60
16	30-40
20	0-10
Pan	0-1

A1.2 For Fine Carbon

TABLE A2—SUPPLIERS FINE CARBON ROUGH SCREEN ANALYSIS SPECIFICATION

Retained on U.S. Standard Sieve No.	Mass %
14	0-0.5
16	10-30
20	50-70
30	0-25
Pan	0-5

A2. Carbon Final Screening

A2.1 Coarse Carbon—A quantity of rough screened material may be sieved through a system of U.S. Standard Sieves 8, 12, and 16 (or their equivalent). Material retained on the No. 12 and No. 16 sieves is combined in the proportions specified under 3.3.2.

A2.2 Fine Carbon—A quantity of rough screened material may be sieved through a system of U.S. Standard Sieves 16, 20, and 30 (or their equivalent). Material retained on the No. 30 and No. 20 sieve is combined in the proportions specified under 3.3.1.

A2.3 Shaking—If a motorized vibrator shaker is used, material should be shaken in small quantities for approximately 10 min. If hand shaken, a longer shake period should be observed to assure that size segregation is complete.

A3. Alternate Strength Determination Method—One alternate method, which has been used to determine strength, utilizes a cyclonic separator. This is acceptable if it provides equivalent strength, ±2%, to the method described in 4.2.

A4. Used Carbon—Current practice forbids the use of carbon more than once, though this requirement is not contained in this document. Used carbon may be reused provided it is rescreened and meets all the requirements of this document, and in addition, at least three replicate tests are made using new and used carbon with spark arresters with the same flow rating, ±10%. The results of these replicate tests shall yield spark arrester efficiencies that are identical, ±5%, or the used carbon shall be rejected. Used carbon may be mixed with new carbon in a proportion up to one-third, without replicate testing, provided that the mixture meets all of the requirements of this document.

A5. Reference Materials—Refer to SAE J335, J342, and J350 for spark arrester test techniques. The standard screens used are described in ASTM E-11.

ENGINE PREHEATERS—SAE J226 JAN86 — SAE Recommended Practice

Report of Engine Committee, approved January 1971, and reaffirmed January 1986.

Purpose—Coolant heaters are used to prevent freezing damage and facilitate engine starting under low temperature conditions. This recommended practice provides a guideline for coolant heaters, including adequate clearances and service accessibility.

Scope—This recommended practice includes information and accommodation of both the electric immersion and tank or side arm type external heaters.

Electric Immersion Type Heaters

Straight Adaptor Design—This type is used for installation through covers, core hole plugs, etc. The heating element loop should be located to permit free convective circulation of the engine coolant.

See Fig. 1 and Table 1.

Pipe Thread Adaptor Design—This type is used where space within the cooling jacket allows clearance for a loop. (See Fig. 2 and Table 2.) Where the heater is screwed into a removable flange, the loop may be shaped as shown in Fig. 1.

Tank Type Heater—Where external or tank type heaters are required, suitable coolant connection must be provided at engine locations assuring adequate fluid circulation. Engines of less than approximately 6.5 ℓ (400 in³) displacement require ¼ NPT minimum size connections. For larger engines, use ⅜, ½, or ¾ NPT according to increasing engine ϕ size, with ¾ NPT being used for all engines of 13.0 ℓ (793 in³) displacement and over.

TABLE 1—STRAIGHT ADAPTOR DESIGN

SAE Type	Rating	Volts	A mm	A in	B Wire
1A	500 W	120	76.20	3.000	16/3
1B	500 W	240	76.20	3.000	16/3
2A	1000 W	120	114.30	4.500	16/3
2B	1000 W	240	114.30	4.500	16/3
3A	1500 W	120	152.40	6.000	16/3
3B	1500 W	240	152.40	6.000	16/3
4A[a]	2000 W	120[a]	190.50	7.500	16/3
4B	2000 W	240	190.50	7.500	16/3

[a] Not recommended.

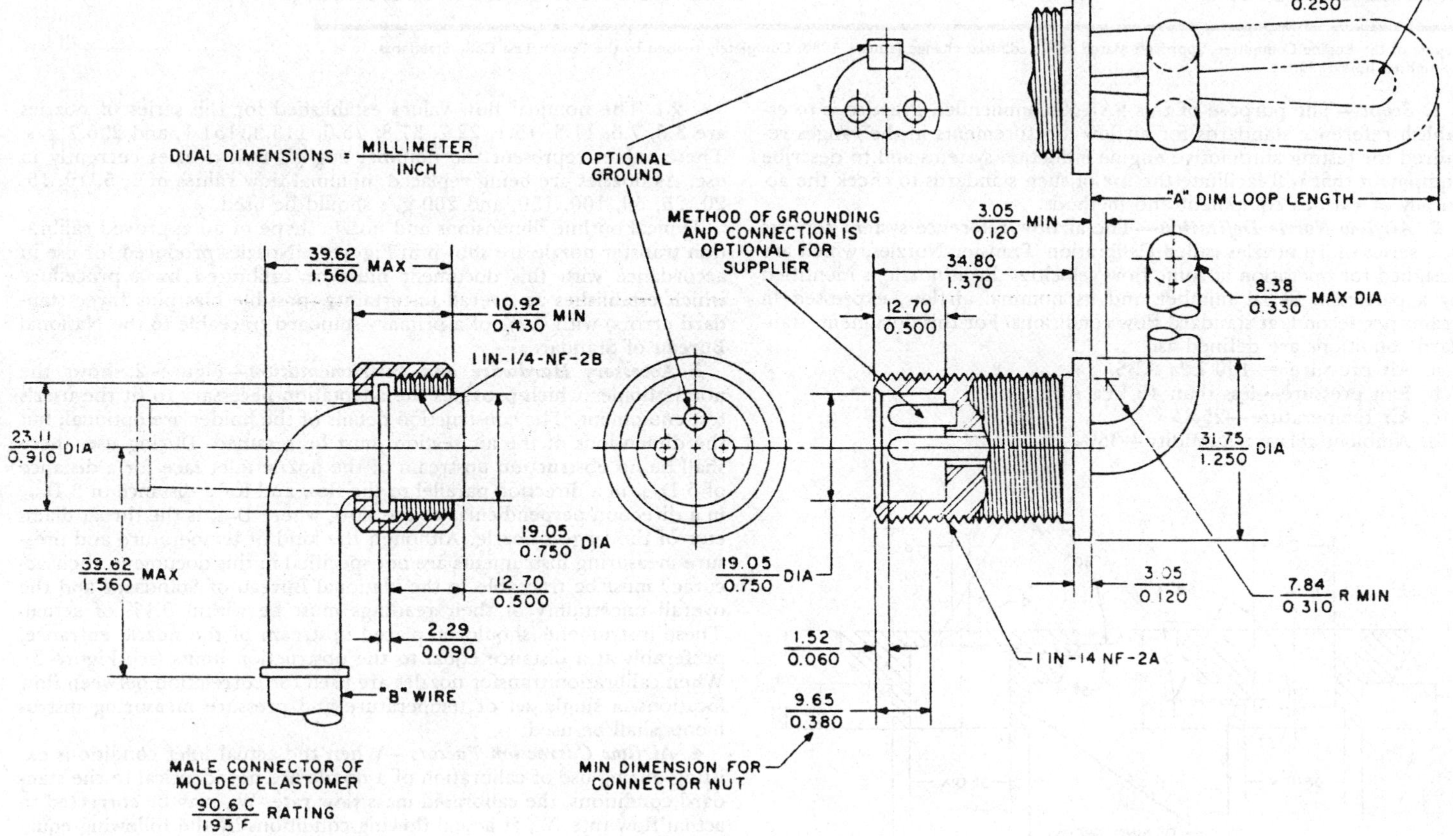

26.371

FIG. 1

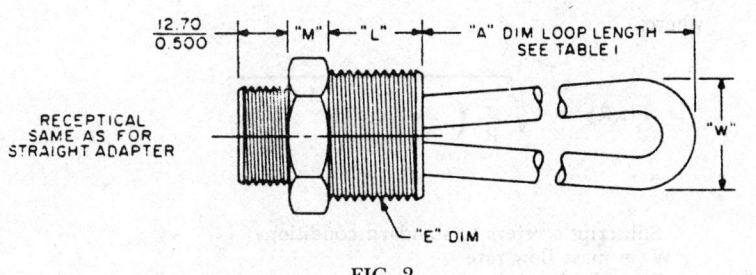

FIG. 2

RECEPTICAL SAME AS FOR STRAIGHT ADAPTER

TABLE 2—PIPE THREAD ADAPTOR DESIGN

E, in NPT	L		M		Max Hex Size (across flat)		W, max	
	mm	in	mm	in	mm	in	mm	in
3/4	19.05	0.750	8.89	0.350	28.45	1.120	23.11	0.910
1	23.88	0.940	8.89	0.350	34.80	1.370	28.45	1.120

(R) AIRFLOW REFERENCE STANDARDS
SAE J228 JUN90
SAE Recommended Practice

Report of the Engine Committee, approved March 1971, editorial change January 1980. Completely revised by the Power Test Code Standards Committee June 1990.

1. Scope—The purpose of this SAE Recommended Practice is to establish reference standards for airflow measurements in the ranges required for testing automotive engine induction systems and to describe equipment that will facilitate the use of such standards to check the accuracy of various equipment and methods.

2. Airflow Nozzle Definition—The airflow reference system consists of a series of 10 nozzles called Calibration Transfer Nozzles, which are designed for operation at sonic flow velocities. Each nozzle is identified by a particular serial number and its nominal airflow, expressed in grams per second, at standard flow conditions. For this document, standard conditions are defined as:
  a. Air pressure = 100 kPa ABS.
  b. Exit pressure—less than 45 kPa ABS.
  c. Air temperature—25°C
  d. Ambient relative humidity—35%.

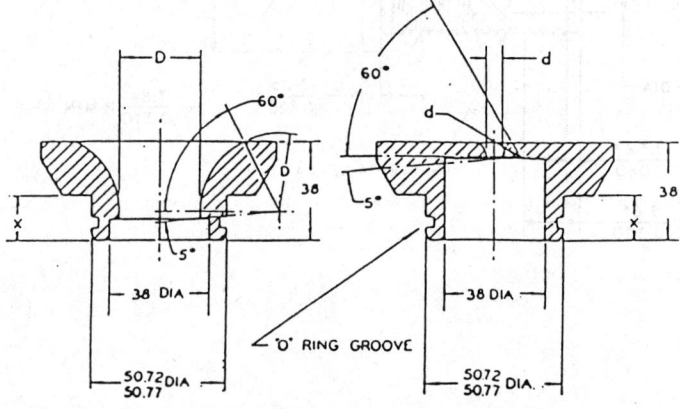

FIG. 1—TYPICAL OUTLINES OF AIRFLOW CALIBRATION TRANSFER NOZZLE

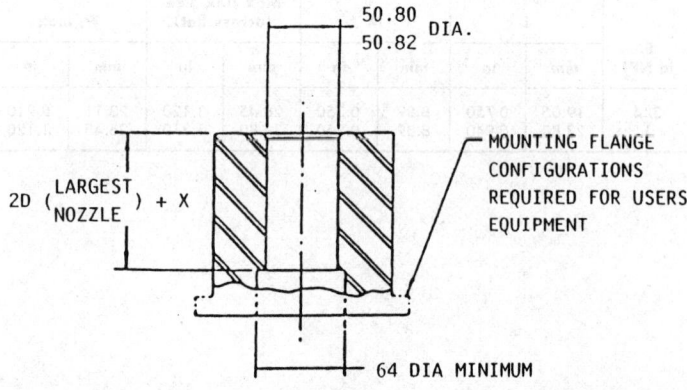

FIG. 2—NOZZLE HOLDER

2.1 The nominal flow values established for the series of nozzles are 3.8, 7.6, 11.3, 15.1, 22.7, 37.8, 75.6, 113.3, 151.1, and 226.7 g/s. These values represent the nominal english unit nozzles currently in use. As nozzles are being replaced, nominal flow values of 1, 5, 10, 15, 20, 25, 50, 100, 150, and 200 g/s should be used.

Typical outline dimensions and nozzle shape of an approved calibration transfer nozzle are shown in Figure 1. Nozzles produced for use in accordance with this document must be calibrated by a procedure which establishes an overall uncertainty (possible bias plus three standard errors) with 0.2% of a primary standard traceable to the National Bureau of Standards.

3. Accessory Hardware and Instrumentation—Figure 2 shows the nozzle holder which provides the adaptation necessary to fit the user's test equipment. The construction details of the holder are optional, but the dimensions of the air section must be retained. During use, there shall be no obstruction upstream of the nozzle inlet face for a distance of 5 D_{max} in a direction parallel to the flow and for a distance of 3 D_{max} in a direction perpendicular to the flow, where D_{max} is the throat diameter of the largest nozzle. Although the kind of temperature and pressure measuring instruments are not specified in this document, their accuracy must be traceable to the National Bureau of Standards and the overall uncertainty of their readings must be within 0.1% of actual. These instruments should be placed upstream of the nozzle entrance, preferably at a distance equal to the obstruction limits (see Figure 3). When calibration transfer nozzles are used for correlation between flow locations, a single set of temperature and pressure measuring instruments shall be used.

4. Airflow Correction Factors—When the actual inlet conditions existing during use of calibration of a nozzle are not identical to the standard conditions, the calibrated mass flow rate, W_s, may be corrected to actual flow rate, W, at actual flowing conditions by the following equation:

$$W = W_s \sqrt{\frac{T_s}{T}} \left(\frac{P}{P_s}\right) \left[\frac{F(\tau, R)}{F(\tau_s, R_s)}\right] \quad \text{(Eq. 1)}$$

where:

$$F(\tau, R) = \sqrt{\frac{\tau}{R}\left(\frac{2}{\tau + 1}\right)^{\frac{\tau + 1}{\tau - 1}}}$$

Subscript s refers to standard conditions
W = mass flow rate
P = absolute pressure at inlet
T = absolute stagnation temperature at inlet
R = gas constant for moist air
τ = specific heat ratio for moist air

Values of $F(\tau,R)/F(\tau_s,R_s)$ may be obtained from Figure 4 and are valid for actual operating conditions over the following ranges:
  a. Inlet pressure, P—96 to 103 kPa ABS
  b. Inlet temperature, T—293 to 303 K
  c. Ambient relative humidity—0 to 50%
  d. Nozzle exit pressure—less than 45 kPa ABS

When the use conditions are not within the limits given above, special nozzle calculations or calibrations must be obtained from the nozzle manufacturer.

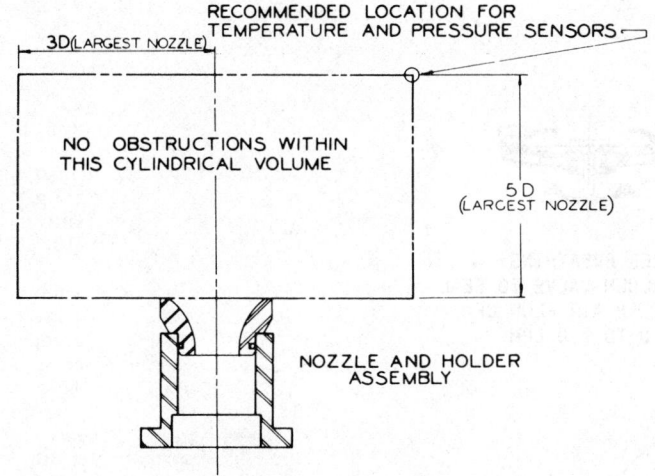

FIG. 3—NOZZLE ENTRANCE OBSTRUCTION LIMITS

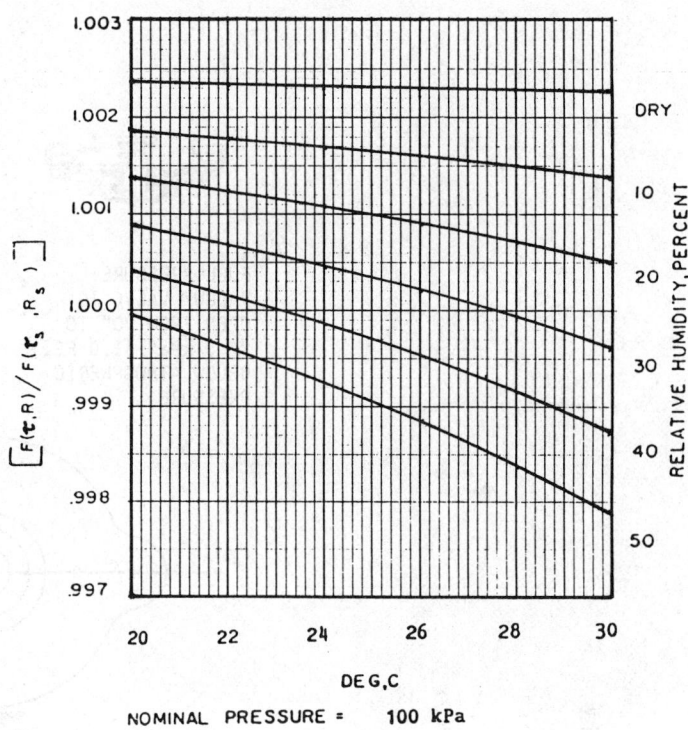

FIG. 4—CORRECTION FACTOR FOR FLOW THROUGH CRITICAL FLOW NOZZLES WITH MOIST AIR

RADIATOR CAPS AND FILLER NECKS—SAE J164 JUN91

SAE Standard

Report of the Engine Committee, approved April 1970, last revised March 1982. Completely revised by the SAE Cooling Systems Standards Committee June 1991.

1. Scope—This SAE Standard was developed primarily for passenger car and truck application, but may be used in marine, industrial, and similar applications.

1.1 Purpose—The purpose of this document is to provide dimensions for the different pressure ratings of boyonet type radiator pressure caps and filler necks. Use of these dimensions will preclude the use of a high pressure rated cap on low pressure systems for the small and medium size applications. (See Figures 1 to 4 and Tables 1 to 4).

This document will allow use of low pressure rated cap on a nominal high pressure system.

2. References

2.1 Applicable Documents—The following publications form a part of this specification to the extent specified herein. The latest issue of SAE publications shall apply.

2.1.1 SAE PUBLICATIONS—Available from SAE, 400 Commonwealth Drive, Warrendale, PA 15096-0001.

SAE J151—Pressure Relief for Cooling System

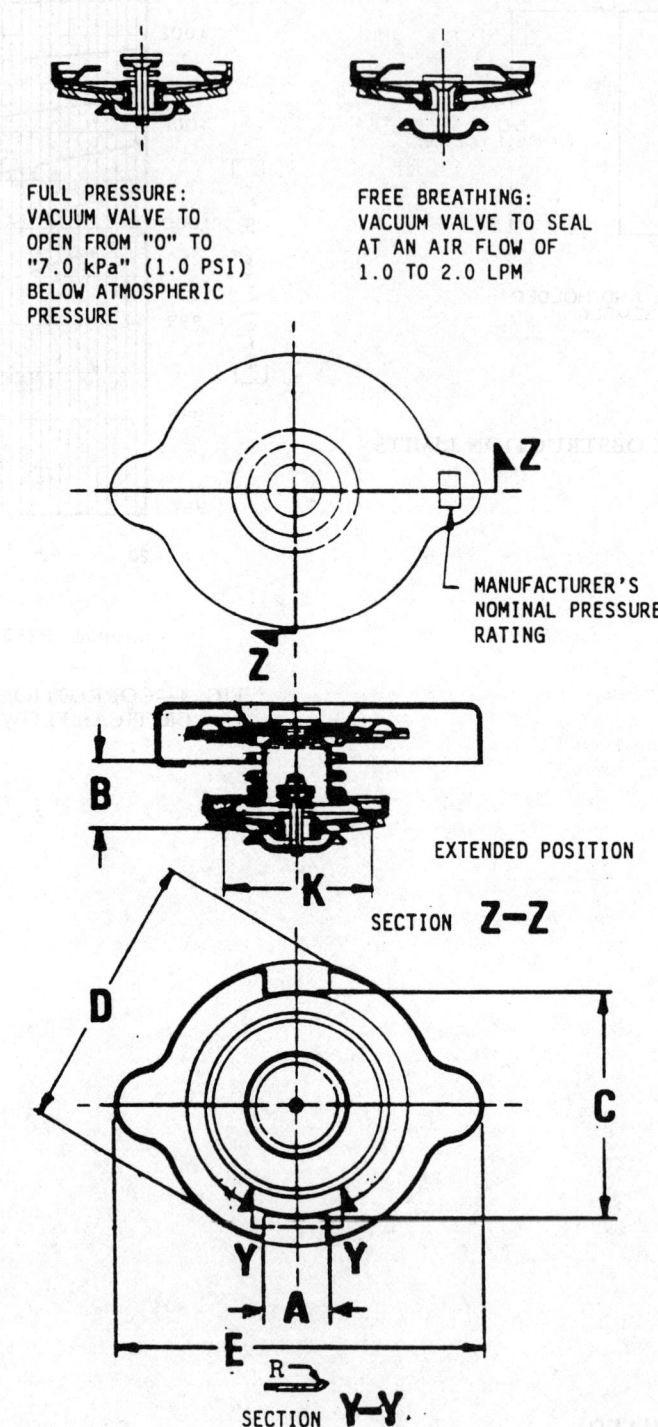

FIGURE 1—PRESSURE TYPE RADIATOR CAP

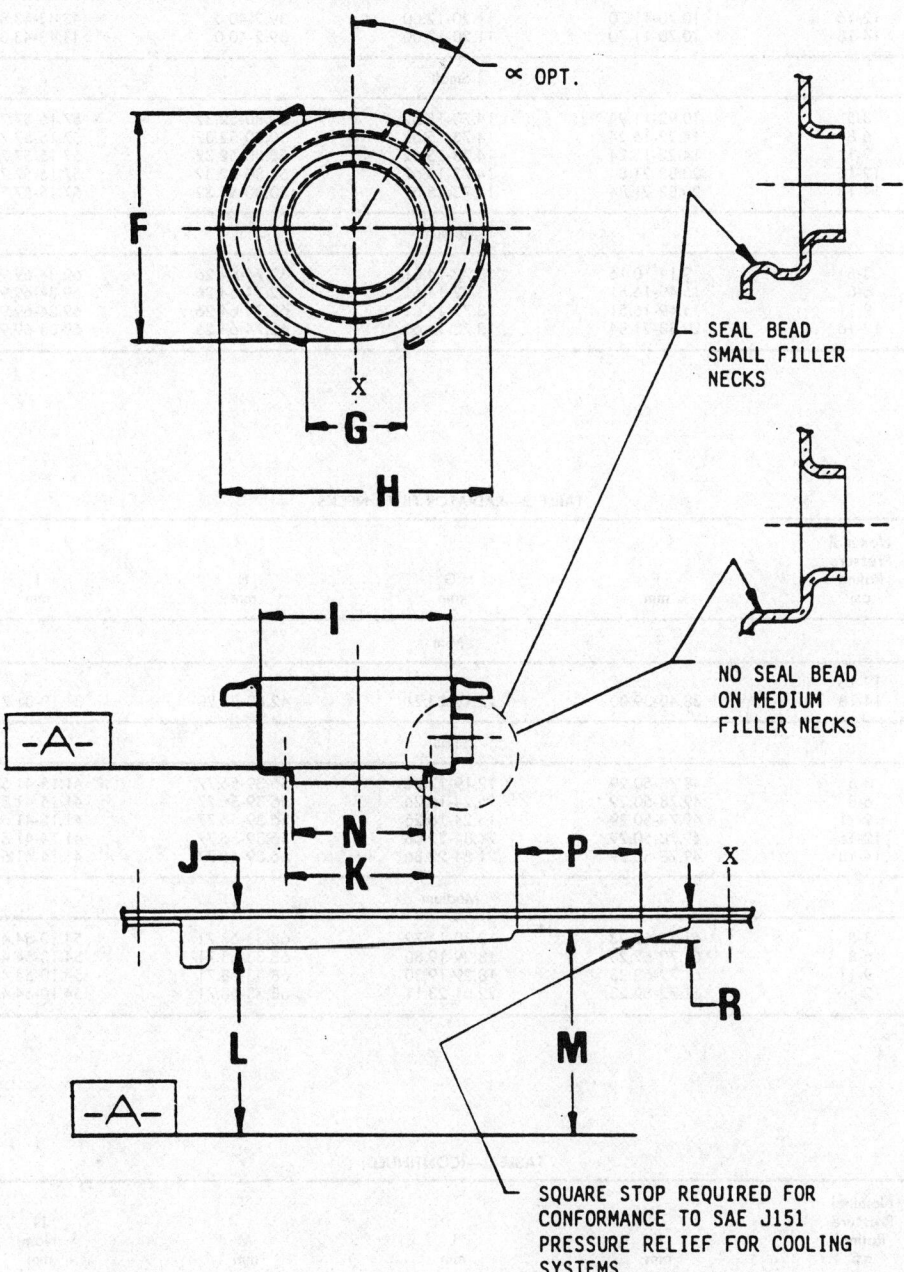

FIGURE 2—RADIATOR FILLER NECKS

TABLE 1—PRESSURE TYPE RADIATOR CAP

Nominal Pressure Rating kPa	Nominal Pressure Rating psi	A mm	B mm	C mm	D mm	E mm
			Mini			
83-110	12-16	10.70-11.70	11.20-12.00	39.2-40.0	43.43-43.80	66.0
97-124	14-18	10.70-11.70	11.20-12.00	39.2-40.0	43.43-43.80	66.0
			Small			
21-34	3-5	10.92-11.94	14.73-15.62	50.80-52.32	57.15-57.79	80.26
41-55	6-8	14.22-15.24	14.73-15.62	50.80-52.32	57.15-57.79	80.26
62-76	9-11	14.22-15.24	14.73-15.62	50.80-52.32	57.15-57.79	80.26
83-110	12-16	20.83-21.84	14.73-15.62	50.80-52.32	57.15-57.79	80.26
97-124	14-18	20.83-21.84	14.73-15.62	50.80-52.32	57.15-57.79	80.26
			Medium			
21-34	3-5	9.14-10.16	13.72-14.61	62.74-64.26	69.34-69.98	92.71
41-55	6-8	15.49-16.51	13.72-14.61	62.74-64.26	69.34-69.98	92.71
62-76	9-11	15.49-16.51	13.72-14.61	62.74-64.26	69.34-69.98	92.71
83-110	12-16	20.83-21.84	13.72-14.61	62.74-64.26	69.34-69.98	92.71

TABLE 2—RADIATOR FILLER NECKS

Nominal Pressure Rating kPa	Nominal Pressure Rating psi	F mm	G mm	H mm	I mm	J mm
			Mini			
83-110	12-16					
97-124	14-18	38.40-39.00	12.19-13.21	42.80-43.20	31.10-31.7	5.08-5.33
			Small			
21-34	3-5	49.78-50.29	12.19-13.21	56.39-56.77	41.15-41.53	5.46-5.72
41-55	6-8	49.78-50.29	15.24-16.26	56.39-56.77	41.15-41.53	5.46-5.72
62-76	9-11	49.78-50.29	15.24-16.26	56.39-56.77	41.15-41.53	5.46-5.72
83-110	12-16	49.78-50.29	21.84-22.86	56.39-56.77	41.14-41.53	5.46-5.72
97-124	14-18	49.78-50.29	21.84-22.86	56.39-56.77	41.15-41.53	5.46-5.72
			Medium			
21-34	3-5	61.72-62.23	12.70-13.72	68.33-68.71	54.10-54.48	5.46-5.72
41-55	6-8	61.72-62.23	18.29-19.30	68.33-68.71	54.10-54.48	5.46-5.72
62-76	9-11	61.72-62.23	18.29-19.30	68.33-68.71	54.10-54.48	5.46-5.72
83-110	12-16	61.72-62.23	22.61-23.11	68.33-68.71	54.10-54.48	5.46-5.72

TABLE 2—(CONTINUED)

Nominal Pressure Rating kPa	Nominal Pressure Rating psi	K mm	L mm	M mm	N Nom mm	P Min mm	R mm
			Mini				
83-110	12-16						
97-124	14-18	22.4-22.7	10.45-10.7	12.43-12.68	19.7	12.04	.70-5.0
			Small				
21-34	3-5	30.7-31.0	13.75-14.0	16.05-16.3	28.3	12.45	4.90-5.16
41-55	6-8	30.7-31.0	13.75-14.0	16.05-16.3	28.3	15.75	4.90-5.16
62-76	9-11	30.7-31.0	13.75-14.0	16.05-16.3	28.3	15.75	4.90-5.16
83-110	12-16	30.7-31.0	13.75-14.0	16.05-16.3	28.3	22.35	4.90-5.16
97-124	14-18	30.7-31.0	13.75-14.0	16.05-16.3	28.3	22.35	4.90-5.16
			Medium				
21-34	3-5	———	13.05-13.3	15.35-15.6	40.70	10.67	4.90-5.16
41-55	6-8	———	13.05-13.3	15.35-15.6	40.70	17.02	4.90-5.16
62-76	9-11	———	13.05-13.3	15.35-15.6	40.70	17.02	4.90-5.16
83-110	12-16	———	13.05-13.3	15.35-15.6	40.70	22.35	4.90-5.16

26.377

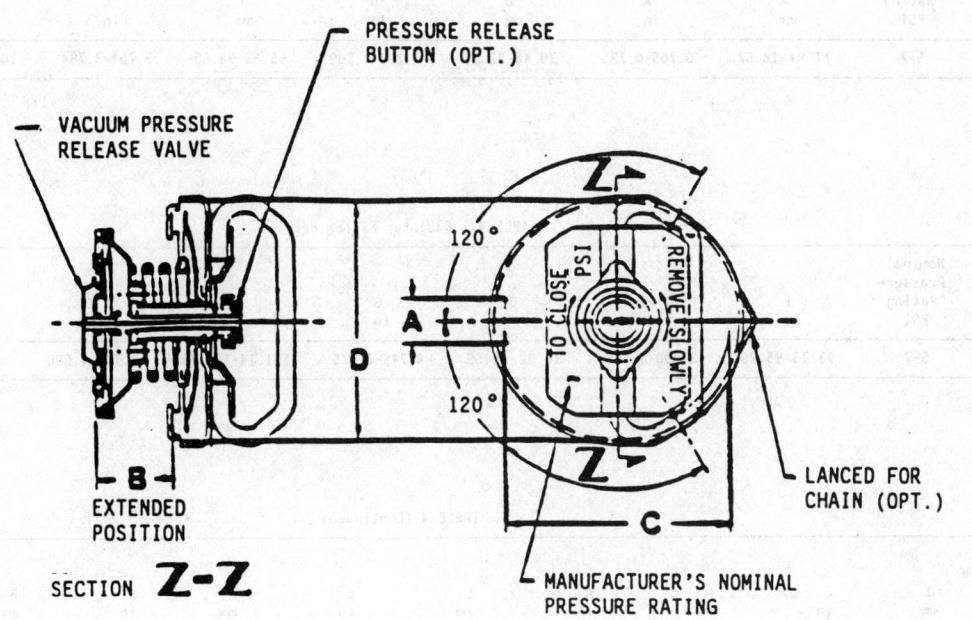

FIGURE 3—PRESSURE TYPE RADIATOR CAP

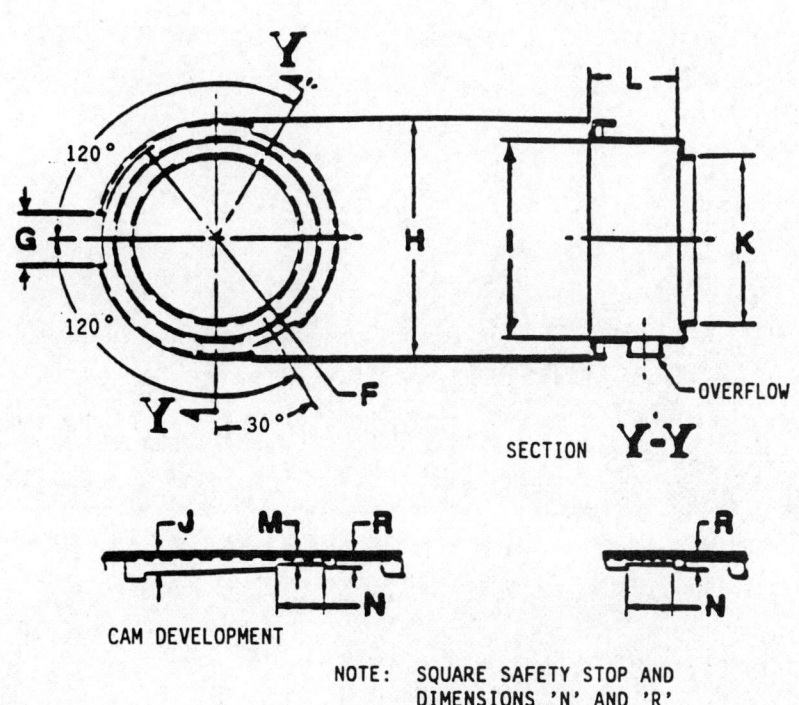

FIGURE 4—RADIATOR FILLER NECK

TABLE 3 - Pressure Type Radiator Cap

Nominal Pressure Rating kPa	Nominal Pressure Rating PSI	A mm	A in	B mm	B in	C mm	C in	D mm	D in
34-48	5-7	17.91-18.67	0.705-0.735	30.48-32.51	1.200-1.280	95.63-96.38	3.764-3.794	102.24-103.00	4.025-4.055

TABLE 4 - Radiator Filler Neck

Nominal Pressure Rating kPa	Nominal Pressure Rating PSI	F mm	F in	G mm	G in	H mm	H in	I mm	I in
34-48	5-7	93.73-95.25	3.690-3.750	18.92-19.69	0.745-0.775	101.14-101.75	3.982-4.006	83.97-84.53	3.306-3.328

TABLE 4 (Continued)

Nominal Pressure Rating kPa	Nominal Pressure Rating PSI	J mm	J in	K mm	K in	L mm	L in	M mm	M in	N mm	N in	R mm	R in
34-48	5-7	7.49-7.75	0.295-0.305	72.90-73.16	2.870-2.880	37.97-38.23	1.496-1.505	4.45-4.70	0.175-0.185	19.43	0.765	5.46-6.22	0.215-0.245

REQUIREMENTS FOR ENGINE COOLING SYSTEM FILLING, DEAERATION AND DRAWDOWN TESTS—SAE J1436 JUN89

SAE Information Report

Report of the Engine Committee, approved September 1983 and reaffirmed by the Cooling Systems Standards Committee 3 June 1989.

1. Purpose—The purpose of this information report is to list the requirements which are in general use for filling, deaeration, and drawdown of engine cooling systems for heavy-duty and industrial applications. The material presented in this document is for information purposes only, and does not constitute an SAE Standard.

2. Levelling—Before starting any test, the vehicle or the industrial equipment must be level.

3. Filling—With the engine off, thermostats closed, a completely drained system (including heater, other accessories and their lines) must fill with cold water at 5 gpm (19 L/min) with hose until the filler neck overflows. Close shutters or block the airflow to the radiator with cardboard segments and run the engine at approximately rated speed without radiator cap until the thermostats open. The opening of the thermostats may be detected by observing the flow in the radiator inlet line sight glass, by noting a sudden rise in inlet line or top tank temperature, or by noting when the coolant temperature exceeds the thermostat rating by 5°F (3°C). Continue running the engine for 5 min, and then stop the engine and measure the amount of water required to refill the system to the 100% full point, which is defined to be at the bottom of the filler neck extension (cold-fill level) or to the recommended cold-fill "Full" mark if there is no filler neck extension.

The quantity of water added shall not exceed 10% of the total system capacity, defined in this information report.

The test applies in general to systems of up to 100 qt (95 L) capacity. However, lower fill rates [for example, 3 gpm (11 L/min)] may be required in special instances and will be specified by the engine manufacturer. For systems over 100 qt (95 L) capacity, the engine manufacturer may call for a higher fill rate in order to keep the total fill time to a reasonable period.

In addition, in certain instances where air entrapment in the fill line (shunt line) may be suspected, the engine manufacturer may call for a bucket fill test. In larger systems, particularly those with remotely mounted heaters, manual air bleed vent valves may be required.

4. Expansion Volume—The radiator must provide an expansion volume equal to a minimum of 6% of the total system capacity. This expansion volume will remain empty during a cold fast fill, but provision shall be made to vent the air from this space to the filler neck during gradual engine warm up. The amount of water required to slow fill the radiator from the bottom of the filler neck extension to the breather hole [usually 0.12 in (3.2 mm) diameter] in the filler neck extension, expressed as a percent of the total system capacity, is the percent expansion volume.

Provision for a filler neck extension tube is the preferred construction for radiators for heavy-duty and industrial applications. However, if a filler neck extension is not provided, the amount of water required to fill the radiator from the recommended "Full" mark to the bottom of the filler neck, expressed as a percent of the total system capacity, is the percent expansion volume. Total system capacity is defined following the requirements for drawdown testing.

5. Deaeration Tests—The engine manufacturers require tests of the cooling system deaeration capability to remove gases from the coolant during operation. These gases may originate from air entrainment during filling, from vortexing at the fill or shunt line connection when a vehicle is not operating on a level surface or due to centrifugal forces in a prolonged turn, or from combustion gases leaking across cylinder head seals.

Because of the differences in the approach and the test methods of the various engine manufacturers, it is important that these tests be performed strictly in accordance with the engine manufacturer's requirements.

A brief description of some of the deaeration tests required by various manufacturers is given in the information report for general information only. Refer to the engine manufacturer's requirements for details.

Deaeration tests are to be performed after determination of the expansion volume.

After determination of the expansion volume, replace the thermostats with blocked-open thermostats and refill the system with a hose until the filler neck overflows. Run the engine at approximately rated speed for 5 min and refill the system to the bottom of the filler neck extension or other "Full" mark. One or more of the following deaeration tests may then be required:

5.1 Test 1 (Deaeration of Fill-Entrained Air)—Run the engine at an approximately rated speed with the shutters closed or the airflow to the radiator blocked with cardboard segments to maintain a top tank temperature of 150°F ± 10 (65°C ± 6) without a radiator cap. Run until a sight glass in the engine outlet (radiator inlet) runs clear of air bubbles. The time from refilling to the bottom of the filler neck extension until the sight glass runs clear of air bubbles shall not exceed 25 min.

5.2 Test 2 (Continuous Deaeration)—Using a special vented radiator cap with the hose led to an inverted water-filled bottle set in a bucket of water for purposes of measuring the volume of vented air, run the engine at an approximately rated speed with blocked-open thermostats and with the shutters closed or the airflow to the radiator blocked with cardboard segments to maintain a top tank temperature of 150°F ± 10 (65°C ± 6).

Inject air into the system and measure the volume of air vented by the deaeration system while monitoring the pump pressure rise. The rate of air venting when a 35% loss in pump pressure rise occurs must equal or exceed an amount specified for each engine model.

5.3 Test 3 (Continuous Deaeration)—With the radiator cap on, run the engine at an approximately rated speed with blocked open thermostats and with the shutters closed or the airflow to the radiator blocked with cardboard segments to maintain a top tank temperature of 150°F ± 10 (65°C ± 6).

Inject air at a rate specified for each engine model [approximately 0.1 cfm (2.8 L/min) per cylinder] and monitor the water pump flow. At a specified air injection rate, the water pump flow must not fall below 50% of the original value, and the coolant loss through the overflow line must not be more than the drawdown rating (determined in Section 6).

5.4 Test 4 (Continuous Deaeration)—Under the conditions of Test 3, inject air at increasing rates until bubbles are seen in a sight glass in the radiator inlet line. This rate of air injection is called the deaeration capacity and the minimum values are specified by the manufacturer for each engine model.

6 Drawdown—After the deaeration tests, run the engine at a governed no load speed, with blocked-open thermostats, and with the shutters closed or the radiator blocked with cardboard segments to maintain top tank temperature at 150°F ± 10 (65°C ± 6) without the radiator cap. When the temperature is reached, add or draw off water until the system is filled to the bottom of the filler neck extension or the other "Full" mark. Then draw off water slowly in 1 qt (1 L) increments from system at a point of positive pressure and measure until air is seen in engine outlet sight glass. The amount of water drawn off, expressed as a percent of total system capacity, is the drawdown rating. This must be equal to or greater than 11% of the total system capacity, but not less than a specified minimum, for systems up to 100 qt (95 L) capacity. For most manufacturers, this specified minimum drawdown rating is 3 qt (3 L). However, some engine manufacturers require higher minimums. For cooling systems with capacity above 100 qt (95 L), the required drawdown rating shall be 11 qt (10.5 L) plus 4% of the system capacity in excess of 100 qt (95 L).

7. Total System Capacity—Following the drawdown test, drain and measure the water from the entire system, being careful that no fluid is trapped in the system. This volume of fluid drained, added to the amount drawn off during the drawdown test, is the total system capacity.

8. Other Requirements—Individual engine manufacturers may have additional cooling system tests or system parameters that they require. Refer to the engine manufacturer's requirements. A sampling of some of these requirements follows.

8.1 Pump Cavitation—Under the rated engine load and an engine outlet temperature of 210°F (99°C), which will provide a temperature equal to or greater than the highest coolant temperature expected at the pump, the loss of water pump pressure rise shall not exceed a specified amount, usually 10%.

8.2 Water Pump Suction—Suction at the inlet to the water pump shall not exceed 3 in Hg (10.2 kPa) at engine high idle, without a radiator cap, and with the thermostats open. Or, alternatively,

8.3 Water Pump Inlet Pressure—There shall be a positive pressure at the inlet to the water pump at all times.

9. Special Considerations for Systems with Surge Tanks or Coolant Recovery Systems—An engine cooling system which has a surge tank can be considered to have a remote mounted radiator top tank. For purposes of these tests, the surge tank shall be considered to be the radiator top tank. Filling should be accomplished through the filler neck on the surge tank. The surge tank will be provided with a filler neck extension or the other cold-fill "Full" mark. The expansion volume for the system is provided in the surge tank in the same manner as in the usual radiator top tank. The total system capacity includes the volume of the surge tank to the bottom of the filler neck extension or the other cold-fill "Full" mark.

An engine cooling system which has a coolant recovery system can be considered to have a remote mounted expansion volume only. For purposes of this test, filling should be accomplished through the radiator filler neck, filling to the bottom of the filler neck, and through the coolant recovery system tank inlet, filling to the recommended cold-fill level. The expansion volume for the system is provided in the coolant recovery system tank, and is equal to the volume from the recommended cold-fill level to the top of the tank.

GLOSSARY OF ENGINE COOLING SYSTEM TERMS—SAE J1004 APR93

SAE Recommended Practice

Report of the Engine Committee approved February 1975, revised August 1988. Reaffirmed by the Cooling Systems Committee April 1993. Rationale statement available.

Foreword—This reaffirmed document has been changed only to reflect the new SAE Technical Standards Board format.

1. Scope—The objective of this glossary is to establish uniform definitions of parts and terminology for engine cooling systems.

2. References—There are no referenced publications specified herein.

3. Definitions

3.1 Aeration—The entrainment of gas (air or combustion gas, or both) in the coolant.

3.2 Afterboil—Boiling of the coolant after engine shutdown caused by residual heat in the engine.

3.3 Afterboil Volume—The quantity of coolant forced from the engine by afterboil. This may or may not be displaced from the system depending upon the system design or coolant level, or both.

3.4 Air-to-Boil Temperature—The ambient temperature at which the coolant at the radiator inlet reaches its boiling point. The coolant boiling point is a function of the absolute pressure and the characteristics of the coolant. However, the term is also quite commonly used without consideration for the absolute pressure or the coolant characteristics. In this case, an open radiator at sea level pressure with water, boiling point 100 °C (212 °F), as the coolant is assumed. The Air-to-Boil Temperature is computed as follows: Boiling temperature (or design temperature, which may be less than the boiling temperature) of the system coolant minus the highest coolant temperature plus ambient temperature.

3.5 Air Recirculation—Fan—The recycling of air already discharged by the fan.

3.6 Air Recirculation—Radiator—The recycling of hot air already passed through the radiator, usually caused by engine enclosure pressure.

3.7 Ambient Temperature—The environmental air temperature in which a unit is operating. In general, the temperature is measured in the shade (no solar radiation) and represents the air temperature for engine cooling performance measurement purposes. Air entering the radiator may or may not be the same as ambient, due to possible heating from other sources or recirculation.

3.8 Approach Temperature Differential—The temperature difference between hot fluid leaving and cold fluid entering the heat exchanger.

3.9 Auxiliary Tank—A separate tank in the cooling system provided to perform one or more of the following functions: (a) filling, (b) coolant reservoir, (c) de-aeration, (d) retention of coolant expelled from radiator by expansion or afterboil, or both, and (e) visible fluid level indication.

3.10 Blocked Open Thermostat—A normal thermostat mechanically blocked open to the position representing its maximum open position; usually used during cooling tests.

3.11 Blower Fan—A fan positioned in a cooling system such that the air passes through the fan before entering the radiator.

3.12 Coolant—A liquid used to transport heat from one point to another.

3.13 Cooling Differential (Inlet Temperature Differential)—The temperature differential existing between the ambient air temperature and the coolant temperature at the radiator inlet.

3.14 Cooling System—A group of inter-related components to effect the transfer of heat.

3.15 Cooling System Capacity (Volume)—The amount of coolant designated in liters (quarts) to completely fill a cooling system to its designated cold level mark.

3.16 De-aeration—The removal or purging of gases (air or combustion gas, or both) which have been entrained in the coolant.

3.17 De-aeration Baffle—A barrier used to separate chambers in a top tank or auxiliary tank to form a de-aerating tank.

3.18 De-aeration Capability—The ability of the cooling system to de-aerate the coolant expressed in terms of time or performance, or both, under specified test procedures.

3.19 De-aerating Tank—A specially designed tank capable of removing entrained air or combustion gas, or both, from the circulating coolant.

3.20 Drawdown—The quantity of coolant which can be lost before impairing the cooling system performance, or grade cooling level, under normal operating conditions.

3.21 Expansion Volume—The volume of space in a cooling system (such as in the radiator top tank or auxiliary tank) which allows for the expansion of coolant resulting from temperature rise.

3.22 Fan Air Flow—The rate of air flow usually in units of cubic meters (cubic feet) per minute that a fan can deliver at standard air conditions and a specified static pressure and speed.

3.23 Fan Drive—Temperature Controlled—A fan drive which can be turned on or off or whose speed can be modulated in accord with temperature conditions of either the coolant or the circulated air. The purpose of the drive is to operate the fan as required for cooling, but when cooling demands permit, allows the fan speed to be reduced or the fan to free-wheel to reduce fan horsepower or fan noise, or both. The clutch may be any of the following types: (a) Dry Clutch, (b) Wet Clutch, (c) Viscous Shear Coupling, (d) Dump and Fill, or (e) Electronically Controlled.

3.24 Fan Drive—Torque Limiting—A drive, usually of a viscous shear type, not controlled directly by coolant or air temperature, which is used to limit maximum fan speed and power absorption.

3.25 Fill Rate—The coolant flow rate usually in liters (gallons) per minute that an empty cooling system will accept up to the full mark without overflowing.

3.26 Flow Rate, Coolant—The rate of flow of coolant through a cooling system component or group of components under specified conditions in liters (gallons) per minute.

3.27 Free Flow—Water flow rate, in liters (gallons) per minute, at a specified water temperature, through a radiator tested in its installed position while maintaining water at atmospheric pressure at seat of pressure cap (or pressure relief valve) for downflow radiator, or at highest point on inlet tank for a crossflow radiator. This rate is used for manufacturing quality control purposes, and for checking suspect radiators which have been in service (and which may be clogged internally).

3.28 Grade Cooling Level—The cooling differential or air-to-boil value obtained while maintaining a specified speed of ascent on a particular percent and length grade.

3.29 Heat Dissipation—The quantity of heat, usually expressed in British thermal units per kilowatts (minute), that a heat transfer component can dissipate under specified conditions.

3.30 Idle Time to Boil—The time required to boil the coolant while idling following other specified conditions.

3.31 Inlet Restriction (Coolant Pump)—The difference between the pressure of the coolant at the pump inlet under no flow conditions with the radiator filler cap removed, and the pressure under full flow conditions with the thermostat blocked fully open and the radiator filler cap removed.

3.32 Inlet Tank Temperature—The temperature of coolant entering the radiator inlet.

3.33 Low Flow Cooling System—A cooling system which under normal conditions operates at a coolant flow through the radiator of significantly less than full engine coolant flow rate, thus increasing the temperature drop of the coolant for a charge air cooler.

3.34 Multi Pass Radiator—A radiator configuration where the core is divided into two or more sections which the coolant passes through increasing the velocity of the coolant and the distance it travels through the radiator.

3.35 Open Radiator Tank—A radiator tank that is open to atmospheric pressure.

3.36 Outlet Tank Temperature—The temperature of the coolant leaving the radiator outlet (this would include any temperature gain or loss from any device located in the radiator outlet tank).

3.37 Outlet Temperature Differential—The temperature difference between the ambient air temperature and the coolant temperatures at the radiator outlet.

3.38 Overflow Bottle—See "Auxiliary Tank."

3.39 Overheating—An operating condition where coolant temperature exceeds design intent. This may be caused by a deficiency in the cooling system or by abnormal operating conditions.

3.40 Pump Cavitation—The formation of gas or vapor bubbles, or both, which reduce the pump delivery.

3.41 Pump Cavitation Pressure—The pressure on the coolant existing at the point where a specified degree of cavitation occurs under defined temperature conditions.

3.42 Pump Cavitation Temperature—Temperature of coolant at the point where a specified degree of cavitation occurs under defined pressure conditions.

3.43 Radiator Air Restriction—The air pressure drop, usually expressed in Pa (inches of water), across a radiator at a specified rate of air flow and air density.

3.44 Radiator Air Baffle—Various types of barriers used to enclose or direct air through the radiator to minimize air recirculation.

3.45 Radiator Cooling Potential—The temperature differential between the air temperature entering the radiator and the average temperature of coolant in the radiator, usually under stabilized conditions.

3.46 Radiator Shutters—Automatic or manually operated devices positioned in front or rear of radiators to control engine coolant temperatures by regulating air flow through the radiators.

3.47 Radiator Tank Baffle—A partition or divider in a radiator tank to direct coolant to a particular section of that radiator.

3.48 Ram Air Flow—The amount of air passing through the radiator as a result of vehicle motion or wind.

3.49 Specific Heat Rejection—The heat rejection of the engine expressed essentially in British thermal units per kilowatts (minute) per brake kilowatts (horsepower). It should be further qualified by engine load, speed, and ambient air temperature.

3.50 Stabilization—A condition attained during specified conditions of engine operation when the temperatures of air, oil, and coolant have reached values which will not change regardless of the length of time the unit is run.

3.51 Standard Air—Air at 21.1 °C (70 °F) and 760 Pa (29.921 in of mercury) and weighing 1.200 kg/m^3 (0.07488 lb/ft^3).

3.52 Suction Fan—A fan positioned in a cooling system so that air passes through the radiator before entering the fan.

3.53 Surge Tank—See "Auxiliary Tank."

3.54 System Restriction (Air)—The static pressure differential which occurs at a given air flow from air entrance through air exit in a system, generally measured in Pa (inches of water).

3.55 Temperature Stability on Drift—The ability of an engine cooling system to maintain coolant temperature on a long vehicular drift (coasting). This is an important characteristic for good heater operation during winter months.

3.56 Thermostat—A self-contained modulating valve in the engine coolant circuit for the purpose of controlling the flow of coolant. The amount of thermostat opening is dependent on coolant temperature; normally it is closed below a specified "start-to-open" temperature and full open at some higher specified temperature.

3.57 Top Tank Temperatures—Same as Inlet Tank Temperature.

3.58 Winter Front—A cover tailored to the front of a radiator for winter use to control air flow through it. Zippered or snap flaps allow manual adjustments to be made as needed.

LABORATORY TESTING OF VEHICLE AND INDUSTRIAL HEAT EXCHANGERS FOR THERMAL CYCLE DURABILITY—SAE J1542 JAN89

SAE Recommended Practice

Report of the Cooling Systems Standards Committee approved January 1989.

1. Purpose—This recommended practice is to provide a test guideline for determining the durability of a heat exchanger under thermal cycle conditions.

2. Scope—This recommended practice is applicable to all liquid-to-air, liquid-to-liquid, air-to-liquid, and air-to-air heat exchangers used in vehicle and industrial cooling systems. This document outlines the tests to determine durability characteristics of the heat exchanger under thermal cycling.

3. Objective—To verify compliance with established criteria that insures durability in a specific application. This document describes a system to thermally induce stresses into a heat exchanger at specified rates and temperatures. The process is accomplished by heating and cooling the unit in a specified manner.

4. Facility Requirement—The facility should provide the following as required:

4.1 Source of "HOT" fluid capable of delivering the fluid to the test unit at specified temperature and pressure.

4.2 Source of "COLD" fluid capable of delivering the fluid to the test unit at specified temperature and pressure.

4.3 Compressed air at specified pressure and temperature.

4.3.1 Charge air cooler test media

4.3.2 Determination of product integrity

4.3.3 Evacuation of steam condensate after heating cycle if required

4.4 Means of moving these fluids to and from the test unit in a specified manner.

4.5 Means of checking heat exchanger integrity.

4.6 A control system adequate to control the thermal cycle as specified. It may include the following:

Flow measuring devices
Pressure gauges
Pressure switches
Fluid filters
Temperature indicators
Pressure regulators for each fluid
Relief and back pressure regulators
Automatic emergency shut-downs
Electrical relays, timers, switches, indicator lights, and related items as required
Thermal cycle counters
Safety features as specified by regulatory codes and common practices
Heaters and controls
Automatic data logging equipment

(See Fig. 1 for typical control system for reference.)

5. Testing

5.1 Check heat exchanger for integrity.

5.2 Install test unit as specified (orientation as in service, if possible) in a safe manner.

5.3 Set up control system to obtain specified cycle (see Fig. 2 for typical thermal cycle).

5.3.1 TYPICAL THERMAL CYCLE TEST TEMPERATURE DIFFERENTIAL

Type of Heat Exchanger	Temperature Differential
Radiator (Liquid-to-Air)	80°C (144°F)
Air Cooled Oil Cooler (Oil-to-Air)	110°C (198°F)
Liquid Cooled Oil Cooler (Oil-to-Liquid)	110°C (198°F)
Air Cooled Charge Air Cooler (Air-to-Air)	140°C (252°F)
Liquid Cooled Charge Air Cooler (Air-to-Liquid)	140°C (252°F)

5.3.2 "HOT" temperature is specified based on specific application. This temperature to be at least equal to normal operating conditions. The system shall be capable of maintaining this temperature.

5.3.3 "HOT-COLD" test temperature differential is based on individual application. A controlled temperature should be maintained within ±6°C (±11°F) by controlling either hot or cold source to take advantage of available resources.

5.3.4 "COLD" temperature is established by test temperature differential.

5.3.5 The test cycle and its segments are to be based on specific application and are measured in seconds. The transition time from HOT to COLD will vary depending on test media, test equipment, and test method. One cycle consists of a heating segment and a cooling segment. A segment is that time required to bring the outlet temperature to within 5% of inlet temperature. It is desirable to minimize segment duration in order to expedite testing and maximize thermal cycle loading. To this end, and to insure compliance with outlet temperature goals, it may be required to circulate ambient media.

5.3.6 Test ambient temperature is to be specified based on specific application.

5.3.7 Nominal pressures and pressure ranges observed within the test should be recorded and maintained.

5.4 Run test to specified conditions mutually established between supplier and user or documented on part print. Types of tests are as follows:

5.4.1 New heat exchanger (no previous testing) for comparison to previously established criteria

5.4.2 Used heat exchanger if specified (document previous history)

5.5 Remove and check heat exchanger for leaks and structural damage.

6. Test Documentation

6.1 During and after test, document leakage rate and location of leaks as well as compare to acceptance criteria.

6.2 Document the following variables:
Unit Orientation (Ref. 5.2)
Maximum Temperature (Ref. 5.3.2)
Temperature Differential (Ref. 5.3.3)
Fluid Type: Hot (Ref. 4.1)
          Cold (Ref. 4.2)
Test Ambient (Ref. 5.3.6)
Unit Pressure Range (Ref. 5.3.7)
Cycle & Segment Duration (Ref. 5.3.5)
Number of Cycles to Completion (Ref. 5.4)

7. This recommended practice is valid for durability comparison of thermal cycle conditions only. Correlation to field results must be developed on individual basis. Other tests affecting heat exchanger durability are SAE J1597 (pressure cycle) and J1598 (vibration). These tests can be run in combination as well as independently.

NOTE: Combination testing may make it difficult to determine cause of failure.

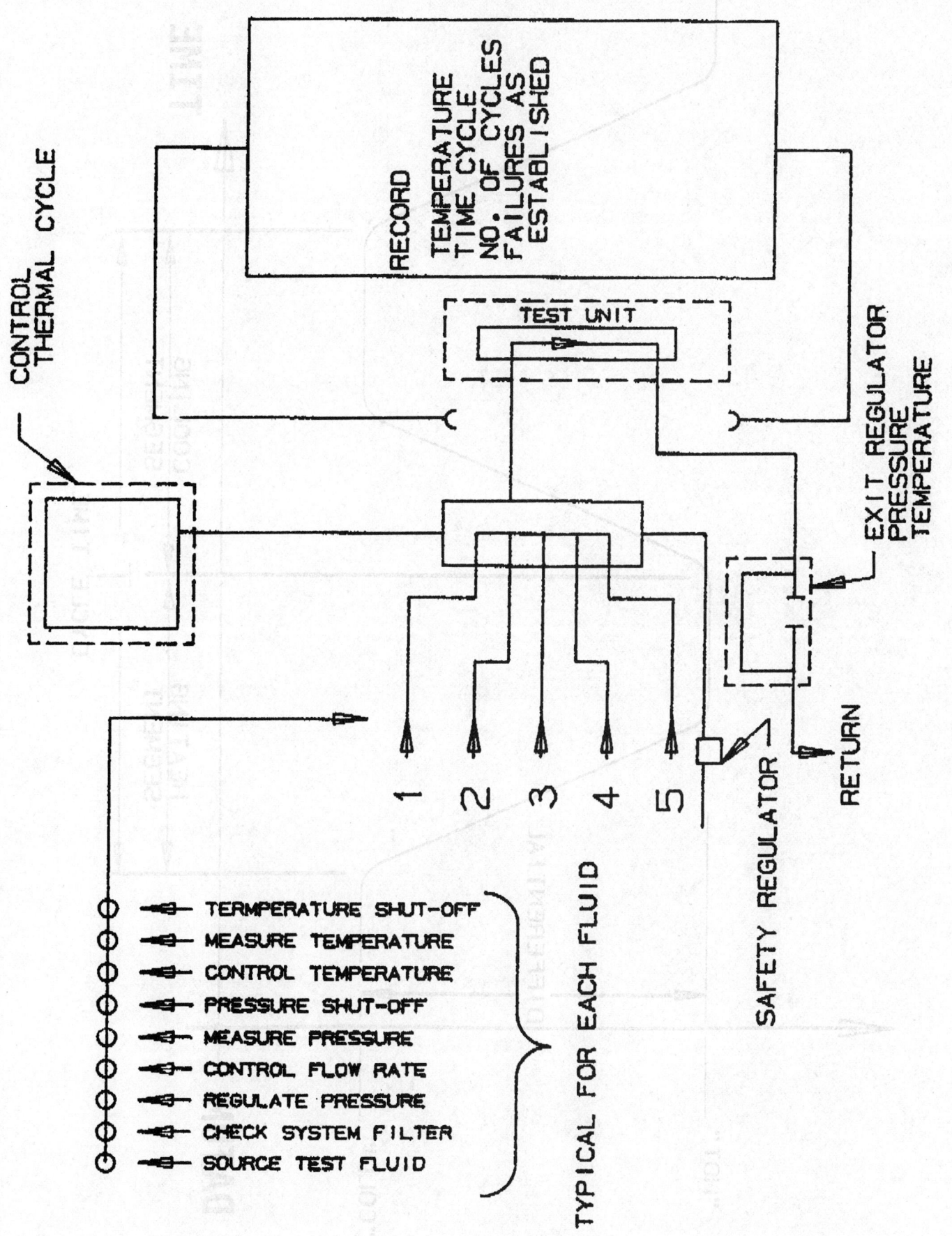

FIG. 1—TYPICAL CONTROL SYSTEM

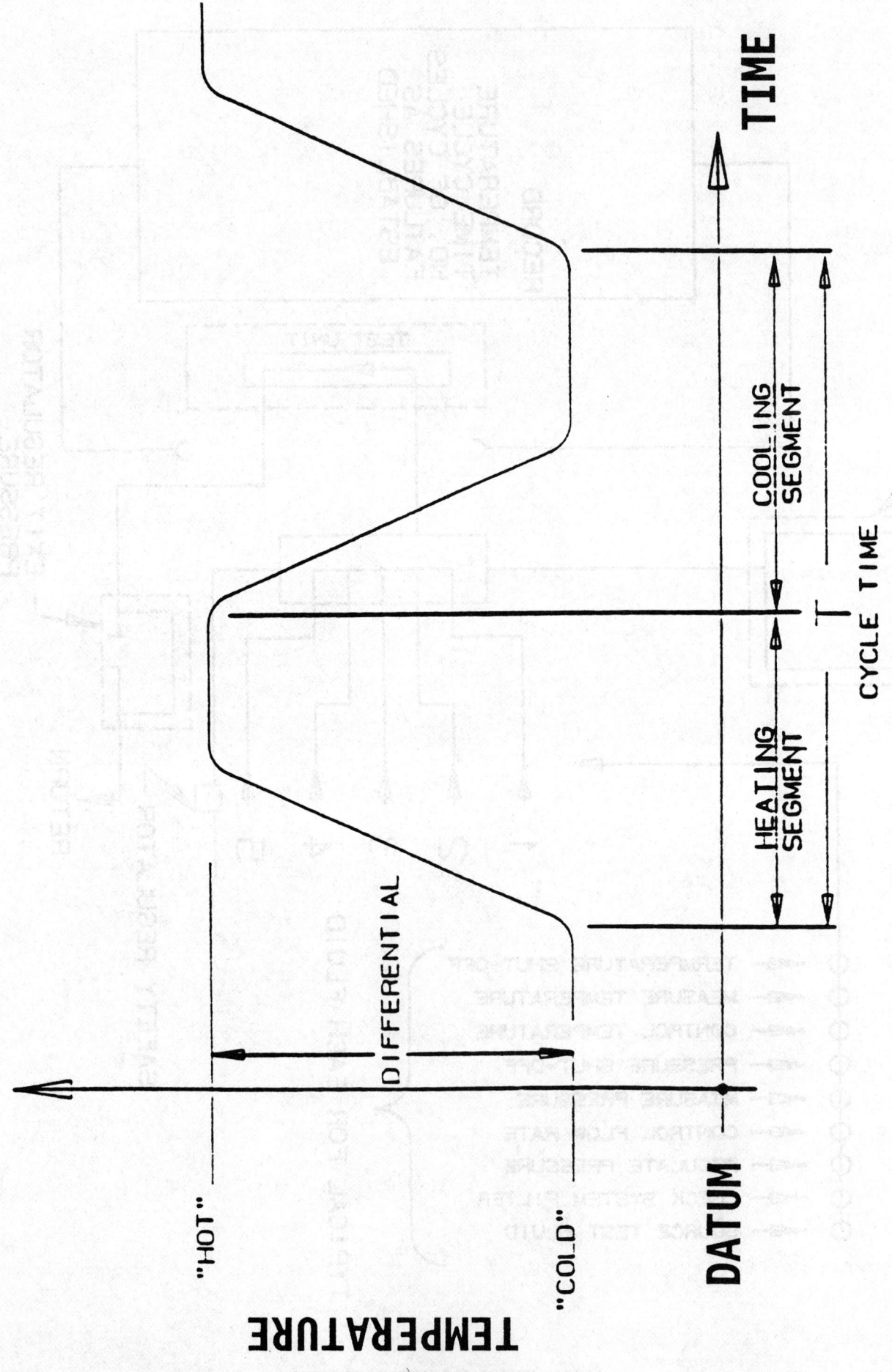

FIG. 2—TYPICAL THERMAL CYCLE

RADIATOR NOMENCLATURE—SAE J631 APR93

SAE Standard

Report of the Truck and Bus Technical Committee, approved May 1949, and revised by the Engine Committee August 1988. Revised by the SAE Cooling Systems Committee April 1993.

1. **Scope**—This SAE Standard documents standard nomenclature in common use for various types of radiator and radiator core construction, as well as for various radiator-related accessories.

2. *References*

   2.1 **Applicable Documents**—The following publications form a part of this specification to the extent specified herein. The latest issue of SAE publications shall apply.

   2.1.1 SAE PUBLICATIONS—Available from SAE, 400 Commonwealth Drive, Warrendale, PA 15096-0001.

   SAE J151—Pressure Relief for Cooling System
   SAE J164—Radiator Caps and Filler Necks

3. *Radiator Cores*

   3.1 **Tube and Plate Fin Core**—An assembly of fluid-carrying tubes of any cross-sectional form, the tubes being joined by heat-conducting fins or plates common to all tubes or groups of tubes. See Figure 1.

   3.2 **Cellular Core**—A number of fluid passages made by joining metal ribbons at the edges and grouped to form a cellular structure. Parts of the cellular structure may be of formed or flat ribbon which is not a part of the fluid passages. See Figure 2.

   3.3 **Tube and Serpentine Fin Core**—An assembly of fluid tubes in line in the direction of airflow and joined in heat transfer relation by corrugated conduction fins inserted between adjacent line of tubes. See Figure 3.

   3.4 **Sheet Metal Radiators**—Figures 4 to 7 illustrate the nomenclature for sheet metal radiators.

   3.5 **Plastic Tank Radiators**—Figures 6 and 7 can also represent radiators having plastic inlet and/or outlet tanks. Inlet and outlet are integral with the appropriate tank. The filler neck and other parts and/or fittings may also be integrated with a tank. Tanks are gasketed and secured to the headers.

   3.6 **Bolted Radiators**—Figures 8 to 11 give the nomenclature for bolted radiators.

4. *Accessories*

   4.1 **Fan Shroud**—An enclosure to duct air between the radiator and the fan. It provides increased fan efficiency by reducing or eliminating fan and/or radiator air recirculation. See Figures 4, 5, 9, and 11.

   4.2 **LFC (Low Flow Cooling) Valve**—A valve to vent the radiator to the surge tank of a low flow cooling radiator to allow air (gas) to escape during system fill and start up, and to allow deaeration, but prevent or minimize coolant by-pass during normal operation.

   4.3 **Radiator Cap, Solid**—A removable device which closes the cooling system fill opening (filler neck). When installed, it permits no leakage under any cooling system operating condition. It must be used in combination with a radiator pressure relief valve. See "Filler Cap" in Figures 5 and 8.

   4.4 **Radiator Pressure Cap**—A removable device which closes the cooling system fill opening (filler neck) and which incorporates both pressure and vacuum relief valves. Refer to SAE J151 and SAE J164. See "Filler Cap" in Figures 4, 6, 7, 9, 10, and 11.

   4.5 **Radiator Pressure Relief Valve**—A device which provides the same features and functions as a radiator pressure cap except that it is not used for system filling. (Some hand tools are required for the removal of this valve.) It must be used in combination with a radiator cap, solid. See "Pressure Valve" in Figures 5 and 8.

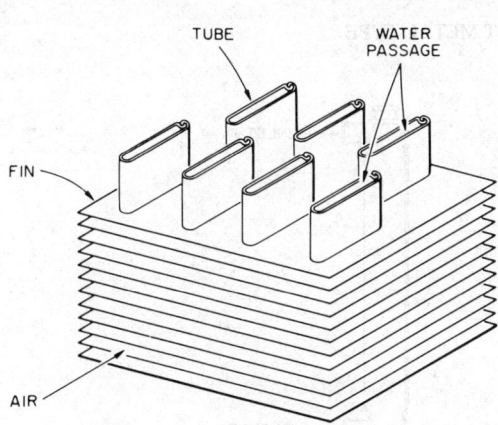

FIGURE 1—TUBE AND PLATE FIN CORE

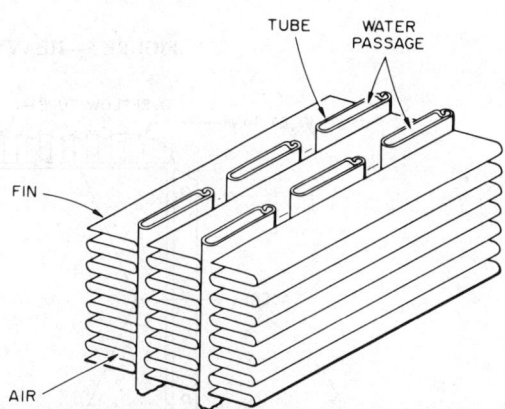

FIGURE 3—TYPICAL TUBE AND SERPENTINE FIN CORE (OFTEN REFERRED TO AS TUBE AND CENTER CORE AND CT CORE)

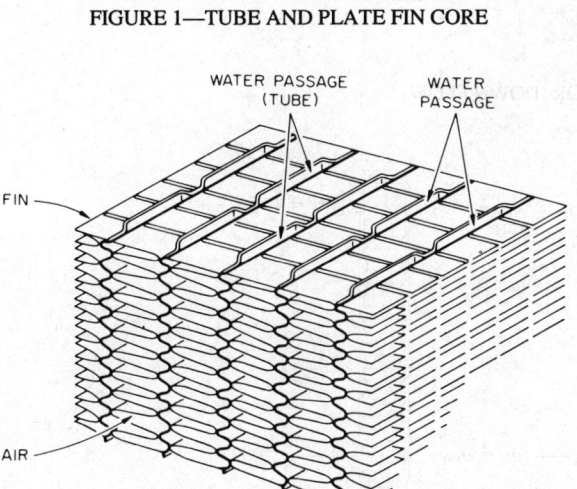

FIGURE 2—CELLULAR CORE

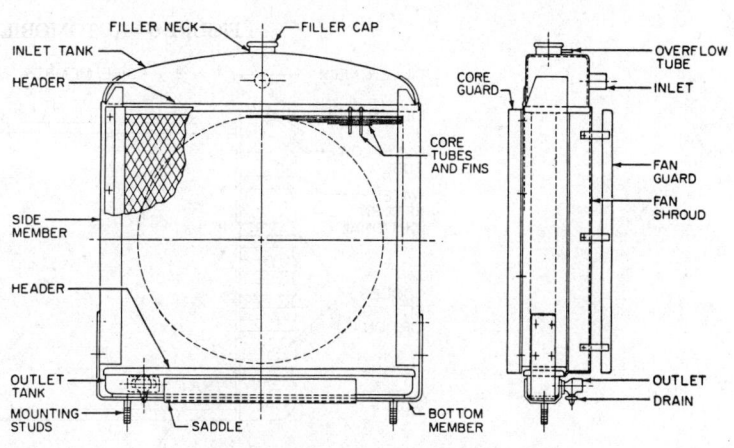

FIGURE 4—SHEET METAL RADIATOR

4.6 Sand Grid—A device between the radiator and the blower fan on certain construction equipment. This device protects the core tubes from damage and erosion on equipment which operates in an environment such that sand, grit, and/or small stones may be picked up in the cooling air stream and hurled at the radiator with great force by the blower fan.

4.7 Tube Protector—A metal shield installed on all radiator core tubes, in the row facing a blower fan, in certain construction equipment. The tube protector prevents tube damage and erosion on equipment which operates in an environment such that sand, grit, and/or small stones may be picked up in the cooling air stream and hurled at the radiator with great force by the blower fan.

4.8 Turbulated Tube—Dimples or impressions on a core tube to increase turbulence and thereby enhance core heat rejection at low coolant flow rates.

4.9 Turbulator—An insert placed inside core tubes to cause coolant flow to be turbulent, and thereby enhance core heat rejection at low coolant flow rates.

NOTE—Nomenclature sketches are typical and not a definitive design.

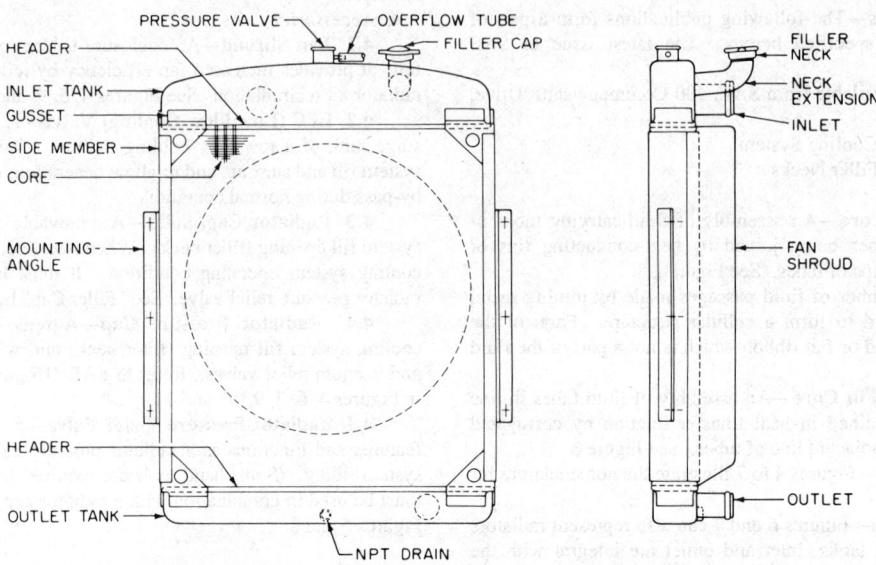

FIGURE 5—HEAVY-DUTY RADIATOR, SHEET METAL TYPE

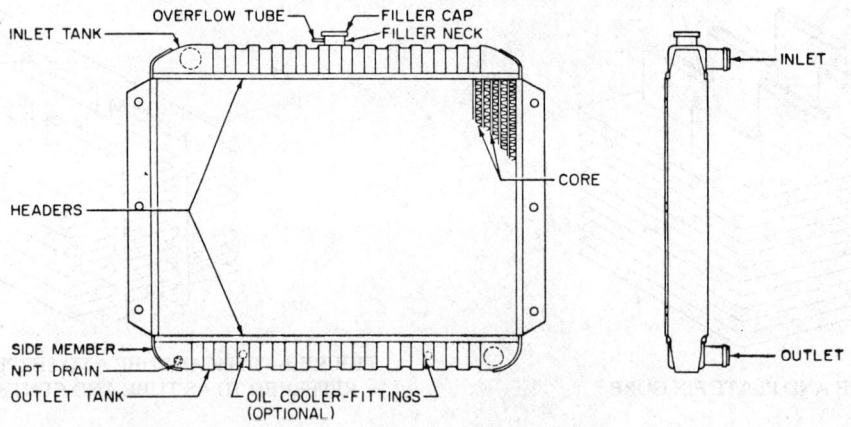

FIGURE 6—AUTOMOBILE RADIATOR (DOWN FLOW)

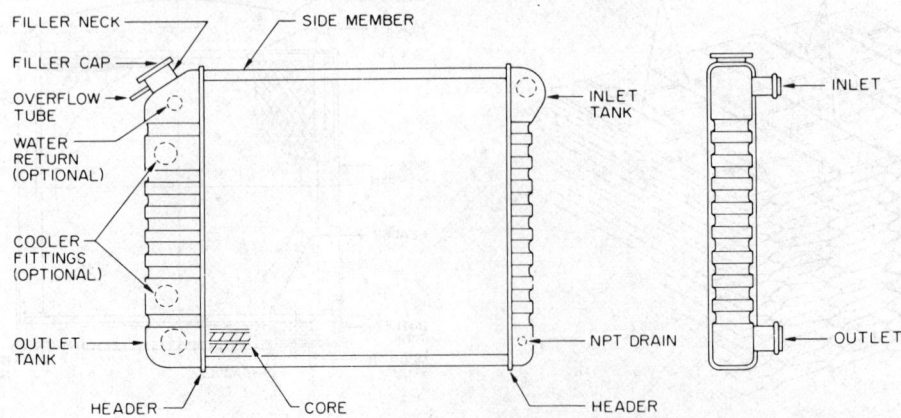

FIGURE 7—AUTOMOBILE RADIATOR (CROSS FLOW)

26.387

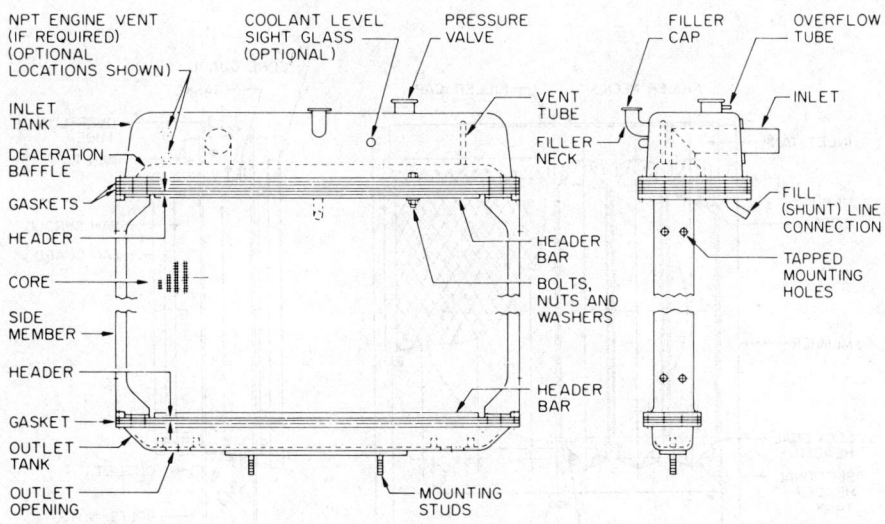

FIGURE 8—BOLTED TYPE RADIATOR, DRAWN TANKS (TYPICAL DEAERATION SYSTEM SHOWN)

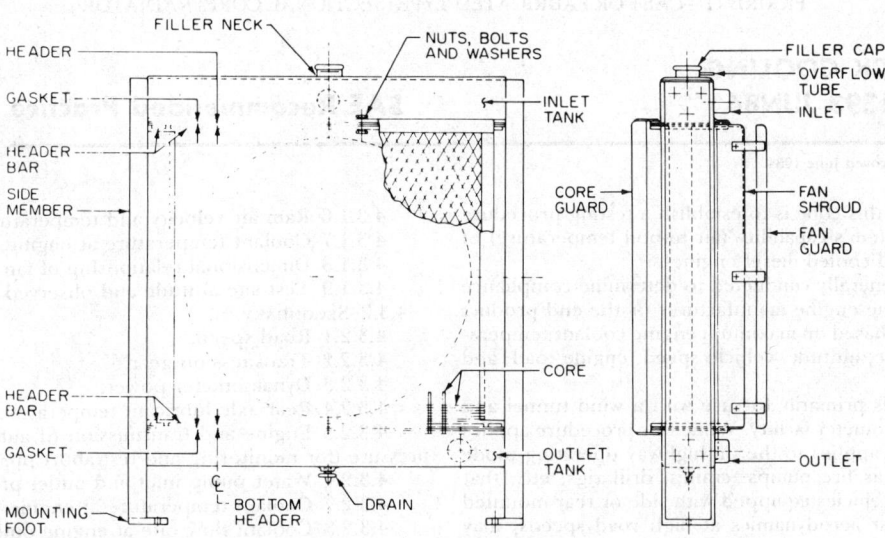

FIGURE 9—FABRICATED TYPE (ONE-PIECE CORE) RADIATOR

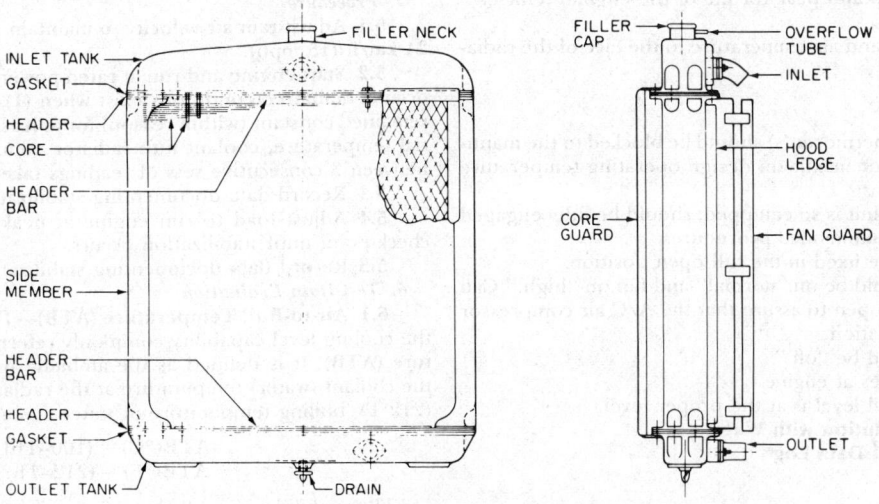

FIGURE 10—CAST TYPE (ONE-PIECE CORE) RADIATOR

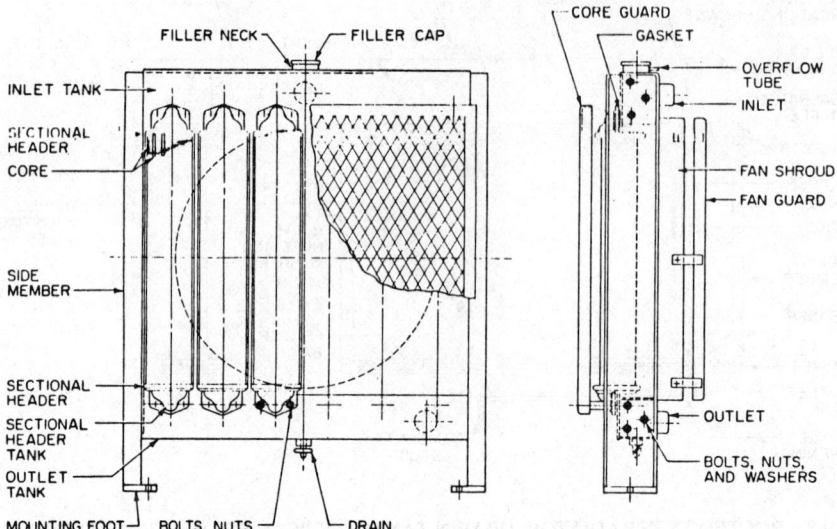

FIGURE 11—CAST OR FABRICATED TYPE (SECTIONAL CORE) RADIATOR

ON-HIGHWAY TRUCK COOLING TEST CODE—SAE J1393 JUN84

SAE Recommended Practice

Report of the Engine Committee, approved June 1984.

1. Scope—The purpose of this code is to establish a testing procedure to determine the cooling system's capability (air-to-boil temperature) of on-highway trucks with liquid cooled diesel engines.

2. Objective—Testing is generally conducted to determine compliance with criteria established by the engine manufacturer or the end product user. Usually, the criteria are based on maximum engine coolant temperature, ambient air temperature, altitude, vehicle speed, engine load, and engine speed.

This test code procedure is primarily for use with a wind tunnel and a chassis or drive train dynamometer facility. While this procedure applies to on-highway trucks, it also applies to the on-highway operating mode of specialized vehicles such as fire pumps, cranes, drill rigs, etc., that have dual purpose engines. Vehicles equipped with side or rear mounted radiators (because of peculiar aerodynamics at high road speeds) may require an alternate procedure.

3. Facility Requirements—The test facility should provide the following features:

3.1 Power Absorbing Device—A chassis or drive train dynamometer capable of absorbing rated hp and peak torque of the engine/vehicle.

3.2 Cooling Air

3.2.1 Constant air velocity and air temperature to the face of the radiator.

4. Test Preparation

4.1 Engine and Vehicle

4.1.1 The engine coolant thermostat(s) should be blocked to the manufacturer's specified opening for maximum design operating temperature condition.

4.1.2 The fan drive, if the unit is so equipped, should be fully engaged using the manufacturer's recommended procedure.

4.1.3 All shutters should be fixed in the full open position.

4.1.4 Air conditioning should be on "normal" and fan on "high." Cab windows and vents are to be open to assure that the A/C air compressor does not cycle during stabilization.

4.1.5 Cab heater fan should be "off."

4.1.6 Block cab heater hoses at engine.

4.1.7 Determine that the oil level is at the proper level.

4.2 Replace Coolant Solution with Water

4.3 Instrumentation and Data Log

4.3.1 PRIMARY

4.3.1.1 Time interval.

4.3.1.2 Engine speed.

4.3.1.3 Fan speed.

4.3.1.4 Fuel flow rate measurement and temperature. Follow engine manufacturer's recommended procedure.

4.3.1.5 Ambient air temperature.

4.3.1.6 Ram air velocity and temperature.

4.3.1.7 Coolant temperature at engine outlet to radiator.

4.3.1.8 Dimensional relationship of fan to core, shroud, and engine.

4.3.1.9 Test site altitude and observed barometric pressure.

4.3.2 SECONDARY

4.3.2.1 Road speed.

4.3.2.2 Transmission gear.

4.3.2.3 Dynamometer power.

4.3.2.4 Rear axle lubricant temperature.

4.3.2.5 Engine and transmission (if automatic) oil temperature and pressure (for monitoring and test abort purposes).

4.3.2.6 Water pump inlet and outlet pressures.

4.3.2.7 Coolant temperature at engine inlet from the radiator.

4.3.2.8 Coolant flow rate at engine outlet to radiator.

4.3.2.9 Air temperature at engine inlet.

4.3.2.10 Intake manifold pressure (turbocharged engine only).

4.3.2.11 Air temperature behind radiator core at the center and at each corner.

5. Procedure

5.1 Adjust ram air velocity to maintain specified test speed, typically 24 km/h (15 mph).

5.2 Start engine and run at rated power and speed until stabilization occurs. Stabilized conditions exist when (1) engine speed and load have remained constant (within reason) for at least 15 min and (2) the differential temperature, coolant into radiator minus ambient, does not change between 3 consecutive sets of readings taken at 5 min intervals.

5.3 Record data documenting stabilization.

5.4 Adjust load to run engine at peak torque and speed or other check point until stabilization occurs.

5.5 Record data documenting stabilization.

6. Test Data Evaluation

6.1 Air-to-Boil Temperature (ATB)—Test data is used to determine the cooling level capability commonly referred to as air-to-boil temperature (ATB). It is defined as the ambient air temperature (T_a) at which the coolant (water) temperature at the radiator inlet (T_b) reaches 100°C (212°F), boiling temperature of water at standard conditions.

$$ATB(°C) = (100 - T_b) + T_a$$
$$ATB(°F) = (212 - T_b) + T_a$$

ATB capabilities are calculated for each combination of ram air velocity, engine load, and engine speed specified by the engine manufacturer/vehicular manufacturer or the end user.

7. Test Results Extrapolation and Adjustment Guidelines

7.1 Due to facility limitations or the existing ambient air conditions, actual test conditions may deviate from those specified by the acceptance criteria. ATB temperature based on test results can be adjusted for variations.

7.1.1 ALTITUDE—When the test is conducted at an altitude above the base of 152 m (500 ft), the calculated ATB temperature should be adjusted upward 1—2°C (2—4°F) per 305 m (1000 ft).

7.1.2 COOLANT—When a test is conducted with a 50/50 ethylene glycol-water mixture instead of the usual water coolant base, the calculated ATB temperature should be adjusted upward 3—4°C (5—7°F). Greater ATB temperature differences may be encountered in a cooling system where the coolant velocity through the radiator tubes is extremely low.

7.1.3 FUEL RATE VARIATION—When the test is conducted at a fuel rate different than the specified value, the ATB temperature should be corrected.

$$ATB\ (°C)_{CORR.} = 100 - (100 - ATB_{TEST}) \frac{(\text{Specified Fuel Rate})}{(\text{Test Fuel Rate})}$$

$$ATB\ (°F)_{CORR.} = 212 - (212 - ATB_{TEST}) \frac{(\text{Specified Fuel Rate})}{(\text{Test Fuel Rate})}$$

7.1.4 AMBIENT AIR TEMPERATURE—Most facilities provide control of ambient air temperature in the 27—38°C (80—100°F) range. Ordinarily, in this range, no adjustment of resulting ATB temperature is required. When the facility's ambient air temperature is below 27°C (80°F), the calculated ATB temperature is adjusted downward 0.5—1°C (1—2°F). Conversely, when the facility's ambient air temperature is above 38°C (100°F), the calculated ATB temperature is adjusted upward 0.5—1°C (1—2°F).

7.1.5 RAM AIR—In general, the ATB temperature increases with increasing ram air velocity. Consequently, facilities without ram air capacity will produce a reduced ATB temperature. Approximately a 2°C (4°F) ATB temperature increase is considered for a 24 km/h (15 mph) change. Above 24 km/h (15 mph), the ATB temperature increase is not predictable.

7.1.6 AIR CONDITIONING—In general, when a radiator mounted condenser core is installed, but the air conditioning unit is not operating during the test, the ATB temperature should be reduced 2°C (4°F).

When no radiator mounted condenser core is installed on the test unit, but will be offered on the model being tested, the ATB temperature should be reduced 4°C (7°F).

8. Analysis of Cooling System ATB Tests—The analysis of a cooling system test presumes that the desired system performance requirements have been previously determined and that the test procedures have been properly followed; fault in these areas is not considered. The analysis of a cooling system test often provides information which can be used to improve the system's (1) performance and (2) production cost effectiveness.

8.1 **Radiator Air Temperature**—The difference between the temperature of the air entering and leaving a radiator core is affected by the performance of all the components in the entire system; the radiator core, fan, fan shroud, and their enclosure, the vehicle itself.

If each of the cooling system components performs at its potential, unencumbered by excessive vehicle restrictions, the difference between core inlet and outlet air temperature is usually in the range of 22—33°C (40—60°F). Deviations (up and down) are indicative of low or high air flow respectively. Whether or not this should be considered normal and/or expected should be determined through analysis with the radiator core manufacturer. It is not meant to imply failure to meet the predetermined system performance criteria, but it does mean that there may be potential for improvement in either the ATB and/or the effectiveness of the system.

8.1.1 CAUSE OF HIGH RADIATOR AIR TEMPERATURE RISE
8.1.1.1 High Core Air Restriction

Change	Potential Result
Decrease fins per inch (core density)	Decreased ATB, fouling tendency, first cost, and operating costs

8.1.1.2 Inadequate Air Flow Performance

Change	Potential Result
Increase fan speed	Increased ATB, fan horsepower, first cost, and operating costs
Increase fan diameter, number of blades and/or pitch	Increased ATB, fan horsepower, first cost, and increased operating costs

Fan Shrouding
  None to ring
  Ring to box
  Box to venturi
	Increased ATB and first cost. Decreased operating costs

Decrease fan tip clearance	Increased ATB

8.1.1.3 Air Recirculation

Change	Potential Result
Reduce air flow restriction, i.e. grill, bumper, etc.	Increased ATB and decreased operating costs
Add baffles, and/or fan shroud	Increased ATB and first costs
Make changes to the vehicle	Decreased operating costs

8.1.2 CAUSE OF LOW RADIATOR TEMPERATURE RISE—A low air temperature rise across the core is caused by the inverse of the above and generally provides opportunities to effect savings for the engine/vehicle manufacturer and/or the end product user.

8.2 **Radiator Coolant Temperature**—The difference between the temperature of the coolant entering and leaving the radiator core is usually in the range of 4—8°C (7—14°F). It is equal to the coolant temperature rise across the engine if no bypass system permits part of the coolant to recirculate without passing through the radiator core. If a substantial portion of the engine coolant is bypassing the core, improved cooling may result from a change which will increase the flow i.e., full blocking bypass thermostat in place of a partial or non blocking thermostat.

Other factors relating to the transfer of heat from the coolant to the core, such as core tube velocity and proper distribution of coolant through all core tubes, are beyond the scope of this analysis; they should be reviewed with the radiator manufacturers.

8.3 **Engine Air Inlet Temperature**—Some increase in the engine combustion air temperature usually occurs as the air flows from the point where it enters the vehicle air intake system to where it enters the engine at the turbocharger or other inlet point. A certain amount of increase is expected and is considered normal. In practice, this increase can usually be kept to 15°C (27°F) or less. However, if the temperature of the air entering the vehicle air intake system is different than that being admitted to the radiator, a condition differing from actual operating conditions may be created. If the temperature of the air entering the vehicle air intake system is higher than the temperature of the air entering the radiator, the ATB temperature will be lowered. For every 4°C (7°F) that the temperature of the air entering the vehicle air intake system is increased over the temperature of the air entering the radiator, the ATB temperature will be reduced 0.5°C (1°F).

LABORATORY TESTING OF VEHICLE AND INDUSTRIAL HEAT EXCHANGERS FOR DURABILITY UNDER VIBRATION INDUCED LOADING—SAE J1598 JUL88

SAE Recommended Practice

Report of the Engine Committee approved July 1988

1. **Purpose**—This recommended practice provides a test guideline for determining the durability of a heat exchanger under specified vibration loading.

2. **Scope**—This recommended practice is applicable to all liquid-to-gas, liquid-to-liquid, gas-to-gas, and gas-to-liquid heat exchangers used in vehicle and industrial cooling systems. This document outlines the test to determine durability characteristics of the heat exchanger from vibration induced loading.

3. **Objective**—To verify compliance with established criteria that insures durability in a specific application. This document describes a system to induce stresses in a heat exchanger resulting from vibration loading. The process is accomplished by vibrating the unit at specified frequencies, and amplitudes or acceleration.

4. **Facility Requirement**—The facility should provide the following as required:

4.1 Vibration equipment with controls on frequency, amplitude, and acceleration (for 50 Hz and below, electro hydraulic is suggested, higher frequencies may require electrodynamic).

4.2 Test Fixture
4.2.1 Actual Installation
4.2.2 If actual installation is not available, a fixture should support the heat exchanger with orientation as in service or as specified. The fixture must be rigid so as not to induce any additional dynamic inputs.
4.3 Means of checking heat exchanger integrity.
4.4 Equipment to monitor motion of unit during test.
4.5 Source for pressurizing and heating test unit if required.
4.6 Additional equipment may include but not be limited to the following:
Accelerometers
Automatic Data Logging Equipment
Automatic Emergency Shutdown
Cycle Counters
Digital Signal Analyzer
Function Generator (sine wave, random noise, square wave, ramp function, etc.)
Load Transducers
Pressure Gauges
Pressure Regulators
Road Data Recording Equipment
Safety features as specified by regulatory codes and common practices
Strain Measurement Equipment
Temperature Monitoring and Control Equipment

5. *Testing*
5.1 Test heat exchanger for integrity.
5.2 Install test unit on vibration test fixture, fill with specified test fluid, orientation the same as in service or as specified, and to include all mounting hardware: isolators, tie rods, mounting brackets, etc. All heat exchanger mounted masses will affect the response characteristics of the test unit and should be attached. Examples of mounted masses for a radiator are: charge air cooler, condensers, oil cooler, air dryers, shrouds, shutter, etc.

5.3 Set up control system to obtain the specified vibration cycle under one of the following modes:
5.3.1 Road or duty cycle (simulation or replication).
5.3.2 Failure mode simulation.
5.3.3 Sine sweep at specified acceleration or amplitude.
5.3.4 Resonant frequency at specified energy input.
5.3.5 Random noise at specified acceleration levels.
5.3.6 Per established specification.
5.4 Pressurize and heat the test unit if required.
5.5 Run test to specified duration or component failure.
5.6 Remove and test heat exchanger for leaks and structural damage.

6. *Test Documentation*
6.1 During and after test, document leakage rate(s), location(s), and structural failures for comparison to acceptance criteria.
6.2 Document the following:
a) Condition of unit prior to test (new or previous history)
b) Unit Orientation (Ref. paragraph 5.2)
c) Test Mode (Ref. paragraph 5.3)
d) Pressure and Temperature (Ref. paragraph 5.4)
e) Duration of Test (Ref. paragraph 5.5)
f) Location of leaks and structural damage (Ref. paragraph 5.6)
6.3 Testing should be adequately documented to allow test reproduction. Documentation to include a test log of the complete test, recording any changes in the heat exchanger and fixture. Test log to include cycles or hours, time of day for all test starts and stops with reasons for stops.

7. This recommended practice is valid for durability comparison of vibration induced loadings. Correlation to field results must be developed on an individual basis. Other tests affecting heat exchanger durability are SAE J1542 (Thermal Cycle) and SAE J1597 (Pressure Cycle). These tests can be run in combination as well as independently.

NOTE: Combination testing may make it difficult to determine cause of failure.

LABORATORY TESTING OF VEHICLE AND INDUSTRIAL HEAT EXCHANGERS FOR PRESSURE CYCLE DURABILITY—SAE J1597 JUL88

SAE Recommended Practice

Report of the Engine Committee approved July 1988.

1. *Purpose* — This recommended practice provides a test guideline for determining the durability of a heat exchanger under specified pressure cycle conditions.

2. *Scope* — This recommended practice is applicable to all liquid-to-gas, liquid-to-liquid, gas-to-gas, and gas-to-liquid heat exchangers used in vehicle and industrial cooling systems. This document outlines the tests to determine durability characteristics of the heat exchanger under pressure cycling.

3. *Objective* — To verify compliance with established criteria that insures durability in a specific application. This document describes a system to induce stresses in a heat exchanger resulting from specified pressure variations. The process is accomplished by varying the pressure in the unit at a specified frequency and temperature.

4. *Facility Requirement* — The facility should provide the following as required:
4.1 Source of pressurized fluid at specified temperature and pressure range.
4.2 Means of moving these fluids to and from the test unit in a specified manner.
4.3 Means of checking heat exchange integrity.
4.4 A system adequate to control the pressure cycle as specified. It may include but not be limited to the following:
Automatic data logging equipment
Automatic emergency shutdowns
Electrical relays, timers, switches, indicator lights, and related items as required
Flow measuring devices
Fluid filters
Heaters and controls
Leak detector
Pressure cycle counters
Pressure gauges
Pressure regulators for each fluid
Pressure switches
Pumps
Relief and back pressure regulators
Safety features as specified by regulatory codes and common practices
Tanks
Temperature indicators
See Fig. 1 for typical control system as reference.

5. *Testing*
5.1 Test heat exchanger for integrity.
5.2 Install test unit as specified (orientation as in service) in a safe manner.
5.3 Set up control system to obtain the specified pressure cycle (see Fig. 2 for typical pressure cycle).
5.3.1 High pressure and pressure range to be specified based on specific application. High pressure and pressure range to be at least equal to or greater than normal operating conditions. CAUTION—Pressures exceeding specified value can cause significant structural damage which will invalidate the test results. The system shall be capable of maintaining this pressure within 5% absolute pressure unless otherwise specified and should be measured within or as near the test unit as possible.

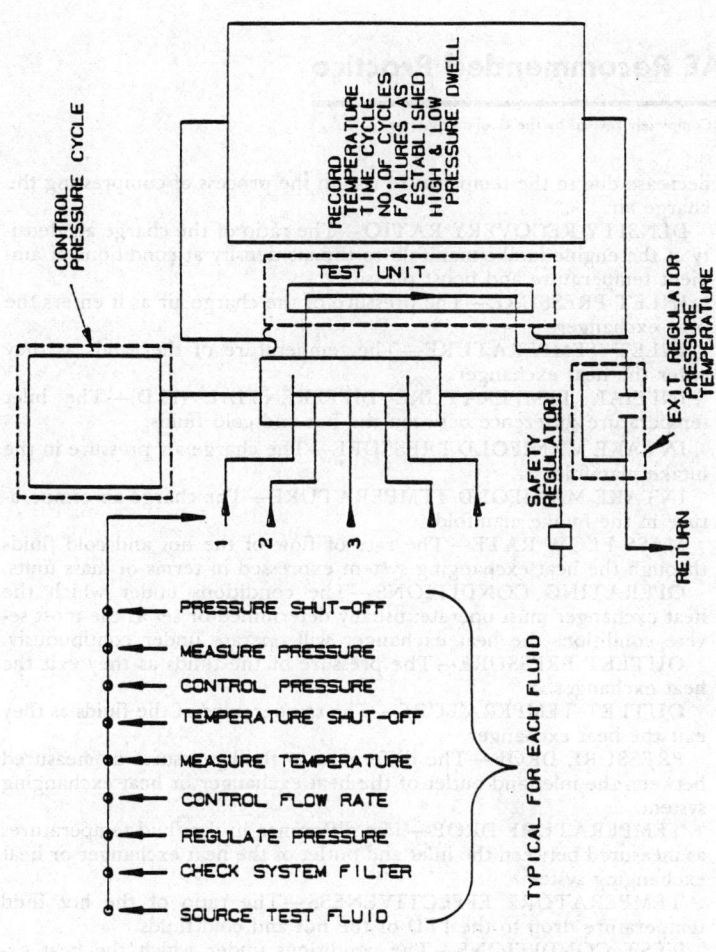

FIG. 1—TYPICAL CONTROL SYSTEM

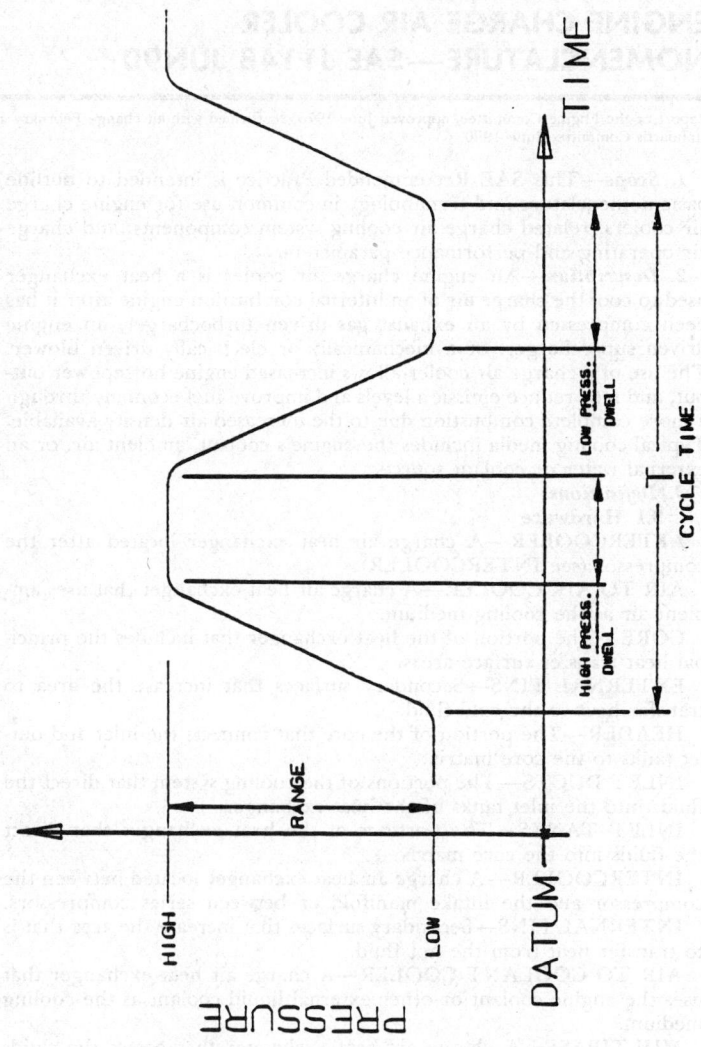

FIG. 2—TYPICAL PRESSURE CYCLE

5.3.2 High and low pressure dwell, and rate of pressure change, are to be as specified.

5.3.3 Fluid temperature to be as specified and controlled within 1% absolute temperature unless otherwise specified. The specified temperature should be near the maximum value expected in service.

5.4 Run test to specified cycle duration or heat exchanger failure.

5.5 Remove and test heat exchanger for leaks and structural damage.

6. Test Documentation

6.1 During and after the test, document leakage rate and location(s) at specified pressure (gage or absolute) and compare to acceptance criteria.

6.2 Document the following:
a) Condition of unit prior to test (new or previous history)
b) Unit Orientation (Ref. paragraph 5.2)
c) Pressure Range (Ref. paragraph 5.3.1)
d) High Pressure (Ref. paragraph 5.3.1)
e) Frequency, Pressure Dwell, High & Low, Rate of Pressure Change (Ref. paragraph 5.3.2)
f) Fluid Temperature (Ref. paragraph 5.3.3)
g) Number of Cycles to Completion (Ref. paragraph 5.4)
h) Location of Leaks and Structural Damage (Ref. paragraph 5.5)
i) If testing is continuous, or in block units

7. This recommended practice is valid for durability comparison of pressure cycle conditions only. Correlation to field results must be developed on an individual basis. Other tests affecting heat exchanger durability are SAE J1542 (Thermal Cycle) and SAE J1598 (Vibration). These tests can be run in combination as well as independently.

NOTE: Combination testing may make it difficult to determine cause of failure.

(R) ENGINE CHARGE AIR COOLER NOMENCLATURE—SAE J1148 JUN90

SAE Recommended Practice

Report of the Engine Committee, approved June 1976, reaffirmed without change February 1984. Completely revised by the Cooling Systems Standards Committee June 1990.

1. Scope—This SAE Recommended Practice is intended to outline basic nomenclature and terminology in common use for engine charge air coolers, related charge air cooling system components, and charge air operating and performance parameters.

2. Description—An engine charge air cooler is a heat exchanger used to cool the charge air of an internal combustion engine after it has been compressed by an exhaust gas driven turbocharger, an engine driven supercharger, or a mechanically or electrically driven blower. The use of a charge air cooler allows increased engine horsepower output, and may reduce emission levels and improve fuel economy through a more complete combustion due to the increased air density available. Typical cooling media includes the engine's coolant, ambient air, or an external water or coolant source.

3. Definitions
3.1 Hardware
AFTERCOOLER—A charge air heat exchanger located after the compressor (see INTERCOOLER).

AIR TO AIR COOLER—A charge air heat exchanger that uses ambient air as the cooling medium.

CORE—The portion of the heat exchanger that includes the principal heat transfer surface areas.

EXTERNAL FINS—Secondary surfaces that increase the area to transfer heat to the cold fluid.

HEADER—The portion of the core that connects the inlet and outlet tanks to the core matrix.

INLET DUCTS—The portions of the cooling system that direct the fluids into the inlet tanks of the heat exchanger.

INLET TANKS—The portions of the heat exchanger that direct the fluids into the core matrix.

INTERCOOLER—A charge air heat exchanger located between the compressor and the intake manifold or between series compressors.

INTERNAL FINS—Secondary surfaces that increase the area that is to transfer heat from the hot fluid.

AIR TO COOLANT COOLER—A charge air heat exchanger that uses the engine coolant or other external liquid coolant as the cooling medium.

MULTIPASS—A charge air heat exchanger that passes the fluids through the core matrix more than once.

OUTLET DUCTS—The portions of the cooling system that direct the fluids out of the outlet tanks of the heat exchanger.

OUTLET TANK—The portion of the heat exchanger that direct the fluids out of the core matrix.

REMOTE MOUNTED—A charge air heat exchanger that is located (mounted) in an area not normally associated with or convenient to the cooling medium.

SINGLE PASS—A charge air heat exchanger that passes the fluids through the core only once.

TUBES—The portions of the heat exchanger core matrix that are used to separate the fluids and are also the primary heat transfer surface areas.

TURBULATOR—Secondary surfaces that increase the turbulence and mixing of the cold or hot fluids, not metallurgically bonded to the primary heat transfer surfaces.

3.2 Operating and Performance Parameters
AMBIENT TEMPERATURE—The temperature of the area surrounding the heat exchanger.

BOOST PRESSURE—The pressure of the charge air as it leaves the turbocharger, supercharger, or other compressor.

DENSITY RECOVERY EFFICIENCY—The ratio of the charge air density increase achieved from cooling the charged air, to the density decrease due to the temperature rise in the process of compressing the charge air.

DENSITY RECOVERY RATIO—The ratio of the charge air density at the engine intake manifold to the air density at conditions of ambient temperature and boost pressure.

INLET PRESSURE—The pressure of the charge air as it enters the heat exchanger.

INLET TEMPERATURE—The temperature of the fluids as they enter the heat exchanger.

INITIAL TEMPERATURE DIFFERENTIAL (ITD)—The inlet temperature difference between the hot and cold fluids.

INTAKE MANIFOLD PRESSURE—The charge air pressure in the intake manifold.

INTAKE MANIFOLD TEMPERATURE—The charge air temperature in the intake manifold.

MASS FLOW RATE—The rate of flow of the hot and cold fluids through the heat exchanging system expressed in terms of mass units.

OPERATING CONDITIONS—The conditions under which the heat exchanger must operate; usually determined or set as the most severe conditions the heat exchanger will operate under continuously.

OUTLET PRESSURE—The pressure of the fluids as they exit the heat exchanger.

OUTLET TEMPERATURE—The temperature of the fluids as they exit the heat exchanger.

PRESSURE DROP—The difference in fluid pressures as measured between the inlet and outlet of the heat exchanger or heat exchanging system.

TEMPERATURE DROP—The difference in the fluid temperatures as measured between the inlet and outlet of the heat exchanger or heat exchanging system.

TEMPERATURE EFFECTIVENESS—The ratio of the hot fluid temperature drop to the ITD of the hot and cold fluids.

TEST CONDITIONS—The conditions under which the heat exchanger is tested to determine its effectiveness and pressure drop, usually the same as the operating conditions.

4. Typical Examples of the Induced Air Flow Path
a. Air > Turbocharger > Charge Air Cooler > Intake Manifold
b. Air > Turbocharger > Charge Air Cooler > Mechanically Driven Blower > Intake Manifold
c. Air > Turbocharger > Mechanically Driven Blower > Charge Air Cooler > Intake Manifold
d. Air > Turbocharger > Charge Air Cooler > Mechanically Driven Blower > Charge Air Cooler > Intake Manifold
e. Air > Mechanically Driven Blower > Charge Air Cooler > Intake Manifold
f. Air > Mechanically Driven Blower > Turbocharger > Charge Air Cooler > Intake Manifold
g. Air > Mechanically Driven Blower > Charge Air Cooler > Turbocharger > Intake Manifold
h. Air > Mechanically Driven Blower > Charge Air Cooler > Turbocharger > Charge Air Cooler > Intake Manifold
i. Air > Turbocharger > Charge Air Cooler > Turbocharger > Charge Air Cooler > Intake Manifold

5. Schematics of Typical Charge Air Coolers
5.1 Air to Coolant Heat Exchangers
a. Mounted in the intake manifold (see Figures 1 and 2)
b. Mounted Remotely (see Figures 3 and 4)
5.2 Air to Air Heat Exchangers
a. Engine fan cooled (see Figure 5)
b. Auxiliary Blower Cooled (See Figure 6)

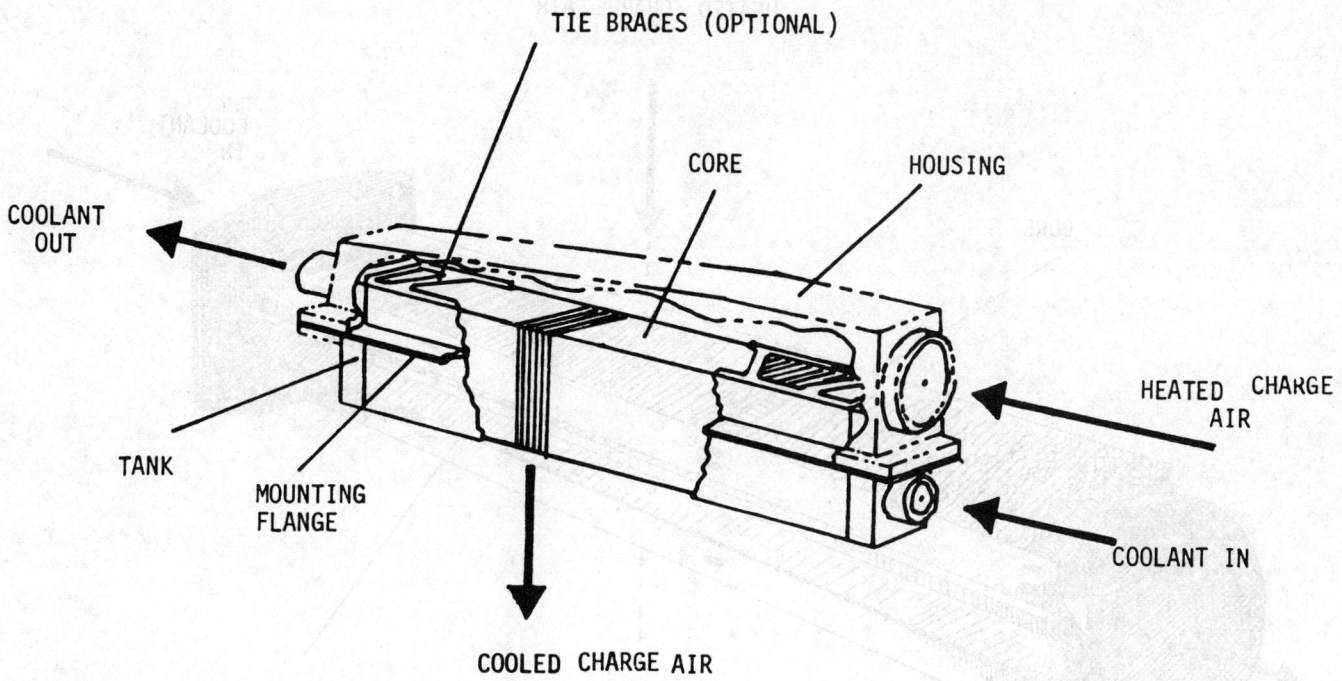

NOTE: Coolant sources can be varied.
Materials have to be compatible with the type of coolant and environment.
Coolant traverses may be a single pass or a multipass arrangement.

FIG. 1—AIR TO COOLANT—MOUNTED IN INTAKE MANIFOLD

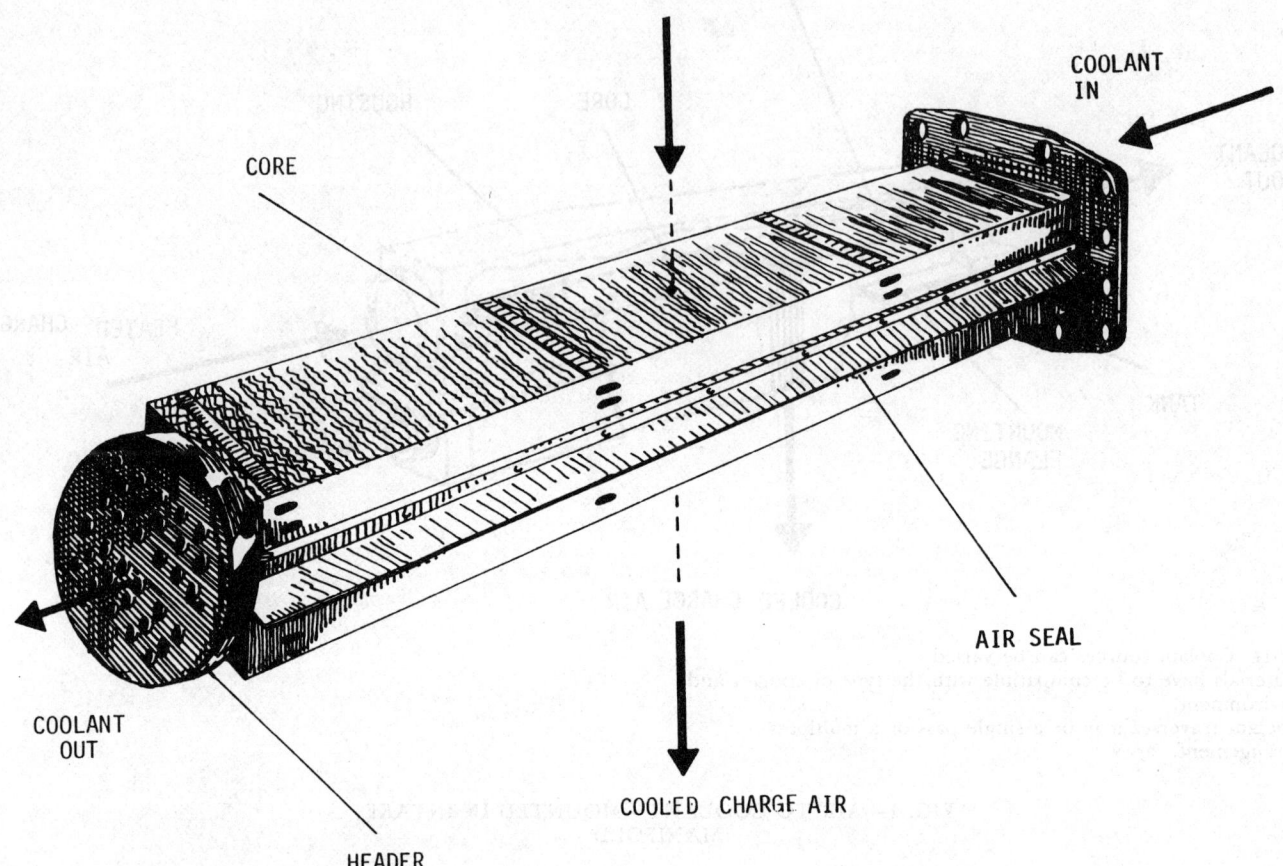

NOTE: Coolant sources can be varied.
Materials have to be compatible with the type of coolant and environment.
Coolant traverses may be a single pass or a multipass arrangement.

FIG. 2—AIR TO COOLANT—MOUNTED IN INTAKE MANIFOLD

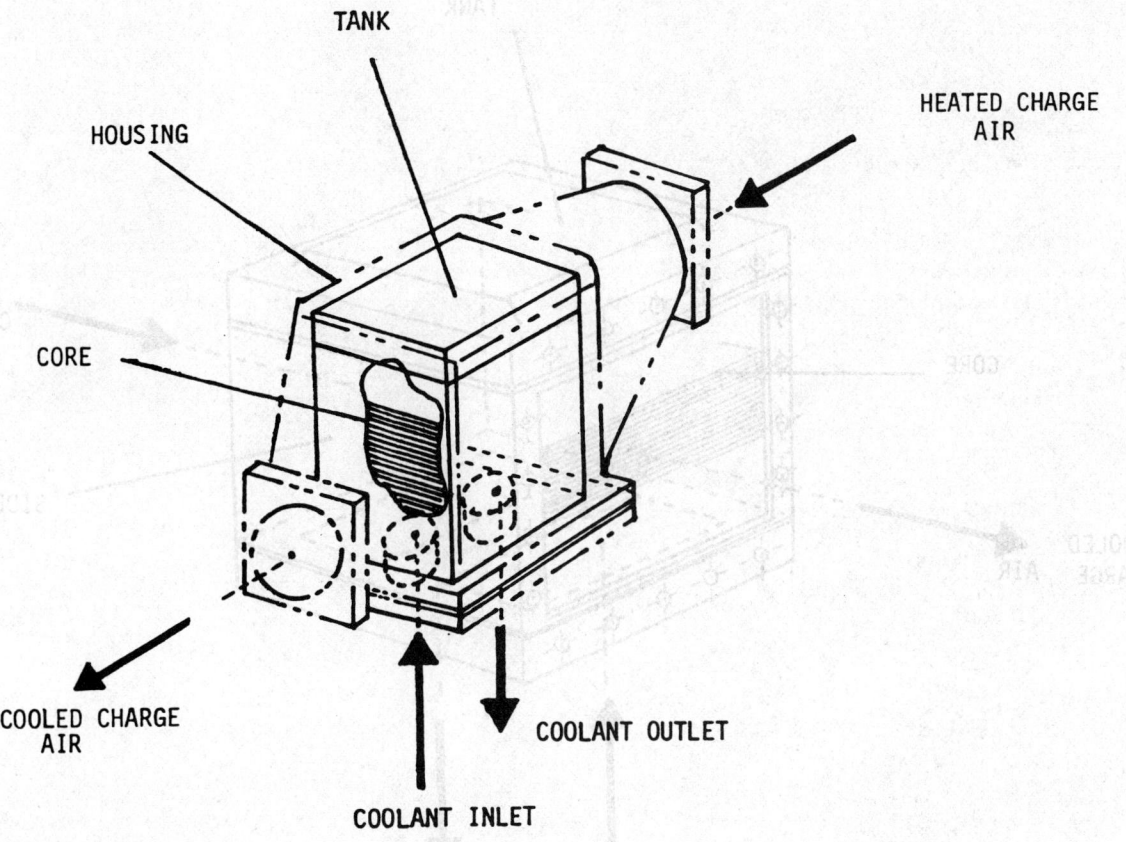

NOTE: Coolant sources can be varied.
Materials have to be compatible with the type of coolant and environment.
Coolant traverses may be a single pass or a multipass arrangement.

FIG. 3—AIR TO COOLANT—MOUNTED REMOTELY

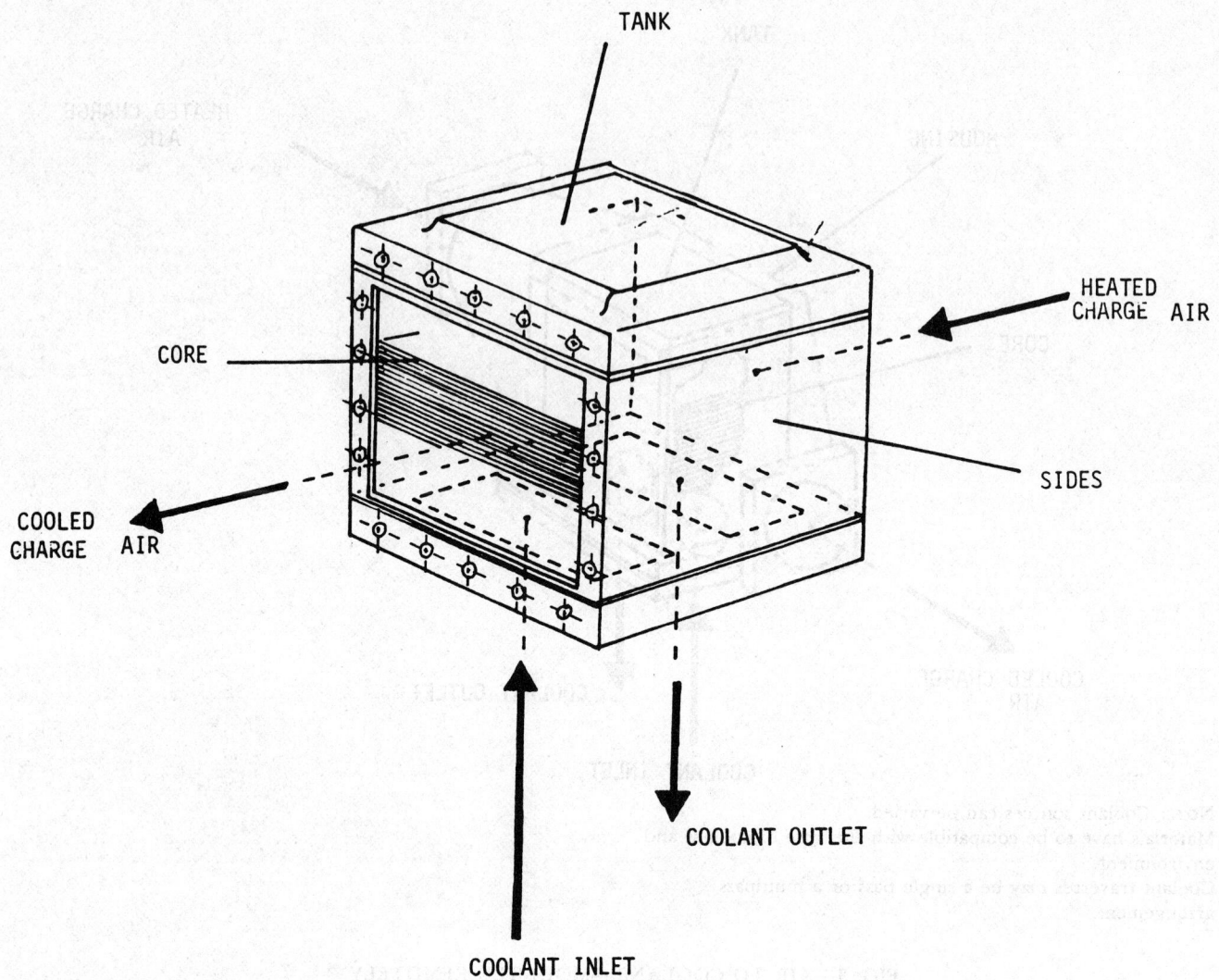

NOTE: Coolant sources can be varied.
Materials have to be compatible with the type of coolant and environment.
Coolant traverses may be a single pass or a multipass arrangement.

FIG. 4—AIR TO COOLANT—MOUNTED REMOTELY

26.397

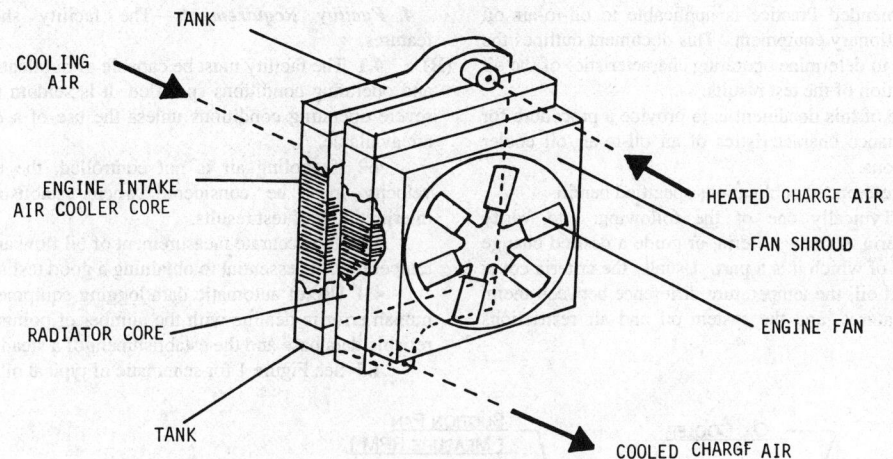

FIG. 5—AIR TO AIR—ENGINE-FAN COOLED—MOUNTED REMOTELY

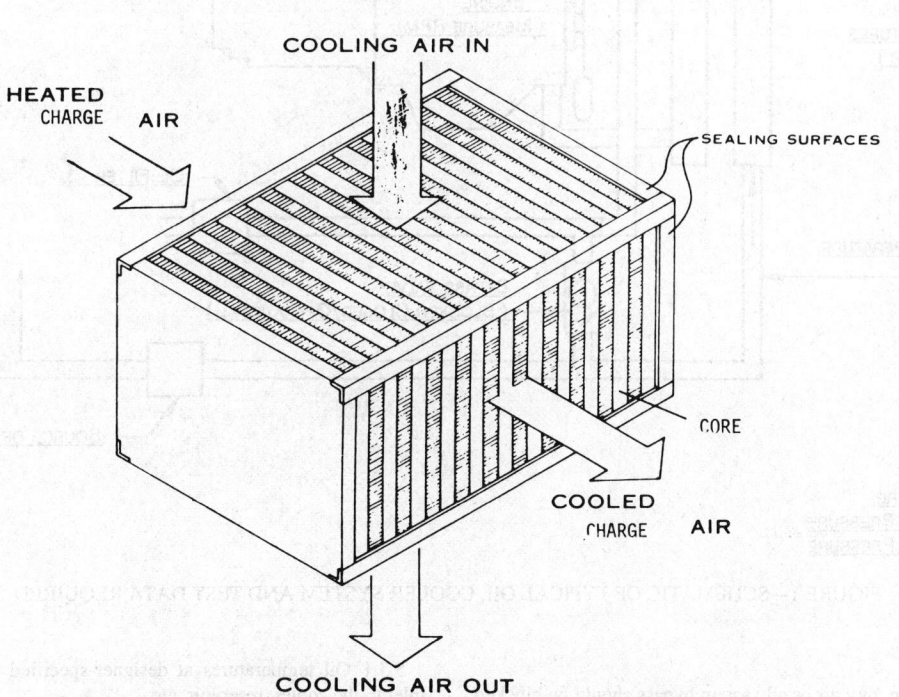

FIG. 6—AIR TO AIR AUXILIARY BLOWER COOLED—MOUNTED REMOTELY

APPLICATION TESTING OF OIL-TO-AIR OIL COOLERS FOR COOLING PERFORMANCE —SAE J1468 MAY93

SAE Recommended Practice

Report of the Engine Committee approved November 1985. Revised by the SAE Cooling Systems Standards Committee May 1993.

(R) **1. Scope**—This SAE Recommended Practice is applicable to oil-to-air oil coolers installed on mobile or stationary equipment. This document outlines the method of procuring the test data to determine operating characteristics of the oil cooling system and the interpretation of the test results.

(R) **1.1 Purpose**—The purpose of this document is to provide a procedure for determining the cooling performance characteristics of an oil-to-air oil cooler under specified operating conditions.

(R) **2. References**—There are no referenced publications specified herein.

(R) **3. Objective (of the test)**—Typically one of the following: to verify compliance with established criteria, set new criteria, or guide a desired change of either the cooler or the system of which it is a part. Usually the criteria cover the mass flows of cooling air and oil, the temperature difference between them, the maximum allowable temperatures, and the system oil and air restrictions imposed by the heat exchanger.

4. Facility Requirements—The facility should provide the following features:

(R) **4.1** The facility must be capable of duplicating the most severe duty cycles and operating conditions specified. It is seldom practical to duplicate the most severe operating conditions unless the use of a dynamometer and wind tunnel are available.

4.2 If cooling air is not controlled, the effects of wind direction and velocity must be considered when establishing vehicle orientation and interpretation of test results.

4.3 The accurate measurement of oil flow and oil pressures plus oil and air temperatures is essential to obtaining a good test result.

4.4 Use of automatic data logging equipment is preferred as it minimizes human error in dealing with the number of points necessary to accumulate for a reliable data base and the establishment of a steady-state operating condition.

4.5 See Figure 1 for schematic of typical oil cooler system.

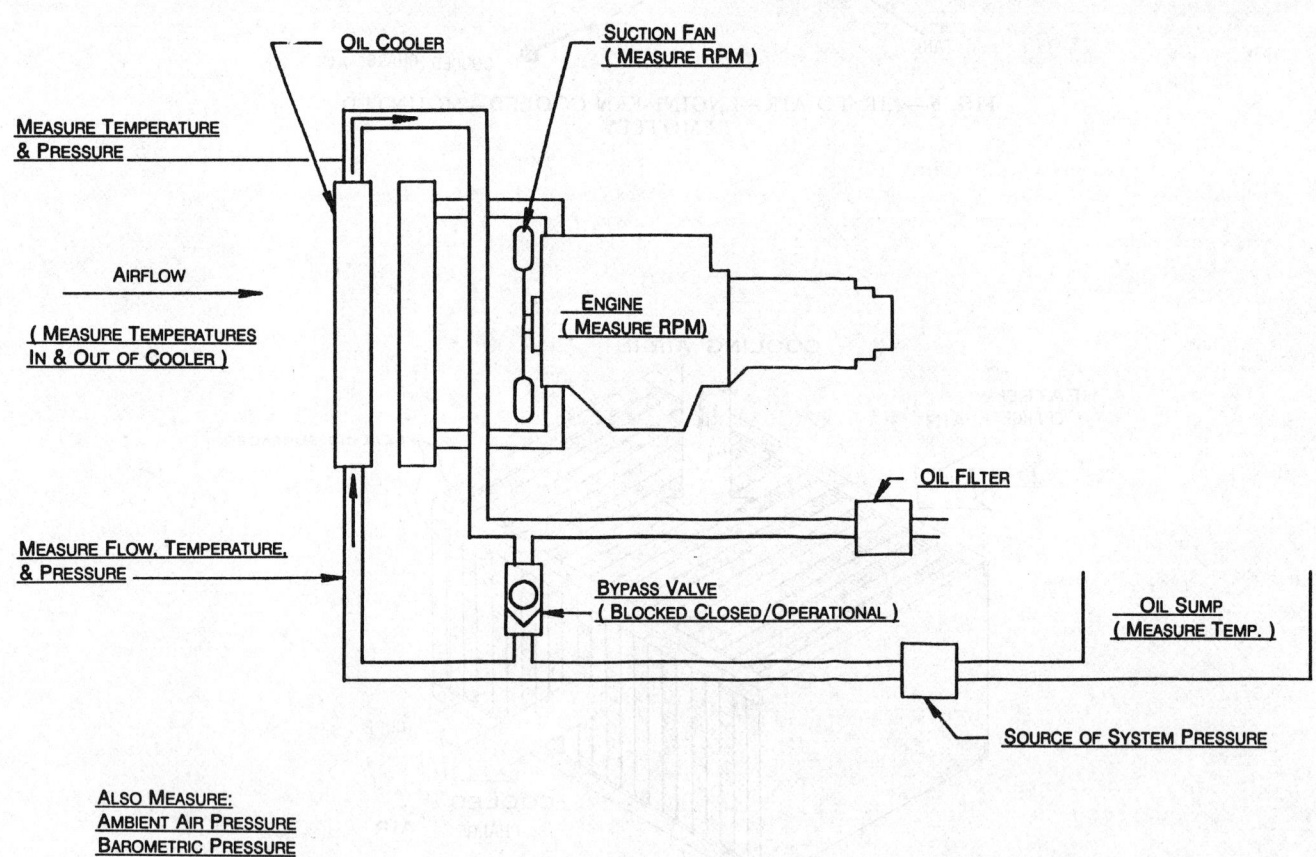

FIGURE 1—SCHEMATIC OF TYPICAL OIL COOLER SYSTEM AND TEST DATA REQUIRED

5. Test Preparation

(R) **5.1** For component testing, any air or oil system bypass should be blocked closed to insure full measured flow of the fluids through the heat exchanger. For system testing, the bypass should be left in the normal operating condition.

(R) **5.2** For component testing, the fan drive, if the unit is so equipped, should be fully engaged using the manufacturer's recommended procedure. For system testing, the drive should be left in the normal operating condition.

5.3 All shutters or other air directional control devices should be fixed for the test in the full open position.

(R) **5.4** All other heat-producing equipment that may adversely affect the air temperature to the oil cooler and fan should be operated during the test in a specified manner.

5.5 Instrumentation and data to be recorded includes the following:

5.5.1 Oil temperatures at designer-specified critical locations, for example, inlet to the cooler, reservoir, etc.

5.5.2 Oil temperature at oil cooler inlet (if not already specified in 5.5.1).

5.5.3 Oil temperature at oil cooler outlet (if not already specified in 5.5.1).

5.5.4 Average air temperature at oil cooler air inlet (multipoint grid normally required).

5.5.5 Average air temperature at oil cooler air outlet (multipoint grid normally required).

(R) 5.5.6 Oil flow (net through the cooler).

NOTE—Pressure drop across flow meter should be kept to a minimum. If extensive plumbing is required to incorporate flow meter, lines to and from meter should be insulated.

5.5.7 Barometric pressure at test site.

5.5.8 Test fluid shall be as specified.
5.5.9 Operating pressure at oil cooler inlet.
5.5.10 Operating pressure at oil cooler outlet.

(R) NOTE—Pressure measurement devices should be installed to eliminate any possible source of error due to turbulence at the point of measurement. Direct massflow measurements are preferred. For structural information, these devices should be capable of measuring millisecond pressure spikes.

5.5.11 Engine or motor operating speeds.
5.5.12 Ambient air temperature.
5.5.13 Actual fan speed and/or vehicle velocity.

5.6 Verify that the oil cooler is mounted in its designated location with proper inlet and outlet connections.

6. Procedure

6.1 Operate test unit in its specified and verified work cycle until practical stabilized thermal conditions have been achieved.

6.2 Collect data for a total of 10 complete work cycles, or for a time span of no less than 10 min.

7. Test Data Evaluation

7.1 Calculate oil cooler heat rejection from the test data.
7.1.1 Oil flow rate.
7.1.2 Oil cooler inlet temperature.
7.1.3 Oil cooler outlet temperature.
7.1.4 Obtain manufacturer's specific heat and density of oil to establish oil thermal characteristics at average oil temperature.

7.1.4.1 Power (kW) = specific heat of oil (kJ/kg · °C) × oil density (kg/L) × oil flow (L/s) × oil cooler inlet to outlet temperature differential (°C).

7.1.4.2 BTU/min = oil specific heat (BTU/lb/°F) × oil density (lb/gal) × oil flow (gal/min) × oil cooler inlet to outlet temperature differential (°F).

7.2 Determine oil stabilization temperature above ambient at the critical location.

7.2.1 Oil stabilization temperature above ambient = oil temperature measured at critical location minus ambient air temperature.

7.3 Compare oil stabilization temperature above ambient with the specification.

7.4 Analyze the test data. Unsatisfactory results could be due to one or more of the following.

7.4.1 Other than expected oil cooler heat load. (Is the oil system rejecting more or less heat than the cooler was designed for?)

(R) 7.4.2 Oil cooler heat rejection performance is not to the manufacturer's specifications.

7.4.3 Other than expected oil flow through the cooler. (Does the measured oil flow match the design value?)

7.4.4 Other than expected airflow through the oil cooler. (Does the measured air temperature difference across the oil cooler core indicate other than expected air flow?) Estimate oil cooler airflow by performing the following calculation:

7.4.4.1 Air flow—See Equation 1.

$$\frac{kg}{s} = \frac{\text{oil cooler power (kW) (from paragraph 7.1.4.1)}}{1.005 \times \text{oil cooler air } \Delta T \text{ (°C)}} \quad \text{(Eq.1)}$$

7.4.4.2 Air flow—See Equation 2.

$$\frac{lb}{min} = \frac{\text{oil cooler BTU/min (from paragraph 7.1.4.2)}}{0.240 \times \text{oil cooler air } \Delta T \text{ (°F)}} \quad \text{(Eq.2)}$$

(Is the oil cooler core too restrictive to airflow? Is air flowing around the cooler rather than through it?)

7.4.5 Poor airflow distribution across the oil cooler core. (Are upstream or downstream obstructions causing poor airflow through portions of the oil cooler core?) Detect by performing anemometer survey or similar technique.

7.4.6 Preheating of air into the oil cooler. (Is ambient air being inordinately heated before entering the oil cooler core? Is hot air discharge recirculating into the oil cooler air inlet?)

7.4.7 Other than expected oil pressure differential between oil cooler inlet and outlet. (Is oil cooler pressure drop compatible with the system design? Is excessive pressure causing pumps to create more heat, oil flow to dump over relief valve, etc.?)

7.4.8 Temperature gradient or difference between critical oil temperature and temperature of oil into the cooler. (Are improvements required in system design? Is more oil flow required in a part of the system? Is one part of the system overheating because the hottest oil is not being circulated directly to the oil cooler?)

7.4.9 Ambient temperature too low. If the test was run substantially below the specified ambient temperature, the oil stabilization temperature above ambient may be significantly different than if the test were run at a higher ambient temperature.

ENGINE COOLING FAN STRUCTURAL ANALYSIS—SAE J1390 APR82

SAE Recommended Practice

Report of the Engine Committee, approved April 1982.

"This SAE Recommended Practice is intended as a guide toward standard practice but may be subject to frequent change to keep pace with experience and technical advances, and this should be kept in mind when considering its use."

1. Purpose—The purpose of this recommended practice is to identify the general methodology for the structural analysis of engine cooling fans, and to provide expanded information on subset practices within the general methodology, such that a user of this practice can adapt specific subsets related to a vehicle[1] class.

2. Overall Scope—Three levels of fan structural analysis are included in this practice:

(a) Initial Structural Integrity
(b) In-vehicle Testing
(c) Durability Test Methods

The Initial Structural Integrity section describes analytical and test methods used to predict potential resonance and therefore possible fatigue accumulation.

The In-vehicle (or machine) section enumerates the general procedure used to conduct a fan strain gage test. Various considerations that may affect the outcome of strain gage data have been described for the user of this procedure to adapt/discard depending on the particular application.

The Durability Test Methods section describes the detailed test procedures that may be used depending on type of fan, equipment availability, and end objective.

Each of the above levels builds upon information derived from the previous level. Engineering judgment is required as to the applicability of each level to a different vehicle environment or a new fan design.

This recommended practice is applicable to heavy duty trucks, construction equipment, industrial and agricultural equipment. It does not necessarily include passenger cars and light trucks.

3. Initial Structural Integrity

3.1 Scope—It is necessary to identify and attempt to evaluate the characteristics of an application which can have an effect on fan durability. Failures almost always occur in fatigue, so careful attention should be paid to avoid resonance or forced vibration of the fan. This section considers vibrational inputs, fan natural frequencies, and operating speed as part of the initial structural integrity analysis. A fan application fact sheet (Table 1) is recommended as a form to communicate between user and fan supplier.

3.1.1 A resonant condition may occur when the natural frequencies of the fan as determined from paragraph 3.3 is coincident with either of the vibrational input frequencies or their harmonics, as calculated in paragraph 3.2. Therefore, it is desirable to separate these frequencies, including the upper and lower harmonics, whenever possible. In addition, torsional vibrations, auxiliary equipment, driveline vibrations, etc., may cause vibratory inputs to the fan which cannot be predicted in equation form.

3.2 Vibrational Inputs

3.2.1 Without verification by in-vehicle testing, the potential critical speeds and frequencies that can affect the fan can only be predicted. Equations 1 and 2 are used to calculate the piston firing and blade pass frequencies (in Hz). The blade pass frequency is associated with obstructions which cause a non-uniform or "pulsating" flow as the fan rotates. These equations are more applicable to fans whose blades are evenly spaced.

$$F_{pf} = \frac{(\text{No. of Cylinders})\ (\text{Engine rpm})}{(30)\ (\text{No. of Cycles})} \quad (1)$$

where number of cycles refers to a 2 or 4 stroke engine

$$F_{bp} = \frac{(\text{Fan rpm})\ (\text{No. of Blades})}{(60)} \quad (2)$$

3.3 Fan Natural Frequencies

3.3.1 The resonant frequencies and location of the node lines of a fan should be measured. This can be done by using a shaker table, accelerometers, strain gages, or by visual observation. The natural frequency of a fan can, and usually will, be affected by the following:

(a) The shape of the fan mounting surface, which should be duplicated as closely as possible;
(b) Torque of the mounting bolts;
(c) Level of excitation when using a shaker; and
(d) Dimensional variations from fan to fan, including material tolerances.

3.3.2 Most natural frequencies of a fan will change slightly as the rotational speed of the fan is varied. The following equation may be used to estimate the change in frequency with rotation (Hz):

$$F_{rotating} = \sqrt{F^2_{static} + \frac{rpm^2}{3600}} \quad (3)$$

3.3.3 Finite element analysis methods may be used as an alternative or adjunct to test methods. Mode shapes and frequency, and relative stress gradients can be numerically predicted and compared to vehicle vibrational inputs.

3.4 Operating Speed

3.4.1 It is desirable to keep the fan operating speed as low as possible to minimize stresses due to centrifugal force.

3.5 Log Sheet

3.5.1 The fan user and supplier may wish to establish a fan application fact sheet. This can be used to identify design parameters which affect fan performance for their particular application and to record the basic fan characteristics. It must be remembered, however, that one cannot anticipate all of the factors that will affect fan durability while early in the design stage. No itemized list will be all-inclusive, and no algorithm is possible. A sample fact sheet is provided as Table 1. Users should feel free to modify it as mutually desired.

4. In-Vehicle Test

4.1 Scope—In-vehicle tests can be used to complement the Initial Structural Integrity section of this practice because the forces on a fan are determined by a combination of the fan's operating environment and the engine/vehicle operational duty cycle conditions. A fan designed for a specific vehicle application and having satisfactory structural strength in its intended application, may have unsatisfactory durability in a different application. The following sections will aid in the selection of a suitable test vehicle, in the preparation and instrumentation of the vehicle for test, and in conducting the actual test.

4.2 Test Vehicle Selection—This subsection considers the kinds of standard and optional vehicle equipment which can impose forces, vibrations, and frequencies on the subject fan. The vehicle equipment environment influences the fan by three different methods. Equipment obstructions ahead of, behind, and around the fan disturb the air flow passing through the fan. This air flow disturbance imposes forces and vibrations on the fan. The engine and other equipment are sources of mechanical vibrations. These vibrations can propagate and be transmitted to the subject fan through its mounting attachment. The fan drive which determines the plane-of-rotation, the center-of-rotation, and the rotational speed directly induces inertial forces and vibrations in the subject fan. Consideration of these factors will permit the tester to specify the appropriate optional equipment for the test vehicle.

4.2.1 PACKAGE OBSTRUCTIONS—Equipment located in the near vicinity of the fan that restricts and obstructs the fan air flow.

4.2.1.1 *Radiator Core*—Consider the variation in air flow restriction due to optional radiator cores with a different number of tube rows, fin densities, and fin designs. Radiator tanks, side structures, and cross-reinforcing rods can be significant obstructions if the fan partially "sweeps" any of these items.

4.2.1.2 *Heat Exchanger Core*—Any finned core located ahead or behind the fan will restrict air flow. Examples are air conditioning freon condenser and coolers for engine lubricating oil, power steering fluid, automatic transmission fluid, and hydraulic PTO fluid.

4.2.1.3 *Shroud*—Optional fan shroud or hand guards.

4.2.1.4 *Radiator Shutters*—Mounting location (in front or behind radiator) and type (modulating or full open-full closed).

4.2.1.5 *Winter Front*—Optional grille mounted winter fronts.

4.2.1.6 *Blockage*—Frame member or body sheet metal obstructions. Also belt pulleys and crankshaft damper can present circular section blockages in close proximity.

---

[1] The term "vehicle" as used in this practice, is defined as an all-inclusive term. "Vehicle" includes the SAE J687 definition of vehicle and the SAE J1057, J1116, and J1234 definitions of machine.

26.401

TABLE 1—FAN APPLICATION FACT SHEET

PART ONE—FAN USER INFORMATION

Customer:
Location:
Contact:
Phone No.:

Performance Reqd.
_____ CFM at Speed _____ rpm
_____ in H₂O SP
Air Density: _____ Est. hp _____

Limitations
Max. Dia. _____ Overspd.: _____
Max. P.W. _____ Max. hp _____

Other
_____
_____

Reqd. Testing
___ Airflow   ___ Spin Pit
___ Shaker Scan   ___ Strain Gage

Application Description
Model No.
Engine:
Rated hp: _____ at rpm _____
Drive Type:
Fan Drive Ratio:
Fan Mtg. Loc.:
Fan Rotation:
Pilot and Bolt Circle:
Max. Runout:
Belt Tension:

System
Shroud: _____ Dia. _____ Type _____
Fan Position:
Clearanceª: _____ Fr _____ Rr _____

___ Endurance ("Bench Test")
___ Other:

PART TWO—FAN SUPPLIER INFORMATION

Manufacturer:
Location:
Contact:
Phone No.:

Proposal
Mfr. P/N
Drawing Attached:
Fan Wt.: _____ Inertia _____
Max. Fan Unbalance
Other

Test Results Attached
___ Airflow Performance
___ Shaker Scan
___ Spin Pit Burst Speed (Act. or Calc.)
___ Strain Gage Data
___ Material Properties
___ Endurance ("Bench Test")
___ Other:

ª Clearance from fan mounting face to nearest obstructions. The proximity of the blades to obstructions should be discussed with the fan vendor.

4.2.1.7 *PTO Shafts*—Front crankshaft driven power take-off shaft.

4.2.1.8 *Deflector*—Optional fan blast deflector.

4.2.2 VIBRATION SOURCES—Equipment that generates mechanical vibrations. Because these equipment items are commonly located in the immediate vicinity of the fan, they also cause fan air flow obstruction effects.

4.2.2.1 *Engine*—Design parameters and/or data permit the tester to consider engine vibration magnitudes and fundamental forcing frequencies.

4.2.2.1.1 Type—Gasoline or diesel, two-stroke cycle or four-stroke cycle, number of cylinders, in-line or Vee, turbocharged or naturally aspirated, uniform or uneven cylinder firing interval.

4.2.2.1.2 Size—Displacement, bore, stroke, and horsepower rating.

4.2.2.1.3 Crankshaft Damper—Determine if a damper is provided, and location of fan drive pulley with respect to damper (is fan driven from crank or damper inertia member?).

4.2.2.1.4 Speed—Governed or rated engine speed; fast idle or subidle.

4.2.2.1.5 Vibration Data—Review any available vibration test data for magnitudes and frequency. Crankshaft torsional data is important.

4.2.2.2 *Accessories*—Power steering pump, air brake compressor, freon compressor, alternator, emissions air pump, vacuum brake pump, and hydraulic brake booster pump.

4.2.2.3 *Driveline*—Optional sizes, lengths, and types (one-piece or two-piece).

4.2.3 FAN DRIVE—The drive ratio and clutch, if used, control fan rotational speed. Additionally, the drive acts as the path to transmit mechanical vibration to the fan.

4.2.3.1 *Type*—Gear or belt, fixed or clutch (on-off, modulated engagement or viscous).

4.2.3.2 *Ratio*—Fan speed to engine speed ratio.

4.2.3.3 *Position*—Fan mounting position on end of crankshaft, on water pump shaft, or remote mounted fan spindle.

4.2.3.4 *Fan Mounting*—Fan mounting surface or spacer configuration.

4.3 **Instrumentation**—This section considers the preparation and instrumentation of the subject fan and the selected test vehicle.

4.3.1 STRAIN GAGE LOCATION—It is important that strain gages are placed at all of the highly stressed locations on the subject fan. Several methods exist for gage location. The best procedure is to employ multiple methods of gage location and to use a generous number of gages. This will reduce the risk that a highly stressed location has been overlooked. The fully strained gaged fan becomes in effect a "master fan." It should be handled carefully during the in-vehicle test and should be safely stored after test completion. Retention of the "master fan" will allow for its use in future in-vehicle tests or to correlate bench durability tests.

4.3.1.1 *Brittle Lacquer*—Strain gage locations are determined by crack patterns developed in the brittle lacquer coating. The brittle coating can be patterned in the test vehicle and/or on a bench test stand. The bench procedure can use rotational and/or axial excitation.

4.3.1.2 *Judgement*—Gage locations determined by past experience with similar fan designs or by analytical analysis of the subject fan design. Contact fan supplier for gage location information.

4.3.1.3 *Modal Analysis*—Gage locations are along the node lines of the principal modes that are expected to receive the largest amounts of vibrational energy.

4.3.1.4 *Finite Element*—Gage locations are determined from the computer analysis of a finite element model of the subject fan.

4.3.1.5 *Photoelastic*—Gage locations are determined from the bi-refringent pattern in a photoelastic coating on the subject fan.

4.3.1.6 *Failures*—Gage locations determined by the fatigue crack pattern in a failed fan.

4.3.2 RECORDED DATA—Suitable instrumentation is required to record engine speed, fan speed when a viscous drive is used, and strain gage output magnitude and frequency. For a viscous drive, sensor air temperature and clutch housing temperature may be recorded. For a plastic fan, the adjacent air temperature and the plastic material temperatures are required.

4.3.3 VEHICLES' MEASUREMENTS—Specific measurements may be made on the selected test vehicle so that the following factors will be known.

4.3.3.1 *Driveline*—Record out-of-balance and verify correct alignment and universal joint phasing.

4.3.3.2 *V-Belts*—Adjust belt tension on all accessory drives including the fan drive to the specified maximum. New V-belts require "break in." Tension new belts 50% higher than normal maximum specification and run for ½ h. Readjust tension to specified maximum.

4.3.3.3 *Fan Mounting*—Record the axial runout of the fan drive mounting surface and the radial runout of the fan drive pilot.

4.3.3.4 *Fan to Shroud*—Record radial tip clearances. Record axial depth of fan penetration into shroud.

4.4 **Vehicle Test**—The intent of the in-vehicle fan test is to record fan strains while vehicle operational duty cycle conditions and the fan operating environment are varied in a systematic way. This will determine the combination of factors that cause the highest fan strains.

4.4.1 VIBRATION SOURCES—Determine the effect on fan strains for different vibration input conditions.

4.4.1.1 *Engine Load*—Test with engine at full load, partial load, and no load conditions.

4.4.1.2 *Engine Speed*—Test condition of maximum speed (governed speed or rated speed and a specific overspeed) and speed transients due to automatic or manual transmission shifting. Test engine start-up, shutdown, idle, and specific below idle speed conditions. Scan the engine operating speed range in a continuous sweep mode in each transmission gear position to identify fan resonances and maximum fan strain amplitudes.

4.4.1.3 *Accessories*—Test the effect of the various accessories on-off duty cycle.

4.4.1.4 *Fan Clutch*—Determine the effect of the fan clutch operating modes. For non-viscous clutches—lockup operation and on-off transients. For viscous clutches—disengaged, fully engaged, and mechanically locked operation.

4.4.1.5 *Fan Drive*—Test with fan mounting components (water pump, spacer, fan drive) that produce the maximum allowable radial and axial runout.

4.4.1.6 *Driveline*—Test with a known added amount of out-of-balance or out-of-phase universal joints.

4.4.1.7 *V-Belts*—Test with both a "loose" belt tension condition and with an over-tightened belt tension condition. This is applicable to the fan drive belts and to all accessories.

4.4.1.8 *Fan*—Test with the maximum allowable out-of-balance.

4.4.2 OBSTRUCTIONS—Determine the effect on fan strains caused by obstruction changes.

4.4.2.1 *Accessories*—Selectively remove individual accessories to determine their obstruction effect on fan strains.

4.4.2.2 *Radiator Core*—Determine effects of change in air flow restriction caused by an optional core. Simulate the restriction increase due to core clogging in service.

4.4.2.3 *Radiator Shutters*—Effects of shutters open, closed, and partially open if the modulating shutter type is used.

4.4.2.4 *Winter Front*—Test the effect on fan strains caused by the use of winter fronts or cardboard.

4.4.2.5 *Shroud*—Test for the effects of an optional fan shroud.

4.4.2.6 *Heat Exchanger Core*—Test for the obstruction effect on fan strains by selectively removing optional heat exchanger cores.

4.4.3 IN-SERVICE EVALUATION—Simulate special conditions occurring during typical vehicle operation.

4.4.3.1 *Shock*—Effects on road, cargo loading, or auxiliary equipment caused shock inputs.

4.4.3.2 *Water Splash*

4.4.3.3 *Dirt build-up on fan blades.*

4.5 Analysis—The strain data gathered during in-vehicle testing must be analyzed to be useful to the engineer. SAE J1099, "Technical Report on Fatigue Properties" provides information that is useful in the analysis of service load and/or strain data. It also contains a list of references that have proven useful in the analysis of fatigue data.

5. Durability Test Methods

5.1 Scope—Durability tests can be used to complement in-vehicle strain gage testing. Testing can also evaluate areas unaccessible to conventional strain gage measurements. Durability testing, with the ability to control and increase strain levels, can demonstrate design factors at loads greater than anticipated service. Bench testing allows evaluation of the fan at a greatly accelerated rate compared to most in-vehicle field testing.

5.2 Methods—A variety of bench tests are used to evaluate fans. It is difficult for any one test to accurately reproduce every operating characteristic. The user must select the method(s) that satisfy his/her durability criteria. Since it is accepted that fan durability considerations are associated with fatigue, the methods outlined are generally designed to evaluate endurance characteristics.

5.2.1 NON-ROTATIONAL RESONANT TEST—The test fan is mounted at the pilot and bolt circle and loaded axially through the range of amplitude and frequencies of interest. This method may not accurately reproduce all modal conditions and will not include strains produced by centrifugal loading.

5.2.1.1 *Equipment (Shaker Table)*—The vibrator should be capable of accepting various fan mounting arrangements with normal input loading from 10–100 lb and a frequency range of 25–500 Hz.

5.2.1.2 *Instrumentation*—A method of monitoring frequency and strain amplitudes that will accurately relate the test conditions to the in-vehicle conditions is required. Additionally, the use of a strobe light and frequency analysis may aid in the test evaluation.

5.2.1.3 *Procedure*—The fan must be mounted using an actual attachment component from the vehicle (that is, hub, spacer, clutch, etc.). This assembly is mounted to the driving member of the loading source. If the test assembly must be suspended, use caution to minimize the frequency effects on the test results.

5.2.1.4 *Durability Test*—Mount the "master fan" used in the vehicle analysis and set the test stand to simulate the frequency mode of vibrating and the highest dynamic strain level observed during the vehicle test. Because the minor variations of physical conditions of nominally identical parts can cause small changes in resonant frequencies, each test fan should be gaged to insure operation at the required test condition. Replace the "master fan" with the test fan and adjust frequency if necessary to maintain peak resonance. Other than frequency, all test conditions must remain as set up with the master fan. Monitor equipment to insure that conditions remain stable during testing. Strain levels higher than anticipated can also be tested.

5.2.2 RAPID CYCLE TEST—This test rotates the fan from zero or a low speed to some predetermined maximum speed and back again. The test has limited applications, but can be used for flex fans and to simulate some unusual in-vehicle situations primarily encountered with viscous drives and other add-on clutches.

5.2.2.1 *Equipment*—Test stand can be driven with an electric motor and should be capable of adjusting the acceleration, maximum speed, deceleration, and dwell time. Test stand power requirements may be substantially higher than the in-vehicle rating of the fan.

5.2.2.2 *Instrumentation*—A method of monitoring frequency and strain amplitudes that will accurately relate the test conditions to the in-vehicle conditions is required. The use of a strobe light and frequency analyzer may aid in the test evaluation. Additionally, methods of measuring speed and acceleration/deceleration will be required.

5.2.2.3 *Procedure*—The fan must be mounted using an actual attachment component from the vehicle (that is, hub, spacer, clutch, etc.). This assembly is mounted to the drive shaft. Use caution to minimize the frequency effects of the slip ring or adapters on the test results.

5.2.2.4 *Durability Test*—Mount the "master fan" that was used in the vehicle analysis and adjust the test stand to duplicate both frequencies and highest strain levels observed during the vehicle test. Replace the "master fan" with a test fan before testing. Monitor acceleration, speeds, and deceleration to insure that conditions remain stable during the test.

5.2.3 TORSIONALLY LOADED ROTATING TEST—This test rotates the fan at some constant operating speed while applying a torsional vibration to the drive shaft. The test can accurately duplicate centrifugal loads while accumulating stress cycles very rapidly. Durability of the test equipment can be a problem.

5.2.3.1 *Equipment*—The test stand can be driven with an electric motor. The fan shaft may be driven through "U" joints or off-center pulleys or other methods to produce the torsional loads. Hydraulically driven, torsional actuators are also available with load and frequency capabilities suitable for fan testing. Speed and power capabilities should reproduce all possible vehicle conditions. It should be noted the "U" joint excited tests may be limited to lower fan speeds because of joint durability problems.

5.2.3.2 *Instrumentation*—A method of measuring speed, frequency, and strain amplitude that will accurately relate the test conditions to the in-vehicle conditions is required.

5.2.3.3 *Procedure*—The fan must be mounted using an actual attachment component from the vehicle (that is, hub, spacer, clutch, etc.). This assembly is mounted to the drive shaft. Use caution to minimize the frequency effects of the slip ring or adapters on the test results.

5.2.3.4 *Durability Test*—Mount the "master fan" used in the vehicle analysis and set the test stand to simulate the frequency and speed at the highest strain levels observed during the in-vehicle test. Because the minor variations of physical conditions of nominally identical parts can cause small changes in resonant frequencies, each test fan should be gaged to insure operation at the required test conditions. Normally one to four gages are sufficient. Replace the "master fan" with a test fan and adjust the speed, if necessary, to maintain peak resonance. All other conditions must remain as established with the "master fan." Monitor equipment to maintain stability during testing. Strain levels higher than anticipated service can also be tested.

5.2.4 OBSTRUCTION LOADED ROTATING TEST—This test rotates the fan at some constant operating speed while applying air loaded vibrations by placing obstructions in front and/or behind the fan. The test can accurately duplicate centrifugal loads while accumulating stress cycles very rapidly.

5.2.4.1 *Equipment*—The test stand can be driven with an electric motor. Radiators, shutters, shrouds, and other vehicle components can be used as obstructions. Additional obstructions extended radially can be evenly or unevenly spaced and used to further increase strain levels. Speed and power capabilities should reproduce all possible vehicle conditions.

5.2.4.2 *Instrumentation*—A method of measuring speed, frequency, and strain amplitude that will accurately relate the test conditions to the in-vehicle conditions is required.

5.2.4.3 *Procedure*—The fan must be mounted using an actual attachment component from the vehicle (that is, hub, spacer, clutch, etc.). This assembly is mounted to the drive shaft. Use caution to minimize the frequency effects of the slip ring or adapters on the test results.

5.2.4.4 *Durability Test*—Mount the "master fan" that was used in the vehicle analysis and adjust the test stand to simulate both frequencies and highest strain levels observed during the vehicle test. Replace the "master fan" with a test fan before testing. Strain levels higher than anticipated can also be tested.

5.2.5 ROTATIONAL RESONANT TEST—In this test, the fan is rotated at some constant operating speed while applying an axial vibration. The test can reproduce a large range of centrifugal and vibratory strains simultaneously in a near infinitely variable manner.

5.2.5.1 *Equipment*—The test stand can be driven with an electric motor. Speed and power capabilities should reproduce all possible vehicle conditions. The vibrator should be capable of a loading in excess of 1000 lb and a frequency range of 25–500 Hz.

5.2.5.2 *Instrumentation*—A method of measuring speed, frequency, and strain amplitude that will accurately relate the test conditions to the in-vehicle conditions is required.

5.2.5.3 *Procedure*—The fan must be mounted using an actual attachment component from the vehicle (that is, hub, spacer, clutch, etc.). This assembly is mounted to the drive shaft.

5.2.5.4 *Durability Test*—Mount the "master fan" used in the vehicle analysis and set the test stand to duplicate the frequency and speed at the highest strain levels observed during the in-vehicle test. Because the minor variations of physical conditions of nominally identical parts can cause small changes in resonant frequencies, each test fan should be gaged to insure operation at the required test conditions. Normally one to four gages are sufficient. Replace the "master fan" with a test fan and adjust the frequency, if necessary, to maintain peak resonance. All other conditions must remain as established with the "master fan." Monitor equipment to maintain stability during testing.

5.3 Durability Test Acceptance Criteria—Since the purpose of bench testing is to evaluate a fan relative to in-vehicle operating conditions, a variety of criteria can be used.

5.3.1 TEST COMPLETION—A test is defined as complete when the fan either reaches the cycle life specified or when any crack is detected. It may be of interest to continue testing beyond detection of a crack or the specified cycle life.

5.3.2 INFINITE LIFE—Durability testing can be used to demonstrate infinite life. Testing can be run at the in-vehicle conditions for a number of cycles that is judged to represent infinite life. Normally accepted values for steel are 10^7 cycles. Caution must be used in evaluating materials other than steel where the fatigue characteristics are not as well defined. Design factors at loads greater than anticipated service can also be tested. Because interpretation of the test data depends on known material characteristics, it is strongly recommended that the actual mechanical and chemical properties of the material in the test fans be available. This test can only demonstrate infinite life at the values tested and does not provide definition of actual design factors. Fan designs can be optimized by using the data available from destructive testing.

5.3.3 FINITE LIFE—The purpose of this test is to establish an actual strength-life diagram of the fan. It can also verify gage locations by determining the actual point of crack initiation. Any of the test methods can be used to develop the necessary data. Generally, the first test point is at a strain level that will easily generate a fatigue crack. Subsequent tests are run at lower strain levels until enough data (usually two or four points) are available to establish a strength-life diagram.

5.4 Other Considerations

5.4.1 NUMBER OF TEST SAMPLES—Because of the nature of fatigue data, enough samples must be run to generate confidence in the test results.

5.4.2 NUMBER OF TEST POINTS—The number of test points will be determined by the resonant conditions found in the vehicle.

5.4.3 GAGE COORDINATION—Some test methods may not duplicate strain levels observed at all gage locations during the in-vehicle evaluation. Some experience with each of the test methods will aid in properly assessing the validity of the results.

5.4.4 TEMPERATURE—Temperature considerations are critical when testing non-metallic fans. Since large changes in mechanical properties can occur, it is often necessary to conduct durability tests at the extremes of the in-vehicle conditions. Temperature effects can be studied with all of the test methods.

5.4.5 TEST FAN FABRICATION—The prototyping method can have an effect on the mechanical properties of the part. If the production process and/or material is different than the method used to fabricate the prototype fan, production parts should be rechecked for durability characteristics.

HEAVY DUTY NON-METALLIC ENGINE COOLING FANS—MATERIAL, MANUFACTURING AND TEST CONSIDERATIONS—SAE J1474 JAN85

SAE Information Report

Report of the Engine Committee, approved January 1985.

This SAE Information Report is intended to serve as a supplement to SAE J1390, Engine Cooling Fan Structural Analysis Recommended Practice.

1. Foreword—Nonmetallic fans can be constructed from a variety of thermoplastic and thermosetting resins, with or without any of the following: reinforcing fibers, fillers, stabilizers, modifiers, and pigments. Among the engine cooling applications commonly described as heavy duty (trucks, buses, construction equipment, industrial equipment, and agricultural equipment), the most widely used combination has been injection molded, glass fiber reinforced nylon. Due to limited experience with other processes and/or materials, this report will address only that combination, hereinafter referred to simply as "non-metallic" or "nylon" fans.

The advent of non-metallic construction has introduced fan manufacturers and fan users to areas of evaluation not required by metal designs. Chief among these are temperature extremes, moisture content, impact resistance, chemical attack, material purity/homogeneity, and aging/weathering. Areas of evaluation affecting both metallic and nonmetallic fans, but requiring somewhat different approaches with nonmetallic parts, include manufacturing quality assurance, dimensional consistency, assembly integrity, natural frequency determination, and durability testing.

2. Purpose—This report exists to identify general methodology which addresses the areas of evaluation listed above for injection molded nylon fans. It is envisioned that those working with other processes and/or materials can use J1390 and this information report as starting points in the development of structural analysis methodology pertinent to their particular combination(s).

3. Overall Scope—The following topics are included in this report:
Section 5—Material Selection
Section 6—Production Considerations
Section 7—Initial Structural Integrity
Section 8—In-Vehicle Testing
Section 9—Laboratory Testing

The Material Selection section lists environmental factors and material properties which should be considered when determining appropriate fan material(s) for a given application.

The Production Considerations section covers various aspects of machine selection, mold design, and process control.

The Initial Structural Integrity section lists factors which should be considered in addition to those covered by Section 3 of J1390.

The In-Vehicle Testing section lists factors which should be considered in addition to those covered by Section 4 of J1390.

The Laboratory Testing section addresses some test considerations and methods for non-metallic fans which differ from those used with metallic fans or which were not included in Section 5 of J1390.

4. Definition of Terms—The following terms relating to injection molding of plastics are used:

Barrel, Liner, and Screw—Components of the plasticizing cylinder of an injection molding machine in which the material is melted and moved forward to the injection nozzle.

Weld or Knit-Line—The area of a molded plastic part, formed by the union of two or more streams of plastic flowing together.

Cold Slug Flow—Condition where insufficient heating of plasticizing cylinder results in unmelted material pellets appearing in the molded part.

Flash—The excess plastic material that is forced from a mold cavity during the molding operation. Flash may also occur between worn mold sections.

Gate—An orifice or opening through which the melted plastic material enters the cavity.

Surface Drag—Skidding of plastic resin along surface of mold due to improper mold temperatures, injection pressure, or injection speed.

Stress Whitening—An affect noted in nylon (crystaline materials) under stress loading occuring as a result of molecular orientation visible as a white area due to the change in the refraction index of the material.

5. Material Selection

5.1 Scope—All materials classified as nylons share certain basic characteristics. However, even within the seemingly limited realm of glass fiber reinforced nylons, many levels and combinations of physical, thermal, and environmental resistance properties are available. These properties result from such factors as the following:

1. Type of nylon (6, 6/6, 6/10, 6/12, etc.)
2. Percentage (by weight) of glass fibers
3. Diameter and length of glass fibers
4. Wetting agent used to promote adhesion between resin and fibers (if present)
5. Heat stabilizer (if present)
6. Impact modifier (if present)
7. Pigmentation (if present)

In order to select a material with appropriate characteristics, the prospective manufacturer and end user should investigate the proposed application thoroughly, and maintain open communication with the various material suppliers.

5.2 Application Factors—Evaluate the following characteristics, as applicable, for all materials used in the fan assembly:

5.2.1 Temperature extremes–highest and lowest
5.2.2 Relative humidity (relative to material moisture content)
5.2.3 PHYSICAL PROPERTIES
    5.2.3.1 Tensile strength
    5.2.3.2 Flexural strength
    5.2.3.3 Flexural modulus
    5.2.3.4 Impact resistance
    5.2.3.5 Elongation
    5.2.3.6 Any other depending on application
5.2.4 THERMAL PROPERTIES
    5.2.4.1 Heat distortion temperature
    5.2.4.2 Expansion characteristic
5.2.5 CHEMICAL RESISTANCE
    5.2.5.1 Petroleum products
    5.2.5.2 Coolant
    5.2.5.3 Other vehicular fluids
    5.2.5.4 Cleaning solvents
    5.2.5.5 Paint(s)
    5.2.5.6 Molded-in component(s)
    5.2.5.7 Salt spray
    5.2.5.8 Any others depending on application
5.2.6 AGING/WEATHERING
    5.2.6.1 Exposure to sunlight
    5.2.6.2 Exposure to ozone
    5.2.6.3 Heat cycling
    5.2.6.4 Abrasion
    5.2.6.5 Storage and shipping

5.3 Manufacturing Factors—Evaluate the following characteristics of the material and process:

5.3.1 Dimensional stability
5.3.2 Melt temperature
5.3.3 Abrasiveness (tool wear)
5.3.4 Sensitivity to moisture content
5.3.5 Molded-in stresses
5.3.6 Any others of concern to manufacturer

6. Production Considerations

6.1 Scope—The fan mold and injection molding machine used must be properly designed or selected to manufacture a consistent product. Similarly, the manufacturing process must be capable of yielding a consistent product. Unless this consistency can be assured, further testing would be fruitless.

6.2 Machine Selection—The type of material to be processed will determine certain machine characteristics which are critical to consistent production. These include:

6.2.1 MECHANICAL OPERATION OF THE MACHINE
    6.2.1.1 Proper barrel, liner, and screw for material type
    6.2.1.2 Temperature control
    6.2.1.3 Injection pressure, stroke, rate, and clamping force
    6.2.1.4 Shot capacity versus shot size
6.2.2 SUPPORT EQUIPMENT
    6.2.2.1 Material driers
    6.2.2.2 Chillers or heaters to regulate mold temperature

6.3 Mold Design—Part design and material selection will determine mold design. Factors considered include:

6.3.1 Physical stability of the mold is essential.
    6.3.1.1 No deflection in the mold base during operation
    6.3.1.2 No shifting inserts during molding
    6.3.1.3 Proper ejector pin operation and location
    6.3.1.4 General fit and proper shut off in all areas
    6.3.1.5 Proper construction and hardening to reduce wear
6.3.2 Gate location can be important due to several factors:
    6.3.2.1 Material flow and fiber orientation
    6.3.2.2 Knit lines
    6.3.2.3 Cold slug flow
    6.3.2.4 Type of gate and size for material and reinforcement
    6.3.2.5 Aesthetics
6.3.3 Proper venting
6.3.4 Material shrinkage characteristics
6.3.5 Even temperature and control in all areas of the mold
6.3.6 Cavity pressure and temperature sensors can be installed for process control.

6.4 Process Control—Every step of the manufacturing process should be closely monitored.

6.4.1 MATERIAL CONTROL
    6.4.1.1 Inspect containers for shipping damage and possible contamination.
    6.4.1.2 Obtain material lot certification.
    6.4.1.3 Infrared spectrophotometry can supply a quick check of material purity.
    6.4.1.4 Consider molding test bars for evaluation of physical properties.
6.4.2 Establish a policy regarding the use of reground material.
6.4.3 Assure proper drying of materials.
6.4.4 Assure proper purging of machines to eliminate contamination of materials in the machine.
6.4.5 Assure proper screw/liner clearance to minimize destruction of reinforcing material (glass fibers).
6.4.6 Assure proper barrel and nozzle temperature,
    6.4.6.1 For proper cavity fill.
    6.4.6.2 To avoid material degradation.
6.4.7 Monitor material residence time at high temperatures to avoid degradation.
6.4.8 Monitor material flow indications on the molded part.
    6.4.8.1 Note slug flow.
    6.4.8.2 Check for surface drag.
    6.4.8.3 Check for glass fiber dispersion and orientation.
    6.4.8.4 Check for burned or scorched material.
    6.4.8.5 Check for flash on edges, ejector pins, or molded in holes.
    6.4.8.6 Check knit or weld lines.
    6.4.8.7 Check match to molded-in components.
6.4.9 Assure proper gate trim.
6.4.10 Monitor any secondary part trimming.
6.4.11 Assure proper date coding.
6.4.12 Assure molded-in components are properly prepared.
    6.4.12.1 Precise location in the mold
    6.4.12.2 Fit
    6.4.12.3 Surface preparation
    6.4.12.4 Temperature compatability
    6.4.12.5 No other detrimental effects on the molded part
6.4.13 Monitor part weight and maintain within acceptable variation. Part weight is a good indicator of process stability.
6.4.14 Monitor part balance and maintain within drawing tolerance. Balance is a good indicator of process stability.
6.4.15 Penetrating dyes, x-rays, sectioning, etc. can be used to check for voids or fit of molded-in components.
6.4.16 Dimensional checks should be performed at the start of each production run, and thereafter on a sampling basis in accordance with customer requirements.
6.4.17 Statistical process control techniques are encouraged wherever applicable.

7. Initial Integrity

7.1 Same as J1390 with the following additions:

7.1.1 ADDED TO PARAGRAPH 3.3.1—Natural frequencies of a non-metallic fan can also be affected by variations in the following:

(e) Ambient temperature;
(f) Moisture content;
(g) Orientation of reinforcing fibers (if present); and
(h) Material homogeneity due to molding process variations.

8. In-Vehicle Testing

8.1 Same as J1390 with the following additions:

8.1.1 ADDED TO PARAGRAPH 4.3.2—Relative humidity and temperature of air at the fan inlet and fan material temperature should be recorded for nonmetallic fans.

8.1.2 ADD PARAGRAPH 4.4.3.4—Varying degrees of moisture content of nonmetallic test fan.

8.1.3 ADD PARAGRAPH 4.4.3.5—Varying operating temperature, which affects natural frequencies and mechanical properties of nonmetallic materials.

8.1.4 ADDED TO PARAGRAPH 4.5—Particular attention should be paid to the fact that properties of nonmetallic materials vary significantly with temperature, humidity, environmental factors, and manufacturing conditions (Reference Section 5). Any analysis should consider the effect of these variations on both measured strain levels and product suitability.

8.2 Strain gage life is an important consideration when materials exhibit strain levels much higher than those for which the gage is designed to measure. Typically, a gage can give a good indication of strain levels several times its rated high cycle life, but only for a limited number of cycles, and then typically with a zero shift and a gage factor shift. It is important to understand these characteristics of the gage in use.

9. Laboratory Testing

9.1 Scope—Laboratory testing can serve several purposes in the validation of a fan for an application. Historically, it has been used to "prove" that a fan will survive for some acceptable length of time under the worst conditions recorded while testing in the application. Laboratory tests can be configured to subject the fan to loads above anticipated use.

Conditions which might be impossible to obtain during testing in the application can be simulated somewhat more easily in laboratory tests. Finally, laboratory testing usually allows evaluation of fan characteristics in a shorter time than is possible with field testing.

9.2 General Test Concerns

9.2.1 Conditioning of nonmetallic fans prior to a test can be as important as the test procedure itself. In the Dry as Molded (D.A.M.) state, some properties are at their maximum value while others are at their minimum. At the other extreme, when saturated with moisture, these physical property values can change by large percentages. In actual fan applications, neither of these conditions is likely to exist.

Depending on the particulars of a given operating environment (ambient temperature, average relative humidity, duty cycle, etc.), a nonmetallic fan will tend to stabilize somewhere between these extremes. It should be recognized, when designing a laboratory test, that conditions which subject the test fan to very high or very low relative humidity (hence moisture content) add severity to the test.

9.2.2 Temperature also plays an important role, since important physical properties of the material vary greatly with test ambient. Percent changes in property values can be several times as large as those due to moisture content changes. Test severity is sensitive to small changes in test temperature, particularly at elevated temperatures.

9.2.3 Test length must also be considered, since the deformation characteristics of plastics include elastic, viscoelastic, and viscous components, the latter two of which are time dependent.

9.2.4 Strain rate dependence of the material in question and how the strain is induced to the part and its magnitude with respect to material fatigue characteristics are very important to understand prior to undertaking any accelerated testing. Accelerated testing may give erroneous results by understating material life.

9.2.5 Combination of the factors outlined above must be considered.

9.2.6 Sequencing of tests may be considered.

9.3 Humidity Soak—Dynamic testing of nonmetallic fans should not be conducted with D.A.M. parts. Conditioning in accordance with (or equivalent to) ASTM D618A is suggested as an absolute minimum. Longer exposure to ambient humidity, or conditioning in water (such as ASTM D618D), is preferred. Prior to using an alternate moisture conditioning routine, it is necessary to understand the potential hydrolysis effects on the material in question. Saturation moisture content, for example, can be obtained by boiling in water until part weight stabilizes. The rate of moisture absorption is temperature dependent, with hot water being absorbed more quickly than cool water. Similarly, hydrolysis effects (if present) vary with conditioning temperature and rate.

9.4 Temperature Soak—Constant exposure to a fixed temperature can be a test itself, or can serve as the initial segment of a larger test. Unless a fan design has unusually thick sections, continuous exposure to a temperature for four hours should be sufficient for equilibrium to be reached.

Testing can involve more than one temperature in a cycling arrangement. Such a test could simulate a duty cycle in an extreme environment. It could also be used to test for loosening of a molded-in plate or mounting bolt "bosses" due to expansion and contraction of the plastic.

9.5 Impact Resistance—An impact test can simulate circumstances where a rock (thrown by tire) or tool (dropped by mechanic) strikes the fan. Such a test generally consists of a uniform missile dropped from a specified height onto a representative section of the fan. The fan would, of course, be stationary, and bolted to a rigid mount. A temperature soak would commonly precede the actual impact(s). Acceptance criteria would be no evidence of cracking, stress whiting, or permanent deformation.

9.6 Spin Burst—Burst testing can serve as a general indicator of the quality of design and/or construction of a molded fan. It can be done in air, but is generally done in vacuum due to equipment power limitations. As suggested previously, burst testing can be done in sequence with other tests which may have caused undetectable damage to a test fan. Procedures identical to those used with assembled metal fans can be followed.

9.7 Cyclic Endurance—"Durability" testing of nonmetallic fans can be accomplished in at least two formats: long-term and accelerated. Both require essentially the same test equipment, but differ in procedure.

9.7.1 Test equipment would consist of a "closed loop" air path with means for heating and cooling the contained air, variable speed drive system for the test fan, and a temperature monitor and control system. Facility to maintain moisture level in the test part may be required.

9.7.2 Instrumentation as used for in-application strain gage tests is required to "set up" the test, using a strain gaged "master" fan.

9.7.3 Procedure for a long-term test would be very similar to that used for assembled metal fans. In most cases, the goal of the test setup would be to reproduce the "worse case" in-application strain condition (magnitude, direction, frequency). This should be accomplished while reasonably duplicating the in-application test conditions of speed, temperature, and humidity. Given the lack of engine torsionals and the natural tendency of nylon material to damp vibration, this may not be possible with the commonly used aerodynamic obstructions. Use of overspeed and overtemperature may be required. If an overload condition is desired, speed and/or temperature can be increased, or obstructions used, to raise strain level in the master fan until a predetermined number is reached. The master fan would then be removed and test fan(s) mounted.

9.7.4 Procedure for an accelerated test is similar. Overspeed and/or overtemperature conditions can be chosen based on particulars of the application. The test fan(s) commonly would be soaked at the selected temperature, then subjected to a number of speed cycles from rest to the overspeed RPM.

9.7.5 Acceptance criteria would generally require no structural failure, and variation of selected "critical" dimensions within a predetermined tolerance. "Cosmetic" damage should be differentiated from structural damage.

OIL COOLER NOMENCLATURE AND GLOSSARY—SAE J1244 AUG88

SAE Recommended Practice

Report of the Engine Committee, approved March 1979, revised June 1981, and reaffirmed August 1988.

1. Scope—This report presents general nomenclature and glossary of terms for oil coolers.

2. Type of Applications—Oil coolers covered in this report are used for cooling of hydrostatics, bearings, differentials, transmissions, engines, power steering, aircraft, and stationary systems.

3. Types of Unit—The basic types are oil to air and oil to water or other liquids. See Nomenclature for examples.

4. Glossary of Oil Cooler Terms

4.1 Baffle—A partition which directs flow of fluids across the core. See Fig. 2 in Nomenclature.

4.2 Baffle Cut—Expresses percentage of area removed from baffle. See Fig. 2 in Nomenclature.

4.3 Baffle Spacing—Distance between adjacent baffles.

4.4 Bonnet—Collector or manifold on end of shell and tube heat exchanger which directs flow of tube-side fluid.

4.5 Core—That section of an oil cooler assembly which is comprised of the heat transfer surfaces.

4.6 End Zone—Space between first or last baffle and adjacent tube sheet (or header) of a shell and tube oil cooler.

4.7 Face Area—Area defined by the core width times core height. (Oil to air coolers.)

4.8 Face Velocity—The velocity of air approaching the core.

NOTE: Volume per unit time divided by face area.

4.9 Fin—Extended Heat Transfer Surface—See Figs. 1, 4, and 6 in Nomenclature.

NOTE: Shell and tube oil coolers may have fins or other extended surface.

4.10 Fixed Bundle Oil Cooler—A shell and tube heat exchanger with the tube bundle permanently installed in the shell. See Fig. 3 in Nomenclature.

4.11 Fouling Factor—See Fouling Resistance.

4.12 Fouling Resistance—The resistance to heat transfer resulting from accumulation of foreign material on the heat transfer surfaces of an oil cooler.

4.13 Header—This term has a dual meaning. It is sometimes used synonymously with tube sheet or tank. See Figs. 2, 3, and 6 in Nomenclature.

4.14 Heat Dissipation—The quality of heat, usually expressed in British Thermal Units per minute (kilowatts), that an oil cooler can dissipate under specified conditions.

4.15 Inlet Temperature Differential—The difference in temperature between the fluid being cooled and the cooling medium at the point each enters the heat exchanger.

4.16 Manifold—See Tank or Bonnet. Refer to Figs. 1, 6, 7, and 8 in Nomenclature.

4.17 Multi-Pass Oil Cooler—An oil cooler that is so circuited that either fluid passes across or through the core more than once.

4.18 Oil Cooler Pressure Relief Valve—A pressure differential activated device which allows oil to bypass the oil cooler.

NOTE: Commonly used for protection under low temperature, high viscosity conditions, or any pressure surge condition where inlet pressure can become excessive.

4.19 Oil Cooler Thermostat—A temperature activated device in the oil cooler circuit, which can either by-pass oil around or modulate oil flow through the cooler.

NOTE: This device regulates oil cooler heat transfer to allow rapid heating of oil at start up or prevent excessive cooling under light load conditions.

4.20 Operating Pressure—That fluid pressure to which the oil cooler is normally exposed during operation.

4.21 Partition—A device that is installed in a manifold, header, bonnet, or tank to create multiple pass of fluids through the core.

4.22 Peak Pressure—The highest pressure to which the oil cooler is intermittently subjected.

4.23 Pressure Drop—The pressure differential between inlet and outlet at a specified fluid flow rate and viscosity.

NOTE: 1. Air side is measured in inches (millimeters) of water.
2. Oil side is measured in psi (kPa).
3. Water side is measured in psi (kPa).

4.24 Removable Tube Bundle Oil Cooler—A shell and tube heat exchanger utilizing seals between the shell and tube fluids allowing the tube bundle to be removed from the shell.

NOTE: Normally used to provide for disassembly and/or thermal expansion. See Fig. 2 in Nomenclature.

4.25 Tank—An enclosure, located at the inlet and/or outlet of an oil cooler, which is sealed against the tube sheet or individual tubes and distributes the tube side fluid into the tubes or collects the tube side fluid as it exits the tubes. See Fig. 6 in Nomenclature.

4.26 Tube Sheet—See Figs. 2 and 3 in Nomenclature.

4.27 Turbulator—A device that increases fluid turbulence for the purpose of increasing heat transfer.

NOTE: For typical configurations, see Figs. 1, 5, 6, and 8 in Nomenclature.

5. Nomenclature

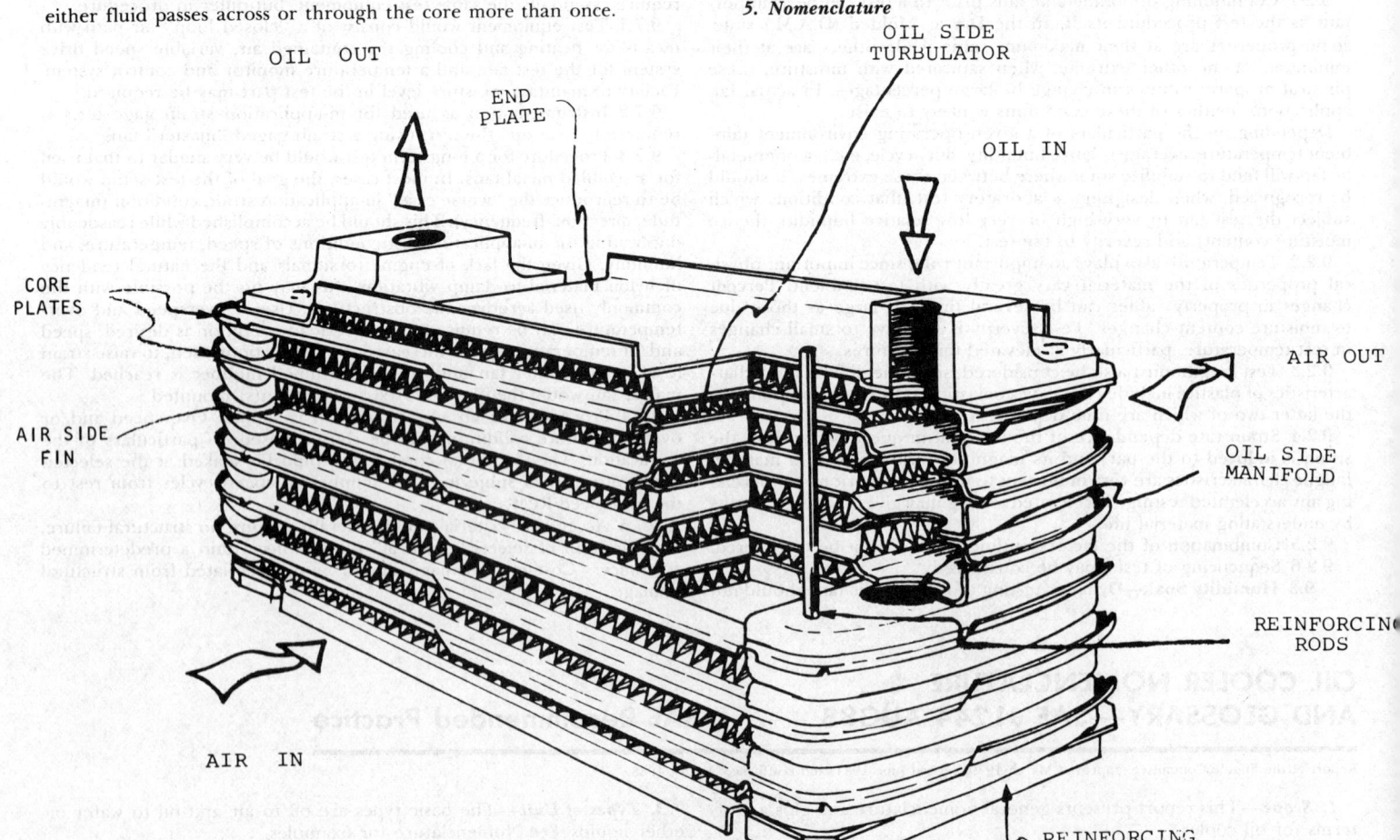

FIG. 1—PLATE FIN SEPARATOR OIL TO AIR COOLER

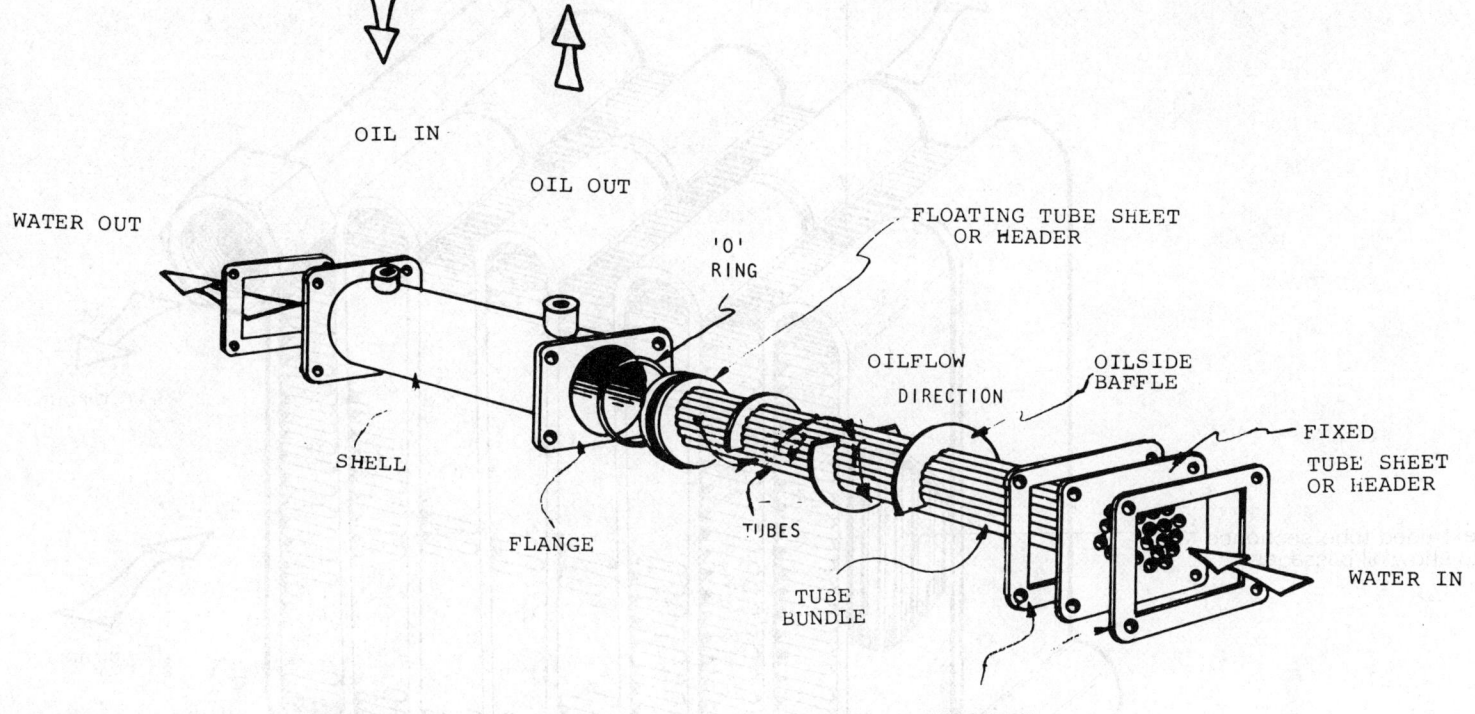

FIG. 2—SHELL AND TUBE REMOVABLE BUNDLE OIL TO WATER COOLER

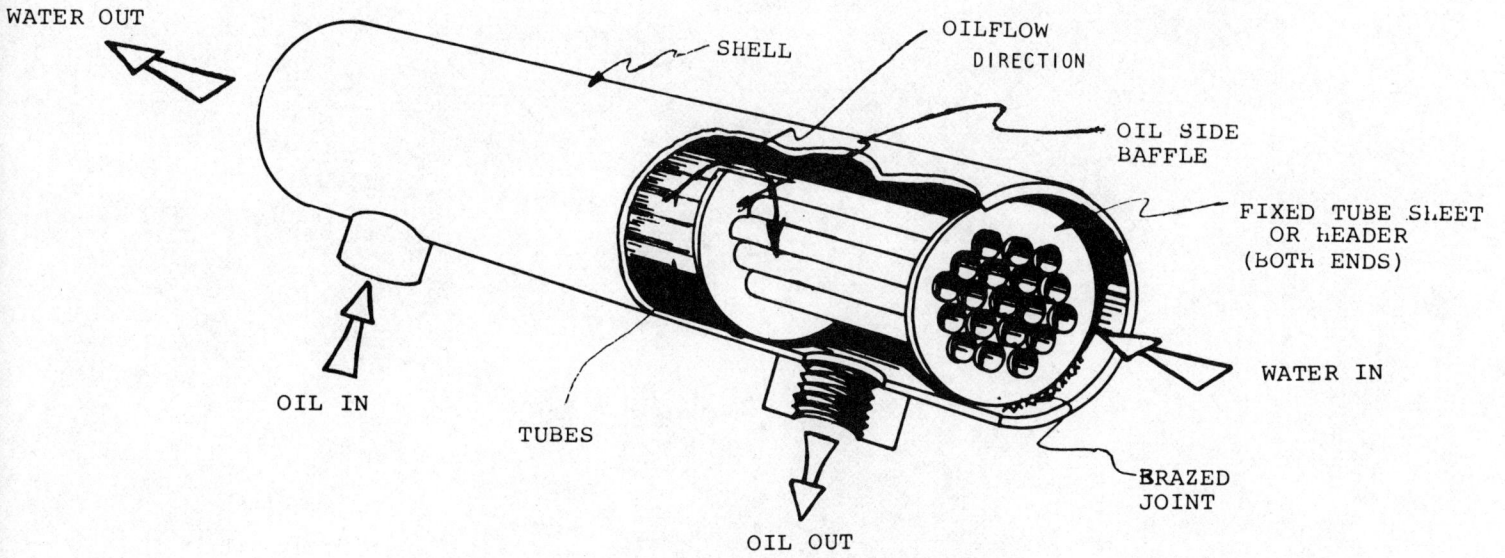

FIG. 3—SHELL AND TUBE OIL TO WATER COOLER-FIXED BUNDLE

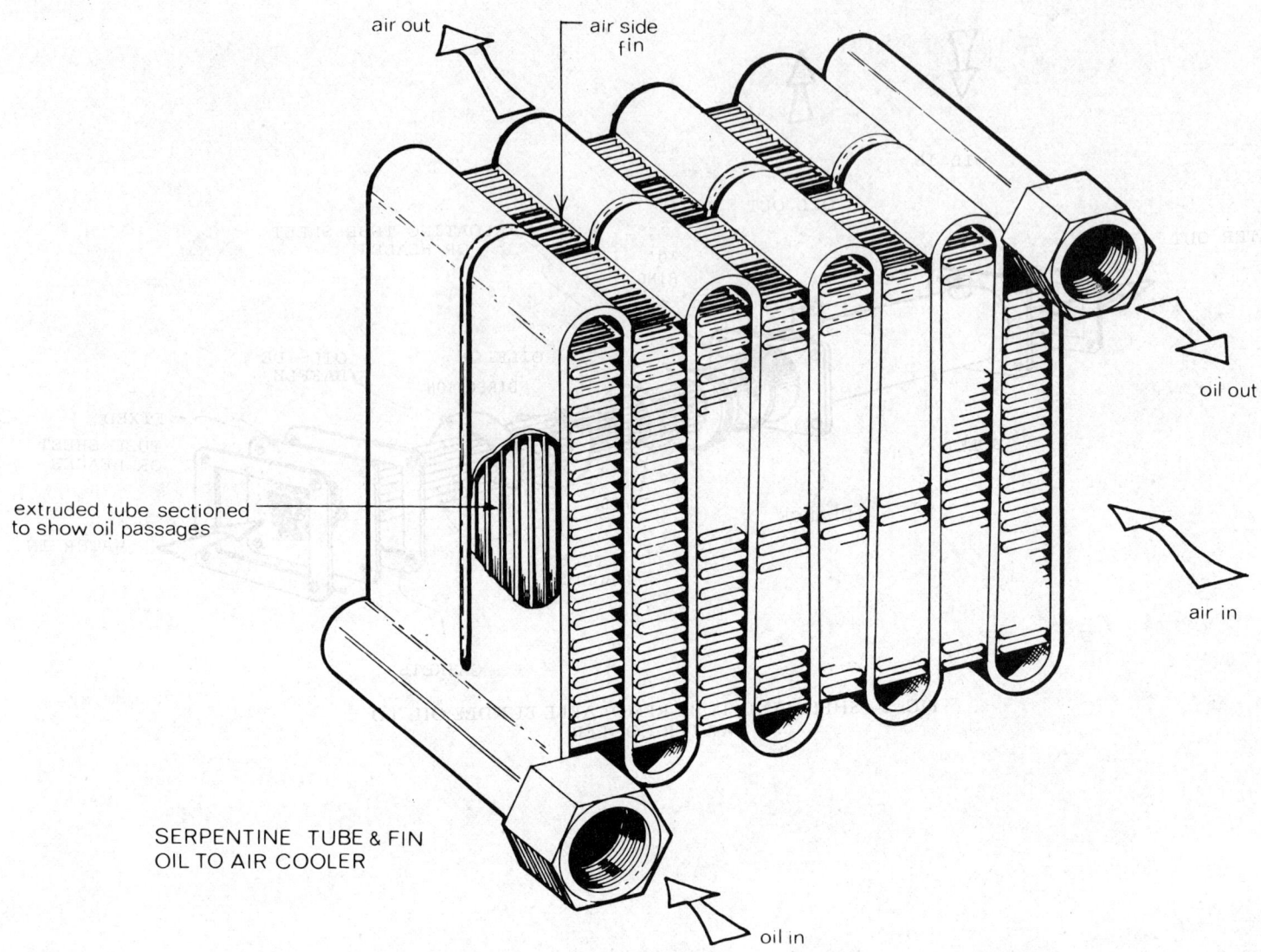

FIG. 4—SERPENTINE TUBE AND FIN OIL TO AIR COOLER

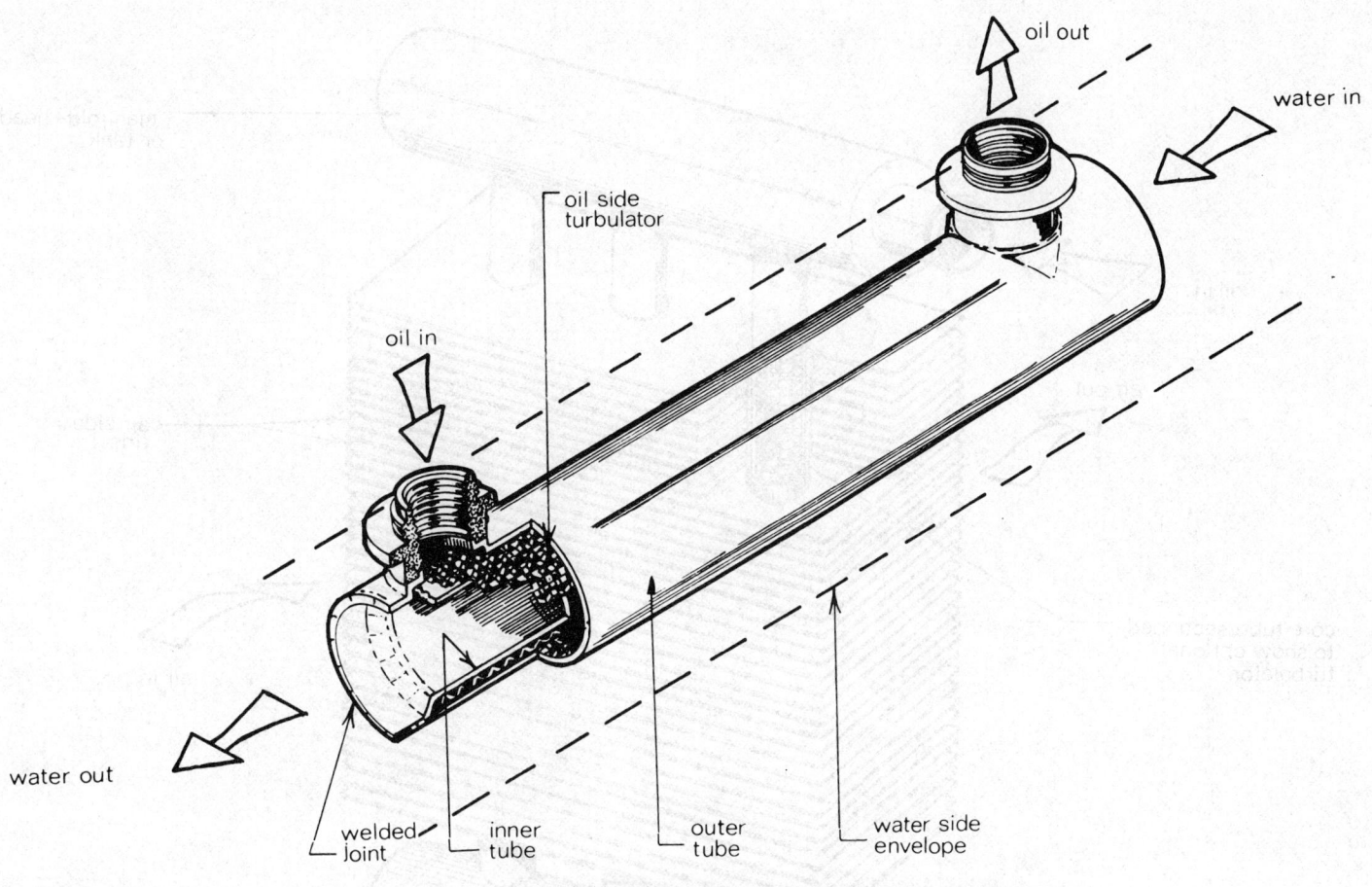

FIG. 5—CONCENTRIC TUBE OIL TO WATER COOLER

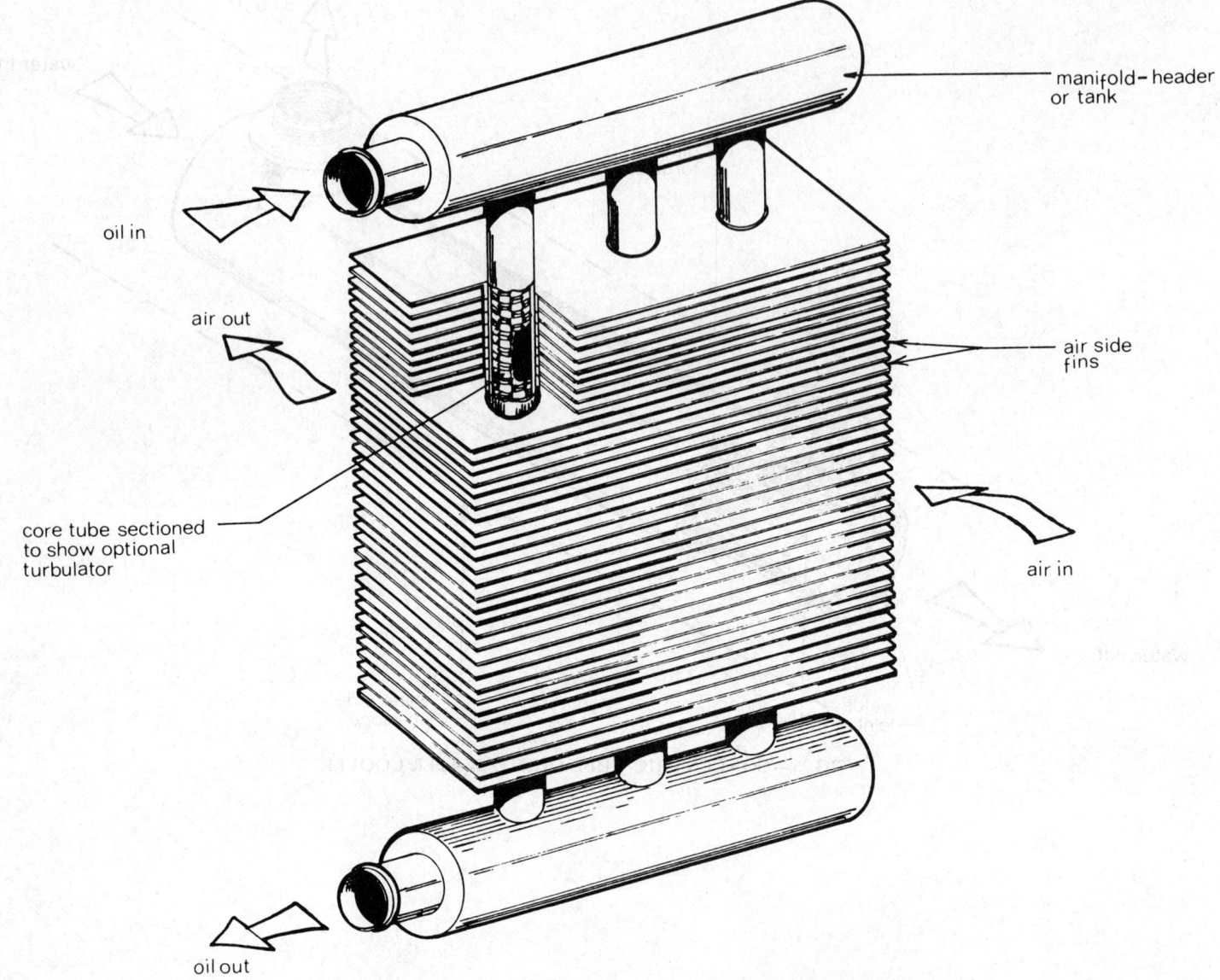

FIG. 6A—ROUND TUBE AND FIN OIL TO AIR COOLER

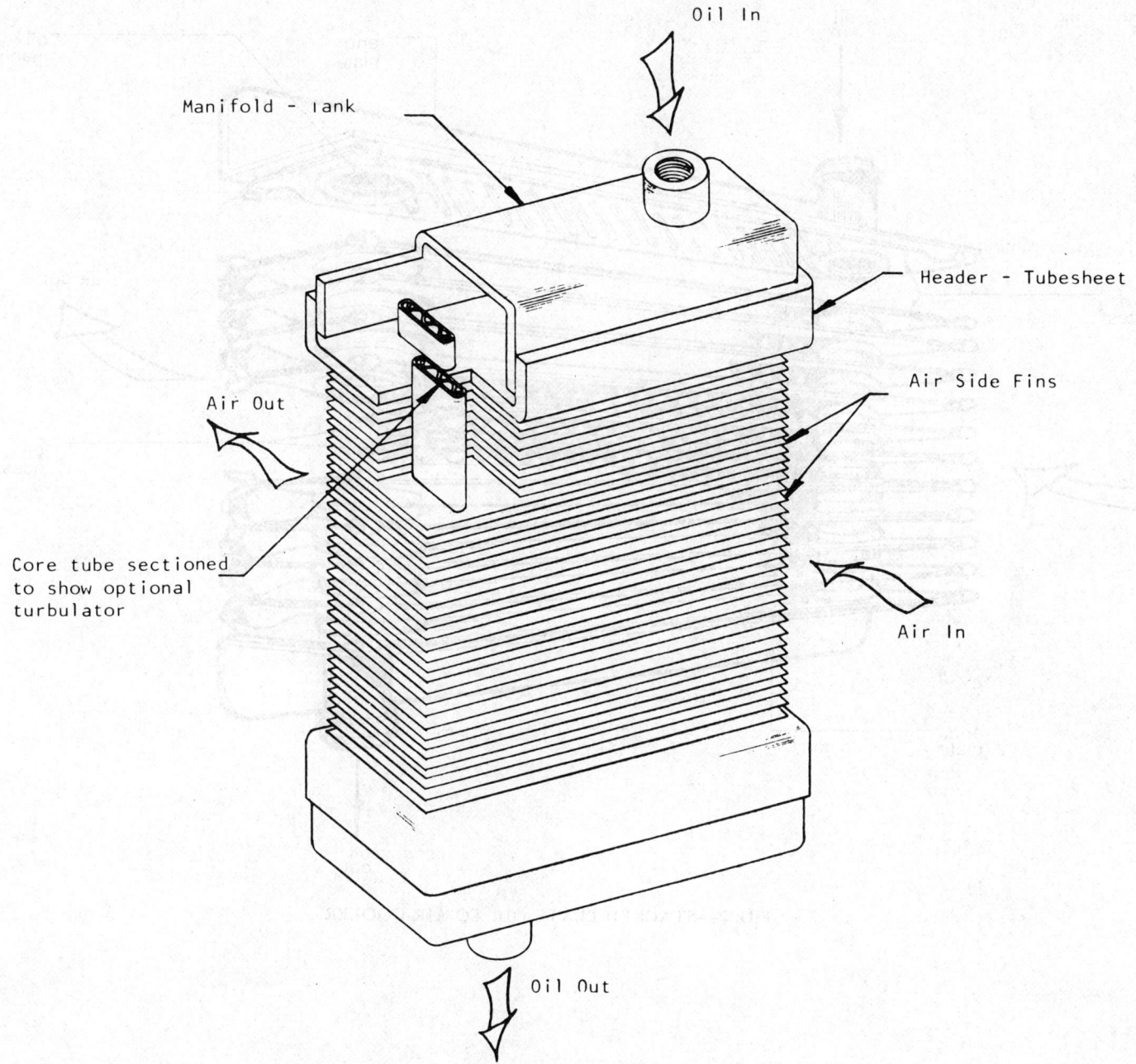

FIG. 6B—OVAL TUBE AND FIN AIR TO OIL COOLER

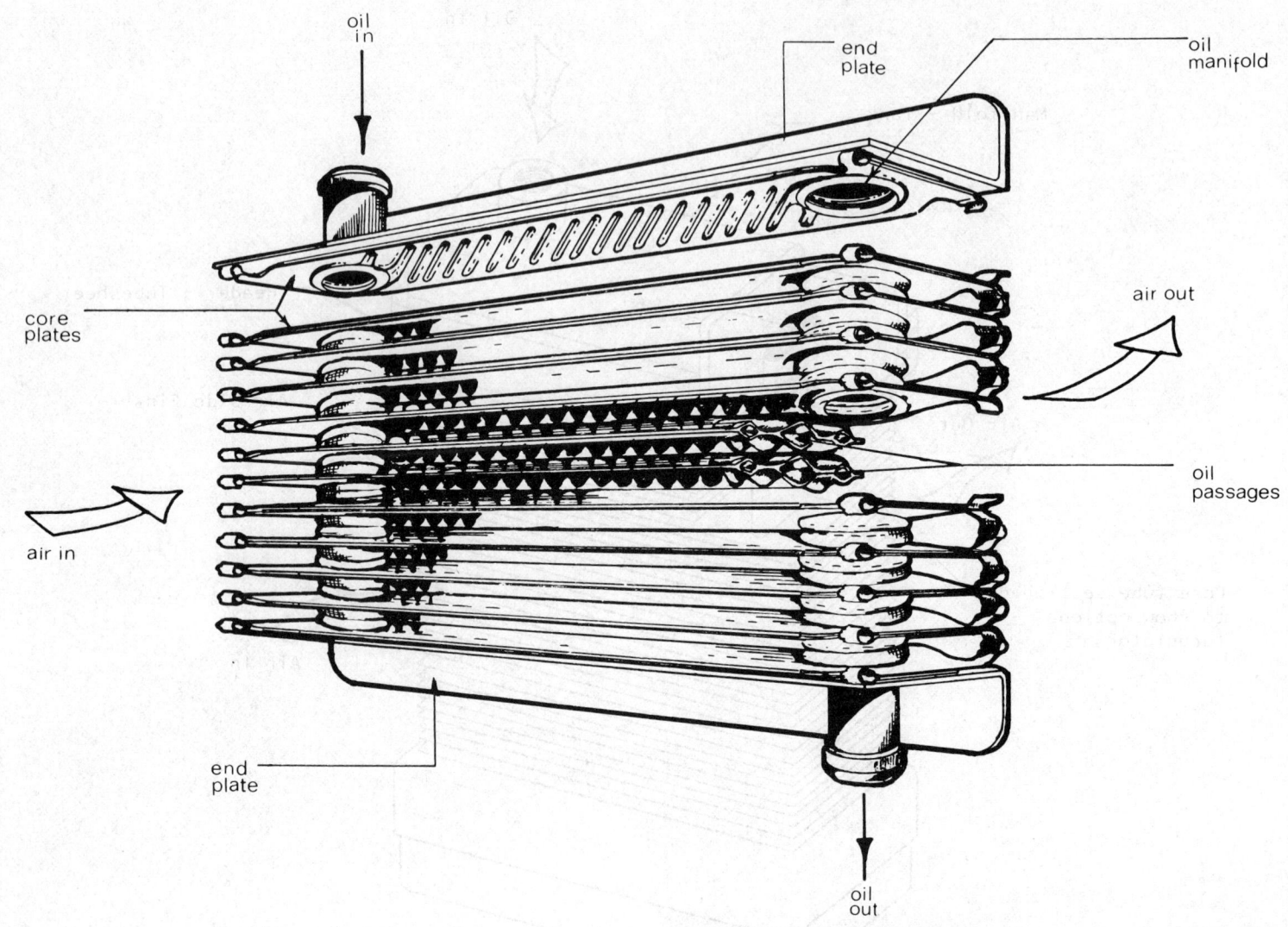

FIG. 7—STACKED PLATE OIL TO AIR COOLER

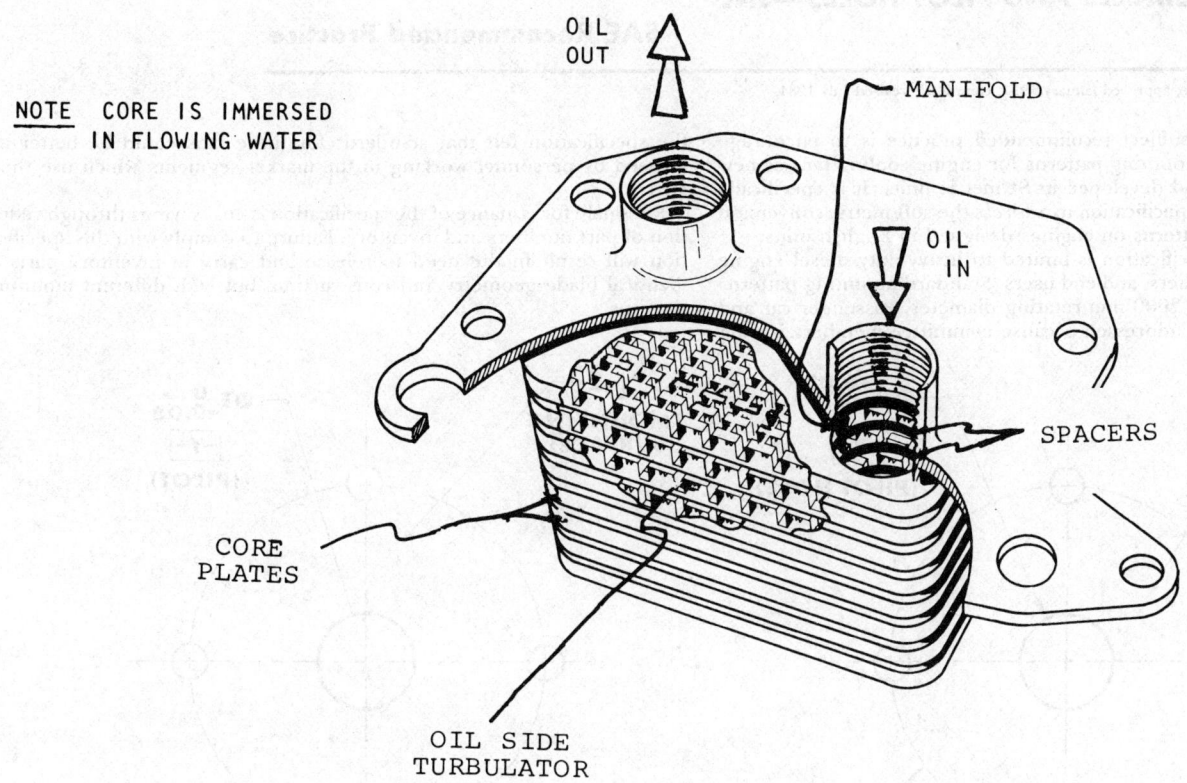

FIG. 8—PLATE TYPE OIL TO WATER COOLER

FAN HUB BOLT CIRCLES AND PILOT HOLES—SAE J635 JUL84

SAE Recommended Practice

Report of the Engine Committee, approved January 1951, completely revised July 1984.

The purpose of the subject recommended practice is to encourage the standardization of mounting patterns for engine cooling fans as new engines are designed and developed in SI metric units. It is specifically not the objective of the specification to address the soft metric conversion of existing mounting patterns on engines designed in English units.

The scope of the specification is limited to heavy-duty diesel engine manufacturers, fan suppliers, and end users. Standard mounting patterns are given for fans up to 2000 mm rotating diameter. Passenger car and light-duty fans were not addressed because committee members issuing the specification felt that standards for these fans could be better addressed by personnel working in the market segments which use those fans.

Rationale for issuance of the specification is cost savings through reduction of part numbers and inventory. Failure to comply with this specification will result in the need to release and carry in inventory parts of identical blade geometry and construction, but with different mounting patterns.

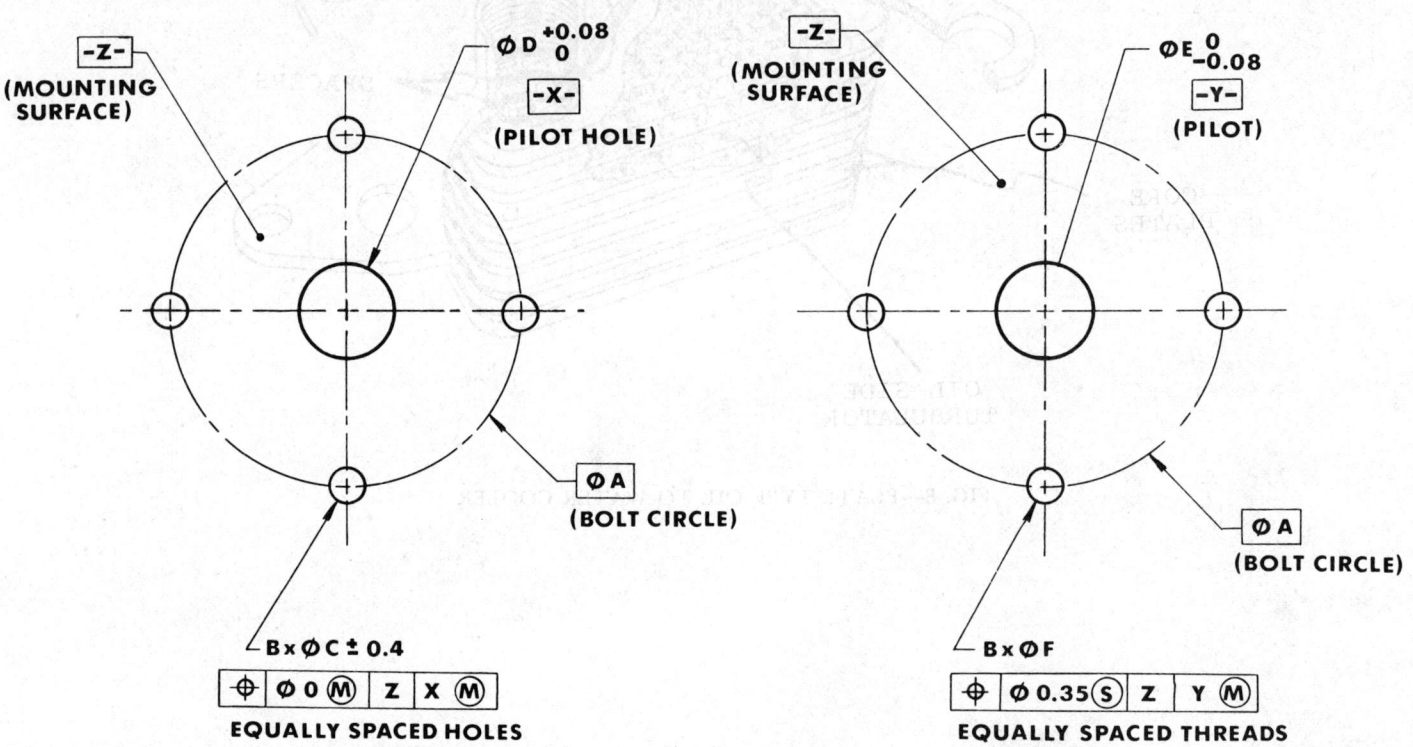

TABLE 1—METRIC FIXED DRIVE MOUNTING PATTERNS (mm)

Fan Diameter	A Bolt Circle Diameter	B Number Bolts	C Bolt Hole Diameter (F + 0.75)	D Fan Pilot Hole Dia.	E Hub Pilot Diameter (D−0.05)	F Bolt Dia.
Thru 600	50.00	4	8.75	25.00	24.95	8
Over 600 Thru 1000	100.00	6	10.75	50.00	49.95	10
Over 1000 Thru 1600	150.00	6	12.75	75.00	74.95	12
Over 1600 Thru 2000	200.00	8	12.75	100.00	99.95	12

R) ENGINE COOLANT PUMP SEALS—SAE J780 JUN90

SAE Standard

26.415

Report of the Engine Committee, approved June 1962, completely revised by the Transmission and Drivertrain Committee November 1978, reaffirmed without change April 1984. Completely revised by the Face Seals Standards Technical Committee June 1990.

1. Scope—This SAE standard outlines physical dimensions and nomenclature for the sizes of seals commonly used in engine coolant pumps of automotive type engines. Its purpose is to define a standard envelope to accommodate installation of various seal designs and to promote uniformity in seal nomenclature.

2. References—For additional information on material combinations, drawing format, qualification and inspection, and quality control data, please refer to SAE J1245, Guide to the Application and Use of Engine Coolant Pump Face Seals.

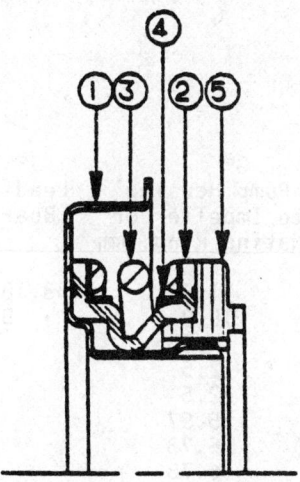

FIG. 1—SPRING-LOADED

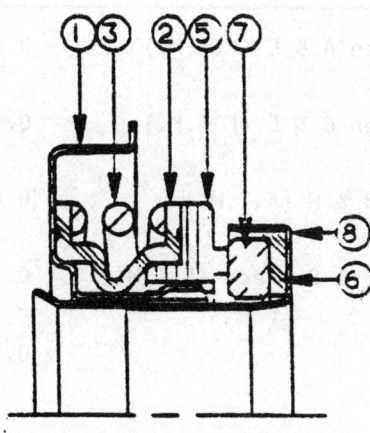

FIG. 2—UNITIZED

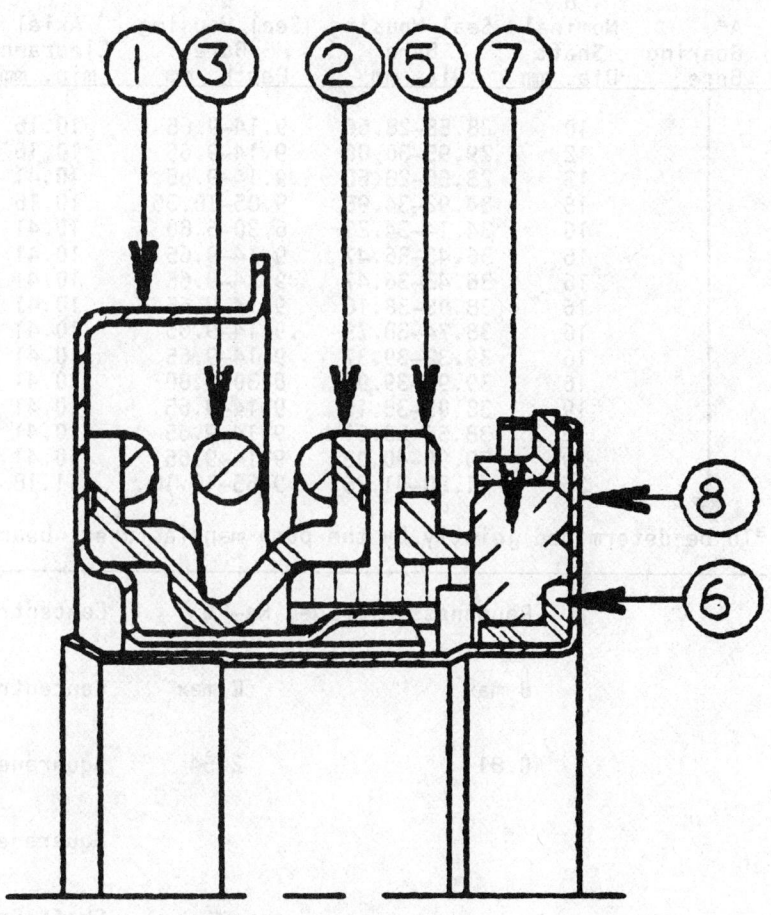

FIG. 3—UNITIZED-POSITIVE DRIVE MATING RING

NOMENCLATURE

1) CARTRIDGE
2) BELLOWS
3) SPRING
4) FERRULE
5) PRIMARY SEAL RING
6) SECONDARY DRIVE SEAL
7) MATING RING
8) UNITIZER

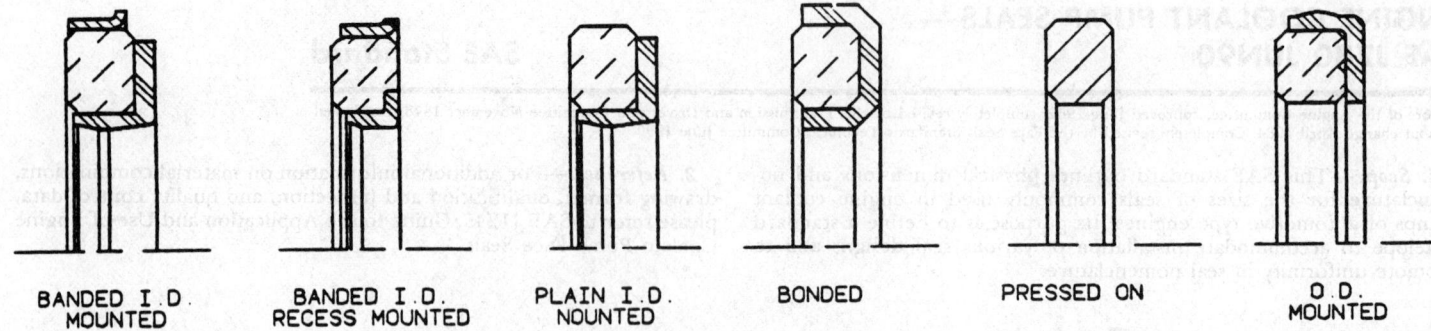

BANDED I.D. MOUNTED   BANDED I.D. RECESS MOUNTED   PLAIN I.D. MOUNTED   BONDED   PRESSED ON   O.D. MOUNTED

FIG. 4—MATING RING TYPES

TABLE 1. REFERENCE DIMENSIONS

A[a] Boaring Bore	B Nominal Shaft Dia, mm	C Seal Housing Bore Dia, mm	D Seal Housing Bore Depth, mm	E Axial Clearance min, mm	F[a] Seal Bore Lead-in Chamfer	G[a] Pump Housing to Impeller or Mating Ring, mm	J[a] Lead-in Chamfer Boaring Shaft End
	10	28.55–28.60	9.14–9.65	10.16	1.02×45°	1.57	3.18 mm × 30° Blended
	12	29.95–30.00	9.14–9.65	10.16	1.02×45°	10.0	
	13	28.55–28.60	9.14–9.65	10.41	1.02×45°	4.75	
	15	34.92–34.95	9.85–10.35	10.16	1.02×45°	12.5	
	16	34.14–34.21	6.30–6.80	10.41	1.02×45°	12.5	
	16	36.43–36.47	9.14–9.65	10.41	1.02×45°	5.97	
	16	36.43–36.47	9.14–9.65	10.41	1.02×45°	6.73	
	16	38.05–38.10	9.14–9.65	10.41	1.02×45°	6.73	
	16	38.74–38.79	9.14–9.65	10.41	1.02×45°	6.73	
	16	39.32–39.37	9.14–9.65	10.41	1.02×45°	6.73	
	16	39.92–39.96	6.30–6.80	10.41	1.02×45°	12.5	
	19	38.05–38.10	9.14–9.65	10.41	1.02×45°	6.73	
	19	38.56–38.61	9.14–9.65	10.41	1.02×45°	12.5	
	19	39.98–40.03	9.14–9.65	10.41	1.02×45°	6.73	
	25	41.20–41.25	9.65–10.16	11.18	1.52×45°	14.68	

[a]To be determined jointly by the pump manufacturer, bearing and seal suppliers.

Roughness Average, Ra- m		Concentricity between A & C (F.I.M.)	0.05 mm max
B max	C max	Concentricity between B & C (F.I.M.)	0.13 mm max
0.81	2.54	Squareness between B & H (F.I.M.)	0.05 mm max
		Squareness between B & Surface (F.I.M.)	0.13 mm max
		Shaft End Play	0.13 mm max

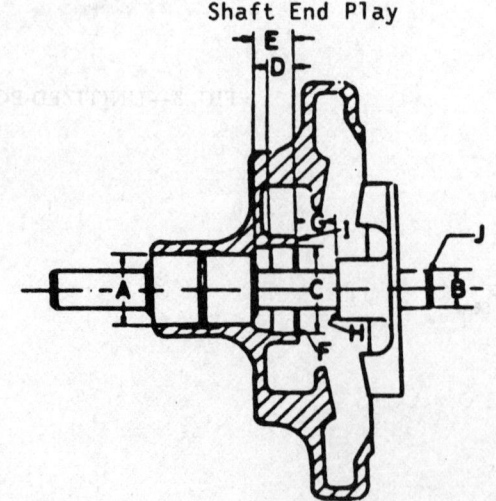

GUIDE TO THE APPLICATION AND USE OF ENGINE COOLANT PUMP FACE SEALS—SAE J1245 JUN82   SAE Recommended Practice

Report of the Transmission and Drivetrain Committee, approved November 1978, revised by the Seals Committee June 1982.

1. Introduction—This recommended practice is intended as a guide in the usage of mechanical face seals for the engine coolant pump application. The main purpose of the guide, is to fill the void caused by the lack of a ready source of practical information on the design and use of the engine coolant pump face seal. Included in the report is a compilation of present practices, viz.; a description of the various types of seals, material combinations, design data, tolerances, drawing format, qualification and inspection information, and quality control data. The terminology used throughout the guide is recommended, and through common usage it is hoped to promote uniformity in seal nomenclature.

2. Seal and Mating Ring Types—The mechanical face seal assemblies utilized in automotive and other heavy duty vehicle engine coolant pumps consist of a seal head assembly and a mating ring. Although many variations of face seal designs exist, two basic concepts are predominantly applied; they are based on the single helical and elastomeric spring preload concepts. Preload is defined as the force applied to the primary seal ring, when located in its normal operating position, to prevent separation of the sealing interfaces during the anticipated duty cycle.

2.1 Single Helical Spring-Loaded Type Seal—This seal type is the most commonly utilized concept for preloading mechanical face seals. The seal head assembly generally consists of a cartridge, bellows, spring, ferrule(s), and primary seal ring (Ref. Fig. 1, SAE J780a (November, 1978)).

Force is applied to the primary seal ring by the compression of a single helical coil spring between the cartridge and primary seal ring during installation of the seal head assembly to its proper operating length. The bellows allow the primary seal ring to move axially, thus compensating for wear without loss of its sealing ability. Ferrules may be used to provide a contact surface for the spring and secondary sealing functions. Under normal operating conditions and with proper materials selection, this design is generally resistant to corrosion, abrasion, time-temperature, and coolant exposure effects.

2.2 Elastomeric Spring Loaded Type Seal—The seal head assembly consists of a primary seal ring and bellows (Ref. Fig. 2, SAE J780a (November, 1978)).

Preload force is applied to the primary seal ring by the axial and/or radial deformation of the elastomeric bellows member. This design is resistant to corrosion and abrasion, but its load characteristics are dependent on elastomeric material properties to resist time-temperature and coolant exposure effects.

2.3 Unitized Type Seal—The unitized seal consists of a seal head assembly of either helical or elastomeric spring preload concept and the mating ring, constructed so as to be handled as a single piece. The unitized seal generally consists of a cartridge, bellows, spring, ferrule(s), primary seal ring, secondary drive seal, mating ring, and unitizer (Ref. Fig. 3, SAE J780a (November, 1978)).

The unitizer is press fitted onto the shaft, thus the unit is not dependent on impeller position to establish operating length.

2.4 Mating Ring Types—There are six widely applied mating ring types, including surfaces of pump components (Ref. Fig. 4, SAE J780a (November, 1978)).

Mating ring types are differentiated by the method of mounting while providing a reliable means of assuring drive, secondary sealing, and to minimize stresses and distortion during the operating cycle; they are as follows: (1) banded, I.D. mounted with a secondary drive seal; (2) plain (unbanded), I.D. mounted with a secondary drive seal; (3) bonded onto the surface or into a cavity of the pump component, e.g., rotating impeller; (4) the surface of the pump component is machined and lapped to provide a mating ring surface; (5) mating ring is press fitted onto the pump shaft; and (6) the mating ring is mounted into the pump component with a secondary drive seal.

NOTE: Type 5 mating ring would result in the most distortion of the mating ring face.

3. Seal Material—Environmental conditions dictate the type of material which should be used in a specific application. Seal materials can be fully evaluated only in terms of specific operating conditions and performance requirements. The following paragraphs give general descriptions of primary seal rings and their mating ring materials, elastomeric compounds, and hardware, also outlining some advantages and disadvantages. It should be recognized that batch to batch variations due to material inconsistencies can occur in all materials listed. Such inconsistencies can alter the performance data given.

3.1 Primary Seal Ring—The primary seal ring is allowed axial motion to permit the sealing face to remain in contact despite shaft endplay, runout, and face wear.

3.1.1 THERMOSET PLASTIC MATERIALS—Thermoset plastic materials with varying amounts of mineral and/or metal fillers are low shrinkage resin materials such as epoxy, phenolic, or polyester, usually molded in the 400–600°F (205–315°C) range. Thermoset plastics are low cost materials.

3.1.1.1 *Advantages:*
(a) Good wear resistance at required temperatures and pressures.
(b) Readily molded to complex geometry and close tolerances.

3.1.1.2 *Disadvantages:*
(a) Poor thermal stability.
(b) Poor thermal conductivity.

3.1.2 CARBON GRAPHITE—Carbon graphite is generally a manufactured product which contains some graphite, natural or synthetic, and which has a rigid, hard structure produced by firing at high temperatures usually ranging between 1650 and 3630°F (900 and 2000°C).

The material can be impregnated with various materials, including metals, to impart a particular carbon mix identity. The material is in the premium cost range.

3.1.2.1 *Advantages:*
(a) Excellent temperature resistance and stability.
(b) Low absorption of coolant and consequent lack of degradation.
(c) Some degree of self-lubricity, thus able to withstand dry runs without galling.
(d) Excellent wear resistance.
(e) Good thermal conductivity (via impregnation).

3.1.2.2 *Disadvantages:*
(a) Difficult to mold relatively complex shapes and maintain close tolerances.
(b) Poor handling characteristics (damages easily).
(c) Possible chemical attack on impregnants.

3.2 Mating Rings—Mating rings are usually of a dissimilar material which is harder than the primary seal ring. The material choice depends upon operating conditions, configuration, costs, and performance requirements.

3.2.1 ALUMINUM OXIDE CERAMICS—Aluminum oxide ceramics generally have an aluminum oxide (Al_2O_3) content ranging from 85–99.9% by weight. The parts are formed by compacting finely ground oxide powders with fluxing agents and inhibitors at high pressures. The formed part is then fired at high temperatures usually ranging between 2550 and 3250°F (1400–1800°C). After firing, the ceramic is a strong, hard, dense material, composed mostly of pure alumina crystals of controlled size. Cost is moderate to premium.

3.2.1.1 *Advantages:*
(a) Excellent wear resistance.
(b) Excellent dimensional stability.
(c) Excellent fluid compatibility.

3.2.1.2 *Disadvantages:*
(a) Mechanical and thermal shock susceptibility.
(b) Difficult to mold complex shapes and maintain tolerances.

3.2.2 POWDERED AND CAST METAL MATERIALS—Metal powders, such as iron, are placed in a die and compressed. The parts are then sintered in a controlled atmosphere, whereas cast metal is poured or injected in its molten state into a mold. These can be supplied in a diversity of alloys for high volume low cost applications.

3.2.2.1 *Advantages:*
(a) Excellent thermal shock resistance.
(b) Good thermal conductivity.

3.2.2.2 *Disadvantages:*
(a) Poor corrosion resistance (compatibility verified).
(b) Moderate wear resistance.

3.2.3 SPRAYED COATINGS—Sprayed coatings can combine various qualities of materials to improve performance and obtain certain economic advantages.

3.2.3.1 *Advantages:*
(a) Good thermal shock resistance.

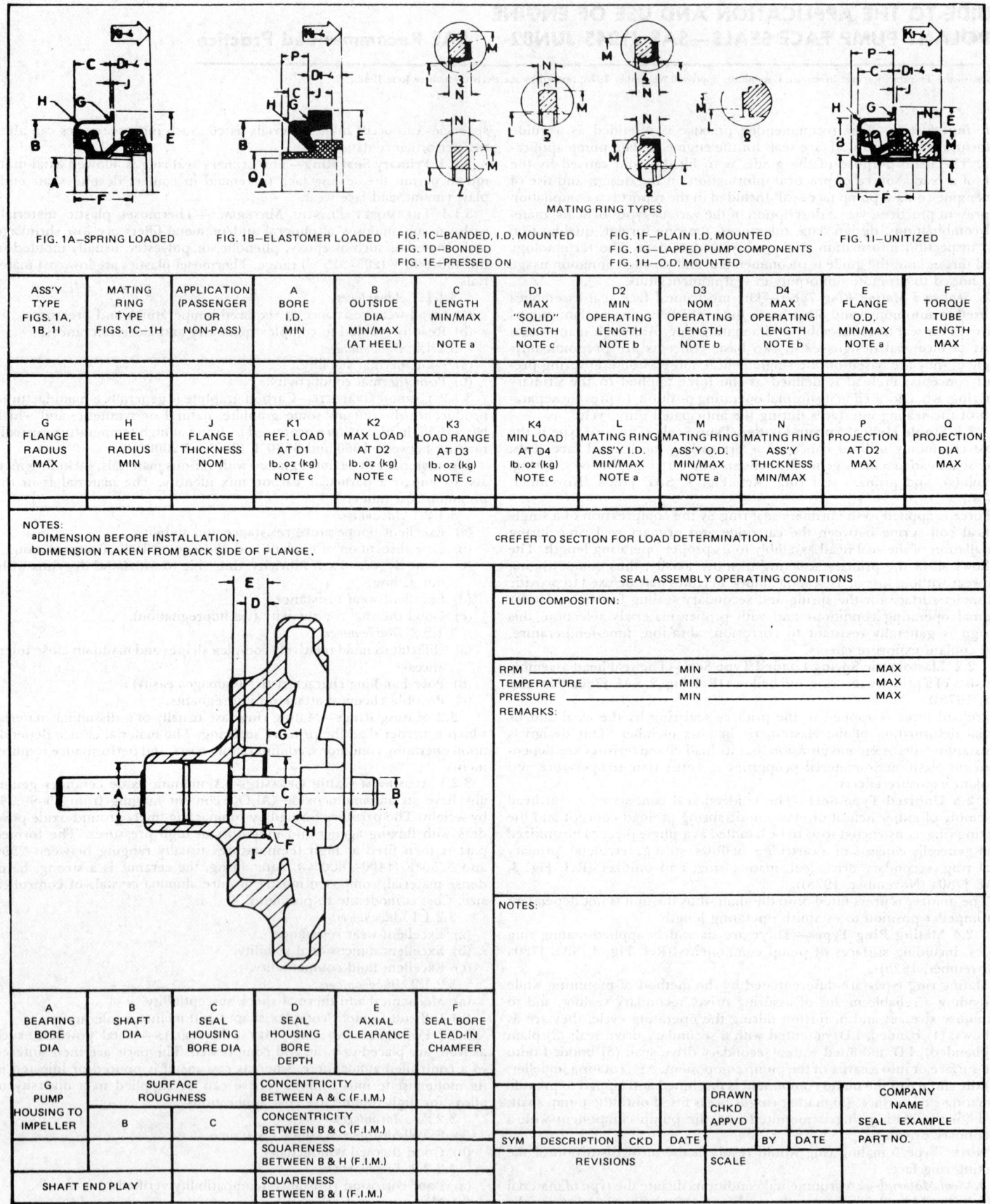

FIG. 1—STANDARD DRAWING FORMAT

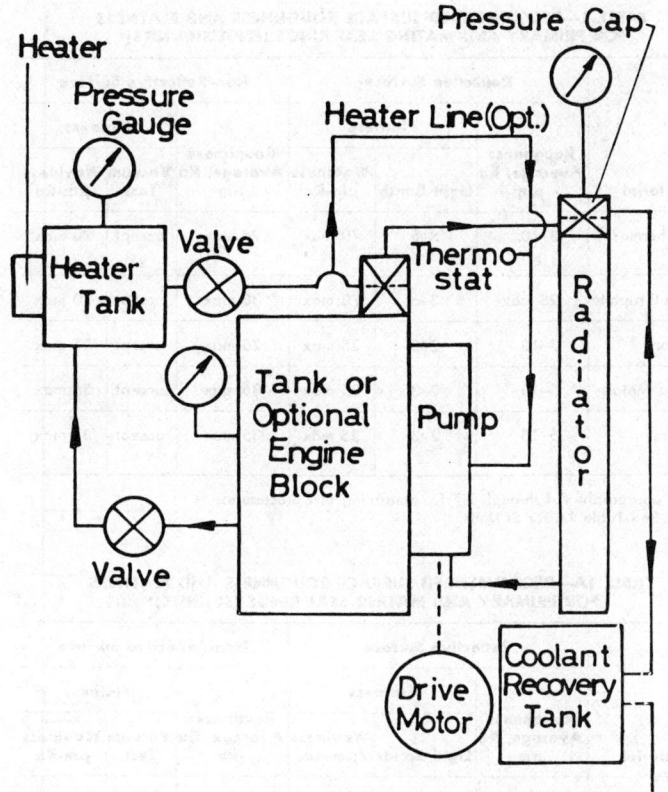

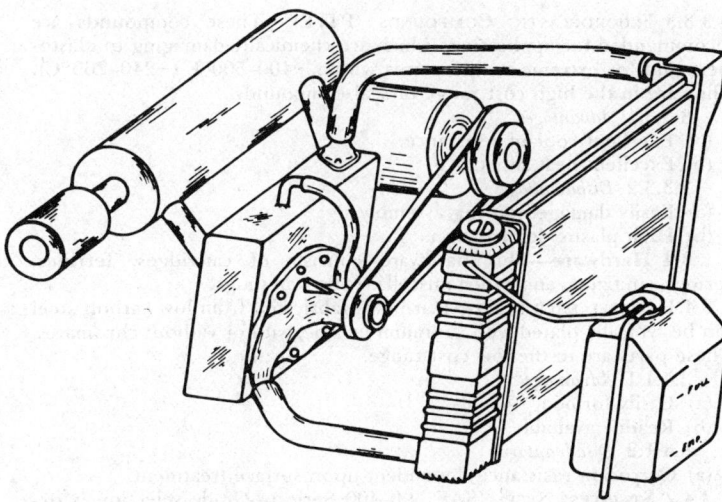

FIG. 3—PICTORIAL ARRANGEMENT OF RECOMMENDED ENGINE COOLANT PUMP SEAL DURABILITY TEST SYSTEM

FIG. 2—SCHEMATIC—ENGINE COOLANT PUMP SEAL DURABILITY TEST SYSTEM

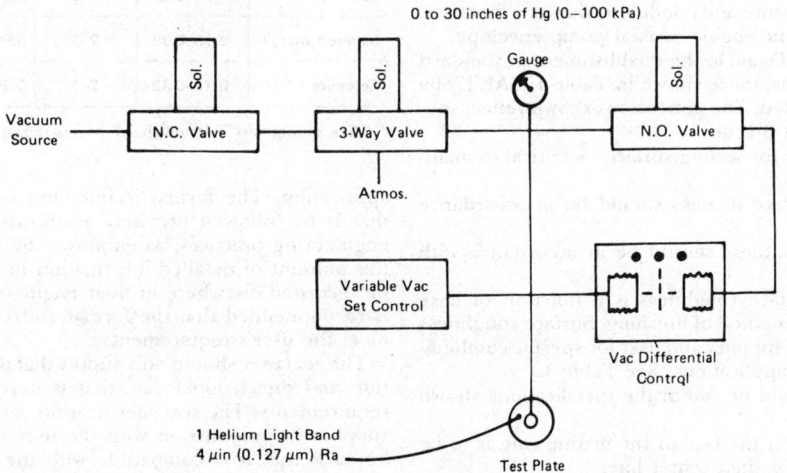

FIG. 4—SCHEMATIC—TYPICAL VACUUM TEST ARRANGEMENT

(b) Excellent wear resistance.
(c) Adaptable to a large variety of sizes and shapes.
  3.2.3.2 *Disadvantages:*
(a) Added cost.
(b) Specialized processing.

3.3 Secondary Seals—Secondary seals are generally elastomers that can be categorized as not-oil or not-solvent resistant, oil or solvent resistant, and heat resistant.

  3.3.1 NITRILE COMPOUNDS (NBR)[1]—This material family's operating range is −65–275°F (−54–135°C). Nitrile is recommended for general use with exposure to coolants. It is in the low cost range.
  3.3.1.1 *Advantages:*

(a) Good processability.
(b) Good oil resistance.
  3.3.1.2 *Disadvantages:*
(a) Limited high temperature life.
(b) Poor to moderate ozone resistance.

  3.3.2 SILICONE COMPOUNDS (V Si)[1]—Silicone compounds are recommended for applications where temperatures are within −65–350°F (−54–177°C). The maximum useable temperature is limited by the decomposition temperatures of the various coolants. Silicone rubbers are in the high cost range of seal compounds.
  3.3.2.1 *Advantages:*
(a) Excellent heat resistance.
(b) Excellent low-temperature properties.
  3.3.2.2 *Disadvantages:*
(a) Poor compatibility in some coolants.
(b) Moderate abrasion resistance.

---

[1] ASTM D 1418, Recommended Practice for Nomenclature for Synthetic Elastomers and Lattices, 1976 Edition, should be used as reference.

3.3.3 Fluoroplastic Compounds (PTFE)[2]—These compounds are recommended for applications which are chemically damaging to elastomers and for extreme temperatures within −400–500°F (−240–260°C). They are in the high cost range of seal compounds.

3.3.3.1 *Advantages:*
(a) Excellent coolant resistance.
(b) Excellent heat resistance.

3.3.3.2 *Disadvantages:*
(a) Easily damaged during assembly.
(b) High plastic deformation.

3.4 Hardware—This hardware consists of cartridges, ferrules, springs, unitizers, and other miscellaneous stampings.

3.4.1 Plated Mild Steel—Hardware fabricated from low carbon steel can be typically plated with cadmium or zinc, with or without chromates. These parts are in the low cost range.

3.4.1.1 *Advantages:*
(a) Easily formed.
(b) Readily available.

3.4.1.2 *Disadvantages:*
(a) Corrosion resistance dependent upon surface treatment.

3.4.2 Stainless Steel (SAE 300–400 Series)—Grade selection is dependent on desired level of corrosion resistance, mechanical properties, method of fabrication, and cost/availability factors.

3.4.2.1 *Advantages:*
(a) High temperature physical property retention.
(b) Excellent corrosion resistance.

3.4.2.2 *Disadvantages:*
(a) Limited workability.
(b) Availability of some grades.

3.4.3 Brass—Generally, annealed 70/30 brass (SAE alloy No. CA 260) is utilized.

3.4.3.1 *Advantages:*
(a) Good corrosion resistance.
(b) Easily formed.

3.4.3.2 *Disadvantages:*
(a) Season (stress corrosion) cracking.
(b) Easily deformed.

4. Application Design Data—The following section is presented to provide guidelines as to specific dimensions and conditions that may functionally affect the applied seal within the engine coolant pump envelope.

4.1 Reference Dimensions—To aid in the establishment of standard pump housing envelope dimensions, those shown in Table 1, SAE J780a (November, 1978), are recommended. The dimensions shown reflect specific dimensions utilized in current practice.

4.2 Flatness—Overall flatness for sealing surfaces is critical to maintain a liquid or gas tight seal.

4.2.1 Primary Seal Ring—Surface flatness should be in accordance with Table 1.

4.2.2 Mating Rings—Surface flatness should be in accordance with Table 1.

4.3 Surface Roughness—Surface roughness is a function of base material, grain size, structure, and method of finishing. Surface roughness is to be evaluated on the part of the supplier and user for specific combinations of materials and for specific applications. See Table 1.

4.4 Waviness—Waviness should be within the specifications shown on Table 1.

4.5 Squareness—Squareness of the face of the mating ring is to be within 0.005 in (0.13 mm) F.I.M. of shaft center line.

4.6 Dynamic Runout—Dynamic runout is defined as twice the distance the center of the shaft is displaced from the center of rotation, expressed in F.I.M., and should not exceed 0.010 in (0.25 mm).

4.7 Eccentricity—Eccentricity is defined as the radial distance the geometric center of a shaft is displaced from the axis of rotation and should be held within 0.005 in (0.13 mm) F.I.M.

4.8 Lead-In Chamfer—A lead-in chamfer is required at the pump housing bore and the bearing shaft end for ease of seal installation and prevention of damage to the secondary seals (see Table 1, SAE J780a (November, 1978)). All corners should be blended smoothly.

4.9 Clearances—All clearances must be large enough to provide sufficient coolant circulation for proper seal functioning.

5. Drawing Designation—It is recommended that the standard SAE seal and housing drawing format be used. This format (Fig. 1) is a composite of the engineering application, seal, and pump housing dimensional data that is required to assure functional compatibility of the seal in a specific application. The format is intended as a guide and it is not required that it be followed precisely as shown. It is understood that standard engineering practices, as employed by some users, will not require that this amount of detailed information be shown on the print since it may be recorded elsewhere in their engineering standards. In those cases, it is recommended that the format and/or sketches be suitably altered to meet the user's requirements.

The seal user should only supply that portion of the engineering application and dimensional data that is necessary for the particular product requirements. The seal specification data should be furnished by the seal supplier in conjunction with the user. This data and information must be such that it is compatible with the engineering application data as supplied by the user.

6. Qualification Test—This test is conducted to determine the durability characteristics of an engine coolant pump seal in a functional engine coolant pump assembly.

6.1 Description of Equipment and Installation—The following equipment and system orientation is recommended (see Figs. 2 and 3).

6.1.1 Tank and heater capacity and pressure drop equivalent to engine cylinder block (engine block equipped with heaters—optional).

6.1.2 Complete production engine coolant pump assembly in which the seal will operate (including heater and bypass lines).

6.1.3 Drive motor capable of driving pump at 6000 rpm.

6.1.4 Production radiator (preferred) or tank with equivalent restriction.

6.1.5 Production radiator cap (check for opening pressure).

6.1.6 Coolant recovery tank.

6.1.7 Production thermostat and housing (checked for opening temperature).

6.1.8 Valves for setting restrictions equivalent to that in production cooling systems.

6.1.9 Pressure gauges to measure system pressures.

---

[2] ASTM D 1600, Tentative Abbreviations of Terms Relating to Plastics, 1975 Edition, should be used as reference.

TABLE 1—RECOMMENDED SURFACE ROUGHNESS AND FLATNESS FOR PRIMARY AND MATING SEAL RINGS (ENGLISH UNITS)[a]

Material	Reflective Surface			Non-Reflective Surface		
	Roughness Average, Ra μin	Flatness		Roughness Average, Ra μin	Flatness	
		Light Bands	Waviness μin-Ra		Vacuum Test	Waviness μin-Ra
Filled Thermoset Plastic	3–10	3–6	70 max	25 max	accept	70 max
Carbon Graphite	25 max	3–6	70 max	30 max	accept	70 max
Cast Iron	3–10	2–3	35 max	20 max	accept	35 max
Sintered Metals	5–15	2–3	35 max	35 max	accept	35 max
Ceramics (Al₂O₃)	5–15	2–3	35 max	35 max	accept	35 max

[a] See paragraphs 7.4 through 7.7 for measuring test procedures.
NOTE: See Table 1A for SI Units.

TABLE 1A—RECOMMENDED SURFACE ROUGHNESS AND FLATNESS FOR PRIMARY AND MATING SEAL RINGS (SI UNITS)[a]

Material	Reflective Surface			Non-Reflective Surface		
	Roughness Average, Ra μm	Flatness		Roughness Average, Ra μm	Flatness	
		Light Bands	Waviness μm-Ra		Vacuum Test	Waviness μm-Ra
Filled Thermoset Plastic	0.08–0.25	3–6	1.78 max	0.64 max	accept	1.78 max
Carbon-Graphite	0.64 max	3–6	1.78 max	0.76 max	accept	1.78 max
Cast Iron	0.08–0.25	2–3	0.89 max	0.51 max	accept	0.89 max
Sintered Metals	0.13–0.38	2–3	0.89 max	0.89 max	accept	0.89 max
Ceramics	0.13–0.38	2–3	0.89 max	0.89 max	accept	0.89 max

[a] See paragraphs 7.4 through 7.7 for measuring test procedures.

6.1.10 Cycle counter and running time meter.
6.1.11 Heater core (optional).
6.1.12 Automatic controls for cycling motor.
6.1.13 Heater(s) (3000 W) with controls.
6.1.14 Coolant hoses and clamps (production parts preferred).

6.2 Procedure—The following procedural outline is provided as a guide; obviously, this procedure should be modified to be compatible with the user's established standard engineering practice.

6.2.1 TEST DURATION AND CONDITIONS—The engine coolant pump assembly should be run under the following conditions commensurate with the user's standards.

6.2.1.1 *Coolant Temperature*—Coolant temperature, at the pump outlet should be maintained at 240–250°F (115–121°C).

6.2.1.2 *System Pressure*—The system pressure should be maintained, at the pump outlet, at a level equivalent to that of a standard production system.

6.2.1.3 *Pump Rotational Speed*—The engine coolant pump rotational speed should be maintained at maximum rated speed. Maximum rated speed is the pump speed attained at the maximum rated engine speed.

Optional: Engine coolant pump speed should be cycled from zero rpm to maximum rated speed in 15 s, held for 30 s, and returned to zero rpm in 15 s. This cycle is to be repeated for the duration of the test.

6.2.1.4 *Coolant Concentration*—The coolant concentration should be maintained in accordance with engine manufacturer's factory specification. However, the coolant boiling point, at test pressure, should be 10°F (5.5°C) greater than test temperature. It should be emphasized here that contaminants in the coolant system, either present originally or developed after a period of operation (Reference #1, #2), will affect seal performance.

The contaminants normally found in the system are soluble and non-soluble in nature. The soluble elements may include constituents from the basic coolant chemistry and various commercial additives. The non-soluble elements may include precipitates from the basic coolant and additives, core sand, and oxides of aluminum, iron and other metallic elements in the coolant system. Soluble and non-soluble oil contaminants may have been introduced during the normal manufacturing process or during system operation and maintenance.

If test data shows non-repeatability, and coolant variations are determined to be the principal contributor, then a standard fluid, such as ASTM 3585–77 can be utilized for test stand qualification.

It is recommended that long term durability and contamination effects be determined by vehicle testing. If it is desirable to establish a contaminated test system fluid, typical fluid composition is as follows:

Fill the cooling system with a mixture of 88% factory fill coolant and 12% Sarasota water (by volume). Sarasota water is made with 82% distilled water and 18% Sarasota Concentrate (by volume). The composition of Sarasota Concentrate is as follows:

Ingredients	Grams/Gallon
Sodium Metasilicate	2.39
Sodium Chloride	5.50
Potassium Chloride	0.64
Sodium Bicarbonate	2.20
Sodium Sulphate	25.40
Distilled Water	Remainder to complete 1 gallon

NOTE: If core sand and/or other non-soluble contaminants are to be purposely added, modifications (such as removal of heater cores, radiators, etc.) to the test system may be necessary to minimize contaminant "drop out" or erosion damage within the system which would change the test conditions.

6.2.1.5 *Heater and Bypass Lines*—All bypass and heater return lines to the engine coolant pump should be connected and operable.

6.2.1.6 *System Pressure Drop*—When an engine block and radiator are not employed, valves on the inlet and discharge sides of the engine coolant pump are to be used to set a restriction equivalent to that found in the production cooling system.

6.2.1.7 *Belt Tension*—Initial engine coolant pump drive belt tension should be set to manufacturer's specifications.

6.2.1.8 *Thermostat*—If an engine is used, a thermostat valve disc position should be mechanically maintained at an opening equal to that achieved at the test temperature.

6.3 Data to be Recorded—To provide meaningful test data, the following minimum data should be recorded:

6.3.1 BEFORE TEST—The following data should be obtained and recorded prior to pump assembly and testing.

6.3.1.1 *Primary Seal Ring*
(a) Material.
(b) Surface roughness.
(c) Surface flatness.
(d) Surface waviness.
(e) Face height.

6.3.1.2 *Mating Ring*
(a) Material.
(b) Surface roughness.
(c) Surface flatness.
(d) Surface waviness.
(e) Hardness.
(f) Squareness to shaft.

6.3.1.3 *Coolant Composition*
(a) Coolant identification.
(b) Coolant formulation.

6.3.1.4 *Seal and Pump Assembly*
(a) Seal operating length.
(b) Seal load at operating length.
(c) Test schedule and allowable leakage.
(d) Shaft speed.

6.3.2 DURING TEST—The following data should be recorded during the test sequence.
6.3.2.1 Pump rpm.
6.3.2.2 Coolant temperature at pump outlet.
6.3.2.3 System pressure.
6.3.2.4 Inlet pressure.
6.3.2.5 Discharge pressure.
6.3.2.6 Seal cavity pressure.

6.3.3 AFTER TEST—The following data or observations should be noted and recorded after completion of the test sequence and the seal has been removed from the pump assembly:

6.3.3.1 *Primary Seal Ring*
(a) Wear pattern.
(b) Surface roughness.
(c) Surface flatness.
(d) Surface waviness.
(e) Face height.

6.3.3.2 *Mating Ring*
(a) Wear pattern.
(b) Surface roughness.
(c) Surface flatness.
(d) Surface waviness.

6.3.3.3 *Bellow Condition*
(a) Time-temperature effect.
(b) Glue joints.
(c) Abrasion.

6.3.3.4 *Spring*
(a) Corrosion.
(b) Fatigue.

6.3.3.5 *Cartridge*
(a) Corrosion.
(b) Pressfit pattern.

7. Inspection and Quality Control Data—The following is presented as an inspection guide and outlines broad general quality control equipment and procedures. These guidelines should be reviewed and modified to be commensurate with the supplier's and user's standard inspection and quality control procedures.

7.1 Concentricity and Squareness Relationships—Concentricity and squareness relationships are defined in accordance with referenced dimensions given in Table 1, SAE J780a (November, 1978), and are:
(a) Concentricity between Bore A and Bore C (F.I.M.).
(b) Concentricity between Shaft B and Bore C (F.I.M.).
(c) Squareness between Shaft B and Surface H (F.I.M.).
(d) Squareness between Shaft B and Surface I (F.I.M.).
(e) Shaft endplay.

Measurement of these relationships shall be made with dial indicators having accuracy within ±0.0001 in (2.5 μm). The use of precision collets and/or expanding mandrels will greatly facilitate measurement of these relationships.

Measurement of concentricity between B and C and squareness between B and I shall be made with bearing and shaft assembly installed in the pump housing. Measurement of squareness between B and H may be made independent of other pump components. See SAE J780a (November, 1978) for recommended limits of concentricity and squareness.

7.2 Operating Length Variation—Operating length variation is defined as dimension (D_4-D_2) given in the standard drawing format Fig. 1, Section 5 (Figs. 1A, 1B, and 1I). Recommended operating length variations are 0.048 in (1.22 mm) for Figs. 1A and 1B and 0.015 in (0.38 mm) for Fig. 1I.

NOTE: Dimension D_2 should exceed dimension D_1 by 0.020 in (0.51 mm) minimum. To determine dimension D_1, compress the seal from its free length to a point where the load reaches 2.5 times the nominal load specified at the nominal operating length, D_3.

7.3 Seal Load Determinations—Seal load determinations are to be performed on spring scale apparatus having the capability of measuring operating length to within ±0.001 in (0.02 mm) and loads to within ±0.1 lb · f (0.4 N). Seal loads are to be determined by compressing the seal to its solid length, D_1, and reading the seal loads in the direction from solid length toward the free length. Repeat the procedure until two consecutive readings are within 0.5 lb · f (2.2 N).

Solid length load is defined as 2.5 times nominal load specified at the nominal operating length.

Maximum load is defined as the load measured at the minimum operating length, D_2. Nominal load is defined as the load measured at nominal operating length, D_3. The minimum load is defined as the load measured at the maximum operating length, D_4. Recommended load tolerance at nominal operating length is ±20%.

7.4 Overall Flatness—It is recommended that overall flatness measurements of reflective primary seal rings be made through the use of an optical flat and helium monochromatic light source (half wave length = 11.6 μin (0.3 μm)). Optical flat reference surface shall be flat to within 1/10 a helium light band. See Table 1 for flatness of various primary face materials. In the case of non-reflective seal rings, stylus or vacuum type equipment shall be used as the recommended means of determining flatness. See paragraphs 7.6 and 7.7 respectively.

7.5 Surface Roughness—Surface roughness of seal rings shall be measured with stylus type surface profiling equipment having either linear or circumferential trace capability. The equipment employed shall be in compliance with ANSI B46.1, Section 4, 1978. Stylus radius shall be 0.0005 in (0.013 mm) and a minimum stroke of 0.125 in (3 mm) shall be used. Stylus load shall be sufficient to maintain contact with the surface without surface destruction and shall not exceed 2.5 g. Wave length cut-off shall be 0.030 in (0.76 mm). It is recommended that a reading (not longer than 3 s) be taken at three different locations and averaged. See Table 1 for surface roughness limits of various seal ring materials.

7.6 Waviness—Waviness of primary seal faces is defined as the deviation from a plane at a given diameter. Primary seal faces generally have two node or *saddle shape* waviness profiles. Waviness shall be measured with stylus type surface profiling equipment having complete circumferential trace capability within an accuracy of ±2.5 μin (±0.06 μm) in the direction vertical to the primary seal ring face. Measurements shall be made on the mean primary seal ring face diameter using a 3/32 in (2 mm) diameter spherical stylus. (NOTE: Care must be taken to eliminate any significant degree of tilt of the primary seal ring face during a waviness measurement.) Surface roughness shall be electronically filtered. See Table 1 for waviness limits of various seal ring materials.

7.7 Vacuum Test—Non-reflective seal rings may be measured for flatness on vacuum-type equipment capable of detecting 0.2 in Hg (0.7 kPa) pressure drop in 10 s at 25 in Hg (85 kPa) test vacuum. The test piece shall be placed against a clean surface lapped to one helium light band and having a surface roughness average of 4 μin (0.1 μm) Ra or less. Figs. 4, 5, and 6 depict typical vacuum test equipment.

8. References:

1. N. E. Payerle, "Engine Coolant Performance in Late Model Passenger Cars"; SAE Paper No. 760631, 1976.
2. E. Beynon, "Fluid Performance of Factory-Installed Antifreeze Coolants in Passenger Car Service"; SAE Paper No. 660128, 1966.

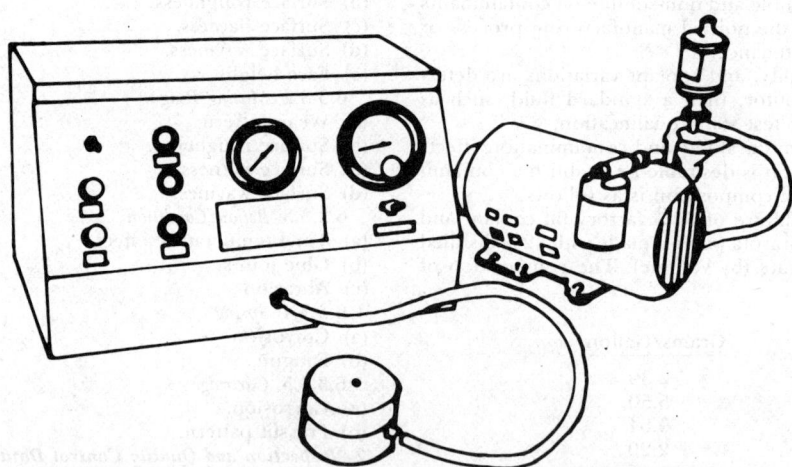

FIG. 5.—PICTORIAL ARRANGEMENT OF ACTUAL VACUUM TEST EQUIPMENT

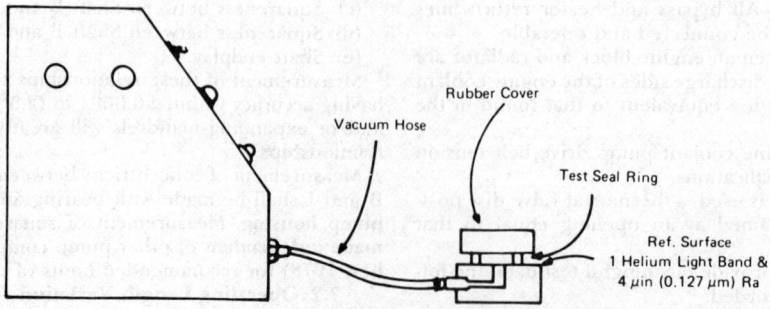

FIG. 6—DETAIL SHOWING TEST SEAL RING

GUIDE TO THE APPLICATION AND USE OF PASSENGER CAR AIR-CONDITIONING COMPRESSOR FACE SEALS—J1954 MAY90 — SAE Recommended Practice

Report of the Face Seal Standards Committee approved May 1990.

1. Scope—This SAE Recommended Practice is intended as a guide in the usage of mechanical face seals for the passenger car air-conditioning compressor application. Included in this guide is a compilation of present practices; for example, a description of various type seals, material combinations, design data, tolerances, drawing format, qualification testing, inspection information and quality control data. The terminology used is recommended to promote uniformity in seal nomenclature.

2. Seal and Mating Ring Types—The mechanical face seal assemblies utilized in passenger car air-conditioning compressors consist of a seal head and a mating ring. Although many variations of face seal designs exist, two basic concepts are predominantly applied with respect to both seal head and mating rings.

2.1 Seal Head—The two basic design classifications are pusher and nonpusher. By definition, pusher seals are mechanical seals employing a secondary sealing element (such as O-rings, V-rings, U-cups, wedges, etc.) that are pushed along the respective sealing surfaces while the primary sealing function at the faces is being performed. A typical pusher seal is shown in Figure 1B. Nonpusher seals shown in Figures 1A, 1C and 1D employ a bellows or diaphragm as a secondary sealing element. Axial displacement of the seal components caused by wear at the seal faces and/or shaft movement causes sliding or "pushing" of the secondary seals in the pusher seals. This same displacement in nonpusher seals is taken up by flexing of the bellows. The selection of seal head design is based primarily on seal environmental conditions.

Independent of classification, the seal head is the rotating element of the seal assembly and is comprised of the following components:

a. Primary Ring (see Item 1, Figure 2) which is in rubbing contact with the stationary mating ring (see Item 5, Figure 2). This interface forms the primary seal.
b. Secondary Seal (see Item 2, Figure 2) as defined in 2.1.
c. Spring (see Item 3, Figure 2) which provides mechanical load to the primary faces.
d. Hardware consisting of retaining and drive devices (see Item 4, Figure 2) which provide reactionary support for the spring and transmit shaft rotation drive to the primary ring. Secondary seal support function is also provided by these components. On some nonpusher seal designs, seal head drive is provided via a positive interference of the elastomeric secondary seal element with the compressor shaft and hardware components.

2.1.1 NONPUSHER SEAL—Nonpusher seals have been preferred for passenger car air-conditioning compressor seals where one or more of the following conditions exist:

a. Axial shaft movement with amplitudes greater than 0.008 in (0.203 mm).
b. Deposition of carbonized oil, foreign material, and/or corrosive products on the compressor shaft in sufficient magnitude to restrict the movement of the pusher-type secondary seal. This restriction prevents the seal from following shaft dynamics or prevents the primary faces from axial movement to accommodate seal face wear. The end result is excess leakage.
c. Distortion of the primary seal faces caused by excessive volume swell of secondary seal elastomer.
d. Excessive face loading due to pressure being sealed. Reduced seal balance is required to minimize heat generation in applications where standard balanced seals cannot be accommodated. Nonpusher seals, although classified as unbalanced, can be designed to have a balance of less than one without a stepped-shaft design normally required for a balanced seal. It should be noted that the balance of nonpusher seals will vary with the pressure sealed. The amount of balance variation is a function of the bellows design and materials. The term "seal balance" as used here is the ratio of hydraulic area closing the seal face to that of hydraulic area opening (Ac/Ao).

2.1.1.1 *Advantages*—In addition to the application conditions listed above, nonpusher seals have the following advantages:

a. Normally do not require a positive mechanical drive with the compressor shaft.
b. Are more tolerant of out-of-square mating ring face conditions.

2.1.1.2 *Disadvantages:*
a. Generally more expensive due to the bellows.
b. Generally more difficult to install into compressor, due to the interference fit between shaft and drive portion of bellows.

2.1.2 PUSHER SEALS—Recent passenger car compressor designs have substantially improved the seal environment, particularly with respect to temperature excursions and shaft end play. These improvements have permitted the use of pusher seals.

2.1.2.1 *Advantages:*
a. Generally lower cost due to the O-ring.
b. Generally reduced exposure of the elastomeric member to interfacial frictional heat.
c. Closer primary ring-to-shaft concentricity.

2.1.2.2 *Disadvantages:*
a. Require positive shaft drive.
b. More vulnerable to seal "hang-up".
c. More vulnerable to primary ring face distortion caused by secondary seal volume change.

2.2 Mating Ring—Two basic mating ring designs are incorporated in air-conditioning compressor seals:

a. End-plate designs as shown in Figure 1G which are an integral part of the compressor.
b. Separate secondary seal-mounted mating rings as shown in Figures 1e, 1f, 1h and 1j.

The selection of the mating ring design is based principally on compressor design, seal operating conditions and mating ring material.

2.2.1 END PLATE DESIGNS—End plate designs are generally machined from fine-grained cast iron with a precision-lapped seal mating surface.

2.2.1.1 *Advantages:*
a. Requires less axial space.
b. Provides better frictional heat transfer capabilities.

2.2.1.2 *Disadvantages:*
a. Geometry generally limits material selection.
b. Subject to face distortion due to bolt stresses. Distortion can be minimized by use of a clamping ring which eliminates localized bolt stresses.
c. More difficult to lap primary seal face due to size and shape.

2.2.2 SEPARATE MATING RINGS—Separate mating rings are fabricated from various materials and are precision lapped.

2.2.2.1 *Advantages:*
a. Due to relatively simple shape, rings can be produced from various materials such as cast iron, sintered iron, high alumina ceramics, etc.
b. Lower cost due to simpler configuration and smaller size.
c. Less face distortion when installed.
d. Easier to lap.

2.2.2.2 *Disadvantages:*
a. Reduced frictional heat dissipation.
b. Secondary seal is often subjected to higher temperatures due to closer proximity to the primary seal faces.
c. More potential for installation damage or misassembly of secondary seals.

2.3 Secondary Seals—Three basic secondary seals are employed. The most universal is an O-ring. Flat gaskets and molded or lathe-cut rings are also incorporated in conjunction with end-plate designs and mating rings. In end-plate designs, the O-ring, molded or lathe-cut ring is preferred over flat gaskets for control of flatness and to improve heat transfer from the end plate.

O-ring configurations are also preferred for separate mating rings, again primarily for simplicity and cost.

3. Seal Materials—Environmental conditions dictate the type of material which should be used in a specific application. Seal materials can be fully evaluated only in terms of specific operating conditions and performance requirements. The following paragraphs describe primary seal ring materials, mating ring materials, secondary seals and the elastomeric compounds used to fabricate them, seal hardware, and springs.

3.1 Primary Ring—The primary ring material must be impervious to environmental pressure, display adequate friction/wear properties, have good thermal conductivity and remain stable under the temperature, pressure and fluid conditions in a compressor.

26.424

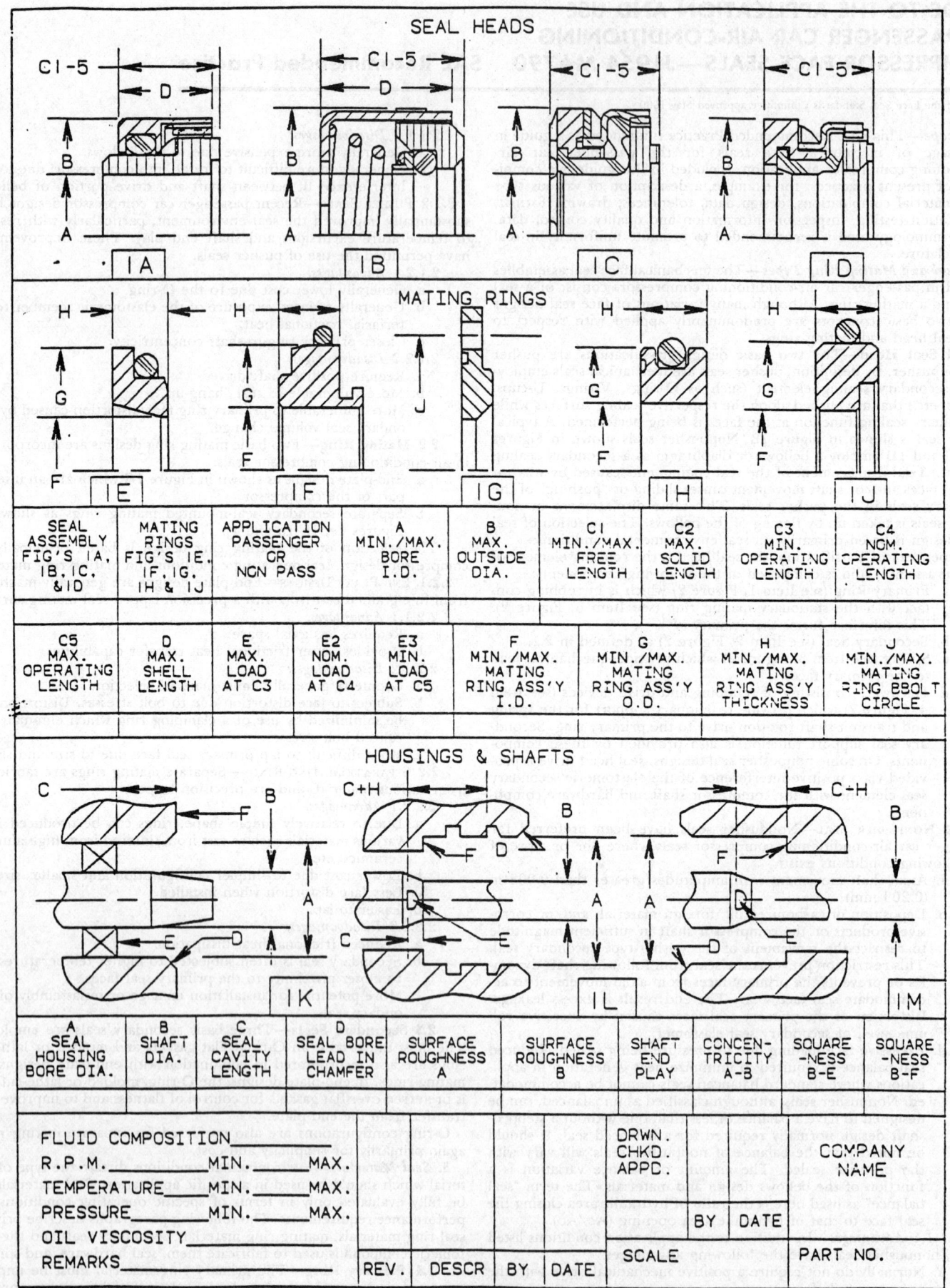

FIG. 1—DRAWING FORMAT

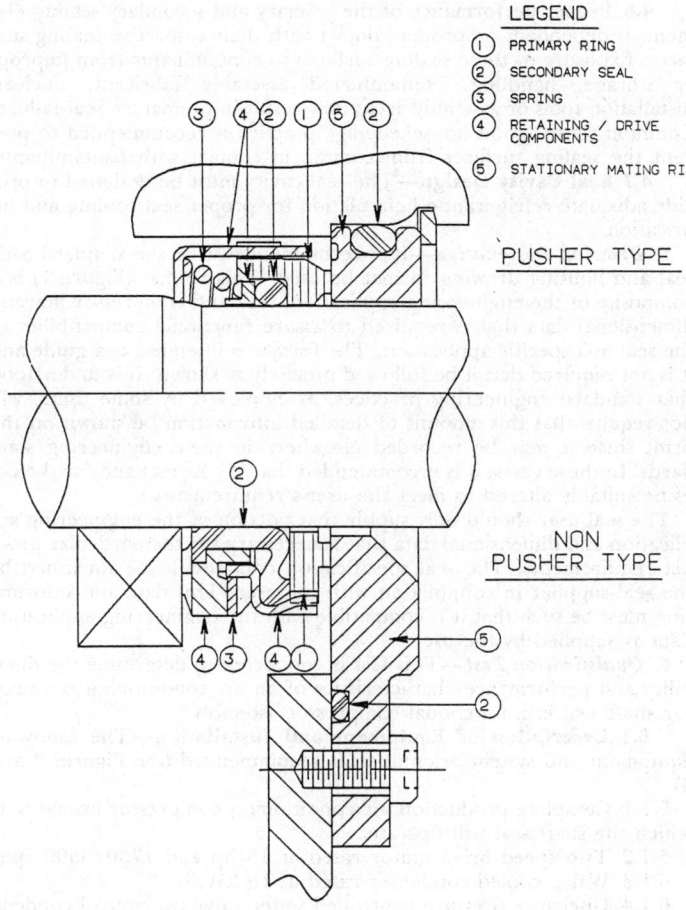

LEGEND
① PRIMARY RING
② SECONDARY SEAL
③ SPRING
④ RETAINING / DRIVE COMPONENTS
⑤ STATIONARY MATING RING

FIG. 2—SEAL COMPONENTS

3.1.1 RESIN BONDED GRAPHITE—Thermoset materials generally made from phenolic resin, graphite and varying amounts of mineral fillers. Usually processed in the 149 to 315°C temperature range.
 3.1.1.1 *Advantages:*
  a. Good wear resistance.
  b. Readily molded to complex geometry and close tolerances.
  c. Good inherent low porosity.
  d. Low cost.
  e. Good handling characteristics (resists chips and cracks).
 3.1.1.2 *Disadvantages:*
  a. Fair thermal stability.
  b. Lower maximum operating temperature.
  c. Fair thermal conductivity.
 3.1.2 CARBON-GRAPHITE—A manufactured product of carbon and graphite in a rigid, hard structure produced by firing at high temperatures usually in the range of 900 to 2000°C. Carbon-graphite is inherently porous after firing and must be made impervious for use as a primary seal ring. Various materials are used to impregnate the carbon-graphite structure.
 3.1.2.1 *Advantages:*
  a. Excellent temperature resistance and stability.
  b. Excellent compatibility with refrigerants and associated lubricants (impregnant must be properly chosen).
  c. Excellent wear resistance.
  d. Good thermal conductivity.
  e. Low thermal distortion coefficient.
 3.1.2.2 *Disadvantages:*
  a. Difficult to mold complex shapes and maintain close tolerances.
  b. Fair handling characteristics (vulnerable to chips and cracks).
  c. High cost.

3.2 Mating Ring—The mating ring is usually a dissimilar material, and harder than the primary seal ring. The material choice depends upon operating conditions, configuration, costs and performance requirements.

3.2.1 CAST IRON—Cast iron is used in various forms and grades. Microstructure and hardness are principal criteria for grade selection.
 3.2.1.1 *Advantages:*
  a. Low cost for complex shapes.
  b. Excellent thermal shock resistance.
  c. Good wear resistance.
  d. Low thermal distortion coefficient.
 3.2.1.2 *Disadvantages:*
  a. Control of microstructure and casting defects critical to insure good performance.
  b. Fair corrosion resistance.
  c. Fair scratch resistance.
3.2.2 CERAMIC—Ceramic materials used are hard, dense aluminum oxides (85 to 99% Al_2O_3 by weight) formed by compacting finely ground oxide powders with fluxing agents and inhibitors at high pressure. The formed part is then fired at temperatures of 1400 to 1800°C. After firing, the part is composed mostly of pure alumina crystals of controlled size.
 3.2.2.1 *Advantages:*
  a. Excellent wear resistance.
  b. Excellent dimensional stability.
  c. Excellent fluid compatibility.
  d. Low cost for simple shapes.
  e. Excellent scratch resistance.
 3.2.2.2 *Disadvantages:*
  a. Mechanical and thermal shock susceptibility.
  b. Difficult to mold complex shapes.
  c. Low thermal conductivity.

3.3 Secondary Seal—Secondary seals are generally made from elastomers principally selected for compatibility with refrigerants and associated lubricants. Other vehicle environmental conditions should also be considered when selecting the secondary seal material.
3.3.1 NEOPRENE COMPOUNDS (CR)—This material's operating temperature range is −40 to 121°C.
 3.3.1.1 *Advantages:*
  a. Good processability.
  b. Fair oil resistance.
  c. Low cost.
  d. Low permeability coefficient.
 3.3.1.2 *Disadvantages:*
  a. Fair temperature resistance.
3.3.2 NITRILE COMPOUNDS (NBR)—This material's operating temperature range is −40 to 121°C.
 3.3.2.1 *Advantages:*
  a. Good processability.
  b. Good oil resistance.
  c. Low cost.
  d. Low permeability coefficient.
 3.3.2.2 *Disadvantages:*
  a. Fair temperature resistance.
3.3.3 FLUOROELASTOMER COMPOUNDS (FKM)—This material's operating temperature range is −32 to 204°C.
 3.3.3.1 *Advantages:*
  a. Excellent high temperature resistance.
  b. Excellent oil resistance.
  c. Low permeability coefficient.
 3.3.3.2 *Disadvantages:*
  a. Poor processability (in terms of shape factor).
  b. High cost.

3.4 Hardware—Hardware components are usually fabricated from low carbon steel or powdered metal. They are easily formed, readily available and low in cost. They may require plating or other treatment to achieve in-process corrosion resistance.

3.5 Springs
3.5.1 COIL SPRINGS—Coil springs are compression springs wound from music wire or stainless steel wire. They are low in cost, and may require plating or other treatment to achieve in-process corrosion resistance.
3.5.2 WAVE SPRINGS—Wave springs are fabricated from carbon steel or stainless steel sheet stock. They are low to moderate cost, and may require plating or other treatment to achieve in-process corrosion protection.

4. Application Design Data—This section is to provide guidelines as to specific dimensions and conditions that may functionally affect the passenger car air-conditioning compressor face seal.

4.1 Flatness—Overall flatness of sealing surfaces is critical to maintain a refrigerant oil or gas tight seal.

4.1.1 PRIMARY SEAL RING AND MATING RINGS—Surface flatness for both should be two helium light bands maximum.

4.2 Surface Roughness—Surface roughness is a function of base material, grain size, structure and method of finishing. Surface roughness is to be evaluated as agreed upon by the supplier and user for specific combinations of materials and for specific applications. See Table I for recommended surface finish roughness for seal rings.

TABLE I—RECOMMENDED SURFACE ROUGHNESS FOR PRIMARY AND MATING SEAL RINGS

ROUGHNESS		RESIN BONDED GRAPHITE	CARBON GRAPHITE	CAST IRON	CERAMICS (Al_2O_3)
µm Ra	MIN	0.025	0.051	0.076	0.10
	MAX	0.38	0.63	0.25	0.45

4.3 Waviness—Waviness should be 0.63 mm maximum for both sealing surfaces.

4.4.1 SQUARENESS—Squareness of the face of the mating ring is to be within 0.05 mm FIM of shaft center line.

4.4.2 SHAFT RUNOUT—Shaft runout is defined as twice the distance the center of the shaft is displaced from the center of rotation, expressed in FIM, and should not exceed 0.05 mm.

4.4.3 CONCENTRICITY—Concentricity is defined as the radial distance the geometric center of the mating ring is displaced from the geometric center of the shaft along the axis of rotation and should be held to within 0.13 mm FIM.

4.4.4 END PLAY—End Play is defined as total axial shaft movement and is a function of bearing selection and mating component fits. The maximum recommended end play is 0.2 mm.

4.5 Installation—Proper installation of seal components is necessary for the seal assembly to function properly.

4.5.1 LEAD-IN CHAMFER—A lead-in chamfer is required on the seal components, compressor housing bore, and the bearing shaft end for ease of seal installation and prevention of damage to the secondary seals. All corners should be blended smoothly and lubricants may be required for seal component assembly.

4.5.2 INSTALLATION TOOLS—Special seal component installation tools may be required to be used as lead-in chamfer or cover sharp edges (keyway, splines, etc.) to protect secondary seals from damage.

4.6 Proper performance of the primary and secondary sealing elements is dependent on proper contact with their respective sealing surfaces. Exposure of these sealing surfaces to contaminants from improper storage, handling, contaminated assembly lubricant, unclean installation tools or assembly areas may result in premature seal failure. Common sense "good housekeeping" practice is recommended to prevent the sealing surfaces from coming in contact with contaminants.

4.7 Seal Cavity Design—The seal cavity must be designed to provide adequate refrigerant-oil circulation for proper seal cooling and lubrication.

5. Drawing Designation—It is recommended that the standard SAE seal and housing drawing format be used. This format (Figure 1) is a composite of the engineering application, seal, and compressor housing dimensional data that is required to assure functional compatibility of the seal in a specific application. The format is intended as a guide and it is not required that it be followed precisely as shown. It is understood that standard engineering practices, as employed by some users, will not require that this amount of detailed information be shown on the print since it may be recorded elsewhere in their engineering standards. In those cases, it is recommended that the format and/or sketches be suitably altered to meet the user's requirements.

The seal user should only supply that portion of the engineering application and dimensional data that is necessary for the particular product requirements. The seal specification data should be furnished by the seal supplier in conjunction with the user. This data and information must be such that it is compatible with the engineering application data as supplied by the user.

6. Qualification Test—This test is conducted to determine the durability and performance characteristics of an air conditioning compressor shaft seal in a functional compressor assembly.

6.1 Description of Equipment and Installation—The following equipment and system orientation is recommended (see Figures 3 and 4):

6.1.1 Complete production air conditioning compressor assembly in which the shaft seal will operate.

6.1.2 Two speed drive motor rated at 15 hp and 1750/3500 rpm.

6.1.3 Water cooled condenser rated at 10 kW/h.

6.1.4 Discharge pressure controlled water valve (to control condenser cooling water flow).

6.1.5 Automotive evaporator and blower motor assembly.

6.1.6 Thermostatic expansion valve (15°SH).

6.1.7 Valves (to add refrigerant charge).

6.1.8 12.8 V power supply (40 amp minimum) for clutch and blower

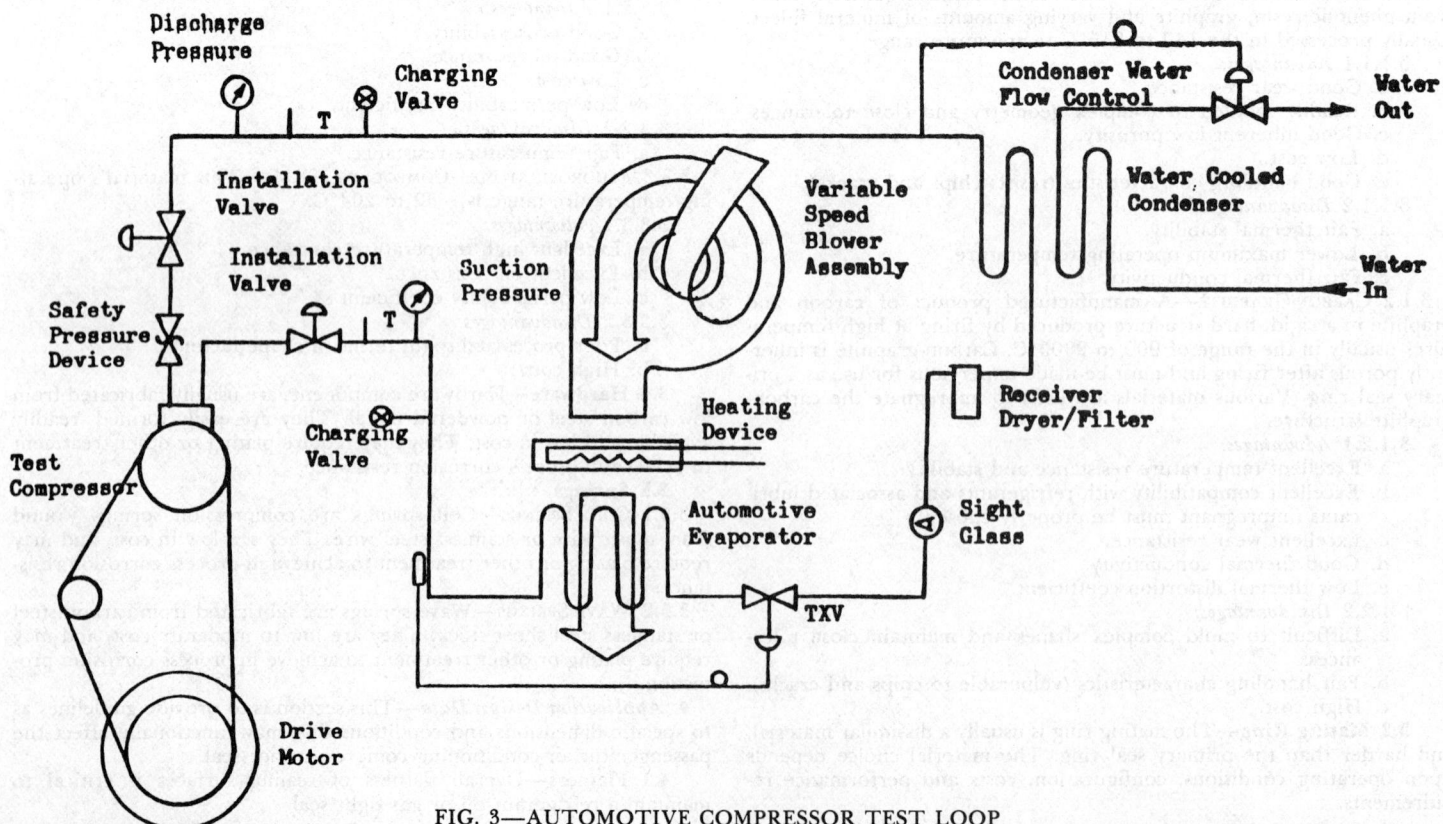

FIG. 3—AUTOMOTIVE COMPRESSOR TEST LOOP

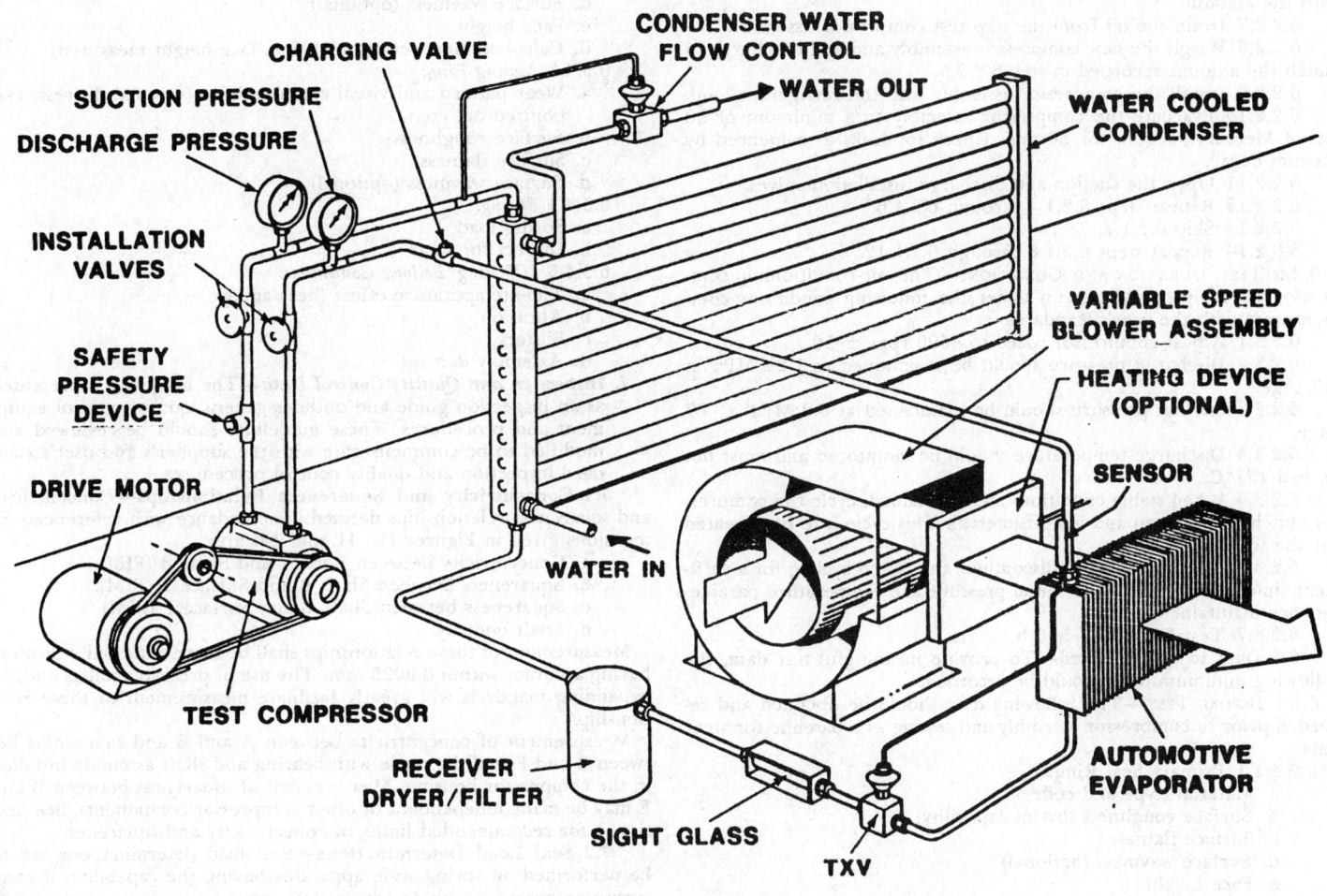

FIG. 4—AUTOMOTIVE COMPRESSOR TEST LOOP

motor assembly.

6.1.9 Cycle counter and running time meter.
6.1.10 Valves for installing and removing the test compressor from the refrigerant loop.
6.1.11 Gauges to measure system pressures.
6.1.12 Thermocouples for measuring system temperatures.
6.1.13 Liquid line sight glass.
6.1.14 Refrigerant hoses and clamps (production parts preferred).
6.1.15 Evacuation pump.
6.1.16 Halogen leak detector capable of measuring 7.1 g/year refrigerant 12 leakage per year and 7.1 g/year calibrated leak master.
6.1.17 Scale accurate to 1.0 mg to determine seal oil weepage.
6.1.18 Scale accurate to 1.77 g, avdp., to determine weight of compressor and oil.
6.1.19 An appropriate pressure relief device must be added to the refrigerant system if not integral to the compressor. All water and refrigerant lines shall have a burst pressure at least two and one half times the operating pressure and conform to all reasonable safe operating practices.
6.1.20 Heating device for evaporator.
6.1.21 Liquid line receiver/dryer and filter.

6.2 Procedure—The following procedural outline is provided as a guide; obviously, this procedure should be modified to be compatible with the user's established standard engineering practice.

6.2.1 INITIAL OPERATING CONDITIONS—The air conditioning compressor assembly should be prepared for test by the following procedure.

6.2.1.1 Drain the oil from the compressor assembly and record the quantity.
6.2.1.2 Weigh the compressor assembly and record.
6.2.1.3 Replace the oil charge and install the compressor assembly into the refrigerant loop.
6.2.1.4 A new felt of known weight is to be inserted at the external side of the seal assembly to absorb oil weepage.

6.2.1.5 Evacuate the system to a minimum of 50 mm of Mercury vacuum for 30 min. Check for leaks as evidenced by vacuum decay.
6.2.1.6 Belt tension should be set to manufacturer's specifications.
6.2.1.7 Weigh in 0.91 kg of refrigerant 12 to the discharge side service valve.
6.2.1.8 With Halogen leak detector calibrated to 14.18 g/year refrigerant 12 leak rate, check all connections for leaks. Measure and record leakage rate at the shaft seal. See Section 8 for recommended seal cavity leakage apparatus and procedure.
6.2.1.9 Operate compressor at 2300 rpm, 1.38 MPa ± .07 gage discharge pressure and 0.2 MPa ± .014 gage suction pressure. Adjust fan speed and/or heat load to the evaporator assembly to obtain desired suction pressure after the discharge pressure is established by adjusting the water control. DO NOT ADD CHARGE TO DISCHARGE SIDE WHILE THE COMPRESSOR IS OPERATING.
6.2.1.10 Add refrigerant 12 vapor to the system through the suction charging valve until the flow in the sight glass just becomes bubble-free.
6.2.1.11 Repeat step 6.2.1.9 if needed.
6.2.1.12 Operate the compressor at the above conditions for 2 h for break-in.

6.2.2 SUBSEQUENT TEST INSTALLATION— Subsequent test compressor installations may be made without discharging the entire refrigerant loop by utilizing the following procedure:

6.2.2.1 Stop the compressor rotation and allow the suction and discharge pressure to equalize.
6.2.2.2 Close the suction and discharge installation valves.
6.2.2.3 Loosen the discharge line connection slightly to allow the refrigerant to bleed off slowly without losing the oil charge.
6.2.2.4 Remove the compressor assembly from the refrigerant loop.
6.2.2.5 Weigh the compressor assembly at end of test and record the weight.
6.2.2.6 Drain the oil from the tested compressor assembly and re-

cord the amount.

6.2.2.7 Drain the oil from the new test compressor assembly.

6.2.2.8 Weigh the new compressor assembly and add oil charge to match the amount recorded in step 6.2.2.5.

6.2.2.9 Install the compressor assembly into the refrigerant loop.

6.2.2.10 Evacuate the compressor assembly to a minimum of 50 μm of Mercury vacuum for 30 min. Check for leaks as evidenced by vacuum decay.

6.2.2.11 Open the suction and discharge installation valves.

6.2.2.12 Repeat steps 6.2.1.4 through 6.2.1.6.

6.2.2.13 Skip 6.2.1.7.

6.2.2.14 Repeat steps 6.2.1.8 through 6.2.1.12.

6.2.3 TEST DURATION AND CONDITIONS—The air-conditioning compressor assembly should be run under the following conditions commensurate with the user's standards:

6.2.3.1 Adjust compressor speed to 4600 rpm ± 50.

6.2.3.2 Discharge pressure should be maintained at 1.24 MPa ± .07 gage.

6.2.3.3 Suction pressure should be maintained at 0.2 MPa ± .07 gage.

6.2.3.4 Discharge temperature should be monitored and must not exceed 121°C.

6.2.3.5 When stable conditions are maintained, cycle the compressor clutch at a 10 s on and 5 s off interval. This cycle is to be repeated for the test duration.

6.2.3.6 At 24 h intervals, discontinue the clutch cycling for a sufficient time to insure that the initial pressure and temperature parameters are maintained.

6.2.3.7 Test Duration—500 h.

6.3 Data to be Recorded—To provide meaningful test data, the following minimum data should be recorded:

6.3.1 BEFORE TEST—The following data should be obtained and recorded prior to compressor assembly and testing as a baseline for wear data:

6.3.1.1 Primary Seal Ring:
  a. Material (type and code)
  b. Surface roughness (by lot capability)
  c. Surface flatness
  d. Surface waviness (optional)
  e. Face height

6.3.1.2 Mating Ring:
  a. Material (type and code)
  b. Surface roughness
  c. Surface flatness
  d. Surface waviness (optional)
  e. Hardness (optional)

6.3.1.3 Seal and Housing Assembly:
  a. Seal head operating length
  b. Seal load at operating length (on an oiled shaft)
  c. Test schedule, refrigerant and vacuum leakage rate
  d. Shaft speed
  e. Squareness of mating ring to shaft
  f. Shaft runout
  g. Shaft end play

6.3.2 DURING TEST—The following data should be recorded during the test sequence at the stabilized condition:
  a. Compressor rpm
  b. System pressures
  c. System temperatures

6.3.3 AFTER TEST—The following data or observations should be noted and recorded after completion of the test sequence prior to removal of the seal from the compressor assembly:
  a. Felt weight
  b. Refrigerant leakage rate at seal housing area
  c. If the refrigerant leakage observed is out of specification, carefully locate the leak source.

6.3.4 AFTER TEST—The following data or observations should be noted and recorded after completion of the test sequence:

6.3.4.1 *Seal and Housing Assembly:*
  a. Seal load at operating length
  b. Shaft end play
  c. Shaft runout
  d. Contamination

6.3.4.2 *Primary Seal Ring:*
  a. Wear pattern and visual characteristics (blisters, voids, cracks, drive wear, etc.)
  b. Surface roughness
  c. Surface flatness
  d. Surface waviness (optional)
  e. Face height
  d. Calculate face wear (initial final face height measured)

6.3.4.3 *Mating Ring:*
  a. Wear pattern and visual characteristics (thermal distress, carbonized oil, etc.)
  b. Surface roughness
  c. Surface flatness
  d. Surface waviness (optional)

6.3.4.4 *Spring:*
  a. Spring load
  b. Inspect for wear

6.3.4.5 *"O" Ring/Bellows Conditions:*
  a. Time-temperature effect (heat aging)
  b. Abrasion
  c. Blisters
  d. Assembly damage

7. Inspection and Quality Control Data—The following is presented as an inspection guide and outlines general quality control equipment and procedures. These guidelines should be reviewed and modified to be commensurate with the supplier's and user's standard inspection and quality control procedures:

7.1 Concentricity and Squareness Relationships—Concentricity and squareness relationships defined in accordance with referenced dimensions given in Figures 1K, 1L and 1M are:
  a. Concentricity between Bore A and Shaft B (FIM).
  b. Squareness between Shaft B and Surface E (FIM).
  c. Squareness between Shaft B and Surface F (FIM).
  d. Shaft endplay.

Measurement of these relationships shall be made with dial indicators having accuracy within 0.0025 mm. The use of precision collets and/or expanding mandrels will greatly facilitate measurement of these relationships.

Measurement of concentricity between A and B and squareness between B and F shall be made with bearing and shaft assembly installed in the compressor housing. Measurement of squareness between B and E may be made independent of other compressor components. See Section 4 for recommended limits of concentricity and squareness.

7.2 Seal Load Determinations—Seal load determinations are to be performed on spring scale apparatus having the capability of measuring operating length to within 0.02 mm and loads to within 0.4N. Seal loads are to be determined by compressing the seal to its solid length, C2, and reading the seal loads in the direction from solid length toward the free length. Repeat the procedure until two consecutive readings are within 2.2n.

Solid length is defined as the dimension at which the seal load is 2.5 times nominal load specified at the nominal operating length.

Maximum load is defined as the load measured at the minimum operating length, C3. Nominal load is defined as the load measured at nominal operating length, C4. The minimum load is defined as the load measured at the maximum operating length, C_5. Recommended load tolerance at nominal operating length is ±20%.

7.3 Operating Length Variation—Operating length variation is defined as dimension (C5-C3) given in Figure 1, (Figures 1A, 1B, 1C, and 1D). Recommended operating length variations are 0.61 mm maximum for Figures 1A, 1B, 1C and 1D.

NOTE: Dimension C3 should exceed dimension C2 by 0.51 mm minimum. To determine dimension C2, compress the seal from its free length to a point where the load reaches 2.5 times the nominal load specified at the nominal operating length, C4.

7.4 Overall Flatness—It is recommended that overall flatness measurements of reflective primary seal face be made through the use of an optical flat and helium monochromatic light source (half wave length = 0.3 μm). Optical flat reference surface shall be flat to within 1/10 a helium light band. See 4.1 for flatness recommendations. In the case of nonreflective seal faces, stylus type equipment shall be used to determine flatness. See 7.6.

7.5 Surface Roughness—Surface roughness of seal faces shall be measured with stylus type surface profiling equipment having either linear or circumferential trace capability. The equipment employed shall be in compliance with ANSI B46.1, Section 4, 1978. Stylus radius shall be 0.013 mm and a minimum stroke of 3.2 mm shall be used. Stylus load shall be sufficient to maintain contact with the surface without surface destruction and shall not exceed 2.5 g. Wave length cutoff shall be 0.76 mm. It is recommended that a data reading be taken at three different locations and averaged. See Table 1 for surface roughness limits of various seal face materials.

7.6 Waviness—Waviness of primary seal faces is defined as the deviation from a plane at a given diameter. Primary seal faces generally have two node or saddle shape waviness profiles. Waviness shall be measured with stylus type surface profiling equipment having complete circumferential trace capability within an accuracy of ±0.06 μm in the direction vertical to the primary seal ring face. Measurements shall be made on the mean primary seal ring face diameter using a 2 mm diameter spherical stylus.

NOTE: Care must be taken to eliminate any significant degree of tilt of the primary seal ring face during a waviness measurement. Surface roughness shall be electronically filtered. See 4.3 for maximum waviness recommendation.

8. Seal Leakage Determination—The following procedure is recommended for quantitatively determining the leak rate of a seal. The physical arrangement of the actual test device should be adjusted for the specific cavity arrangement being tested.

8.1 The intent of this procedure is to have a device which isolates the seal cavity from all surrounding halogen sources. This cavity is continuously flushed with a known halogen-free gas (pure air or nitrogen). Using this arrangement, any halogen detected will be from the seal cavity being evaluated. Calibration of the device is achieved by installing a calibrated leak in a dummy cavity of the same configuration as that being evaluated. By first calibrating the device on the dummy cavity; the values obtained during test can be directly equated to the calibrated leak.

8.2 Equipment—The following equipment is required for this test procedure:
    a. Electronic halogen leak detector.
    b. Calibrated standard leak source (Figure 5)
    c. Halogen-free gas supply (dry air or Nitrogen)
    d. Dummy seal cavity fixture.
    e. Gas supply orifice.
    f. Seal cavity leak test fixture (Figure 6)

8.3 Equipment Arrangement—The equipment is arranged as shown in Figure 7. The gas flow is established by sizing the orifice for thorough and continuous flushing of the seal cavity at a rate greater than 10 cavity vol/s.

8.4 Test Procedure:

8.4.1 CALIBRATION:

8.4.1.1 Set air source for proper flushing.

8.4.1.2 Turn "on" halogen leak detector.

8.4.1.3 Set standard leak source at 14.18 g/year.

8.4.1.4 Install seal cavity leak test device on standard leak source (without probe). (See Figure 5.)

8.4.1.5 Flush the leak source for 15 s.

8.4.1.6 Install leak detector probe in device.

8.4.1.7 After 5 s, calibrate leak detector for a mid-scale reading to the value of the standard leak source.

8.4.1.8 Reading should be stable for 5 s.

8.4.1.9 Remove leak-detector probe.

8.4.2 SEAL TESTING:

8.4.2.1 Remove the dust shield and felt wicking (if used) from compressor.

8.4.2.2 Calibrate device per 8.4.1.

8.4.2.3 Install device in compressor seal cavity. (See Figure 7.)

8.4.2.4 Flush seal cavity for 15 s.

8.4.2.5 Install leak-detector probe in device. If excessive halogen signal is indicated, immediately remove probe to avoid severely contaminated sensing element. Allow flushing to continue for an additional minute and again install the probe. If excessive halogen is again indicated, allow to stabilize for 5 s and record the value indicated as the cavity leak rate in g/year.

8.4.2.6 Remove leak detector probe from device.

8.4.2.7 Recheck calibration per 8.4.1. If calibration is validated, testing is complete. If validation is not achieved, repeat the test.

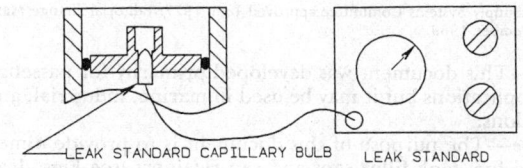

FIG. 5—CALIBRATED STANDARD LEAK SOURCE

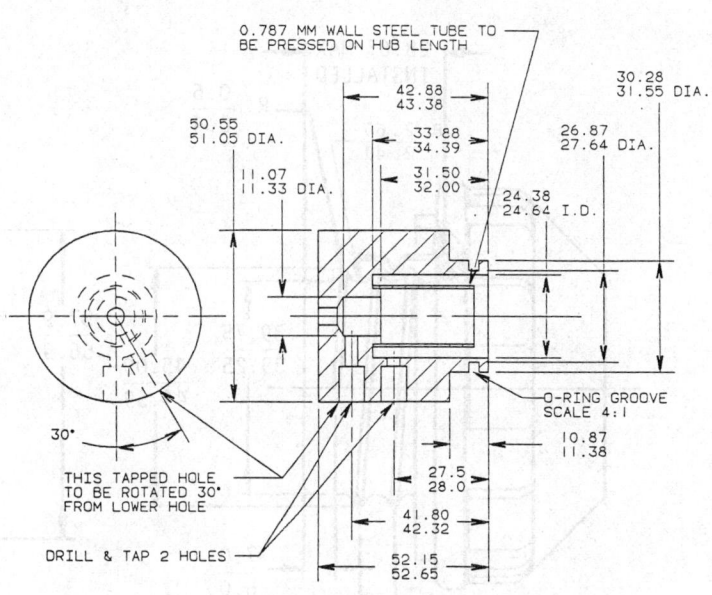

FIG. 6—TYPICAL CAVITY LEAK TEST FIXTURE

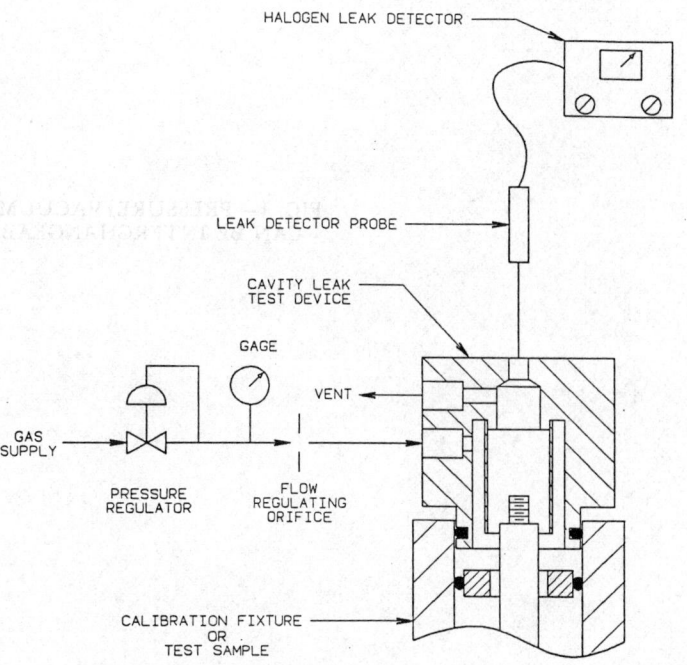

FIG. 7—EQUIPMENT ARRANGEMENT

(R) FUEL TANK FILLER CAP AND CAP RETAINER THREADED—SAE J1114 NOV89 — SAE Recommended Practice

Report of Fuel Supply Systems Committee approved June 1977. Editorial change March 1978. Completely revised by the Fuel Supply Systems Forum Committee November 1989.

1. Scope—This document was developed primarily for passenger car and truck applications but it may be used in marine, industrial, and similar applications.

2. Purpose—The purpose of this document is to provide dimensions for threaded fuel tank filler caps and cap retainers (see Figs. 1 and 2).

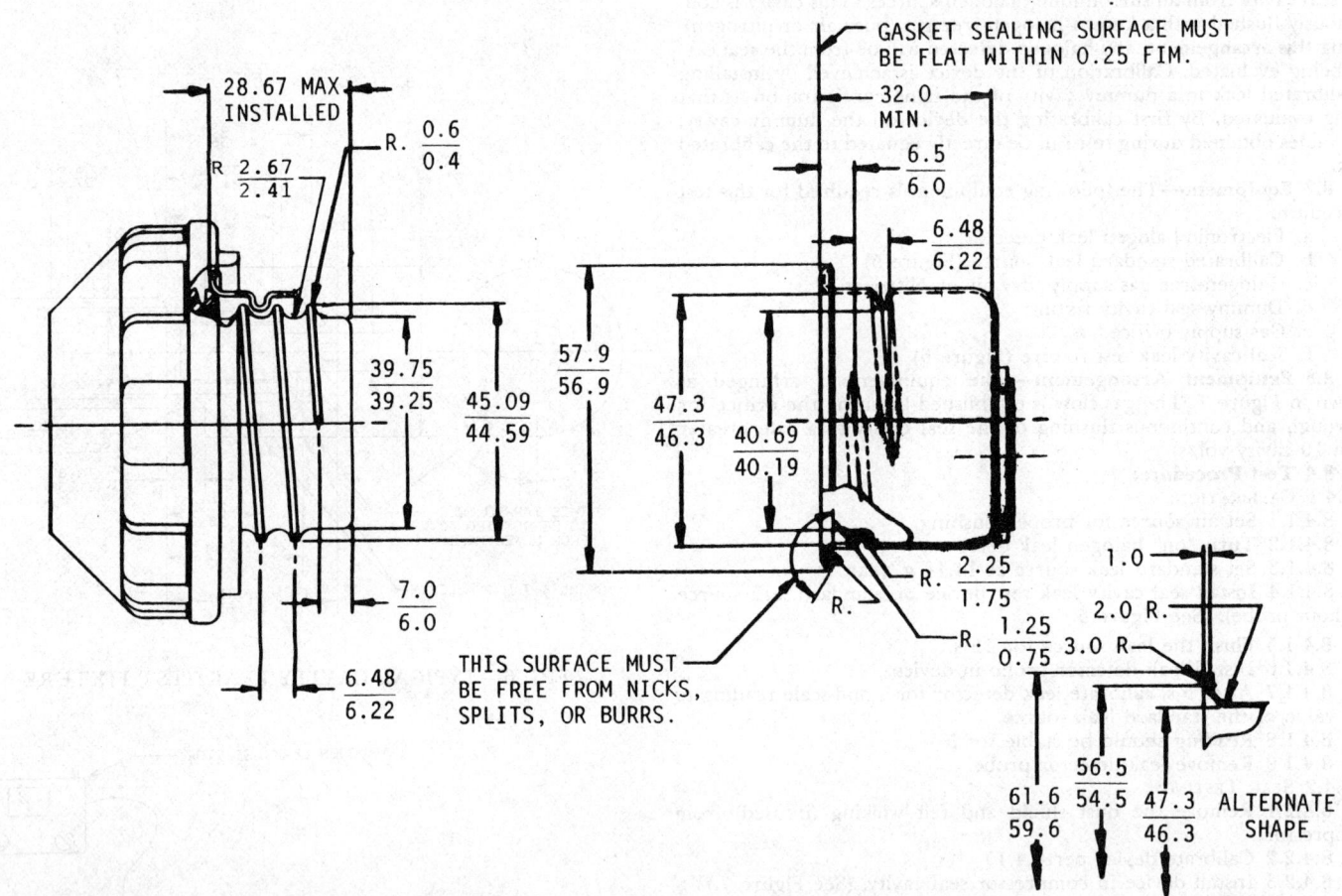

FIG. 1—PRESSURE/VACUUM (CLOCKWISE ROTATION) CAP CAN BE INTERCHANGEABLE WITH FIG. 2 FILLERNECK.

26.431

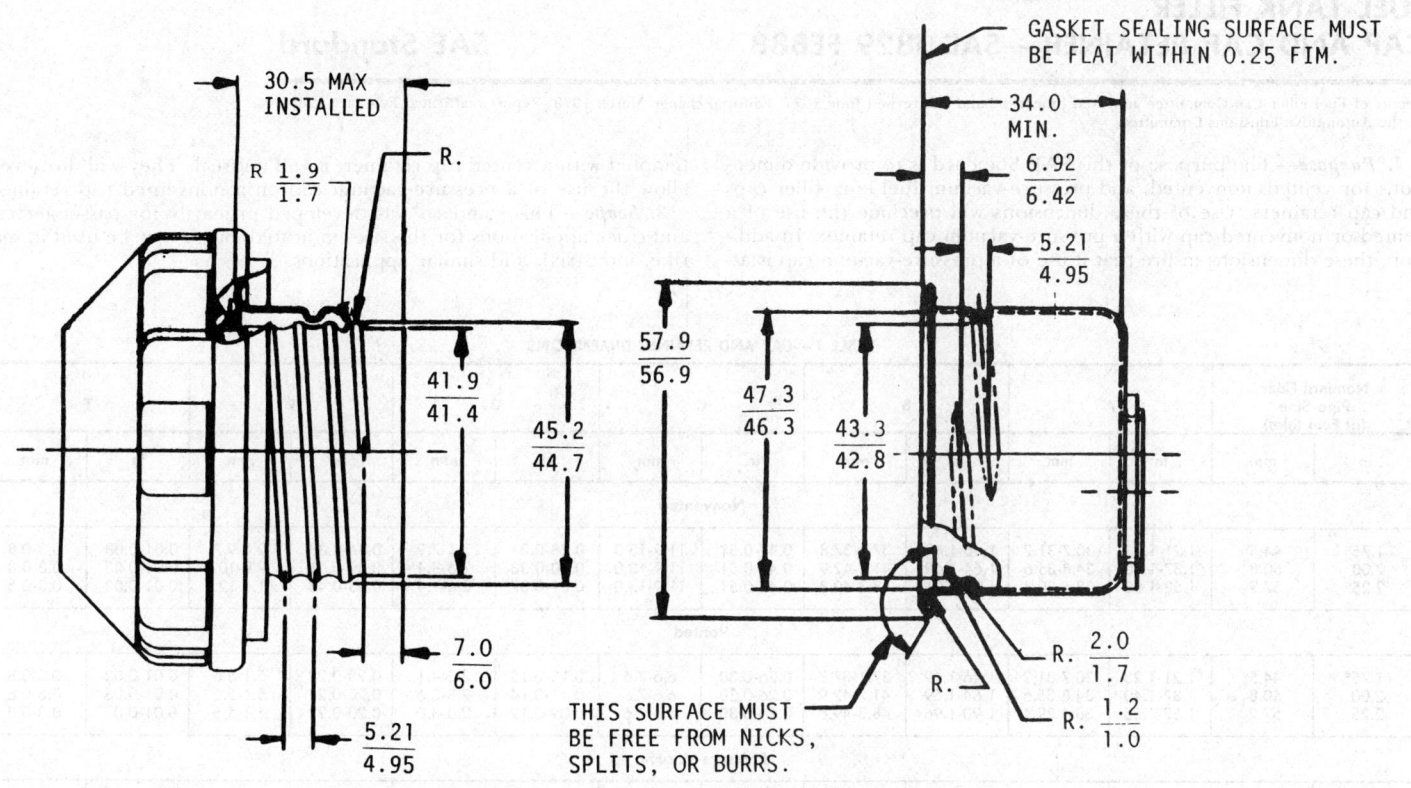

FIG. 2—NONVENTED OR VACUUM ONLY (CLOCKWISE ROTATION) CAP IS NOT INTERCHANGEABLE WITH FIG. 1 FILLERNECK.

FUEL TANK FILLER CAP AND CAP RETAINER—SAE J829 FEB88

SAE Standard

Report of Fuel Filler Cap Committee approved June 1962 and last revised June 1977. Editorial change March 1978. Report reaffirmed February 1988 by the Automotive Emissions Committee.

1. Purpose—The purpose of this SAE Standard is to provide dimensions for vented, nonvented, and pressure-vacuum fuel tank filler caps and cap retainers. Use of these dimensions will preclude the use of a vented or nonvented cap with a pressure-vacuum cap retainer. In addition, these dimensions insure that if use of a pressure-vacuum cap is attempted with a vented cap retainer, it will not seal. They will, however, allow the use of a pressure-vacuum cap on a nonvented cap retainer.

2. Scope—This standard was developed primarily for passenger car and truck applications for the sizes indicated, but it may be used in marine, industrial, and similar applications.

TABLE 1—CAP AND RETAINER DIMENSIONS

Nominal Filler Pipe Size (at Fuel Inlet)		A		B		C		D		E		F	
in	mm	in	mm	in	mm	in	mm	in	mm	in	mm	in	mm
Nonvented													
1.75[a]	44.5[a]	1.21-1.23	30.7-31.2	1.46-1.49	37.1-37.8	0.47-0.51	11.9-13.0	0.28-0.31	7.1-7.9	0.36-0.38	9.1-9.7	0.01-0.03	0.3-0.8
2.00	50.8	1.37-1.40	34.8-35.6	1.64-1.69	41.7-42.9	0.47-0.51	11.9-13.0	0.30-0.33	7.6-8.4	0.39-0.41	9.9-10.4	0.01-0.03	0.3-0.8
2.25	57.2	1.52-1.55	38.6-39.4	1.90-1.96	48.3-49.8	0.47-0.51	11.9-13.0	0.34-0.37	8.6-9.4	0.45-0.48	11.4-12.2	0.01-0.03	0.3-0.8
Vented													
1.75[a]	44.5[a]	1.21-1.23	30.7-31.2	1.46-1.49	37.1-37.8	0.26-0.30	6.6-7.6	0.13-0.16	3.3-4.1	0.20-0.22	5.1-5.6	0.01-0.03	0.3-0.8
2.00	50.8	1.37-1.40	34.8-35.6	1.64-1.69	41.7-42.9	0.26-0.30	6.6-7.6	0.11-0.14	2.8-3.6	0.20-0.22	5.1-5.6	0.01-0.03	0.3-0.8
2.25	57.2	1.52-1.55	38.6-39.4	1.90-1.96	48.3-49.8	0.26-0.30	6.6-7.6	0.09-0.12	2.3-3.0	0.20-0.22	5.1-5.6	0.01-0.03	0.3-0.8
Pressure-Vacuum													
2.00	50.8	1.37-1.40	34.8-35.6	1.64-1.69	41.7-42.9	0.26-0.30	6.6-7.6	0.30-0.33	7.6-8.4	0.39-0.41	9.9-10.4	0.01-0.03	0.3-0.8
2.25	57.2	1.52-1.55	38.6-39.4	1.90-1.96	48.3-49.8	0.26-0.30	6.6-7.6	0.34-0.37	8.6-9.4	0.45-0.48	11.4-12.2	0.01-0.03	0.3-0.8

Nominal Filler Pipe Size (at Fuel Inlet)		G		H. Min		J		K		L		Deg X ±4
in	mm	in	mm	in	mm	in	mm	in	mm	in	mm	
Nonvented												
1.75[a]	44.5[a]	0.48-0.51	12.2-13.0	0.12	3.0	1.40-1.43	35.6-36.3	1.17-1.19	29.7-30.2	0.36-0.40	9.1-10.2	130
2.00	50.8	0.54-0.57	13.7-14.5	0.12	3.0	1.57-1.63	39.9-41.4	1.33-1.36	33.8-34.5	0.36-0.40	9.1-10.2	140
2.25	57.2	0.60-0.63	15.2-16.0	0.12	3.0	1.80-1.86	45.7-47.2	1.48-1.51	37.6-38.4	0.36-0.40	9.1-10.2	140
Vented												
1.75[a]	44.5[a]	0.30-0.33	7.6-8.4	0.12	3.0	1.40-1.43	35.6-36.3	1.17-1.19	29.7-30.2	0.21-0.24	5.3-6.1	130
2.00	50.8	0.30-0.33	7.6-8.4	0.12	3.0	1.57-1.63	39.9-41.4	1.33-1.36	33.8-34.5	0.21-0.24	5.3-6.1	140
2.25	57.2	0.30-0.33	7.6-8.4	0.12	3.0	1.80-1.86	45.7-47.2	1.48-1.51	37.6-38.4	0.21-0.24	5.3-6.1	140
Pressure-Vacuum												
2.00	50.8	0.54-0.57	13.7-14.5	0.12	3.0	1.57-1.63	39.9-41.4	1.33-1.36	33.8-34.5	0.21-0.24	5.3-6.1	140
2.25	57.2	0.60-0.63	15.2-16.0	0.12	3.0	1.80-1.86	45.7-47.2	1.48-1.51	37.6-38.4	0.21-0.24	5.3-6.1	140

[a] The 1.75 in (44.5 mm) filler pipe inlet size is not recommended for passenger cars and trucks. Application refers to SAE J1140 MAR80, Filler Pipe and Openings of Motor Vehicle Fuel Tanks.

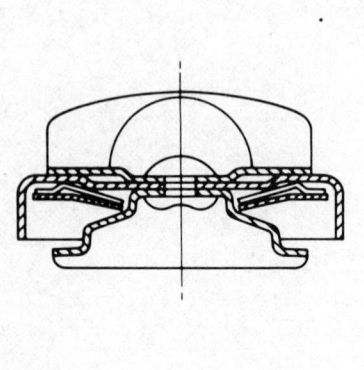
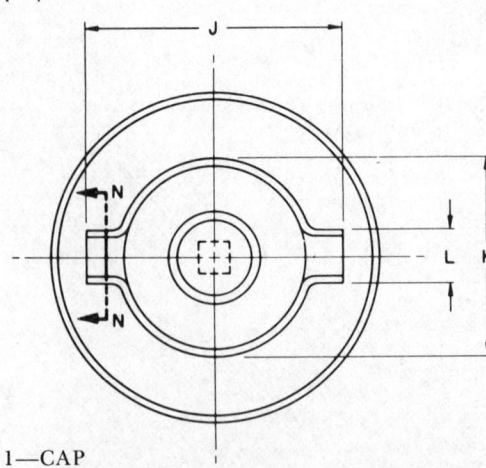

FIG. 1—CAP

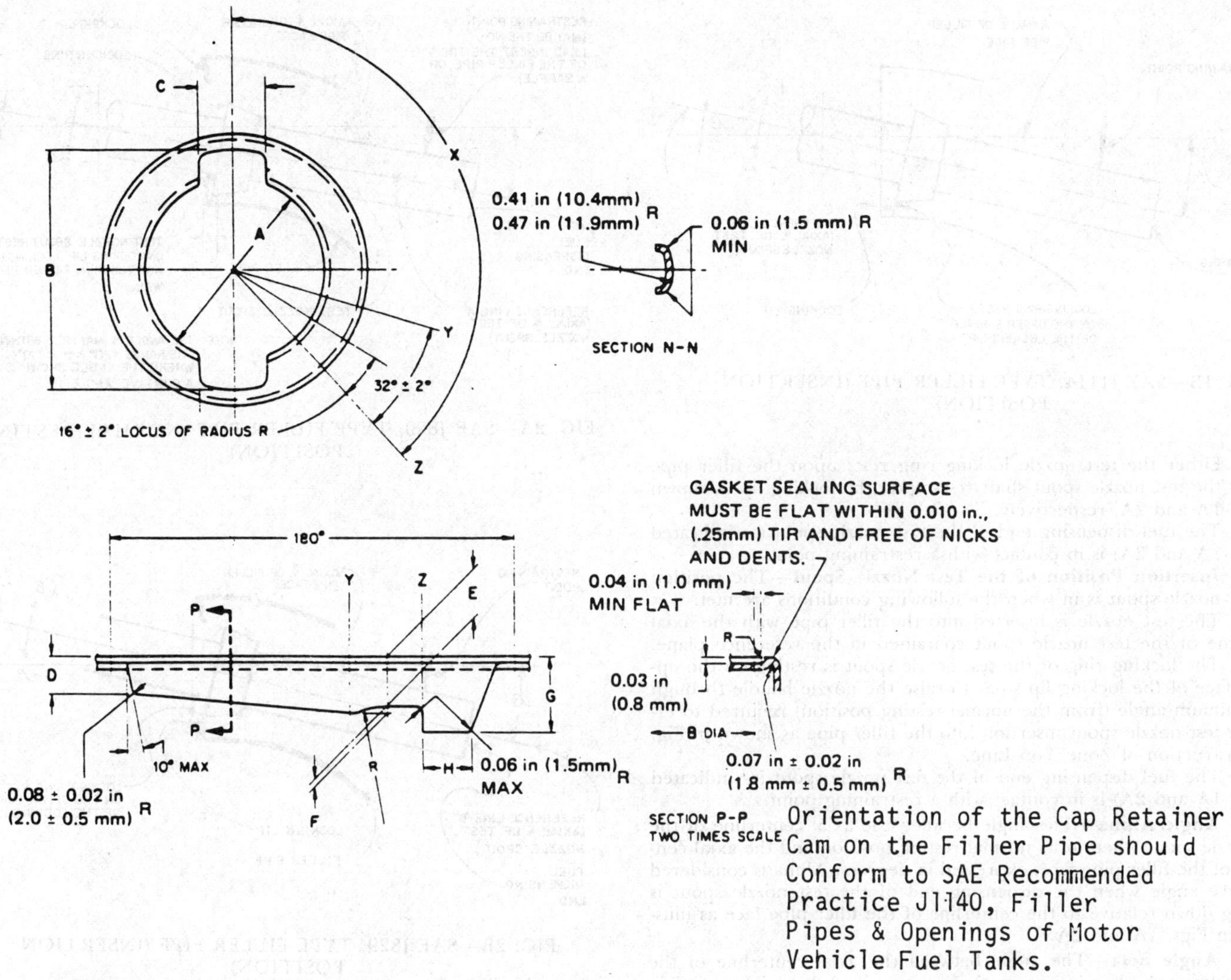

ΦFIG.2—CAP RETAINER

Orientation of the Cap Retainer Cam on the Filler Pipe should Conform to SAE Recommended Practice J1140, Filler Pipes & Openings of Motor Vehicle Fuel Tanks.

FILLER PIPES AND OPENINGS OF MOTOR VEHICLE FUEL TANKS— SAE J1140 FEB88

SAE Recommended Practice

Report of the Fuel Supply Systems Committee approved December 1976, last revised March 1980, and reaffirmed by the SAE Automotive Emissions Committee February 1988.

1. Purpose—The purpose of this recommended practice is to ensure compatibility between new vehicle designs and refueling vapor recovery nozzles by their dimensions and specifications.

2. Scope—This recommended practice was developed primarily for gasoline-powered passenger car and truck applications but may be used in marine, industrial, and similar applications where refueling vapor recovery is required.

3. Definitions

3.1 Filler Pipe Face—The gasket sealing surface as defined in SAE J829 MAR78, Fig. 2 and SAE J1114 MAR78, Fig. 2.

3.2 Test Nozzle Spout—The rod with dimensions as shown in Fig. 4, used to establish the reference lines around which the filler pipe access zone is defined.

3.3 Reference Plane—The plane which contains the axial centerline of the filler pipe face, and is turned in the direction which the manufacturer of the automobile has determined to be the orientation in which the nozzle is to be inserted.

3.4 Normal Resting Position of Test Nozzle Spout—The position the test nozzle spout is in when the following conditions are met:

3.4.1 The test nozzle spout is inserted into the filler pipe, such that the axial centerline of the test nozzle spout lies in the reference plane.

3.4.2 The locking ring of the test nozzle spout is located immediately on the inside (that is, the vehicle tank side) of the locking lip.

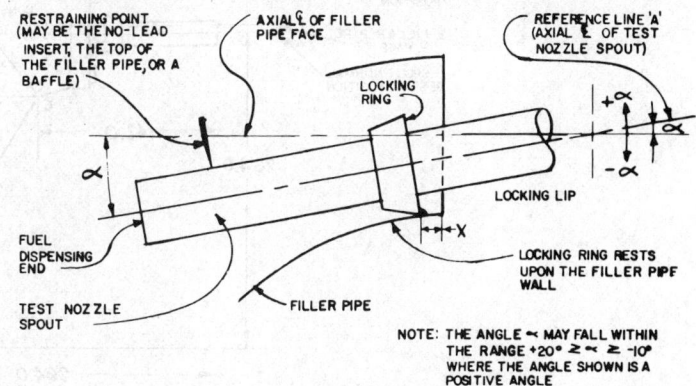

FIG. 1A—SAE J1114, TYPE FILLER PIPE (NORMAL RESTING POSITION)

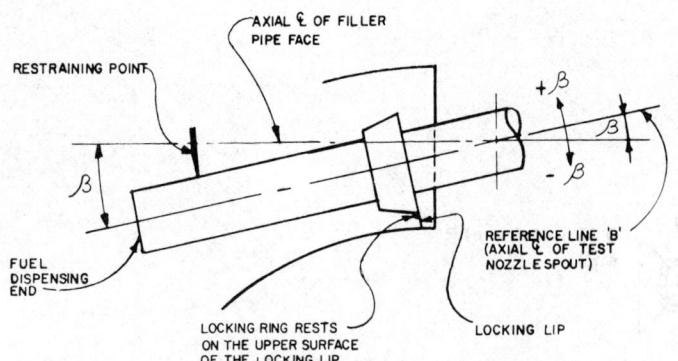

FIG. 1B—SAE J1114, TYPE FILLER PIPE (INSERTION POSITION)

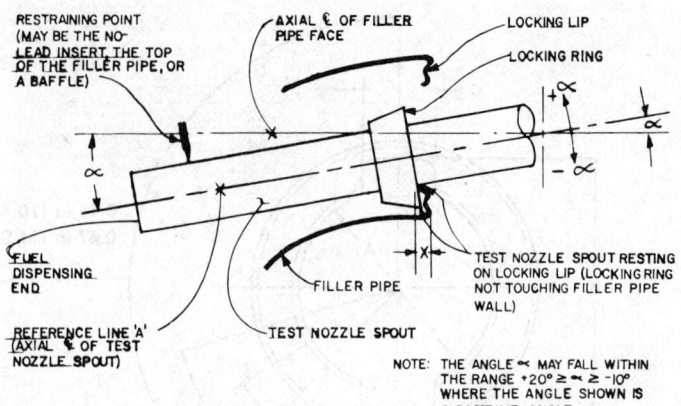

FIG. 2A—SAE J829, TYPE FILLER PIPE (NORMAL RESTING POSITION)

3.4.3 Either the test nozzle locking ring rests upon the filler pipe wall, or the test nozzle spout shaft rests upon the locking lip as shown in Figs. 1A and 2A, respectively.

3.4.4 The fuel dispensing end of the test nozzle spout (as indicated in Figs. 1A and 2A) is in contact with a restraining point.

3.5 Insertion Position of the Test Nozzle Spout—The position the test nozzle spout is in when the following conditions are met:

3.5.1 The test nozzle is inserted into the filler pipe with the axial centerline of the test nozzle spout contained in the reference plane.

3.5.2 The locking ring of the test nozzle spout is resting on the upper surface of the locking lip so as to raise the nozzle handle through the minimum angle (from the normal resting position) required to effect the test nozzle spout insertion into the filler pipe as shown in Fig. 3—Construction of Zone Top Line.

3.5.3 The fuel dispensing end of the test nozzle spout (as indicated in Figs. 1A and 2A) is in contact with a restraining point.

3.6 Angle Alpha—The angle between the axial centerline of the test nozzle spout when in its normal resting position and the axial centerline of the filler pipe face, expressed in degrees. Alpha is considered a positive angle when the dispensing end of the test nozzle spout is pointing down relative to the centerline of the filler pipe face as illustrated in Figs. 1A and 2A.

3.7 Angle Beta—The angle between the axial centerline of the test nozzle spout when in its insertion position, and the axial centerline

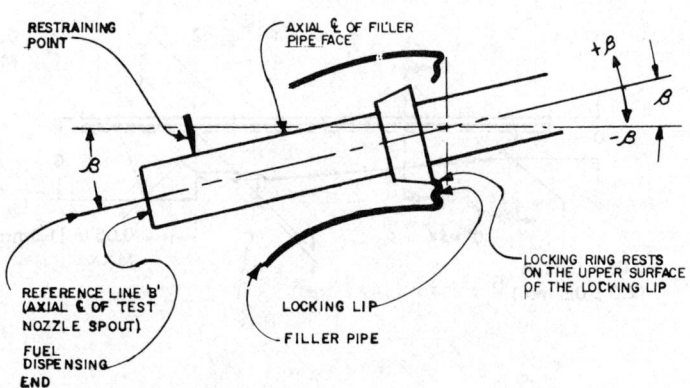

FIG. 2B—SAE J829, TYPE FILLER PIPE (INSERTION POSITION)

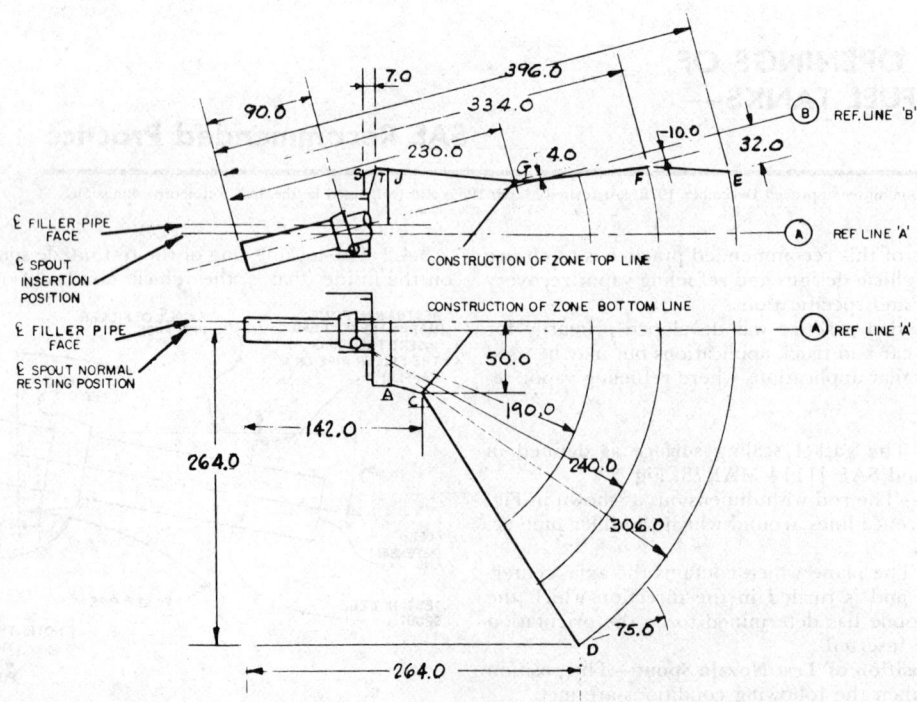

FIG. 3—CONSTRUCTION OF FILLER PIPE ACCESS ZONE

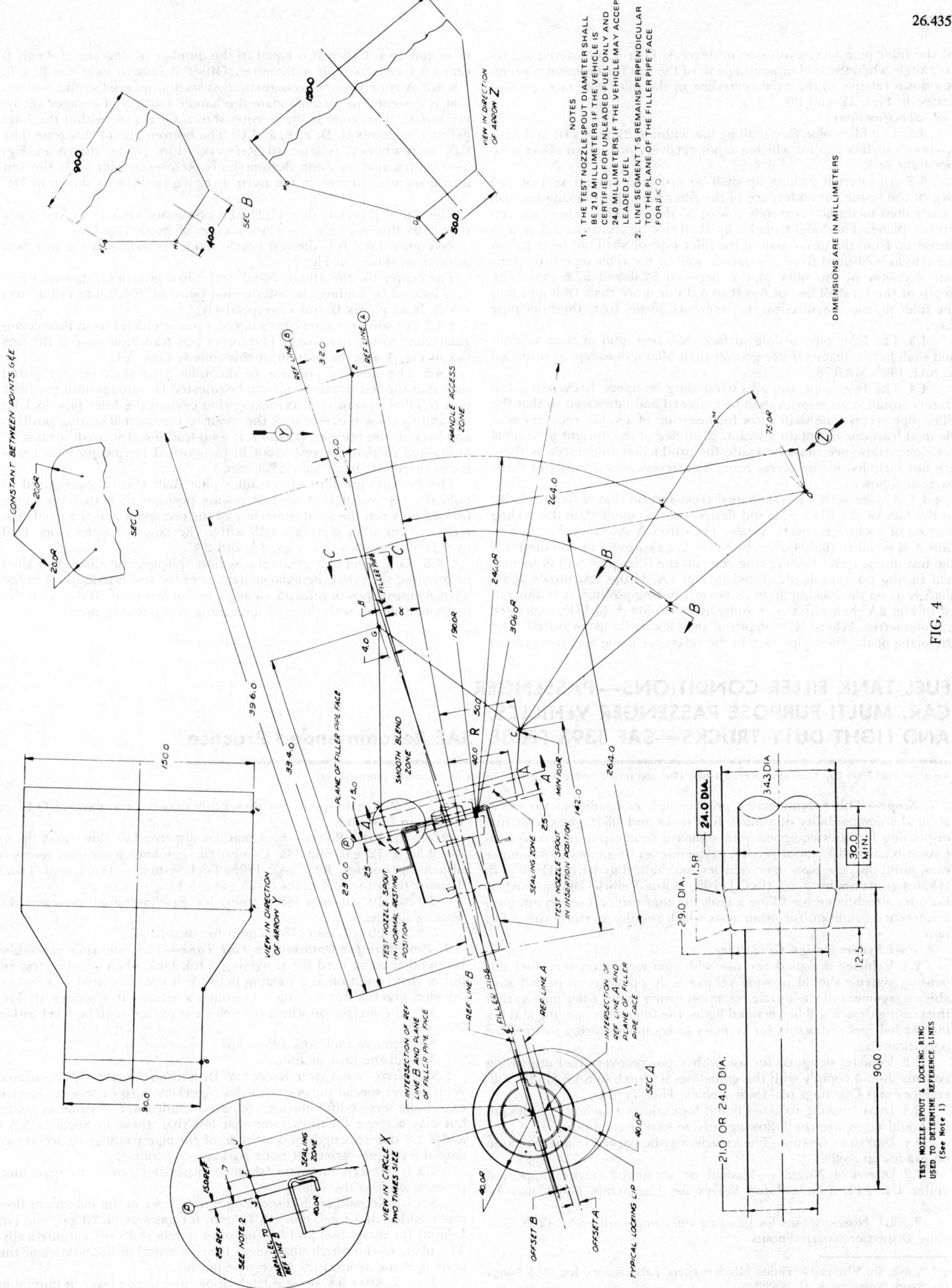

FIG. 4

of the filler pipe face, expressed in degrees. Beta is considered a positive angle when the fuel dispensing end of the test nozzle spout is pointing down relative to the axial centerline of the filler pipe face, as illustrated in Figs. 1B and 2B.

4. Specifications

4.1 The filler pipe face shall be flat within 0.25 mm TIR and have a smooth surface against which a vapor recovery nozzle can effect a vapor-tight seal.

4.2 An internal locking lip shall be provided around at least 100 deg of the inside circumference of the filler pipe. The locking lip shall be oriented such that it extends at least 35 deg to either side of the reference plane. The height of the lip shall not be less than 2.5 mm as measured from the inside wall of the filler pipe or shall not be less than 8.5 mm as measured from the outside wall of the filler pipe if the outside diameter of the filler pipe is between 52.0 and 57.5 mm. The depth of the lip shall be not less than 4.0 nor more than 13.0 mm into the filler pipe as measured in the reference plane, from the filler pipe face.

4.3 The filler pipe sealing surface shall be round in cross section, and shall have a diameter not greater than 50.0 mm except as provided in SAE J829 MAR78.

4.4 The filler pipe and all surrounding bumpers, body parts, and factory installed accessories shall be designed and fabricated so that the filler pipe access zone shall allow for insertion of a vapor recovery nozzle in at least one orientation within ± 90 deg of the upright or vertical position. Allowance must be made for production tolerances as these are not included in the access zone. The access zone consists of three parts as follows:

4.4.1 A zone with an oval shaped cross section that is fixed relative to the face of the filler pipe and designed to accommodate the sealing portion of a vapor recovery nozzle. The offset A shown on Fig. 4, Section A is equal to $(0.004)\alpha^2 - (0.3)\alpha + 2.0$ expressed in millimetres if the test nozzle spout locking ring rests on the filler pipe wall in its normal resting position as illustrated in Fig. 1A. If the test nozzle spout shaft rests on the locking lip in its normal resting position as is illustrated in Fig. 2A then offset $A = (0.004)\alpha^2 - (0.3)\alpha + (0.11)X$ expressed in millimetres. Where X = depth of the locking lip as measured from the plane of the filler pipe face in the reference plane and is expressed in millimetres. Offset B is equal to the number of degrees of Angle β times 0.4 expressed in millimetres. Offset B fails to exist for $\beta \leq 0$.

4.4.2 A zone with a rectangular cross-section tapered at the bottom, that is designed to accommodate the handle portion of a vapor recovery nozzle. This zone is the portion shown on Fig. 4, within the lines defined by points C, D, E, F, and G. The bottom line of this zone (line CD) is positioned relative to Reference Line A as shown in Fig. 3—Construction of Zone Bottom Line. Reference Line A is the test nozzle spout centerline in the normal resting position as shown in Fig. 3.

The top of this zone (line G, F, E) is positioned relative to Reference Line B as shown in Fig. 3—Construction of Zone Top Line.

Reference Line B is the test nozzle spout centerline in the insertion position as shown in Fig. 3.

The center for the 190.0, 240.0, and 306.0 mm radii, shown in Fig. 4, is located by finding the intersection point of 306.0 mm radius arcs struck from points D and E, respectively.

4.4.3 A transition zone consisting of a smooth blend from the rectangular zone to the oval zone. The top of this transition zone is the line G-J in Fig. 4 and the bottom of this zone is Line A-C.

4.5 The internal portions of the filler pipe shall be configured such that the test nozzle spout can be inserted far enough into the filler pipe to allow entrance of its locking ring beyond the filler pipe locking lip, and to allow movement of the spout to the normal resting position and back to the released position. If a no-lead insert is used, it must be positioned so that the test spout in the normal resting position penetrates the insert by at least 22.5 mm.

The internal portions of the filler pipe shall also be configured to hold the test spout in a normal resting position such that the angle formed between the axial centerline of the test spout and the axial centerline of the filler pipe face falls within the range of alpha from +20 to −10 deg as shown in Figs. 1A and 2A.

4.6 As a minimum assurance against spillage, the filler pipe shall be oriented such that the axial centerline of the test nozzle spout in the normal resting position forms an angle of not less than 30 deg with the horizontal plane, with the fuel dispensing end pointing down.

FUEL TANK FILLER CONDITIONS—PASSENGER CAR, MULTI-PURPOSE PASSENGER VEHICLES, AND LIGHT DUTY TRUCKS—SAE J398 FEB88 SAE Recommended Practice

Report of Fuel Filler Cap Committee approved June 1969 and revised by the Automotive Emissions Committee February 1988.

1. Scope

This recommended practice defines conditions for evaluating the compatibility of vehicle fuel tanks and filler pipes with fuel dispensing facilities equipped with standard (non-vapor recovery) configuration as well as vapor recovery type nozzles. It applies to passenger cars, multi-purpose passenger vehicles, and light-duty trucks (10 000 lb (4536 kg) maximum gvw), (Ref. J1100, Motor Vehicle Dimensions). It includes a technique for filling a tank full that can be used to establish a reference condition for other tests which require starting with a full tank.

2. Fuel System Design Guidelines

2.1 Vehicles designed for use with non-vapor recovery fuel dispensing systems should provide vehicle body openings to permit suitable engagement of the nozzle retention spring in the filler pipe so that the sensing device will be retained below the filler pipe opening and below leaded fuel restrictors for vehicles so equipped during unattended operation.

2.2 Vehicles designed for use with vapor recovery fuel dispensing systems should comply with the guidelines set forth in SAE J1140, Filler Pipes and Openings of Motor Vehicle Fuel Tanks.

2.3 In attempting to avoid liquid losses during fueling, the designer should recognize the following service station conditions:

2.3.1 DRIVEWAY GRADE—The vehicle may be parked in any direction on grades up to 6%.

2.3.2 DISPENSER NOZZLES—Manual or automatic nozzles approved under UL 842, Standard for Valves for Flammable Fluids may be used.[1]

2.3.2.1 Nozzle spouts for gasoline will comply with SAE J285, Gasoline Dispenser Nozzle Spouts.

2.3.2.2 Nozzle spouts for diesel fuel may have a nominal O.D. up to 1 3/16 in (30 mm).

2.3.3 DELIVERY RATES—Fuel may be dispensed at flow rates up to:

2.3.3.1 12 gal/min (45 L/min) for gasoline non-vapor recovery dispensing nozzles. Ref. SAE J703, Fuel Systems—Truck and Truck Tractor (F-8) for fuel tanks of 25 gal (95 L).

2.3.3.2 10 gal/min (38 L/min) for gasoline vapor recovery dispensing nozzles.

2.3.3.3 20 gal/min (76 L/min) for diesel fuel.

3. Procedure for Obtaining a Full Tank

The following procedure is intended to be used for providing a full tank when another test requires that condition as a starting point. It is not presented as a recommended practice for fueling a customer's vehicle at a service station.

3.1 The surface on which the vehicle is parked shall be level within 2% grade.

3.2 Remove fuel tank filler cap.

3.3 Fill the tank as follows:

3.3.1 WITH NON-VAPOR RECOVERY DISPENSING NOZZLE—Recreational vehicles and special purpose vehicles (wreckers, airport service, electrical/phone service lift vehicles, etc.) are included in this scope as guide, but may achieve fill rates somewhat less than those in Sections 3.3.1 and 3.3.2 due to length and attitude of fill pipe routings to accommodate the unique variety of body package constraints.

3.3.1.1 Insert automatic shutoff dispensing nozzle to maximum penetration into the filler pipe.

3.3.1.2 Set and latch dispensing nozzle lever in the maximum flow rate position of 12 gal/min (45 L/min) for gasoline or 20 gal/min (76 L/min) for diesel fuel and let run until nozzle shuts off automatically. The plane of the nozzle shall be as nearly vertical as boundaries of the body opening around the filler pipe permit.

3.3.1.3 After 5 s, set and latch dispensing nozzle lever in minimum

---

[1] "Valves for Flammable Fluids," Underwriters' Laboratories, Inc., 333 Pfingsten Road, Northbrook, IL 60062.

flow rate position of 4-5 gal/min (15-19 L/min) and let run until nozzle shuts off automatically.

NOTE: Minimum flow rate to cause automatic shutoff must be maintained.

3.3.1.4 After 5 s, withdraw the nozzle until the tip extends 2 in (50 mm) into the filler pipe neck (sensing orifice must remain below the filler pipe opening), or if the filler neck is equipped with a leaded fuel restrictor, withdraw the nozzle until the sensing orifice is just on the fuel side of the leaded fuel restrictor. Set and hold at a minimum flow rate position of 4-5 gal/min (15-19 L/min) until nozzle shuts off automatically. Allow residual fuel to drain from the nozzle into the filler pipe to avoid spillage and remove the nozzle.

3.3.2 WITH REFUELING VAPOR RECOVERY DISPENSING NOZZLE

3.3.2.1 Insert the refueling vapor recovery nozzle into the filler pipe in the normal resting position as defined in SAE J1140, Filler Pipes and Openings of Motor Vehicle Fuel Tanks. Allow the retaining mechanism to engage on the lip of the filler tube.

3.3.2.2 Set and latch dispensing nozzle lever in the maximum flow rate position of 10 gal/min (38 L/min) for gasoline and let run until nozzle shuts off automatically.

3.3.2.3 Wait 10 s, then set and latch dispensing nozzle lever in minimum flow rate position of 4-5 gal/min (15-19 L/min) and let run until nozzle shuts off automatically.

NOTE: Minimum flow rate to cause automatic shutoff must be maintained.

3.3.2.4 Repeat paragraph 3.3.2.3.

3.3.2.5 Disengage the nozzle retaining mechanism, allow residual fuel to drain from the nozzle into the filler pipe to avoid spillage and remove the nozzle.

3.4 Attach fuel tank filler cap securely.

R) COARSE DROPLET WATER/FUEL SEPARATION TEST PROCEDURE— SAE J1839 MAY90

SAE Information Report

Report of the Engine Committee approved February 1987. Completely revised by the Filter Test Methods Standards Technical Committee 15 May 1990.

Foreword—Water in fuels is one of the major causes of diesel engine maintenance problems. The effects of water in fuel are characterized by corrosion of fuel system parts, plugging of filters and orifices and, in some cases, failure of fuel injection equipment. Water in fuel often dissolves sulfur compounds, becomes acidic, and enhances corrosion in fuel injection systems as well as in the engine itself. The presence of water also encourages microbiological growth, which generates orifice and filter restricting sludge. Further, due to displacement of fuel lubrication in close tolerance injector parts, and rapid expansion of heated water at the fuel injector tip, galling, and more serious failure may also occur.

During transportation, transfer, and storage of fuel, water may become entrained in a variety of ways. The mode and timing of water entry in the handling sequence before use, as well as the chemistry of the fuel itself (additives and surfactants), will determine what form the contaminant takes. In systems where the water and fuel pass through high shear pumps, fuel water interfacial tension is relatively low, and settling time is minimized, fine emulsions may predominate. In systems where water enters before or after low shear pumps, or where there is a prolonged settling time in high interfacial tension fuel, larger water droplets may predominate. In some systems, both fine emulsions and large droplets may be present simultaneously. Generally, fine emulsions are more likely to predominate on the pressure side of high shear pumps, whereas larger water droplets are more likely to predominate on the suction side of pumps. (A water removal test procedure designed for applications where finely dispersed droplets predominate is also recommended. This procedure is given in SAE J1488.)

The following test procedure is relevant to coarsely dispersed water separation devices whether applied on the suction or discharge side of engine fuel transfer pumps.[1] The procedure is well suited to lower flow rates, although it may be applied with due caution to higher flow rates [up to approximately 1.6×10^{-3} m^3/s (~25 gpm)]. It has been designed to approximate field conditions in a practical manner. A water dispersing technique simulating the water droplet sizes experienced drawing fuel/water mixtures through fuel lines and fittings (180 to 260 μm mean droplet size), is used to reproduce field conditions where coarse droplets predominate. The test fuel may be an actual fuel sample (with additives) that is to be used in the field, or it may be No. 2 fuel oil that has been clay treated (conditioned) so as to enable equal and reproducible laboratory comparisons of various test devices. Test fuel conditioning is recommended for laboratory comparisons only, as this treatment may yield water removal efficiency results, which are significantly different from those obtained using water separating devices in untreated fuel. Furthermore, testing unused "clean" water separators may provide water removal efficiencies, which are far superior to those obtained from the same water separators after very short exposure to natural fuel and natural fuel contaminants.

1. Scope—To determine the undissolved water removal performance of a fuel/water separator under controlled laboratory conditions, using water droplets.

2. References

2.1 Applicable Documents

SAE J1488, Emulsified Water/Fuel Separation Test Procedure
ISO R 760
ASTM D 971

3. Test Apparatus—A test system, as illustrated in Figure 1, is to include:

3.1 Operating sump with a flat bottom. Material to be compatible with diesel fuel (polyethylene, etc.). Sump size to be adequate to hold test fluid volume under test conditions. Outlet to be 2.5 cm from bottom of tank or higher.

3.2 A pump, capable of providing test flow rate, under test conditions.

3.3 A water dispersing device, as described in Appendix A.

3.4 Thermometer or temperature readout accurate ±2°C under test conditions.

3.5 The fuel/water separator under test.

3.6 Differential pressure gauge or manometer with 1.0 mm Hg or 0.1 kPa subdivisions, or as required.

3.7 Inline static mixer with at least three internal mixing units, or similar, to provide a representative sample at the sample port.

3.8 Sample port. Provision must be made to allow adequate flushing, immediately prior to sampling.

3.9 Test fuel flow meter, flow range appropriate for the unit under test, accuracy ±5%, repeatability ±2%.

3.10 A final fuel/water separator assembly, such that not more than 30 ppm undissolved water is recycled on an average basis under test conditions.

3.11 Suitable heat exchange and controls for maintaining constant test temperatures as specified.

3.12 Suitable valve or other control for adjusting and controlling test flow.

3.13 Diffuser, to promote uniform mixing in test reservoir.

3.14 Water flow meter with range to suit application capable of measuring flows of 0.25% of fuel flow rate, ±5% accuracy.

3.15 Suitable valve for adjusting and controlling water injection flow.

3.16 A supply of clean deionized or distilled water with a surface tension greater than 6×10^{-2} N/m (60 dynes/cm) at 20°C ± 1.5. Adequate pressure must be available to inject water on the high pressure side of the pump. (Use of a deionizing filter, for example, housed Barnstead D8901, will permit use of local water supply and pressure.)

3.17 Automatic Karl Fischer Titration Apparatus for water content analysis.

3.18 All interconnecting piping should be selected, sized, and oriented to prevent the separation of water and solid contaminants from the fuel or provide traps for these contaminants (13 mm PVC is recom-

---

[1] This procedure recommends pressure side location of the test unit for ease and convenience of testing only. Modification of the procedure to place the test unit on the suction side of the pump should not alter test results as long as the water droplet size distribution remains unaltered.

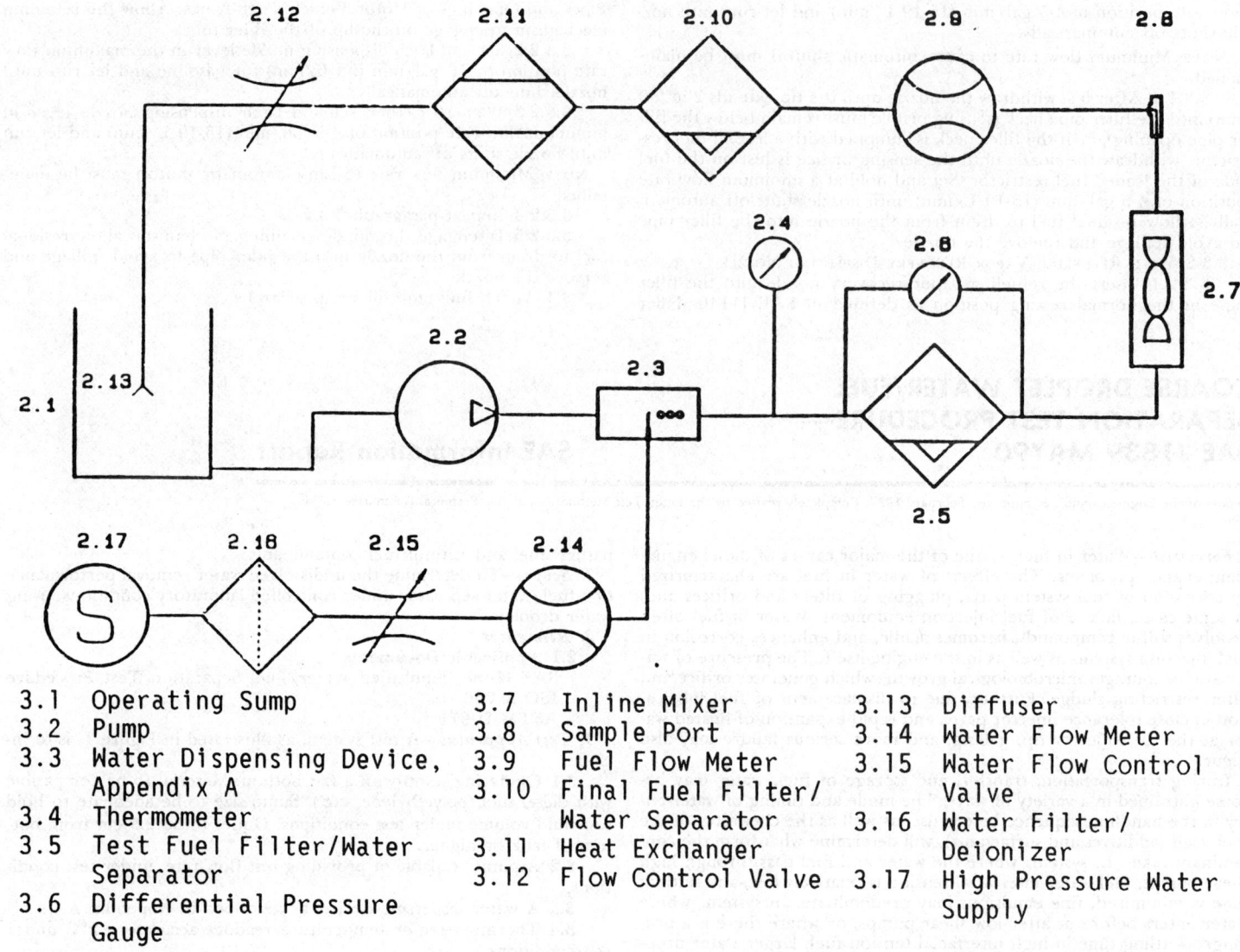

- 3.1 Operating Sump
- 3.2 Pump
- 3.3 Water Dispensing Device, Appendix A
- 3.4 Thermometer
- 3.5 Test Fuel Filter/Water Separator
- 3.6 Differential Pressure Gauge
- 3.7 Inline Mixer
- 3.8 Sample Port
- 3.9 Fuel Flow Meter
- 3.10 Final Fuel Filter/Water Separator
- 3.11 Heat Exchanger
- 3.12 Flow Control Valve
- 3.13 Diffuser
- 3.14 Water Flow Meter
- 3.15 Water Flow Control Valve
- 3.16 Water Filter/Deionizer
- 3.17 High Pressure Water Supply

FIG. 1

mended for 0 to 23 lpm flow rates).

4. Test Materials

4.1 Test Fluid—Since fuel oil contains various constituents, the test oil type should be categorized and recorded as one of the following:

a. A sample of the fluid used in the application
b. No. 2 diesel fuel, locally available
c. Specially treated fluid, per Appendix B
d. A standard reference fluid to be specified

In all these cases, it should be understood that the results are relevant to this fuel and that some amount of variance in performance can be expected with different fuels, depending on the particular design of the test fuel/water separator.

4.2 Clean, distilled or deionized water with a surface tension greater than 6×10^{-2} N/m (60 dynes/cm) at 20°C ± 1.5.

5. Test Conditions

5.1 Volume of Fuel in the Test System (including filters, piping, etc.):

Shall be five times the flow rate, per minute, with a minimum of 3.8×10^{-2} m³ (38 L).

5.2 Temperature:
26.6°C ± 2.5 measured at the test separator inlet.

5.3 Flow Rate of Fuel:
Rated flow of unit to be tested or as specified.

5.4 Water Flow Rate:
0.25% of fuel flow rate.

6. Test Procedures

6.1 If clay-treated fuel (Appendix B) is not the selected fluid, use a fresh quantity of fluid. Retreat clay-treated fuel, if used.

6.2 Determine the water saturation level for each batch of test fluid:
a. According to Appendix C
b. By using tables or charts where available; see Appendix D

6.3 Install cleanup filter (95% efficiency at 5 μm or better suggested) in place of test filter; fill fuel tank; start circulation of fuel at cleanup filter flow rating. Continue system cleanup until a particulate contamination level of 5 mg/L or less and 30 ppm undissolved water or less is obtained. Low cleanup flow rates and/or lower efficiencies will require long cleanup times. The final fuel/water separator may be used as a cleanup filter if suitable.

6.4 Install test fuel/water separator to be tested (see Figure 1).

6.5 With flow set near zero, start pump, adjust to specified flow rate.

6.6 Bleed air from system if necessary, take initial (fuel only) pressure drop reading at the test flow.

6.7 Open the water valve and adjust water flow rate to 0.25% of fuel flow. Start the clock at the same time water begins to flow. Establish proper flow rate within 1 min.

6.8 Without interrupting test flows, periodically drain the water from the water collection sump of the unit under test. Do not let water build up beyond the maximum recommended level of the water sump. DO NOT TAKE ANY SAMPLES WHEN ASSEMBLY IS BEING DRAINED.

6.9 Record test time for each drain.

6.10 After 10 min carefully withdraw a sample, being sure to flush the sample port thoroughly. Be sure that the sample syringe or container is thoroughly dry. Analyze the sample immediately using the Karl Fischer method (see ISO R 760), or similar. Record each reading. Additional sampling is permitted. Repeat this sampling procedure every 20 min thereafter until termination of the test.

6.11 Record the differential pressure across the test fuel/water separator at each effluent sample interval.

6.12 Terminate the test if one or more of the following conditions is met:

a. Water concentration in effluent fuel is above an acceptable level, to be specified by manufacturer or user

b. 2 h and 30 min or more of test time has elapsed and an equilibrium pressure drop is attained. Equilibrium pressure drop has been reached when an increase of no more than 2.5 mm Hg (0.34 kPa) occurs during a 20 min interval

c. Differential pressure exceeds an upper limit specified by the manufacturer or user

7. Presentation of Data

7.1 Plot concentration of undissolved water in effluent (ppm by volume)[2] versus time (minutes) on linear graph paper. Undissolved water = total water minus dissolved water (see 6.2). Indicate drain times on graph.

7.2 Plot pressure drop (mm Hg or kPa) versus time (minutes).

7.3 Record the following:

7.3.1 Test fluid viscosity, type, composition, specific gravity, and source.

7.3.2 Fuel flow rate.

7.3.3 Actual test temperature.

7.3.4 Total test time.

7.3.5 Equilibrium pressure drop.

7.3.6 Dissolved water saturation level, method used to determine level.

7.4 Calculate and report the average free water content of effluent by the following equation:

where:
$$E_{av} = \Sigma \ E_i \times [(t_i - t_{i-1})/t_{total}] \quad \text{(Eq. 1)}$$

E_{av} = Average undissolved water content of effluent, ppm by volume

E_i = Undissolved water content of the "i th" sample = Total water content of the sample-water saturation level (5.2), ppm by volume

$t_i - t_{i-1}$ = Time since previous sample, minutes

t_{total} = Total test time to final sample, minutes

7.5 Calculate and report average undissolved water separation efficiency by using the following equation:

$$\text{Average Efficiency} = (1 - \frac{E_{av}}{2500}) \times 100 \quad \text{(Eq. 3)}$$

APPENDIX A

FIG. A1—WATER DISPERSING DEVICE

(Fuel velocity past nozzle tip should be between 0.75 and 1.50 m/s. Water velocity through nozzle should be between 4.7 and 7.0 m/s.)

Where LPM = Test Flow Rate

D = Maximum diameter (mm) = $20 \sqrt{LPM \times .07074}$ for 0.75 m/s velocity

D = Minimum diameter (mm) = $20 \sqrt{LPM \times .03536}$ for 1.50 m/s velocity

d = Maximum diameter (mm) = $20 \sqrt{LPM \times .0000282}$ for 4.7 m/s velocity

d = Minimum diameter (mm) = $20 \sqrt{LPM \times .0000189}$ for 7.0 m/s velocity

(See Figures A2 and A3 for nozzle I.D. and gauge at various test flow rates.)

This water dispersing method has been found to produce dispersed phase droplet populations with mean diameters of approximately 180 to 260 μm.

[2] Many Karl Fischer titrators will determine micrograms. Convert this to ppm by volume.

$$\text{ppm by vol} = \frac{\text{Karl Fischer reading } \mu gms}{\text{Sample Volume (mL)}} = \frac{\text{Karl Fischer } \mu gms \times \text{Fuel sp. gr.}}{\text{Sample Weight (grams)}} \quad \text{(Eq. 2)}$$

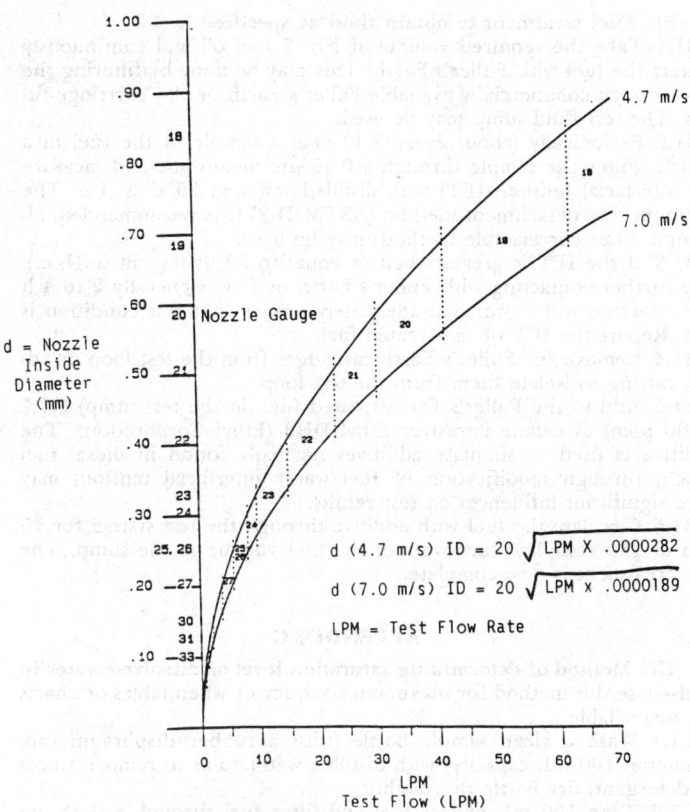

FIG. A2—DISPERSING NOZZLE SIZE AND GAUGE FOR 0 TO 70 LPM TEST FLOW AT 4.7 TO 7.0 M/S VELOCITY THROUGH NOZZLE (ASSUMES 0.25% WATER ADD RATE)

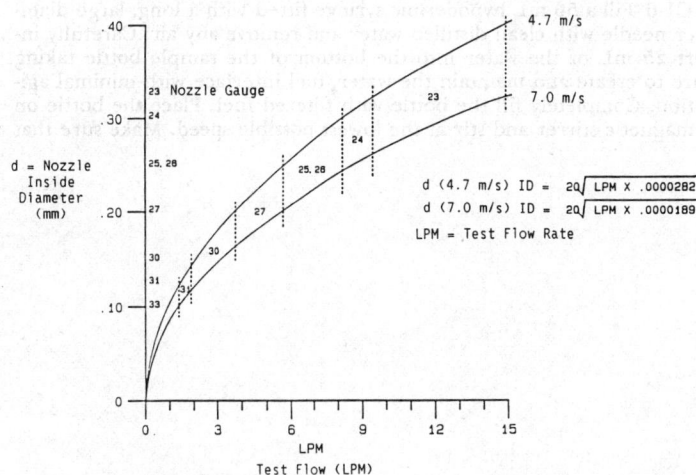

FIG. A3—DISPERSING NOZZLE SIZE AND GAUGE FOR 0 TO 15 LPM TEST FLOW AT 4.7 TO 7.0 M/S VELOCITY THROUGH NOZZLE (ASSUMES 0.25% WATER ADD RATE)

APPENDIX B

B1. Fuel treatment to obtain fluid as specified in 4.1

B1.1 Take the required volume of No. 2 fuel oil and continuously contact the fuel with Fuller's Earth. This may be done by filtering the fuel through commercially available Fuller's Earth or clay cartridge filters. The test fluid sump may be used.

B1.2 Periodically (about every 2 h) take a sample of the fuel in a beaker. Filter the sample through a 0.45 μm membrane and measure the interfacial tension (IFT) with distilled water at 20°C ± 1.5. The platinum ring detachment method (ASTM D 971) is recommended, although other correlatable methods may be used.

B1.3 If the IFT is greater than or equal to 30 dynes/cm (mN/m), stop further contacting with Fuller's Earth or clay. Generally 2 to 4 h of contacting will more than adequately ensure that this condition is met. Report the IFT of the treated fuel.

B1.4 Remove the Fuller's Earth cartridges from the test loop or adjust valving to isolate them from the test loop.

B1.5 Add to the Fuller's Earth treated fuel (in the test sump) 0.1% (1000 ppm) of cetane improver Ethyl DII3 (Ethyl Corporation). The additive is used to simulate additives normally found in diesel fuel which, through modification of fuel/water interfacial tension, may have significant influences on test results.

B1.6 Circulate the fuel with additive through the test system for 15 min or two complete turnovers of the fuel volume in the sump. The fuel treatment is now complete.

APPENDIX C

C1. Method of determining saturation level of dissolved water in fuel
—Use this method for maximum accuracy or when tables or charts are unavailable.

C1.1 Wash a clean sample bottle (with a rubber diaphragm cap; minimum 100 mL capacity) with distilled water so as to remove traces of detergent; dry bottle thoroughly.

C1.2 Take 150 mL of test fuel and filter fuel through a 0.45 μm membrane compatible with fuel oil.

C1.3 Determine total water concentration in the fuel by the Karl Fischer method in ppm by volume.

C1.4 If the water concentration is below or equal to 100 ppm, proceed to C1.5; if not, repeat C1.2 and C1.3. If necessary, cool fluid to −4°C before filtering (step C1.2).

C1.5 Place 75 mL of the filtered fuel into the dried clean 100 mL sample bottle. Insert a PTFE coated magnetic stirrer.

C1.6 Fill a 50 mL hypodermic syringe fitted with a long, large diameter needle with clean distilled water and remove any air. Carefully insert 25 mL of the water into the bottom of the sample bottle taking care to create and maintain the water/fuel interface with minimal agitation. Completely fill the bottle with filtered fuel. Place the bottle on a magnetic stirrer and stir at the lowest possible speed. Make sure that the water/fuel interface is not strongly agitated and that no appreciable vortex develops as a result of mixing.

C1.7 After mixing for approximately 18 h (overnight), place the bottle gently in a water bath at the test temperature for 2 h; insert a clean, dried, hypodermic syringe through the rubber diaphragm in the cap; gently withdraw 2 mL of fuel from the top and analyze for water content using the Karl Fischer method. Take three readings. The average is the water saturation.

C1.8 Convert water saturation to ppm by volume; if the concentration is not between 100 to 150 ppm by volume, then repeat steps C1.6 through C1.8 to confirm; otherwise report this as the dissolved water saturation level in fuel.

APPENDIX D
SOLUBILITY OF WATER IN NO. 2 FUEL OIL

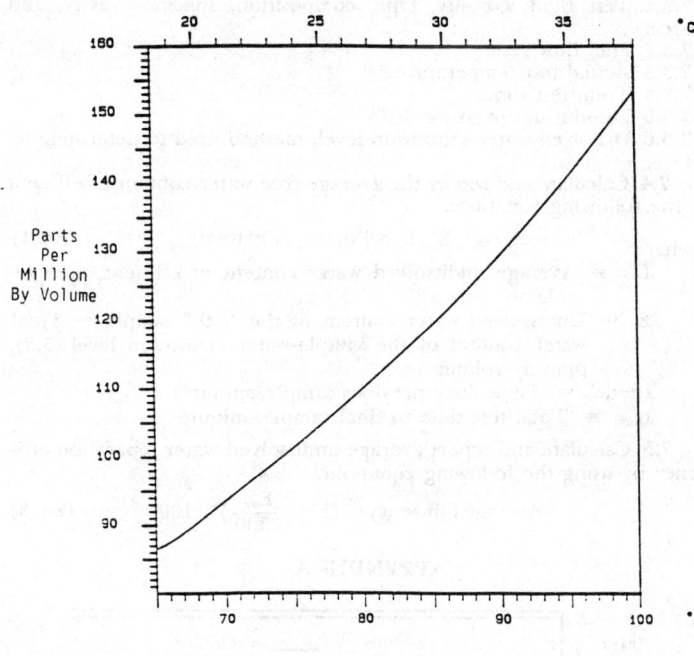

Temperature

R) GASOLINE DISPENSER NOZZLE SPOUTS—SAE J285 SEP92

SAE Recommended Practice

Report of the Fuel Supply Systems Committee, approved 1972, revised August 1974 and reaffirmed by the Automotive Emissions Committee March 1988. Completely revised by the SAE Refueling Interface Standards Committee September 1992.

1. **Scope**—This SAE Recommended Practice provides standardized dimensions for nozzle spouts and a system for differentiating between "unleaded gasoline" nozzle spouts and all other fuel nozzle spouts.

If emission control equipment requires unleaded gasoline exclusively and other fuels not meeting this specification are available, differentiation is accomplished by providing differences between the outside diameter of the nozzle spouts used to dispense "unleaded gasoline" and those used for all other fuels. These differences establish a basis on which fuel filler inlets that will accept only "unleaded gasoline" can be designed.

Spouts used to dispense "unleaded gasoline" should have a nominal OD of 20.6 mm (13/16 in) and be straight for 85 to 95 mm (3.35 to 3.74 in) from the outlet. It is understood that tolerances and normal use may increase the spout up to 21.3 mm (0.84 in) OD.

The spouts for all other fuels should have a nominal OD of 23.8 mm (15/16 in) or more. Tolerances may decrease these spouts to 23.6 mm (0.93 in) OD. An exception is that a 20.6 mm (13/16 in) spout may be used, if such a spout is provided with means that positively prevents the dispensing of leaded fuels into fuel filler inlets designed to accept only "unleaded gasoline."

Identification of spouts is optional. If desired, "unleaded gasoline" spouts may be identified with the letters UN-L stamped or rolled in the spouts and located anywhere on their OD except for the first 63.5 mm (2.5 in) from the outlet. Spouts for all other fuels may be identified with the letters LEAD located anywhere on their OD.

The standardized dimensions are summarized in Table 1 and referenced in Figure 1. These dimensions take into account present nozzle interchangeability in dispenser nozzle boot fit requirements of UL 87 and UL 842. These dimensions minimize the variability of positioning the nozzle spout in the fuel inlets of passenger cars and light-duty vehicles.

2. **References**

2.1 **Applicable Documents**—The following publications form a part of this specification to the extent specified herein. The latest issue of SAE publications shall apply.

2.1.1 SAE PUBLICATION—Available from SAE, 400 Commonwealth Drive, Warrendale, PA 15096-0001.

SAE J1140—Filler Pipes and Openings of Motor Vehicle Fuel Tanks

2.1.2 UL PUBLICATIONS—Available from Underwriters Laboratories, 333 Pfingsten Road, Northbrook, FL 60062-2096.

UL 87—Power Operated Dispensing Devices
UL 842—Valves for Flammable Liquids

TABLE 1—UNLEADED SPOUT DIMENSIONS

Symbol	Description	Unleaded Gasoline	Other Fuels
D_1	Nozzle Outside Diameter	2.13/20.5 mm (0.839/0.807 in)	23.6 mm min. (0.929 in)
L_1	Minimum Length of D_1	75.0 mm (2.95 in)	----
	Roundness of D_1	Within Dia. Limits	----
C_t	Nozzle Tip Chamfer	0.25/0.75 m × 30 degrees (0.010/0.030 in)	----
D_2	Nozzle Anchor Device Outside Diameter	35 mm max. (1.375 in)	----
L_2	Length of Straight Part of Nozzle Spout	85/95 mm (3.35/3.74 in)	----
L_3	Distance Between Nozzle End and First Anchor Position	85/95 mm (3.35/3.74 in)	----
L_4	Distance Between Nozzle End and Sensor Centerline	17.0/18.0 mm (0.669/0.709 in)	----
L_5	Clearance from Fuel Dispensing End and Any Part of Nozzle Body	Consistent with SAE J1140 and UL Reg's	Consistent with SAE J1140 and UL Reg's
R_b	Nozzle Spout Bend Radius	90/120 mm (3.54/4.72 in)	----
a	Bend Angle of Nozzle Spout	22.0 degrees/26.0 degrees	----

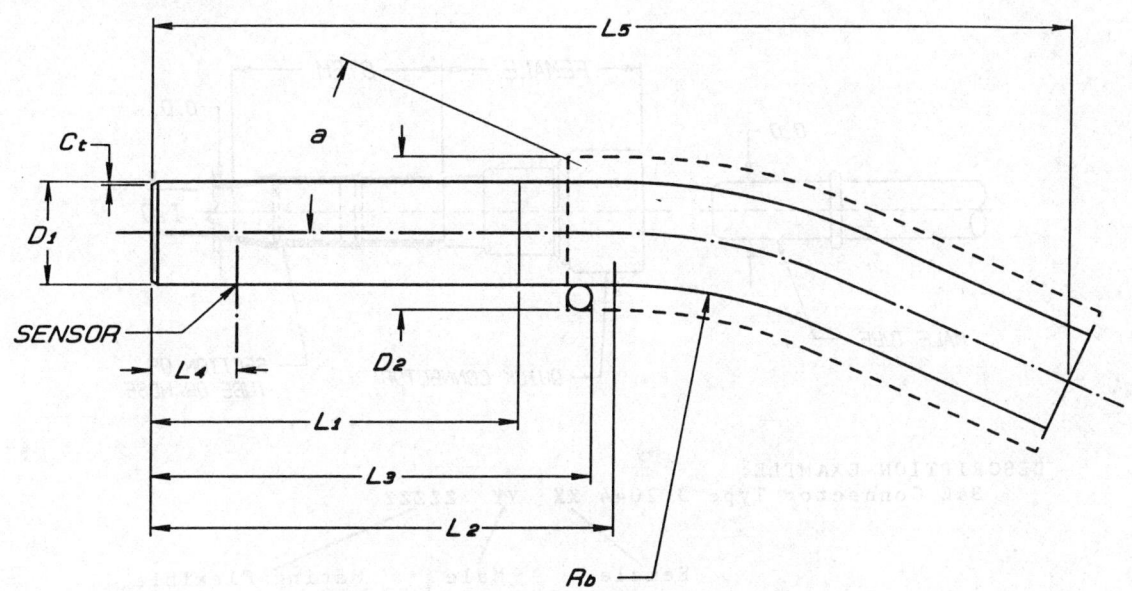

FIGURE 1—UNLEADED SPOUT DIMENSIONS

SAE QUICK CONNECTOR SPECIFICATIONS FOR LIQUID FUEL SYSTEMS—
SAE J2044 JUN92

SAE Recommended Practice

Report of the SAE Fuel Lines and Fittings Standards Committee approved June 1992.

TABLE OF CONTENTS
1. Scope
2. References
2.1 Applicable Documents
3. Size Designation
4. Test Temperatures
5. Functional Requirements
5.1 Leak Test
5.2 Assembly Effort
5.3 Pull Apart Effort
5.4 Side Load Capability
6. Design Verification/Validation Testing
6.1 Corrosion
6.2 Zinc Chloride Resistance
6.3 External Chemical & Environmental Resistance
6.4 Fuel Compatibility
6.5 Life Cycle Tests
6.6 Flow Restriction Requirements
6.7 Elevated Temperature Burst
7. Design Verification/Validation and In-Process Testing Matrix

1. Scope—This SAE Recommended Practice defines minimum functional requirements for quick connectors used in supply and return, liquid lines, for flexible tubing fuel systems. The document applies to automotive and light truck gasoline and diesel fuel systems, with operating pressures up to 500 kPa, 5 Bar, (72.5 psig), and the fuel lines and connectors routed such that continuous operating temperature exposure is less than 115 °C (239 °F). These tests apply to new fittings in assembly operations: for service operations, the male tube should be lubricated with engine oil before coupling.

Vehicle O.E.M. fuel system specifications may impose additional requirements beyond the scope of this general SAE document. In those cases, the O.E.M. specification takes precedence over this document.

2. References

2.1 Applicable Documents—The following publications form a part of this specification to the extent specified herein. The latest issue of SAE publications shall apply.

2.1.1 SAE PUBLICATIONS—Available from SAE, 400 Commonwealth Drive, Warrendale, PA 15096-0001.

SAE J2045—Flexible Hose Assemblies for Gasoline Fuel Injection Fuel Supply Systems

2.1.2 ASTM PUBLICATIONS—Available from ASTM, 1916 Race Street, Philadelphia, PA 19103.

ASTM B 117—Method of Salt Spray (Fog) Testing
ASTM D 1171-68—Test Method for Rubber Deterioration—Surface Ozone Cracking Outdoors or Chamber Triangular Specimens)
ASTM G 23—Practice for Operating Light- and Water-Exposure Apparatus (Carbon-Arc Type) for Exposure of Nonmetallic Materials

3. Size Designation—For this document, the quick connector size designation consists of two numbers. The first number designates the OD of the mating tube end, and the second number designates the tubing size suited to the stem. (See Figures 1 to 5.)

NOTE—On nylon and metal tubing, the OD is used to designate size; rubber and P.T.F.E. hose use the I.D. to designate size.

4. Test Temperatures—Unless otherwise specified all tests will be performed at room temperature 23 °C ± 2 °C (73.4 °F ± 4 °F).

5. Functional Requirements—This section defines the minimum functional requirements for quick connectors used in flexible tubing fuel systems.

5.1 Leak Test—Fuel system quick connectors shall meet the leak test requirements of this document. In order to provide a production compatible process, all leak testing should be performed using compressed air to insure the connectors will not leak fuel.

5.1.1 TEST PROCEDURE—Insert leak test pins, shown in Figure 6, into the quick connectors and pressurize between the O-ring seals with suitable air leak test equipment.

5.1.2 LOW PRESSURE—69 kPa ± 7 kPa, 0.69 bar ± 0.07 bar (10 psig ± 1 psig).

5.1.3 ACCEPTANCE CRITERIA—2 cc/minute, Maximum

5.1.4 HIGH PRESSURE—1034 kPa ± 35 kPa, 10.34 bar ± 0.35 bar (150 psig ± 5 psig)

5.1.5 ACCEPTANCE CRITERIA—5 cc/minute, Maximum

5.2 Connector Assembly Effort—Connector assembly effort is the force required to fully insert (latch or retain) the male tube end into the quick connector. Use a tensile/compression tester suitable to verify con-

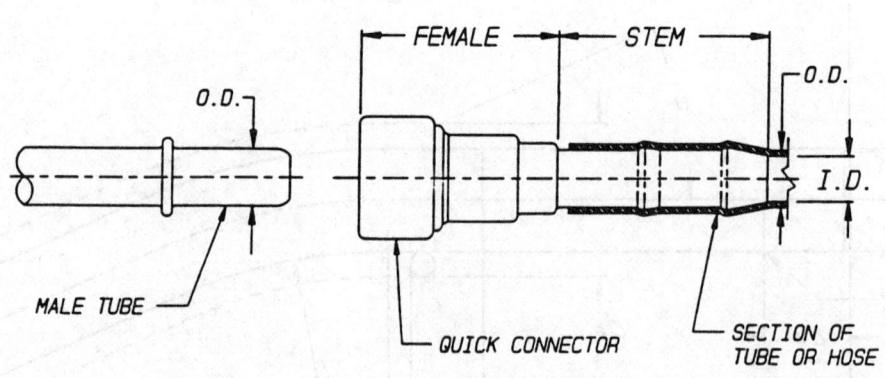

FIGURE 1—SIZE DESIGNATION

formance of this document.

5.2.1 TEST PROCEDURE—For New Parts
a. Test the connector as supplied. Do not add additional lubrication to the connector or test pin.
b. Attach quick connect fitting to suitable test fixture.
c. Wipe the test pins, before each test, with a clean lint free cloth to prevent an accumulation of lubrication.
d. Insert assembly test pin, shown in Figure 7, into the connector at a rate of 51 mm/min ± 5 mm/min (2 in/min ± 0.2 in/min) and measure assembly effort.

5.2.2 TEST PROCEDURE—For Fittings Exposed To Fuel (Per 6.5), or Corrosion (Per 6.1)
a. Allow samples to dry 48 h before insertion testing.
b. Lubricate test pin with SAE 30 weight oil.
c. Insert assembly test pin, shown in Figure 7, into the connector at a rate of 51 mm/min ± 5mm/min (2 in/min ± 0.2 in/min) and measure assembly effort.

5.2.3 TEST REQUIREMENT
a. First time assembly effort must not exceed 67 N (15 lb).
b. Assembly effort after fuel compatibility test must not exceed 111 N (25 lb).
c. Assembly effort after corrosion test must not exceed 111 N (25 lb).

5.3 Quick Connector Pull-Apart Test—Quick connector pull apart is the force required to pull the mating tube end out of the quick connector. Use a suitable tensile tester to verify conformance to this document.

5.3.1 TEST PROCEDURE
a. Attach the connector body stem to a fixture suitable for pulling axially through the centerline of the connector.
b. Use the pull apart test pin shown in Figure 8 of this document.
c. Apply a pull apart load, at a speed of 51 mm/min ± 5 mm/min (2 in/min ± 0.20 in/min), until complete separation occurs.

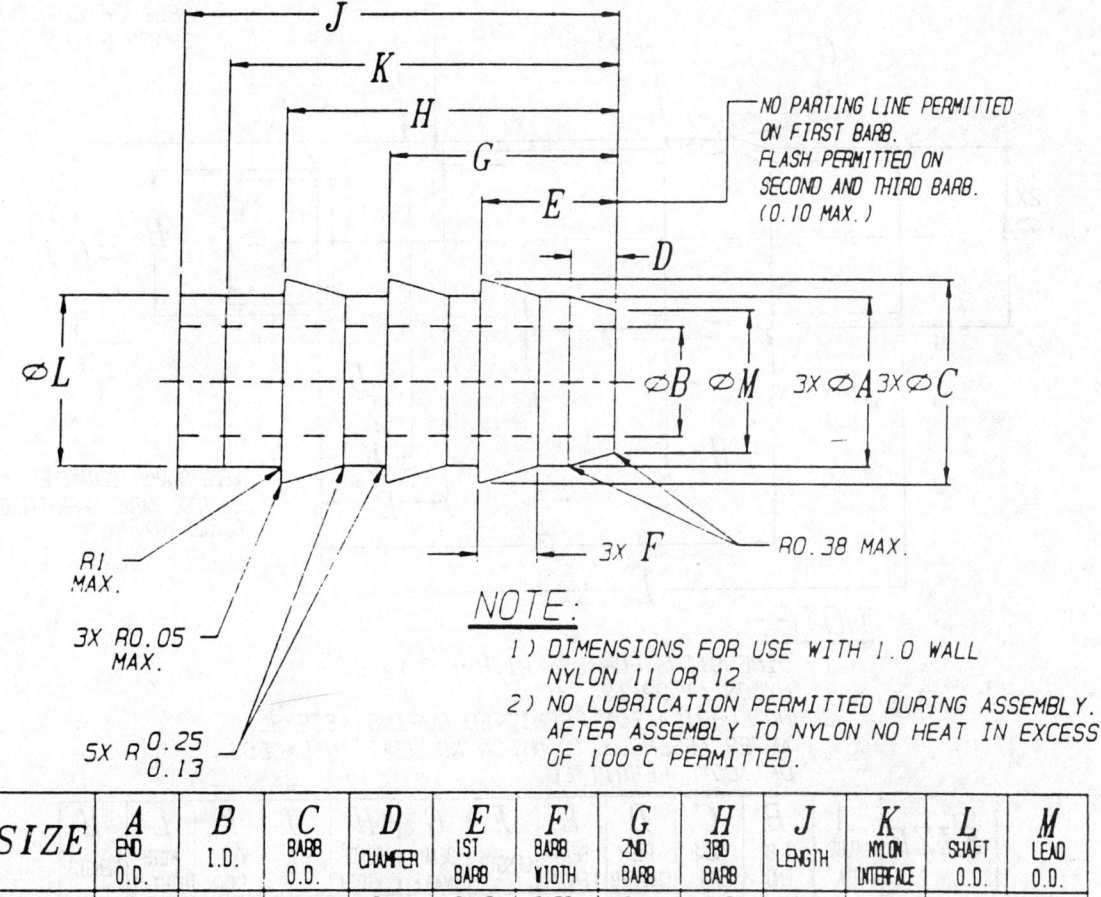

SIZE	A END O.D.	B I.D.	C BARB O.D.	D CHAMFER	E 1ST BARB	F BARB WIDTH	G 2ND BARB	H 3RD BARB	J LENGTH	K NYLON INTERFACE	L SHAFT O.D.	M LEAD O.D.
1/4" (6.35)	N/A	N/A	N/A	3.1 / 2.5	6.13 / 5.86	2.72 / 2.46	10.14 / 9.88	14.64 / 14.37	18.6 MIN.	17 MIN.	N/A	N/A
5/16" (7.93)	7.52 / 7.27	4.83 / 4.67	9.10 / 8.70	2.25 / 1.75	6.25 / 5.75	2.85 / 2.35	10.25 / 9.75	14.75 / 14.25	19.0 MIN.	17 MIN.	8.25 / 7.75	6.25 / 6.09
3/8" (9.53)	9.44 / 9.29	6.08 / 5.88	11.12 / 10.60	2.25 / 1.75	6.25 / 5.75	2.85 / 2.35	10.25 / 9.75	14.75 / 14.25	19.5 MIN.	19 MIN.	9.85 / 9.29	7.87 / 7.67
1/2" (12.7)	12.45 / 12.30	9.34 / 9.08	14.65 / 14.50	2.25 / 1.75	6.25 / 5.75	2.75 / 2.25	10.25 / 9.75	14.75 / 14.25	20.0 MIN.	20 MIN.	N/A	10.70 / 10.55
8mm (.315")	7.52 / 7.27	4.83 / 4.67	9.10 / 8.70	2.25 / 1.75	6.25 / 5.75	2.85 / 2.35	10.25 / 9.75	14.75 / 14.25	19.0 MIN.	17 MIN.	8.25 / 7.75	6.25 / 6.09
10mm (.394")	9.92 / 9.77	6.56 / 6.36	11.60 / 11.08	2.25 / 1.75	6.25 / 5.75	2.85 / 2.35	10.25 / 9.75	14.75 / 14.25	19.5 MIN.	19 MIN.	9.92 / 9.77	8.35 / 8.15

FIGURE 2—TYPICAL MOLDED OR MACHINED STEM CONFIGURATION FOR PLASTIC TUBING

5.3.2 TEST REQUIREMENTS—The force required to separate the tube end from the connector must exceed 450 N (101 lb).

5.4 Connector Side Load Capability—Connector side load capability consists of a side load leak test and a side load fracture test.

The fuel quick connector must meet the side load and air leak test requirements of this document.

5.4.1 TEST PROCEDURE
a. Insert fitting stem into a 200 mm length design intent hose with the opposite end sealed.
b. Attach the connector to a suitable fixture shown in Figure 9 of this document.
c. Internally pressurize the connector between the O-ring seals with 1034 kPa ± 35 kPa, 10.34 bar ± 0.35 bar (150 psig ± 5 psig) air pressure.

NOTE—Appropriate safety precautions should be taken when testing with high pressure air.

d. Side load the connector, at a rate of 12.7 mm/min ± 0.5 mm/min (0.50 in/min ± 0.02 in/min) to the specified force and perform the leak test. At the end of the leak test, increase the side load force until fracture or kinking of the connector occurs. Kinking of the stem is permitted if no rupture results.

5.4.2 TEST REQUIREMENTS
a. Side Load Air Leak Test. (See Table 1.)

5.4.3 ACCEPTANCE CRITERIA
a. No fracture or rupture of the component permitted. Air leak not to exceed 8 cc/min at 10.34 bar ± 0.34 bar (150 psig ± 5 psig).
b. Side load force to fracture or kink test. (See Table 2.)

6. Design Verification/Validation Testing

6.1 Corrosion Test—The corrosion test is performed to assure the quick connector will meet the functional requirements of the fuel system after exposure to the corrosion test.

6.1.1 TEST PROCEDURE
a. Insert a mating tube end, shown in Figure 10, into the quick connectors.

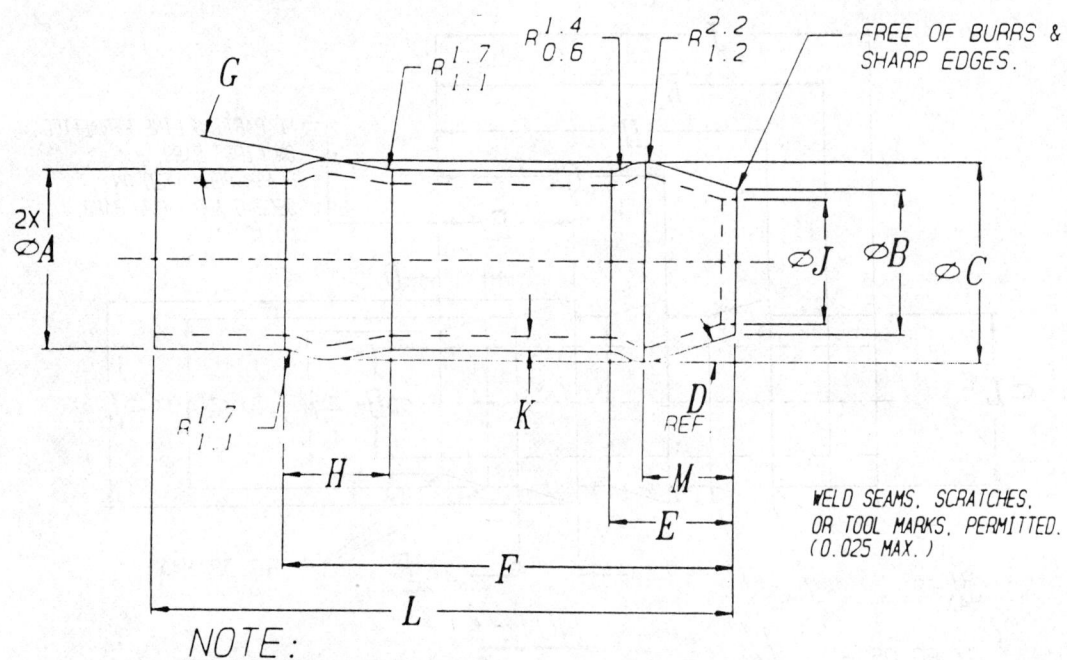

NOTE:
1) DIMENSIONS FOR USE WITH 1.0 WALL NYLON 11 OR 12
2) NO LUBRICATION PERMITTED DURING ASSEMBLY. AFTER ASSEMBLY TO NYLON NO HEAT IN EXCESS OF 100°C PERMITTED.

SIZE	A TUBE DIA.	B TUBE END	C BEAD DIA.	D REF. CHAMFER	E BEAD LENGTH	F LENGTH	G BEAD ANGLE	H BEAD WIDTH	J I.D.	K WALL THICK	L NYLON INTERFACE	M LENGTH
1/4" (6.35)	6.35	N/A	N/A	N/A	N/A	19.15 17.65	9°-11°	5.0 4.5	3.84 MIN.	.62 MIN.	30.0 MIN.	N/A
5/16" (7.92)	7.92	6.60 6.10	8.89 8.64	15°	7.11 5.84	20.05 19.55	9°-11°	5.0 4.5	4.67 MIN.	.62 MIN.	30.0 MIN.	4.0
3/8" (9.53)	9.53	8.12 7.62	10.41 10.16	17°	7.36 6.09	21.35 20.85	9°-11°	5.0 4.5	5.90 MIN.	.62 MIN.	33.0 MIN.	4.5
1/2" (12.7)	12.70	N/A	N/A	N/A	N/A	19.15 17.65	9°-11°	5.0 4.5	9.33 MIN.	.62 MIN.	33.0 MIN.	N/A
8mm (.315")	7.57	6.25 5.75	8.54 8.29	15°	7.11 5.84	20.05 19.55	9°-11°	5.0 4.5	4.32 MIN.	.62 MIN.	30.0 MIN.	4.0
10mm (.394")	10.00	8.60 8.10	10.89 10.64	17°	7.36 6.09	21.35 21.85	9°-11°	5.0 4.5	6.38 MIN.	.62 MIN.	33.0 MIN.	4.5

FIGURE 3—TYPICAL STEEL TUBE STEM CONFIGURATION FOR PLASTIC TUBING

b. Cap the mating tube end and stem end of the connector so that internal surfaces remain free of water and corrosion.
c. Perform salt spray test per ASTM B 117.

6.1.2 TEST REQUIREMENTS

a. The connectors shall be capable of meeting the functional requirements of Section 5 after 500 h salt spray. Appearance is not a functional requirement.

6.2 Zinc Chloride Resistance—The zinc chloride resistance test is performed to assure the quick connector meets the functional requirements of the fuel system after exposure to the zinc chloride test.

6.2.1 TEST PROCEDURE

a. Insert a mating tube end, shown in Figure 10, into the quick connectors.
b. Cap the mating tube end and stem end of the connector so that internal surfaces remain free of water and corrosion.
c. The connector shall be immersed in a 50% aqueous solution (by weight) of zinc chloride for 200 h at 60 °C (140 °F). The solution may change percentage during the exposure.
d. The connectors then must be removed from solution, do not rinse or clean.
e. The connectors must then be exposed to heated air at 60 °C (140 °F) for 200 h.
f. Samples are to be inspected after each exposure sequence and any evidence of cracking recorded.

6.2.2 ACCEPTANCE CRITERIA

a. No cracks or fractures of connector components permitted.
b. The connectors shall be capable of meeting the functional requirements of Section 5 after exposure to zinc chloride.

6.3 External Chemical and Environmental Resistance—The external chemical resistance test is performed to assure the quick connectors will meet the functional requirements of the fuel system after exposure to typical automotive fluids.

6.3.1 TEST PROCEDURE

a. Insert a mating tube end, shown in Figure 10, into the quick connector.
b. Cap the mating tube end and stem end of the connector.
c. The connector shall be immersed in the fluids, specified in 6.3.2, at room temperature, unless otherwise specified.
d. Perform functional tests specified in Section 5.

6.3.2 FLUIDS OR MEDIUM—(See Table 3.)

TABLE 1—SIDE LOAD AIR LEAK TEST

Connector Size (Stem OD)	Side Load Force
1/4 in	111 N ± 9 N (25 lb ± 2 lb)
5/16 in	138 N ± 9 N (31 lb ± 2 lb)
3/8 in	187 N ± 9 N (42 lb ± 2 lb)
1/2 in	249 N ± 9 N (56 lb ± 2 lb)
8 mm	138 N ± 9 N (31 lb ± 2 lb)
10 mm	200 N ± 9 N (45 lb ± 2 lb)

TABLE 2—SIDE LOAD FORCE TO FRACTURE OR KINK TEST

Connector Size (Stem OD)	Side Load Force Minimum
1/4 in	200 N (45 lb)
5/16 in	245 N (55 lb)
3/8 in	311 N (70 lb)
1/2 in	403 N (90 lb)
8 mm	245 N (55 lb)
10 mm	350 N (79 lb)

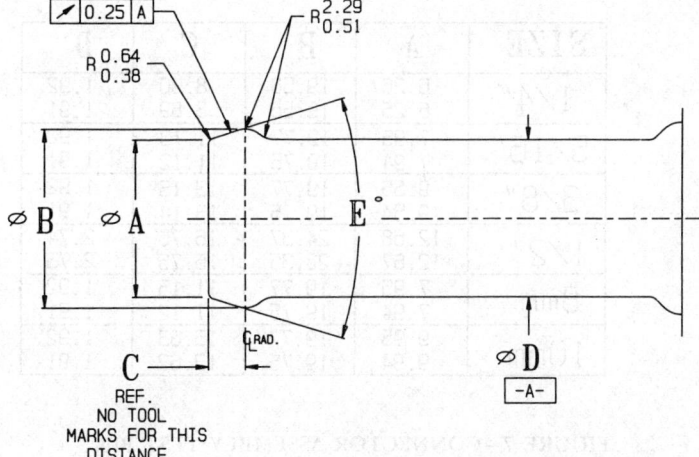

SIZE	A END O.D.	B BARB O.D.	C BARB CENTER LINE	D STEM O.D.	E BARB ANGLE
1/4" (6.35)	6.20	7.37	2.54	6.35	30° 26°
5/16" (7.93)	7.78	9.14	2.54	7.93	30° 26°
3/8" (9.53)	9.38	10.92	2.54	9.53	30° 26°
1/2" (12.70)	12.55	14.22	3.05	12.70	30° 26°
8mm (.315")	7.85	9.14	2.54	8.0	30° 26°
10mm (.394")	9.85	11.14	2.54	10.0	30° 26°

FIGURE 4—TYPICAL MOLDED OR MACHINED END CONFIGURATION FOR RUBBER HOSE

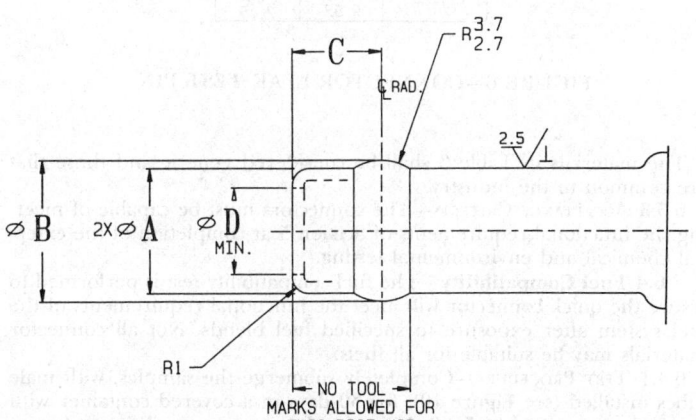

SIZE	A END O.D.	B BARB O.D.	C BARB CENTER LINE	D MIN. I.D.
1/4" (6.35)	6.35	7.34	5.08	4.11
5/16" (7.94)	7.94	8.92	5.46	5.60
3/8" (9.53)	9.53	10.52	5.84	7.11
1/2" (12.70)	12.70	13.72	6.22	10.49
8mm (.315")	8.00	8.56	5.46	5.13
10mm (.394")	10.00	10.99	5.84	7.51

FIGURE 5—TYPICAL MOLDED OR MACHINED END CONFIGURATION FOR P.T.F.E. TUBE

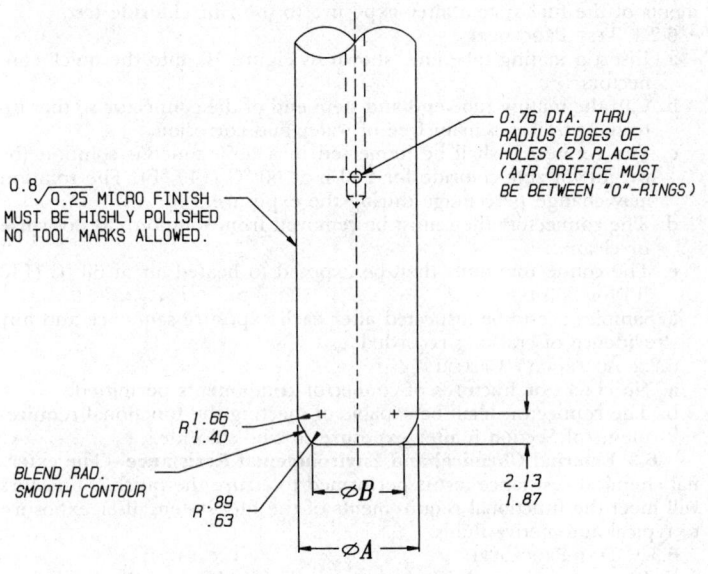

FIGURE 6 — CONNECTOR LEAK TEST PIN

The materials in Table 3 shall be considered generic and those that are common to the industry.

6.3.3 Acceptance Criteria — The connectors must be capable of meeting the functional requirements of Section 5 at completion of the external chemical and environmental testing.

6.4 Fuel Compatibility — The fuel compatibility test is performed to assure the quick connector will meet the functional requirements of the fuel system after exposure to specified fuel blends. Not all connector materials may be suitable for all fuels.

6.4.1 Test Procedure — Completely submerge the samples, with male tubes installed (see Figure 10), for 30 days in a covered container with the fuel as specified in 5.4.2. New samples must be used for each test. Agitate the container every 10 days.

6.4.2 Test Fuels (Ref. SAE Gasoline/Methanol Mixtures for Materials Testing #90-0868EG)

  a. ASTM reference fuel "C"
  b. SAE CE15 (ASTM reference fuel "C" plus 15% ethyl alcohol)

TABLE 3 — FLUIDS OR MEDIUM

Fluid or Medium	Exposure Time	Procedure
Automatic Transmission Fluid	30 day	Soak
Brake Fluid	30 day	Soak
Ethylene Glycol (Coolant)	30 day	Soak
Ozone	144 hr	ASTM D 1171-68
Ultraviolet Rays	30 day	ASTM G 23
Diesel Fuels	30 day	Soak
Motor Oil	30 day	Soak
Windshield Washer Fluid	30 day	Soak
Lacquer Thinner - Used in Assembly Plants	30 day	Soak
Car Wash Detergent	30 day	Soak
Engine Degreasers	30 day	Soak
Vehicle Undercoating	30 day	Soak

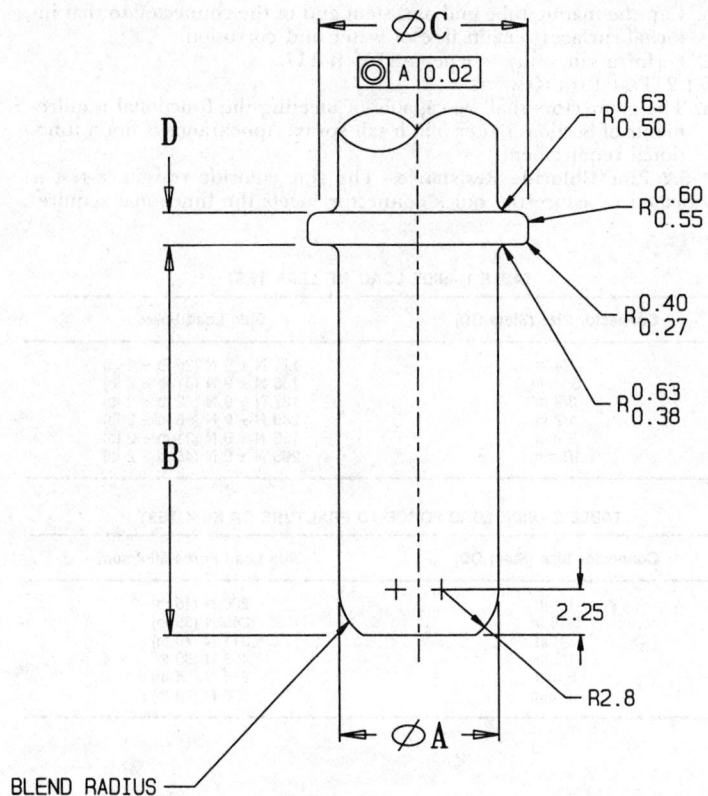

FIGURE 7 — CONNECTOR ASSEMBLY TEST PIN

  c. SAE CM30 (ASTM reference fuel "C" plus 30% methyl alcohol)
  d. SAE CP (Auto-Oxidized Fuel)

6.4.3 Test Requirements — The quick connector shall meet the functional requirements of Section 5 after completion of the fuel compatibility test, half of the samples tested immediately and the remaining half tested after 48 h dry-out.

6.5 Life Cycle Test — The life cycle test is performed to assure the connector will meet the functional requirements of the fuel system when exposed to pressure/vibration/temperature cycles.

TABLE 4 — G LOAD

Maintain G Load	From	To
2 G	7 HZ	25 HZ
10	25	50
20	50	75
18	75	100
16	100	125
14	125	150
12	150	175
10	175	200

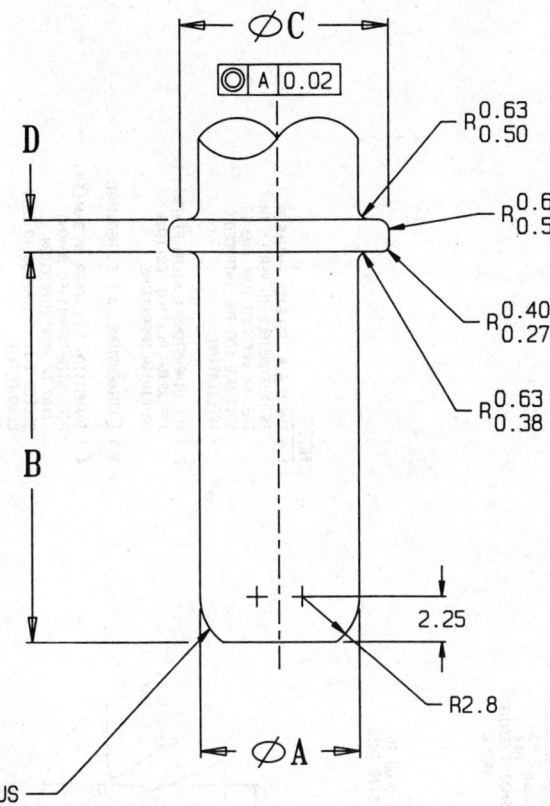

SIZE	A	B	C	D
1/4"	6.26 6.25	19.10 19.06	8.61 8.60	1.61 1.60
5/16"	7.84 7.83	19.31 19.27	10.84 10.83	1.61 1.60
3/8"	9.44 9.43	19.31 19.27	12.74 12.73	1.61 1.60
1/2"	12.55 12.54	24.90 23.86	16.27 16.26	2.51 2.50
8mm	7.84 7.83	19.31 19.27	10.84 10.83	1.61 1.60
10mm	9.84 9.83	19.31 19.27	13.22 13.21	1.61 1.60

FIGURE 8 – CONNECTOR PULL APART TEST PIN

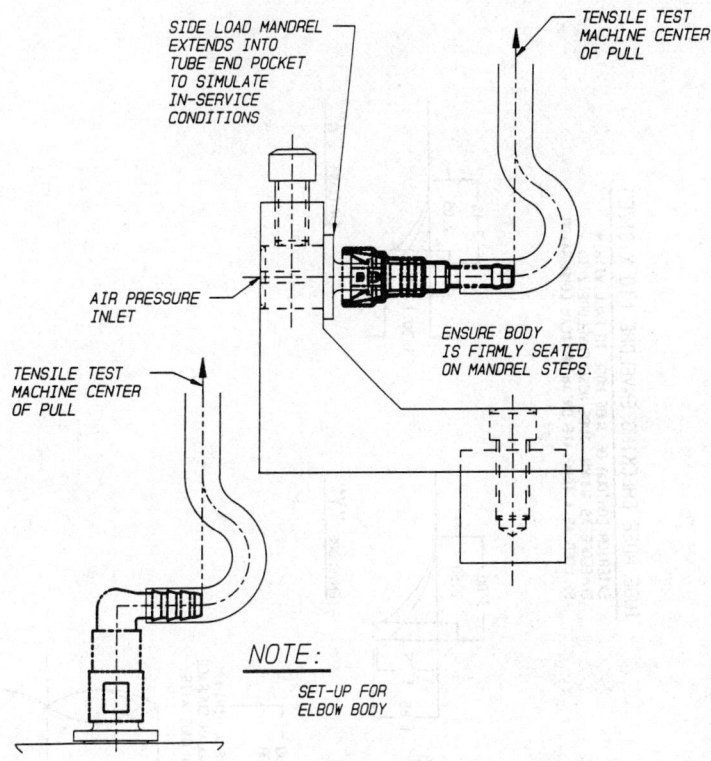

FIGURE 9 – SIDE LOAD TEST FIXTURE

6.5.1 TEST PROCEDURE
a. Attach a quick connector to each end of a 500 mm (19.69 in) length of suitable tubing.
b. Air pressure leak test the assembly per 5.1, except use a male tube as shown in Figure 10.
c. Connect the assembly to a test fixture as shown in Figure 11.
d. Test Fluid-Mobile Arctic 155 refrigerant oil or equivalent.
NOTE – USE OF FLAMMABLE MATERIALS IS NOT RECOMMENDED.

6.5.2 VIBRATION FREQUENCY – Continuously sweep the frequency from 7 HZ to 200 HZ at a rate of 0.3 HZ/s (approximately 3 sweeps/h).

6.5.3 G LOAD – (See Table 4.)

6.5.4 VIBRATION DURATION – Maintain vibration as specified in the test cycles.

6.5.5 FLUID PRESSURE – During pressure portions of the test alternate pressure between 0 and 413 kPa ± 34 kPa, (60 psig ± 5 psig) alternate pressure one time per minute. (One minute at each pressure.)

6.5.6 FLUID FLOW – Flow rate during the specified test cycle is 6.65 LPM ± 1 LPM (1.75 GPM ± 0.26 GPM).

6.5.7 TEST DURATION – 336 h (14 test cycles) (14 days)

NOTE – The test may be interrupted or shut down for weekends at the end of 6.5.8.5.

6.5.8 TEST CYCLE – The test cycle consists of five sections to simulate hot operation, hot soak, hot operation after hot soak, cold soak, and cold operation. See Table 5.

Included in the beginning hot and cold test sections, are temperature transition times of 1 h maximum.

6.5.8.1 Hot Operation Test
a. Length of Time – 7 h
b. Fluid Pressure – yes
c. Fluid Flown – yes
d. Vibration – yes
e. Air Temperature – 125 °C ± 5 °C (257 °F ± 9 °F)
f. Fluid Temperature – 66 °C ± 5 °C (151 °F ± 9 °F)

6.5.8.2 Hot Soak
a. Length of Time – 2 h
b. Fluid Pressure – yes
c. Fluid Flow – no
d. Vibration – no
e. Air Temperature – 125 °C ± 5 °C (257 °F ± 9 °F)
f. Fluid Temperature – (Heat to air temperature)

6.5.8.3 Hot Operation After Hot Soak
a. Length of Time – 7 h
b. Fluid Pressure – yes
c. Fluid Flow – yes
d. Vibration – yes
e. Air Temperature – 125 °C ± 5 °C (257 °F ± 9 °F)
f. Fluid Temperature – 66 °C ± 5 °C (151 °F ± 9 °F)

6.5.8.4 Cold Soak
a. Length of Time – 7 h
b. Fluid Pressure – yes
c. Fluid Flow – no
d. Vibration – no
e. Air Temperature -40 °C (-40 °F)
f. Fluid Temperature Cools to Air Temperature

6.5.8.5 Cold Operation
a. Length of cycle – 1 h
b. Fluid Pressure – yes
c. Fluid Flow – yes
d. Vibration – yes
e. Air Temperature -40 °C (-40 °F)

FIGURE 10 — FUEL QUICK CONNECTOR STANDARD MALE TUBE END

f. Fluid Temperature Cooled by Air Temperature
   6.5.9 TEST REQUIREMENTS
   a. No fluid leaks permitted during or at completion of test.
   b. At completion of the life cycle test, the connector shall be capable of meeting the functional requirements of Section 5.
   c. Perform visual inspection of connector components. No failures, cracks, or unusual wear permitted.

6.6 Flow Restriction—All the fuel quick connectors shall be designed to provide minimal flow restriction.
   6.6.1 TEST PROCEDURE—Measure ΔP across the fitting with a flow rate of 100 Lph (26.4 gph) using a 50/50 glycol and water mixture.
   6.6.2 ACCEPTANCE CRITERIA
   a. To be determined.

6.7 Elevated Temperature Burst Test—The connector burst test is performed to assure the quick connectors will withstand the pressure requirements of the fuel system at maximum operating temperature.

This test can be performed as part of the tube/hose assembly requirements of SAE J2045 or as follows.
   6.7.1 TEST PROCEDURE
   a. Insert a fuel quick connector in each end of a short length, 150 mm (6 in), of reinforced fuel hose. Secure each end with a hose clamp.
   b. Insert a male tube end, shown in Figure 10, into the connectors.
   c. Attach hose assembly to a suitable, air or fluid, burst pressure source.
   d. Place the hose assembly in a suitable environment chamber and hold at 115 °C (239 °F) for 1 h.
   e. Perform burst by pressurizing the hose assembly at a rate of 3450 kPa (500 psig) per minute until burst or rupture occurs.
   6.7.2 ACCEPTANCE CRITERIA
   a. Burst pressure must exceed 2000 kPa (290 psig).
   b. This burst test is for the quick connector only. Leakage or rupture of the hose is not a failure. If the hose fails, the test must be rerun with hose capable of higher pressure, if practical.

7. Design Verification/Validation and In-Process Testing Matrix—See Table 6. This section describes the minimum In-Process Testing requirements for fuel system quick connectors, mating tube ends, and attachment joints.

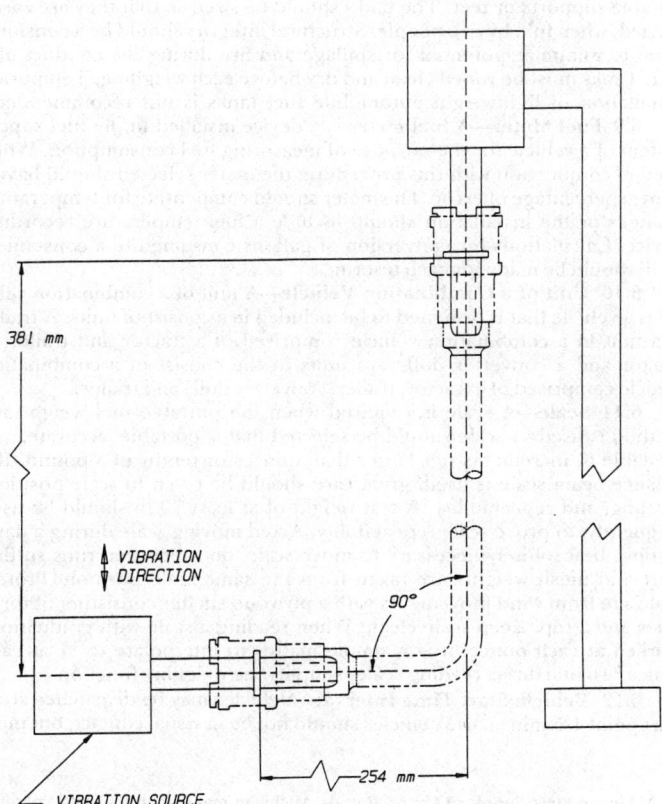

FIGURE 11—LIFE CYCLE TEST FIXTURE

Lot is a quantity of a product produced under similar conditions so that the product within the lot is expected to be homogeneous in all significant attributes. Unless otherwise stated, a lot consists of 8 h production produced within 1 day.

TABLE 5—LIFE CYCLE TEST SCHEDULE

Section	Time	Air Temperature	Fluid Temperature	Pressure	Flow	Vibration
6.5.8.1	1st h	125 °C[1]	66 °C[1]	Yes	Yes	Yes
	2nd h	125 °C	66 °C	Yes	Yes	Yes
	3rd h	125 °C	66 °C	Yes	Yes	Yes
	4th h	125 °C	66 °C	Yes	Yes	Yes
	5th h	125 °C	66 °C	Yes	Yes	Yes
	6th h	125 °C	66 °C	Yes	Yes	Yes
	7th h	125 °C	66 °C	Yes	Yes	Yes
6.5.8.2	8th h	125 °C	125 °C[1]	Yes	No	No
	9th h	125 °C	125 °C	Yes	No	No
6.5.8.3	10th h	125 °C	66 °C[1]	Yes	Yes	Yes
	11th h	125 °C	66 °C	Yes	Yes	Yes
	12th h	125 °C	66 °C	Yes	Yes	Yes
	13th h	125 °C	66 °C	Yes	Yes	Yes
	14th h	125 °C	66 °C	Yes	Yes	Yes
	15th h	125 °C	66 °C	Yes	Yes	Yes
	16th h	125 °C	66 °C	Yes	Yes	Yes
6.5.8.4	17th h	-40 °C[1]	-40 °C	Yes	No	No
	18th h	-40 °C	-40 °C	Yes	No	No
	19th h	-40 °C	-40 °C	Yes	No	No
	20th h	-40 °C	-40 °C	Yes	No	No
	21st h	-40 °C	-40 °C	Yes	No	No
	22nd h	-40 °C	-40 °C	Yes	No	No
	23rd h	-40 °C	-40 °C	Yes	No	No
6.5.8.5	24th h	-40 °C	-40 °C	Yes	Yes	Yes

[1] Temperature may be in transition.

TABLE 6—DESIGN VERIFICATION/VALIDATION AND IN-PROCESS TESTING MATRIX

Test	Section	Design Verification Minimum Sample Size	Design Verification Acceptance Criteria	In Process Minimum Sample Size	In Process Acceptance Criteria
Quick Connector Leak	5.1	50	No Failures	100%	No Failures
Connector Assembly Effort	5.2	50	$\bar{X}$ + 3 Sigma	5/lot	$\bar{X}$ + 3 Sigma
Connector Pull-Apart	5.3	50	$\bar{X}$ − 3 Sigma	5/lot	$\bar{X}$ − 3 Sigma
Connector Side-Load Capability	5.4	50	$\bar{X}$ − 3 Sigma	5/lot	$\bar{X}$ − 3 Sigma
Corrosion Test	6.1	10	No Failures	10/year	No Failures
Zinc Chloride Resistance	6.2	10	No Failures	10/year	No Failures
External Chemical Resistance	6.3	10	No Failures	10/year	No Failures
Fuel Compatibility	6.4	10	No Failures	10/year	No Failures
Life Cycle	6.5	10	No Failures	10/year	No Failures
Flow Restriction	6.6	10	T.B.D.	10/year	No Failures
Elevated Temperature Burst	6.7	10	$\bar{X}$ − 3 Sigma	10/year	$\bar{X}$ − 3 Sigma

JOINT TMC/SAE FUEL CONSUMPTION IN-SERVICE TEST PROCEDURE TYPE III—SAE J1526 JUN87

SAE Recommended Practice

Report of the Truck and Bus In-Service Test Procedure Subcommittee approved June 1987.

1. *Purpose*—This recommended practice provides a standard test procedure for comparing the fuel economy of components or systems of the type which can be switched from one vehicle to another in a short period of time. This test procedure is also ideally suited for comparing the fuel consumption of one vehicle to another, and one component of a combination vehicle to the same component in another. This procedure is specifically designed to be completed in one (1) day.

2. *Scope*—The test utilizes two medium to heavy duty in-service vehicles operated over interstate type highways. The relative fuel economy of the component, system, or vehicle under test is expressed as a percentage improvement or as a percentage of fuel saved. This factor is calculated using relative fuel consumption while operating with and without the test component, system, or vehicle under evaluation. Accuracy obtained from the use of this test procedure can be ± 1% when property executed. This procedure is not intended to replace SAE J1264 OCT86, Joint RCCC/SAE Fuel Consumption Test Procedure or SAE J1321 OCT86, Joint TMC/SAE Fuel Consumption Test Procedure Type II, but will enhance a fleet's or manufacturer's ability to do a wide variety of fuel consumption tests on highway.

The following basic rules must be applied to this procedure to ensure valid test results:

 a. A single test is inconclusive. A single test result may be an indicator. Test results must be repeatable to have validity.

 b. The more variables controlled, the more conclusive the results.

 c. All test procedures are accurate within prescribed limits. If the component, system or vehicle being tested by a given procedure shows a degree of improvement which is equal to or less than the accuracy limit of the procedure, an additional number of tests should be conducted to prove validity.

3. *Summary of Equipment Required*
 a. Two medium to heavy duty vehicles.
 b. A device, system, component of a combination vehicle or vehicle to be tested. If vehicle is to be tested, it is one of the two required by Item a. above.
 c. A minimum of two portable fuel tanks fitted with quick disconnect supply and return lines and suitable retainers; for example, chain binders.
 d. A scale calibrated in increments of 0.1 lb (45 g) or 1 oz (28.4 g) or temperature compensating fuel flow meters.
 e. A test weight of about 70 lbs.
 f. A wind screen to shield scale.
 g. On-site fuel storage; for example, three barrels on a pickup truck and a hand pump, or a portable pump drawing fuel from the tank of a vehicle not being tested.
 h. Compass, thermometer and wind speed indicator.

4. *Summary of the Procedure*
 a. Two test vehicles are warmed up.
 b. The two vehicles, one of which is or contains the object of the test, are driven over a test route at a designated speed a minimum of three times. Fuel consumption and elapsed time are measured for each vehicle on each test run. When the ratios of the fuel consumed by both vehicles are within 2%, and when the elapsed times of each vehicle on each test run are within 1/2%, half of the test procedure is complete.
 c. The test device, system, or component of a combination vehicle is switched from one vehicle to the other. If an entire vehicle is being tested, the drivers switch vehicles.
 d. The two vehicles are re-warmed, if required (see paragraph 8.1).
 e. The two vehicles are again driven over the test route a minimum of three times, or until the fuel ratio and time requirements of Item b. above have been met.
 f. To determine if the subject of the test is more fuel efficient, add the fuel consumed by the vehicles when they were operated with the device, system, or component of a combination vehicle and the fuel consumed by the vehicles when they were operated without the device, system or component of a combination vehicle and compare the sums. If an entire vehicle is being tested, add the fuel consumed by the vehicles when driven by Driver "A" and the fuel consumed by the vehicles when driven by Driver "B" and compare the sums.

5. *Identification*—Sufficient information should be recorded to identify the vehicles under test and the route over which the test is conducted. Recommended minimum information can be recorded on Type III Test Data Form #1.

6. *Definitions*

 6.1 Complete Test—The complete test is comprised of two test segments (paragraph 6.2).

 6.2 Test Segments— A test segment is a minimum of three valid A and B data points (paragraph 6.3). If more than three valid test runs are used in calculations for one segment, an equal number of valid test runs will be required in the other segment.

 6.3 Data Point—A data point is the quantity of fuel consumed by Vehicle "A" or "B" on one test run. Two data points from the same run form an A/B ratio (paragraph 6.4).

 6.4 A/B Ratio—An A/B ratio is the quantity of fuel consumed by Vehicle "A" during a test run divided by the quantity of fuel consumed by Vehicle "B" on the same test run (paragraph 6.5).

 6.5 Test Run—A test run is a complete circuit of the test route (paragraph 6.6). A test run always starts and ends at a common point. Vehicles "A" and "B" negotiate each test run simultaneously.

 6.6 Test Route—A test route should be selected that is typical of the general direction and terrain of the user's operation. A test route should be a minimum of 28 miles and should consume a minimum of 3 gal, or 20% of portable tank volume, of fuel in each test vehicle (paragraph 6.7). A test route longer than 50 miles may preclude test completion in one day.

 6.7 Test Vehicles—Test vehicles are two vehicles selected for comparison and operated over a prescribed route, as required by this procedure. The vehicles are to be properly prepared for safe and normal highway operation and must be fitted with a portable fuel tank (paragraph 6.8) or a fuel meter (paragraph 6.9). Use of either or both is optional.

 6.8 Portable Fuel Tank—Each test vehicle is equipped with a portable fuel tank. The tank should be designed so that it can be readily mounted close to the truck engine and fuel supply system. The tank should be equipped with quick disconnect fittings attached to pipes that can accommodate the fuel flow required by the engine and fuel return from certain engines. The tanks should also be equipped with a vent, handles, and suitable supports or feet. The tanks should be sized so that they are easily carried, when full, by two people. Structural integrity should be a consideration to minimize potential for spillage and fire during the conduct of a test. Tanks must be wiped clean and dry before each weighing. Temporary installation of lightweight automobile fuel tanks is not recommended.[1]

 6.9 Fuel Meter—A fuel meter is a device installed in the fuel supply system of a vehicle for the purpose of measuring fuel consumption. When used in conjunction with this procedure, the meter selected should have a known percentage of error. The meter should compensate for temperature changes or the installation should include a fuel temperature recording device. Calculations for conversion of gallons consumed to a constant of 60°F should be made for each test run.

 6.10 Unit of a Combination Vehicle—A unit of a combination vehicle is a vehicle that is designed to be included in a consist of units. A trailer is a unit in a combination vehicle comprised of a tractor and trailer. A tractor and a converter dolly are units in the consist of a combination vehicle comprised of a tractor, trailer, converter dolly and trailer.

 6.11 Scale—A scale is required when the portable fuel weigh tank method is used. A scale should be selected that is portable, accurate, and readable to increments not larger than ounces or tenths of a pound. If a balance beam scale is used, great care should be given to scale position, leveling, and repeatability. A test weight of at least 70 lb should be used frequently, to prove scale repeatability. Avoid moving scale during a day's testing. If absolutely necessary to move scale, do it between runs so that start and finish weights are taken from the same scale position. Protect scale site from wind movements with a plywood shelter consisting of three sides and a top. Keep scale clean. When reading a scale with graduations marked at each ounce, it is a simple matter to interpolate to ¼ at each ounce.[2] Digital direct reading scales simplify the weighing function.

 6.12 Vehicle Start Time Interval—Vehicles may be dispatched from start point 1-5 min apart. Vehicles should not be in visual contact, but must

---

[1] Michigan Fleet Supply of Grand Rapids, Michigan manufacturers an 18 gal tank, Model M24, which meets these requirements and has been used extensively for on-highway testing.

[2] A good scale for this purpose is Acu-Weigh, Model 200 or equivalent.

TYPE III TEST DATA FORM #1 (VEHICLE IDENTIFICATION)

Power Unit

Fleet _____ Date _____ Test # _____

	Vehicle A	Vehicle B
Unit Number	_____	_____
Make	_____	_____
Model	_____	_____
Year	_____	_____
Number of Axles	_____	_____
Number of Drive Axles	_____	_____
Engine Make/Model	_____	_____
Governed Speed @ No Load (High Idle)	_____ rpm	_____ rpm
Rated Power (bhp)	_____ hp (kw)	_____ hp (kw)
Rated Speed	_____ rpm	_____ rpm
Peak torque	_____ lb-ft	_____ lb-ft
Peak Torque Speed	_____ rpm	_____ rpm
Transmission Make/Model	_____	_____
Geared For	_____ mph (km/h)	_____ mph (km/h)
	at _____ rpm	at _____ rpm
	in _____ gear	in _____ gear
Differential Make/Model	_____	_____
Differential Ratio	_____	_____
Tire Size/Type/Make/Model	____ / ____ / ____	____ / ____ / ____
Tire Pressure (Cold)	_____ psi (kPa)	_____ psi (kPa)
5th Wheel Setting (express in in (mm) the distance 5th wheel fulcrum is ahead or behind the center line of bogie)	_____ in (mm)	_____ in (mm)

NOTE: In areas where two units are shown [i.e., hp (kw)] circle the unit used.

---

be close enough to maximize the potential of experiencing the same atmospheric ambients. Interval selected must be the same for each test run.

6.13 Vehicle Stopped Time Interval—Vehicle detention time between test runs should be as short as possible and is usually less than 5 min. The detention or stopped time should be the same between all runs of each segment. This requirement assures equal cool-down.

6.14 Vehicle Stopped Time Between Segments—If the time required to switch the component, system, or component of a combination vehicle being tested from one vehicle to the other, between test segments, is more than 20 min, it is required that the vehicles be re-warmed.

7. Test Preparation

7.1 Test Route Selection—Experience has shown that a four lane, limited access, divided highway or a test track should be selected for a test site. If a public highway is to be used, the site selected should not include the possibility of an increase in traffic that would cause either test vehicle to change speed or impede maneuvering. Each test run must be operationally identical to all other test runs. Test sites can be found readily anywhere in the United States. The interstate system is recommended. Avoid using a trucking terminal truck stop or a manufacturer's facility as a stop and start point if they are congested areas which cause test delays and unusable data.

If the distance from a truck domicile to a test site is planned to be 50 or 60 miles, that distance traveled qualifies for warm-up time and usually locates the start-stop point in a low traffic volume area. The test route should have a very low propensity for traffic-induced slow downs. The test route should include a parking area with easy, uncontrolled access to the highway. A rest area on an interstate highway is ideally suited for parking and for the start and stop point of each test run. The test route should accommodate direction changes without interference by traffic control signals or other vehicles. Test route turn arounds are best accomplished by an off ramp, stop sign, overpass, an on ramp, or a cloverleaf. Traffic control signals should not be included in a test route. When testing aerodynamic devices, an ideal test route would include nearly equal distances of North-South and East-West highways having a cloverleaf for direction changes or a closed loop, such as a beltway. A test route should be a minimum of 28 miles and should consume in each vehicle a minimum of 3 gal of fuel, or 20% of portable tank volume.

7.2 Drive Cycle Selection—On-highway fuel economy testing readily accommodates a "long haul" or interstate cycle test. A "short haul" or

TYPE III TEST DATA FORM #1 (VEHICLE IDENTIFICATION) (CONTINUED)

Devices, Components, or Systems That Are Incorporated into Test Vehicle Specifications

Fleet _____   Date _____   Test # _____

	Vehicle A			Vehicle B		
	No	Yes	Type	No	Yes	Type
Radiator Shutters (on-off or modulating)						
Engine Cooling Fan Sys. (Describe below-A)						
Aerodynamic Device (Describe below-B)						
Engine Oil						
Transmission Lube						
Differential Lube						
Fuel Heater						
Oil Cooler						
Tag Axle						
Air Lift Axle(s)						
Low Back Pressure Exhaust System						
Other:						

A _____ / _____

B _____ / _____

suburban cycle test can be accomplished on the same stretch of highway by using the intermediate on off ramps and including speed changes between 40 and 55 mph. "Local cycle" or city stop and go testing should be accomplished in a closed test area. A drive cycle should be selected that is indicative of the user's typical vehicles in-service experience and include average gross load, average speed, average stops per mile and idling. For transit bus testing, the ADB driving cycle should be considered.

7.3 Test Speed Selection—The test speeds selected should be representative of average operation, as determined by the operator conducting the test, and be within the capability of the test vehicles. Vehicles are to be operated according to vehicle, engine, and transmission manufacturer's recommendations (engine speeds and shift points). If the test vehicles can be operated in more than one transmission or differential ratio over any part of the test route at the speed selected, a predetermined driving procedure must be specified. At no time during the test cycle should one vehicle control the speed or performance of the other vehicle; however, they should be run at basically the same time in order to experience the same ambient operating conditions.

7.4 Test Vehicle Specification and Configuration—If a device or system is to be tested, the test vehicles must be identical except for the device or system being evaluated.

If a component of a combination vehicle is being tested, all components of both combination vehicles must be the same except the component being tested.

If two complete vehicles are being compared, they need not be the same. The differences in specifications or configurations are what is being compared.

Gross weight of both vehicles should be the same, unless the equipment being tested changes the gross weight, in which case the cargo weight should be the same. If cargo weight differential is being tested, results should be expressed in ton-miles/gal saved or ton-miles/gal improvement.

Vehicles and tires should have a minimum of 2,500 miles of use before being included in a two truck test. (10,000 miles are recommended.) If an operational change occurs, the test supervisor must be advised at the completion of that run.

7.5 Drivers—Driver technique, consistency in shifting, braking, acceleration, and speed/rpm maintenance is crucial to the quality of the data points and the accuracy of the test results. The drivers who start the test must complete the test. Driver substitution is not permitted.

7.6 Observers—Complex driving cycles require observers; simple driving cycles do not require observers. If observers are used, they should have a contributing function such as advising driver of gear change points, speed change points, and elapsed time between mile markers. Observers, if needed, should be assigned to a driver and assist that driver on every test run of both segments. Observers should avoid distracting driver's attention. Noncontributing passengers should be avoided in order to reduce distraction and mistakes.

7.7 Measurement of Fuel, Weather and Speed

7.7.1 PORTABLE WEIGH TANK METHOD—This method of fuel consumption measurement requires that a portable tank be installed on each vehicle. The tank is filled with the fuel and weighed before each test run and after each test run. The difference in weight is the amount of fuel consumed on that test run. Before each weighing, the tank should be wiped clean of road dirt. Fill caps should be in place. If an inadvertent spill takes place before or during installation, the tank must be weighed again. If a spill or leakage takes place during the test run or prior to weighing at end of test run, that test run data cannot be used. Hoses must be disconnected from the portable tanks within 5 sec of engine shutdown at the end of each test run. Disconnect the return line first. Failure to do this may result in unequal drain back, particularly from the fuel return line.

7.7.2 FLOW METER METHOD—If vehicles are fitted with on-board flow meters, these meters should be capable of temperature/density compensation and be calibrated to a minimum accuracy of ± 0.5% at a flow rate consistent with the vehicle being tested. If a float controlled mixing tank is

26.453

TYPE III TEST DATA FORM #1 (VEHICLE IDENTIFICATION) (CONTINUED)

Fleet_____ Date_____ Test #_____

Detailed Description of Vehicle, Component, or System Modification Being Tested:

_____
_____
_____
_____
_____
_____
_____

Length of Test Route from Start to Stop Point:_____ miles (km)

Test Route: (Describe in detail number of lanes; type of road surface; type of turnarounds; type, if any, of traffic control devices; type of terrain, hills, cuts, curves; special driving instructions; etc.)

_____
_____
_____
_____
_____

Driver(s) Interview

Handling, Power, and Braking Characteristics of Vehicle(s) during Test (see paragraph 7.5):

Vehicle A_____
_____
_____
_____

Vehicle B_____
_____
_____

a component of the meter system, it is necessary to measure the volume of fuel before and after each test run and adjust meter reading accordingly. Failure to include variations in float levels with this type of meter in total fuel consumed on each test run will result in incorrect data.

7.7.3 Fuel Temperature—The fuel temperature in the portable weighing tanks should be kept below 160°F (71°C). Fuel coolers can be used to maintain the temperature below that value, but positioning the portable weigh tank in an area of good air flow is an easier solution. Fuel heaters should not be used unless required by low ambient temperatures. At least 3 gal of fuel should remain in the portable tank at the end of each test run to prevent fuel heat-up.

7.8 Weather Measurement—Complete test summaries should include conditions of weather that prevailed during the conduct of the test. This can be accomplished by contacting the weather service at a nearby airport or by using inexpensive hand held instruments. Data recorded should include temperature, wind speed and direction, and relative humidity. Instruments for measuring these factors are available in outdoor supply or marine stores.

7.9 Speed Measurement—It is imperative that both drivers have a speedometer or a tachometer that is very easy to read and has a steady needle. It is also important that the speedometer be checked for accuracy and any error be known by the drivers and accounted for when monitoring test vehicle speed. Test route mileage should be measured with an odometer or hubometer that is known to be accurate and that distance be used when making calculations for both trucks. If cruise control is used, it is important that the exact same speed is selected and maintained each test run. It is also important that the same points on the test route are used for engaging and disengaging cruise.

If drivers are required to switch from one vehicle to the other vehicle between test segments, dashboard instrument error should be clearly displayed in front of driver to assure that equal vehicle speed is maintained on both segments. To further assist drivers in equal speed control of both vehicles, the use of observers timing the passing of mile markers and calling out speeds, or the use of accurate digital tachometers, is recommended. This test is sensitive to speed and elapsed time, and failure to maintain equal speeds when drivers change vehicles will produce false results.

8. Test Procedure

8.1 Vehicle Warm-up—Vehicles "A" and "B" should follow the same start and warm-up procedures. Warm-up speeds should be at or near test

TYPE III TEST DATA FORM #1 (VEHICLE IDENTIFICATION) (CONTINUED)

Trailer/Body

Fleet _____ Date _____ Test # _____

	Vehicle A	Vehicle B
Unit Number	_____	_____
Make	_____	_____
Model	_____	_____
Year	_____	_____
Type (Van, Flatbed, Tank, Etc.)	_____	_____
Type of Side	_____	_____
Type of Corner	_____	_____
Height	_____	_____
Length	_____	_____
Tire Size/Type/Make/Model	_____	_____
Tire Pressure (Cold)	___/___/___ psi (kPa)	___/___/___ psi (kPa)
Number of Axles on Trailer(s)	_____	_____
G.V.W. (Measured on Scale)	_____	_____
Kingpin Setting	_____ in (mm)	_____ in (mm)
Cab-to-Trailer Gap	_____ in (mm)	_____ in (mm)

speeds. The time of warm-up must not be less than 1 h. Longer warm-up periods may be required at colder temperatures. Warm-up and driver familiarization with the test route and establishing speed control, can be accomplished at the same time.

If the switch between test segments of the component, system, or component of a combination vehicle being tested requires less than 20 min, re-warming of the vehicles may be waived.

This test procedure is structured to measure fuel consumption differences of warmed-up vehicles. A vehicle is considered to be sufficiently warmed up when temperatures are near stabilization in tires, hubs, bearings, differentials, and transmissions.

8.2 Test Segment 1—When Vehicle "A" (modified) and Vehicle "B" (unmodified) are sufficiently warmed up, and the drivers (and observers, when required) are sufficiently familiar with the vehicles and test route, they are brought to the test route stop/start point and parked, with engines stopped. Pre-weighed portable tanks are mounted on each vehicle, all required data is recorded (including fuel meter reading, if used) and Vehicle "A" is dispatched on Test Run I. This is accomplished by starting the engine and immediately moving out, the trip time starting when the wheels begin to roll. A 30-60 sec idle period used for final equipment checks is permissible, but, if used, must be the same for each test run. The start time interval between vehicles should be 1-5 min. The total time each vehicle is stopped between warm-up and Test Run I, and between all subsequent test runs must be the same and need not be over 5 min. The interval between vehicles during test runs is maintained to ensure that one vehicle will not impose an artificial performance limit on the following vehicle, and the vehicles should not be in visual contact. The purpose of maintaining equal time intervals between test runs is to control and assure equal cool-down of both test vehicles. Test Vehicle "B" is dispatched at the proper interval and the trip start time recorded when the wheels begin to roll.

As each vehicle returns, they are stopped and the engine speed reduced to idle for 1 min minus total stopped time at turn arounds or other pre-designated stops on the test course. The trip stop time is recorded when the engine stops. Also, immediately after the engine is stopped, the quick disconnects are uncoupled, fuel return first, and the portable tank is dismounted, weighed, recorded, refilled, weighed, recorded, and re-mounted (if fuel meters are used, they are read and recorded). The quick disconnects are reconnected and the vehicles are dispatched on Test Run II exactly as they were the previous test run. If four or more tanks are used, pre-weighed tanks can be installed and the trucks started on the next run, saving a considerable amount of time.

Test run elapsed time should be timed and recorded by the person responsible for the conduct of the test. Drivers (or observers, if required) should time the amount of idle time that may occur at turn around stop signs and advise person recording so that it can be subtracted from the idle time at end of each test run. This is usually not more than 5-20 s and is subtracted from total trip time since the engine at idle burns an insignificant amount of fuel in that amount of time. Drivers should not accelerate engine above idle speed when stopped. Another very practical and easy method of determining total trip time is to have an observer measure and record the maneuvering times at start, turn arounds, and finish. The sum of these four times, subtracted from total trip time, would be "highway only time". This should be done with a stop watch and should be measured from the same points each trip. Use Form 2-5. Using this method will usually result in fewer voided test runs due to inconsistent times. However, if this alternate method is used, it is very important that the drivers start, stop and make the turn arounds exactly alike each test run. The elapsed test run times of each vehicle on each test run must not vary more than 0.5%. Test runs that do not meet this time constraint cannot be accepted and must be repeated.

Sufficient successful test runs must be completed to comprise a minimum of three valid A/B ratios, each fitting within a 2% band (see example on following test forms). See paragraphs 7.8 and 9.18.

8.3 Switch of Device, System or Vehicle Between Test Segments—After the completion of Segment I, the component, system, unit of a combination vehicle, or the entire vehicle under test is switched or removed from Vehicle "A" and is installed on, or included in, or becomes Vehicle "B". If a component or system is being tested, Driver "A" stays with Vehicle "A" and Driver 'B" stays with Vehicle "B" for the entire test. If a unit of a combination vehicle (tractor, trailer(s) or converter dolly) is being tested, Driver "A" stays with Vehicle "A" and Driver "B" stays with Vehicle "B" for the entire test. If a complete vehicle is being tested (straight truck or combination vehicle), Driver "A" drives Vehicle "A"

SEGMENT 1

This Side for Data Related To Vehicles WITH TEST Device, System or Component of A Combination Vehicle or Data Related to the Performance of VEHICLE A driven by DRIVER A				This Side for Data Related To Vehicles WITHOUT TEST Device, System or Component of A Combination Vehicle or Data Related to the Performance of VEHICLE B driven by DRIVER B				Fleet _____ Date _____ Test Number and Description _____ Test Site and Route and Exit # _____			
Run #	Tank #	Fuel Start	Elapsed Time	Driver	Run #	Tank #	Fuel Start	Elapsed Time	Fuel Consumed	2% Fuel and 1/2% Time	amb. temp
Driver	Run Start Time	Fuel Stop	Stopped Time		Run Start Time		Fuel Stop	Stopped Time	T/C or A/B Ratio	Requirements Met-Mark OK	wind dir
	Run Stop Time				Run Stop Time						wind speed
Truck #		Total	Run Time	Truck #			Total	Run Time		Fuel | Time	time

Comments Re Vehicle Condition - Highway Condition - Traffic - Weather - Mistakes

```
After 3 runs:   Highest A/C or T/C ratio  _____ x .98 = _____ min. acceptable ratio
After 4 runs:   Highest A/B or T/C ratio  _____ x .98 = _____ min. acceptable ratio
                Second highest ratio      _____ x .98 = _____ min. acceptable ratio
After 5 runs:   Highest ratio             _____ x .98 = _____ min. acceptable ratio
                Second highest ratio      _____ x .98 = _____ min. acceptable ratio
                Third highest ratio       _____ x .98 = _____ min. acceptable ratio
```

during the first segment and Vehicle "B" during the second segment, and Driver "B" drives Vehicle "B" during the first segment and Vehicle "A" during the second segment. This will result in each driver being involved with the test item (component, system, unit of a combination vehicle or vehicles) for one complete segment. (See cautionary note 9.17.)

No modifications, other than the test, are to be done to either vehicle during segments. Test modifications should be of short time duration to enable both segments to be completed in one day. If the test modifications require enough time to prevent completion in one day, the Type II procedure should be used.

A typical test would require 1 h for warm-up, 3-4 h to complete Segment I, 1-2 h to complete modifications, 1 h of warm-up and 3-4 h to complete Segment II, for a total elapsed time of 9-12 h. A test utilizing the minimum number of test runs in each segment (3), run over the shortest allowable test route, and requiring only a switch of vehicles (10 min) can be executed without the warm-up between segments, thus reducing overall time to less than 5 h.

8.4 Test Segment II—When Vehicle "A" (unmodified) and Vehicle "B" (modified) are sufficiently warmed up, Segment II can be started and is finished when three valid A/B ratios are completed in the same manner as Segment I.

8.5 Fuel Consumption Calculations—Each vehicle's data points must meet the 0.5% time constraint to be eligible for inclusion in an A/B ratio. As each pair of qualifying "A" and "B" data points are recorded at the completion of each Segment I Test Run, an A/B ratio is calculated. A calculation to determine A/B ratios within 2% should be made after the third test run and then after each succeeding test run that is required. Calculations usually require less than 5 min; therefore, a short delay in starting next test run is permitted. Avoid longer cool down periods.

After the switch of the item (component, system, unit of a combination vehicle, or complete vehicle) being tested, and with re-warmed up vehicles, unless switch takes less than 20 min, Segment II Test Runs are made until, using the same parameter as in Segment I, three valid A/B ratios have been included within a 2% band (multiplying highest numerical ratio by 0.98 will result in minimum acceptable A/B ratio). Using only test runs that resulted in valid A/B ratios, add the pounds of fuel consumed by Vehicle "A"

SEGMENT 2

This Side for Data Related To Vehicles **WITH TEST** Device, System or Component of A Combination Vehicle or Data Related to the Performance of VEHICLE A driven by DRIVER B					This Side for Data Related To Vehicles **WITHOUT TEST** Device, System or Component of A Combination Vehicle or Data Related to the Performance of VEHICLE B driven by DRIVER A				Fleet _____ Date _____ Test Number and Description _____ Test Site and Route and Exit # _____		
Run # / Driver / Truck #	Tank # / Run Start Time / Run Stop Time	Fuel Start / Fuel Stop / Total	Elapsed Time / Stopped Time / Run Time	Driver	Run Start Time / Run Stop Time	Tank # / Fuel Stop	Fuel Start / Total	Elapsed Time / Stopped Time / Run Time	Fuel Consumed T/C or A/B Ratio	2% Fuel and 1/2% Time Requirements Met-Mark OK / Fuel / Time	amb. temp / wind dir / wind speed / time

Comments Re Vehicle Condition - Highway Condition - Traffic - Weather - Mistakes

After 3 runs: Highest A/C or T/C ratio _____ x .98 = _____ min. acceptable ratio
After 4 runs: Highest A/B or T/C ratio _____ x .98 = _____ min. acceptable ratio
 Second highest ratio _____ x .98 = _____ min. acceptable ratio
After 5 runs: Highest ratio _____ x .98 = _____ min. acceptable ratio
 Second highest ratio _____ x .98 = _____ min. acceptable ratio
 Third highest ratio _____ x .98 = _____ min. acceptable ratio

modified, Segment I, and the pounds of fuel consumed by Vehicle "B" modified, Segment II. Then, using the same test runs, add the fuel consumed by Vehicle "B" unmodified, Segment I and the fuel consumed by Vehicle "A" unmodified, Segment II. The number of valid test runs in each segment must be the same.

By subtracting the lowest amount of total fuel consumed as determined above, from the highest amount of fuel consumed, the effect of the tested item is shown to be either more fuel efficient or less.

If fuel was saved as a result of both trucks operating with the modification, the percent of fuel saved can be determined by subtracting A_1 and B_2 modified from $B_1 + A_2$ unmodified and dividing the result by $B_1 + A_2$ unmodified. Multiply this answer by 100.

If percent improvement is required, subtract $A_1 + B_2$ modified from $B_1 + A_2$ unmodified and divide the result by $A_1 + B_2$ modified. Multiply this answer by 100.

Formula:
% Fuel Saved = (A+B unmodified − A+B modified) ÷ A+B unmodified × 100.

% Improvement = (A+B unmodified − A+B modified) ÷ A+B modified × 100.

Where: A+B modified = sum of the runs made by Vehicle A, Segment I, modified and the sum of the runs made by Vehicle B, Segment II, modified.

A+B Unmodified = the sum of the runs made by Vehicle B, Segment I, unmodified and the sum of the runs made by Vehicle A, Segment II, unmodified.

If one component is tested by alternative modification of two vehicles, it is more fuel efficient by the percentage calculated above.

If two vehicles or two components of a combination vehicle are tested, the one with the lowest total fuel usage becomes "modified" in the above calculations and is more fuel efficient, by the percentages calculated.

9. Cautionary Notes

9.1 It has been determined that the optimum long haul test route is one that starts and stops at a common point, and has easy access to the test route. The turn around should be either the cloverleaf type or an off ramp with a stop sign, overhead (or underneath) crossover, and an on ramp. A turn around point with traffic control lights must be avoided. A test route that has had mile (km) markers installed is recommended. For other test routes (P&D, construction, transit, buses, etc.), experience has shown that

**CALCULATION FORM
FOR
FUEL CONSUMPTION COMPARISON
OF TWO VEHICLES
WHEN A DEVICE, SYSTEM OR
COMPONENT OF A COMBINATION VEHICLE
IS SWITCHED FROM ONE VEHICLE TO THE
OTHER BETWEEN TEST SEGMENTS.**

Fleet	Test Number and Description	Test Route	Date

SEGMENT 1

Test Runs	WITH VEHICLE A DRIVER A Check Valid Runs	Lbs (gals) Fuel Consumed	Test Runs	WITHOUT VEHICLE B DRIVER B Check Valid Runs	Lbs (gals) Fuel Consumed
1			1		
2			2		
3			3		
4			4		
5			5		

SEGMENT 2

Test Runs	WITH VEHICLE B DRIVER B Check Valid Runs	Lbs (gals) Fuel Consumed	Test Runs	WITHOUT VEHICLE A DRIVER A Check Valid Runs	Lbs (gals) Fuel Consumed
1			1		
2			2		
3			3		
4			4		
5			5		

Total of valid runs _____   Total of valid runs _____
Both Segments   Both Segments

The terms "with" and "without" on this form refer to the vehicles fuel consumption when they were operated over the test route with and without the device, system or component of a combination vehicle being tested or compared.

1. Without − with = lbs. fuel saved
   _____ − _____ = _____ lbs fuel saved
2. Lbs fuel saved ÷ lbs/gal. (or 7.01) = gallons fuel saved
   _____ ÷ _____ = _____ gls fuel saved
3. (Without − with) ÷ without = percent fuel saved
   (_____ − _____) ÷ _____ = _____ % fuel saved
4. (Without − with) ÷ with = percent improvement
   (_____ − _____) ÷ _____ = _____ % improvement

---

this test procedure is acceptable. However, care must be taken in establishing routes and their inherent driving cycles to ensure they are representative of the operating parameters of the equipment under test and they are repeatable by average drivers.

9.1.1 For transit buses, the Transit Coach Operating Profile Duty Cycle may be used.[3]

9.2 If trailers are used, the trailers included in the consist of Vehicles "A" and "B" must stay with their respective tractors throughout both test segments, unless a tractor or trailer is the item being tested, in which case a switch will take place between segments. If it cannot be assured that revenue equipment can be held out of revenue service for the duration of the test period, consideration should be given to renting trailers. The use of revenue cargo for test weight should be avoided to prevent delay of freight or an untimely and costly termination of the test procedure due to cargo delivery commitments. Experience has determined that renting baled, used newspaper is inexpensive, easy and quick to handle, weight adjustable, and available in every city.

9.3 Experience has determined when using portable weigh tanks, that two tanks for each test vehicle will save a considerable amount of time since the weighing and filling process is done after the vehicles leave the start point. This reduces the stopped time between test runs to 2 or 3 min while tanks are switched. Tanks should be color coded and/or numbered to prevent error.

9.4 Portable tanks, if used, should be weighed on the same portable scales. Scales should be checked with a known, dead weight of approximately 70 lb (45.4 ct.) before each series of readings. Experience has shown that placement of the portable tank on the scale in the same position will improve accuracy. Once a scale is in place for a day's testing, it should not be removed. Experience has also shown that errors in weighing can be detected and corrected without loss of valuable data by weighing each tank twice and by having a second person verify the weight and watch the recording as it is put on the record.

9.5 It is strongly recommended that all drivers and observers of Vehicles "A" and "B" be required to drive and ride over the test route at least twice before testing. Familiarity with grades, required shifting, braking, speed maintenance, etc., will lead to greater accuracy and repeatability.

9.6 To minimize test variability, when driving the warm-up run it is recommended that each driver mentally note the precise location on the test route where he applies the brakes and for how long, where he shifts gears, and where he accelerates and decelerates. Each subsequent run should be an exact duplicate of the previous run and no attempt to improve should be made.

The use of stop watches by observers and/or drivers to facilitate the measurement of time and speed between mile (km) markers has been found to be a valuable aid in meeting the time requirements of this test procedure. The use of mechanical or electronic ground speed or cruise control has been found to be very beneficial.

It has also been found useful to select mile (km) marker checkpoints along the route and record the time between markers, the time to negotiate a cloverleaf, and the time elapsed from interstate ramp to ramp. The selected checkpoints should remain the same for each test run. No attempt should be made to compensate for a fast or slow elapsed time between two previous checkpoints.

9.7 To minimize test variability, it is recommended that all vehicles being tested be in similar mechanical conditions and have (except in the

---

[3] Baseline Advanced Design Transit Coach Specification, Part II, paragraph 1.2 (17). Guidelines procurement document for a new 30 and 40 ft (10.4 and 12.2 m) coach design. Published by DOT and UMTA.

CALCULATION FORM
FOR
FUEL CONSUMPTION COMPARISON
OF TWO VEHICLES
WHEN DRIVERS SWITCH VEHICLES
BETWEEN TEST SEGMENTS

Fleet	Test Number and Description	Test Route	Date

SEGMENT 1

Test Runs	WITH VEHICLE A _____ DRIVER A _____ Check Valid Runs	Lbs (gals) Fuel Consumed	Test Runs	WITHOUT VEHICLE B _____ DRIVER B _____ Check Valid Runs	Lbs (gals) Fuel Consumed
1			1		
2			2		
3			3		
4			4		
5			5		

SEGMENT 2

	WITH VEHICLE A _____ DRIVER B _____			WITHOUT VEHICLE B _____ DRIVER A _____	
1			1		
2			2		
3			3		
4			4		
5			5		
Total of valid runs Both Segments			Total of valid runs Both Segments		

1. Fuel consumed by vehicle B minus fuel consumed by Vehicle A = lbs fuel saved
   _____ - _____ = _____ lbs. fuel saved

2. Lbs. fuel saved divided by weight of fuel per lb. (or 7.01) = _____ gals saved
   _____ - _____ (or 7.01) = _____ gals saved.

3. (Vehicle B - Vehicle A) ÷ Vehicle B = % fuel saved.
   (_____ - _____) ÷ _____ = _____ % fuel saved.

4. (Vehicle B - Vehicle A) ÷ Vehicle A = % improvement
   (_____ - _____) ÷ _____ = _____ % improvement

case where this is the item being evaluated):

a. Each engine governor set to manufacturer's recommendation or the operator's standard.

b. New air cleaner element and new fuel filters. Installation of new air cleaner element can be waived if vehicle's inlet restriction does not exceed 15 in H$_2$O (3.7 kPa).

c. Each vehicle reasonably clean and free of sheet metal dents, tears, or missing body parts. Fiberglass hoods should be intact.

d. Cab side window openings the same in each vehicle, open or closed, for the entire test. For transit buses, all windows should stay closed for entire test, due to the drag caused by cross winds.

e. Accessory load for each vehicle as consistent as possible, for example, by turning air conditioning off, defroster off, heat switch at the same position, and lights on. Be aware that on many vehicles the air conditioner runs when the defroster is turned on. If in doubt, disconnect air conditioner.

f. Trailer free of damage to exterior surfaces that would affect aerodynamic drag.

g. Truck/tractor alignment checked and proper. Trailer axle alignment checked and proper.

h. Each vehicle properly lubricated prior to test. All fluid levels should be checked and be at prescribed levels.

i. Temperature controlled fan drives locked in the same operating mode throughout the test, either on or off. Shutters blocked open unless being tested.

j. Cold tire pressures measured and inflated to operator's standards.

k. A stall check made and correct shift points verified on vehicles equipped with automatic transmissions and torque converters.

l. Exhaust back pressure below engine manufacturer's maximum recommended limit and within 0.5 in Hg (1.7 kPa) between test vehicle engines of the same make and model.

m. Proper brake adjustment. Be aware that proper brake adjustment for a test requires that there be no shoe to drum contact with brakes released. Automatic slack adjusters should be disarmed to prevent actuation and adjustment.

9.8 At the end of each warm-up and at the end of each test run, all vehicles must be checked for mechanical changes that would affect test results. Typical checks would include:

a. Oil pressure and leaks.
b. Coolant temperature and leaks.
c. Exhaust gas temperature.
d. Engine air filter restriction.
e. Electrical load.
f. Tire condition (for example, audible leaks, etc.).
g. Brake dragging (for example, temperatures, odors, etc.).
h. Exhaust smoke.
i. Observed ability to maintain selected test speed.
j. Fluid or air leaks.
k. Intake manifold pressures (turbocharger boost).
l. Fuel leaks—if any fuel leaks from truck plumbing, portable tank, or meter, the leak must be corrected and the prior test runs must be considered suspect. A wet road surface will mask a small fuel leak; caution is advised.

The above items are monitored by the driver, who should be instructed to advise the test supervisor of any changes in the operation of the vehicle at the end of each test run.

9.9 Drivers should be interviewed between test runs to ascertain any differences in the apparent handling, power, and braking characteristics of their respective vehicles. If changes occur between the test runs of either test segment, the test data should be discarded and the test rerun after correction of the problem.

9.10 In order to obtain results which may be considered representative of actual service conditions, it is important to reproduce typical service conditions during the test. This applies to load weights, routes, grades, vehicle speeds, weather, wind conditions, drivers, etc. For example, if the actual service vehicles generally operate in a part of the country where hills exist over a substantial portion of the routes, the test should be conducted on similar terrain in order to obtain the most representative results.

The procedure is route or terrain sensitive and that fact must always temper the use of results. For example, the use of flat test tracks produces flat test track answers, which may not be representative of similar equipment used in a severe positive and negative grade application.

9.11 Because of the special nature of aerodynamic drag reduction equipment (deflectors, body fairings, roof fairings, vortex stabilizers, etc.), when testing aerodynamic devices, route selection should include near equal distances of North-South East-West highway, with a cloverleaf centered to accommodate direction changes, or a circular route, such as a beltway around a city (make sure traffic is always loose enough to maintain the predetermined test speed), to minimize the effects of high or low yaw angle winds. The entire range of results may be either higher or lower than average conditions, depending upon the weather (wind velocity and direction) on the days during which the tests were conducted.

9.12 The accuracy of odometers and speedometers should be determined and compensations made for error during actual test runs. If odometer readings (total miles (km)), between two vehicles differ, it is recommended that the two elapsed mileage (km) readings be averaged and this value be used for calculation purposes. Another acceptable method would be to measure the test course with a vehicle with known speedometer and odometer accuracy and use that distance for calculations of mpg (km/L) conversions.

9.13 If test participants are extremely careful and pay attention to all details of the procedure, it has been found that it is highly unusual that more than five test runs are required to complete a segment. It has also been found that, almost without exception, a procedural error or a mechanical problem can be identified when it is necessary to throw out a test run.

9.14 If more than five test runs are required, a mechanical or operational (procedural) problem exists. The best recommendation, based on experience, would be to continue making test runs and recording data, the express purpose being to determine the cause of the data variable and correct it. Complete the day practicing procedure technique and start over the next day. Under normal circumstances, if the test supervisor has carefully read the procedure and is using it as written, he can train uninitiated drivers and start/stop site helpers in a day, and produce three valid A/B data points within three or four test runs thereafter. An experienced test supervisor can frequently eliminate the first day lost to practicing. A casual approach to the details required for homogenous or "mirror image" test runs by one member of the test crew will cause as many data problems as an intermittent dragging brake or an intermittent air leak causing erratic air compressor operation.

9.15 Any power-consuming accessories that are not used eliminate a variable that may affect results - lock out of service clutch fans, air conditioners, etc. and correct air leaks to reduce air compressor action. Instruct drivers regarding the use of windows, lights, fans, etc. and have an agreement regarding usage prior to start of test, the main point being that whatever is used in the first test run must be used for all test runs of that segment and for the same duration. If wipers are required, they should be turned on or off simultaneously by CB radio or at some prearranged point such as the next turn around stop sign after the rain starts. Compressors affect fuel economy by approximately 2% and should, therefore, be managed carefully through control of leaks, use of air-operated accessories and similar brake operations.

9.16 Experience has disclosed that quick-disconnect O-rings should be available to replace those affected by fuel exposure. Failure to do this can cause lost time and possible fuel spillage which may result in bad data for a test run.

9.17 The speed that each vehicle maintains during the test runs of the second segment must be exactly the same as the speeds that were maintained during the first segment. If drivers switch vehicles between segments, special care may be required to assure that the dash-mounted speedometers and/or tachometers are accurate and the same in each vehicle, or accurate digital tachometers should be installed, or an observer used to measure time between mile posts.

9.18 When two vehicles with horsepower differences of fifty or more are being tested together, great care must be exercised by both drivers to assure that the potentially higher performance capabilities of one truck at the turnarounds do not cause problems meeting the 0.05% time constraint requirement.

φFUEL ECONOMY MEASUREMENT ROAD TEST PROCEDURE—SAE J1082 JAN89

SAE Standard

Report of the Fuel Economy Measurement Procedures Task Force, approved April 1974 and completely revised by the Light Duty Fuel Economy Measurement Committee January 1989.

1. Purpose—This SAE Standard provides uniform testing procedures for measuring the fuel economy of light duty vehicles (motor vehicles designed primarily for transportation of persons or property and rated at 10 000 lb (4500 kg) or less) on suitable roads. (The development of the new SAE Motor Vehicle Fuel Economy Measurement Procedures–SAE 75000b.)

2. Scope—This procedure incorporates driving cycles that produce fuel consumption data relating to Urban, Suburban, and Interstate driving patterns and is intended to be used to determine the relative fuel economy among vehicles and driving patterns under warmed-up conditions on test tracks, suitable roads or chassis dynamometers.[1] The urban driving cycle forms the basis of a Cold-Start Test Procedure described in SAE J1256.

3. Definitions

3.1 Driving Cycles

3.1.1 URBAN CYCLE—Driving pattern defined by paragraph 8.3.4 which is similar to driving conditions in the central business district of a large city.

3.1.2 SUBURBAN CYCLE—Driving pattern defined by paragraph 8.3.5 which is similar to driving conditions in suburban areas of a large city.

3.1.3 INTERSTATE CYCLE—Driving patterns defined by paragraphs 8.3.6 and 8.3.7 which are similar to driving conditions on expressways.

3.2 Test Vehicle—Passenger car or light truck prepared for test according to Section 7.

3.3 Test Vehicle Weight—Unloaded vehicle weight plus 300 lb (136 kg).

3.3.1 UNLOADED VEHICLE WEIGHT (CURB WEIGHT)—The weight of the vehicle as built to production parts list with maximum capacity of all fluids necessary for operation of the vehicle.

3.3.2 DRIVER AND PASSENGER OR BALLAST WEIGHT—300 lb (136 kg) includes occupants, instrumentation, and ballast, if necessary.

NOTE: This weight will be distributed to properly simulate passenger locations and vehicle attitude (one passenger in driver's position, and one passenger or equivalent weight in front seat passenger position).

3.3.3 CHASSIS DYNAMOMETER INERTIA WEIGHT AND HORSEPOWER SETTINGS—These settings should be established and set in accordance with SAE J1263–Road Load Measurement and Dynamometer Simulation Using Coastdown Techniques.

3.4 Observed Economy—Observed economy is the fuel economy measured during a driving cycle. It is determined by dividing the actual miles (kilometers) driven on the cycle by the number of gallons (liters) consumed. Economy should be expressed as miles per gallon (kilometers per liter).

3.5 Corrected Economy—Corrected economy is the observed economy multiplied by the correction factors listed in Section 10. The corrected fuel economy should be expressed as miles per gallon (kilometers per liter).[2]

3.6 Correction Factors—Factors which are used to adjust data to the standard ambient condition of 60°F (15.6°C) and 29.00 in Hg (98.2 kPa) and reference fuel properties.

3.7 Average Fuel Economy—Average fuel economy is the total distance driven divided by the total volume of fuel consumed in a series of replicate tests. When the distance driven in each of the tests is identical, as may be assumed for this procedure, the average fuel economy is determined by taking the harmonic average of the individual economies.

---

[1] Though these tests cycles can be run on a chassis dynamometer, this procedure cannot be used for compliance with mandatory fuel economy standards or fuel economy labelling for light duty vehicles first established by the "Energy Policy and Conservation Act," Public Law 94-163, 94th Congress S. 622, December 22, 1975. Details of the mandatory dynamometer procedure can be obtained by contacting Environmental Protection Agency, Fourth and M Street, S.W., Washington, DC 20460. It should be noted that correlation between chassis dynamometer and road test results has not been established.

[2] The corrected economy and average fuel economy may be expressed in terms of fuel consumption, for example, L/100 km, if the appropriate conversions are made. When average fuel economy is expressed as consumption, the average fuel consumption is the arithmetic average of the individual consumptions.

$$\text{Average Fuel Economy} = \frac{n}{1/MPG_1 + 1/MPG_2 + 1/MPG_n}$$

n = the number of replicate tests

3.8 Test Repeatability Guidelines—These guidelines are intended to provide an estimate of repeatability of test data for replicate tests and are based on a standard deviation equal to 1.9% of the mean.

3.8.1 ESTIMATE OF THE 95TH PERCENTILE RANGE FOR REPLICATE TESTS—The 95th percentile range (R) equals 0.019Q times the average fuel economy, where Q equals the critical value obtained from a table for the Studentized[3] range and the average fuel economy for n tests. Selected values for 0.019Q are:

n	0.019Q
2	0.053
3	0.063
4	0.069
5	0.073
10	0.085

Example 1 (U.S. units)—If a vehicle obtains 14.5 mile/gal and 15.5 mile/gal on two tests on the same cycle, the average fuel economy would be 14.98 mile/gal and the 95th percentile range would be:

$$R = 0.053 \times 14.98 = 0.79 \text{ mile/gal}$$

The difference between the two tests is 1.0 mile/gal which is greater than the difference that would be expected for 95% of the cases in which two tests were conducted. Consequently, additional tests should be conducted to provide more confidence in the average fuel economy.

Example 2 (SI units)—If a vehicle obtains 6.20 km/L and 6.60 km/L on two tests on the same cycle, the average fuel economy would be 6.39 km/L and the 95th percentile range would be:

$$R = 0.053 \times 6.39 = 0.34 \text{ km/L}$$

The difference between the two tests is 0.4 km/L which is greater than the difference that would be expected for 95% of the cases in which two tests were conducted. Consequently, additional tests should be conducted to provide more confidence in the average fuel economy.

3.8.2 ESTIMATE OF THE AVERAGE FUEL ECONOMY AT A 90% CONFIDENCE INTERVAL.

$$\frac{\text{Average at 90\%}}{\text{Confidence Interval}} = \text{Average Fuel Economy} \pm \left[\frac{0.031}{\sqrt{n}} \times \text{(Average Fuel Economy)}\right]$$

Example 1 (U.S. units)—If a vehicle obtained 14.8 mile/gal and 15.2 mile/gal on two tests on the same cycle, the average fuel economy would be 15.0 and the 90% confidence interval would be:

$$15.0 \pm \left[\frac{0.031}{\sqrt{2}} \times 15.0\right] = 15.0 \pm 0.3 \text{ mile/gal}$$

Example 2 (SI units)—If a vehicle obtained 6.29 km/L and 6.46 km/L on two tests on the same cycle, the average fuel economy would be 6.37 km/L and the 90% confidence interval would be:

$$6.37 \pm \left[\frac{0.031}{\sqrt{2}} \times 6.37\right] = 6.37 \pm 0.14 \text{ km/L}$$

4. Instrumentation—All instrumentation shall be calibrated.

4.1 Fuel—The fuel measurement device must be compatible with the vehicle fuel system and should alter the fuel temperature and pressure as little as practical. The fuel measurement system must be accurate to within 0.5% of the fuel used during a driving cycle.

4.2 Speed—The speed indicating device shall indicate vehicle speed in miles per hour (kilometers per hour) and be accurate within 0.5 mile/h (1 km/h).

4.3 Acceleration—The acceleration indicating device must be capable of indicating both positive and negative acceleration. It shall indicate acceleration/deceleration in ft/s^2 (m/s^2) and be accurate within 0.5 ft/s^2 (0.2 m/s^2). (Refer to paragraph 6.5 for Chassis Dynamometer Testing.)

4.4 Time—The time measuring instrument must be capable of measuring the time interval to 0.1 s and be accurate within 0.1 s in 1 min.

4.5 Temperature—The temperature indicating devices must be capable of measuring to the nearest 2°F or 1°C. Accuracy must be within ±2°F or ±1°C. The sensing element shall be shielded from radiant heat sources.

4.6 Absolute Barometric Pressure—An aneroid or mercury barometer should be used. This device should be accurate within 0.1 in Hg or 0.3 kPa.

4.7 Wind—Wind speed should be measured with a device that provides an indication of wind speed that is accurate within 2 mile/h (3 km/h). Wind direction should also be indicated.

4.8 Distance—A distance indicating device is required if the tests are not conducted on a premarked course. This device must be capable of indicating distance to within 15 ft (5 m) and must be capable of accuracy within 30 ft in 1 mile (6 m in 1 km).

4.9 Vehicle Weight—Vehicle weight should be measured with a device that is accurate within ±0.5% with minimum resolution of 10 lb (5 kg).

4.10 Dynamometer Inertia Weight—The dynamometer inertia weight is established reflecting the inertia of the nonrotating tires and the vehicle test weight. The inertia weight should be set to the nearest flywheel increment for mechanical inertia dynamometers or within 10 lb for electrical inertia dynamometers.

5. Test Material

5.1 Test Vehicle—The test vehicle shall be completely defined on the Test Vehicle Specifications and Preparation Form. (The test vehicle will normally be representative of a production built vehicle - any exceptions must be properly noted.)

5.2 Test Fuel—Normally, service station fuel will be satisfactory for test purposes, provided that it is consistent with the manufacturer's recommendations for the vehicle and with the ASTM D 439 volatility guidelines. Specific gravity or API gravity for both gasoline and diesels shall be recorded.

Also gasoline octane rating $\left(\frac{R + M}{2}\right)$ shall be recorded and other properties such as distillation and Reid vapor pressure should be recorded when available.

5.3 Lubricants—Lubricants used shall conform to the manufacturer's recommendation for the predominant weather condition in which the vehicle is being tested.

6. Test Conditions

6.1 Ambient Temperature—Tests should be conducted at ambient temperatures between 30 °F(−1°C) and 90°F (32°C).

6.2 Wind Velocity—Urban cycle tests must not be conducted when average wind speed exceeds 15 mile/h (24 km/h) or when gusts exceed 20 mile/h (32 km/h). For the Suburban and Interstate Cycle tests, these limits should be reduced to 10 mile/h (16 km/h) average and 15 mile/h (24 km/h) gusts.

6.3 Road Conditions—Roads must be dry, clean, smooth, and not exceed 1.0% grade. If operating on a closed track, the start and stop points should be selected such that the schedule elevation difference is 10 ft (3 m) or less.

6.4 It is recommended that roadside markers be used to indicate the points at which speed changes are to be made as indicated in paragraph 8.3.

6.5 A driver's aid is recommended for dynamometer operations reflecting the test cycles described in paragraph 8.3.

7. Test Vehicle Preparation

7.1 Break-In—The vehicle should have accumulated a minimum of 2000 miles (3200 km) of operation prior to test. At least 1000 miles (1600 km) must have been driven at cycling speeds between 40 mile/h (64 km/h) and maximum legal highway speeds. If a closed track is available for break-in, the maximum speed should not exceed 100 mile/h (160 km/h). Unless the testing is specifically evaluating lubricant effects of fuel economy, care should be taken to ensure that lubricant changes or additions do not take place over the duration of the test, and that engine oil has a minimum of 2000 miles use prior to testing. Chassis dynamometer break-in is acceptable. All of the tires must have operated at least 100 road or track miles (160 km) prior to the test. Tires must have at least 75% of the tread remaining and tread must be in good condition. For dynamometer testing, the vehicle should have experienced at least 500 miles of cyclic break-in for the tires and brakes.

7.2 Inspection—The vehicle must be inspected and adjusted where necessary to meet manufacturer's specifications. Checks are specified on the Test Vehicle Specifications and Preparation Form.

7.3 Instrumentation—The fuel measuring device and other instrumentation, as necessary, must be installed in a manner not to hinder the vehicle operation or operating characteristics.

7.4 Test Weight—The vehicle weight must be adjusted to provide the test weight indicated in paragraph 3.3 (this test weight includes instrumentation and operator).

7.5 Tire Pressure—The cold tire pressure should be the minimum recommended by the manufacturer for the vehicle test weight

---

[3] D. B. Owen, "Handbook of Statistical Tables" Reading, MA: Addison Wesley Publishing Co., Inc. 1962. 1962, pp 144-148

and should be set before vehicle operation immediately prior to the vehicle warm up at the beginning of the test.

8. Test Procedure

8.1 Warm-Up—The vehicle must be driven a minimum of 20 miles (32 km) at 55 mile/h (90 km/h) or maximum legal highway speed to stabilize engine and driveline operating temperatures immediately before running the first driving cycle.

8.2 Vehicle Controls

8.2.1 Air conditioning compressor, headlamps, and other accessories that consume power should be turned off unless required for safe vehicle operation. The battery should be fully charged to minimize alternator loading.

8.2.2 Vehicle windows must remain closed while fuel consumption is being measured during the Suburban and Interstate Cycles.

8.3 Driving Schedules

8.3.1 GENERAL DRIVING INSTRUCTIONS

8.3.1.1 Vehicles incapable of attaining acceleration rates specified by the driving schedules will be driven at maximum acceleration until specified schedule speed is reached.

8.3.1.2 Vehicles with automatic transmissions should be driven with the transmission in a range that ensures all forward gears can be automatically engaged. If transmission hunting is encountered at a specified acceleration, the acceleration should be increased to maintain the transmission in the lower gear and this departure from the schedule noted on the data form.

8.3.1.3 Vehicles equipped with manual transmissions will be operated in the following manner: Idles will be made in gear, clutch disengaged. Decelerations will be made in gear, and the clutch will be disengaged at 15 mile/h (24 km/h) on a stop. All cruise operation should be in the highest gear that will prevent engine lugging. Downshifts will be permitted to obtain specified acceleration rates after a deceleration or to obtain a smooth engine operation at a slow speed. The following manual transmission shift speeds are guidelines only and may be modified up or down as necessary to ensure that the specified acceleration rates are attained and to avoid engine lugging or overspeed. Departure from shift speeds specified below should be noted on the data form. Manufacturer's recommended shift speed/shift lights may be used providing their use is noted on the data form.

8.3.1.4 Vehicles with truck-type manual transmissions containing a creeper gear will not use the creeper gear during the driving cycle.

8.3.1.5 Vehicles with manual transmissions will be shifted during accelerations at the specified speeds (mile/h) shown below.

Shifts	TRANSMISSION TYPE		
	3	4	5
1-2	15	15	15
2-3	25	25	25
3-4	—	35	40
4-5	—	—	45

Note any deviations from this schedule on the data form.

8.3.1.6 Shift into the highest possible gear whenever a specified cruise speed is reached. For example, the 20 mile/h (32 km/h) cruise after accelerating at the 0.5, 0.7, and 0.8 mile (0.80, 1.13, and 1.29 km) markers in the urban cycle would be conducted in the highest gear that will prevent engine lugging.

8.3.1.7 Vehicles with overdrive transmissions where the overdrive unit engages automatically are to be driven with the actuator switch in a position which ensures engagement when conditions for operation are reached. On vehicles where overdrive is engaged manually (such as designated overdrive gear), upshift to overdrive at the manufacturer's recommended speed for smooth operation. Where specified accelerations cannot be maintained in overdrive, make the complete acceleration in the conventional gear and engage overdrive upon reaching the specified speed.

8.3.1.8 On vehicles with automatic transmission, brakes should be applied to maintain the schedule speed if the engine idle results in vehicle speed above that specified. For manual transmission vehicles, the transmission should be downshifted.

8.3.2 GENERAL CYCLE INSTRUCTIONS

8.3.2.1 The Urban Cycle will normally be run on a 2 mile (3.2 km) straightaway. The Suburban and Interstate Cycles may be run on either a closed track or on a straightaway. For tests on a straightaway less than 2 mile (3.2 km) long, turn-arounds may be made at normal stop intervals. A test on a straightaway shall consist of successive cycles run in opposite directions to minimize wind and grade effects. A test on a closed track shall consist of one cycle.

8.3.2.2 Effort should be made to perform the Interstate Schedule acceleration and decelerations as specified. The Urban and Suburban acceleration and decelerations should be maintained within ±1 ft/s² (0.3 m/s²) of that specified. Vehicle speeds should be maintained within ±1 mile/h (1.6 km/h).

8.3.2.3 Driving cycle maneuvers are initiated at the points indicated, except for the stop at the end of the Urban Cycle, which is to be completed by the point indicated.

8.3.2.4 Fuel temperature will be recorded on the data form during all idle periods or at the beginning and end of the cycle on the Interstate Schedules.

8.3.2.5 Record weather data for each test cycle.

8.3.2.6 Ambient conditions should be such that repeatability may be attained in as few cycles as possible.

8.3.2.7 Fuel consumed for each schedule, as indicated by a fuel meter, should be the average of at least two consecutive tests that repeat within 2%. If the measured fuel readings are not within 2%, additional tests are required until this criteria is met before calculating the fuel economy. Elapsed time should repeat within 1%.

8.3.2.8 The driving cycles are to be conducted on warmed-up vehicles (refer to initial warm-up procedure in paragraph 8.1).

8.3.3 GENERAL CYCLE SUMMARY

Cycle	Average Speed		Nominal Test Time	Test Distance		Time	Stops
	mile/h	(km/h)	s	mile	(km)	s	
Urban	15.6	(25.1)	463	2.0	(3.22)	60	8
Suburban	41.1	(66.1)	455	5.2	(8.37)	14	2
55 mile/h Interstate	55.0	(88.5)	308	4.7	(7.56)	—	—
70 mile/h Interstate	70.0	(112.6)	242	4.7	(7.56)	—	—

8.3.4 Urban Driving Cycle

Distance		Operation
mile	(km)	

0.0 (0.0) Start fuel meter and timing device, idle 15 s, accelerate to 15 mile/h (24 km/h) at 7 ft/s² (2.1 m/s²). Proceed at 15 mile/h (24 km/h) to the 0.2 mile (0.32 km) marker.

0.2 (0.32) Stop at 4 ft/s² (1.2 m/s²), accelerate to 15 mile/h (24/km/h) at 7 ft/s² (2.1 m/s²). Proceed at 15 mile/h (24/km/h) to the 0.3 mile (0.48 km) marker.

0.3 (0.48) Decelerate to 5 mile/h (8 km/h) at 4 ft/s² (1.2 m/s²), accelerate to 15 mile/h (24 km/h) at 7 ft/s² (2.1 m/s²). Proceed at 15 mile/h (24 km/h) to the 0.5 mile (0.80 km) marker.

0.5 (0.80) Stop at 4 ft/s² (1.2 m/s²), record fuel temperature and idle 15 s, accelerate to 20 mile/h (32 km/h) at 7 ft/s² (2.1 m/s²). Proceed at 20 mile/h (32 km/h) to the 0.7 mile (1.13 km) marker.

0.7 (1.13) Stop at 4 ft/s² (1.2 m/s²), accelerate to 20 mile/h (32 km/h) at 7 ft/s² (2.1 m/s²). Proceed at 20 mile/h (32 km/h) to the 0.8 mile (1.29 km) marker.

0.8 (1.29) Decelerate to 10 mile/h (16 km/h) at 4 ft/s² (1.2 m/s²), accelerate to 20 mile/h (32 km/h) at 5 ft/s² (1.5 m/s²). Proceed at 20 mile/h (32 km/h) to the 1.0 mile (1.61 km) marker.

1.0 (1.61) Stop at 4 ft/s² (1.2 m/s²), record fuel temperature and idle 15 s, accelerate to 15 mile/h (24 km/h) at 7 ft/s² (2.1 m/s²), then to 25 mile/h (40 km/h) at 5 ft/s² (1.5 m/s²). Proceed at 25 mile/h (40 km/h) to the 1.2 mile (1.93 km) marker.

1.2 (1.93) Stop at 4 ft/s² (1.2 m/s²), accelerate to 15 mile/h (24 km/h) at 7 ft/s² (2.1 m/s²), then to 25 mile/h (40 km/h) at 5 ft/s² (1.5 m/s²). Proceed at 25 mile/h (40 km/h) to the 1.3 mile (2.09 km) marker.

1.3 (2.09) Decelerate to 15 mile/h (24 km/h) at 4 ft/s² (1.2 m/s²), accelerate to 25 mile/h (40 km/h) at 5 ft/s² (1.5 m/s²). Proceed at 25 mile/h (40 km/h) to the 1.5 mile (2.41 km) marker.

1.5 (2.41) Stop at 4 ft/s² (1.2 m/s²), record fuel temperature and idle 15 s, accelerate to 15 mile/h (24 km/h) at 7 ft/s² (2.1 m/s²), then to 30 mile/h (48 km/h) at 5 ft/s² (1.5 m/s²). Proceed at 30 mile/h (48 km/h) to the 1.7 mile (2.74 km) marker.

1.7 (2.74) Stop at 4 ft/s² (1.2 m/s²), accelerate to 15 mile/h (24 km/h) at 7 ft/s² (2.1 m/s²) and then to 30 mile/h (48 km/h) at 5 ft/s² (1.5 m/s²). Proceed at 30 mile/h (48 km/h) to the 1.8 mile (2.90 km) marker.

1.8 (2.90) Decelerate to 20 mile/h (32 km/h) at 4 ft/s² (1.2 m/s²), accelerate to 30 mile/h (48 km/h) at 5 ft/s² (1.5 m/s²). Proceed at 30 mile/h (48 km/h).

2.0 (3.22) Begin braking at 4 ft/s² (1.2 m/s²) to arrive at stop at 2.0 mile (3.22 km) marker. Stop fuel meter and timing device at stop, record fuel consumed, elapsed time, and fuel temperature.

0.0 (0.0) Run recheck cycle.

8.3.5 Suburban Driving Cycle

Distance		Operation
mile	(km)	

0.0 (0.0) Approach starting line at 40 mile/h (64 km/h). At line, start fuel measuring and timing devices, accelerate to 60 mile/h (97 km/h) at 3 ft/s² (0.9 m/s²). Proceed at 60 mile/h (97 km/h) to the 0.7 mile (1.13 km) marker.

0.7 (1.13) Decelerate to 30 mile/h (48 km/h) at 4 ft/s² (1.2 m/s²). Accelerate to 50 mile/h (80 km/h) at 3 ft/s² (0.9 m/s²). Proceed at 50 mile/h (80 km/h) to the 2.0 mile (3.22 km) marker.

2.00 (3.22) Stop at 4 ft/s² (1.2 m/s²), record fuel temperature and idle 7 s, accelerate to 15 mile/h (24 km/h) at 7 ft/s² (2.1 m/s²). Continue accelerating to 25 mile/h (40 km/h) at 5 ft/s² (1.5 m/s²). Continue accelerating to 40 mile/h (64 km/h) at 3 ft/s² (0.9 m/s²). Proceed at 40 mile/h (64 km/h) to the 2.6 mile (4.18 km) marker.

2.60 (4.18) Accelerate to 50 mile/h (80 km/h) at 3 ft/s² (0.9 m/s²). Proceed at 50 mile/h (80 km/h) to the 3.3 mile (5.31 km) marker.

3.30 (5.31) Stop at 4 ft/s² (1.2 m/s²), record fuel temperature and idle 7 s, accelerate to 15 mile/h (24 km/h) at 7 ft/s² (2.1 m/s²). Continue accelerating to 25 mile/h (40 km/h) at 5 ft/s² (1.5 m/s²). Continue accelerating to 40 mile/h (64 km/h) at 3 ft/s² (0.9 m/s²). Proceed at 40 mile/h (64 km/h) to the 5.2 mile (8.37 km) marker.

5.2 (8.37) Stop fuel measuring and timing devices while driving at 40 mile/h (64 km/h) at 5.2 mile (8.37 km). Record fuel consumed, elapsed time, and fuel temperature.

0.0 (0.0) Run recheck cycle.

8.3.6 Interstate Cycle 55 mile/h (89 km/h)

Distance		Operation
mile	(km)	

0.0 (0.0) Aproach the starting line at 55 mile/h (89 km/h). Record fuel temperature at line, start fuel measuring and timing devices. Proceed at 55 mile/h (89 km/h) to the 0.2 mile (0.32 km) marker.

0.20 (0.32) Accelerate to 60 mile/h (97 km/h) at 1 ft/s² (0.3 m/s²). Immediately decelerate to 50 mile/h (80 km/h) at 1 ft/s² (0.3 m/s²). Immediately accelerate to 55 mile/h (89 km/h) at 1 ft/s² (0.3 m/s²). Proceed at 55 mile/h (89 km/h) to the 1.2 mile (1.93 km) marker.

1.2 (1.93) Repeat accelerations and decelerations as at 0.20 mile (0.32 km). Proceed to the 2.2 mile (3.54 km) marker.

2.2 (3.54) Repeat accelerations and decelerations as to 0.20 mile (0.32 km). Proceed to the 3.2 mile (5.15 km) marker.

3.2 (5.15) Repeat accelerations and decelerations as to 0.20 mile (0.32 km). Proceed to the 4.7 mile (7.56 km) marker.

4.7 (7.56) Stop fuel measuring and timing device while driving at 55 mile/h (89 km/h) at 4.7 mile (7.56 km). Record fuel consumed, elapsed time, and fuel temperature.

0.0 (0.0) Run recheck cycle.

8.3.7 Interstate Cycle 70 mile/h (113 km/h)

Distance		Operation
mile	(km)	

0.0 (0.0) Approach the starting line at 70 mile/h (113 km/h). Record fuel temperature at line, start fuel measuring and timing devices. Proceed at 70 mile/h (113 km/h) to the 0.2 mile (0.32 km) marker.

0.20 (0.32) Accelerate to 75 mile/h (121 km/h) at 1 ft/s² (0.3 m/s²). Immediately decelerate to 65 mile/h (105 km/h) at 1 ft/s² (0.3 m/s²). Immediately accelerate to 70 mile/h (113 km/h) at 1 ft/s² (0.3 m/s²). Proceed at 70 mile/h (113 km/h) to the 1.2 mile (1.93 km) marker.

1.2 (1.93) Repeat accelerations and decelerations as at 0.20 mile (0.32 km). Proceed to the 2.2 mile (3.54 km) marker.

2.2 (3.54) Repeat accelerations and decelerations as at 0.20 mile (0.32 km). Proceed to the 3.2 mile (5.15 km) marker.

3.2 (5.15) Repeat accelerations and decelerations as at 0.20 mile (0.32 km). Proceed to the 4.7 mile (7.56 km) marker.

4.7 (7.56) Stop fuel measuring and timing device while driving at 70 mile/h (113 km/h) at 4.7 mile (7.56 km). Record fuel consumed, elapsed time, and fuel temperature.

0.0 (0.0) Run recheck cycle.

9. Data Recording—Data shall be entered as required on test data forms.

9.1 Test Vehicle Specifications and Preparation Form
9.2 *Data Form
9.3 *Summary Sheet

10. Data Correction (U.S. Units)
10.1 Reference Conditions

Ambient Temperature	60°F
Fuel Temperature	60°F
Barometric Pressure	29.00 in Hg (wet)
Fuel Gravity (Gasoline)	0.737 Specific Gravity, 60.5° API Gravity
Fuel Gravity (ASTM 1D)	0.820 Specific Gravity, 41.0° API Gravity
(ASTM 2D)	0.845 Specific Gravity, 36.0° API Gravity
Fuel Net Heating Value	
(ASTM 1D)	126 700 Btu/gal
(ASTM 2D)	129 900 Btu/gal

10.2 Fuel Economy Correcton (Gasoline)

10.2.1 DEFINITIONS (UNITS)
- T_A – Average ambient temperature during test cycle (°F)
- T_f – Average fuel temperature at measuring instrument during test cycle (°F)
- P – Average barometric pressure during test cycle (in Hg)
- G_s – Specific gravity of test fuel at 60°F
- G_A – API gravity of test fuel at 60°F
- FE_o – Observed fuel economy (mile/gal)
- FE_c – Corrected fuel economy (mile/gal)

10.2.2 CORRECTION FORMULA
$$FE_c = FE_o \cdot C_1 \cdot C_2 \cdot C_3 \cdot C_4$$

10.2.3 CORRECTION FACTORS
- $C_1 = 1.0 + 0.0014 (60 - T_A)$
- $C_2 = 1.0$ Urban Cycle
  - $= 1.0 + 0.0072 (P - 29.00)$ Suburban Cycle
  - $= 1.0 + 0.0084 (P - 29.00)$ 55 mile/h Interstate Cycle
  - $= 1.0 + 0.0144 (P - 29.00)$ 70 mile/h Interstate Cycle
- $C_3 = 1.0 + 0.8 (0.737 - G_s)$
- $C_3 = 1.0 + 0.0032 (G_A - 60.5)$

C_4† is derived from Table 1 based on gravity of fuel at 60°F and T_f or from the following analytical equation:
$$C_4 = a + bT_f + cT_f^2$$
where the coefficients a, b, c, are:

	ASTM FUEL GROUP			
	1	2	3	4
Spec Gravity Range	0.8499–0.9659	0.7754–0.8498	0.7239–0.7753	0.6723–0.7238
Coefficient				
a	$9.7645 (10)^{-1}$	$9.7108 (10)^{-1}$	$9.6513 (10)^{-1}$	$9.5982 (10)^{-1}$
b	$3.8674 (10)^{-4}$	$4.6590 (10)^{-4}$	$5.5473 (10)^{-4}$	$6.3156 (10)^{-4}$
c	$9.3735 (10)^{-8}$	$2.6156 (10)^{-7}$	$4.3541 (10)^{-7}$	$6.2624 (10)^{-7}$

10.3 Fuel Economy Correction (Diesel)*

NOTE: The method for correcting observed fuel economy for vehicles with diesel engines has not been investigated to the same degree that it has for gasoline powered vehicles. However, the ambient temperature and barometric pressure corrections are primarily for changes in air density and its effect on aerodynamic drag. Hence, correction factors for gasoline powered vehicles are recommended for use.

10.3.1 DEFINITIONS—See paragraph 10.2.1.
- H = Volumetric heating value of diesel test fuel (Btu/gal)

10.3.2 CORRECTION FORMULA
$$FE_c = FE_o \cdot C_1 \cdot C_2 \cdot C_3 \cdot C_4$$

10.3.3 CORRECTION FACTORS
- $C_1 = 1.0 + 0.0014 (60 - T_A)$
- $C_2 = 1.0$ Urban Cycle
  - $= 1.0 + 0.0072 (P - 29.00)$ Suburban Cycle
  - $= 1.0 + 0.0084 (P - 29.00)$ 55 mile/h Interstate Cycle
  - $= 1.0 + 0.0144 (P - 29.00)$ 70 mile/h Interstate Cycle
- $C_3 = K/H$
- K = 126 700 (ASTM 1D type fuels)
  - = 129 900 (ASTM 2D type fuels)

H – shall be obtained from Fig. 1 by using the observed API gravity at 60°F and the 50% distillation point or from calorimeter tests.

C_4 is derived from Table 1 based on gravity of fuel at 60°F and T_f or from the following analytical equation
$$C_4 = a + bT_f + cT_f^2$$
where the coefficients a, b, c are as shown in paragraph 10.2.3.

11. Data Correction (SI Units)

11.1 Reference Conditions
Ambient Temperature	15.6°C
Fuel Temperature	15.6°C
Barometric pressure	98 kPa
Fuel Gravity (gasoline)	0.737 specific gravity
Fuel Gravity (ASTM 1D)	0.820 specific gravity
(ASTM 2D)	0.845 specific gravity
Fuel Net Heating Value	
(ASTM 1D)	35.31 MJ/L
(ASTM 2D)	36.21 MJ/L

11.2 Fuel Economy Correction (Gasoline)

11.2.1 DEFINITIONS (UNITS)
- T_A – Average ambient temperature during test cycle (°C)
- T_f – Average fuel temperature during test cycle (°C)
- P – Average barometric pressure during test cycle (kPa)
- G_s – Specific gravity of test fuel at 15.6°C
- FE_o – Observed fuel economy (km/L)
- FE_c – Corrected fuel economy (km/L)

11.2.2 CORRECTION FORMULA
$$FE_c = FE_o \cdot C_1 \cdot C_2 \cdot C_3 \cdot C_4$$

11.2.3 CORRECTION FACTORS
- $C_1 = 1.0 + 0.0025 (15.6 - T_A)$
- $C_2 = 1.0$ Urban Cycle
  - $= 1.0 + 0.0021 (P - 98)$ Suburban Cycle
  - $= 1.0 + 0.0025 (P - 98)$ 89 km/h Interstate Cycle
  - $= 1.0 + 0.0043 (P - 98)$ 113 km/h Interstate Cycle
- $C_3 = 1.0 + 0.8 (0.737 - G_s)$

C_4 is deried from Table 1A based on gravity of fuel at 15.6°C and T_f or from the following analytical equation:
$$C_4 = a' + b'T_f + c'T_f^2$$
where the coefficients a', b', c' are:

	ASTM FUEL GROUP			
Coefficient	1	2	3	4
a'	$9.8892 (10)^{-1}$	$9.8626 (10)^{-1}$	$9.8333 (10)^{-1}$	$9.8067 (10)^{-1}$
b'	$7.0693 (10)^{-4}$	$8.6875 (10)^{-4}$	$1.0487 (10)^{-3}$	$1.2090 (10)^{-3}$
c'	$3.0370 (10)^{-7}$	$8.4745 (10)^{-7}$	$1.4107 (10)^{-6}$	$2.0290 (10)^{-6}$

11.3 Fuel Economy Correction (Diesel)

NOTE: The method for correcting observed fuel economy for vehicles with diesel engines has not been investigated to the same degree that it has for gasoline powered vehicles. However, the ambient temperature and barometric pressure corrections are primarily for changes in air density and its effect on aerodynamic drag. Hence, the correction factors for gasoline powered vehicles are recommended for use until additional data become available.

11.3.1 DEFINITIONS (See paragraph 11.2.1):
- H = Volumetric heating value of test fuel (MJ/L)

11.3.2 CORRECTION FORMULA
$$FE_c = FE_o \cdot C_1 \cdot C_2 \cdot C_3 \cdot C_4$$

11.3.3 CORRECTION FACTORS
- $C_1 = 1.0 + 0.0025 (15.6 - T_A)$
- $C_2 = 1.0$ Urban Cycle
  - $= 1.0 + 0.0021 (P - 98)$ Suburban Cycle
  - $= 1.0 + 0.0025 (P - 98)$ 89 km/h Interstate Cycle
  - $= 1.0 + 0.0043 (P - 98)$ 113 km/h Interstate Cycle
- $C_3 = K/H$
- K = 35.31 mJ/L for ASTM 1D type fuel
  - = 36.21 mJ/L for ASTM 2D type fuel

H shall be determined from Fig. 1 by using the API gravity at 15.6°C and 50% distillation point or from calorimeter tests.

C_4 is derived from Table 1A based on gravity of fuel at 15.6°F and T_f or from the following analytical equation:
$$C_4 = a' + b'T_f + c'T_f^2$$
where the coefficients a', b', c' are

	ASTM FUEL GROUP	
Coefficient	1	2
a'	$9.8892 (10)^{-1}$	$9.8626 (10)^{-1}$
b'	$7.0693 (10)^{-4}$	$8.6875 (10)^{-4}$
c'	$3.0370 (10)^{-7}$	$8.4745 (10)^{-7}$

12. Data Presentation

- **12.1** Test Vehicle Specifications and Preparation List
- **12.2** Fuel Economy Test-Data Form
- **12.3** Fuel Economy Test-Data Summary Sheet

---

*Data Form & Summary Sheet will be on separate pages for clarity.

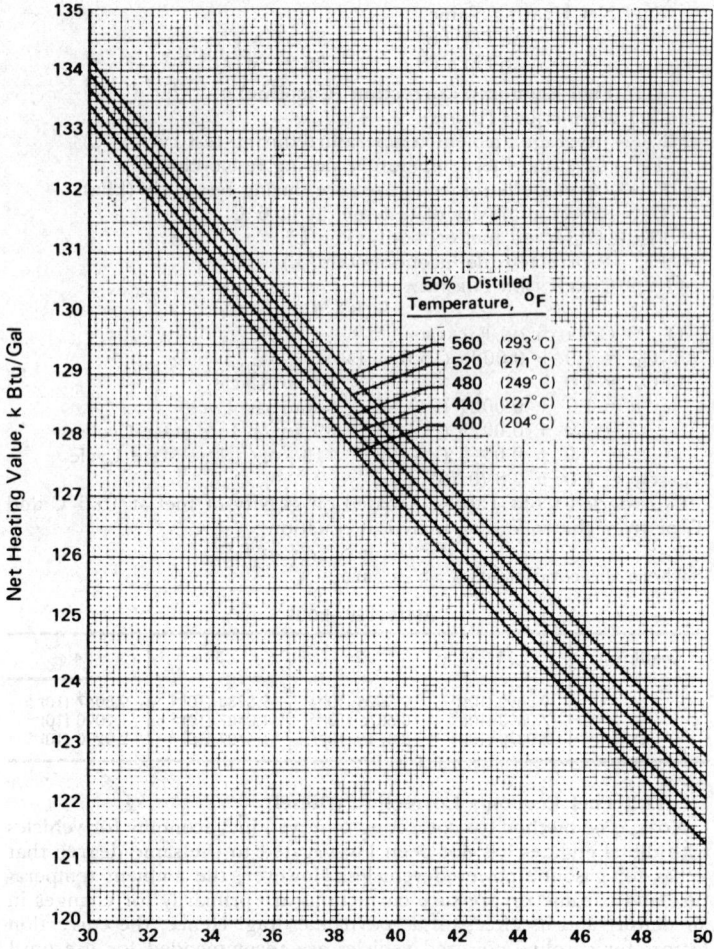

Note: 1 Btu/gal = 278.7 J/L
k Btu/gal = 1000 Btu/gal

FIG. 1—VOLUMETRIC NET HEAT CONTENT OF DIESEL FUELS

TABLE 1—C₄—FUEL TEMPERATURE CORRECTION FACTOR^a (U.S. UNITS)

ASTM Group Number	Specific Gravity Range	API Gravity Range, °API
1	0.8499–0.9659	15.0–34.9
2	0.7754–0.8498	35.0–50.9
3	0.7239–0.7753	51.0–63.9
4	0.6723–0.7238	64.0–78.9

Fuel Temp., °F	C₄				Fuel Temp., °F	C₄			
	Group 1	Group 2	Group 3	Group 4		Group 1	Group 2	Group 3	Group 4
0	0.9765	0.9711	0.9651	0.9598	75	1.0060	1.0076	1.0092	1.0107
1	0.9768	0.9715	0.9656	0.9604	76	1.0064	1.0081	1.0098	1.0114
2	0.9772	0.9720	0.9662	0.9611	77	1.0068	1.0085	1.0104	1.0121
3	0.9776	0.9725	0.9667	0.9616	78	1.0071	1.0090	1.0110	1.0129
4	0.9780	0.9730	0.9673	0.9623	79	1.0076	1.0095	1.0116	1.0136
5	0.9784	0.9734	0.9679	0.9629	80	1.0080	1.0100	1.0122	1.0143
6	0.9788	0.9739	0.9684	0.9636	81	1.0084	1.0105	1.0129	1.0151
7	0.9791	0.9743	0.9690	0.9642	82	1.0088	1.0110	1.0136	1.0158
8	0.9795	0.9748	0.9696	0.9649	83	1.0092	1.0115	1.0142	1.0166
9	0.9799	0.9753	0.9701	0.9655	84	1.0096	1.0120	1.0148	1.0173
10	0.9803	0.9758	0.9707	0.9661	85	1.0100	1.0126	1.0154	1.0180
11	0.9807	0.9763	0.9713	0.9668	86	1.0104	1.0131	1.0160	1.0187
12	0.9811	0.9768	0.9718	0.9675	87	1.0108	1.0136	1.0167	1.0195
13	0.9814	0.9772	0.9724	0.9681	88	1.0112	1.0141	1.0173	1.0202
14	0.9818	0.9777	0.9730	0.9688	89	1.0116	1.0146	1.0179	1.0209
15	0.9822	0.9782	0.9735	0.9695	90	1.0120	1.0151	1.0185	1.0218
16	0.9826	0.9787	0.9741	0.9701	91	1.0124	1.0156	1.0192	1.0225
17	0.9830	0.9790	0.9747	0.9708	92	1.0129	1.0162	1.0198	1.0232
18	0.9835	0.9795	0.9753	0.9714	93	1.0133	1.0167	1.0205	1.0240
19	0.9839	0.9800	0.9759	0.9721	94	1.0137	1.0172	1.0211	1.0247
20	0.9843	0.9805	0.9765	0.9728	95	1.0141	1.0177	1.0218	1.0254
21	0.9846	0.9810	0.9770	0.9734	96	1.0145	1.0182	1.0224	1.0262
22	0.9850	0.9814	0.9776	0.9741	97	1.0148	1.0187	1.0230	1.0269
23	0.9854	0.9819	0.9782	0.9748	98	1.0152	1.0193	1.0236	1.0276
24	0.9858	0.9824	0.9788	0.9753	99	1.0156	1.0198	1.0243	1.0285
25	0.9862	0.9829	0.9793	0.9760	100	1.0160	1.0203	1.0249	1.0292
26	0.9866	0.9834	0.9799	0.9767	101	1.0165	1.0208	1.0255	1.0300
27	0.9870	0.9839	0.9805	0.9773	102	1.0169	1.0213	1.0262	1.0307
28	0.9874	0.9843	0.9811	0.9780	103	1.0173	1.0219	1.0269	1.0314
29	0.9878	0.9848	0.9816	0.9787	104	1.0177	1.0224	1.0275	1.0322
30	0.9881	0.9853	0.9822	0.9793	105	1.0181	1.0229	1.0282	1.0330
31	0.9885	0.9858	0.9828	0.9800	106	1.0185	1.0234	1.0288	1.0338
32	0.9889	0.9863	0.9834	0.9807	107	1.0190	1.0240	1.0294	1.0346
33	0.9893	0.9868	0.9840	0.9814	108	1.0194	1.0245	1.0301	1.0353
34	0.9897	0.9873	0.9845	0.9820	109	1.0198	1.0250	1.0307	1.0360
35	0.9901	0.9878	0.9851	0.9827	110	1.0201	1.0255	1.0314	1.0368
36	0.9905	0.9882	0.9857	0.9834	111	1.0205	1.0261	1.0320	1.0376
37	0.9909	0.9887	0.9863	0.9841	112	1.0209	1.0266	1.0327	1.0384
38	0.9913	0.9892	0.9869	0.9847	113	1.0213	1.0271	1.0334	1.0392
39	0.9917	0.9897	0.9875	0.9854	114	1.0218	1.0276	1.0340	1.0399
40	0.9921	0.9902	0.9880	0.9861	115	1.0222	1.0282	1.0347	1.0407
41	0.9925	0.9907	0.9886	0.9868	116	1.0226	1.0287	1.0353	1.0414
42	0.9928	0.9912	0.9892	0.9875	117	1.0230	1.0291	1.0359	1.0422
43	0.9932	0.9917	0.9898	0.9881	118	1.0234	1.0296	1.0366	1.0431
44	0.9936	0.9922	0.9904	0.9888	119	1.0238	1.0302	1.0372	1.0438
45	0.9940	0.9926	0.9910	0.9895	120	1.0243	1.0307	1.0380	1.0446
46	0.9944	0.9930	0.9916	0.9902	121	1.0246	1.0312	1.0386	1.0454
47	0.9948	0.9935	0.9922	0.9910	122	1.0250	1.0318	1.0393	1.0461
48	0.9952	0.9940	0.9928	0.9917	123	1.0254	1.0323	1.0399	1.0469
49	0.9956	0.9945	0.9933	0.9924	124	1.0258	1.0328	1.0406	1.0478
50	0.9960	0.9950	0.9939	0.9930	125	1.0263	1.0334	1.0412	1.0485
51	0.9964	0.9955	0.9946	0.9937	126	1.0267	1.0339	1.0419	1.0493
52	0.9968	0.9960	0.9952	0.9944	127	1.0271	1.0344	1.0425	1.0501
53	0.9973	0.9965	0.9958	0.9951	128	1.0275	1.0350	1.0433	1.0509
54	0.9977	0.9971	0.9964	0.9958	129	1.0279	1.0355	1.0440	1.0517
55	0.9981	0.9976	0.9971	0.9965	130	1.0283	1.0360	1.0446	1.0525
56	0.9984	0.9980	0.9977	0.9973	131	1.0287	1.0366	1.0452	1.0533
57	0.9988	0.9985	0.9983	0.9980	132	1.0291	1.0371	1.0459	1.0541
58	0.9992	0.9990	0.9988	0.9986	133	1.0295	1.0377	1.0466	1.0548
59	0.9996	0.9995	0.9994	0.9993	134	1.0300	1.0382	1.0473	1.0557
60	1.0000	1.0000	1.0000	1.0000	135	1.0304	1.0387	1.0480	1.0565
61	1.0004	1.0005	1.0006	1.0007	136	1.0308	1.0393	1.0486	1.0573
62	1.0008	1.0010	1.0012	1.0014	137	1.0312	1.0398	1.0493	1.0581
63	1.0012	1.0015	1.0018	1.0021	138	1.0316	1.0404	1.0500	1.0589
64	1.0016	1.0020	1.0024	1.0028	139	1.0320	1.0409	1.0506	1.0598
65	1.0020	1.0025	1.0030	1.0035	140	1.0324	1.0414	1.0513	1.0606
66	1.0024	1.0030	1.0036	1.0042	141	1.0328	1.0420	1.0520	1.0613
67	1.0028	1.0035	1.0042	1.0049	142	1.0333	1.0425	1.0527	1.0621
68	1.0032	1.0040	1.0049	1.0056	143	1.0337	1.0431	1.0534	1.0630
69	1.0036	1.0045	1.0055	1.0064	144	1.0341	1.0436	1.0541	1.0638
70	1.0040	1.0050	1.0061	1.0072	145	1.0345	1.0442	1.0547	1.0646
71	1.0044	1.0055	1.0067	1.0079	146	1.0350	1.0447	1.0554	1.0654
72	1.0048	1.0060	1.0074	1.0086	147	1.0353	1.0452	1.0561	1.0662
73	1.0052	1.0065	1.0080	1.0093	148	1.0357	1.0458	1.0568	1.0671
74	1.0056	1.0070	1.0086	1.0100	149	1.0362	1.0464	1.0575	1.0679
					150	1.0366	1.0469	1.0582	1.0687

ᵃThis table is based on Tables 25 and 7 of "Petroleum Measurement Tables" published by the American Society for Testing and Materials, 1916 Race St., Philadelphia, PA, 19103. Values given are reciprocals of the multiplier values in Tables 25 and 7 as:

$$C_4 = \frac{1}{\text{multiplier for volume reduction to } 15.6°C}$$

TABLE 1A—C₄—FUEL TEMPERATURE CORRECTION FACTOR[a] (SI UNITS)

ASTM Group Number	Specific Gravity Range	API Gravity Range, °API
1	0.8499–0.9659	15.0–34.9
2	0.7754–0.8498	35.0–50.9
3	0.7239–0.7753	51.0–63.9
4	0.6723–0.7238	64.0–78.9

Fuel Temp., °C	C_4 Group 1	C_4 Group 2	C_4 Group 3	C_4 Group 4	Fuel Temp., °C	C_4 Group 1	C_4 Group 2	C_4 Group 3	C_4 Group 4
−15	0.9784	0.9734	0.9679	0.9629	25	1.0068	1.0085	1.0104	1.0121
−14	0.9791	0.9742	0.9689	0.9641	26	1.0075	1.0095	1.0115	1.0134
−13	0.9798	0.9751	0.9699	0.9653	27	1.0083	1.0104	1.0127	1.0147
−12	0.9805	0.9760	0.9710	0.9664	28	1.0090	1.0113	1.0138	1.0161
−11	0.9812	0.9768	0.9720	0.9676	29	1.0097	1.0122	1.0150	1.0174
−10	0.9818	0.9777	0.9730	0.9688	30	1.0104	1.0131	1.0161	1.0187
−9	0.9826	0.9785	0.9741	0.9700	31	1.0112	1.0141	1.0172	1.0201
−8	0.9833	0.9794	0.9751	0.9712	32	1.0119	1.0150	1.0184	1.0214
−7	0.9840	0.9803	0.9761	0.9724	33	1.0126	1.0159	1.0195	1.0228
−6	0.9847	0.9811	0.9772	0.9736	34	1.0133	1.0168	1.0207	1.0241
−5	0.9854	0.9819	0.9782	0.9748	35	1.0141	1.0177	1.0218	1.0254
−4	0.9861	0.9828	0.9792	0.9760	36	1.0148	1.0187	1.0229	1.0268
−3	0.9868	0.9837	0.9803	0.9771	37	1.0155	1.0196	1.0241	1.0281
−2	0.9875	0.9845	0.9813	0.9783	38	1.0162	1.0205	1.0252	1.0295
−1	0.9882	0.9854	0.9823	0.9795	39	1.0170	1.0214	1.0264	1.0308
0	0.9889	0.9863	0.9834	0.9807	40	1.0177	1.0224	1.0275	1.0322
1	0.9896	0.9871	0.9844	0.9820	41	1.0184	1.0233	1.0287	1.0336
2	0.9903	0.9880	0.9854	0.9832	42	1.0191	1.0243	1.0299	1.0350
3	0.9910	0.9889	0.9865	0.9844	43	1.0199	1.0252	1.0310	1.0364
4	0.9918	0.9898	0.9876	0.9856	44	1.0206	1.0262	1.0322	1.0378
5	0.9925	0.9907	0.9886	0.9868	45	1.0213	1.0271	1.0334	1.0392
6	0.9932	0.9916	0.9897	0.9881	46	1.0221	1.0281	1.0346	1.0406
7	0.9939	0.9924	0.9908	0.9893	47	1.0228	1.0290	1.0358	1.0419
8	0.9946	0.9933	0.9918	0.9906	48	1.0235	1.0300	1.0370	1.0433
9	0.9953	0.9942	0.9929	0.9918	49	1.0243	1.0309	1.0381	1.0447
10	0.9960	0.9950	0.9939	0.9930	50	1.0250	1.0318	1.0393	1.0461
11	0.9968	0.9960	0.9950	0.9943	51	1.0258	1.0328	1.0405	1.0475
12	0.9975	0.9969	0.9961	0.9956	52	1.0265	1.0338	1.0417	1.0490
13	0.9982	0.9978	0.9972	0.9968	53	1.0272	1.0347	1.0429	1.0504
14	0.9989	0.9986	0.9983	0.9981	54	1.0280	1.0357	1.0441	1.0519
15	0.9996	0.9995	0.9994	0.9993	55	1.0287	1.0366	1.0453	1.0533
16	1.0003	1.0004	1.0005	1.0005	56	1.0294	1.0376	1.0465	1.0548
17	1.0010	1.0013	1.0016	1.0018	57	1.0302	1.0385	1.0477	1.0562
18	1.0018	1.0022	1.0027	1.0031	58	1.0309	1.0395	1.0489	1.0577
19	1.0025	1.0031	1.0038	1.0044	59	1.0316	1.0404	1.0501	1.0592
20	1.0032	1.0040	1.0049	1.0056	60	1.0324	1.0414	1.0513	1.0606
21	1.0039	1.0049	1.0060	1.0069	61	1.0332	1.0424	1.0526	1.0621
22	1.0046	1.0058	1.0071	1.0082	62	1.0339	1.0434	1.0538	1.0635
23	1.0054	1.0067	1.0082	1.0095	63	1.0347	1.0444	1.0550	1.0650
24	1.0061	1.0076	1.0093	1.0108	64	1.0354	1.0454	1.0562	1.0664

[a]This table is based on Tables 25 and 7 of "Petroleum Measurement Tables" published by the American Society for Testing and Materials, 1916 Race St., Philadelphia, PA, 19103. Values given are reciprocals of the multiplier values as:

$$C_4 = \frac{1}{\text{multiplier for volume reduction to } 60°F}$$

SUMMARY SHEET
FUEL ECONOMY TEST DATA - LIGHT DUTY VEHICLES

DATE _____

VEHICLE MAKE _____ ENGINE _____ CAR NO. _____

VIN _____ TRANSMISSION _____ AXLE RATIO _____

TIRE MAKE _____ SIZE _____ PRESSURE _____

VEHICLE TEST WEIGHT _____

TEST ROAD _____ SURFACE _____

FUEL ECONOMY[a]

	CORRECTED[b] mile/gal (km/L)	OBSERVED mile/gal (km/L)	FUEL TEMP. °F(°C)	AIR TEMP. °F(°C)	BARO. PRESS. in Hg (kPa)	WIND SPEED-DIRECTION mile/h (km/h)
URBAN CYCLE						
SUBURBAN CYCLE						
55 mile/h (89 km/h) INTERSTATE CYCLE						
70 mile/h (113 km/h) INTERSTATE CYCLE						

[a]Cross out those units not used.
[b]The Corrected Fuel Economy may be expressed in terms of fuel as consumed litres per 100km.

COMMENTS: _____
_____
_____
_____
_____

TESTED BY: _____

26.468

FUEL ECONOMY TEST - LIGHT DUTY VEHICLES
TEST VEHICLE SPECIFICATIONS AND PREPARATION

SPECIFICATION LIST

DATE _____
CAR NO. _____
YEAR AND MAKE _____
MODEL AND BODY _____
VEHICLE INDENT. NO. _____
PRODUCTION _____ OTHER _____
ENGINE TYPE _____ DISP. _____
NET hp (U.S.) _____ COMP. RATIO _____
ENGINE NO. _____
ENGINE EMISSION CALIBRATION NO. _____
CARB. TYPE AND NO. _____
DISTRIBUTOR NO. _____
EXHAUST SYSTEM TYPE _____
TRANSMISSION _____
DRIVE AXLE TYPE AND RATIO _____
BRAKES, (DISC OR DRUM) F_____ R_____
STEERING _____
*TIRE MAKE _____ SIZE _____
   LOAD RANGE _____ TYPE _____
   % TREAD _____
COLD INFLATION-TIRE PRESSURE
   LF_____ RF_____
   LR_____ RR_____
TEST WEIGHT_____

*TIRES MUST HAVE A MINIMUM OF 100 miles (160 km) BREAK-IN ON ROAD OR TRACK

CHECK LIST

_____ENGINE OIL LEVEL OK
_____COOLANT LEVEL OK
_____TRANSMISSION FLUID LEVEL OK
_____BELTS AND HOSES-TIGHT
_____EMISSION CONTROLS-FUNCTIONAL
_____CHOKE OPERATION-FUNCTIONAL
_____THROTTLE OPERATION-FUNCTIONAL
_____PUMP AND NOZZLES-NO LEAKS
_____IGNITION WIRES-TIGHT
_____BRAKE DRAG NOT EXCESSIVE
_____TRANSMISSION OPERATION
_____TIRE PRESSURE AND CONDITION
_____ENGINE TUNE-PERFORMED
_____IGNITION TIMING_____
_____IDLE rpm _____ IDLE CO _____
_____IGNITION POINT DWELL_____
_____WHEEL ALIGNMENT-PERFORMED
_____AIR CLEANER-CLEAN

_____A/C COMPRESSOR LOAD-REMOVED
_____NO FUEL LEAKS
_____MANIFOLD HEAT VALVE-FUNCTIONAL
_____FAN CLUTCH-FUNCTIONAL

$$\% \text{ TREAD} = \frac{\text{AVERAGE TREAD DEPTH OF TEST TIRES}}{\text{AVERAGE TREAD DEPTH OF IDENTICAL NEW TIRE}} \cdot 100$$

EQUIPPED WITH OPTIONAL EQUIPMENT
_____POWER BRAKES
_____POWER STEERING
_____AIR CONDITIONING
_____RADIO
_____POWER SEATS
_____POWER WINDOWS
_____POWER DOOR LOCKS
_____
_____

COMMENTS:_____
_____
_____
_____
CAR CHECKED BY:_____
DATE:_____

TEST FUEL SPECIFICATIONS
FUEL TYPE AND GRADE _____
GRAVITY (API OR SPECIFIC)_____ AT 60°F (15.6°C)
REID VAPOR PRESSURE_____ psi (kPa)
DISTILLATION   10%_____ °F (°C)
               50%_____ °F (°C)
               90%_____ °F (°C)
Φ OCTANE $\frac{R+M}{2}$ _____

26.469

DATA FORM
FUEL ECONOMY TEST-LIGHT DUTY VEHICLES

CAR NO. _____ VIN _____ TEST VEHICLE WEIGHT _____
YEAR AND MAKE _____ TEST ROAD _____
MODEL AND BODY _____ TYPE OF SURFACE _____
ENGINE TYPE _____ DISP. _____ INSTRUMENTATION
DRIVE AXLE RATIO _____ 1. _____ NO. _____
TRANSMISSION _____ 2. _____ NO. _____
TIRE MAKE _____ SIZE _____ 3. _____ NO. _____
TIRE PRESSURE: F _____ R _____ 4. _____ NO. _____
FUEL TYPE _____ SPECIFIC GRAVITY _____

URBAN CYCLE (2.00 mile) (3.22 km)　　DATE _____ TIME _____
ODOMETER: START _____ FINISH _____

	FUEL TEMPERATURE					ACCUM	ACCUM	AMBIENT CONDITIONS					
	0.5	1.0	1.5	2.0	mile			TEMPERATURE		BAROMETRIC		WIND	
DIRECTION	(0.8)	(1.6)	(2.4)	(3.2)	km	Time	Fuel	Start	Fin.	Start	Fin.	Start	Fin.

SUBURBAN CYCLE (5.2 mile) (8.37 km)　　DATE _____ TIME _____
ODOMETER: START _____ FINISH _____

	FUEL TEMPERATURE				ACCUM	ACCUM	AMBIENT CONDITIONS					
	2.0	3.3	5.2	mile			TEMPERATURE		BAROMETRIC		WIND	
DIRECTION	(3.2)	(5.3)	(8.4)	km	Time	Fuel	Start	Fin.	Start	Fin.	Start	Fin.

_____ mile/h (km/h) INTERSTATE CYCLE (4.7 miles) (7.56 km) DATE _____ TIME _____
ODOMETER: START _____ FINISH _____

	FUEL TEMPERATURE		ACCUM	ACCUM	AMBIENT CONDITIONS					
	0.0	4.7 mile			TEMPERATURE		BAROMETRIC		WIND	
DIRECTION	(0.0)	(7.6) km	Time	Fuel	Start	Fin.	Start	Fin.	Start	Fin.

TESTED BY: _____

(R) ROAD LOAD MEASUREMENT AND DYNAMOMETER SIMULATION USING COASTDOWN TECHNIQUES—SAE J1263 JUN91

SAE Recommended Practice

Report of the Passenger Car Light Duty Truck Fuel Economy Measurement Committee approved June 1979, revised by the Passenger Car and Light Truck Fuel Economy Measurement Committee May 1984. Completely revised by the SAE Light Duty Vehicle Performance and Economy Measurements Standards Committee June 1991. Rationale statement available.

Foreword—This procedure was originally designed in the English units system and has been converted to the metric (MKS) system of units. This will explain the somewhat awkward values used in this procedure.

1. Scope—This procedure covers measurement of vehicle road load on a straight, level road at speeds less than 113 km/h (70 mph).

1.1 Purpose—This SAE Recommended Practice provides uniform testing procedures for measuring the road load force on a vehicle as a function of vehicle velocity and for simulation of that road load force at 80 km/h (50 mph) on a chassis dynamometer.

2. References

2.1 Applicable Documents—The following publications form a part of this specification to the extent specified herein. The latest issue of SAE publications shall apply.

2.1.1 SAE PUBLICATIONS—Available from SAE, 400 Commonwealth Drive, Warrendale, PA 15096-0001.

SAE J1100—Motor Vehicle Dimensions

2.1.2 FEDERAL REGULATIONS—Available from the Superintendent of Documents, U. S. Government Printing Office, Washington, DC 20402.

Code of Federal Regulations, 40 CFR 600, Appendix A

2.2 Related Publications—The following publications are provided for information purposes only and are not a required part of this document.

R. A. White and H. H. Korst, "The Determination of Vehicle Drag Contribution from Coast-Down Test," SAE Paper 720099, 1972.

T. P. Yasin, "The Analytical Basis of Automobile Coastdown Testing," SAE Paper 780334, 1978.

B. Dayman, Jr., "Tire Rolling Resistance Measurements from Coast-Down Tests," SAE Paper 760153, 1976.

G. D. Thompson, "Prediction of Dynamometer Power Absorption to Simulate Light Duty Truck Road Load," SAE Paper 770844, 1977.

J. R. Smith, J. C. Tracy, and D. S. Potter, "Tire Rolling Resistance—A Speed Dependent Contribution," SAE Paper 780255, 1978.

B. Dayman, Jr., "Realistic Effects of Winds on the Aerodynamics Resistance of Automobiles," SAE Paper 780337, 1978.

2.3 Definitions

2.3.1 TEST WEIGHT—Is the weight of the vehicle as tested; including driver, operator (if necessary), and all instrumentation.

2.3.2 TEST MASS—Is the mass of the vehicle as tested; including driver, operator (if necessary), and all instrumentation.

2.3.3 EFFECTIVE MASS—Is equal to the sum of the test mass and the effective inertias of the driven and nondriven axles.

2.3.4 FRONTAL AREA—Is the area of the orthogonal projection of the vehicle including tires and suspension components onto a plane perpendicular to the longitudinal axis of the vehicle.

2.4 Symbols

a = Slope of the actual dynamometer horsepower $-1/\delta$ t curve (kW-s or hp-s).

A = Vehicle frontal area (see 6.6) (m² or ft²).

b = Zero intercept of the actual dynamometer horsepower $-1/\delta$ t curve (kW or hp).

C_D = Aerodynamic drag coefficient (dimensionless).

C_D' = Aerodynamic drag coefficient for nonzero yaw (dimensionless) (see 8.2).

C_{DY} = Crosswind aerodynamic drag coefficient (dimensionless).

D = Denominator of expressions to determine a and b from vehicle-dynamometer coastdown data.

f_0, f_2 = Coefficients of the zeroth and second order terms (respectively) in the road load force equation, (N or lb and N/[km/h²] or lb/mph²).

f_0', f_2' = Coefficients of the zeroth and second order terms (respectively) in the road load force equation, (N or lb and N/[km/h²] or lb/mph²) corrected to standard conditions.

g = Acceleration of gravity 9.80 m/s² or 32.174 ft/s².

I = Moment of inertia of tire, wheel, and brake rotor or drum kg-m² or ft-lb-s².

IW = Inertia weight or equivalent weight of inertia simulation mechanism (N or lb).

k = Drag coefficient dependence on yaw angle ψ.

k_t = Temperature coefficient of rolling resistance (°C⁻¹ or °F⁻¹).

M = Vehicle test mass (kg or slugs).

m_e = Effective mass of a wheel, tire, and brake assembly (kg or slugs).

M_{DLC} = Total equivalent mass of drivetrain components (kg or slugs).

M_{IW} = Equivalent mass of dynamometer inertia simulation mechanism (IWC/g) (kg or slugs).

M_e = Total effective vehicle mass (kg or slugs).

N = Number of data points used in Part 2.

P_o = Reference atmospheric pressure 736.6 mm Hg (29.00 in Hg).

r = Tire rolling radius (m or ft).

S = +1 or −1, depending on vehicle coastdown direction.

t_o = Time (seconds).

$t-t_o$ = Coastdown time interval (seconds).

δt = Vehicle coastdown time on the chassis dynamometer(s).

T = Ambient temperature of test area (K or °R).

T_o = Standard temperature (20 °C = 293.15 K, 68 °F = 527.67 °R).

V = Vehicle speed (km/h or mph).

V_1, V_o = Final and initial speeds in the calculation of the coastdown time interval (km/h or mph).

V_w = Wind velocity (km/h or mph).

v_x = Component of wind parallel to track (km/h or mph).

v_y = Component of wind perpendicular to track (km/h or mph).

W = Vehicle test weight (N or lb).

W_w = Weight of tire, wheel, and brake rotor or drum (N or lb).

μ = Coefficient of rolling resistance (dimensionless).

μ_o = Velocity independent coefficient of rolling resistance (dimensionless).

μ' = Velocity coefficient of rolling resistance ([km/h]⁻² or [mph]⁻²).

ρ = Air mass density (kg/m³ or slugs/ft³).

ψ = Aerodynamic yaw angle (radians).

3. Part 1—Vehicle Road Load Measurement

3.1 Instrumentation—All instrumentation must be calibrated.

3.2 Speed Time—An instrument to measure vehicle speed as a function of elapsed time is used in this procedure. The device must meet the following specifications:

a. Time:
  (1) Accuracy ±0.1% of total coastdown time interval
  (2) Resolution 0.1 s
b. Speed:
  (1) Accuracy ±0.4 km/h (±0.25 mph)
  (2) Resolution ±0.2 km/h (0.1 mph)

3.3 Temperature—The temperature indicating devices must have a resolution of 2 °F or 1 °C and an accuracy of ±2 °F or ±1 °C. The sensing element must be shielded from radiant heat sources.

3.4 Atmospheric Pressure—A barometer with an accuracy of ±0.7 kPa or ±0.2 in Hg is necessary.

3.5 Wind—Wind speed and direction during the test should be continuously monitored. Wind measurements should permit the determination of average longitudinal and crosswind components to within ±1.6 km/h (±1 mph).

3.6 Vehicle Weight—Vehicle weight should be measured to an accuracy of ±5 kg (±10 lb) per axle.

3.7 Tire Pressure—Should be measured to an accuracy of ±3 kPa (±0.5 psi).

4. Test Material

4.1 Test Vehicle—The test vehicle should be uniquely described on the Vehicle Road Test Data Sheet (see Figure 1). In particular, any modifications from the normal configuration of the vehicle should be noted.

5. Test Conditions

5.1 Ambient Temperature—Tests may be conducted at ambient temperatures between −1 °C (30 °F) and 32 °C (90 °F). The recommended temperature range is from 5 °C (41 °F) to 32 °C (90 °F). Data ob-

26.471

Date _____
Vehicle No. _____
Odometer Reading _____

VEHICLE DESCRIPTION:

Model year _____
Make _____
Model _____
Inertia Weight Class _____
Ballast –

   Amount   Location

_____   _____
_____   _____
_____   _____

Weight Removed _____
Trim Heights –
   LF _____ RF _____
   LR _____ RR _____

FINAL WEIGHT (TEST WEIGHT)

Fifth Wheel Down and Off Scale –
   LF _____ RF _____
   Total Front _____
Fifth Wheel Down and Off Scale –
   LR _____ RF _____
   Total Rear _____
   Total _____

Fifth Wheel Included –
   LF _____ RF _____
   Total Front _____
Fifth Wheel Included –
   LR _____ RR _____
   Total Rear _____
   Total _____

EQUIVALENT INERTIA:

Front Wheel/Brake/Tire (each) _____
Rear Wheel/Brake/Tire (each) _____
Total Non-Driven Axle _____
Total Drive Line _____
   Total _____
[ ] Check if estimated

AMBIENT CONDITIONS:

Temperature–Start _____ Finish _____ Average _____ (°C/°F)

Wind –

   Average Wind _____ (km/h/mph) PREPARATION AREA:
   Direction _____ (deg)
   Peak Gusts _____ (km/h/mph) Temperature _____ (°C/°F)

Atmospheric Pressure – Start _____ Finish _____ Average _____ (mm Hg/in Hg)

NO RAIN OR FOG – ROADS MUST BE DRY

TIRES:

Size _____
Make _____
Tread Depth >75% Yes [ ] No [ ]
Pressures –

                Specified   Actual
   Front   _____   _____
   Rear   _____   _____
Dynamometer Test Drive Wheel Tire
Pressure _____

WHEELS:

Size _____
Wheel Covers _____

CHASSIS:

Axle Ratio _____
Transmission Type _____
Brake Type –
   Front _____
   Rear _____
Toe-In (+=in) _____

FRONTAL AREA DETERMINATION:

Area _____
[ ] Check if estimated

NOTE: Check Brake Drag and Fluid Levels

COMMENTS: _____
_____
_____
_____
_____
_____
_____

FIGURE 1 – VEHICLE ROAD TEST DATA SHEET

tained at temperatures outside this range cannot be reliably adjusted to standard conditions by 8.5.

5.2 Fog—Tests may not be run during foggy conditions.

5.3 Winds—Tests may not be conducted when wind speeds average more than 16 km/h (10 mph) (or when peak wind speeds are more than 20 km/h [12.3 mph]). The average of the component of the wind velocity perpendicular to the test road may not exceed 8 km/h (5 mph).

5.4 Road Conditions—Roads must be dry, clean, smooth, and must not exceed 0.5% grade. In addition, the grade should be constant and the road should be straight since variations in grade or straightness can significantly affect results. (The road surface should be concrete or rolled asphalt (or equivalent) in good condition since rough roads can significantly affect rolling resistance.)

5.5 Coastdown Speed Range—The range of speeds over which the vehicle is coasted should be as long as possible considering the length of the straightaway. The speed interval must include 80 km/h (50 mph) and should include the range of 100 to 40 km/h (60 to 25 mph).

6. Vehicle Preparation

6.1 Break-In—The test vehicle should have accumulated a minimum of 500 km (300 miles) prior to testing. The tires should have accumulated a minimum of 160 km (100 miles) and should have at least 75% of the original tread depth remaining. In addition, if a twin roll dynamometer is to be used, the drive axle tires should have a minimum of 1 h at 80 km/h (50 mph) on the dynamometer rolls before conducting the dynamometer road load simulation portion of this procedure. All tire break-in should be performed on the test vehicle or under load conditions similar to those imposed by the test vehicle.

6.2 Vehicle Check-In—The following items should be compared to manufacturer's recommendation and recorded on the Vehicle Road Test Data Sheet (see Figure 1) prior to the test:
  a. Tire type, size, and cold inflation pressure (see 6.5)
  b. Wheel size, conditions, and presence of wheel covers
  c. Brake adjustment
  d. Lubricants in the drivetrain and in the nondriving wheel bearings
  e. Vehicle suspension heights

6.3 Instrumentation—The speed-time measuring device and other necessary equipment must be installed so that they do not hinder vehicle operation or alter the operating characteristics of the vehicle.

6.4 Pretest Weight—The weight of the vehicle prior to testing should be appropriate for the vehicle represented; for example, consideration should be given to the effect of the added weight of the test instrumentation.

6.5 Tire Pressure—Inflate the tires of the test vehicle to the manufacturer's recommended cold inflation pressure, corrected for the temperature difference (if any) between the vehicle tires and the test area. The tire pressure should be increased 1 psi for each 13 °F that the vehicle preparation area temperature is above the test area temperature or 1 kPa for each Celsius degree. Record the actual inflation pressure and preparation area temperature on the Vehicle Road Test Data Sheet (see Figure 1).

6.6 Vehicle Frontal Area—The vehicle frontal area must be known, measured, or estimated and the value recorded on the Vehicle Road Test Data Sheet (see Figure 1). The frontal area may be estimated by the following equation:

$$A = 0.8 \, (H101) \times (W103) \quad \text{(Eq.1)}$$

where:

$H101$ = Body height (meters [feet]) measured according to SAE J1100
$W103$ = Body width (meters [feet]) measured according to SAE J1100

6.7 Vehicle Warm-Up—The vehicle must be driven a minimum of 30 min at an average speed of 80 km/h (50 mph) immediately prior to the test.

7. Coastdown Test

7.1 Alternating Directions—A minimum of 10 runs are made in alternating directions. The runs must be paired for the data reduction process in order to reduce error.

7.2 Procedure—The vehicle windows must be closed. At the start of each run, accelerate the vehicle to a speed 8 km/h (5 mph) above the high point of the coastdown speed range, start the recording equipment, and shift into neutral and let the engine idle. The vehicle clutch must be engaged. At a speed less than the lower point of the coastdown speed range, stop the recording equipment, engage the transmission, and prepare for the next run.

7.3 Lane Changes—While coasting, lane changes should be avoided if at all possible. If necessary, they should be done as slowly as possible and over a distance of at least a half kilometer (a quarter mile). If such a gradual change cannot be made, abort the run.

7.4 Data to be Recorded—Record the direction and number of each run (including aborted runs) in such a way that the speed time data can be separated by run number. Record the ambient temperature and atmospheric pressure after the warm-up and after the test. Average the two values to determine the value to be used in the data reduction. The total wind and either the wind direction or the crosswind component of the total wind must be recorded. The wind quantities should be recorded, screened for gusts exceeding the ambient condition limits in 5.3, and averaged. Record the results on the Vehicle Road Test Data Sheet (see Figure 1).

7.5 Vehicle Test Weight or Mass—After the coastdown run, weigh the vehicle to determine the vehicle test weight or mass. Include the weight of the fifth wheel, driver, and all instrumentation used. Record on the Vehicle Road Test Data Sheet (see Figure 1).

7.6 Axle Weights—Measure each axle weight as in 7.5 except with the fifth wheel (if used) in the operating position and off the scale. Record on the Vehicle Road Test Data Sheet (see Figure 1).

8. Data Reduction

This section prescribes the technique for analyzing a set of coastdown data and the correction factors employed in the determination of the coefficients of the road load force equation. These corrected coefficients are used to calculate the time required to freely decelerate from 88 to 72 km/h (55 to 45 mph) on a chassis dynamometer.

8.1 Road Load Force—A two-parameter road load force equation is fitted to the $V(t)$ data of Section 7 according to the technique of White and Korst (SAE Paper 720099). The coefficients of the road load force equation are determined for each individual $V(t)$ coastdown and are then averaged over all pairs of coastdowns in each data set. Corrections are applied for wind (both parallel and perpendicular to the coastdown path), for the temperature dependence of rolling resistance, and for the density dependence of aerodynamic drag. The corrected coefficients are then used to construct the vehicle force-velocity equation characteristics of the vehicle under standard ambient conditions with no wind. This force is then corrected for inertial difference between the road test configuration and the dynamometer test configuration and the resultant force is used to calculate the time to coast from 88 to 72 km/h (55 to 45 mph) on a chassis dynamometer (see 8.6).

8.1.1 ADDITIONAL APPLICATIONS—For dynamometers with more than one adjustable load parameter, the force—velocity and coastdown time—speed interval equations (Eqs. 5 and 17) may be used to simulate the vehicle road load in several speed intervals within the experimental data range.

8.2 Solution to Force Equations—The road load force is related to the coastdown deceleration rate by:

$$-M_e \frac{dV}{dt} = \mu W + 1/2 \rho C_D' A [(V + Sv_x)^2 + V_y^2] \quad \text{(Eq.2)}$$

where:

$$\mu = \mu_o(1 + \mu' V^2)$$

The drag coefficient C_D' is the sum of the drag coefficient at zero yaw (C_D) and a coefficient k times the square of the sine of the yaw angle.

$$C_D' = C_D + k \sin^2 \psi \quad \text{(Eq.3)}$$
$$= C_D + k v_y^2/[(V + Sv_x)^2 + v_y^2]$$

The road load force can now be expressed as:

$$-M_e \frac{dV}{dt} = \mu_o W(1 + \mu' V^2) + 1/2 \rho C_D A(V^2 + v_x^2)$$
$$+ 1/2 \rho (C_D + k) A v_y^2 \quad \text{(Eq.4)}$$

where:

The term linear in V has been ignored. The error introduced by ignoring this term (and the road grade) is minimized by the averaging process subsequently applied because these terms change sign for each change in coastdown direction.

Collecting terms, (Eq. 4) may be rewritten as:

$$-M_e \frac{dV}{dt} = f_o + f_2 V^2 \quad \text{(Eq.5)}$$

where:

$$f_2 = \mu_o \mu' W + 1/2 \rho C_D A$$
$$f_o = \mu_o W + (f_2 - \mu_o \mu' W) v_x^2 + 1/2 \rho C_{DY} A v_y^2$$

The crosswind aerodynamic drag coefficient C_{DY} is a measure of the response of the vehicle to the crosswind component of the wind at small yaw angles. It may be calculated by:

$$C_{DY} = C_D + k \quad \text{(Eq.6)}$$

The integral solution to this equation is:

$$T - T_0 = \frac{M_e}{\sqrt{f_0 f_2}} \left(\tan^{-1}\left(\sqrt{\frac{f_2}{f_0}} V_1\right) - \tan^{-1}\left(\sqrt{\frac{f_2}{f_0}} V_2\right) \right) \quad (Eq.7)$$

This is the equation which, after correction of the coefficients determined by the White and Korst technique or equivalent, is used to calculate the coastdown time interval. The units for M_e, f_0, f_2, V, and V_0 must be chosen so that the argument of the inverse tangent function is dimensionless and the resultant coastdown time is in seconds. The individual terms and their corrections are described in the following paragraphs.

8.3 Effective Vehicle Mass — The effective vehicle mass (M_e) is the sum of the final vehicle test mass ($M = W/g$) and the effective mass of the rotating components. The effective mass of the drivetrain components other than the wheels, tires, and brakes may be ignored. For each tire, wheel, and brake rotor or drum, the effective mass, m_e is:

$$m_e = \frac{I}{r^2} \quad (Eq.8)$$

where:
r is the rolling radius of the tire and I is the polar moment of inertia of the assembly. The polar moment of inertia may be measured or may be estimated by circular disk expression:

$$I = \frac{W_w}{g} \frac{r^2}{2} \quad (Eq.9)$$

where:
W_w is the weight of the tire, wheel, and brake rotor or drum. If no measurements are available, the effective inertia of all the rotating components may be estimated as 3.0% of the vehicle test mass.

$$M_e = 1.03 \frac{W}{g} = 1.03M \text{ (For use in Eq. 7)} \quad (Eq.10)$$

For chassis dynamometer calculations the equivalent inertia of the driveline components, M_{DLC} is the effective inertia of two tire and wheel assemblies plus the effective inertia of two drive wheel brake rotors or drums. If no measurements are available, M_{DLC} may be estimated as 1.5% of the test mass.

$$M_{DLC} = 0.015 \frac{W}{g} = 0.015M \text{ (For use in Eqs. 17 and 18)} \quad (Eq.11)$$

8.4 Data Acceptability Criteria — Experience has shown that the criteria of this section are necessary and sufficient to provide accurate and precise test results. Data which exceed these criteria generally arise from wind gusts or driver inputs, which violate the assumption that the forces on the vehicle are depicted by Eq. 5.

8.4.1 Analyze each individual coastdown V(t) in the set of paired runs obtained in Section 7 by the White and Korst method, or equivalent, to obtain the coefficients f_0 and f_2. A computer program that performs this task is supplied in Appendix A along with a sample data set (see Figures A1 and A2) and the corresponding output. Using (Eq. 7) of 8.2, compare each individual V(t) trace and its analytical counterpart V(f_0, f_2, t). If the root-mean-square difference (error) exceeds 0.40 km/h (0.25 mph) on any individual run, discard that run and the paired run in the opposite direction. If less than three pairs comply with this criterion, the test run is invalid.

8.4.2 Of the paired runs meeting the criterion of 8.4.1, those which fail to satisfy the following criteria regarding f_0 and f_2 must also be discarded.
  a. The standard deviation of the f_0's must be less than 11 N (2.5 lb) or 5% of the mean. If this value is exceeded, discard the run and its pair with f_0 farthest from the mean and recompute the standard deviation until compliance is obtained or until the remaining pairs number less than three.
  b. The standard deviation of all the f_2's must be less than 0.011 N/(km/h)² (0.001 lb/[mile/h]²) or 3% of the mean. If this value is exceeded, discard the run and its pair with f_2 farthest from the mean and recompute the standard deviation until compliance is obtained or until the remaining pairs number less than three. If less than three pairs remain, the test run is invalid. Average f_0's and f_2's of all remaining runs to determine an f_0 and f_2.

8.5 Data Correction — The average f_0 and f_2 values must now be corrected to a standard set of ambient conditions. The standard conditions are:

  a. Temperature — 20 °C (68 °F)
  b. Atmospheric pressure — 736.6 mm Hg (29.00 in Hg)
  c. Zero wind
  d. The effect of humidity on air density may be ignored

8.5.1 Wind Corrections to f_0 — Separate the rolling resistance from wind effects as follows:

$$\mu_0 W = (f_0 - f_2 v_x^2 - 1/2 \rho C_{DY} A v_y^2)/(1 - \mu' v_x^2) \quad (Eq.12)$$

where:
ρ = mass density of ambient air.
μ' = velocity coefficient of rolling resistance.
C_{DY} = crosswind aerodynamic drag coefficient.
v_x = component of wind parallel to track.
v_y = component of wind perpendicular to the track.

Unless specific information about the test vehicle is available, use the following values for the coefficients:

$$\mu' = 19 \times 10^{-6} \text{ (km/h)}^{-2} \text{ } (50 \times 10^{-6} \text{(mph)}^{-2}) \quad (Eq.13)$$
$$C_{DY} = 0.8181 \text{ (m/s)}^2/\text{(km/h)}^2 \text{ } (3.4 \text{ (ft/s)}^2/\text{(mph)}^2)$$

8.5.2 Temperature Correction to f_0 — The temperature dependence of rolling resistance shall be corrected by:

$$f_0' = \mu_0 W [1 + k_t (T - T_0)] \quad (Eq.14)$$

Unless specific information about the test vehicle is available, use:

$$k_t = 8.6 \times 10^{-3}/°C = 4.8 \times 10^{-3}/°F \quad (Eq.15)$$

NOTE: Significant changes in sun load may affect rolling resistance and, consequently, contribute to test variations.

8.5.3 Weight Correction — Eq. 14 may be used to correct the test results for small weight differences (less than 50 kg or 100 lb) between the actual vehicle test weight and a desired vehicle test weight (see 7.5).

8.5.4 Air Density Correction to f_2 — Adjust the coefficient of the V^2 term to standard ambient conditions ($\rho = 1.1678$ kg/m³ [$\rho = 0.002266$ slugs/ft³]) by the equation:

$$f_2' = (p_0 T / p T_0) [f_2 - \mu'(\mu_0 W)] + \mu' f_0' \quad (Eq.16)$$

where:
p = barometric pressure
p_0 = 736.6 mm Hg (29.00 in)
T = absolute temperature of ambient air (K or °R)
T_0 = 293.16 K (527.69°R)

8.6 Calculation of Coastdown Time

8.6.1 Calculation of 55 to 45 mph Coastdown Time — The 55 to 45 mph coastdown time on a chassis dynamometer, which duplicates the force at 50 mph, measured on the road is given by:

$$dt = t - t_0 = \frac{1.467(M_{IW} + M_{DLC})}{\sqrt{f_0' f_2'}} \left(\tan^{-1}\left(\sqrt{\frac{f_2'}{f_0'}} 55\right) - \tan^{-1}\left(\sqrt{\frac{f_2'}{f_0'}} 45\right) \right) \quad (Eq.17)$$

8.6.2 Calculation of 88 to 72 km/h Coastdown Time — The 88 to 72 km/h coastdown time on a chassis dynamometer, which duplicates the force at 80 km/h, measured on the road is given by:

$$dt = t - t_0 = \frac{(M_{IW} + M_{DLC})}{3.6\sqrt{f_0' f_2'}} \left(\tan^{-1}\left(\sqrt{\frac{f_2'}{f_0'}} 88\right) - \tan^{-1}\left(\sqrt{\frac{f_2'}{f_0'}} 72\right) \right)$$

9. Part 2 — Chassis Dynamometer Simulation of Vehicle Road Load — This portion of the procedure covers the measurement of total load on the vehicle as a function of dynamometer absorber horsepower when the vehicle is mounted on a chassis dynamometer. This procedure is written to determine the total load on the vehicle at 80 km/h (50 mph). It may also be extended to determine the load on the vehicle at other speeds.

10. Definitions

10.1 Indicated Horsepower — The indicated horsepower is the load horsepower which is set on the dynamometer and does not include the friction in the dynamometer bearings or inertia simulation mechanism.

10.2 Actual Horsepower — The actual horsepower is the load horsepower which includes the friction in the dynamometer bearings and inertia simulation mechanism.

11. Equipment

11.1 Test Site — A chassis dynamometer, which has an adjustable power absorber setting capability, is necessary. The dynamometer speed during this test must be measured from the shaft, which drives the power absorber and weights. The dynamometer must be calibrated over a range of horsepower settings and at the inertia weight used in the test.

11.2 Test Instrumentation — An instrument that can measure the time to coastdown between two speeds on a chassis dynamometer will be

used. The specifications for this instrument are as follows:
   a. Speed:
      Accuracy—±0.040 km/h (±0.25 mph)
      Resolution—0.16 km/h (0.1 mph)
   b. Time Base:
      Accuracy—±0.1%
      Resolution—0.1 s

11.3 Test Vehicle—The test vehicle should be the same as the vehicle used during the coastdown tests on the road in Part 1. No differences in the vehicle's transmission, axle, tire size, and construction, or vehicle load on the drive axle can exist between the vehicle used on the road and the vehicle to be tested on the dynamometer. The vehicle must meet the requirements of 6.1, and must not have been operated on the dynamometer or driven during the previous 4 h to provide stabilized tire pressure.

12. Procedure

12.1 Preparation

12.1.1 Preselect the actual dynamometer horsepower values at which the coastdown time will be measured (see Figure 2). A minimum of three values may be chosen using engineering judgment or the horsepowers discussed in 12.1.5, 12.1.6, 12.2.5, and 12.2.7 may be used. Enter the values under A, B, and C on the Dynamometer Test Worksheet (see Figure 3).

12.1.2 For small, twin roll dynamometer, set the drive wheel tire pressures at 310 kPa (45 psi) or lower cold setting. If greater pressures are necessary in order to prolong tire life, or if such pressures are specified by the manufacturer, they may be used and recorded in the Vehicle Test Data Sheet (see Figure 1). The tire pressure used for this road load simulation should also be used for any subsequent emission or fuel economy tests. For large roll dynamometers, set the drive wheel tire pressure to the manufacturer's recommended value.

12.1.3 Adjust the vehicle total drive wheel weight to the desired drive axle test weight (see 7.6) within ±7 kg (±15 lb) as indicated on the Dynamometer Test Worksheet (see Figure 4). The test weight can be adjusted by adding ballast to the trunk or by removing the spare tire and jack.

12.1.4 Set the vehicle on the dynamometer and restrain it. The vehicle should be restrained with a horizontal force in such a manner that there is no change in the vertical load on the drive axle. Place the cooling fan in front of the vehicle and record the odometer reading on the Dynamometer Test Worksheet (see Figure 4). Additional fans for cooling tires should not be used unless absolutely necessary to ensure stabilized tire temperature or to prevent tire damage.

12.1.5 Select the dynamometer inertia weight requested on the Dynamometer Test Worksheet (see Figure 4). Adjust the 80 km/h (50 mph) actual dynamometer horsepower (warm-up load) to Setting A or the value listed in Table 1 for the selected inertia weight class.

12.1.6 Warm the test vehicle and the dynamometer at 80 km/h (50 mph) for 30 min. Near the end of this warm-up period, reset the dynamometer absorber load to the value set in 12.1.5 (Setting A). Record the indicated horsepower on the Dynamometer Test Worksheet (see Figure 3).

12.1.7 Alternately, the test vehicle and the dynamometer may be warmed up by driving two Highway Fuel Economy Driving Schedules (Reference, Code of Federal Regulations, 40 CFR 600, Appendix I). Following this warm-up, immediately accelerate the vehicle to a speed of 80 km/h (50 mph), reset the dynamometer absorber load to the value set in 12.1.5 (Setting A), and record the indicated horsepower on the Dynamometer Test Worksheet (see Figure 3).

12.1.8 After the warm-up, insure that the speed measurement device is connected to the dynamometer roll, which is coupled to the inertia simulation mechanism.

12.2 Coastdown Time Measurement

12.2.1 After the warm-up, accelerate to 105 km/h (65 mph). When the vehicle reaches 105 km/h (65 mph), place the transmission in the NEUTRAL position, engage the clutch, and allow the vehicle to lose speed. Do not touch the brakes or disengage the clutch during the coastdown. If the vehicle speed drops below 65 km/h (40 mph) for more than 1 min during the test, the test is invalid.

12.2.2 At a speed less than 72 km/h (45 mph), but not below 64 km/h (40 mph), engage the transmission, accelerate to 55 mph, and decelerate to 80 km/h (50 mph). Allow the indicated absorber load to stabilize at 80 km/h (50 mph).

12.2.3 Record the time required to coast from 88 to 72 km/h (55 to 45 mph) and the indicated 80 km/h (50 mph) horsepower on the Dynamometer Test Worksheet (see Figure 3).

12.2.4 Repeat 12.2.1 to 12.2.3 until the times from three coastdowns are within ±0.10 s of each other. If the coastdown times are not consistent, the dynamometer may be malfunctioning. Repeat 12.2.1 to 12.2.3 until consistent times are established or terminate the test.

12.2.5 Increase the absorber setting to give an actual dynamometer horsepower setting of 2 hp more than the absorber horsepower listed on the Dynamometer Test Worksheet (Setting B).

12.2.6 Repeat 12.2.1 to 12.2.3 until the times from three coastdowns are within ±0.10 s of each other. If the coastdown times are not consistent, the dynamometer may be malfunctioning. Repeat 12.2.1 to 12.2.3 until consistent times are established or terminate the test.

12.2.7 Decrease the absorber setting until the actual dynamometer horsepower setting is approximately 50% of the horsepower listed on the Dynamometer Test Worksheet (Setting C) (see Figure 3). Repeat 12.2.6.

12.2.8 The intent of the above is to insure that the road coastdown is centrally located within the bracket found by the dynamometer coastdown times.

NOTE: A retest will be necessary if the coastdown time at this lower absorber setting is less than 88 to 72 km/h (55 to 45 mph) coastdown time obtained on the road test procedure. See the Dynamometer Test Worksheet (see Figure 3) for that time, if available.

12.2.9 If more than three values were chosen in 12.1.1, repeat 12.2.1 to 12.2.4 until all data is obtained.

12.2.10 Remove the vehicle from the dynamometer, measure and record the drive axle test weight (with driver) on the Dynamometer Test Worksheet (see Figure 3).

13. Coastdown Time—Actual Horsepower Calculation

13.1 Data Reduction—Fill in the data on the Dynamometer Coastdown Worksheet (see Figure 4). Enter all data points and calculate the inverse of 88 to 72 km/h (55 to 45 mph) coastdown time ($1/\delta t$) for each actual horsepower value. Refer to the dynamometer calibration table and enter the actual horsepower values for each setting. Alternately, if the dynamometer calibration coefficients are available, they should be entered on the Dynamometer Coastdown Worksheet (see Figure 4) along with the indicated horsepower values.

13.1.1 Calculate the best fitting straight line through the data points using the Least Squares method. Use $1/\delta t$ as the independent variable (x), and the actual absorber horsepower as the dependent variable (y).

$$\text{Actual horsepower} = \frac{a}{\delta t} - b \qquad (Eq.19)$$

$$D = N\Sigma(1/\delta t)^2 - (\Sigma 1/\delta t)^2 \qquad (Eq.20)$$

$$a = \frac{[N\Sigma((1/\delta t) \times \text{actual hp})] - [\Sigma(1/\delta t)\Sigma(\text{actual hp})]}{D} \qquad (Eq.21)$$

$$b = \frac{[\Sigma(1/\delta t)\Sigma((1/\delta t) \times \text{actual hp})] - [\Sigma(1/\delta t)^2 \Sigma(\text{actual hp})]}{D} \qquad (Eq.22)$$

13.1.2 Enter the values for a and b on both the Dynamometer Test Worksheet (see Figure 3) and the Dynamometer Coastdown Worksheet (see Figure 4).

13.1.3 Plot the data on the Dynamometer Coastdown Worksheet (see Figure 4) and draw in the straight line represented by Eq. 19.

13.1.4 Experience has shown that for a properly operating dynamometer if any measured actual horsepower differs from the actual horsepower calculated by Eq. 19 by more than 0.5 hp, the data point is invalid and must be removed and the straight line recalculated using 13.1.1 to 13.2.3. If less than six data points are valid, the test is invalid. This determination is easily made from the $1/\delta t$ versus horsepower graph (see Figure 2) on the Dynamometer Coastdown Worksheet (see Figure 4).

13.2 Horsepower Determination

13.2.1 Determine the 88 to 72 km/h (55 to 45 mph) coastdown time (δt) from the road test data (Part 1). δt must be between data points.

13.2.2 Calculate the absorber horsepower for the vehicle using Eq. 19. This is the actual dynamometer horsepower that duplicates the vehicle road load at 80 km/h (50 mph) on a chassis dynamometer. This actu-

TABLE 1—ACTUAL DYNAMOMETER HORSEPOWER (WARM-UP LOAD)

Inertia Weight Class (lb)	Actual hp
≤2000	7.5
2125–2500	8.8
2625–3000	9.9
3125–3500	10.8
3750–4000	11.6
4250–4500	12.4
4750–5000	13.0
5000–5500+	13.6

al dynamometer horsepower value is applicable to dynamometers with identical roll size, number of rolls, roll spacing, surface finish, and method of restraining the vehicle.

13.2.3 When the dynamometer is being adjusted preparatory to operating a test vehicle at the road load determination in Part 2, the load setting should be made with the rolls connected to the power absorption unit running at exactly 80 km/h (50 mph).

13.2.4 When adjusting the indicated horsepower on a chassis roll dynamometer to set up the actual horsepower determined by this procedure, operate the dynamometer roll attached to the power absorption unit at the speed used for measuring indicated horsepower under 12.2.3.

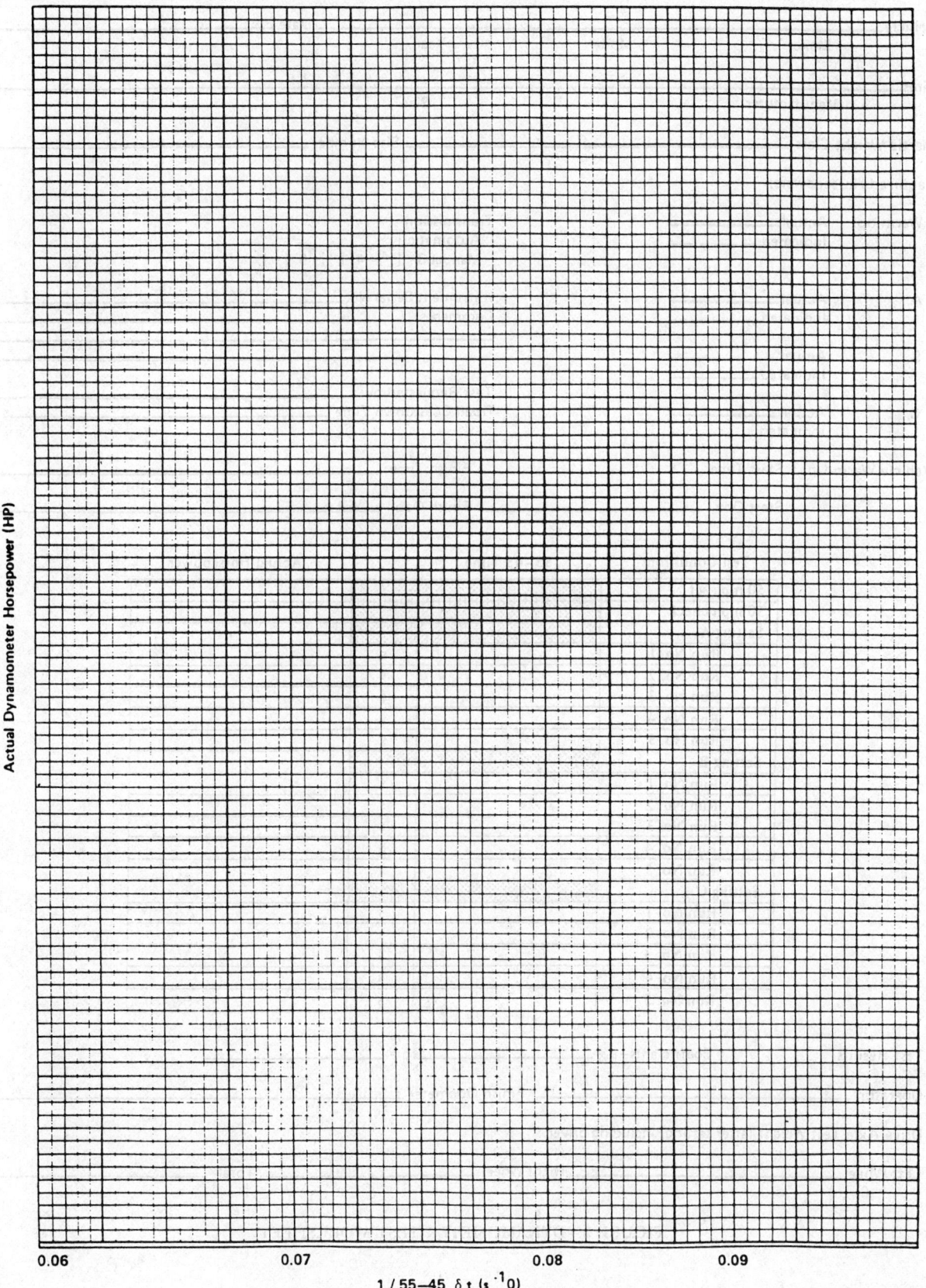

FIGURE 2—ACTUAL DYNAMOMETER HORSEPOWER

Vehicle No. _____

DYNAMOMETER TEST WORKSHEET

Vehicle _____ Date _____
        Make   Model   Year

Tires _____
      Manufacturer   Size

Inertia Weight Class _____ Site Number _____

Required Horsepowers:

Warm-up: Actual _____   Transmission
             Indicated _____   Automatic [ ]
                                        Manual 3 [ ] 4 [ ] 5 [ ]

A.   Actual _____   Instrument Serial No. _____
           Indicated _____   Comments _____

B.   Actual _____
           Indicated _____
                                        Test Requested by: _____
C.   Actual _____   Road Coastdown Time _____ s
           Indicated _____

Vehicle Warm-Up: Start Time _____ Finish Time _____

                    Start Odo. _____ Finish Odo. _____

Test Conditions	55–45 Time		Actual Horsepower
50 mph set	░░░░░	░░░░░	
50 mph check	░░░░░	░░░░░	
Setting A	░░░░░	░░░░░	
Run No. 1		s	
Run No. 2		s	
Run No. 3		s	
Run No. 4		s	
Run No. 5		s	
Setting B	░░░░░	░░░░░	
Run No. 1		s	
Run No. 2		s	
Run No. 3		s	
Run No. 4		s	
Run No. 5		s	
Setting C	░░░░░	░░░░░	
Run No. 1		s	
Run No. 2		s	
Run No. 3		s	
Run No. 4		s	
Run No. 5		s	

Test Results   Coefficients – a = _____ b = _____

Operator _____ Finish Odometer _____

Drive Axle Test Weight (after testing) including driver.

Left Wheel _____ Right Wheel _____ Total _____

FIGURE 3 – DYNAMOMETER TEST WORKSHEET

DYNAMOMETER COASTDOWN WORKSHEET

Test Date _____ Site _____

Dynamometer Calibration Coefficients

slope = _____

zero intercept = _____

Actual hp = Slope × indicated hp + zero intercept

Actual Horsepower = $\frac{a}{(\delta t)} - b$ (Eq. 10)

Output values:

a = _____

b = _____ (include minus sign)

Vehicle Identifiers:

Car No. _____

Make _____

Model Year _____

Data Point Number	δt	Indicated hp	1/δt	Actual hp	(1/δt)²	(1/δt × Actual hp)
1						
2						
3						
4						
5						
6						
7						
8						
9						
10						
11						
12						
13						
14						
15						
N =		Σ	Σ1/δt	Σhp	Σ(1/δt)²	Σ(1/δt × hp)

FIGURE 4 – DYNAMOMETER COASTDOWN WORKSHEET

APPENDIX A

A.1 COASTSAE

```
100 * THIS PROGRAM REDUCES COASTDOWN SPEED-TIME DATA AND APPLIES THE APPROPRIATE
110 * CORRECTION FACTORS ACCORDING TO THE SAE RECOMMENDED PRACTICE FOR
120 * ROAD LOAD MEASUREMENT AND DYNAMOMETER SIMULATION USING COASTDOWN TECHNIQUES.
130 *
160 *
170 * THIS PROGRAM USES VELOCITY VS TIME DATA PAIRS TO COMPUTE STATISTICALLY
180 * OPTIMIZED COEFFICIENTS OF THE FORCE-VELOCITY EQUATION. THE FORM OF THE
190 * INPUT DATA IS DESCRIBED BELOW. YOU MUST CREATE A SYSTEM FILE WITH A
200 * UNIQUE NAME TO CONTAIN THE DATA. THIS FILE CONTAINS BASIC VEHICLE AND
210 * ATMOSPHERIC DATA AND THE VELOCITY DATA POINTS. THE FILE MUST BE
220 * ORGANIZED AS FOLLOWS:
230 *
240 *
250 * LINE 1 - A 72 CHARACTER (OR LESS) TITLE WITH DESCRIPTIVE INFO.
260 * LINE 2 - EFFECTIVE WEIGHT,AREA,AIR,TEMP,NUMBER OF RUNS
270 *          AND THE TIME INCREMENT BETWEEN VELOCITY POINTS.
280 * LINE 3 - NO. OF POINTS THIS RUN,V1,V2,V3,V4,.......VN
290 * LINE 4 - NO. OF POINTS NEXT RUN,V1,V2,V3,V4,.......VN
300 * LINE 5 - ETC.
310 * USE COMMAS TO CONTINUE YOUR DATA ON THE NEXT LINE IF NECESSARY.
320 * LINE N-1 - ACTUAL TEST WEIGHT,WIND SPEED,WIND DIRECTION FROM THE NORTH (DEG),
330 *           TEST TRACK SITE(0+TRACK A,1=TRACK B,2=TRACK C,3=TRACK D), AND CDY (OPTIONAL)
331 *        ############################################################
332 *        #                                                          #
333 *        #   NOTE: As written, Tracks A and D have an angle of 0    #
334 *        #         degrees. Track B is at 349 degrees and Track C   #
335 *        #         is at 329 degrees. This information is needed in #
336 *        #         order to resolve the wind into parallel and cross-#
337 *        #         wind components. (See lines 4080 to 4150).        #
338 *        #         Alternately, a user may enter wind in degrees    #
339 *        #         relative to the track and use Track A (code 0)   #
340 *        #         as input.                                        #
341 *        #                                                          #
342 *        ############################################################
343 * LINE N - DYNAMOMETER TEST WEIGHT AND DRIVELINE INERTIA
350 * LINE N+1 - MODEL YEAR,DIVISION OR MANUFACTURER,MODEL,VEHICLE NUMBER,TRANSMISSION TYPE,
360 *            TIRE SIZE,TIRE MANUFACTURER (3 LETTER CODE),TIRE PRESSURE-FRONT,
370 *            TIRE PRESSURE-REAR,TEST DATE,DRIVER
380 *
390 * THEN SAVE THIS FILE AND TYPE   EXE COASTSAE
400 * NOTE THAT YOUR CURRENT TERMINAL FILE REMAINS AS THE FOREGROUND
410 * FILE  AND THAT A LIST COMMAND SHOULD SHOW YOU YOUR DATA AND NOT
420 * THE PROGRAM.
430 * THE PROGRAM WILL ASK YOU FOR THE NAME OF YOUR DATA FILE.
440 * SOMETIMES THE PROGRAM HAS TROUBLE BRACKETTING THE ROOTS OF THE
450 * FUNCTION. THIS USUALLY MEANS THAT THE GUESS WAS
460 * EITHER TOO LARGE OR TOO SMALL (OBVIOUSLY)
470 * OR ELSE YOUR DATA IS MOT ACCURATE ENOUGH AT THE LAST TWO
480 * POINTS THAT WERE ENTERED. ADD MORE DATA.
570 *
```

COASTSAE (continued)

```
580 *      CHARACTER VARIABLES FOR OUTPUT SHEET
590 *
600 CHARACTER*8 FNAME
610 CHARACTER*4 MODYR
620 CHARACTER*10 DIVISN
630 CHARACTER*10 MODEL
640 CHARACTER*5 VEHNO
650 CHARACTER *2 TRANS
660 CHARACTER*10 TIRE
670 CHARACTER*3 TMFG
680 CHARACTER*8 TDATE,PDATE
690 CHARACTER*8 DRIVER
700 CHARACTER*72 TITLE
710 INTEGER UNIT1,UNIT2
720 COMMON V(150),TAU(150),TIME(150),N
730 COMMON BETA
740 DIMENSION IRUNS(26),CDS(26),RS(26),BS(26),TOS(26)
750 DIMENSION RMSS(26),PAIR(13),PAIRR(13),RSMET(26),BSMET(26),RMSSM(26)
760 INTEGER DELTAT
770 INTEGER TPF,TPR
780 CALL DATE(PDATE)
790 *
800 *      PROGRAM STARTUP. ENTER DATA FILE AND DECIDE ON FORM OF
810 *      OUTPUT INCLUDING UNITS.
820 *
830 PRINT,"COAST-DOWN TEST DATA REDUCTION PROGRAM: COASTSAE"
840 PRINT
850 PRINT,"LIST THE SOURCE FILE FOR DATA FILE REQUIREMENTS"
870 PRINT
880 PRINT,"ENTER DATA FILE NAME"
890 INPUT'(A8)',FNAME
900 OPEN(1,FNAME)
910 PRINT,"DO YOU WISH INDIVIDUAL RUN SUMMARIES? (2=QUICKRUN, 1=YES, 0=NO)"
920 INPUT'(I1)',NOTYPE
930 PRINT,"IS INPUT IN ENGLISH OR METRIC UNITS (1=METRIC, 0=ENGLISH)"
940 INPUT'(I1)',UNIT1
950 PRINT,"DO YOU WISH ENGLISH OF METRIC OUTPUT (1=METRIC, 0=ENGLISH)"
960 INPUT'(I1)',UNIT2
970 READ(1,'(#,A72)') TITLE
975 PRINT
980 PRINT,TITLE
985 PRINT
990 READ(1,'(#,V)') W,A,T,PRESS,NRUNS,DELTAT
1000 *
1010 *    IF INPUT UNITS ARE METRIC, TRANSFORM TO ENGLISH
1020 *
1030 IF(UNIT1.EQ.0) GO TO 851
1040 W = W / 0.4536
1050 A = A / 0.0929
1060 T = 1.8 * T + 32.0
```

COASTSAE (continued)

```
1070 PRESS = PRESS / 25.4
1080 851 CONTINUE
1090 IGAIN=1
1100 IF(DELTAT.LT.5)IGAIN=6/DELTAT
1110 IF(NRUNS.LT.2) NOTYPE=1
1120 T=T+459.67
1130 RHO=32.174*.002378*PRESS/29.92*520/T
1140 13 FORMAT(1H ,"FOR RUN NUMBER ",I2," SPEED DATA POINTS ARE :")
1150 1 FORMAT(24A4)
1170 DO 1000 J=1,NRUNS
1180  69 CONTINUE
1190 *
1200 *     PARROT BACK THE SPEED-TIME DATA (UNLESS NOTYPE=2)
1210 *
1220 IF(NOTYPE.EQ.1) PRINT11
1230 IF(NOTYPE.NE.2) PRINT 13,J
1240 READ(1,'(#,V)') N,(V(I),I=1,N)
1250 IF(NOTYPE.NE.2) PRINT '(13F6.1)',(V(I),I=1,N)
1260 IF(UNIT1.EQ.0) GO TO 852
1270 DO 853 I=1,N
1280 853 V(I)=V(I)/1.609
1290 852 CONTINUE
1300 *
1310 *              MAKE ALL THE INITIAL GUESSES TO FIT THE DATA.
1320 *     --->NOTE THAT ALL OF THE DATA REDUCTION IS ACTUALLY DONE    <---
1330 *     --->IN ENGLISH UNITS. THIS IS A HOLDOVER FROM AN OLDER PROGRAM.<---
1340 *
1350 VO=V(1)
1360 VOMET=VO*1.609
1370 DO 10 I=1,N
1380 TIME(I)=(I-1)*DELTAT
1390 10 V(I)=V(I)/VO
1400 TO=(V(N)*IGAIN*DELTAT/(V(N-IGAIN)-V(N)))+TIME(N)
1410 TL=TO
1420 TR=TO
1430 IF(NOTYPE.EQ.1) PRINT17,TO
1440 17 FORMAT(1H0,"INITIAL APPROXIMATION OF TO",T40,F8.4)
1450 600 TL=TL-2.
1460 TR=TR+5.
1470 *
1480 *     FIT THE DATA FOR THIS RUN
1490 *
1500 CALL YTER8T(TO,DVDTO,TL,TR,.0001,50,IER)
1510 IF(IER-1)50,51,52
1520 51 PRINT,"NO CONVERGENCE"
1530 GO TO 1001
1540 52 IF(TL.GT.TO/2.) GO TO 600
1550 PRINT,"COULD NOT BRACKET ROOT OF TO - CHECK INPUT DATA !!!"
1560 1001 RS(J)=0.
1570 IRUNS(J)=J
```

COASTSAE (continued)

```
1580 CDS(J)=0.
1600 RMSS(J)=0.
1610 GO TO 1000
1620 50 CONTINUE
1630 *
1640 *     FROM THE SUBROUTINE OUTPUT CONSTRUCT THE COEFFICIENTS F0 AND F2.
1650 *     THE COEFFICIENTS ARE LABELED R AND B HERE IN THE LISTING, AGAIN
1660 *     AS A HOLDOVER FROM AN OLDER PROGRAM. THE METRIC COEFFICIENTS ARE ALSO CALCULATED.
1670 *
1680 B=ATAN(BETA)
1690 CD=B*W*2.*BETA/(V0*T0*RHO*A*88./60.)
1700 R=W*B*V0*88./(BETA*T0*32.2*60.)
1710 RMET=R*4.448
1720 BCOEFF=RHO*A*CD*88.*88./(60.*60.*2.*32.174)
1730 BMET=BCOEFF*1.7181
1740 CALL SPEEDS(V0,RMS,UNIT1)
1750 RMSMET=RMS*1.609
1760 *
1770 *     IF NOTYPE=1, PRINT OUT ALL THE VARIOUS PARAMETERS IN BOTH SYSTEMS OF UNITS.
1780 *
1790 IF(NOTYPE.EQ.1) PRINT 6,R,RMET
1800 6 FORMAT(" F0 CONSTANT COEFFICIENT:",T40,F8.4," LBS."T60,F8.4," NEWTONS")
1810 IF(NOTYPE.EQ.1) PRINT 7,BCOEFF,BMET
1820 7 FORMAT(" V**2 COEFFICIENT;",T40,F8.4," LBS/MPH**2",T60,F8.4," N/KPH**2")
1830 IF(NOTYPE.EQ.1) PRINT 8,BETA
1840 IF(NOTYPE.EQ.1) PRINT 12,V0,V0MET
1850 12 FORMAT(" INITIAL VELOCITY:",T40,F5.1,"   MPH",T60,F5.1,"   KPH")
1860 8 FORMAT(1H0,"PARAMETER BETA",T40,F8.4)
1870 IF(NOTYPE.EQ.1) PRINT 9,T0
1880 9 FORMAT(1H0,"TOTAL COASTDOWN TIME IN SECONDS",T40,F8.4)
1890 IF(NOTYPE.EQ.1) PRINT,RMS,RMSMET
1900 15 FORMAT(" TOTAL RMS ERROR IN VELOCITY FIT:",T40,F8.4," MPH",T60,F8.4," KPH")
1910 BS(J)=BCOEFF
1920 BSMET(J)=BCOEFF*1.7181
1930 IRUNS(J)=J
1940 CDS(J)=CD
1950 RS(J)=R
1960 RSMET(J)=R*4.448
1970 * BETAS(J)=BETA
1980 RMSS(J)=RMS
1990 RMSSM(J)=RMS*1.609
2000 IF(NOTYPE.EQ.1)PRINT 11
2010 11 FORMAT(1H0)
2020 1000 CONTINUE
2030 *
2040 *     READ IN THE REST OF THE DATA AND PARROT IT BACK IN THE APPROPIATE UNITS.
2050 *     IF CDY IS NOT READ, SET IF EQUAL TO THE STANDARD VALUE.
2052 *
2060 PRINT,"CORRECTION FACTOR DATA INPUTS BEING READ ARE :"
2070 READ(1,'(#,V)') WACT,W0,ALPHA,L1,CDY
```

COASTSAE (continued)

```
2080 IF(UNIT1.EQ.0)GOTO854
2090 PRINT'(T1,"ACTUAL MASS",F8.2," KG",/,T1,"WIND SPEED ",F4.1," KPH",/,
2100 &T1,"WIND DIRECTION ",F6.1," DEGREES FROM NORTH")',WACT,WO,ALPHA
2110 WO=WO/1.609
2120 WACT=WACT/0.4536
2130 854 CONTINUE
2140 PRINT'(T1,"ACTUAL WEIGHT ",F8.2," LBS",
2150 &/,T1,"WIND SPEED ",F3.1," MPH",/,
2160 &T1,"WIND DIRECTION ",F6.1," DEGREES FROM NORTH")',WACT,WO,ALPHA
2170 867 IF(L1.EQ.0)PRINT,"TRACK A"
2180 IF(L1.EQ.1) PRINT,"TRACK B"
2190 IF(L1.EQ.2) PRINT,"TRACK C"
2200 IF(L1.EQ.3) PRINT,"TRACK D"
2210 IF(L1.LT.0.OR.L1.GT.3) PRINT,"TEST TRACK CODE ERROR -- BUT WILL CONTINUE"
2212 PRINT'(T1,"CROSSWIND FACTOR (CDY) = ",F5.3)',CDY
2220 PRINT
2230 PRINT,"DYNAMOMETER PARAMETERS BEING READ ARE :"
2240 READ(1,'(#,V)') XID,XIDLC
2250 IF(UNIT1.EQ.1)DLCMET=XIDLC
2260 IF(UNIT1.EQ.1)XIDLC=XIDLC/0.4536
2270 IF(UNIT1.EQ.0)DLCMET=XIDLC*0.4536
2280 PRINT'(T1,"DYNO INERTIA WEIGHT",F8.2," LBS")',XID
2290 PRINT'(T1,"DRIVELINE EFFECTIVE INERTIA:",F8.2," LBS",F10.2," KG")',XIDLC,DLCMET
2300 READ(1,'(#,V)') MODYR,DIVISN,MODEL,VEHNO,TRANS,TIRE,TMFG,TPF,TPR,TDATE,DRIVER
2310 PRINT,".................................................."
2320 IF(NOTYPE.EQ.2)GOTO901
2330 *
2340 *     BEGIN THE PRINTING BY SKIPPING TO THE TOP OF THE NEXT PAGE.
2350 *
2360 PRINT 99
2370 99 FORMAT(1H1)
2380 PRINT,"           COASTDOWN TEST DATA PROCESSING PROGRAM: COASTSAE"
2390 IF(NRUNS.EQ.1)GOTO1003
2400 PRINT
2410 PRINT,"MODEL YEAR _____    TIRE SIZE _____    AXLE RATIO _____"
2420 PRINT'(1H+,11X,A4,21X,A10)',MODYR,TIRE
2430 PRINT,"MAKE (DIVISION) _____   TIRE MAKE _____   TRANS. TYPE _____"
2440 PRINT'(1H+,15X,A10,11X,A3,27X,A2)',DIVISN,TMFG,TRANS
2450 PRINT,"MODEL TYPE _____   TIRE PRESSURE  F ___ R ___  BRAKE TYPE F ____ R ____"
2460 PRINT'(1H+,11X,A10,23X,I3,3X,I3)',MODEL,TPF,TPR
2470 PRINT,"CAR NO. _____        TEST WEIGHT F ____ R ____   AERO DEVICES _____"
2480 PRINT'(1H+,8X,A5)',VEHNO
2490 PRINT,"ODOMETER _____    TRIM HEIGHT LF ____ RF ____   TEST DATE _____"
2500 PRINT'(1H+,64X,A8)',TDATE
2510 PRINT,"VIN _____              LR ____ RR ____  TESTED BY _____"
2520 PRINT'(1H+,64X,A8)',DRIVER
2530 PRINT,"TOE-IN (+=IN) _____  COMMENTS _____"
2540 PRINT
2550 PRINT,TITLE
2560 PRINT
```

COASTSAE (continued)

```
2570 PRINT 20,
2580 20 FORMAT(1H ,T12,"SUMMARY OF RESULTS FOR UNCORRECTED INDIVIDUAL RUNS")
2590 PRINT,"                       ROAD LOAD FORCE = F0 + F2*(V**2)"
2600 62 CONTINUE
2610 PRINT 22,
2620 22 FORMAT(1H0,T1,"RUN",T9,"F2 COEFFICIENT",T28,"F0 COEFFICIENT",
2630 &T48,"R.M.S. ERROR")
2640 *
2650 *     PRINT OUT F0, F2 AND RMS ERROR FOR EACH RUN
2660 *
2670 IF(UNIT2.EQ.0)PRINT 23
2680 IF(UNIT2.EQ.1)PRINT 855
2690 855 FORMAT(1H ,T10," N/KPH**2",T28,"CONSTANT ( N )",
2700 &T48,"VELOCITY FIT(KPH)")
2710 23 FORMAT(1H ,T10,"LBS/MPH**2"T28,"CONSTANT (LBS.)",
2720 &T48,"VELOCITY FIT(MPH)")
2730 901 CONTINUE
2740 DO 25 I=1,NRUNS
2750 IF(NOTYPE.NE.2.AND.UNIT2.EQ.0)PRINT 26,IRUNS(I),BS(I),RS(I),RMSS(I)
2760 IF(NOTYPE.NE.2.AND.UNIT2.EQ.1)PRINT 26,IRUNS(I),BSMET(I),RSMET(I),RMSSM(I)
2770 26 FORMAT(1H ,T1,I2,T12,F6.4,T31,F6.1,T49,F7.4,T66,F6.3)
2780 IF(UNIT2.EQ.0)GOTO856
2790 *
2800 *     IF THE RMS ERROR IS EXCESSIVE, FLAG IT.
2810 *
2820 IF(RMSSM(I).GT.0.40)PRINT'("+",T56,"<---EXCESSIVE")'
2830 IF(RMSSM(I).GT.0.40)RS(I)=0.0
2840 IF(RMSSM(I).GT.0.40)EXCESS=1.0
2850 GOTO25
2860 856 CONTINUE
2870 IF(RMSS(I).GT.0.25)PRINT'("+",T56,"<---EXCESSIVE")'
2880 IF(RMSS(I).GT.0.25)RS(I)=0
2890 IF(RMSS(I).GT.0.25)EXCESS=1.
2900 25 CONTINUE
2910 24 PRINT
2920 J=-1
2930 *
2940 *     AVERAGE THE COEFFICIENTS, LEAVING OUT THE RUNS WITH EXCESSIVE RMS.
2950 *
2960 DO 96 I=1,NRUNS,2
2970 IF(RS(I).EQ.0.OR.RS(I+1).EQ.0)GOTO96
2980 J=J+2
2990 IRUNS(J)=IRUNS(I)
3000 IRUNS(J+1)=IRUNS(1+1)
3010 BS(J)=BS(I)
3020 BS(J+1)=BS(I+1)
3030 RMSS(J)=RMSS(I)
3040 RMSS(J+1)=RMSS(I+1)
3050 RS(J)=RS(I)
3060 RS(J+1)=RS(I+1)
```

COASTSAE (continued)

```
3070 CDS(J)=CDS(I)
3080 CDS(J+1)=CDS(I+1)
3090 96 CONTINUE
3100 NPAIRS=(J+1)/2
3110 NRUNS=NPAIRS*2
3120 IF(NPAIRS.LT.3)PRINT,"   INSUFFICIENT DATA REMAINING FOR VALID STATISTICAL PROCESSING"
3130 IF(NPAIRS.LT.3)PRINT'(1H ,///,"*** DATA NOT BOUNDED BY CONSTRAINTS ON INTERNAL CONSISTANCY ***",///)'
3150 IF(NPAIRS.LE.0)GOTO63
3160 IF(LIST.EQ.0)GOTO66
3170 IF(NOTYPE.NE.2)PRINT22,
3180 IF(NOTYPE.NE.2.AND.UNIT2.EQ.0)PRINT23,
3190 IF(NOTYPE.NE.2.AND.UNIT2.EQ.1)PRINT855,
3200 DO 67 I=1,NRUNS
3210 IF(NOTYPE.NE.2.AND.UNIT2.EQ.0)PRINT26,IRUNS(I),BS(I),RS(I),RMSS(I)
3220 67 IF(NOTYPE.NE.2.AND.UNIT2.EQ.1)PRINT26,IRUNS(I),BSMET(I),RSMET(I),RMSSM(I)
3230 66 CONTINUE
3240 DO 30 I=1,NPAIRS
3250 PAIRR(I)=(RS(2*I-1)+RS(2*I))/2.
3260 30 PAIR(I)=(BS(2*I-1)+BS(2*I))/2.
3270 CDMEAN=0.
3280 RMEAN=0.
3290 DO 31 I=1,NPAIRS
3300 RMEAN=RMEAN+PAIRR(I)
3310 31 CDMEAN=CDMEAN+PAIR(I)
3320 RMEAN=RMEAN/NPAIRS
3330 CDMEAN=CDMEAN/NPAIRS
3340 SDR=0.
3350 SDCD=0.
3360 DO 32 I=1,NPAIRS
3370 SDR=SDR+(PAIRR(I)-RMEAN)**2
3380 32 SDCD=SDCD+(PAIR(I)-CDMEAN)**2
3390 SDR=SQRT(SDR/NPAIRS)
3400 SDRMET=SDR*4.448
3410 RMET=RMEAN*4.448
3420 PERCR=SDR/RMEAN
3430 SDCD=SQRT(SDCD/NPAIRS)
3440 IF(NOTYPE.NE.2)PRINT58,NPAIRS
3450 58 FORMAT(1H ,T20,"SUMMARY OF RESULTS FOR ",I1," PAIRED RUNS")
3460 IF(EXCESS.EQ.1.0.AND.UNIT2.EQ.0)PRINT,"RUNS WITH EXCESSIVE ERROR (>.25 MPH) WERE DELETED"
3470 IF(EXCESS.EQ.1.0.AND.UNIT2.EQ.1)PRINT,"RUNS WITH EXCESSIVE ERROR (>.40 KPH) WERE DELETED"
3480 PRINT
3490 BMEAN=CDMEAN
3500 BMET=CDMEAN*1.7181
3510 BSD=SDCD
3520 BSDMET=SDCD*1.7181
3530 PERCB=BSDMET/BMET
3540 *
3550 *    CHECK THAT THE S.D. OF F0 AND F2 ARE WITHIN LIMITS. IF NOT, THROW OUT
3560 *    THE RUN AND ITS PAIR FURTHEST FROM THE MEAN AND RE-COMPUTE.
3570 *
```

COASTSAE (continued)

```
3580 IF(NOTYPE.NE.2.AND.UNIT2.EQ.0)PRINT48,BMEAN,BSD
3590 IF(NOTYPE.NE.2.AND.UNIT2.EQ.1)PRINT857,BMET,BSDMET
3600 IF(UNIT2.EQ.0)GOTO858
3610 IF(BSDMET.GE.0.011.OR.PERCB.GE.0.03)PRINT3
3620 IF(BSDMET.GE.0.011.OR.PERCB.GE.0.03)DELETE=1.0
3630 GOTO859
3640 858 CONTINUE
3650 IF(BSD.GE.0.001.OR.PERCB.GE.0.03)PRINT3
3660 IF(BSD.GE.0.001.OR.PERCB.GE.0.03)DELETE=1.0
3670 3 FORMAT(1H+,T66,"<---EXCESSIVE")
3680 48 FORMAT(1H ,"OBSERVED F2 COEFFICIENT (LBS/MPH**2)",T40,F7.4,T60,F6.4)
3690 857 FORMAT(1H ,"OBSERVED F2 COEFFICIENT (  N/KPH**2)",T40,F7.4,T60,F6.4)
3700 859 CONTINUE
3710 IF(NOTYPE.NE.2.AND.UNIT2.EQ.0)PRINT,"OBSERVED F0 CONSTANT (LBS)"
3720 IF(NOTYPE.NE.2.AND.UNIT2.EQ.0)PRINT41,RMEAN,SDR
3730 IF(NOTYPE.NE.2.AND.UNIT2.EQ.1)PRINT,"OBSERVED F0 CONSTANT (N)"
3740 IF(NOTYPE.NE.2.AND.UNIT2.EQ.1)PRINT860,RMET,SDRMET
3750 860 FORMAT(1H+,T33,"MEAN",T40,F7.1,T50,"STD. DEV."T60,F6.1)
3760 IF(UNIT2.EQ.0)GOTO861
3770 IF(SDRMET.GE.11..OR.PERCR.GE.0.05)PRINT3
3780 IF(SDRMET.GE.11..OR.PERCR.GE.0.05)DELETE=2.0
3790 GOTO200
3800 861 CONTINUE
3810 IF(SDR.GE.2.5.OR.PERCR.GE.0.05)PRINT3
3820 IF(SDR.GE.2.5.OR.PERCR.GE.0.05)DELETE=2.0
3830 41 FORMAT(1H+,T33,"MEAN",T40,F7.1,T50,"STD. DEV.",T60,F6.1)
3840 200 CONTINUE
3850 IF(DELETE.EQ.0.)GOTO61
3860 TEMP=0.
3870 NGRUN=0
3880 ERROR=0.
3890 DO 60 I=1,NPAIRS
3900 TEMP=ABS(RMEAN-PAIRR(I))
3910 IF(TEMP.GT.ERROR)NGRUN=I
3920 60 IF(TEMP.GT.ERROR)ERROR=TEMP
3930 NGRUN=NGRUN*2
3940 RS(NGRUN)=0.
3950 IF(NOTYPE.NE.2)PRINT
3960 PRINT,"          SUMMARY OF REMAINING UNCORRECTED RUNS"
3970 PRINT,"          WITH WORST ERROR RUN PAIRS DELETED"
3980 DELETE=0.
3990 LIST=1
4000 EXCESS=0.
4010 GOTO24
4020 61 CONTINUE
4030 *
4040 *     CORRECTION FACTOR APPLICATION
4050 *
4060 RHO=RHO/32.174
4070 *
```

COASTSAE (continued)

```
4080 *       RESOLVE WIND INTO PARALLEL AND PERPENDICULAR COMPONENTS. TRACK ANGLES ARE:
4090 *       TRACKS A AND D = 0 DEGREES, TRACK B = 349 DEGREES, TRACK C = 329 DEGREES.
4100 *
4110 X=W0*SIN((ALPHA+11*L1)/57.2958)
4120 IF(L1.EQ.2)X=W0*SIN((ALPHA+31)/57.2958)
4130 IF(L1.EQ.3)X=W0*SIN(ALPHA/57.2958)
4133 VX2=W0*W0-X*X
4135 IF(VX2.LT.0.0)VX2=0.0
4140 VX=SQRT(VX2)
4150 VY=X
4160 *
4170 *       APPLY CORRECTION FACTORS TO F0 FOR PARALLEL WIND (VX), CROSSWIND (VY)
4180 *       CDY IS THE CROSSWIND DRAG COEFFICIENT WHICH CAN BE EITHER THE DEFAULT VALUE
4190 *       OF 3.40 OR THE USER MAY INPUT HIS OR HER OWN IN THE DATA (LINE 130).
4200 *
4210 IF(CDY.EQ.0.0)CDY=3.40
4220 F0=(RMEAN-BMEAN*VX*VX-0.5*RHO*A*CDY*VY*VY)/(1-0.00005*VX*VX)
4230 *
4240 *       APPLY THE TEMPERATURE CORRECTION TO F0
4250 *
4260 TERM1=F0*(1.0+0.0048*(T-527.67))
4270 *
4280 *       CONSTRUCT THE AIR DENSITY CORRECTION FACTOR (CF) AND APPLY IF TO F2.
4290 *
4300 CF=29.00*T/PRESS/527.67
4310 *
4320 *       APPLY THE AIR DENSITY CORRECTION ONLY TO THAT PART OF F2 WHICH IS DUE
4330 *       TO AIR RESISTANCE. SUBTRACT OFF THE TIRE CONTRIBUTION. APPLY THE
4340 *       CORRECTION FACTOR AND THEN ADD BACK IN THE CORRECTED TIRE CONTRIBUTION.
4350 *
4360 TERM2=CF*(BMEAN-(0.00005*F0))+0.00005*TERM1
4370 *
4380 *       CONVERT THE RESULTS TO METRIC UNITS IF NECESSARY.
4390 *
4400 IF(UNIT2.EQ.1)TERM1=TERM1*4.448
4410 IF(UNIT2.EQ.1)TERM2=TERM2*1.7181
4420 IF(NOTYPE.NE.2)PRINT
4430 IF(NOTYPE.NE.2)PRINT,"CORRECTION TO REFERENCE CONDITIONS OF"
4440 IF(NOTYPE.NE.2)PRINT,"    29.00 IN HG. (736.6 MM HG.)"
4450 IF(NOTYPE.NE.2)PRINT,"    WIND = 0.0 MPH"
4460 IF(NOTYPE.NE.2)PRINT,"    AIR TEMPERATURE = 68 DEG F. (20 DEG C.)"
4465 PRINT
4470 PRINT,"TOTAL CORRECTED ROAD LOAD FORCE IS GIVEN BY :"
4480 PRINT
4490 PRINT'(T10,"F = ",T14,F5.1,T20,"+ ",T22,F5.4,T28,"* V**2")',TERM1,TERM2
4500 PRINT
4510 *
4520 *       CONSTRUCT THE VARIOUS USEFUL OUTPUT VALUES AND PRINT THEM OUT.
4530 *
4540 F=TERM1+TERM2*50.0*50.0
```

COASTSAE (continued)

```
4550 IF(UNIT2.EQ.1)F=TERM1+TERM2*80.0*80.0
4570 IF(UNIT2.EQ.0)PRINT'(" ROAD LOAD AT 50 MPH =",F6.1," LBS (CORRECTED)")',F
4580 IF(UNIT2.EQ.1)PRINT'(" ROAD LOAD AT 80 KPH +",F6.1," N (CORRECTED)")',F
4590 TRLHP=F*50.*88./60./550.
4600 TRLKW=T*0.02222
4610 IF(UNIT2.EQ.0)PRINT'(T1,"TOTAL ROAD LOAD H.P. = ",F6.2)',TRLHP
4620 IF(UNIT2.EQ.1)PRINT'(T1,"TOTAL ROAD LOAD POWER = ",F6.2," KW")')',TRLKW
4630 PRINT
4640 F2X=TERM2*50.*50.
4650 IF(UNIT2.EQ.1)F2X=TERM2*80.*80.
4660 IF(NOTYPE.NE.2.AND.UNIT2.EQ.0)PRINT'(T1,"F2 X 2500 = ",F6.2,".....F2 UNROUNDED")',F2X
4670 IF(NOTYPE.NE.2.AND.UNIT2.EQ.1)PRINT'(T1,"F2 X 6400 = ",F6.2," N...F2 UNROUNDED")',F2X
4690 IF(UNIT2.EQ.0)GOTO862
4700 CLARK1=SQRT(TERM2/TERM1)*88.
4710 CLARK2=SQRT(TERM2/TERM1)*72.
4720 DT=(XID+XIDLC)*0.4536/(SQRT(TERM1*TERM2))*(ATAN(CLARK1)-ATAN(CLARK2))*5.0/18.0
4725 PRINT
4740 PRINT'(T1,"CALCULATED 88-72 KPH DYNO COASTDOWN TIME =",F6.2," SEC")',DT
4750 GOTO863
4760 862 CLARK1=SQRT(TERM2/TERM1)*55.0
4770 CLARK2=SQRT(TERM2/TERM1)*45.0
4780 DT=(XID+XIDLC)/32.174/(SQRT(TERM1*TERM2))*(ATAN(CLARK1)-ATAN(CLARK2))*1.466667
4785 PRINT
4790 PRINT'(T1,"CALCULATED 55-45 MPH DYNO COASTDOWN TIME = ",F6.2," SEC")',DT
4800 863 CONTINUE
4810 T=T-459.67
4820 63 CONTINUE
4830 PRINT,
4850 IF(NPAIRS.LT.3)PRINT,"*** HOWEVER INSUFFICIENT DATA ARE PRESENT FOR THESE TO BE ACCEPTABLE VALUES ***"
4870 PRINT," INPUT DATA :"
4880 WKG=W*0.4536
4890 WACTKG=WACT*0.4536
4900 AM3=A*0.0929
4910 TMET=(T-32.0)/1.8
4920 PBAR=PRESS*25.4
4930 WOKPH=WO*1.609
4940 *
4950 *     PRINT OUT THE INPUT DATA, BUT IN THE OUTPUT UNIT SYSTEM.
4960 *
4970 IF(UNIT2.EQ.1)PRINT 864,WKG,WACTKG,AM3,TMET,PBAR,WOKPH,ALPHA
4980 864 FORMAT(T1,"EFFECTIVE WEIGHT (KG)",T30,F8.2,T43,"ACTUAL WEIGHT (KG)",T67,F8.2,/,
5000 &T1,"FRONTAL AREA (M**2)",T30,F8.3,T43,"AIR TEMPERATURE (DEG C)",T67,F8.1,/,
5010 &T1,"ATM. PRESSURE (MM HG)",T30,F8.0,T43,"WIND SPEED (KPH)",T67,F8.1,/,
5020 &T1,"WIND DIRECTION FROM NORTH (DEG)",T30,F8.1)
5030 IF(UNIT2.EQ.0)PRINT 46,W,WACT,A,T,PRESS,WO,ALPHA
5040 46 FORMAT(T1,"EFFECTIVE WEIGHT (LBS)",T30,F8.2,T43,"ACTUAL WEIGHT (LBS)",T67,F8.2,/,
5060 &T1,"FRONTAL AREA (SQ. FT.)",T30,F8.2,T43,"AIR TEMPERATURE (DEG F)",T67,F8.0,/,
5070 &T1,"ATM. PRESSURE (IN.HG.)",T30,F8.2,T43,"WIND SPEED (MPH)",T67,F8.1,/,
5080 &T1,"WIND DIRECTION FROM NORTH (DEG)",T33,F5.1)
5090 IF(L1.EQ.0)PRINT'(1H+,T43,"TRACK A")'
```

COASTSAE (continued)

```
5100 IF(L1.EQ.1)PRINT'(1H+,T43,"TRACK B")'
5110 IF(L1.EQ.2)PRINT'(1H+,T43,"TRACK C")'
5120 IF(L1.EQ.3)PRINT'(1H+,T43,"TRACK D")'
5130 IF(L1.GT.3)PRINT,"UNKOWN TEST SITE"
5140 PRINT'(T1,"DYNAMOMETER INERTIA WEIGHT",T30,F6.0," LBS.",T43,"CROSSWIND FACTOR (CDY)",T70,F5.3)',XID,CDY
5150 IF(UNIT2.EQ.0)PRINT'(T1,"DRIVELINE INERTIA WEIGHT",T30,F8.2)',XIDLC
5160 IF(UNIT2.EQ.1)PRINT'(T1,"DRIVELINE EFFECTIVE INERTIA (KG)",T30,F8.2)',DLCMET
5170 PRINT'(T1,"PROCESSING DATE",T30,A8,T43,"DATA FILE NAME",T67,A8)',PDATE,FNAME
5190 PRINT
5200 PRINT
5210 1003 CONTINUE
5220 CLOSE(1)
5230 PRINT11
5240 PRINT11
5250 PRINT11
5260 84 STOP
5270 END
5280 *
5290 *      SUBROUTINE YTER8T(X,F,XL1,EPS,IEND,IER)
5300 *
5310 SUBROUTINE YTER8T(X,F,XLI,XRI,EPS,IEND,IER)
5320 IER=0
5330 XL=XLI
5340 XR=XRI
5350 X=XL
5360 TOL=X
5370 CALL VALU8T(TOL,F)
5380 IF(F)1,16,1
5390 1 FL=F
5400 X=XR
5410 TOL=X
5420 CALL VALU8T(TOL,F)
5430 IF(F)2,16,2
5440 2 FR=F
5450 IF(SIGN(1.,FL)+SIGN(1.,FR)) 25,3,25
5460 3 I=0
5470 TOLF=100.*EPS
5480 4 I=I+1
5490 DO 13 K=1,IEND
5500 X=.5*(XL+XR)
5510 TOL=X
5520 CALL VALU8T(TOL,F)
5530 IF(F) 5,16,5
5540 5 IF(SIGN(1.,F)+SIGN(1.,FR))7,6,7
5550 6 TOL=XL
5560 XL=XR
5570 XR=TOL
5580 TOL=FL
5590 FL=FR
5600 FR=TOL
```

COASTSAE (continued)

```
5610 7 TOL=F-FL
5620 A=F*TOL
5630 A=A+A
5640 IF(A-FR*(FR-FL))8,9,9
5650 8 IF(I-IEND)17,17,9
5660 9 XR=X
5670 FR=F
5680 TOL=EPS
5690 A=ABS(XR)
5700 IF(A-1.)11,11,10
5710 10 TOL=TOL*A
5720 11 IF(ABS(XR-XL)-TOL)12,12,13
5730 12 IF(ABS(FR-FL)-TOLF)14,14,13
5740 13 CONTINUE
5750 IER=1
5760 14 IF(ABS(FR)-ABS(FL))16,16,15
5770 15 X=XL
5780 F=FL
5790 16 RETURN
5800 17 A=FR-F
5810 DX=(X-XL)*FL*(1.+F*(A-TOL)/(A*(FR-FL)))/TOL
5820 XM=X
5830 FM=F
5840 X=XL-DX
5850 TOL=X
5860 CALL VALU8T(TOL,F)
5870 IF(F)18,16,18
5880 18 TOL=EPS
5890 A=ABS(X)
5900 IF(A-1.) 20,20,19
5910 19 TOL=TOL*A
5920 20 IF(ABS(DX)-TOL)21,21,22
5930 21 IF(ABS(F)-TOLF)16,16,22
5940 22 IF(SIGN(1.,F)+SIGN(1.,FL)) 24,23,24
5950 23 XR=X
5960 FR=F
5970 GO TO 4
5980 24 XL=X
5990 FL=F
6000 XR=XM
6010 FR=FM
6020 GO TO 4
6030 25 IER=2
6040 RETURN
6050 END
6060 *
6070 *     SUBROUTINE VALU8T(TO,DVDTO)
6080 *
6090 SUBROUTINE VALU8T (TO,DVDTO)
6100 COMMON V(150),TAU(150),TIME(150),N
```

COASTSAE (continued)

```
6110 COMMON BETA
6120 DO 10 I=1,N
6130 10 TAU(I)=TIME(I)/TO
6140 BL=.5
6150 BR=5.
6160 CALL YTER8B(ABETA,DVDB,BL,BR,.0001,50,IER)
6170 IF (IER-1)50,51,52
6180 50 BETA=ABETA
6190 SIGMA1=0.
6200 SIGMA2=0
6210  DO 20 I=1,N
6220 A=1.-TAU(I)
6230 B=ATAN(BETA)
6240 C=A*B
6250 SIGMA1=SIGMA1+V(I)*TIME(I)/COS(C)**2.
6260 20 SIGMA2=SIGMA2+TIME(I)*SIN(C)/(BETA*COS(C)**3)
6270 DVDTO=SIGMA1-SIGMA2
6280 RETURN
6290 51 PRINT," NO ROOT IN GIVEN RANGE OF BETA "
6300 RETURN
6310 52 PRINT," COULD NOT BRACKET THE ROOT OF BETA BETWEEN ",BL,BR
6320 RETURN
6330 END
6340 *
6350 *     SUBROUTINE YTER8B(X,F,XLI,XRI,EPS,IEND,IER)
6360 *
6370 SUBROUTINE YTER8B(X,F,XLI,XRI,EPS,IEND,IER)
6380 IER=0
6390 XL=XLI
6400 XR=XRI
6410 X=XL
6420 TOL=X
6430 CALL VALU8B(TOL,F)
6440 IF(F)1,16,1
6450 1 FL=F
6460 X=XR
6470 TOL=X
6480 CALL VALU8B(TOL,F)
6490 IF(F)2,16,2
6500 2 FR=F
6510 IF(SIGN(1.,FL)+SIGN(1.,FR)) 25,3,25
6520 3 I=0
6530 TOLF=100.*EPS
6540 4 I=I+1
6550 DO 13 K=1,IEND
6560 X=.5*(XL+XR)
6570 TOL=X
6580 CALL VALU8B(TOL,F)
6590 IF(F) 5,16,5
6600 5 IF(SIGN(1.,F)+SIGN(1.,FR))7,6,7
```

COASTSAE (continued)

```
6610  6 TOL=XL
6620    XL=XR
6630    XR=TOL
6640    TOL=FL
6650    FL=FR
6660    FR=TOL
6670  7 TOL=F-FL
6680    A=F*TOL
6690    A=A+A
6700    IF(A-FR*(FR-FL))8,9,9
6710  8 IF(I-IEND)17,17,9
6720  9 XR=X
6730    FR=F
6740    TOL=EPS
6750    A=ABS(XR)
6760    IF(A-1.)11,11,10
6770 10 TOL=TOL*A
6780 11 IF(ABS(XR-XL)-TOL)12,12,13
6790 12 IF(ABS(FR-FL)-TOLF)14,14,13
6800 13 CONTINUE
6810    IER=1
6820 14 IF(ABS(FR)-ABS(FL))16,16,15
6830 15 X=XL
6840    F=FL
6850 16 RETURN
6860 17 A=FR-F
6870    DX=(X-XL)*FL*(1.+F*(A-TOL)/(A*(FR-FL)))/TOL
6880    XM=X
6890    FM=F
6900    X=XL-DX
6910    TOL=X
6950    CALL VALU8B(TOL,F)
6960    IF(F)18,16,18
6970 18 TOL=EPS
6980    A=ABS(X)
6990    IF(A-1.) 20,20,19
7000 19 TOL=TOL*A
7010 20 IF(ABS(DX)-TOL)21,21,22
7020 21 IF(ABS(F)-TOLF)16,16,22
7030 22 IF(SIGN(1.,F)+SIGN(1.,FL)) 24,23,24
7040 23 XR=X
7050    FR=F
7060    GO TO 4
7070 24 XL=X
7080    FL=F
7090    XR=XM
7100    FR=FM
7110    GO TO 4
7120 25 IER=2
7130    RETURN
```

COASTSAE (continued)

```
7140 END
7150 *
7160 *        SUBROUTINE VALU8B(ABETA,DVDB)
7170 *
7180 SUBROUTINE VALU8B (ABETA,DVDB)
7190 COMMON V(150),TAU(150),TIME(150),N
7200 COMMON BETA
7210 BETA=ABETA
7220 SIGMA1=0.
7230 SIGMA2=0.
7240 SIGMA3=0.
7250 SIGMA4=0.
7260 B=ATAN(BETA)
7270 DO 10 I=1,N
7280 A=1-TAU(I)
7290 C=A*B
7300 D=SIN(C)/COS(C)
7310 SIGMA1=SIGMA1+V(I)*D
7320 SIGMA2=SIGMA2+D**2.
7330 SIGMA3=SIGMA3+V(I)*A/COS(C)**2.
7340 SIGMA4=SIGMA4+A*D/COS(C)**2.
7350 10 CONTINUE
7360 DVDB=SIGMA3*BETA**2.
7370 &   -BETA*(1.+BETA**2.)*SIGMA1
7380 &   -BETA*SIGMA4+(1.+BETA**2.)*SIGMA2
7390 RETURN
7400 END
7405 *
7406 *        SUBROUTINE SPEEDS(V0,RMS,UNIT1)
7407 *
7410 SUBROUTINE SPEEDS(V0,RMS,UNIT1)
7420 *
7430 *     SUBROUTINE SPEEDS(V0,RMS,UNIT1) TO CALCULATE THE FITTED SPEEDS FOR EACH DATA
7440 *     POINT AND CALCULATE THE RMS ERROR IN THE VELOCITY FIT.  IF THE ERROR IS
7450 *     EXCESSIVE, BOTH THE FITTED SPEEDS AND THE INDIVIDUAL ERRORS ARE PRINTED
7460 *     AS AN AID IN FINDING KEYPUNCH-TYPE ERRORS.  THE RMS LIMITS ARE 0.25 MPH
7470 *     OR 0.40 KPH DEPENDING ON THE INPUT UNITS CHOSEN.
7480 *
7490 COMMON V(150),TAU(150),TIME(150),N
7500 COMMON BETA
7505 INTEGER UNIT1
7510 DIMENSION VC(150),VDIF(150)
7520 SIGMA=0.
7530 DO 10 I=1,N
7540 A=1.-TAU(I)
7550 B=ATAN(BETA)
7560 C=A*B
7570 VCALC=V0*SIN(C)/(BETA*COS(C))
7580 VC(I) = VCALC
7590 VDIF(I) = V(I)*V0 - VCALC
```

COASTSAE (continued)

```
7600 SIGMA=SIGMA+(V(I)*V0-VCALC)**2
7610 10 CONTINUE
7620 RMS = SQRT(SIGMA/N)
7630 RMSMET = RMS * 1.609
7640 IF(UNIT1.EQ.1)GO TO 865
7650 IF(RMS.LE.0.25) GO TO 902
7660 PRINT," RMS ERROR GREATER THAN 0.25 MPH.   INDIVIDUAL ERRORS ARE:"
7670 PRINT'(13F6.1)',(VC(I),I=1,N)
7680 PRINT,"INDIVIDUAL ERROR ARE AS FOLLOWS:"
7690 PRINT'(13F6.2)',(VDIF(I),I=1,N)
7700 GO TO 902
7710 865 IF(RMSMET.LE.0.40)GO TO 902
7720 DO 866 I=1,N
7730 VC(I) = VC(I) * 1.609
7740 866 VDIF(I) = VDIF(I) * 1.609
7750 PRINT," RMS ERROR GREATER THAN 0.40 KPH.  INDIVIDUAL ERRORS ARE:"
7760 PRINT'(13F6.1)',(VC(I),I=1,N)
7770 PRINT," INDIVIDUAL ERRORS ARE AS FOLLOWS:"
7780 PRINT'(13F9.4)',(VDIF(I),I=1,N)
7790 902 CONTINUE
7800 RETURN
7810 END
```

26.494

COAST DOWN TEST DATA FORMATTED FOR COASTSAE

CAR NO. _____ TIRES _____ DATE _____
BODY STYLE _____ SIZE _____ TIME _____
V.I.N. _____ PRESSURE: FRONT _____ REAR _____ TEMP: °F _____
ODOMETER _____ - _____ TEST WEIGHT: BAROMETER _____
 FRONT _____ + REAR _____ = _____ WIND _____

File Name	Line Number	Car Number, Model, Style; Tire Size, Mfr; Test Date; Run Number												
Title Line	10												
		Effective Weight		Frontal Area		Ambient Temp.		Barometer		No. of Runs		Time Increment	
Test Condition	20												
Run Number													
1 N	30												
	35												
2 S	40												
	45												
3 N	50												
	55												
4 S	60												
	65												
5 N	70												
	75												
6 S	80												
	85												
7 N	90												
	95												
8 S	100												
	105												
9 N	110												
	115												
10 S	120												
	125												
Correction Factor		Actual Weight		Wind Speed		Wind Direction		Test Track		CDY (Optional)			
	130												
Dyno Parameters		Dyno Inertia Weight		Drive Line Inertia Wt.									
	135												
Vehicle Information		Model Year		Division or Manufacturer		Model		Vehicle Number		Transmission Type		Tire Size		Tire Manufacturer		Tire Pressure F R
	140												
		Test Date / /		Driver									

FIGURE A1 — COASTDOWN TEST DATA FORMATTED FOR COASTSAE

MODEL YEAR 1979
MAKE (DIVISION) AJAX
MODEL TYPE TURKEY
CAR NO. SAE-0
ODOMETER
VIN
TOE-IN (+=IN)

TIRE SIZE P185/80R13
TIRE MAKE UNI
TIRE PRESSURES F 26 R 26
TEST WEIGHT F ____ R ____
TRIM HEIGHT LF ____ RF ____
              LR ____ RR ____
COMMENTS

AXLE RATIO
TRANS. TYPE M4
BRAKE TYPE F ____ R ____
AERO DEVICES
TEST DATE 9/19/78
TESTED BY QUINTON

SAMPLE COASTDOWN DATA SET USING NONSTANDARD CDY VALUE

SUMMARY OF RESULTS FOR UNCORRECTED INDIVIDUAL RUNS

ROAD LOAD FORCE = F0 + F2*(V**2)

RUN	F2 COEFFICIENT LBS/MPH**2	F0 COEFFICIENT CONSTANT (LBS)	R.M.S. ERROR VELOCITY FIT (MPH)
1	0.0254	29.4	0.1529
2	0.0224	21.5	0.1209
3	0.0242	28.9	0.1135
4	0.0225	21.4	0.1193
5	0.0246	27.9	0.1017
6	0.0222	21.3	0.1028
7	0.0252	25.5	0.1001
8	0.0224	23.6	0.0741
9	0.0244	25.5	0.1038
10	0.0224	24.8	0.0910

SUMMARY OF RESULTS FOR 5 PAIRED RUNS

OBSERVED F2 COEFFICIENT (LBS/MPH**2)   0.0236   0.0001
OBSERVED F0 CONSTANT (LBS)   MEAN   25.0   STD. DEV.   0.3

CORRECTION TO REFERENCE CONDITIONS OF
    29.00 IN HG (736.6 MM HG)
    WIND = 0.0 MPH
    AIR TEMPERATURE = 68 DEG F (20 DEG C)
TOTAL CORRECTED ROAD LOAD FORCE IS GIVEN BY:
    F = 26.1 + .0237 * V**2

ROAD LOAD AT 50 MPH = 85.4 LBS (CORRECTED)
TOTAL ROAD LOAD H.P. = 11.39

F2 X 2500 = 59.25.....F2 UNROUNDED
CALCULATED 55-45 MPH DYNO COAST DOWN TIME = 16.28 SEC

INPUT DATA:

EFFECTIVE WEIGHT (LBS)	2926.00	ACTUAL WEIGHT (LBS)	2854.00
FRONTAL AREA (SQ. FT.)	21.10	AIR TEMPERATURE (DEG F)	79.00
ATM. PRESSURE (IN HG)	29.50	WIND SPEED (MPH)	2.5
WIND DIRECTION FROM NORTH (DEG)	192.00	MILFORD TEST SITE	
DYNAMOMETER INERTIA WEIGHT	3000 LBS	CROSSWIND FACTOR (CDY)	3000.
DRIVELINE INERTIA WEIGHT	36.70		
PROCESSING DATE	10/16/78	DATA FILE NAME	SAEDATA

FIGURE A2—COASTDOWN TEST DATA PROCESSING PROGRAM: COASTSAE

```
EXE COASTSAE

COAST-DOWN TEST DATA REDUCTION PROGRAM:  COASTSAE

LIST THE SOURCE FILE FOR DTSS DATA FILE REQUIREMENTS
     REVISED 8/21/78 CEC
ENTER DATA FILE NAME? SAEDATA
DO YOU WISH INDIVIDUAL RUN SUMMARIES (2=QUICKRUN, 1=YES, 0=NO)? 0
IS INPUT IN ENGLISH OR METRIC UNITS (1=METRIC, 0=ENGLISH)? 0
DO YOU WISH ENGLISH OR METRIC OUTPUT (1=METRIC, 0=ENGLISH)? 0
SAMPLE COASTDOWN DATA SET USING NONSTANDARD CDY VALUE
FOR RUN NUMBER 1 SPEED DATA POINTS ARE:
   70.0  64.5  59.6  55.4  51.6  48.4  45.3  42.3  39.6  37.1  34.7  32.4  30.2
   28.3  26.6  25.0  23.1  21.6  20.4  18.9
FOR RUN NUMBER 2 SPEED DATA POINTS ARE:
   70.0  65.5  61.1  57.4  53.8  51.1  48.1  45.4  43.1  40.7  38.6  36.7  34.7
   32.8  31.2  29.5  28.2  26.8  25.2  23.9  22.6  21.7  20.5  19.4
FOR RUN NUMBER 3 SPEED DATA POINTS ARE:
   70.0  64.9  60.1  56.0  52.2  48.7  45.7  43.1  40.3  37.9  35.3  33.4  31.2
   29.3  27.5  25.7  24.0  22.6  21.2  19.4  18.3
FOR RUN NUMBER 4 SPEED DATA POINTS ARE:
   70.0  65.4  60.9  57.2  53.8  51.0  48.1  45.3  43.0  40.6  38.5  36.5  34.6
   33.1  31.2  29.7  28.1  26.7  25.5  23.9  22.9  21.4  20.3  19.1
FOR RUN NUMBER 5 SPEED DATA POINTS ARE:
   70.0  64.7  60.3  55.9  52.4  48.8  45.8  43.0  40.3  37.8  35.4  33.4  31.4
   29.5  27.6  25.7  24.4  22.7  21.4  19.9  18.6
FOR RUN NUMBER 6 SPEED DATA POINTS ARE:
   70.0  65.3  61.2  57.4  54.0  51.0  48.4  45.5  43.2  41.0  38.9  36.9  34.8
   33.1  31.4  29.9  28.5  27.0  25.7  24.2  22.9  21.7  20.8  19.5  18.3
FOR RUN NUMBER 7 SPEED DATA POINTS ARE:
   70.0  64.6  60.1  56.0  52.1  48.8  45.7  43.0  40.5  38.2  35.9  33.8  31.6
   29.9  28.2  26.4  24.9  23.3  21.8  20.5  19.2
FOR RUN NUMBER 8 SPEED DATA POINTS ARE:
   70.0  65.3  61.0  57.2  53.8  50.6  47.7  45.1  42.5  40.3  37.9  35.9  34.0
   32.3  30.7  28.8  27.3  25.9  24.4  23.1  21.8  20.7  19.3  18.1
FOR RUN NUMBER 9 SPEED DATA POINTS ARE:
   70.0  64.9  60.3  56.2  52.5  49.3  46.3  43.6  41.0  38.6  36.3  34.1  32.2
   30.3  28.6  26.9  25.4  23.9  22.2  20.8  19.9  18.5
FOR RUN NUMBER 10 SPEED DATA POINTS ARE:
   70.0  65.3  60.9  57.1  53.8  50.5  47.3  44.8  42.3  39.8  37.6  35.7  33.6
   31.9  30.2  28.6  26.8  25.5  24.1  22.5  21.3  19.9  18.7
CORRECTION FACTOR DATA INPUTS BEING READ ARE:
ACTUAL WEIGHT 2854.00 LBS
WIND SPEED 2.5 MPH
WIND DIRECTION 192.0 DEGREES FROM NORTH
MILFORD TEST TRACK
CROSSWIND FACTOR (CDY) = 3.000

DYNAMOMETER PARAMETERS BEING READ ARE:
DYNO INERTIA WEIGHT 3000.00 LBS
DRIVELINE EFFECTIVE INERTIA:   36.70 LBS   16.65 KG
```

FIGURE A-2 (CONTINUED)

```
LIS

SAEDATA  16 OCT78  10:05

10 SAMPLE COASTDOWN DATA SET USING NONSTANDARD CDY VALUE
20 2926,21.1,79,29.50,10,5
30 20,70,64.5,59.6,55.4,51.6,48.4,45.3,42.3,39.6,37.1,34.7,32.4,30.2,28.3,26.6,25,23.1,21.6,20.4,18.9
35 24,70,65.5,61.1,57.4,53.8,51.1,48.1,45.4,43.1,40.7,38.6,36.7,34.7,32.8,31.2,29.5,28.2,26.8,25.2,23.9,22.6,21.7,20.5,19.4
40 21,70,64.9,60.1,56.0,52.2,48.7,45.7,43.1,40.3,37.9,35.3,33.4,31.2,29.3,27.5,25.7,24.0,22.6,21.2,19.4,18.3
45 24,70,55.4,60.9,57.2,53.8,51,48.1,45.3,43,40.6,38.5,36.5,34.6,33.1,31.2,29.7,28.1,26.7,25.5,23.9,22.9,21.4,20.3,19.1
50 21,70,64.7,60.3,55.9,52.4,48.8,45.8,43,40.3,37.8,35.4,33.4,31.4,29.5,27.6,25.7,24.4,22.7,21.4,19.9,18.6
55 25,70,65.3,61.2,57.4,54,51,48.4,45.5,43.2,41,38.9,36.8,34.8,33.1,31.4,29.9,28.5,27,25.7,24.2,22.9,21.7,20.8,19.5,18.3
60 21,70,64.6,60.1,56,52.1,48.8,45.7,43.0,40.5,38.2,35.9,33.8,31.6,29.9,28.2,26.4,24.9,23.3,21.8,20.5,19.2
65 24,70,65.3,61.0,57.2,53.8,50.6,47.7,45.1,42.5,40.3,37.9,35.9,34,32.3,30.7,28.8,27.3,25.9,24.4,23.1,21.8,20.7,19.3,18.1
70 22,70,64.9,60.3,56.2,52.5,49.3,46.3,43.6,41,38.6,36.3,34.1,32.2,30.3,28.6,26.9,25.4,23.9,22.2,20.8,19.9,18.5
75 23,70,65.3,60.9,57.1,53.8,50.5,47.3,44.8,42.3,39.8,37.6,35.7,33.6,31.9,30.2,28.6,26.8,25.5,24.1,22.5,21.3,19.9,18.7
80 2854,2.5,192,0,3.0
85 3000,36.7
90 1979,AJAX,TURKEY,SAE-0,M4,P185/80R13,UNI,26,26,9/19/78,QUINTON
READY
```

FIGURE A-2 (CONTINUED)

SAE COLD START AND DRIVEABILITY PROCEDURE—SAE J1635 MAY93

SAE Recommended Practice

Report of the SAE Light-Duty Vehicle Performance and Economy Measurement Standards Committee approved May 1993.

1. Scope

1.1 To subjectively evaluate engine starting behavior and driveability characteristics of a motor vehicle which has been soaked at ambient temperature for a given time period after attaining a stabilized engine coolant temperature. This SAE Recommended Practice also defines driveability defects and the rating system.

1.2 This evaluation may be affected by ambient temperature, altitude, fuel, and the road system.

1.2.1 The vehicle should be evaluated with all fuels recommended by the manufacturer. A partial list comprises reformulated gasoline, ethanol/gasoline and methanol/gasoline blends of various proportions, diesel #1, and diesel #2.

2. References
There are no referenced publications specified herein.

3. Test Conditions

3.1 The test road should be paved, level, smooth, and dry. If the vehicle is evaluated on a public road, traffic should be light.

3.2 Tests may be conducted at any ambient temperature, but it is recommended that average winds not exceed 25 km/h (15 mph) or gusts not exceed 40 km/h (25 mph). Winds mask defects such as surge, hesitation, and bucking. Precipitation or fog may affect traction and must be avoided.

3.3 Nomenclature for transmission gear shall be as follows:

3.3.1 Automatic transmission is designated as A; the gear is represented by the first letter of the appropriate word (i.e., A-D is Drive).

3.3.2 Manual transmission is designated as M; the gear is represented by the appropriate number (i.e., M-1 is First). For idle rating purposes, the clutch is to be disengaged.

4. Equipment

4.1 Stop watch
4.2 Engine tachometer
4.3 Intake manifold absolute pressure (MAP) gauge

5. Vehicle Preparation

5.1 The vehicle must be properly equipped, maintained, and adjusted for ambient temperature operation. This includes the use of recommended:

5.1.1 Fuel (at least 1/4 tank)
5.1.2 Weight and viscosity of engine oil
5.1.3 Coolant composition
5.1.4 Thermostat
5.1.5 Spark plug type and gap
5.1.6 Tires, at proper cold inflation pressure

5.1.6.1 When the ambient temperature at which the test is to be conducted differs more than 14 °C (26 °F) from that at which the vehicle is being prepared, it is recommended that the inflation pressure be altered by the following pressure increment:

$$P = k \text{ (Prep Temp} - \text{Test Temp)} \quad \text{(Eq.1)}$$

where:

P = Change in pressure
k = 1 kPa/°C or 1 psi/13 °F

5.1.6.2 If the ambient test temperature is higher than the preparation area temperature, then subtract the tire pressure correction factor from the specified tire pressure.

5.1.6.3 If the ambient test temperature is lower than the preparation area temperature, then add the tire pressure correction factor to the specified tire pressure, e.g.,

a. Prep Temp 21 °C (70 °F)
b. Test Temp 13 °C (56 °F)
c. (21 °C to 13 °C) 1 kPa/°C = +8 kPa or (70 °F to 56 °F) 1 psi/13 °F = +1.1 psi

5.1.7 Battery and battery cable sizes
5.1.8 Cold weather starting aids
5.1.9 Transmission fluid
5.1.10 Air cleaner
5.1.11 Air cleaner hot-air stove

5.2 The vehicle is to be loaded 136 kg ± 11.4 kg (300 lb ± 25 lb) above curb weight, inclusive of weight of driver and test equipment.

5.3 Prepare the vehicle for a cold/partial cooldown/hot start as follows:

5.3.1 If the engine coolant temperature is not at a stable value when the vehicle is delivered, then prepare the vehicle by driving at a steady state. At a time equal to the desired soak period before the start test, bring engine to stabilized coolant temperature by driving for at least 16 km (10 mile) at 70 to 90 km/h (45 to 55 mph). At temperatures lower than -18 °C (0 °F), drive for a minimum of 32 km (20 mile). See Table 1.

TABLE 1—LENGTH OF SOAK PERIOD

Type of Start Test	Soak Period
Cold	8 to 36 h (12 h nominal)
Partial cooldown	1.5 to 4 h
Hot	5 to 30 min

5.3.2 Record manifold absolute pressure (kPa) for stabilized engine idle (for use in 7.3) and RPM (gears A-D, A-P or M-1, M-0) at conclusion of warm-up for use with Table 2.

5.3.3 Place gear selector in PARK (A-P) or NEUTRAL (M-0) and turn off engine. Set the parking brake.

5.3.4 Set the heater/air conditioner controls as follows, with the blower switch in HIGH (if selectable) and the temperature set to a maximum tolerable level.

a. Non A/C—Defrost, LOW fan, maximum tolerable temperature (MTT)
b. Manual A/C—Defrost, LOW fan, MTT
c. Semi-auto A/C—Defrost, LOW fan, MTT
d. Automatic A/C—Automatic LOW fan, MTT

5.3.5 Allow the vehicle to soak at ambient temperature for the desired period.

TABLE 2—DEFINITION OF ACCELERATION

Description	Manifold Pressure (DRIVE)	Fraction of Full Pedal Travel
Light accel	40.4 + 0.60 * Idle kPa	1/4
Moderate accel	60.6 + 0.40 * Idle kPa	1/4 < Pedal < 1/2
Heavy accel	85.8 + 0.15 * Idle kPa	1/2 < Pedal < 7/8
Wide Open Throttle (WOT)	101 kPa	Full

6. Cold Start Procedure

6.1 Record all test information on the Cold Start and Driveability data sheet. See Figure 1.

6.2 For diesel engines, turn ignition key to RUN position and record "WAIT" light on-time.

6.3 Start engine per Owner's Manual/visor information procedure. Record start time.

6.4 If engine fails to start after 15 s, stop cranking. Follow Owner's Manual procedure for a no start. Begin cranking and record total cranking time until engine starts. After three failures to start, the vehicle should be withdrawn from testing for diagnosis and repair.

6.5 As soon as the engine starts, turn on headlights and backlight defroster (if so equipped). The defroster is not to be reset if it shuts off automatically.

6.5.1 Fan is kept at low setting throughout test to provide a quiet ambient for assessing engine roughness, knock, and engine run-on.

6.5.2 If vehicle is tested with heated windshield activated, note this condition on the data sheet. See Figure 1.

6.6 Record the following readings (as applicable) in gear A-P or M-0 immediately after start. Do not force the idle; engine is to idle at closed throttle. For a manual transmission, record data with clutch engaged and disengaged.

6.6.1 Idle speed, RPM
6.6.2 Manifold absolute pressure, kPa
6.6.3 Idle quality (see Appendix A).

6.7 If engine stalls, repeat Steps 6.2 through 6.4. Record number of stalls and starting times. After three failures to start, the vehicle should be withdrawn from testing for diagnosis and repair.

6.8 Allow engine to idle 15 s in gear A-P or M-0. Apply brake, shift to gear A-D or M-1 (clutch disengaged) range, idle for 10 s and record information in 6.6. If engine stalls, restart immediately. Do not record restart time. Record number of stalls.

VEHICLE INFORMATION

Vehicle Identification _____ (VIN)

Model Year _____ Manufacturer _____

Model _____ Odometer _____ km

Body Style _____ (2D, 4D, SW, TR = truck)

Engine Displacement: _____ (Liters)   Configuration _____ (L4, V6, V8)

Induction System _____ (Normal, Turbocharged, Supercharged)

Fuel Delivery System _____ (CARB, TBI, DIE, PFI)

Transmission _____ (A3, A4, A5, M5, M6)

Final Drive Ratio _____   Drive Axle _____ (Front, Rear, AWD)

Test Weight: _____ kg   Fuel Type _____

Tire Brand & Model _____

Tire Size _____ Cold Tire Press ___ kPa front ___ kPa rear

Comments _____

TEST INFORMATION

Date _____   Driver _____

Observer _____

Soak Time _____   Soak Temperature _____ °C.

Acceleration based on ___ Manifold Pressure ___ Throttle Pedal travel

Manifold Pressure at Idle _____ kPa

Condition	Start Rating	Idle Rating			Drive Rating	Ambient Temp (C)	Soak Time (hr)
		Idle Neutral Reverse	Idle Drive	Idle			
Cold Drive		/	/				
Comments							
Warm Drive	--------	/	/				------
Comments							
Hot Drive		/	/				
Comments							
Partial Cool Dwn		/	/				
Comments							

FIGURE 1—COLD START AND DRIVEABILITY DATA SHEET

7. Driveability Procedure

7.1 After the cold start evaluation, the vehicle is driven according to Cycle 1, then Cycle 2, followed by a repeat of Cycle 1 and Cycle 2.

7.2 Record the frequency and severity of the following defects:
7.2.1 Backfire (note whether induction or exhaust)
7.2.2 Bucking
7.2.3 Detonation
7.2.4 Harshness
7.2.5 Hesitation/Stumble
7.2.6 Stall
7.2.7 Surge
7.2.8 Vibration

7.3 Accelerations are calculated as a function of:

7.3.1 Manifold absolute pressure at idle (gear A-D or M-1, clutch disengaged). See 5.3.2.

OR

7.3.2 Full pedal travel (Diesel engine or when no MAP gauge available/erratic MAP at idle).

7.4 All automatic transmissions are to be operated in highest drive/overdrive position (D/D3/D4/OD). Manual transmissions are to be shifted when signaled by a Shift Indicator Light (SIL) or at speeds specified in the Owner's Manual for normal driving. Upshifts may be lower than the recommended speeds if the vehicle will be cruising at a speed below the recommended acceleration shift points. For accelerations from a rest with a manual transmission, clutch should be released simultaneously with throttle application. Idle stops are to be made at a normal rate, with the clutch depressed on manual transmission vehicles.

7.5 Place transmission in REVERSE, and back vehicle for 15 m (50 ft). Evaluate driveability for smoothness and note any stalls (see Appendix A for rating methods). Idle for 10 s in gear A-R or M-R. Record idle parameters (see 6.6). Select gear A-D or M-1. Idle for 10 s. Record idle parameters (see 6.6).

7.6 After 10 s in gear A-D or M-1, drive schedule (Cycle 1) in Table 3 and Figure 2. Record idle parameters at the 0.5 km (0.3 mile) stop. See Table 2 for definition of throttle setting.

TABLE 3—SCHEDULE DRIVEABILITY—CYCLE 1

Distance km	Distance (mile)	Operation and Throttle Setting	Speed km/h	Speed (mph)
0.0-0.2	0.0-0.1	Light accel	0-40	0-25
0.2-0.3	0.1-0.2	Steady state	40	25
0.3	0.2	Heavy accel	40-55	25-35
0.3-0.5	0.2-0.3	Steady state	55	35
0.5	0.3	BRAKE	55-0	35-0
0.5	0.3	10 s idle	0	0
0.5	0.3	WOT accel	0-55	0-35
0.5	0.3	Closed-throttle decel	55-15	35-10
0.5-0.6	0.3-0.4	Steady state	15	10
0.6	0.4	Moderate accel	15-40	10-25
0.6-0.8	0.4-0.5	Steady state	40	25
0.8	0.5	BRAKE	40-0	25-0
0.8	0.5	30 s IDLE	0	0
0.8	0.5	Lock-to-lock steering maneuver		

7.7 At the 0.8 km (0.5 mile) marker, brake to a stop off the roadway. Idle for 30 s in gear A-D or M-1. Record idle parameters.

7.7.1 If vehicle has power steering, at end of idle period, turn steering wheel to full left steering stop, then to full right steering stop, and return to center. Note any stalls or changes in engine RPM.

7.8 Cycle 2 is illustrated in Table 4 and Figure 3. Rate and record defects in these maneuvers as in 7.5. Idle 30 s in gear A-D or M-1 at end of Cycle 2. See Table 2 for definition of throttle setting.

7.8.1 CROWD ACCELERATION—Maintain a constant intake manifold absolute pressure by continually increasing the throttle opening with increasing engine speed.

7.8.2 INTERRUPTED ACCELERATION—Moderate throttle acceleration, followed immediately by a quick brake stop. Total vehicle travel must be 0.5 to 1.5 m (1.5 to 5.0 ft).

7.9 Repeat Cycle 1 (Step 7.6).

7.10 Repeat Cycle 2 (Step 7.8).

7.11 If the warm driving characteristics are to be evaluated, return to 7.5.

7.12 Shift the transmission into gear A-P or M-0 and shut off engine. Check for engine run-on. If the engine does run-on, place left foot on brake and shift transmission into gear A-D or M-5 (let clutch out). Note if run-on continues.

TABLE 4—SCHEDULE DRIVEABILITY—CYCLE 2

Distance km	Distance (mile)	Operation and Throttle Setting	Speed km/h	Speed (mph)
0.8-1.1	0.5-0.7	Crowd accel	0-70	0-45
1.1-1.4	0.7-0.9	Steady state	70	45
1.4	0.9	Closed-throttle decel	70-40	45-25
1.4	0.9	Heavy accel	40-55	25-35
1.4-1.6	0.9-1.0	Steady state	55	35
1.6	1.0	BRAKE	55-0	35-0
1.6	1.0	5 s IDLE	0	0
1.6	1.0	Interrupted accel	N/A	N/A
1.6-1.7	1.0-1.05	Moderate accel	0-40	0-25
1.7	1.05	BRAKE	40-0	25-0
1.7	1.05	30 s IDLE	0	0

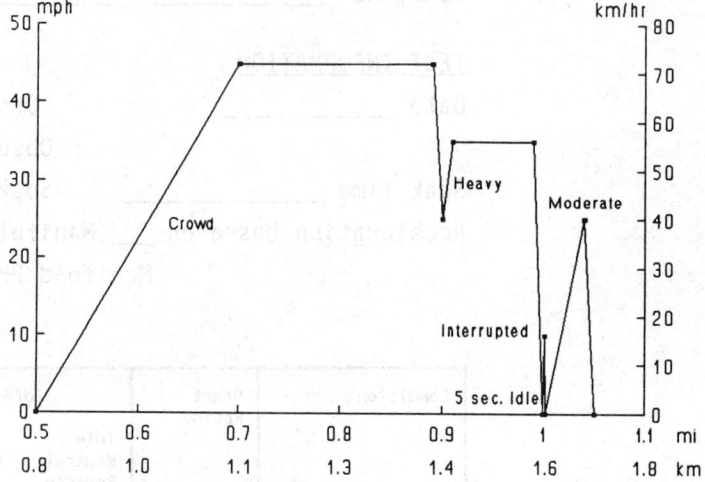

FIGURE 3—DRIVEABILITY SCHEDULE—CYCLE 2

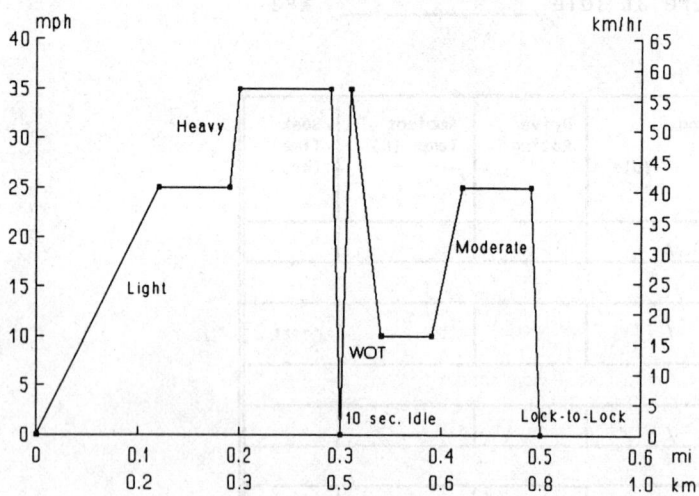

FIGURE 2—DRIVEABILITY SCHEDULE—CYCLE 1

APPENDIX A
RATING SYSTEM

A.1 The "SID" (Start, Idle, Drive) Rating System

A.1.1 A "1" to "9" (9 is the best) numerical rating system is used in rating the following vehicle functions under specific operating conditions:

A.1.1.1 FUNCTIONS
 a. Engine starting (S)
 b. Idle Quality Rating (I)
 c. Vehicle Driveability Rating (D)

A.1.1.2 CONDITIONS
 a. Cold Drive, including warm-up (CD)
 b. Warm Drive (WD)
 c. Hot Drive (HD)
 d. Partial Cooldown Drive (PCD)

A.1.2 Rating Index and Criteria—See Table A1.

TABLE A1—RATING INDEX AND CRITERIA

1	2	3	4	5	6	7	8	9
Unreliable	Unreliable	Lack of Confidence	Disturbing	Borderline	Marginal	Fair	Very Good	Excellent

A.1.3 Rating Calculation—Final Rating = 9 − (Rating Downgrades)

A.1.4 Rating Matrix

A.2 Determination of the START Rating

A.2.1 The definition of engine STARTING is found in the Section A.5.

A.2.2 Vehicle start ratings are based on the actual start time (seconds) of a gasoline-fueled engine under specific ambient temperature conditions. The following matrix is used to determine start ratings when a stall has not occurred. All start times will be to the closest whole second.

A.2.3 Cold Start Ratings—See Table A3.

TABLE A2—RATING MATRIX

Condition	Start Rating	Idle Rating Idle Neutral	Idle Rating Idle Drive	Idle Rating Idle Rev	Drive Rating	Ambient Temp (°C)	Soak Time (h)
Cold Drive		/	/				
Warm Drive	--------						------
Hot Drive		/	/				
Partial Cooldown		/	/				

NOTE—Warm drive is conducted immediately following the cold drive.

TABLE A3—COLD START RATING AT AMBIENT TEMPERATURE

Start Time (s)	Below -18 °C (0 °F)	At or Above -18 °C (0 °F)
1.0 or less	9	9
2.0	9	7
3.0	8	5
4.0	7	4
5.0	6	3
6.0	5	2
7.0	4	1
8.0	3	1
9.0	2	1
10.0 or more	1	1

A.2.3.1 HOT AND PARTIAL COOLDOWN START RATINGS—See Table A4.

TABLE A4—HOT AND PARTIAL COOLDOWN START RATINGS

Start Time	Start Rating
1.0 or less	9
2.0	7
3.0	5
4.0	3
5.0	2
6.0 or more	1

A.2.4 Start Rating Downgrades Due to Stall
A.2.4.1 1 STALL—2.0 point downgrade
A.2.4.2 2 STALLS—5.0 point downgrade
A.2.4.3 OVER 2 STALLS—Automatic rating of "1"
A.2.4.4 FALSE START—(Same as Stall)

A.3 Determination of the IDLE Rating

A.3.1 The definition of engine IDLE is found in the Section A.5.

A.3.2 Engine idle quality is rated subjectively in idle neutral (IN), idle drive (ID), and idle reverse (IR) shift lever positions. Idle ratings are determined using the criteria in Table A5.

A.3.3 IDLE Rating Downgrades due to Specific Defects (note gear in which stall occurs, such as A-D, A-R, or A-P):

A.3.4.1 1 STALL AT IDLE—Automatic "3" rating
A.3.4.2 ≥2 STALLS AT IDLE—Automatic "1" rating
A.3.4.3 IDLE FLUCTUATION
 a. Heavy—1.0 point downgrade
 b. Trace/Light—0.5 point downgrade
A.3.4.4 AFTER-RUN—2.0 point downgrade

A.4 Determination of the DRIVEABILITY Rating

A.4.1 The definition of specific DRIVEABILITY defect terms are found in Section A.5.

TABLE A5—IDLE RATINGS

Rating	Definition
9	Excellent idle quality, cannot feel engine running.
8	Engine operation smooth, flawless, barely perceptible.
7	Engine vibration noticeable, but unobjectionable.
6	Slight engine roughness, but speed remains relatively constant.
5	Moderately rough engine, irritating condition.
4	Disturbing engine roughness, but still confident of continual operation.
3	Uncertainty that engine will stay running; heavy roughness.
2	Frequent stalls, will not operate consistently.
1	Multiple stalls, uncontrolled operation, throttle manipulation required to keep running.

A.4.2 Vehicle driveability is a subjective "worst case" judgment of the vehicle's ability to perform under all styles of driving maneuvers. Specific driveability defects, if found during the driveability evaluation, will result in rating degradation (detract from "9" rating). Driveability ratings are determined using the criteria in Table A6:

TABLE A6—DRIVEABILITY RATINGS

Rating	Definition
9	Excellent driveability, no trace of defects, solid/responsive.
8	No noticeable defects, less responsive or flat performance.
7	One or more slight defects present; barely noticeable.
6	One or more defects present; very noticeable, not objectionable.
5	Obvious defects present; irritating, will probably generate complaints.
4	Disturbing defects present, but still confident of continual operation.
3	Undermines driver confidence, not reliable.
2	Failure to stay running, will not operate consistently.
1	Uncontrollable, unpredictable operation.

A.4.3 Drive Rating Downgrades Due to Specific Defects
A.4.3.1 EACH OCCURRENCE
 a. 1 driving Stall—Automatic "3" rating
 b. ≥2 driving Stalls—Automatic "1" rating
 c. Backfire—2.0 point downgrade
A.4.3.2 COLLECTIVE OCCURRENCES
 a. Surge
  (1) Heavy—1.0 point downgrade
  (2) Trace/Light—0.5 point downgrade
 b. Hesitation
  (1) Severe—1.0 point downgrade
  (2) Trace/Light—0.5 point downgrade
 c. Bucking—1.0 point downgrade
 d. Detonation/Spark Knock
  (1) Heavy—2.0 point downgrade
  (2) Trace—0.5 point downgrade
 e. Closed-Throttle Engine Braking (Automatic Transmission only)
  (1) Excessively impedes—2.0 point downgrade
  (2) Noticeably impedes—1.0 point downgrade

A.5 Glossary of Driveability Terms

A.5.1 Standardization of Driveability Terms and Definitions
A.5.2 Terminology for Driveability Defects
A.5.2.1 After-run
A.5.2.2 Backfire (Induction/Exhaust)
A.5.2.3 Bucking
A.5.2.4 Closed-Throttle Engine Braking
A.5.2.5 Detonation/Spark Knock
A.5.2.6 False Start
A.5.2.7 Harshness
A.5.2.8 Hesitation/Stumble
A.5.2.9 Idle Fluctuation
A.5.2.10 Idle Roughness
A.5.2.11 Stall
A.5.2.11 Surge
A.5.2.12 Vibration
A.5.3 Definition of Terms

Acceleration—The process of increasing vehicle speed (providing a positive change in velocity) at various constant accelerator pedal positions. Acceleration is measured by magnitude (force of gravity, g) and relative change over time.

Accessory-Induced Variations—Accessory-induced variations in engine operation can be generated by activation of various accessory drive systems, such as:
 a. Air conditioning compressor
 b. Power steering pump
 c. Alternator (defroster, headlamps, heated windshield/backlight)

After-run—Engine continuing to run after the ignition key is turned to "Off" position (also "run-on" or "dieseling").

Backfire—Explosion of the air/fuel mixture in either the exhaust system or induction system.

Bucking—A series of engine- and/or transmission-related torque fluctuations sufficiently abrupt to load and unload the drivetrain, causing a jerky fore/aft motion.

Closed-Throttle Deceleration—Reduction in vehicle speed induced by returning throttle pedal to idle/no load position.

Closed-Throttle Engine Braking—The ability of the engine/powertrain to smoothly and consistently decelerate a moving vehicle following accelerator pedal release.

Cold Start—Follows a soak period of a minimum of 8 h and a maximum of 36 h for a fully warmed-up vehicle.

Crowd Acceleration—Acceleration at a constant manifold absolute pressure.

Deceleration—The process of reducing vehicle speed (providing a negative change in velocity) at various constant accelerator pedal positions less than "road load" position, or a fully released accelerator pedal position. Deceleration is measured by magnitude (force of gravity, g) and relative change over time.

Detonation/Spark Knock—A metallic noise, such as a "pinging," "knocking," or "rattling" sound, caused by abnormal combustion.

False Start—The failure of the engine to "run up" or to generate adequate power during cranking to achieve self-sustaining operation.

Harshness—An unpleasant disturbance resulting in loss of smoothness in engine, vehicle, or transmission operation.

Heavy Tip-in (HTI)—Accelerator pedal depression greater than 1/2 but less than 7/8 travel.

Hesitation/Stumble—A momentary lack of acceleration during or following accelerator pedal depression. Hesitation can occur either before acceleration begins after the accelerator pedal has been opened, or in the rate of acceleration once it has begun.

Hot Start—Follows a soak period of a minimum of 5 min and a maximum of 40 min for a fully warmed-up vehicle.

Idle—An engine running condition with the accelerator pedal fully released and the vehicle stationary.

Idle Fluctuation—An idle quality condition characterized by excessive variations in engine speed.

Idle Roughness—A partial or continuous engine-related disturbance at idle resulting in loss of smooth operation. It may be characterized by engine misfire, steering wheel shake, body vibration, etc.

Light Tip-in (LTI)—Less than 1/4 accelerator pedal travel.

Lugging—The vehicle's inability to smoothly accelerate from a low engine speed, high load condition. Lugging is more likely on vehicles equipped with manual transmissions if upshifts to lower numerical gear ratios occur too early. Lugging can also occur on vehicles equipped with automatic transmissions, if they incorporate an aggressive "overdrive" upshift schedule.

Medium Tip-in (MTI)—Accelerator pedal depression greater than 1/4 but less than 1/2 travel.

Partial Cooldown Start—Follows a soak period of a minimum of 1.5 h and a maximum of 4 h for a fully warmed-up vehicle.

Part-Throttle Acceleration—Any acceleration at a constant accelerator pedal position less than wide open throttle position.

Road Load—Maintenance of a constant vehicle speed in any given gear on a smooth/level road.

Stall—Engine stoppage, other than intentional shutoff or clumsy clutch operation, experienced after self-sustaining engine operation has occurred.

Starting—The ability to bring the engine RPM up to a stable idle speed from an engine "off" condition. Start time is defined as the total elapsed crank time between starter motor first engagement and the point when the starter is disengaged because the engine has reached self-sustaining operation. The engine must be started in accordance with the Manufacturer's Recommended Starting Procedure.

Surge—An engine or vehicle condition consisting of periodic variations in velocity or rate of acceleration.

Vibration—Periodic motion of a system or component transmitted into the passenger compartment and sensed by the driver.

Warm Drive—Following cold start, after thermostat has opened and engine attains normal operating temperature.

Wide Open Throttle—Any acceleration at maximum accelerator pedal travel.

VEHICLE ACCELERATION MEASUREMENT—
SAE J1491 JUN90

SAE Recommended Practice

Report of the Light Duty Vehicle fuel Economy Accelerations Measurement Committee, approved June 1985. Completely revised by the Light Duty Vehicle Economy Acceleration Measurement Standards Committee June 1990.

1. Scope—To define a test procedure that when conducted will provide a repeatable measure of a vehicle's maximum acceleration performance.

1.1 Purpose—This SAE Recommended Practice provides a standardized means of measuring acceleration performance of passenger cars and light duty trucks.

2. References
No ISO comparable.

2.1 Definition

2.1.1 UNLOADED VEHICLE WEIGHT—The weight of the vehicle as built with production parts with maximum capacity of all fluids necessary for operation of the vehicle.

3. Instrumentation—(All instrumentation must be calibrated.)

3.1 Speed-Time—An instrument to measure vehicle speed as a function of elapsed time is used in this procedure. The device must meet the following specifications:
   a. Time:
      (1) Accuracy ±0.1 s
      (2) Resolution 0.1 s
   b. Vehicle Speed:
      (1) Accuracy ±0.50 mph (0.8 km/h)
      (2) Resolution 0.25 mph (0.4 km/h)
   c. Engine Speed (tachometer):
      (1) Accuracy ±50 rpm
      (2) Resolution 25 rpm

3.2 Temperature—The ambient temperature indicating devices must have a resolution of 2°F or 1°C and an accuracy of ±2°F or ±1°C. The sensing elements must be shielded from radiant heat sources.

3.3 Atmospheric Pressure—A barometer with an accuracy of ±0.2 in Hg or 0.7 kPa.

3.4 Wind—Wind speed and direction during the test should be continuously monitored. Wind measurements should permit the determination of average longitudinal and crosswind components to within ±1 mph (±2 km/h).

3.5 Vehicle Weight—Vehicle weight should be measured to an accuracy of ±10 lb (±5 kg) per axle.

3.6 Tire Pressure—Should be measured to an accuracy of ±1 psi (±7 kPa).

3.7 Distance—A distance indicating device is required. This device must be capable of indicating distance to within 1 ft and must be capable of accuracy within 5 ft in 1 mile.

4. Test Material—(See Figure 1)

4.1 Test Vehicle—The test vehicle shall be completely defined on the test vehicle specifications and preparation list. The test vehicle will normally be representative of a standard production built vehicle; any optional or nonstandard equipment must be noted (i.e., roof racks, optional mirrors, fog lamps, spoilers, optional axle ratio, etc.). Record any equipment that is removed for test.

4.2 Test Fuel—Commercially available fuel as recommended by the manufacturer will normally be used for test purposes. If the information is available or if a special test fuel is used, the fuel specifications should be recorded, such as—fuel generic type, gasoline octane rating or diesel cetane rating, brand name, specific gravity, Reid vapor pressure.

4.3 Lubricants—Lubricants used shall conform to the manufacturer's recommendation for the predominant weather condition in which the vehicle is being tested.

5. Test Conditions

5.1 Ambient Temperature—The test should be conducted at ambient temperatures between 30 and 90°F (−1 and 32°C).

5.2 Adverse Weather Conditions—The tests may not be run during foggy, rainy, or snowy conditions.

5.3 Wind Velocity—The tests may not be conducted when wind speeds average more than 15 mph (24 km/h) (or when peak wind speeds are more than 20 mph (32 km/h)).

5.4 Road Conditions—The roads must be dry, clean, smooth, and must not exceed 0.5% grade. In addition, the grade should be constant and the road should be straight. The road surface should be concrete or rolled asphalt (or equivalent) and in good condition; testing should not be conducted on slippery roads.

5.5 Speed Limitation—These tests should be run on a controlled track or proving grounds. If run on public roads or highways, speed should not exceed posted speed limit, and vehicle should not interfere with traffic flow or otherwise operate in a manner that would be hazardous.

6. Vehicle Preparation

6.1 Break-In—The vehicle should have at least 2000 miles of operation before test. Tires must have at least 75% of the tread remaining and tread must be in good condition. All tires must have at least 100 miles of run in before test.

6.2 Vehicle Check List

6.2.1 The vehicle must be inspected and adjusted where necessary to meet manufacturer's specifications, particularly if vehicle is exhibiting abnormal performance characteristics during acceleration. Tune and time engine, and make all other adjustments, such as front end alignment, and functional checks in accordance with manufacturer's published procedures.

6.2.2 Operate, observe, and reset, if necessary, the throttle linkage to ensure wide open throttle occurs.

6.2.3 If the vehicle is equipped with automatic transmission, ensure that automatic transmission shift points are within manufacturer's published specifications.

6.2.4 Ensure that brake drag is not excessive.

6.3 Instrumentation—The speed time measurement device and other necessary test equipment must be installed so that they do not hinder vehicle operation or alter the operating characteristics of the vehicle.

6.4 Test Weight—The unloaded vehicle weight +300 lb (includes driver and all instrumentation) and the fifth wheel in the raised position.

6.5 Tire Pressure—The cold tire pressure should be the standard recommended by the manufacturer for the vehicle test weight and installed tires.

6.6 Vehicle Warm-Up—The vehicle must be driven a minimum of 20 miles at an average speed of 55 mph ± 5 (88 km/h) immediately prior to the test. Alternative schedules that provide equivalent vehicle warm-up can be substituted. There should not be more than a 5 min time lapse between the warm-up and the start of test.

6.7 Vehicle Data—Record all information as specified on the attached Vehicle Specification Sheet. (See Figure 1).

7. Test Procedure

7.1 Test Schedules

7.1.1 Perform wide open throttle (WOT) accelerations from a standing start and record the following:
   a. 0 to 30 mph—Record Elapsed Time
   b. 0 to 50 mph—Record Elapsed Time
   c. 0 to 60 mph—Record Elapsed Time
   d. 1/4 mile—Record Elapsed Time and Terminal Speed
   e. 0 to 5 s—Record Distance Covered and Terminal Speed

7.1.2 Also perform the following test at wide open throttle:
   a. 40 to 60 mph: Record Elapsed Time

7.2 Automatic Transmission Operating Procedure—From a standing start with engine at idle (braked if necessary), with the shift selector in the "drive" position, accelerate with wide open throttle. The vehicle should be operated to achieve maximum performance with minimum wheel spin. Time zero starts at the instant the driver's foot moves the accelerator pedal.

7.3 Manual Transmission Operating Procedure—From a standing start, the vehicle should be operated to achieve maximum performance with minimum wheel spin. Clutch operation, as well as shift point selection, should be optimized for performance without exceeding the maximum specified engine rpm. Time zero starts at the instant of clutch pedal movement.

7.4 40 to 60 Test Procedure—Starting from a stabilized 40 mph, accelerate with wide open throttle to 60 mph. Manual transmissions should be run both in top gear and top gear less one, with 4- or 5-speed transmissions. Three speed, manual transmission should be run in top gear only. Manual transmissions should not be downshifted during this test.

Automatic transmissions will be allowed to downshift as determined by the vehicle transmission controls.

7.5 Test Data

7.5.1 Run a minimum of six individual runs, three in each direction. When difficulty is experienced in one run, the pair is excluded.

7.5.2 Record all data specified on the Vehicle Data Sheet. (See Figure 2).

7.6 Operation of Accessories

7.6.1 The headlamps are to be off. If the vehicle is equipped with pop-up lamps, the lamp pods should be in the down position. The lights should be on if required for safe vehicle operation, and so noted under remarks on the Vehicle Data Sheet. (See Figure 2).

7.6.2 The heater blower motor shall be used in the "low" position only.

7.6.3 The vehicles equipped with air-conditioning should have the compressor clutch wire disconnected before the start of test.

7.6.4 Radio operation is optional.

7.6.5 All other electrical accessories must be in the off position.

7.6.6 Windows should remain closed during test runs.

8. Data Reduction

8.1 Data Calculation—Simple averages will be calculated for all valid multiple test observations (pairs of data).

8.2 Data Presentation—Data should be presented in accordance with the Vehicle Data Sheet. (See Figure 2). Alternatively, continuous plots may be charted as follows: speed vs. time, time vs. distance, or other data considered appropriate.

8.3 General

8.3.1 DATA VARIABILITY—Because of unpredictable effects of wind on vehicle performance, the following guideline is suggested as a criterion for test acceptability.

The coefficient of variation (standard deviation of the paired runs)/mean of individual runs should not be greater than 3%. On the 40 to 60 mph acceleration, the coefficient of variation should not exceed 6%.

8.3.2 WEATHER CORRECTION—No provision for weather correction is made in this procedure. Variables such as temperature, humidity, barometric pressure, wind speed, and direction should be considered by the tester in evaluating the test results.

LEVEL ROAD W.O.T. ACCELERATION PERFORMANCE
VEHICLE SPECIFICATION SHEET

Report No. _____
Date _____

Vehicle: Make_____ Model_____ Year_____ Car No._____
Odometer_____ VIN_____
Test Weight_____lb RF_____lb LF_____lb RR_____lb LR_____lb

EQUIPPED WITH

Engine:
  Type_____ Displacement_____ Compression Ratio_____
  No. of Cylinders_____ Rated SAE Horsepower_____
  Fuel System_____
  Engine Fan Type_____
  Electronic Engine Control Yes____ No____ Knock Sensor Yes____ No____
  Idle Speed_____rpm (Drive) _____rpm (Neutral)
  Redline_____rpm Initial Timing_____
  Boost_____ Number of Valves_____

Transmission:
  Type_____
  Forward Ratios_____

Axle Ratio:
  Front Wheel Drive[1]_____
  Rear Wheel Drive[2]_____
  [1]Overall Top Gear Ratio = Transfer Drive Ratio X Final Reduction Gear Ratio
  [2]Rear Wheel Drive = Rear Axle Ratio

Tires:
  Manufacturer_____ Model_____
  Type_____ Size_____
  Pressure (Cold) Front_____psi Rear_____psi

Brake Type: Front_____ Rear_____
Exhaust System
Type: _____

Operational Checklist:
Choke Operation_____
Throttle Operation_____
Transmission Operation_____

Brake Drag_____
Parking Brake_____
Wheel Alignment_____

Fluid Level Checklist:
Engine Oil_____
Battery_____
Transmission_____
Coolant_____
Brake_____
Differential_____
Power Steering_____

Test Fuel Specifications:
Test Fuel Type and Grade_____
Gravity (API or Specific)_____ @60°F (15.6°C)
Reid Vapor Pressure_____psi (kPa)
Distillation 10%_____ °F (°C)
            50%_____ °F (°C)
            90%_____ °F (°C)
Test Fuel Octane No. RON_____ MON_____
            $\frac{R + M}{2}$
Test Fuel Cetane No._____
Test Fuel Viscosity_____

Optional Equipment: _____

Equipment Removed
for Test: _____

Notes: _____

FIG. 1—VEHICLE SPECIFICATION SHEET

26.505

LEVEL ROAD W.O.T. ACCELERATION PERFORMANCE
VEHICLE DATA SHEET

Report No._____
Date_____

Vehicle: Make_____ Model_____ Year_____ Car No._____
Odometer_____

Test Location_____ Track Orientation_____
Start of Test: Date_____ Time_____
Completion of Test: Date_____ Time_____
Driver_____
Transmission: Automatic Shift Mode_____ Manual Launch rpm_____
Shift rpm (1-2)_____ (2-3)_____ (3-4)_____
Remarks:_____

Ambient Conditions for Test:
Temp._____°F _____°C   Barometric Pressure_____in Hg (kPa)   Relative Humidity_____%[a]
Wind Velocity_____mph   Direction_____   Peak Wind Velocity_____mph

0-30 mph (Elapsed Time)[b]
1_____ 2_____ 3_____ 4_____ 5_____ 6_____   Avg._____s   Variability[c]_____%

0-50 mph (Elapsed Time)[b]
1_____ 2_____ 3_____ 4_____ 5_____ 6_____   Avg._____s   Variability[c]_____%

0-60 mph (Elapsed Time)[b]
1_____ 2_____ 3_____ 4_____ 5_____ 6_____   Avg._____s   Variability[c]_____%

40-60 mph (Elapsed Time) Top Gear[b]
1_____ 2_____ 3_____ 4_____ 5_____ 6_____   Avg._____s   Variability[c]_____%

40-60 mph (Elapsed Time) Top Gear—Less One[b]
1_____ 2_____ 3_____ 4_____ 5_____ 6_____   Avg._____s   Variability[c]_____%

0-5 s (Distance Covered)[b]
1_____ 2_____ 3_____ 4_____ 5_____ 6_____   Avg._____ft   Variability[c]_____%

0-5 s (Terminal Speed)[b]
1_____ 2_____ 3_____ 4_____ 5_____ 6_____   Avg._____mph   Variability[c]_____%

1/4 mile (Elapsed Time)[b]
1_____ 2_____ 3_____ 4_____ 5_____ 6_____   Avg._____s   Variability[c]_____%

1/4 mile (Terminal Speed)[b]
1_____ 2_____ 3_____ 4_____ 5_____ 6_____   Avg._____mph   Variability[c]_____%

[a]Record, if available.
[b]Tests must be conducted in alternate directions then collectively averaged. Two valid paired runs are considered adequate. When difficulty is experienced in one run, the pair should be excluded.
[c]Variability equals standard deviation of tests divided by test average multiplied by 100.

FIG. 2—VEHICLE DATA SHEET

FUEL ECONOMY MEASUREMENT—ROAD TEST PROCEDURE—COLD START AND WARM-UP FUEL ECONOMY—SAE J1256 OCT88

SAE Recommended Practice

Report of the Passenger Car and Light Duty Truck Fuel Economy Measurement Committee, approved May 1979, completely revised by the Light Duty Vehicle Fuel Economy Acceleration Measurement Standards Committee October 1988.

1. Purpose—This recommended practice provides a uniform test procedure for measuring the fuel economy of light duty vehicles during cold start and warm-up operation (motor vehicles designed primarily for transportation of persons or property and rated at 10 000 lb GVW (4500 kg) or less on suitable roads.

2. Scope—This procedure is a modification of the urban driving cycles noted in SAE J1082 and which is run on a suitable road or test track. The procedure yields cold start/warm-up fuel economy values indicative of consumer level at the ambient condition of the test. Within referenced limitations, the procedure can be utilized to determine the fuel economy differential among vehicles or between vehicle changes.

3. Definitions

3.1 Driving Cycle—The urban driving pattern is defined by paragraph 8.3.4.2. This driving pattern is a modification for cold starting of the SAE J1082 urban driving cycle.[1]

3.2 Test Vehicle—Passenger car or light truck prepared for test according to Section 7.

3.3 Test Vehicle Weight—Unloaded vehicle weight (curb weight) plus 300 lb (136 kg).

3.3.1 CHASSIS DYNAMOMETER INERTIA WEIGHT AND HORSE POWER SETTINGS—Should be established and set in accordance with SAE J1263, Road Load Measurement and Dynamometer Simulation Using Coastdown Techniques.

3.4 Observed Cold Fuel Economy—Observed cold fuel economy will be reported for both 0-2 mile (0-3.2 km) and 0-4 mile (0-6.4 km) cycles. The 0-2 mile (0-3.2 km) observed fuel economy will be determined by dividing the 2.0 mile (3.2 km) of the urban cycle by the fuel consumed during Cycle 1. The 0-4 mile (0-6.4 km) observed fuel economy will be determined by dividing 4.0 mile (6.4 km) by the total fuel consumed during Cycle 1 and Cycle 2. Economy will be expressed in miles per gallon (kilometers per liter).

3.5 Corrected Cold Fuel Economy—Corrected economy is the observed economy multiplied by the correction factors listed in Sections 11 and 12. Corrected fuel economy will be expressed as miles per gallon (kilometers per liter).

3.6 Correction Factors—Factors which are used to adjust data to the standard ambient condition of 60°F (15.6°C) and reference fuel properties.

3.7 Average Fuel Economy—Average fuel economy is the total distance driven divided by the total volume of fuel consumed in a series of replicate tests. When the distance driven in each of the tests is identical, as may be assumed for this procedure, the average fuel economy is determined by taking the harmonic average of the individual economies.

$$\text{Average Fuel Economy} = \frac{n}{1/MPG_1 + 1/MPG_2 + 1/MPG_n}$$

3.8 Test Repeatability Guidelines—These guidelines are intended to provide an estimate of repeatability of test data for replicate tests (assumes a standard deviation equal to 1.9% of the mean).

3.8.1 ESTIMATE OF THE 95TH PERCENTILE RANGE FOR REPLICATE TESTS—The 95th percentile range (R) equals 0.019Q times the average fuel economy, where Q equals the critical value obtained from a table for the student t range and the average fuel economy for n tests.

Selected values for 0.019Q are:

n	0.019Q
2	0.053
3	0.063
4	0.069
5	0.073
10	0.085

Example 1 (U.S. units) - If a vehicle obtains 14.5 mile/gal and 15.5 mile/gal on two tests on the same cycle, the corrected average fuel economy would be 15.0 mile/gal and the 95th percentile range would be:

$$R = 0.053 \times 15.0 = 0.80 \text{ mile/gal}$$

The difference between the two tests is 1.0 mile/gal which is greater than the difference that would be expected for 95% of the cases in which two tests were conducted. Consequently, additional tests should be conducted to provide more confidence in the average fuel economy.

Example 2 (SI units) - If a vehicle obtains 6.20 km/L and 6.60 km/L on two tests on the same cycle, the corrected average fuel economy would be 6.39 km/L and the 95th percentile range would be:

$$R = 0.053 \times 6.39 = 0.34 \text{ km/L}$$

The difference between the two tests is 0.4 km/L which is greater than the difference that would be expected for 95% of the cases in which two tests were conducted. Consequently, additional tests should be conducted to provide more confidence in the average fuel economy.

3.8.2 Estimate of the average fuel economy at a 90% confidence interval.

$$\text{Average at 90\% Confidence Interval} = \text{Average Fuel Economy} \pm \left[\frac{0.031}{\sqrt{n}} \times (\text{Average Fuel Economy})\right]$$

n is the number of replicate tests.

Example 1 (U.S. units) - If a vehicle obtained 14.8 mile/gal and 15.2 mile/gal on two tests on the same cycle, the average fuel economy would be 15.0 and the 90% confidence interval would be:

$$15.0 \pm \left[\frac{0.031}{\sqrt{2}} \times 15.0\right] = 15.0 \pm 0.3 \text{ mile/gal}$$

Example 2 (SI units) - If a vehicle obtained 6.29 km/L and 6.46 km/L on two tests on the same cycle, the average fuel economy would be 6.37 km/L and the 90% confidence interval would be:

$$6.37 \pm \left[\frac{0.031}{\sqrt{2}} \times 6.37\right] = 6.37 \pm 0.14 \text{ km/L}$$

4. Instrumentation

All instrumentation shall be calibrated.

4.1 Fuel—The fuel measurement device must be compatible with the vehicle fuel system and should alter the fuel temperature and pressure as little as practical. The fuel measurement system must be accurate to within 0.5% of the fuel used during a driving cycle.

4.2 Speed—The speed indicating device shall indicate vehicle speed in miles per hour (kilometers per hour) and be accurate within 0.5 mile/h (1 km/h).

4.3 Acceleration—The acceleration indicating device must be capable of indicating both positive and negative acceleration. It shall indicate acceleration/deceleration in ft/s^2 (m/s^2) and be accurate within 0.5 ft/s^2 (0.2 m/s^2). (Refer to paragraph 6.5 for Chassis Dynamometer Testing.)

4.4 Time—The time measuring instrument must be capable of measuring the time interval to 0.1 s and be accurate within 0.1 s in 1 min.

4.5 Temperature—The temperature indicating devices must be capable of measuring to the nearest 2°F or 1°C. Accuracy must be within ±2°F or ±1°C. The sensing element shall be shielded from radiant heat sources.

4.6 Absolute Barometric Pressure—An aneroid or mercury barometer should be used. This device should be accurate within 0.1 in Hg or 0.3 kPa.

---

[1] Though this test cycle can be run on a chassis dynamometer, it does not apply to mandatory fuel economy standards or fuel economy labeling requirements for passenger vehicles first established by the "Energy Policy and Conservation Act", Public Law 94-163, 94th Congress S.622, December 22, 1975. Details of the mandatory dynamometer procedure can be obtained by contacting Environmental Protection Agency, Fourth and M Streets, S.W., Washington, DC 20460. It should be noted that correlation of this test cycle between chassis dynamometers and road test results has not been established.

4.7 Wind—Wind speed should be measured with a device that provides an indication of wind speed that is accurate within 2 mile/h (3 km/h). Wind direction should also be indicated.

4.8 Distance—A distance indicating device is required if the tests are not conducted on a premarked course. This device must be capable of indicating distance to within 15 ft (5 m) and must be capable of accuracy within 30 ft in 1 mile (6 m in 1 km).

4.9 Vehicle Weight—Vehicle weight should be measured with a device that is accurate within 0.5% with a resolution of 10 lb (5 kg) or less.

4.10 Dynamometer Inertia Weight—The dynamometer inertia weight is established reflecting the inertia of the non-rotating tires and the vehicle test weight. The inertia weight should be set to the nearest flywheel increment for mechanical inertia dynamometers, or within 10 lb for electrical inertia dynamometers.

5. Test Material

5.1 Test Vehicle—The test vehicle shall be completely defined on the Test Vehicle Specifications and Preparation Form. (The test vehicle will normally be representative of a production built vehicle - any exceptions must be properly noted.)

5.2 Test Fuel—Normally, service station fuel will be satisfactory for test purposes, provided that it is consistent with the manufacturer's recommendations for the vehicle and with the ASTM D 439 Volatility Guidelines. Gasoline octane rating $\left(\frac{R+M}{2}\right)$, specific gravity or API gravity for diesel fuel or gasoline used shall be recorded. Other properties such as distillation and Reid vapor pressure should be recorded when available.

5.3 Lubricants—Lubricants used shall conform to the manufacturer's recommendation for the predominant weather condition in which the vehicle is being tested.

6. Test Conditions

6.1 Ambient Temperature—Tests and soaks should be conducted at ambient temperatures between 30 and 90°F (−1 and 32°C). For temperatures below 30°F (−1°C), the temperature correction factors may not be valid.

6.2 Wind Speed—Tests must not be conducted when average wind speed exceeds 15 mile/h (24 km/h) or when gusts exceed 20 mile/h (32 km/h). During the soak period, higher wind speed is allowable.

6.3 Road Conditions—Roads must be dry, clean, smooth, and not exceed 1.0% grade. If operating on a closed track, the start and stop points should be selected such that the schedule elevation difference is 10 ft (3 m) or less.

6.4 It is recommended that roadside markers be used to indicate the points at which speed changes are to be made as indicated in paragraph 8.3.

6.5 Driver's Aid—A driver's aid is recommended for dynamometer operation reflecting the schedule in paragraph 8.3.

7. Test Vehicle Preparation

7.1 Break-In—The vehicle should have accumulated a minimum of 2000 miles (3200 km) of operation prior to test. At least 1000 miles (1600 km) must have been driven at cycling speeds between 40 mile/h (64 km/h) and maximum legal highway speeds. If a closed track is available for break-in, the maximum speed should not exceed 100 mile/h (160 km/h). Chassis dynamometer break-in is acceptable. All of the tires must have operated at least 100 road or track miles (160 km) prior to the test. Tires must have at least 75% of the tread remaining and tread must be in good condition. Vehicles for dynamometer testing should have experienced at least 500 miles of road cyclic operation.

7.2 Inspection—The vehicle must be inspected and adjusted where necessary to meet manufacturer's specifications. Checks are specified on the Test Vehicle Specifications and Preparation Form.

7.3 Instrumentation—The fuel measuring device, temperature indicating devices, and other instrumentation, as necessary, must be installed in a manner not to hinder the vehicle operation or operating characteristics.

7.4 Test Weight—The vehicle weight must be adjusted to provide the test weight indicated in paragraph 3.3 (this test weight includes instrumentation and operator).

7.5 Tire Pressure—The cold tire pressure should be the minimum recommended by the manufacturer for the vehicle test weight.

8. Test Procedure

8.1 Preconditioning—The vehicle must be driven a minimum of 20 miles (32 km) at 55 mile/h (90 km/h) or maximum legal highway speed to stabilize engine and driveline operating temperatures immediately prior to the cold soak.

8.1.1 COLD SOAK—The vehicle should be parked close enough to the start of the test course to fulfill the requirements of paragraph 8.3.4 and exposed to ambient weather for a period of not less than 12 h nor more than 24 h prior to the beginning of a cold urban test. The temperature of the engine oil, engine block coolant, and/or ambient air is to be recorded just prior to the start of the test. In areas of rapidly changing ambient temperature, all three temperatures should be recorded and should not vary by more than 10°F (6°C) for a valid test.

8.1.2 START PROCEDURES - SPARK IGNITION ENGINES

8.1.2.1 *Manufacturer's Recommended Procedure*—Where the manufacturer has specified a vehicle starting procedure, that procedure should be followed.

8.1.2.2 Where no recommended starting procedure is specified, the following procedure is recommended. Deviations or other starting procedures should be documented.

8.1.2.2.1 Manual Choke Equipped Vehicles—Vehicles equipped with manual chokes will be started by slowly depressing the accelerator pedal fully to the floor while simultaneously adjusting the choke to the fully closed position. Release the accelerator pedal and start the engine. When the engine is running, release the choke only to the point where the engine continues to run smoothly. The choke should continue to be released during the driving cycle as rapidly as possible while maintaining smooth operation.

8.1.2.2.2 Automatic Choke Equipped Vehicles—Vehicles equipped with automatic chokes will be started by turning the key to the ON position, then fully depressing the accelerator and releasing, and starting the vehicle. After the vehicle has been moved to the starting position, tap the throttle lightly to reduce the idle speed.

8.1.2.2.3 Fuel Injection Equipped Vehicles—Vehicles equipped with an FI system will be started by turning the key to the ON position, then starting the vehicle. No effort to reduce idle speed should be made.

8.1.3 START PROCEDURES - DIESEL ENGINES

8.1.3.1 *Manufacturer's Recommended Procedure*—Where the manufacturer has specified a vehicle starting procedure, that procedure should be followed.

8.1.3.2 Where no starting procedures are recommended, the starting procedure used should be documented.

8.1.3.2.1 Vehicles Equipped with Glow Plug Lights or Other Preheat Warning Devices—Glow plugs are normally preheated via a separate ignition switch position or a separate actuation switch. Actuate the glow plug. When the glow plug light goes out or otherwise indicates the preheat cycle is completed, depress the accelerator pedal to at least halfway to the floor and hold, and turn the ignition key to the start position. Release the ignition switch and throttle pedal when the engine has reached a smooth running condition.

8.1.3.2.2 Vehicles NOT Equipped with Glow Plug Lights or Other Preheat Warning Devices—Glow plug preheaters are required to start automotive type diesel engines. Sixty-second preheats will generally be satisfactory at 30°F (−1°C). Fifteen seconds of preheat will be sufficient at 90°F (32°C) ambient. After the appropriate preheat cycle, the starting procedure will be followed as above.

8.2 Vehicle Controls—Air conditioning compressor, headlamps, and other switched accessories that consume power should be turned off unless required for safe vehicle operation. Air conditioning or heater/defroster operation required to clear windows should be turned off as soon as safe vehicle operation is possible. Previous to the preconditioning cycle, the vehicle's battery should be in a fully charged state.

8.3 Driving Schedules

8.3.1 GENERAL DRIVING INSTRUCTIONS

8.3.1.1 Vehicles incapable of attaining acceleration rates specified by the driving schedules will be driven at maximum acceleration until specified schedule speed is reached.

8.3.1.2 Vehicles with automatic transmissions should be driven with the transmission in a range that insures all forward gears can be automatically engaged. If transmission hunting is encountered at a specified acceleration, the acceleration should be increased to maintain the transmission in the lower gear and this departure from the schedule noted on the data form.

8.3.1.3 Vehicles equipped with manual transmissions will be operated in the following manner: Idles will be made in gear, clutch disengaged. Decelerations will be made in gear, and the clutch will be disengaged at 15 mile/h (24 km/h) on a stop. All cruise operation should be in the highest gear that will prevent engine lugging. Downshifts will be permitted to obtain specified acceleration rates after a deceleration or to obtain a smooth engine operation at a slow speed. The following manual transmission shift speeds are guidelines only and may be modified up or down as necessary to ensure that the specified acceleration rates are attained and to avoid engine lugging or overspeed. Departure from the shift speeds specified below should be noted on the data form. Manufacturer's recommended shift speeds/shift indicator lights may be used providing their use is noted on the data form.

8.3.1.4 Vehicle with truck-type manual transmissions containing a creeper gear will not use the creeper gear during the driving cycle.

8.3.1.5 Vehicles with manual transmissions will be shifted during accelerations at the specified speed (mile/h) shown below. Manufacturer's alternative shift schedules or shift indicator light shifting can be used if the specified driving schedule can be maintained.

ACCELERATION UPSHIFTS	TRANSMISSION TYPE		
	3	4	5
1 – 2	15	15	15
2 – 3	25	25	25

8.3.1.6 During cruise modes, use the highest possible gear consistent with smooth engine operation.

8.3.1.7 On vehicles with an automatic transmission, brakes should be applied to maintain the schedule speed if the engine idle results in vehicle speed above that specified. For manual transmission vehicles, the transmission should be downshifted.

8.3.2 GENERAL CYCLE INSTRUCTIONS

8.3.2.1 The cold start cycle follows the same driving cycle as the J1082 urban cycle with special instructions for starting and movement prior to the start of the test. It should be noted on the test data sheet if defroster, heater, or air conditioning usage were required for safety; otherwise, they should not be used during this test.

8.3.2.2 The cold start tests are conducted as a sequence of 2 urban cycles of 2.0 mile (3.2 km) each. The 0-2.0 mile (0-3.2 km) test is Cycle 1 and the 2.0-4.0 mile (3.2-6.4 km) test is Cycle 2. Fuel economy is reported for both the 0-2.0 mile (0-3.2 km) Cycle 1 and the 0-4.0 mile (0-6.4 km) Cycle 1 plus Cycle 2 test distances. (Additional cycles may be run to meet the warmed-up urban cycle requirements of SAE J1082.)

8.3.2.3 The urban acceleration and deceleration should be maintained within 1 ft/s^2 (0.3 m/s^2) of that specified. Vehicle speeds should be maintained within 1 mile/h (1.6 km/h).

8.3.2.4 Driving cycle maneuvers are initiated at the points indicated except for the stop at the end of the urban cycle, which is to be completed by the point indicated.

8.3.2.5 Fuel temperature will be recorded on the data form during all idle periods.

8.3.2.6 Record weather data for each test cycle.

8.3.2.7 At least two cold start tests will be conducted, each preceded by a 12-24 h soak period. Record the number of stalls which occur during the test and the total fuel used, including starts. Total elasped time for each test, with time lost due to starting and stalls subtracted, must repeat within 1%. Additional tests are required until this criteria is met before calculating fuel economy. Observed cold fuel economy will be calculated for each test in accordance with paragraph 3.4.

8.3.3 GENERAL CYCLE SUMMARY TABLE—This cycle summary is exclusive of start, stall, and restart time or distance required to move the vehicle from the cold soak area to the driving course.

Cycle	Average Speed		Test Time	Test Distance		Idle Time	Stops
	mile/h	(km/h)	s	mile	(km)	s	
Cold Start Cycle 0-2 mile (0-3.2 km)	15.6	(25.1)	463	2.0	(3.22)	60	8
Cold Start Cycle 0-4 mile (0-6.4 km)	15.6	(25.1)	926	4.0	(6.44)	120	16

8.3.4 COLD START DRIVING CYCLE

8.3.4.1 *Pre-Starting Instructions*—Fuel consumption for the Cycle 1 test will include all fuel used to start the vehicle and run the cycle including that fuel used for the carburetor bowl fill, if required, but excluding fuel line fill. Immediately prior to starting the vehicle, fill the fuel line and filter but not the carburetor float bowl. One method of accomplishing this would require the installation of a three-way valve as close to the carburetor as possible. A purge line, with a clear section of hose in it, could then be tapped off and an observer could tell when the fuel line was void of air. The fuel lines would then be filled through the three-way valve.

Fuel injected vehicles will require specialized fuel measurement instrumentation and should be addressed on an individual basis.

Note that this driving cycle commences with a 15-s idle. If it is necessary to move the vehicle from a cold soak parking area to the driving course, it must be accomplished during the initial 15-s idle period and while the engine is operating on fast idle. As an alternative, the test vehicle may be pushed from a cold soak area to the start of the driving course. This movement, however, should not be greater than the distance that the vehicle would travel under its own power as described above.

8.3.4.2 *Driving Cycle (0-2 mile (0-3.2 km) Cycle)*

Distance		Operation
mi	(km)	
0.0	(0.0)	Start fuel meter, timing device and engine, idle 15 s moving vehicle to the "0" mile marker if needed, accelerate to 15 mile/h (24 km/h) at 7 ft/s^2 (2.1 m/s^2). Proceed at 15 mile/h (24 km/h) to the 0.2 mile (0.32 km) marker.
0.2	(0.32)	Stop at 4 ft/s^2 (1.2 m/s^2), accelerate to 15 mile/h (24km/h) at 7 ft/s^2 (2.1 m/s^2). Proceed at 15 mile/h (24km/h) to the 0.3 mile (0.48 km) marker.
0.3	(0.48)	Decelerate to 5 mile/h (8 km/h) at 4 ft/s^2 (1.2 m/s^2), accelerate to 15 mile/h (24 km/h) at 7 ft/s^2 (2.1 m/s^2). Proceed at 15 mile/h (24 km/h) to the 0.5 mile (0.80 km) marker.
0.5	(0.80)	Stop at 4 ft/s^2 (1.2 m/s^2), record fuel temperature and idle 15 s, accelerate to 20 mile/h (32 km/h) at 7 ft/s^2 (2.1 m/s^2). Proceed at 20 mile/h (32 km/h) to the 0.7 mile (1.13 km) marker.
0.7	(1.13)	Stop at 4 ft/s^2 (1.2 m/s^2), accelerate to 20 mile/h (32 km/h) at 7 ft/s^2 (2.1 m/s^2). Proceed at 20 mile/h (32 km/h) to the 0.8 mile (1.29 km) marker.
0.8	(1.29)	Decelerate to 10 mile/h (16 km/h) at 4 ft/s^2 (1.2 m/s^2), accelerate to 20 mile/h (32 km/h) at 5 ft/s^2 (1.5 m/s^2). Proceed at 20 mile/h (32 km/h) to the 1.0 mile (1.61 km) marker.
1.0	(1.61)	Stop at 4 ft/s^2 (1.2 m/s^2), record fuel temperature and idle 15 s, accelerate to 15 mile/h (24 km/h) at 7 ft/s^2 (2.1 m/s^2), then to 25 mile/h (40 km/h) at 5 ft/s^2 (1.5 m/s^2). Proceed at 25 mile/h (40 km/h) to the 1.2 mile (1.93 km) marker.
1.2	(1.93)	Stop at 4 ft/s^2 (1.2 m/s^2), accelerate to 15 mile/h (24 km/h) at 7 ft/s^2 (2.1 m/s^2), then to 25 mile/h (40 km/h) at 5 ft/s^2 (1.5 m/s^2). Proceed at 25 mile/h (40 km/h) to the 1.3 mile (2.09 km) marker.
1.3	(2.09)	Decelerate to 15 mile/h (24 km/h) at 4 ft/s^2 (1.2 m/s^2), accelerate to 25 mile/h (40 km/h) at 5 ft/s^2 (1.5 m/s^2). Proceed at 25 mile/h (40 km/h) to the 1.5 mile (2.41 km) marker.
1.5	(2.41)	Stop at 4 ft/s^2 (1.2 m/s^2), record fuel temperature and idle 15 s, accelerate to 15 mile/h (24 km/h) at 7 ft/s^2 (2.1 m/s^2), then to 30 mile/h (48 km/h) at 5 ft/s^2 (1.5 m/s^2). Proceed at 30 mile/h (48 km/h) to the 1.7 mile (2.74 km) marker.
1.7	(2.74)	Stop at 4 ft/s^2 (1.2 m/s^2), accelerate to 15 mile/h (24 km/h) at 7 ft/s^2 (2.1 m/s^2), then to 30 mile/h (48 km/h) at 5 ft/s^2 (1.5 m/s^2). Proceed at 30 mile/h (48 km/h) to the 1.8 mile (2.90 km) marker.
1.8	(2.90)	Decelerate to 20 mile/h (32 km/h) at 4 ft/s^2 (1.2 m/s^2), accelerate to 30 mile/h (48 km/h) at 5 ft/s^2 (1.5 m/s^2). Proceed at 30 mile/h (48 km/h).
2.0	(3.22)	Begin braking at 4 ft/s^2 (1.2 m/s^2) to arrive at stop at the 2.0 mile (3.22 km) marker. Stop fuel meter and timing device, record fuel consumed, elapsed time, and fuel temperature.

8.3.4.3 *Driving Cycle (0-4 mile (0-6.4 km) Cycle)*—The 2.0-4.0 mile (3.2-6.4 km) test is the second half of the 0-4.0 mile (0-6.4 km) test, and is a repetition of the test cycle shown in paragraph 8.3.4.2. The 2.0 and 4.0 mile (3.2-6.4 km) test begins be re-starting the fuel meter and timing device but not the engine, followed by the 15-s idle, and continues with the remaining driving cycle instructions. The 0.2 miles (0.32 km) stop, thus, begins at 2.2 mile (3.54) and the 0.3 mile (0.48 km) deceleration begins at 2.3 mile (3.70 km), etc.

9. Data Recording—Data shall be entered as required on test data forms.
  9.1 Test Vehicle Specifications and Preparation List
  9.2 Data Form
  9.3 Summary Sheet

10. Data Correction - General—Fuel economy data is corrected with three factors: C_1, C_2, and C_3. Factors C_2 and C_3 account for differentials in fuel properties and should be applied to all data. C_1 corrects fuel economy on the basis of average vehicle temperature response.

NOTE: If fuel economy results are to be quoted as results indicative of values at the ambient temperature condition of the test, correction factor C_1 should not be applied. When comparing relative fuel economy, it is rare that tests will be run at the same ambient condition. For this comparison, correction factor C_1 can be applied. Where a vehicle is suspected to respond differently than as indicated by the supplied correction factor or where extreme accuracy is needed, comparative tests should be run over a temperature span and an independent correction factor, C_1, be developed. Where the supplied correction factor C_1 is used, it should be used with caution to correct for small changes in test temperature conditions. Correction factor C_1 represents an accumulation of test results on about 25 1975-1977 model year vehicles with as many as 60 tests on individual cars. This correction factor will not precisely correct any specific vehicle, over-correcting some and under-correcting others, and can induce error with attempts to correct over large temperature spans. The extreme of the data available indicates that one of the 25 test vehicles was over-corrected by 1.8% while another was undercorrected by 5.8% for a 10°F (6°C) change in ambient. It is, therefore, recommended that data for comparison purposes be run within a temperature span of 10°F (6°C) to minimize the likelihood of induced error.

11. Data Correction (U.S. Units)
  11.1 Reference Conditions
  Reference Temperature 60°F
  Fuel Temperature 60°F
  Fuel Gravity (Gasoline) 0.737 Specific Gravity, 60.5° API Gravity
  Fuel Gravity (ASTM 1D) 0.820 Specific Gravity, 41.0° API Gravity
  Fuel Gravity (ASTM 2D) 0.845 Specific Gravity, 36.0° API Gravity
  Fuel Net Heating Value
    (ASTM 1D) 126 700 Btu/gal
    (ASTM 2D) 129 900 Btu/gal

  11.2 Fuel Economy Correction (Gasoline)
  11.2.1 Definitions (Units)
  T_A - Average ambient temperature during cold urban test cycle (°F)
  T_f - Average fuel temperature at measuring instrument during test cycle (°F)
  G_s - Specific gravity of test fuel at 60°F
  G_A - API gravity of test fuel at 60°F
  FE_o - Observed fuel economy (mile/gal)
  FE_c - Corrected fuel economy (mile/gal)
  11.2.2 Correction Formula
  $$FE_c = FE_o \cdot C_1 \cdot C_2 \cdot C_3$$
  11.2.3 Correction Factors
  $C_1 = 1 + 0.0025 (15.6 - T_A)$
  $C_2 = 1.0$ Urban Cycle
     $= 1.0 + 0.0021 (P - 98)$ Suburban Cycle
     $= 1.0 + 0.0025 (P - 98)$ 89 km/h Interstate Cycle
     $= 1.0 + 0.0043 (P - 98)$ 113 km/h Interstate Cycle
  $C_3 = 1.0 + 0.8 (0.737 - G_s)$

C_4 is derived from Table 1 based on gravity of fuel at 60°F and T_f or from the following analytical equation:
$$C_4 = a + bT_f + cT_f^2$$
where the coefficients a, b, c, are:

ASTM FUEL GROUP

Spec Gravity Range	1 0.8499–0.9659	2 0.7754–0.8498	3 0.7239–0.7753	4 0.6723–0.7238
Coefficient				
a	9.7645 (10)⁻¹	9.7108 (10)⁻¹	9.6513 (10)⁻¹	9.5982 (10)⁻¹
b	3.8674 (10)⁻⁴	4.6590 (10)⁻⁴	5.5473 (10)⁻⁴	6.3156 (10)⁻⁴
c	9.3735 (10)⁻⁸	2.6156 (10)⁻⁷	4.3541 (10)⁻⁷	6.2624 (10)⁻⁷

11.3 Fuel Economy Correction (Diesel)
NOTE: The method for correcting observed fuel economy for vehicles with diesel engines has not been investigated to the same degree that it has for gasoline-powered vehicles. The correction factors for gasoline-powered vehicles are recommended for use.

11.3.1 Definitions—See paragraph 11.2.1.
H = Volumetric heating value of diesel test fuel (Btu/gal)
11.3.2 Correction Formula
$$FE_c = FE_o \cdot C_1 \cdot C_2 \cdot C_3$$
11.3.3 Correction Factors
$C_1 = 1 + 0.0050 (60 - T_A)$ 0-2 mile (See Note in Section 10.)
   $= 1 + 0.0044 (60 - T_A)$ 0-4 mile
$C_2 = K/H$
K = 126 700 (ASTM 1D type fuels)
   = 129 900 (ASTM 2D type fuels)

H – shall be obtained from Fig. 1 by using the observed API gravity at 60°F and the 50% distillation point or from calorimeter tests.
C_3 is derived from Table 1 based on gravity of fuel at 60°F and T_f or from the following analytical equation:
$$C_3 = a + bT_f + cT_f^2$$
where the coefficients a, b, c are noted in paragraph 11.2.3.

12. Data Correction (SI) Units
  12.1 Reference Conditions
  Reference Temperature 15.6°C
  Fuel Temperature 15.6°C
  Fuel Gravity (Gasoline) 0.737 Specific Gravity
  Fuel Gravity (ASTM 1D) 0.820 Specific Gravity
  Fuel Gravity (ASTM 2D) 0.845 Specific Gravity
  Fuel Net Heating Value
    (ASTM 1D) 35.31 mJ/L
    (ASTM 2D) 36.21 mJ/L

  12.2 Fuel Economy Correction (Gasoline)
  12.2.1 Definitions (Units)
  T_A - Average ambient temperature during test cycle (°C)
  T_f - Average fuel temperature during test cycle (°C)
  G_s - Specific gravity of test fuel at 15.6°C
  G_A - API gravity of test fuel at 15.6°C
  FE_o - Observed fuel economy (km/L)
  FE_c - Corrected fuel economy (km/L)
  12.2.2 Correction Formula
  $$FE_c = FE_o \cdot C_1 \cdot C_2 \cdot C_3$$
  12.2.3 Correction Factors
  $C_1 = 1 + 0.0090 (15.6 - T_A)$ 0-2 mile (See Note in Section 10.)
     $= 1 + 0.0079 (15.6 - T_A)$ 0-4 mile
  $C_2 = 1.0 + 0.8 (0.737 - G_s)$
     or
  $C_2 = 1.0 + 0.0032 (G_A - 60.5)$

C_3 is derived from Table 1A based on gravity of fuel at 15.6°C and T_f or from the following analytical equation:
$$C_3 = a' + b'T_f + c'T_f^2$$
where the coefficients a', b', c' are:

ASTM FUEL GROUP

Spec Gravity Range	1 0.8499–0.9659	2 0.7754–0.8498	3 0.7239–0.7753	4 0.6723–0.7238
Coefficient				
a'	9.8892 (10)⁻¹	9.8626 (10)⁻¹	9.8333 (10)⁻¹	9.8067 (10)⁻¹
b'	7.0693 (10)⁻⁴	8.6875 (10)⁻⁴	1.0487 (10)⁻³	1.2090 (10)⁻³
c'	3.0370 (10)⁻⁷	8.4745 (10)⁻⁷	1.4107 (10)⁻⁶	2.0290 (10)⁻⁶

12.3 Fuel Economy Correction (Diesel)

NOTE: The method for correcting observed fuel economy for vehicles with diesel engines has not been investigated to the same degree that it has for gasoline powered vehicles.

The correction factors for gasoline powered vehicles are recommended for use.

12.3.1 DEFINITIONS—See paragraph 12.2.1.

H = Volumetric heating value of test fuel (mJ/L)

12.3.2 CORRECTION FORMULA

$$FE_c = FE_o \cdot C_1 \cdot C_2 \cdot C_3$$

12.3.3 CORRECTION FACTORS

$C_1 = 1.0 + 0.0090 (15.6 - T_A)$ 0-2 miles (see Note in Section 10.)
$ = 1 + 0.0079 (15.6 - T_A)$ 0-4 miles

$C_2 = K/H$
$K = 35.31$ MJ/L for ASTM 1D type fuel
$ = 36.21$ MJ/L for ASTM 2D type fuel

H - shall be determined from Fig. 1 by using the API gravity at 15.6°C and 50% distillation point or from calorimeter tests.

C_3 is derived from Table 1A based on gravity of fuel at 15.6°F and T_f or from the following analytical equation:

$$C_3 = a' + b'T_f + c'T_f^2$$

where the coefficients a', b', c' are noted in paragraph 12.2.3.

13. Data Presentation

13.1 Test Vehicle Specifications and Preparation List
13.2 Fuel Economy Test-Data Form
13.3 Fuel Economy Test-Data Summary Sheet

Fuel Temp., °F	C₃				Fuel Temp., °F	C₃			
	Group 1	Group 2	Group 3	Group 4		Group 1	Group 2	Group 3	Group 4
0	0.9765	0.9711	0.9651	0.9598	75	1.0060	1.0076	1.0092	1.0107
1	0.9768	0.9715	0.9656	0.9604	76	1.0064	1.0081	1.0098	1.0114
2	0.9772	0.9720	0.9662	0.9611	77	1.0068	1.0085	1.0104	1.0121
3	0.9776	0.9725	0.9667	0.9616	78	1.0071	1.0090	1.0110	1.0129
4	0.9780	0.9730	0.9673	0.9623	79	1.0076	1.0095	1.0116	1.0136
5	0.9784	0.9734	0.9679	0.9629	80	1.0080	1.0100	1.0122	1.0143
6	0.9788	0.9739	0.9684	0.9636	81	1.0084	1.0105	1.0129	1.0151
7	0.9791	0.9743	0.9690	0.9642	82	1.0088	1.0110	1.0136	1.0158
8	0.9795	0.9748	0.9696	0.9649	83	1.0092	1.0115	1.0142	1.0166
9	0.9799	0.9753	0.9701	0.9655	84	1.0096	1.0120	1.0148	1.0173
10	0.9803	0.9758	0.9707	0.9661	85	1.0100	1.0126	1.0154	1.0180
11	0.9807	0.9763	0.9713	0.9668	86	1.0104	1.0131	1.0160	1.0187
12	0.9811	0.9768	0.9718	0.9675	87	1.0108	1.0136	1.0167	1.0195
13	0.9814	0.9772	0.9724	0.9681	88	1.0112	1.0141	1.0173	1.0202
14	0.9818	0.9777	0.9730	0.9688	89	1.0116	1.0146	1.0179	1.0209
15	0.9822	0.9782	0.9735	0.9695	90	1.0120	1.0151	1.0185	1.0218
16	0.9826	0.9787	0.9741	0.9701	91	1.0124	1.0156	1.0192	1.0225
17	0.9830	0.9790	0.9747	0.9708	92	1.0129	1.0162	1.0198	1.0232
18	0.9835	0.9795	0.9753	0.9714	93	1.0133	1.0167	1.0205	1.0240
19	0.9839	0.9800	0.9759	0.9721	94	1.0137	1.0172	1.0211	1.0247
20	0.9843	0.9805	0.9765	0.9728	95	1.0141	1.0177	1.0218	1.0254
21	0.9846	0.9810	0.9770	0.9734	96	1.0145	1.0182	1.0224	1.0262
22	0.9850	0.9814	0.9776	0.9741	97	1.0148	1.0187	1.0230	1.0269
23	0.9854	0.9819	0.9782	0.9748	98	1.0152	1.0193	1.0236	1.0276
24	0.9858	0.9824	0.9788	0.9753	99	1.0156	1.0198	1.0243	1.0285
25	0.9862	0.9829	0.9793	0.9760	100	1.0160	1.0203	1.0249	1.0292
26	0.9866	0.9834	0.9799	0.9767	101	1.0165	1.0208	1.0255	1.0300
27	0.9870	0.9839	0.9805	0.9773	102	1.0169	1.0213	1.0262	1.0307
28	0.9874	0.9843	0.9811	0.9780	103	1.0173	1.0219	1.0269	1.0314
29	0.9878	0.9848	0.9816	0.9787	104	1.0177	1.0224	1.0275	1.0322
30	0.9881	0.9853	0.9822	0.9793	105	1.0181	1.0229	1.0282	1.0330
31	0.9885	0.9858	0.9828	0.9800	106	1.0185	1.0234	1.0288	1.0338
32	0.9889	0.9863	0.9834	0.9807	107	1.0190	1.0240	1.0294	1.0346
33	0.9893	0.9868	0.9840	0.9814	108	1.0194	1.0245	1.0301	1.0353
34	0.9897	0.9873	0.9845	0.9820	109	1.0198	1.0250	1.0307	1.0360
35	0.9901	0.9878	0.9851	0.9827	110	1.0201	1.0255	1.0314	1.0368
36	0.9905	0.9882	0.9857	0.9834	111	1.0205	1.0261	1.0320	1.0376
37	0.9909	0.9887	0.9863	0.9841	112	1.0209	1.0266	1.0327	1.0384
38	0.9913	0.9892	0.9869	0.9847	113	1.0213	1.0271	1.0334	1.0392
39	0.9917	0.9897	0.9875	0.9854	114	1.0218	1.0276	1.0340	1.0399
40	0.9921	0.9902	0.9880	0.9861	115	1.0222	1.0282	1.0347	1.0407
41	0.9925	0.9907	0.9886	0.9868	116	1.0226	1.0287	1.0353	1.0414
42	0.9928	0.9912	0.9892	0.9875	117	1.0230	1.0291	1.0359	1.0422
43	0.9932	0.9917	0.9898	0.9881	118	1.0234	1.0296	1.0366	1.0431
44	0.9936	0.9922	0.9904	0.9888	119	1.0238	1.0302	1.0372	1.0438
45	0.9940	0.9926	0.9910	0.9895	120	1.0243	1.0307	1.0380	1.0446
46	0.9944	0.9930	0.9916	0.9902	121	1.0246	1.0312	1.0386	1.0454
47	0.9948	0.9935	0.9922	0.9910	122	1.0250	1.0318	1.0393	1.0461
48	0.9952	0.9940	0.9928	0.9917	123	1.0254	1.0323	1.0399	1.0469
49	0.9956	0.9945	0.9933	0.9924	124	1.0258	1.0328	1.0406	1.0478
50	0.9960	0.9950	0.9939	0.9930	125	1.0263	1.0334	1.0412	1.0485
51	0.9964	0.9955	0.9946	0.9937	126	1.0267	1.0339	1.0419	1.0493
52	0.9968	0.9960	0.9952	0.9944	127	1.0271	1.0344	1.0425	1.0501
53	0.9973	0.9965	0.9958	0.9951	128	1.0275	1.0350	1.0433	1.0509
54	0.9977	0.9971	0.9964	0.9958	129	1.0279	1.0355	1.0440	1.0517
55	0.9981	0.9976	0.9971	0.9965	130	1.0283	1.0360	1.0446	1.0525
56	0.9984	0.9980	0.9977	0.9973	131	1.0287	1.0366	1.0452	1.0533
57	0.9988	0.9985	0.9983	0.9980	132	1.0291	1.0371	1.0459	1.0541
58	0.9992	0.9990	0.9988	0.9986	133	1.0295	1.0377	1.0466	1.0548
59	0.9996	0.9995	0.9994	0.9993	134	1.0300	1.0382	1.0473	1.0557
60	1.0000	1.0000	1.0000	1.0000	135	1.0304	1.0387	1.0480	1.0565
61	1.0004	1.0005	1.0006	1.0007	136	1.0308	1.0393	1.0486	1.0573
62	1.0008	1.0010	1.0012	1.0014	137	1.0312	1.0398	1.0493	1.0581
63	1.0012	1.0015	1.0018	1.0021	138	1.0316	1.0404	1.0500	1.0589
64	1.0016	1.0020	1.0024	1.0028	139	1.0320	1.0409	1.0506	1.0598
65	1.0020	1.0025	1.0030	1.0035	140	1.0324	1.0414	1.0513	1.0606
66	1.0024	1.0030	1.0036	1.0042	141	1.0328	1.0420	1.0520	1.0613
67	1.0028	1.0035	1.0042	1.0049	142	1.0333	1.0425	1.0527	1.0621
68	1.0032	1.0040	1.0049	1.0056	143	1.0337	1.0431	1.0534	1.0630
69	1.0036	1.0045	1.0055	1.0064	144	1.0341	1.0436	1.0541	1.0638
70	1.0040	1.0050	1.0061	1.0072	145	1.0345	1.0442	1.0547	1.0646
71	1.0044	1.0055	1.0067	1.0079	146	1.0350	1.0447	1.0554	1.0654
72	1.0048	1.0060	1.0074	1.0086	147	1.0353	1.0452	1.0561	1.0662
73	1.0052	1.0065	1.0080	1.0093	148	1.0357	1.0458	1.0568	1.0671
74	1.0056	1.0070	1.0086	1.0100	149	1.0362	1.0464	1.0575	1.0679
					150	1.0366	1.0469	1.0582	1.0687

ᵃThis table is based on Tables 25 and 7 of "Petroleum Measurement Tables" published by the American Society for Testing and Materials, 1916 Race St., Philadelphia, PA, 19103. Values given are reciprocals of the multiplier values in Tables 25 and 7 as:

C_3 = 1/multiplier for volume reduction to 60°F

TABLE 1A—C₃—FUEL TEMPERATURE CORRECTION FACTOR[a] (SI UNITS)

ASTM Group Number	Specific Gravity Range	API Gravity Range, °API
1	0.8499-0.9659	15.0-34.9
2	0.7754-0.8498	35.0-50.9
3	0.7239-0.7753	51.0-63.9
4	0.6723-0.7238	64.0-78.9

Fuel Temp., °C	C_3 Group 1	C_3 Group 2	C_3 Group 3	C_3 Group 4	Fuel Temp., °C	C_3 Group 1	C_3 Group 2	C_3 Group 3	C_3 Group 4
−15	0.9784	0.9734	0.9679	0.9629	25	1.0068	1.0085	1.0104	1.0121
−14	0.9791	0.9742	0.9689	0.9641	26	1.0075	1.0095	1.0115	1.0134
−13	0.9798	0.9751	0.9699	0.9653	27	1.0083	1.0104	1.0127	1.0147
−12	0.9805	0.9760	0.9710	0.9664	28	1.0090	1.0113	1.0138	1.0161
−11	0.9812	0.9768	0.9720	0.9676	29	1.0097	1.0122	1.0150	1.0174
−10	0.9818	0.9777	0.9730	0.9688	30	1.0104	1.0131	1.0161	1.0187
−9	0.9826	0.9785	0.9741	0.9700	31	1.0112	1.0141	1.0172	1.0201
−8	0.9833	0.9794	0.9751	0.9712	32	1.0119	1.0150	1.0184	1.0214
−7	0.9840	0.9803	0.9761	0.9724	33	1.0126	1.0159	1.0195	1.0228
−6	0.9847	0.9811	0.9772	0.9736	34	1.0133	1.0168	1.0207	1.0241
−5	0.9854	0.9819	0.9782	0.9748	35	1.0141	1.0177	1.0218	1.0254
−4	0.9861	0.9828	0.9792	0.9760	36	1.0148	1.0187	1.0229	1.0268
−3	0.9868	0.9837	0.9803	0.9771	37	1.0155	1.0196	1.0241	1.0281
−2	0.9875	0.9845	0.9813	0.9783	38	1.0162	1.0205	1.0252	1.0295
−1	0.9882	0.9854	0.9823	0.9795	39	1.0170	1.0214	1.0264	1.0308
0	0.9889	0.9863	0.9834	0.9807	40	1.0177	1.0224	1.0275	1.0322
1	0.9896	0.9871	0.9844	0.9820	41	1.0184	1.0233	1.0287	1.0336
2	0.9903	0.9880	0.9854	0.9832	42	1.0191	1.0243	1.0299	1.0350
3	0.9910	0.9889	0.9865	0.9844	43	1.0199	1.0252	1.0310	1.0364
4	0.9918	0.9898	0.9876	0.9856	44	1.0206	1.0262	1.0322	1.0378
5	0.9925	0.9907	0.9886	0.9868	45	1.0213	1.0271	1.0334	1.0392
6	0.9932	0.9916	0.9897	0.9881	46	1.0221	1.0281	1.0346	1.0406
7	0.9939	0.9924	0.9908	0.9893	47	1.0228	1.0290	1.0358	1.0419
8	0.9946	0.9933	0.9918	0.9906	48	1.0235	1.0300	1.0370	1.0433
9	0.9953	0.9942	0.9929	0.9918	49	1.0243	1.0309	1.0381	1.0447
10	0.9960	0.9950	0.9939	0.9930	50	1.0250	1.0318	1.0393	1.0461
11	0.9968	0.9960	0.9950	0.9943	51	1.0258	1.0328	1.0405	1.0475
12	0.9975	0.9969	0.9961	0.9956	52	1.0265	1.0338	1.0417	1.0490
13	0.9982	0.9978	0.9972	0.9968	53	1.0272	1.0347	1.0429	1.0504
14	0.9989	0.9986	0.9983	0.9981	54	1.0280	1.0357	1.0441	1.0519
15	0.9996	0.9995	0.9994	0.9993	55	1.0287	1.0366	1.0453	1.0533
16	1.0003	1.0004	1.0005	1.0005	56	1.0294	1.0376	1.0465	1.0548
17	1.0010	1.0013	1.0016	1.0018	57	1.0302	1.0385	1.0477	1.0562
18	1.0018	1.0022	1.0027	1.0031	58	1.0309	1.0395	1.0489	1.0577
19	1.0025	1.0031	1.0038	1.0044	59	1.0316	1.0404	1.0501	1.0592
20	1.0032	1.0040	1.0049	1.0056	60	1.0324	1.0414	1.0513	1.0606
21	1.0039	1.0049	1.0060	1.0069	61	1.0332	1.0424	1.0526	1.0621
22	1.0046	1.0058	1.0071	1.0082	62	1.0339	1.0434	1.0538	1.0635
23	1.0054	1.0067	1.0082	1.0095	63	1.0347	1.0444	1.0550	1.0650
24	1.0061	1.0076	1.0093	1.0108	64	1.0354	1.0454	1.0562	1.0664

[a]This table is based on Tables 25 and 7 of "Petroleum Measurement Tables" published by the American Society for Testing and Materials, 1916 Race St., Philadelphia, PA, 19103. Values given are reciprocals of the multiplier values as:

C_3 = 1/multiplier for volume reduction to 15.6°C

26.513

SUMMARY SHEET
FUEL ECONOMY TEST DATA - LIGHT DUTY VEHICLES

DATE _____

VEHICLE MAKE _____ ENGINE _____ CAR NO. _____

VIN _____ TRANSMISSION _____ AXLE RATIO _____

TIRE MAKE _____ SIZE _____ PRESSURE _____

VEHICLE TEST WEIGHT _____

TEST ROAD _____ SURFACE _____

	FUEL ECONOMY		FUEL TEMP. °F (°C)	AIR TEMP. °F (°C)	BARO. PRESS. in Hg (kPa)	WIND SPEED-DIRECTION mi/h (km/h)
	CORRECTED mi/gal (km/L)	OBSERVED mi/gal (km/L)				
COLD START (0-2 mi) (0-3.2 km)	_____	_____	_____	_____	_____	_____
COLD START (0-4 mi) (0-6.4 km)	_____	_____	_____	_____	_____	_____

COMMENTS: _____
_____
_____

TESTED BY: _____

26.514

FUEL ECONOMY TEST - LIGHT DUTY VEHICLES
TEST VEHICLE SPECIFICATIONS AND PREPARATION

SPECIFICATION LIST

DATE _____
CAR NO. _____
YEAR AND MAKE _____
MODEL AND BODY _____
VEHICLE INDENT. NO. _____
PRODUCTION _____ OTHER _____
ENGINE TYPE _____ DISP. _____
NET hp (U.S.) _____ COMP. RATIO _____
ENGINE NO. _____
ENGINE EMISSION CALIBRATION NO. _____
CARB. TYPE AND NO. _____
DISTRIBUTOR NO. _____
EXHAUST SYSTEM TYPE _____
TRANSMISSION _____
DRIVE AXLE TYPE AND RATIO _____
BRAKES, (DISC OR DRUM) F_____ R_____
STEERING _____
*TIRE MAKE _____ SIZE _____
    LOAD RANGE _____ TYPE _____
    % TREAD _____
COLD INFLATION-TIRE PRESSURE
    LF_____ RF_____
    LR_____ RR_____
TEST WEIGHT_____

*TIRES MUST HAVE A MINIMUM OF 100 miles (160 km)
BREAK-IN ON ROAD OR TRACK

CHECK LIST

_____ ENGINE OIL LEVEL OK
_____ COOLANT LEVEL OK
_____ TRANSMISSION FLUID LEVEL OK
_____ BELTS AND HOSES-TIGHT
_____ EMISSION CONTROLS-FUNCTIONAL
_____ CHOKE OPERATION-FUNCTIONAL
_____ THROTTLE OPERATION-FUNCTIONAL
_____ PUMP AND NOZZLES-NO LEAKS
_____ IGNITION WIRES-TIGHT
_____ BRAKE DRAG NOT EXCESSIVE
_____ TRANSMISSION OPERATION
_____ TIRE PRESSURE AND CONDITION
_____ ENGINE TUNE-PERFORMED
_____ IGNITION TIMING_____
_____ IDLE rpm _____ IDLE CO _____
_____ IGNITION POINT DWELL_____
_____ WHEEL ALIGNMENT-PERFORMED
_____ AIR CLEANER-CLEAN
_____ A/C COMPRESSOR LOAD-REMOVED
_____ NO FUEL LEAKS
_____ MANIFOLD HEAT VALVE-FUNCTIONAL
_____ FAN CLUTCH-FUNCTIONAL

$$\% \text{ TREAD} = \frac{\text{AVERAGE TREAD DEPTH OF TEST TIRES}}{\text{AVERAGE TREAD DEPTH OF IDENTICAL NEW TIRE}} \cdot 100$$

EQUIPPED WITH OPTIONAL EQUIPMENT
_____ POWER BRAKES
_____ POWER STEERING
_____ AIR CONDITIONING
_____ RADIO
_____ POWER SEATS
_____ POWER WINDOWS
_____ POWER DOOR LOCKS
_____
_____

COMMENTS:_____
_____
_____
_____
CAR CHECKED BY:_____
DATE:_____

TEST FUEL SPECIFICATIONS
FUEL TYPE AND GRADE _____
GRAVITY (API OR SPECIFIC) _____ AT 60°F (15.6°C)
REID VAPOR PRESSURE_____ psi (kPa)
DISTILLATION   10% _____ °F (°C)
                  50% _____ °F (°C)
                  90% _____ °F (°C)
Φ OCTANE $\frac{R+M}{2}$ _____

26.515

DATA FORM
FUEL ECONOMY TEST-LIGHT DUTY VEHICLES

CAR NO. _____ VIN _____ TEST VEHICLE WEIGHT_____
YEAR AND MAKE _____ TEST ROAD_____
MODEL AND BODY _____ TYPE OF SURFACE_____
ENGINE TYPE _____ DISP. _____ INSTRUMENTATION
DRIVE AXLE RATIO _____ 1. _____ NO. _____
TRANSMISSION_____ 2. _____ NO. _____
TIRE MAKE _____ SIZE _____ 3. _____ NO. _____
TIRE PRESSURE: F_____ R_____ 4. _____ NO. _____
FUEL TYPE _____ SPECIFIC GRAVITY _____

URBAN CYCLE (2.00 mile) (3.22 km) DATE _____ TIME _____
ODOMETER: START _____ FINISH _____

DIRECTION	FUEL TEMPERATURE					ACCUM Time	ACCUM Fuel	AMBIENT CONDITIONS					
	0.5 (0.8)	1.0 (1.6)	1.5 (2.4)	2.0 (3.2)	mile km			TEMPERATURE Start	Fin.	BAROMETRIC Start	Fin.	WIND Start	Fin.
___	___	___	___	___	___	___	___	___	___	___	___	___	___
___	___	___	___	___	___	___	___	___	___	___	___	___	___
___	___	___	___	___	___	___	___	___	___	___	___	___	___
___	___	___	___	___	___	___	___	___	___	___	___	___	___

SUBURBAN CYCLE (5.2 mile) (8.37 km) DATE _____ TIME _____
ODOMETER: START _____ FINISH _____

DIRECTION	FUEL TEMPERATURE				ACCUM Time	ACCUM Fuel	AMBIENT CONDITIONS					
	2.0 (3.2)	3.3 (5.3)	5.2 (8.4)	mile km			TEMPERATURE Start	Fin.	BAROMETRIC Start	Fin.	WIND Start	Fin.
___	___	___	___	___	___	___	___	___	___	___	___	___
___	___	___	___	___	___	___	___	___	___	___	___	___
___	___	___	___	___	___	___	___	___	___	___	___	___
___	___	___	___	___	___	___	___	___	___	___	___	___

_____ mile/h (km/h) INTERSTATE CYCLE (4.7 miles) (7.56 km) DATE _____ TIME _____
ODOMETER: START _____ FINISH _____

DIRECTION	FUEL TEMPERATURE			ACCUM Time	ACCUM Fuel	AMBIENT CONDITIONS					
	0.0 (0.0)	4.7 (7.6)	mile km			TEMPERATURE Start	Fin.	BAROMETRIC Start	Fin.	WIND Start	Fin.
___	___	___	___	___	___	___	___	___	___	___	___
___	___	___	___	___	___	___	___	___	___	___	___
___	___	___	___	___	___	___	___	___	___	___	___
___	___	___	___	___	___	___	___	___	___	___	___

TESTED BY:_____